6TH EDITION

Algebra & Trigonometry

Graphs and Models

Marvin L. Bittinger
Indiana University Purdue University Indianapolis

Judith A. Beecher

David J. Ellenbogen
Community College of Vermont

Judith A. Penna

PEARSON

Boston Columbus Indianapolis New York San Francisco
Amsterdam Cape Town Dubai London Madrid Milan Munich Paris Montréal Toronto
Delhi Mexico City São Paulo Sydney Hong Kong Seoul Singapore Taipei Tokyo

Editorial Director	Chris Hoag
Editor in Chief	Anne Kelly
Acquisitions Editor	Chelsea Kharakozova
Editorial Assistant	Ashley Gordon
Project Manager	Rachel S. Reeve
Program Management Team Lead	Karen Wernholm
Project Management Team Lead	Christina Lepre
Media Producer	Erica Lange
TestGen Content Manager	Marty Wright
MathXL Content Manager	Kristina Evans
Product Marketing Manager	Rachel Ross
Field Marketing Manager	Peggy Lucas
Marketing Assistant	Fiona Murray
Senior Author Support/Technology Specialist	Joe Vetere
Rights and Permissions Project Manager	Gina M. Cheselka
Procurement Specialist	Carol Melville
Associate Director of Design	Andrea Nix
Program Design Lead	Barbara T. Atkinson
Text Design, Art Editing, and Photo Research	The Davis Group, Inc.
Editorial and Production Coordination	Martha Morong/Quadrata, Inc.
Composition	Lumina Datamatics, Inc.
Illustrations	Network Graphics, William Melvin, and Lumina Datamatics, Inc.
Cover Design	Jenny Willingham/Infiniti
Cover Image	hxdbzxy/Shutterstock

Library of Congress Cataloging-in-Publication Data

Algebra and trigonometry : graphs and models / Marvin L. Bittinger, Indiana University Purdue University Indianapolis [and three others].—Sixth edition.
pages cm
Includes index.
ISBN 978-0-13-417904-9 (hardcover)—ISBN 0-13-417904-8 (hardcover)—ISBN 978-0-13-418752-5 (hardcover)—ISBN 0-13-418752-0 (hardcover)
1. Algebra—Textbooks. 2. Algebra—Graphic methods—Textbooks. 3. Trigonometry—Textbooks. 4. Functional analysis—Textbooks. I. Bittinger, Marvin L.
QA152.3.A49 2017
512'.13—dc23

2015035065

2 17

www.pearsonhighered.com

ISBN 13: 978-0-13-417904-9
ISBN 10: 0-13-417904-8

Contents

Preface

Algebra and Trigonometry: Graphs and Models is known for enabling students to "see the math" through its

- focus on visualization,
- early introduction of functions,
- integration of technology, and
- connections between math concepts and the real world.

New!

With the new edition, we continue to innovate by positioning the review material as a more effective tool for teachers and students. Chapter R from the previous edition has been condensed into **28 Just-in-Time review topics** that are placed before Chapter 1. This new review feature is designed to give each student the opportunity to be successful in this course by providing a quick review of topics from intermediate algebra that will be built upon in new algebra and trigonometry topics. The review can be used in an individualized instruction format since some students will require more review than others. Treating the review in this manner will allow more time to cover the algebra and trigonometry topics in the syllabus.

On the other hand, some instructors might choose to review some or all of the topics with the entire class at the beginning of the course or in a just-in-time format as each is needed. We think that instructors will appreciate the flexibility that the Just-in-Time feature offers.

Additional resources in the MyMathLab course reflect the themes of just-in-time review and concept retention. For example, new Cumulative Review assignments allow students to synthesize and retain concepts learned throughout the course. The Just-in-Time review topics within MyMathLab allow for assignable Getting Ready review quizzes that lead to personalized Getting Ready homework focused on areas in which students need additional practice.

Our overarching goal is to provide students with a learning experience that will not only lead to success in this course, but also prepare them to be successful in the mathematics courses that they take in the future.

Content Changes to the Sixth Edition

- **Just-in-Time Review** Review of prerequisite algebra topics is now presented when students need it most.

New!

  - A set of 28 numbered, short review topics creates an efficient review of intermediate algebra topics:

1. Real Numbers
2. Properties of Real Numbers
3. Order on the Number Line
4. Absolute Value
5. Operations with Real Numbers
6. Interval Notation
7. Integers as Exponents
8. Scientific Notation
9. Order of Operations
10. Introduction to Polynomials
11. Add and Subtract Polynomials
12. Multiply Polynomials
13. Special Products of Binomials
14. Factor Polynomials; The FOIL Method

15. Factor Polynomials: The *ac*-Method
16. Special Factorizations
17. Equation-Solving Principles
18. Inequality-Solving Principles
19. The Principle of Zero Products
20. The Principle of Square Roots
21. Simplify Rational Expressions
22. Multiply and Divide Rational Expressions
23. Add and Subtract Rational Expressions
24. Simplify Complex Rational Expressions
25. Simplify Radical Expressions
26. Rationalize Denominators
27. Rational Exponents
28. The Pythagorean Theorem

- This feature is placed before Chapter 1. Just-in-Time icons are positioned throughout the text next to the appropriate example where review of an intermediate algebra topic would be helpful.

- **Cumulative Reviews** For enhanced concept review, cumulative reviews, assignable in MyMathLab, allow students to synthesize and retain concepts learned throughout the course.
- **Informed Exercises** We have analyzed the MyMathLab usage data which has informed our revision of the exercises for this new edition. The goal is to ultimately improve the quality and quantity of exercises that are most relevant.

Emphasis on Functions

Functions are the core of this course and are presented as a thread that runs throughout the course rather than as an isolated topic. We introduce functions in Chapter 1, whereas many traditional algebra and trigonometry textbooks cover equation-solving in Chapter 1. Our approach of introducing students to a relatively new concept at the beginning of the course, rather than requiring them to begin with a review of material that was previously covered in intermediate algebra, immediately engages them and serves to help them avoid the temptation to neglect studying early in the course because "I already know this."

The concept of a function can be challenging for students. By repeatedly exposing them to the language, notation, and use of functions, demonstrating visually how functions relate to equations and graphs, and also showing how functions can be used to model real data, we hope to ensure that students not only become comfortable with functions but also come to understand and appreciate them. You will see this emphasis on functions woven throughout the other themes that follow.

Classify the Function Exercises With a focus on conceptual understanding, students are asked periodically to identify a number of functions by their type (linear, quadratic, rational, and so on). As students progress through the text, the variety of functions with which they are familiar increases and these exercises become more challenging. The "classifying the function" exercises appear with the review exercises in the Skill Maintenance portion of an exercise set. (See pp. 262 and 353–354.)

Visual Emphasis

Our early introduction of functions allows graphs to be used to provide a visual aspect to solving equations and inequalities. For example, we are able to show students both algebraically and visually that the solutions of a quadratic equation $ax^2 + bx + c = 0$ are the zeros of the quadratic function $f(x) = ax^2 + bx + c$, as well as the first coordinates of the x-intercepts of the graph of that function. This makes it possible for students, particularly visual learners, to gain a quick understanding of these concepts. (See pp. 178, 181, 221, 281, and 342.)

Visualizing the Graph Appearing at least once in every chapter, this feature provides students with an opportunity to match an equation with its graph by focusing on the characteristics of the equation and the corresponding attributes of the graph. (See pp. 138, 194, and 276.) In MyMathLab, animated Visualizing the Graph features for each chapter allow students to interact with graphs on an entirely new level. In addition to this full-page feature, many of the exercise sets include exercises in which the student is asked to match an equation with its graph or to find an equation of a function from its graph. (See pp. 140, 141, 230, and 326.)

Side-by-Side Examples Many examples are presented in a side-by-side, two-column format in which the algebraic solution of an equation appears in the left column and a graphical solution appears in the right column. (See pp. 244, 284, and 357.) This enables students to visualize and comprehend the connections among the solutions of an equation, the zeros of a function, and the x-intercepts of the graph of a function.

New!

Guided Visualizations These new figures help bring mathematical concepts to life. They are included in MyMathLab as both a teaching and a learning tool. Used as a lecture tool, the figures help engage students more fully and save the time that would otherwise be spent drawing figures by hand. Questions pertaining to each guided visualization are assignable in MyMathLab and reinforce active learning, critical thinking, and conceptual learning.

Integrated Technology In order to increase students' understanding of the course content through a visual means, we integrate graphing calculator technology throughout. The use of the graphing calculator is woven throughout the text's exposition, exercise sets, and testing program without sacrificing algebraic skills. Graphing calculator technology is included in order to enhance—not replace—students' mathematical skills, and to alleviate the tedium associated with certain procedures. (See pp. 176, 273–274, and 355.) The graphing calculator windows enhance the visual element of the text, providing graphical interpretations of solutions of equations, zeros of functions, and x-intercepts of graphs of functions.

› Making Connections

Zeros, Solutions, and x-Intercepts We find that when students understand the connections among the real zeros of a function, the solutions of its associated equation, and the first coordinates of the x-intercepts of its graph, a door opens to a new level of mathematical comprehension that increases the probability of success in this course. We emphasize zeros, solutions, and x-intercepts throughout the text by using consistent, precise terminology and including exceptional graphics. Seeing this theme repeated in different contexts leads to a better understanding and retention of these concepts. (See pp. 171 and 181.)

Connecting the Concepts This feature highlights the importance of connecting concepts. When students are presented with concepts in visual form—using graphs, an outline, or a chart—rather than merely in paragraphs of text, comprehension is streamlined and retention is enhanced. The visual aspect of this feature invites students to stop and check their understanding of how concepts work together in one section or in several sections. This check in turn enhances student performance on homework assignments and exams. (See pp. 69, 181, and 249.)

Annotated Examples We have included over 1070 annotated examples designed to fully prepare the student to work the exercises. Learning is carefully guided with the use of numerous color-coded art pieces and step-by-step annotations. Substitutions and annotations are highlighted in red for emphasis. (See pp. 175 and 349–350.)

Now Try Exercises Now Try Exercises are found after nearly every example. This feature encourages active learning by asking students to do an exercise in the exercise set that is similar to the example that the student has just read. (See pp. 173, 268, and 322.)

Synthesis Exercises These exercises appear at the end of each exercise set and encourage critical thinking by requiring students to synthesize concepts from several sections or to take a concept a step further than in the general exercises. For the Sixth Edition, these exercises are assignable in MyMathLab. (See pp. 251–252, 330, and 380.)

Real-Data Applications We encourage students to see and interpret the mathematics that appears every day in the world around them. Throughout the writing process, we conducted an energetic search for real-data applications, and the result is a variety of examples and exercises that connect the mathematical content with everyday life. Most of these applications feature source lines and many include charts and graphs. Many are drawn from the fields of health, business and economics, life and physical sciences, social science, and areas of general interest such as sports and travel. (See pp. 37 ("Food Stamp Program"), 63 ("Industrial Robots"), 141 ("Medical Care Abroad"), 184 ("Funding for Afghan Security"), 231 ("Vinyl Album Sales"), 328 ("Alfalfa Imported by China"), 560 ("Vietnam Veterans Memorial"), 649 ("Cosmetic Surgery"), 658 ("Top Art Auction Sales"), 735 ("The Ellipse at the White House"), and 813 ("The Economic Multiplier; Super Bowl XLVII").)

Ongoing Review

The most significant change to the Sixth Edition is the new Just-in-Time Review feature, designed to provide students with efficient and effective review of basic algebra skills.

New! Just-in-Time Review Chapter R has been condensed into 28 numbered, short review topics to create an efficient review of intermediate algebra topics. This feature is placed before Chapter 1.

- Just-In-Time icons are placed throughout the text next to the example where review of an intermediate algebra topic would be helpful. (See pp. 33, 95, 164, 220, and 315.)
- The coverage of each topic contains worked-out examples and a short exercise set. Answers to all exercises appear at the end of the answers at the back of the book.
- Worked-out solutions to all Just-in-Time exercises are included in the *Student Solutions Manual.*

Just in Time 20

THE PRINCIPLE OF SQUARE ROOTS

The principle of square roots can be used to solve some quadratic equations.

THE PRINCIPLE OF SQUARE ROOTS

If $x^2 = k$, then $x = -\sqrt{k}$ *or* $x = \sqrt{k}$.

EXAMPLES Solve.

1.
$$s^2 - 144 = 0$$
$$s^2 = 144$$
$$s = -\sqrt{144} \quad or \quad s = \sqrt{144}$$
$$s = -12 \quad or \quad s = 12$$

The solutions are -12 and 12, or ± 12.

2.
$$3x^2 - 21 = 0$$
$$3x^2 = 21$$
$$x^2 = 7$$
$$x = -\sqrt{7} \quad or \quad x = \sqrt{7}$$

The solutions are $-\sqrt{7}$ and $\sqrt{7}$, or $\pm\sqrt{7}$.

Do Exercises 1–6.

Solve.

1. $x^2 - 36 = 0$

2. $2y^2 - 20 = 0$

3. $6z^2 = 18$

4. $3t^2 - 15 = 0$

5. $z^2 - 1 = 24$

6. $5x^2 - 75 = 0$

Mid-Chapter Mixed Review This review reinforces understanding of the mathematical concepts and skills covered in the first half of the chapter before students move on to new material in the second half of the chapter. Each review begins with at least three true/false exercises that require students to consider the concepts they have studied and also contains exercises that drill the skills from all prior sections of the chapter. They are available as assignments in MyMathLab. (See pp. 121–122 and 346–347.)

Collaborative Discussion and Writing Exercises appear in the Mid-Chapter Mixed Review as well. These exercises can be discussed in small groups or by the class as a whole to encourage students to talk about the key mathematical concepts in the chapter. They can also be assigned to individual students to give them an opportunity to write about mathematics. (See pp. 199 and 253).

A section reference (shown in red) is provided for each exercise in the Mid-Chapter Mixed Review. This tells the student which section to refer to if help is needed to work the exercise. Answers to all exercises in the Mid-Chapter Mixed Review are given at the back of the book.

Study Guide This feature is found at the beginning of the **Summary and Review** near the end of each chapter. Presented in a two-column format and organized by section, this feature gives key concepts and terms in the left column and a worked-out example in the right column. It provides students with a concise and effective review of the chapter that is a solid basis for studying for a test. In MyMathLab, these Study Guides are accompanied by narrated examples to reinforce the key concepts and ideas. (See pp. 210–215 and 381–387.)

Exercise Sets There are over 7060 exercises in this text. The exercise sets are enhanced with real-data applications and source lines, detailed art pieces, tables, graphs, and photographs. In addition to the exercises that provide students with concepts presented in the section, the exercise sets feature the following elements to provide ongoing review of topics presented earlier:

- **Skill Maintenance Exercises.** These exercises provide an ongoing review of concepts previously presented in the course, enhancing students' retention of these concepts. They include **Vocabulary Reinforcement**, described next, and **Classifying the Function** exercises, described earlier in the section "Emphasis on Functions." A section reference (shown in red) is provided for each exercise. This tells the student which section to refer to if help is needed to work the exercise. Answers to all Skill Maintenance exercises appear in the answer section at the back of the book. (See pp. 128, 206, 279–280, and 345.)
- **Enhanced Vocabulary Reinforcement Exercises.** This feature checks and reviews students' understanding of the vocabulary introduced throughout the text. It appears once in every chapter, in the Skill Maintenance portion of an exercise set, and is intended to provide a continuing review of the terms that students must know in order to be able to communicate effectively in the language of mathematics. (See pp. 149–150, 209, and 279.)
- **Enhanced Synthesis Exercises.** These exercises appear at the end of each exercise set and encourage critical thinking by requiring students to synthesize concepts from several sections or to take a concept a step further than in the general exercises. For the Sixth Edition, these exercises are assignable in MyMathLab.

Review Exercises These exercises in the **Summary and Review** supplement the Study Guide by providing a thorough and comprehensive review of the skills taught in the chapter. A group of true/false exercises appears first, followed by a large number of exercises that drill the skills and concepts taught in the chapter. In addition, three multiple-choice exercises, one of which involves identifying the graph of a function, are included in the Review Exercises for every chapter. Each review exercise is accompanied by a section reference that, as in the Mid-Chapter Mixed Review, directs students to the section in which the material being reviewed can be found. Also included are Collaborative Discussion and Writing exercises. These exercises are described under the Mid-Chapter Mixed Review heading on p. xv. (See pp. 215–217 and 388–391.)

Chapter Test The test at the end of each chapter allows students to test themselves and target areas that need further study before taking the in-class test. Each Chapter Test includes a multiple-choice exercise involving identifying the graph of a function. Answers to all questions in the Chapter Tests appear in the answer section at the back of the book, along with corresponding section references. (See pp. 218 and 391–392.)

DOMAIN

REVIEW SECTION 1.2.

Review Icons Placed next to the concept that a student is currently studying, a review icon references a section of the text in which the student can find and review topics on which the current concept is built. (See pp. 263 and 308.)

Acknowledgments

We wish to express our heartfelt thanks to a number of people who have contributed in special ways to the development of this textbook. Our editor, Chelsea Kharakozova, and Editor in Chief, Anne Kelly, encouraged and supported our vision. We are very appreciative of the marketing insight provided by Peggy Lucas and Rachel Ross, our marketing managers, and of the support that we received from the entire Pearson team, including Rachel Reeve, project manager, Barbara Atkinson, cover designer,

Ashley Gordon, editorial assistant, and Fiona Murray, marketing assistant. We also thank Erica Lange, media producer, for her creative work on the media products that accompany this text. And we are immensely grateful to Martha Morong for her editorial and production services, and to Geri Davis for her text design and art editing, and for the endless hours of hard work they have done to make this a book of which we are proud. We also thank Laurie Hurley and Holly Martinez for their meticulous accuracy checking and proofreading of the text.

The following reviewers made invaluable contributions to the development of the recent editions and we thank them for that:

Gerald Allen, *Howard College*
Robin Ayers, *Western Kentucky University*
Heidi Barrett, *Arapahoe Community College*
George Behr, *Coastline Community College*
Kimberly Bennekin, *Georgia Perimeter College*
*Nadine Bluett, *Front Range Community College*
Marc Campbell, *Daytona Beach Community College*
*Shawn Clift, *Eastern Kentucky University*
Mark A. Crawford, Jr., *Waubonsee Community College*
Brad Feldser, *Kennesaw State University*
Homa Ghaussi-Mujtaba, *Lansing Community College*
Bob Gravelle, *Colorado Technical University, Colorado Springs*
*Mako E. Haruta, *University of Hartford*
Judy Hayes, *Lake-Sumter Community College*
Michelle Hollis, *Bowling Green Community College*
*Patricia Ann Hussey, *Triton College*
Glenn Jablonski, *Triton College*
Bridgette Jacob, *Onondaga Community College*
Symon Kimitei, *Kennesaw State University*
Deanna Kindhart, *Illinois Central College*
Pamela Krompak, *Owens Community College*
Laud Kwaku, *Owens Community College*
Carol A. Lucas, *University of Central Oklahoma*
*Claude Moore, *Cape Fear Community College*
Daniel Olson, *Purdue University, North Central*
*Priti Patel, *Tarrant County Community College*
Cloyd A. Payne, *Owens Community College*
Randy K. Ross, *Morehead State University*
Daniel Russow, *Arizona Western College*
Brian Schworm, *Morehead State University*
*Nicholas Sedlock, *Framingham State University*
*Pavel Sikorskii, *Michigan State University*
Judith Staver, *Florida Community College at Jacksonville, South Campus*
*Laura Taylor, *Cape Fear Community College*
Jean Hunt Thorton, *Western Kentucky University*
*Pat Velicky, *Florence–Darlington Technical College*
*Jim Voss, *Front Range Community College*
Douglas Windham, *Tallahassee Community College*
Weicheng Xuan, *Arizona Western College*
*Cathleen Zucco-Teveloff, *Rider University*

** Reviewers of the Sixth Edition*

M.L.B.
J.A.B.
D.J.E.
J.A.P.

BREAK THROUGH
To improving results

Resources for Success

MyMathLab® Online Course for *Algebra & Trigonometry: Graphs and Models* by Bittinger/Beecher/Ellenbogen/Penna

(access code required)

MyMathLab is available to accompany Pearson's market-leading text offerings. To give students a consistent tone, voice, and teaching method, each text's flavor and approach is tightly integrated throughout the accompanying MyMathLab course, making learning the material as seamless as possible.

With this new edition, instructors will find even more assignable exercises in MyMathLab to match many of the in-text features so many instructors have grown to rely on, as well as a new Skills for Success Module.

Skill Maintenance Quizzes and Cumulative Reviews

Instructors can now assign MyMathLab quizzes generated from the *Skill Maintenance Exercises* found in the text. These quizzes support ongoing review to help students maintain essential skills.

For enhanced concept review throughout the course, *Cumulative Reviews* are now also assignable in MyMathLab. These assignments allow students to synthesize and retain concepts.

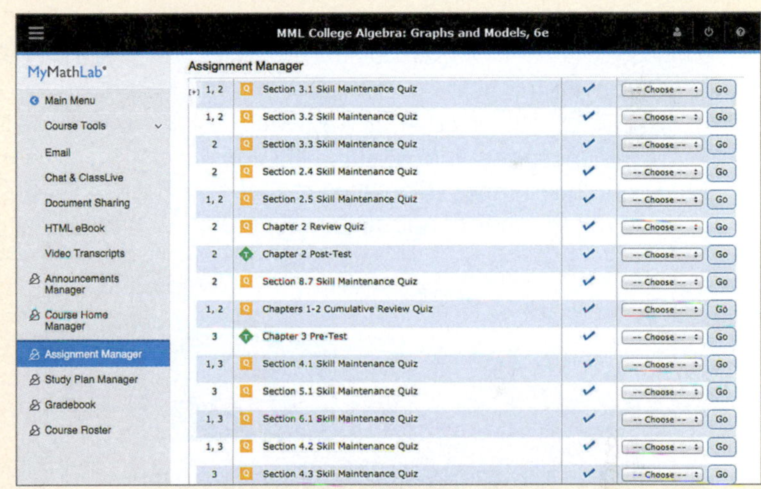

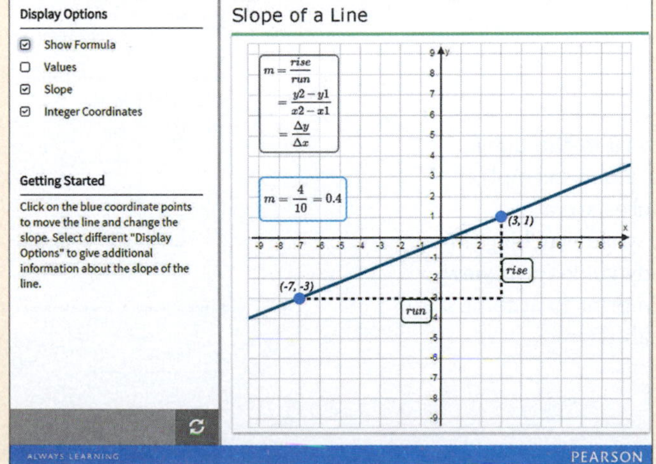

NEW! Guided Interactive Figures

These engaging interactive figures bring mathematical concepts to life, helping students visualize the concepts through directed explorations and purposeful manipulation. Guided Visualizations are assignable in MyMathLab and encourage active learning, critical thinking, and conceptual learning.

www.mymathlab.com

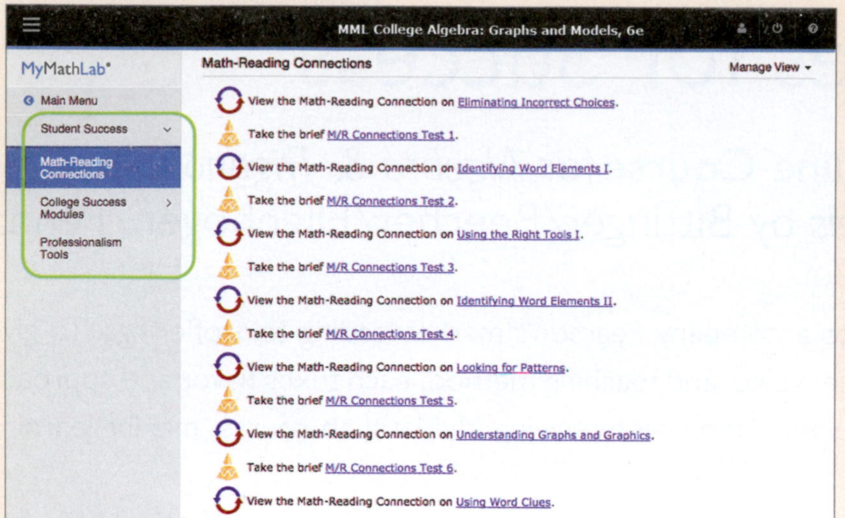

Skills for Success Modules

Skills for Success Modules help foster success in collegiate courses and prepare students for future professions. Topics such as "Time Management," "Stress Management" and "Financial Literacy" are available within the MyMathLab course. Instructors can integrate these media-rich activities with their traditional MyMathLab assignments.

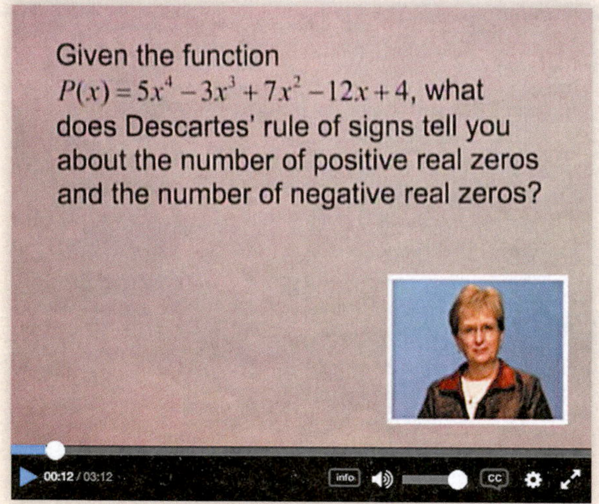

Video Assessment Exercises

Featuring authors Judy Beecher and Judy Penna, these Example Solutions videos walk students through the detailed solution process for nearly all examples from the text. Corresponding exercises check for conceptual understanding. Videos and exercises are assignable in MyMathLab.

NEW! Video Notebook

This author-created notebook contains fill-in-the-blank worksheets to accompany the authors' video examples. Key definitions, theorems, and procedures are also included. After filling in the worksheet while watching the video, the student has an excellent study guide for review and test preparation. This is available in print to accompany the text or as a download in MyMathLab.

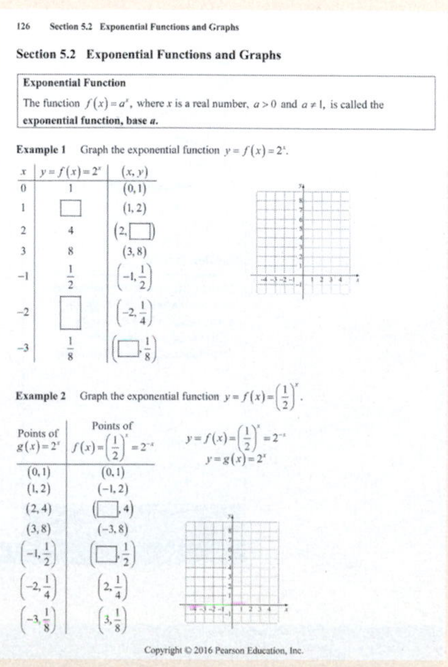

126 Section 5.2 Exponential Functions and Graphs

Section 5.2 Exponential Functions and Graphs

Exponential Function

The function $f(x) = a^x$, where x is a real number, $a > 0$ and $a \neq 1$, is called the **exponential function, base *a*.**

Example 1 Graph the exponential function $y = f(x) = 2^x$.

x	$y = f(x) = 2^x$	(x, y)
0	1	$(0, 1)$
1	$\square$	$(1, 2)$
2	4	$(2, \square)$
3	8	$(3, 8)$
−1	$\frac{1}{2}$	$\left(-1, \frac{1}{2}\right)$
−2	$\square$	$\left(-2, \frac{1}{4}\right)$
−3	$\frac{1}{8}$	$\left(\square, \frac{1}{8}\right)$

Example 2 Graph the exponential function $y = f(x) = \left(\frac{1}{2}\right)^x$.

Points of $g(x) = 2^x$	Points of $f(x) = \left(\frac{1}{2}\right)^x = 2^{-x}$
$(0, 1)$	$(0, 1)$
$(1, 2)$	$(-1, 2)$
$(2, 4)$	$(\square, 4)$
$(3, 8)$	$(-3, 8)$
$\left(-1, \frac{1}{2}\right)$	$\left(\square, \frac{1}{2}\right)$
$\left(-2, \frac{1}{4}\right)$	$\left(2, \frac{1}{4}\right)$
$\left(-3, \frac{1}{8}\right)$	$\left(3, \frac{1}{8}\right)$

$y = f(x) = \left(\frac{1}{2}\right)^x = 2^{-x}$

$y = g(x) = 2^x$

Instructor Resources

Additional resources can be downloaded from within MyMathLab and at www.pearsonhighered.com, or hardcopy resources can be ordered from your Pearson sales representative.

Annotated Instructor's Edition

The instructor's edition includes all answers to the exercise sets. Shorter answers are presented on the same page as the exercise; longer answers are in the back of the text. Sample homework assignments are indicated by a blue underline and may be assigned in MyMathLab®. Available upon request from your Pearson sales representative.

Instructor's Solutions Manual (Download Only)

Written by author Judy Penna, this resource contains worked-out solutions to all exercises in the exercise sets, Mid-Chapter Mixed Reviews, Chapter Reviews, and Chapter Tests, as well as solutions for all the Just-In-Time exercises.

PowerPoint® Lecture Slides

Feature presentations written and designed specifically for this text. These lecture slides provide an outline for presenting definitions, figures, and key examples from the text.

Online Test Bank (Download Only)

Contains four free-response test forms for each chapter following the same format and having the same level of difficulty as the test in the main text and two multiple-choice test forms for each chapter. It also provides six forms of the final examination, four with free-response questions and two with multiple-choice questions.

TestGen®

TestGen® (www.pearsoned.com/testgen) enables instructors to build, edit, print, and administer tests using a computerized bank of questions developed to cover all the objectives of the text.

Learning Catalytics Integration

MyMathLab now provides **Learning Catalytics**, an interactive student response tool that uses students' smartphones, tablets, or laptops to engage them in more sophisticated tasks and thinking. Learning Catalytics contains Pearson-created content for algebra and trigonometry that allows instructors to take advantage of this exciting technology immediately.

Student Resources

Additional resources to promote student success.

Author Example Videos

Ideal for distance learning or supplemental instruction, these videos feature authors Judy Beecher and Judy Penna working through and explaining examples in the text. Assignable in MyMathLab with new Video Assessment exercises.

Video Notebook

This notebook can accompany the text and/or MyMathLab course. It contains fill-in-the-blank worksheets to accompany the video examples presented by the authors. Key definitions, theorems, and procedures are also included. After filling in the worksheet while watching the video, the student has an excellent study guide for review and test preparation. This is available electronically for download within MyMathLab or as a printed resource.

Student's Solutions Manual

Written by author Judy Penna, this resource contains completely worked-out solutions with step-by-step annotations for all the odd-numbered exercises in the exercise sets, Mid-Chapter Mixed Reviews, and Chapter Reviews, as well as solutions for all the Chapter Test exercises and the Just-In-Time exercises.

Chapter R: Basic Concepts of Algebra

Available within MyMathLab, Chapter R supplements the prerequisite topics of the Just-in-Time review with more in-depth coverage and exercises.

- R.1 The Real-Number System
- R.2 Integer Exponents, Scientific Notation, and Order of Operations
- R.3 Addition, Subtraction, and Multiplication of Polynomials
- R.4 Factoring
- R.5 The Basics of Equation Solving
- R.6 Rational Expressions
- R.7 Radical Notation and Rational Exponents

To the Student

Guide to Success

Success can be planned. Combine goals and good study habits to create a plan for success that works for you. The following list contains study tips that your authors consider most helpful.

Skills for Success

- **Set goals and expect success.** Approach this class experience with a positive attitude.
- **Communicate with your instructor** when you need extra help.
- **Take your text with you to class and lab.** Each section in the text is designed with headings and boxed information that provide an outline for easy reference.
- **Ask questions in class, lab, and tutoring sessions.** Instructors encourage them, and other students probably have the same questions.
- **Begin each homework assignment as soon as possible.** If you have difficulty, you will then have the time to access supplementary resources.
- **Carefully read the instructions** before working homework exercises **and include all steps**.
- **Form a study group** with fellow students. Verbalizing questions about topics that you do not understand can clarify the material for you.
- After each quiz or test, **write out corrected step-by step solutions** to all missed questions. They will provide a valuable study guide for the midterm exam and the final exam.
- **MyMathLab has numerous tools to help you succeed.** Use MyMathLab to create a personalized study plan and practice skills with sample quizzes and tests.
- **Knowing math vocabulary is an important step toward success.** Review vocabulary using the Vocabulary Reinforcement exercises in the text and in MyMathLab.
- If you miss a lecture, **watch the video in the Multimedia Library** of MyMathLab that explains the concepts you missed.

In writing this textbook, we challenged ourselves to do everything possible to help you learn the concepts and skills contained between its covers so that you will be successful in this course and in the mathematics courses you take in the future. We realize that your time is both valuable and limited, so we communicate in a highly visual way that allows you to learn quickly and efficiently. We are confident that, if you invest an adequate amount of time in the learning process, this text will be of great value to you. We wish you a positive learning experience.

Marv Bittinger
Judy Beecher
David Ellenbogen
Judy Penna

JUST-IN-TIME Review

Throughout this text, there are Just-in-Time icons, numbered 1–28, that refer to brief reviews of the following 28 intermediate-algebra topics. Each mini-review lesson is accompanied by several exercises. All answers are provided in the answer section at the back of the text.

1. Real Numbers
2. Properties of Real Numbers
3. Order on the Number Line
4. Absolute Value
5. Operations with Real Numbers
6. Interval Notation
7. Integers as Exponents
8. Scientific Notation
9. Order of Operations
10. Introduction to Polynomials
11. Add and Subtract Polynomials
12. Multiply Polynomials
13. Special Products of Binomials
14. Factor Polynomials; The FOIL Method
15. Factor Polynomials: The *ac*-Method
16. Special Factorizations
17. Equation-Solving Principles
18. Inequality-Solving Principles
19. The Principle of Zero Products
20. The Principle of Square Roots
21. Simplify Rational Expressions
22. Multiply and Divide Rational Expressions
23. Add and Subtract Rational Expressions
24. Simplify Complex Rational Expressions
25. Simplify Radical Expressions
26. Rationalize Denominators
27. Rational Exponents
28. The Pythagorean Theorem

Just in Time 1

REAL NUMBERS

Some frequently used sets of real numbers and the relationships among them are shown below.

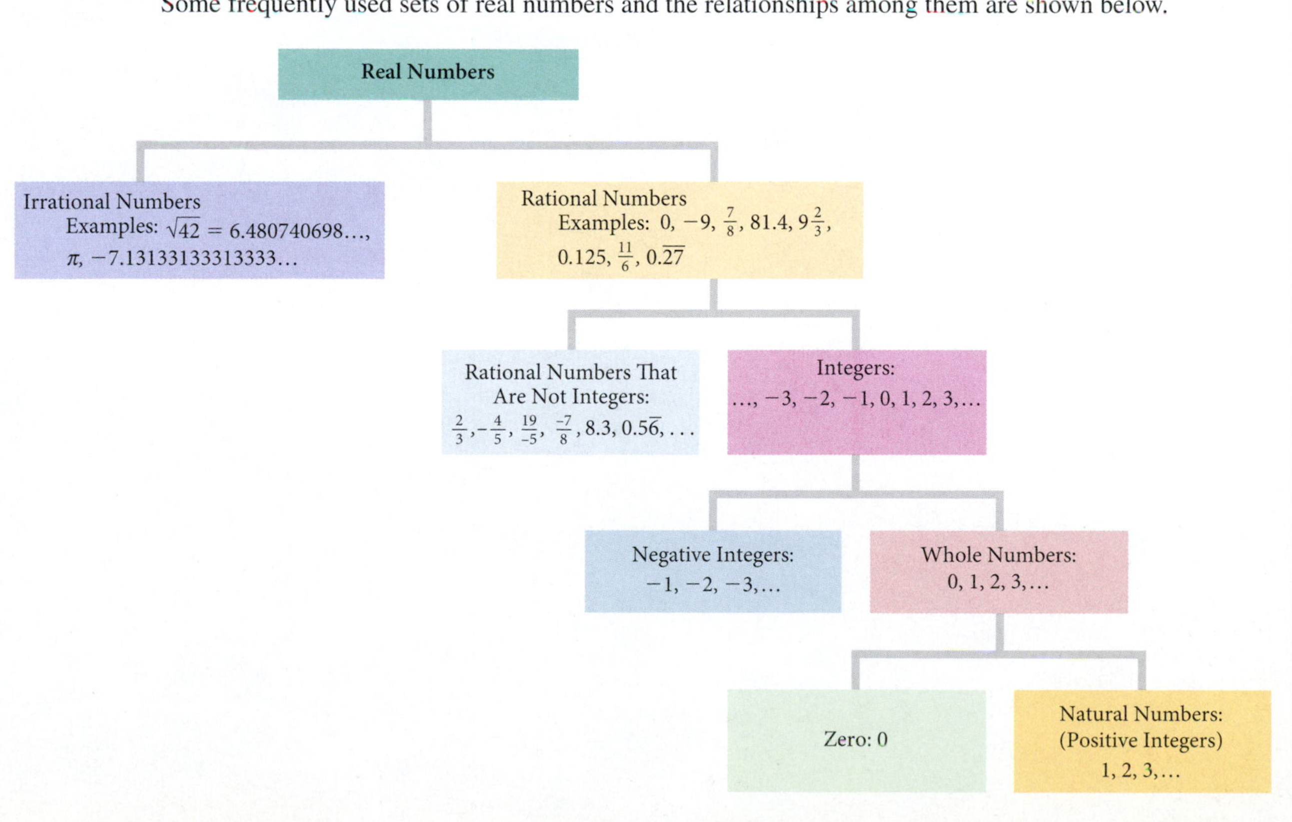

(continued)

JUST-IN-TIME Review

Numbers that can be expressed in the form p/q, where p and q are integers and $q \neq 0$, are **rational numbers**. Decimal notation for rational numbers either *terminates* (ends) or *repeats*. Each of the following is a rational number:

$$0, \quad -17, \quad \frac{13}{4}, \quad \sqrt{25} = 5;$$

$$\frac{1}{4} = 0.25 \quad \text{(terminating decimal);}$$

$$-\frac{5}{11} = -0.454545\ldots = -0.\overline{45} \quad \text{(repeating decimal);}$$

$$\frac{5}{6} = 0.8333\ldots = 0.8\overline{3} \quad \text{(repeating decimal).}$$

The real numbers that are not rational are **irrational numbers**. Decimal notation for irrational numbers neither terminates nor repeats. Each of the following is an irrational number. Note in each case that there is no repeating block of digits.

$\sqrt{2} = 1.414213562\ldots,$

$\sqrt{65} = 8.062257748\ldots,$

$-6.12122122212222\ldots,$

$\pi = 3.1415926535\ldots$

($\frac{22}{7}$ and 3.14 are *rational approximations* of the irrational number π.)

The set of all rational numbers combined with the set of all irrational numbers gives us the set of **real numbers.**

Do Exercises 1–6.

In Exercises 1–6, consider the numbers

$\frac{2}{3}$, 6, $\sqrt{3}$, -2.45, $\sqrt[6]{26}$, $18.\overline{4}$, -11, $\sqrt[3]{27}$, $5\frac{1}{6}$, $7.151551555\ldots$, $-\sqrt{35}$, $\sqrt[5]{3}$, $-\frac{8}{7}$, 0, $\sqrt{16}$.

1. Which are rational numbers?
2. Which are rational numbers but not integers?
3. Which are irrational numbers?
4. Which are integers?
5. Which are whole numbers?
6. Which are real numbers?

All answers to the Just-in-Time exercises appear at the end of the answers at the back of the book.

Just in Time 2

PROPERTIES OF REAL NUMBERS

PROPERTIES OF REAL NUMBERS

For any real numbers a, b, and c:

$a + b = b + a$ and $ab = ba$	Commutative properties of addition and multiplication
$a + (b + c) = (a + b) + c$ and $a(bc) = (ab)c$	Associative properties of addition and multiplication
$a + 0 = 0 + a = a$	Additive identity property
$-a + a = a + (-a) = 0$	Additive inverse property
$a \cdot 1 = 1 \cdot a = a$	Multiplicative identity property
$a \cdot \frac{1}{a} = \frac{1}{a} \cdot a = 1 \; (a \neq 0)$	Multiplicative inverse property
$a(b + c) = ab + ac$ and $a(b - c) = ab - ac$	Distributive property

EXAMPLES Name the property illustrated in each sentence.

1. $8 \cdot 5 = 5 \cdot 8$	Commutative property of multiplication
2. $14 + (-14) = 0$	Additive inverse property
3. $2(a - b) = 2a - 2b$	Distributive property
4. $5 + (m + n) = (5 + m) + n$	Associative property of addition
5. $6 \cdot 1 = 1 \cdot 6 = 6$	Multiplicative identity property
6. $q + t = t + q$	Commutative property of addition
7. $\frac{7}{9} + 0 = \frac{7}{9}$	Additive identity property
8. $-2(cd) = (-2c)d$	Associative property of multiplication
9. $3(x + y) = 3x + 3y$	Distributive property
10. $8 \cdot \frac{1}{8} = 1$	Multiplicative inverse property

Do Exercises 1–10.

Name the property illustrated by the sentence.

1. $-24 + 24 = 0$

2. $7(xy) = (7x)y$

3. $9(r - s) = 9r - 9s$

4. $11 + z = z + 11$

5. $-20 \cdot 1 = -20$

6. $5(x + y) = (x + y)5$

7. $q + 0 = q$

8. $75 \cdot \frac{1}{75} = 1$

9. $(x + y) + w = x + (y + w)$

10. $8(a + b) = 8a + 8b$

All answers to the Just-in-Time exercises appear at the end of the answers at the back of the book.

Just in Time 3 ORDER ON THE NUMBER LINE

The real numbers are modeled using a **number line,** as shown below. Each point on the line represents a real number, and every real number is represented by a point on the line.

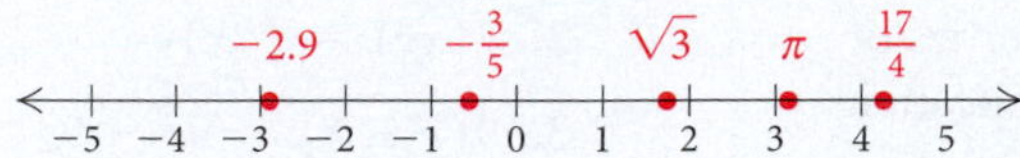

The order of the real numbers can be determined from the number line. If a number a is to the left of a number b, then a **is less than** b $(a < b)$. Similarly, a **is greater than** b $(a > b)$ if a is to the right of b on the number line. For example, we see from the number line above that $-2.9 < -\frac{3}{5}$, because -2.9 is to the left of $-\frac{3}{5}$. Also, $\frac{17}{4} > \sqrt{3}$, because $\frac{17}{4}$ is to the right of $\sqrt{3}$.

The statement $a \le b$, read "a is less than or equal to b," is true if either $a < b$ is true or $a = b$ is true. A similar statement holds for $a \ge b$.

Do Exercises 1–6.

Classify the inequality as true or false.

1. $9 < -9$
2. $-10 \le -1$
3. $-\sqrt{26} < -5$
4. $\sqrt{6} \ge \sqrt{6}$
5. $-30 > -25$
6. $-\frac{4}{5} > -\frac{5}{4}$

Just in Time 4 ABSOLUTE VALUE

The **absolute value** of a number a, denoted $|a|$, is its distance from 0 on the number line. For example, $|-5| = 5$, because the distance of -5 from 0 is 5. For any real number a,

$$|a| = a \quad \text{if } a \ge 0 \qquad \text{and} \qquad |a| = -a \quad \text{if } a < 0.$$

EXAMPLES Simplify.

1. $|-10| = 10$
2. $|0| = 0$
3. $\left|\frac{4}{9}\right| = \frac{4}{9}$

Absolute value can be used to find the distance between two points on the number line. For any real numbers a and b, the distance between a and b is $|a - b|$ or, equivalently, $|b - a|$.

EXAMPLE 4 Find the distance between -2 and 3.

$$|-2 - 3| = |-5| = 5,$$

or equivalently,

$$|3 - (-2)| = |3 + 2| = |5| = 5$$

Do Exercises 1–8.

Simplify.

1. $|-98|$
2. $|0|$
3. $|4.7|$
4. $\left|-\frac{2}{3}\right|$

Find the distance between the given pair of points on the number line.

5. $-7,\ 13$
6. $2,\ 14.6$
7. $-39,\ -28$
8. $-\frac{3}{4},\ \frac{15}{8}$

Just in Time 5

OPERATIONS WITH REAL NUMBERS

RULES FOR OPERATIONS WITH REAL NUMBERS

Addition

- *Positive Numbers*: Add the same way that we add arithmetic numbers. The answer is positive.
- *Negative Numbers*: Add absolute values. The answer is negative.
- *A Positive Number and a Negative Number:* If the numbers have the same absolute value, the answer is 0. If the numbers have different absolute values, subtract the smaller absolute value from the larger. If the positive number has the greater absolute value, the answer is positive. If the negative number has the greater absolute value, the answer is negative.

Subtraction

- To subtract, add the opposite, or additive inverse, of the number being subtracted.

Multiplication and Division, where the divisor is nonzero

- Multiply or divide the absolute values. If the signs are the same, the answer is positive. If the signs are different, the answer is negative.

EXAMPLES Add.

1. $9 + (-29) = -20$
2. $-9 + (-29) = -38$
3. $-9 + 29 = 20$

EXAMPLES Subtract.

4. $15 - 6 = 15 + (-6) = 9$
5. $15 - (-6) = 15 + 6 = 21$
6. $-15 - 6 = -15 + (-6) = -21$
7. $-15 - (-6) = -15 + 6 = -9$

EXAMPLES Multiply or divide.

8. $-5 \cdot 20 = -100$
9. $32 \div (-4) = -8$
10. $-32 \div 4 = -8$
11. $-32 \div (-4) = 8$
12. $-5 \cdot (-20) = 100$
13. $5 \cdot 20 = 100$

Do Exercises 1–15.

Compute and simplify.

1. $8 - (-11)$
2. $-\frac{3}{10} \cdot \left(-\frac{1}{3}\right)$
3. $15 \div (-3)$
4. $-4 - (-1)$
5. $7 \cdot (-50)$
6. $-0.5 - 5$
7. $-3 + 27$
8. $-400 \div (-40)$
9. $4.2 \cdot (-3)$
10. $-13 - (-33)$
11. $-60 + 45$
12. $\frac{1}{2} - \frac{2}{3}$
13. $-24 \div 3$
14. $-6 + (-16)$
15. $-\frac{1}{2} \div \left(-\frac{5}{8}\right)$

Just in Time 6 — INTERVAL NOTATION

Sets of real numbers can be expressed using **interval notation**. For example, for real numbers a and b such that $a < b$, the **open interval** (a, b) is the set of real numbers between, but not including, a and b.

Some intervals extend without bound in one or both directions. The interval $[a, \infty)$, for example, begins at a and extends to the right without bound. The bracket indicates that a is included in the interval.

SET NOTATION	INTERVAL NOTATION	GRAPH
$\{x \mid a < x < b\}$	(a, b)	
$\{x \mid a \le x \le b\}$	$[a, b]$	
$\{x \mid a \le x < b\}$	$[a, b)$	
$\{x \mid a < x \le b\}$	$(a, b]$	
$\{x \mid x > a\}$	(a, ∞)	
$\{x \mid x \ge a\}$	$[a, \infty)$	
$\{x \mid x < b\}$	$(-\infty, b)$	
$\{x \mid x \le b\}$	$(-\infty, b]$	
$\{x \mid x \text{ is a real number}\}$	$(-\infty, \infty)$	

Do Exercises 1–10.

Write interval notation.

1. $\{x \mid -5 \le x \le 5\}$
2. $\{x \mid -3 < x \le -1\}$
3. $\{x \mid x \le -2\}$
4. $\{x \mid x > 3.8\}$
5. $\{x \mid 7 < x\}$
6. $\{x \mid -2 < x < 2\}$

Write interval notation for the graph.

7. −5 −4 −3 −2 −1 0 1 2 3 4 5
8. 1.7 −5 −4 −3 −2 −1 0 1 2 3 4 5
9. −5 −4 −3 −2 −1 0 1 2 3 4 5
10. $\sqrt{5}$ −5 −4 −3 −2 −1 0 1 2 3 4 5

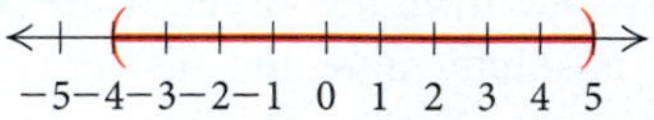

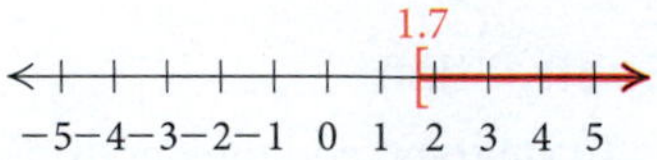

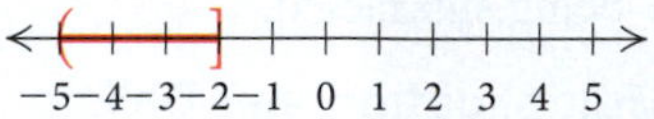

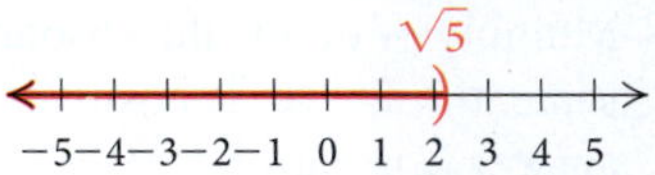

Just in Time 7

INTEGERS AS EXPONENTS

When a positive integer is used as an *exponent*, it indicates the number of times that a factor appears in a product. For example, 7^3 means $7 \cdot 7 \cdot 7$, where 7 is the **base** and 3 is the **exponent**.

For any nonzero numbers a and b and any integers m and n,

$$a^0 = 1, \qquad a^{-m} = \frac{1}{a^m}, \quad \text{and} \quad \frac{a^{-m}}{b^{-n}} = \frac{b^n}{a^m}.$$

PROPERTIES OF EXPONENTS

For any real numbers a and b and any integers m and n, assuming 0 is not raised to a nonpositive power:

$a^m \cdot a^n = a^{m+n}$	Product rule
$\frac{a^m}{a^n} = a^{m-n}\ (a \neq 0)$	Quotient rule
$(a^m)^n = a^{mn}$	Power rule
$(ab)^m = a^m b^m$	Raising a product to a power
$\left(\frac{a}{b}\right)^m = \frac{a^m}{b^m}\ (b \neq 0)$	Raising a quotient to a power

EXAMPLES Simplify each of the following.

1. $4^2 \cdot 4^{-5} = 4^{2+(-5)} = 4^{-3}$, or $\frac{1}{4^3}$
2. $\left(\frac{7}{9}\right)^0 = 1$
3. $(8^2)^{-5} = 8^{2(-5)} = 8^{-10}$, or $\frac{1}{8^{10}}$
4. $\frac{x^{11}}{x^4} = x^{11-4} = x^7$
5. $\left(\frac{a}{b}\right)^3 = \frac{a^3}{b^3}$
6. $(cd)^{-2} = c^{-2}d^{-2}$, or $\frac{1}{c^2d^2}$

Do Exercises 1–10.

Simplify.

1. 3^{-6}
2. $\frac{1}{(0.2)^{-5}}$
3. $\frac{w^{-4}}{z^{-9}}$
4. $\left(\frac{z}{y}\right)^2$
5. 100^0
6. $\frac{a^5}{a^{-3}}$
7. $(2xy^3)(-3x^{-5}y)$
8. $x^{-4} \cdot x^{-7}$
9. $(mn)^{-6}$
10. $(t^{-5})^4$

Just in Time 8

SCIENTIFIC NOTATION

We can use scientific notation to name both very large and very small positive numbers and to perform computations.

SCIENTIFIC NOTATION

Scientific notation for a number is an expression of the type

$$N \times 10^m,$$

where $1 \leq N < 10$, N is in decimal notation, and m is an integer.

EXAMPLES

Convert to scientific notation.

1. 9,460,000,000,000

Since 9,460,000,000,000 is a large number, the exponent on 10 will be positive. We want to move the decimal point between 9 and 4. This is a move of 12 places, so the exponent will be 12.

$$9{,}460{,}000{,}000{,}000 = 9.46 \times 10^{12}$$

2. 0.000073

Since 0.000073 is a small number, the exponent on 10 will be negative. We want to move the decimal point between 7 and 3. This is a move of 5 places, so the exponent will be -5.

$$0.000073 = 7.3 \times 10^{-5}$$

Convert to decimal notation.

3. 5.4×10^7

The exponent is positive. We will move the decimal point 7 places to the right.

$$5.4 \times 10^7 = 54{,}000{,}000$$

4. 3.819×10^{-3}

The exponent is negative. We will move the decimal point 3 places to the left.

$$3.819 \times 10^{-3} = 0.003819$$

Do Exercises 1–8.

Convert to scientific notation.

1. 18,500,000

2. 0.000786

3. 0.0000000023

4. 8,927,000,000

Convert to decimal notation.

5. 4.3×10^{-8}

6. 5.17×10^6

7. 6.203×10^{11}

8. 2.94×10^{-5}

Just in Time 9

ORDER OF OPERATIONS

Recall that to simplify the expression $3 + 4 \cdot 5$, first we multiply 4 and 5 to get 20 and then we add 3 to get 23. Mathematicians have agreed on the following procedure, or rules for order of operations.

RULES FOR ORDER OF OPERATIONS

1. Do all calculations within grouping symbols before operations outside. When nested grouping symbols are present, work from the inside out.
2. Evaluate all exponential expressions.
3. Do all multiplications and divisions in order from left to right.
4. Do all additions and subtractions in order from left to right.

EXAMPLES

1. $8(5 - 3)^3 - 20$
$= 8 \cdot 2^3 - 20$
$= 8 \cdot 8 - 20$
$= 64 - 20$
$= 44$

2. $10[7 - 4(8 - 5)]$
$= 10[7 - 4 \cdot 3]$
$= 10[7 - 12]$
$= 10[-5]$
$= -50$

3. $-32 \div 2 \times (-4) \div (-2)$
$= -16 \times (-4) \div (-2)$
$= 64 \div (-2)$
$= -32$

4. $\dfrac{10 \div (8 - 6) + 9 \cdot 4}{2^5 + 3^2}$
$= \dfrac{10 \div 2 + 9 \cdot 4}{32 + 9}$
$= \dfrac{5 + 36}{41}$
$= \dfrac{41}{41}$
$= 1$

5. $3^{12} \cdot 3^{-4} \div 3^3 \cdot 3^{-1}$
$= 3^8 \div 3^3 \cdot 3^{-1}$
$= 3^5 \cdot 3^{-1}$
$= 3^4$
$= 81$

Do Exercises 1–8.

Calculate.

1. $3 + 18 \div 6 - 3$
2. $5 \cdot 3 + 8 \cdot 3^2 + 4(6 - 2)$
3. $5(3 - 8 \cdot 3^2 + 4 \cdot 6 - 2)$
4. $16 \div 4 \cdot 4 \div 2 \cdot 256$
5. $2^6 \cdot 2^{-3} \div 2^{10} \div 2^{-8}$
6. $\dfrac{4(8 - 6)^2 - 4 \cdot 3 + 2 \cdot 8}{3^1 + 19^0}$
7. $64 \div [(-4) \div (-2)]$
8. $6[9 - (3 - 2)] + 4(2 - 3)$

Just in Time 10 — INTRODUCTION TO POLYNOMIALS

Polynomials are a type of algebraic expression that you will often encounter in your study of algebra. Some examples of polynomials are

$3x - 4y$, $5y^3 - \frac{7}{3}y^2 + 3y - 2$, $-2.3a^4$, 16, and $z^6 - \sqrt{5}$.

Algebraic expressions like $8x - 13$, $x^2 + 3x - 4$, and $3a^5 - 11 + a$ are **polynomials in one variable.** Algebraic expressions like $3ab^3 - 8$ and $5x^4y^2 - 3x^3y^8 + 7xy^2 + 6$ are **polynomials in several variables.** The terms of a polynomial are separated by + signs. The terms of $2x^2 - 8x + 3 = 2x^2 + (-8x) + 6$ are $2x^2$, $-8x$, and 3. The **degree of a term** is the sum of the exponents of the variables in that term. The **degree of a polynomial** is the degree of the term of highest degree.

A polynomial with just one term, like $-9y^6$, is a **monomial.** If a polynomial has two terms, like $x^2 + 4$, it is a **binomial.** A polynomial with three terms, like $4x^2 - 4xy + 1$, is a **trinomial.**

EXAMPLES Determine the degree of the polynomial.

1. $2x^3 - 1$ Degree: 3
2. -5 $(-5 = -5x^0)$ Degree: 0
3. $w^2 - 3.5 + 4w^5 = 4w^5 + w^2 - 3.5$ Degree: 5
4. $7xy^3 - 16x^2y^4$ Degree: $2 + 4$, or 6

Do Exercises 1–8.

Determine the degree of the polynomial.

1. $5 - x^6$
2. $x^2y^5 - x^7y + 4$
3. $2a^4 - 3 + a^2$
4. -41
5. $4x - x^3 + 0.1x^8 - 2x^5$

Classify the polynomial as a monomial, a binomial, or a trinomial.

6. $x - 3$
7. $14y^5$
8. $2y - \frac{1}{4}y^2 + 8$

Just in Time 11 — ADD AND SUBTRACT POLYNOMIALS

If two terms of an expression have the same variables raised to the same powers, they are called **like terms**, or **similar terms**. We can **combine**, or **collect, like terms** using the distributive property. For example, $3y^2$ and $5y^2$ are like terms and $3y^2 + 5y^2 = (3 + 5)y^2 = 8y^2$. We add or subtract polynomials by combining like terms.

EXAMPLES Add or subtract each of the following.

1. $(-5x^3 + 3x^2 - x) + (12x^3 - 7x^2 + 3)$
   $= (-5x^3 + 12x^3) + (3x^2 - 7x^2) - x + 3$
   $= (-5 + 12)x^3 + (3 - 7)x^2 - x + 3$
   $= 7x^3 - 4x^2 - x + 3$
2. $(6x^2y^3 - 9xy) - (5x^2y^3 - 4xy)$
   $= 6x^2y^3 - 9xy - 5x^2y^3 + 4xy$
   $= x^2y^3 - 5xy$

Do Exercises 1–5.

Add or subtract.

1. $(8y - 1) - (3 - y)$
2. $(3x^2 - 2x - x^3 + 2) - (5x^2 - 8x - x^3 + 4)$
3. $(2x + 3y + z - 7) + (4x - 2y - z + 8) + (-3x + y - 2z - 4)$
4. $(3ab^2 + 4a^2b - 2ab + 6) + (-ab^2 - 5a^2b + 8ab + 4)$
5. $(5x^2 + 4xy - 3y^2 + 2) - (9x^2 - 4xy + 2y^2 - 1)$

Just in Time 12 — MULTIPLY POLYNOMIALS

To multiply monomials, we first multiply their coefficients, and then we multiply their variables.

EXAMPLES

1. $(-2x^3)(5x^4) = (-2\cdot 5)(x^3\cdot x^4) = -10x^7$
2. $(3yz^2)(8y^3z^5) = (3\cdot 8)(y\cdot y^3)(z^2\cdot z^5) = 24y^4z^7$

We can find the product of two binomials by multiplying the **F**irst terms, then the **O**uter terms, then the **I**nner terms, then the **L**ast terms. Then we combine like terms, if possible. This procedure is sometimes called **FOIL**.

EXAMPLE 3 Multiply: $(2x - 7)(3x + 4)$.

F L F O I L

$$(2x - 7)(3x + 4) = 6x^2 + 8x - 21x - 28$$
$$= 6x^2 - 13x - 28$$

I O

EXAMPLE 4 Multiply: $(5c - d)(4c - 9d)$.

$$(5c - d)(4c - 9d) = 20c^2 - 45cd - 4cd + 9d^2$$
$$= 20c^2 - 49cd + 9d^2$$

Do Exercises 1–6.

Multiply.

1. $(3a^2)(-7a^4)$
2. $(y - 3)(y + 5)$
3. $(x + 6)(x + 3)$
4. $(2a + 3)(a + 5)$
5. $(2x + 3y)(2x + y)$
6. $(11t - 1)(3t + 4)$

Just in Time 13 — SPECIAL PRODUCTS OF BINOMIALS

SPECIAL PRODUCTS OF BINOMIALS

$(A + B)^2 = A^2 + 2AB + B^2$	Square of a sum
$(A - B)^2 = A^2 - 2AB + B^2$	Square of a difference
$(A + B)(A - B) = A^2 - B^2$	Product of a sum and a difference

EXAMPLES

1. $(4x + 1)^2 = (4x)^2 + 2\cdot 4x\cdot 1 + 1^2$
   $= 16x^2 + 8x + 1$
2. $(3y^2 - 2)^2 = (3y^2)^2 - 2\cdot 3y^2\cdot 2 + 2^2$
   $= 9y^4 - 12y^2 + 4$
3. $(x^2 + 3y)(x^2 - 3y) = (x^2)^2 - (3y)^2$
   $= x^4 - 9y^2$

Do Exercises 1–6.

Multiply.

1. $(x + 3)^2$
2. $(5x - 3)^2$
3. $(2x + 3y)^2$
4. $(a - 5b)^2$
5. $(n + 6)(n - 6)$
6. $(3y + 4)(3y - 4)$

Just in Time 14

FACTOR POLYNOMIALS; THE FOIL METHOD

When a polynomial is to be factored, we should always look first to factor out a factor that is common to all the terms using the distributive property. We generally look for the constant common factor with the largest absolute value and for variables with the largest exponent common to all the terms.

EXAMPLE 1 Factor: $15 + 10x - 5x^2$.

$$15 + 10x - 5x^2 = 5 \cdot 3 + 5 \cdot 2x - 5 \cdot x^2 = 5(3 + 2x - x^2)$$

In some polynomials, pairs of terms have a common binomial factor that can be removed in a process called **factoring by grouping**.

EXAMPLE 2 Factor: $x^3 + 3x^2 - 5x - 15$.

$$\begin{aligned} x^3 + 3x^2 - 5x - 15 &= (x^3 + 3x^2) + (-5x - 15) \\ &= x^2(x + 3) - 5(x + 3) \\ &= (x + 3)(x^2 - 5) \end{aligned}$$

Some trinomials can be factored into the product of two binomials. To factor a trinomial of the form $x^2 + bx + c$, we look for binomial factors of the form $(x + p)(x + q)$, where $p \cdot q = c$ and $p + q = b$. That is, we look for two numbers p and q whose sum is the coefficient of the middle term of the polynomial, b, and whose product is the constant term, c.

EXAMPLES Factor.

3. $x^2 + 5x + 6 = (x + 2)(x + 3)$

4. $x^4 - 6x^3 + 8x^2 = x^2(x^2 - 6x + 8) = x^2(x - 2)(x - 4)$

To factor trinomials of the type $ax^2 + bx + c, a \neq 1$, using the **FOIL method:**

1. Factor out the largest common factor.
2. Find two First terms whose product is ax^2:

$$(\square x + \quad)(\square x + \quad) = ax^2 + bx + c.$$

FOIL

3. Find two Last terms whose product is c:

$$(\quad x + \square)(\quad x + \square) = ax^2 + bx + c$$

FOI**L**

4. Repeat steps (2) and (3) until a combination is found for which the sum of the Outer product and the Inner product is bx:

$$(\square x + \square)(\square x + \square) = ax^2 + bx + c.$$

I, O — F**OI**L

(*continued*)

EXAMPLES Factor.

5. $3x^2 - 10x - 8 = (3x + 2)(x - 4)$
6. $12y^2 + 44y - 45 = (2y + 9)(6y - 5)$
7. $r^2 - 7rs + 6s^2 = (r - 6s)(r - s)$
8. $y^4 - 3y^2 - 40 = (y^2 + 5)(y^2 - 8)$

Do Exercises 1–12.

Factor out the largest common factor.

1. $3x + 18$
2. $2z^3 - 8z^2$

Factor by grouping.

3. $3x^3 - x^2 + 18x - 6$
4. $t^3 + 6t^2 - 2t - 12$

Factor the trinomial.

5. $w^2 - 7w + 10$
6. $t^2 + 8t + 15$
7. $2n^2 - 20n - 48$
8. $y^4 - 9y^3 + 14y^2$
9. $2n^2 + 9n - 56$
10. $2y^2 + y - 6$
11. $b^2 - 6bt + 5t^2$
12. $x^4 - 7x^2 - 30$

Just in Time 15

FACTOR POLYNOMIALS: THE *ac*-METHOD

A second method of factoring trinomials of the type $ax^2 + bx + c, a \neq 1$, is known as the ***ac*-method**, or the **grouping method**.

THE *ac*-METHOD FOR FACTORING TRINOMIALS

1. Factor out the largest common factor. The remaining trinomial is $ax^2 + bx + c$.
2. Multiply the leading coefficient a and the constant c.
3. Try to factor the product ac so that the sum of the factors is b. That is, find integers p and q such that $pq = ac$ and $p + q = b$.
4. Split the middle term, writing it as a sum using the factors found in step (3).
5. Factor by grouping.

EXAMPLE Factor: $6x^2 + 23x + 20$.

There is no common factor other than 1 or -1. We multiply the leading coefficient, 6, and the constant, 20: $6 \cdot 20 = 120$. Then we look for a factorization of 120 in which the sum of the factors is the coefficient of the middle term, 23. That factorization is $8 \cdot 15$.

(*continued*)

Split the middle term: $23x = 8x + 15x$.

Factor by grouping:

$$\begin{aligned} 6x^2 + 23x + 20 &= 6x^2 + 8x + 15x + 20 \\ &= 2x(3x + 4) + 5(3x + 4) \\ &= (3x + 4)(2x + 5). \end{aligned}$$

Do Exercises 1–3.

Factor.

1. $8x^2 - 6x - 9$
2. $10t^2 + 4t - 6$
3. $18a^2 - 51a + 15$

SPECIAL FACTORIZATIONS

SPECIAL FACTORIZATIONS

- Trinomial Squares:

$$A^2 + 2AB + B^2 = (A + B)^2;$$
$$A^2 - 2AB + B^2 = (A - B)^2$$

- Difference of Squares:

$$A^2 - B^2 = (A + B)(A - B)$$

- Sum or Difference of Cubes:

$$A^3 + B^3 = (A + B)(A^2 - AB + B^2);$$
$$A^3 - B^3 = (A - B)(A^2 + AB + B^2)$$

EXAMPLES

1. $x^2 - 16 = x^2 - 4^2 = (x + 4)(x - 4)$
2. $x^2 + 8x + 16 = x^2 + 2 \cdot x \cdot 4 + 4^2 = (x + 4)^2$
3. $25y^2 - 30y + 9 = (5y)^2 - 2 \cdot 5y \cdot 3 + 3^2 = (5y - 3)^2$
4. $x^3 + 27 = x^3 + 3^3 = (x + 3)(x^2 - 3x + 9)$
5. $16y^3 - 250 = 2(8y^3 - 125) = 2[(2y)^3 - 5^3] = 2(2y - 5)(4y^2 + 10y + 25)$

Do Exercises 1–10.

Factor the difference of squares.

1. $z^2 - 81$
2. $16x^2 - 9$
3. $7pq^4 - 7py^4$

Factor the square of a binomial.

4. $x^2 + 12x + 36$
5. $9z^2 - 12z + 4$
6. $a^3 + 24a^2 + 144a$

Factor the sum or the difference of cubes.

7. $x^3 + 64$
8. $m^3 - 216$
9. $3a^5 - 24a^2$
10. $t^6 + 1$

Just in Time 17

EQUATION-SOLVING PRINCIPLES

For any real numbers a, b, and c,

THE ADDITION PRINCIPLE

If $a = b$ is true, then $a + c = b + c$ is true.

THE MULTIPLICATION PRINCIPLE

If $a = b$ is true, then $ac = bc$ is true.

EXAMPLES Solve.

1.
$$\begin{aligned} y - 11 &= 12 \\ y - 11 + 11 &= 12 + 11 \\ y &= 23 \end{aligned}$$

Check:
$$\begin{array}{c|c} \multicolumn{2}{c}{y - 11 = 12} \\ \hline 23 - 11 \; ? & 12 \\ 12 & 12 \end{array} \quad \text{TRUE}$$

The solution is 23.

2.
$$\begin{aligned} 15c &= 90 \\ \frac{1}{15} \cdot 15c &= \frac{1}{15} \cdot 90 \\ c &= 6 \end{aligned}$$

Check:
$$\begin{array}{c|c} \multicolumn{2}{c}{15c = 90} \\ \hline 15 \cdot 6 \; ? & 90 \\ 90 & 90 \end{array} \quad \text{TRUE}$$

The solution is 6.

3.
$$\begin{aligned} \frac{1}{4}x + 5 &= 8 \\ \frac{1}{4}x + 5 - 5 &= 8 - 5 \\ \frac{1}{4}x &= 3 \\ 4 \cdot \frac{1}{4}x &= 4 \cdot 3 \\ x &= 12 \end{aligned}$$

Check:
$$\begin{array}{c|c} \multicolumn{2}{c}{\frac{1}{4}x + 5 = 8} \\ \hline \frac{1}{4} \cdot 12 + 5 \; ? & 8 \\ 3 + 5 & \\ 8 & 8 \end{array} \quad \text{TRUE}$$

The solution is 12.

4.
$$\begin{aligned} 2x + 3 &= 1 - 6(x - 1) \\ 2x + 3 &= 1 - 6x + 6 \\ 2x + 3 &= 7 - 6x \\ 2x + 3 + 6x &= 7 - 6x + 6x \\ 8x + 3 &= 7 \\ 8x + 3 - 3 &= 7 - 3 \\ 8x &= 4 \\ \frac{8x}{8} &= \frac{4}{8} \\ x &= \frac{1}{2} \end{aligned}$$

Check:
$$\begin{array}{c|c} \multicolumn{2}{c}{2x + 3 = 1 - 6(x - 1)} \\ \hline 2\left(\frac{1}{2}\right) + 3 \; ? & 1 - 6\left(\frac{1}{2} - 1\right) \\ 1 + 3 & 1 + 3 \\ 4 & 4 \end{array} \quad \text{TRUE}$$

The solution is $\frac{1}{2}$.

Do Exercises 1–8.

Solve.

1. $7t = 70$

2. $x - 5 = 7$

3. $3x + 4 = -8$

4. $6x - 15 = 45$

5. $7y - 1 = 23 - 5y$

6. $3m - 7 = -13 + m$

7. $2(x + 7) = 5x + 14$

8. $5y - 4(2y - 10) = 25$

Just in Time 18

INEQUALITY-SOLVING PRINCIPLES

For any real numbers a, b, and c:

THE ADDITION PRINCIPLE FOR INEQUALITIES

If $a < b$ is true, then $a + c < b + c$ is true.

THE MULTIPLICATION PRINCIPLE FOR INEQUALITIES

a) If $a < b$ and $c > 0$ are true, then $ac < bc$ is true.

b) If $a < b$ and $c < 0$ are true, then $ac > bc$ is true.

(When both sides of an inequality are multiplied by a negative number, the inequality symbol must be reversed.)

Similar statements hold for $a \leq b$.

EXAMPLES Solve.

1. $a + 9 \leq -50$

$a + 9 - 9 \leq -50 - 9$

$a \leq -59$

The solution set is $(-\infty, -59]$.

2. $-5x < 4$

$-\frac{1}{5}(-5x) > -\frac{1}{5}(4)$

$x > -\frac{4}{5}$

The solution set is $\left(-\frac{4}{5}, \infty\right)$.

3. $2y - 1 < 5$

$2y - 1 + 1 < 5 + 1$

$2y < 6$

$\frac{1}{2} \cdot 2y < \frac{1}{2} \cdot 6$

$y < 3$

The solution set is $(-\infty, 3)$.

4. $4 - 3x \geq 13$

$-4 + 4 - 3x \geq -4 + 13$

$-3x \geq 9$

$\frac{-3x}{-3} \leq \frac{9}{-3}$

$x \leq -3$

The solution set is $(-\infty, -3]$.

Do Exercises 1–6.

Solve.

1. $p + 25 \geq -100$

2. $-\frac{2}{3}x > 6$

3. $9x - 1 < 17$

4. $-x - 16 \geq 40$

5. $\frac{1}{3}y - 6 < 3$

6. $8 - 2w \leq -14$

Just in Time 19

THE PRINCIPLE OF ZERO PRODUCTS

The product of two numbers is 0 if one or both of the numbers is 0. Furthermore, *if any product is* 0, *then a factor must be* 0. For example:

If $7x = 0$, then we know that $x = 0$.

If $x(2x - 9) = 0$, then we know that $x = 0$ *or* $2x - 9 = 0$.

If $(x + 3)(x - 2) = 0$, then we know that $x + 3 = 0$ *or* $x - 2 = 0$.

> **THE PRINCIPLE OF ZERO PRODUCTS**
>
> If $ab = 0$ is true, then $a = 0$ or $b = 0$, and if $a = 0$ or $b = 0$, then $ab = 0$.

Some quadratic equations can be solved using the principle of zero products.

EXAMPLES Solve.

1. $(2q - 7)(q + 4) = 0$

$$2q - 7 = 0 \quad or \quad q + 4 = 0$$
$$2q = 7 \quad or \quad q = -4$$
$$q = \frac{7}{2} \quad or \quad q = -4$$

The solutions are -4 and $\frac{7}{2}$.

2. $5x^2 - 75x = 0$

$$5x(x - 15) = 0$$
$$5x = 0 \quad or \quad x - 15 = 0$$
$$x = 0 \quad or \quad x = 15$$

The solutions are 0 and 15.

3. $x^2 - 3x - 4 = 0$

$$(x + 1)(x - 4) = 0$$
$$x + 1 = 0 \quad or \quad x - 4 = 0$$
$$x = -1 \quad or \quad x = 4$$

The solutions are -1 and 4.

4. $2x^2 + 5x - 3 = 0$

$$(x + 3)(2x - 1) = 0$$
$$x + 3 = 0 \quad or \quad 2x - 1 = 0$$
$$x = -3 \quad or \quad 2x = 1$$
$$x = -3 \quad or \quad x = \frac{1}{2}$$

The solutions are -3 and $\frac{1}{2}$.

Do Exercises 1–8.

Solve.

1. $2y^2 + 42y = 0$

2. $(a + 7)(a - 1) = 0$

3. $(5y + 3)(y - 4) = 0$

4. $6x^2 + 7x - 5 = 0$

5. $t(t - 8) = 0$

6. $x^2 - 8x - 33 = 0$

7. $x^2 + 13x = 30$

8. $12x^2 - 7x - 12 = 0$

Just in Time 20 THE PRINCIPLE OF SQUARE ROOTS

The principle of square roots can be used to solve some quadratic equations.

THE PRINCIPLE OF SQUARE ROOTS

If $x^2 = k$, then $x = -\sqrt{k}$ *or* $x = \sqrt{k}$.

EXAMPLES Solve.

1. $s^2 - 144 = 0$

$s^2 = 144$

$s = -\sqrt{144}$ *or* $s = \sqrt{144}$

$s = -12$ *or* $s = 12$

The solutions are -12 and 12, or ± 12.

2. $3x^2 - 21 = 0$

$3x^2 = 21$

$x^2 = 7$

$x = -\sqrt{7}$ *or* $x = \sqrt{7}$

The solutions are $-\sqrt{7}$ and $\sqrt{7}$, or $\pm\sqrt{7}$.

Do Exercises 1–6.

Solve.

1. $x^2 - 36 = 0$

2. $2y^2 - 20 = 0$

3. $6z^2 = 18$

4. $3t^2 - 15 = 0$

5. $z^2 - 1 = 24$

6. $5x^2 - 75 = 0$

Just in Time 21 — SIMPLIFY RATIONAL EXPRESSIONS

A **rational expression** is the quotient of two polynomials. The **domain** of an algebraic expression is the set of all real numbers for which the expression is defined. Since division by 0 is not defined, any number that makes the denominator 0 is not in the domain of a rational expression.

EXAMPLE 1 Find the domain of

$$\frac{x^2 - 4}{x^2 - 4x - 5}.$$

We solve the equation $x^2 - 4x - 5 = 0$, or $(x + 1)(x - 5) = 0$, to find the numbers that are not in the domain. The solutions are -1 and 5. Since the denominator is 0 when $x = -1$ *or* $x = 5$, the domain is the set of all real numbers except -1 and 5.

EXAMPLE 2 Simplify: $\dfrac{9x^2 + 6x - 3}{12x^2 - 12}$.

$$\begin{aligned}\frac{9x^2 + 6x - 3}{12x^2 - 12} &= \frac{3(3x^2 + 2x - 1)}{12(x^2 - 1)}\\ &= \frac{3(x + 1)(3x - 1)}{3\cdot 4(x + 1)(x - 1)}\\ &= \frac{3(x + 1)}{3(x + 1)}\cdot\frac{3x - 1}{4(x - 1)}\\ &= 1\cdot\frac{3x - 1}{4(x - 1)}\\ &= \frac{3x - 1}{4(x - 1)}\end{aligned}$$

Canceling is a shortcut that is often used to remove a factor of 1.

EXAMPLE 3 Simplify: $\dfrac{2 - x}{x^2 + x - 6}$.

$$\begin{aligned}\frac{2 - x}{x^2 + x - 6} &= \frac{2 - x}{(x + 3)(x - 2)}\\ &= \frac{-1(x - 2)}{(x + 3)(x - 2)}\\ &= \frac{-1\cancel{(x - 2)}}{(x + 3)\cancel{(x - 2)}}\\ &= \frac{-1}{x + 3}, \text{ or } -\frac{1}{x + 3}\end{aligned}$$

Do Exercises 1–6.

Find the domain of the rational expression.

1. $\dfrac{3x - 3}{x(x - 1)}$

2. $\dfrac{y + 6}{y^2 + 4y - 21}$

Simplify.

3. $\dfrac{x^2 - 4}{x^2 - 4x + 4}$

4. $\dfrac{x^2 + 2x - 3}{x^2 - 9}$

5. $\dfrac{x^3 - 6x^2 + 9x}{x^3 - 3x^2}$

6. $\dfrac{6y^2 + 12y - 48}{3y^2 - 9y + 6}$

Just in Time 22

MULTIPLY AND DIVIDE RATIONAL EXPRESSIONS

To multiply rational expressions, we multiply numerators and multiply denominators and, if possible, simplify the result. To divide rational expressions, we multiply the dividend by the reciprocal of the divisor and, if possible, simplify the result; that is,

$$\frac{a}{b}\cdot\frac{c}{d}=\frac{ac}{bd} \quad \text{and} \quad \frac{a}{b}\div\frac{c}{d}=\frac{a}{b}\cdot\frac{d}{c}=\frac{ad}{bc}.$$

EXAMPLES Multiply or divide.

1. $$\frac{a^2-4}{16a}\cdot\frac{20a^2}{a+2}=\frac{(a^2-4)(20a^2)}{16a(a+2)}$$
$$=\frac{\cancel{(a+2)}(a-2)\cdot\cancel{4}\cdot 5\cdot a\cdot\cancel{a}}{\cancel{4}\cdot 4\cdot\cancel{a}\cdot\cancel{(a+2)}}$$
$$=\frac{5a(a-2)}{4}$$

2. $$\frac{x-2}{12}\div\frac{x^2-4x+4}{3x^3+15x^2}=\frac{x-2}{12}\cdot\frac{3x^3+15x^2}{x^2-4x+4}$$
$$=\frac{(x-2)(3x^3+15x^2)}{12(x^2-4x+4)}$$
$$=\frac{\cancel{(x-2)}(\cancel{3})(x^2)(x+5)}{\cancel{3}\cdot 4\,\cancel{(x-2)}(x-2)}$$
$$=\frac{x^2(x+5)}{4(x-2)}$$

Do Exercises 1–6.

Multiply or divide and, if possible, simplify.

1. $\dfrac{r-s}{r+s}\cdot\dfrac{r^2-s^2}{(r-s)^2}$

2. $\dfrac{m^2-n^2}{r+s}\div\dfrac{m-n}{r+s}$

3. $\dfrac{4x^2+9x+2}{x^2+x-2}\cdot\dfrac{x^2-1}{3x^2+x-2}$

4. $\dfrac{a^2-a-2}{a^2-a-6}\div\dfrac{a^2-2a}{2a+a^2}$

5. $\dfrac{3x+12}{2x-8}\div\dfrac{(x+4)^2}{(x-4)^2}$

6. $\dfrac{x^2-y^2}{x^3-y^3}\cdot\dfrac{x^2+xy+y^2}{x^2+2xy+y^2}$

Just in Time 23 ADD AND SUBTRACT RATIONAL EXPRESSIONS

When rational expressions have the same denominator, we can add or subtract by adding or subtracting the numerators and retaining the common denominator. If the denominators differ, we must find equivalent rational expressions that have a common denominator before we can add or subtract. In general, it is most efficient to find the **least common denominator (LCD)** of the expressions.

To find the least common denominator of rational expressions, factor each denominator and form the product that uses each factor the greatest number of times it occurs in any factorization.

EXAMPLE 1 Add.

$$\frac{x^2 - 4x + 4}{2x^2 - 3x + 1} + \frac{x + 4}{2x - 2}$$

$$= \frac{x^2 - 4x + 4}{(2x - 1)(x - 1)} + \frac{x + 4}{2(x - 1)}$$

The LCD is $(2x - 1)(x - 1)(2)$, or $2(2x - 1)(x - 1)$.

$$= \frac{x^2 - 4x + 4}{(2x - 1)(x - 1)} \cdot \frac{2}{2} + \frac{x + 4}{2(x - 1)} \cdot \frac{2x - 1}{2x - 1}$$

$$= \frac{2x^2 - 8x + 8}{(2x - 1)(x - 1)(2)} + \frac{2x^2 + 7x - 4}{2(x - 1)(2x - 1)}$$

$$= \frac{4x^2 - x + 4}{2(2x - 1)(x - 1)}$$

EXAMPLE 2 Subtract.

$$\frac{x}{x^2 + 11x + 30} - \frac{5}{x^2 + 9x + 20}$$

$$= \frac{x}{(x + 5)(x + 6)} - \frac{5}{(x + 5)(x + 4)}$$

The LCD is $(x + 5)(x + 6)(x + 4)$.

$$= \frac{x}{(x + 5)(x + 6)} \cdot \frac{x + 4}{x + 4} - \frac{5}{(x + 5)(x + 4)} \cdot \frac{x + 6}{x + 6}$$

$$= \frac{x^2 + 4x}{(x + 5)(x + 6)(x + 4)} - \frac{5x + 30}{(x + 5)(x + 4)(x + 6)}$$

$$= \frac{x^2 + 4x - (5x + 30)}{(x + 5)(x + 6)(x + 4)} = \frac{x^2 + 4x - 5x - 30}{(x + 5)(x + 6)(x + 4)}$$

$$= \frac{x^2 - x - 30}{(x + 5)(x + 6)(x + 4)} = \frac{\cancel{(x + 5)}(x - 6)}{\cancel{(x + 5)}(x + 6)(x + 4)}$$

$$= \frac{x - 6}{(x + 6)(x + 4)}$$

Do Exercises 1–6.

Add or subtract and, if possible, simplify.

1. $\dfrac{a - 3b}{a + b} + \dfrac{a + 5b}{a + b}$

2. $\dfrac{x^2 - 5}{3x^2 - 5x - 2} + \dfrac{x + 1}{3x - 6}$

3. $\dfrac{a^2 + 1}{a^2 - 1} - \dfrac{a - 1}{a + 1}$

4. $\dfrac{9x + 2}{3x^2 - 2x - 8} + \dfrac{7}{3x^2 + x - 4}$

5. $\dfrac{y}{y^2 - y - 20} - \dfrac{2}{y + 4}$

6. $\dfrac{3y}{y^2 - 7y + 10} - \dfrac{2y}{y^2 - 8y + 15}$

Just in Time 24

SIMPLIFY COMPLEX RATIONAL EXPRESSIONS

A **complex rational expression** has rational expressions in its numerator or its denominator or both.

EXAMPLE Simplify: $\dfrac{\frac{1}{a}+\frac{1}{b}}{\frac{1}{a^3}+\frac{1}{b^3}}$.

Method 1:

$$\frac{\frac{1}{a}+\frac{1}{b}}{\frac{1}{a^3}+\frac{1}{b^3}} = \frac{\frac{1}{a}+\frac{1}{b}}{\frac{1}{a^3}+\frac{1}{b^3}} \cdot \frac{a^3b^3}{a^3b^3}$$

The LCD of the four rational expressions in the numerator and the denominator is a^3b^3.

$$= \frac{\left(\frac{1}{a}+\frac{1}{b}\right)(a^3b^3)}{\left(\frac{1}{a^3}+\frac{1}{b^3}\right)(a^3b^3)} = \frac{a^2b^3 + a^3b^2}{b^3 + a^3}$$

$$= \frac{a^2b^2(b+a)}{(b+a)(b^2 - ba + a^2)} = \frac{a^2b^2}{b^2 - ba + a^2}$$

Method 2:

$$\frac{\frac{1}{a}+\frac{1}{b}}{\frac{1}{a^3}+\frac{1}{b^3}} = \frac{\frac{1}{a}\cdot\frac{b}{b}+\frac{1}{b}\cdot\frac{a}{a}}{\frac{1}{a^3}\cdot\frac{b^3}{b^3}+\frac{1}{b^3}\cdot\frac{a^3}{a^3}}$$

The LCD is ab. (numerator)

The LCD is a^3b^3. (denominator)

$$= \frac{\frac{b}{ab}+\frac{a}{ab}}{\frac{b^3}{a^3b^3}+\frac{a^3}{a^3b^3}}$$

$$= \frac{\frac{b+a}{ab}}{\frac{b^3+a^3}{a^3b^3}} = \frac{b+a}{ab}\cdot\frac{a^3b^3}{b^3+a^3}$$

$$= \frac{(b+a)(ab)(a^2b^2)}{(ab)(b+a)(b^2 - ba + a^2)}$$

$$= \frac{a^2b^2}{b^2 - ba + a^2}$$

Do Exercises 1–5.

Simplify.

1. $\dfrac{\frac{x}{y}-\frac{y}{x}}{\frac{1}{y}+\frac{1}{x}}$

2. $\dfrac{\frac{a-b}{b}}{\frac{a^2-b^2}{ab}}$

3. $\dfrac{w+\frac{8}{w^2}}{1+\frac{2}{w}}$

4. $\dfrac{\frac{x^2-y^2}{xy}}{\frac{x-y}{y}}$

5. $\dfrac{\frac{a}{b}-\frac{b}{a}}{\frac{1}{a}-\frac{1}{b}}$

Note: $b - a = -1(a - b)$.

Just in Time 25

SIMPLIFY RADICAL EXPRESSIONS

The symbol $\sqrt{a}$ denotes the nonnegative square root of a, and the symbol $\sqrt[3]{a}$ denotes the real-number cube root of a. The symbol $\sqrt[n]{a}$ denotes the nth root of a; that is, a number whose nth power is a. The symbol $\sqrt[n]{\ }$ is called a **radical**, and the expression under the radical is called the **radicand**. The number n (which is omitted when it is 2) is called the **index**.

Any positive number has two square roots, one positive and one negative. Similarly, for any even index, a positive number has two real-number roots. The positive root is called the **principal root**. Any real number has only one real-number odd root.

EXAMPLES Simplify.

1. $\sqrt{36} = 6$ because $6 \cdot 6 = 36$.
2. $-\sqrt{36} = -6$
3. $\sqrt[3]{-8} = -2$
4. $\sqrt[5]{\dfrac{32}{243}} = \dfrac{2}{3}$
5. $\sqrt[4]{-16}$ is not a real number.

PROPERTIES OF RADICALS

Let a and b be any real numbers or expressions for which the given roots exist. For any natural numbers m and n $(n \neq 1)$:

1. If n is even, $\sqrt[n]{a^n} = |a|$.
2. If n is odd, $\sqrt[n]{a^n} = a$.
3. $\sqrt[n]{a} \cdot \sqrt[n]{b} = \sqrt[n]{ab}$.
4. $\sqrt[n]{\dfrac{a}{b}} = \dfrac{\sqrt[n]{a}}{\sqrt[n]{b}}$ $(b \neq 0)$.
5. $\sqrt[n]{a^m} = (\sqrt[n]{a})^m$.

Here, we assume that no radicands are formed by raising negative quantities to even powers and, consequently, we will not use absolute-value notation when we simplify radical expressions involving variables.

EXAMPLES Simplify.

6. $\sqrt{(-5)^2} = |-5| = 5$
7. $\sqrt[3]{(-5)^3} = -5$
8. $\sqrt[4]{4} \cdot \sqrt[4]{5} = \sqrt[4]{4 \cdot 5} = \sqrt[4]{20}$
9. $\sqrt[3]{8^5} = (\sqrt[3]{8})^5 = 2^5 = 32$
10. $\sqrt{50} = \sqrt{25 \cdot 2} = \sqrt{25} \cdot \sqrt{2} = 5\sqrt{2}$
11. $\dfrac{\sqrt{72}}{\sqrt{6}} = \sqrt{\dfrac{72}{6}} = \sqrt{12} = \sqrt{4 \cdot 3} = \sqrt{4} \cdot \sqrt{3} = 2\sqrt{3}$
12. $\sqrt{216x^5y^3} = \sqrt{36 \cdot 6 \cdot x^4 \cdot x \cdot y^2 \cdot y} = \sqrt{36x^4y^2}\sqrt{6xy}$
 $= 6x^2y\sqrt{6xy}$
13. $8\sqrt{50} - 3\sqrt{8} = 8\sqrt{25 \cdot 2} - 3\sqrt{4 \cdot 2} = 8 \cdot 5\sqrt{2} - 3 \cdot 2\sqrt{2}$
 $= 40\sqrt{2} - 6\sqrt{2} = (40 - 6)\sqrt{2} = 34\sqrt{2}$
14. $(5 - \sqrt{2})(4 + 3\sqrt{2}) = 20 + 15\sqrt{2} - 4\sqrt{2} - 3(\sqrt{2})^2$
 $= 20 + 11\sqrt{2} - 6 = 14 + 11\sqrt{2}$

Do Exercises 1–20.

Simplify. Assume that no radicands were formed by raising negative quantities to even powers.

1. $\sqrt{(-21)^2}$
2. $\sqrt{9y^2}$
3. $\sqrt{(a-2)^2}$
4. $\sqrt[3]{-27x^3}$
5. $\sqrt[4]{81x^8}$
6. $\sqrt[5]{32}$
7. $\sqrt[4]{48x^6y^4}$
8. $\sqrt{15}\sqrt{35}$
9. $\dfrac{\sqrt{40xy}}{\sqrt{8x}}$
10. $\dfrac{\sqrt[3]{3x^2}}{\sqrt[3]{24x^5}}$
11. $\sqrt{x^2 - 4x + 4}$
12. $\sqrt{2x^3y}\sqrt{12xy}$
13. $\sqrt[3]{3x^2y}\sqrt[3]{36x}$
14. $5\sqrt{2} + 3\sqrt{32}$
15. $7\sqrt{12} - 2\sqrt{3}$
16. $2\sqrt{32} + 3\sqrt{8} - 4\sqrt{18}$
17. $6\sqrt{20} - 4\sqrt{45} + \sqrt{80}$
18. $(2 + \sqrt{3})(5 + 2\sqrt{3})$
19. $(\sqrt{8} + 2\sqrt{5})(\sqrt{8} - 2\sqrt{5})$
20. $(1 + \sqrt{3})^2$

Just in Time 26

RATIONALIZE DENOMINATORS

There are times when we need to remove the radicals in a denominator. This procedure is called **rationalizing the denominator**. It is done by multiplying by 1 in such a way as to obtain a perfect *n*th power in the denominator.

EXAMPLES Rationalize the denominator.

1. $\sqrt{\frac{3}{2}} = \sqrt{\frac{3}{2} \cdot \frac{2}{2}} = \sqrt{\frac{6}{4}} = \frac{\sqrt{6}}{\sqrt{4}} = \frac{\sqrt{6}}{2}$

2. $\frac{2}{\sqrt{3}} = \frac{2}{\sqrt{3}} \cdot \frac{\sqrt{3}}{\sqrt{3}} = \frac{2\sqrt{3}}{3}$

3. $\frac{\sqrt[3]{7}}{\sqrt[3]{9}} = \frac{\sqrt[3]{7}}{\sqrt[3]{9}} \cdot \frac{\sqrt[3]{3}}{\sqrt[3]{3}} = \frac{\sqrt[3]{21}}{\sqrt[3]{27}} = \frac{\sqrt[3]{21}}{3}$

Pairs of expressions of the form $a\sqrt{b} + c\sqrt{d}$ and $a\sqrt{b} - c\sqrt{d}$ are called **conjugates**. The product of such a pair contains no radicals and can be used to rationalize a denominator or a numerator.

EXAMPLE 4 Rationalize the denominator: $\frac{7}{3 + \sqrt{5}}$.

$$\begin{aligned}\frac{7}{3 + \sqrt{5}} &= \frac{7}{3 + \sqrt{5}} \cdot \frac{3 - \sqrt{5}}{3 - \sqrt{5}} \\ &= \frac{21 - 7\sqrt{5}}{3^2 - 3\sqrt{5} + 3\sqrt{5} - (\sqrt{5})^2} \\ &= \frac{21 - 7\sqrt{5}}{9 - 5} = \frac{21 - 7\sqrt{5}}{4}\end{aligned}$$

EXAMPLE 5 Rationalize the denominator: $\frac{1 + \sqrt{2}}{\sqrt{5} + \sqrt{10}}$.

$$\begin{aligned}\frac{1 + \sqrt{2}}{\sqrt{5} + \sqrt{10}} &= \frac{1 + \sqrt{2}}{\sqrt{5} + \sqrt{10}} \cdot \frac{\sqrt{5} - \sqrt{10}}{\sqrt{5} - \sqrt{10}} \\ &= \frac{\sqrt{5} - \sqrt{10} + \sqrt{10} - \sqrt{20}}{(\sqrt{5})^2 - (\sqrt{10})^2} \\ &= \frac{\sqrt{5} - \sqrt{20}}{5 - 10} \\ &= \frac{\sqrt{5} - 2\sqrt{5}}{-5} \\ &= \frac{-\sqrt{5}}{-5} \\ &= \frac{\sqrt{5}}{5}\end{aligned}$$

Do Exercises 1–8.

Rationalize the denominator.

1. $\frac{4}{\sqrt{11}}$

2. $\sqrt{\frac{3}{7}}$

3. $\frac{\sqrt[3]{7}}{\sqrt[3]{2}}$

4. $\sqrt[3]{\frac{16}{9}}$

5. $\frac{3}{\sqrt{30} - 4}$

6. $\frac{4}{\sqrt{7} - \sqrt{3}}$

7. $\frac{6}{\sqrt{m} - \sqrt{n}}$

8. $\frac{1 - \sqrt{2}}{\sqrt{3} - \sqrt{6}}$

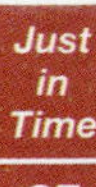

Just in Time 27

RATIONAL EXPONENTS

For any real number a and any natural numbers m and n, $n \neq 1$, for which $\sqrt[n]{a}$ exists:

$$a^{1/n} = \sqrt[n]{a}, \quad a^{m/n} = (\sqrt[n]{a})^m = \sqrt[n]{a^m}, \quad \text{and} \quad a^{-m/n} = \frac{1}{a^{m/n}}.$$

EXAMPLES Convert to radical notation and, if possible, simplify.

1. $m^{1/6} = \sqrt[6]{m}$
2. $7^{3/4} = \sqrt[4]{7^3}$, or $(\sqrt[4]{7})^3$
3. $8^{-5/3} = \dfrac{1}{8^{5/3}} = \dfrac{1}{(\sqrt[3]{8})^5} = \dfrac{1}{2^5} = \dfrac{1}{32}$

EXAMPLES Convert to exponential notation.

4. $(\sqrt[4]{7xy})^5 = (7xy)^{5/4}$
5. $\sqrt[6]{x^3} = x^{3/6} = x^{1/2}$

EXAMPLES Simplify and then, if appropriate, write radical notation.

6. $x^{5/6} \cdot x^{2/3} = x^{5/6+2/3} = x^{9/6} = x^{3/2} = \sqrt{x^3}$
   $= \sqrt{x^2}\sqrt{x} = x\sqrt{x}$
7. $(x+3)^{5/2}(x+3)^{-1/2} = (x+3)^{5/2-1/2} = (x+3)^2$

Do Exercises 1–11.

Convert to radical notation and, if possible, simplify.

1. $y^{5/6}$
2. $x^{2/3}$
3. $16^{3/4}$
4. $4^{7/2}$
5. $125^{-1/3}$
6. $32^{-4/5}$

Convert to exponential notation.

7. $\sqrt[12]{y^4}$
8. $\sqrt{x^5}$

Simplify and then, if appropriate, write radical notation.

9. $x^{1/2} \cdot x^{2/3}$
10. $(a-2)^{9/4}(a-2)^{-1/4}$
11. $(m^{1/2}n^{5/2})^{2/3}$

Just in Time 28

THE PYTHAGOREAN THEOREM

A **right triangle** is a triangle with a 90° angle, as shown in the following figure. The small square in the corner indicates the 90° angle.

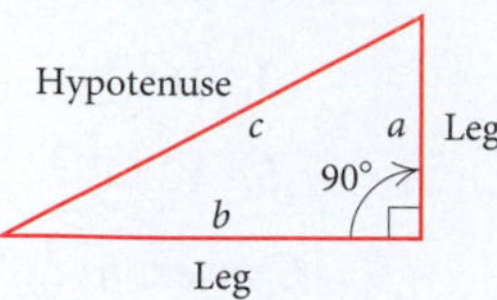

In a right triangle, the longest side is called the **hypotenuse**. It is also the side opposite the right angle. The other two sides are called **legs**. We generally use the letters a and b for the lengths of the legs and c for the length of the hypotenuse. They are related as follows.

THE PYTHAGOREAN THEOREM

In any right triangle, if a and b are the lengths of the legs and c is the length of the hypotenuse, then

$$a^2 + b^2 = c^2.$$

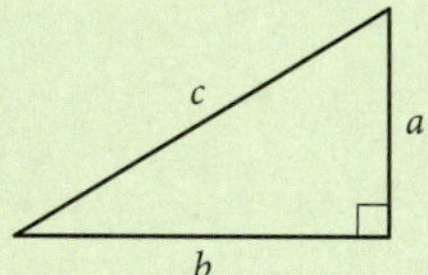

The equation $a^2 + b^2 = c^2$ is called the **Pythagorean equation**.

EXAMPLE 1 Find the length of the hypotenuse of this right triangle. Give an exact answer and an approximation to three decimal places.

$$4^2 + 5^2 = c^2$$
$$16 + 25 = c^2$$
$$41 = c^2$$
$$c = \sqrt{41}$$
$$c \approx 6.403$$

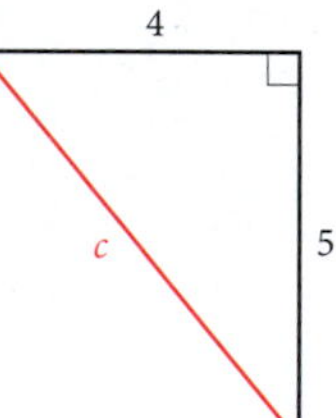

EXAMPLE 2 Find the length of leg b of this right triangle. Give an exact answer and an approximation to three decimal places.

$$10^2 + b^2 = 12^2$$
$$100 + b^2 = 144$$
$$b^2 = 144 - 100$$
$$b^2 = 44$$
$$b = \sqrt{44}$$
$$b \approx 6.633$$

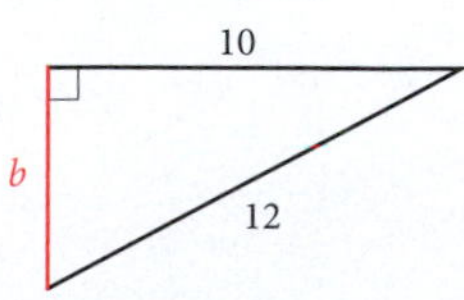

Do Exercises 1–5.

Find the length of the third side of each right triangle. Where appropriate, give both an exact answer and an approximation to three decimal places.

1.

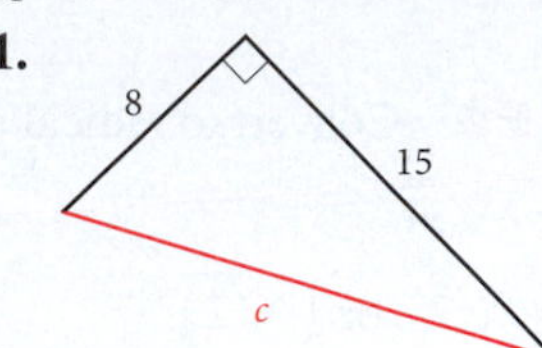

2.

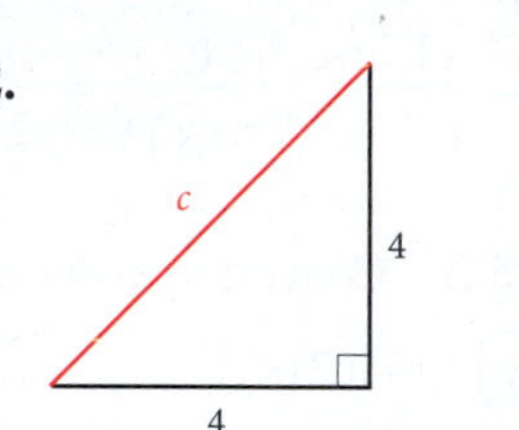

3.

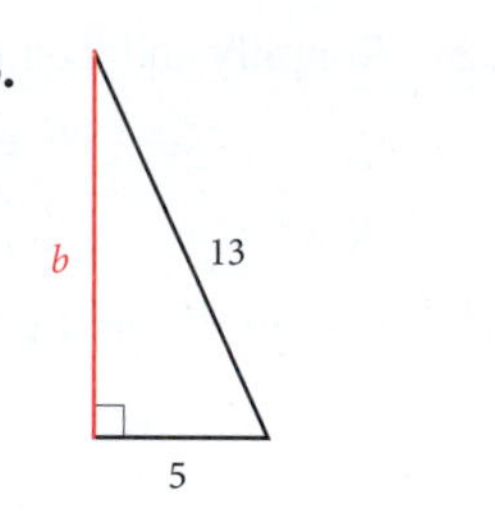

4.

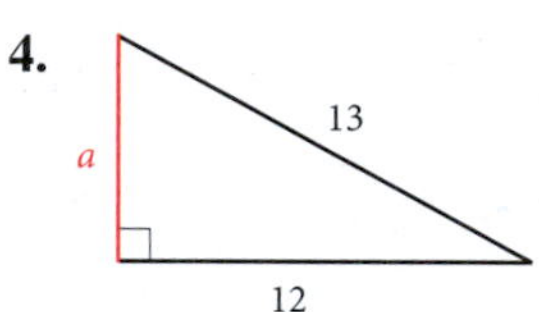

5.

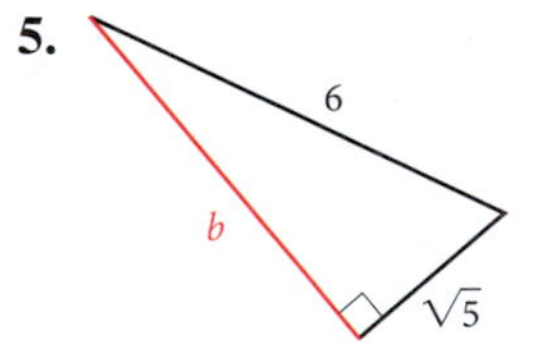

Graphs, Functions, and Models

CHAPTER

1

APPLICATION

This problem appears as Exercise 60 in the Review Exercises.

The number of female medical school graduates has increased each year from 2006 to 2014. Model the data shown on p. 88 with a linear function where the number of female medical school graduates W is a function of the year x and where x is the number of years after 2006. Then, using this function, estimate the number of female graduates in 2013.

1.1 Introduction to Graphing

- Plot points.
- Determine whether an ordered pair is a solution of an equation.
- Find the x- and y-intercepts of an equation of the form $Ax + By = C$.
- Graph equations.
- Find the distance between two points in the plane, and find the midpoint of a segment.
- Find an equation of a circle with a given center and radius, and given an equation of a circle in standard form, find the center and the radius.
- Graph equations of circles.

Graphs

Graphs provide a means of displaying, interpreting, and analyzing data in a visual format. It is not uncommon to open a newspaper or a magazine and encounter graphs. Examples of bar, line, and circle graphs are shown below.

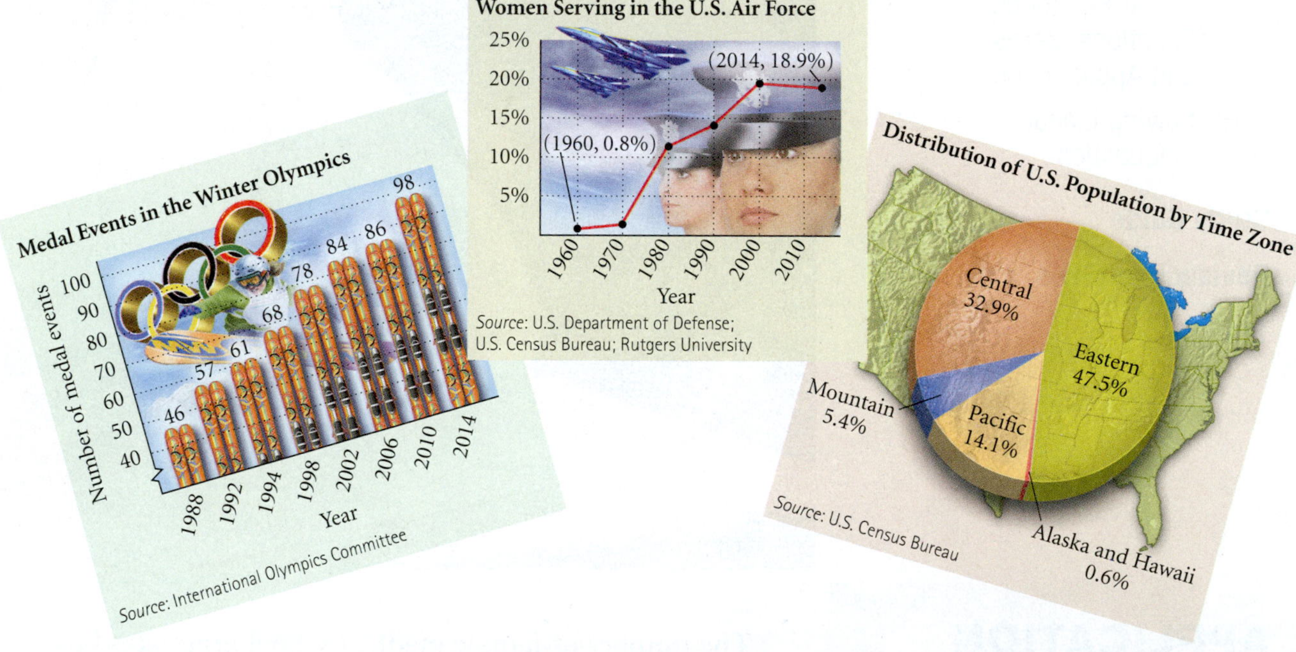

Many real-world situations can be modeled, or described mathematically, using equations in which two variables appear. We use a plane to graph a pair of numbers. To locate points on a plane, we use two perpendicular number lines, called **axes**, that intersect at $(0, 0)$. We call this point the **origin**. The horizontal axis is called the ***x*-axis**, and the vertical axis is called the ***y*-axis**. (Other variables, such as a and b, can also be used.) The axes divide the plane into four regions, called **quadrants**, denoted by Roman numerals and numbered counterclockwise from the upper right. Arrows show the positive direction of each axis.

Each point (x, y) in the plane is described by an **ordered pair**. The first number, x, indicates the point's horizontal location with respect to the y-axis, and the

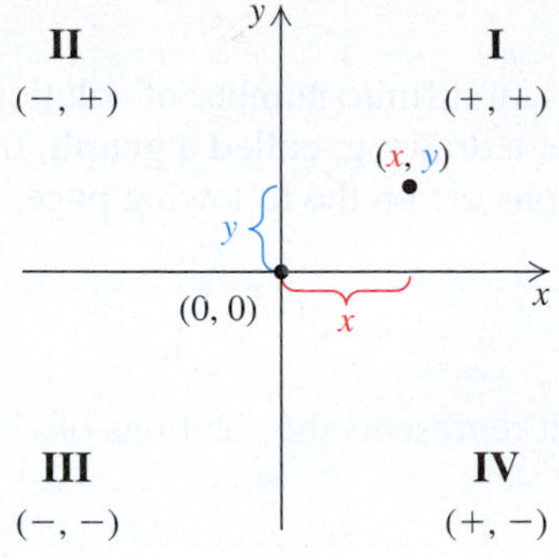

second number, y, indicates the point's vertical location with respect to the x-axis. We call x the **first coordinate,** the **x-coordinate**, or the **abscissa**. We call y the **second coordinate,** the **y-coordinate**, or the **ordinate**. Such a representation is called the **Cartesian coordinate system** in honor of the French mathematician and philosopher René Descartes (1596–1650).

In the first quadrant, both coordinates of a point are positive. In the second quadrant, the first coordinate is negative and the second is positive. In the third quadrant, both coordinates are negative, and in the fourth quadrant, the first coordinate is positive and the second is negative.

EXAMPLE 1 Graph and label the points $(-3, 5)$, $(4, 3)$, $(3, 4)$, $(-4, -2)$, $(3, -4)$, $(0, 4)$, $(-3, 0)$, and $(0, 0)$.

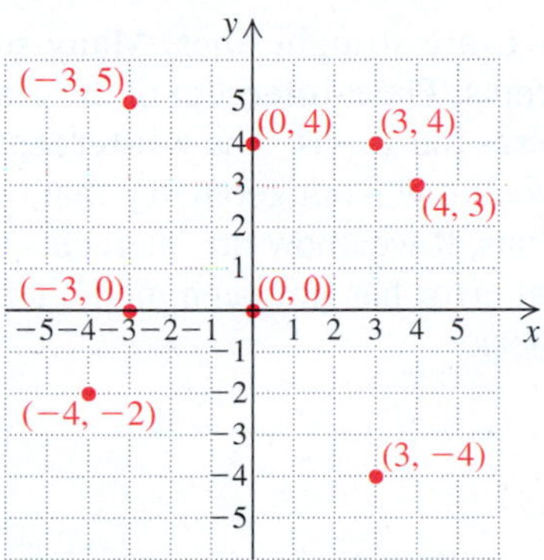

Solution To graph or **plot** $(-3, 5)$, we note that the x-coordinate, -3, tells us to move from the origin 3 units horizontally in the negative direction, or 3 units to the left of the y-axis. Then we move 5 units up from the x-axis.* To graph the other points, we proceed in a similar manner. (See the graph at left.) Note that the point $(4, 3)$ is different from the point $(3, 4)$.

Now Try Exercise 3.

Solutions of Equations

Equations in two variables, like $2x + 3y = 18$, have solutions (x, y) that are ordered pairs such that when the first coordinate is substituted for x and the second coordinate is substituted for y, the result is a true equation. The first coordinate in an ordered pair generally represents the variable that occurs first alphabetically.

EXAMPLE 2 Determine whether each ordered pair is a solution of the equation $2x + 3y = 18$.

a) $(-5, 7)$ **b)** $(3, 4)$

Solution We substitute the ordered pair into the equation and determine whether the resulting equation is true.

a)
$$\begin{array}{r|l} \multicolumn{2}{c}{2x + 3y = 18} \\ \hline 2(-5) + 3(7) & ?\ 18 \\ -10 + 21 & \\ 11 & 18 \end{array}$$
FALSE

We substitute −5 for x and 7 for y (alphabetical order).

The equation $11 = 18$ is false, so $(-5, 7)$ is not a solution.

b)
$$\begin{array}{r|l} \multicolumn{2}{c}{2x + 3y = 18} \\ \hline 2(3) + 3(4) & ?\ 18 \\ 6 + 12 & \\ 18 & 18 \end{array}$$
TRUE

We substitute 3 for x and 4 for y.

The equation $18 = 18$ is true, so $(3, 4)$ is a solution.

```
2(−5)+3*7
                11
2*3+3*4
                18
```

We can also perform these substitutions on a graphing calculator. When we substitute -5 for x and 7 for y, we get 11. Since $11 \neq 18$, $(-5, 7)$ is not a solution of the equation. When we substitute 3 for x and 4 for y, we get 18, so $(3, 4)$ is a solution.

Now Try Exercise 11.

*Here the notation $(-3, 5)$ represents an ordered pair. This notation can also represent an open interval. See Just-in-Time 6 review on p. J-6. The context in which the notation appears usually makes the meaning clear.

Graphs of Equations

The equation considered in Example 2 actually has an infinite number of solutions. Since we cannot list all the solutions, we will make a drawing, called a **graph**, that represents them. Some suggestions for drawing graphs are on the following page.

> **TO GRAPH AN EQUATION**
>
> To **graph an equation** is to make a drawing that represents the solutions of that equation.

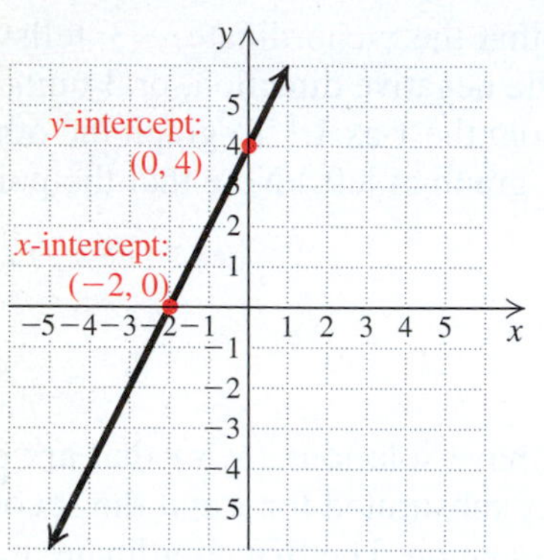

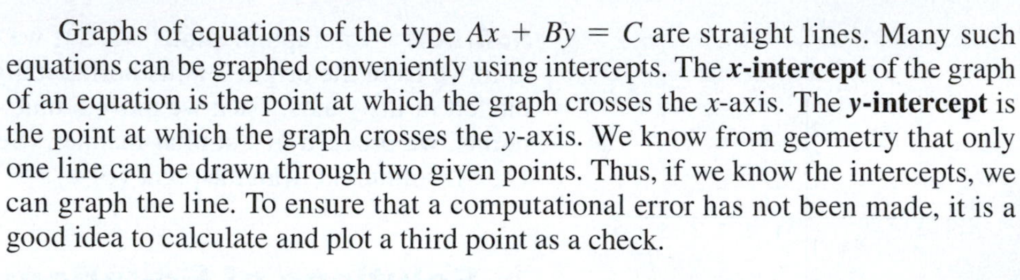

Graphs of equations of the type $Ax + By = C$ are straight lines. Many such equations can be graphed conveniently using intercepts. The ***x*-intercept** of the graph of an equation is the point at which the graph crosses the x-axis. The ***y*-intercept** is the point at which the graph crosses the y-axis. We know from geometry that only one line can be drawn through two given points. Thus, if we know the intercepts, we can graph the line. To ensure that a computational error has not been made, it is a good idea to calculate and plot a third point as a check.

> ***x*-INTERCEPT AND *y*-INTERCEPT**
>
> An ***x*-intercept** is a point $(a, 0)$. To find a, let $y = 0$ and solve for x.
>
> A ***y*-intercept** is a point $(0, b)$. To find b, let $x = 0$ and solve for y.

EXAMPLE 3 Graph: $2x + 3y = 18$.

Solution The graph is a line. To find ordered pairs that are solutions of this equation, we can replace either x or y with any number and then solve for the other variable. In this case, it is convenient to find the intercepts of the graph. For instance, if x is replaced with 0, then

$$\begin{aligned} 2 \cdot 0 + 3y &= 18 \\ 3y &= 18 \\ y &= 6. \quad \text{Dividing by 3 on both sides} \end{aligned}$$

Thus, $(0, 6)$ is a solution. It is the *y-intercept* of the graph. If y is replaced with 0, then

$$\begin{aligned} 2x + 3 \cdot 0 &= 18 \\ 2x &= 18 \\ x &= 9. \quad \text{Dividing by 2 on both sides} \end{aligned}$$

Thus, $(9, 0)$ is a solution. It is the *x-intercept* of the graph. We find a third solution as a check. If x is replaced with 3, then

$$\begin{aligned} 2 \cdot 3 + 3y &= 18 \\ 6 + 3y &= 18 \\ 3y &= 12 \quad \text{Subtracting 6 on both sides} \\ y &= 4. \quad \text{Dividing by 3 on both sides} \end{aligned}$$

Thus, $(3, 4)$ is a solution.

> **STUDY TIPS**
>
> Success can be planned. Combine goals and good study habits to create a plan for success that works for you. A list of study tips that your authors consider most helpful are included in the Guide to Success in the front of the text before Chapter 1.

> **SUGGESTIONS FOR DRAWING GRAPHS**
>
> 1. Calculate solutions and list the ordered pairs in a table.
> 2. Use graph paper.
> 3. Draw axes and label them with the variables.
> 4. Use arrows on the axes to indicate positive directions.
> 5. Scale the axes; that is, label the tick marks on the axes. Consider the ordered pairs found in part (1) above when choosing the scale.
> 6. Plot the ordered pairs, look for patterns, and complete the graph. Label the graph with the equation being graphed.

We list the solutions in a table and then plot the points. Note that the points appear to lie on a straight line.

x	y	(x, y)
0	6	(0, 6)
9	0	(9, 0)
3	4	(3, 4)

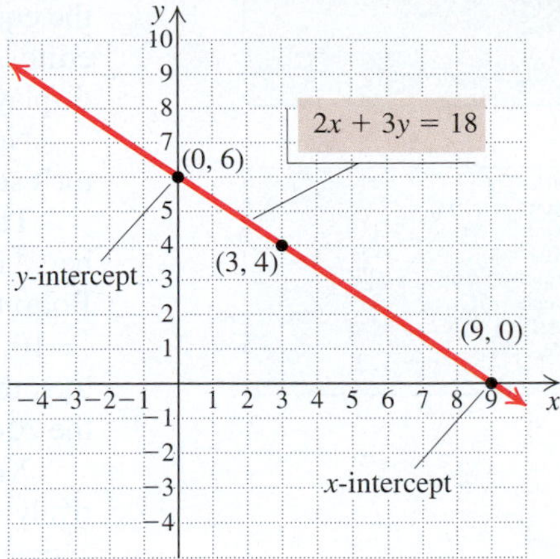

Were we to graph additional solutions of $2x + 3y = 18$, they would lie on the same straight line. Thus, to complete the graph, we use a straightedge to draw a line, as shown in the figure. This line represents all solutions of the equation. Every point on the line represents a solution; every solution is represented by a point on the line.

Now Try Exercise 17.

When graphing some equations, it is convenient to first solve for y and then find ordered pairs. We can use the addition and multiplication principles to solve for y.

EXAMPLE 4 Graph: $3x - 5y = -10$.

Solution We first solve for y:

Just in Time **17**

$$3x - 5y = -10$$
$$-5y = -3x - 10 \quad \text{Subtracting } 3x \text{ on both sides}$$
$$y = \tfrac{3}{5}x + 2. \quad \text{Multiplying by } -\tfrac{1}{5} \text{ on both sides}$$

By choosing multiples of 5 for x, we can avoid adding and subtracting fraction values when calculating y. For example, if we choose -5 for x, we get

$$y = \tfrac{3}{5}x + 2 = \tfrac{3}{5}(-5) + 2 = -3 + 2 = -1.$$

The following table lists a few points. We plot the points and draw the graph.

x	y	(x, y)
−5	−1	(−5, −1)
0	2	(0, 2)
5	5	(5, 5)

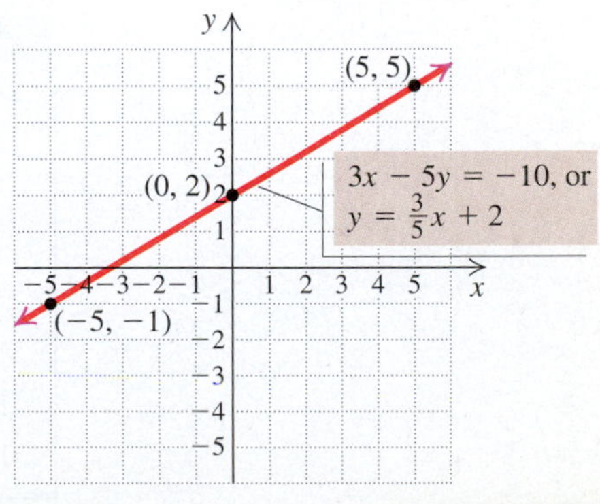

Now Try Exercise 29.

In the equation $y = \frac{3}{5}x + 2$ in Example 4, the value of y *depends* on the value chosen for x, so x is said to be the **independent variable** and y the **dependent variable**.

Plot1 Plot2 Plot3
\Y1 = (3/5)X+2
\Y2=
\Y3=
\Y4=
\Y5=
\Y6=
\Y7=

We can graph an equation on a graphing calculator. Many calculators require an equation to be entered in the form $y =$. In such a case, if the equation is not initially given in this form, it must be solved for y before it is entered in the calculator. For the equation $3x - 5y = -10$ in Example 4, we enter $y = \frac{3}{5}x + 2$ on the equation-editor, or $y =$, screen in the form $y = (3/5)x + 2$, which some calculators will display as shown in the window at left.

WINDOW
Xmin = −10
Xmax = 10
Xscl = 1
Ymin = −10
Ymax = 10
Yscl = 1
Xres = 1

Next, we determine the portion of the xy-plane that will appear on the calculator's screen. That portion of the plane is called the **viewing window**.

The notation used in this text to denote a window setting consists of four numbers [L, R, B, T], which represent the **L**eft and **R**ight endpoints of the x-axis and the **B**ottom and **T**op endpoints of the y-axis, respectively. The window with the settings $[-10, 10, -10, 10]$ is the **standard viewing window**. On some graphing calculators, the standard window can be selected quickly using the ZSTANDARD feature from the ZOOM menu.

Xmin and Xmax are used to set the left and right endpoints of the x-axis, respectively; Ymin and Ymax are used to set the bottom and top endpoints of the y-axis, respectively. The settings Xscl and Yscl give the scales for the axes. For example, Xscl = 1 and Yscl = 1 means that there is 1 unit between tick marks on each of the axes. In this text, scaling factors other than 1 will be listed by the window unless they are readily apparent.

After entering the equation $y = (3/5)x + 2$ and choosing a viewing window, we can then draw the graph.

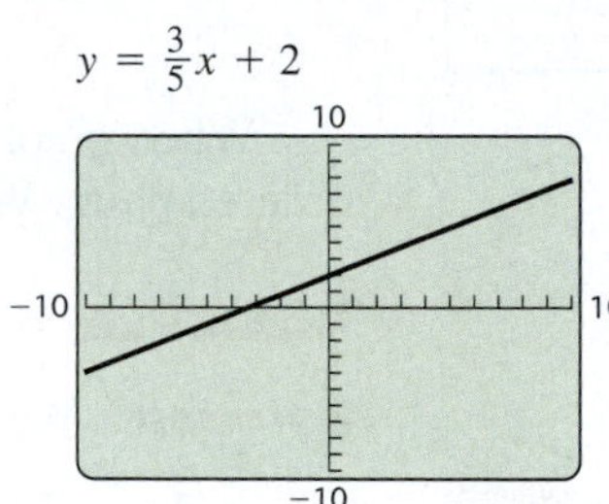

EXAMPLE 5 Graph: $y = x^2 - 9x - 12$.

Just in Time 7

Solution Note that since this equation is not of the form $Ax + By = C$, its graph is not a straight line. We make a table of values, plot enough points to obtain an idea of the shape of the curve, and connect the points with a smooth curve. It is important to scale the axes to include most of the ordered pairs listed in the table. Here it is appropriate to use a larger scale on the y-axis than on the x-axis.

x	y	(x, y)
−3	24	(−3, 24)
−1	−2	(−1, −2)
0	−12	(0, −12)
2	−26	(2, −26)
4	−32	(4, −32)
5	−32	(5, −32)
10	−2	(10, −2)
12	24	(12, 24)

① **Select values for x.**
② **Compute values for y.**

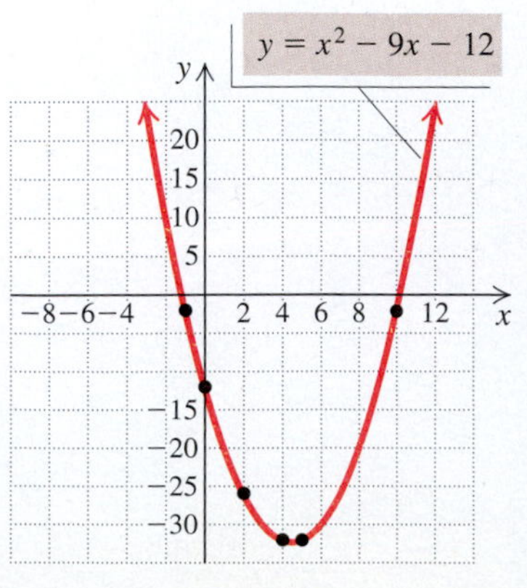

Now Try Exercise 39.

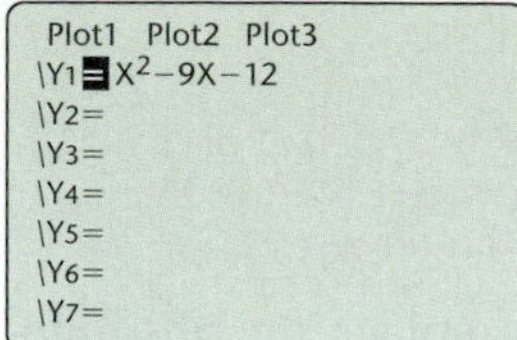

FIGURE 1.

TABLE SETUP
TblStart = −3
ΔTbl = 1
Indpnt: Auto Ask
Depend: Auto Ask

FIGURE 2.

A graphing calculator can be used to create a table of ordered pairs that are solutions of an equation. For the equation in Example 5, $y = x^2 - 9x - 12$, we first enter the equation on the equation-editor screen. (See Fig. 1.) Then we set up a table in AUTO mode by designating a value for TBLSTART and a value for ΔTBL. The calculator will produce a table beginning with the value of TBLSTART and continuing by adding ΔTBL to supply succeeding x-values. For the equation $y = x^2 - 9x - 12$, we let TBLSTART $= -3$ and ΔTBL $= 1$. (See Fig. 2.)

We can scroll up and down in the table to find values other than those shown in Fig. 3. We can also graph this equation on the graphing calculator, as shown in Fig. 4.

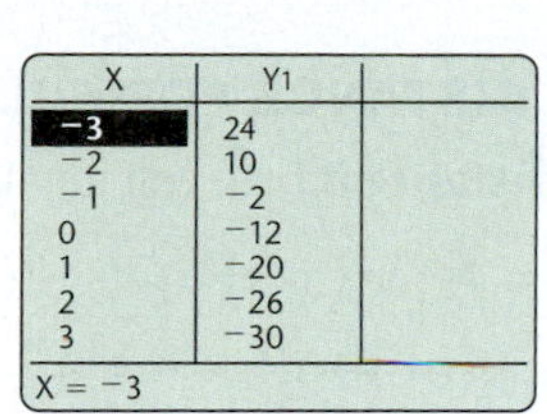

FIGURE 3.

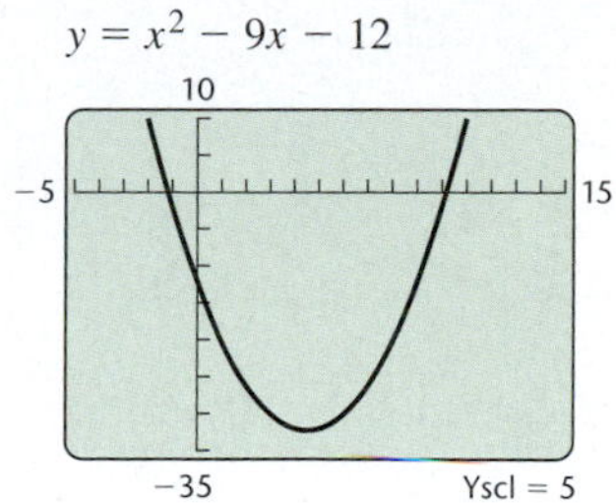

FIGURE 4.

The Distance Formula

Suppose that a photographer must determine the distance between two points, A and B, on opposite sides of a lane of the Panama Canal. One way in which he or she might proceed is to measure two legs of a right triangle that is situated as shown below. The Pythagorean equation, $c^2 = a^2 + b^2$, where c is the length of the hypotenuse and a and b are the lengths of the legs, can then be used to find the length of the hypotenuse, which is the distance from A to B.

The \$5.25 billion expansion of the Panama Canal doubled its capacity. A third canal lane is scheduled to open in 2016. (Source: *Panama Canal Authority*)

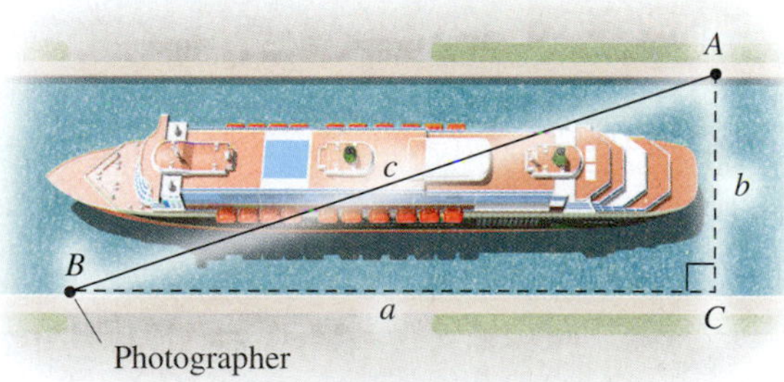

A similar strategy is used to find the distance between two points in a plane. For two points (x_1, y_1) and (x_2, y_2), we can draw a right triangle in which the legs have lengths $|x_2 - x_1|$ and $|y_2 - y_1|$.

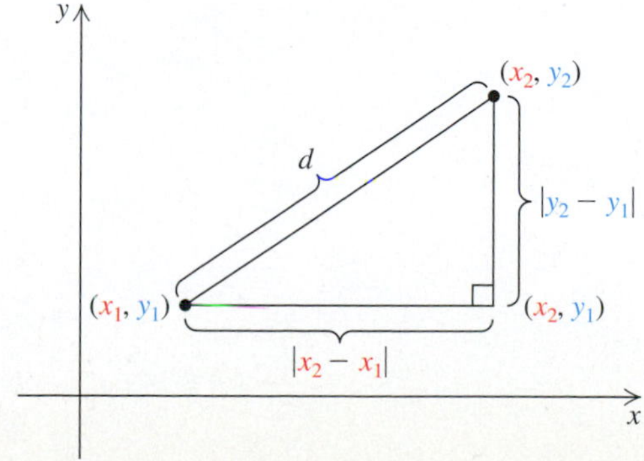

Just in Time 4, 28

Using the Pythagorean equation, $c^2 = a^2 + b^2$, we have

$$d^2 = |x_2 - x_1|^2 + |y_2 - y_1|^2.$$

Substituting d for c, $|x_2 - x_1|$ for a, and $|y_2 - y_1|$ for b in the Pythagorean equation

Because we are squaring, we can use parentheses to replace the absolute-value symbols:

$$d^2 = (x_2 - x_1)^2 + (y_2 - y_1)^2.$$

Taking the principal square root, we obtain the distance formula.

THE DISTANCE FORMULA

The **distance d** between any two points (x_1, y_1) and (x_2, y_2) is given by

$$d = \sqrt{(x_2 - x_1)^2 + (y_2 - y_1)^2}.$$

The subtraction of the x-coordinates can be done in any order, as can the subtraction of the y-coordinates. Although we derived the distance formula by considering two points not on a horizontal line or a vertical line, the distance formula holds for *any* two points.

Just in Time 5, 25

EXAMPLE 6 Find the distance between each pair of points.

a) $(-2, 2)$ and $(3, -6)$ **b)** $(-1, -5)$ and $(-1, 2)$

Solution We substitute into the distance formula.

a) $d = \sqrt{(x_2 - x_1)^2 + (y_2 - y_1)^2}$

$= \sqrt{[3 - (-2)]^2 + (-6 - 2)^2}$

$= \sqrt{5^2 + (-8)^2} = \sqrt{25 + 64}$

$= \sqrt{89} \approx 9.4$

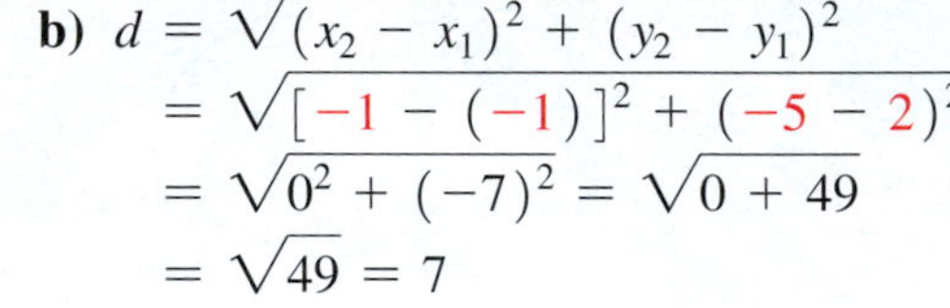

b) $d = \sqrt{(x_2 - x_1)^2 + (y_2 - y_1)^2}$

$= \sqrt{[-1 - (-1)]^2 + (-5 - 2)^2}$

$= \sqrt{0^2 + (-7)^2} = \sqrt{0 + 49}$

$= \sqrt{49} = 7$

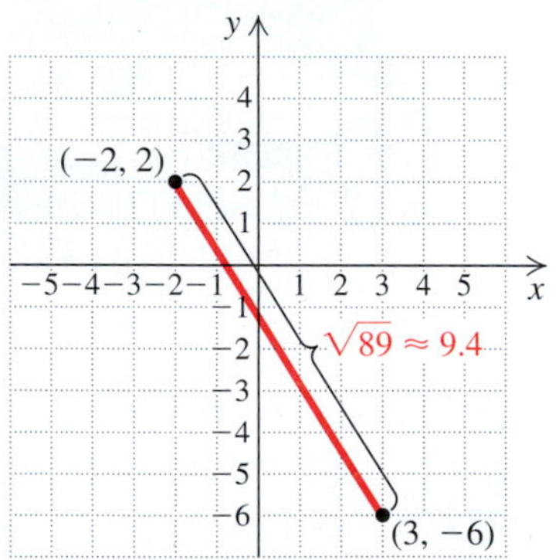

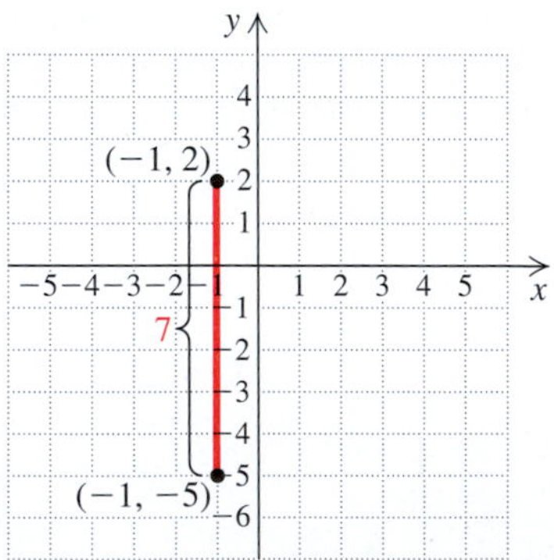

Now Try Exercises 63 and 71.

EXAMPLE 7 The point $(-2, 5)$ is on a circle that has $(3, -1)$ as its center. Find the length of the radius of the circle.

Solution Since the length of the radius is the distance from the center to a point on the circle, we substitute into the distance formula:

$$d = \sqrt{(x_2 - x_1)^2 + (y_2 - y_1)^2};$$

$$r = \sqrt{[3 - (-2)]^2 + (-1 - 5)^2}$$ **Substituting r for d, $(3, -1)$ for (x_2, y_2), and $(-2, 5)$ for (x_1, y_1). Either point can serve as (x_1, y_1).**

$$= \sqrt{5^2 + (-6)^2} = \sqrt{25 + 36}$$

$$= \sqrt{61} \approx 7.8.$$ **Rounding to the nearest tenth**

The radius of the circle is approximately 7.8.

Now Try Exercise 77.

Midpoints of Segments

The distance formula can be used to develop a method of determining the *midpoint* of a segment when the endpoints are known. We state the formula and leave its proof to the exercises.

THE MIDPOINT FORMULA

If the endpoints of a segment are (x_1, y_1) and (x_2, y_2), then the coordinates of the **midpoint** of the segment are

$$\left(\frac{x_1 + x_2}{2}, \frac{y_1 + y_2}{2}\right).$$

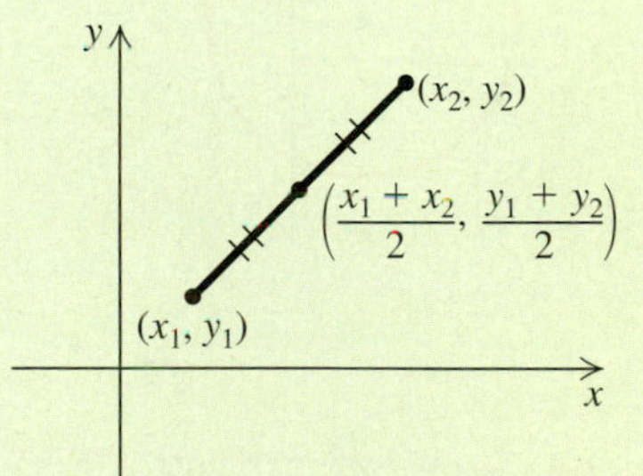

Note that we obtain the coordinates of the midpoint by averaging the coordinates of the endpoints. This is a good way to remember the midpoint formula.

EXAMPLE 8 Find the midpoint of the segment whose endpoints are $(-4, -2)$ and $(2, 5)$.

Solution Using the midpoint formula, we obtain

$$\left(\frac{-4 + 2}{2}, \frac{-2 + 5}{2}\right) = \left(\frac{-2}{2}, \frac{3}{2}\right) = \left(-1, \frac{3}{2}\right).$$

Now Try Exercise 83.

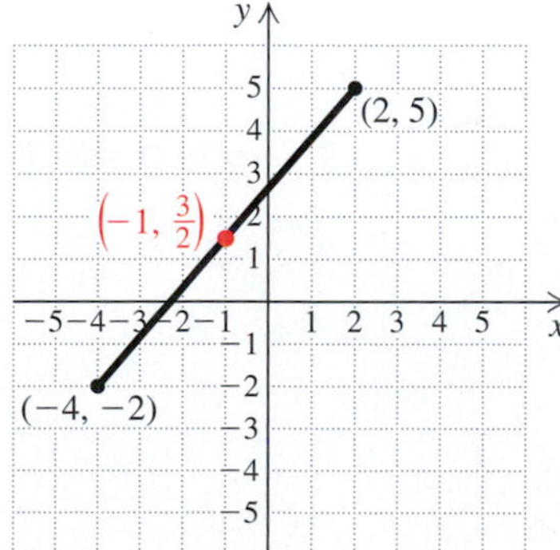

EXAMPLE 9 The diameter of a circle connects the points $(2, -3)$ and $(6, 4)$ on the circle. Find the coordinates of the center of the circle.

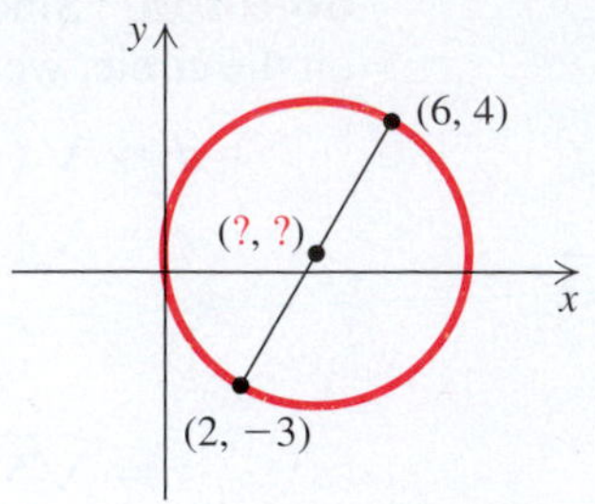

Solution Since the center of the circle is the midpoint of the diameter, we use the midpoint formula:

$$\left(\frac{2+6}{2}, \frac{-3+4}{2}\right), \quad \text{or} \quad \left(\frac{8}{2}, \frac{1}{2}\right), \quad \text{or} \quad \left(4, \frac{1}{2}\right).$$

The coordinates of the center are $\left(4, \frac{1}{2}\right)$.

Now Try Exercise 95.

Circles

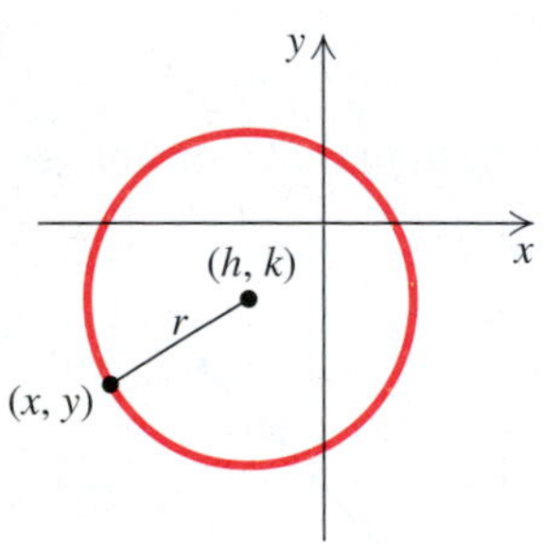

A **circle** is the set of all points in a plane that are a fixed distance r from a *center* (h, k). Thus if a point (x, y) is to be r units from the center, we must have

$$r = \sqrt{(x-h)^2 + (y-k)^2}.$$ Using the distance formula, $d = \sqrt{(x_2 - x_1)^2 + (y_2 - y_1)^2}$

Squaring both sides gives an equation of a circle. The distance r is the length of a *radius* of the circle.

THE EQUATION OF A CIRCLE

The standard form of the equation of a circle having center (h, k) and radius r is

$$(x-h)^2 + (y-k)^2 = r^2.$$

EXAMPLE 10 Find an equation of the circle having radius 5 and center $(3, -7)$.

Solution Using the standard form, we have

$$[x-3]^2 + [y-(-7)]^2 = 5^2 \quad \text{Substituting}$$
$$(x-3)^2 + (y+7)^2 = 25.$$

Now Try Exercise 99.

EXAMPLE 11 Graph the circle $(x + 5)^2 + (y - 2)^2 = 16$.

Solution We write the equation in standard form to determine the center and the radius:

$$[x - (-5)]^2 + [y - 2]^2 = 4^2.$$

The center is $(-5, 2)$ and the radius is 4. We locate the center and draw the circle using a compass.

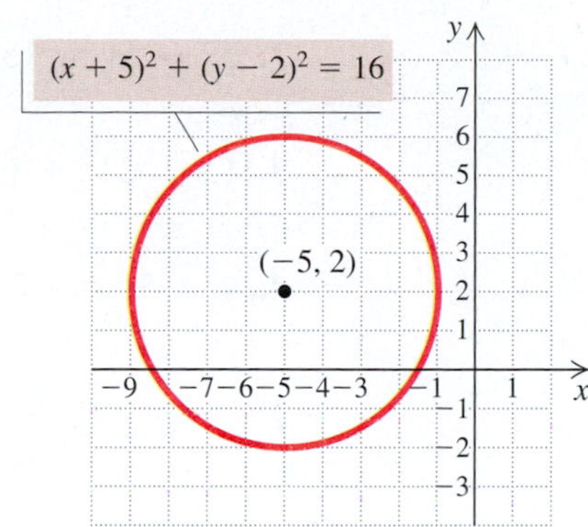

Now Try Exercise 111.

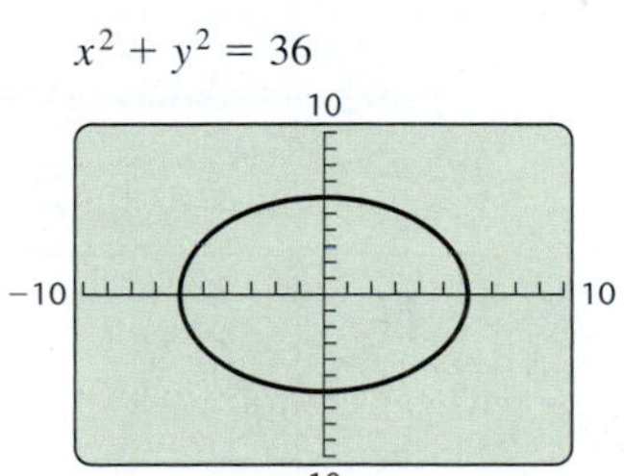

FIGURE 1.

$x^2 + y^2 = 36$

FIGURE 2.

Circles can also be graphed using a graphing calculator. We show one method of doing so here.

When we graph a circle, we select a viewing window in which the distance between units is visually the same on both axes. This procedure is called **squaring the viewing window**. We do this so that the graph will not be distorted. A graph of the circle $x^2 + y^2 = 36$ in a nonsquared window is shown in Fig. 1.

On many graphing calculators, the ratio of the height to the width of the viewing screen is $\frac{2}{3}$. When we choose a window in which Xscl = Yscl and the length of the y-axis is $\frac{2}{3}$ the length of the x-axis, the window will be squared. The windows with dimensions $[-6, 6, -4, 4]$, $[-9, 9, -6, 6]$, and $[-12, 12, -8, 8]$ are examples of squared windows. A graph of the circle $x^2 + y^2 = 36$ in a squared window is shown in Fig. 2. Many graphing calculators have an option on the ZOOM menu that squares the window automatically.

EXAMPLE 12 Graph the circle $(x - 2)^2 + (y + 1)^2 = 16$.

Solution The circle $(x - 2)^2 + (y + 1)^2 = 16$ has center $(2, -1)$ and radius 4, so the viewing window $[-9, 9, -6, 6]$ is a good choice for the graph.

To graph a circle, we select the CIRCLE feature from the DRAW menu and enter the coordinates of the center and the length of the radius. The graph of the circle $(x - 2)^2 + (y + 1)^2 = 16$ is shown here. For more on graphing circles with a graphing calculator, see the material on conic sections in a later chapter.

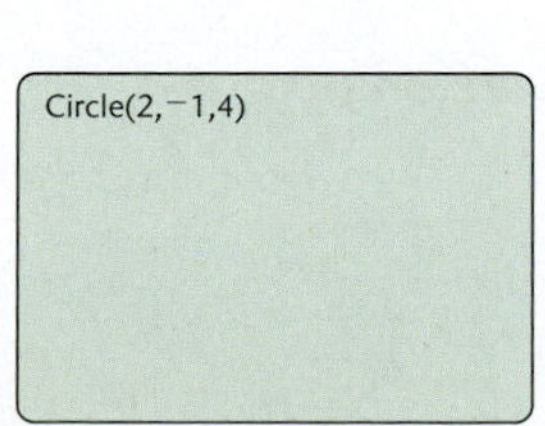

$(x - 2)^2 + (y + 1)^2 = 16$

Now Try Exercise 113.

Visualizing the Graph

Match the equation with its graph.

1. $y = -x^2 + 5x - 3$

2. $3x - 5y = 15$

3. $(x - 2)^2 + (y - 4)^2 = 36$

4. $y - 5x = -3$

5. $x^2 + y^2 = \frac{25}{4}$

6. $15y - 6x = 90$

7. $y = -\frac{2}{3}x - 2$

8. $(x + 3)^2 + (y - 1)^2 = 16$

9. $3x + 5y = 15$

10. $y = x^2 - x - 4$

Answers on page A-1

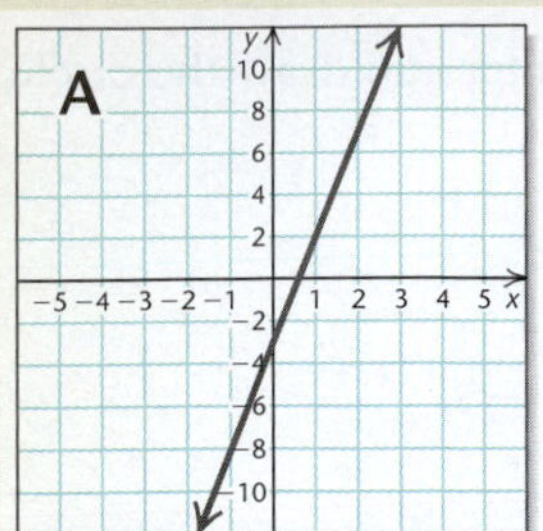

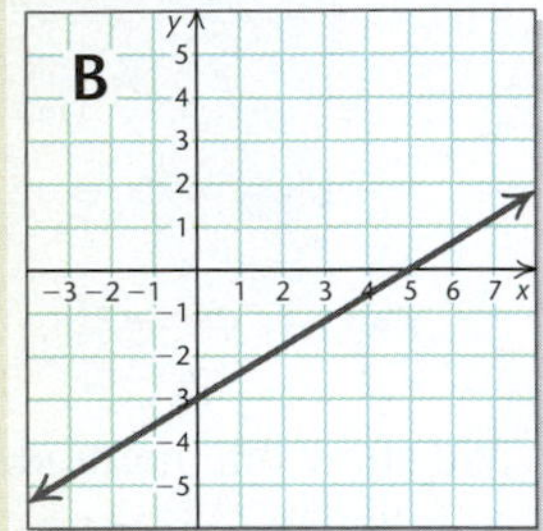

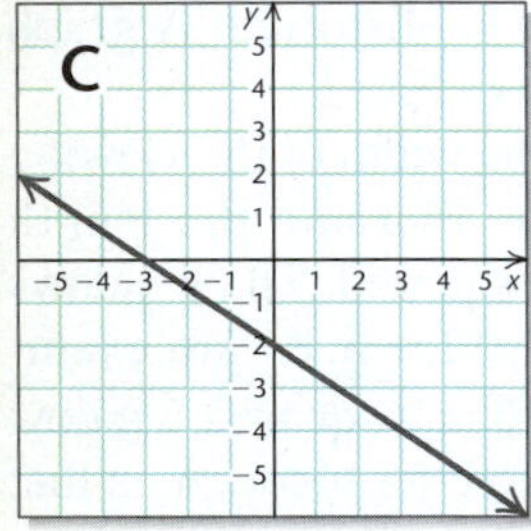

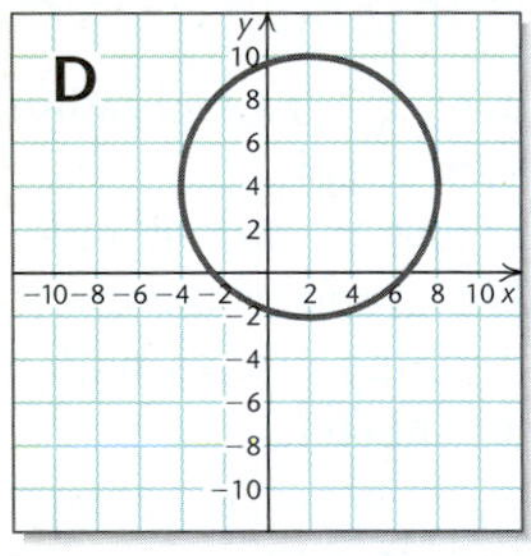

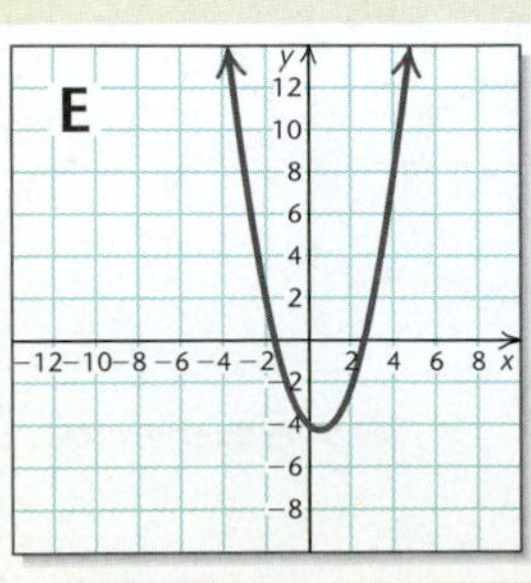

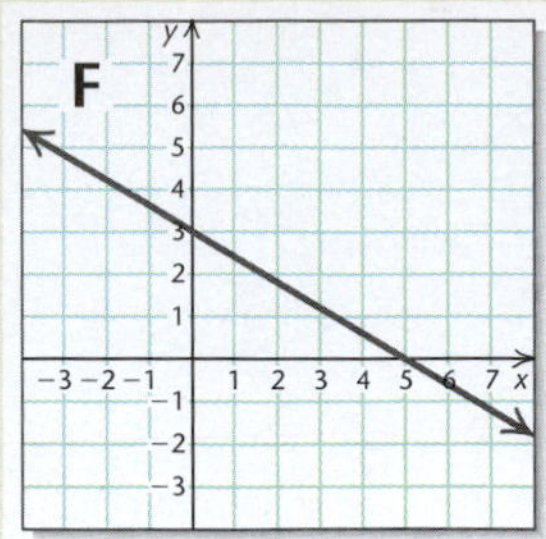

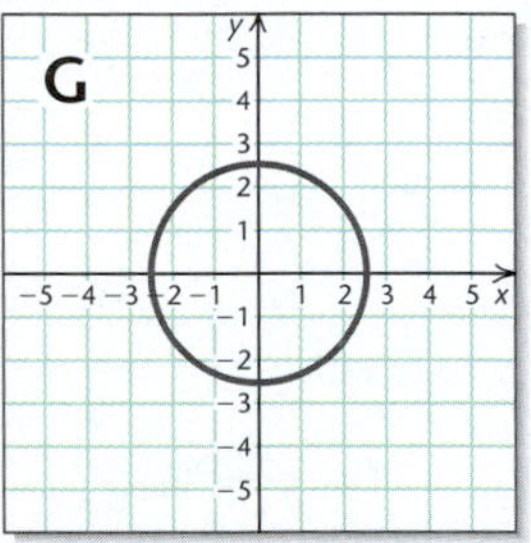

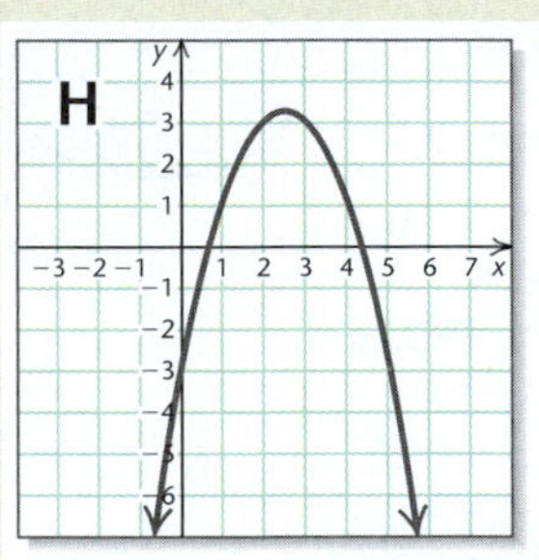

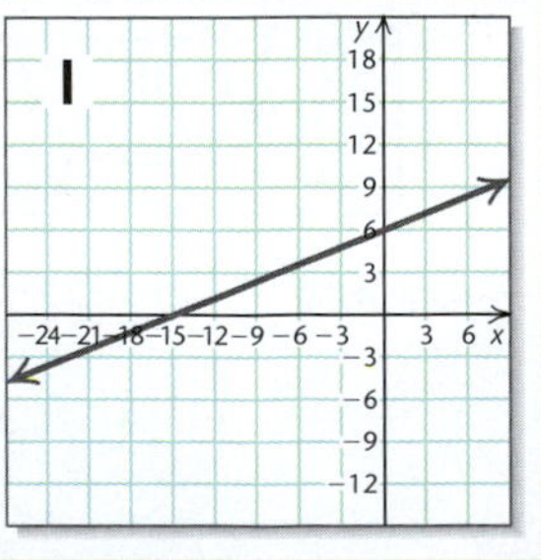

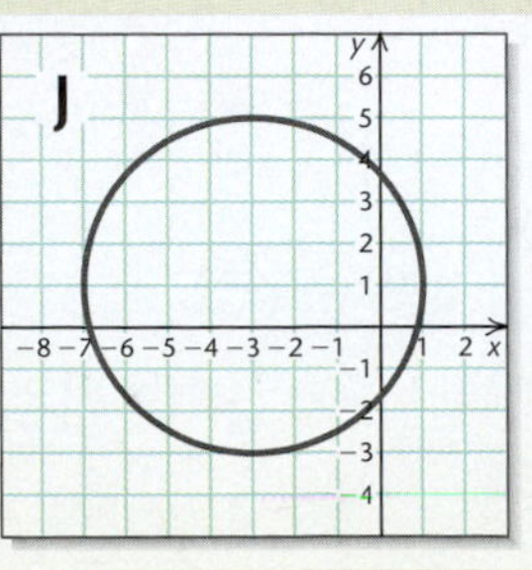

1.1 Exercise Set

Use the following graph for Exercises 1 and 2.

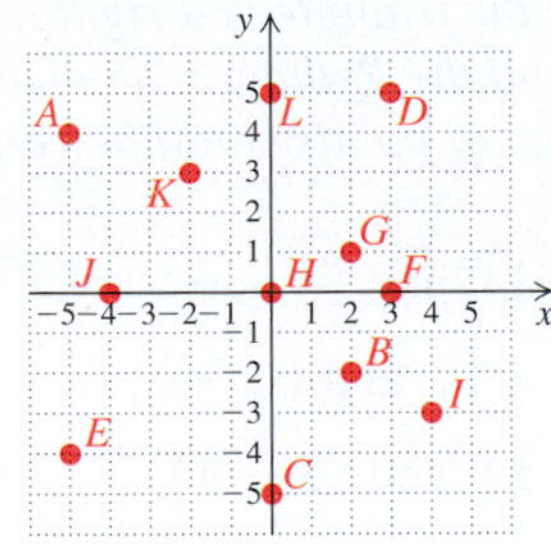

1. Find the coordinates of points A, B, C, D, E, and F.

2. Find the coordinates of points G, H, I, J, K, and L.

Graph and label the given points.

3. $(4, 0)$, $(-3, -5)$, $(-1, 4)$, $(0, 2)$, $(2, -2)$
4. $(1, 4)$, $(-4, -2)$, $(-5, 0)$, $(2, -4)$, $(0, 3)$
5. $(-5, 1)$, $(5, 1)$, $(2, 3)$, $(2, -1)$, $(0, 1)$
6. $(0, -1)$, $(4, -3)$, $(-5, 2)$, $(2, 0)$, $(-1, -5)$

Express the data pictured in the graph as ordered pairs, letting the first coordinate represent the year and the second coordinate the amount or percent.

7. **Southwest Airlines: Number of Cities Served**

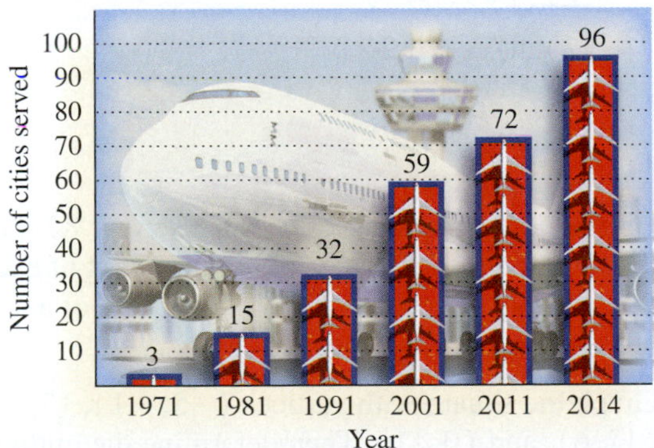

8. **Women Serving in the Marines**

Source: U. S. Department of Veterans Affairs, Rutgers University

Use substitution to determine whether the given ordered pairs are solutions of the given equation.

9. $(-1, -9)$, $(0, 2)$; $y = 7x - 2$
10. $\left(\frac{1}{2}, 8\right)$, $(-1, 6)$; $y = -4x + 10$
11. $\left(\frac{2}{3}, \frac{3}{4}\right)$, $\left(1, \frac{3}{2}\right)$; $6x - 4y = 1$
12. $(1.5, 2.6)$, $(-3, 0)$; $x^2 + y^2 = 9$
13. $\left(-\frac{1}{2}, -\frac{4}{5}\right)$, $\left(0, \frac{3}{5}\right)$; $2a + 5b = 3$
14. $\left(0, \frac{3}{2}\right)$, $\left(\frac{2}{3}, 1\right)$; $3m + 4n = 6$
15. $(-0.75, 2.75)$, $(2, -1)$; $x^2 - y^2 = 3$
16. $(2, -4)$, $(4, -5)$; $5x + 2y^2 = 70$

Find the intercepts and then graph the line.

17. $5x - 3y = -15$
18. $2x - 4y = 8$
19. $2x + y = 4$
20. $3x + y = 6$
21. $4y - 3x = 12$
22. $3y + 2x = -6$

Graph the equation.

23. $y = 3x + 5$
24. $y = -2x - 1$
25. $x - y = 3$
26. $x + y = 4$
27. $y = -\frac{3}{4}x + 3$
28. $3y - 2x = 3$
29. $5x - 2y = 8$
30. $y = 2 - \frac{4}{3}x$
31. $x - 4y = 5$
32. $6x - y = 4$
33. $2x + 5y = -10$
34. $4x - 3y = 12$
35. $y = -x^2$
36. $y = x^2$
37. $y = x^2 - 3$
38. $y = 4 - x^2$
39. $y = -x^2 + 2x + 3$
40. $y = x^2 + 2x - 1$

In Exercises 41–44, use a graphing calculator to match the equation with one of the graphs (a)–(d) that follow.

a)
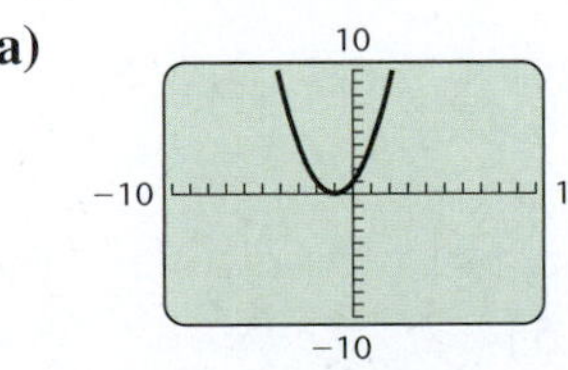

b)
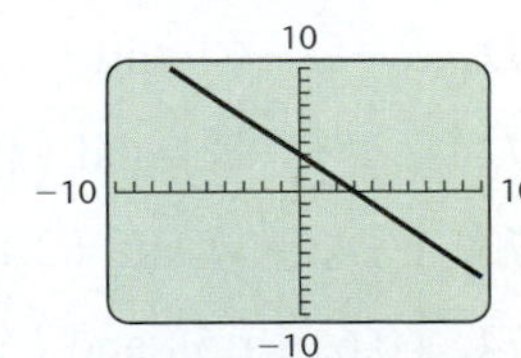

c)
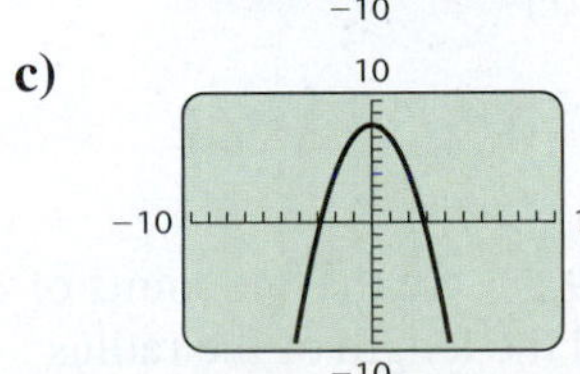

d)
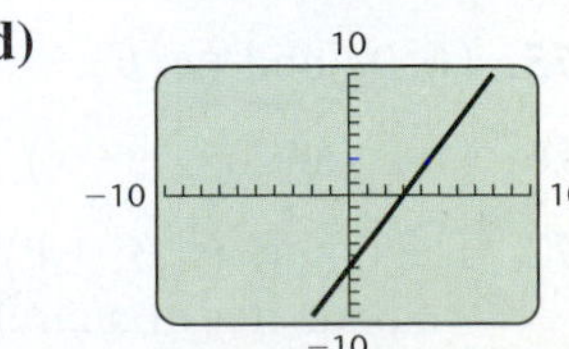

41. $y = 3 - x$
42. $2x - y = 6$
43. $y = x^2 + 2x + 1$
44. $y = 8 - x^2$

Use a graphing calculator to graph the equation in the standard window.

45. $y = 2x + 1$

46. $y = 3x - 4$

47. $4x + y = 7$

48. $5x + y = -8$

49. $y = \frac{1}{3}x + 2$

50. $y = \frac{3}{2}x - 4$

51. $2x + 3y = -5$

52. $3x + 4y = 1$

53. $y = x^2 + 6$

54. $y = x^2 - 8$

55. $y = 2 - x^2$

56. $y = 5 - x^2$

57. $y = x^2 + 4x - 2$

58. $y = x^2 - 5x + 3$

Graph the equation in the standard window and in the given window. Determine which window better shows the shape of the graph and the x- and y-intercepts.

59. $y = 3x^2 - 6$
$[-4, 4, -4, 4]$

60. $y = -2x + 24$
$[-15, 15, -10, 30]$, with Xscl = 3 and Yscl = 5

61. $y = -\frac{1}{6}x^2 + \frac{1}{12}$
$[-1, 1, -0.3, 0.3]$, with Xscl = 0.1 and Yscl = 0.1

62. $y = 6 - x^2$
$[-3, 3, -3, 3]$

Find the distance between the pair of points. Give an exact answer and, where appropriate, an approximation to three decimal places.

63. $(4, 6)$ and $(5, 9)$

64. $(-3, 7)$ and $(2, 11)$

65. $(-11, -8)$ and $(1, -13)$

66. $(-60, 5)$ and $(-20, 35)$

67. $(6, -1)$ and $(9, 5)$

68. $(-4, -7)$ and $(-1, 3)$

69. $\left(-8, \frac{7}{11}\right)$ and $\left(8, \frac{7}{11}\right)$

70. $\left(\frac{1}{2}, -\frac{4}{25}\right)$ and $\left(\frac{1}{2}, -\frac{13}{25}\right)$

71. $\left(-\frac{3}{5}, -4\right)$ and $\left(-\frac{3}{5}, \frac{2}{3}\right)$

72. $\left(-\frac{11}{3}, -\frac{1}{2}\right)$ and $\left(\frac{1}{3}, \frac{5}{2}\right)$

73. $(-4.2, 3)$ and $(2.1, -6.4)$

74. $(0.6, -1.5)$ and $(-8.1, -1.5)$

75. $(0, 0)$ and (a, b)

76. (r, s) and $(-r, -s)$

77. The points $(-3, -1)$ and $(9, 4)$ are the endpoints of the diameter of a circle. Find the length of the radius of the circle.

78. The point $(0, 1)$ is on a circle that has center $(-3, 5)$. Find the length of the diameter of the circle.

The converse of the Pythagorean theorem is also a true statement: If the sum of the squares of the lengths of two sides of a triangle is equal to the square of the length of the third side, then the triangle is a right triangle. Use the distance formula and the Pythagorean theorem to determine whether the set of points could be vertices of a right triangle.

79. $(-4, 5)$, $(6, 1)$, and $(-8, -5)$

80. $(-3, 1)$, $(2, -1)$, and $(6, 9)$

81. $(-4, 3)$, $(0, 5)$, and $(3, -4)$

82. The points $(-3, 4)$, $(2, -1)$, $(5, 2)$, and $(0, 7)$ are vertices of a quadrilateral. Show that the quadrilateral is a rectangle. (*Hint*: Show that the opposite sides of the quadrilateral are the same length and that the two diagonals are the same length.)

Find the midpoint of the segment having the given endpoints.

83. $(4, -9)$ and $(-12, -3)$

84. $(7, -2)$ and $(9, 5)$

85. $\left(0, \frac{1}{2}\right)$ and $\left(-\frac{2}{5}, 0\right)$

86. $(0, 0)$ and $\left(-\frac{7}{13}, \frac{2}{7}\right)$

87. $(6.1, -3.8)$ and $(3.8, -6.1)$

88. $(-0.5, -2.7)$ and $(4.8, -0.3)$

89. $(-6, 5)$ and $(-6, 8)$

90. $(1, -2)$ and $(-1, 2)$

91. $\left(-\frac{1}{6}, -\frac{3}{5}\right)$ and $\left(-\frac{2}{3}, \frac{5}{4}\right)$

92. $\left(\frac{2}{9}, \frac{1}{3}\right)$ and $\left(-\frac{2}{5}, \frac{4}{5}\right)$

93. Graph the rectangle described in Exercise 82. Then determine the coordinates of the midpoint of each of the four sides. Are the midpoints vertices of a rectangle?

94. Graph the square with vertices $(-5, -1)$, $(7, -6)$, $(12, 6)$, and $(0, 11)$. Then determine the midpoint of each of the four sides. Are the midpoints vertices of a square?

95. The points $(\sqrt{7}, -4)$ and $(\sqrt{2}, 3)$ are endpoints of the diameter of a circle. Determine the center of the circle.

96. The points $(-3, \sqrt{5})$ and $(1, \sqrt{2})$ are endpoints of the diagonal of a square. Determine the center of the square.

In Exercises 97 and 98, how would you change the window so that the circle is not distorted? Answers may vary.

97. $(x + 3)^2 + (y - 2)^2 = 36$

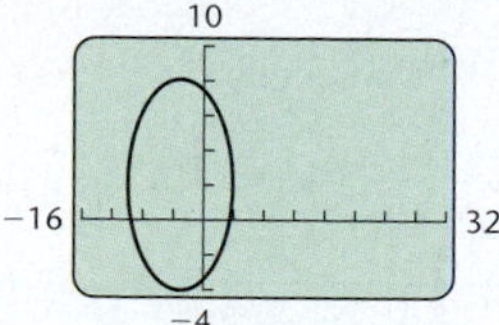

98. $(x - 4)^2 + (y + 5)^2 = 49$

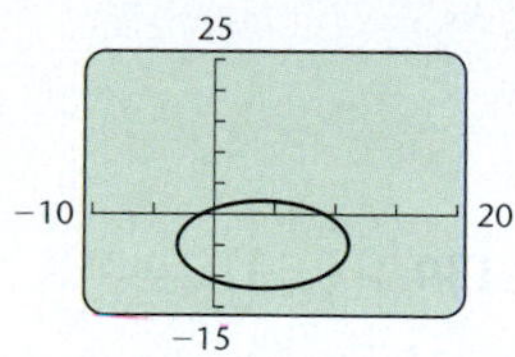

Find an equation for a circle satisfying the given conditions.

99. Center $(2, 3)$, radius of length $\frac{5}{3}$

100. Center $(4, 5)$, diameter of length 8.2

101. Center $(-1, 4)$, passes through $(3, 7)$

102. Center $(6, -5)$, passes through $(1, 7)$

103. The points $(7, 13)$ and $(-3, -11)$ are at the ends of a diameter.

104. The points $(-9, 4)$, $(-2, 5)$, $(-8, -3)$, and $(-1, -2)$ are vertices of an inscribed square.

105. Center $(-2, 3)$, tangent (touching at one point) to the y-axis

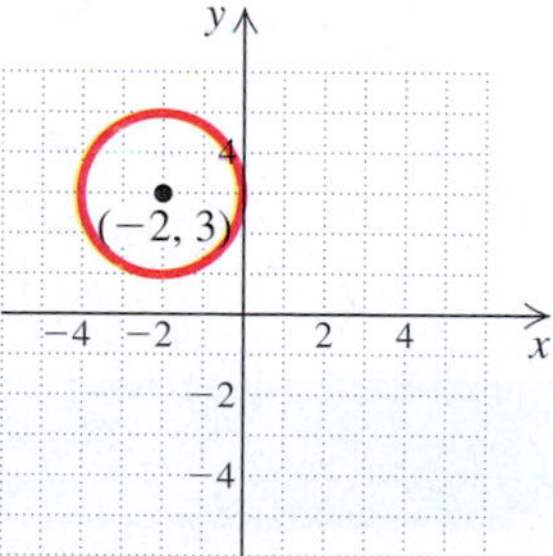

106. Center $(4, -5)$, tangent to the x-axis

Find the center and the radius of the circle. Then graph the circle by hand. Check your graph with a graphing calculator.

107. $x^2 + y^2 = 4$

108. $x^2 + y^2 = 81$

109. $x^2 + (y - 3)^2 = 16$

110. $(x + 2)^2 + y^2 = 100$

111. $(x - 1)^2 + (y - 5)^2 = 36$

112. $(x - 7)^2 + (y + 2)^2 = 25$

113. $(x + 4)^2 + (y + 5)^2 = 9$

114. $(x + 1)^2 + (y - 2)^2 = 64$

Find the equation of the circle.

115.

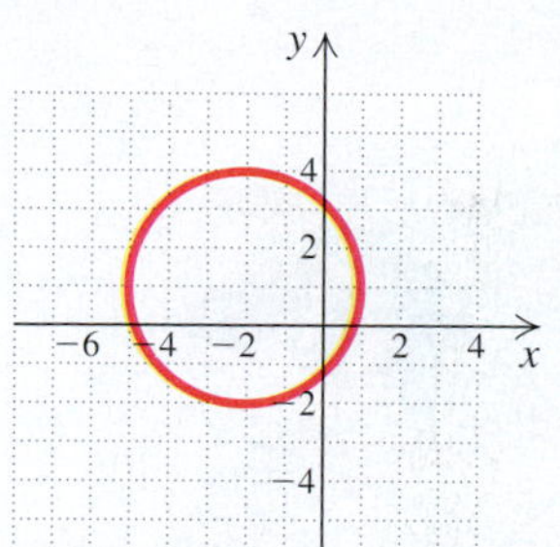

116.

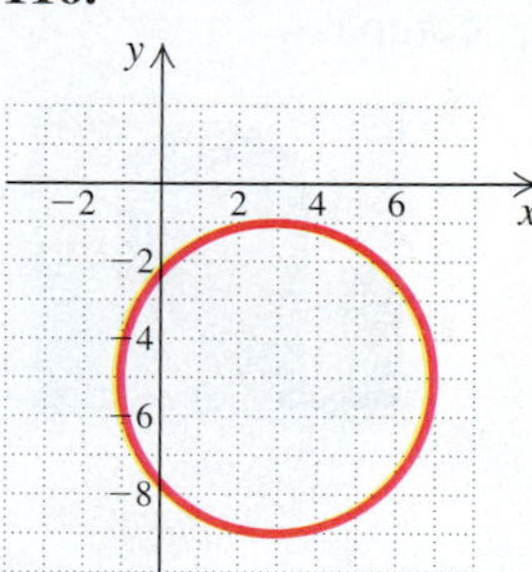

117.

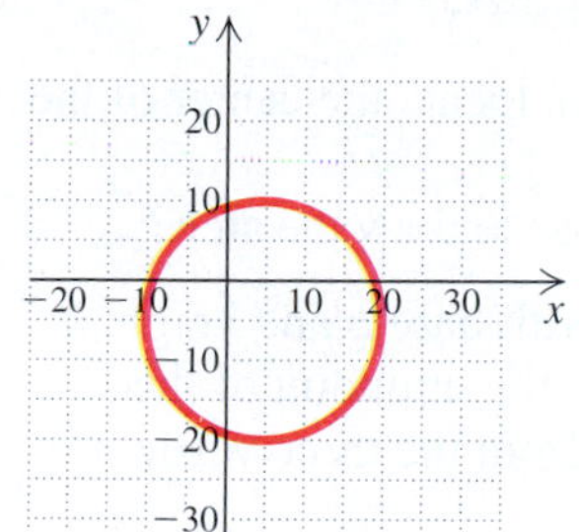

118.

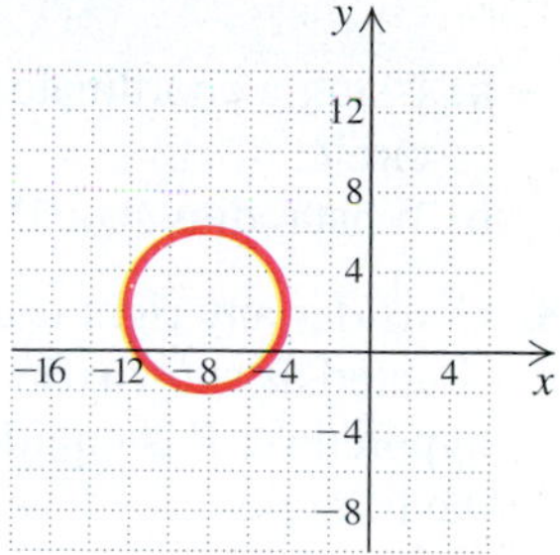

Synthesis

To the student and the instructor: *The Synthesis exercises found at the end of every exercise set challenge students to combine concepts or skills studied in that section or in preceding parts of the text.*

119. If the point (p, q) is in the fourth quadrant, in which quadrant is the point $(q, -p)$?

Find the distance between the pair of points and find the midpoint of the segment having the given points as endpoints.

120. $\left(a, \frac{1}{a}\right)$ and $\left(a + h, \frac{1}{a + h}\right)$

121. $(a, \sqrt{a})$ and $(a + h, \sqrt{a + h})$

Find an equation of a circle satisfying the given conditions.

122. Center $(-5, 8)$ with a circumference of 10π units

123. Center $(2, -7)$ with an area of 36π square units

124. Find the point on the x-axis that is equidistant from the points $(-4, -3)$ and $(-1, 5)$.

125. Find the point on the y-axis that is equidistant from the points $(-2, 0)$ and $(4, 6)$.

126. Determine whether the points $(-1, -3)$, $(-4, -9)$, and $(2, 3)$ are collinear.

127. *An Arch of a Circle in Carpentry.* Matt is remodeling the front entrance to his home and needs to cut an arch for the top of an entranceway. The arch must be 8 ft wide and 2 ft high. To draw the arch, he will use a stretched string with chalk attached at one end as a compass.

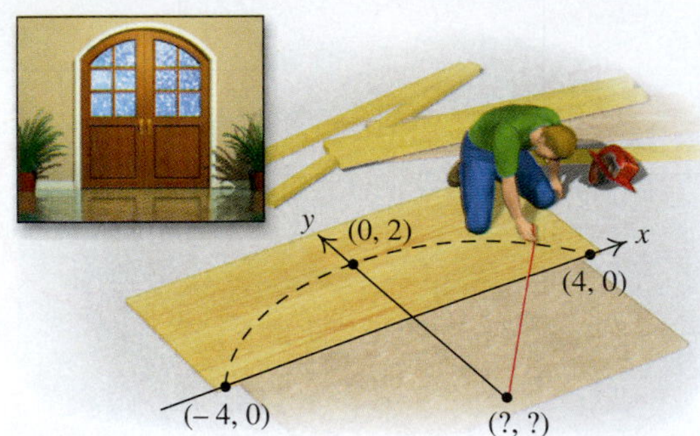

a) Using a coordinate system, locate the center of the circle.

b) What radius should Matt use to draw the arch?

128. Consider any right triangle with base b and height h, situated as shown. Show that the midpoint of the hypotenuse P is equidistant from the three vertices of the triangle.

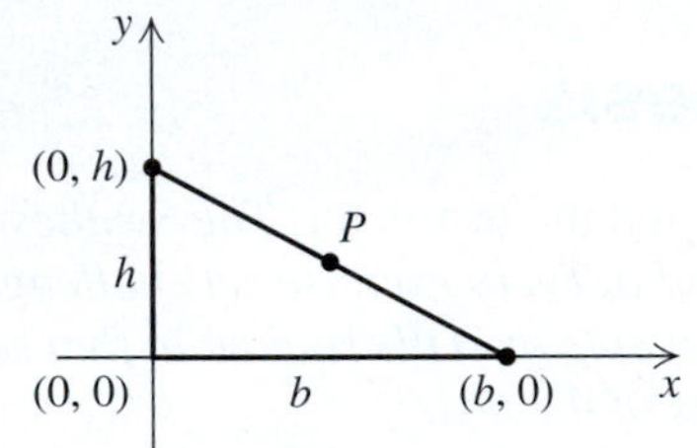

Determine whether each of the following points lies on the ***unit circle***, $x^2 + y^2 = 1$.

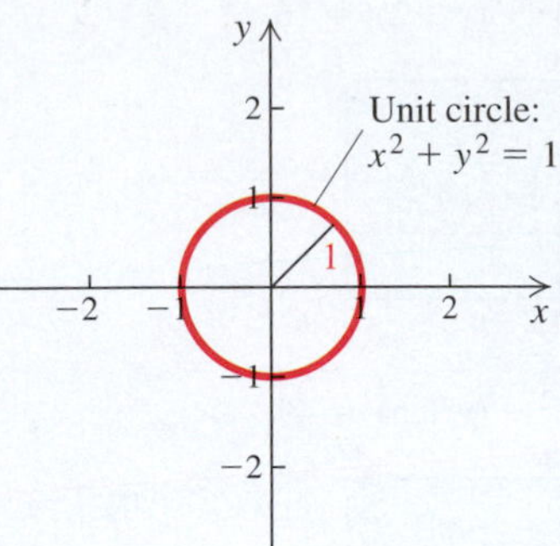

129. $\left(\frac{\sqrt{3}}{2}, -\frac{1}{2}\right)$

130. $(0, -1)$

131. $\left(-\frac{\sqrt{2}}{2}, \frac{\sqrt{2}}{2}\right)$

132. $\left(\frac{1}{2}, -\frac{\sqrt{3}}{2}\right)$

133. Prove the midpoint formula by showing that $\left(\frac{x_1 + x_2}{2}, \frac{y_1 + y_2}{2}\right)$ is equidistant from the points (x_1, y_1) and (x_2, y_2).

1.2 Functions and Graphs

- Determine whether a correspondence or a relation is a function.
- Find function values, or outputs, using a formula or a graph.
- Graph functions.
- Determine whether a graph is that of a function.
- Find the domain and the range of a function.
- Solve applied problems using functions.

We now focus our attention on a concept that is fundamental to many areas of mathematics—the idea of a *function*.

Functions

We first consider an application.

Pay-As You-Go Exercise Program. A community center offers a pay-as you-go exercise program. The total cost is a membership fee of \$20 plus \$7.50 for every class attended. If a person attends 7 classes, the total cost is

$$\$7.50(7) + \$20, \quad \text{or} \quad \$72.50.$$

We can express this relationship with a set of ordered pairs, a graph, and an equation. A few ordered pairs are listed in the following table.

x	y	Ordered Pairs: (x, y)	Correspondence
1	27.50	$(1, 27.50)$	$1 \rightarrow 27.50$
2	35	$(2, 35)$	$2 \rightarrow 35$
4	50	$(4, 50)$	$4 \rightarrow 50$
7	72.50	$(7, 72.50)$	$7 \rightarrow 72.50$
10	95	$(10, 95)$	$10 \rightarrow 95$

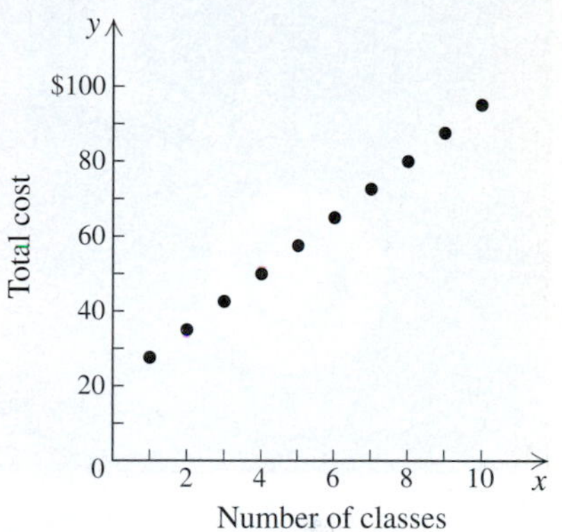

The ordered pairs express a relationship, or a correspondence, between the first coordinate and the second coordinate. We can see this relationship in the graph as well. The equation that describes the correspondence is

$$y = 7.50x + 20, \quad \text{where } x \text{ is a natural number.}$$

This is an example of a *function*. In this case, the total cost of the exercise classes y is a function of the number of classes attended x; that is, y is a function of x, where x is the independent variable and y is the dependent variable.

Let's consider some other correspondences before giving the definition of a function.

First Set	Correspondence	Second Set
To each person	there corresponds	that person's DNA.
To each truck sold	there corresponds	its price.
To each real number	there corresponds	the square of that number.

In each correspondence, the first set is called the **domain** and the second set is called the **range**. For each member, or **element**, in the domain, there is *exactly one* member in the range to which it corresponds. That is, each person has exactly *one* DNA, each truck has exactly *one* price, and each real number has exactly *one* square. Each correspondence is a *function*.

FUNCTION

A **function** is a correspondence between a first set, called the **domain**, and a second set, called the **range**, such that each member of the domain corresponds to *exactly one* member of the range.

It is important to note that not every correspondence between two sets is a function.

EXAMPLE 1 Determine whether each of the following correspondences is a function.

a) 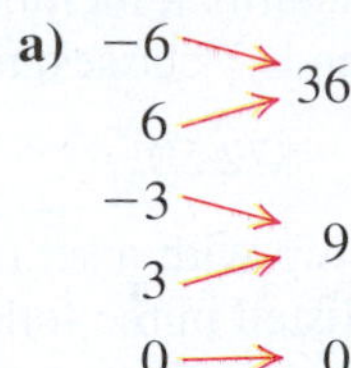

b) 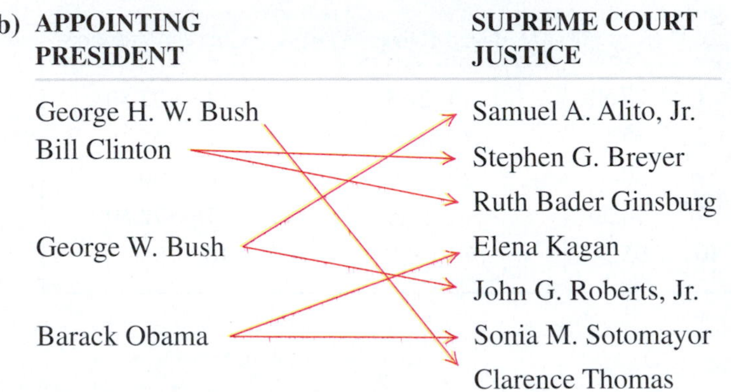

Solution

a) This correspondence *is* a function because each member of the domain corresponds to exactly one member of the range. Note that the definition of a function allows more than one member of the domain to correspond to the same member of the range.

b) This correspondence *is not* a function because there is at least one member of the domain that is paired with more than one member of the range (Bill Clinton with Stephen G. Breyer and Ruth Bader Ginsburg; George W. Bush with Samuel A. Alito, Jr., and John G. Roberts, Jr.; Barack Obama with Elena Kagan and Sonia M. Sotomayor).

Now Try Exercises 5 and 7.

EXAMPLE 2 Determine whether each of the following correspondences is a function.

	DOMAIN	CORRESPONDENCE	RANGE
a)	Years in which a presidential election occurs	The person elected	A set of presidents
b)	All automobiles produced in 2016	Each automobile's VIN	A set of VINs
c)	The set of all professional golfers who won a PGA tournament in 2014	The tournament won	The set of all PGA tournaments in 2014
d)	The set of all PGA tournaments in 2014	The winner of the tournament	The set of all golfers who won a PGA tournament in 2014

Solution

a) This correspondence *is* a function because in each presidential election *exactly one* president is elected.

b) This correspondence *is* a function because each automobile has *exactly one* VIN.

c) This correspondence *is not* a function because a winning golfer could be paired with more than one tournament.

d) This correspondence *is* a function because each tournament has only one winning golfer.

Now Try Exercises 11 and 13.

When a correspondence between two sets is not a function, it may still be an example of a **relation**.

> **RELATION**
>
> A **relation** is a correspondence between a first set, called the **domain**, and a second set, called the **range**, such that each member of the domain corresponds to *at least one* member of the range.

All the correspondences in Examples 1 and 2 are relations, but, as we have seen, not all are functions. Relations are sometimes written as sets of ordered pairs (as we saw earlier in the example on the total cost of attending a number of exercise classes) in which elements of the domain are the first coordinates of the ordered pairs and elements of the range are the second coordinates. For example, instead of writing $-3 \rightarrow 9$, as we did in Example 1(a), we could write the ordered pair $(-3, 9)$.

EXAMPLE 3 Determine whether each of the following relations is a function. Identify the domain and the range.

a) $\{(9, -5), (9, 5), (2, 4)\}$

b) $\{(-2, 5), (5, 7), (0, 1), (4, -2)\}$

c) $\{(-5, 3), (0, 3), (6, 3)\}$

Solution

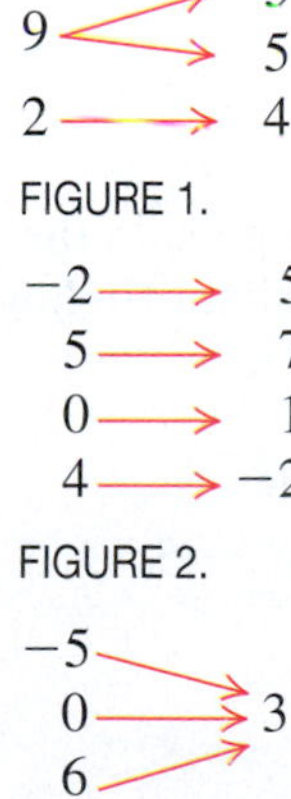

FIGURE 1.
FIGURE 2.
FIGURE 3.

a) The relation *is not* a function because the ordered pairs $(9, -5)$ and $(9, 5)$ have the same first coordinate and different second coordinates. (See Fig. 1.)

The domain is the set of all first coordinates: $\{9, 2\}$.

The range is the set of all second coordinates: $\{-5, 5, 4\}$.

b) The relation *is* a function because *no* two ordered pairs have the same first coordinate and different second coordinates. (See Fig. 2.)

The domain is the set of all first coordinates: $\{-2, 5, 0, 4\}$.

The range is the set of all second coordinates: $\{5, 7, 1, -2\}$.

c) The relation *is* a function because *no* two ordered pairs have the same first coordinate and different second coordinates. (See Fig. 3.)

The domain is $\{-5, 0, 6\}$.

The range is $\{3\}$.

Now Try Exercises 17 and 19.

Notation for Functions

Functions used in mathematics are often given by equations. They generally require that certain calculations be performed in order to determine which member of the range is paired with each member of the domain. For example, in Section 1.1, we graphed the function $y = x^2 - 9x - 12$ by doing calculations like the following:

$$\text{for } x = -2,\ y = (-2)^2 - 9(-2) - 12 = 10,$$
$$\text{for } x = 0,\ y = 0^2 - 9 \cdot 0 - 12 = -12, \quad \text{and}$$
$$\text{for } x = 1,\ y = 1^2 - 9 \cdot 1 - 12 = -20.$$

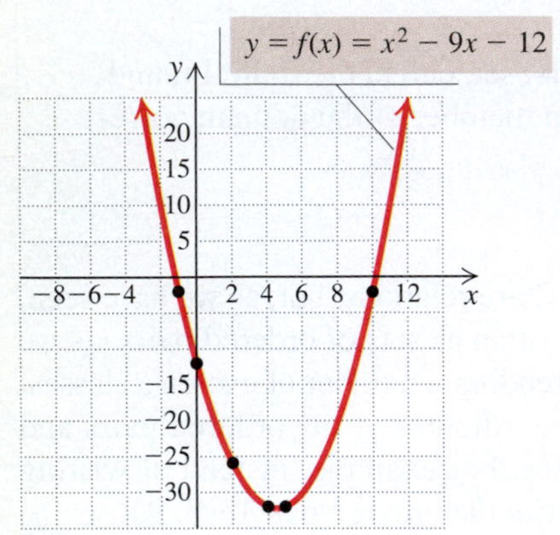

A more concise notation is often used. For $y = x^2 - 9x - 12$, the **inputs** (members of the domain) are values of x substituted into the equation. The **outputs** (members of the range) are the resulting values of y. If we call the function f, we can use x to represent an arbitrary *input* and $f(x)$—read "f of x," or "f at x," or "the value of f at x"—to represent the corresponding *output*. In this notation, the function given by $y = x^2 - 9x - 12$ is written as $f(x) = x^2 - 9x - 12$, and the above calculations would be

$$f(-2) = (-2)^2 - 9(-2) - 12 = 10,$$
$$f(0) = 0^2 - 9 \cdot 0 - 12 = -12,$$
$$f(1) = 1^2 - 9 \cdot 1 - 12 = -20.$$

Keep in mind that $f(x)$ *does not* mean $f \cdot x$.

Thus, instead of writing "when $x = -2$, the value of y is 10," we can simply write "$f(-2) = 10$," which can be read as "f of -2 is 10" or "for the input -2, the output of f is 10." The letters g and h are also often used to name functions.

EXAMPLE 4 A function f is given by $f(x) = 2x^2 - x + 3$. Find each of the following.

a) $f(0)$ **b)** $f(-7)$

c) $f(5a)$ **d)** $f(a - 4)$

Solution We can think of this formula as follows:

$$f(\square) = 2(\square)^2 - (\square) + 3.$$

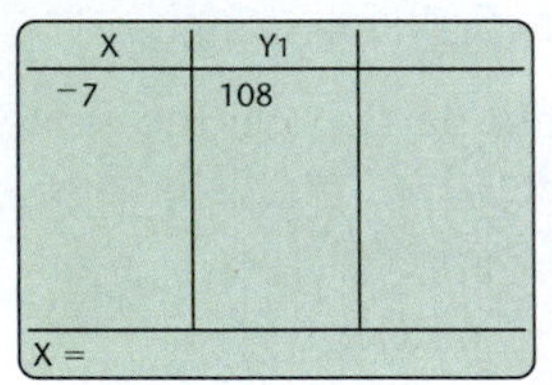

Thus to find an output for a given input we think: "Whatever goes in the blank on the left goes in the blank(s) on the right." This gives us a "recipe" for finding outputs.

a) $f(0) = 2(0)^2 - 0 + 3 = 0 - 0 + 3 = 3$

b) $f(-7) = 2(-7)^2 - (-7) + 3 = 2 \cdot 49 + 7 + 3 = 108$

c) $f(5a) = 2(5a)^2 - 5a + 3 = 2 \cdot 25a^2 - 5a + 3 = 50a^2 - 5a + 3$

d) $f(a - 4) = 2(a - 4)^2 - (a - 4) + 3$
$= 2(a^2 - 8a + 16) - a + 4 + 3$
$= 2a^2 - 16a + 32 - a + 4 + 3$
$= 2a^2 - 17a + 39$

Just in Time 13

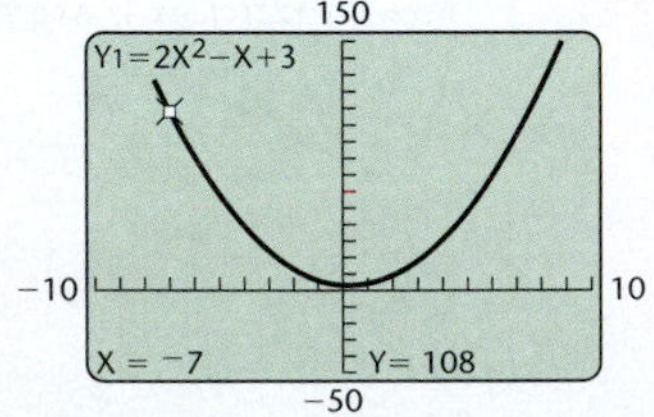

We can also find function values with a graphing calculator. Most calculators do not use function notation "$f(x) = \ldots$" to enter a function formula. Instead, we must enter the function using "$y = \ldots$." At left, we illustrate finding $f(-7)$ from part (b), first with the TABLE feature set in ASK mode and then with the VALUE feature from the CALC menu. We see on both screens that $f(-7) = 108$.

Now Try Exercise 21.

Graphs of Functions

We graph functions in the same way that we graph equations. We find ordered pairs (x, y), or $(x, f(x))$, plot points, and complete the graph.

EXAMPLE 5 Graph each of the following functions.

a) $f(x) = x^2 - 5$ **b)** $f(x) = x^3 - x$ **c)** $f(x) = \sqrt{x + 4}$

Solution We select values for x and find the corresponding values of $f(x)$. Then we plot the points and connect them with a smooth curve.

a) $f(x) = x^2 - 5$

x	$f(x)$	$(x, f(x))$
-3	4	$(-3, 4)$
-2	-1	$(-2, -1)$
-1	-4	$(-1, -4)$
0	-5	$(0, -5)$
1	-4	$(1, -4)$
2	-1	$(2, -1)$
3	4	$(3, 4)$

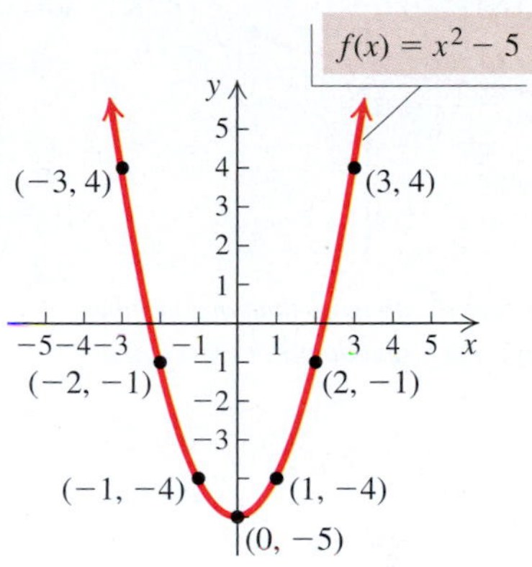

b) $f(x) = x^3 - x$

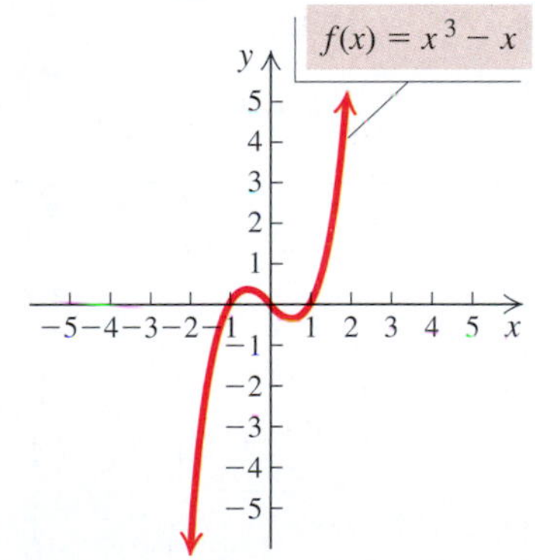

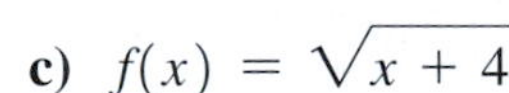

c) $f(x) = \sqrt{x + 4}$

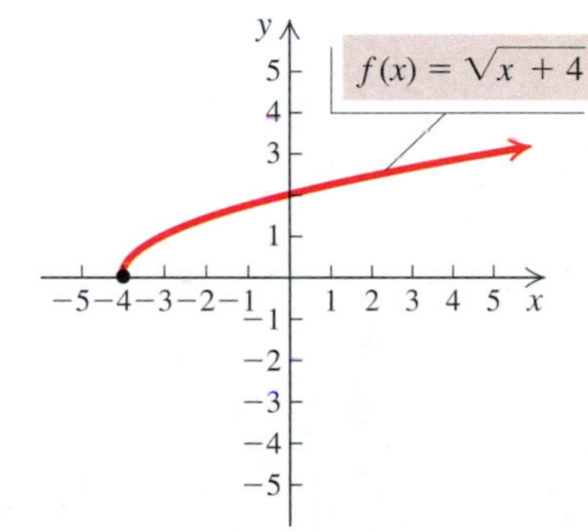

$y = x^3 - x$

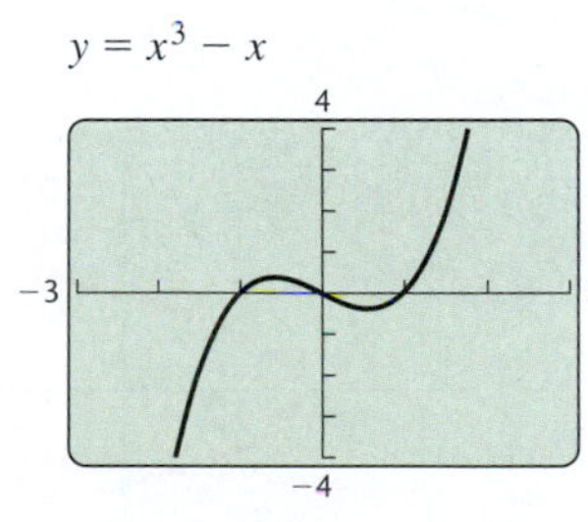

$y = \sqrt{x + 4}$

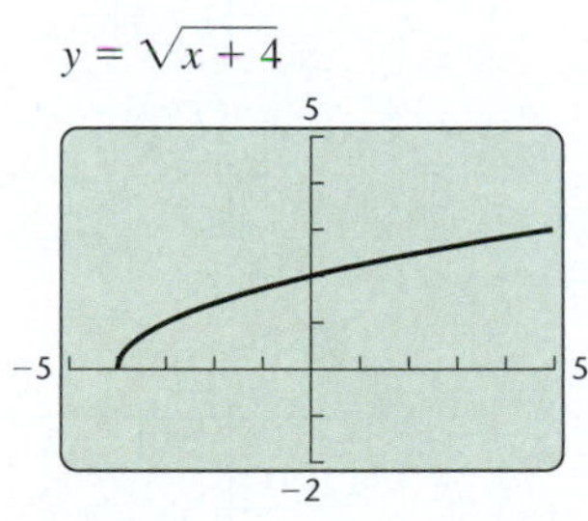

We can check the graphs with a graphing calculator. The checks for parts (b) and (c) are shown at left.

Now Try Exercise 33.

Function values can also be determined from a graph.

EXAMPLE 6 For the function $f(x) = x^2 - 6$, use the graph at left to find each of the following function values.

a) $f(-3)$ **b)** $f(1)$

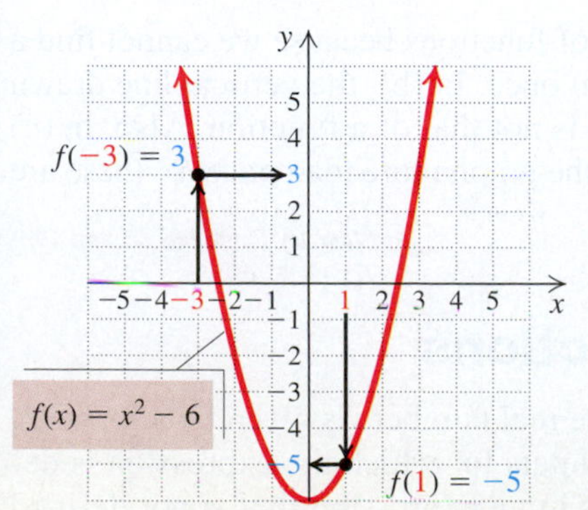

Solution

a) To find the function value $f(-3)$ from the graph, we locate the input -3 on the horizontal axis, move vertically to the graph of the function, and then move horizontally to find the output on the vertical axis. We see that $f(-3) = 3$.

b) To find the function value $f(1)$, we locate the input 1 on the horizontal axis, move vertically to the graph, and then move horizontally to find the output on the vertical axis. We see that $f(1) = -5$.

Now Try Exercise 37.

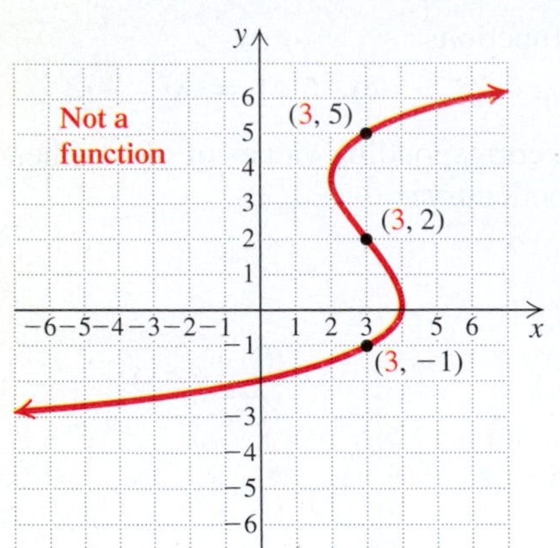

Since 3 is paired with more than one member of the range, the graph does not represent a function.

We know that when one member of the domain is paired with two or more different members of the range, the correspondence *is not* a function. Thus, when a graph contains two or more different points with the same first coordinate, the graph cannot represent a function. (See the graph at left. Note that 3 is paired with -1, 2, and 5.) Points sharing a common first coordinate are vertically above or below each other. This leads us to the *vertical-line test*.

THE VERTICAL-LINE TEST

If it is possible for a vertical line to cross a graph more than once, then the graph *is not* the graph of a function.

To apply the vertical-line test, we try to find a vertical line that crosses the graph more than once. If we succeed, then the graph is not that of a function. If we do not, then the graph is that of a function.

EXAMPLE 7 Which of graphs (a)–(f) (in red) are graphs of functions? In graph (f), the solid dot shows that $(-1, 1)$ belongs to the graph. The open circle shows that $(-1, -2)$ does *not* belong to the graph.

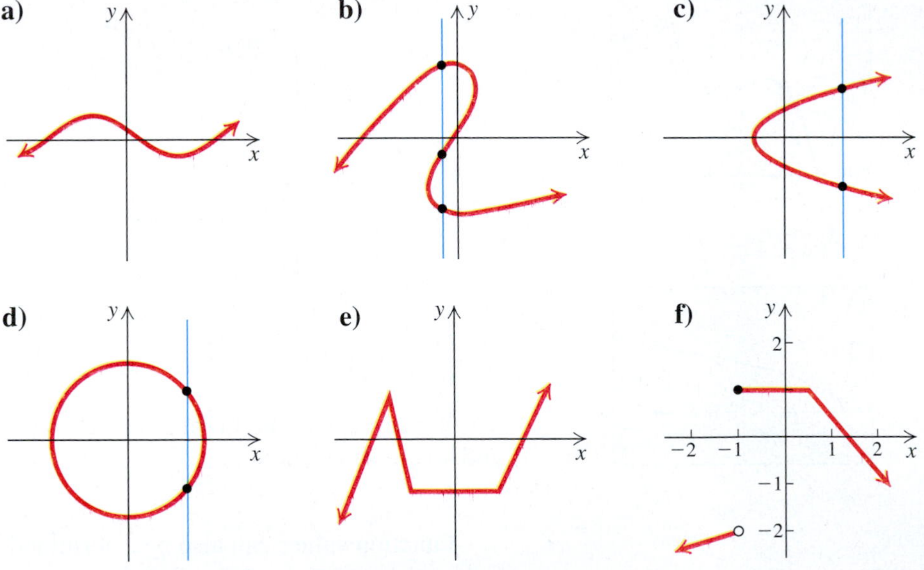

Solution Graphs (a), (e), and (f) are graphs of functions because we cannot find a vertical line that crosses any of them more than once. In (b), the vertical line drawn crosses the graph at three points, so graph (b) is not that of a function. Also, in (c) and (d), we can find a vertical line that crosses the graph more than once, so these are not graphs of functions.

Now Try Exercises 45 and 49.

Finding Domains of Functions

When a function f whose inputs and outputs are real numbers is given by a formula, the *domain* is understood to be the set of all inputs for which the expression is defined as a real number. When an input results in an expression that is not defined as a real number, we say that the function value *does not exist* and that the number being substituted *is not* in the domain of the function.

EXAMPLE 8 Find the indicated function values, if possible, and determine whether the given values are in the domain of the function.

a) $f(1)$ and $f(3)$, for $f(x) = \dfrac{1}{x - 3}$

b) $g(16)$ and $g(-7)$, for $g(x) = \sqrt{x} + 5$

Solution

$y = 1/(x - 3)$

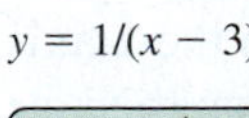

X	Y1	
1	−.5	
3	ERROR	

X =

a) $f(1) = \dfrac{1}{1 - 3} = \dfrac{1}{-2} = -\dfrac{1}{2}$

Since $f(1)$ is defined, 1 is in the domain of f.

$$f(3) = \frac{1}{3 - 3} = \frac{1}{0}$$

Since division by 0 is not defined, $f(3)$ does not exist and the number 3 is not in the domain of f. In a table from a graphing calculator, this is indicated with an ERROR message.

$y = \sqrt{x} + 5$

X	Y1	
16	9	
−7	ERROR	

X =

b) $g(16) = \sqrt{16} + 5 = 4 + 5 = 9$

Since $g(16)$ is defined, 16 is in the domain of g.

$$g(-7) = \sqrt{-7} + 5$$

Since $\sqrt{-7}$ is not defined as a real number, $g(-7)$ does not exist and the number -7 is not in the domain of g. Note the ERROR message in the table at left.

As we see in Example 8, inputs that make a denominator 0 or that yield a negative radicand in an even root are not in the domain of a function.

EXAMPLE 9 Find the domain of each of the following functions.

a) $f(x) = \dfrac{1}{x - 7}$

b) $h(x) = \dfrac{3x^2 - x + 7}{x^2 + 2x - 3}$

c) $f(x) = x^3 + |x|$

d) $g(x) = \sqrt[3]{x - 1}$

Solution

a) Because $x - 7 = 0$ when $x = 7$, the only input that results in a denominator of 0 is 7. The domain is $\{x \mid x \neq 7\}$. We can also write the solution using interval notation and the symbol $\cup$ for the **union**, or inclusion, of both sets: $(-\infty, 7) \cup (7, \infty)$.

b) We can substitute any real number in the numerator, but we must avoid inputs that make the denominator 0. To find those inputs, we solve $x^2 + 2x - 3 = 0$, or $(x + 3)(x - 1) = 0$. Since $x^2 + 2x - 3$ is 0 for -3 and 1, the domain consists of the set of all real numbers except -3 and 1, or $\{x \mid x \neq -3 \textit{ and } x \neq 1\}$, or $(-\infty, -3) \cup (-3, 1) \cup (1, \infty)$.

c) We can substitute any real number for x. The domain is the set of all real numbers, or $(-\infty, \infty)$.

d) Because the index is odd, the radicand, $x - 1$, can be any real number. Thus x can be any real number. The domain is all real numbers, or $(-\infty, \infty)$.

Now Try Exercises 53, 55, and 59.

Visualizing Domain and Range

Keep the following in mind regarding the *graph* of a function:

Domain = the set of a function's inputs, found on the horizontal x-axis;

Range = the set of a function's outputs, found on the vertical y-axis.

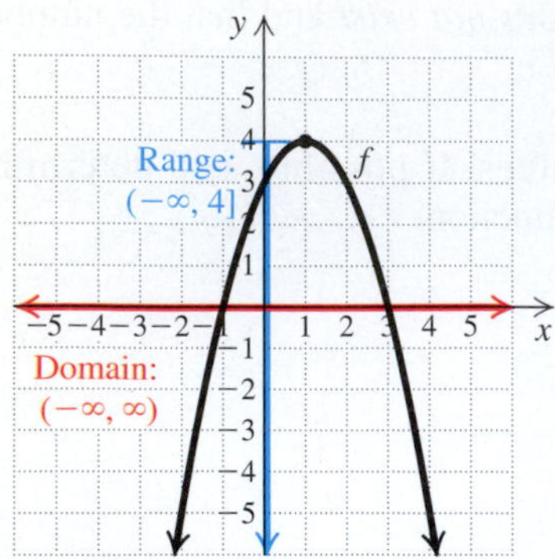

Consider the graph of function f, shown at left. To determine the domain of f, we look for the inputs on the x-axis that correspond to a point on the graph. We see that they include the entire set of real numbers, illustrated in red on the x-axis. Thus the domain is $(-\infty, \infty)$. To find the range, we look for the outputs on the y-axis that correspond to a point on the graph. We see that they include 4 and all real numbers less than 4, illustrated in blue on the y-axis. The bracket at 4 indicates that 4 is included in the interval. The range is $\{y \mid y \leq 4\}$, or $(-\infty, 4]$.

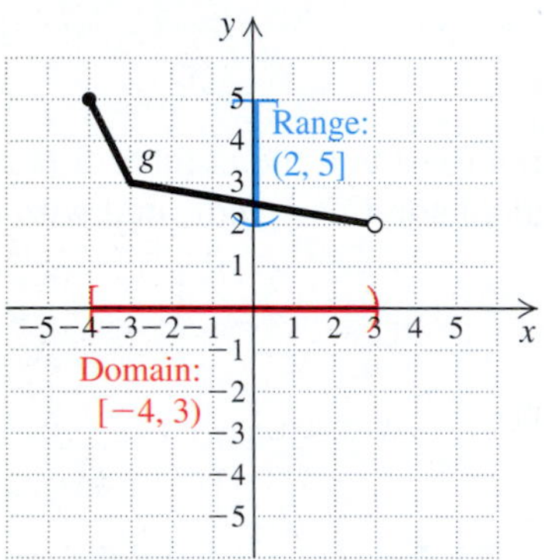

Let's now consider the graph of function g, shown at left. The solid dot shows that $(-4, 5)$ belongs to the graph. The open circle shows that $(3, 2)$ does *not* belong to the graph.

We see that the inputs of the function are -4 and all real numbers between -4 and 3, illustrated in red on the x-axis. The bracket at -4 indicates that -4 is included in the domain. The parenthesis at 3 indicates that 3 is not included in the domain. The domain is $\{x \mid -4 \leq x < 3\}$, or $[-4, 3)$. The outputs of the function are 5 and all real numbers between 2 and 5, illustrated in blue on the y-axis. The parenthesis at 2 indicates that 2 is not included in the range. The bracket at 5 indicates that 5 is included in the range. The range is $\{y \mid 2 < y \leq 5\}$, or $(2, 5]$.

EXAMPLE 10 Using the graph of the function, find the domain and the range of each of the following functions.

a) $f(x) = \frac{1}{2}x + 1$

b) $f(x) = \sqrt{x + 4}$

c) $f(x) = x^3 - x$

d) $f(x) = \dfrac{1}{x - 2}$

e) $f(x) = x^4 - 2x^2 - 3$

f) $f(x) = \sqrt{4 - (x - 3)^2}$

Solution

a)

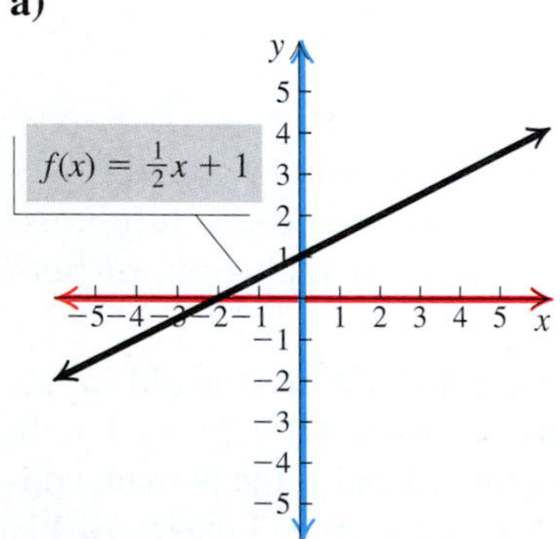

Domain = all real numbers, $(-\infty, \infty)$; range = all real numbers, $(-\infty, \infty)$

b)

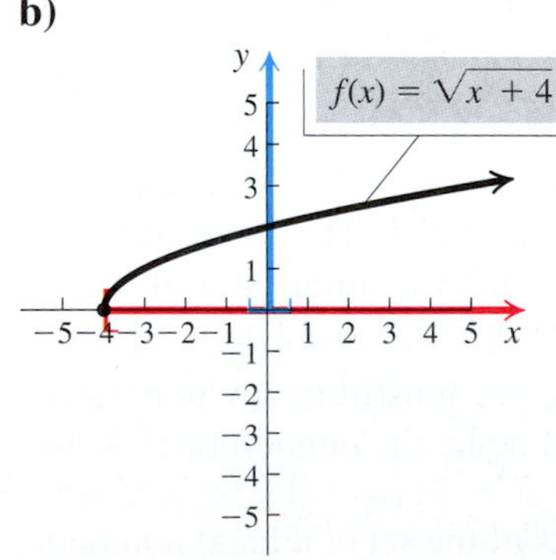

Domain = $[-4, \infty)$; range = $[0, \infty)$

c)

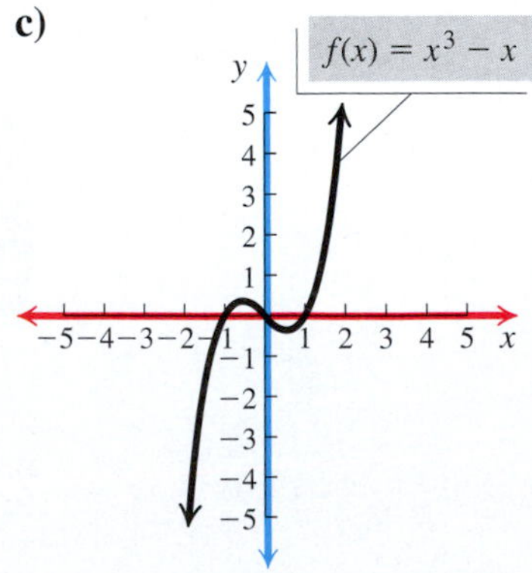

Domain = all real numbers, $(-\infty, \infty)$; range = all real numbers, $(-\infty, \infty)$

d)

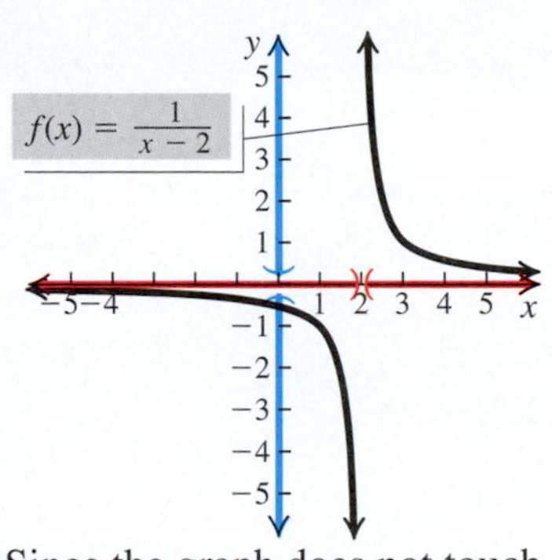

Since the graph does not touch or cross either the vertical line $x = 2$ or the x-axis $y = 0$, 2 is excluded from the domain and 0 is excluded from the range. Domain $= (-\infty, 2) \cup (2, \infty)$; range $= (-\infty, 0) \cup (0, \infty)$

e)

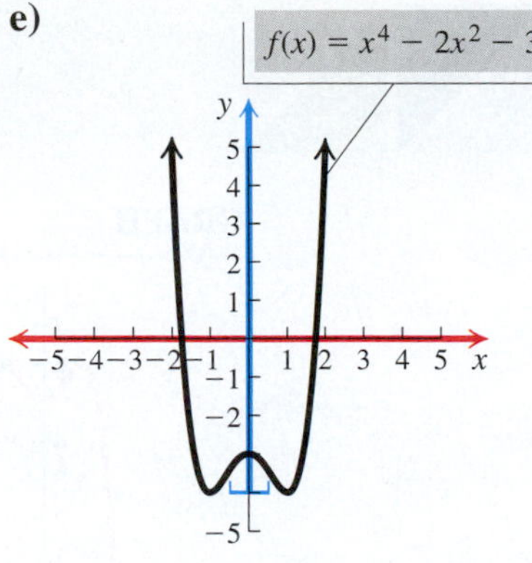

Domain = all real numbers, $(-\infty, \infty)$; range $= [-4, \infty)$

f)

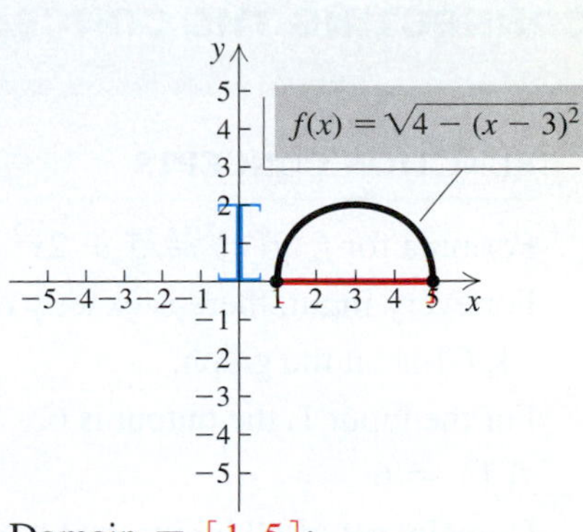

Domain $= [1, 5]$; range $= [0, 2]$

Now Try Exercises 75 and 83.

Always consider adding the reasoning of Example 9 to a graphical analysis. Think, "What can I input?" to find the domain. Think, "What do I get out?" to find the range. Thus, in Examples 10(c) and 10(e), it might not appear as though the domain is all real numbers because the graph rises steeply, but by examining the equation we see that we can indeed substitute any real number for x.

Applications of Functions

EXAMPLE 11 ***Linear Expansion of a Bridge.*** The linear expansion L of the steel center span of a suspension bridge that is 1420 m long is a function of the change in temperature t, in degrees Celsius, from winter to summer and is given by

$$L(t) = 0.000013 \cdot 1420 \cdot t,$$

where 0.000013 is the coefficient of linear expansion for steel and L is in meters. Find the linear expansion of the steel center span when the change in temperature from winter to summer is 30°, 42°, 50°, and 56° Celsius.

Solution We use a graphing calculator with the TABLE feature set in ASK mode to compute the function values. We find that

X	Y1	
30	.5538	
42	.77532	
50	.923	
56	1.0338	
X =		

$$L(30) = 0.5538 \text{ m},$$
$$L(42) = 0.77532 \text{ m},$$
$$L(50) = 0.923 \text{ m}, \quad \text{and}$$
$$L(56) = 1.0338 \text{ m}.$$

Now Try Exercise 87.

CONNECTING THE CONCEPTS

FUNCTION CONCEPTS

Formula for f: $f(x) = 5 + 2x^2 - x^4$.
For every input, there is exactly one output.
$(1, 6)$ is on the graph.
For the input 1, the output is 6.
$f(1) = 6$
Domain: set of all inputs $= (-\infty, \infty)$
Range: set of all outputs $= (-\infty, 6]$

GRAPH

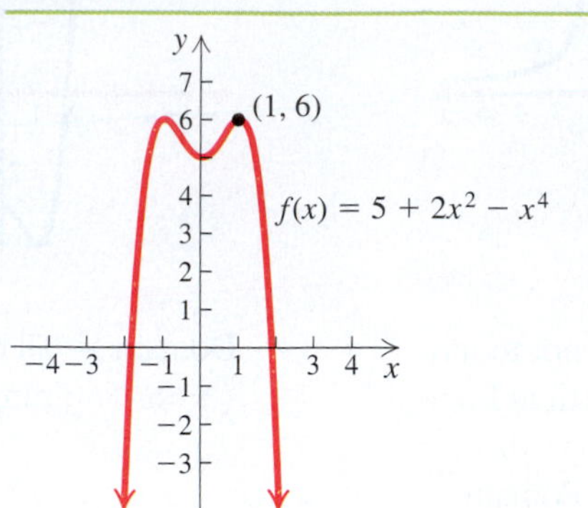

1.2 Exercise Set

In Exercises 1–14, determine whether the correspondence is a function.

1. $a \to w$, $b \to y$, $c \to z$

2. $m \to q$, $n \to s$, $o \to r$

3. $-6 \to 36$, $-2 \to 4$, $2 \to 4$

4. $-3 \to 2$, $1 \to 4$, $1 \to 6$, $5 \to 6$, $9 \to 8$

5. $m \to A$, $m \to B$, $n \to C$, $r \to D$, $s \to C$

6. $a \to r$, $b \to t$, $c \to s$, $d \to s$

7.

PAINTING	ARTIST
Night Watch	Vincent van Gogh
Old Guitarist	Claude Monet
Irises, Saint-Remy	Pablo Picasso
Starry Night	Rembrandt van Rijn
The Water-Lily Pond	Leonardo da Vinci
Sunflowers	
Mona Lisa	
Woman with a Parasol	
An Elephant	

8.

COMPOSER	BROADWAY MUSICAL
Marvin Hamlisch	*Billy Elliot the Musical*
Elton John	*Cats*
Claude-Michel Schönberg	*A Chorus Line*
Andrew Lloyd Webber	*Evita*
	Les Misérables
	The Lion King
	The Phantom of the Opera
	Miss Saigon

	DOMAIN	CORRESPONDENCE	RANGE
9.	A set of cars in a parking lot	Each car's license number	A set of letters and numbers
10.	A set of people in a town	A doctor a person uses	A set of doctors
11.	The integers less than 9	Five times the integer	A subset of integers
12.	A set of members of a rock band	An instrument each person plays	A set of instruments
13.	A set of students in a class	A student sitting in a neighboring seat	A set of students

14. A set of bags of chips on a shelf | Each bag's weight | A set of weights

Determine whether the relation is a function. Identify the domain and the range.

15. $\{(2, 10), (3, 15), (4, 20)\}$

16. $\{(3, 1), (5, 1), (7, 1)\}$

17. $\{(-7, 3), (-2, 1), (-2, 4), (0, 7)\}$

18. $\{(1, 3), (1, 5), (1, 7), (1, 9)\}$

19. $\{(-2, 1), (0, 1), (2, 1), (4, 1), (-3, 1)\}$

20. $\{(5, 0), (3, -1), (0, 0), (5, -1), (3, -2)\}$

21. Given that $g(x) = 3x^2 - 2x + 1$, find each of the following.

a) $g(0)$ **b)** $g(-1)$
c) $g(3)$ **d)** $g(-x)$
e) $g(1 - t)$

22. Given that $f(x) = 5x^2 + 4x$, find each of the following.

a) $f(0)$ **b)** $f(-1)$
c) $f(3)$ **d)** $f(t)$
e) $f(t - 1)$

23. Given that $g(x) = x^3$, find each of the following.

a) $g(2)$ **b)** $g(-2)$
c) $g(-x)$ **d)** $g(3y)$
e) $g(2 + h)$

24. Given that $f(x) = 2|x| + 3x$, find each of the following.

a) $f(1)$ **b)** $f(-2)$
c) $f(-x)$ **d)** $f(2y)$
e) $f(2 - h)$

25. Given that

$$g(x) = \frac{x - 4}{x + 3},$$

find each of the following.

a) $g(5)$ **b)** $g(4)$
c) $g(-3)$ **d)** $g(-16.25)$
e) $g(x + h)$

26. Given that

$$f(x) = \frac{x}{2 - x},$$

find each of the following.

a) $f(2)$ **b)** $f(1)$
c) $f(-16)$ **d)** $f(-x)$
e) $f\left(-\frac{2}{3}\right)$

27. Find $g(0)$, $g(-1)$, $g(5)$, and $g\left(\frac{1}{2}\right)$ for

$$g(x) = \frac{x}{\sqrt{1 - x^2}}.$$

28. Find $h(0)$, $h(2)$, and $h(-x)$ for

$$h(x) = x + \sqrt{x^2 - 1}.$$

In Exercises 29 and 30, use a graphing calculator and the TABLE *feature set in* ASK *mode.*

29. Given that

$$g(x) = 0.06x^3 - 5.2x^2 - 0.8x,$$

find $g(-2.1)$, $g(5.08)$, and $g(10.003)$. Round answers to the nearest tenth.

30. Given that

$$h(x) = 3x^4 - 10x^3 + 5x^2 - x + 6,$$

find $h(-11)$, $h(7)$, and $h(15)$.

Graph the function.

31. $f(x) = \frac{1}{2}x + 3$

32. $f(x) = \sqrt{x} - 1$

33. $f(x) = -x^2 + 4$

34. $f(x) = x^2 + 1$

35. $f(x) = \sqrt{x - 1}$

36. $f(x) = x - \frac{1}{2}x^3$

In Exercises 37–42, a graph of a function is shown. Using the graph, find the indicated function values; that is, given the inputs, find the outputs.

37. $h(1)$, $h(3)$, and $h(4)$

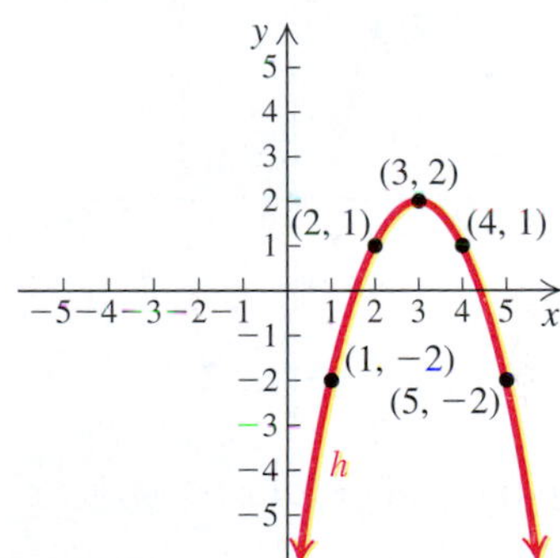

38. $t(-4)$, $t(0)$, and $t(3)$

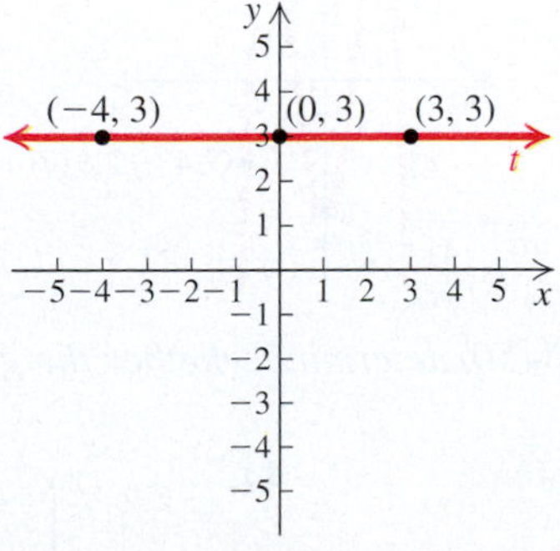

39. $s(-4)$, $s(-2)$, and $s(0)$

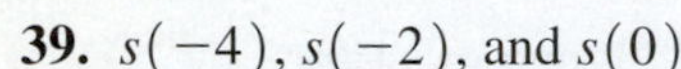

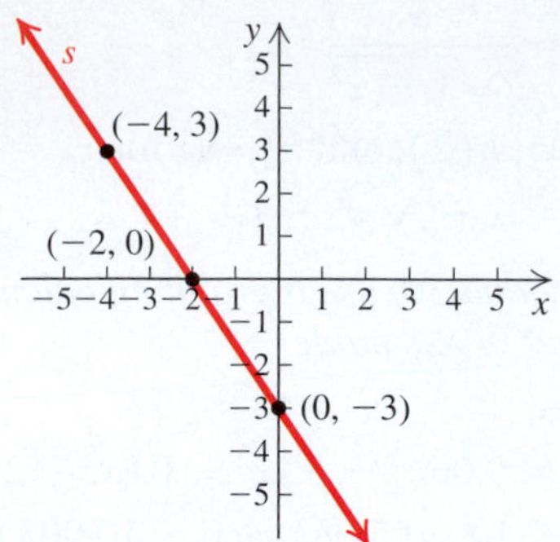

40. $g(-4)$, $g(-1)$, and $g(0)$

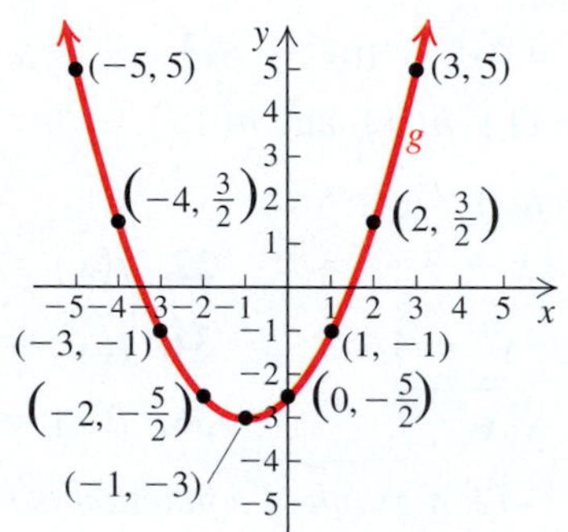

41. $f(-1)$, $f(0)$, and $f(1)$

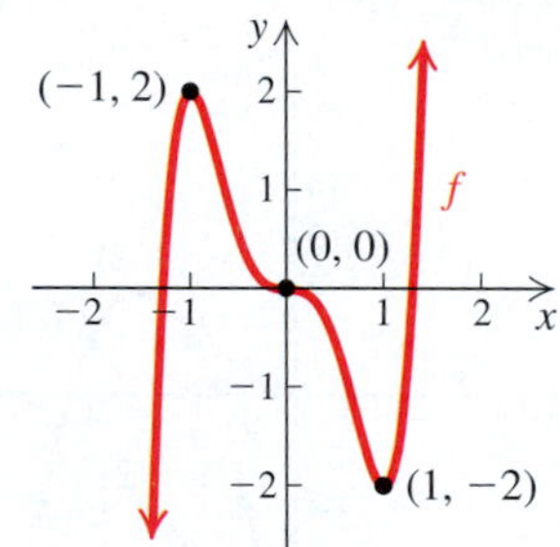

42. $g(-2)$, $g(0)$, and $g(2.4)$

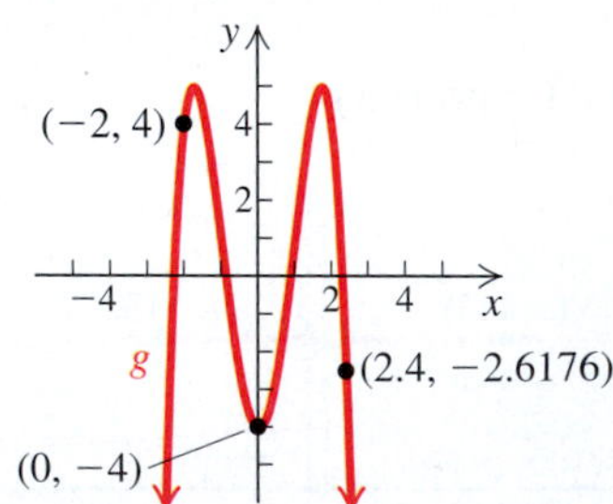

In Exercises 43–50, determine whether the graph is that of a function.

43.

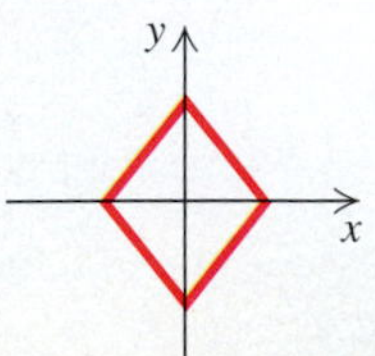

44.

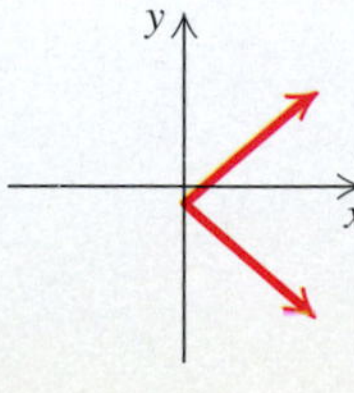

45.

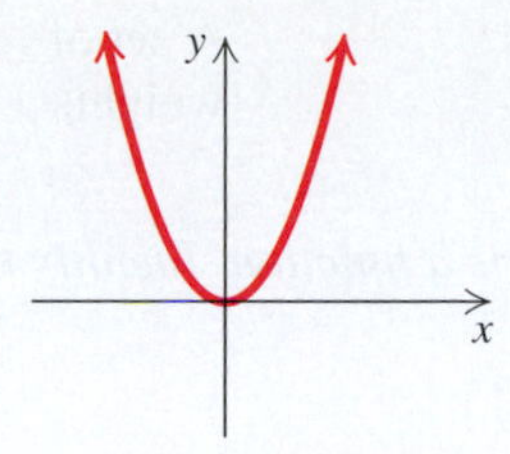

46.

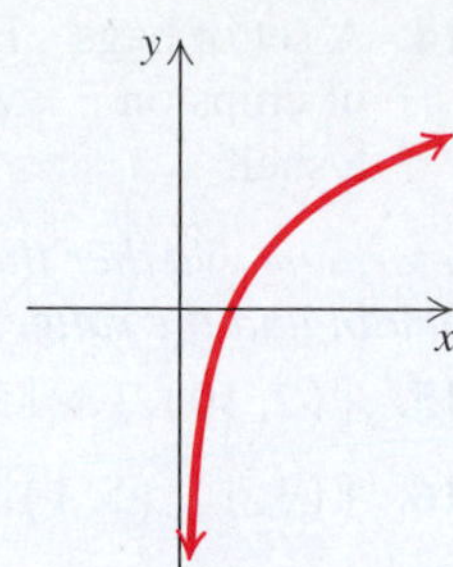

47.

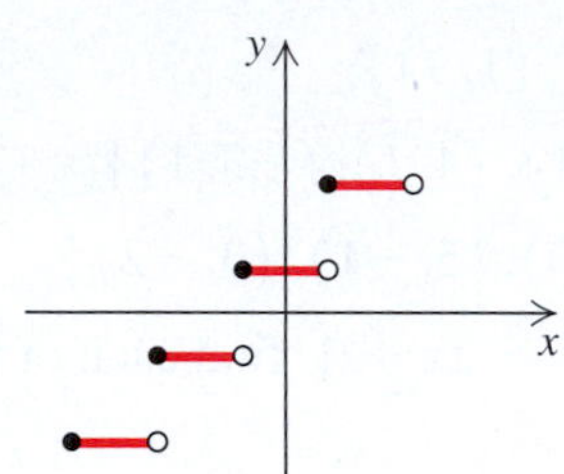

48.

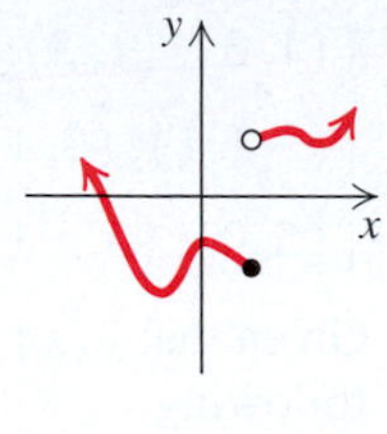

49.

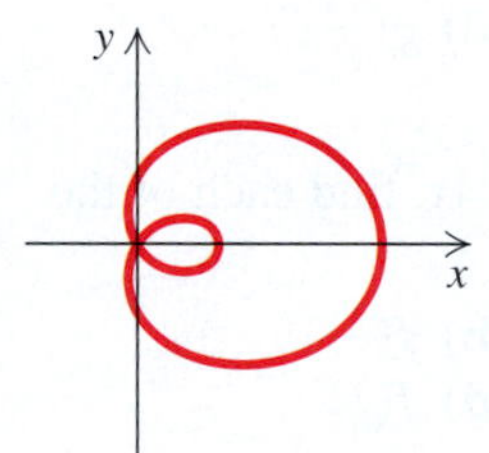

50.

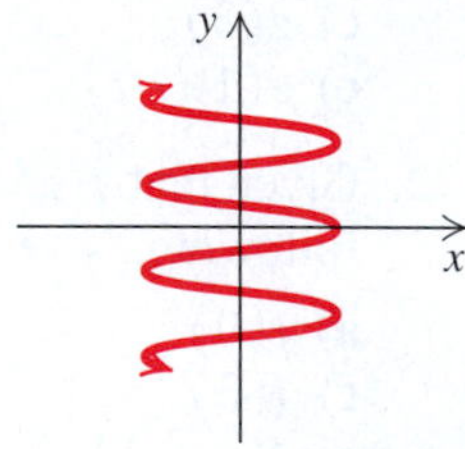

Find the domain of the function. Do not use a graphing calculator.

51. $f(x) = 7x + 4$

52. $f(x) = |3x - 2|$

53. $f(x) = |6 - x|$

54. $f(x) = \dfrac{1}{x^4}$

55. $f(x) = 4 - \dfrac{2}{x}$

56. $f(x) = \dfrac{1}{5}x^2 - 5$

57. $f(x) = \dfrac{x + 5}{2 - x}$

58. $f(x) = \dfrac{8}{x + 4}$

59. $f(x) = \dfrac{1}{x^2 - 4x - 5}$

60. $f(x) = \dfrac{(x - 2)(x + 9)}{x^3}$

61. $f(x) = \sqrt[3]{x + 10} - 1$

62. $f(x) = \sqrt[3]{4 - x}$

63. $f(x) = \dfrac{8 - x}{x^2 - 7x}$

64. $f(x) = \dfrac{x^4 - 2x^3 + 7}{3x^2 - 10x - 8}$

65. $f(x) = \frac{1}{10}|x|$

66. $f(x) = x^2 - 2x$

In Exercises 67–74, determine the domain and the range of the function.

67.

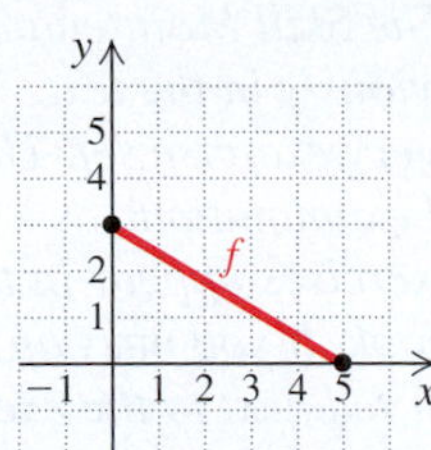

68.

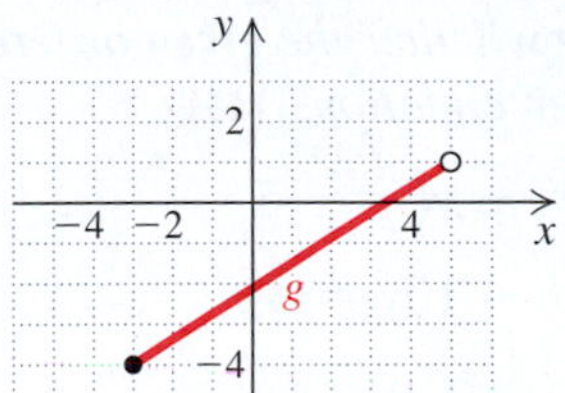

69.

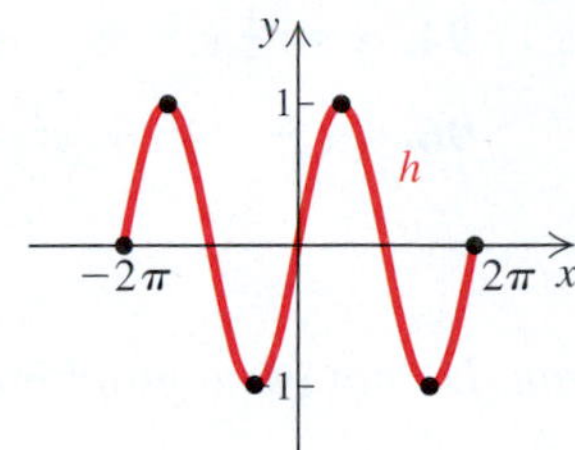

70.

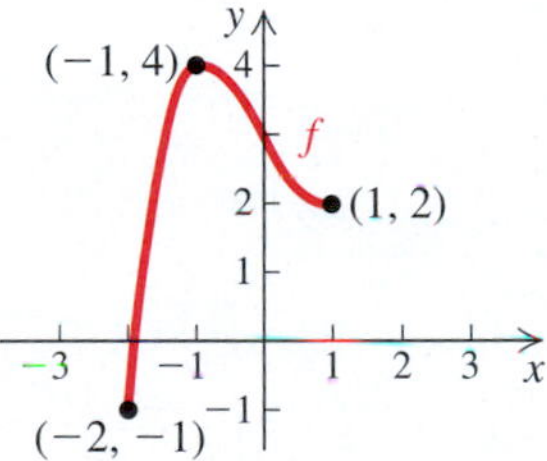

71.

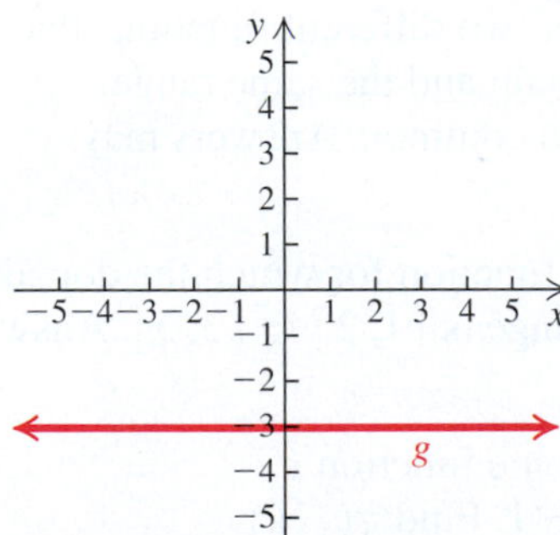

72.

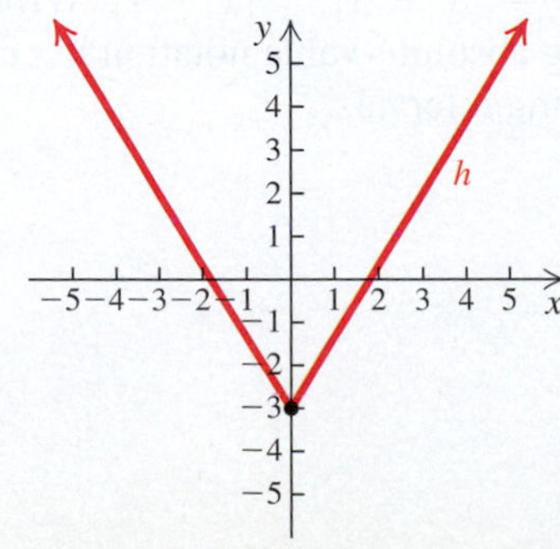

73.

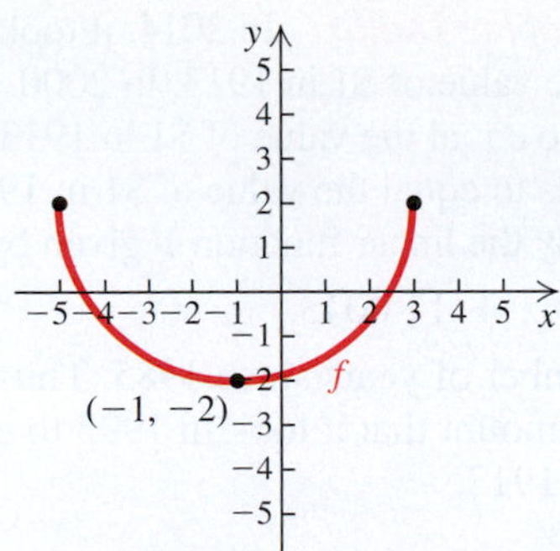

74.

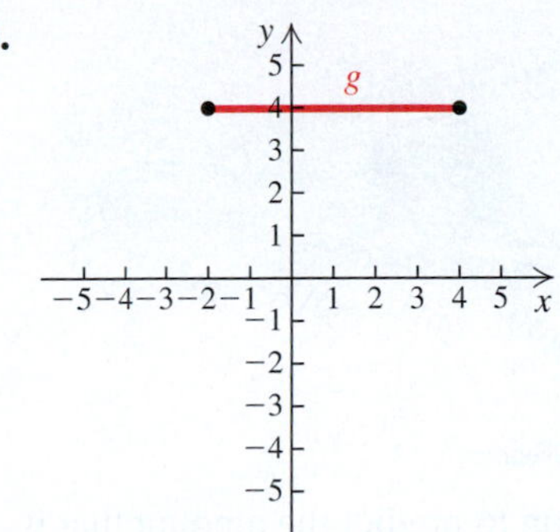

Graph the function with a graphing calculator. Then visually estimate the domain and the range.

75. $f(x) = |x|$

76. $f(x) = |x| - 2$

77. $f(x) = 3x - 2$

78. $f(x) = 5 - 3x$

79. $f(x) = \dfrac{1}{x - 3}$

80. $f(x) = \dfrac{1}{x + 1}$

81. $f(x) = (x - 1)^3 + 2$

82. $f(x) = (x - 2)^4 + 1$

83. $f(x) = \sqrt{7 - x}$

84. $f(x) = \sqrt{x + 8}$

85. $f(x) = -x^2 + 4x - 1$

86. $f(x) = 2x^2 - x^4 + 5$

87. *Boiling Point and Elevation.* The elevation E, in meters, above sea level at which the boiling point of water is t degrees Celsius is given by the function

$$E(t) = 1000(100 - t) + 580(100 - t)^2.$$

At what elevation is the boiling point 99.5°? 100°?

88. *Windmill Power.* Under certain conditions, the power P, in watts per hour, generated by a windmill with winds blowing v miles per hour is given by

$$P(v) = 0.015v^3.$$

Find the power generated by 15-mph winds and by 35-mph winds.

89. *Decreasing Value of the Dollar.* In 2014, it took \$23.63 to equal the value of \$1 in 1913. In 2000, it took only \$17.39 to equal the value of \$1 in 1913. The amount that it takes to equal the value of \$1 in 1913 can be estimated by the linear function V given by

$$V(x) = 0.4306x + 11.0043,$$

where x is the number of years since 1985. Thus, $V(10)$ gives the amount that it took in 1995 to equal the value of \$1 in 1913.

Source: usinflationcalculator.com

a) Use this function to predict the amount that it will take in 2018 and in 2025 to equal the value of \$1 in 1913.
b) When will it take approximately \$32 to equal the value of \$1 in 1913?

90. *Population of the United States.* The population P of the United States in 1960 was 179,323,175. In 2015, the population was 320,400,215. The population of the United States can be estimated by the linear function P given by

$$P(x) = 2{,}511{,}040x + 151{,}143{,}509,$$

where x is the number of years after 1950. Thus, $P(20)$ gives the population in 1970.

a) Use this function to estimate the population in 1980 and in 2020.
b) When will the population be approximately 400,000,000?

Skill Maintenance

To the student and the instructor: *The Skill Maintenance exercises review skills covered previously in the text. You can expect such exercises in every exercise set. They provide excellent review for a final examination. Answers to all skill maintenance exercises appear in the answer section at the back of the book. If you miss an exercise, restudy the objective shown in red next to the exercise or the instruction line that precedes it.*

Use substitution to determine whether the given ordered pairs are solutions of the given equation. **[1.1]**

91. $(-3, -2), (2, -3);\ y^2 - x^2 = -5$

92. $\left(\frac{4}{5}, -2\right), \left(\frac{11}{5}, \frac{1}{10}\right);\ 15x - 10y = 32$

Graph the equation. **[1.1]**

93. $y = (x - 1)^2$

94. $y = \frac{1}{3}x - 6$

95. $-2x - 5y = 10$

96. $(x - 3)^2 + y^2 = 4$

Synthesis

Find the domain of the function. Do not use a graphing calculator.

97. $f(x) = \sqrt[4]{2x + 5} + 3$

98. $f(x) = \dfrac{\sqrt{x + 1}}{x}$

99. $f(x) = \dfrac{\sqrt{x + 6}}{(x + 2)(x - 3)}$

100. $f(x) = \sqrt{x} - \sqrt{4 - x}$

101. Give an example of two different functions that have the same domain and the same range, but have no pairs in common. Answers may vary.

102. Draw a graph of a function for which the domain is $[-4, 4]$ and the range is $[1, 2] \cup [3, 5]$. Answers may vary.

103. Suppose that for some function g, $g(x + 3) = 2x + 1$. Find $g(-1)$.

104. Suppose that $f(x) = |x + 3| - |x - 4|$. Write $f(x)$ without using absolute-value notation if x is in each of the following intervals.

a) $(-\infty, -3)$
b) $[-3, 4)$
c) $[4, \infty)$

1.3 Linear Functions, Slope, and Applications

- Determine the slope of a line given two points on the line.
- Solve applied problems involving slope, or average rate of change.
- Find the slope and the y-intercept of a line given the equation $y = mx + b$, or $f(x) = mx + b$.
- Graph a linear equation using the slope and the y-intercept.
- Solve applied problems involving linear functions.

In real-life situations, we often need to make decisions on the basis of limited information. When the given information is used to formulate an equation or an inequality that at least approximates the situation mathematically, we have created a **model**. One of the most frequently used mathematical models is *linear*. The graph of a linear model is a straight line.

Linear Functions

Let's examine the connections among equations, functions, and graphs that are *straight lines*. First, examine the graphs of linear functions and nonlinear functions shown here. Note that the graphs of the two types of functions are quite different.

Linear Functions

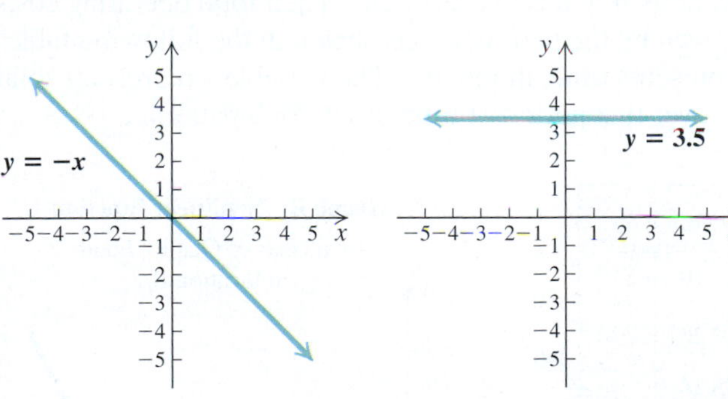

Nonlinear Functions

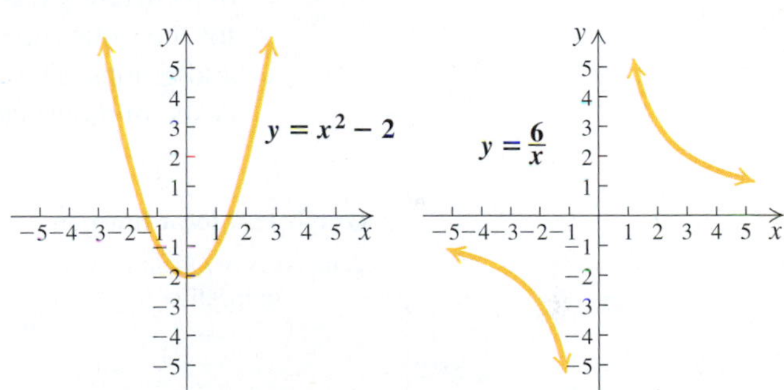

We begin with the definition of a linear function and related terminology, which are illustrated with graphs on the following page.

LINEAR FUNCTION

A function f is a **linear function** if it can be written as

$$f(x) = mx + b,$$

where m and b are constants.

If $m = 1$ and $b = 0$, then the function is the **identity function** $f(x) = x$.
If $m = 0$, then the function is a **constant function** $f(x) = b$.

Linear function:
$y = mx + b$

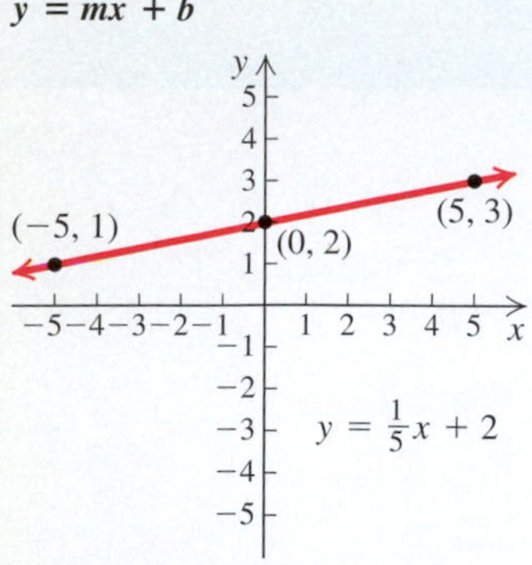

Identity function:
$y = 1 \cdot x + 0$, **or** $y = x$

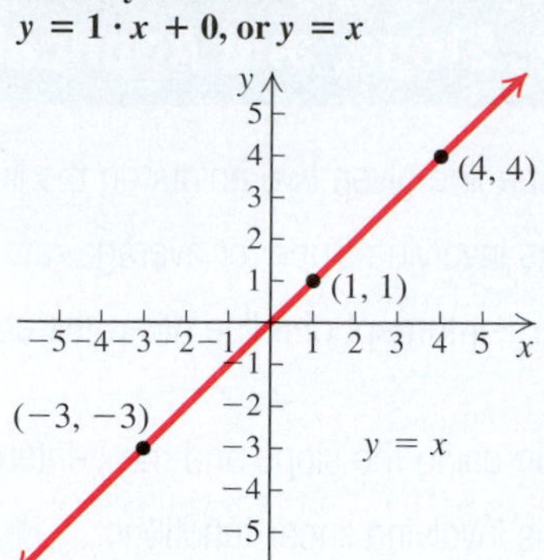

Constant function:
$y = 0 \cdot x + b$, **or** $y = b$ **(Horizontal line)**

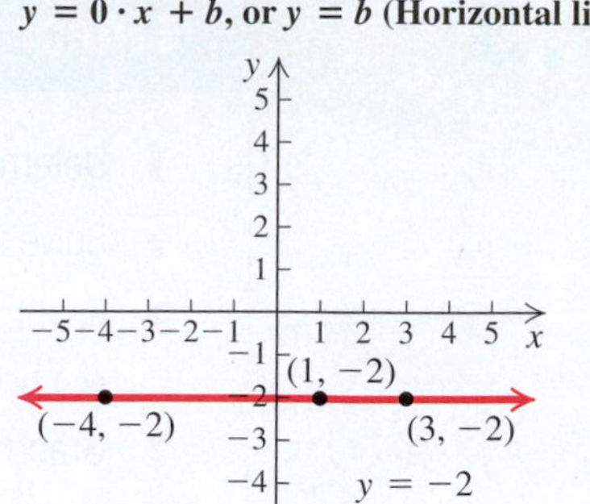

Vertical line: $x = a$
(*not* a function)

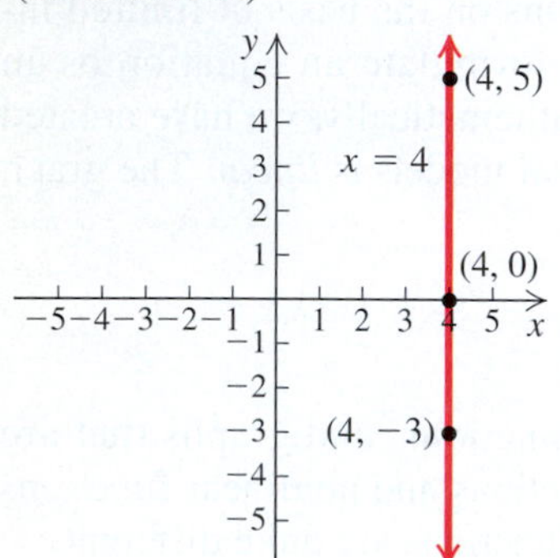

> **HORIZONTAL LINES AND VERTICAL LINES**
>
> **Horizontal lines** are given by equations of the type $y = b$ or $f(x) = b$. (They *are* functions.)
>
> **Vertical lines** are given by equations of the type $x = a$. (They *are not* functions.)

The Linear Function $f(x) = mx + b$ and Slope

To attach meaning to the constant m in the equation $f(x) = mx + b$, we first consider an application. Suppose that Quality Foods is a wholesale supplier to restaurants that currently has stores in locations A and B in a large city. Their total operating costs for the same time period are given by the two functions shown in the following tables and graphs. The variable x represents time, in months. The variable y represents total costs, in thousands of dollars, over that period of time. Look for a pattern.

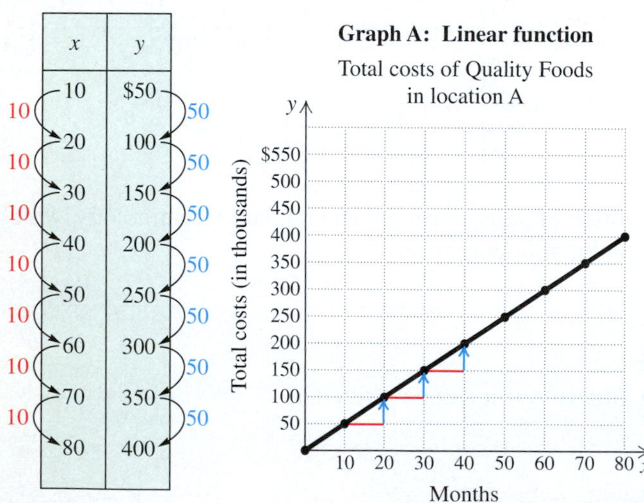

x	y
10	$50
20	100
30	150
40	200
50	250
60	300
70	350
80	400

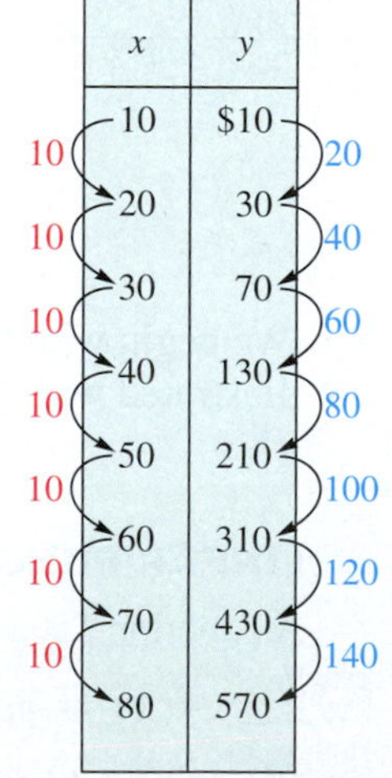

x	y
10	$10
20	30
30	70
40	130
50	210
60	310
70	430
80	570

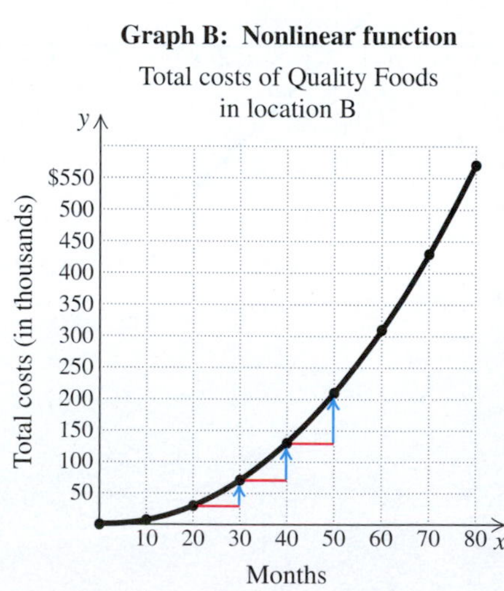

We see in graph A that *every* change of 10 months results in a $50 thousand change in total costs. But in graph B, changes of 10 months do *not* result in constant changes in total costs. This is a way to distinguish linear functions from nonlinear functions. The rate at which a linear function changes, or the steepness of its graph, is constant.

Mathematically, we define the steepness, or the **slope**, of a line as the ratio of its vertical change (*rise*) to the corresponding horizontal change (*run*). Slope represents the **rate of change** of y with respect to x.

SLOPE

The **slope** m of a line containing points (x_1, y_1) and (x_2, y_2) is given by

$$m = \frac{\text{rise}}{\text{run}} = \frac{\text{the change in } y}{\text{the change in } x} = \frac{y_2 - y_1}{x_2 - x_1} = \frac{y_1 - y_2}{x_1 - x_2}.$$

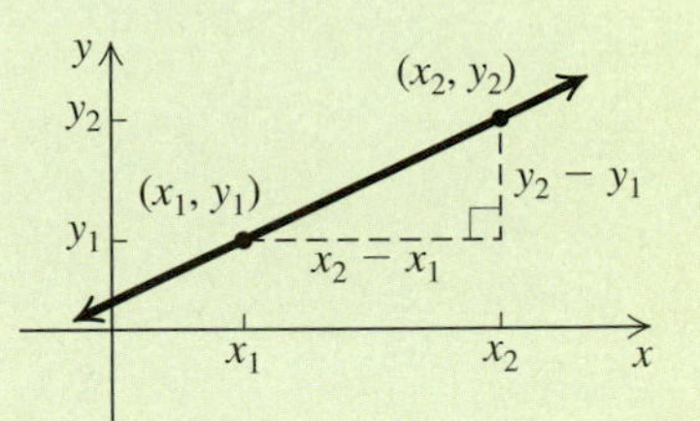

Just in Time 5

EXAMPLE 1 Graph the function $f(x) = -\frac{2}{3}x + 1$ and determine its slope.

Solution Since the equation for f is in the form $f(x) = mx + b$, we know that it is a linear function. We can graph it by connecting two points on the graph with a straight line. We calculate two ordered pairs, plot the points, graph the function, and determine the slope:

$$f(3) = -\frac{2}{3}\cdot 3 + 1 = -2 + 1 = -1;$$

$$f(9) = -\frac{2}{3}\cdot 9 + 1 = -6 + 1 = -5;$$

Pairs: $(3, -1)$, $(9, -5)$;

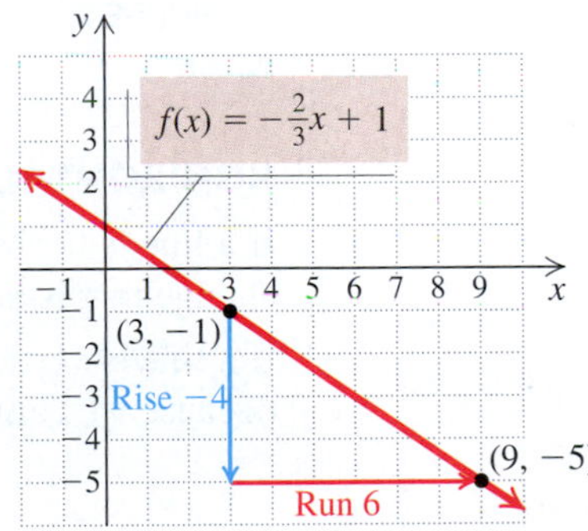

$$\text{Slope} = m = \frac{f(x_2) - f(x_1)}{x_2 - x_1} = \frac{y_2 - y_1}{x_2 - x_1}$$

$$= \frac{-5 - (-1)}{9 - 3} = \frac{-4}{6} = -\frac{2}{3}.$$

The slope is the same for any two points on a line. Thus, to check our work, note that $f(6) = -\frac{2}{3}\cdot 6 + 1 = -4 + 1 = -3$. Using the points $(6, -3)$ and $(3, -1)$, we have

$$m = \frac{-1 - (-3)}{3 - 6} = \frac{2}{-3} = -\frac{2}{3}.$$

EXPLORING WITH TECHNOLOGY

We can animate the effect of the slope m in linear functions of the type $f(x) = mx$ with a graphing calculator. Graph the equations

$$y_1 = x, \quad y_2 = 2x, \quad y_3 = 5x, \text{ and } y_4 = 10x$$

by entering them as $y_1 = \{1, 2, 5, 10\}x$. What do you think the graph of $y = 128x$ will look like?

Clear the screen and graph the equations

$$y_1 = -x, \quad y_2 = -2x, \quad y_3 = -4x, \text{ and } y_4 = -10x.$$

What do you think the graph of $y = -200x$ will look like?

We can also use the points in the opposite order when computing slope:

$$m = \frac{-3 - (-1)}{6 - 3} = \frac{-2}{3} = -\frac{2}{3}.$$

Note too that the slope of the line is the number m in the equation for the function $f(x) = -\frac{2}{3}x + 1$.

Now Try Exercises 7 and 31.

The *slope* of the line given by $f(x) = mx + b$ is m.

If a line slants up from left to right, the change in x and the change in y have the same sign, so the line has a positive slope. The larger the slope, the steeper the line, as shown in Fig. 1. If a line slants down from left to right, the change in x and the change in y are of opposite signs, so the line has a negative slope. The larger the absolute value of the slope, the steeper the line, as shown in Fig. 2. Considering $y = mx$ when $m = 0$, we have $y = 0x$, or $y = 0$. Note that this horizontal line is the x-axis, as shown in Fig. 3.

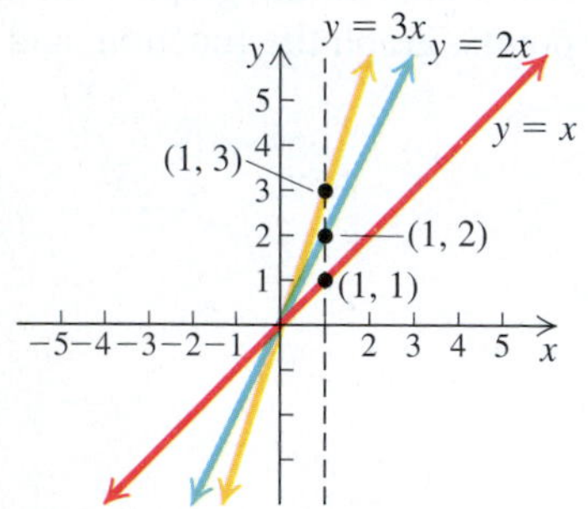

FIGURE 1.

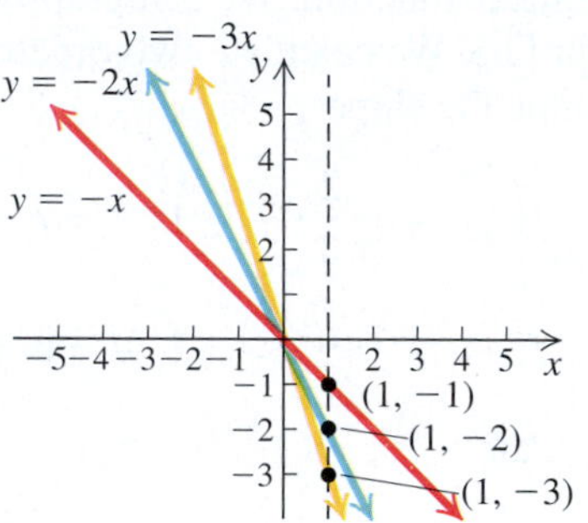

FIGURE 2.

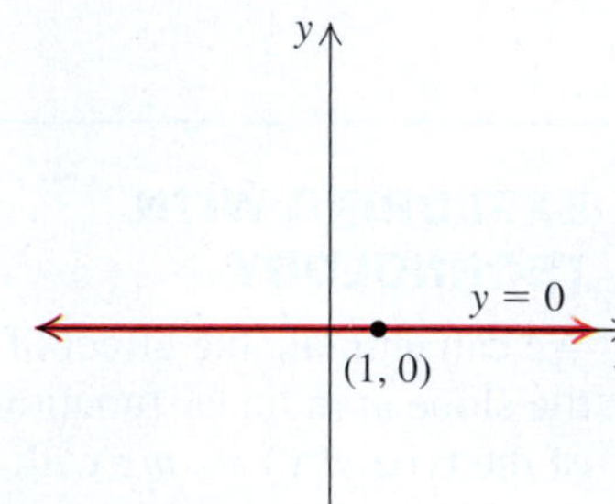

FIGURE 3.

HORIZONTAL LINES AND VERTICAL LINES

If a line is horizontal, the change in y for any two points is 0 and the change in x is nonzero. Thus a horizontal line has slope 0. (See Fig. 4.)

If a line is vertical, the change in x is 0. Thus the slope is *not defined* because we cannot divide by 0. (See Fig. 5.)

Horizontal lines

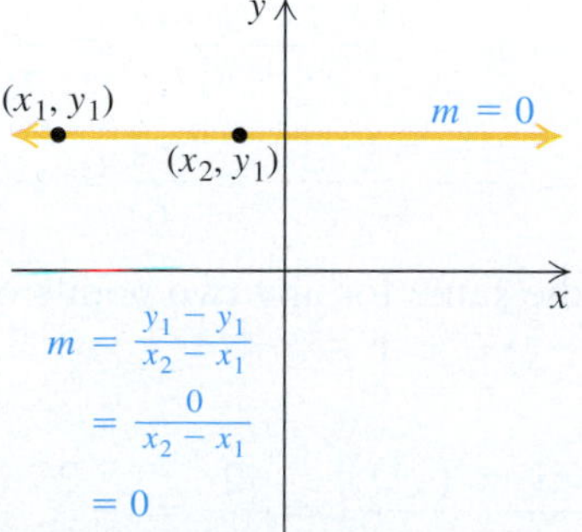

FIGURE 4.

Vertical lines

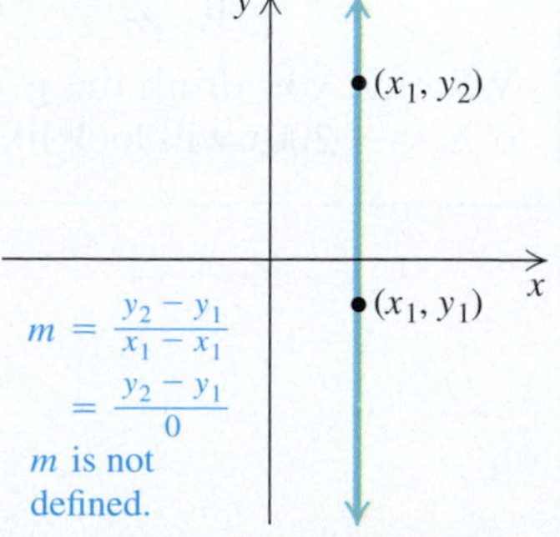

FIGURE 5.

Note that zero slope and an undefined slope are two very different concepts.

EXAMPLE 2 Graph each linear equation and determine its slope.

a) $x = -2$ **b)** $y = \frac{5}{2}$

Solution

a) Since y is missing in $x = -2$, any value for y will do.

x	y
-2	0
-2	3
-2	-4

Choose any number for y; x must be -2.

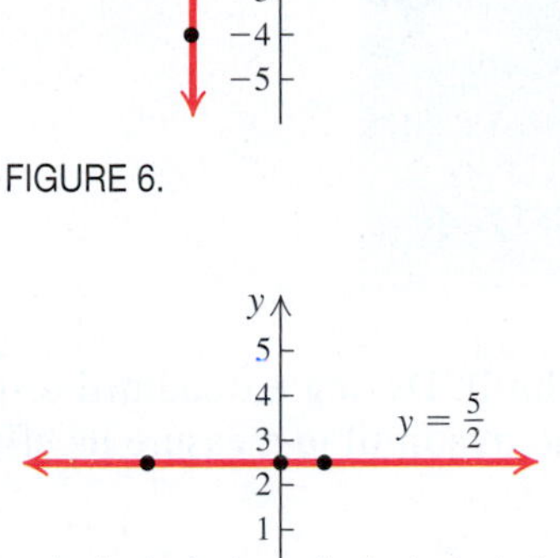

FIGURE 6.

The graph is a *vertical line* 2 units to the left of the y-axis. (See Fig. 6.) The slope is not defined. The graph is *not* the graph of a function.

b) Since x is missing in $y = \frac{5}{2}$, any value for x will do.

x	y
0	$\frac{5}{2}$
-3	$\frac{5}{2}$
1	$\frac{5}{2}$

Choose any number for x; y must be $\frac{5}{2}$.

FIGURE 7.

The graph is a *horizontal line* $\frac{5}{2}$, or $2\frac{1}{2}$, units above the x-axis. (See Fig. 7.) The slope is 0. The graph is the graph of a constant function.

Now Try Exercises 17 and 23.

Applications of Slope

Slope has many real-world applications. Numbers like 2%, 4%, and 7% are often used to represent the **grade** of a road. Such a number is meant to tell how steep a road is on a hill or a mountain. For example, a 4% grade means that the road rises (or falls) 4 ft for every horizontal distance of 100 ft.

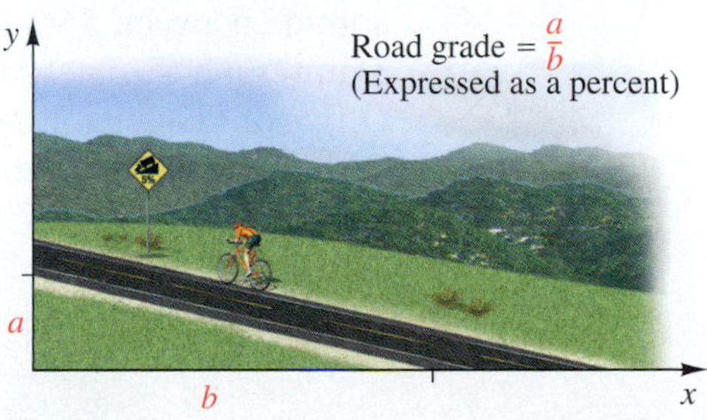

The 2014 Olympic downhill course at Rosa Khutor Alpine Resort, located 40 km from Sochi, Russia, has the largest vertical drop ever built for an Olympic event. With a run of nearly 3500 m and a vertical drop of over 1075 m, the resulting grade, or slope, is approximately 31%. (*Source*: "Sochi's Gold Medal Ski Resort," by Brian Pinella, sochimagazine.com)

The concept of grade is also used with a treadmill. During a treadmill test, a cardiologist might change the slope, or grade, of the treadmill to measure its effect on a person's heart rate.

Another example occurs in hydrology. The strength or force of a river depends on how far the river falls vertically compared to how far it flows horizontally.

EXAMPLE 3 ***Curb Ramps.*** Curb ramps provide independent access to sidewalks for those who use wheelchairs. Guidelines for the grade of a curb ramp suggest a grade between 5.9% and 8.3%. A federal law states that every vertical rise of 1 ft requires a horizontal run of at least 12 ft. (*Source*: Federal Highway Administration, Office of Planning, Environment, and Realty) Find the grade of the curb ramp shown in the following figure.

Solution The grade, or slope, is given by $m = \dfrac{3 \text{ in.}}{42 \text{ in.}} = \dfrac{1}{14} \approx 7.1\%.$

AVERAGE RATE OF CHANGE

Slope can also be considered as an **average rate of change**. To find the average rate of change between any two data points on a graph, we determine the slope of the line that passes through the two points.

EXAMPLE 4 ***Food Stamp Program.*** The number of people participating in the federal Supplemental Nutrition Assistance Program has increased from 17.2 million in 2000 to 47.6 million in 2013. The following graph illustrates this upward trend. Find the average rate of change in the number of people using food stamps from 2000 to 2013.

Enrollment in the Federal Supplemental Nutrition Assistance Program

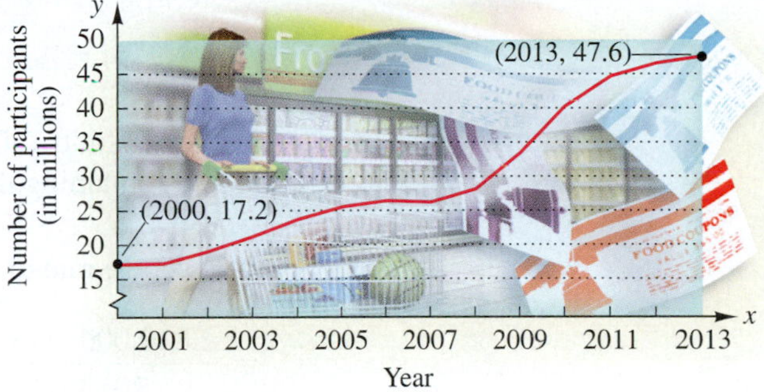

Source: U.S. Department of Agriculture

Solution We use the coordinates of two points on the graph. In this case, we use (2000, 17.2) and (2013, 47.6). Then we compute the slope, or the average rate of change, as follows:

$$\begin{aligned} \text{Slope} &= \text{Average rate of change} \\ &= \frac{\text{Change in } y}{\text{Change in } x} = \frac{47.6 - 17.2}{2013 - 2000} \\ &= \frac{30.4}{13} \approx 2.3. \end{aligned}$$

The result tells us that each year from 2000 to 2013, the number of participants in the federal Supplemental Nutrition Assistance Program increased an average of 2.3 million. The average rate of change over this 13-year period was an increase of 2.3 million participants per year.

Now Try Exercise 41.

EXAMPLE 5 ***Oil Imports.*** Increased oil production in the United States has resulted in decreased imports of crude oil. The total number of barrels imported in 2008 was 3,590,000. This number had decreased to 2,810,000 barrels in 2013. (*Source*: U.S. Census Bureau) Find the average rate of change in crude oil imports from 2008 to 2013.

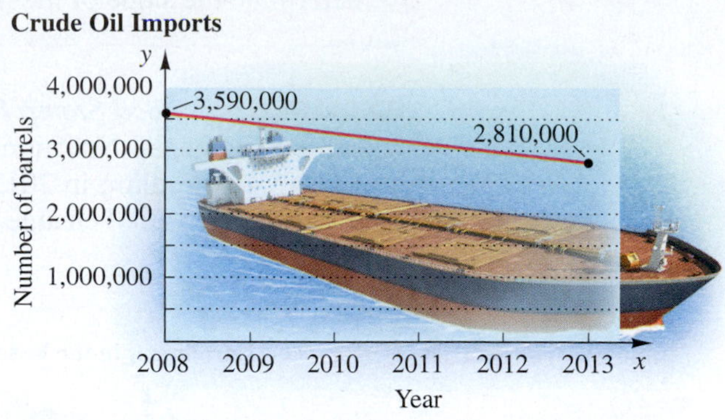

Source: U.S. Census Bureau

Solution Using the points (2008, 3,590,000) and (2013, 2,810,000), we compute the slope of the line containing these two points:

$$\text{Slope} = \text{Average rate of change} = \frac{\text{Change in } y}{\text{Change in } x}$$

$$= \frac{2{,}810{,}000 - 3{,}590{,}000}{2013 - 2008} = \frac{-780{,}000}{5} = -156{,}000.$$

The result tells us that each year from 2008 to 2013, the number of barrels of imported crude oil decreased on average 156,000 barrels. The average rate of change over the 5-year period was a decrease of 156,000 barrels per year.

Now Try Exercise 47.

Slope–Intercept Equations of Lines

***y*-INTERCEPT**

REVIEW SECTION 1.1.

Compare the graphs of the equations

$$y = 3x \quad \text{and} \quad y = 3x - 2.$$

Note that the graph of $y = 3x - 2$ is a shift of the graph of $y = 3x$ down 2 units and that $y = 3x - 2$ has y-intercept $(0, -2)$. That is, the graph is parallel to $y = 3x$ and it crosses the y-axis at $(0, -2)$. The point $(0, -2)$ is the *y-intercept* of the graph.

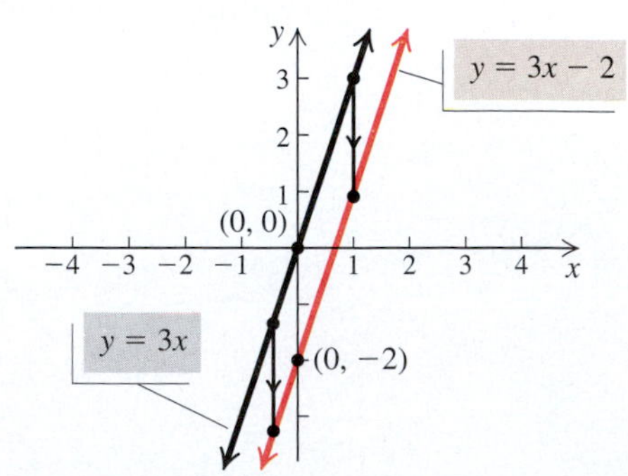

EXPLORING WITH TECHNOLOGY

We can use a graphing calculator to explore the effect of the constant b in linear equations of the type $f(x) = mx + b$. Begin with the graph of $y = x$. Now graph the lines $y = x + 3$ and $y = x - 4$ in the same viewing window. Try entering these equations as $y = x + \{0, 3, -4\}$ and compare the graphs. How do the last two lines differ from $y = x$? What do you think the line $y = x - 6$ will look like?

Clear the first set of equations and graph $y = -0.5x$, $y = -0.5x - 4$, and $y = -0.5x + 3$ in the same viewing window. Describe what happens to the graph of $y = -0.5x$ when a number b is added.

THE SLOPE–INTERCEPT EQUATION

The linear function f given by

$$f(x) = mx + b$$

is written in slope–intercept form. The graph of an equation in this form is a straight line parallel to $f(x) = mx$. The constant m is called the slope, and the y-intercept is $(0, b)$.

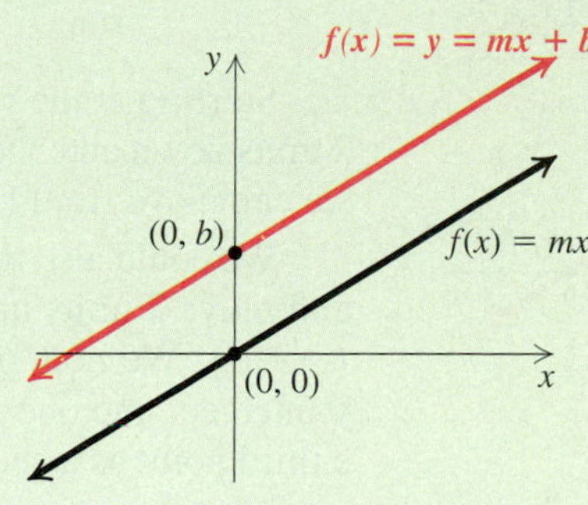

We can read the slope m and the y-intercept $(0, b)$ directly from the equation of a line written in slope–intercept form $y = mx + b$.

EXAMPLE 6 Find the slope and the y-intercept of the line with equation $y = -0.25x - 3.8$.

Solution

$$y = -0.25x - 3.8$$

Slope $= -0.25$; y-intercept $= (0, -3.8)$

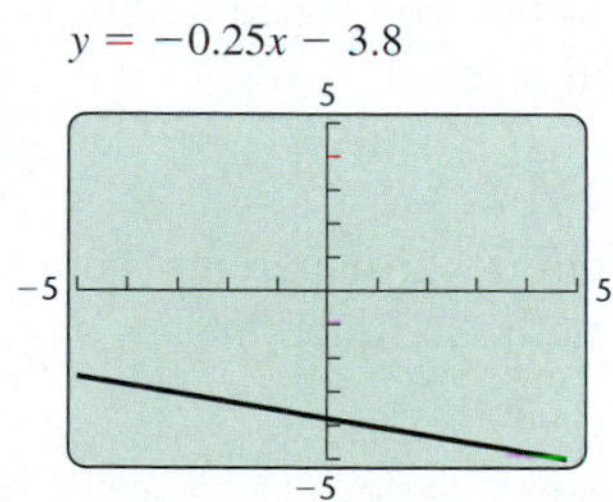

Now Try Exercise 49.

Any equation whose graph is a straight line is a **linear equation**. To find the slope and the y-intercept of the graph of a linear equation, we can solve for y, and then read the information from the equation.

Just in Time 2

EXAMPLE 7 Find the slope and the y-intercept of the line with equation $3x - 6y - 7 = 0$.

Solution We solve for y:

$$3x - 6y - 7 = 0$$
$$-6y = -3x + 7 \quad \text{Adding } -3x \text{ and } 7 \text{ on both sides}$$
$$-\tfrac{1}{6}(-6y) = -\tfrac{1}{6}(-3x + 7) \quad \text{Multiplying by } -\tfrac{1}{6} \text{ on both sides}$$
$$y = \tfrac{1}{2}x - \tfrac{7}{6}. \quad \text{Using a distributive law}$$

Thus the slope is $\frac{1}{2}$, and the y-intercept is $\left(0, -\frac{7}{6}\right)$.

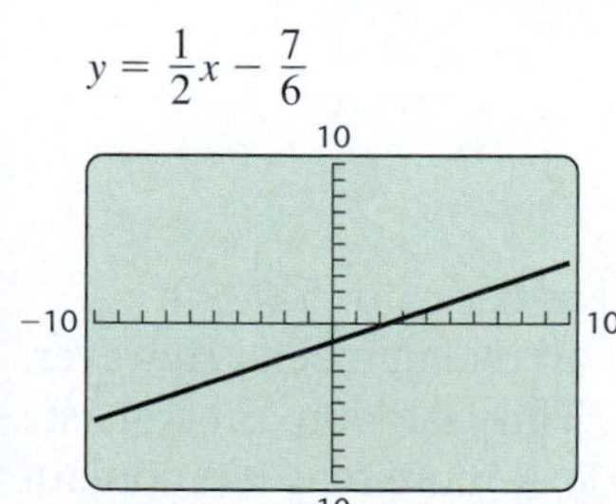

Now Try Exercise 61.

Graphing $f(x) = mx + b$ Using m and b

We can also graph a linear equation using its slope and y-intercept.

EXAMPLE 8 Graph: $y = -\frac{2}{3}x + 4$.

Solution This equation is in slope–intercept form, $y = mx + b$. The y-intercept is $(0, 4)$. We plot this point. We can think of the slope $\left(m = -\frac{2}{3}\right)$ as $\frac{-2}{3}$.

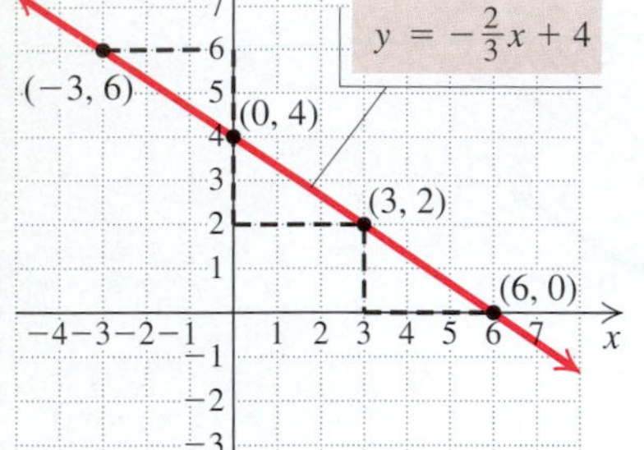

$$m = \frac{\text{rise}}{\text{run}} = \frac{\text{change in } y}{\text{change in } x} = \frac{-2}{3} \begin{array}{l} \leftarrow \text{Move 2 units down.} \\ \leftarrow \text{Move 3 units to the right.} \end{array}$$

Starting at the y-intercept and using the slope, we find another point by moving 2 units down and 3 units to the right. We get a new point $(3, 2)$. In a similar manner, we can move from $(3, 2)$ to find another point, $(6, 0)$.

We could also think of the slope $\left(m = -\frac{2}{3}\right)$ as $\frac{2}{-3}$. Then we can start at $(0, 4)$ and move 2 units up and 3 units to the left. We get to another point on the graph, $(-3, 6)$. We now plot the points and draw the line. Note that we need only the y-intercept and one other point in order to graph the line, but it's a good idea to find a third point as a check that the first two points are correct.

Now Try Exercise 63.

Applications of Linear Functions

We now consider an application of linear functions.

EXAMPLE 9 *Estimating Adult Height.* There is no *proven* way to predict a child's adult height, but a linear function can be used to *estimate* it, given the sum of the heights of the child's parents. The adult height M, in inches, of a male child whose parents' combined height is x, in inches, can be estimated with the function

$$M(x) = 0.5x + 2.5.$$

The adult height F, in inches, of a female child whose parents' combined height is x, in inches, can be estimated with the function

$$F(x) = 0.5x - 2.5.$$

(*Source*: Jay L. Hoecker, M.D., MayoClinic.com) Estimate the height of a female child whose parents' combined height is 135 in. What is the domain of this function?

Solution We substitute in the function:

$$F(135) = 0.5(135) - 2.5 = 65.$$

Thus we can estimate the adult height of the female child as 65 in., or 5 ft 5 in.

Theoretically, the domain of the function is the set of all real numbers. However, the context of the problem dictates a different domain. Thus the domain consists of all positive real numbers—that is, the interval $(0, \infty)$. A more realistic domain might be 100 in. to 110 in.—that is, the interval $[100, 110]$.

Now Try Exercise 73.

Visualizing the Graph

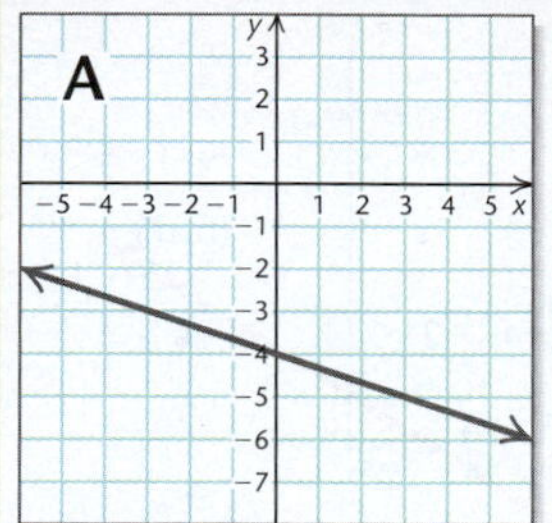

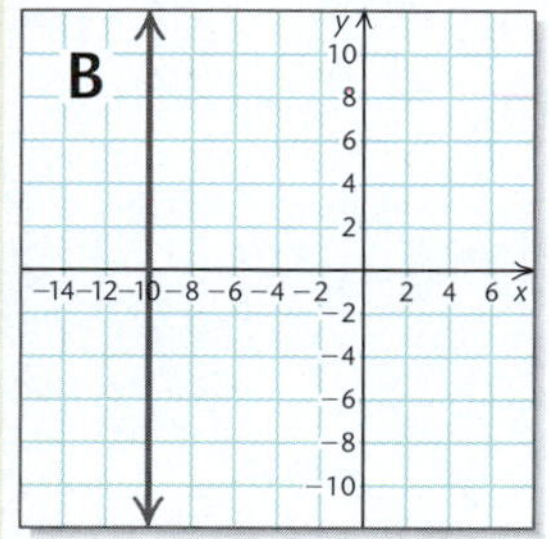

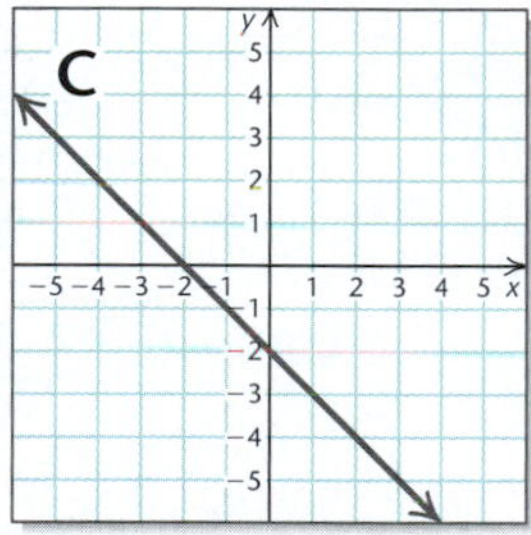

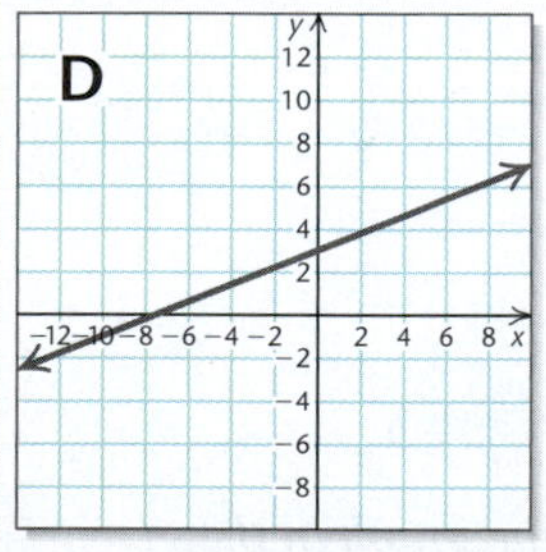

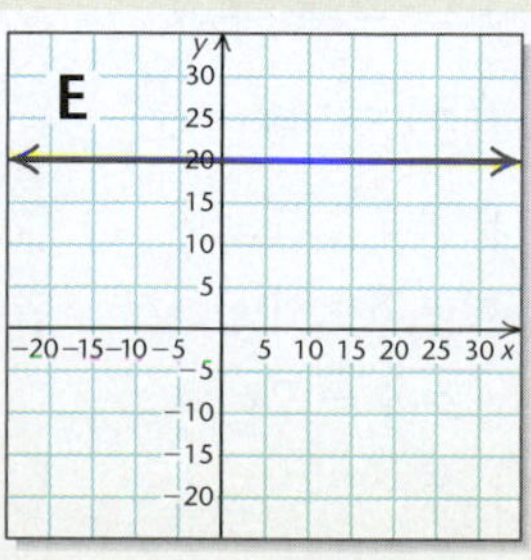

Match the equation with its graph.

1. $y = 20$

2. $5y = 2x + 15$

3. $y = -\frac{1}{3}x - 4$

4. $x = \frac{5}{3}$

5. $y = -x - 2$

6. $y = 2x$

7. $y = -3$

8. $3y = -4x$

9. $x = -10$

10. $y = x + \frac{7}{2}$

Answers on page A-3

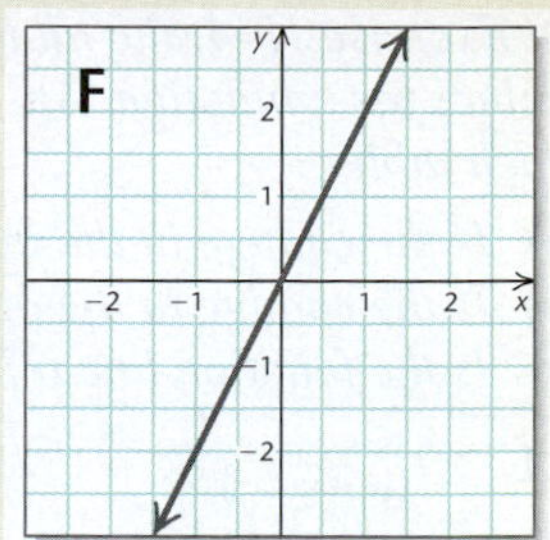

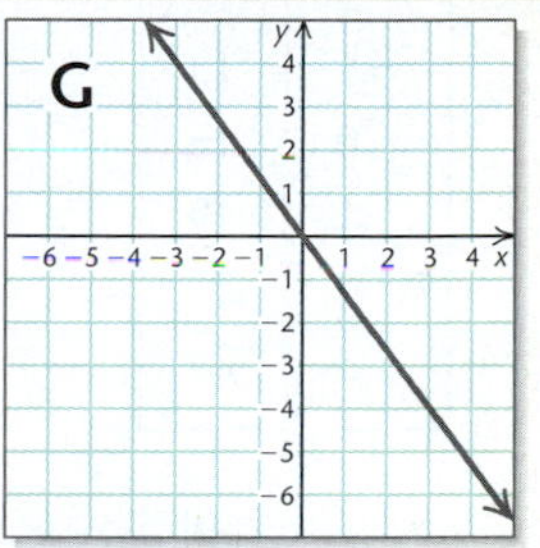

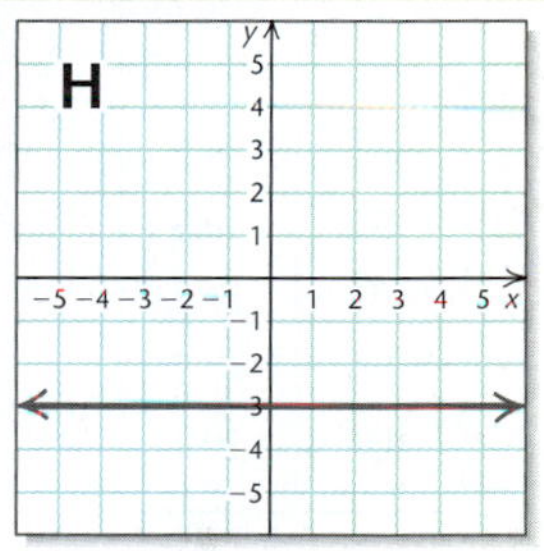

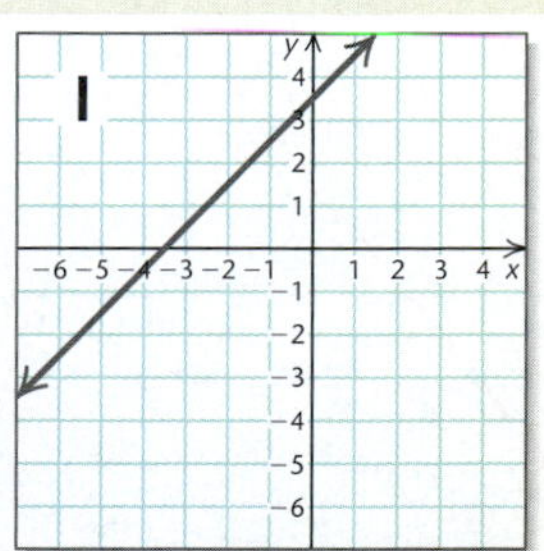

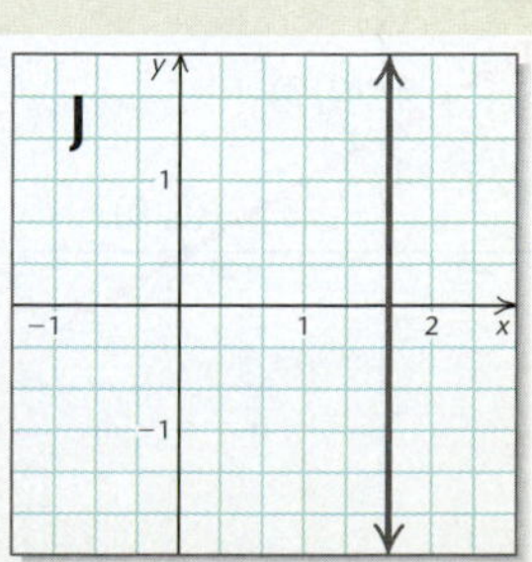

1.3 Exercise Set

In Exercises 1–4, the table of data contains input–output values for a function. Answer the following questions for each table.

a) *Is the change in the inputs x the same?*
b) *Is the change in the outputs y the same?*
c) *Is the function linear?*

1.

x	y
−3	7
−2	10
−1	13
0	16
1	19
2	22
3	25

2.

x	y
20	12.4
30	24.8
40	49.6
50	99.2
60	198.4
70	396.8
80	793.6

3.

x	y
11	3.2
26	5.7
41	8.2
56	9.3
71	11.3
86	13.7
101	19.1

4.

x	y
2	−8
4	−12
6	−16
8	−20
10	−24
12	−28
14	−32

Find the slope of the line containing the given points.

5.

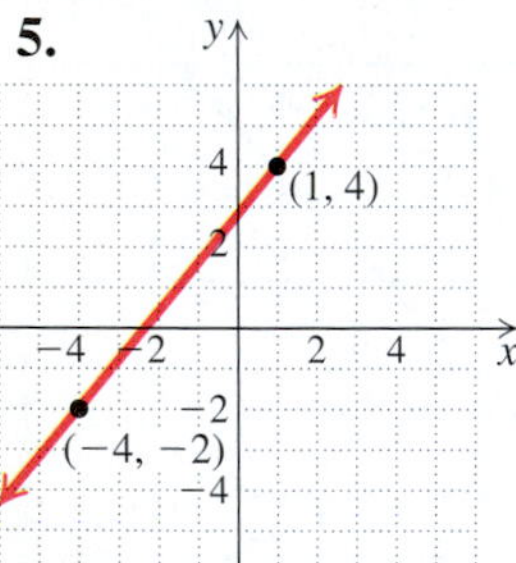

6.

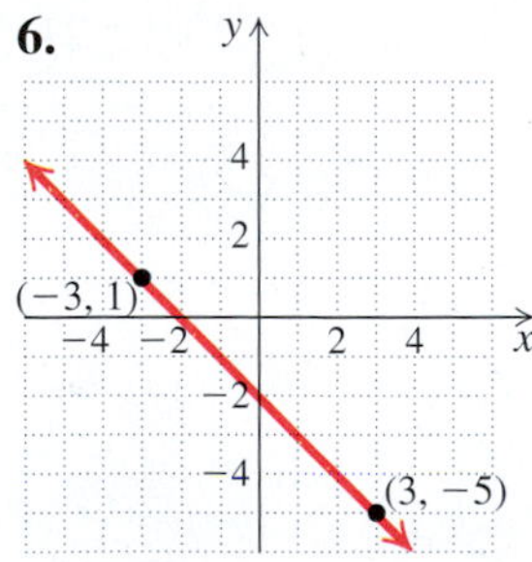

7.

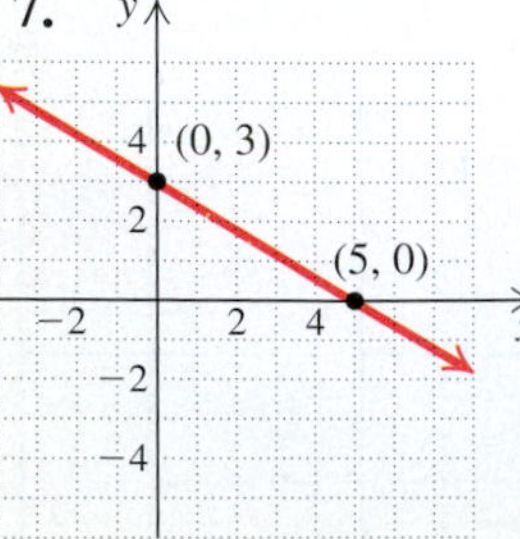

8.

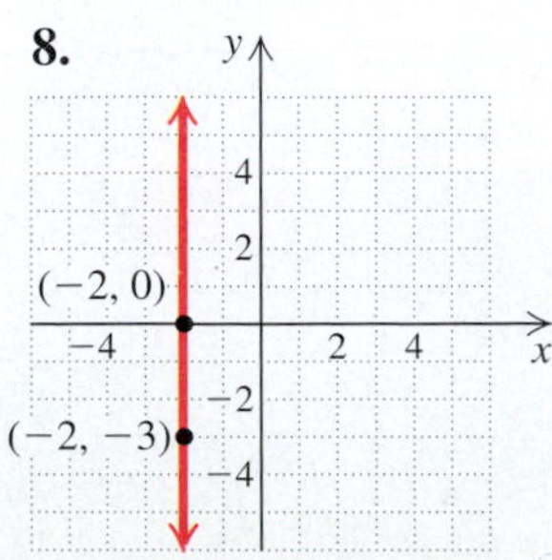

9.

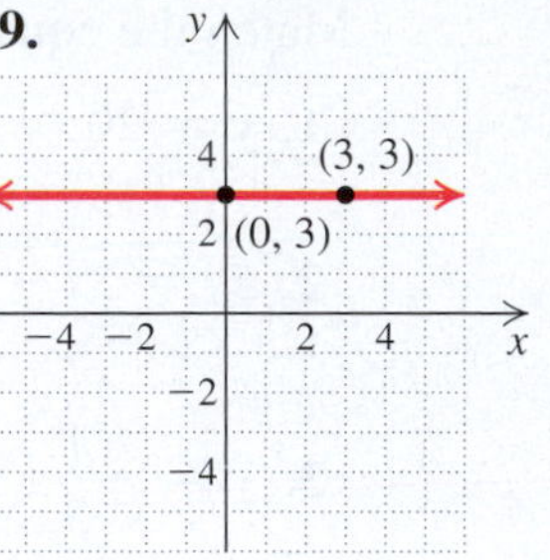

10.

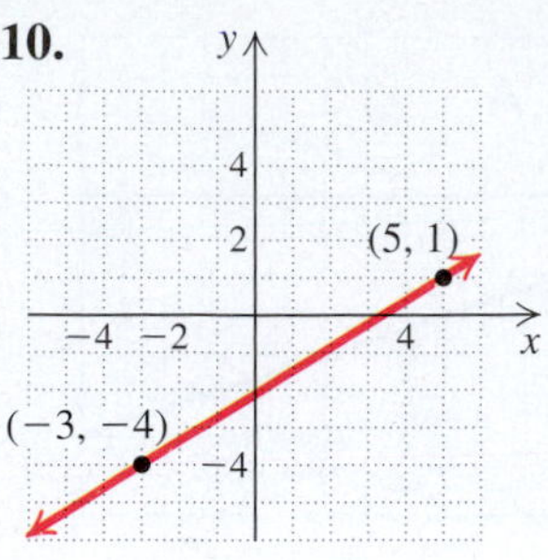

11. $(9, 4)$ and $(-1, 2)$

12. $(-3, 7)$ and $(5, -1)$

13. $(4, -9)$ and $(4, 6)$

14. $(-6, -1)$ and $(2, -13)$

15. $(0.7, -0.1)$ and $(-0.3, -0.4)$

16. $\left(-\frac{3}{4}, -\frac{1}{4}\right)$ and $\left(\frac{2}{7}, -\frac{5}{7}\right)$

17. $(2, -2)$ and $(4, -2)$

18. $(-9, 8)$ and $(7, -6)$

19. $\left(\frac{1}{2}, -\frac{3}{5}\right)$ and $\left(-\frac{1}{2}, \frac{3}{5}\right)$

20. $(-8.26, 4.04)$ and $(3.14, -2.16)$

21. $(16, -13)$ and $(-8, -5)$

22. $(\pi, -3)$ and $(\pi, 2)$

23. $(-10, -7)$ and $(-10, 7)$

24. $(\sqrt{2}, -4)$ and $(0.56, -4)$

25. $f(4) = 3$ and $f(-2) = 15$

26. $f(-4) = -5$ and $f(4) = 1$

27. $f\left(\frac{1}{5}\right) = \frac{1}{2}$ and $f(-1) = -\frac{11}{2}$

28. $f(8) = -1$ and $f\left(-\frac{2}{3}\right) = \frac{10}{3}$

29. $f(-6) = \frac{4}{5}$ and $f(0) = \frac{4}{5}$

30. $g\left(-\frac{9}{2}\right) = \frac{2}{9}$ and $g\left(\frac{2}{5}\right) = -\frac{5}{2}$

Determine the slope, if it exists, of the graph of the given linear equation.

31. $y = 1.3x - 5$

32. $y = -\frac{2}{5}x + 7$

33. $x = -2$

34. $f(x) = 4x - \frac{1}{4}$

35. $f(x) = -\frac{1}{2}x + 3$

36. $y = \frac{3}{4}$

37. $y = 9 - x$

38. $x = 8$

39. $y = 0.7$

40. $y = \frac{4}{5} - 2x$

41. *Price of a World Series Ticket.* In 1946, the lowest price of a World Series ticket was \$1.20. By 2012, the lowest price of a ticket had increased to \$110. (*Source*: AARP.com) Find the average rate of change in the lowest price of a World Series ticket from 1946 to 2012.

42. *Population Loss.* The population of Detroit, Michigan, decreased from 1,027,974 in 1990 to 688,701 in 2013 (*Source*: U.S. Census Bureau). Find the average rate of change in the population of Detroit, Michigan, over the 23-year period.

43. *Population Loss.* The population of Cleveland, Ohio, decreased from 478,403 in 2000 to 390,113 in 2013 (*Source*: U.S. Census Bureau). Find the average rate of change in the population of Cleveland, Ohio, over the 13-year period.

44. *Decreasing Size of Cattle Herd.* Drought has been the major reason for the decrease in the U.S. cattle herd in recent years. The number of cattle is at its lowest level since 1952. In 2006, there were 96.6 million head of cattle. This number had fallen to 87.7 million by 2014. (*Source*: U.S. Department of Agriculture) Find the average rate of change in the number of cattle from 2006 to 2014.

45. *Whole-Milk Consumption.* The annual per-capita consumption of whole milk in the United States was 25.3 gal in 1970. By 2011, this amount had decreased to 5.5 gal. Find the average rate of change in per-capita consumption of whole milk from 1970 to 2011.

46. *Chicken Consumption.* The annual per-capita consumption of chicken in the United States was 42.5 lb in 1990. By 2011, this amount had increased to 58.4 lb. (*Source*: Economic Research Service, U.S. Department of Agriculture) Find the average rate of change in per-capita consumption of chicken from 1990 to 2011.

47. *Growing Almonds.* In 2003, 550,000 acres of farmland in California were devoted to growing almonds. By 2012, the number of acres used to grow almonds had increased to 810,000. (*Source*: USDA National Agricultural Statistics Service) Find the average rate of change in the number of acres in California used to grow almonds from 2003 to 2012.

48. *ATM Fees.* The average fee to use an out-of-network ATM was \$2.66 in 2004. By 2014, this fee had increased to \$4.35. (*Source*: Bankrate.com) Find the average rate of change in out-of-network ATM fees from 2004 to 2014.

Find the slope and the y-intercept of the line with the given equation.

49. $y = \frac{3}{5}x - 7$

50. $f(x) = -2x + 3$

51. $x = -\frac{2}{5}$

52. $y = \frac{4}{7}$

53. $f(x) = 5 - \frac{1}{2}x$

54. $y = 2 + \frac{3}{7}x$

55. $3x + 2y = 10$

56. $2x - 3y = 12$

57. $y = -6$

58. $x = 10$

59. $5y - 4x = 8$

60. $5x - 2y + 9 = 0$

61. $4y - x + 2 = 0$

62. $f(x) = 0.3 + x$

Graph the equation using the slope and the y-intercept.

63. $y = -\frac{1}{2}x - 3$

64. $y = \frac{3}{2}x + 1$

65. $f(x) = 3x - 1$

66. $f(x) = -2x + 5$

67. $3x - 4y = 20$

68. $2x + 3y = 15$

69. $x + 3y = 18$

70. $5y - 2x = -20$

71. *Whales and Pressure at Sea Depth.* Whales can withstand extreme changes in atmospheric pressure because their bodies are flexible. Their rib cages and lungs can collapse safely under pressure. Sperm whales can hunt for squid at depths of 7000 ft or more. (*Sources*: National Ocean Service; National Oceanic and Atmospheric Administration) The function P, given by

$$P(d) = \frac{1}{33}d + 1,$$

gives the pressure, in atmospheres (atm), at a given depth d, in feet, under the sea.

a) Graph P.
b) Find $P(0)$, $P(33)$, $P(1000)$, $P(5000)$, and $P(7000)$.

72. *Stopping Distance on Glare Ice.* The stopping distance (at some fixed speed) of regular tires on glare ice is a function of the air temperature F, in degrees Fahrenheit. This function is estimated by

$$D(F) = 2F + 115,$$

where $D(F)$ is the stopping distance, in feet, when the air temperature is F, in degrees Fahrenheit.

a) Graph D.
b) Find $D(-20°)$, $D(0°)$, $D(10°)$, and $D(32°)$.
c) Explain why the domain should be restricted to $[-57.5°, 32°]$.

73. *Reaction Time.* Suppose that while driving a car, you suddenly see a deer standing in the road. Your brain registers the information and sends a signal to your foot to hit the brake. The car travels a distance D, in feet, during this time, where D is a function of the speed r, in miles per hour, of the car when you see the deer. That reaction distance is a linear function given by

$$D(r) = \frac{11}{10}r + \frac{1}{2}.$$

a) Find the slope of this line and interpret its meaning in this application.
b) Graph D.
c) Find $D(5)$, $D(10)$, $D(20)$, $D(50)$, and $D(65)$.
d) What is the domain of this function? Explain.

74. *Straight-Line Depreciation.* A contractor buys a new truck for \$38,000. The truck is purchased on January 1 and is expected to last 5 years, at the end of which time its *trade-in*, or *salvage, value* will be \$16,500. If the company figures the decline or depreciation in value to be the same each year, then the salvage value V, after t years, is given by the linear function

$$V(t) = \$38{,}000 - \$4300t, \quad \text{for } 0 \le t \le 5.$$

a) Find $V(0)$, $V(1)$, $V(2)$, $V(3)$, and $V(5)$.
b) Find the domain and the range of this function.

75. *Total Cost.* Richard is considering relocating to an assisted living facility. He learns that there is an initial community fee of \$2250 and a monthly charge of \$3380 for level-one care. Write an equation that can be used to determine the total cost $C(t)$ for t months of level-one care. Then find the total cost for 20 months.

76. *Total Cost.* Superior Cable Television charges a \$95 installation fee and \$125 per month for the Star plan. Write an equation that can be used to determine the total cost $C(t)$ for t months of the Star plan. Then find the total cost for 18 months of service.

*In Exercises 77 and 78, the term **fixed costs** refers to the start-up costs of operating a business. This includes machinery and building costs. The term **variable costs** refers to what it costs a business to produce or service one item.*

77. Max's Custom Lacrosse Stringing experienced fixed costs of \$750 and variable costs of \$15 for each lacrosse stick that was restrung. Write an equation that can be used to determine the total cost when x sticks are restrung. Then determine the total cost of restringing 32 lacrosse sticks.

78. Soosie's Cookie Company had fixed costs of \$1250 and variable costs of \$4.25 per dozen gourmet cookies that were baked and packaged for sale. Write an equation that can be used to determine the total cost when x dozens of cookies are baked and sold. Then determine the total cost of baking and selling 85 dozen gourmet cookies.

Skill Maintenance

If $f(x) = x^2 - 3x$, find each of the following. **[1.2]**

79. $f\left(\frac{1}{2}\right)$

80. $f(5)$

81. $f(-5)$

82. $f(-a)$

83. $f(a + h)$

Synthesis

84. *Grade of Treadmills.* A treadmill is 5 ft long and is set at an 8% grade. How high is the end of the treadmill?

Find the slope of the line containing the given points.

85. (a, a^2) and $(a + h, (a + h)^2)$

86. $(r, s + t)$ and (r, s)

Suppose that f is a linear function. Determine whether each of the following statements is true or false.

87. $f(c - d) = f(c) - f(d)$

88. $f(kx) = kf(x)$

Let $f(x) = mx + b$. Find a formula for $f(x)$ given each of the following.

89. $f(x + 2) = f(x) + 2$

90. $f(3x) = 3f(x)$

Mid-Chapter Mixed Review

Determine whether each of the following statements is true or false.

1. The x-intercept of the line that passes through $\left(-\frac{2}{3}, \frac{3}{2}\right)$ and the origin is $\left(-\frac{2}{3}, 0\right)$. **[1.1]**

2. All functions are relations, but not all relations are functions. **[1.2]**

3. The line parallel to the y-axis that passes through $(-5, 25)$ is $y = -5$. **[1.3]**

4. Find the intercepts of the graph of the line $-8x + 5y = -40$. **[1.1]**

For each pair of points, find the distance between the points and the midpoint of the segment having the points as endpoints. **[1.1]**

5. $(-8, -15)$ and $(3, 7)$

6. $\left(-\frac{3}{4}, \frac{1}{5}\right)$ and $\left(\frac{1}{4}, -\frac{4}{5}\right)$

7. Find an equation for a circle having center $(-5, 2)$ and radius 13. **[1.1]**

8. Find the center and the radius of the circle given by the equation $(x - 3)^2 + (y + 1)^2 = 4$. **[1.1]**

Graph the equation.

9. $3x - 6y = 6$ [1.1]

10. $y = -\frac{1}{2}x + 3$ [1.3]

11. $y = 2 - x^2$ [1.1]

12. $(x + 4)^2 + y^2 = 4$ [1.1]

13. Given that $f(x) = x - 2x^2$, find $f(-4)$, $f(0)$, and $f(1)$. [1.2]

14. Given that $g(x) = \dfrac{x + 6}{x - 3}$, find $g(-6)$, $g(0)$, and $g(3)$. [1.2]

Find the domain of the function. [1.2]

15. $g(x) = x + 9$

16. $f(x) = \dfrac{-5}{x + 5}$

17. $h(x) = \dfrac{1}{x^2 + 2x - 3}$

Graph the function. [1.2]

18. $f(x) = -2x$

19. $g(x) = x^2 - 1$

20. Determine the domain and the range of the function shown in the figure at right. [1.2]

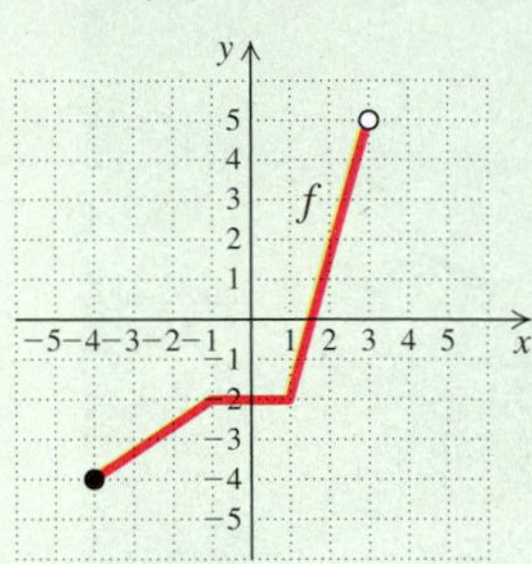

Find the slope of the line containing the given points. [1.3]

21. $(-2, 13)$ and $(-2, -5)$

22. $(10, -1)$ and $(-6, 3)$

23. $\left(\frac{5}{7}, \frac{1}{3}\right)$ and $\left(\frac{2}{7}, \frac{1}{3}\right)$

Determine the slope, if it exists, and the y-intercept of the line with the given equation. [1.3]

24. $f(x) = -\frac{1}{9}x + 12$

25. $y = -6$

26. $x = 2$

27. $3x - 16y + 1 = 0$

COLLABORATIVE DISCUSSION AND WRITING

To the student and the instructor: *The Collaborative Discussion and Writing exercises are meant to be answered with one or more sentences. They can be discussed and answered collaboratively by the entire class or by small groups.*

28. Explain as you would to a fellow student how the numerical value of the slope of a line can be used to describe the slant and the steepness of that line. [1.3]

29. Discuss why the graph of a vertical line $x = a$ cannot represent a function. [1.3]

30. Explain in your own words the difference between the domain of a function and the range of a function. [1.2]

31. Explain how you could find the coordinates of a point $\frac{7}{8}$ of the way from point A to point B. [1.1]

1.4 Equations of Lines and Modeling

- Determine equations of lines.
- Given the equations of two lines, determine whether their graphs are parallel or perpendicular.
- Model a set of data with a linear function.
- Fit a regression line to a set of data. Then use the linear model to make predictions.

Slope–Intercept Equations of Lines

In Section 1.3, we developed the slope–intercept equation $y = mx + b$, or $f(x) = mx + b$. If we know the slope and the y-intercept of a line, we can find an equation of the line using the slope–intercept equation.

EXAMPLE 1 A line has slope $-\frac{7}{9}$ and y-intercept $(0, 16)$. Find an equation of the line.

Solution We use the slope–intercept equation and substitute $-\frac{7}{9}$ for m and 16 for b:

$$\begin{aligned} y &= mx + b \\ y &= -\tfrac{7}{9}x + 16, \quad \text{or} \\ f(x) &= -\tfrac{7}{9}x + 16. \end{aligned}$$

Now Try Exercise 7.

EXAMPLE 2 A line has slope $-\frac{2}{3}$ and contains the point $(-3, 6)$. Find an equation of the line.

Solution We use the slope–intercept equation, $y = mx + b$, and substitute $-\frac{2}{3}$ for m: $y = -\frac{2}{3}x + b$. Then, using the point $(-3, 6)$, we substitute -3 for x and 6 for y in $y = -\frac{2}{3}x + b$. Finally, we solve for b.

$$\begin{aligned} y &= mx + b \\ y &= -\tfrac{2}{3}x + b && \text{Substituting } -\tfrac{2}{3} \text{ for } m \\ 6 &= -\tfrac{2}{3}(-3) + b && \text{Substituting } -3 \text{ for } x \text{ and } 6 \text{ for } y \\ 6 &= 2 + b \\ 4 &= b && \text{Solving for } b. \text{ The } y\text{-intercept is } (0, b). \end{aligned}$$

The equation of the line is $y = -\frac{2}{3}x + 4$, or $f(x) = -\frac{2}{3}x + 4$.

Now Try Exercise 13.

Point–Slope Equations of Lines

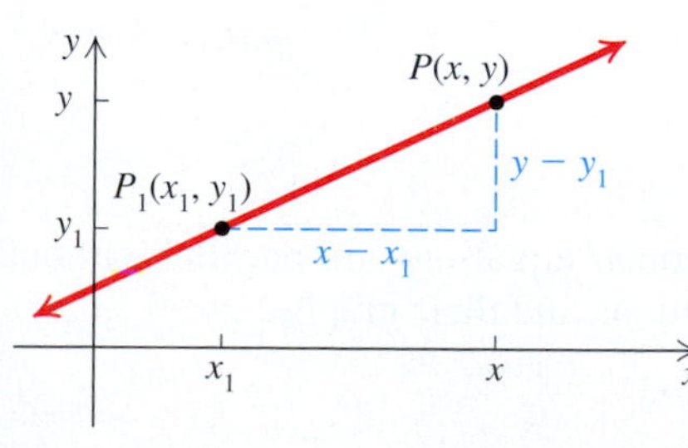

Another formula that can be used to determine an equation of a line is the *point–slope equation*. Suppose that we have a nonvertical line and that the coordinates of point P_1 on the line are (x_1, y_1). We can think of P_1 as fixed and imagine another point P on the line with coordinates (x, y). Thus the slope is given by

$$\frac{y - y_1}{x - x_1} = m.$$

Multiplying by $x - x_1$ on both sides, we get the *point–slope equation* of the line:

$$\begin{aligned} (x - x_1) \cdot \frac{y - y_1}{x - x_1} &= m \cdot (x - x_1) \\ y - y_1 &= m(x - x_1). \end{aligned}$$

POINT–SLOPE EQUATION

The **point–slope equation** of the line with slope m passing through (x_1, y_1) is

$$y - y_1 = m(x - x_1).$$

If we know the slope of a line and the coordinates of one point on the line, we can find an equation of the line using either the point–slope equation,

$$y - y_1 = m(x - x_1),$$

or the slope–intercept equation,

$$y = mx + b.$$

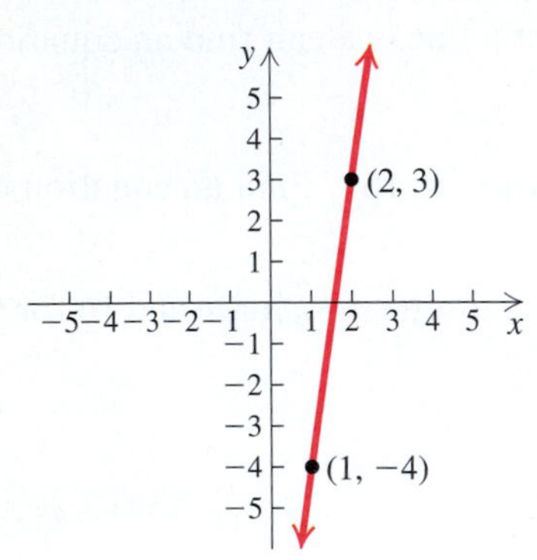

EXAMPLE 3 Find an equation of the line containing the points $(2, 3)$ and $(1, -4)$.

Solution We first determine the slope:

$$m = \frac{-4 - 3}{1 - 2} = \frac{-7}{-1} = 7.$$

Using the Point–Slope Equation: We substitute 7 for m and either of the points $(2, 3)$ or $(1, -4)$ for (x_1, y_1) in the point–slope equation. In this case, we use $(2, 3)$.

$$y - y_1 = m(x - x_1) \quad \text{Point–slope equation}$$
$$y - 3 = 7(x - 2) \quad \text{Substituting}$$
$$y - 3 = 7x - 14$$
$$y = 7x - 11, \quad \text{or}$$
$$f(x) = 7x - 11$$

Using the Slope–Intercept Equation: We substitute 7 for m and either of the points $(2, 3)$ or $(1, -4)$ for (x, y) in the slope–intercept equation and solve for b. Here we use $(1, -4)$.

$$y = mx + b \quad \text{Slope–intercept equation}$$
$$-4 = 7 \cdot 1 + b \quad \text{Substituting}$$
$$-4 = 7 + b$$
$$-11 = b \quad \text{Solving for } b$$

We substitute 7 for m and -11 for b in $y = mx + b$ to get

$$y = 7x - 11, \quad \text{or}$$
$$f(x) = 7x - 11.$$

Now Try Exercise 19.

Parallel Lines

Can we determine whether the graphs of two linear equations are parallel without graphing them? Let's look at three pairs of equations and their graphs.

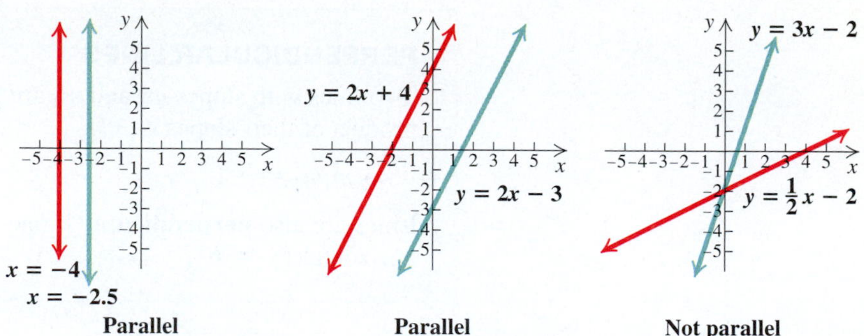

If two different lines, such as $x = -4$ and $x = -2.5$, are vertical, then they are parallel. Thus two equations such as $x = a_1$ and $x = a_2$, where $a_1 \neq a_2$, have graphs that are *parallel lines*. Two nonvertical lines, such as $y = 2x + 4$ and $y = 2x - 3$, or, in general, $y = mx + b_1$ and $y = mx + b_2$, where the slopes are the *same* and $b_1 \neq b_2$, also have graphs that are *parallel lines.*

PARALLEL LINES

Vertical lines are **parallel**. Nonvertical lines are **parallel** if and only if they have the same slope and different y-intercepts.

Perpendicular Lines

Can we examine a pair of equations to determine whether their graphs are perpendicular without graphing the equations? Let's look at the following pairs of equations and their graphs.

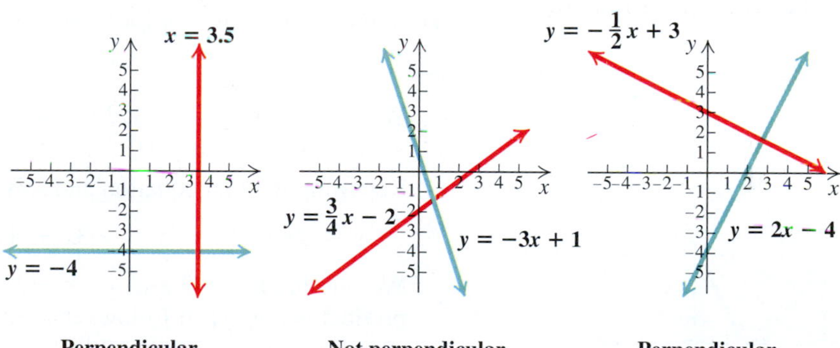

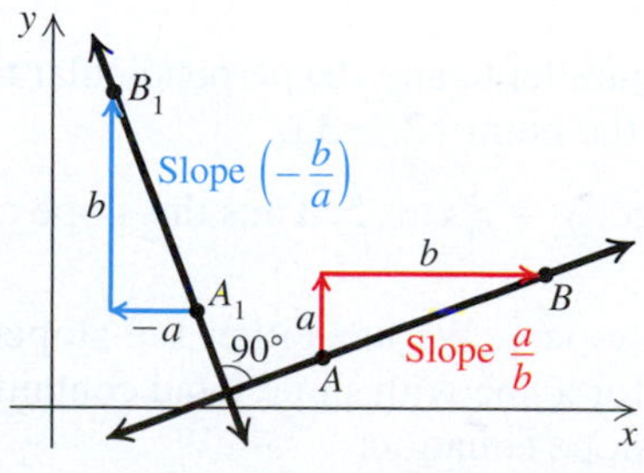

If one line is vertical and another is horizontal, they are perpendicular. For example, the lines $x = 3.5$ and $y = -4$ are perpendicular. Otherwise, how can we tell whether two lines are perpendicular? Consider a line $\overleftrightarrow{AB}$, as shown in the figure at left, with slope a/b. Then think of rotating the line 90° to get a line $\overleftrightarrow{A_1B_1}$ perpendicular to $\overleftrightarrow{AB}$. For the new line, the rise and the run are interchanged, but the run is now negative. Thus the slope of the new line is $-b/a$, which is the opposite of the reciprocal of the slope of the first line. Also note that when we multiply the slopes, we get

$$\frac{a}{b}\left(-\frac{b}{a}\right) = -1.$$

This is the condition under which lines will be perpendicular.

PERPENDICULAR LINES

Two lines with slopes m_1 and m_2 are **perpendicular** if and only if the product of their slopes is -1:

$$m_1m_2 = -1.$$

Lines are also **perpendicular** if one is vertical $(x = a)$ and the other is horizontal $(y = b)$.

If a line has slope m_1, the slope m_2 of a line perpendicular to it is $-1/m_1$. The slope of one line is the *opposite of the reciprocal* of the other:

$$m_2 = -\frac{1}{m_1}, \quad \text{or} \quad m_1 = -\frac{1}{m_2}.$$

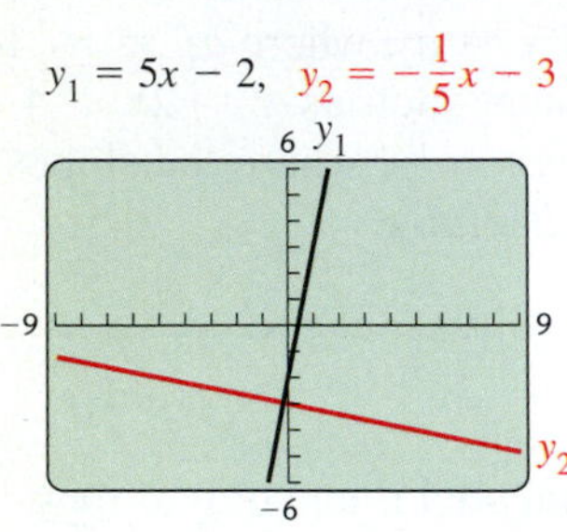

FIGURE 1.

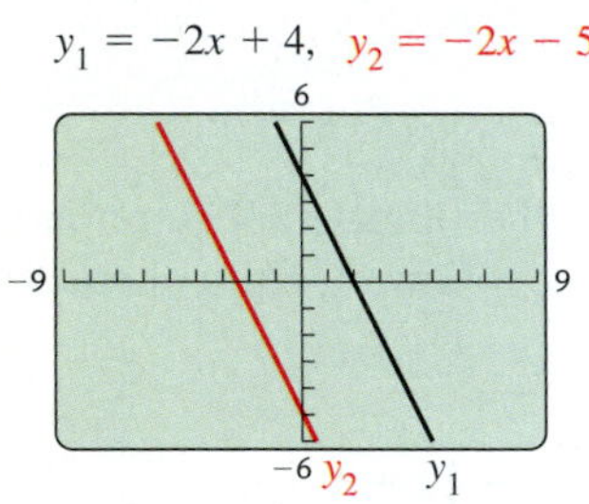

FIGURE 2.

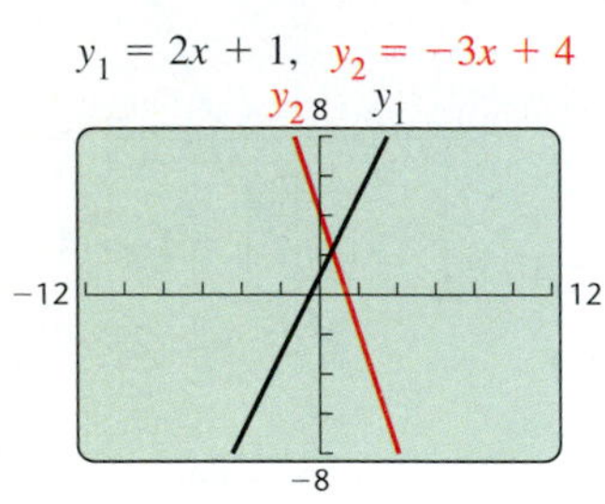

FIGURE 3.

EXAMPLE 4 Determine whether each of the following pairs of lines is parallel, perpendicular, or neither.

a) $y + 2 = 5x,\ 5y + x = -15$ **b)** $2y + 4x = 8,\ 5 + 2x = -y$

c) $2x + 1 = y,\ y + 3x = 4$

Solution We use the slopes of the lines to determine whether the lines are parallel or perpendicular.

a) We solve each equation for y:

$$y = 5x - 2, \qquad y = -\tfrac{1}{5}x - 3.$$

The slopes are 5 and $-\frac{1}{5}$. Their product is -1, so the lines are perpendicular. (See Fig. 1.) Note in the graphs at left that the graphing calculator windows have been squared to avoid distortion. (Review squaring windows in Section 1.1.)

b) Solving each equation for y, we get

$$y = -2x + 4, \qquad y = -2x - 5.$$

We see that $m_1 = -2$ and $m_2 = -2$. Since the slopes are the same and the y-intercepts, $(0, 4)$ and $(0, -5)$, are different, the lines are parallel. (See Fig. 2.)

c) Rewriting the first equation and solving the second equation for y, we have

$$y = 2x + 1, \qquad y = -3x + 4.$$

We see that $m_1 = 2$ and $m_2 = -3$. Since the slopes are not the same and their product is not -1, it follows that the lines are neither parallel nor perpendicular. (See Fig. 3.)

Now Try Exercises 35 and 39.

EXAMPLE 5 Write equations of the lines **(a)** parallel to and **(b)** perpendicular to the graph of the line $4y - x = 20$ and containing the point $(2, -3)$.

Solution We first solve $4y - x = 20$ for y to get $y = \frac{1}{4}x + 5$. Thus the slope of the given line is $\frac{1}{4}$.

a) The line parallel to the given line will have slope $\frac{1}{4}$. We use either the slope–intercept equation or the point–slope equation for a line with slope $\frac{1}{4}$ and containing the point $(2, -3)$. Here we use the point–slope equation:

$$\begin{aligned} y - y_1 &= m(x - x_1) \\ y - (-3) &= \tfrac{1}{4}(x - 2) \\ y + 3 &= \tfrac{1}{4}x - \tfrac{1}{2} \\ y &= \tfrac{1}{4}x - \tfrac{7}{2}. \end{aligned}$$

$y_1 = \frac{1}{4}x + 5$, $y_2 = \frac{1}{4}x - \frac{7}{2}$,
$y_3 = -4x + 5$

b) The slope of the perpendicular line is the opposite of the reciprocal of $\frac{1}{4}$, or -4. Again, we use the point–slope equation to write an equation for a line with slope -4 and containing the point $(2, -3)$:

$$\begin{aligned} y - y_1 &= m(x - x_1) \\ y - (-3) &= -4(x - 2) \\ y + 3 &= -4x + 8 \\ y &= -4x + 5. \end{aligned}$$

Now Try Exercise 43.

SUMMARY OF TERMINOLOGY ABOUT LINES

TERMINOLOGY	MATHEMATICAL INTERPRETATION
Slope	$m = \frac{y_2 - y_1}{x_2 - x_1}$, or $\frac{y_1 - y_2}{x_1 - x_2}$
Slope–intercept equation	$y = mx + b$
Point–slope equation	$y - y_1 = m(x - x_1)$
Horizontal line	$y = b$
Vertical line	$x = a$
Parallel lines	$m_1 = m_2, b_1 \neq b_2$; or $x = a_1, x = a_2, a_1 \neq a_2$
Perpendicular lines	$m_1 m_2 = -1$, or $m_2 = -\frac{1}{m_1}$; or $x = a, y = b$

Creating a Mathematical Model

1. Recognize real-world problem.

2. Collect data.

3. Analyze data.

4. Construct model.

5. Test and refine model.

6. Explain and predict.

Mathematical Models

When a real-world problem can be described in mathematical language, we have a **mathematical model**. For example, the natural numbers constitute a mathematical model for situations in which counting is essential. Situations in which algebra can be brought to bear often require the use of functions as models.

Mathematical models are abstracted from real-world situations. The mathematical model gives results that allow one to predict what will happen in that real-world situation. If the predictions are inaccurate or the results of experimentation do not conform to the model, the model must be changed or discarded.

Mathematical modeling can be an ongoing process. For example, finding a mathematical model that will provide an accurate prediction of population growth is not a simple task. Any population model that one might devise would need to be reshaped as further information is acquired.

Curve Fitting

We will develop and use many kinds of mathematical models in this text. In this chapter, we have used *linear* functions as models. Other types of functions, such as quadratic, cubic, and exponential functions, can also model data. These functions are *nonlinear*. Modeling with quadratic functions and cubic functions is discussed in Chapter 4. Modeling with exponential functions is discussed in Chapter 5.

Quadratic function:
$y = ax^2 + bx + c, a > 0$

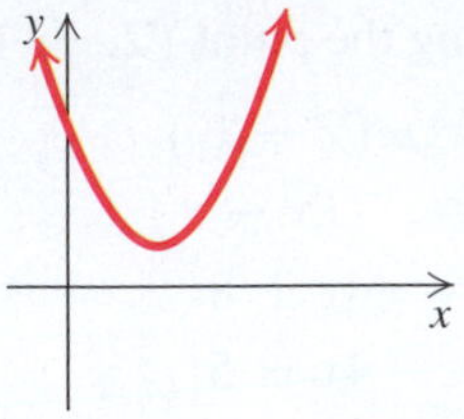

Cubic function:
$y = ax^3 + bx^2 + cx + d, a > 0$

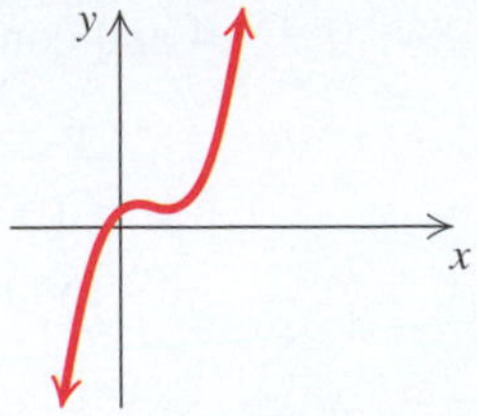

Exponential function:
$y = ab^x, a, b > 0, b \neq 1$

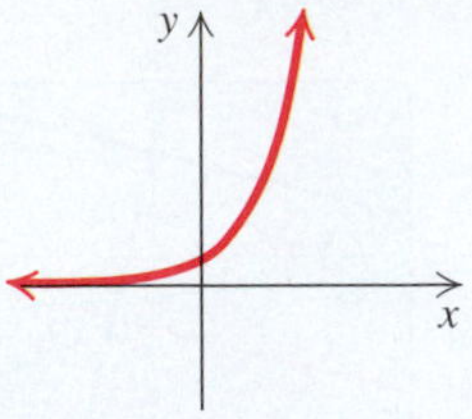

In general, we try to find a function that fits, as well as possible, observations (data), theoretical reasoning, and common sense. We call this **curve fitting**; it is one aspect of mathematical modeling.

Let's look at some data and related graphs or **scatterplots** and determine whether a linear function seems to fit the set of data.

Year, x	Cost of a 30-Second Super Bowl Commercial (in millions)	Scatterplot
2010, 0	$2.9	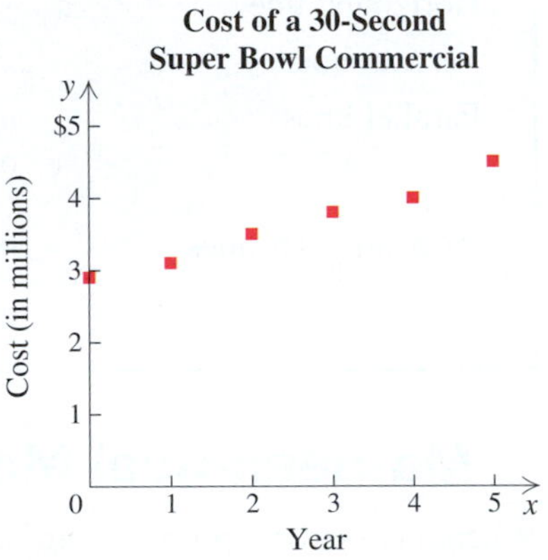 It appears that the data points can be represented, or modeled, by a linear function. The graph is **linear**.
2011, 1	3.1	
2012, 2	3.5	
2013, 3	3.8	
2014, 4	4.0	
2015, 5	4.5	

Source: huffingtonpost.com

Year, x	Deposits in All U.S. FDIC-Insured Commercial and Savings Banks (in trillions)	Scatterplot
1960, 0	$ 0.310	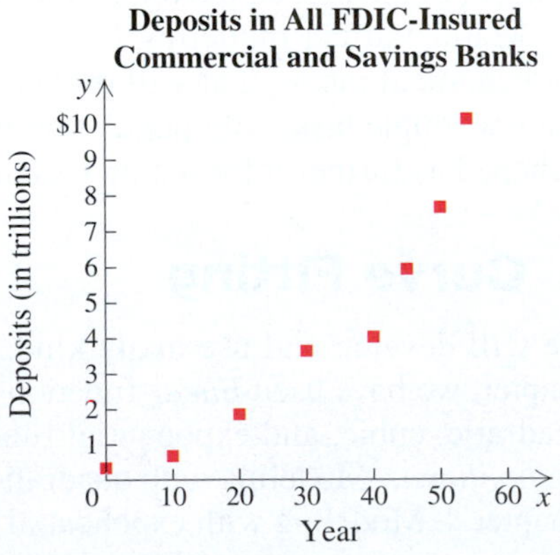 It appears that the data points cannot be modeled accurately by a linear function. The graph is **nonlinear**.
1970, 10	0.687	
1980, 20	1.833	
1990, 30	3.637	
2000, 40	4.004	
2005, 45	5.934	
2010, 50	7.677	
2014, 54	10.113	

Source: *Summary of Deposits*, Federal Deposit Insurance Corporation

Looking at the scatterplots, we see that the data on the cost of a Super Bowl commercial seem to be rising in a manner to suggest that a *linear function* might fit, although a "perfect" straight line cannot be drawn through the data points. A linear function does not seem to fit the data on deposits in U.S. banks.

EXAMPLE 6 ***Cost of a 30-Sec Super Bowl Commercial.*** The cost of a 30-sec Super Bowl commercial has increased \$1.6 million from 2010 to 2015. Model the data in the table on p. 52 with a linear function. Then estimate the cost of a 30-sec commercial in 2018.

Solution We can choose any two of the data points to determine an equation. Note that the first coordinate is the number of years since 2010 and the second coordinate is the corresponding cost of a 30-sec Super Bowl commercial in millions of dollars. Let's use (1, 3.1) and (4, 4.0).

We first determine the slope of the line:

$$m = \frac{4.0 - 3.1}{4 - 1} = \frac{0.9}{3} = 0.3.$$

Then we substitute 0.3 for m and either of the points $(1, 3.1)$ or $(4, 4.0)$ for (x_1, y_1) in the point–slope equation. In this case, let's use $(1, 3.1)$. We get

$$y - y_1 = m(x - x_1) \qquad \textbf{Point–slope equation}$$
$$y - 3.1 = 0.3(x - 1), \qquad \textbf{Substituting}$$

which simplifies to

$$y = 0.3x + 2.8,$$

where x is the number of years after 2010 and y is in millions of dollars.

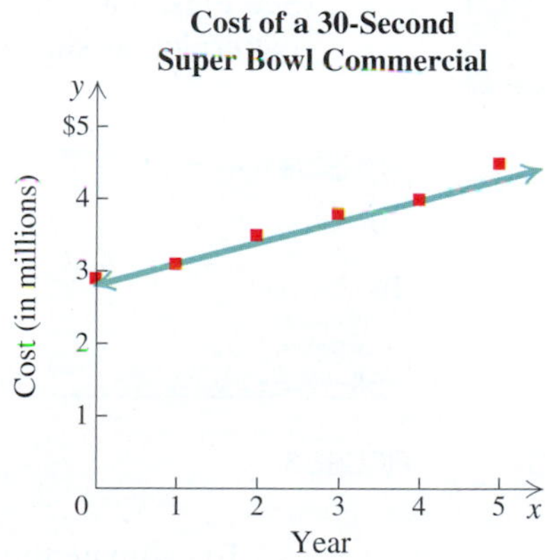

Next, we estimate the cost of a 30-sec commercial in 2018 by substituting 8 $(2018 - 2010 = 8)$ for x in the model:

$$\begin{aligned} y &= 0.3x + 2.8 && \textbf{Model} \\ &= 0.3(8) + 2.8 && \textbf{Substituting} \\ &= 5.2. \end{aligned}$$

We estimate that the cost of a 30-sec Super Bowl commercial will be \$5.2 million in 2018.

Now Try Exercise 61.

A model and the estimates it produces are dependent on the data points used. In Example 6, if we were to use the data points $(0, 2.9)$ and $(5, 4.5)$, our model would be

$$y = 0.32x + 2.9,$$

and our estimate for the cost of a 30-sec commercial in 2018 would be \$5.46 million, about \$0.26 million more than the estimate provided by the first model.

Models that consider all the data points, not just two, are generally better models. The linear model that best fits the data can be found using a graphing calculator and a procedure called **linear regression**.

Linear Regression

Although discussion leading to a complete understanding of linear regression belongs in a statistics course, we present the procedure here because we can carry it out easily using technology. The graphing calculator gives us the powerful capability to find linear models and to make predictions using them.

Consider the data presented before Example 6 on the cost of a 30-sec Super Bowl commercial. We can fit a regression line of the form $y = mx + b$ to the data using the LINEAR REGRESSION feature on a graphing calculator.

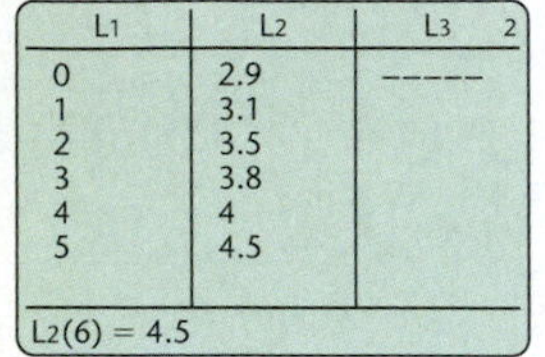

FIGURE 1.

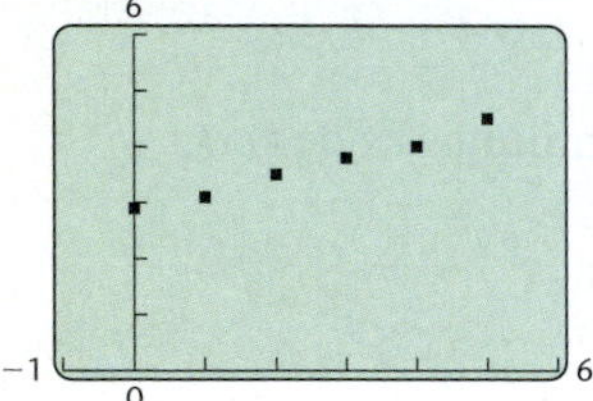

FIGURE 2.

EXAMPLE 7 ***Cost of a 30-Sec Super Bowl Commercial.*** Fit a regression line to the data on the cost of a 30-sec Super Bowl commercial given in the table on p. 52. Then use the function to predict the cost of a 30-sec commercial in 2018.

Solution First, we enter the data in lists on the calculator. We enter the values of the independent variable x in list L1 and the corresponding values of the dependent variable y in list L2. (See Fig. 1.) The graphing calculator can then create a scatterplot of the data, as shown in Fig. 2.

When we select the LINEAR REGRESSION feature from the STAT CALC menu, we find the linear equation that best models the data. It is

$$y = 0.3142857143x + 2.847619048. \quad \text{Regression line}$$

(See Figs. 3 and 4.) We can then graph the regression line on the same graph as the scatterplot, as shown in Fig. 5.

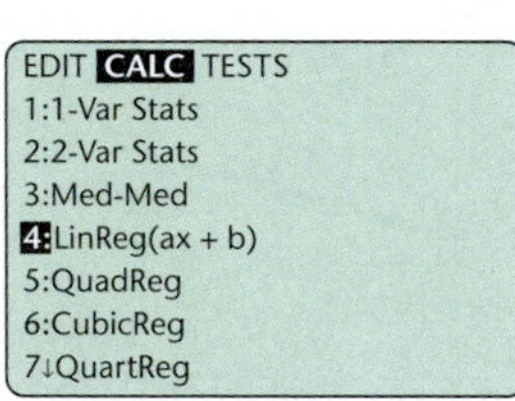

FIGURE 3.

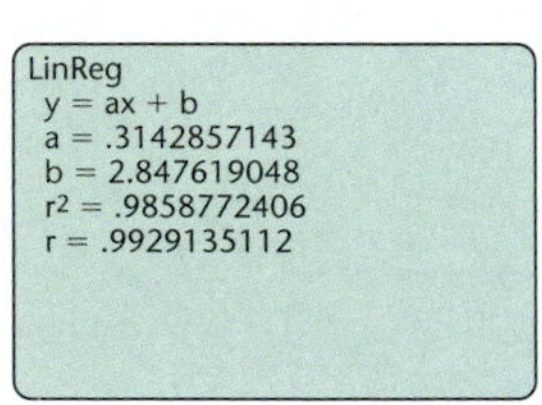

FIGURE 4.

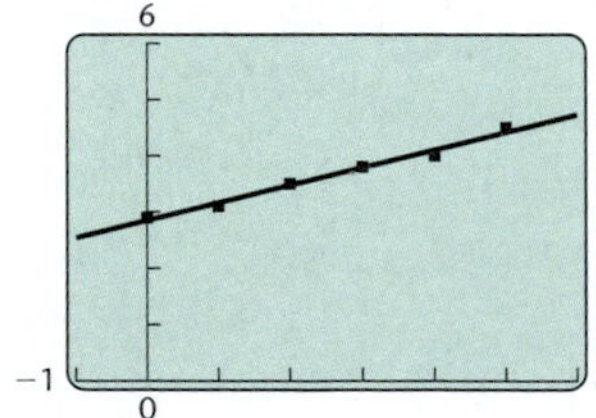

FIGURE 5.

To estimate the cost of a 30-sec Super Bowl commercial in 2018, we substitute 8 for x in the regression equation, which has been saved as Y1. Using this model, we see that the cost of a 30-sec commercial is predicted to be about \$5.36 million in 2018. (See Fig. 6.)

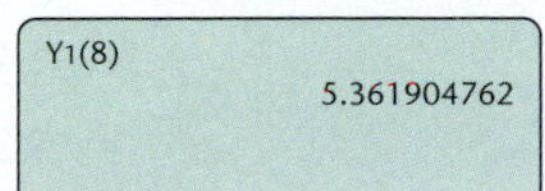

FIGURE 6.

Note that \$5.36 million is closer to the value \$5.46 million that we found with data points $(0, 2.9)$ and $(5, 4.5)$ following Example 6 than to the value \$5.2 million that we found with the data points $(1, 3.1)$ and $(4, 4.0)$ in Example 6.

Now Try Exercises 69(a) and 69(b).

The Correlation Coefficient

On some graphing calculators with the DIAGNOSTIC feature turned on, a constant r between -1 and 1, called the **coefficient of linear correlation**, appears with the equation of the regression line. Though we cannot develop a formula for calculating r in this text, keep in mind that it is used to describe the strength of the linear relationship between x and y. The closer $|r|$ is to 1, the better the correlation. A positive value of r also indicates that the regression line has a positive slope, and a negative value of r indicates that the regression line has a negative slope. As shown in Fig. 4, for the data on the cost of a 30-sec Super Bowl commercial just discussed, $r = 0.9929135112$, which indicates a very good linear correlation.

The following scatterplots summarize the interpretation of a correlation coefficient.

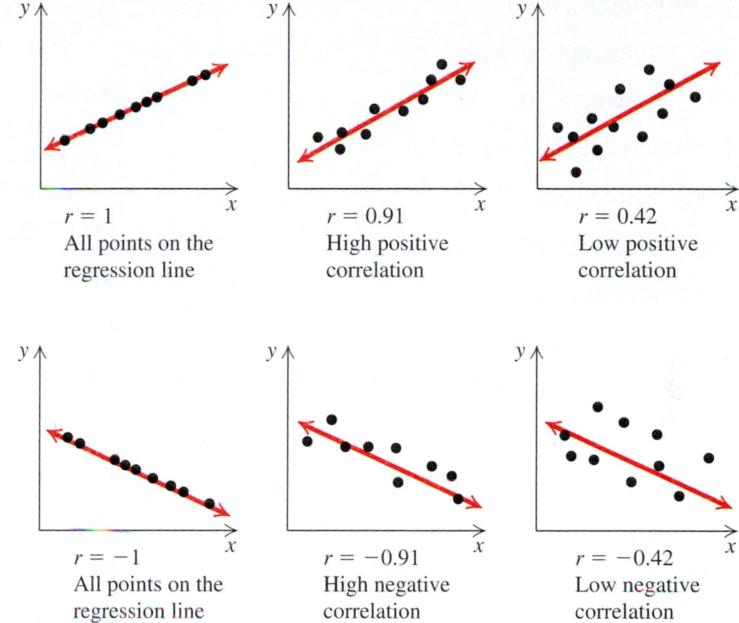

1.4 Exercise Set

Find the slope and the y-intercept of the graph of the linear equation. Then write the equation of the line in slope–intercept form.

1.

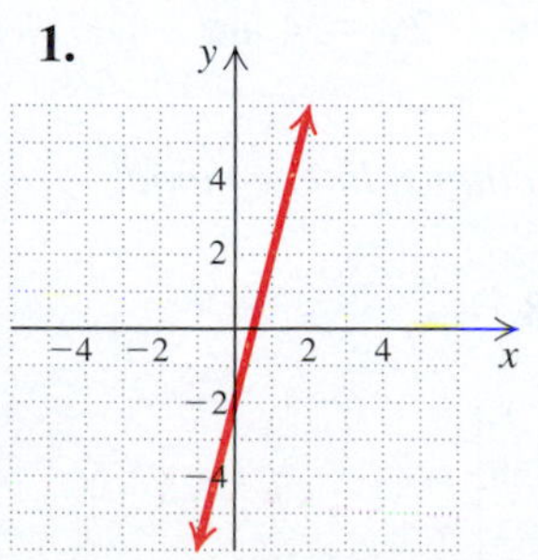

2.

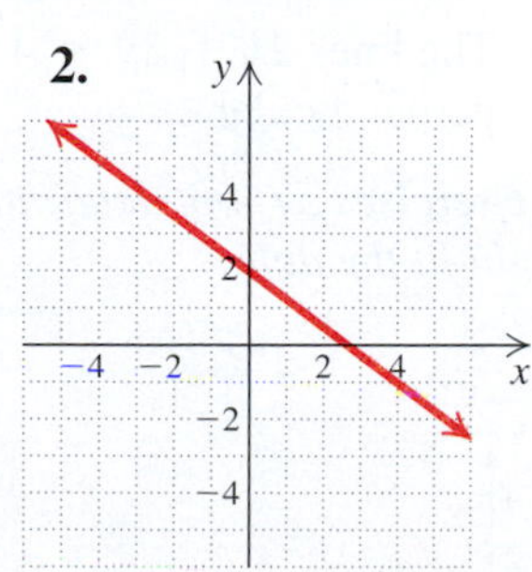

3.

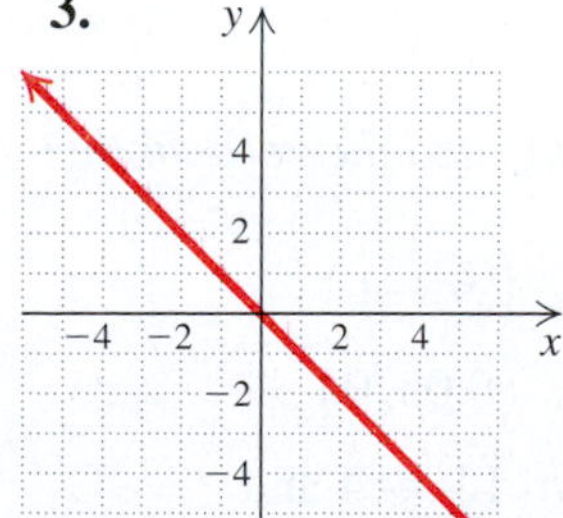

4.

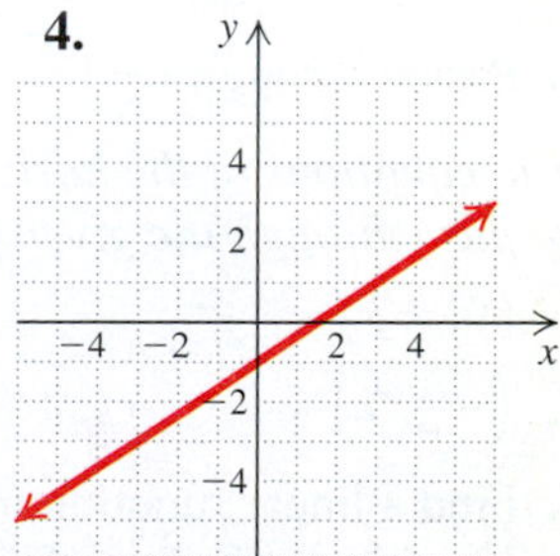

5.

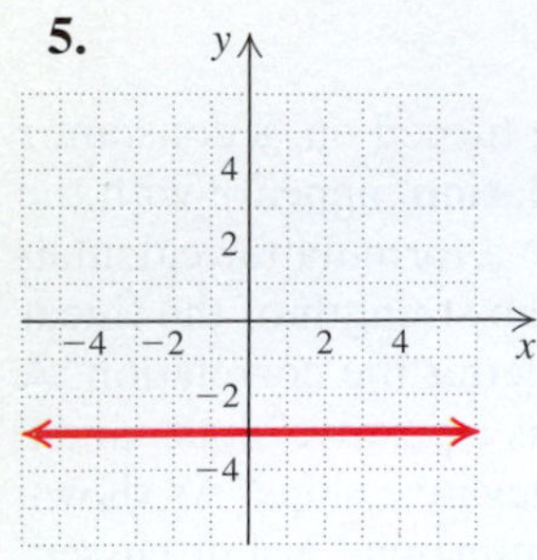

6.

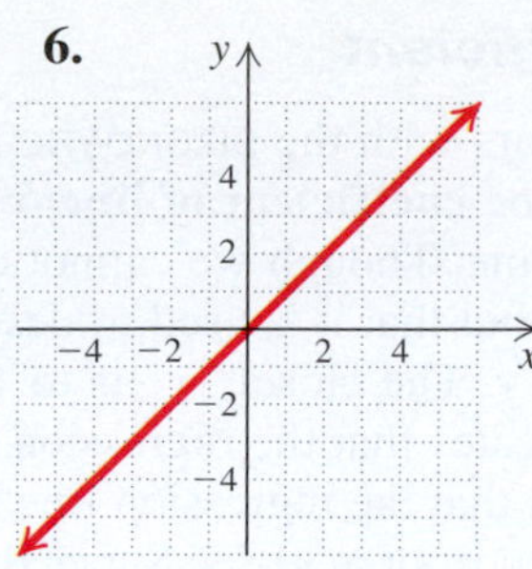

Write a slope–intercept equation for a line with the given characteristics.

7. $m = \frac{2}{9}$, y-intercept $(0, 4)$

8. $m = -\frac{3}{8}$, y-intercept $(0, 5)$

9. $m = -4$, y-intercept $(0, -7)$

10. $m = \frac{2}{7}$, y-intercept $(0, -6)$

11. $m = -4.2$, y-intercept $\left(0, \frac{3}{4}\right)$

12. $m = -4$, y-intercept $\left(0, -\frac{3}{2}\right)$

13. $m = \frac{2}{9}$, passes through $(3, 7)$

14. $m = -\frac{3}{8}$, passes through $(5, 6)$

15. $m = 0$, passes through $(-2, 8)$

16. $m = -2$, passes through $(-5, 1)$

17. $m = -\frac{3}{5}$, passes through $(-4, -1)$

18. $m = \frac{2}{3}$, passes through $(-4, -5)$

19. Passes through $(-1, 5)$ and $(2, -4)$

20. Passes through $\left(-3, \frac{1}{2}\right)$ and $\left(1, \frac{1}{2}\right)$

21. Passes through $(7, 0)$ and $(-1, 4)$

22. Passes through $(-3, 7)$ and $(-1, -5)$

23. Passes through $(0, -6)$ and $(3, -4)$

24. Passes through $(-5, 0)$ and $\left(0, \frac{4}{5}\right)$

25. Passes through $(-4, 7.3)$ and $(0, 7.3)$

26. Passes through $(-13, -5)$ and $(0, 0)$

Write equations of the horizontal line and the vertical line that pass through the given point.

27. $(0, -3)$

28. $\left(-\frac{1}{4}, 7\right)$

29. $\left(\frac{2}{11}, -1\right)$

30. $(0.03, 0)$

31. Find a linear function h given $h(1) = 4$ and $h(-2) = 13$. Then find $h(2)$.

32. Find a linear function g given $g\left(-\frac{1}{4}\right) = -6$ and $g(2) = 3$. Then find $g(-3)$.

33. Find a linear function f given $f(5) = 1$ and $f(-5) = -3$. Then find $f(0)$.

34. Find a linear function h given $h(-3) = 3$ and $h(0) = 2$. Then find $h(-6)$.

Determine whether the pair of lines is parallel, perpendicular, or neither.

35. $y = \frac{26}{3}x - 11$, $y = -\frac{3}{26}x - 11$

36. $y = -3x + 1$, $y = -\frac{1}{3}x + 1$

37. $y = \frac{2}{5}x - 4$, $y = -\frac{2}{5}x + 4$

38. $y = \frac{3}{2}x - 8$, $y = 8 + 1.5x$

39. $x + 2y = 5$, $2x + 4y = 8$

40. $2x - 5y = -3$, $2x + 5y = 4$

41. $y = 4x - 5$, $4y = 8 - x$

42. $y = 7 - x$, $y = x + 3$

Write a slope–intercept equation for a line passing through the given point that is parallel to the given line. Then write a second equation for a line passing through the given point that is perpendicular to the given line.

43. $(3, 5)$, $y = \frac{2}{7}x + 1$

44. $(-1, 6)$, $f(x) = 2x + 9$

45. $(-7, 0)$, $y = -0.3x + 4.3$

46. $(-4, -5)$, $2x + y = -4$

47. $(3, -2)$, $3x + 4y = 5$

48. $(8, -2)$, $y = 4.2(x - 3) + 1$

49. $(3, -3)$, $x = -1$

50. $(4, -5)$, $y = -1$

Determine whether the statement is true or false.

51. The lines $x = -3$ and $y = 5$ are perpendicular.

52. The lines $y = 2x - 3$ and $y = -2x - 3$ are perpendicular.

53. The lines $y = \frac{2}{5}x + 4$ and $y = \frac{2}{5}x - 4$ are parallel.

54. The intersection of the lines $y = 2$ and $x = -\frac{3}{4}$ is $\left(-\frac{3}{4}, 2\right)$.

55. The lines $x = -1$ and $x = 1$ are perpendicular.

56. The lines $2x + 3y = 4$ and $3x - 2y = 4$ are perpendicular.

In Exercises 57–60, determine whether a linear model might fit the data.

57.

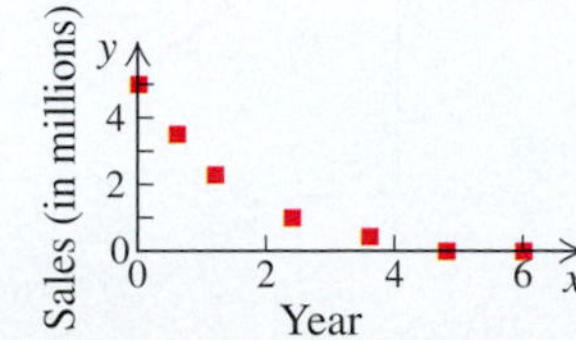

58.

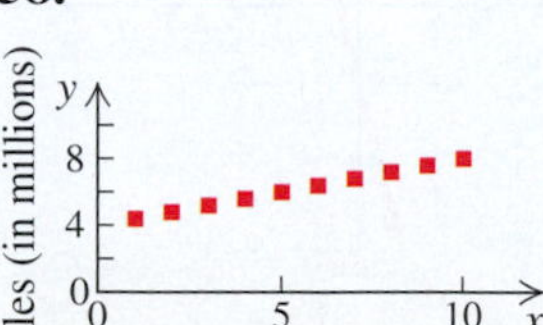

59. **60.**

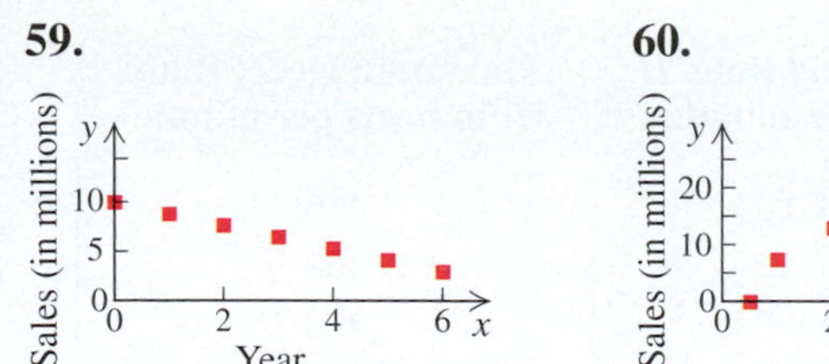

61. *Cost of Health Insurance.* The following table lists the average monthly cost to workers for family health insurance for various years.

Year, x	Average Monthly Cost to Workers for Family Health Insurance
2009, 0	$293
2010, 1	333
2011, 2	344
2012, 3	360
2013, 4	380
2014, 5	402

Source: Kaiser Family Foundation

a) Model the data with a linear function. Let the independent variable represent the number of years after 2009; that is, the data points are $(0, 293)$, $(3, 360)$, and so on. Answers may vary depending on the data points used.

b) Using the function found in part (a), predict the average cost to workers for family health insurance in 2018 and in 2023.

62. *U.S. Banks.* The following table lists the numbers of commercial and savings banks in the United States for various years.

Year, x	Number of U.S. Banks
2007, 0	8605
2008, 1	8441
2009, 2	8185
2010, 3	7821
2011, 4	7523
2012, 5	7255
2013, 6	6950
2014, 7	6669

Source: Federal Deposit Insurance Corporation

a) Model the data with a linear function. Let the independent variable represent the number of years after 2007. Answers may vary depending on the data points used.

b) Using the function found in part (a), predict the number of banks in the United States in 2017 and in 2020.

63. *Airline Add-On Fees.* Data on airline revenue from add-on fees are listed in the following table. Model the data with a linear function. Then, using that function, predict the revenue from airline add-on fees in 2019. Answers may vary depending on the data points used.

Year, x	Airline Revenue from Add-On Fees (in billions)
2010, 0	$22.6
2011, 1	32.5
2012, 2	36.1
2013, 3	42.6
2014, 4	49.9

Sources: U.S. airline consultancy, IdeaWorks Company; rental firm CarTrawler

64. *Licensed Drivers.* Data on the number of licensed drivers in the United States in selected years are listed in the following table. Model the data with a linear function, estimate the number of licensed drivers in 2005, and predict the number of licensed drivers in 2021. Answers may vary depending on the data points used.

Year, x	Number of Licensed Drivers (in millions)
1980, 0	145.295
1990, 10	167.015
2000, 20	190.625
2010, 30	210.115
2012, 32	211.815

Sources: Federal Highway Administration; U.S. Department of Transportation

65. *Bottled Water.* Data on the per-capita consumption, in gallons, of bottled water in the United States are given in the following table. Model the data with a linear function and predict the per-capita consumption of bottled water in 2017. Answers may vary depending on the data points used.

Year, x	Per-Capita Consumption of Bottled Water (in gallons)
2009, 0	27.6
2010, 1	28.3
2011, 2	29.2
2012, 3	30.8
2013, 4	32.0

Source: Beverage Marketing Corporation

66. *Electricity Use.* Data on the average annual household use of electricity, in kilowatt-hours, are listed in the following table. Model the data with a linear function and predict the average annual household electricity use in 2019. Answers may vary depending on the data points used.

Year, x	Annual Electricity Use (in kilowatt-hours)
2010, 0	11,504
2011, 1	11,280
2012, 2	10,837
2013, 3	10,819

Source: Energy Information Administration

67. **a)** Use a graphing calculator to fit a regression line to the data in Exercise 61.
b) Predict the average monthly cost to workers for family health insurance in 2018 and compare the value with the result found in Exercise 61.
c) Find the correlation coefficient for the regression line and determine whether the line fits the data closely.

68. **a)** Use a graphing calculator to fit a regression line to the data in Exercise 62.
b) Predict the number of banks in the United States in 2020 and compare the result with the result found with the model in Exercise 62.
c) Find the correlation coefficient for the regression line and determine whether the line fits the data closely.

69. **a)** Use a graphing calculator to fit a regression line to the data in Exercise 63.
b) Predict airline revenue from add-on fees in 2019 and compare the value with the result found in Exercise 63.
c) Find the correlation coefficient for the regression line and determine whether the line fits the data closely.

70. **a)** Use a graphing calculator to fit a regression line to the data in Exercise 64.
b) Predict the number of licensed drivers in the United States in 2021 and compare the result with the prediction found with the model in Exercise 64.
c) Find the correlation coefficient for the regression line and determine whether the line fits the data closely.

71. *Maximum Heart Rate.* A person who is exercising should not exceed his or her maximum heart rate, which is determined on the basis of that person's gender, age, and resting heart rate. The following table relates resting heart rate and maximum heart rate for a 20-year-old man.

Resting Heart Rate, H (in beats per minute)	Maximum Heart Rate, M (in beats per minute)
50	166
60	168
70	170
80	172

Source: American Heart Association

a) Use a graphing calculator to model the data with a linear function.
b) Estimate the maximum heart rate if the resting heart rate is 40, 65, 76, and 84.
c) What is the correlation coefficient? How confident are you about using the regression line to estimate function values?

72. *Study Time versus Grades.* A math instructor asked her students to keep track of how much time each spent studying a chapter on functions in her algebra–trigonometry course. She collected the information together with test scores from that chapter's test. The data are listed in the following table.

Study Time, x (in hours)	Test Grade, y (in percent)
23	81%
15	85
17	80
9	75
21	86
13	80
16	85
11	93

a) Use a graphing calculator to model the data with a linear function.
b) Predict a student's score if he or she studies 24 hr, 6 hr, and 18 hr.
c) What is the correlation coefficient? How confident are you about using the regression line to predict function values?

› Skill Maintenance

Find the slope of the line containing the given points. **[1.3]**

73. $(2, -8)$ and $(-5, -1)$

74. $(5, 7)$ and $(5, -7)$

Find an equation for a circle satisfying the given conditions. [1.1]

75. Center $(-7, -1)$, radius of length $\frac{9}{5}$

76. Center $(0, 3)$, diameter of length 5

Synthesis

77. Find k so that the line containing the points $(-3, k)$ and $(4, 8)$ is parallel to the line containing the points $(5, 3)$ and $(1, -6)$.

78. *Road Grade.* Using the following figure, find the road grade and an equation giving the height y as a function of the horizontal distance x.

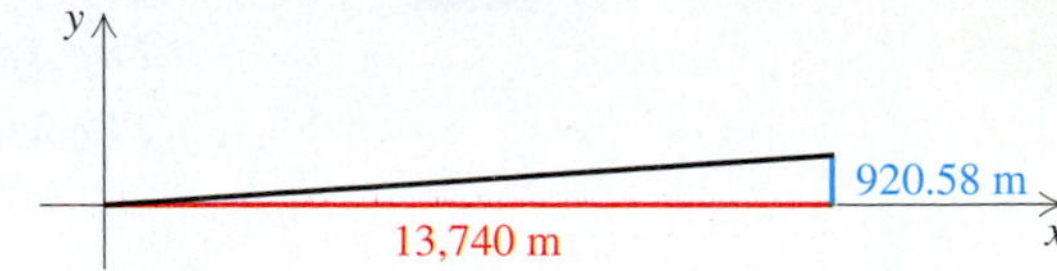

79. Find an equation of the line passing through the point $(4, 5)$ and perpendicular to the line passing through the points $(-1, 3)$ and $(2, 9)$.

1.5 Linear Equations, Functions, Zeros, and Applications

- Solve linear equations.
- Solve applied problems using linear models.
- Find zeros of linear functions.

An **equation** is a statement that two expressions are equal. To **solve** an equation in one variable is to find all the values of the variable that make the equation true. Each of these values is a **solution** of the equation. The set of all solutions of an equation is its **solution set**. Some examples of **equations in one variable** are

$$2x + 3 = 5, \quad 3(x - 1) = 4x + 5,$$

$$x^2 - 3x + 2 = 0, \quad \text{and} \quad \frac{x - 3}{x + 4} = 1.$$

Linear Equations

The first two equations above are *linear equations* in one variable. We define such equations as follows.

A **linear equation in one variable** is an equation that can be expressed in the form $mx + b = 0$, where m and b are real numbers and $m \neq 0$.

Equations that have the same solution set are **equivalent equations**. For example, $2x + 3 = 5$ and $x = 1$ are equivalent equations because 1 is the solution of each equation. On the other hand, $x^2 - 3x + 2 = 0$ and $x = 1$ are not equivalent equations because 1 and 2 are both solutions of $x^2 - 3x + 2 = 0$ but 2 is not a solution of $x = 1$.

To solve an equation, we find an equivalent equation in which the variable is isolated. The following principles allow us to solve linear equations.

EQUATION-SOLVING PRINCIPLES

For any real numbers a, b, and c:

***The Addition Principle*:** If $a = b$ is true, then $a + c = b + c$ is true.

***The Multiplication Principle*:** If $a = b$ is true, then $ac = bc$ is true.

Note to the student and the instructor: We assume that students come to a College Algebra course with some equation-solving skills from their study of Intermediate Algebra. Thus a portion of the material in this section might be considered by some to be review in nature. We present this material here in order to use linear functions, with which students are familiar, to lay the groundwork for zeros of higher-order polynomial functions and their connection to solutions of equations and x-intercepts of graphs.

EXAMPLE 1 Solve: $\frac{3}{4}x - 1 = \frac{7}{5}$.

Solution When we have an equation that contains fractions, it is often convenient to multiply on both sides of the equation by the least common denominator (LCD) of the fractions in order to clear the equation of fractions. We have

$$\frac{3}{4}x - 1 = \frac{7}{5} \qquad \text{The LCD is } 4 \cdot 5, \text{ or } 20.$$

$$20\left(\frac{3}{4}x - 1\right) = 20 \cdot \frac{7}{5} \qquad \text{Multiplying by the LCD on both sides to clear fractions}$$

$$20 \cdot \frac{3}{4}x - 20 \cdot 1 = 28$$

$$15x - 20 = 28$$

$$15x - 20 + 20 = 28 + 20 \qquad \text{Using the addition principle to add 20 on both sides}$$

$$15x = 48$$

$$\frac{15x}{15} = \frac{48}{15} \qquad \text{Using the multiplication principle to multiply by } \tfrac{1}{15}, \text{ or divide by 15, on both sides}$$

$$x = \frac{48}{15}$$

$$x = \frac{16}{5}. \qquad \text{Simplifying. Note that } \tfrac{3}{4}x - 1 = \tfrac{7}{5} \text{ and } x = \tfrac{16}{5} \text{ are equivalent equations.}$$

Check:

$\frac{3}{4}x - 1 = \frac{7}{5}$		
$\frac{3}{4} \cdot \frac{16}{5} - 1$?	$\frac{7}{5}$	Substituting $\frac{16}{5}$ for x
$\frac{12}{5} - \frac{5}{5}$		
$\frac{7}{5}$	$\frac{7}{5}$	TRUE

The solution is $\frac{16}{5}$.

Now Try Exercise 15.

$y_1 = \frac{3}{4}x - 1,\ y_2 = \frac{7}{5}$

Intersection X = 3.2 Y = 1.4

X▸Frac 16/5

We can use the INTERSECT feature on a graphing calculator to solve equations. We call this the **Intersect method**. To use the Intersect method to solve the equation in Example 1, for instance, we graph $y_1 = \frac{3}{4}x - 1$ and $y_2 = \frac{7}{5}$. The value of x for which $y_1 = y_2$ is the solution of the equation $\frac{3}{4}x - 1 = \frac{7}{5}$. This value of x is the first coordinate of the point of intersection of the graphs of y_1 and y_2. Using the INTERSECT feature, we find that the first coordinate of this point is 3.2. We can find fraction notation for the solution by using the ▶FRAC feature. The solution is 3.2, or $\frac{16}{5}$.

EXAMPLE 2 Solve: $2(5 - 3x) = 8 - 3(x + 2)$.

Algebraic Solution

We have

$$\begin{aligned} 2(5 - 3x) &= 8 - 3(x + 2) \\ 10 - 6x &= 8 - 3x - 6 && \text{Using the distributive property} \\ 10 - 6x &= 2 - 3x && \text{Collecting like terms} \\ 10 - 6x + 6x &= 2 - 3x + 6x && \text{Using the addition principle to add } 6x \text{ on both sides} \\ 10 &= 2 + 3x \\ 10 - 2 &= 2 + 3x - 2 && \text{Using the addition principle to add } -2\text{, or subtract 2, on both sides} \\ 8 &= 3x \\ \frac{8}{3} &= \frac{3x}{3} && \text{Using the multiplication principle to multiply by } \tfrac{1}{3}\text{, or divide by 3, on both sides} \\ \frac{8}{3} &= x. \end{aligned}$$

Check:

$$\begin{array}{r|l} 2(5 - 3x) & = 8 - 3(x + 2) \\ \hline 2\left(5 - 3\cdot\frac{8}{3}\right) & ?\ 8 - 3\left(\frac{8}{3} + 2\right) \quad \text{Substituting } \tfrac{8}{3} \text{ for } x \\ 2(5 - 8) & 8 - 3\left(\frac{14}{3}\right) \\ 2(-3) & 8 - 14 \\ -6 & -6 \quad \text{TRUE} \end{array}$$

The solution is $\frac{8}{3}$.

Graphical Solution

We graph $y_1 = 2(5 - 3x)$ and $y_2 = 8 - 3(x + 2)$. The first coordinate of the point of intersection of the graphs is the value of x for which $2(5 - 3x) = 8 - 3(x + 2)$ and is thus the solution of the equation.

$y_1 = 2(5 - 3x)$, $y_2 = 8 - 3(x + 2)$

Intersection X = 2.6666667 Y = −6

The solution is approximately 2.6666667.

We can find fraction notation for the exact solution by using the ▶FRAC feature. The solution is $\frac{8}{3}$.

Now Try Exercise 27.

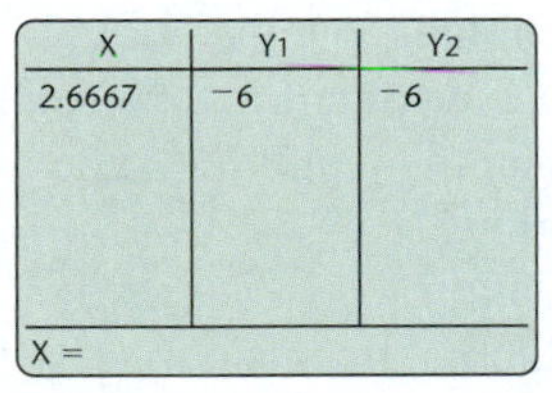

X	Y1	Y2
2.6667	−6	−6

X =

We can use the TABLE feature on a graphing calculator, set in ASK mode, to check the solutions of equations. In Example 2, for instance, we let $y_1 = 2(5 - 3x)$ and $y_2 = 8 - 3(x + 2)$. When $\frac{8}{3}$ is entered for x, we see that $y_1 = y_2$, or $2(5 - 3x) = 8 - 3(x + 2)$. Thus, $\frac{8}{3}$ is the solution of the equation. (Note that the calculator converts $\frac{8}{3}$ to decimal notation in the table.)

Special Cases

Some equations have *no* solution.

EXAMPLE 3 Solve: $-24x + 7 = 17 - 24x$.

Solution We have

$$\begin{aligned} -24x + 7 &= 17 - 24x \\ 24x - 24x + 7 &= 24x + 17 - 24x && \text{Adding } 24x \\ 7 &= 17. && \text{We get a false equation.} \end{aligned}$$

No matter what number we substitute for x, we get a false equation. Thus the equation has *no* solution.

Now Try Exercise 11.

There are some equations for which *any* real number is a solution.

EXAMPLE 4 Solve: $3 - \frac{1}{3}x = -\frac{1}{3}x + 3$.

Solution We have

$$3 - \tfrac{1}{3}x = -\tfrac{1}{3}x + 3$$
$$\tfrac{1}{3}x + 3 - \tfrac{1}{3}x = \tfrac{1}{3}x - \tfrac{1}{3}x + 3 \qquad \text{Adding } \tfrac{1}{3}x$$
$$3 = 3. \qquad \text{We get a true equation.}$$

Replacing x with any real number gives a true equation. Thus *any* real number is a solution. This equation has *infinitely* many solutions. The solution set is the set of real numbers, $\{x \mid x \text{ is a real number}\}$, or $(-\infty, \infty)$.

Now Try Exercise 3.

Applications Using Linear Models

Mathematical techniques can be used to answer questions arising from real-world situations. Linear equations and linear functions *model* many of these situations.

The following strategy is of great assistance in problem solving.

FIVE STEPS FOR PROBLEM SOLVING

1. **Familiarize** yourself with the problem situation. If the problem is presented in words, this means to read carefully. Some or all of the following can also be helpful.
   a) Make a drawing, if it makes sense to do so.
   b) Make a written list of the known facts and a list of what you wish to find out.
   c) Assign variables to represent unknown quantities.
   d) Organize the information in a chart or a table, if appropriate.
   e) Find further information. Look up a formula, consult a reference book or an expert in the field, or do research on the Internet.
   f) Guess or estimate the answer and check your guess or estimate.
2. **Translate** the problem situation to mathematical language or symbolism. For most of the problems you will encounter in algebra, this means to write one or more equations, but sometimes an inequality or some other mathematical symbolism may be appropriate.
3. **Carry out** some type of mathematical manipulation. Use your mathematical skills to find a possible solution. In algebra, this usually means to solve an equation, an inequality, or a system of equations or inequalities.
4. **Check** to see whether your possible solution actually fits the problem situation and is thus really a solution of the problem. Although you may have solved an equation, the solution(s) of the equation might not be solution(s) of the original problem.
5. **State** the answer clearly using a complete sentence.

EXAMPLE 5 ***Industrial Robots.*** Companies in China purchased 36,560 industrial robots in 2013. This is 54% more than the number of industrial robots purchased by companies in the United States. (*Source*: International Federation of Robotics Statistical Department) How many industrial robots were bought by companies in the United States in 2013?

Solution

1. **Familiarize.** Let's estimate that 20,000 industrial robots were purchased by U.S. companies. Then the number of robots purchased by Chinese companies would be

$$20{,}000 + 54\% \cdot 20{,}000 = 1(20{,}000) + 0.54(20{,}000) = 1.54(20{,}000) = 30{,}800.$$

   Since we know that Chinese companies actually bought 36,560 robots, our estimate of 20,000 is too low. Nevertheless, the calculations performed indicate how we can translate the problem to an equation. We let $x =$ the number of industrial robots purchased by U.S. companies in 2013. Then $x + 54\%x$, or $1 \cdot x + 0.54x$, or $1.54x$, is the number of robots purchased by Chinese companies.

2. **Translate.** We translate to an equation:

$$\underbrace{\text{Number of robots purchased by Chinese companies}}_{1.54x} \ \underbrace{\text{is}}_{=} \ \underbrace{36{,}560}_{36{,}560.}$$

3. **Carry out.** We solve the equation, as follows:

$$1.54x = 36{,}560$$
$$x = \frac{36{,}560}{1.54} \approx 23{,}740.$$

4. **Check.** 54% of 23,740 is about 12,820, and $23{,}740 + 12{,}820 = 36{,}560$.

5. **State.** Companies in the United States purchased approximately 23,740 industrial robots in 2013.

Now Try Exercise 33.

EXAMPLE 6 ***Studying Abroad.*** In the 2012–2013 school year 47,058 U.S. students studied abroad in Italy and in France. There were 12,638 more students studying in Italy than in France. (*Source*: Institute of International Education 2014 Report) Find the number of U.S. students studying abroad in Italy and in France.

Solution

1. **Familiarize.** The number of U.S. students studying in Italy is described in terms of the number of students studying in France, so we let $x =$ the number of students studying in France. Then $x + 12{,}638 =$ the number of students studying in Italy.

2. **Translate.** We translate to an equation:

$$\underbrace{\text{Number of students studying in France}}_{x} \ \underbrace{\text{plus}}_{+} \ \underbrace{\text{Number of students studying in Italy}}_{x + 12{,}638} \ \underbrace{\text{is}}_{=} \ \underbrace{47{,}058}_{47{,}058.}$$

3. **Carry out.** We solve the equation, as follows:

$$x + x + 12{,}638 = 47{,}058$$
$$2x = 34{,}420 \quad \text{Collecting like terms and subtracting 12,638 on both sides}$$
$$x = 17{,}210. \quad \text{Dividing by 2 on both sides}$$

   If $x = 17{,}210$, then $x + 12{,}638 = 17{,}210 + 12{,}638 = 29{,}848$.

4. **Check.** If there were 29,848 U.S. students studying abroad in Italy and 17,210 in France, then the total number of U.S. students studying abroad in Italy and in France was 29,848 + 17,210, or 47,058. Also, 29,848 is 12,638 more than 17,210. The answer checks.
5. **State.** In the 2012–2013 school year, there were 29,848 U.S. students studying abroad in Italy and 17,210 students studying in France.

Now Try Exercise 57.

In some applications, we need to use a formula that describes the relationships among variables. When a situation involves distance, rate (also called speed or velocity), and time, for example, we use the following formula.

THE MOTION FORMULA

The distance d traveled by an object moving at rate r in time t is given by

$$d = r \cdot t.$$

EXAMPLE 7 ***Airplane Speed.*** Delta Airlines' fleet includes B737/800's, each with a cruising speed of 517 mph, and Saab 340B's, each with a cruising speed of 290 mph (*Source*: Delta Airlines). Suppose that a Saab 340B takes off and travels at its cruising speed. One hour later, a B737/800 takes off and follows the same route, traveling at its cruising speed. How long will it take the B737/800 to overtake the Saab 340B?

Solution

1. **Familiarize.** We make a drawing showing both the known information and the unknown information. We let $t =$ the time, in hours, that the B737/800 travels before it overtakes the Saab 340B. Since the Saab 340B takes off 1 hr before the 737, it will travel for $t + 1$ hr before being overtaken. The planes will have traveled the same distance, d, when one overtakes the other.

We can also organize the information in a table, as follows.

$d \quad = \quad r \quad \cdot \quad t$

	Distance	Rate	Time	
B737/800	d	517	t	→ $d = 517t$
Saab 340B	d	290	$t + 1$	→ $d = 290(t + 1)$

2. **Translate.** Using the formula $d = rt$ in each row of the table, we get two expressions for d:

$$d = 517t \quad \text{and} \quad d = 290(t + 1).$$

Since the distances are the same, we have the following equation:

$$517t = 290(t + 1).$$

3. **Carry out.** We solve the equation, as follows:

$$517t = 290(t + 1)$$

$$517t = 290t + 290 \qquad \text{Using the distributive property}$$

$$227t = 290 \qquad \text{Subtracting } 290t \text{ on both sides}$$

$$t \approx 1.28. \qquad \text{Dividing by 227 on both sides and rounding to the nearest hundredth}$$

4. **Check.** If the B737/800 travels for about 1.28 hr, then the Saab 340B travels for about 1.28 + 1, or 2.28 hr. In 2.28 hr, the Saab 340B travels $290(2.28)$, or 661.2 mi; and in 1.28 hr, the B737/800 travels $517(1.28)$, or 661.76 mi. Since 661.76 mi $\approx$ 661.2 mi, the answer checks. (Remember: We rounded the value of t.)

5. **State.** About 1.28 hr after the B737/800 has taken off, it will overtake the Saab 340B.

Now Try Exercise 41.

For some applications, we need to use a formula to find the amount of interest earned by an investment or the amount of interest due on a loan.

THE SIMPLE-INTEREST FORMULA

The **simple interest** I on a principal of P dollars at interest rate r for t years is given by

$$I = Prt.$$

EXAMPLE 8 ***Student Loans.*** Damarion's two student loans total \$28,000. One loan is at 5% simple interest, and the other is at 3% simple interest. After 1 year, Damarion owes \$1040 in interest. What is the amount of each loan?

Solution

1. **Familiarize.** We let x = the amount borrowed at 5% interest. Then the remainder of the \$28,000, or $28,000 - x$, is borrowed at 3%. We organize the information in a table, keeping in mind the formula $I = Prt$.

	Amount Borrowed	Interest Rate	Time	Amount of Interest
5% Loan	x	5%, or 0.05	1 year	$x(0.05)(1)$, or $0.05x$
3% Loan	$28,000 - x$	3%, or 0.03	1 year	$(28,000 - x)(0.03)(1)$, or $0.03(28,000 - x)$
Total	28,000			1040

2. **Translate.** The total amount of interest on the two loans is \$1040. Thus we can translate to the following equation:

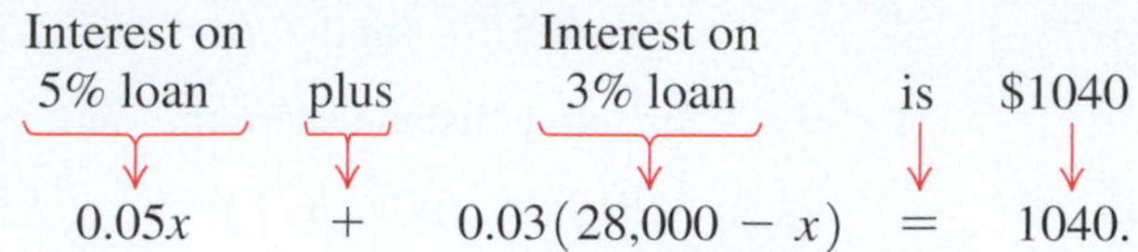

3. **Carry out.** We solve the equation, as follows:

$$0.05x + 0.03(28{,}000 - x) = 1040$$
$$0.05x + 840 - 0.03x = 1040 \quad \text{Using the distributive property}$$
$$0.02x + 840 = 1040 \quad \text{Collecting like terms}$$
$$0.02x = 200 \quad \text{Subtracting 840 on both sides}$$
$$x = 10{,}000. \quad \text{Dividing by 0.02 on both sides}$$

If $x = 10{,}000$, then $28{,}000 - x = 28{,}000 - 10{,}000 = 18{,}000$.

4. **Check.** The interest on \$10,000 at 5% for 1 year is \$10,000(0.05)(1), or \$500. The interest on \$18,000 at 3% for 1 year is \$18,000(0.03)(1), or \$540. Since \$500 + \$540 = \$1040, the answer checks.

5. **State.** Damarion borrowed \$10,000 at 5% interest and \$18,000 at 3% interest.

Now Try Exercise 55.

Sometimes we use formulas from geometry when solving applied problems. In the following example, we use the formula for the perimeter P of a rectangle with length l and width w: $P = 2l + 2w$.

EXAMPLE 9 ***Solar Panels.*** In December 2009, a solar energy farm was completed at the Denver International Airport. More than 9200 rectangular solar panels were installed (*Sources*: Woods Allee, Denver International Airport; www.solarpanelstore.com; *The Denver Post*). A solar panel, or photovoltaic panel, converts sunlight into electricity. The length of a panel is 13.6 in. less than twice the width, and the perimeter is 207.4 in. Find the length and the width.

Solution

1. **Familiarize.** We first make a drawing. Since the length of the panel is described in terms of the width, we let w = the width, in inches. Then $2w - 13.6$ = the length, in inches.

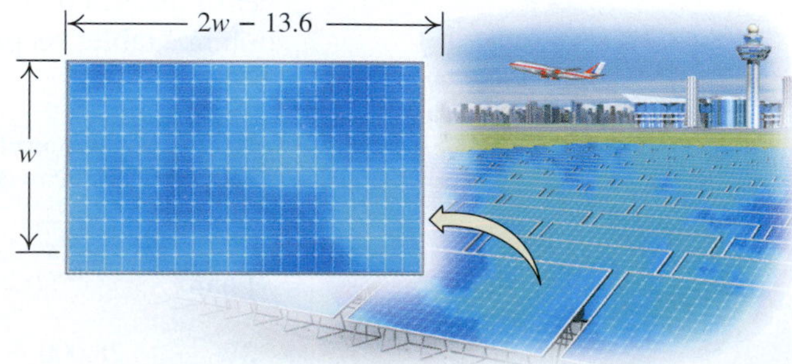

2. **Translate.** We use the formula for the perimeter of a rectangle:

$$P = 2l + 2w$$
$$207.4 = 2(2w - 13.6) + 2w. \quad \text{Substituting 207.4 for } P \text{ and } 2w - 13.6 \text{ for } l$$

3. **Carry out.** We solve the equation:

$$\begin{aligned} 207.4 &= 2(2w - 13.6) + 2w \\ 207.4 &= 4w - 27.2 + 2w && \text{Using the distributive property} \\ 207.4 &= 6w - 27.2 && \text{Collecting like terms} \\ 234.6 &= 6w && \text{Adding 27.2 on both sides} \\ 39.1 &= w. && \text{Dividing by 6 on both sides} \end{aligned}$$

If $w = 39.1$, then $2w - 13.6 = 2(39.1) - 13.6 = 78.2 - 13.6 = 64.6$.

4. **Check.** The length, 64.6 in., is 13.6 in. less than twice the width, 39.1 in. Also

$$2 \cdot 64.6 \text{ in.} + 2 \cdot 39.1 \text{ in.} = 129.2 \text{ in.} + 78.2 \text{ in.} = 207.4 \text{ in.}$$

The answer checks.

5. **State.** The length of the solar panel is 64.6 in., and the width is 39.1 in.

Now Try Exercise 49.

EXAMPLE 10 *Cab Fare.* Metro Taxi charges a \$2.50 pickup fee and \$2 per mile traveled. Grayson's cab fare from the airport to his hotel is \$32.50. How many miles did he travel in the cab?

Solution

1. **Familiarize.** Let's guess that Grayson traveled 12 mi in the cab. Then his fare would be

$$\$2.50 + \$2 \cdot 12 = \$2.50 + \$24 = \$26.50.$$

We see that our guess is low, but the calculation shows us how to translate the problem to an equation. We let $m =$ the number of miles that Grayson traveled in the cab.

2. **Translate.** We translate to an equation:

Pickup fee	plus	Cost per mile	times	Number of miles traveled	is	Total charge
↓	↓	↓	↓	↓	↓	↓
2.50	+	2	·	m	=	32.50

3. **Carry out.** We solve the equation:

$$\begin{aligned} 2.50 + 2 \cdot m &= 32.50 \\ 2m &= 30 && \text{Subtracting 2.50 on both sides} \\ m &= 15. && \text{Dividing by 2 on both sides} \end{aligned}$$

4. **Check.** If Grayson travels 15 mi in the cab, the mileage charge is \$2 · 15, or \$30. Then, with the \$2.50 pickup fee included, his total charge is \$2.50 + \$30, or \$32.50. The answer checks.

5. **State.** Grayson traveled 15 mi in the cab.

Now Try Exercise 65.

Zeros of Linear Functions

An input for which a function's output is 0 is called a **zero** of the function. We will restrict our attention in this section to zeros of linear functions. This allows us to become familiar with the concept of a zero, and it lays the groundwork for working with zeros of other types of functions in succeeding chapters.

LINEAR FUNCTIONS

REVIEW SECTION 1.3.

ZEROS OF FUNCTIONS

An input c of a function f is called a **zero** of the function if the output for the function is 0 when the input is c. That is, c is a zero of f if $f(c) = 0$.

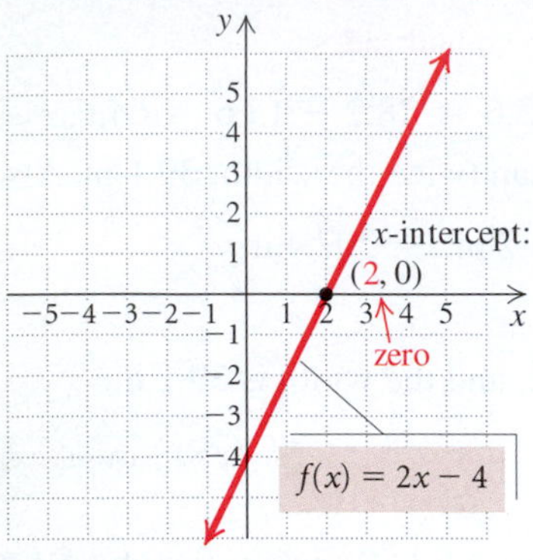

Recall that a linear function is given by $f(x) = mx + b$, where m and b are constants. For the linear function $f(x) = 2x - 4$, we have $f(2) = 2 \cdot 2 - 4 = 0$, so 2 is a **zero** of the function. In fact, 2 is the *only* zero of this function. In general, a **linear function $f(x) = mx + b$, with $m \neq 0$, has exactly one zero.**

The zero, 2, is the first coordinate of the point at which the graph crosses the x-axis. This point, $(2, 0)$, is the *x-intercept* of the graph. Thus when we find the zero of a linear function, we are also finding the first coordinate of the x-intercept of the graph of the function.

For every linear function $f(x) = mx + b$, there is an associated linear equation $mx + b = 0$. When we find the zero of a function $f(x) = mx + b$, we are also finding the solution of the equation $mx + b = 0$.

EXAMPLE 11 Find the zero of $f(x) = 5x - 9$.

Algebraic Solution

We find the value of x for which $f(x) = 0$:

$5x - 9 = 0$ **Setting $f(x) = 0$**

$5x = 9$ **Adding 9 on both sides**

$x = \frac{9}{5}$, or 1.8. **Dividing by 5 on both sides**

Using a table, set in ASK mode, we can check the solution. We enter $y = 5x - 9$ on the equation-editor screen and then enter the value $x = \frac{9}{5}$, or 1.8, in the table.

X	Y1
1.8	0

X =

We see that $y = 0$ when $x = 1.8$, so the number 1.8 checks. The zero is $\frac{9}{5}$, or 1.8. This means that $f\left(\frac{9}{5}\right) = 0$, or $f(1.8) = 0$. Note that the *zero* of the function $f(x) = 5x - 9$ is the *solution* of the equation $5x - 9 = 0$.

Graphical Solution

The solution of $5x - 9 = 0$ is also the zero of $f(x) = 5x - 9$. Thus we can solve an equation by finding the zeros of the function associated with it. We call this the **Zero method**.

We graph $y = 5x - 9$ in the standard window and use the ZERO feature from the CALC menu to find the zero of $f(x) = 5x - 9$, as shown in the following figure. Note that the x-intercept must appear in the window when the ZERO feature is used.

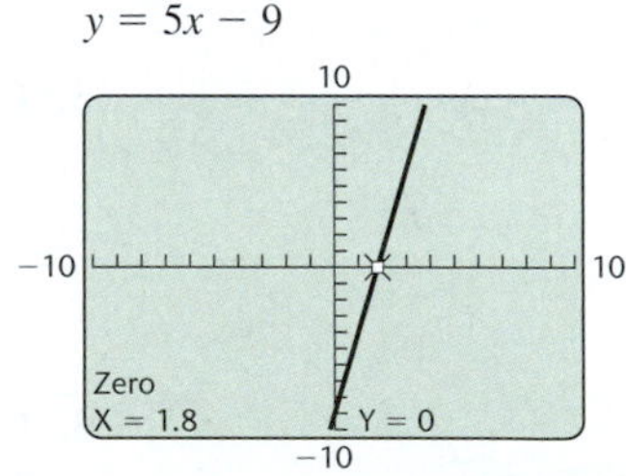

We can check algebraically by substituting 1.8 for x:

$$f(1.8) = 5(1.8) - 9 = 9 - 9 = 0.$$

The zero of $f(x) = 5x - 9$ is 1.8, or $\frac{9}{5}$.

Now Try Exercise 73.

CONNECTING THE CONCEPTS

The Intersect Method and the Zero Method

An equation such as $x - 1 = 2x - 6$ can be solved using the Intersect method by graphing $y_1 = x - 1$ and $y_2 = 2x - 6$ and using the INTERSECT feature to find the first coordinate of the point of intersection of the graphs.

The equation can also be solved using the Zero method by writing it with 0 on one side of the equals sign and then using the ZERO feature.

Solve: $x - 1 = 2x - 6$.

THE INTERSECT METHOD

Graph $y_1 = x - 1$ and $y_2 = 2x - 6$.

Point of intersection: $(5, 4)$

Solution: 5

THE ZERO METHOD

First, add $-2x$ and 6 on both sides of the equation to get 0 on one side:

$$x - 1 = 2x - 6$$
$$x - 1 - 2x + 6 = 0.$$

Graph

$$y_3 = x - 1 - 2x + 6.$$

Zero: 5

Solution: 5

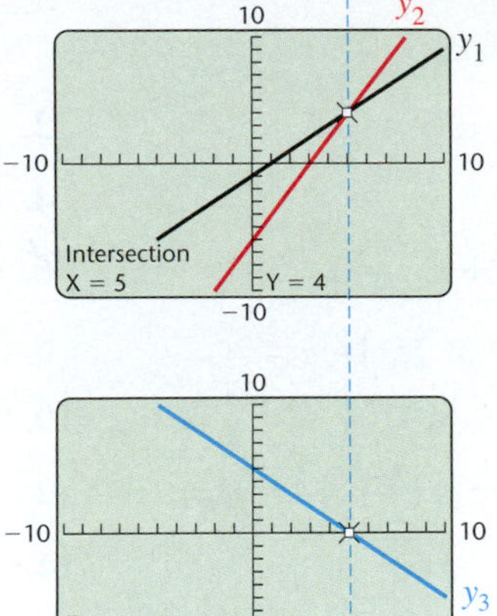

CONNECTING THE CONCEPTS

Zeros, Solutions, and Intercepts

The zero of a linear function $f(x) = mx + b$, with $m \neq 0$, is the solution of the linear equation $mx + b = 0$ and is the first coordinate of the x-intercept of the graph of $f(x) = mx + b$. To find the zero of $f(x) = mx + b$, we solve $f(x) = 0$, or $mx + b = 0$.

FUNCTION

Linear Function

$$f(x) = 2x - 4, \text{ or}$$
$$y = 2x - 4$$

ZERO OF THE FUNCTION; SOLUTION OF THE EQUATION

To find the **zero** of $f(x)$, we solve $f(x) = 0$:

$$2x - 4 = 0$$
$$2x = 4$$
$$x = 2.$$

The **solution** of $2x - 4 = 0$ is 2. This is the zero of the function $f(x) = 2x - 4$. That is, $f(2) = 0$.

ZERO OF THE FUNCTION; *x*-INTERCEPT OF THE GRAPH

The zero of $f(x)$ is the first coordinate of the ***x*-intercept** of the graph of $y = f(x)$.

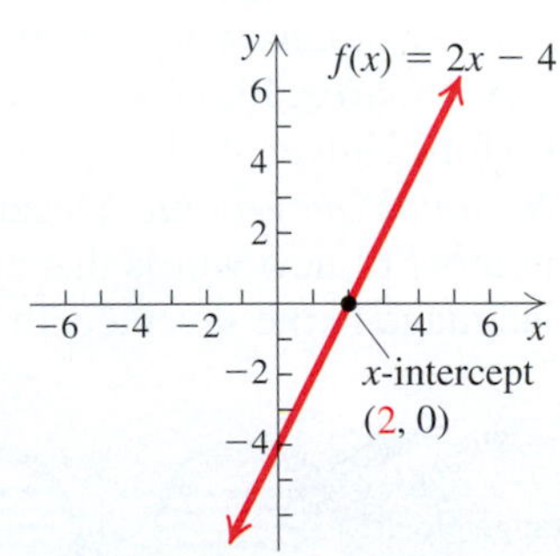

1.5 Exercise Set

Solve.

1. $4x + 5 = 21$
2. $2y - 1 = 3$
3. $23 - \frac{2}{5}x = -\frac{2}{5}x + 23$
4. $\frac{6}{5}y + 3 = \frac{3}{10}$
5. $4x + 3 = 0$
6. $3x - 16 = 0$
7. $3 - x = 12$
8. $4 - x = -5$
9. $3 - \frac{1}{4}x = \frac{3}{2}$
10. $10x - 3 = 8 + 10x$
11. $\frac{2}{11} - 4x = -4x + \frac{9}{11}$
12. $8 - \frac{2}{9}x = \frac{5}{6}$
13. $8 = 5x - 3$
14. $9 = 4x - 8$
15. $\frac{2}{5}y - 2 = \frac{1}{3}$
16. $-x + 1 = 1 - x$
17. $y + 1 = 2y - 7$
18. $5 - 4x = x - 13$
19. $2x + 7 = x + 3$
20. $5x - 4 = 2x + 5$
21. $3x - 5 = 2x + 1$
22. $4x + 3 = 2x - 7$
23. $4x - 5 = 7x - 2$
24. $5x + 1 = 9x - 7$
25. $5x - 2 + 3x = 2x + 6 - 4x$
26. $5x - 17 - 2x = 6x - 1 - x$
27. $7(3x + 6) = 11 - (x + 2)$
28. $4(5y + 3) = 3(2y - 5)$
29. $3(x + 1) = 5 - 2(3x + 4)$
30. $4(3x + 2) - 7 = 3(x - 2)$
31. $2(x - 4) = 3 - 5(2x + 1)$
32. $3(2x - 5) + 4 = 2(4x + 3)$

33. *New Words in the English Language.* During the nineteenth century, 75,029 new words entered the English language. This is about 46.9% more than the number of new words entered in the seventeenth century. (*Source*: Philip Durkin and Katherine Martin, Oxford University Press; "English by the Book," *National Geographic*, December 2013) Find the number of new words that appeared in the English language in the seventeenth century.

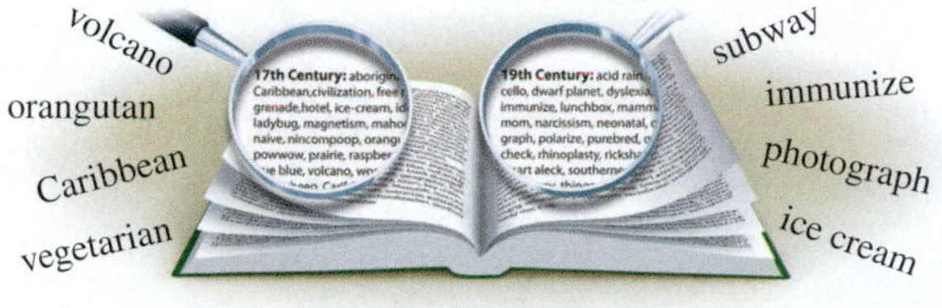

34. *Calorie Intake.* The average worldwide daily calorie intake per person has increased from 2200 to 2800 calories since the early 1960s. The average daily calorie intake per person in the United States is 3688. This is about 86.4% more than the average daily calorie intake per person in Haiti. (*Sources*: UN Food and Agriculture Organization; World Health Organization) Find the average daily calorie intake per person in Haiti.

35. *Amount Borrowed.* Kea borrowed money from her father at 5% simple interest to help pay her tuition at Wellington Community College. At the end of 1 year, she owed a total of $1365 in principal and interest. How much did she borrow?

36. *Amount of an Investment.* Miles makes an investment at 4% simple interest. At the end of 1 year, the total value of the investment is $1560. How much was originally invested?

37. *Angle Measure.* In triangle *ABC*, angle *B* is five times as large as angle *A*. The measure of angle *C* is 2° less than that of angle *A*. Find the measures of the angles. (*Hint*: The sum of the angle measures is 180°.)

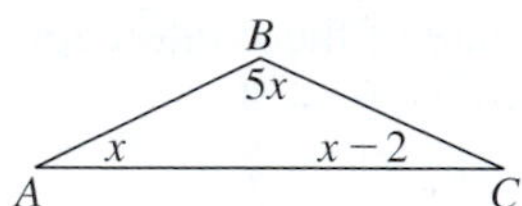

38. *Angle Measure.* In triangle *ABC*, angle *B* is twice as large as angle *A*. Angle *C* measures 20° more than angle *A*. Find the measures of the angles.

39. *Clothing Trade Deficit.* Imports of apparel and clothing accessories to the United States totaled $87.924 billion in 2013. This amount was $2.299 billion less than twenty-five times the apparel and clothing accessories exports that year. (*Source*: Bureau of Economic Analysis, U.S. Department of Commerce) Find the amount of apparel and clothing accessories exports from the United States in 2013.

40. *Foreign Trade.* In 2013, the total value of exports from the United States was $1,579,593,000,000. That year, exports were $445,432,000,000 more than half of the U.S. imports. (*Source*: U.S. Bureau of Economic Analysis, U.S. Department of Commerce) Find the value of imports to the United States in 2013.

41. *Train Speeds.* A Central Railway freight train leaves a station and travels due north at a speed of 60 mph. One hour later, an Amtrak passenger train leaves the same station and travels due north on a parallel track at a speed of 80 mph. How long will it take the passenger train to overtake the freight train?

42. *Distance Traveled.* A private airplane leaves Midway Airport and flies due east at a speed of 180 km/h. Two hours later, a jet leaves Midway and flies due east at a speed of 900 km/h. How far from the airport will the jet overtake the private plane?

43. *TV Tweets.* In 2014, there were 13.8 million tweets about the 56th Annual Grammy Awards. This number was 11.7 million more than the number of tweets about the State of the Union address. (*Source*: Nielsen) Find the number of tweets about the State of the Union address.

44. *Hourly Earnings.* The average hourly earnings of a medical records and health information technician is $33.05 less than the average hourly earnings of a purchasing manager. The average hourly earnings of a medical records and health information technician is $16.81. (*Sources*: Bureau of Labor Statistics; Economic Modeling Specialists International; Debra Auerbach, CareerBuilder.com) Find the average hourly earnings of a purchasing manager.

45. *Commission vs. Salary.* Samantha has a choice between receiving either a monthly salary of $1800 from Furniture by Design or a base salary of $1600 and a 4% commission on the amount of furniture that she sells during the month. For what amount of sales will the two choices be equal?

46. *Sales Commission.* Edward, a consumer electronics salesperson, earns a base salary of $1270 per month and a commission of 6% on the amount of sales that he makes. One month Edward received a paycheck for $3154. Find the amount of his sales for the month.

47. *Studying Abroad.* In the 2012–2013 school year, approximately 820,000 foreign students studied in the United States. The number of U.S. students who studied abroad that same year was about seven-twentieths of the number of foreign students who studied in the United States. (*Source*: Pew Research Center) Find the number of U.S. students who studied abroad during the 2012–2013 school year.

48. *Population Density.* The population density in China is 365.3 persons per square mile. The population density in the United States is approximately one-fourth of the density in China. (*Source*: *The World Almanac* 2014) Find the population density in the United States.

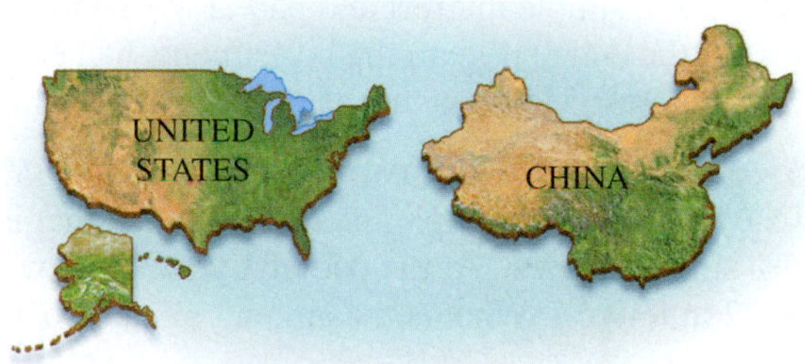

49. *Soccer-Field Dimensions.* The width of the soccer field recommended for players under the age of 12 is 35 yd less than the length. The perimeter of the field is 330 yd. (*Source*: U.S. Youth Soccer) Find the dimensions of the field.

50. *Poster Dimensions.* Marissa is designing a poster to promote the Talbot Street Art Fair. The width of the poster will be two-thirds of its height, and its perimeter will be 100 in. Find the dimensions of the poster.

51. *Test-Plot Dimensions.* Morgan's Seeds has a rectangular test plot with a perimeter of 322 m. The length is 25 m more than the width. Find the dimensions of the plot.

52. *Garden Dimensions.* The children at Tiny Tots Day Care plant a rectangular vegetable garden with a perimeter of 39 m. The length is twice the width. Find the dimensions of the garden.

53. *Flying into a Headwind.* An airplane that travels 450 mph in still air encounters a 30-mph headwind. How long will it take the plane to travel 1050 mi into the wind?

54. *Flying with a Tailwind.* An airplane that can travel 375 mph in still air is flying with a 25-mph tailwind. How long will it take the plane to travel 700 mi with the wind?

55. *Investment Income.* Anton invested a total of \$5000, part at 3% simple interest and part at 4% simple interest. At the end of 1 year, the investments had earned \$176 interest. How much was invested at each rate?

56. *Student Loans.* Vera's two student loans total \$9000. One loan is at 5% simple interest, and the other is at 6% simple interest. At the end of 1 year, Vera owes \$492 in interest. What is the amount of each loan?

57. *Patents.* In 2013, IBM (International Business Machines) received 2133 more patents than Samsung. Together, they received 11,485 patents. (*Source*: IFI Claims Patent Services) How many patents did each company receive?

58. *Books about Presidents.* There are 5493 print and e-books written about both George Washington and Abraham Lincoln. There are 1675 more books about Lincoln than about Washington. (*Source*: Bowker Books in Print) How many books have been written about each president?

59. *Estates of Celebrities.* From October 2013 to October 2014, the estates of Michael Jackson and Elvis Presley together took in \$195 million. The income to the estate of Michael Jackson was \$30 million more than twice the income to the estate of Elvis Presley. (*Source*: 'Forbes' list of Top-Earning Dead Celebrities) Find the income to each estate.

60. *Ocean Depth.* The average depth of the Pacific Ocean is 14,040 ft, and its depth is 8890 ft less than the sum of the average depths of the Atlantic Ocean and the Indian Ocean. The average depth of the Indian Ocean is 272 ft less than four-fifths of the average depth of the Atlantic Ocean. (*Source*: *Time Almanac* 2010) Find the average depth of the Indian Ocean.

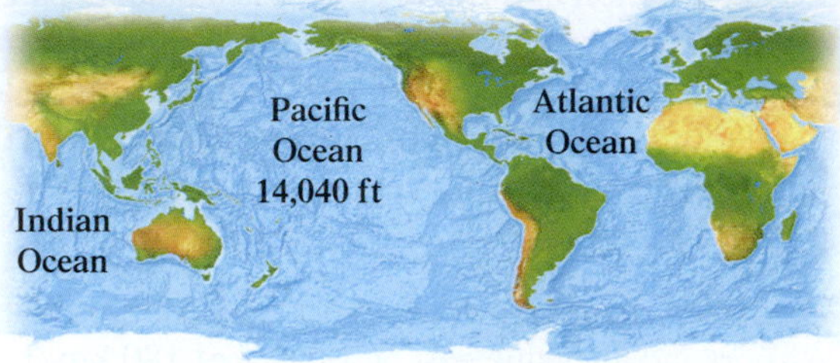

61. *Water Weight.* Water accounts for 55% of a woman's weight (*Source*: ga.water.usgs.gov/edu). Lily weighs 135 lb. How much of her body weight is water?

62. *Water Weight.* Water accounts for 60% of a man's weight (*Source*: ga.water.usgs.gov/edu). Jake weighs 186 lb. How much of his body weight is water?

63. *Traveling Upstream.* A kayak moves at a rate of 12 mph in still water. If the river's current flows at a rate of 4 mph, how long does it take the boat to travel 36 mi upstream?

64. *Traveling Downstream.* Angelo's kayak travels 14 km/h in still water. If the river's current flows at a rate of 2 km/h, how long will it take him to travel 20 km downstream?

65. *Cab Fare.* City Cabs charges a \$1.75 pickup fee and \$1.50 per mile traveled. Diego's fare for a cross-town ride is \$19.75. How far did he travel in the cab?

66. *Hourly Wage.* Rosalyn worked 48 hr one week and earned a \$1066 paycheck. She earns time and a half (1.5 times her regular hourly wage) for the number of hours that she works in excess of 40. What is Rosalyn's regular hourly wage?

67. *Olive Oil.* Together, Italy, Spain, and the United States consume 58% of the world's olive oil. The percentage consumed in Italy is $3\frac{3}{4}$ times the percentage consumed in the United States. The percentage consumed in Spain is $\frac{2}{3}$ of the percentage consumed in Italy. (*Source*: www.OliveOilEmporium.com) Find the percent of the world's olive oil consumed in each country.

68. *NFL Stadium Elevation.* The elevations of the 31 NFL stadiums range from 3 ft at Mercedes-Benz Superdome, New Orleans, Louisiana, to 5280 ft at Sports Authority Field at Mile High, Denver, Colorado. The elevation of Sports Authority Field at Mile High is 275 ft higher than seven times the elevation of Lucas Oil Stadium in Indianapolis, Indiana. What is the elevation of Lucas Oil Stadium?

Find the zero of the linear function.

69. $f(x) = x + 5$

70. $f(x) = 5x + 20$

71. $f(x) = -2x + 11$

72. $f(x) = 8 + x$

73. $f(x) = 16 - x$

74. $f(x) = -2x + 7$

75. $f(x) = x + 12$

76. $f(x) = 8x + 2$

77. $f(x) = -x + 6$

78. $f(x) = 4 + x$

79. $f(x) = 20 - x$

80. $f(x) = -3x + 13$

81. $f(x) = \frac{2}{5}x - 10$

82. $f(x) = 3x - 9$

83. $f(x) = -x + 15$

84. $f(x) = 4 - x$

In Exercises 85–90, use the given graph to find each of the following: **(a)** *the x-intercept and* **(b)** *the zero of the function.*

85.

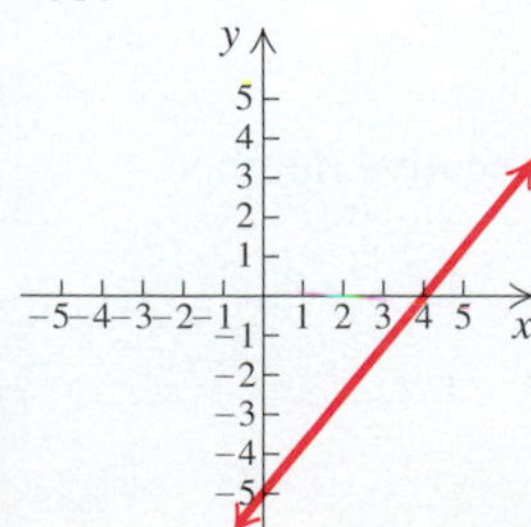

86.

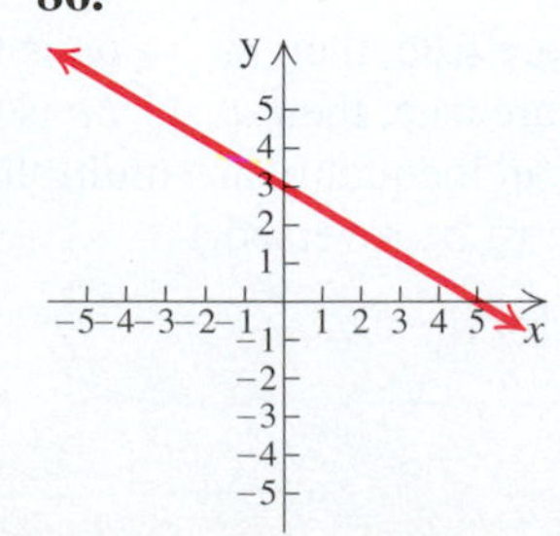

87.

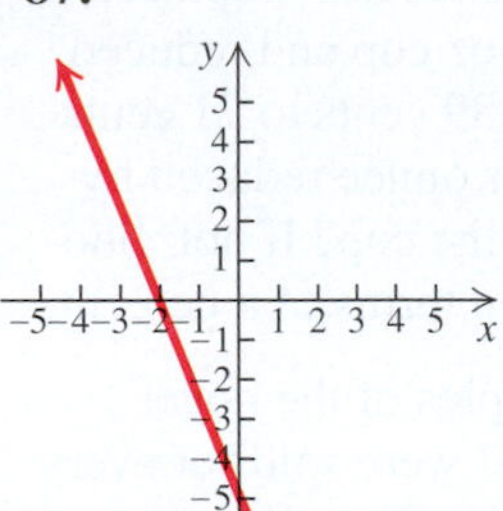

88.

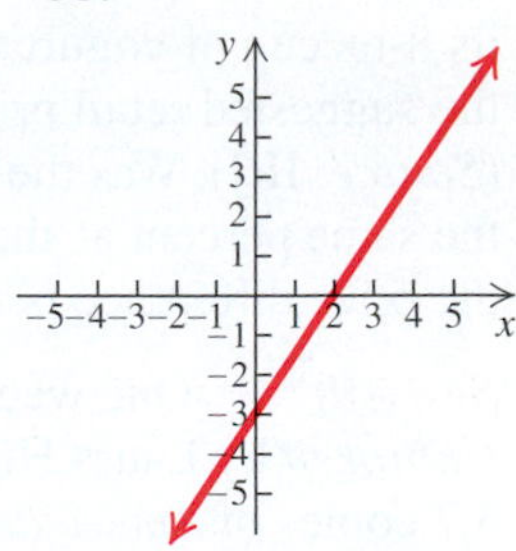

89.

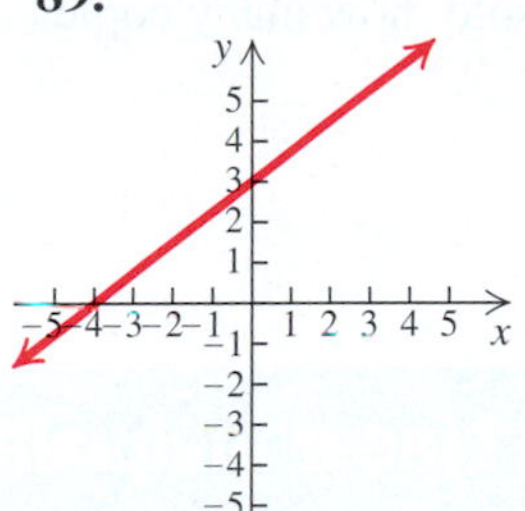

90.

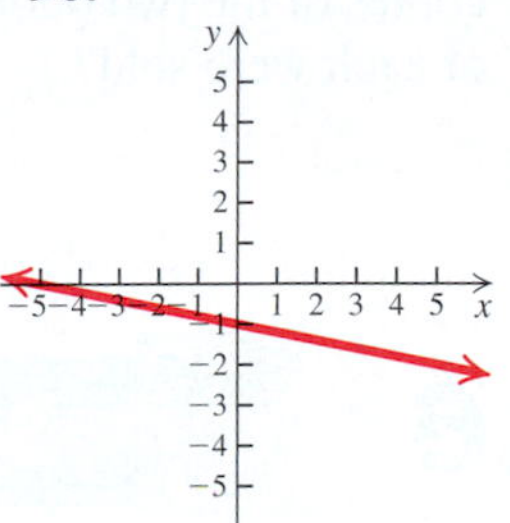

Skill Maintenance

91. Write a slope–intercept equation for the line containing the point $(-1, 4)$ and parallel to the line $3x + 4y = 7$. **[1.4]**

92. Write an equation of the line containing the points $(-5, 4)$ and $(3, -2)$. **[1.4]**

93. Find the distance between $(2, 2)$ and $(-3, -10)$. **[1.1]**

94. Find the midpoint of the segment with endpoints $\left(-\frac{1}{2}, \frac{2}{5}\right)$ and $\left(-\frac{3}{2}, \frac{3}{5}\right)$. **[1.1]**

95. Given that $f(x) = \dfrac{x}{x - 3}$, find $f(-3), f(0)$, and $f(3)$.

96. Find the slope and the y-intercept of the line with the equation $7x - y = \frac{1}{2}$.

Synthesis

State whether each of the following is a linear function.

97. $f(x) = 7 - \frac{3}{2}x$

98. $f(x) = \dfrac{3}{2x} + 5$

99. $f(x) = x^2 + 1$

100. $f(x) = \frac{3}{4}x - (2.4)^2$

Solve.

101. $2x - \{x - [3x - (6x + 5)]\} = 4x - 1$

102. $14 - 2[3 + 5(x - 1)] = 3\{x - 4[1 + 6(2 - x)]\}$

103. *Packaging and Price.* Dannon recently replaced its 8-oz cup of yogurt with a 6-oz cup and reduced the suggested retail price from 89 cents to 71 cents (*Source*: IRI). Was the price per ounce reduced by the same percent as the size of the cup? If not, find the price difference per ounce in terms of a percent.

104. *Bestsellers.* One week, 10 copies of the novel *Unbroken* by Laura Hillenbrand were sold for every 3.7 copies of *American Sniper* by Chris Kyle, Scott McEwen, and Jim DeFelice that were sold (*Source*: *USA Today* Best-Selling Books). If a total of 11,371 copies of the two books were sold, how many copies of each were sold?

105. *Running vs. Walking.* A 150-lb person who runs at 6 mph for 1 hr burns about 720 calories. The same person, walking at 4 mph for 90 min, burns about 480 calories. (*Source*: FitSmart, *USA Weekend*, July 19–21, 2002) Suppose that a 150-lb person runs at 6 mph for 75 min. How far would the person have to walk at 4 mph in order to burn the same number of calories used running?

1.6 Solving Linear Inequalities

- Solve linear inequalities.
- Solve compound inequalities.
- Solve applied problems using inequalities.

An **inequality** is a sentence with $<$, $>$, $\leq$, or $\geq$ as its verb. An example is $3x - 5 < 6 - 2x$. To **solve** an inequality is to find all values of the variable that make the inequality true. Each of these values is a **solution** of the inequality, and the set of all such solutions is its **solution set**. Inequalities that have the same solution set are called **equivalent inequalities**.

Linear Inequalities

The principles for solving inequalities are similar to those for solving equations.

PRINCIPLES FOR SOLVING INEQUALITIES

For any real numbers a, b, and c:

The Addition Principle for Inequalities:

If $a < b$ is true, then $a + c < b + c$ is true.

The Multiplication Principle for Inequalities:

a) If $a < b$ and $c > 0$ are true, then $ac < bc$ is true.
b) If $a < b$ and $c < 0$ are true, then $ac > bc$ is true.
(When both sides of an inequality are multiplied by a negative number, the inequality sign must be reversed.)

Similar statements hold for $a \leq b$.

First-degree inequalities with one variable, like those in Example 1 below, are **linear inequalities**.

Just in Time 3,6

EXAMPLE 1 Solve the inequality. Then graph the solution set.

a) $3x - 5 < 6 - 2x$ **b)** $13 - 7x \geq 10x - 4$

Solution

a) $3x - 5 < 6 - 2x$

$5x - 5 < 6$ **Using the addition principle for inequalities; adding $2x$**

$5x < 11$ **Using the addition principle for inequalities; adding 5**

$x < \frac{11}{5}$ **Using the multiplication principle for inequalities; multiplying by $\frac{1}{5}$, or dividing by 5**

Any number less than $\frac{11}{5}$ is a solution. The solution set is $\{x|x < \frac{11}{5}\}$, or $(-\infty, \frac{11}{5})$. The graph of the solution set is shown below.

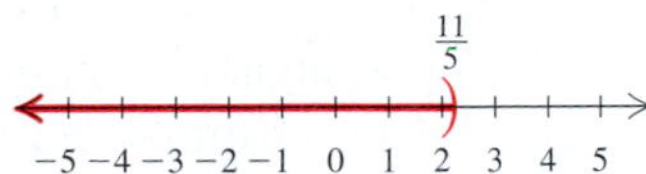

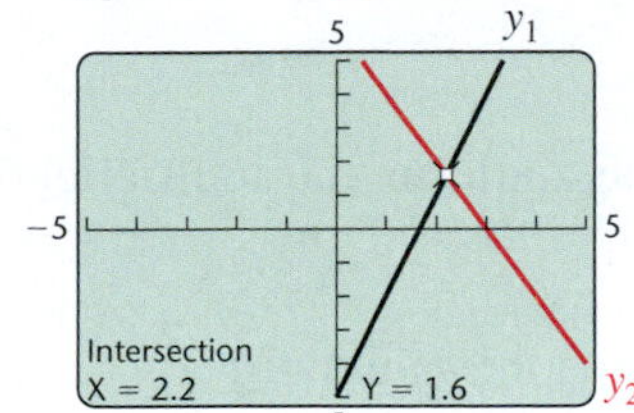

To check, we can graph $y_1 = 3x - 5$ and $y_2 = 6 - 2x$. The graph at left shows that for $x < 2.2$, or $x < \frac{11}{5}$, the graph of y_1 lies below the graph of y_2, or $y_1 < y_2$.

b) $13 - 7x \geq 10x - 4$

$13 - 17x \geq -4$ **Subtracting $10x$**

$-17x \geq -17$ **Subtracting 13**

$x \leq 1$ **Dividing by −17 and reversing the inequality sign**

The solution set is $\{x|x \leq 1\}$, or $(-\infty, 1]$. The graph of the solution set is shown below.

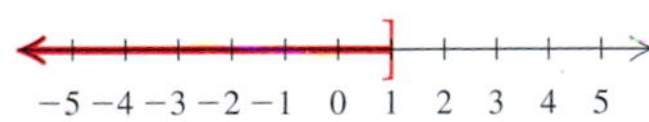

Now Try Exercises 1 and 3.

EXAMPLE 2 Find the domain of the function.

a) $f(x) = \sqrt{x - 6}$ **b)** $h(x) = \dfrac{x}{\sqrt{3 - x}}$

Solution

a) The radicand, $x - 6$, must be greater than or equal to 0. We solve the inequality $x - 6 \geq 0$:

$$x - 6 \geq 0$$
$$x \geq 6.$$

The domain is $\{x|x \geq 6\}$, or $[6, \infty)$.

b) Any real number can be an input for x in the numerator, but inputs for x must be restricted in the denominator. We must have $3 - x \geq 0$ and $\sqrt{3 - x} \neq 0$. Thus, $3 - x > 0$. We solve for x:

$$3 - x > 0$$
$$-x > -3 \quad \text{Subtracting 3}$$
$$x < 3. \quad \text{Multiplying by } -1 \text{ and reversing the inequality sign}$$

The domain is $\{x | x < 3\}$, or $(-\infty, 3)$.

Now Try Exercises 17 and 21.

Compound Inequalities

When two inequalities are joined by the word *and* or the word *or*, a **compound inequality** is formed. A compound inequality like

$$-3 < 2x + 5 \quad \textit{and} \quad 2x + 5 \leq 7$$

is called a **conjunction**, because it uses the word *and*. The sentence

$$-3 < 2x + 5 \leq 7$$

is an abbreviation for the preceding conjunction.

Compound inequalities can be solved using the addition and multiplication principles for inequalities.

EXAMPLE 3 Solve $-3 < 2x + 5 \leq 7$. Then graph the solution set.

Solution We have

$$-3 < 2x + 5 \leq 7$$
$$-8 < 2x \leq 2 \quad \text{Subtracting 5}$$
$$-4 < x \leq 1. \quad \text{Dividing by 2}$$

The solution set is $\{x | -4 < x \leq 1\}$, or $(-4, 1]$. The graph of the solution set is shown below.

Now Try Exercise 23.

A compound inequality like $2x - 5 \leq -7$ *or* $2x - 5 > 1$ is called a **disjunction**, because it contains the word *or*. Unlike some conjunctions, it cannot be abbreviated; that is, it cannot be written without the word *or*.

EXAMPLE 4 Solve: $2x - 5 \leq -7$ *or* $2x - 5 > 1$. Then graph the solution set.

Solution We have

$$2x - 5 \leq -7 \quad \textit{or} \quad 2x - 5 > 1$$
$$2x \leq -2 \quad \textit{or} \quad 2x > 6 \quad \text{Adding 5}$$
$$x \leq -1 \quad \textit{or} \quad x > 3. \quad \text{Dividing by 2}$$

The solution set is $\{x | x \leq -1 \textit{ or } x > 3\}$. We can also write the solution set using interval notation and the symbol $\cup$ for the **union** or inclusion of both sets: $(-\infty, -1] \cup (3, \infty)$. The graph of the solution set is shown below.

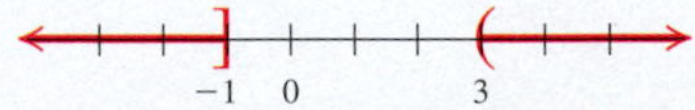

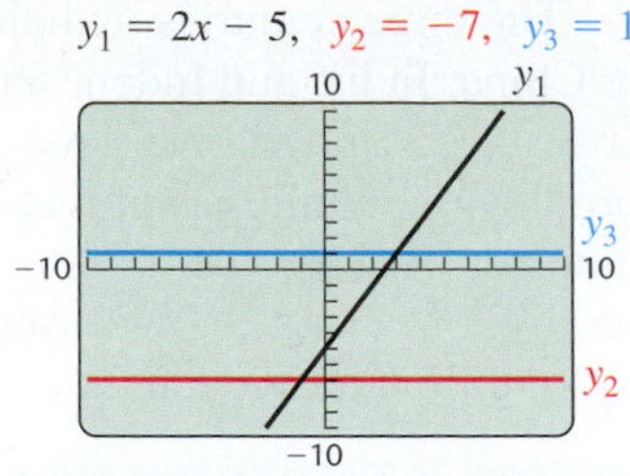

To check, we graph $y_1 = 2x - 5$, $y_2 = -7$, and $y_3 = 1$. Note that for $\{x \mid x \le -1 \text{ or } x > 3\}$, $y_1 \le y_2$ or $y_1 > y_3$.

Now Try Exercise 35.

An Application

EXAMPLE 5 ***Income Plans.*** For her interior decorating job, Natália can be paid in one of two ways:

Plan A: \$250 plus \$10 per hour;

Plan B: \$20 per hour.

Suppose that a job takes n hours. For what values of n is plan B better for Natália?

Solution

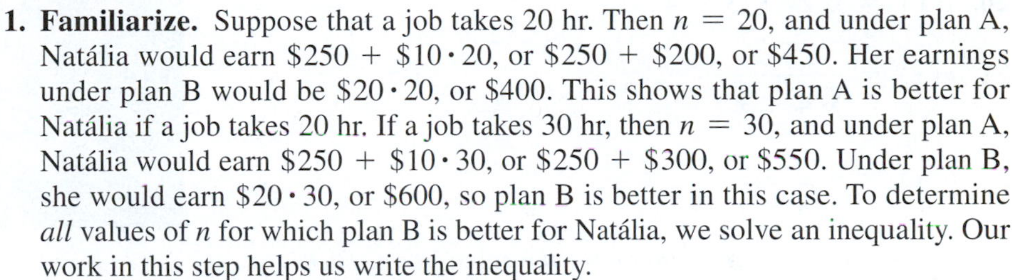

1. **Familiarize.** Suppose that a job takes 20 hr. Then $n = 20$, and under plan A, Natália would earn $\$250 + \$10 \cdot 20$, or $\$250 + \$200$, or \$450. Her earnings under plan B would be $\$20 \cdot 20$, or \$400. This shows that plan A is better for Natália if a job takes 20 hr. If a job takes 30 hr, then $n = 30$, and under plan A, Natália would earn $\$250 + \$10 \cdot 30$, or $\$250 + \$300$, or \$550. Under plan B, she would earn $\$20 \cdot 30$, or \$600, so plan B is better in this case. To determine *all* values of n for which plan B is better for Natália, we solve an inequality. Our work in this step helps us write the inequality.

2. **Translate.** We translate to an inequality:

Income from plan B | is greater than | income from plan A

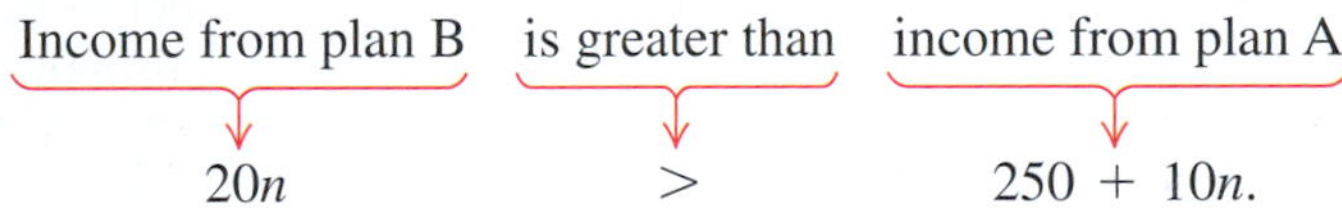

$20n \quad > \quad 250 + 10n.$

2. **Carry out.** We solve the inequality:

$$20n > 250 + 10n$$
$$10n > 250 \qquad \text{Subtracting } 10n \text{ on both sides}$$
$$n > 25. \qquad \text{Dividing by 10 on both sides}$$

4. **Check.** For $n = 25$, the income from plan A is $\$250 + \$10 \cdot 25$, or $\$250 + \$250$, or \$500, and the income from plan B is $\$20 \cdot 25$, or \$500. This shows that for a job that takes 25 hr to complete, the income is the same under either plan. In the *Familiarize* step, we saw that plan B pays more for a 30-hr job. Since $30 > 25$, this provides a partial check of the result. We cannot check all values of n.

5. **State.** For values of n greater than 25 hr, plan B is better for Natália.

Now Try Exercise 45.

1.6 Exercise Set

Solve and graph the solution set.

1. $4x - 3 > 2x + 7$
2. $8x + 1 \ge 5x - 5$
3. $x + 6 < 5x - 6$
4. $3 - x < 4x + 7$
5. $4 - 2x \le 2x + 16$
6. $3x - 1 > 6x + 5$
7. $14 - 5y \le 8y - 8$
8. $8x - 7 < 6x + 3$
9. $7x - 7 > 5x + 5$
10. $12 - 8y \ge 10y - 6$
11. $3x - 3 + 2x \ge 1 - 7x - 9$

12. $5y - 5 + y \le 2 - 6y - 8$

13. $-\frac{3}{4}x \ge -\frac{5}{8} + \frac{2}{3}x$

14. $-\frac{5}{6}x \le \frac{3}{4} + \frac{8}{3}x$

15. $4x(x - 2) < 2(2x - 1)(x - 3)$

16. $(x + 1)(x + 2) > x(x + 1)$

Find the domain of the function.

17. $h(x) = \sqrt{x - 7}$

18. $g(x) = \sqrt{x + 8}$

19. $f(x) = \sqrt{1 - 5x} + 2$

20. $f(x) = \sqrt{2x + 3} - 4$

21. $g(x) = \dfrac{5}{\sqrt{4 + x}}$

22. $h(x) = \dfrac{x}{\sqrt{8 - x}}$

Solve and write interval notation for the solution set. Then graph the solution set.

23. $-2 \le x + 1 < 4$

24. $-3 < x + 2 \le 5$

25. $5 \le x - 3 \le 7$

26. $-1 < x - 4 < 7$

27. $-3 \le x + 4 \le 3$

28. $-5 < x + 2 < 15$

29. $-2 < 2x + 1 < 5$

30. $-3 \le 5x + 1 \le 3$

31. $-4 \le 6 - 2x < 4$

32. $-3 < 1 - 2x \le 3$

33. $-5 < \frac{1}{2}(3x + 1) < 7$

34. $\frac{2}{3} \le -\frac{4}{5}(x - 3) < 1$

35. $3x \le -6$ *or* $x - 1 > 0$

36. $2x < 8$ *or* $x + 3 \ge 10$

37. $2x + 3 \le -4$ *or* $2x + 3 \ge 4$

38. $3x - 1 < -5$ *or* $3x - 1 > 5$

39. $2x - 20 < -0.8$ *or* $2x - 20 > 0.8$

40. $5x + 11 \le -4$ *or* $5x + 11 \ge 4$

41. $x + 14 \le -\frac{1}{4}$ *or* $x + 14 \ge \frac{1}{4}$

42. $x - 9 < -\frac{1}{2}$ *or* $x - 9 > \frac{1}{2}$

43. *World Rice Production.* The three countries with the most rice production are China, India, and Indonesia. The equation $y = 9.06x + 410.81$ provides a good estimate of world rice production in millions of metric tons, where x is the number of years after 1980. (*Source*: www.geohive.com) For what years will world rice production exceed 820 million metric tons?

44. *Social Security Disability.* The equation $y = 0.326x + 7.148$ can be used to estimate the number of people, in millions, collecting Social Security disability payments, where x is the number of years after 2007 (*Source*: Social Security Administration). For what years will the number of people collecting disability payments be more than 12 million?

45. *Moving Costs.* Acme Movers charges \$200 plus \$45 per hour to move a household across town. Leo's Movers charges \$65 per hour. For what lengths of time does it cost less to hire Leo's Movers?

46. *Investment Income.* Jalyn plans to invest \$12,000, part at 4% simple interest and the rest at 6% simple interest. What is the most that she can invest at 4% and still be guaranteed at least \$650 in interest per year?

47. *Investment Income.* Dillon plans to invest \$7500, part at 4% simple interest and the rest at 5% simple interest. What is the most that he can invest at 4% and still be guaranteed at least \$325 in interest per year?

48. *Investment Income.* A foundation invests \$150,000 at simple interest, part at 7%, twice that amount at 4%, and the rest at 5.5%. What is the most that the foundation can invest at 4% and still be guaranteed at least \$7575 in interest per year?

49. *Investment Income.* A university invests \$1,400,000 at simple interest, part at 5%, half that amount at 3.5%, and the rest at 5.5%. What is the most that the university can invest at 3.5% and still be guaranteed at least \$68,000 in interest per year?

50. *Income Plans.* Karen can be paid in one of two ways for selling insurance policies:

Plan A: A salary of \$750 per month, plus a commission of 10% of sales;

Plan B: A salary of \$1000 per month, plus a commission of 8% of sales in excess of \$2000.

For what amount of monthly sales is plan A better than plan B if we can assume that sales are always more than \$2000?

51. *Income Plans.* Curt can be paid in one of two ways for the furniture he sells:

Plan A: A salary of \$900 per month, plus a commission of 10% of sales;

Plan B: A salary of \$1200 per month, plus a commission of 15% of sales in excess of \$8000.

For what amount of monthly sales is plan B better than plan A if we can assume that Curt's sales are always more than \$8000?

52. *Income Plans.* Jeanette can be paid in one of two ways for painting a house:

Plan A: \$200 plus \$12 per hour;

Plan B: \$20 per hour.

Suppose that a job takes n hours to complete. For what values of n is plan A better for Jeanette?

Skill Maintenance

Vocabulary Reinforcement

In each of Exercises 53–56, fill in the blank(s) with the correct term(s). Some of the given choices will not be used; others will be used more than once.

constant	domain
function	distance formula
any	exactly one
midpoint formula	identity
y-intercept	x-intercept
range	

53. A(n) ________ is a correspondence between a first set, called the ________, and a second set, called the ________, such that each member of the ________ corresponds to ________ member of the ________. [1.2]

54. The ________ is $\left(\frac{x_1 + x_2}{2}, \frac{y_1 + y_2}{2}\right)$. [1.1]

55. A(n) ________ is a point $(a, 0)$. [1.1]

56. A function f is a linear function if it can be written as $f(x) = mx + b$, where m and b are constants. If $m = 0$, the function is a(n) ________ function $f(x) = b$. If $m = 1$ and $b = 0$, the function is the ________ function $f(x) = x$. [1.3]

Synthesis

Solve.

57. $2x \le 5 - 7x < 7 + x$

58. $x \le 3x - 2 \le 2 - x$

59. $3y < 4 - 5y < 5 + 3y$

60. $y - 10 < 5y + 6 \le y + 10$

Chapter 1 Summary and Review

STUDY GUIDE

KEY TERMS AND CONCEPTS | **EXAMPLES**

SECTION 1.1: INTRODUCTION TO GRAPHING

Graphing Equations

To **graph** an equation is to make a drawing that represents the solutions of that equation. We can graph an equation by selecting values for one variable and finding the corresponding values for the other variable. We list the solutions (ordered pairs) in a table, plot the points, and draw the graph.

Graph: $y = 4 - x^2$.

x	$y = 4 - x^2$	(x, y)
0	4	$(0, 4)$
-1	3	$(-1, 3)$
1	3	$(1, 3)$
-2	0	$(-2, 0)$
2	0	$(2, 0)$

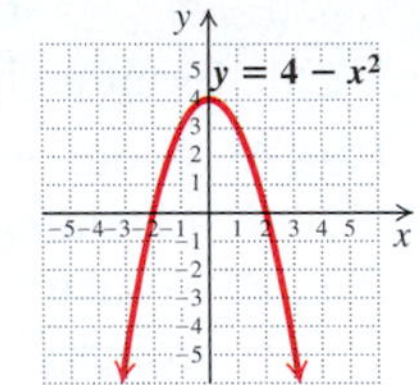

Intercepts

An ***x*-intercept** is a point $(a, 0)$.
To find a, let $y = 0$ and solve for x.

A ***y*-intercept** is a point $(0, b)$.
To find b, let $x = 0$ and solve for y.

We can graph a straight line by plotting the intercepts and drawing the line containing them.

Graph using intercepts: $2x - y = 4$.

We let $y = 0$:

$$2x - 0 = 4$$
$$2x = 4$$
$$x = 2.$$

The x-intercept is $(2, 0)$.

We let $x = 0$:

$$2 \cdot 0 - y = 4$$
$$-y = 4$$
$$y = -4.$$

The y-intercept is $(0, -4)$.

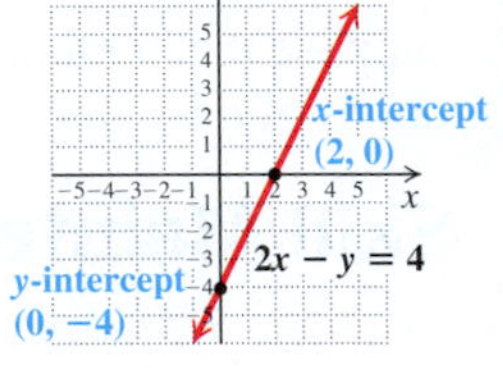

Distance Formula

The **distance** d between any two points (x_1, y_1) and (x_2, y_2) is given by

$$d = \sqrt{(x_2 - x_1)^2 + (y_2 - y_1)^2}.$$

Find the distance between $(-5, 7)$ and $(2, -3)$.

$$d = \sqrt{[2 - (-5)]^2 + (-3 - 7)^2}$$
$$= \sqrt{7^2 + (-10)^2}$$
$$= \sqrt{49 + 100}$$
$$= \sqrt{149} \approx 12.2$$

Midpoint Formula

If the endpoints of a segment are (x_1, y_1) and (x_2, y_2), then the coordinates of the **midpoint** of the segment are

$$\left(\frac{x_1 + x_2}{2}, \frac{y_1 + y_2}{2}\right).$$

Find the midpoint of the segment whose endpoints are $(-10, 4)$ and $(3, 8)$.

$$\left(\frac{x_1 + x_2}{2}, \frac{y_1 + y_2}{2}\right) = \left(\frac{-10 + 3}{2}, \frac{4 + 8}{2}\right)$$
$$= \left(-\frac{7}{2}, 6\right)$$

Circles

The **standard form** of the equation of a circle with center (h, k) and radius r is

$$(x - h)^2 + (y - k)^2 = r^2.$$

Find an equation of a circle with center $(1, -6)$ and radius 8.

$$(x - h)^2 + (y - k)^2 = r^2$$
$$(x - 1)^2 + [y - (-6)]^2 = 8^2$$
$$(x - 1)^2 + (y + 6)^2 = 64$$

Given the circle

$$(x + 9)^2 + (y - 2)^2 = 121,$$

determine the center and the radius.

Writing in standard form, we have

$$[x - (-9)]^2 + (y - 2)^2 = 11^2.$$

The center is $(-9, 2)$, and the radius is 11.

SECTION 1.2: FUNCTIONS AND GRAPHS

Functions

A **function** is a correspondence between a first set, called the **domain**, and a second set, called the **range**, such that each member of the domain corresponds to *exactly one* member of the range.

Consider the function given by

$$g(x) = |x| - 1.$$
$$g(-3) = |-3| - 1$$
$$= 3 - 1$$
$$= 2$$

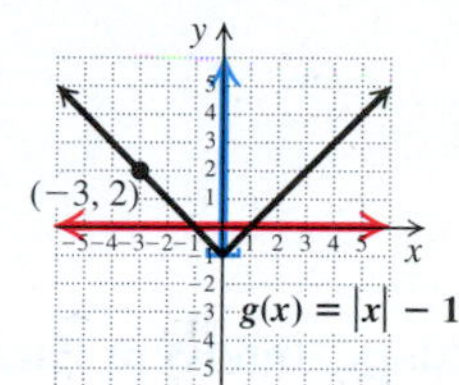

For the input -3, the output is 2: $g(-3) = 2$.
The point $(-3, 2)$ is on the graph.
Domain: Set of all inputs $= \{x \mid x \text{ is a real number}\}$, or $(-\infty, \infty)$.
Range: Set of all outputs: $\{y \mid y \geq -1\}$, or $[-1, \infty)$.

The Vertical-Line Test

If it is possible for a vertical line to cross a graph more than once, then the graph *is not* the graph of a function.

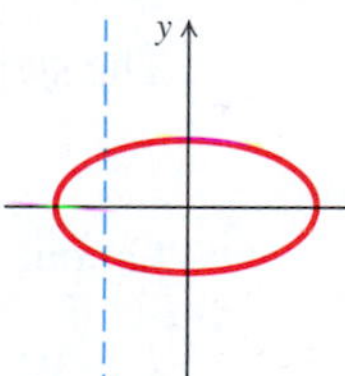

This *is not* the graph of a function because a vertical line can cross it more than once, as shown.

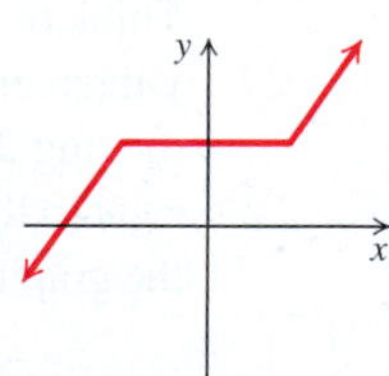

This *is* the graph of a function because no vertical line can cross it more than once.

Domain

When a function f whose inputs and outputs are real numbers is given by a formula, the **domain** is the set of all inputs for which the expression is defined as a real number.

Find the domain of the function given by

$$h(x) = \frac{x - 1}{(x + 5)(x - 10)}.$$

Division by 0 is not defined. Since $x + 5 = 0$ when $x = -5$ and $x - 10 = 0$ when $x = 10$, the domain of h is

$\{x \mid x \text{ is a real number } \textit{and } x \neq -5 \textit{ and } x \neq 10\}$,
or $(-\infty, -5) \cup (-5, 10) \cup (10, \infty)$.

SECTION 1.3: LINEAR FUNCTIONS, SLOPE, AND APPLICATIONS

Slope

$$m = \frac{\text{rise}}{\text{run}} = \frac{y_2 - y_1}{x_2 - x_1} = \frac{y_1 - y_2}{x_1 - x_2}$$

Slope can also be considered as an **average rate of change**. To find the average rate of change between two data points on a graph, determine the slope of the line that passes through the points.

The slope of the line containing the points $(3, -10)$ and $(-2, 6)$ is

$$m = \frac{y_2 - y_1}{x_2 - x_1} = \frac{6 - (-10)}{-2 - 3}$$
$$= \frac{16}{-5} = -\frac{16}{5}.$$

In 2000, the population of Flint, Michigan, was 124,943. By 2012, the population had decreased to 100,515. Find the average rate of change in population from 2000 to 2012.

$$\text{Average rate of change} = m = \frac{100{,}515 - 124{,}943}{2012 - 2000}$$
$$= \frac{-24{,}428}{12} \approx -2036$$

The average rate of change in population over the 12-year period was a decrease of about 2036 people per year.

Slope–Intercept Form of an Equation

$$f(x) = mx + b$$

The slope of the line is m.

The y-intercept of the line is $(0, b)$.

Determine the slope and the y-intercept of the line given by $5x - 7y = 14$.

We first find the slope–intercept form:

$$5x - 7y = 14$$
$$-7y = -5x + 14 \qquad \text{Adding } -5x$$
$$y = \frac{5}{7}x - 2. \qquad \text{Multiplying by } -\frac{1}{7}$$

The slope is $\frac{5}{7}$, and the y-intercept is $(0, -2)$.

To graph an equation written in slope–intercept form, plot the y-intercept and use the slope to find another point. Then draw the line.

Graph: $f(x) = -\frac{2}{3}x + 4$.

We plot the y-intercept, $(0, 4)$. Think of the slope as $\frac{-2}{3}$. From the y-intercept, we find another point by moving 2 units down and 3 units to the right to the point $(3, 2)$. We then draw the graph.

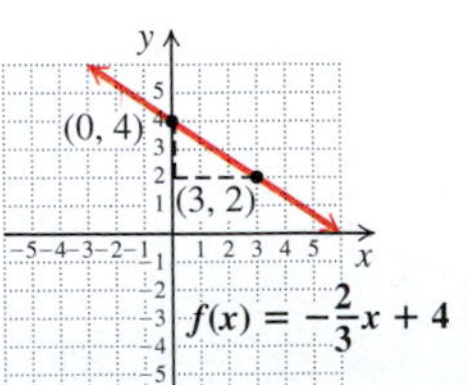

Horizontal Lines

The graph of $y = b$, or $f(x) = b$, is a horizontal line with y-intercept $(0, b)$. The slope of a horizontal line is 0.

Vertical Lines

The graph of $x = a$ is a vertical line with x-intercept $(a, 0)$. The slope of a vertical line is *not* defined.

Graph $y = -4$ and determine its slope.

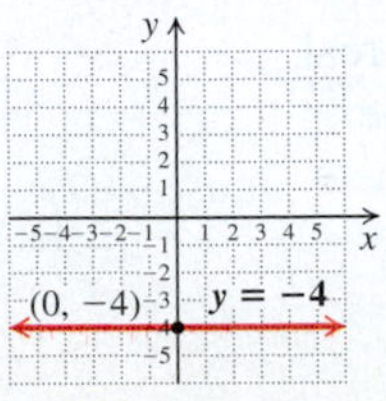

The slope is 0.

Graph $x = 3$ and determine its slope.

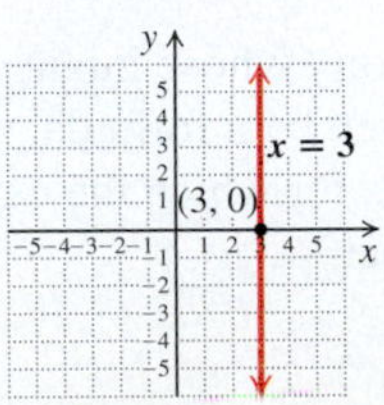

The slope is not defined.

SECTION 1.4: EQUATIONS OF LINES AND MODELING

Slope–Intercept Form of an Equation

$$y = mx + b, \text{ or } f(x) = mx + b$$

The slope of the line is m.

The y-intercept of the line is $(0, b)$.

Point–Slope Form of an Equation

$$y - y_1 = m(x - x_1)$$

The slope of the line is m.

The line passes through (x_1, y_1).

Write the slope–intercept equation for a line with slope $-\frac{2}{9}$ and y-intercept $(0, 4)$.

$y = mx + b$ **Using the slope–intercept form**

$y = -\frac{2}{9}x + 4$ **Substituting $-\frac{2}{9}$ for m and 4 for b**

Write the slope–intercept equation for a line that passes through $(-5, 7)$ and $(3, -9)$.

We first determine the slope:

$$m = \frac{-9 - 7}{3 - (-5)} = \frac{-16}{8} = -2.$$

Using the slope–intercept form: We substitute -2 for m and either $(-5, 7)$ or $(3, -9)$ for (x, y) and solve for b:

$y = mx + b$

$7 = -2 \cdot (-5) + b$ **Using $(-5, 7)$**

$7 = 10 + b$

$-3 = b.$

The slope–intercept equation is $y = -2x - 3$.

Using the point–slope equation: We substitute -2 for m and either $(-5, 7)$ or $(3, -9)$ for (x_1, y_1):

$y - y_1 = m(x - x_1)$

$y - (-9) = -2(x - 3)$ **Using $(3, -9)$**

$y + 9 = -2x + 6$

$y = -2x - 3.$

The slope–intercept equation is $y = -2x - 3$.

Parallel Lines

Vertical lines are parallel. Nonvertical lines are **parallel** if and only if they have the same slope and different y-intercepts.

Write the slope–intercept equation for a line passing through $(-3, 1)$ that is parallel to the line $y = \frac{2}{3}x + 5$.

The slope of $y = \frac{2}{3}x + 5$ is $\frac{2}{3}$, so the slope of a line parallel to this line is also $\frac{2}{3}$. We use either the slope–intercept equation or the point–slope equation for a line with slope $\frac{2}{3}$ and containing the point $(-3, 1)$. Here we use the point–slope equation and substitute $\frac{2}{3}$ for m, -3 for x_1, and 1 for y_1.

$y - y_1 = m(x - x_1)$

$y - 1 = \frac{2}{3}[x - (-3)]$

$y - 1 = \frac{2}{3}(x + 3)$

$y - 1 = \frac{2}{3}x + 2$

$y = \frac{2}{3}x + 3$ **Slope–intercept form**

Perpendicular Lines

Two lines are **perpendicular** if and only if the product of their slopes is -1 or if one line is vertical $(x = a)$ and the other is horizontal $(y = b)$.

Write the slope–intercept equation for a line that passes through $(-3, 1)$ and is perpendicular to the line $y = \frac{2}{3}x + 5$.

The slope of $y = \frac{2}{3}x + 5$ is $\frac{2}{3}$, so the slope of a line perpendicular to this line is the opposite of the reciprocal of $\frac{2}{3}$, or $-\frac{3}{2}$. Here we use the point–slope equation and substitute $-\frac{3}{2}$ for m, -3 for x_1, and 1 for y_1.

$$y - y_1 = m(x - x_1)$$
$$y - 1 = -\frac{3}{2}[x - (-3)]$$
$$y - 1 = -\frac{3}{2}(x + 3)$$
$$y - 1 = -\frac{3}{2}x - \frac{9}{2}$$
$$y = -\frac{3}{2}x - \frac{7}{2}$$

SECTION 1.5: LINEAR EQUATIONS, FUNCTIONS, ZEROS, AND APPLICATIONS

Equation–Solving Principles

Addition Principle: If $a = b$ is true, then $a + c = b + c$ is true.

Multiplication Principle: If $a = b$ is true, then $ac = bc$ is true.

Solve: $2(3x - 7) = 15 - (x + 1)$.

$$2(3x - 7) = 15 - (x + 1)$$
$$6x - 14 = 15 - x - 1 \quad \text{Using the distributive property}$$
$$6x - 14 = 14 - x \quad \text{Collecting like terms}$$
$$6x - 14 + x = 14 - x + x \quad \text{Adding } x \text{ on both sides}$$
$$7x - 14 = 14$$
$$7x - 14 + 14 = 14 + 14 \quad \text{Adding 14 on both sides}$$
$$7x = 28$$
$$\frac{7x}{7} = \frac{28}{7} \quad \text{Dividing by 7 on both sides}$$
$$x = 4$$

Check:

$2(3x - 7)$	$= 15 - (x + 1)$
$2(3 \cdot 4 - 7)$?	$15 - (4 + 1)$
$2(12 - 7)$	$15 - 5$
$2 \cdot 5$	10
10	10 TRUE

The solution is 4.

Special Cases

Some equations have *no* solution.

Solve: $2 + 17x = 17x - 9$.

$$2 + 17x = 17x - 9$$
$$2 + 17x - 17x = 17x - 9 - 17x \quad \text{Subtracting } 17x \text{ on both sides}$$
$$2 = -9 \quad \text{False equation}$$

We get a false equation; thus the equation has *no* solution.

(continued)

There are some equations for which *any* real number is a solution.

Solve: $5 - \frac{1}{2}x = -\frac{1}{2}x + 5.$

$$5 - \frac{1}{2}x = -\frac{1}{2}x + 5$$

$$5 - \frac{1}{2}x + \frac{1}{2}x = -\frac{1}{2}x + 5 + \frac{1}{2}x \quad \text{Adding } \tfrac{1}{2}x \text{ on both sides}$$

$$5 = 5 \quad \text{True equation}$$

We get a true equation. Thus any real number is a solution. The solution set is

$$\{x|x \text{ is a real number}\}, \quad \text{or} \quad (-\infty, \infty).$$

Zeros of Functions

An input c of a function f is called a **zero** of the function if the output for the function is 0 when the input is c. That is,

c is a zero of f if $f(c) = 0$.

A linear function $f(x) = mx + b$, with $m \neq 0$, has exactly one zero.

Find the zero of the linear function

$$f(x) = \frac{5}{8}x - 40.$$

We find the value of x for which $f(x) = 0$:

$$\frac{5}{8}x - 40 = 0 \quad \text{Setting } f(x) = 0$$

$$\frac{5}{8}x = 40 \quad \text{Adding 40 on both sides}$$

$$\frac{8}{5} \cdot \frac{5}{8}x = \frac{8}{5} \cdot 40 \quad \text{Multiplying by } \tfrac{8}{5} \text{ on both sides}$$

$$x = 64.$$

We can check by substituting 64 for x:

$$f(64) = \tfrac{5}{8} \cdot 64 - 40 = 40 - 40 = 0.$$

The zero of $f(x) = \frac{5}{8}x - 40$ is 64.

SECTION 1.6: SOLVING LINEAR INEQUALITIES

Principles for Solving Linear Inequalities

Addition Principle:
If $a < b$ is true, then $a + c < b + c$ is true.

Multiplication Principle:
If $a < b$ and $c > 0$ are true, then $ac < bc$ is true.
If $a < b$ and $c < 0$ are true, then $ac > bc$ is true.

Similar statements hold for $a \leq b$.

Solve $3x - 2 \leq 22 - 5x$ and graph the solution set.

$$3x - 2 \leq 22 - 5x$$

$$3x - 2 + 5x \leq 22 - 5x + 5x \quad \text{Adding } 5x \text{ on both sides}$$

$$8x - 2 \leq 22$$

$$8x - 2 + 2 \leq 22 + 2 \quad \text{Adding 2 on both sides}$$

$$8x \leq 24$$

$$\frac{8x}{8} \leq \frac{24}{8} \quad \text{Dividing by 8 on both sides}$$

$$x \leq 3$$

The solution set is

$$\{x|x \leq 3\}, \quad \text{or} \quad (-\infty, 3].$$

The graph of the solution set is as follows.

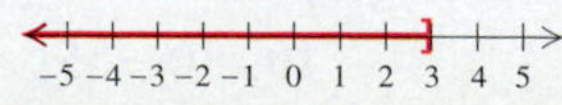

Compound Inequalities When two inequalities are joined by the word *and* or the word *or*, a compound inequality is formed. *A Conjunction*: $1 < 3x - 20$ *and* $3x - 20 \le 40$, or $1 < 3x - 20 \le 40$ *A Disjunction*: $8x - 1 \le -17$ *or* $8x - 1 > 7$	Solve: $1 < 3x - 20 \le 40$. $1 < 3x - 20 \le 40$ $21 < 3x \le 60$ **Adding 20** $7 < x \le 20$ **Dividing by 3** The solution set is $\{x \mid 7 < x \le 20\}$, or $(7, 20]$. Solve: $8x - 1 \le -17$ *or* $8x - 1 > 7$. $8x - 1 \le -17$ *or* $8x - 1 > 7$ $8x \le -16$ *or* $8x > 8$ **Adding 1** $x \le -2$ *or* $x > 1$ **Dividing by 8** The solution set is $\{x \mid x \le -2$ *or* $x > 1\}$, or $(-\infty, -2] \cup (1, \infty)$.

REVIEW EXERCISES

Answers to all of the review exercises appear in the answer section at the back of the book. If you get an incorrect answer, restudy the objective indicated in red next to the exercise or the instruction line that precedes it.

Determine whether the statement is true or false.

1. If the line $ax + y = c$ is perpendicular to the line $x - by = d$, then $\frac{a}{b} = 1$. [1.4]

2. The intersection of the lines $y = \frac{1}{2}$ and $x = -5$ is $(-5, \frac{1}{2})$. [1.3]

3. The domain of the function $f(x) = \dfrac{\sqrt{3 - x}}{x}$ does not contain -3 and 0. [1.2]

4. The line parallel to the x-axis that passes through $\left(-\frac{1}{4}, 7\right)$ is $x = -\frac{1}{4}$. [1.3]

5. The zero of a linear function f is the first coordinate of the x-intercept of the graph of $y = f(x)$. [1.5]

6. If $a < b$ is true and $c \ne 0$, then $ac < bc$ is true. [1.6]

Use substitution to determine whether the given ordered pairs are solutions of the given equation. [1.1]

7. $\left(3, \frac{24}{9}\right), (0, -9)$; $2x - 9y = -18$

8. $(0, 7), (7, 1)$; $y = 7$

Find the intercepts and then graph the line. [1.1]

9. $2x - 3y = 6$

10. $10 - 5x = 2y$

Graph the equation. [1.1]

11. $y = -\frac{2}{3}x + 1$

12. $2x - 4y = 8$

13. $y = 2 - x^2$

14. Find the distance between $(3, 7)$ and $(-2, 4)$. [1.1]

15. Find the midpoint of the segment with endpoints $(3, 7)$ and $(-2, 4)$. [1.1]

16. Find the center and the radius of the circle with equation $(x + 1)^2 + (y - 3)^2 = 9$. Then graph the circle. [1.1]

Find an equation for a circle satisfying the given conditions. [1.1]

17. Center: $(0, -4)$, radius of length $\frac{3}{2}$

18. Center: $(-2, 6)$, radius of length $\sqrt{13}$

19. Diameter with endpoints $(-3, 5)$ and $(7, 3)$

Determine whether the correspondence is a function. [1.2]

20.
$-6 \to 1$
$-1 \to 3$
$2 \to 10$
$2 \to 3$
$7 \to 12$

21.
h → r
i → s
j → t
k → t

Determine whether the relation is a function. Identify the domain and the range. [1.2]

22. $\{(3, 1), (5, 3), (7, 7), (3, 5)\}$

23. $\{(2, 7), (-2, -7), (7, -2), (0, 2), (1, -4)\}$

24. Given that $f(x) = x^2 - x - 3$, find each of the following. [1.2]

a) $f(0)$ **b)** $f(-3)$
c) $f(a - 1)$ **d)** $f(-x)$

25. Given that $f(x) = \dfrac{x - 7}{x + 5}$, find each of the following. [1.2]

a) $f(7)$ **b)** $f(x + 1)$
c) $f(-5)$ **d)** $f\left(-\frac{1}{2}\right)$

26. A graph of a function is shown below. Find $f(2)$, $f(-4)$, and $f(0)$. [1.2]

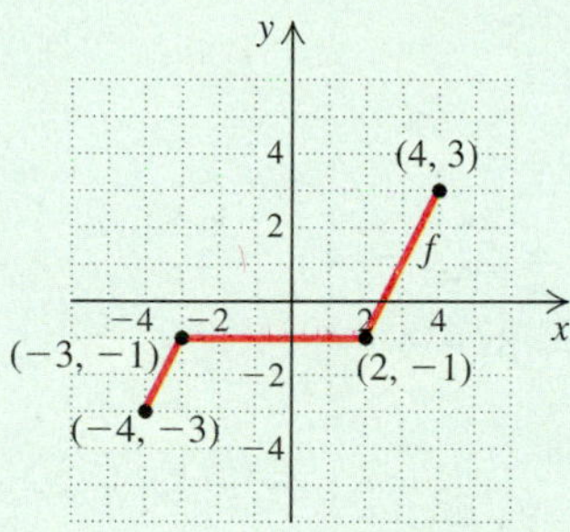

Determine whether the graph is that of a function. [1.2]

27.

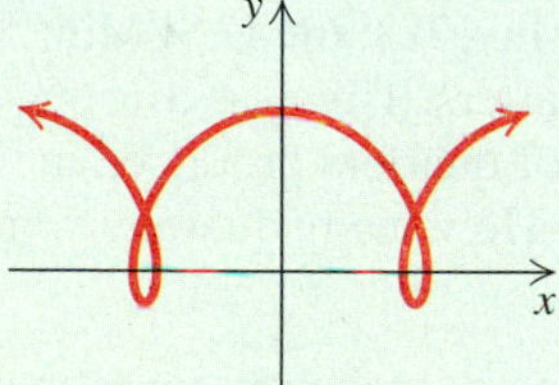

28.

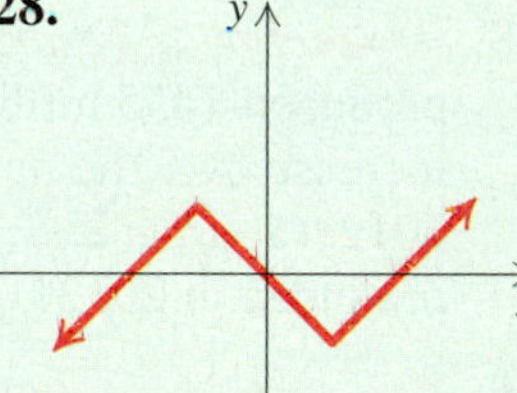

29.

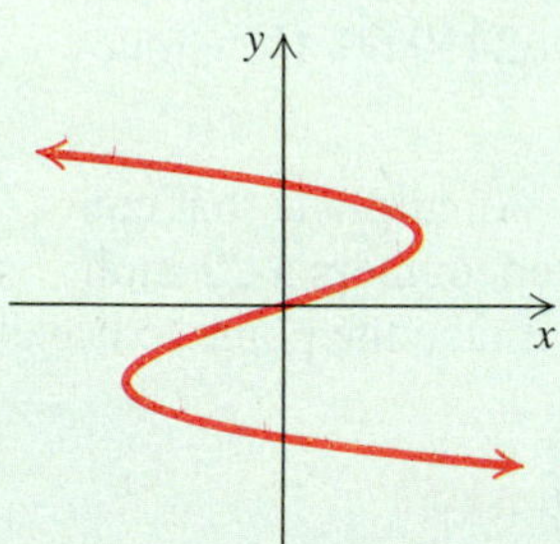

30.

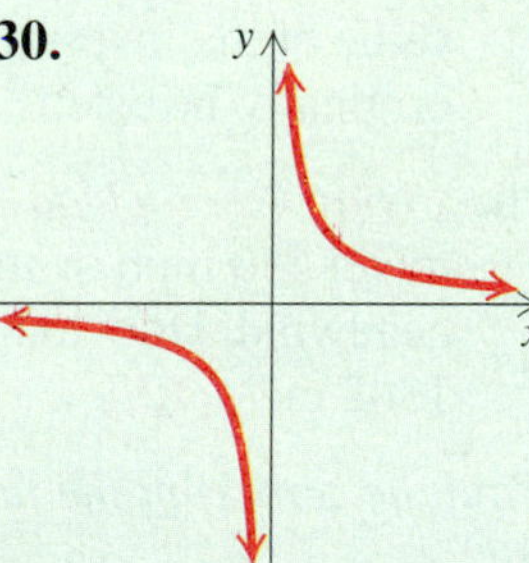

Find the domain of the function. [1.2]

31. $f(x) = 4 - 5x + x^2$

32. $f(x) = \dfrac{3}{x} + 2$

33. $f(x) = \dfrac{1}{x^2 - 6x + 5}$ **34.** $f(x) = \dfrac{-5x}{|16 - x^2|}$

Graph the function. Then visually estimate the domain and the range. [1.2]

35. $f(x) = \sqrt{16 - x^2}$

36. $g(x) = |x - 5|$

37. $f(x) = x^3 - 7$

38. $h(x) = x^4 + x^2$

In Exercises 39 and 40, the table of data contains input–output values for a function. Answer the following questions. [1.3]

a) *Is the change in the inputs, x, the same?*
b) *Is the change in the outputs, y, the same?*
c) *Is the function linear?*

39.

x	y
−3	8
−2	11
−1	14
0	17
1	20
2	22
3	26

40.

x	y
20	11.8
30	24.2
40	36.6
50	49.0
60	61.4
70	73.8
80	86.2

Find the slope of the line containing the given points. [1.3]

41. $(2, -11), (5, -6)$

42. $(5, 4), (-3, 4)$

43. $\left(\frac{1}{2}, 3\right), \left(\frac{1}{2}, 0\right)$

44. *Coffee Consumption.* The U.S. annual per-capita consumption of coffee was 26.8 gal in 1990. By 2012, this amount had decreased to 24.7 gal. (*Source*: Economic Research Service, U.S. Department of Agriculture) Find the average rate of change in per-capita coffee consumption from 1990 to 2012. [1.3]

Find the slope and the y-intercept of the line with the given equation. [1.3]

45. $y = -\frac{7}{11}x - 6$ **46.** $-2x - y = 7$

47. Graph $y = -\frac{1}{4}x + 3$ using the slope and the y-intercept. [1.3]

48. *Total Cost.* Clear County Cable Television charges a \$110 installation fee and \$85 per month for basic service. Write an equation that can be used to determine the total cost $C(t)$ of t months of basic cable television service. Find the total cost of 1 year of service. [1.3]

49. *Temperature and Depth of the Earth.* The function T given by $T(d) = 10d + 20$ can be used to determine the temperature T, in degrees Celsius, at a depth d, in kilometers, inside the earth.

a) Find $T(5)$, $T(20)$, and $T(1000)$. [1.3]
b) The radius of the earth is about 5600 km. Use this fact to determine the domain of the function. [1.3]

Write a slope–intercept equation for a line with the following characteristics. [1.4]

50. $m = -\frac{2}{3}$, y-intercept $(0, -4)$

51. $m = 3$, passes through $(-2, -1)$

52. Passes through $(4, 1)$ and $(-2, -1)$

53. Write equations of the horizontal line and the vertical line that pass through $\left(-4, \frac{2}{5}\right)$. [1.4]

54. Find a linear function h given $h(-2) = -9$ and $h(4) = 3$. Then find $h(0)$. [1.4]

Determine whether the lines are parallel, perpendicular, or neither. [1.4]

55. $3x - 2y = 8$,
$6x - 4y = 2$

56. $y - 2x = 4$,
$2y - 3x = -7$

57. $y = \frac{3}{2}x + 7$,
$y = -\frac{2}{3}x - 4$

Given the point $(1, -1)$ *and the line* $2x + 3y = 4$:

58. Find an equation of the line containing the given point and parallel to the given line. [1.4]

59. Find an equation of the line containing the given point and perpendicular to the given line. [1.4]

60. *Female Medical School Graduates.* Data in the following table list the numbers of female medical school graduates for selected years 2006–2014.

Year, x	Female Medical School Graduates in the United States, W
2006, 0	7748
2008, 2	7969
2010, 4	8133
2012, 6	8291
2014, 8	8576

Source: Association of American Medical Colleges

a) Without using the regression feature on a graphing calculator, model the data with a linear function where the number of female medical school graduates W is a function of the year x and where x is the number of years after 2006. Then using this function, estimate the number of female graduates in 2013. Answers may vary depending on the data points used. [1.4]

b) Using a graphing calculator, fit a regression line to the data and use it to estimate the number of female medical school graduates in 2013. What is the correlation coefficient for the regression line? How close a fit is the regression line? [1.4]

Solve. [1.5]

61. $4y - 5 = 1$

62. $3x - 4 = 5x + 8$

63. $5(3x + 1) = 2(x - 4)$

64. $2(n - 3) = 3(n + 5)$

65. $\frac{3}{5}y - 2 = \frac{3}{8}$

66. $5 - 2x = -2x + 3$

67. $x - 13 = -13 + x$

68. *Production of Quarters.* In 2013, the U.S. Mint produced 1455 million quarters. This was a 156% increase over the number of quarters produced in 2012. (*Source*: U.S. Mint) How many quarters were produced in 2012? [1.5]

69. *Amount of Investment.* James makes an investment at 5.2% simple interest. At the end of 1 year, the total value of the investment is \$2419.60. How much was originally invested? [1.5]

70. *Flying into a Headwind.* An airplane that can travel 550 mph in still air encounters a 20-mph headwind. How long will it take the plane to travel 1802 mi? [1.5]

Find the zero(s) of the function. [1.5]

71. $f(x) = 6x - 18$

72. $f(x) = x - 4$

73. $f(x) = 2 - 10x$

74. $f(x) = 8 - 2x$

Solve and write interval notation for the solution set. Then graph the solution set. [1.6]

75. $2x - 5 < x + 7$

76. $3x + 1 \geq 5x + 9$

77. $-3 \leq 3x + 1 \leq 5$

78. $-2 < 5x - 4 \leq 6$

79. $2x < -1$ *or* $x - 3 > 0$

80. $3x + 7 \leq 2$ *or* $2x + 3 \geq 5$

81. *Homeschooled Children in the United States.* The equation $y = 0.073x + 0.848$ can be used to estimate the number of homeschooled children in the United States, in millions, where x is the number of years after 1999 (*Source*: Department of Education's National Center for Education Statistics). For what years will the number of homeschooled children exceed 2.3 million? **[1.6]**

82. *Temperature Conversion.* The formula $C = \frac{5}{9}(F - 32)$ can be used to convert Fahrenheit temperatures F to Celsius temperatures C. For what Fahrenheit temperatures is the Celsius temperature lower than 45°C? **[1.6]**

83. The domain of the function

$$f(x) = \frac{x + 3}{8 - 4x}$$

is which of the following? **[1.2]**

A. $(-3, 2)$
B. $(-\infty, 2) \cup (2, \infty)$
C. $(-\infty, -3) \cup (-3, 2) \cup (2, \infty)$
D. $(-\infty, -3) \cup (-3, \infty)$

84. The center of the circle described by the equation $(x - 1)^2 + y^2 = 9$ is which of the following? **[1.1]**

A. $(-1, 0)$
B. $(1, 0)$
C. $(0, -3)$
D. $(-1, 3)$

85. The graph of $f(x) = -\frac{1}{2}x - 2$ is which of the following? **[1.3]**

A.

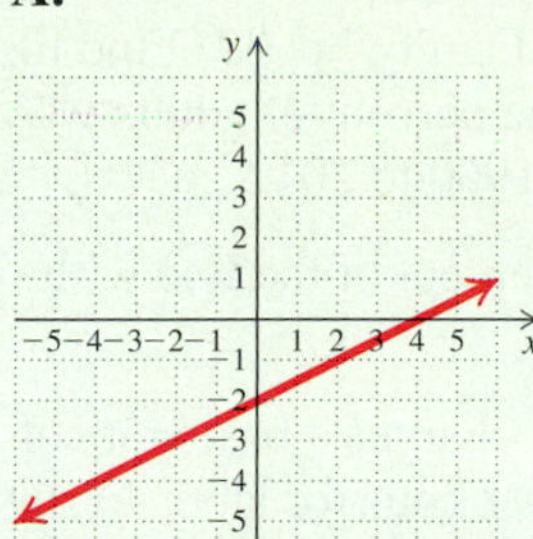

B.

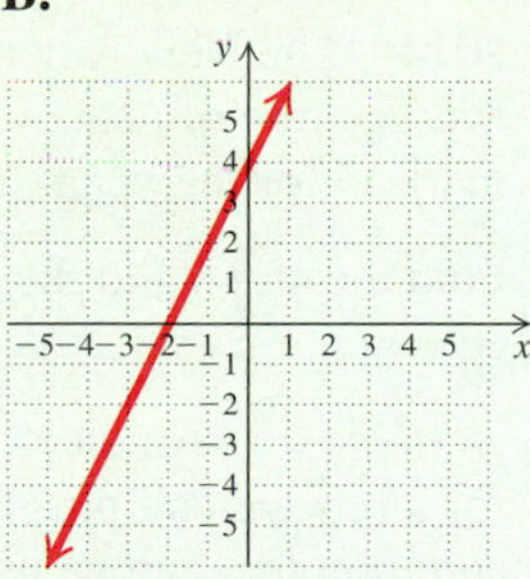

C.

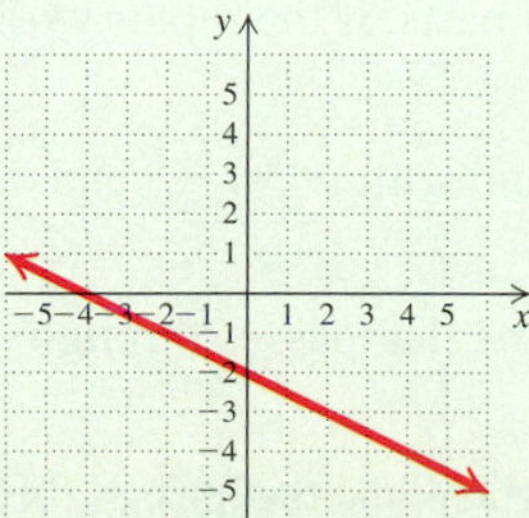

D.

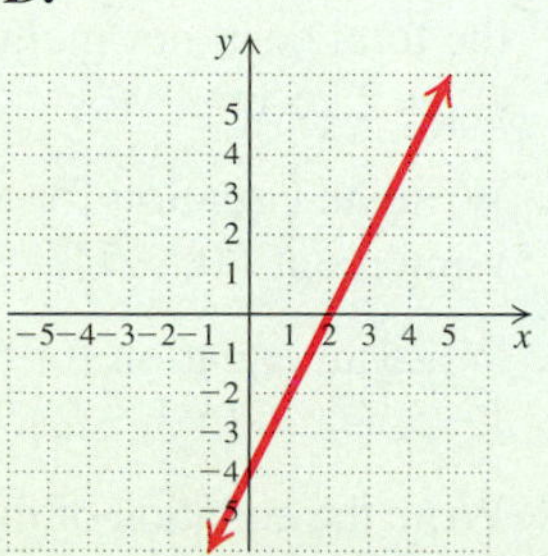

Synthesis

86. Find the point on the x-axis that is equidistant from the points $(1, 3)$ and $(4, -3)$. **[1.1]**

Find the domain. **[1.2]**

87. $f(x) = \dfrac{\sqrt{1 - x}}{x - |x|}$

88. $f(x) = (x - 9x^{-1})^{-1}$

Collaborative Discussion and Writing

89. Discuss why the graph of $f(x) = -\frac{3}{5}x + 4$ is steeper than the graph of $g(x) = \frac{1}{2}x - 6$. **[1.3]**

90. As the first step in solving

$$3x - 1 = 8,$$

Stella multiplies by $\frac{1}{3}$ on both sides. What advice would you give her about the procedure for solving equations? **[1.5]**

91. Is it possible for a disjunction to have no solution? Why or why not? **[1.6]**

92. Explain in your own words why a linear function $f(x) = mx + b$, with $m \neq 0$, has exactly one zero. **[1.5]**

93. Why can the conjunction $3 < x$ *and* $x < 4$ be written as $3 < x < 4$, but the disjunction $x < 3$ *or* $x > 4$ cannot be written as $3 > x > 4$? **[1.6]**

94. Explain in your own words what a function is. **[1.2]**

1 Chapter Test

1. Determine whether the ordered pair $\left(\frac{1}{2}, \frac{9}{10}\right)$ is a solution of the equation $5y - 4 = x$.

2. Find the intercepts of $5x - 2y = -10$ and graph the line.

3. Find the distance between $(5, 8)$ and $(-1, 5)$.

4. Find the midpoint of the segment with endpoints $(-2, 6)$ and $(-4, 3)$.

5. Find the center and the radius of the circle
$$(x + 4)^2 + (y - 5)^2 = 36.$$

6. Find an equation of the circle with center $(-1, 2)$ and radius $\sqrt{5}$.

7. **a)** Determine whether the relation
$$\{(-4, 7), (3, 0), (1, 5), (0, 7)\}$$
is a function. Answer yes or no.
b) Find the domain of the relation.
c) Find the range of the relation.

8. Given that $f(x) = 2x^2 - x + 5$, find each of the following.
a) $f(-1)$
b) $f(a + 2)$

9. Given that $f(x) = \dfrac{1 - x}{x}$, find each of the following.
a) $f(0)$ **b)** $f(1)$

10. Using the graph below, find $f(-3)$.

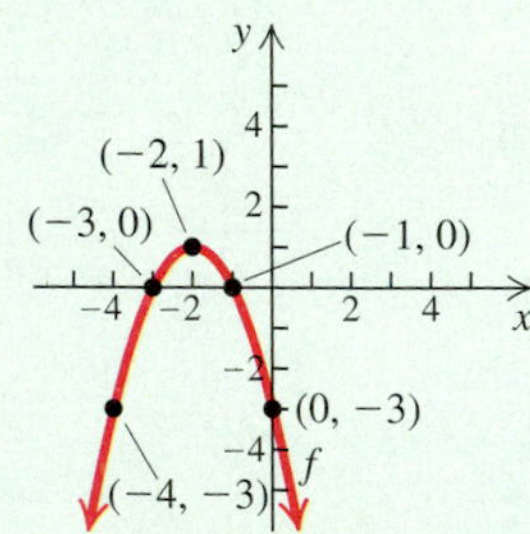

11. Determine whether the graph is that of a function. Answer yes or no.

a)

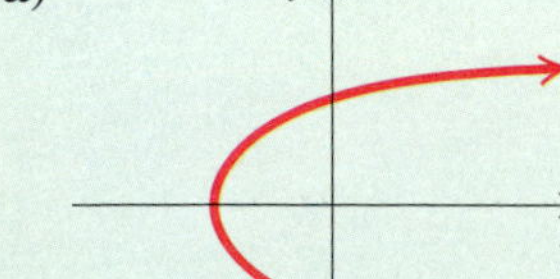

b)

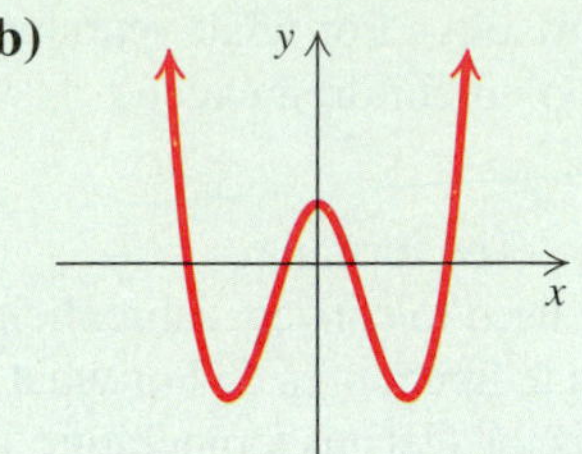

Find the domain of the function.

12. $f(x) = \dfrac{1}{x - 4}$

13. $g(x) = x^3 + 2$

14. $h(x) = \sqrt{25 - x^2}$

15. **a)** Graph: $f(x) = |x - 2| + 3$.
b) Visually estimate the domain of $f(x)$.
c) Visually estimate the range of $f(x)$.

Find the slope of the line containing the given points.

16. $\left(-2, \frac{2}{3}\right), (-2, 5)$

17. $(4, -10), (-8, 12)$

18. $(-5, 6), \left(\frac{3}{4}, 6\right)$

19. *Number of Married Adults.* The number of married adults in the United States is declining. In 1960, 72% of adults were married. This number had decreased to 51% by 2012. (*Source*: *AARP Magazine*, June–July 2014, "The New American Family," p. 32) Find the average rate of change in the percent of adults who are married for the years from 1960 to 2012.

20. Find the slope and the y-intercept of the line with equation $-3x + 2y = 5$.

21. *Total Cost.* An electrician charges a basic rate of \$65 for a service call plus \$48 per hour for labor. Write an equation that can be used to determine the cost $C(t)$ of hiring an electrician to do repair work. Then find the total cost, not including parts, if the repair work takes 2.25 hr.

22. Write an equation for the line with $m = -\frac{5}{8}$ and y-intercept $(0, -5)$.

23. Write an equation for the line that passes through $(-5, 4)$ and $(3, -2)$.

24. Write the equation of the vertical line that passes through $\left(-\frac{3}{8}, 11\right)$.

25. Determine whether the lines are parallel, perpendicular, or neither.

$$2x + 3y = -12,$$
$$2y - 3x = 8$$

26. Find an equation of the line containing the point $(-1, 3)$ and parallel to the line $x + 2y = -6$.

27. Find an equation of the line containing the point $(-1, 3)$ and perpendicular to the line $x + 2y = -6$.

28. *Weekly Earnings.* Data in the following table list the average weekly earnings of U.S. production workers from 2003 to 2013.

Year, x	Average Weekly Earnings of U.S. Production Workers
2003, 0	$517.82
2005, 2	544.05
2007, 4	589.27
2009, 6	616.01
2011, 8	653.19
2013, 10	677.67

Source: Bureau of Labor Statistics, U.S. Department of Labor

a) Without using the regression feature on a graphing calculator, model the data with a linear function. Then using this function, predict the average weekly earnings of U.S. production workers in 2017. Answers may vary depending on the data points used.

b) Using a graphing calculator, fit a regression line to the data and use it to predict the average weekly earnings of production workers in 2017. What is the correlation coefficient for the regression line?

Solve.

29. $6x + 7 = 1$

30. $2.5 - x = -x + 2.5$

31. $\frac{3}{2}y - 4 = \frac{5}{3}y + 6$

32. $2(4x + 1) = 8 - 3(x - 5)$

33. *Parking-Lot Dimensions.* The parking lot behind Kai's Kafé has a perimeter of 210 m. The width is three-fourths of the length. What are the dimensions of the parking lot?

34. *Pricing.* Kokona's Juice Bar prices its bottled juices by raising the wholesale price 50% and then adding 25¢. What is the wholesale price of a bottle of juice that sells for $2.95?

35. Find the zero(s) of the function

$$f(x) = 3x + 9.$$

Solve and write interval notation for the solution set. Then graph the solution set.

36. $5 - x \geq 4x + 20$

37. $-7 < 2x + 3 < 9$

38. $2x - 1 \leq 3$ *or* $5x + 6 \geq 26$

39. *Moving Costs.* Morgan Movers charges $90 plus $25 per hour to move households across town. McKinley Movers charges $40 per hour for crosstown moves. For what lengths of time does it cost less to hire Morgan Movers?

40. The graph of $g(x) = 1 - \frac{1}{2}x$ is which of the following ?

A.

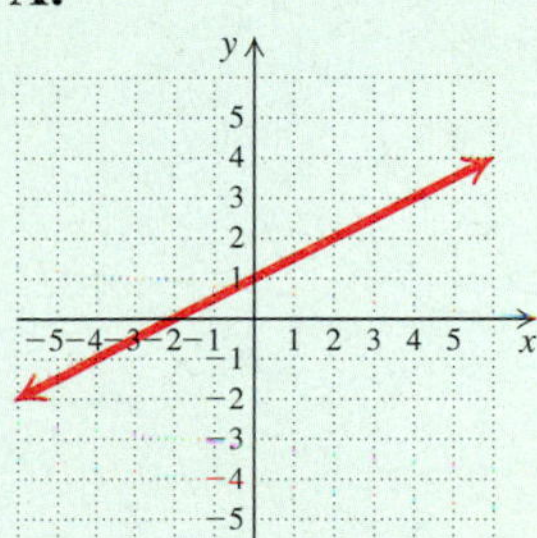

B.

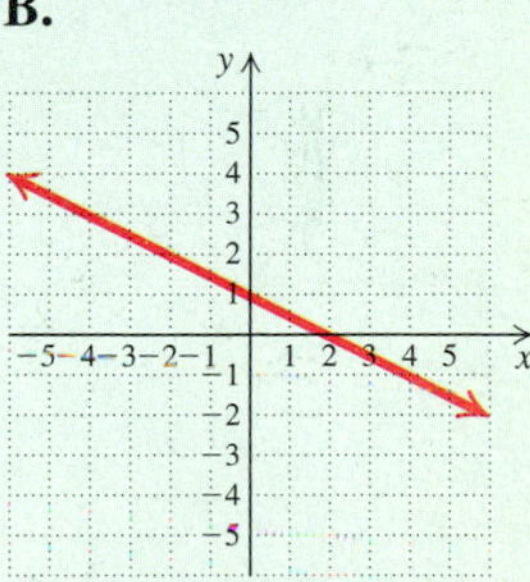

C.

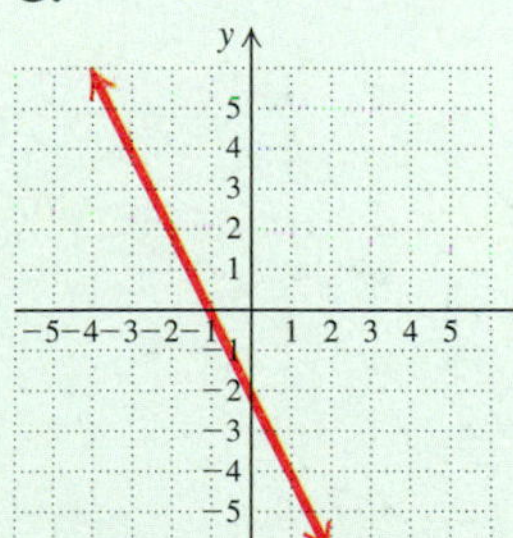

D.

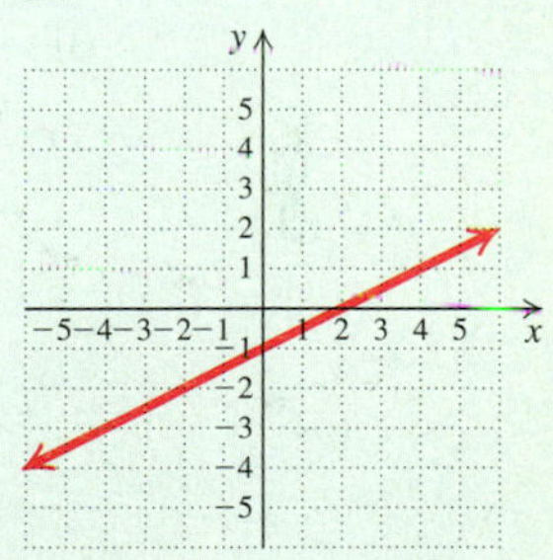

Synthesis

41. Suppose that for some function h, $h(x + 2) = \frac{1}{2}x$. Find $h(-2)$.

More on Functions

CHAPTER 2

APPLICATION

This problem appears as Exercise 27 in Section 2.1.

A wholesale nursery estimates that it will sell N fruit trees after spending a dollars on advertising, where

$$N(a) = -a^2 + 300a + 6, \quad 0 \le a \le 300,$$

and a is measured in thousands of dollars. For what advertising expenditure will the greatest number of fruit trees be sold? How many fruit trees will be sold for that amount?

2.1 Increasing, Decreasing, and Piecewise Functions; Applications

- Graph functions, looking for intervals on which the function is increasing, decreasing, or constant, and estimate relative maxima and minima.
- Given an application, find a function that models the application. Find the domain of the function and function values, and then graph the function.
- Graph functions defined piecewise.

Because functions occur in so many real-world situations, it is important to be able to analyze them carefully.

Increasing, Decreasing, and Constant Functions

On a given interval, if the graph of a function rises from left to right, it is said to be **increasing** on that interval. If the graph drops from left to right, it is said to be **decreasing**. If the function values stay the same on the interval, the function is said to be **constant**.

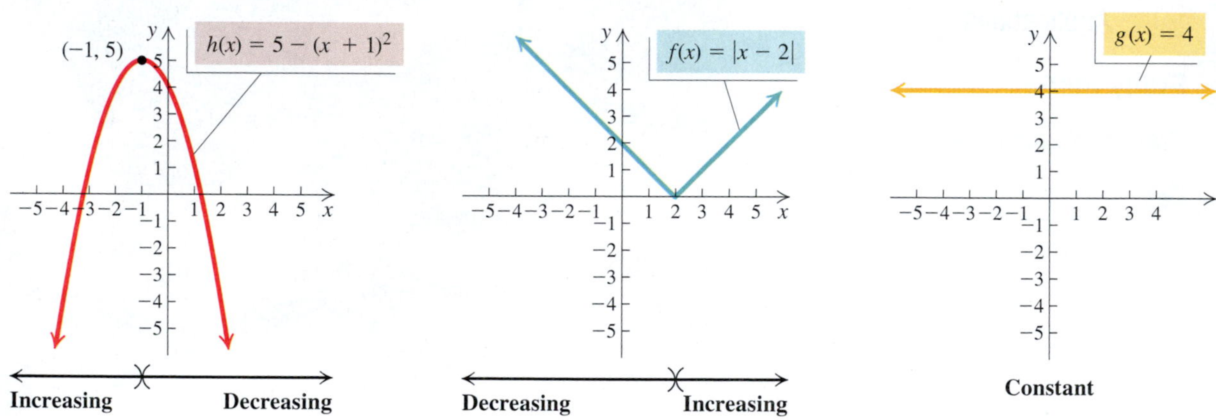

We are led to the following definitions.

> **INCREASING, DECREASING, AND CONSTANT FUNCTIONS**
>
> A function f is said to be **increasing** on an *open* interval I, if for all a and b in that interval, $a < b$ implies $f(a) < f(b)$. (See Fig. 1 on the following page.)
>
> A function f is said to be **decreasing** on an *open* interval I, if for all a and b in that interval, $a < b$ implies $f(a) > f(b)$. (See Fig. 2.)
>
> A function f is said to be **constant** on an *open* interval I, if for all a and b in that interval, $f(a) = f(b)$. (See Fig. 3.)

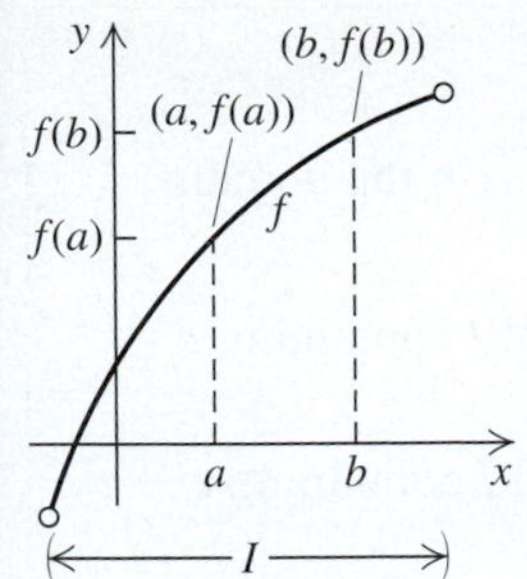

For $a < b$ in I, $f(a) < f(b)$; f is ***increasing*** on I.

FIGURE 1.

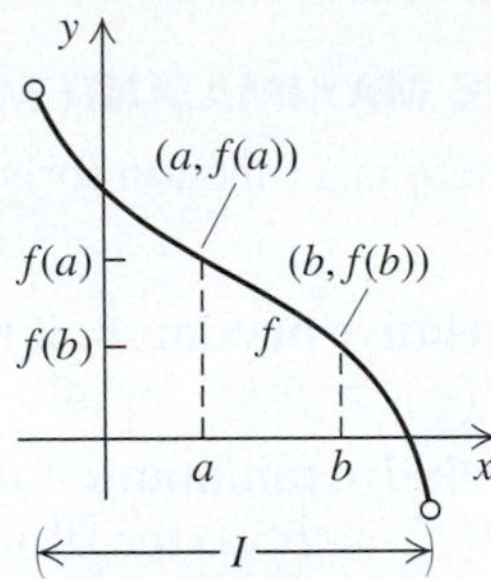

For $a < b$ in I, $f(a) > f(b)$; f is ***decreasing*** on I.

FIGURE 2.

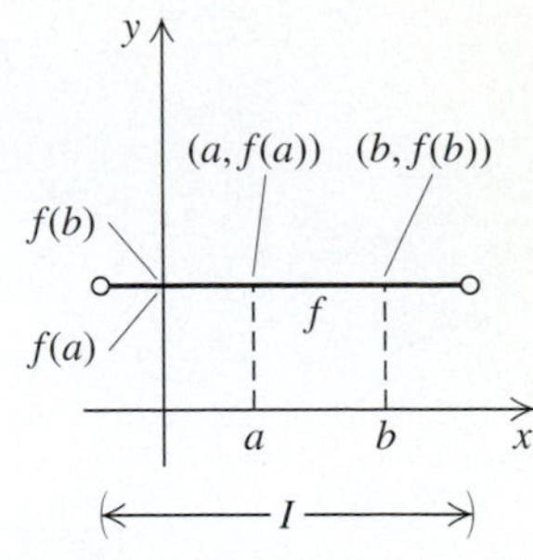

For all a and b in I, $f(a) = f(b)$; f is ***constant*** on I.

FIGURE 3.

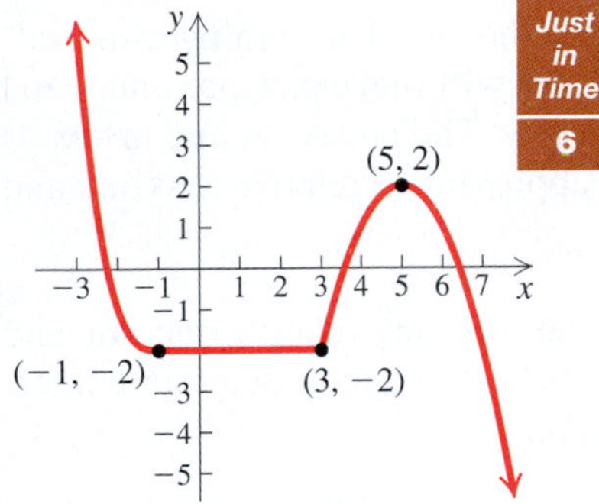

Just in Time 6

EXAMPLE 1 Determine the intervals on which the function in the figure at left is **(a)** increasing; **(b)** decreasing; **(c)** constant.

Solution When expressing interval(s) on which a function is increasing, decreasing, or constant, we consider only values in the *domain* of the function. Since the domain of this function is $(-\infty, \infty)$, we consider all real values of x.

a) As x-values (that is, values in the domain) increase from $x = 3$ to $x = 5$, y-values (that is, values in the range) increase from -2 to 2. Thus the function is increasing on the interval $(3, 5)$.

b) As x-values increase from negative infinity to -1, y-values decrease; y-values also decrease as x-values increase from 5 to positive infinity. Thus the function is decreasing on the intervals $(-\infty, -1)$ and $(5, \infty)$.

c) As x-values increase from -1 to 3, y remains -2. The function is constant on the interval $(-1, 3)$.

Now Try Exercise 5.

In calculus, the slope of a line tangent to the graph of a function at a particular point is used to determine whether the function is increasing, decreasing, or neither. If the slope is positive, the function is increasing; if the slope is negative, the function is decreasing; if the slope is 0 over an interval, the function is constant. Since slope cannot be both positive and negative at the same point, a function cannot be both increasing and decreasing at a specific point. For this reason, increasing, decreasing, and constant intervals are expressed in *open interval* notation. In Example 1, if $[3, 5]$ had been used for the increasing interval and $[5, \infty)$ for a decreasing interval, the function would be both increasing and decreasing at $x = 5$. This is not possible.

Relative Maximum and Minimum Values

Consider the graph shown below. Note the "peaks" and "valleys" at the x-values c_1, c_2, and c_3. The function value $f(c_2)$ is called a **relative maximum** (plural, **maxima**). Each of the function values $f(c_1)$ and $f(c_3)$ is called a **relative minimum** (plural, **minima**).

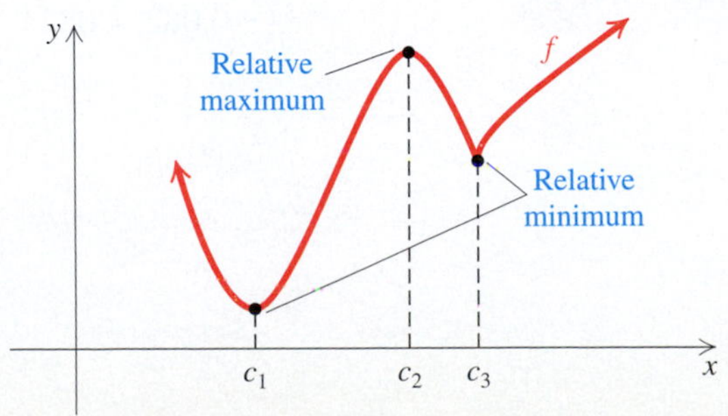

> **RELATIVE MAXIMA AND MINIMA**
>
> Suppose that f is a function for which $f(c)$ exists for some c in the domain of f. Then:
>
> $f(c)$ is a **relative maximum** if there exists an *open* interval I containing c such that $f(c) > f(x)$, for all x in I where $x \neq c$; and
>
> $f(c)$ is a **relative minimum** if there exists an *open* interval I containing c such that $f(c) < f(x)$, for all x in I where $x \neq c$.

Simply stated, $f(c)$ is a *relative maximum* if $(c, f(c))$ is the highest point in some *open* interval, and $f(c)$ is a *relative minimum* if $(c, f(c))$ is the lowest point in some *open* interval.

If you take a calculus course, you will learn a method for determining exact values of relative maxima and minima. In Section 3.3, we will find exact maximum and minimum values of quadratic functions algebraically. The MAXIMUM and MINIMUM features on a graphing calculator can be used to approximate relative maxima and minima.

EXAMPLE 2 Use a graphing calculator to determine any relative maxima and minima of the function $f(x) = 0.1x^3 - 0.6x^2 - 0.1x + 2$ and to determine intervals on which the function is increasing or decreasing.

Solution We first graph the function, experimenting with the window dimensions as needed. The curvature is seen fairly well with window settings of $[-4, 6, -3, 3]$. Using the MAXIMUM and MINIMUM features, we determine the relative maximum value and the relative minimum value of the function.

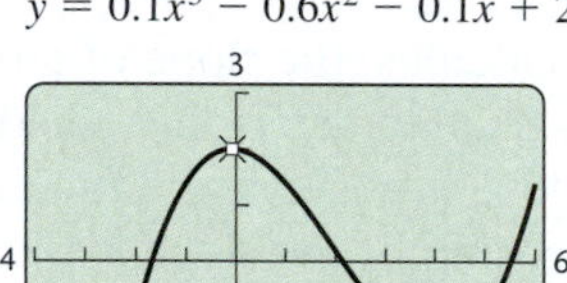

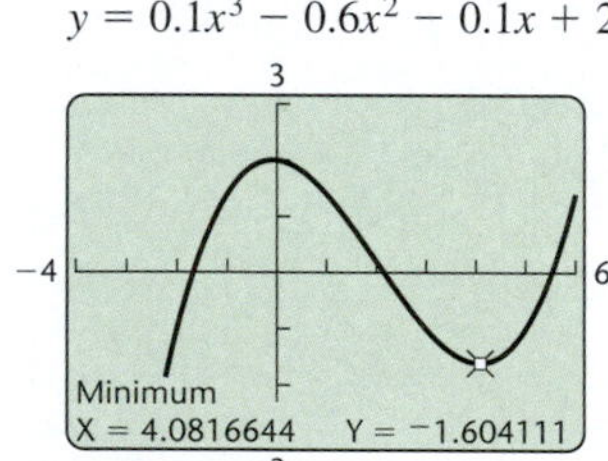

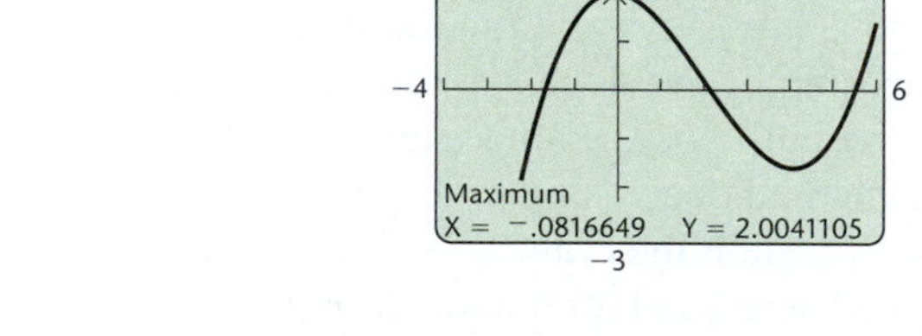

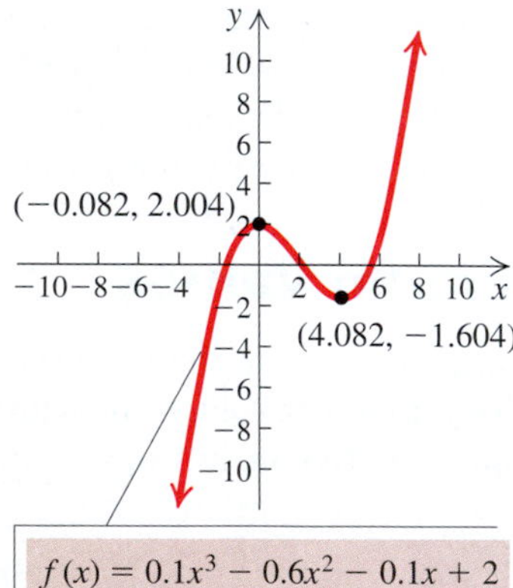

We see that the relative maximum value of the function is about 2.004. It occurs when $x \approx -0.082$. We also approximate the relative minimum, -1.604 at $x \approx 4.082$.

We note that the graph rises, or increases, from the left and stops increasing at the relative maximum. From this point, the graph decreases to the relative minimum and then begins to rise again. Thus the function is *increasing* on the intervals

$$(-\infty, -0.082) \quad \text{and} \quad (4.082, \infty)$$

and *decreasing* on the interval

$$(-0.082, 4.082).$$

Now Try Exercise 23.

Applications of Functions

Many real-world situations can be modeled by functions.

EXAMPLE 3 *Car Distance.* Two nurses, Kiara and Matias, drive away from a hospital at right angles to each other. Kiara's speed is 35 mph and Matias's is 40 mph.

a) Express the distance between the cars as a function of time, $d(t)$.

b) Find the domain of the function.

Solution

a) Suppose that 1 hr goes by. At that time, Kiara has traveled 35 mi and Matias has traveled 40 mi. We can use the Pythagorean theorem to find the distance between them. This distance would be the length of the hypotenuse of a right triangle with legs measuring 35 mi and 40 mi. After 2 hr, the triangle's legs would measure $2 \cdot 35$, or 70 mi, and $2 \cdot 40$, or 80 mi. Noting that the distances will always be changing, we make a drawing and let $t =$ the time, in hours, that Kiara and Matias have been driving since leaving the hospital.

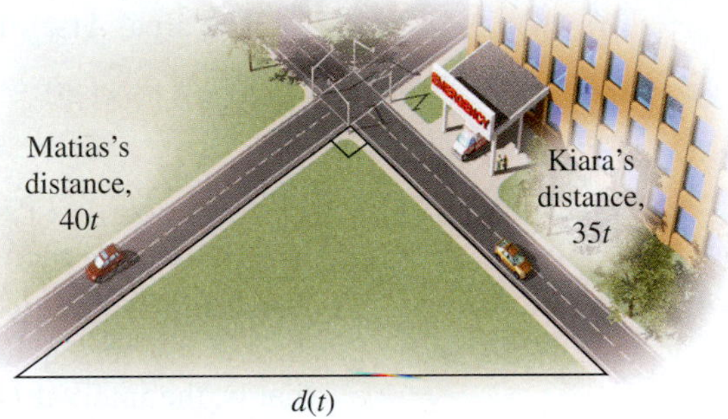

After t hours, Kiara has traveled $35t$ miles and Matias $40t$ miles. We now use the Pythagorean theorem:

$$[d(t)]^2 = (35t)^2 + (40t)^2.$$

Because distance must be nonnegative, we need consider only the positive square root when solving for $d(t)$:

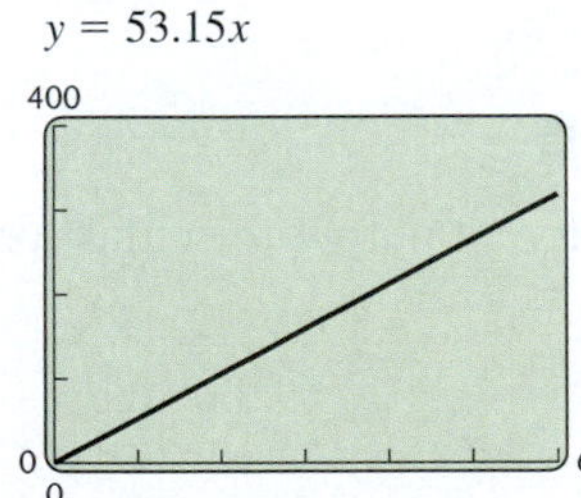

$$\begin{aligned} d(t) &= \sqrt{(35t)^2 + (40t)^2} \\ &= \sqrt{1225t^2 + 1600t^2} \\ &= \sqrt{2825t^2} \\ &\approx 53.15|t| && \text{Approximating the root to two decimal places} \\ &\approx 53.15t. && \text{Since } t \geq 0, |t| = t. \end{aligned}$$

Thus, $d(t) = 53.15t$, $t \geq 0$.

b) Since the time traveled, t, must be nonnegative, the domain is the set of nonnegative real numbers $[0, \infty)$.

Now Try Exercise 35.

EXAMPLE 4 *Area of Office Space.* A community college has 30 ft of dividers with which to set off a rectangular area for a student testing center. If a corner of the math lab is used for the testing center, the partition need only form two sides of a rectangle.

a) Express the floor area of the testing center as a function of the length of the partition.

b) Find the domain of the function.

c) Graph the function.

d) Find the dimensions that maximize the floor area.

Solution

a) Note that the dividers will form two sides of a rectangle. If, for example, 14 ft of dividers are used for the length of the rectangle, that would leave $30 - 14$, or 16 ft of dividers for the width. Thus if $x =$ the length, in feet, of the rectangle, then $30 - x =$ the width. We represent this information in a drawing, as shown below.

The area, $A(x)$, is given by

$$A(x) = x(30 - x) \quad \text{Area = length · width.}$$
$$= 30x - x^2.$$

The function $A(x) = 30x - x^2$ can be used to express the area of the rectangle as a function of the length.

b) Because the length and the width of the rectangle must be positive and only 30 ft of dividers are available, we restrict the domain of A to $\{x \mid 0 < x < 30\}$—that is, the interval $(0, 30)$.

c) The graph is shown at left.

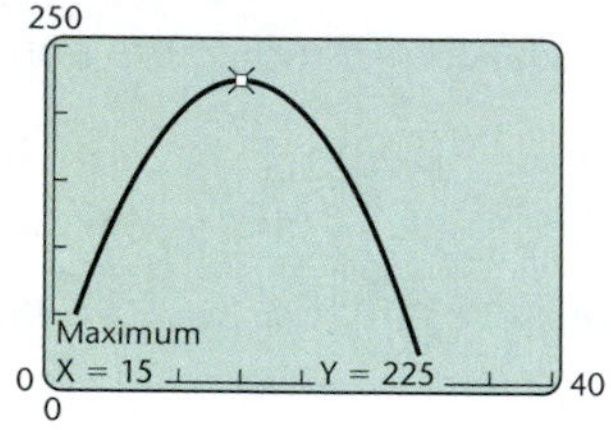

d) We use the MAXIMUM feature as shown on the graph at left. The maximum value of the area function on the interval $(0, 30)$ is 225 when $x = 15$. Thus the dimensions that maximize the area are

$$\text{Length} = x = 15 \text{ ft} \quad \text{and}$$
$$\text{Width} = 30 - x = 30 - 15 = 15 \text{ ft}.$$

Now Try Exercise 33.

Functions Defined Piecewise

Sometimes functions are defined **piecewise** using different output formulas for different pieces, or parts, of the domain.

Just in Time 5

EXAMPLE 5 For the function defined as

$$f(x) = \begin{cases} x + 1, & \text{for } x < -2, \\ 5, & \text{for } -2 \le x \le 3, \\ x^2, & \text{for } x > 3, \end{cases}$$

find $f(-5)$, $f(-3)$, $f(0)$, $f(3)$, $f(4)$, and $f(10)$.

Solution First, we determine which part of the domain contains the given input. Then we use the corresponding formula to find the output.

Since $-5 < -2$, we use the formula $f(x) = x + 1$:

$$f(-5) = -5 + 1 = -4.$$

Since $-3 < -2$, we use the formula $f(x) = x + 1$ again:

$$f(-3) = -3 + 1 = -2.$$

Since $-2 \leq 0 \leq 3$, we use the formula $f(x) = 5$:

$f(0) = 5$.

Since $-2 \leq 3 \leq 3$, we use the formula $f(x) = 5$ a second time:

$f(3) = 5$.

Since $4 > 3$, we use the formula $f(x) = x^2$:

$f(4) = 4^2 = 16$.

Since $10 > 3$, we once again use the formula $f(x) = x^2$:

$f(10) = 10^2 = 100$.

Now Try Exercise 47.

EXAMPLE 6 Graph the function defined as

$$g(x) = \begin{cases} \frac{1}{3}x + 3, & \text{for } x < 3, \\ -x, & \text{for } x \geq 3. \end{cases}$$

Solution Since the function is defined in two pieces, or parts, we create the graph in two parts.

a) We graph $g(x) = \frac{1}{3}x + 3$ *only* for inputs x less than 3. That is, we use $g(x) = \frac{1}{3}x + 3$ only for x-values in the interval $(-\infty, 3)$. Some ordered pairs that are solutions of this piece of the function are shown in Table 1.

b) We graph $g(x) = -x$ *only* for inputs x greater than or equal to 3. That is, we use $g(x) = -x$ only for x-values in the interval $[3, \infty)$. Some ordered pairs that are solutions of this piece of the function are shown in Table 2.

TABLE 1

x $(x < 3)$	$g(x) = \frac{1}{3}x + 3$
-3	2
0	3
2	$3\frac{2}{3}$

TABLE 2

x $(x \geq 3)$	$g(x) = -x$
3	-3
4	-4
6	-6

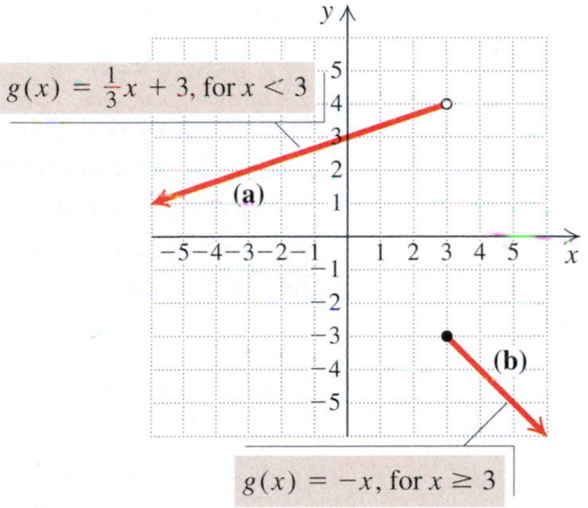

Now Try Exercise 51.

EXAMPLE 7 Graph the function defined as

$$f(x) = \begin{cases} 4, & \text{for } x \leq 0, \\ 4 - x^2, & \text{for } 0 < x \leq 2, \\ 2x - 6, & \text{for } x > 2. \end{cases}$$

Solution We create the graph in three pieces, or parts.

a) We graph $f(x) = 4$ *only* for inputs x less than or equal to 0. That is, we use $f(x) = 4$ only for x-values in the interval $(-\infty, 0]$. Some ordered pairs that are solutions of this piece of the function are shown in Table 3.

b) We graph $f(x) = 4 - x^2$ *only* for inputs x greater than 0 and less than or equal to 2. That is, we use $f(x) = 4 - x^2$ only for x-values in the interval $(0, 2]$. Some ordered pairs that are solutions of this piece of the function are shown in Table 4.

TABLE 3

x $(x \leq 0)$	$f(x) = 4$
-5	4
-2	4
0	4

TABLE 4

x $(0 < x \leq 2)$	$f(x) = 4 - x^2$
$\frac{1}{2}$	$3\frac{3}{4}$
1	3
2	0

TABLE 5

x $(x > 2)$	$f(x) = 2x - 6$
$2\frac{1}{2}$	-1
3	0
5	4

c) We graph $f(x) = 2x - 6$ *only* for inputs x greater than 2. That is, we use $f(x) = 2x - 6$ only for x-values in the interval $(2, \infty)$. Some ordered pairs that are solutions of this piece of the function are shown in Table 5.

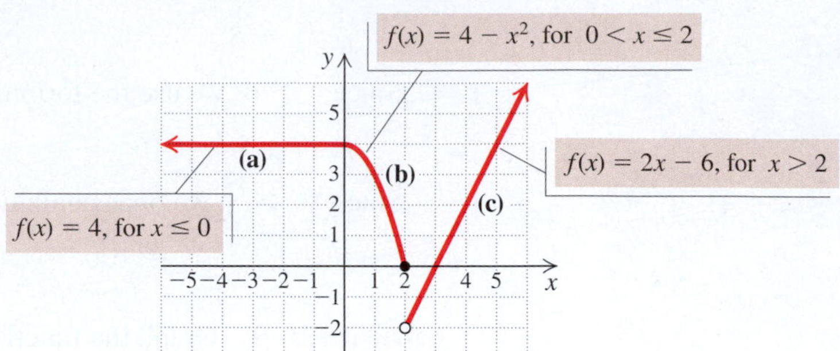

Now Try Exercise 55.

Just in Time 21

EXAMPLE 8 Graph the function defined as

$$f(x) = \begin{cases} \dfrac{x^2 - 4}{x + 2}, & \text{for } x \neq -2, \\ 3, & \text{for } x = -2. \end{cases}$$

Solution When $x \neq -2$, the denominator of $(x^2 - 4)/(x + 2)$ is nonzero, so we can simplify:

$$\frac{x^2 - 4}{x + 2} = \frac{(x + 2)(x - 2)}{x + 2} = x - 2.$$

Thus,

$$f(x) = x - 2, \quad \text{for } x \neq -2.$$

The graph of this part of the function consists of a line with a "hole" at the point $(-2, -4)$, indicated by the open circle. The hole occurs because the piece of the function represented by $(x^2 - 4)/(x + 2)$ is not defined for $x = -2$. By the definition of the function, we see that $f(-2) = 3$, so we plot the point $(-2, 3)$ above the open circle.

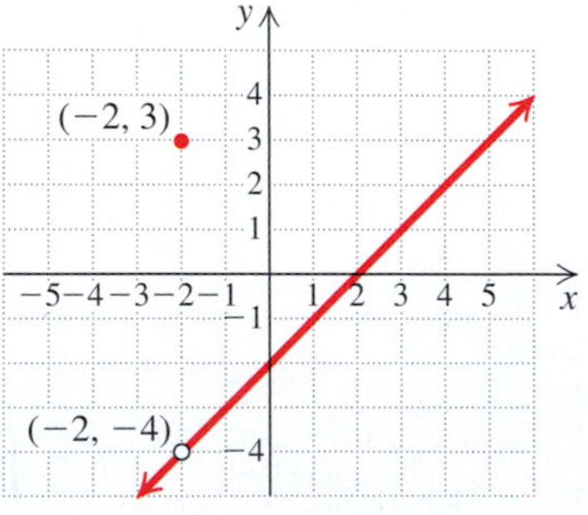

Now Try Exercise 59.

$y = (x^2 - 4)/(x + 2)$

3.1, −4.7, 4.7, −6.2

X	Y1
−2.3	−4.3
−2.2	−4.2
−2.1	−4.1
−2	ERROR
−1.9	−3.9
−1.8	−3.8
−1.7	−3.7

X = −2

The hand-drawn graph of the piece of the function in Example 8 represented by $y = (x^2 - 4)/(x + 2)$, $x \neq 2$, can be checked using a graphing calculator. When $y = (x^2 - 4)/(x + 2)$ is graphed, the hole may or may not be visible, depending on the window dimensions chosen. When we use the ZDECIMAL feature from the ZOOM menu, note that the hole will appear, as shown in the graph at left. If we examine a table of values for this function, we see that an ERROR message corresponds to the x-value -2. This indicates that -2 is *not* in the domain of the function $y = (x^2 - 4)/(x + 2)$. However, -2 *is* in the domain of the function f in Example 8 because $f(-2)$ is defined to be 3.

A piecewise function with importance in calculus and computer programming is the **greatest integer function**, denoted $f(x) = [\![x]\!]$, or $\text{int}(x)$.

> **GREATEST INTEGER FUNCTION**
>
> $f(x) = [\![x]\!] =$ the greatest integer *less than or equal to x.*

The greatest integer function pairs each input with the greatest integer *less than or equal to* that input. Thus, x-values 1, $1\frac{1}{2}$, and 1.8 are all paired with the y-value 1. Other pairings are shown below.

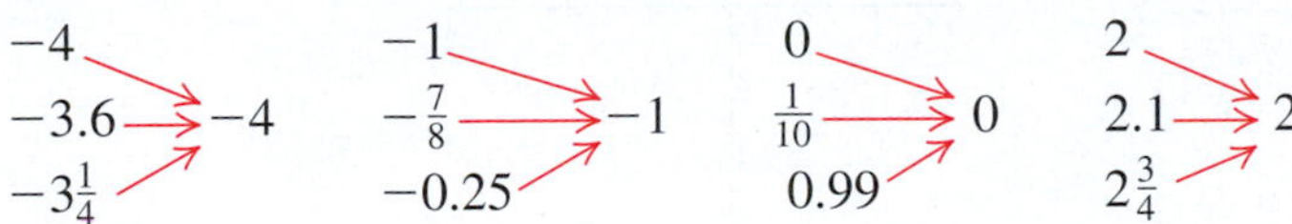

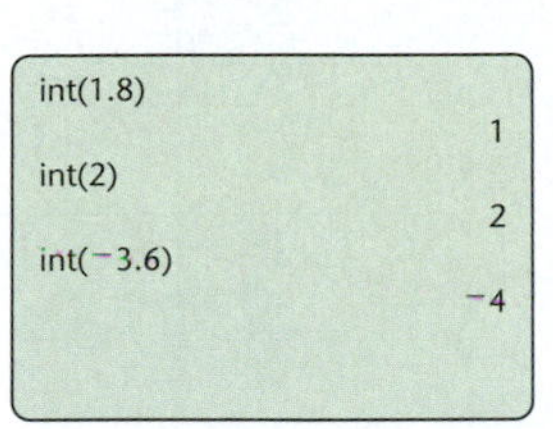

These values can be checked with a graphing calculator using the int(feature from the NUM submenu in the MATH menu.

EXAMPLE 9 Graph $f(x) = [\![x]\!]$ and determine its domain and range.

Solution The greatest integer function can also be defined as a piecewise function with an infinite number of statements. When plotting points by hand, it can be helpful to use the TABLE feature on a graphing calculator to find ordered pairs.

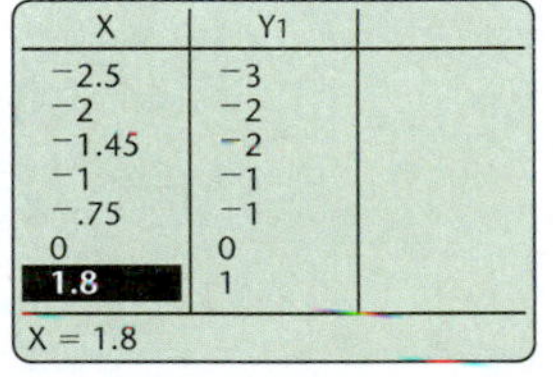

X	Y1
−2.5	−3
−2	−2
−1.45	−2
−1	−1
−.75	−1
0	0
1.8	1

X = 1.8

Just in Time 3

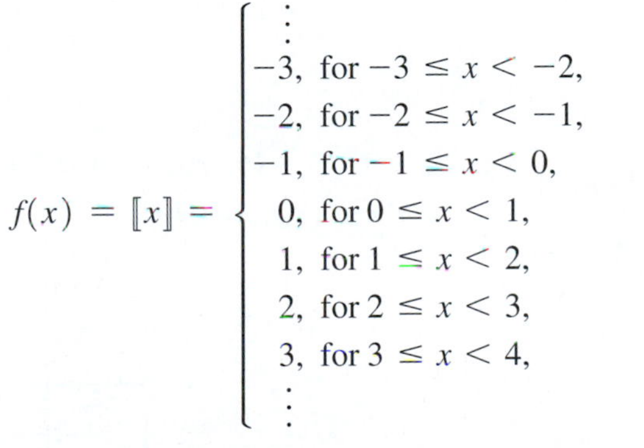

$$f(x) = [\![x]\!] = \begin{cases} \vdots \\ -3, & \text{for } -3 \le x < -2, \\ -2, & \text{for } -2 \le x < -1, \\ -1, & \text{for } -1 \le x < 0, \\ 0, & \text{for } 0 \le x < 1, \\ 1, & \text{for } 1 \le x < 2, \\ 2, & \text{for } 2 \le x < 3, \\ 3, & \text{for } 3 \le x < 4, \\ \vdots \end{cases}$$

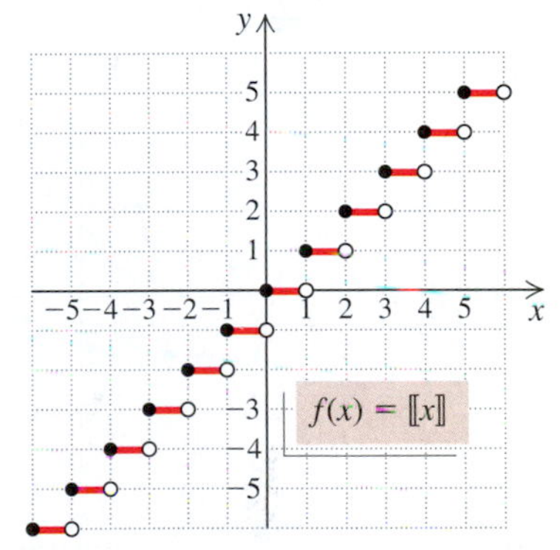

We see that the domain of this function is the set of all real numbers, $(-\infty, \infty)$, and the range is the set of all integers, $\{\ldots, -3, -2, -1, 0, 1, 2, 3, \ldots\}$.

Now Try Exercise 63.

If we had used a calculator for Example 9, we would see the graph shown below.

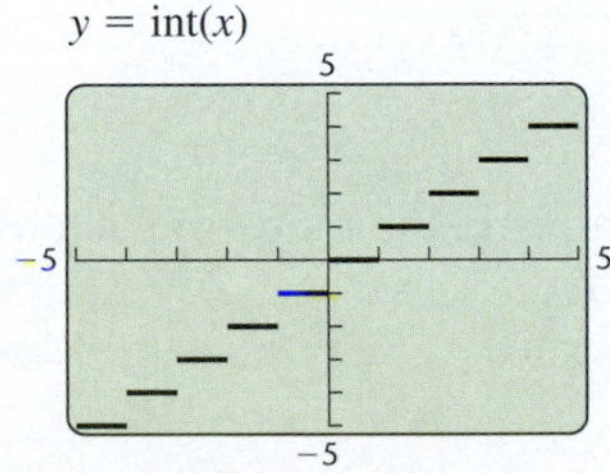

2.1 Exercise Set

Determine the intervals on which the function is **(a)** *increasing,* **(b)** *decreasing, and* **(c)** *constant.*

1.

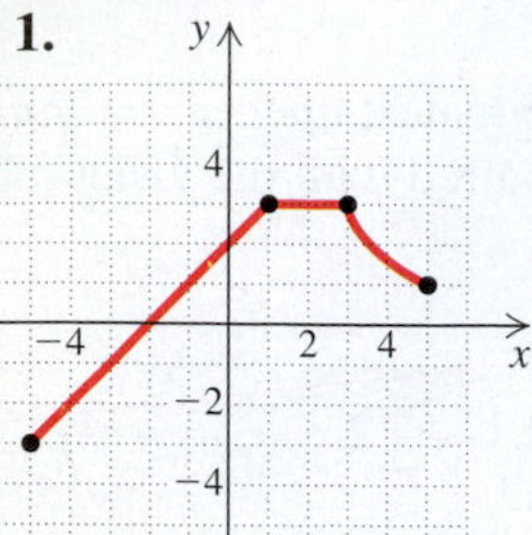

2.

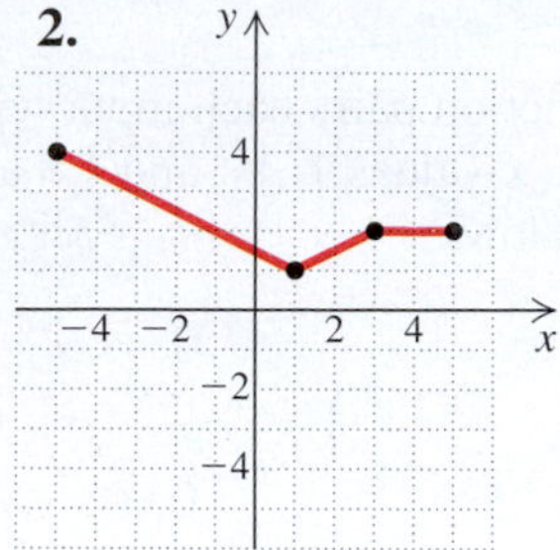

3.

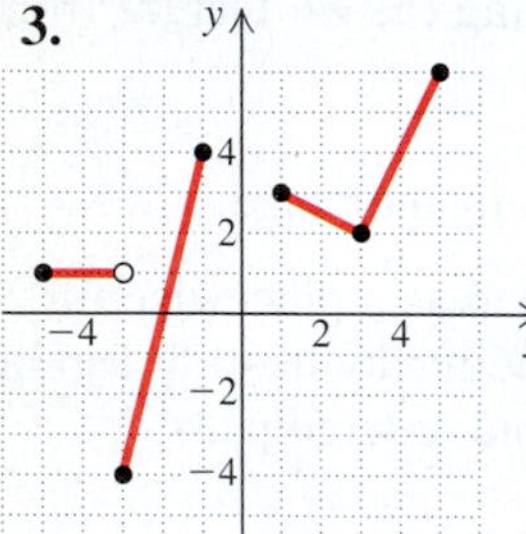

4.

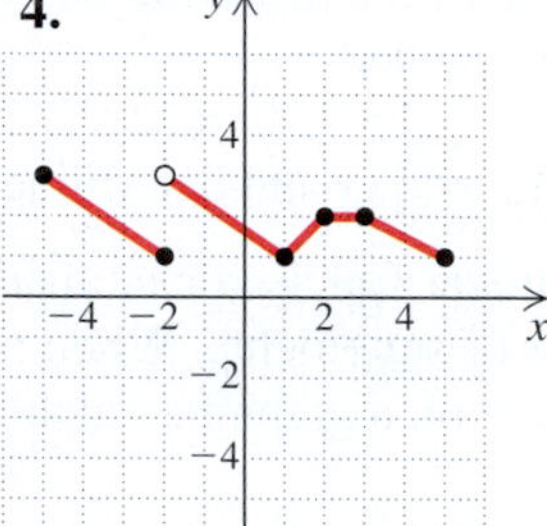

5.

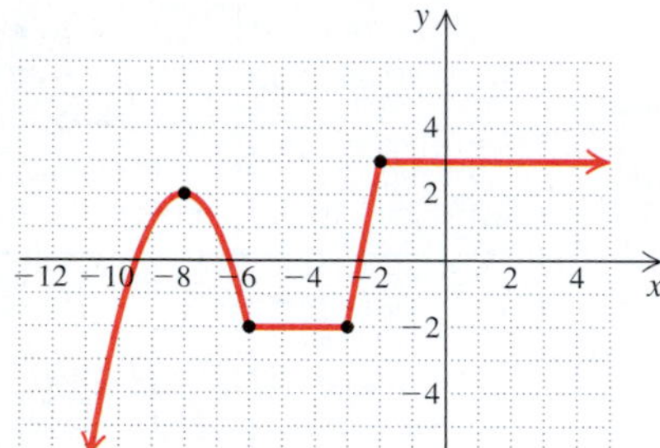

6.

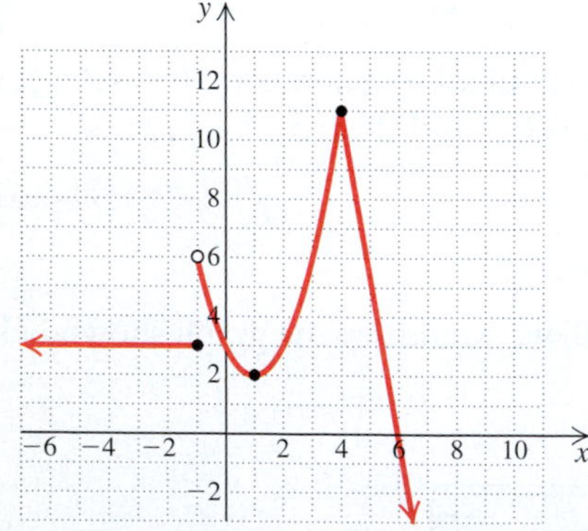

7.–12. Determine the domain and the range of each of the functions graphed in Exercises 1–6.

Using the graph, determine any relative maxima or minima of the function and the intervals on which the function is increasing or decreasing.

13. $f(x) = -x^2 + 5x - 3$

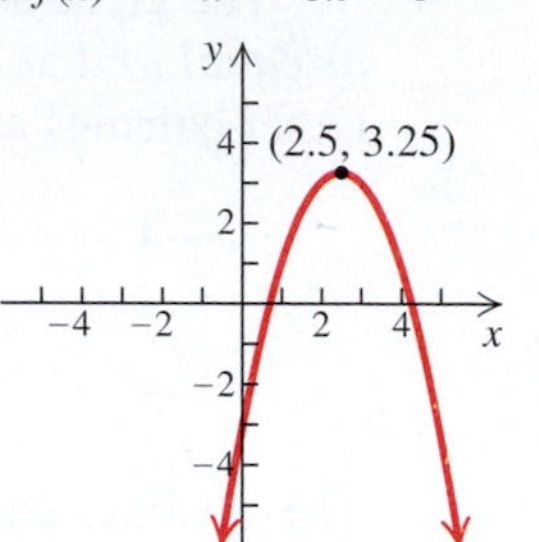

14. $f(x) = x^2 - 2x + 3$

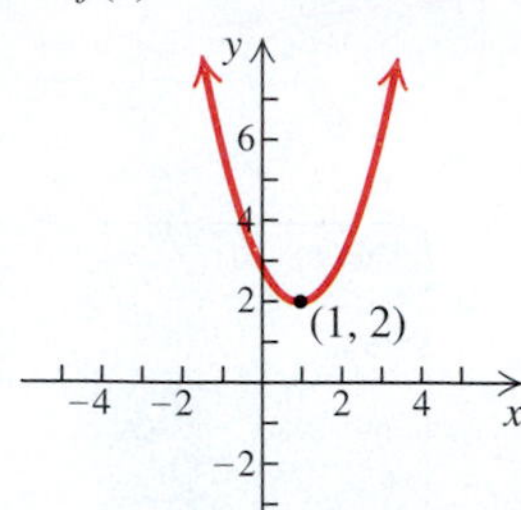

15. $f(x) = \frac{1}{4}x^3 - \frac{1}{2}x^2 - x + 2$

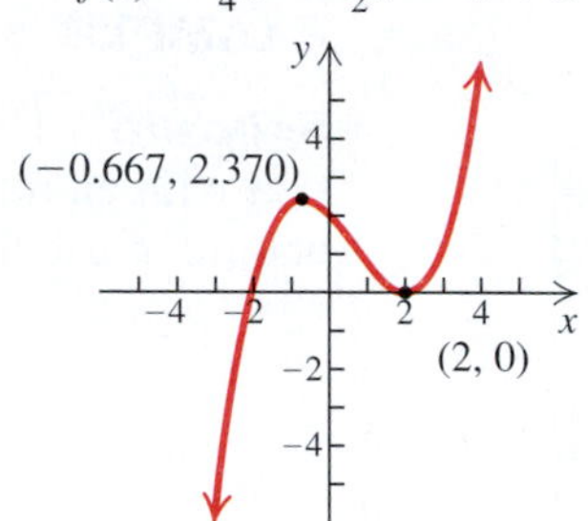

16. $f(x) = -0.09x^3 + 0.5x^2 - 0.1x + 1$

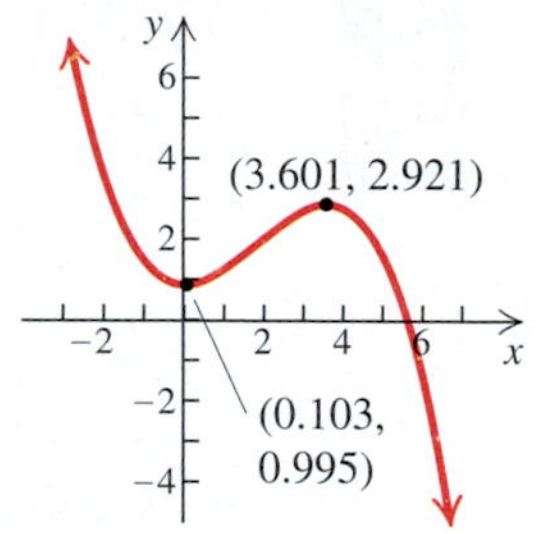

Graph the function. Estimate the intervals on which the function is increasing or decreasing and any relative maxima or minima.

17. $f(x) = x^2$

18. $f(x) = 4 - x^2$

19. $f(x) = 5 - |x|$

20. $f(x) = |x + 3| - 5$

21. $f(x) = x^2 - 6x + 10$

22. $f(x) = -x^2 - 8x - 9$

Graph the function using the given viewing window. Find the intervals on which the function is increasing or decreasing and find any relative maxima or minima. Change the viewing window if it seems appropriate for further analysis.

23. $f(x) = -x^3 + 6x^2 - 9x - 4$,
$[-3, 7, -20, 15]$

24. $f(x) = 0.2x^3 - 0.2x^2 - 5x - 4$,
$[-10, 10, -30, 20]$

25. $f(x) = 1.1x^4 - 5.3x^2 + 4.07$,
$[-4, 4, -4, 8]$

26. $f(x) = 1.2(x + 3)^4 + 10.3(x + 3)^2 + 9.78$,
$[-9, 3, -40, 100]$

27. *Advertising Effect.* A wholesale nursery estimates that it will sell N fruit trees after spending a dollars on advertising, where

$$N(a) = -a^2 + 300a + 6, \quad 0 \le a \le 300,$$

and a is measured in thousands of dollars.

a) Graph the function using a graphing calculator.
b) Use the MAXIMUM feature to find the relative maximum.
c) For what advertising expenditure will the greatest number of fruit trees be sold? How many fruit trees will be sold for that amount?

28. *Temperature During an Illness.* The temperature of a patient during an illness is given by the function

$$T(t) = -0.1t^2 + 1.2t + 98.6, \quad 0 \le t \le 12,$$

where T is the temperature, in degrees Fahrenheit, at time t, in days, after the onset of the illness.

a) Graph the function using a graphing calculator.
b) Use the MAXIMUM feature to determine at what time the patient's temperature was the highest. What was the highest temperature?

Use a graphing calculator to find the intervals on which the function is increasing or decreasing. Consider the entire set of real numbers if no domain is given.

29. $f(x) = \dfrac{8x}{x^2 + 1}$

30. $f(x) = \dfrac{-4}{x^2 + 1}$

31. $f(x) = x\sqrt{4 - x^2}$, for $-2 \le x \le 2$

32. $f(x) = -0.8x\sqrt{9 - x^2}$, for $-3 \le x \le 3$

33. *Lumberyard.* Rick's lumberyard has 480 yd of fencing with which to enclose a rectangular area. If the enclosed area is x yards long, express its area as a function of its length.

34. *Triangular Flag.* A seamstress is designing a triangular flag so that the length of the base of the triangle, in inches, is 7 less than twice the height h. Express the area of the flag as a function of the height.

35. *Blimp Distance.* The Goodyear Blimp can be seen flying at an altitude of 3500 ft above the Motor Speedway during the Indianapolis 500 race. The slanted distance directly to the Pagoda at the start–finish line is d feet. Express the horizontal distance h as a function of d.

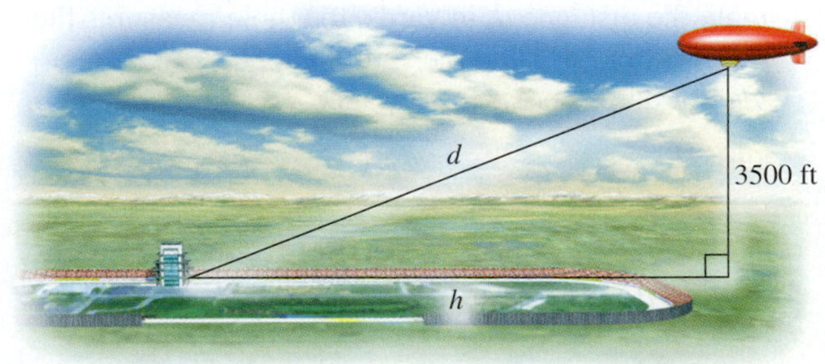

36. *Rising Balloon.* A hot-air balloon rises straight up from the ground at a rate of 120 ft/min. The balloon is tracked from a rangefinder on the ground at point P, which is 400 ft from the release point Q of the balloon. Let $d =$ the distance from the balloon to the rangefinder and $t =$ the time, in minutes, since the balloon was released. Express d as a function of t.

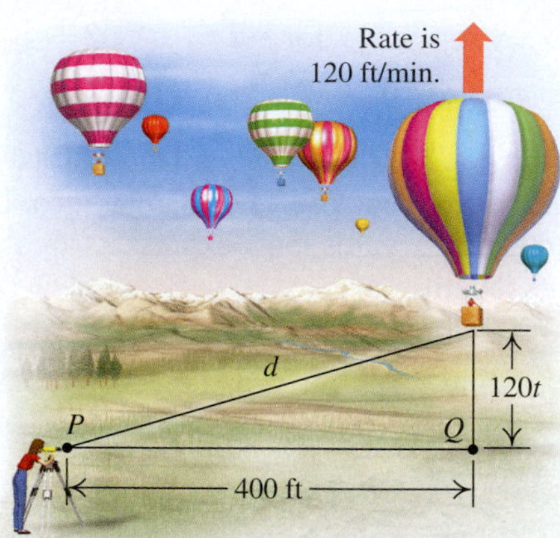

37. *Inscribed Rhombus.* A rhombus is inscribed in a rectangle that is w meters wide with a perimeter of 40 m. Each vertex of the rhombus is a midpoint of a side of the rectangle. Express the area of the rhombus as a function of the width of the rectangle.

38. *Carpet Area.* A carpet installer uses 46 ft of linen tape to bind the edges of a rectangular hall runner. If the runner is w feet wide, express its area as a function of the width.

39. *Golf Distance Finder.* A device used in golf to estimate the distance d, in yards, to a hole measures the size s, in inches, that the 7-ft pin appears to be in a viewfinder. Express the distance d as a function of s.

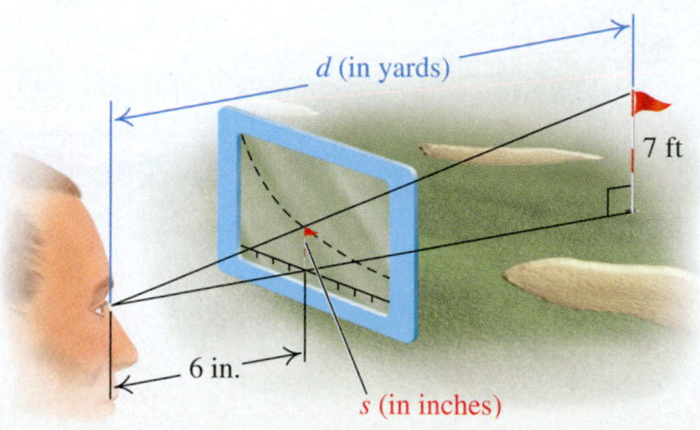

40. *Gas Tank Volume.* A gas tank has ends that are hemispheres of radius r feet. The cylindrical midsection is 6 ft long. Express the volume of the tank as a function of r.

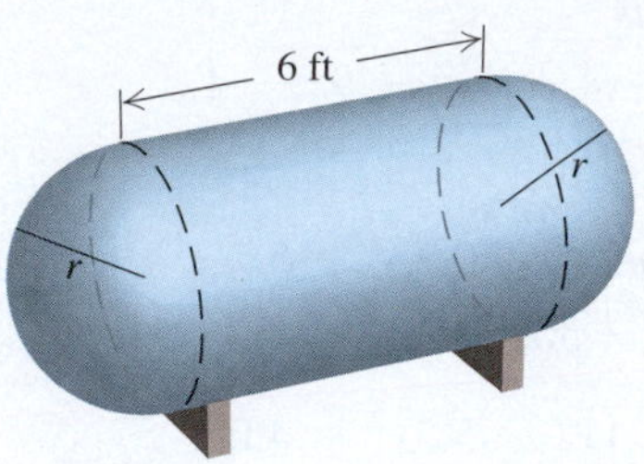

41. *Swimming Areas.* A summer camp has 240 ft of float line with which to rope off three adjacent rectangular areas of a lake for swimming lessons, one for each of three levels of swimming ability. A beach forms one side of the swimming areas. Suppose that the width of each area is x yards.

a) Express the total area of the three swimming areas as a function of x.
b) Find the domain of the function.
c) Using the graph of the function shown below, determine the dimensions that yield the maximum area.

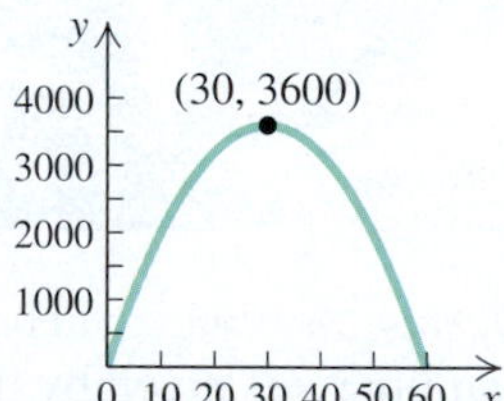

42. *Play Space.* A car dealership has 24 ft of dividers with which to enclose a rectangular play space in a corner of a customer lounge. The sides against the wall require no partition. Suppose the play space is x feet long.

a) Express the area of the play space as a function of x.
b) Find the domain of the function.
c) Using the graph shown below, determine the dimensions that yield the maximum area.

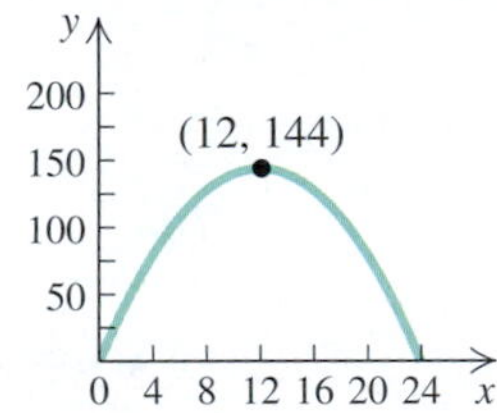

43. *Volume of a Box.* From a 12-cm by 12-cm piece of cardboard, square corners are cut out so that the sides can be folded up to make a box.

a) Express the volume of the box as a function of the side x, in centimeters, of a cut-out square.
b) Find the domain of the function.
c) Graph the function with a graphing calculator.
d) What dimensions yield the maximum volume?

44. *Office File.* Designs Unlimited plans to produce a one-component vertical file by bending the long side of an 8-in. by 14-in. sheet of plastic along two lines to form a ⊔ shape.

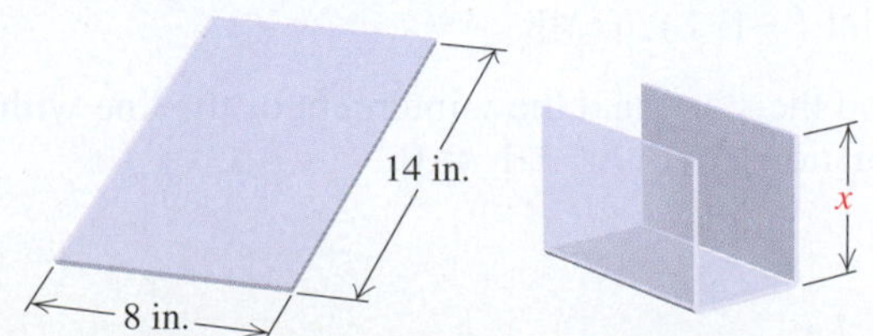

a) Express the volume of the file as a function of the height x, in inches, of the file.
b) Find the domain of the function.
c) Graph the function with a graphing calculator.
d) How tall should the file be in order to maximize the volume that the file can hold?

45. *Area of an Inscribed Rectangle.* A rectangle that is x feet wide is inscribed in a circle of radius 8 ft.

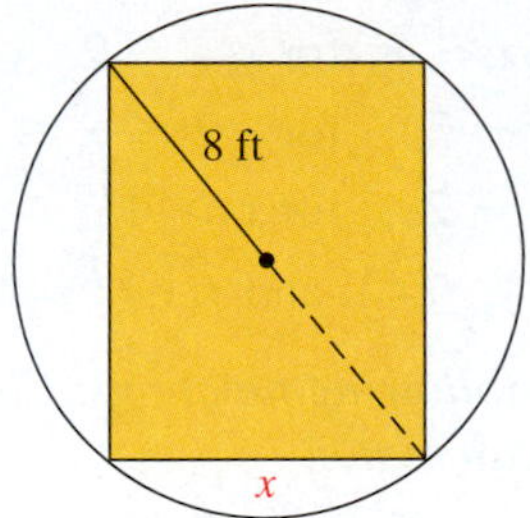

a) Express the area of the rectangle as a function of x.
b) Find the domain of the function.
c) Graph the function with a graphing calculator.
d) What dimensions maximize the area of the rectangle?

46. *Cost of Material.* A rectangular box with volume 320 ft^3 is built with a square base and top. The cost is $\$1.50/\text{ft}^2$ for the bottom, $\$2.50/\text{ft}^2$ for the sides, and $\$1/\text{ft}^2$ for the top. Let $x =$ the length of the base, in feet.

a) Express the cost of the box as a function of x.
b) Find the domain of the function.
c) Graph the function with a graphing calculator.
d) What dimensions minimize the cost of the box?

For each piecewise function, find the specified function values.

47. $g(x) = \begin{cases} x + 4, & \text{for } x \leq 1, \\ 8 - x, & \text{for } x > 1 \end{cases}$

$g(-4)$, $g(0)$, $g(1)$, and $g(3)$

48. $f(x) = \begin{cases} 3, & \text{for } x \le -2, \\ \frac{1}{2}x + 6, & \text{for } x > -2 \end{cases}$

$f(-5), f(-2), f(0)$, and $f(2)$

49. $h(x) = \begin{cases} -3x - 18, & \text{for } x < -5, \\ 1, & \text{for } -5 \le x < 1, \\ x + 2, & \text{for } x \ge 1 \end{cases}$

$h(-6), h(0), h(1)$, and $h(4)$

50. $f(x) = \begin{cases} -5x - 8, & \text{for } x < -2, \\ \frac{1}{2}x + 5, & \text{for } -2 \le x \le 4, \\ 10 - 2x, & \text{for } x > 4 \end{cases}$

$f(-4), f(-2), f(4)$, and $f(6)$

Graph each of the following functions. Check your results using a graphing calculator.

51. $f(x) = \begin{cases} \frac{1}{2}x, & \text{for } x < 0, \\ x + 3, & \text{for } x \ge 0 \end{cases}$

52. $f(x) = \begin{cases} -\frac{1}{3}x + 2, & \text{for } x \le 0, \\ x - 5, & \text{for } x > 0 \end{cases}$

53. $f(x) = \begin{cases} -\frac{3}{4}x + 2, & \text{for } x < 4, \\ -1, & \text{for } x \ge 4 \end{cases}$

54. $h(x) = \begin{cases} 2x - 1, & \text{for } x < 2, \\ 2 - x, & \text{for } x \ge 2 \end{cases}$

55. $f(x) = \begin{cases} x + 1, & \text{for } x \le -3, \\ -1, & \text{for } -3 < x < 4, \\ \frac{1}{2}x, & \text{for } x \ge 4 \end{cases}$

56. $f(x) = \begin{cases} 4, & \text{for } x \le -2, \\ x + 1, & \text{for } -2 < x < 3, \\ -x, & \text{for } x \ge 3 \end{cases}$

57. $g(x) = \begin{cases} \frac{1}{2}x - 1, & \text{for } x < 0, \\ 3, & \text{for } 0 \le x \le 1, \\ -2x, & \text{for } x > 1 \end{cases}$

58. $f(x) = \begin{cases} \dfrac{x^2 - 9}{x + 3}, & \text{for } x \ne -3, \\ 5, & \text{for } x = -3 \end{cases}$

59. $f(x) = \begin{cases} 2, & \text{for } x = 5, \\ \dfrac{x^2 - 25}{x - 5}, & \text{for } x \ne 5 \end{cases}$

60. $f(x) = \begin{cases} \dfrac{x^2 + 3x + 2}{x + 1}, & \text{for } x \ne -1, \\ 7, & \text{for } x = -1 \end{cases}$

61. $f(x) = [\![x]\!]$

62. $f(x) = 2[\![x]\!]$

63. $g(x) = 1 + [\![x]\!]$

64. $h(x) = \frac{1}{2}[\![x]\!] - 2$

65.–70. Find the domain and the range of each of the functions defined in Exercises 51–56.

Determine the domain and the range of the piecewise function. Then write an equation for the function.

71.

72.

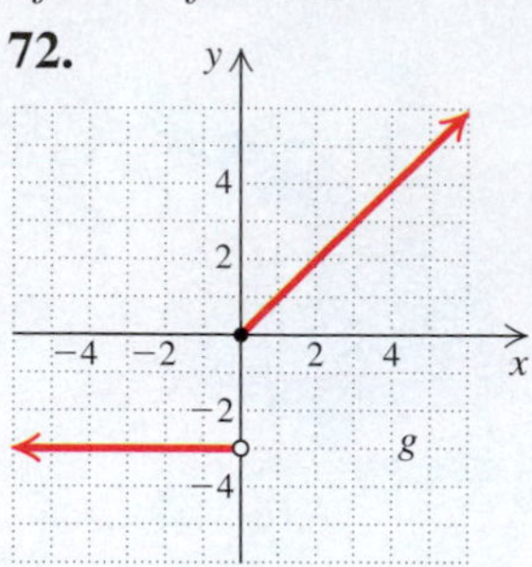

73.

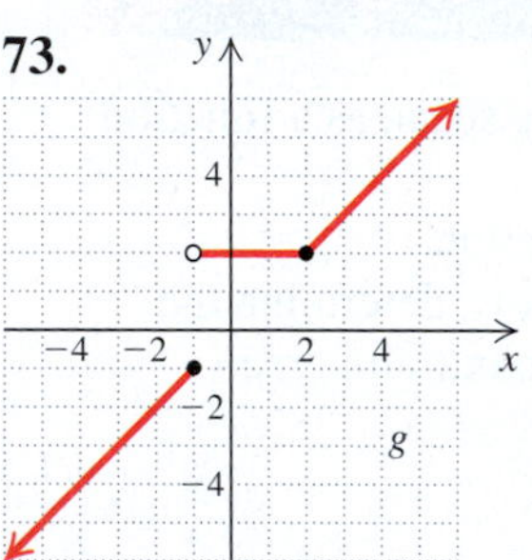

74.

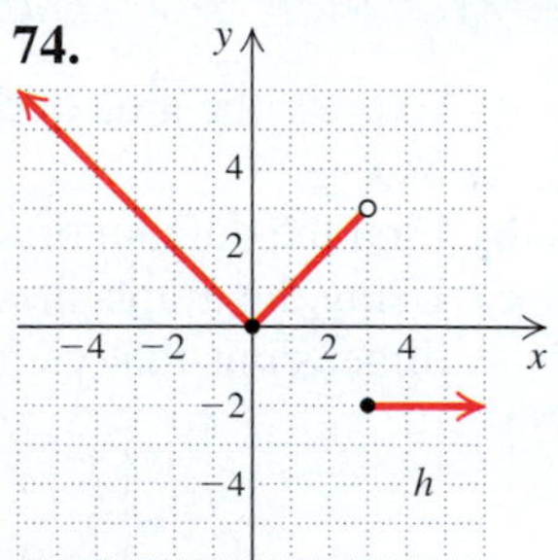

75.

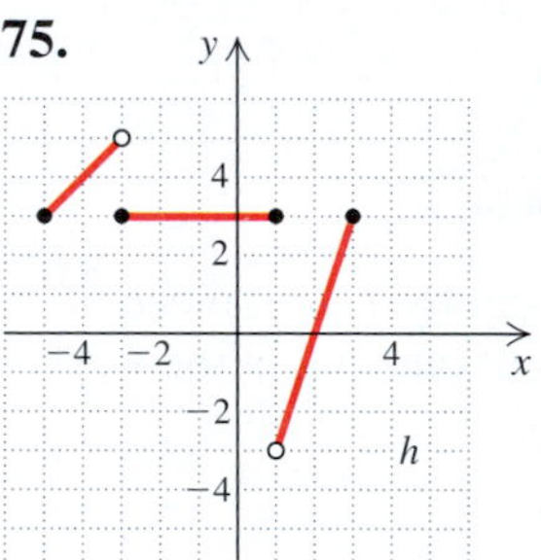

76.

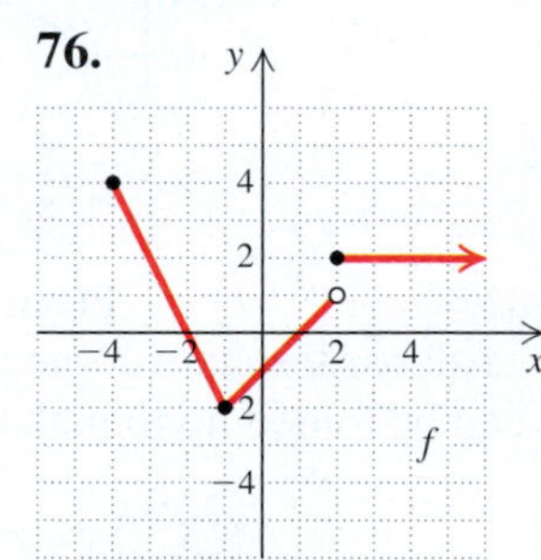

Skill Maintenance

77. Given $f(x) = 5x^2 - 7$, find each of the following. **[1.2]**

a) $f(-3)$
b) $f(3)$
c) $f(a)$
d) $f(-a)$

78. Given $f(x) = 4x^3 - 5x$, find each of the following. **[1.2]**

a) $f(2)$
b) $f(-2)$
c) $f(a)$
d) $f(-a)$

79. Write an equation of the line perpendicular to the graph of the line $8x - y = 10$ and containing the point $(-1, 1)$. **[1.4]**

80. Find the slope and the y-intercept of the line with equation $2x - 9y + 1 = 0$. **[1.4]**

› Synthesis

Using a graphing calculator, estimate the interval on which the function is increasing or decreasing and any relative maxima or minima.

81. $f(x) = x^4 + 4x^3 - 36x^2 - 160x + 400$

82. $f(x) = 3.22x^5 - 5.208x^3 - 11$

83. *Parking Costs.* A parking garage charges \$3 for up to (but not including) 1 hr of parking, \$6 for up to 2 hr of parking, \$9 for up to 3 hr of parking, and so on. Let $C(t)$ = the cost of parking for t hours.

a) Graph the function.

b) Write an equation for $C(t)$ using the greatest integer notation $[\![t]\!]$.

84. If $[\![x + 2]\!] = -3$, what are the possible inputs for x?

85. If $([\![x]\!])^2 = 25$, what are the possible inputs for x?

86. *Minimizing Power Line Costs.* A power line is constructed from a power station at point A to an island at point I, which is 1 mi directly out in the water from a point B on the shore. Point B is 4 mi downshore from the power station at A. It costs \$5000 per mile to lay the power line under water and \$3000 per mile to lay the power line under ground. The line comes to the shore at point S downshore from A. Let x = the distance from B to S.

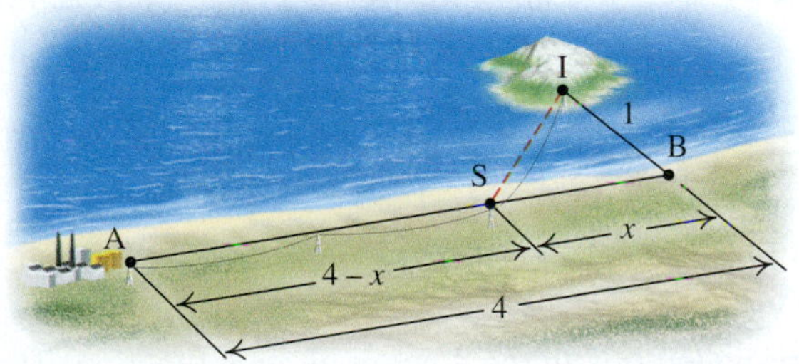

a) Express the cost C of laying the line as a function of x.

b) At what distance x from point B should the line come to shore in order to minimize cost?

87. *Volume of an Inscribed Cylinder.* A right circular cylinder of height h and radius r is inscribed in a right circular cone with a height of 10 ft and a base with radius 6 ft.

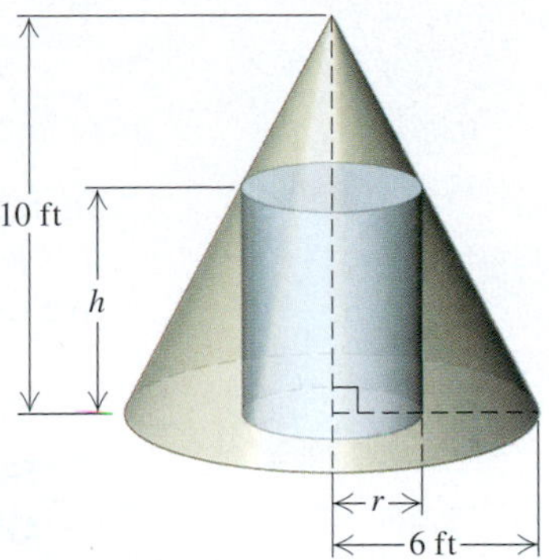

a) Express the height h of the cylinder as a function of r.

b) Express the volume V of the cylinder as a function of r.

c) Express the volume V of the cylinder as a function of h.

2.2 The Algebra of Functions

› Find the sum, the difference, the product, and the quotient of two functions, and determine the domains of the resulting functions.

› Find the difference quotient for a function.

› The Algebra of Functions: Sums, Differences, Products, and Quotients

We now use addition, subtraction, multiplication, and division to combine functions and obtain new functions.

Consider the following two functions f and g:

$$f(x) = x + 2 \quad \text{and} \quad g(x) = x^2 + 1.$$

Since $f(3) = 3 + 2 = 5$ and $g(3) = 3^2 + 1 = 10$, we have

$$f(3) + g(3) = 5 + 10 = 15,$$
$$f(3) - g(3) = 5 - 10 = -5,$$
$$f(3) \cdot g(3) = 5 \cdot 10 = 50,$$

and

$$\frac{f(3)}{g(3)} = \frac{5}{10} = \frac{1}{2}.$$

In fact, so long as x is in the domain of *both* f and g, we can easily compute $f(x) + g(x)$, $f(x) - g(x)$, $f(x) \cdot g(x)$, and, assuming $g(x) \neq 0$, $f(x)/g(x)$. We use the notation shown below.

SUMS, DIFFERENCES, PRODUCTS, AND QUOTIENTS OF FUNCTIONS

If f and g are functions and x is in the domain of each function, then:

$$(f + g)(x) = f(x) + g(x),$$
$$(f - g)(x) = f(x) - g(x),$$
$$(fg)(x) = f(x) \cdot g(x),$$
$$(f/g)(x) = f(x)/g(x), \text{ provided } g(x) \neq 0.$$

EXAMPLE 1 Given that $f(x) = x + 1$ and $g(x) = \sqrt{x + 3}$, find each of the following.

a) $(f + g)(x)$ **b)** $(f + g)(6)$ **c)** $(f + g)(-4)$

Solution

a) $(f + g)(x) = f(x) + g(x)$
$= x + 1 + \sqrt{x + 3}$ **This cannot be simplified.**

b) We can find $(f + g)(6)$ provided 6 is in the domain of *each* function. The domain of f is all real numbers. The domain of g is all real numbers x for which $x + 3 \geq 0$, or $x \geq -3$. This is the interval $[-3, \infty)$. We see that 6 is in both domains, so we have

$$f(6) = 6 + 1 = 7, \qquad g(6) = \sqrt{6 + 3} = \sqrt{9} = 3,$$
$$(f + g)(6) = f(6) + g(6) = 7 + 3 = 10.$$

Another method is to use the formula found in part (a):

$$(f + g)(6) = 6 + 1 + \sqrt{6 + 3} = 7 + \sqrt{9} = 7 + 3 = 10.$$

We can check our work using a graphing calculator by entering

$$y_1 = x + 1, \qquad y_2 = \sqrt{x + 3}, \quad \text{and} \quad y_3 = y_1 + y_2$$

on the $y =$ screen. Then on the home screen, we find Y3(6).

```
Plot1  Plot2  Plot3
\Y1= X+1
\Y2= √(X+3)
\Y3= Y1+Y2
\Y4=
\Y5=
\Y6=
\Y7=
```

```
Y3(6)
                10
```

c) To find $(f + g)(-4)$, we must first determine whether -4 is in the domain of each function. We note that -4 is not in the domain of g, $[-3, \infty)$. That is, $\sqrt{-4 + 3}$ is not a real number. Thus, $(f + g)(-4)$ does not exist.

Now Try Exercise 15.

It is useful to view the concept of the sum of two functions graphically. In the following graph, we see the graphs of two functions f and g and their sum, $f + g$. Consider finding $(f + g)(4)$, or $f(4) + g(4)$. We can locate $g(4)$ on the graph of g and measure it. Then we add that length on top of $f(4)$ on the graph of f. The sum gives us $(f + g)(4)$.

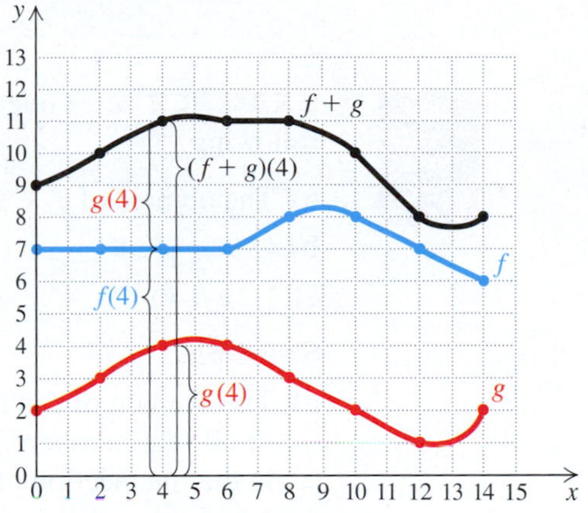

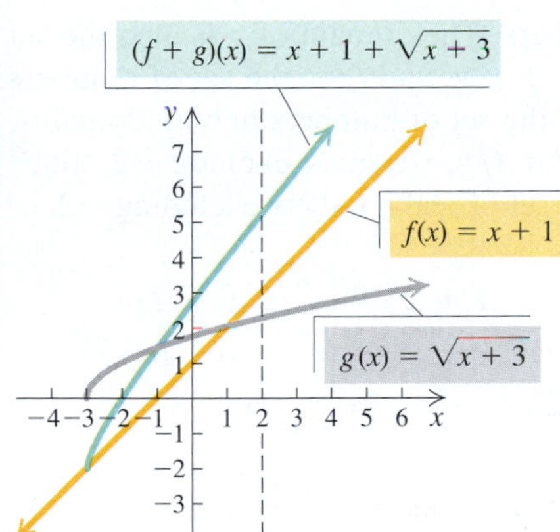

With this in mind, let's view Example 1 from a graphical perspective. Let's look at the graphs of

$$f(x) = x + 1, \qquad g(x) = \sqrt{x + 3}, \quad \text{and}$$
$$(f + g)(x) = x + 1 + \sqrt{x + 3}.$$

See the graph at left. Note that the domain of f is the set of all real numbers. The domain of g is $[-3, \infty)$. The domain of $f + g$ is the set of numbers in the intersection of the domains. This is the set of numbers in both domains.

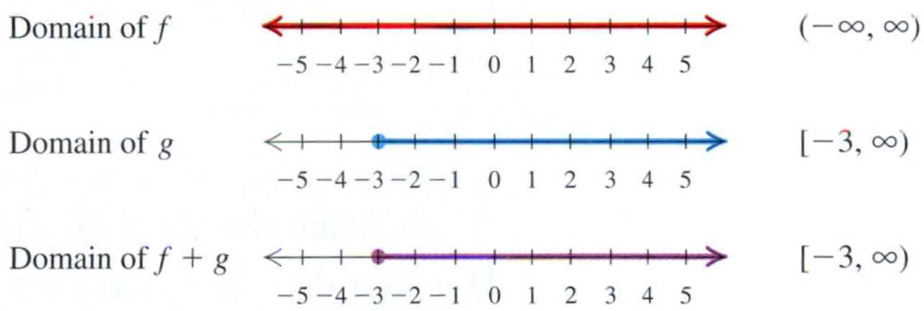

Thus the domain of $f + g$ is $[-3, \infty)$.

We can confirm that the y-coordinates of the graph of $(f + g)(x)$ are the sums of the corresponding y-coordinates of the graphs of $f(x)$ and $g(x)$. Here we confirm it for $x = 2$.

$$f(x) = x + 1 \qquad\qquad g(x) = \sqrt{x + 3}$$
$$f(2) = 2 + 1 = 3; \qquad g(2) = \sqrt{2 + 3} = \sqrt{5};$$

$$(f + g)(x) = x + 1 + \sqrt{x + 3}$$
$$(f + g)(2) = 2 + 1 + \sqrt{2 + 3}$$
$$= 3 + \sqrt{5} = f(2) + g(2).$$

DOMAIN

REVIEW SECTION 1.2.

Let's also examine the domains of $f - g$, fg, and f/g for the functions $f(x) = x + 1$ and $g(x) = \sqrt{x + 3}$ of Example 1. The domains of $f - g$ and fg are the same as the domain of $f + g$, $[-3, \infty)$, because numbers in this interval are in the domains of *both* functions. For f/g, $g(x)$ cannot be 0. Since $\sqrt{x + 3} = 0$ when $x = -3$, we must exclude -3 so the domain of f/g is $(-3, \infty)$.

DOMAINS OF $f + g$, $f - g$, fg, AND f/g

If f and g are functions, then the domain of the functions $f + g$, $f - g$, and fg is the intersection of the domain of f and the domain of g. The domain of f/g is also the intersection of the domains of f and g with the exclusion of any x-values for which $g(x) = 0$.

Just in Time 11, 12

EXAMPLE 2 Given that $f(x) = x^2 - 4$ and $g(x) = x + 2$, find each of the following.

a) The domain of $f + g$, $f - g$, fg, and f/g

b) $(f + g)(x)$ **c)** $(f - g)(x)$ **d)** $(fg)(x)$

e) $(f/g)(x)$ **f)** $(gg)(x)$

Solution

a) The domain of f is the set of all real numbers. The domain of g is also the set of all real numbers. The domain of $f + g$, $f - g$, and fg is the set of numbers in the intersection of the domains—that is, the set of numbers in both domains, which is again the set of real numbers. For f/g, we must exclude -2, since $g(-2) = 0$. Thus the domain of f/g is the set of real numbers excluding -2, or $(-\infty, -2) \cup (-2, \infty)$.

b) $(f + g)(x) = f(x) + g(x) = (x^2 - 4) + (x + 2) = x^2 + x - 2$

c) $(f - g)(x) = f(x) - g(x) = (x^2 - 4) - (x + 2) = x^2 - x - 6$

d) $(fg)(x) = f(x) \cdot g(x) = (x^2 - 4)(x + 2) = x^3 + 2x^2 - 4x - 8$

e) $(f/g)(x) = \dfrac{f(x)}{g(x)} = \dfrac{x^2 - 4}{x + 2}$ Note that $g(x) = 0$ when $x = -2$, so $(f/g)(x)$ is *not defined* when $x = -2$.

$= \dfrac{(x + 2)(x - 2)}{x + 2}$ Factoring

$= x - 2$ Removing a factor of 1: $\dfrac{x + 2}{x + 2} = 1$

Thus, $(f/g)(x) = x - 2$ with the added stipulation that $x \neq -2$ since -2 is not in the domain of $(f/g)(x)$.

f) $(gg)(x) = g(x) \cdot g(x) = [g(x)]^2 = (x + 2)^2 = x^2 + 4x + 4$

Now Try Exercise 21.

Difference Quotients

In Section 1.3, we learned that the slope of a line can be considered as an average *rate of change*. Here let's consider a nonlinear function f and draw a line through two points $(x, f(x))$ and $(x + h, f(x + h))$ as shown at left.

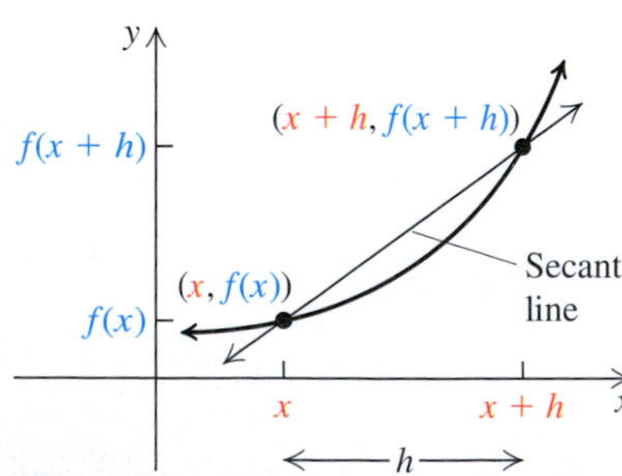

The slope of the line, called a **secant line**, is

$$\frac{f(x + h) - f(x)}{x + h - x},$$

which simplifies to

$$\frac{f(x + h) - f(x)}{h}.$$ Difference quotient

This ratio is called the **difference quotient**, or the **average rate of change**. In calculus, it is important to be able to find and simplify difference quotients.

EXAMPLE 3 For the function f given by $f(x) = 2x - 3$, find and simplify the difference quotient

$$\frac{f(x + h) - f(x)}{h}.$$

Solution

$$\frac{f(x + h) - f(x)}{h} = \frac{2(x + h) - 3 - (2x - 3)}{h} \quad \text{Substituting}$$

$$= \frac{2x + 2h - 3 - 2x + 3}{h} \quad \text{Removing parentheses}$$

$$= \frac{2h}{h} = 2 \quad \text{Simplifying}$$

Now Try Exercise 49.

Just in Time 24

EXAMPLE 4 For the function f given by $f(x) = \frac{1}{x}$, find and simplify the difference quotient

$$\frac{f(x + h) - f(x)}{h}.$$

Solution

$$\frac{f(x + h) - f(x)}{h} = \frac{\frac{1}{x + h} - \frac{1}{x}}{h} \quad \text{Substituting}$$

$$= \frac{\frac{1}{x + h} \cdot \frac{x}{x} - \frac{1}{x} \cdot \frac{x + h}{x + h}}{h} \quad \text{The LCD of } \frac{1}{x + h} \text{ and } \frac{1}{x} \text{ is } x(x + h).$$

$$= \frac{\frac{x}{x(x + h)} - \frac{x + h}{x(x + h)}}{h}$$

$$= \frac{\frac{x - (x + h)}{x(x + h)}}{h} \quad \text{Subtracting in the numerator}$$

$$= \frac{\frac{x - x - h}{x(x + h)}}{h} \quad \text{Removing parentheses}$$

$$= \frac{\frac{-h}{x(x + h)}}{h} \quad \text{Simplifying the numerator}$$

$$= \frac{-h}{x(x + h)} \cdot \frac{1}{h} \quad \text{Multiplying by the reciprocal of the divisor}$$

$$= \frac{-h \cdot 1}{x \cdot (x + h) \cdot h}$$

$$= \frac{-1 \cdot \cancel{h}}{x \cdot (x + h) \cdot \cancel{h}} \quad \text{Rewriting } -h \cdot 1 \text{ as } -1 \cdot h$$

$$= \frac{-1}{x(x + h)}, \text{ or } -\frac{1}{x(x + h)}$$

Now Try Exercise 55.

EXAMPLE 5 For the function f given by $f(x) = 2x^2 - x - 3$, find and simplify the difference quotient

$$\frac{f(x + h) - f(x)}{h}.$$

Solution We first find $f(x + h)$:

$$f(x + h) = 2(x + h)^2 - (x + h) - 3$$ Substituting $x + h$ for x in $f(x) = 2x^2 - x - 3$

$$= 2[x^2 + 2xh + h^2] - (x + h) - 3$$
$$= 2x^2 + 4xh + 2h^2 - x - h - 3.$$

Then

$$\frac{f(x + h) - f(x)}{h} = \frac{[2x^2 + 4xh + 2h^2 - x - h - 3] - [2x^2 - x - 3]}{h}$$
$$= \frac{2x^2 + 4xh + 2h^2 - x - h - 3 - 2x^2 + x + 3}{h}$$
$$= \frac{4xh + 2h^2 - h}{h}$$
$$= \frac{h(4x + 2h - 1)}{h \cdot 1} = \frac{h}{h} \cdot \frac{4x + 2h - 1}{1} = 4x + 2h - 1.$$

Now Try Exercise 63.

2.2 Exercise Set

Given that $f(x) = x^2 - 3$ and $g(x) = 2x + 1$, find each of the following, if it exists.

1. $(f + g)(5)$
2. $(fg)(0)$
3. $(f - g)(-1)$
4. $(fg)(2)$
5. $(f/g)\left(-\frac{1}{2}\right)$
6. $(f - g)(0)$
7. $(fg)\left(-\frac{1}{2}\right)$
8. $(f/g)(-\sqrt{3})$
9. $(g - f)(-1)$
10. $(g/f)\left(-\frac{1}{2}\right)$

Given that $h(x) = x + 4$ and $g(x) = \sqrt{x - 1}$, find each of the following, if it exists.

11. $(h - g)(-4)$
12. $(gh)(10)$
13. $(g/h)(1)$
14. $(h/g)(1)$
15. $(g + h)(1)$
16. $(hg)(3)$

For each pair of functions in Exercises 17–34:

a) *Find the domain of f, g, $f + g$, $f - g$, fg, ff, f/g, and g/f.*
b) *Find $(f + g)(x)$, $(f - g)(x)$, $(fg)(x)$, $(ff)(x)$, $(f/g)(x)$, and $(g/f)(x)$.*

17. $f(x) = 2x + 3$, $g(x) = 3 - 5x$
18. $f(x) = -x + 1$, $g(x) = 4x - 2$
19. $f(x) = x - 3$, $g(x) = \sqrt{x + 4}$
20. $f(x) = x + 2$, $g(x) = \sqrt{x - 1}$
21. $f(x) = 2x - 1$, $g(x) = -2x^2$
22. $f(x) = x^2 - 1$, $g(x) = 2x + 5$
23. $f(x) = \sqrt{x - 3}$, $g(x) = \sqrt{x + 3}$
24. $f(x) = \sqrt{x}$, $g(x) = \sqrt{2 - x}$
25. $f(x) = x + 1$, $g(x) = |x|$
26. $f(x) = 4|x|$, $g(x) = 1 - x$
27. $f(x) = x^3$, $g(x) = 2x^2 + 5x - 3$
28. $f(x) = x^2 - 4$, $g(x) = x^3$
29. $f(x) = \frac{4}{x + 1}$, $g(x) = \frac{1}{6 - x}$
30. $f(x) = 2x^2$, $g(x) = \frac{2}{x - 5}$
31. $f(x) = \frac{1}{x}$, $g(x) = x - 3$

32. $f(x) = \sqrt{x + 6},\ g(x) = \frac{1}{x}$

33. $f(x) = \frac{3}{x - 2},\ g(x) = \sqrt{x - 1}$

34. $f(x) = \frac{2}{4 - x},\ g(x) = \frac{5}{x - 1}$

In Exercises 35–40, consider the functions F and G as shown in the following graph.

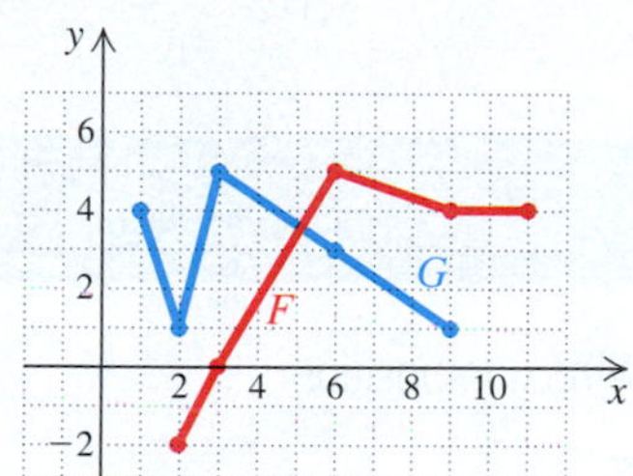

35. Find the domain of F, the domain of G, and the domain of $F + G$.

36. Find the domain of $F - G$, FG, and F/G.

37. Find the domain of G/F.

38. Graph $F + G$.

39. Graph $G - F$.

40. Graph $F - G$.

In Exercises 41–46, consider the functions F and G as shown in the following graph.

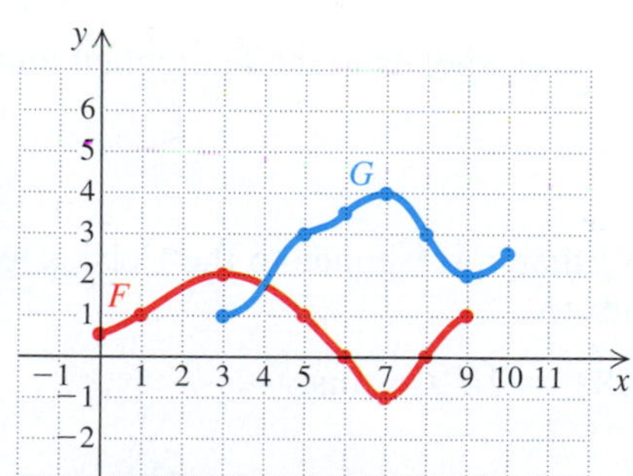

41. Find the domain of F, the domain of G, and the domain of $F + G$.

42. Find the domain of $F - G$, FG, and F/G.

43. Find the domain of G/F.

44. Graph $F + G$.

45. Graph $G - F$.

46. Graph $F - G$.

47. *Total Cost, Revenue, and Profit.* In economics, functions that involve revenue, cost, and profit are used. For example, suppose that $R(x)$ and $C(x)$ denote the total revenue and the total cost, respectively, of producing a new grocery cart for Ogata Wholesalers. Then the difference

$$P(x) = R(x) - C(x)$$

represents the total profit for producing x carts. Given

$$R(x) = 60x - 0.4x^2 \quad \text{and} \quad C(x) = 3x + 13,$$

find each of the following.

a) $P(x)$
b) $R(100)$, $C(100)$, and $P(100)$
c) Using a graphing calculator, graph the three functions in the viewing window $[0, 160, 0, 3000]$.

48. *Total Cost, Revenue, and Profit.* Given that

$$R(x) = 200x - x^2 \quad \text{and} \quad C(x) = 5000 + 8x$$

for a new tablet produced by Visual Communications, find each of the following. (See Exercise 47.)

a) $P(x)$
b) $R(175)$, $C(175)$, and $P(175)$
c) Using a graphing calculator, graph the three functions in the viewing window $[0, 200, 0, 10{,}000]$.

For each function f, construct and simplify the difference quotient

$$\frac{f(x + h) - f(x)}{h}.$$

49. $f(x) = 3x - 5$

50. $f(x) = 4x - 1$

51. $f(x) = 6x + 2$

52. $f(x) = 5x + 3$

53. $f(x) = \frac{1}{3}x + 1$

54. $f(x) = -\frac{1}{2}x + 7$

55. $f(x) = \frac{1}{3x}$

56. $f(x) = \frac{1}{2x}$

57. $f(x) = -\frac{1}{4x}$

58. $f(x) = -\frac{1}{x}$

59. $f(x) = x^2 + 1$

60. $f(x) = x^2 - 3$

61. $f(x) = 4 - x^2$

62. $f(x) = 2 - x^2$

63. $f(x) = 3x^2 - 2x + 1$

64. $f(x) = 5x^2 + 4x$

65. $f(x) = 4 + 5|x|$

66. $f(x) = 2|x| + 3x$

67. $f(x) = x^3$

68. $f(x) = x^3 - 2x$

69. $f(x) = \frac{x - 4}{x + 3}$

70. $f(x) = \frac{x}{2 - x}$

Skill Maintenance

Graph the equation. **[1.1], [1.3]**

71. $y = 3x - 1$

72. $2x + y = 4$

73. $x - 3y = 3$

74. $y = x^2 + 1$

Synthesis

75. Write equations for two functions f and g such that the domain of $f - g$ is

$$\{x \mid x \neq -7 \text{ and } x \neq 3\}.$$

76. For functions h and f, find the domain of $h + f$, $h - f$, hf, and h/f if

$h = \{(-4, 13), (-1, 7), (0, 5), (\frac{5}{2}, 0), (3, -5)\}$, and

$f = \{(-4, -7), (-2, -5), (0, -3), (3, 0), (5, 2), (9, 6)\}$.

77. Find the domain of $(h/g)(x)$ given that

$$h(x) = \frac{5x}{3x - 7} \quad \text{and} \quad g(x) = \frac{x^4 - 1}{5x - 15}.$$

2.3 The Composition of Functions

- Find the composition of two functions and the domain of the composition.
- Decompose a function as a composition of two functions.

The Composition of Functions

In real-world situations, it is not uncommon for the output of a function to depend on some input that is itself an output of another function. For instance, the amount that a person pays as state income tax usually depends on the amount of adjusted gross income on the person's federal tax return, which, in turn, depends on his or her annual earnings. Such functions are called **composite functions**.

To see how composite functions work, suppose that a chemistry student needs a formula to convert Fahrenheit temperatures to Kelvin units. The formula

$$c(t) = \tfrac{5}{9}(t - 32)$$

gives the Celsius temperature $c(t)$ that corresponds to the Fahrenheit temperature t. The formula

$$k(c(t)) = c(t) + 273$$

gives the Kelvin temperature $k(c(t))$ that corresponds to the Celsius temperature $c(t)$. Thus, 50° Fahrenheit corresponds to

$$c(50) = \tfrac{5}{9}(50 - 32) = \tfrac{5}{9}(18) = 10° \text{ Celsius},$$

and 10° Celsius corresponds to

$$k(c(50)) = k(10) = 10 + 273 = 283 \text{ Kelvin units},$$

which is usually written 283K. We see that 50° Fahrenheit is the same as 283K. This two-step procedure can be used to convert any Fahrenheit temperature to Kelvin units.

	°F Fahrenheit	°C Celsius	K Kelvin
Boiling point of water	212°	100°	373 K
	50° →	10° →	283 K
Freezing point of water	32°	0°	273 K
Absolute zero	−460°	−273°	0 K

$y_1 = \frac{5}{9}(x - 32)$, $y_2 = y_1 + 273$

X	Y1	Y2
50	10	283
59	15	288
68	20	293
77	25	298
86	30	303
95	35	308
104	40	313

X = 50

In the table shown at left, we use a graphing calculator to convert Fahrenheit temperatures x to Celsius temperatures y_1, using $y_1 = \frac{5}{9}(x - 32)$. We also convert Celsius temperatures to Kelvin units y_2, using $y_2 = y_1 + 273$.

A student making numerous conversions might look for a formula that converts directly from Fahrenheit to Kelvin. Such a formula can be found by substitution:

$$\begin{aligned} k(c(t)) &= c(t) + 273 \\ &= \frac{5}{9}(t - 32) + 273 && \text{Substituting } \frac{5}{9}(t - 32) \text{ for } c(t) \\ &= \frac{5}{9}t - \frac{160}{9} + 273 \\ &= \frac{5}{9}t - \frac{160}{9} + \frac{2457}{9} \\ &= \frac{5t + 2297}{9}. && \text{Simplifying} \end{aligned}$$

We can show on a graphing calculator that the same values that appear in the table for y_2 will appear when y_2 is entered as

$$y_2 = \frac{5x + 2297}{9}.$$

Since the formula found above expresses the Kelvin temperature as a new function K of the Fahrenheit temperature t, we can write

$$K(t) = \frac{5t + 2297}{9},$$

where $K(t)$ is the Kelvin temperature corresponding to the Fahrenheit temperature, t. Here we have $K(t) = k(c(t))$. The new function K is called the **composition** of k and c and can be denoted $k \circ c$ (read "k composed with c," "the composition of k and c," or "k circle c").

COMPOSITION OF FUNCTIONS

The **composite function** $f \circ g$, the **composition** of f and g, is defined as

$$(f \circ g)(x) = f(g(x)),$$

where x is in the domain of g and $g(x)$ is in the domain of f.

EXAMPLE 1 Given that $f(x) = 2x - 5$ and $g(x) = x^2 - 3x + 8$, find each of the following.

a) $(f \circ g)(x)$ and $(g \circ f)(x)$ **b)** $(f \circ g)(7)$ and $(g \circ f)(7)$
c) $(g \circ g)(1)$ **d)** $(f \circ f)(x)$

Solution Consider each function separately:

$$f(x) = 2x - 5 \qquad \text{This function multiplies each input by 2 and then subtracts 5.}$$

and

$$g(x) = x^2 - 3x + 8. \qquad \text{This function squares an input, subtracts three times the input from the result, and then adds 8.}$$

a) To find $(f \circ g)(x)$, we substitute $g(x)$ for x in the equation for $f(x)$:

$$(f \circ g)(x) = f(g(x)) = f(x^2 - 3x + 8)$$

$x^2 - 3x + 8$ is the input for f.

$$= 2(x^2 - 3x + 8) - 5$$

f multiplies the input by 2 and then subtracts 5.

$$= 2x^2 - 6x + 16 - 5$$
$$= 2x^2 - 6x + 11.$$

To find $(g \circ f)(x)$, we substitute $f(x)$ for x in the equation for $g(x)$:

$$(g \circ f)(x) = g(f(x)) = g(2x - 5)$$

$2x - 5$ is the input for g.

$$= (2x - 5)^2 - 3(2x - 5) + 8$$

g squares the input, subtracts three times the input, and then adds 8.

$$= 4x^2 - 20x + 25 - 6x + 15 + 8$$
$$= 4x^2 - 26x + 48.$$

b) To find $(f \circ g)(7)$, we first find $g(7)$. Then we use $g(7)$ as an input for f:

$$(f \circ g)(7) = f(g(7)) = f(7^2 - 3 \cdot 7 + 8)$$
$$= f(36) = 2 \cdot 36 - 5$$
$$= 72 - 5 = 67.$$

To find $(g \circ f)(7)$, we first find $f(7)$. Then we use $f(7)$ as an input for g:

$$(g \circ f)(7) = g(f(7)) = g(2 \cdot 7 - 5)$$
$$= g(9) = 9^2 - 3 \cdot 9 + 8$$
$$= 81 - 27 + 8 = 62.$$

We could also find $(f \circ g)(7)$ and $(g \circ f)(7)$ by substituting 7 for x in the equations that we found in part (a):

$$(f \circ g)(x) = 2x^2 - 6x + 11$$
$$(f \circ g)(7) = 2 \cdot 7^2 - 6 \cdot 7 + 11 = 98 - 42 + 11 = 67;$$
$$(g \circ f)(x) = 4x^2 - 26x + 48$$
$$(g \circ f)(7) = 4 \cdot 7^2 - 26 \cdot 7 + 48 = 196 - 182 + 48 = 62.$$

$y_1 = 2x - 5,\ y_2 = x^2 - 3x + 8$

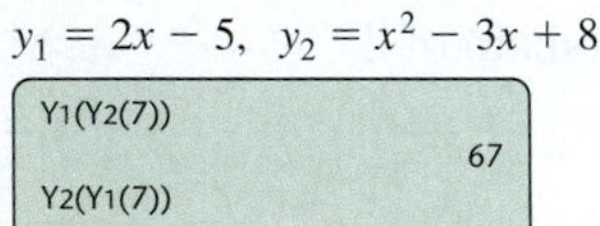

We can check our work using a graphing calculator. On the equation-editor screen, we enter $f(x)$ as $y_1 = 2x - 5$ and $g(x)$ as $y_2 = x^2 - 3x + 8$. Then, on the home screen, we find $(f \circ g)(7)$ and $(g \circ f)(7)$ using the function notations Y1(Y2(7)) and Y2(Y1(7)), respectively.

c) $(g \circ g)(1) = g(g(1)) = g(1^2 - 3 \cdot 1 + 8)$

$$= g(1 - 3 + 8) = g(6)$$
$$= 6^2 - 3 \cdot 6 + 8$$
$$= 36 - 18 + 8 = 26$$

d) $(f \circ f)(x) = f(f(x)) = f(2x - 5)$

$$= 2(2x - 5) - 5$$
$$= 4x - 10 - 5 = 4x - 15$$

Now Try Exercises 1 and 15.

$y_1 = (f \circ g)(x) = 2x^2 - 6x + 11,$
$y_2 = (g \circ f)(x) = 4x^2 - 26x + 48$

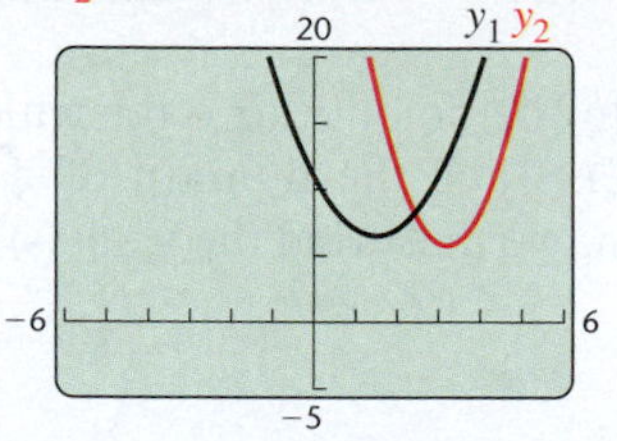

Example 1 illustrates that, as a rule, $(f \circ g)(x) \neq (g \circ f)(x)$. We can see this graphically, as shown in the graphs at left.

EXAMPLE 2 Given that $f(x) = \sqrt{x}$ and $g(x) = x - 3$:

a) Find $f \circ g$ and $g \circ f$.

b) Find the domain of $f \circ g$ and the domain of $g \circ f$.

Solution

a) $(f \circ g)(x) = f(g(x)) = f(x - 3) = \sqrt{x - 3}$
$(g \circ f)(x) = g(f(x)) = g(\sqrt{x}) = \sqrt{x} - 3$

b) The domain of $f(x)$ is $\{x | x \geq 0\}$, or $[0, \infty)$. Any real number can be an input for $g(x)$, so the domain of $g(x)$ is $(-\infty, \infty)$.

The domain of $f \circ g$ consists of the values of x in the domain of g, $(-\infty, \infty)$, for which $g(x)$ is nonnegative. (Recall that the inputs of $f(x)$ must be nonnegative.) Thus we have

$$g(x) \geq 0$$
$$x - 3 \geq 0 \qquad \text{Substituting } x - 3 \text{ for } g(x)$$
$$x \geq 3.$$

We see that the domain of $f \circ g$ is $\{x | x \geq 3\}$, or $[3, \infty)$.

We can also find the domain of $f \circ g$ by examining the composite function itself, $(f \circ g)(x) = \sqrt{x - 3}$, keeping in mind any restrictions on the domain of g. Since any real number can be an input for g, the only restriction on $f \circ g$ is that the radicand must be nonnegative. We have

$$x - 3 \geq 0$$
$$x \geq 3.$$

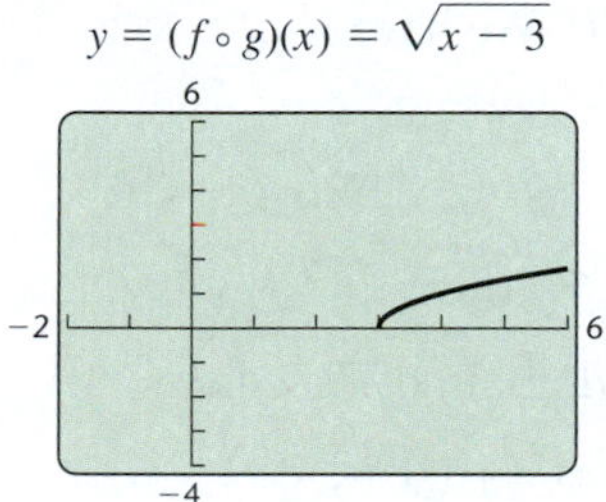
FIGURE 1.

Again, we see that the domain of $f \circ g$ is $\{x | x \geq 3\}$, or $[3, \infty)$. The graph in Fig. 1 confirms this.

The domain of $g \circ f$ consists of the values of x in the domain of f, $[0, \infty)$, for which $g(x)$ is defined. Since g can accept *any* real number as an input, any output from f is acceptable, so the entire domain of f is the domain of $g \circ f$. That is, the domain of $g \circ f$ is $\{x | x \geq 0\}$, or $[0, \infty)$.

We can also examine the composite function itself to find its domain, keeping in mind any restrictions on the domain of f. First, recall that the domain of f is $\{x | x \geq 0\}$, or $[0, \infty)$. Then consider $(g \circ f)(x) = \sqrt{x} - 3$. The radicand cannot be negative, so we have $x \geq 0$. As above, we see that the domain of $g \circ f$ is the domain of f, $\{x | x \geq 0\}$, or $[0, \infty)$. The graph in Fig. 2 confirms this.

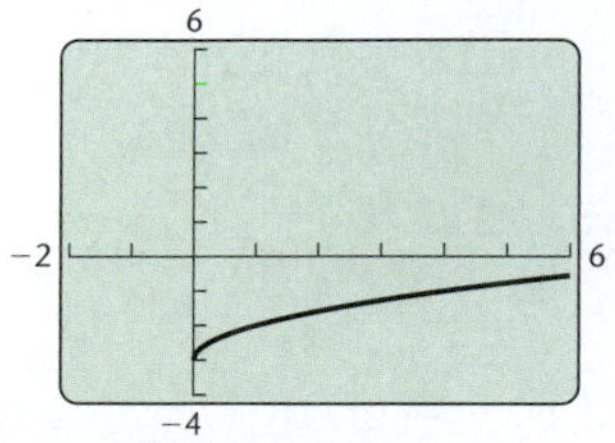

$y = (g \circ f)(x) = \sqrt{x} - 3$

FIGURE 2.

Now Try Exercise 27.

Just in Time 24

EXAMPLE 3 Given that $f(x) = \dfrac{1}{x - 2}$ and $g(x) = \dfrac{5}{x}$, find $f \circ g$ and $g \circ f$ and the domain of each.

Solution We have

$$(f \circ g)(x) = f(g(x)) = f\left(\frac{5}{x}\right) = \frac{1}{\frac{5}{x} - 2} = \frac{1}{\frac{5 - 2x}{x}} = \frac{x}{5 - 2x};$$

$$(g \circ f)(x) = g(f(x)) = g\left(\frac{1}{x - 2}\right) = \frac{5}{\frac{1}{x - 2}} = 5(x - 2).$$

Values of x that make the denominator 0 are not in the domains of these functions. Since $x - 2 = 0$ when $x = 2$, the domain of f is $\{x|x \neq 2\}$. The denominator of g is x, so the domain of g is $\{x|x \neq 0\}$.

The domain of $f \circ g$ consists of the values of x in the domain of g for which $g(x) \neq 2$. (Recall that 2 cannot be an input of f.) Since the domain of g is $\{x|x \neq 0\}$, 0 is not in the domain of $f \circ g$. In addition, we must find the value(s) of x for which $g(x) = 2$. We have

$$g(x) = 2$$

$$\frac{5}{x} = 2 \qquad \textbf{Substituting } \frac{5}{x} \textbf{ for } g(x)$$

$$5 = 2x$$

$$\frac{5}{2} = x.$$

This tells us that $\frac{5}{2}$ is also *not* in the domain of $f \circ g$. Then the domain of $f \circ g$ is

$$\{x|x \neq 0 \text{ and } x \neq \tfrac{5}{2}\}, \quad \text{or} \quad (-\infty, 0) \cup (0, \tfrac{5}{2}) \cup (\tfrac{5}{2}, \infty).$$

We can also examine the composite function $f \circ g$ to find its domain, keeping in mind any restrictions on the domain of g. First, recall that 0 is not in the domain of g, so it cannot be in the domain of $(f \circ g)(x) = x/(5 - 2x)$. We must also exclude the value(s) of x for which the denominator of $f \circ g$ is 0. We have

$$5 - 2x = 0$$

$$5 = 2x$$

$$\tfrac{5}{2} = x.$$

Again, we see that $\frac{5}{2}$ is also not in the domain, so the domain of $f \circ g$ is

$$\{x|x \neq 0 \text{ and } x \neq \tfrac{5}{2}\}, \quad \text{or} \quad (-\infty, 0) \cup (0, \tfrac{5}{2}) \cup (\tfrac{5}{2}, \infty).$$

The domain of $g \circ f$ consists of the values of x in the domain of f for which $f(x) \neq 0$. (Recall that 0 cannot be an input of g.) The domain of f is $\{x|x \neq 2\}$, so 2 is not in the domain of $g \circ f$. Next, we determine whether there are values of x for which $f(x) = 0$:

$$f(x) = 0$$

$$\frac{1}{x - 2} = 0 \qquad \textbf{Substituting } \frac{1}{x - 2} \textbf{ for } f(x)$$

$$(x - 2) \cdot \frac{1}{x - 2} = (x - 2) \cdot 0 \qquad \textbf{Multiplying by } x - 2$$

$$1 = 0. \qquad \textbf{False equation}$$

We see that there are no values of x for which $f(x) = 0$, so there are no additional restrictions on the domain of $g \circ f$. Thus the domain of $g \circ f$ is

$$\{x|x \neq 2\}, \quad \text{or} \quad (-\infty, 2) \cup (2, \infty).$$

We can also examine $g \circ f$ to find its domain. First, recall that 2 is not in the domain of f, so it cannot be in the domain of $(g \circ f)(x) = 5(x - 2)$. Since $5(x - 2)$ is defined for all real numbers, there are no additional restrictions on the domain of $g \circ f$. The domain is

$$\{x|x \neq 2\}, \quad \text{or} \quad (-\infty, 2) \cup (2, \infty).$$

Now Try Exercise 23.

Decomposing a Function as a Composition

In calculus, we often need to recognize how a function can be expressed as the composition of two functions. In this way, we are "decomposing" the function.

EXAMPLE 4 If $h(x) = (2x - 3)^5$, find $f(x)$ and $g(x)$ such that $h(x) = (f \circ g)(x)$.

Solution The function $h(x)$ raises $(2x - 3)$ to the 5th power. Two functions that can be used for the composition are

$$f(x) = x^5 \quad \text{and} \quad g(x) = 2x - 3.$$

We can check by forming the composition:

$$h(x) = (f \circ g)(x) = f(g(x)) = f(2x - 3) = (2x - 3)^5.$$

This is the most "obvious" solution. There can be other less obvious solutions. For example, if

$$f(x) = (x + 7)^5 \quad \text{and} \quad g(x) = 2x - 10,$$

then

$$\begin{aligned} h(x) = (f \circ g)(x) &= f(g(x)) \\ &= f(2x - 10) \\ &= [2x - 10 + 7]^5 = (2x - 3)^5. \end{aligned}$$

Now Try Exercise 39.

EXAMPLE 5 If $h(x) = \frac{1}{(x + 3)^3}$, find $f(x)$ and $g(x)$ such that $h(x) = (f \circ g)(x)$.

Solution Two functions that can be used are

$$f(x) = \frac{1}{x^3} \quad \text{and} \quad g(x) = x + 3.$$

We check by forming the composition:

$$h(x) = (f \circ g)(x) = f(g(x)) = f(x + 3) = \frac{1}{(x + 3)^3}.$$

There are other functions that can be used as well. For example, if

$$f(x) = \frac{1}{x} \quad \text{and} \quad g(x) = (x + 3)^3,$$

then

$$h(x) = (f \circ g)(x) = f(g(x)) = f((x + 3)^3) = \frac{1}{(x + 3)^3}.$$

Now Try Exercise 41.

2.3 Exercise Set

Given that $f(x) = 3x + 1$, $g(x) = x^2 - 2x - 6$, and $h(x) = x^3$, find each of the following.

1. $(f \circ g)(-1)$

2. $(g \circ f)(-2)$

3. $(h \circ f)(1)$

4. $(g \circ h)\left(\frac{1}{2}\right)$

5. $(g \circ f)(5)$

6. $(f \circ g)\left(\frac{1}{3}\right)$

7. $(f \circ h)(-3)$

8. $(h \circ g)(3)$

9. $(g \circ g)(-2)$

10. $(g \circ g)(3)$

11. $(h \circ h)(2)$

12. $(h \circ h)(-1)$

13. $(f \circ f)(-4)$

14. $(f \circ f)(1)$

15. $(h \circ h)(x)$

16. $(f \circ f)(x)$

Find $(f \circ g)(x)$ and $(g \circ f)(x)$ and the domain of each.

17. $f(x) = x + 3$, $g(x) = x - 3$

18. $f(x) = \frac{4}{5}x$, $g(x) = \frac{5}{4}x$

19. $f(x) = x + 1$, $g(x) = 3x^2 - 2x - 1$

20. $f(x) = 3x - 2$, $g(x) = x^2 + 5$

21. $f(x) = x^2 - 3$, $g(x) = 4x - 3$

22. $f(x) = 4x^2 - x + 10$, $g(x) = 2x - 7$

23. $f(x) = \dfrac{4}{1 - 5x}$, $g(x) = \dfrac{1}{x}$

24. $f(x) = \dfrac{6}{x}$, $g(x) = \dfrac{1}{2x + 1}$

25. $f(x) = 3x - 7$, $g(x) = \dfrac{x + 7}{3}$

26. $f(x) = \frac{2}{3}x - \frac{4}{5}$, $g(x) = 1.5x + 1.2$

27. $f(x) = 2x + 1$, $g(x) = \sqrt{x}$

28. $f(x) = \sqrt{x}$, $g(x) = 2 - 3x$

29. $f(x) = 20$, $g(x) = 0.05$

30. $f(x) = x^4$, $g(x) = \sqrt[4]{x}$

31. $f(x) = \sqrt{x + 5}$, $g(x) = x^2 - 5$

32. $f(x) = x^5 - 2$, $g(x) = \sqrt[5]{x + 2}$

33. $f(x) = x^2 + 2$, $g(x) = \sqrt{3 - x}$

34. $f(x) = 1 - x^2$, $g(x) = \sqrt{x^2 - 25}$

35. $f(x) = \dfrac{1 - x}{x}$, $g(x) = \dfrac{1}{1 + x}$

36. $f(x) = \dfrac{1}{x - 2}$, $g(x) = \dfrac{x + 2}{x}$

37. $f(x) = x^3 - 5x^2 + 3x + 7$, $g(x) = x + 1$

38. $f(x) = x - 1$, $g(x) = x^3 + 2x^2 - 3x - 9$

Find $f(x)$ and $g(x)$ such that $h(x) = (f \circ g)(x)$. Answers may vary.

39. $h(x) = (4 + 3x)^5$

40. $h(x) = \sqrt[3]{x^2 - 8}$

41. $h(x) = \dfrac{1}{(x - 2)^4}$

42. $h(x) = \dfrac{1}{\sqrt{3x + 7}}$

43. $h(x) = \dfrac{x^3 - 1}{x^3 + 1}$

44. $h(x) = |9x^2 - 4|$

45. $h(x) = \left(\dfrac{2 + x^3}{2 - x^3}\right)^6$

46. $h(x) = \left(\sqrt{x} - 3\right)^4$

47. $h(x) = \sqrt{\dfrac{x - 5}{x + 2}}$

48. $h(x) = \sqrt{1 + \sqrt{1 + x}}$

49. $h(x) = (x + 2)^3 - 5(x + 2)^2 + 3(x + 2) - 1$

50. $h(x) = 2(x - 1)^{5/3} + 5(x - 1)^{2/3}$

51. *Ripple Spread.* A stone is thrown into a pond, creating a circular ripple that spreads over the pond in such a way that the radius is increasing at a rate of 3 ft/sec.

a) Find a function $r(t)$ for the radius in terms of t.

b) Find a function $A(r)$ for the area of the ripple in terms of the radius r.

c) Find $(A \circ r)(t)$. Explain the meaning of this function.

52. The surface area S of a right circular cylinder is given by the formula $S = 2\pi rh + 2\pi r^2$. If the height is twice the radius, find each of the following.

a) A function $S(r)$ for the surface area as a function of r

b) A function $S(h)$ for the surface area as a function of h

53. *Blouse Sizes.* A blouse that is size x in Japan is size $s(x)$ in the United States, where $s(x) = x - 3$. A blouse that is size x in the United States is size $t(x)$ in Australia, where $t(x) = x + 4$. (*Source:* www.onlineconversion.com) Find a function that will convert blouse sizes in Japan to blouse sizes in Australia.

54. A manufacturer of tools, selling rechargeable drills to a chain of home improvement stores, charges $6 more per drill than its manufacturing cost m. The stores then sell each drill for 150% of the price that it paid the manufacturer. Find a function $P(m)$ for the price at the home improvement stores.

Skill Maintenance

Consider the following linear equations. Without graphing them, answer the questions below. [1.3], [1.4]

a) $y = x$ **b)** $y = -5x + 4$
c) $y = \frac{2}{3}x + 1$ **d)** $y = -0.1x + 6$
e) $y = 3x - 5$ **f)** $y = -x - 1$
g) $2x - 3y = 6$ **h)** $6x + 3y = 9$

55. Which, if any, have y-intercept $(0, 1)$?

56. Which, if any, have the same y-intercept?

57. Which slope down from left to right?

58. Which has the steepest slope?

59. Which pass(es) through the origin?

60. Which, if any, have the same slope?

61. Which, if any, are parallel?

62. Which, if any, are perpendicular?

Synthesis

63. Let $p(a)$ represent the number of pounds of grass seed required to seed a lawn with area a. Let $c(s)$ represent the cost of s pounds of grass seed. Which composition makes sense: $(c \circ p)(a)$ or $(p \circ c)(s)$? What does it represent?

64. Write equations of two functions f and g such that $f \circ g = g \circ f = x$. (In Section 5.1, we will study inverse functions. If $f \circ g = g \circ f = x$, functions f and g are *inverses* of each other.)

Mid-Chapter Mixed Review

Determine whether the statement is true or false.

1. $f(c)$ is a relative maximum if $(c, f(c))$ is the highest point in some open interval containing c. [2.1]

2. If f and g are functions, then the domain of the functions $f + g$, $f - g$, fg, and f/g is the intersection of the domain of f and the domain of g. [2.2]

3. In general, $(f \circ g)(x) \neq (g \circ f)(x)$. [2.3]

4. Determine the intervals on which the function is (a) increasing; (b) decreasing; (c) constant. [2.1]

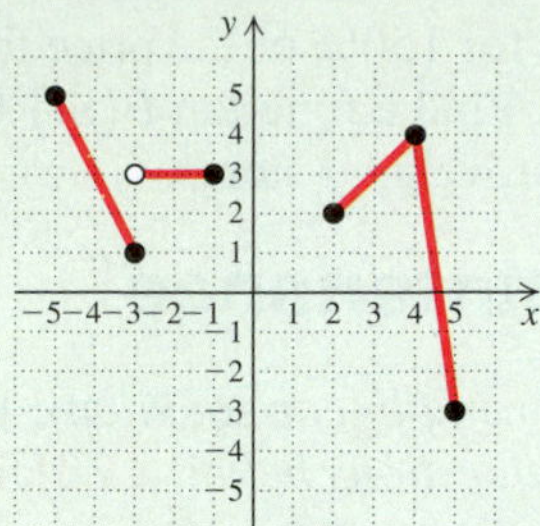

5. Using the graph, determine any relative maxima or minima of the function and the intervals on which the function is increasing or decreasing. [2.1]

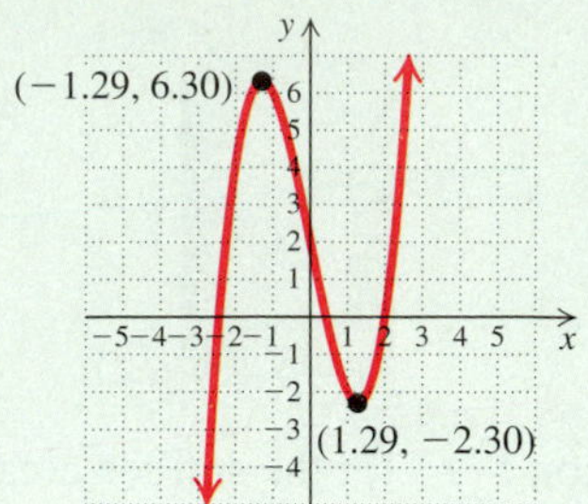

6. Determine the domain and the range of the function graphed in Exercise 4. [2.1]

7. *Window Design.* Lucas is designing a window for the peak of an A-frame house. The base is 4 ft more than the height h. Express the area of the window as a function of the height. [2.1]

8. For the function defined as

$$f(x) = \begin{cases} x - 5, & \text{for } x \leq -3, \\ 2x + 3, & \text{for } -3 < x \leq 0, \\ \frac{1}{2}x, & \text{for } x > 0, \end{cases}$$

find $f(-5), f(-3), f(-1)$, and $f(6)$. [2.1]

9. Graph the function defined as

$$g(x) = \begin{cases} x + 2, & \text{for } x < -4, \\ -x, & \text{for } x \geq -4. \end{cases}$$ [2.1]

Given that $f(x) = 3x - 1$ and $g(x) = x^2 + 4$, find each of the following, if it exists. [2.2]

10. $(f + g)(-1)$

11. $(fg)(0)$

12. $(g - f)(3)$

13. $(g/f)\left(\frac{1}{3}\right)$

For each pair of functions in Exercises 14 and 15:

a) *Find the domains of $f, g, f + g, f - g, fg, ff, f/g$, and g/f.*
b) *Find $(f + g)(x)$, $(f - g)(x)$, $(fg)(x)$, $(ff)(x)$, $(f/g)(x)$, and $(g/f)(x)$.* [2.2]

14. $f(x) = 2x + 5,\ g(x) = -x - 4$

15. $f(x) = x - 1,\ g(x) = \sqrt{x + 2}$

For each function f in Exercises 16 and 17, construct and simplify the difference quotient

$$\frac{f(x + h) - f(x)}{h}.$$ [2.2]

16. $f(x) = 4x - 3$

17. $f(x) = 6 - x^2$

Given that $f(x) = 5x - 4$, $g(x) = x^3 + 1$, and $h(x) = x^2 - 2x + 3$, find each of the following. [2.3]

18. $(f \circ g)(1)$

19. $(g \circ h)(2)$

20. $(f \circ f)(0)$

21. $(h \circ f)(-1)$

Find $(f \circ g)(x)$ and $(g \circ f)(x)$ and the domain of each. [2.3]

22. $f(x) = \frac{1}{2}x,\ g(x) = 6x + 4$

23. $f(x) = 3x + 2,\ g(x) = \sqrt{x}$

COLLABORATIVE DISCUSSION AND WRITING

24. If $g(x) = b$, where b is a positive constant, describe how the graphs of $y = h(x)$ and $y = (h - g)(x)$ will differ. [2.2]

25. If the domain of a function f is the set of real numbers and the domain of a function g is also the set of real numbers, under what circumstances do $(f + g)(x)$ and $(f/g)(x)$ have different domains? [2.2]

26. If f and g are linear functions, what can you say about the domain of $f \circ g$ and the domain of $g \circ f$? [2.3]

27. Nora determines the domain of $f \circ g$ by examining only the formula for $(f \circ g)(x)$. Is her approach valid? Why or why not? [2.3]

2.4 Symmetry

- Determine whether a graph is symmetric with respect to the x-axis, the y-axis, and the origin.
- Determine whether a function is even, odd, or neither even nor odd.

Symmetry

Symmetry occurs often in nature and in art. For example, when viewed from the front, the bodies of most animals are at least approximately symmetric. This means that each eye is the same distance from the center of the bridge of the nose, each shoulder is the same distance from the center of the chest, and so on. Architects have used symmetry for thousands of years to enhance the beauty of buildings.

A knowledge of symmetry in mathematics helps us graph and analyze equations and functions.

Consider the points $(4, 2)$ and $(4, -2)$ that appear on the graph of $x = y^2$. Points like these have the same x-value but opposite y-values and are **reflections** of each other across the x-axis. If, for any point (x, y) on a graph, the point $(x, -y)$ is also on the graph, then the graph is said to be **symmetric with respect to the *x*-axis.** If we fold the graph on the x-axis, the parts above and below the x-axis will coincide.

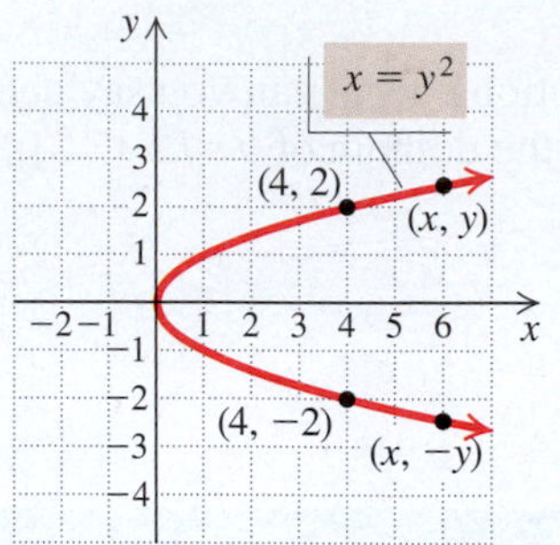

Consider the points $(3, 4)$ and $(-3, 4)$ that appear on the graph of $y = x^2 - 5$ shown below. Points like these have the same y-value but opposite x-values and are **reflections** of each other across the y-axis. If, for any point (x, y) on a graph, the point $(-x, y)$ is also on the graph, then the graph is said to be **symmetric with respect to the *y*-axis**. If we fold the graph on the y-axis, the parts to the left and to the right of the y-axis will coincide.

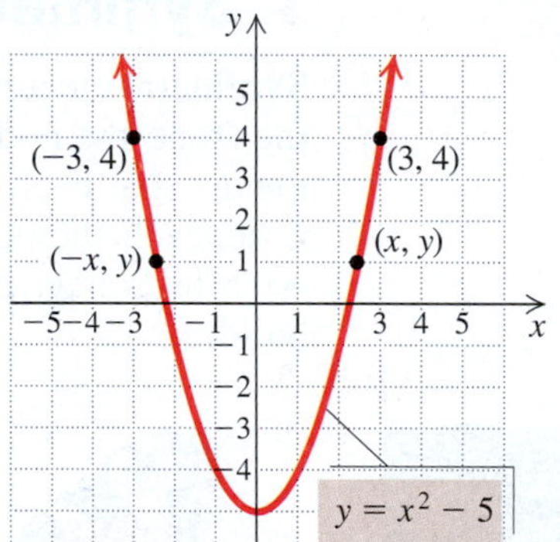

Consider the points $(-3, \sqrt{7})$ and $(3, -\sqrt{7})$ that appear on the graph of $x^2 = y^2 + 2$ shown below. Note that if we take the opposites of the coordinates of one pair, we get the other pair. If, for any point (x, y) on a graph, the point $(-x, -y)$ is also on the graph, then the graph is said to be **symmetric with respect to the origin**. Visually, if we rotate the graph 180° about the origin, the resulting figure coincides with the original.

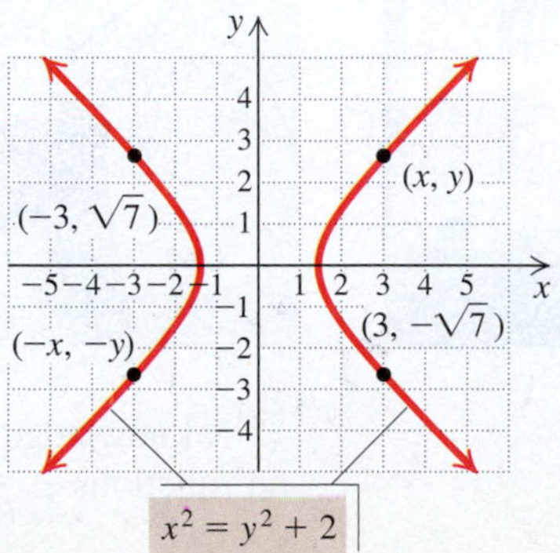

ALGEBRAIC TESTS OF SYMMETRY

x-axis: If replacing y with $-y$ produces an equivalent equation, then the graph is *symmetric with respect to the x-axis.*

y-axis: If replacing x with $-x$ produces an equivalent equation, then the graph is *symmetric with respect to the y-axis.*

Origin: If replacing x with $-x$ and y with $-y$ produces an equivalent equation, then the graph is *symmetric with respect to the origin.*

EXAMPLE 1 Test $y = x^2 + 2$ for symmetry with respect to the x-axis, the y-axis, and the origin.

Algebraic Solution

***x*-Axis:**
We replace y with $-y$:

$$y = x^2 + 2$$
$$-y = x^2 + 2$$
$$y = -x^2 - 2. \quad \text{Multiplying by } -1 \text{ on both sides}$$

The resulting equation *is not* equivalent to the original equation, so the graph *is not* symmetric with respect to the x-axis.

***y*-Axis:**
We replace x with $-x$:

$$y = x^2 + 2$$
$$y = (-x)^2 + 2$$
$$y = x^2 + 2. \quad \text{Simplifying}$$

The resulting equation *is* equivalent to the original equation, so the graph *is* symmetric with respect to the y-axis.

Origin:
We replace x with $-x$ and y with $-y$:

$$y = x^2 + 2$$
$$-y = (-x)^2 + 2$$
$$-y = x^2 + 2 \quad \text{Simplifying}$$
$$y = -x^2 - 2.$$

The resulting equation *is not* equivalent to the original equation, so the graph *is not* symmetric with respect to the origin.

Graphical Solution

We use a graphing calculator to graph the equation.

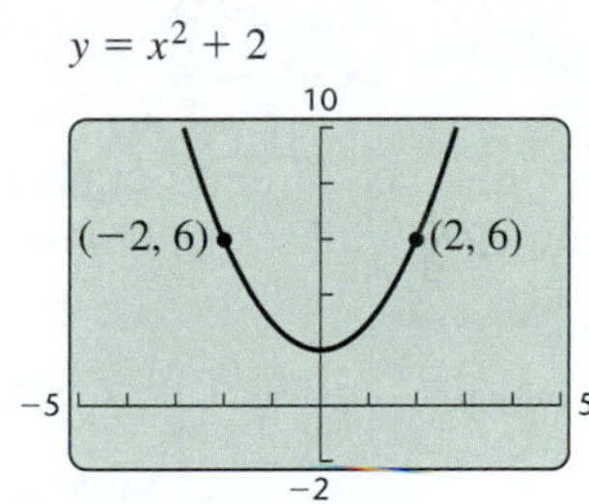

Note that if the graph were folded on the x-axis, the parts above and below the x-axis would not coincide so the graph *is not* symmetric with respect to the x-axis. If it were folded on the y-axis, the parts to the left and to the right of the y-axis would coincide so the graph *is* symmetric with respect to the y-axis. If we rotated it 180° around the origin, the resulting graph would not coincide with the original graph so the graph *is not* symmetric with respect to the origin.

Now Try Exercise 11.

The algebraic method is often easier to apply than the graphical method, especially with equations that we may not be able to graph easily. It is also often more precise.

EXAMPLE 2 Test $x^2 + y^4 = 5$ for symmetry with respect to the x-axis, the y-axis, and the origin.

Algebraic Solution

***x*-Axis:**
We replace y with $-y$:

$$x^2 + y^4 = 5$$
$$x^2 + (-y)^4 = 5$$
$$x^2 + y^4 = 5.$$

The resulting equation *is* equivalent to the original equation. Thus the graph *is* symmetric with respect to the x-axis.

***y*-Axis:**
We replace x with $-x$:

$$x^2 + y^4 = 5$$
$$(-x)^2 + y^4 = 5$$
$$x^2 + y^4 = 5.$$

The resulting equation *is* equivalent to the original equation, so the graph *is* symmetric with respect to the y-axis.

Origin:
We replace x with $-x$ and y with $-y$:

$$x^2 + y^4 = 5$$
$$(-x)^2 + (-y)^4 = 5$$
$$x^2 + y^4 = 5.$$

The resulting equation *is* equivalent to the original equation, so the graph *is* symmetric with respect to the origin.

Graphical Solution

To graph $x^2 + y^4 = 5$ using a graphing calculator, we first solve the equation for y:

$$y = \pm\sqrt[4]{5 - x^2}.$$

Then on the Y= screen we enter the equations

$$y_1 = \sqrt[4]{5 - x^2} \quad \text{and}$$
$$y_2 = -\sqrt[4]{5 - x^2}.$$

$x^2 + y^4 = 5$

$y_1 = \sqrt[4]{5 - x^2}$ and $y_2 = -\sqrt[4]{5 - x^2}$

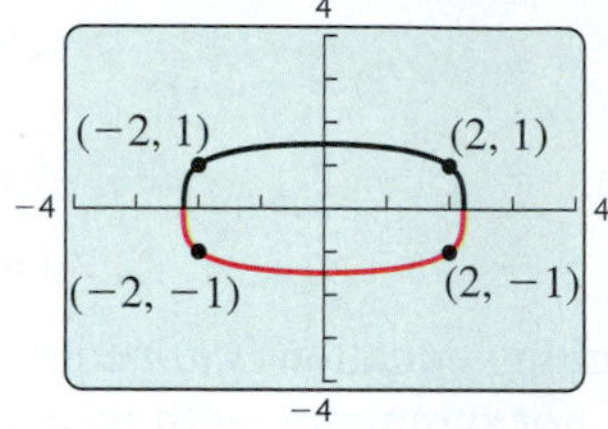

From the graph of the equation, we see symmetry with respect to both axes and with respect to the origin.

Now Try Exercise 21.

DETERMINING EVEN AND ODD FUNCTIONS

Given the function $f(x)$:

1. Find $f(-x)$ and simplify. If $f(x) = f(-x)$, then f is even.
2. Find $-f(x)$, simplify, and compare with $f(-x)$ from step (1). If $f(-x) = -f(x)$, then f is odd.

Except for the function $f(x) = 0$, a function cannot be *both* even and odd. Thus if $f(x) \neq 0$ and we see in step (1) that $f(x) = f(-x)$ (that is, f is even), we need not continue.

Even Functions and Odd Functions

Now we relate symmetry to graphs of functions.

EVEN FUNCTIONS AND ODD FUNCTIONS

If the graph of a function f is symmetric with respect to the y-axis, we say that it is an **even function**. That is, for each x in the domain of f, $f(x) = f(-x)$.

If the graph of a function f is symmetric with respect to the origin, we say that it is an **odd function**. That is, for each x in the domain of f, $f(-x) = -f(x)$.

An algebraic procedure for determining even functions and odd functions is shown at left. In the next example, we show an even function and an odd function. Many functions are neither even nor odd.

EXAMPLE 3 Determine whether each of the following functions is even, odd, or neither.

a) $f(x) = 5x^7 - 6x^3 - 2x$ b) $h(x) = 5x^6 - 3x^2 - 7$

Algebraic Solution

a) $f(x) = 5x^7 - 6x^3 - 2x$

1. $f(-x) = 5(-x)^7 - 6(-x)^3 - 2(-x)$
$= 5(-x^7) - 6(-x^3) + 2x$
$(-x)^7 = (-1 \cdot x)^7 = (-1)^7 x^7 = -x^7;$
$(-x)^3 = (-1 \cdot x)^3 = (-1)^3 x^3 = -x^3$
$= -5x^7 + 6x^3 + 2x$

We see that $f(x) \neq f(-x)$. Thus, f is *not* even.

2. $-f(x) = -(5x^7 - 6x^3 - 2x)$
$= -5x^7 + 6x^3 + 2x$

We see that $f(-x) = -f(x)$. Thus, f is odd.

b) $h(x) = 5x^6 - 3x^2 - 7$

1. $h(-x) = 5(-x)^6 - 3(-x)^2 - 7$
$= 5x^6 - 3x^2 - 7$

We see that $h(x) = h(-x)$. Thus the function is even.

Graphical Solution

a) We see that the graph appears to be symmetric with respect to the origin. The function is odd.

$y = 5x^7 - 6x^3 - 2x$

6
−3 3
−6

b) We see that the graph appears to be symmetric with respect to the y-axis. The function is even.

$y = 5x^6 - 3x^2 - 7$

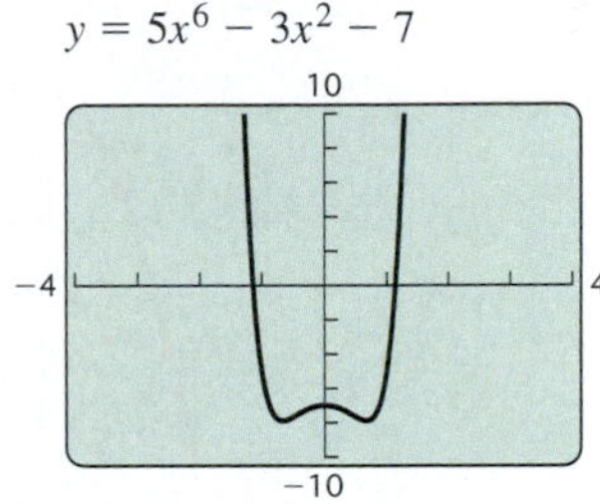

Now Try Exercises 39 and 41.

2.4 Exercise Set

Determine visually whether the graph is symmetric with respect to the x-axis, the y-axis, and the origin.

1.
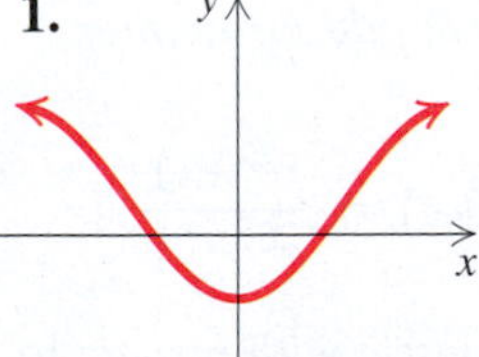

2.
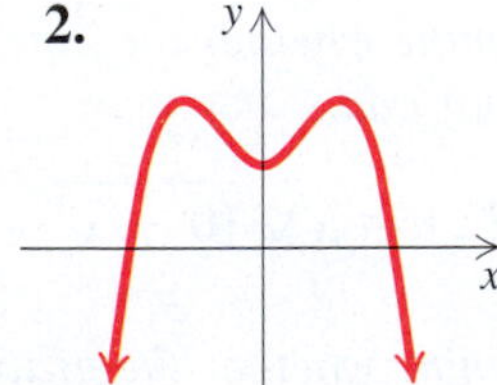

3.
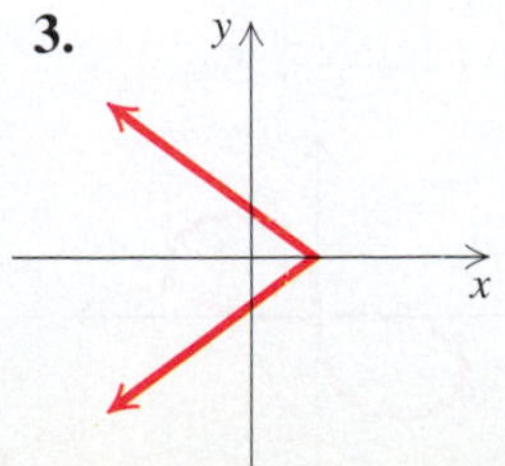

4.
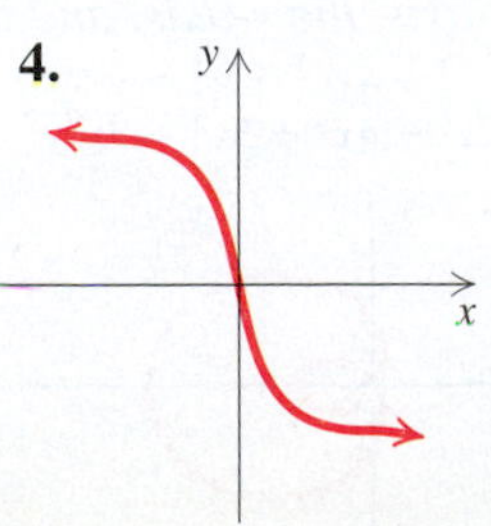

5.
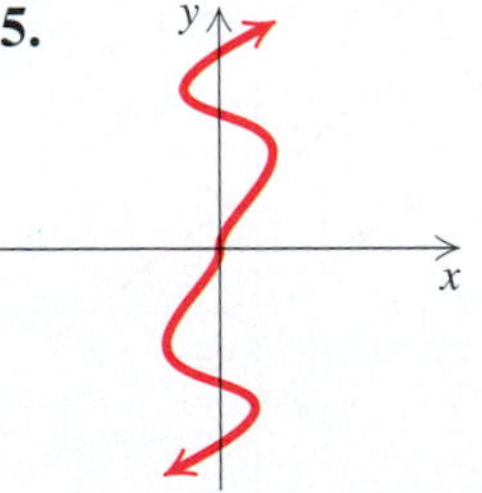

6.
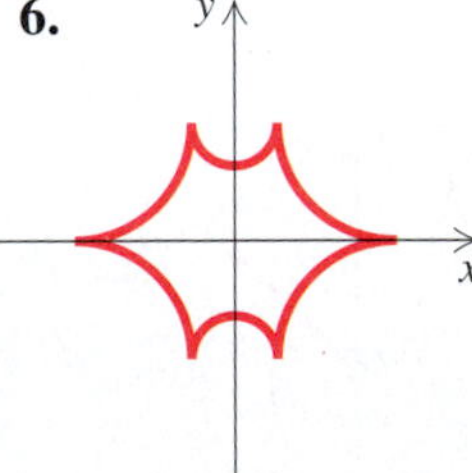

First, graph the equation and determine visually whether it is symmetric with respect to the x-axis, the y-axis, and the origin. Then verify your assertion algebraically.

7. $y = |x| - 2$
8. $y = |x + 5|$
9. $5y = 4x + 5$
10. $2x - 5 = 3y$
11. $5y = 2x^2 - 3$
12. $x^2 + 4 = 3y$
13. $y = 1/x$
14. $y = -(4/x)$

Test algebraically whether the graph is symmetric with respect to the x-axis, the y-axis, and the origin. Then check your work graphically, if possible, using a graphing calculator.

15. $5x - 5y = 0$

16. $6x + 7y = 0$

17. $3x^2 - 2y^2 = 3$

18. $5y = 7x^2 - 2x$

19. $y = |2x|$

20. $y^3 = 2x^2$

21. $2x^4 + 3 = y^2$

22. $2y^2 = 5x^2 + 12$

23. $3y^3 = 4x^3 + 2$

24. $3x = |y|$

25. $xy = 12$

26. $xy - x^2 = 3$

Find the point that is symmetric to the given point with respect to the x-axis, the y-axis, and the origin.

27. $(-5, 6)$

28. $\left(\frac{7}{2}, 0\right)$

29. $(-10, -7)$

30. $\left(1, \frac{3}{8}\right)$

31. $(0, -4)$

32. $(8, -3)$

Determine visually whether the function is even, odd, or neither even nor odd.

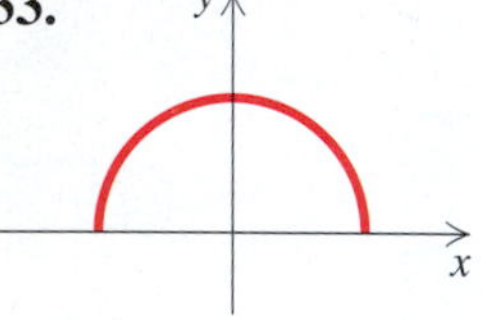

34.

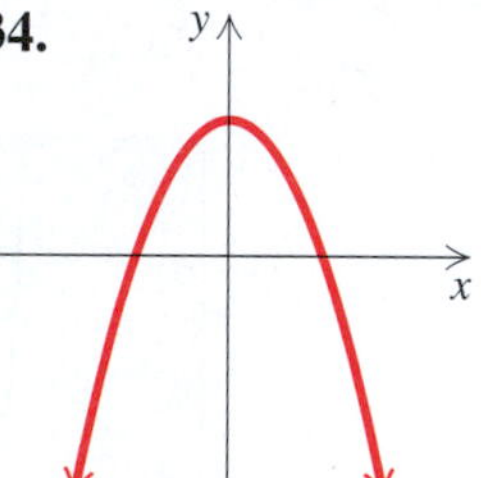

35.

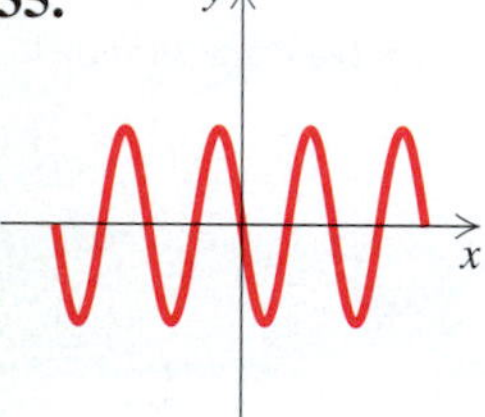

36.

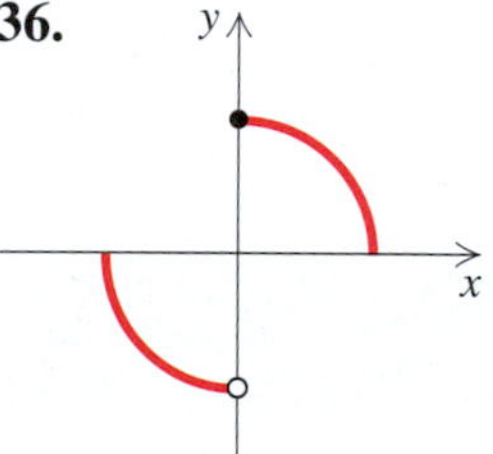

37.

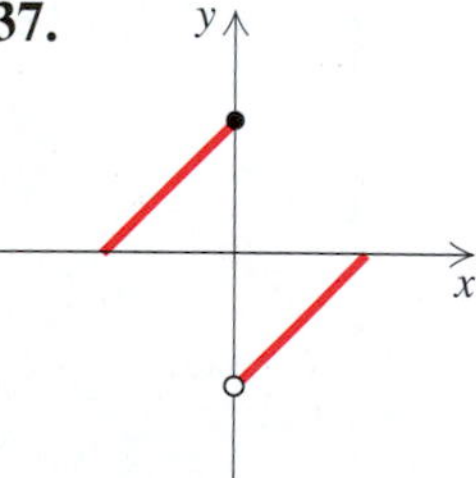

38.

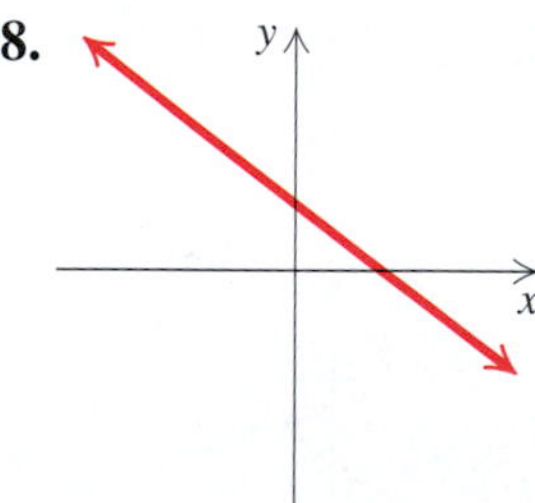

Determine algebraically whether the function is even, odd, or neither even nor odd. Then check your work graphically, where possible, using a graphing calculator.

39. $f(x) = -3x^3 + 2x$

40. $f(x) = 7x^3 + 4x - 2$

41. $f(x) = 5x^2 + 2x^4 - 1$

42. $f(x) = x + \dfrac{1}{x}$

43. $f(x) = x^{17}$

44. $f(x) = \sqrt[3]{x}$

45. $f(x) = x - |x|$

46. $f(x) = \dfrac{1}{x^2}$

47. $f(x) = 8$

48. $f(x) = \sqrt{x^2 + 1}$

Skill Maintenance

49. Graph: $f(x) = \begin{cases} x - 2, & \text{for } x \le -1, \\ 3, & \text{for } -1 < x \le 2, \\ x, & \text{for } x > 2. \end{cases}$ **[2.1]**

50. *Peace Corps Volunteers.* Since 1961, there has been a total of 6688 Peace Corps volunteers from the University of California–Berkeley and the University of Wisconsin–Madison. The number of volunteers from the University of California–Berkeley is 464 more than the number of volunteers from the University of Wisconsin–Madison. (*Source:* Peace Corps 2014) Find the number of Peace Corps volunteers from each university. **[1.5]**

Synthesis

Determine whether the function is even, odd, or neither even nor odd.

51. $f(x) = x\sqrt{10 - x^2}$

52. $f(x) = \dfrac{x^2 + 1}{x^3 - 1}$

Determine whether the graph is symmetric with respect to the x-axis, the y-axis, and the origin.

53. $y^2 + 4xy^2 - y^4 = x^4 - 4x^3 + 3x^2 + 2x^2y^2$

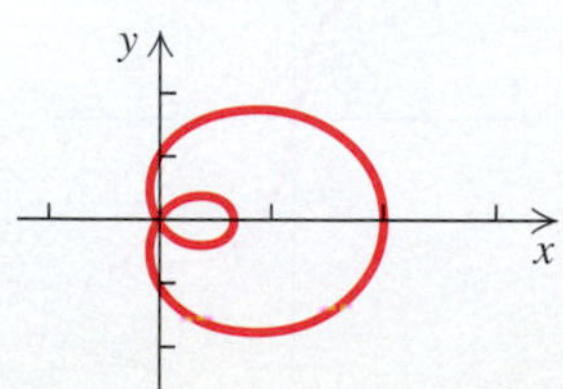

54. $(x^2 + y^2)^2 = 2xy$

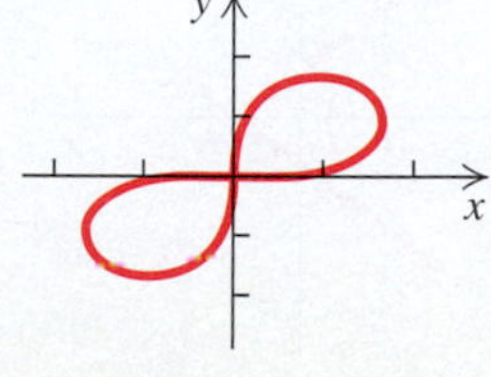

55. Show that if f is *any* function, then the function E defined by

$$E(x) = \frac{f(x) + f(-x)}{2}$$

is even.

56. Show that if f is *any* function, then the function O defined by

$$O(x) = \frac{f(x) - f(-x)}{2}$$

is odd.

57. Consider the functions E and O of Exercises 55 and 56.

a) Show that $f(x) = E(x) + O(x)$. This means that every function can be expressed as the sum of an even function and an odd function.

b) Let $f(x) = 4x^3 - 11x^2 + \sqrt{x} - 10$. Express f as a sum of an even function and an odd function.

Determine whether the statement is true or false.

58. The product of two odd functions is odd.

59. The sum of two even functions is even.

60. The product of an even function and an odd function is odd.

2.5 Transformations

› Given the graph of a function, graph its transformation under translations, reflections, stretchings, and shrinkings.

› Transformations of Functions

The graphs of some basic functions are shown below. Others can be seen on the inside back cover.

Identity function: $y = x$

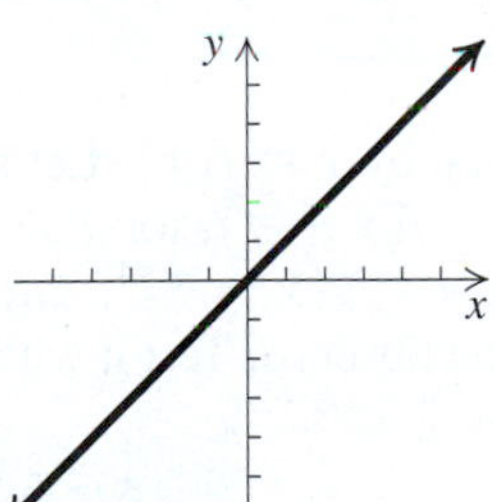

Squaring function: $y = x^2$

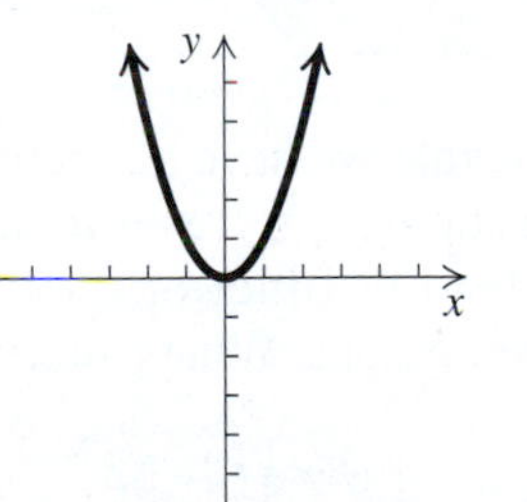

Square root function: $y = \sqrt{x}$

Cubing function: $y = x^3$

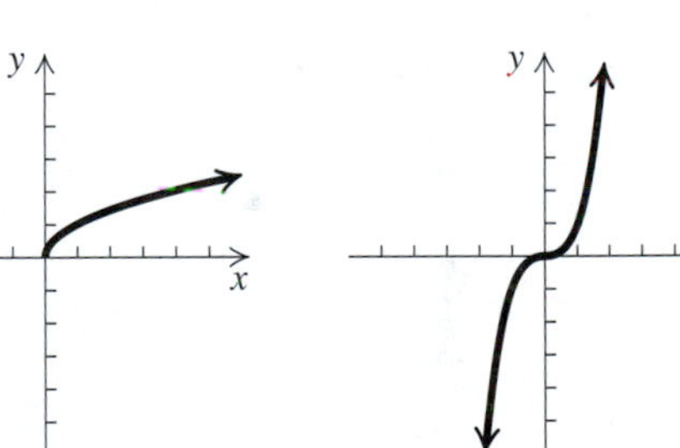

Cube root function: $y = \sqrt[3]{x}$

Reciprocal function: $y = \frac{1}{x}$

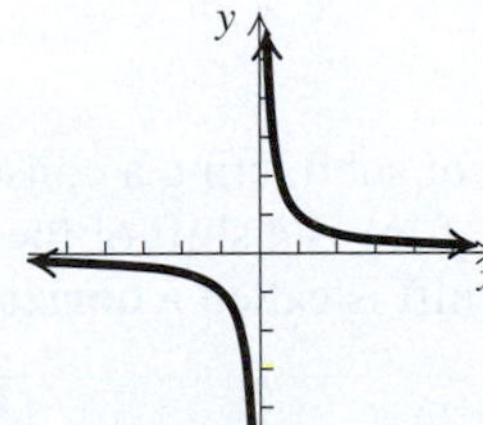

Absolute-value function: $y = |x|$

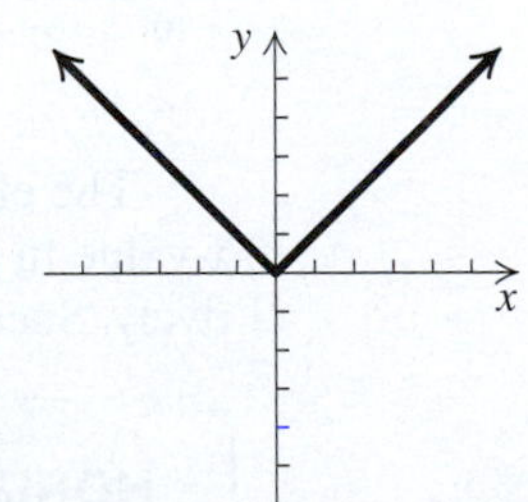

These functions can be considered building blocks for many other functions. We can create graphs of new functions by shifting them horizontally or vertically, stretching or shrinking them, and reflecting them across an axis. We now consider these **transformations**.

Vertical Translations and Horizontal Translations

Suppose that we have a function given by $y = f(x)$. Let's explore the graphs of the new functions $y = f(x) + b$ and $y = f(x) - b$, for $b > 0$.

Consider the functions $y = \frac{1}{5}x^4$, $y = \frac{1}{5}x^4 + 5$, and $y = \frac{1}{5}x^4 - 3$ and compare their graphs. What pattern do you see? Test it with some other functions.

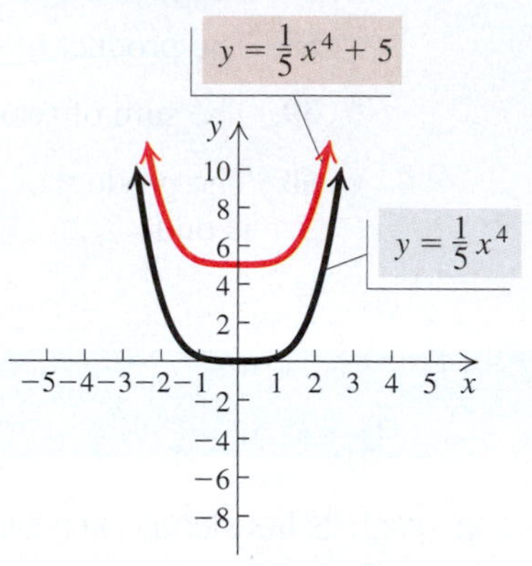

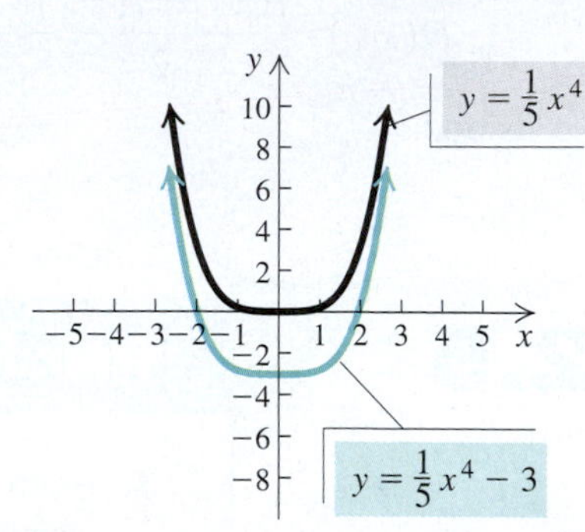

The effect of adding a constant to or subtracting a constant from $f(x)$ in $y = f(x)$ is a shift of the graph of $f(x)$ up or down, respectively. Such a shift is called a **vertical translation**.

VERTICAL TRANSLATION

For $b > 0$:

the graph of $y = f(x) + b$ is the graph of $y = f(x)$ shifted *up b* units;

the graph of $y = f(x) - b$ is the graph of $y = f(x)$ shifted *down b* units.

Suppose that we have a function given by $y = f(x)$. Let's explore the graphs of the new functions $y = f(x - d)$ and $y = f(x + d)$, for $d > 0$.

Consider the functions $y = \frac{1}{5}x^4$, $y = \frac{1}{5}(x - 3)^4$, and $y = \frac{1}{5}(x + 7)^4$ and compare their graphs. What pattern do you observe? Test it with some other functions.

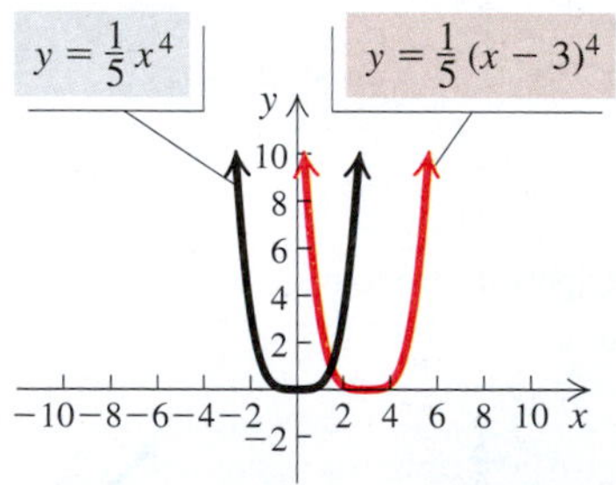

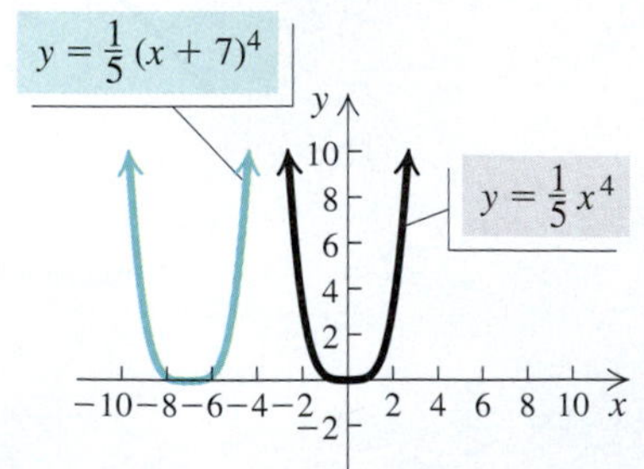

The effect of subtracting a constant from the x-value or adding a constant to the x-value in $y = f(x)$ is a shift of the graph of $f(x)$ to the right or to the left, respectively. Such a shift is called a **horizontal translation**.

HORIZONTAL TRANSLATION

For $d > 0$:

the graph of $y = f(x - d)$ is the graph of $y = f(x)$ shifted *to the right d* units;

the graph of $y = f(x + d)$ is the graph of $y = f(x)$ shifted *to the left d* units.

EXAMPLE 1 Graph each of the following. Before doing so, describe how each graph can be obtained from one of the basic graphs shown on the preceding pages.

a) $g(x) = x^2 - 6$ b) $h(x) = |x - 4|$

c) $g(x) = \sqrt{x + 2}$ d) $h(x) = \sqrt{x + 2} - 3$

Solution

a) To graph $g(x) = x^2 - 6$, first think of the graph of $f(x) = x^2$. Since $g(x) = f(x) - 6$, the graph of $g(x) = x^2 - 6$ is the graph of $f(x) = x^2$ shifted, or translated, *down* 6 units. (See Fig. 1.)

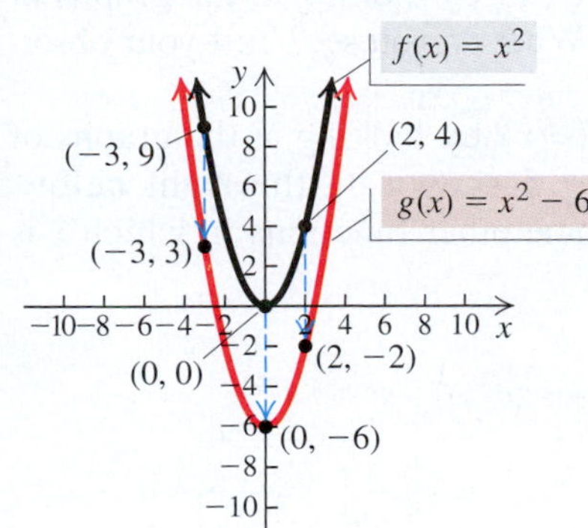

FIGURE 1.

Let's compare some points on the graphs of f and g.

Points on f:	$(-3, 9)$,	$(0, 0)$,	$(2, 4)$
	↓	↓	↓
Corresponding points on g:	$(-3, 3)$,	$(0, -6)$,	$(2, -2)$

We note that the y-coordinate of a point on the graph of g is 6 less than the corresponding y-coordinate on the graph of f.

b) To graph $h(x) = |x - 4|$, first think of the graph of $f(x) = |x|$. Since $h(x) = f(x - 4)$, the graph of $h(x) = |x - 4|$ is the graph of $f(x) = |x|$ shifted *right* 4 units. (See Fig. 2.)

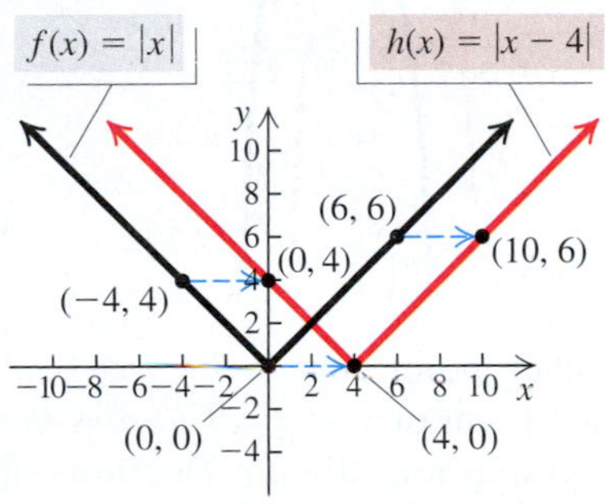

FIGURE 2.

Let's again compare points on the two graphs.

Points on f:	$(-4, 4)$,	$(0, 4)$,	$(6, 6)$
	↓	↓	↓
Corresponding points on h:	$(0, 4)$,	$(4, 0)$,	$(10, 6)$

Noting points on f and h, we see that the x-coordinate of a point on the graph of h is 4 more than the x-coordinate of the corresponding point on f.

c) To graph $g(x) = \sqrt{x + 2}$, first think of the graph of $f(x) = \sqrt{x}$. Since $g(x) = f(x + 2)$, the graph of $g(x) = \sqrt{x + 2}$ is the graph of $f(x) = \sqrt{x}$ shifted *left* 2 units. (See Fig. 3.)

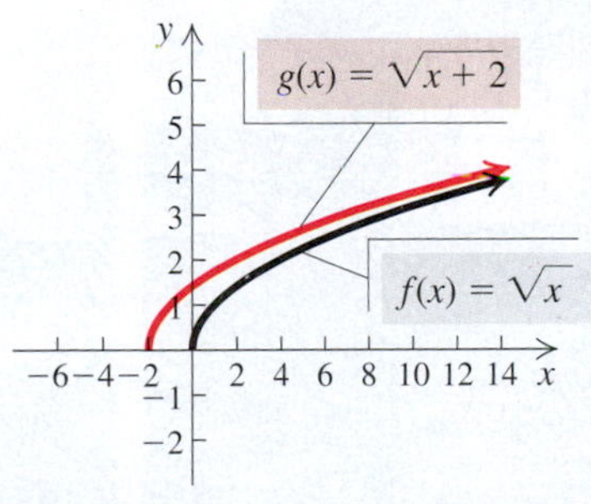

FIGURE 3.

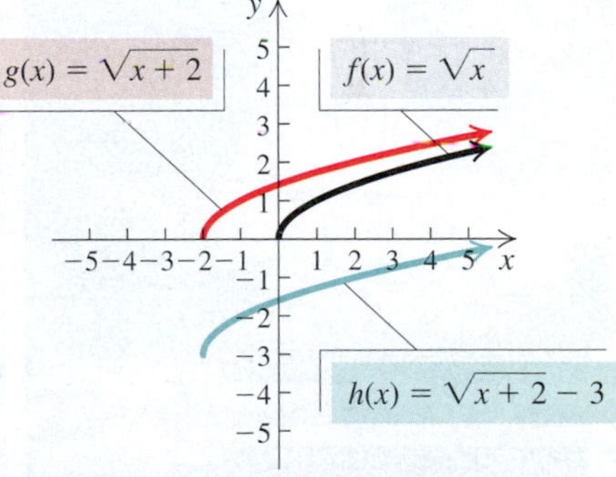

FIGURE 4.

d) To graph $h(x) = \sqrt{x + 2} - 3$, first think of the graph of $f(x) = \sqrt{x}$. In part (c), we found that the graph of $g(x) = \sqrt{x + 2}$ is the graph of $f(x) = \sqrt{x}$ shifted *left* 2 units. Since $h(x) = g(x) - 3$, we shift the graph of $g(x) = \sqrt{x + 2}$ *down* 3 units. Together, the graph of $f(x) = \sqrt{x}$ is shifted *left* 2 units and *down* 3 units. (See Fig. 4.)

Now Try Exercises 3 and 15.

Reflections

Suppose that we have a function given by $y = f(x)$. Let's explore the graphs of the new functions $y = -f(x)$ and $y = f(-x)$.

Compare the functions $y = f(x)$ and $y = -f(x)$ by looking at the graphs of $y = \frac{1}{5}x^4$ and $y = -\frac{1}{5}x^4$ shown on the left below. What do you see? Test your observation with some other functions y_1 and y_2, where $y_2 = -y_1$.

Compare the functions $y = f(x)$ and $y = f(-x)$ by looking at the graphs of $y = 2x^3 - x^4 + 5$ and $y = 2(-x)^3 - (-x)^4 + 5$ shown on the right below. What do you see? Test your observation with some other functions in which x is replaced with $-x$.

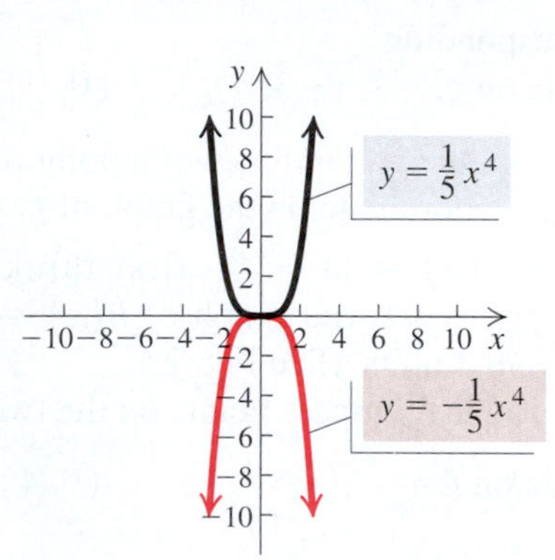

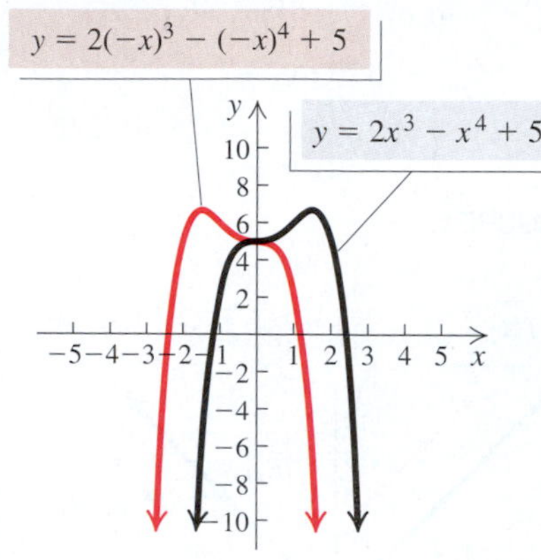

Given the graph of $y = f(x)$, we can reflect each point *across the x-axis* to obtain the graph of $y = -f(x)$. We can reflect each point of $y = f(x)$ *across the y-axis* to obtain the graph of $y = f(-x)$. The new graphs are called **reflections** of $y = f(x)$. The following photos illustrate reflection.

> **REFLECTIONS**
>
> The graph of $y = -f(x)$ is the **reflection** of the graph of $y = f(x)$ across the x-axis.
>
> The graph of $y = f(-x)$ is the **reflection** of the graph of $y = f(x)$ across the y-axis.
>
> If a point (x, y) is on the graph of $y = f(x)$, then $(x, -y)$ is on the graph of $y = -f(x)$, and $(-x, y)$ is on the graph of $y = f(-x)$.

$y_1 = x^3 - 4x^2$,
$y_2 = (-x)^3 - 4(-x)^2$

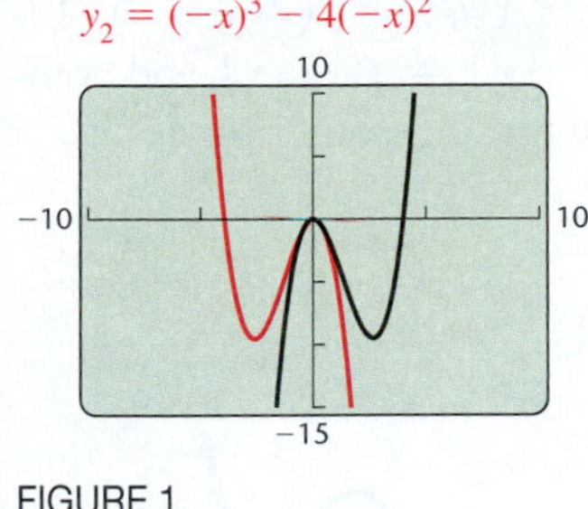

FIGURE 1.

$y_1 = x^3 - 4x^2$, $y_2 = -x^3 + 4x^2$

20
−10
10
−20

FIGURE 2.

EXAMPLE 2 Graph each of the following. Before doing so, describe how each graph can be obtained from the graph of $f(x) = x^3 - 4x^2$.

a) $g(x) = (-x)^3 - 4(-x)^2$ **b)** $h(x) = 4x^2 - x^3$

Solution

a) We first note that

$$f(-x) = (-x)^3 - 4(-x)^2 = g(x).$$

Thus the graph of g is a *reflection* of the graph of f across the y-axis. (See Fig. 1.) If (x, y) is on the graph of f, then $(-x, y)$ is on the graph of g. For example, $(2, -8)$ is on f and $(-2, -8)$ is on g.

b) We first note that

$$\begin{aligned} -f(x) &= -(x^3 - 4x^2) \\ &= -x^3 + 4x^2 \\ &= h(x). \end{aligned}$$

Thus the graph of h is a *reflection* of the graph of f across the x-axis. (See Fig. 2.) If (x, y) is on the graph of f, then $(x, -y)$ is on the graph of h. For example, $(2, -8)$ is on f and $(2, 8)$ is on h.

Vertical and Horizontal Stretchings and Shrinkings

Suppose that we have a function given by $y = f(x)$. Let's explore the graphs of the new functions $y = af(x)$ and $y = f(cx)$.

Let's consider the functions $y = f(x) = x^3 - x$, $y = \frac{1}{10}(x^3 - x) = \frac{1}{10}f(x)$, $y = 2(x^3 - x) = 2f(x)$, and $y = -2(x^3 - x) = -2f(x)$ and compare their graphs. What pattern do you observe? Test it with some other functions.

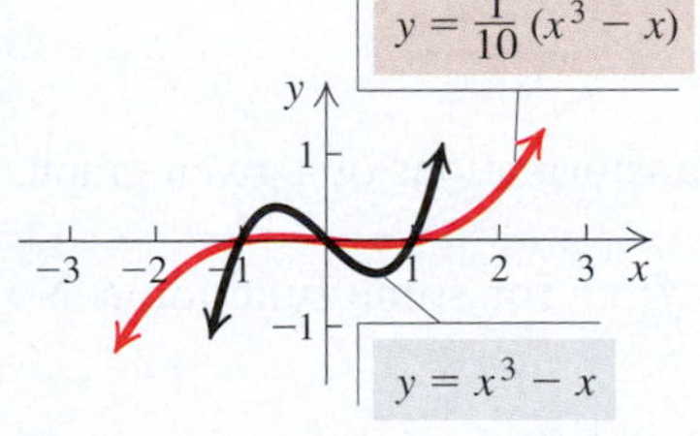

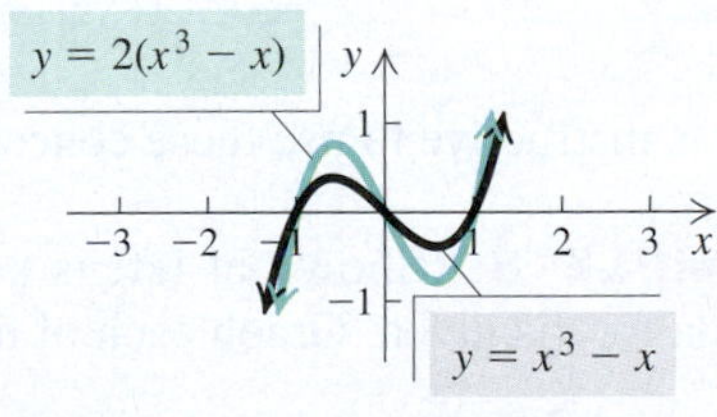

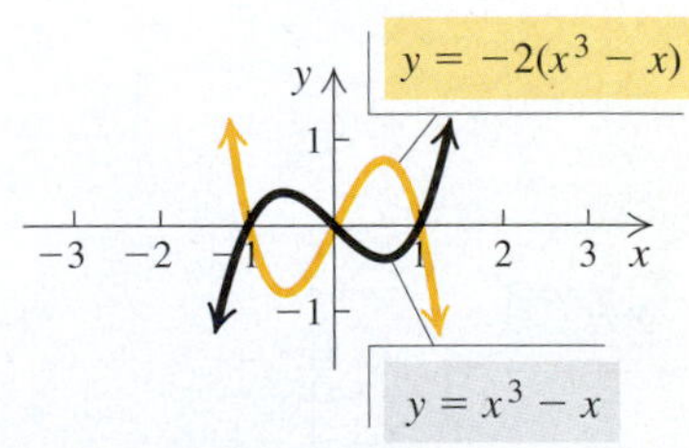

Consider any function f given by $y = f(x)$. Multiplying $f(x)$ by any constant a, where $|a| > 1$, to obtain $g(x) = af(x)$ will *stretch* the graph vertically away from the x-axis. If $0 < |a| < 1$, then the graph will be flattened or *shrunk* vertically toward the x-axis. If $a < 0$, the graph is also reflected across the x-axis.

Just in Time 4

VERTICAL STRETCHING AND SHRINKING

The graph of $y = af(x)$ can be obtained from the graph of $y = f(x)$ by

stretching vertically for $|a| > 1$, or
shrinking vertically for $0 < |a| < 1$.

For $a < 0$, the graph is also reflected across the x-axis.
(The y-coordinates of the graph of $y = af(x)$ can be obtained by multiplying the y-coordinates of $y = f(x)$ by a.)

Let's consider the functions $y = f(x) = x^3 - x$, $y = (2x)^3 - (2x) = f(2x)$, $y = \left(\frac{1}{2}x\right)^3 - \left(\frac{1}{2}x\right) = f\left(\frac{1}{2}x\right)$, and $y = \left(-\frac{1}{2}x\right)^3 - \left(-\frac{1}{2}x\right) = f\left(-\frac{1}{2}x\right)$ and compare their graphs. What pattern do you observe? Test it with some other functions.

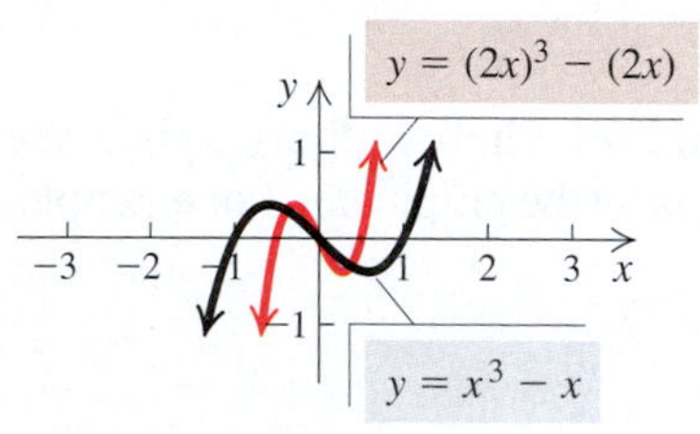

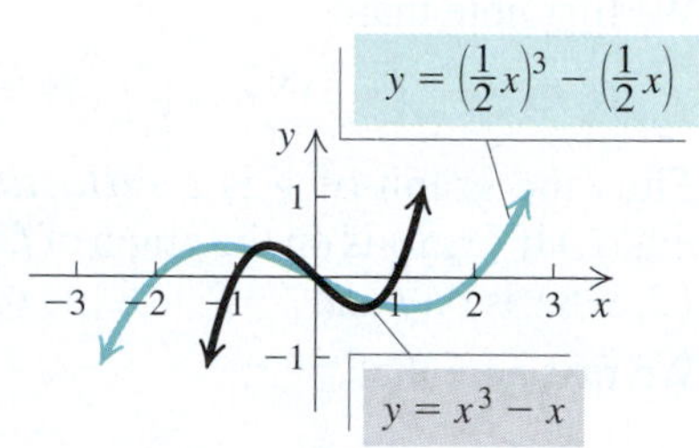

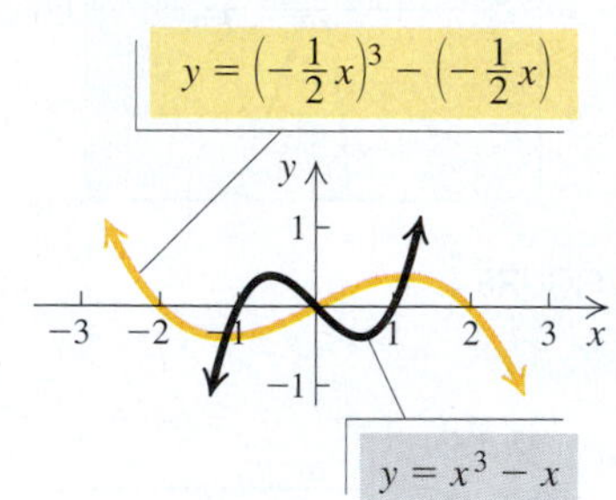

The constant c in the equation $g(x) = f(cx)$ will *shrink* the graph of $y = f(x)$ horizontally toward the y-axis if $|c| > 1$. If $0 < |c| < 1$, the graph will be *stretched* horizontally away from the y-axis. If $c < 0$, the graph is also reflected across the y-axis.

HORIZONTAL STRETCHING AND SHRINKING

The graph of $y = f(cx)$ can be obtained from the graph of $y = f(x)$ by

shrinking horizontally for $|c| > 1$, or
stretching horizontally for $0 < |c| < 1$.

For $c < 0$, the graph is also reflected across the y-axis.
(The x-coordinates of the graph of $y = f(cx)$ can be obtained by dividing the x-coordinates of the graph of $y = f(x)$ by c.)

It is instructive to use these concepts to create transformations of a given graph.

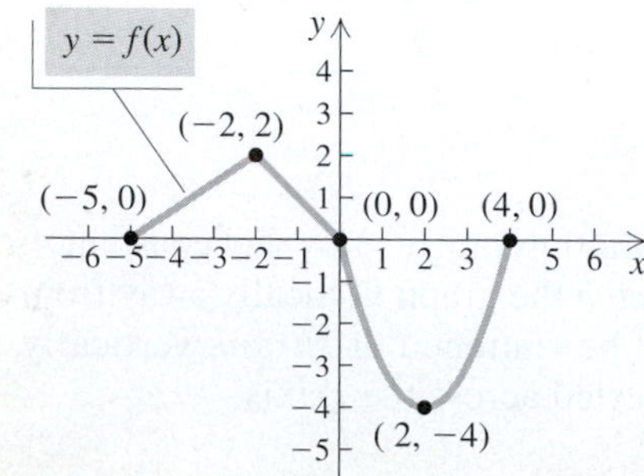

EXAMPLE 3 Shown at left is a graph of $y = f(x)$ for some function f. No formula for f is given. Graph each of the following.

a) $g(x) = 2f(x)$ **b)** $h(x) = \frac{1}{2}f(x)$
c) $r(x) = f(2x)$ **d)** $s(x) = f\left(\frac{1}{2}x\right)$
e) $t(x) = f\left(-\frac{1}{2}x\right)$

Solution

a) Since $|2| > 1$, the graph of $g(x) = 2f(x)$ is a vertical stretching of the graph of $y = f(x)$ by a factor of 2. We can consider the key points $(-5, 0)$, $(-2, 2)$, $(0, 0)$, $(2, -4)$, and $(4, 0)$ on the graph of $y = f(x)$. The transformation multiplies each

y-coordinate by 2 to obtain the key points $(-5, 0)$, $(-2, 4)$, $(0, 0)$, $(2, -8)$, and $(4, 0)$ on the graph of $g(x) = 2f(x)$. The graph is shown below.

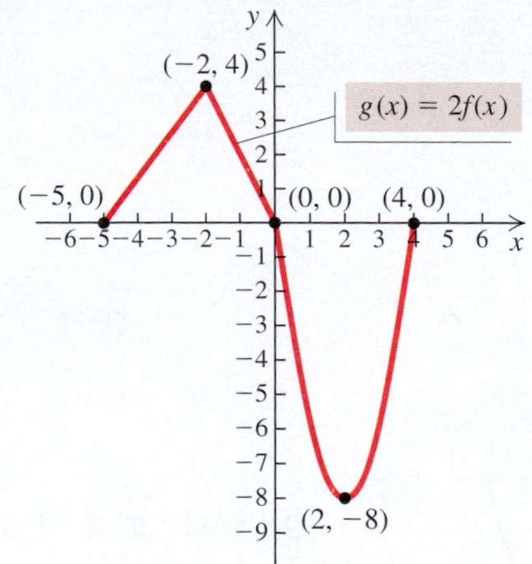

b) Since $\left|\frac{1}{2}\right| < 1$, the graph of $h(x) = \frac{1}{2}f(x)$ is a vertical shrinking of the graph of $y = f(x)$ by a factor of $\frac{1}{2}$. We again consider the key points $(-5, 0)$, $(-2, 2)$, $(0, 0)$, $(2, -4)$, and $(4, 0)$ on the graph of $y = f(x)$. The transformation multiplies each y-coordinate by $\frac{1}{2}$ to obtain the key points $(-5, 0)$, $(-2, 1)$, $(0, 0)$, $(2, -2)$, and $(4, 0)$ on the graph of $h(x) = \frac{1}{2}f(x)$. The graph is shown on the left below.

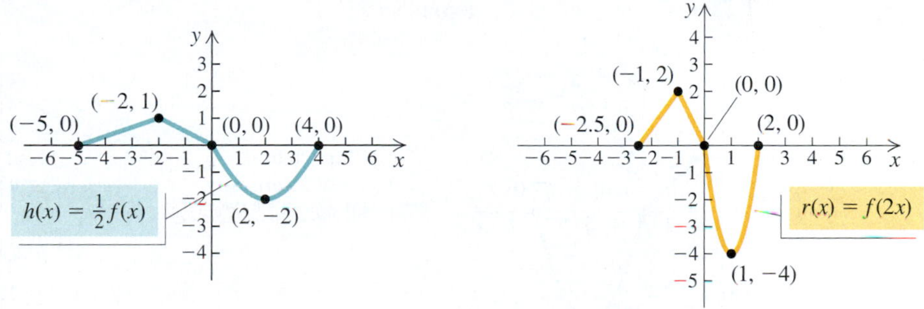

c) Since $|2| > 1$, the graph of $r(x) = f(2x)$ is a horizontal shrinking of the graph of $y = f(x)$. We consider the key points $(-5, 0)$, $(-2, 2)$, $(0, 0)$, $(2, -4)$, and $(4, 0)$ on the graph of $y = f(x)$. The transformation divides each x-coordinate by 2 to obtain the key points $(-2.5, 0)$, $(-1, 2)$, $(0, 0)$, $(1, -4)$, and $(2, 0)$ on the graph of $r(x) = f(2x)$. The graph is shown on the right above.

d) Since $\left|\frac{1}{2}\right| < 1$, the graph of $s(x) = f\left(\frac{1}{2}x\right)$ is a horizontal stretching of the graph of $y = f(x)$. We consider the key points $(-5, 0)$, $(-2, 2)$, $(0, 0)$, $(2, -4)$, and $(4, 0)$ on the graph of $y = f(x)$. The transformation divides each x-coordinate by $\frac{1}{2}$ (which is the same as multiplying by 2) to obtain the key points $(-10, 0)$, $(-4, 2)$, $(0, 0)$, $(4, -4)$, and $(8, 0)$ on the graph of $s(x) = f\left(\frac{1}{2}x\right)$. The graph is shown below.

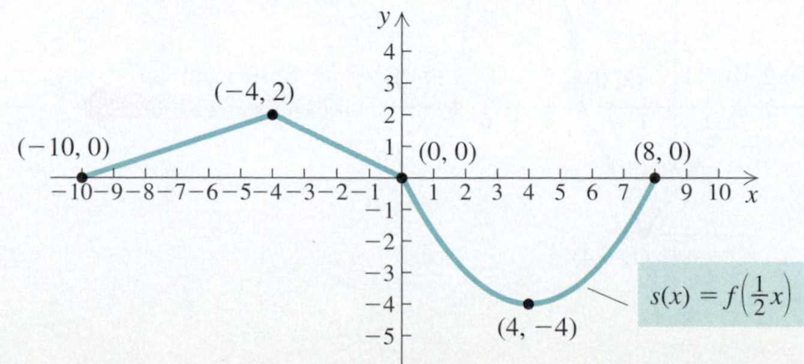

e) The graph of $t(x) = f\left(-\frac{1}{2}x\right)$ can be obtained by reflecting the graph in part (d) across the y-axis, as shown below.

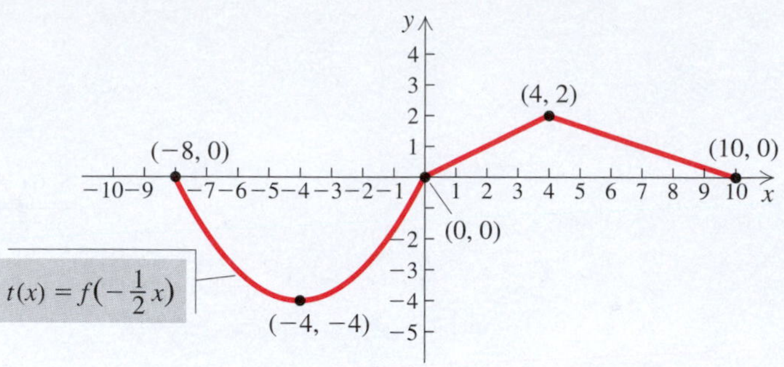

> Now Try Exercises 59 and 61.

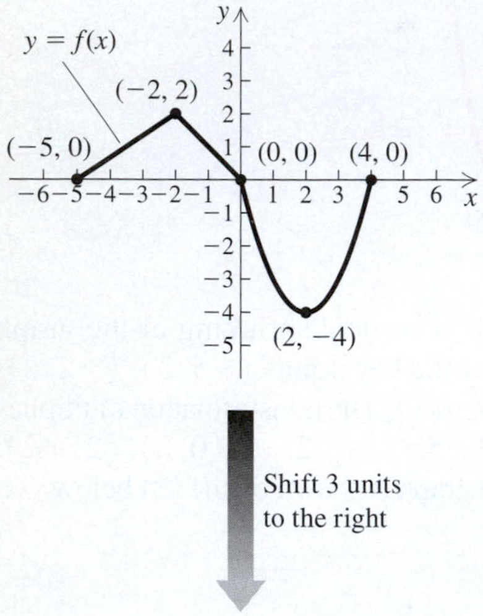

EXAMPLE 4 Use the graph of $y = f(x)$ shown at left to graph

$$y = -2f(x - 3) + 1.$$

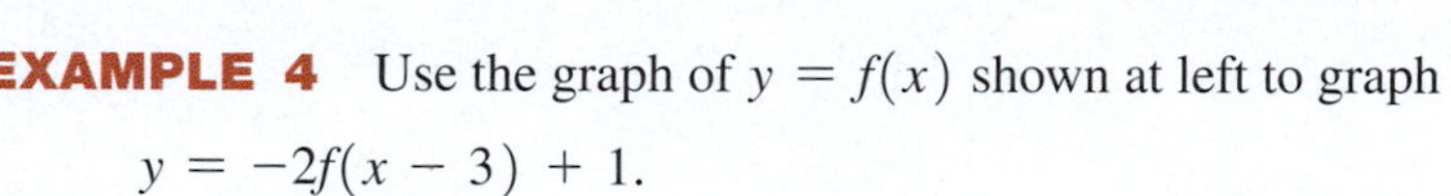

Solution

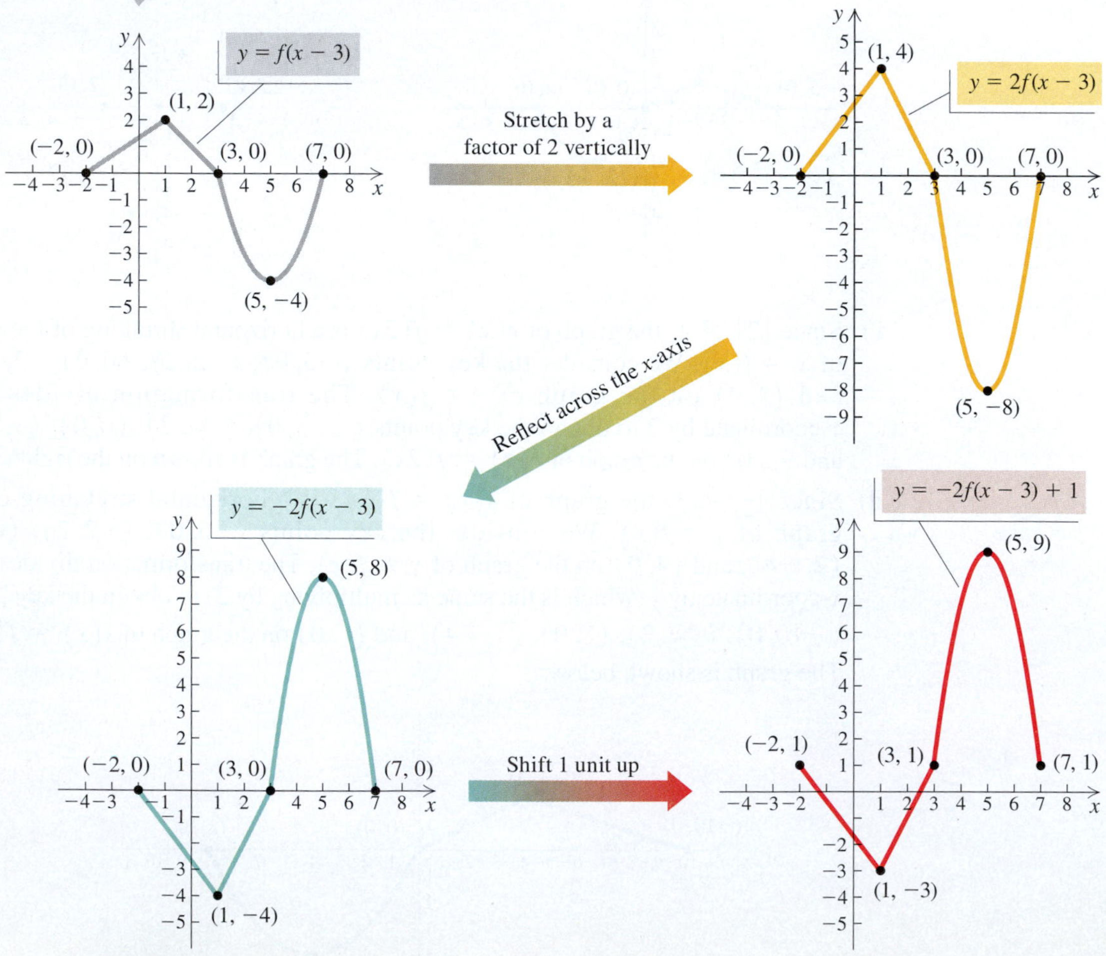

Shift 3 units to the right

Stretch by a factor of 2 vertically

Reflect across the x-axis

Shift 1 unit up

> Now Try Exercise 63.

SUMMARY OF TRANSFORMATIONS OF $y = f(x)$

***Vertical Translation:* $y = f(x) \pm b$**

For $b > 0$:

the graph of $y = f(x) + b$ is the graph of $y = f(x)$ shifted *up* b units;

the graph of $y = f(x) - b$ is the graph of $y = f(x)$ shifted *down* b units.

***Horizontal Translation:* $y = f(x \mp d)$**

For $d > 0$:

the graph of $y = f(x - d)$ is the graph of $y = f(x)$ shifted *right* d units;

the graph of $y = f(x + d)$ is the graph of $y = f(x)$ shifted *left* d units.

Reflections

Across the x-axis:

The graph of $y = -f(x)$ is the reflection of the graph of $y = f(x)$ across the x-axis.

Across the y-axis:

The graph of $y = f(-x)$ is the reflection of the graph of $y = f(x)$ across the y-axis.

***Vertical Stretching or Shrinking:* $y = af(x)$**

The graph of $y = af(x)$ can be obtained from the graph of $y = f(x)$ by

stretching vertically for $|a| > 1$, or
shrinking vertically for $0 < |a| < 1$.

For $a < 0$, the graph is also reflected across the x-axis.

***Horizontal Stretching or Shrinking:* $y = f(cx)$**

The graph of $y = f(cx)$ can be obtained from the graph of $y = f(x)$ by

shrinking horizontally for $|c| > 1$, or
stretching horizontally for $0 < |c| < 1$.

For $c < 0$, the graph is also reflected across the y-axis.

Visualizing the Graph

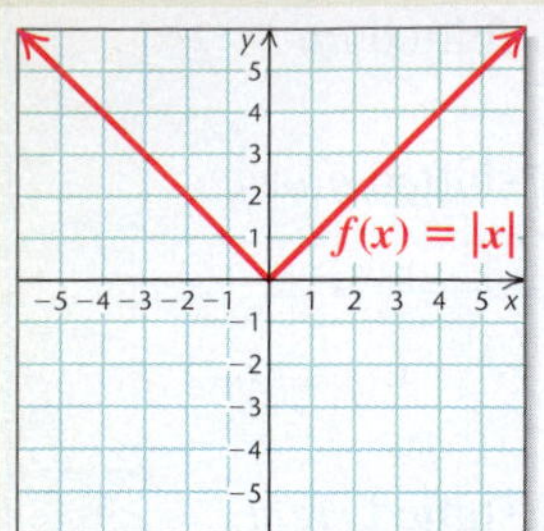

Match the function with its graph. Use transformation graphing techniques to obtain the graph of g from the basic function $f(x) = |x|$ shown at top left.

1. $g(x) = -2|x|$

2. $g(x) = |x - 1| + 1$

3. $g(x) = -\left|\frac{1}{3}x\right|$

4. $g(x) = |2x|$

5. $g(x) = |x + 2|$

6. $g(x) = |x| + 3$

7. $g(x) = -\frac{1}{2}|x - 4|$

8. $g(x) = \frac{1}{2}|x| - 3$

9. $g(x) = -|x| - 2$

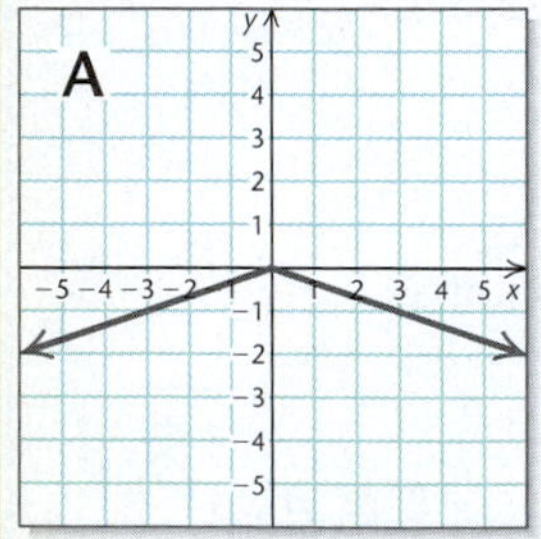

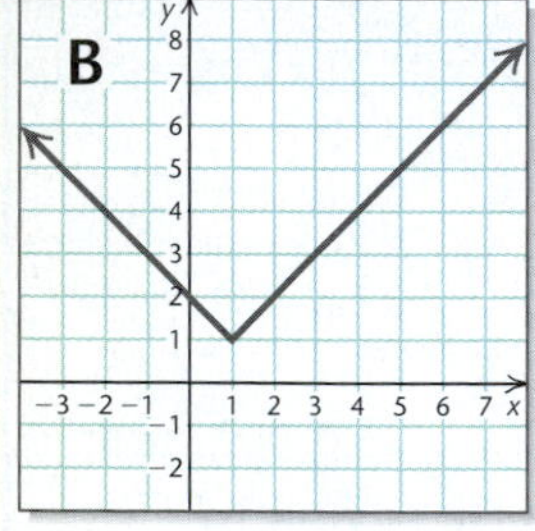

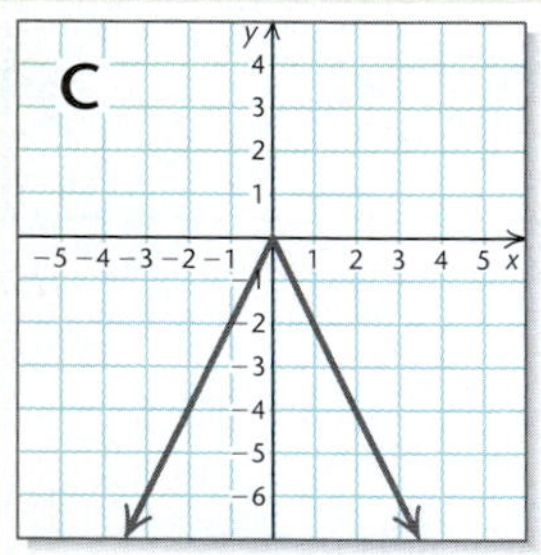

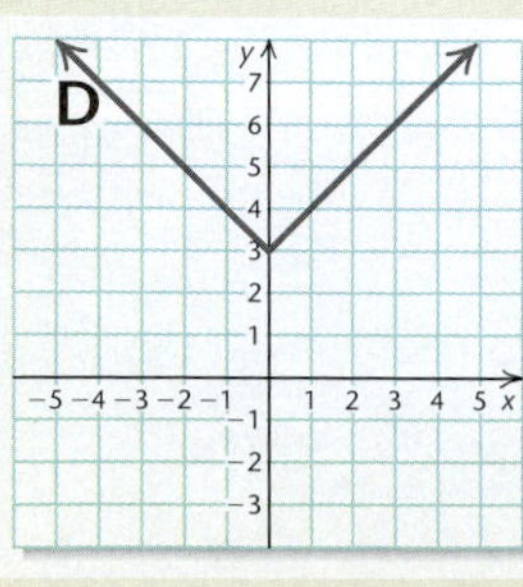

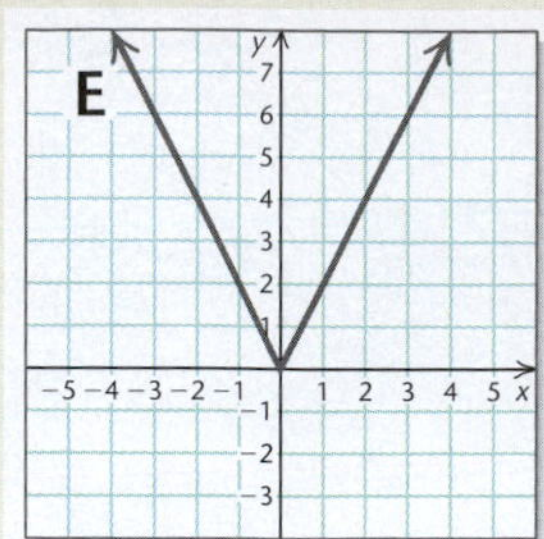

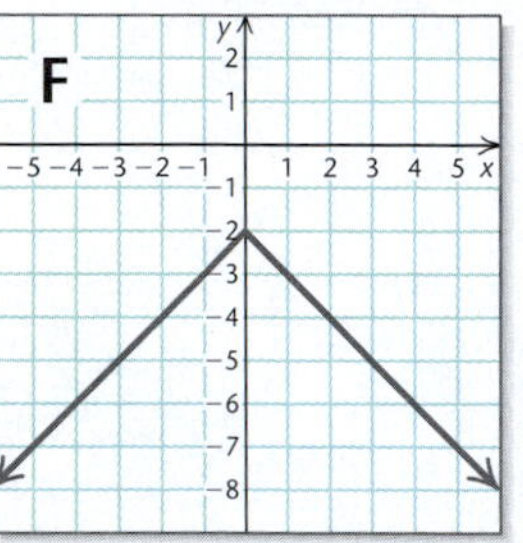

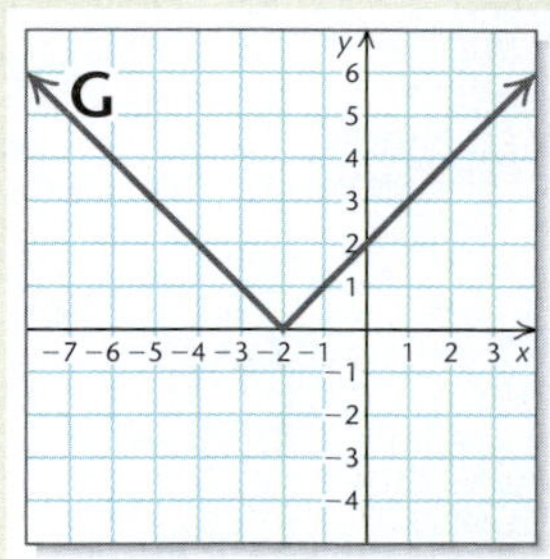

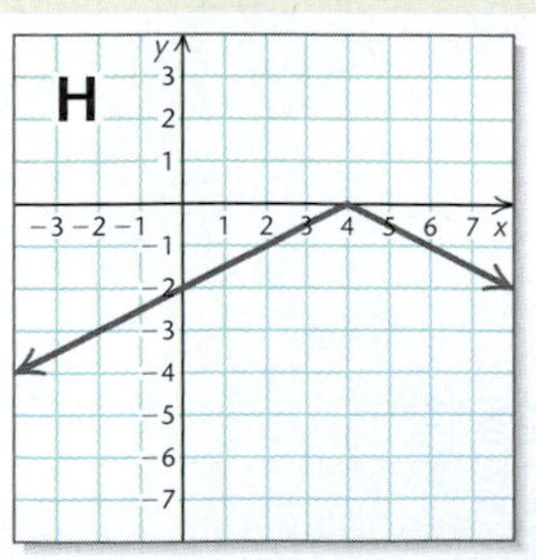

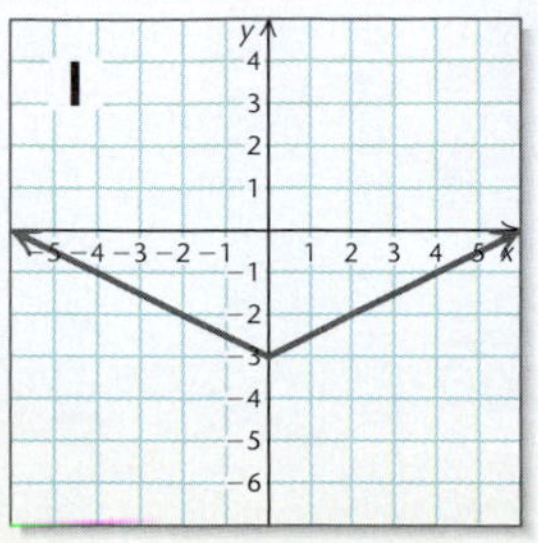

Answers on page A-10

2.5 Exercise Set

Describe how the graph of the function can be obtained from one of the basic graphs on p. 129. Then graph the function by hand or with a graphing calculator.

1. $f(x) = (x - 3)^2$
2. $g(x) = x^2 + \frac{1}{2}$
3. $g(x) = x - 3$
4. $g(x) = -x - 2$
5. $h(x) = -\sqrt{x}$
6. $g(x) = \sqrt{x - 1}$
7. $h(x) = \frac{1}{x} + 4$
8. $g(x) = \frac{1}{x - 2}$
9. $h(x) = -3x + 3$
10. $f(x) = 2x + 1$
11. $h(x) = \frac{1}{2}|x| - 2$
12. $g(x) = -|x| + 2$
13. $g(x) = -(x - 2)^3$
14. $f(x) = (x + 1)^3$
15. $g(x) = (x + 1)^2 - 1$
16. $h(x) = -x^2 - 4$
17. $g(x) = \frac{1}{3}x^3 + 2$
18. $h(x) = (-x)^3$
19. $f(x) = \sqrt{x + 2}$
20. $f(x) = -\frac{1}{2}\sqrt{x - 1}$
21. $f(x) = \sqrt[3]{x} - 2$
22. $h(x) = \sqrt[3]{x + 1}$

Describe how the graph of the function can be obtained from one of the basic graphs on p. 129.

23. $g(x) = |3x|$
24. $f(x) = \frac{1}{2}\sqrt[3]{x}$
25. $h(x) = \frac{2}{x}$
26. $f(x) = |x - 3| - 4$
27. $f(x) = 3\sqrt{x} - 5$
28. $f(x) = 5 - \frac{1}{x}$
29. $g(x) = \left|\frac{1}{3}x\right| - 4$
30. $f(x) = \frac{2}{3}x^3 - 4$
31. $f(x) = -\frac{1}{4}(x - 5)^2$
32. $f(x) = (-x)^3 - 5$
33. $f(x) = \frac{1}{x + 3} + 2$
34. $g(x) = \sqrt{-x} + 5$
35. $h(x) = -(x - 3)^2 + 5$
36. $f(x) = 3(x + 4)^2 - 3$

The point $(-12, 4)$ is on the graph of $y = f(x)$. Find the corresponding point on the graph of $y = g(x)$.

37. $g(x) = \frac{1}{2}f(x)$
38. $g(x) = f(x - 2)$
39. $g(x) = f(-x)$
40. $g(x) = f(4x)$
41. $g(x) = f(x) - 2$
42. $g(x) = f\left(\frac{1}{2}x\right)$
43. $g(x) = 4f(x)$
44. $g(x) = -f(x)$

Given that $f(x) = x^2 + 3$, match the function g with a transformation of f from one of A–D.

45. $g(x) = x^2 + 4$ — A. $f(x - 2)$
46. $g(x) = 9x^2 + 3$ — B. $f(x) + 1$
47. $g(x) = (x - 2)^2 + 3$ — C. $2f(x)$
48. $g(x) = 2x^2 + 6$ — D. $f(3x)$

Write an equation for a function that has a graph with the given characteristics.

49. The shape of $y = x^2$, but reflected across the x-axis and shifted right 8 units
50. The shape of $y = \sqrt{x}$, but shifted left 6 units and down 5 units
51. The shape of $y = |x|$, but shifted left 7 units and up 2 units
52. The shape of $y = x^3$, but reflected across the x-axis and shifted right 5 units
53. The shape of $y = 1/x$, but shrunk horizontally by a factor of 2 and shifted down 3 units
54. The shape of $y = x^2$, but shifted right 6 units and up 2 units
55. The shape of $y = x^2$, but reflected across the x-axis and shifted right 3 units and up 4 units
56. The shape of $y = |x|$, but stretched horizontally by a factor of 2 and shifted down 5 units
57. The shape of $y = \sqrt{x}$, but reflected across the y-axis and shifted left 2 units and down 1 unit
58. The shape of $y = 1/x$, but reflected across the x-axis and shifted up 1 unit

A graph of $y = f(x)$ follows. No formula for f is given. In Exercises 59–66, graph the given equation.

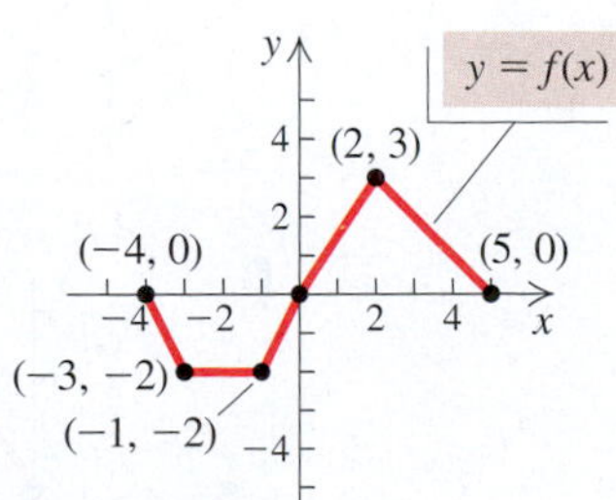

59. $g(x) = -2f(x)$
60. $g(x) = \frac{1}{2}f(x)$
61. $g(x) = f\left(-\frac{1}{2}x\right)$
62. $g(x) = f(2x)$

63. $g(x) = -\frac{1}{2}f(x - 1) + 3$

64. $g(x) = -3f(x + 1) - 4$

65. $g(x) = f(-x)$ **66.** $g(x) = -f(x)$

A graph of $y = g(x)$ follows. No formula for g is given. In Exercises 67–70, graph the given equation.

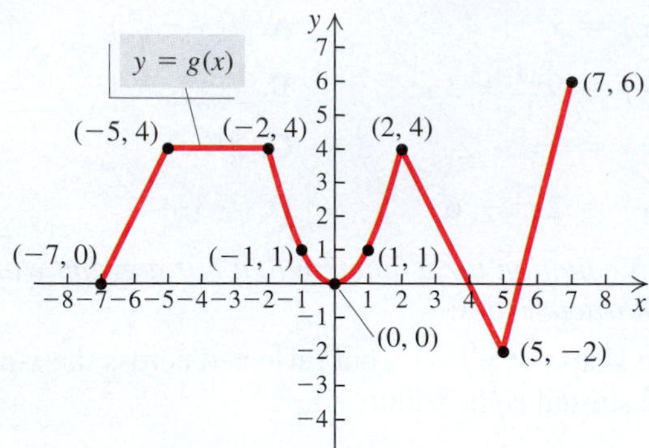

67. $h(x) = -g(x + 2) + 1$

68. $h(x) = \frac{1}{2}g(-x)$ **69.** $h(x) = g(2x)$

70. $h(x) = 2g(x - 1) - 3$

The graph of the function f is shown in figure (a). In Exercises 71–78, match the function g with one of the graphs (a)–(h) that follow. Some graphs may be used more than once and some may not be used at all.

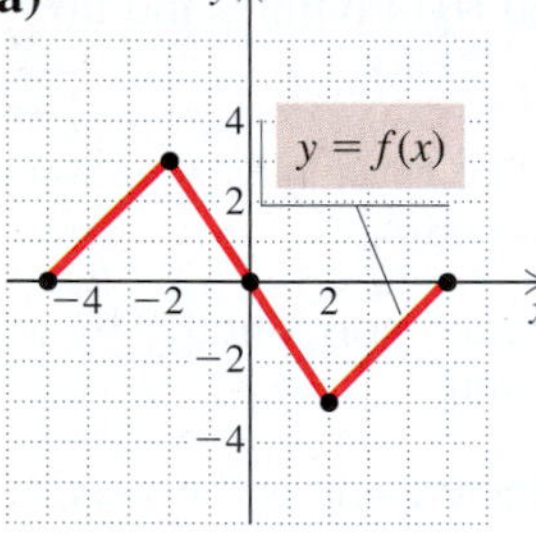

b)

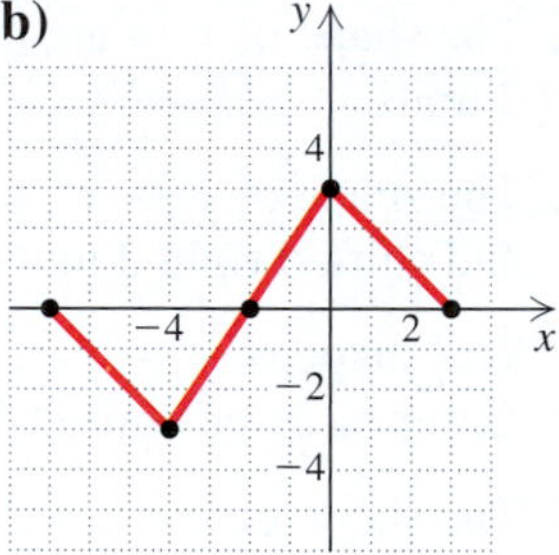

c)

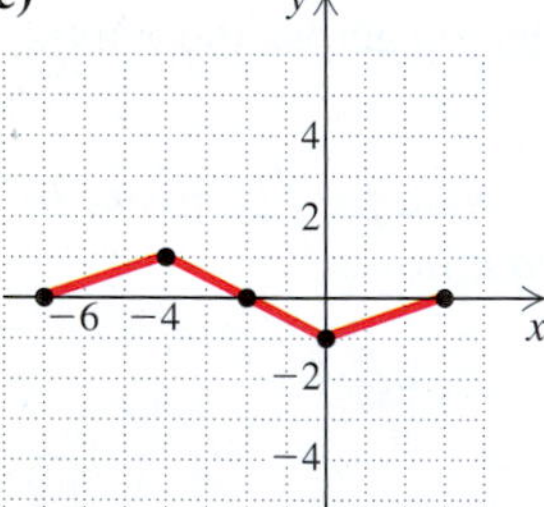

d)

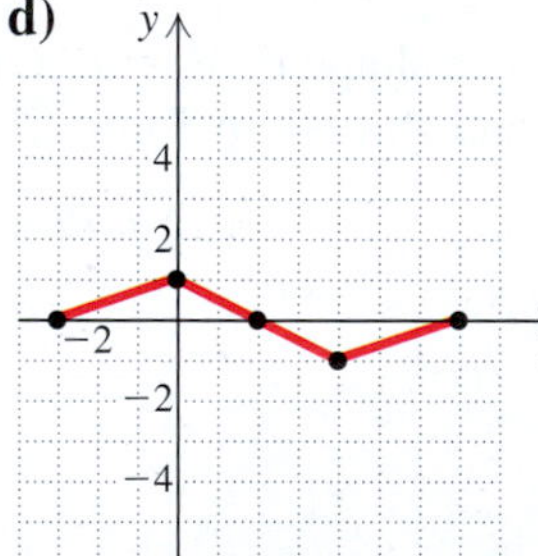

e)

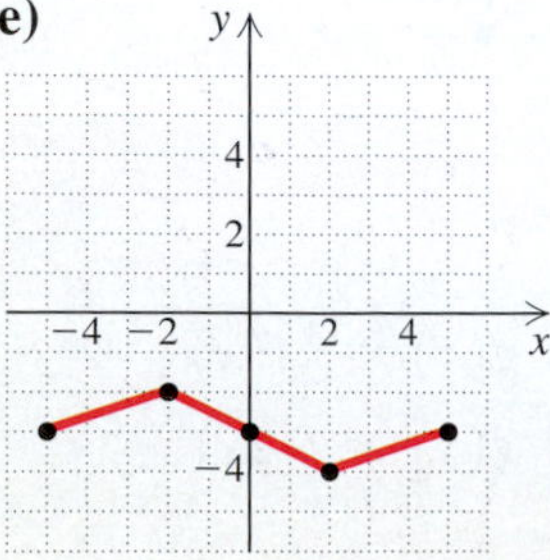

f)

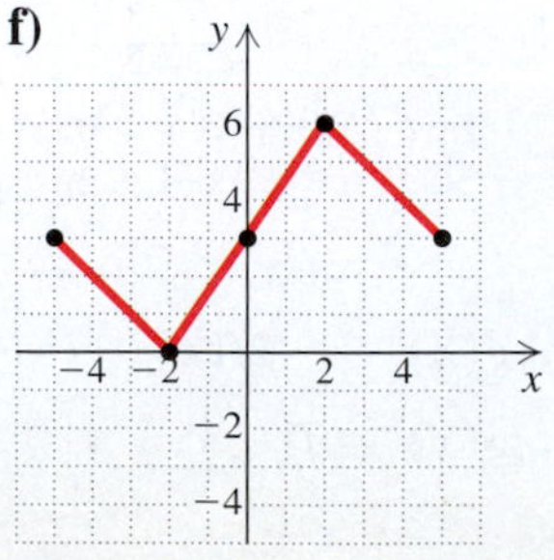

g)

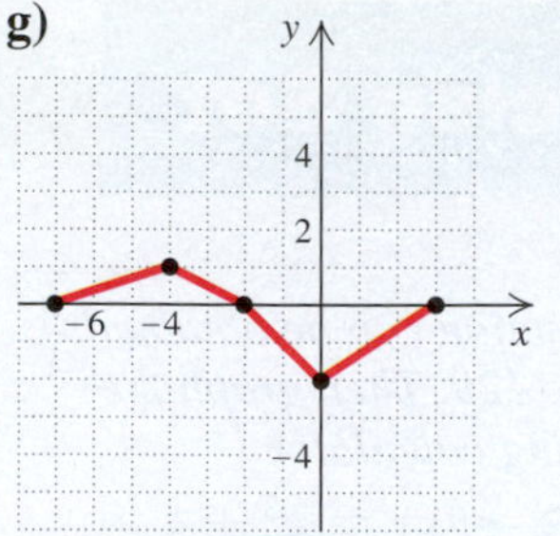

h)

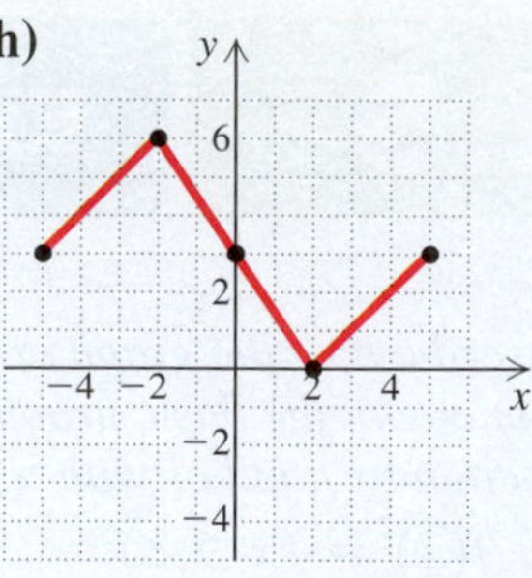

71. $g(x) = f(-x) + 3$

72. $g(x) = f(x) + 3$

73. $g(x) = -f(x) + 3$

74. $g(x) = -f(-x)$

75. $g(x) = \frac{1}{3}f(x - 2)$

76. $g(x) = \frac{1}{3}f(x) - 3$

77. $g(x) = \frac{1}{3}f(x + 2)$

78. $g(x) = -f(x + 2)$

For each pair of functions, determine whether $g(x) = f(-x)$.

79. $f(x) = 2x^4 - 35x^3 + 3x - 5$,
$g(x) = 2x^4 + 35x^3 - 3x - 5$

80. $f(x) = \frac{1}{4}x^4 + \frac{1}{5}x^3 - 81x^2 - 17$,
$g(x) = \frac{1}{4}x^4 + \frac{1}{5}x^3 + 81x^2 - 17$

A graph of the function $f(x) = x^3 - 3x^2$ is shown below. Exercises 81–84 show graphs of functions transformed from this one. Find a formula for each function.

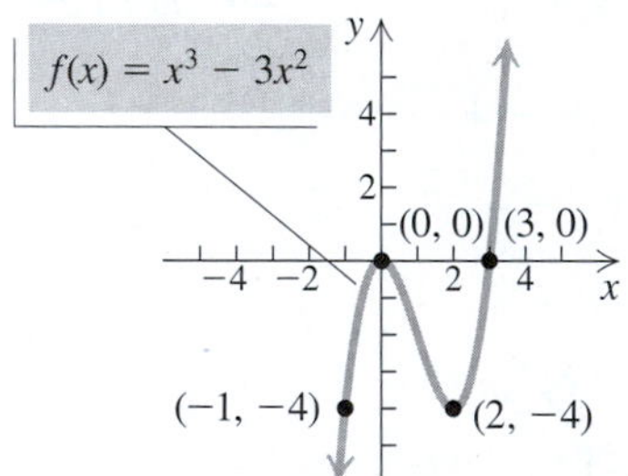

81.

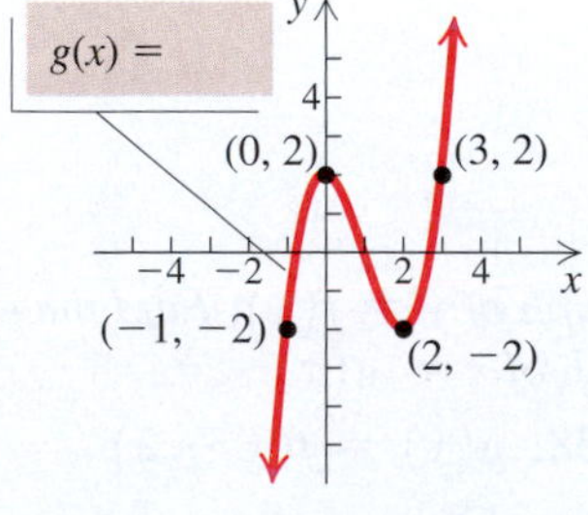

82.

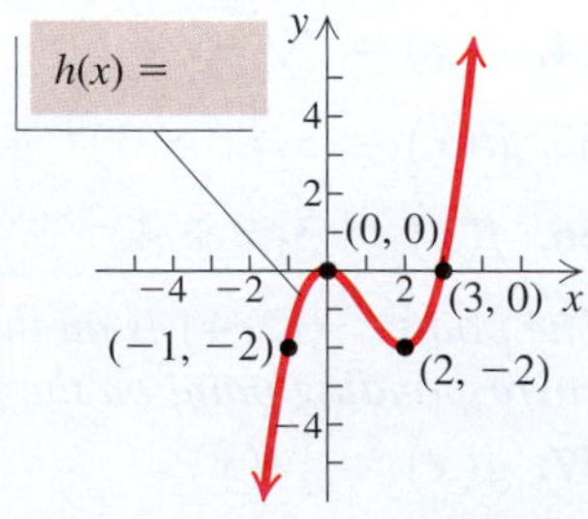

83.

84.

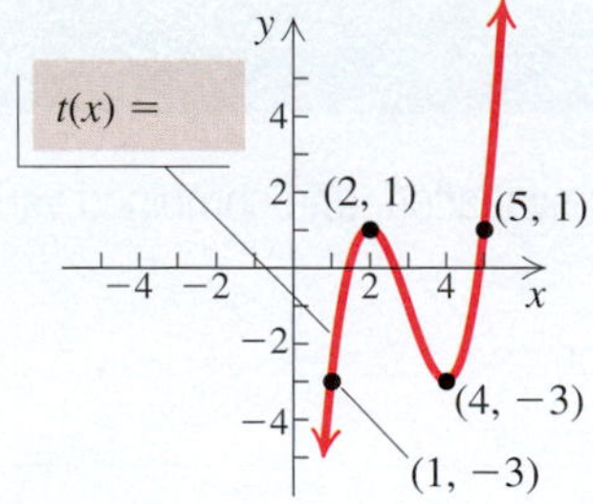

› Skill Maintenance

Determine algebraically whether the graph is symmetric, with respect to the x-axis, the y-axis, and the origin. **[2.4]**

85. $y = 3x^4 - 3$

86. $y^2 = x$

87. $2x - 5y = 0$

88. *Educational Level and Income.* The average annual wages of a 64-year-old person with a bachelor's degree is $67,735. This amount is approximately 53.7% more than the average wages of a 64-year-old person with only a high school diploma. (*Sources:* U.S. Census Bureau; Bureau of Labor Statistics, Current Population Survey, March Supplement) Find the average annual wages of a 64-year-old person with only a high school diploma. **[1.5]**

89. *Medical Care Abroad.* The cost of medical care abroad is often significantly less than in the United States. In 2014, the cost of knee replacement surgery in the United States was $34,000. This cost was $4000 more than four times the cost of knee replacement surgery in India. (*Sources:* aarp.org, Sarah Barchus and Beth Howard; Center for Medical Tourism Research in San Antonio) Find the cost of knee replacement surgery in India. **[1.5]**

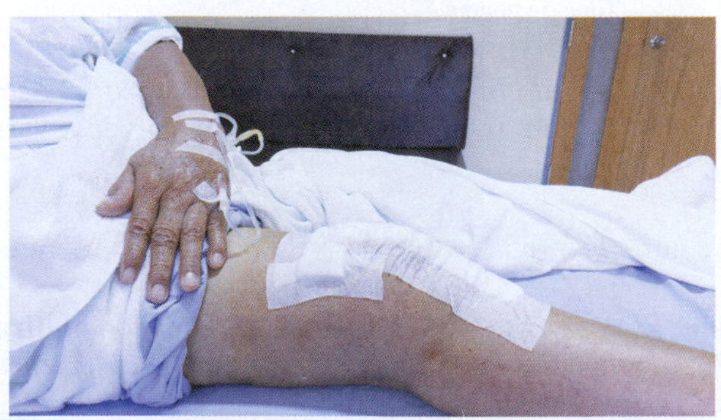

90. *Undergraduate Students from Foreign Countries.* In 2013–2014, Saudi Arabia and Canada sent 82,223 undergraduate students to the United States to study in universities. Saudi Arabia sent 25,615 more students than Canada sent. (*Source:* Institute for International Educators' "Open Doors" report) How many students did each country send to the United States to study? **[1.5]**

› Synthesis

Use the following graph of the function f for Exercises 91 and 92.

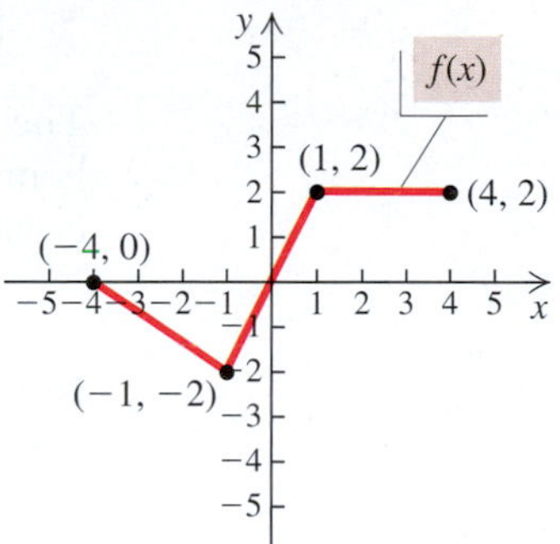

91. Graph: $y = |f(x)|$.

92. Graph: $y = f(|x|)$.

Use the following graph of the function g for Exercises 93 and 94.

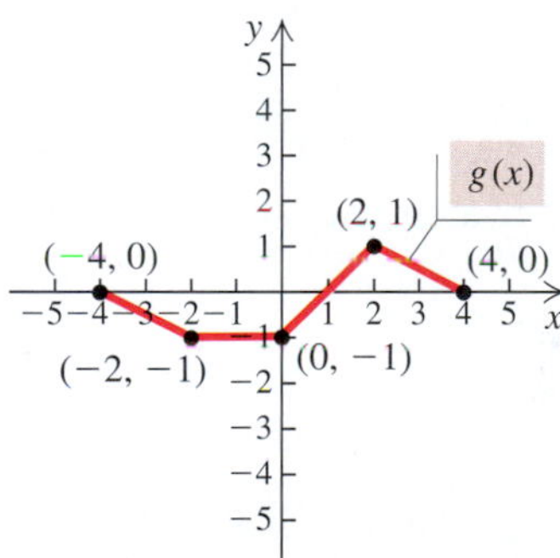

93. Graph: $y = g(|x|)$.

94. Graph: $y = |g(x)|$.

Graph each of the following using a graphing calculator. Before doing so, describe how the graph can be obtained from a more basic graph. Give the domain and the range of the function.

95. $f(x) = [\![x - \frac{1}{2}]\!]$

96. $f(x) = |\sqrt{x} - 1|$

97. If $(3, 4)$ is a point on the graph of $y = f(x)$, what point do you know is on the graph of $y = 2f(x)$? of $y = 2 + f(x)$? of $y = f(2x)$?

98. Find the zeros of $f(x) = 3x^5 - 20x^3$. Then, without using a graphing calculator, state the zeros of $f(x - 3)$ and $f(x + 8)$.

2.6 Variation and Applications

- Find equations of direct variation, inverse variation, and combined variation given values of the variables.
- Solve applied problems involving variation.

We now extend our study of formulas and functions by considering applications involving variation.

Direct Variation

The median hourly wage for an elevator and escalator installer/repairer is \$35 per hour (*Source:* U.S. Bureau of Labor Statistics). In 1 hr, \$35 is earned; in 2 hr, \$70 is earned; in 3 hr, \$105 is earned; and so on. This gives rise to a set of ordered pairs:

$$(1, 35), \quad (2, 70), \quad (3, 105), \quad (4, 140), \quad \text{and so on.}$$

Note that the ratio of the second coordinate to the first coordinate is the same number for each pair:

$$\frac{35}{1} = 35, \quad \frac{70}{2} = 35, \quad \frac{105}{3} = 35, \quad \frac{140}{4} = 35, \quad \text{and so on.}$$

Earnings for Elevator and Escalator Installer/Repairer

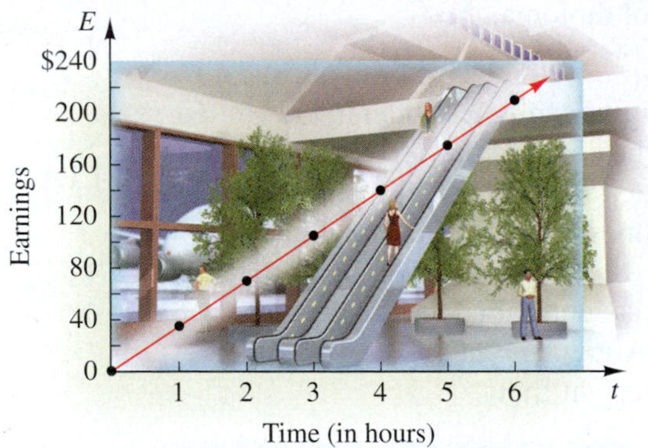

Whenever a situation produces pairs of numbers in which the *ratio is constant*, we say that there is **direct variation**. In this case, the amount earned E varies directly as the time worked t:

$$\frac{E}{t} = 35 \text{ (a constant)}, \quad \text{or} \quad E = 35t,$$

or, if we use function notation, $E(t) = 35t$. This equation is an equation of **direct variation**. The coefficient, 35, is called the **variation constant**. In this case, it is the rate of change of earnings with respect to time.

The graph of $y = kx, k > 0$, always goes through the origin and rises from left to right. Note that as x increases, y increases; that is, the function is increasing on the interval $(0, \infty)$. The constant k is also the slope of the line.

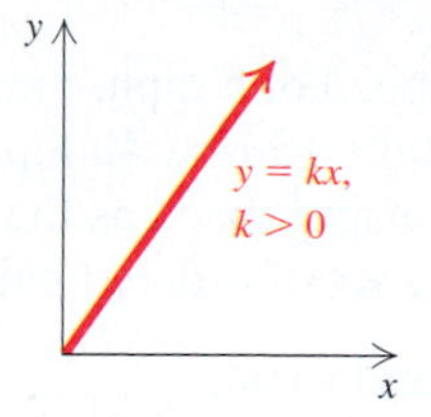

DIRECT VARIATION

If a situation gives rise to a linear function $f(x) = kx$, or $y = kx$, where k is a positive constant, we say that we have **direct variation**, or that **y varies directly as x**, or that **y is directly proportional to x**. The number k is called the **variation constant**, or the **constant of proportionality**.

EXAMPLE 1 Find the variation constant and an equation of variation in which y varies directly as x, and $y = 32$ when $x = 2$.

Solution We know that $(2, 32)$ is a solution of $y = kx$. Thus,

$$y = kx$$
$$32 = k \cdot 2 \quad \text{Substituting}$$
$$\frac{32}{2} = k \quad \text{Solving for } k$$
$$16 = k. \quad \text{Simplifying}$$

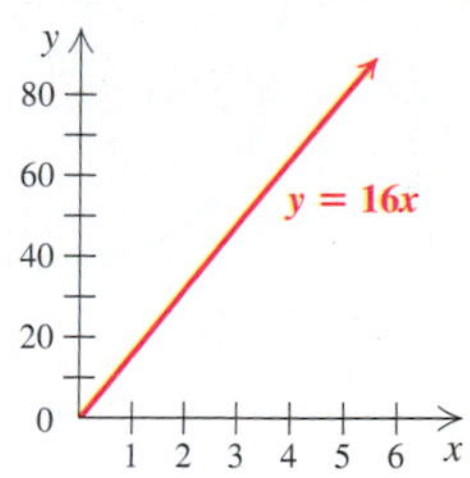

The variation constant, 16, is the rate of change of y with respect to x. The equation of variation is $y = 16x$.

Now Try Exercise 1.

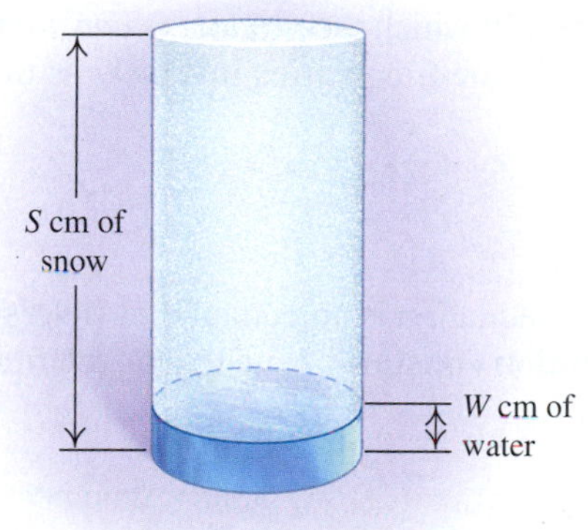

EXAMPLE 2 *Water from Melting Snow.* The number of centimeters of water W produced from melting snow varies directly as S, the number of centimeters of snow. Meteorologists have found that under certain conditions 150 cm of snow will melt to 16.8 cm of water. To how many centimeters of water will 200 cm of snow melt under the same conditions?

Solution We can express the amount of water as a function of the amount of snow. Thus, $W(S) = kS$, where k is the variation constant. We first find k using the given data and then find an equation of variation:

$$W(S) = kS \quad W \text{ varies directly as } S.$$
$$W(150) = k \cdot 150 \quad \text{Substituting 150 for } S$$
$$16.8 = k \cdot 150 \quad \text{Replacing } W(150) \text{ with } 16.8$$
$$\frac{16.8}{150} = k \quad \text{Solving for } k$$
$$0.112 = k. \quad \text{This is the variation constant.}$$

The equation of variation is $W(S) = 0.112S$.

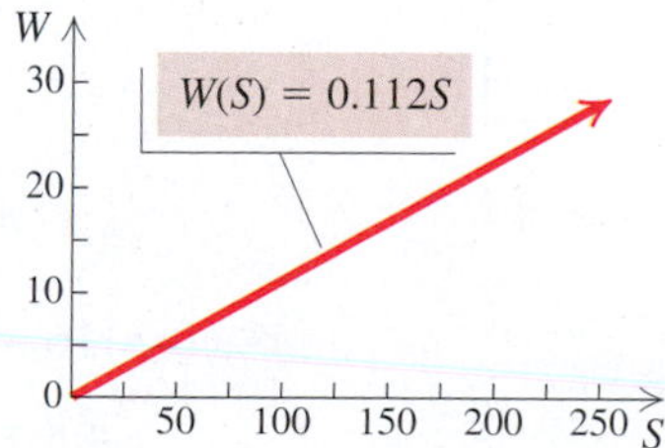

Next, we use the equation to find how many centimeters of water will result from melting 200 cm of snow:

$$W(S) = 0.112S$$
$$W(200) = 0.112(200) \quad \textbf{Substituting}$$
$$W = 22.4.$$

Thus, 200 cm of snow will melt to 22.4 cm of water.

Now Try Exercise 17.

Inverse Variation

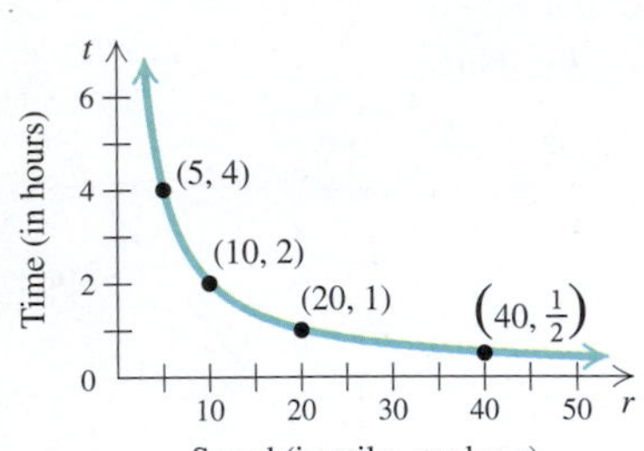

Suppose that a bus is traveling a distance of 20 mi. At a speed of 5 mph, the trip will take 4 hr; at 10 mph, it will take 2 hr; at 20 mph, it will take 1 hr; at 40 mph, it will take $\frac{1}{2}$ hr; and so on. We plot this information on a graph, using speed as the first coordinate and time as the second coordinate to determine a set of ordered pairs:

$$(5, 4), \quad (10, 2), \quad (20, 1), \quad \left(40, \tfrac{1}{2}\right), \quad \text{and so on.}$$

Note that the products of the coordinates are all the same number:

$$5 \cdot 4 = 20, \quad 10 \cdot 2 = 20, \quad 20 \cdot 1 = 20, \quad 40 \cdot \tfrac{1}{2} = 20, \quad \text{and so on.}$$

Whenever a situation produces pairs of numbers in which the *product is constant*, we say that there is **inverse variation**. In this case, the time varies inversely as the speed, or rate:

$$rt = 20 \text{ (a constant)}, \quad \text{or} \quad t = \frac{20}{r},$$

or, if we use function notation, $t(r) = 20/r$. This equation is an equation of **inverse variation**. The coefficient, 20, is called the **variation constant**. Note that as the first number increases, the second number decreases.

The graph of $y = k/x$, $k > 0$, is like the one shown below. Note that as x increases, y decreases; that is, the function is decreasing on the interval $(0, \infty)$.

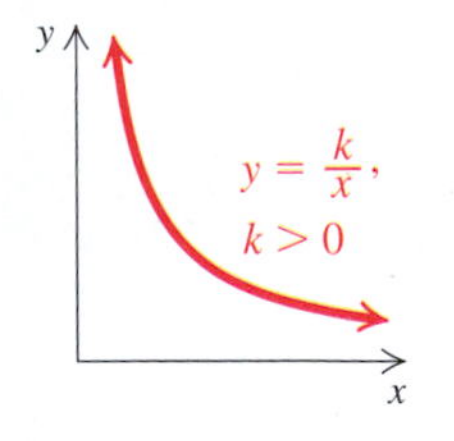

INVERSE VARIATION

If a situation gives rise to a function $f(x) = k/x$, or $y = k/x$, where k is a positive constant, we say that we have **inverse variation**, or that ***y* varies inversely as *x***, or that ***y* is inversely proportional to *x***. The number k is called the **variation constant**, or the **constant of proportionality**.

EXAMPLE 3 Find the variation constant and an equation of variation in which y varies inversely as x, and $y = 16$ when $x = 0.3$.

Solution We know that $(0.3, 16)$ is a solution of $y = k/x$. We substitute:

$$y = \frac{k}{x}$$
$$16 = \frac{k}{0.3} \quad \textbf{Substituting}$$
$$(0.3)16 = k \quad \textbf{Solving for } k$$
$$4.8 = k. \quad \textbf{Simplifying}$$

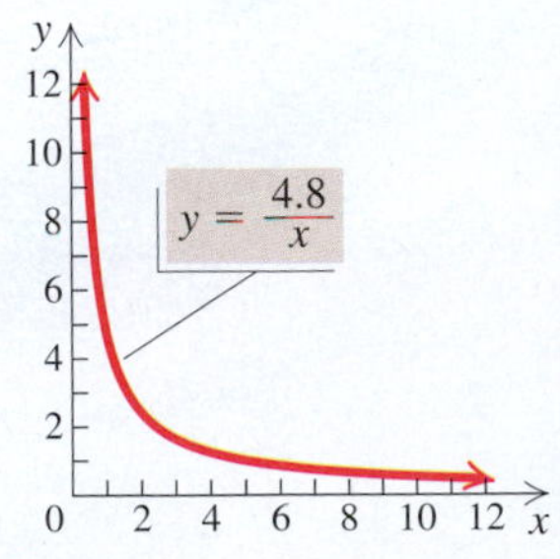

The variation constant is 4.8. The equation of variation is $y = 4.8/x$.

Now Try Exercise 3.

There are many real-world problems that translate to an equation of inverse variation.

EXAMPLE 4 ***Filling a Swimming Pool.*** The time t required to fill a swimming pool varies inversely as the rate of flow r of water into the pool. A tank truck can fill a pool in 90 min at a rate of 1500 L/min. How long would it take to fill the pool at a rate of 1800 L/min?

Solution We can express the amount of time required as a function of the rate of flow. Thus we have $t(r) = k/r$. We first find k using the given information and then find an equation of variation:

$$t(r) = \frac{k}{r} \qquad t \text{ varies inversely as } r.$$

$$t(1500) = \frac{k}{1500} \qquad \text{Substituting 1500 for } r$$

$$90 = \frac{k}{1500} \qquad \text{Replacing } t(1500) \text{ with 90}$$

$$90 \cdot 1500 = k \qquad \text{Solving for } k$$

$$135{,}000 = k. \qquad \text{This is the variation constant.}$$

The equation of variation is

$$t(r) = \frac{135{,}000}{r}.$$

Next, we use the equation to find the time that it would take to fill the pool at a rate of 1800 L/min:

$$t(r) = \frac{135{,}000}{r}$$

$$t(1800) = \frac{135{,}000}{1800} \qquad \text{Substituting}$$

$$t = 75.$$

Thus it would take 75 min to fill the pool at a rate of 1800 L/min.

Now Try Exercise 15.

Let's summarize the procedure for solving variation problems.

SOLVING VARIATION PROBLEMS

1. Determine whether direct variation or inverse variation applies.
2. Write an equation of the form $y = kx$ (for direct variation) or $y = k/x$ (for inverse variation), substitute the known values, and solve for k.
3. Write the equation of variation, and use it to find the unknown value(s) in the problem.

Combined Variation

We now look at other kinds of variation.

y varies **directly as the nth power of x** if there is some positive constant k such that

$$y = kx^n.$$

y varies **inversely as the nth power of x** if there is some positive constant k such that

$$y = \frac{k}{x^n}.$$

y varies **jointly as x and z** if there is some positive constant k such that

$$y = kxz.$$

There are other types of combined variation as well. Consider the formula for the volume of a right circular cylinder, $V = \pi r^2 h$, in which V, r, and h are variables and π is a constant. We say that V varies jointly as h and the square of r. In this formula, π is the variation constant.

EXAMPLE 5 Find an equation of variation in which y varies directly as the square of x, and $y = 12$ when $x = 2$.

Solution We write an equation of variation and find k:

$$\begin{aligned} y &= kx^2 \\ 12 &= k \cdot 2^2 \quad \text{Substituting} \\ 12 &= k \cdot 4 \\ 3 &= k. \end{aligned}$$

Thus, $y = 3x^2$.

Now Try Exercise 27.

EXAMPLE 6 Find an equation of variation in which y varies jointly as x and z, and $y = 42$ when $x = 2$ and $z = 3$.

Solution We have

$$\begin{aligned} y &= kxz \\ 42 &= k \cdot 2 \cdot 3 \quad \text{Substituting} \\ 42 &= k \cdot 6 \\ 7 &= k. \end{aligned}$$

Thus, $y = 7xz$.

Now Try Exercise 29.

EXAMPLE 7 Find an equation of variation in which y varies jointly as x and z and inversely as the square of w, and $y = 105$ when $x = 3$, $z = 20$, and $w = 2$.

Solution We have

$$\begin{aligned} y &= k \cdot \frac{xz}{w^2} \\ 105 &= k \cdot \frac{3 \cdot 20}{2^2} \quad \text{Substituting} \\ 105 &= k \cdot 15 \\ 7 &= k. \end{aligned}$$

Thus, $y = 7\frac{xz}{w^2}$, or $y = \frac{7xz}{w^2}$.

Now Try Exercise 33.

Many applied problems can be modeled using equations of combined variation.

EXAMPLE 8 *Volume of a Tree.* The volume of wood V in a tree varies jointly as the height h and the square of the girth g. (Girth is distance around.) If the volume of a redwood tree is 216 m^3 when the height is 30 m and the girth is 1.5 m, what is the height of a tree whose volume is 344 m^3 and whose girth is 1.6 m?

Solution We first find k using the first set of data. Then we solve for h using the second set of data.

$$V = khg^2$$
$$216 = k \cdot 30 \cdot 1.5^2$$
$$216 = k \cdot 30 \cdot 2.25$$
$$216 = k \cdot 67.5$$
$$3.2 = k$$

Thus the equation of variation is $V = 3.2hg^2$. We substitute the second set of data into the equation:

$$344 = 3.2 \cdot h \cdot 1.6^2$$
$$344 = 3.2 \cdot h \cdot 2.56$$
$$344 = 8.192 \cdot h$$
$$42 \approx h.$$

The height of the tree is about 42 m.

Now Try Exercise 35.

2.6 Exercise Set

Find the variation constant and an equation of variation for the given situation.

1. y varies directly as x, and $y = 54$ when $x = 12$
2. y varies directly as x, and $y = 0.1$ when $x = 0.2$
3. y varies inversely as x, and $y = 3$ when $x = 12$
4. y varies inversely as x, and $y = 12$ when $x = 5$
5. y varies directly as x, and $y = 1$ when $x = \frac{1}{4}$
6. y varies inversely as x, and $y = 0.1$ when $x = 0.5$
7. y varies inversely as x, and $y = 32$ when $x = \frac{1}{8}$
8. y varies directly as x, and $y = 3$ when $x = 33$
9. y varies directly as x, and $y = \frac{3}{4}$ when $x = 2$
10. y varies inversely as x, and $y = \frac{1}{5}$ when $x = 35$
11. y varies inversely as x, and $y = 1.8$ when $x = 0.3$
12. y varies directly as x, and $y = 0.9$ when $x = 0.4$

13. *Child's Allowance.* The Harrisons decide to give their children a weekly allowance that is directly proportional to each child's age. Their 6-year-old daughter receives an allowance of \$5.50. What is their 9-year-old son's allowance?

14. *Sales Tax.* The amount of sales tax paid on a product is directly proportional to the purchase price. In Iowa, the sales tax on a Nook Glowlight™ that sells for \$119 is \$7.14. What is the sales tax on an e-book that sells for \$21?

15. *Rate of Travel.* The time t required to drive a fixed distance varies inversely as the speed r. It takes 5 hr at a speed of 80 km/h to drive a fixed distance. How long will it take to drive the same distance at a speed of 70 km/h?

16. *Beam Weight.* The weight W that a horizontal beam can support varies inversely as the length L of the beam. Suppose that an 8-m beam can support 1200 kg. How many kilograms can a 14-m beam support?

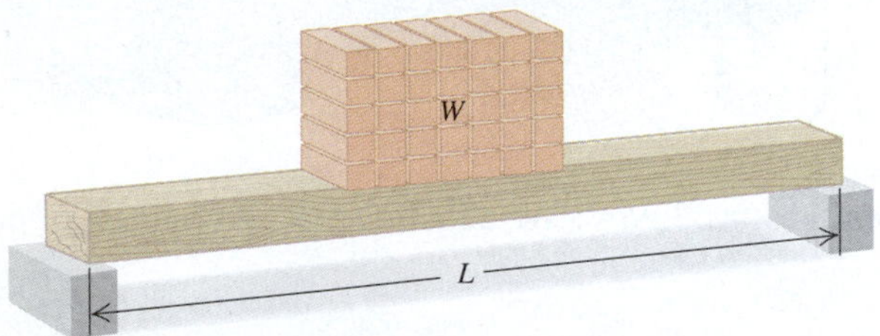

17. *Fat Intake.* The maximum number of grams of fat that should be in a diet varies directly as a person's weight. A person weighing 120 lb should consume no more than 60 g of fat per day. What is the maximum daily fat intake for a person weighing 180 lb?

18. *House of Representatives.* The number of representatives N that each state has varies directly as the number of people P living in the state. If California, with 38,333,000 residents, has 53 representatives, how many representatives does Texas, with a population of 26,448,000, have?

19. *Work Rate.* The time T required to do a job varies inversely as the number of people P working. It takes 5 hr for 7 bricklayers to build a park wall. How long will it take 10 bricklayers to complete the job?

20. *Pumping Rate.* The time t required to empty a tank varies inversely as the rate r of pumping. If a pump can empty a tank in 45 min at the rate of 600 kL/min, how long will it take the pump to empty the same tank at the rate of 1000 kL/min?

21. *Hooke's Law.* Hooke's law states that the distance d that a spring will stretch varies directly as the mass m of an object hanging from the spring. If a 3-kg mass stretches a spring 40 cm, how far will a 5-kg mass stretch the spring?

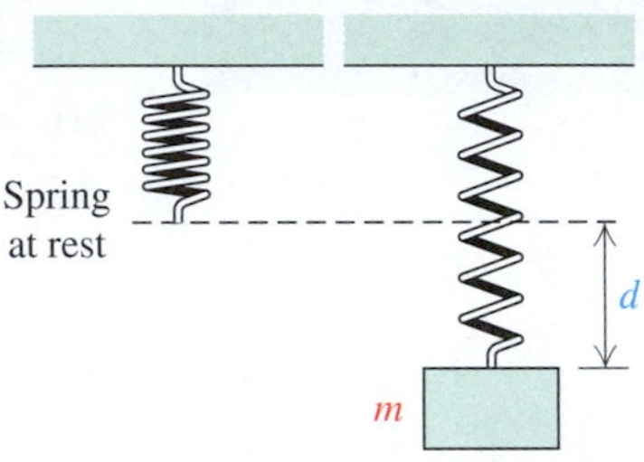

22. *Relative Aperture.* The relative aperture, or f-stop, of a 23.5-mm diameter lens is directly proportional to the focal length F of the lens. If a 150-mm focal length has an f-stop of 6.3, find the f-stop of a 23.5-mm diameter lens with a focal length of 80 mm.

23. *Musical Pitch.* The pitch P of a musical tone varies inversely as its wavelength W. One tone has a pitch of 330 vibrations per second and a wavelength of 3.2 ft. Find the wavelength of another tone that has a pitch of 550 vibrations per second.

24. *Weight on Mars.* The weight M of an object on Mars varies directly as its weight E on Earth. A person who weighs 95 lb on Earth weighs 35.9 lb on Mars. How much would a 100-lb person weigh on Mars?

Find an equation of variation for the given situation.

25. y varies inversely as the square of x, and $y = 0.15$ when $x = 0.1$

26. y varies inversely as the square of x, and $y = 6$ when $x = 3$

27. y varies directly as the square of x, and $y = 0.15$ when $x = 0.1$

28. y varies directly as the square of x, and $y = 6$ when $x = 3$

29. y varies jointly as x and z, and $y = 56$ when $x = 7$ and $z = 8$

30. y varies directly as x and inversely as z, and $y = 4$ when $x = 12$ and $z = 15$

31. y varies jointly as x and the square of z, and $y = 105$ when $x = 14$ and $z = 5$

32. y varies jointly as x and z and inversely as w, and $y = \frac{3}{2}$ when $x = 2$, $z = 3$, and $w = 4$

33. y varies jointly as x and z and inversely as the product of w and p, and $y = \frac{3}{28}$ when $x = 3$, $z = 10$, $w = 7$, and $p = 8$

34. y varies jointly as x and z and inversely as the square of w, and $y = \frac{12}{5}$ when $x = 16$, $z = 3$, and $w = 5$

35. *Intensity of Light.* The intensity I of light from a light bulb varies inversely as the square of the distance d from the bulb. Suppose that I is 90 W/m^2 (watts per square meter) when the distance is 5 m. How much *farther* would it be to a point where the intensity is 40 W/m^2?

36. *Atmospheric Drag.* Wind resistance, or atmospheric drag, tends to slow down moving objects. Atmospheric drag varies jointly as an object's surface area A and velocity v. If a car traveling at a speed of 40 mph with a surface area of 37.8 ft^2 experiences a drag of 222 N (Newtons), how fast must a car with 51 ft^2 of surface area travel in order to experience a drag force of 430 N?

37. *Braking Distance of a Car.* The braking distance d of a car after the brakes have been applied varies directly as the square of the speed r. If a car traveling 60 mph can stop in 200 ft, how fast can a car travel and still stop in 72 ft?

38. *Weight of an Astronaut.* The weight W of an object varies inversely as the square of the distance d from the center of the earth. At sea level (3978 mi from the center of the earth), an astronaut weighs 220 lb. Find his weight when he is 200 mi above the surface of the earth.

39. *Earned-Run Average.* A pitcher's earned-run average E varies directly as the number R of earned runs allowed and inversely as the number I of innings pitched. In 2014, Clayton Kershaw of the Los Angeles Dodgers had an earned-run average of 1.77. He gave up 39 earned runs in 198.1 innings. How many earned runs would he have given up had he pitched 220 innings with the same average? Round to the nearest whole number.

40. *Boyle's Law.* The volume V of a given mass of a gas varies directly as the temperature T and inversely as the pressure P. If $V = 231$ cm^3 when $T = 42°$ and $P = 20$ kg/cm^2, what is the volume when $T = 30°$ and $P = 15$ kg/cm^2?

Skill Maintenance

Vocabulary Review

In each of Exercises 41–45, fill in the blank with the correct term. Some of the given choices will not be used.

even function	relative maximum
odd function	relative minimum
constant function	solution
composite function	zero
direct variation	perpendicular
inverse variation	parallel

41. Nonvertical lines are ______________ if and only if they have the same slope and different y-intercepts. **[1.4]**

42. An input c of a function f is a(n) ______________ of the function if $f(c) = 0$. **[1.5]**

43. For a function f for which $f(c)$ exists, $f(c)$ is a(n) ______________ if $f(c)$ is the lowest point in some open interval. **[2.1]**

44. If the graph of a function is symmetric with respect to the origin, then f is a(n) ______________. **[2.4]**

45. An equation $y = k/x$ is an equation of ______________. **[2.6]**

› Synthesis

46. In each of the following equations, state whether y varies directly as x, inversely as x, or neither directly nor inversely as x.

a) $7xy = 14$
b) $x - 2y = 12$
c) $-2x + 3y = 0$
d) $x = \frac{3}{4}y$
e) $\frac{x}{y} = 2$

47. *Volume and Cost.* An 18-oz jar of peanut butter in the shape of a right circular cylinder is 5 in. high and 3 in. in diameter and sells for \$2.89. In the same store, a 22-oz jar of the same brand is $5\frac{1}{4}$ in. high and $3\frac{1}{4}$ in. in diameter. If the cost is directly proportional to volume, what should the price of the larger jar be? If the cost is directly proportional to weight, what should the price of the larger jar be?

48. Describe in words the variation given by the equation

$$Q = \frac{kp^2}{q^3}.$$

49. *Area of a Circle.* The area of a circle varies directly as the square of the length of a diameter. What is the variation constant?

Chapter 2 Summary and Review

STUDY GUIDE

KEY TERMS AND CONCEPTS | **EXAMPLES**

SECTION 2.1: INCREASING, DECREASING, AND PIECEWISE FUNCTIONS; APPLICATIONS

Increasing, Decreasing, and Constant Functions

A function f is said to be **increasing** on an *open* interval I if for all a and b in that interval, $a < b$ implies $f(a) < f(b)$.

A function f is said to be **decreasing** on an *open* interval I if for all a and b in that interval, $a < b$ implies $f(a) > f(b)$.

A function f is said to be **constant** on an *open* interval I if for all a and b in that interval, $f(a) = f(b)$.

Determine the intervals on which the function is **(a)** increasing; **(b)** decreasing; **(c)** constant.

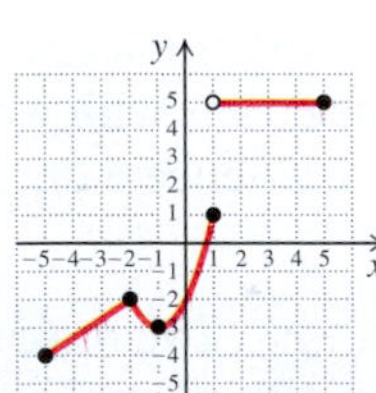

a) As x-values increase from -5 to -2, y-values increase from -4 to -2; y-values also increase as x-values increase from -1 to 1. Thus the function is increasing on the intervals $(-5, -2)$ and $(-1, 1)$.

b) As x-values increase from -2 to -1, y-values decrease from -2 to -3, so the function is decreasing on the interval $(-2, -1)$.

c) As x-values increase from 1 to 5, y remains 5, so the function is constant on the interval $(1, 5)$.

Relative Maxima and Minima

Suppose that f is a function for which $f(c)$ exists for some c in the domain of f. Then:

$f(c)$ is a **relative maximum** if there exists an *open* interval I containing c such that $f(c) > f(x)$ for all x in I, where $x \neq c$; and

$f(c)$ is a **relative minimum** if there exists an *open* interval I containing c such that $f(c) < f(x)$ for all x in I, where $x \neq c$.

Determine any relative maxima or minima of the function.

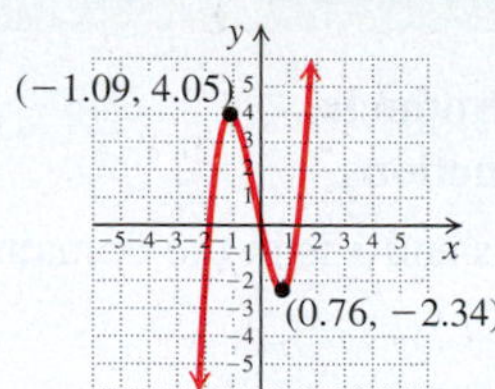

We see from the graph that the function has one relative maximum, 4.05. It occurs when $x = -1.09$. We also see that there is one relative minimum, -2.34. It occurs when $x = 0.76$.

Some applied problems can be modeled by functions.

See Examples 3 and 4 on pp. 97 and 98.

To graph a function that is defined **piecewise**, graph the function in parts as defined by its output formulas.

Graph the function defined as

$$f(x) = \begin{cases} 2x - 3, & \text{for } x < 1, \\ x + 1, & \text{for } x \geq 1. \end{cases}$$

We create the graph in two parts. First, we graph $f(x) = 2x - 3$ for inputs x less than 1. Then we graph $f(x) = x + 1$ for inputs x greater than or equal to 1.

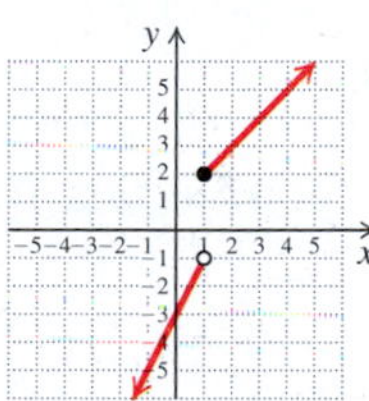

Greatest Integer Function

$f(x) = [\![x]\!] =$ the greatest integer less than or equal to x.

The graph of the greatest integer function is shown below. Each input is paired with the greatest integer less than or equal to that input.

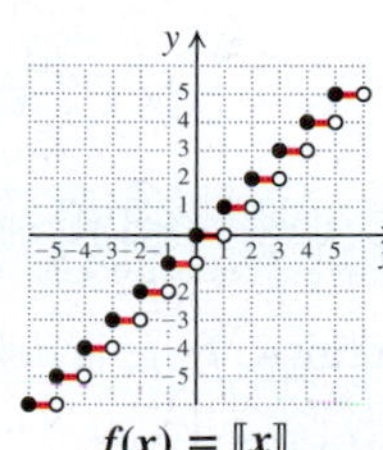

$f(x) = [\![x]\!]$

SECTION 2.2: THE ALGEBRA OF FUNCTIONS

Sums, Differences, Products, and Quotients of Functions

If f and g are functions and x is in the domain of each function, then:

$$(f+g)(x) = f(x) + g(x),$$
$$(f-g)(x) = f(x) - g(x),$$
$$(fg)(x) = f(x) \cdot g(x),$$
$$(f/g)(x) = f(x)/g(x),$$
provided $g(x) \neq 0$.

Given that $f(x) = x - 4$ and $g(x) = \sqrt{x+5}$, find each of the following.

a) $(f+g)(x)$ **b)** $(f-g)(x)$

c) $(fg)(x)$ **d)** $(f/g)(x)$

a) $(f+g)(x) = f(x) + g(x) = x - 4 + \sqrt{x+5}$

b) $(f-g)(x) = f(x) - g(x) = x - 4 - \sqrt{x+5}$

c) $(fg)(x) = f(x) \cdot g(x) = (x-4)\sqrt{x+5}$

d) $(f/g)(x) = f(x)/g(x) = \dfrac{x-4}{\sqrt{x+5}}, x > -5$

Domains of $f+g$, $f-g$, fg, and f/g

If f and g are functions, then the domain of the functions $f+g$, $f-g$, and fg is the intersection of the domain of f and the domain of g. The domain of f/g is the intersection of the domain of f and the domain of g, with the exclusion of any x-values for which $g(x) = 0$.

For the functions f and g above, find the domains of $f+g$, $f-g$, fg, and f/g.

The domain of $f(x) = x - 4$ is the set of all real numbers. The domain of $g(x) = \sqrt{x+5}$ is the set of all real numbers for which $x + 5 \geq 0$, or $x \geq -5$, or $[-5, \infty)$. Then the domain of $f+g$, $f-g$, and fg is the set of numbers in the intersection of these domains, or $[-5, \infty)$.

Since $g(-5) = 0$, we must exclude -5 from the domain of f/g. Thus the domain of f/g is $(-5, \infty)$.

The **difference quotient** for a function $f(x)$ is the ratio

$$\frac{f(x+h) - f(x)}{h}.$$

For the function $f(x) = x^2 - 4$, construct and simplify the difference quotient.

$$\begin{aligned}\frac{f(x+h) - f(x)}{h} &= \frac{[(x+h)^2 - 4] - (x^2 - 4)}{h} \\ &= \frac{x^2 + 2xh + h^2 - 4 - x^2 + 4}{h} \\ &= \frac{2xh + h^2}{h} = \frac{h(2x+h)}{h} \\ &= 2x + h\end{aligned}$$

SECTION 2.3: THE COMPOSITION OF FUNCTIONS

The **composition of functions**, $f \circ g$, is defined as

$$(f \circ g)(x) = f(g(x)),$$

where x is in the domain of g and $g(x)$ is in the domain of f.

Given that $f(x) = 2x - 1$ and $g(x) = \sqrt{x}$, find each of the following.

a) $(f \circ g)(4)$ **b)** $(g \circ g)(625)$

c) $(f \circ g)(x)$ **d)** $(g \circ f)(x)$

e) The domain of $f \circ g$ and the domain of $g \circ f$

a) $(f \circ g)(4) = f(g(4)) = f(\sqrt{4}) = f(2) = 2 \cdot 2 - 1 = 4 - 1 = 3$

b) $(g \circ g)(625) = g(g(625)) = g(\sqrt{625}) = g(25) = \sqrt{25} = 5$

c) $(f \circ g)(x) = f(g(x)) = f(\sqrt{x}) = 2\sqrt{x} - 1$

(continued)

	d) $(g \circ f)(x) = g(f(x)) = g(2x - 1) = \sqrt{2x - 1}$ **e)** The domain and the range of $f(x)$ are both $(-\infty, \infty)$, and the domain and the range of $g(x)$ are both $[0, \infty)$. Since f can accept any real number as an input, the domain of $f \circ g$ consists of all real numbers that are outputs of g, or $[0, \infty)$. Any real number can be an input for f. Since $g(x)$ is not defined for negative radicands, we must have $2x - 1 \geq 0$, or $x \geq \frac{1}{2}$, so the domain of g and thus the domain of $g \circ f$ is $\left[\frac{1}{2}, \infty\right)$.
When we **decompose** a function, we write it as the composition of two functions.	If $h(x) = \sqrt{3x + 7}$, find $f(x)$ and $g(x)$ such that $h(x) = (f \circ g)(x)$. This function finds the square root of $3x + 7$, so one decomposition is $f(x) = \sqrt{x}$ and $g(x) = 3x + 7$. There are other correct answers, but this one is probably the most obvious.

SECTION 2.4: SYMMETRY

Algebraic Tests of Symmetry *x-axis*: If replacing y with $-y$ produces an equivalent equation, then the graph is *symmetric with respect to the x-axis.* *y-axis*: If replacing x with $-x$ produces an equivalent equation, then the graph is *symmetric with respect to the y-axis.* *Origin*: If replacing x with $-x$ and y with $-y$ produces an equivalent equation, then the graph is *symmetric with respect to the origin.*	Test $y = 2x^3$ for symmetry with respect to the x-axis, the y-axis, and the origin. *x-axis*: We replace y with $-y$: $-y = 2x^3$ $y = -2x^3$. **Multiplying by −1** The resulting equation *is not* equivalent to the original equation, so the graph *is not* symmetric with respect to the x-axis. *y-axis*: We replace x with $-x$: $y = 2(-x)^3 = -2x^3$. The resulting equation *is not* equivalent to the original equation, so the graph *is not* symmetric with respect to the y-axis. *Origin*: We replace x with $-x$ and y with $-y$: $-y = 2(-x)^3$ $-y = -2x^3$ $y = 2x^3$. The resulting equation *is* equivalent to the original equation, so the graph *is* symmetric with respect to the origin.
Even Functions and Odd Functions If the graph of a function is symmetric with respect to the y-axis, we say that it is an **even function**. That is, for each x in the domain of f, $f(x) = f(-x)$. If the graph of a function is symmetric with respect to the origin, we say that it is an **odd function**. That is, for each x in the domain of f, $f(-x) = -f(x)$.	Determine whether each function is even, odd, or neither. **a)** $g(x) = 2x^2 - 4$ **b)** $h(x) = x^5 - 3x^3 - x$ **a)** We first find $g(-x)$ and simplify: $g(-x) = 2(-x)^2 - 4 = 2x^2 - 4$. $g(x) = g(-x)$, so g *is* even. Since a function other than $f(x) = 0$ cannot be *both* even and odd and g is even, we need not test to see if it is an odd function. *(continued)*

b) We first find $h(-x)$ and simplify:

$$h(-x) = (-x)^5 - 3(-x)^3 - (-x)$$
$$= -x^5 + 3x^3 + x.$$

$h(x) \neq h(-x)$, so h *is not* even.

Next, we find $-h(x)$ and simplify:

$$-h(x) = -(x^5 - 3x^3 - x)$$
$$= -x^5 + 3x^3 + x.$$

$h(-x) = -h(x)$, so h *is* odd.

SECTION 2.5: TRANSFORMATIONS

Vertical Translation

For $b > 0$:

the graph of $y = f(x) + b$ is the graph of $y = f(x)$ shifted *up b* units;

the graph of $y = f(x) - b$ is the graph of $y = f(x)$ shifted *down b* units.

Horizontal Translation

For $d > 0$:

the graph of $y = f(x - d)$ is the graph of $y = f(x)$ shifted *right d* units;

the graph of $y = f(x + d)$ is the graph of $y = f(x)$ shifted *left d* units.

Graph $g(x) = (x - 2)^2 + 1$. Before doing so, describe how the graph can be obtained from the graph of $f(x) = x^2$.

First, note that the graph of $h(x) = (x - 2)^2$ is the graph of $f(x) = x^2$ shifted right 2 units. Then the graph of $g(x) = (x - 2)^2 + 1$ is the graph of $h(x) = (x - 2)^2$ shifted up 1 unit. Thus the graph of g is obtained by shifting the graph of $f(x) = x^2$ right 2 units and up 1 unit.

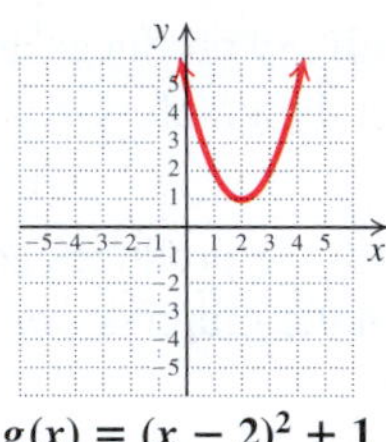

$g(x) = (x - 2)^2 + 1$

Reflections

The graph of $y = -f(x)$ is the **reflection** of $y = f(x)$ across the x-axis.

The graph of $y = f(-x)$ is the **reflection** of $y = f(x)$ across the y-axis.

If a point (x, y) is on the graph of $y = f(x)$, then $(x, -y)$ is on the graph of $y = -f(x)$, and $(-x, y)$ is on the graph of $y = f(-x)$.

Graph each of the following. Before doing so, describe how each graph can be obtained from the graph of $f(x) = x^2 - x$.

a) $g(x) = x - x^2$ **b)** $h(x) = (-x)^2 - (-x)$

a) Note that

$$-f(x) = -(x^2 - x)$$
$$= -x^2 + x$$
$$= x - x^2$$
$$= g(x).$$

Thus the graph is a reflection of the graph of $f(x) = x^2 - x$ across the x-axis.

$f(x) = x^2 - x$

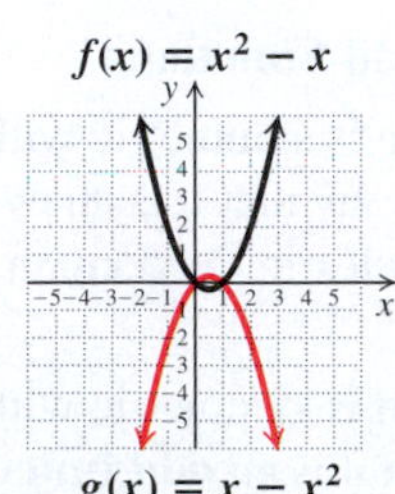

$g(x) = x - x^2$

(continued)

b) Note that

$$f(-x) = (-x)^2 - (-x) = h(x).$$

Thus the graph of $h(x) = (-x)^2 - (-x)$ is a reflection of the graph of $f(x) = x^2 - x$ across the y-axis.

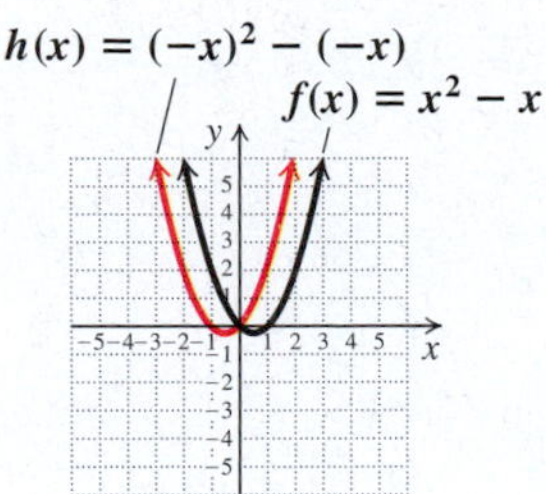

Vertical Stretching and Shrinking

The graph of $y = af(x)$ can be obtained from the graph of $y = f(x)$ by:

stretching vertically for $|a| > 1$, or

shrinking vertically for $0 < |a| < 1$.

For $a < 0$, the graph is also reflected across the x-axis.

(The y-coordinates of the graph of $y = af(x)$ can be obtained by multiplying the y-coordinates of $y = f(x)$ by a.)

Horizontal Stretching and Shrinking

The graph of $y = f(cx)$ can be obtained from the graph of $y = f(x)$ by:

shrinking horizontally for $|c| > 1$,

or

stretching horizontally for $0 < |c| < 1$.

For $c < 0$, the graph is also reflected across the y-axis.

(The x-coordinates of the graph of $y = f(cx)$ can be obtained by dividing the x-coordinates of $y = f(x)$ by c.)

A graph of $y = g(x)$ is shown below. Use this graph to graph each of the given equations.

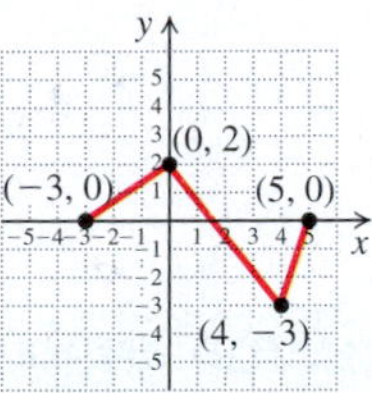

a) $f(x) = g(2x)$ **b)** $f(x) = -2g(x)$

c) $f(x) = \frac{1}{2}g(x)$ **d)** $f(x) = g\left(\frac{1}{2}x\right)$

a) Since $|2| > 1$, the graph of $f(x) = g(2x)$ is a horizontal shrinking of the graph of $y = g(x)$. The transformation divides each x-coordinate of g by 2.

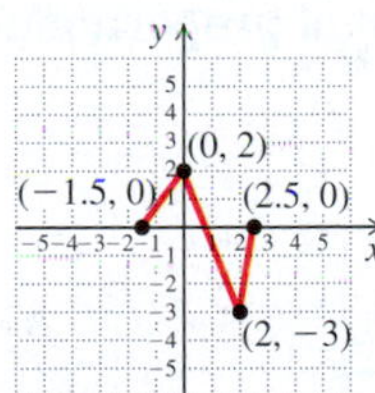

b) Since $|-2| > 1$, the graph of $f(x) = -2g(x)$ is a vertical stretching of the graph of $y = g(x)$. Since $-2 < 0$, the graph is also reflected across the x-axis. The transformation multiplies each y-coordinate of g by 2.

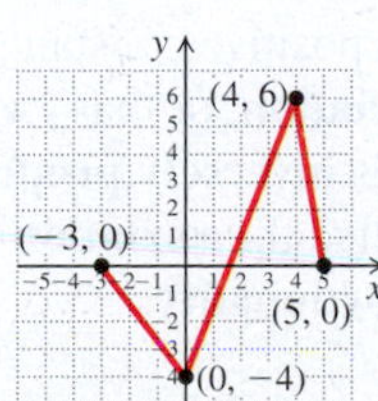

(continued)

c) Since $\left|\frac{1}{2}\right| < 1$, the graph of $f(x) = \frac{1}{2}g(x)$ is a vertical shrinking of the graph of $y = g(x)$. The transformation multiplies each y-coordinate of g by $\frac{1}{2}$.

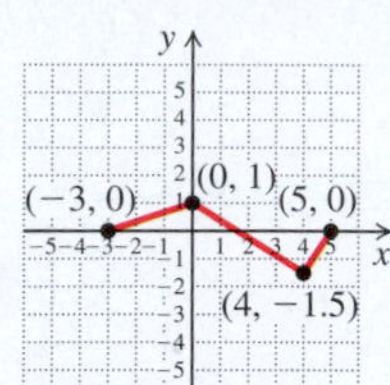

d) Since $\left|\frac{1}{2}\right| < 1$, the graph of $f(x) = g\left(\frac{1}{2}x\right)$ is a horizontal stretching of the graph of $y = g(x)$. The transformation divides each x-coordinate of g by $\frac{1}{2}$ (which is the same as multiplying by 2).

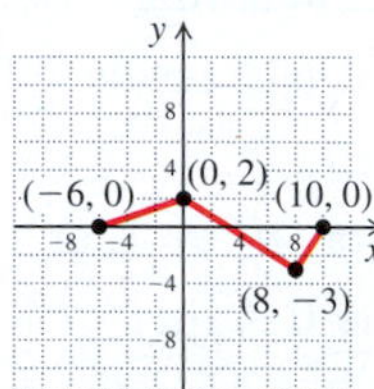

SECTION 2.6: VARIATION AND APPLICATIONS

Direct Variation

If a situation gives rise to a linear function $f(x) = kx$, or $y = kx$, where k is a positive constant, we say that we have **direct variation**, or that y **varies directly as** x, or that y **is directly proportional to** x. The number k is called the **variation constant**, or the **constant of proportionality**.

Find an equation of variation in which y varies directly as x, and $y = 24$ when $x = 8$. Then find the value of y when $x = 5$.

First, we have

$$y = kx \qquad \textbf{y varies directly as x.}$$
$$24 = k \cdot 8 \qquad \textbf{Substituting}$$
$$3 = k \qquad \textbf{Variation constant}$$

The equation of variation is $y = 3x$. Now we use the equation to find the value of y when $x = 5$:

$$y = 3x$$
$$= 3 \cdot 5 \qquad \textbf{Substituting}$$
$$= 15.$$

When $x = 5$, the value of y is 15.

Inverse Variation

If a situation gives rise to a function $f(x) = k/x$, or $y = k/x$, where k is a positive constant, we say that we have **inverse variation**, or that y **varies inversely as** x, or that y **is inversely proportional to** x. The number k is called the **variation constant**, or the **constant of proportionality**.

Find an equation of variation in which y varies inversely as x, and $y = 5$ when $x = 0.1$. Then find the value of y when $x = 10$.

First, we have

$$y = \frac{k}{x} \qquad \textbf{y varies inversely as x.}$$
$$5 = \frac{k}{0.1} \qquad \textbf{Substituting}$$
$$0.5 = k. \qquad \textbf{Variation constant}$$

The equation of variation is $y = \frac{0.5}{x}$.

(continued)

	Now we use the equation to find the value of y when $x = 10$: $y = \frac{0.5}{x}$ $= \frac{0.5}{10}$ **Substituting** $= 0.05.$ When $x = 10$, the value of y is 0.05. **Combined Variation**
y varies **directly as the *n*th power of *x*** if there is some positive constant k such that $y = kx^n.$ y varies **inversely as the *n*th power of *x*** if there is some positive constant k such that $y = \frac{k}{x^n}.$ y varies **jointly as *x* and *z*** if there is some positive constant k such that $y = kxz.$	Find an equation of variation in which y varies jointly as w and the square of x and inversely as z, and $y = 8$ when $w = 3, x = 2$, and $z = 6$. First, we have $y = k \cdot \frac{wx^2}{z}$ $8 = k \cdot \frac{3 \cdot 2^2}{6}$ **Substituting** $8 = k \cdot \frac{3 \cdot 4}{6}$ $8 = 2k$ $4 = k.$ **Variation constant** The equation of variation is $y = 4\frac{wx^2}{z}$, or $y = \frac{4wx^2}{z}$.

REVIEW EXERCISES

Determine whether the statement is true or false.

1. The greatest integer function pairs each input with the greatest integer less than or equal to that input. [2.1]
2. In general, for functions f and g, the domain of $f \circ g =$ the domain of $g \circ f$. [2.3]
3. The graph of $y = (x - 2)^2$ is the graph of $y = x^2$ shifted right 2 units. [2.5]
4. The graph of $y = -x^2$ is the reflection of the graph of $y = x^2$ across the x-axis. [2.5]

Determine the intervals on which the function is **(a)** *increasing;* **(b)** *decreasing; and* **(c)** *constant.* [2.1]

5.

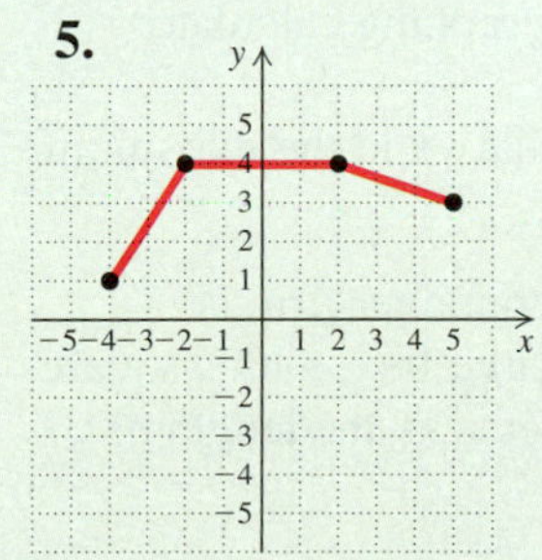

6.

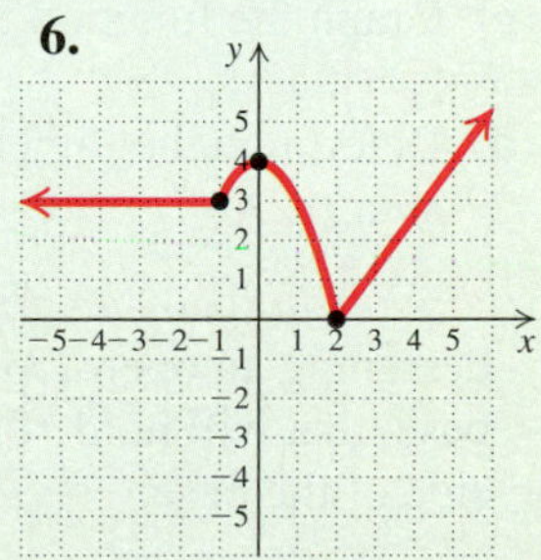

Graph the function. Estimate the intervals on which the function is increasing or decreasing, and estimate any relative maxima or minima. [2.1]

7. $f(x) = x^2 - 1$

8. $f(x) = 2 - |x|$

Use a graphing calculator to find the intervals on which the function is increasing or decreasing, and find any relative maxima or minima. [2.1]

9. $f(x) = x^2 - 4x + 3$

10. $f(x) = -x^2 + x + 6$

11. $f(x) = x^3 - 4x$

12. $f(x) = 2x - 0.5x^3$

13. *Fenced Patio.* Syd has 48 ft of rolled bamboo fence with which to enclose a rectangular patio. The house forms one side of the patio. Suppose that two sides of the patio each measures x feet. Express the area of the patio as a function of x. [2.1]

14. *Inscribed Rectangle.* A rectangle is inscribed in a semicircle of radius 2, as shown. The variable x = half the length of the rectangle. Express the area of the rectangle as a function of x. [2.1]

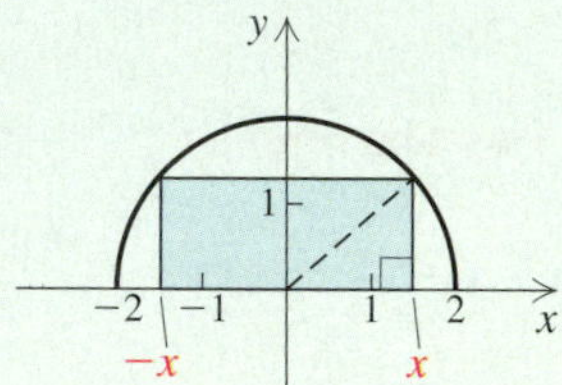

15. *Dog Pen.* Mamie has 66 ft of fencing with which to enclose a rectangular dog pen. The side of her garage forms one side of the pen. Suppose that the side of the pen parallel to the garage is x feet long.

a) Express the area of the dog pen as a function of x. [2.1]
b) Find the domain of the function. [2.1]
c) Graph the function using a graphing calculator. [2.1]
d) Determine the dimensions that yield the maximum area. [2.1]

16. *Minimizing Surface Area.* A container firm is designing an open-top rectangular box, with a square base, that will hold 108 in^3. Let x = the length of a side of the base.

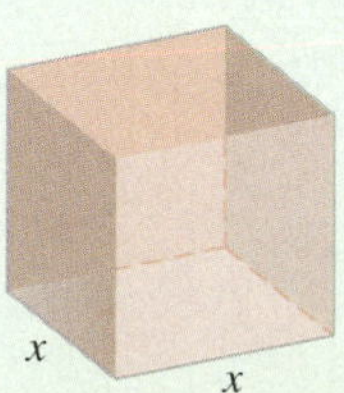

a) Express the surface area as a function of x. [2.1]
b) Find the domain of the function. [2.1]
c) Using the following graph, determine the dimensions that will minimize the surface area of the box. [2.1]

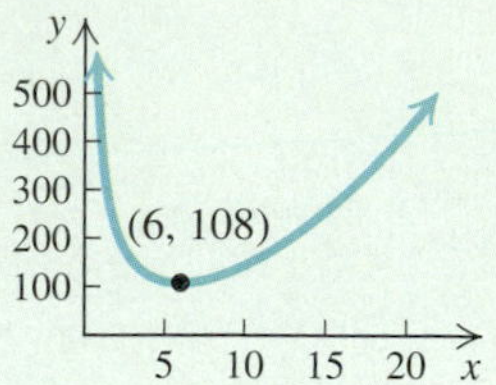

Graph each of the following. [2.1]

17. $f(x) = \begin{cases} -x, & \text{for } x \le -4, \\ \frac{1}{2}x + 1, & \text{for } x > -4 \end{cases}$

18. $f(x) = \begin{cases} x^3, & \text{for } x < -2, \\ |x|, & \text{for } -2 \le x \le 2, \\ \sqrt{x-1}, & \text{for } x > 2 \end{cases}$

19. $f(x) = \begin{cases} \dfrac{x^2 - 1}{x + 1}, & \text{for } x \ne -1, \\ 3, & \text{for } x = -1 \end{cases}$

20. $f(x) = [\![x]\!]$

21. $f(x) = [\![x - 3]\!]$

22. For the function in Exercise 18, find $f(-1)$, $f(5)$, $f(-2)$, and $f(-3)$. [2.1]

23. For the function in Exercise 19, find $f(-2)$, $f(-1)$, $f(0)$, and $f(4)$. [2.1]

Given that $f(x) = \sqrt{x - 2}$ *and* $g(x) = x^2 - 1$, *find each of the following, if it exists.* [2.2]

24. $(f - g)(6)$

25. $(fg)(2)$

26. $(f + g)(-1)$

For each pair of functions in Exercises 27 and 28:

a) *Find the domain of* f, g, $f + g$, $f - g$, fg, *and* f/g. [2.2]
b) *Find* $(f + g)(x)$, $(f - g)(x)$, $(fg)(x)$, *and* $(f/g)(x)$. [2.2]

27. $f(x) = \dfrac{4}{x^2}$, $g(x) = 3 - 2x$

28. $f(x) = 3x^2 + 4x$, $g(x) = 2x - 1$

29. Given the total-revenue and total-cost functions $R(x) = 120x - 0.5x^2$ and $C(x) = 15x + 6$, find the total-profit function $P(x)$. [2.2]

For each function f, construct and simplify the difference quotient. [2.2]

30. $f(x) = 2x + 7$

31. $f(x) = 3 - x^2$

32. $f(x) = \frac{4}{x}$

Given that $f(x) = 2x - 1$, $g(x) = x^2 + 4$, and $h(x) = 3 - x^3$, find each of the following. [2.3]

33. $(f \circ g)(1)$

34. $(g \circ f)(1)$

35. $(h \circ f)(-2)$

36. $(g \circ h)(3)$

37. $(f \circ h)(-1)$

38. $(h \circ g)(2)$

39. $(f \circ f)(x)$

40. $(h \circ h)(x)$

For each pair of functions in Exercises 41 and 42:

a) *Find $(f \circ g)(x)$ and $(g \circ f)(x)$.* [2.3]

b) *Find the domain of $f \circ g$ and the domain of $g \circ f$.* [2.3]

41. $f(x) = \frac{4}{x^2}$, $g(x) = 3 - 2x$

42. $f(x) = 3x^2 + 4x$, $g(x) = 2x - 1$

Find $f(x)$ and $g(x)$ such that $h(x) = (f \circ g)(x)$. [2.3]

43. $h(x) = \sqrt{5x + 2}$

44. $h(x) = 4(5x - 1)^2 + 9$

Graph the given equation and determine visually whether it is symmetric with respect to the x-axis, the y-axis, and the origin. Then verify your assertion algebraically. [2.4]

45. $x^2 + y^2 = 4$

46. $y^2 = x^2 + 3$

47. $x + y = 3$

48. $y = x^2$

49. $y = x^3$

50. $y = x^4 - x^2$

Determine visually whether the function is even, odd, or neither even nor odd. [2.4]

51.

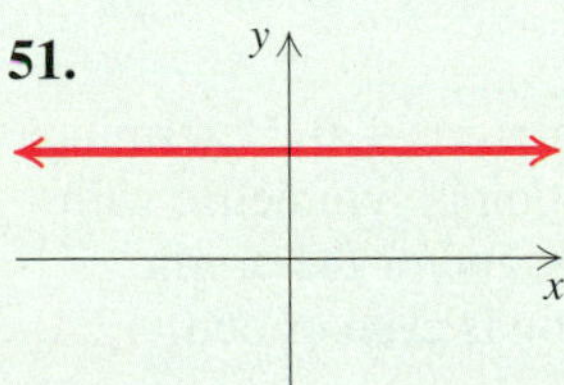

52.

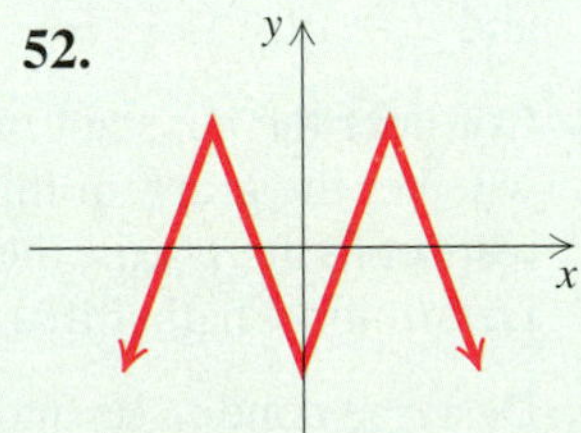

53.

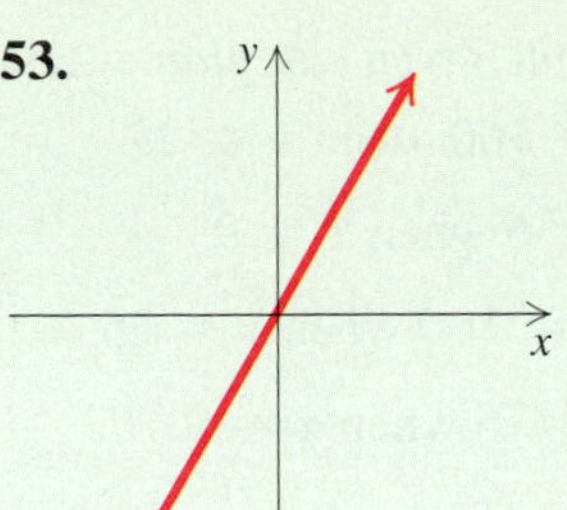

54.

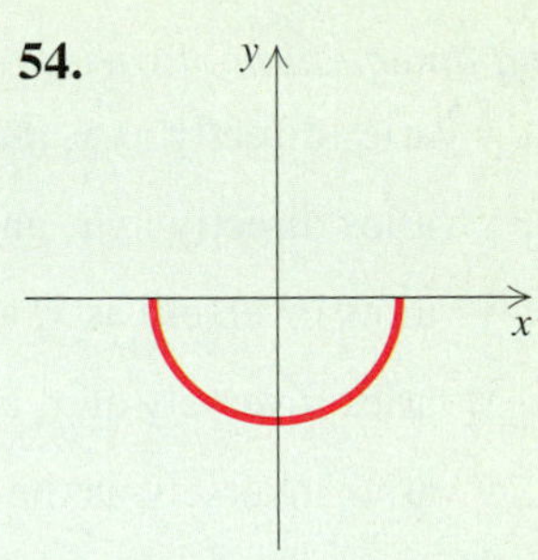

In Exercises 55–60, test whether the function is even, odd, or neither even nor odd. [2.4]

55. $f(x) = 9 - x^2$

56. $f(x) = x^3 - 2x + 4$

57. $f(x) = x^7 - x^5$

58. $f(x) = |x|$

59. $f(x) = \sqrt{16 - x^2}$

60. $f(x) = \frac{10x}{x^2 + 1}$

Write an equation for a function that has a graph with the given characteristics. [2.5]

61. The shape of $y = x^2$, but shifted left 3 units

62. The shape of $y = \sqrt{x}$, but reflected across the x-axis and shifted right 3 units and up 4 units

63. The shape of $y = |x|$, but stretched vertically by a factor of 2 and shifted right 3 units

A graph of $y = f(x)$ is shown below. No formula for f is given. Graph each of the following. [2.5]

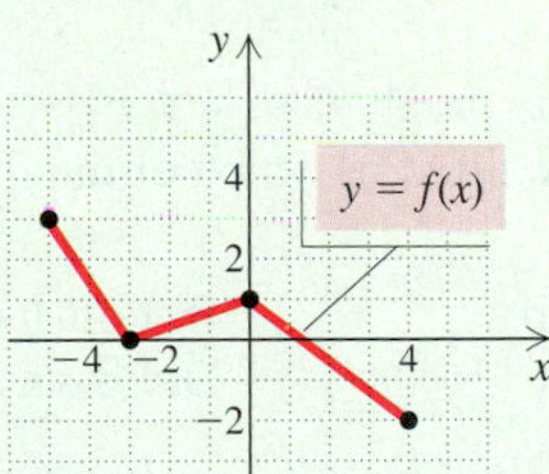

64. $y = f(x - 1)$

65. $y = f(2x)$

66. $y = -2f(x)$

67. $y = 3 + f(x)$

Find an equation of variation for the given situation. **[2.6]**

68. y varies directly as x, and $y = 100$ when $x = 25$.

69. y varies directly as x, and $y = 6$ when $x = 9$.

70. y varies inversely as x, and $y = 100$ when $x = 25$.

71. y varies inversely as x, and $y = 6$ when $x = 9$.

72. y varies inversely as the square of x, and $y = 12$ when $x = 2$.

73. y varies jointly as x and the square of z and inversely as w, and $y = 2$ when $x = 16$, $w = 0.2$, and $z = \frac{1}{2}$.

74. *Pumping Time.* The time t required to empty a tank varies inversely as the rate r of pumping. If a pump can empty a tank in 35 min at the rate of 800 kL/min, how long will it take the pump to empty the same tank at the rate of 1400 kL/min? **[2.6]**

75. *Test Score.* The score N on a test varies directly as the number of correct responses a. Sam answers 29 questions correctly and earns a score of 87. What would Sam's score have been if he had answered 25 questions correctly? **[2.6]**

76. *Power of Electric Current.* The power P expended by heat in an electric circuit of fixed resistance varies directly as the square of the current C in the circuit. A circuit expends 180 watts when a current of 6 amperes is flowing. What is the amount of heat expended when the current is 10 amperes? **[2.6]**

77. For $f(x) = x + 1$ and $g(x) = \sqrt{x}$, the domain of $(g \circ f)(x)$ is which of the following? **[2.3]**

A. $[-1, \infty)$ **B.** $[-1, 0)$
C. $[0, \infty)$ **D.** $(-\infty, \infty)$

78. For $b > 0$, the graph of $y = f(x) + b$ is the graph of $y = f(x)$ shifted in which of the following ways? **[2.5]**

A. Right b units **B.** Left b units
C. Up b units **D.** Down b units

79. The graph of the function f is shown below.

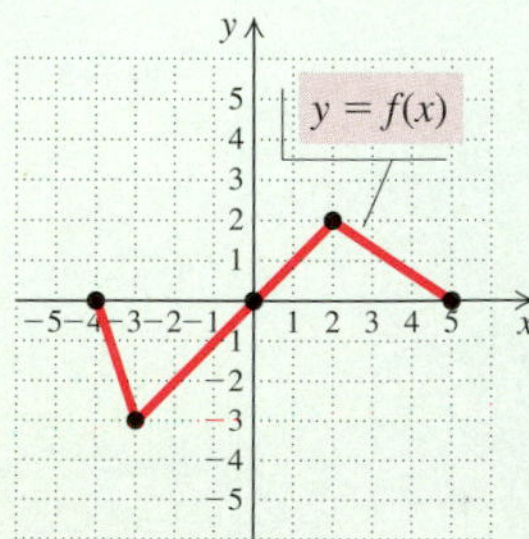

The graph of $g(x) = -\frac{1}{2}f(x) + 1$ is which of the following? **[2.5]**

A.

B.

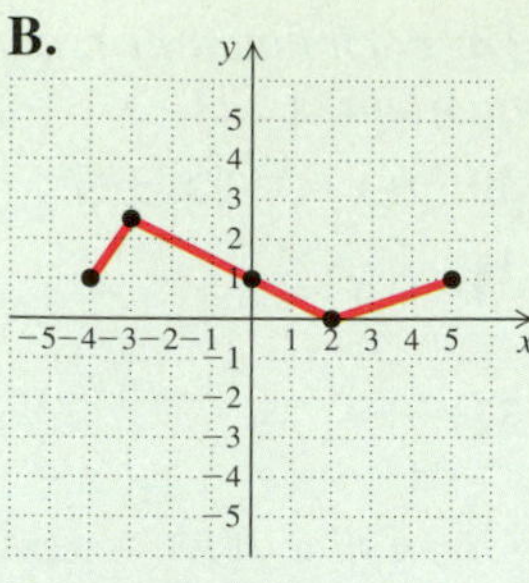

C.

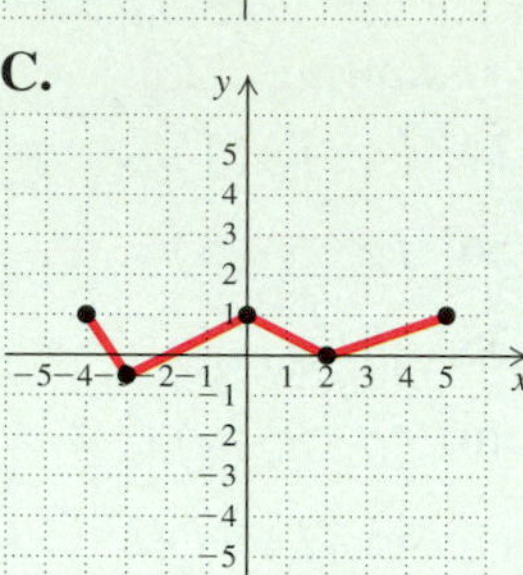

D.

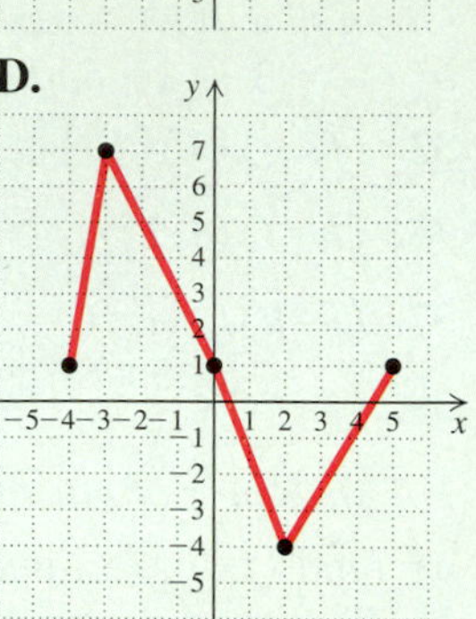

Synthesis

80. Prove that the sum of two odd functions is odd. **[2.2], [2.4]**

81. Describe how the graph of $y = -f(-x)$ is obtained from the graph of $y = f(x)$. **[2.5]**

Collaborative Discussion and Writing

82. Given that $f(x) = 4x^3 - 2x + 7$, find each of the following. Then discuss how each expression differs from the other. **[1.2], [2.5]**

a) $f(x) + 2$
b) $f(x + 2)$
c) $f(x) + f(2)$

83. Given the graph of $y = f(x)$, explain and contrast the effect of the constant c on the graphs of $y = f(cx)$ and $y = cf(x)$. **[2.5]**

84. Consider the constant function $f(x) = 0$. Determine whether the graph of this function is symmetric with respect to the x-axis, the y-axis, and/or the origin. Determine whether this function is even or odd. **[2.4]**

85. Describe conditions under which you would know whether a polynomial function

$$f(x) = a_nx^n + a_{n-1}x^{n-1} + \cdots + a_2x^2 + a_1x + a_0$$

is even or odd without using an algebraic procedure. Explain. **[2.4]**

86. If y varies directly as x^2, explain why doubling x would not cause y to be doubled as well. **[2.6]**

87. If y varies directly as x and x varies inversely as z, how does y vary with regard to z? Why? **[2.6]**

2 Chapter Test

1. Determine the intervals on which the function is **(a)** increasing; **(b)** decreasing; and **(c)** constant.

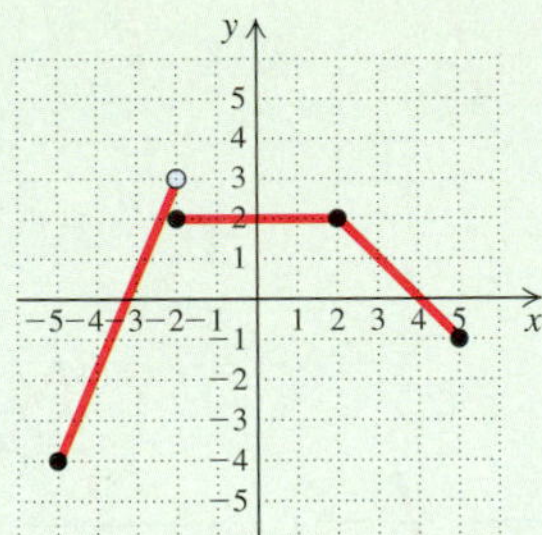

2. Graph the function $f(x) = 2 - x^2$. Estimate the intervals on which the function is increasing or decreasing, and estimate any relative maxima or minima.

3. Use a graphing calculator to find the intervals on which the function $f(x) = x^3 + 4x^2$ is increasing or decreasing, and find any relative maxima or minima.

4. *Triangular Pennant.* A softball team is designing a triangular pennant such that the height is 6 in. less than four times the length of the base b. Express the area of the pennant as a function of b.

5. Graph:

$$f(x) = \begin{cases} x^2, & \text{for } x < -1, \\ |x|, & \text{for } -1 \le x \le 1, \\ \sqrt{x-1}, & \text{for } x > 1. \end{cases}$$

6. For the function in Exercise 5, find $f\left(-\frac{7}{8}\right)$, $f(5)$, and $f(-4)$.

Given that $f(x) = x^2 - 4x + 3$ and $g(x) = \sqrt{3 - x}$, find each of the following, if it exists.

7. $(f + g)(-6)$

8. $(f - g)(-1)$

9. $(fg)(2)$

10. $(f/g)(1)$

For $f(x) = x^2$ and $g(x) = \sqrt{x - 3}$, find each of the following.

11. The domain of f

12. The domain of g

13. The domain of $f + g$

14. The domain of $f - g$

15. The domain of fg

16. The domain of f/g

17. $(f + g)(x)$

18. $(f - g)(x)$

19. $(fg)(x)$

20. $(f/g)(x)$

For each function, construct and simplify the difference quotient.

21. $f(x) = \frac{1}{2}x + 4$

22. $f(x) = 2x^2 - x + 3$

Given that $f(x) = x^2 - 1$, $g(x) = 4x + 3$, and $h(x) = 3x^2 + 2x + 4$, find each of the following.

23. $(g \circ h)(2)$

24. $(f \circ g)(-1)$

25. $(h \circ f)(1)$

26. $(g \circ g)(x)$

For $f(x) = \sqrt{x - 5}$ and $g(x) = x^2 + 1$:

27. Find $(f \circ g)(x)$ and $(g \circ f)(x)$.

28. Find the domain of $(f \circ g)(x)$ and the domain of $(g \circ f)(x)$.

29. Find $f(x)$ and $g(x)$ such that

$$h(x) = (f \circ g)(x) = (2x - 7)^4.$$

30. Determine whether the graph of $y = x^4 - 2x^2$ is symmetric with respect to the x-axis, the y-axis, and the origin.

31. Determine whether the function

$$f(x) = \frac{2x}{x^2 + 1}$$

is even, odd, or neither even nor odd. Show your work.

32. Write an equation for a function that has the shape of $y = x^2$, but shifted right 2 units and down 1 unit.

33. Write an equation for a function that has the shape of $y = x^2$, but reflected across the x-axis and shifted left 2 units and up 3 units.

34. The graph of a function $y = f(x)$ is shown below. No formula for f is given. Graph $y = -\frac{1}{2}f(x)$.

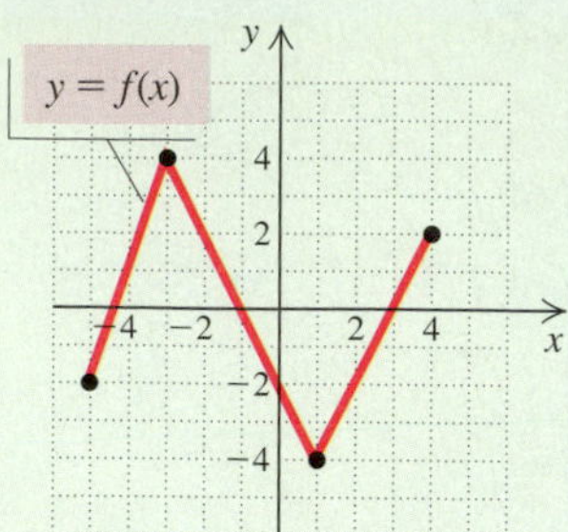

35. Find an equation of variation in which y varies inversely as x, and $y = 5$ when $x = 6$.

36. Find an equation of variation in which y varies directly as x, and $y = 60$ when $x = 12$.

37. Find an equation of variation in which y varies jointly as x and the square of z and inversely as w, and $y = 100$ when $x = 0.1$, $z = 10$, and $w = 5$.

38. The stopping distance d of a car after the brakes have been applied varies directly as the square of the speed r. If a car traveling 60 mph can stop in 200 ft, how long will it take a car traveling 30 mph to stop?

39. The graph of the function f is shown below.

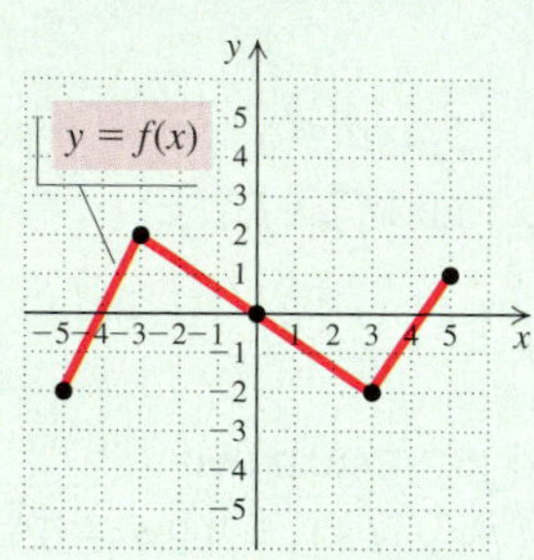

The graph of $g(x) = 2f(x) - 1$ is which of the following?

A.

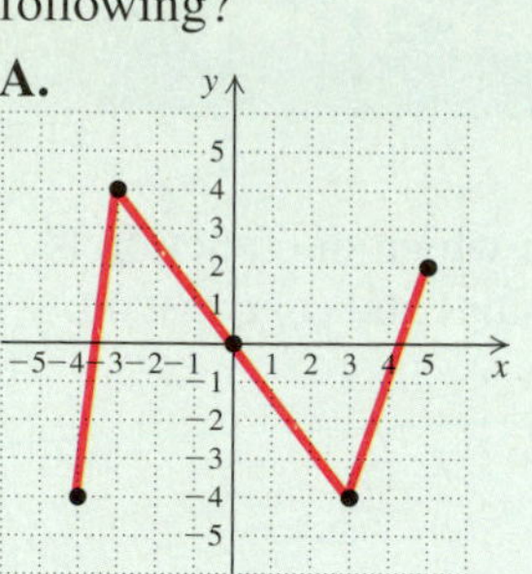

B.

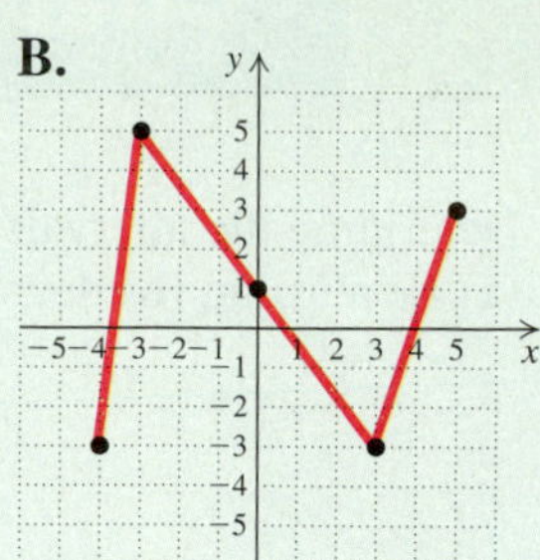

C.

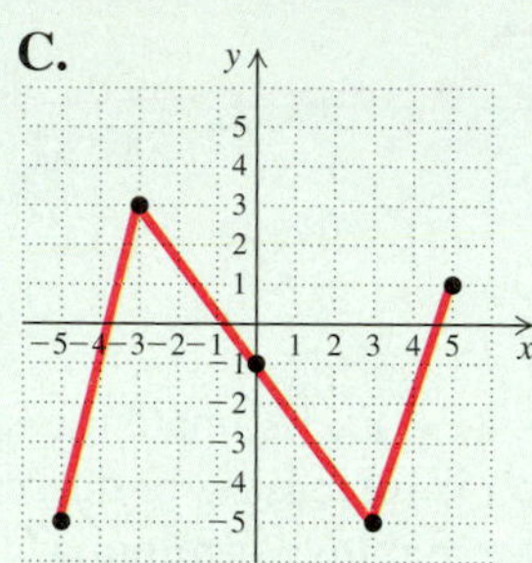

D.

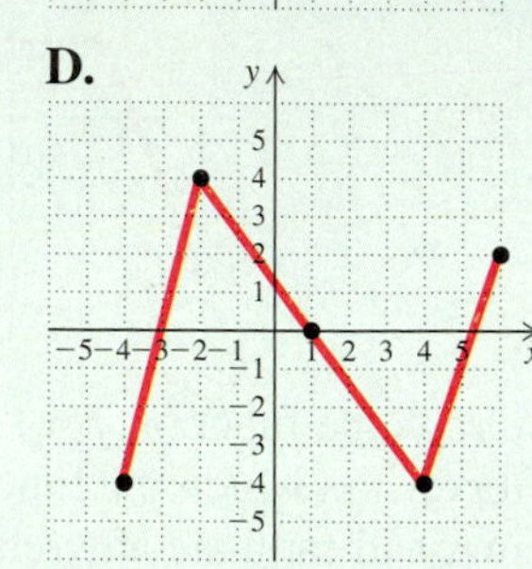

Synthesis

40. If $(-3, 1)$ is a point on the graph of $y = f(x)$, what point do you know is on the graph of $y = f(3x)$?

CHAPTER 3

Quadratic Functions and Equations; Inequalities

APPLICATION

This problem appears as Example 10 in Section 3.2.

The numbers of both magazine launches and magazine closures have increased in recent years. The function $m(x) = 34x^2 - 59x + 81$ can be used to estimate the number of magazine closures after 2012 (*Source*: MediaFinder.com, online database of U.S. and Canadian print and digital publications). In what year was the number of magazine closures about 99?

3.1 The Complex Numbers

› Perform computations involving complex numbers.

Some functions have zeros that are not real numbers. In order to find the zeros of such functions, we must consider the **complex-number system**.

› The Complex-Number System

Just in Time 1

We know that the square root of a negative number is not a real number. For example, $\sqrt{-1}$ is not a real number because there is no real number x such that $x^2 = -1$. This means that certain equations, like $x^2 = -1$ or $x^2 + 1 = 0$, do not have real-number solutions, and certain functions, like $f(x) = x^2 + 1$, do not have real-number zeros. Consider the graph of $f(x) = x^2 + 1$.

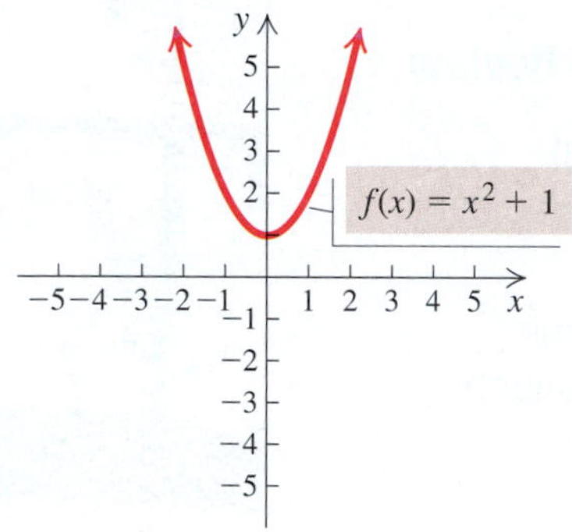

We see that the graph does not cross the x-axis and thus has no x-intercepts. This illustrates that the function $f(x) = x^2 + 1$ has no real-number zeros. Thus there are no real-number solutions of the corresponding equation $x^2 + 1 = 0$.

We can define a nonreal number that is a solution of the equation $x^2 + 1 = 0$.

THE NUMBER *i*

The number i is defined such that

$$i = \sqrt{-1} \quad \text{and} \quad i^2 = -1.$$

To express roots of negative numbers in terms of i, we can use the fact that

$$\sqrt{-p} = \sqrt{-1 \cdot p} = \sqrt{-1} \cdot \sqrt{p} = i\sqrt{p}$$

when p is a positive real number.

Just in Time 25

EXAMPLE 1 Express each number in terms of i.

a) $\sqrt{-7}$ **b)** $\sqrt{-16}$ **c)** $-\sqrt{-13}$

d) $-\sqrt{-64}$ **e)** $\sqrt{-48}$

Solution

a) $\sqrt{-7} = \sqrt{-1 \cdot 7} = \sqrt{-1} \cdot \sqrt{7}$
$= i\sqrt{7}$, or $\sqrt{7}i$

i is *not* under the radical.

b) $\sqrt{-16} = \sqrt{-1 \cdot 16} = \sqrt{-1} \cdot \sqrt{16}$
$= i \cdot 4 = 4i$

c) $-\sqrt{-13} = -\sqrt{-1 \cdot 13} = -\sqrt{-1} \cdot \sqrt{13}$
$= -i\sqrt{13}$, or $-\sqrt{13}i$

d) $-\sqrt{-64} = -\sqrt{-1 \cdot 64} = -\sqrt{-1} \cdot \sqrt{64}$
$= -i \cdot 8 = -8i$

e) $\sqrt{-48} = \sqrt{-1 \cdot 48} = \sqrt{-1} \cdot \sqrt{48}$
$= i\sqrt{16 \cdot 3}$
$= i \cdot 4\sqrt{3}$
$= 4i\sqrt{3}$, or $4\sqrt{3}i$

Now Try Exercises 1, 7, and 9.

The complex numbers are formed by adding real numbers and multiples of i.

COMPLEX NUMBERS

A **complex number** is a number of the form $a + bi$, where a and b are real numbers. The number a is said to be the **real part** of $a + bi$, and the number b is said to be the **imaginary part** of $a + bi$.*

Note that either a or b or both can be 0. When $b = 0$, $a + bi = a + 0i = a$, so every real number is a complex number. A complex number like $3 + 4i$ or $17i$, in which $b \neq 0$, is called an **imaginary number**. A complex number like $17i$ or $-4i$, in which $a = 0$ and $b \neq 0$, is sometimes called a **pure imaginary number**. The relationships among various types of complex numbers are shown in the figure below.

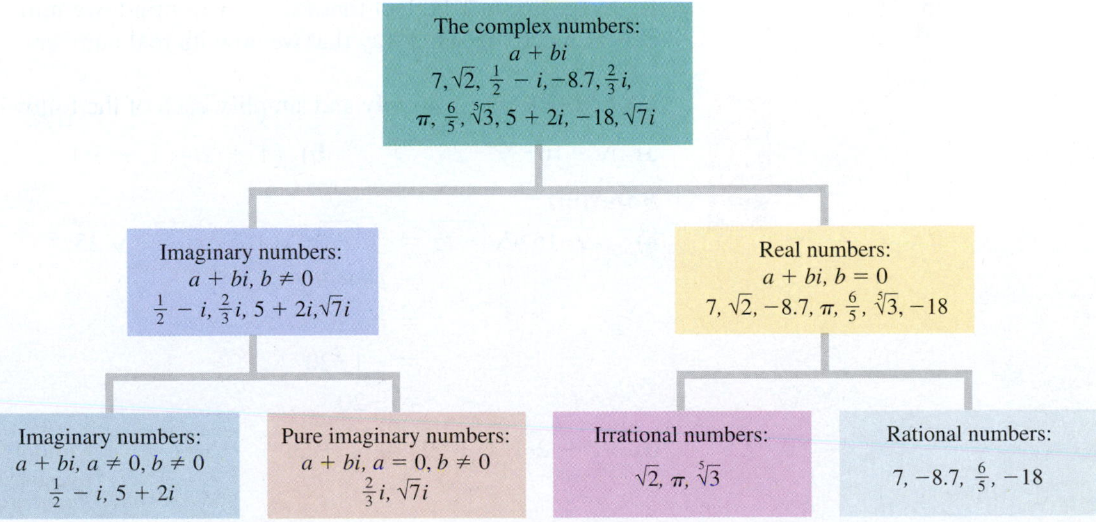

*Sometimes bi is considered to be the imaginary part.

Addition and Subtraction

The complex numbers obey the commutative, associative, and distributive laws. Thus we can add and subtract them as we do binomials. We collect the real parts and the imaginary parts of complex numbers just as we collect like terms in binomials.

EXAMPLE 2 Add or subtract and simplify each of the following.

a) $(8 + 6i) + (3 + 2i)$ **b)** $(4 + 5i) - (6 - 3i)$

Solution

a) $(8 + 6i) + (3 + 2i) = (8 + 3) + (6i + 2i)$ Collecting the real parts and the imaginary parts

$= 11 + (6 + 2)i = 11 + 8i$

b) $(4 + 5i) - (6 - 3i) = (4 - 6) + [5i - (-3i)]$ Note that 6 and $-3i$ are both being subtracted.

$= -2 + 8i$

Now Try Exercises 11 and 21.

(8+6i)+(3+2i)	
	11+8i
(4+5i)−(6−3i)	
	−2+8i

When set in $a + bi$ mode, most graphing calculators can perform operations on complex numbers. The operations in Example 2 are shown in the window at left. Some calculators will express a complex number in the form (a, b) rather than $a + bi$.

Multiplication

When $\sqrt{a}$ and $\sqrt{b}$ are real numbers, $\sqrt{a} \cdot \sqrt{b} = \sqrt{ab}$, but this is not true when $\sqrt{a}$ and $\sqrt{b}$ are not real numbers. Thus,

$$\sqrt{-2} \cdot \sqrt{-5} = \sqrt{-1} \cdot \sqrt{2} \cdot \sqrt{-1} \cdot \sqrt{5}$$
$$= i\sqrt{2} \cdot i\sqrt{5}$$
$$= i^2\sqrt{10} = -1\sqrt{10} = -\sqrt{10} \quad \text{is correct!}$$

But

$$\sqrt{-2} \cdot \sqrt{-5} = \sqrt{(-2)(-5)} = \sqrt{10} \quad \text{is wrong!}$$

Keeping this and the fact that $i^2 = -1$ in mind, we multiply with imaginary numbers in much the same way that we do with real numbers.

Just in Time 12, 13

EXAMPLE 3 Multiply and simplify each of the following.

a) $\sqrt{-16} \cdot \sqrt{-25}$ **b)** $(1 + 2i)(1 + 3i)$ **c)** $(3 - 7i)^2$

Solution

a) $\sqrt{-16} \cdot \sqrt{-25} = \sqrt{-1} \cdot \sqrt{16} \cdot \sqrt{-1} \cdot \sqrt{25}$

$= i \cdot 4 \cdot i \cdot 5$

$= i^2 \cdot 20$

$= -1 \cdot 20$ $i^2 = -1$

$= -20$

b) $(1 + 2i)(1 + 3i) = 1 + 3i + 2i + 6i^2$ Multiplying each term of one number by every term of the other (FOIL)

$= 1 + 3i + 2i - 6$ $i^2 = -1$

$= -5 + 5i$ Collecting like terms

c) $(3 - 7i)^2 = 3^2 - 2 \cdot 3 \cdot 7i + (7i)^2$ Recall that $(A - B)^2 = A^2 - 2AB + B^2$.

$= 9 - 42i + 49i^2$

$= 9 - 42i - 49$ $\quad i^2 = -1$

$= -40 - 42i$

Now Try Exercises 31, 39, and 55.

We can multiply complex numbers on a graphing calculator set in $a + bi$ mode. The products found in Example 3 are shown below.

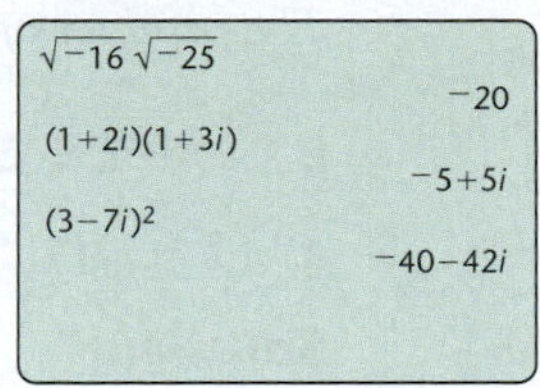

Just in Time 7

Recall that -1 raised to an *even* power is 1, and -1 raised to an *odd* power is -1. Simplifying powers of i can then be done by using the fact that $i^2 = -1$ and expressing the given power of i in terms of i^2. Consider the following:

$$i = \sqrt{-1},$$
$$i^2 = -1,$$
$$i^3 = i^2 \cdot i = (-1)i = -i,$$
$$i^4 = (i^2)^2 = (-1)^2 = 1,$$
$$i^5 = i^4 \cdot i = (i^2)^2 \cdot i = (-1)^2 \cdot i = 1 \cdot i = i,$$
$$i^6 = (i^2)^3 = (-1)^3 = -1,$$
$$i^7 = i^6 \cdot i = (i^2)^3 \cdot i = (-1)^3 \cdot i = -1 \cdot i = -i,$$
$$i^8 = (i^2)^4 = (-1)^4 = 1.$$

Note that the powers of i cycle through the values i, -1, $-i$, and 1.

EXAMPLE 4 Simplify each of the following.

a) i^{37} b) i^{58}

c) i^{75} d) i^{80}

Solution

a) $i^{37} = i^{36} \cdot i = (i^2)^{18} \cdot i = (-1)^{18} \cdot i = 1 \cdot i = i$

b) $i^{58} = (i^2)^{29} = (-1)^{29} = -1$

c) $i^{75} = i^{74} \cdot i = (i^2)^{37} \cdot i = (-1)^{37} \cdot i = -1 \cdot i = -i$

d) $i^{80} = (i^2)^{40} = (-1)^{40} = 1$

Now Try Exercises 79 and 83.

These powers of i can also be simplified in terms of i^4 rather than i^2. Consider i^{37} in Example 4(a), for instance. When we divide 37 by 4, we get 9 with a remainder of 1. Then $37 = 4 \cdot 9 + 1$, so

$$i^{37} = (i^4)^9 \cdot i = 1^9 \cdot i = 1 \cdot i = i.$$

The other examples shown above can be done in a similar manner.

Conjugates and Division

Conjugates of complex numbers are defined as follows.

CONJUGATE OF A COMPLEX NUMBER

The **conjugate** of a complex number $a + bi$ is $a - bi$. The numbers $a + bi$ and $a - bi$ are **complex conjugates**.

Each of the following pairs of numbers are complex conjugates:

$-3 + 7i$ and $-3 - 7i$; $14 - 5i$ and $14 + 5i$; and $8i$ and $-8i$.

The product of a complex number and its conjugate is a real number.

EXAMPLE 5 Multiply each of the following.

a) $(5 + 7i)(5 - 7i)$ **b)** $(8i)(-8i)$

Solution

(5+7i)(5−7i) 74
(8i)(−8i) 64

a) $$\begin{aligned}(5 + 7i)(5 - 7i) &= 5^2 - (7i)^2 \quad \text{Using } (A + B)(A - B) = A^2 - B^2\\ &= 25 - 49i^2\\ &= 25 - 49(-1)\\ &= 25 + 49\\ &= 74\end{aligned}$$

b) $$\begin{aligned}(8i)(-8i) &= -64i^2\\ &= -64(-1)\\ &= 64\end{aligned}$$

Now Try Exercise 49.

Conjugates are used when we divide complex numbers.

EXAMPLE 6 Divide $2 - 5i$ by $1 - 6i$.

Solution We write fraction notation and then multiply by 1, using the conjugate of the denominator to form the symbol for 1.

$$\begin{aligned}\frac{2 - 5i}{1 - 6i} &= \frac{2 - 5i}{1 - 6i} \cdot \frac{1 + 6i}{1 + 6i} \quad \text{Note that } 1 + 6i \text{ is the conjugate of the divisor, } 1 - 6i.\\ &= \frac{(2 - 5i)(1 + 6i)}{(1 - 6i)(1 + 6i)}\\ &= \frac{2 + 12i - 5i - 30i^2}{1 - 36i^2}\\ &= \frac{2 + 7i + 30}{1 + 36} \quad i^2 = -1\\ &= \frac{32 + 7i}{37}\\ &= \frac{32}{37} + \frac{7}{37}i. \quad \text{Writing the quotient in the form } a + bi\end{aligned}$$

Now Try Exercise 69.

(2−5i)/(1−6i)▸Frac $\frac{32}{37}+\frac{7}{37}i$

With a graphing calculator set in $a + bi$ mode, we can divide complex numbers and express the real parts and the imaginary parts in fraction form, just as we did in Example 6.

3.1 Exercise Set

Express the number in terms of i.

1. $\sqrt{-3}$
2. $\sqrt{-21}$
3. $\sqrt{-25}$
4. $\sqrt{-100}$
5. $-\sqrt{-33}$
6. $-\sqrt{-59}$
7. $-\sqrt{-81}$
8. $-\sqrt{-9}$
9. $\sqrt{-98}$
10. $\sqrt{-28}$

Simplify. Write answers in the form a + bi, where a and b are real numbers.

11. $(-5 + 3i) + (7 + 8i)$
12. $(-6 - 5i) + (9 + 2i)$
13. $(4 - 9i) + (1 - 3i)$
14. $(7 - 2i) + (4 - 5i)$
15. $(12 + 3i) + (-8 + 5i)$
16. $(-11 + 4i) + (6 + 8i)$
17. $(-1 - i) + (-3 - i)$
18. $(-5 - i) + (6 + 2i)$
19. $(3 + \sqrt{-16}) + (2 + \sqrt{-25})$
20. $(7 - \sqrt{-36}) + (2 + \sqrt{-9})$
21. $(10 + 7i) - (5 + 3i)$
22. $(-3 - 4i) - (8 - i)$
23. $(13 + 9i) - (8 + 2i)$
24. $(-7 + 12i) - (3 - 6i)$
25. $(6 - 4i) - (-5 + i)$
26. $(8 - 3i) - (9 - i)$
27. $(-5 + 2i) - (-4 - 3i)$
28. $(-6 + 7i) - (-5 - 2i)$
29. $(4 - 9i) - (2 + 3i)$
30. $(10 - 4i) - (8 + 2i)$
31. $\sqrt{-4} \cdot \sqrt{-36}$
32. $\sqrt{-49} \cdot \sqrt{-9}$
33. $\sqrt{-81} \cdot \sqrt{-25}$
34. $\sqrt{-16} \cdot \sqrt{-100}$
35. $7i(2 - 5i)$
36. $3i(6 + 4i)$
37. $-2i(-8 + 3i)$
38. $-6i(-5 + i)$
39. $(1 + 3i)(1 - 4i)$
40. $(1 - 2i)(1 + 3i)$
41. $(2 + 3i)(2 + 5i)$
42. $(3 - 5i)(8 - 2i)$
43. $(-4 + i)(3 - 2i)$
44. $(5 - 2i)(-1 + i)$
45. $(8 - 3i)(-2 - 5i)$
46. $(7 - 4i)(-3 - 3i)$
47. $(3 + \sqrt{-16})(2 + \sqrt{-25})$
48. $(7 - \sqrt{-16})(2 + \sqrt{-9})$
49. $(5 - 4i)(5 + 4i)$
50. $(5 + 9i)(5 - 9i)$
51. $(3 + 2i)(3 - 2i)$
52. $(8 + i)(8 - i)$
53. $(7 - 5i)(7 + 5i)$
54. $(6 - 8i)(6 + 8i)$
55. $(4 + 2i)^2$
56. $(5 - 4i)^2$
57. $(-2 + 7i)^2$
58. $(-3 + 2i)^2$
59. $(1 - 3i)^2$
60. $(2 - 5i)^2$
61. $(-1 - i)^2$
62. $(-4 - 2i)^2$
63. $(3 + 4i)^2$
64. $(6 + 5i)^2$
65. $\dfrac{3}{5 - 11i}$
66. $\dfrac{i}{2 + i}$
67. $\dfrac{5}{2 + 3i}$
68. $\dfrac{-3}{4 - 5i}$
69. $\dfrac{4 + i}{-3 - 2i}$
70. $\dfrac{5 - i}{-7 + 2i}$
71. $\dfrac{5 - 3i}{4 + 3i}$
72. $\dfrac{6 + 5i}{3 - 4i}$
73. $\dfrac{2 + \sqrt{3}i}{5 - 4i}$
74. $\dfrac{\sqrt{5} + 3i}{1 - i}$
75. $\dfrac{1 + i}{(1 - i)^2}$
76. $\dfrac{1 - i}{(1 + i)^2}$
77. $\dfrac{4 - 2i}{1 + i} + \dfrac{2 - 5i}{1 + i}$
78. $\dfrac{3 + 2i}{1 - i} + \dfrac{6 + 2i}{1 - i}$

Simplify.

79. i^{11}
80. i^{7}
81. i^{35}
82. i^{24}
83. i^{64}
84. i^{42}
85. $(-i)^{71}$
86. $(-i)^{6}$
87. $(5i)^{4}$
88. $(2i)^{5}$

Skill Maintenance

89. Write a slope–intercept equation for the line containing the point $(3, -5)$ and perpendicular to the line $3x - 6y = 7$. [1.4]

Given that $f(x) = x^2 + 4$ and $g(x) = 3x + 5$, find each of the following. [2.2]

90. The domain of $f - g$

91. The domain of f/g

92. $(f - g)(x)$

93. $(f/g)(2)$

94. For the function $f(x) = x^2 - 3x + 4$, construct and simplify the difference quotient
$$\frac{f(x+h) - f(x)}{h}.$$ [2.2]

Synthesis

Determine whether the statement is true or false.

95. The sum of two numbers that are complex conjugates of each other is always a real number.

96. The conjugate of a sum is the sum of the conjugates of the individual complex numbers.

97. The conjugate of a product is the product of the conjugates of the individual complex numbers.

Let $z = a + bi$ and $\bar{z} = a - bi$.

98. Find a general expression for $1/z$.

99. Find a general expression for $z\bar{z}$.

100. Solve $z + 6\bar{z} = 7$ for z.

101. Multiply and simplify:
$$[x - (3 + 4i)][x - (3 - 4i)].$$

3.2 Quadratic Equations, Functions, Zeros, and Models

- Find zeros of quadratic functions and solve quadratic equations by using the principle of zero products, by using the principle of square roots, by completing the square, and by using the quadratic formula.
- Solve equations that are reducible to quadratic.
- Solve applied problems using quadratic equations.

Quadratic Equations and Quadratic Functions

In this section, we will explore the relationship between the solutions of quadratic equations and the zeros of quadratic functions. We define quadratic equations and quadratic functions as follows.

QUADRATIC EQUATIONS

A **quadratic equation** is an equation that can be written in the form
$$ax^2 + bx + c = 0, \quad a \neq 0,$$
where a, b, and c are real numbers.

QUADRATIC FUNCTIONS

A **quadratic function** f is a function that can be written in the form
$$f(x) = ax^2 + bx + c, \quad a \neq 0,$$
where a, b, and c are real numbers.

A quadratic equation written in the form $ax^2 + bx + c = 0$ is said to be in **standard form**.

ZEROS OF A FUNCTION
REVIEW SECTION 1.5.

The *zeros* of a quadratic function $f(x) = ax^2 + bx + c$ are the *solutions* of the associated quadratic equation $ax^2 + bx + c = 0$. (These solutions are sometimes called *roots* of the equation.) Quadratic functions can have real-number zeros or imaginary-number zeros and quadratic equations can have real-number solutions or imaginary-number solutions. If the zeros or solutions are real numbers, they are also the first coordinates of the x-intercepts of the graph of the quadratic function.

The following principles allow us to solve many quadratic equations.

EQUATION-SOLVING PRINCIPLES

The Principle of Zero Products: If $ab = 0$ is true, then $a = 0$ or $b = 0$, and if $a = 0$ or $b = 0$, then $ab = 0$.

The Principle of Square Roots: If $x^2 = k$, then $x = \sqrt{k}$ *or* $x = -\sqrt{k}$.

Just in Time 14, 19

EXAMPLE 1 Solve: $2x^2 - x = 3$.

Algebraic Solution

We factor and use the principle of zero products:

$$\begin{aligned} 2x^2 - x &= 3 \\ 2x^2 - x - 3 &= 0 \\ (x+1)(2x-3) &= 0 \\ x + 1 = 0 \quad &or \quad 2x - 3 = 0 \\ x = -1 \quad &or \quad 2x = 3 \\ x = -1 \quad &or \quad x = \tfrac{3}{2}. \end{aligned}$$

Check:

For $x = -1$:

$$\begin{array}{c|c} 2x^2 - x & 3 \\ \hline 2(-1)^2 - (-1) \ ? & 3 \\ 2 \cdot 1 + 1 & \\ 2 + 1 & \\ 3 & 3 \quad \text{TRUE} \end{array}$$

For $x = \frac{3}{2}$:

$$\begin{array}{c|c} 2x^2 - x & 3 \\ \hline 2\left(\frac{3}{2}\right)^2 - \frac{3}{2} \ ? & 3 \\ 2 \cdot \frac{9}{4} - \frac{3}{2} & \\ \frac{9}{2} - \frac{3}{2} & \\ \frac{6}{2} & \\ 3 & 3 \quad \text{TRUE} \end{array}$$

The solutions are -1 and $\frac{3}{2}$.

Graphical Solution

The solutions of the equation $2x^2 - x = 3$, or the equivalent equation $2x^2 - x - 3 = 0$, are the zeros of the function $f(x) = 2x^2 - x - 3$. They are also the first coordinates of the x-intercepts of the graph of $f(x) = 2x^2 - x - 3$.

$y = 2x^2 - x - 3$

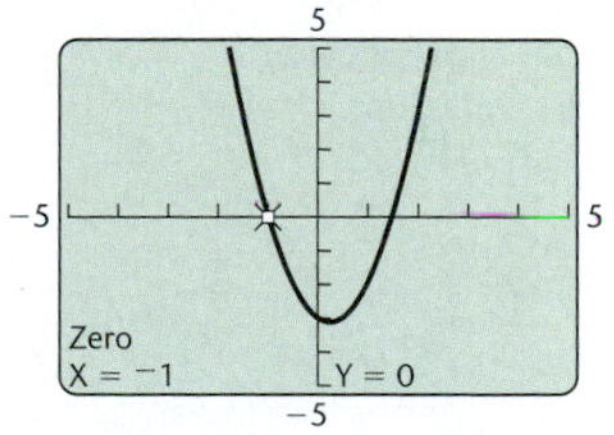

$y = 2x^2 - x - 3$

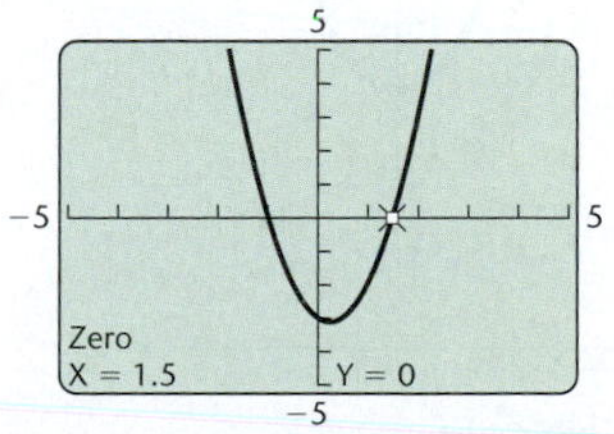

The solutions are -1 and 1.5, or -1 and $\frac{3}{2}$.

Now Try Exercise 3.

Just in Time 20

EXAMPLE 2 Solve: $2x^2 - 10 = 0$.

Solution We have

$$
\begin{aligned}
2x^2 - 10 &= 0 \\
2x^2 &= 10 && \text{Adding 10 on both sides} \\
x^2 &= 5 && \text{Dividing by 2 on both sides} \\
x = \sqrt{5} \quad or \quad x &= -\sqrt{5}. && \text{Using the principle of square roots}
\end{aligned}
$$

Check:

$$
\begin{array}{c|c}
2x^2 - 10 = 0 & \\
\hline
2(\pm\sqrt{5})^2 - 10 \ ? & 0 \\
2 \cdot 5 - 10 & \\
10 - 10 & \\
0 & 0 \quad \text{TRUE}
\end{array}
$$

We can check both solutions at once.

The solutions are $\sqrt{5}$ and $-\sqrt{5}$, or $\pm\sqrt{5}$.

Now Try Exercise 7.

$f(x) = x^2 - 3x - 4$

$(-1, 0)$ $(4, 0)$

Two real-number zeros
Two x-intercepts

FIGURE 1.

We have seen that some quadratic equations can be solved by factoring and using the principle of zero products. For example, consider the equation $x^2 - 3x - 4 = 0$:

$$
\begin{aligned}
x^2 - 3x - 4 &= 0 \\
(x + 1)(x - 4) &= 0 && \text{Factoring} \\
x + 1 = 0 \quad or \quad x - 4 &= 0 && \text{Using the principle of zero products} \\
x = -1 \quad or \quad x &= 4.
\end{aligned}
$$

The equation $x^2 - 3x - 4 = 0$ has *two real-number* solutions, -1 and 4. These are the zeros of the associated quadratic function $f(x) = x^2 - 3x - 4$ and the first coordinates of the x-intercepts of the graph of this function. (See Fig. 1.)

Next, consider the equation $x^2 - 6x + 9 = 0$. Again, we factor and use the principle of zero products:

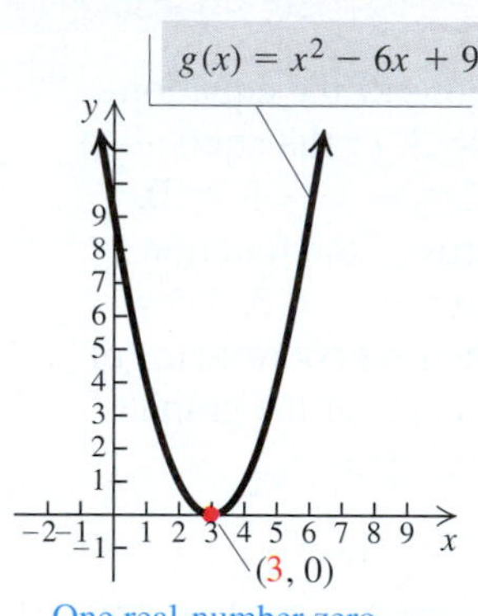

One real-number zero
One x-intercept

FIGURE 2.

$$
\begin{aligned}
x^2 - 6x + 9 &= 0 \\
(x - 3)(x - 3) &= 0 && \text{Factoring} \\
x - 3 = 0 \quad or \quad x - 3 &= 0 && \text{Using the principle of zero products} \\
x = 3 \quad or \quad x &= 3.
\end{aligned}
$$

The equation $x^2 - 6x + 9 = 0$ has *one real-number* solution, 3. It is the zero of the quadratic function $g(x) = x^2 - 6x + 9$ and the first coordinate of the x-intercept of the graph of this function. (See Fig. 2.)

The principle of square roots can be used to solve quadratic equations like $x^2 + 13 = 0$:

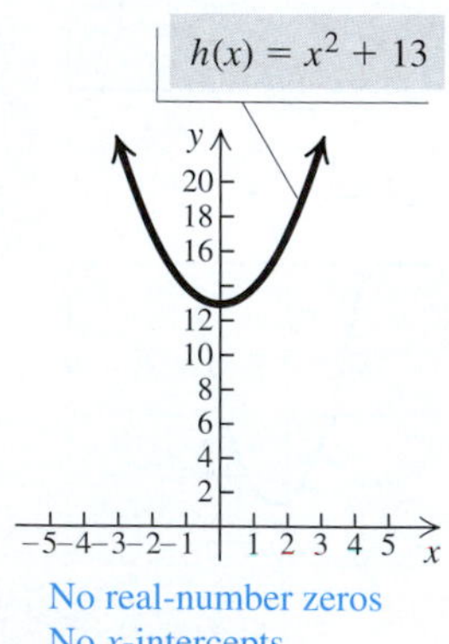

No real-number zeros
No x-intercepts

FIGURE 3.

$$
\begin{aligned}
x^2 + 13 &= 0 \\
x^2 &= -13 \\
x &= \pm\sqrt{-13} && \text{Using the principle of square roots} \\
x &= \pm\sqrt{13}i. && \sqrt{-13} = \sqrt{-1} \cdot \sqrt{13} = i \cdot \sqrt{13} = \sqrt{13}i
\end{aligned}
$$

The equation has *two imaginary-number* solutions, $-\sqrt{13}i$ and $\sqrt{13}i$. These are the zeros of the associated quadratic function $h(x) = x^2 + 13$. Since the zeros are not real numbers, the graph of the function has no x-intercepts. (See Fig. 3.)

Completing the Square

Neither the principle of zero products nor the principle of square roots would yield the *exact* zeros of a function like $f(x) = x^2 - 6x - 10$ or the *exact* solutions of the associated equation $x^2 - 6x - 10 = 0$. If we wish to find exact zeros or solutions, we can use a procedure called **completing the square** and then use the principle of square roots.

EXAMPLE 3 Find the zeros of $f(x) = x^2 - 6x - 10$ by completing the square.

Solution We find the values of x for which $f(x) = 0$; that is, we solve the associated equation $x^2 - 6x - 10 = 0$. Our goal is to find an equivalent equation of the form $x^2 + bx + c = d$ in which $x^2 + bx + c$ is a perfect square. Since

$$x^2 + bx + \left(\frac{b}{2}\right)^2 = \left(x + \frac{b}{2}\right)^2,$$

the number c is found by taking half the coefficient of the x-term and squaring it. Thus for the equation $x^2 - 6x - 10 = 0$, we have

$$x^2 - 6x - 10 = 0$$

$$x^2 - 6x = 10 \qquad \text{Adding 10}$$

$$x^2 - 6x + 9 = 10 + 9 \qquad \text{Adding 9 on both sides to complete the square: } \left(\frac{b}{2}\right)^2 = \left(\frac{-6}{2}\right)^2 = (-3)^2 = 9$$

$$x^2 - 6x + 9 = 19.$$

Because $x^2 - 6x + 9$ is a perfect square, we are able to write it as $(x - 3)^2$, the square of a binomial. We can then use the principle of square roots to finish the solution:

$$(x - 3)^2 = 19 \qquad \text{Factoring}$$

$$x - 3 = \pm\sqrt{19} \qquad \text{Using the principle of square roots}$$

$$x = 3 \pm \sqrt{19}. \qquad \text{Adding 3}$$

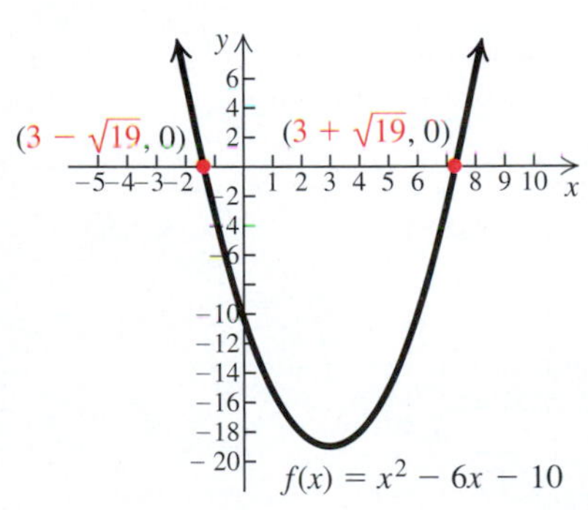

Therefore, the solutions of the equation are $3 + \sqrt{19}$ and $3 - \sqrt{19}$, or simply $3 \pm \sqrt{19}$. The zeros of $f(x) = x^2 - 6x - 10$ are also $3 + \sqrt{19}$ and $3 - \sqrt{19}$, or $3 \pm \sqrt{19}$.

We can find decimal approximations for $3 \pm \sqrt{19}$ using a calculator:

$$3 + \sqrt{19} \approx 7.359 \quad \text{and} \quad 3 - \sqrt{19} \approx -1.359.$$

The zeros are approximately 7.359 and -1.359.

Now Try Exercise 31.

Approximations for the zeros of the quadratic function $f(x) = x^2 - 6x - 10$ in Example 3 can be found using the Zero method.

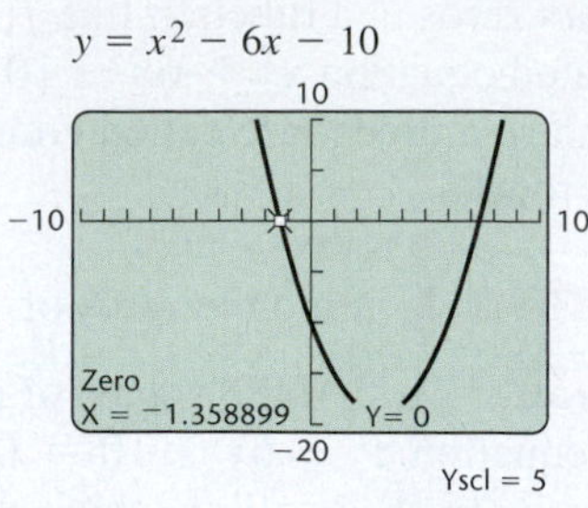

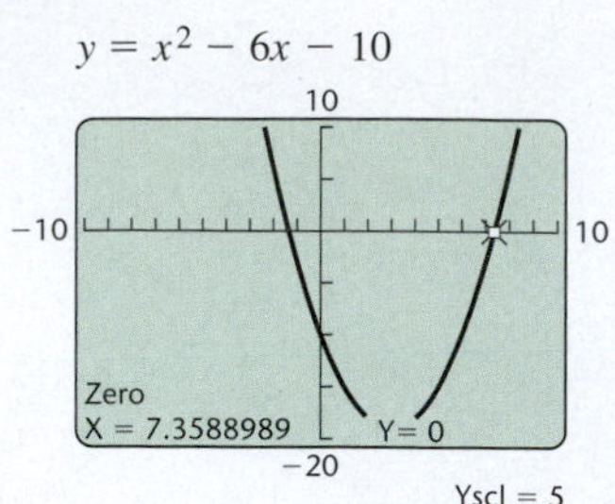

Before we can complete the square, the coefficient of the x^2-term must be 1. When it is not, we divide by the x^2-coefficient on both sides of the equation.

EXAMPLE 4 Solve: $2x^2 - 1 = 3x$.

Solution We have

$$2x^2 - 1 = 3x$$

$$2x^2 - 3x - 1 = 0 \qquad \text{Subtracting } 3x\text{. We are unable to factor the result.}$$

$$2x^2 - 3x = 1 \qquad \text{Adding 1}$$

$$x^2 - \frac{3}{2}x = \frac{1}{2} \qquad \text{Dividing by 2 to make the } x^2\text{-coefficient 1}$$

$$x^2 - \frac{3}{2}x + \frac{9}{16} = \frac{1}{2} + \frac{9}{16} \qquad \text{Completing the square: } \tfrac{1}{2}\left(-\tfrac{3}{2}\right) = -\tfrac{3}{4} \text{ and } \left(-\tfrac{3}{4}\right)^2 = \tfrac{9}{16}\text{; adding } \tfrac{9}{16}$$

$$\left(x - \frac{3}{4}\right)^2 = \frac{17}{16} \qquad \text{Factoring and simplifying}$$

$$x - \frac{3}{4} = \pm\frac{\sqrt{17}}{4} \qquad \text{Using the principle of square roots and the quotient rule for radicals}$$

$$x = \frac{3}{4} \pm \frac{\sqrt{17}}{4} \qquad \text{Adding } \tfrac{3}{4}$$

$$x = \frac{3 \pm \sqrt{17}}{4}.$$

The solutions are

$$\frac{3 + \sqrt{17}}{4} \text{ and } \frac{3 - \sqrt{17}}{4}, \text{ or } \frac{3 \pm \sqrt{17}}{4}.$$

Now Try Exercise 35.

To solve a quadratic equation by completing the square:

1. Isolate the terms with variables on one side of the equation and arrange them in descending order.
2. Divide by the coefficient of the squared term if that coefficient is not 1.
3. Complete the square by finding half the coefficient of the first-degree term and adding its square on both sides of the equation.
4. Express one side of the equation as the square of a binomial.
5. Use the principle of square roots.
6. Solve for the variable.

Using the Quadratic Formula

Because completing the square works for *any* quadratic equation, it can be used to solve the general quadratic equation $ax^2 + bx + c = 0$ for x. The result will be a formula that can be used to solve any quadratic equation quickly.

Consider any quadratic equation in standard form:

$$ax^2 + bx + c = 0, \quad a \neq 0.$$

For now, we assume that $a > 0$ and solve by completing the square. As the steps are carried out, compare them with those of Example 4.

$$ax^2 + bx + c = 0 \qquad \textbf{Standard form}$$

$$ax^2 + bx = -c \qquad \textbf{Adding } -c$$

$$x^2 + \frac{b}{a}x = -\frac{c}{a} \qquad \textbf{Dividing by } a$$

Half of $\frac{b}{a}$ is $\frac{b}{2a}$, and $\left(\frac{b}{2a}\right)^2 = \frac{b^2}{4a^2}$. Thus we add $\frac{b^2}{4a^2}$:

$$x^2 + \frac{b}{a}x + \frac{b^2}{4a^2} = -\frac{c}{a} + \frac{b^2}{4a^2} \qquad \textbf{Adding } \frac{b^2}{4a^2} \textbf{ to complete the square}$$

$$\left(x + \frac{b}{2a}\right)^2 = -\frac{4ac}{4a^2} + \frac{b^2}{4a^2}$$

Factoring on the left; finding a common denominator on the right: $-\frac{c}{a} = -\frac{c}{a} \cdot \frac{4a}{4a} = -\frac{4ac}{4a^2}$

$$\left(x + \frac{b}{2a}\right)^2 = \frac{b^2 - 4ac}{4a^2}$$

$$x + \frac{b}{2a} = \pm\frac{\sqrt{b^2 - 4ac}}{2a}$$

Using the principle of square roots and the quotient rule for radicals. Since $a > 0$, $\sqrt{4a^2} = 2a$.

$$x = -\frac{b}{2a} \pm \frac{\sqrt{b^2 - 4ac}}{2a} \qquad \textbf{Adding } -\frac{b}{2a}$$

$$x = \frac{-b \pm \sqrt{b^2 - 4ac}}{2a}.$$

It can also be shown that this result holds if $a < 0$.

THE QUADRATIC FORMULA

The solutions of $ax^2 + bx + c = 0$, $a \neq 0$, are given by

$$x = \frac{-b \pm \sqrt{b^2 - 4ac}}{2a}.$$

EXAMPLE 5 Solve $3x^2 + 2x = 7$. Find exact solutions and approximate solutions rounded to three decimal places.

Algebraic Solution

After writing the equation in standard form, we are unable to factor, so we identify a, b, and c in order to use the quadratic formula:

$$3x^2 + 2x = 7$$
$$3x^2 + 2x - 7 = 0;$$
$$a = 3, \quad b = 2, \quad c = -7.$$

We then use the quadratic formula:

$$x = \frac{-b \pm \sqrt{b^2 - 4ac}}{2a}$$
$$= \frac{-2 \pm \sqrt{2^2 - 4(3)(-7)}}{2(3)} \quad \text{Substituting}$$
$$= \frac{-2 \pm \sqrt{4 + 84}}{6} = \frac{-2 \pm \sqrt{88}}{6}$$
$$= \frac{-2 \pm \sqrt{4 \cdot 22}}{6} = \frac{-2 \pm 2\sqrt{22}}{6}$$
$$= \frac{2(-1 \pm \sqrt{22})}{2 \cdot 3}$$
$$= \frac{2}{2} \cdot \frac{-1 \pm \sqrt{22}}{3}$$
$$= \frac{-1 \pm \sqrt{22}}{3}.$$

The exact solutions are

$$\frac{-1 - \sqrt{22}}{3} \quad \text{and} \quad \frac{-1 + \sqrt{22}}{3}.$$

Using a calculator, we approximate the solutions to be -1.897 and 1.230.

Graphical Solution

Using the Intersect method, we graph $y_1 = 3x^2 + 2x$ and $y_2 = 7$ and use the INTERSECT feature to find the coordinates of the points of intersection. The first coordinates of these points are the solutions of the equation $y_1 = y_2$, or $3x^2 + 2x = 7$.

$y_1 = 3x^2 + 2x, \quad y_2 = 7$

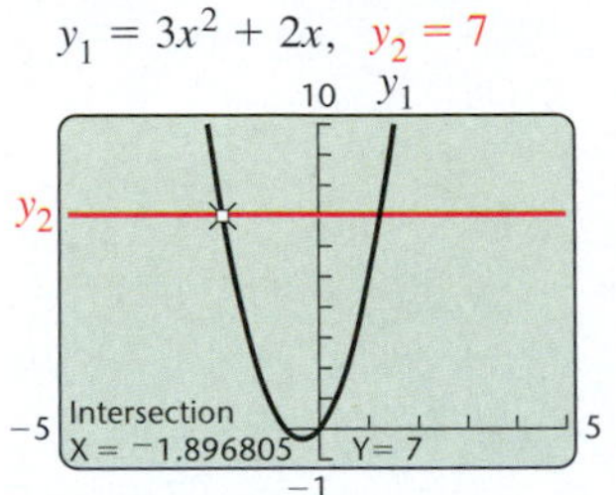

$y_1 = 3x^2 + 2x, \quad y_2 = 7$

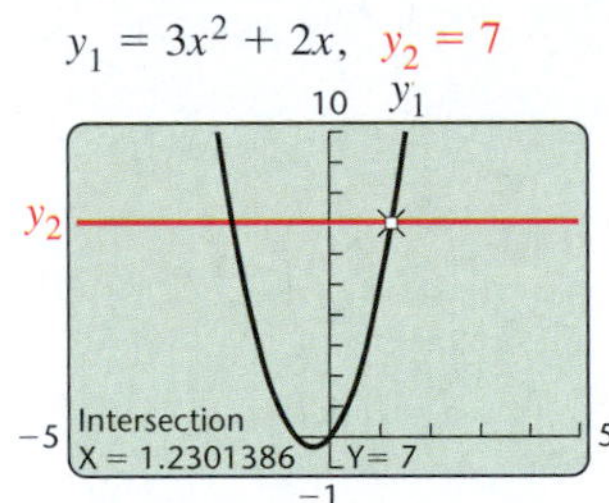

The solutions of $3x^2 + 2x = 7$ are approximately -1.897 and 1.230. We could also write the equation in standard form, $3x^2 + 2x - 7 = 0$, and use the Zero method.

Now Try Exercise 41.

Not all quadratic equations can be solved graphically.

EXAMPLE 6 Solve: $x^2 + 5x + 8 = 0$.

Algebraic Solution

To find the solutions, we use the quadratic formula. For $x^2 + 5x + 8 = 0$, we have

$$a = 1, \quad b = 5, \quad c = 8;$$

$$x = \frac{-b \pm \sqrt{b^2 - 4ac}}{2a}$$

$$= \frac{-5 \pm \sqrt{5^2 - 4(1)(8)}}{2 \cdot 1} \qquad \text{Substituting}$$

$$= \frac{-5 \pm \sqrt{25 - 32}}{2}$$

$$= \frac{-5 \pm \sqrt{-7}}{2} = \frac{-5 \pm \sqrt{7}i}{2}.$$

The solutions are $-\frac{5}{2} - \frac{\sqrt{7}}{2}i$ and $-\frac{5}{2} + \frac{\sqrt{7}}{2}i$.

Graphical Solution

The graph of the function $f(x) = x^2 + 5x + 8$ shows no x-intercepts.

$y = x^2 + 5x + 8$

10, −8, 8, −2

Thus the function has no real-number zeros and there are no real-number solutions of the associated equation $x^2 + 5x + 8 = 0$. This is a quadratic equation that cannot be solved graphically.

Now Try Exercise 47.

The Discriminant

From the quadratic formula, we know that the solutions x_1 and x_2 of a quadratic equation are given by

$$x_1 = \frac{-b + \sqrt{b^2 - 4ac}}{2a} \quad \text{and} \quad x_2 = \frac{-b - \sqrt{b^2 - 4ac}}{2a}.$$

The expression $b^2 - 4ac$ shows the nature of the solutions. This expression is called the **discriminant**. If it is 0, then it makes no difference whether we choose the plus sign or the minus sign in the formula. That is, $x_1 = -\frac{b}{2a} = x_2$, so there is just one solution. In this case, we sometimes say that there is one repeated real solution. If the discriminant is positive, there will be two different real solutions. If it is negative, we will be taking the square root of a negative number; hence there will be two imaginary-number solutions, and they will be complex conjugates.

DISCRIMINANT

For $ax^2 + bx + c = 0$, where a, b, and c are real numbers, $a \neq 0$:

$b^2 - 4ac = 0 \longrightarrow$ One real-number solution;

$b^2 - 4ac > 0 \longrightarrow$ Two different real-number solutions;

$b^2 - 4ac < 0 \longrightarrow$ Two different imaginary-number solutions, complex conjugates.

In Example 5, the discriminant, 88, is positive, indicating that there are two different real-number solutions. The negative discriminant, −7, in Example 6 indicates that there are two different imaginary-number solutions.

Equations Reducible to Quadratic

Some equations can be treated as quadratic, provided we make a suitable substitution. For example, consider the following:

$$x^4 - 5x^2 + 4 = 0$$
$$(x^2)^2 - 5x^2 + 4 = 0 \qquad x^4 = (x^2)^2$$
$$u^2 - 5u + 4 = 0. \qquad \text{Substituting } u \text{ for } x^2$$

The equation $u^2 - 5u + 4 = 0$ can be solved for u by factoring or using the quadratic formula. Then we can reverse the substitution, replacing u with x^2, and solve for x. Equations like the one above are said to be **reducible to quadratic**, or **quadratic in form**.

EXAMPLE 7 Solve: $x^4 - 5x^2 + 4 = 0$.

Algebraic Solution

We let $u = x^2$ and substitute:

$$u^2 - 5u + 4 = 0 \qquad \text{Substituting } u \text{ for } x^2$$
$$(u - 1)(u - 4) = 0 \qquad \text{Factoring}$$
$$u - 1 = 0 \quad or \quad u - 4 = 0 \qquad \text{Using the principle of zero products}$$
$$u = 1 \quad or \quad u = 4.$$

Don't stop here! We must solve for the original variable. We substitute x^2 for u and solve for x:

$$x^2 = 1 \quad or \quad x^2 = 4$$
$$x = \pm 1 \quad or \quad x = \pm 2. \qquad \text{Using the principle of square roots}$$

The solutions are -1, 1, -2, and 2.

Graphical Solution

Using the Zero method, we graph the function $y = x^4 - 5x^2 + 4$ and use the ZERO feature to find the zeros.

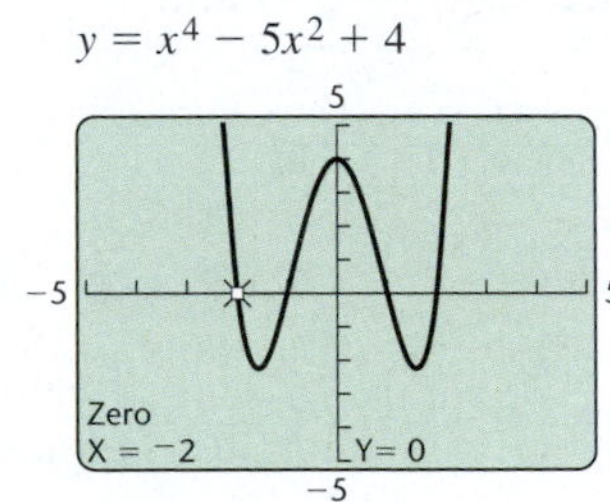

The leftmost zero is -2. Using the ZERO feature three more times, we find that the other zeros are -1, 1, and 2. Thus the solutions of $x^4 - 5x^2 + 4 = 0$ are -2, -1, 1, and 2.

Now Try Exercise 91.

Just in Time 27

EXAMPLE 8 Solve: $t^{2/3} - 2t^{1/3} - 3 = 0$.

Solution We let $u = t^{1/3}$ and substitute:

$$t^{2/3} - 2t^{1/3} - 3 = 0$$
$$(t^{1/3})^2 - 2t^{1/3} - 3 = 0$$
$$u^2 - 2u - 3 = 0 \qquad \text{Substituting } u \text{ for } t^{1/3}$$
$$(u + 1)(u - 3) = 0 \qquad \text{Factoring}$$
$$u + 1 = 0 \quad or \quad u - 3 = 0 \qquad \text{Using the principle of zero products}$$
$$u = -1 \quad or \quad u = 3.$$

Now we must solve for the original variable, t. We substitute $t^{1/3}$ for u and solve for t:

$$t^{1/3} = -1 \quad or \quad t^{1/3} = 3$$
$$(t^{1/3})^3 = (-1)^3 \quad or \quad (t^{1/3})^3 = 3^3 \qquad \text{Cubing on both sides}$$
$$t = -1 \quad or \quad t = 27.$$

The solutions are -1 and 27.

Now Try Exercise 99.

Applications

Some applied problems can be translated to quadratic equations.

EXAMPLE 9 ***Museums in China.*** The number of museums in China increased from approximately 2000 in the year 2000 to over 3500 by the end of 2012. In 2012, a record 451 new museums opened. For comparison, in the United States, only 20–40 new museums were opened per year from 2000 to 2008. The function

$$h(x) = 30.992x^2 + 4.108x + 2294.594$$

can be used to estimate the number of museums in China, x years after 2005. (*Source:* The Economist/www.economist.com).

a) Estimate the number of museums that will be in China in 2017 if the number of new museums that open per year continues growing at the same rate.

b) In what year was the number of museums in China 2600?

Solution

a) For 2017, $x = 2017 - 2005 = 12$. We substitute 12 for x and find $h(12)$:

$$\begin{aligned} h(x) &= 30.992x^2 + 4.108x + 2294.594 \\ h(12) &= 30.992(12)^2 + 4.108(12) + 2294.594 \\ h(12) &= 4462.848 + 49.296 + 2294.594 \approx 6807. \end{aligned}$$

In 2017, there will be approximately 6807 museums in China.

b) We substitute 2600 for $h(x)$ and solve for x:

$$\begin{aligned} h(x) &= 30.992x^2 + 4.108x + 2294.594 \\ 2600 &= 30.992x^2 + 4.108x + 2294.594 \\ 0 &= 30.992x^2 + 4.108x - 305.406. \end{aligned}$$

We then use the quadratic formula, with $a = 30.992$, $b = 4.108$, and $c = -305.406$:

$$\begin{aligned} x &= \frac{-b \pm \sqrt{b^2 - 4ac}}{2a} \\ x &= \frac{-4.108 \pm \sqrt{(4.108)^2 - 4(30.992)(-305.406)}}{2(30.992)} \\ x &= \frac{-4.108 \pm \sqrt{37{,}877.44667}}{61.984} \\ x &= 3.074 \quad \textit{or} \quad x = -3.206. \end{aligned}$$

Because we are looking for a year after 2005, we use the positive solution. Thus there were about 2600 museums in China 3 years after 2005, or in 2008.

Now Try Exercise 107.

EXAMPLE 10 ***Magazine Closures.*** The numbers of both magazine launches and magazine closures have increased in recent years. The function

$$m(x) = 34x^2 - 59x + 81$$

can be used to estimate the number of magazine closures after 2012 (*Source*: Media Finder.com, online database of U.S. and Canadian print and digital publications). In what year was the number of magazine closures 99?

Solution We substitute 99 for $m(x)$ and solve for x:

$$99 = 34x^2 - 59x + 81$$
$$0 = 34x^2 - 59x - 18.$$

We then use the quadratic formula, with $a = 34$, $b = -59$, and $c = -18$:

$$x = \frac{-(-59) \pm \sqrt{(-59)^2 - 4 \cdot 34 \cdot (-18)}}{2 \cdot 34} \quad \text{Substituting}$$
$$x = \frac{59 \pm \sqrt{5929}}{68}$$
$$x = 2 \quad or \quad x \approx -0.3.$$

Because we are looking for a year after 2012, we use the positive solution. Thus there were 99 magazine closures 2 years after 2012, or in 2014.

Now Try Exercise 109.

EXAMPLE 11 *Train Speeds.* Two trains leave a station at the same time. One train travels due west, and the other travels due south. The train traveling west travels 20 km/h faster than the train traveling south. After 2 hr, the trains are 200 km apart. Find the speed of each train.

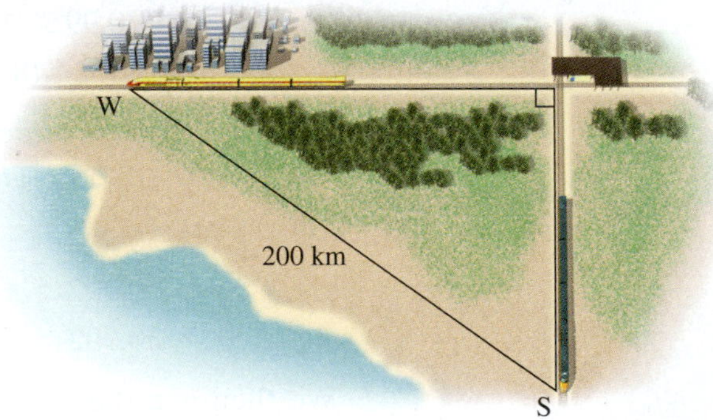

Solution

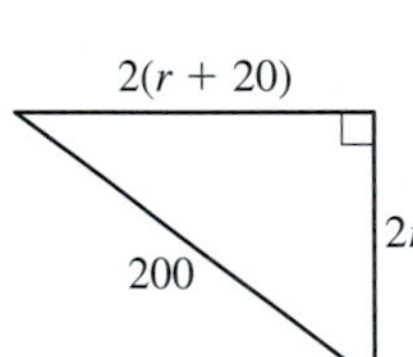

1. **Familiarize.** First, we make a drawing. We let $r =$ the speed of the train traveling south, in kilometers per hour. Then $r + 20 =$ the speed of the train traveling west, in kilometers per hour. We use the motion formula $d = rt$, where d is the distance, r is the rate (or speed), and t is the time. After 2 hr, the train traveling south has traveled $2r$ kilometers, and the train traveling west has traveled $2(r + 20)$ kilometers. We add these distances to the drawing.

Just in Time 28

2. **Translate.** We use the Pythagorean theorem, $a^2 + b^2 = c^2$, where a and b are the lengths of the legs of a right triangle and c is the length of the hypotenuse:

$$[2(r + 20)]^2 + (2r)^2 = 200^2.$$

3. **Carry out.** We solve the equation:

$$[2(r + 20)]^2 + (2r)^2 = 200^2$$
$$4(r^2 + 40r + 400) + 4r^2 = 40{,}000$$
$$4r^2 + 160r + 1600 + 4r^2 = 40{,}000$$
$$8r^2 + 160r + 1600 = 40{,}000 \quad \text{Collecting like terms}$$
$$8r^2 + 160r - 38{,}400 = 0 \quad \text{Subtracting 40,000}$$
$$r^2 + 20r - 4800 = 0 \quad \text{Dividing by 8}$$
$$(r + 80)(r - 60) = 0 \quad \text{Factoring}$$
$$r + 80 = 0 \quad or \quad r - 60 = 0 \quad \text{Principle of zero products}$$
$$r = -80 \quad or \quad r = 60.$$

$y_1 = [2(x + 20)]^2 + (2x)^2$,
$y_2 = 200^2$

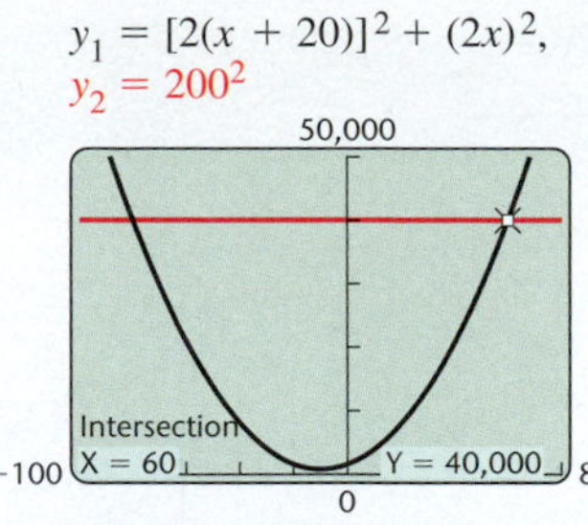

We also can solve this equation with a graphing calculator using the Intersect method, as shown in the window at left.

4. **Check.** Since speed cannot be negative, we check only 60. If the speed of the train traveling south is 60 km/h, then the speed of the train traveling west is $60 + 20$, or 80 km/h. In 2 hr, the train heading south travels $60 \cdot 2$, or 120 km, and the train heading west travels $80 \cdot 2$, or 160 km. Then they are $\sqrt{120^2 + 160^2}$, or $\sqrt{40{,}000}$, or 200 km apart. The answer checks.

5. **State.** The speed of the train heading south is 60 km/h, and the speed of the train heading west is 80 km/h.

Now Try Exercise 113.

CONNECTING THE CONCEPTS — Zeros, Solutions, and Intercepts

The zeros of a function $y = f(x)$ are also the solutions of the equation $f(x) = 0$, and the real-number zeros are the first coordinates of the x-intercepts of the graph of the function.

FUNCTION

Linear Function

$f(x) = 2x - 4$, or
$y = 2x - 4$

ZEROS OF THE FUNCTION; SOLUTIONS OF THE EQUATION

To find the **zero** of $f(x)$, we solve $f(x) = 0$:

$$2x - 4 = 0$$
$$2x = 4$$
$$x = 2.$$

The **solution** of the equation $2x - 4 = 0$ is 2. This is the zero of the function $f(x) = 2x - 4$; that is, $f(2) = 0$.

x-INTERCEPTS OF THE GRAPH

The zero of $f(x)$ is the first coordinate of the **x-intercept** of the graph of $y = f(x)$.

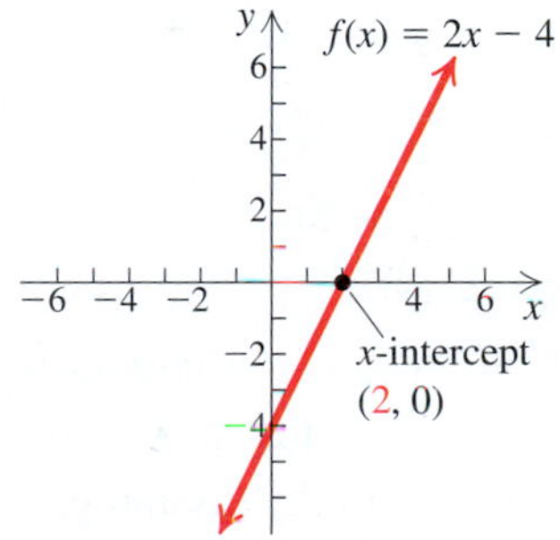

Quadratic Function

$g(x) = x^2 - 3x - 4$, or
$y = x^2 - 3x - 4$

To find the **zeros** of $g(x)$, we solve $g(x) = 0$:

$$x^2 - 3x - 4 = 0$$
$$(x + 1)(x - 4) = 0$$
$$x + 1 = 0 \quad or \quad x - 4 = 0$$
$$x = -1 \quad or \quad x = 4.$$

The **solutions** of the equation $x^2 - 3x - 4 = 0$ are -1 and 4. They are the zeros of the function $g(x) = x^2 - 3x - 4$; that is, $g(-1) = 0$ and $g(4) = 0$.

The real-number zeros of $g(x)$ are the first coordinates of the **x-intercepts** of the graph of $y = g(x)$.

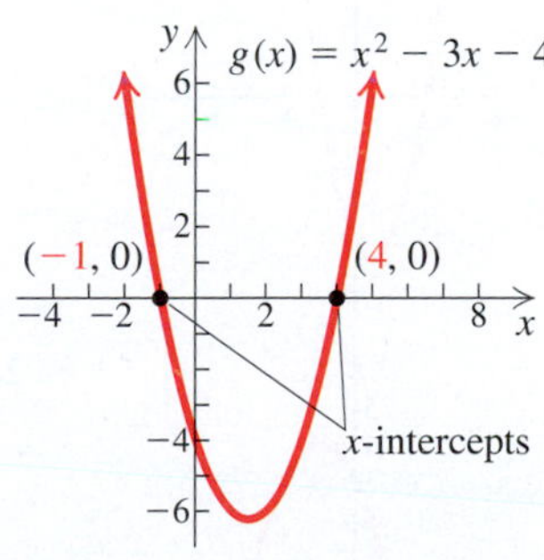

3.2 Exercise Set

Solve.

1. $(2x - 3)(3x - 2) = 0$
2. $(5x - 2)(2x + 3) = 0$
3. $x^2 - 8x - 20 = 0$
4. $x^2 + 6x + 8 = 0$
5. $3x^2 + x - 2 = 0$
6. $10x^2 - 16x + 6 = 0$
7. $4x^2 - 12 = 0$
8. $6x^2 = 36$
9. $3x^2 = 21$
10. $2x^2 - 20 = 0$
11. $5x^2 + 10 = 0$
12. $4x^2 + 12 = 0$
13. $x^2 + 16 = 0$
14. $x^2 + 25 = 0$
15. $2x^2 = 6x$
16. $18x + 9x^2 = 0$
17. $3y^3 - 5y^2 - 2y = 0$
18. $3t^3 + 2t = 5t^2$
19. $7x^3 + x^2 - 7x - 1 = 0$
(*Hint*: Factor by grouping.)
20. $3x^3 + x^2 - 12x - 4 = 0$
(*Hint*: Factor by grouping.)

In Exercises 21–28, use the given graph to find **(a)** *the x-intercepts and* **(b)** *the zeros of the function.*

21.
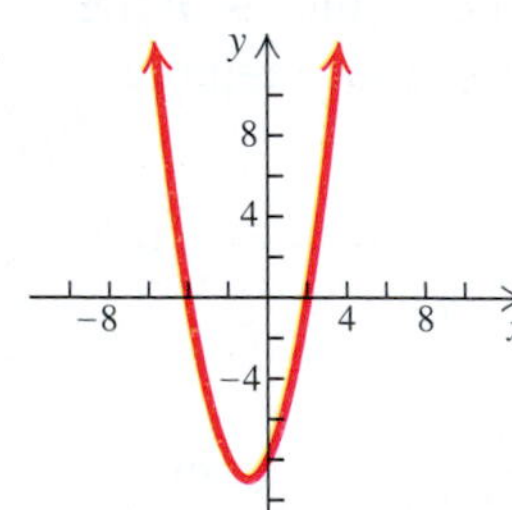

22.
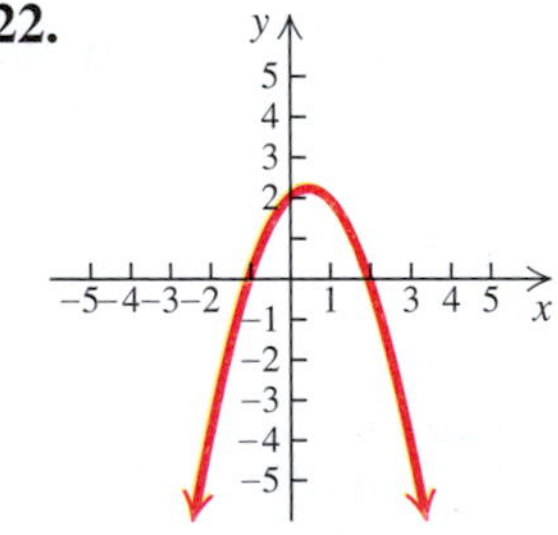

23.
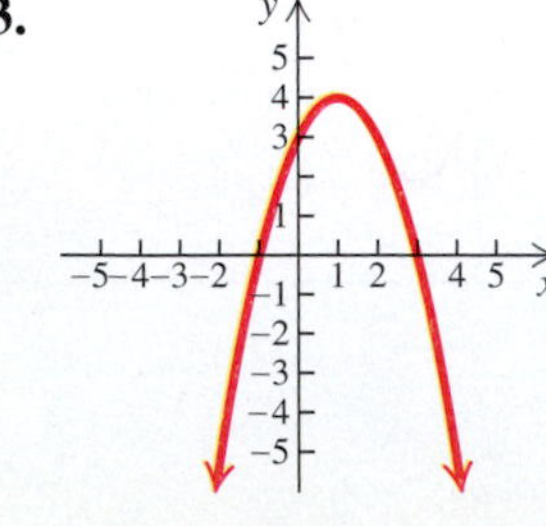

24.
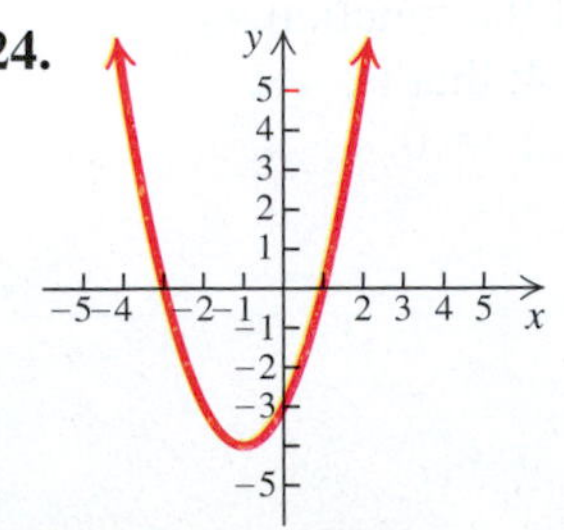

25.
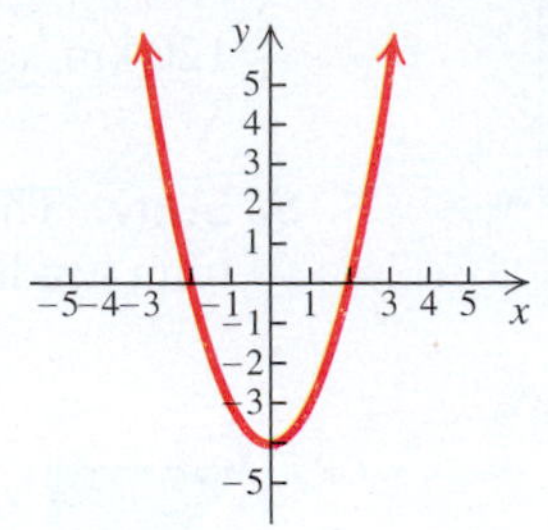

26.
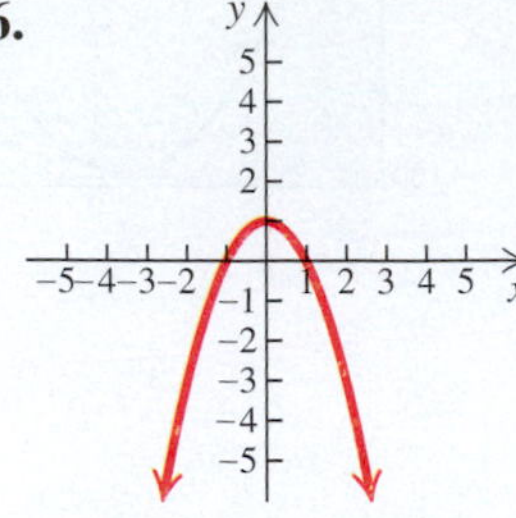

27.
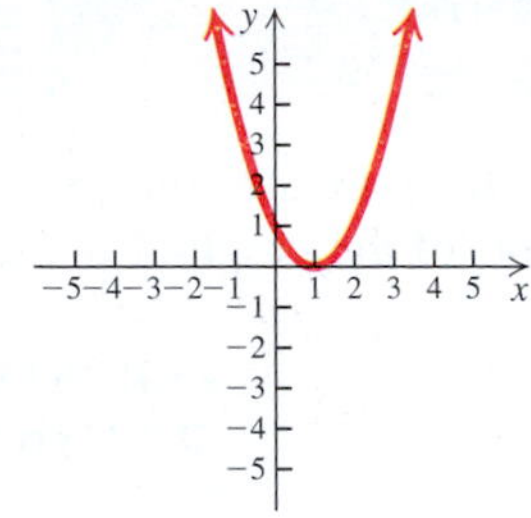

28.
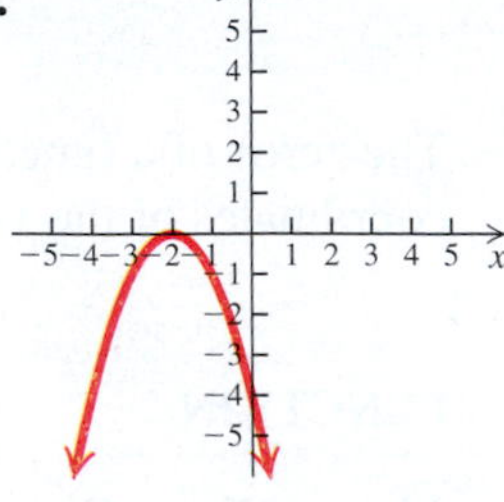

Solve by completing the square to obtain exact solutions.

29. $x^2 + 6x = 7$
30. $x^2 + 8x = -15$
31. $x^2 = 8x - 9$
32. $x^2 = 22 + 10x$
33. $x^2 + 8x + 25 = 0$
34. $x^2 + 6x + 13 = 0$
35. $3x^2 + 5x - 2 = 0$
36. $2x^2 - 5x - 3 = 0$

Use the quadratic formula to find exact solutions.

37. $x^2 - 2x = 15$
38. $x^2 + 4x = 5$
39. $5m^2 + 3m = 2$
40. $2y^2 - 3y - 2 = 0$
41. $3x^2 + 6 = 10x$
42. $3t^2 + 8t + 3 = 0$
43. $x^2 + x + 2 = 0$
44. $x^2 + 1 = x$
45. $5t^2 - 8t = 3$
46. $5x^2 + 2 = x$
47. $3x^2 + 4 = 5x$
48. $2t^2 - 5t = 1$
49. $x^2 - 8x + 5 = 0$
50. $x^2 - 6x + 3 = 0$
51. $3x^2 + x = 5$
52. $5x^2 + 3x = 1$
53. $2x^2 + 1 = 5x$
54. $4x^2 + 3 = x$
55. $5x^2 + 2x = -2$
56. $3x^2 + 3x = -4$

For each of the following, find the discriminant, $b^2 - 4ac$, and then determine whether one real-number solution, two different real-number solutions, or two different imaginary-number solutions exist.

57. $4x^2 = 8x + 5$
58. $4x^2 - 12x + 9 = 0$
59. $x^2 + 3x + 4 = 0$
60. $x^2 - 2x + 4 = 0$
61. $5t^2 - 7t = 0$
62. $5t^2 - 4t = 11$

Solve graphically. Round solutions to three decimal places, where appropriate.

63. $x^2 - 8x + 12 = 0$

64. $5x^2 + 42x + 16 = 0$

65. $7x^2 - 43x + 6 = 0$

66. $10x^2 - 23x + 12 = 0$

67. $6x + 1 = 4x^2$

68. $3x^2 + 5x = 3$

69. $2x^2 - 4 = 5x$

70. $4x^2 - 2 = 3x$

Find the zeros of the function algebraically. Give exact answers.

71. $f(x) = x^2 + 6x + 5$

72. $f(x) = x^2 - x - 2$

73. $f(x) = x^2 - 3x - 3$

74. $f(x) = 3x^2 + 8x + 2$

75. $f(x) = x^2 - 5x + 1$

76. $f(x) = x^2 - 3x - 7$

77. $f(x) = x^2 + 2x - 5$

78. $f(x) = x^2 - x - 4$

79. $f(x) = 2x^2 - x + 4$

80. $f(x) = 2x^2 + 3x + 2$

81. $f(x) = 3x^2 - x - 1$

82. $f(x) = 3x^2 + 5x + 1$

83. $f(x) = 5x^2 - 2x - 1$

84. $f(x) = 4x^2 - 4x - 5$

85. $f(x) = 4x^2 + 3x - 3$

86. $f(x) = x^2 + 6x - 3$

Use a graphing calculator to find the zeros of the function. Round to three decimal places.

87. $f(x) = 3x^2 + 2x - 4$

88. $f(x) = 9x^2 - 8x - 7$

89. $f(x) = 5.02x^2 - 4.19x - 2.057$

90. $f(x) = 1.21x^2 - 2.34x - 5.63$

Solve.

91. $x^4 - 3x^2 + 2 = 0$

92. $x^4 + 3 = 4x^2$

93. $x^4 + 3x^2 = 10$

94. $x^4 - 8x^2 = 9$

95. $y^4 + 4y^2 - 5 = 0$

96. $y^4 - 15y^2 - 16 = 0$

97. $x - 3\sqrt{x} - 4 = 0$
(*Hint*: Let $u = \sqrt{x}$.)

98. $2x - 9\sqrt{x} + 4 = 0$

99. $m^{2/3} - 2m^{1/3} - 8 = 0$
(*Hint*: Let $u = m^{1/3}$.)

100. $t^{2/3} + t^{1/3} - 6 = 0$

101. $x^{1/2} - 3x^{1/4} + 2 = 0$

102. $x^{1/2} - 4x^{1/4} = -3$

103. $(2x - 3)^2 - 5(2x - 3) + 6 = 0$
(*Hint*: Let $u = 2x - 3$.)

104. $(3x + 2)^2 + 7(3x + 2) - 8 = 0$

105. $(2t^2 + t)^2 - 4(2t^2 + t) + 3 = 0$

106. $12 = (m^2 - 5m)^2 + (m^2 - 5m)$

Multigenerational Households. After declining between 1940 and 1980, the number of multigenerational American households has been increasing since 1980. The function

$$h(x) = 0.012x^2 - 0.583x + 35.727$$

can be used to estimate the number of multigenerational households in the United States, in millions, x years after 1940 (Source: Pew Research Center*). Use this function for Exercises 107 and 108.*

107. In what year were there 40 million multigenerational households?

108. In what year were there 55 million multigenerational households?

Funding for Afghan Security. *The number of U.S. forces in Afghanistan decreased to approximately 34,000 in 2014 from a high of about 100,000 in 2010. The amount of U.S. funding for Afghan security forces also decreased during this period. The function*

$$f(x) = -1.321x^2 + 5.156x + 5.517$$

can be used to estimate the amount of U.S. funding for Afghan security forces, in billions of dollars, x years after 2009 (Sources: U.S. Department of Defense; Brookings Institution; International Security Assistance Force; ESRI). Use this function for Exercises 109 and 110.

109. In what year was the amount of U.S. funding for Afghan security forces about \$10.5 billion?

110. In what year was the amount of U.S. funding for Afghan security forces about \$5.0 billion?

Time of a Free Fall. *The formula $s = 16t^2$ is used to approximate the distance s, in feet, that an object falls freely from rest in t seconds. Use this formula for Exercises 111 and 112.*

111. The Taipei 101 Tower, also known as the Taipei Financial Center, in Taipei, Taiwan, is 1670 ft tall. How long would it take an object dropped from the top to reach the ground?

112. At 630 ft, the Gateway Arch in St. Louis is the tallest man-made monument in the United States. How long would it take an object dropped from the top to reach the ground?

113. The length of a rectangular poster is 1 ft more than the width, and a diagonal of the poster is 5 ft. Find the length and the width.

114. The length of one leg of a right triangle is 7 cm less than the length of the other leg. The length of the hypotenuse is 13 cm. Find the lengths of the legs.

115. One number is 5 greater than another. The product of the numbers is 36. Find the numbers.

116. One number is 6 less than another. The product of the numbers is 72. Find the numbers.

117. *Box Construction.* An open box is made from a 10-cm by 20-cm piece of tin by cutting a square from each corner and folding up the edges. The area of the resulting base is 96 cm^2. What is the length of the sides of the squares?

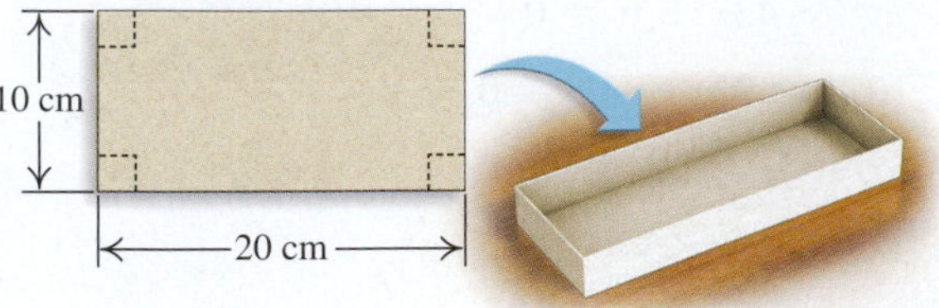

118. *Petting Zoo Dimensions.* At the Glen Island Zoo, 170 m of fencing was used to enclose a rectangular petting area of 1750 m^2. Find the dimensions of the petting area.

119. *Dimensions of a Rug.* Find the dimensions of a rectangular Persian rug whose perimeter is 28 ft and whose area is 48 ft^2.

120. *Picture Frame Dimensions.* The rectangular frame on a picture is 8 in. by 10 in. outside and is of uniform width. What is the width of the frame if 48 in^2 of the picture shows?

State whether the function is linear or quadratic.

121. $f(x) = 4 - 5x$

122. $f(x) = 4 - 5x^2$

123. $f(x) = 7x^2$

124. $f(x) = 23x + 6$

125. $f(x) = 1.2x - (3.6)^2$

126. $f(x) = 2 - x - x^2$

Skill Maintenance

Cost of a Super Bowl Ad. *The cost of a 30-sec Super Bowl ad has increased more than 70% since 2004. The function*

$$C(x) = 0.17x + 2.25$$

can be used to estimate the cost of a 30-sec ad, in millions of dollars, x years after 2004 (Source: Katar Media). Use this function for Exercises 127 and 128. **[1.2]**

127. Estimate the cost of a 30-sec Super Bowl ad in 2014.

128. When will the cost of a 30-sec Super Bowl ad reach \$5.0 million?

Determine whether the graph is symmetric with respect to the x-axis, the y-axis, and the origin. [2.4]

129. $3x^2 + 4y^2 = 5$ **130.** $y^3 = 6x^2$

Determine whether the function is even, odd, or neither even nor odd. [2.4]

131. $f(x) = 2x^3 - x$

132. $f(x) = 4x^2 + 2x - 3$

Synthesis

For each equation in Exercises 133–136, under the given condition: **(a)** *Find k and* **(b)** *find a second solution.*

133. $kx^2 - 17x + 33 = 0$; one solution is 3

134. $kx^2 - 2x + k = 0$; one solution is -3

135. $x^2 - kx + 2 = 0$; one solution is $1 + i$

136. $x^2 - (6 + 3i)x + k = 0$; one solution is 3

Solve.

137. $(x - 2)^3 = x^3 - 2$

138. $(x + 1)^3 = (x - 1)^3 + 26$

139. $(6x^3 + 7x^2 - 3x)(x^2 - 7) = 0$

140. $\left(x - \frac{1}{5}\right)\left(x^2 - \frac{1}{4}\right) + \left(x - \frac{1}{5}\right)\left(x^2 + \frac{1}{8}\right) = 0$

141. $x^2 + x - \sqrt{2} = 0$

142. $x^2 + \sqrt{5}x - \sqrt{3} = 0$

143. $2t^2 + (t - 4)^2 = 5t(t - 4) + 24$

144. $9t(t + 2) - 3t(t - 2) = 2(t + 4)(t + 6)$

145. $\sqrt{x - 3} - \sqrt[4]{x - 3} = 2$

146. $x^2 + 3x + 1 - \sqrt{x^2 + 3x + 1} = 8$

147. $\left(y + \frac{2}{y}\right)^2 + 3y + \frac{6}{y} = 4$

148. Solve $\frac{1}{2}at^2 + v_0t + x_0 = 0$ for t.

3.3 Analyzing Graphs of Quadratic Functions

- Find the vertex, the axis of symmetry, and the maximum or minimum value of a quadratic function using the method of completing the square.
- Graph quadratic functions.
- Solve applied problems involving maximum and minimum function values.

Graphing Quadratic Functions of the Type $f(x) = a(x - h)^2 + k$

The graph of a quadratic function is called a **parabola**. The graph of every parabola evolves from the graph of the squaring function $f(x) = x^2$ using transformations.

Exploring with Technology

Think of transformations and look for patterns. Consider the following functions:

$$y_1 = x^2, \quad y_2 = -0.4x^2,$$
$$y_3 = -0.4(x - 2)^2, \quad y_4 = -0.4(x - 2)^2 + 3.$$

Graph y_1 and y_2. How do you get from the graph of y_1 to y_2?
Graph y_2 and y_3. How do you get from the graph of y_2 to y_3?
Graph y_3 and y_4. How do you get from the graph of y_3 to y_4?

Consider the following functions:

$$y_1 = x^2, \quad y_2 = 2x^2,$$
$$y_3 = 2(x + 3)^2, \quad y_4 = 2(x + 3)^2 - 5.$$

Graph y_1 and y_2. How do you get from the graph of y_1 to y_2?

Graph y_2 and y_3. How do you get from the graph of y_2 to y_3?

Graph y_3 and y_4. How do you get from the graph of y_3 to y_4?

TRANSFORMATIONS

REVIEW SECTION 2.5.

We get the graph of $f(x) = a(x - h)^2 + k$ from the graph of $f(x) = x^2$ as follows:

$$f(x) = x^2$$
$$\downarrow$$
$$f(x) = ax^2 \qquad \text{Vertical stretching or shrinking with a reflection across the } x\text{-axis if } a < 0$$
$$\downarrow$$
$$f(x) = a(x - h)^2 \qquad \text{Horizontal translation}$$
$$\downarrow$$
$$f(x) = a(x - h)^2 + k. \qquad \text{Vertical translation}$$

Consider the following graphs of the form $f(x) = a(x - h)^2 + k$. The point (h, k) at which the graph turns is called the **vertex**. The maximum or minimum value of $f(x)$ occurs at the vertex. Each graph has a line $x = h$ that is called the **axis of symmetry**.

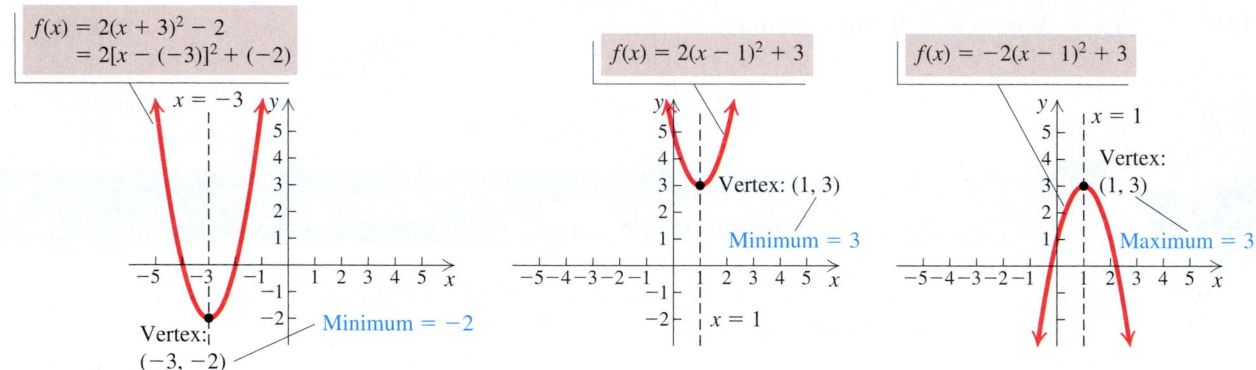

CONNECTING THE CONCEPTS — Graphing Quadratic Functions

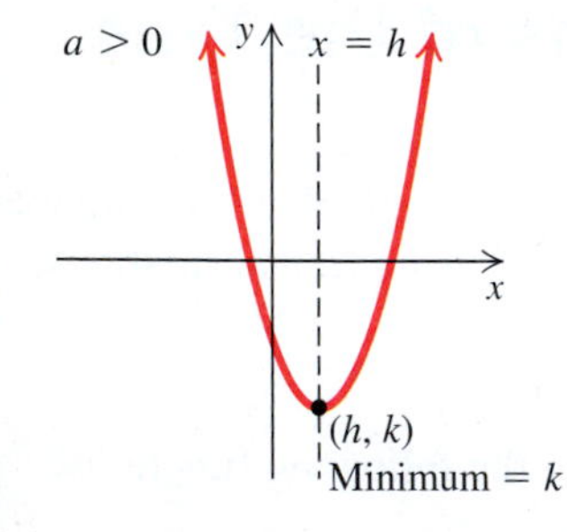

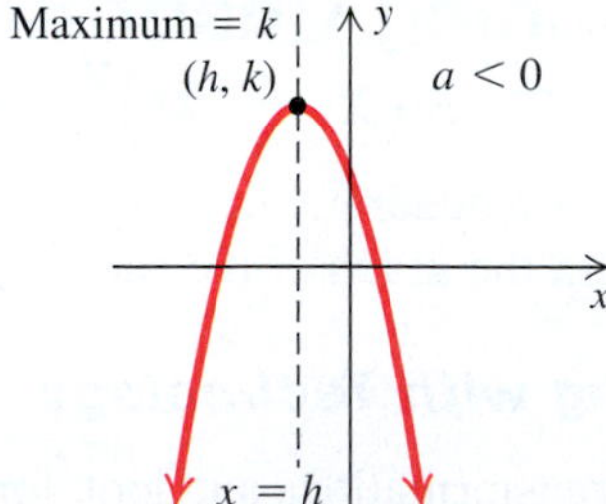

The graph of the function $f(x) = a(x - h)^2 + k$ is a parabola that

- opens up if $a > 0$ and down if $a < 0$;
- has (h, k) as the vertex;
- has $x = h$ as the axis of symmetry;
- has k as a minimum value (output) if $a > 0$;
- has k as a maximum value if $a < 0$.

As we saw in Section 2.5, the constant a serves to stretch or shrink the graph vertically. As a parabola is stretched vertically, it becomes narrower, and as it is shrunk vertically, it becomes wider. That is, as $|a|$ increases, the graph becomes narrower, and as $|a|$ gets close to 0, the graph becomes wider.

If the equation is in the form $f(x) = a(x - h)^2 + k$, we can learn a great deal about the graph without actually graphing the function.

Function	$f(x) = 3(x - \frac{1}{4})^2 - 2$ $= 3(x - \frac{1}{4})^2 + (-2)$	$g(x) = -3(x + 5)^2 + 7$ $= -3[x - (-5)]^2 + 7$
Vertex	$(\frac{1}{4}, -2)$	$(-5, 7)$
Axis of Symmetry	$x = \frac{1}{4}$	$x = -5$
Maximum	None ($3 > 0$, so the graph opens up.)	7 ($-3 < 0$, so the graph opens down.)
Minimum	-2 ($3 > 0$, so the graph opens up.)	None ($-3 < 0$, so the graph opens down.)

Note that the vertex (h, k) is used to find the maximum or minimum value of the function. The maximum or minimum value is the number k, *not* the ordered pair (h, k).

Graphing Quadratic Functions of the Type $f(x) = ax^2 + bx + c, a \neq 0$

We now use a modification of the method of completing the square as an aid in graphing and analyzing quadratic functions of the form $f(x) = ax^2 + bx + c, a \neq 0$.

EXAMPLE 1 Find the vertex, the axis of symmetry, and the maximum or minimum value of $f(x) = x^2 + 10x + 23$. Then graph the function.

Solution To express

$$f(x) = x^2 + 10x + 23$$

in the form

$$f(x) = a(x - h)^2 + k,$$

we complete the square on the terms involving x. To do so, we take half the coefficient of x and square it, obtaining $(10/2)^2$, or 25. We now add and subtract that number on the *right side*:

$$f(x) = x^2 + 10x + 23 = x^2 + 10x + 25 - 25 + 23.$$

Since $25 - 25 = 0$, the new expression for the function is equivalent to the original expression. Note that this process differs from the one we used to complete the square in order to solve a quadratic equation, where we added the same number on both sides of the equation to obtain an equivalent equation. Instead, when we complete the square to write a function in the form $f(x) = a(x - h)^2 + k$, we add and subtract the same number on one side. The entire process is shown below:

$$f(x) = x^2 + 10x + 23$$ Note that 25 completes the square for $x^2 + 10x$.

$$= x^2 + 10x + 25 - 25 + 23$$ Adding $25 - 25$, or 0, to the right side

$$= (x^2 + 10x + 25) - 25 + 23$$ Regrouping

$$= (x + 5)^2 - 2$$ Factoring and simplifying

$$= [x - (-5)]^2 + (-2).$$ Writing in the form $f(x) = a(x - h)^2 + k$

Keeping in mind that this function will have a minimum value since $a > 0$ $(a = 1)$, from this form of the function we know the following:

Vertex: $(-5, -2)$;

Axis of symmetry: $x = -5$;

Minimum value of the function: -2.

To graph the function by hand, we first plot the vertex, $(-5, -2)$, and find several points on either side of $x = -5$. Then we plot these points and connect them with a smooth curve. We see that the points $(-4, -1)$ and $(-3, 2)$ are reflections of the points $(-6, -1)$ and $(-7, 2)$, respectively, across the axis of symmetry, $x = -5$.

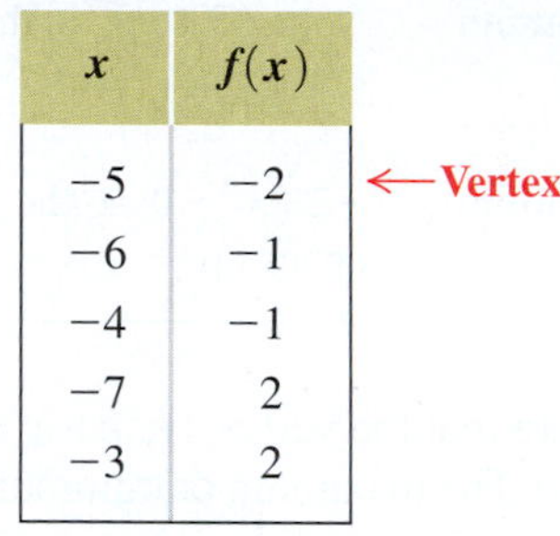

x	$f(x)$
-5	-2 ← Vertex
-6	-1
-4	-1
-7	2
-3	2

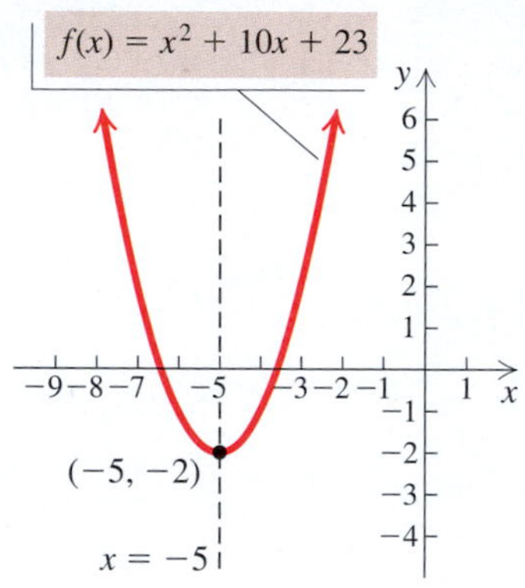

The graph of $f(x) = x^2 + 10x + 23$, or

$$f(x) = [x - (-5)]^2 + (-2),$$

shown above, is a shift of the graph of $y = x^2$ left 5 units and down 2 units.

Now Try Exercise 3.

Keep in mind that the axis of symmetry is not part of the graph; it is a characteristic of the graph. If you fold the graph on its axis of symmetry, the two halves of the graph will coincide.

EXAMPLE 2 Find the vertex, the axis of symmetry, and the maximum or minimum value of $g(x) = x^2/2 - 4x + 8$. Then graph the function.

Solution We complete the square in order to write the function in the form $g(x) = a(x - h)^2 + k$. First, we factor $\frac{1}{2}$ out of the first two terms. This makes the coefficient of x^2 within the parentheses 1:

$$g(x) = \frac{x^2}{2} - 4x + 8$$

$$= \frac{1}{2}(x^2 - 8x) + 8. \quad \text{Factoring } \tfrac{1}{2} \text{ out of the first two terms: } x^2/2 - 4x = \tfrac{1}{2}\cdot x^2 - \tfrac{1}{2}\cdot 8x$$

Next, we complete the square inside the parentheses: Half of -8 is -4, and $(-4)^2 = 16$. We add and subtract 16 inside the parentheses:

$$g(x) = \tfrac{1}{2}(x^2 - 8x + 16 - 16) + 8$$

$$= \tfrac{1}{2}(x^2 - 8x + 16) - \tfrac{1}{2}\cdot 16 + 8 \quad \text{Using the distributive law to remove } -16 \text{ from within the parentheses}$$

$$= \tfrac{1}{2}(x^2 - 8x + 16) - 8 + 8$$

$$= \tfrac{1}{2}(x - 4)^2 + 0, \text{ or } \tfrac{1}{2}(x - 4)^2. \quad \text{Factoring and simplifying}$$

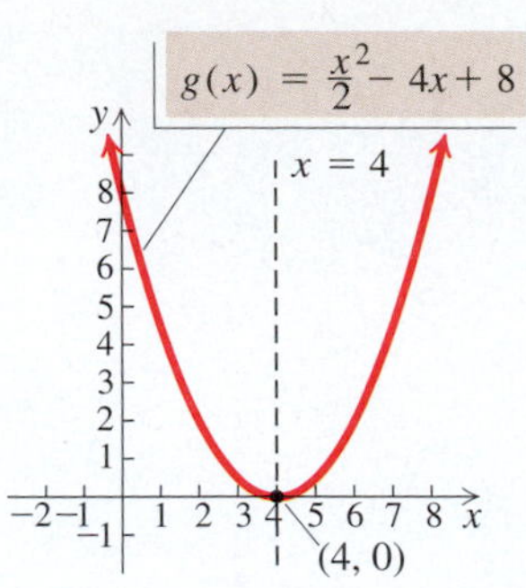

We know the following:

Vertex: $(4, 0)$;

Axis of symmetry: $x = 4$;

Minimum value of the function: 0.

Finally, we plot the vertex and several points on either side of it and draw the graph of the function. The graph of g is a vertical shrinking of the graph of $y = x^2$ along with a shift right 4 units.

Now Try Exercise 9.

EXAMPLE 3 Find the vertex, the axis of symmetry, and the maximum or minimum value of $f(x) = -2x^2 + 10x - \frac{23}{2}$. Then graph the function.

Solution We have

$$\begin{aligned} f(x) &= -2x^2 + 10x - \tfrac{23}{2} \\ &= -2(x^2 - 5x) - \tfrac{23}{2} && \text{Factoring } -2 \text{ out of the first two terms} \\ &= -2\left(x^2 - 5x + \tfrac{25}{4} - \tfrac{25}{4}\right) - \tfrac{23}{2} && \text{Completing the square inside the parentheses} \\ &= -2\left(x^2 - 5x + \tfrac{25}{4}\right) - 2\left(-\tfrac{25}{4}\right) - \tfrac{23}{2} && \text{Using the distributive law to remove } -\tfrac{25}{4} \text{ from within the parentheses} \\ &= -2\left(x^2 - 5x + \tfrac{25}{4}\right) + \tfrac{25}{2} - \tfrac{23}{2} \\ &= -2\left(x - \tfrac{5}{2}\right)^2 + 1. \end{aligned}$$

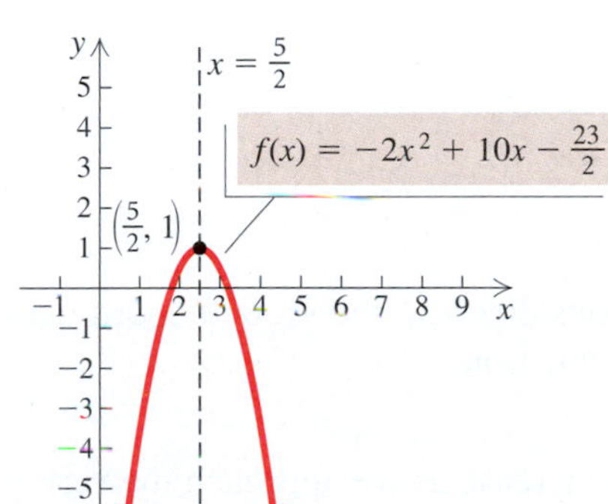

This form of the function yields the following:

Vertex: $\left(\frac{5}{2}, 1\right)$;

Axis of symmetry: $x = \frac{5}{2}$;

Maximum value of the function: 1.

The graph is found by shifting the graph of $f(x) = x^2$ right $\frac{5}{2}$ units, reflecting it across the x-axis, stretching it vertically, and shifting it up 1 unit.

Now Try Exercise 13.

Just in Time 24

In many situations, we want to use a formula to find the coordinates of the vertex directly from the equation $f(x) = ax^2 + bx + c$. One way to develop such a formula is to first note that the x-coordinate of the vertex is centered between the x-intercepts, or zeros, of the function. By averaging the two solutions of $ax^2 + bx + c = 0$, we find a formula for the x-coordinate of the vertex:

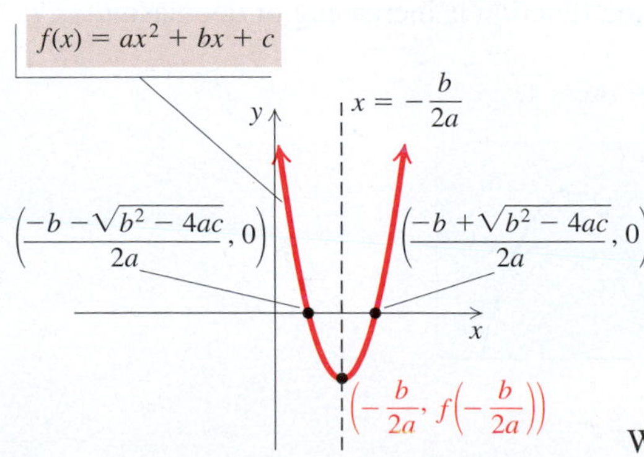

$$\begin{aligned} x\text{-coordinate of vertex} &= \frac{\dfrac{-b - \sqrt{b^2 - 4ac}}{2a} + \dfrac{-b + \sqrt{b^2 - 4ac}}{2a}}{2} \\ &= \frac{\dfrac{-2b}{2a}}{2} = \frac{-\dfrac{b}{a}}{2} \\ &= -\frac{b}{a} \cdot \frac{1}{2} = -\frac{b}{2a}. \end{aligned}$$

We use this value of x to find the y-coordinate of the vertex, $f\left(-\dfrac{b}{2a}\right)$.

THE VERTEX OF A PARABOLA

The **vertex** of the graph of $f(x) = ax^2 + bx + c$ is

$$\left(-\frac{b}{2a}, f\left(-\frac{b}{2a}\right)\right).$$

↑ We calculate the x-coordinate.

↑ We substitute to find the y-coordinate.

EXAMPLE 4 For the function $f(x) = -x^2 + 14x - 47$:

a) Find the vertex.

b) Determine whether there is a maximum or a minimum value and find that value.

c) Find the range.

d) On what intervals is the function increasing? decreasing?

Solution There is no need to graph the function.

a) The x-coordinate of the vertex is

$$-\frac{b}{2a} = -\frac{14}{2(-1)} = -\frac{14}{-2} = 7.$$

Since

$$f(7) = -7^2 + 14 \cdot 7 - 47 = -49 + 98 - 47 = 2,$$

the vertex is $(7, 2)$.

b) Since a is negative $(a = -1)$, the graph opens down so the second coordinate of the vertex, 2, is the maximum value of the function.

c) The range is $(-\infty, 2]$.

d) Since the graph opens down, function values increase as we approach the vertex from the left and decrease as we move away from the vertex on the right. Thus the function is increasing on the interval $(-\infty, 7)$ and decreasing on $(7, \infty)$.

Now Try Exercise 31.

We can use a graphing calculator to work Example 4. Once we have graphed $y = -x^2 + 14x - 47$, we see that the graph opens down and thus has a maximum value. We can use the MAXIMUM feature to find the coordinates of the vertex. Using these coordinates, we can then find the maximum value and the range of the function along with the intervals on which the function is increasing or decreasing.

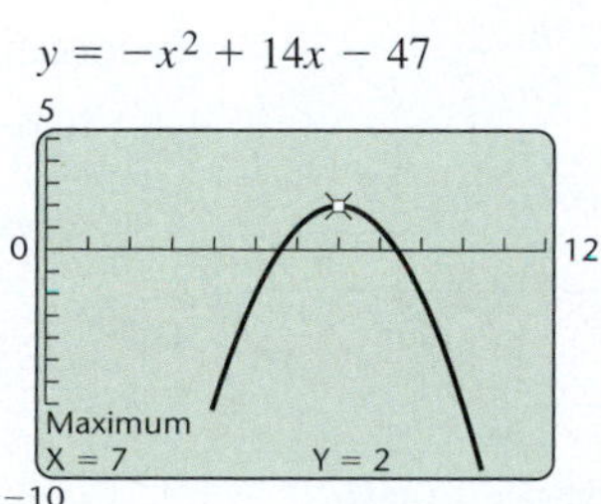

Applications

Many real-world situations involve finding the maximum or the minimum value of a quadratic function.

EXAMPLE 5 ***Maximizing Area.*** A landscaper has enough stone to enclose a rectangular koi pond next to an existing garden wall of the Englemans' house with 24 ft of stone wall. If the garden wall forms one side of the rectangle, what is the maximum area that the landscaper can enclose? What dimensions of the koi pond will yield this area?

PROBLEM-SOLVING STRATEGY

REVIEW SECTION 1.5.

Solution We will use the five-step problem-solving strategy.

1. **Familiarize.** We first make a drawing of the situation, using w to represent the width of the koi pond, in feet. Then $(24 - 2w)$ feet of stone is available for the length. Suppose that the koi pond were 1 ft wide. Then its length would be $24 - 2 \cdot 1 = 22$ ft, and its area would be $(22 \text{ ft})(1 \text{ ft}) = 22 \text{ ft}^2$. If the koi pond were 2 ft wide, its length would be $24 - 2 \cdot 2 = 20$ ft, and its area would be $(20 \text{ ft})(2 \text{ ft}) = 40 \text{ ft}^2$. This is larger than the first area we found, but we do not know if it is the maximum possible area. To find the maximum area, we will find a function that represents the area and then determine its maximum value.

2. **Translate.** We write a function for the area of the koi pond. We have

$$A(w) = (24 - 2w)w \qquad \mathbf{A = lw;\ l = 24 - 2w}$$
$$= -2w^2 + 24w,$$

where $A(w)$ is the area of the koi pond, in square feet, as a function of the width w.

3. **Carry out.** To solve this problem, we need to determine the maximum value of $A(w)$ and find the dimensions for which that maximum occurs. Since A is a quadratic function and w^2 has a negative coefficient, we know that the function has a maximum value that occurs at the vertex of the graph of the function. The first coordinate of the vertex, $(w, A(w))$, is

$$w = -\frac{b}{2a} = -\frac{24}{2(-2)} = -\frac{24}{-4} = 6.$$

Thus, if $w = 6$ ft, then the length $l = 24 - 2 \cdot 6 = 12$ ft, and the area is $(12 \text{ ft})(6 \text{ ft}) = 72 \text{ ft}^2$.

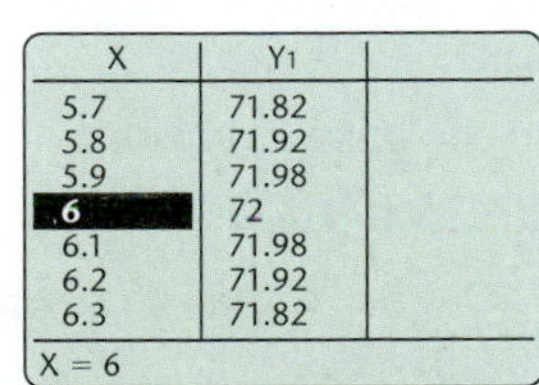

X	Y1	
5.7	71.82	
5.8	71.92	
5.9	71.98	
6	72	
6.1	71.98	
6.2	71.92	
6.3	71.82	

X = 6

$y = (24 - 2x)x$

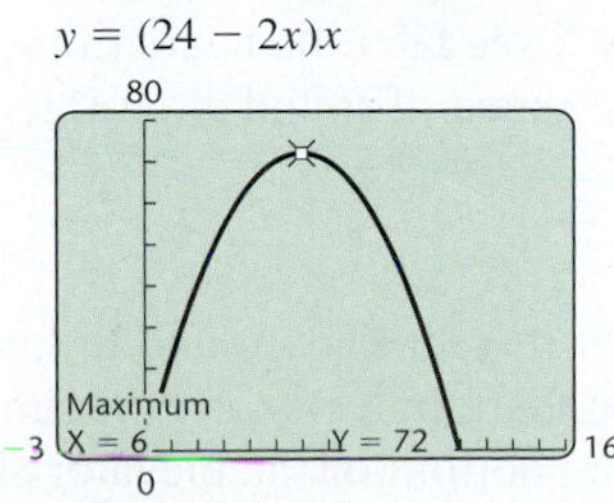

4. **Check.** As a partial check, we note that $72 \text{ ft}^2 > 40 \text{ ft}^2$, which is the larger area that we found in a guess in the *Familiarize* step. As a more complete check, assuming that the function $A(W)$ is correct, we could examine a table of values for $A(w) = (24 - 2w)w$ and/or examine its graph.

5. **State.** The maximum possible area is 72 ft^2 when the koi pond is 6 ft wide and 12 ft long.

Now Try Exercise 45.

EXAMPLE 6 ***Height of a Rocket.*** A model rocket is launched with an initial velocity of 100 ft/sec from the top of a hill that is 20 ft high. Its height, in feet, t seconds after it has been launched is given by the function $s(t) = -16t^2 + 100t + 20$. Determine the time at which the rocket reaches its maximum height and find the maximum height.

Solution

1., 2. Familiarize and **Translate.** We are given the function in the statement of the problem: $s(t) = -16t^2 + 100t + 20$.

3. Carry out. We need to find the maximum value of the function and the value of t for which it occurs. Since $s(t)$ is a quadratic function and t^2 has a negative coefficient, we know that the maximum value of the function occurs at the vertex of the graph of the function. The first coordinate of the vertex gives the time t at which the rocket reaches its maximum height. It is

$$t = -\frac{b}{2a} = -\frac{100}{2(-16)} = -\frac{100}{-32} = 3.125.$$

The second coordinate of the vertex gives the maximum height of the rocket. We substitute in the function to find it:

$$s(3.125) = -16(3.125)^2 + 100(3.125) + 20 = 176.25.$$

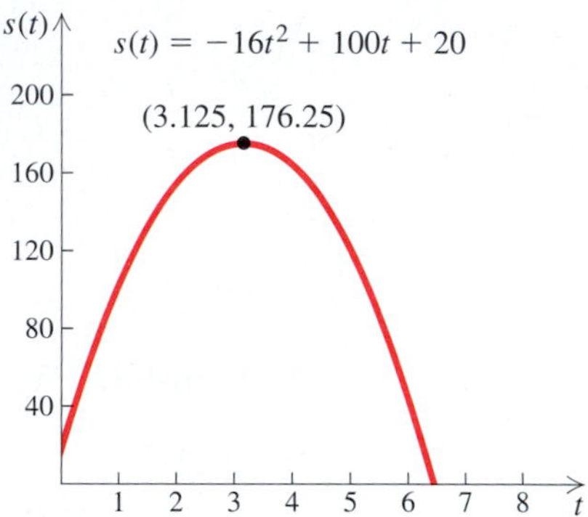

4. Check. As a check, we can complete the square to write the function in the form $s(t) = a(t - h)^2 + k$ and determine the coordinates of the vertex from this form of the function. We get

$$s(t) = -16(t - 3.125)^2 + 176.25.$$

This confirms that the vertex is $(3.125, 176.25)$, so the answer checks.

5. State. The rocket reaches a maximum height of 176.25 ft 3.125 sec after it has been launched.

Now Try Exercise 41.

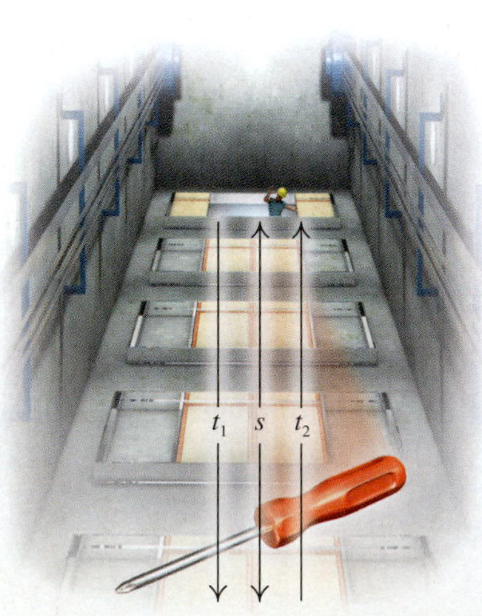

EXAMPLE 7 ***Determining the Height of an Elevator Shaft.*** Jared drops a screwdriver from the top of an elevator shaft. Exactly 5 sec later, he hears the sound of the screwdriver hitting the bottom of the shaft. The speed of sound is 1100 ft/sec. How tall is the elevator shaft?

Solution

1. Familiarize. We first make a drawing and label it with known and unknown information. We let s = the height of the elevator shaft, in feet, t_1 = the time, in seconds, that it takes for the screwdriver to hit the bottom of the elevator shaft, and t_2 = the time, in seconds, that it takes for the sound to reach the top of the elevator shaft. This gives us the equation

$$t_1 + t_2 = 5. \qquad (1)$$

2. **Translate.** Can we find any relationship between the two times and the distance s? Often in problem solving you may need to look up related formulas in a physics book, in another mathematics book, or on the Internet. We find that the formula

$$s = 16t^2$$

gives the distance, in feet, that a dropped object falls in t seconds. The time t_1 that it takes the screwdriver to hit the bottom of the elevator shaft can be found as follows:

$$s = 16t_1^2, \quad \text{or} \quad \frac{s}{16} = t_1^2, \quad \text{so} \quad t_1 = \frac{\sqrt{s}}{4}. \qquad \text{Taking the positive square root} \qquad (2)$$

To find an expression for t_2, the time that it takes the sound to travel to the top of the well, recall that *Distance* = *Rate* · *Time*. Thus,

$$s = 1100t_2, \quad \text{or} \quad t_2 = \frac{s}{1100}. \qquad (3)$$

We now have expressions for t_1 and t_2, both in terms of s. Substituting into equation (1), we obtain

$$t_1 + t_2 = 5, \quad \text{or} \quad \frac{\sqrt{s}}{4} + \frac{s}{1100} = 5. \qquad (4)$$

3. **Carry out.**

Algebraic Solution

We solve equation (4) for s. Multiplying by 1100, we get

$$275\sqrt{s} + s = 5500, \quad \text{or} \quad s + 275\sqrt{s} - 5500 = 0.$$

This equation is reducible to quadratic with $u = \sqrt{s}$. Substituting, we get

$$u^2 + 275u - 5500 = 0.$$

Using the quadratic formula, we can solve for u:

$$u = \frac{-b \pm \sqrt{b^2 - 4ac}}{2a}$$

$$= \frac{-275 + \sqrt{275^2 - 4 \cdot 1 \cdot (-5500)}}{2 \cdot 1} \qquad \text{We want only the positive solution.}$$

$$= \frac{-275 + \sqrt{97{,}625}}{2} \approx 18.725.$$

Since $u \approx 18.725$, we have

$$\sqrt{s} \approx 18.725$$

$$s \approx 350.6. \qquad \text{Squaring both sides and rounding to the nearest tenth}$$

Graphical Solution

We use the Intersect method. It will probably require some trial and error to determine an appropriate window.

$$y_1 = \frac{\sqrt{x}}{4} + \frac{x}{1100}, \quad y_2 = 5$$

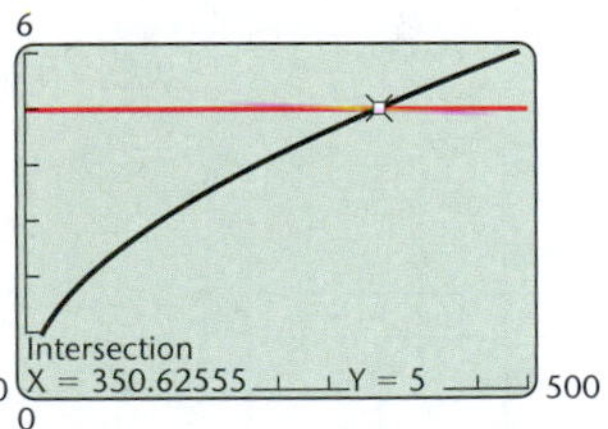

4. **Check.** To check, we can substitute 350.6 for s in equation (4) and see that $t_1 + t_2 \approx 5$. We leave the computation to the student.

5. **State.** The height of the elevator shaft is about 350.6 ft.

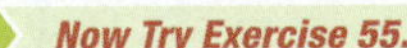

Visualizing the Graph

Match the equation with its graph.

1. $y = 3x$
2. $y = -(x - 1)^2 + 3$
3. $(x + 2)^2 + (y - 2)^2 = 9$
4. $y = 3$
5. $2x - 3y = 6$
6. $(x - 1)^2 + (y + 3)^2 = 4$
7. $y = -2x + 1$
8. $y = 2x^2 - x - 4$
9. $x = -2$
10. $y = -3x^2 + 6x - 2$

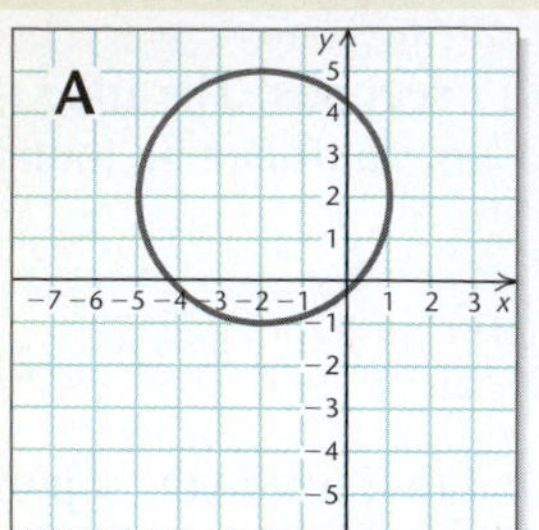

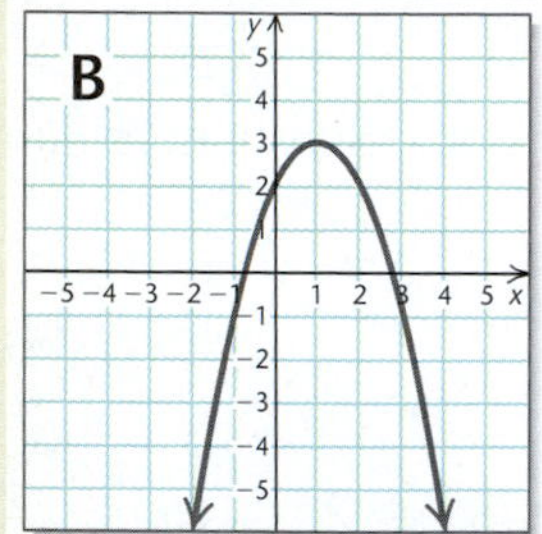

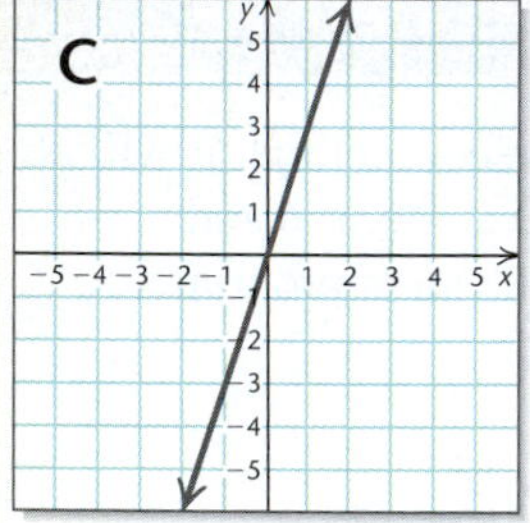

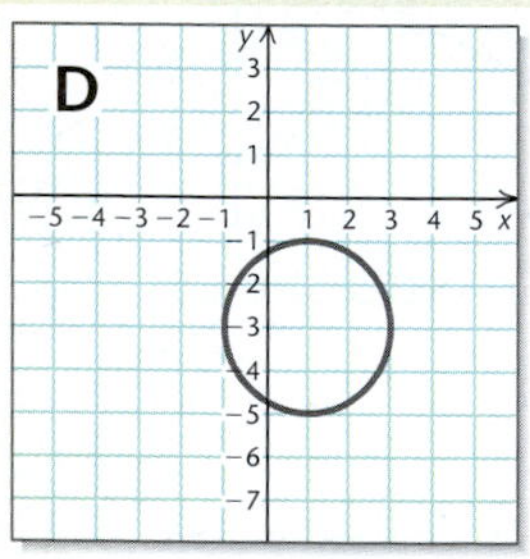

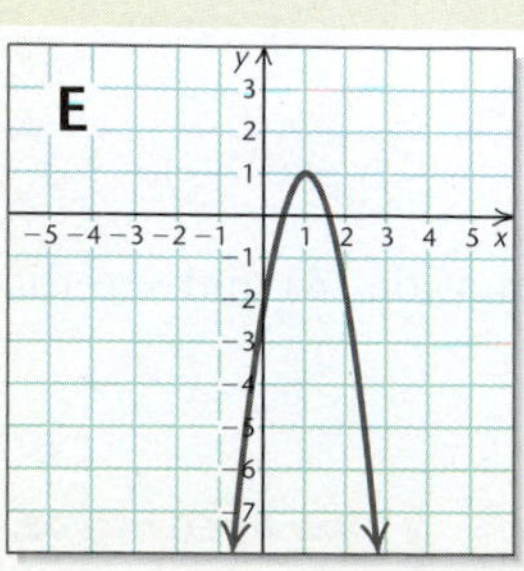

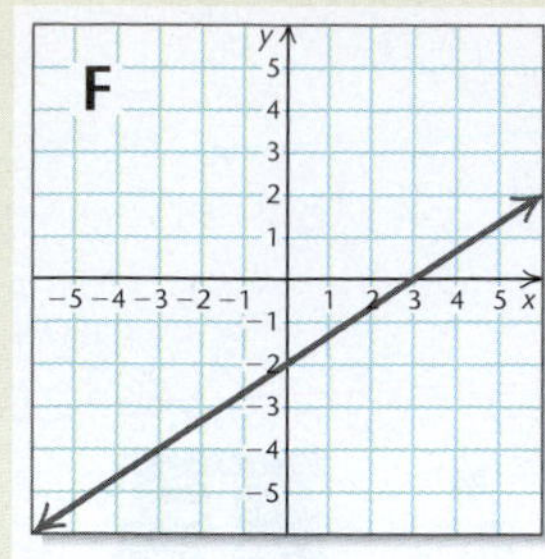

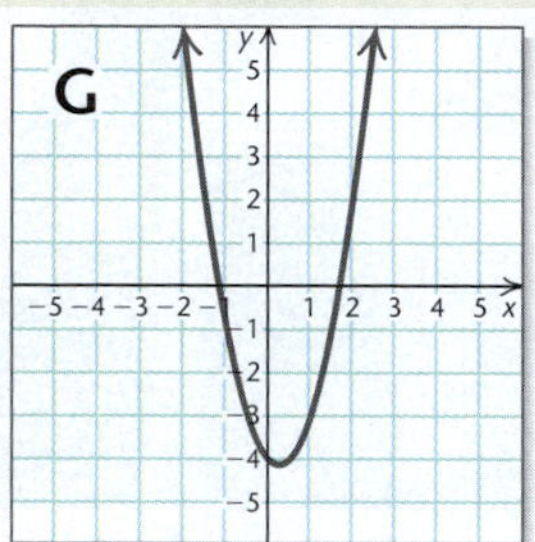

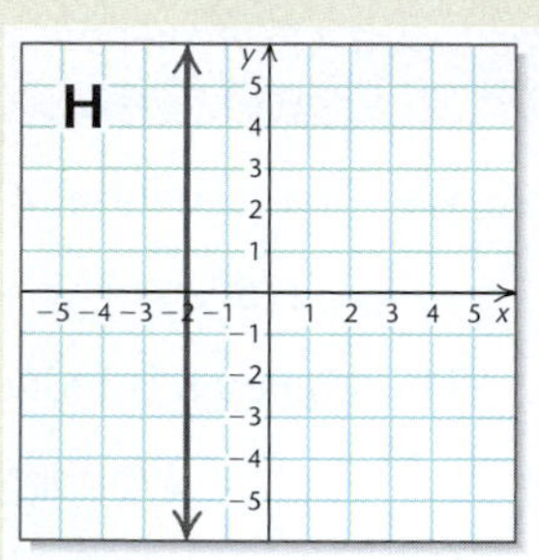

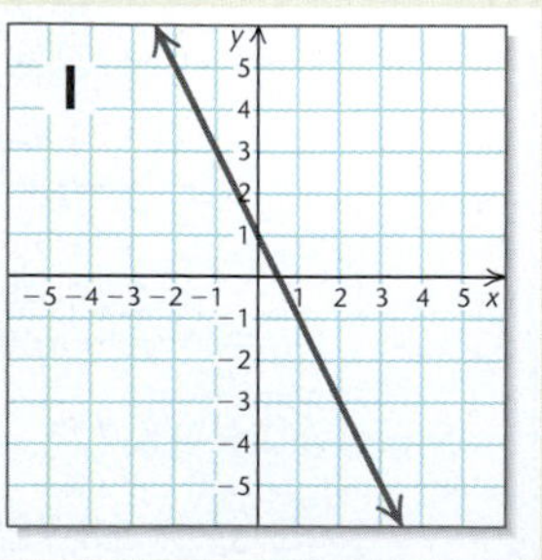

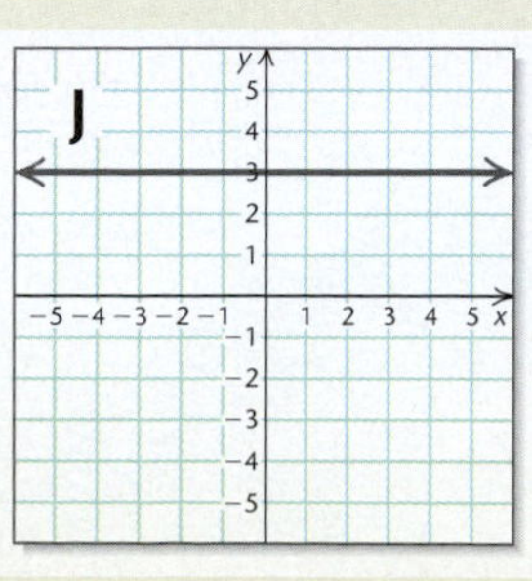

Answers on page A-14

3.3 Exercise Set

In Exercises 1 and 2, use the given graph to find the following.

a) *The vertex.*
b) *The axis of symmetry.*
c) *The maximum or the minimum value of the function.*

1.

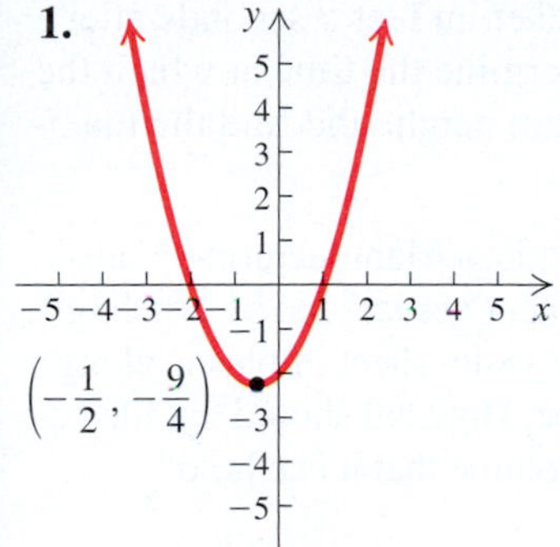

2.

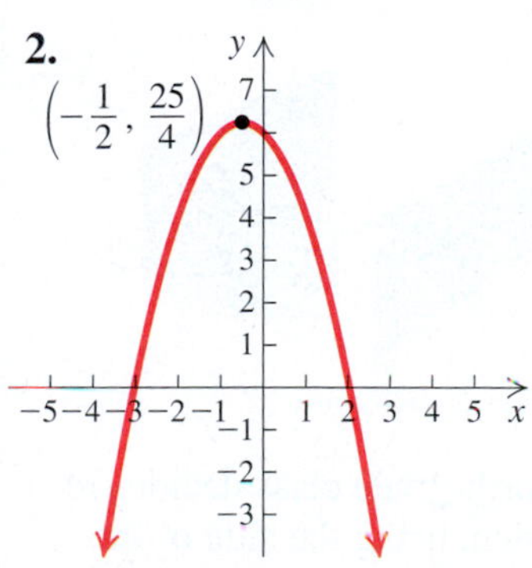

In Exercises 3–16:

a) *Find the vertex.*
b) *Find the axis of symmetry.*
c) *Determine whether there is a maximum or a minimum value and find that value.*
d) *Graph the function.*

3. $f(x) = x^2 - 8x + 12$

4. $g(x) = x^2 + 7x - 8$

5. $f(x) = x^2 - 7x + 12$

6. $g(x) = x^2 - 5x + 6$

7. $f(x) = x^2 + 4x + 5$

8. $f(x) = x^2 + 2x + 6$

9. $g(x) = \dfrac{x^2}{2} + 4x + 6$

10. $g(x) = \dfrac{x^2}{3} - 2x + 1$

11. $g(x) = 2x^2 + 6x + 8$

12. $f(x) = 2x^2 - 10x + 14$

13. $f(x) = -x^2 - 6x + 3$

14. $f(x) = -x^2 - 8x + 5$

15. $g(x) = -2x^2 + 2x + 1$

16. $f(x) = -3x^2 - 3x + 1$

In Exercises 17–24, match the equation with one of the graphs (a)–(h) that follow.

a)

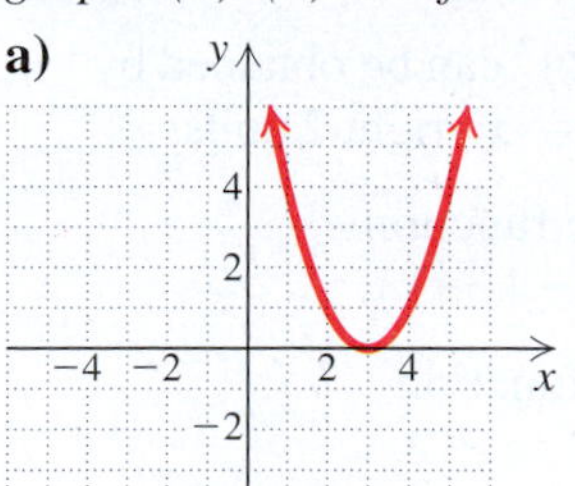

b)

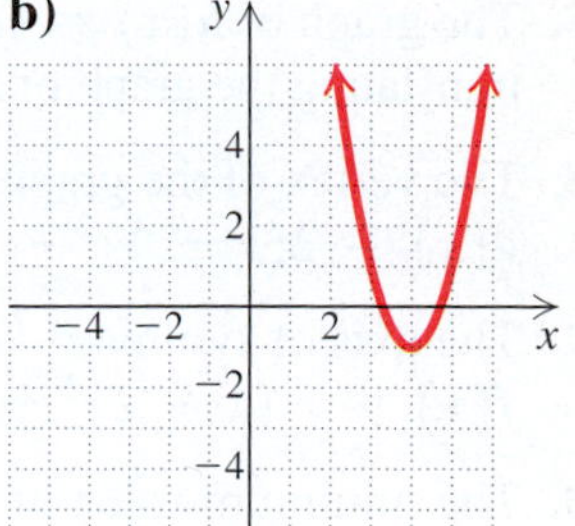

c)

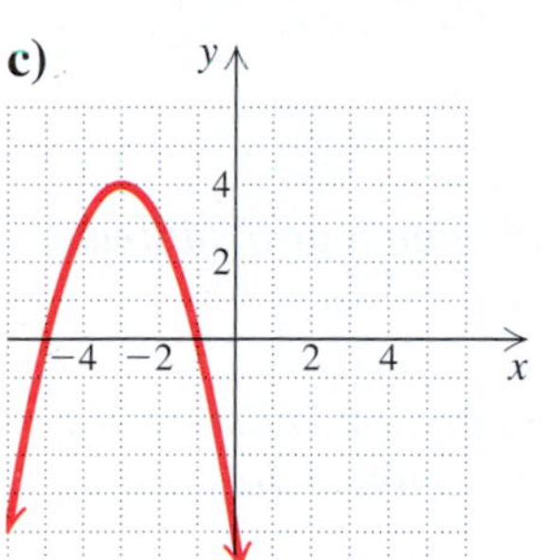

d)

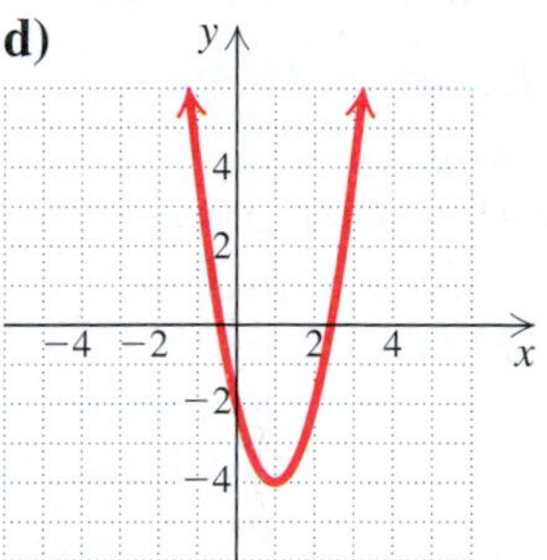

e)

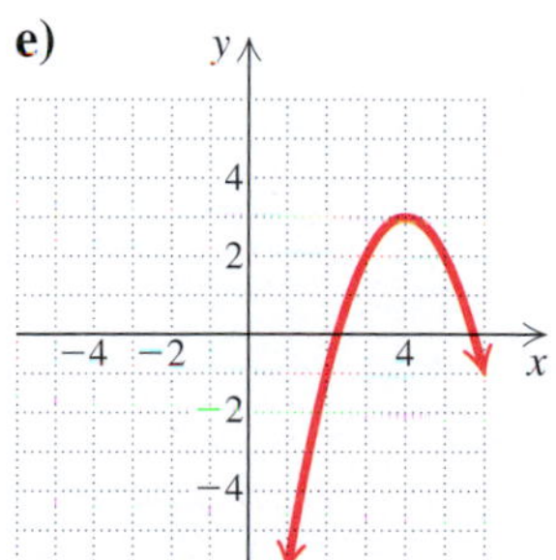

f)

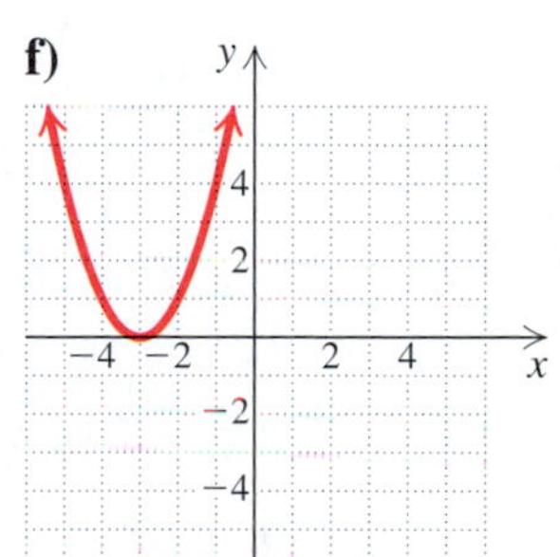

g)

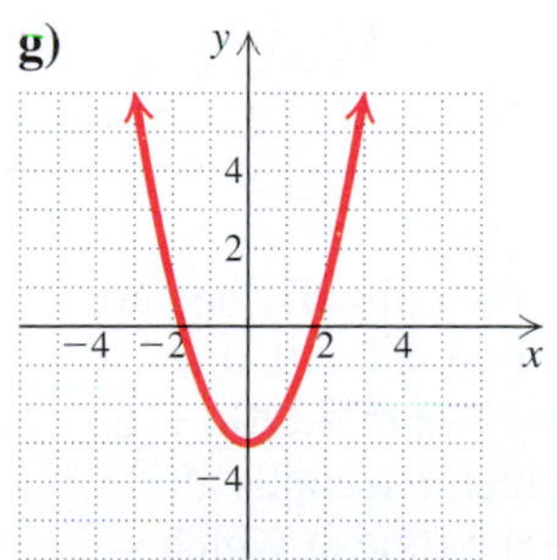

h)

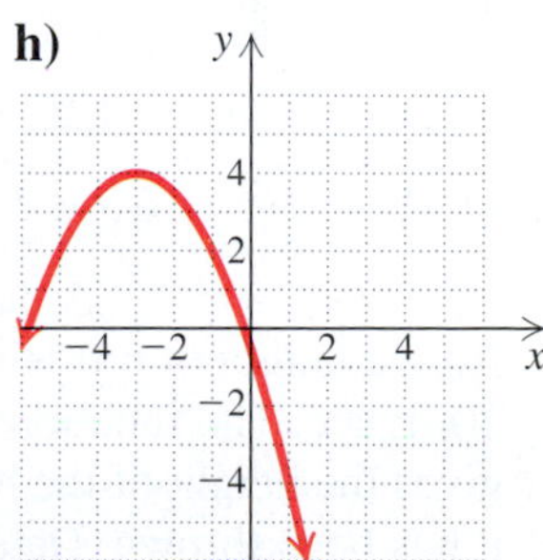

17. $y = (x + 3)^2$

18. $y = -(x - 4)^2 + 3$

19. $y = 2(x - 4)^2 - 1$

20. $y = x^2 - 3$

21. $y = -\frac{1}{2}(x + 3)^2 + 4$

22. $y = (x - 3)^2$

23. $y = -(x + 3)^2 + 4$

24. $y = 2(x - 1)^2 - 4$

Determine whether the statement is true or false.

25. The function $f(x) = -3x^2 + 2x + 5$ has a maximum value.

26. The vertex of the graph of $f(x) = ax^2 + bx + c$ is $-\frac{b}{2a}$.

27. The graph of $h(x) = (x + 2)^2$ can be obtained by translating the graph of $h(x) = x^2$ right 2 units.

28. The vertex of the graph of the function $g(x) = 2(x - 4)^2 - 1$ is $(-4, -1)$.

29. The axis of symmetry of the function $f(x) = -(x + 2)^2 - 4$ is $x = -2$.

30. The minimum value of the function $f(x) = 3(x - 1)^2 + 5$ is 5.

In Exercises 31–40:

a) *Find the vertex.*
b) *Determine whether there is a maximum or a minimum value and find that value.*
c) *Find the range.*
d) *Find the intervals on which the function is increasing and the intervals on which the function is decreasing.*

31. $f(x) = x^2 - 6x + 5$

32. $f(x) = x^2 + 4x - 5$

33. $f(x) = 2x^2 + 4x - 16$

34. $f(x) = \frac{1}{2}x^2 - 3x + \frac{5}{2}$

35. $f(x) = -\frac{1}{2}x^2 + 5x - 8$

36. $f(x) = -2x^2 - 24x - 64$

37. $f(x) = 3x^2 + 6x + 5$

38. $f(x) = -3x^2 + 24x - 49$

39. $g(x) = -4x^2 - 12x + 9$

40. $g(x) = 2x^2 - 6x + 5$

41. *Height of a Ball.* A ball is thrown directly upward from a height of 6 ft with an initial velocity of 20 ft/sec. The function $s(t) = -16t^2 + 20t + 6$ gives the height of the ball, in feet, t seconds after it has been thrown. Determine the time at which the ball reaches its maximum height and find the maximum height.

42. *Height of a Projectile.* A stone is thrown directly upward from a height of 30 ft with an initial velocity of 60 ft/sec. The height of the stone, in feet, t seconds after it has been thrown is given by the function $s(t) = -16t^2 + 60t + 30$. Determine the time at which the stone reaches its maximum height and find the maximum height.

43. *Height of a Rocket.* A model rocket is launched with an initial velocity of 120 ft/sec from a height of 80 ft. The height of the rocket, in feet, t seconds after it has been launched is given by the function $s(t) = -16t^2 + 120t + 80$. Determine the time at which the rocket reaches its maximum height and find the maximum height.

44. *Height of a Rocket.* A model rocket is launched with an initial velocity of 150 ft/sec from a height of 40 ft. The function $s(t) = -16t^2 + 150t + 40$ gives the height of the rocket, in feet, t seconds after it has been launched. Determine the time at which the rocket reaches its maximum height and find the maximum height.

45. *Maximizing Volume.* Mendoza Manufacturing plans to produce a one-compartment vertical file by bending the long side of a 10-in. by 18-in. sheet of plastic along two lines to form a ⊔-shape. How tall should the file be in order to maximize the volume that it can hold?

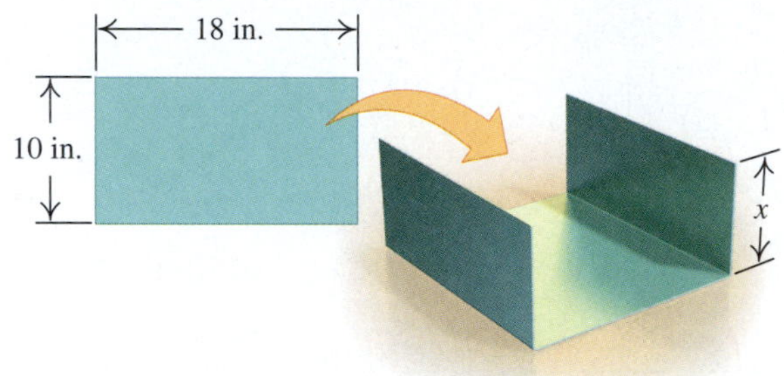

46. *Maximizing Area.* A fourth-grade class decides to enclose a rectangular garden, using the side of the school as one side of the rectangle. What is the maximum area that the class can enclose using 32 ft of fence? What should the dimensions of the garden be in order to yield this area?

47. *Maximizing Area.* The sum of the base and the height of a triangle is 20 cm. Find the dimensions for which the area is a maximum.

48. *Maximizing Area.* The sum of the base and the height of a parallelogram is 69 cm. Find the dimensions for which the area is a maximum.

49. *Minimizing Cost.* Designs for #1 Canines has determined that when x hundred portable doghouses are produced, the average cost per doghouse is given by

$$C(x) = 0.1x^2 - 4.2x + 72.4,$$

where $C(x)$ is in hundreds of dollars. How many doghouses should be produced in order to minimize the average cost per doghouse?

Maximizing Profit. In business, profit is the difference between revenue and cost; that is,

$$\textit{Total profit} = \textit{Total revenue} - \textit{Total cost},$$
$$P(x) = R(x) - C(x),$$

where x is the number of units sold. Find the maximum profit and the number of units that must be sold in order to yield the maximum profit for each of the following.

50. $R(x) = 5x,\ C(x) = 0.001x^2 + 1.2x + 60$

51. $R(x) = 50x - 0.5x^2,\ C(x) = 10x + 3$

52. $R(x) = 20x - 0.1x^2,\ C(x) = 4x + 2$

53. *Maximizing Area.* A berry farmer needs to separate and enclose two adjacent rectangular fields, one for blueberries and one for strawberries. If a lake forms one side of the fields and 240 yd of fencing is available, what is the largest total area that can be enclosed?

54. *Norman Window.* A Norman window is a rectangle with a semicircle on top. Sky Blue Windows is designing a Norman window that will require 24 ft of trim on the outer edges. What dimensions will allow the maximum amount of light to enter a house?

55. *Finding the Depth of a Well.* Two seconds after a chlorine tablet has been dropped into a well, a splash is heard. The speed of sound is 1100 ft/sec. How far is the top of the well from the water? (*Hint*: See Example 7.)

56. *Finding the Height of a Cliff.* A water balloon is dropped from a cliff. Exactly 3 sec later, the sound of the balloon hitting the ground reaches the top of the cliff. How high is the cliff? (*Hint*: See Example 7.)

Skill Maintenance

For each function f, construct and simplify the difference quotient

$$\frac{f(x+h) - f(x)}{h}. \quad [2.2]$$

57. $f(x) = 3x - 7$

58. $f(x) = 2x^2 - x + 4$

A graph of $y = f(x)$ follows. No formula is given for f. Make a hand-drawn graph of each of the following. [2.5]

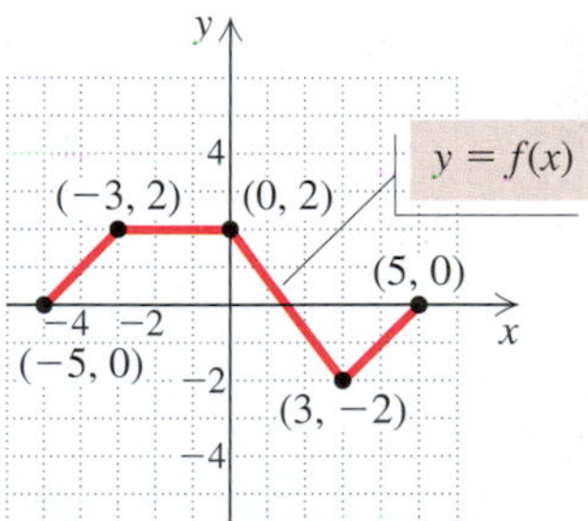

59. $g(x) = -2f(x)$

60. $g(x) = f(2x)$

Synthesis

61. Find c such that

$$f(x) = -0.2x^2 - 3x + c$$

has a maximum value of -225.

62. Find b such that
$$f(x) = -4x^2 + bx + 3$$
has a maximum value of 50.

63. Graph: $f(x) = (|x| - 5)^2 - 3$.

64. Find a quadratic function with vertex $(4, -5)$ and containing the point $(-3, 1)$.

65. *Minimizing Area.* A 24-in. piece of string is cut into two pieces. One piece is used to form a circle while the other is used to form a square. How should the string be cut so that the sum of the areas is a minimum?

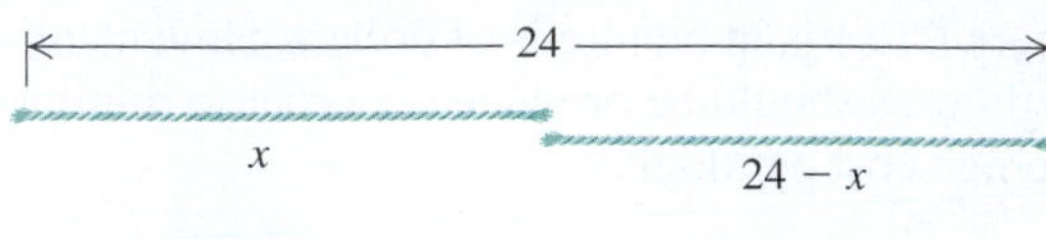

Mid-Chapter Mixed Review

Determine whether the statement is true or false.

1. The product of a complex number and its conjugate is a real number. [3.1]

2. Every quadratic equation has at least one x-intercept. [3.2]

3. If a quadratic equation has two different real-number solutions, then its discriminant is positive. [3.2]

4. The vertex of the graph of the function $f(x) = 3(x + 4)^2 + 5$ is $(4, 5)$. [3.3]

Express the number in terms of i. [3.1]

5. $\sqrt{-36}$

6. $\sqrt{-5}$

7. $-\sqrt{-16}$

8. $\sqrt{-32}$

Simplify. Write answers in the form $a + bi$, where a and b are real numbers. [3.1]

9. $(3 - 2i) + (-4 + 3i)$

10. $(-5 + i) - (2 - 4i)$

11. $(2 + 3i)(4 - 5i)$

12. $\dfrac{3 + i}{-2 + 5i}$

Simplify. [3.1]

13. i^{13}

14. i^{44}

15. $(-i)^5$

16. $(2i)^6$

Solve. [3.2]

17. $x^2 + 3x - 4 = 0$

18. $2x^2 + 6 = -7x$

19. $4x^2 = 24$

20. $x^2 + 100 = 0$

21. Find the zeros of $f(x) = 4x^2 - 8x - 3$ by completing the square. Show your work. [3.2]

In Exercises 22–24:

a) *Find the discriminant $b^2 - 4ac$, and then determine whether one real-number solution, two different real-number solutions, or two different imaginary-number solutions exist.* [3.2]

b) *Solve the equation, finding exact solutions and approximate solutions rounded to three decimal places, where appropriate.* [3.2]

22. $x^2 - 3x - 5 = 0$

23. $4x^2 - 12x + 9 = 0$

24. $3x^2 + 2x = -1$

Solve. [3.2]

25. $x^4 + 5x^2 - 6 = 0$

26. $2x - 5\sqrt{x} + 2 = 0$

27. One number is 2 more than another. The product of the numbers is 35. Find the numbers. [3.2]

In Exercises 28 and 29:

a) *Find the vertex.* [3.3]
b) *Find the axis of symmetry.* [3.3]
c) *Determine whether there is a maximum or a minimum value, and find that value.* [3.3]
d) *Find the range.* [3.3]
e) *Find the intervals on which the function is increasing and the intervals on which the function is decreasing.* [3.3]
f) *Graph the function.* [3.3]

28. $f(x) = x^2 - 6x + 7$

29. $f(x) = -2x^2 - 4x - 5$

30. The sum of the base and the height of a triangle is 16 in. Find the dimensions for which the area is a maximum. [3.3]

COLLABORATIVE DISCUSSION AND WRITING

31. Is the sum of two imaginary numbers always an imaginary number? Explain your answer. [3.1]

32. The graph of a quadratic function can have 0, 1, or 2 x-intercepts. How can you predict the number of x-intercepts without drawing the graph or (completely) solving an equation? [3.2]

33. Discuss two ways in which we used completing the square in this chapter. [3.2], [3.3]

34. Suppose that the graph of $f(x) = ax^2 + bx + c$ has x-intercepts $(x_1, 0)$ and $(x_2, 0)$. What are the x-intercepts of $g(x) = -ax^2 - bx - c$? Explain. [3.3]

3.4 Solving Rational Equations and Radical Equations

- Solve rational equations.
- Solve radical equations.

Rational Equations

Equations containing rational expressions are called **rational equations**. Solving such equations involves multiplying on both sides by the least common denominator (LCD) of all the rational expressions to *clear the equation of fractions*.

EXAMPLE 1 Solve: $\dfrac{x - 8}{3} + \dfrac{x - 3}{2} = 0.$

Algebraic Solution

We have

$$\frac{x-8}{3}+\frac{x-3}{2}=0 \qquad \textbf{The LCD is } 3\cdot 2\textbf{, or } 6.$$

$$6\left(\frac{x-8}{3}+\frac{x-3}{2}\right)=6\cdot 0 \qquad \textbf{Multiplying by the LCD on both sides to clear fractions}$$

$$6\cdot\left(\frac{x-8}{3}\right)+6\cdot\left(\frac{x-3}{2}\right)=0$$

$$2(x-8)+3(x-3)=0$$

$$2x-16+3x-9=0$$

$$5x-25=0$$

$$5x=25$$

$$x=5.$$

The possible solution is 5. We check using a table in ASK mode.

$$y=\frac{x-8}{3}+\frac{x-3}{2}$$

X	Y1	
5	0	

X =

Since the value of $\frac{x-8}{3}+\frac{x-3}{2}$ is 0 when $x=5$, the number 5 is the solution.

Graphical Solution

We use the Zero method. The solution of the equation

$$\frac{x-8}{3}+\frac{x-3}{2}=0$$

is the zero of the function

$$f(x)=\frac{x-8}{3}+\frac{x-3}{2}.$$

$$y=\frac{x-8}{3}+\frac{x-3}{2}$$

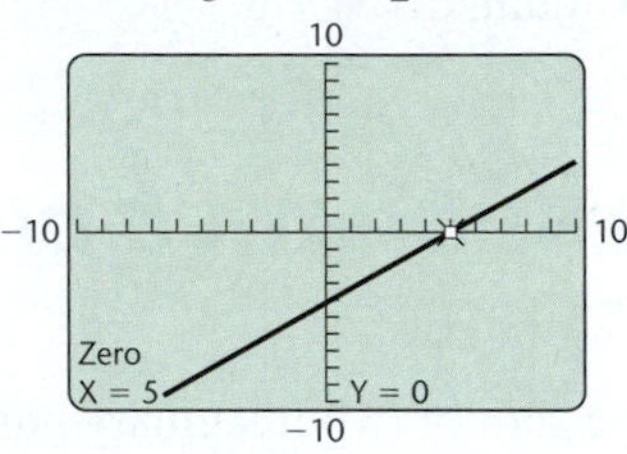

The zero of the function is 5. Thus the solution of the equation is 5.

Now Try Exercise 3.

CAUTION! Clearing fractions is a valid procedure when solving rational equations but not when adding, subtracting, multiplying, or dividing rational expressions. A rational expression may have operation signs but it will have no equals sign. A rational equation *always* has an equals sign. For example, $\frac{x-8}{3}+\frac{x-3}{2}$ is a rational expression but $\frac{x-8}{3}+\frac{x-3}{2}=0$ is a rational equation.

To *simplify* the rational *expression* $\frac{x-8}{3}+\frac{x-3}{2}$, we first find the LCD and write each fraction with that denominator. The final result is usually a rational expression.

To *solve* the rational *equation* $\frac{x-8}{3}+\frac{x-3}{2}=0$, we first multiply on both sides by the LCD to clear fractions. The final result is one or more numbers. As we will see in Example 2, these numbers must be checked in the original equation.

When we use the multiplication principle to multiply (or divide) on both sides of an equation by an expression with a variable, we might not obtain an equivalent equation. We must check the possible solutions obtained in this manner by substituting them in the original equation. The next example illustrates this.

EXAMPLE 2 Solve: $\dfrac{x^2}{x-3} = \dfrac{9}{x-3}$.

Solution The LCD is $x - 3$.

$$(x-3)\cdot\frac{x^2}{x-3} = (x-3)\cdot\frac{9}{x-3}$$
$$x^2 = 9$$
$$x = -3 \quad or \quad x = 3 \qquad \textbf{Using the principle of square roots}$$

The possible solutions are -3 and 3. We check.

Check: For -3:

$$\frac{x^2}{x-3} = \frac{9}{x-3}$$
$$\frac{(-3)^2}{-3-3} \stackrel{?}{=} \frac{9}{-3-3}$$
$$\frac{9}{-6} \;\Big|\; \frac{9}{-6} \qquad \text{TRUE}$$

For 3:

$$\frac{x^2}{x-3} = \frac{9}{x-3}$$
$$\frac{3^2}{3-3} \stackrel{?}{=} \frac{9}{3-3}$$
$$\frac{9}{0} \;\Big|\; \frac{9}{0} \qquad \text{NOT DEFINED}$$

$y_1 = \dfrac{x^2}{x-3}$, $y_2 = \dfrac{9}{x-3}$

X	Y1	Y2
−3	−1.5	−1.5
3	ERROR	ERROR

X =

The number -3 checks, so it is a solution. Since division by 0 is not defined, 3 is not a solution. Note that 3 is not in the domain of either $x^2/(x-3)$ or $9/(x-3)$.

We can also use a table on a graphing calculator to check the possible solutions. When $x = -3$, we see that $y_1 = -1.5 = y_2$, so -3 is a solution. When $x = 3$, we get ERROR messages. This indicates that 3 is not in the domain of y_1 or y_2 and thus is not a solution.

Now Try Exercise 9.

EXAMPLE 3 Solve: $\dfrac{2}{3x+6} + \dfrac{1}{x^2-4} = \dfrac{4}{x-2}$.

Solution We first factor the denominators in order to determine the LCD:

$$\frac{2}{3(x+2)} + \frac{1}{(x+2)(x-2)} = \frac{4}{x-2} \qquad \textbf{The LCD is } \mathbf{3(x+2)(x-2).}$$

$$3(x+2)(x-2)\left(\frac{2}{3(x+2)} + \frac{1}{(x+2)(x-2)}\right) = 3(x+2)(x-2)\cdot\frac{4}{x-2}$$

Multiplying by the LCD to clear fractions

$$2(x-2) + 3 = 3\cdot 4(x+2)$$
$$2x - 4 + 3 = 12x + 24$$
$$2x - 1 = 12x + 24$$
$$-10x = 25$$
$$x = -\tfrac{5}{2}.$$

The possible solution is $-\frac{5}{2}$. We check this on a graphing calculator.

$$y_1 = \frac{2}{3x+6} + \frac{1}{x^2-4}, \quad y_2 = \frac{4}{x-2}$$

X	Y1	Y2
−2.5	−.8889	−.8889

X =

We see that $y_1 = y_2$ when $x = -\frac{5}{2}$, or -2.5, so $-\frac{5}{2}$ is the solution.

Now Try Exercise 21.

Radical Equations

A **radical equation** is an equation in which variables appear in one or more radicands. For example,

$$\sqrt{2x-5} - \sqrt{x-3} = 1$$

is a radical equation. The following principle is used to solve such equations.

THE PRINCIPLE OF POWERS

For any positive integer n:

If $a = b$ is true, then $a^n = b^n$ is true.

EXAMPLE 4 Solve: $\sqrt{3x+1} = 4$.

Algebraic Solution

We use the principle of powers and square both sides:

$$\begin{aligned} \sqrt{3x+1} &= 4 \\ \left(\sqrt{3x+1}\right)^2 &= 4^2 \\ 3x+1 &= 16 \\ 3x &= 15 \\ x &= 5. \end{aligned}$$

Check:

$$\begin{array}{c|c} \sqrt{3x+1} & 4 \\ \hline \sqrt{3 \cdot 5 + 1} & ?\ 4 \\ \sqrt{15+1} & \\ \sqrt{16} & \\ 4 & 4 \quad \text{TRUE} \end{array}$$

The solution is 5.

Graphical Solution

We graph $y_1 = \sqrt{3x+1}$ and $y_2 = 4$ and then use the INTERSECT feature. We see that the solution is 5. The check shown in the following table confirms that the solution is 5.

$y_1 = \sqrt{3x+1}$, $y_2 = 4$

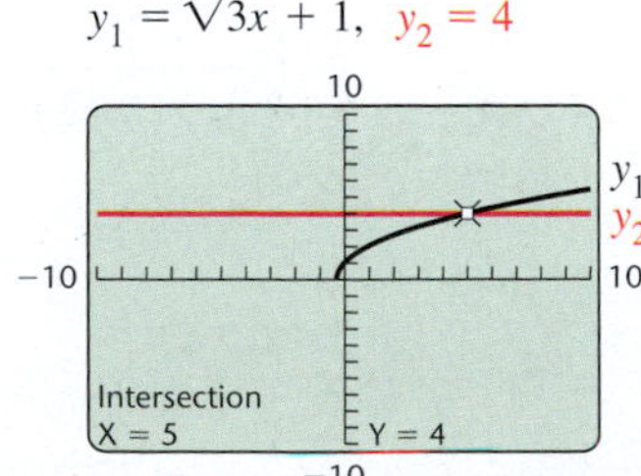

X	Y1	Y2
5	4	4

X =

Now Try Exercise 31.

In Example 4, the radical was isolated on one side of the equation. If this had not been the case, our first step would have been to isolate the radical. We do so in the next example.

EXAMPLE 5 Solve: $5 + \sqrt{x + 7} = x$.

Algebraic Solution

We first isolate the radical and then use the principle of powers:

$$
\begin{aligned}
5 + \sqrt{x + 7} &= x \\
\sqrt{x + 7} &= x - 5 && \text{Subtracting 5 on both sides to isolate the radical} \\
(\sqrt{x + 7})^2 &= (x - 5)^2 && \text{Using the principle of powers; squaring both sides} \\
x + 7 &= x^2 - 10x + 25 \\
0 &= x^2 - 11x + 18 && \text{Subtracting } x \text{ and 7} \\
0 &= (x - 9)(x - 2) && \text{Factoring}
\end{aligned}
$$

$$x - 9 = 0 \quad or \quad x - 2 = 0$$

$$x = 9 \quad or \quad x = 2.$$

The possible solutions are 9 and 2.

Check: For 9:

$$
\begin{array}{c|c}
\multicolumn{2}{c}{5 + \sqrt{x + 7} = x} \\
\hline
5 + \sqrt{9 + 7} \ ? & 9 \\
5 + \sqrt{16} & \\
5 + 4 & \\
9 & 9 \quad \text{TRUE}
\end{array}
$$

For 2:

$$
\begin{array}{c|c}
\multicolumn{2}{c}{5 + \sqrt{x + 7} = x} \\
\hline
5 + \sqrt{2 + 7} \ ? & 2 \\
5 + \sqrt{9} & \\
5 + 3 & \\
8 & 2 \quad \text{FALSE}
\end{array}
$$

Since 9 checks but 2 does not, the only solution is 9.

Graphical Solution

We graph $y_1 = 5 + \sqrt{x + 7}$ and $y_2 = x$. Using the INTERSECT feature, we see that the solution is 9.

$y_1 = 5 + \sqrt{x + 7}$, $y_2 = x$

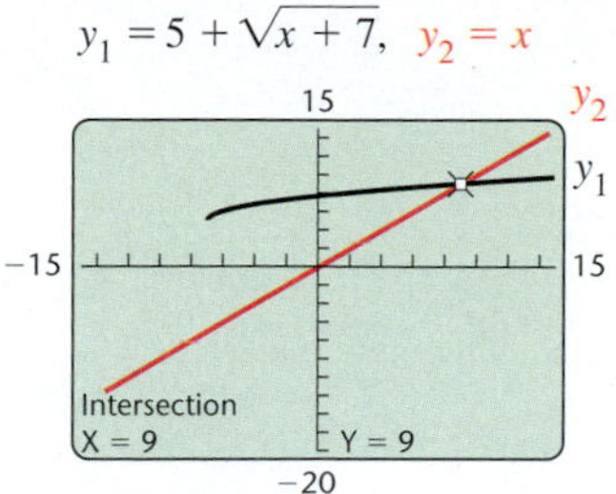

We can also use the ZERO feature to get this result. To do so, we first write the equivalent equation $5 + \sqrt{x + 7} - x = 0$. The zero of the function $f(x) = 5 + \sqrt{x + 7} - x$ is 9, so the solution of the original equation is 9.

$y = 5 + \sqrt{x + 7} - x$

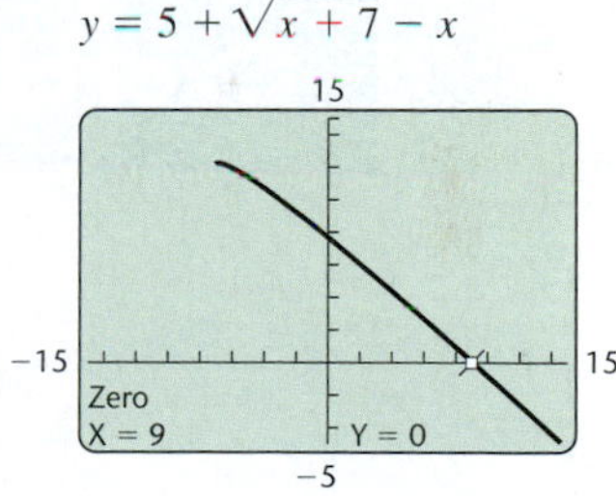

Note that the graphs show that the equation has only one solution.

Now Try Exercise 55.

When we raise both sides of an equation to an even power, the resulting equation can have solutions that the original equation does not. This is because the converse of the principle of powers is not necessarily true. That is, if $a^n = b^n$ is true, we do not know that $a = b$ is true. For example, $(-2)^2 = 2^2$, but $-2 \neq 2$. Thus, as we saw in Example 5, it is necessary to check the possible solutions in the original equation when the principle of powers is used to raise both sides of an equation to an even power.

When a radical equation has two radical terms on one side, we isolate one of them and then use the principle of powers. If, after doing so, a radical term remains, we repeat these steps.

EXAMPLE 6 Solve: $\sqrt{x-3} + \sqrt{x+5} = 4$.

Solution We have

$$\sqrt{x-3} = 4 - \sqrt{x+5}$$ Isolating one radical

$$(\sqrt{x-3})^2 = (4 - \sqrt{x+5})^2$$ Using the principle of powers; squaring both sides

$$x - 3 = 16 - 8\sqrt{x+5} + (x+5)$$

$$x - 3 = 21 - 8\sqrt{x+5} + x$$ Collecting like terms

$$-24 = -8\sqrt{x+5}$$ Isolating the remaining radical; subtracting x and 21 on both sides

$$3 = \sqrt{x+5}$$ Dividing by -8 on both sides

$$3^2 = (\sqrt{x+5})^2$$ Using the principle of powers; squaring both sides

$$9 = x + 5$$

$$4 = x.$$ Subtracting 5 on both sides

We check the possible solution, 4, on a graphing calculator.

$y_1 = \sqrt{x-3} + \sqrt{x+5},\ y_2 = 4$

X	Y1	Y2
4	4	4

X =

Since $y_1 = y_2$ when $x = 4$, the number 4 checks. It is the solution.

Now Try Exercise 65.

3.4 Exercise Set

Solve.

1. $\frac{1}{4} + \frac{1}{5} = \frac{1}{t}$
2. $\frac{1}{3} - \frac{5}{6} = \frac{1}{x}$
3. $\frac{x+2}{4} - \frac{x-1}{5} = 15$
4. $\frac{t+1}{3} - \frac{t-1}{2} = 1$
5. $\frac{1}{2} + \frac{2}{x} = \frac{1}{3} + \frac{3}{x}$
6. $\frac{1}{t} + \frac{1}{2t} + \frac{1}{3t} = 5$
7. $\frac{5}{3x+2} = \frac{3}{2x}$
8. $\frac{2}{x-1} = \frac{3}{x+2}$
9. $\frac{y^2}{y+4} = \frac{16}{y+4}$
10. $\frac{49}{w-7} = \frac{w^2}{w-7}$
11. $x + \frac{6}{x} = 5$
12. $x - \frac{12}{x} = 1$
13. $\frac{6}{y+3} + \frac{2}{y} = \frac{5y-3}{y^2-9}$
14. $\frac{3}{m+2} + \frac{2}{m} = \frac{4m-4}{m^2-4}$
15. $\frac{2x}{x-1} = \frac{5}{x-3}$
16. $\frac{2x}{x+7} = \frac{5}{x+1}$
17. $\frac{2}{x+5} + \frac{1}{x-5} = \frac{16}{x^2-25}$
18. $\frac{2}{x^2-9} + \frac{5}{x-3} = \frac{3}{x+3}$
19. $\frac{3x}{x+2} + \frac{6}{x} = \frac{12}{x^2+2x}$
20. $\frac{3y+5}{y^2+5y} + \frac{y+4}{y+5} = \frac{y+1}{y}$

21. $\frac{1}{5x+20} - \frac{1}{x^2-16} = \frac{3}{x-4}$

22. $\frac{1}{4x+12} - \frac{1}{x^2-9} = \frac{5}{x-3}$

23. $\frac{2}{5x+5} - \frac{3}{x^2-1} = \frac{4}{x-1}$

24. $\frac{1}{3x+6} - \frac{1}{x^2-4} = \frac{3}{x-2}$

25. $\frac{8}{x^2-2x+4} = \frac{x}{x+2} + \frac{24}{x^3+8}$

26. $\frac{18}{x^2-3x+9} - \frac{x}{x+3} = \frac{81}{x^3+27}$

27. $\frac{x}{x-4} - \frac{4}{x+4} = \frac{32}{x^2-16}$

28. $\frac{x}{x-1} - \frac{1}{x+1} = \frac{2}{x^2-1}$

29. $\frac{1}{x-6} - \frac{1}{x} = \frac{6}{x^2-6x}$

30. $\frac{1}{x-15} - \frac{1}{x} = \frac{15}{x^2-15x}$

31. $\sqrt{3x-4} = 1$

32. $\sqrt{4x+1} = 3$

33. $\sqrt{2x-5} = 2$

34. $\sqrt{3x+2} = 6$

35. $\sqrt{7-x} = 2$

36. $\sqrt{5-x} = 1$

37. $\sqrt{1-2x} = 3$

38. $\sqrt{2-7x} = 2$

39. $\sqrt[3]{5x-2} = -3$

40. $\sqrt[3]{2x+1} = -5$

41. $\sqrt[4]{x^2-1} = 1$

42. $\sqrt[5]{3x+4} = 2$

43. $\sqrt{y-1} + 4 = 0$

44. $\sqrt{m+1} - 5 = 8$

45. $\sqrt{b+3} - 2 = 1$

46. $\sqrt{x-4} + 1 = 5$

47. $\sqrt{z+2} + 3 = 4$

48. $\sqrt{y-5} - 2 = 3$

49. $\sqrt{2x+1} - 3 = 3$

50. $\sqrt{3x-1} + 2 = 7$

51. $\sqrt{2-x} - 4 = 6$

52. $\sqrt{5-x} + 2 = 8$

53. $\sqrt[3]{6x+9} + 8 = 5$

54. $\sqrt[5]{2x-3} - 1 = 1$

55. $\sqrt{x+4} + 2 = x$

56. $\sqrt{x+1} + 1 = x$

57. $\sqrt{x-3} + 5 = x$

58. $\sqrt{x+3} - 1 = x$

59. $\sqrt{x+7} = x+1$

60. $\sqrt{6x+7} = x+2$

61. $\sqrt{3x+3} = x+1$

62. $\sqrt{2x+5} = x-5$

63. $\sqrt{5x+1} = x-1$

64. $\sqrt{7x+4} = x+2$

65. $\sqrt{x-3} + \sqrt{x+2} = 5$

66. $\sqrt{x} - \sqrt{x-5} = 1$

67. $\sqrt{3x-5} + \sqrt{2x+3} + 1 = 0$

68. $\sqrt{2m-3} = \sqrt{m+7} - 2$

69. $\sqrt{x} - \sqrt{3x-3} = 1$

70. $\sqrt{2x+1} - \sqrt{x} = 1$

71. $\sqrt{2y-5} - \sqrt{y-3} = 1$

72. $\sqrt{4p+5} + \sqrt{p+5} = 3$

73. $\sqrt{y+4} - \sqrt{y-1} = 1$

74. $\sqrt{y+7} + \sqrt{y+16} = 9$

75. $\sqrt{x+5} + \sqrt{x+2} = 3$

76. $\sqrt{6x+6} = 5 + \sqrt{21-4x}$

77. $x^{1/3} = -2$

78. $t^{1/5} = 2$

79. $t^{1/4} = 3$

80. $m^{1/2} = -7$

Solve.

81. $\frac{P_1V_1}{T_1} = \frac{P_2V_2}{T_2}$, for T_1
(A chemistry formula for gases)

82. $\frac{1}{F} = \frac{1}{m} + \frac{1}{p}$, for F
(A formula from optics)

83. $W = \sqrt{\frac{1}{LC}}$, for C
(An electricity formula)

84. $s = \sqrt{\frac{A}{6}}$, for A
(A geometry formula)

85. $\frac{1}{R} = \frac{1}{R_1} + \frac{1}{R_2}$, for R_2
(A formula for resistance)

86. $\frac{1}{t} = \frac{1}{a} + \frac{1}{b}$, for t
(A formula for work rate)

87. $I = \sqrt{\frac{A}{P}} - 1$, for P
(A compound-interest formula)

88. $T = 2\pi\sqrt{\frac{1}{g}}$, for g
(A pendulum formula)

89. $\frac{1}{F} = \frac{1}{m} + \frac{1}{p}$, for p
(A formula from optics)

90. $\frac{V^2}{R^2} = \frac{2g}{R+h}$, for h
(A formula for escape velocity)

Skill Maintenance

Find the zero of the function. **[1.5]**

91. $f(x) = 15 - 2x$

92. $f(x) = -3x + 9$

Solve. **[1.5]**

93. *Pork Production.* Together, China and the United States, the top two pork producers worldwide, produced 64,308,000 metric tons of pork in 2013. China produced 1,260,000 metric tons more than five times the number of metric tons produced by the United States. (*Source*: United Nations Food and Agriculture Organization) How many metric tons of pork did each country produce in 2013?

94. *Student Loan Debt.* In 2014, the average student loan debt per college graduate was about \$33,050. This was about 77.6% more than the average student loan debt per college graduate in 2004. (*Source*: Analysis by Mark Kantrowitz, publisher of Edvisors.com) What was the average student loan debt per college graduate in 2004?

Synthesis

Solve.

95. $(x - 3)^{2/3} = 2$

96. $\dfrac{x + 3}{x + 2} - \dfrac{x + 4}{x + 3} = \dfrac{x + 5}{x + 4} - \dfrac{x + 6}{x + 5}$

97. $\sqrt{x + 5} + 1 = \dfrac{6}{\sqrt{x + 5}}$

98. $\sqrt{15 + \sqrt{2x + 80}} = 5$

99. $x^{2/3} = x$

3.5 Solving Equations and Inequalities with Absolute Value

- Solve equations with absolute value.
- Solve inequalities with absolute value.

Equations with Absolute Value

Recall that the absolute value of a number is its distance from 0 on the number line. We use this concept to solve equations with absolute value.

For $a > 0$ and an algebraic expression X:

$$|X| = a \quad \text{is equivalent to} \quad X = -a \quad or \quad X = a.$$

EXAMPLE 1 Solve: $|x| = 5$.

Algebraic Solution	Graphical Solution
We have $|x| = 5$ $x = -5$ *or* $x = 5$. **Writing an equivalent statement** The solutions are -5 and 5. To check, note that -5 and 5 are both 5 units from 0 on the number line. 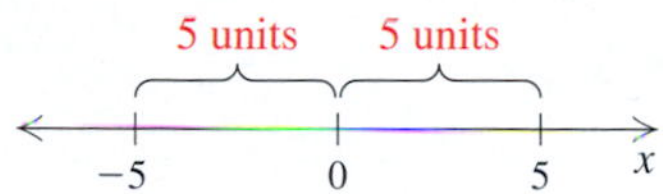	Using the Intersect method, we graph $y_1 = |x|$ and $y_2 = 5$ and find the first coordinates of the points of intersection. 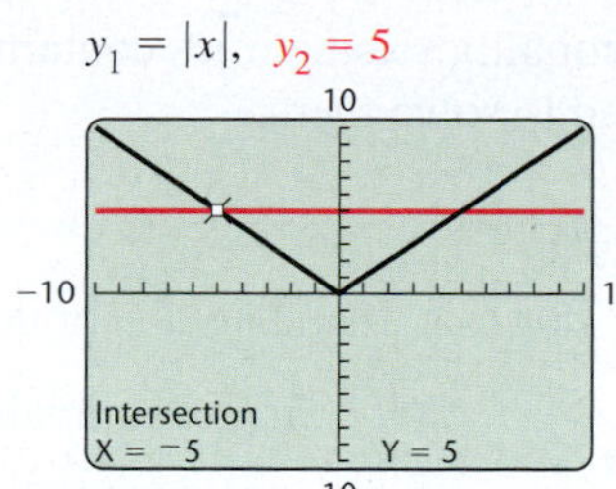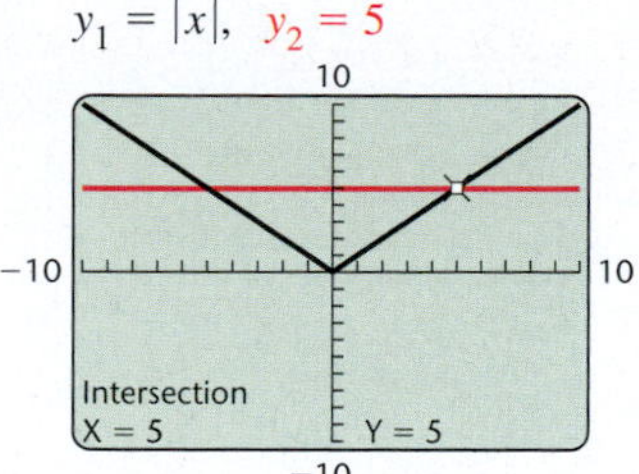 The solutions are -5 and 5. We could also have used the Zero method to get this result, graphing $y = |x| - 5$ and using the ZERO feature twice. 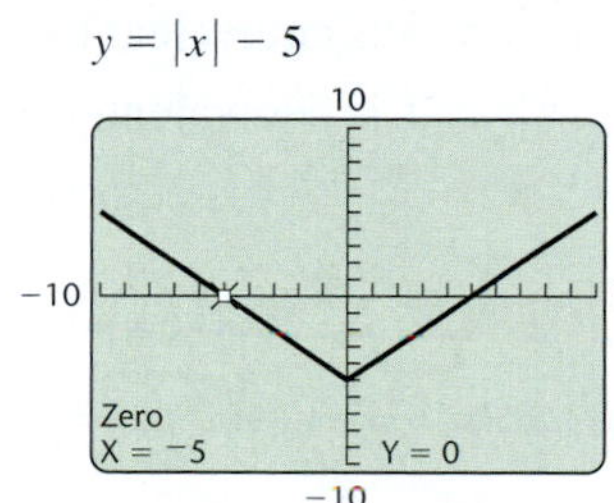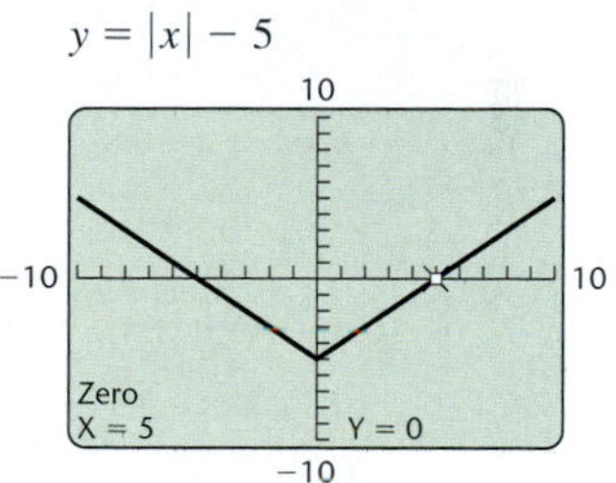 The zeros of $f(x) = |x| - 5$ are -5 and 5, so the solutions of the original equation are -5 and 5.

Now Try Exercise 1.

EXAMPLE 2 Solve: $|x - 3| - 1 = 4$.

Solution First, we add 1 on both sides to get an expression of the form $|X| = a$:

$$|x - 3| - 1 = 4$$
$$|x - 3| = 5$$
$$x - 3 = -5 \quad or \quad x - 3 = 5$$ **$|X| = a$ is equivalent to $X = -a$ or $X = a$.**
$$x = -2 \quad or \quad x = 8.$$ **Adding 3**

Check:

For -2:

$|x - 3| - 1 = 4$	
$|-2 - 3| - 1$	? 4
$|-5| - 1$	
$5 - 1$	
4	4 TRUE

For 8:

$|x - 3| - 1 = 4$	
$|8 - 3| - 1$	? 4
$|5| - 1$	
$5 - 1$	
4	4 TRUE

The solutions are -2 and 8.

Now Try Exercise 21.

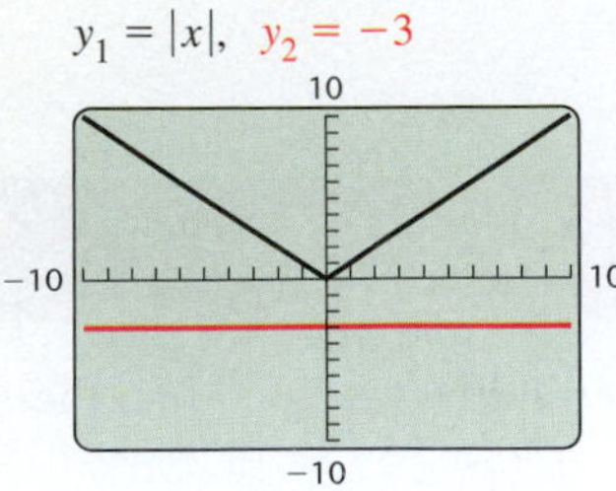

When $a = 0$, $|X| = a$ is equivalent to $X = 0$. Note that for $a < 0$, $|X| = a$ has no solution, because the absolute value of an expression is never negative. We can use a graph to illustrate the last statement for a specific value of a. For example, if we let $a = -3$ and graph $y_1 = |x|$ and $y_2 = -3$, we see that the graphs do not intersect, as shown at left. Thus the equation $|x| = -3$ has no solution. The solution set is the **empty set**, denoted $\varnothing$.

Inequalities with Absolute Value

Inequalities sometimes contain absolute-value notation. The following properties are used to solve them.

> For $a > 0$ and an algebraic expression X:
>
> $|X| < a$ is equivalent to $-a < X < a$.
>
> $|X| > a$ is equivalent to $X < -a$ *or* $X > a$.
>
> Similar statements hold for $|X| \le a$ and $|X| \ge a$.

For example,

$|x| < 3$ is equivalent to $-3 < x < 3$;
$|y| \ge 1$ is equivalent to $y \le -1$ *or* $y \ge 1$; and
$|2x + 3| \le 4$ is equivalent to $-4 \le 2x + 3 \le 4$.

EXAMPLE 3 Solve and graph the solution set: $|3x + 2| < 5$.

Solution We have

$$|3x + 2| < 5$$
$$-5 < 3x + 2 < 5 \quad \text{Writing an equivalent inequality}$$
$$-7 < 3x < 3 \quad \text{Subtracting 2}$$
$$-\tfrac{7}{3} < x < 1. \quad \text{Dividing by 3}$$

The solution set is $\left\{x \middle| -\frac{7}{3} < x < 1\right\}$, or $\left(-\frac{7}{3}, 1\right)$. The graph of the solution set is shown below.

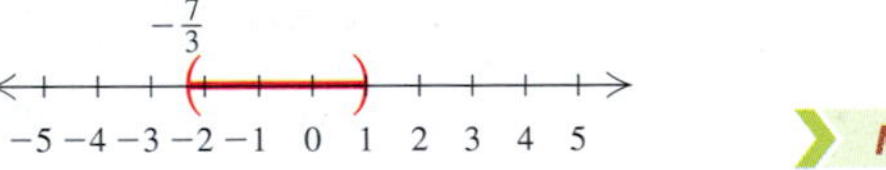

Now Try Exercise 45.

EXAMPLE 4 Solve and graph the solution set: $|5 - 2x| \ge 1$.

Solution We have

$$|5 - 2x| \ge 1$$
$$5 - 2x \le -1 \quad or \quad 5 - 2x \ge 1 \quad \text{Writing an equivalent inequality}$$
$$-2x \le -6 \quad or \quad -2x \ge -4 \quad \text{Subtracting 5}$$
$$x \ge 3 \quad or \quad x \le 2. \quad \text{Dividing by } -2 \text{ and reversing the inequality signs}$$

The solution set is $\{x|x \le 2 \; or \; x \ge 3\}$, or $(-\infty, 2] \cup [3, \infty)$. The graph of the solution set is shown below.

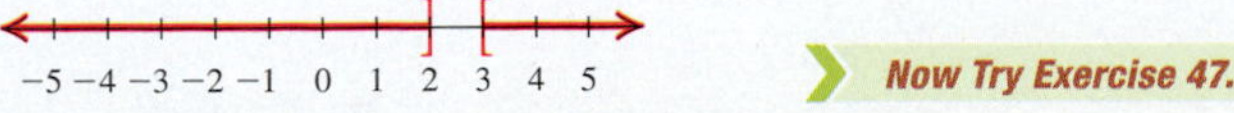

Now Try Exercise 47.

3.5 Exercise Set

Solve.

1. $|x| = 7$
2. $|x| = 4.5$
3. $|x| = 0$
4. $|x| = \frac{3}{2}$
5. $|x| = \frac{5}{6}$
6. $|x| = -\frac{3}{5}$
7. $|x| = -10.7$
8. $|x| = 12$
9. $|3x| = 1$
10. $|5x| = 4$
11. $|8x| = 24$
12. $|6x| = 0$
13. $|x - 1| = 4$
14. $|x - 7| = 5$
15. $|x + 2| = 6$
16. $|x + 5| = 1$
17. $|3x + 2| = 1$
18. $|7x - 4| = 8$
19. $|\frac{1}{2}x - 5| = 17$
20. $|\frac{1}{3}x - 4| = 13$
21. $|x - 1| + 3 = 6$
22. $|x + 2| - 5 = 9$
23. $|x + 3| - 2 = 8$
24. $|x - 4| + 3 = 9$
25. $|3x + 1| - 4 = -1$
26. $|2x - 1| - 5 = -3$
27. $|4x - 3| + 1 = 7$
28. $|5x + 4| + 2 = 5$
29. $12 - |x + 6| = 5$
30. $9 - |x - 2| = 7$
31. $7 - |2x - 1| = 6$
32. $5 - |4x + 3| = 2$

Solve and write interval notation for the solution set. Then graph the solution set.

33. $|x| < 7$
34. $|x| \leq 4.5$
35. $|x| \leq 2$
36. $|x| < 3$
37. $|x| \geq 4.5$
38. $|x| > 7$
39. $|x| > 3$
40. $|x| \geq 2$
41. $|3x| < 1$
42. $|5x| \leq 4$
43. $|2x| \geq 6$
44. $|4x| > 20$
45. $|x + 8| < 9$
46. $|x + 6| \leq 10$
47. $|x + 8| \geq 9$
48. $|x + 6| > 10$
49. $|x - \frac{1}{4}| < \frac{1}{2}$
50. $|x - 0.5| \leq 0.2$
51. $|2x + 3| \leq 9$
52. $|3x + 4| < 13$
53. $|x - 5| > 0.1$
54. $|x - 7| \geq 0.4$
55. $|6 - 4x| \geq 8$
56. $|5 - 2x| > 10$
57. $|x + \frac{2}{3}| \leq \frac{5}{3}$
58. $|x + \frac{3}{4}| < \frac{1}{4}$
59. $\left|\frac{2x + 1}{3}\right| > 5$
60. $\left|\frac{2x - 1}{3}\right| \geq \frac{5}{6}$
61. $|2x - 4| < -5$
62. $|3x + 5| < 0$
63. $|7 - x| \geq -4$
64. $|2x + 1| > -\frac{1}{2}$

Skill Maintenance

Vocabulary Reinforcement

In each of Exercises 65–72, fill in the blank with the correct term. Some of the given choices will not be used.

distance formula
midpoint formula
function
relation
x-intercept
y-intercept
perpendicular
parallel
horizontal lines
vertical lines
symmetric with respect to the x-axis
symmetric with respect to the y-axis
symmetric with respect to the origin
increasing
decreasing
constant

65. A(n) ______________ is a point $(0, b)$. [1.1]
66. The ______________ is $d = \sqrt{(x_2 - x_1)^2 + (y_2 - y_1)^2}$. [1.1]
67. A(n) ______________ is a correspondence such that each member of the domain corresponds to at least one member of the range. [1.2]
68. A(n) ______________ is a correspondence such that each member of the domain corresponds to exactly one member of the range. [1.2]
69. ______________ are given by equations of the type $y = b$, or $f(x) = b$. [1.3]
70. Nonvertical lines are ______________ if and only if they have the same slope and different y-intercepts. [1.4]
71. A function f is said to be ______________ on an open interval I if, for all a and b in that interval, $a < b$ implies $f(a) > f(b)$. [2.1]
72. For an equation $y = f(x)$, if replacing x with $-x$ produces an equivalent equation, then the graph is ______________. [2.4]

Synthesis

Solve.

73. $|3x - 1| > 5x - 2$
74. $|x + 2| \leq |x - 5|$
75. $|p - 4| + |p + 4| < 8$
76. $|x| + |x + 1| < 10$
77. $|x - 3| + |2x + 5| > 6$

Chapter 3 Summary and Review

STUDY GUIDE

KEY TERMS AND CONCEPTS	EXAMPLE
SECTION 3.1: THE COMPLEX NUMBERS	
The number i is defined such that $i = \sqrt{-1}$ and $i^2 = -1$.	Express each number in terms of i. $\sqrt{-5} = \sqrt{-1 \cdot 5} = \sqrt{-1} \cdot \sqrt{5} = i\sqrt{5}$, or $\sqrt{5}i$; $-\sqrt{-36} = -\sqrt{-1 \cdot 36} = -\sqrt{-1} \cdot \sqrt{36} = -i \cdot 6 = -6i$
A **complex number** is a number of the form $a + bi$, where a and b are real numbers. The number a is said to be the **real part** of $a + bi$, and the number b is said to be the **imaginary part** of $a + bi$. To **add** or **subtract complex numbers**, we add or subtract the real parts, and we add or subtract the imaginary parts.	Add or subtract. $(-3 + 4i) + (5 - 8i) = (-3 + 5) + (4i - 8i)$ $= 2 - 4i$; $(6 - 7i) - (10 + 3i) = (6 - 10) + (-7i - 3i)$ $= -4 - 10i$
When we **multiply complex numbers**, we must keep in mind the fact that $i^2 = -1$. Note that $\sqrt{a} \cdot \sqrt{b} \neq \sqrt{ab}$ when $\sqrt{a}$ and $\sqrt{b}$ are not real numbers.	Multiply. $\sqrt{-4} \cdot \sqrt{-100} = \sqrt{-1} \cdot \sqrt{4} \cdot \sqrt{-1} \cdot \sqrt{100}$ $= i \cdot 2 \cdot i \cdot 10$ $= i^2 \cdot 20$ $= -1 \cdot 20$ $i^2 = -1$ $= -20$; $(2 - 5i)(3 + i) = 6 + 2i - 15i - 5i^2$ $= 6 - 13i - 5(-1)$ $= 6 - 13i + 5$ $= 11 - 13i$
The **conjugate of a complex number** $a + bi$ is $a - bi$. The numbers $a + bi$ and $a - bi$ are **complex conjugates**. Conjugates are used when we **divide complex numbers**.	Divide. $\frac{5 - 2i}{3 + i} = \frac{5 - 2i}{3 + i} \cdot \frac{3 - i}{3 - i}$ **$3 - i$ is the conjugate of the divisor, $3 + i$.** $= \frac{15 - 5i - 6i + 2i^2}{9 - i^2}$ $= \frac{15 - 11i - 2}{9 + 1}$ $i^2 = -1$ $= \frac{13 - 11i}{10}$ $= \frac{13}{10} - \frac{11}{10}i$

SECTION 3.2: QUADRATIC EQUATIONS, FUNCTIONS, ZEROS, AND MODELS

A **quadratic equation** is an equation that can be written in the form

$$ax^2 + bx + c = 0, \quad a \neq 0,$$

where a, b, and c are real numbers.

A **quadratic function** f is a function that can be written in the form

$$f(x) = ax^2 + bx + c, \quad a \neq 0,$$

where a, b, and c are real numbers.

The **zeros** of a quadratic function $f(x) = ax^2 + bx + c$ are the *solutions* of the associated quadratic equation $ax^2 + bx + c = 0$.

$3x^2 - 2x + 4 = 0$ and $5 - 4x = x^2$ are examples of quadratic equations. The equation $3x^2 - 2x + 4 = 0$ is written in **standard form**.

The functions $f(x) = 2x^2 + x + 1$ and $f(x) = 5x^2 - 4$ are examples of quadratic functions.

The Principle of Zero Products

If $ab = 0$ is true, then $a = 0$ *or* $b = 0$, and if $a = 0$ *or* $b = 0$, then $ab = 0$.

Solve: $3x^2 - 4 = 11x$.

$$3x^2 - 4 = 11x$$

$$3x^2 - 11x - 4 = 0$$ **Subtracting $11x$ on both sides to get 0 on one side of the equation**

$$(3x + 1)(x - 4) = 0$$ **Factoring**

$$3x + 1 = 0 \quad or \quad x - 4 = 0$$ **Using the principle of zero products**

$$3x = -1 \quad or \quad x = 4$$

$$x = -\frac{1}{3} \quad or \quad x = 4$$

The solutions are $-\frac{1}{3}$ and 4.

The Principle of Square Roots

If $x^2 = k$, then $x = \sqrt{k}$ *or* $x = -\sqrt{k}$.

Solve: $3x^2 - 18 = 0$.

$$3x^2 - 18 = 0$$

$$3x^2 = 18$$ **Adding 18 on both sides**

$$x^2 = 6$$ **Dividing by 3 on both sides**

$$x = \sqrt{6} \quad or \quad x = -\sqrt{6}$$ **Using the principle of square roots**

The solutions are $\sqrt{6}$ and $-\sqrt{6}$, or $\pm\sqrt{6}$.

To solve a quadratic equation by **completing the square**:

1. Isolate the terms with variables on one side of the equation and arrange them in descending order.
2. Divide by the coefficient of the squared term if that coefficient is not 1.
3. Complete the square by taking half the coefficient of the first-degree term and adding its square on both sides of the equation.
4. Express one side of the equation as the square of a binomial.
5. Use the principle of square roots.
6. Solve for the variable.

Solve: $2x^2 - 3 = 6x$.

$$2x^2 - 3 = 6x$$

$$2x^2 - 6x - 3 = 0 \quad \text{Subtracting } 6x$$

$$2x^2 - 6x = 3 \quad \text{Adding 3}$$

$$x^2 - 3x = \frac{3}{2} \quad \text{Dividing by 2 to make the } x^2\text{-coefficient 1}$$

$$x^2 - 3x + \frac{9}{4} = \frac{3}{2} + \frac{9}{4} \quad \text{Completing the square: } \tfrac{1}{2}(-3) = -\tfrac{3}{2} \text{ and } (-\tfrac{3}{2})^2 = \tfrac{9}{4}\text{; adding } \tfrac{9}{4}$$

$$\left(x - \frac{3}{2}\right)^2 = \frac{15}{4} \quad \text{Factoring and simplifying}$$

$$x - \frac{3}{2} = \pm\frac{\sqrt{15}}{2} \quad \text{Using the principle of square roots and the quotient rule for radicals}$$

$$x = \frac{3}{2} \pm \frac{\sqrt{15}}{2} = \frac{3 \pm \sqrt{15}}{2}$$

The solutions are $\frac{3 + \sqrt{15}}{2}$ and $\frac{3 - \sqrt{15}}{2}$, or $\frac{3 \pm \sqrt{15}}{2}$.

The solutions of $ax^2 + bx + c = 0$, $a \neq 0$, can be found using the **quadratic formula**:

$$x = \frac{-b \pm \sqrt{b^2 - 4ac}}{2a}.$$

Solve: $x^2 - 6 = 3x$.

$$x^2 - 6 = 3x$$

$$x^2 - 3x - 6 = 0 \quad \text{Standard form}$$

$$a = 1, b = -3, c = -6$$

$$x = \frac{-b \pm \sqrt{b^2 - 4ac}}{2a}$$

$$= \frac{-(-3) \pm \sqrt{(-3)^2 - 4(1)(-6)}}{2 \cdot 1}$$

$$= \frac{3 \pm \sqrt{9 + 24}}{2}$$

$$= \frac{3 \pm \sqrt{33}}{2} \quad \text{Exact solutions}$$

Using a calculator, we approximate the solutions to be 4.372 and −1.372.

Discriminant

For $ax^2 + bx + c = 0$, where a, b, and c are real numbers:

$b^2 - 4ac = 0 \longrightarrow$ One real-number solution;

$b^2 - 4ac > 0 \longrightarrow$ Two different real-number solutions;

$b^2 - 4ac < 0 \longrightarrow$ Two different imaginary-number solutions, complex conjugates.

For the equation above, $x^2 - 6 = 3x$, we see that $b^2 - 4ac$ is 33. Since 33 is positive, there are two different real-number solutions.

For $2x^2 - x + 4 = 0$, with $a = 2$, $b = -1$, and $c = 4$, the discriminant, $(-1)^2 - 4 \cdot 2 \cdot 4 = 1 - 32 = -31$, is negative, so there are two different imaginary-number (or nonreal) solutions.

For $x^2 - 6x + 9 = 0$, with $a = 1$, $b = -6$, and $c = 9$, the discriminant, $(-6)^2 - 4 \cdot 1 \cdot 9 = 36 - 36 = 0$, is 0 so there is one real-number solution.

Equations **reducible to quadratic**, or **quadratic in form**, can be treated as quadratic equations if a suitable substitution is made.

Solve: $x^4 - x^2 - 12 = 0$.

$x^4 - x^2 - 12 = 0$ Let $u = x^2$. Then $u^2 = (x^2)^2 = x^4$.

$u^2 - u - 12 = 0$ Substituting

$(u - 4)(u + 3) = 0$

$u - 4 = 0$ *or* $u + 3 = 0$

$u = 4$ *or* $u = -3$ Solving for u

$x^2 = 4$ *or* $x^2 = -3$

$x = \pm 2$ *or* $x = \pm\sqrt{3}i$ Solving for x

The solutions are 2, -2, $\sqrt{3}i$, and $-\sqrt{3}i$.

SECTION 3.3: ANALYZING GRAPHS OF QUADRATIC FUNCTIONS

Graphing Quadratic Equations

The graph of the function $f(x) = a(x - h)^2 + k$ is a parabola that:

- opens up if $a > 0$ and down if $a < 0$;
- has (h, k) as the vertex;
- has $x = h$ as the axis of symmetry;
- has k as a minimum value (output) if $a > 0$;
- has k as a maximum value if $a < 0$.

We can use a modification of the technique of completing the square as an aid in analyzing and graphing quadratic functions.

Find the vertex, the axis of symmetry, and the maximum or minimum value of $f(x) = 2x^2 + 12x + 12$.

$f(x) = 2x^2 + 12x + 12$

$= 2(x^2 + 6x) + 12$ Note that 9 completes the square for $x^2 + 6x$.

$= 2(x^2 + 6x + 9 - 9) + 12$ Adding $9 - 9$, or 0, inside the parentheses

$= 2(x^2 + 6x + 9) - 2 \cdot 9 + 12$ Using the distributive law to remove -9 from within the parentheses

$= 2(x + 3)^2 - 6$

$= 2[x - (-3)]^2 + (-6)$

The function is now written in the form

$f(x) = a(x - h)^2 + k$

with $a = 2$, $h = -3$, and $k = -6$. Because $a > 0$, we know the graph opens up and thus the function has a minimum value. We also know the following:

Vertex (h, k): $(-3, -6)$;

Axis of symmetry $x = h$: $x = -3$;

Minimum value of the function k: -6.

To graph the function, we first plot the vertex and then find several points on either side of it. We plot these points and connect them with a smooth curve.

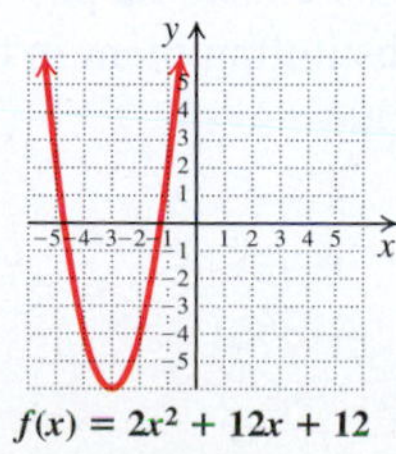

$f(x) = 2x^2 + 12x + 12$

The Vertex of a Parabola The **vertex** of the graph of $f(x) = ax^2 + bx + c$ is $\left(-\frac{b}{2a}, f\left(-\frac{b}{2a}\right)\right)$. We calculate the x-coordinate. We substitute to find the y-coordinate.	Find the vertex of the function $f(x) = -3x^2 + 6x + 1$. $-\frac{b}{2a} = -\frac{6}{2(-3)} = 1$ $f(1) = -3 \cdot 1^2 + 6 \cdot 1 + 1 = 4$ The vertex is $(1, 4)$.
Some applied problems can be solved by finding the maximum or minimum value of a quadratic function.	See Examples 5–7 on pp. 191–193.

SECTION 3.4: SOLVING RATIONAL EQUATIONS AND RADICAL EQUATIONS

A **rational equation** is an equation containing one or more rational expressions. When we solve a rational equation, we usually first multiply by the least common denominator (LCD) of all the rational expressions to clear the fractions.

CAUTION! When we multiply by an expression containing a variable, we might not obtain an equation equivalent to the original equation, so we *must* check the possible solutions obtained by substituting them in the original equation.

Solve: $\frac{5}{x+2} - \frac{4}{x^2-4} = \frac{x-3}{x-2}$.

$$\frac{5}{x+2} - \frac{4}{x^2-4} = \frac{x-3}{x-2}$$

$$\frac{5}{x+2} - \frac{4}{(x+2)(x-2)} = \frac{x-3}{x-2} \quad \text{The LCD is } (x+2)(x-2).$$

$$(x+2)(x-2)\left(\frac{5}{x+2} - \frac{4}{(x+2)(x-2)}\right) = (x+2)(x-2)\cdot\frac{x-3}{x-2}$$

$$5(x-2) - 4 = (x+2)(x-3)$$
$$5x - 10 - 4 = x^2 - x - 6$$
$$5x - 14 = x^2 - x - 6$$
$$0 = x^2 - 6x + 8$$
$$0 = (x-2)(x-4)$$
$$x - 2 = 0 \quad or \quad x - 4 = 0$$
$$x = 2 \quad or \quad x = 4$$

The number 2 does not check, but 4 does. The solution is 4.

A **radical equation** is an equation that contains one or more radicals. We use the **principle of powers** to solve radical equations.

For any positive integer n:

If $a = b$ is true, then $a^n = b^n$ is true.

CAUTION! If $a^n = b^n$ is true, it is not necessarily true that $a = b$, so we *must* check the possible solutions obtained by substituting them in the original equation.

Solve: $\sqrt{x+2} + \sqrt{x-1} = 3$.

$$\sqrt{x+2} + \sqrt{x-1} = 3$$
$$\sqrt{x+2} = 3 - \sqrt{x-1} \quad \text{Isolating one radical}$$
$$(\sqrt{x+2})^2 = (3 - \sqrt{x-1})^2$$
$$x + 2 = 9 - 6\sqrt{x-1} + (x-1)$$
$$x + 2 = 8 - 6\sqrt{x-1} + x$$
$$-6 = -6\sqrt{x-1} \quad \text{Isolating the remaining radical}$$
$$1 = \sqrt{x-1} \quad \text{Dividing by } -6$$
$$1^2 = (\sqrt{x-1})^2$$
$$1 = x - 1$$
$$2 = x$$

The number 2 checks. It is the solution.

SECTION 3.5: SOLVING EQUATIONS AND INEQUALITIES WITH ABSOLUTE VALUE

We use the following property to **solve equations with absolute value**. For $a > 0$ and an algebraic expression X: $\|X\| = a$ is equivalent to $X = -a$ *or* $X = a$.	Solve: $\|x + 1\| = 4$. $\|x + 1\| = 4$ $x + 1 = -4$ *or* $x + 1 = 4$ $x = -5$ *or* $x = 3$ Both numbers check. The solutions are -5 and 3.
The following properties are used to **solve inequalities with absolute value**. For $a > 0$ and an algebraic expression X: $\|X\| < a$ is equivalent to $-a < X < a$. $\|X\| > a$ is equivalent to $X < -a$ *or* $X > a$. Similar statements hold for $\|X\| \leq a$ and $\|X\| \geq a$.	Solve: $\|x - 2\| < 3$. $\|x - 2\| < 3$ $-3 < x - 2 < 3$ $-1 < x < 5$ **Adding 2** The solution set is $\{x \| -1 < x < 5\}$, or $(-1, 5)$. $\|3x\| \geq 6$ $3x \leq -6$ *or* $3x \geq 6$ $x \leq -2$ *or* $x \geq 2$ **Dividing by 3** The solution set is $\{x \| x \leq -2 \text{ or } x \geq 2\}$, or $(-\infty, -2] \cup [2, \infty)$.

REVIEW EXERCISES

Determine whether the statement is true or false.

1. We can use the quadratic formula to solve any quadratic equation. [3.2]
2. The function $f(x) = -3(x + 4)^2 - 1$ has a maximum value. [3.3]
3. For any positive integer n, if $a^n = b^n$ is true, then $a = b$ is true. [3.4]
4. An equation with absolute value cannot have two negative-number solutions. [3.5]

Solve. [3.2]

5. $(2y + 5)(3y - 1) = 0$
6. $x^2 + 4x - 5 = 0$
7. $3x^2 + 2x = 8$
8. $5x^2 = 15$
9. $x^2 + 10 = 0$

Find the zero(s) of the function. [3.2]

10. $f(x) = x^2 - 2x + 1$
11. $f(x) = x^2 + 2x - 15$
12. $f(x) = 2x^2 - x - 5$
13. $f(x) = 3x^2 + 2x + 3$

Solve.

14. $\dfrac{5}{2x + 3} + \dfrac{1}{x - 6} = 0$ [3.4]
15. $\dfrac{3}{8x + 1} + \dfrac{8}{2x + 5} = 1$ [3.4]
16. $\sqrt{5x + 1} - 1 = \sqrt{3x}$ [3.4]
17. $\sqrt{x - 1} - \sqrt{x - 4} = 1$ [3.4]
18. $|x - 4| = 3$ [3.5]
19. $|2y + 7| = 9$ [3.5]

Solve and write interval notation for the solution set. Then graph the solution set. [3.5]

20. $|5x| \geq 15$
21. $|3x + 4| < 10$
22. $|6x - 1| < 5$
23. $|x + 4| \geq 2$
24. Solve $\dfrac{1}{M} + \dfrac{1}{N} = \dfrac{1}{P}$ for P. [3.4]

Express in terms of i. [3.1]

25. $-\sqrt{-40}$

26. $\sqrt{-12} \cdot \sqrt{-20}$

27. $\dfrac{\sqrt{-49}}{-\sqrt{-64}}$

Simplify each of the following. Write the answer in the form a + bi, where a and b are real numbers. [3.1]

28. $(6 + 2i) + (-4 - 3i)$

29. $(3 - 5i) - (2 - i)$

30. $(6 + 2i)(-4 - 3i)$

31. $\dfrac{2 - 3i}{1 - 3i}$

32. i^{23}

Solve by completing the square to obtain exact solutions. Show your work. [3.2]

33. $x^2 - 3x = 18$

34. $3x^2 - 12x - 6 = 0$

Solve. Give exact solutions. [3.2]

35. $3x^2 + 10x = 8$

36. $r^2 - 2r + 10 = 0$

37. $x^2 = 10 + 3x$

38. $x = 2\sqrt{x} - 1$

39. $y^4 - 3y^2 + 1 = 0$

40. $(x^2 - 1)^2 - (x^2 - 1) - 2 = 0$

41. $(p - 3)(3p + 2)(p + 2) = 0$

42. $x^3 + 5x^2 - 4x - 20 = 0$

In Exercises 43 and 44, complete the square to:

a) *find the vertex;*
b) *find the axis of symmetry;*
c) *determine whether there is a maximum or a minimum value and find that value;*
d) *find the range; and*
e) *graph the function.* [3.3]

43. $f(x) = -4x^2 + 3x - 1$

44. $f(x) = 5x^2 - 10x + 3$

In Exercises 45–48, match the equation with one of the figures (a)–(d) that follow. [3.3]

a)

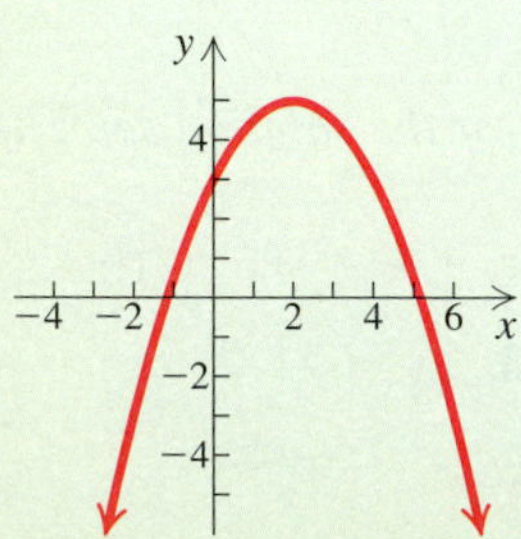

b)

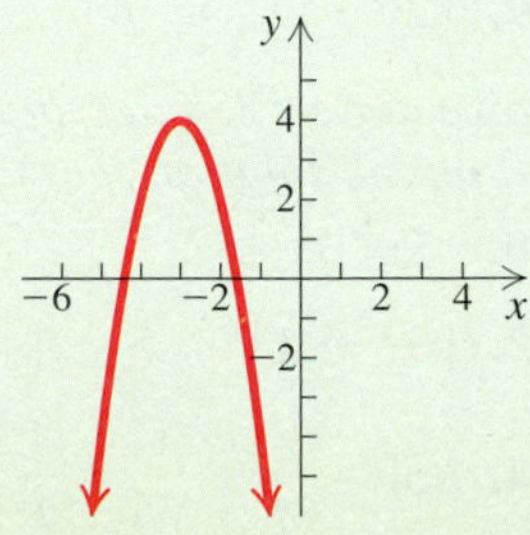

c)

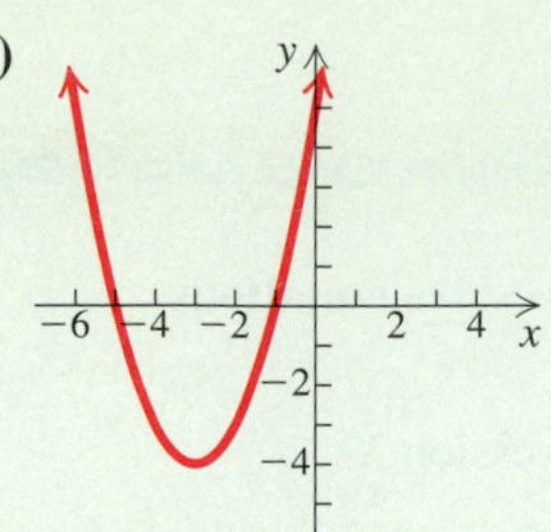

d)

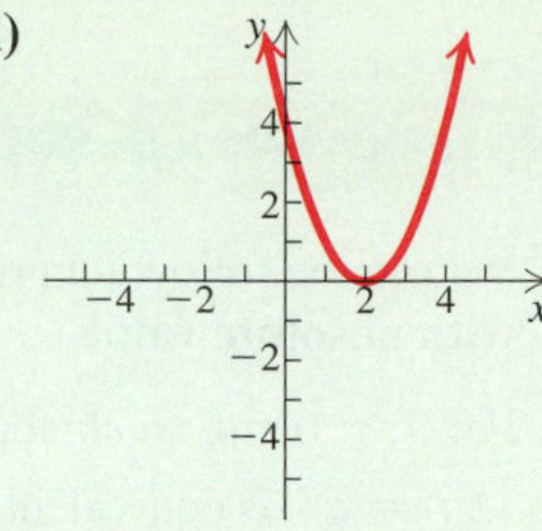

45. $y = (x - 2)^2$

46. $y = (x + 3)^2 - 4$

47. $y = -2(x + 3)^2 + 4$

48. $y = -\frac{1}{2}(x - 2)^2 + 5$

49. *Legs of a Right Triangle.* The hypotenuse of a right triangle is 50 ft. One leg is 10 ft longer than the other. What are the lengths of the legs? [3.2]

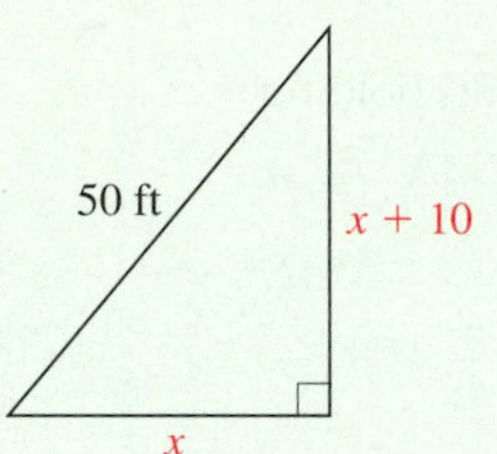

50. *Bicycling Speed.* Harry and Rebecca leave a campsite, Harry biking due north and Rebecca biking due east. Harry bikes 7 km/h slower than Rebecca. After 4 hr, they are 68 km apart. Find the speed of each bicyclist. [3.2]

51. *Sidewalk Width.* A 60-ft by 80-ft parking lot is torn up to install a sidewalk of uniform width around its perimeter. The new area of the parking lot is two-thirds of the old area. How wide is the sidewalk? [3.2]

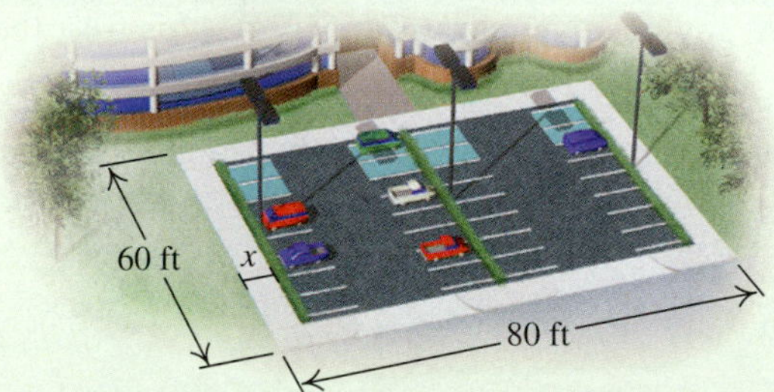

52. *Maximizing Volume.* The Garcias have 24 ft of flexible fencing with which to build a rectangular "toy corral." If the fencing is 2 ft high, what dimensions should the corral have in order to maximize its volume? [3.3]

53. *Dimensions of a Box.* An open box is made from a 10-cm by 20-cm piece of aluminum by cutting a square from each corner and folding up the edges. The area of the resulting base is 90 cm^2. What is the length of the sides of the squares? [3.2]

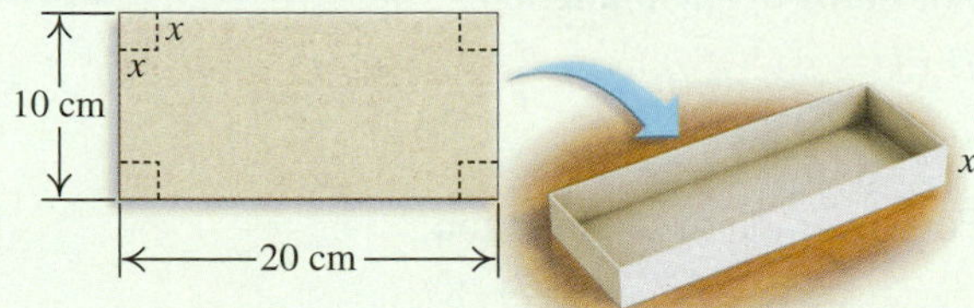

54. What are the zeros of $f(x) = 2x^2 - 5x + 1$? [3.2]

A. $\dfrac{5 \pm \sqrt{17}}{2}$ **B.** $\dfrac{5 \pm \sqrt{17}}{4}$

C. $\dfrac{5 \pm \sqrt{33}}{4}$ **D.** $\dfrac{-5 \pm \sqrt{17}}{4}$

55. Solve: $\sqrt{4x + 1} + \sqrt{2x} = 1$. [3.4]

A. There are two solutions.
B. There is only one solution. It is less than 1.
C. There is only one solution. It is greater than 1.
D. There is no solution.

56. The graph of $f(x) = (x - 2)^2 - 3$ is which of the following? [3.3]

A.

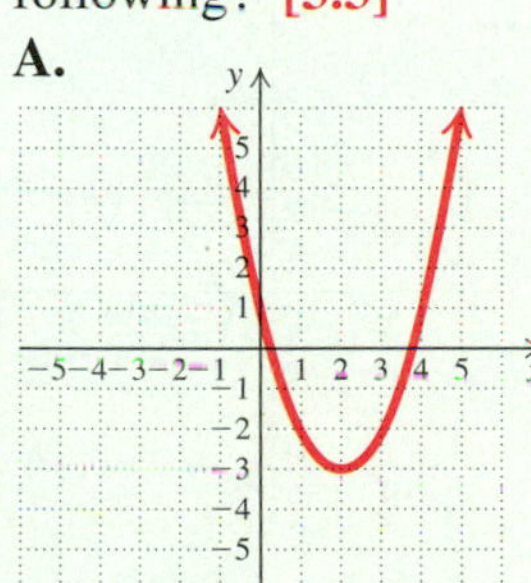

B.

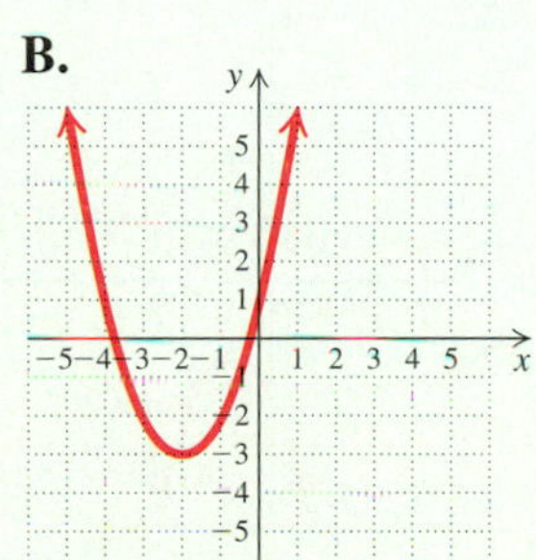

C.

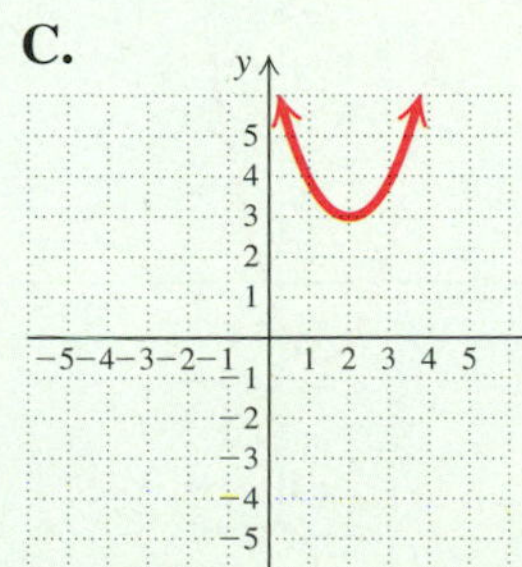

D.

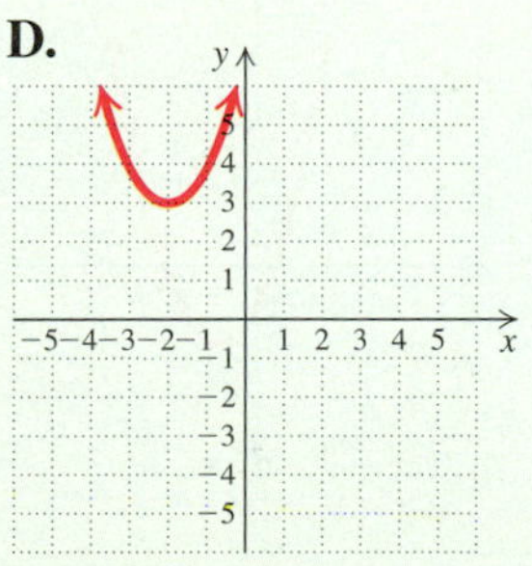

Synthesis

Solve.

57. $\sqrt{\sqrt{\sqrt{x}}} = 2$ [3.4]

58. $(t - 4)^{4/5} = 3$ [3.4]

59. $(x - 1)^{2/3} = 4$ [3.4]

60. $(2y - 2)^2 + y - 1 = 5$ [3.2]

61. $\sqrt{x + 2} + \sqrt[4]{x + 2} - 2 = 0$ [3.2]

62. At the beginning of the year, \$3500 was deposited in a savings account. One year later, \$4000 was deposited in another account. The interest rate was the same for both accounts. At the end of the second year, there was a total of \$8518.35 in the accounts. What was the annual interest rate? [3.2]

63. Find b such that $f(x) = -3x^2 + bx - 1$ has a maximum value of 2. [3.3]

Collaborative Discussion and Writing

64. Is the product of two imaginary numbers always an imaginary number? Explain your answer. [3.1]

65. Is it possible for a quadratic function to have one real zero and one imaginary zero? Why or why not? [3.2]

66. If the graphs of

$$f(x) = a_1(x - h_1)^2 + k_1$$

and

$$g(x) = a_2(x - h_2)^2 + k_2$$

have the same shape, what, if anything, can you conclude about the a's, the h's, and the k's? Explain your answer. [3.3]

67. Explain why it is necessary to check the possible solutions of a rational equation. [3.4]

68. Explain why it is necessary to check the possible solutions when the principle of powers is used to solve an equation. [3.4]

69. Explain why $|x| < p$ has no solution for $p \le 0$. [3.5]

70. Explain why all real numbers are solutions of $|x| > p$, for $p < 0$. [3.5]

3 Chapter Test

Solve. Find exact solutions.

1. $(2x - 1)(x + 5) = 0$

2. $6x^2 - 36 = 0$

3. $x^2 + 4 = 0$

4. $x^2 - 2x - 3 = 0$

5. $x^2 - 5x + 3 = 0$

6. $2t^2 - 3t + 4 = 0$

7. $x + 5\sqrt{x} - 36 = 0$

8. $\dfrac{3}{3x + 4} + \dfrac{2}{x - 1} = 2$

9. $\sqrt{x + 4} - 2 = 1$

10. $\sqrt{x + 4} - \sqrt{x - 4} = 2$

11. $|x + 4| = 7$

12. $|4y - 3| = 5$

Solve and write interval notation for the solution set. Then graph the solution set.

13. $|x + 3| \le 4$

14. $|2x - 1| < 5$

15. $|x + 5| > 2$

16. $|3x - 5| \ge 7$

17. Solve $\dfrac{1}{A} + \dfrac{1}{B} = \dfrac{1}{C}$ for B.

18. Solve $R = \sqrt{3np}$ for n.

19. Solve $x^2 + 4x = 1$ by completing the square. Find the exact solutions. Show your work.

20. The tallest structure in the United States, at 2063 ft, is the KTHI-TV tower in Blanchard, North Dakota (*Source*: *The Cambridge Fact Finder*). How long would it take an object falling freely from the top to reach the ground? (Use the formula $s = 16t^2$, where s is the distance, in feet, that an object falls freely from rest in t seconds.)

Express in terms of i.

21. $\sqrt{-43}$

22. $-\sqrt{-25}$

Simplify.

23. $(5 - 2i) - (2 + 3i)$

24. $(3 + 4i)(2 - i)$

25. $\dfrac{1 - i}{6 + 2i}$

26. i^{33}

Find the zeros of each function.

27. $f(x) = 4x^2 - 11x - 3$

28. $f(x) = 2x^2 - x - 7$

29. For the graph of the function

$$f(x) = -x^2 + 2x + 8:$$

a) Find the vertex.
b) Find the axis of symmetry.
c) State whether there is a maximum or a minimum value and find that value.
d) Find the range.
e) Graph the function.

30. *Maximizing Area.* A homeowner wants to fence a rectangular play yard using 80 ft of fencing. The side of the house will be used as one side of the rectangle. Find the dimensions for which the area is a maximum.

31. The graph of $f(x) = (x - 1)^2 - 2$ is which of the following?

A.

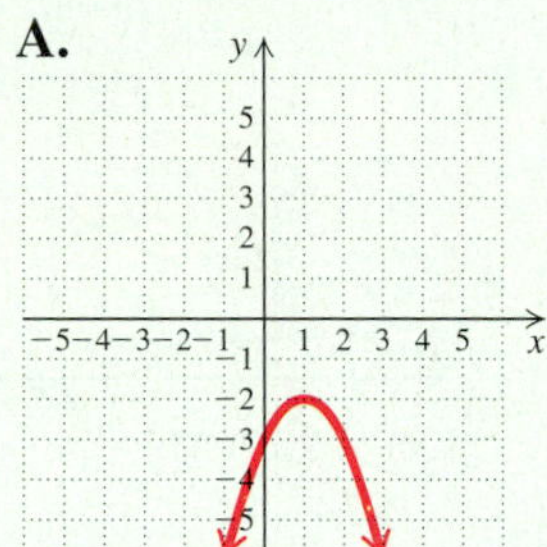

B.

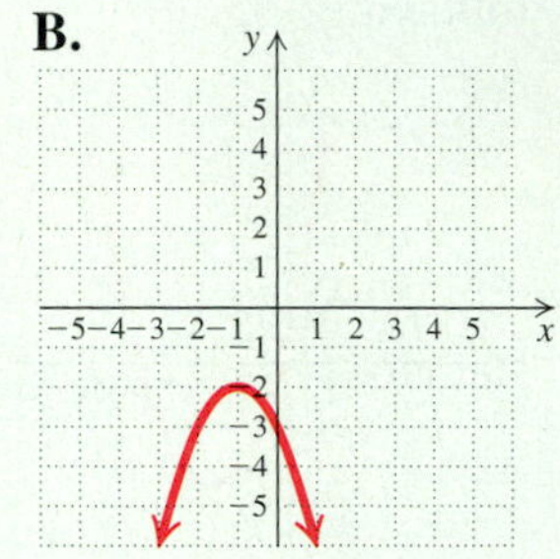

C.

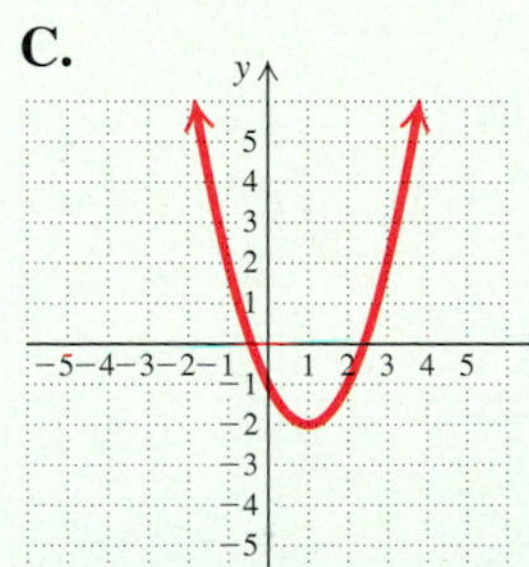

D.

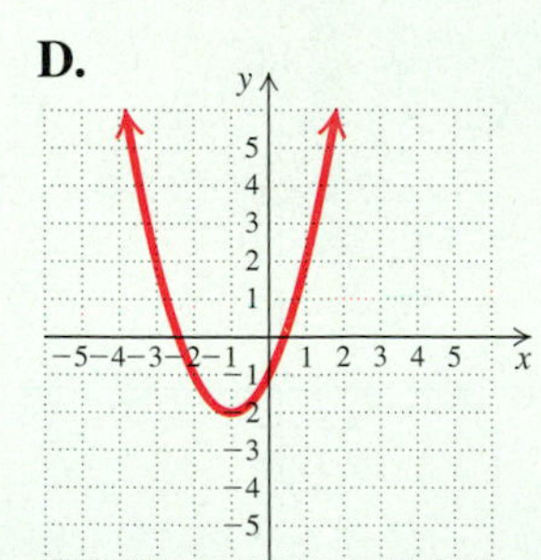

Synthesis

32. Find a such that $f(x) = ax^2 - 4x + 3$ has a maximum value of 12.

CHAPTER

4

Polynomial Functions and Rational Functions

APPLICATION

This problem appears as Example 9 in Section 4.1.

U.S. Army personnel on active duty make up approximately 37% of the total number in the U.S. military. Midyear numbers, in thousands, of U.S. Army personnel for selected years are listed in the table on p. 228.

Model the data with a quadratic function, a cubic function, and a quartic function. Then using R^2, the coefficient of determination, decide which function is the best fit.

4.1 Polynomial Functions and Modeling

- Determine the behavior of the graph of a polynomial function using the leading-term test.
- Factor polynomial functions and find their zeros and their multiplicities.
- Use a graphing calculator to graph a polynomial function and find its real-number zeros, its relative maximum and minimum values, and its domain and range.
- Solve applied problems using polynomial models; fit quadratic, cubic, and quartic polynomial functions to data.

There are many different kinds of functions. The constant, linear, and quadratic functions that we studied in Chapters 1 and 3 are part of a larger group of functions called *polynomial functions.*

POLYNOMIAL FUNCTION

A **polynomial function** P is given by

$$P(x) = a_n x^n + a_{n-1}x^{n-1} + a_{n-2}x^{n-2} + \cdots + a_1 x + a_0,$$

where the coefficients $a_n, a_{n-1}, \ldots, a_1, a_0$ are real numbers and the exponents are whole numbers.

The first nonzero coefficient, a_n, is called the **leading coefficient**. The term $a_n x^n$ is called the **leading term**. The **degree** of the polynomial function is n. Some examples of polynomial functions follow.

POLYNOMIAL FUNCTION	EXAMPLE	DEGREE	LEADING TERM	LEADING COEFFICIENT
Constant	$f(x) = 3 \quad (f(x) = 3 = 3x^0)$	0	3	3
Linear	$f(x) = \frac{2}{3}x + 5 \quad \left(f(x) = \frac{2}{3}x + 5 = \frac{2}{3}x^1 + 5\right)$	1	$\frac{2}{3}x$	$\frac{2}{3}$
Quadratic	$f(x) = 4x^2 - x + 3$	2	$4x^2$	4
Cubic	$f(x) = x^3 + 2x^2 + x - 5$	3	x^3	1
Quartic	$f(x) = -x^4 - 1.1x^3 + 0.3x^2 - 2.8x - 1.7$	4	$-x^4$	-1

The function $f(x) = 0$ can be described in many ways:

$$f(x) = 0 = 0x^2 = 0x^{15} = 0x^{48},$$

and so on. For this reason, we say that the constant function $f(x) = 0$ has no degree.

Functions such as

$$f(x) = \frac{2}{x} + 5, \text{ or } 2x^{-1} + 5, \quad \text{and} \quad g(x) = \sqrt{x} - 6, \text{ or } x^{1/2} - 6,$$

are *not* polynomial functions because the exponents -1 and $\frac{1}{2}$ are *not* whole numbers.

From our study of functions in Chapters 1–3, we know how to find or at least estimate many characteristics of a polynomial function. Let's consider two examples for review.

Quadratic Function

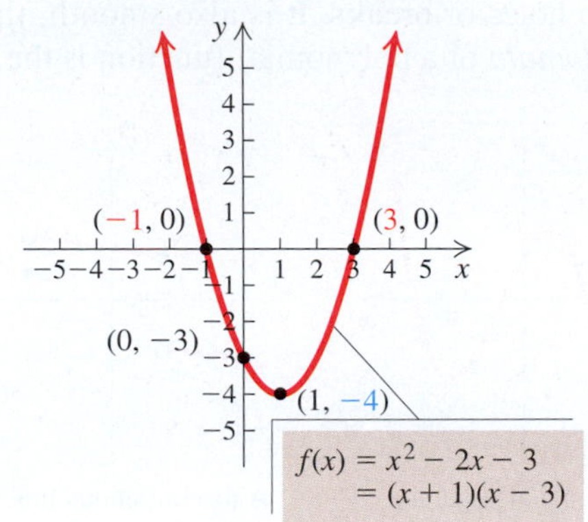

Function: $f(x) = x^2 - 2x - 3$
$= (x + 1)(x - 3)$

Zeros: $-1, 3$

x-intercepts: $(-1, 0), (3, 0)$

y-intercept: $(0, -3)$

Minimum: -4 at $x = 1$

Maximum: None

Domain: All real numbers, $(-\infty, \infty)$

Range: $[-4, \infty)$

Cubic Function

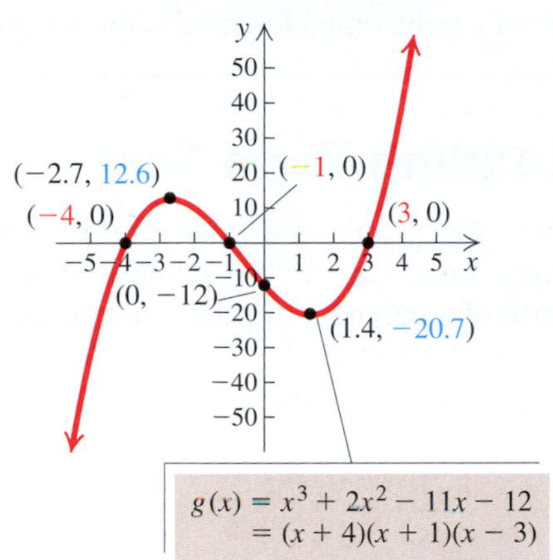

Function: $g(x) = x^3 + 2x^2 - 11x - 12$
$= (x + 4)(x + 1)(x - 3)$

Zeros: $-4, -1, 3$

x-intercepts: $(-4, 0), (-1, 0), (3, 0)$

y-intercept: $(0, -12)$

Relative minimum: -20.7 at $x = 1.4$

Relative maximum: 12.6 at $x = -2.7$

Domain: All real numbers, $(-\infty, \infty)$

Range: All real numbers, $(-\infty, \infty)$

All graphs of polynomial functions have some characteristics in common. Compare the following graphs. How do the graphs of polynomial functions differ from the graphs of nonpolynomial functions? Describe some characteristics of the graphs of polynomial functions that you observe.

Polynomial Functions

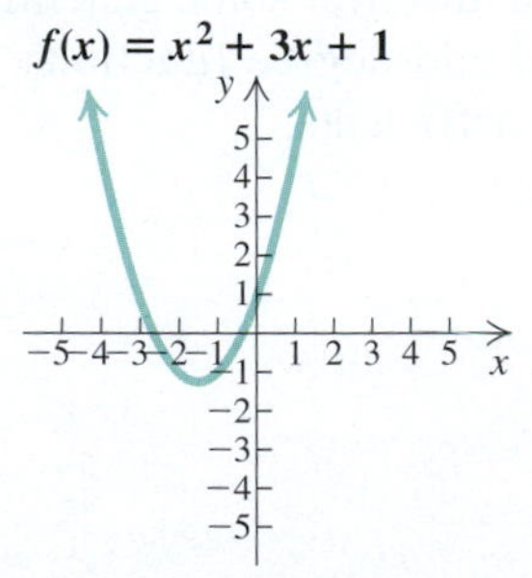

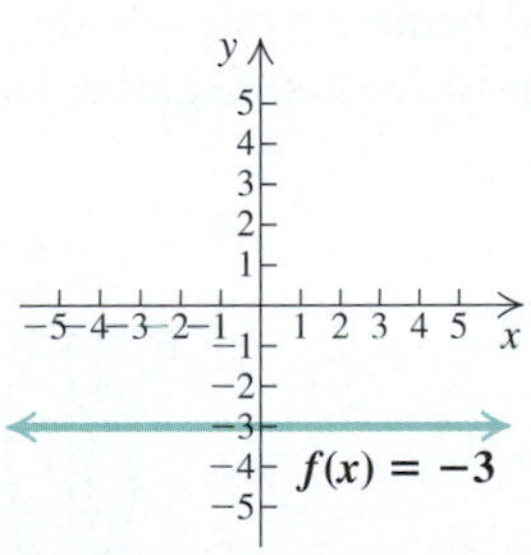

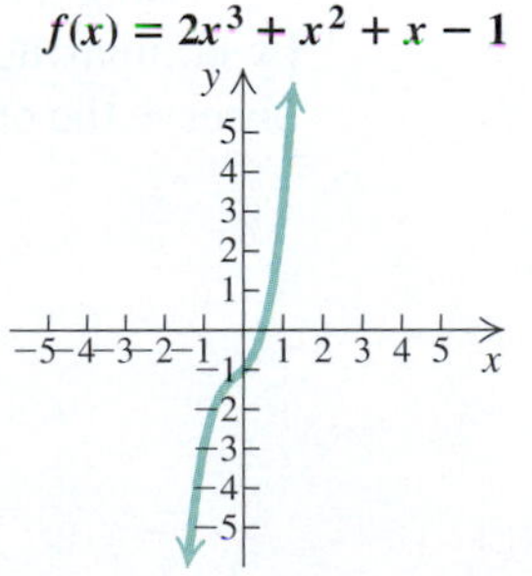

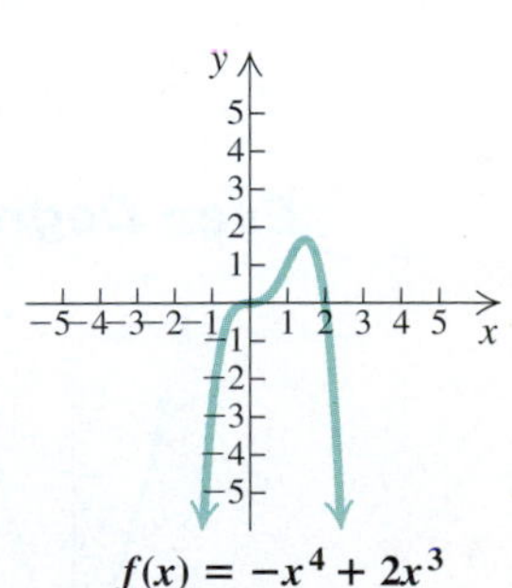

Nonpolynomial Functions

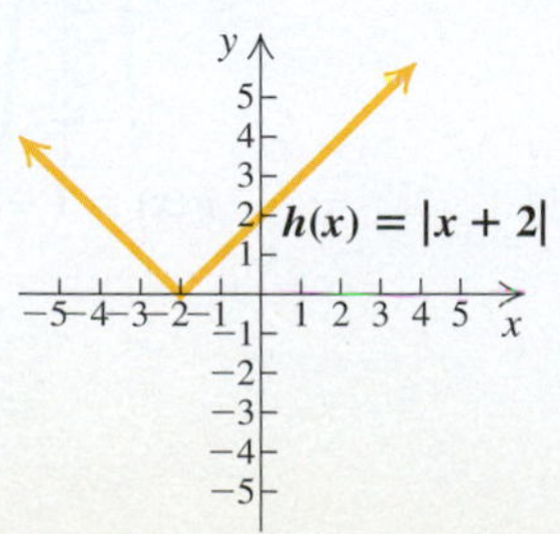

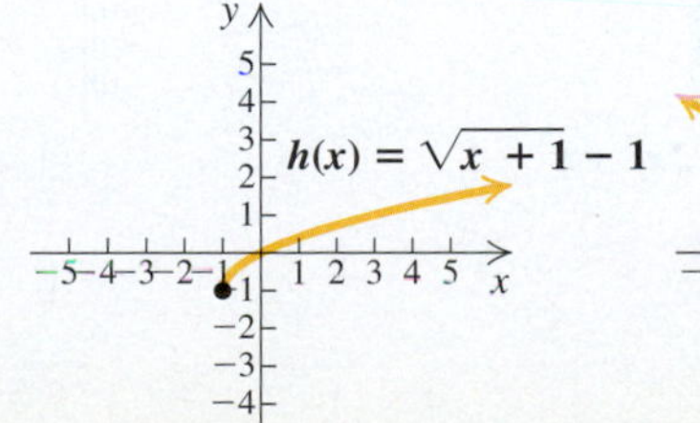

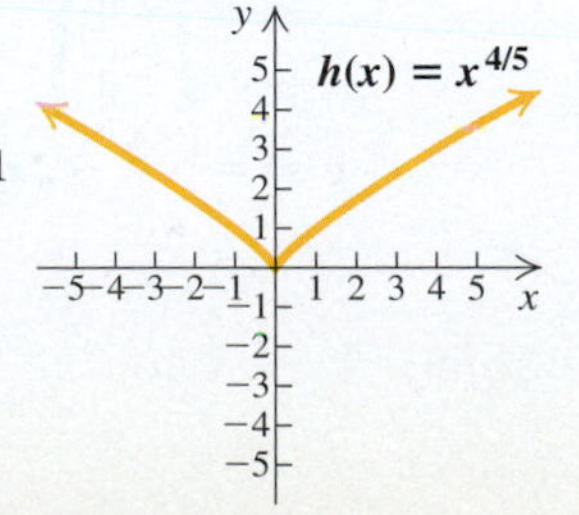

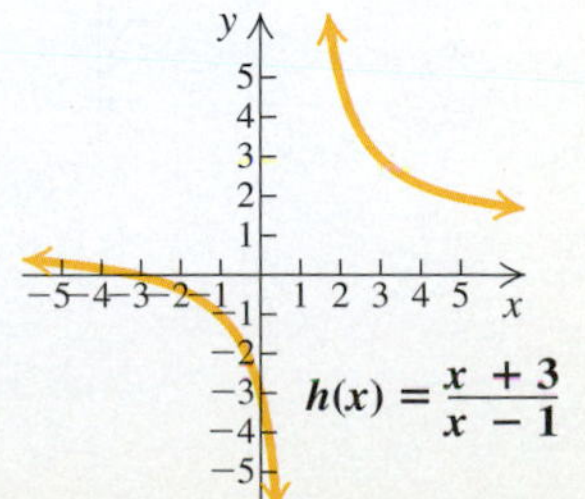

DOMAIN OF A FUNCTION

REVIEW SECTION 1.2.

You probably noted that the graph of a polynomial function is *continuous*; that is, it has no holes or breaks. It is also smooth; there are no sharp corners. Furthermore, the *domain* of a polynomial function is the set of all real numbers, $(-\infty, \infty)$.

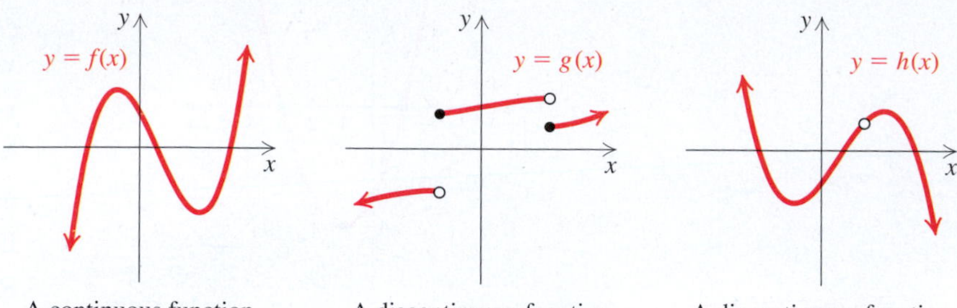

The *domain* of a polynomial function is the set of all real numbers, $(-\infty, \infty)$.

The Leading-Term Test

The behavior of the graph of a polynomial function as x becomes very large $(x \to \infty)$ or very small $(x \to -\infty)$ is referred to as the end behavior of the graph. The leading term of a polynomial function determines its end behavior.

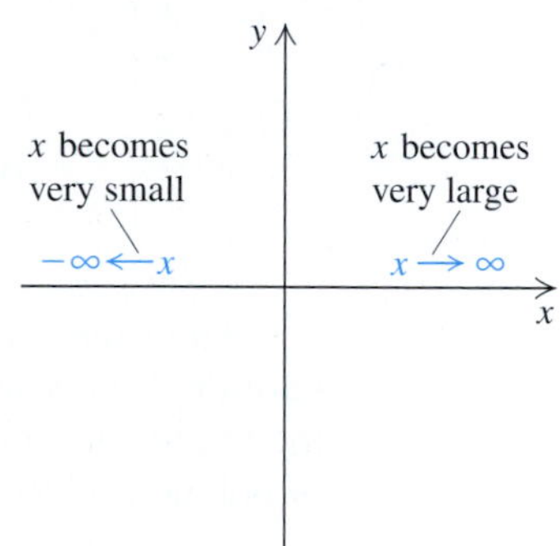

Using the graphs shown below, let's see if we can discover some general patterns by comparing the end behavior of even-degree and odd-degree functions. We also observe the effect of positive and negative leading coefficients.

Even Degree

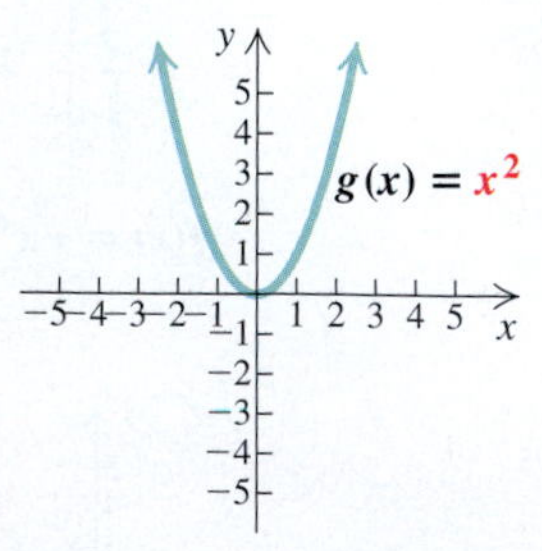

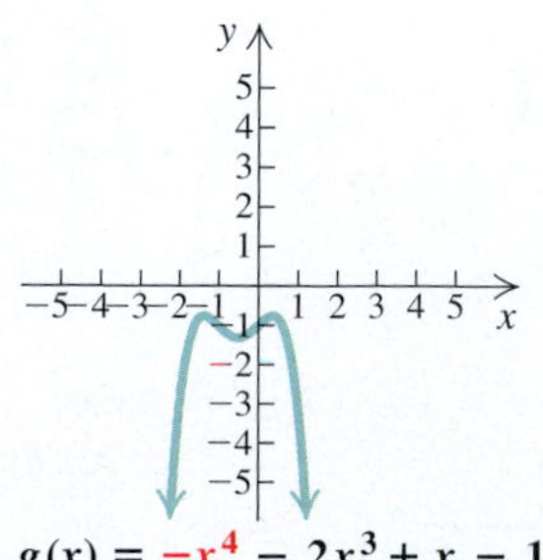

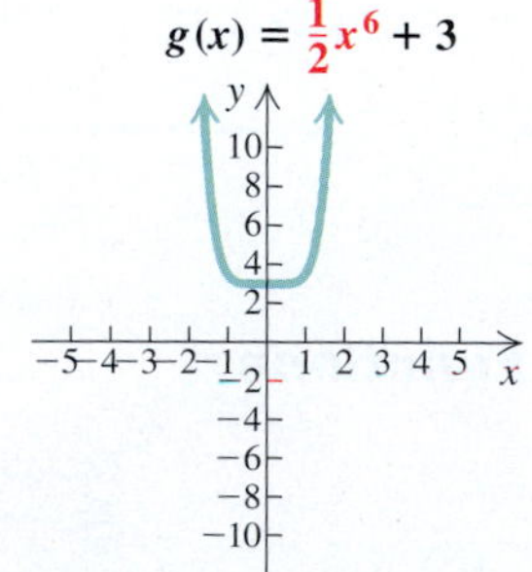

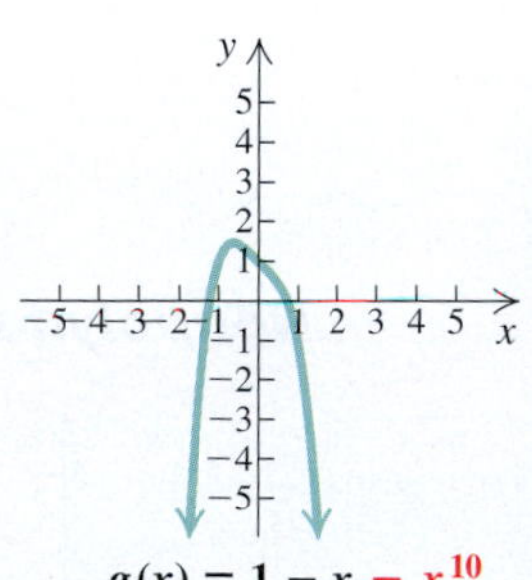

Odd Degree

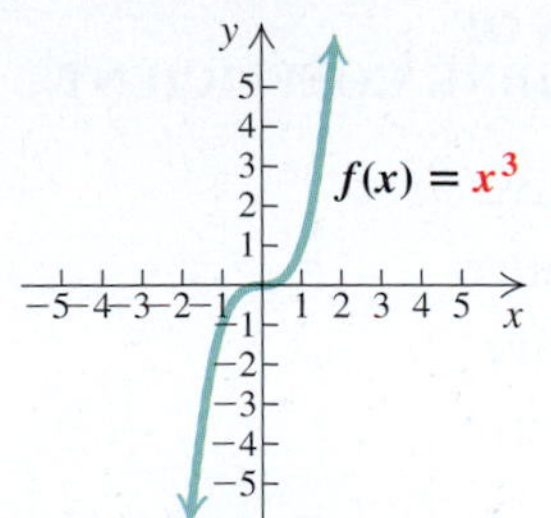

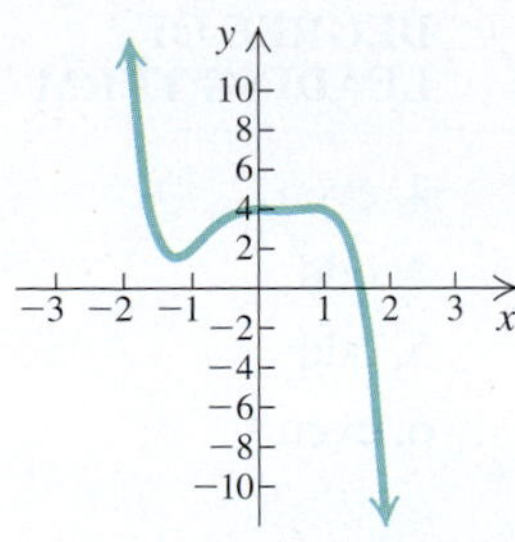

$f(x) = -x^5 + 2x^3 - x^2 + 4$

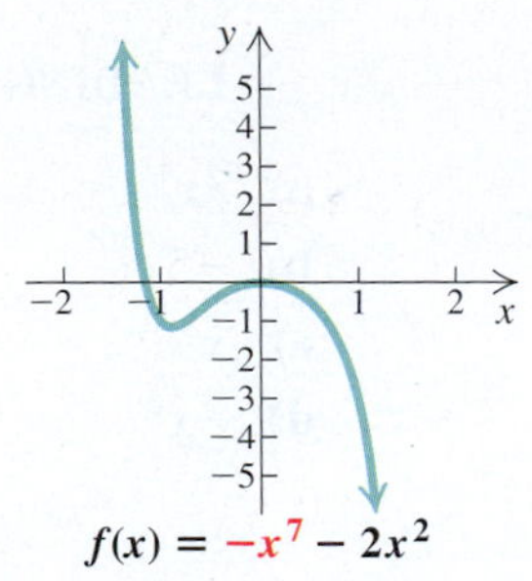

$f(x) = -x^7 - 2x^2$

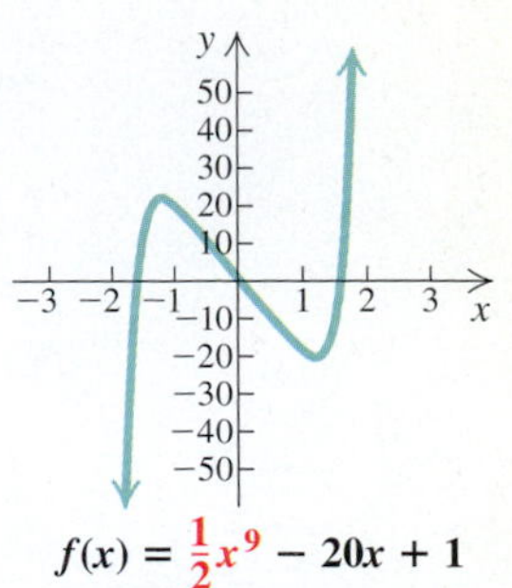

$f(x) = \frac{1}{2}x^9 - 20x + 1$

We can summarize our observations as follows.

THE LEADING-TERM TEST

If a_nx^n is the leading term of a polynomial function, then the behavior of the graph as $x \to \infty$ or as $x \to -\infty$ can be described in one of the four following ways.

n	$a_n > 0$	$a_n < 0$
Even		
Odd		

The portion of the graph is not determined by this test.

EXAMPLE 1 Using the leading-term test, match each of the following functions with one of the graphs A–D that follow.

a) $f(x) = 3x^4 - 2x^3 + 3$

b) $f(x) = -5x^3 - x^2 + 4x + 2$

c) $f(x) = x^5 + \frac{1}{4}x + 1$

d) $f(x) = -x^6 + x^5 - 4x^3$

A.

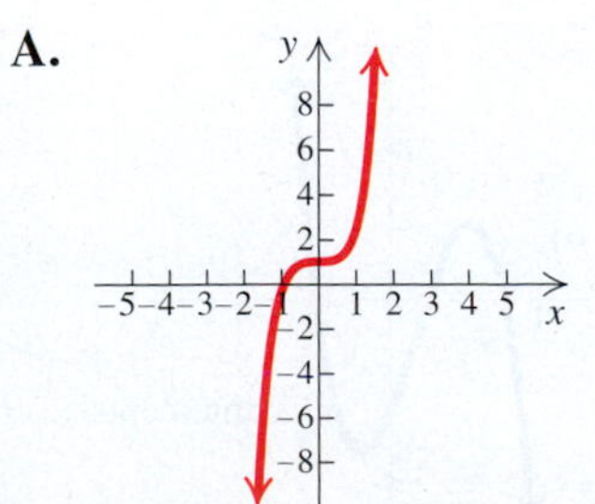

B.

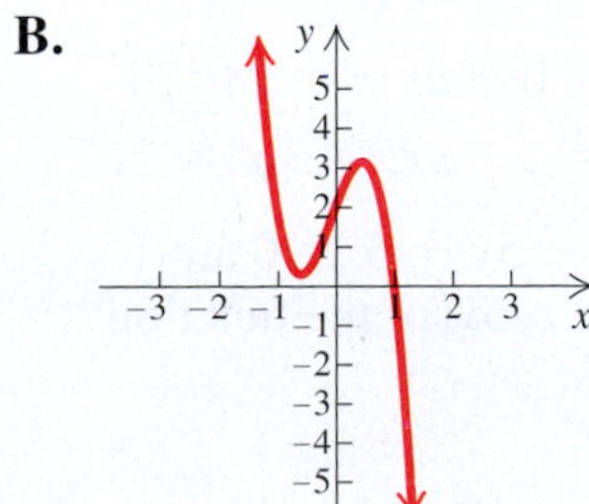

C.

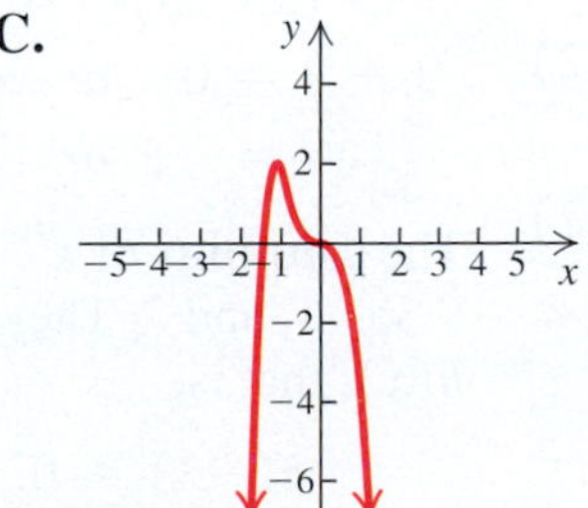

D.

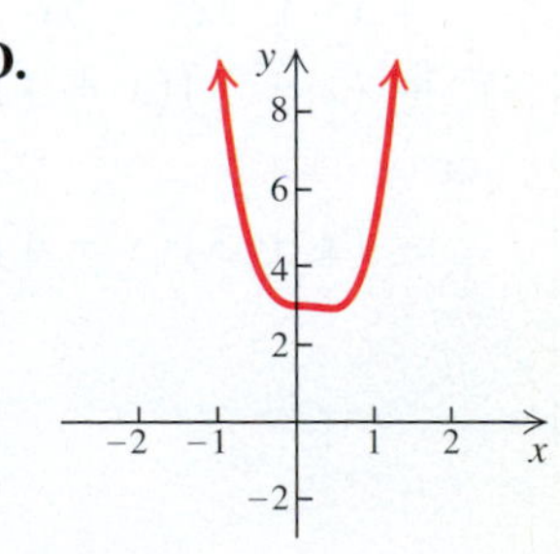

Solution

	LEADING TERM	DEGREE OF LEADING TERM	SIGN OF LEADING COEFFICIENT	GRAPH
a)	$3x^4$	4, even	Positive	D
b)	$-5x^3$	3, odd	Negative	B
c)	x^5	5, odd	Positive	A
d)	$-x^6$	6, even	Negative	C

Now Try Exercise 19.

Finding Zeros of Factored Polynomial Functions

Let's review the meaning of the real zeros of a function and their connection to the x-intercepts of the function's graph.

CONNECTING THE CONCEPTS — Zeros, Solutions, and Intercepts

FUNCTION	ZEROS OF THE FUNCTION; SOLUTIONS OF THE EQUATION	ZEROS OF THE FUNCTION; x-INTERCEPTS OF THE GRAPH
Quadratic Polynomial $g(x) = x^2 - 2x - 8$ $= (x + 2)(x - 4)$, or $y = (x + 2)(x - 4)$	To find the **zeros** of $g(x)$, we solve $g(x) = 0$: $x^2 - 2x - 8 = 0$; $(x + 2)(x - 4) = 0$; $x + 2 = 0$ *or* $x - 4 = 0$; $x = -2$ *or* $x = 4$. The **solutions** of $x^2 - 2x - 8 = 0$ are -2 and 4. They are the zeros of the function $g(x)$; that is, $g(-2) = 0$ and $g(4) = 0$.	The real-number zeros of $g(x)$ are the x-coordinates of the **x-intercepts** of the graph of $y = g(x)$. 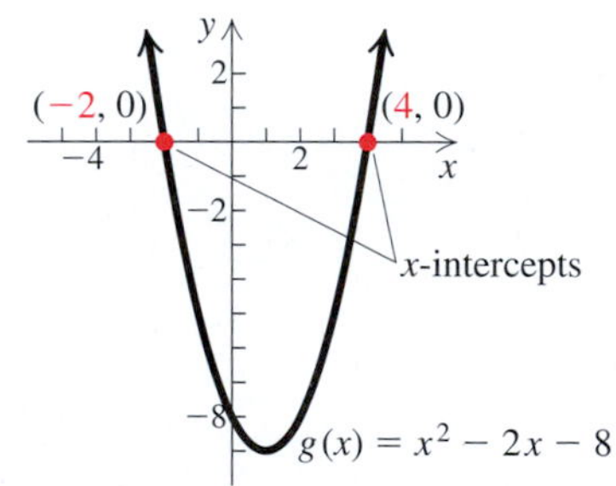
Cubic Polynomial $h(x)$ $= x^3 + 2x^2 - 5x - 6$ $= (x + 3)(x + 1)(x - 2)$, or $y = (x + 3)(x + 1)(x - 2)$	To find the **zeros** of $h(x)$, we solve $h(x) = 0$: $x^3 + 2x^2 - 5x - 6 = 0$; $(x + 3)(x + 1)(x - 2) = 0$; $x + 3 = 0$ *or* $x + 1 = 0$ *or* $x - 2 = 0$; $x = -3$ *or* $x = -1$ *or* $x = 2$. The **solutions** of $x^3 + 2x^2 - 5x - 6 = 0$ are -3, -1, and 2. They are the zeros of the function $h(x)$; that is, $h(-3) = 0$, $h(-1) = 0$, and $h(2) = 0$.	The real-number zeros of $h(x)$ are the x-coordinates of the **x-intercepts** of the graph of $y = h(x)$. 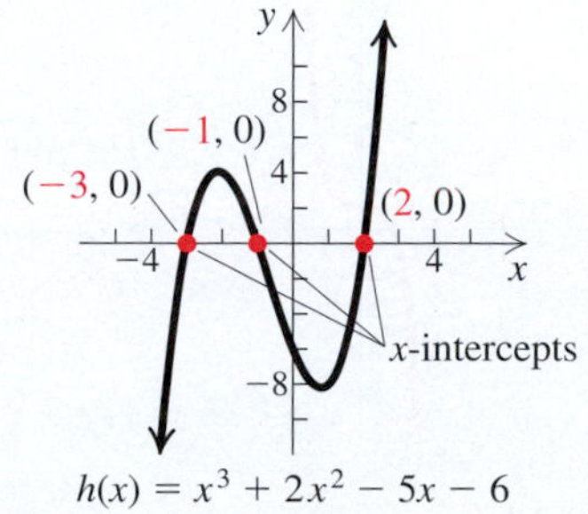

The connection between the real-number zeros of a function and the x-intercepts of the graph of the function is easily seen in the preceding examples. If c is a real zero of a function (that is, if $f(c) = 0$), then $(c, 0)$ is an x-intercept of the graph of the function.

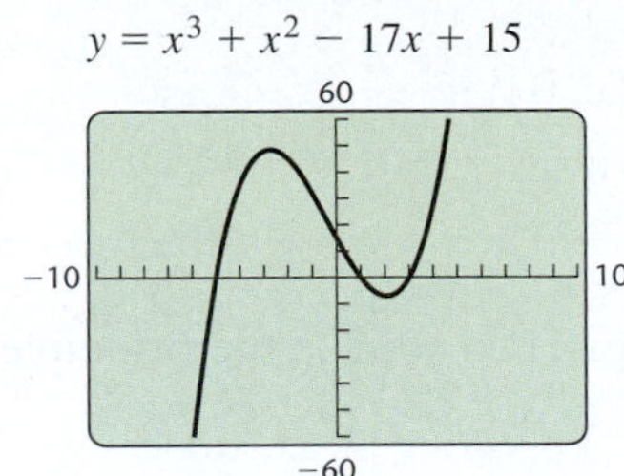

EXAMPLE 2 Consider $P(x) = x^3 + x^2 - 17x + 15$. Determine whether each of the numbers 2 and -5 is a zero of $P(x)$.

Solution We first evaluate $P(2)$:

$$P(2) = (2)^3 + (2)^2 - 17(2) + 15 = -7. \quad \text{Substituting 2 into the polynomial}$$

Since $P(2) \neq 0$, we know that 2 is *not* a zero of the polynomial function.

We then evaluate $P(-5)$:

$$P(-5) = (-5)^3 + (-5)^2 - 17(-5) + 15 = 0. \quad \text{Substituting } -5 \text{ into the polynomial}$$

Since $P(-5) = 0$, we know that -5 is a zero of $P(x)$.

Now Try Exercise 23.

Let's take a closer look at the polynomial function

$$h(x) = x^3 + 2x^2 - 5x - 6$$

Just in Time 19

(see Connecting the Concepts on p. 224). The factors of $h(x)$ are

$$x + 3, \qquad x + 1, \quad \text{and} \quad x - 2,$$

and the zeros are

$$-3, \qquad -1, \quad \text{and} \quad 2.$$

We note that when the polynomial is expressed as a product of linear factors, each factor determines a zero of the function. Thus if we know the linear factors of a polynomial function $f(x)$, we can easily find the zeros of $f(x)$ by solving the equation $f(x) = 0$ using the principle of zero products.

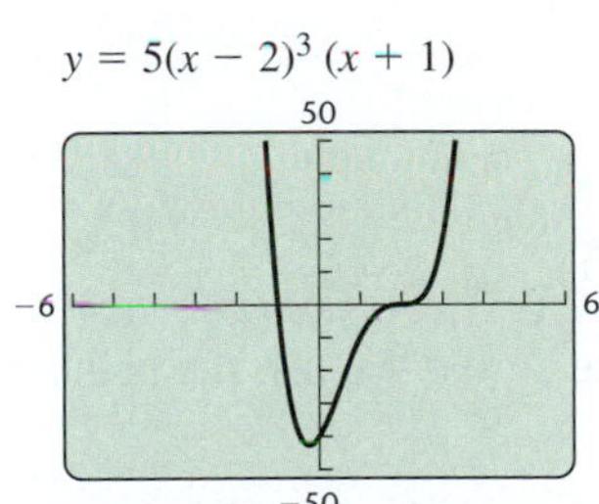

FIGURE 1

EXAMPLE 3 Find the zeros of

$$f(x) = 5(x - 2)(x - 2)(x - 2)(x + 1) = 5(x - 2)^3(x + 1).$$

Solution To solve the equation $f(x) = 0$, we use the principle of zero products, solving $x - 2 = 0$ and $x + 1 = 0$. The zeros of $f(x)$ are 2 and -1. (See Fig. 1.)

$y = -(x - 1)^2 (x + 2)^2$

12

−5 5

−16

FIGURE 2

EXAMPLE 4 Find the zeros of

$$g(x) = -(x - 1)(x - 1)(x + 2)(x + 2) = -(x - 1)^2(x + 2)^2.$$

Solution To solve the equation $g(x) = 0$, we use the principle of zero products, solving $x - 1 = 0$ and $x + 2 = 0$. The zeros of $g(x)$ are 1 and -2. (See Fig. 2.)

Let's consider the occurrences of the zeros in the functions in Examples 3 and 4 and their relationship to the graphs of those functions. In Example 3, the factor $x - 2$ occurs three times. In a case like this, we say that the zero we obtain from this factor, 2, has a **multiplicity** of 3. The factor $x + 1$ occurs one time. The zero we obtain from this factor, -1, has a *multiplicity* of 1.

In Example 4, the factors $x - 1$ and $x + 2$ each occur two times. Thus both zeros, 1 and -2, have a *multiplicity* of 2.

Note, in Example 3, that the zeros have odd multiplicities and the graph crosses the x-axis at both -1 and 2. But in Example 4, the zeros have even multiplicities and the graph is tangent to (touches but does not cross) the x-axis at -2 and 1. This leads us to the following generalization.

> **EVEN MULTIPLICITY AND ODD MULTIPLICITY**
>
> If $(x - c)^k$, $k \geq 1$, is a factor of a polynomial function $P(x)$ and $(x - c)^{k+1}$ is not a factor and:
>
> - k is odd, then the graph crosses the x-axis at $(c, 0)$;
> - k is even, then the graph is tangent to the x-axis at $(c, 0)$.

Just in Time 14

Some polynomials can be factored by grouping. Then we use the principle of zero products to find their zeros.

EXAMPLE 5 Find the zeros of

$$f(x) = x^3 - 2x^2 - 9x + 18.$$

Solution We factor by grouping, as follows:

$y = x^3 - 2x^2 - 9x + 18$

35, −5, 5, −25

$$\begin{aligned} f(x) &= x^3 - 2x^2 - 9x + 18 \\ &= x^2(x - 2) - 9(x - 2) && \text{Grouping } x^3 \text{ with } -2x^2 \text{ and } -9x \text{ with 18 and factoring each group} \\ &= (x - 2)(x^2 - 9) && \text{Factoring out } x - 2 \\ &= (x - 2)(x + 3)(x - 3). && \text{Factoring } x^2 - 9 \end{aligned}$$

Then, by the principle of zero products, the solutions of the equation $f(x) = 0$ are 2, −3, and 3. These are the zeros of $f(x)$.

Now Try Exercise 39.

Other factoring techniques can also be used.

EXAMPLE 6 Find the zeros of

$$f(x) = x^4 + 4x^2 - 45.$$

Solution We factor as follows:

$$f(x) = x^4 + 4x^2 - 45 = (x^2 - 5)(x^2 + 9).$$

We now solve the equation $f(x) = 0$ to determine the zeros. We use the principle of zero products:

$$\begin{aligned} (x^2 - 5)(x^2 + 9) &= 0 \\ x^2 - 5 = 0 \quad &or \quad x^2 + 9 = 0 \\ x^2 = 5 \quad &or \quad x^2 = -9 \\ x = \pm\sqrt{5} \quad &or \quad x = \pm\sqrt{-9} = \pm 3i. \end{aligned}$$

The solutions are $\pm\sqrt{5}$ and $\pm 3i$. These are the zeros of $f(x)$.

Now Try Exercise 37.

$y = x^4 + 4x^2 - 45$

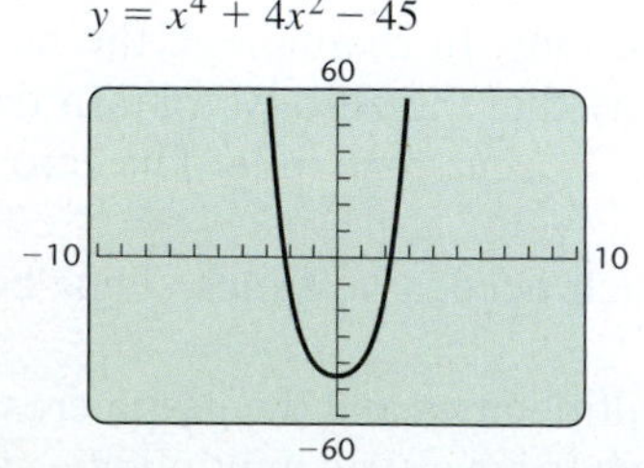

Only the real-number zeros of a function correspond to the x-intercepts of its graph. For instance, the real-number zeros of the function in Example 6, $-\sqrt{5}$ and $\sqrt{5}$, can be seen on the graph of the function at left, but the nonreal zeros, $-3i$ and $3i$, cannot.

> Every polynomial function of degree n, with $n \geq 1$, has at least one zero and at most n zeros.

This is often stated as follows: "Every polynomial function of degree n, with $n \geq 1$, has *exactly* n zeros." This statement is compatible with the preceding statement, if one takes multiplicities into account.

Finding Real Zeros on a Calculator

Finding exact values of the real zeros of a function can be difficult. We can find approximations using a graphing calculator.

EXAMPLE 7 Find the real zeros of the function f given by

$$f(x) = 0.1x^3 - 0.6x^2 - 0.1x + 2.$$

Approximate the zeros to three decimal places.

Solution We use a graphing calculator, trying to create a graph that clearly shows the curvature and the intercepts. Then we look for points where the graph crosses the x-axis. We know that there are no more than 3 because the degree of the polynomial is 3. It appears that there are three different zeros, one near -2, one near 2, and one near 6. We use the ZERO feature to find them.

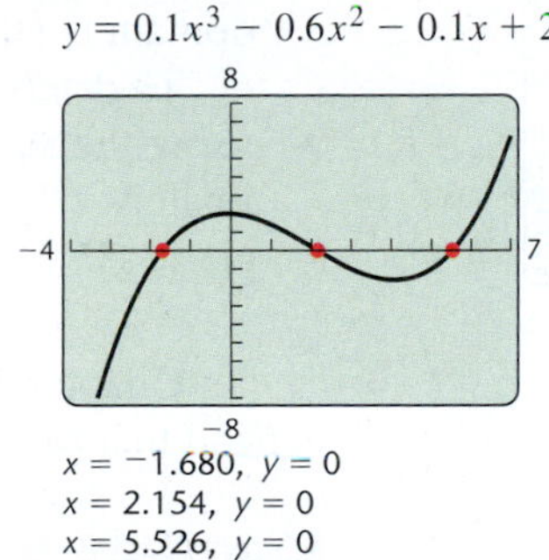

The zeros are approximately -1.680, 2.154, and 5.526. Now Try Exercise 43.

Polynomial Models

Polynomial functions have many uses as models in science, engineering, and business. The simplest use of polynomial functions in applied problems occurs when we merely evaluate a polynomial function. In such cases, a model has already been developed.

EXAMPLE 8 ***Ibuprofen in the Bloodstream.*** The polynomial function

$$M(t) = 0.5t^4 + 3.45t^3 - 96.65t^2 + 347.7t$$

can be used to estimate the number of milligrams of the pain relief medication ibuprofen in the bloodstream t hours after 400 mg of the medication has been taken.

a) Find the number of milligrams in the bloodstream at $t = 0, 0.5, 1, 1.5$, and so on, up to 6 hr. Round the function values to the nearest tenth.

b) Find the domain, the relative maximum and where it occurs, and the range.

Solution

a) We can evaluate the function with the TABLE feature of a graphing calculator set in AUTO mode. We start at 0 and use a step-value of 0.5.

X	Y_1
0	0
.5	150.15
1	255
1.5	318.26
2	344.4
2.5	338.63
3	306.9

X =

$M(0) = 0$, $\quad M(3.5) = 255.9$,
$M(0.5) = 150.2$, $\quad M(4) = 193.2$,
$M(1) = 255$, $\quad M(4.5) = 126.9$,
$M(1.5) = 318.3$, $\quad M(5) = 66$,
$M(2) = 344.4$, $\quad M(5.5) = 20.2$,
$M(2.5) = 338.6$, $\quad M(6) = 0$.
$M(3) = 306.9$,

$y = 0.5x^4 + 3.45x^3 - 96.65x^2 + 347.7x$

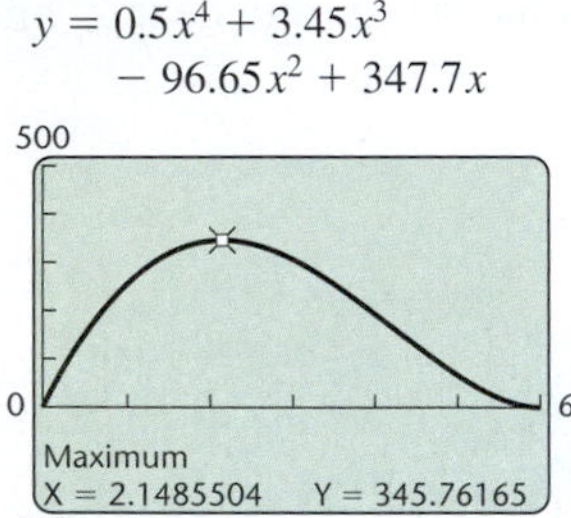

b) Recall that the domain of a polynomial function, unless restricted by a statement of the function, is $(-\infty, \infty)$. The implications of this application restrict the domain of the function. If we assume that a patient had not taken any of the medication before, it seems reasonable that $M(0) = 0$; that is, at time 0, there is 0 mg of the medication in the bloodstream. After the medication has been taken, $M(t)$ will be positive for a period of time and eventually decrease back to 0 when $t = 6$ and not increase again (unless another dose is taken). Thus the restricted domain is $[0, 6]$.

To determine the range, we find the relative maximum value of the function using the MAXIMUM feature. The maximum is about 345.8 mg. It occurs approximately 2.15 hr, or 2 hr 9 min, after the initial dose has been taken. The range is about $[0, 345.8]$.

Now Try Exercise 63.

In Chapter 1, we used regression to model data with linear functions. We now expand that procedure to include quadratic, cubic, and quartic models.

EXAMPLE 9 ***Army Personnel on Active Duty.*** U.S. Army personnel on active duty make up approximately 37% of the total number in the U.S. military. Midyear numbers, in thousands, of U.S. Army personnel for selected years are listed in the following table.

Year, x	Number of Active-Duty Army Personnel (in thousands)
2005, 0	492.7
2007, 2	522.0
2008, 3	539.2
2010, 5	566.0
2012, 7	550.1
2014, 9	512.1

Source: Department of the Army, U.S. Department of Defense

Looking at the table above, we note that the data, in thousands of active-duty Army personnel, could be modeled with a quadratic function, a cubic function, or a quartic function.

a) Model the data with a quadratic function, a cubic function, and a quartic function. Let the first coordinate of each data point be the number of years after 2005; that is, enter the data as $(0, 492.7)$, $(2, 522.0)$, $(3, 539.2)$, and so on. Then using R^2, the **coefficient of determination**, decide which function is the best fit.

b) Graph the function with the scatterplot of the data.

c) Use the answer to part (a) to estimate the number of active-duty Army personnel in 2006, in 2009, in 2013, and in 2020.

Solution

a) Using the REGRESSION feature with DIAGNOSTIC turned on, we get the following.

QuadReg
y = ax2 + bx + c
a = −2.695463547
b = 27.48096918
c = 486.7387796
R2 = .9281065984

CubicReg
y = ax3 + bx2 + cx + d
a = −.2824686269
b = 1.063162809
c = 15.42806538
d = 491.7267161
R2 = .9810056029

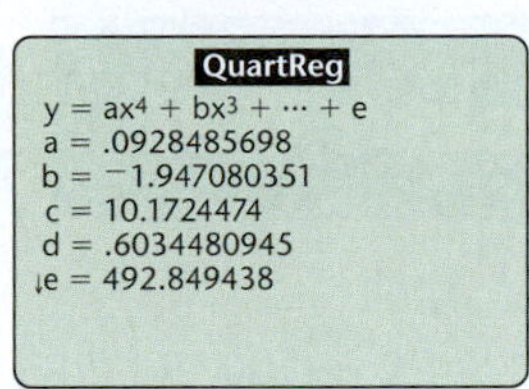

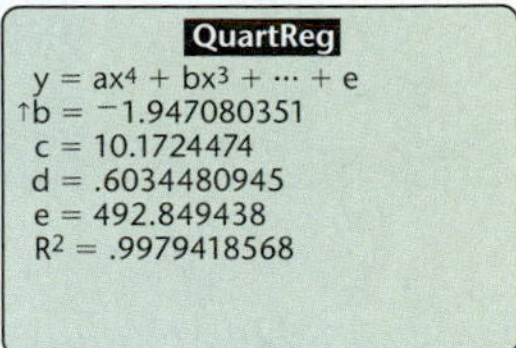

The R^2-value for the quartic function is closer to 1 than the R^2-values for the quadratic function and the cubic function. Thus the quartic function is the best fit:

$$f(x) = 0.0928485698x^4 - 1.947080351x^3 + 10.1724474x^2 + 0.6034480945x + 492.849438.$$

b) The scatterplot and the graph are shown below.

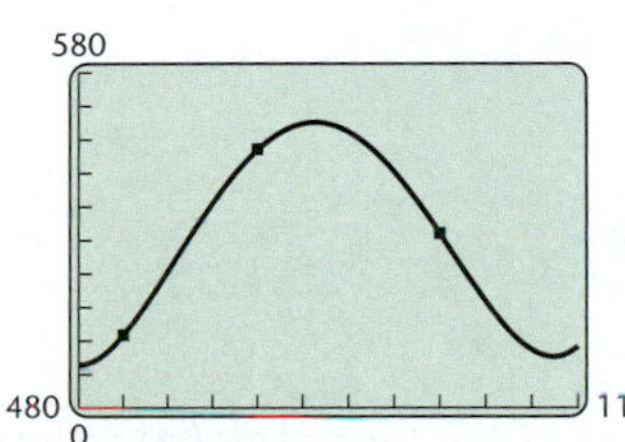

c) We evaluate the function found in part (a).

X	Y1
1	501.77
4	557.18
8	532.12
15	919.76

X =

Using this function, we can estimate the number of active-duty Army personnel as 501,800 in 2006, 557,200 in 2009, and 532,100 in 2013. Looking at the given data in the table above, we see that these estimates appear to be fairly accurate.

If we use the function to estimate the number of active-duty Army personnel in 2020, we find an answer of about 919,800. This estimate is not realistic since it is not reasonable to expect the number of active-duty Army personnel to increase by 407,700 from 2014 to 2020.

The quartic model has a higher value for R^2 than the quadratic model or the cubic model over the domain of the data, but this number does not reflect the degree of accuracy for extended values. It is always important when using regression to analyze predictions with common sense and knowledge of current trends.

Now Try Exercise 77.

4.1 Exercise Set

Determine the leading term, the leading coefficient, and the degree of the polynomial. Then classify the polynomial function as constant, linear, quadratic, cubic, or quartic.

1. $g(x) = \frac{1}{2}x^3 - 10x + 8$
2. $f(x) = 15x^2 - 10 + 0.11x^4 - 7x^3$
3. $h(x) = 0.9x - 0.13$
4. $f(x) = -6$
5. $g(x) = 305x^4 + 4021$
6. $h(x) = 2.4x^3 + 5x^2 - x + \frac{7}{8}$
7. $h(x) = -5x^2 + 7x^3 + x^4$
8. $f(x) = 2 - x^2$
9. $g(x) = 4x^3 - \frac{1}{2}x^2 + 8$
10. $f(x) = 12 + x$

In Exercises 11–18, select one of the four sketches (a)–(d) that follow to describe the end behavior of the graph of the function.

a) 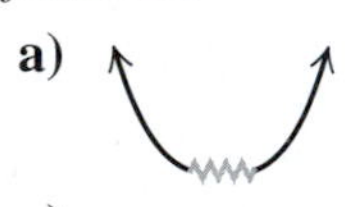

b) 

c) 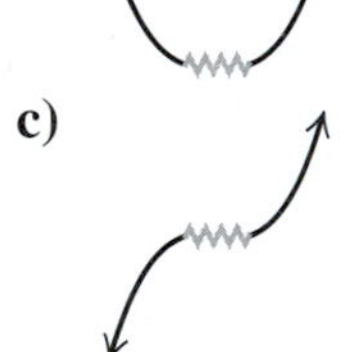

d) 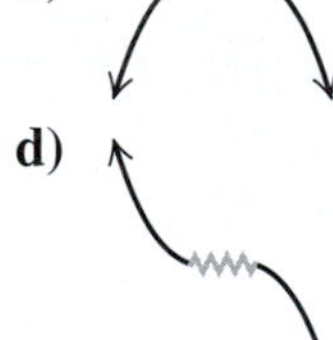

11. $f(x) = -3x^3 - x + 4$
12. $f(x) = \frac{1}{4}x^4 + \frac{1}{2}x^3 - 6x^2 + x - 5$
13. $f(x) = -x^6 + \frac{3}{4}x^4$
14. $f(x) = \frac{2}{5}x^5 - 2x^4 + x^3 - \frac{1}{2}x + 3$
15. $f(x) = -3.5x^4 + x^6 + 0.1x^7$
16. $f(x) = -x^3 + x^5 - 0.5x^6$
17. $f(x) = 10 + \frac{1}{10}x^4 - \frac{2}{5}x^3$
18. $f(x) = 2x + x^3 - x^5$

In Exercises 19–22, use the leading-term test to match the function with one of the graphs (a)–(d) that follow.

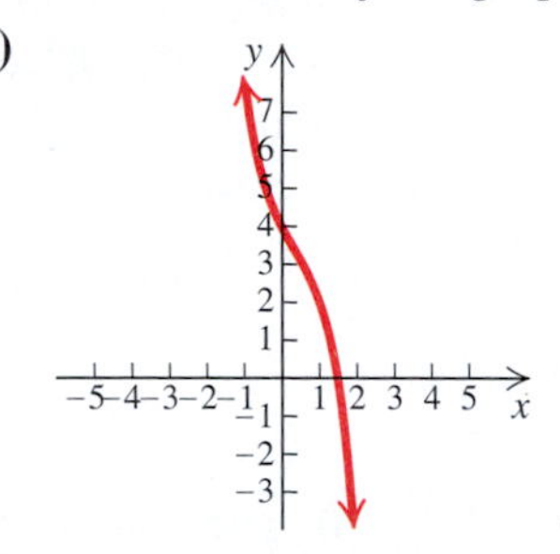

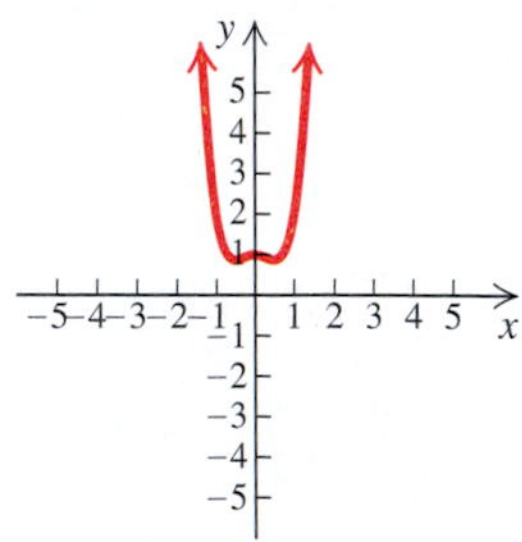

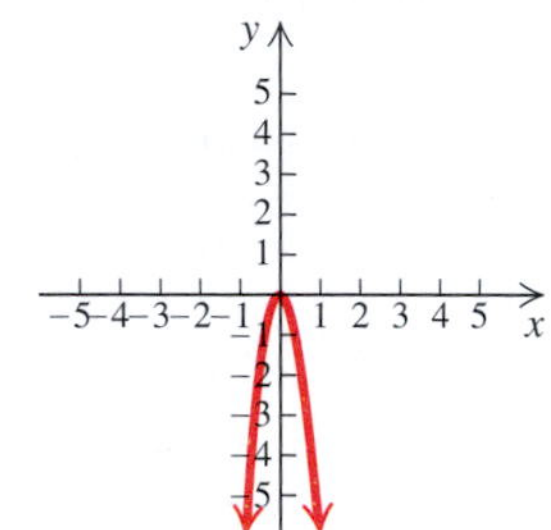

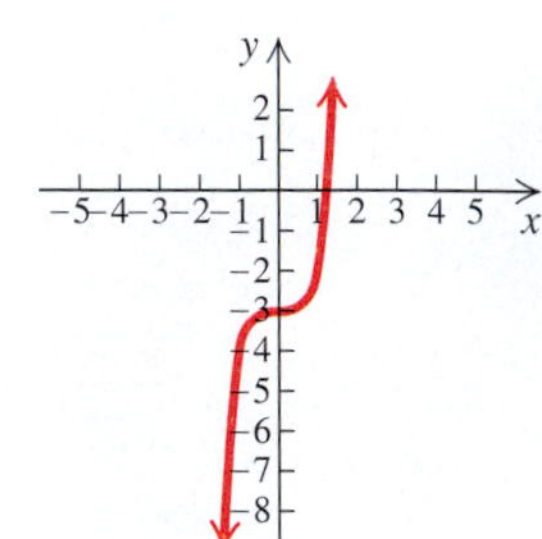

19. $f(x) = -x^6 + 2x^5 - 7x^2$
20. $f(x) = 2x^4 - x^2 + 1$
21. $f(x) = x^5 + \frac{1}{10}x - 3$
22. $f(x) = -x^3 + x^2 - 2x + 4$
23. Use substitution to determine whether 4, 5, and -2 are zeros of
$$f(x) = x^3 - 9x^2 + 14x + 24.$$
24. Use substitution to determine whether 2, 3, and -1 are zeros of
$$f(x) = 2x^3 - 3x^2 + x + 6.$$
25. Use substitution to determine whether 2, 3, and -1 are zeros of
$$g(x) = x^4 - 6x^3 + 8x^2 + 6x - 9.$$
26. Use substitution to determine whether 1, -2, and 3 are zeros of
$$g(x) = x^4 - x^3 - 3x^2 + 5x - 2.$$

Find the zeros of the polynomial function and state the multiplicity of each.

27. $f(x) = (x + 3)^2(x - 1)$

28. $f(x) = (x + 5)^3(x - 4)(x + 1)^2$

29. $f(x) = -2(x - 4)(x - 4)(x - 4)(x + 6)$

30. $f(x) = \left(x + \frac{1}{2}\right)(x + 7)(x + 7)(x + 5)$

31. $f(x) = (x^2 - 9)^3$

32. $f(x) = (x^2 - 4)^2$

33. $f(x) = x^3(x - 1)^2(x + 4)$

34. $f(x) = x^2(x + 3)^2(x - 4)(x + 1)^4$

35. $f(x) = -8(x - 3)^2(x + 4)^3x^4$

36. $f(x) = (x^2 - 5x + 6)^2$

37. $f(x) = x^4 - 4x^2 + 3$

38. $f(x) = x^4 - 10x^2 + 9$

39. $f(x) = x^3 + 3x^2 - x - 3$

40. $f(x) = x^3 - x^2 - 2x + 2$

41. $f(x) = 2x^3 - x^2 - 8x + 4$

42. $f(x) = 3x^3 + x^2 - 48x - 16$

Using a graphing calculator, find the real zeros of the function. Approximate the zeros to three decimal places.

43. $f(x) = x^3 - 3x - 1$

44. $f(x) = x^3 + 3x^2 - 9x - 13$

45. $f(x) = x^4 - 2x^2$

46. $f(x) = x^4 - 2x^3 - 5.6$

47. $f(x) = x^3 - x$

48. $f(x) = 2x^3 - x^2 - 14x - 10$

49. $f(x) = x^8 + 8x^7 - 28x^6 - 56x^5 + 70x^4 + 56x^3 - 28x^2 - 8x + 1$

50. $f(x) = x^6 - 10x^5 + 13x^3 - 4x^2 - 5$

Using a graphing calculator, estimate the real zeros, the relative maxima and minima, and the range of the polynomial function.

51. $g(x) = x^3 - 1.2x + 1$

52. $h(x) = -\frac{1}{2}x^4 + 3x^3 - 5x^2 + 3x + 6$

53. $f(x) = x^6 - 3.8$

54. $h(x) = 2x^3 - x^4 + 20$

55. $f(x) = x^2 + 10x - x^5$

56. $f(x) = 2x^4 - 5.6x^2 + 10$

Determine whether the statement is true or false.

57. If $P(x) = (x - 3)^4(x + 1)^3$, then the graph of the polynomial function $y = P(x)$ crosses the x-axis at $(3, 0)$.

58. If $P(x) = (x + 2)^2\left(x - \frac{1}{4}\right)^5$, then the graph of the polynomial function $y = P(x)$ crosses the x-axis at $\left(\frac{1}{4}, 0\right)$.

59. If $P(x) = (x - 2)^3(x + 5)^6$, then the graph of $y = P(x)$ is tangent to the x-axis at $(-5, 0)$.

60. If $P(x) = (x + 4)^2(x - 1)^2$, then the graph of $y = P(x)$ is tangent to the x-axis at $(4, 0)$.

61. *Vinyl Album Sales.* Vinyl record albums are making a comeback. Sales of vinyl albums rose 51% from 2013 to 2014. The sales data over the years 2001 to 2014 are modeled by the quartic function

$$f(x) = 0.000553x^4 - 0.00735x^3 + 0.06496x^2 - 0.24441x + 1.41345,$$

where x is the number of years after 2001 and $f(x)$ is the number of albums in millions (*Source:* Nielsen SoundScan). Find the number of vinyl albums sold in 2008, in 2012, and in 2016.

62. *Railroad Miles.* The greatest combined length of U.S.-owned operating railroad track existed in 1916, when industrial activity increased during World War I. The total length has decreased ever since. The data over the years 1900 to 2011 are modeled by the quartic function

$$f(x) = -0.002391x^4 + 0.949686x^3 - 123.648199x^2 + 4729.3635x + 198{,}846.4097,$$

where x is the number of years after 1900 and $f(x)$ is in miles (*Source:* Association of American Railroads). Find the number of miles of operating railroad track in the United States in 1916, in 1960, in 2000, and in 2016.

63. *Print Book Sales.* The sales of print books increased about 7% from 2012 to 2014. The sales data over the years 2008 to 2014 are modeled by the cubic function

$$s(x) = 4.958x^3 - 38.125x^2 + 26.417x + 778,$$

where x is the number of years after 2008 and $s(x)$ is the number of print books, in millions (*Source:* Nielsen). Find the number of print books sold in 2009, in 2014, and in 2015.

64. *Dog Years.* A dog's life span is typically much shorter than that of a human. The cubic function

$$d(x) = 0.010255x^3 - 0.340119x^2 + 7.397499x + 6.618361,$$

where x is the dog's age, in years, approximates the equivalent human age in years. Estimate the equivalent human age for dogs that are 3, 12, and 16 years old.

65. *Projectile Motion.* A stone thrown downward with an initial velocity of 34.3 m/sec will travel a distance of s meters, where

$$s(t) = 4.9t^2 + 34.3t$$

and t is in seconds. If a stone is thrown downward at 34.3 m/sec from a height of 294 m, how long will it take the stone to hit the ground?

66. *Games in a Sports League.* If there are x teams in a sports league and all the teams play each other twice, a total of $N(x)$ games are played, where

$$N(x) = x^2 - x.$$

A softball league has 9 teams, each of which plays the others twice. If the league pays $110 per game for the field and the umpires, how much will it cost to play the entire schedule?

67. *Prison Admissions.* Since 2006, total admissions to state and federal prisons have been declining (*Source:* Bureau of Justice Statistics). The quartic function

$$p(x) = 6.213x^4 - 432.347x^3 + 1922.987x^2 + 20{,}503.912x + 638{,}684.984,$$

where x is the number of years after 2001, can be used to estimate the number of admissions to state and federal prisons from 2001 to 2012. Estimate the number of prison admissions in 2003, in 2006, and in 2011.

68. *Obesity.* The percentage of adults in the United States who are obese is increasing (*Source:* Gallup–Healthways Well-Being Index). The cubic function

$$f(x) = 0.102x^3 - 0.764x^2 + 1.595x + 25.494,$$

where x is the number of years after 2008, can be used to estimate the percentage of adults who are obese. Using this function, estimate the percentage of adults who were obese in 2009 and in 2013.

69. *Interest Compounded Annually.* When P dollars is invested at interest rate i, compounded annually, for t years, the investment grows to A dollars, where

$$A = P(1 + i)^t.$$

Trevor's parents deposit $8000 in a savings account when Trevor is 16 years old. The principal plus interest is to be used for a truck when Trevor is 18 years old. Find the interest rate i if the $8000 grows to $9039.75 in 2 years.

70. *Interest Compounded Annually.* When P dollars is invested at interest rate i, compounded annually, for t years, the investment grows to A dollars, where

$$A = P(1 + i)^t.$$

When Sara enters the 11th grade, her grandparents deposit $10,000 in a college savings account. Find the interest rate i if the $10,000 grows to $11,193.64 in 2 years.

For the scatterplots and graphs in Exercises 71–76, determine which, if any, of the following functions might be used as a model for the data.

a) *Linear,* $f(x) = mx + b$
b) *Quadratic,* $f(x) = ax^2 + bx + c, a > 0$
c) *Quadratic,* $f(x) = ax^2 + bx + c, a < 0$
d) *Polynomial, not linear or quadratic*

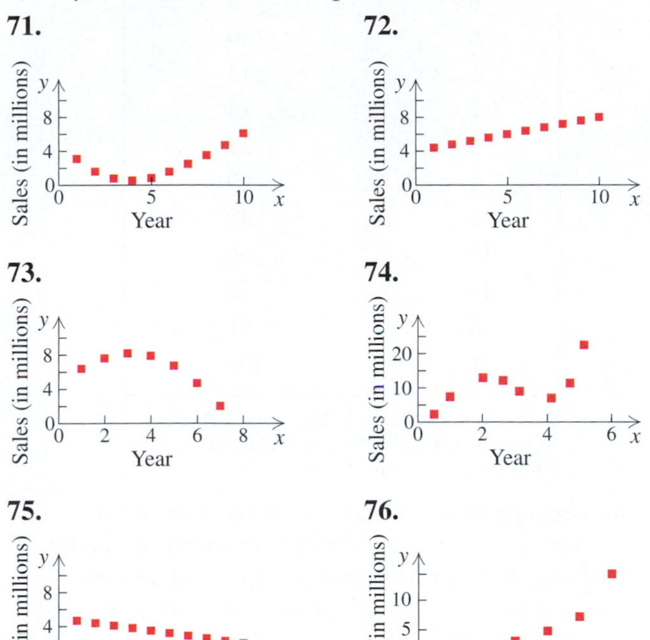

77. *Foreign Adoptions.* The number of foreign adoptions in the United States has declined in recent years, as shown in the following table.

Year, x	Number of U.S. Foreign Adoptions from Top 15 Countries, y
2003, 0	21,320
2004, 1	22,911
2005, 2	22,710
2006, 3	20,705
2007, 4	19,741
2008, 5	17,229
2009, 6	12,782
2010, 7	11,059
2011, 8	9,320
2012, 9	8,668
2013, 10	7,094

Sources: Office of Immigration Statistics; Department of Homeland Security

a) Use a graphing calculator to fit quadratic, cubic, and quartic functions to the data. Let x represent the number of years after 2003. Using R^2-values, determine which function is the best fit.

b) Using the function found in part (a), estimate the number of U.S. foreign adoptions in 2014.

78. *U.S. Farm Acreage.* As the number of farms has decreased in the United States, the average size of the remaining farms has grown larger, as shown in the following table.

Year, x	Average Acreage per Farm, y
1900, 0	147
1910, 10	139
1920, 20	149
1930, 30	157
1940, 40	174
1950, 50	213
1960, 60	297
1970, 70	374
1980, 80	426
1990, 90	460
1995, 95	438
2000, 100	436
2003, 103	441
2005, 105	445
2009, 109	418
2013, 113	435

Sources: National Agricultural Statistics Service; U.S. Department of Agriculture

a) Use a graphing calculator to fit quadratic, cubic, and quartic functions to the data. Let x represent the number of years since 1900. Using R^2-values, determine which function is the best fit.

b) Using the function found in part (a), estimate the average acreage in 1955, in 1998, and in 2011.

79. *Classified Ad Revenue.* The following table lists the newspaper revenue, in billions of dollars, from classified ads for selected years from 1975 to 2013.

Year, x	Newspaper Revenue from Classified Ads, y (in billions of dollars)
1975, 0	$ 2.159
1980, 5	4.222
1985, 10	8.375
1990, 15	11.506
1995, 20	13.742
2000, 25	19.608
2005, 30	17.312
2006, 31	16.986
2007, 32	14.186
2008, 33	9.975
2009, 34	6.179
2010, 35	5.648
2011, 36	5.028
2012, 37	4.626
2013, 38	4.140

Source: Research Department, Newspaper Association of America

a) Use a graphing calculator to fit cubic and quartic functions to the data. Let x represent the number of years after 1975. Using R^2-values, determine which function is the better fit.

b) Using the function found in part (a), estimate the newspaper revenue from classified ads in 1988 and in 2002.

80. *Dog Years.* A dog's life span is typically much shorter than that of a human. Age equivalents for dogs and humans are listed in the following table.

Age of Dog, x (in years)	Human Age, $h(x)$ (in years)
0.25	5
0.5	10
1	15
2	24
4	32
6	40
8	48
10	56
14	72
18	91
21	106

Source: Based on "How to Determine a Dog's Age in Human Years," by Melissa Maroff, eHow Contributor, www.ehow.com

a) Use a graphing calculator to fit linear and cubic functions to the data. Which function has the better fit?

b) Using the function from part (a), estimate the equivalent human age for dogs that are 5, 11, and 15 years old.

Skill Maintenance

Find the distance between the pair of points. **[1.1]**

81. $(3, -5)$ and $(0, -1)$

82. $(4, 2)$ and $(-2, -4)$

83. Find the center and the radius of the circle

$$(x - 3)^2 + (y + 5)^2 = 49.$$ **[1.1]**

84. The diameter of a circle connects the points $(-6, 5)$ and $(-2, 1)$ on the circle. Find the coordinates of the center of the circle and the length of the radius. **[1.1]**

Solve.

85. $2y - 3 \geq 1 - y + 5$ **[1.6]**

86. $(x - 2)(x + 5) > x(x - 3)$ **[1.6]**

87. $|x + 6| \geq 7$ **[3.5]**

88. $|x + \frac{1}{4}| \leq \frac{2}{3}$ **[3.5]**

Synthesis

Determine the degree and the leading term of the polynomial function.

89. $f(x) = (x^5 - 1)^2(x^2 + 2)^3$

90. $f(x) = (10 - 3x^5)^2(5 - x^4)^3(x + 4)$

4.2 Graphing Polynomial Functions

- Graph polynomial functions.
- Use the intermediate value theorem to determine whether a function has a real zero between two given real numbers.

Graphing Polynomial Functions

In addition to using the leading-term test and finding the zeros of a polynomial function, it is helpful to consider the following facts when graphing the function.

If $P(x)$ is a polynomial function of degree n, then the graph of the function has:

- at most n real zeros, and thus at most n x-intercepts;
- at most $n - 1$ turning points.

(Turning points on a graph, also called relative maxima and minima, occur when the function changes from decreasing to increasing or from increasing to decreasing.)

EXAMPLE 1 Graph the polynomial function $h(x) = -2x^4 + 3x^3$.

Solution

1. First, we use the leading-term test to determine the end behavior of the graph. The leading term is $-2x^4$. The degree, 4, is even, and the coefficient, -2, is negative. Thus the end behavior of the graph as $x \to \infty$ and as $x \to -\infty$ can be sketched as follows.

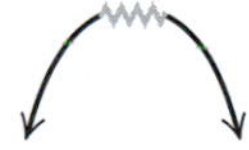

2. The zeros of the function are the first coordinates of the x-intercepts of the graph. To find the zeros, we solve $h(x) = 0$ by factoring and using the principle of zero products.

$$-2x^4 + 3x^3 = 0$$

$$-x^3(2x - 3) = 0 \qquad \text{Factoring}$$

$$-x^3 = 0 \quad or \quad 2x - 3 = 0 \qquad \text{Using the principle of zero products}$$

$$x = 0 \quad or \quad x = \tfrac{3}{2}.$$

The zeros of the function are 0 and $\frac{3}{2}$. Note that the multiplicity of 0 is 3 and the multiplicity of $\frac{3}{2}$ is 1. The x-intercepts are $(0, 0)$ and $\left(\frac{3}{2}, 0\right)$.

3. The zeros divide the x-axis into three intervals:

$$(-\infty, 0), \qquad \left(0, \tfrac{3}{2}\right), \quad \text{and} \quad \left(\tfrac{3}{2}, \infty\right).$$

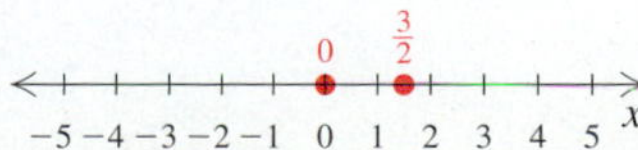

The sign of $h(x)$ is the same for all values of x in an interval. That is, $h(x)$ is positive for all x-values in an interval or $h(x)$ is negative for all x-values in an interval. To determine which, we choose a test value for x from each interval and find $h(x)$.

$y_1 = -2x^4 + 3x^3$

X	Y1
−1	−5
1	1
2	−8

X =

Interval	$(-\infty, 0)$	$\left(0, \frac{3}{2}\right)$	$\left(\frac{3}{2}, \infty\right)$
Test Value, x	-1	1	2
Function Value, $h(x)$	-5	1	-8
Sign of $h(x)$	$-$	$+$	$-$
Location of Points on Graph	Below x-axis	Above x-axis	Below x-axis

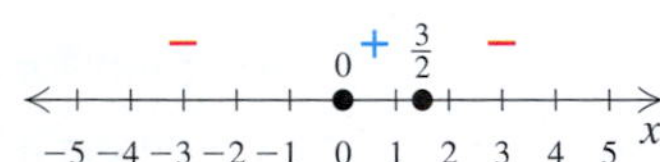

This test-point procedure also gives us three points to plot. In this case, we have $(-1, -5)$, $(1, 1)$, and $(2, -8)$.

4. We found the y-intercept in step (2) but if we hadn't, we would find $h(0)$ to determine it:

$$h(x) = -2x^4 + 3x^3$$
$$h(0) = -2 \cdot 0^4 + 3 \cdot 0^3 = 0.$$

The y-intercept is $(0, 0)$.

5. A few additional points are helpful when completing the graph.

x	$h(x)$
−1.5	−20.25
−0.5	−0.5
0.5	0.25
2.5	−31.25

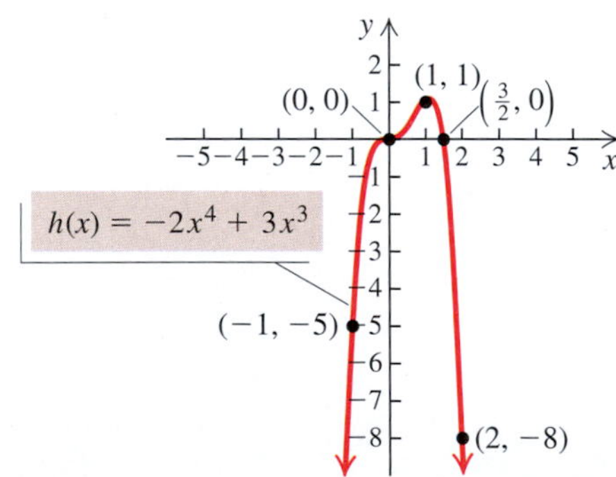

6. The degree of h is 4. The graph of h can have at most 4 x-intercepts and at most 3 turning points. In fact, it has 2 x-intercepts and 1 turning point. The zeros, 0 and $\frac{3}{2}$, each have odd multiplicities: 3 for 0 and 1 for $\frac{3}{2}$. Since the multiplicities are odd, the graph crosses the x-axis at 0 and $\frac{3}{2}$. The end behavior of the graph is what we described in step (1). As $x \to \infty$ and also as $x \to -\infty$, $h(x) \to -\infty$. The graph appears to be correct.

> Now Try Exercise 23.

The following is a procedure for graphing polynomial functions.

> To graph a polynomial function:
>
> 1. Use the leading-term test to determine the end behavior.
> 2. Find the zeros of the function by solving $f(x) = 0$. Any real zeros are the first coordinates of the x-intercepts.
> 3. Use the x-intercepts (zeros) to divide the x-axis into intervals, and choose a test point in each interval to determine the sign of all function values in that interval.
> 4. Find $f(0)$. This gives the y-intercept of the function.
> 5. If necessary, find additional function values to determine the general shape of the graph and then draw the graph.
> 6. As a partial check, use the facts that the graph has at most n x-intercepts and at most $n - 1$ turning points. Multiplicity of zeros can also be considered in order to check where the graph crosses or is tangent to the x-axis. We can also check the graph with a graphing calculator.

EXAMPLE 2 Graph the polynomial function

$$f(x) = 2x^3 + x^2 - 8x - 4.$$

Solution

1. The leading term is $2x^3$. The degree, 3, is odd, and the coefficient, 2, is positive. Thus the end behavior of the graph will appear as follows.

Just in Time 14

2. To find the zeros, we solve $f(x) = 0$. Here we can use factoring by grouping.

$$2x^3 + x^2 - 8x - 4 = 0$$
$$x^2(2x + 1) - 4(2x + 1) = 0 \quad \text{Factoring by grouping}$$
$$(2x + 1)(x^2 - 4) = 0$$
$$(2x + 1)(x + 2)(x - 2) = 0 \quad \text{Factoring a difference of squares}$$

The zeros are $-\frac{1}{2}$, -2, and 2. Each is of multiplicity 1. The x-intercepts are $(-2, 0)$, $\left(-\frac{1}{2}, 0\right)$, and $(2, 0)$.

3. The zeros divide the x-axis into four intervals:

$$(-\infty, -2), \quad \left(-2, -\tfrac{1}{2}\right), \quad \left(-\tfrac{1}{2}, 2\right), \quad \text{and} \quad (2, \infty).$$

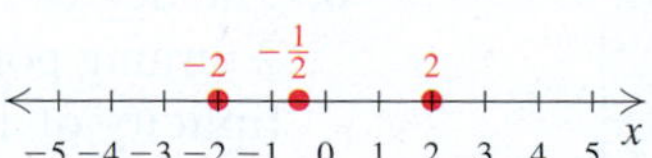

We choose a test value for x from each interval and find $f(x)$.

$y_1 = 2x^3 + x^2 - 8x - 4$

X	Y1
−3	−25
−1	3
1	−9
3	35

X =

Interval	$(-\infty, -2)$	$\left(-2, -\frac{1}{2}\right)$	$\left(-\frac{1}{2}, 2\right)$	$(2, \infty)$
Test Value, x	-3	-1	1	3
Function Value, $f(x)$	-25	3	-9	35
Sign of $f(x)$	$-$	$+$	$-$	$+$
Location of Points on Graph	Below x-axis	Above x-axis	Below x-axis	Above x-axis

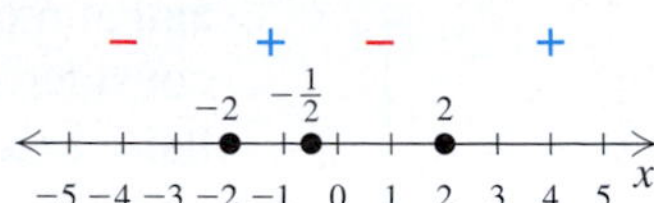

The test values and the corresponding function values also give us four points on the graph: $(-3, -25)$, $(-1, 3)$, $(1, -9)$, and $(3, 35)$.

4. To determine the y-intercept, we find $f(0)$:

$$f(x) = 2x^3 + x^2 - 8x - 4$$
$$f(0) = 2 \cdot 0^3 + 0^2 - 8 \cdot 0 - 4 = -4.$$

The y-intercept is $(0, -4)$.

5. We find a few additional points and complete the graph.

x	$f(x)$
-2.5	-9
-1.5	3.5
0.5	-7.5
1.5	-7

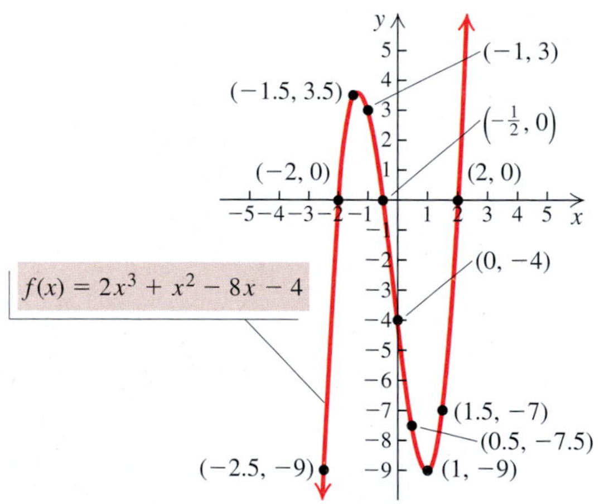

$y = 2x^3 + x^2 - 8x - 4$

6. The degree of f is 3. The graph of f can have at most 3 x-intercepts and at most 2 turning points. It has 3 x-intercepts and 2 turning points. Each zero has a multiplicity of 1. Thus the graph crosses the x-axis at -2, $-\frac{1}{2}$, and 2. The graph has the end behavior described in step (1). As $x \to -\infty$, $f(x) \to -\infty$, and as $x \to \infty$, $f(x) \to \infty$. The graph appears to be correct.

Now Try Exercise 33.

Some polynomials are difficult to factor. In the next example, the polynomial is given in factored form. In Sections 4.3 and 4.4, we will learn methods that facilitate determining factors of such polynomials.

EXAMPLE 3 Graph the polynomial function

$$g(x) = x^4 - 7x^3 + 12x^2 + 4x - 16$$
$$= (x + 1)(x - 2)^2(x - 4).$$

Solution

1. The leading term is x^4. The degree, 4, is even, and the coefficient, 1, is positive. The following sketch shows the end behavior.

2. To find the zeros, we solve $g(x) = 0$:

$$(x + 1)(x - 2)^2(x - 4) = 0.$$

The zeros are -1, 2, and 4; 2 is of multiplicity 2; the others are of multiplicity 1. The x-intercepts are $(-1, 0)$, $(2, 0)$, and $(4, 0)$.

3. The zeros divide the x-axis into four intervals:

$$(-\infty, -1), \quad (-1, 2), \quad (2, 4), \quad \text{and} \quad (4, \infty).$$

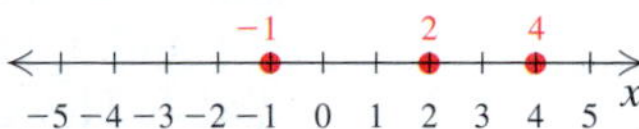

We choose a test value for x from each interval and find $g(x)$.

$y_1 = x^4 - 7x^3 + 12x^2 + 4x - 16$

X	Y1
-1.25	13.863
1	-6
3	-4
4.25	6.6445
X =	

Interval	$(-\infty, -1)$	$(-1, 2)$	$(2, 4)$	$(4, \infty)$
Test Value, x	-1.25	1	3	4.25
Function Value, $g(x)$	≈ 13.9	-6	-4	≈ 6.6
Sign of $g(x)$	+	−	−	+
Location of Points on Graph	Above x-axis	Below x-axis	Below x-axis	Above x-axis

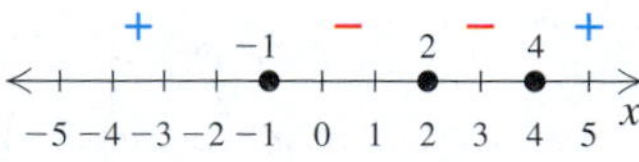

The test values and the corresponding function values also give us four points on the graph: $(-1.25, 13.9)$, $(1, -6)$, $(3, -4)$, and $(4.25, 6.6)$.

4. To determine the y-intercept, we find $g(0)$:

$$g(x) = x^4 - 7x^3 + 12x^2 + 4x - 16$$
$$g(0) = 0^4 - 7 \cdot 0^3 + 12 \cdot 0^2 + 4 \cdot 0 - 16 = -16.$$

The y-intercept is $(0, -16)$.

5. We find a few additional points and draw the graph.

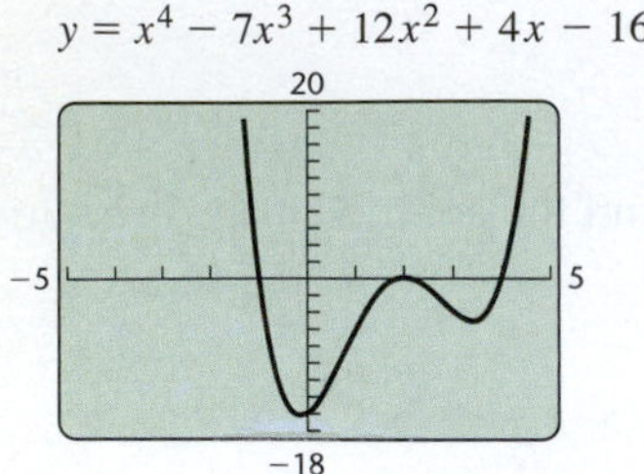

x	$g(x)$
−0.5	−14.1
0.5	−11.8
1.5	−1.6
2.5	−1.3
3.5	−5.1

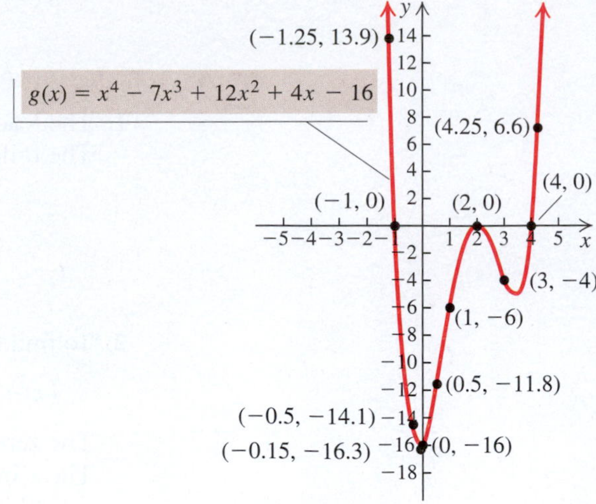

6. The degree of g is 4. The graph of g can have at most 4 x-intercepts and at most 3 turning points. It has 3 x-intercepts and 3 turning points. One of the zeros, 2, has a multiplicity of 2, so the graph is tangent to the x-axis at 2. The other zeros, -1 and 4, each have a multiplicity of 1 so the graph crosses the x-axis at -1 and 4. The graph has the end behavior described in step (1). As $x \to \infty$ and as $x \to -\infty$, $g(x) \to \infty$. The graph appears to be correct.

Now Try Exercise 19.

The Intermediate Value Theorem

Polynomial functions are continuous, hence their graphs are unbroken. The domain of a polynomial function, unless restricted by the statement of the function, is $(-\infty, \infty)$. Suppose that two polynomial function values $P(a)$ and $P(b)$ have opposite signs. Since P is continuous, its graph must be a curve from $(a, P(a))$ to $(b, P(b))$ without a break. Then it follows that the curve must cross the x-axis at some point c between a and b; that is, the function has a zero at c between a and b.

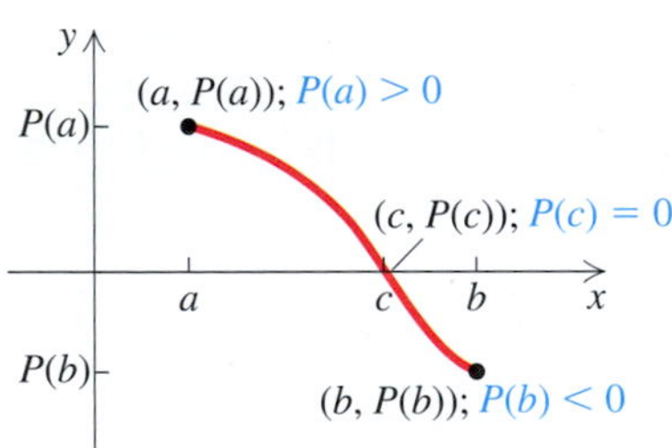

THE INTERMEDIATE VALUE THEOREM

For any polynomial function $P(x)$ with real coefficients, suppose that for $a \neq b$, $P(a)$ and $P(b)$ are of opposite signs. Then the function has a real zero between a and b.

The intermediate value theorem *cannot* be used to determine whether there is or is not a real zero between a and b when $P(a)$ and $P(b)$ have the *same* sign.

EXAMPLE 4 Using the intermediate value theorem, determine, if possible, whether the function has a real zero between a and b.

a) $f(x) = x^3 + x^2 - 6x;\ a = -4, b = -2$

b) $f(x) = x^3 + x^2 - 6x;\ a = -1, b = 3$

c) $g(x) = \frac{1}{3}x^4 - x^3;\ a = -\frac{1}{2}, b = \frac{1}{2}$

d) $g(x) = \frac{1}{3}x^4 - x^3;\ a = 1, b = 2$

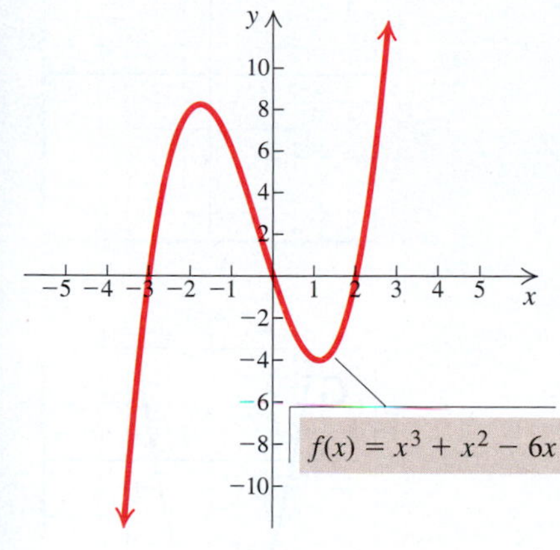

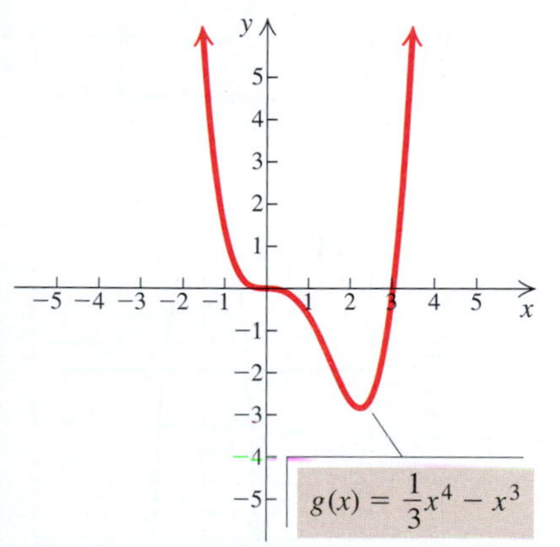

Solution We find $f(a)$ and $f(b)$ or $g(a)$ and $g(b)$ and determine whether they differ in sign. The graphs of $f(x)$ and $g(x)$ at left provide a visual check of the conclusions.

a) $f(-4) = (-4)^3 + (-4)^2 - 6(-4) = -24,$
$f(-2) = (-2)^3 + (-2)^2 - 6(-2) = 8$

Note that $f(-4)$ is negative and $f(-2)$ is positive. By the intermediate value theorem, since $f(-4)$ and $f(-2)$ have opposite signs, then $f(x)$ has a real zero between -4 and -2. The graph at left confirms this.

b) $f(-1) = (-1)^3 + (-1)^2 - 6(-1) = 6,$
$f(3) = 3^3 + 3^2 - 6(3) = 18$

Both $f(-1)$ and $f(3)$ are positive. Thus the intermediate value theorem *does not allow* us to determine whether there is a real zero between -1 and 3. Note that the graph of $f(x)$ shows that there are two zeros between -1 and 3.

c) $g\left(-\frac{1}{2}\right) = \frac{1}{3}\left(-\frac{1}{2}\right)^4 - \left(-\frac{1}{2}\right)^3 = \frac{7}{48},$
$g\left(\frac{1}{2}\right) = \frac{1}{3}\left(\frac{1}{2}\right)^4 - \left(\frac{1}{2}\right)^3 = -\frac{5}{48}$

Since $g\left(-\frac{1}{2}\right)$ and $g\left(\frac{1}{2}\right)$ have opposite signs, $g(x)$ has a real zero between $-\frac{1}{2}$ and $\frac{1}{2}$. The graph at left confirms this.

d) $g(1) = \frac{1}{3}(1)^4 - 1^3 = -\frac{2}{3},$
$g(2) = \frac{1}{3}(2)^4 - 2^3 = -\frac{8}{3}$

Both $g(1)$ and $g(2)$ are negative. Thus the intermediate value theorem *does not allow* us to determine whether there is a real zero between 1 and 2. Note that the graph of $g(x)$ at left shows that there are no real zeros between 1 and 2.

Now Try Exercises 39 and 43.

Visualizing the Graph

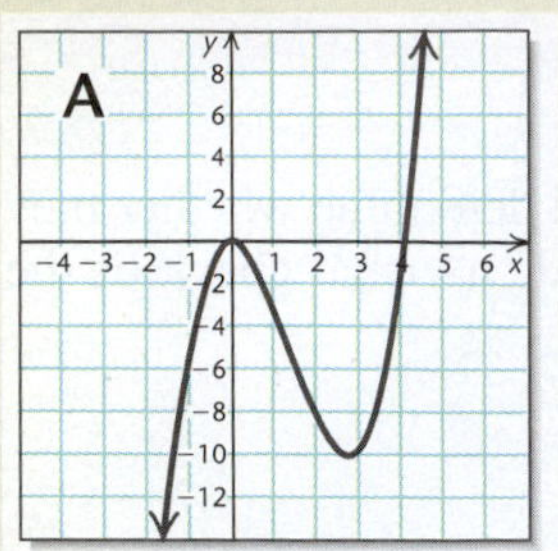

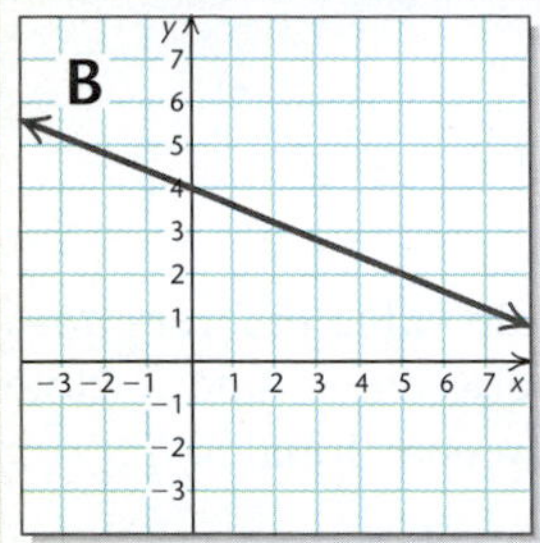

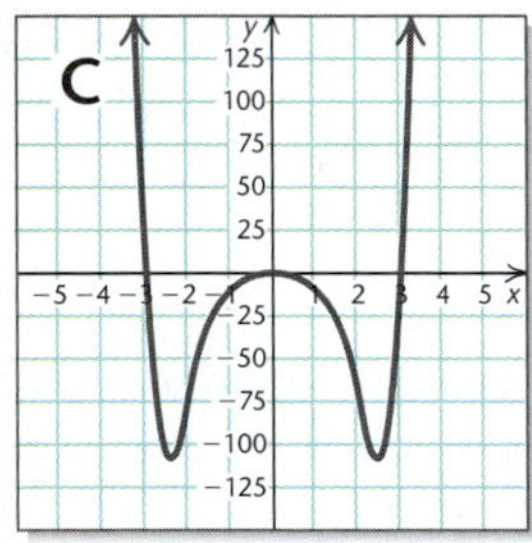

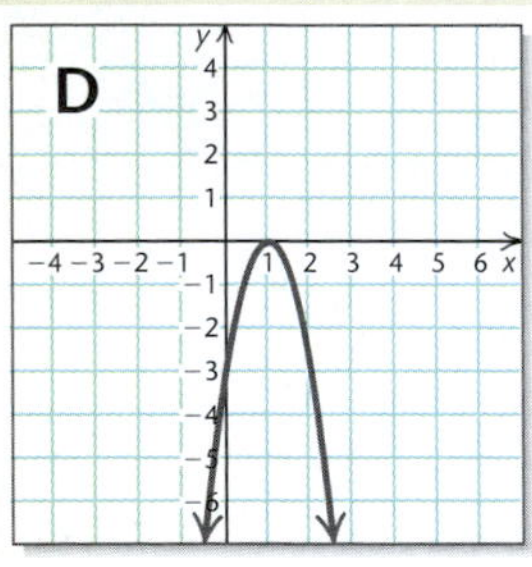

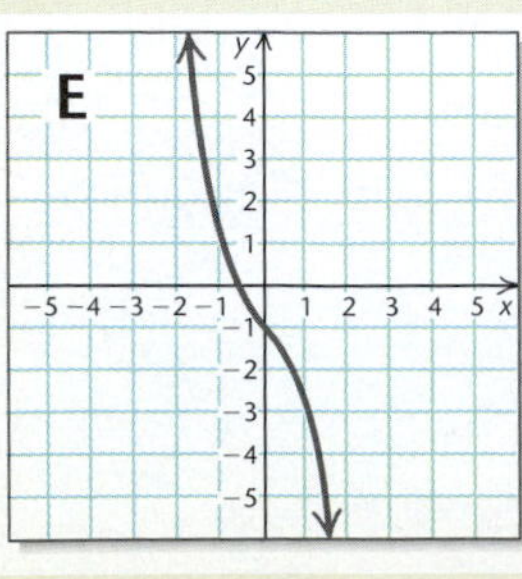

Match the function with its graph.

1. $f(x) = -x^4 - x + 5$

2. $f(x) = -3x^2 + 6x - 3$

3. $f(x) = x^4 - 4x^3 + 3x^2 + 4x - 4$

4. $f(x) = -\frac{2}{5}x + 4$

5. $f(x) = x^3 - 4x^2$

6. $f(x) = x^6 - 9x^4$

7. $f(x) = x^5 - 3x^3 + 2$

8. $f(x) = -x^3 - x - 1$

9. $f(x) = x^2 + 7x + 6$

10. $f(x) = \frac{7}{2}$

Answers on page A-18

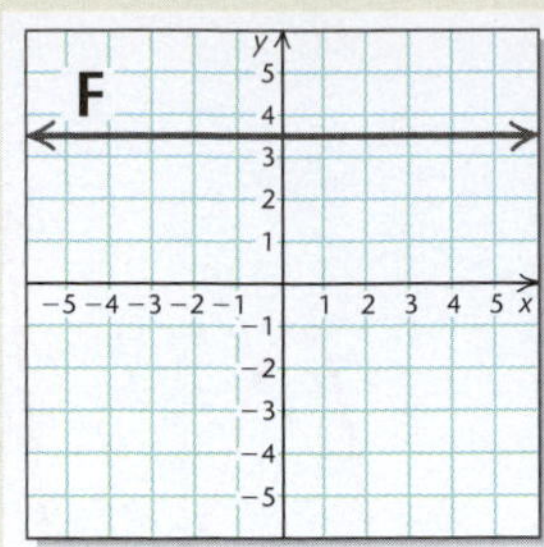

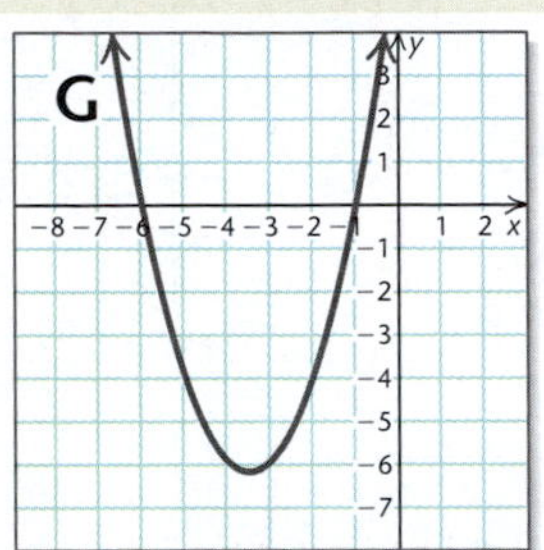

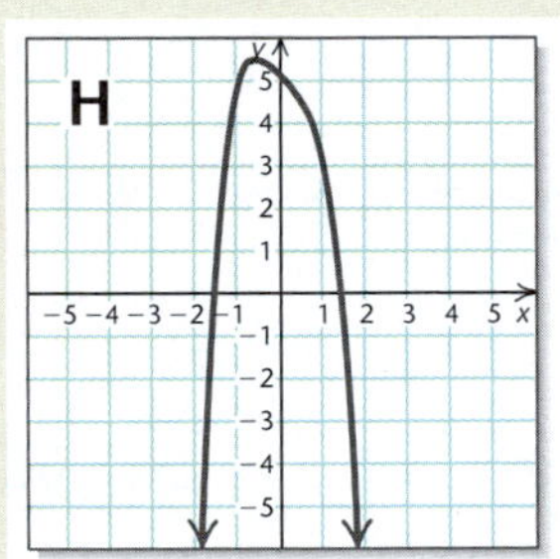

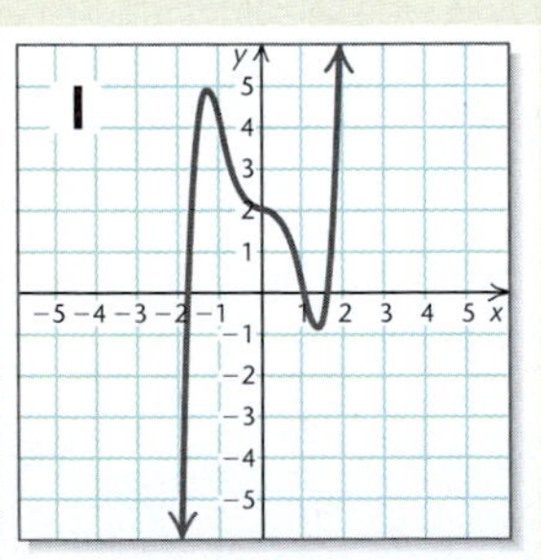

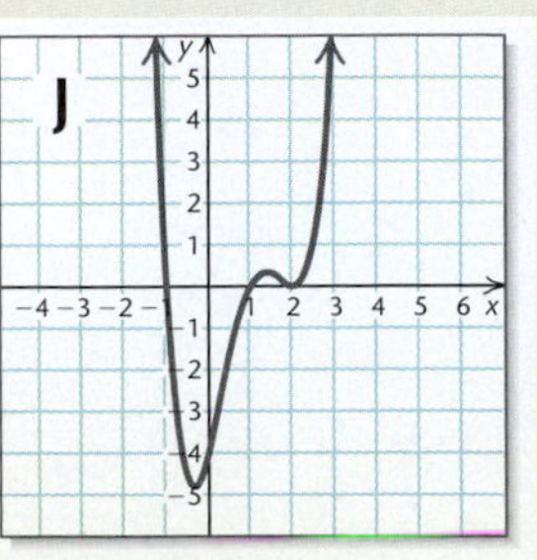

4.2 Exercise Set

For each function in Exercises 1–6, state:

a) *the maximum number of real zeros that the function can have;*

b) *the maximum number of x-intercepts that the graph of the function can have; and*

c) *the maximum number of turning points that the graph of the function can have.*

1. $f(x) = x^5 - x^2 + 6$

2. $f(x) = -x^2 + x^4 - x^6 + 3$

3. $f(x) = x^{10} - 2x^5 + 4x - 2$

4. $f(x) = \frac{1}{4}x^3 + 2x^2$

5. $f(x) = -x - x^3$

6. $f(x) = -3x^4 + 2x^3 - x - 4$

In Exercises 7–12, use the leading-term test and your knowledge of y-intercepts to match the function with one of the graphs (a)–(f) that follow.

a)

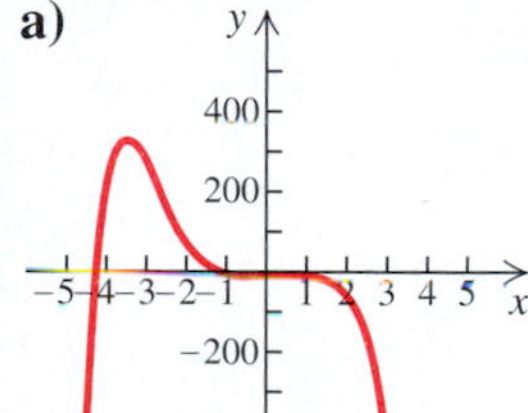

b)

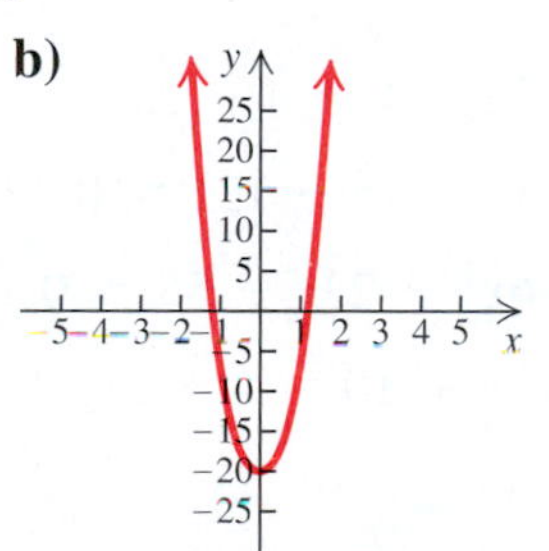

c)

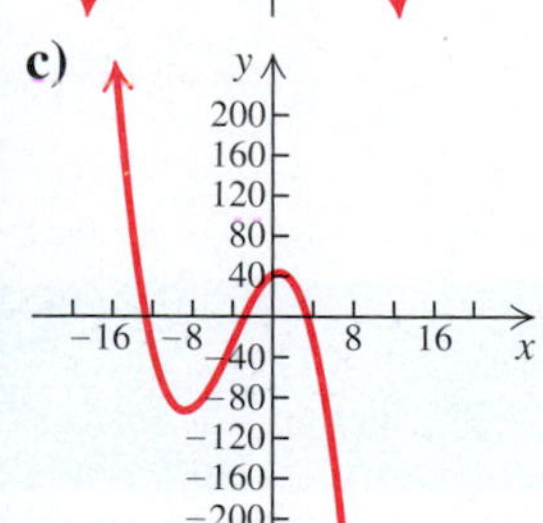

d)

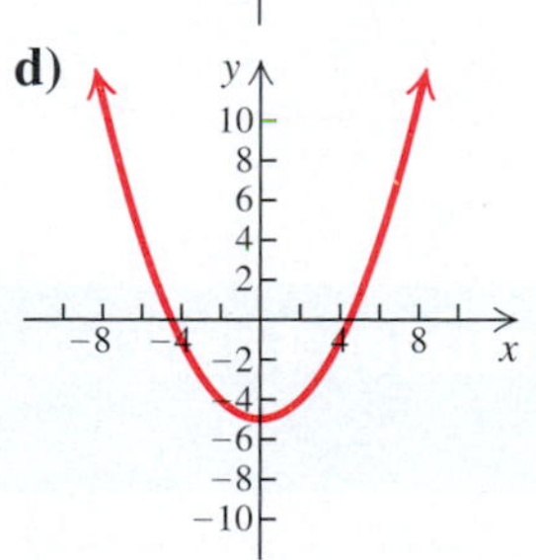

e)

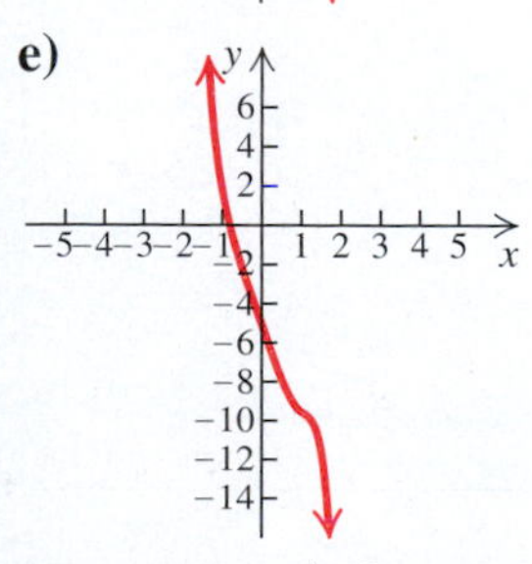

f)

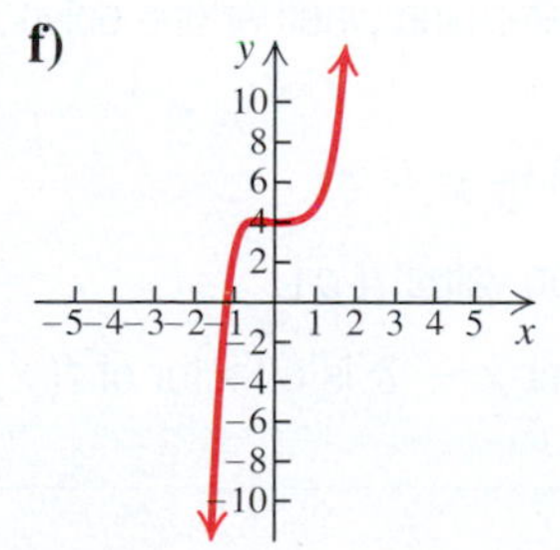

7. $f(x) = \frac{1}{4}x^2 - 5$

8. $f(x) = -0.5x^6 - x^5 + 4x^4 - 5x^3 - 7x^2 + x - 3$

9. $f(x) = x^5 - x^4 + x^2 + 4$

10. $f(x) = -\frac{1}{3}x^3 - 4x^2 + 6x + 42$

11. $f(x) = x^4 - 2x^3 + 12x^2 + x - 20$

12. $f(x) = -0.3x^7 + 0.11x^6 - 0.25x^5 + x^4 + x^3 - 6x - 5$

Graph the polynomial function. Follow the steps outlined in the procedure on p. 237.

13. $f(x) = -x^3 - 2x^2$

14. $g(x) = x^4 - 4x^3 + 3x^2$

15. $h(x) = x^2 + 2x - 3$

16. $f(x) = x^2 - 5x + 4$

17. $h(x) = x^5 - 4x^3$

18. $f(x) = x^3 - x$

19. $h(x) = x(x - 4)(x + 1)(x - 2)$

20. $f(x) = x(x - 1)(x + 3)(x + 5)$

21. $g(x) = -\frac{1}{4}x^3 - \frac{3}{4}x^2$

22. $f(x) = \frac{1}{2}x^3 + \frac{5}{2}x^2$

23. $g(x) = -x^4 - 2x^3$

24. $h(x) = x^3 - 3x^2$

25. $f(x) = -\frac{1}{2}(x - 2)(x + 1)^2(x - 1)$

26. $g(x) = (x - 2)^3(x + 3)$

27. $g(x) = -x(x - 1)^2(x + 4)^2$

28. $h(x) = -x(x - 3)(x - 3)(x + 2)$

29. $f(x) = (x - 2)^2(x + 1)^4$

30. $g(x) = x^4 - 9x^2$

31. $g(x) = -(x - 1)^4$

32. $h(x) = (x + 2)^3$

33. $h(x) = x^3 + 3x^2 - x - 3$

34. $g(x) = -x^3 + 2x^2 + 4x - 8$

35. $f(x) = 6x^3 - 8x^2 - 54x + 72$

36. $h(x) = x^5 - 5x^3 + 4x$

Graph the piecewise function.

37. $g(x) = \begin{cases} -x + 3, & \text{for } x \le -2, \\ 4, & \text{for } -2 < x < 1, \\ \frac{1}{2}x^3, & \text{for } x \ge 1 \end{cases}$

38. $h(x) = \begin{cases} -x^2, & \text{for } x < -2, \\ x + 1, & \text{for } -2 \le x < 0, \\ x^3 - 1, & \text{for } x \ge 0 \end{cases}$

Using the intermediate value theorem, determine, if possible, whether the function f has a real zero between a and b.

39. $f(x) = x^3 + 3x^2 - 9x - 13;\ a = -5, b = -4$

40. $f(x) = x^3 + 3x^2 - 9x - 13;\ a = 1, b = 2$

41. $f(x) = 3x^2 - 2x - 11;\ a = -3, b = -2$

42. $f(x) = 3x^2 - 2x - 11;\ a = 2, b = 3$

43. $f(x) = x^4 - 2x^2 - 6;\ a = 2, b = 3$

44. $f(x) = 2x^5 - 7x + 1;\ a = 1, b = 2$

45. $f(x) = x^3 - 5x^2 + 4;\ a = 4, b = 5$

46. $f(x) = x^4 - 3x^2 + x - 1;\ a = -3, b = -2$

Skill Maintenance

Match the equation with one of the graphs (a)–(f) that follow.

a)

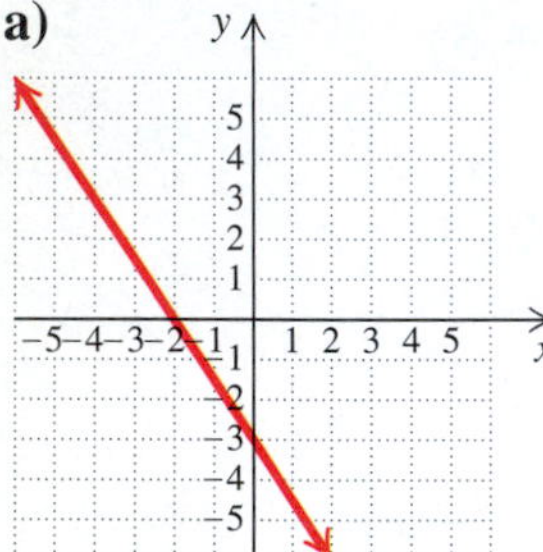

b)

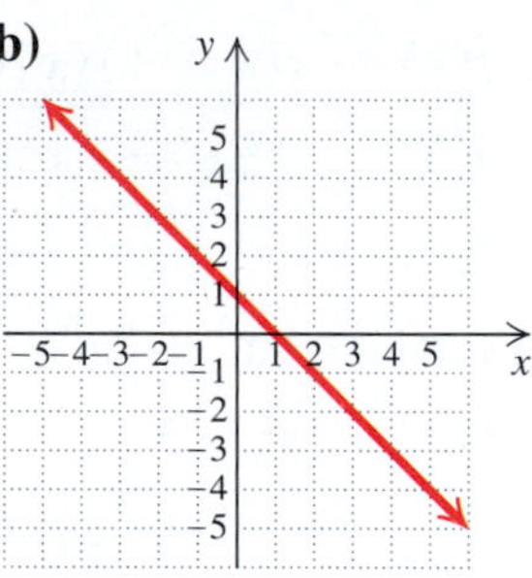

c)

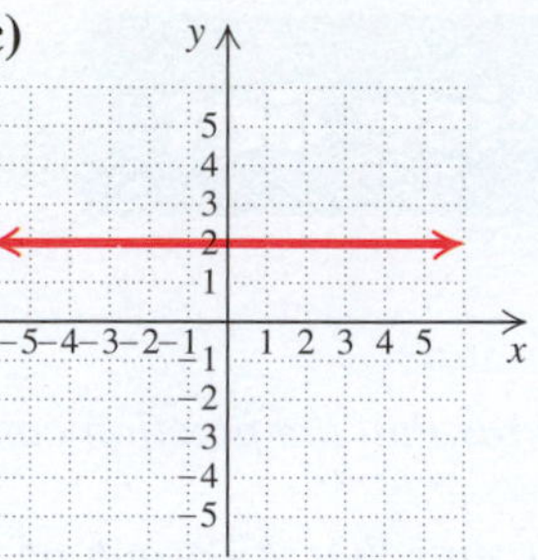

d)

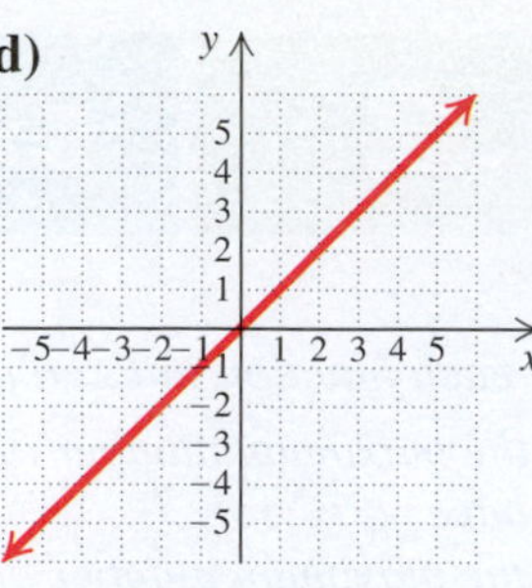

e)

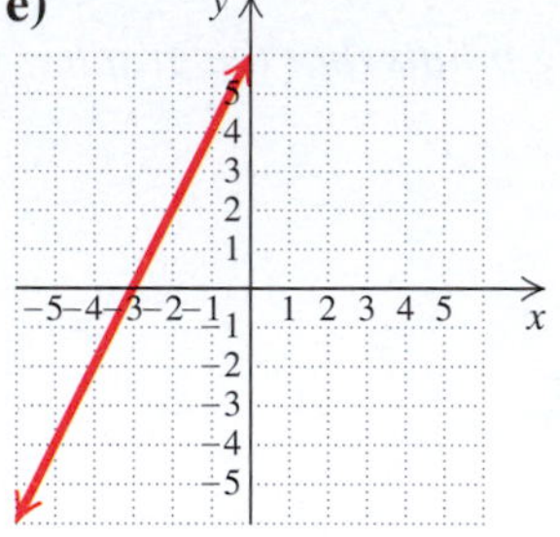

f)

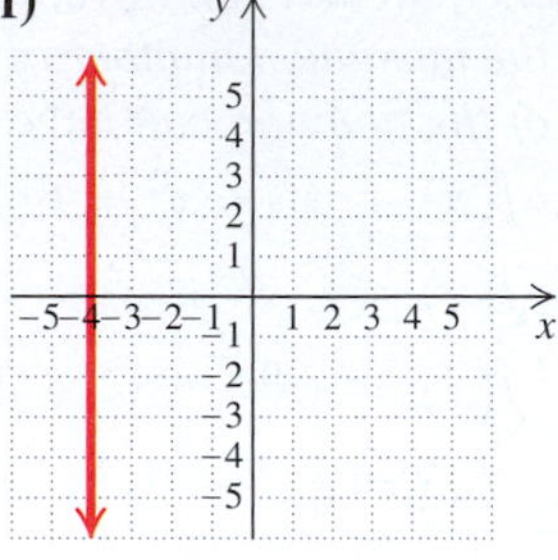

47. $y = x$ [1.1]

48. $x = -4$ [1.3]

49. $y - 2x = 6$ [1.1]

50. $3x + 2y = -6$ [1.1]

51. $y = 1 - x$ [1.1]

52. $y = 2$ [1.3]

Solve.

53. $2x - \frac{1}{2} = 4 - 3x$ [1.5]

54. $x^3 - x^2 - 12x = 0$ [4.1]

55. $6x^2 - 23x - 55 = 0$ [3.2]

56. $\frac{3}{4}x + 10 = \frac{1}{5} + 2x$ [1.5]

4.3 Polynomial Division; The Remainder Theorem and the Factor Theorem

- Perform long division with polynomials and determine whether one polynomial is a factor of another.
- Use synthetic division to divide a polynomial by $x - c$.
- Use the remainder theorem to find a function value $f(c)$.
- Use the factor theorem to determine whether $x - c$ is a factor of $f(x)$.

In general, finding exact zeros of many polynomial functions is neither easy nor straightforward. We now develop concepts that help us find exact zeros of certain polynomial functions with degree 3 or greater. Consider the polynomial

$$h(x) = x^3 + 2x^2 - 5x - 6 = (x + 3)(x + 1)(x - 2).$$

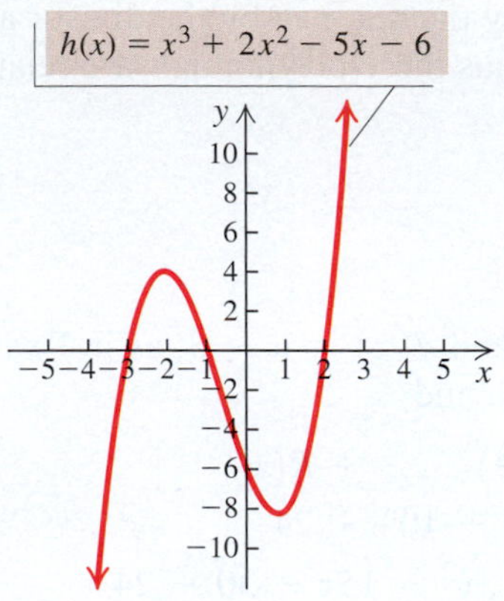

The factors are

$$x + 3, \quad x + 1, \quad \text{and} \quad x - 2,$$

and the zeros are

$$-3, \quad -1, \quad \text{and} \quad 2.$$

When a polynomial is expressed in factored form, each factor determines a zero of the function. Thus if we know the factors of a polynomial, we can easily find the zeros. The "reverse" is also true: If we know the zeros of a polynomial function, we can find the factors of the polynomial.

Division and Factors

When we divide one polynomial by another, we obtain a quotient and a remainder. If the remainder is 0, then the divisor is a factor of the dividend.

EXAMPLE 1 Divide to determine whether $x + 1$ and $x - 3$ are factors of

$$x^3 + 2x^2 - 5x - 6.$$

Solution We divide $x^3 + 2x^2 - 5x - 6$ by $x + 1$.

$$\begin{array}{r l}
 & \text{Quotient} \\
 & x^2 + x - 6 \\
\text{Divisor} \rightarrow x + 1 \,\big) & x^3 + 2x^2 - 5x - 6 \leftarrow \text{Dividend} \\
 & \underline{x^3 + x^2} \\
 & x^2 - 5x \\
 & \underline{x^2 + x} \\
 & -6x - 6 \\
 & \underline{-6x - 6} \\
 & 0 \leftarrow \text{Remainder}
\end{array}$$

Since the remainder is 0, we know that $x + 1$ is a factor of $x^3 + 2x^2 - 5x - 6$. In fact, we know that

$$x^3 + 2x^2 - 5x - 6 = (x + 1)(x^2 + x - 6).$$

We divide $x^3 + 2x^2 - 5x - 6$ by $x - 3$.

$$\begin{array}{r l}
 & x^2 + 5x + 10 \\
x - 3 \,\big) & x^3 + 2x^2 - 5x - 6 \\
 & \underline{x^3 - 3x^2} \\
 & 5x^2 - 5x \\
 & \underline{5x^2 - 15x} \\
 & 10x - 6 \\
 & \underline{10x - 30} \\
 & 24 \leftarrow \text{Remainder}
\end{array}$$

The remainder, 24, is not 0, so we know that $x - 3$ is *not* a factor of $x^3 + 2x^2 - 5x - 6$.

Now Try Exercise 3.

When we divide a polynomial $P(x)$ by a divisor $d(x)$, a polynomial $Q(x)$ is the quotient and a polynomial $R(x)$ is the remainder. The quotient $Q(x)$ must have degree less than that of the dividend $P(x)$. The remainder $R(x)$ must either be 0 or have degree less than that of the divisor $d(x)$.

As in arithmetic, to check division, we multiply the quotient by the divisor and add the remainder, to see if we get the dividend. Thus these polynomials are related as follows:

$$P(x) = d(x) \cdot Q(x) + R(x)$$

Dividend **Divisor** **Quotient** **Remainder**

For instance, if $P(x) = x^3 + 2x^2 - 5x - 6$ and $d(x) = x - 3$, as in Example 1, then $Q(x) = x^2 + 5x + 10$ and $R(x) = 24$, and

$$\begin{aligned} P(x) \qquad &= \quad d(x) \quad \cdot \qquad Q(x) \qquad + R(x) \\ x^3 + 2x^2 - 5x - 6 &= (x - 3) \cdot (x^2 + 5x + 10) + 24 \\ &= x^3 + 5x^2 + 10x - 3x^2 - 15x - 30 + 24 \\ &= x^3 + 2x^2 - 5x - 6. \end{aligned}$$

The Remainder Theorem and Synthetic Division

Consider the function

$$h(x) = x^3 + 2x^2 - 5x - 6.$$

When we divided $h(x)$ by $x + 1$ and $x - 3$ in Example 1, the remainders were 0 and 24, respectively. Let's now find the function values $h(-1)$ and $h(3)$:

$$h(-1) = (-1)^3 + 2(-1)^2 - 5(-1) - 6 = 0;$$
$$h(3) = (3)^3 + 2(3)^2 - 5(3) - 6 = 24.$$

Note that the function values are the same as the remainders. This suggests the following theorem.

THE REMAINDER THEOREM

If a number c is substituted for x in the polynomial $f(x)$, then the result $f(c)$ is the remainder that would be obtained by dividing $f(x)$ by $x - c$. That is, if $f(x) = (x - c) \cdot Q(x) + R$, then $f(c) = R$.

Proof (Optional). The equation $f(x) = d(x) \cdot Q(x) + R(x)$, where $d(x) = x - c$, is the basis of this proof. If we divide $f(x)$ by $x - c$, we obtain a quotient $Q(x)$ and a remainder $R(x)$ related as follows:

$$f(x) = (x - c) \cdot Q(x) + R(x).$$

The remainder $R(x)$ must either be 0 or have degree less than $x - c$. Thus, $R(x)$ must be a constant. Let's call this constant R. The equation above is true for any replacement of x, so we replace x with c. We get

$$\begin{aligned} f(c) &= (c - c) \cdot Q(c) + R \\ &= 0 \cdot Q(c) + R \\ &= R. \end{aligned}$$

Thus the function value $f(c)$ is the remainder obtained when we divide $f(x)$ by $x - c$. ■

The remainder theorem motivates us to find a rapid way of dividing by $x - c$ in order to find function values. To streamline division, we can arrange the work so that duplicate and unnecessary writing is avoided. Consider the following:

$$(4x^3 - 3x^2 + x + 7) \div (x - 2).$$

A.

$$\begin{array}{r} 4x^2 + 5x + 11 \\ x - 2\overline{)\,4x^3 - 3x^2 + x + 7} \\ \underline{4x^3 - 8x^2} \\ 5x^2 + x \\ \underline{5x^2 - 10x} \\ 11x + 7 \\ \underline{11x - 22} \\ 29 \end{array}$$

B.

$$\begin{array}{r} 4 \quad 5 \quad 11 \\ 1 - 2\overline{)\,4 - 3 + 1 + 7} \\ \underline{4 - 8} \\ 5 + 1 \\ \underline{5 - 10} \\ 11 + 7 \\ \underline{11 - 22} \\ 29 \end{array}$$

The division in (B) is the same as that in (A), but we wrote only the coefficients. The red numerals are duplicated, so we look for an arrangement in which they are not duplicated. In place of the divisor in the form $x - c$, we can simply use c and then add rather than subtract. When the procedure is "collapsed," we have the algorithm known as **synthetic division**.

C. *Synthetic Division*

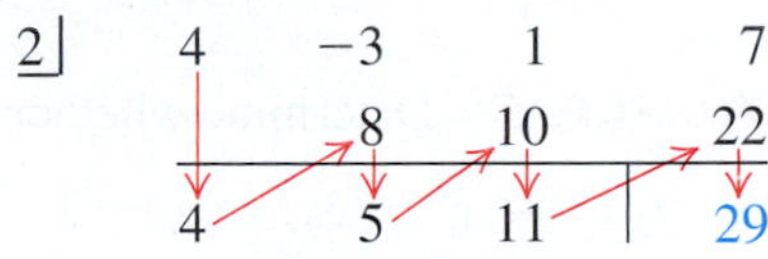

The divisor is $x - 2$; thus we use 2 in synthetic division.

We "bring down" the 4. Then we multiply it by the 2 to get 8 and add to get 5. We then multiply 5 by 2 to get 10, add, and so on. The last number, 29, is the remainder. The others, 4, 5, and 11, are the coefficients of the quotient, $4x^2 + 5x + 11$. (Note that the degree of the quotient is 1 less than the degree of the dividend when the degree of the divisor is 1.)

When using synthetic division, we write a 0 for a missing term in the dividend.

Just in Time 5

EXAMPLE 2 Use synthetic division to find the quotient and the remainder:

$$(2x^3 + 7x^2 - 5) \div (x + 3).$$

Solution First, we note that $x + 3 = x - (-3)$.

$$\begin{array}{r|rrrr} -3 & 2 & 7 & 0 & -5 \\ & & -6 & -3 & 9 \\ \hline & 2 & 1 & -3 & 4 \end{array}$$

***Note:* We must write a 0 for the missing x-term.**

The quotient is $2x^2 + x - 3$. The remainder is 4.

Now Try Exercise 13.

We can now use synthetic division to find polynomial function values.

EXAMPLE 3 Given that $f(x) = 2x^5 - 3x^4 + x^3 - 2x^2 + x - 8$, find $f(10)$.

Solution By the remainder theorem, $f(10)$ is the remainder when $f(x)$ is divided by $x - 10$. We use synthetic division to find that remainder.

$$\begin{array}{r|rrrrrr} 10 & 2 & -3 & 1 & -2 & 1 & -8 \\ & & 20 & 170 & 1710 & 17{,}080 & 170{,}810 \\ \hline & 2 & 17 & 171 & 1708 & 17{,}081 & 170{,}802 \end{array}$$

Thus, $f(10) = 170{,}802$.

Now Try Exercise 25.

Compare the computations in Example 3 with those in a direct substitution:

$$\begin{aligned} f(10) &= 2(10)^5 - 3(10)^4 + (10)^3 - 2(10)^2 + 10 - 8 \\ &= 2 \cdot 100{,}000 - 3 \cdot 10{,}000 + 1000 - 2 \cdot 100 + 10 - 8 \\ &= 200{,}000 - 30{,}000 + 1000 - 200 + 10 - 8 \\ &= 170{,}802. \end{aligned}$$

The computations in synthetic division are less complicated than those involved in substituting. The easiest way to find $f(10)$ is to use one of the methods for evaluating a function on a graphing calculator. In the figure at left, we show the result when we enter $y_1 = 2x^5 - 3x^4 + x^3 - 2x^2 + x - 8$ and then use function notation on the home screen.

Y1(10)
170802

EXAMPLE 4 Determine whether 5 is a zero of $g(x)$, where

$$g(x) = x^4 - 26x^2 + 25.$$

Solution We use synthetic division and the remainder theorem to find $g(5)$.

$$\begin{array}{r|rrrrr} 5 & 1 & 0 & -26 & 0 & 25 \\ & & 5 & 25 & -5 & -25 \\ \hline & 1 & 5 & -1 & -5 & 0 \end{array}$$

Writing 0's for missing terms: $x^4 + 0x^3 - 26x^2 + 0x + 25$

Since $g(5) = 0$, the number 5 is a zero of $g(x)$.

Now Try Exercise 31.

EXAMPLE 5 Determine whether i is a zero of $f(x)$, where

$$f(x) = x^3 - 3x^2 + x - 3.$$

Solution We use synthetic division and the remainder theorem to find $f(i)$.

$$\begin{array}{r|rrrr} i & 1 & -3 & 1 & -3 \\ & & i & -3i - 1 & 3 \\ \hline & 1 & -3 + i & -3i & 0 \end{array}$$

$i(-3 + i) = -3i + i^2 = -3i - 1$

$i(-3i) = -3i^2 = 3$

Since $f(i) = 0$, the number i is a zero of $f(x)$.

Now Try Exercise 35.

Finding Factors of Polynomials

We now consider a useful result that follows from the remainder theorem.

THE FACTOR THEOREM

For a polynomial $f(x)$, if $f(c) = 0$, then $x - c$ is a factor of $f(x)$.

Proof (Optional). If we divide $f(x)$ by $x - c$, we obtain a quotient and a remainder, related as follows:

$$f(x) = (x - c) \cdot Q(x) + f(c).$$

Then if $f(c) = 0$, we have

$$f(x) = (x - c) \cdot Q(x),$$

so $x - c$ is a factor of $f(x)$. ■

The factor theorem is very useful in factoring polynomials and hence in solving polynomial equations and finding zeros of polynomial functions. If we know a zero of a polynomial function, we know a factor.

EXAMPLE 6 Let $f(x) = x^3 - 3x^2 - 6x + 8$. Factor $f(x)$ and solve the equation $f(x) = 0$.

Solution We look for linear factors of the form $x - c$. Let's try $x + 1$, or $x - (-1)$. (In the next section, we will learn a method for choosing the numbers to try for c.) We use synthetic division to determine whether $f(-1) = 0$.

$$\begin{array}{r|rrrr} -1 & 1 & -3 & -6 & 8 \\ & & -1 & 4 & 2 \\ \hline & 1 & -4 & -2 & 10 \end{array}$$

Since $f(-1) \neq 0$, we know that $x + 1$ *is not a factor* of $f(x)$. We now try $x - 1$.

$$\begin{array}{r|rrrr} 1 & 1 & -3 & -6 & 8 \\ & & 1 & -2 & -8 \\ \hline & 1 & -2 & -8 & 0 \end{array}$$

Since $f(1) = 0$, we know that $x - 1$ *is one factor* of $f(x)$ and the quotient, $x^2 - 2x - 8$, is another. Thus,

$$f(x) = (x - 1)(x^2 - 2x - 8).$$

The trinomial $x^2 - 2x - 8$ is easily factored, so we have

$$f(x) = (x - 1)(x - 4)(x + 2).$$

We now solve the equation $f(x) = 0$. To do so, we use the principle of zero products:

$$(x - 1)(x - 4)(x + 2) = 0$$
$$x - 1 = 0 \quad or \quad x - 4 = 0 \quad or \quad x + 2 = 0$$
$$x = 1 \quad or \quad x = 4 \quad or \quad x = -2.$$

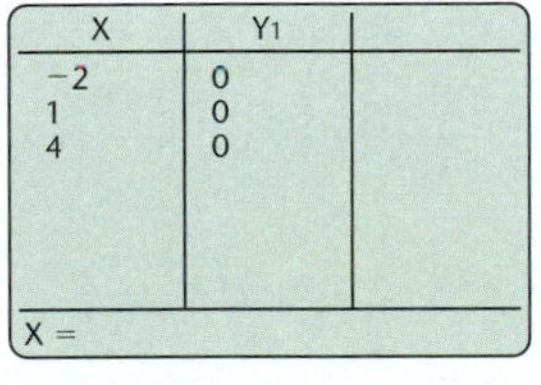

The solutions of the equation $x^3 - 3x^2 - 6x + 8 = 0$ are -2, 1, and 4. They are also the zeros of the function $f(x) = x^3 - 3x^2 - 6x + 8$. We can use a table set in ASK mode to check the solutions.

Now Try Exercise 41.

CONNECTING THE CONCEPTS

Consider the function

$$f(x) = (x - 2)(x + 3)(x + 1), \quad \text{or} \quad f(x) = x^3 + 2x^2 - 5x - 6,$$

and its graph.

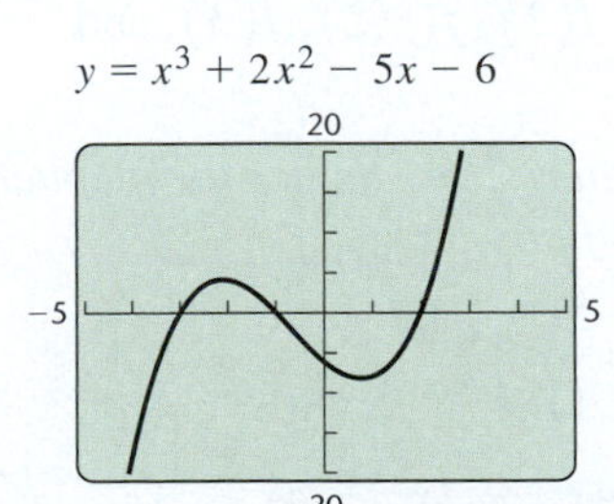

We can make the following statements:

- -3 is a zero of f.
- $f(-3) = 0$.
- -3 is a solution of $f(x) = 0$.
- $(-3, 0)$ is an x-intercept of the graph of f.
- 0 is the remainder when $f(x)$ is divided by $x + 3$, or $x - (-3)$.
- $x - (-3)$, or $x + 3$, is a factor of f.

Similar statements are also true for -1 and 2.

4.3 Exercise Set

1. For the function
$$f(x) = x^4 - 6x^3 + x^2 + 24x - 20,$$
use long division to determine whether each of the following is a factor of $f(x)$.

a) $x + 1$ **b)** $x - 2$ **c)** $x + 5$

2. For the function
$$h(x) = x^3 - x^2 - 17x - 15,$$
use long division to determine whether each of the following is a factor of $h(x)$.

a) $x + 5$ **b)** $x + 1$ **c)** $x + 3$

3. For the function
$$g(x) = x^3 - 2x^2 - 11x + 12,$$
use long division to determine whether each of the following is a factor of $g(x)$.

a) $x - 4$ **b)** $x - 3$ **c)** $x - 1$

4. For the function
$$f(x) = x^4 + 8x^3 + 5x^2 - 38x + 24,$$
use long division to determine whether each of the following is a factor of $f(x)$.

a) $x + 6$ **b)** $x + 1$ **c)** $x - 4$

In each of the following, a polynomial $P(x)$ and a divisor $d(x)$ are given. Use long division to find the quotient $Q(x)$ and the remainder $R(x)$ when $P(x)$ is divided by $d(x)$. Express $P(x)$ in the form $d(x) \cdot Q(x) + R(x)$.

5. $P(x) = x^3 - 8$,
$d(x) = x + 2$

6. $P(x) = 2x^3 - 3x^2 + x - 1$,
$d(x) = x - 3$

7. $P(x) = x^3 + 6x^2 - 25x + 18$,
$d(x) = x + 9$

8. $P(x) = x^3 - 9x^2 + 15x + 25$,
$d(x) = x - 5$

9. $P(x) = x^4 - 2x^2 + 3$,
$d(x) = x + 2$

10. $P(x) = x^4 + 6x^3$,
$d(x) = x - 1$

Use synthetic division to find the quotient and the remainder.

11. $(2x^4 + 7x^3 + x - 12) \div (x + 3)$

12. $(x^3 - 7x^2 + 13x + 3) \div (x - 2)$

13. $(x^3 - 2x^2 - 8) \div (x + 2)$

14. $(x^3 - 3x + 10) \div (x - 2)$

15. $(3x^3 - x^2 + 4x - 10) \div (x + 1)$

16. $(4x^4 - 2x + 5) \div (x + 3)$

17. $(x^5 + x^3 - x) \div (x - 3)$

18. $(x^7 - x^6 + x^5 - x^4 + 2) \div (x + 1)$

19. $(x^4 - 1) \div (x - 1)$

20. $(x^5 + 32) \div (x + 2)$

21. $(2x^4 + 3x^2 - 1) \div \left(x - \frac{1}{2}\right)$

22. $(3x^4 - 2x^2 + 2) \div \left(x - \frac{1}{4}\right)$

Use synthetic division to find the function values. Then check your work using a graphing calculator.

23. $f(x) = x^3 - 6x^2 + 11x - 6$; find $f(1)$, $f(-2)$, and $f(3)$.

24. $f(x) = x^3 + 7x^2 - 12x - 3$; find $f(-3)$, $f(-2)$, and $f(1)$.

25. $f(x) = x^4 - 3x^3 + 2x + 8$; find $f(-1)$, $f(4)$, and $f(-5)$.

26. $f(x) = 2x^4 + x^2 - 10x + 1$; find $f(-10)$, $f(2)$, and $f(3)$.

27. $f(x) = 2x^5 - 3x^4 + 2x^3 - x + 8$; find $f(20)$ and $f(-3)$.

28. $f(x) = x^5 - 10x^4 + 20x^3 - 5x - 100$; find $f(-10)$ and $f(5)$.

29. $f(x) = x^4 - 16$; find $f(2)$, $f(-2)$, $f(3)$, and $f\left(1 - \sqrt{2}\right)$.

30. $f(x) = x^5 + 32$; find $f(2)$, $f(-2)$, $f(3)$, and $f(2 + 3i)$.

Using synthetic division, determine whether the numbers are zeros of the polynomial function.

31. $-3, 2$; $f(x) = 3x^3 + 5x^2 - 6x + 18$

32. $-4, 2$; $f(x) = 3x^3 + 11x^2 - 2x + 8$

33. $-3, 1$; $h(x) = x^4 + 4x^3 + 2x^2 - 4x - 3$

34. $2, -1$; $g(x) = x^4 - 6x^3 + x^2 + 24x - 20$

35. $i, -2i$; $g(x) = x^3 - 4x^2 + 4x - 16$

36. $\frac{1}{3}, 2$; $h(x) = x^3 - x^2 - \frac{1}{9}x + \frac{1}{9}$

37. $-3, \frac{1}{2}$; $f(x) = x^3 - \frac{7}{2}x^2 + x - \frac{3}{2}$

38. $i, -i, -2$; $f(x) = x^3 + 2x^2 + x + 2$

Factor the polynomial function $f(x)$. Then solve the equation $f(x) = 0$.

39. $f(x) = x^3 + 4x^2 + x - 6$

40. $f(x) = x^3 + 5x^2 - 2x - 24$

41. $f(x) = x^3 - 6x^2 + 3x + 10$

42. $f(x) = x^3 + 2x^2 - 13x + 10$

43. $f(x) = x^3 - x^2 - 14x + 24$

44. $f(x) = x^3 - 3x^2 - 10x + 24$

45. $f(x) = x^4 - 7x^3 + 9x^2 + 27x - 54$

46. $f(x) = x^4 - 4x^3 - 7x^2 + 34x - 24$

47. $f(x) = x^4 - x^3 - 19x^2 + 49x - 30$

48. $f(x) = x^4 + 11x^3 + 41x^2 + 61x + 30$

Sketch the graph of the polynomial function. Follow the procedure outlined on p. 237. Use synthetic division and the remainder theorem to find the zeros.

49. $f(x) = x^4 - x^3 - 7x^2 + x + 6$

50. $f(x) = x^4 + x^3 - 3x^2 - 5x - 2$

51. $f(x) = x^3 - 7x + 6$

52. $f(x) = x^3 - 12x + 16$

53. $f(x) = -x^3 + 3x^2 + 6x - 8$

54. $f(x) = -x^4 + 2x^3 + 3x^2 - 4x - 4$

Skill Maintenance

Solve. Find exact solutions. [3.2]

55. $2x^2 + 12 = 5x$

56. $7x^2 + 4x = 3$

In Exercises 57–59, consider the function

$$g(x) = x^2 + 5x - 14.$$

57. What are the inputs if the output is -14? [3.2]

58. What is the output if the input is 3? [1.2]

59. Given an output of -20, find the corresponding inputs. [3.2]

60. *Disruptive Airline Passengers.* The number of reported disruptive airline passengers has increased linearly over the years, rising from 339 cases in 2007 to 8217 cases in 2013 (*Source:* International Air Transportation Association). Using these two data points, find a linear function, $f(x) = mx + b$, that models the data. Let x represent the number of years after 2007. Then use this function to estimate the number of reported cases of disruptive airline passengers in 2011 and in 2017. [1.4]

61. The sum of the base and the height of a triangle is 30 in. Find the dimensions for which the area is a maximum. [3.3]

Synthesis

In Exercises 62 and 63, a graph of a polynomial function is given. On the basis of the graph:

a) *Find as many factors of the polynomial as you can.*

b) *Construct a polynomial function with the zeros shown in the graph.*

c) *Can you find any other polynomial functions with the given zeros?*

d) *Can you find more than one polynomial function with the given zeros and the same graph?*

62.

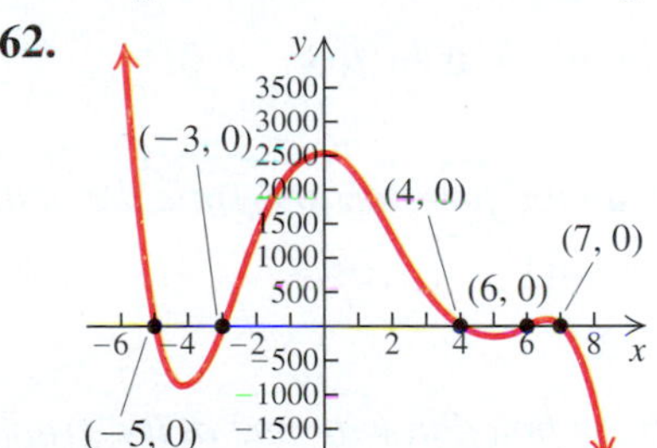

63.

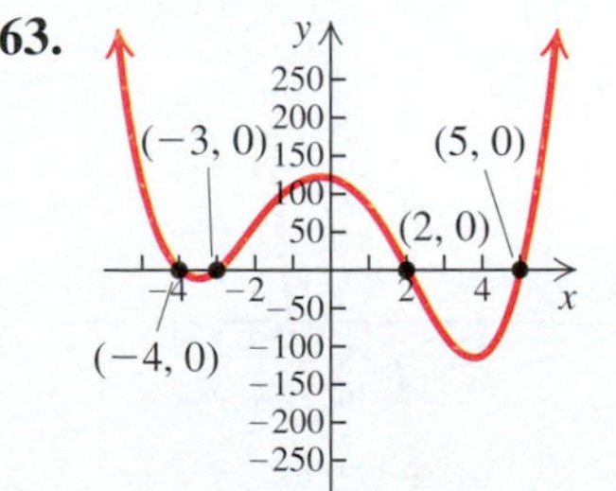

64. For what values of k will the remainder be the same when $x^2 + kx + 4$ is divided by $x - 1$ and by $x + 1$?

65. Find k such that $x + 2$ is a factor of $x^3 - kx^2 + 3x + 7k$.

66. *Beam Deflection.* A beam rests at two points A and B and has a concentrated load applied to its center. Let $y =$ the deflection, in feet, of the beam at a distance of x feet from A. Under certain conditions, this deflection is given by

$$y = \tfrac{1}{13}x^3 - \tfrac{1}{14}x.$$

Find the zeros of the polynomial on the interval $[0, 2]$.

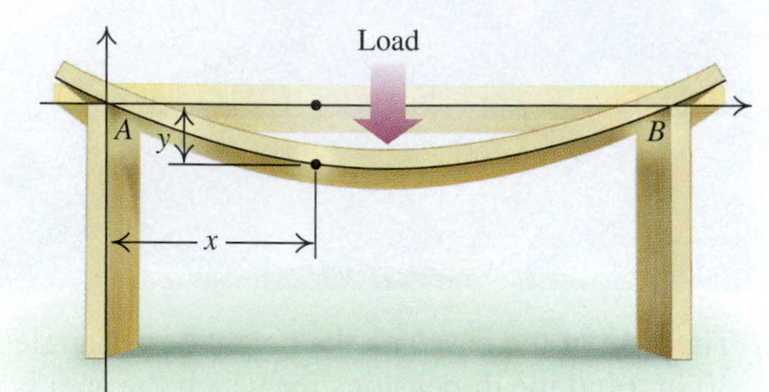

Solve.

67. $\dfrac{2x^2}{x^2 - 1} + \dfrac{4}{x + 3} = \dfrac{12x - 4}{x^3 + 3x^2 - x - 3}$

68. $\dfrac{6x^2}{x^2 + 11} + \dfrac{60}{x^3 - 7x^2 + 11x - 77} = \dfrac{1}{x - 7}$

69. Find a 15th-degree polynomial for which $x - 1$ is a factor. Answers may vary.

Use synthetic division to divide.

70. $(x^4 - y^4) \div (x - y)$

71. $(x^3 + 3ix^2 - 4ix - 2) \div (x + i)$

72. $(x^2 - 4x - 2) \div [x - (3 + 2i)]$

73. $(x^2 - 3x + 7) \div (x - i)$

Mid-Chapter Mixed Review

Determine whether the statement is true or false.

1. The y-intercept of the graph of the function $P(x) = 5 - 2x^3$ is $(5, 0)$. **[4.2]**

2. The degree of the polynomial $x - \frac{1}{2}x^4 - 3x^6 + x^5$ is 6. **[4.1]**

3. If $f(x) = (x + 7)(x - 8)$, then $f(8) = 0$. **[4.3]**

4. If $f(12) = 0$, then $x + 12$ is a factor of $f(x)$. **[4.3]**

Find the zeros of the polynomial function and state the multiplicity of each. **[4.1]**

5. $f(x) = (x^2 - 10x + 25)^3$

6. $h(x) = 2x^3 + x^2 - 50x - 25$

7. $g(x) = x^4 - 3x^2 + 2$

8. $f(x) = -6(x - 3)^2(x + 4)$

In Exercises 9–12, match the function with one of the graphs (a)–(d) that follow. **[4.2]**

a)

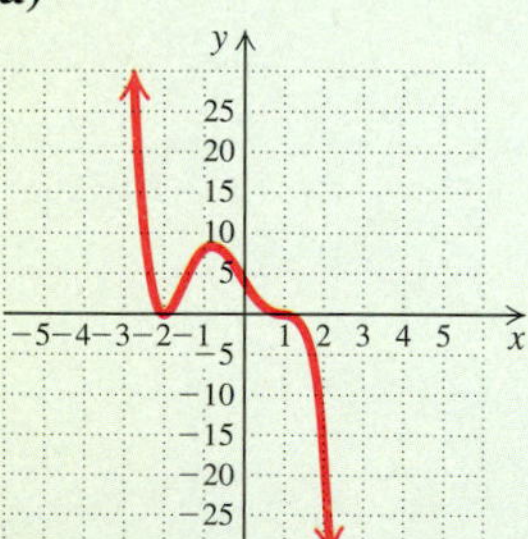

b)

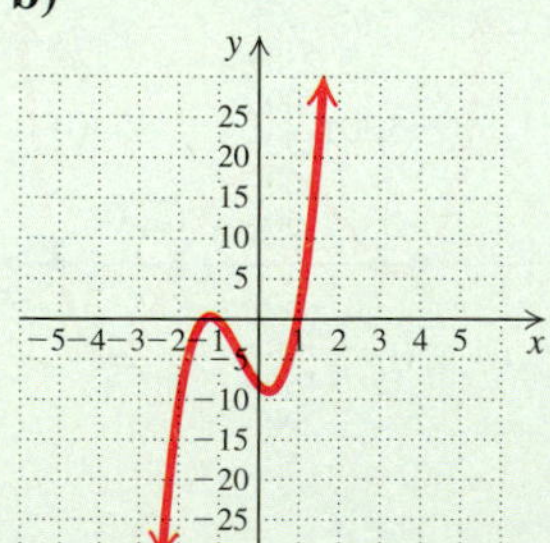

c)

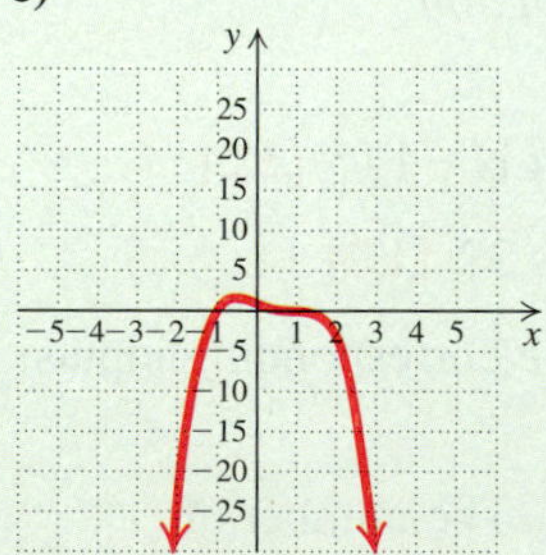

d)

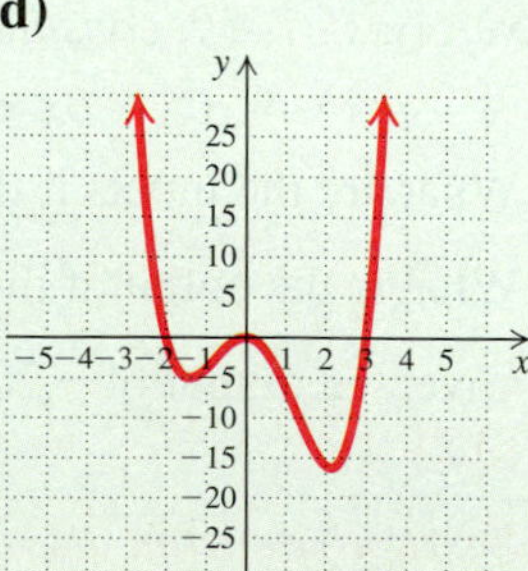

9. $f(x) = x^4 - x^3 - 6x^2$

10. $f(x) = -(x - 1)^3(x + 2)^2$

11. $f(x) = 6x^3 + 8x^2 - 6x - 8$

12. $f(x) = -(x - 1)^3(x + 1)$

Using the intermediate value theorem, determine, if possible, whether the function has at least one real zero between a and b. **[4.2]**

13. $f(x) = x^3 - 2x^2 + 3;\ a = -2, b = 0$

14. $f(x) = x^3 - 2x^2 + 3;\ a = -\frac{1}{2}, b = 1$

15. For the polynomial $P(x) = x^4 - 6x^3 + x - 2$ and the divisor $d(x) = x - 1$, use long division to find the quotient $Q(x)$ and the remainder $R(x)$ when $P(x)$ is divided by $d(x)$. Express $P(x)$ in the form $d(x) \cdot Q(x) + R(x)$. [4.3]

Use synthetic division to find the quotient and the remainder. [4.3]

16. $(3x^4 - x^3 + 2x^2 - 6x + 6) \div (x - 2)$

17. $(x^5 - 5) \div (x + 1)$

Use synthetic division to find the function values. [4.3]

18. $g(x) = x^3 - 9x^2 + 4x - 10$; find $g(-5)$

19. $f(x) = 20x^2 - 40x$; find $f\left(\frac{1}{2}\right)$

20. $f(x) = 5x^4 + x^3 - x$; find $f(-\sqrt{2})$

Using synthetic division, determine whether the numbers are zeros of the polynomial function. [4.3]

21. $-3i, 3$; $f(x) = x^3 - 4x^2 + 9x - 36$

22. $-1, 5$; $f(x) = x^6 - 35x^4 + 259x^2 - 225$

Factor the polynomial function $f(x)$. Then solve the equation $f(x) = 0$. [4.3]

23. $f(x) = x^3 - 2x^2 - 55x + 56$

24. $f(x) = x^4 - 2x^3 - 13x^2 + 14x + 24$

COLLABORATIVE DISCUSSION AND WRITING

25. How is the range of a polynomial function related to the degree of the polynomial? [4.1]

26. Is it possible for the graph of a polynomial function to have no y-intercept? no x-intercepts? Explain your answer. [4.2]

27. Explain why values of a function must be all positive or all negative between consecutive zeros. [4.2]

28. In synthetic division, why is the degree of the quotient 1 less than that of the dividend? [4.3]

4.4 Theorems about Zeros of Polynomial Functions

- Find a polynomial with specified zeros.
- For a polynomial function with integer coefficients, find the rational zeros and the other zeros, if possible.
- Use Descartes' rule of signs to find information about the number of real zeros of a polynomial function with real coefficients.

We will now allow the coefficients of a polynomial to be complex numbers. In certain cases, we will restrict the coefficients to be real numbers, rational numbers, or integers, as shown in the following examples.

Polynomial	Type of Coefficient
$5x^3 - 3x^2 + (2 + 4i)x + i$	Complex
$5x^3 - 3x^2 + \sqrt{2}x - \pi$	Real
$5x^3 - 3x^2 + \frac{2}{3}x - \frac{7}{4}$	Rational
$5x^3 - 3x^2 + 8x - 11$	Integer

The Fundamental Theorem of Algebra

A linear, or first-degree, polynomial function $f(x) = mx + b$ (where $m \neq 0$) has just one zero, $-b/m$. It can be shown that any quadratic polynomial function $f(x) = ax^2 + bx + c$ with complex numbers for coefficients has at least one, and at most two, complex zeros. The following theorem is a generalization. No proof is given in this text.

> **THE FUNDAMENTAL THEOREM OF ALGEBRA**
>
> Every polynomial function of degree n, with $n \geq 1$, has at least one zero in the set of complex numbers.

Note that although the fundamental theorem of algebra guarantees that a zero exists, it does not tell how to find it. Recall that the zeros of a polynomial function $f(x)$ are the solutions of the polynomial equation $f(x) = 0$. We now develop some concepts that can help in finding zeros. First, we consider one of the results of the fundamental theorem of algebra.

> Every polynomial function f of degree n, with $n \geq 1$, can be factored into n linear factors (not necessarily unique); that is,
>
> $$f(x) = a_n(x - c_1)(x - c_2) \cdots (x - c_n).$$

Finding Polynomials with Given Zeros

Given several numbers, we can find a polynomial function with those numbers as its zeros.

$y = x^3 - x^2 + 9x - 9$

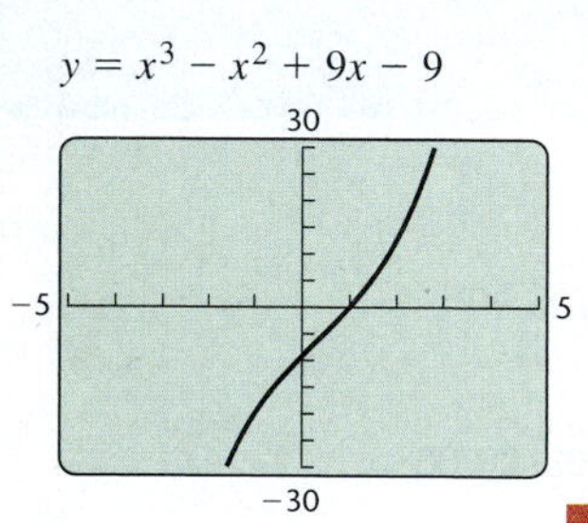

EXAMPLE 1 Find a polynomial function of degree 3, having the zeros 1, $3i$, and $-3i$.

Solution Such a function has factors $x - 1$, $x - 3i$, and $x + 3i$, so we have

$$f(x) = a_n(x - 1)(x - 3i)(x + 3i).$$

The number a_n can be any nonzero number. The simplest polynomial function will be obtained if we let it be 1. If we then multiply the factors, we obtain

Just in Time **12**

$$\begin{aligned} f(x) &= (x - 1)(x^2 - 9i^2) && \text{Multiplying } (x - 3i)(x + 3i) \\ &= (x - 1)(x^2 + 9) && -9i^2 = -9(-1) = 9 \\ &= x^3 - x^2 + 9x - 9. \end{aligned}$$

Now Try Exercise 3.

$y = x^5 - x^4 - 9x^3 - 11x^2 - 4x$

100
-3
5
-200

EXAMPLE 2 Find a polynomial function of degree 5 with -1 as a zero of multiplicity 3, 4 as a zero of multiplicity 1, and 0 as a zero of multiplicity 1.

Solution Proceeding as in Example 1, letting $a_n = 1$, we obtain

$$\begin{aligned} f(x) &= [x - (-1)]^3(x - 4)(x - 0) \\ &= (x + 1)^3(x - 4)x \\ &= (x^3 + 3x^2 + 3x + 1)(x^2 - 4x) \\ &= x^5 - x^4 - 9x^3 - 11x^2 - 4x. \end{aligned}$$

Zeros of Polynomial Functions with Real Coefficients

Consider the quadratic equation $x^2 - 2x + 2 = 0$, with real coefficients. Its solutions are $1 + i$ and $1 - i$. Note that they are complex conjugates. This generalizes to any polynomial equation with real coefficients.

NONREAL ZEROS: $a + bi$ AND $a - bi$, $b \neq 0$

If a complex number $a + bi$, $b \neq 0$, is a zero of a polynomial function $f(x)$ with *real* coefficients, then its conjugate, $a - bi$, is also a zero. For example, if $2 + 7i$ is a zero of a polynomial function $f(x)$ with real coefficients, then its conjugate, $2 - 7i$, is also a zero. (Nonreal zeros occur in conjugate pairs.)

In order for the preceding to be true, it is essential that the coefficients be *real* numbers.

Rational Coefficients

When a polynomial function has rational numbers for coefficients, certain irrational zeros also occur in pairs, as described in the following theorem.

IRRATIONAL ZEROS: $a + c\sqrt{b}$ AND $a - c\sqrt{b}$, b IS NOT A PERFECT SQUARE

If $a + c\sqrt{b}$, where a, b, and c are rational and b is not a perfect square, is a zero of a polynomial function $f(x)$ with *rational* coefficients, then its conjugate, $a - c\sqrt{b}$, is also a zero. For example, if $-3 + 5\sqrt{2}$ is a zero of a polynomial function $f(x)$ with rational coefficients, then its conjugate, $-3 - 5\sqrt{2}$, is also a zero. (Irrational zeros occur in conjugate pairs.)

EXAMPLE 3 Suppose that a polynomial function of degree 6 with rational coefficients has

$$-2 + 5i, \quad -2i, \quad \text{and} \quad 1 - \sqrt{3}$$

as three of its zeros. Find the other zeros.

Solution Since the coefficients are rational, the other zeros are the conjugates of the given zeros,

$$-2 - 5i, \quad 2i, \quad \text{and} \quad 1 + \sqrt{3}.$$

There are no other zeros because a polynomial function of degree 6 can have at most 6 zeros.

Now Try Exercise 19.

EXAMPLE 4 Find a polynomial function of lowest degree with rational coefficients that has $-\sqrt{3}$ and $1 + i$ as two of its zeros.

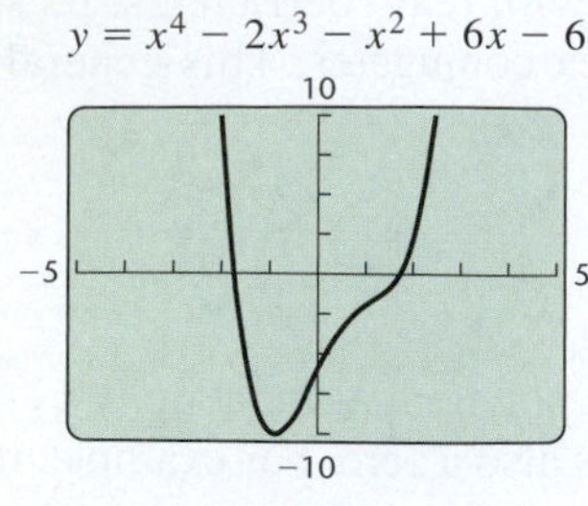

Solution The function must also have the zeros $\sqrt{3}$ and $1 - i$. Because we want to find the polynomial function of lowest degree with the given zeros, we will not include additional zeros; that is, we will write a polynomial function of degree 4. Thus if we let $a_n = 1$, the polynomial function is

$$\begin{aligned} f(x) &= [x - (-\sqrt{3})][x - \sqrt{3}][x - (1 + i)][x - (1 - i)] \\ &= (x + \sqrt{3})(x - \sqrt{3})[(x - 1) - i][(x - 1) + i] \\ &= (x^2 - 3)[(x - 1)^2 - i^2] \\ &= (x^2 - 3)[x^2 - 2x + 1 + 1] \\ &= (x^2 - 3)(x^2 - 2x + 2) \\ &= x^4 - 2x^3 - x^2 + 6x - 6. \end{aligned}$$

Now Try Exercise 39.

Integer Coefficients and the Rational Zeros Theorem

It is not always easy to find the zeros of a polynomial function. However, if a polynomial function has integer coefficients, there is a procedure that will yield all the rational zeros.

THE RATIONAL ZEROS THEOREM

Let

$$P(x) = a_n x^n + a_{n-1}x^{n-1} + \cdots + a_1 x + a_0,$$

where all the coefficients are integers. Consider a rational number denoted by p/q, where p and q are relatively prime (having no common factor besides -1 and 1). If p/q is a zero of $P(x)$, then p is a factor of a_0 and q is a factor of a_n.

EXAMPLE 5 Given $f(x) = 3x^4 - 11x^3 + 10x - 4$:

a) Find the rational zeros and then the other zeros; that is, solve $f(x) = 0$.

b) Factor $f(x)$ into linear factors.

Solution

a) Because the degree of $f(x)$ is 4, there are at most 4 distinct zeros. All of the coefficients are integers. The rational zeros theorem says that if a rational number p/q is a zero of $f(x)$, then p must be a factor of -4 and q must be a factor of 3. Thus the possibilities for p/q are

$$\frac{\textit{Possibilities for } p}{\textit{Possibilities for } q}: \quad \frac{\pm 1, \pm 2, \pm 4}{\pm 1, \pm 3};$$

Possibilities for p/q: $1, -1, 2, -2, 4, -4, \frac{1}{3}, -\frac{1}{3}, \frac{2}{3}, -\frac{2}{3}, \frac{4}{3}, -\frac{4}{3}$.

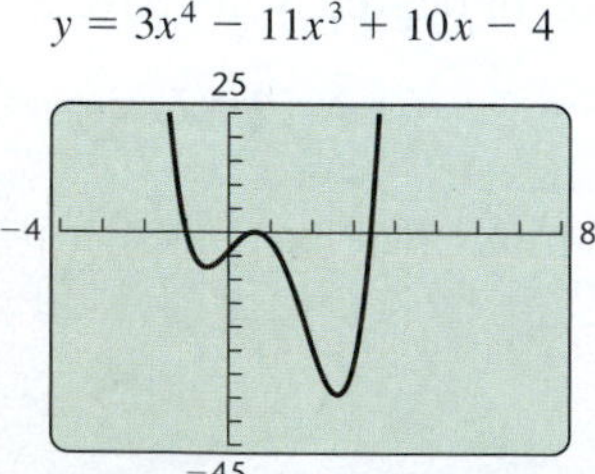

We could use the TABLE feature or some other method to find function values. However, if we use synthetic division, the quotient polynomial becomes a beneficial by-product if a zero is found. Rather than use synthetic division to check *each* of these possibilities, we graph the function and inspect the graph for zeros that appear to be near any of the possible rational zeros. (See the graph at left.)

From the graph, we see that of the possibilities in the list, only the numbers -1, $\frac{1}{3}$, and $\frac{2}{3}$ might be rational zeros.

We try -1.

−1	3	−11	0	10	−4
		−3	14	−14	4
	3	−14	14	−4	0

We have $f(-1) = 0$, so -1 is a zero. Thus, $x + 1$ is a factor of $f(x)$. Using the results of the synthetic division, we can express $f(x)$ as

$$f(x) = (x + 1)(3x^3 - 14x^2 + 14x - 4).$$

We now consider the factor $3x^3 - 14x^2 + 14x - 4$ and check the other possible zeros. We try $\frac{1}{3}$.

$$\begin{array}{r|rrrr} 1/3 & 3 & -14 & 14 & -4 \\ & & 1 & -\frac{13}{3} & \frac{29}{9} \\ \hline & 3 & -13 & \frac{29}{3} & \Big| -\frac{7}{9} \end{array}$$

Since $f\left(\frac{1}{3}\right) \neq 0$, we know that $\frac{1}{3}$ is not a zero.

Let's now try $\frac{2}{3}$.

$$\begin{array}{r|rrrr} 2/3 & 3 & -14 & 14 & -4 \\ & & 2 & -8 & 4 \\ \hline & 3 & -12 & 6 & \Big| \; 0 \end{array}$$

Since the remainder is 0, we know that $x - \frac{2}{3}$ is a factor of $3x^3 - 14x^2 + 14x - 4$ and is also a factor of $f(x)$. Thus, $\frac{2}{3}$ is a zero of $f(x)$.

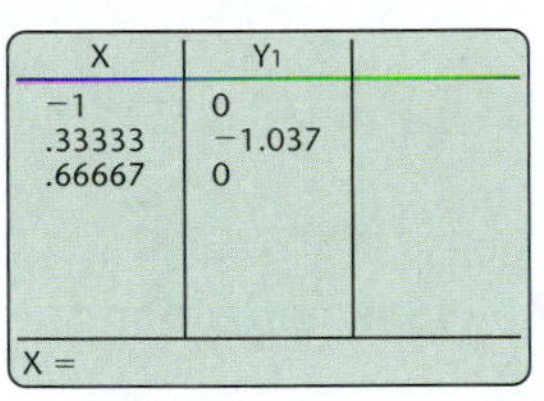

X	Y1
−1	0
.33333	−1.037
.66667	0

X =

We can check the zeros with the TABLE feature. (See the window at left.) Note that the graphing calculator converts $\frac{1}{3}$ and $\frac{2}{3}$ to decimal notation. Since $f(-1) = 0$ and $f\left(\frac{2}{3}\right) = 0$, -1 and $\frac{2}{3}$ are zeros. Since $f\left(\frac{1}{3}\right) \neq 0$, $\frac{1}{3}$ is not a zero.

Using the results of the synthetic division, we can factor further:

$$f(x) = (x + 1)\left(x - \tfrac{2}{3}\right)(3x^2 - 12x + 6) \quad \text{Using the results of the last synthetic division}$$

$$= (x + 1)\left(x - \tfrac{2}{3}\right) \cdot 3 \cdot (x^2 - 4x + 2). \quad \text{Removing a factor of 3}$$

The quadratic formula can be used to find the values of x for which $x^2 - 4x + 2 = 0$. Those values are also zeros of $f(x)$:

$$x = \frac{-b \pm \sqrt{b^2 - 4ac}}{2a}$$

$$= \frac{-(-4) \pm \sqrt{(-4)^2 - 4 \cdot 1 \cdot 2}}{2 \cdot 1} \quad a = 1, b = -4, \text{ and } c = 2$$

$$= \frac{4 \pm \sqrt{8}}{2} = \frac{4 \pm 2\sqrt{2}}{2} = \frac{2(2 \pm \sqrt{2})}{2}$$

$$= 2 \pm \sqrt{2}.$$

The rational zeros are -1 and $\frac{2}{3}$. The other zeros are $2 \pm \sqrt{2}$.

b) The complete factorization of $f(x)$ is

$$f(x) = 3(x + 1)\left(x - \tfrac{2}{3}\right)[x - (2 - \sqrt{2})][x - (2 + \sqrt{2})], \quad \text{or}$$
$$f(x) = (x + 1)(3x - 2)[x - (2 - \sqrt{2})][x - (2 + \sqrt{2})].$$

Replacing $3\left(x - \frac{2}{3}\right)$ with $(3x - 2)$

Now Try Exercise 55.

EXAMPLE 6 Given $f(x) = 2x^5 - x^4 - 4x^3 + 2x^2 - 30x + 15$:

a) Find the rational zeros and then the other zeros; that is, solve $f(x) = 0$.

b) Factor $f(x)$ into linear factors.

Solution

a) Because the degree of $f(x)$ is 5, there are at most 5 distinct zeros. All of the coefficients are integers. According to the rational zeros theorem, any rational zero of f must be of the form p/q, where p is a factor of 15 and q is a factor of 2. The possibilities are

$$\frac{\textit{Possibilities for } p}{\textit{Possibilities for } q}: \quad \frac{\pm 1, \pm 3, \pm 5, \pm 15}{\pm 1, \pm 2};$$

$$\textit{Possibilities for } p/q: \quad 1, -1, 3, -3, 5, -5, 15, -15, \tfrac{1}{2}, -\tfrac{1}{2}, \tfrac{3}{2}, -\tfrac{3}{2}, \tfrac{5}{2}, -\tfrac{5}{2}, \tfrac{15}{2}, -\tfrac{15}{2}.$$

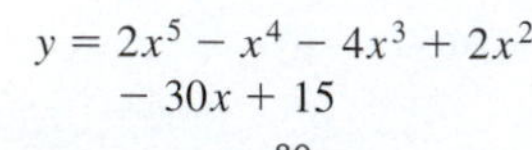

$y = 2x^5 - x^4 - 4x^3 + 2x^2 - 30x + 15$

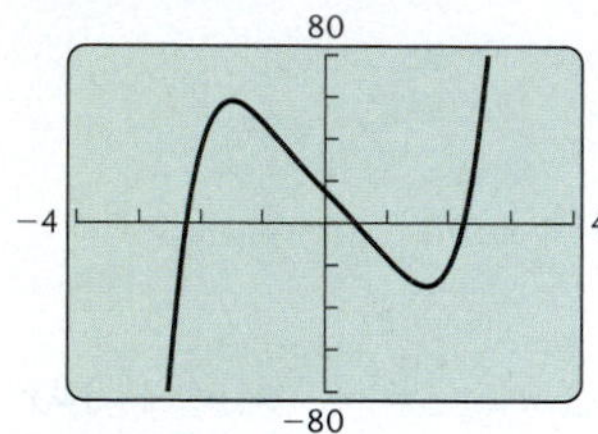

X	Y1
−2.5	−69.38
.5	0
2.5	46.25
X =	

Rather than use synthetic division to check each of these possibilities, we graph $y = 2x^5 - x^4 - 4x^3 + 2x^2 - 30x + 15$. (See the graph at left.) We can then inspect the graph for zeros that appear to be near any of the possible rational zeros.

From the graph, we see that of the possibilities in the list, only the numbers $-\frac{5}{2}, \frac{1}{2}$, and $\frac{5}{2}$ might be rational zeros. By synthetic division or using the TABLE feature of a graphing calculator (see the table at left), we see that only $\frac{1}{2}$ is actually a rational zero.

$$\begin{array}{r|rrrrrr} 1/2 & 2 & -1 & -4 & 2 & -30 & 15 \\ & & 1 & 0 & -2 & 0 & -15 \\ \hline & 2 & 0 & -4 & 0 & -30 & 0 \end{array}$$

This means that $x - \frac{1}{2}$ is a factor of $f(x)$. We write the factorization and try to factor further:

$$\begin{aligned} f(x) &= \left(x - \tfrac{1}{2}\right)(2x^4 - 4x^2 - 30) \\ &= \left(x - \tfrac{1}{2}\right)\cdot 2\cdot(x^4 - 2x^2 - 15) \qquad \text{Factoring out the 2} \\ &= \left(x - \tfrac{1}{2}\right)\cdot 2\cdot(x^2 - 5)(x^2 + 3). \qquad \text{Factoring the trinomial} \end{aligned}$$

We now solve the equation $f(x) = 0$ to determine the zeros. We use the principle of zero products:

$$\begin{aligned} &\left(x - \tfrac{1}{2}\right)\cdot 2\cdot(x^2 - 5)(x^2 + 3) = 0 \\ x - \tfrac{1}{2} = 0 \quad &or \quad x^2 - 5 = 0 \quad or \quad x^2 + 3 = 0 \\ x = \tfrac{1}{2} \quad &or \quad x^2 = 5 \quad or \quad x^2 = -3 \\ x = \tfrac{1}{2} \quad &or \quad x = \pm\sqrt{5} \quad or \quad x = \pm\sqrt{3}i. \end{aligned}$$

There is only one rational zero, $\frac{1}{2}$. The other zeros are $\pm\sqrt{5}$ and $\pm\sqrt{3}i$.

b) The factorization into linear factors is

$$f(x) = 2\left(x - \tfrac{1}{2}\right)(x + \sqrt{5})(x - \sqrt{5})(x + \sqrt{3}i)(x - \sqrt{3}i), \quad \text{or}$$
$$f(x) = (2x - 1)(x + \sqrt{5})(x - \sqrt{5})(x + \sqrt{3}i)(x - \sqrt{3}i).$$

Replacing $2(x - \frac{1}{2})$ with $(2x - 1)$

Now Try Exercise 61.

Descartes' Rule of Signs

The development of a rule that helps determine the number of positive real zeros and the number of negative real zeros of a polynomial function is credited to the French mathematician René Descartes. To use the rule, we must have the polynomial arranged in descending order or ascending order, with no zero terms written in and the constant term not 0. Then we determine the number of *variations of sign*, that is, the number of times, in reading through the polynomial, that successive coefficients are of different signs.

EXAMPLE 7 Determine the number of variations of sign in the polynomial function $P(x) = 2x^5 - 3x^2 + x + 4$.

Solution We have

$$P(x) = 2x^5 - 3x^2 + x + 4$$

From positive to negative; a variation

From negative to positive; a variation

Both positive; no variation

The number of variations of sign is 2.

Just in Time 7

Note the following:

$$P(-x) = 2(-x)^5 - 3(-x)^2 + (-x) + 4$$
$$= -2x^5 - 3x^2 - x + 4.$$

We see that the number of variations of sign in $P(-x)$ is 1. It occurs as we go from $-x$ to 4.

We now state Descartes' rule, without proof.

DESCARTES' RULE OF SIGNS

Let $P(x)$, written in descending order or ascending order, be a polynomial function with real coefficients and a nonzero constant term. The number of positive real zeros of $P(x)$ is either:

1. The same as the number of variations of sign in $P(x)$, or
2. Less than the number of variations of sign in $P(x)$ by a positive even integer.

The number of negative real zeros of $P(x)$ is either:

3. The same as the number of variations of sign in $P(-x)$, or
4. Less than the number of variations of sign in $P(-x)$ by a positive even integer.

A zero of multiplicity m must be counted m times.

In each of Examples 8–10, what does Descartes' rule of signs tell you about the number of positive real zeros and the number of negative real zeros?

EXAMPLE 8 $P(x) = 2x^5 - 5x^2 - 3x + 6$

Solution The number of variations of sign in $P(x)$ is 2. Therefore, the number of positive real zeros is either 2 or less than 2 by 2, 4, 6, and so on. Thus the number of positive real zeros is either 2 or 0, since a negative number of zeros has no meaning.

$$P(-x) = -2x^5 - 5x^2 + 3x + 6$$

The number of variations of sign in $P(-x)$ is 1. Thus there is exactly 1 negative real zero. Since nonreal, complex conjugates occur in pairs, we also know the possible ways in which nonreal zeros might occur. The table at left summarizes all the possibilities for real zeros and nonreal zeros of $P(x)$.

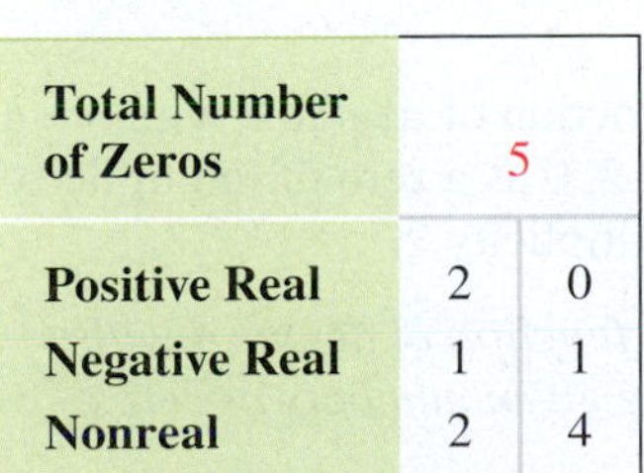

Total Number of Zeros	5	
Positive Real	2	0
Negative Real	1	1
Nonreal	2	4

Now Try Exercise 93.

EXAMPLE 9 $P(x) = 5x^4 - 3x^3 + 7x^2 - 12x + 4$

Solution There are 4 variations of sign. Thus the number of positive real zeros is either

$$4 \quad or \quad 4 - 2 \quad or \quad 4 - 4;$$

that is, the number of positive real zeros is 4, 2, or 0.

$$P(-x) = 5x^4 + 3x^3 + 7x^2 + 12x + 4$$

There are 0 variations of sign, so there are no negative real zeros. The table at left summarizes all the possibilities for real zeros and nonreal zeros of $P(x)$.

Total Number of Zeros	4		
Positive Real	4	2	0
Negative Real	0	0	0
Nonreal	0	2	4

Now Try Exercise 81.

EXAMPLE 10 $P(x) = 6x^6 - 2x^2 - 5x$

Solution As written, the polynomial does not satisfy the conditions of Descartes' rule of signs because the constant term is 0. But because x is a factor of every term, we know that the polynomial has 0 as a zero. We can then factor as follows:

$$P(x) = x(6x^5 - 2x - 5).$$

Now we analyze $Q(x) = 6x^5 - 2x - 5$ and $Q(-x) = -6x^5 + 2x - 5$. The number of variations of sign in $Q(x)$ is 1. Therefore, there is exactly 1 positive real zero. The number of variations of sign in $Q(-x)$ is 2. Thus the number of negative real zeros is 2 or 0. The same results apply to $P(x)$. Since nonreal, complex conjugates occur in pairs, we know the possible ways in which nonreal zeros might occur. The table at left summarizes all the possibilities for real zeros and nonreal zeros of $P(x)$.

Total Number of Zeros	6	
0 as a Zero	1	1
Positive Real	1	1
Negative Real	2	0
Nonreal	2	4

Now Try Exercise 95.

4.4 Exercise Set

Find a polynomial function of degree 3 with the given numbers as zeros.

1. $-2, 3, 5$
2. $-1, 0, 4$
3. $-3, 2i, -2i$
4. $2, i, -i$
5. $\sqrt{2}, -\sqrt{2}, 3$
6. $-5, \sqrt{3}, -\sqrt{3}$
7. $1 - \sqrt{3}, 1 + \sqrt{3}, -2$
8. $-4, 1 - \sqrt{5}, 1 + \sqrt{5}$
9. $1 + 6i, 1 - 6i, -4$
10. $1 + 4i, 1 - 4i, -1$
11. $-\frac{1}{3}, 0, 2$
12. $-3, 0, \frac{1}{2}$
13. Find a polynomial function of degree 5 with -1 as a zero of multiplicity 3, 0 as a zero of multiplicity 1, and 1 as a zero of multiplicity 1.
14. Find a polynomial function of degree 4 with -2 as a zero of multiplicity 1, 3 as a zero of multiplicity 2, and -1 as a zero of multiplicity 1.
15. Find a polynomial function of degree 4 with -1 as a zero of multiplicity 3 and 0 as a zero of multiplicity 1.
16. Find a polynomial function of degree 5 with $-\frac{1}{2}$ as a zero of multiplicity 2, 0 as a zero of multiplicity 1, and 1 as a zero of multiplicity 2.

Suppose that a polynomial function of degree 4 with rational coefficients has the given numbers as zeros. Find the other zero(s).

17. $-1, \sqrt{3}, \frac{11}{3}$
18. $-\sqrt{2}, -1, \frac{4}{5}$
19. $-i, 2 - \sqrt{5}$
20. $i, -3 + \sqrt{3}$

21. $3i, 0, -5$

22. $3, 0, -2i$

23. $-4 - 3i, 2 - \sqrt{3}$

24. $6 - 5i, -1 + \sqrt{7}$

Suppose that a polynomial function of degree 5 with rational coefficients has the given numbers as zeros. Find the other zero(s).

25. $-\frac{1}{2}, \sqrt{5}, -4i$

26. $\frac{3}{4}, -\sqrt{3}, 2i$

27. $-5, 0, 2 - i, 4$

28. $-2, 3, 4, 1 - i$

29. $6, -3 + 4i, 4 - \sqrt{5}$

30. $-3 - 3i, 2 + \sqrt{13}, 6$

31. $-\frac{3}{4}, \frac{3}{4}, 0, 4 - i$

32. $-0.6, 0, 0.6, -3 + \sqrt{2}$

Find a polynomial function of lowest degree with rational coefficients that has the given numbers as some of its zeros.

33. $1 + i, 2$

34. $2 - i, -1$

35. $4i$

36. $-5i$

37. $-4i, 5$

38. $3, -i$

39. $1 - i, -\sqrt{5}$

40. $2 - \sqrt{3}, 1 + i$

41. $\sqrt{5}, -3i$

42. $-\sqrt{2}, 4i$

Given that the polynomial function has the given zero, find the other zeros.

43. $f(x) = x^3 + 5x^2 - 2x - 10;\ -5$

44. $f(x) = x^3 - x^2 + x - 1;\ 1$

45. $f(x) = x^4 - 5x^3 + 7x^2 - 5x + 6;\ -i$

46. $f(x) = x^4 - 16;\ 2i$

47. $f(x) = x^3 - 6x^2 + 13x - 20;\ 4$

48. $f(x) = x^3 - 8;\ 2$

List all possible rational zeros of the function.

49. $f(x) = x^5 - 3x^2 + 1$

50. $f(x) = x^7 + 37x^5 - 6x^2 + 12$

51. $f(x) = 2x^4 - 3x^3 - x + 8$

52. $f(x) = 3x^3 - x^2 + 6x - 9$

53. $f(x) = 15x^6 + 47x^2 + 2$

54. $f(x) = 10x^{25} + 3x^{17} - 35x + 6$

For each polynomial function:

a) *Find the rational zeros and then the other zeros; that is, solve $f(x) = 0$.*

b) *Factor $f(x)$ into linear factors.*

55. $f(x) = x^3 + 3x^2 - 2x - 6$

56. $f(x) = x^3 - x^2 - 3x + 3$

57. $f(x) = 3x^3 - x^2 - 15x + 5$

58. $f(x) = 4x^3 - 4x^2 - 3x + 3$

59. $f(x) = x^3 - 3x + 2$

60. $f(x) = x^3 - 2x + 4$

61. $f(x) = 2x^3 + 3x^2 + 18x + 27$

62. $f(x) = 2x^3 + 7x^2 + 2x - 8$

63. $f(x) = 5x^4 - 4x^3 + 19x^2 - 16x - 4$

64. $f(x) = 3x^4 - 4x^3 + x^2 + 6x - 2$

65. $f(x) = x^4 - 3x^3 - 20x^2 - 24x - 8$

66. $f(x) = x^4 + 5x^3 - 27x^2 + 31x - 10$

67. $f(x) = x^3 - 4x^2 + 2x + 4$

68. $f(x) = x^3 - 8x^2 + 17x - 4$

69. $f(x) = x^3 + 8$

70. $f(x) = x^3 - 8$

71. $f(x) = \frac{1}{3}x^3 - \frac{1}{2}x^2 - \frac{1}{6}x + \frac{1}{6}$

72. $f(x) = \frac{2}{3}x^3 - \frac{1}{2}x^2 + \frac{2}{3}x - \frac{1}{2}$

Find only the rational zeros of the function. If there are none, state this.

73. $f(x) = x^4 + 2x^3 - 5x^2 - 4x + 6$

74. $f(x) = x^4 - 3x^3 - 9x^2 - 3x - 10$

75. $f(x) = x^3 - x^2 - 4x + 3$

76. $f(x) = 2x^3 + 3x^2 + 2x + 3$

77. $f(x) = x^4 + 2x^3 + 2x^2 - 4x - 8$

78. $f(x) = x^4 + 6x^3 + 17x^2 + 36x + 66$

79. $f(x) = x^5 - 5x^4 + 5x^3 + 15x^2 - 36x + 20$

80. $f(x) = x^5 - 3x^4 - 3x^3 + 9x^2 - 4x + 12$

What does Descartes' rule of signs tell you about the number of positive real zeros and the number of negative real zeros of the function?

81. $f(x) = 3x^5 - 2x^2 + x - 1$

82. $g(x) = 5x^6 - 3x^3 + x^2 - x$

83. $h(x) = 6x^7 + 2x^2 + 5x + 4$

84. $P(x) = -3x^5 - 7x^3 - 4x - 5$

85. $F(p) = 3p^{18} + 2p^4 - 5p^2 + p + 3$

86. $H(t) = 5t^{12} - 7t^4 + 3t^2 + t + 1$

87. $C(x) = 7x^6 + 3x^4 - x - 10$

88. $g(z) = -z^{10} + 8z^7 + z^3 + 6z - 1$

89. $h(t) = -4t^5 - t^3 + 2t^2 + 1$

90. $P(x) = x^6 + 2x^4 - 9x^3 - 4$

91. $f(y) = y^4 + 13y^3 - y + 5$

92. $Q(x) = x^4 - 2x^2 + 12x - 8$

93. $r(x) = x^4 - 6x^2 + 20x - 24$

94. $f(x) = x^5 - 2x^3 - 8x$

95. $R(x) = 3x^5 - 5x^3 - 4x$

96. $f(x) = x^4 - 9x^2 - 6x + 4$

Sketch the graph of the polynomial function. Follow the procedure outlined on p. 237. Use the rational zeros theorem when finding the zeros.

97. $f(x) = 4x^3 + x^2 - 8x - 2$

98. $f(x) = 3x^3 - 4x^2 - 5x + 2$

99. $f(x) = 2x^4 - 3x^3 - 2x^2 + 3x$

100. $f(x) = 4x^4 - 37x^2 + 9$

Skill Maintenance

For Exercises 101 and 102, complete the square to:

a) *find the vertex;*
b) *find the axis of symmetry; and*
c) *determine whether there is a maximum or a minimum function value and find that value.* **[3.3]**

101. $f(x) = x^2 - 8x + 10$

102. $f(x) = 3x^2 - 6x - 1$

Find the zeros of the function.

103. $f(x) = -\frac{4}{5}x + 8$ **[1.5]**

104. $g(x) = x^2 - 8x - 33$ **[3.2]**

Determine the leading term, the leading coefficient, and the degree of the polynomial. Then describe the end behavior of the function's graph and classify the polynomial function as constant, linear, quadratic, cubic, or quartic.

105. $g(x) = -x^3 - 2x^2$ **[4.1]**

106. $f(x) = -x^2 - 3x + 6$ **[3.3]**

107. $f(x) = -\frac{4}{9}$ **[1.3], [4.1]**

108. $h(x) = x - 2$ **[1.3], [4.1]**

109. $g(x) = x^4 - 2x^3 + x^2 - x + 2$ **[4.1]**

110. $h(x) = x^3 + \frac{1}{2}x^2 - 4x - 3$ **[4.1]**

Synthesis

111. Consider $f(x) = 2x^3 - 5x^2 - 4x + 3$. Find the solutions of each equation.

a) $f(x) = 0$
b) $f(x - 1) = 0$
c) $f(x + 2) = 0$
d) $f(2x) = 0$

112. Use the rational zeros theorem and the equation $x^4 - 12 = 0$ to show that $\sqrt[4]{12}$ is irrational.

Find the rational zeros of the function.

113. $P(x) = 2x^5 - 33x^4 - 84x^3 + 2203x^2 - 3348x - 10{,}080$

114. $P(x) = x^6 - 6x^5 - 72x^4 - 81x^2 + 486x + 5832$

4.5 Rational Functions

- For a rational function, find the domain and graph the function, identifying all of the asymptotes.
- Solve applied problems involving rational functions.

Now we turn our attention to functions that represent the quotient of two polynomials. Whereas the sum, the difference, or the product of two polynomials is a polynomial, in general the quotient of two polynomials is *not* itself a polynomial.

A *rational number* can be expressed as the quotient of two integers, p/q, where $q \neq 0$. A *rational function* is formed by the quotient of two polynomials, $p(x)/q(x)$, where $q(x) \neq 0$. Here are some examples of rational functions and their graphs.

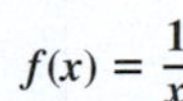

$f(x) = \frac{1}{x}$

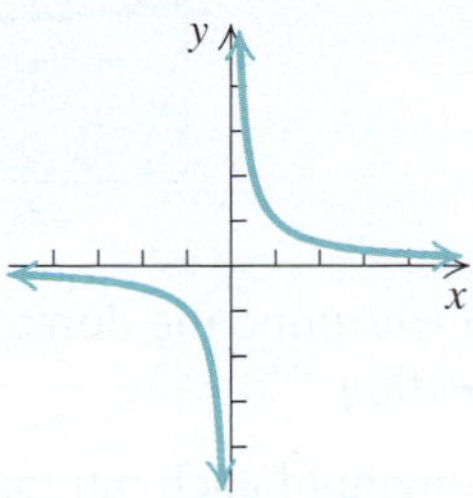

$f(x) = \frac{1}{x^2}$

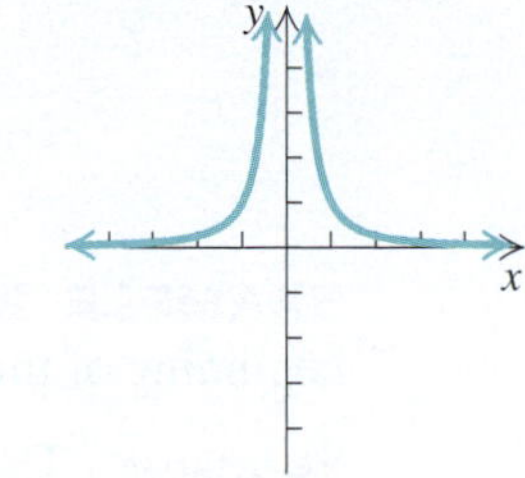

$f(x) = \frac{x - 3}{x^2 + x - 2}$

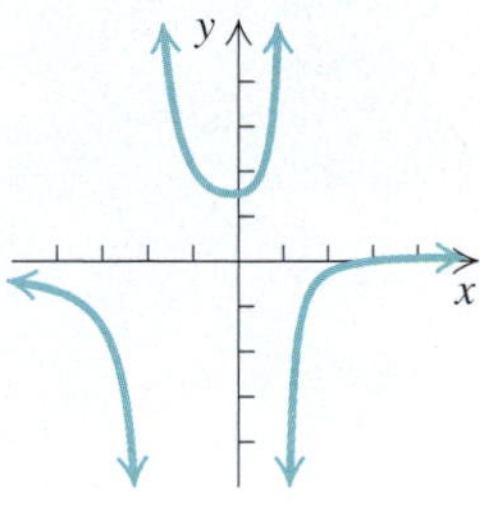

$f(x) = \frac{2x + 5}{2x - 6}$

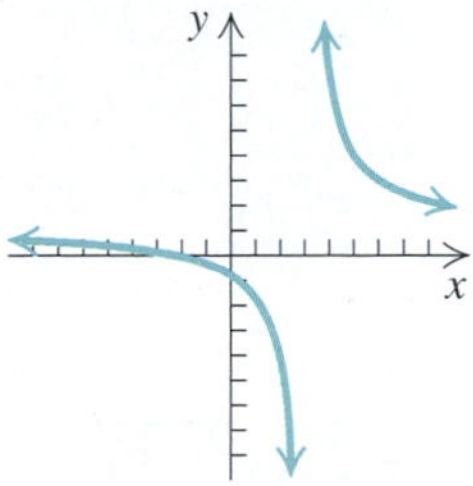

$f(x) = \frac{x^2 + 2x - 3}{x^2 - x - 2}$

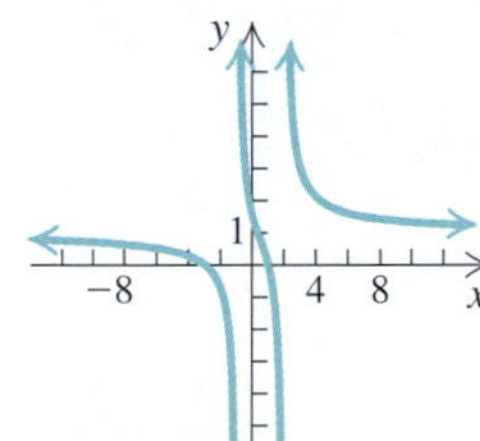

$f(x) = \frac{-x^2}{x + 1}$

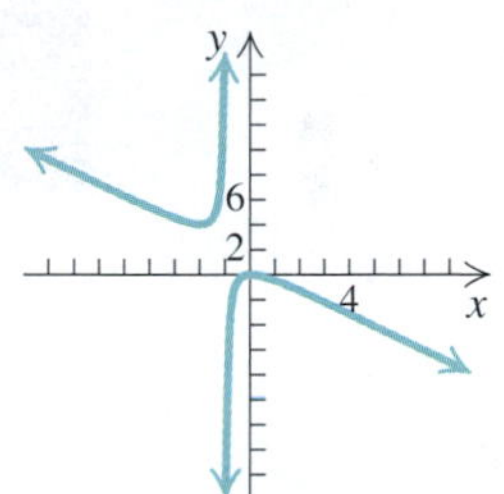

RATIONAL FUNCTION

A **rational function** is a function *f* that is a quotient of two polynomials. That is,

$$f(x) = \frac{p(x)}{q(x)},$$

where $p(x)$ and $q(x)$ are polynomials and where $q(x)$ is not the zero polynomial. The domain of *f* consists of all inputs *x* for which $q(x) \neq 0$.

› The Domain of a Rational Function

EXAMPLE 1 Consider

$$f(x) = \frac{1}{x - 3}.$$

Find the domain and graph *f*.

DOMAINS OF FUNCTIONS

REVIEW SECTION 1.2.

Solution When the denominator $x - 3$ is 0, we have $x = 3$, so the only input that results in a denominator of 0 is 3. Thus the domain is

$$\{x | x \neq 3\}, \text{ or } (-\infty, 3) \cup (3, \infty).$$

The graph of this function is the graph of $y = 1/x$ translated 3 units to the right.

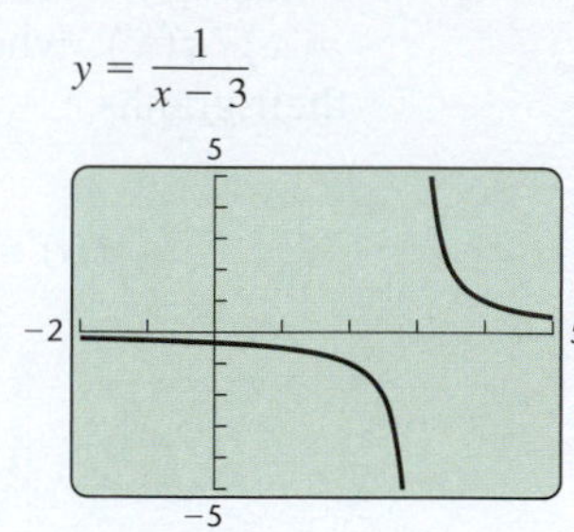

EXAMPLE 2 Determine the domain of each of the functions illustrated at the beginning of this section.

Solution The domain of each rational function will be the set of all real numbers except those values that make the denominator 0. To determine those exceptions, we set the denominator equal to 0 and solve for x.

Just in Time 6

FUNCTION	DOMAIN
$f(x) = \frac{1}{x}$	$\{x \mid x \neq 0\}$, or $(-\infty, 0) \cup (0, \infty)$
$f(x) = \frac{1}{x^2}$	$\{x \mid x \neq 0\}$, or $(-\infty, 0) \cup (0, \infty)$
$f(x) = \frac{x-3}{x^2+x-2} = \frac{x-3}{(x+2)(x-1)}$	$\{x \mid x \neq -2 \text{ and } x \neq 1\}$, or $(-\infty, -2) \cup (-2, 1) \cup (1, \infty)$
$f(x) = \frac{2x+5}{2x-6} = \frac{2x+5}{2(x-3)}$	$\{x \mid x \neq 3\}$, or $(-\infty, 3) \cup (3, \infty)$
$f(x) = \frac{x^2+2x-3}{x^2-x-2} = \frac{x^2+2x-3}{(x+1)(x-2)}$	$\{x \mid x \neq -1 \text{ and } x \neq 2\}$, or $(-\infty, -1) \cup (-1, 2) \cup (2, \infty)$
$f(x) = \frac{-x^2}{x+1}$	$\{x \mid x \neq -1\}$, or $(-\infty, -1) \cup (-1, \infty)$

As a partial check of the domains, we can observe the discontinuities (breaks) in the graphs of these functions. (See p. 263.)

Asymptotes

Vertical Asymptotes

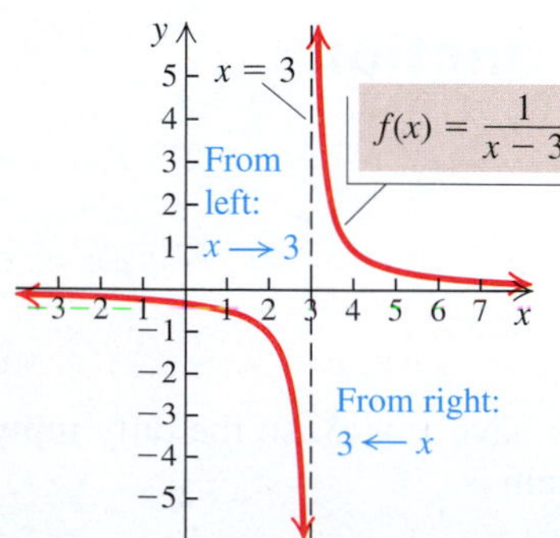

Vertical asymptote: $x = 3$

Look at the graph of $f(x) = 1/(x - 3)$, shown at left. (Also see Example 1.) Let's explore what happens as x-values get closer and closer to 3 from the left. We then explore what happens as x-values get closer and closer to 3 from the right.

From the left:

x	2	$2\frac{1}{2}$	$2\frac{99}{100}$	$2\frac{9999}{10,000}$	$2\frac{999,999}{1,000,000}$	$\rightarrow 3$
$f(x)$	−1	−2	−100	−10,000	−1,000,000	$\rightarrow -\infty$

From the right:

x	4	$3\frac{1}{2}$	$3\frac{1}{100}$	$3\frac{1}{10{,}000}$	$3\frac{1}{1{,}000{,}000}$	$\rightarrow 3$
$f(x)$	1	2	100	10,000	1,000,000	$\rightarrow \infty$

We see that as x-values get closer and closer to 3 from the left, the function values (y-values) decrease without bound (that is, they approach negative infinity, $-\infty$). Similarly, as the x-values approach 3 from the right, the function values increase without bound (that is, they approach positive infinity, ∞). We write this as

$$f(x) \rightarrow -\infty \text{ as } x \rightarrow 3^- \quad \text{and} \quad f(x) \rightarrow \infty \text{ as } x \rightarrow 3^+.$$

We read "$f(x) \rightarrow -\infty$ as $x \rightarrow 3^-$" as "$f(x)$ decreases without bound as x approaches 3 from the left." We read "$f(x) \rightarrow \infty$ as $x \rightarrow 3^+$" as "$f(x)$ increases without bound as x approaches 3 from the right." The notation $x \rightarrow 3$ means that x gets as close to 3 as possible without being equal to 3. The vertical line $x = 3$ is said to be a *vertical asymptote* for this curve.

In general, the line $x = a$ is a **vertical asymptote** for the graph of f if any of the following is true:

$$f(x) \rightarrow \infty \text{ as } x \rightarrow a^-, \quad \text{or} \quad f(x) \rightarrow -\infty \text{ as } x \rightarrow a^-, \quad \text{or}$$
$$f(x) \rightarrow \infty \text{ as } x \rightarrow a^+, \quad \text{or} \quad f(x) \rightarrow -\infty \text{ as } x \rightarrow a^+.$$

The following figures show the four ways in which a vertical asymptote can occur.

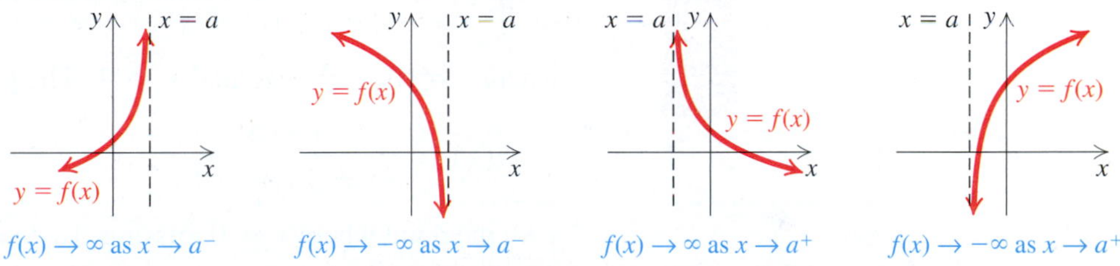

The vertical asymptotes of a rational function $f(x) = p(x)/q(x)$ are found by determining the zeros of $q(x)$ that are not also zeros of $p(x)$. If $p(x)$ and $q(x)$ are polynomials with no common factors other than constants, we need determine only the zeros of the denominator $q(x)$.

DETERMINING VERTICAL ASYMPTOTES

For a rational function $f(x) = p(x)/q(x)$, where $p(x)$ and $q(x)$ are polynomials with no common factors other than constants, if a is a zero of the denominator, then the line $x = a$ is a vertical asymptote for the graph of the function.

EXAMPLE 3 Determine the vertical asymptotes for the graph of each of the following functions.

a) $f(x) = \dfrac{2x - 11}{x^2 + 2x - 8}$

b) $h(x) = \dfrac{x^2 - 4x}{x^3 - x}$

c) $g(x) = \dfrac{x - 2}{x^3 - 5x}$

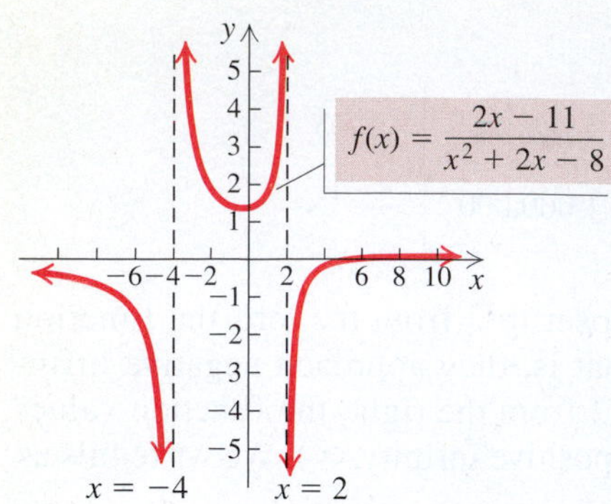

FIGURE 1.

Solution

a) First, we factor the denominator:

$$f(x) = \frac{2x - 11}{x^2 + 2x - 8} = \frac{2x - 11}{(x + 4)(x - 2)}.$$

The numerator and the denominator have no common factors. The zeros of the denominator are -4 and 2. Thus the vertical asymptotes for the graph of $f(x)$ are the lines $x = -4$ and $x = 2$. (See Fig. 1.)

b) We factor the numerator and the denominator:

$$h(x) = \frac{x^2 - 4x}{x^3 - x} = \frac{x(x - 4)}{x(x^2 - 1)} = \frac{x(x - 4)}{x(x + 1)(x - 1)}.$$

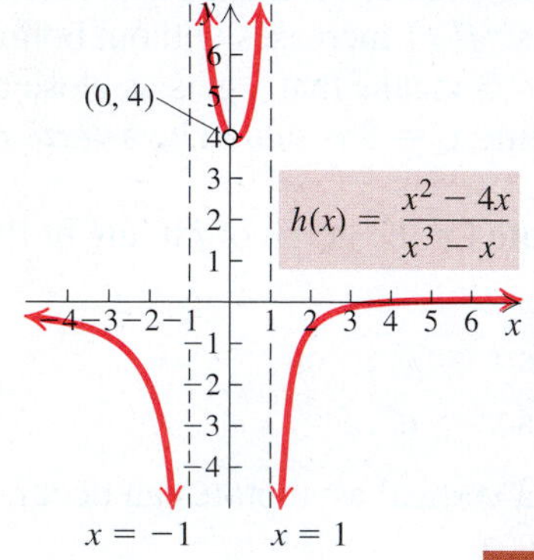

FIGURE 2.

The domain of the function is $\{x | x \neq -1 \text{ and } x \neq 0 \text{ and } x \neq 1\}$, or $(-\infty, -1) \cup (-1, 0) \cup (0, 1) \cup (1, \infty)$. Note that the numerator and the denominator share a common factor, x. The vertical asymptotes of $h(x)$ are found by determining the zeros of the denominator, $x(x + 1)(x - 1)$, that are *not* also zeros of the numerator, $x(x - 4)$. The zeros of $x(x + 1)(x - 1)$ are 0, -1, and 1. The zeros of $x(x - 4)$ are 0 and 4. Thus, although the denominator has three zeros, the graph of $h(x)$ has only two vertical asymptotes, $x = -1$ and $x = 1$. (See Fig. 2.)

The rational expression $[x(x - 4)]/[x(x + 1)(x - 1)]$ can be simplified. Thus,

$$h(x) = \frac{x(x - 4)}{x(x + 1)(x - 1)} = \frac{x - 4}{(x + 1)(x - 1)},$$

where $x \neq 0$, $x \neq -1$, and $x \neq 1$. The graph of $h(x)$ is the graph of

$$h(x) = \frac{x - 4}{(x + 1)(x - 1)}$$

with the point where $x = 0$ missing. To determine the y-coordinate of the "hole," we substitute 0 for x:

$$h(0) = \frac{0 - 4}{(0 + 1)(0 - 1)} = \frac{-4}{1 \cdot (-1)} = 4.$$

Thus the "hole" is located at $(0, 4)$.

c) We factor the denominator:

$$g(x) = \frac{x - 2}{x^3 - 5x} = \frac{x - 2}{x(x^2 - 5)}.$$

The numerator and the denominator have no common factors. We find the zeros of the denominator, $x(x^2 - 5)$. Solving $x(x^2 - 5) = 0$, we get

$$x = 0 \quad \text{or} \quad x^2 - 5 = 0$$
$$x = 0 \quad \text{or} \quad x^2 = 5$$
$$x = 0 \quad \text{or} \quad x = \pm\sqrt{5}.$$

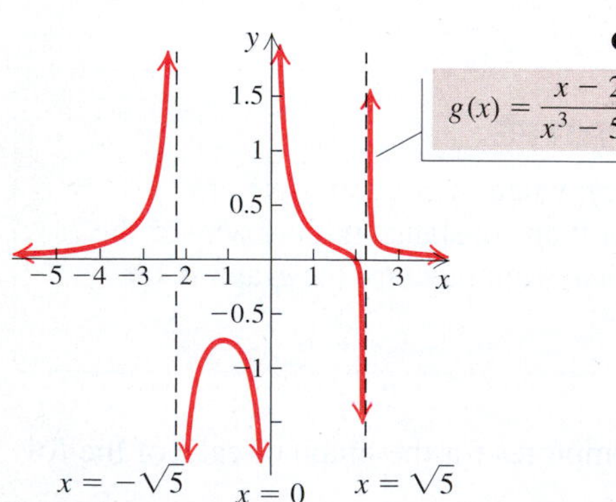

FIGURE 3.

The zeros of the denominator are 0, $\sqrt{5}$, and $-\sqrt{5}$. Thus the vertical asymptotes are the lines $x = 0$, $x = \sqrt{5}$, and $x = -\sqrt{5}$. (See Fig. 3.)

Now Try Exercises 15 and 19.

Horizontal Asymptotes

Looking again at the graph of $f(x) = 1/(x - 3)$ (also see Example 1), let's explore what happens to $f(x) = 1/(x - 3)$ as x increases without bound (approaches positive infinity, ∞) and as x decreases without bound (approaches negative infinity, $-\infty$).

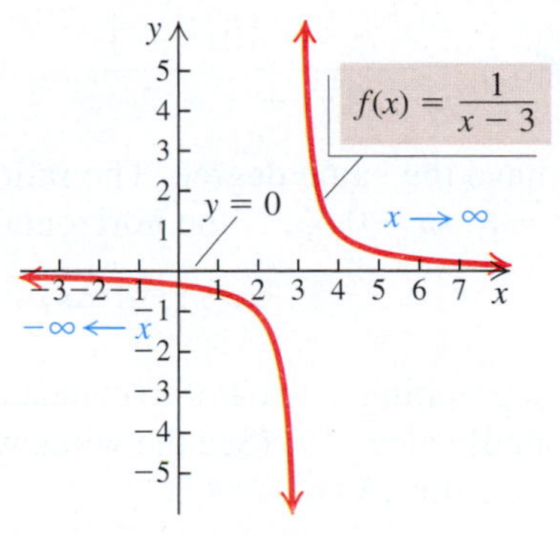

Horizontal asymptote: $y = 0$

x increases without bound:

x	100	5000	1,000,000	$\to \infty$
$f(x)$	≈ 0.0103	≈ 0.0002	≈ 0.000001	$\to 0$

x decreases without bound:

x	-300	-8000	$-1,000,000$	$\to -\infty$
$f(x)$	≈ -0.0033	≈ -0.0001	≈ -0.000001	$\to 0$

We see that

$$\frac{1}{x-3} \to 0 \text{ as } x \to \infty \quad \text{and} \quad \frac{1}{x-3} \to 0 \text{ as } x \to -\infty.$$

Since $y = 0$ is the equation of the x-axis, we say that the curve approaches the x-axis asymptotically and that the x-axis is a *horizontal asymptote* for the curve.

In general, the line $y = b$ is a **horizontal asymptote** for the graph of f if either or both of the following are true:

$$f(x) \to b \text{ as } x \to \infty \quad \text{or} \quad f(x) \to b \text{ as } x \to -\infty.$$

The following figures illustrate four ways in which horizontal asymptotes can occur. In each case, the curve gets close to the line $y = b$ either as $x \to \infty$ or as $x \to -\infty$. Keep in mind that the symbols ∞ and $-\infty$ convey the idea of increasing without bound and decreasing without bound, respectively.

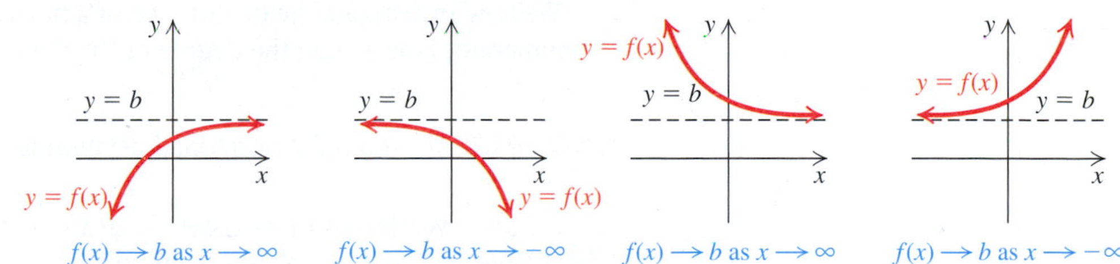

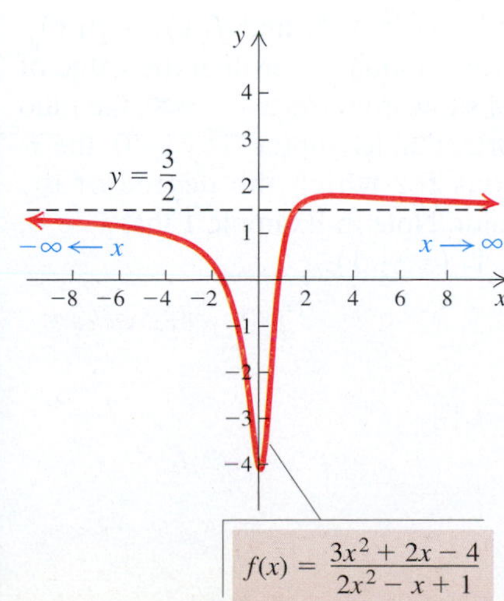

How can we determine a horizontal asymptote? As x gets very large or very small, the value of a polynomial function $p(x)$ is dominated by the function's leading term. Because of this, if $p(x)$ and $q(x)$ have the *same* degree, the value of $p(x)/q(x)$ as $x \to \infty$ or as $x \to -\infty$ is dominated by the ratio of the numerator's leading coefficient to the denominator's leading coefficient.

For $f(x) = (3x^2 + 2x - 4)/(2x^2 - x + 1)$, we see that the numerator, $3x^2 + 2x - 4$, is dominated by $3x^2$ and the denominator, $2x^2 - x + 1$, is dominated by $2x^2$, so $f(x)$ approaches $3x^2/2x^2$, or $3/2$ as x gets very large or very small:

$$\frac{3x^2 + 2x - 4}{2x^2 - x + 1} \to \frac{3}{2}, \text{ or } 1.5, \text{ as } x \to \infty, \quad \text{and}$$

$$\frac{3x^2 + 2x - 4}{2x^2 - x + 1} \to \frac{3}{2}, \text{ or } 1.5, \text{ as } x \to -\infty.$$

We say that the curve approaches the horizontal line $y = \frac{3}{2}$ asymptotically and that $y = \frac{3}{2}$ is a *horizontal asymptote* for the curve.

It follows that when the numerator and the denominator of a rational function have the same degree, the line $y = a/b$ is the horizontal asymptote, where a and b are the leading coefficients of the numerator and the denominator, respectively.

EXAMPLE 4 Find the horizontal asymptote: $f(x) = \dfrac{-7x^4 - 10x^2 + 1}{11x^4 + x - 2}$.

Solution The numerator and the denominator have the same degree. The ratio of the leading coefficients is $-\frac{7}{11}$, so the line $y = -\frac{7}{11}$, or $-0.\overline{63}$, is the horizontal asymptote.

Now Try Exercise 21.

X	Y1	
100000	−.6364	
−80000	−.6364	

X =

As a partial check of Example 4, we could use a graphing calculator to evaluate the function for a very large value of x and a very small value of x. (See the window at left.) It is useful in calculus to multiply by 1. Here we use $(1/x^4)/(1/x^4)$:

$$f(x) = \frac{-7x^4 - 10x^2 + 1}{11x^4 + x - 2} \cdot \frac{\frac{1}{x^4}}{\frac{1}{x^4}} = \frac{\frac{-7x^4}{x^4} - \frac{10x^2}{x^4} + \frac{1}{x^4}}{\frac{11x^4}{x^4} + \frac{x}{x^4} - \frac{2}{x^4}}$$

$$= \frac{-7 - \frac{10}{x^2} + \frac{1}{x^4}}{11 + \frac{1}{x^3} - \frac{2}{x^4}}.$$

As $|x|$ becomes very large, each expression whose denominator is a power of x tends toward 0. Specifically, as $x \rightarrow \infty$ or as $x \rightarrow -\infty$, we have

$$f(x) \rightarrow \frac{-7 - 0 + 0}{11 + 0 - 0}, \quad \text{or} \quad f(x) \rightarrow -\frac{7}{11}.$$

The horizontal asymptote is $y = -\frac{7}{11}$, or $-0.\overline{63}$.

We now investigate the occurrence of a horizontal asymptote when the degree of the numerator is less than the degree of the denominator.

EXAMPLE 5 Find the horizontal asymptote: $f(x) = \dfrac{2x + 3}{x^3 - 2x^2 + 4}$.

Solution We let $p(x) = 2x + 3$, $q(x) = x^3 - 2x^2 + 4$, and $f(x) = p(x)/q(x)$. Note that as $x \rightarrow \infty$, the value of $q(x)$ grows much faster than the value of $p(x)$. Because of this, the ratio $p(x)/q(x)$ shrinks toward 0. As $x \rightarrow -\infty$, the ratio $p(x)/q(x)$ behaves in a similar manner. The horizontal asymptote is $y = 0$, the x-axis. This is the case for all rational functions for which the degree of the numerator is less than the degree of the denominator. Note in Example 1 that $y = 0$, the x-axis, is the horizontal asymptote of $f(x) = 1/(x - 3)$.

Now Try Exercise 23.

The following statements describe the two ways in which a horizontal asymptote occurs.

DETERMINING A HORIZONTAL ASYMPTOTE

- When the numerator and the denominator of a rational function have the same degree, the line $y = a/b$ is the horizontal asymptote, where a and b are the leading coefficients of the numerator and the denominator, respectively.
- When the degree of the numerator of a rational function is less than the degree of the denominator, the x-axis, or $y = 0$, is the horizontal asymptote.
- When the degree of the numerator of a rational function is greater than the degree of the denominator, there is no horizontal asymptote.

EXAMPLE 6 Graph

$$g(x) = \frac{2x^2 + 1}{x^2}.$$

Include and label all asymptotes.

Solution Since 0 is the zero of the denominator and is not a zero of the numerator, the y-axis, $x = 0$, is the vertical asymptote. Note also that the degree of the numerator is the same as the degree of the denominator. Thus, $y = 2/1$, or 2, is the horizontal asymptote.

To draw the graph, we first draw the asymptotes with dashed lines. Then we compute and plot some ordered pairs and draw the two branches of the curve. We can check the graph with a graphing calculator.

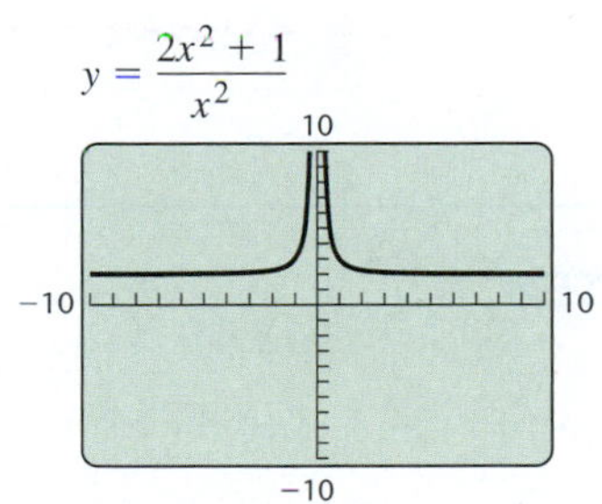

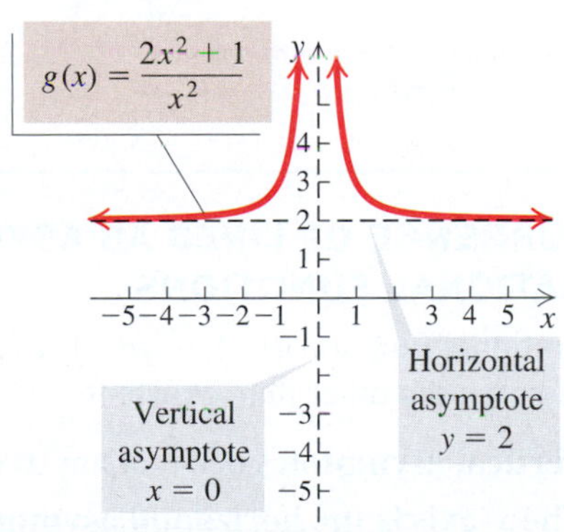

x	$g(x)$
−2	2.25
−1.5	$2.\overline{4}$
−1	3
−0.5	6
0.5	6
1	3
1.5	$2.\overline{4}$
2	2.25

Now Try Exercise 41.

Oblique Asymptotes

Sometimes a line that is neither horizontal nor vertical is an asymptote. Such a line is called an **oblique asymptote**, or a **slant asymptote**.

EXAMPLE 7 Find all the asymptotes of

$$f(x) = \frac{2x^2 - 3x - 1}{x - 2}.$$

Solution The line $x = 2$ is the vertical asymptote because 2 is the zero of the denominator and is not a zero of the numerator. There is no horizontal asymptote because the degree of the numerator is greater than the degree of the denominator. When the degree of the numerator is 1 greater than the degree of the denominator, we divide to find an equivalent expression:

$$\frac{2x^2 - 3x - 1}{x - 2} = (2x + 1) + \frac{1}{x - 2}.$$

$$\begin{array}{r} 2x + 1 \\ x - 2\overline{)2x^2 - 3x - 1} \\ \underline{2x^2 - 4x} \\ x - 1 \\ \underline{x - 2} \\ 1 \end{array}$$

We see that when $x \to \infty$ or $x \to -\infty$, $1/(x - 2) \to 0$ and the value of $f(x) \to 2x + 1$. This means that as $|x|$ becomes very large, the graph of $f(x)$ gets very close to the graph of $y = 2x + 1$. Thus the line $y = 2x + 1$ is the oblique asymptote.

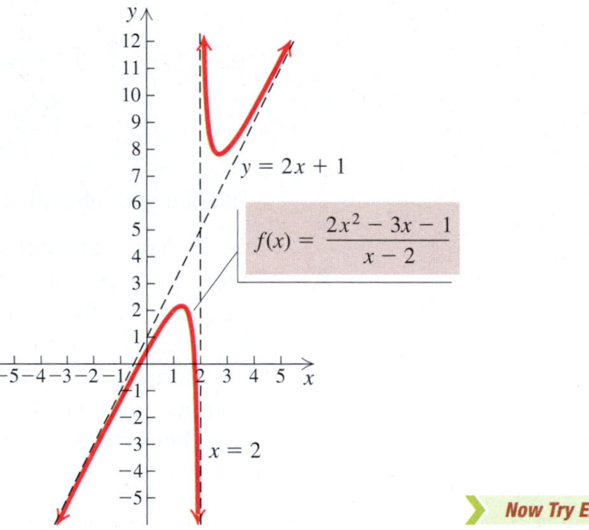

Now Try Exercise 59.

OCCURRENCE OF LINES AS ASYMPTOTES OF RATIONAL FUNCTIONS

For a rational function $f(x) = p(x)/q(x)$, where $p(x)$ and $q(x)$ have no common factors other than constants:

Vertical asymptotes occur at any x-values that make the denominator 0.

The x-axis is the horizontal asymptote when the degree of the numerator is less than the degree of the denominator.

A horizontal asymptote other than the x-axis occurs when the numerator and the denominator have the same degree.

An oblique asymptote occurs when the degree of the numerator is 1 greater than the degree of the denominator.

There can be only one horizontal asymptote or one oblique asymptote and never both.

An asymptote is *not* part of the graph of the function.

The following statements are also true.

CROSSING AN ASYMPTOTE

- The graph of a rational function *never crosses* a vertical asymptote.
- The graph of a rational function *might cross* a horizontal asymptote but does not necessarily do so.

Shown below is an outline of a procedure that we can follow to create accurate graphs of rational functions.

To graph a rational function $f(x) = p(x)/q(x)$, where $p(x)$ and $q(x)$ have no common factor other than constants:

1. Find any real zeros of the denominator. Determine the domain of the function and sketch any vertical asymptotes.
2. Find the horizontal asymptote or the oblique asymptote, if there is one, and sketch it.
3. Find any zeros of the function. The zeros are found by determining the zeros of the numerator. These are the first coordinates of the x-intercepts of the graph.
4. Find $f(0)$. This gives the y-intercept, $(0, f(0))$, of the function.
5. Find other function values to determine the general shape. Then draw the graph.

EXAMPLE 8 Graph: $f(x) = \dfrac{2x + 3}{3x^2 + 7x - 6}$.

Solution

1. We find the zeros of the denominator by solving $3x^2 + 7x - 6 = 0$. Since
$$3x^2 + 7x - 6 = (3x - 2)(x + 3),$$
the zeros are $\frac{2}{3}$ and -3. Thus the domain excludes $\frac{2}{3}$ and -3 and is
$$(-\infty, -3) \cup \left(-3, \tfrac{2}{3}\right) \cup \left(\tfrac{2}{3}, \infty\right).$$
Since neither zero of the denominator is a zero of the numerator, the graph has vertical asymptotes $x = -3$ and $x = \frac{2}{3}$. We sketch these as dashed lines.
2. Because the degree of the numerator is less than the degree of the denominator, the x-axis, $y = 0$, is the horizontal asymptote.
3. To find the zeros of the numerator, we solve $2x + 3 = 0$ and get $x = -\frac{3}{2}$. Thus, $-\frac{3}{2}$ is the zero of the function, and the pair $\left(-\frac{3}{2}, 0\right)$ is the x-intercept.
4. We find $f(0)$:
$$f(0) = \frac{2 \cdot 0 + 3}{3 \cdot 0^2 + 7 \cdot 0 - 6} = \frac{3}{-6} = -\frac{1}{2}.$$
The point $\left(0, -\frac{1}{2}\right)$ is the y-intercept.

5. We find other function values to determine the general shape. We choose values in each interval of the domain, as shown in the following table, and then draw the graph. Note that the graph of this function crosses its horizontal asymptote at $x = -\frac{3}{2}$.

x	y
−4.5	−0.26
−3.25	−1.19
−2.5	0.42
−0.5	−0.23
0.5	−2.29
0.75	4.8
1.5	0.53
3.5	0.18

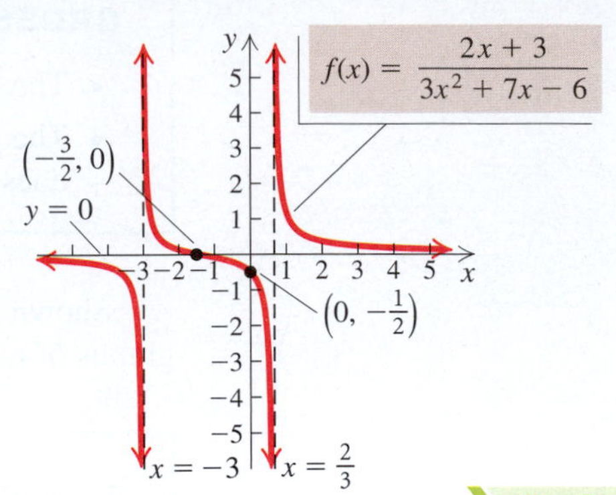

Now Try Exercise 65.

EXAMPLE 9 Graph: $g(x) = \dfrac{x^2 - 1}{x^2 + x - 6}$.

Solution

1. We find the zeros of the denominator by solving $x^2 + x - 6 = 0$. Since

$$x^2 + x - 6 = (x + 3)(x - 2),$$

the zeros are -3 and 2. Thus the domain excludes the x-values -3 and 2 and is

$$(-\infty, -3) \cup (-3, 2) \cup (2, \infty).$$

Since neither zero of the denominator is a zero of the numerator, the graph has vertical asymptotes $x = -3$ and $x = 2$. We sketch these as dashed lines.

2. The numerator and the denominator have the same degree, so the horizontal asymptote is determined by the ratio of the leading coefficients: $1/1$, or 1. Thus, $y = 1$ is the horizontal asymptote. We sketch it with a dashed line.

3. To find the zeros of the numerator, we solve $x^2 - 1 = 0$. The solutions are -1 and 1. Thus, -1 and 1 are the zeros of the function and the pairs $(-1, 0)$ and $(1, 0)$ are the x-intercepts.

4. We find $g(0)$:

$$g(0) = \frac{0^2 - 1}{0^2 + 0 - 6} = \frac{-1}{-6} = \frac{1}{6}.$$

Thus, $\left(0, \frac{1}{6}\right)$ is the y-intercept.

5. We find other function values to determine the general shape and draw the graph.

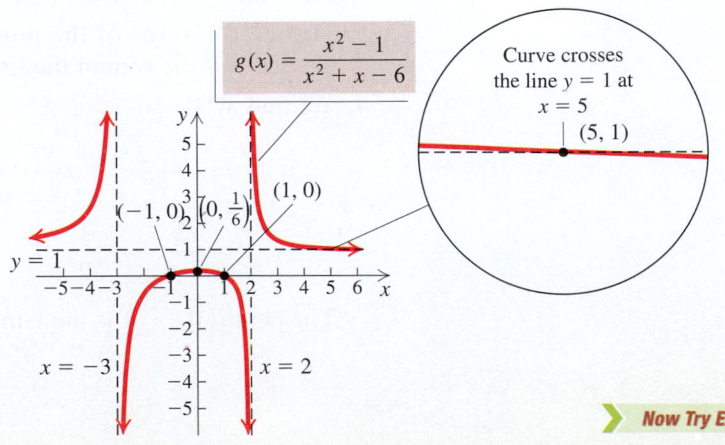

Now Try Exercise 77.

The magnified portion of the graph in Example 9 above shows another situation in which a graph can cross its horizontal asymptote. The point where $g(x)$ crosses $y = 1$ can be found by setting $g(x) = 1$ and solving for x:

$$\begin{aligned} \frac{x^2 - 1}{x^2 + x - 6} &= 1 \\ x^2 - 1 &= x^2 + x - 6 \\ -1 &= x - 6 \qquad \textbf{Subtracting } x^2 \\ 5 &= x. \qquad \textbf{Adding 6} \end{aligned}$$

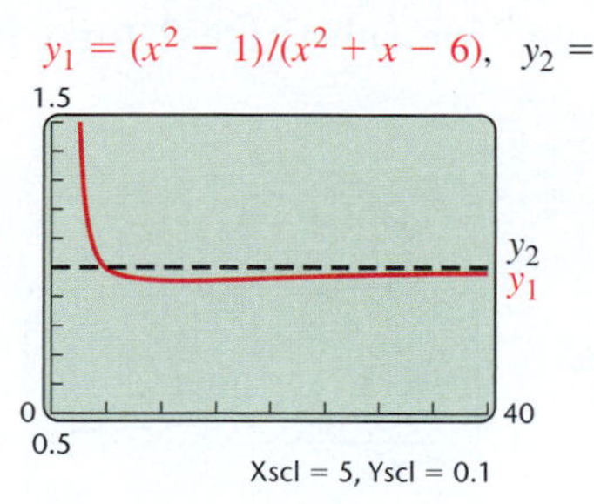

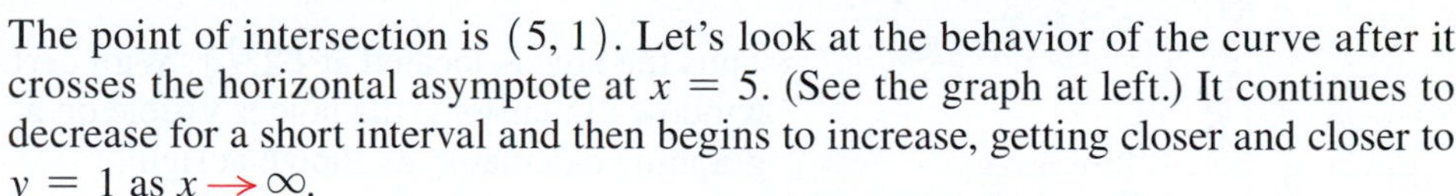

The point of intersection is $(5, 1)$. Let's look at the behavior of the curve after it crosses the horizontal asymptote at $x = 5$. (See the graph at left.) It continues to decrease for a short interval and then begins to increase, getting closer and closer to $y = 1$ as $x \rightarrow \infty$.

Graphs of rational functions can also cross an oblique asymptote. The graph of

$$f(x) = \frac{2x^3}{x^2 + 1}$$

shown below crosses its oblique asymptote $y = 2x$. **Remember: Graphs can cross horizontal asymptotes or oblique asymptotes, but they cannot cross vertical asymptotes.**

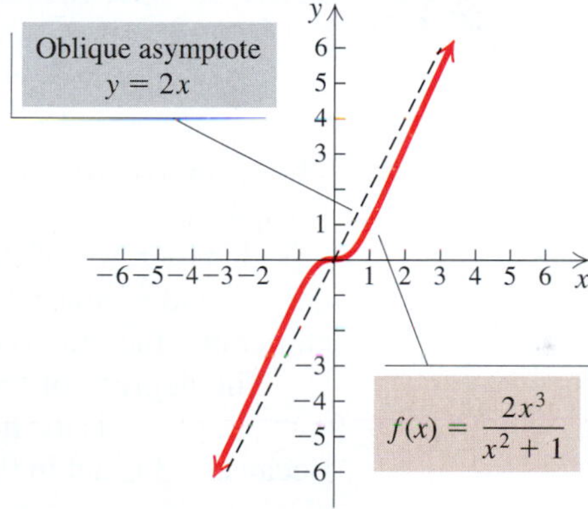

Let's now graph a rational function $f(x) = p(x)/q(x)$, where $p(x)$ and $q(x)$ have a common factor, $x - c$. The graph of such a function has a "hole" in it. We first saw this situation in Example 3(b), where the common factor was x.

EXAMPLE 10 Graph: $g(x) = \dfrac{x - 2}{x^2 - x - 2}$.

Solution We first express the denominator in factored form:

$$g(x) = \frac{x - 2}{x^2 - x - 2} = \frac{x - 2}{(x + 1)(x - 2)}.$$

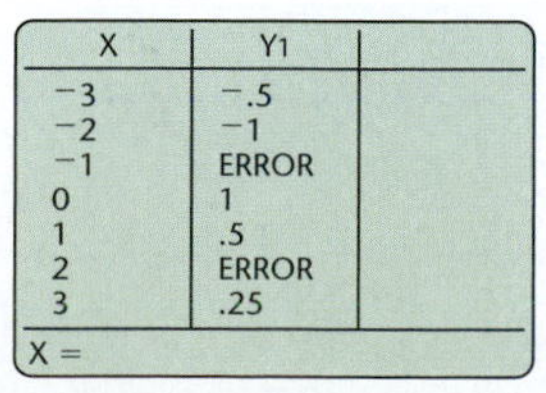

X	Y1
−3	−.5
−2	−1
−1	ERROR
0	1
1	.5
2	ERROR
3	.25

X =

The domain of the function is $\{x \mid x \neq -1 \text{ and } x \neq 2\}$, or $(-\infty, -1) \cup (-1, 2) \cup (2, \infty)$. Note that both the numerator and the denominator have the factor $x - 2$. The zeros of the denominator are -1 and 2, and the zero of the numerator is 2. Since -1 is the only zero of the denominator that is *not* a zero of the numerator, the graph of the function has $x = -1$ as its only vertical asymptote. The degree of the numerator is less than the degree of the denominator, so $y = 0$ is the horizontal asymptote. There are no zeros of the function and thus no x-intercepts, because 2 is the only zero of the numerator and 2 is not in the domain of the function. Since $g(0) = 1$, $(0, 1)$ is the y-intercept. We draw the graph indicating the "hole" when $x = 2$ with an open circle.

The rational expression $(x-2)/[(x+1)(x-2)]$ can be simplified. Thus,

$$g(x) = \frac{x-2}{(x+1)(x-2)} = \frac{1}{x+1}, \quad \text{where } x \neq -1 \text{ and } x \neq 2.$$

The graph of $g(x)$ is the graph of $y = 1/(x+1)$ with the point where $x = 2$ missing. To determine the coordinates of the "hole," we substitute 2 for x in $g(x) = 1/(x+1)$:

$$g(2) = \frac{1}{2+1} = \frac{1}{3}.$$

Thus the hole is located at $\left(2, \frac{1}{3}\right)$. With certain window dimensions, the hole is visible on a graphing calculator, as shown at right.

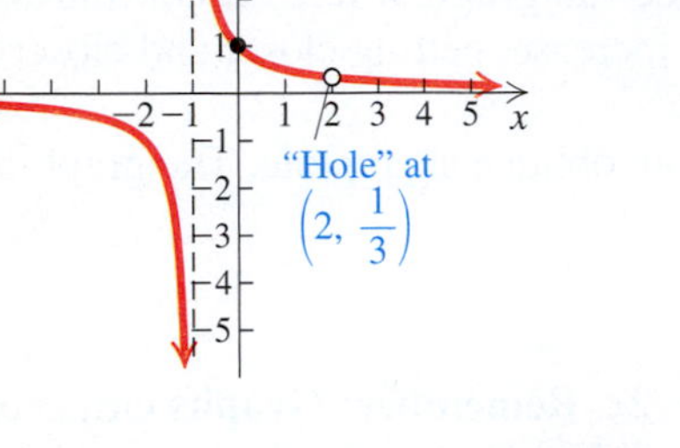

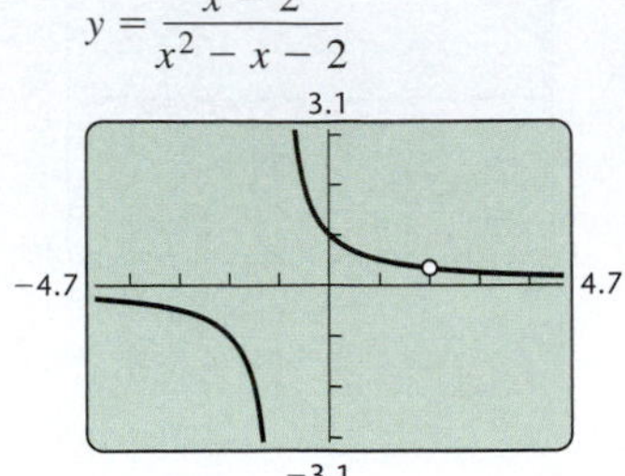

Now Try Exercise 49.

Just in Time 21

EXAMPLE 11 Graph: $f(x) = \dfrac{-2x^2 - x + 15}{x^2 - x - 12}$.

Solution We first express the numerator and the denominator in factored form:

$$f(x) = \frac{-2x^2 - x + 15}{x^2 - x - 12} = \frac{-(2x^2 + x - 15)}{x^2 - x - 12} = \frac{-(2x-5)(x+3)}{(x-4)(x+3)}.$$

The domain of the function is $\{x \mid x \neq -3 \text{ and } x \neq 4\}$, or $(-\infty, -3) \cup (-3, 4) \cup (4, \infty)$. The numerator and the denominator have the common factor $x + 3$. The zeros of the denominator are -3 and 4, and the zeros of the numerator are -3 and $\frac{5}{2}$. Since 4 is the only zero of the denominator that is *not* a zero of the numerator, the graph of the function has $x = 4$ as its *only* vertical asymptote.

The degrees of the numerator and the denominator are the same, so the line $y = \frac{-2}{1} = -2$ is the horizontal asymptote. The zeros of the numerator are $\frac{5}{2}$ and -3. Because -3 is not in the domain of the function, the only x-intercept is $\left(\frac{5}{2}, 0\right)$. Since $f(0) = \frac{15}{-12} = -\frac{5}{4}$, then $\left(0, -\frac{5}{4}\right)$ is the y-intercept.

The rational function

$$\frac{-(2x-5)(x+3)}{(x-4)(x+3)}$$

can be simplified. Thus,

$$f(x) = \frac{-(2x-5)(x+3)}{(x-4)(x+3)} = \frac{-(2x-5)}{x-4}, \quad \text{where } x \neq -3 \text{ and } x \neq 4.$$

The graph of $f(x)$ is the graph of $y = -(2x-5)/(x-4)$ with the point where $x = -3$ missing. To determine the coordinates of the hole, we substitute -3 for x in $f(x) = -(2x-5)/(x-4)$:

$$f(-3) = \frac{-[2(-3) - 5]}{-3 - 4}$$

$$= \frac{-[-11]}{-7} = \frac{11}{-7} = -\frac{11}{7}.$$

Thus the hole is located at $\left(-3, -\frac{11}{7}\right)$. We draw the graph indicating the hole when $x = -3$ with an open circle.

x	y
−5	−1.67
−4	−1.63
−3	Not defined
−2	−1.5
−1	−1.4
0	−1.25
1	−1
2	−0.5

x	y
3	1
3.5	4
4	Not defined
5	−5
6	−3.5
7	−3
8	−2.75

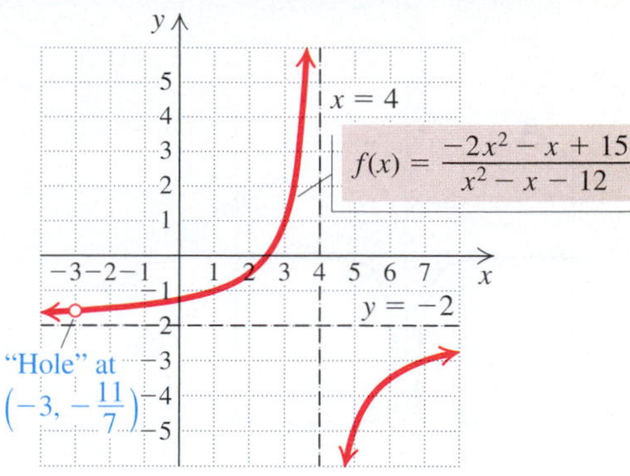

Now Try Exercise 67.

Applications

EXAMPLE 12 ***Temperature During an Illness.*** A person's temperature T, in degrees Fahrenheit, during an illness is given by the function

$$T(t) = \frac{4t}{t^2 + 1} + 98.6,$$

where time t is given in hours since the onset of the illness.

a) Graph the function on the interval $[0, 48]$.

b) Find the temperature at $t = 0, 1, 2, 5, 12$, and 24.

c) Find the horizontal asymptote of the graph of $T(t)$. Complete:

$$T(t) \to \square \text{ as } t \to \infty.$$

d) Give the meaning of the answer to part (c) in terms of the application.

e) Find the maximum temperature during the illness.

Solution

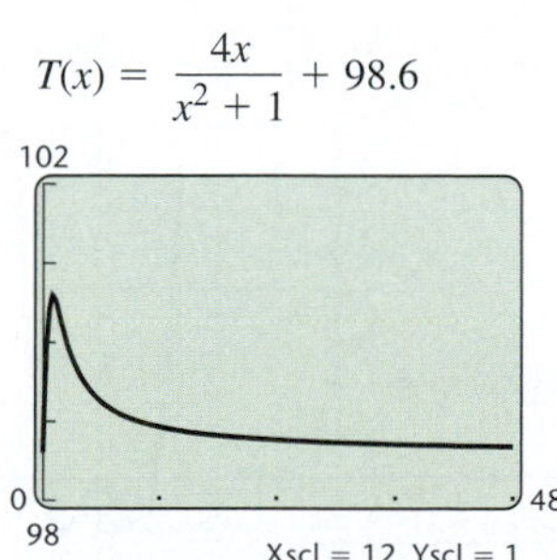

a) The graph is shown at left.

b) We have

$$T(0) = 98.6, \quad T(1) = 100.6, \quad T(2) = 100.2,$$
$$T(5) \approx 99.369, \quad T(12) \approx 98.931, \quad \text{and} \quad T(24) \approx 98.766.$$

c) Since

$$T(t) = \frac{4t}{t^2 + 1} + 98.6,$$
$$= \frac{98.6t^2 + 4t + 98.6}{t^2 + 1},$$

the horizontal asymptote is $y = 98.6/1$, or 98.6. Then it follows that $T(t) \to 98.6$ as $t \to \infty$.

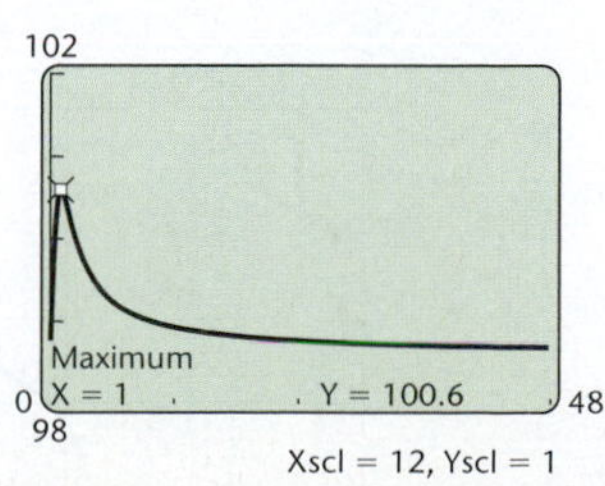

d) As time goes on, the temperature returns to "normal," which is 98.6°F.

e) Using the MAXIMUM feature on a graphing calculator, we find the maximum temperature to be 100.6°F at $t = 1$ hr.

Now Try Exercise 83.

Visualizing the Graph

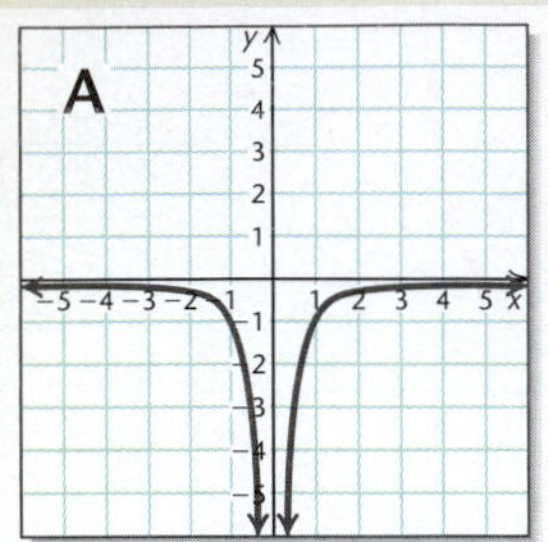

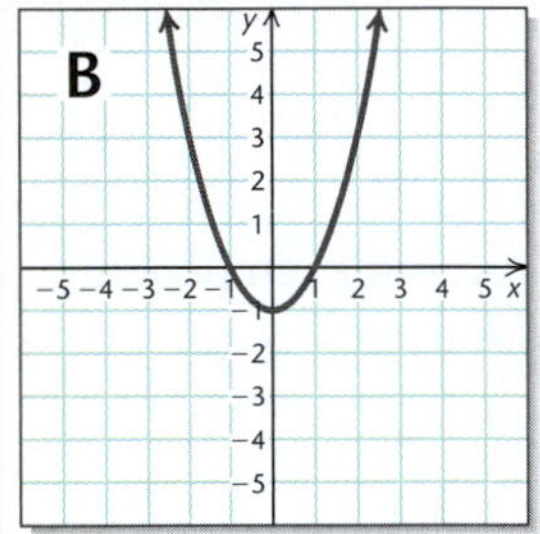

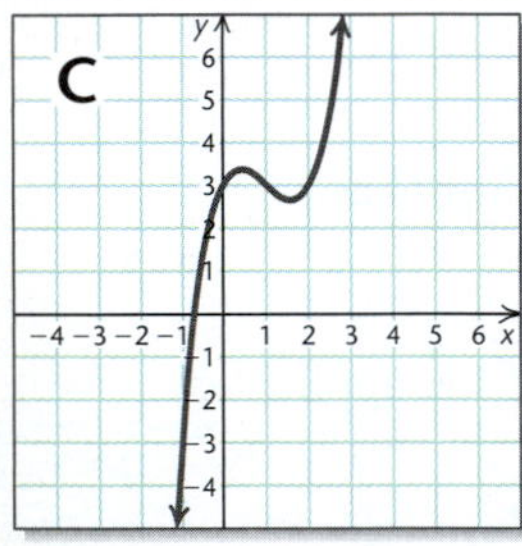

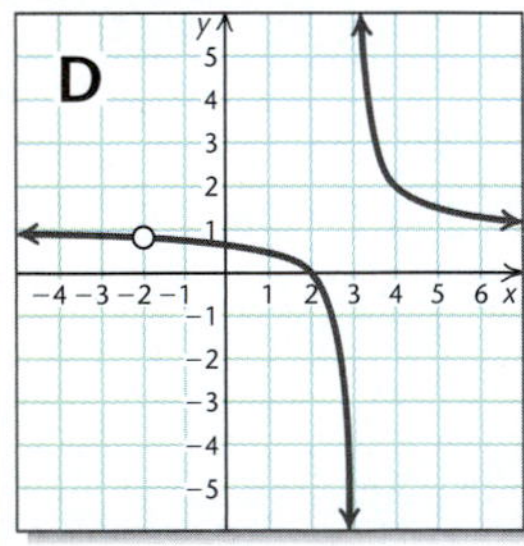

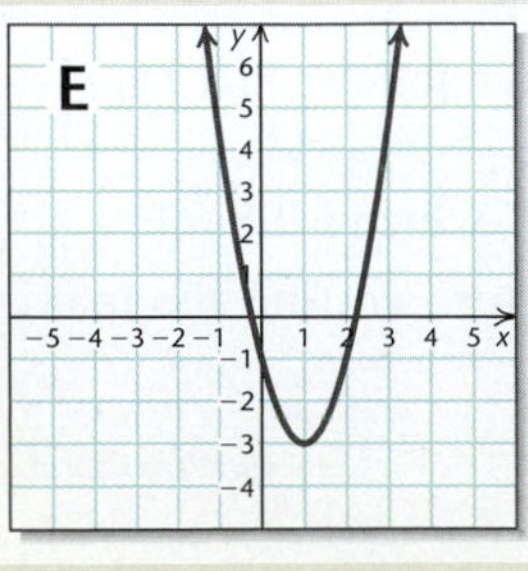

Match the function with its graph.

1. $f(x) = -\dfrac{1}{x^2}$

2. $f(x) = x^3 - 3x^2 + 2x + 3$

3. $f(x) = \dfrac{x^2 - 4}{x^2 - x - 6}$

4. $f(x) = -x^2 + 4x - 1$

5. $f(x) = \dfrac{x - 3}{x^2 + x - 6}$

6. $f(x) = \dfrac{3}{4}x + 2$

7. $f(x) = x^2 - 1$

8. $f(x) = x^4 - 2x^2 - 5$

9. $f(x) = \dfrac{8x - 4}{3x + 6}$

10. $f(x) = 2x^2 - 4x - 1$

Answers on page A-20

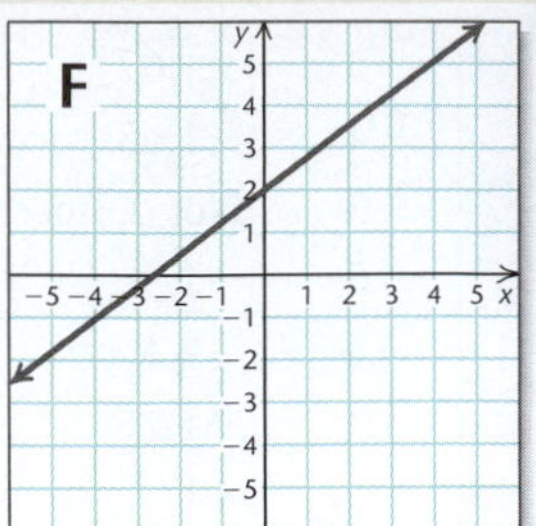

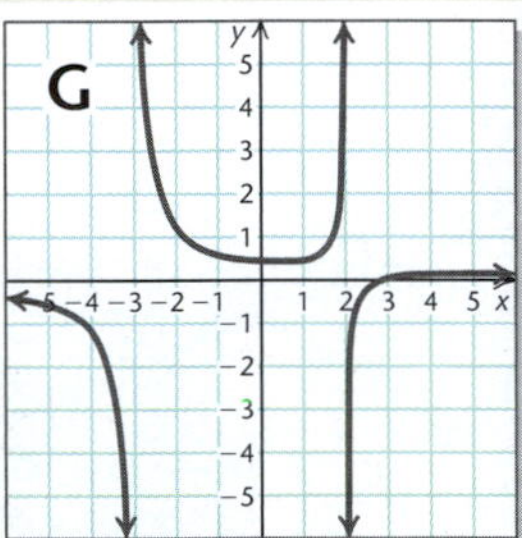

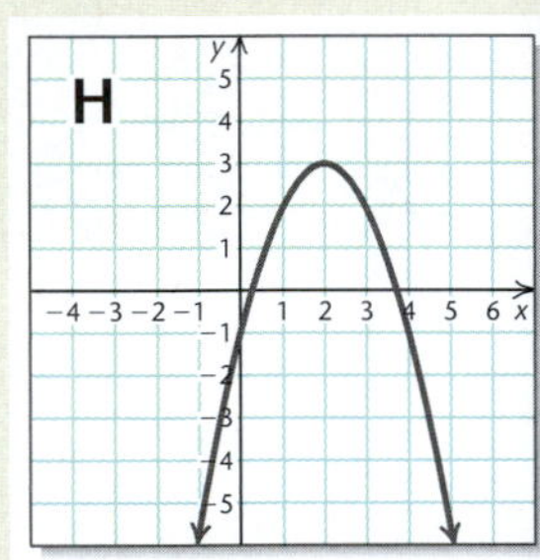

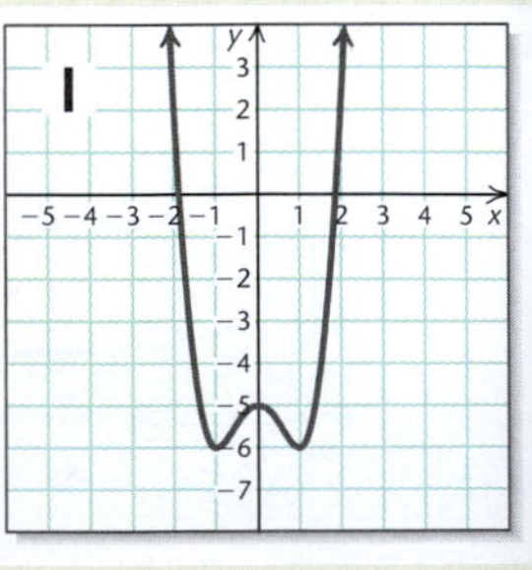

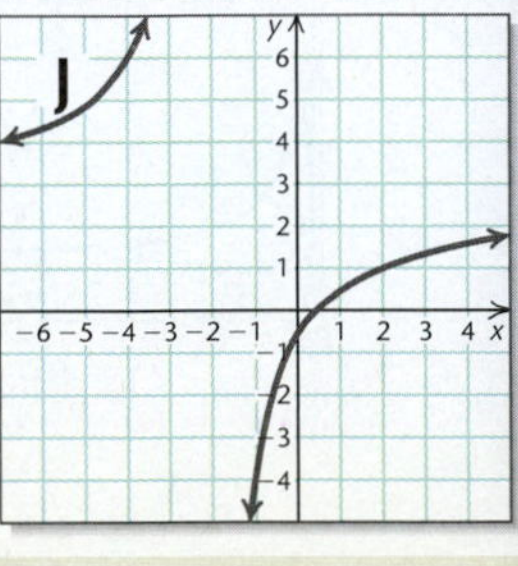

4.5 Exercise Set

Determine the domain of the function.

1. $f(x) = \dfrac{x^2}{2 - x}$

2. $f(x) = \dfrac{1}{x^3}$

3. $f(x) = \dfrac{x + 1}{x^2 - 6x + 5}$

4. $f(x) = \dfrac{(x + 4)^2}{4x - 3}$

5. $f(x) = \dfrac{3x - 4}{3x + 15}$

6. $f(x) = \dfrac{x^2 + 3x - 10}{x^2 + 2x}$

In Exercises 7–12, use your knowledge of asymptotes and intercepts to match the equation with one of the graphs (a)–(f) that follow. List all asymptotes. Check your work using a graphing calculator.

a)
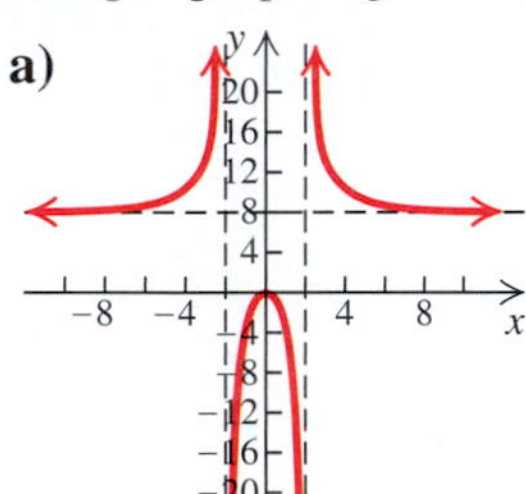

b)
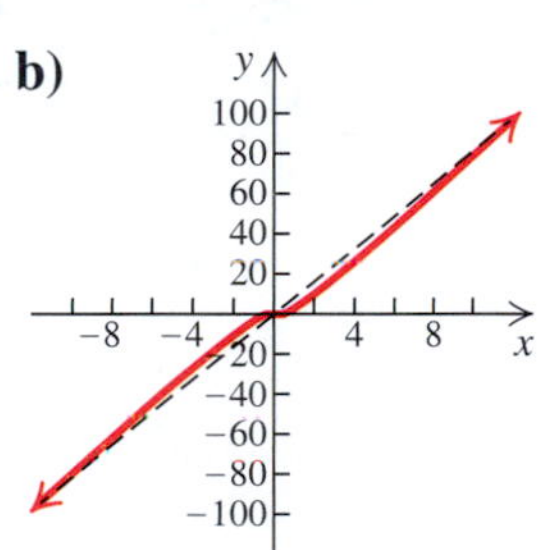

c)
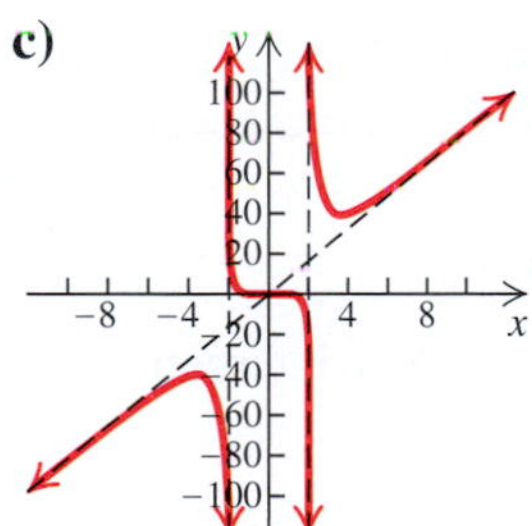

d)
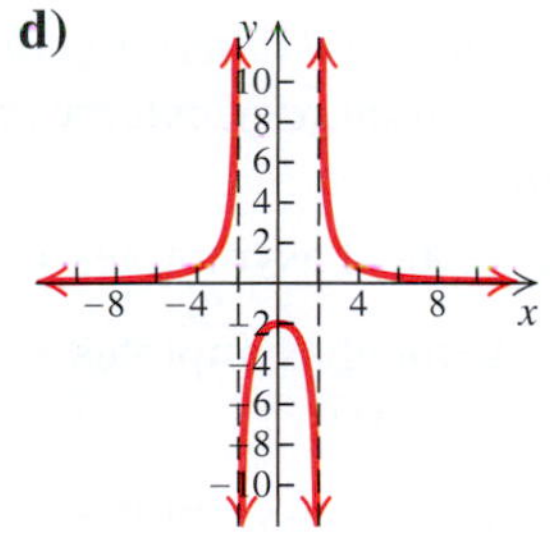

e)
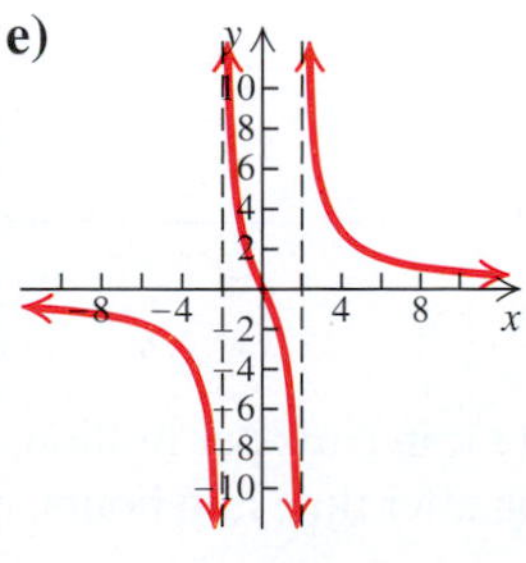

f)
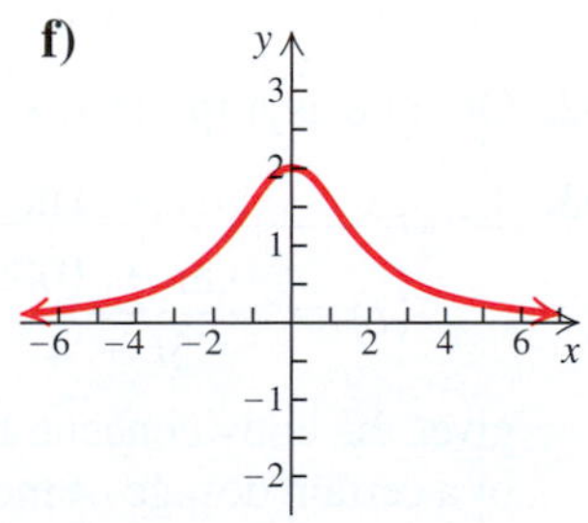

7. $f(x) = \dfrac{8}{x^2 - 4}$

8. $f(x) = \dfrac{8}{x^2 + 4}$

9. $f(x) = \dfrac{8x}{x^2 - 4}$

10. $f(x) = \dfrac{8x^2}{x^2 - 4}$

11. $f(x) = \dfrac{8x^3}{x^2 - 4}$

12. $f(x) = \dfrac{8x^3}{x^2 + 4}$

Determine the vertical asymptotes of the graph of the function.

13. $g(x) = \dfrac{1}{x^2}$

14. $f(x) = \dfrac{4x}{x^2 + 10x}$

15. $h(x) = \dfrac{x + 7}{2 - x}$

16. $g(x) = \dfrac{x^4 + 2}{x}$

17. $f(x) = \dfrac{3 - x}{(x - 4)(x + 6)}$

18. $h(x) = \dfrac{x^2 - 4}{x(x + 5)(x - 2)}$

19. $g(x) = \dfrac{x^3}{2x^3 - x^2 - 3x}$

20. $f(x) = \dfrac{x + 5}{x^2 + 4x - 32}$

Determine the horizontal asymptote of the graph of the function.

21. $f(x) = \dfrac{3x^2 + 5}{4x^2 - 3}$

22. $g(x) = \dfrac{x + 6}{x^3 + 2x^2}$

23. $h(x) = \dfrac{x^2 - 4}{2x^4 + 3}$

24. $f(x) = \dfrac{x^5}{x^5 + x}$

25. $g(x) = \dfrac{x^3 - 2x^2 + x - 1}{x^2 - 16}$

26. $h(x) = \dfrac{8x^4 + x - 2}{2x^4 - 10}$

Determine the oblique asymptote of the graph of the function.

27. $g(x) = \dfrac{x^2 + 4x - 1}{x + 3}$

28. $f(x) = \dfrac{x^2 - 6x}{x - 5}$

29. $h(x) = \dfrac{x^4 - 2}{x^3 + 1}$

30. $g(x) = \dfrac{12x^3 - x}{6x^2 + 4}$

31. $f(x) = \dfrac{x^3 - x^2 + x - 4}{x^2 + 2x - 1}$

32. $h(x) = \dfrac{5x^3 - x^2 + x - 1}{x^2 - x + 2}$

Make a hand-drawn graph for each of Exercises 33–78. Be sure to label all the asymptotes. List the domain and the x-intercepts and the y-intercepts. Check your work using a graphing calculator.

33. $f(x) = \frac{1}{x}$

34. $g(x) = \frac{1}{x^2}$

35. $h(x) = -\frac{4}{x^2}$

36. $f(x) = -\frac{6}{x}$

37. $g(x) = \frac{x^2 - 4x + 3}{x + 1}$

38. $h(x) = \frac{2x^2 - x - 3}{x - 1}$

39. $f(x) = \frac{-2}{x - 5}$

40. $f(x) = \frac{1}{x - 5}$

41. $f(x) = \frac{2x + 1}{x}$

42. $f(x) = \frac{3x - 1}{x}$

43. $f(x) = \frac{x + 3}{x^2 - 9}$

44. $f(x) = \frac{x - 1}{x^2 - 1}$

45. $f(x) = \frac{x}{x^2 + 3x}$

46. $f(x) = \frac{3x}{3x - x^2}$

47. $f(x) = \frac{1}{(x - 2)^2}$

48. $f(x) = \frac{-2}{(x - 3)^2}$

49. $f(x) = \frac{x^2 + 2x - 3}{x^2 + 4x + 3}$

50. $f(x) = \frac{x^2 - x - 2}{x^2 - 5x - 6}$

51. $f(x) = \frac{1}{x^2 + 3}$

52. $f(x) = \frac{-1}{x^2 + 2}$

53. $f(x) = \frac{x^2 - 4}{x - 2}$

54. $f(x) = \frac{x^2 - 9}{x + 3}$

55. $f(x) = \frac{x - 1}{x + 2}$

56. $f(x) = \frac{x - 2}{x + 1}$

57. $f(x) = \frac{x^2 + 3x}{2x^3 - 5x^2 - 3x}$

58. $f(x) = \frac{3x}{x^2 + 5x + 4}$

59. $f(x) = \frac{x^2 - 9}{x + 1}$

60. $f(x) = \frac{x^3 - 4x}{x^2 - x}$

61. $f(x) = \frac{x^2 + x - 2}{2x^2 + 1}$

62. $f(x) = \frac{x^2 - 2x - 3}{3x^2 + 2}$

63. $g(x) = \frac{3x^2 - x - 2}{x - 1}$

64. $f(x) = \frac{2x^2 - 5x - 3}{2x + 1}$

65. $f(x) = \frac{x - 1}{x^2 - 2x - 3}$

66. $f(x) = \frac{x + 2}{x^2 + 2x - 15}$

67. $f(x) = \frac{3x^2 + 11x - 4}{x^2 + 2x - 8}$

68. $f(x) = \frac{2x^2 - 3x - 9}{x^2 - 2x - 3}$

69. $f(x) = \frac{x - 3}{(x + 1)^3}$

70. $f(x) = \frac{x + 2}{(x - 1)^3}$

71. $f(x) = \frac{x^3 + 1}{x}$

72. $f(x) = \frac{x^3 - 1}{x}$

73. $f(x) = \frac{x^3 + 2x^2 - 15x}{x^2 - 5x - 14}$

74. $f(x) = \frac{x^3 + 2x^2 - 3x}{x^2 - 25}$

75. $f(x) = \frac{5x^4}{x^4 + 1}$

76. $f(x) = \frac{x + 1}{x^2 + x - 6}$

77. $f(x) = \frac{x^2}{x^2 - x - 2}$

78. $f(x) = \frac{x^2 - x - 2}{x + 2}$

Find a rational function that satisfies the given conditions. Answers may vary, but try to give the simplest answer possible.

79. Vertical asymptotes $x = -4$, $x = 5$

80. Vertical asymptotes $x = -4$, $x = 5$; x-intercept $(-2, 0)$

81. Vertical asymptotes $x = -4$, $x = 5$; horizontal asymptote $y = \frac{3}{2}$; x-intercept $(-2, 0)$

82. Oblique asymptote $y = x - 1$

83. *Medical Dosage.* The function

$$N(t) = \frac{0.8t + 1000}{5t + 4}, \quad t \geq 15,$$

gives the body concentration $N(t)$, in parts per million, of a certain dosage of medication after time t, in hours.

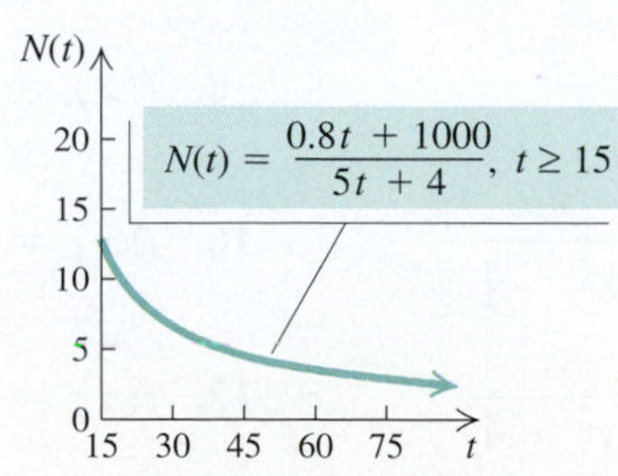

a) Find the horizontal asymptote of the graph and complete the following:

$N(t) \to \square$ as $t \to \infty$.

b) Explain the meaning of the answer to part (a) in terms of the application.

84. *Average Cost.* The average cost per light, in dollars, for a company to produce x roadside emergency lights is given by the function

$$A(x) = \frac{2x + 100}{x}, \quad x > 0.$$

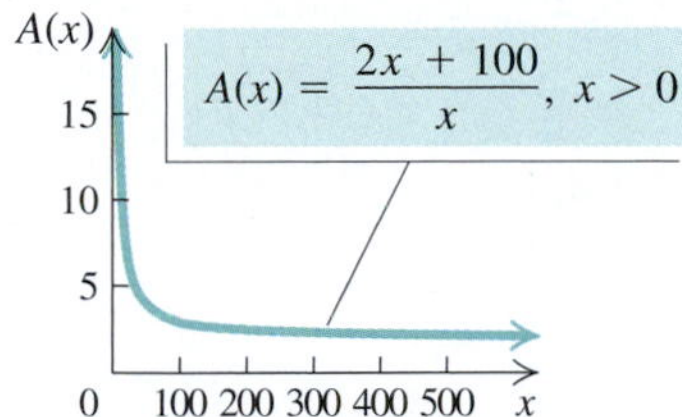

a) Find the horizontal asymptote of the graph and complete the following:

$A(x) \to \square$ as $x \to \infty$.

b) Explain the meaning of the answer to part (a) in terms of the application.

85. *Population Growth.* The population P, in thousands, of a resort community is given by

$$P(t) = \frac{500t}{2t^2 + 9},$$

where t is the time, in months, since the city council raised the property taxes.

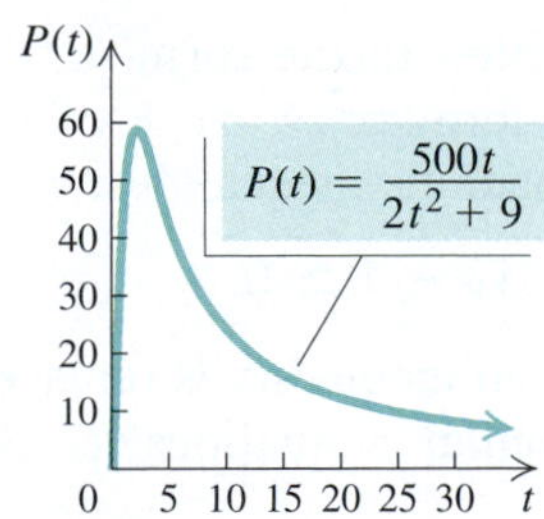

a) Find the population at $t = 0$, 1, 3, and 8 months.

b) Find the horizontal asymptote of the graph and complete the following:

$P(t) \to \square$ as $t \to \infty$.

c) Explain the meaning of the answer to part (b) in terms of the application.

86. *Minimizing Surface Area.* The Hold-It Container Co. is designing an open-top rectangular box, with a square base, that will hold 108 cubic centimeters.

a) Express the surface area S as a function of the length x of a side of the base.

b) Use a graphing calculator to graph the function on the interval $(0, \infty)$.

c) Estimate the minimum surface area and the value of x that will yield it.

Skill Maintenance

Vocabulary Reinforcement

In each of Exercises 87–95, fill in the blank with the correct term. Some of the given choices will not be used. Others will be used more than once.

x-intercept	midpoint formula
y-intercept	horizontal lines
odd function	vertical lines
even function	point–slope equation
domain	slope–intercept equation
range	difference quotient
slope	$f(x) = f(-x)$
distance formula	$f(-x) = -f(x)$

87. A function is a correspondence between a first set, called the ______________, and a second set, called the ______________, such that each member of the ______________ corresponds to exactly one member of the ______________. **[1.2]**

88. The ______________ of a line containing (x_1, y_1) and (x_2, y_2) is given by $(y_2 - y_1)/(x_2 - x_1)$. **[1.3]**

89. The ______________ of the line with slope m and y-intercept $(0, b)$ is $y = mx + b$. **[1.3]**

90. The ______________ of the line with slope m passing through (x_1, y_1) is $y - y_1 = m(x - x_1)$. **[1.4]**

91. A(n) ______________ is a point $(a, 0)$. **[1.1]**

92. For each x in the domain of an odd function f, ______________. **[2.4]**

93. ______________ are given by equations of the type $x = a$. **[1.3]**

94. The ______________ is $\left(\frac{x_1 + x_2}{2}, \frac{y_1 + y_2}{2}\right)$. [1.1]

95. A(n) ______________ is a point $(0, b)$. [1.1]

› Synthesis

96. Graph

$$y_1 = \frac{x^3 + 4}{x} \quad \text{and} \quad y_2 = x^2$$

using the same viewing window. Explain how the parabola $y_2 = x^2$ can be thought of as a nonlinear asymptote for y_1.

Find the nonlinear asymptote of the function.

97. $f(x) = \dfrac{x^5 + 2x^3 + 4x^2}{x^2 + 2}$

98. $f(x) = \dfrac{x^4 + 3x^2}{x^2 + 1}$

Graph the function.

99. $f(x) = \dfrac{2x^3 + x^2 - 8x - 4}{x^3 + x^2 - 9x - 9}$

100. $f(x) = \dfrac{x^3 + 4x^2 + x - 6}{x^2 - x - 2}$

4.6 Polynomial Inequalities and Rational Inequalities

- › Solve polynomial inequalities.
- › Solve rational inequalities.

We will use a combination of algebraic methods and graphical methods to solve polynomial inequalities and rational inequalities.

› Polynomial Inequalities

Just as a quadratic equation can be written in the form $ax^2 + bx + c = 0$, a **quadratic inequality** can be written in the form $ax^2 + bx + c \;\square\; 0$, where $\square$ is $<$, $>$, $\leq$, or $\geq$. Here are some examples of quadratic inequalities:

$$x^2 - 4x - 5 < 0 \quad \text{and} \quad -\tfrac{1}{2}x^2 + 4x - 7 \geq 0.$$

When the inequality symbol in a polynomial inequality is replaced with an equals sign, a **related equation** is formed. Polynomial inequalities can be solved once the related equation has been solved.

EXAMPLE 1 Solve: $x^2 - 4x - 5 > 0$.

Solution We are asked to find all x-values for which $x^2 - 4x - 5 > 0$. To locate these values, we graph $f(x) = x^2 - 4x - 5$. Then we note that whenever the graph passes through an x-intercept, the function changes sign. Thus to solve $x^2 - 4x - 5 > 0$, we first solve the *related equation* $x^2 - 4x - 5 = 0$ to find all zeros of the function:

$$x^2 - 4x - 5 = 0$$
$$(x + 1)(x - 5) = 0.$$

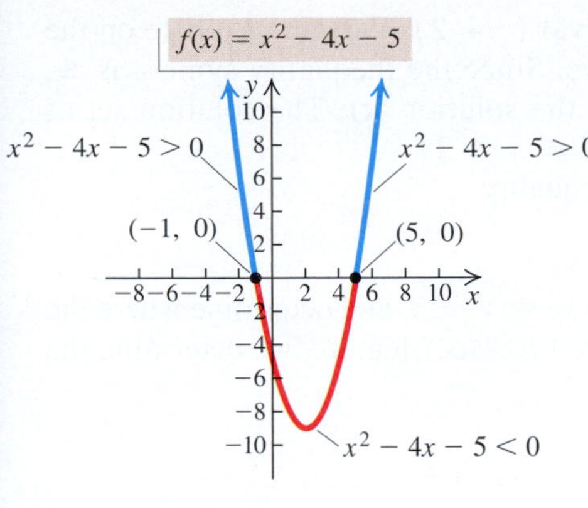

The zeros are -1 and 5. Thus the x-intercepts of the graph are $(-1, 0)$ and $(5, 0)$, as shown at left. The zeros divide the x-axis into three intervals:

$$(-\infty, -1), \quad (-1, 5), \quad \text{and} \quad (5, \infty).$$

The sign of $x^2 - 4x - 5$ is the same for all values of x in a given interval. Thus we choose a test value for x from each interval and find $f(x)$. We can also determine the sign of $f(x)$ in each interval by simply looking at the graph of the function.

Interval	$(-\infty, -1)$	$(-1, 5)$	$(5, \infty)$
Test Value	$f(-2) = 7$	$f(0) = -5$	$f(7) = 16$
Sign of $f(x)$	Positive	Negative	Positive

Just in Time 3

Since we are solving $x^2 - 4x - 5 > 0$, the solution set consists of only two of the three intervals, those in which the sign of $f(x)$ is positive. Since the inequality sign is $>$, we do not include the endpoints of the intervals in the solution set. The solution set is

$$(-\infty, -1) \cup (5, \infty), \quad \text{or} \quad \{x \mid x < -1 \text{ or } x > 5\}.$$

Now Try Exercise 27.

EXAMPLE 2 Solve: $x^2 + 3x - 5 \le x + 3$.

Solution By subtracting $x + 3$ on both sides, we form an equivalent inequality:

$$x^2 + 3x - 5 - x - 3 \le 0$$
$$x^2 + 2x - 8 \le 0.$$

We need to find all x-values for which $x^2 + 2x - 8 \le 0$. To visualize these values, we first graph $f(x) = x^2 + 2x - 8$ and then determine the zeros of the function.

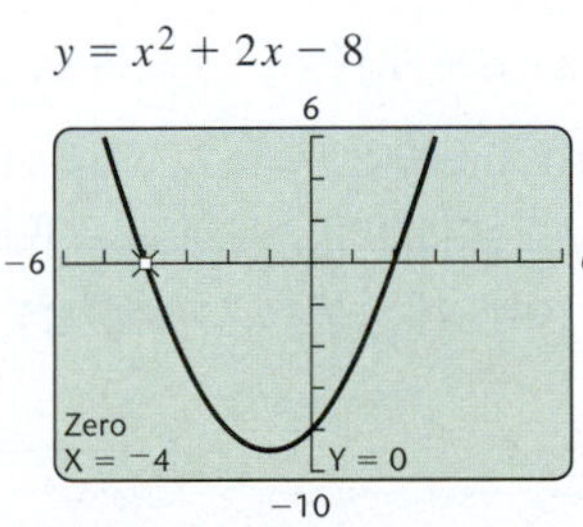

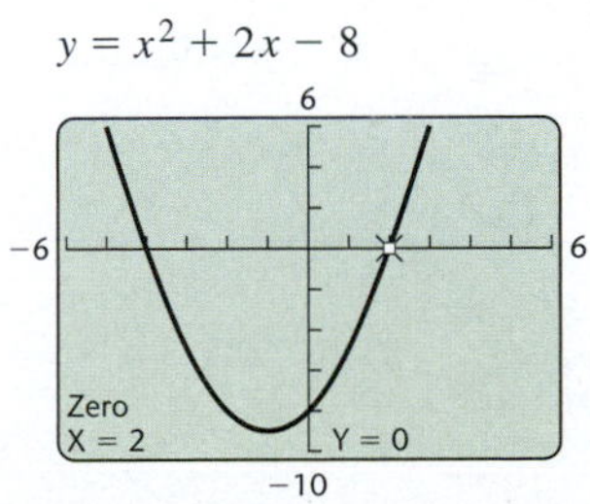

X	Y1
−5	7
0	−8
4	16

X =

Using the ZERO feature, we see that the zeros are -4 and 2.

The intervals to be considered are $(-\infty, -4)$, $(-4, 2)$, and $(2, \infty)$. Using test values for $f(x)$, we determine the sign of $f(x)$ in each interval. (See the table at left.)

Function values are negative in the interval $(-4, 2)$. We can also note on the graph where the function values are negative. Since the inequality symbol is $\leq$, we include the endpoints of the interval in the solution set. The solution set of $x^2 + 3x - 5 \leq x + 3$ is $[-4, 2]$, or $\{x \mid -4 \leq x \leq 2\}$.

An alternative approach to solving the inequality

$$x^2 + 3x - 5 \leq x + 3$$

is to graph both sides, $y_1 = x^2 + 3x - 5$ and $y_2 = x + 3$, and determine where the graph of y_1 is below the graph of y_2. Using the INTERSECT feature, we determine the points of intersection, $(-4, -1)$ and $(2, 5)$.

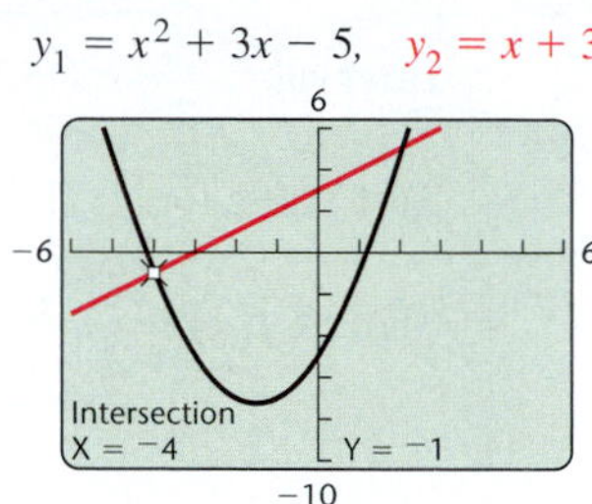

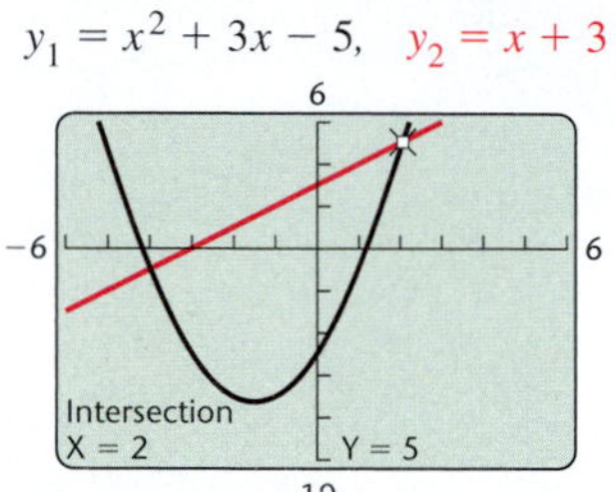

The graph of y_1 is below the graph of y_2 over the interval $(-4, 2)$. Since the inequality symbol is $\leq$, we include the endpoints of the interval in the solution. Thus the solution set of $x^2 + 3x - 5 \leq x + 3$ is $[-4, 2]$.

Now Try Exercise 29.

Quadratic inequalities are one type of **polynomial inequality**. Other examples of polynomial inequalities are

$$-2x^4 + x^2 - 3 < 7, \qquad \tfrac{2}{3}x + 4 \geq 0, \quad \text{and} \quad 4x^3 - 2x^2 > 5x + 7.$$

EXAMPLE 3 Solve: $x^3 - x > 0$.

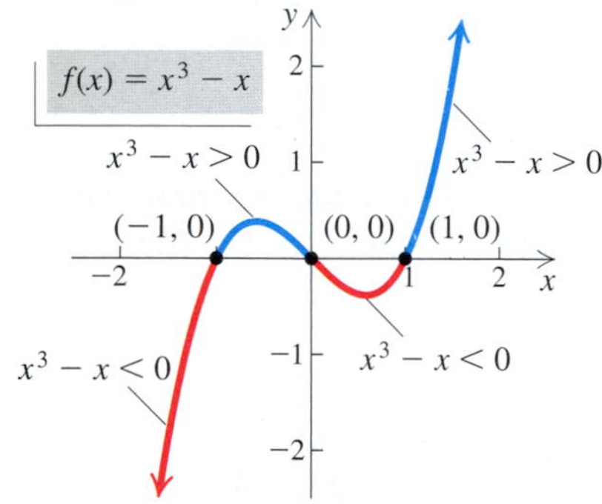

Solution We are asked to find all x-values for which $x^3 - x > 0$. To locate these values, we graph $f(x) = x^3 - x$. Then we note that whenever the function changes sign, its graph passes through an x-intercept. Thus to solve $x^3 - x > 0$, we first solve the related equation $x^3 - x = 0$ to find all zeros of the function:

$$x^3 - x = 0$$
$$x(x^2 - 1) = 0$$
$$x(x + 1)(x - 1) = 0.$$

The zeros are -1, 0, and 1. Thus the x-intercepts of the graph are $(-1, 0)$, $(0, 0)$, and $(1, 0)$, as shown in the figure at left. The zeros divide the x-axis into four intervals:

$$(-\infty, -1), \qquad (-1, 0), \qquad (0, 1), \quad \text{and} \quad (1, \infty).$$

−1 0 1 x

The sign of $x^3 - x$ is the same for all values of x in a given interval. Thus we choose a test value for x from each interval and find $f(x)$. We can use the TABLE feature set in ASK mode to determine the sign of $f(x)$ in each interval.

X	Y1
−2	−6
−.5	.375
.5	−.375
2	6

X =

We can also determine the sign of $f(x)$ in each interval by simply looking at the graph of the function.

Interval	$(-\infty, -1)$	$(-1, 0)$	$(0, 1)$	$(1, \infty)$
Test Value	$f(-2) = -6$	$f(-0.5) = 0.375$	$f(0.5) = -0.375$	$f(2) = 6$
Sign of $f(x)$	Negative	Positive	Negative	Positive

Since we are solving $x^3 - x > 0$, the solution set consists of only two of the four intervals, those in which the sign of $f(x)$ is *positive*. We see that the solution set is $(-1, 0) \cup (1, \infty)$, or $\{x \mid -1 < x < 0 \text{ or } x > 1\}$.

Now Try Exercise 39.

To solve a polynomial inequality:

1. Find an equivalent inequality with $P(x)$ on one side and 0 on the other.
2. Change the inequality symbol to an equals sign and solve the related equation; that is, solve $P(x) = 0$.
3. Use the solutions to divide the x-axis into intervals. Then select a test value from each interval and determine the sign of the polynomial on the interval.
4. Determine the intervals for which the inequality is satisfied and write interval notation or set-builder notation for the solution set. Include the endpoints of the intervals in the solution set if the inequality symbol is $\leq$ or $\geq$.

EXAMPLE 4 Solve: $3x^4 + 10x \le 11x^3 + 4$.

Solution By subtracting $11x^3 + 4$, we form the equivalent inequality

$$3x^4 - 11x^3 + 10x - 4 \le 0.$$

Algebraic Solution

To solve the related equation

$$3x^4 - 11x^3 + 10x - 4 = 0,$$

we need to use the theorems of Section 4.4. We solved this equation in Example 5 in Section 4.4. The solutions are

$$-1, \quad 2 - \sqrt{2}, \quad \tfrac{2}{3}, \quad \text{and} \quad 2 + \sqrt{2},$$

or approximately

$$-1, \quad 0.586, \quad 0.667, \quad \text{and} \quad 3.414.$$

These numbers divide the x-axis into five intervals: $(-\infty, -1)$, $(-1, 2 - \sqrt{2})$, $(2 - \sqrt{2}, \frac{2}{3})$, $(\frac{2}{3}, 2 + \sqrt{2})$, and $(2 + \sqrt{2}, \infty)$.

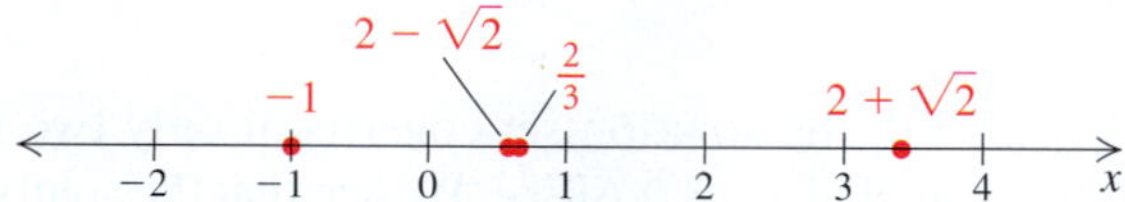

We then let $f(x) = 3x^4 - 11x^3 + 10x - 4$ and, using test values for x, determine the sign of $f(x)$ in each interval.

X	Y1
−2	112
0	−4
.6	.0128
1	−2
4	100

X =

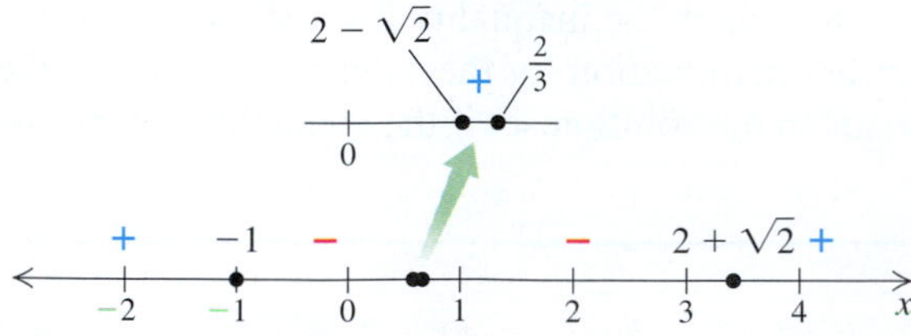

Function values are negative in the intervals $(-1, 2 - \sqrt{2})$ and $(\frac{2}{3}, 2 + \sqrt{2})$. Since the inequality sign is $\le$, we include the endpoints of the intervals in the solution set. The solution set is

$$[-1, 2 - \sqrt{2}] \cup [\tfrac{2}{3}, 2 + \sqrt{2}], \quad \text{or}$$

$$\{x \mid -1 \le x \le 2 - \sqrt{2} \ or \ \tfrac{2}{3} \le x \le 2 + \sqrt{2}\}.$$

Graphical Solution

We graph $y = 3x^4 - 11x^3 + 10x - 4$ using a viewing window that reveals the curvature of the graph.

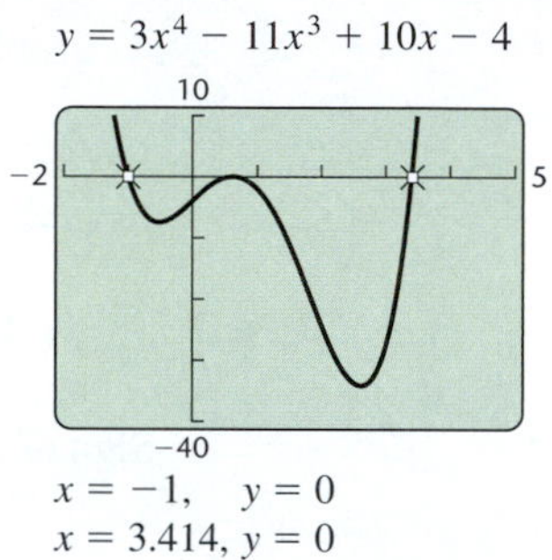

Using the ZERO feature, we see that two of the zeros are -1 and approximately 3.414 $(2 + \sqrt{2} \approx 3.414)$. However, this window leaves us uncertain about the number of zeros of the function on the interval $[0, 1]$. The following window shows another view of the zeros on the interval $[0, 1]$. Those zeros are about 0.586 and 0.667 $(2 - \sqrt{2} \approx 0.586; \frac{2}{3} \approx 0.667)$.

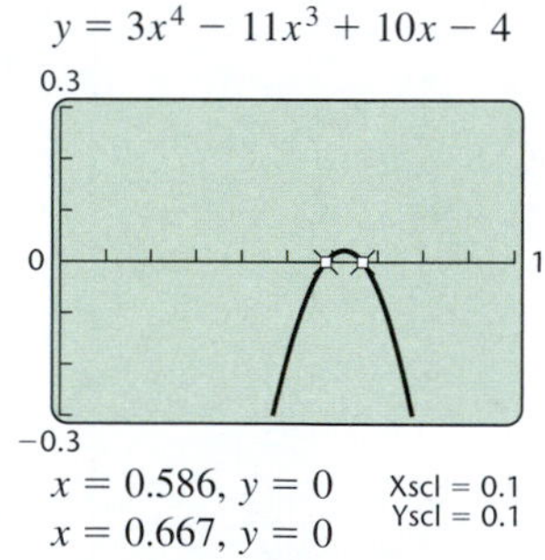

The intervals to be considered are $(-\infty, -1)$, $(-1, 0.586)$, $(0.586, 0.667)$, $(0.667, 3.414)$, and $(3.414, \infty)$. We note on the graph where the function is negative. Then including appropriate endpoints, we find that the solution set is approximately

$$[-1, 0.586] \cup [0.667, 3.414], \quad \text{or}$$

$$\{x \mid -1 \le x \le 0.586 \ or \ 0.667 \le x \le 3.414\}.$$

Now Try Exercise 45.

Rational Inequalities

Some inequalities involve rational expressions and functions. These are called **rational inequalities**. To solve rational inequalities, we need to make some adjustments to the preceding method.

EXAMPLE 5 Solve: $\dfrac{3x}{x+6} < 0$.

Solution We look for all values of x for which the related function

$$f(x) = \frac{3x}{x+6}$$

is not defined or is 0. These are called **critical values**.

The denominator tells us that $f(x)$ is not defined when $x = -6$. Next, we solve $f(x) = 0$:

$$\frac{3x}{x+6} = 0$$

$$(x+6) \cdot \frac{3x}{x+6} = (x+6) \cdot 0 \quad \text{Multiplying by } x + 6$$

$$3x = 0$$

$$x = 0.$$

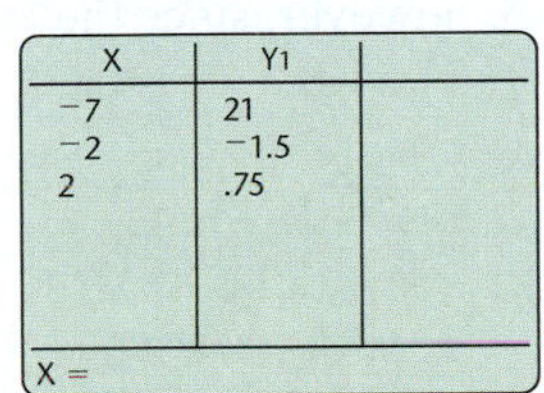

X	Y1
−7	21
−2	−1.5
2	.75

X =

The critical values are −6 and 0. These values divide the x-axis into three intervals:

$$(-\infty, -6), \qquad (-6, 0), \quad \text{and} \quad (0, \infty).$$

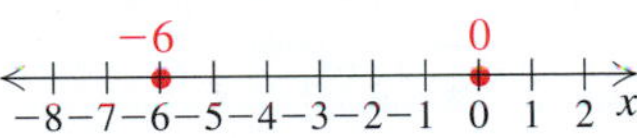

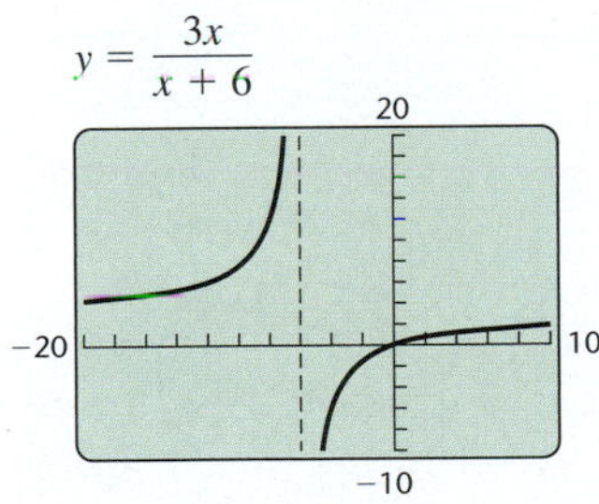

We then use a test value to determine the sign of $f(x)$ on each interval.

Function values are negative on only the interval $(-6, 0)$. Although $f(0) = 0$, the inequality symbol is $<$, so we know that 0 is not included in the solution set. Note that since −6 is not in the domain of f, −6 cannot be part of the solution set. The solution set is

$$(-6, 0), \quad \text{or} \quad \{x \mid -6 < x < 0\}.$$

The graph of $f(x)$ shows where $f(x)$ is positive and where it is negative.

Now Try Exercise 53.

EXAMPLE 6 Solve: $\dfrac{x+1}{2x-4} \le 1$.

Solution We first subtract 1 on both sides in order to find an equivalent inequality with 0 on one side:

$$\frac{x+1}{2x-4} - 1 \le 0.$$

Algebraic Solution

We look for all values of x for which the related function

$$f(x) = \frac{x + 1}{2x - 4} - 1$$

is not defined or is 0. These are called **critical values**.

A look at the denominator shows that $f(x)$ is not defined for $x = 2$. Next, we solve $f(x) = 0$:

$$\frac{x + 1}{2x - 4} - 1 = 0$$

$$(2x - 4)\left(\frac{x + 1}{2x - 4} - 1\right) = (2x - 4)\cdot 0$$

$$x + 1 - (2x - 4)\cdot 1 = 0$$

$$x + 1 - 2x + 4 = 0$$

$$-x = -5$$

$$x = 5.$$

The critical values are 2 and 5. These values divide the x-axis into three intervals:

$$(-\infty, 2), \quad (2, 5), \quad \text{and} \quad (5, \infty).$$

We then use a test value to determine the sign of $f(x)$ in each interval.

X	Y1
0	−1.25
3	1
6	−.125

X =

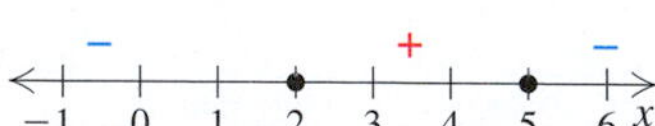

Function values are negative on the intervals $(-\infty, 2)$ and $(5, \infty)$. Since $f(5) = 0$ and the inequality symbol is $\leq$, we know that 5 is in the solution set. Note that since 2 is not in the domain of f, it cannot be part of the solution set.

The solution set is $(-\infty, 2) \cup [5, \infty)$.

Graphical Solution

The graph of

$$y = \frac{x + 1}{2x - 4} - 1$$

confirms the two critical values found algebraically: 2 where $f(x)$ is not defined and 5 where $f(x) = 0$.

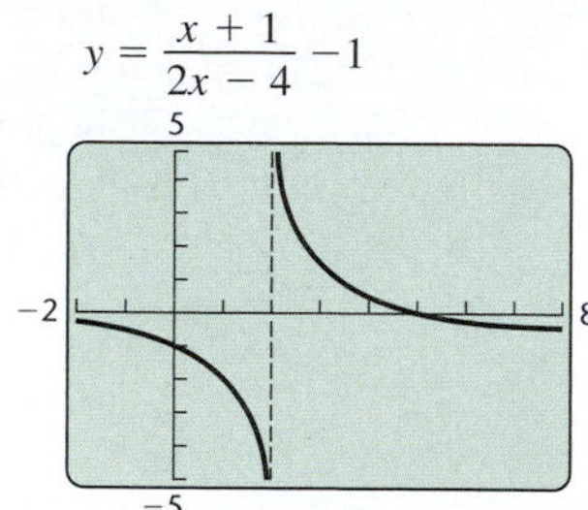

The graph shows where y is negative. Note that 2 cannot be in the solution set since y is not defined for this value. We do include 5, however, since the inequality symbol is $\leq$ and $f(5) = 0$.

The solution set is

$$(-\infty, 2) \cup [5, \infty).$$

Now Try Exercise 59.

EXAMPLE 7 Solve: $\dfrac{x - 3}{x + 4} \geq \dfrac{x + 2}{x - 5}$.

Solution We first subtract $(x + 2)/(x - 5)$ on both sides in order to find an equivalent inequality with 0 on one side:

$$\frac{x - 3}{x + 4} - \frac{x + 2}{x - 5} \geq 0.$$

Algebraic Solution

We look for all values of x for which the related function

$$f(x) = \frac{x-3}{x+4} - \frac{x+2}{x-5}$$

is not defined or is 0. These are the critical values.

A look at the denominators shows that $f(x)$ is not defined for $x = -4$ and $x = 5$. Next, we solve $f(x) = 0$:

$$\frac{x-3}{x+4} - \frac{x+2}{x-5} = 0$$

$$(x+4)(x-5)\left(\frac{x-3}{x+4} - \frac{x+2}{x-5}\right) = (x+4)(x-5)\cdot 0$$

$$(x-5)(x-3) - (x+4)(x+2) = 0$$

$$x^2 - 8x + 15 - (x^2 + 6x + 8) = 0$$

$$x^2 - 8x + 15 - x^2 - 6x - 8 = 0$$

$$-14x + 7 = 0$$

$$-14x = -7$$

$$x = \tfrac{1}{2}.$$

The critical values are -4, $\frac{1}{2}$, and 5. These values divide the x-axis into four intervals:

$$(-\infty, -4), \quad \left(-4, \tfrac{1}{2}\right), \quad \left(\tfrac{1}{2}, 5\right), \quad \text{and} \quad (5, \infty).$$

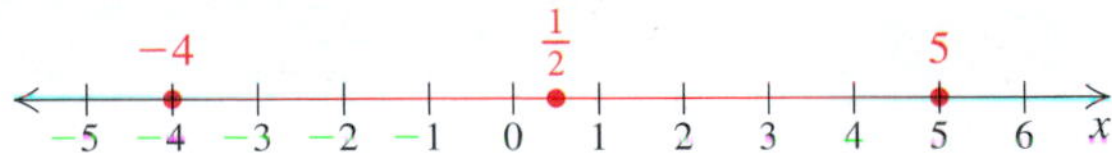

We then use a test value to determine the sign of $f(x)$ in each interval.

X	Y1
−5	7.7
−2	−2.5
3	2.5
6	−7.7

X =

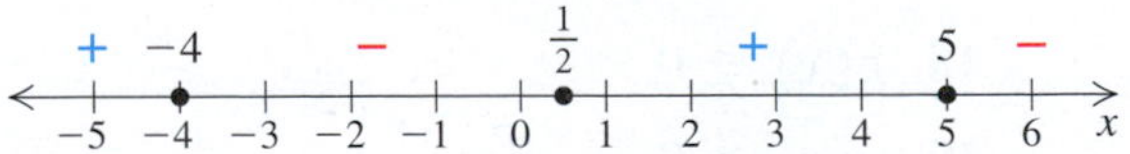

Function values are positive on the intervals $(-\infty, -4)$ and $\left(\frac{1}{2}, 5\right)$. Since $f\left(\frac{1}{2}\right) = 0$ and the inequality symbol is $\geq$, we know that $\frac{1}{2}$ must be in the solution set. Note that since neither -4 nor 5 is in the domain of f, they cannot be part of the solution set.

The solution set is $(-\infty, -4) \cup \left[\frac{1}{2}, 5\right)$.

Graphical Solution

We graph

$$y = \frac{x-3}{x+4} - \frac{x+2}{x-5}$$

in the standard window, which shows the curvature of the function.

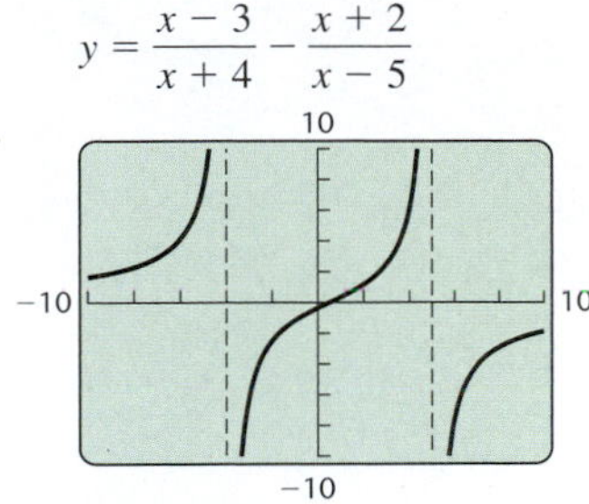

Using the ZERO feature, we find that 0.5 is a zero.

We then look for values where the function is not defined. By examining the denominators $x + 4$ and $x - 5$, we see that $f(x)$ is not defined for $x = -4$ and $x = 5$.

The critical values are -4, 0.5, and 5.

The graph shows where y is positive and where it is negative. Note that -4 and 5 cannot be in the solution set since y is not defined for these values. We do include 0.5, however, since the inequality symbol is $\geq$ and $f(0.5) = 0$. The solution set is

$$(-\infty, -4) \cup [0.5, 5).$$

Now Try Exercise 61.

The following is a method for solving rational inequalities.

To solve a rational inequality:

1. Find an equivalent inequality with 0 on one side.
2. Change the inequality symbol to an equals sign and solve the related equation.
3. Find values of the variable for which the related rational function is not defined.
4. The numbers found in steps (2) and (3) are called critical values. Use the critical values to divide the x-axis into intervals. Then determine the function's sign in each interval using an x-value from the interval or using the graph of the equation.
5. Select the intervals for which the inequality is satisfied and write interval notation or set-builder notation for the solution set. If the inequality symbol is $\leq$ or $\geq$, then the solutions to step (2) should be included in the solution set. The x-values found in step (3) are never included in the solution set.

It works well to use a combination of algebraic methods and graphical methods to solve polynomial inequalities and rational inequalities. The algebraic methods give exact numbers for the critical values, and the graphical methods usually allow us to see easily what intervals satisfy the inequality.

4.6 Exercise Set

For the function $f(x) = x^2 + 2x - 15$, solve each of the following.

1. $f(x) = 0$
2. $f(x) < 0$
3. $f(x) \leq 0$
4. $f(x) > 0$
5. $f(x) \geq 0$

For the function

$$g(x) = \frac{x-2}{x+4},$$

solve each of the following.

6. $g(x) = 0$
7. $g(x) > 0$
8. $g(x) \leq 0$
9. $g(x) \geq 0$
10. $g(x) < 0$

For the function

$$h(x) = \frac{7x}{(x-1)(x+5)},$$

solve each of the following.

11. $h(x) = 0$
12. $h(x) \leq 0$
13. $h(x) \geq 0$
14. $h(x) > 0$
15. $h(x) < 0$

For the function $g(x) = x^5 - 9x^3$, solve each of the following.

16. $g(x) = 0$
17. $g(x) < 0$
18. $g(x) \leq 0$
19. $g(x) > 0$
20. $g(x) \geq 0$

In Exercises 21–24, a related function is graphed. Solve the given inequality.

21. $x^3 + 6x^2 < x + 30$

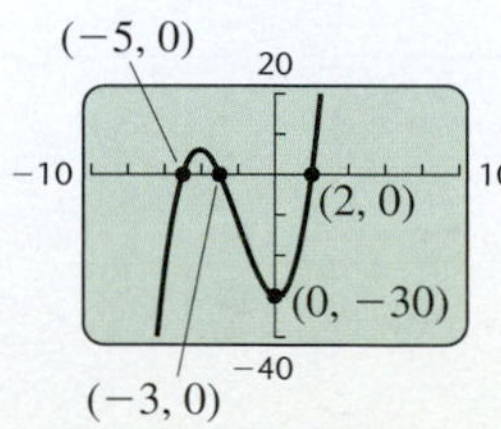

22. $x^4 - 27x^2 - 14x + 120 \geq 0$

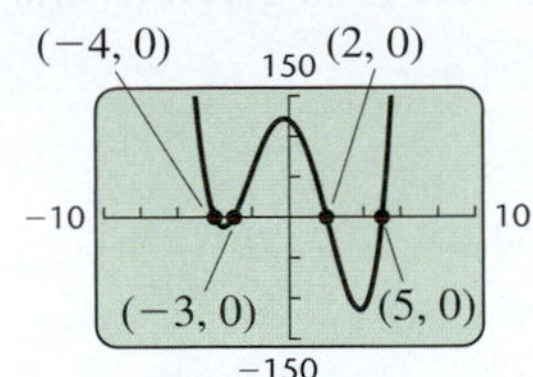

23. $\dfrac{8x}{x^2 - 4} \geq 0$

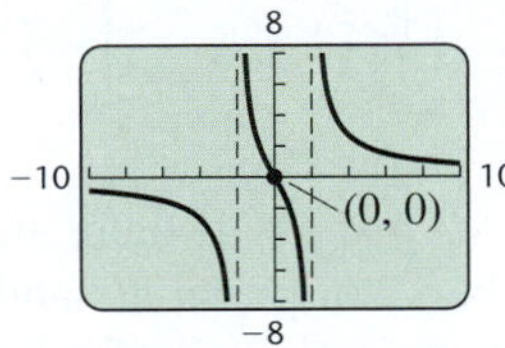

24. $\dfrac{8}{x^2 - 4} < 0$

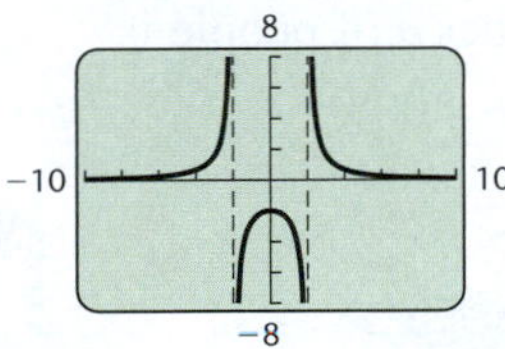

Solve.

25. $(x - 1)(x + 4) < 0$
26. $(x + 3)(x - 5) < 0$
27. $x^2 + x - 2 > 0$
28. $x^2 - x - 6 > 0$
29. $x^2 - x - 5 \geq x - 2$
30. $x^2 + 4x + 7 \geq 5x + 9$
31. $x^2 > 25$
32. $x^2 \leq 1$
33. $4 - x^2 \leq 0$
34. $11 - x^2 \geq 0$
35. $6x - 9 - x^2 < 0$
36. $x^2 + 2x + 1 \leq 0$
37. $x^2 + 12 < 4x$
38. $x^2 - 8 > 6x$
39. $4x^3 - 7x^2 \leq 15x$
40. $2x^3 - x^2 < 5x$
41. $x^3 + 3x^2 - x - 3 \geq 0$
42. $x^3 + x^2 - 4x - 4 \geq 0$
43. $x^3 - 2x^2 < 5x - 6$
44. $x^3 + x \leq 6 - 4x^2$
45. $x^5 + x^2 \geq 2x^3 + 2$
46. $x^5 + 24 > 3x^3 + 8x^2$
47. $2x^3 + 6 \leq 5x^2 + x$
48. $2x^3 + x^2 < 10 + 11x$
49. $x^3 + 5x^2 - 25x \leq 125$
50. $x^3 - 9x + 27 \geq 3x^2$
51. $0.1x^3 - 0.6x^2 - 0.1x + 2 < 0$
52. $19.2x^3 + 12.8x^2 + 144 \geq 172.8x + 3.2x^4$

List the critical values of the related function. Then solve the inequality.

53. $\dfrac{1}{x + 4} > 0$
54. $\dfrac{1}{x - 3} \leq 0$
55. $\dfrac{-4}{2x + 5} < 0$
56. $\dfrac{-2}{5 - x} \geq 0$
57. $\dfrac{2x}{x - 4} \geq 0$
58. $\dfrac{5x}{x + 1} < 0$
59. $\dfrac{x + 1}{x - 2} \geq 3$
60. $\dfrac{x}{x - 5} < 2$
61. $\dfrac{x - 4}{x + 3} - \dfrac{x + 2}{x - 1} \leq 0$
62. $\dfrac{x + 1}{x - 2} - \dfrac{x - 3}{x - 1} < 0$
63. $\dfrac{x + 6}{x - 2} > \dfrac{x - 8}{x - 5}$
64. $\dfrac{x - 7}{x + 2} \geq \dfrac{x - 9}{x + 3}$
65. $x - 2 > \dfrac{1}{x}$
66. $4 \geq \dfrac{4}{x} + x$
67. $\dfrac{2}{x^2 - 4x + 3} \leq \dfrac{5}{x^2 - 9}$
68. $\dfrac{3}{x^2 - 4} \leq \dfrac{5}{x^2 + 7x + 10}$
69. $\dfrac{3}{x^2 + 1} \geq \dfrac{6}{5x^2 + 2}$
70. $\dfrac{4}{x^2 - 9} < \dfrac{3}{x^2 - 25}$
71. $\dfrac{5}{x^2 + 3x} < \dfrac{3}{2x + 1}$
72. $\dfrac{2}{x^2 + 3} > \dfrac{3}{5 + 4x^2}$
73. $\dfrac{5x}{7x - 2} > \dfrac{x}{x + 1}$
74. $\dfrac{x^2 - x - 2}{x^2 + 5x + 6} < 0$

75. $\dfrac{x}{x^2 + 4x - 5} + \dfrac{3}{x^2 - 25} \le \dfrac{2x}{x^2 - 6x + 5}$

76. $\dfrac{2x}{x^2 - 9} + \dfrac{x}{x^2 + x - 12} \ge \dfrac{3x}{x^2 + 7x + 12}$

77. *Temperature During an Illness.* A person's temperature T, in degrees Fahrenheit, during an illness is given by the function

$$T(t) = \frac{4t}{t^2 + 1} + 98.6,$$

where t is the time since the onset of the illness, in hours. Find the interval on which the temperature was over 100°F. (See Example 12 in Section 4.5.)

78. *Population Growth.* The population P, in thousands, of a resort community is given by

$$P(t) = \frac{500t}{2t^2 + 9},$$

where t is the time, in months, since the city council raised the property taxes. Find the interval on which the population was 40,000 or greater. (See Exercise 85 in Exercise Set 4.5.)

79. *Total Profit.* Flexl, Inc., determines that its total profit is given by the function

$$P(x) = -3x^2 + 630x - 6000.$$

a) Flexl makes a profit for those nonnegative values of x for which $P(x) > 0$. Find the values of x for which Flexl makes a profit.

b) Flexl loses money for those nonnegative values of x for which $P(x) < 0$. Find the values of x for which Flexl loses money.

80. *Height of a Thrown Object.* The function

$$S(t) = -16t^2 + 32t + 1920$$

gives the height S, in feet, of an object thrown upward with a velocity of 32 ft/sec from a cliff that is 1920 ft high. Here t is the time, in seconds, that the object is in the air.

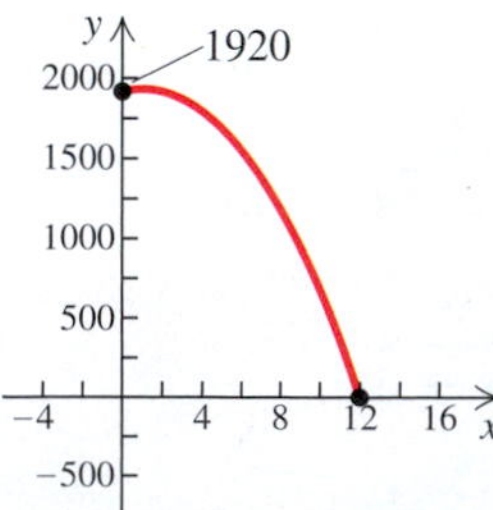

a) For what times is the height greater than 1920 ft?

b) For what times is the height less than 640 ft?

81. *Number of Diagonals.* A polygon with n sides has D diagonals, where D is given by the function

$$D(n) = \frac{n(n - 3)}{2}.$$

Find the number of sides n if

$$27 \le D \le 230.$$

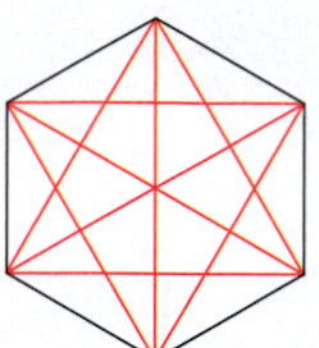

82. *Number of Handshakes.* If there are n people in a room, the number N of possible handshakes by all the people in the room is given by the function

$$N(n) = \frac{n(n - 1)}{2}.$$

For what number n of people is

$$66 \le N \le 300?$$

Skill Maintenance

Find an equation for a circle satisfying the given conditions. **[1.1]**

83. Center: $(-2, 4)$; radius of length 3

84. Center: $(0, -3)$; diameter of length $\frac{7}{2}$

In Exercises 85 and 86:

a) *Find the vertex.*

b) *Determine whether there is a maximum or a minimum value and find that value.*

c) *Find the range.* **[3.3]**

85. $h(x) = -2x^2 + 3x - 8$

86. $g(x) = x^2 - 10x + 2$

Synthesis

Solve.

87. $|x^2 - 5| = 5 - x^2$

88. $x^4 - 6x^2 + 5 > 0$

89. $2|x|^2 - |x| + 2 \leq 5$

90. $(7 - x)^{-2} < 0$

91. $\left|1 + \frac{1}{x}\right| < 3$

92. $\left|2 - \frac{1}{x}\right| \leq 2 + \left|\frac{1}{x}\right|$

93. Write a quadratic inequality for which the solution set is $(-4, 3)$.

94. Write a polynomial inequality for which the solution set is $[-4, 3] \cup [7, \infty)$.

Find the domain of the function.

95. $f(x) = \sqrt{\frac{72}{x^2 - 4x - 21}}$

96. $f(x) = \sqrt{x^2 - 4x - 21}$

Chapter 4 Summary and Review

STUDY GUIDE

KEY TERMS AND CONCEPTS | **EXAMPLES**

SECTION 4.1: POLYNOMIAL FUNCTIONS AND MODELING

Polynomial Function

$$P(x) = a_n x^n + a_{n-1}x^{n-1} + a_{n-2}x^{n-2} + \cdots + a_1 x + a_0,$$

where the coefficients $a_n, a_{n-1}, \ldots, a_1, a_0$ are real numbers and the exponents are whole numbers.

The first nonzero coefficient, a_n, is called the **leading coefficient**. The term $a_n x^n$ is called the **leading term**. The **degree** of the polynomial function is n.

Classifying polynomial functions by degree:

Type	Degree
Constant	0
Linear	1
Quadratic	2
Cubic	3
Quartic	4

Consider the polynomial

$$P(x) = \frac{1}{3}x^2 + x - 4x^5 + 2.$$

Leading term: $-4x^5$

Leading coefficient: -4

Degree of polynomial: 5

Classify the following polynomial functions:

Function	Type
$f(x) = -2$	Constant
$f(x) = 0.6x - 11$	Linear
$f(x) = 5x^2 + x - 4$	Quadratic
$f(x) = 5x^3 - x + 10$	Cubic
$f(x) = -x^4 + 8x^3 + x$	Quartic

The Leading-Term Test

If $a_n x^n$ is the leading term of a polynomial function, then the behavior of the graph as $x \to \infty$ and as $x \to -\infty$ can be described in one of the following four ways.

a) If n is even, and $a_n > 0$:

b) If n is even, and $a_n < 0$:

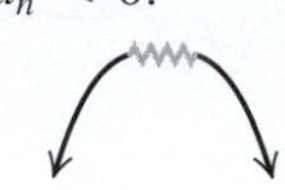

c) If n is odd, and $a_n > 0$:

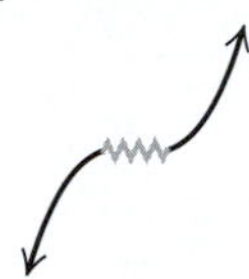

d) If n is odd, and $a_n < 0$:

Using the leading-term test, describe the end behavior of the graph of each of the following functions by selecting one of (a)–(d) shown at left.

$$h(x) = -2x^6 + x^4 - 3x^2 + x$$

The leading term $a_n x^n$ is $-2x^6$. Since 6 is even and $-2 < 0$, the shape is shown in (b).

$$g(x) = 4x^3 - 8x + 1$$

The leading term, $a_n x^n$, is $4x^3$. Since 3 is odd and $4 > 0$, the shape is shown in (c).

Zeros of Functions

If c is a real zero of a function $f(x)$ (that is, $f(c) = 0$), then $x - c$ is a factor of $f(x)$ and $(c, 0)$ is an x-intercept of the graph of the function.

If we know the linear factors of a polynomial function $f(x)$, we can find the zeros of $f(x)$ by solving the equation $f(x) = 0$ using the principle of zero products.

Every function of degree n, with $n \geq 1$, has at least one zero and at most n zeros.

To find the zeros of

$$f(x) = -2(x - 3)(x + 8)^2,$$

solve $-2(x - 3)(x + 8)^2 = 0$ using the principle of zero products:

$$x - 3 = 0 \quad or \quad x + 8 = 0$$
$$x = 3 \quad or \quad x = -8.$$

The zeros of $f(x)$ are 3 and -8.

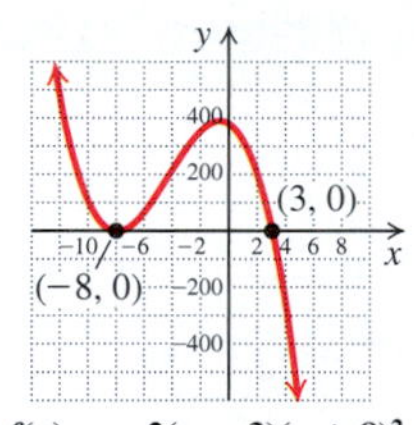

$f(x) = -2(x - 3)(x + 8)^2$

To find the zeros of

$$h(x) = x^4 - 12x^2 - 64,$$

solve $h(x) = 0$:

$$x^4 - 12x^2 - 64 = 0$$
$$(x^2 - 16)(x^2 + 4) = 0$$
$$(x + 4)(x - 4)(x^2 + 4) = 0$$
$$x + 4 = 0 \quad or \quad x - 4 = 0 \quad or \quad x^2 + 4 = 0$$
$$x = -4 \quad or \quad x = 4 \quad or \quad x^2 = -4$$
$$x = -4 \quad or \quad x = 4 \quad or \quad x = \pm\sqrt{-4}$$
$$x = -4 \quad or \quad x = 4 \quad or \quad x = \pm 2i.$$

The zeros of $h(x)$ are -4, 4, $-2i$, and $2i$.

Even and Odd Multiplicity

If $(x - c)^k$, $k \geq 1$, is a factor of a polynomial function $P(x)$ and $(x - c)^{k+1}$ is not a factor and:

- k is odd, then the graph crosses the x-axis at $(c, 0)$;
- k is even, then the graph is tangent to the x-axis at $(c, 0)$.

For $f(x) = -2(x - 3)(x + 8)^2$ graphed above, note that for the factor $x - 3$, or $(x - 3)^1$, the exponent 1 is odd and the graph crosses the x-axis at $(3, 0)$. For the factor $(x + 8)^2$, the exponent 2 is even and the graph is tangent to the x-axis at $(-8, 0)$.

SECTION 4.2: GRAPHING POLYNOMIAL FUNCTIONS

If $P(x)$ is a polynomial function of degree n, the graph of the function has:

- at most n real zeros, and thus at most n x-intercepts, and
- at most $n - 1$ turning points.

To Graph a Polynomial Function

1. Use the leading-term test to determine the end behavior.
2. Find the zeros of the function by solving $f(x) = 0$. Any real zeros are the first coordinates of the x-intercepts.
3. Use the x-intercepts (zeros) to divide the x-axis into intervals and choose a test point in each interval to determine the sign of all function values in that interval. For all x-values in an interval, $f(x)$ is either always positive for all values or always negative for all values.
4. Find $f(0)$. This gives the y-intercept of the function.
5. If necessary, find additional function values to determine the general shape of the graph and then draw the graph.

Graph: $h(x) = x^4 - 12x^2 - 16x = x(x - 4)(x + 2)^2$.

1. The leading term is x^4. Since 4 is even and $1 > 0$, the end behavior of the graph can be sketched as follows.

2. Solve $x(x - 4)(x + 2)^2 = 0$. The solutions are 0, 4, and -2. The zeros of $h(x)$ are 0, 4, and -2. The x-intercepts are $(0, 0)$, $(4, 0)$, and $(-2, 0)$. The multiplicity of 0 and 4 is 1. The graph will cross the x-axis at 0 and 4. The multiplicity of -2 is 2. The graph is tangent to the x-axis at -2.
3. The zeros divide the x-axis into four intervals.

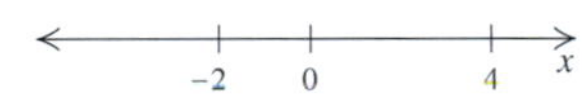

Interval	$(-\infty, -2)$	$(-2, 0)$	$(0, 4)$	$(4, \infty)$
Test Value	-3	-1	1	5
Function Value, $h(x)$	21	5	-27	245
Sign of $h(x)$	+	+	−	+
Location of Points on Graph	Above x-axis	Above x-axis	Below x-axis	Above x-axis

Four points on the graph are $(-3, 21)$, $(-1, 5)$, $(1, -27)$, and $(5, 245)$.

4. Find $h(0)$:

$$h(0) = 0(0 - 4)(0 + 2)^2 = 0.$$

The y-intercept is $(0, 0)$.

(continued)

5. Find additional points and draw the graph.

x	$h(x)$
-2.5	4.1
-1.5	2.1
-0.5	5.1
0.5	-10.9
2	-64
3	-75

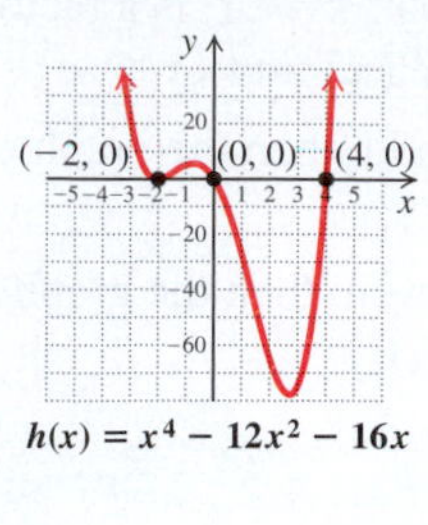

The Intermediate Value Theorem

For any polynomial function $P(x)$ with real coefficients, suppose that for $a \neq b$, $P(a)$ and $P(b)$ are of opposite signs. Then the function has at least one real zero between a and b.

The intermediate value theorem *cannot* be used to determine whether there is a real zero between a and b when $P(a)$ and $P(b)$ have the *same* sign.

Use the intermediate value theorem to determine, if possible, whether each of the following functions has a real zero between a and b.

$$f(x) = 2x^3 - 5x^2 + x - 2, \quad a = 2, b = 3;$$
$$f(2) = -4; \; f(3) = 10$$

Since $f(2)$ and $f(3)$ have opposite signs, $f(x)$ has at least one real zero between 2 and 3.

$$f(x) = 2x^3 - 5x^2 + x - 2, \quad a = -2, b = -1;$$
$$f(-2) = -40; \; f(-1) = -10$$

Both $f(-2)$ and $f(-1)$ are negative. Thus the intermediate value theorem does not allow us to determine whether there is a real zero between -2 and -1.

SECTION 4.3: POLYNOMIAL DIVISION; THE REMAINDER THEOREM AND THE FACTOR THEOREM

Polynomial Division

$$P(x) = d(x) \cdot Q(x) + R(x)$$

Dividend Divisor Quotient Remainder

When we divide a polynomial $P(x)$ by a divisor $d(x)$, a polynomial $Q(x)$ is the quotient and a polynomial $R(x)$ is the remainder. The quotient $Q(x)$ must have degree less than that of the dividend $P(x)$. The remainder $R(x)$ must either be 0 or have degree less than that of the divisor $d(x)$. If $R(x) = 0$, then the divisor $d(x)$ is a factor of the dividend.

Given $P(x) = x^4 - 6x^3 + 9x^2 + 4x - 12$ and $d(x) = x + 2$, use long division to find the quotient and the remainder when $P(x)$ is divided by $d(x)$. Express $P(x)$ in the form $d(x) \cdot Q(x) + R(x)$.

$$\begin{array}{r} x^3 - 8x^2 + 25x - 46 \\ x + 2 \overline{) x^4 - 6x^3 + 9x^2 + 4x - 12} \\ \underline{x^4 + 2x^3} \qquad\qquad\qquad\quad \\ -8x^3 + 9x^2 \qquad\qquad \\ \underline{-8x^3 - 16x^2} \qquad\qquad \\ 25x^2 + 4x \qquad \\ \underline{25x^2 + 50x} \qquad \\ -46x - 12 \\ \underline{-46x - 92} \\ 80 \end{array}$$

$Q(x) = x^3 - 8x^2 + 25x - 46$ and $R(x) = 80$. Thus, $P(x) = (x + 2)(x^3 - 8x^2 + 25x - 46) + 80$. Since $R(x) \neq 0$, $x + 2$ is not a factor of $P(x)$.

The Remainder Theorem

If a number c is substituted for x in the polynomial $f(x)$, then the result $f(c)$ is the remainder that would be obtained by dividing $f(x)$ by $x - c$. That is, if $f(x) = (x - c) \cdot Q(x) + R$, then $f(c) = R$.

The long-division process can be streamlined with synthetic division. Synthetic division can also be used to find polynomial function values.

Repeat the division shown above using synthetic division. Note that the divisor $x + 2 = x - (-2)$.

$$\begin{array}{r|rrrrr} -2 & 1 & -6 & 9 & 4 & -12 \\ & & -2 & 16 & -50 & 92 \\ \hline & 1 & -8 & 25 & -46 & 80 \end{array}$$

Again, note that $Q(x) = x^3 - 8x^2 + 25x - 46$ and $R(x) = 80$. Since $R(x) \neq 0$, $x - (-2)$, or $x + 2$, is not a factor of $P(x)$.

Now divide $P(x)$ by $x - 3$.

$$\begin{array}{r|rrrrr} 3 & 1 & -6 & 9 & 4 & -12 \\ & & 3 & -9 & 0 & 12 \\ \hline & 1 & -3 & 0 & 4 & 0 \end{array}$$

$Q(x) = x^3 - 3x^2 + 4$ and $R(x) = 0$. Since $R(x) = 0$, $x - 3$ is a factor of $P(x)$.

For $f(x) = 2x^5 - x^3 - 3x^2 - 4x + 15$, find $f(-2)$.

$$\begin{array}{r|rrrrrr} -2 & 2 & 0 & -1 & -3 & -4 & 15 \\ & & -4 & 8 & -14 & 34 & -60 \\ \hline & 2 & -4 & 7 & -17 & 30 & -45 \end{array}$$

Thus, $f(-2) = -45$.

The Factor Theorem

For a polynomial $f(x)$, if $f(c) = 0$, then $x - c$ is a factor of $f(x)$.

Let $g(x) = x^4 + 8x^3 + 6x^2 - 40x + 25$. Factor $g(x)$ and solve $g(x) = 0$.

Use synthetic division to look for factors of the form $x - c$. Let's try $x + 5$.

$$\begin{array}{r|rrrrr} -5 & 1 & 8 & 6 & -40 & 25 \\ & & -5 & -15 & 45 & -25 \\ \hline & 1 & 3 & -9 & 5 & 0 \end{array}$$

Since $g(-5) = 0$, the number -5 is a zero of $g(x)$ and $x - (-5)$, or $x + 5$, is a factor of $g(x)$. This gives us

$$g(x) = (x + 5)(x^3 + 3x^2 - 9x + 5).$$

Let's try $x + 5$ again with the factor $x^3 + 3x^2 - 9x + 5$.

$$\begin{array}{r|rrrr} -5 & 1 & 3 & -9 & 5 \\ & & -5 & 10 & -5 \\ \hline & 1 & -2 & 1 & 0 \end{array}$$

Now we have

$$g(x) = (x + 5)^2(x^2 - 2x + 1).$$

The trinomial $x^2 - 2x + 1$ easily factors, so

$$g(x) = (x + 5)^2(x - 1)^2.$$

Solve $g(x) = 0$. The solutions of $(x + 5)^2(x - 1)^2 = 0$ are -5 and 1. They are also the zeros of $g(x)$.

SECTION 4.4: THEOREMS ABOUT ZEROS OF POLYNOMIAL FUNCTIONS

The Fundamental Theorem of Algebra

Every polynomial function of degree n, $n \geq 1$, with complex coefficients has at least one zero in the system of complex numbers.

Every polynomial function f of degree n, with $n \geq 1$, can be factored into n linear factors (not necessarily unique); that is,

$$f(x) = a_n(x - c_1)(x - c_2)\cdots(x - c_n).$$

Find a polynomial function of degree 5 with -4 and 2 as zeros of multiplicity 1, and -1 as a zero of multiplicity 3.

$$\begin{aligned} f(x) &= [x - (-4)][x - 2][x - (-1)]^3 \\ &= (x + 4)(x - 2)(x + 1)^3 \\ &= x^5 + 5x^4 + x^3 - 17x^2 - 22x - 8 \end{aligned}$$

Nonreal Zeros

$\boldsymbol{a + bi}$ **and** $\boldsymbol{a - bi,\ b \neq 0}$

If a complex number $a + bi$, $b \neq 0$, is a zero of a polynomial function $f(x)$ with *real* coefficients, then its conjugate, $a - bi$, is also a zero. (Nonreal zeros occur in conjugate pairs.)

Irrational Zeros

$\boldsymbol{a + c\sqrt{b}}$ **and** $\boldsymbol{a - c\sqrt{b}}$**,**
$\boldsymbol{b}$ **not a perfect square**

If $a + c\sqrt{b}$, where a, b, and c are rational and b is not a perfect square, is a zero of a polynomial function $f(x)$ with *rational* coefficients, then its conjugate, $a - c\sqrt{b}$, is also a zero. (Irrational zeros occur in conjugate pairs.)

Find a polynomial function with rational coefficients of lowest degree with $1 - i$ and $\sqrt{7}$ as two of its zeros.

If $1 - i$ is a zero, then $1 + i$ is also a zero. If $\sqrt{7}$ is a zero, then $-\sqrt{7}$ is also a zero.

$$\begin{aligned} f(x) &= [x - (1 - i)][x - (1 + i)][x - \sqrt{7}] \times \\ &\qquad [x - (-\sqrt{7})] \\ &= [(x - 1) + i][(x - 1) - i](x - \sqrt{7})(x + \sqrt{7}) \\ &= [(x - 1)^2 - i^2](x^2 - 7) \\ &= (x^2 - 2x + 1 + 1)(x^2 - 7) \\ &= (x^2 - 2x + 2)(x^2 - 7) \\ &= x^4 - 2x^3 - 5x^2 + 14x - 14 \end{aligned}$$

The Rational Zeros Theorem

Consider the polynomial function

$$P(x) = a_nx^n + a_{n-1}x^{n-1} + a_{n-2}x^{n-2} + \cdots + a_1x + a_0,$$

where all the coefficients are integers and $n \geq 1$. Also, consider a rational number p/q, where p and q have no common factor other than -1 and 1. If p/q is a zero of $P(x)$, then p is a factor of a_0 and q is a factor of a_n.

For $f(x) = 2x^4 - 9x^3 - 16x^2 - 9x - 18$, solve $f(x) = 0$ and factor $f(x)$ into linear factors.

There are at most 4 distinct zeros. Any rational zeros of f must be of the form p/q, where p is a factor of -18 and q is a factor of 2.

Possibilities for p / *Possibilities for q*: $\dfrac{\pm 1, \pm 2, \pm 3, \pm 6, \pm 9, \pm 18}{\pm 1, \pm 2}$

Possibilities for p/q: $1, -1, 2, -2, 3, -3, 6, -6, 9, -9, 18, -18, \frac{1}{2}, -\frac{1}{2}, \frac{3}{2}, -\frac{3}{2}, \frac{9}{2}, -\frac{9}{2}$

Use synthetic division to check the possibilities. We leave it to the student to verify that ± 1, ± 2, and ± 3 are not zeros. Let's try 6.

$$\begin{array}{r|rrrrr} 6 & 2 & -9 & -16 & -9 & -18 \\ & & 12 & 18 & 12 & 18 \\ \hline & 2 & 3 & 2 & 3 & 0 \end{array}$$

Since $f(6) = 0$, 6 is a zero and $x - 6$ is a factor of $f(x)$. Now we express $f(x)$ as

$$f(x) = (x - 6)(2x^3 + 3x^2 + 2x + 3).$$

(continued)

Let's consider the factor $2x^3 + 3x^2 + 2x + 3$ and check the other possibilities. Let's try $-\frac{3}{2}$.

$$\begin{array}{r|rrrr} -\frac{3}{2} & 2 & 3 & 2 & 3 \\ & & -3 & 0 & -3 \\ \hline & 2 & 0 & 2 & 0 \end{array}$$

Since $f\left(-\frac{3}{2}\right) = 0$, $-\frac{3}{2}$ is also a zero and $x + \frac{3}{2}$ is a factor of $f(x)$. We express $f(x)$ as

$$f(x) = (x - 6)\left(x + \tfrac{3}{2}\right)(2x^2 + 2)$$
$$= 2(x - 6)\left(x + \tfrac{3}{2}\right)(x^2 + 1).$$

Now we solve the equation $f(x) = 0$ to determine the zeros. We see that the only rational zeros are 6 and $-\frac{3}{2}$. The other zeros are $\pm i$.

The factorization into linear factors is

$$f(x) = 2(x - 6)\left(x + \tfrac{3}{2}\right)(x - i)(x + i), \quad \text{or}$$
$$(x - 6)(2x + 3)(x - i)(x + i).$$

Descartes' Rule of Signs

Let $P(x)$, written in descending order or ascending order, be a polynomial function with real coefficients and a nonzero constant term. The number of positive real zeros of $P(x)$ is either:

1. The same as the number of variations of sign in $P(x)$, or
2. Less than the number of variations of sign in $P(x)$ by a positive even integer.

The number of negative real zeros of $P(x)$ is either:

3. The same as the number of variations of sign in $P(-x)$, or
4. Less than the number of variations of sign in $P(-x)$ by a positive even integer.

A zero of multiplicity m must be counted m times.

Determine the number of positive real zeros and the number of negative real zeros of

$$P(x) = 4x^5 - x^4 - 2x^3 + 8x - 10.$$

There are 3 variations of sign in $P(x)$. Thus the number of positive real zeros is 3 or 1.

$$P(-x) = -4x^5 - x^4 + 2x^3 - 8x - 10$$

There are 2 variations of sign in $P(-x)$. Thus the number of negative real zeros is 2 or 0.

SECTION 4.5: RATIONAL FUNCTIONS

Rational Function

A **rational function** is a function f that is a quotient of two polynomials. That is,

$$f(x) = \frac{p(x)}{q(x)},$$

where $p(x)$ and $q(x)$ are polynomials and $q(x)$ is not the zero polynomial. The domain of $f(x)$ consists of all x for which $q(x) \neq 0$.

Determine the domain of each function.

FUNCTION	DOMAIN
$f(x) = \frac{1}{x^5}$	$(-\infty, 0) \cup (0, \infty)$
$f(x) = \frac{x+6}{x^2+2x-8} = \frac{x+6}{(x-2)(x+4)}$	$(-\infty, -4) \cup (-4, 2) \cup (2, \infty)$

Vertical Asymptotes

For a rational function $f(x) = p(x)/q(x)$, where $p(x)$ and $q(x)$ are polynomials with *no common factors* other than constants, if a is a zero of the denominator, then the line $x = a$ is a vertical asymptote for the graph of the function.

Horizontal Asymptotes

When the numerator and the denominator of a rational function have the same degree, the line $y = a/b$ is the horizontal asymptote, where a and b are the leading coefficients of the numerator and the denominator, respectively.

When the degree of the numerator is less than the degree of the denominator, the x-axis, or $y = 0$, is the horizontal asymptote.

When the degree of the numerator is greater than the degree of the denominator, there is *no* horizontal asymptote.

Oblique Asymptotes

When the degree of the numerator is 1 greater than the degree of the denominator, there is an oblique asymptote.

Determine the vertical, horizontal, and oblique asymptotes of the graph of the function.

FUNCTION	ASYMPTOTES
$f(x) = \frac{x^2-2}{x-1}$	Vertical: $x = 1$ Horizontal: None Oblique: $y = x + 1$
$f(x) = \frac{3x-4}{x^2+6x-7}$	Vertical: $x = -7; x = 1$ Horizontal: $y = 0$ Oblique: None
$f(x) = \frac{2x^2+9x-5}{3x^2+13x+12}$	Vertical: $x = -\frac{4}{3}; x = -3$ Horizontal: $y = \frac{2}{3}$ Oblique: None

To Graph a Rational Function

$f(x) = p(x)/q(x)$, where $p(x)$ and $q(x)$ have no common factor other than constants:

1. Find any real zeros of the denominator. Determine the domain of the function and sketch any vertical asymptotes.
2. Find the horizontal asymptote or the oblique asymptote, if there is one, and sketch it.

(continued)

Graph: $g(x) = \frac{x^2-4}{x^2+4x-5}$.

Domain: The zeros of the denominator are -5 and 1. The domain is $(-\infty, -5) \cup (-5, 1) \cup (1, \infty)$.

Vertical asymptotes: Since neither zero of the denominator is a zero of the numerator, the graph has vertical asymptotes at $x = -5$ and $x = 1$.

Horizontal asymptote: The degree of the numerator is the same as the degree of the denominator, so the horizontal asymptote is determined by the ratio of the leading coefficients: $1/1$, or 1. The horizontal asymptote is $y = 1$.

(continued)

3. Find any zeros of the function. The zeros are found by determining the zeros of the numerator. These are the first coordinates of the x-intercepts of the graph.
4. Find $f(0)$. This gives the y-intercept, $(0, f(0))$, of the graph.
5. Find other function values to determine the general shape. Then draw the graph.

Crossing an Asymptote

The graph of a rational function never crosses a vertical asymptote.

The graph of a rational function might cross a horizontal asymptote but does not necessarily do so.

Oblique asymptote: None

Zeros of g: Solving $g(x) = 0$ gives us -2 and 2, so the zeros are -2 and 2.

x-intercepts: $(-2, 0)$ and $(2, 0)$

y-intercept: $\left(0, \frac{4}{5}\right)$, because $g(0) = \frac{4}{5}$

Other values:

x	y
−8	2.22
−6	4.57
−4	−2.4
−3	−0.63
−1	0.38
0.5	1.36
1.5	−0.54
3	0.31
4	0.44

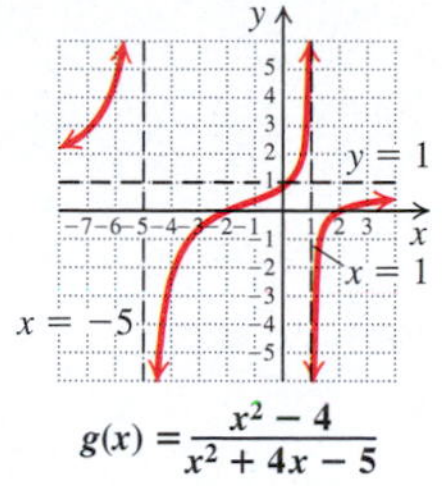

Graph: $f(x) = \dfrac{x + 3}{x^2 - x - 12}$.

Domain: The zeros of the denominator are -3 and 4. The domain is $(-\infty, -3) \cup (-3, 4) \cup (4, \infty)$.

Vertical asymptote: Since 4 is the only zero of the denominator that is not a zero of the numerator, the only vertical asymptote is $x = 4$.

Horizontal asymptote: Because the degree of the numerator is less than the degree of the denominator, the x-axis, $y = 0$, is the horizontal asymptote.

Oblique asymptote: None

Zeros of f: The equation $f(x) = 0$ has no solutions because -3 is not in the domain of the function. Thus there are no zeros of f.

x-intercepts: None

y-intercept: $\left(0, -\frac{1}{4}\right)$ because $f(0) = -\frac{1}{4}$

Hole in the graph:

$$f(x) = \frac{x + 3}{(x + 3)(x - 4)} = \frac{1}{x - 4},$$

where $x \neq -3$ and $x \neq 4$.

(continued)

To determine the coordinates of the hole, substitute -3 for x in $f(x) = 1/(x - 4)$:

$$f(-3) = \frac{1}{-3 - 4} = -\frac{1}{7}.$$

The hole is located at $\left(-3, -\frac{1}{7}\right)$.

Other values:

x	y
-4	-0.13
-2	-0.17
1	-0.33
3	-1
5	1
7	0.33

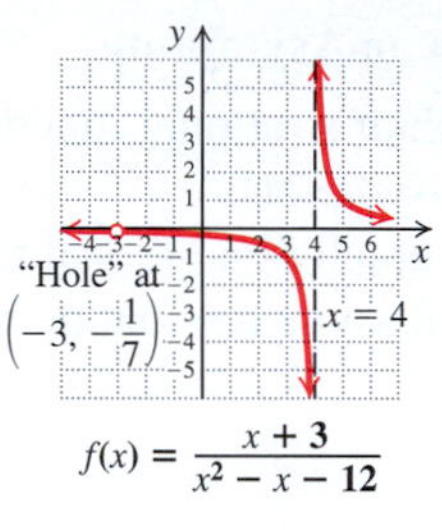

SECTION 4.6: POLYNOMIAL INEQUALITIES AND RATIONAL INEQUALITIES

To Solve a Polynomial Inequality

1. Find an equivalent inequality with 0 on one side.
2. Change the inequality symbol to an equals sign and solve the related equation.
3. Use the solutions to divide the x-axis into intervals. Then select a test value from each interval and determine the polynomial's sign on the interval.
4. Determine the intervals for which the inequality is satisfied and write interval notation or set-builder notation for the solution set. Include the endpoints of the intervals in the solution set if the inequality symbol is $\leq$ or $\geq$.

Solve: $x^3 - 3x^2 \leq 6x - 8$.

Equivalent inequality: $x^3 - 3x^2 - 6x + 8 \leq 0$.

First, we solve the related equation:

$$x^3 - 3x^2 - 6x + 8 = 0.$$

The solutions are -2, 1, and 4. These numbers divide the x-axis into 4 intervals. Next, we let $f(x) = x^3 - 3x^2 - 6x + 8$ and, using test values for $f(x)$, determine the sign of $f(x)$ in each interval.

INTERVAL	TEST VALUE	SIGN OF $f(x)$
$(-\infty, -2)$	$f(-3) = -28$	$-$
$(-2, 1)$	$f(0) = 8$	$+$
$(1, 4)$	$f(2) = -8$	$-$
$(4, \infty)$	$f(6) = 80$	$+$

$-$ $+$ $-$ $+$
-2 1 4 x

Test values are negative in the intervals $(-\infty, -2)$ and $(1, 4)$. Since the inequality sign is $\leq$, we include the endpoints of the intervals in the solution set.

The solution set is

$$(-\infty, -2] \cup [1, 4].$$

To Solve a Rational Inequality

1. Find an equivalent inequality with 0 on one side.
2. Change the inequality symbol to an equals sign and solve the related equation.
3. Find values of the variable for which the related rational function is not defined.
4. The numbers found in steps (2) and (3) are called *critical values*. Use the critical values to divide the x-axis into intervals. Then determine the function's sign in each interval using an x-value from the interval or using the graph of the equation.
5. Select the intervals for which the inequality is satisfied and write interval notation or set-builder notation for the solution set. If the inequality symbol is $\leq$ or $\geq$, then the solutions to step (2) should be included in the solution set. The x-values found in step (3) are never included in the solution set.

Solve: $\dfrac{x-1}{x+5} > \dfrac{x+3}{x-2}$.

Equivalent inequality: $\dfrac{x-1}{x+5} - \dfrac{x+3}{x-2} > 0$

Related function: $f(x) = \dfrac{x-1}{x+5} - \dfrac{x+3}{x-2}$

The function is not defined for $x = -5$ and $x = 2$. Solving $f(x) = 0$, we get $x = -\frac{13}{11}$. The critical values are -5, $-\frac{13}{11}$, and 2. These divide the x-axis into four intervals.

INTERVAL	TEST VALUE	SIGN OF $f(x)$
$(-\infty, -5)$	$f(-6) = 6.63$	+
$\left(-5, -\frac{13}{11}\right)$	$f(-2) = -0.75$	−
$\left(-\frac{13}{11}, 2\right)$	$f(0) = 1.3$	+
$(2, \infty)$	$f(3) = -5.75$	−

+ − + −
−5 $-\frac{13}{11}$ 2 x

Test values are positive in the intervals $(-\infty, -5)$ and $\left(-\frac{13}{11}, 2\right)$. Since $f\left(-\frac{13}{11}\right) = 0$ and -5 and 2 are not in the domain of f, -5, $-\frac{13}{11}$, and 2 cannot be part of the solution set.

The solution set is

$$(-\infty, -5) \cup \left(-\tfrac{13}{11}, 2\right).$$

REVIEW EXERCISES

Determine whether the statement is true or false.

1. If $f(x) = (x+a)(x+b)(x-c)$, then $f(-b) = 0$. [4.3]
2. The graph of a rational function never crosses a vertical asymptote. [4.5]
3. For the function $g(x) = x^4 - 8x^2 - 9$, the only possible rational zeros are 1, -1, 3, and -3. [4.4]
4. The graph of $P(x) = x^6 - x^8$ has at most 6 x-intercepts. [4.2]
5. The domain of the function
$$f(x) = \frac{x-4}{(x+2)(x-3)}$$
is $(-\infty, -2) \cup (3, \infty)$. [4.5]

Use a graphing calculator to graph the polynomial function. Then estimate the function's **(a)** *zeros,* **(b)** *relative maxima,* **(c)** *relative minima, and* **(d)** *domain and range.* [4.1]

6. $f(x) = -2x^2 - 3x + 6$
7. $f(x) = x^3 + 3x^2 - 2x - 6$
8. $f(x) = x^4 - 3x^3 + 2x^2$

Determine the leading term, the leading coefficient, and the degree of the polynomial. Then classify the polynomial function as constant, linear, quadratic, cubic, or quartic. [4.1]

9. $f(x) = 7x^2 - 5 + 0.45x^4 - 3x^3$
10. $h(x) = -25$
11. $g(x) = 6 - 0.5x$
12. $f(x) = \frac{1}{3}x^3 - 2x + 3$

Use the leading-term test to describe the end behavior of the graph of the function. **[4.1]**

13. $f(x) = -\frac{1}{2}x^4 + 3x^2 + x - 6$

14. $f(x) = x^5 + 2x^3 - x^2 + 5x + 4$

Find the zeros of the polynomial function and state the multiplicity of each. **[4.1]**

15. $g(x) = \left(x - \frac{2}{3}\right)(x + 2)^3(x - 5)^2$

16. $f(x) = x^4 - 26x^2 + 25$

17. $h(x) = x^3 + 4x^2 - 9x - 36$

18. *Interest Compounded Annually.* When P dollars is invested at interest rate i, compounded annually, for t years, the investment grows to A dollars, where

$$A = P(1 + i)^t.$$

a) Find the interest rate i if \$6250 grows to \$6760 in 2 years. **[4.1]**

b) Find the interest rate i if \$1,000,000 grows to \$1,215,506.25 in 4 years. **[4.1]**

19. *Cholesterol Level and the Risk of Heart Attack.* The following table lists data concerning the relationship of cholesterol level in men to the risk of a heart attack.

Cholesterol Level	Number of Men per 100,000 Who Suffer a Heart Attack
100	30
200	65
250	100
275	130

Source: Nutrition Action Newsletter

a) Use regression on a graphing calculator to fit linear, quadratic, and cubic functions to the data. **[4.1]**

b) It is also known that 180 of 100,000 men with a cholesterol level of 300 have a heart attack. Which function in part (a) would best make this prediction? **[4.1]**

c) Use the answer to part (b) to predict the heart attack rate for men with cholesterol levels of 350 and of 400. **[4.1]**

Sketch the graph of the polynomial function.

20. $f(x) = -x^4 + 2x^3$ **[4.2]**

21. $g(x) = (x - 1)^3(x + 2)^2$ **[4.2]**

22. $h(x) = x^3 + 3x^2 - x - 3$ **[4.2]**

23. $f(x) = x^4 - 5x^3 + 6x^2 + 4x - 8$ **[4.2], [4.3], [4.4]**

24. $g(x) = 2x^3 + 7x^2 - 14x + 5$ **[4.2], [4.4]**

Using the intermediate value theorem, determine, if possible, whether the function f has a zero between a and b. **[4.2]**

25. $f(x) = 4x^2 - 5x - 3;\ a = 1, b = 2$

26. $f(x) = x^3 - 4x^2 + \frac{1}{2}x + 2;\ a = -1, b = 1$

In each of the following, a polynomial $P(x)$ and a divisor $d(x)$ are given. Use long division to find the quotient $Q(x)$ and the remainder $R(x)$ when $P(x)$ is divided by $d(x)$. Express $P(x)$ in the form $d(x) \cdot Q(x) + R(x)$. **[4.3]**

27. $P(x) = 6x^3 - 2x^2 + 4x - 1,$
$d(x) = x - 3$

28. $P(x) = x^4 - 2x^3 + x + 5,$
$d(x) = x + 1$

Use synthetic division to find the quotient and the remainder. **[4.3]**

29. $(x^3 + 2x^2 - 13x + 10) \div (x - 5)$

30. $(x^4 + 3x^3 + 3x^2 + 3x + 2) \div (x + 2)$

31. $(x^5 - 2x) \div (x + 1)$

Use synthetic division to find the indicated function value. **[4.3]**

32. $f(x) = x^3 + 2x^2 - 13x + 10;\ f(-2)$

33. $f(x) = x^4 - 16;\ f(-2)$

34. $f(x) = x^5 - 4x^4 + x^3 - x^2 + 2x - 100;$
$f(-10)$

Using synthetic division, determine whether the given numbers are zeros of the polynomial function. **[4.3]**

35. $-i, -5;\ f(x) = x^3 - 5x^2 + x - 5$

36. $-1, -2;\ f(x) = x^4 - 4x^3 - 3x^2 + 14x - 8$

37. $\frac{1}{3}, 1;\ f(x) = x^3 - \frac{4}{3}x^2 - \frac{5}{3}x + \frac{2}{3}$

38. $2, -\sqrt{3};\ f(x) = x^4 - 5x^2 + 6$

Factor the polynomial $f(x)$. Then solve the equation $f(x) = 0$. **[4.3], [4.4]**

39. $f(x) = x^3 + 2x^2 - 7x + 4$

40. $f(x) = x^3 + 4x^2 - 3x - 18$

41. $f(x) = x^4 - 4x^3 - 21x^2 + 100x - 100$

42. $f(x) = x^4 - 3x^2 + 2$

Find a polynomial function of degree 3 with the given numbers as zeros. **[4.4]**

43. $-4, -1, 2$

44. $-3,\ 1 - i,\ 1 + i$

45. $\frac{1}{2}, 1 - \sqrt{2}, 1 + \sqrt{2}$

46. Find a polynomial function of degree 4 with -5 as a zero of multiplicity 3 and $\frac{1}{2}$ as a zero of multiplicity 1. **[4.4]**

47. Find a polynomial function of degree 5 with -3 as a zero of multiplicity 2, 2 as a zero of multiplicity 1, and 0 as a zero of multiplicity 2. [4.4]

Suppose that a polynomial function of degree 5 with rational coefficients has the given zeros. Find the other zero(s). [4.4]

48. $-\frac{2}{3}, \sqrt{5}, 4 + i$

49. $0, 1 + \sqrt{3}, -\sqrt{3}$

50. $-\sqrt{2}, \frac{1}{2}, 1, 2$

Find a polynomial function of lowest degree with rational coefficients and the following as some of its zeros. [4.4]

51. $\sqrt{11}$

52. $-i, 6$

53. $-1, 4, 1 + i$

54. $\sqrt{5}, -2i$

55. $\frac{1}{3}, 0, -3$

List all possible rational zeros. [4.4]

56. $h(x) = 4x^5 - 2x^3 + 6x - 12$

57. $g(x) = 3x^4 - x^3 + 5x^2 - x + 1$

58. $f(x) = x^3 - 2x^2 + x - 24$

For each polynomial function:

a) *Find the rational zeros and then the other zeros; that is, solve $f(x) = 0$.* [4.4]

b) *Factor $f(x)$ into linear factors.* [4.4]

59. $f(x) = 3x^5 + 2x^4 - 25x^3 - 28x^2 + 12x$

60. $f(x) = x^3 - 2x^2 - 3x + 6$

61. $f(x) = x^4 - 6x^3 + 9x^2 + 6x - 10$

62. $f(x) = x^3 + 3x^2 - 11x - 5$

63. $f(x) = 3x^3 - 8x^2 + 7x - 2$

64. $f(x) = x^5 - 8x^4 + 20x^3 - 8x^2 - 32x + 32$

65. $f(x) = x^6 + x^5 - 28x^4 - 16x^3 + 192x^2$

66. $f(x) = 2x^5 - 13x^4 + 32x^3 - 38x^2 + 22x - 5$

What does Descartes' rule of signs tell you about the number of positive real zeros and the number of negative real zeros of each of the following polynomial functions? [4.4]

67. $f(x) = 2x^6 - 7x^3 + x^2 - x$

68. $h(x) = -x^8 + 6x^5 - x^3 + 2x - 2$

69. $g(x) = 5x^5 - 4x^2 + x - 1$

Graph the function. Be sure to label all the asymptotes. List the domain and the x- and y-intercepts. [4.5]

70. $f(x) = \dfrac{x^2 - 5}{x + 2}$

71. $f(x) = \dfrac{5}{(x - 2)^2}$

72. $f(x) = \dfrac{x^2 + x - 6}{x^2 - x - 20}$

73. $f(x) = \dfrac{x - 2}{x^2 - 2x - 15}$

In Exercises 74 and 75, find a rational function that satisfies the given conditions. Answers may vary, but try to give the simplest answer possible. [4.5]

74. Vertical asymptotes $x = -2, x = 3$

75. Vertical asymptotes $x = -2, x = 3$; horizontal asymptote $y = 4$; x-intercept $(-3, 0)$

76. *Medical Dosage.* The function

$$N(t) = \frac{0.7t + 2000}{8t + 9}, \quad t \geq 5,$$

gives the body concentration $N(t)$, in parts per million, of a certain dosage of medication after time t, in hours.

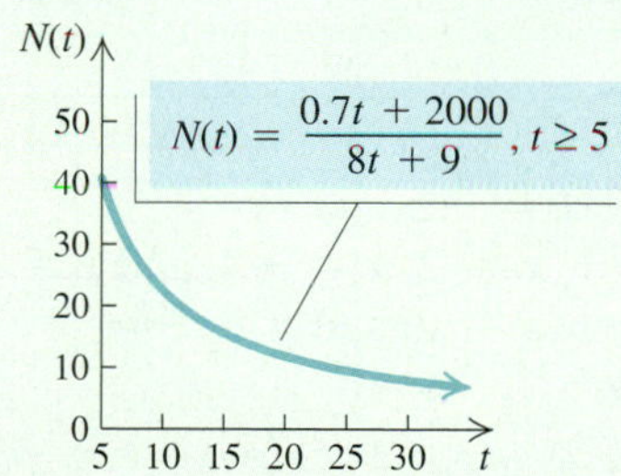

a) Find the horizontal asymptote of the graph and complete the following:

$N(t) \rightarrow$ ______ as $t \rightarrow \infty$. [4.5]

b) Explain the meaning of the answer to part (a) in terms of the application. [4.5]

Solve. [4.6]

77. $x^2 - 9 < 0$

78. $2x^2 > 3x + 2$

79. $(1 - x)(x + 4)(x - 2) \leq 0$

80. $\dfrac{x - 2}{x + 3} < 4$

81. *Height of a Rocket.* The function

$$S(t) = -16t^2 + 80t + 224$$

gives the height S, in feet, of a model rocket launched with a velocity of 80 ft/sec from a hill that is 224 ft high, where t is the time, in seconds.

a) Determine when the rocket reaches the ground. [4.1]

b) On what interval is the height greater than 320 ft? [4.1], [4.6]

82. *Population Growth.* The population P, in thousands, of Novi is given by

$$P(t) = \frac{8000t}{4t^2 + 10},$$

where t is the time, in months. Find the interval on which the population was 400,000 or greater. [4.6]

83. Which of the following is the domain of the function

$$g(x) = \frac{x^2 + 2x - 3}{x^2 - 5x + 6}? \text{ [4.5]}$$

A. $(-\infty, 2) \cup (2, 3) \cup (3, \infty)$
B. $(-\infty, -3) \cup (-3, 1) \cup (1, \infty)$
C. $(-\infty, 2) \cup (3, \infty)$
D. $(-\infty, -3) \cup (1, \infty)$

84. Which of the following are the vertical asymptotes of the function

$$f(x) = \frac{x - 4}{(x + 1)(x - 2)(x + 4)}? \text{ [4.5]}$$

A. $x = 1$, $x = -2$, and $x = 4$
B. $x = -1$, $x = 2$, $x = -4$, and $x = 4$
C. $x = -1$, $x = 2$, and $x = -4$
D. $x = 4$

85. The graph of $f(x) = -\frac{1}{2}x^4 + x^3 + 1$ is which of the following? [4.2]

A.

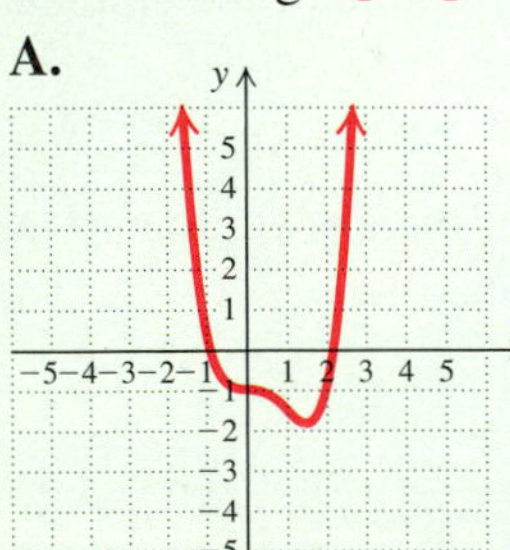

B.

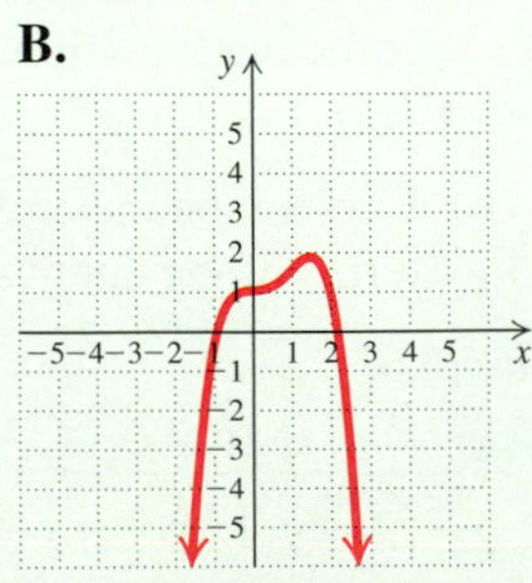

C.

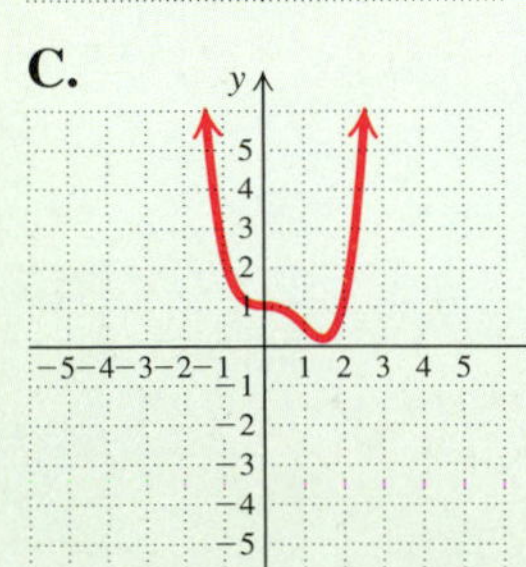

D.

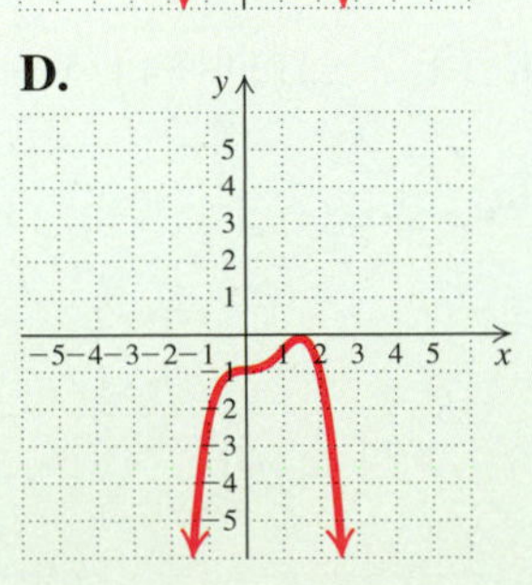

Synthesis

Solve.

86. $x^2 \geq 5 - 2x$ [4.6]

87. $\left|1 - \frac{1}{x^2}\right| < 3$ [4.6]

88. $x^4 - 2x^3 + 3x^2 - 2x + 2 = 0$ [4.4]

89. $(x - 2)^{-3} < 0$ [4.6]

90. Express $x^3 - 1$ as a product of linear factors. [4.4]

91. Find k such that $x + 3$ is a factor of $x^3 + kx^2 + kx - 15$. [4.3]

92. When $x^2 - 4x + 3k$ is divided by $x + 5$, the remainder is 33. Find the value of k. [4.3]

Find the domain of the function. [4.6]

93. $f(x) = \sqrt{x^2 + 3x - 10}$

94. $f(x) = \sqrt{x^2 - 3.1x + 2.2} + 1.75$

95. $f(x) = \dfrac{1}{\sqrt{5 - |7x + 2|}}$

Collaborative Discussion and Writing

96. Explain the difference between a polynomial function and a rational function. [4.1], [4.5]

97. Is it possible for a third-degree polynomial with rational coefficients to have no real zeros? Why or why not? [4.4]

98. Explain and contrast the three types of asymptotes considered for rational functions. [4.5]

99. If $P(x)$ is an even function, and by Descartes' rule of signs, $P(x)$ has one positive real zero, how many negative real zeros does $P(x)$ have? Explain. [4.4]

100. Explain why the graph of a rational function cannot have both a horizontal asymptote and an oblique asymptote. [4.5]

101. Under what circumstances would a quadratic inequality have a solution set that is a closed interval? [4.6]

4 Chapter Test

Determine the leading term, the leading coefficient, and the degree of the polynomial. Then classify the polynomial as constant, linear, quadratic, cubic, or quartic.

1. $f(x) = 2x^3 + 6x^2 - x^4 + 11$

2. $h(x) = -4.7x + 29$

3. Find the zeros of the polynomial function and state the multiplicity of each:
$$f(x) = x(3x - 5)(x - 3)^2(x + 1)^3.$$

4. *Hybrid Automobiles.* In 2005, only 205,828 hybrid automobiles were sold, while in 2013, 498,054 were sold (*Source*: WardsAuto Group, a division of Penton). The cubic function
$$f(x) = 3707.968x^3 - 40{,}437.526x^2 + 126{,}421.240x + 197{,}407.131,$$
where x is the number of years after 2005, can be used to estimate the number of hybrid automobiles sold in years 2005 to 2013. Use this function to estimate the number of hybrid automobiles sold in 2008 and in 2012.

Sketch the graph of the polynomial function.

5. $f(x) = x^3 - 5x^2 + 2x + 8$

6. $f(x) = -2x^4 + x^3 + 11x^2 - 4x - 12$

Using the intermediate value theorem, determine, if possible, whether the function has a zero between a and b.

7. $f(x) = -5x^2 + 3;\ a = 0, b = 2$

8. $g(x) = 2x^3 + 6x^2 - 3;\ a = -2, b = -1$

9. Use long division to find the quotient $Q(x)$ and the remainder $R(x)$ when $P(x)$ is divided by $d(x)$. Express $P(x)$ in the form $d(x) \cdot Q(x) + R(x)$. Show your work.
$$P(x) = x^4 + 3x^3 + 2x - 5,$$
$$d(x) = x - 1$$

10. Use synthetic division to find the quotient and the remainder. Show your work.
$$(3x^3 - 12x + 7) \div (x - 5)$$

11. Use synthetic division to find $P(-3)$ for $P(x) = 2x^3 - 6x^2 + x - 4$. Show your work.

12. Use synthetic division to determine whether -2 is a zero of $f(x) = x^3 + 4x^2 + x - 6$. Answer yes or no. Show your work.

13. Find a polynomial function of degree 4 with -3 as a zero of multiplicity 2 and 0 and 6 as zeros of multiplicity 1.

14. Suppose that a polynomial function of degree 5 with rational coefficients has 1, $\sqrt{3}$, and $2 - i$ as zeros. Find the other zeros.

Find a polynomial function of lowest degree with rational coefficients and the following as some of its zeros.

15. $-10,\ 3i$

16. $0,\ -\sqrt{3},\ 1 - i$

List all possible rational zeros.

17. $f(x) = 2x^3 + x^2 - 2x + 12$

18. $h(x) = 10x^4 - x^3 + 2x - 5$

For each polynomial function:

a) *Find the rational zeros and then the other zeros; that is, solve $f(x) = 0$.*
b) *Factor $f(x)$ into linear factors.*

19. $f(x) = x^3 + x^2 - 5x - 5$

20. $f(x) = 2x^4 - 11x^3 + 16x^2 - x - 6$

21. $f(x) = x^3 + 4x^2 + 4x + 16$

22. $f(x) = 3x^4 - 11x^3 + 15x^2 - 9x + 2$

23. What does Descartes' rule of signs tell you about the number of positive real zeros and the number of negative real zeros of the following function?
$$g(x) = -x^8 + 2x^6 - 4x^3 - 1$$

Graph the function. Be sure to label all the asymptotes. List the domain and the x- and y-intercepts.

24. $f(x) = \dfrac{2}{(x - 3)^2}$

25. $f(x) = \dfrac{x + 3}{x^2 - 3x - 4}$

26. Find a rational function that has vertical asymptotes $x = -1$ and $x = 2$ and x-intercept $(-4, 0)$.

Solve.

27. $2x^2 > 5x + 3$

28. $\dfrac{x + 1}{x - 4} \le 3$

29. The function $S(t) = -16t^2 + 64t + 192$ gives the height S, in feet, of a model rocket launched with a velocity of 64 ft/sec from a hill that is 192 ft high.

a) Determine how long it will take the rocket to reach the ground.

b) Find the interval on which the height of the rocket is greater than 240 ft.

30. The graph of $f(x) = x^3 - x^2 - 2$ is which of the following?

A.

B.

C.

D.

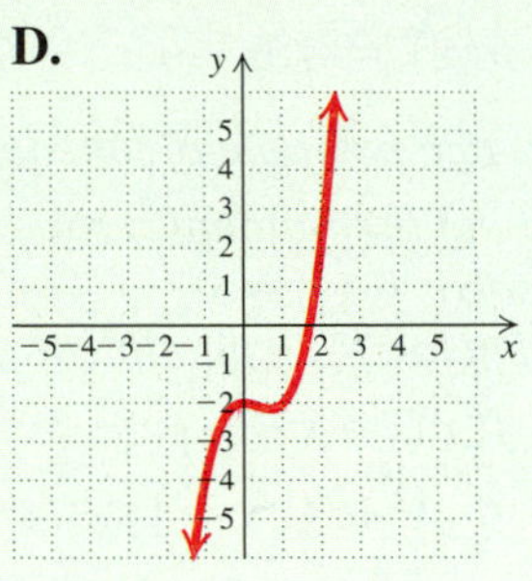

Synthesis

31. Find the domain of $f(x) = \sqrt{x^2 + x - 12}$.

CHAPTER

5

Exponential Functions and Logarithmic Functions

APPLICATION

This problem appears as Exercise 65 in Section 5.2.

Annual milk consumption per capita in China has increased greatly in recent years. Consequently, the demand for imported alfalfa has increased exponentially. In 2008, approximately 20,000 tons of U.S. alfalfa were imported by China. This number increased to approximately 650,000 tons by 2013. (*Source*: *National Geographic*, May 2015). The increase in imported U.S. alfalfa can be modeled by the exponential function $A(x) = 22{,}611.008(1.992)^x$, where x is the number of years after 2008. Find the number of tons of U.S. alfalfa imported by China in 2012 and in 2016.

5.1 Inverse Functions

- Determine whether a function is one-to-one, and if it is, find a formula for its inverse.
- Simplify expressions of the type $(f \circ f^{-1})(x)$ and $(f^{-1} \circ f)(x)$.

Inverses

When we go from an output of a function back to its input or inputs, we get an inverse relation. When that relation is a function, we have an inverse function.

Consider the relation h given as follows:

$$h = \{(-8, 5), (4, -2), (-7, 1), (3.8, 6.2)\}.$$

Suppose that we *interchange* the first and second coordinates. The relation we obtain is called the **inverse** of the relation h and is given as follows:

RELATIONS

REVIEW SECTION 1.2.

$$\text{Inverse of } h = \{(5, -8), (-2, 4), (1, -7), (6.2, 3.8)\}.$$

INVERSE RELATION

Interchanging the first and second coordinates of each ordered pair in a relation produces the **inverse relation**.

EXAMPLE 1 Consider the relation g given by

$$g = \{(2, 4), (-1, 3), (-2, 0)\}.$$

Graph the relation in blue. Find the inverse and graph it in red.

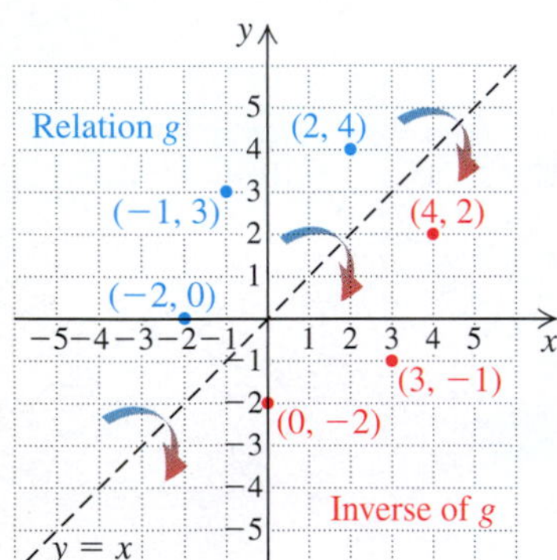

Solution The relation g is shown in blue in the figure at left. The inverse of the relation is

$$\{(4, 2), (3, -1), (0, -2)\}$$

and is shown in red. The pairs in the inverse are reflections of the pairs in g across the line $y = x$.

Now Try Exercise 1.

INVERSE RELATION

If a relation is defined by an equation, interchanging the variables produces an equation of the **inverse relation.**

EXAMPLE 2 Find an equation for the inverse of the relation

$$y = x^2 - 5x.$$

Solution We interchange x and y and obtain an equation of the inverse:

$$x = y^2 - 5y.$$

Now Try Exercise 9.

If a relation is given by an equation, then the solutions of the inverse can be found from those of the original equation by interchanging the first and second coordinates of each ordered pair. Thus the graphs of a relation and its inverse are always reflections of each other across the line $y = x$. This is illustrated with the equations of Example 2 in the following tables and graph. We will explore inverses and their graphs later in this section.

$x = y^2 - 5y$	y
6	−1
0	0
−6	2
−4	4

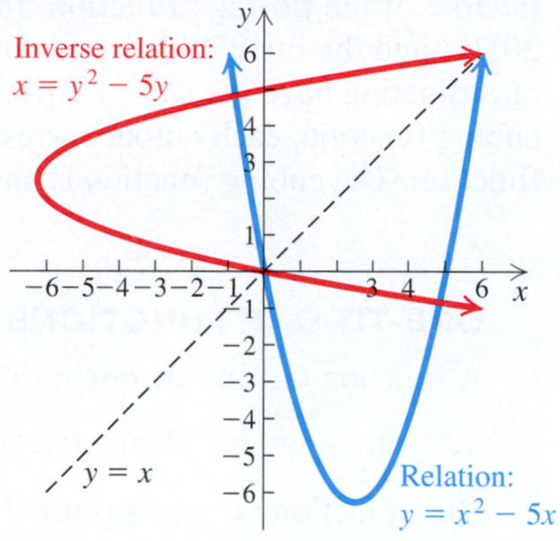

x	$y = x^2 - 5x$
−1	6
0	0
2	−6
4	−4

Inverses and One-to-One Functions

Let's consider the following two functions.

Year (domain)	First-Class Postage Cost, in cents (range)
2006	→ 39
2007	→ 41
2008	→ 42
2009	→ 44
2010	→ 44
2011	→ 44
2012	→ 45
2013	→ 46
2014	→ 49
2015	→ 49

Source: U.S. Postal Service

Number (domain)	Cube (range)
−3	→ −27
−2	→ −8
−1	→ −1
0	→ 0
1	→ 1
2	→ 8
3	→ 27

Suppose that we reverse the arrows. Are these inverse relations functions?

Year (range)	First-Class Postage Cost, in cents (domain)
2006 ←	39
2007 ←	41
2008 ←	42
2009, 2010, 2011 ←	44
2012 ←	45
2013 ←	46
2014, 2015 ←	49

Source: U.S. Postal Service

Number (range)	Cube (domain)
−3 ←	−27
−2 ←	−8
−1 ←	−1
0 ←	0
1 ←	1
2 ←	8
3 ←	27

We see that the inverse of the postage function is not a function. Like all functions, each input in the postage function has exactly one output. However, the output for 2009, 2010, and 2011 is 44. Also, the output for 2014 and 2015 is 49. Thus in the inverse of the postage function, the input 44 has *three* outputs—2009, 2010, and 2011—and the input 49 has *two* outputs—2014 and 2015. When two or more inputs of a function have the same output, the inverse relation cannot be a function. In the cubing function, each output corresponds to exactly one input, so its inverse is also a function. The cubing function is an example of a **one-to-one function**.

ONE-TO-ONE FUNCTIONS

A function f is **one-to-one** if different inputs have different outputs—that is,

$$\text{if} \quad a \neq b, \quad \text{then} \quad f(a) \neq f(b).$$

Or, a function f is **one-to-one** if when the outputs are the same, the inputs are the same—that is,

$$\text{if} \quad f(a) = f(b), \quad \text{then} \quad a = b.$$

If the inverse of a function f is also a function, it is named f^{-1} (read "f-inverse").

The -1 in f^{-1} is *not* an exponent!

Do *not* misinterpret the -1 in f^{-1} as a negative exponent: f^{-1} does *not* mean the reciprocal of f and $f^{-1}(x)$ is *not* equal to $\dfrac{1}{f(x)}$.

ONE-TO-ONE FUNCTIONS AND INVERSES

- If a function f is one-to-one, then its inverse f^{-1} is a function.
- The domain of a one-to-one function f is the range of the inverse f^{-1}.
- The range of a one-to-one function f is the domain of the inverse f^{-1}.

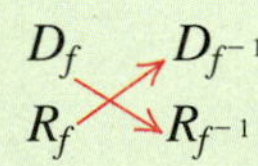

- A function that is increasing over its entire domain or is decreasing over its entire domain is a one-to-one function.

EXAMPLE 3 Given the function f described by $f(x) = 2x - 3$, prove that f is one-to-one (that is, it has an inverse that is a function).

Solution To show that f is one-to-one, we show that if $f(a) = f(b)$, then $a = b$. Assume that $f(a) = f(b)$ for a and b in the domain of f. Since $f(a) = 2a - 3$ and $f(b) = 2b - 3$, we have

$$\begin{aligned} 2a - 3 &= 2b - 3 \\ 2a &= 2b && \textbf{Adding 3} \\ a &= b. && \textbf{Dividing by 2} \end{aligned}$$

Thus, if $f(a) = f(b)$, then $a = b$. This shows that f is one-to-one.

Now Try Exercise 17.

EXAMPLE 4 Given the function g described by $g(x) = x^2$, prove that g is not one-to-one.

Solution We can prove that g is not one-to-one by finding two numbers a and b for which $a \neq b$ and $g(a) = g(b)$. Two such numbers are -3 and 3, because $-3 \neq 3$ and $g(-3) = g(3) = 9$. Thus g is not one-to-one.

Now Try Exercise 21.

The following graphs show a function, in blue, and its inverse, in red. To determine whether the inverse is a function, we can apply the vertical-line test to its graph. By reflecting each such vertical line across the line $y = x$, we obtain an equivalent **horizontal-line test** for the original function.

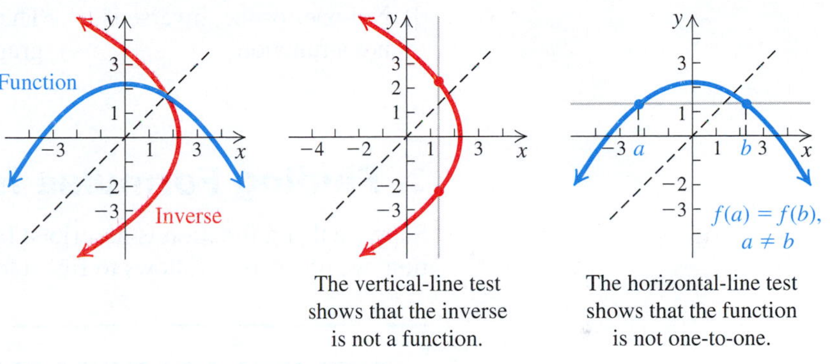

The vertical-line test shows that the inverse is not a function.

The horizontal-line test shows that the function is not one-to-one.

> **HORIZONTAL-LINE TEST**
>
> If it is possible for a horizontal line to intersect the graph of a function more than once, then the function is *not* one-to-one and its inverse is *not* a function.

EXAMPLE 5 From the graph shown in each of (a)–(d), determine whether the function is one-to-one and thus has an inverse that is a function.

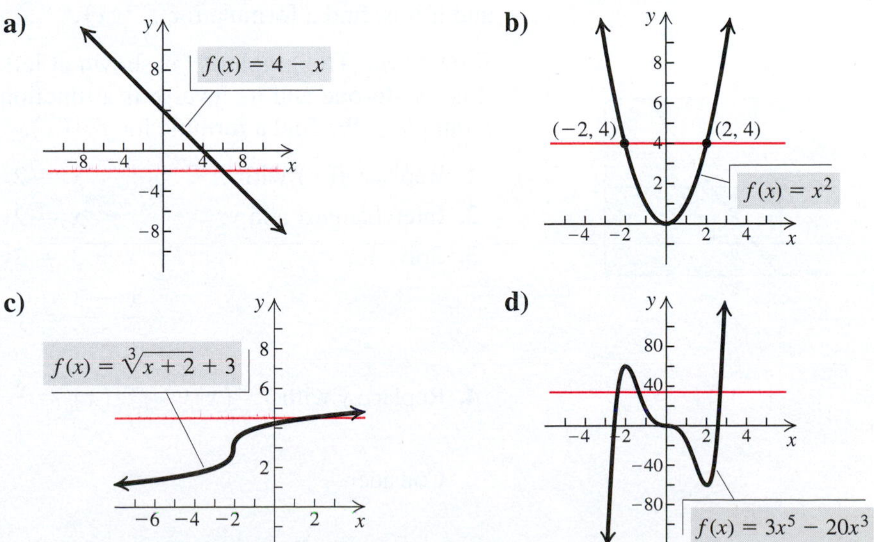

Solution For each function, we apply the horizontal-line test.

	RESULT	REASON
a)	One-to-one; inverse is a function	No horizontal line intersects the graph more than once.
b)	Not one-to-one; inverse is not a function	There are many horizontal lines that intersect the graph more than once. Note that where the line $y = 4$ intersects the graph, the first coordinates are -2 and 2. Although these are different inputs, they have the same output, 4.
c)	One-to-one; inverse is a function	No horizontal line intersects the graph more than once.
d)	Not one-to-one; inverse is not a function	There are many horizontal lines that intersect the graph more than once.

Now Try Exercises 25 and 27.

Finding Formulas for Inverses

Suppose that a function is described by a formula. If it has an inverse that is a function, we proceed as follows to find a formula for f^{-1}.

OBTAINING A FORMULA FOR AN INVERSE

If a function f is one-to-one, a formula for its inverse can generally be found as follows:

1. Replace $f(x)$ with y.
2. Interchange x and y.
3. Solve for y.
4. Replace y with $f^{-1}(x)$.

EXAMPLE 6 Determine whether the function $f(x) = 2x - 3$ is one-to-one, and if it is, find a formula for $f^{-1}(x)$.

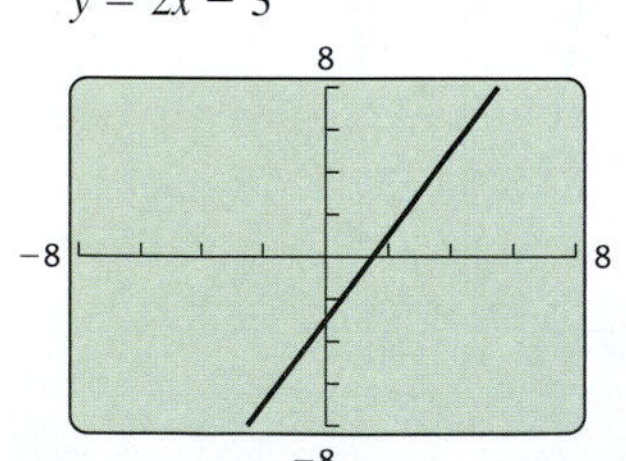

Solution The graph of f is shown at left. It passes the horizontal-line test. Thus it is one-to-one and its inverse is a function. We also proved that f is one-to-one in Example 3. We find a formula for $f^{-1}(x)$.

1. Replace $f(x)$ with y: $\quad y = 2x - 3$
2. Interchange x and y: $\quad x = 2y - 3$
3. Solve for y: $\quad x + 3 = 2y$

$$\frac{x+3}{2} = y$$

4. Replace y with $f^{-1}(x)$: $\quad f^{-1}(x) = \dfrac{x+3}{2}$.

Now Try Exercise 57.

Consider

$$f(x) = 2x - 3 \quad \text{and} \quad f^{-1}(x) = \frac{x+3}{2}$$

from Example 6. For the input 5, we have

$$f(5) = 2 \cdot 5 - 3 = 10 - 3 = 7.$$

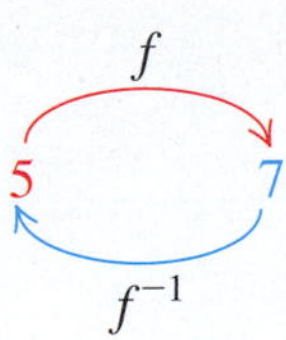

The output is 7. Now we use 7 for the input in the inverse:

$$f^{-1}(7) = \frac{7+3}{2} = \frac{10}{2} = 5.$$

The function f takes the number 5 to 7. The inverse function f^{-1} takes the number 7 back to 5.

EXAMPLE 7 Graph

$$f(x) = 2x - 3 \quad \text{and} \quad f^{-1}(x) = \frac{x+3}{2}$$

using the same set of axes. Then compare the two graphs.

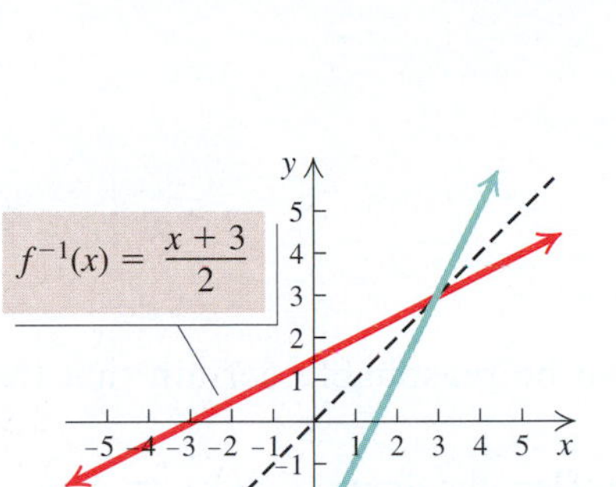

Solution The graphs of f and f^{-1} are shown at left. The solutions of the inverse function can be found from those of the original function by interchanging the first and second coordinates of each ordered pair.

x	$f(x) = 2x - 3$
-1	-5
0	-3 ← y-intercept
2	1
3	3

x	$f^{-1}(x) = \frac{x+3}{2}$
-5	-1
-3	0 ← x-intercept
1	2
3	3

On some graphing calculators, we can graph the inverse of a function after graphing the function itself by accessing a drawing feature.

When we interchange x and y in finding a formula for the inverse of $f(x) = 2x - 3$, we are in effect reflecting the graph of that function across the line $y = x$. For example, when the coordinates of the y-intercept of the graph of f, $(0, -3)$, are reversed, we get the x-intercept of the graph of f^{-1}, $(-3, 0)$. If we were to graph $f(x) = 2x - 3$ in wet ink and fold along the line $y = x$, the graph of $f^{-1}(x) = (x + 3)/2$ would be formed by the ink transferred from f.

> The graph of f^{-1} is a reflection of the graph of f across the line $y = x$.

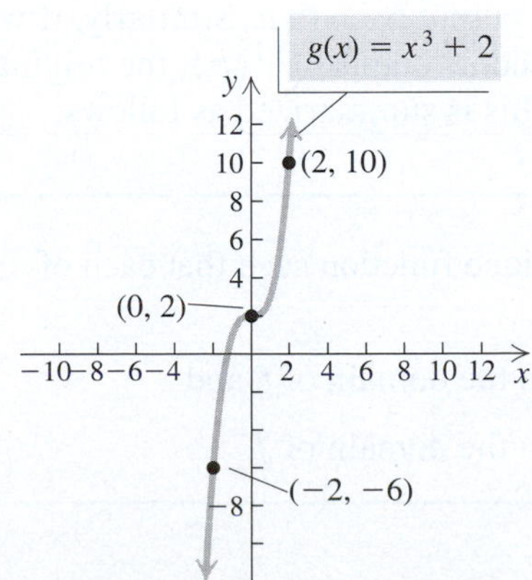

EXAMPLE 8 Consider $g(x) = x^3 + 2$.

a) Determine whether the function is one-to-one.

b) If it is one-to-one, find a formula for its inverse.

c) Graph the function and its inverse.

Solution

a) The graph of $g(x) = x^3 + 2$ is shown at left. It passes the horizontal-line test and thus has an inverse that is a function. We also know that $g(x)$ is one-to-one because it is an increasing function over its entire domain.

b) We follow the procedure for finding an inverse.

1. Replace $g(x)$ with y: $\quad y = x^3 + 2$

2. Interchange x and y: $\quad x = y^3 + 2$

3. Solve for y: $\quad x - 2 = y^3$

$$\sqrt[3]{x - 2} = y$$

4. Replace y with $g^{-1}(x)$: $\quad g^{-1}(x) = \sqrt[3]{x - 2}$.

We can test a point as a partial check:

$$g(x) = x^3 + 2$$
$$g(3) = 3^3 + 2 = 27 + 2 = 29.$$

Will $g^{-1}(29) = 3$? We have

$$g^{-1}(x) = \sqrt[3]{x - 2}$$
$$g^{-1}(29) = \sqrt[3]{29 - 2} = \sqrt[3]{27} = 3.$$

x	$g(x)$
-2	-6
-1	1
0	2
1	3
2	10

Since $g(3) = 29$ and $g^{-1}(29) = 3$, we can be reasonably certain that the formula for $g^{-1}(x)$ is correct.

c) To find the graph of the inverse function, we reflect the graph of $g(x) = x^3 + 2$ across the line $y = x$. This can be done by plotting points or by using a graphing calculator.

x	$g^{-1}(x)$
-6	-2
1	-1
2	0
3	1
10	2

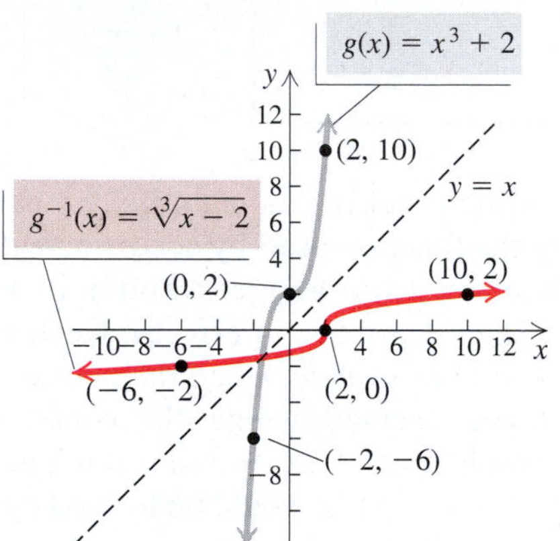

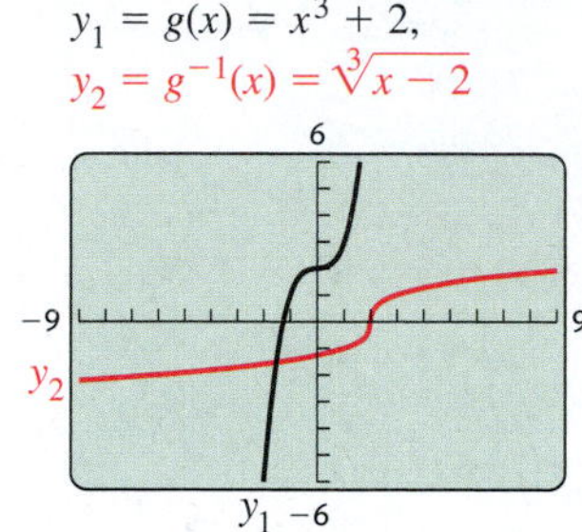

Now Try Exercises 63 and 79.

Inverse Functions and Composition

Suppose that we were to use some input a for a one-to-one function f and find its output, $f(a)$. The function f^{-1} would then take that output back to a. Similarly, if we began with an input b for the function f^{-1} and found its output, $f^{-1}(b)$, the original function f would then take that output back to b. This is summarized as follows.

> If a function f is one-to-one, then f^{-1} is the unique function such that each of the following holds:
>
> $(f^{-1} \circ f)(x) = f^{-1}(f(x)) = x$, for each x in the domain of f, and
>
> $(f \circ f^{-1})(x) = f(f^{-1}(x)) = x$, for each x in the domain of f^{-1}.

EXAMPLE 9 Given that $f(x) = 5x + 8$, use composition of functions to show that

$$f^{-1}(x) = \frac{x - 8}{5}.$$

Solution We find $(f^{-1} \circ f)(x)$ and $(f \circ f^{-1})(x)$ and check to see that each is x:

$$\begin{aligned}(f^{-1} \circ f)(x) &= f^{-1}(f(x)) \\ &= f^{-1}(5x + 8) = \frac{(5x + 8) - 8}{5} = \frac{5x}{5} = x; \\ (f \circ f^{-1})(x) &= f(f^{-1}(x)) \\ &= f\left(\frac{x - 8}{5}\right) = 5\left(\frac{x - 8}{5}\right) + 8 = x - 8 + 8 = x.\end{aligned}$$

Now Try Exercise 87.

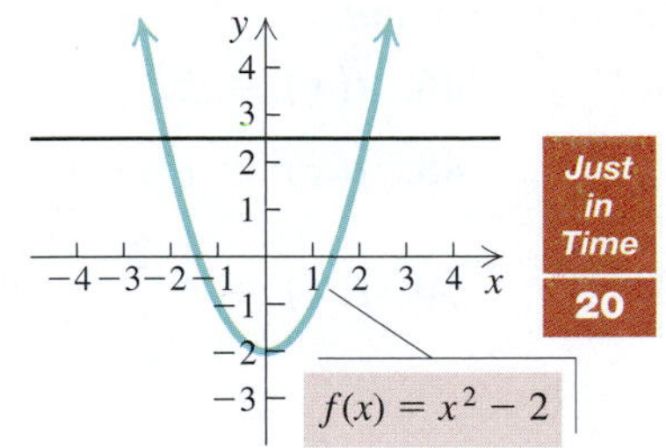

Just in Time 20

$f^{-1}(x) = \sqrt{x + 2}$

$f(x) = x^2 - 2$, for inputs $x \geq 0$

Restricting a Domain

In the case in which the inverse of a function is not a function, the domain of the function can be restricted to allow the inverse to be a function. Let's consider the function $f(x) = x^2 - 2$. It is not one-to-one. The graph is shown at left.

Suppose that we had tried to find a formula for the inverse as follows:

$$\begin{aligned} y &= x^2 - 2 && \text{Replacing } f(x) \text{ with } y \\ x &= y^2 - 2 && \text{Interchanging } x \text{ and } y \\ x + 2 &= y^2 \\ \pm\sqrt{x + 2} &= y. && \text{Solving for } y \end{aligned}$$

This is not the equation of a function. An input of, say, 2 would yield two outputs, -2 and 2. In such cases, it is convenient to consider "part" of the function by restricting the domain of $f(x)$. For example, if we restrict the domain of $f(x) = x^2 - 2$ to nonnegative numbers, then its inverse is a function, as shown at left by the graphs of $f(x) = x^2 - 2$, $x \geq 0$, and $f^{-1}(x) = \sqrt{x + 2}$.

5.1 Exercise Set

Find the inverse of the relation.

1. $\{(7, 8), (-2, 8), (3, -4), (8, -8)\}$
2. $\{(0, 1), (5, 6), (-2, -4)\}$
3. $\{(-1, -1), (-3, 4)\}$
4. $\{(-1, 3), (2, 5), (-3, 5), (2, 0)\}$

Find an equation of the inverse relation.

5. $y = 4x - 5$
6. $2x^2 + 5y^2 = 4$
7. $x^3y = -5$
8. $y = 3x^2 - 5x + 9$
9. $x = y^2 - 2y$
10. $x = \frac{1}{2}y + 4$

Graph the equation by substituting and plotting points. Then reflect the graph across the line $y = x$ to obtain the graph of its inverse.

11. $x = y^2 - 3$
12. $y = x^2 + 1$
13. $y = 3x - 2$
14. $x = -y + 4$
15. $y = |x|$
16. $x + 2 = |y|$

Given the function f, prove that f is one-to-one using the definition of a one-to-one function on p. 310.

17. $f(x) = \frac{1}{3}x - 6$

18. $f(x) = 4 - 2x$

19. $f(x) = x^3 + \frac{1}{2}$

20. $f(x) = \sqrt[3]{x}$

Given the function g, prove that g is not one-to-one using the definition of a one-to-one function on p. 310.

21. $g(x) = 1 - x^2$

22. $g(x) = 3x^2 + 1$

23. $g(x) = x^4 - x^2$

24. $g(x) = \dfrac{1}{x^6}$

Using the horizontal-line test, determine whether the function is one-to-one.

25. $f(x) = 2.7^x$

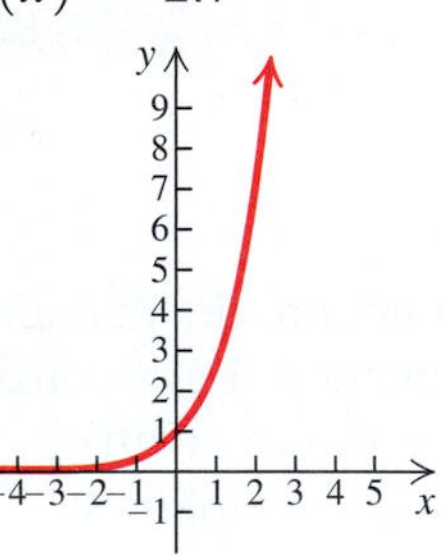

26. $f(x) = 2^{-x}$

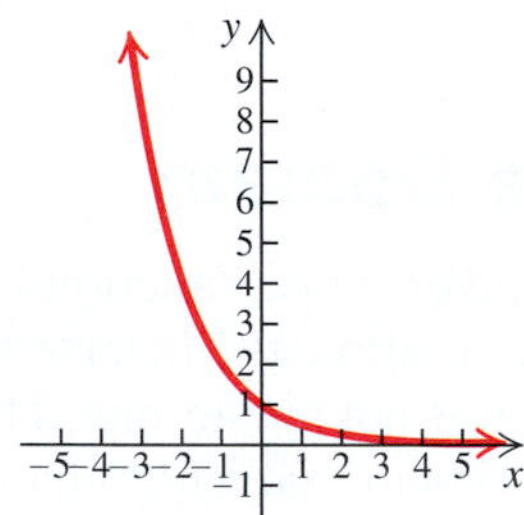

27. $f(x) = 4 - x^2$

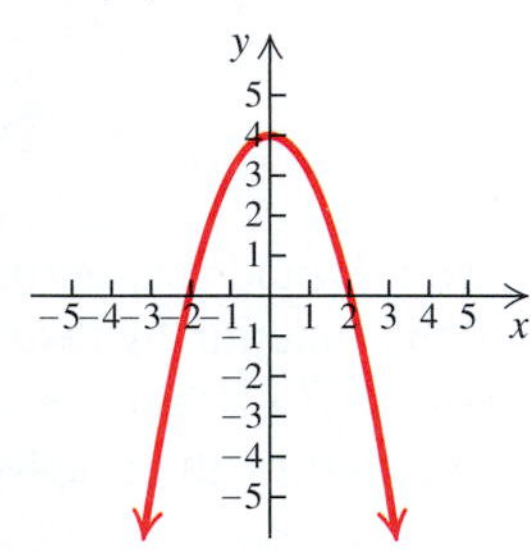

28. $f(x) = x^3 - 3x + 1$

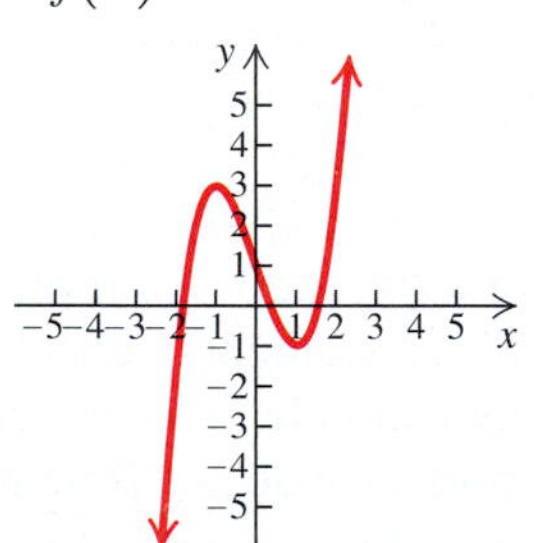

29. $f(x) = \dfrac{8}{x^2 - 4}$

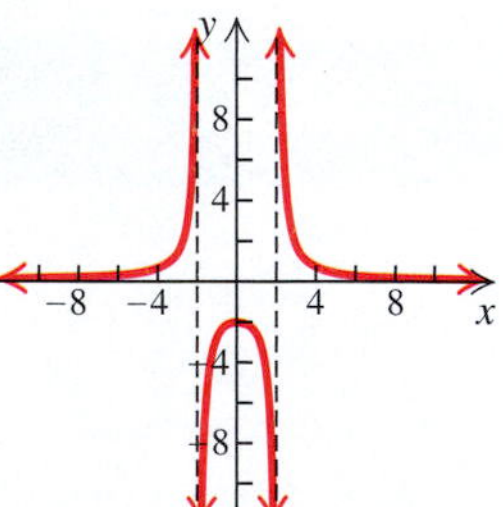

30. $f(x) = \sqrt{\dfrac{10}{4 + x}}$

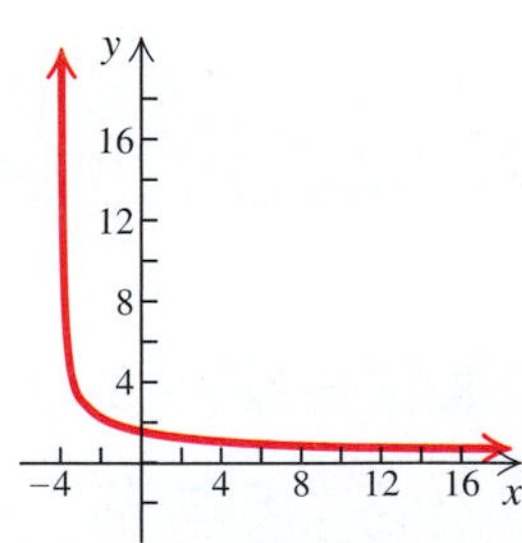

31. $f(x) = \sqrt[3]{x + 2} - 2$

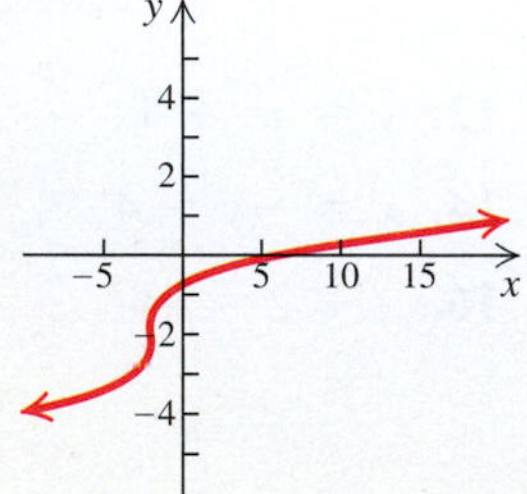

32. $f(x) = \dfrac{8}{x}$

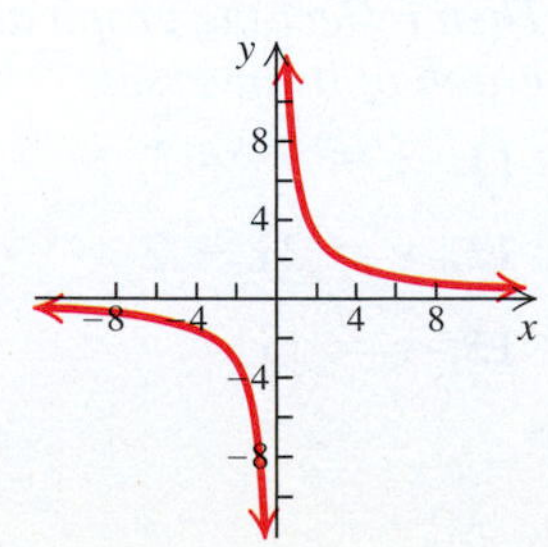

Graph the function and determine whether the function is one-to-one using the horizontal-line test.

33. $f(x) = 5x - 8$

34. $f(x) = 3 + 4x$

35. $f(x) = 1 - x^2$

36. $f(x) = |x| - 2$

37. $f(x) = |x + 2|$

38. $f(x) = -0.8$

39. $f(x) = -\dfrac{4}{x}$

40. $f(x) = \dfrac{2}{x + 3}$

41. $f(x) = \frac{2}{3}$

42. $f(x) = \frac{1}{2}x^2 + 3$

43. $f(x) = \sqrt{25 - x^2}$

44. $f(x) = -x^3 + 2$

Graph the function and its inverse using a graphing calculator. Use an inverse drawing feature, if available. Find the domain and the range of f and of f^{-1}.

45. $f(x) = 0.8x + 1.7$

46. $f(x) = 2.7 - 1.08x$

47. $f(x) = \frac{1}{2}x - 4$

48. $f(x) = x^3 - 1$

49. $f(x) = \sqrt{x - 3}$

50. $f(x) = -\dfrac{2}{x}$

51. $f(x) = x^2 - 4, x \geq 0$

52. $f(x) = 3 - x^2, x \geq 0$

53. $f(x) = (3x - 9)^3$

54. $f(x) = \sqrt[3]{\dfrac{x - 3.2}{1.4}}$

In Exercises 55–70, for each function:

a) *Determine whether it is one-to-one.*

b) *If the function is one-to-one, find a formula for the inverse.*

55. $f(x) = x + 4$

56. $f(x) = 7 - x$

57. $f(x) = 2x - 1$

58. $f(x) = 5x + 8$

59. $f(x) = \dfrac{4}{x + 7}$

60. $f(x) = -\dfrac{3}{x}$

61. $f(x) = \dfrac{x + 4}{x - 3}$

62. $f(x) = \dfrac{5x - 3}{2x + 1}$

63. $f(x) = x^3 - 1$

64. $f(x) = (x + 5)^3$

65. $f(x) = x\sqrt{4 - x^2}$

66. $f(x) = 2x^2 - x - 1$

67. $f(x) = 5x^2 - 2, x \geq 0$

68. $f(x) = 4x^2 + 3, x \geq 0$

69. $f(x) = \sqrt{x + 1}$

70. $f(x) = \sqrt[3]{x - 8}$

Find the inverse by thinking about the operations of the function and then reversing, or undoing, them. Check your work algebraically.

	FUNCTION	INVERSE
71.	$f(x) = 3x$	$f^{-1}(x) = \square$
72.	$f(x) = \frac{1}{4}x + 7$	$f^{-1}(x) = \square$
73.	$f(x) = -x$	$f^{-1}(x) = \square$
74.	$f(x) = \sqrt[3]{x} - 5$	$f^{-1}(x) = \square$
75.	$f(x) = \sqrt[3]{x - 5}$	$f^{-1}(x) = \square$
76.	$f(x) = x^{-1}$	$f^{-1}(x) = \square$

Each graph in Exercises 77–82 is the graph of a one-to-one function f. Sketch the graph of the inverse function f^{-1}.

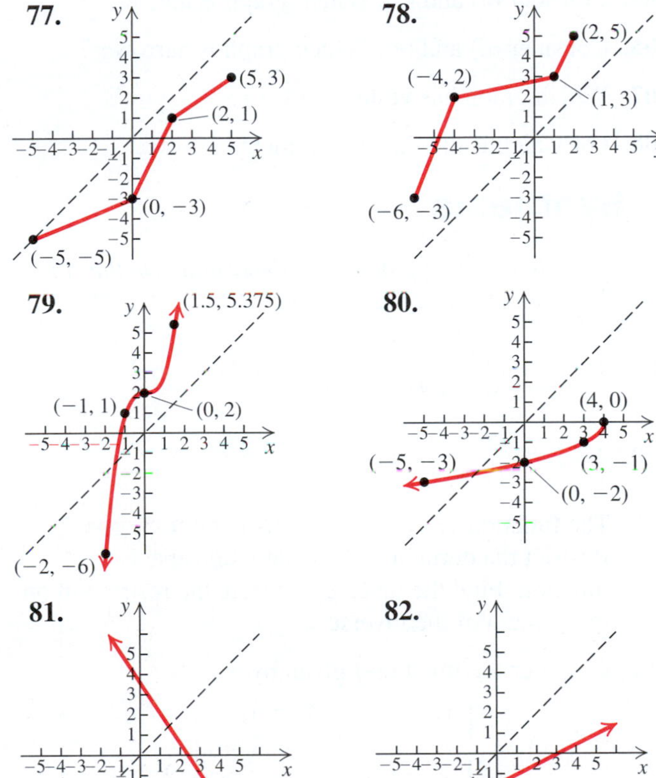

For the function f, use composition of functions to show that f^{-1} is as given.

83. $f(x) = \frac{7}{8}x,\ f^{-1}(x) = \frac{8}{7}x$

84. $f(x) = \dfrac{x + 5}{4},\ f^{-1}(x) = 4x - 5$

85. $f(x) = \dfrac{1 - x}{x},\ f^{-1}(x) = \dfrac{1}{x + 1}$

86. $f(x) = \sqrt[3]{x + 4},\ f^{-1}(x) = x^3 - 4$

87. $f(x) = \dfrac{2}{5}x + 1,\ f^{-1}(x) = \dfrac{5x - 5}{2}$

88. $f(x) = \dfrac{x + 6}{3x - 4},\ f^{-1}(x) = \dfrac{4x + 6}{3x - 1}$

Find the inverse of the given one-to-one function f. Give the domain and the range of f and of f^{-1}, and then graph both f and f^{-1} on the same set of axes.

89. $f(x) = 5x - 3$

90. $f(x) = 2 - x$

91. $f(x) = \dfrac{2}{x}$

92. $f(x) = -\dfrac{3}{x + 1}$

93. $f(x) = \frac{1}{3}x^3 - 2$

94. $f(x) = \sqrt[3]{x} - 1$

95. $f(x) = \dfrac{x + 1}{x - 3}$

96. $f(x) = \dfrac{x - 1}{x + 2}$

97. Find $f(f^{-1}(5))$ and $f^{-1}(f(a))$:
$f(x) = x^3 - 4.$

98. Find $f^{-1}(f(p))$ and $f(f^{-1}(1253))$:
$$f(x) = \sqrt[5]{\frac{2x - 7}{3x + 4}}.$$

99. *Hitting Lessons.* A summer little-league baseball team determines that the cost per player of a group hitting lesson is given by the formula
$$C(x) = \frac{72 + 2x}{x},$$
where x is the number of players in the group and $C(x)$ is in dollars.

a) Determine the cost per player of a group hitting lesson when there are 2, 5, and 8 players in the group.

b) Find a formula for the inverse of the function and explain what it represents.

c) Use the inverse function to determine the number of players in the group lesson when the cost per player is \$74, \$20, and \$11.

100. *Women's Shoe Sizes.* A function that will convert women's shoe sizes in the United States to those in Australia is

$$s(x) = \frac{2x - 3}{2}$$

(*Source*: OnlineConversion.com).

a) Determine the women's shoe sizes in Australia that correspond to sizes 5, $7\frac{1}{2}$, and 8 in the United States.
b) Find a formula for the inverse of the function and explain what it represents.
c) Use the inverse function to determine the women's shoe sizes in the United States that correspond to sizes 3, $5\frac{1}{2}$, and 7 in Australia.

101. *E-Commerce Holiday Sales.* Retail e-commerce holiday season sales (November and December), in billions of dollars, x years after 2009 is given by the function

$$H(x) = 4.798x + 28.1778$$

(*Source*: statista.com).

a) Determine the total amount of e-commerce holiday sales in 2012 and in 2016.
b) Find $H^{-1}(x)$ and explain what it represents.

102. *Converting Temperatures.* The following formula can be used to convert Fahrenheit temperatures x to Celsius temperatures $T(x)$:

$$T(x) = \frac{5}{9}(x - 32).$$

a) Find $T(-13°)$ and $T(86°)$.
b) Find $T^{-1}(x)$ and explain what it represents.

Skill Maintenance

Consider quadratic functions (a)–(h) that follow. Without graphing them, answer the questions below. [3.3]

a) $f(x) = 2x^2$
b) $f(x) = -x^2$
c) $f(x) = \frac{1}{4}x^2$
d) $f(x) = -5x^2 + 3$
e) $f(x) = \frac{2}{3}(x - 1)^2 - 3$
f) $f(x) = -2(x + 3)^2 + 1$
g) $f(x) = (x - 3)^2 + 1$
h) $f(x) = -4(x + 1)^2 - 3$

103. Which functions have a maximum value?

104. Which graphs open up?

105. Consider (a) and (c). Which graph is narrower?

106. Consider (d) and (e). Which graph is narrower?

107. Which graph has vertex $(-3, 1)$?

108. For which is the line of symmetry $x = 0$?

Synthesis

Using only a graphing calculator, determine whether the functions are inverses of each other.

109. $f(x) = \sqrt[3]{\frac{x - 3.2}{1.4}}$, $g(x) = 1.4x^3 + 3.2$

110. $f(x) = \frac{2x - 5}{4x + 7}$, $g(x) = \frac{7x - 4}{5x + 2}$

111. The function $f(x) = x^2 - 3$ is not one-to-one. Restrict the domain of f so that its inverse is a function. Find the inverse and state the restriction on the domain of the inverse.

112. Consider the function f given by

$$f(x) = \begin{cases} x^3 + 2, & x \le -1, \\ x^2, & -1 < x < 1, \\ x + 1, & x \ge 1. \end{cases}$$

Does f have an inverse that is a function? Why or why not?

113. Find three examples of functions that are their own inverses; that is, $f = f^{-1}$.

114. Given the function $f(x) = ax + b$, $a \neq 0$, find the values of a and b for which $f^{-1}(x) = f(x)$.

5.2 Exponential Functions and Graphs

- Graph exponential equations and exponential functions.
- Solve applied problems involving exponential functions and their graphs.

We now turn our attention to the study of a set of functions that are very rich in application. Consider the following graphs. Each one illustrates an *exponential function*. In this section, we consider such functions and some important applications.

Skype Users Online at Same Time

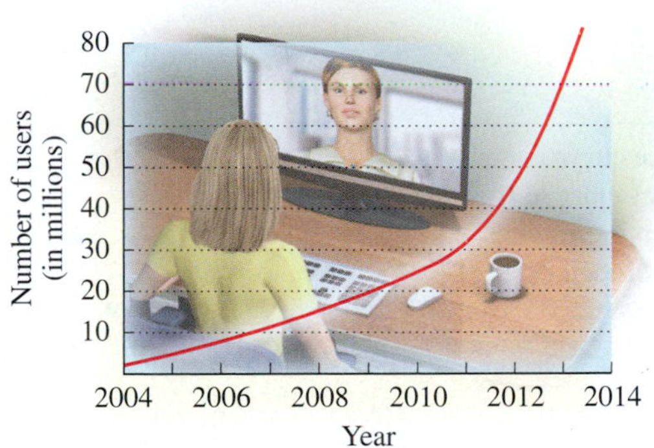

Source: Skype Numerology Blog

Postseason Bowl Games

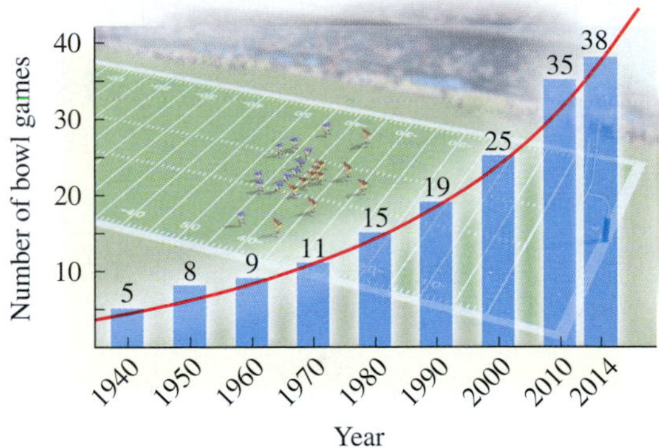

Source: USA TODAY research, College Football Data Warehouse

Graphing Exponential Functions

We now define exponential functions. We assume that a^x has meaning for any real number x and any positive real number a and that the laws of exponents still hold, though we will not prove them here.

> **EXPONENTIAL FUNCTION**
>
> The function $f(x) = a^x$, where x is a real number, $a > 0$ and $a \neq 1$, is called the **exponential function, base *a*.**

We require the **base** to be positive in order to avoid the imaginary numbers that would occur by taking even roots of negative numbers: An example is $(-1)^{1/2}$, the square root of -1, which is not a real number. The restriction $a \neq 1$ is made to exclude the constant function $f(x) = 1^x = 1$, which does not have an inverse that is a function because it is not one-to-one.

The following are examples of exponential functions:

$$f(x) = 2^x, \qquad f(x) = \left(\tfrac{1}{2}\right)^x, \qquad f(x) = 3.57^{(x-2)}.$$

Note that, in contrast to functions like $f(x) = x^5$ and $f(x) = x^{1/2}$ in which the variable is the base of an exponential expression, the variable in an exponential function is *in the exponent*.

Let's now consider graphs of exponential functions.

Just in Time 7

EXAMPLE 1 Graph the exponential function

$$y = f(x) = 2^x.$$

Solution We compute some function values and list the results in a table.

x	y $y = f(x) = 2^x$	(x, y)
0	1	$(0, 1)$
1	2	$(1, 2)$
2	4	$(2, 4)$
3	8	$(3, 8)$
-1	$\frac{1}{2}$	$\left(-1, \frac{1}{2}\right)$
-2	$\frac{1}{4}$	$\left(-2, \frac{1}{4}\right)$
-3	$\frac{1}{8}$	$\left(-3, \frac{1}{8}\right)$

$$f(0) = 2^0 = 1; \qquad f(-1) = 2^{-1} = \frac{1}{2^1} = \frac{1}{2};$$
$$f(1) = 2^1 = 2;$$
$$f(2) = 2^2 = 4; \qquad f(-2) = 2^{-2} = \frac{1}{2^2} = \frac{1}{4};$$
$$f(3) = 2^3 = 8; \qquad f(-3) = 2^{-3} = \frac{1}{2^3} = \frac{1}{8}.$$

Next, we plot these points and connect them with a smooth curve. Be sure to plot enough points to determine how steeply the curve rises.

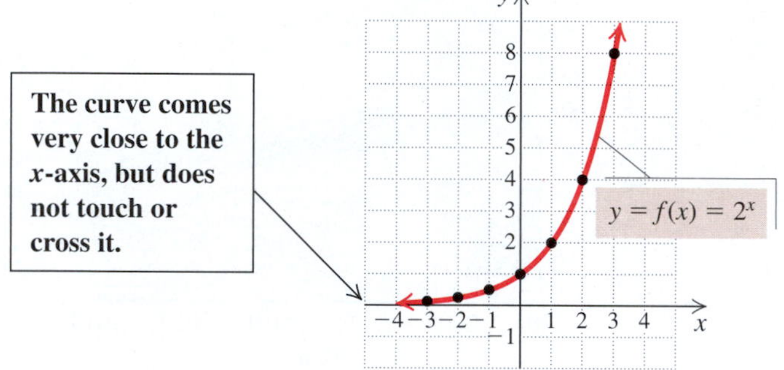

HORIZONTAL ASYMPTOTES

REVIEW SECTION 4.5.

Note that as x increases, the function values increase without bound. As x decreases, the function values decrease, getting close to 0. That is, as $x \to -\infty$, $y \to 0$. Thus the x-axis, or the line $y = 0$, is a horizontal asymptote. As the x-inputs decrease, the curve gets closer and closer to this line, but does not cross it.

Now Try Exercise 11.

EXAMPLE 2 Graph the exponential function $y = f(x) = \left(\frac{1}{2}\right)^x$.

Solution Before we plot points and draw the curve, note that

$$y = f(x) = \left(\tfrac{1}{2}\right)^x = (2^{-1})^x = 2^{-x}.$$

This tells us that this graph is a reflection of the graph of $y = 2^x$ across the y-axis. For example, if $(3, 8)$ is a point of the graph of $f(x) = 2^x$, then $(-3, 8)$ is a point of the graph of $f(x) = 2^{-x}$. Selected points are listed in the table at left.

Points of $f(x) = 2^x$	Points of $f(x) = \left(\frac{1}{2}\right)^x = 2^{-x}$
$(0, 1)$	$(0, 1)$
$(1, 2)$	$(-1, 2)$
$(2, 4)$	$(-2, 4)$
$(3, 8)$	$(-3, 8)$
$\left(-1, \frac{1}{2}\right)$	$\left(1, \frac{1}{2}\right)$
$\left(-2, \frac{1}{4}\right)$	$\left(2, \frac{1}{4}\right)$
$\left(-3, \frac{1}{8}\right)$	$\left(3, \frac{1}{8}\right)$

Next, we plot points and connect them with a smooth curve.

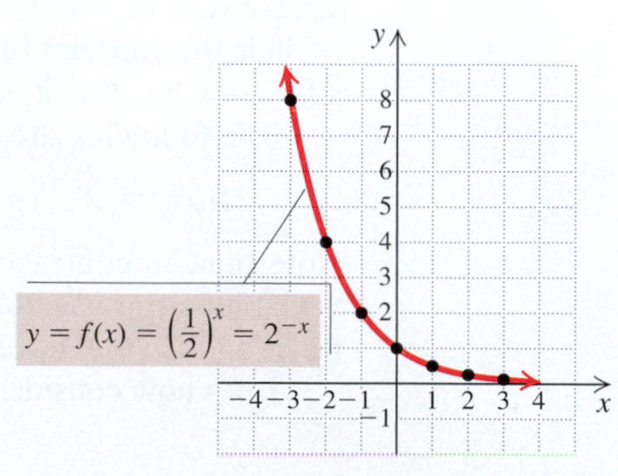

EXPLORING WITH TECHNOLOGY

Use the TRACE and TABLE features to confirm that the graphs of $y = f(x) = 2^x$ and $y = f(x) = \left(\frac{1}{2}\right)^x$ never cross the x-axis.

Note that as x increases, the function values decrease, getting close to 0. The x-axis, $y = 0$, is the horizontal asymptote. As x decreases, the function values increase without bound.

Now Try Exercise 15.

Note the patterns in the following graphs of exponential functions.

CONNECTING THE CONCEPTS — Properties of Exponential Functions

Let's list and compare some characteristics of exponential functions, keeping in mind that the definition of an exponential function, $f(x) = a^x$, requires that a be positive and different from 1.

$f(x) = a^x, a > 0, a \neq 1$

Continuous

One-to-one

Domain: $(-\infty, \infty)$

Range: $(0, \infty)$

Increasing if $a > 1$

Decreasing if $0 < a < 1$

Horizontal asymptote is x-axis

y-intercept: $(0, 1)$

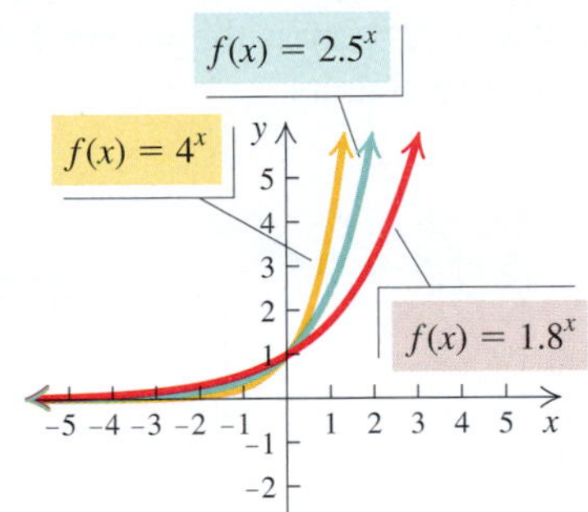

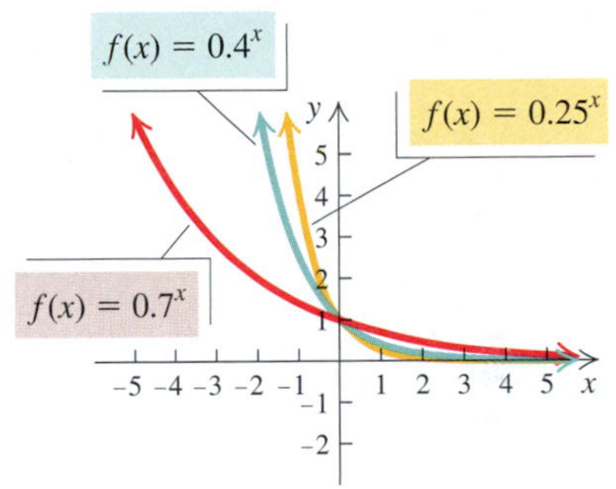

TRANSFORMATIONS

REVIEW SECTION 2.5.

To graph other types of exponential functions, keep in mind the ideas of translation, stretching, and reflection. All these concepts allow us to visualize the graph before drawing it.

EXAMPLE 3 Graph each of the following by hand. Before doing so, describe how each graph can be obtained from the graph of $f(x) = 2^x$. Then check your graph with a graphing calculator.

a) $f(x) = 2^{x-2}$ **b)** $f(x) = 2^x - 4$ **c)** $f(x) = 5 - 0.5^x$

Solution

a) The graph of $f(x) = 2^{x-2}$ is the graph of $y = 2^x$ shifted *right* 2 units.

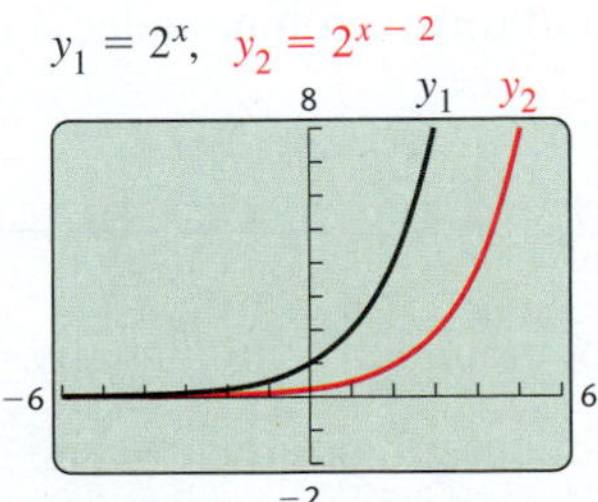

x	$f(x)$
-1	$\frac{1}{8}$
0	$\frac{1}{4}$
1	$\frac{1}{2}$
2	1
3	2
4	4
5	8

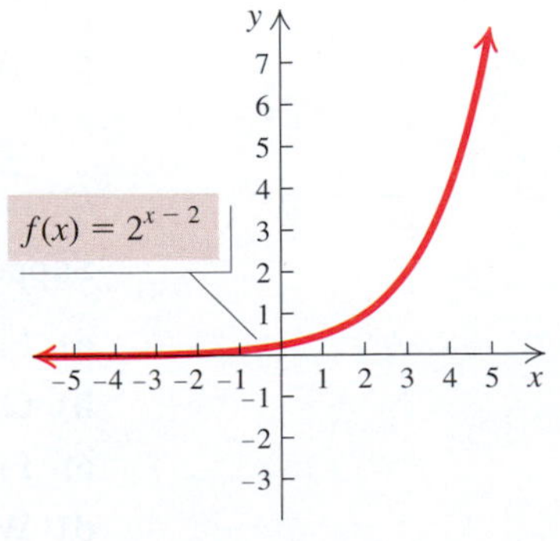

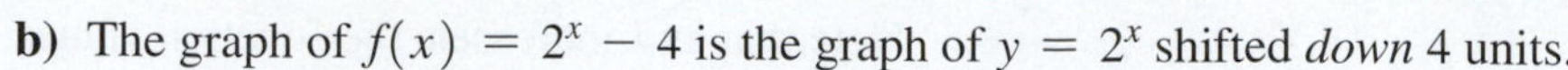

b) The graph of $f(x) = 2^x - 4$ is the graph of $y = 2^x$ shifted *down* 4 units.

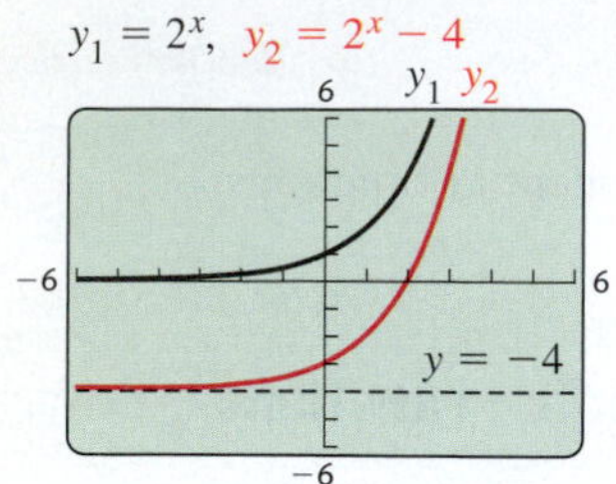

x	$f(x)$
−2	$-3\frac{3}{4}$
−1	$-3\frac{1}{2}$
0	−3
1	−2
2	0
3	4

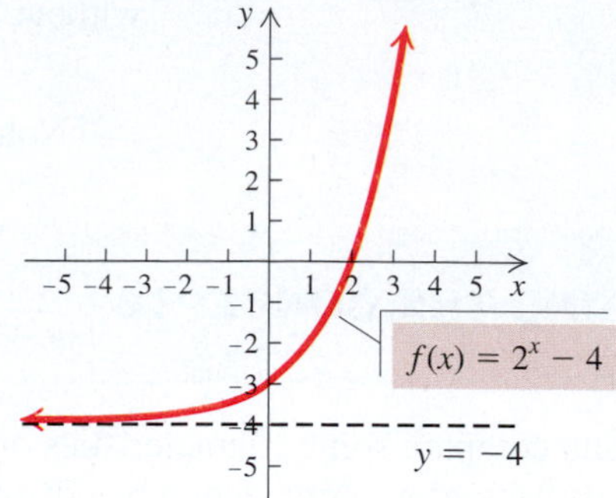

c) The graph of $f(x) = 5 - 0.5^x = 5 - \left(\frac{1}{2}\right)^x = 5 - 2^{-x}$ is a reflection of the graph of $y = 2^x$ across the y-axis, followed by a reflection across the x-axis and then a shift *up* 5 units.

$y_1 = 2^x$, $y_2 = 5 - 0.5^x$

x	$f(x)$
−3	−3
−2	1
−1	3
0	4
1	$4\frac{1}{2}$
2	$4\frac{3}{4}$

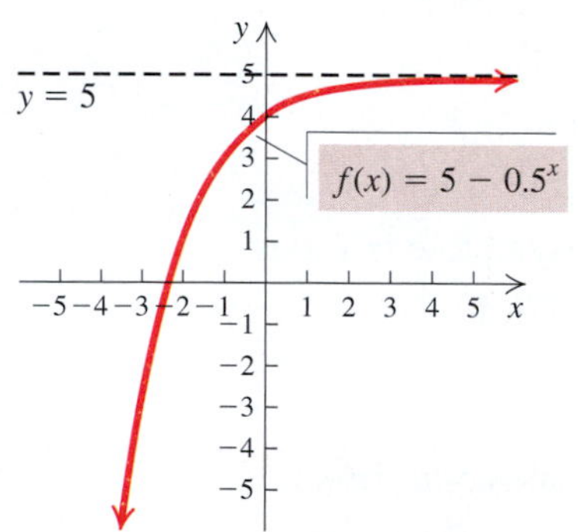

Now Try Exercises 27 and 33.

Applications

Graphing calculators are especially helpful when we are working with exponential functions. They not only facilitate computations but they also allow us to visualize the functions. One of the most frequent applications of exponential functions occurs with compound interest.

EXAMPLE 4 ***Compound Interest.*** The amount of money A to which a principal P will grow after t years at interest rate r (in decimal form), compounded n times per year, is given by the function

$$A(t) = P\left(1 + \frac{r}{n}\right)^{nt}.$$

Suppose that \$100,000 is invested at 6.5% interest, compounded semiannually.

a) Find a function for the amount to which the investment grows after t years.

b) Graph the function.

c) Find the amount of money in the account at $t = 0, 4, 8,$ and 10 years.

d) When will the amount of money in the account reach \$400,000?

Solution

$y = 100{,}000(1.0325)^{2x}$

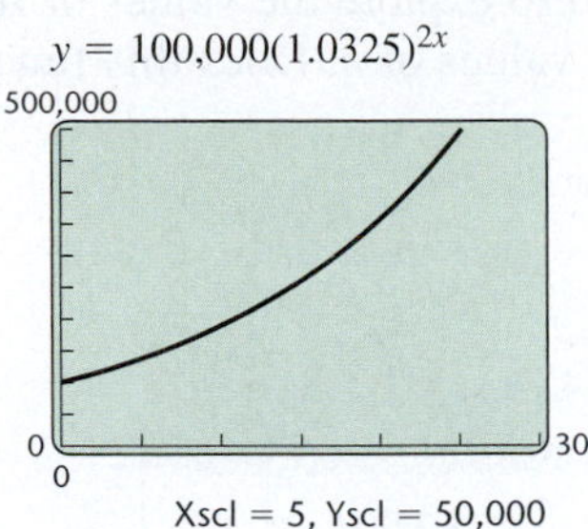

Xscl = 5, Yscl = 50,000

$y = 100{,}000(1.0325)^{2x}$

Y1(0)
100000
Y1(4)
129157.7535
Y1(8)
166817.253

a) Since $P = \$100{,}000$, $r = 6.5\% = 0.065$, and $n = 2$, we can substitute these values and write the following function:

$$A(t) = 100{,}000\left(1 + \frac{0.065}{2}\right)^{2 \cdot t} = \$100{,}000(1.0325)^{2t}.$$

b) For the graph shown at left, we use the viewing window $[0, 30, 0, 500{,}000]$ because of the large numbers and the fact that negative time values have no meaning in this application.

c) We can compute function values using function notation on the home screen of a graphing calculator. (See the window at left.) We can also calculate the values directly on a graphing calculator by substituting in the expression for $A(t)$:

$$A(0) = 100{,}000(1.0325)^{2 \cdot 0} = \$100{,}000;$$
$$A(4) = 100{,}000(1.0325)^{2 \cdot 4} \approx \$129{,}157.75;$$
$$A(8) = 100{,}000(1.0325)^{2 \cdot 8} \approx \$166{,}817.25;$$
$$A(10) = 100{,}000(1.0325)^{2 \cdot 10} \approx \$189{,}583.79.$$

d) To find the amount of time it takes for the account to grow to \$400,000, we set

$$100{,}000(1.0325)^{2t} = 400{,}000$$

and solve for t. One way we can do this is by graphing the equations

$$y_1 = 100{,}000(1.0325)^{2x} \quad \text{and} \quad y_2 = 400{,}000.$$

Then we can use the Intersect method to estimate the first coordinate of the point of intersection. (See Fig. 1 below.)

We can also use the Zero method to estimate the zero of the function $y = 100{,}000(1.0325)^{2x} - 400{,}000$. (See Fig. 2 below.)

Regardless of the method we use, we see that the account grows to \$400,000 after about 21.67 years, or about 21 years and 8 months.

INTERSECT METHOD
REVIEW SECTION 1.5.

ZERO METHOD
REVIEW SECTION 1.5.

$y_1 = 100{,}000(1.0325)^{2x}$,
$y_2 = 400{,}000$

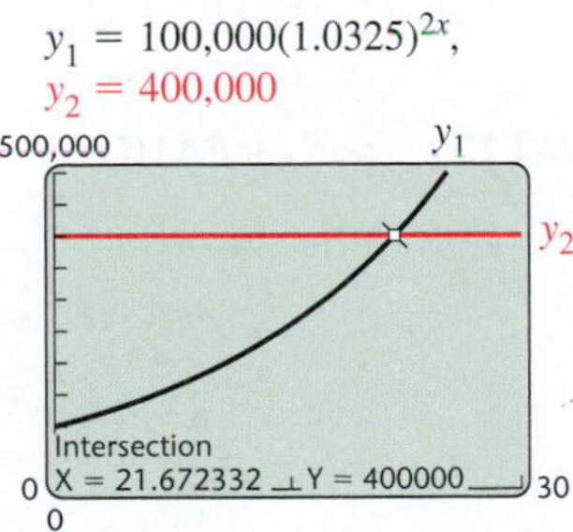

FIGURE 1.

$y_1 = 100{,}000(1.0325)^{2x} - 400{,}000$

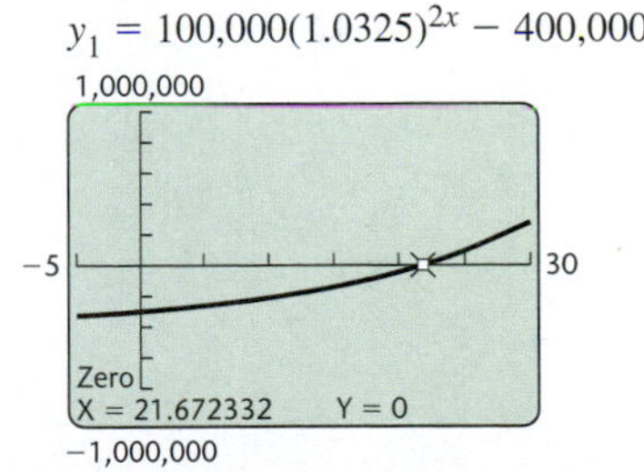

FIGURE 2.

Now Try Exercise 51.

The Number e

We now consider a very special number in mathematics. In 1741, Leonhard Euler named this number e. Though you may not have encountered it before, you will see here and in future mathematics courses that it has many important applications. To explain this number, we use the compound interest formula $A = P(1 + r/n)^{nt}$ discussed in Example 4. Suppose that \$1 is invested at 100% interest for 1 year. Since $P = 1$, $r = 100\% = 1$, and $t = 1$, the formula above becomes a function A defined in terms of the number of compounding periods n:

$$A = P\left(1 + \frac{r}{n}\right)^{nt} = 1\left(1 + \frac{1}{n}\right)^{n \cdot 1} = \left(1 + \frac{1}{n}\right)^{n}.$$

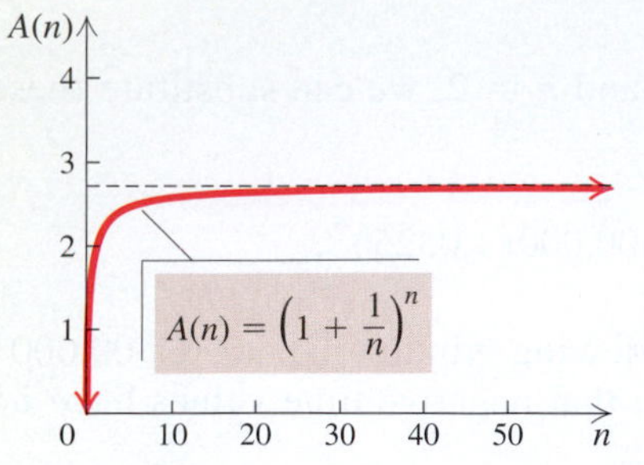

Let's use the graph of this function shown at left to explore the values of $A(n)$ as $n \rightarrow \infty$. Consider the graph for larger and larger values of n. Does this function have a horizontal asymptote?

Let's find some function values using a calculator.

n, Number of Compounding Periods	$A(n) = \left(1 + \frac{1}{n}\right)^n$
1 (compounded annually)	$2.00
2 (compounded semiannually)	2.25
3	2.3704
4 (compounded quarterly)	2.4414
5	2.4883
100	2.7048
365 (compounded daily)	2.7146
8760 (compounded hourly)	2.7181

It appears from these values that the graph does have a horizontal asymptote, $y \approx 2.7$. As the values of n get larger and larger, the function values get closer and closer to the number Euler named e. Its decimal representation does not terminate or repeat; it is irrational.

$$e = 2.7182818284\ldots$$

EXAMPLE 5 Find each value of e^x, to four decimal places, using the e^x key on a calculator.

a) e^3 **b)** $e^{-0.23}$ **c)** e^0

Solution

FUNCTION VALUE	READOUT	ROUNDED
a) e^3	e3 20.08553692	20.0855
b) $e^{-0.23}$	e^-.23 .7945336025	0.7945
c) e^0	e0 1	1

Now Try Exercises 1 and 3.

Graphs of Exponential Functions, Base e

We demonstrate ways in which to graph exponential functions.

EXAMPLE 6 Graph $f(x) = e^x$ and $g(x) = e^{-x}$.

Solution We can compute points for each equation using a table or the e^x key on a calculator. Then we plot these points and draw the graphs of the functions.

$y_1 = e^x$, $y_2 = e^{-x}$

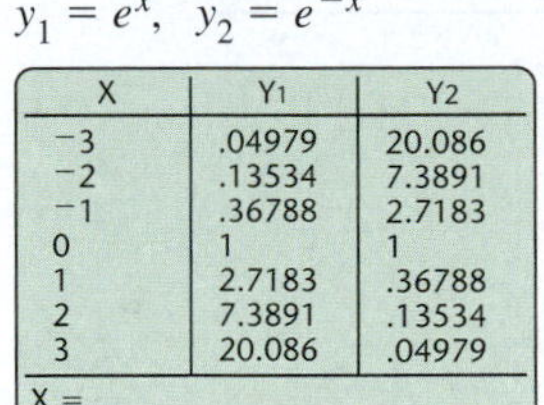

X	Y1	Y2
−3	.04979	20.086
−2	.13534	7.3891
−1	.36788	2.7183
0	1	1
1	2.7183	.36788
2	7.3891	.13534
3	20.086	.04979

X =

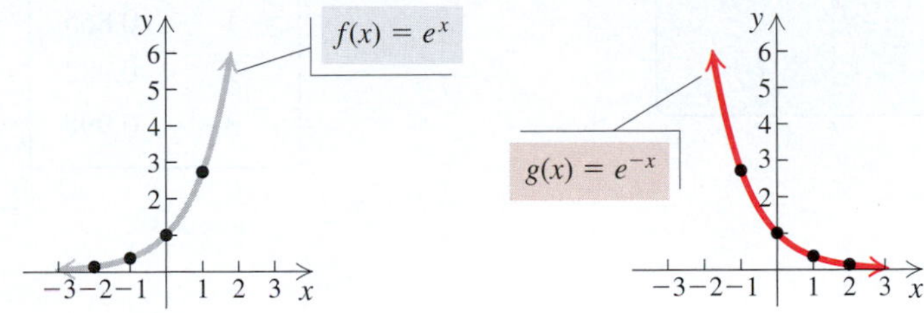

Note that the graphs are reflections of each other across the y-axis.

Now Try Exercise 23.

EXAMPLE 7 Graph each of the following by hand. Before doing so, describe how each graph can be obtained from the graph of $y = e^x$.

a) $f(x) = e^{x+3}$ **b)** $f(x) = e^{-0.5x}$ **c)** $f(x) = 1 - e^{-2x}$

Solution

a) The graph of $f(x) = e^{x+3}$ is a translation of the graph of $y = e^x$ left 3 units.

$y_1 = e^x$, $y_2 = e^{x+3}$

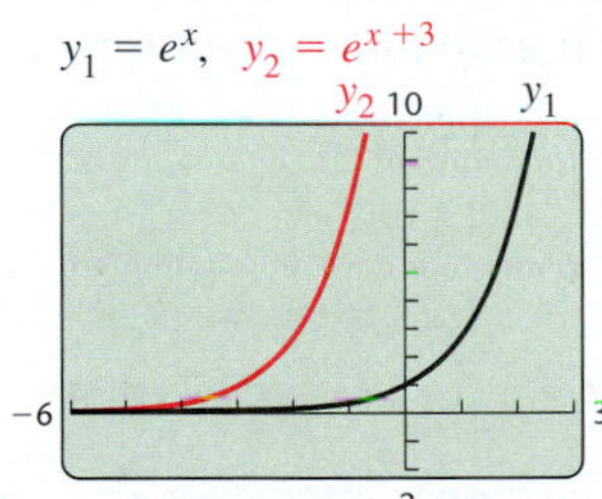

x	$f(x)$
−7	0.018
−5	0.135
−3	1
−1	7.389
0	20.086

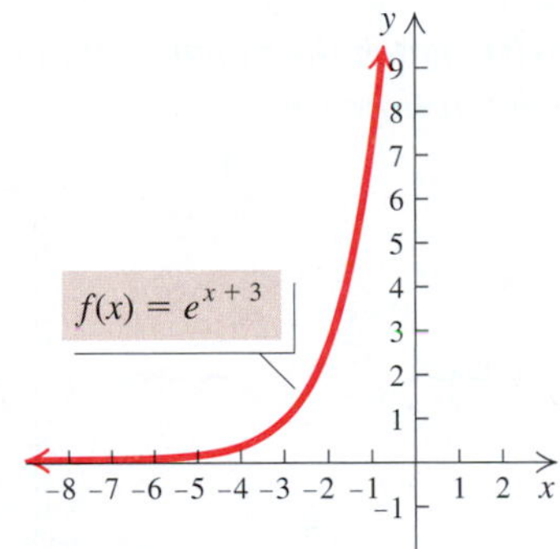

b) We note that the graph of $f(x) = e^{-0.5x}$ is a horizontal stretching of the graph of $y = e^x$ followed by a reflection across the y-axis.

$y_1 = e^x$, $y_2 = e^{-0.5x}$

x	$f(x)$
−2	2.718
−1	1.649
0	1
1	0.607
2	0.368

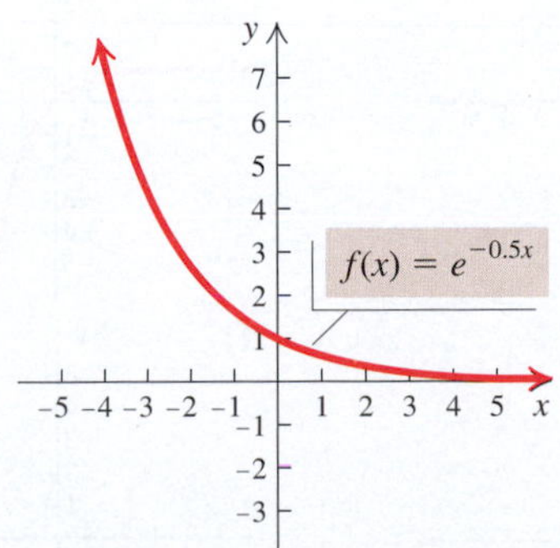

c) The graph of $f(x) = 1 - e^{-2x}$ is a horizontal shrinking of the graph of $y = e^x$, followed by a reflection across the y-axis, then across the x-axis, followed by a translation up 1 unit.

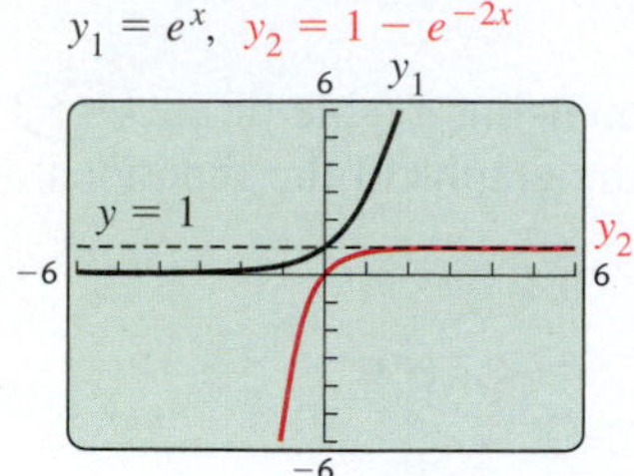

x	$f(x)$
−1	−6.389
0	0
1	0.865
2	0.982
3	0.998

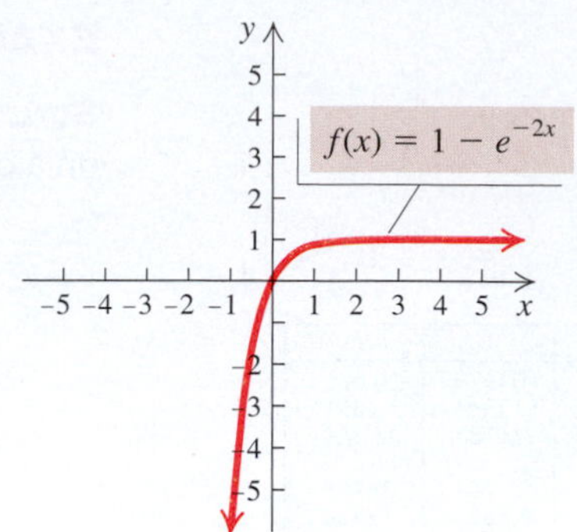

Now Try Exercises 41 and 47.

5.2 Exercise Set

Find each of the following, to four decimal places, using a calculator.

1. e^4

2. e^{10}

3. $e^{-2.458}$

4. $\left(\dfrac{1}{e^3}\right)^2$

In Exercises 5–10, match the function with one of the graphs (a)–(f) that follow.

a)

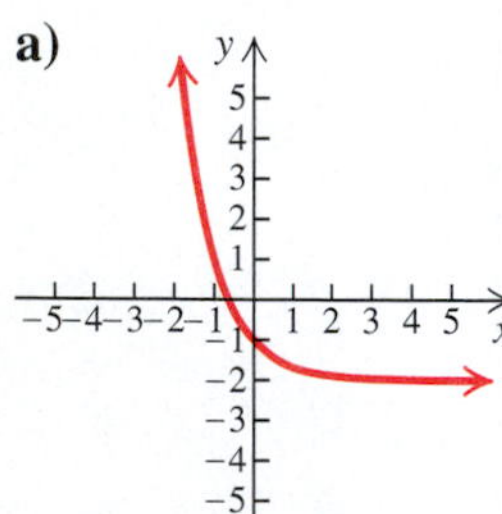

b)

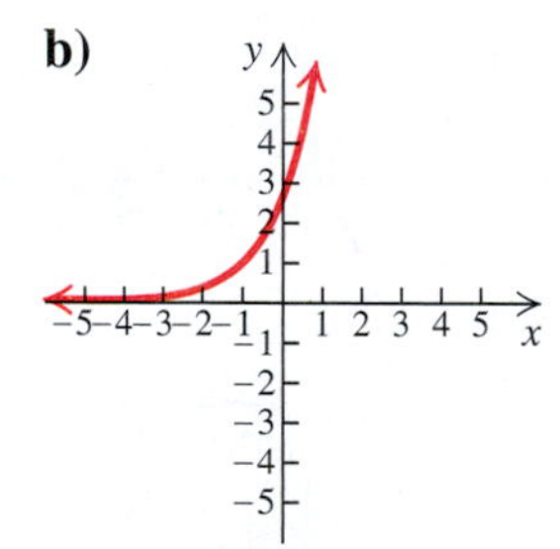

c)

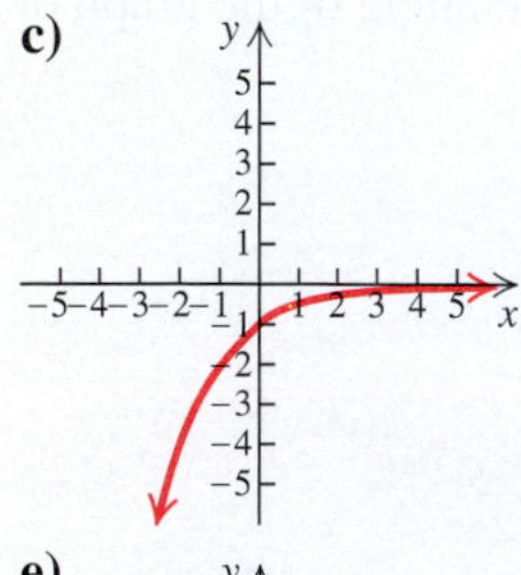

d)

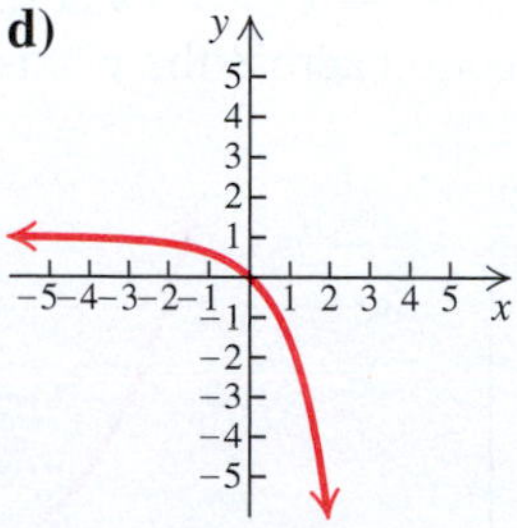

e)

f)

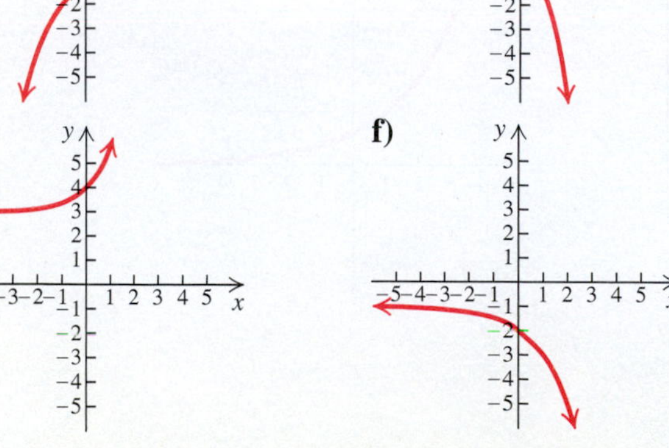

5. $f(x) = -2^x - 1$

6. $f(x) = -\left(\frac{1}{2}\right)^x$

7. $f(x) = e^x + 3$

8. $f(x) = e^{x+1}$

9. $f(x) = 3^{-x} - 2$

10. $f(x) = 1 - e^x$

Graph the function by substituting and plotting points. Then check your work using a graphing calculator.

11. $f(x) = 3^x$

12. $f(x) = 5^x$

13. $f(x) = 6^x$

14. $f(x) = 3^{-x}$

15. $f(x) = \left(\frac{1}{4}\right)^x$

16. $f(x) = \left(\frac{2}{3}\right)^x$

17. $y = -2^x$

18. $y = 3 - 3^x$

19. $f(x) = -0.25^x + 4$

20. $f(x) = 0.6^x - 3$

21. $f(x) = 1 + e^{-x}$

22. $f(x) = 2 - e^{-x}$

23. $y = \frac{1}{4}e^x$

24. $y = 2e^{-x}$

25. $f(x) = 1 - e^{-x}$

26. $f(x) = e^x - 2$

Sketch the graph of the function and check the graph with a graphing calculator. Describe how each graph can be obtained from the graph of a basic exponential function.

27. $f(x) = 2^{x+1}$

28. $f(x) = 2^{x-1}$

29. $f(x) = 2^x - 3$

30. $f(x) = 2^x + 1$

31. $f(x) = 2^{1-x} + 2$

32. $f(x) = 5 - 2^{-x}$

33. $f(x) = 4 - 3^{-x}$

34. $f(x) = 2^{x-1} - 3$

35. $f(x) = \left(\frac{3}{2}\right)^{x-1}$

36. $f(x) = 3^{4-x}$

37. $f(x) = 2^{x+3} - 5$

38. $f(x) = -3^{x-2}$

39. $f(x) = 3 \cdot 2^{x-1} + 1$

40. $f(x) = 2 \cdot 3^{x+1} - 2$

41. $f(x) = e^{2x}$

42. $f(x) = e^{-0.2x}$

43. $f(x) = \frac{1}{2}(1 - e^x)$

44. $f(x) = 3(1 + e^x) - 2$

45. $y = e^{-x+1}$

46. $y = e^{2x} + 1$

47. $f(x) = 2(1 - e^{-x})$

48. $f(x) = 1 - e^{-0.01x}$

Graph the piecewise function.

49. $f(x) = \begin{cases} e^{-x} - 4, & \text{for } x < -2, \\ x + 3, & \text{for } -2 \le x < 1, \\ x^2, & \text{for } x \ge 1 \end{cases}$

50. $g(x) = \begin{cases} 4, & \text{for } x \le -3, \\ x^2 - 6, & \text{for } -3 < x < 0, \\ e^x, & \text{for } x \ge 0 \end{cases}$

51. *Compound Interest.* Suppose that \$82,000 is invested at $4\frac{1}{2}\%$ interest, compounded quarterly.
   a) Find the function for the amount to which the investment grows after t years.
   b) Graph the function.
   c) Find the amount of money in the account at $t = 0$, 2, 5, and 10 years.
   d) When will the amount of money in the account reach \$100,000?

52. *Compound Interest.* Suppose that \$750 is invested at 7% interest, compounded semiannually.
   a) Find the function for the amount to which the investment grows after t years.
   b) Graph the function.
   c) Find the amount of money in the account at $t = 1$, 6, 10, 15, and 25 years.
   d) When will the amount of money in the account reach \$3000?

53. *Interest on a CD.* On Elizabeth's sixth birthday, her grandparents present her with a \$3000 certificate of deposit (CD) that earns 5% interest, compounded quarterly. If the CD matures on her sixteenth birthday, what amount will be available then?

54. *Interest in a College Trust Fund.* Following the birth of his child, Benjamin deposits \$10,000 in a college trust fund where interest is 3.9%, compounded semiannually.
   a) Find a function for the amount in the account after t years.
   b) Find the amount of money in the account at $t = 0$, 4, 8, 10, 18, and 21 years.

In Exercises 55–62, use the compound-interest formula to find the account balance A with the given conditions:

P = principal,
r = interest rate,
n = number of compounding periods per year,
t = time, in years,
A = account balance.

	P	r	Compounded	n	t	A
55.	\$3,000	4%	Semiannually		2	
56.	\$12,500	3%	Quarterly		3	
57.	\$120,000	2.5%	Annually		10	
58.	\$120,000	2.5%	Quarterly		10	
59.	\$53,500	$5\frac{1}{2}\%$	Quarterly		$6\frac{1}{2}$	
60.	\$6,250	$6\frac{3}{4}\%$	Semiannually		$4\frac{1}{2}$	
61.	\$17,400	8.1%	Daily		5	
62.	\$900	7.3%	Daily		$7\frac{1}{4}$	

63. *Alternative-Fuel Vehicles.* The sales of alternative-fuel vehicles have more than tripled since 1995 (*Source*: Energy Information Administration). The exponential function

$$A(x) = 246{,}855(1.0931)^x,$$

where x is the number of years after 1995, can be used to estimate the number of alternative-fuel vehicles sold in a given year. Find the number of alternative-fuel vehicles sold in 2000 and in 2013. Then project the number of alternative-fuel vehicles sold in 2018.

64. *Increasing CPU Power.* The central processing unit (CPU) power in computers has increased significantly over the years. The CPU power in Macintosh computers has grown exponentially from 8 MHz in 1984 to 3400 MHz in 2013 (*Source*: Apple). The exponential function

$$M(t) = 7.91477(1.26698)^t,$$

where t is the number of years after 1984, can be used to estimate the CPU power in a Macintosh computer in a given year. Find the CPU power of a Macintosh Performa 5320CD in 1995 and of an iMac G6 in 2009. Round to the nearest one MHz.

65. *Alfalfa Imported by China.* The amount and quality of milk produced increases when dairy cows are fed high-quality alfalfa. Although annual milk consumption per capita in China has been one of the lowest in the world, it has increased greatly in recent years. Consequently, the demand for imported alfalfa has increased exponentially. In 2008, approximately 20,000 tons of U.S. alfalfa were imported by China. This number increased to approximately 650,000 tons by 2013. (*Sources*: *National Geographic*, May 2015, Arjen Hoekstra, University of Twente; USDA Economic Research Service; Shefali Sharma and Zhang Rou, Institute for Agriculture and Trade Policy; FAO; Ministry of Agriculture, People's Republic of China) The increase in imported U.S. alfalfa can be modeled by the exponential function

$$A(x) = 22{,}611.008(1.992)^x,$$

where x is the number of years after 2008. Find the number of tons of U.S. alfalfa imported by China in 2012 and in 2016.

66. *U.S. Imports.* The amount of imports to the United States has increased exponentially since 1980 (*Sources*: U.S. Census Bureau; U.S. Bureau of Economic Analysis; U.S. Department of Commerce). The exponential function

$$I(x) = 297.539(1.075)^x,$$

where x is the number of years after 1980, can be used to estimate the total amount of U.S. imports, in billions of dollars. Find the total amount of imports to the United States in 1995, in 2005, in 2010, and in 2013. Round to the nearest billion dollars.

67. *E-Cigarette Sales.* The electronic cigarette was launched in 2007, and since then sales have increased from about \$20 million in 2008 to about \$500 million in 2012 (*Sources*: UBS; forbes.com). The exponential function

$$S(x) = 20.913(2.236)^x,$$

where x is the number of years after 2008, models the sales, in millions of dollars. Use this function to estimate the sales of e-cigarettes in 2011 and in 2015. Round to the nearest million dollars.

68. *Earthquakes in Oklahoma.* The number of earthquakes in Oklahoma has increased dramatically since 2008. Researchers are noting that as the number of wastewater wells that have resulted from oil and gas operations increases, the number of earthquakes also increases (*Sources*: U.S. Geological Survey; Oklahoma Geological Survey). The number of earthquakes with a 3.0 or greater magnitude in Oklahoma can be modeled by the exponential function

$$E(x) = 20.279(1.785)^x,$$

where x is the number of years after 2009. Find the number of earthquakes with a magnitude greater than 3.0 in 2010, in 2013, and in 2016.

69. *Centenarian Population.* The centenarian population in the United States has grown over 65% in the last 30 years. In 1980, there were only 32,194 residents ages 100 and over. This number had grown to 53,364 by 2010. (*Sources*: Population Projections Program; U.S. Census Bureau; U.S. Department of Commerce; "What People Who Live to 100 Have in Common," by Emily Brandon, *U.S. News and World Report*, January 7, 2013) The exponential function

$$H(t) = 80{,}040.68(1.0481)^t,$$

where t is the number of years after 2015, can be used to project the number of centenarians. Use this function to project the centenarian population in 2020 and in 2050.

70. *Bachelor's Degrees Earned.* The exponential function

$$D(t) = 347(1.024)^t$$

gives the number of bachelor's degrees, in thousands, earned in the United States t years after 1970 (*Sources*: National Center for Educational Statistics; U.S. Department of Education). Find the number of bachelor's degrees earned in 1985, in 2000, and in 2014. Then estimate the number of bachelor's degrees that will be earned in 2020. Round to the nearest thousand degrees.

71. *Charitable Giving.* Over the last four decades, the amount of charitable giving in the United States has grown exponentially from approximately \$20.7 billion in 1969 to approximately \$316.2 billion in 2012 (*Sources*: Giving USA Foundation; Volunteering in America by the Corporation for National & Community Service; National Philanthropic Trust; School of Philanthropy, Indiana University Purdue University Indianapolis). The exponential function

$$G(x) = 20.7(1.066)^x,$$

where x is the number of years after 1969, can be used to estimate the amount of charitable giving, in billions of dollars, in a given year. Find the amount of charitable giving in 1982, in 1995, and in 2010. Then use this function to estimate the amount of charitable giving in 2017. Round to the nearest billion dollars.

72. *Price of Admission to the Magic Kingdom.* In 2015, the price of a one-day, one-park admission to Disney's Magic Kingdom in Florida rose to \$105. The exponential function

$$D(x) = 4.532(1.078)^x,$$

where x is the number of years after 1971, models the price of a ticket. (*Source*: AllEars.net, an independent Disney consumer website) Find the price of a ticket in 1980, in 2000, and in 2012. Then use the function to project the price of a ticket in 2020.

73. *Salvage Value.* A restaurant purchased a 72-in. range with six burners for \$6982. The value of the range each year is 85% of the value of the preceding year. After t years, its value, in dollars, is given by the exponential function

$$V(t) = 6982(0.85)^t.$$

a) Graph the function.
b) Find the value of the range after 0, 1, 2, 5, and 8 years.
c) The restaurant decides to replace the range when its value has declined to \$1000. After how long will the range be replaced?

74. *Salvage Value.* A landscape company purchased a backhoe for \$56,395. The value of the backhoe each year is 90% of the value of the preceding year. After t years, its value, in dollars, is given by the exponential function

$$V(t) = 56{,}395(0.9)^t.$$

a) Graph the function.
b) Find the value of the backhoe after 0, 1, 3, 6, and 10 years. Round to the nearest dollar.

75. *Advertising.* A company begins an Internet advertising campaign to market a new telephone. The percentage of the target market that buys a product is generally a function of the length of the advertising campaign. The estimated percentage is given by

$$f(t) = 100(1 - e^{-0.04t}),$$

where t is the number of days of the campaign.

a) Graph the function.
b) Find $f(25)$, the percentage of the target market that has bought the phone after a 25-day advertising campaign.
c) After how long will 90% of the target market have bought the phone?

76. *Growth of a Stock.* The value of a stock is given by the function

$$V(t) = 58(1 - e^{-1.1t}) + 20,$$

where V is the value of the stock after time t, in months.

a) Graph the function.
b) Find $V(1)$, $V(2)$, $V(4)$, $V(6)$, and $V(12)$.
c) After how long will the value of the stock be \$75?

In Exercises 77–90, use a graphing calculator to match the equation with one of the figures (a)–(n) that follow.

a)

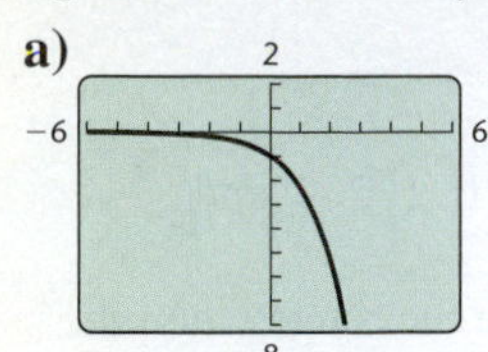

b)

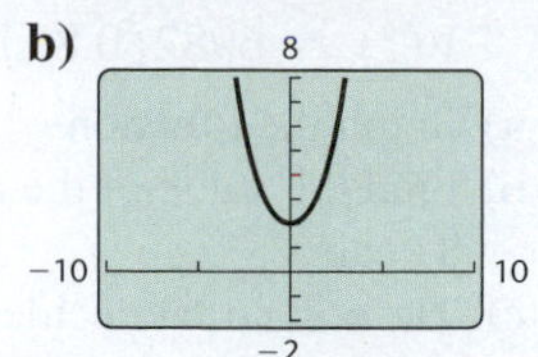

c)

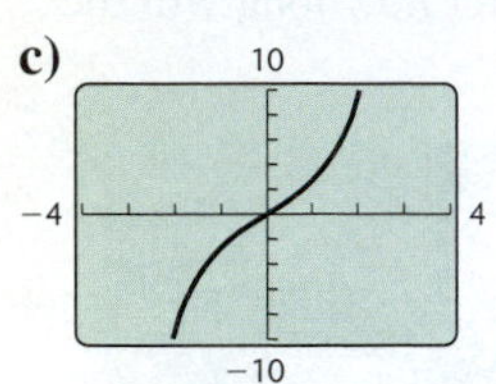

d)

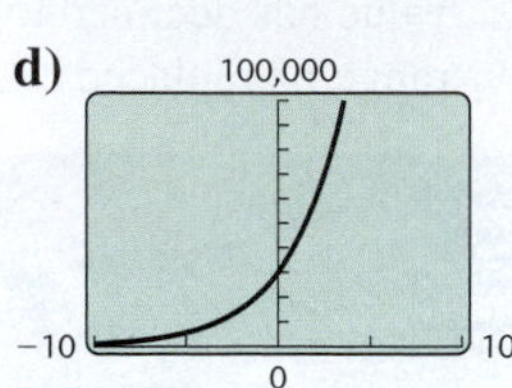

e)

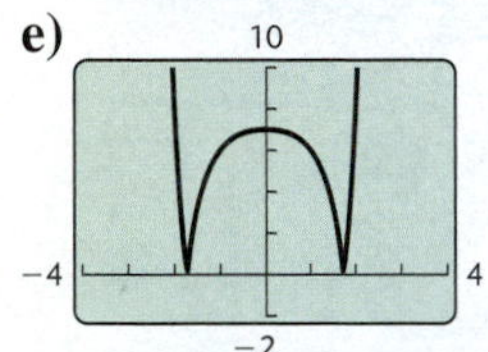

f)

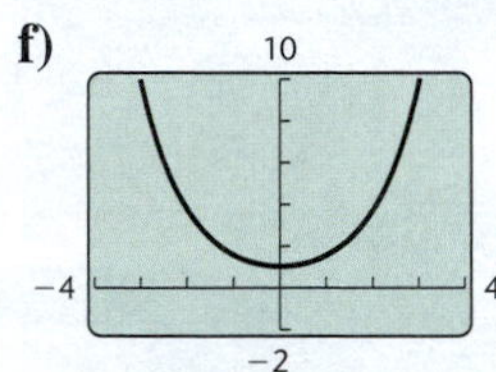

g)

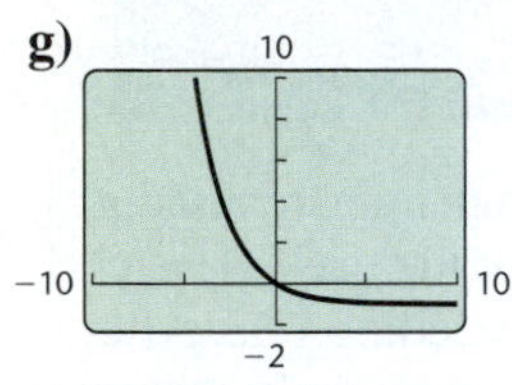

h)

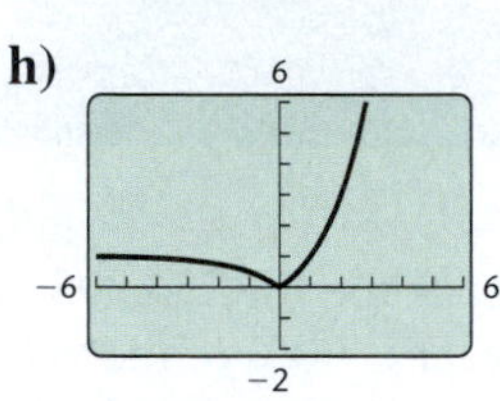

i)

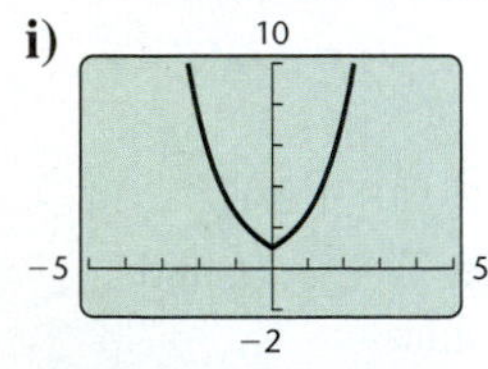

j)

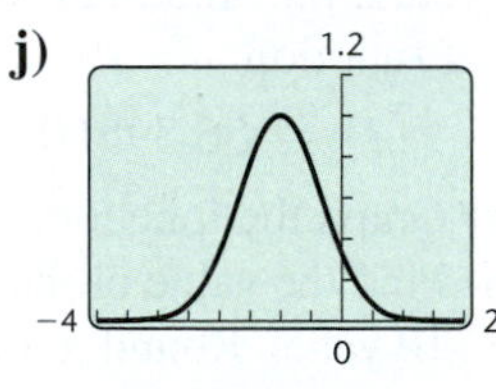

k)

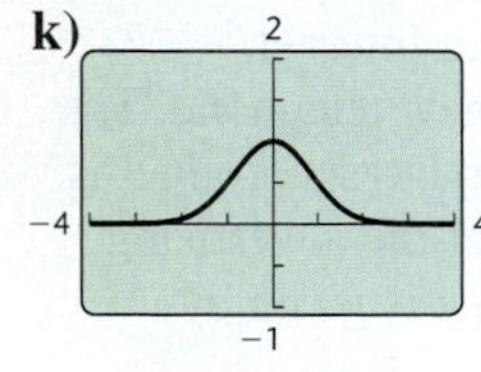

l)

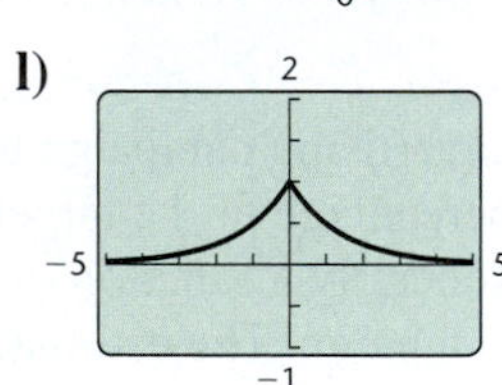

m)

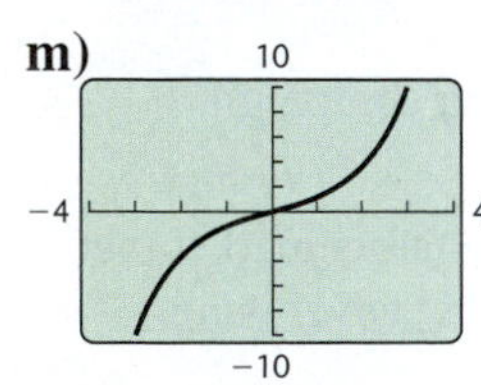

n)

10 −10 10 −2

77. $y = 3^x - 3^{-x}$

78. $y = 3^{-(x+1)^2}$

79. $f(x) = -2.3^x$

80. $f(x) = 30{,}000(1.4)^x$

81. $y = 2^{-|x|}$

82. $y = 2^{-(x-1)}$

83. $f(x) = (0.58)^x - 1$

84. $y = 2^x + 2^{-x}$

85. $g(x) = e^{|x|}$

86. $f(x) = |2^x - 1|$

87. $y = 2^{-x^2}$

88. $y = |2^{x^2} - 8|$

89. $g(x) = \dfrac{e^x - e^{-x}}{2}$

90. $f(x) = \dfrac{e^x + e^{-x}}{2}$

Use a graphing calculator to find the point(s) of intersection of the graphs of each of the following pairs of equations.

91. $y = |1 - 3^x|$, $y = 4 + 3^{-x^2}$

92. $y = 4^x + 4^{-x}$, $y = 8 - 2x - x^2$

93. $y = 2e^x - 3$, $y = \dfrac{e^x}{x}$

94. $y = \dfrac{1}{e^x + 1}$, $y = 0.3x + \dfrac{7}{9}$

Solve graphically.

95. $5.3^x - 4.2^x = 1073$

96. $e^x = x^3$

97. $2^x > 1$

98. $3^x \leq 1$

99. $2^x + 3^x = x^2 + x^3$

100. $31{,}245e^{-3x} = 523{,}467$

› Skill Maintenance

Simplify. [3.1]

101. $(1 - 4i)(7 + 6i)$

102. $\dfrac{2 - i}{3 + i}$

Find the x-intercepts and the zeros of the function.

103. $f(x) = 2x^2 - 13x - 7$ [3.2]

104. $h(x) = x^3 - 3x^2 + 3x - 1$ [4.3]

105. $h(x) = x^4 - x^2$ [4.1]

106. $g(x) = x^3 + x^2 - 12x$ [4.1]

Solve.

107. $x^3 + 6x^2 - 16x = 0$ [4.1]

108. $3x^2 - 6 = 5x$ [3.2]

› Synthesis

109. Which is larger, 7^π or π^7? 70^{80} or 80^{70}?

In Exercises 110 and 111:

a) *Graph using a graphing calculator.*
b) *Approximate the zeros.*
c) *Approximate the relative maximum and minimum values. If your graphing calculator has a* MAX–MIN *feature, use it.*

110. $f(x) = x^2e^{-x}$

111. $f(x) = e^{-x^2}$

112. Graph $f(x) = x^{1/(x-1)}$ for $x > 0$. Use a graphing calculator and the TABLE feature to identify the horizontal asymptote.

113. For the function f, construct and simplify the difference quotient. (See Section 2.2 for review.)

$$f(x) = 2e^x - 3$$

5.3 Logarithmic Functions and Graphs

- Find common logarithms and natural logarithms with and without a calculator.
- Convert between exponential equations and logarithmic equations.
- Change logarithm bases.
- Graph logarithmic functions.
- Solve applied problems involving logarithmic functions.

We now consider *logarithmic*, or *logarithm, functions*. These functions are inverses of exponential functions and have many applications.

Logarithmic Functions

We have noted that every exponential function (with $a > 0$ and $a \neq 1$) is one-to-one. Thus such a function has an inverse that is a function. In this section, we will study these inverse functions, called logarithmic functions, and use them in applications. We can draw the graph of the inverse of an exponential function by interchanging x and y.

EXAMPLE 1 Graph: $x = 2^y$.

Solution Note that x is alone on one side of the equation. We can find ordered pairs that are solutions by choosing values for y and then computing the corresponding x-values.

For $y = 0$, $x = 2^0 = 1$.
For $y = 1$, $x = 2^1 = 2$.
For $y = 2$, $x = 2^2 = 4$.
For $y = 3$, $x = 2^3 = 8$.
For $y = -1$, $x = 2^{-1} = \dfrac{1}{2^1} = \dfrac{1}{2}$.
For $y = -2$, $x = 2^{-2} = \dfrac{1}{2^2} = \dfrac{1}{4}$.
For $y = -3$, $x = 2^{-3} = \dfrac{1}{2^3} = \dfrac{1}{8}$.

x $x = 2^y$	y	(x, y)
1	0	$(1, 0)$
2	1	$(2, 1)$
4	2	$(4, 2)$
8	3	$(8, 3)$
$\frac{1}{2}$	-1	$\left(\frac{1}{2}, -1\right)$
$\frac{1}{4}$	-2	$\left(\frac{1}{4}, -2\right)$
$\frac{1}{8}$	-3	$\left(\frac{1}{8}, -3\right)$

(1) Choose values for y.
(2) Compute values for x.

We plot the points and connect them with a smooth curve. Note that the curve does not touch or cross the y-axis. The y-axis is a vertical asymptote.

Note too that this curve is the graph of $y = 2^x$ reflected across the line $y = x$, as we would expect for an inverse. The inverse of $y = 2^x$ is $x = 2^y$.

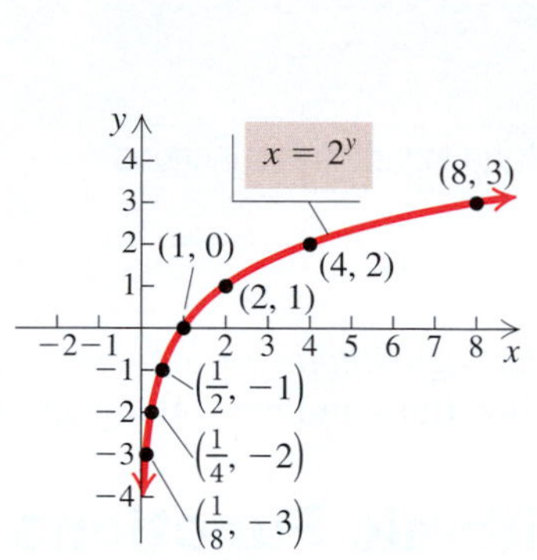

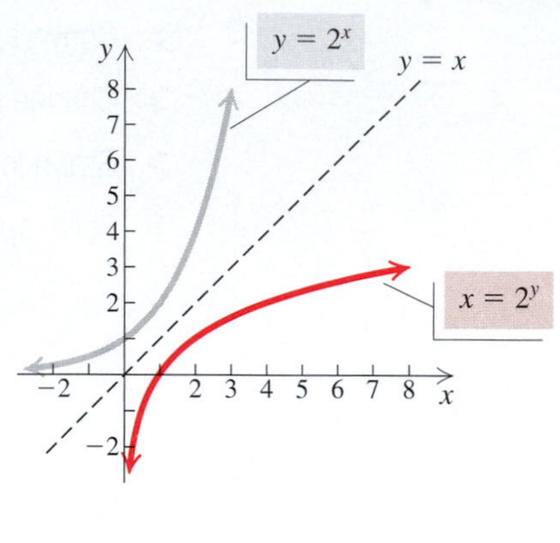

Now Try Exercise 1.

To find a formula for f^{-1} when $f(x) = 2^x$, we use the method discussed in Section 5.1:

1. Replace $f(x)$ with y: $\quad y = 2^x$
2. Interchange x and y: $\quad x = 2^y$
3. Solve for y: $\quad y =$ the power to which we raise 2 to get x
4. Replace y with $f^{-1}(x)$: $\quad f^{-1}(x) =$ the power to which we raise 2 to get x.

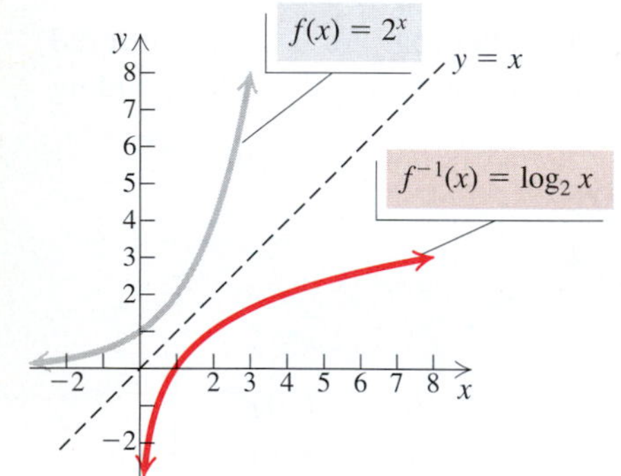

Mathematicians have defined a new symbol to replace the words "the power to which we raise 2 to get x." That symbol is "$\log_2 x$," read "the logarithm, base 2, of x."

LOGARITHMIC FUNCTION, BASE 2

"$\log_2 x$," read "the logarithm, base 2, of x," means "the power to which we raise 2 to get x."

Thus if $f(x) = 2^x$, then $f^{-1}(x) = \log_2 x$. For example,

$$f^{-1}(8) = \log_2 8 = 3,$$

because

3 is the power to which we raise 2 to get 8.

Similarly, $\log_2 13$ is the power to which we raise 2 to get 13. As yet, we have no simpler way to say this other than

"$\log_2 13$ is the power to which we raise 2 to get 13."

Later, however, we will learn how to approximate this expression using a calculator.

For any exponential function $f(x) = a^x$, its inverse is called a **logarithmic function, base *a***. The graph of the inverse can be obtained by reflecting the graph of $y = a^x$ across the line $y = x$, to obtain $x = a^y$. Then $x = a^y$ is equivalent to $y = \log_a x$. We read $\log_a x$ as "the logarithm, base a, of x."

The inverse of $f(x) = a^x$ is given by $f^{-1}(x) = \log_a x$.

LOGARITHMIC FUNCTION, BASE a

We define $y = \log_a x$ as that number y such that $x = a^y$, where $x > 0$ and a is a positive constant other than 1.

Let's look at the graphs of $f(x) = a^x$ and $f^{-1}(x) = \log_a x$ for $a > 1$ and for $0 < a < 1$.

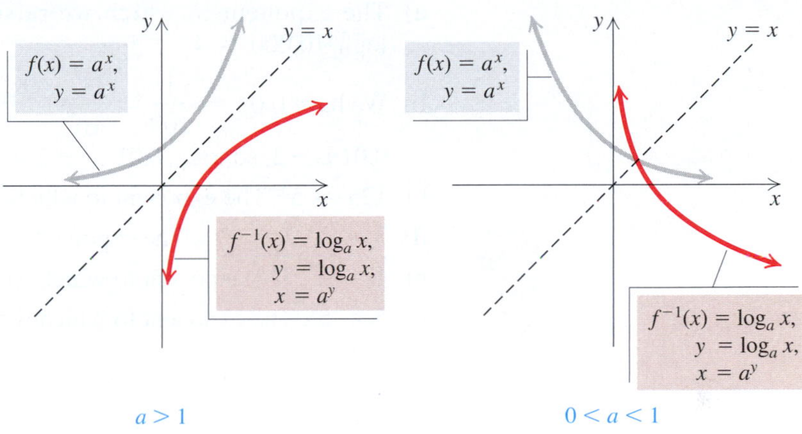

$a > 1$ $\qquad$ $0 < a < 1$

Note that the graphs of $f(x)$ and $f^{-1}(x)$ are reflections of each other across the line $y = x$.

CONNECTING THE CONCEPTS

Comparing Exponential Functions and Logarithmic Functions

In the following table, we compare exponential functions and logarithmic functions with bases a greater than 1. Similar statements could be made for a, where $0 < a < 1$. It is helpful to visualize the differences by carefully observing the graphs.

EXPONENTIAL FUNCTION

$y = a^x$
$f(x) = a^x$
$a > 1$
Continuous
One-to-one
Domain: All real numbers, $(-\infty, \infty)$
Range: All positive real numbers, $(0, \infty)$
Increasing
Horizontal asymptote is x-axis: $(a^x \to 0$ as $x \to -\infty)$
y-intercept: $(0, 1)$
There is no x-intercept.

LOGARITHMIC FUNCTION

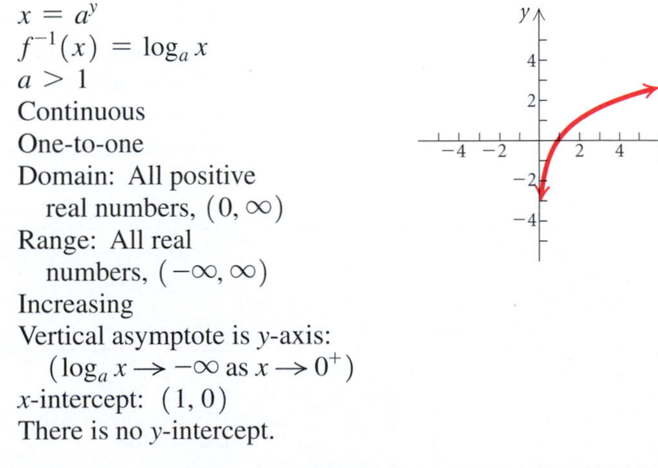

$x = a^y$
$f^{-1}(x) = \log_a x$
$a > 1$
Continuous
One-to-one
Domain: All positive real numbers, $(0, \infty)$
Range: All real numbers, $(-\infty, \infty)$
Increasing
Vertical asymptote is y-axis: $(\log_a x \to -\infty$ as $x \to 0^+)$
x-intercept: $(1, 0)$
There is no y-intercept.

Finding Certain Logarithms

Let's use the definition of logarithms to find some logarithmic values.

EXAMPLE 2 Find each of the following logarithms.

a) $\log_{10} 10{,}000$ **b)** $\log_{10} 0.01$ **c)** $\log_5 125$

d) $\log_9 3$ **e)** $\log_6 1$ **f)** $\log_5 5$

Solution

a) The exponent to which we raise 10 to obtain 10,000 is 4; thus, we have $\log_{10} 10{,}000 = 4$.

b) We have $0.01 = \frac{1}{100} = \frac{1}{10^2} = 10^{-2}$. The exponent to which we raise 10 to get 0.01 is -2, so $\log_{10} 0.01 = -2$.

c) $125 = 5^3$. The exponent to which we raise 5 to get 125 is 3, so $\log_5 125 = 3$.

Just in Time 27

d) $3 = \sqrt{9} = 9^{1/2}$. The exponent to which we raise 9 to get 3 is $\frac{1}{2}$, so $\log_9 3 = \frac{1}{2}$.

e) $1 = 6^0$. The exponent to which we raise 6 to get 1 is 0, so $\log_6 1 = 0$.

f) $5 = 5^1$. The exponent to which we raise 5 to get 5 is 1, so $\log_5 5 = 1$.

Now Try Exercises 9 and 15.

Examples 2(e) and 2(f) illustrate two important properties of logarithms. The property $\log_a 1 = 0$ follows from the fact that $a^0 = 1$. Thus, $\log_5 1 = 0$, $\log_{10} 1 = 0$, and so on. The property $\log_a a = 1$ follows from the fact that $a^1 = a$. Thus, $\log_5 5 = 1$, $\log_{10} 10 = 1$, and so on.

> $\log_a 1 = 0$ and $\log_a a = 1$, for any logarithmic base a.

Converting Between Exponential Equations and Logarithmic Equations

It is helpful in dealing with logarithmic functions to remember that a logarithm of a number is an *exponent*. It is the exponent y in $x = a^y$. You might think to yourself, "the logarithm, base a, of a number x is the power to which a must be raised to get x."

We are led to the following. (The symbol $\longleftrightarrow$ means that the two statements are equivalent; that is, when one is true, the other is true. The words "if and only if" can be used in place of $\longleftrightarrow$.)

> $\log_a x = y \longleftrightarrow x = a^y$ A logarithm is an exponent!

EXAMPLE 3 Convert each of the following to a logarithmic equation.

a) $16 = 2^x$ **b)** $10^{-3} = 0.001$ **c)** $e^t = 70$

Solution

a) $16 = 2^x$ $\log_2 16 = x$

The exponent is the logarithm.

The base remains the same.

b) $10^{-3} = 0.001 \rightarrow \log_{10} 0.001 = -3$

c) $e^t = 70 \rightarrow \log_e 70 = t$

Now Try Exercise 37.

EXAMPLE 4 Convert each of the following to an exponential equation.

a) $\log_2 32 = 5$ b) $\log_a Q = 8$ c) $x = \log_t M$

Solution

The logarithm is the exponent.

a) $\log_2 32 = 5 \quad 2^5 = 32$

The base remains the same.

b) $\log_a Q = 8 \rightarrow a^8 = Q$

c) $x = \log_t M \rightarrow t^x = M$

Now Try Exercise 45.

Finding Logarithms on a Calculator

Before calculators became so widely available, base-10 logarithms, or **common logarithms**, were used extensively to simplify complicated calculations. In fact, that is why logarithms were invented. The abbreviation **log**, with no base written, is used to represent common logarithms, or base-10 logarithms. Thus,

$\log x$ means $\log_{10} x$.

For example, log 29 means $\log_{10} 29$. Let's compare log 29 with log 10 and log 100:

$$\left.\begin{aligned} \log 10 = \log_{10} 10 &= 1 \\ \log 29 &= ? \\ \log 100 = \log_{10} 100 &= 2 \end{aligned}\right\}$$

Since 29 is between 10 and 100, it seems reasonable that log 29 is between 1 and 2.

On a calculator, the key for common logarithms is generally marked LOG. Using that key, we find that

$$\log 29 \approx 1.462397998 \approx 1.4624$$

rounded to four decimal places. Since $1 < 1.4624 < 2$, our answer seems reasonable. This also tells us that $10^{1.4624} \approx 29$.

EXAMPLE 5 Find each of the following common logarithms on a calculator. If you are using a graphing calculator, set the calculator in REAL mode. Round to four decimal places.

a) log 645,778 b) log 0.0000239 c) log (−3)

Solution

FUNCTION VALUE	READOUT	ROUNDED
a) log 645,778	log(645778) 5.810083246	5.8101
b) log 0.0000239	log(0.0000239) −4.621602099	−4.6216
c) log (−3)	ERR:NONREAL ANS 1:Quit 2:Goto *	Does not exist as a real number

*If the graphing calculator is set in $a + bi$ mode, the readout is .4771212547 + 1.364376354i.

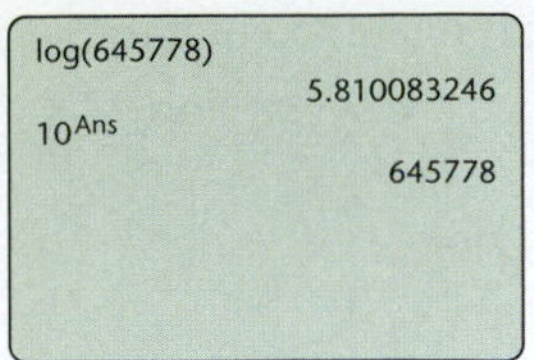

A check for part (a) is shown at left. Since 5.810083246 is the power to which we raise 10 to get 645,778, we can check part (a) by finding $10^{5.810083246}$. We can check part (b) in a similar manner. In part (c), $\log(-3)$ does not exist as a real number because there is no real-number power to which we can raise 10 to get -3. The number 10 raised to any real-number power is positive. The common logarithm of a negative number does not exist as a real number. Recall that logarithmic functions are inverses of exponential functions, and since the range of an exponential function is $(0, \infty)$, the domain of $f(x) = \log_a x$ is $(0, \infty)$.

Now Try Exercises 57 and 61.

Natural Logarithms

Logarithms, base e, are called **natural logarithms**. The abbreviation "ln" is generally used for natural logarithms. Thus,

$$\ln x \quad \text{means} \quad \log_e x.$$

For example, ln 53 means $\log_e 53$. On a calculator, the key for natural logarithms is generally marked LN. Using that key, we find that

$$\ln 53 \approx 3.970291914$$
$$\approx 3.9703$$

rounded to four decimal places. This also tells us that $e^{3.9703} \approx 53$.

EXAMPLE 6 Find each of the following natural logarithms on a calculator. If you are using a graphing calculator, set the calculator in REAL mode. Round to four decimal places.

a) ln 645,778 **b)** ln 0.0000239 **c)** $\ln(-5)$
d) $\ln e$ **e)** ln 1

Solution

	FUNCTION VALUE	READOUT	ROUNDED
a)	ln 645,778	ln(645778) 13.37821107	13.3782
b)	ln 0.0000239	ln(0.0000239) −10.6416321	−10.6416
c)	$\ln(-5)$	ERR:NONREAL ANS 1:Quit 2:Goto *	Does not exist as a real number
d)	$\ln e$	ln(e) 1	1
e)	ln 1	ln(1) 0	0

*If the graphing calculator is set in $a + bi$ mode, the readout is $1.609437912 + 3.141592654i$.

ln(0.0000239)
−10.6416321
e^{Ans}
2.39E−5

Since 13.37821107 is the power to which we raise e to get 645,778, we can check part (a) by finding $e^{13.37821107}$. We can check parts (b), (d), and (e) in a similar manner. A check for part (b) is shown at left. In parts (d) and (e), note that $\ln e = \log_e e = 1$ and $\ln 1 = \log_e 1 = 0$.

Now Try Exercises 65 and 67.

$\ln 1 = 0$ and $\ln e = 1$, for the logarithmic base e.

With some calculators, it is possible to find a logarithm with any logarithmic base using the logBASE operation from the MATH MATH menu. The computation of $\log_4 15$ is shown in the window below.

$\log_4(15)$
1.953445298

Changing Logarithmic Bases

Most calculators give the values of both common logarithms and natural logarithms. To find a logarithm with a base other than 10 or e, we can use the following conversion formula.

THE CHANGE-OF-BASE FORMULA

For any logarithmic bases a and b, and any positive number M,

$$\log_b M = \frac{\log_a M}{\log_a b}.$$

We will prove this result in the next section.

EXAMPLE 7 Find $\log_5 8$ using common logarithms.

Solution First, we let $a = 10$, $b = 5$, and $M = 8$. Then we substitute into the change-of-base formula:

$$\log_5 8 = \frac{\log_{10} 8}{\log_{10} 5} \quad \text{Substituting}$$

$$\approx 1.2920. \quad \text{Using a calculator}$$

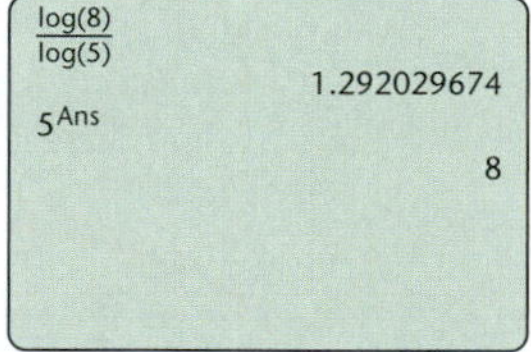

Since $\log_5 8$ is the power to which we raise 5 to get 8, we would expect this power to be greater than 1 $(5^1 = 5)$ and less than 2 $(5^2 = 25)$, so the result is reasonable. The check is shown in the window at left.

Now Try Exercise 69.

We can also use base e for a conversion.

EXAMPLE 8 Find $\log_5 8$ using natural logarithms.

Solution Substituting e for a, 5 for b, and 8 for M, we have

$$\log_5 8 = \frac{\log_e 8}{\log_e 5}$$

$$= \frac{\ln 8}{\ln 5} \approx 1.2920.$$

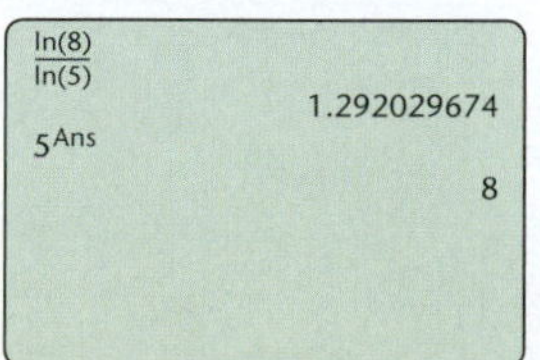

The check is shown at left. Note that we get the same value using base e for the conversion that we did using base 10 in Example 7.

Now Try Exercise 75.

Graphs of Logarithmic Functions

We demonstrate several ways to graph logarithmic functions.

EXAMPLE 9 Graph: $y = f(x) = \log_5 x$.

Solution The equation $y = \log_5 x$ is equivalent to $x = 5^y$. We can find ordered pairs that are solutions by choosing values for y and computing the corresponding x-values. We then plot points, remembering that x is still the first coordinate.

For $y = 0$, $x = 5^0 = 1$.

For $y = 1$, $x = 5^1 = 5$.

For $y = 2$, $x = 5^2 = 25$.

For $y = 3$, $x = 5^3 = 125$.

For $y = -1$, $x = 5^{-1} = \dfrac{1}{5}$.

For $y = -2$, $x = 5^{-2} = \dfrac{1}{25}$.

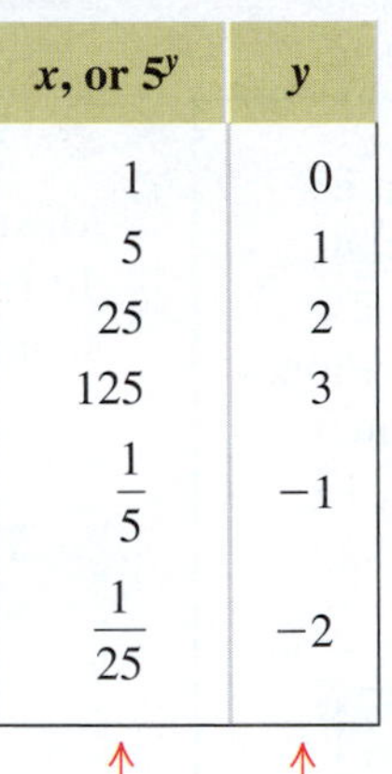

x, or 5^y	y
1	0
5	1
25	2
125	3
$\frac{1}{5}$	−1
$\frac{1}{25}$	−2

(1) Select y.

(2) Compute x.

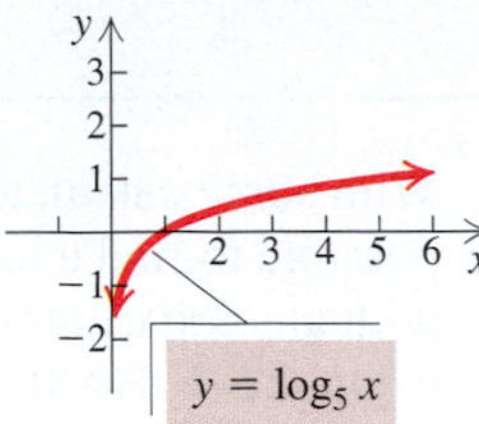

Two other methods for graphing this function are shown at left.

Now Try Exercise 5.

To use a graphing calculator to graph $y = \log_5 x$ in Example 9, we first change the base or use the logBASE operation. Here we change from base 5 to base e:

$$y = \log_5 x = \frac{\ln x}{\ln 5}.$$

$y = \log_5 x = \dfrac{\ln x}{\ln 5} = \dfrac{\ln (x)}{\ln (5)}$

Some graphing calculators have a feature that graphs inverses automatically. If we begin with $y_1 = 5^x$, the graphs of both y_1 and its inverse $y_2 = \log_5 x$ will be drawn.

$y_1 = 5^x$, $y_2 = \log_5 x$

EXAMPLE 10 Graph: $g(x) = \ln x$.

Solution To graph $y = g(x) = \ln x$, we select values for x and use a calculator to find the corresponding values of $\ln x$. We then plot points and draw the curve. We can graph the function $y = \ln x$ directly using a graphing calculator.

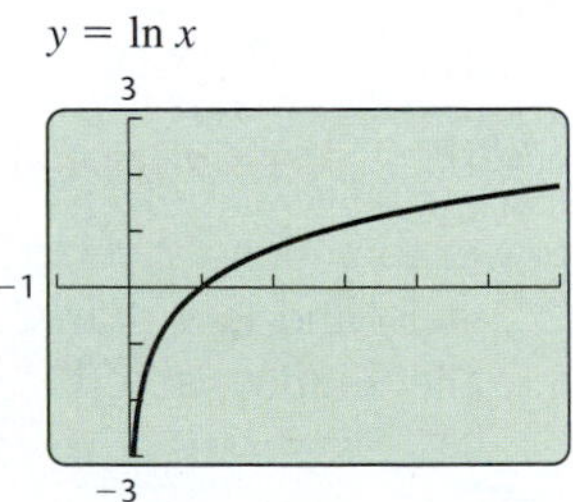

x	$g(x)$
0.5	−0.7
1	0
2	0.7
3	1.1
4	1.4
5	1.6

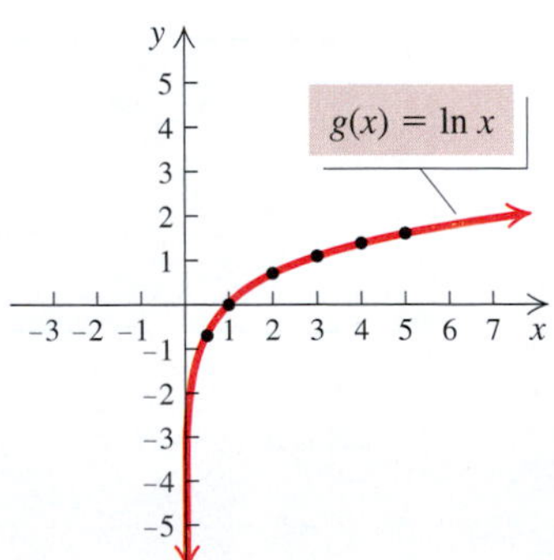

We could also write $g(x) = \ln x$, or $y = \ln x$, as $x = e^y$, select values for y, and use a calculator to find the corresponding values of x.

Now Try Exercise 7.

Recall that the graph of $f(x) = \log_a x$, for any base a, has the x-intercept $(1, 0)$. The domain is the set of positive real numbers, and the range is the set of all real numbers. The y-axis is the vertical asymptote.

EXAMPLE 11 Graph each of the following by hand. Before doing so, describe how each graph can be obtained from the graph of $y = \ln x$. Give the domain and the vertical asymptote of each function. Then check your graph with a graphing calculator.

a) $f(x) = \ln(x + 3)$

b) $f(x) = 3 - \frac{1}{2}\ln x$

c) $f(x) = |\ln(x - 1)|$

Solution

a) The graph of $f(x) = \ln(x + 3)$ is a shift of the graph of $y = \ln x$ left 3 units. The domain is the set of all real numbers greater than -3, $(-3, \infty)$. The line $x = -3$ is the vertical asymptote.

$y_1 = \ln x$, $y_2 = \ln(x + 3)$

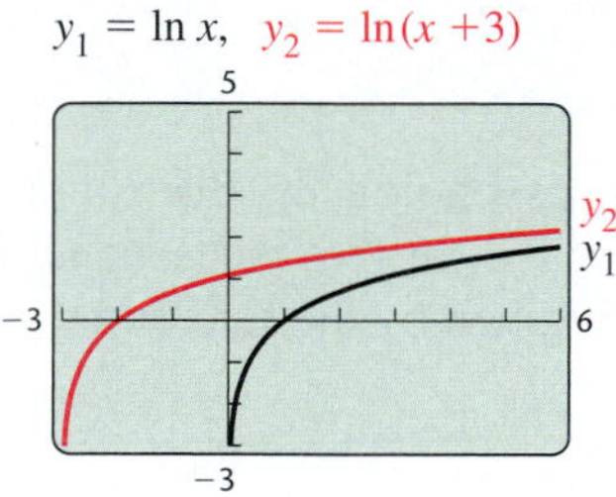

x	$f(x)$
−2.9	−2.303
−2	0
0	1.099
2	1.609
4	1.946

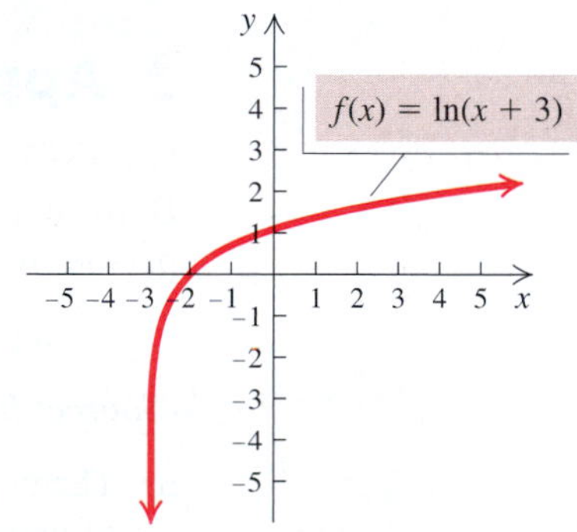

b) The graph of $f(x) = 3 - \frac{1}{2}\ln x$ is a vertical shrinking of the graph of $y = \ln x$, followed by a reflection across the x-axis, and then a translation up 3 units. The domain is the set of all positive real numbers, $(0, \infty)$. The y-axis is the vertical asymptote.

$y_1 = \ln x$, $y_2 = 3 - \frac{1}{2}\ln x$

x	$f(x)$
0.1	4.151
1	3
3	2.451
6	2.104
9	1.901

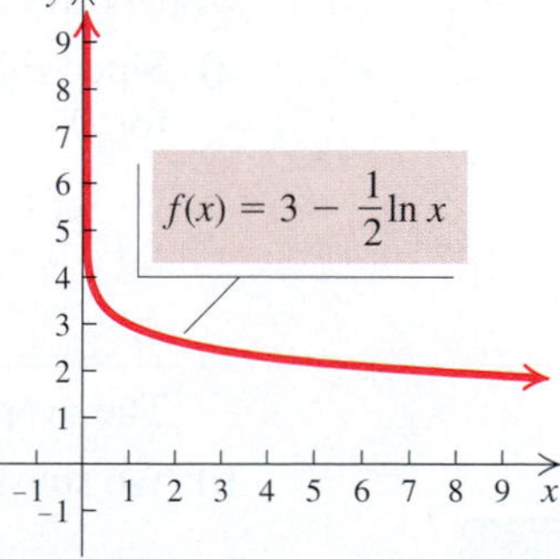

c) The graph of $f(x) = |\ln(x - 1)|$ is a translation of the graph of $y = \ln x$ right 1 unit. Then the absolute value has the effect of reflecting negative outputs across the x-axis. The domain is the set of all real numbers greater than 1, $(1, \infty)$. The line $x = 1$ is the vertical asymptote.

x	$f(x)$
1.1	2.303
2	0
4	1.099
6	1.609
8	1.946

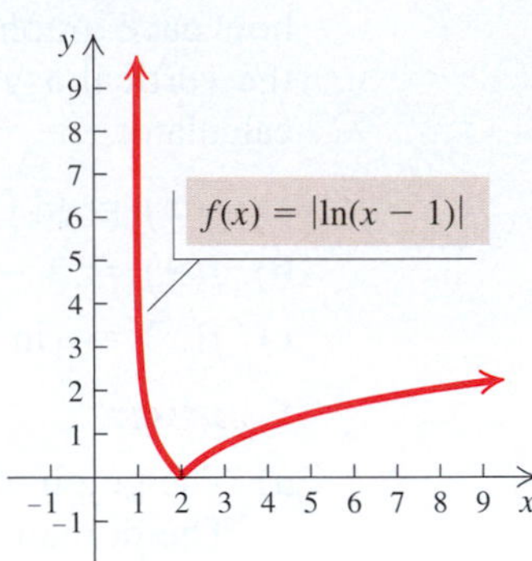

Now Try Exercise 89.

› Applications

EXAMPLE 12 ***Walking Speed.*** In a study by psychologists Bornstein and Bornstein, it was found that the average walking speed w, in feet per second, of a person living in a city of population P, in thousands, is given by the function

$$w(P) = 0.37 \ln P + 0.05$$

(*Source*: *International Journal of Psychology*).

a) The population of Billings, Montana, is 106,954. Find the average walking speed of people living in Billings.

b) The population of Chicago, Illinois, is 2,714,856. Find the average walking speed of people living in Chicago.

c) Graph the function.

d) A sociologist computes the average walking speed in a city to be approximately 2.0 ft/sec. Use this information to estimate the population of the city.

Solution

a) Since P is in thousands and $106{,}954 = 106.954$ thousand, we substitute 106.954 for P:

$$w(106.954) = 0.37 \ln 106.954 + 0.05 \quad \text{Substituting}$$
$$\approx 1.8. \quad \text{Finding the natural logarithm and simplifying}$$

The average walking speed of people living in Billings is about 1.8 ft/sec.

b) We substitute 2714.856 for P:

$$w(2714.856) = 0.37 \ln 2714.856 + 0.05 \quad \text{Substituting}$$
$$\approx 3.0.$$

The average walking speed of people living in Chicago is about 3.0 ft/sec.

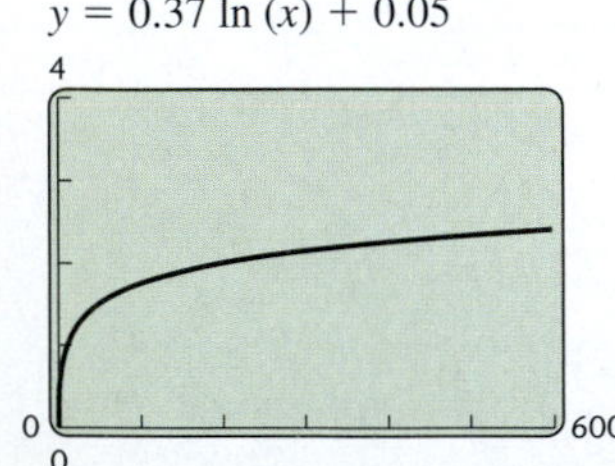

c) We graph with a viewing window of $[0, 600, 0, 4]$ because inputs are very large and outputs are very small by comparison.

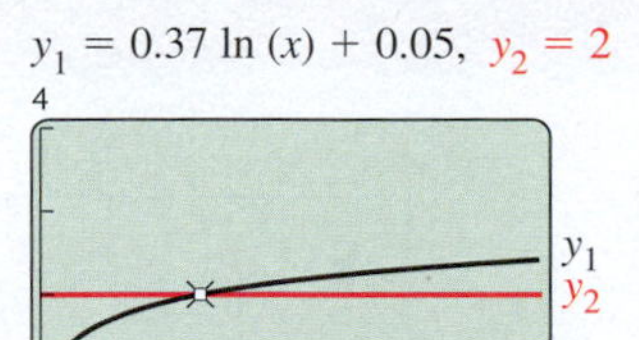

d) To find the population for which the average walking speed is 2.0 ft/sec, we substitute 2.0 for $w(P)$,

$$2.0 = 0.37 \ln P + 0.05,$$

and solve for P.

We will use the Intersect method. We graph the equations $y_1 = 0.37 \ln x + 0.05$ and $y_2 = 2$ and use the INTERSECT feature to approximate the point of intersection.

We see that in a city in which the average walking speed is 2.0 ft/sec, the population is about 194.5 thousand, or 194,500.

Now Try Exercise 95(d).

EXAMPLE 13 ***Earthquake Magnitude.*** Measured on the Richter scale, the magnitude R of an earthquake of intensity I is defined as

$$R = \log \frac{I}{I_0},$$

where I_0 is a minimum intensity used for comparison. We can think of I_0 as a threshold intensity that is the weakest earthquake that can be recorded on a seismograph. If one earthquake is 10 times as intense as another, its magnitude on the Richter scale is 1 greater than that of the other. If one earthquake is 100 times as intense as another, its magnitude on the Richter scale is 2 higher, and so on. Thus an earthquake whose magnitude is 5 on the Richter scale is 10 times as intense as an earthquake whose magnitude is 4. Earthquake intensities can be interpreted as multiples of the minimum intensity I_0.

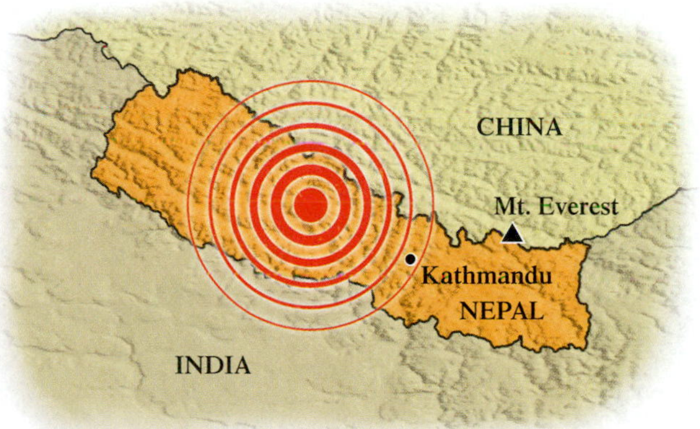

The Nepal region earthquake in South Asia on April 25, 2015, had an intensity of $10^{7.8} \cdot I_0$ (*Source*: earthquake.usgs.gov). It caused extensive loss of life and severe structural damage to buildings, railways, and roads. What was the magnitude of this earthquake on the Richter scale?

Solution We substitute into the formula:

$$R = \log \frac{I}{I_0} = \log \frac{10^{7.8} \cdot I_0}{I_0} = \log 10^{7.8} = 7.8.$$

The magnitude of the earthquake was 7.8 on the Richter scale.

Now Try Exercise 97(a).

Visualizing the Graph

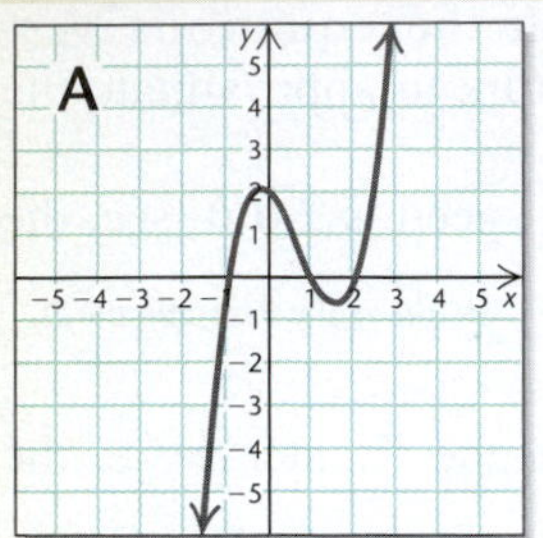

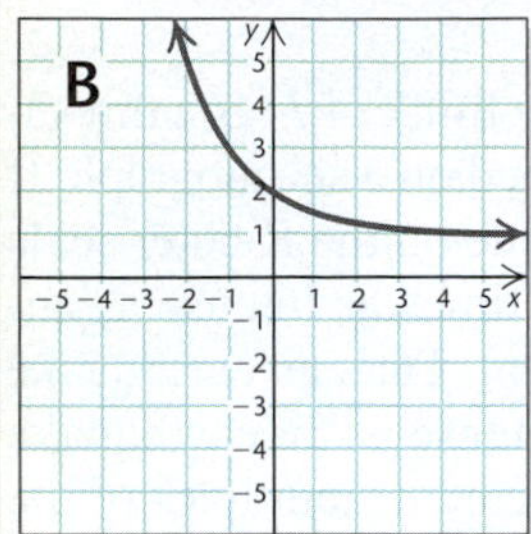

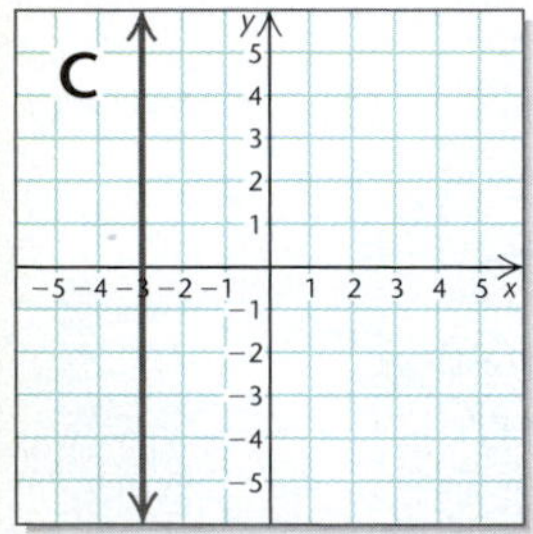

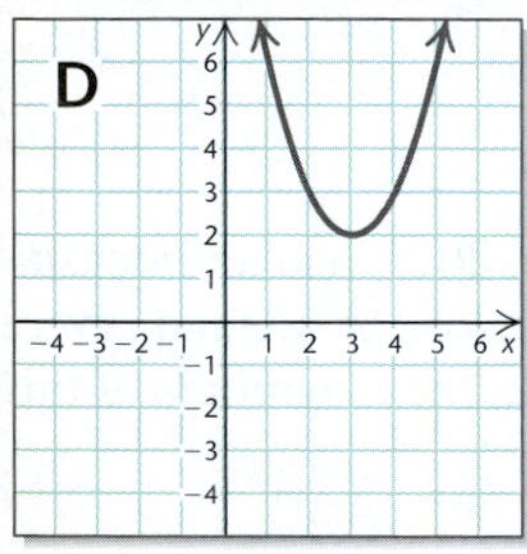

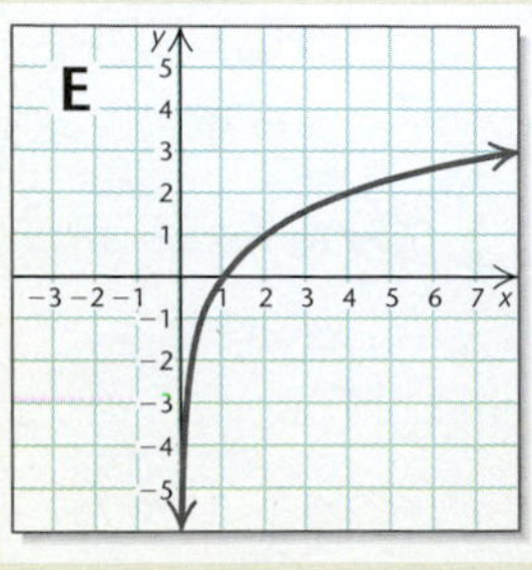

Match the equation or function with its graph.

1. $f(x) = 4^x$
2. $f(x) = \ln x - 3$
3. $(x + 3)^2 + y^2 = 9$
4. $f(x) = 2^{-x} + 1$
5. $f(x) = \log_2 x$
6. $f(x) = x^3 - 2x^2 - x + 2$
7. $x = -3$
8. $f(x) = e^x - 4$
9. $f(x) = (x - 3)^2 + 2$
10. $3x = 6 + y$

Answers on page A-28

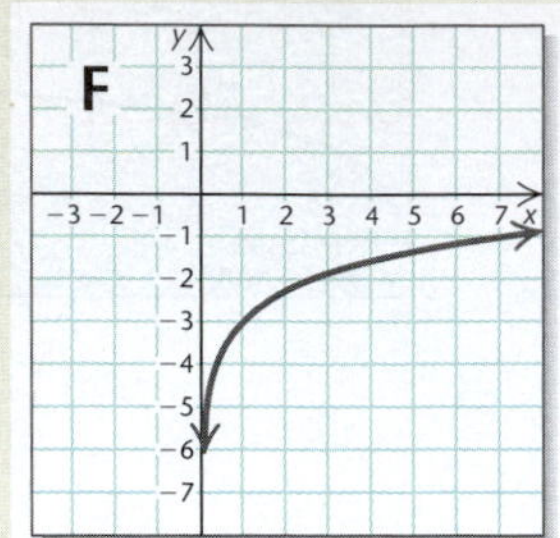

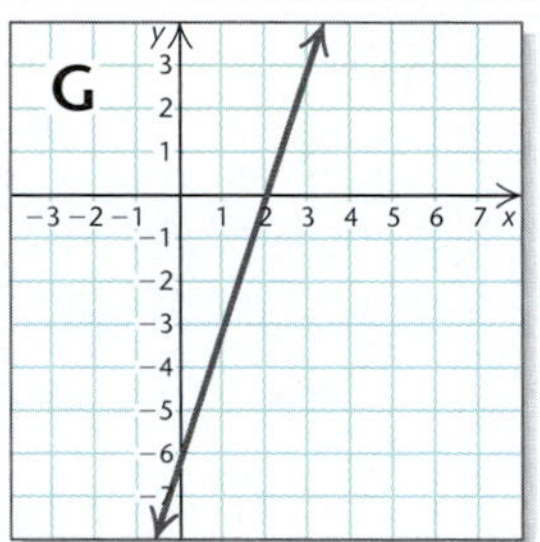

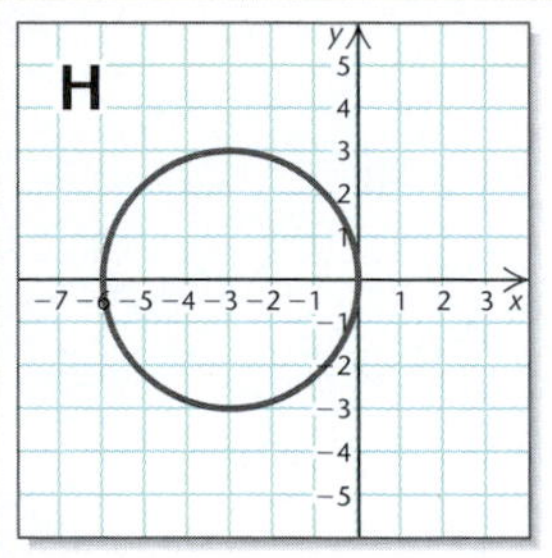

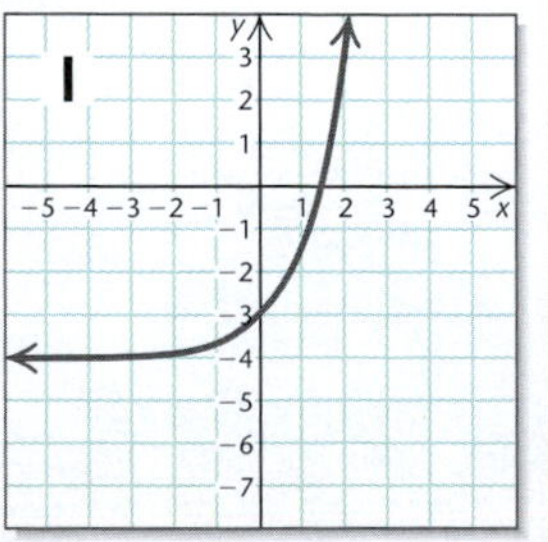

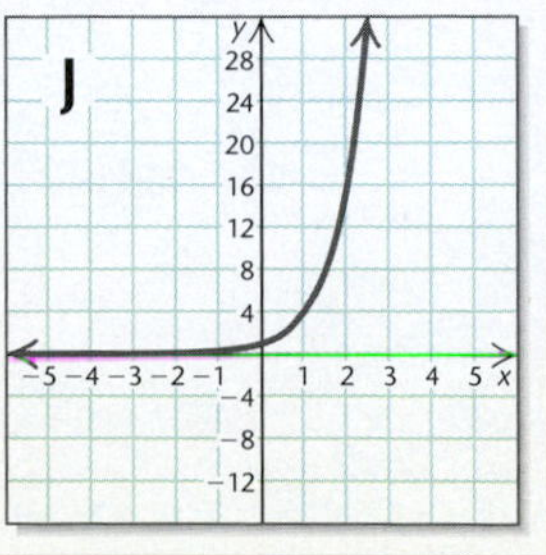

5.3 Exercise Set

Make a hand-drawn graph of each of the following. Then check your work using a graphing calculator.

1. $x = 3^y$
2. $x = 4^y$
3. $x = \left(\frac{1}{2}\right)^y$
4. $x = \left(\frac{4}{3}\right)^y$
5. $y = \log_3 x$
6. $y = \log_4 x$
7. $f(x) = \log x$
8. $f(x) = \ln x$

Find each of the following. Do not use a calculator.

9. $\log_2 16$
10. $\log_3 9$
11. $\log_5 125$
12. $\log_2 64$
13. $\log 0.001$
14. $\log 100$
15. $\log_2 \frac{1}{4}$
16. $\log_8 2$
17. $\ln 1$
18. $\ln e$
19. $\log 10$
20. $\log 1$
21. $\log_5 5^4$
22. $\log \sqrt{10}$
23. $\log_3 \sqrt[4]{3}$
24. $\log 10^{8/5}$
25. $\log 10^{-7}$
26. $\log_5 1$
27. $\log_{49} 7$
28. $\log_3 3^{-2}$
29. $\ln e^{3/4}$
30. $\log_2 \sqrt{2}$
31. $\log_4 1$
32. $\ln e^{-5}$
33. $\ln \sqrt{e}$
34. $\log_{64} 4$

Convert to a logarithmic equation.

35. $10^3 = 1000$
36. $5^{-3} = \frac{1}{125}$
37. $8^{1/3} = 2$
38. $10^{0.3010} = 2$
39. $e^3 = t$
40. $Q^t = x$
41. $e^2 = 7.3891$
42. $e^{-1} = 0.3679$
43. $p^k = 3$
44. $e^{-t} = 4000$

Convert to an exponential equation.

45. $\log_5 5 = 1$
46. $t = \log_4 7$
47. $\log 0.01 = -2$
48. $\log 7 = 0.845$
49. $\ln 30 = 3.4012$
50. $\ln 0.38 = -0.9676$
51. $\log_a M = -x$
52. $\log_t Q = k$
53. $\log_a T^3 = x$
54. $\ln W^5 = t$

Find each of the following using a calculator. Round to four decimal places.

55. $\log 3$
56. $\log 8$
57. $\log 532$
58. $\log 93{,}100$
59. $\log 0.57$
60. $\log 0.082$
61. $\log(-2)$
62. $\ln 50$
63. $\ln 2$
64. $\ln(-4)$
65. $\ln 809.3$
66. $\ln 0.00037$
67. $\ln(-1.32)$
68. $\ln 0$

Find the logarithm using common logarithms and the change-of-base formula.

69. $\log_4 100$
70. $\log_3 20$
71. $\log_{100} 0.3$
72. $\log_\pi 100$
73. $\log_{200} 50$
74. $\log_{5.3} 1700$

Find the logarithm using natural logarithms and the change-of-base formula.

75. $\log_3 12$
76. $\log_4 25$
77. $\log_{100} 15$
78. $\log_9 100$

Graph the function and its inverse using the same set of axes. Use any method.

79. $f(x) = 3^x,\ f^{-1}(x) = \log_3 x$
80. $f(x) = \log_4 x,\ f^{-1}(x) = 4^x$
81. $f(x) = \log x,\ f^{-1}(x) = 10^x$
82. $f(x) = e^x,\ f^{-1}(x) = \ln x$

For each of the following functions, briefly describe how the graph can be obtained from the graph of a basic logarithmic function. Then graph the function using a graphing calculator. Give the domain and the vertical asymptote of each function.

83. $f(x) = \log_2(x + 3)$
84. $f(x) = \log_3(x - 2)$
85. $y = \log_3 x - 1$
86. $y = 3 + \log_2 x$
87. $f(x) = 4 \ln x$
88. $f(x) = \frac{1}{2} \ln x$
89. $y = 2 - \ln x$
90. $y = \ln(x + 1)$
91. $f(x) = \frac{1}{2} \log(x - 1) - 2$
92. $f(x) = 5 - 2 \log(x + 1)$

Graph the piecewise function.

93. $g(x) = \begin{cases} 5, & \text{for } x \le 0, \\ \log x + 1, & \text{for } x > 0 \end{cases}$

94. $f(x) = \begin{cases} 1 - x, & \text{for } x \le -1, \\ \ln(x + 1), & \text{for } x > -1 \end{cases}$

95. *Walking Speed.* Refer to Example 12. The data on various cities and their populations are given below. Find the average walking speed in each city. Round to the nearest tenth of a foot per second.

a) El Paso, Texas: 672,538
b) Phoenix, Arizona: 1,488,750
c) Birmingham, Alabama: 212,038
d) Milwaukee, Wisconsin: 598,916
e) Honolulu, Hawaii: 345,610
f) Charlotte, North Carolina: 775,202
g) Omaha, Nebraska: 421,570
h) Sydney, Australia: 3,908,643

96. *Forgetting.* Students in a political science class took a final exam and then took equivalent forms of the exam at monthly intervals thereafter. The average score $S(t)$, as a percent, after t months was found to be given by the function

$$S(t) = 78 - 15 \log(t + 1), \quad t \geq 0.$$

a) What was the average score when the students initially took the test, $t = 0$?
b) What was the average score after 4 months? after 24 months?
c) Graph the function.
d) After what time t was the average score 50%?

97. *Earthquake Magnitude.* Refer to Example 13. The data on various locations of earthquakes and their intensities are given below. Find the magnitude of each earthquake on the Richter scale.

a) San Francisco, California, 1906: $10^{7.7} \cdot I_0$
b) Chile, 1960: $10^{9.5} \cdot I_0$
c) Iran, 2003: $10^{6.6} \cdot I_0$
d) Turkey, 1999: $10^{7.6} \cdot I_0$
e) Chile, 2014: $10^{8.2} \cdot I_0$
f) China, 2008: $10^{7.9} \cdot I_0$
g) Spain, 2011: $10^{5.1} \cdot I_0$
h) Sumatra, 2004: $10^{9.3} \cdot I_0$

98. *pH of Substances in Chemistry.* In chemistry, the pH of a substance is defined as

$$\text{pH} = -\log[\text{H}^+],$$

where H^+ is the hydrogen ion concentration, in moles per liter. Find the pH of each substance.

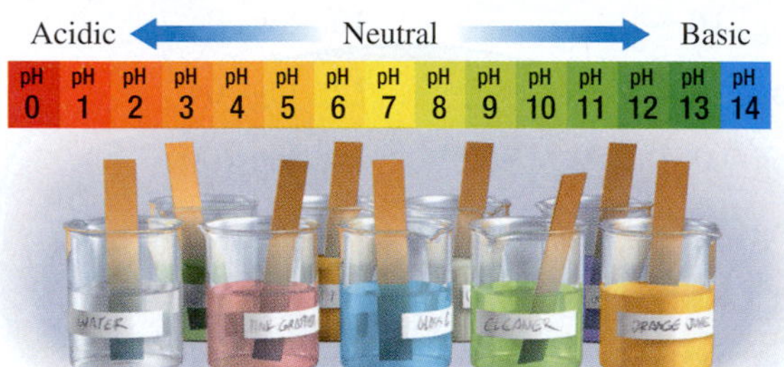

SUBSTANCE	HYDROGEN ION CONCENTRATION
a) Pineapple juice	1.6×10^{-4}
b) Hair conditioner	0.0013
c) Mouthwash	6.3×10^{-7}
d) Eggs	1.6×10^{-8}
e) Tomatoes	6.3×10^{-5}

99. Find the hydrogen ion concentration of each substance, given the pH. (See Exercise 98.) Express the answer in scientific notation.

SUBSTANCE	pH
a) Tap water	7
b) Rainwater	5.4
c) Orange juice	3.2
d) Wine	4.8

100. *Advertising.* A model for advertising response is given by the function

$$N(a) = 1000 + 200 \ln a, \quad a \geq 1,$$

where $N(a)$ is the number of units sold when a is the amount spent on advertising, in thousands of dollars.

a) How many units were sold after spending \$1000 $(a = 1)$ on advertising?
b) How many units were sold after spending \$5000?
c) Graph the function.
d) How much would have to be spent in order to sell 2000 units?

101. *Loudness of Sound.* The **loudness** L, in bels (after Alexander Graham Bell), of a sound of intensity I is defined to be

$$L = \log \frac{I}{I_0},$$

where I_0 is the minimum intensity detectable by the human ear (such as the tick of a watch at 20 ft under quiet conditions). If a sound is 10 times as intense as another, its loudness is 1 bel greater than that of the other. If a sound is 100 times as intense as another, its loudness is 2 bels greater, and so on. The bel is a large unit, so a subunit, the **decibel**, is generally used. For L, in decibels, the formula is

$$L = 10 \log \frac{I}{I_0}.$$

Find the loudness, in decibels, of each sound with the given intensity.

SOUND	INTENSITY
a) Jet engine at 100 ft	$10^{14} \cdot I_0$
b) Loud rock concert	$10^{11.5} \cdot I_0$
c) Bird calls	$10^{4} \cdot I_0$
d) Normal conversation	$10^{6.5} \cdot I_0$
e) Thunder	$10^{12} \cdot I_0$
f) Loudest sound possible	$10^{19.4} \cdot I_0$

Skill Maintenance

Find the slope and the y-intercept of the line. [1.4]

102. $3x - 10y = 14$

103. $y = 6$

104. $x = -4$

Use synthetic division to find the function values. [4.3]

105. $g(x) = x^3 - 6x^2 + 3x + 10$; find $g(-5)$

106. $f(x) = x^4 - 2x^3 + x - 6$; find $f(-1)$

Find a polynomial function of degree 3 with the given numbers as zeros. Answers may vary. [4.4]

107. $\sqrt{7}, -\sqrt{7}, 0$

108. $4i, -4i, 1$

Synthesis

Simplify.

109. $\dfrac{\log_5 8}{\log_5 2}$

110. $\dfrac{\log_3 64}{\log_3 16}$

Find the domain of the function.

111. $f(x) = \log_5 x^3$

112. $f(x) = \log_4 x^2$

113. $f(x) = \ln |x|$

114. $f(x) = \log(3x - 4)$

Solve.

115. $\log_2(2x + 5) < 0$

116. $\log_2(x - 3) \geq 4$

In Exercises 117–120, match the equation with one of the figures (a)–(d) that follow.

a)

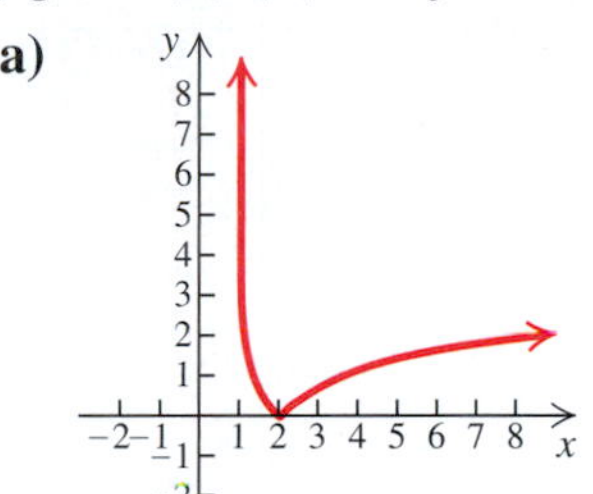

b)

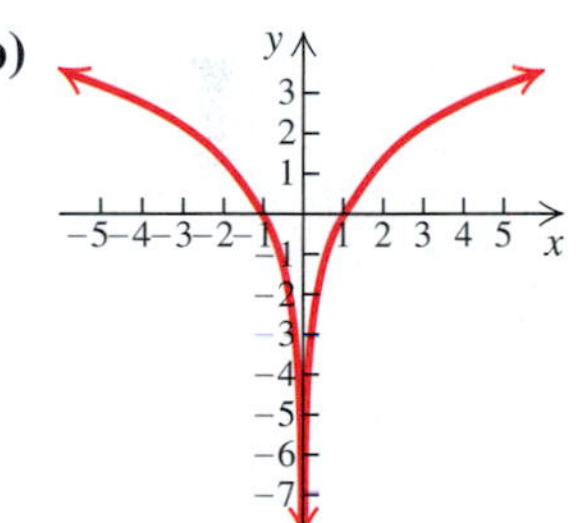

c)

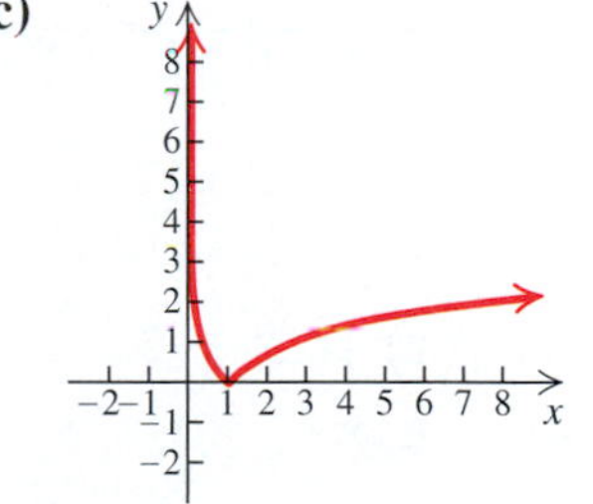

d)

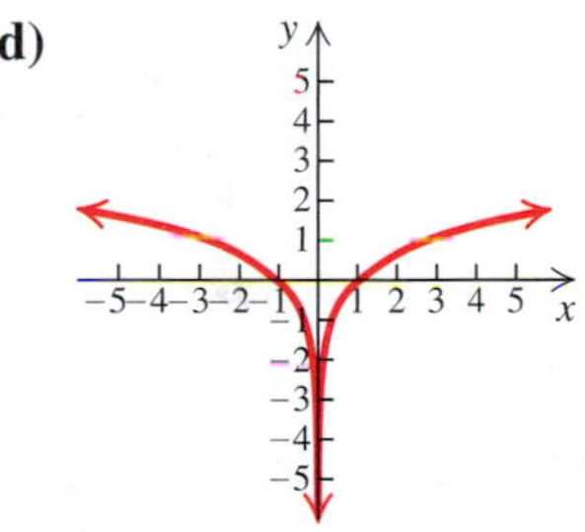

117. $f(x) = \ln |x|$

118. $f(x) = |\ln x|$

119. $f(x) = \ln x^2$

120. $g(x) = |\ln(x - 1)|$

For Exercises 121–124:

a) *Graph the function.*
b) *Estimate the zeros.*
c) *Estimate the relative maximum values and the relative minimum values.*

121. $f(x) = x \ln x$

122. $f(x) = x^2 \ln x$

123. $f(x) = \dfrac{\ln x}{x^2}$

124. $f(x) = e^{-x} \ln x$

125. Using a graphing calculator, find the point(s) of intersection of the graphs of the following.

$$y = 4 \ln x, \qquad y = \frac{4}{e^x + 1}$$

Mid-Chapter Mixed Review

Determine whether the statement is true or false.

1. The domain of all logarithmic functions is $[1, \infty)$. [5.3]

2. The range of a one-to-one function f is the domain of its inverse f^{-1}. [5.1]

3. The y-intercept of $f(x) = e^{-x}$ is $(0, -1)$. [5.2]

For each function, determine whether it is one-to-one, and if the function is one-to-one, find a formula for its inverse. [5.1]

4. $f(x) = -\dfrac{2}{x}$

5. $f(x) = 3 + x^2$

6. $f(x) = \dfrac{5}{x - 2}$

7. Given the function $f(x) = \sqrt{x - 5}$, use composition of functions to show that $f^{-1}(x) = x^2 + 5$. [5.1]

8. Given the one-to-one function $f(x) = x^3 + 2$, find the inverse and give the domain and the range of f and of f^{-1}. [5.1]

Match the function with one of the graphs (a)–(h) that follow. [5.2], [5.3]

a)

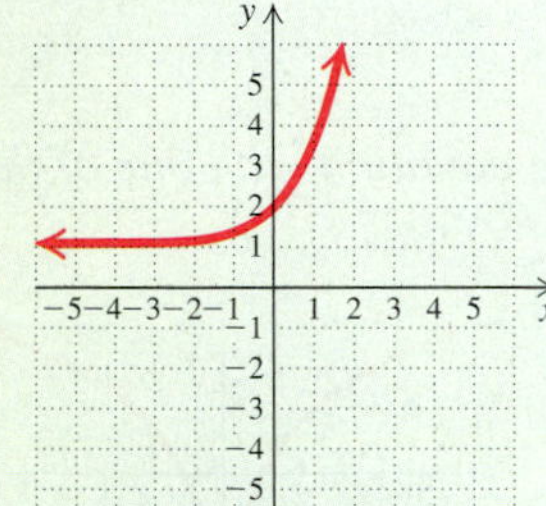

b)

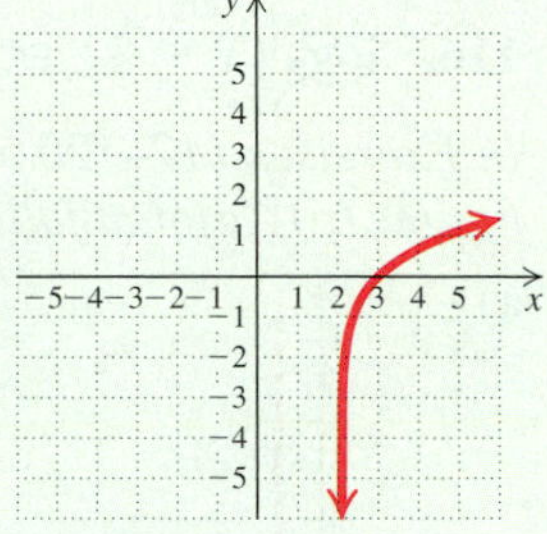

c)

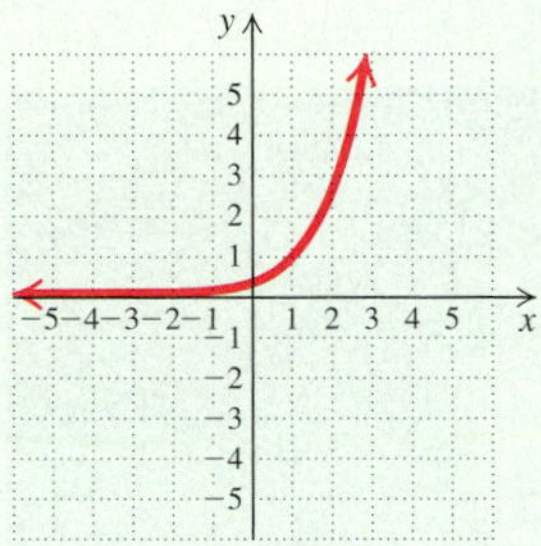

d)

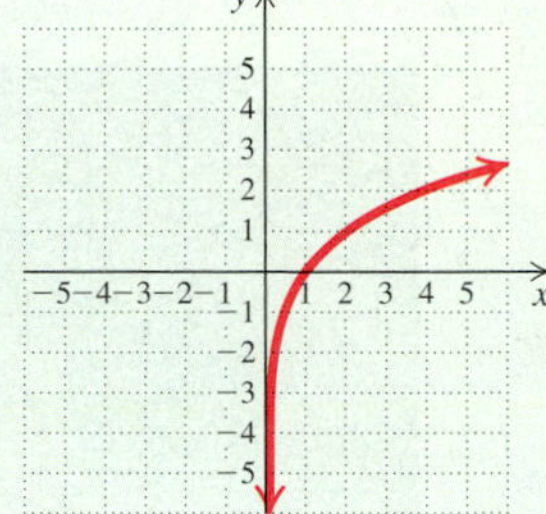

e)

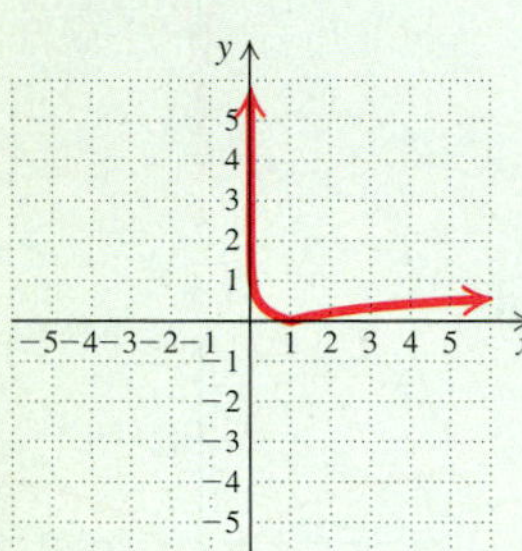

f)

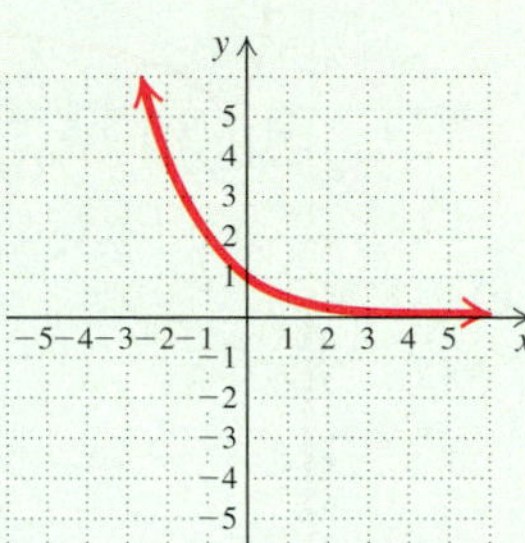

g)

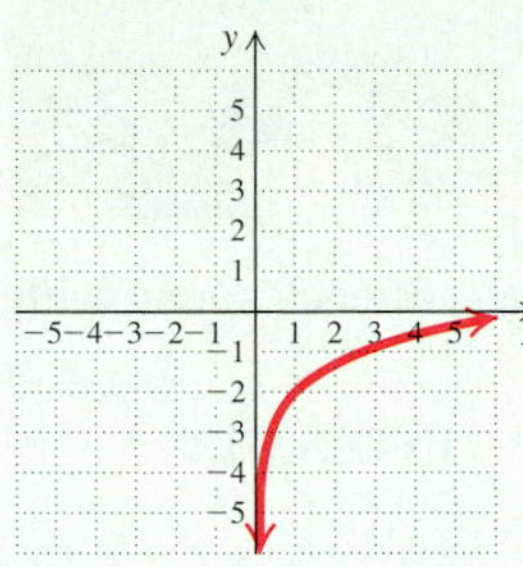

h)

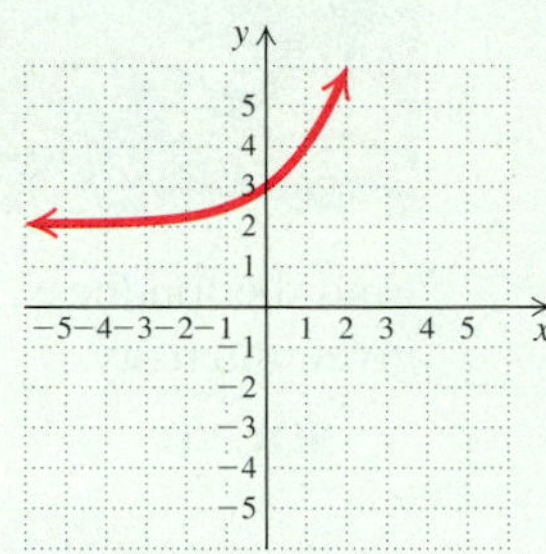

9. $y = \log_2 x$

10. $f(x) = 2^x + 2$

11. $f(x) = e^{x-1}$

12. $f(x) = \ln x - 2$

13. $f(x) = \ln(x - 2)$

14. $y = 2^{-x}$

15. $f(x) = |\log x|$

16. $f(x) = e^x + 1$

17. Suppose that \$3200 is invested at $4\frac{1}{2}\%$ interest, compounded quarterly. Find the amount of money in the account in 6 years. [5.2]

Find each of the following without a calculator. [5.3]

18. $\log_4 1$

19. $\ln e^{-4/5}$

20. $\log 0.01$

21. $\ln e^2$

22. $\ln 1$

23. $\log_2 \dfrac{1}{16}$

24. $\log 1$

25. $\log_3 27$

26. $\log \sqrt[4]{10}$

27. $\ln e$

28. Convert $e^{-6} = 0.0025$ to a logarithmic equation. [5.3]

29. Convert $\log T = r$ to an exponential equation. [5.3]

Find the logarithm using the change-of-base formula. [5.3]

30. $\log_3 20$

31. $\log_\pi 10$

COLLABORATIVE DISCUSSION AND WRITING

32. Explain why an even function f does not have an inverse f^{-1} that is a function. [5.1]

33. Suppose that \$10,000 is invested for 8 years at 6.4% interest, compounded annually. In what year will the most interest be earned? Why? [5.2]

34. Describe the differences between the graphs of $f(x) = x^3$ and $g(x) = 3^x$. [5.2]

35. If $\log b < 0$, what can you say about b? [5.3]

5.4 Properties of Logarithmic Functions

- Convert from logarithms of products, powers, and quotients to expressions in terms of individual logarithms, and conversely.
- Simplify expressions of the type $\log_a a^x$ and $a^{\log_a x}$.

We now establish some properties of logarithmic functions. These properties are based on the corresponding rules for exponents.

Just in Time 7

Logarithms of Products

The first property of logarithms corresponds to the product rule for exponents: $a^m \cdot a^n = a^{m+n}$.

THE PRODUCT RULE

For any positive numbers M and N and any logarithmic base a,

$$\log_a MN = \log_a M + \log_a N.$$

(The logarithm of a product is the sum of the logarithms of the factors.)

EXAMPLE 1 Express as a sum of logarithms: $\log_3 (9 \cdot 27)$.

Solution We have

$$\log_3 (9 \cdot 27) = \log_3 9 + \log_3 27. \quad \text{Using the product rule}$$

As a check, note that

$$\log_3 (9 \cdot 27) = \log_3 243 = 5 \quad 3^5 = 243$$

and $\log_3 9 + \log_3 27 = 2 + 3 = 5.$ $\quad 3^2 = 9;\ 3^3 = 27$

Now Try Exercise 1.

EXAMPLE 2 Express as a single logarithm: $\log_2 p^3 + \log_2 q$.

Solution We have

$$\log_2 p^3 + \log_2 q = \log_2 (p^3 q).$$

Now Try Exercise 35.

A Proof of the Product Rule: Let $\log_a M = x$ and $\log_a N = y$. Converting to exponential equations, we have $a^x = M$ and $a^y = N$. Then

$$MN = a^x \cdot a^y = a^{x+y}.$$

Converting back to a logarithmic equation, we get

$$\log_a MN = x + y.$$

Remembering what x and y represent, we know it follows that

$$\log_a MN = \log_a M + \log_a N. \quad \blacksquare$$

Logarithms of Powers

The second property of logarithms corresponds to the power rule for exponents: $(a^m)^n = a^{mn}$.

THE POWER RULE

For any positive number M, any logarithmic base a, and any real number p,

$$\log_a M^p = p \log_a M.$$

(The logarithm of a power of M is the exponent times the logarithm of M.)

EXAMPLE 3 Express each of the following as a product.

a) $\log_a 11^{-3}$

b) $\log_a \sqrt[4]{7}$

c) $\ln x^6$

Solution

a) $\log_a 11^{-3} = -3 \log_a 11$ **Using the power rule**

b) $\log_a \sqrt[4]{7} = \log_a 7^{1/4}$ **Writing exponential notation**

$= \frac{1}{4} \log_a 7$ **Using the power rule**

c) $\ln x^6 = 6 \ln x$ **Using the power rule**

Now Try Exercises 13 and 15.

A Proof of the Power Rule: Let $x = \log_a M$. The equivalent exponential equation is $a^x = M$. Raising both sides to the power p, we obtain

$$(a^x)^p = M^p, \quad \text{or} \quad a^{xp} = M^p.$$

Converting back to a logarithmic equation, we get

$$\log_a M^p = xp.$$

But $x = \log_a M$, so substituting gives us

$$\log_a M^p = (\log_a M)p = p \log_a M. \quad \blacksquare$$

Logarithms of Quotients

The third property of logarithms corresponds to the quotient rule for exponents: $a^m/a^n = a^{m-n}$.

THE QUOTIENT RULE

For any positive numbers M and N, and any logarithmic base a,

$$\log_a \frac{M}{N} = \log_a M - \log_a N.$$

(The logarithm of a quotient is the logarithm of the numerator minus the logarithm of the denominator.)

EXAMPLE 4 Express as a difference of logarithms: $\log_t \frac{8}{w}$.

Solution We have

$$\log_t \frac{8}{w} = \log_t 8 - \log_t w. \quad \text{Using the quotient rule}$$

Now Try Exercise 17.

EXAMPLE 5 Express as a single logarithm: $\log_b 64 - \log_b 16$.

Solution We have

$$\log_b 64 - \log_b 16 = \log_b \frac{64}{16} = \log_b 4.$$

Now Try Exercise 37.

A Proof of the Quotient Rule: The proof follows from both the product rule and the power rule:

$$\begin{aligned}\log_a \frac{M}{N} &= \log_a MN^{-1} = \log_a M + \log_a N^{-1} && \text{Using the product rule}\\ &= \log_a M + (-1)\log_a N && \text{Using the power rule}\\ &= \log_a M - \log_a N.\end{aligned}$$

■

COMMON ERRORS

$\log_a MN \neq (\log_a M)(\log_a N)$	The logarithm of a product is *not* the product of the logarithms.
$\log_a (M + N) \neq \log_a M + \log_a N$	The logarithm of a sum is *not* the sum of the logarithms.
$\log_a \frac{M}{N} \neq \frac{\log_a M}{\log_a N}$	The logarithm of a quotient is *not* the quotient of the logarithms.
$(\log_a M)^p \neq p \log_a M$	The power of a logarithm is *not* the exponent times the logarithm.

Applying the Properties

EXAMPLE 6 Express each of the following in terms of sums and differences of logarithms.

a) $\log_a \frac{x^2y^5}{z^4}$ **b)** $\log_a \sqrt[3]{\frac{a^2b}{c^5}}$ **c)** $\log_b \frac{ay^5}{m^3n^4}$

Solution

$$\begin{aligned}\textbf{a) } \log_a \frac{x^2y^5}{z^4} &= \log_a (x^2y^5) - \log_a z^4 && \text{Using the quotient rule}\\ &= \log_a x^2 + \log_a y^5 - \log_a z^4 && \text{Using the product rule}\\ &= 2\log_a x + 5\log_a y - 4\log_a z && \text{Using the power rule}\end{aligned}$$

b) $\log_a \sqrt[3]{\frac{a^2 b}{c^5}} = \log_a \left(\frac{a^2 b}{c^5}\right)^{1/3}$ — Writing exponential notation

$= \frac{1}{3} \log_a \frac{a^2 b}{c^5}$ — Using the power rule

$= \frac{1}{3} (\log_a a^2 b - \log_a c^5)$ — Using the quotient rule. The parentheses are necessary.

$= \frac{1}{3} (2 \log_a a + \log_a b - 5 \log_a c)$ — Using the product rule and the power rule

$= \frac{1}{3} (2 + \log_a b - 5 \log_a c)$ — $\log_a a = 1$

$= \frac{2}{3} + \frac{1}{3} \log_a b - \frac{5}{3} \log_a c$ — Multiplying to remove parentheses

c) $\log_b \frac{ay^5}{m^3 n^4} = \log_b ay^5 - \log_b m^3 n^4$ — Using the quotient rule

$= (\log_b a + \log_b y^5) - (\log_b m^3 + \log_b n^4)$ — Using the product rule

$= \log_b a + \log_b y^5 - \log_b m^3 - \log_b n^4$ — Removing parentheses

$= \log_b a + 5 \log_b y - 3 \log_b m - 4 \log_b n$ — Using the power rule

Now Try Exercises 25 and 31.

EXAMPLE 7 Express as a single logarithm:

$$5 \log_b x - \log_b y + \tfrac{1}{4} \log_b z.$$

Solution We have

$5 \log_b x - \log_b y + \frac{1}{4} \log_b z = \log_b x^5 - \log_b y + \log_b z^{1/4}$ — Using the power rule

$= \log_b \frac{x^5}{y} + \log_b z^{1/4}$ — Using the quotient rule

$= \log_b \frac{x^5 z^{1/4}}{y}$, or $\log_b \frac{x^5 \sqrt[4]{z}}{y}$. — Using the product rule

Now Try Exercise 41.

EXAMPLE 8 Express as a single logarithm:

$$\ln (3x + 1) - \ln (3x^2 - 5x - 2).$$

Solution We have

$\ln (3x + 1) - \ln (3x^2 - 5x - 2) = \ln \frac{3x + 1}{3x^2 - 5x - 2}$ — Using the quotient rule

$= \ln \frac{3x + 1}{(3x + 1)(x - 2)}$ — Factoring

$= \ln \frac{1}{x - 2}$. — Simplifying

Now Try Exercise 45.

EXAMPLE 9 Given that $\log_a 2 \approx 0.301$ and $\log_a 3 \approx 0.477$, find each of the following, if possible.

a) $\log_a 6$ **b)** $\log_a \frac{2}{3}$ **c)** $\log_a 81$

d) $\log_a \frac{1}{4}$ **e)** $\log_a 5$ **f)** $\dfrac{\log_a 3}{\log_a 2}$

Solution

a) $\log_a 6 = \log_a (2 \cdot 3) = \log_a 2 + \log_a 3$ Using the product rule
$\approx 0.301 + 0.477$
≈ 0.778

b) $\log_a \frac{2}{3} = \log_a 2 - \log_a 3$ Using the quotient rule
$\approx 0.301 - 0.477 \approx -0.176$

c) $\log_a 81 = \log_a 3^4 = 4 \log_a 3$ Using the power rule
$\approx 4(0.477) \approx 1.908$

d) $\log_a \frac{1}{4} = \log_a 1 - \log_a 4$ Using the quotient rule
$= 0 - \log_a 2^2$ $\log_a 1 = 0$; $4 = 2^2$
$= -2 \log_a 2$ Using the power rule
$\approx -2(0.301) \approx -0.602$

e) $\log_a 5$ *cannot* be found using these properties and the given information.

$\log_a 5 \neq \log_a 2 + \log_a 3$ $\log_a 2 + \log_a 3 = \log_a (2 \cdot 3) = \log_a 6$

f) $\dfrac{\log_a 3}{\log_a 2} \approx \dfrac{0.477}{0.301} \approx 1.585$ We simply divide, not using any of the properties.

Now Try Exercises 53 and 55.

Simplifying Expressions of the Type $\log_a a^x$ and $a^{\log_a x}$

We have two final properties of logarithms to consider. The first follows from the product rule: Since $\log_a a^x = x \log_a a = x \cdot 1 = x$, we have $\log_a a^x = x$. This property also follows from the definition of a logarithm: x is the power to which we raise a in order to get a^x.

THE LOGARITHM OF A BASE TO A POWER

For any logarithmic base a and any real number x,

$$\log_a a^x = x.$$

(The logarithm, base a, of a to a power is the power.)

EXAMPLE 10 Simplify each of the following.

a) $\log_a a^8$ **b)** $\ln e^{-t}$ **c)** $\log 10^{3k}$

Solution

a) $\log_a a^8 = 8$ 8 is the power to which we raise a in order to get a^8.

b) $\ln e^{-t} = \log_e e^{-t} = -t$ $\ln e^x = x$

c) $\log 10^{3k} = \log_{10} 10^{3k} = 3k$

Now Try Exercises 65 and 73.

Let $M = \log_a x$. Then $a^M = x$. Substituting $\log_a x$ for M, we obtain $a^{\log_a x} = x$. This also follows from the definition of a logarithm: $\log_a x$ is the power to which a is raised in order to get x.

A BASE TO A LOGARITHMIC POWER

For any logarithmic base a and any positive real number x,

$$a^{\log_a x} = x.$$

(The number a raised to the power $\log_a x$ is x.)

EXAMPLE 11 Simplify each of the following.

a) $4^{\log_4 k}$ **b)** $e^{\ln 5}$ **c)** $10^{\log 7t}$

Solution

a) $4^{\log_4 k} = k$

b) $e^{\ln 5} = e^{\log_e 5} = 5$

c) $10^{\log 7t} = 10^{\log_{10} 7t} = 7t$

Now Try Exercises 69 and 71.

CHANGE-OF-BASE FORMULA

REVIEW SECTION 5.3.

A Proof of the Change-of-Base Formula: We close this section by proving the change-of-base formula and summarizing the properties of logarithms considered thus far in this chapter. In Section 5.3, we used the change-of-base formula,

$$\log_b M = \frac{\log_a M}{\log_a b},$$

to make base conversions in order to find logarithmic values using a calculator. Let $x = \log_b M$. Then

$$b^x = M \quad \text{Definition of logarithm}$$

$$\log_a b^x = \log_a M \quad \text{Taking the logarithm on both sides}$$

$$x \log_a b = \log_a M \quad \text{Using the power rule}$$

$$x = \frac{\log_a M}{\log_a b}, \quad \text{Dividing by } \log_a b$$

so

$$x = \log_b M = \frac{\log_a M}{\log_a b}.$$

■

Following is a summary of the properties of logarithms.

SUMMARY OF THE PROPERTIES OF LOGARITHMS

The Product Rule:	$\log_a MN = \log_a M + \log_a N$
The Power Rule:	$\log_a M^p = p \log_a M$
The Quotient Rule:	$\log_a \frac{M}{N} = \log_a M - \log_a N$
The Change-of-Base Formula:	$\log_b M = \frac{\log_a M}{\log_a b}$
Other Properties:	$\log_a a = 1, \quad \log_a 1 = 0,$ $\log_a a^x = x, \quad a^{\log_a x} = x$

5.4 Exercise Set

Express as a sum of logarithms and simplify, if possible.

1. $\log_3 (81 \cdot 27)$
2. $\log_2 (8 \cdot 64)$
3. $\log_5 (5 \cdot 125)$
4. $\log_4 (64 \cdot 4)$
5. $\log_t 8Y$
6. $\log 0.2x$
7. $\ln xy$
8. $\ln ab$

Express as a product.

9. $\log_b t^3$
10. $\log_a x^4$
11. $\log y^8$
12. $\ln y^5$
13. $\log_c K^{-6}$
14. $\log_b Q^{-8}$
15. $\ln \sqrt[3]{4}$
16. $\ln \sqrt{a}$

Express as a difference of logarithms.

17. $\log_t \dfrac{M}{8}$
18. $\log_a \dfrac{76}{13}$
19. $\log \dfrac{x}{y}$
20. $\ln \dfrac{a}{b}$
21. $\ln \dfrac{r}{s}$
22. $\log_b \dfrac{3}{w}$

Express in terms of sums and differences of logarithms.

23. $\log_a 6xy^5z^4$
24. $\log_a x^3y^2z$
25. $\log_b \dfrac{p^2q^5}{m^4b^9}$
26. $\log_b \dfrac{x^2y}{b^3}$
27. $\ln \dfrac{2}{3x^3y}$
28. $\log \dfrac{5a}{4b^2}$
29. $\log \sqrt{r^3t}$
30. $\ln \sqrt[3]{5x^5}$
31. $\log_a \sqrt{\dfrac{x^6}{p^5q^8}}$
32. $\log_c \sqrt[3]{\dfrac{y^3z^2}{x^4}}$
33. $\log_a \sqrt[4]{\dfrac{m^8n^{12}}{a^3b^5}}$
34. $\log_a \sqrt{\dfrac{a^6b^8}{a^2b^5}}$

Express as a single logarithm and, if possible, simplify.

35. $\log_a 75 + \log_a 2$
36. $\log 0.01 + \log 1000$
37. $\log 10{,}000 - \log 100$
38. $\ln 54 - \ln 6$
39. $\frac{1}{2} \log n + 3 \log m$
40. $\frac{1}{2} \log a - \log 2$
41. $\frac{1}{2} \log_a x + 4 \log_a y - 3 \log_a x$
42. $\frac{2}{5} \log_a x - \frac{1}{3} \log_a y$
43. $\ln x^2 - 2 \ln \sqrt{x}$
44. $\ln 2x + 3(\ln x - \ln y)$
45. $\ln (x^2 - 4) - \ln (x + 2)$
46. $\log (x^3 - 8) - \log (x - 2)$
47. $\log (x^2 - 5x - 14) - \log (x^2 - 4)$
48. $\log_a \dfrac{a}{\sqrt{x}} - \log_a \sqrt{ax}$
49. $\ln x - 3[\ln (x - 5) + \ln (x + 5)]$
50. $\frac{2}{3}[\ln (x^2 - 9) - \ln (x + 3)] + \ln (x + y)$
51. $\frac{3}{2} \ln 4x^6 - \frac{4}{5} \ln 2y^{10}$
52. $120(\ln \sqrt[5]{x^3} + \ln \sqrt[3]{y^2} - \ln \sqrt[4]{16z^5})$

Given that $\log_a 2 \approx 0.301$, $\log_a 7 \approx 0.845$, *and* $\log_a 11 \approx 1.041$, *find each of the following, if possible. Round the answer to the nearest thousandth.*

53. $\log_a \frac{2}{11}$
54. $\log_a 14$
55. $\log_a 98$
56. $\log_a \frac{1}{7}$
57. $\dfrac{\log_a 2}{\log_a 7}$
58. $\log_a 9$

Given that $\log_b 2 \approx 0.693$, $\log_b 3 \approx 1.099$, *and* $\log_b 5 \approx 1.609$, *find each of the following, if possible. Round the answer to the nearest thousandth.*

59. $\log_b 125$
60. $\log_b \frac{5}{3}$
61. $\log_b \frac{1}{6}$
62. $\log_b 30$
63. $\log_b \dfrac{3}{b}$
64. $\log_b 15b$

Simplify.

65. $\log_p p^3$
66. $\log_t t^{2713}$
67. $\log_e e^{|x-4|}$
68. $\log_q q^{\sqrt{3}}$
69. $3^{\log_3 4x}$
70. $5^{\log_5 (4x-3)}$
71. $10^{\log w}$
72. $e^{\ln x^3}$
73. $\ln e^{8t}$
74. $\log 10^{-k}$
75. $\log_b \sqrt{b}$
76. $\log_b \sqrt{b^3}$

Skill Maintenance

In each of Exercises 77–86, classify the function as linear, quadratic, cubic, quartic, rational, exponential, or logarithmic.

77. $f(x) = 5 - x^2 + x^4$ [4.1]
78. $f(x) = 2^x$ [5.2]

79. $f(x) = -\frac{3}{4}$ [1.3]

80. $f(x) = 4^x - 8$ [5.2]

81. $f(x) = -\frac{3}{x}$ [4.5]

82. $f(x) = \log x + 6$ [5.3]

83. $f(x) = -\frac{1}{3}x^3 - 4x^2 + 6x + 42$ [4.1]

84. $f(x) = \frac{x^2 - 1}{x^2 + x - 6}$ [4.5]

85. $f(x) = \frac{1}{2}x + 3$ [1.4]

86. $f(x) = 2x^2 - 6x + 3$ [3.2]

Synthesis

Solve for x.

87. $5^{\log_5 8} = 2x$

88. $\ln e^{3x-5} = -8$

Express as a single logarithm and, if possible, simplify.

89. $\log_a (x^2 + xy + y^2) + \log_a (x - y)$

90. $\log_a (a^{10} - b^{10}) - \log_a (a + b)$

Express as a sum or a difference of logarithms.

91. $\log_a \frac{x - y}{\sqrt{x^2 - y^2}}$

92. $\log_a \sqrt{9 - x^2}$

93. Given that $\log_a x = 2$, $\log_a y = 3$, and $\log_a z = 4$, find

$$\log_a \frac{\sqrt[4]{y^2z^5}}{\sqrt[4]{x^3z^{-2}}}.$$

Determine whether each of the following is true or false. Assume that a, x, M, and N are positive.

94. $\log_a M + \log_a N = \log_a (M + N)$

95. $\log_a M - \log_a N = \log_a \frac{M}{N}$

96. $\frac{\log_a M}{\log_a N} = \log_a M - \log_a N$

97. $\frac{\log_a M}{x} = \log_a M^{1/x}$

98. $\log_a x^3 = 3 \log_a x$

99. $\log_a 8x = \log_a x + \log_a 8$

100. $\log_N (MN)^x = x \log_N M + x$

Suppose that $\log_a x = 2$. Find each of the following.

101. $\log_a \left(\frac{1}{x}\right)$

102. $\log_{1/a} x$

103. Simplify:

$\log_{10} 11 \cdot \log_{11} 12 \cdot \log_{12} 13 \cdots \log_{998} 999 \cdot \log_{999} 1000.$

Write each of the following without using logarithms.

104. $\log_a x + \log_a y - mz = 0$

105. $\ln a - \ln b + xy = 0$

Prove each of the following for any base a and any positive number x.

106. $\log_a \left(\frac{1}{x}\right) = -\log_a x = \log_{1/a} x$

107. $\log_a \left(\frac{x + \sqrt{x^2 - 5}}{5}\right) = -\log_a (x - \sqrt{x^2 - 5})$

5.5 Solving Exponential Equations and Logarithmic Equations

- Solve exponential equations.
- Solve logarithmic equations.

Solving Exponential Equations

Equations with variables in the exponents, such as

$$3^x = 20 \quad \text{and} \quad 2^{5x} = 64,$$

are called **exponential equations**.

Sometimes, as is the case with the equation $2^{5x} = 64$, we can write each side as a power of the same number:

$$2^{5x} = 2^6.$$

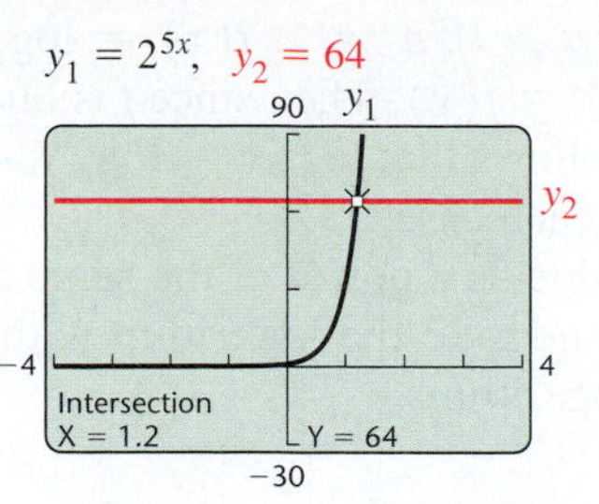

We can then set the exponents equal and solve:

$$5x = 6$$
$$x = \tfrac{6}{5}, \text{ or } 1.2.$$

We use the following property to solve exponential equations.

BASE–EXPONENT PROPERTY

For any $a > 0$, $a \neq 1$,

$$a^x = a^y \longleftrightarrow x = y.$$

ONE-TO-ONE FUNCTIONS

REVIEW SECTION 5.1.

This property follows from the fact that for any $a > 0$, $a \neq 1$, $f(x) = a^x$ is a one-to-one function. If $a^x = a^y$, then $f(x) = f(y)$. Then since f is one-to-one, it follows that $x = y$. Conversely, if $x = y$, it follows that $a^x = a^y$, since we are raising a to the same power in each case.

EXAMPLE 1 Solve: $2^{3x-7} = 32$.

Algebraic Solution

Note that $32 = 2^5$. Thus we can write each side as a power of the same number:

$$2^{3x-7} = 2^5.$$

Since the bases are the same number, 2, we can use the base–exponent property and set the exponents equal:

$$3x - 7 = 5$$
$$3x = 12$$
$$x = 4.$$

Check:

$2^{3x-7} = 32$	
$2^{3(4)-7}$?	32
2^{12-7}	
2^5	
32	32 TRUE

The solution is 4.

Graphical Solution

We will use the Intersect method. We graph

$$y_1 = 2^{3x-7} \quad \text{and} \quad y_2 = 32$$

to find the coordinates of the point of intersection. The first coordinate of this point is the solution of the equation $y_1 = y_2$, or $2^{3x-7} = 32$.

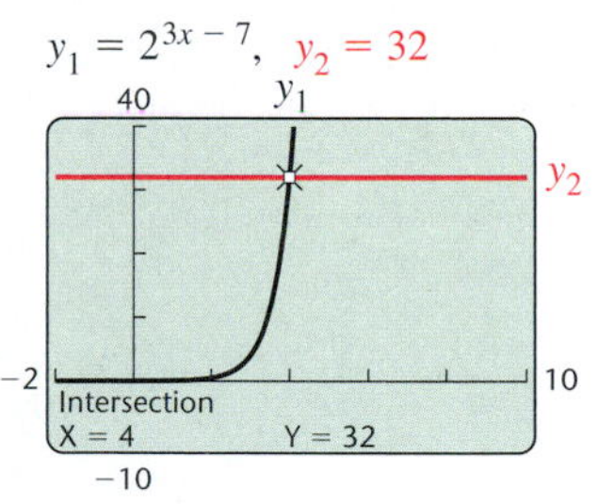

The solution is 4.

We could also write the equation in the form $2^{3x-7} - 32 = 0$ and use the Zero method.

Now Try Exercise 7.

Another property that is used when solving some exponential equations and logarithmic equations is as follows.

PROPERTY OF LOGARITHMIC EQUALITY

For any $M > 0$, $N > 0$, $a > 0$, and $a \neq 1$,

$$\log_a M = \log_a N \longleftrightarrow M = N.$$

This property follows from the fact that for any $a > 0, a \neq 1, f(x) = \log_a x$ is a one-to-one function. If $\log_a x = \log_a y$, then $f(x) = f(y)$. Then since f is one-to-one, it follows that $x = y$. Conversely, if $x = y$, it follows that $\log_a x = \log_a y$, since we are taking the logarithm of the same number in each case.

When it does not seem possible to write each side as a power of the same base, we can use the property of logarithmic equality and take the logarithm with any base on each side and then use the power rule for logarithms.

EXAMPLE 2 Solve: $3^x = 20$.

Algebraic Solution

We have

$$3^x = 20$$

$$\log 3^x = \log 20 \qquad \text{Taking the common logarithm on both sides}$$

$$x \log 3 = \log 20 \qquad \text{Using the power rule}$$

$$x = \frac{\log 20}{\log 3}. \qquad \text{Dividing by } \log 3$$

This is an exact answer. We cannot simplify further, but we can approximate using a calculator:

$$x = \frac{\log 20}{\log 3} \approx 2.7268.$$

We can check this by finding $3^{2.7268}$:

$$3^{2.7268} \approx 20.$$

The solution is about 2.7268.

Graphical Solution

We will use the Intersect method. We graph

$$y_1 = 3^x \quad \text{and} \quad y_2 = 20$$

to find the x-coordinate of the point of intersection. That x-coordinate is the value of x for which $3^x = 20$ and is thus the solution of the equation.

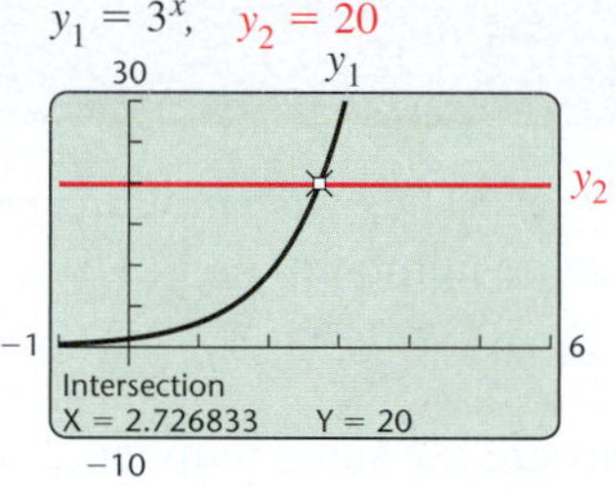

The solution is approximately 2.7268.

We could also write the equation in the form $3^x - 20 = 0$ and use the Zero method.

Now Try Exercise 11.

In Example 2, we took the common logarithm on both sides of the equation. Any base will give the same result. Let's try base 3. We have

$$3^x = 20$$

$$\log_3 3^x = \log_3 20$$

$$x = \log_3 20 \qquad \log_a a^x = x$$

$$x = \frac{\log 20}{\log 3} \qquad \text{Using the change-of-base formula}$$

$$x \approx 2.7268.$$

$\log_3(20)$

2.726833028

Note that we changed the base to do the final calculation. We also could find $\log_3 20$ directly using the logBASE operation.

EXAMPLE 3 Solve: $100e^{0.08t} = 2500$.

Algebraic Solution

It will make our work easier if we take the natural logarithm when working with equations that have e as a base.

We have

$$100e^{0.08t} = 2500$$

$$e^{0.08t} = 25 \quad \text{Dividing by 100}$$

$$\ln e^{0.08t} = \ln 25 \quad \text{Taking the natural logarithm on both sides}$$

$$0.08t = \ln 25 \quad \text{Finding the logarithm of a base to a power: } \log_a a^x = x$$

$$t = \frac{\ln 25}{0.08} \quad \text{Dividing by 0.08}$$

$$\approx 40.2.$$

The solution is about 40.2.

Graphical Solution

Using the Intersect method, we graph the equations

$$y_1 = 100e^{0.08x} \quad \text{and} \quad y_2 = 2500$$

and determine the point of intersection. The first coordinate of the point of intersection is the solution of the equation $100e^{0.08x} = 2500$.

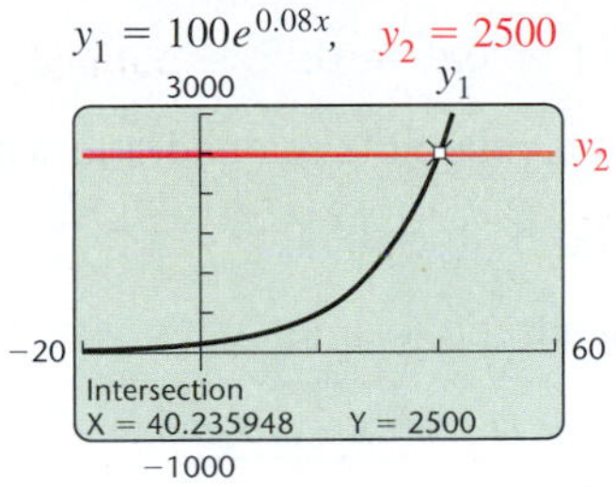

The solution is approximately 40.2.

Now Try Exercise 19.

EXAMPLE 4 Solve: $4^{x+3} = 3^{-x}$.

Algebraic Solution

We have

$$4^{x+3} = 3^{-x}$$

$$\log 4^{x+3} = \log 3^{-x} \quad \text{Taking the common logarithm on both sides}$$

$$(x+3)\log 4 = -x\log 3 \quad \text{Using the power rule}$$

$$x\log 4 + 3\log 4 = -x\log 3 \quad \text{Removing parentheses}$$

$$x\log 4 + x\log 3 = -3\log 4 \quad \text{Adding } x\log 3 \text{ and subtracting } 3\log 4$$

$$x(\log 4 + \log 3) = -3\log 4 \quad \text{Factoring on the left}$$

$$x = \frac{-3\log 4}{\log 4 + \log 3} \quad \text{Dividing by } \log 4 + \log 3$$

$$\approx -1.6737.$$

The solution is about -1.6737.

Graphical Solution

We will use the Intersect method. We graph

$$y_1 = 4^{x+3} \quad \text{and} \quad y_2 = 3^{-x}$$

to find the x-coordinate of the point of intersection. That x-coordinate is the value of x for which $4^{x+3} = 3^{-x}$ and is the solution of the equation.

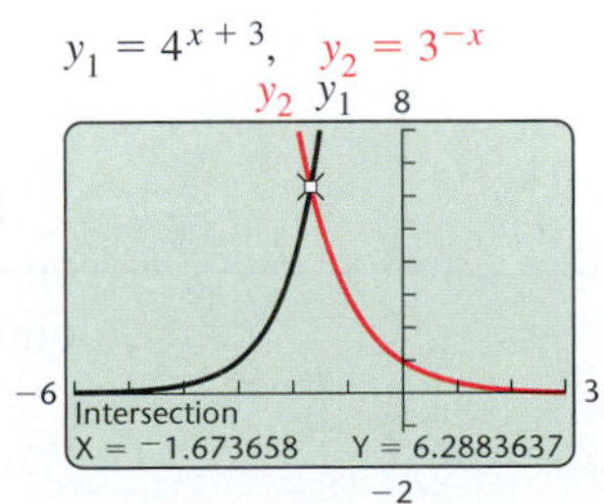

The solution is approximately -1.6737.

We could also write the equation in the form $4^{x+3} - 3^{-x} = 0$ and use the Zero method.

Now Try Exercise 21.

EQUATIONS REDUCIBLE TO QUADRATIC

REVIEW SECTION 3.2.

EXAMPLE 5 Solve: $e^x + e^{-x} - 6 = 0$.

Algebraic Solution

In this case, we have more than one term with x in the exponent:

$$e^x + e^{-x} - 6 = 0$$

$$e^x + \frac{1}{e^x} - 6 = 0 \quad \text{Rewriting } e^{-x} \text{ with a positive exponent}$$

$$e^{2x} + 1 - 6e^x = 0. \quad \text{Multiplying by } e^x \text{ on both sides}$$

This equation is reducible to quadratic with $u = e^x$:

$$u^2 - 6u + 1 = 0.$$

We use the quadratic formula with $a = 1$, $b = -6$, and $c = 1$:

$$u = \frac{-b \pm \sqrt{b^2 - 4ac}}{2a}$$

$$u = \frac{-(-6) \pm \sqrt{(-6)^2 - 4 \cdot 1 \cdot 1}}{2 \cdot 1}$$

$$u = \frac{6 \pm \sqrt{32}}{2} = \frac{6 \pm 4\sqrt{2}}{2}$$

$$u = \frac{2(3 \pm 2\sqrt{2})}{2}$$

$$u = 3 \pm 2\sqrt{2}$$

$$e^x = 3 \pm 2\sqrt{2}. \quad \text{Replacing } u \text{ with } e^x$$

We now take the natural logarithm on both sides:

$$\ln e^x = \ln(3 \pm 2\sqrt{2})$$

$$x = \ln(3 \pm 2\sqrt{2}). \quad \text{Using } \ln e^x = x$$

Approximating each of the solutions, we obtain 1.76 and -1.76.

Graphical Solution

Using the Zero method, we begin by graphing the function

$$y = e^x + e^{-x} - 6.$$

Then we find the zeros of the function.

The leftmost zero is about -1.76. Using the ZERO feature one more time, we find that the other zero is about 1.76.

The solutions are about -1.76 and 1.76.

Now Try Exercise 25.

It is possible that when encountering an equation like the one in Example 5, you might not recognize that it could be solved in the algebraic manner shown. This points out the value of the graphical solution.

Solving Logarithmic Equations

Equations containing variables in logarithmic expressions, such as $\log_2 x = 4$ and $\log x + \log(x + 3) = 1$, are called **logarithmic equations**. To solve logarithmic equations algebraically, we first try to obtain a single logarithmic expression on one side and then write an equivalent exponential equation.

EXAMPLE 6 Solve: $\log_3 x = -2$.

Algebraic Solution

We have

$$\log_3 x = -2$$

$$3^{-2} = x \quad \text{Converting to an exponential equation}$$

$$\frac{1}{3^2} = x$$

$$\frac{1}{9} = x.$$

Check:

$\log_3 x = -2$	
$\log_3 \frac{1}{9}$?	-2
$\log_3 3^{-2}$	
-2	-2 TRUE

The solution is $\frac{1}{9}$.

Graphical Solution

We use the change-of-base formula and graph the equations

$$y_1 = \log_3 x = \frac{\ln x}{\ln 3} \quad \text{and} \quad y_2 = -2.$$

We could also graph $y_1 = \log_3 x$ directly with the logBASE operation. Then we use the Intersect method with a graphing calculator.

$y_1 = \log_3 x = \frac{\ln x}{\ln 3}$, $y_2 = -2$

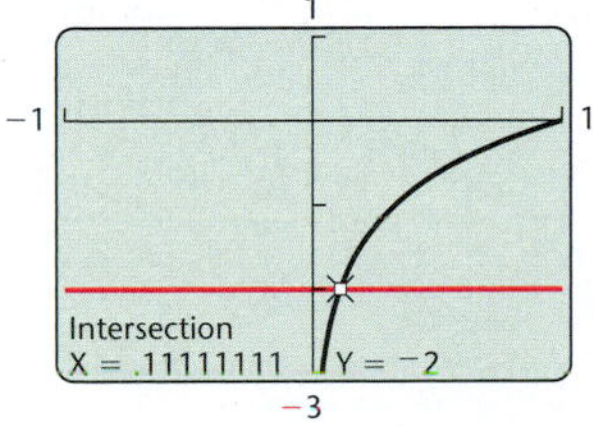

If the solution is a rational number, we can usually find fraction notation for the exact solution by using the FRAC feature from the MATH submenu of the MATH menu.

The solution is $\frac{1}{9}$.

Now Try Exercise 33.

Just in Time 19

EXAMPLE 7 Solve: $\log x + \log(x + 3) = 1$.

Algebraic Solution

In this case, we have common logarithms. Writing the base of 10 will help us understand the problem:

$$\log_{10} x + \log_{10}(x + 3) = 1$$

$$\log_{10}[x(x + 3)] = 1 \qquad \text{Using the product rule to obtain a single logarithm}$$

$$x(x + 3) = 10^1 \qquad \text{Writing an equivalent exponential equation}$$

$$x^2 + 3x = 10$$

$$x^2 + 3x - 10 = 0$$

$$(x - 2)(x + 5) = 0 \qquad \text{Factoring}$$

$$x - 2 = 0 \quad or \quad x + 5 = 0$$

$$x = 2 \quad or \quad x = -5.$$

Check: For 2:

$$\begin{array}{r|l} \log x + \log(x + 3) & = 1 \\ \hline \log 2 + \log(2 + 3) & ?\ 1 \\ \log 2 + \log 5 & \\ \log(2 \cdot 5) & \\ \log 10 & \\ 1 & 1 \quad \text{TRUE} \end{array}$$

For −5:

$$\begin{array}{c} \log x + \log(x + 3) = 1 \\ \hline \log(-5) + \log(-5 + 3)\ ?\ 1 \quad \text{FALSE} \end{array}$$

The number −5 is not a solution because negative numbers do not have real-number logarithms. The solution is 2.

Graphical Solution

We can graph the equations

$$y_1 = \log x + \log(x + 3)$$

and

$$y_2 = 1$$

and use the Intersect method. The first coordinate of the point of intersection is the solution of the equation.

$y_1 = \log x + \log(x + 3)$, $y_2 = 1$

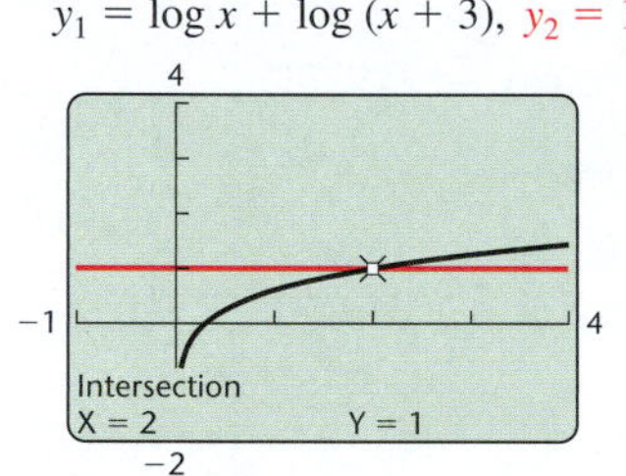

We could also graph the function

$$y = \log x + \log(x + 3) - 1$$

and use the Zero method. The zero of the function is the solution of the equation.

$y_1 = \log x + \log(x + 3) - 1$

The solution of the equation is 2. From the graph, we can easily see that there is only one solution.

Now Try Exercise 41.

EXAMPLE 8 Solve: $\log_3(2x - 1) - \log_3(x - 4) = 2$.

Algebraic Solution

We have

$$\log_3(2x - 1) - \log_3(x - 4) = 2$$

$$\log_3 \frac{2x - 1}{x - 4} = 2 \quad \text{Using the quotient rule}$$

$$\frac{2x - 1}{x - 4} = 3^2 \quad \text{Writing an equivalent exponential equation}$$

$$\frac{2x - 1}{x - 4} = 9$$

$$(x - 4) \cdot \frac{2x - 1}{x - 4} = (x - 4) \cdot 9 \quad \text{Multiplying by the LCD, } x - 4$$

$$2x - 1 = 9x - 36$$

$$35 = 7x$$

$$5 = x.$$

Check:

$\log_3(2x - 1) - \log_3(x - 4) = 2$	
$\log_3(2 \cdot 5 - 1) - \log_3(5 - 4)$?	2
$\log_3 9 - \log_3 1$	
$2 - 0$	
2	2 TRUE

The solution is 5.

Graphical Solution

Here we use the Intersect method to find the solution of the equation. We use the change-of-base formula and graph the equations

$$y_1 = \frac{\ln(2x - 1)}{\ln 3} - \frac{\ln(x - 4)}{\ln 3}$$

and

$$y_2 = 2.$$

We could also graph y_1 directly with the logBASE operation.

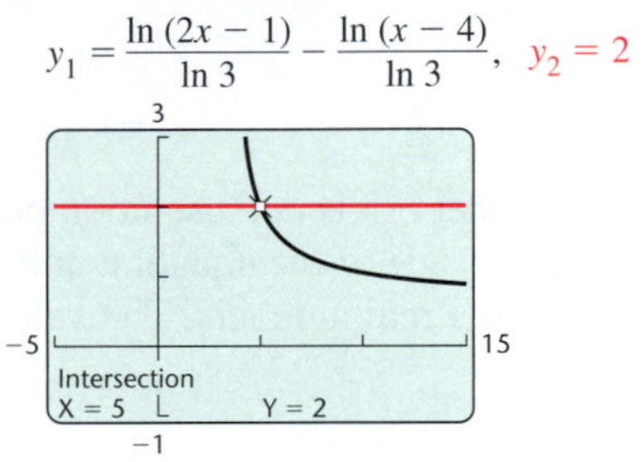

$y_1 = \frac{\ln(2x - 1)}{\ln 3} - \frac{\ln(x - 4)}{\ln 3}$, $y_2 = 2$

The solution is 5.

Now Try Exercise 45.

EXAMPLE 9 Solve: $\ln(4x+6) - \ln(x+5) = \ln x$.

Algebraic Solution

We have

$$\ln(4x+6) - \ln(x+5) = \ln x$$

$$\ln\frac{4x+6}{x+5} = \ln x \qquad \text{Using the quotient rule}$$

$$\frac{4x+6}{x+5} = x \qquad \text{Using the property of logarithmic equality}$$

$$(x+5)\cdot\frac{4x+6}{x+5} = (x+5)x \qquad \text{Multiplying by } x+5$$

$$4x+6 = x^2+5x$$

$$0 = x^2+x-6$$

$$0 = (x+3)(x-2) \qquad \text{Factoring}$$

$$x+3=0 \quad or \quad x-2=0$$

$$x=-3 \quad or \quad x=2.$$

The number -3 is not a solution because $\ln[4(-3)+6]$, or $\ln(-6)$, on the left side of the equation and $\ln(-3)$ on the right side of the equation are not real numbers. The value 2 checks and is the solution.

Graphical Solution

The solution of the equation

$$\ln(4x+6) - \ln(x+5) = \ln x$$

is the zero of the function

$$f(x) = \ln(4x+6) - \ln(x+5) - \ln x.$$

The solution is also the first coordinate of the x-intercept of the graph of the function. Here we use the Zero method.

$y_1 = \ln(4x+6) - \ln(x+5) - \ln x$

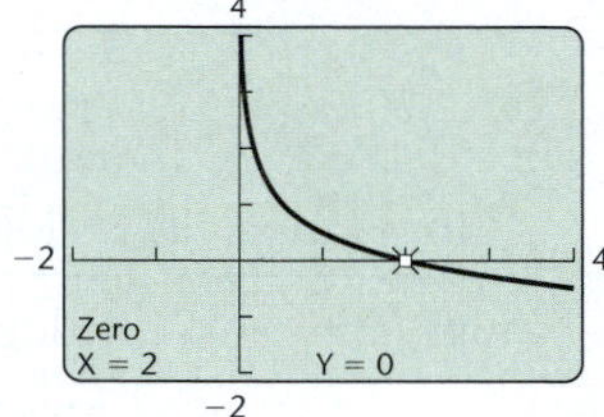

The solution of the equation is 2. From the graph, we can easily see that there is only one solution.

Now Try Exercise 43.

Sometimes we encounter equations for which an algebraic solution seems difficult or impossible.

$y_1 = e^{0.5x} - 7.3$, $y_2 = 2.08x + 6.2$

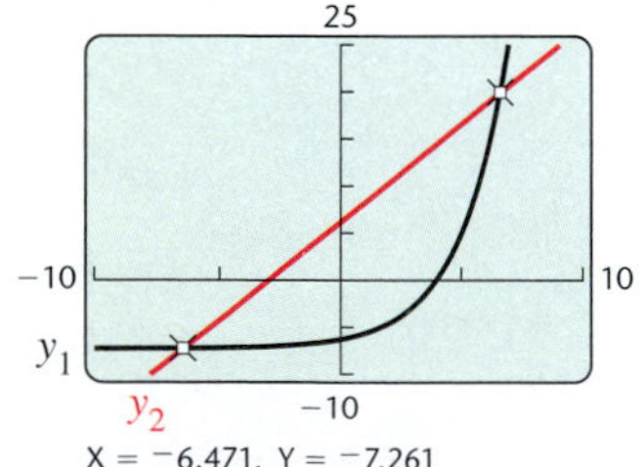

EXAMPLE 10 Solve: $e^{0.5x} - 7.3 = 2.08x + 6.2$.

Graphical Solution We graph the equations

$$y_1 = e^{0.5x} - 7.3 \quad \text{and} \quad y_2 = 2.08x + 6.2$$

on a graphing calculator and use the Intersect method. (See the window at left.)

We can also consider the equation

$$y = e^{0.5x} - 7.3 - 2.08x - 6.2, \quad \text{or} \quad y = e^{0.5x} - 2.08x - 13.5,$$

and use the Zero method. The approximate solutions are -6.471 and 6.610.

Now Try Exercise 67.

5.5 Exercise Set

Solve the exponential equation algebraically. Then check using a graphing calculator. Round to three decimal places, if appropriate.

1. $3^x = 81$

2. $2^x = 32$

3. $2^{2x} = 8$

4. $3^{7x} = 27$

5. $2^x = 33$

6. $2^x = 40$

7. $5^{4x-7} = 125$

8. $4^{3x-5} = 16$

9. $27 = 3^{5x} \cdot 9^{x^2}$

10. $3^{x^2+4x} = \frac{1}{27}$

11. $84^x = 70$

12. $28^x = 10^{-3x}$

13. $10^{-x} = 5^{2x}$

14. $15^x = 30$

15. $e^{-c} = 5^{2c}$

16. $e^{4t} = 200$

17. $e^t = 1000$

18. $e^{-t} = 0.04$

19. $e^{-0.03t} = 0.08$

20. $1000e^{0.09t} = 5000$

21. $3^x = 2^{x-1}$

22. $5^{x+2} = 4^{1-x}$

23. $(3.9)^x = 48$

24. $250 - (1.87)^x = 0$

25. $e^x + e^{-x} = 5$

26. $e^x - 6e^{-x} = 1$

27. $3^{2x-1} = 5^x$

28. $2^{x+1} = 5^{2x}$

29. $2e^x = 5 - e^{-x}$

30. $e^x + e^{-x} = 4$

Solve the logarithmic equation algebraically. Then check using a graphing calculator.

31. $\log_5 x = 4$

32. $\log_2 x = -3$

33. $\log x = -4$

34. $\log x = 1$

35. $\ln x = 1$

36. $\ln x = -2$

37. $\log_{64} \frac{1}{4} = x$

38. $\log_{125} \frac{1}{25} = x$

39. $\log_2 (10 + 3x) = 5$

40. $\log_5 (8 - 7x) = 3$

41. $\log x + \log (x - 9) = 1$

42. $\log_2 (x + 1) + \log_2 (x - 1) = 3$

43. $\log_2 (x + 20) - \log_2 (x + 2) = \log_2 x$

44. $\log (x + 5) - \log (x - 3) = \log 2$

45. $\log_8 (x + 1) - \log_8 x = 2$

46. $\log x - \log (x + 3) = -1$

47. $\log x + \log (x + 4) = \log 12$

48. $\log_3 (x + 14) - \log_3 (x + 6) = \log_3 x$

49. $\log (x + 8) - \log (x + 1) = \log 6$

50. $\ln x - \ln (x - 4) = \ln 3$

51. $\log_4 (x + 3) + \log_4 (x - 3) = 2$

52. $\ln (x + 1) - \ln x = \ln 4$

53. $\log (2x + 1) - \log (x - 2) = 1$

54. $\log_5 (x + 4) + \log_5 (x - 4) = 2$

55. $\ln (x + 8) + \ln (x - 1) = 2 \ln x$

56. $\log_3 x + \log_3 (x + 1) = \log_3 2 + \log_3 (x + 3)$

Solve.

57. $\log_6 x = 1 - \log_6 (x - 5)$

58. $2^{x^2-9x} = \frac{1}{256}$

59. $9^{x-1} = 100(3^x)$

60. $2 \ln x - \ln 5 = \ln (x + 10)$

61. $e^x - 2 = -e^{-x}$

62. $2 \log 50 = 3 \log 25 + \log (x - 2)$

Use a graphing calculator to find the approximate solutions of the equation.

63. $2^x - 5 = 3x + 1$

64. $0.082e^{0.05x} = 0.034$

65. $xe^{3x} - 1 = 3$

66. $4x - 3^x = -6$

67. $5e^{5x} + 10 = 3x + 40$

68. $4 \ln (x + 3.4) = 2.5$

69. $\log_8 x + \log_8 (x + 2) = 2$

70. $\ln x^2 = -x^2$

71. $\log_5 (x + 7) - \log_5 (2x - 3) = 1$

72. $\log_3 x + 7 = 4 - \log_5 x$

Approximate the point(s) of intersection of the pair of equations.

73. $2.3x + 3.8y = 12.4,\ y = 1.1 \ln (x - 2.05)$

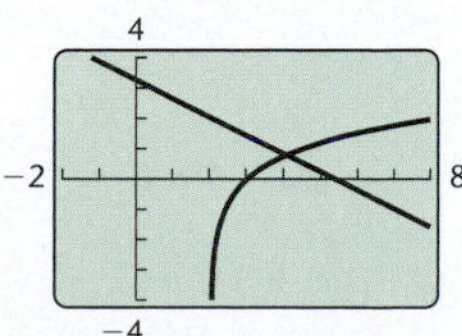

74. $y = \ln 3x,\ y = 3x - 8$

75. $y = 2.3 \ln (x + 10.7),\ y = 10e^{-0.007x^2}$

76. $y = 2.3 \ln (x + 10.7),\ y = 10e^{-0.07x^2}$

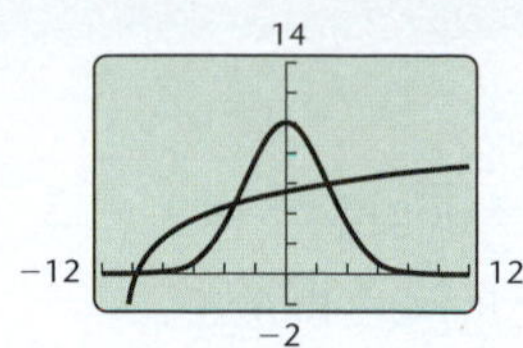

Skill Maintenance

In Exercises 77–80:

a) *Find the vertex.*
b) *Find the axis of symmetry.*
c) *Determine whether there is a maximum or a minimum value and find that value.* [3.3]

77. $g(x) = x^2 - 6$

78. $f(x) = -x^2 + 6x - 8$

79. $G(x) = -2x^2 - 4x - 7$

80. $H(x) = 3x^2 - 12x + 16$

Synthesis

Solve using any method.

81. $\dfrac{e^x + e^{-x}}{e^x - e^{-x}} = 3$

82. $\ln (\ln x) = 2$

83. $\sqrt{\ln x} = \ln \sqrt{x}$

84. $\ln \sqrt[4]{x} = \sqrt{\ln x}$

85. $(\log_3 x)^2 - \log_3 x^2 = 3$

86. $\log_3 (\log_4 x) = 0$

87. $\ln x^2 = (\ln x)^2$

88. $x\left(\ln \frac{1}{6}\right) = \ln 6$

89. $5^{2x} - 3 \cdot 5^x + 2 = 0$

90. $x^{\log x} = \dfrac{x^3}{100}$

91. $\ln x^{\ln x} = 4$

92. $|2^{x^2} - 8| = 3$

93. $\dfrac{\sqrt{(e^{2x} \cdot e^{-5x})^{-4}}}{e^x \div e^{-x}} = e^7$

94. Given that $a = (\log_{125} 5)^{\log_5 125}$, find the value of $\log_3 a$.

95. Given that $a = \log_8 225$ and $b = \log_2 15$, express a as a function of b.

96. Given that $f(x) = e^x - e^{-x}$, find $f^{-1}(x)$ if it exists.

5.6 Applications and Models: Growth and Decay; Compound Interest

- Solve applied problems involving exponential growth and exponential decay.
- Solve applied problems involving compound interest.
- Find models involving exponential functions and logarithmic functions.

Exponential functions and logarithmic functions with base e are rich in applications to many fields such as business, science, psychology, and sociology.

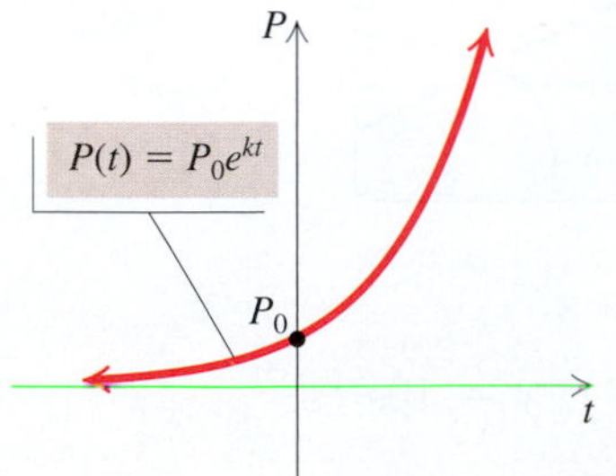

Population Growth

The function

$$P(t) = P_0 e^{kt}, \quad k > 0,$$

is a model of many kinds of population growth, whether it be a population of people, bacteria, smartphones, or money. In this function, P_0 is the population at time 0, P is the population after time t, and k is called the **exponential growth rate**. The graph of such an equation is shown at left.

EXAMPLE 1 ***Population Growth of Ghana.*** In 2014, the population of Ghana, located on the west coast of Africa, was about 25.8 million, and the exponential growth rate was 2.19% per year (*Source*: *CIA World Factbook*, 2015).

a) Find the exponential growth function.

b) Graph the exponential growth function.

c) Estimate the population in 2018.

d) At this growth rate, when will the population be 40 million?

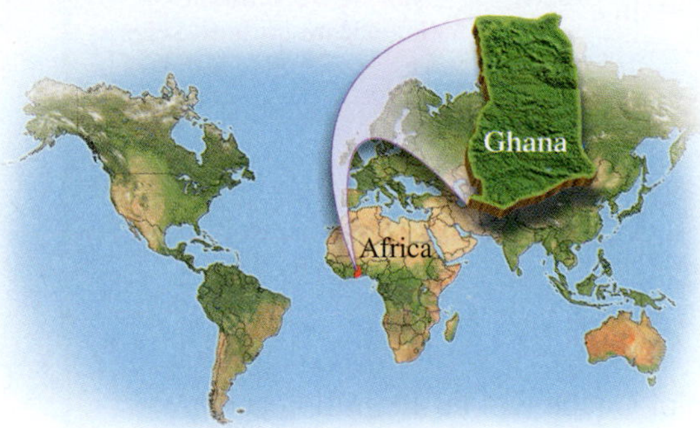

Solution

a) At $t = 0$ (2014), the population was 25.8 million, and the exponential growth rate was 2.19% per year. We substitute 25.8 for P_0 and 2.19%, or 0.0219, for k to obtain the exponential growth function

$$P(t) = 25.8e^{0.0219t},$$

where t is the number of years after 2014 and $P(t)$ is in millions.

b) Using a graphing calculator, we obtain the graph of the exponential growth function.

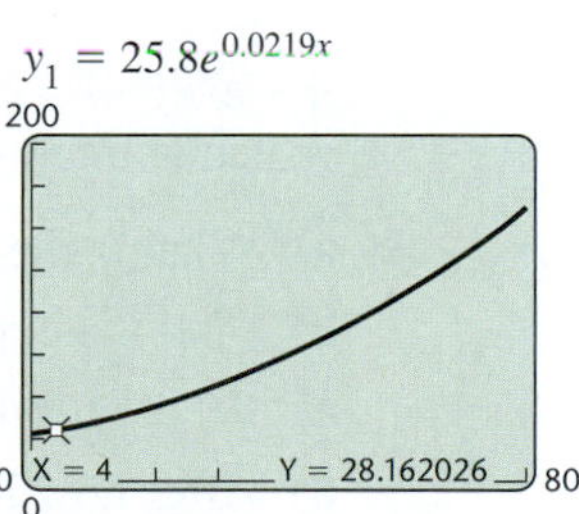

c) In 2018, $t = 4$; that is, 4 years have passed since 2014. To find the population in 2018, we substitute 4 for t:

$$P(4) = 25.8e^{0.0219(4)} = 25.8e^{0.0876} \approx 28.2.$$

We can also use the VALUE feature from the CALC menu on a graphing calculator to find $P(4)$. We estimate the population to be about 28.2 million, or 28,200,000, in 2018.

d) We are looking for the time t for which $P(t) = 40$. To find t, we solve the equation

$$40 = 25.8e^{0.0219t}$$

using both an algebraic method and a graphical method.

Algebraic Solution

We have

$$40 = 25.8e^{0.0219t} \quad \text{Substituting 40 for } P(t)$$

$$\frac{40}{25.8} = e^{0.0219t} \quad \text{Dividing by 25.8}$$

$$\ln \frac{40}{25.8} = \ln e^{0.0219t} \quad \text{Taking the natural logarithm on both sides}$$

$$\ln \frac{40}{25.8} = 0.0219t \quad \ln e^x = x$$

$$\frac{\ln \frac{40}{25.8}}{0.0219} = t \quad \text{Dividing by 0.0219}$$

$$20 \approx t.$$

The population of Ghana will be 40 million about 20 years after 2014.

Graphical Solution

Using the Intersect method, we graph the equations

$$y_1 = 25.8e^{0.0219t} \quad \text{and} \quad y_2 = 40$$

and find the first coordinate of their point of intersection.

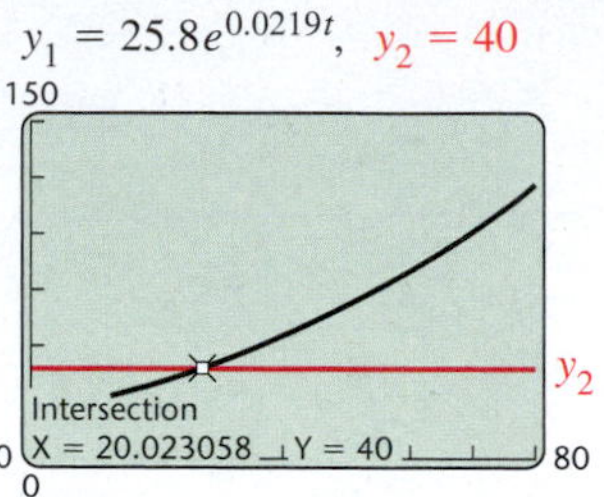

The solution is about 20, so the population of Ghana will be 40 million about 20 years after 2014.

Now Try Exercise 1.

Interest Compounded Continuously

When interest is paid on interest, we call it **compound interest**. Suppose that an amount P_0 is invested in a savings account at interest rate k **compounded continuously**. The amount $P(t)$ in the account after t years is given by the exponential function

$$P(t) = P_0e^{kt}.$$

EXAMPLE 2 ***Interest Compounded Continuously.*** Suppose that \$2000 is invested at interest rate k, compounded continuously, and grows to \$2504.65 in 5 years.

a) What is the interest rate?

b) Find the exponential growth function.

c) What will the balance be after 10 years?

d) After how long will the \$2000 have doubled?

Solution

a) At $t = 0$, $P(0) = P_0 = \$2000$. Thus the exponential growth function is of the form

$$P(t) = 2000e^{kt}.$$

We know that $P(5) = \$2504.65$. We substitute and solve for k:

$$2504.65 = 2000e^{k(5)} \quad \text{Substituting 2504.65 for } P(t) \text{ and 5 for } t$$

$$2504.65 = 2000e^{5k}$$

$$\frac{2504.65}{2000} = e^{5k}. \quad \text{Dividing by 2000}$$

Then

$$\ln \frac{2504.65}{2000} = \ln e^{5k} \qquad \text{Taking the natural logarithm}$$

$$\ln \frac{2504.65}{2000} = 5k \qquad \text{Using } \ln e^x = x$$

$$\frac{\ln \frac{2504.65}{2000}}{5} = k \qquad \text{Dividing by 5}$$

$$0.045 \approx k.$$

$y_1 = 2000e^{5x}$, $y_2 = 2504.65$

4000

Intersection
X = .04500036 Y = 2504.65

−0.5 0.5

0

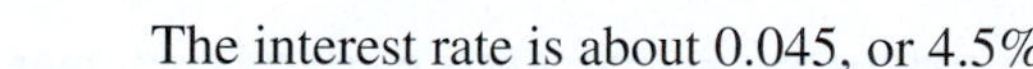

The interest rate is about 0.045, or 4.5%.

We can also find k by graphing the equations

$$y_1 = 2000e^{5x} \quad \text{and} \quad y_2 = 2504.65$$

and using the Intersect feature to approximate the first coordinate of the point of intersection. The interest rate is about 0.045, or 4.5%.

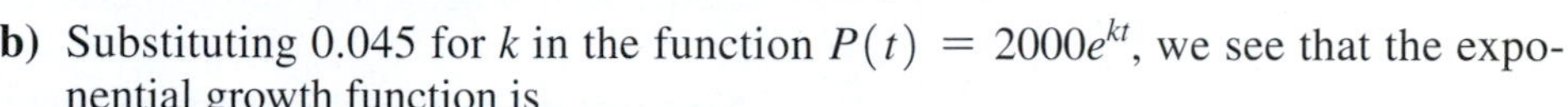

b) Substituting 0.045 for k in the function $P(t) = 2000e^{kt}$, we see that the exponential growth function is

$$P(t) = 2000e^{0.045t}.$$

c) The balance after 10 years is

$$P(10) = 2000e^{0.045(10)} = 2000e^{0.45} \approx \$3136.62.$$

d) To find the doubling time T, we set $P(T) = 2 \cdot P_0 = 2 \cdot \$2000 = \$4000$ and solve for T. We solve

$$4000 = 2000e^{0.045T}$$

using both an algebraic method and a graphical method.

Algebraic Solution

We have

$$4000 = 2000e^{0.045T}$$

$$2 = e^{0.045T} \qquad \text{Dividing by 2000}$$

$$\ln 2 = \ln e^{0.045T} \qquad \text{Taking the natural logarithm}$$

$$\ln 2 = 0.045T \qquad \ln e^x = x$$

$$\frac{\ln 2}{0.045} = T \qquad \text{Dividing by 0.045}$$

$$15.4 \approx T.$$

Thus the original investment of \$2000 will double in about 15.4 years.

Graphical Solution

We use the Zero method. We graph the function

$$y = 2000e^{0.045x} - 4000$$

and find the zero of the function. The zero of the function is the solution of the equation.

$y = 2000e^{0.045x} - 4000$

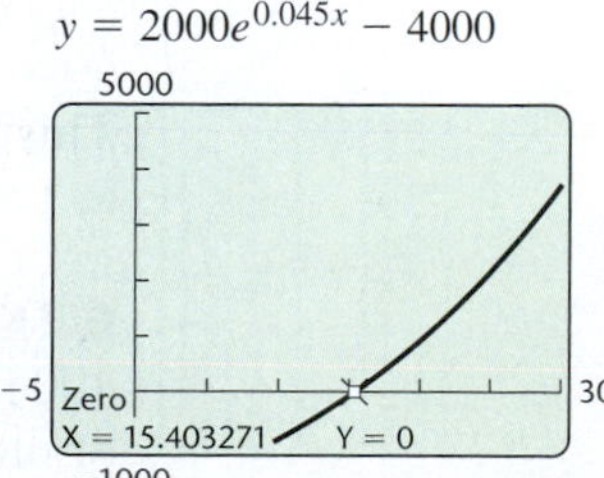

The solution is about 15.4, so the original investment of \$2000 will double in about 15.4 years.

Now Try Exercise 7.

We can find a general expression relating the growth rate k and the doubling time T by solving the following equation:

$$
\begin{aligned}
2P_0 &= P_0e^{kT} && \textbf{Substituting } 2P_0 \textbf{ for } P \textbf{ and } T \textbf{ for } t\\
2 &= e^{kT} && \textbf{Dividing by } P_0\\
\ln 2 &= \ln e^{kT} && \textbf{Taking the natural logarithm}\\
\ln 2 &= kT && \textbf{Using } \ln e^x = x\\
\frac{\ln 2}{k} &= T.
\end{aligned}
$$

GROWTH RATE AND DOUBLING TIME

The **growth rate** k and the **doubling time** T are related by

$$kT = \ln 2, \quad \text{or} \quad k = \frac{\ln 2}{T}, \quad \text{or} \quad T = \frac{\ln 2}{k}.$$

Note that the relationship between k and T does not depend on P_0.

EXAMPLE 3 ***Population Growth.*** The population of the Philippines is now doubling every 37.7 years (*Source*: *CIA World Factbook*, 2014). What is the exponential growth rate?

Solution We have

$$k = \frac{\ln 2}{T} = \frac{\ln 2}{37.7} \approx 0.0184 \approx 1.84\%.$$

The growth rate of the population of the Philippines is about 1.84% per year.

Now Try Exercise 3(e).

Models of Limited Growth

The model $P(t) = P_0e^{kt}$, $k > 0$, has many applications involving unlimited population growth. However, in some populations, there can be factors that prevent a population from exceeding some limiting value—perhaps a limitation on food, living space, or other natural resources. One model of such growth is

$$P(t) = \frac{a}{1 + be^{-kt}}.$$

This is called a **logistic function**. This function increases toward a *limiting value a* as $t \to \infty$. Thus, $y = a$ is the horizontal asymptote of the graph of $P(t)$.

EXAMPLE 4 ***Limited Population Growth in a Lake.*** A lake is stocked with 400 fish of a new variety. The size of the lake, the availability of food, and the number of other fish restrict the growth of that type of fish in the lake to a limiting value of 2500. The population gets closer and closer to this limiting value, but never reaches it. The population of fish in the lake after time t, in months, is given by the logistic function

$$P(t) = \frac{2500}{1 + 5.25e^{-0.32t}}.$$

a) Graph the function.

b) Find the population after 0, 1, 5, 10, 15, and 20 months.

Solution

a) We use a graphing calculator to graph the function. The graph is shown below. Note that this function increases toward a limiting value of 2500. The graph has $y = 2500$ as a horizontal asymptote.

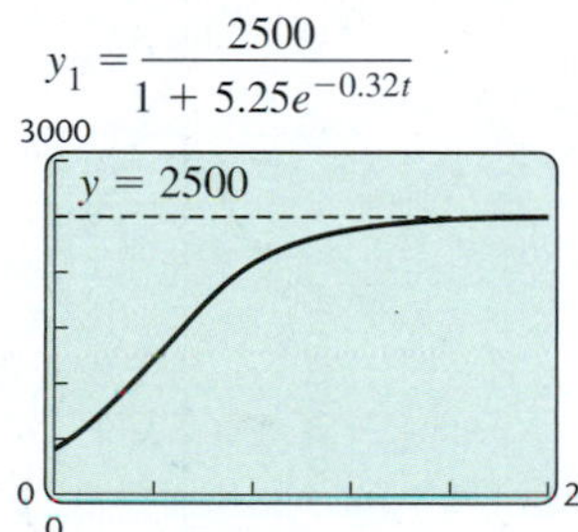

X	Y1
0	400
1	519.5
5	1213.6
10	2059.3
15	2396.5
20	2478.4

X =

b) We can use the TABLE feature on a graphing calculator set in ASK mode to find the function values. (See the window on the right above.) Thus the population will be about 400 after 0 months, 520 after 1 month, 1214 after 5 months, 2059 after 10 months, 2397 after 15 months, and 2478 after 20 months.

Now Try Exercise 17.

Another model of limited growth is provided by the function

$$P(t) = L(1 - e^{-kt}), \quad k > 0,$$

which is shown graphed at right. This function also increases toward a limiting value L as $t \to \infty$, so $y = L$ is the horizontal asymptote of the graph of $P(t)$.

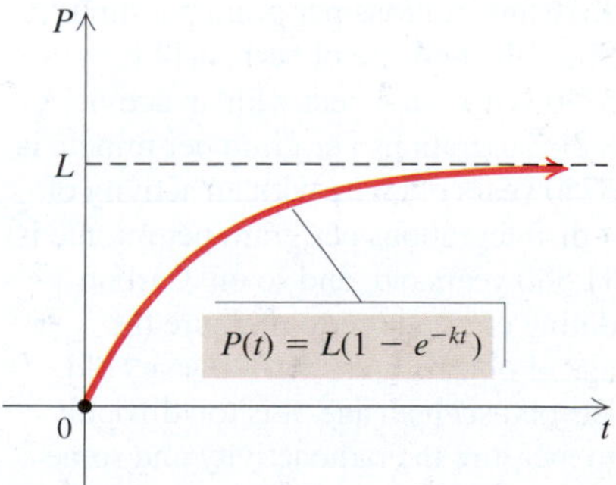

Exponential Decay

The function

$$P(t) = P_0 e^{-kt}, \quad k > 0,$$

is an effective model of the decline, or decay, of a population. An example is the decay of a radioactive substance. In this case, P_0 is the amount of the substance at time $t = 0$, and $P(t)$ is the amount of the substance left after time t, where k is a positive constant that depends on the situation. The constant k is called the **decay rate**.

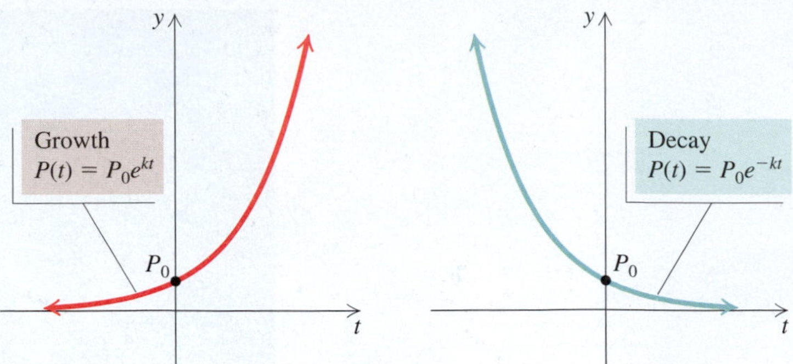

The **half-life** of bismuth (Bi-210) is 5 days. This means that half of an amount of radioactive bismuth will cease to be radioactive in 5 days. The effect of half-life T for nonnegative inputs is shown in the following graph. The exponential function gets close to 0, but never reaches 0, as t gets very large. Thus, according to an exponential decay model, a radioactive substance never completely decays.

How can scientists determine that an animal bone has lost 30% of its carbon-14? The assumption is that the percentage of carbon-14 in the atmosphere is the same as that in living plants and animals. When a plant or an animal dies, the amount of carbon-14 that it contains decays exponentially. A scientist can burn an animal bone and use a Geiger counter to determine the percentage of the smoke that is carbon-14. The amount by which this varies from the percentage in the atmosphere tells how much carbon-14 has been lost.

The process of carbon-14 dating was developed by the American chemist Willard E. Libby in 1952. It is known that the radioactivity in a living plant is 16 disintegrations per gram per minute. Since the half-life of carbon-14 is 5750 years, an object with an activity of 8 disintegrations per gram per minute is 5750 years old, one with an activity of 4 disintegrations per gram per minute is 11,500 years old, and so on. Carbon-14 dating can be used to measure the age of objects up to 40,000 years old. Beyond such an age, it is too difficult to measure the radioactivity and some other method would need to be used.

Carbon-14 dating was used to find the age of the Dead Sea Scrolls. It was also used to refute the authenticity of the Shroud of Turin, presumed to have covered the body of Christ.

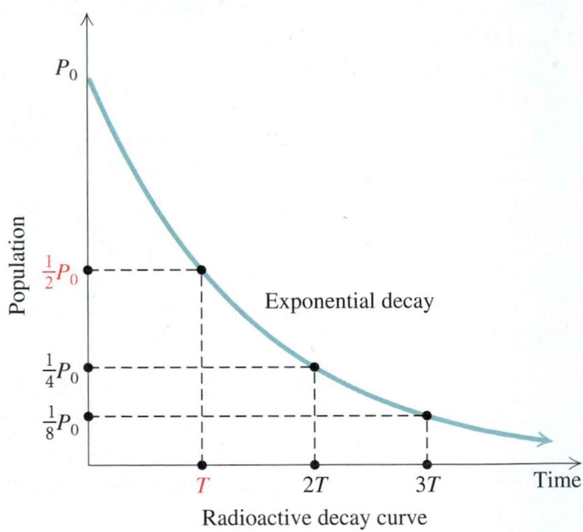

Radioactive decay curve

We can find a general expression relating the decay rate k and the half-life time T by solving the following equation:

$$\frac{1}{2}P_0 = P_0e^{-kT} \qquad \text{Substituting } \tfrac{1}{2}P_0 \text{ for } P \text{ and } T \text{ for } t$$

$$\frac{1}{2} = e^{-kT} \qquad \text{Dividing by } P_0$$

$$\ln\frac{1}{2} = \ln e^{-kT} \qquad \text{Taking the natural logarithm on both sides}$$

$$\ln 2^{-1} = -kT \qquad \frac{1}{2} = 2^{-1};\ \ln e^x = x$$

$$-\ln 2 = -kT \qquad \text{Using the power rule}$$

$$\frac{\ln 2}{k} = T. \qquad \text{Dividing by } -k$$

DECAY RATE AND HALF-LIFE

The **decay rate k** and the **half-life T** are related by

$$kT = \ln 2, \quad \text{or} \quad k = \frac{\ln 2}{T}, \quad \text{or} \quad T = \frac{\ln 2}{k}.$$

Note that the relationship between decay rate and half-life is the same as that between growth rate and doubling time.

EXAMPLE 5 *Carbon Dating.* The radioactive element carbon-14 has a half-life of 5750 years. The percentage of carbon-14 present in the remains of organic matter can be used to determine the age of that organic matter. Archaeologists discovered that the linen wrapping from one of the Dead Sea Scrolls had lost 22.3% of its carbon-14 at the time it was found. How old was the linen wrapping?

In 1947, a Bedouin youth looking for a stray goat climbed into a cave at Kirbet Qumran on the shores of the Dead Sea near Jericho and came upon earthenware jars containing an incalculable treasure of ancient manuscripts. Shown here are fragments of those Dead Sea Scrolls, a portion of some 600 or so texts found so far and which concern the Jewish books of the Bible. Officials date them before 70 A.D., making them the oldest Biblical manuscripts by 1000 years.

Solution We first find k when the half-life T is 5750 years:

$$k = \frac{\ln 2}{T} = \frac{\ln 2}{5750} = 0.00012.$$

Now we have the function

$$P(t) = P_0 e^{-0.00012t}.$$

(This function can be used for any subsequent carbon-dating problem.) If the linen wrapping has lost 22.3% of its carbon-14 from an initial amount P_0, then $77.7\%P_0$ is the amount present. To find the age t of the wrapping, we solve the equation for t:

$$77.7\%P_0 = P_0 e^{-0.00012t} \qquad \text{Substituting } 77.7\%P_0 \text{ for } P$$

$$0.777 = e^{-0.00012t} \qquad \text{Dividing by } P_0 \text{ and writing } 77.7\% \text{ as } 0.777$$

$$\ln 0.777 = \ln e^{-0.00012t} \qquad \text{Taking the natural logarithm on both sides}$$

$$\ln 0.777 = -0.00012t \qquad \ln e^x = x$$

$$\frac{\ln 0.777}{-0.00012} = t \qquad \text{Dividing by } -0.00012$$

$$2103 \approx t.$$

Thus the linen wrapping on the Dead Sea Scrolls was about 2103 years old when it was found.

Now Try Exercise 9.

Exponential and Logarithmic Curve Fitting

We have added several new functions that can be considered when we fit curves to data. Let's review some of them.

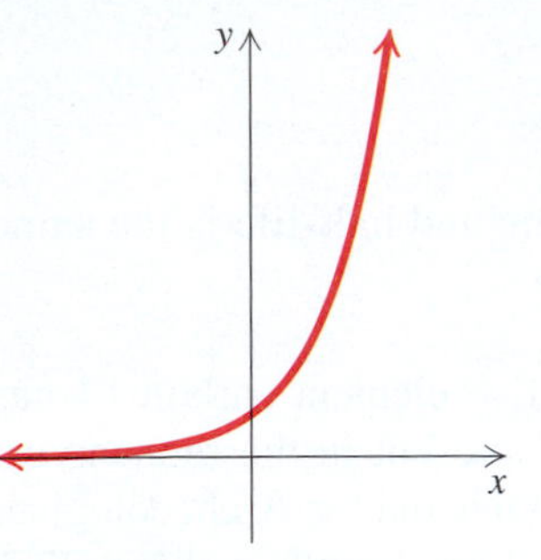

Exponential:
$f(x) = ab^x$, or ae^{kx}
$a > 0, b > 1, k > 0$

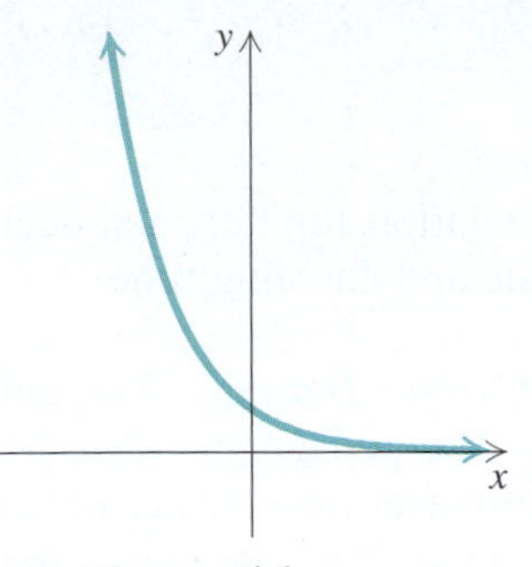

Exponential:
$f(x) = ab^{-x}$, or ae^{-kx}
$a > 0, b > 1, k > 0$

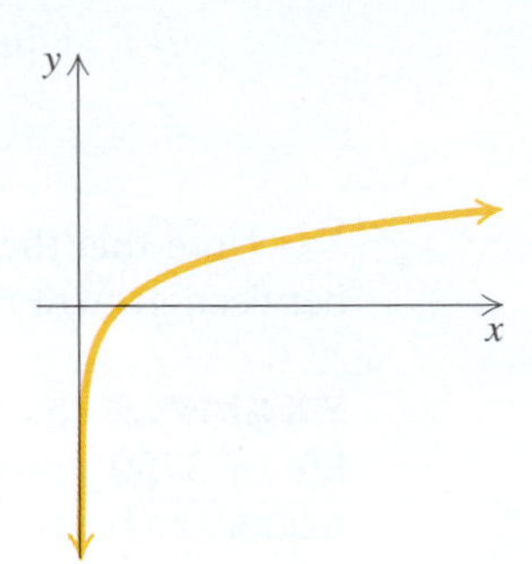

Logarithmic:
$f(x) = a + b \ln x$
$b > 0$

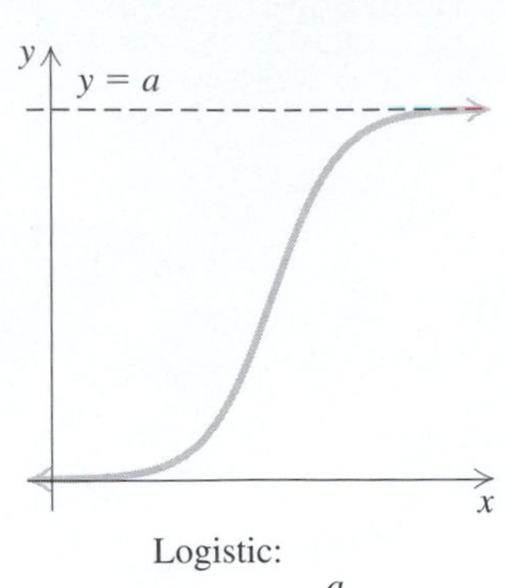

Logistic:
$f(x) = \dfrac{a}{1 + be^{-kx}}$
$a, b, k > 0$

Now, when we analyze a set of data for curve fitting, these models can be considered as well as polynomial functions (such as linear, quadratic, cubic, and quartic functions) and rational functions.

EXAMPLE 6 ***Health Expenditures.*** Health costs have increased exponentially over the past few decades. The following table lists national health expenditures in selected years from 1960 to 2012.

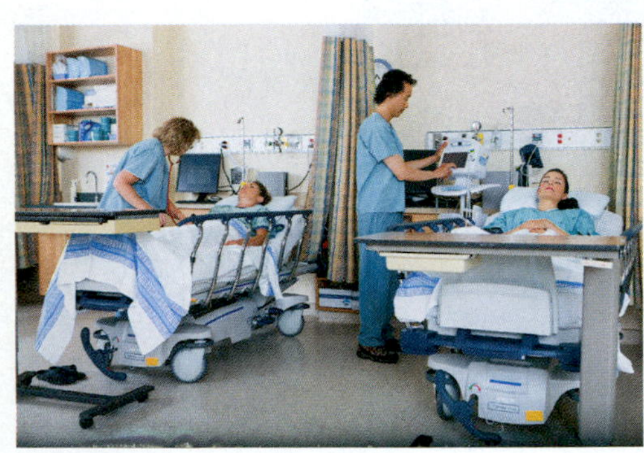

Year, x	National Health Expenditures (in billions)
1960, 0	\$ 27
1970, 10	75
1980, 20	256
1990, 30	724
2000, 40	1400
2010, 50	2600
2012, 52	2800

Sources: Centers for Medicare and Medicaid Services

a) Use a graphing calculator to fit an exponential function to the data.

b) Graph the function with the scatterplot of the data.

c) Use the function to estimate national health expenditures in 1985, in 2003, and in 2011. Then use the function to estimate health expenditures in 2016.

Solution

a) We will fit an equation of the type $y = a \cdot b^x$ to the data, where x is the number of years after 1960. Entering the data into the calculator and carrying out the regression procedure, we find that the equation is

$$y = 34.11812301(1.093629292)^x.$$

(See Fig. 1.) The correlation coefficient, r, is very close to 1. This gives us an indication that the exponential function fits the data well.

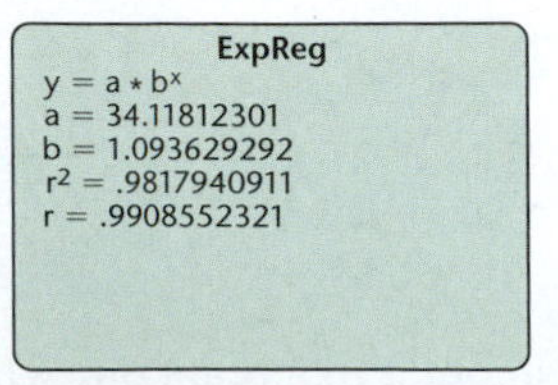

FIGURE 1.

b) The scatterplot and the graph are shown in Fig. 2 on the left below.

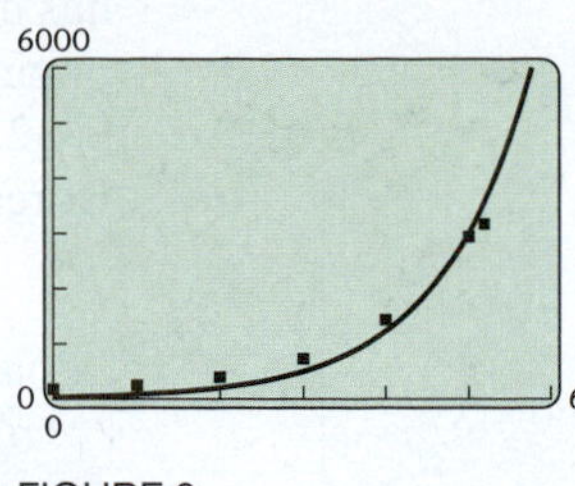

FIGURE 2.

6000
Y1 = 34.118123008385 * 1.09
X = 25
Y = 319.69694
0
0
60

FIGURE 3.

c) Using the VALUE feature in the CALC menu (see Fig. 3 on the right above), we evaluate the function found in part (a) for $x = 25$ $(1985 - 1960 = 25)$ and estimate national health expenditures in 1985 to be about \$320 billion. We continue using the VALUE feature to find health expenditures in additional years—2003: \$1601 billion, or \$1.601 trillion; 2011: \$3276 billion, or \$3.276 trillion; 2016: \$5125 billion, or \$5.125 trillion.

Now Try Exercise 29.

On some graphing calculators, there may be a REGRESSION feature that yields an exponential function, base *e*. If not, and you wish to find such a function, a conversion can be done using the following.

CONVERTING FROM BASE *b* TO BASE *e*

$$b^x = e^{x(\ln b)}$$

Then, for the equation in Example 6, we have

$$\begin{aligned} y &= 34.11812301(1.093629292)^x \\ &= 34.11812301e^{x(\ln 1.093629292)} \\ &= 34.11812301e^{0.089501791x}. \end{aligned}$$

We can prove this conversion formula using properties of logarithms, as follows:

$$e^{x(\ln b)} = e^{\ln b^x} = b^x.$$

5.6 Exercise Set

1. *Population Growth of Raleigh.* Raleigh, North Carolina, is one of the fastest growing cities in the United States. In 2014, the population was approximately 438,000, and the growth rate was about 3.4% per year.

   a) Find the exponential growth function.
   b) Estimate the population of Raleigh in 2017.
   c) When will the population of Raleigh be 500,000?
   d) Find the doubling time.

2. *Population Growth of Rabbits.* Under ideal conditions, a population of rabbits has an exponential growth rate of 11.7% per day. Consider an initial population of 100 rabbits.
   a) Find the exponential growth function.
   b) Graph the function.
   c) What will the population be after 7 days? after 2 weeks?
   d) Find the doubling time.

3. *Population Growth.* Complete the following table.

Country	Growth Rate, k	Doubling Time, T
a) United States		99.0 years
b) Bolivia		42.5 years
c) Uganda	3.32%	
d) Australia	1.11%	
e) Sweden		385 years
f) Laos	2.32%	
g) India	1.28%	
h) China		150.7 years
i) Guinea		26.3 years
j) Hong Kong	0.39%	

4. *E-Book Sales.* The revenue from e-book sales (not including educational textbooks) accounted for only 0.5% of U.S. publishing sales in 2006. This percentage grew to 22.6% in 2012. (*Source*: Association of American Publishers) Assuming that the exponential growth model applies:
   a) Find the value of k and write the function.
   b) Estimate the percentage of U.S. publishing sales that were e-book sales in 2009 and in 2010. Round to the nearest tenth.

5. *Population Growth of Haiti.* The population of Haiti has a growth rate of 1.08% per year. In 2015, the population was 9,996,731, and the land area of Haiti is 32,961,561,600 square yards. (*Source*: U.S. Census Bureau)

Assuming that this growth rate continues and is exponential, after how long will there be one person for every square yard of land?

6. *Georgia O'Keeffe Painting.* On November 20, 2014, Georgia O'Keeffe's painting, "Jimson Weed/White Flower No. 1," sold at Sotheby's auction in New York City for $44.4 million, a record price for any female artist's work that was sold at auction. The painting had previously sold for $1 million in 1994. (*Sources*: theguardian.com; sothebys.com) Assuming that the value A of the painting has grown exponentially:
   a) Find the value of k, and determine the exponential growth function, assuming that $A_0 = \$1$ million and t is the number of years after 1994.
   b) Estimate the value of the painting in 2019.
   c) What is the doubling time for the value of the painting?
   d) After how long will the value of the painting be $280 million, assuming that there is no change in the growth rate?

7. *Interest Compounded Continuously.* Suppose that \$10,000 is invested at an interest rate of 5.4% per year, compounded continuously.

a) Find the exponential function that describes the amount in the account after time t, in years.
b) What is the balance after 1 year? 2 years? 5 years? 10 years?
c) What is the doubling time?

8. *Interest Compounded Continuously.* Complete the following table.

Initial Investment at $t = 0, P_0$	Interest Rate, k	Doubling Time, T	Amount After 5 Years
a) \$35,000	3.2%		
b) \$5000			\$7,130.90
c)	5.6%		\$9,923.47
d)		11 years	\$17,539.32
e) \$109,000			\$136,503.18
f)		46.2 years	\$19,552.82

9. *Carbon Dating.* In 1970, Amos Flora of Flora, Indiana, discovered teeth and jawbones while dredging a creek. Scientists discovered that the bones were from a mastodon and that they had lost 77.2% of their carbon-14. How old were the bones at the time that they were discovered? (*Sources*: "Farm Yields Bones Thousands of Years Old," by Dan McFeely, *Indianapolis Star*, October 20, 2008; Field Museum of Chicago, Bill Turnbull, anthropologist)

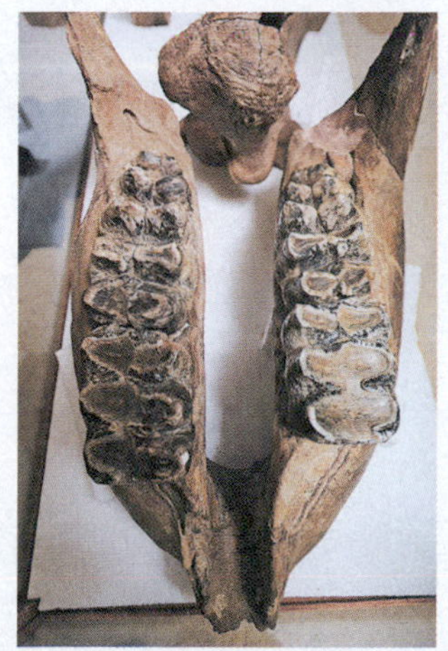

10. *Tomb in the Valley of the Kings.* In February 2006, in the Valley of the Kings in Egypt, a team of archaeologists uncovered the first tomb since King Tut's tomb was found in 1922. The tomb contained five wooden sarcophagi that contained mummies. The archaeologists believe that the mummies are from the 18th Dynasty, about 3300 to 3500 years ago. Determine the amount of carbon-14 that the mummies have lost.

11. *Radioactive Decay.* Complete the following table.

Radioactive Substance	Decay Rate, k	Half-Life T
a) Polonium (Po-218)		3.1 min
b) Lead (Pb-210)		22.3 years
c) Iodine (I-125)	1.15% per day	
d) Krypton (Kr-85)	6.5% per year	
e) Strontium (Sr-90)		29.1 years
f) Uranium (U-232)		70.0 years
g) Plutonium (Pu-239)		24,100 years

12. *Advertising Revenue.* The amount of advertising revenue in U.S. newspapers has declined continually since 2006. In 2006, the advertising revenue was \$49.3 billion, and by 2013 that amount had decreased to \$20.7 billion (*Source*: Newspaper Association of America). Assuming that the amount of newspaper advertising revenue decreased according to the exponential decay model:

a) Find the value of k, and write an exponential function that describes the advertising revenue after time t, in years, where t is the number of years after 2006.
b) Estimate the advertising revenue in 2008 and in 2012.

c) At this decay rate, when will the advertising revenue be $16 billion?

13. *Porsche 928.* The market value of the 1993–1995 Porsche 928 has had a recent upswing, increasing from $8000 in 2000 to $15,400 in 2015 (*Sources*: Haggerty; *Hemmings Motor News*, May 2015, p. 36) Assuming that the value V of the car has grown exponentially:

a) Find the value of k, and determine the exponential growth function, assuming that $V_0 = 8000$ and t is the number of years after 2000.
b) Estimate the value of the car in 2011.
c) What is the doubling time for the value of the car?
d) After how long will the value of the car be $25,000, assuming that there is no change in the growth rate?

14. *Married Adults.* The data in the following table show that the percentage of adults in the United States who are currently married is declining.

Year	Percent of Adults Who Are Married
1960	72.2%
1980	62.3
2000	57.4
2010	51.4
2012	50.5

Sources: Pew Research Center; U.S. Census Bureau

Assuming that the percentage of adults who are married will continue to decrease according to the exponential decay model:

a) Use the data for 1960 and 2012 to find the value of k and to write an exponential function that describes the percent of adults married after time t, in years, where t is the number of years after 1960.
b) Estimate the percent of adults who are married in 2015 and in 2018.
c) At this decay rate, in which year will the percent of adults who are married be 40%?

15. *British Guiana 1c Magenta Stamp.* The British Guiana 1c magenta stamp is considered the world's most famous stamp. It was issued in 1856, and only one of these stamps is known to exist. The sale of this stamp has broken the record for a single auction price four times. In 1980, John E. DuPont bought the stamp for $935,000. On June 17, 2014, the DuPont estate sold the stamp at Sotheby's auction in New York for a record price of $9,480,000 (*Source*: www.sothebys.com). Assuming that the value S of the stamp has grown exponentially:

a) Find the value of k, and determine the exponential growth function, assuming that $S_0 = \$935{,}000$ and t is the number of years after 1980.
b) Estimate the value of the stamp in 2000.
c) What is the doubling time for the value of the stamp?
d) After how long will the value of the stamp be $12 million, assuming that there is no change in growth rate?

16. *Oil Consumption.* In 1980, China consumed 1.85 million barrels of oil per day. By 2012, that consumption had grown to 10.28 million barrels per day. (*Sources*: U.S. Energy Information Administration; NextBigThingInvestor.com) Assuming that the consumption of oil C_0 in China has grown exponentially:

a) Find the value of k, and determine the exponential growth function, assuming that $C_0 = 1.85$ and t is the number of years after 1980.
b) Estimate the consumption of oil in 2005.
c) What is the doubling time for the consumption of oil in China?
d) After how long will the consumption of oil in China be 13 million barrels per day, assuming that there is no change in the growth rate?

17. *Spread of an Epidemic.* In a town whose population is 3500, a disease creates an epidemic. The number of people N infected t days after the disease has begun is given by the function

$$N(t) = \frac{3500}{1 + 19.9e^{-0.6t}}.$$

a) Graph the function.
b) How many are initially infected with the disease $(t = 0)$?
c) Find the number infected after 2 days, 5 days, 8 days, 12 days, and 16 days.
d) Using this model, can you say whether all 3500 people will ever be infected? Explain.

18. *Limited Population Growth in a Lake.* A lake is stocked with 640 fish of a new variety. The size of the lake, the availability of food, and the number of other fish restrict the growth of that type of fish in the lake to a limiting value of 3040. The population of fish in the lake after time t, in months, is given by the function

$$P(t) = \frac{3040}{1 + 3.75e^{-0.32t}}.$$

a) Graph the function.
b) Find the population after 0, 1, 5, 10, 15, and 20 months.

Newton's Law of Cooling. Suppose that a body with temperature T_1 is placed in surroundings with temperature T_0 different from that of T_1. The body will either cool or warm to temperature $T(t)$ after time t, in minutes, where

$$T(t) = T_0 + (T_1 - T_0)e^{-kt}.$$

Use this law in Exercises 19–22.

19. A cup of coffee with temperature 105°F is placed in a freezer with temperature 0°F. After 5 min, the temperature of the coffee is 70°F. What will its temperature be after 10 min?

20. A dish of lasagna baked at 375°F is taken out of the oven at 11:15 A.M. into a kitchen that is 72°F. After 3 min, the temperature of the lasagna is 365°F. What will the temperature of the lasagna be at 11:30 A.M.?

21. A chilled gelatin salad that has a temperature of 43°F is taken from the refrigerator and placed on the dining room table in a room that is 68°F. After 12 min, the temperature of the salad is 55°F. What will the temperature of the salad be after 20 min?

22. *When Was the Murder Committed?* The police discover the body of a murder victim. Critical to solving the crime is determining when the murder was committed. The coroner arrives at the murder scene at 12:00 P.M. She immediately takes the temperature of the body and finds it to be 94.6°F. She then takes the temperature 1 hr later and finds it to be 93.4°F. The temperature of the room is 70°F. When was the murder committed?

In Exercises 23–28, determine which, if any, of these functions might be used as a model for the data in the scatterplot.

a) *Quadratic,* $f(x) = ax^2 + bx + c$
b) *Polynomial, not quadratic*
c) *Exponential,* $f(x) = ab^x$, *or* P_0e^{kx}, $k > 0$
d) *Exponential,* $f(x) = ab^{-x}$, *or* P_0e^{-kx}, $k > 0$
e) *Logarithmic,* $f(x) = a + b\ln x$
f) *Logistic,* $f(x) = \dfrac{a}{1 + be^{-kx}}$

23.

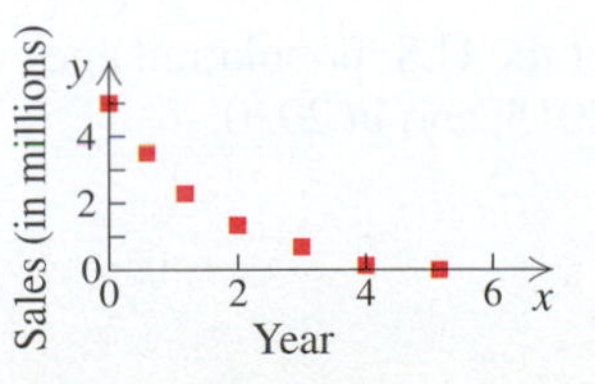

24.

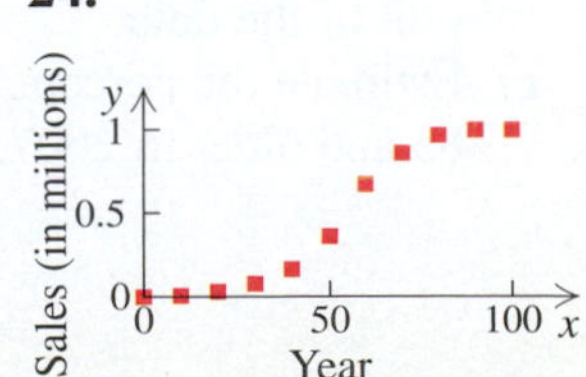

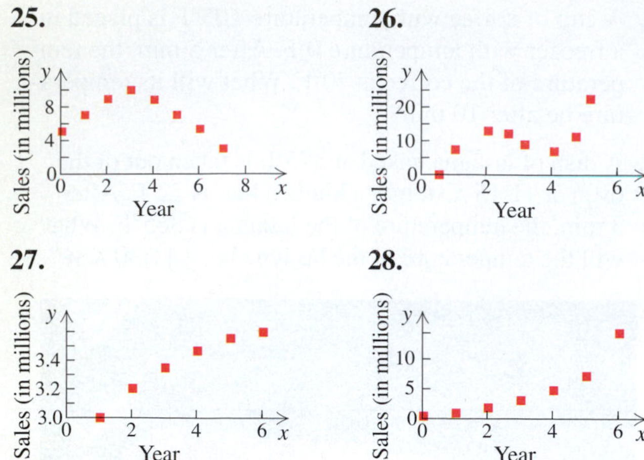

29. *Percent of Americans Ages 85 and Older.* In 1900, 0.2% of the U.S. population, or 122,000 people, were ages 85 and older. This number grew to 5,751,000 in 2010. The following table lists data regarding the percentage of the U.S. population ages 85 and older in selected years from 1900 to 2010.

Year, x	Percent of U.S. Population Ages 85 and Older, y
1900, 0	0.2%
1910, 10	0.2
1920, 20	0.2
1930, 30	0.2
1940, 40	0.3
1950, 50	0.4
1960, 60	0.5
1970, 70	0.7
1980, 80	1.0
1990, 90	1.2
1995, 95	1.4
2000, 100	1.5
2005, 105	1.7
2010, 110	1.9

Sources: U.S. Census Bureau; U.S. Department of Commerce

a) Use a graphing calculator to fit an exponential function to the data, where x is the number of years after 1900. Determine whether the function is a good fit.

b) Graph the function found in part (a) with a scatterplot of the data.

c) Estimate the percentage of the U.S. population ages 85 and older in 2007, in 2015, and in 2020.

30. *Forgetting.* In an economics class, students were given a final exam at the end of the course. Then they were retested with an equivalent test at subsequent time intervals. Their scores after time x, in months, are listed in the following table.

Time, x (in months)	Score, y
1	84.9%
2	84.6
3	84.4
4	84.2
5	84.1
6	83.9

a) Use a graphing calculator to fit a logarithmic function $y = a + b \ln x$ to the data.

b) Use the function to predict test scores after 8, 10, 24, and 36 months.

c) After how long will the test scores fall below 82%?

31. *Costs of Alzheimer's Disease.* The costs of caring for Americans with Alzheimer's disease have surpassed the costs of caring for cancer patients. The following table lists the costs and projected costs of caring for Alzheimer's patients in selected years.

Year, x	Costs of Alzheimer's Disease (in billions)
2010, 0	\$ 172
2014, 4	214
2020, 10	276
2025, 15	363
2030, 20	486
2035, 25	640
2040, 30	824
2045, 35	1009
2050, 40	1205

Sources: aarp.org/bulletin, "Where's the War on Alzheimer's," by T. R. Reid, January–February 2015; Alzheimer's Association

a) Use a graphing calculator to fit an exponential function to the data, where x is the number of years after 2010.

b) Graph the function found in part (a).

c) Use the function found in part (a) to estimate the costs of Alzheimer's disease in 2016 and in 2023.

32. *Wind Power Capacity.* U.S. wind power capacity has been increasing exponentially in recent years. The following table lists, for selected years, the U.S. wind power capacity, in megawatts (MW).

Year, x	U.S. Wind Power Capacity (in megawatts, MW)
2002, 0	4,557
2004, 2	6,619
2006, 4	11,450
2008, 6	25,065
2010, 8	40,283
2012, 10	60,012
2014, 12	65,879

Source: American Wind Energy Association

a) Use a graphing calculator to fit an exponential function to the data, where x is the number of years after 2002.
b) Graph the function in part (a).
c) Use the function in part (a) to estimate the wind capacity in 2011 and in 2015.

33. *Architects in Indiana.* The number of architects employed in Indiana in 2013 is more than 500 less than the number employed in 2007. The following table lists the number of architects employed in selected years.

Year, x	Number of Architects Employed In Indiana
2007, 0	1460
2009, 2	1280
2011, 4	1010
2013, 6	950

Source: Indiana Department of Workforce Development

a) Use a graphing calculator to model the data with an exponential function, where x is the number of years after 2007.
b) Use the function found in part (a) to estimate the number of architects employed in Indiana in 2008 and in 2012.
c) If the number of architects continues to decrease at the same rate, in what year will the number of architects employed in Indiana be 900?

34. *Effect of Advertising.* A company introduced a new software product on a trial run in a city. They advertised the product on television and found the following data regarding the percent P of people who bought the product after x ads were run.

Number of Ads, x	Percentage Who Bought, P
0	0.2%
10	0.7
20	2.7
30	9.2
40	27.0
50	57.6
60	83.3
70	94.8
80	98.5
90	99.6

a) Use a graphing calculator to fit a logistic function

$$P(x) = \frac{a}{1 + be^{-kx}}$$

to the data.
b) What percent of people bought the product when 55 ads were run? 100 ads?
c) Find the horizontal asymptote for the graph. Interpret the asymptote in terms of the advertising situation.

Skill Maintenance

Vocabulary Reinforcement

In Exercises 35–40, choose the correct name of the principle or rule from the given choices.

principle of zero products
multiplication principle for equations
product rule
addition principle for inequalities
power rule
multiplication principle for inequalities
principle of square roots
quotient rule

35. For any real numbers a, b, and c: If $a < b$ and $c > 0$ are true, then $ac < bc$ is true. If $a < b$ and $c < 0$ are true, then $ac > bc$ is true. **[1.6]**

____

36. For any positive numbers M and N and any logarithm base a, $\log_a MN = \log_a M + \log_a N$. **[5.4]**

____

37. If $ab = 0$ is true, then $a = 0$ or $b = 0$, and if $a = 0$ or $b = 0$, then $ab = 0$. **[3.2]**

____

38. If $x^2 = k$, then $x = \sqrt{k}$ or $x = -\sqrt{k}$. **[3.2]**

____

39. For any positive number M, any logarithm base a, and any real number p, $\log_a M^p = p \log_a M$. **[5.4]**

____

40. For any real numbers a, b, and c: If $a = b$ is true, then $ac = bc$ is true. **[1.5]**

____

Synthesis

41. *Supply and Demand.* The supply function and the demand function for the sale of a certain type of DVD player are given by

$$S(p) = 150e^{0.004p} \quad \text{and} \quad D(p) = 480e^{-0.003p},$$

respectively, where $S(p)$ is the number of DVD players that the company is willing to sell at price p and $D(p)$ is the quantity that the public is willing to buy at price p. Find p such that $D(p) = S(p)$. This is called the **equilibrium price**.

42. *Carbon Dating.* Recently, while digging in Chaco Canyon, New Mexico, archaeologists found corn pollen that was 4000 years old (*Source*: *American Anthropologist*). This was evidence that Native Americans had been cultivating crops in the Southwest centuries earlier than scientists had thought. What percent of the carbon-14 had been lost from the pollen?

43. *Present Value.* Following the birth of a child, a grandparent wants to make an initial investment P_0 that will grow to \$50,000 for the child's education at age 18. Interest is compounded continuously at 5.2%. What should the initial investment be? Such an amount is called the **present value** of \$50,000 due 18 years from now.

44. *Present Value.*

a) Solve $P = P_0 e^{kt}$ for P_0.

b) Referring to Exercise 43, find the present value of \$50,000 due 18 years from now at interest rate 6.4%, compounded continuously.

45. *The Beer–Lambert Law.* A beam of light enters a medium such as water or smog with initial intensity I_0. Its intensity decreases depending on the thickness (or concentration) of the medium. The intensity I at a depth (or concentration) of x units is given by

$$I = I_0 e^{-\mu x}.$$

The constant μ (the Greek letter "mu") is called the **coefficient of absorption**, and it varies with the medium. For sea water, $\mu = 1.4$.

a) What percentage of light intensity I_0 remains in sea water at a depth of 1 m? 3 m? 5 m? 50 m?

b) Plant life cannot exist below 10 m. What percentage of I_0 remains at 10 m?

46. Given that $y = ax^b$, take the natural logarithm on both sides. Let $Y = \ln y$ and $X = \ln x$. Consider Y as a function of X. What kind of function is Y?

47. Given that $y = ae^x$, take the natural logarithm on both sides. Let $Y = \ln y$. Consider Y as a function of x. What kind of function is Y?

Chapter 5 Summary and Review

STUDY GUIDE

KEY TERMS AND CONCEPTS	EXAMPLES
SECTION 5.1: INVERSE FUNCTIONS	
Inverse Relation If a relation is defined by an equation, then interchanging the variables produces an equation of the inverse relation.	Given $y = -5x + 7$, find an equation of the inverse relation. $y = -5x + 7$ **Relation** $\downarrow \quad \downarrow$ $x = -5y + 7$ **Inverse relation**
One-to-One Functions A function f is one-to-one if different inputs have different outputs—that is, if $a \neq b$, then $f(a) \neq f(b)$. Or a function f is one-to-one if when the outputs are the same, the inputs are the same—that is, if $f(a) = f(b)$, then $a = b$.	Prove that $f(x) = 16 - 3x$ is one-to-one. Show that if $f(a) = f(b)$, then $a = b$. Assume that $f(a) = f(b)$. Since $f(a) = 16 - 3a$ and $f(b) = 16 - 3b$, $16 - 3a = 16 - 3b$ $-3a = -3b$ $a = b$. Thus, if $f(a) = f(b)$, then $a = b$ and f is one-to-one.
Horizontal-Line Test If it is possible for a horizontal line to intersect the graph of a function more than once, then the function is *not* one-to-one and its inverse is *not* a function. **One-to-One Functions and Inverses** • If a function f is one-to-one, then its inverse f^{-1} is a function. • The domain of a one-to-one function f is the range of the inverse f^{-1}. • The range of a one-to-one function f is the domain of the inverse f^{-1}. • A function that is increasing over its entire domain or is decreasing over its entire domain is a one-to-one function. The -1 in f^{-1} is *not* an exponent.	Using its graph, determine whether each function is one-to-one. **a)** 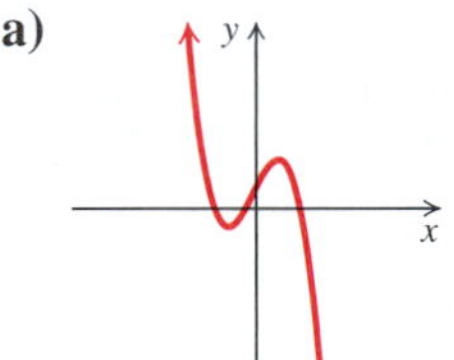 **b)** 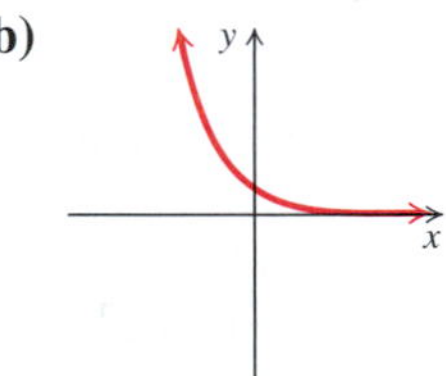 **a)** There are many horizontal lines that intersect the graph more than once. Thus the function is *not* one-to-one and its inverse is *not* a function. **b)** No horizontal line intersects the graph more than once. Thus the function is one-to-one and its inverse is a function.

(*continued*)

Obtaining a Formula for an Inverse

If a function f is one-to-one, a formula for its inverse can generally be found as follows:

1. Replace $f(x)$ with y.
2. Interchange x and y.
3. Solve for y.
4. Replace y with $f^{-1}(x)$.

The graph of f^{-1} is a reflection of the graph of f across the line $y = x$.

Given the one-to-one function $f(x) = 2 - x^3$, find a formula for its inverse. Then graph the function and its inverse on the same set of axes.

$f(x) = 2 - x^3$

1. $y = 2 - x^3$ **Replacing $f(x)$ with y**
2. $x = 2 - y^3$ **Interchanging x and y**
3. Solve for y:

   $y^3 = 2 - x$ **Adding y^3 and subtracting x**

   $y = \sqrt[3]{2 - x}$.
4. $f^{-1}(x) = \sqrt[3]{2 - x}$ **Replacing y with $f^{-1}(x)$**

$f(x) = 2 - x^3$ and $f^{-1}(x) = \sqrt[3]{2 - x}$

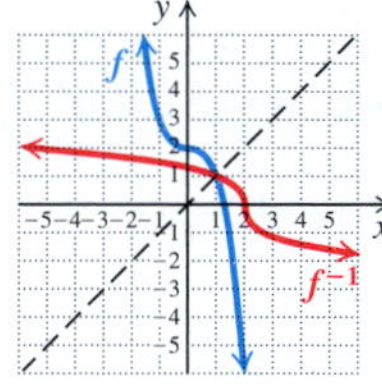

If a function f is one-to-one, then f^{-1} is the unique function such that each of the following holds:

$(f^{-1} \circ f)(x) = f^{-1}(f(x)) = x,$
for each x in the domain of f, and

$(f \circ f^{-1})(x) = f(f^{-1}(x)) = x,$
for each x in the domain of f^{-1}.

Given $f(x) = \dfrac{3 + x}{x}$, use composition of functions to show that $f^{-1}(x) = \dfrac{3}{x - 1}$.

$$(f^{-1} \circ f)(x) = f^{-1}(f(x)) = f^{-1}\left(\frac{3 + x}{x}\right) = \frac{3}{\frac{3 + x}{x} - 1}$$

$$= \frac{3}{\frac{3 + x - x}{x}} = \frac{3}{\frac{3}{x}} = 3 \cdot \frac{x}{3} = x;$$

$$(f \circ f^{-1})(x) = f(f^{-1}(x)) = f\left(\frac{3}{x - 1}\right)$$

$$= \frac{3 + \frac{3}{x - 1}}{\frac{3}{x - 1}} = \frac{\frac{3(x - 1) + 3}{x - 1}}{\frac{3}{x - 1}}$$

$$= \frac{3(x - 1) + 3}{x - 1} \cdot \frac{x - 1}{3}$$

$$= \frac{3x - 3 + 3}{3} = \frac{3x}{3} = x$$

SECTION 5.2: EXPONENTIAL FUNCTIONS AND GRAPHS

Exponential Function

$y = a^x$, or $f(x) = a^x$, $\quad a > 0, a \neq 1$

Continuous

One-to-one

Domain: $(-\infty, \infty)$

Range: $(0, \infty)$

Increasing if $a > 1$

Decreasing if $0 < a < 1$

Horizontal asymptote is x-axis

y-intercept: $(0, 1)$

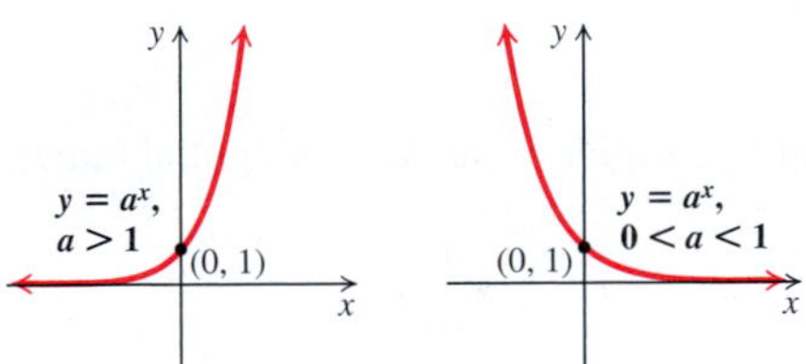

Graph: $f(x) = 2^x$, $g(x) = 2^{-x}$, $h(x) = 2^{x-1}$, and $t(x) = 2^x - 1$.

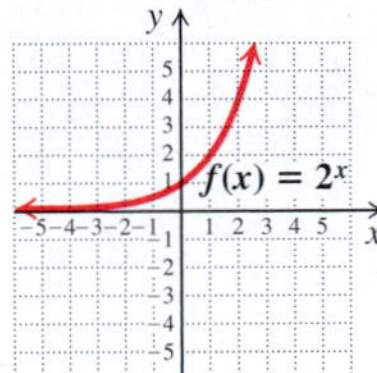

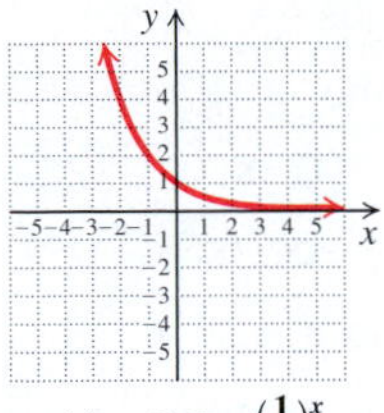

$g(x) = 2^{-x} = \left(\frac{1}{2}\right)^x$

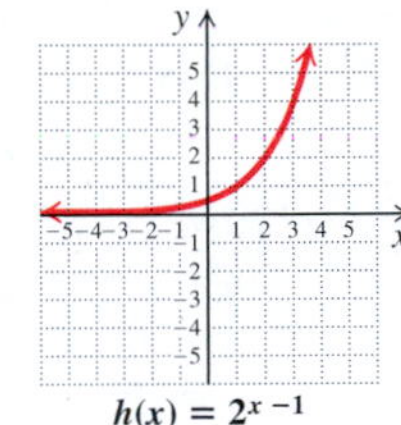

$h(x) = 2^{x-1}$

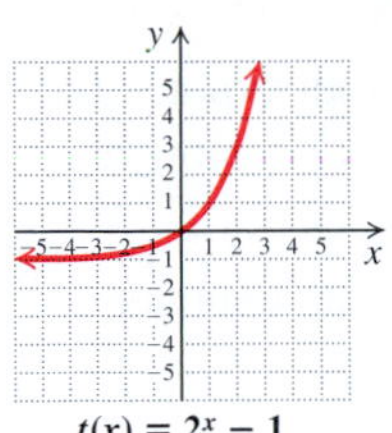

$t(x) = 2^x - 1$

Compound Interest

The amount of money A to which a principal P will grow after t years at interest rate r (in decimal form), compounded n times per year, is given by the formula

$$A = P\left(1 + \frac{r}{n}\right)^{nt}.$$

Suppose that \$5000 is invested at 3.5% interest, compounded quarterly. Find the money in the account after 3 years.

$$A = P\left(1 + \frac{r}{n}\right)^{nt} = 5000\left(1 + \frac{0.035}{4}\right)^{4\cdot 3}$$

$$\approx \$5551.02$$

The Number *e*

$e = 2.7182818284\ldots$

Find each of the following, to four decimal places, using a calculator.

$e^{-3} \approx 0.0498;$

$e^{4.5} \approx 90.0171$

Graph: $f(x) = e^x$ and $g(x) = e^{-x+2} - 4$.

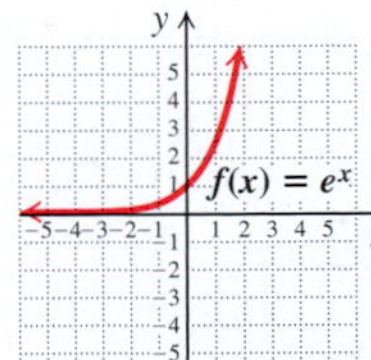

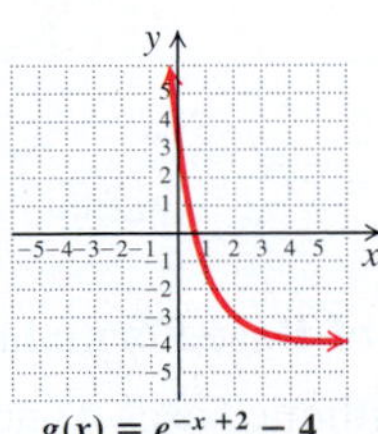

$g(x) = e^{-x+2} - 4$

SECTION 5.3: LOGARITHMIC FUNCTIONS AND GRAPHS

Logarithmic Function

$y = \log_a x, \quad x > 0, a > 0, a \neq 1$

- Continuous
- One-to-one
- Domain: $(0, \infty)$
- Range: $(-\infty, \infty)$
- Increasing if $a > 1$
- Vertical asymptote is y-axis
- x-intercept: $(1, 0)$

The inverse of an exponential function $f(x) = a^x$ is given by $f^{-1}(x) = \log_a x$.

Graph: $f(x) = \log_2 x$ and $g(x) = \ln(x - 1) + 2$.

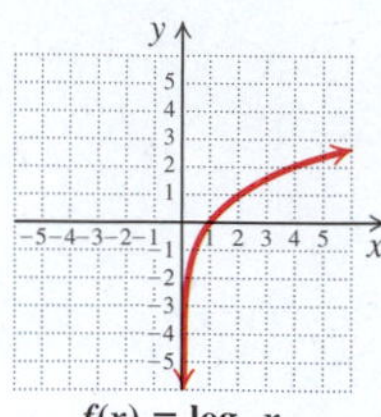

$f(x) = \log_2 x$

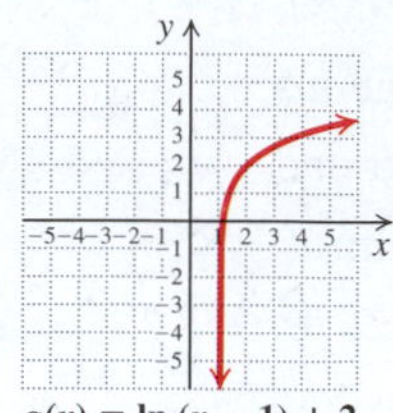

$g(x) = \ln(x - 1) + 2$

A logarithm is an exponent:

$$\log_a x = y \longleftrightarrow x = a^y.$$

$\log x$ means $\log_{10} x$ **Common logarithms**

$\ln x$ means $\log_e x$ **Natural logarithms**

Convert each logarithmic equation to an exponential equation.

$$\log_4 \frac{1}{16} = -2 \longleftrightarrow 4^{-2} = \frac{1}{16};$$

$$\ln R = 3 \longleftrightarrow e^3 = R$$

Convert each exponential equation to a logarithmic equation.

$$e^{-5} = 0.0067 \longleftrightarrow \ln 0.0067 = -5;$$

$$7^2 = 49 \longleftrightarrow \log_7 49 = 2$$

For any logarithm base a,

$$\log_a 1 = 0 \quad \text{and} \quad \log_a a = 1.$$

For the logarithm base e,

$$\ln 1 = 0 \quad \text{and} \quad \ln e = 1.$$

Find each of the following without using a calculator.

$\log 100 = 2$; $\quad \log 10^{-5} = -5$;

$\ln 1 = 0$; $\quad \log_9 9 = 1$;

$\ln \sqrt[3]{e} = \frac{1}{3}$; $\quad \log_2 64 = 6$;

$\log_8 1 = 0$; $\quad \ln e = 1$

Find each of the following using a calculator and rounding to four decimal places.

$\ln 223 = 5.4072$; $\quad \log \frac{2}{9} = -0.6532$;

$\log(-8)$ Does not exist; $\quad \ln 0.06 = -2.8134$

The Change-of-Base Formula

For any logarithmic bases a and b, and any positive number M,

$$\log_b M = \frac{\log_a M}{\log_a b}.$$

Find $\log_3 11$ using common logarithms:

$$\log_3 11 = \frac{\log 11}{\log 3} \approx 2.1827.$$

Find $\log_3 11$ using natural logarithms:

$$\log_3 11 = \frac{\ln 11}{\ln 3} \approx 2.1827.$$

Earthquake Magnitude

The magnitude R, measured on the Richter scale, of an earthquake of intensity I is defined as

$$R = \log \frac{I}{I_0},$$

where I_0 is a minimum intensity used for comparison.

What is the magnitude on the Richter scale of an earthquake of intensity $10^{6.8} \cdot I_0$?

$$R = \log \frac{I}{I_0} = \log \frac{10^{6.8} \cdot I_0}{I_0} = \log 10^{6.8} = 6.8$$

SECTION 5.4: PROPERTIES OF LOGARITHMIC FUNCTIONS

The Product Rule

For any positive numbers M and N, and any logarithmic base a,

$$\log_a MN = \log_a M + \log_a N.$$

The Power Rule

For any positive number M, any logarithmic base a, and any real number p,

$$\log_a M^p = p \log_a M.$$

The Quotient Rule

For any positive numbers M and N, and any logarithmic base a,

$$\log_a \frac{M}{N} = \log_a M - \log_a N.$$

Express $\log_c \sqrt{\frac{c^2 r}{b^3}}$ in terms of sums and differences of logarithms.

$$\begin{aligned}\log_c \sqrt{\frac{c^2 r}{b^3}} &= \log_c \left(\frac{c^2 r}{b^3}\right)^{1/2} \\ &= \tfrac{1}{2} \log_c \left(\frac{c^2 r}{b^3}\right) \\ &= \tfrac{1}{2}(\log_c c^2 r - \log_c b^3) \\ &= \tfrac{1}{2}(\log_c c^2 + \log_c r - \log_c b^3) \\ &= \tfrac{1}{2}(2 + \log_c r - 3 \log_c b) \\ &= 1 + \tfrac{1}{2} \log_c r - \tfrac{3}{2} \log_c b\end{aligned}$$

Express $\ln(3x^2 + 5x - 2) - \ln(x + 2)$ as a single logarithm.

$$\begin{aligned}\ln(3x^2 + 5x - 2) - \ln(x + 2) &= \ln \frac{3x^2 + 5x - 2}{x + 2} \\ &= \ln \frac{(3x - 1)(x + 2)}{x + 2} \\ &= \ln(3x - 1)\end{aligned}$$

Given $\log_a 7 \approx 0.8451$ and $\log_a 5 \approx 0.6990$, find $\log_a \frac{1}{7}$ and $\log_a 35$.

$$\log_a \frac{1}{7} = \log_a 1 - \log_a 7 \approx 0 - 0.8451 \approx -0.8451;$$

$$\begin{aligned}\log_a 35 = \log_a (7 \cdot 5) &= \log_a 7 + \log_a 5 \\ &\approx 0.8451 + 0.6990 \\ &\approx 1.5441\end{aligned}$$

For any base a and any real number x,

$$\log_a a^x = x.$$

For any base a and any positive real number x,

$$a^{\log_a x} = x.$$

Simplify each of the following.

$8^{\log_8 k} = k$; $\quad \log 10^{43} = 43$;

$\log_a a^4 = 4$; $\quad e^{\ln 2} = 2$

SECTION 5.5: SOLVING EXPONENTIAL EQUATIONS AND LOGARITHMIC EQUATIONS

The Base–Exponent Property

For any $a > 0, a \neq 1$,

$$a^x = a^y \longleftrightarrow x = y.$$

Solve: $3^{2x-3} = 81$.

$$\begin{aligned} 3^{2x-3} &= 3^4 \qquad 81 = 3^4 \\ 2x - 3 &= 4 \\ 2x &= 7 \\ x &= \tfrac{7}{2} \end{aligned}$$

The solution is $\frac{7}{2}$.

The Property of Logarithmic Equality

For any $M > 0, N > 0, a > 0$, and $a \neq 1$,

$$\log_a M = \log_a N \longleftrightarrow M = N.$$

Solve: $6^{x-2} = 2^{-3x}$.

$$\begin{aligned} \log 6^{x-2} &= \log 2^{-3x} \\ (x - 2)\log 6 &= -3x \log 2 \\ x \log 6 - 2 \log 6 &= -3x \log 2 \\ x \log 6 + 3x \log 2 &= 2 \log 6 \\ x(\log 6 + 3 \log 2) &= 2 \log 6 \\ x &= \frac{2 \log 6}{\log 6 + 3 \log 2} \\ x &\approx 0.9257 \end{aligned}$$

Solve: $\log_3 (x - 2) + \log_3 x = 1$.

$$\begin{aligned} \log_3 [x(x - 2)] &= 1 \\ x(x - 2) &= 3^1 \\ x^2 - 2x - 3 &= 0 \\ (x - 3)(x + 1) &= 0 \\ x - 3 = 0 \quad &or \quad x + 1 = 0 \\ x = 3 \quad &or \quad x = -1 \end{aligned}$$

The number -1 is not a solution because negative numbers do not have real-number logarithms. The value 3 checks and is the solution.

Solve: $\ln (x + 10) - \ln (x + 4) = \ln x$.

$$\begin{aligned} \ln \frac{x + 10}{x + 4} &= \ln x \\ \frac{x + 10}{x + 4} &= x \\ x + 10 &= x(x + 4) \\ x + 10 &= x^2 + 4x \\ 0 &= x^2 + 3x - 10 \\ 0 &= (x + 5)(x - 2) \\ x + 5 = 0 \quad &or \quad x - 2 = 0 \\ x = -5 \quad &or \quad x = 2 \end{aligned}$$

The number -5 is not a solution because $-5 + 4 = -1$ and $\ln(-1)$ is not a real number. The value 2 checks and is the solution.

SECTION 5.6: APPLICATIONS AND MODELS: GROWTH AND DECAY; COMPOUND INTEREST

Exponential Growth Model $P(t) = P_0e^{kt}, \quad k > 0$ **Doubling Time** $kT = \ln 2, \quad \text{or} \quad k = \frac{\ln 2}{T},$ $\text{or} \quad T = \frac{\ln 2}{k}$	In April 2015, the population of the United States was 320.8 million, and the exponential growth rate was 0.7% per year (*Source*: *CIA World Factbook* 2015). After how long will the population be double what it was in 2015? Estimate the population in 2020. With a population growth rate of 0.7%, or 0.007, the doubling time T is $T = \frac{\ln 2}{k} = \frac{\ln 2}{0.007} \approx 99.$ The population of the United States will be double what it was in 2015 in about 99 years after 2015. The exponential growth function is $P(t) = 320.8e^{0.007t},$ where t is the number of years after 2015 and $P(t)$ is in millions. Since $t = 5$ in 2020, we substitute 5 for t: $P(5) = 320.8e^{0.007(5)} = 320.8e^{0.035} \approx 332.2.$ The population will be about 332.2 million, or 332,200,000 in 2020.
Interest Compounded Continuously $P(t) = P_0e^{kt}, \quad k > 0$	Suppose that \$20,000 is invested at interest rate k, compounded continuously, and grows to \$23,236.68 in 3 years. What is the interest rate? What will the balance be in 8 years? The exponential growth function is of the form $P(t) = 20{,}000e^{kt}$. Given that $P(3) = \$23{,}236.68$, substituting 3 for t and 23,236.68 for $P(t)$ gives $23{,}236.68 = 20{,}000e^{k(3)}$ to get $k \approx 0.05$, or 5%. We then substitute 0.05 for k and 8 for t and determine $P(8)$: $P(8) = 20{,}000e^{0.05(8)} = 20{,}000e^{0.4} \approx \$29{,}836.49.$
Exponential Decay Model $P(t) = P_0e^{-kt}, \quad k > 0$ **Half-Life** $kT = \ln 2, \quad \text{or} \quad k = \frac{\ln 2}{T},$ $\text{or} \quad T = \frac{\ln 2}{k}$	Archaeologists discovered an animal bone that had lost 65.2% of its carbon-14 at the time it was found. How old was the bone? The decay rate for carbon-14 is 0.012%, or 0.00012. If the bone has lost 65.2% of its carbon-14 from an initial amount P_0, then $34.8\%P_0$ is the amount present. We substitute $34.8\%P_0$ for $P(t)$ and solve: $34.8\%P_0 = P_0e^{-0.00012t}$ $0.348 = e^{-0.00012t}$ $\ln 0.348 = -0.00012t$ $\frac{\ln 0.348}{-0.00012} = t$ $8796 \approx t.$ The bone was about 8796 years old when it was found.

REVIEW EXERCISES

Determine whether the statement is true or false.

1. The domain of a one-to-one function f is the range of the inverse f^{-1}. [5.1]
2. The x-intercept of $f(x) = \log x$ is $(0, 1)$. [5.3]
3. The graph of f^{-1} is a reflection of the graph of f across $y = 0$. [5.1]
4. If it is not possible for a horizontal line to intersect the graph of a function more than once, then the function is one-to-one and its inverse is a function. [5.1]
5. The range of all exponential functions is $[0, \infty)$. [5.2]
6. The horizontal asymptote of $y = 2^x$ is $y = 0$. [5.2]
7. Find the inverse of the relation $\{(1.3, -2.7), (8, -3), (-5, 3), (6, -3), (7, -5)\}$. [5.1]
8. Find an equation of the inverse relation. [5.1]
   a) $y = -2x + 3$
   b) $y = 3x^2 + 2x - 1$
   c) $0.8x^3 - 5.4y^2 = 3x$

Graph the function and determine whether the function is one-to-one using the horizontal-line test. [5.1]

9. $f(x) = -|x| + 3$
10. $f(x) = x^2 + 1$
11. $f(x) = 2x - \frac{3}{4}$
12. $f(x) = -\dfrac{6}{x + 1}$

In Exercises 13–18, given the function:

a) *Sketch the graph and determine whether the function is one-to-one.* [5.1], [5.3]
b) *If it is one-to-one, find a formula for the inverse.* [5.1], [5.3]

13. $f(x) = 2 - 3x$
14. $f(x) = \dfrac{x + 2}{x - 1}$
15. $f(x) = \sqrt{x - 6}$
16. $f(x) = x^3 - 8$
17. $f(x) = 3x^2 + 2x - 1$
18. $f(x) = e^x$

For the function f, use composition of functions to show that f^{-1} is as given. [5.1]

19. $f(x) = 6x - 5$, $f^{-1}(x) = \dfrac{x + 5}{6}$
20. $f(x) = \dfrac{x + 1}{x}$, $f^{-1}(x) = \dfrac{1}{x - 1}$

Find the inverse of the given one-to-one function f. Give the domain and the range of f and of f^{-1} and then graph both f and f^{-1} on the same set of axes. [5.1]

21. $f(x) = 2 - 5x$
22. $f(x) = \dfrac{x - 3}{x + 2}$
23. Find $f(f^{-1}(657))$:
$$f(x) = \frac{4x^5 - 16x^{37}}{119x}, \quad x > 1.$$ [5.1]
24. Find $f(f^{-1}(a))$: $f(x) = \sqrt[3]{3x - 4}$. [5.1]

Graph the function.

25. $f(x) = \left(\frac{1}{3}\right)^x$ [5.2]
26. $f(x) = 1 + e^x$ [5.2]
27. $f(x) = -e^{-x}$ [5.2]
28. $f(x) = \log_2 x$ [5.3]
29. $f(x) = \frac{1}{2} \ln x$ [5.3]
30. $f(x) = \log x - 2$ [5.3]

In Exercises 31–36, match the equation with one of the figures (a)–(f) that follow.

a) 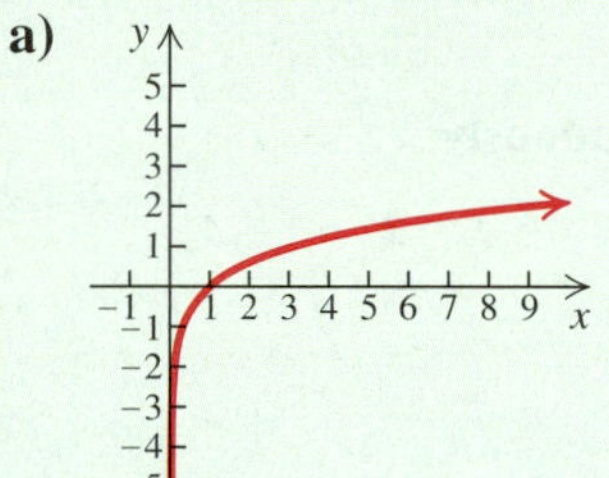

b) 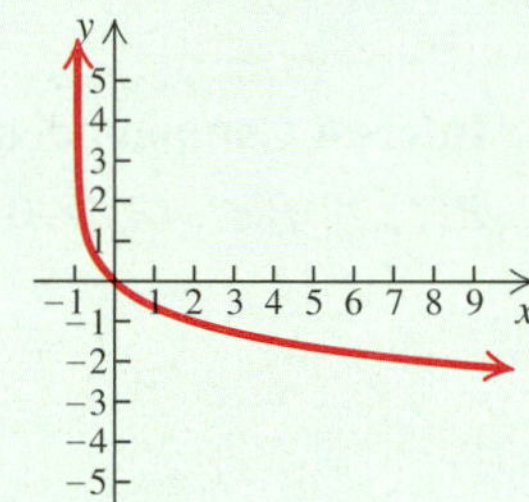

c) 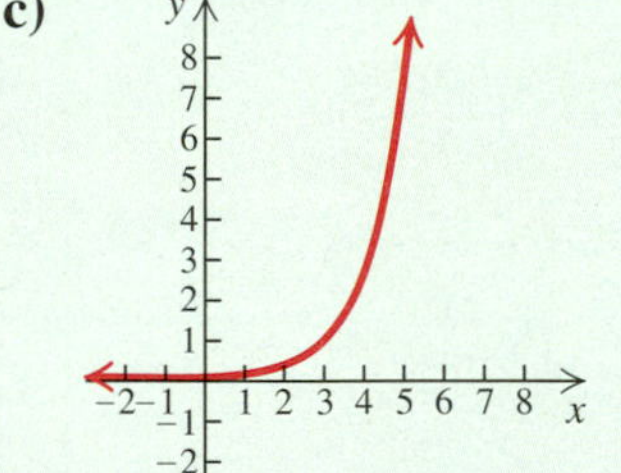

d) 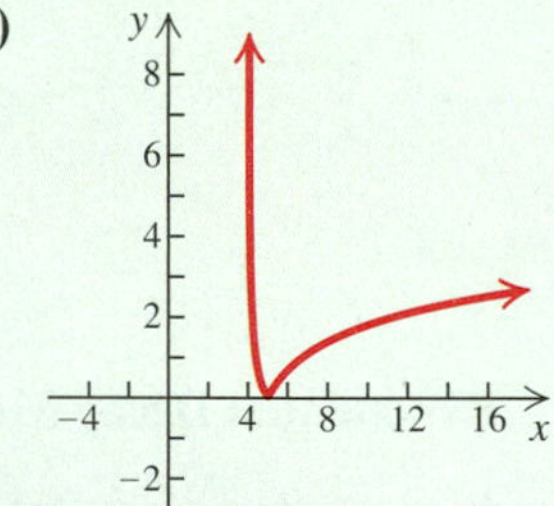

e) 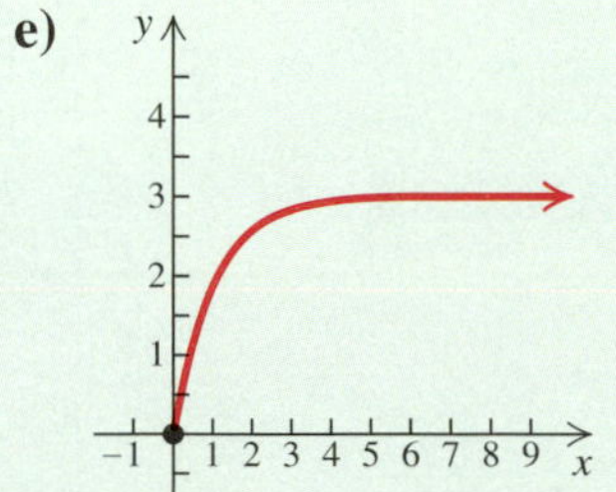

f) 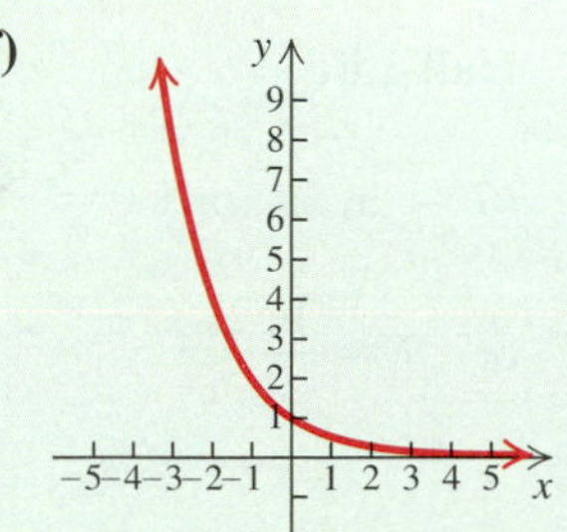

31. $f(x) = e^{x-3}$ [5.2]
32. $f(x) = \log_3 x$ [5.3]
33. $y = -\log_3 (x + 1)$ [5.3]

34. $y = \left(\frac{1}{2}\right)^x$ [5.2]

35. $f(x) = 3(1 - e^{-x}), x \geq 0$ [5.2]

36. $f(x) = |\ln(x - 4)|$ [5.3]

Find each of the following. Do not use a calculator. [5.3]

37. $\log_5 125$

38. $\log 100{,}000$

39. $\ln e$

40. $\ln 1$

41. $\log 10^{1/4}$

42. $\log_3 \sqrt{3}$

43. $\log 1$

44. $\log 10$

45. $\log_2 \sqrt[3]{2}$

46. $\log 0.01$

Convert to an exponential equation. [5.3]

47. $\log_4 x = 2$

48. $\log_a Q = k$

Convert to a logarithmic equation. [5.3]

49. $4^{-3} = \frac{1}{64}$

50. $e^x = 80$

Find each of the following using a calculator. Round to four decimal places. [5.3]

51. $\log 11$

52. $\log 0.234$

53. $\ln 3$

54. $\ln 0.027$

55. $\log(-3)$

56. $\ln 0$

Find the logarithm using the change-of-base formula. [5.3]

57. $\log_5 24$

58. $\log_8 3$

Express as a single logarithm and, if possible, simplify. [5.4]

59. $3 \log_b x - 4 \log_b y + \frac{1}{2} \log_b z$

60. $\ln(x^3 - 8) - \ln(x^2 + 2x + 4) + \ln(x + 2)$

Express in terms of sums and differences of logarithms. [5.4]

61. $\ln \sqrt[4]{wr^2}$

62. $\log \sqrt[3]{\frac{M^2}{N}}$

Given that $\log_a 2 = 0.301$, $\log_a 5 = 0.699$, *and* $\log_a 6 = 0.778$, *find each of the following.* [5.4]

63. $\log_a 3$

64. $\log_a 50$

65. $\log_a \frac{1}{5}$

66. $\log_a \sqrt[3]{5}$

Simplify. [5.4]

67. $\ln e^{-5k}$

68. $\log_5 5^{-6t}$

Solve. [5.5]

69. $\log_4 x = 2$

70. $3^{1-x} = 9^{2x}$

71. $e^x = 80$

72. $4^{2x-1} - 3 = 61$

73. $\log_{16} 4 = x$

74. $\log_x 125 = 3$

75. $\log_2 x + \log_2(x - 2) = 3$

76. $\log(x^2 - 1) - \log(x - 1) = 1$

77. $\log x^2 = \log x$

78. $e^{-x} = 0.02$

79. *Saving for College.* Following the birth of triplets, the grandparents deposit \$30,000 in a college trust fund that earns 4.2% interest, compounded quarterly.

a) Find a function for the amount in the account after t years. [5.2]

b) Find the amount in the account at $t = 0, 6, 12,$ and 18 years. [5.2]

80. *Breweries.* The number of breweries in the United States has increased in recent years. The total number of breweries can be estimated using the exponential function

$$B(t) = 2456.1(1.188)^t,$$

where t is the number of years after 2012 (*Source*: Brewers Association). Find the number of breweries in 2014. Then use this function to estimate the number of breweries in 2018. [5.2]

81. How long will it take an investment to double if it is invested at 4.5%, compounded continuously? [5.6]

82. The population of a metropolitan area consisting of 8 counties doubled in 26 years. What was the exponential growth rate? [5.6]

83. How old is a skeleton that has lost 27% of its carbon-14? [5.6]

84. The hydrogen ion concentration of milk is 2.3×10^{-6}. What is the pH? (See Exercise 98 in Exercise Set 5.3.) [5.3]

85. *Earthquake Magnitude.* The earthquake in Haiti on January 12, 2010, had an intensity of $10^{7.0} \cdot I_0$ (*Source*: U.S. Geological Survey). What is the magnitude of the earthquake on the Richter scale? [5.3]

86. What is the loudness, in decibels, of a sound whose intensity is $1000I_0$? (See Exercise 101 in Exercise Set 5.3.) [5.3]

87. *Walking Speed.* The average walking speed w, in feet per second, of a person living in a city of population P, in thousands, is given by the function

$$w(P) = 0.37 \ln P + 0.05.$$

a) The population of Wichita, Kansas, is 353,823. Find the average walking speed. [5.3]

b) A city's population has an average walking speed of 3.4 ft/sec. Find the population. [5.6]

88. *Social Security Distributions.* Cash Social Security distributions were \$35 million, or \$0.035 billion, in 1940. This amount has increased exponentially to \$786 billion in 2012. (*Source*: Pew Research Center) Assuming that the exponential growth model applies:

a) Find the exponential growth rate k. [5.6]
b) Find the exponential growth function. [5.6]
c) Estimate the total cash distributions in 1970, in 2000, and in 2015. [5.6]
d) In what year will the cash benefits reach \$2 trillion? [5.6]

89. *The Population of Cambodia.* The population of Cambodia was 15.2 million in 2013, and the exponential growth rate was 1.67% per year (*Source*: U.S. Census Bureau, World Population Profile).

a) Find the exponential growth function. [5.6]
b) What will the population be in 2017? in 2020? [5.6]
c) When will the population be 18 million? [5.6]
d) What is the doubling time? [5.6]

90. *Atmospheric Carbon as Carbon Dioxide (CO_2).* Scientists are of the opinion that the world's cumulative emissions of carbon as CO_2 should not exceed one trillion metric tons. The following table lists the cumulative atmospheric carbon as CO_2, in billions of metric tons, in selected years.

Year, x	Cumulative Atmospheric Carbon as CO_2 (in billions of metric tons)
1933, 0	100
1967, 34	200
1983, 50	300
1997, 64	400
2008, 75	500
2012, 79	545

Sources: *National Geographic*, April 2014, "Can Goal Ever Be Clean?," p. 38, by Michelle Nijhuis; Thomas Boden, Carbon Dioxide Information Analysis Center/Oak Ridge Laboratory, U.S. Department of Energy; EPA

a) Use a graphing calculator to fit an exponential function to the data, where x is the number of years after 1933. [5.6]
b) Graph the function with a scatterplot of the data. [5.6]
c) Estimate the cumulative atmospheric carbon in 1970, in 2000, and in 2015. [5.6]
d) In what year will the cumulative atmospheric carbon as CO_2 reach one trillion metric tons? [5.6]

91. Using only a graphing calculator, determine whether the following functions are inverses of each other.

$$f(x) = \frac{4 + 3x}{x - 2}, \quad g(x) = \frac{x + 4}{x - 3} \quad [5.1]$$

92. a) Use a graphing calculator to graph

$$f(x) = 5e^{-x} \ln x$$

in the viewing window $[-1, 10, -5, 5]$. [5.2], [5.3]

b) Estimate the relative maximum and minimum values of the function. [5.2], [5.3]

93. Which of the following is the horizontal asymptote of the graph of $f(x) = e^{x-3} + 2$? [5.2]

A. $y = -2$ **B.** $y = -3$
C. $y = 3$ **D.** $y = 2$

94. Which of the following is the domain of the logarithmic function $f(x) = \log(2x - 3)$? [5.3]

A. $\left(\frac{3}{2}, \infty\right)$ **B.** $\left(-\infty, \frac{3}{2}\right)$
C. $(3, \infty)$ **D.** $(-\infty, \infty)$

95. The graph of $f(x) = 2^{x-2}$ is which of the following? [5.2]

A.

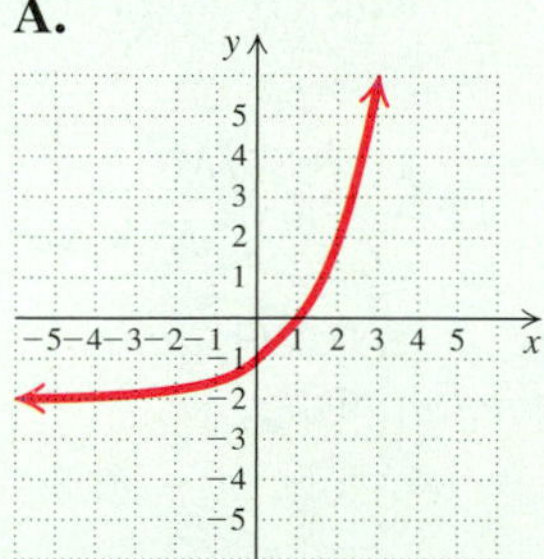

B.

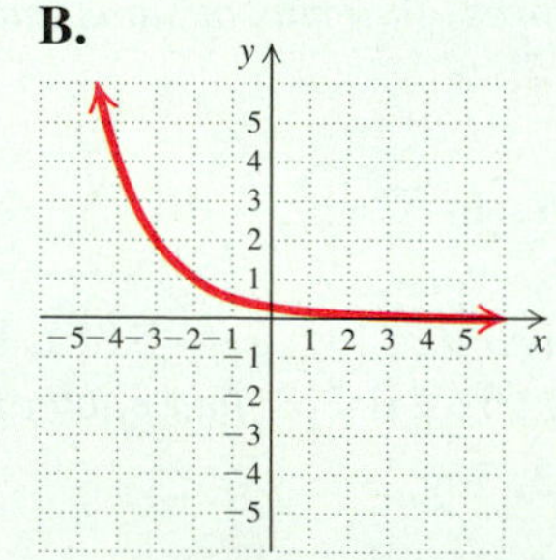

C.

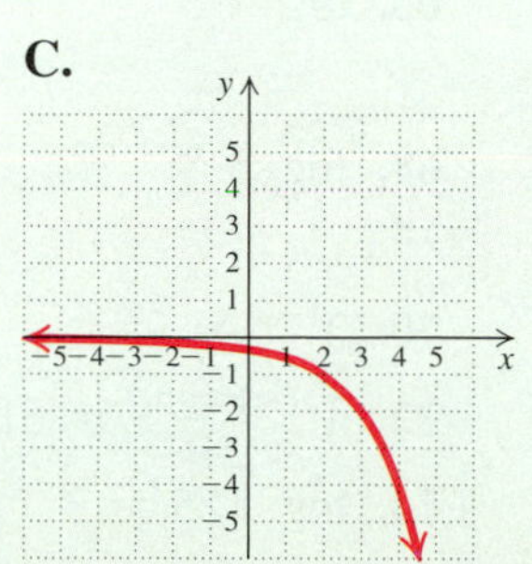

D.

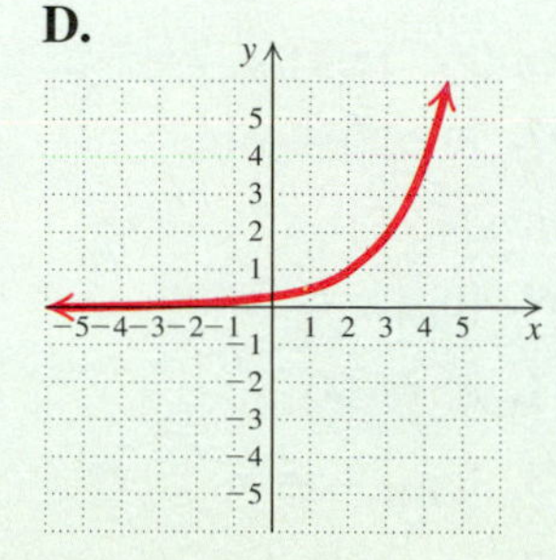

96. The graph of $f(x) = \log_2 x$ is which of the following? [5.3]

A.

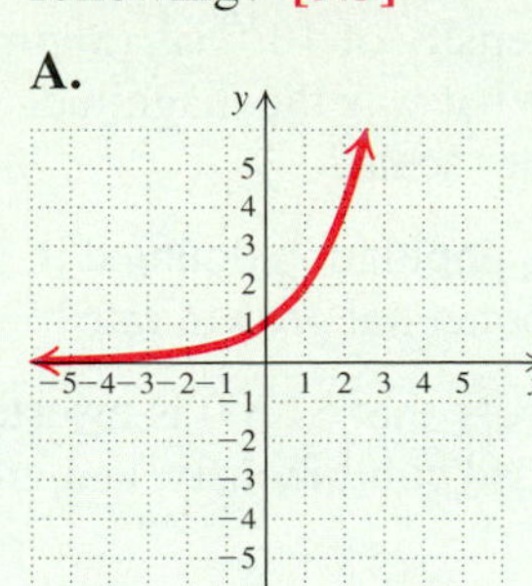

B.

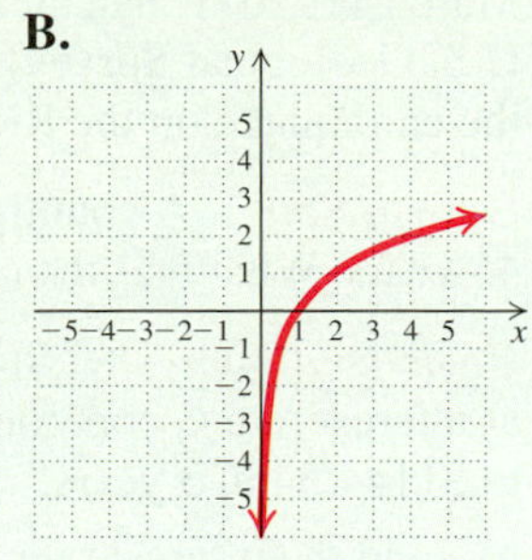

C.

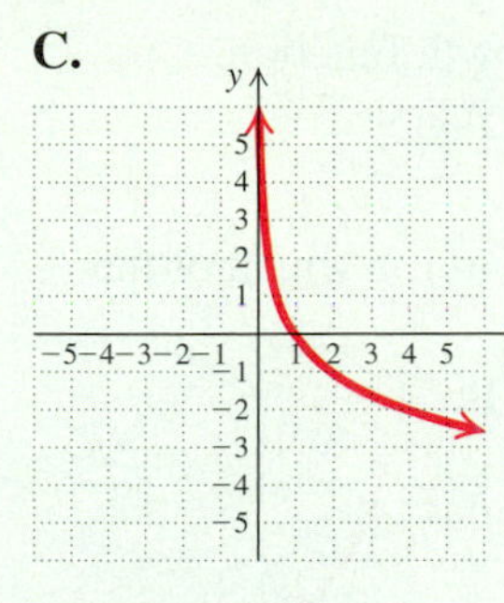

D.

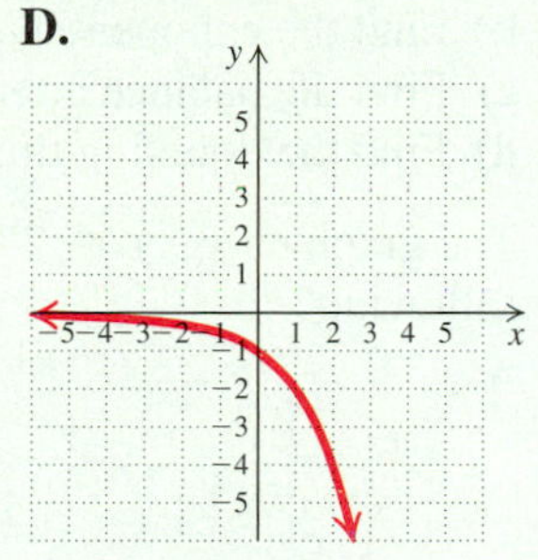

Synthesis

Solve. [5.5]

97. $|\log_4 x| = 3$

98. $\log x = \ln x$

99. $5^{\sqrt{x}} = 625$

100. Find the domain: $f(x) = \log_3 (\ln x)$. [5.3]

Collaborative Discussion and Writing

101. Explain how the graph of $f(x) = \ln x$ can be used to obtain the graph of $g(x) = e^{x-2}$. [5.3]

102. *Atmospheric Pressure.* Atmospheric pressure P at an altitude a is given by

$$P = P_0 e^{-0.00005a},$$

where P_0 is the pressure at sea level, approximately 14.7 lb/in^2 (pounds per square inch). Explain how a barometer, or some device for measuring atmospheric pressure, can be used to find the height of a skyscraper. [5.6]

103. Explain the errors, if any, in the following: [5.4]

$$\log_a ab^3 = (\log_a a)(\log_a b^3) = 3 \log_a b.$$

104. Describe the difference between $f^{-1}(x)$ and $[f(x)]^{-1}$. [5.1]

5 Chapter Test

1. Find the inverse of the relation

$$\{(-2, 5), (4, 3), (0, -1), (-6, -3)\}.$$

Determine whether the function is one-to-one. Answer yes or no.

2.

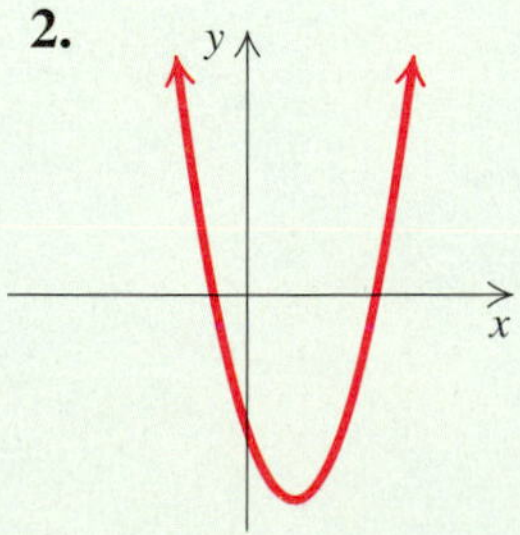

3.

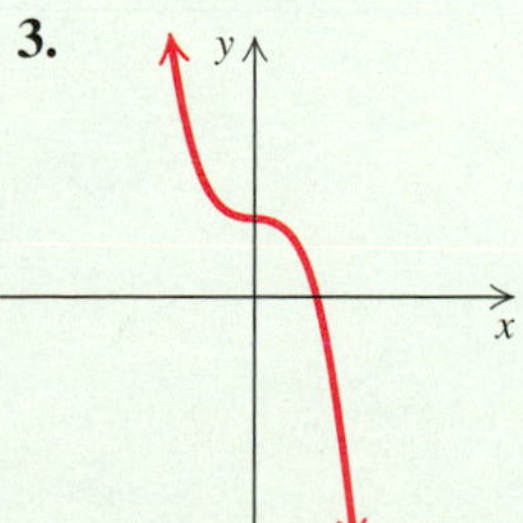

In Exercises 4–7, given the function:

a) *Sketch the graph and determine whether the function is one-to-one.*

b) *If it is one-to-one, find a formula for the inverse.*

4. $f(x) = x^3 + 1$

5. $f(x) = 1 - x$

6. $f(x) = \dfrac{x}{2 - x}$

7. $f(x) = x^2 + x - 3$

8. Use composition of functions to show that f^{-1} is as given:

$$f(x) = -4x + 3, \qquad f^{-1}(x) = \frac{3 - x}{4}.$$

9. Find the inverse of the one-to-one function

$$f(x) = \frac{1}{x - 4}.$$

Give the domain and the range of f and of f^{-1} and then graph both f and f^{-1} on the same set of axes.

Graph the function.

10. $f(x) = 4^{-x}$

11. $f(x) = \log x$

12. $f(x) = e^x - 3$

13. $f(x) = \ln (x + 2)$

Find each of the following. Do not use a calculator.

14. $\log 0.00001$

15. $\ln e$

16. $\ln 1$

17. $\log_4 \sqrt[5]{4}$

18. Convert to an exponential equation: $\ln x = 4$.

19. Convert to a logarithmic equation: $3^x = 5.4$.

Find each of the following using a calculator. Round to four decimal places.

20. $\ln 16$

21. $\log 0.293$

22. Find $\log_6 10$ using the change-of-base formula.

23. Express as a single logarithm:

$$2 \log_a x - \log_a y + \tfrac{1}{2} \log_a z.$$

24. Express $\ln \sqrt[5]{x^2 y}$ in terms of sums and differences of logarithms.

25. Given that $\log_a 3 = 1.585$ and $\log_a 15 = 3.907$, find $\log_a 5$.

26. Simplify: $\ln e^{-4t}$.

Solve.

27. $\log_{25} 5 = x$

28. $\log_3 x + \log_3 (x + 8) = 2$

29. $3^{4-x} = 27^x$

30. $e^x = 65$

31. *Earthquake Magnitude.* The Tohoku earthquake near the northeast coast of Honshu, Japan, on March 11, 2011, had an intensity of $10^{9.0} \cdot I_0$ (*Source*: U.S. Geological Survey). What was the magnitude of the earthquake on the Richter scale?

32. *Growth Rate.* A country's population doubled in 45 years. What was the exponential growth rate?

33. *Compound Interest.* Suppose that \$1000 is invested at interest rate k, compounded continuously, and grows to \$1144.54 in 3 years.

a) Find the interest rate.
b) Find the exponential growth function.
c) Find the balance after 8 years.
d) Find the doubling time.

34. The graph of $f(x) = 2^{x-1} + 1$ is which of the following?

A.

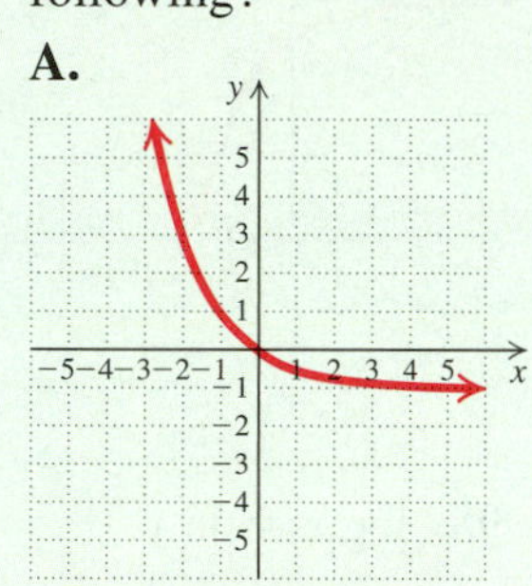

B.

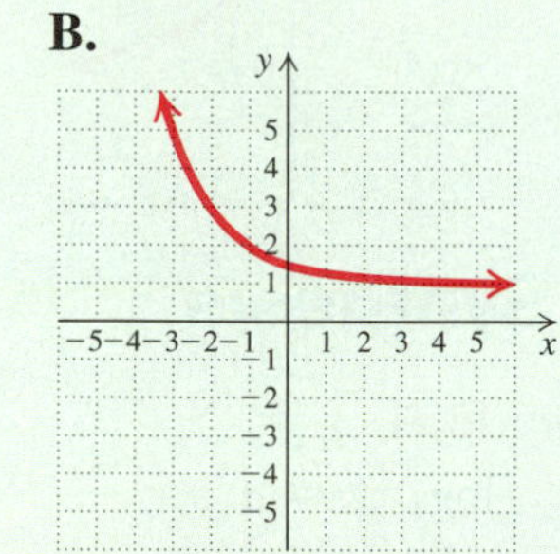

C.

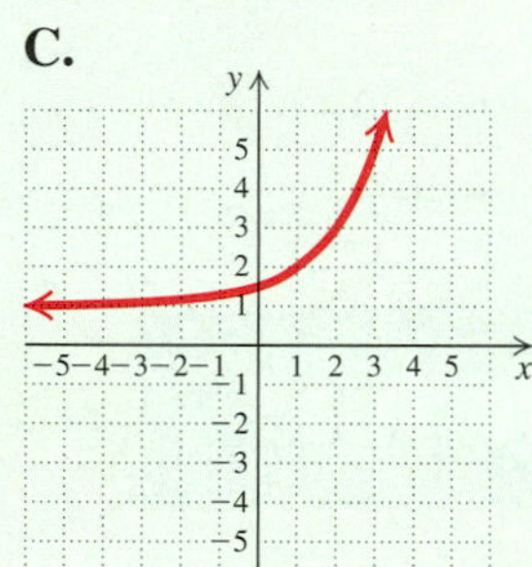

D.

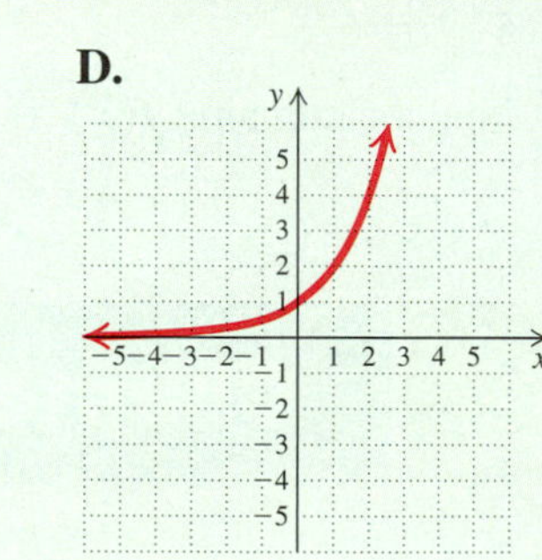

› Synthesis

35. Solve: $4^{\sqrt[3]{x}} = 8$.

CHAPTER

6

The Trigonometric Functions

APPLICATION

This problem appears as Exercise 23 in Section 6.2.

The longest escalator in the world is in the subway system in St. Petersburg, Russia. The escalator is 1084.6 ft long and drops a vertical distance of 195.8 ft. What is its angle of depression?

6.1 Trigonometric Functions of Acute Angles

- Determine the six trigonometric ratios for a given acute angle of a right triangle.
- Determine the trigonometric function values of 30°, 45°, and 60°.
- Using a calculator, find function values for any acute angle, and given a function value of an acute angle, find the angle.
- Given the function values of an acute angle, find the function values of its complement.

The Trigonometric Ratios

STUDY TIPS

Success can be planned. Combine goals and good study habits to create a plan for success that works for you. A list of study tips that your authors consider most helpful are included in the Guide to Success in the front of the text.

We begin our study of trigonometry by considering right triangles and acute angles measured in degrees. An **acute angle** is an angle with measure greater than 0° and less than 90°. Greek letters such as α (alpha), β (beta), γ (gamma), θ (theta), and ϕ (phi) are often used to denote an angle. Consider a right triangle with one of its acute angles labeled θ. The side opposite the right angle is called the **hypotenuse**. The other sides of the triangle are referenced by their position relative to the acute angle θ. One side is opposite θ and one is adjacent to θ.

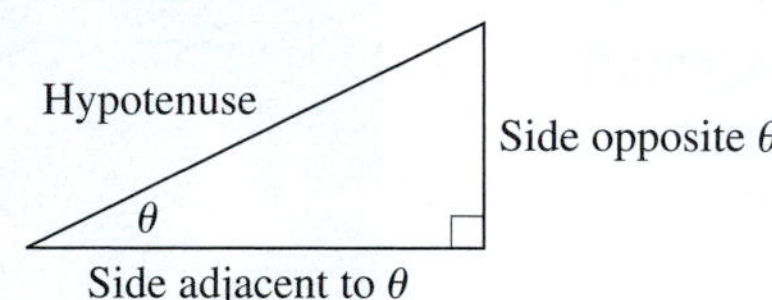

The *lengths* of the sides of the triangle are used to define the six trigonometric ratios:

sine (sin), cosecant (csc),
cosine (cos), secant (sec),
tangent (tan), cotangent (cot).

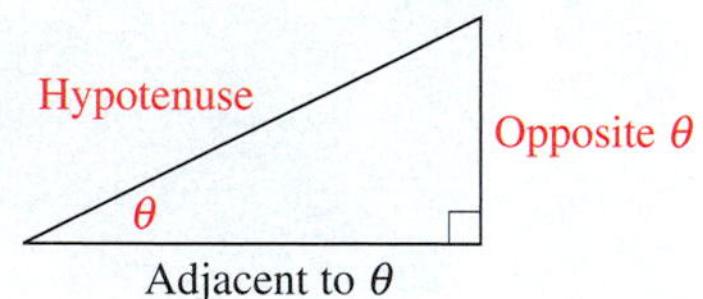

FIGURE 1.

The **sine of θ** is the length of the side opposite θ divided by the length of the hypotenuse (see Fig. 1):

$$\sin \theta = \frac{\text{length of side opposite } \theta}{\text{length of hypotenuse}}.$$

The ratio depends on the measure of angle θ and thus the ratio is a function of θ. The notation $\sin \theta$ actually means $\sin(\theta)$, where sin, or sine, is the name of the function.

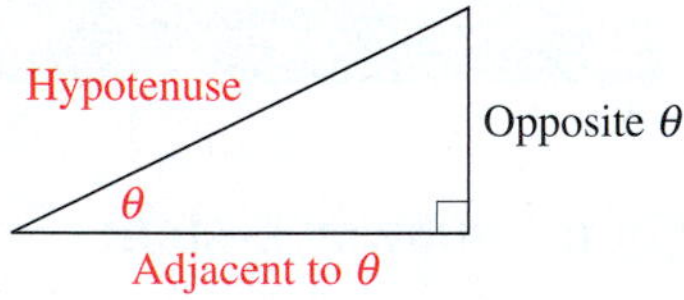

FIGURE 2.

The **cosine of θ** is the length of the side adjacent to θ divided by the length of the hypotenuse (see Fig. 2):

$$\cos \theta = \frac{\text{length of side adjacent to } \theta}{\text{length of hypotenuse}}.$$

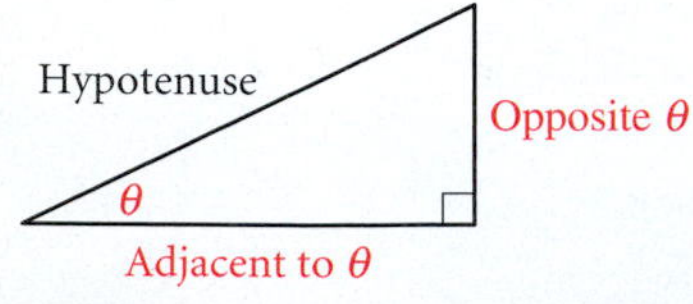

FIGURE 3.

The **tangent of θ** is the length of the side opposite θ divided by the length of the side adjacent to θ (see Fig. 3):

$$\tan \theta = \frac{\text{length of side opposite } \theta}{\text{length of side adjacent to } \theta}.$$

The six trigonometric ratios, or trigonometric functions, are defined as follows. Here the *domain* of each function is the *set of acute angles*. Later in this chapter, the domain will be extended first to the set of all angles, or rotations, and then to the set of real numbers.

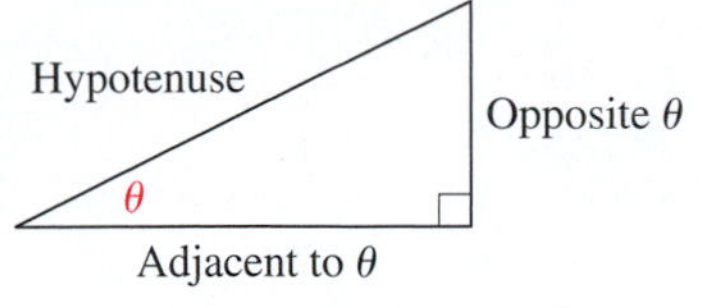

TRIGONOMETRIC FUNCTION VALUES OF AN ACUTE ANGLE θ

Let θ be an acute angle of a right triangle. Then the six trigonometric functions of θ are as follows:

$$\sin\theta = \frac{\text{side opposite }\theta}{\text{hypotenuse}}, \qquad \csc\theta = \frac{\text{hypotenuse}}{\text{side opposite }\theta},$$

$$\cos\theta = \frac{\text{side adjacent to }\theta}{\text{hypotenuse}}, \qquad \sec\theta = \frac{\text{hypotenuse}}{\text{side adjacent to }\theta},$$

$$\tan\theta = \frac{\text{side opposite }\theta}{\text{side adjacent to }\theta}, \qquad \cot\theta = \frac{\text{side adjacent to }\theta}{\text{side opposite }\theta}.$$

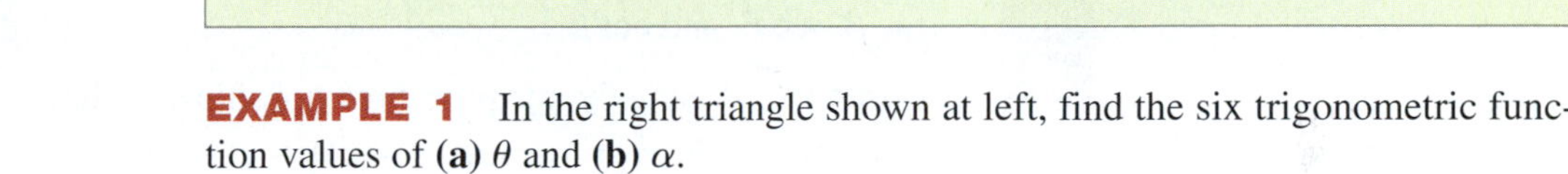

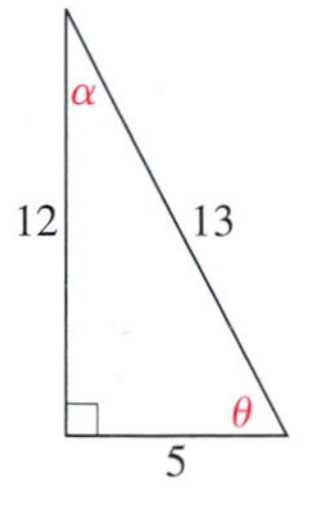

EXAMPLE 1 In the right triangle shown at left, find the six trigonometric function values of **(a)** θ and **(b)** α.

Solution We use the definitions.

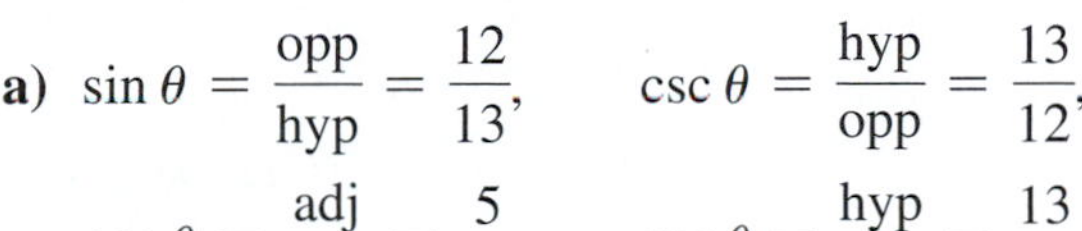

a) $\sin\theta = \frac{\text{opp}}{\text{hyp}} = \frac{12}{13}$, $\quad\csc\theta = \frac{\text{hyp}}{\text{opp}} = \frac{13}{12}$,

$\cos\theta = \frac{\text{adj}}{\text{hyp}} = \frac{5}{13}$, $\quad\sec\theta = \frac{\text{hyp}}{\text{adj}} = \frac{13}{5}$,

$\tan\theta = \frac{\text{opp}}{\text{adj}} = \frac{12}{5}$, $\quad\cot\theta = \frac{\text{adj}}{\text{opp}} = \frac{5}{12}$

The references to opposite, adjacent, and hypotenuse are relative to θ.

b) $\sin\alpha = \frac{\text{opp}}{\text{hyp}} = \frac{5}{13}$, $\quad\csc\alpha = \frac{\text{hyp}}{\text{opp}} = \frac{13}{5}$,

$\cos\alpha = \frac{\text{adj}}{\text{hyp}} = \frac{12}{13}$, $\quad\sec\alpha = \frac{\text{hyp}}{\text{adj}} = \frac{13}{12}$,

$\tan\alpha = \frac{\text{opp}}{\text{adj}} = \frac{5}{12}$, $\quad\cot\alpha = \frac{\text{adj}}{\text{opp}} = \frac{12}{5}$

The references to opposite, adjacent, and hypotenuse are relative to α.

Now Try Exercise 1.

In Example 1(a), we note that the value of $\csc\theta$, $\frac{13}{12}$, is the reciprocal of $\frac{12}{13}$, the value of $\sin\theta$. Likewise, we see the same reciprocal relationship between the values of $\sec\theta$ and $\cos\theta$ and between the values of $\cot\theta$ and $\tan\theta$. For any angle, the cosecant, secant, and cotangent function values are the reciprocals of the sine, cosine, and tangent function values, respectively.

RECIPROCAL FUNCTIONS

$$\csc\theta = \frac{1}{\sin\theta}, \qquad \sec\theta = \frac{1}{\cos\theta}, \qquad \cot\theta = \frac{1}{\tan\theta}$$

If we know the values of the sine, cosine, and tangent functions of an angle, we can use these reciprocal relationships to find the values of the cosecant, secant, and cotangent functions of that angle.

EXAMPLE 2 Given that $\sin\phi = \frac{4}{5}$, $\cos\phi = \frac{3}{5}$, and $\tan\phi = \frac{4}{3}$, find $\csc\phi$, $\sec\phi$, and $\cot\phi$.

Solution Using the reciprocal relationships, we have

$$\csc\phi = \frac{1}{\sin\phi} = \frac{1}{\frac{4}{5}} = \frac{5}{4}, \quad \sec\phi = \frac{1}{\cos\phi} = \frac{1}{\frac{3}{5}} = \frac{5}{3},$$

and $$\cot\phi = \frac{1}{\tan\phi} = \frac{1}{\frac{4}{3}} = \frac{3}{4}.$$

Just in Time 26

EXAMPLE 3 Given that $\sin\beta = \frac{\sqrt{21}}{5}$, $\cos\beta = \frac{2}{5}$, and $\tan\beta = \frac{\sqrt{21}}{2}$, find $\csc\beta$, $\sec\beta$, and $\cot\beta$.

Solution Using the reciprocal relationships, we have

$$\csc\beta = \frac{1}{\sin\beta} = \frac{1}{\frac{\sqrt{21}}{5}} = \frac{5}{\sqrt{21}}$$

$$= \frac{5}{\sqrt{21}} \cdot \frac{\sqrt{21}}{\sqrt{21}} = \frac{5\sqrt{21}}{21}, \quad \text{Rationalizing the denominator}$$

$$\sec\beta = \frac{1}{\cos\beta} = \frac{1}{\frac{2}{5}} = \frac{5}{2},$$

and $$\cot\beta = \frac{1}{\tan\beta} = \frac{1}{\frac{\sqrt{21}}{2}} = \frac{2}{\sqrt{21}} = \frac{2}{\sqrt{21}} \cdot \frac{\sqrt{21}}{\sqrt{21}} = \frac{2\sqrt{21}}{21}.$$

Now Try Exercise 7.

Triangles are said to be **similar** if their corresponding angles have the *same* measure. In similar triangles, the lengths of corresponding sides are in the same ratio. The right triangles shown below are similar. Note that the corresponding angles are equal, and the length of each side of the second triangle is four times the length of the corresponding side of the first triangle.

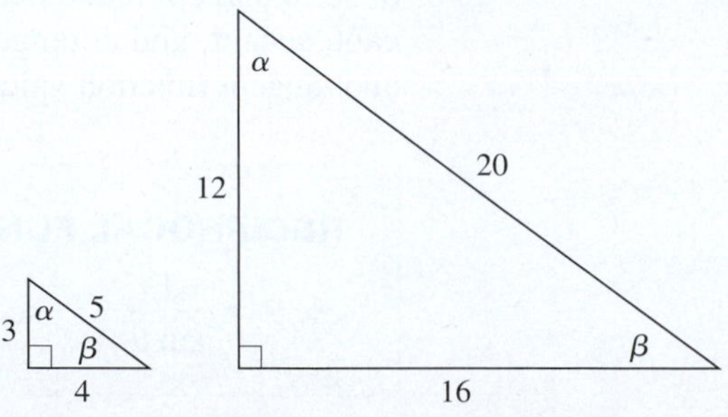

Let's take a look at the sine, cosine, and tangent values of β in each triangle. Can we expect corresponding function values to be the same?

FIRST TRIANGLE	SECOND TRIANGLE
$\sin\beta = \frac{3}{5}$	$\sin\beta = \frac{12}{20} = \frac{3}{5}$
$\cos\beta = \frac{4}{5}$	$\cos\beta = \frac{16}{20} = \frac{4}{5}$
$\tan\beta = \frac{3}{4}$	$\tan\beta = \frac{12}{16} = \frac{3}{4}$

For the two triangles, the values of $\sin\beta$, $\cos\beta$, and $\tan\beta$ are the same. The lengths of the sides are proportional—thus the *ratios* are the same. This must be the case because in order for the sine, cosine, and tangent to be functions, there must be only one output (the ratio) for each input (the angle β).

> The trigonometric function values of θ depend only on the measure of the angle, *not* on the size of the triangle.

The Six Related Functions

We can find the other five trigonometric function values of an acute angle when one of the function-value ratios is known.

EXAMPLE 4 If $\sin\beta = \frac{6}{7}$ and β is an acute angle, find the other five trigonometric function values of β.

Solution We know from the definition of the sine function that the ratio

$$\frac{6}{7} \text{ is } \frac{\text{opp}}{\text{hyp}}.$$

PYTHAGOREAN EQUATION

REVIEW SECTION 1.1.

Using this information, let's consider a right triangle in which the hypotenuse has length 7 and the side opposite β has length 6. To find the length of the side adjacent to β, we use the *Pythagorean equation*:

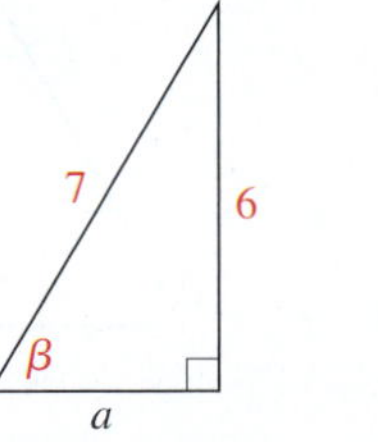

$$\begin{aligned} a^2 + b^2 &= c^2 \\ a^2 + 6^2 &= 7^2 \\ a^2 + 36 &= 49 \\ a^2 &= 49 - 36 = 13 \\ a &= \sqrt{13}. \end{aligned}$$

We now use the lengths of the three sides to find the other five ratios:

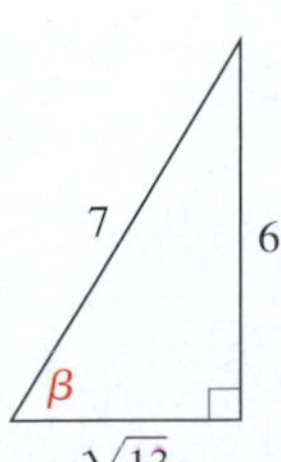

$$\sin\beta = \frac{6}{7}, \qquad \csc\beta = \frac{7}{6},$$

$$\cos\beta = \frac{\sqrt{13}}{7}, \qquad \sec\beta = \frac{7}{\sqrt{13}}, \text{ or } \frac{7\sqrt{13}}{13},$$

$$\tan\beta = \frac{6}{\sqrt{13}}, \text{ or } \frac{6\sqrt{13}}{13}, \qquad \cot\beta = \frac{\sqrt{13}}{6}.$$

Now Try Exercise 9.

Function Values of 30°, 45°, and 60°

In Examples 1 and 4, we found the trigonometric function values of an acute angle of a right triangle when the lengths of the three sides were known. In most situations, we are asked to find the function values when the measure of the acute angle is given. For certain special angles such as 30°, 45°, and 60°, which are frequently seen in applications, we can use geometry to determine the function values.

A right triangle with a 45° angle actually has two 45° angles. Thus the triangle is *isosceles*, and the legs are the same length. Let's consider such a triangle whose legs have length 1. Then we can find the length of its hypotenuse, c, using the Pythagorean equation as follows:

$$1^2 + 1^2 = c^2, \quad \text{or} \quad c^2 = 2, \quad \text{or} \quad c = \sqrt{2}.$$

Such a triangle is shown below. From this diagram, we can easily determine the trigonometric function values of 45°.

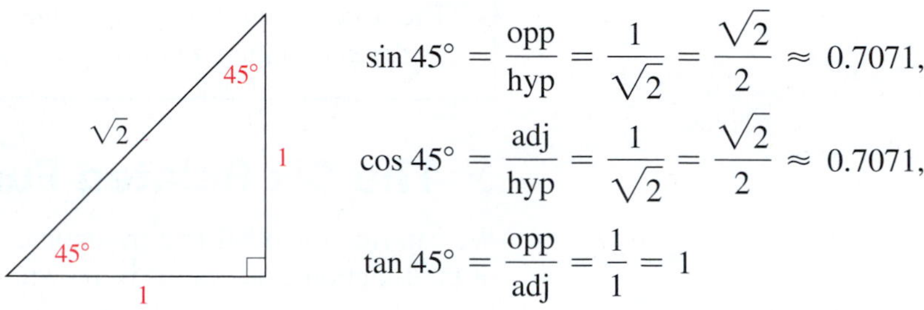

$$\sin 45° = \frac{\text{opp}}{\text{hyp}} = \frac{1}{\sqrt{2}} = \frac{\sqrt{2}}{2} \approx 0.7071,$$

$$\cos 45° = \frac{\text{adj}}{\text{hyp}} = \frac{1}{\sqrt{2}} = \frac{\sqrt{2}}{2} \approx 0.7071,$$

$$\tan 45° = \frac{\text{opp}}{\text{adj}} = \frac{1}{1} = 1$$

It is sufficient to find only the function values of the sine, cosine, and tangent, since the other three function values are their reciprocals.

It is also possible to determine the function values of 30° and 60°. A right triangle with 30° and 60° acute angles is half of an equilateral triangle, as shown in the following figure. Thus if we choose an equilateral triangle whose sides have length 2 and take half of it, we obtain a right triangle that has a hypotenuse of length 2 and a leg of length 1. The other leg has length a, which can be found as follows:

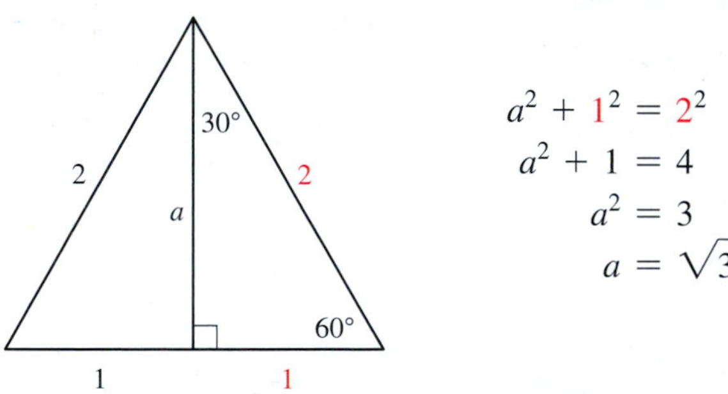

$$a^2 + 1^2 = 2^2$$
$$a^2 + 1 = 4$$
$$a^2 = 3$$
$$a = \sqrt{3}.$$

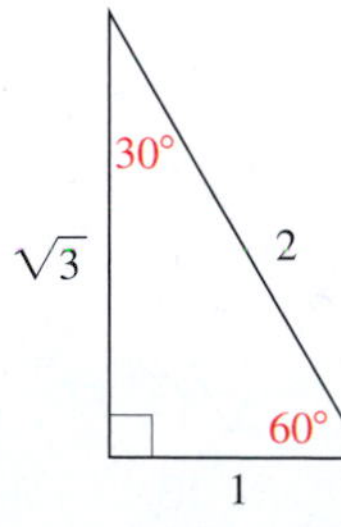

We can now determine the function values of 30° and 60°:

$$\sin 30° = \frac{1}{2} = 0.5, \qquad \sin 60° = \frac{\sqrt{3}}{2} \approx 0.8660,$$

$$\cos 30° = \frac{\sqrt{3}}{2} \approx 0.8660, \qquad \cos 60° = \frac{1}{2} = 0.5,$$

$$\tan 30° = \frac{1}{\sqrt{3}} = \frac{\sqrt{3}}{3} \approx 0.5774, \qquad \tan 60° = \frac{\sqrt{3}}{1} = \sqrt{3} \approx 1.7321.$$

Since we will often use the function values of 30°, 45°, and 60°, either the triangles that yield them or the values themselves should be memorized.

	30°	45°	60°
sin	$1/2$	$\sqrt{2}/2$	$\sqrt{3}/2$
cos	$\sqrt{3}/2$	$\sqrt{2}/2$	$1/2$
tan	$\sqrt{3}/3$	1	$\sqrt{3}$

Let's now use what we have learned about trigonometric functions of special angles to solve problems. We will consider such applications in greater detail in Section 6.2.

EXAMPLE 5 ***Height of a Fireworks Display.*** Trajectories for fireworks involve variables such as the launch angle, the launch velocity, and the size of the shell. Using physics to calculate the trajectory, fireworks technicians know exactly the path of the shell. Launch angles vary from 45° to 90°. For every inch of the diameter of the shell, a firework can travel about 100 ft vertically and 70 ft horizontally. These distances can vary depending on air resistance. (*Sources*: www.thinkquest.org, Pyrotechnico; www.dispatch.com, Aaron Harden) A 6-in. shell launched at an angle of 60° travels a horizontal distance of 390 ft. Approximate the height of the fireworks display. Round the answer to the nearest foot.

Solution We begin with a diagram of the situation. We know the measure of an acute angle and the length of the adjacent side.

Since we want to determine the length of the opposite side, we can use either the tangent ratio or the cotangent ratio. In this case, we use the tangent ratio:

$$\tan 60° = \frac{\text{opp}}{\text{adj}} = \frac{h}{390}$$

$$390 \cdot \tan 60° = h \quad \textbf{Multiplying by 390 on both sides}$$

$$390 \cdot \sqrt{3} = h \quad \textbf{Substituting; } \tan 60° = \sqrt{3}$$

$$675 \approx h.$$

The fireworks display is approximately 675 ft high.

Now Try Exercise 29.

Function Values of Any Acute Angle

Historically, the measure of an angle has been expressed in degrees, minutes, and seconds. One minute, denoted $1'$, is such that $60' = 1°$, or $1' = \frac{1}{60} \cdot (1°)$. One second, denoted $1''$, is such that $60'' = 1'$, or $1'' = \frac{1}{60} \cdot (1')$. Then 61 degrees, 27 minutes, 4 seconds can be written as $61°27'4''$. The use of this **D°M′S″ form** was common before the widespread use of calculators. Now the preferred notation is to express fraction parts of degrees in **decimal degree form**. For example, $61°27'4'' \approx 61.45°$ when written in decimal degree form. Although the D°M′S″ notation is still widely used in navigation, we will most often use the decimal form in this text.

Most calculators can convert D°M′S″ notation to decimal degree notation and vice versa. Procedures among calculators vary.

EXAMPLE 6 Convert $5°42'30''$ to decimal degree notation.

Solution We can use a graphing calculator set in DEGREE mode (see the window at left) to convert between D°M′S″ form and decimal degree form.

NORMAL SCI ENG
FLOAT 0123456789
RADIAN DEGREE
FUNC PAR POL SEQ
CONNECTED DOT
SEQUENTIAL SIMUL
REAL a+bi re^θi
FULL HORIZ G-T

To convert D°M′S″ form to decimal degree form, we enter $5°42'30''$ using the ANGLE menu for the degree and minute symbols and ALPHA + for the third symbol. Pressing ENTER gives us

$$5°42'30'' \approx 5.71°,$$

rounded to the nearest hundredth of a degree.

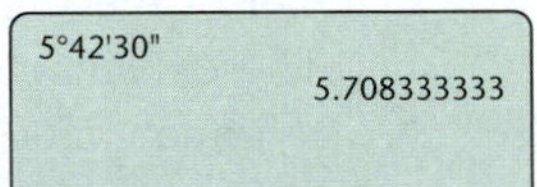

Without a calculator, we can convert as follows:

$$\begin{aligned} 5°42'30'' &= 5° + 42' + 30'' \\ &= 5° + 42' + \frac{30'}{60} && 1'' = \frac{1}{60}'; \; 30'' = \frac{30}{60}' \\ &= 5° + 42.5' && \frac{30}{60}' = 0.5' \\ &= 5° + \frac{42.5°}{60} && 1' = \frac{1}{60}°; \; 42.5' = \frac{42.5°}{60} \\ &\approx 5.71°. && \frac{42.5°}{60} \approx 0.71° \end{aligned}$$

Now Try Exercise 37.

EXAMPLE 7 Convert $72.18°$ to D°M′S″ notation.

Solution To convert decimal degree form to D°M′S″ form, we enter 72.18 and access the ▶DMS feature in the ANGLE menu.

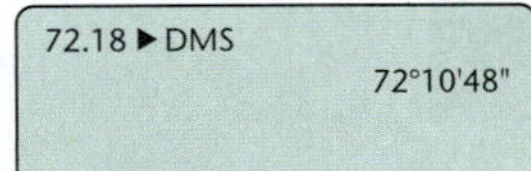

The result is

$$72.18° = 72°10'48''.$$

Without a calculator, we can convert as follows:

$$\begin{aligned} 72.18° &= 72° + 0.18 \times 1° \\ &= 72° + 0.18 \times 60' && 1° = 60' \\ &= 72° + 10.8' \\ &= 72° + 10' + 0.8 \times 1' \\ &= 72° + 10' + 0.8 \times 60'' && 1' = 60'' \\ &= 72° + 10' + 48'' \\ &= 72°10'48''. \end{aligned}$$

Now Try Exercise 45.

So far we have measured angles using degrees. Another useful unit for angle measure is the radian, which we will study in Section 6.4. Most calculators work with either degrees or radians. Be sure to use whichever mode is appropriate. In this section, we use the DEGREE mode.

Keep in mind the difference between an exact answer and an approximation. For example,

$$\sin 60^\circ = \frac{\sqrt{3}}{2}. \quad \textbf{This is exact.}$$

But using a calculator, you get an answer like

$$\sin 60^\circ \approx 0.8660254038. \quad \textbf{This is an approximation.}$$

Calculators generally provide values only of the sine, cosine, and tangent functions. You can find values of the cosecant, secant, and cotangent by taking reciprocals of the sine, cosine, and tangent functions, respectively.

EXAMPLE 8 Using a calculator, find the trigonometric function value, rounded to four decimal places, of each of the following.

a) $\tan 29.7^\circ$ **b)** $\sec 48^\circ$ **c)** $\sin 84^\circ 10' 39''$

Solution

a) We check to be sure that the calculator is in DEGREE mode. The function value is

$$\tan 29.7^\circ \approx 0.5703899297$$
$$\approx 0.5704. \quad \text{Rounded to four decimal places}$$

b) The secant function value can be found by taking the reciprocal of the cosine function value:

$$\sec 48^\circ = \frac{1}{\cos 48^\circ} \approx 1.49447655 \approx 1.4945.$$

c) We enter $\sin 84^\circ 10' 39''$. The result is

$$\sin 84^\circ 10' 39'' \approx 0.9948409474 \approx 0.9948.$$

Now Try Exercises 61 and 69.

We can use the TABLE feature on a graphing calculator to find an angle for which we know a trigonometric function value.

EXAMPLE 9 Find the acute angle, to the nearest tenth of a degree, whose sine value is approximately 0.20113—that is, given $\sin \theta = 0.20113$, find θ.

Solution With a graphing calculator set in DEGREE mode, we first enter the equation $y = \sin x$. With a minimum value of 0 and a step-value of 0.1, we scroll through the table of values looking for the y-value closest to 0.20113.

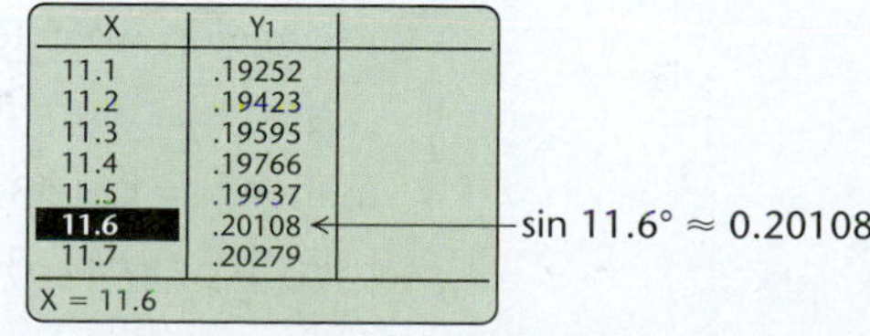

X	Y1
11.1	.19252
11.2	.19423
11.3	.19595
11.4	.19766
11.5	.19937
11.6	.20108
11.7	.20279

X = 11.6

We find that 11.6° is the angle whose sine value is about 0.20113.

The quickest way to find the angle with a calculator is to use an inverse function key. (We first studied inverse functions in the chapter on exponential functions and logarithmic functions.) First, check to be sure that your calculator is in DEGREE mode. Usually two keys must be pressed in sequence. For this example, if we press

2ND SIN .20113 ENTER,

we find that the acute angle whose sine is 0.20113 is approximately 11.60304613°, or 11.6°.

Now Try Exercise 75.

EXAMPLE 10 *Ladder Safety.* A window-washing crew has purchased new 30-ft extension ladders. The manufacturer states that the safest placement on a wall is to extend the ladder to 25 ft and to position the base 6.5 ft from the wall (*Source*: R. D. Werner Co., Inc.). What angle does the ladder make with the ground in this position?

Solution We make a drawing and then use the most convenient trigonometric function. Because we know the length of the side adjacent to θ and the length of the hypotenuse, we choose the cosine function.

From the definition of the cosine function, we have

$$\cos\theta = \frac{\text{adj}}{\text{hyp}} = \frac{6.5\text{ ft}}{25\text{ ft}} = 0.26.$$

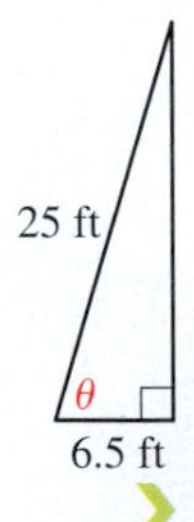

Using a calculator, we find the acute angle whose cosine is 0.26:

$$\theta \approx 74.92993786°. \quad \text{Pressing 2ND COS 0.26 ENTER}$$

Thus when the ladder is in its safest position, it makes an angle of about 75° with the ground.

Cofunctions and Complements

Two angles are **complementary** whenever the sum of their measures is 90°. Each is the complement of the other. In a right triangle, the acute angles are complementary, because the sum of all three angle measures is 180° and the right angle accounts for 90° of this total. Thus if one acute angle of a right triangle is θ, the other is $90° - \theta$.

The six trigonometric function values of each of the acute angles in the following triangle are listed on the right. Note that 53° and 37° are complementary angles because $53° + 37° = 90°$.

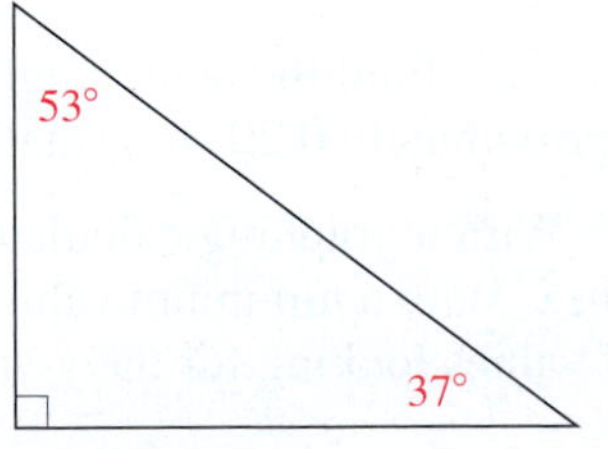

$\sin 37° \approx 0.6018$ $\csc 37° \approx 1.6616$
$\cos 37° \approx 0.7986$ $\sec 37° \approx 1.2521$
$\tan 37° \approx 0.7536$ $\cot 37° \approx 1.3270$

$\sin 53° \approx 0.7986$ $\csc 53° \approx 1.2521$
$\cos 53° \approx 0.6018$ $\sec 53° \approx 1.6616$
$\tan 53° \approx 1.3270$ $\cot 53° \approx 0.7536$

For these angles, we note that

$\sin 37° = \cos 53°$, $\cos 37° = \sin 53°$,
$\tan 37° = \cot 53°$, $\cot 37° = \tan 53°$,
$\sec 37° = \csc 53°$, $\csc 37° = \sec 53°$.

The sine of an angle is also the cosine of the angle's complement. Similarly, the tangent of an angle is the cotangent of the angle's complement, and the secant of an angle is the cosecant of the angle's complement. These pairs of functions are called **cofunctions**. A list of cofunction identities follows.

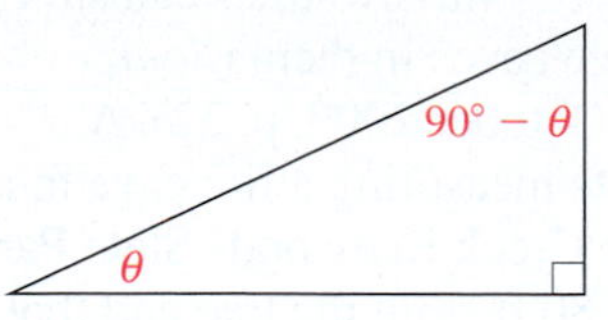

COFUNCTION IDENTITIES

$\sin\theta = \cos(90° - \theta)$, $\quad\cos\theta = \sin(90° - \theta)$,
$\tan\theta = \cot(90° - \theta)$, $\quad\cot\theta = \tan(90° - \theta)$,
$\sec\theta = \csc(90° - \theta)$, $\quad\csc\theta = \sec(90° - \theta)$

EXAMPLE 11 Given that $\sin 18° \approx 0.3090$, $\cos 18° \approx 0.9511$, and $\tan 18° \approx 0.3249$, find the six trigonometric function values of 72°.

Solution Using reciprocal relationships, we know that

$$\csc 18° = \frac{1}{\sin 18°} \approx 3.2361,$$

$$\sec 18° = \frac{1}{\cos 18°} \approx 1.0515,$$

and $$\cot 18° = \frac{1}{\tan 18°} \approx 3.0777.$$

Since 72° and 18° are complementary, we have

$$\sin 72° = \cos 18° \approx 0.9511, \quad \cos 72° = \sin 18° \approx 0.3090,$$
$$\tan 72° = \cot 18° \approx 3.0777, \quad \cot 72° = \tan 18° \approx 0.3249,$$
$$\sec 72° = \csc 18° \approx 3.2361, \quad \csc 72° = \sec 18° \approx 1.0515.$$

Now Try Exercise 97.

6.1 Exercise Set

In Exercises 1–6, find the six trigonometric function values of the specified angle.

1.

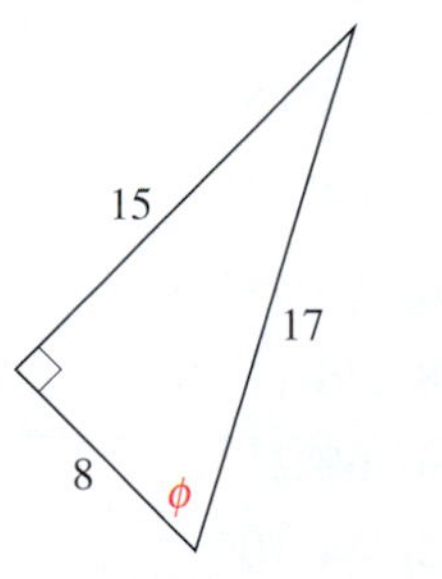

2.

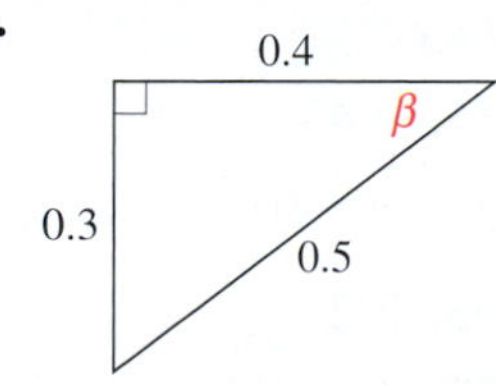

3.

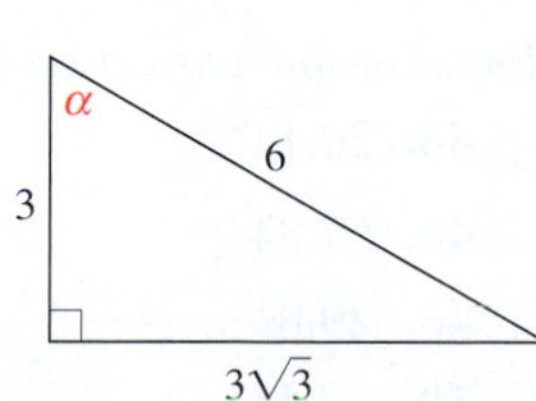

4.

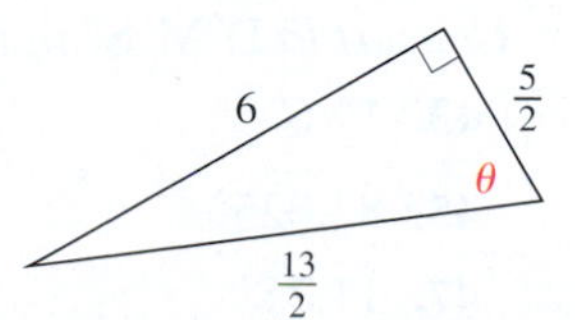

5.

6.

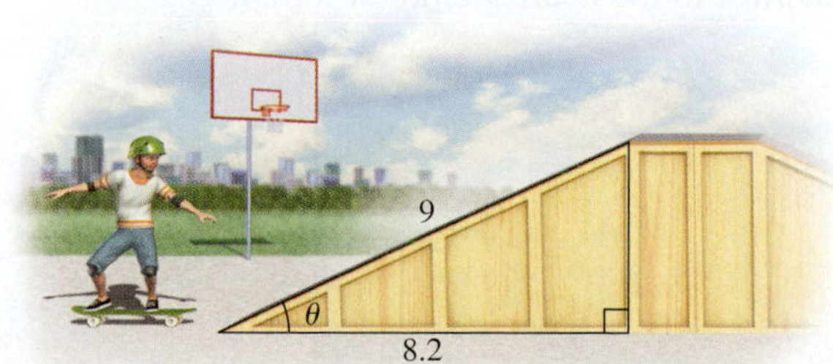

7. Given that $\sin \alpha = \frac{\sqrt{5}}{3}$, $\cos \alpha = \frac{2}{3}$, and $\tan \alpha = \frac{\sqrt{5}}{2}$, find $\csc \alpha$, $\sec \alpha$, and $\cot \alpha$.

8. Given that $\sin \beta = \frac{2\sqrt{2}}{3}$, $\cos \beta = \frac{1}{3}$, and $\tan \beta = 2\sqrt{2}$, find $\csc \beta$, $\sec \beta$, and $\cot \beta$.

Given a function value of an acute angle, find the other five trigonometric function values.

9. $\sin \theta = \frac{24}{25}$
10. $\cos \sigma = 0.7$
11. $\tan \phi = 2$
12. $\cot \theta = \frac{1}{3}$
13. $\csc \theta = 1.5$
14. $\sec \beta = \sqrt{17}$
15. $\cos \beta = \frac{\sqrt{5}}{5}$
16. $\sin \sigma = \frac{10}{11}$

Find the exact function value.

17. $\cos 45°$
18. $\tan 30°$
19. $\sec 60°$
20. $\sin 45°$
21. $\cot 60°$
22. $\csc 45°$
23. $\sin 30°$
24. $\cos 60°$
25. $\tan 45°$
26. $\sec 30°$
27. $\csc 30°$
28. $\tan 60°$

29. *Four Square.* The game Four Square is making a comeback on college campuses. The game is played on a 16-ft square court divided into four smaller squares that meet in the center (*Source*: www.squarefour.org/rules).

If a line is drawn diagonally from one corner to another corner, then a right triangle QTS is formed, where $\angle QTS$ is 45°. Using the cosecant function, find the length of the diagonal. Round the answer to the nearest tenth of a foot.

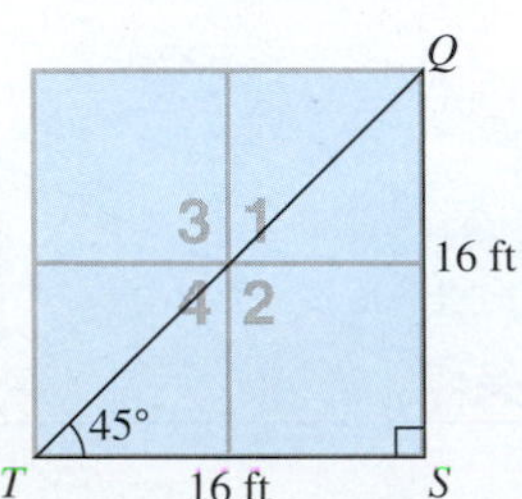

30. *Distance to a Fire Cave.* Massive trees can survive wildfires that leave large caves in them (*Source*: *National Geographic*, October 2009, p. 32). A hiker observes scientists measuring a fire cave in a redwood tree in Prairie Creek Redwoods State Park. He estimates that he is 80 ft from the tree and that the angle between the ground and the line of sight to the scientists is 60°. Approximate how high the fire cave is. Round the answer to the nearest foot.

Convert to decimal degree notation. Round to two decimal places.

31. 9°43′
32. 52°15′
33. 35°50″
34. 64°53′
35. 3°2′
36. 19°47′23″
37. 49°38′46″
38. 76°11′34″
39. 15′5″
40. 68°2″
41. 5°53″
42. 44′10″

Convert to D°M′S″ *notation. Round to the nearest second.*

43. 17.6°
44. 20.14°
45. 83.025°
46. 67.84°
47. 11.75°
48. 29.8°
49. 47.8268°
50. 0.253°
51. 0.9°
52. 30.2505°
53. 39.45°
54. 2.4°

Find the function value. Round to four decimal places.

55. $\cos 51^\circ$

56. $\cot 17^\circ$

57. $\tan 4^\circ 13'$

58. $\sin 26.1^\circ$

59. $\sec 38.43^\circ$

60. $\cos 74^\circ 10' 40''$

61. $\cos 40.35^\circ$

62. $\csc 45.2^\circ$

63. $\sin 69^\circ$

64. $\tan 63^\circ 48'$

65. $\tan 85.4^\circ$

66. $\cos 4^\circ$

67. $\csc 89.5^\circ$

68. $\sec 35.28^\circ$

69. $\cot 30^\circ 25' 6''$

70. $\sin 59.2^\circ$

Find the acute angle θ, to the nearest tenth of a degree, for the given function value.

71. $\sin \theta = 0.5125$

72. $\tan \theta = 2.032$

73. $\tan \theta = 0.2226$

74. $\cos \theta = 0.3842$

75. $\sin \theta = 0.9022$

76. $\tan \theta = 3.056$

77. $\cos \theta = 0.6879$

78. $\sin \theta = 0.4005$

79. $\cot \theta = 2.127$

$\left(\textit{Hint}: \tan \theta = \dfrac{1}{\cot \theta}.\right)$

80. $\csc \theta = 1.147$

81. $\sec \theta = 1.279$

82. $\cot \theta = 1.351$

Find the exact acute angle θ for the given function value.

83. $\sin \theta = \dfrac{\sqrt{2}}{2}$

84. $\cot \theta = \dfrac{\sqrt{3}}{3}$

85. $\cos \theta = \dfrac{1}{2}$

86. $\sin \theta = \dfrac{1}{2}$

87. $\tan \theta = 1$

88. $\cos \theta = \dfrac{\sqrt{3}}{2}$

89. $\csc \theta = \dfrac{2\sqrt{3}}{3}$

90. $\tan \theta = \sqrt{3}$

91. $\cot \theta = \sqrt{3}$

92. $\sec \theta = \sqrt{2}$

Use the cofunction and reciprocal identities to complete each of the following.

93. $\cos 20^\circ = \_\_\_\_\_ 70^\circ = \dfrac{1}{\_\_\_\_\_ 20^\circ}$

94. $\sin 64^\circ = \_\_\_\_\_ 26^\circ = \dfrac{1}{\_\_\_\_\_ 64^\circ}$

95. $\tan 52^\circ = \cot \_\_\_\_\_ = \dfrac{1}{\_\_\_\_\_ 52^\circ}$

96. $\sec 13^\circ = \csc \_\_\_\_\_ = \dfrac{1}{\_\_\_\_\_ 13^\circ}$

97. Given that

$\sin 65^\circ \approx 0.9063$, $\cos 65^\circ \approx 0.4226$,
$\tan 65^\circ \approx 2.1445$, $\cot 65^\circ \approx 0.4663$,
$\sec 65^\circ \approx 2.3662$, $\csc 65^\circ \approx 1.1034$,

find the six function values of 25°.

98. Given that

$\sin 8^\circ \approx 0.1392$, $\cos 8^\circ \approx 0.9903$,
$\tan 8^\circ \approx 0.1405$, $\cot 8^\circ \approx 7.1154$,
$\sec 8^\circ \approx 1.0098$, $\csc 8^\circ \approx 7.1853$,

find the six function values of 82°.

99. Given that $\sin 71^\circ 10' 5'' \approx 0.9465$, $\cos 71^\circ 10' 5'' \approx 0.3228$, and $\tan 71^\circ 10' 5'' \approx 2.9321$, find the six function values of $18^\circ 49' 55''$.

100. Given that $\sin 38.7^\circ \approx 0.6252$, $\cos 38.7^\circ \approx 0.7804$, and $\tan 38.7^\circ \approx 0.8012$, find the six function values of 51.3°.

101. Given that $\sin 82^\circ = p$, $\cos 82^\circ = q$, and $\tan 82^\circ = r$, find the six function values of 8° in terms of p, q, and r.

Skill Maintenance

Make a hand-drawn graph of the function. Then check your work using a graphing calculator.

102. $f(x) = 2^{-x}$ [5.2]

103. $f(x) = e^{x/2}$ [5.2]

104. $g(x) = \log_2 x$ [5.3]

105. $h(x) = \ln x$ [5.3]

Solve. [5.5]

106. $e^t = 10{,}000$

107. $5^x = 625$

108. $\log(3x + 1) - \log(x - 1) = 2$

109. $\log_7 x = 3$

Synthesis

110. Given that $\cos \theta = 0.9651$, find $\csc(90^\circ - \theta)$.

111. Given that $\sec \beta = 1.5304$, find $\sin(90^\circ - \beta)$.

112. Find the six trigonometric function values of α.

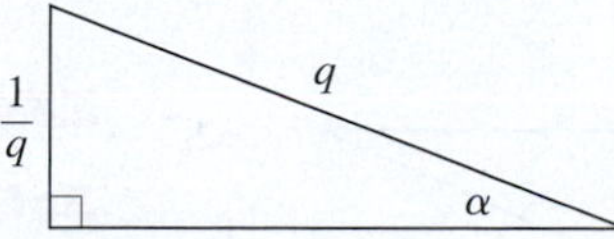

113. Show that the area of this triangle is $\frac{1}{2}ab \sin \theta$.

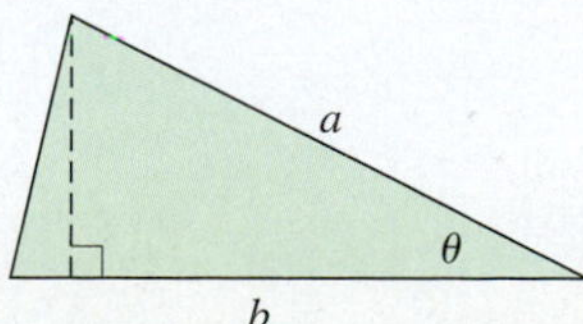

6.2 Applications of Right Triangles

- Solve right triangles.
- Solve applied problems involving right triangles and trigonometric functions.

Solving Right Triangles

Now that we can find function values for any acute angle, it is possible to *solve* right triangles. To **solve** a triangle means to find the lengths of *all* sides and the measures of *all* angles.

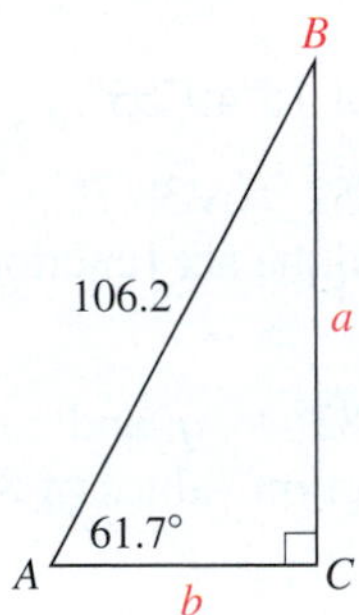

EXAMPLE 1 In $\triangle ABC$ (shown at left), find a, b, and B, where a and b represent lengths of sides and B represents the measure of $\angle B$. Here we use standard lettering for naming the sides and angles of a right triangle: Side a is opposite angle A, side b is opposite angle B, where a and b are the legs, and side c, the hypotenuse, is opposite angle C, the right angle.

Solution In $\triangle ABC$, we know three of the measures:

$$\begin{aligned} A &= 61.7^\circ, & a &= ?, \\ B &= ?, & b &= ?, \\ C &= 90^\circ, & c &= 106.2. \end{aligned}$$

Since the sum of the angle measures of any triangle is 180° and $C = 90°$, the sum of A and B is 90°. Thus,

$$B = 90^\circ - A = 90^\circ - 61.7^\circ = 28.3^\circ.$$

We are given an acute angle and the hypotenuse. This suggests that we can use the sine and cosine ratios to find a and b, respectively:

$$\sin 61.7^\circ = \frac{\text{opp}}{\text{hyp}} = \frac{a}{106.2} \quad \text{and} \quad \cos 61.7^\circ = \frac{\text{adj}}{\text{hyp}} = \frac{b}{106.2}.$$

Solving for a and b, we get

$$\begin{aligned} a &= 106.2 \sin 61.7^\circ & \text{and} \quad b &= 106.2 \cos 61.7^\circ \\ a &\approx 93.5 & b &\approx 50.3. \end{aligned}$$

Thus,

$$\begin{aligned} A &= 61.7^\circ, & a &\approx 93.5, \\ B &= 28.3^\circ, & b &\approx 50.3, \\ C &= 90^\circ, & c &= 106.2. \end{aligned}$$

Now Try Exercise 1.

EXAMPLE 2 In $\triangle DEF$ (shown at left), find D and F. Then find d.

Solution In $\triangle DEF$, we know three of the measures:

$$\begin{aligned} D &= ?, & d &= ?, \\ E &= 90^\circ, & e &= 23, \\ F &= ?, & f &= 13. \end{aligned}$$

We know the side adjacent to D and the hypotenuse. This suggests the use of the cosine ratio:

$$\cos D = \frac{\text{adj}}{\text{hyp}} = \frac{13}{23}.$$

We now find the angle whose cosine is $\frac{13}{23}$. To the nearest hundredth of a degree,

$D \approx 55.58°$. Pressing 2ND COS (13/23) ENTER

Since the sum of D and F is 90°, we can find F by subtracting:

$$F = 90° - D \approx 90° - 55.58° \approx 34.42°.$$

We could use the Pythagorean equation to find d, but we will use a trigonometric function here. We could use $\cos F$, $\sin D$, or the tangent or cotangent ratios for either D or F. Let's use $\tan D$:

$$\tan D = \frac{\text{opp}}{\text{adj}} = \frac{d}{13}, \quad \text{or} \quad \tan 55.58° \approx \frac{d}{13}.$$

Then

$$d \approx 13 \tan 55.58° \approx 19.$$

The six measures are

$$D \approx 55.58°, \qquad d \approx 19,$$
$$E = 90°, \qquad e = 23,$$
$$F \approx 34.42°, \qquad f = 13.$$

Now Try Exercise 9.

Applications

Right triangles can be used to model and solve many applied problems in the real world.

EXAMPLE 3 ***Walking at Niagara Falls.*** While visiting Niagara Falls, a tourist walking toward Horseshoe Falls on a walkway next to Niagara Parkway notices the entrance to the Cave of the Winds attraction directly across the Niagara River. She continues walking for another 1000 ft and finds that the entrance is still visible but at approximately a 50° angle to the walkway.

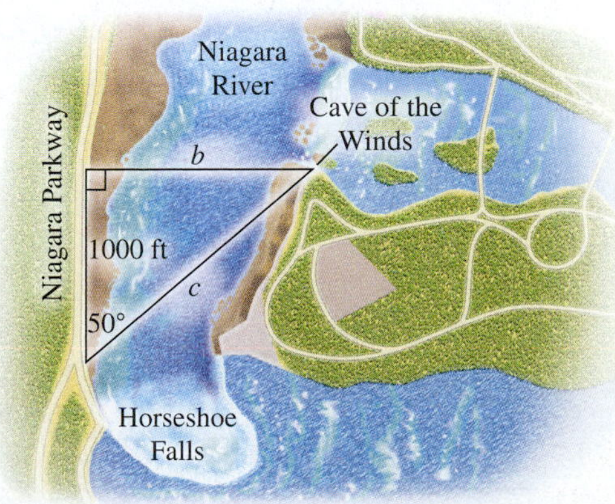

a) How many feet is she from the entrance to the Cave of the Winds?

b) What is the approximate width of the Niagara River at the point directly across from the entrance to the Cave of the Winds?

Solution

a) We know the side adjacent to the 50° angle and want to find the hypotenuse. We can use the cosine function:

$$\cos 50^\circ = \frac{1000 \text{ ft}}{c}$$

$$c \cos 50^\circ = 1000 \text{ ft} \qquad \textbf{Multiplying by } c$$

$$c = \frac{1000 \text{ ft}}{\cos 50^\circ} \qquad \textbf{Dividing by } \cos 50^\circ$$

$$c \approx 1556 \text{ ft}.$$

After walking 1000 ft, she is approximately 1556 ft from the entrance to the Cave of the Winds.

b) We know the side adjacent to the 50° angle and want to find the opposite side. We can use the tangent function:

$$\tan 50^\circ = \frac{b}{1000 \text{ ft}}$$

$$b = 1000 \text{ ft} \cdot \tan 50^\circ \approx 1192 \text{ ft}.$$

The width of the Niagara River directly across from the Cave of the Winds is approximately 1192 ft.

Now Try Exercise 21.

EXAMPLE 4 ***Rafters for a House.*** House framers can use trigonometric functions to determine the lengths of rafters for a house. They first choose the pitch of the roof, or the ratio of the rise over the run. Then using a triangle with that ratio, they calculate the length of the rafter needed for the house. José is constructing rafters for a roof with a 10/12 pitch on a house that is 42 ft wide. Find the length x of the rafter of the house to the nearest tenth of a foot.

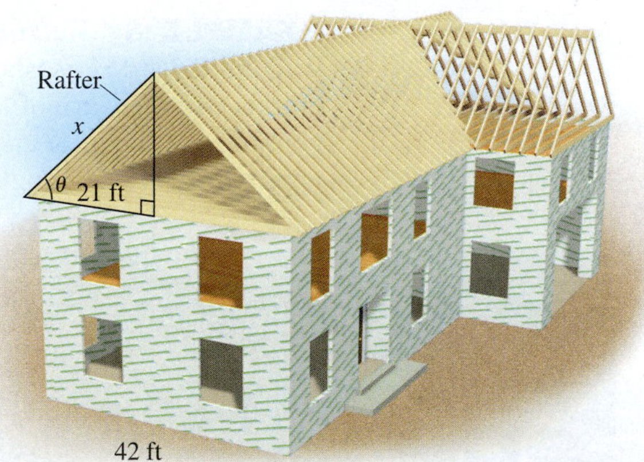

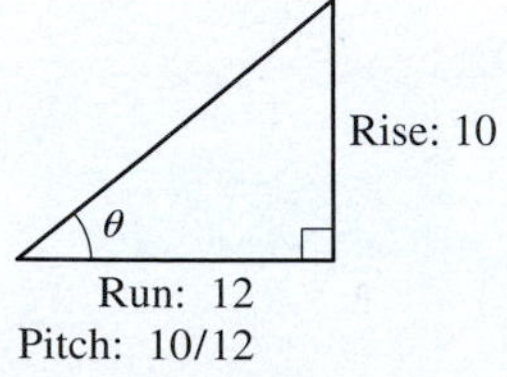

Solution We first find the angle θ that the rafter makes with the side wall. We know the rise, 10, and the run, 12, so we can use the tangent function to determine the angle that corresponds to the pitch of 10/12:

$$\tan \theta = \frac{10}{12} \approx 0.8333.$$

Using a calculator, we find that $\theta \approx 39.8^\circ$. Since trigonometric function values of θ depend only on the measure of the angle and not on the size of the triangle, the angle for the rafter is also 39.8°.

To determine the length x of the rafter, we can use the cosine function. (See the figure at left.) Note that the width of the house is 42 ft, and a leg of this triangle is half that length, 21 ft.

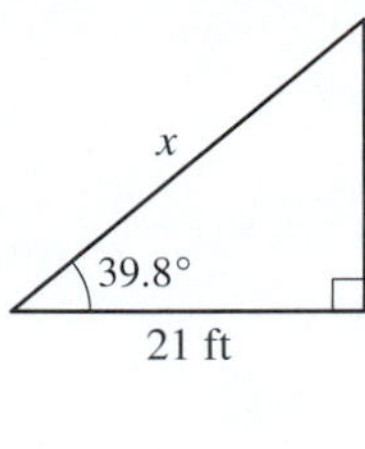

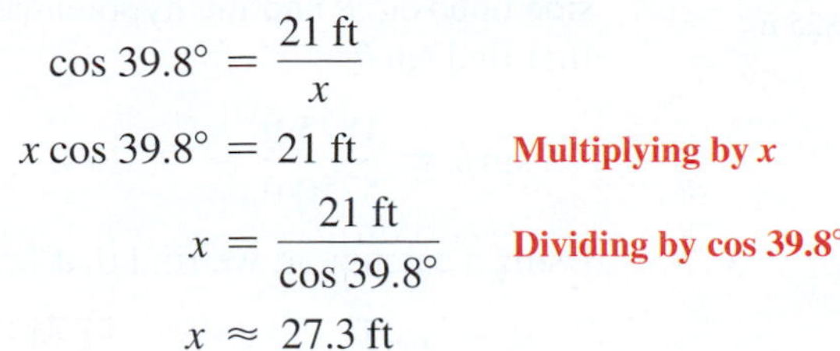

$$\cos 39.8° = \frac{21 \text{ ft}}{x}$$

$$x \cos 39.8° = 21 \text{ ft} \qquad \textbf{Multiplying by } x$$

$$x = \frac{21 \text{ ft}}{\cos 39.8°} \qquad \textbf{Dividing by } \cos 39.8°$$

$$x \approx 27.3 \text{ ft}$$

The length of the rafter for this house is approximately 27.3 ft.

Now Try Exercise 33.

Many applications with right triangles involve an *angle of elevation* or an *angle of depression*. The angle between the horizontal and a line of sight above the horizontal is called an **angle of elevation**. The angle between the horizontal and a line of sight below the horizontal is called an **angle of depression**. For example, suppose that you are looking straight ahead and then you move your eyes up to look at an approaching airplane. The angle that your eyes pass through is an angle of elevation. If the pilot of the plane is looking forward and then looks down, the pilot's eyes pass through an angle of depression.

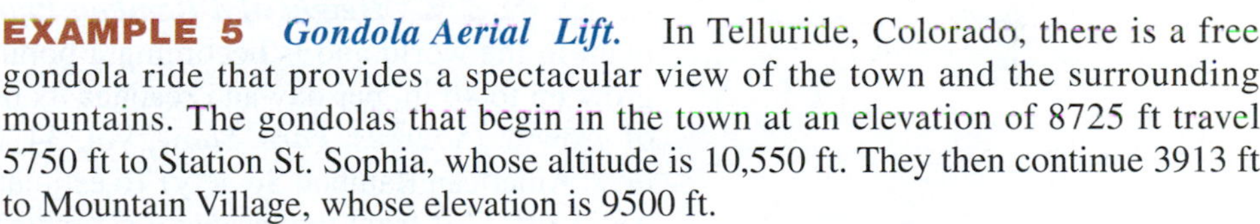

EXAMPLE 5 ***Gondola Aerial Lift.*** In Telluride, Colorado, there is a free gondola ride that provides a spectacular view of the town and the surrounding mountains. The gondolas that begin in the town at an elevation of 8725 ft travel 5750 ft to Station St. Sophia, whose altitude is 10,550 ft. They then continue 3913 ft to Mountain Village, whose elevation is 9500 ft.

a) What is the angle of elevation from the town to Station St. Sophia?

b) What is the angle of depression from Station St. Sophia to Mountain Village?

Solution We begin by labeling a drawing with the given information.

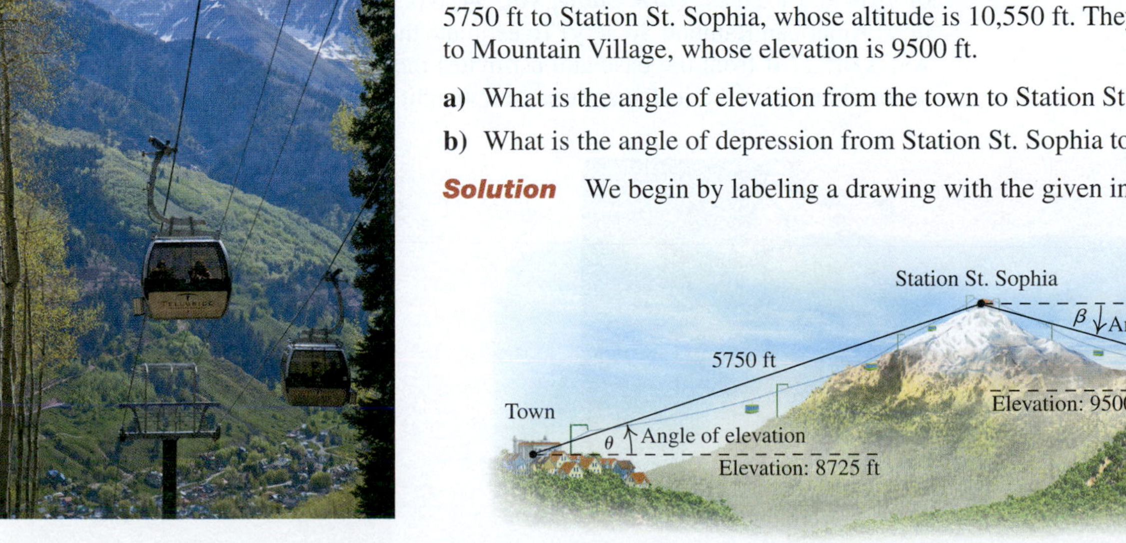

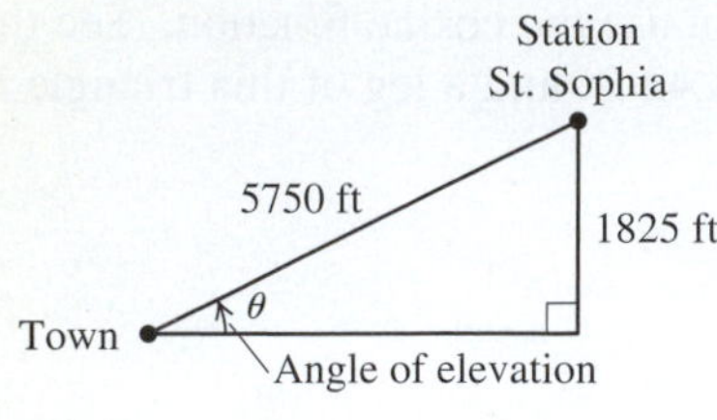

a) The difference in the elevation of Station St. Sophia and the elevation of the town is 10,550 ft − 8725 ft, or 1825 ft. This measure is the length of the side opposite the angle of elevation, θ, in the right triangle shown at left. Since we know the side opposite θ and the hypotenuse, we can find θ by using the sine function. We first find $\sin\theta$:

$$\sin\theta = \frac{1825 \text{ ft}}{5750 \text{ ft}} \approx 0.3174.$$

Using a calculator, we find that

$\theta \approx 18.5°$. Pressing 2ND SIN 0.3174 ENTER

Thus the angle of elevation from the town to Station St. Sophia is approximately 18.5°.

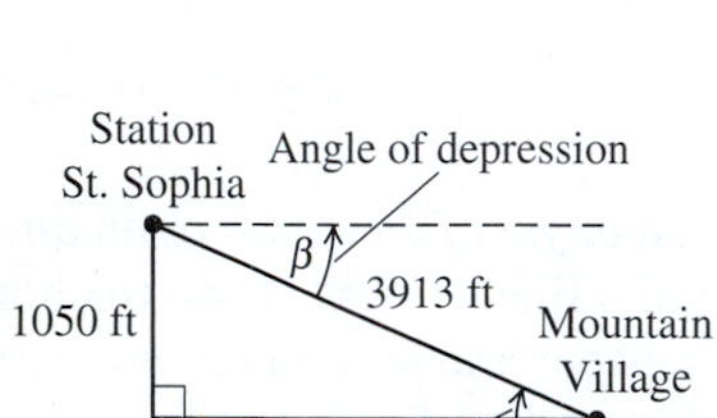

b) When parallel lines are cut by a transversal, alternate interior angles are equal. Thus the angle of depression, β, from Station St. Sophia to Mountain Village is equal to the angle of elevation from Mountain Village to Station St. Sophia, so we can use the right triangle shown at left.

The difference in the elevation of Station St. Sophia and the elevation of Mountain Village is 10,550 ft − 9500 ft, or 1050 ft. Since we know the side opposite the angle of elevation and the hypotenuse, we can again use the sine function:

$$\sin\beta = \frac{1050 \text{ ft}}{3913 \text{ ft}} \approx 0.2683.$$

Using a calculator, we find that

$$\beta \approx 15.6°.$$

The angle of depression from Station St. Sophia to Mountain Village is approximately 15.6°.

Now Try Exercise 17.

EXAMPLE 6 ***Height of a Bamboo Plant.*** Bamboo is the fastest growing land plant in the world and is becoming a popular wood for hardwood flooring. It can grow up to 46 in. per day and reaches its maximum height and girth in one season of growth. (*Sources*: *Farm Show*, Vol. 34, No. 4, 2010, p. 7; *U-Cut Bamboo Business*; American Bamboo Society) To estimate the height of a bamboo shoot, a farmer walks off 27 ft from the base and estimates the angle of elevation to the top of the shoot to be 70°. What is the approximate height h of the bamboo shoot?

Solution From the figure, we have

$$\tan 70^\circ = \frac{h}{27 \text{ ft}}$$

$$h = 27 \text{ ft} \cdot \tan 70^\circ \approx 74 \text{ ft}.$$

The height of the bamboo shoot is approximately 74 ft.

Some applications of trigonometry involve the concept of direction, or bearing. In this text, we present two ways of giving direction, the first below and the second in Section 6.3.

Bearing: First-Type

One method of giving direction, or **bearing**, involves reference to a north–south line using an acute angle. For example, N55°W means 55° west of north and S67°E means 67° east of south.

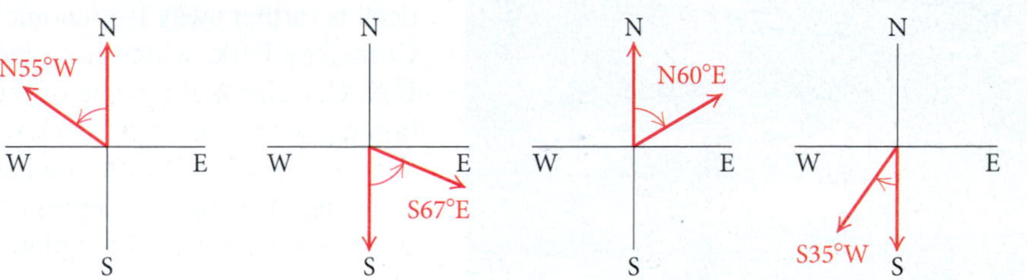

A second-type of bearing that gives directions in degrees from north is covered in Section 6.3.

EXAMPLE 7 ***Distance to a Forest Fire.*** A forest ranger at point A sights a fire directly south. A second ranger at point B, 7.5 mi east of the first ranger, sights the same fire at a bearing of S27°23′W. How far from A is the fire?

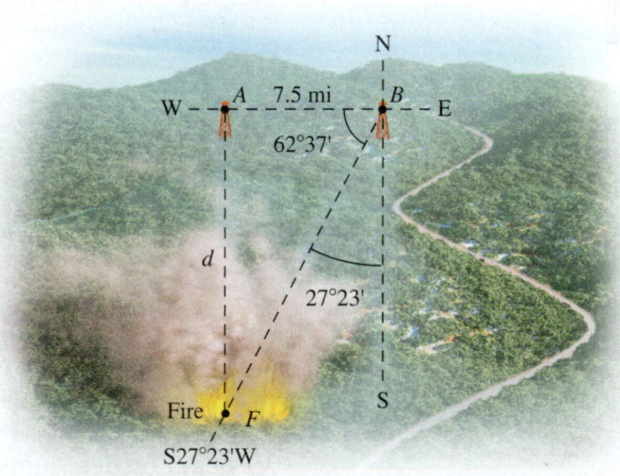

Solution We first find the complement of $27°23'$:

$$\begin{aligned} B &= 90° - 27°23' \qquad \text{Angle } B \text{ is opposite side } d \text{ in triangle } BAF. \\ &= 62°37' \\ &\approx 62.62°. \end{aligned}$$

From the figure shown on the preceding page, we see that the desired distance d is a side of right triangle BAF. We have

$$\frac{d}{7.5 \text{ mi}} \approx \tan 62.62°$$

$$d \approx 7.5 \text{ mi} \cdot \tan 62.62° \approx 14.5 \text{ mi}.$$

The forest ranger at point A is about 14.5 mi from the fire.

Now Try Exercise 37.

EXAMPLE 8 ***U.S. Cellular Field.*** In U.S. Cellular Field, the home of the Chicago White Sox baseball team, the first row of seats in the upper deck is farther away from home plate than the last row of seats in the original Comiskey Park, which it replaced. Although there is no obstructed view in U.S. Cellular Field, some of the fans still complain about the distance from home plate to the upper deck of seats. From a seat in the last row of the upper deck directly behind the batter, the angle of depression to home plate is 29.9°, and the angle of depression to the pitcher's mound is 24.2°. Find **(a)** the viewing distance to home plate and **(b)** the viewing distance to the pitcher's mound.

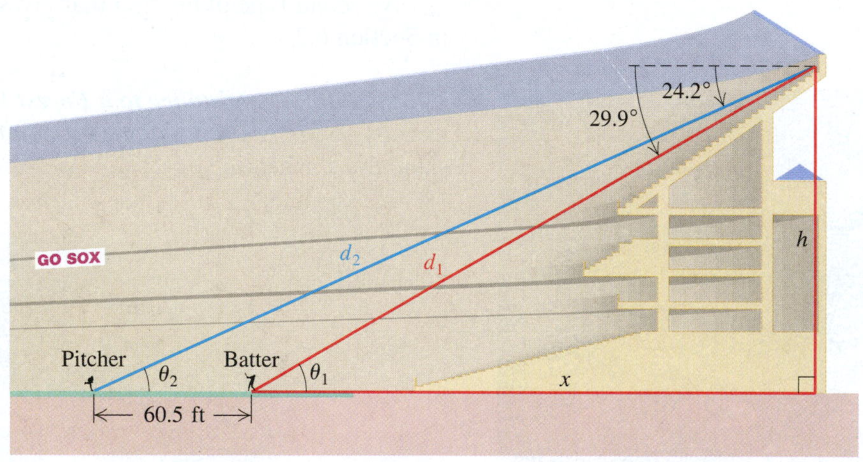

Solution From geometry, we know that $\theta_1 = 29.9°$ and $\theta_2 = 24.2°$. The standard distance from home plate to the pitcher's mound is 60.5 ft. In the drawing, we let d_1 be the viewing distance to home plate, d_2 the viewing distance to the pitcher's mound, h the elevation of the last row, and x the horizontal distance from the batter to a point directly below the seat in the last row of the upper deck.

We begin by determining the distance x. We use the tangent function with $\theta_1 = 29.9°$ and $\theta_2 = 24.2°$:

$$\tan 29.9° = \frac{h}{x} \qquad \text{and} \qquad \tan 24.2° = \frac{h}{x + 60.5}$$

or

$$h = x \tan 29.9° \qquad \text{and} \qquad h = (x + 60.5) \tan 24.2°.$$

Then substituting $x \tan 29.9°$ for h in the second equation, we obtain

$$x \tan 29.9° = (x + 60.5) \tan 24.2°.$$

Just in Time 17

Solving for x, we get

$$x \tan 29.9° = x \tan 24.2° + 60.5 \tan 24.2°$$
$$x \tan 29.9° - x \tan 24.2° = x \tan 24.2° + 60.5 \tan 24.2° - x \tan 24.2°$$
$$x(\tan 29.9° - \tan 24.2°) = 60.5 \tan 24.2°$$
$$x = \frac{60.5 \tan 24.2°}{\tan 29.9° - \tan 24.2°}$$
$$x \approx 216.5.$$

We can then find d_1 and d_2 using the cosine function:

$$\cos 29.9° = \frac{216.5}{d_1} \quad \text{and} \quad \cos 24.2° = \frac{216.5 + 60.5}{d_2}$$

or

$$d_1 = \frac{216.5}{\cos 29.9°} \quad \text{and} \quad d_2 = \frac{277}{\cos 24.2°}$$
$$d_1 \approx 249.7 \qquad d_2 \approx 303.7.$$

The distance to home plate is about 250 ft,* and the distance to the pitcher's mound is about 304 ft.

Now Try Exercise 25.

*In the original Comiskey Park, the distance to home plate was only 150 ft.

6.2 Exercise Set

In Exercises 1–6, solve the right triangle.

1.

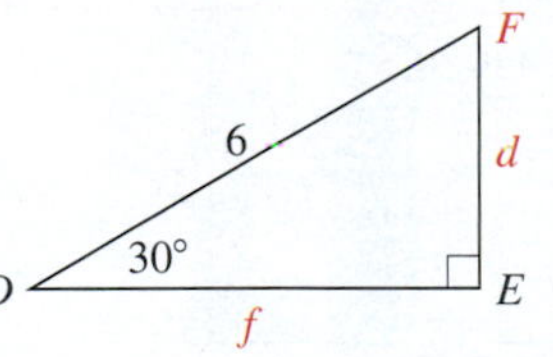

2.

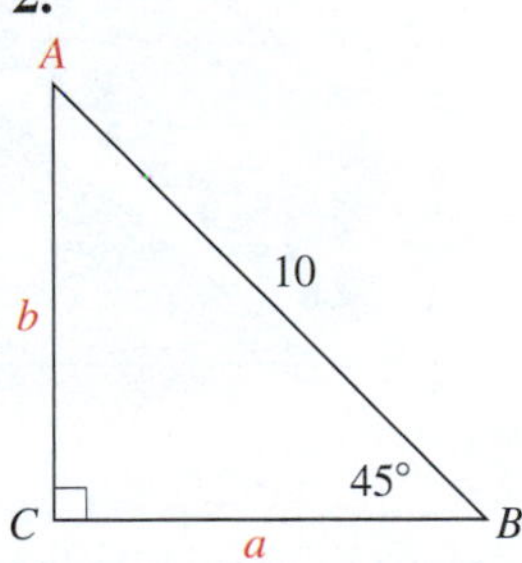

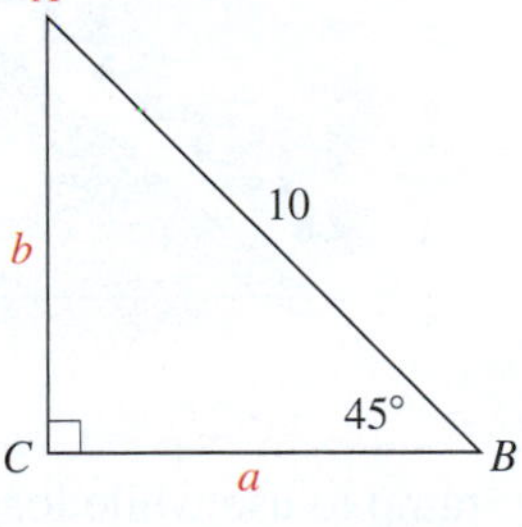

3.

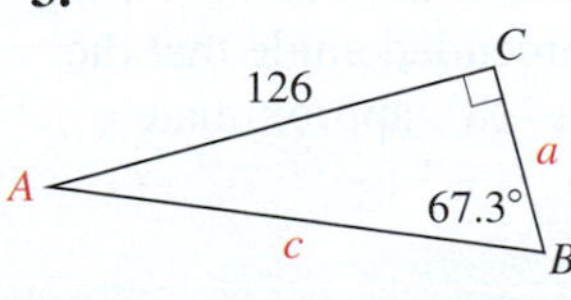

4.

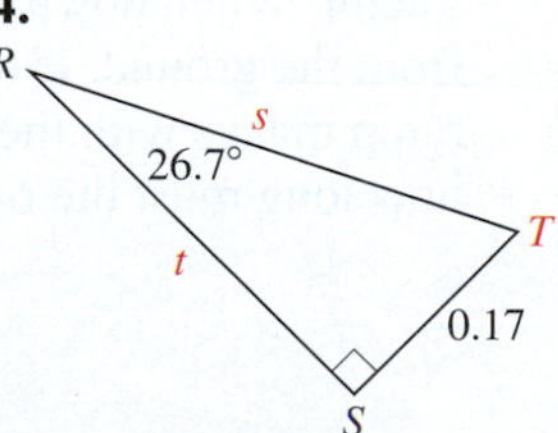

5.

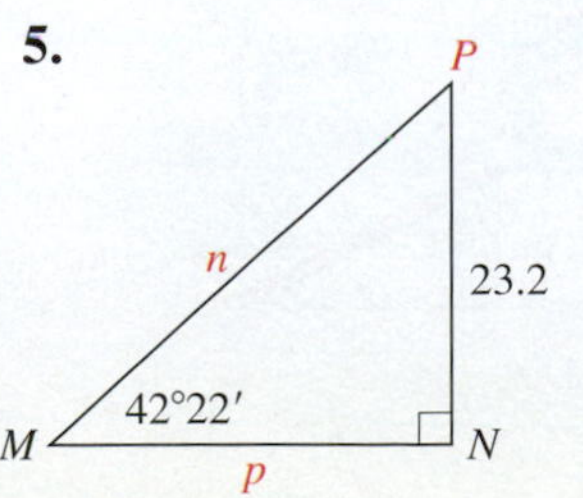

6.

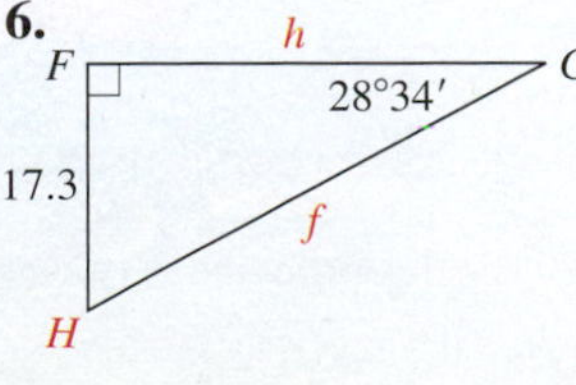

In Exercises 7–16, solve the right triangle. (Standard lettering has been used.)

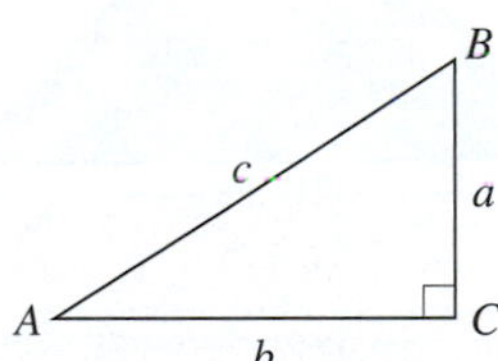

7. $A = 87°43'$, $a = 9.73$
8. $a = 12.5$, $b = 18.3$
9. $b = 100$, $c = 450$
10. $B = 56.5°$, $c = 0.0447$
11. $A = 47.58°$, $c = 48.3$
12. $B = 20.6°$, $a = 7.5$
13. $A = 35°$, $b = 40$
14. $B = 69.3°$, $b = 93.4$
15. $b = 1.86$, $c = 4.02$
16. $a = 10.2$, $c = 20.4$

17. *Aerial Photography.* An aerial photographer who photographs farm properties for a real estate company has determined from experience that the best photo is taken at a height of approximately 475 ft and a distance of 850 ft from the farmhouse. What is the angle of depression from the plane to the house?

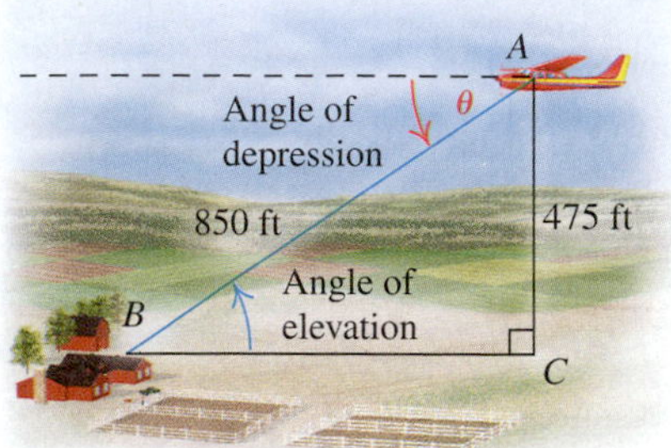

18. *Memorial Flag Case.* A tradition in the United States is to drape an American flag over the casket of a deceased U.S. Forces veteran. At the burial, the flag is removed, folded into a triangle, and presented to the family. The folded flag will fit in a case in the form of an isosceles right triangle, as shown below. The inside dimension across the bottom of the case is $21\frac{1}{2}$ in. (*Source*: Bruce Kieffer, *Woodworker's Journal,* August 2006). Using trigonometric functions, find the length x and round the answer to the nearest tenth of an inch.

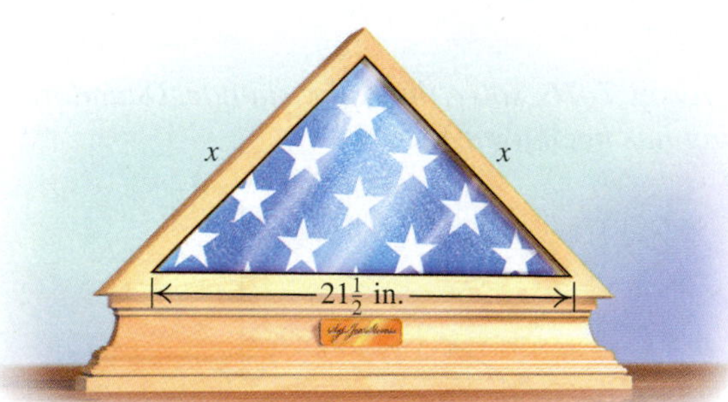

19. *Zip Line.* The ZipRider®, a zip line at Icy Straight Point, Alaska, is 5495 ft long, and has a vertical drop of 1320 ft (*Source*: www.ziprider.com). Find its angle of depression.

20. *Setting a Fishing Reel Line Counter.* A fisherman who is fishing 50 ft directly out from a visible tree stump near the shore wants to position his line and bait approximately N35°W of the boat and west of the stump. Using the right triangle shown in the drawing, determine the reel's line counter setting, to the nearest foot, to position the line directly west of the stump.

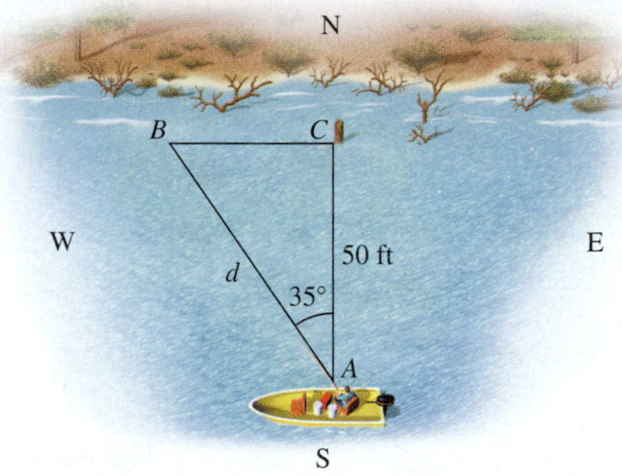

21. *Framing a Closet.* Sam is framing a closet under a stairway. The stairway is 16 ft 3 in. long, and its angle of elevation is 38°. Find the depth of the closet to the nearest inch.

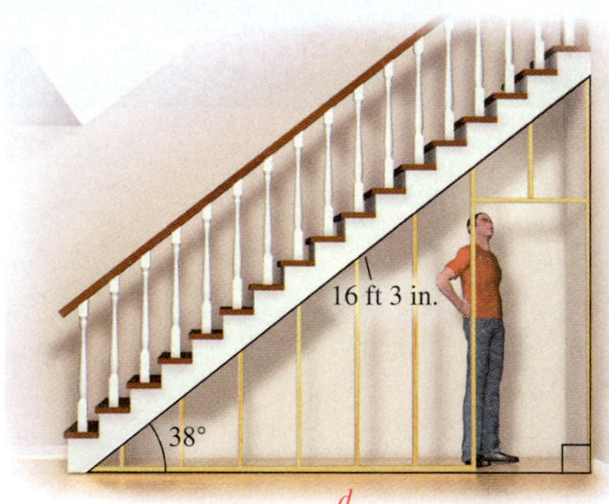

22. *Loading Ramp.* Charles needs to purchase a custom ramp to use while loading and unloading a garden tractor. When down, the tailgate of his truck is 38 in. from the ground. If the recommended angle that the ramp makes with the ground is 28°, approximately how long must the ramp be?

23. *Longest Escalator.* The longest escalator in the world is in the subway system in St. Petersburg, Russia. The escalator is 1084.6 ft long and drops a vertical distance of 195.8 ft. What is its angle of depression?

24. *Cloud Height.* To measure cloud height at night, a vertical beam of light is directed on a spot on the cloud. From a point 135 ft away from the light source, the angle of elevation to the spot is found to be 67.35°. Find the height of the cloud to the nearest foot.

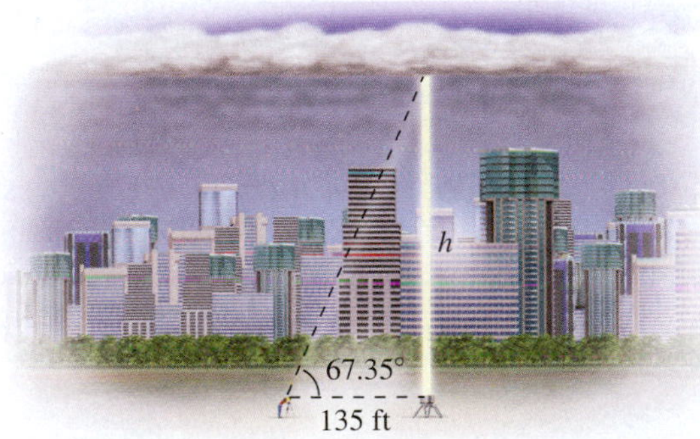

25. *Mount Rushmore National Memorial.* While visiting Mount Rushmore in Rapid City, South Dakota, Landon approximated the angle of elevation to the top of George Washington's head to be 35°. After walking 250 ft closer, he guessed that the angle of elevation had increased by 15°. Approximate the height of the Mount Rushmore memorial, to the top of George Washington's head. Round the answer to the nearest foot.

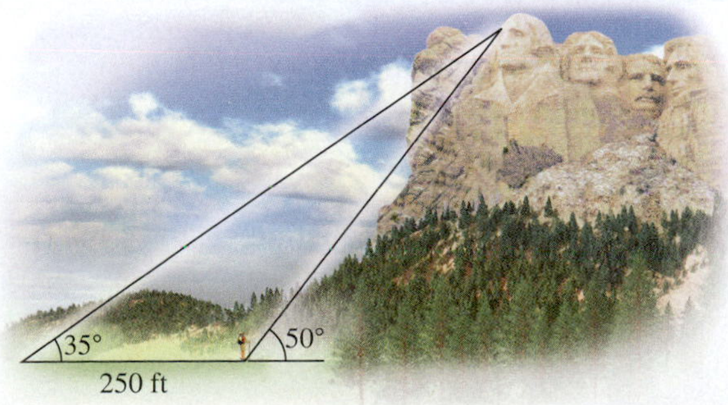

26. *Golden Gate Bridge.* The Golden Gate Bridge has two main towers of equal height that support the two main cables. A visitor on a tour boat passing through San Francisco Bay views the top of one of the towers and estimates the angle of elevation to be 30°. After sailing 670 ft closer, he estimates the angle of elevation to this same tower to be 50°. Approximate the height of the tower to the nearest foot.

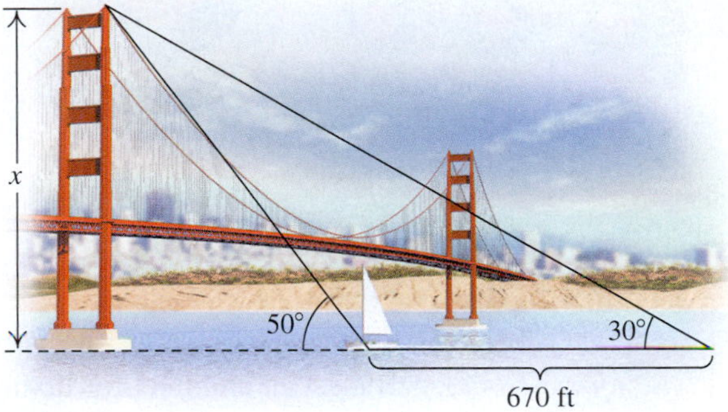

27. *Inscribed Pentagon.* A regular pentagon is inscribed in a circle of radius 15.8 cm. Find the perimeter of the pentagon.

28. *Height of a Weather Balloon.* A weather balloon is directly west of two observing stations that are 10 mi apart. The angles of elevation of the balloon from the two stations are 17.6° and 78.2°. How high is the balloon?

29. *Height of a Building.* A window washer on a ladder looks at a nearby building 100 ft away, noting that the angle of elevation to the top of the building is 18.7° and the angle of depression to the bottom of the building is 6.5°. How tall is the nearby building?

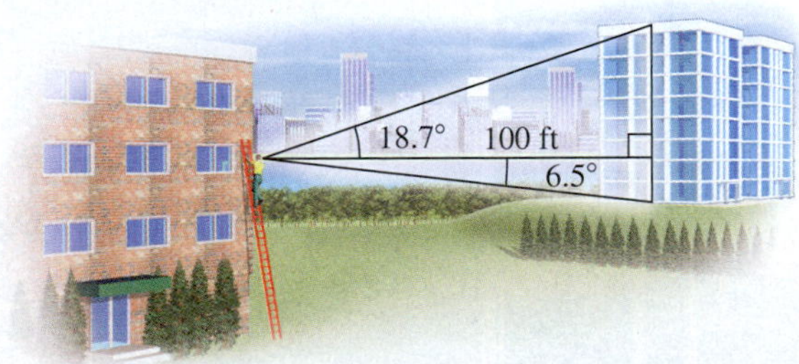

30. *Height of a Kite.* For a science fair project, a group of students tested different materials used to construct kites. Their instructor provided an instrument that accurately measures the angle of elevation. In one of the tests, the angle of elevation was 63.4° with 670 ft of string out. Assuming the string was taut, how high was the kite?

31. *Quilt Design.* Nancy is designing a quilt that she will enter in the quilt competition at the State Fair. The quilt consists of twelve identical squares with 4 rows of 3 squares each. Each square is to have a regular octagon inscribed in a circle, as shown in the figure.

Each side of the octagon is to be 7 in. long. Find the radius of the circumscribed circle and the dimensions of the quilt. Round the answers to the nearest hundredth of an inch.

32. *Rafters for a House.* Blaise, an architect for luxury homes, is designing a house that is 46 ft wide with a roof whose pitch is 11/12. Determine the length of the rafters needed for this house. Round the answer to the nearest tenth of a foot.

33. *Rafters for a Medical Office.* The pitch of the roof for a medical office needs to be 5/12. If the building is 33 ft wide, how long must the rafters be?

34. *Angle of Elevation.* The Millau Viaduct in southern France is the tallest cable-stayed bridge in the world (*Source*: www.abelard.org). What is the angle of elevation of the sun when a pylon with height 343 m casts a shadow of 186 m?

35. *Distance Between Towns.* From a hot-air balloon 2 km high, the angles of depression to two towns in line with the balloon and on the same side of the balloon are 81.2° and 13.5°. How far apart are the towns?

36. *Distance from a Lighthouse.* From the top of a lighthouse 55 ft above sea level, the angle of depression to a small boat is 11.3°. How far from the foot of the lighthouse is the boat?

37. *Lightning Detection.* In extremely large forests, it is not cost-effective to position forest rangers in towers or to use small aircraft to continually watch for fires. Since lightning is a frequent cause of fire, lightning detectors are now commonly used instead. These devices not only give a bearing on the location but also measure the intensity of the lightning. A detector at point Q is situated 15 mi west of a central fire station at point R. The bearing from Q to where lightning hits due south of R is S37.6°E. How far is the hit from point R?

38. *Length of an Antenna.* A vertical antenna is mounted atop a 50-ft pole. From a point on level ground 75 ft from the base of the pole, the antenna subtends an angle of 10.5°. Find the length of the antenna.

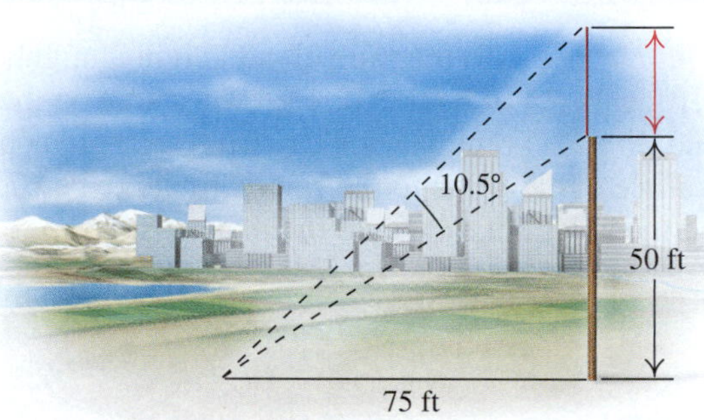

39. *Lobster Boat.* A lobster boat is situated due west of a lighthouse. A barge is 12 km south of the lobster boat. From the barge, the bearing to the lighthouse is N63°20′E. How far is the lobster boat from the lighthouse?

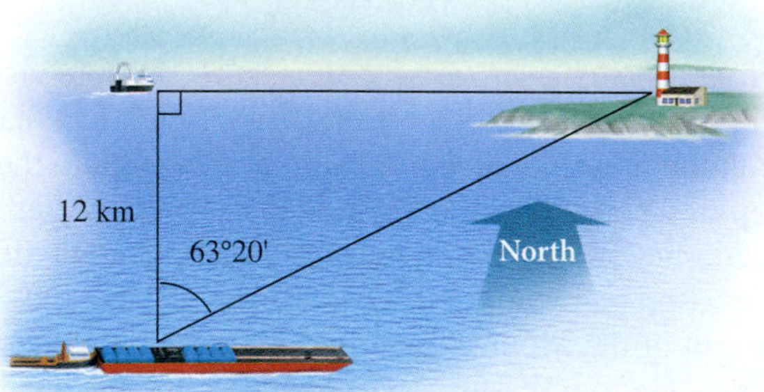

Skill Maintenance

Find the distance between the points. **[1.1]**

40. $(-9, 3)$ and $(0, 0)$

41. $(8, -2)$ and $(-6, -4)$

42. Convert to a logarithmic equation: $e^4 = t$. **[5.3]**

43. Convert to an exponential equation:
$\log 0.001 = -3$. **[5.3]**

› Synthesis

44. *Diameter of a Pipe.* A V-gauge is used to find the diameter of a pipe. The advantage of such a device is that it is rugged, it is accurate, and it has no moving parts to break down. In the figure, the measure of angle AVB is 54°. A pipe is placed in the V-shaped slot and the distance VP is used to estimate the diameter. The line VP is calibrated by listing as its units the corresponding diameters. This, in effect, establishes a function between VP and d.

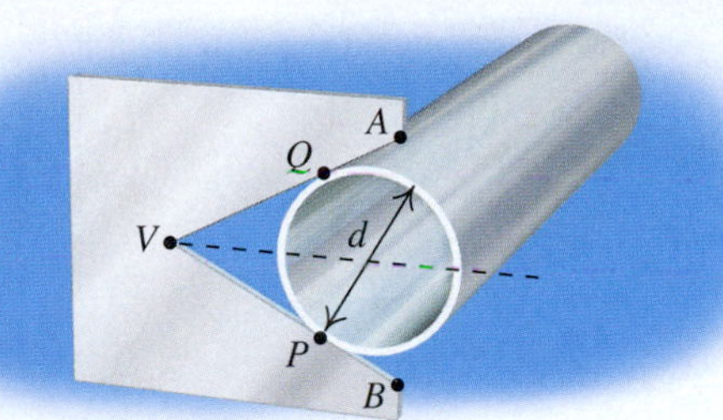

a) Suppose that the diameter of a pipe is 2 cm. What is the distance VP?
b) Suppose that the distance VP is 3.93 cm. What is the diameter of the pipe?
c) Find a formula for d in terms of VP.
d) Find a formula for VP in terms of d.

45. Find h, to the nearest tenth.

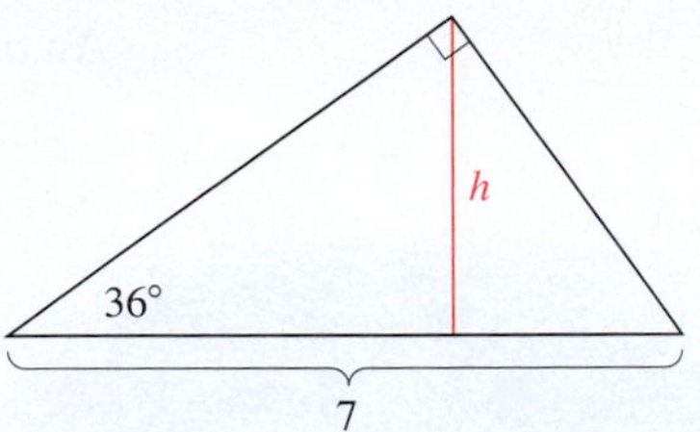

46. *Sound of an Airplane.* It is common experience to hear the sound of a low-flying airplane and look at the wrong place in the sky to see the plane. Suppose that a plane is traveling directly at you at a speed of 200 mph and an altitude of 3000 ft, and you hear the sound at what seems to be an angle of elevation of 20°. At what angle θ should you actually look in order to see the plane? Consider the speed of sound to be 1100 ft/sec.

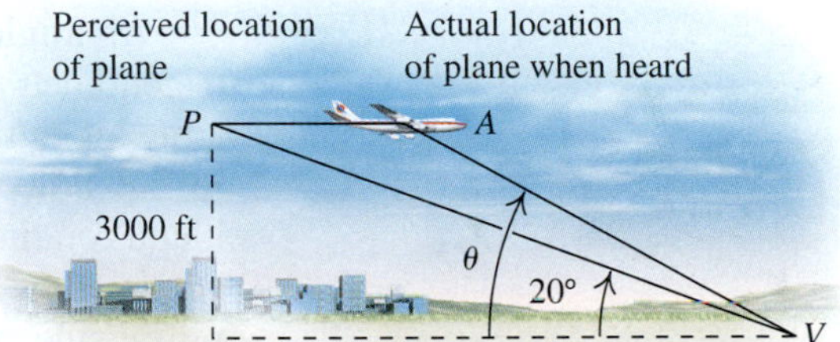

6.3 Trigonometric Functions of Any Angle

- › Find angles that are coterminal with a given angle, and find the complement and the supplement of a given angle.
- › Determine the six trigonometric function values for any angle in standard position when the coordinates of a point on the terminal side are given.
- › Find the function values for any angle whose terminal side lies on an axis.
- › Find the function values for an angle whose terminal side makes an angle of 30°, 45°, or 60° with the x-axis.
- › Use a calculator to find function values and angles.

Angles, Rotations, and Degree Measure

An *angle* is a familiar figure in the world around us.

An **angle** is the union of two rays with a common endpoint called the **vertex**. In trigonometry, we often think of an angle as a **rotation**. To do so, think of locating a ray along the positive x-axis with its endpoint at the origin. This ray is called the **initial side** of the angle. Though we leave that ray fixed, think of making a copy of it and rotating it. A rotation *counterclockwise* is a **positive rotation**, and a rotation *clockwise* is a **negative rotation**. The ray at the end of the rotation is called the **terminal side** of the angle. The angle formed is said to be in **standard position**.

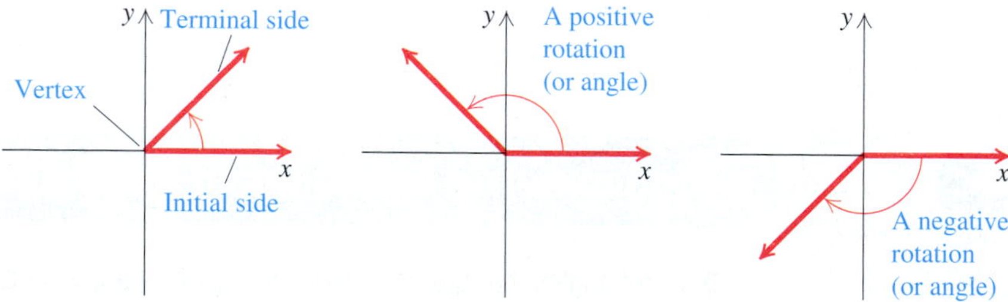

The measure of an angle or rotation may be given in degrees. The Babylonians developed the idea of dividing the circumference of a circle into 360 equal parts, or degrees. If we let the measure of one of these parts be 1°, then one complete positive revolution or rotation has a measure of 360°. One half of a revolution has a measure of 180°, one fourth of a revolution has a measure of 90°, and so on. We can also speak of an angle of measure 60°, 135°, 330°, or 420°. The terminal sides of these angles lie in quadrants I, II, IV, and I, respectively. The negative rotations −30°, −110°, and −225° represent angles with terminal sides in quadrants IV, III, and II, respectively.

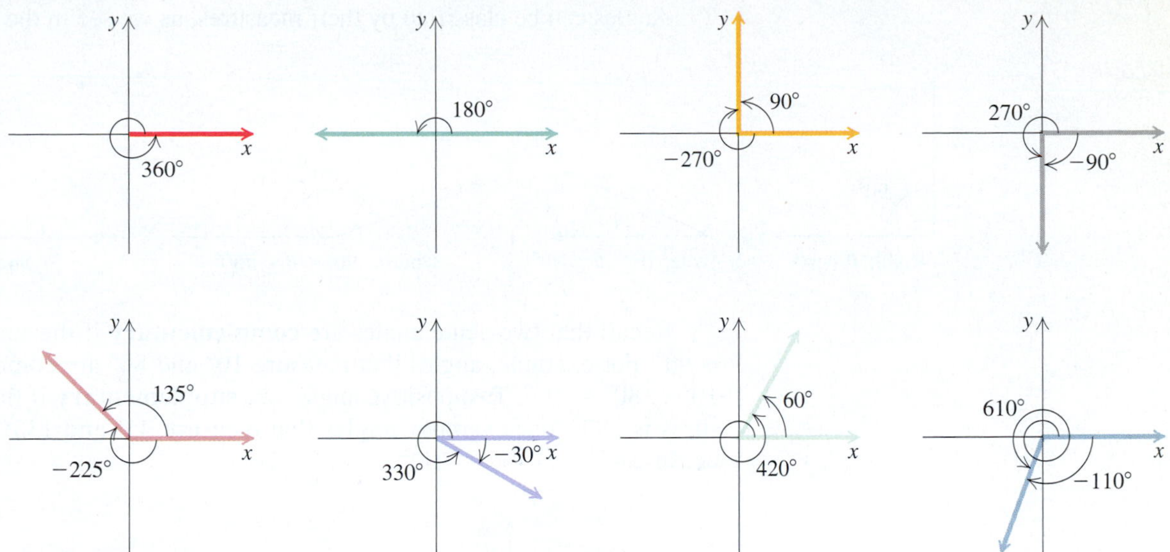

If two or more angles have the same terminal side, the angles are said to be **coterminal**. To find angles coterminal with a given angle, we add or subtract multiples of 360°. For example, 420°, shown above, has the same terminal side as 60°, since $420° = 360° + 60°$. Thus we say that angles of measure 60° and 420° are coterminal. The negative rotation that measures −300° is also coterminal with 60° because $60° - 360° = -300°$. The set of all angles coterminal with 60° can be expressed as $60° + n \cdot 360°$, where n is an integer. Other examples of coterminal angles shown above are 90° and −270°, −90° and 270°, 135° and −225°, −30° and 330°, and −110° and 610°.

Just in Time 5

EXAMPLE 1 Find two positive angles and two negative angles that are coterminal with **(a)** 51° and **(b)** −7°.

Solution

a) We add and subtract multiples of 360°. Many answers are possible.

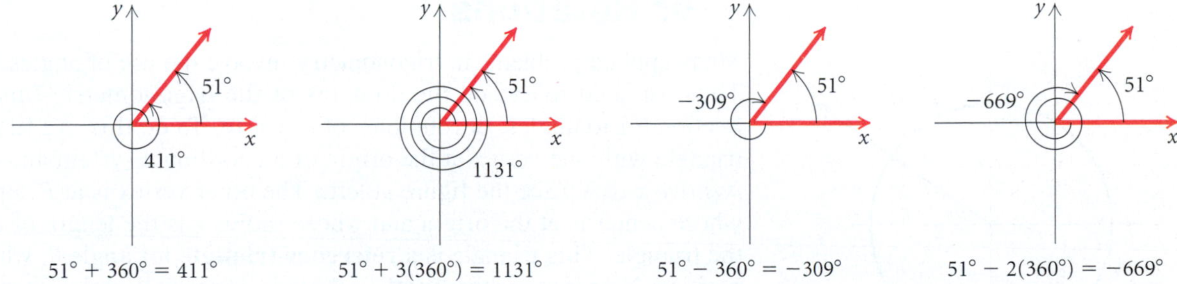

Thus angles of measure 411°, 1131°, −309°, and −669° are coterminal with 51°.

b) We have the following:

$$-7° + 360° = 353°, \qquad -7° + 2(360°) = 713°,$$
$$-7° - 360° = -367°, \qquad -7° - 10(360°) = -3607°.$$

Thus angles of measure 353°, 713°, −367°, and −3607° are coterminal with −7°.

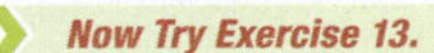

Angles can be classified by their measures, as we see in the following figures.

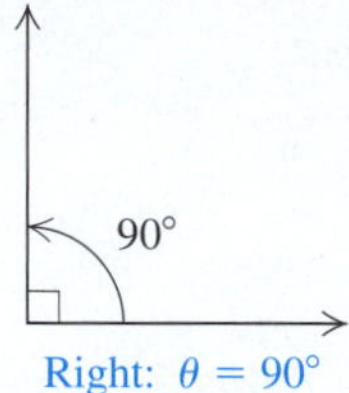

Right: $\theta = 90°$

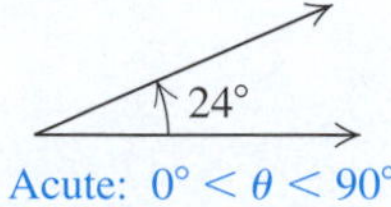

Acute: $0° < \theta < 90°$

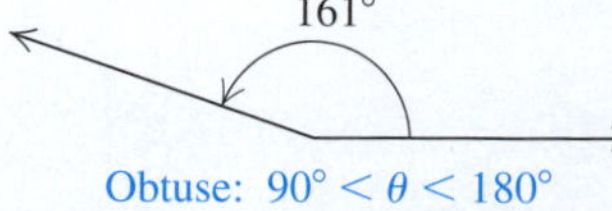

Obtuse: $90° < \theta < 180°$

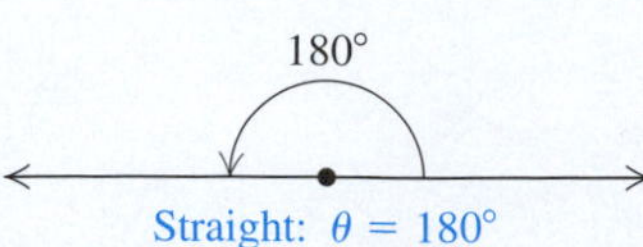

Straight: $\theta = 180°$

Recall that two acute angles are **complementary** if the sum of their measures is 90°. For example, angles that measure 10° and 80° are complementary because $10° + 80° = 90°$. Two positive angles are **supplementary** if the sum of their measures is 180°. For example, angles that measure 45° and 135° are supplementary because $45° + 135° = 180°$.

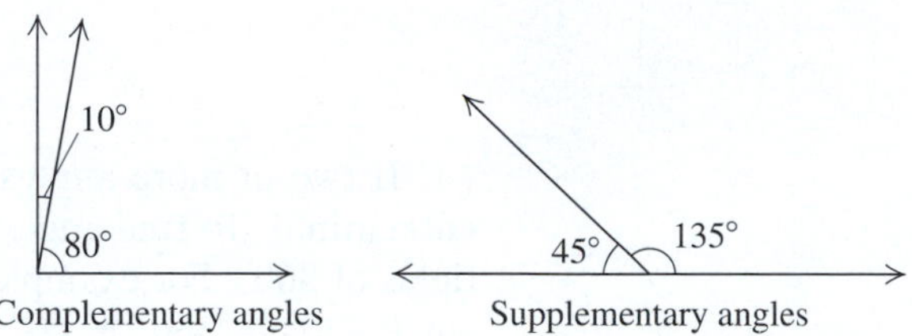

Complementary angles   Supplementary angles

EXAMPLE 2 Find the complement and the supplement of 71.46°.

Solution We have

$$90° - 71.46° = 18.54°,$$
$$180° - 71.46° = 108.54°.$$

Thus the complement of 71.46° is 18.54° and the supplement is 108.54°.

Now Try Exercise 19.

Trigonometric Functions of Angles or Rotations

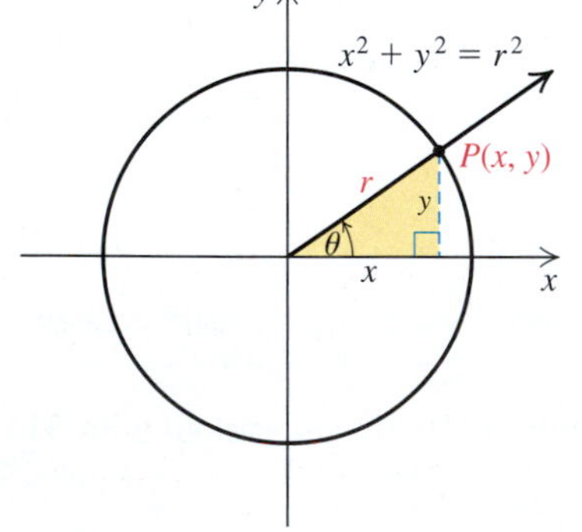

Many applied problems in trigonometry involve the use of angles that are not acute. Thus we need to extend the domains of the trigonometric functions defined in Section 6.1 to angles, or rotations, of *any* size. To do this, we first consider a right triangle with one vertex at the origin of a coordinate system and one vertex *on the positive x-axis*. (See the figure at left.) The other vertex is at P, a point on the circle whose center is at the origin and whose radius r is the length of the hypotenuse of the triangle. This triangle is a **reference triangle** for angle θ, which is in standard position. Note that y is the length of the side opposite θ, and x is the length of the side adjacent to θ.

Recalling the definitions from Section 6.1, we note that three of the trigonometric functions of angle θ are defined as follows:

$$\sin\theta = \frac{\text{opp}}{\text{hyp}} = \frac{y}{r}, \qquad \cos\theta = \frac{\text{adj}}{\text{hyp}} = \frac{x}{r}, \qquad \tan\theta = \frac{\text{opp}}{\text{adj}} = \frac{y}{x}.$$

Since x and y are the coordinates of the point P and the length of the radius is the length of the hypotenuse, we can also define these functions as follows:

$$\sin\theta = \frac{y\text{-coordinate}}{\text{radius}}, \qquad \cos\theta = \frac{x\text{-coordinate}}{\text{radius}}, \qquad \tan\theta = \frac{y\text{-coordinate}}{x\text{-coordinate}}.$$

We will use these definitions for functions of angles of any measure. The following figures show angles whose terminal sides lie in quadrants II, III, and IV.

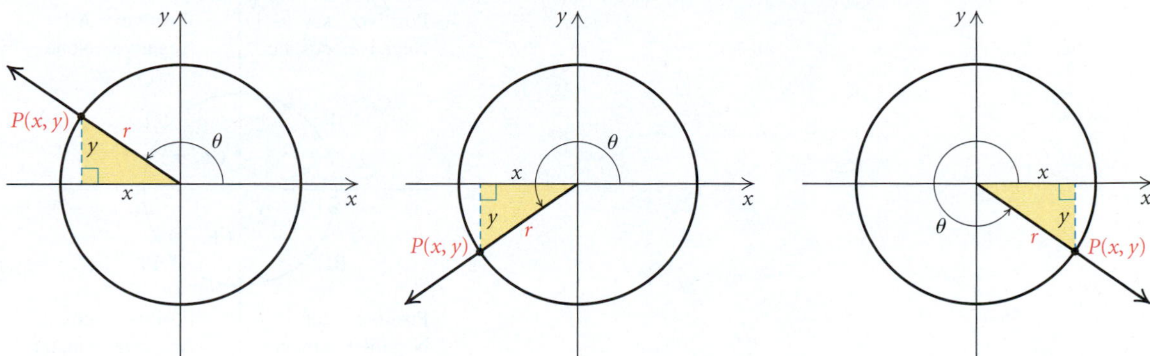

A reference triangle can be drawn for angles in any quadrant, as shown. Note that the angle is in standard position; that is, it is always measured from the positive half of the x-axis. The point $P(x, y)$ is a point, other than the vertex, on the terminal side of the angle. Each of its two coordinates may be positive, negative, or zero, depending on the location of the terminal side. *The length of the radius, which is also the length of the hypotenuse of the reference triangle, is always considered positive.* (Note that $x^2 + y^2 = r^2$, or $r = \sqrt{x^2 + y^2}$.) Regardless of the location of P, we have the following definitions. We now extend the *domain* of the six trigonometric functions from acute angles (see p. 395) to include *all angles, or rotations.*

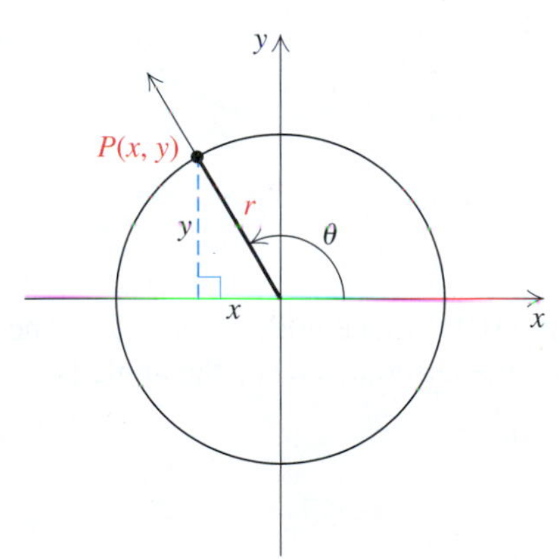

TRIGONOMETRIC FUNCTIONS OF ANY ANGLE θ

Suppose that $P(x, y)$ is any point other than the vertex on the terminal side of any angle θ in standard position, and r is the radius, or distance from the origin to $P(x, y)$. Then the trigonometric functions are defined as follows:

$$\sin\theta = \frac{y\text{-coordinate}}{\text{radius}} = \frac{y}{r}, \qquad \csc\theta = \frac{\text{radius}}{y\text{-coordinate}} = \frac{r}{y}\ (y \neq 0),$$

$$\cos\theta = \frac{x\text{-coordinate}}{\text{radius}} = \frac{x}{r}, \qquad \sec\theta = \frac{\text{radius}}{x\text{-coordinate}} = \frac{r}{x}\ (x \neq 0),$$

$$\tan\theta = \frac{y\text{-coordinate}}{x\text{-coordinate}} = \frac{y}{x}\ (x \neq 0), \qquad \cot\theta = \frac{x\text{-coordinate}}{y\text{-coordinate}} = \frac{x}{y}\ (y \neq 0).$$

Values of the trigonometric functions can be positive, negative, or zero, depending on where the terminal side of the angle lies. Since the length of the radius is always positive, the signs of the function values depend only on the coordinates of the point P on the terminal side of the angle. In the first quadrant, all function values are positive because both coordinates are positive. In the second quadrant, first coordinates are negative and second coordinates are positive; thus only the sine and the cosecant values are positive. Similarly, we can determine the signs of the function values in the third and the fourth quadrants. *Because of the reciprocal relationships, we need to learn only the signs for the sine, cosine, and tangent functions.*

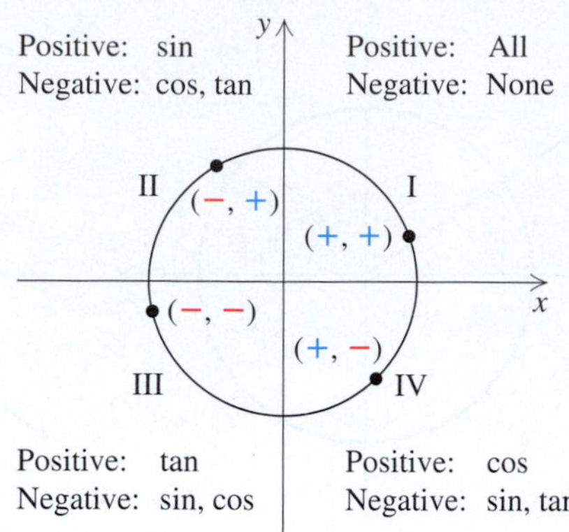

EXAMPLE 3 Find the six trigonometric function values for each angle shown.

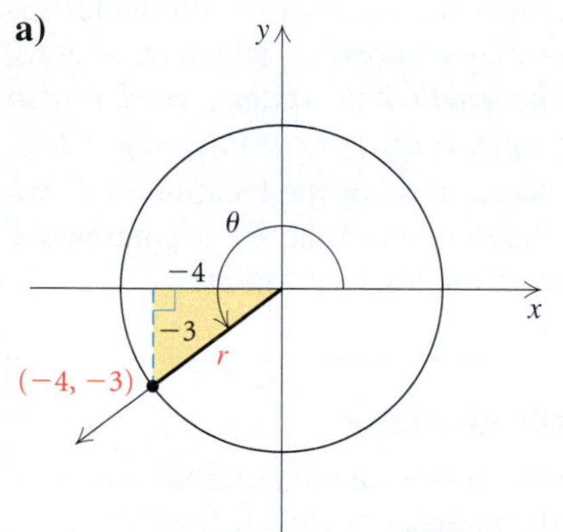

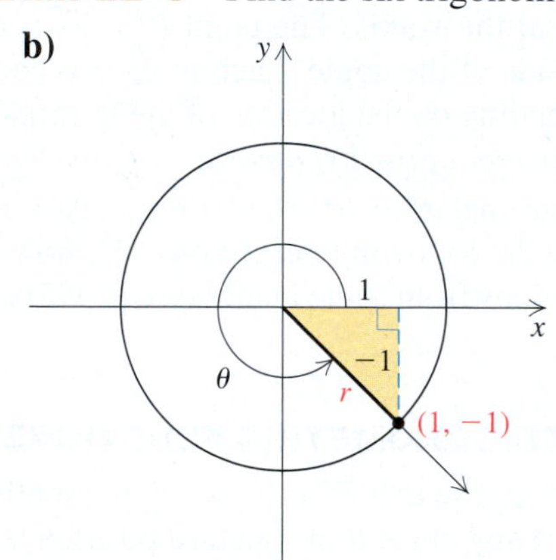

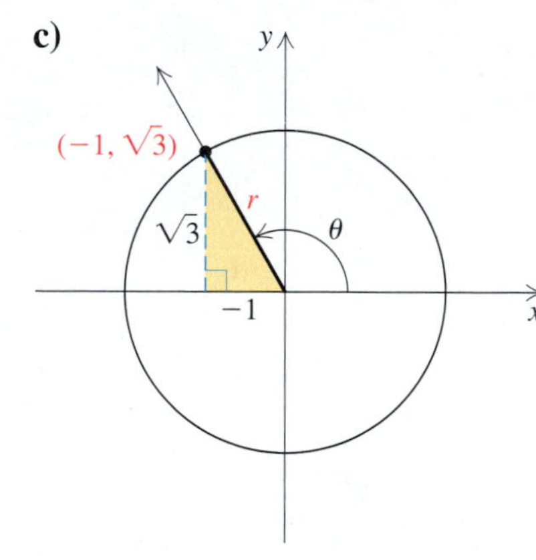

Solution

a) We first determine r, the distance from the origin $(0, 0)$ to the point $(-4, -3)$. The distance between $(0, 0)$ and any point (x, y) on the terminal side of the angle is

$$r = \sqrt{(x-0)^2 + (y-0)^2} = \sqrt{x^2 + y^2}.$$

Just in Time 25

Substituting -4 for x and -3 for y, we find

$$r = \sqrt{(-4)^2 + (-3)^2} = \sqrt{16+9} = \sqrt{25} = 5.$$

Using the definitions of the trigonometric functions, we can now find the function values of θ. We substitute -4 for x, -3 for y, and 5 for r:

$$\sin\theta = \frac{y}{r} = \frac{-3}{5} = -\frac{3}{5}, \qquad \csc\theta = \frac{r}{y} = \frac{5}{-3} = -\frac{5}{3},$$

$$\cos\theta = \frac{x}{r} = \frac{-4}{5} = -\frac{4}{5}, \qquad \sec\theta = \frac{r}{x} = \frac{5}{-4} = -\frac{5}{4},$$

$$\tan\theta = \frac{y}{x} = \frac{-3}{-4} = \frac{3}{4}, \qquad \cot\theta = \frac{x}{y} = \frac{-4}{-3} = \frac{4}{3}.$$

As expected, the tangent value and the cotangent value are positive and the other four values are negative. This is true for all angles in quadrant III.

b) We first determine r, the distance from the origin to the point $(1, -1)$:

$$r = \sqrt{1^2 + (-1)^2} = \sqrt{1+1} = \sqrt{2}.$$

Substituting 1 for x, -1 for y, and $\sqrt{2}$ for r, we find

$$\sin\theta = \frac{y}{r} = \frac{-1}{\sqrt{2}} = -\frac{\sqrt{2}}{2}, \qquad \csc\theta = \frac{r}{y} = \frac{\sqrt{2}}{-1} = -\sqrt{2},$$

$$\cos\theta = \frac{x}{r} = \frac{1}{\sqrt{2}} = \frac{\sqrt{2}}{2}, \qquad \sec\theta = \frac{r}{x} = \frac{\sqrt{2}}{1} = \sqrt{2},$$

$$\tan\theta = \frac{y}{x} = \frac{-1}{1} = -1, \qquad \cot\theta = \frac{x}{y} = \frac{1}{-1} = -1.$$

c) We determine r, the distance from the origin to the point $(-1, \sqrt{3})$:

$$r = \sqrt{(-1)^2 + (\sqrt{3})^2} = \sqrt{1+3} = \sqrt{4} = 2.$$

Substituting -1 for x, $\sqrt{3}$ for y, and 2 for r, we find the trigonometric function values of θ are

$$\sin\theta = \frac{\sqrt{3}}{2}, \qquad \csc\theta = \frac{2}{\sqrt{3}} = \frac{2\sqrt{3}}{3},$$

$$\cos\theta = \frac{-1}{2} = -\frac{1}{2}, \qquad \sec\theta = \frac{2}{-1} = -2,$$

$$\tan\theta = \frac{\sqrt{3}}{-1} = -\sqrt{3}, \qquad \cot\theta = \frac{-1}{\sqrt{3}} = -\frac{\sqrt{3}}{3}.$$

Now Try Exercise 29.

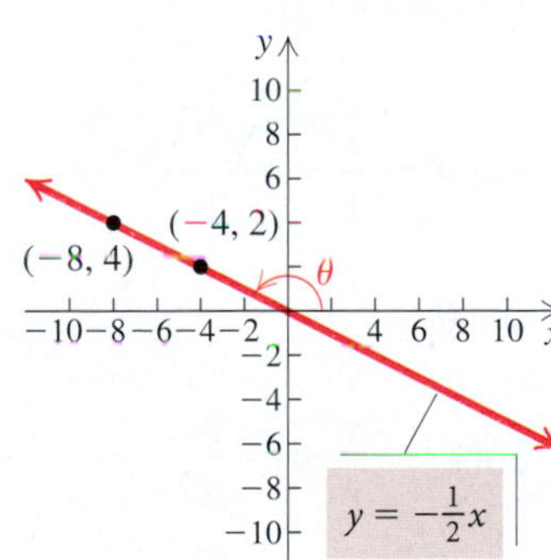

Any point other than the origin on the terminal side of an angle in standard position can be used to determine the trigonometric function values of that angle. The function values are the same regardless of which point is used. To illustrate this, let's consider an angle θ in standard position whose terminal side lies on the line $y = -\frac{1}{2}x$. We can determine two second-quadrant solutions of the equation, find the length r for each point, and then compare the sine, cosine, and tangent function values using each point.

If $x = -4$, then $y = -\frac{1}{2}(-4) = 2$.
If $x = -8$, then $y = -\frac{1}{2}(-8) = 4$.
For $(-4, 2)$, $r = \sqrt{(-4)^2 + 2^2} = \sqrt{20} = 2\sqrt{5}$.
For $(-8, 4)$, $r = \sqrt{(-8)^2 + 4^2} = \sqrt{80} = 4\sqrt{5}$.

Using $(-4, 2)$ and $r = 2\sqrt{5}$, we find that

$$\sin\theta = \frac{y}{r} = \frac{2}{2\sqrt{5}} = \frac{1}{\sqrt{5}} = \frac{\sqrt{5}}{5},$$

$$\cos\theta = \frac{x}{r} = \frac{-4}{2\sqrt{5}} = \frac{-2}{\sqrt{5}} = -\frac{2\sqrt{5}}{5},$$

and $\tan\theta = \frac{y}{x} = \frac{2}{-4} = -\frac{1}{2}.$

Using $(-8, 4)$ and $r = 4\sqrt{5}$, we find that

$$\sin\theta = \frac{y}{r} = \frac{4}{4\sqrt{5}} = \frac{1}{\sqrt{5}} = \frac{\sqrt{5}}{5},$$

$$\cos\theta = \frac{x}{r} = \frac{-8}{4\sqrt{5}} = \frac{-2}{\sqrt{5}} = -\frac{2\sqrt{5}}{5},$$

and $\tan\theta = \frac{y}{x} = \frac{4}{-8} = -\frac{1}{2}.$

We see that the function values are the same using either point. This illustrates that any point other than the origin on the terminal side of an angle can be used to determine the trigonometric function values.

> The trigonometric function values of θ depend only on the angle, not on the choice of the point on the terminal side that is used to compute them.

The Six Related Functions

When we know one of the function values of an angle, we can find the other five if we know the quadrant in which the terminal side lies. The procedure is to sketch a reference triangle in the appropriate quadrant, use the Pythagorean equation as needed to find the lengths of its sides, and then find the ratios of the sides.

EXAMPLE 4 Given that $\tan\theta = -\frac{2}{3}$ and θ is in the second quadrant, find the other function values.

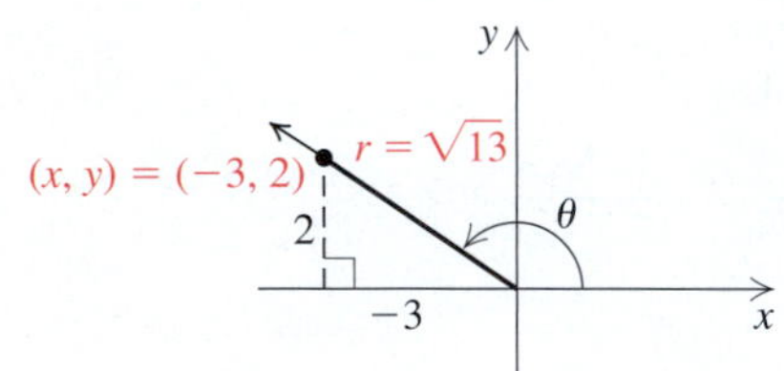

Solution We first sketch a second-quadrant angle. Since

$$\tan\theta = \frac{y}{x} = -\frac{2}{3} = \frac{2}{-3},$$ **Expressing $-\frac{2}{3}$ as $\frac{2}{-3}$ since θ is in quadrant II**

we make the legs' lengths 3 and 2. We measure off the 3 units in the negative direction since θ is in quadrant II. The hypotenuse must then have length $\sqrt{3^2 + 2^2}$, or $\sqrt{13}$. Now we read off the appropriate ratios:

$$\sin\theta = \frac{y}{r} = \frac{2}{\sqrt{13}}, \quad \text{or} \quad \frac{2\sqrt{13}}{13}, \qquad \csc\theta = \frac{r}{y} = \frac{\sqrt{13}}{2},$$

$$\cos\theta = \frac{x}{r} = -\frac{3}{\sqrt{13}}, \quad \text{or} \quad -\frac{3\sqrt{13}}{13}, \qquad \sec\theta = \frac{r}{x} = -\frac{\sqrt{13}}{3},$$

$$\tan\theta = \frac{y}{x} = -\frac{2}{3}, \qquad \cot\theta = \frac{x}{y} = -\frac{3}{2}.$$

Now Try Exercise 35.

Terminal Side on an Axis

An angle whose terminal side falls on one of the axes is a **quadrantal angle**. One of the coordinates of any point on that side is 0. The definitions of the trigonometric functions still apply, but in some cases, function values will not be defined because a denominator will be 0.

EXAMPLE 5 Find the sine, cosine, and tangent values for 90°, 180°, 270°, and 360°.

Solution We first make a drawing of each angle in standard position and label a point on the terminal side. Since the function values are the same for all points on the terminal side, we choose $(0, 1)$, $(-1, 0)$, $(0, -1)$, and $(1, 0)$ for convenience. Note that $r = 1$ for each choice.

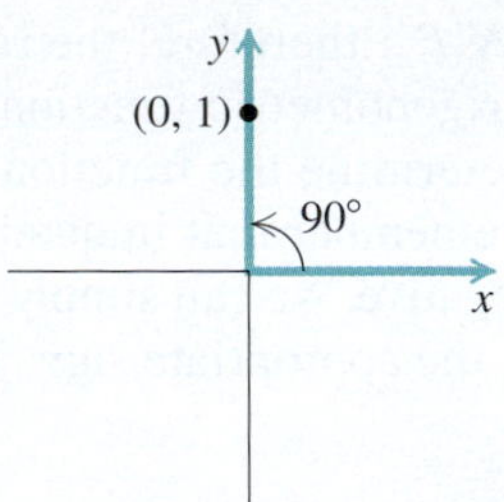

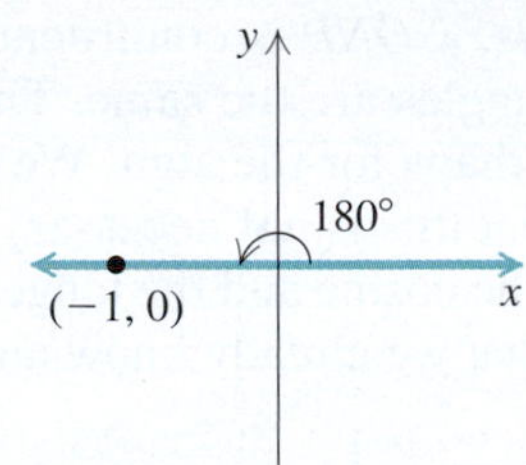

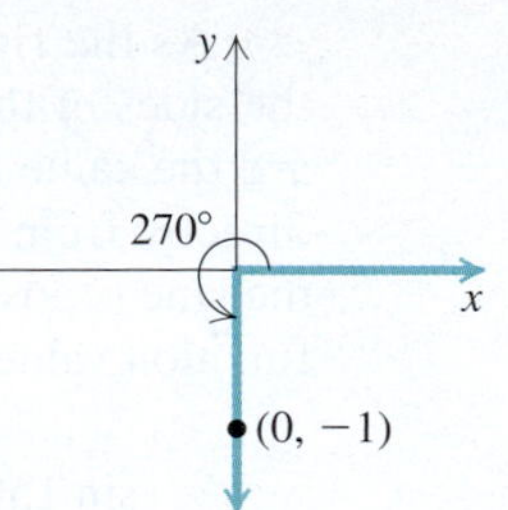

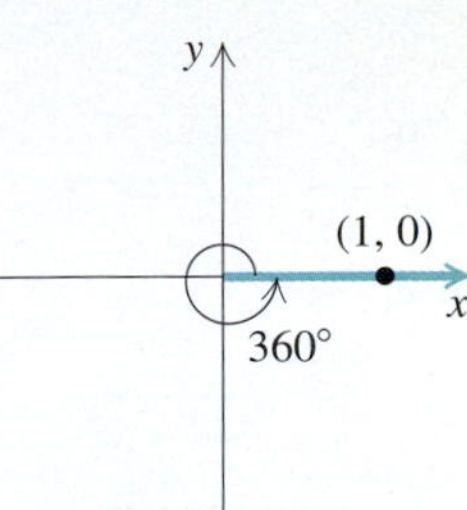

Then by the definitions we get

$\sin 90° = \frac{1}{1} = 1,$ $\quad\sin 180° = \frac{0}{1} = 0,$ $\quad\sin 270° = \frac{-1}{1} = -1,$ $\quad\sin 360° = \frac{0}{1} = 0,$

$\cos 90° = \frac{0}{1} = 0,$ $\quad\cos 180° = \frac{-1}{1} = -1,$ $\quad\cos 270° = \frac{0}{1} = 0,$ $\quad\cos 360° = \frac{1}{1} = 1,$

$\tan 90° = \frac{1}{0},$ **Not defined** $\quad\tan 180° = \frac{0}{-1} = 0,$ $\quad\tan 270° = \frac{-1}{0},$ **Not defined** $\quad\tan 360° = \frac{0}{1} = 0.$

In Example 5, all the values can be found using a calculator, but you will find that it is convenient to be able to compute them mentally. It is also helpful to note that coterminal angles have the same function values. For example, 0° and 360° are coterminal; thus, $\sin 0° = 0$, $\cos 0° = 1$, and $\tan 0° = 0$.

EXAMPLE 6 Find each of the following.

a) $\sin(-90°)$ **b)** $\csc 540°$

Solution

a) We note that $-90°$ is coterminal with 270°. Thus,

$$\sin(-90°) = \sin 270° = -1.$$

b) Since $540° = 180° + 360°$, 540° and 180° are coterminal. Thus,

$$\csc 540° = \csc 180° = \frac{1}{\sin 180°} = \frac{1}{0}, \text{ which is not defined.}$$

Now Try Exercises 45 and 55.

ERR: DIVIDE BY 0
1: Quit
2: Goto

or

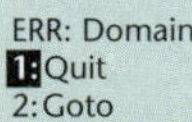

ERR: Domain
1: Quit
2: Goto

Trigonometric values can always be checked using a calculator. When the value is not defined, the calculator will display an ERROR message, as shown at left.

Reference Angles: 30°, 45°, and 60°

We can also mentally determine trigonometric function values whenever the terminal side makes a 30°, 45°, or 60° angle with the x-axis. Consider, for example, an angle of 150°. The terminal side makes a 30° angle with the x-axis, since $180° - 150° = 30°$.

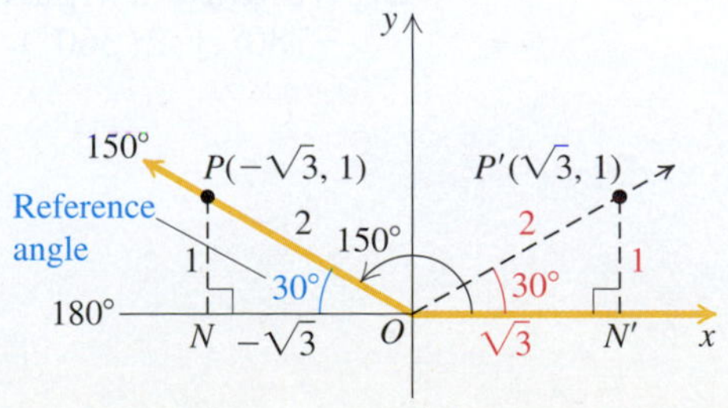

As the figure shows, $\triangle ONP$ is congruent to $\triangle ON'P'$; therefore, the ratios of the sides of the two triangles are the same. Thus the trigonometric function values are the same except perhaps for the sign. We could determine the function values directly from $\triangle ONP$, but this is not necessary. If we remember that in quadrant II, the sine is positive and the cosine and the tangent are negative, we can simply use the function values of 30° that we already know and prefix the appropriate sign. Thus,

$$\sin 150^\circ = \sin 30^\circ = \frac{1}{2},$$

$$\cos 150^\circ = -\cos 30^\circ = -\frac{\sqrt{3}}{2},$$

and $$\tan 150^\circ = -\tan 30^\circ = -\frac{1}{\sqrt{3}}, \quad \text{or} \quad -\frac{\sqrt{3}}{3}.$$

Triangle ONP is the reference triangle and the acute angle $\angle NOP$ is called a *reference angle.*

REFERENCE ANGLE

The **reference angle** for an angle is the acute angle formed by the terminal side of the angle and the x-axis.

EXAMPLE 7 Find the sine, cosine, and tangent function values for each of the following.

a) 225° **b)** −780°

Solution

a) We draw a figure showing the terminal side of a 225° angle. The reference angle is 225° − 180°, or 45°.

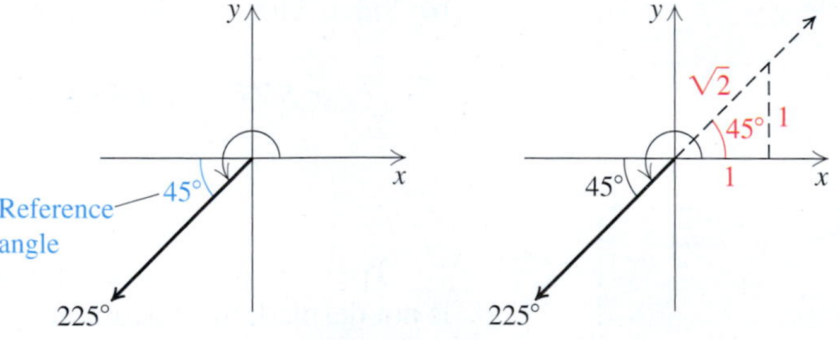

Recall from Section 6.1 that $\sin 45^\circ = \sqrt{2}/2$, $\cos 45^\circ = \sqrt{2}/2$, and $\tan 45^\circ = 1$. Also note that in the third quadrant, the sine and the cosine are negative and the tangent is positive. Thus we have

$$\sin 225^\circ = -\frac{\sqrt{2}}{2}, \quad \cos 225^\circ = -\frac{\sqrt{2}}{2}, \quad \text{and} \quad \tan 225^\circ = 1.$$

b) We draw a figure showing the terminal side of a −780° angle. Since $-780^\circ + 2(360^\circ) = -60^\circ$, we know that −780° and −60° are coterminal.

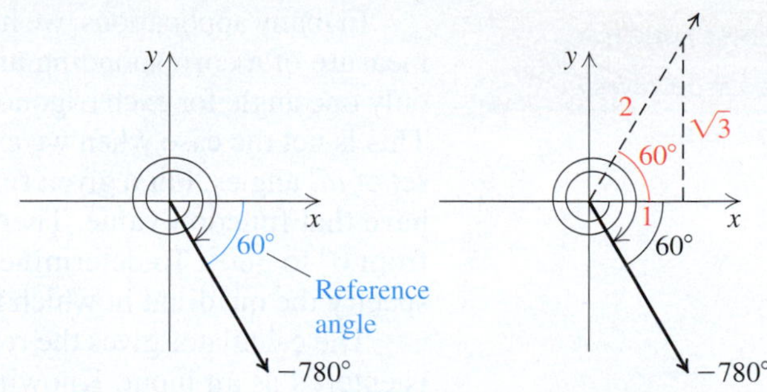

The reference angle for $-60°$ is the acute angle formed by the terminal side of the angle and the x-axis. Thus the reference angle for $-60°$ is $60°$. We know that since $-780°$ is a fourth-quadrant angle, the cosine is positive and the sine and the tangent are negative. Recalling that $\sin 60° = \sqrt{3}/2$, $\cos 60° = 1/2$, and $\tan 60° = \sqrt{3}$, we have

$$\sin(-780°) = -\frac{\sqrt{3}}{2},$$

$$\cos(-780°) = \frac{1}{2},$$

and $\tan(-780°) = -\sqrt{3}.$

Now Try Exercises 47 and 51.

Function Values for Any Angle

When the terminal side of an angle falls on one of the axes or makes a 30°, 45°, or 60° angle with the x-axis, we can find exact function values without the use of a calculator. But this group is only a small subset of *all* angles. Using a calculator, we can approximate the trigonometric function values of *any* angle. In fact, we can approximate or find exact function values of all angles without using a reference angle.

EXAMPLE 8 Find each of the following function values using a calculator and round the answer to four decimal places, where appropriate.

a) $\cos 112°$
b) $\sec 500°$
c) $\tan(-83.4°)$
d) $\csc 351.75°$
e) $\cos 2400°$
f) $\sin 175°40'9''$
g) $\cot(-135°)$

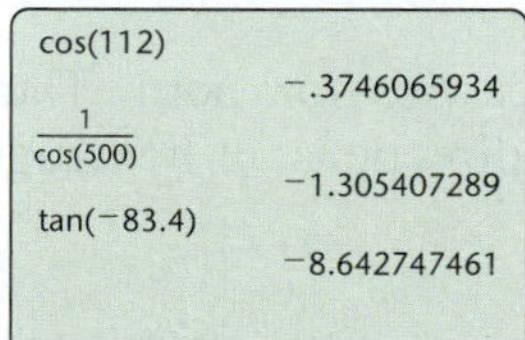

Solution Using a calculator set in DEGREE mode, we find the values.

a) $\cos 112° \approx -0.3746$

b) $\sec 500° = \dfrac{1}{\cos 500°} \approx -1.3054$

c) $\tan(-83.4°) \approx -8.6427$

```
1/sin(351.75)
                 −6.968999424
cos(2400)
                           −.5
sin(175°40'9")
                  .0755153443
```

d) $\csc 351.75° = \dfrac{1}{\sin 351.75°} \approx -6.9690$

e) $\cos 2400° = -0.5$

f) $\sin 175°40'9'' \approx 0.0755$

g) $\cot(-135°) = \dfrac{1}{\tan(-135°)} = 1$

Now Try Exercises 85 and 91.

INVERSE FUNCTIONS

REVIEW SECTION 5.1.

In many applications, we have a trigonometric function value and want to find the measure of a corresponding angle. When only acute angles are considered, there is only one angle for each trigonometric function value. (See Example 9 in Section 6.1.) This is not the case when we extend the domain of the trigonometric functions to the set of *all* angles. For a given function value, there is an infinite number of angles that have that function value. There can be two such angles for each value in the range from 0° to 360°. To determine a unique answer in the interval $(0°, 360°)$, we must specify the quadrant in which the terminal side lies.

The calculator gives the reference angle as an output for each function value that is entered as an input. Knowing the reference angle and the quadrant in which the terminal side lies, we can find the specified angle.

EXAMPLE 9 Given the function value and the quadrant restriction, find θ.

a) $\sin \theta = 0.2812,\ 90° < \theta < 180°$

b) $\cot \theta = -0.1611,\ 270° < \theta < 360°$

Solution

a) We first sketch the angle in the second quadrant. We use the calculator to find the acute angle (reference angle) whose sine is 0.2812. The reference angle is approximately 16.33°. We find the angle θ by subtracting 16.33° from 180°:

$$180° - 16.33° = 163.67°.$$

Thus, $\theta \approx 163.67°$.

b) We begin by sketching the angle in the fourth quadrant. Because the tangent and cotangent values are reciprocals, we know that

$$\tan \theta \approx \frac{1}{-0.1611} \approx -6.2073.$$

We use the calculator to find the acute angle (reference angle) whose tangent is 6.2073, ignoring the fact that $\tan \theta$ is negative. The reference angle is approximately 80.85°. We find angle θ by subtracting 80.85° from 360°:

$$360° - 80.85° = 279.15°.$$

Thus, $\theta \approx 279.15°$.

Now Try Exercise 97.

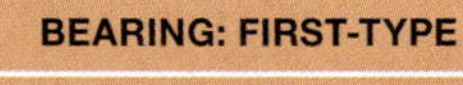

BEARING: FIRST-TYPE

REVIEW SECTION 6.2.

Bearing: Second-Type

In aerial navigation, directions are given in degrees clockwise from north. Thus east is 90°, south is 180°, and west is 270°. Several aerial directions, or **bearings**, are given below.

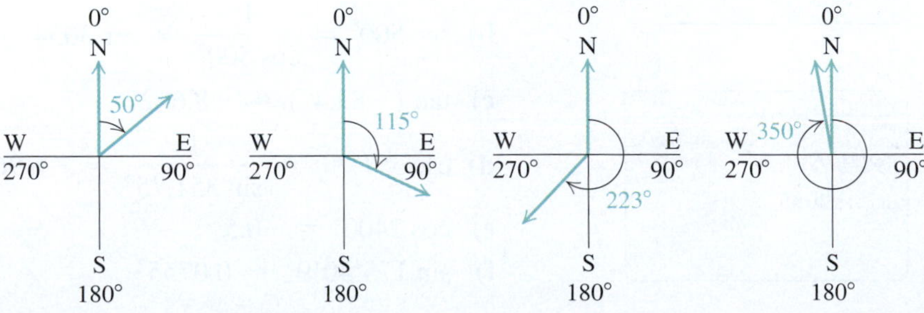

EXAMPLE 10 ***Aerial Navigation.*** An airplane flies 218 mi from an airport in a direction of 245°. How far south of the airport is the plane then? How far west?

Solution We first find the measure of $\angle ABC$:

$$B = 270° - 245° = 25°.$$ **Angle *B* is opposite side *b* in the right triangle.**

From the figure shown at left, we see that the distance south of the airport b and the distance west of the airport a are legs of a right triangle. We have

$$\frac{b}{218} = \sin 25^\circ$$

$$b = 218 \sin 25^\circ \approx 92 \text{ mi}$$

and

$$\frac{a}{218} = \cos 25^\circ$$

$$a = 218 \cos 25^\circ \approx 198 \text{ mi.}$$

The airplane is about 92 mi south and about 198 mi west of the airport.

Now Try Exercise 105.

6.3 Exercise Set

For angles of the following measures, state in which quadrant the terminal side lies. It helps to sketch the angle in standard position.

1. 187° **2.** -14.3°

3. $245^\circ 15'$ **4.** -120°

5. 800° **6.** 1075°

7. -460.5° **8.** 315°

9. -912° **10.** $13^\circ 15' 58''$

11. 537° **12.** -345.14°

Find two positive angles and two negative angles that are coterminal with the given angle. Answers may vary.

13. 74° **14.** -81° **15.** 115.3°

16. $275^\circ 10'$ **17.** -180° **18.** -310°

Find the complement and the supplement.

19. 17.11° **20.** $47^\circ 38'$

21. $12^\circ 3' 14''$ **22.** 9.038°

23. 45.2° **24.** 67.31°

Find the six trigonometric function values for the angle shown.

25.

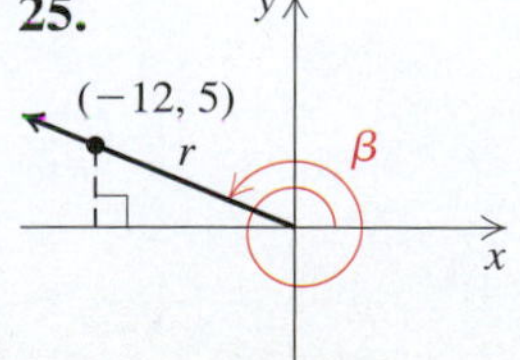

26.

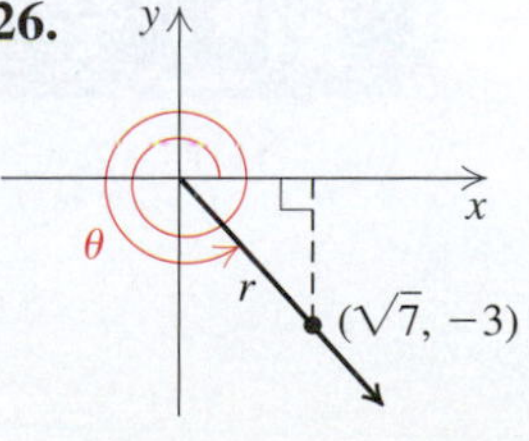

27.

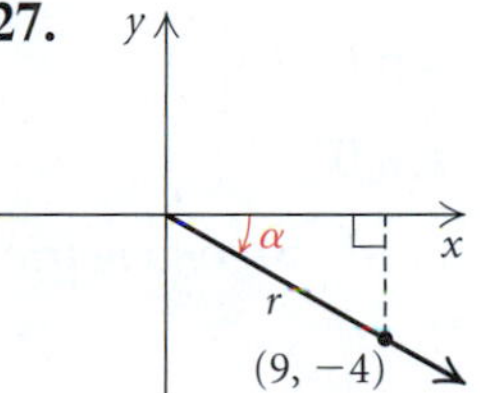

28.

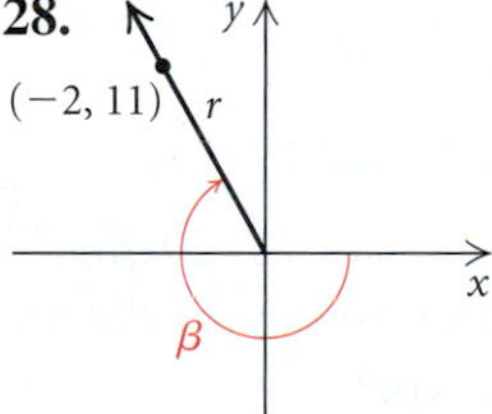

29.

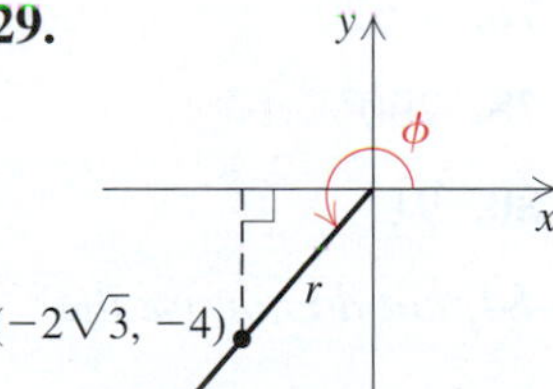

30.

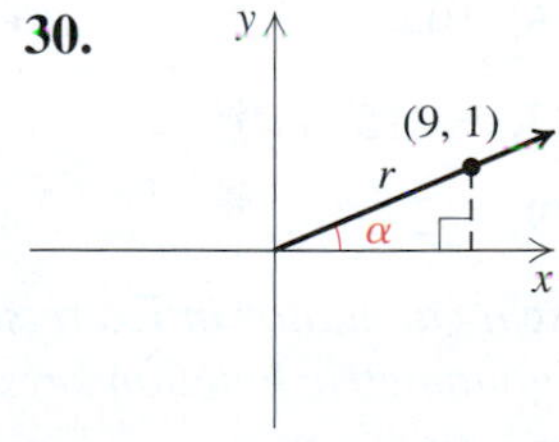

The terminal side of angle θ in standard position lies on the given line in the given quadrant. Find $\sin\theta$, $\cos\theta$, and $\tan\theta$.

31. $2x + 3y = 0$; quadrant IV

32. $4x + y = 0$; quadrant II

33. $5x - 4y = 0$; quadrant I

34. $y = 0.8x$; quadrant III

A function value and a quadrant are given. Find the other five trigonometric function values. Give exact answers.

35. $\sin\theta = -\frac{1}{3}$; quadrant III

36. $\tan\beta = 5$; quadrant I

37. $\cot\theta = -2$; quadrant IV

38. $\cos\alpha = -\frac{4}{5}$; quadrant II

39. $\cos\phi = \frac{3}{5}$; quadrant IV

40. $\sin\theta = -\frac{5}{13}$; quadrant III

Find the reference angle and the exact function value if they exist.

41. $\cos 150°$
42. $\sec(-225°)$
43. $\tan(-135°)$
44. $\sin(-45°)$
45. $\sin 7560°$
46. $\tan 270°$
47. $\cos 495°$
48. $\tan 675°$
49. $\csc(-210°)$
50. $\sin 300°$
51. $\cot 570°$
52. $\cos(-120°)$
53. $\tan 330°$
54. $\cot 855°$
55. $\sec(-90°)$
56. $\sin 90°$
57. $\cos(-180°)$
58. $\csc 90°$
59. $\tan 240°$
60. $\cot(-180°)$
61. $\sin 495°$
62. $\sin 1050°$
63. $\csc 225°$
64. $\sin(-450°)$
65. $\cos 0°$
66. $\tan 480°$
67. $\cot(-90°)$
68. $\sec 315°$
69. $\cos 90°$
70. $\sin(-135°)$
71. $\cos 270°$
72. $\tan 0°$

In Exercises 73–80, find the signs of the six trigonometric function values for the given angles.

73. $319°$
74. $-57°$
75. $194°$
76. $-620°$
77. $-215°$
78. $290°$
79. $-272°$
80. $91°$

Use a calculator in Exercises 81–84, but do not use the trigonometric function keys.

81. Given that
$\sin 41° = 0.6561,$
$\cos 41° = 0.7547,$
$\tan 41° = 0.8693,$
find the trigonometric function values for 319°.

82. Given that
$\sin 27° = 0.4540,$
$\cos 27° = 0.8910,$
$\tan 27° = 0.5095,$
find the trigonometric function values for 333°.

83. Given that
$\sin 65° = 0.9063,$
$\cos 65° = 0.4226,$
$\tan 65° = 2.1445,$
find the trigonometric function values for 115°.

84. Given that
$\sin 35° = 0.5736,$
$\cos 35° = 0.8192,$
$\tan 35° = 0.7002,$
find the trigonometric function values for 215°.

Find the function value. Round to four decimal places.

85. $\tan 310.8°$
86. $\cos 205.5°$
87. $\cot 146.15°$
88. $\sin(-16.4°)$
89. $\sin 118°42'$
90. $\cos 273°45'$
91. $\cos(-295.8°)$
92. $\tan 1086.2°$
93. $\cos 5417°$
94. $\sec 240°55'$
95. $\csc 520°$
96. $\sin 3824°$

Given the function value and the quadrant restriction, find θ.

	FUNCTION VALUE	INTERVAL	θ
97.	$\sin\theta = -0.9956$	$(270°, 360°)$	____
98.	$\tan\theta = 0.2460$	$(180°, 270°)$	____
99.	$\cos\theta = -0.9388$	$(180°, 270°)$	____
100.	$\sec\theta = -1.0485$	$(90°, 180°)$	____
101.	$\tan\theta = -3.0545$	$(270°, 360°)$	____
102.	$\sin\theta = -0.4313$	$(180°, 270°)$	____
103.	$\csc\theta = 1.0480$	$(0°, 90°)$	____
104.	$\cos\theta = -0.0990$	$(90°, 180°)$	____

105. *Aerial Navigation.* An airplane flies 150 km from an airport in a direction of 120°. How far east of the airport is the plane then? How far south?

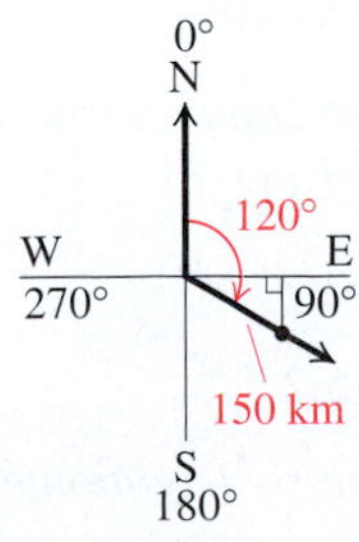

106. *Aerial Navigation.* An airplane leaves an airport and travels for 100 mi in a direction of 300°. How far north of the airport is the plane then? How far west?

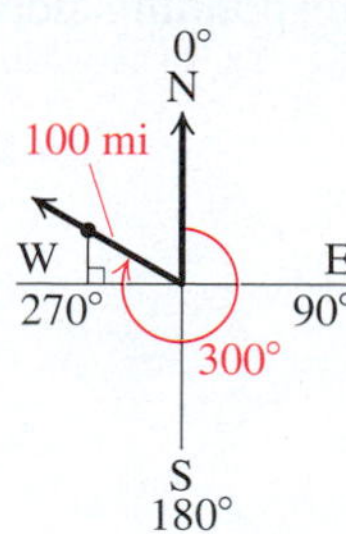

107. *Aerial Navigation.* An airplane travels at 150 km/h for 2 hr in a direction of 138° from Omaha. At the end of this time, how far south of Omaha is the plane?

108. *Aerial Navigation.* An airplane travels at 120 km/h for 2 hr in a direction of 319° from Chicago. At the end of this time, how far north of Chicago is the plane?

Skill Maintenance

Graph the function. Sketch and label any vertical asymptotes.

109. $f(x) = \dfrac{1}{x^2 - 25}$ **[4.5]**

110. $g(x) = x^3 - 2x + 1$ **[4.2]**

Determine the domain and the range of the function.

111. $f(x) = \dfrac{x - 4}{x + 2}$ **[1.2], [4.5]**

112. $g(x) = \dfrac{x^2 - 9}{2x^2 - 7x - 15}$ **[1.2], [4.1], [4.5]**

Find the zeros of the function.

113. $f(x) = 12 - x$ **[1.5]**

114. $g(x) = x^2 - x - 6$ **[3.2]**

Find the x-intercept(s) of the graph of the function.

115. $f(x) = 12 - x$ **[1.5]**

116. $g(x) = x^2 - x - 6$ **[3.2]**

Synthesis

117. *Tallest Ferris Wheel.* On March 31, 2014, the world's tallest ferris wheel, the High Roller, opened in Las Vegas at the LINQ. Each cabin, which can fit up to 40 passengers, is 220 ft from the center of the wheel. The wheel makes one revolution in approximately 30 min. (*Source*: Nancy Trejos, *USA Today*, April 7, 2014). When you board, you are 6 ft above the ground. After you have rotated through an angle of 315°, how far above the ground are you?

118. *Valve Cap on a Bicycle.* The valve cap on a bicycle wheel is 12.5 in. from the center of the wheel. From the position shown, the wheel starts to roll. After the wheel has turned 390°, how far above the ground is the valve cap? Assume that the outer radius of the tire is 13.375 in.

Mid-Chapter Mixed Review

Determine whether the statement is true or false.

1. If $\sin \alpha > 0$ and $\cot \alpha > 0$, then α is in the first quadrant. [6.3]

2. The lengths of corresponding sides in similar triangles are in the same ratio. [6.1]

3. If θ is an acute angle and $\csc \theta \approx 1.5539$, then $\cos(90° - \theta) \approx 0.6435$. [6.1]

Solve the right triangle. [6.2]

4. 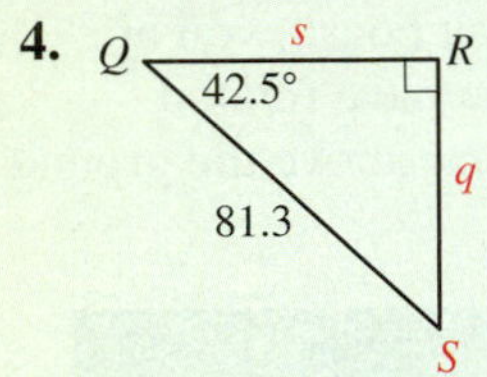

5. 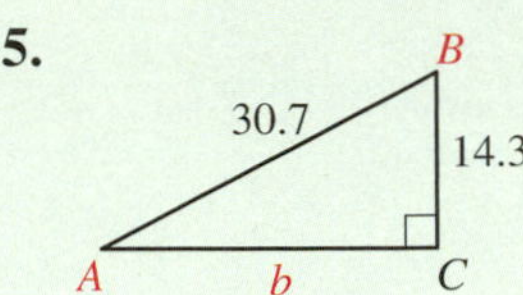

Find two positive angles and two negative angles that are coterminal with the given angle. Answers may vary. [6.3]

6. $-75°$

7. $214°30'$

Find the complement and the supplement of the given angle. [6.3]

8. $18.2°$

9. $87°15'10''$

10. Given that $\sin 25° = 0.4226$, $\cos 25° = 0.9063$, and $\tan 25° = 0.4663$, find the six trigonometric function values for 155°. Use a calculator, but do not use the trigonometric function keys. [6.3]

11. Find the six trigonometric function values for the angle shown. [6.3]

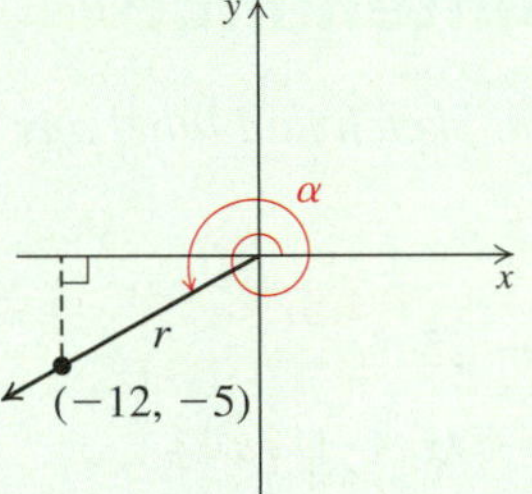

12. Given $\cot \theta = 2$ and θ in quadrant III, find the other five trigonometric function values. [6.3]

13. Given $\cos \alpha = \frac{2}{9}$ and $0° < \alpha < 90°$, find the other five trigonometric function values. [6.1]

14. Convert $42°8'50''$ to decimal degree notation. Round to four decimal places. [6.1]

15. Convert 51.18° to degrees, minutes, and seconds. [6.1]

16. Given that $\sin 9° \approx 0.1564$, $\cos 9° \approx 0.9877$, and $\tan 9° \approx 0.1584$, find the six trigonometric function values of 81°. [6.1]

17. If $\tan \theta = 2.412$ and θ is acute, find the angle to the nearest tenth of a degree. [6.1]

18. *Aerial Navigation.* An airplane travels at 200 mph for $1\frac{1}{2}$ hr in a direction of 285° from Atlanta. At the end of this time, how far west of Atlanta is the plane? [6.3]

Without a calculator, find the exact function value. [6.1], [6.3]

19. $\tan 210°$
20. $\sin 45°$
21. $\cot 30°$
22. $\sec 135°$
23. $\cos 45°$
24. $\csc(-30°)$
25. $\sin 90°$
26. $\cos 270°$
27. $\sin 120°$
28. $\sec 180°$
29. $\tan(-240°)$
30. $\cot(-315°)$
31. $\sin 750°$
32. $\csc 45°$
33. $\cos 210°$
34. $\cot 0°$
35. $\csc 150°$
36. $\tan 90°$
37. $\sec 3600°$
38. $\cos 495°$

Find the function value. Round the answer to four decimal places. [6.1], [6.3]

39. $\cos 39.8°$
40. $\sec 50°$
41. $\tan 2183°$
42. $\sin 10°28'3''$
43. $\csc(-74°)$
44. $\cot 142.7°$
45. $\sin(-40.1°)$
46. $\cos 87°15'$

COLLABORATIVE DISCUSSION AND WRITING

47. Why do the function values of θ depend only on the angle and not on the choice of a point on the terminal side? [6.3]

48. Explain the difference between reciprocal functions and cofunctions. [6.1]

49. In Section 6.1, the trigonometric functions are defined as functions of acute angles. What appear to be the ranges for the sine, cosine, and tangent functions given the restricted domain as the set of angles whose measures are greater than 0° and less than 90°? [6.1]

50. Why is the domain of the tangent function different from the domains of the sine function and the cosine function? [6.3]

6.4 Radians, Arc Length, and Angular Speed

- Find points on the unit circle that are determined by real numbers.
- Convert between radian measure and degree measure; find coterminal, complementary, and supplementary angles.
- Find the length of an arc of a circle; find the measure of a central angle of a circle.
- Convert between linear speed and angular speed.

Another useful unit of angle measure is called a **radian**. To introduce radian measure, we use a circle centered at the origin with a radius of length 1. Such a circle is called a **unit circle**. Its equation is $x^2 + y^2 = 1$.

CIRCLES

REVIEW SECTION 1.1.

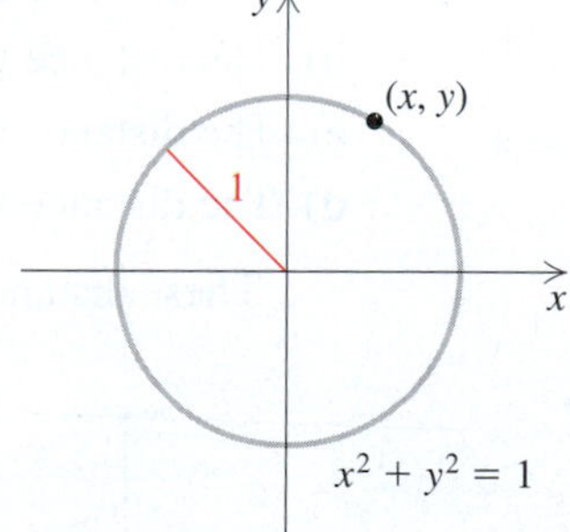

Distances on the Unit Circle

The circumference of a circle of radius r is $2\pi r$. Thus for the unit circle, where $r = 1$, the circumference is 2π. If a point starts at A and travels around the circle (Fig. 1 on the following page), it will travel a distance of 2π. If it travels halfway around the circle (Fig. 2), it will travel a distance of $\frac{1}{2} \cdot 2\pi$, or π.

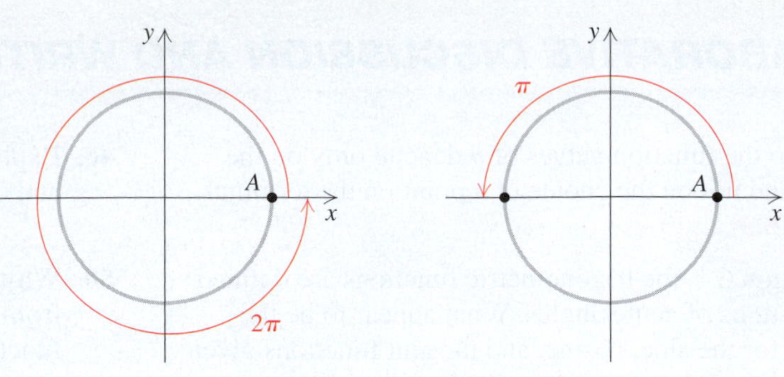

FIGURE 1. FIGURE 2.

If a point C travels $\frac{1}{8}$ of the way around the circle (Fig. 3), it will travel a distance of $\frac{1}{8} \cdot 2\pi$, or $\pi/4$. Note that C is $\frac{1}{4}$ of π the distance from A to B. If a point D travels $\frac{1}{6}$ of the way around the circle (Fig. 4), it will travel a distance of $\frac{1}{6} \cdot 2\pi$, or $\pi/3$. Note that D is $\frac{1}{3}$ of π the distance from A to B.

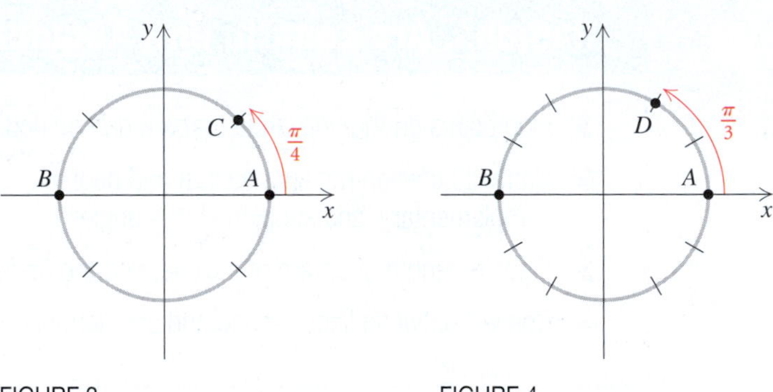

FIGURE 3. FIGURE 4.

EXAMPLE 1 How far will a point travel if it goes **(a)** $\frac{1}{4}$, **(b)** $\frac{1}{12}$, **(c)** $\frac{3}{8}$, and **(d)** $\frac{5}{6}$ of the way around the unit circle?

Solution

a) $\frac{1}{4}$ of the total distance around the circle is $\frac{1}{4} \cdot 2\pi$, which is $\frac{1}{2}\pi$, or $\pi/2$.

b) The distance will be $\frac{1}{12} \cdot 2\pi$, which is $\frac{1}{6}\pi$, or $\pi/6$.

c) The distance will be $\frac{3}{8} \cdot 2\pi$, which is $\frac{3}{4}\pi$, or $3\pi/4$.

d) The distance will be $\frac{5}{6} \cdot 2\pi$, which is $\frac{5}{3}\pi$, or $5\pi/3$. Think of $5\pi/3$ as $\pi + \frac{2}{3}\pi$.

These distances are illustrated in the following figures.

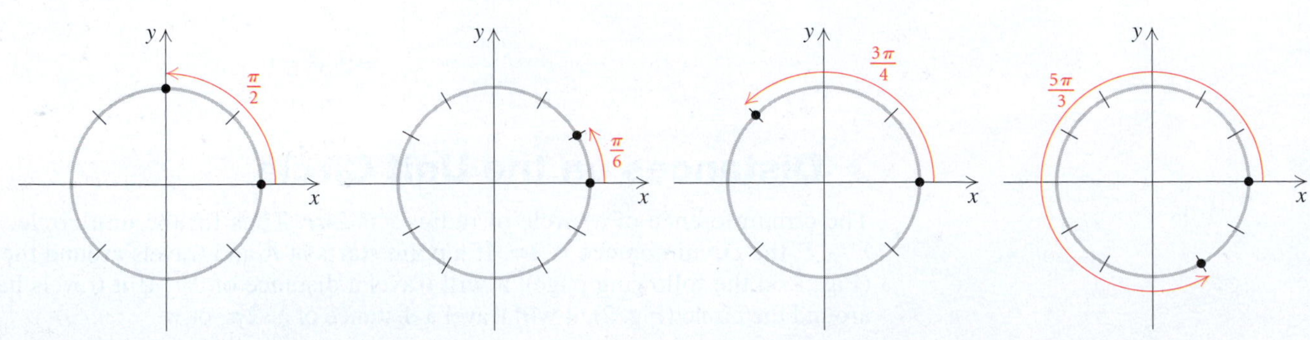

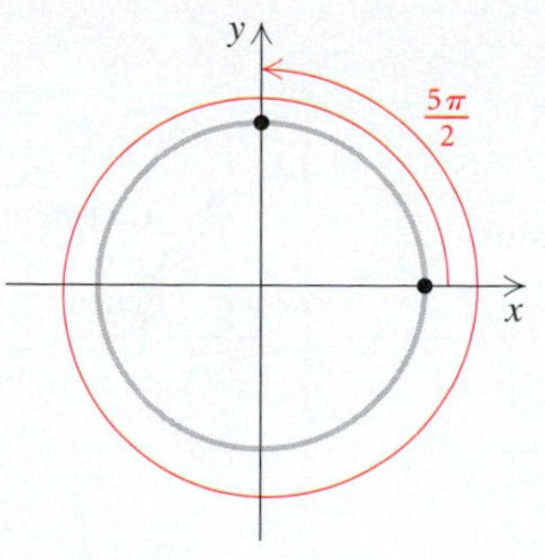

FIGURE 5.

A point may travel completely around the circle and then continue. For example, if it goes around once and then continues $\frac{1}{4}$ of the way around, it will have traveled a distance of $2\pi + \frac{1}{4} \cdot 2\pi$, or $5\pi/2$ (Fig. 5). *Every* real number determines a point on the unit circle. For the positive number 10, for example, we start at A and travel counterclockwise a distance of 10. The point at which we stop is the point "determined" by the number 10. Note that $2\pi \approx 6.28$ and that $10 \approx 1.6(2\pi)$. Thus the point for 10 travels around the unit circle about 1.6, or $1\frac{3}{5}$, times (Fig. 6).

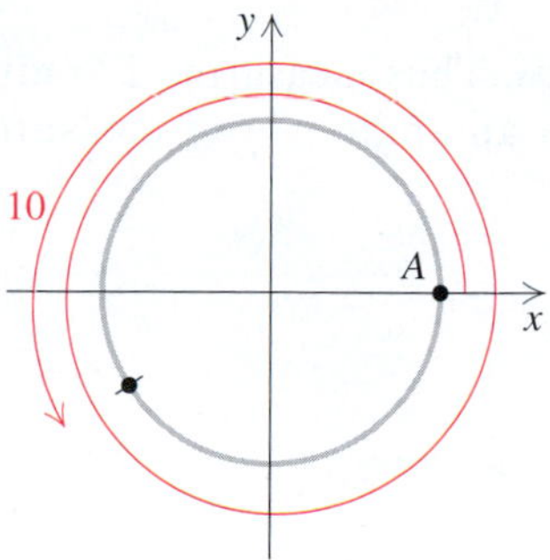

FIGURE 6.

For a negative number, we move clockwise around the circle. Points for $-\pi/4$ and $-3\pi/2$ are shown in the following figures. The number 0 determines the point A.

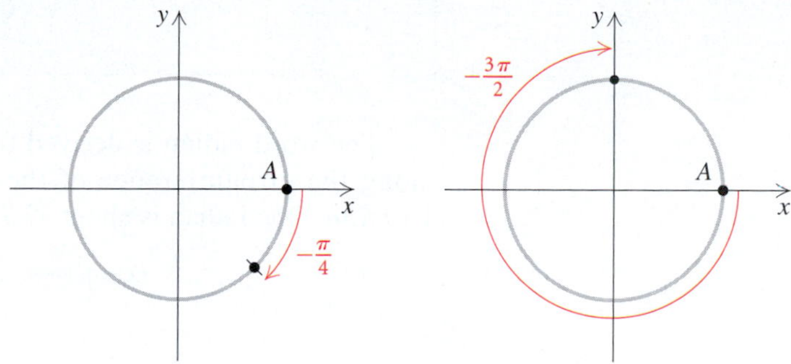

EXAMPLE 2 On the unit circle, mark the point determined by each of the following real numbers.

a) $\frac{9\pi}{4}$ **b)** $-\frac{7\pi}{6}$

Solution

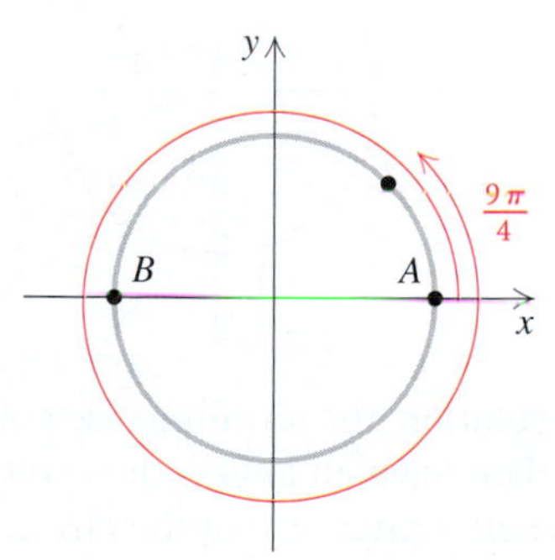

a) Think of $9\pi/4$ as $2\pi + \frac{1}{4}\pi$. (See the figure at left.) Since $9\pi/4 > 0$, the point moves counterclockwise. The point goes completely around once and then continues $\frac{1}{4}$ of the way from A to B.

b) The number $-7\pi/6$ is negative, so the point moves clockwise. From A to B, the distance is π, or $\frac{6}{6}\pi$, so we need to go beyond B another distance of $\pi/6$, clockwise. (See the figure below.)

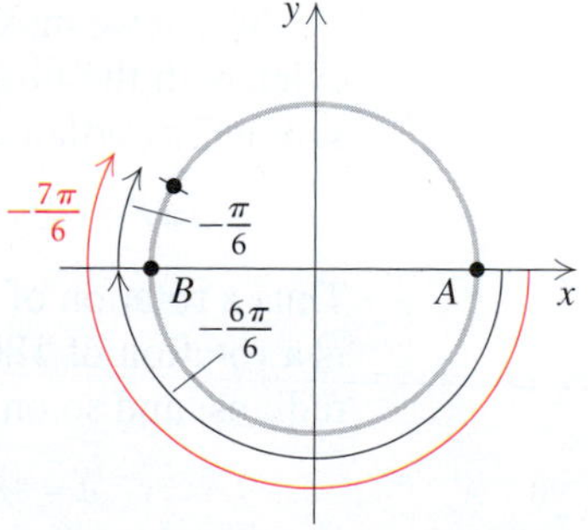

Now Try Exercise 1.

Radian Measure

Degree measure is a common unit of angle measure in many everyday applications. But in many scientific fields and in mathematics (calculus, in particular), there is another commonly used unit of measure called the *radian*.

RADIANS

Consider the unit circle. Recall that this circle has radius 1. Suppose we measure, moving counterclockwise, an arc of length 1, and mark a point T on the circle. If we draw a ray from the origin through T, we have formed an angle. The measure of that angle is 1 **radian**.

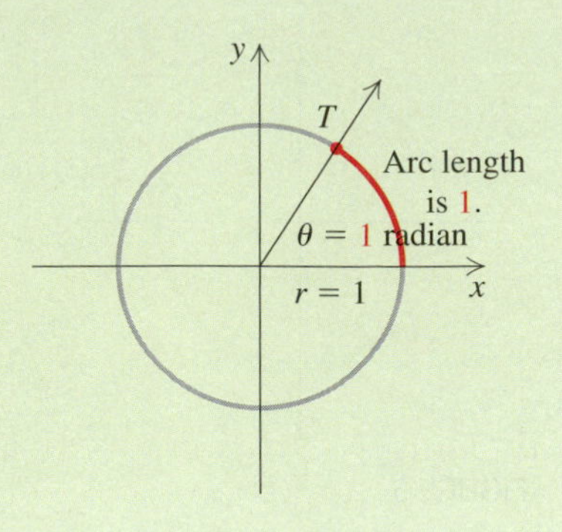

The word radian is derived from the word *radius*. Thus measuring 1 "radius" along the circumference of the circle determines an angle whose measure is 1 *radian*. One radian is about 57.3°.

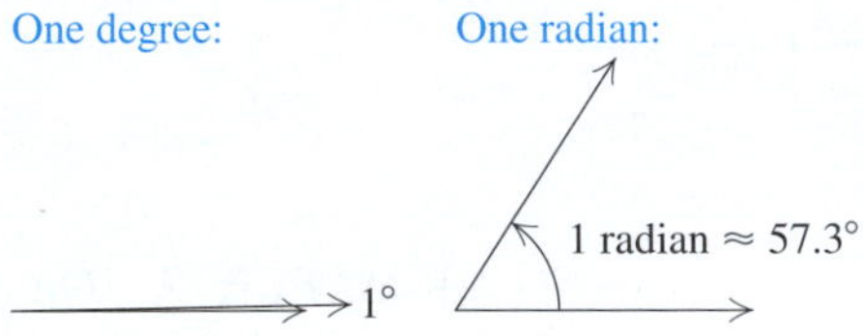

Angles that measure 2 radians, 3 radians, and 6 radians are shown below.

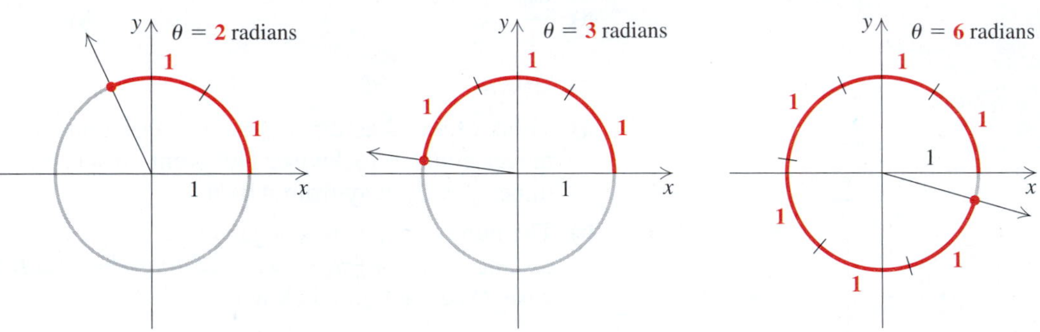

When we make a complete (counterclockwise) revolution, the terminal side coincides with the initial side on the positive x-axis. We then have an angle whose measure is 2π radians, or about 6.28 radians, which is the circumference of the circle:

$$2\pi r = 2\pi(1) = 2\pi.$$

Thus a rotation of 360° (1 revolution) has a measure of 2π radians. A half revolution is a rotation of 180°, or π radians. A quarter revolution is a rotation of 90°, or $\pi/2$ radians, and so on.

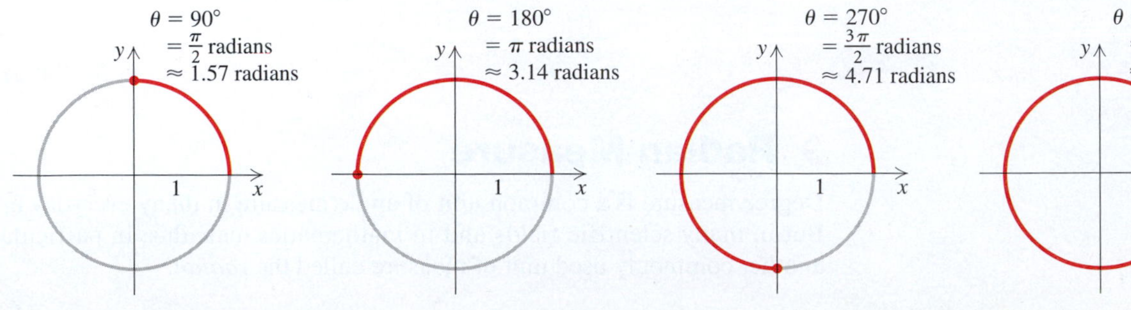

To convert between degrees and radians, we first note that

$$360° = 2\pi \text{ radians}.$$

It follows that

$$180° = \pi \text{ radians}.$$

To make conversions, we multiply by 1, noting that:

CONVERTING BETWEEN DEGREE MEASURE AND RADIAN MEASURE

$$\frac{\pi \text{ radians}}{180°} = \frac{180°}{\pi \text{ radians}} = 1.$$

To convert from degree to radian measure, multiply by $\frac{\pi \text{ radians}}{180°}$.

To convert from radian to degree measure, multiply by $\frac{180°}{\pi \text{ radians}}$.

EXAMPLE 3 Convert each of the following to radians.

a) 120° **b)** −297.25°

Solution

a)
$$\begin{aligned} 120° &= 120° \cdot \frac{\pi \text{ radians}}{180°} \quad \textbf{Multiplying by 1} \\ &= \frac{120°}{180°}\pi \text{ radians} \\ &= \frac{2\pi}{3} \text{ radians, or about 2.09 radians} \end{aligned}$$

b)
$$\begin{aligned} -297.25° &= -297.25° \cdot \frac{\pi \text{ radians}}{180°} \\ &= -\frac{297.25°}{180°}\pi \text{ radians} \\ &= -\frac{297.25\pi}{180} \text{ radians} \\ &\approx -5.19 \text{ radians} \end{aligned}$$

We also can use a calculator set in RADIAN mode to convert the angle measures. We enter the angle measure followed by ° (degrees) from the ANGLE menu. Finally, we press ENTER to see radian measure.

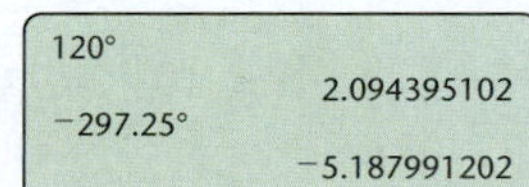

Now Try Exercises 11 and 23.

EXAMPLE 4 Convert each of the following to degrees.

a) $\frac{3\pi}{4}$ radians **b)** 8.5 radians

Solution

a) $$\frac{3\pi}{4}\text{ radians} = \frac{3\pi}{4}\text{ radians}\cdot\frac{180°}{\pi\text{ radians}} \qquad \textbf{Multiplying by 1}$$
$$= \frac{3\pi}{4\pi}\cdot 180° = \frac{3}{4}\cdot 180° = 135°$$

b) $$8.5\text{ radians} = 8.5\text{ radians}\cdot\frac{180°}{\pi\text{ radians}}$$
$$= \frac{8.5(180°)}{\pi} \approx 487.01°$$

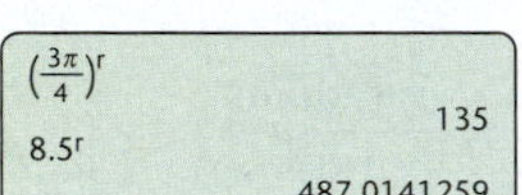

With a calculator set in DEGREE mode, we can enter the angle measure followed by ^{r} (radians) from the ANGLE menu. Finally, we press ENTER to see degree measure.

> Now Try Exercises 35 and 43.

The radian–degree equivalents of the most commonly used angle measures are illustrated in the following figures.

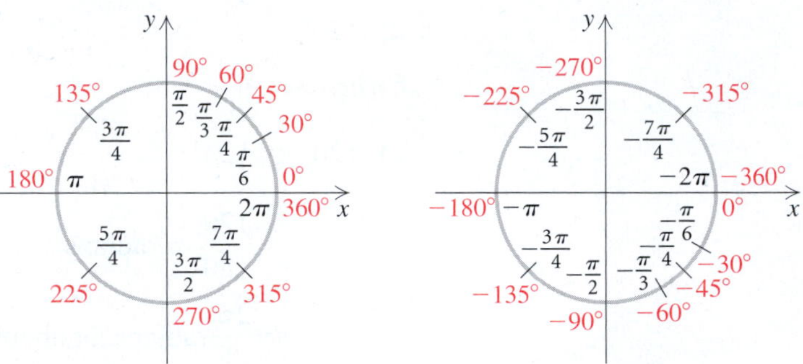

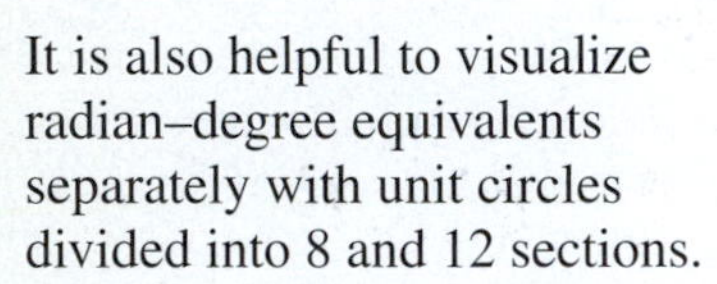

It is also helpful to visualize radian–degree equivalents separately with unit circles divided into 8 and 12 sections.

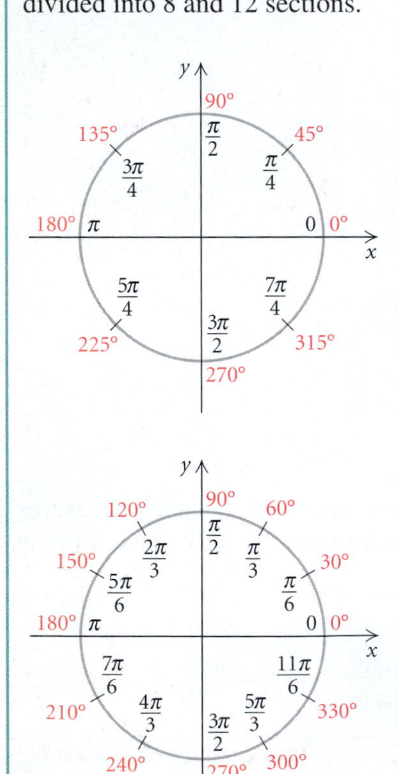

When a rotation is given in radians, the word "radians" is optional and is most often omitted. **Thus if no unit is given for a rotation, the rotation is understood to be in radians.**

We can also find coterminal, complementary, and supplementary angles in radian measure just as we did for degree measure in Section 6.3.

EXAMPLE 5 Find a positive angle and a negative angle that are coterminal with $2\pi/3$. Many answers are possible.

Solution To find angles coterminal with a given angle, we add or subtract multiples of 2π:

$$\frac{2\pi}{3} + 2\pi = \frac{2\pi}{3} + \frac{6\pi}{3} = \frac{8\pi}{3},$$
$$\frac{2\pi}{3} - 3(2\pi) = \frac{2\pi}{3} - \frac{18\pi}{3} = -\frac{16\pi}{3}.$$

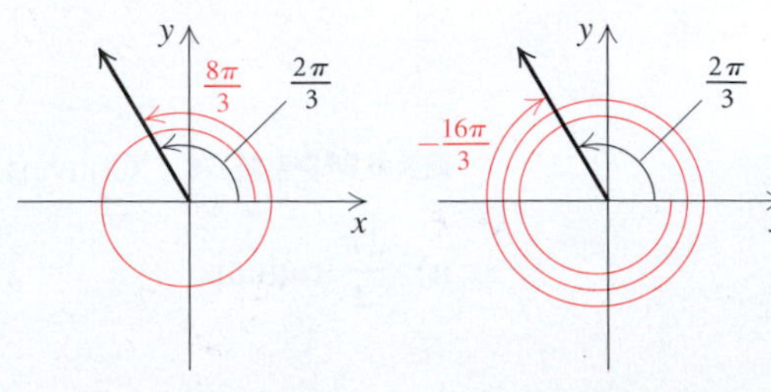

Thus, $8\pi/3$ and $-16\pi/3$ are two of the many angles coterminal with $2\pi/3$.

> Now Try Exercise 51.

EXAMPLE 6 Find the complement and the supplement of $\pi/6$.

Solution Since 90° equals $\pi/2$ radians, the complement of $\pi/6$ is

$$\frac{\pi}{2} - \frac{\pi}{6} = \frac{3\pi}{6} - \frac{\pi}{6} = \frac{2\pi}{6}, \quad \text{or} \quad \frac{\pi}{3}.$$

Since 180° equals π radians, the supplement of $\pi/6$ is

$$\pi - \frac{\pi}{6} = \frac{6\pi}{6} - \frac{\pi}{6} = \frac{5\pi}{6}.$$

Thus the complement of $\pi/6$ is $\pi/3$ and the supplement is $5\pi/6$.

Now Try Exercise 55.

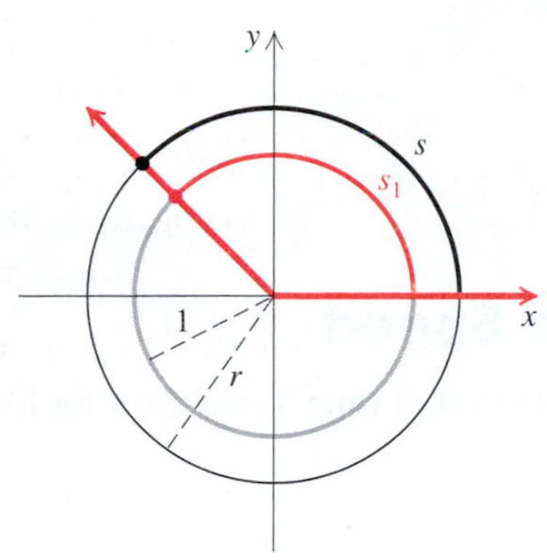

Arc Length and Central Angles

Radian measure can be determined using a circle other than a unit circle. In the figure at left, a unit circle (with radius 1) is shown along with another circle (with radius r, $r \neq 1$). The angle shown is a **central angle** of both circles.

From geometry, we know that the arcs that the angle subtends have their lengths in the same ratio as the radii of the circles. The radii of the circles are r and 1. The corresponding arc lengths are s and s_1. Thus we have the proportion

$$\frac{s}{s_1} = \frac{r}{1},$$

which also can be written as

$$\frac{s_1}{1} = \frac{s}{r}.$$

Now s_1 is the *radian measure* of the rotation in question. It is common to use a Greek letter, such as θ, for the measure of an angle or a rotation and the letter s for arc length. Adopting this convention, we rewrite the proportion above as

$$\theta = \frac{s}{r}.$$

In any circle, the measure (in radians) of a central angle, the arc length the angle subtends, and the length of the radius are related in this fashion. Or, in general, the following is true.

RADIAN MEASURE

The **radian measure** θ of a rotation is the **ratio** of the distance s, traveled by a point at a radius r from the center of rotation, to the length of the radius r:

$$\theta = \frac{s}{r}.$$

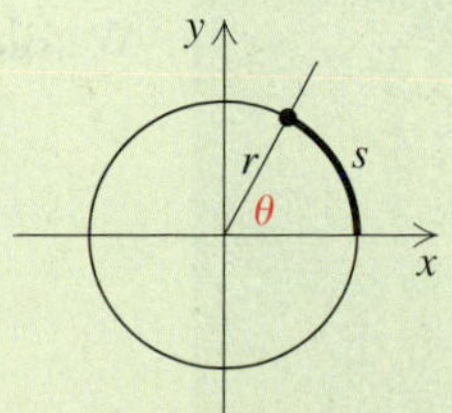

When we are using the formula $\theta = s/r$, θ must be in radians and s and r must be expressed in the same unit.

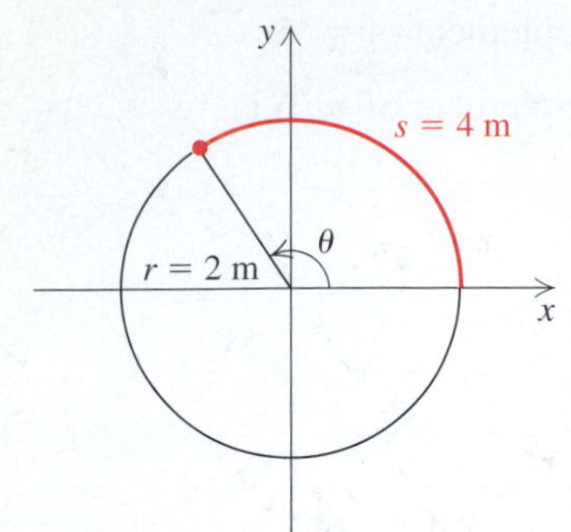

EXAMPLE 7 Find the measure of a rotation in radians when a point 2 m from the center of rotation travels 4 m.

Solution We have

$$\theta = \frac{s}{r}$$
$$= \frac{4\text{ m}}{2\text{ m}} = 2. \quad \textbf{The unit is understood to be radians.}$$

Now Try Exercise 65.

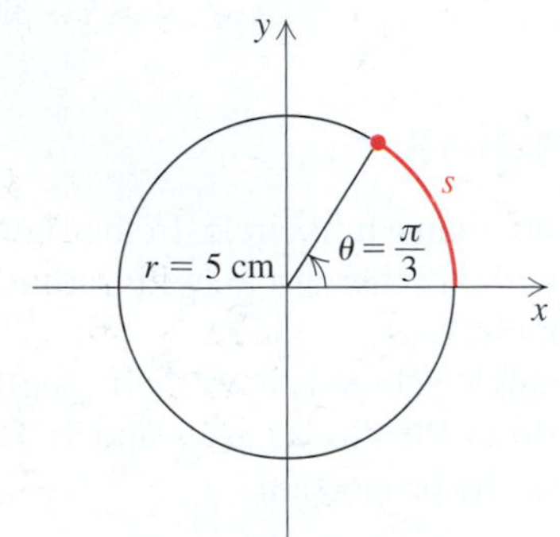

EXAMPLE 8 Find the length of an arc of a circle of radius 5 cm associated with an angle of $\pi/3$ radians.

Solution We have

$$\theta = \frac{s}{r}, \quad \text{or} \quad s = r\theta.$$

Thus, $s = 5\text{ cm} \cdot \pi/3$, or about 5.24 cm.

Now Try Exercise 63.

Linear Speed and Angular Speed

Linear speed is defined to be distance traveled per unit of time. If we use v for linear speed, s for distance, and t for time, then

$$v = \frac{s}{t}.$$

Similarly, **angular speed** is defined to be amount of rotation per unit of time. For example, we might speak of the angular speed of a bicycle wheel as 150 revolutions per minute or the angular speed of the earth as 2π radians per day. The Greek letter ω (omega) is generally used for angular speed. Thus for a rotation θ and time t, angular speed is defined as

$$\omega = \frac{\theta}{t}.$$

© 2006, The Children's Museum of Indianapolis

As an example of how these definitions can be applied, let's consider the refurbished carousel at the Children's Museum in Indianapolis, Indiana. It consists of three circular rows of animals. All animals, regardless of the row, travel at the same angular speed. But the animals in the outer row travel at a greater linear speed than those in the inner rows. What is the relationship between the linear speed v and the angular speed ω?

To develop the relationship we seek, recall that, for rotations measured in radians, $\theta = s/r$. This is equivalent to

$$s = r\theta.$$

We divide by time, t, to obtain

$$\frac{s}{t} = \frac{r\theta}{t} \quad \textbf{Dividing by } t$$

$$\underset{\substack{\downarrow \\ v}}{\frac{s}{t}} = r \cdot \underset{\substack{\downarrow \\ \omega.}}{\frac{\theta}{t}}$$

Now s/t is linear speed v, and θ/t is angular speed ω. Thus we have the relationship we seek,

$$v = r\omega.$$

LINEAR SPEED IN TERMS OF ANGULAR SPEED

The **linear speed** v of a point a distance r from the center of rotation is given by

$$v = r\omega,$$

where ω is the **angular speed**, in radians, per unit of time.

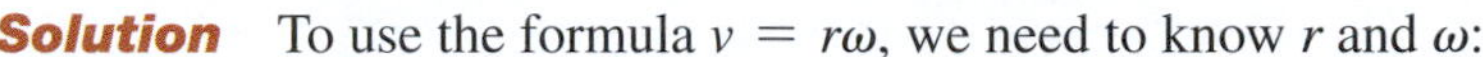

For the formula $v = r\omega$, the units of distance for v and r must be the same, ω must be in radians per unit of time, and the units of time for v and ω must be the same.

EXAMPLE 9 ***Linear Speed of an Earth Satellite.*** An earth satellite in circular orbit 1200 km high makes one complete revolution every 90 min. What is its linear speed? Use 6400 km for the length of a radius of the earth.

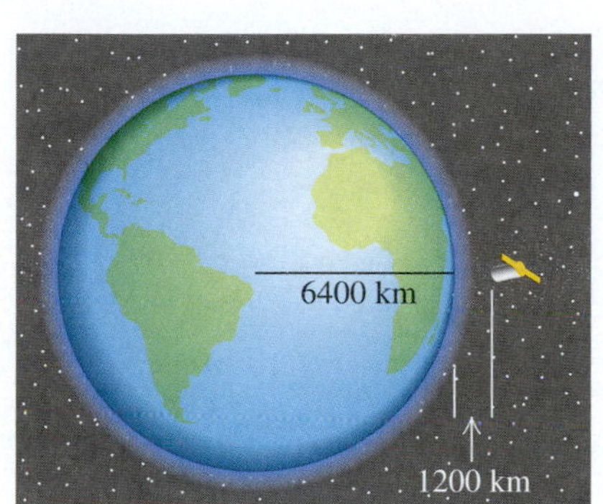

Solution To use the formula $v = r\omega$, we need to know r and ω:

$$\begin{aligned} r &= 6400 \text{ km} + 1200 \text{ km} \qquad \text{Radius of earth plus height of satellite} \\ &= 7600 \text{ km}; \end{aligned}$$

$$\omega = \frac{\theta}{t} = \frac{2\pi}{90 \text{ min}} = \frac{\pi}{45 \text{ min}}. \qquad \text{We have, as usual, omitted the word radians.}$$

Now, using $v = r\omega$, we have

$$v = 7600 \text{ km} \cdot \frac{\pi}{45 \text{ min}} = \frac{7600\pi}{45} \cdot \frac{\text{km}}{\text{min}} \approx 531 \frac{\text{km}}{\text{min}}.$$

Thus the linear speed of the satellite is approximately 531 km/min.

Now Try Exercise 71.

EXAMPLE 10 ***Angular Speed of a Capstan.*** An anchor on a Navy vessel is hoisted at a rate of 2 ft/sec as the line is wound around a capstan with a 1.4-yd diameter. What is the angular speed of the capstan?

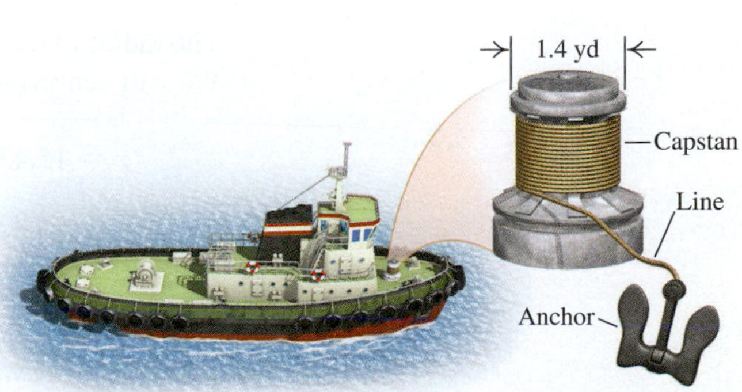

Solution We will use the formula $v = r\omega$ in the form $\omega = v/r$, taking care to use the proper units. Since v is given in feet per second, we need to give r in feet:

$$r = \frac{d}{2} = \frac{1.4}{2}\text{yd} \cdot \frac{3\text{ ft}}{1\text{ yd}} = 2.1\text{ ft}.$$

Then ω will be in radians per second:

$$\omega = \frac{v}{r} = \frac{2\text{ ft/sec}}{2.1\text{ ft}} = \frac{2\text{ ft}}{\text{sec}} \cdot \frac{1}{2.1\text{ ft}} \approx 0.952/\text{sec}.$$

Thus the angular speed is approximately 0.952 radian/sec. **Now Try Exercise 77.**

The formulas $\theta = \omega t$ and $v = r\omega$ can be used in combination to find distances and angles in various situations involving rotational motion.

EXAMPLE 11 ***Angle of Revolution.*** A 2014 Toyota FJ Cruiser is traveling at a speed of 70 mph. Its tires have an outside diameter of 30.875 in. Find the angle through which a tire turns in 10 sec.

Solution Recall that $\omega = \theta/t$, or $\theta = \omega t$. Thus we can find θ if we know ω and t. To find ω, we use the formula $v = r\omega$. The linear speed v of a point on the outside of the tire is the speed of the FJ Cruiser, 70 mph. For convenience, we first convert 70 mph to feet per second:

$$\begin{aligned} v &= 70\frac{\text{mi}}{\text{hr}} \cdot \frac{1\text{ hr}}{60\text{ min}} \cdot \frac{1\text{ min}}{60\text{ sec}} \cdot \frac{5280\text{ ft}}{1\text{ mi}} \\ &\approx 102.667\frac{\text{ft}}{\text{sec}}. \end{aligned}$$

The radius of the tire is half the diameter. Now $r = d/2 = 30.875/2 = 15.4375$ in. We will convert to feet, since v is in feet per second:

$$\begin{aligned} r &= 15.4375\text{ in.} \cdot \frac{1\text{ ft}}{12\text{ in.}} \\ &= \frac{15.4375}{12}\text{ft} \\ &\approx 1.29\text{ ft}. \end{aligned}$$

Using $v = r\omega$, we have

$$102.667\frac{\text{ft}}{\text{sec}} = 1.29\text{ ft} \cdot \omega,$$

so

$$\omega = \frac{102.667 \text{ ft/sec}}{1.29 \text{ ft}} \approx \frac{79.59}{\text{sec}}.$$

Then in 10 sec,

$$\theta = \omega t = \frac{79.59}{\text{sec}} \cdot 10 \text{ sec} \approx 796.$$

Thus the angle, in radians, through which a tire turns in 10 sec is 796.

Now Try Exercise 79.

EXAMPLE 12 ***Angular Speed of a Gear Wheel.*** One gear wheel turns another, the teeth being on the rims. The wheels have 9-in. and 5-in. radii, and the smaller wheel rotates at 48 rpm (revolutions per minute). Find the angular speed of the larger wheel, in radians per second.

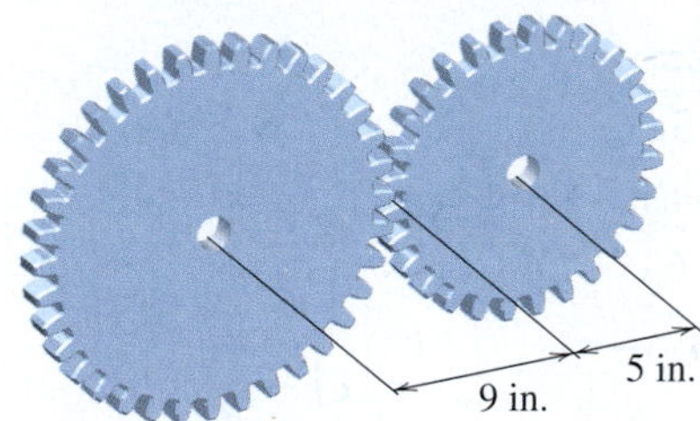

Solution Let ω_1 = the angular speed of the smaller wheel and ω_2 = the angular speed of the larger wheel. The wheels have the same linear speed, so we have

$$v = 5\omega_1 = 9\omega_2.$$

We first convert the angular speed of the smaller wheel, 48 rpm, to radians per second:

$$\begin{aligned}\omega_1 = 48 \text{ rpm} &= \frac{48 \cdot 2\pi}{1 \text{ min}} \qquad \textbf{1 revolution} = \mathbf{2\pi}; \; \textbf{48 revolutions} = \mathbf{48 \cdot 2\pi = 96\pi} \\ &= \frac{96\pi}{1 \text{ min}} \cdot \frac{1 \text{ min}}{60 \text{ sec}} \\ &= 1.6\pi/\text{sec} \\ &\approx 5.027/\text{sec}.\end{aligned}$$

Next, we substitute 5.027/sec for ω_1 and solve for ω_2:

$$\begin{aligned}5\omega_1 &= 9\omega_2 \\ 5(5.027/\text{sec}) &= 9\omega_2 \\ 25.135/\text{sec} &= 9\omega_2 \\ 2.793/\text{sec} &\approx \omega_2.\end{aligned}$$

The angular speed of the larger wheel is about 2.793 radians/sec, or 2.793/sec.

Now Try Exercise 81.

6.4 Exercise Set

For each of Exercises 1–4, sketch a unit circle and mark the points determined by the given real numbers.

1. **a)** $\frac{\pi}{4}$ **b)** $\frac{3\pi}{2}$ **c)** $\frac{3\pi}{4}$
d) π **e)** $\frac{11\pi}{4}$ **f)** $\frac{17\pi}{4}$

2. **a)** $\frac{\pi}{2}$ **b)** $\frac{5\pi}{4}$ **c)** 2π
d) $\frac{9\pi}{4}$ **e)** $\frac{13\pi}{4}$ **f)** $\frac{23\pi}{4}$

3. **a)** $\frac{\pi}{6}$ **b)** $\frac{2\pi}{3}$ **c)** $\frac{7\pi}{6}$
d) $\frac{10\pi}{6}$ **e)** $\frac{14\pi}{6}$ **f)** $\frac{23\pi}{4}$

4. **a)** $-\frac{\pi}{2}$ **b)** $-\frac{3\pi}{4}$ **c)** $-\frac{5\pi}{6}$
d) $-\frac{5\pi}{2}$ **e)** $-\frac{17\pi}{6}$ **f)** $-\frac{9\pi}{4}$

Find two real numbers between -2π and 2π that determine each of the points on the unit circle.

5.

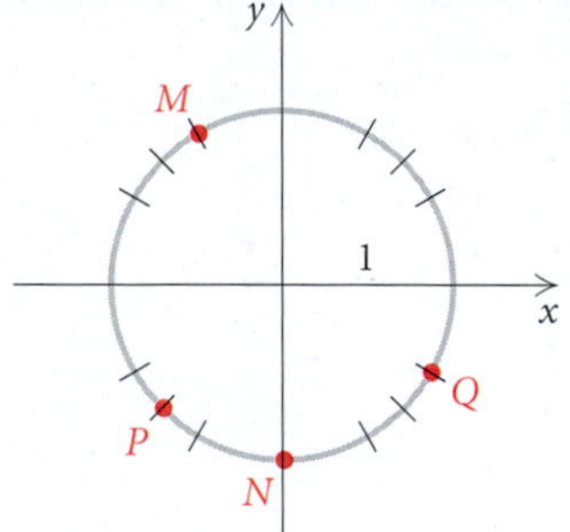

6.

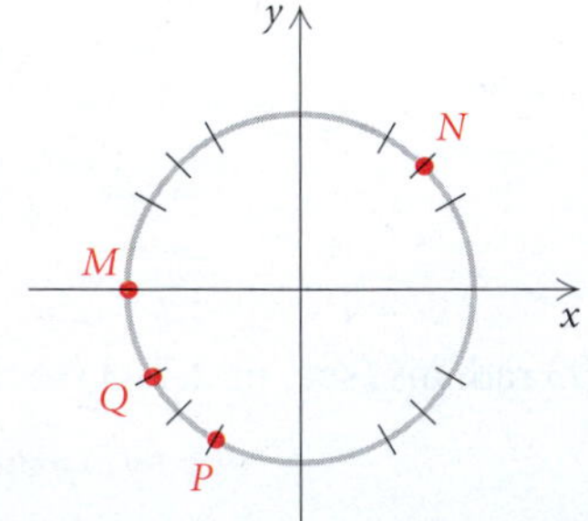

For Exercises 7 and 8, sketch a unit circle and mark the approximate location of the point determined by the given real number.

7. **a)** 2.4 **b)** 7.5
c) 32 **d)** 320

8. **a)** 0.25 **b)** 1.8
c) 47 **d)** 500

Convert to radian measure. Leave the answer in terms of π.

9. 75° **10.** 30°

11. 200° **12.** −135°

13. −214.6° **14.** 37.71°

15. −180° **16.** 90°

17. 12.5° **18.** 6.3°

19. −340° **20.** −60°

Convert to radian measure. Round the answer to two decimal places.

21. 240° **22.** 15°

23. −60° **24.** 145°

25. 117.8° **26.** −231.2°

27. 1.354° **28.** 584°

29. 345° **30.** −75°

31. 95° **32.** 24.8°

Convert to degree measure. Round the answer to two decimal places where appropriate.

33. $-\frac{3\pi}{4}$ **34.** $\frac{7\pi}{6}$

35. 8π **36.** $-\frac{\pi}{3}$

37. 1 **38.** −17.6

39. 2.347 **40.** 25

41. $\frac{5\pi}{4}$ **42.** -6π

43. −90 **44.** 37.12

45. $\frac{2\pi}{7}$ **46.** $\frac{\pi}{9}$

47. Certain positive angles are marked here in degrees. Find the corresponding radian measures.

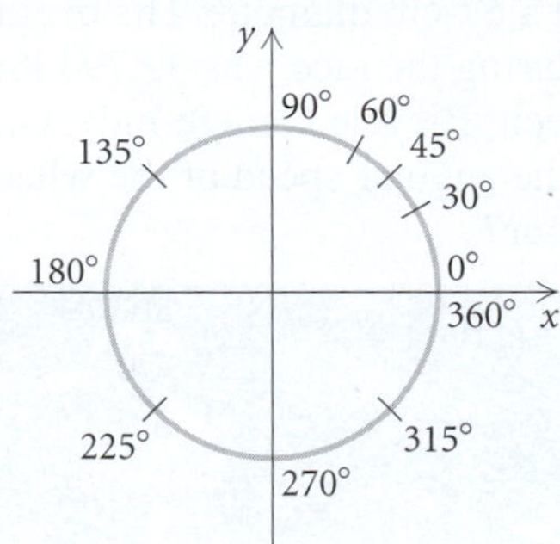

48. Certain negative angles are marked here in degrees. Find the corresponding radian measures.

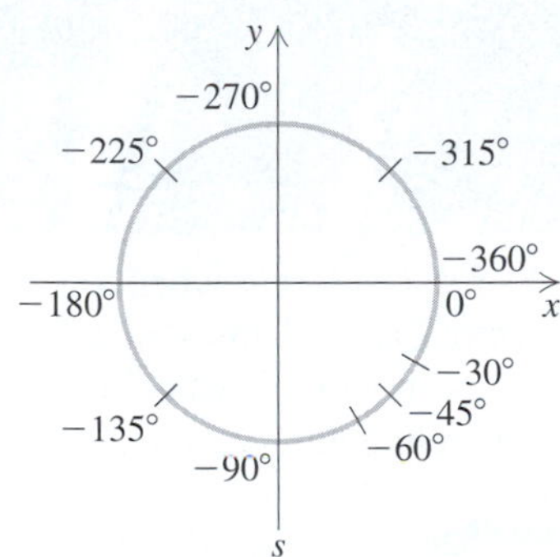

Find a positive angle and a negative angle that are coterminal with the given angle. Answers may vary.

49. $\frac{\pi}{4}$

50. $\frac{5\pi}{3}$

51. $\frac{7\pi}{6}$

52. π

53. $-\frac{2\pi}{3}$

54. $-\frac{3\pi}{4}$

Find the complement and the supplement.

55. $\frac{\pi}{3}$

56. $\frac{5\pi}{12}$

57. $\frac{3\pi}{8}$

58. $\frac{\pi}{4}$

59. $\frac{\pi}{12}$

60. $\frac{\pi}{6}$

Complete the following table. Round the answers to two decimal places where appropriate.

	DISTANCE, s (ARC LENGTH)	RADIUS, r	ANGLE, θ
61.	8 ft	$3\frac{1}{2}$ ft	______
62.	200 cm	______	45°
63.	______	4.2 in.	$\frac{5\pi}{12}$
64.	16 yd	______	5

65. In a circle with a 120-cm radius, an arc 132 cm long subtends an angle of how many radians? how many degrees, to the nearest degree?

66. In a circle with a 10-ft diameter, an arc 20 ft long subtends an angle of how many radians? how many degrees, to the nearest degree?

67. In a circle with a 2-yd radius, how long is an arc associated with an angle of 1.6 radians?

68. In a circle with a 5-m radius, how long is an arc associated with an angle of 2.1 radians?

69. *Angle of Revolution.* A tire on a 2014 Dodge Durango SUV has an outside diameter of 36.32 in. Through what angle (in radians) does the tire turn while traveling 1 mi?

70. *Angle of Revolution.* Through how many radians does the minute hand of a wristwatch rotate from 12:40 P.M. to 1:30 P.M.?

71. *Linear Speed.* A flywheel with a 15-cm diameter is rotating at a rate of 7 radians/sec. What is the linear speed of a point on its rim, in centimeters per minute?

72. *Linear Speed.* A wheel with a 30-cm radius is rotating at a rate of 3 radians/sec. What is the linear speed of a point on its rim, in meters per minute?

73. *Linear Speeds on a Carousel.* When Brett and Will ride the carousel described earlier in this section, Brett always selects a horse on the outside row, whereas Will prefers the row closest to the center. These rows are 19 ft 3 in. and 13 ft 11 in. from the center, respectively. The angular speed of the carousel is 2.4 revolutions per minute. (*Source*: The Children's Museum,

Indianapolis, IN) What is the difference, in miles per hour, in the linear speeds of Brett and Will?

74. *Angular Speed of a Printing Press.* This text was printed on a four-color web heatset offset press. A cylinder on this press has a 21-in. diameter. The linear speed of a point on the cylinder's surface is 18.33 feet per second. What is the angular speed of the cylinder, in revolutions per hour? Printers often refer to the angular speed as impressions per hour (IPH). (*Source*: R. R. Donnelley, Willard, Ohio)

75. *Linear Speed at the Equator.* The earth has a 4000-mi radius and rotates one revolution every 24 hr. What is the linear speed of a point on the equator, in miles per hour?

76. *Linear Speed of the Earth.* The earth is about 93,000,000 mi from the sun and traverses its orbit, which is nearly circular, every 365.25 days. What is the linear velocity of the earth in its orbit, in miles per hour?

77. *Determining the Speed of a River.* A water wheel has a 10-ft radius. To get a good approximation of the speed of the river, you count the revolutions of the wheel and find that it makes 14 revolutions per minute (rpm). What is the speed of the river, in miles per hour?

78. *The Tour de France.* Cadel Evans of Australia won the 2015 Tour de France bicycle race. The wheel of his bicycle had a 67-cm diameter. His overall average linear speed during the race was 39.794 km/h (*Source*: Preston Green, Bicycle Garage Indy, Greenwood, IN). What was the angular speed of the wheel, in revolutions per hour?

79. *John Deere Tractor.* A rear wheel on a John Deere 8300 farm tractor has a 23-in. radius. Find the angle (in radians) through which a wheel rotates in 12 sec if the tractor is traveling at a speed of 22 mph.

80. *Angular Speed of a Pulley.* Two pulleys, 50 cm and 30 cm in diameter, respectively, are connected by a belt. The larger pulley makes 12 revolutions per minute. Find the angular speed of the smaller pulley, in radians per second.

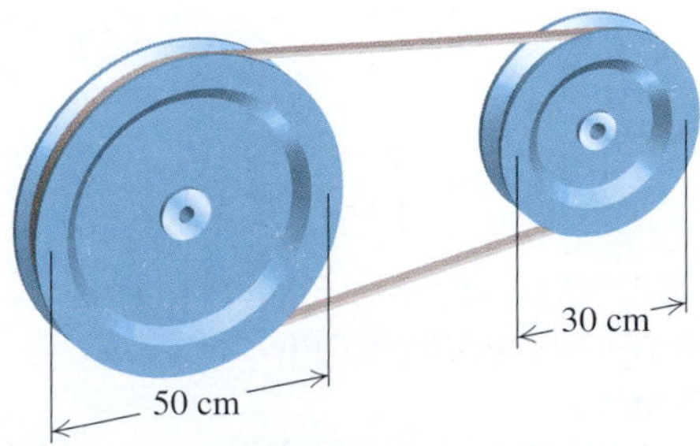

81. *Angular Speed of a Gear Wheel.* One gear wheel turns another, the teeth being on the rims. The wheels have 40-cm and 50-cm radii, and the smaller wheel rotates at 20 rpm. Find the angular speed of the larger wheel, in radians per second.

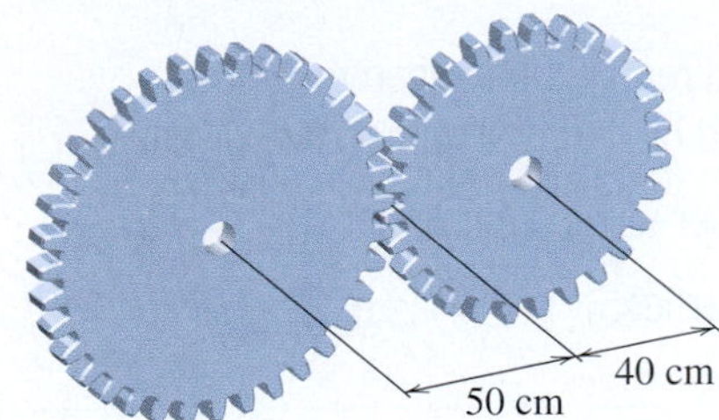

Skill Maintenance

Vocabulary Reinforcement

In each of Exercises 82–89, fill in the blanks with the correct terms. Some of the given choices will not be used.

inverse	relation
horizontal line	vertical asymptote
vertical line	horizontal asymptote
exponential function	even function
logarithmic function	odd function
natural	sine of θ
common	cosine of θ
logarithm	tangent of θ
one-to-one	

82. The domain of a(n) ______________ function f is the range of the inverse f^{-1}. [5.1]

83. The ______________ is the length of the side adjacent to θ divided by the length of the hypotenuse. [6.1]

84. The function $f(x) = a^x$, where x is a real number, $a > 0$ and $a \neq 1$, is called the ______________, base a. [5.2]

85. The graph of a rational function may or may not cross a(n) ______________. [4.5]

86. If the graph of a function f is symmetric with respect to the origin, we say that it is a(n) ______________. [2.4]

87. Logarithms, base e, are called ______________ logarithms. [5.3]

88. If it is possible for a(n) ______________ to intersect the graph of a function more than once, then the function is not one-to-one and its ______________ is not a function. [5.1]

89. A(n) ______________ is an exponent. [5.3]

Synthesis

90. A point on the unit circle has y-coordinate $-\sqrt{21}/5$. What is its x-coordinate? Check using a calculator.

91. On the earth, one degree of latitude is how many kilometers? how many miles? (Assume that the radius of the earth is 6400 km, or 4000 mi, approximately.)

92. A **grad** is a unit of angle measure similar to a degree. A right angle has a measure of 100 grads. Convert each of the following to grads.

a) $48°$ **b)** $\dfrac{5\pi}{7}$

93. A **mil** is a unit of angle measure. A right angle has a measure of 1600 mils. Convert each of the following to degrees, minutes, and seconds.

a) 100 mils **b)** 350 mils

94. *Hands of a Clock.* At what time between noon and 1:00 P.M. are the hands of a clock perpendicular?

95. *Distance Between Points on the Earth.* To find the distance between two points on the earth when their latitude and longitude are known, we can use a right triangle for an excellent approximation if the points are not too far apart. Point A is at latitude $38°27'30''$ N, longitude $82°57'15''$ W, and point B is at latitude $38°28'45''$ N, longitude $82°56'30''$ W. Find the distance from A to B in nautical miles. (One minute of latitude is one nautical mile.)

6.5 Circular Functions: Graphs and Properties

- Given the coordinates of a point on the unit circle, find its reflections across the x-axis, the y-axis, and the origin.
- Determine the six trigonometric function values for a real number when the coordinates of the point on the unit circle determined by that real number are given.
- Find trigonometric function values for any real number using a calculator.
- Graph the six circular functions and state their properties.

The domains of the trigonometric functions, defined in Sections 6.1 and 6.3, have been sets of angles or rotations measured in a *real number* of degree units. We can also consider the domains to be sets of **real numbers**, or **radians**, introduced in Section 6.4. Many applications in calculus that use the trigonometric functions refer only to radians.

Let's again consider radian measure and the unit circle. We defined radian measure for θ as

$$\theta = \frac{s}{r}. \quad \textbf{θ is a real number without units.}$$

When $r = 1$,

$$\theta = \frac{s}{1}, \quad \text{or} \quad \theta = s.$$

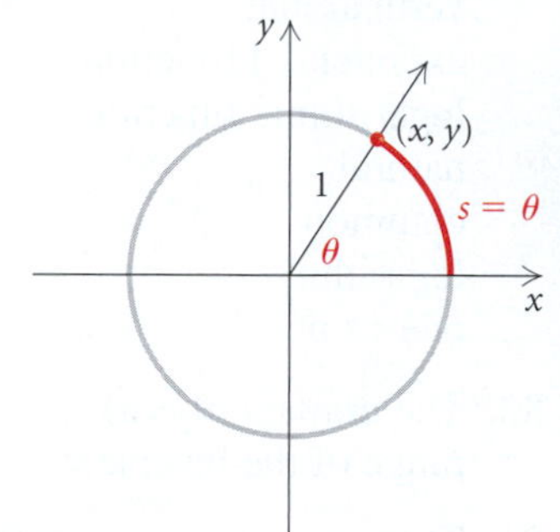

The arc length s on the unit circle is the same as the radian measure of the angle θ.

In the figure above, the point (x, y) is the point where the terminal side of the angle with radian measure s intersects the *unit circle*. We can now extend our definitions of the trigonometric functions using *domains* composed of *real numbers*, or *radians*. Trigonometric functions with domains composed of real numbers are called **circular functions**.

In the definitions,

- *s can be considered the radian measure of an angle or*
- *the measure of an arc length on the unit circle.*

Either way, s is a real number. To each real number s, there corresponds an arc length s on the unit circle.

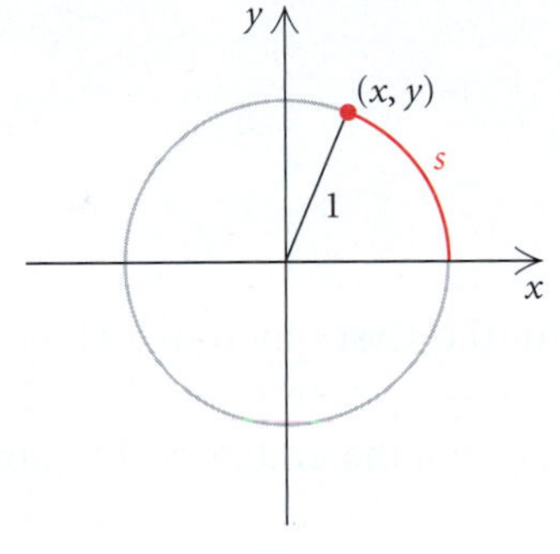

BASIC CIRCULAR FUNCTIONS

For a real number s that determines a point (x, y) on the unit circle:

$$\sin s = \text{second coordinate} = \frac{y}{1} = y,$$

$$\cos s = \text{first coordinate} = \frac{x}{1} = x,$$

$$\tan s = \frac{\text{second coordinate}}{\text{first coordinate}} = \frac{y}{x} \quad (x \neq 0),$$

$$\csc s = \frac{1}{\text{second coordinate}} = \frac{1}{y} \quad (y \neq 0),$$

$$\sec s = \frac{1}{\text{first coordinate}} = \frac{1}{x} \quad (x \neq 0),$$

$$\cot s = \frac{\text{first coordinate}}{\text{second coordinate}} = \frac{x}{y} \quad (y \neq 0).$$

We can consider the domains of trigonometric functions to be real numbers rather than angles. We can determine these values for a specific real number if we know the coordinates of the point on the unit circle determined by that number. As with degree measure, we can also find these function values directly using a calculator.

Reflections on the Unit Circle

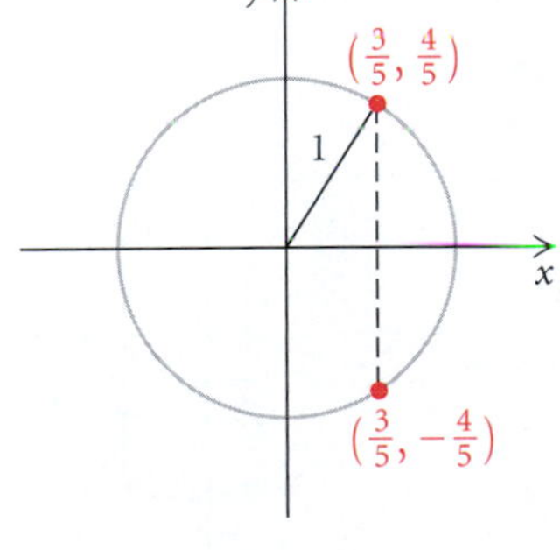

Let's consider the unit circle and a few of its points. For any point (x, y) on the unit circle, $x^2 + y^2 = 1$, we know that $-1 \le x \le 1$ and $-1 \le y \le 1$. If we know the x- or y-coordinate of a point on the unit circle, we can find the other coordinate. If $x = \frac{3}{5}$, then

$$\left(\tfrac{3}{5}\right)^2 + y^2 = 1$$
$$y^2 = 1 - \tfrac{9}{25} = \tfrac{16}{25}$$
$$y = \pm\tfrac{4}{5}.$$

Thus, $\left(\frac{3}{5}, \frac{4}{5}\right)$ and $\left(\frac{3}{5}, -\frac{4}{5}\right)$ are points on the unit circle. There are two points with an x-coordinate of $\frac{3}{5}$.

Now let's consider the radian measure $\pi/3$ and determine the coordinates of the point on the unit circle determined by $\pi/3$. We construct a right triangle by dropping a perpendicular segment from the point to the x-axis.

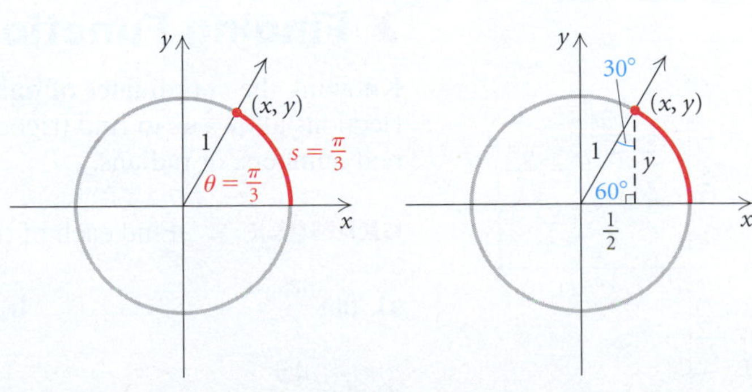

Since $\pi/3 = 60°$, we have a 30°–60° right triangle in which the side opposite the 30° angle is one half of the hypotenuse. The hypotenuse, or radius, is 1, so the side opposite the 30° angle is $\frac{1}{2} \cdot 1$, or $\frac{1}{2}$. Using the Pythagorean equation, we can find the other side:

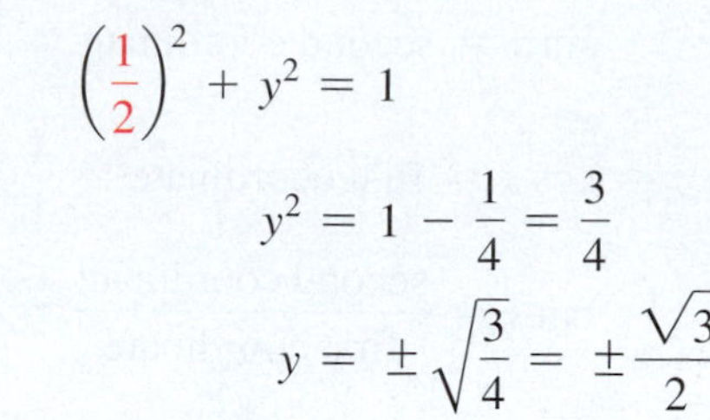

$$\left(\frac{1}{2}\right)^2 + y^2 = 1$$

$$y^2 = 1 - \frac{1}{4} = \frac{3}{4}$$

$$y = \pm\sqrt{\frac{3}{4}} = \pm\frac{\sqrt{3}}{2}.$$

We know that y is positive since the point is in the first quadrant. Thus the coordinates of the point determined by $\pi/3$ are $x = 1/2$ and $y = \sqrt{3}/2$, or $\left(1/2, \sqrt{3}/2\right)$. We can always check to see if a point is on the unit circle by substituting into the equation $x^2 + y^2 = 1$:

$$\left(\frac{1}{2}\right)^2 + \left(\frac{\sqrt{3}}{2}\right)^2 = \frac{1}{4} + \frac{3}{4} = 1.$$

Because a unit circle is symmetric with respect to the x-axis, the y-axis, and the origin, we can use the coordinates of one point on the unit circle to find coordinates of its reflections.

EXAMPLE 1 Each of the following points lies on the unit circle. Find their reflections across the x-axis, the y-axis, and the origin.

a) $\left(\frac{3}{5}, \frac{4}{5}\right)$ **b)** $\left(\frac{\sqrt{2}}{2}, \frac{\sqrt{2}}{2}\right)$ **c)** $\left(\frac{1}{2}, \frac{\sqrt{3}}{2}\right)$

Solution

a)

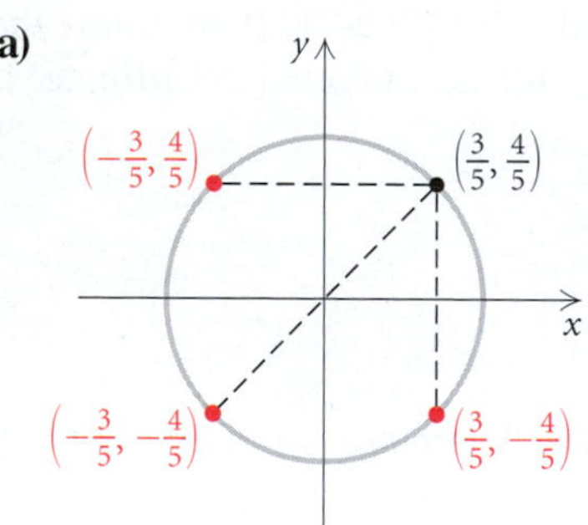

b)

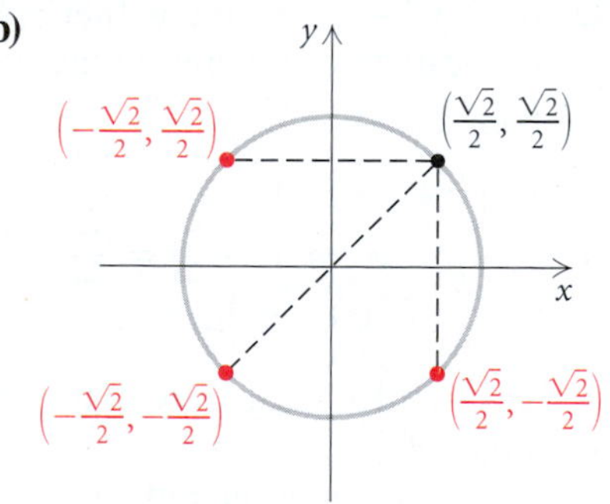

c)

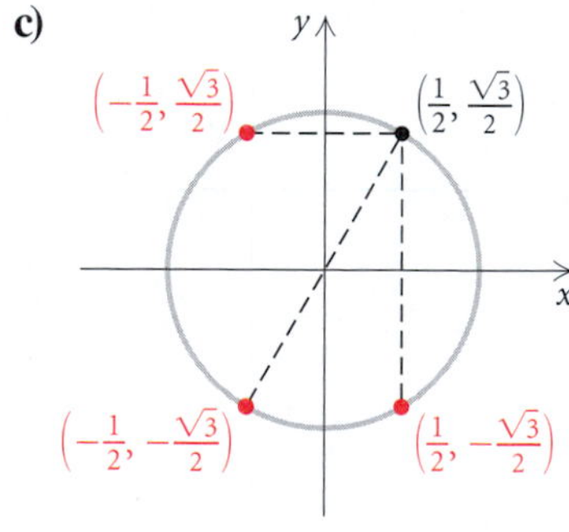

Now Try Exercise 1.

Finding Function Values

Knowing the coordinates of only a few points on the unit circle along with their reflections allows us to find trigonometric function values of the most frequently used real numbers, or radians.

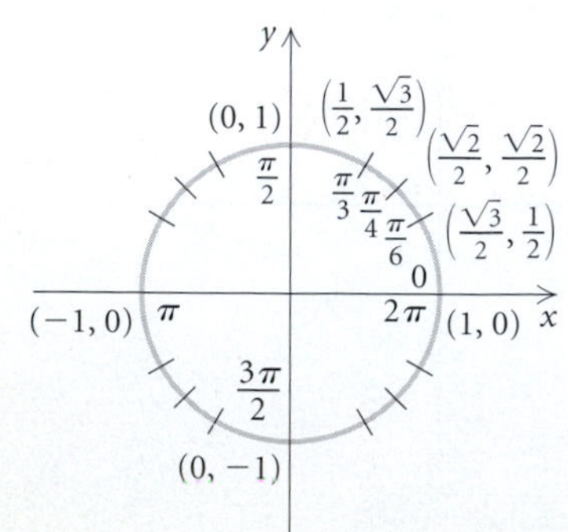

EXAMPLE 2 Find each of the following function values.

a) $\tan\frac{\pi}{3}$ **b)** $\cos\frac{3\pi}{4}$ **c)** $\sin\left(-\frac{\pi}{6}\right)$

d) $\cos\frac{4\pi}{3}$ **e)** $\cot\pi$ **f)** $\csc\left(-\frac{7\pi}{2}\right)$

Solution We locate the point on the unit circle determined by the rotation, and then find its coordinates using reflection if necessary.

a) The coordinates of the point determined by $\pi/3$ are $(1/2, \sqrt{3}/2)$.

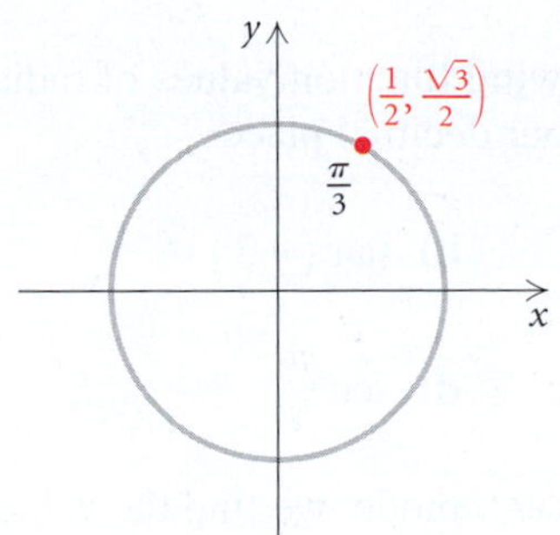

Thus, $\tan \dfrac{\pi}{3} = \dfrac{y}{x} = \dfrac{\sqrt{3}/2}{1/2} = \sqrt{3}$.

b) The reflection of $(\sqrt{2}/2, \sqrt{2}/2)$ across the y-axis is $(-\sqrt{2}/2, \sqrt{2}/2)$.

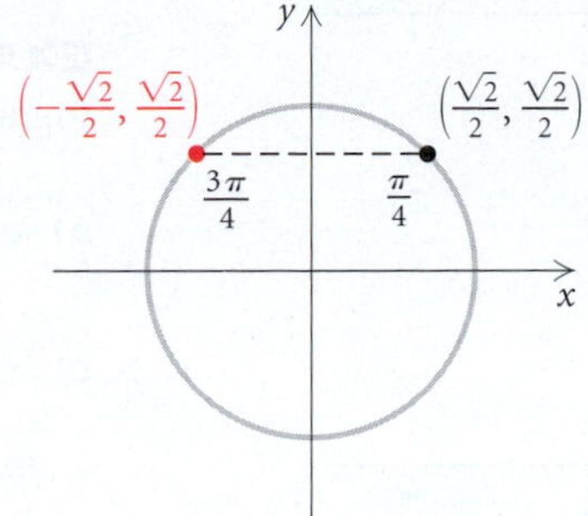

Thus, $\cos \dfrac{3\pi}{4} = x = -\dfrac{\sqrt{2}}{2}$.

c) The reflection of $(\sqrt{3}/2, 1/2)$ across the x-axis is $(\sqrt{3}/2, -1/2)$.

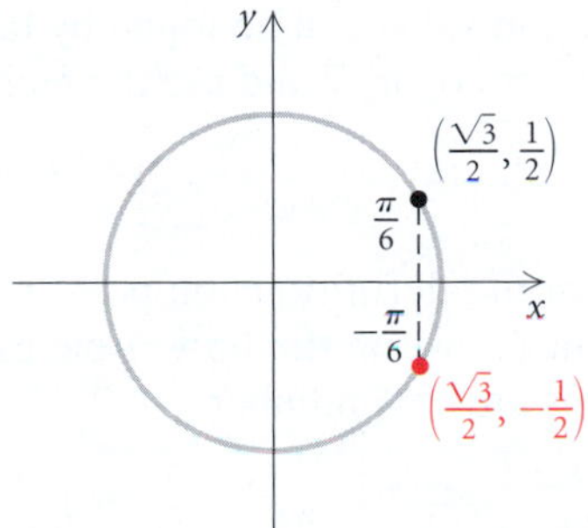

Thus, $\sin\left(-\dfrac{\pi}{6}\right) = y = -\dfrac{1}{2}$.

d) The reflection of $(1/2, \sqrt{3}/2)$ across the origin is $(-1/2, -\sqrt{3}/2)$.

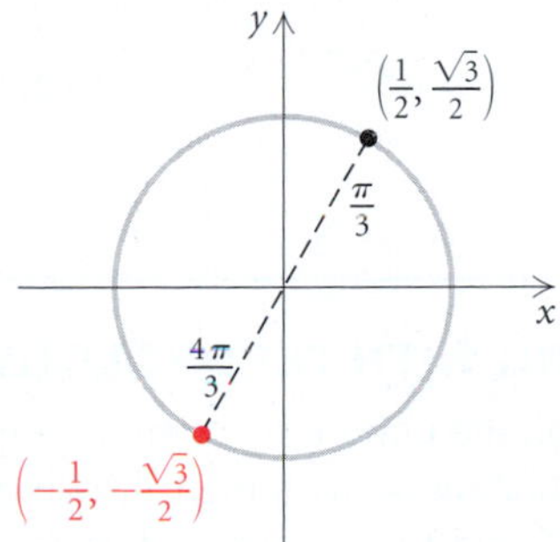

Thus, $\cos \dfrac{4\pi}{3} = x = -\dfrac{1}{2}$.

e) The coordinates of the point determined by π are $(-1, 0)$.

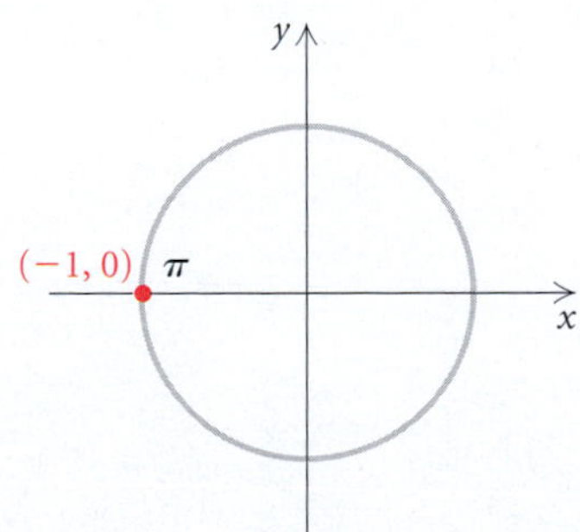

Thus, $\cot \pi = \dfrac{x}{y} = \dfrac{-1}{0}$, which is not defined.

We can also think of $\cot \pi$ as the reciprocal of $\tan \pi$. Since $\tan \pi = y/x = 0/-1 = 0$ and the reciprocal of 0 is not defined, we know that $\cot \pi$ is not defined.

f) The coordinates of the point determined by $-7\pi/2$ are $(0, 1)$.

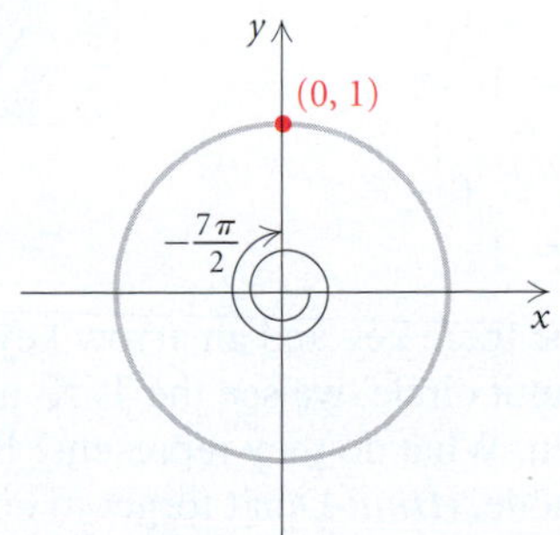

Thus, $\csc\left(-\dfrac{7\pi}{2}\right) = \dfrac{1}{y} = \dfrac{1}{1} = 1$.

Now Try Exercises 9 and 11.

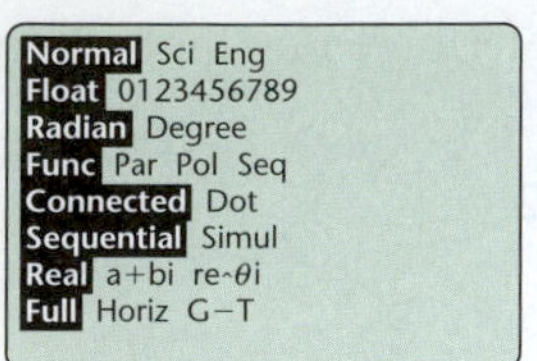

Using a calculator, we can find trigonometric function values of any real number without knowing the coordinates of the point that it determines on the unit circle. Most calculators have both degree mode and radian mode. When finding function values of radian measures, or real numbers, we *must* set the calculator in RADIAN mode. (See the window at left.)

EXAMPLE 3 Find each of the following function values of radian measures using a calculator. Round the answers to four decimal places.

a) $\cos \frac{2\pi}{5}$ **b)** $\tan(-3)$

c) $\sin 24.9$ **d)** $\sec \frac{\pi}{7}$

cos(2π/5)
.3090169944
tan(−3)
.1425465431
sin(24.9)
−.2306457059

Solution Using a calculator set in RADIAN mode, we find the values.

a) $\cos \frac{2\pi}{5} \approx 0.3090$ **b)** $\tan(-3) \approx 0.1425$

c) $\sin 24.9 \approx -0.2306$ **d)** $\sec \frac{\pi}{7} = \frac{1}{\cos \frac{\pi}{7}} \approx 1.1099$

Note in part (d) that the secant function value can be found by taking the reciprocal of the cosine value. Thus we can enter $\cos \pi/7$ and use the reciprocal key.

Now Try Exercises 25 and 33.

EXPLORING WITH TECHNOLOGY

We can graph the unit circle using a graphing calculator. We use PARAMETRIC mode with the following window and let $X_{1T} = \cos T$ and $Y_{1T} = \sin T$. Here we use DEGREE mode.

WINDOW

Tmin = 0
Tmax = 360
Tstep = 15
Xmin = −1.5
Xmax = 1.5
Xscl = 1
Ymin = −1
Ymax = 1
Yscl = 1

X1T = cos(T) Y1T = sin(T)
T = 30
X = .8660254 Y = .5

Using the trace key and an arrow key to move the cursor around the unit circle, we see the T, X, and Y values appear on the screen. What do they represent? Repeat this exercise in RADIAN mode. (*Hint*: Don't forget to convert Tmin, Tmax, and Tstep to radians.) What do the T, X, and Y values represent? (For more on parametric equations, see Section 10.7.)

From the definitions on p. 449, we can relabel any point (x, y) on the unit circle as $(\cos s, \sin s)$, where s is any real number.

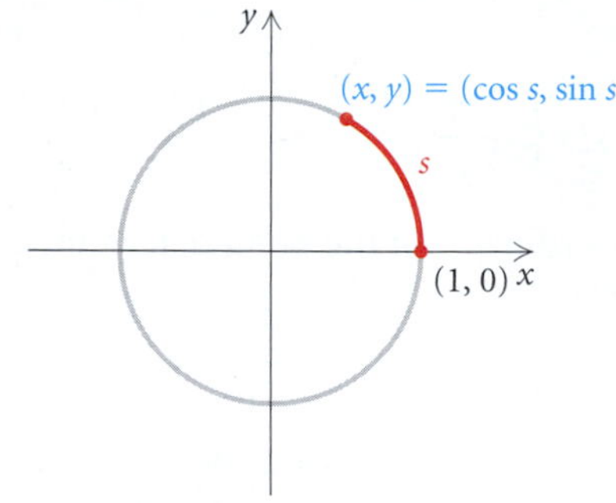

Graphs of the Sine and Cosine Functions

Properties of functions can be observed from their graphs. We begin by graphing the sine and cosine functions. We make a table of values, plot the points, and then connect those points with a smooth curve. It is helpful to first draw a unit circle and label a few points with coordinates. We can either use the coordinates as the function values or find approximate sine and cosine values directly with a calculator.

s	$\sin s$	$\cos s$
0	0	1
$\pi/6$	0.5	0.8660
$\pi/4$	0.7071	0.7071
$\pi/3$	0.8660	0.5
$\pi/2$	1	0
$3\pi/4$	0.7071	−0.7071
π	0	−1
$5\pi/4$	−0.7071	−0.7071
$3\pi/2$	−1	0
$7\pi/4$	−0.7071	0.7071
2π	0	1

s	$\sin s$	$\cos s$
0	0	1
$-\pi/6$	−0.5	0.8660
$-\pi/4$	−0.7071	0.7071
$-\pi/3$	−0.8660	0.5
$-\pi/2$	−1	0
$-3\pi/4$	−0.7071	−0.7071
$-\pi$	0	−1
$-5\pi/4$	0.7071	−0.7071
$-3\pi/2$	1	0
$-7\pi/4$	0.7071	0.7071
-2π	0	1

The graphs are as follows.

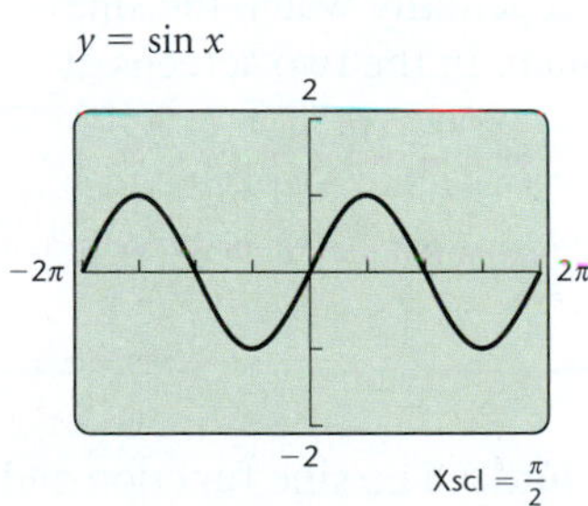

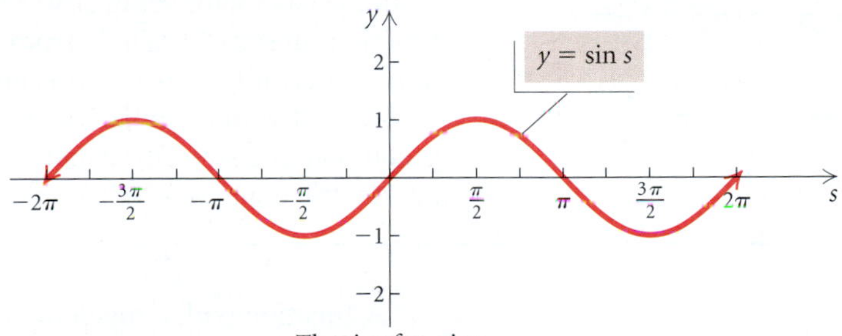

The sine function

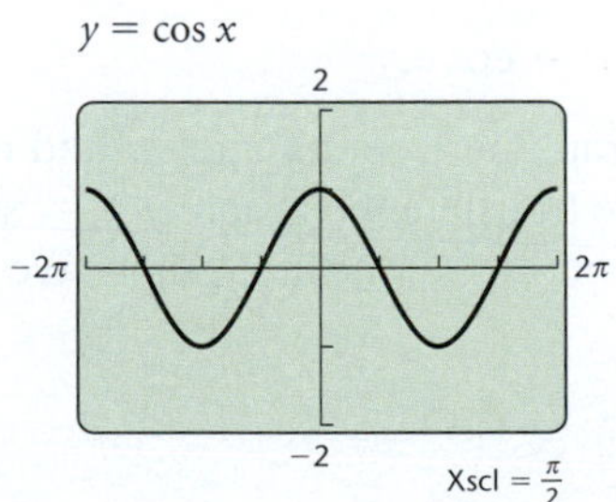

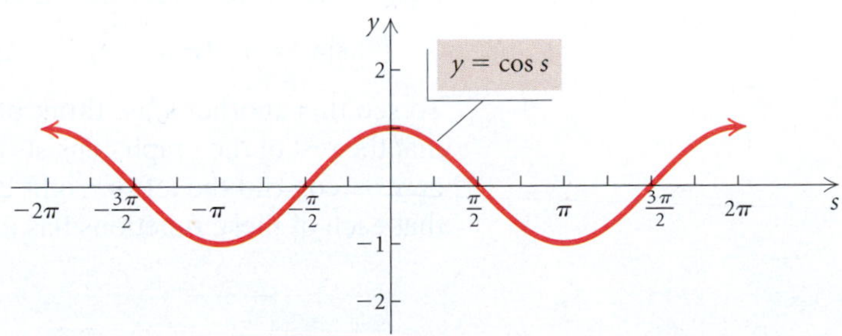

The cosine function

The sine and cosine functions are continuous functions. Note in the graph of the sine function that function values increase from 0 at $s = 0$ to 1 at $s = \pi/2$, then decrease to 0 at $s = \pi$, decrease further to -1 at $s = 3\pi/2$, and increase to 0 at 2π. The reverse pattern follows when s decreases from 0 to -2π. Note in the graph of the cosine function that function values start at 1 when $s = 0$, and decrease to 0 at $s = \pi/2$. They decrease further to -1 at $s = \pi$, then increase to 0 at $s = 3\pi/2$, and increase further to 1 at $s = 2\pi$. An identical pattern follows when s decreases from 0 to -2π.

From the unit circle and the graphs of the functions, we know that the domain of both the sine and cosine functions is the entire set of real numbers, $(-\infty, \infty)$. The range of each function is the set of all real numbers from -1 to 1, $[-1, 1]$.

DOMAIN AND RANGE OF THE SINE FUNCTION AND THE COSINE FUNCTION

The *domain* of the sine function and the cosine function is $(-\infty, \infty)$.

The *range* of the sine function and the cosine function is $[-1, 1]$.

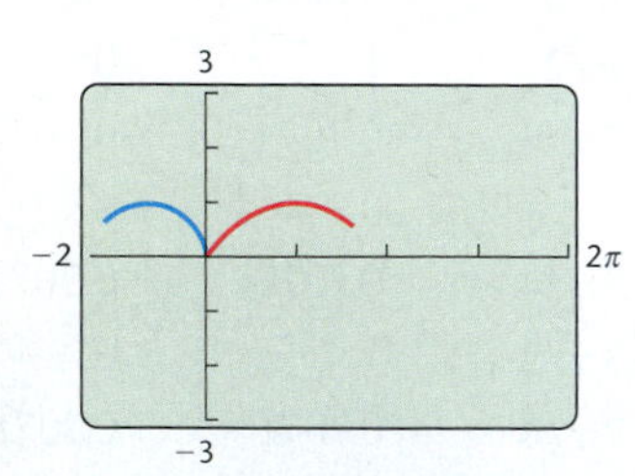

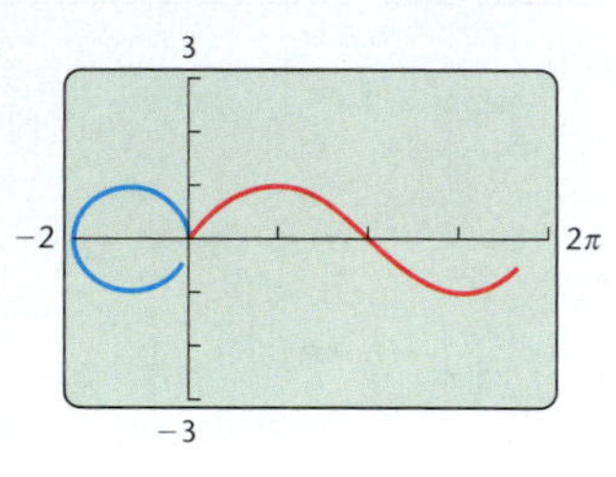

EXPLORING WITH TECHNOLOGY

Another way to construct the sine and cosine graphs is by considering the unit circle and transferring vertical distances for the sine function and horizontal distances for the cosine function. Using a graphing calculator, we can visualize the transfer of these distances. We use the calculator set in PARAMETRIC mode and RADIAN mode and let $X_{1T} = \cos T - 1$ and $Y_{1T} = \sin T$ for the unit circle centered at $(-1, 0)$ and $X_{2T} = T$ and $Y_{2T} = \sin T$ for the sine curve. Use the following window settings.

Tmin $= 0$	Xmin $= -2$	Ymin $= -3$
Tmax $= 2\pi$	Xmax $= 2\pi$	Ymax $= 3$
Tstep $= .1$	Xscl $= \pi/2$	Yscl $= 1$

With the calculator set in SIMULTANEOUS mode, we can actually watch the sine function (in red) "unwind" from the unit circle (in blue). In the two screens at left, we partially illustrate this animated procedure.

Consult your calculator's instruction manual for specific keystrokes and graph both the sine curve and the cosine curve in this manner. (For more on parametric equations, see Section 10.7.)

A function with a repeating pattern is called **periodic**. The sine function and the cosine function are examples of periodic functions. The values of these functions repeat themselves every 2π units. In other words, for any s, we have

$$\sin(s + 2\pi) = \sin s \quad \text{and} \quad \cos(s + 2\pi) = \cos s.$$

To see this another way, think of the part of the graph between 0 and 2π and note that the rest of the graph consists of copies of it. If we translate the graph of $y = \sin x$ or $y = \cos x$ to the left or right 2π units, we will obtain the original graph. We say that each of these functions has a period of 2π.

PERIODIC FUNCTION

A function f is said to be **periodic** if there exists a positive constant p such that

$$f(s + p) = f(s)$$

for all s in the domain of f. The smallest such positive number p is called the period of the function.

The period p can be thought of as the length of the shortest recurring interval.

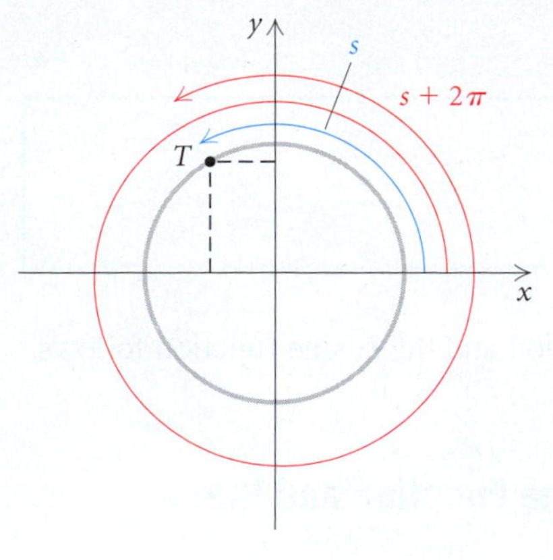

We can also use the unit circle to verify that the period of the sine and cosine functions is 2π. Consider any real number s and the point T that it determines on a unit circle, as shown at left. If we increase s by 2π, the point determined by $s + 2\pi$ is again the point T. Hence for any real number s,

$$\sin(s + 2\pi) = \sin s \quad \text{and} \quad \cos(s + 2\pi) = \cos s.$$

It is also true that $\sin(s + 4\pi) = \sin s$, $\sin(s + 6\pi) = \sin s$, and so on. In fact, for *any* integer k, the following equations are identities:

$$\sin[s + k(2\pi)] = \sin s \quad \text{and} \quad \cos[s + k(2\pi)] = \cos s,$$

or $$\sin s = \sin(s + 2k\pi) \quad \text{and} \quad \cos s = \cos(s + 2k\pi).$$

The **amplitude** of a periodic function is defined as one half of the distance between its maximum and minimum function values. It is always positive. Both the graphs and the unit circle verify that the maximum value of the sine and cosine functions is 1, whereas the minimum value of each is -1. Thus,

Just in Time 4

$$\text{the amplitude of the sine function} = \tfrac{1}{2}|1 - (-1)| = 1$$

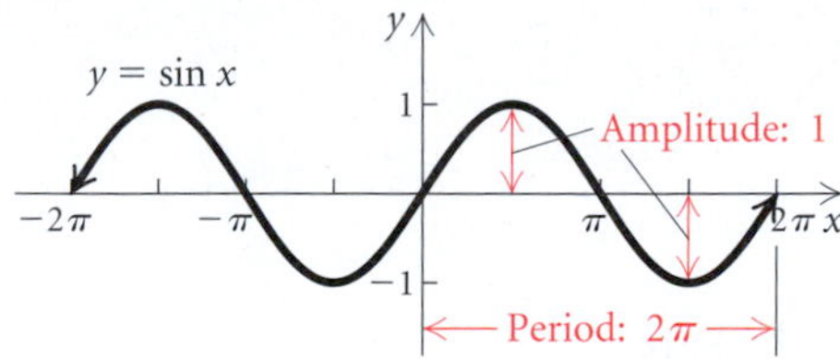

and $$\text{the amplitude of the cosine function} = \tfrac{1}{2}|1 - (-1)| = 1.$$

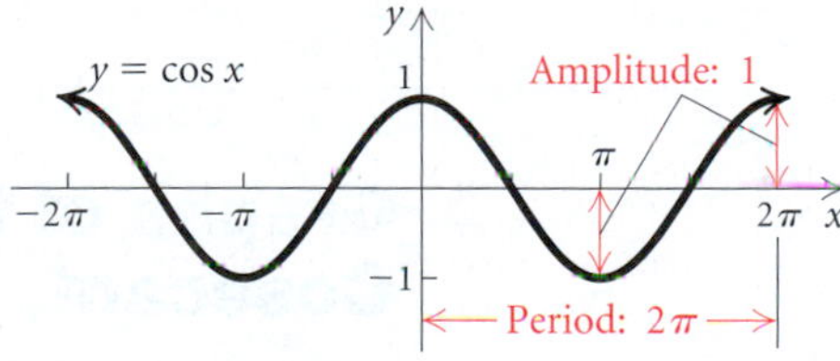

Consider any real number s and its opposite, $-s$. These numbers determine points T and T_1 on a unit circle that are symmetric with respect to the x-axis.

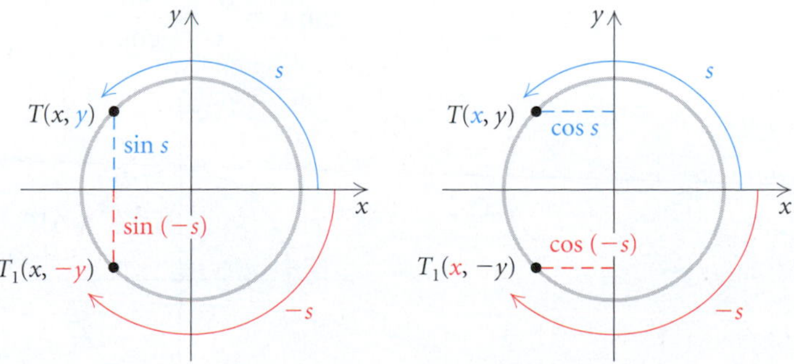

Because their second coordinates are opposites of each other, we know that for any number s,

$$\sin(-s) = -\sin s.$$

Because their first coordinates are the same, we know that for any number s,

$$\cos(-s) = \cos s.$$

EXPLORING WITH TECHNOLOGY

Using the TABLE feature on a graphing calculator, compare the y-values for $y_1 = \sin x$ and $y_2 = \sin(-x)$ and for $y_3 = \cos x$ and $y_4 = \cos(-x)$. Set TblStart $= 0$ and ΔTbl $= \pi/12$.

X	Y1	Y2
0	0	0
.2618	.25882	−.2588
.5236	.5	−.5
.7854	.70711	−.7071
1.0472	.86603	−.866
1.309	.96593	−.9659
1.5708	1	−1

X = 0

X	Y3	Y4
0	1	1
.2618	.96593	.96593
.5236	.86603	.86603
.7854	.70711	.70711
1.0472	.5	.5
1.309	.25882	.25882
1.5708	0	0

X = 0

What appears to be the relationship between $\sin x$ and $\sin(-x)$ and that between $\cos x$ and $\cos(-x)$?

Thus we have shown the following.

EVEN AND ODD FUNCTIONS

REVIEW SECTION 2.4.

The sine function is *odd*.
The cosine function is *even*.

A summary of the properties of the sine function and the cosine function follows.

CONNECTING THE CONCEPTS — **Comparing the Sine Function and the Cosine Function**

SINE FUNCTION

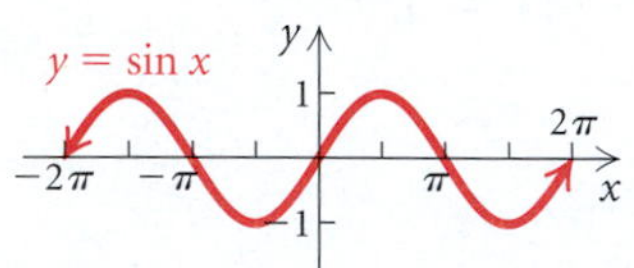

1. Continuous
2. Period: 2π
3. Domain: All real numbers
4. Range: $[-1, 1]$
5. Amplitude: 1
6. Odd: $\sin(-s) = -\sin s$

COSINE FUNCTION

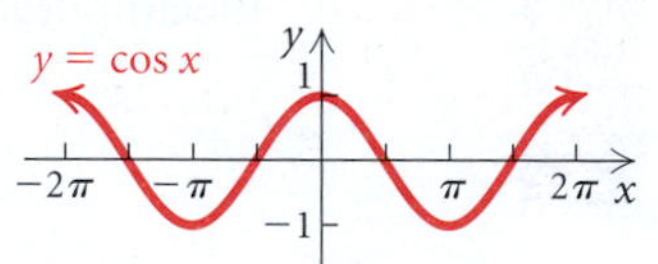

1. Continuous
2. Period: 2π
3. Domain: All real numbers
4. Range: $[-1, 1]$
5. Amplitude: 1
6. Even: $\cos(-s) = \cos s$

Graphs of the Tangent, Cotangent, Cosecant, and Secant Functions

To graph the tangent function, we could make a table of values using a calculator, but in this case it is easier to begin with the definition of tangent and the coordinates of a few points on the unit circle. We recall that

$$\tan s = \frac{y}{x} = \frac{\sin s}{\cos s}.$$

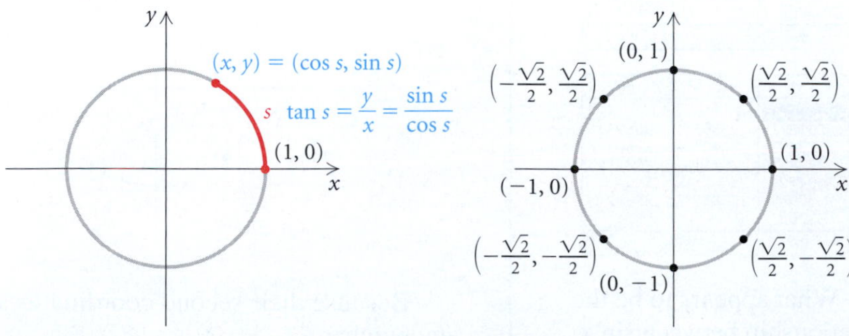

The tangent function is not defined when x, the first coordinate, is 0. That is, it is not defined for any number s whose cosine is 0:

$$s = \pm\frac{\pi}{2}, \pm\frac{3\pi}{2}, \pm\frac{5\pi}{2}, \ldots.$$

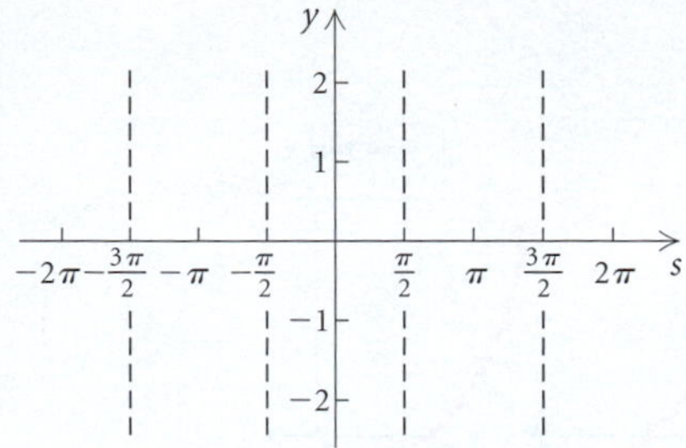
FIGURE 1.

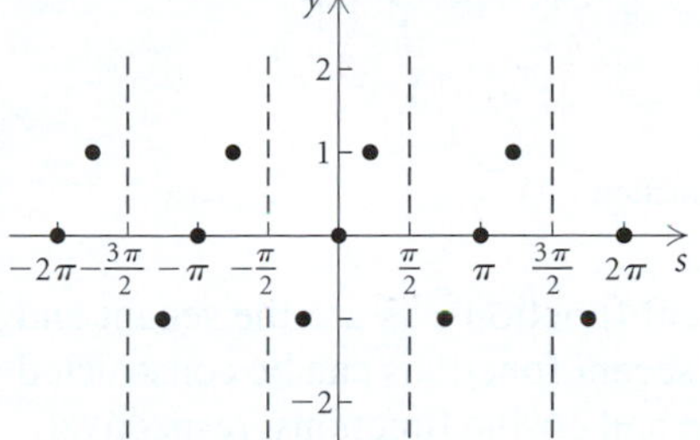
FIGURE 2.

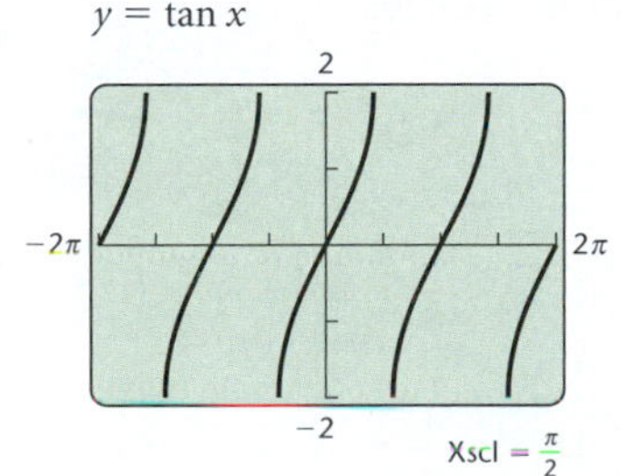

We draw vertical asymptotes at these locations (see Fig. 1 at left).

We also note that

$$\tan s = 0 \text{ at } s = 0, \pm\pi, \pm 2\pi, \pm 3\pi, \dots,$$

$$\tan s = 1 \text{ at } s = \dots -\frac{7\pi}{4}, -\frac{3\pi}{4}, \frac{\pi}{4}, \frac{5\pi}{4}, \frac{9\pi}{4}, \dots,$$

$$\tan s = -1 \text{ at } s = \dots -\frac{9\pi}{4}, -\frac{5\pi}{4}, -\frac{\pi}{4}, \frac{3\pi}{4}, \frac{7\pi}{4}, \dots.$$

We can add these ordered pairs to the graph (see Fig. 2 at left) and investigate the values in $(-\pi/2, \pi/2)$ using a calculator. Note that the function value is 0 when $s = 0$, and the values increase without bound as s increases toward $\pi/2$. The graph gets closer and closer to the vertical asymptote as s gets closer to $\pi/2$, but it never touches the line. As s decreases from 0 to $-\pi/2$, the values decrease without bound. Again the graph gets closer and closer to a vertical asymptote, but it never touches it. We now complete the graph.

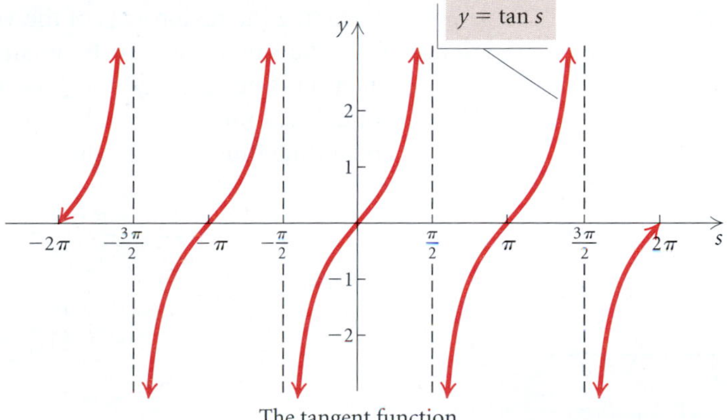

The tangent function

From the graph, we see that the tangent function is continuous except where it is not defined. The period of the tangent function is π. Note that although there is a period, there is no amplitude because there are no maximum and minimum values. When $\cos s = 0$, $\tan s$ is not defined $(\tan s = \sin s/\cos s)$. Thus the domain of the tangent function is the set of all real numbers except $(\pi/2) + k\pi$, where k is an integer. The range of the function is the set of all real numbers.

THE DOMAIN AND THE RANGE OF THE TANGENT FUNCTION

The domain of the tangent function is all real numbers except $\frac{\pi}{2} + k\pi$, where k is an integer.

The range of the tangent function is $(-\infty, \infty)$.

The cotangent function $(\cot s = \cos s/\sin s)$ is not defined when y, the second coordinate, is 0—that is, it is not defined for any number s whose sine is 0. Thus the cotangent is not defined for $s = 0, \pm\pi, \pm 2\pi, \pm 3\pi, \dots.$

The graph of the function is shown below.

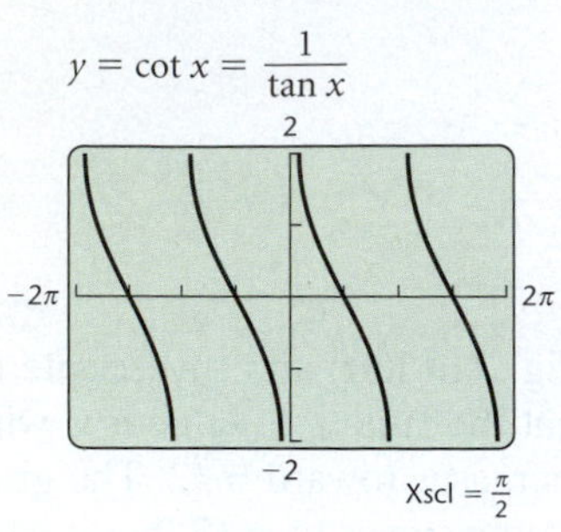

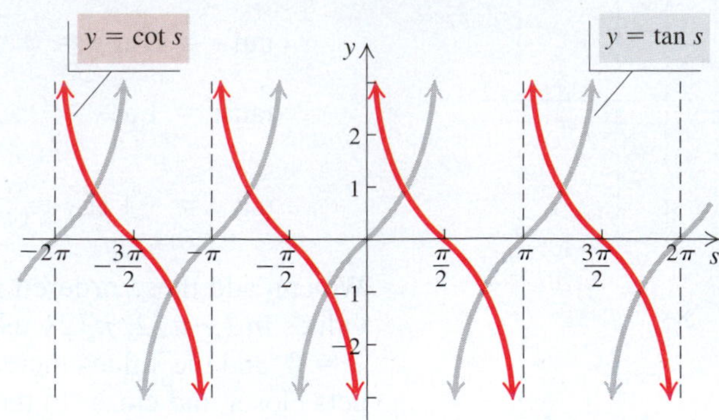

The cotangent function

The cosecant and sine functions are reciprocal functions, as are the secant and cosine functions. The graphs of the cosecant and secant functions can be constructed by finding the reciprocals of the values of the sine and cosine functions, respectively. Thus the functions will be positive together and negative together. The cosecant function is not defined for those numbers s whose sine is 0. The secant function is not defined for those numbers s whose cosine is 0. In the following graphs, the sine and cosine functions are shown by the gray curves for reference.

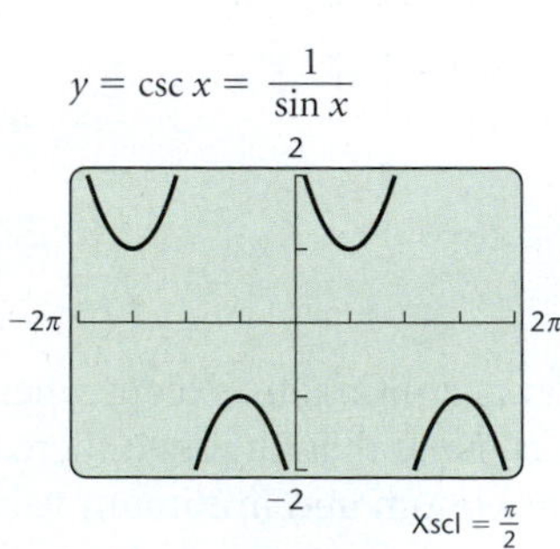

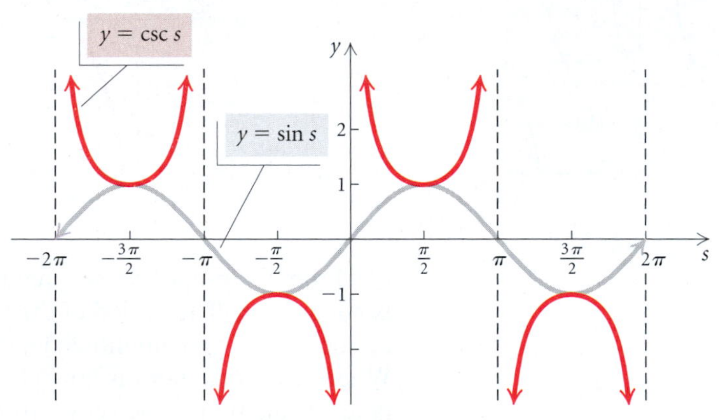

The cosecant function

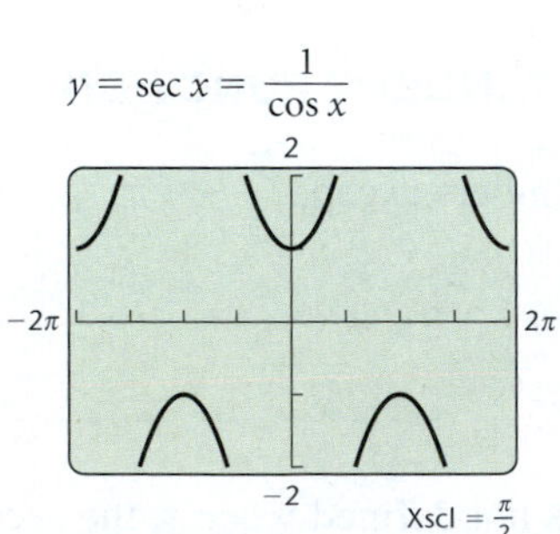

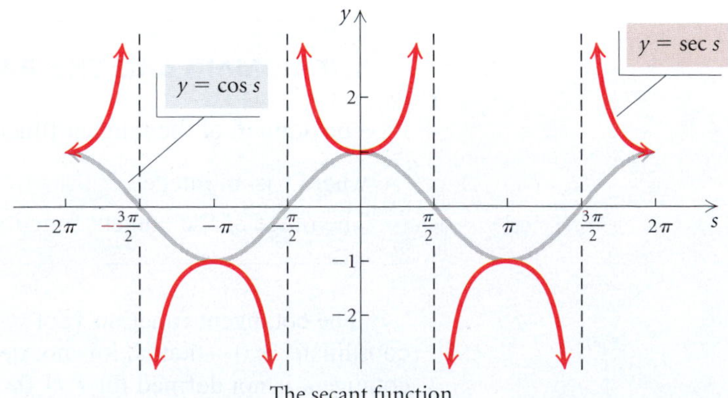

The secant function

The following is a summary of the basic properties of the tangent, cotangent, cosecant, and secant functions. These functions are continuous except where they are not defined.

Now Try Exercises 49.

CONNECTING THE CONCEPTS

Comparing the Tangent, Cotangent, Cosecant, and Secant Functions

TANGENT FUNCTION

1. Period: π
2. Domain: All real numbers except $(\pi/2) + k\pi$, where k is an integer
3. Range: All real numbers

COTANGENT FUNCTION

1. Period: π
2. Domain: All real numbers except $k\pi$, where k is an integer
3. Range: All real numbers

COSECANT FUNCTION

1. Period: 2π
2. Domain: All real numbers except $k\pi$, where k is an integer
3. Range: $(-\infty, -1] \cup [1, \infty)$

SECANT FUNCTION

1. Period: 2π
2. Domain: All real numbers except $(\pi/2) + k\pi$, where k is an integer
3. Range: $(-\infty, -1] \cup [1, \infty)$

In this chapter, we have used the letter s for arc length and have avoided the letters x and y, which generally represent first and second coordinates, respectively. Nevertheless, we can represent the arc length on a unit circle by any variable, such as s, t, x, or θ. Each arc length determines a point that can be labeled with an ordered pair. The first coordinate of that ordered pair is the cosine of the arc length, and the second coordinate is the sine of the arc length. The identities we have developed hold no matter what symbols are used for variables—for example, $\cos(-s) = \cos s$, $\cos(-x) = \cos x$, $\cos(-\theta) = \cos\theta$, and $\cos(-t) = \cos t$.

6.5 Exercise Set

The following points are on the unit circle. Find the coordinates of their reflections across **(a)** *the x-axis,* **(b)** *the y-axis, and* **(c)** *the origin.*

1. $\left(-\frac{3}{4}, \frac{\sqrt{7}}{4}\right)$

2. $\left(\frac{2}{3}, \frac{\sqrt{5}}{3}\right)$

3. $\left(\frac{2}{5}, -\frac{\sqrt{21}}{5}\right)$

4. $\left(-\frac{\sqrt{3}}{2}, -\frac{1}{2}\right)$

5. The number $\pi/4$ determines a point on the unit circle with coordinates $\left(\sqrt{2}/2, \sqrt{2}/2\right)$. What are the coordinates of the point determined by $-\pi/4$?

6. A number β determines a point on the unit circle with coordinates $\left(-2/3, \sqrt{5}/3\right)$. What are the coordinates of the point determined by $-\beta$?

Find the function value using coordinates of points on the unit circle. Give exact answers.

7. $\sin \pi$

8. $\cos\left(-\frac{\pi}{3}\right)$

9. $\cot \frac{7\pi}{6}$

10. $\tan \frac{11\pi}{4}$

11. $\sin(-3\pi)$

12. $\csc \frac{3\pi}{4}$

13. $\cos \frac{5\pi}{6}$

14. $\tan\left(-\frac{\pi}{4}\right)$

15. $\sec \frac{\pi}{2}$

16. $\cos 10\pi$

17. $\cos \frac{\pi}{6}$

18. $\sin \frac{2\pi}{3}$

19. $\sin \frac{5\pi}{4}$

20. $\cos \frac{11\pi}{6}$

21. $\sin(-5\pi)$

22. $\tan \frac{3\pi}{2}$

23. $\cot \frac{5\pi}{2}$

24. $\tan \frac{5\pi}{3}$

Find the function value using a calculator set in RADIAN *mode. Round the answer to four decimal places, where appropriate.*

25. $\tan \frac{\pi}{7}$

26. $\cos\left(-\frac{2\pi}{5}\right)$

27. $\sec 37$

28. $\sin 11.7$

29. $\cot 342$

30. $\tan 1.3$

31. $\cos 6\pi$

32. $\sin \frac{\pi}{10}$

33. $\csc 4.16$

34. $\sec \frac{10\pi}{7}$

35. $\tan \frac{7\pi}{4}$

36. $\cos 2000$

37. $\sin\left(-\frac{\pi}{4}\right)$

38. $\cot 7\pi$

39. $\sin 0$

40. $\cos(-29)$

41. $\tan \frac{2\pi}{9}$

42. $\sin \frac{8\pi}{3}$

In Exercises 43–46, recall that the graph of $y = f(-x)$ is a reflection of the graph of $y = f(x)$ across the y-axis and that the graph of $y = -f(x)$ is a reflection of the graph of $y = f(x)$ across the x-axis.

43. a) Sketch a graph of $y = \sin x$.
b) By reflecting the graph in part (a), sketch a graph of $y = \sin(-x)$.
c) By reflecting the graph in part (a), sketch a graph of $y = -\sin x$.
d) How do the graphs in parts (b) and (c) compare?

44. a) Sketch a graph of $y = \cos x$.
b) By reflecting the graph in part (a), sketch a graph of $y = \cos(-x)$.
c) By reflecting the graph in part (a), sketch a graph of $y = -\cos x$.
d) How do the graphs in parts (a) and (b) compare?

45. a) Sketch a graph of $y = \tan x$.
b) By reflecting the graph of part (a), sketch a graph of $y = \tan(-x)$.
c) By reflecting the graph of part (a), sketch a graph of $y = -\tan x$.
d) How do the graphs of parts (b) and (c) compare?

46. a) Sketch a graph of $y = \sec x$.
b) By reflecting the graph of part (a), sketch a graph of $y = \sec(-x)$.
c) By reflecting the graph of part (a), sketch a graph of $y = -\sec x$.
d) How do the graphs of parts (a) and (b) compare?

In Exercises 47–50, recall that the graph of $y = f(x - d)$ is the graph of $y = f(x)$ shifted right d units and that the graph of $y = f(x + d)$ is the graph of $y = f(x)$ shifted left d units.

47. a) Sketch a graph of $y = \sin x$.
b) By translating, sketch a graph of $y = \sin(x + \pi)$.
c) By reflecting the graph of part (a), sketch a graph of $y = -\sin x$.
d) How do the graphs of parts (b) and (c) compare?

48. a) Sketch a graph of $y = \sin x$.
b) By translating, sketch a graph of $y = \sin(x - \pi)$.
c) By reflecting the graph of part (a), sketch a graph of $y = -\sin x$.
d) How do the graphs of parts (b) and (c) compare?

49. a) Sketch a graph of $y = \cos x$.
b) By translating, sketch a graph of $y = \cos(x + \pi)$.
c) By reflecting the graph of part (a), sketch a graph of $y = -\cos x$.
d) How do the graphs of parts (b) and (c) compare?

50. a) Sketch a graph of $y = \cos x$.
b) By translating, sketch a graph of $y = \cos(x - \pi)$.
c) By reflecting the graph of part (a), sketch a graph of $y = -\cos x$.
d) How do the graphs of parts (b) and (c) compare?

51. Of the six circular functions, which are even? Which are odd?

52. Of the six circular functions, which have period π? Which have period 2π?

Consider the coordinates on the unit circle for Exercises 53–56.

53. In which quadrants is the tangent function positive? negative?

54. In which quadrants is the sine function positive? negative?

55. In which quadrants is the cosine function positive? negative?

56. In which quadrants is the cosecant function positive? negative?

Skill Maintenance

Graph both functions in the same viewing window and describe how g is a transformation of f. [2.4]

57. $f(x) = x^2,\ g(x) = 2x^2 - 3$

58. $f(x) = x^2,\ g(x) = (x - 2)^2$

59. $f(x) = |x|,\ g(x) = \frac{1}{2}|x - 4| + 1$

60. $f(x) = x^3,\ g(x) = -x^3$

Write an equation for a function that has a graph with the given characteristics. Check using a graphing calculator.

61. The shape of $y = x^3$, but reflected across the x-axis, shifted right 2 units, and shifted down 1 unit [2.4]

62. The shape of $y = 1/x$, but shrunk vertically by a factor of $\frac{1}{4}$ and shifted up 3 units [2.4]

Synthesis

Complete. (For example, $\sin(x + 2\pi) = \sin x$.*)*

63. $\sin(x + 2k\pi)$, k an integer $=$ ______

64. $\cos(x + 2k\pi)$, k an integer $=$ ______

65. $\sin(\pi - x) =$ ______

66. $\cos(\pi - x) =$ ______

67. $\cos(x - \pi) =$ ______

68. $\cos(x + \pi) =$ ______

69. $\sin(x + \pi) =$ ______

70. $\sin(x - \pi) =$ ______

71. Find all numbers x that satisfy the following. Check using a graphing calculator.

a) $\sin x = 1$
b) $\cos x = -1$
c) $\sin x = 0$

72. Find $f \circ g$ and $g \circ f$, where $f(x) = x^2 + 2x$ and $g(x) = \cos x$.

Use a graphing calculator to determine the domain, the range, the period, and the amplitude of the function.

73. $y = (\sin x)^2$

74. $y = |\cos x| + 1$

Determine the domain of the function.

75. $f(x) = \sqrt{\cos x}$

76. $g(x) = \dfrac{1}{\sin x}$

77. $f(x) = \dfrac{\sin x}{\cos x}$

78. $g(x) = \log(\sin x)$

Graph.

79. $y = 3 \sin x$

80. $y = \sin |x|$

81. $y = \sin x + \cos x$

82. $y = |\cos x|$

83. One of the motivations for developing trigonometry with a unit circle is that you can actually "see" $\sin\theta$ and $\cos\theta$ on the circle. Note in the following figure that $AP = \sin\theta$ and $OA = \cos\theta$. It turns out that you can also "see" the other four trigonometric functions. Prove each of the following.

a) $BD = \tan\theta$ **b)** $OD = \sec\theta$
c) $OE = \csc\theta$ **d)** $CE = \cot\theta$

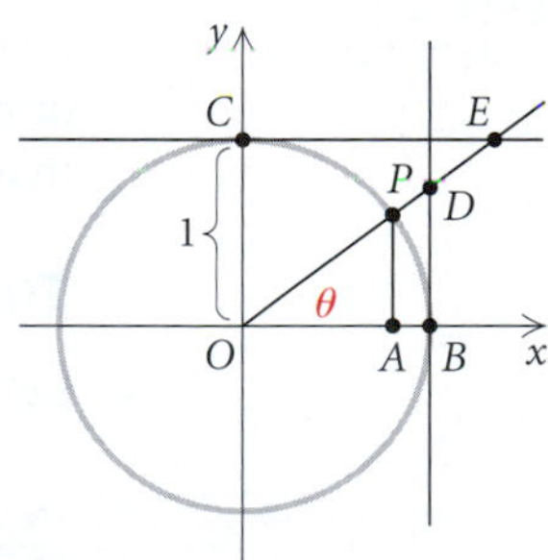

84. Using graphs, determine all numbers x that satisfy

$$\sin x < \cos x.$$

85. Using a calculator, consider $(\sin x)/x$, where x is between 0 and $\pi/2$. As x approaches 0, this function approaches a limiting value. What is it?

6.6 Graphs of Transformed Sine Functions and Cosine Functions

- Graph transformations of $y = \sin x$ and $y = \cos x$ in the form

  $$y = A \sin (Bx - C) + D$$

  and

  $$y = A \cos (Bx - C) + D$$

  and determine the amplitude, the period, and the phase shift.
- Graph sums of functions.
- Graph functions (damped oscillations) found by multiplying trignometric functions by other functions.

Variations of Basic Graphs

In Section 6.5, we graphed all six trigonometric functions. In this section, we will consider variations of the graphs of the sine and the cosine functions. For example, we will graph equations like the following:

$$y = 5 \sin \tfrac{1}{2}x, \qquad y = \cos (2x - \pi), \quad \text{and} \quad y = \tfrac{1}{2}\sin x - 3.$$

In particular, we are interested in graphs of functions in the form

$$y = A \sin (Bx - C) + D$$

and

$$y = A \cos (Bx - C) + D,$$

TRANSFORMATIONS OF FUNCTIONS

REVIEW SECTION 2.5.

where A, B, C, and D are constants. These constants have the effect of translating, reflecting, stretching, and shrinking the basic graphs. Let's first examine the effect of each constant individually. Then we will consider the combined effects of more than one constant.

The Constant D

Let's observe the effect of the constant D in the following graphs.

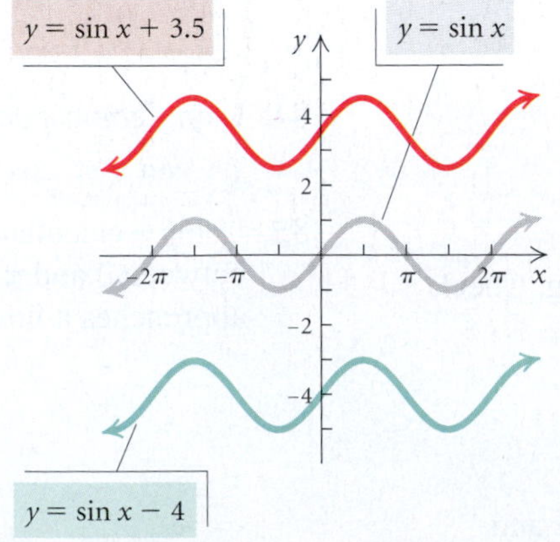

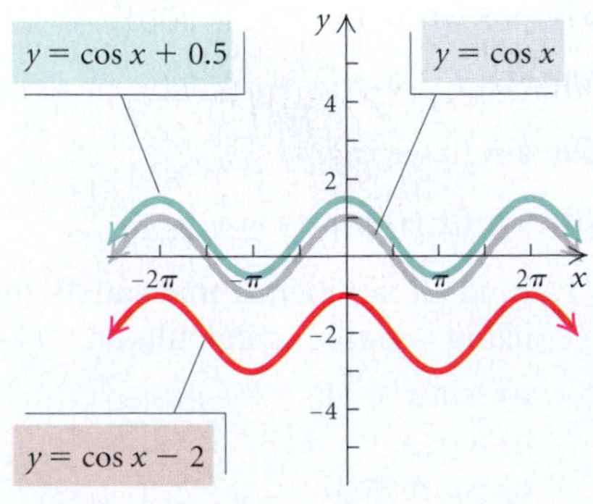

Just in Time 4

THE CONSTANT D: TRANSLATING VERTICALLY

The constant D in $y = A \sin(Bx - C) + D$ and $y = A \cos(Bx - C) + D$ *translates* the graphs *vertically* up D units if $D > 0$ or down $|D|$ units if $D < 0$.

EXAMPLE 1 Sketch a graph of $y = \sin x + 3$.

Solution The graph of $y = \sin x + 3$ is a *vertical* translation of the graph of $y = \sin x$ up 3 units. One way to sketch the graph is to first consider $y = \sin x$ on an interval of length 2π, say, $[0, 2\pi]$. The zeros of the function and the maximum and minimum values can be considered key points. These are

$$(0, 0), \quad \left(\frac{\pi}{2}, 1\right), \quad (\pi, 0), \quad \left(\frac{3\pi}{2}, -1\right), \quad (2\pi, 0).$$

These key points are translated up 3 units to obtain the key points of the graph of $y = \sin x + 3$. These are

$$(0, 3), \quad \left(\frac{\pi}{2}, 4\right), \quad (\pi, 3), \quad \left(\frac{3\pi}{2}, 2\right), \quad (2\pi, 3).$$

The graph of $y = \sin x + 3$ can be sketched on the interval $[0, 2\pi]$ and extended to obtain the rest of the graph by repeating the graph on intervals of length 2π.

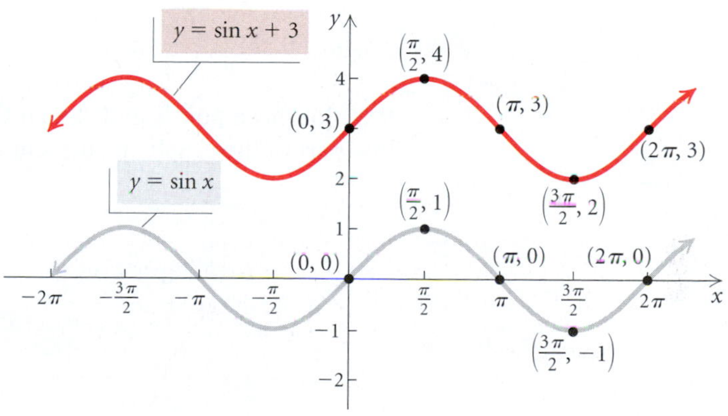

The Constant A

Next, we consider the effect of the constant A. What can we observe in the following graphs? What is the effect of the constant A on the graph of the basic function when **(a)** $0 < A < 1$? **(b)** $A > 1$? **(c)** $-1 < A < 0$? **(d)** $A < -1$?

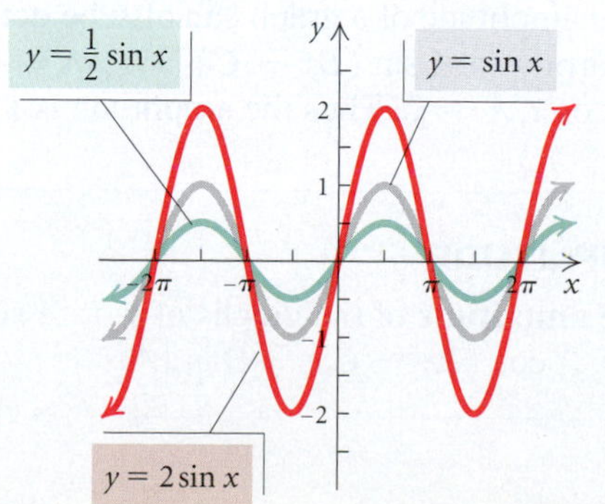

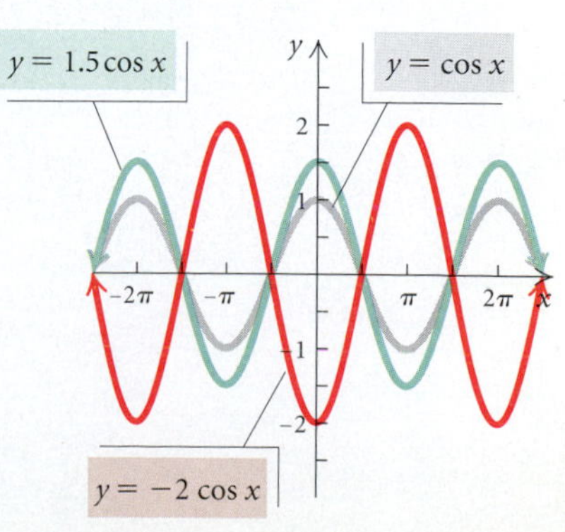

THE CONSTANT A: VERTICAL STRETCHING AND SHRINKING

The constant A in $y = A \sin(Bx - C) + D$ and $y = A \cos(Bx - C) + D$ *stretches* or *shrinks* the graphs *vertically*. If $|A| > 1$, then there will be a vertical stretching. If $|A| < 1$, then there will be a vertical shrinking. If $A < 0$, the graph is also *reflected* across the x-axis.

EXAMPLE 2 Sketch a graph of $y = 2 \cos x$. What is the amplitude?

Solution The constant 2 in $y = 2 \cos x$ has the effect of stretching the graph of $y = \cos x$ vertically by a factor of 2. Since the function values of $y = \cos x$ are such that $-1 \le \cos x \le 1$, the function values of $y = 2 \cos x$ are such that $-2 \le 2 \cos x \le 2$. The maximum value of $y = 2 \cos x$ is 2, and the minimum value is -2. Thus the **amplitude** is

AMPLITUDE

REVIEW SECTION 6.5.

$$\tfrac{1}{2}|2 - (-2)|, \text{ or } 2.$$

We draw the graph of $y = \cos x$ and consider its key points,

$$(0, 1), \quad \left(\frac{\pi}{2}, 0\right), \quad (\pi, -1), \quad \left(\frac{3\pi}{2}, 0\right), \quad (2\pi, 1),$$

on the interval $[0, 2\pi]$.

We then multiply the second coordinates by 2 to obtain the key points of $y = 2 \cos x$. These are

$$(0, 2), \quad \left(\frac{\pi}{2}, 0\right), \quad (\pi, -2), \quad \left(\frac{3\pi}{2}, 0\right), \quad (2\pi, 2).$$

We plot these points and sketch the graph on the interval $[0, 2\pi]$. Then we repeat this part of the graph on adjacent intervals of length 2π.

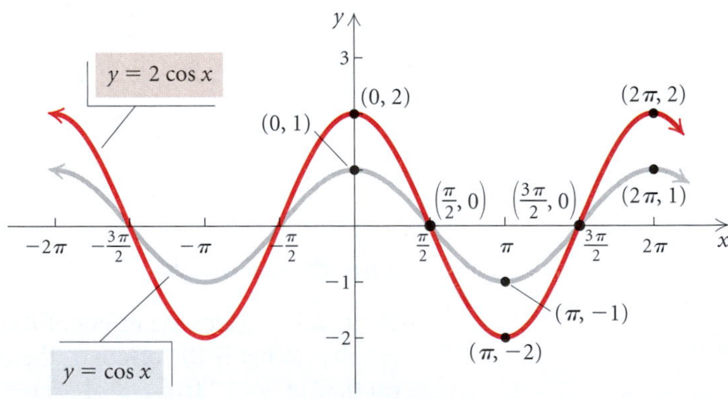

The amplitude of a graph can also be determined by finding $|A|$ from equations of the form $y = A \sin(Bx - C) + D$ or $y = A \cos(Bx - C) + D$. In Example 2, $y = 2 \cos x$, $A = 2$. Thus the amplitude is $|2|$, or 2.

AMPLITUDE

The **amplitude** of the graphs of $y = A \sin(Bx - C) + D$ and $y = A \cos(Bx - C) + D$ is $|A|$.

EXAMPLE 3 Sketch a graph of $y = -\frac{1}{2}\sin x$. What is the amplitude?

Solution The amplitude of the graph is $\left|-\frac{1}{2}\right|$, or $\frac{1}{2}$. The graph of $y = -\frac{1}{2}\sin x$ is a vertical shrinking and a reflection of the graph of $y = \sin x$ across the x-axis. In graphing, the key points of $y = \sin x$,

$$(0, 0),\quad \left(\frac{\pi}{2}, 1\right),\quad (\pi, 0),\quad \left(\frac{3\pi}{2}, -1\right),\quad (2\pi, 0),$$

are transformed to

$$(0, 0),\quad \left(\frac{\pi}{2}, -\frac{1}{2}\right),\quad (\pi, 0),\quad \left(\frac{3\pi}{2}, \frac{1}{2}\right),\quad (2\pi, 0).$$

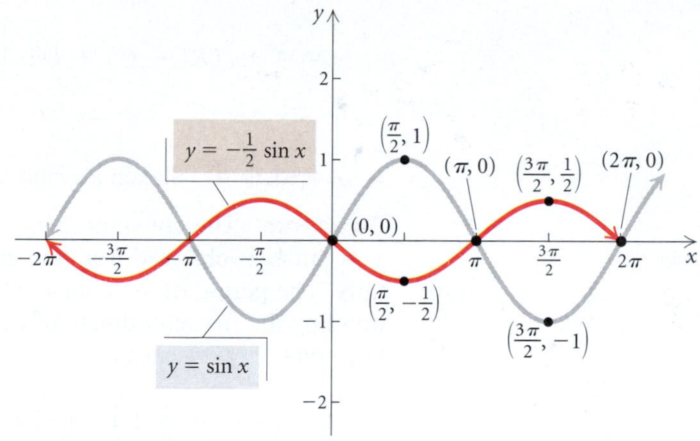

The Constant B

Next, let's consider the effect of the constant B. Changes in the constants A and D *do not* change the period. But what effect, if any, does a change in B have on the period of the function? Let's observe the period of each of the following graphs.

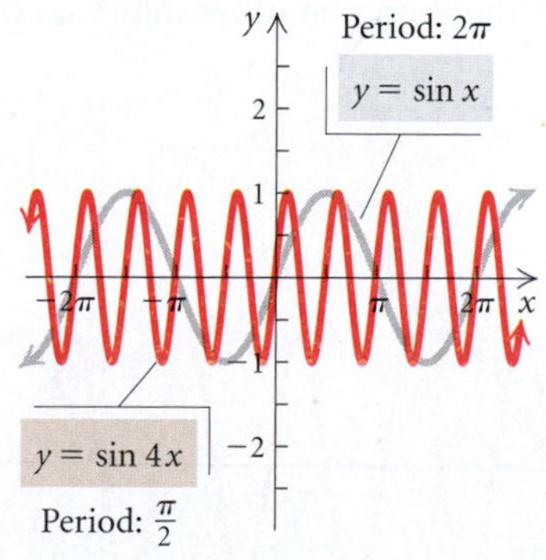

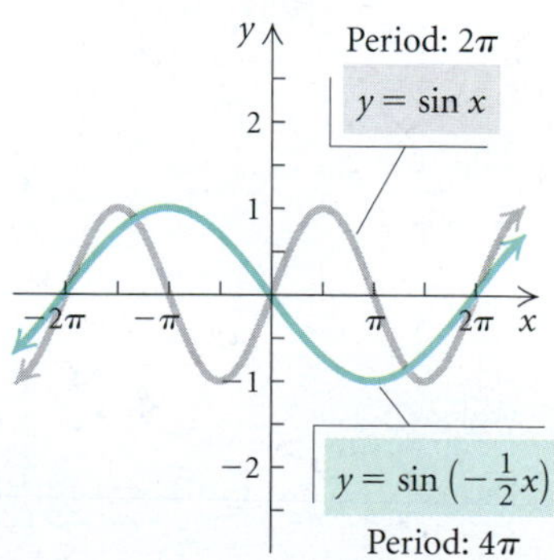

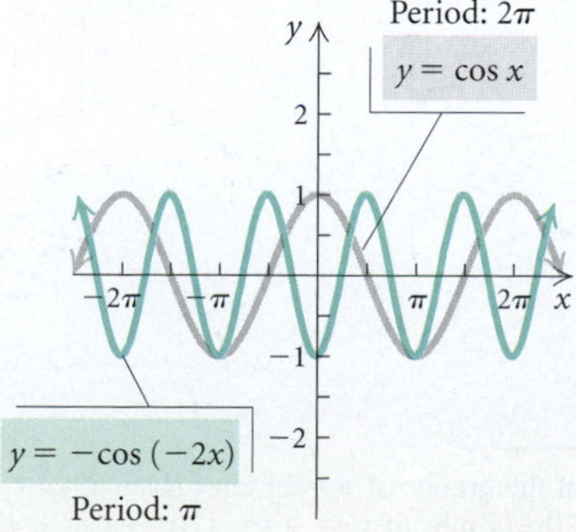

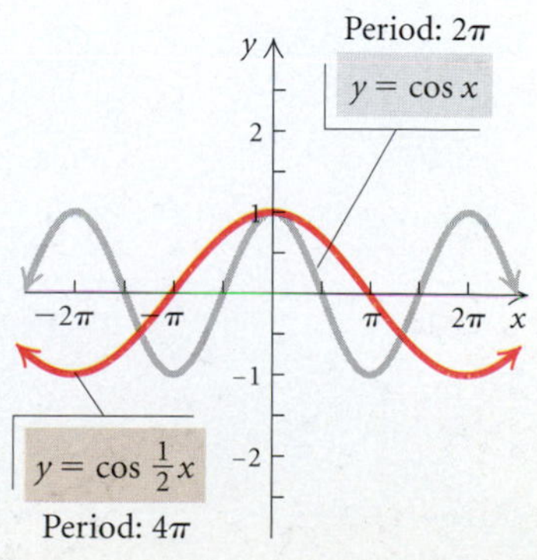

THE CONSTANT B: STRETCHING OR SHRINKING HORIZONTALLY

The constant B in $y = A \sin (Bx - C) + D$ and $y = A \cos (Bx - C) + D$ *stretches* or *shrinks* the graph *horizontally*. If $|B| < 1$, then there will be a horizontal stretching. If $|B| > 1$, then there will be a horizontal shrinking. If $B < 0$, the graph is also *reflected* across the y-axis.

PERIOD

The **period** of the graphs of $y = A \sin (Bx - C) + D$ and $y = A \cos (Bx - C) + D$ is $\left|\frac{2\pi}{B}\right|$.*

EXAMPLE 4 Sketch a graph of $y = \sin 4x$. What is the period?

Solution The constant B has the effect of changing the period. The graph of $y = \sin 4x$ is obtained from the graph of $y = \sin x$ by shrinking the graph horizontally. The period of $y = \sin 4x$ is $|2\pi/4|$, or $\pi/2$. The new graph is obtained by dividing the first coordinate of each ordered-pair solution of $y = \sin 4x$ by 4. The key points of $y = \sin x$ are

$$(0, 0),\quad \left(\frac{\pi}{2}, 1\right),\quad (\pi, 0),\quad \left(\frac{3\pi}{2}, -1\right),\quad (2\pi, 0).$$

These are transformed to the key points of $y = \sin 4x$, which are

$$(0, 0),\quad \left(\frac{\pi}{8}, 1\right),\quad \left(\frac{\pi}{4}, 0\right),\quad \left(\frac{3\pi}{8}, -1\right),\quad \left(\frac{\pi}{2}, 0\right).$$

We plot these key points and sketch in the graph on the shortened interval $[0, \pi/2]$. Then we repeat the graph on other intervals of length $\pi/2$.

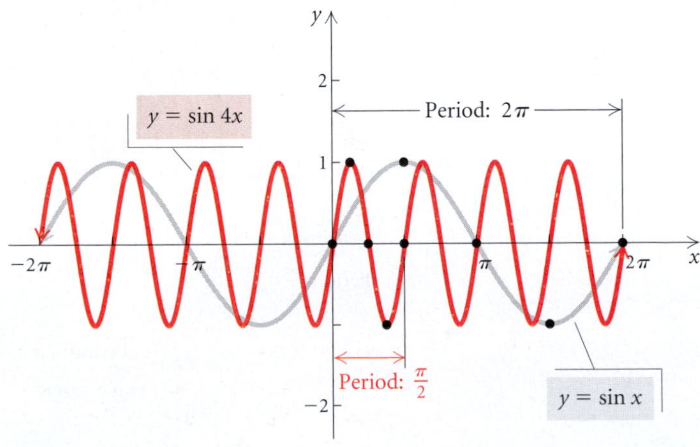

*The period of the graphs of $y = A \tan (Bx - C) + D$ and $y = A \cot (Bx - C) + D$ is $|\pi/B|$. The period of the graphs of $y = A \sec (Bx - C) + D$ and $y = A \csc (Bx - C) + D$ is $|2\pi/B|$.

The Constant C

Next, we examine the effect of the constant C. The curve in each of the following graphs has an amplitude of 1 and a period of 2π, but there are six distinct graphs. What is the effect of the constant C?

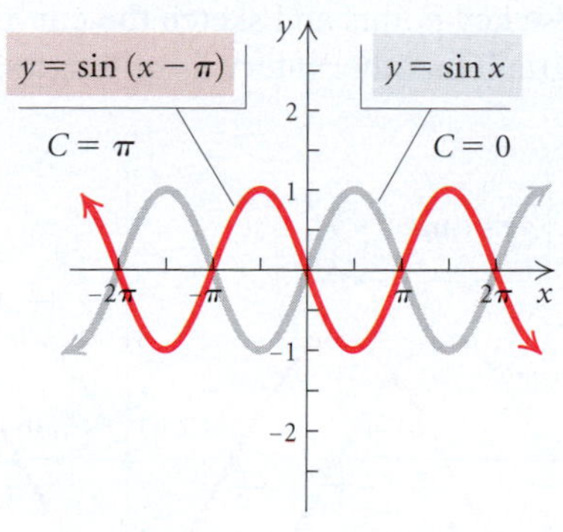

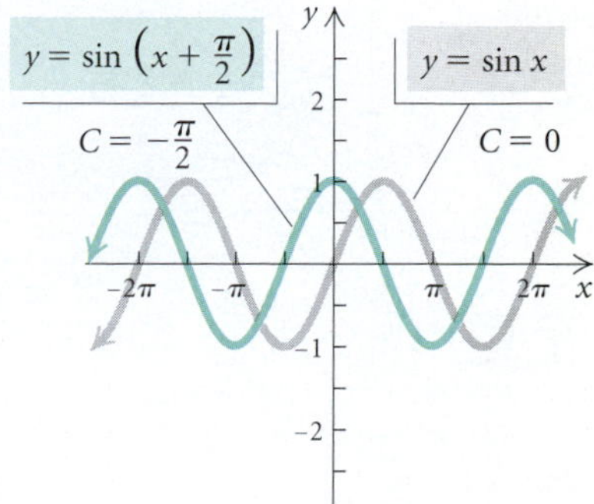

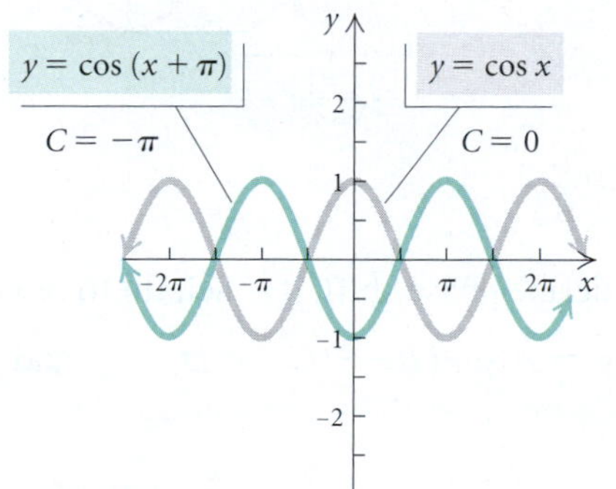

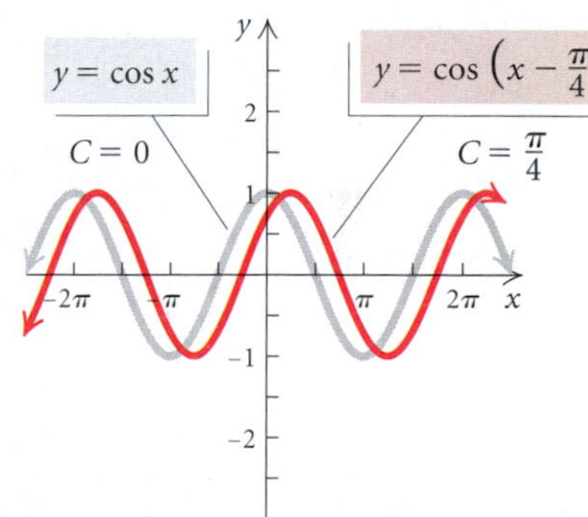

THE CONSTANT C: TRANSLATING HORIZONTALLY $\frac{C}{B}$

For each of the functions of the form

$$y = A\sin(Bx - C) + D \quad \text{and} \quad y = A\cos(Bx - C) + D:$$

- If $B = 1$, then the constant C translates the graph horizontally to the right C units if $C > 0$ and to the left $|C|$ units if $C < 0$.
- If $B \neq 1$ and $B > 0$, then $\frac{C}{B}$ translates the graph horizontally to the right $\frac{C}{B}$ units if $\frac{C}{B} > 0$ and to the left $\left|\frac{C}{B}\right|$ units if $\frac{C}{B} < 0$.

EXAMPLE 5 Sketch a graph of $y = \sin\left(x - \frac{\pi}{2}\right)$.

Solution The amplitude is 1, and the period is 2π. The graph of $y = \sin(x - \pi/2)$ is obtained from the graph of $y = \sin x$ by translating the graph horizontally—to the right C units if $C > 0$ and to the left $|C|$ units if $C < 0$. The graph of $y = \sin(x - \pi/2)$ is a translation of the graph of $y = \sin x$ to the right $\pi/2$ units. The value $\pi/2$ is called the **phase shift**. The key points of $y = \sin x$,

$$(0, 0), \quad \left(\frac{\pi}{2}, 1\right), \quad (\pi, 0), \quad \left(\frac{3\pi}{2}, -1\right), \quad (2\pi, 0),$$

are transformed by adding $\pi/2$ to each of the first coordinates to obtain the following key points of $y = \sin(x - \pi/2)$:

$$\left(\frac{\pi}{2}, 0\right), \quad (\pi, 1), \quad \left(\frac{3\pi}{2}, 0\right), \quad (2\pi, -1), \quad \left(\frac{5\pi}{2}, 0\right).$$

We plot these key points and sketch the curve on the interval $[\pi/2, 5\pi/2]$. Then we repeat the graph on other intervals of length 2π.

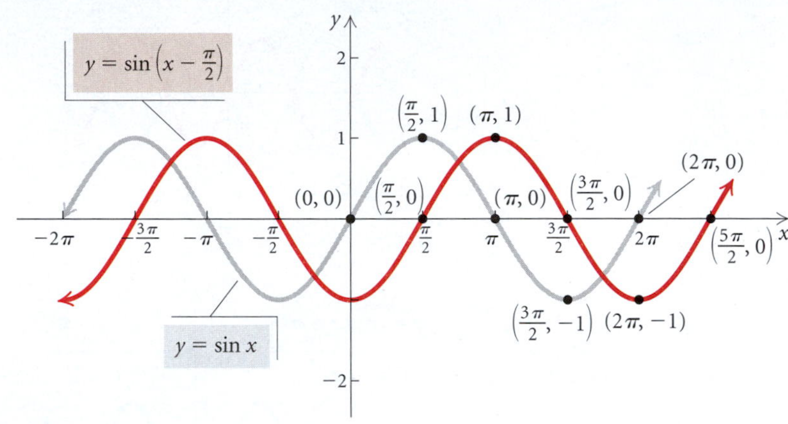

To define phase shift, it is helpful to rewrite

$$y = A\sin(Bx - C) + D \quad \text{and} \quad y = A\cos(Bx - C) + D$$

as

$$y = A\sin\left[B\left(x - \frac{C}{B}\right)\right] + D \quad \text{and} \quad y = A\cos\left[B\left(x - \frac{C}{B}\right)\right] + D.$$

PHASE SHIFT

The **phase shift** of the graphs

$$y = A\sin(Bx - C) + D = A\sin\left[B\left(x - \frac{C}{B}\right)\right] + D$$

and

$$y = A\cos(Bx - C) + D = A\cos\left[B\left(x - \frac{C}{B}\right)\right] + D \text{ is } \frac{C}{B}.$$

- If $\frac{C}{B} > 0$, then the graph is translated to the right $\frac{C}{B}$ units.
- If $\frac{C}{B} < 0$, then the graph is translated to the left $\left|\frac{C}{B}\right|$ units.

EXAMPLE 6 Sketch a graph of $y = \cos(2x - \pi)$.

Solution The graph of

$$y = \cos(2x - \pi)$$

is the same as the graph of

$$y = 1 \cdot \cos\left[2\left(x - \frac{\pi}{2}\right)\right] + 0.$$

The amplitude is 1. The factor 2 shrinks the period by half, making the period $|2\pi/2|$, or π. The phase shift $\pi/2$ translates the graph of $y = \cos 2x$ to the right $\pi/2$ units. Because $D = 0$, there is no vertical translation. Thus, to form the graph, we first graph $y = \cos x$, followed by $y = \cos 2x$ and then $y = \cos[2(x - \pi/2)]$.

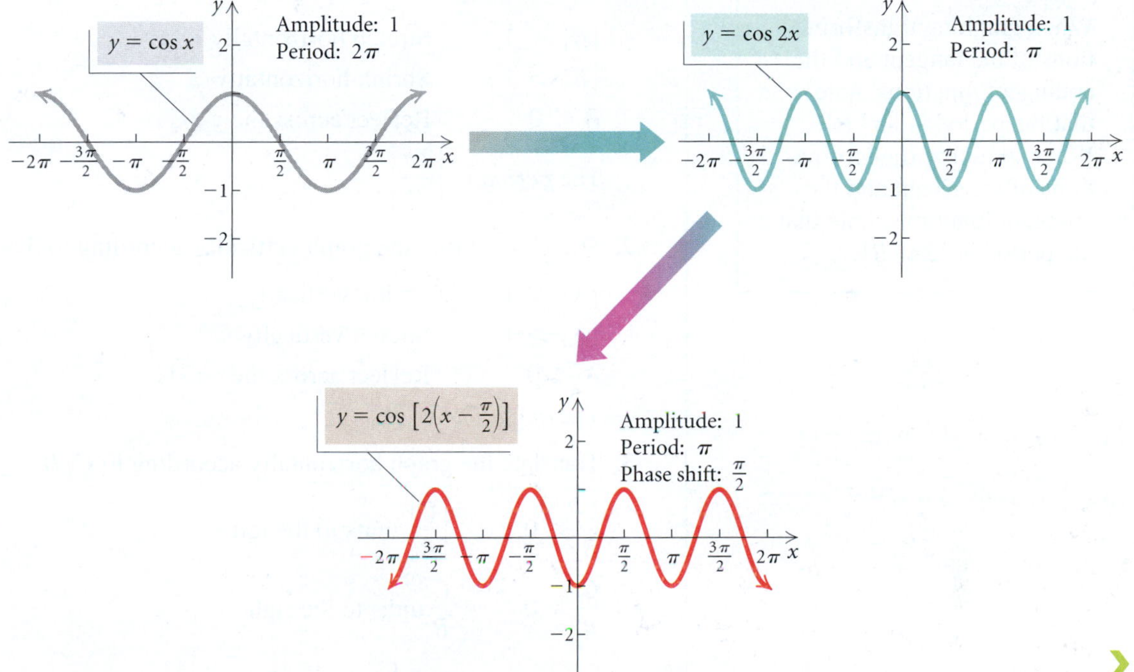

Let's now summarize the effect of the constants. When graphing, we carry out the procedures *in the order* listed. Be sure that the horizontal stretching or shrinking based on the constant B is done before the translation based on the phase shift C/B.

TRANSFORMATIONS OF SINE FUNCTIONS AND COSINE FUNCTIONS

To graph

$$y = A\sin(Bx - C) + D = A\sin\left[B\left(x - \frac{C}{B}\right)\right] + D$$

and

$$y = A\cos(Bx - C) + D = A\cos\left[B\left(x - \frac{C}{B}\right)\right] + D,$$

follow these steps in the order in which they are listed.

1. Stretch or shrink the graph horizontally according to B.

$\lvert B\rvert < 1$	Stretch horizontally
$\lvert B\rvert > 1$	Shrink horizontally
$B < 0$	Reflect across the y-axis

The *period* is $\left|\frac{2\pi}{B}\right|$.

2. Stretch or shrink the graph vertically according to A.

$\lvert A\rvert < 1$	Shrink vertically
$\lvert A\rvert > 1$	Stretch vertically
$A < 0$	Reflect across the x-axis

The *amplitude* is $|A|$.

3. Translate the graph horizontally according to C/B.

$\frac{C}{B} < 0$	$\left\lvert\frac{C}{B}\right\rvert$ units to the left
$\frac{C}{B} > 0$	$\frac{C}{B}$ units to the right

The *phase shift* is $\frac{C}{B}$.

4. Translate the graph vertically according to D.

$D < 0$	$\lvert D\rvert$ units down
$D > 0$	D units up

When graphing transformations of the tangent and the cotangent functions, note that the period is $|\pi/B|$. When graphing transformations of the secant and the cosecant functions, note that the period is $|2\pi/B|$.

EXAMPLE 7 Sketch a graph of $y = 3\sin(2x + \pi/2) + 1$. Find the amplitude, the period, and the phase shift.

Solution We first note that

$$y = 3\sin\left(2x + \frac{\pi}{2}\right) + 1$$

$$= 3\sin\left[2x - \left(-\frac{\pi}{2}\right)\right] + 1 \qquad y = A\sin(Bx - C) + D;\ C = -\pi/2$$

$$= 3\sin\left[2\left(x - \left(-\frac{\pi}{4}\right)\right)\right] + 1. \qquad y = A\sin\left[B\left(x - \frac{C}{B}\right)\right] + D;\ \frac{C}{B} = \frac{-\pi/2}{2} = -\frac{\pi}{4}$$

Then we have the following:

$$\text{Amplitude} = |A| = |3| = 3,$$

$$\text{Period} = \left|\frac{2\pi}{B}\right| = \left|\frac{2\pi}{2}\right| = \pi,$$

$$\text{Phase shift} = \frac{C}{B} = \frac{-\pi/2}{2} = -\frac{\pi}{4}.$$

To create the final graph, we begin with the basic sine curve, $y = \sin x$. Then we sketch graphs of each of the following equations in sequence.

1. $y = \sin 2x$
2. $y = 3 \sin 2x$
3. $y = 3 \sin\left[2\left(x - \left(-\frac{\pi}{4}\right)\right)\right]$
4. $y = 3 \sin\left[2\left(x - \left(-\frac{\pi}{4}\right)\right)\right] + 1$

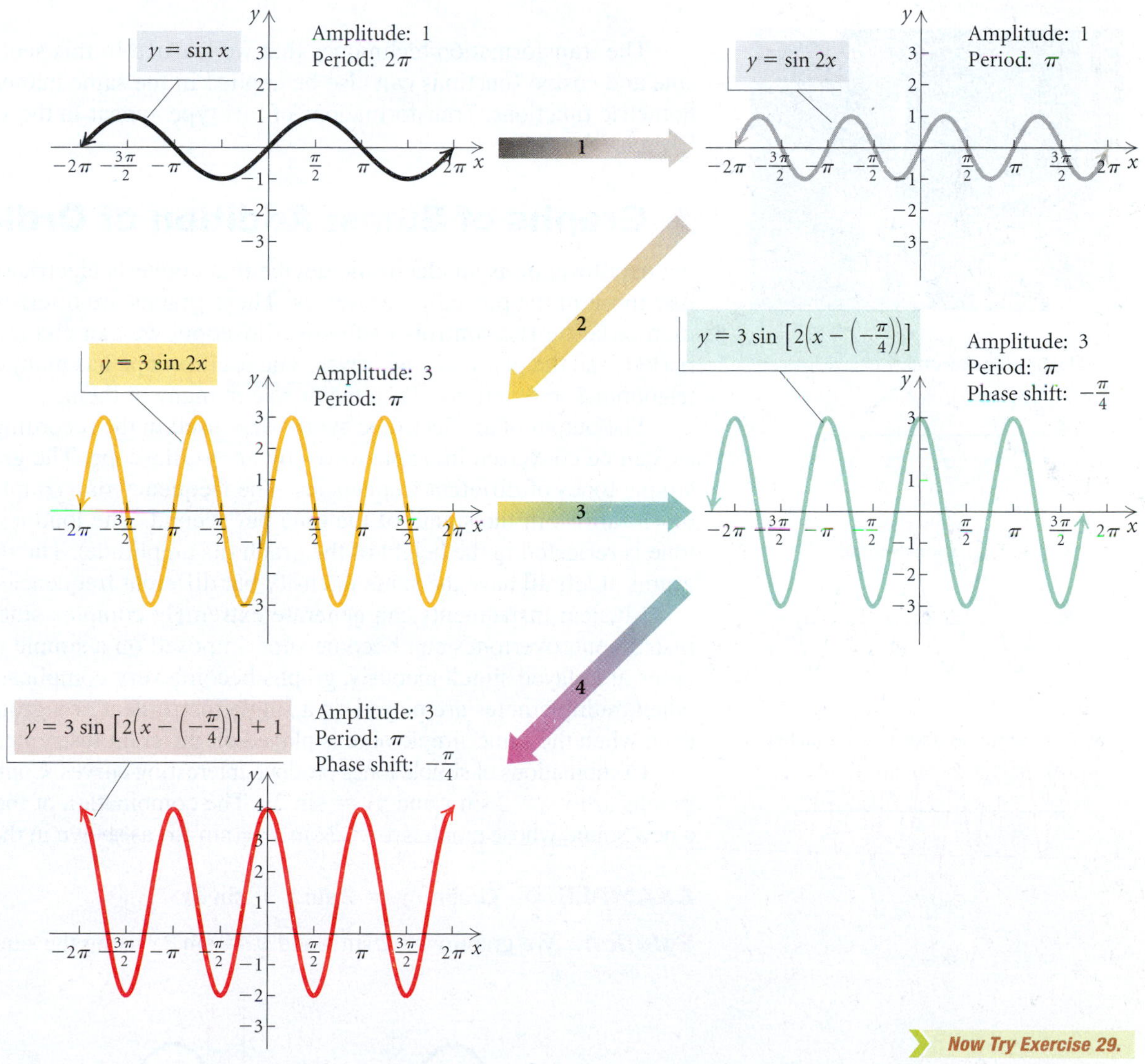

Now Try Exercise 29.

All the graphs in Examples 1–7 can be checked using a graphing calculator. Even though it is faster and more accurate to graph using a calculator, graphing by hand gives us a greater understanding of the effect of changing the constants A, B, C, and D.

Graphing calculators are especially convenient when a period or a phase shift is not a multiple of $\pi/4$.

EXAMPLE 8 Graph $y = 3\cos 2\pi x - 1$. Find the amplitude, the period, and the phase shift.

Solution First we note the following:

$$\text{Amplitude} = |A| = |3| = 3,$$

$$\text{Period} = \left|\frac{2\pi}{B}\right| = \left|\frac{2\pi}{2\pi}\right| = |1| = 1,$$

$$\text{Phase shift} = \frac{C}{B} = \frac{0}{2\pi} = 0.$$

$y = 3\cos(2\pi x) - 1$

There is no phase shift in this case because the constant $C = 0$. The graph has a vertical translation of the graph of the cosine function down 1 unit, an amplitude of 3, and a period of 1, so we can use $[-4, 4, -5, 5]$ as the viewing window.

Now Try Exercise 31.

The transformation techniques that we learned in this section for graphing the sine and cosine functions can also be applied in the same manner to the other trigonometric functions. Transformations of this type appear in the synthesis exercises in Exercise Set 6.6.

Graphs of Sums: Addition of Ordinates

An **oscilloscope** is an electronic device that converts electrical signals into graphs like those in the preceding examples. These graphs are often called sine waves. By manipulating the controls of the oscilloscope, we can change the amplitude, the period, and the phase of sine waves. The oscilloscope has many applications, and the trigonometric functions play a major role in many of them.

The output of an electronic synthesizer used in the recording and playing of music can be converted into sine waves by an oscilloscope. The graphs at left illustrate simple tones of different frequencies. The frequency of a simple tone is the number of vibrations in the signal of the tone per second. The loudness or intensity of the tone is reflected in the height of the graph (its amplitude). The three tones in the diagrams at left all have the same intensity but different frequencies.

220 vibrations/second: 1 octave lower

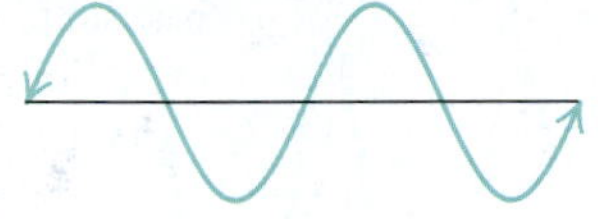

440 vibrations/second

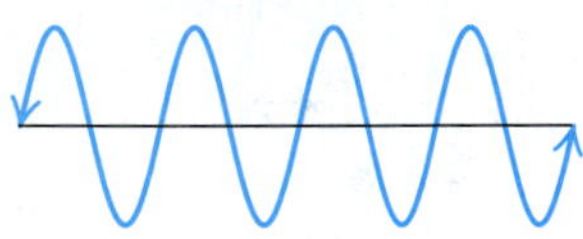

880 vibrations/second: 1 octave higher

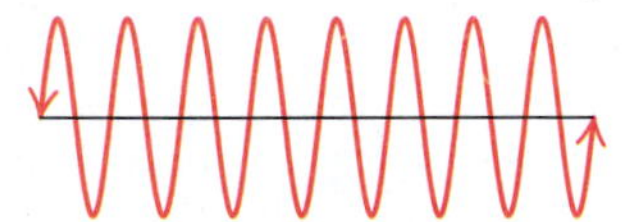

Musical instruments can generate extremely complex sine waves. On a single instrument, overtones can become superimposed on a simple tone. When multiple notes are played simultaneously, graphs become very complicated. This can happen when multiple notes are played on a single instrument or a group of instruments, or even when the same simple note is played on different instruments.

Combinations of simple tones produce interesting curves. Consider two tones whose graphs are $y_1 = 2\sin x$ and $y_2 = \sin 2x$. The combination of the two tones produces a new sound whose graph is $y = 2\sin x + \sin 2x$, as shown in the following example.

EXAMPLE 9 Graph: $y = 2\sin x + \sin 2x$.

Solution We graph $y = 2\sin x$ and $y = \sin 2x$ using the same set of axes.

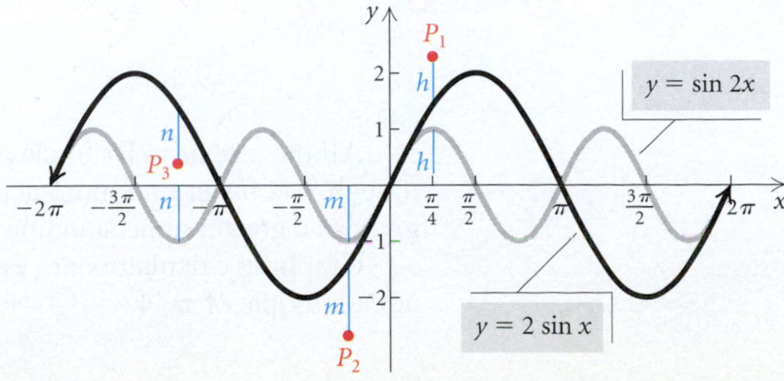

Next, we graphically add some y-coordinates, or ordinates, to obtain points on the graph that we seek. At $x = \pi/4$, we transfer the distance h, which is the value of $\sin 2x$, up to add it to the value of $2 \sin x$. Point P_1 is on the graph that we seek. At $x = -\pi/4$, we use a similar procedure, but this time both ordinates are negative. Point P_2 is on the graph. At $x = -5\pi/4$, we add the negative ordinate of $\sin 2x$ to the positive ordinate of $2 \sin x$. Point P_3 is also on the graph. We continue to plot points in this fashion and then connect them to get the desired graph, shown below. This method is called **addition of ordinates**, because we add the y-values (ordinates) of $y = \sin 2x$ to the y-values (ordinates) of $y = 2 \sin x$. Note that the period of $2 \sin x$ is 2π and the period of $\sin 2x$ is π. The period of the sum $2 \sin x + \sin 2x$ is 2π, the least common multiple of 2π and π.

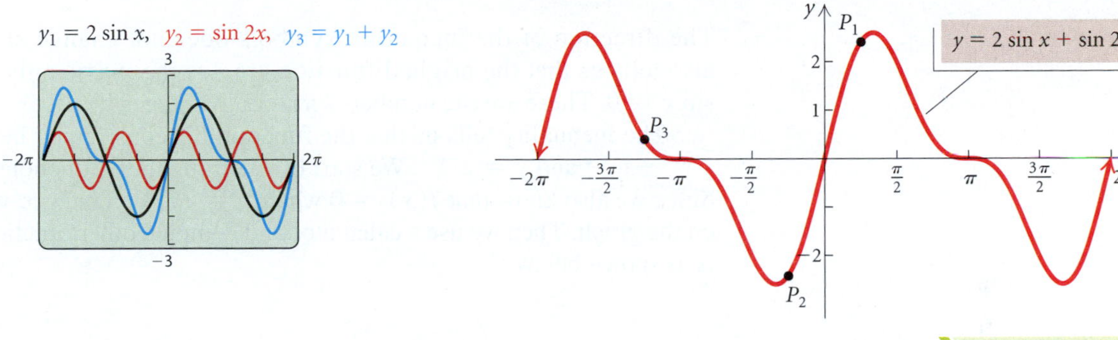

Now Try Exercise 47.

Using a graphing calculator, we can quickly determine the period of a trigonometric function that is a combination of sine and cosine functions.

EXAMPLE 10 Graph $y = 2 \cos x - \sin 3x$ and determine its period.

Solution We graph $y = 2 \cos x - \sin 3x$ with appropriate dimensions. The period appears to be 2π.

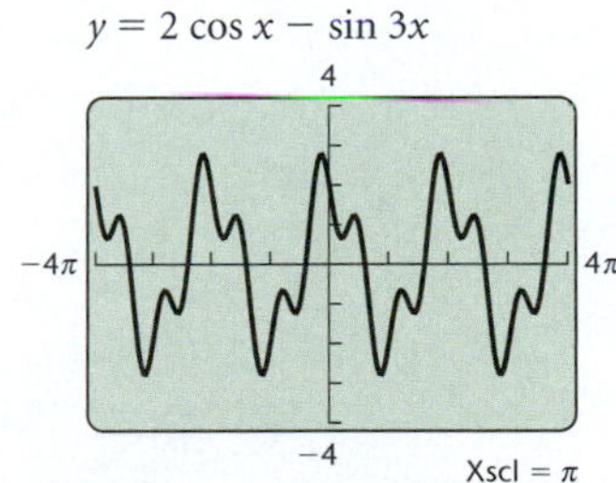

Now Try Exercise 55.

Damped Oscillation: Multiplication of Ordinates

Suppose that a weight is attached to a spring and the spring is stretched and put into motion. The weight oscillates up and down. If we could assume falsely that the weight will bob up and down forever, then its height h after time t, in seconds, might be approximated by a function like

$$h(t) = 5 + 2 \sin (6\pi t).$$

Over a short time period, this might be a valid model, but experience tells us that eventually the spring will come to rest. A more appropriate model is provided by the following example, which illustrates **damped oscillation**.

EXAMPLE 11 Sketch a graph of $f(x) = e^{-x/2}\sin x$.

Solution The function f is the product of two functions g and h, where

$$g(x) = e^{-x/2} \quad \text{and} \quad h(x) = \sin x.$$

Thus, to find function values, we can **multiply ordinates.** Let's do more analysis before graphing. Note that for any real number x,

$$-1 \le \sin x \le 1.$$

Recall from Chapter 5 that all values of the exponential function are positive. Thus we can multiply by $e^{-x/2}$ and obtain the inequality

$$-e^{-x/2} \le e^{-x/2}\sin x \le e^{-x/2}.$$

The direction of the inequality symbols does not change since $e^{-x/2} > 0$. This also tells us that the original function crosses the x-axis only at values for which $\sin x = 0$. These are the numbers $k\pi$, for any integer k.

The inequality tells us that the function f is constrained between the graphs of $y = -e^{-x/2}$ and $y = e^{-x/2}$. We start by graphing these functions using dashed lines. Since we also know that $f(x) = 0$ when $x = k\pi$, k an integer, we mark these points on the graph. Then we use a calculator and compute other function values. The graph is as shown below.

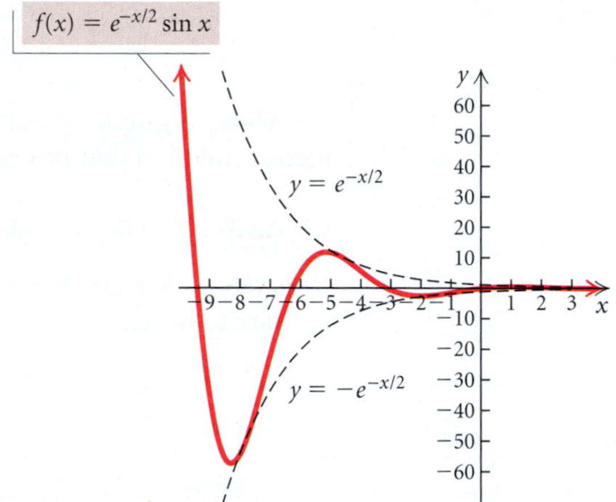

Now Try Exercise 63.

Visualizing the Graph

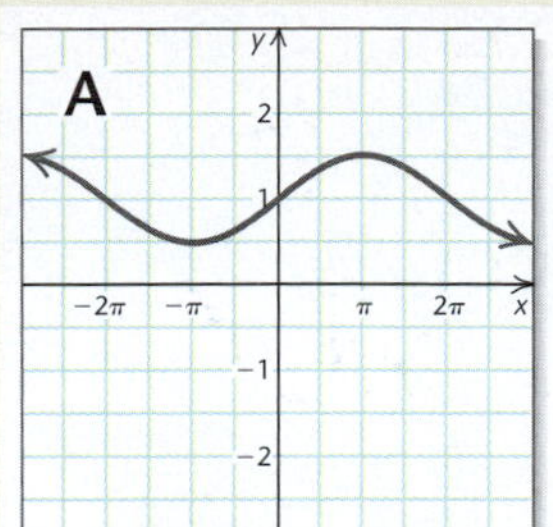

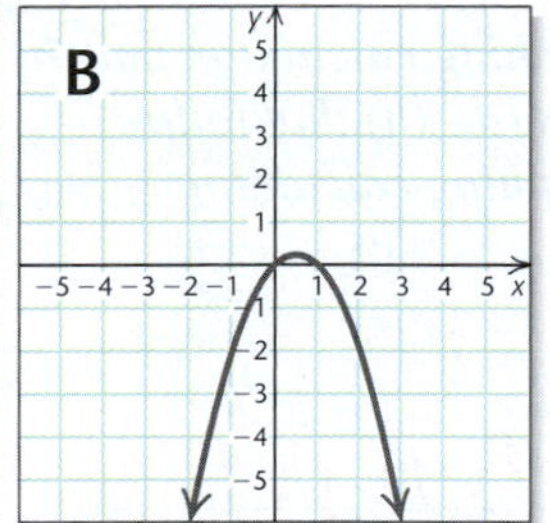

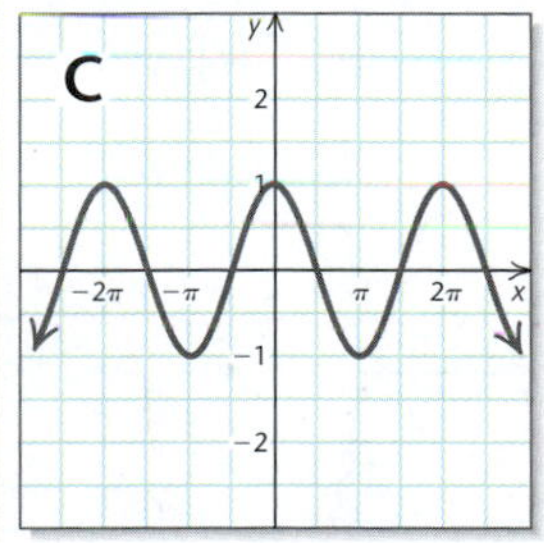

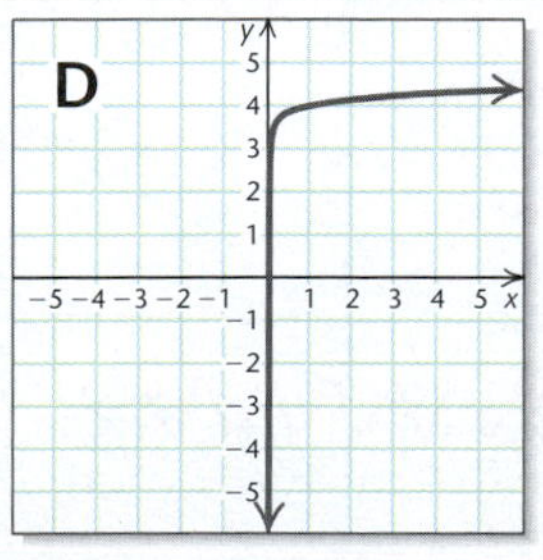

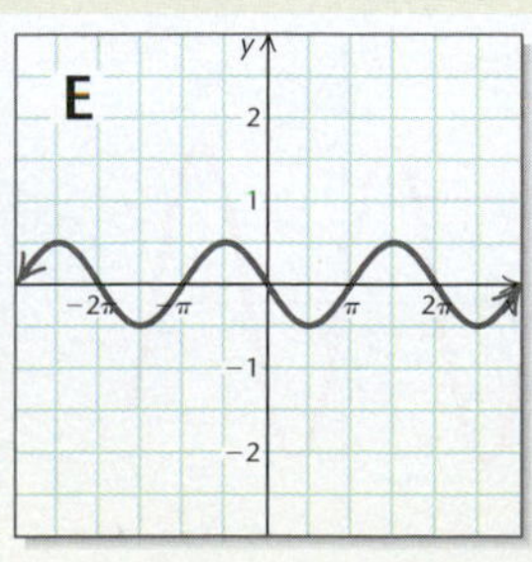

Match the function with its graph.

1. $f(x) = -\sin x$

2. $f(x) = 2x^3 - x + 1$

3. $y = \frac{1}{2}\cos\left(x + \frac{\pi}{2}\right)$

4. $f(x) = \cos\left(\frac{1}{2}x\right)$

5. $y = -x^2 + x$

6. $y = \frac{1}{2}\log x + 4$

7. $f(x) = 2^{x-1}$

8. $f(x) = \frac{1}{2}\sin\left(\frac{1}{2}x\right) + 1$

9. $f(x) = -\cos(x - \pi)$

10. $f(x) = -\frac{1}{2}x^4$

Answers on page A-36

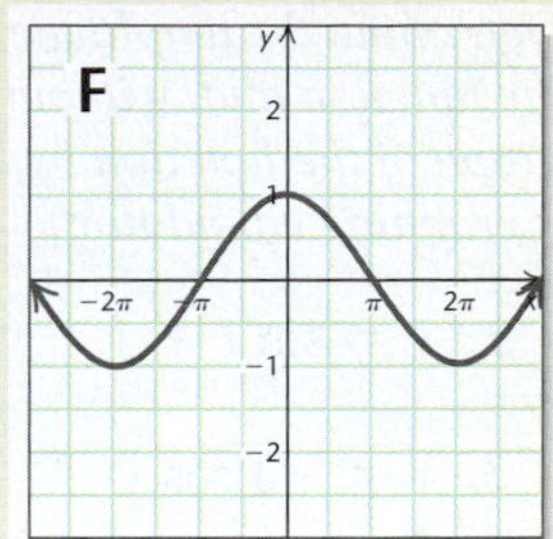

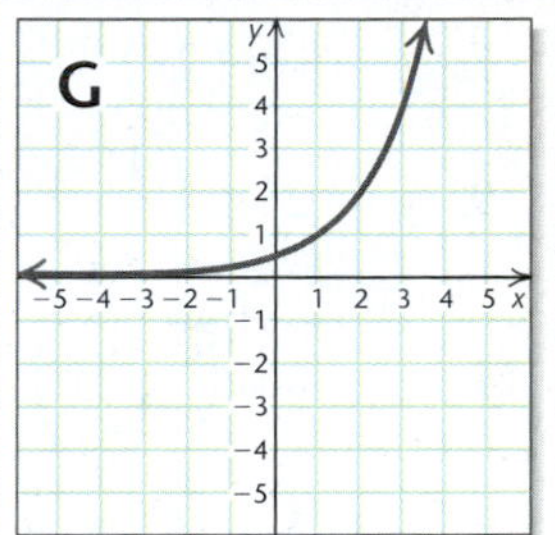

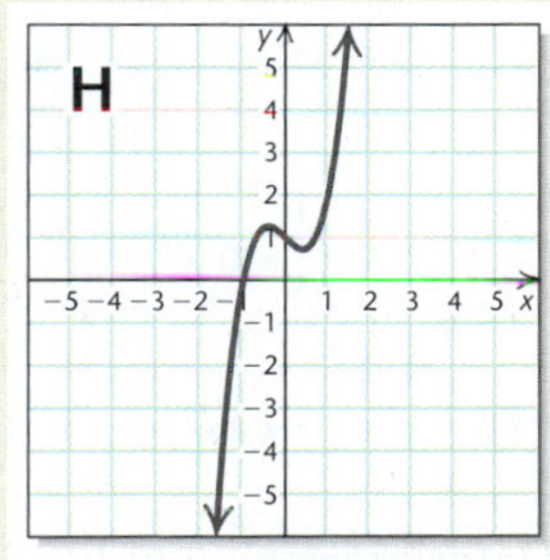

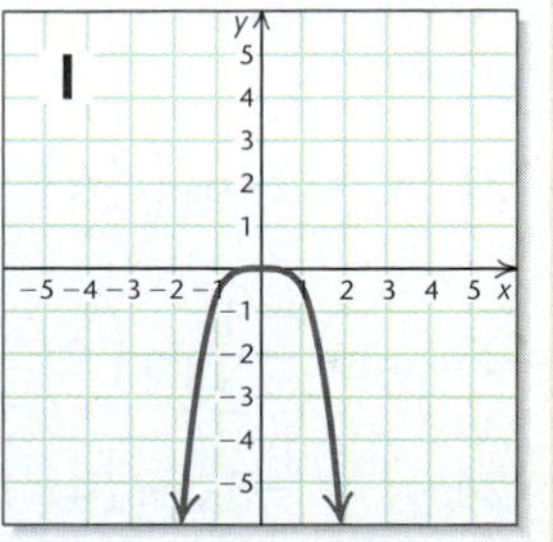

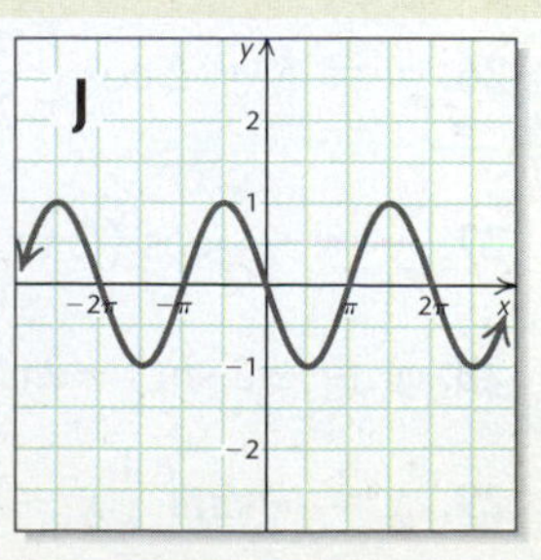

6.6 Exercise Set

Determine the amplitude, the period, and the phase shift of the function and, without a graphing calculator, sketch the graph of the function by hand. Then check the graph using a graphing calculator.

1. $y = \sin x + 1$

2. $y = \frac{1}{4}\cos x$

3. $y = -3\cos x$

4. $y = \sin(-2x)$

5. $y = \frac{1}{2}\cos x$

6. $y = \sin\left(\frac{1}{2}x\right)$

7. $y = \sin(2x)$

8. $y = \cos x - 1$

9. $y = 2\sin\left(\frac{1}{2}x\right)$

10. $y = \cos\left(x - \frac{\pi}{2}\right)$

11. $y = \cos\left(-\frac{1}{2}x\right)$

12. $y = \sin\left(-\frac{1}{4}x\right)$

13. $y = \frac{1}{2}\sin\left(x + \frac{\pi}{2}\right)$

14. $y = \cos x - \frac{1}{2}$

15. $y = 3\cos(x - \pi)$

16. $y = -\sin\left(\frac{1}{4}x\right) + 1$

17. $y = \frac{1}{3}\sin x - 4$

18. $y = \cos\left(\frac{1}{2}x + \frac{\pi}{2}\right)$

19. $y = -\cos(-x) + 2$

20. $y = \frac{1}{2}\sin\left(2x - \frac{\pi}{4}\right)$

Determine the amplitude, the period, and the phase shift of the function. Then check by graphing the function using a graphing calculator. Try to visualize the graph before creating it.

21. $y = 2\cos\left(\frac{1}{2}x - \frac{\pi}{2}\right)$

22. $y = 4\sin\left(\frac{1}{4}x + \frac{\pi}{8}\right)$

23. $y = -\frac{1}{2}\sin\left(2x + \frac{\pi}{2}\right)$

24. $y = -3\cos(4x - \pi) + 2$

25. $y = 2 + 3\cos(\pi x - 3)$

26. $y = 5 - 2\cos\left(\frac{\pi}{2}x + \frac{\pi}{2}\right)$

27. $y = -\frac{1}{2}\cos(2\pi x) + 2$

28. $y = -2\sin(-2x + \pi) - 2$

29. $y = -\sin\left(\frac{1}{2}x - \frac{\pi}{2}\right) + \frac{1}{2}$

30. $y = \frac{1}{3}\cos(-3x) + 1$

31. $y = \cos(-2\pi x) + 2$

32. $y = \frac{1}{2}\sin(2\pi x + \pi)$

33. $y = -\frac{1}{4}\cos(\pi x - 4)$

34. $y = 2\sin(2\pi x + 1)$

In Exercises 35–42, without a graphing calculator, match the function with one of the graphs (a)–(h) that follow. Then check your work using a graphing calculator.

a)

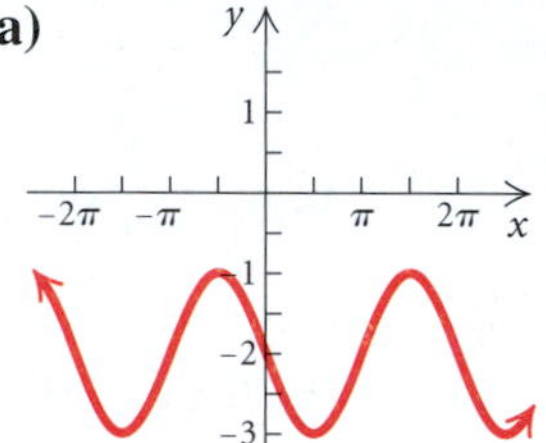

b)

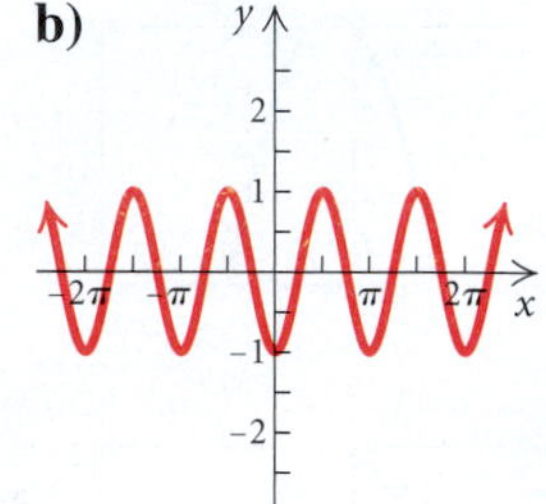

c)

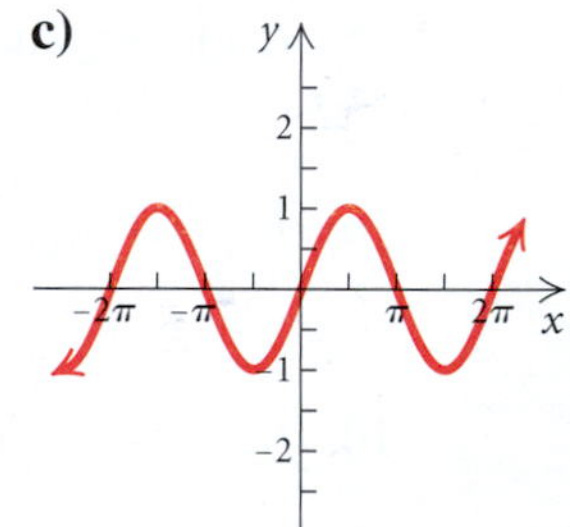

d)

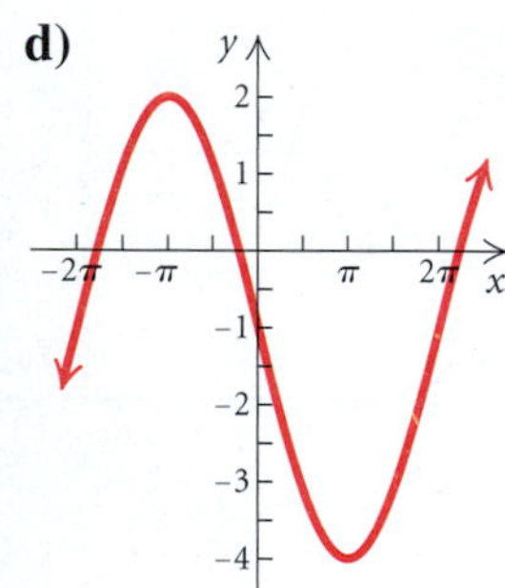

e)

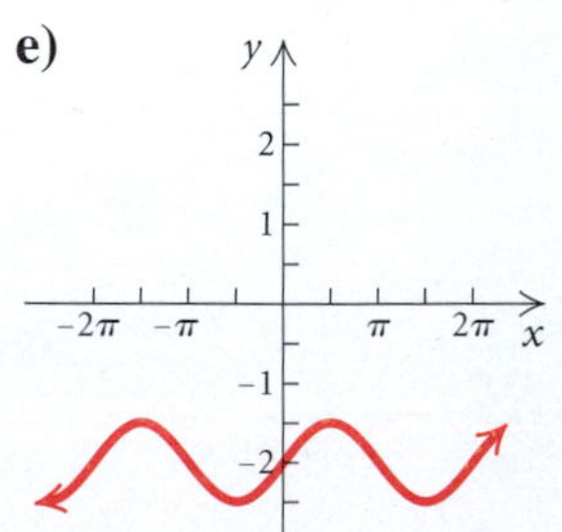

f)

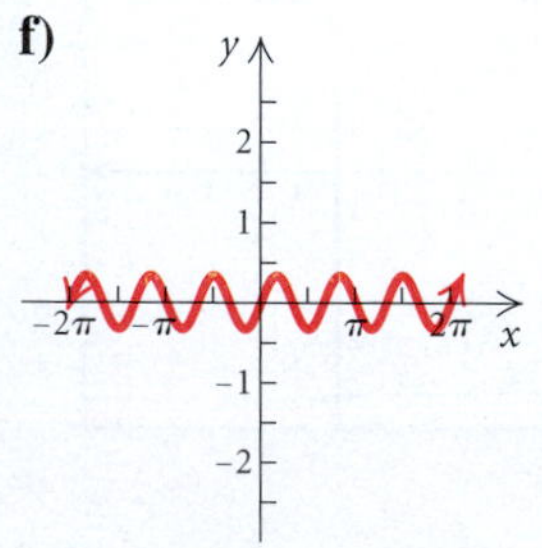

g)

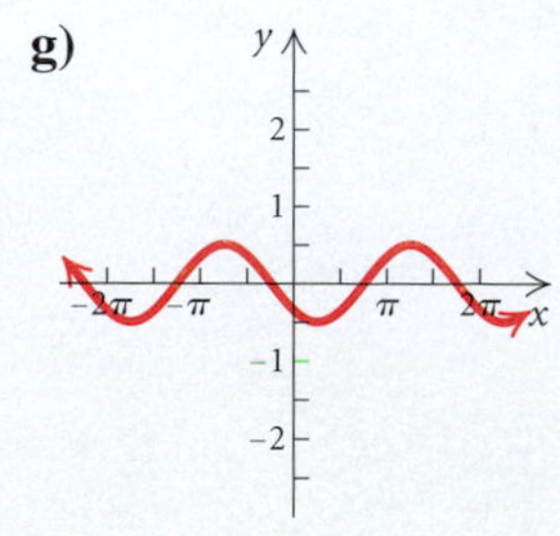

h)

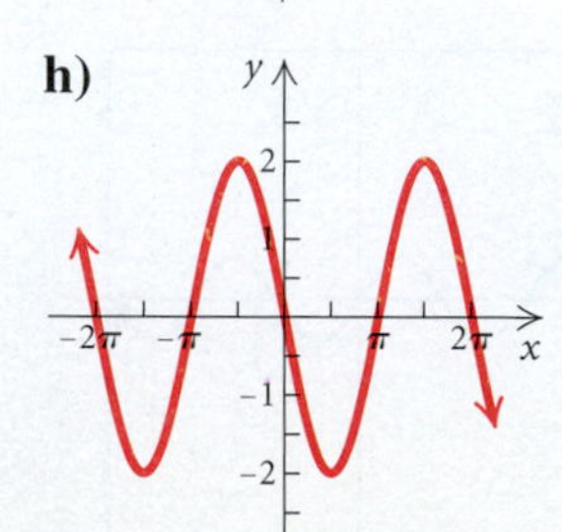

35. $y = -\cos 2x$

36. $y = \frac{1}{2}\sin x - 2$

37. $y = 2\cos\left(x + \frac{\pi}{2}\right)$

38. $y = -3\sin\frac{1}{2}x - 1$

39. $y = \sin(x - \pi) - 2$

40. $y = -\frac{1}{2}\cos\left(x - \frac{\pi}{4}\right)$

41. $y = \frac{1}{3}\sin 3x$

42. $y = \cos\left(x - \frac{\pi}{2}\right)$

In Exercises 43–46, determine the equation of the function that is graphed. Answers may vary.

43.

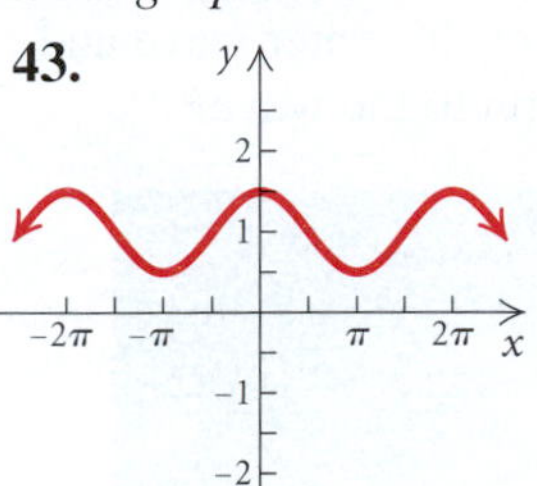

44.

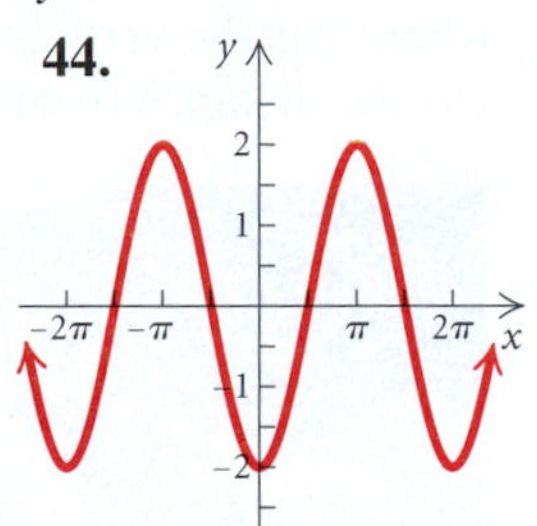

45.

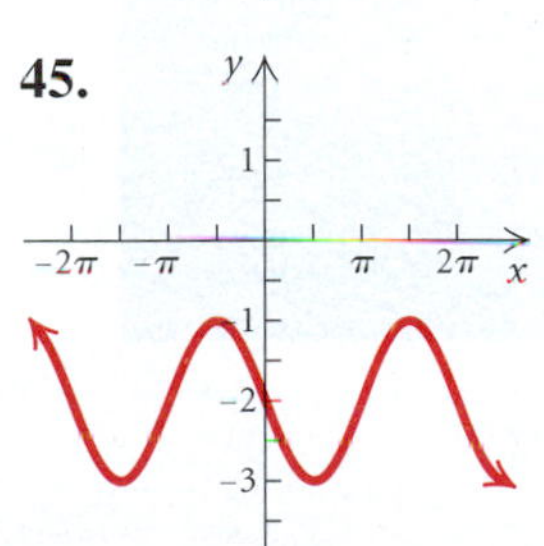

46.

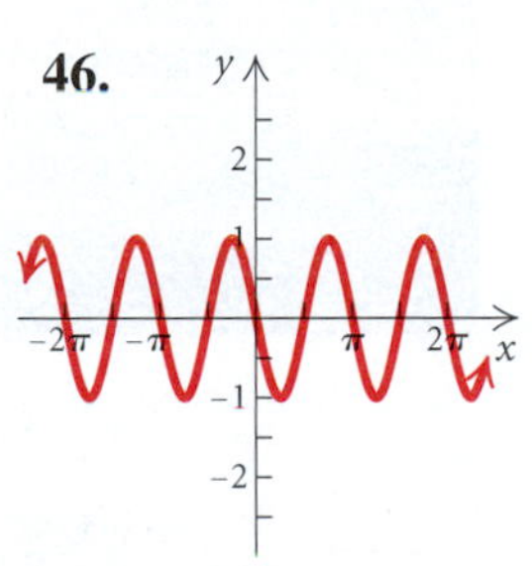

Graph using addition of ordinates. Then check your work using a graphing calculator.

47. $y = 2\cos x + \cos 2x$

48. $y = 3\cos x + \cos 3x$

49. $y = \sin x + \cos 2x$

50. $y = 2\sin x + \cos 2x$

51. $y = \sin x - \cos x$

52. $y = 3\cos x - \sin x$

53. $y = 3\cos x + \sin 2x$

54. $y = 3\sin x - \cos 2x$

Use a graphing calculator to graph the function.

55. $y = x + \sin x$

56. $y = -x - \sin x$

57. $y = \cos x - x$

58. $y = -(\cos x - x)$

59. $y = \cos 2x + 2x$

60. $y = \cos 3x + \sin 3x$

61. $y = 4\cos 2x - 2\sin x$

62. $y = 7.5\cos x + \sin 2x$

Graph each of the following.

63. $f(x) = e^{-x/2}\cos x$

64. $f(x) = e^{-0.4x}\sin x$

65. $f(x) = 0.6x^2\cos x$

66. $f(x) = e^{-x/4}\sin x$

67. $f(x) = x\sin x$

68. $f(x) = |x|\cos x$

69. $f(x) = 2^{-x}\sin x$

70. $f(x) = 2^{-x}\cos x$

Skill Maintenance

Classify the function as linear, quadratic, cubic, quartic, rational, exponential, logarithmic, or trigonometric.

71. $f(x) = \frac{x + 4}{x}$ **[4.5]**

72. $y = \frac{1}{2}\log x - 4$ **[5.3]**

73. $y = x^4 - x - 2$ **[4.1]**

74. $\frac{3}{4}x + \frac{1}{2}y = -5$ **[1.3]**

75. $f(x) = \sin x - 3$ **[6.6]**

76. $f(x) = 0.5e^{x-2}$ **[5.2]**

77. $y = \frac{2}{5}$ **[1.3]**

78. $y = \sin x + \cos x$ **[6.6]**

79. $y = x^2 - x^3$ **[4.1]**

80. $f(x) = \left(\frac{1}{2}\right)^x$ **[5.2]**

Synthesis

Find the maximum and minimum values of the function.

81. $y = 2\cos\left[3\left(x - \frac{\pi}{2}\right)\right] + 6$

82. $y = \frac{1}{2}\sin(2x - 6\pi) - 4$

The transformation techniques that we learned in this section for graphing the sine and cosine functions can also be applied to the other trigonometric functions. Sketch a graph of each of the following. Then check your work using a graphing calculator.

83. $y = -\tan x$

84. $y = \tan(-x)$

85. $y = \csc(-x)$

86. $y = -\cot x$

87. $y = \frac{1}{2}\sec\left(\frac{1}{2}x\right)$

88. $y = -\frac{3}{2}\csc x$

89. $y = -2 + \cot x$

90. $y = -\sec x + 2$

91. $y = 2\tan\left(\frac{1}{2}x\right)$

92. $y = \cot(2x)$

93. $y = 2\sec(x - \pi)$

94. $y = 4\tan\left(\frac{1}{4}x + \frac{\pi}{8}\right)$

95. $y = \cot\left(x + \frac{\pi}{2}\right) - 1$

96. $y = \sec(x + \pi) + 2$

97. $y = 2\csc\left(\frac{1}{2}x - \frac{3\pi}{4}\right)$

98. $y = 4\sec(2x - \pi)$

Use a graphing calculator to graph each of the following on the given interval and approximate the zeros.

99. $f(x) = \dfrac{\sin x}{x}$; $[-12, 12]$

100. $f(x) = \dfrac{\cos x - 1}{x}$; $[-12, 12]$

101. $f(x) = x^3 \sin x$; $[-5, 5]$

102. $f(x) = \dfrac{(\sin x)^2}{x}$; $[-4, 4]$

103. *Temperature During an Illness.* The temperature T of a patient during a 12-day illness is given by

$$T(t) = 101.6° + 3° \sin\left(\frac{\pi}{8}t\right).$$

a) Graph the function on the interval $[0, 12]$.
b) What are the maximum and minimum temperatures during the illness?

104. *Periodic Sales.* A company in a northern climate has sales of skis as given by

$$S(t) = 10\left(1 - \cos\frac{\pi}{6}t\right),$$

where t is the time, in months ($t = 0$ corresponds to July 1), and $S(t)$ is in thousands of dollars.

a) Graph the function on a 12-month interval $[0, 12]$.
b) What is the period of the function?
c) What is the minimum amount of sales and when does it occur?
d) What is the maximum amount of sales and when does it occur?

105. *Satellite Location.* A satellite circles the earth in such a way that it is y miles from the equator (north or south, height not considered) t minutes after its launch, where

$$y(t) = 3000\left[\cos\frac{\pi}{45}(t - 10)\right].$$

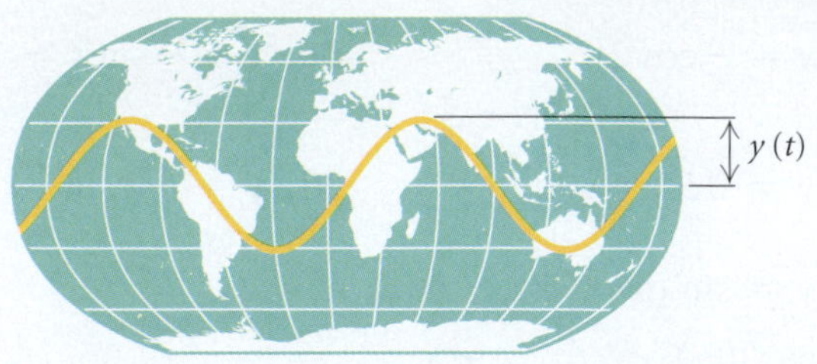

What are the amplitude, the period, and the phase shift?

106. *Water Wave.* The cross-section of a water wave is given by

$$y = 3 \sin\left(\frac{\pi}{4}x + \frac{\pi}{4}\right),$$

where y is the vertical height of the water wave and x is the distance from the origin to the wave.

$y = 3\sin\left(\frac{\pi}{4}x + \frac{\pi}{4}\right)$

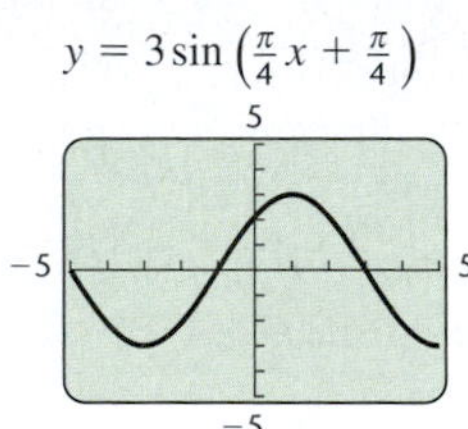

What are the amplitude, the period, and the phase shift?

107. *Damped Oscillations.* Suppose that the motion of a spring is given by

$$d(t) = 6e^{-0.8t}\cos(6\pi t) + 4,$$

where d is the distance, in inches, of a weight from the point at which the spring is attached to a ceiling, after t seconds. How far do you think the spring is from the ceiling when the spring stops bobbing?

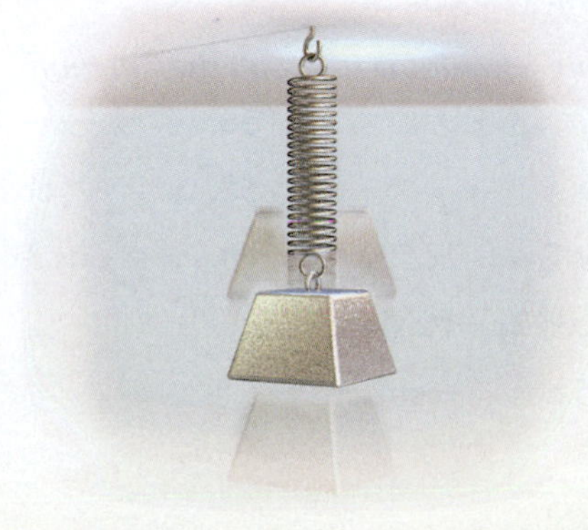

Chapter 6 Summary and Review

STUDY GUIDE

KEY TERMS AND CONCEPTS | **EXAMPLES**

SECTION 6.1: TRIGONOMETRIC FUNCTIONS OF ACUTE ANGLES

Trigonometric Function Values of an Acute Angle θ

Let θ be an acute angle of a right triangle. The six trigonometric functions of θ are as follows:

$$\sin\theta = \frac{\text{opp}}{\text{hyp}}, \qquad \csc\theta = \frac{\text{hyp}}{\text{opp}},$$

$$\cos\theta = \frac{\text{adj}}{\text{hyp}}, \qquad \sec\theta = \frac{\text{hyp}}{\text{adj}},$$

$$\tan\theta = \frac{\text{opp}}{\text{adj}}, \qquad \cot\theta = \frac{\text{adj}}{\text{opp}}.$$

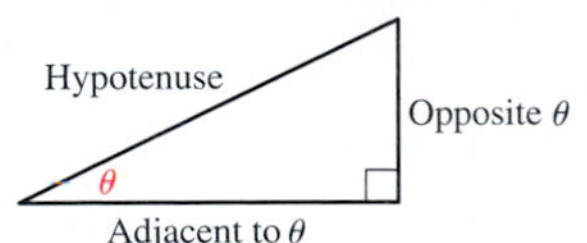

If $\cos\alpha = \frac{3}{8}$ and α is an acute angle, find the other five trigonometric function values of α.

$$\cos\alpha = \frac{3}{8} \begin{array}{l}\longleftarrow \text{adj} \\ \longleftarrow \text{hyp}\end{array}$$

We find the missing length using the Pythagorean equation: $a^2 + b^2 = c^2$.

$$a^2 + 3^2 = 8^2$$
$$a^2 = 64 - 9$$
$$a = \sqrt{55}$$

$$\sin\alpha = \frac{\sqrt{55}}{8}, \qquad \csc\alpha = \frac{8}{\sqrt{55}}, \text{ or } \frac{8\sqrt{55}}{55},$$

$$\cos\alpha = \frac{3}{8}, \qquad \sec\alpha = \frac{8}{3},$$

$$\tan\alpha = \frac{\sqrt{55}}{3}, \qquad \cot\alpha = \frac{3}{\sqrt{55}}, \text{ or } \frac{3\sqrt{55}}{55}$$

Reciprocal Functions

$$\csc\theta = \frac{1}{\sin\theta}, \qquad \sec\theta = \frac{1}{\cos\theta}, \qquad \cot\theta = \frac{1}{\tan\theta}$$

Given that $\sin\beta = \frac{5}{13}$, $\cos\beta = \frac{12}{13}$, and $\tan\beta = \frac{5}{12}$, find $\csc\beta$, $\sec\beta$, and $\cot\beta$.

$$\csc\beta = \frac{13}{5}, \qquad \sec\beta = \frac{13}{12}, \qquad \cot\beta = \frac{12}{5}$$

Function Values of Special Angles

We often use the function values of 30°, 45°, and 60°. Either the following triangles or the values themselves should be memorized.

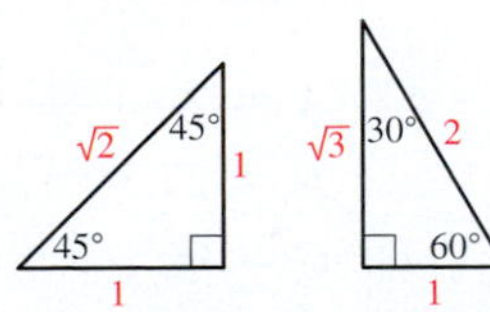

	30°	45°	60°
sin	$1/2$	$\sqrt{2}/2$	$\sqrt{3}/2$
cos	$\sqrt{3}/2$	$\sqrt{2}/2$	$1/2$
tan	$\sqrt{3}/3$	1	$\sqrt{3}$

Find the exact function value.

$$\csc 45^\circ = \frac{2}{\sqrt{2}}, \text{ or } \sqrt{2}, \qquad \tan 60^\circ = \sqrt{3},$$

$$\sin 30^\circ = \frac{1}{2}, \qquad \cot 45^\circ = 1,$$

$$\cos 60^\circ = \frac{1}{2}, \qquad \sec 30^\circ = \frac{2}{\sqrt{3}}, \text{ or } \frac{2\sqrt{3}}{3},$$

$$\cot 30^\circ = \frac{3}{\sqrt{3}}, \text{ or } \sqrt{3}, \qquad \sin 60^\circ = \frac{\sqrt{3}}{2},$$

$$\tan 45^\circ = 1, \qquad \sec 45^\circ = \frac{2}{\sqrt{2}}, \text{ or } \sqrt{2},$$

$$\csc 60^\circ = \frac{2}{\sqrt{3}}, \text{ or } \frac{2\sqrt{3}}{3}, \qquad \cos 45^\circ = \frac{\sqrt{2}}{2}$$

Most calculators can convert D°M′S″ notation to decimal degree notation and vice versa. Procedures among calculators vary. We also can convert without using a calculator.

Convert 17°42′35″ to decimal degree notation, rounding the answer to the nearest hundredth of a degree.

$$17°42'35'' = 17° + 42' + \frac{35'}{60} \approx 17° + 42.5833'$$
$$\approx 17° + \frac{42.5833°}{60} \approx 17.71°$$

Convert 23.12° to D°M′S″ notation.

$$\begin{aligned}23.12° &= 23° + 0.12 \times 1° = 23° + 0.12 \times 60' \\ &= 23° + 7.2' = 23° + 7' + 0.2 \times 1' \\ &= 23° + 7' + 0.2 \times 60'' = 23° + 7' + 12'' \\ &= 23°\,7'12''\end{aligned}$$

Cofunction Identities

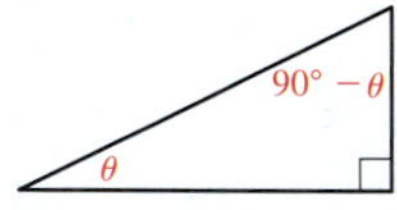

$\sin\theta = \cos(90° - \theta)$,
$\cos\theta = \sin(90° - \theta)$,
$\tan\theta = \cot(90° - \theta)$,
$\cot\theta = \tan(90° - \theta)$,
$\sec\theta = \csc(90° - \theta)$,
$\csc\theta = \sec(90° - \theta)$

Given that $\sin 47° \approx 0.7314$, $\cos 47° \approx 0.6820$, and $\tan 47° \approx 1.0724$, find the six trigonometric function values for 43°.

First, we find csc 47°, sec 47°, and cot 47°:

$$\csc 47° = \frac{1}{\sin 47°} \approx 1.3672,$$
$$\sec 47° = \frac{1}{\cos 47°} \approx 1.4663,$$
$$\cot 47° = \frac{1}{\tan 47°} \approx 0.9325.$$

We know that $43° = 90° - 47°$, so we have

$$\begin{aligned}\sin 43° &= \cos 47° \approx 0.6820, \\ \cos 43° &= \sin 47° \approx 0.7314, \\ \tan 43° &= \cot 47° \approx 0.9325, \\ \csc 43° &= \sec 47° \approx 1.4663, \\ \sec 43° &= \csc 47° \approx 1.3672, \\ \cot 43° &= \tan 47° \approx 1.0724.\end{aligned}$$

SECTION 6.2: APPLICATIONS OF RIGHT TRIANGLES

Solving a Triangle

To **solve** a triangle means to find the lengths of *all* sides and the measures of *all* angles.

Solve this right triangle.

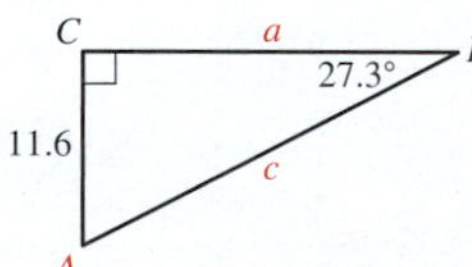

$A = ?$, $a = ?$,
$B = 27.3°$, $b = 11.6$,
$C = 90°$, $c = ?$

First, we find A: $A = 90° - 27.3° = 62.7°$.
Then we use the tangent and the cosine functions to find a and c:

$$\tan 62.7° = \frac{a}{11.6} \qquad \cos 62.7° = \frac{11.6}{c}$$
$$11.6 \tan 62.7° = a$$
$$22.5 \approx a, \qquad c = \frac{11.6}{\cos 62.7°}$$
$$c \approx 25.3.$$

SECTION 6.3: TRIGONOMETRIC FUNCTIONS OF ANY ANGLE

Coterminal Angles

If two or more angles have the same terminal side, the angles are said to be **coterminal**.

To find angles coterminal with a given angle, we add or subtract multiples of 360°.

Find two positive angles and two negative angles that are coterminal with 123°.

$$123^\circ + 360^\circ = 483^\circ,$$
$$123^\circ + 3(360^\circ) = 1203^\circ,$$
$$123^\circ - 360^\circ = -237^\circ,$$
$$123^\circ - 2(360^\circ) = -597^\circ$$

The angles 483°, 1203°, −237°, and −597° are coterminal with 123°.

Complementary Angles and Supplementary Angles

Two acute angles are **complementary** if the sum of their measures is 90°.

Two positive angles are **supplementary** if the sum of their measures is 180°.

Find the complement and the supplement of 83.5°.

$$90^\circ - 83.5^\circ = 6.5^\circ,$$
$$180^\circ - 83.5^\circ = 96.5^\circ$$

The complement of 83.5° is 6.5°, and the supplement of 83.5° is 96.5°.

Trigonometric Functions of Any Angle θ

If $P(x, y)$ is any point on the terminal side of any angle θ in standard position, and r is the distance from the origin to $P(x, y)$, where $r = \sqrt{x^2 + y^2}$, then

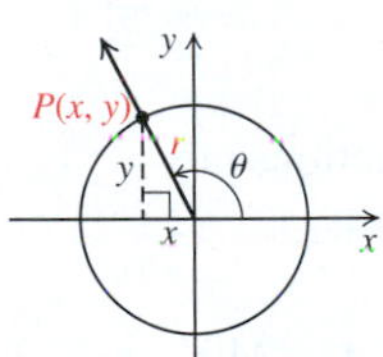

$$\sin\theta = \frac{y}{r}, \quad \csc\theta = \frac{r}{y},$$
$$\cos\theta = \frac{x}{r}, \quad \sec\theta = \frac{r}{x},$$
$$\tan\theta = \frac{y}{x}, \quad \cot\theta = \frac{x}{y}.$$

The trigonometric function values of θ depend only on the angle, not on the choice of the point on the terminal side that is used to compute them.

Signs of Function Values

The signs of the function values depend only on the coordinates of the point P on the terminal side of an angle.

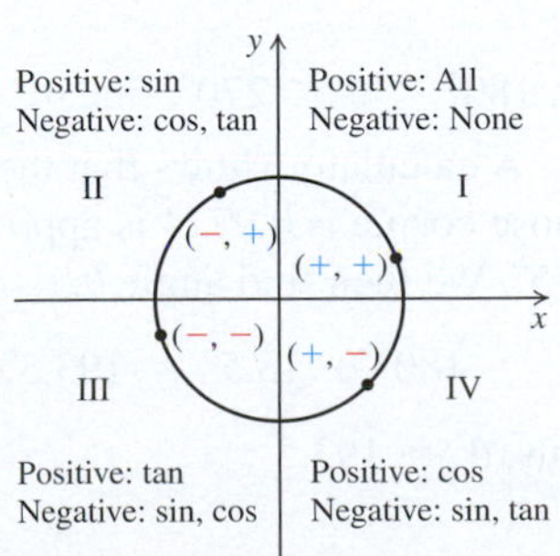

Find the six trigonometric function values for the angle shown.

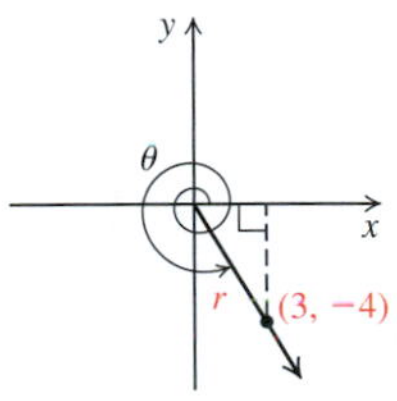

We first determine r:

$$r = \sqrt{x^2 + y^2} = \sqrt{3^2 + (-4)^2} = \sqrt{25} = 5.$$

$$\sin\theta = -\frac{4}{5}, \quad \csc\theta = -\frac{5}{4},$$
$$\cos\theta = \frac{3}{5}, \quad \sec\theta = \frac{5}{3},$$
$$\tan\theta = -\frac{4}{3}, \quad \cot\theta = -\frac{3}{4}$$

Given that $\cos\alpha = -\frac{1}{5}$ and α is in the third quadrant, find the other function values.

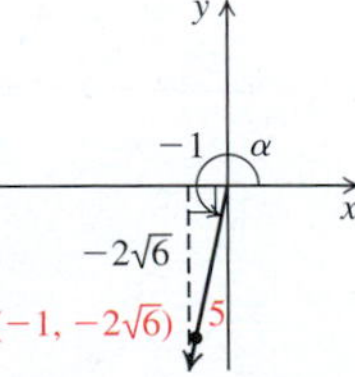

One leg of the reference triangle has length 1, and the length of the hypotenuse is 5. The length of the other leg is $\sqrt{5^2 - 1^2}$, or $\sqrt{24}$, or $2\sqrt{6}$.

$$\sin\alpha = -\frac{2\sqrt{6}}{5}, \quad \csc\alpha = -\frac{5}{2\sqrt{6}}, \text{ or } -\frac{5\sqrt{6}}{12},$$
$$\cos\alpha = -\frac{1}{5}, \quad \sec\alpha = -5,$$
$$\tan\alpha = 2\sqrt{6}, \quad \cot\alpha = \frac{1}{2\sqrt{6}}, \text{ or } \frac{\sqrt{6}}{12}$$

Trigonometric Function Values of Quadrantal Angles

An angle whose terminal side falls on one of the axes is a **quadrantal angle**.

	0° 360°	90°	180°	270°
sin	0	1	0	−1
cos	1	0	−1	0
tan	0	Not defined	0	Not defined

Find the exact function value.

$\tan(-90°)$ is not defined,
$\sin 450° = 1$,
$\csc 270° = -1$,
$\cos 720° = 1$,
$\sec(-180°) = -1$,
$\cot(-360°)$ is not defined

Reference Angles

The **reference angle** for an angle is the acute angle formed by the terminal side of the angle and the x-axis.

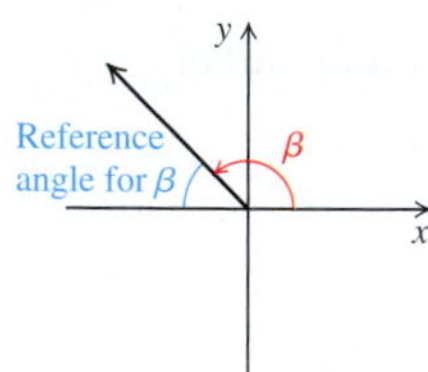

When the reference angle is 30°, 45°, or 60°, we can mentally determine trigonometric function values.

Find the sine, cosine, and tangent values for 240°.

The reference angle is $240° - 180°$, or 60°. Recall that $\sin 60° = \frac{\sqrt{3}}{2}$, $\cos 60° = \frac{1}{2}$, and $\tan 60° = \sqrt{3}$.

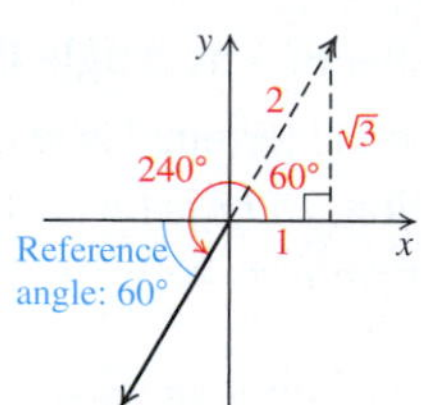

In the third quadrant, the sine and the cosine functions are negative, and the tangent function is positive. Thus,

$$\sin 240° = -\frac{\sqrt{3}}{2}, \quad \cos 240° = -\frac{1}{2}, \quad \text{and} \quad \tan 240° = \sqrt{3}.$$

Trigonometric Function Values of Any Angle

Using a calculator, we can approximate the trigonometric function values of any angle.

Find each of the following function values using a calculator set in DEGREE mode. Round the values to four decimal places, where appropriate.

$\csc 285° \approx -1.0353$, $\quad \cos 51° \approx 0.6293$,
$\sin 25°14'38'' \approx 0.4265$, $\quad \sec(-45°) \approx 1.4142$,
$\tan(-1020°) \approx 1.7321$, $\quad \sin 810° = 1$

Given $\cos\theta \approx -0.9724$, $180° < \theta < 270°$, find θ.

A calculator shows that the acute angle whose cosine is 0.9724 is approximately 13.5°. We then find angle θ:

$$180° + 13.5° = 193.5°.$$

Thus, $\theta \approx 193.5°$.

Aerial Navigation

In aerial navigation, directions are given in degrees clockwise from *north*. For example, a direction, or bearing, of 195° is shown below.

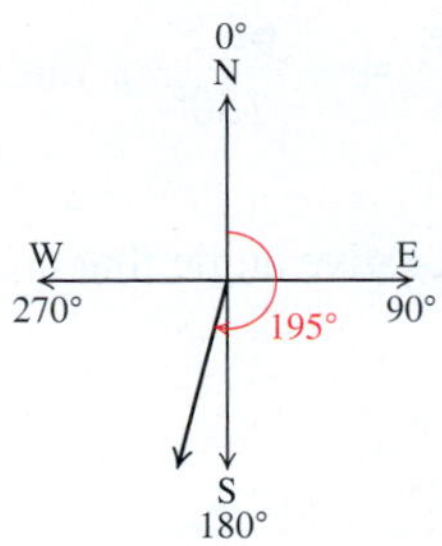

An airplane flies 320 mi from an airport in a direction of 305°. How far north of the airport is the plane then? How far west?

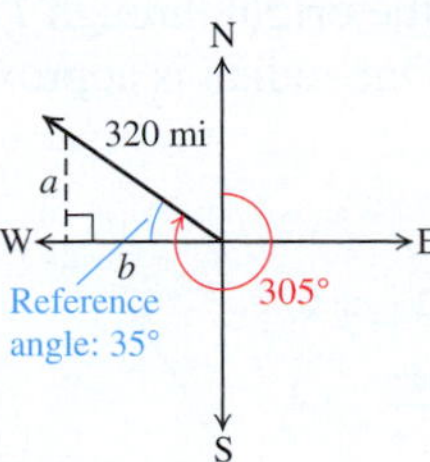

The distance north of the airport a and the distance west of the airport b are legs of a right triangle. The reference angle is $305° - 270° = 35°$. Thus,

$$\frac{a}{320} = \sin 35°$$

$$a = 320 \sin 35° \approx 184;$$

$$\frac{b}{320} = \cos 35°$$

$$b = 320 \cos 35° \approx 262.$$

The airplane is about 184 mi north and about 262 mi west of the airport.

SECTION 6.4: RADIANS, ARC LENGTH, AND ANGULAR SPEED

The Unit Circle

A circle centered at the origin with a radius of length 1 is called a **unit circle**. Its equation is $x^2 + y^2 = 1$.

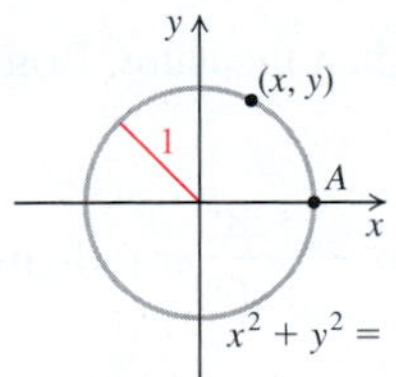

The circumference of a circle of radius r is $2\pi r$. For a unit circle, where $r = 1$, the circumference is 2π. If a point starts at A and travels around the circle, it travels a distance of 2π.

Find two real numbers between -2π and 2π that determine each of the labeled points.

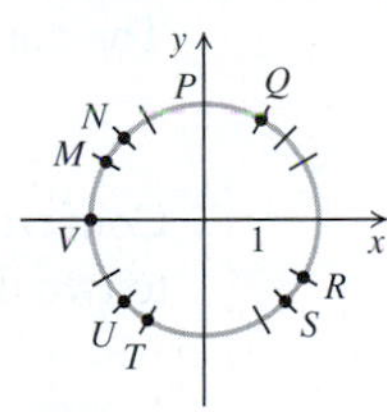

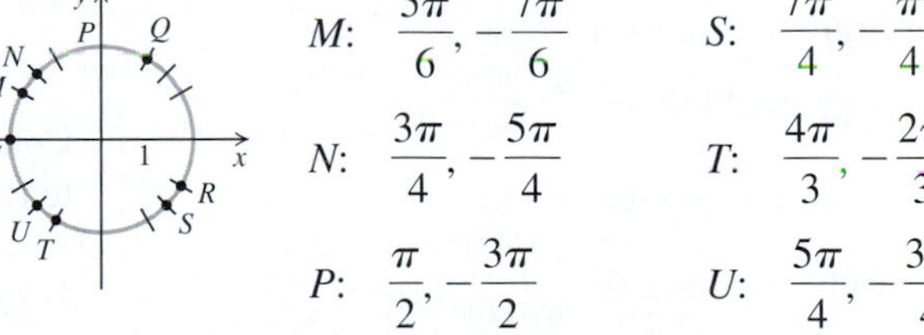

M: $\frac{5\pi}{6}, -\frac{7\pi}{6}$ S: $\frac{7\pi}{4}, -\frac{\pi}{4}$

N: $\frac{3\pi}{4}, -\frac{5\pi}{4}$ T: $\frac{4\pi}{3}, -\frac{2\pi}{3}$

P: $\frac{\pi}{2}, -\frac{3\pi}{2}$ U: $\frac{5\pi}{4}, -\frac{3\pi}{4}$

Q: $\frac{\pi}{3}, -\frac{5\pi}{3}$ V: $\pi, -\pi$

R: $\frac{11\pi}{6}, -\frac{\pi}{6}$

Radian Measure

Consider the unit circle $(r = 1)$ and arc length 1. If a ray is drawn from the origin through T, an angle of 1 radian is formed. One radian is approximately 57.3°.

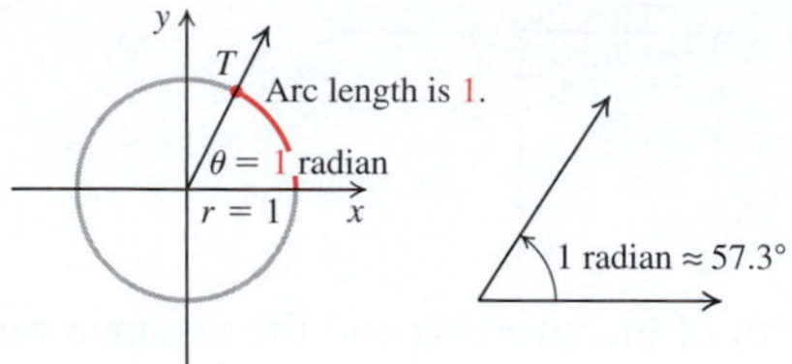

A complete counterclockwise revolution is an angle whose measure is 2π radians, or about 6.28 radians. Thus a rotation of 360° (1 revolution) has a measure of 2π radians.

Radian–Degree Equivalents

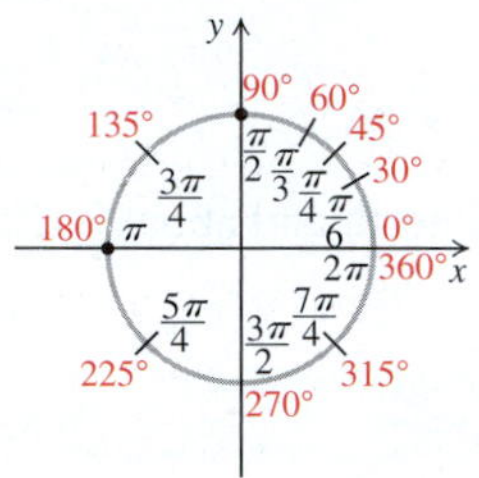

Converting Between Degree Measure and Radian Measure

To convert from degree measure to radian measure, multiply by $\dfrac{\pi \text{ radians}}{180°}$.

To convert from radian measure to degree measure, multiply by $\dfrac{180°}{\pi \text{ radians}}$.

If no unit is given for a rotation, the rotation is understood to be in radians.

Convert 150° and −63.5° to radian measure. Leave answers in terms of π.

$$150° = 150° \cdot \frac{\pi \text{ radians}}{180°} = \frac{150°}{180°}\pi \text{ radians} = \frac{5\pi}{6};$$

$$-63.5° = -63.5° \cdot \frac{\pi \text{ radians}}{180°} = -\frac{63.5°}{180°}\pi \text{ radians} \approx -0.35\pi$$

Find a positive angle and a negative angle that are coterminal with $\dfrac{7\pi}{4}$.

$$\frac{7\pi}{4} + 2\pi = \frac{7\pi}{4} + \frac{8\pi}{4} = \frac{15\pi}{4};$$

$$\frac{7\pi}{4} - 3(2\pi) = \frac{7\pi}{4} - 6\pi = \frac{7\pi}{4} - \frac{24\pi}{4} = -\frac{17\pi}{4}$$

Two angles coterminal with $\dfrac{7\pi}{4}$ are $\dfrac{15\pi}{4}$ and $-\dfrac{17\pi}{4}$.

Find the complement and the supplement of $\dfrac{\pi}{8}$.

$$\frac{\pi}{2} - \frac{\pi}{8} = \frac{4\pi}{8} - \frac{\pi}{8} = \frac{3\pi}{8}; \qquad \mathbf{90° = \frac{\pi}{2}}$$

$$\pi - \frac{\pi}{8} = \frac{8\pi}{8} - \frac{\pi}{8} = \frac{7\pi}{8} \qquad \mathbf{180° = \pi}$$

The complement of $\dfrac{\pi}{8}$ is $\dfrac{3\pi}{8}$, and the supplement of $\dfrac{\pi}{8}$ is $\dfrac{7\pi}{8}$.

Convert −328° and 29.2° to radian measure. Round the answers to two decimal places.

$$-328° = -328° \cdot \frac{\pi \text{ radians}}{180°} = -\frac{328°}{180°}\pi \text{ radians} \approx -5.72;$$

$$29.2° = 29.2° \cdot \frac{\pi \text{ radians}}{180°} = \frac{29.2°}{180°}\pi \text{ radians} \approx 0.51$$

Convert $-\dfrac{2\pi}{3}$, 5π, and -1.3 to degree measure. Round the answers to two decimal places.

$$-\frac{2\pi}{3} = -\frac{2\pi}{3} \cdot \frac{180°}{\pi \text{ radians}} = -\frac{2}{3} \cdot 180° = -120°;$$

$$5\pi = 5\pi \cdot \frac{180°}{\pi \text{ radians}} = 5 \cdot 180° = 900°;$$

$$-1.3 = -1.3 \cdot \frac{180°}{\pi \text{ radians}} = \frac{-1.3(180°)}{\pi} \approx -74.48°$$

Radian Measure

The **radian measure θ** of a rotation is the ratio of the distance s traveled by a point at a radius r from the center of rotation to the length of the radius r:

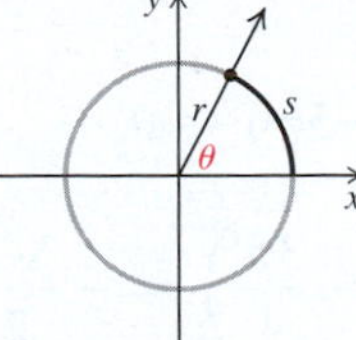

$$\theta = \frac{s}{r}.$$

When the formula $\theta = s/r$ is used, θ must be in radians and s and r must be expressed in the same unit.

Find the measure of a rotation in radians when a point 6 cm from the center of rotation travels 13 cm.

$$\theta = \frac{s}{r} = \frac{13 \text{ cm}}{6 \text{ cm}} = \frac{13}{6} \text{ radians}$$

Find the length of an arc of a circle of radius 10 yd associated with an angle of $5\pi/4$ radians.

$$\theta = \frac{s}{r}, \quad \text{or} \quad s = r\theta;$$

$$s = r\theta = 10 \text{ yd} \cdot \frac{5\pi}{4} \approx 39.3 \text{ yd}$$

Linear Speed and Angular Speed

Linear speed v is the distance s traveled per unit of time t:

$$v = \frac{s}{t}.$$

Angular speed ω is the amount of rotation θ per unit of time t:

$$\omega = \frac{\theta}{t}.$$

Linear Speed in Terms of Angular Speed

The linear speed v of a point a distance r from the center of rotation is given by

$$v = r\omega,$$

where ω is the angular speed, in radians, per unit of time. The unit of distance for v and r must be the same, ω must be in radians per unit of time, and v and ω must be expressed in the same unit of time.

A wheel with a 40-cm radius is rotating at a rate of 2.5 radians/sec. What is the linear speed of a point on its rim, in meters per minute?

We first express r in meters and w in radians per minute.

$$r = 40 \text{ cm} \cdot \frac{1 \text{ m}}{100 \text{ cm}} = 0.4 \text{ m};$$

$$\omega = \frac{2.5 \text{ radians}}{1 \text{ sec}} \cdot \frac{60 \text{ sec}}{1 \text{ min}} = \frac{150 \text{ radians}}{1 \text{ min}}$$

Then we find linear speed.

$$v = r\omega = 0.4 \text{ m} \cdot \frac{150}{1 \text{ min}} = \frac{60 \text{ m}}{\text{min}}$$

SECTION 6.5: CIRCULAR FUNCTIONS: GRAPHS AND PROPERTIES

Domains of the Trigonometric Functions

In Sections 6.1 and 6.3, the domains of the trigonometric functions were defined as a set of angles or rotations measured in a real number of degree units. In Section 6.4, the domains were considered to be sets of real numbers, or radians. Radian measure for θ is defined as $\theta = s/r$. When $r = 1$, $\theta = s$. The arc length s on the unit circle is the same as the radian measure of the angle θ.

Basic Circular Functions

On the unit circle, s can be considered the radian measure of an angle or the measure of an arc length. In either case, it is a real number. Trigonometric functions with domains composed of real numbers are called **circular functions.**

For a real number s that determines a point (x, y) on the unit circle:

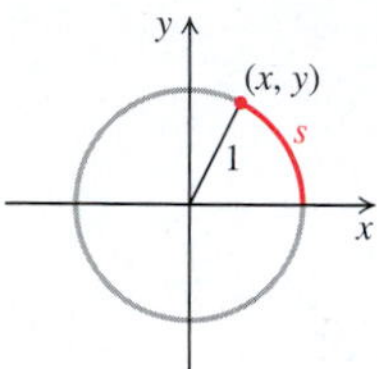

$$\sin s = y,$$

$$\cos s = x,$$

$$\tan s = \frac{y}{x}, x \neq 0,$$

$$\csc s = \frac{1}{y}, y \neq 0,$$

$$\sec s = \frac{1}{x}, x \neq 0,$$

$$\cot s = \frac{x}{y}, y \neq 0.$$

Find each function value using coordinates of a point on the unit circle.

$\sin(-5\pi) = 0,$ $\quad \csc \frac{\pi}{3} = \frac{2}{\sqrt{3}}$, or $\frac{2\sqrt{3}}{3}$,

$\cos\left(-\frac{3\pi}{4}\right) = -\frac{\sqrt{2}}{2},$ $\quad \sec \frac{\pi}{6} = \frac{2}{\sqrt{3}}$, or $\frac{2\sqrt{3}}{3}$,

$\tan \frac{5\pi}{2}$ is not defined, $\quad \cot \frac{23\pi}{6} = -\sqrt{3},$

$\cos\left(-\frac{5\pi}{6}\right) = -\frac{\sqrt{3}}{2},$ $\quad \tan \frac{7\pi}{4} = -1$

Find each function value using a calculator set in RADIAN mode. Round the answers to four decimal places, where appropriate.

$\cos(-14.7) \approx -0.5336,$ $\quad \tan \frac{3\pi}{2}$ is not defined,

$\sin \frac{9\pi}{5} \approx -0.5878,$ $\quad \sec 214 \approx 1.0733$

Reflections

Because a unit circle is symmetric with respect to the x-axis, the y-axis, and the origin, the coordinates of one point on the unit circle can be used to find coordinates of its reflections.

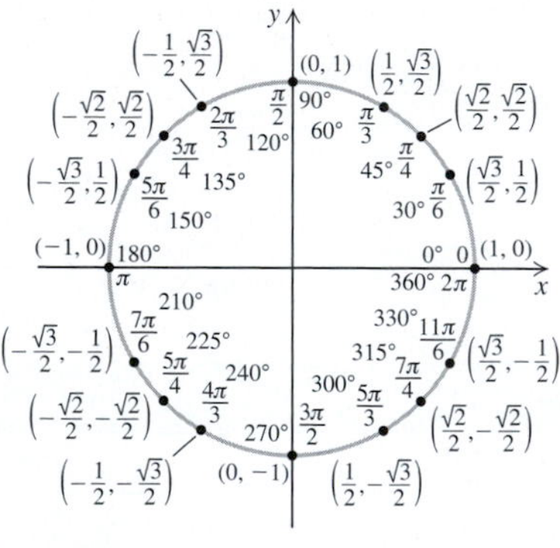

The point $\left(\frac{3}{5}, \frac{4}{5}\right)$ is on the unit circle. Find the coordinates of its reflection across **(a)** the x-axis, **(b)** the y-axis, and **(c)** the origin.

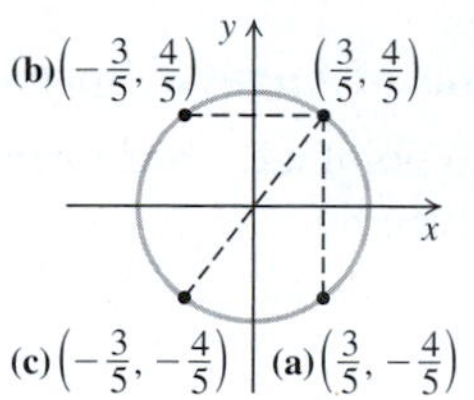

Periodic Function

A function f is said to be **periodic** if there exists a positive constant p such that

$$f(s + p) = f(s)$$

for all s in the domain of f. The smallest such positive number p is called the period of the function.

Graph the sine, the cosine, and the tangent functions. For graphs of the cosecant, the secant, and the cotangent functions, see p. 458.

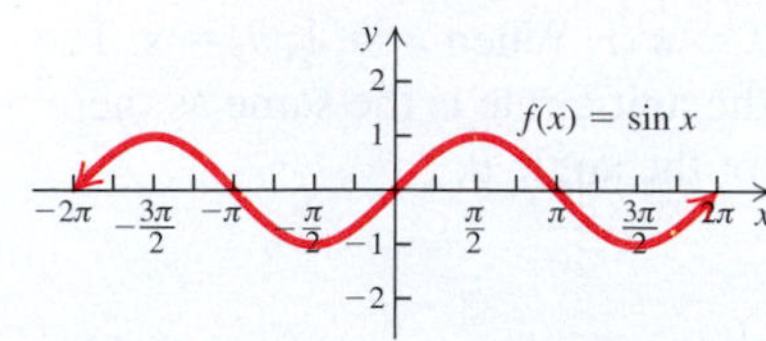

(continued)

Amplitude

The **amplitude** of a periodic function is defined as one half of the distance between its maximum and minimum function values. It is always positive.

Sine Function

1. Continuous
2. Period: 2π
3. Domain: All real numbers
4. Range: $[-1, 1]$
5. Amplitude: 1
6. Odd: $\sin(-s) = -\sin s$

Cosine Function

1. Continuous
2. Period: 2π
3. Domain: All real numbers
4. Range: $[-1, 1]$
5. Amplitude: 1
6. Even: $\cos(-s) = \cos s$

Tangent Function

1. Period: π
2. Domain: All real numbers except $(\pi/2) + k\pi$, where k is an integer
3. Range: All real numbers
4. Odd: $\tan(-s) = -\tan s$

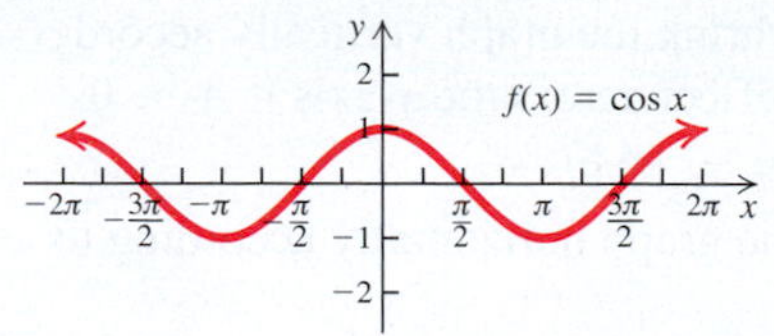

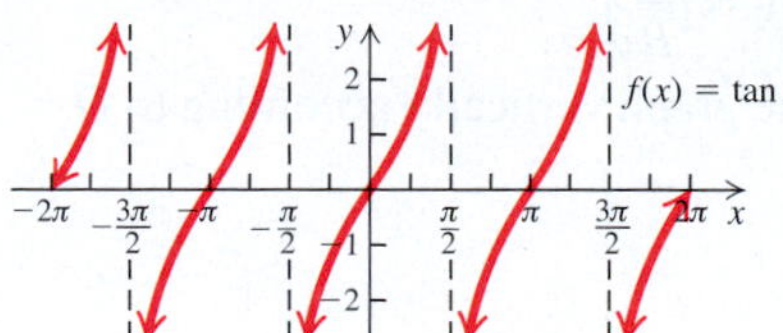

Compare the domains of the sine, the cosine, and the tangent functions.

FUNCTION	DOMAIN
sine	All real numbers
cosine	All real numbers
tangent	All real numbers except $\pi/2 + k\pi$, where k is an integer

Compare the ranges of the sine, the cosine, and the tangent functions.

FUNCTION	RANGE
sine	$[-1, 1]$
cosine	$[-1, 1]$
tangent	All real numbers

Compare the periods of the six trigonometric functions.

FUNCTION	PERIOD
sine, cosine, cosecant, secant	2π
tangent, cotangent	π

SECTION 6.6: GRAPHS OF TRANSFORMED SINE FUNCTIONS AND COSINE FUNCTIONS

Transformations of the Sine Function and the Cosine Function

To graph $y = A\sin(Bx - C) + D$ and $y = A\cos(Bx - C) + D$:

1. Stretch or shrink the graph horizontally according to B. Reflect across the y-axis if $B < 0$. $\left(\text{Period} = \left|\dfrac{2\pi}{B}\right|\right)$

Determine the amplitude, the period, and the phase shift of

$$y = -\frac{1}{2}\sin\left(2x - \frac{\pi}{2}\right) + 1$$

and sketch the graph of the function.

$$y = -\frac{1}{2}\sin\left(2x - \frac{\pi}{2}\right) + 1$$
$$= -\frac{1}{2}\sin\left[2\left(x - \frac{\pi}{4}\right)\right] + 1$$

(continued)

2. Stretch or shrink the graph vertically according to A. Reflect across the x-axis if $A < 0$. (Amplitude $= |A|$)
3. Translate the graph horizontally according to C/B.
$\left(\text{Phase shift} = \frac{C}{B}\right)$
4. Translate the graph vertically according to D.

Period: $\left|\frac{2\pi}{2}\right| = \pi$

Amplitude: $\left|-\frac{1}{2}\right| = \frac{1}{2}$

Phase shift: $\frac{\pi/2}{2} = \frac{\pi}{4}$

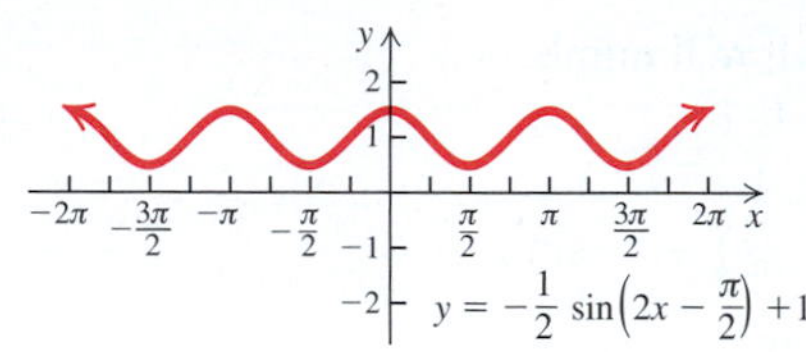

REVIEW EXERCISES

Determine whether the statement is true or false.

1. Given that $(-a, b)$ is a point on the unit circle and θ is in the second quadrant, then $\cos\theta$ is a. **[6.4], [6.5]**

2. Given that $(-c, -d)$ is a point on the unit circle and θ is in the second quadrant, then $\tan\theta = -\frac{c}{d}$. **[6.4], [6.5]**

3. The measure 300° is greater than the measure 5 radians. **[6.4]**

4. If $\sec\theta > 0$ and $\cot\theta < 0$, then θ is in the fourth quadrant. **[6.3]**

5. The amplitude of $y = \frac{1}{2}\sin x$ is twice as large as the amplitude of $y = \sin\frac{1}{2}x$. **[6.6]**

6. The supplement of $\frac{9}{13}\pi$ is greater than the complement of $\frac{\pi}{6}$. **[6.4]**

7. Find the six trigonometric function values of θ. **[6.1]**

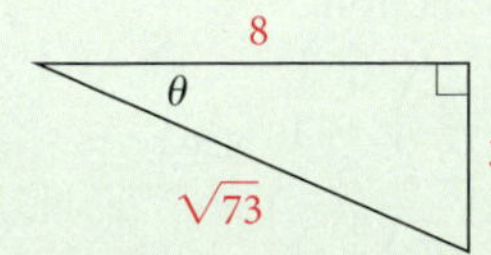

8. Given that β is acute and $\sin\beta = \frac{\sqrt{91}}{10}$, find the other five trigonometric function values. **[6.1]**

Find the exact function value, if it exists.

9. $\cos 45°$ **[6.1]**
10. $\cot 60°$ **[6.1]**
11. $\cos 495°$ **[6.3]**
12. $\sin 150°$ **[6.3]**
13. $\sec(-270°)$ **[6.3]**
14. $\tan(-600°)$ **[6.3]**
15. $\csc 60°$ **[6.1]**
16. $\cot(-45°)$ **[6.3]**

17. Convert 22.27° to degrees, minutes, and seconds. Round to the nearest second. **[6.1]**

18. Convert $47°33'27''$ to decimal degree notation. Round to two decimal places. **[6.1]**

Find the function value. Round to four decimal places. **[6.3]**

19. $\tan 2184°$
20. $\sec 27.9°$
21. $\cos 18°13'42''$
22. $\sin 245°24'$
23. $\cot(-33.2°)$
24. $\sin 556.13°$

Find θ in the interval indicated. Round the answer to the nearest tenth of a degree. **[6.3]**

25. $\cos\theta = -0.9041$, $(180°, 270°)$

26. $\tan\theta = 1.0799$, $(0°, 90°)$

Find the exact acute angle θ, in degrees, given the function value. **[6.1]**

27. $\sin\theta = \frac{\sqrt{3}}{2}$
28. $\tan\theta = \sqrt{3}$

29. $\cos \theta = \dfrac{\sqrt{2}}{2}$

30. $\sec \theta = \dfrac{2\sqrt{3}}{3}$

31. Given that $\sin 59.1° \approx 0.8581$, $\cos 59.1° \approx 0.5135$, and $\tan 59.1° \approx 1.6709$, find the six function values for $30.9°$. [6.1]

Solve each of the following right triangles. Standard lettering has been used. [6.2]

32. $a = 7.3$, $c = 8.6$

33. $a = 30.5$, $B = 51.17°$

34. One leg of a right triangle bears east. The hypotenuse is 734 m long and bears N57°23′E. Find the perimeter of the triangle.

35. An observer's eye is 6 ft above the floor. A mural is being viewed. The bottom of the mural is at floor level. The observer looks down 13° to see the bottom and up 17° to see the top. How tall is the mural?

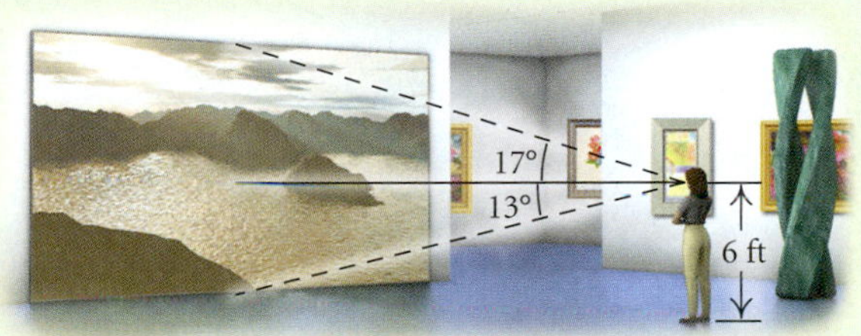

For angles of the following measures, state in which quadrant the terminal side lies. [6.3]

36. $142°11'5''$

37. $-635.2°$

38. $-392°$

Find a positive angle and a negative angle that are coterminal with the given angle. Answers may vary.

39. $65°$ [6.3]

40. $\dfrac{7\pi}{3}$ [6.4]

Find the complement and the supplement.

41. $13.4°$ [6.3]

42. $\dfrac{\pi}{6}$ [6.4]

43. Find the six trigonometric function values for the angle θ shown. [6.3]

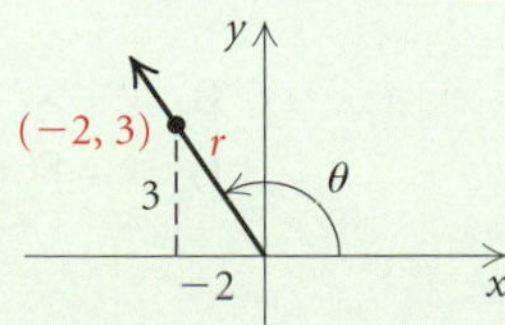

44. Given that $\tan \theta = 2/\sqrt{5}$ and that the terminal side is in quadrant III, find the other five trigonometric function values. [6.3]

45. An airplane travels at 530 mph for $3\frac{1}{2}$ hr in a direction of 160° from Minneapolis, Minnesota. At the end of that time, how far south of Minneapolis is the airplane? [6.3]

46. On a unit circle, mark and label the points determined by $7\pi/6$, $-3\pi/4$, $-\pi/3$, and $9\pi/4$. [6.4]

For angles of the following measures, convert to radian measure in terms of π, and convert to radian measure not in terms of π. Round the answer to two decimal places. [6.4]

47. $145.2°$

48. $-30°$

Convert to degree measure. Round the answer to two decimal places where appropriate. [6.4]

49. $\dfrac{3\pi}{2}$

50. 3

51. -4.5

52. 11π

53. Find the length of an arc of a circle, given a central angle of $\pi/4$ and a radius of 7 cm. [6.4]

54. An arc 18 m long on a circle of radius 8 m subtends an angle of how many radians? how many degrees, to the nearest degree? [6.4]

55. A waterwheel in a watermill has a radius of 7 ft and makes a complete revolution in 70 sec. What is the linear speed, in feet per minute, of a point on the rim? [6.4]

56. An automobile wheel has a diameter of 14 in. If the car travels at a speed of 55 mph, what is the angular velocity, in radians per hour, of a point on the edge of the wheel? [6.4]

57. The point $\left(\frac{3}{5}, -\frac{4}{5}\right)$ is on a unit circle. Find the coordinates of its reflections across the x-axis, the y-axis, and the origin. [6.5]

Find the exact function value, if it exists. [6.5]

58. $\cos \pi$

59. $\tan \dfrac{5\pi}{4}$

60. $\sin \dfrac{5\pi}{3}$

61. $\sin \left(-\dfrac{7\pi}{6}\right)$

62. $\tan \dfrac{\pi}{6}$

63. $\cos (-13\pi)$

Find the function value, if it exists. Round to four decimal places. [6.5]

64. $\sin 24$

65. $\cos (-75)$

66. $\cot 16\pi$

67. $\tan \dfrac{3\pi}{7}$

68. $\sec 14.3$

69. $\cos \left(-\dfrac{\pi}{5}\right)$

70. Graph each of the six trigonometric functions from -2π to 2π. [6.5]

71. What is the period of each of the six trigonometric functions? [6.5]

72. Complete the following table. [6.5]

Function	Domain	Range
sine		
cosine		
tangent		

73. Complete the following table with the sign of the specified trigonometric function value in each of the four quadrants. [6.3]

Function	I	II	III	IV
sine				
cosine				
tangent				

Determine the amplitude, the period, and the phase shift of the function, and sketch the graph of the function. Then check the graph using a graphing calculator. [6.6]

74. $y = \sin \left(x + \dfrac{\pi}{2}\right)$

75. $y = 3 + \dfrac{1}{2}\cos \left(2x - \dfrac{\pi}{2}\right)$

In Exercises 76–79, without using a graphing calculator, match the function with one of the graphs (a)–(d) that follow. Then check your work using a graphing calculator. [6.6]

a)

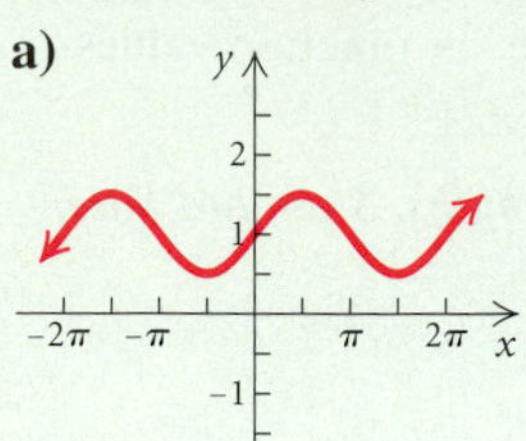

b)

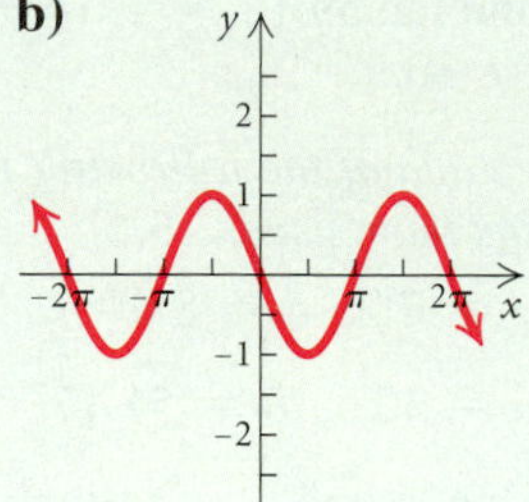

c)

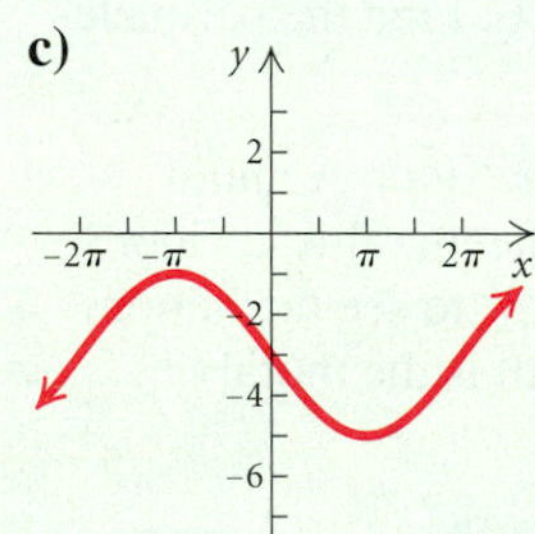

d)

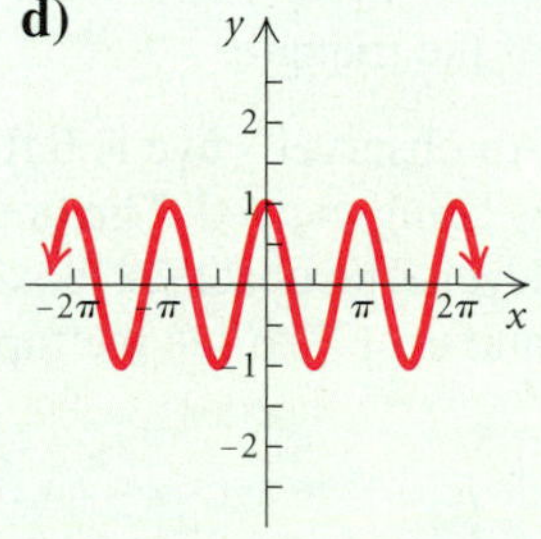

76. $y = \cos 2x$

77. $y = \dfrac{1}{2}\sin x + 1$

78. $y = -2 \sin \dfrac{1}{2}x - 3$

79. $y = -\cos \left(x - \dfrac{\pi}{2}\right)$

80. Sketch a graph of $y = 3\cos x + \sin x$ for values of x between 0 and 2π. [6.6]

81. Graph: $f(x) = e^{-0.7x}\cos x$. [6.6]

82. Which of the following is the reflection of $\left(-\dfrac{1}{2}, \dfrac{\sqrt{3}}{2}\right)$ across the y-axis? [6.5]

A. $\left(\dfrac{1}{2}, -\dfrac{\sqrt{3}}{2}\right)$

B. $\left(\dfrac{\sqrt{3}}{2}, \dfrac{1}{2}\right)$

C. $\left(\dfrac{1}{2}, \dfrac{\sqrt{3}}{2}\right)$

D. $\left(\dfrac{\sqrt{3}}{2}, -\dfrac{1}{2}\right)$

83. Which of the following is the domain of the cosine function? [6.5]

A. $(-1, 1)$

B. $(-\infty, \infty)$

C. $[0, \infty)$

D. $[-1, 1]$

84. The graph of $f(x) = -\cos(-x)$ is which of the following? **[6.6]**

A.

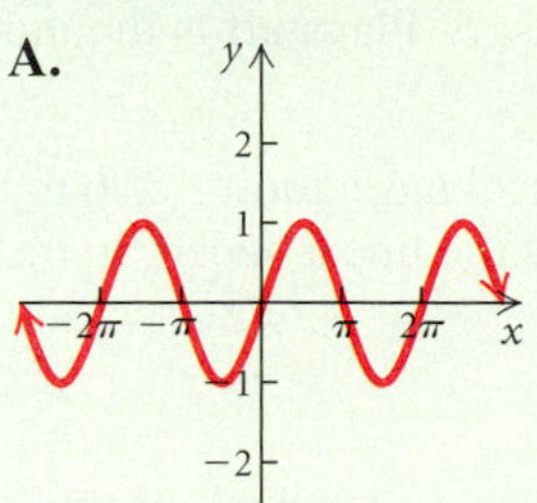

B.

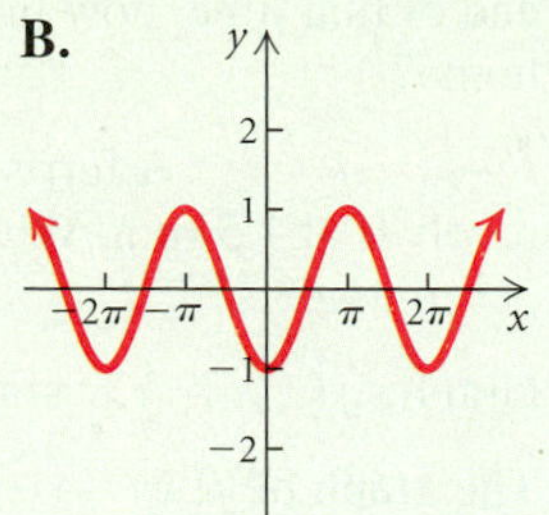

C.

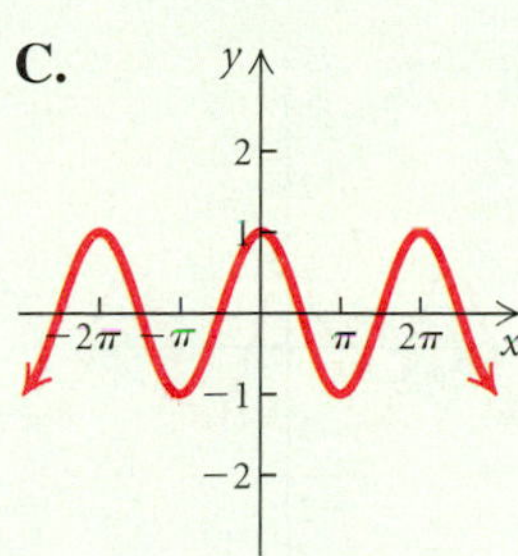

D.

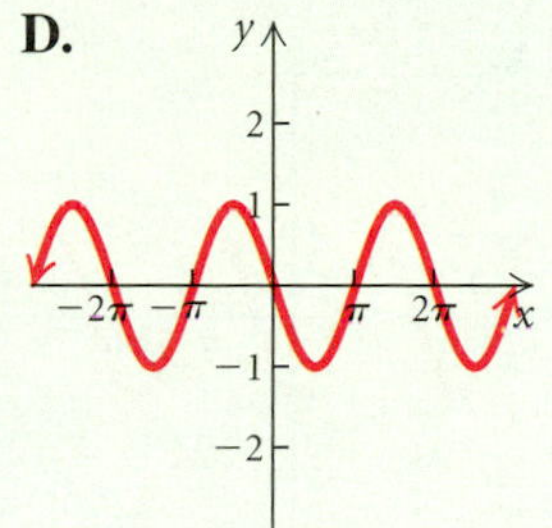

Synthesis

85. Graph $y = 3 \sin(x/2)$, and determine the domain, the range, and the period. **[6.6]**

86. In the following graph, $y_1 = \sin x$ is shown in black and y_2 is shown in red. Express y_2 as a transformation of the graph of y_1. **[6.6]**

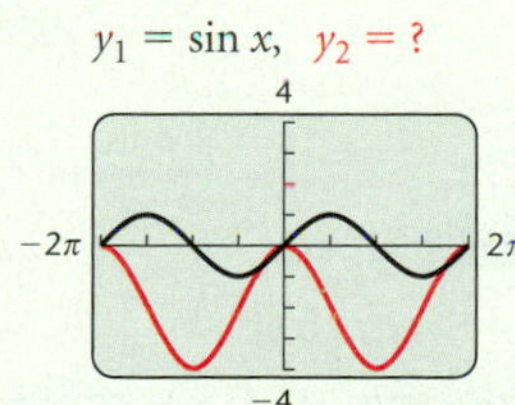

87. Find the domain of $y = \log(\cos x)$. **[6.6]**

Collaborative Discussion and Writing

88. Compare the terms radian and degree. **[6.1], [6.4]**

89. In circular motion with a fixed angular speed, the length of the radius is directly proportional to the linear speed. Explain why using an example. **[6.4]**

90. Explain why both the sine function and the cosine function are continuous, but the tangent function is not continuous. **[6.5]**

91. In the transformation steps listed in Section 6.6, why must step (1) precede step (3)? Give an example that illustrates this. **[6.6]**

92. In the equations $y = A \sin(Bx - C) + D$ and $y = A \cos(Bx - C) + D$, which constants translate the graphs and which constants stretch and shrink the graphs? Describe in your own words the effect of each constant. **[6.6]**

93. Two new cars are each driven at an average speed of 60 mph for an extended highway test drive of 2000 mi. The diameters of the wheels of the two cars are 15 in. and 16 in., respectively. If the cars use tires of equal durability and profile, differing only by the diameter, which car will probably need new tires first? Explain your answer. **[6.4]**

6 Chapter Test

1. Find the six trigonometric function values of θ.

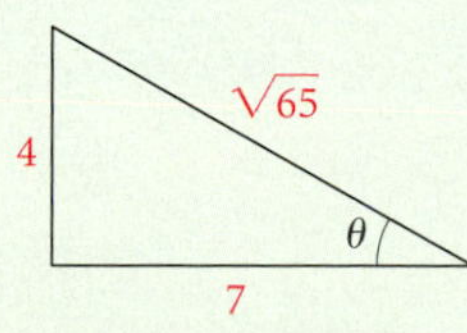

Find the exact function value, if it exists.

2. $\sin 120°$

3. $\tan(-45°)$

4. $\cos 3\pi$

5. $\sec \dfrac{5\pi}{4}$

6. Convert $38°27'56''$ to decimal degree notation. Round to two decimal places.

Find the function values. Round to four decimal places.

7. $\tan 526.4°$

8. $\sin(-12°)$

9. $\sec \dfrac{5\pi}{9}$

10. $\cos 76.07$

11. Find the exact acute angle θ, in degrees, for which $\sin \theta = \frac{1}{2}$.

12. Given that $\sin 28.4° \approx 0.4756$, $\cos 28.4° \approx 0.8796$, and $\tan 28.4° \approx 0.5407$, find the six trigonometric function values for $61.6°$.

13. Solve the right triangle with $b = 45.1$ and $A = 35.9°$. Standard lettering has been used.

14. Find a positive angle and a negative angle coterminal with a $112°$ angle.

15. Find the supplement of $\frac{5\pi}{6}$.

16. Given that $\sin \theta = -4/\sqrt{41}$ and that the terminal side is in quadrant IV, find the other five trigonometric function values.

17. Convert $210°$ to radian measure in terms of π.

18. Convert $\frac{3\pi}{4}$ to degree measure.

19. Find the length of an arc of a circle given a central angle of $\pi/3$ and a radius of 16 cm.

Consider the function $y = -\sin (x - \pi/2) + 1$ *for Exercises 20–23.*

20. Find the amplitude.

21. Find the period.

22. Find the phase shift.

23. Which of the following is the graph of the function?

A.

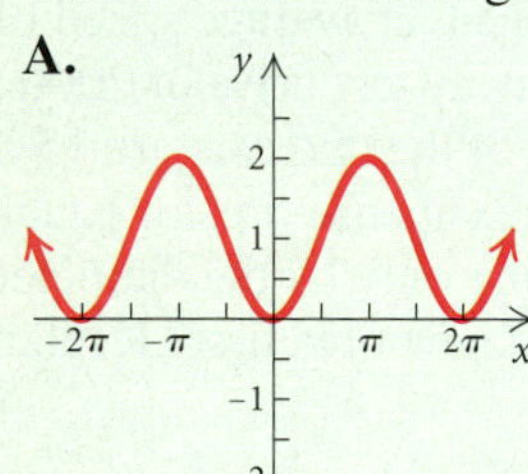

B.

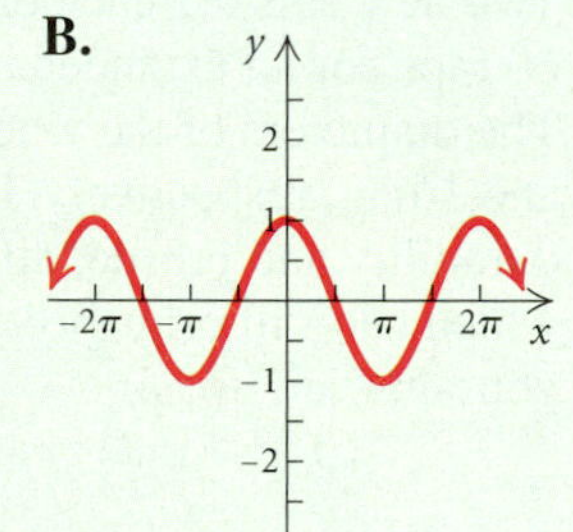

C.

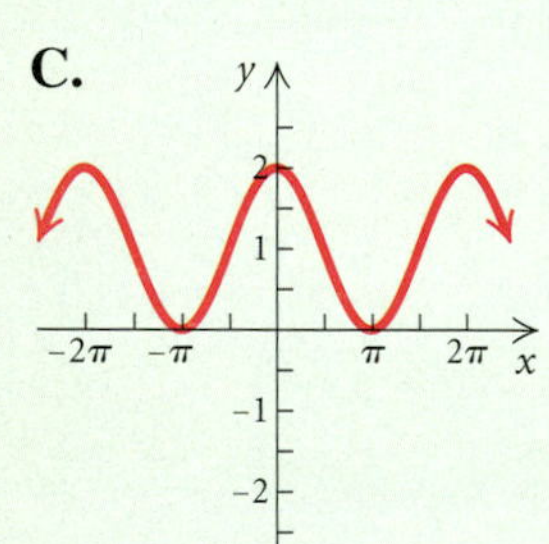

D.

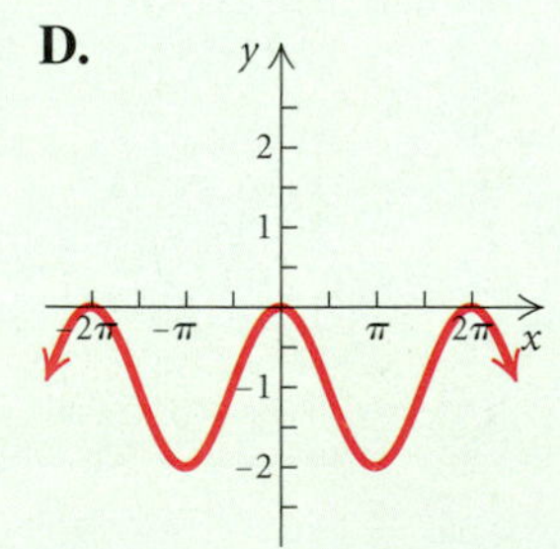

24. *Ski Dubai Resort.* Ski Dubai is the first indoor ski resort in the Middle East. Its longest ski run drops 60 ft and has an angle of depression of approximately $8.6°$ (*Source*: www.SkiDubai.com). Find the length of the ski run. Round the answer to the nearest foot.

25. *Location.* A motor home travels at 50 mph for 6 hr in a direction of $115°$ from Flagstaff, Arizona. At the end of that time, how far east of Flagstaff is the motor home?

26. *Linear Speed.* A ferris wheel has a radius of 6 m and revolves at 1.5 rpm. What is the linear speed, in meters per minute?

27. Graph: $f(x) = \frac{1}{2}x^2 \sin x$.

28. The graph of $f(x) = -\sin(-x)$ is which of the following?

A.

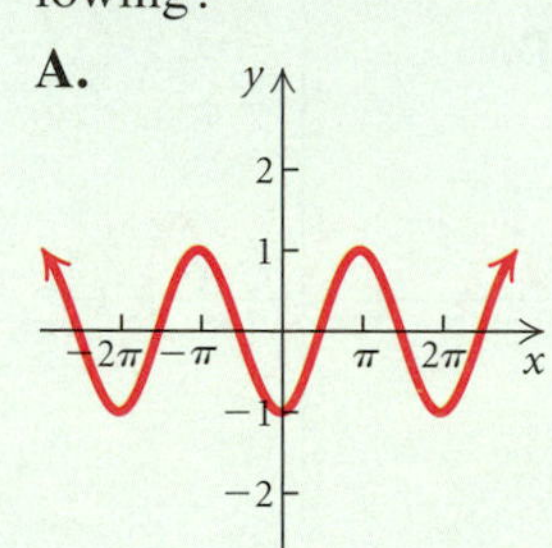

B.

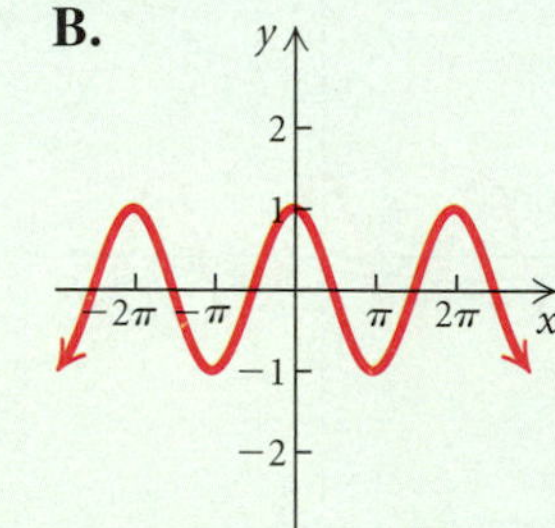

C.

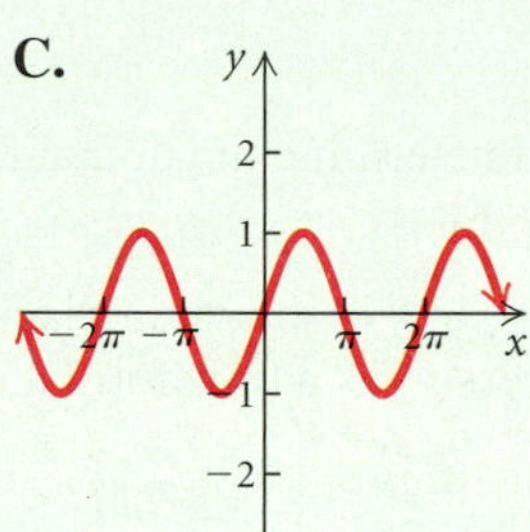

D.

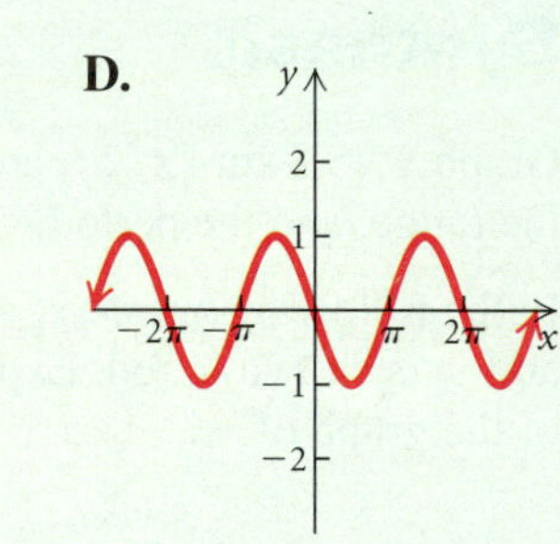

Synthesis

29. Determine the domain of $f(x) = \frac{-3}{\sqrt{\cos x}}$.

CHAPTER 7

Trigonometric Identities, Inverse Functions, and Equations

APPLICATION

This problem appears as Exercise 51 in Section 7.5.

Sales of fishing boats fluctuate in cycles. The following cosine function can be used to estimate the total amount of sales of fishing boats, y, in thousands of dollars, in month x, for a business:

$$y = 15.328 \cos (0.475x - 1.728) + 87.223.$$

Approximate the total amount of sales to the nearest dollar for November and for March. (*Hint*: 1 represents January, 2 represents February, and so on.)

7.1 Identities: Pythagorean and Sum and Difference

- State the Pythagorean identities.
- Simplify and manipulate expressions containing trigonometric expressions.
- Use the sum and difference identities to find function values.

An **identity** is an equation that is true for all *possible* replacements of the variables. The following is a list of the identities studied in Chapter 6.

BASIC IDENTITIES

$$\sin x = \frac{1}{\csc x}, \quad \csc x = \frac{1}{\sin x}, \quad \sin(-x) = -\sin x,$$
$$\cos(-x) = \cos x,$$
$$\cos x = \frac{1}{\sec x}, \quad \sec x = \frac{1}{\cos x}, \quad \tan(-x) = -\tan x,$$
$$\tan x = \frac{1}{\cot x}, \quad \cot x = \frac{1}{\tan x}, \quad \tan x = \frac{\sin x}{\cos x},$$
$$\cot x = \frac{\cos x}{\sin x}$$

In this section, we will develop some other important identities.

Pythagorean Identities

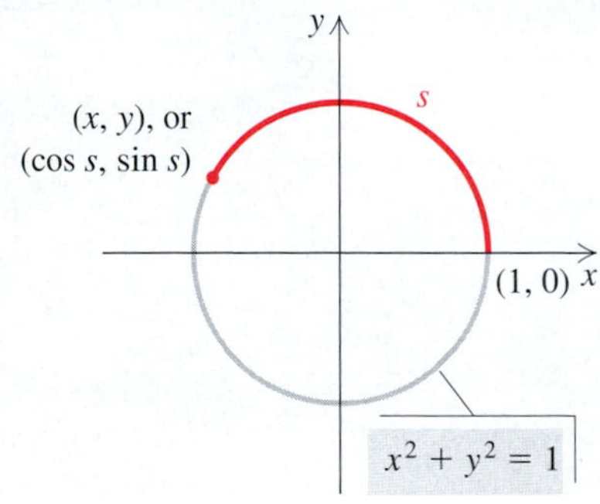

We now consider three other identities that are fundamental to a study of trigonometry. They are called the *Pythagorean identities*. Recall that the equation of a unit circle in the xy-plane is

$$x^2 + y^2 = 1.$$

For any point on the unit circle, the coordinates x and y satisfy this equation. Suppose that a real number s determines a point on the unit circle with coordinates (x, y), or $(\cos s, \sin s)$. Then $x = \cos s$ and $y = \sin s$. Substituting $\cos s$ for x and $\sin s$ for y in the equation of the unit circle gives us the identity

$$(\cos s)^2 + (\sin s)^2 = 1, \qquad \text{Substituting } \cos s \text{ for } x \text{ and } \sin s \text{ for } y$$

which can be expressed as

$$\mathbf{\sin^2 s + \cos^2 s = 1.}$$

It is conventional in trigonometry to use the notation $\sin^2 s$ rather than $(\sin s)^2$. Note that $\sin^2 s \neq \sin s^2$.

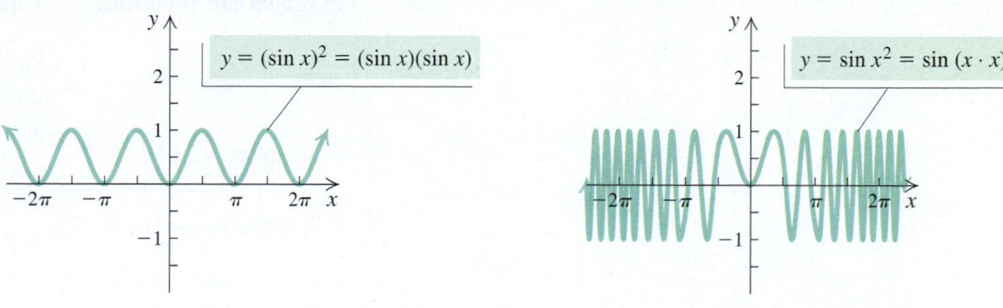

The identity $\sin^2 s + \cos^2 s = 1$ gives a relationship between the sine and the cosine of any real number s. It is an important **Pythagorean identity**.

We can divide by $\sin^2 s$ on both sides of the preceding identity:

$$\frac{\sin^2 s}{\sin^2 s} + \frac{\cos^2 s}{\sin^2 s} = \frac{1}{\sin^2 s}. \qquad \text{Dividing by } \sin^2 s$$

Simplifying gives us a second Pythagorean identity:

$$\mathbf{1 + \cot^2 s = \csc^2 s.}$$

This equation is true for any replacement of s with a real number for which $\sin^2 s \neq 0$, since we divided by $\sin^2 s$. But the numbers for which $\sin^2 s = 0$ (or $\sin s = 0$) are exactly those for which the cotangent function and the cosecant function are not defined. Thus our new equation holds for all real numbers s for which $\cot s$ and $\csc s$ are defined and is thus an identity.

The third Pythagorean identity can be obtained by dividing by $\cos^2 s$ on both sides of the first Pythagorean identity:

$$\frac{\sin^2 s}{\cos^2 s} + \frac{\cos^2 s}{\cos^2 s} = \frac{1}{\cos^2 s} \qquad \text{Dividing by } \cos^2 s$$

$$\mathbf{\tan^2 s + 1 = \sec^2 s.} \qquad \text{Simplifying}$$

This equation is true for any replacement of s with a real number for which $\cos^2 s \neq 0$, since we divided by $\cos^2 s$. But the numbers for which $\cos^2 s = 0$ (or $\cos s = 0$) are exactly those for which the tangent function and the secant function are not defined. Thus our new equation holds for all real numbers s for which $\tan s$ and $\sec s$ are defined and is thus an identity.

The identities that we have developed hold no matter what symbols are used for the variables. For example, we could write

$$\sin^2 s + \cos^2 s = 1, \qquad \sin^2 \theta + \cos^2 \theta = 1, \quad \text{or} \quad \sin^2 x + \cos^2 x = 1.$$

PYTHAGOREAN IDENTITIES

$\sin^2 x + \cos^2 x = 1,$

$1 + \cot^2 x = \csc^2 x,$

$1 + \tan^2 x = \sec^2 x$

It is often helpful to express the Pythagorean identities in equivalent forms.

Pythagorean Identities	Equivalent Forms
$\sin^2 x + \cos^2 x = 1$	$\sin^2 x = 1 - \cos^2 x$ $\cos^2 x = 1 - \sin^2 x$
$1 + \cot^2 x = \csc^2 x$	$1 = \csc^2 x - \cot^2 x$ $\cot^2 x = \csc^2 x - 1$
$1 + \tan^2 x = \sec^2 x$	$1 = \sec^2 x - \tan^2 x$ $\tan^2 x = \sec^2 x - 1$

Simplifying Trigonometric Expressions

We can factor, simplify, and manipulate trigonometric expressions in the same way that we manipulate strictly algebraic expressions.

EXAMPLE 1 Multiply and simplify: $\cos x\,(\tan x - \sec x)$.

Solution

$\cos x(\tan x - \sec x)$

$= \cos x \tan x - \cos x \sec x$ — **Multiplying**

$= \cos x \dfrac{\sin x}{\cos x} - \cos x \dfrac{1}{\cos x}$ — **Recalling the identities $\tan x = \sin x/\cos x$ and $\sec x = 1/\cos x$ and substituting**

$= \sin x - 1$ — **Simplifying**

Now Try Exercise 3.

There is no general procedure for simplifying trigonometric expressions, but it is often helpful to write everything in terms of sines and cosines, as we did in Example 1. We also look for a Pythagorean identity within a trigonometric expression.

EXAMPLE 2 Factor and simplify: $\sin^2 x \cos^2 x + \cos^4 x$.

Solution

$\sin^2 x \cos^2 x + \cos^4 x$

$= \cos^2 x\,(\sin^2 x + \cos^2 x)$ — **Factoring out the common factor**

$= \cos^2 x \cdot (1)$ — **Using $\sin^2 x + \cos^2 x = 1$**

$= \cos^2 x$

Now Try Exercises 9 and 13.

A graphing calculator can be used to perform a partial check of an identity. First, we graph the expression on the left side of the equals sign. Then we graph the expression on the right side using the same screen. If the two graphs are indistinguishable, then we have a partial verification that the equation is an identity. Of course, we can never see the entire graph, so there can always be some doubt. Also, the graphs may not overlap precisely, but you may not be able to tell because the difference between the graphs may be less than the width of a pixel. However, if the graphs are obviously different, we know that a mistake has been made.

Consider the identity in Example 1:

$$\cos x\,(\tan x - \sec x) = \sin x - 1.$$

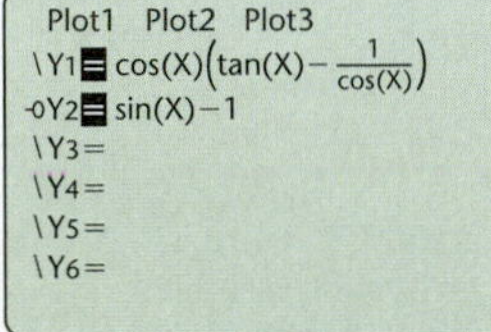

Recalling that $\sec x = 1/\cos x$, we enter

$$y_1 = \cos x\,(\tan x - 1/\cos x) \quad \text{and} \quad y_2 = \sin x - 1.$$

$y_1 = \cos x\left(\tan x - \dfrac{1}{\cos x}\right)$,
$y_2 = \sin x - 1$

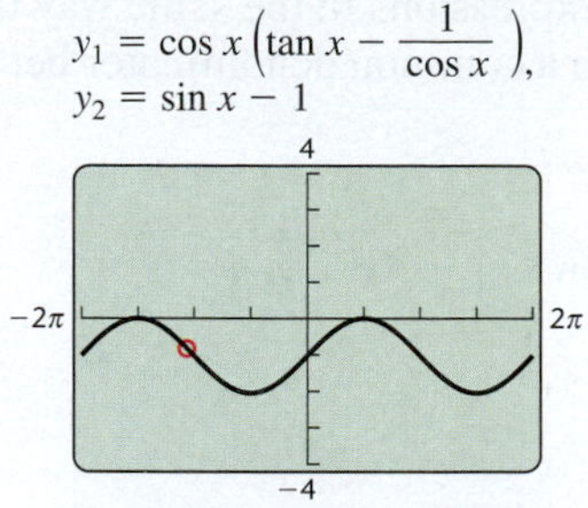

X	Y1	Y2
−6.283	−1	−1
−5.498	−.2929	−.2929
−4.712	ERROR	0
−3.927	−.2929	−.2929
−3.142	−1	−1
−2.356	−1.707	−1.707
−1.571	ERROR	−2

X = −6.28318530718

TblStart = -2π
ΔTbl = $\pi/4$

To graph, we first select SEQUENTIAL mode. Then we select the "line"-graph style for y_1 and the "path"-graph style, denoted by -O, for y_2. The calculator will graph y_1 first. Then it will graph y_2 as the circular cursor traces the leading edge of the graph, allowing us to determine whether the graphs coincide. As you can see in the graph screen at left, the graphs appear to be identical. Thus, $\cos x\,(\tan x - \sec x) = \sin x - 1$ is most likely an identity.

The TABLE feature can also be used to check identities. Note in the table at left that the function values are the same except for those values of x for which $\cos x = 0$. The domain of y_1 excludes these values. The domain of y_2 is the set of all real numbers. Thus all real numbers except $\pm\pi/2, \pm 3\pi/2, \pm 5\pi/2, \ldots$ are possible replacements for x in the identity. Recall that an identity is an equation that is true for all *possible* replacements.

EXAMPLE 3 Simplify each of the following trigonometric expressions.

a) $\dfrac{\cot(-\theta)}{\csc(-\theta)}$ b) $\dfrac{2\sin^2 t + \sin t - 3}{1 - \cos^2 t - \sin t}$

Solution

Just in Time 21

a) $\dfrac{\cot(-\theta)}{\csc(-\theta)} = \dfrac{\dfrac{\cos(-\theta)}{\sin(-\theta)}}{\dfrac{1}{\sin(-\theta)}}$ **Rewriting in terms of sines and cosines**

$= \dfrac{\cos(-\theta)}{\sin(-\theta)} \cdot \dfrac{\sin(-\theta)}{1}$ **Multiplying by the reciprocal, $\sin(-\theta)/1$**

$= \cos(-\theta)$ **Removing a factor of 1, $\sin(-\theta)/\sin(-\theta)$**

$= \cos\theta$ **The cosine function is even.**

Recall that the sine function is odd, $\sin(-\theta) = -\sin\theta$, and the cosine function is even, $\cos(-\theta) = \cos\theta$. It can be shown that the tangent, the cotangent, and the cosecant functions are odd and the secant function is even. We can also simplify this expression using those identities:

$$\frac{\cot(-\theta)}{\csc(-\theta)} = \frac{-\cot\theta}{-\csc\theta} = \frac{\cot\theta}{\csc\theta} = \frac{\dfrac{\cos\theta}{\sin\theta}}{\dfrac{1}{\sin\theta}}$$

$$= \frac{\cos\theta}{\sin\theta} \cdot \frac{\sin\theta}{1} = \cos\theta.$$

Just in Time 14

b) $\dfrac{2\sin^2 t + \sin t - 3}{1 - \cos^2 t - \sin t}$

$= \dfrac{2\sin^2 t + \sin t - 3}{\sin^2 t - \sin t}$ **Substituting $\sin^2 t$ for $1 - \cos^2 t$**

$= \dfrac{(2\sin t + 3)(\sin t - 1)}{\sin t\,(\sin t - 1)}$ **Factoring in both the numerator and the denominator**

$= \dfrac{2\sin t + 3}{\sin t}$ **Simplifying**

$= \dfrac{2\sin t}{\sin t} + \dfrac{3}{\sin t}$

$= 2 + \dfrac{3}{\sin t}$, or $2 + 3\csc t$

Now Try Exercises 17 and 19.

We can add and subtract trigonometric rational expressions in the same way that we do algebraic expressions, writing expressions with a common denominator before adding and subtracting numerators.

Just in Time 23

EXAMPLE 4 Add and simplify: $\dfrac{\cos x}{1 + \sin x} + \tan x$.

Solution

$$\frac{\cos x}{1+\sin x} + \tan x = \frac{\cos x}{1+\sin x} + \frac{\sin x}{\cos x} \qquad \text{Using } \tan x = \frac{\sin x}{\cos x}$$

$$= \frac{\cos x}{1+\sin x}\cdot\frac{\cos x}{\cos x} + \frac{\sin x}{\cos x}\cdot\frac{1+\sin x}{1+\sin x} \qquad \text{Multiplying by forms of 1}$$

$$= \frac{\cos^2 x + \sin x + \sin^2 x}{\cos x\,(1+\sin x)} \qquad \text{Adding}$$

$$= \frac{1+\sin x}{\cos x\,(1+\sin x)} \qquad \text{Using } \sin^2 x + \cos^2 x = 1$$

$$= \frac{1}{\cos x}, \quad \text{or} \quad \sec x \qquad \text{Simplifying}$$

Now Try Exercise 27.

When radicals occur, the use of absolute value is sometimes necessary, but it can be difficult to determine when to use it. In Examples 5 and 6, we will assume that all radicands are nonnegative. This means that the identities are meant to be confined to certain quadrants.

Just in Time 25

EXAMPLE 5 Multiply and simplify: $\sqrt{\sin^3 x \cos x}\cdot\sqrt{\cos x}$.

Solution

$$\sqrt{\sin^3 x\,\cos x}\cdot\sqrt{\cos x} = \sqrt{\sin^3 x\,\cos^2 x}$$

$$= \sqrt{\sin^2 x\,\cos^2 x\,\sin x}$$

$$= \sin x\,\cos x\sqrt{\sin x}$$

Now Try Exercise 31.

Just in Time 26

EXAMPLE 6 Rationalize the denominator: $\sqrt{\dfrac{2}{\tan x}}$.

Solution

$$\sqrt{\frac{2}{\tan x}} = \sqrt{\frac{2}{\tan x}\cdot\frac{\tan x}{\tan x}}$$

$$= \sqrt{\frac{2\tan x}{\tan^2 x}}$$

$$= \frac{\sqrt{2\tan x}}{\tan x}$$

Now Try Exercise 37.

Often in calculus, a substitution is a useful manipulation, as we show in the following example.

EXAMPLE 7 Express $\sqrt{9 + x^2}$ as a trigonometric function of θ without using radicals by letting $x = 3 \tan \theta$. Assume that $0 < \theta < \pi/2$. Then find $\sin \theta$ and $\cos \theta$.

Solution We have

$$\begin{aligned}
\sqrt{9 + x^2} &= \sqrt{9 + (3 \tan \theta)^2} && \text{Substituting } 3 \tan \theta \text{ for } x\\
&= \sqrt{9 + 9 \tan^2 \theta}\\
&= \sqrt{9(1 + \tan^2 \theta)} && \text{Factoring}\\
&= \sqrt{9 \sec^2 \theta} && \text{Using } 1 + \tan^2 x = \sec^2 x\\
&= 3|\sec \theta| = 3 \sec \theta. && \text{For } 0 < \theta < \pi/2, \sec \theta > 0, \text{ so } |\sec \theta| = \sec \theta.
\end{aligned}$$

We can express $\sqrt{9 + x^2} = 3 \sec \theta$ as

$$\sec \theta = \frac{\sqrt{9 + x^2}}{3}.$$

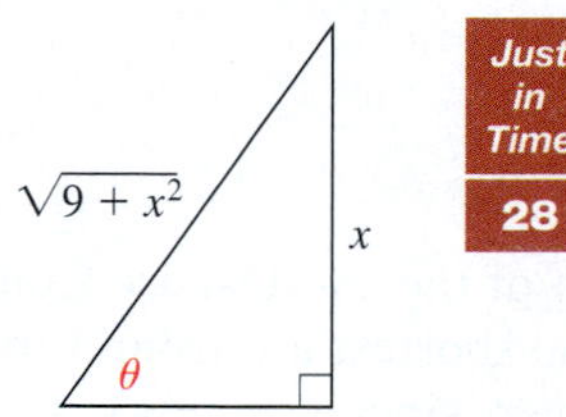

Just in Time 28

In a right triangle, we know that $\sec \theta$ is hypotenuse/adjacent, when θ is one of the acute angles. Using the Pythagorean equation, we can determine that the side opposite θ is x. Then from the right triangle, we see that

$$\sin \theta = \frac{x}{\sqrt{9 + x^2}} \quad \text{and} \quad \cos \theta = \frac{3}{\sqrt{9 + x^2}}.$$

Now Try Exercise 45.

Sum and Difference Identities

We now develop some important identities involving sums or differences of two numbers (or angles), beginning with an identity for the cosine of the difference of two numbers. We use the letters u and v for these numbers.

Let's consider a real number u in the interval $[\pi/2, \pi]$ and a real number v in the interval $[0, \pi/2]$. These determine points A and B on the unit circle, as shown below. The arc length s is $u - v$, and we know that $0 \leq s \leq \pi$. Recall that the coordinates of A are $(\cos u, \sin u)$ and the coordinates of B are $(\cos v, \sin v)$.

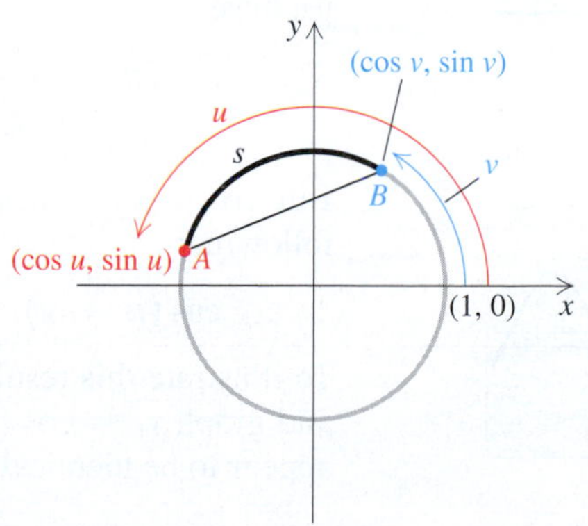

DISTANCE FORMULA

REVIEW SECTION 1.1.

Using the distance formula, we can write an expression for the distance AB:

$$AB = \sqrt{(\cos u - \cos v)^2 + (\sin u - \sin v)^2}.$$

This can be simplified as follows:

$$\begin{aligned}
AB &= \sqrt{\cos^2 u - 2 \cos u \cos v + \cos^2 v + \sin^2 u - 2 \sin u \sin v + \sin^2 v}\\
&= \sqrt{(\sin^2 u + \cos^2 u) + (\sin^2 v + \cos^2 v) - 2(\cos u \cos v + \sin u \sin v)}\\
&= \sqrt{2 - 2(\cos u \cos v + \sin u \sin v)}.
\end{aligned}$$

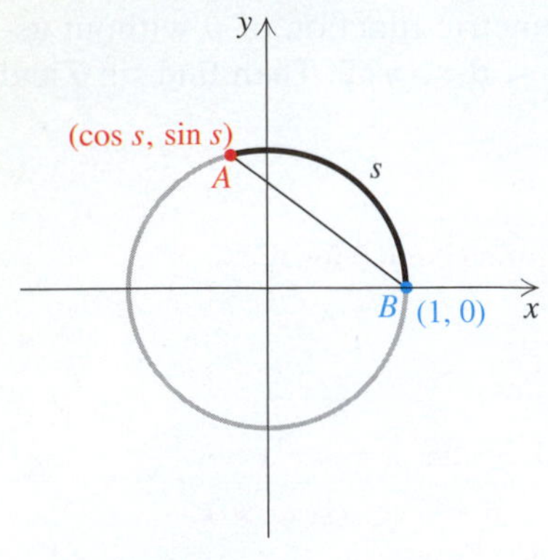

Now let's imagine rotating the circle on p. 499 so that point B is at $(1, 0)$, as shown at left. Although the coordinates of point A are now $(\cos s, \sin s)$, the distance AB has not changed.

Again we use the distance formula to write an expression for the distance AB:

$$AB = \sqrt{(\cos s - 1)^2 + (\sin s - 0)^2}.$$

This can be simplified as follows:

$$\begin{aligned} AB &= \sqrt{\cos^2 s - 2\cos s + 1 + \sin^2 s} \\ &= \sqrt{(\sin^2 s + \cos^2 s) + 1 - 2\cos s} \\ &= \sqrt{2 - 2\cos s}. \end{aligned}$$

Equating our two expressions for AB, we obtain

$$\sqrt{2 - 2(\cos u \cos v + \sin u \sin v)} = \sqrt{2 - 2\cos s}.$$

Solving this equation for $\cos s$ gives

$$\cos s = \cos u \cos v + \sin u \sin v. \quad \textbf{(1)}$$

But $s = u - v$, so we have the equation

$$\cos(u - v) = \cos u \cos v + \sin u \sin v. \quad \textbf{(2)}$$

Formula (1) above holds when s is the length of the shortest arc from A to B. Given any real numbers u and v, the length of the shortest arc from A to B is not always $u - v$. In fact, it could be $v - u$. However, since $\cos(-x) = \cos x$, we know that $\cos(v - u) = \cos(u - v)$. Thus, $\cos s$ is always equal to $\cos(u - v)$. Formula (2) holds for all real numbers u and v. That formula is thus the identity we sought:

$$\mathbf{\cos(u - v) = \cos u \cos v + \sin u \sin v.}$$

$y_1 = \cos(x - 3)$,
$y_2 = \cos x \cos 3 + \sin x \sin 3$

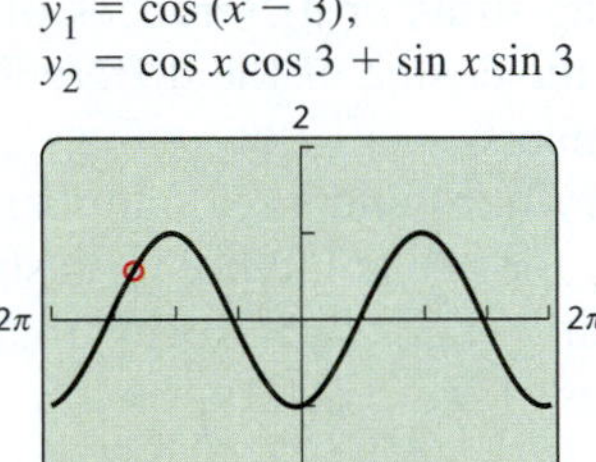

To illustrate this result using a graphing calculator, we replace u with x and v with 3 and graph $y_1 = \cos(x - 3)$ and $y_2 = \cos x \cos 3 + \sin x \sin 3$. The graphs appear to be identical.

The cosine sum formula follows easily from the one we have just derived. Let's consider $\cos(u + v)$. This is equal to $\cos[u - (-v)]$, and by the identity above, we have

$$\begin{aligned} \cos(u + v) &= \cos[u - (-v)] \\ &= \cos u \cos(-v) + \sin u \sin(-v). \end{aligned}$$

But $\cos(-v) = \cos v$ and $\sin(-v) = -\sin v$, so the identity we seek is the following:

$$\mathbf{\cos(u + v) = \cos u \cos v - \sin u \sin v.}$$

$y_1 = \cos(x + 2)$,
$y_2 = \cos x \cos 2 - \sin x \sin 2$

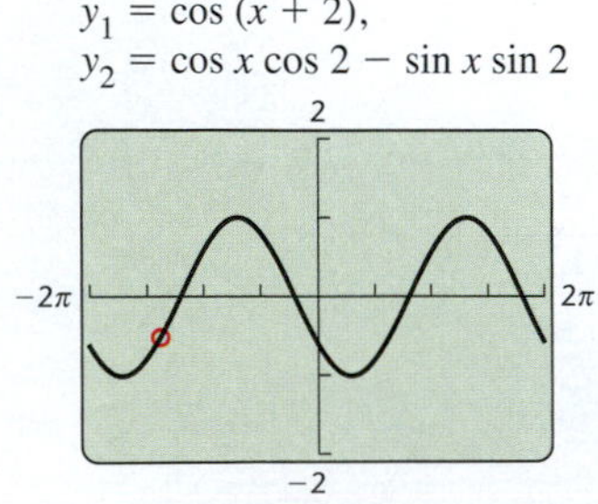

To illustrate this result using a graphing calculator, we replace u with x and v with 2 and graph $y_1 = \cos(x + 2)$ and $y_2 = \cos x \cos 2 - \sin x \sin 2$. Again, the graphs appear to be identical.

EXAMPLE 8 Find $\cos(5\pi/12)$ exactly.

Solution We can express $5\pi/12$ as a difference of two numbers whose exact sine and cosine values are known:

$$\frac{5\pi}{12} = \frac{9\pi}{12} - \frac{4\pi}{12}, \quad \text{or} \quad \frac{3\pi}{4} - \frac{\pi}{3}.$$

Then, using $\cos(u - v) = \cos u \cos v + \sin u \sin v$, we have

$$\begin{aligned}\cos\frac{5\pi}{12} = \cos\left(\frac{3\pi}{4} - \frac{\pi}{3}\right) &= \cos\frac{3\pi}{4}\cos\frac{\pi}{3} + \sin\frac{3\pi}{4}\sin\frac{\pi}{3}\\ &= -\frac{\sqrt{2}}{2}\cdot\frac{1}{2} + \frac{\sqrt{2}}{2}\cdot\frac{\sqrt{3}}{2}\\ &= -\frac{\sqrt{2}}{4} + \frac{\sqrt{6}}{4}\\ &= \frac{\sqrt{6} - \sqrt{2}}{4}.\end{aligned}$$

$\cos\left(\frac{5\pi}{12}\right)$	.2588190451
$\frac{\sqrt{6}-\sqrt{2}}{4}$	.2588190451

We can check using a graphing calculator set in RADIAN mode.

Now Try Exercise 51.

Consider $\cos(\pi/2 - \theta)$. We can use the identity for the cosine of a difference to simplify as follows:

$$\begin{aligned}\cos\left(\frac{\pi}{2} - \theta\right) &= \cos\frac{\pi}{2}\cos\theta + \sin\frac{\pi}{2}\sin\theta\\ &= 0\cdot\cos\theta + 1\cdot\sin\theta\\ &= \sin\theta.\end{aligned}$$

Thus we have developed the identity

$$\boldsymbol{\sin\theta = \cos\left(\frac{\pi}{2} - \theta\right).} \qquad (3)$$

This cofunction identity first appeared in Section 6.1.

This identity holds for any real number θ. From it, we can obtain an identity for the cosine function. We first let α be any real number. Then we replace θ in $\sin\theta = \cos(\pi/2 - \theta)$ with $\pi/2 - \alpha$. This gives us

$$\sin\left(\frac{\pi}{2} - \alpha\right) = \cos\left[\frac{\pi}{2} - \left(\frac{\pi}{2} - \alpha\right)\right] = \cos\alpha,$$

which yields the identity

$$\boldsymbol{\cos\alpha = \sin\left(\frac{\pi}{2} - \alpha\right).} \qquad (4)$$

Using identities (3) and (4) and the identity for the cosine of a difference, we can obtain an identity for the sine of a sum. We start with identity (3) and substitute $u + v$ for θ:

$$\sin\theta = \cos\left(\frac{\pi}{2} - \theta\right) \qquad \textbf{Identity (3)}$$

$$\begin{aligned}\sin(u + v) &= \cos\left[\frac{\pi}{2} - (u + v)\right] && \textbf{Substituting } u + v \textbf{ for } \theta\\ &= \cos\left[\left(\frac{\pi}{2} - u\right) - v\right]\\ &= \cos\left(\frac{\pi}{2} - u\right)\cos v + \sin\left(\frac{\pi}{2} - u\right)\sin v && \textbf{Using the identity for the cosine of a difference}\\ &= \sin u \cos v + \cos u \sin v. && \textbf{Using identities (3) and (4)}\end{aligned}$$

Thus the identity we seek is

$$\sin(u + v) = \sin u \cos v + \cos u \sin v.$$

To find a formula for the sine of a difference, we can use the identity just derived, substituting $-v$ for v:

$$\sin(u + (-v)) = \sin u \cos(-v) + \cos u \sin(-v).$$

Simplifying gives us

$$\sin(u - v) = \sin u \cos v - \cos u \sin v.$$

EXAMPLE 9 Find sin 105° exactly.

Solution We express 105° as the sum of two measures:

$$105° = 45° + 60°.$$

Then

$$\begin{aligned} \sin 105° &= \sin(45° + 60°) \\ &= \sin 45° \cos 60° + \cos 45° \sin 60° \qquad \text{Using } \sin(u+v) = \sin u \cos v + \cos u \sin v \\ &= \frac{\sqrt{2}}{2} \cdot \frac{1}{2} + \frac{\sqrt{2}}{2} \cdot \frac{\sqrt{3}}{2} \\ &= \frac{\sqrt{2} + \sqrt{6}}{4}. \end{aligned}$$

sin(105)
.9659258263
$\frac{\sqrt{2}+\sqrt{6}}{4}$
.9659258263

We can check this result using a graphing calculator set in DEGREE mode.

Now Try Exercise 55.

Formulas for the tangent of a sum or a difference can be derived using identities already established. A summary of the sum and difference identities follows.

SUM AND DIFFERENCE IDENTITIES

$$\sin(u \pm v) = \sin u \cos v \pm \cos u \sin v,$$

$$\cos(u \pm v) = \cos u \cos v \mp \sin u \sin v,$$

$$\tan(u \pm v) = \frac{\tan u \pm \tan v}{1 \mp \tan u \tan v}$$

There are six identities here, half of them obtained by using the signs shown in color.

Just in Time 24

EXAMPLE 10 Find tan 15° exactly.

Solution We rewrite 15° as 45° − 30° and use the identity for the tangent of a difference:

$$\begin{aligned} \tan 15° &= \tan(45° - 30°) \\ &= \frac{\tan 45° - \tan 30°}{1 + \tan 45° \tan 30°} \\ &= \frac{1 - \sqrt{3}/3}{1 + 1 \cdot \sqrt{3}/3} \\ &= \frac{3 - \sqrt{3}}{3 + \sqrt{3}}. \end{aligned}$$

Now Try Exercise 53.

EXAMPLE 11 Assume that $\sin \alpha = \frac{2}{3}$ and $\sin \beta = \frac{1}{3}$ and that α and β are between 0 and $\pi/2$. Then evaluate $\sin(\alpha + \beta)$.

Solution Using the identity for the sine of a sum, we have

$$\begin{aligned}\sin(\alpha + \beta) &= \sin \alpha \cos \beta + \cos \alpha \sin \beta \\ &= \tfrac{2}{3}\cos \beta + \tfrac{1}{3}\cos \alpha.\end{aligned}$$

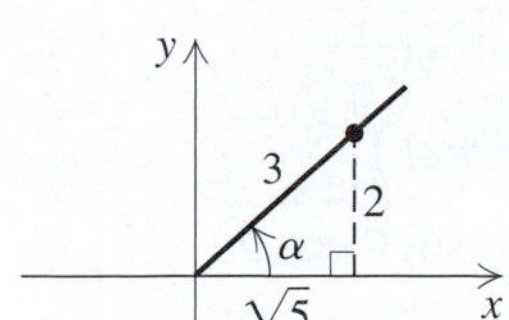

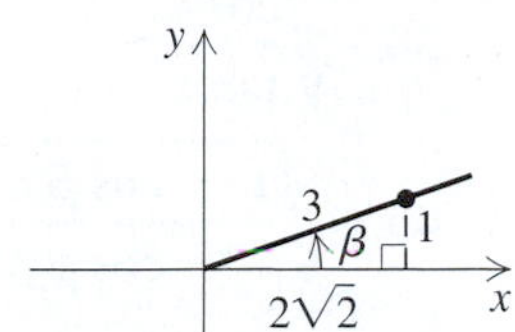

To finish, we need to know the values of $\cos \beta$ and $\cos \alpha$. Using reference triangles and the Pythagorean equation, we can determine these values from the diagrams:

$$\cos \alpha = \frac{\sqrt{5}}{3} \quad \text{and} \quad \cos \beta = \frac{2\sqrt{2}}{3}.$$

Cosine values are positive in the first quadrant.

Substituting these values gives us

$$\begin{aligned}\sin(\alpha + \beta) &= \frac{2}{3}\cdot\frac{2\sqrt{2}}{3} + \frac{1}{3}\cdot\frac{\sqrt{5}}{3} \\ &= \frac{4\sqrt{2}}{9} + \frac{\sqrt{5}}{9}, \quad \text{or} \quad \frac{4\sqrt{2} + \sqrt{5}}{9}.\end{aligned}$$

Now Try Exercise 65.

EXAMPLE 12 Assume that $\cos \alpha = -\frac{4}{5}$ with α between π and $3\pi/2$ and that $\cos \beta = -\frac{2}{5}$ with β between $\pi/2$ and π. Then evaluate $\cos(\alpha - \beta)$.

Solution Using the identity for the cosine of a difference, we have

$$\begin{aligned}\cos(\alpha - \beta) &= \cos \alpha \cos \beta + \sin \alpha \sin \beta \\ &= \left(-\frac{4}{5}\right)\cdot\left(-\frac{2}{5}\right) + \sin \alpha \sin \beta \\ &= \frac{8}{25} + \sin \alpha \sin \beta.\end{aligned}$$

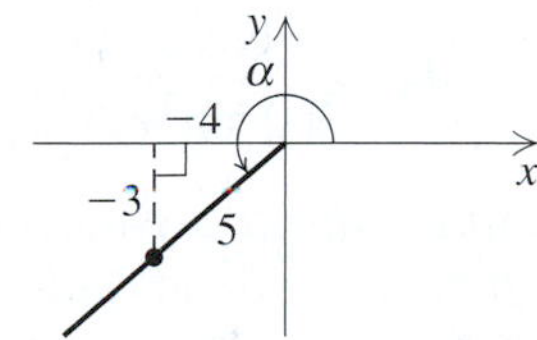

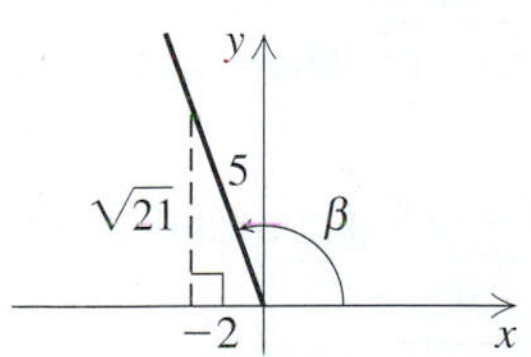

We need to know the values of $\sin \alpha$ and $\sin \beta$. Using reference triangles and the Pythagorean theorem, we can determine these values from the diagrams:

$$\sin \alpha = -\frac{3}{5} \quad \text{and} \quad \sin \beta = \frac{\sqrt{21}}{5}.$$

Substituting these values gives us

$$\begin{aligned}\cos(\alpha - \beta) &= \frac{8}{25} + \left(-\frac{3}{5}\right)\cdot\left(\frac{\sqrt{21}}{5}\right) \\ &= \frac{8}{25} - \frac{3\sqrt{21}}{25} = \frac{8 - 3\sqrt{21}}{25}.\end{aligned}$$

Now Try Exercise 69.

7.1 Exercise Set

Multiply and simplify. Check your result using a graphing calculator.

1. $(\sin x - \cos x)(\sin x + \cos x)$
2. $\tan x(\cos x - \csc x)$
3. $\cos y \sin y(\sec y + \csc y)$
4. $(\sin x + \cos x)(\sec x + \csc x)$
5. $(\sin \phi - \cos \phi)^2$
6. $(1 + \tan x)^2$
7. $(\sin x + \csc x)(\sin^2 x + \csc^2 x - 1)$
8. $(1 - \sin t)(1 + \sin t)$

Factor and simplify, if possible. Check your result using a graphing calculator.

9. $\sin x \cos x + \cos^2 x$

10. $\tan^2\theta - \cot^2\theta$

11. $\sin^4 x - \cos^4 x$

12. $4\sin^2 y + 8\sin y + 4$

13. $2\cos^2 x + \cos x - 3$

14. $3\cot^2\beta + 6\cot\beta + 3$

15. $\sin^3 x + 27$

16. $1 - 125\tan^3 s$

Simplify and check using a graphing calculator.

17. $\dfrac{\sin^2 x \cos x}{\cos^2 x \sin x}$

18. $\dfrac{30\sin^3 x \cos x}{6\cos^2 x \sin x}$

19. $\dfrac{\sin^2 x + 2\sin x + 1}{\sin x + 1}$

20. $\dfrac{\cos^2\alpha - 1}{\cos\alpha + 1}$

21. $\dfrac{4\tan t \sec t + 2\sec t}{6\tan t \sec t + 2\sec t}$

22. $\dfrac{\csc(-x)}{\cot(-x)}$

23. $\dfrac{\sin^4 x - \cos^4 x}{\sin^2 x - \cos^2 x}$

24. $\dfrac{4\cos^3 x}{\sin^2 x}\cdot\left(\dfrac{\sin x}{4\cos x}\right)^2$

25. $\dfrac{5\cos\phi}{\sin^2\phi}\cdot\dfrac{\sin^2\phi - \sin\phi\cos\phi}{\sin^2\phi - \cos^2\phi}$

26. $\dfrac{\tan^2 y}{\sec y} \div \dfrac{3\tan^3 y}{\sec y}$

27. $\dfrac{1}{\sin^2 s - \cos^2 s} - \dfrac{2}{\cos s - \sin s}$

28. $\left(\dfrac{\sin x}{\cos x}\right)^2 - \dfrac{1}{\cos^2 x}$

29. $\dfrac{\sin^2\theta - 9}{2\cos\theta + 1}\cdot\dfrac{10\cos\theta + 5}{3\sin\theta + 9}$

30. $\dfrac{9\cos^2\alpha - 25}{2\cos\alpha - 2}\cdot\dfrac{\cos^2\alpha - 1}{6\cos\alpha - 10}$

Simplify and check using a graphing calculator. Assume that all radicands are nonnegative.

31. $\sqrt{\sin^2 x \cos x}\cdot\sqrt{\cos x}$

32. $\sqrt{\cos^2 x \sin x}\cdot\sqrt{\sin x}$

33. $\sqrt{\cos\alpha\sin^2\alpha} - \sqrt{\cos^3\alpha}$

34. $\sqrt{\tan^2 x - 2\tan x \sin x + \sin^2 x}$

35. $(1 - \sqrt{\sin y})(\sqrt{\sin y} + 1)$

36. $\sqrt{\cos\theta}(\sqrt{2\cos\theta} + \sqrt{\sin\theta\cos\theta})$

Rationalize the denominator.

37. $\sqrt{\dfrac{\sin x}{\cos x}}$

38. $\sqrt{\dfrac{\cos x}{\tan x}}$

39. $\sqrt{\dfrac{\cos^2 y}{2\sin^2 y}}$

40. $\sqrt{\dfrac{1 - \cos\beta}{1 + \cos\beta}}$

Rationalize the numerator.

41. $\sqrt{\dfrac{\cos x}{\sin x}}$

42. $\sqrt{\dfrac{\sin x}{\cot x}}$

43. $\sqrt{\dfrac{1 + \sin y}{1 - \sin y}}$

44. $\sqrt{\dfrac{\cos^2 x}{2\sin^2 x}}$

Use the given substitution to express the given radical expression as a trigonometric function without radicals. Assume that $a > 0$ and $0 < \theta < \pi/2$. Then find expressions for the indicated trigonometric functions.

45. Let $x = a\sin\theta$ in $\sqrt{a^2 - x^2}$. Then find $\cos\theta$ and $\tan\theta$.

46. Let $x = 2\tan\theta$ in $\sqrt{4 + x^2}$. Then find $\sin\theta$ and $\cos\theta$.

47. Let $x = 3\sec\theta$ in $\sqrt{x^2 - 9}$. Then find $\sin\theta$ and $\cos\theta$.

48. Let $x = a\sec\theta$ in $\sqrt{x^2 - a^2}$. Then find $\sin\theta$ and $\cos\theta$.

Use the given substitution to express the given radical expression as a trigonometric function without radicals. Assume that $0 < \theta < \pi/2$.

49. Let $x = \sin\theta$ in $\dfrac{x^2}{\sqrt{1 - x^2}}$.

50. Let $x = 4\sec\theta$ in $\dfrac{\sqrt{x^2 - 16}}{x^2}$.

Use the sum and difference identities to evaluate exactly. Then check using a graphing calculator.

51. $\sin\dfrac{\pi}{12}$

52. $\cos 75^\circ$

53. $\tan 105^\circ$

54. $\tan\dfrac{5\pi}{12}$

55. $\cos 15°$

56. $\sin \frac{7\pi}{12}$

First write each of the following as a trigonometric function of a single angle. Then evaluate.

57. $\sin 37° \cos 22° + \cos 37° \sin 22°$

58. $\cos 83° \cos 53° + \sin 83° \sin 53°$

59. $\cos 19° \cos 5° - \sin 19° \sin 5°$

60. $\sin 40° \cos 15° - \cos 40° \sin 15°$

61. $\dfrac{\tan 20° + \tan 32°}{1 - \tan 20° \tan 32°}$

62. $\dfrac{\tan 35° - \tan 12°}{1 + \tan 35° \tan 12°}$

63. Derive the formula for the tangent of a sum.

64. Derive the formula for the tangent of a difference.

Assuming that $\sin u = \frac{3}{5}$ *and* $\sin v = \frac{4}{5}$ *and that u and v are between 0 and* $\pi/2$*, evaluate each of the following exactly.*

65. $\cos (u + v)$

66. $\tan (u - v)$

67. $\sin (u - v)$

68. $\cos (u - v)$

Assuming that $\cos \alpha = -\frac{3}{7}$ *with* α *between* $\pi/2$ *and* π *and that* $\cos \beta = \frac{8}{9}$ *with* β *between* $3\pi/2$ *and* 2π*, evaluate each of the following exactly.*

69. $\cos (\alpha + \beta)$

70. $\sin (\alpha - \beta)$

Assuming that $\sin \theta = 0.6249$ *and* $\cos \phi = 0.1102$ *and that both* θ *and* ϕ *are first-quadrant angles, evaluate each of the following.*

71. $\tan (\theta + \phi)$

72. $\sin (\theta - \phi)$

73. $\cos (\theta - \phi)$

74. $\cos (\theta + \phi)$

Simplify.

75. $\sin (\alpha + \beta) + \sin (\alpha - \beta)$

76. $\cos (\alpha + \beta) - \cos (\alpha - \beta)$

77. $\cos (u + v) \cos v + \sin (u + v) \sin v$

78. $\sin (u - v) \cos v + \cos (u - v) \sin v$

Skill Maintenance

Solve. [1.5]

79. $2x - 3 = 2\left(x - \frac{3}{2}\right)$

80. $x - 7 = x + 3.4$

Given that $\sin 31° = 0.5150$ *and* $\cos 31° = 0.8572$*, find the specified function value.* [6.1]

81. $\sec 59°$

82. $\tan 59°$

Synthesis

Angles Between Lines. *One of the identities gives an easy way to find an angle formed by two lines. Consider two lines with equations* l_1: $y = m_1x + b_1$ *and* l_2: $y = m_2x + b_2$.

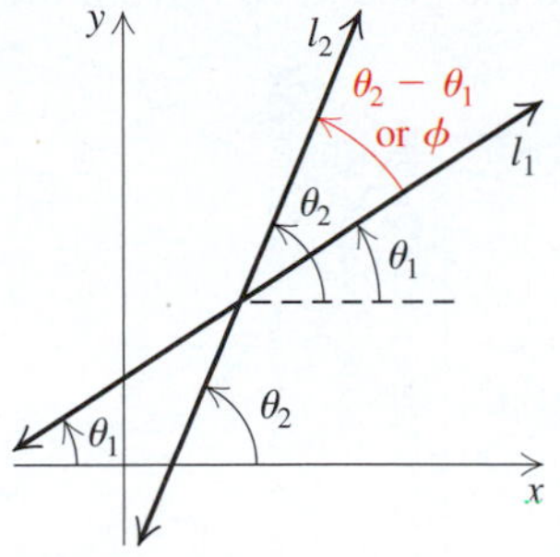

The slopes m_1 *and* m_2 *are the tangents of the angles* θ_1 *and* θ_2 *that the lines form with the positive direction of the x-axis. Thus we have* $m_1 = \tan \theta_1$ *and* $m_2 = \tan \theta_2$*. To find the measure of* $\theta_2 - \theta_1$*, or* ϕ*, we proceed as follows:*

$$\tan \phi = \tan (\theta_2 - \theta_1)$$
$$= \frac{\tan \theta_2 - \tan \theta_1}{1 + \tan \theta_2 \tan \theta_1} = \frac{m_2 - m_1}{1 + m_2m_1}.$$

This formula also holds when the lines are taken in the reverse order. When ϕ *is acute,* $\tan \phi$ *will be positive. When* ϕ *is obtuse,* $\tan \phi$ *will be negative.*

Find the measure of the angle from l_1 *to* l_2.

83. l_1: $2x = 3 - 2y$,
l_2: $x + y = 5$

84. l_1: $3y = \sqrt{3}x + 3$,
l_2: $y = \sqrt{3}x + 2$

85. l_1: $y = 3$,
l_2: $x + y = 5$

86. l_1: $2x + y - 4 = 0$,
l_2: $y - 2x + 5 = 0$

87. *Circus Guy Wire.* In a circus, a guy wire A is attached to the top of a 30-ft pole. Wire B is used for performers to walk up to the tight wire, 10 ft above the ground. Find the angle ϕ between the wires if they are attached to the ground 40 ft from the pole.

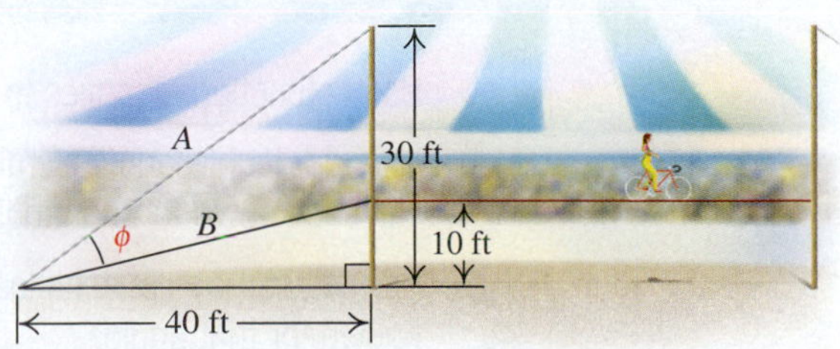

88. *Rope Course and Climbing Wall.* For a rope course and climbing wall, a guy wire R is attached 47 ft high on a vertical pole. Another guy wire S is attached 40 ft above the ground on the same pole. (*Source*: Experiential Resources, Inc., Todd Domeck, Owner) Find the angle α between the wires if they are attached to the ground 50 ft from the pole.

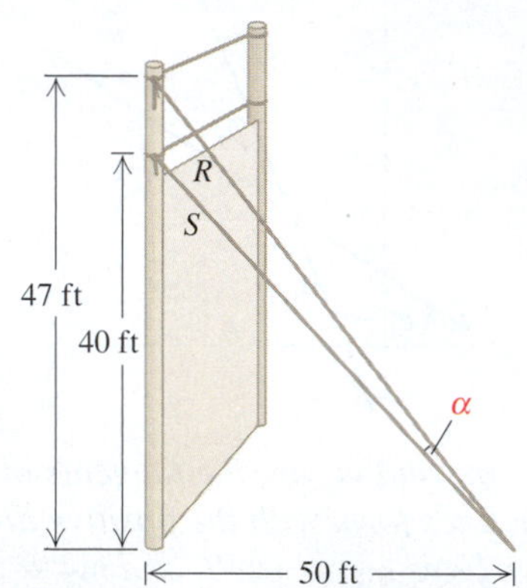

89. Given that $f(x) = \cos x$, show that

$$\frac{f(x+h) - f(x)}{h} = \cos x\left(\frac{\cos h - 1}{h}\right) - \sin x\left(\frac{\sin h}{h}\right).$$

90. Given that $f(x) = \sin x$, show that

$$\frac{f(x+h) - f(x)}{h} = \sin x\left(\frac{\cos h - 1}{h}\right) + \cos x\left(\frac{\sin h}{h}\right).$$

Show that each of the following is not an identity by finding a replacement or replacements for which the sides of the equation do not name the same number. Then use a graphing calculator to show that the equation is not an identity.

91. $\dfrac{\sin 5x}{x} = \sin 5$

92. $\sqrt{\sin^2\theta} = \sin\theta$

93. $\cos(2\alpha) = 2\cos\alpha$

94. $\sin(-x) = \sin x$

95. $\dfrac{\cos 6x}{\cos x} = 6$

96. $\tan^2\theta + \cot^2\theta = 1$

Find the slope of line l_1, where m_2 is the slope of line l_2 and ϕ is the smallest positive angle from l_1 to l_2.

97. $m_2 = \frac{2}{3}$, $\phi = 30°$

98. $m_2 = \frac{4}{3}$, $\phi = 45°$

99. Line l_1 contains the points $(-3, 7)$ and $(-3, -2)$. Line l_2 contains $(0, -4)$ and $(2, 6)$. Find the smallest positive angle from l_1 to l_2.

100. Line l_1 contains the points $(-2, 4)$ and $(5, -1)$. Find the slope of line l_2 such that the angle from l_1 to l_2 is 45°.

101. Find an identity for $\cos 2\theta$. (*Hint*: $2\theta = \theta + \theta$.)

102. Find an identity for $\sin 2\theta$. (*Hint*: $2\theta = \theta + \theta$.)

Derive the identity. Check using a graphing calculator.

103. $\tan\left(x + \dfrac{\pi}{4}\right) = \dfrac{1 + \tan x}{1 - \tan x}$

104. $\sin\left(x - \dfrac{3\pi}{2}\right) = \cos x$

105. $\sin(\alpha + \beta) + \sin(\alpha - \beta) = 2\sin\alpha\cos\beta$

106. $\dfrac{\sin(\alpha + \beta)}{\cos(\alpha - \beta)} = \dfrac{\tan\alpha + \tan\beta}{1 + \tan\alpha\tan\beta}$

7.2 Identities: Cofunction, Double-Angle, and Half-Angle

- Use cofunction identities to derive other identities.
- Use the double-angle identities to find function values of twice an angle when one function value is known for that angle.
- Use the half-angle identities to find function values of half an angle when one function value is known for that angle.
- Simplify trigonometric expressions using the double-angle identities and the half-angle identities.

Cofunction Identities

Each of the identities listed below yields a conversion to a *cofunction*. For this reason, we call them cofunction identities.

COFUNCTION IDENTITIES

$$\sin\left(\frac{\pi}{2} - x\right) = \cos x, \qquad \cos\left(\frac{\pi}{2} - x\right) = \sin x,$$

$$\tan\left(\frac{\pi}{2} - x\right) = \cot x, \qquad \cot\left(\frac{\pi}{2} - x\right) = \tan x,$$

$$\sec\left(\frac{\pi}{2} - x\right) = \csc x, \qquad \csc\left(\frac{\pi}{2} - x\right) = \sec x$$

We verified the first two of these identities in Section 7.1. The other four can be proved using the first two and the definitions of the trigonometric functions. These identities hold for all real numbers, and thus, for all angle measures, but if we restrict θ to values such that $0° < \theta < 90°$, or $0 < \theta < \pi/2$, then we have a special application to the acute angles of a right triangle.

Comparing graphs can lead to possible identities. On the left below, we see that the graph of $y = \sin(x + \pi/2)$ is a translation of the graph of $y = \sin x$ to the left $\pi/2$ units. On the right, we see the graph of $y = \cos x$.

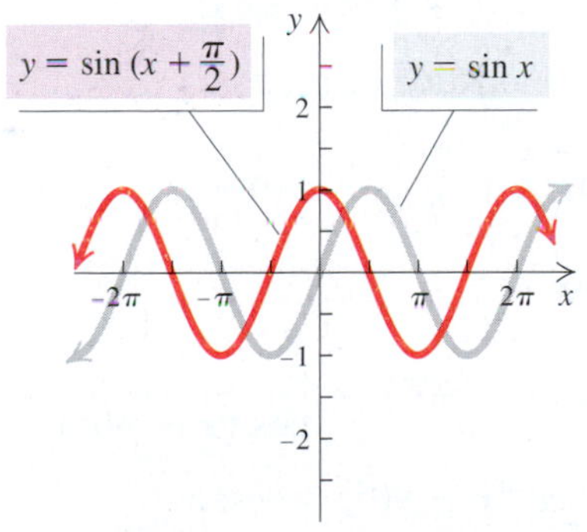

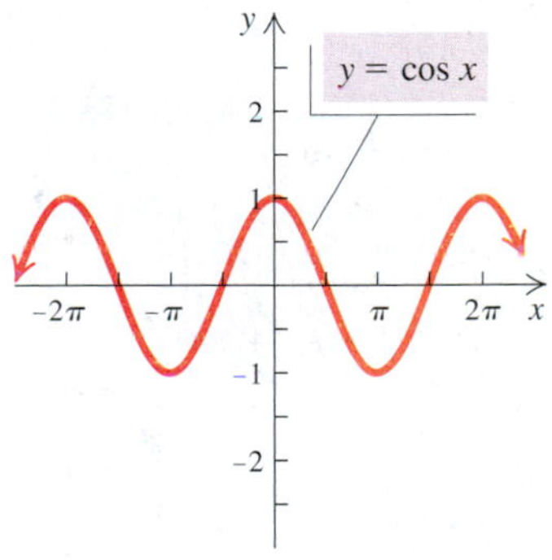

Comparing the graphs, we note a possible identity:

$$\sin\left(x + \frac{\pi}{2}\right) = \cos x.$$

The identity can be proved using the identity for the sine of a sum developed in Section 7.1.

EXAMPLE 1 Prove the identity $\sin(x + \pi/2) = \cos x$.

Solution

$$\begin{aligned}\sin\left(x + \frac{\pi}{2}\right) &= \sin x \cos\frac{\pi}{2} + \cos x \sin\frac{\pi}{2} && \text{Using } \sin(u+v) = \sin u\cos v + \cos u \sin v\\ &= \sin x \cdot 0 + \cos x \cdot 1\\ &= \cos x\end{aligned}$$

We now state four additional cofunction identities. These new identities that involve the sine and cosine functions can be verified using previously established identities as seen in Example 1.

COFUNCTION IDENTITIES FOR THE SINE AND THE COSINE

$$\sin\left(x \pm \frac{\pi}{2}\right) = \pm\cos x, \qquad \cos\left(x \pm \frac{\pi}{2}\right) = \mp\sin x$$

EXAMPLE 2 Find an identity for each of the following.

a) $\tan\left(x + \frac{\pi}{2}\right)$ **b)** $\sec\,(x - 90°)$

Solution

a) We have

$$\tan\left(x + \frac{\pi}{2}\right) = \frac{\sin\left(x + \frac{\pi}{2}\right)}{\cos\left(x + \frac{\pi}{2}\right)} \qquad \text{Using } \tan x = \frac{\sin x}{\cos x}$$

$$= \frac{\cos x}{-\sin x} \qquad \text{Using cofunction identities}$$

$$= -\cot x.$$

Thus the identity we seek is

$$\tan\left(x + \frac{\pi}{2}\right) = -\cot x.$$

b) We have

$$\sec\,(x - 90°) = \frac{1}{\cos\,(x - 90°)} = \frac{1}{\sin x} = \csc x.$$

Thus, $\sec\,(x - 90°) = \csc x$.

Now Try Exercises 5 and 7.

Double-Angle Identities

If we double an angle of measure x, the new angle will have measure $2x$. **Double-angle identities** give trigonometric function values of $2x$ in terms of function values of x. To develop these identities, we will use the sum formulas from the preceding section. We first develop a formula for $\sin 2x$. Recall that

$$\sin\,(u + v) = \sin u \cos v + \cos u \sin v.$$

We will consider a number x and substitute it for both u and v in this identity. Doing so gives us

$$\sin\,(x + x) = \sin 2x$$
$$= \sin x \cos x + \cos x \sin x$$
$$= 2 \sin x \cos x.$$

Our first double-angle identity is thus

$$\mathbf{\sin 2x = 2 \sin x \cos x.}$$

$y_1 = \sin 2x$, $y_2 = 2 \sin x \cos x$

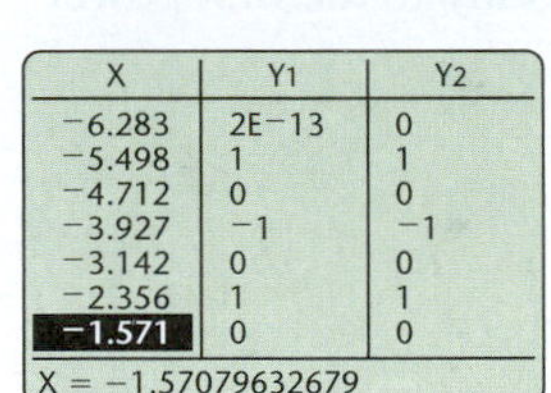

X	Y1	Y2
−6.283	2E−13	0
−5.498	1	1
−4.712	0	0
−3.927	−1	−1
−3.142	0	0
−2.356	1	1
−1.571	0	0

X = −1.57079632679

TblStart = -2π
ΔTbl = $\pi/4$

As a partial verification of this identity, we can graph

$$y_1 = \sin 2x \quad \text{and} \quad y_2 = 2 \sin x \cos x$$

using the "line"-graph style for y_1 and the "path"-graph style for y_2 and see that they appear to have the same graph, as shown at left. We can also use the TABLE feature.

Double-angle identities for the cosine and tangent functions can be derived in much the same way as the identity above:

$$\cos 2x = \cos^2 x - \sin^2 x, \qquad \tan 2x = \frac{2 \tan x}{1 - \tan^2 x}.$$

EXAMPLE 3 Given that $\tan \theta = -\frac{3}{4}$ and θ is in quadrant II, find each of the following.

a) $\sin 2\theta$ **b)** $\cos 2\theta$

c) $\tan 2\theta$ **d)** The quadrant in which 2θ lies

Solution By drawing a reference triangle as shown, we find that

$$\sin \theta = \frac{3}{5}$$

and

$$\cos \theta = -\frac{4}{5}.$$

Thus we have the following.

a) $\sin 2\theta = 2 \sin \theta \cos \theta = 2 \cdot \frac{3}{5} \cdot \left(-\frac{4}{5}\right) = -\frac{24}{25}$

b) $\cos 2\theta = \cos^2 \theta - \sin^2 \theta = \left(-\frac{4}{5}\right)^2 - \left(\frac{3}{5}\right)^2 = \frac{16}{25} - \frac{9}{25} = \frac{7}{25}$

c) $\tan 2\theta = \dfrac{2 \tan \theta}{1 - \tan^2 \theta} = \dfrac{2 \cdot \left(-\frac{3}{4}\right)}{1 - \left(-\frac{3}{4}\right)^2} = \dfrac{-\frac{3}{2}}{1 - \frac{9}{16}} = -\dfrac{3}{2} \cdot \dfrac{16}{7} = -\dfrac{24}{7}$

Note that $\tan 2\theta$ could have been found more easily in this case by simply dividing:

$$\tan 2\theta = \frac{\sin 2\theta}{\cos 2\theta} = \frac{-\frac{24}{25}}{\frac{7}{25}} = -\frac{24}{7}.$$

d) Since $\sin 2\theta$ is negative and $\cos 2\theta$ is positive, we know that 2θ is in quadrant IV.

Now Try Exercise 9.

Two other useful identities for $\cos 2x$ can be derived easily, as follows.

$$\begin{aligned} \cos 2x &= \cos^2 x - \sin^2 x \\ &= (1 - \sin^2 x) - \sin^2 x \\ &= 1 - 2 \sin^2 x \end{aligned}$$

$$\begin{aligned} \cos 2x &= \cos^2 x - \sin^2 x \\ &= \cos^2 x - (1 - \cos^2 x) \\ &= 2 \cos^2 x - 1 \end{aligned}$$

DOUBLE-ANGLE IDENTITIES

$$\sin 2x = 2 \sin x \cos x,$$

$$\tan 2x = \frac{2 \tan x}{1 - \tan^2 x}$$

$$\begin{aligned} \cos 2x &= \cos^2 x - \sin^2 x \\ &= 1 - 2 \sin^2 x \\ &= 2 \cos^2 x - 1 \end{aligned}$$

Solving the last two cosine double-angle identities for $\sin^2 x$ and $\cos^2 x$, respectively, we obtain two more identities:

$$\sin^2 x = \frac{1 - \cos 2x}{2} \quad \text{and} \quad \cos^2 x = \frac{1 + \cos 2x}{2}.$$

Using division and these two identities gives us the following useful identity:

$$\tan^2 x = \frac{1 - \cos 2x}{1 + \cos 2x}.$$

EXAMPLE 4 Find an equivalent expression for each of the following.

a) $\sin 3\theta$ in terms of function values of θ

b) $\cos^3 x$ in terms of function values of x or $2x$, raised only to the first power

Solution

a)
$$\begin{aligned}\sin 3\theta &= \sin (2\theta + \theta)\\ &= \sin 2\theta \cos \theta + \cos 2\theta \sin \theta\\ &= (2 \sin \theta \cos \theta) \cos \theta + (2 \cos^2\theta - 1) \sin \theta\end{aligned}$$

Using $\sin 2\theta = 2 \sin \theta \cos \theta$ and $\cos 2\theta = 2 \cos^2\theta - 1$

$$\begin{aligned}&= 2 \sin \theta \cos^2\theta + 2 \sin \theta \cos^2\theta - \sin \theta\\ &= 4 \sin \theta \cos^2\theta - \sin \theta\end{aligned}$$

We could also substitute $\cos^2\theta - \sin^2\theta$ or $1 - 2 \sin^2\theta$ for $\cos 2\theta$. Each substitution leads to a different result, but all results are equivalent.

b)
$$\begin{aligned}\cos^3 x &= \cos^2 x \cos x\\ &= \frac{1 + \cos 2x}{2} \cos x\\ &= \frac{\cos x + \cos x \cos 2x}{2}\end{aligned}$$

Now Try Exercise 15.

Half-Angle Identities

If we take half of an angle of measure x, the new angle will have measure $x/2$. **Half-angle identities** give trigonometric function values of $x/2$ in terms of function values of x. To develop these identities, we replace x with $x/2$ and take square roots. For example,

$$\sin^2 x = \frac{1 - \cos 2x}{2}$$ Solving the identity $\cos 2x = 1 - 2 \sin^2 x$ for $\sin^2 x$

$$\sin^2 \frac{x}{2} = \frac{1 - \cos 2 \cdot \frac{x}{2}}{2}$$ Substituting $\frac{x}{2}$ for x

$$\sin^2 \frac{x}{2} = \frac{1 - \cos x}{2}$$

$$\sin \frac{x}{2} = \pm\sqrt{\frac{1 - \cos x}{2}}.$$ Taking square roots

The formula is called a *half-angle formula*. The use of + and − depends on the quadrant in which the angle $x/2$ lies. Half-angle identities for the cosine and tangent functions can be derived in a similar manner. Two additional formulas for the half-angle tangent identity are also listed on the following page.

HALF-ANGLE IDENTITIES

$$\sin \frac{x}{2} = \pm\sqrt{\frac{1 - \cos x}{2}},$$

$$\cos \frac{x}{2} = \pm\sqrt{\frac{1 + \cos x}{2}},$$

$$\tan \frac{x}{2} = \pm\sqrt{\frac{1 - \cos x}{1 + \cos x}}$$

$$= \frac{\sin x}{1 + \cos x} = \frac{1 - \cos x}{\sin x}$$

EXAMPLE 5 Find $\tan(\pi/8)$ exactly. Then check the answer using a graphing calculator in RADIAN mode.

Solution We have

$$\tan\frac{\pi}{8} = \tan\frac{\frac{\pi}{4}}{2} = \frac{\sin\frac{\pi}{4}}{1 + \cos\frac{\pi}{4}} = \frac{\frac{\sqrt{2}}{2}}{1 + \frac{\sqrt{2}}{2}} = \frac{\frac{\sqrt{2}}{2}}{\frac{2 + \sqrt{2}}{2}} = \frac{\sqrt{2}}{2 + \sqrt{2}}$$

$$= \frac{\sqrt{2}}{2 + \sqrt{2}} \cdot \frac{2 - \sqrt{2}}{2 - \sqrt{2}} = \frac{2\sqrt{2} - 2}{4 - 2} = \frac{2(\sqrt{2} - 1)}{2} = \sqrt{2} - 1.$$

$\tan\left(\frac{\pi}{8}\right)$
.4142135624
$\sqrt{2}-1$
.4142135624

Now Try Exercise 21.

The identities that we have developed are also useful for simplifying trigonometric expressions.

EXAMPLE 6 Simplify each of the following.

a) $\dfrac{\sin x \cos x}{\frac{1}{2}\cos 2x}$ **b)** $2\sin^2\dfrac{x}{2} + \cos x$

Solution

a) We can obtain $2 \sin x \cos x$ in the numerator by multiplying the expression by $\frac{2}{2}$:

$$\frac{\sin x \cos x}{\frac{1}{2}\cos 2x} = \frac{2}{2} \cdot \frac{\sin x \cos x}{\frac{1}{2}\cos 2x} = \frac{2 \sin x \cos x}{\cos 2x}$$

$$= \frac{\sin 2x}{\cos 2x} \quad \text{Using } \sin 2x = 2 \sin x \cos x$$

$$= \tan 2x.$$

$y_1 = 2\sin^2\frac{x}{2} + \cos x, \; y_2 = 1$

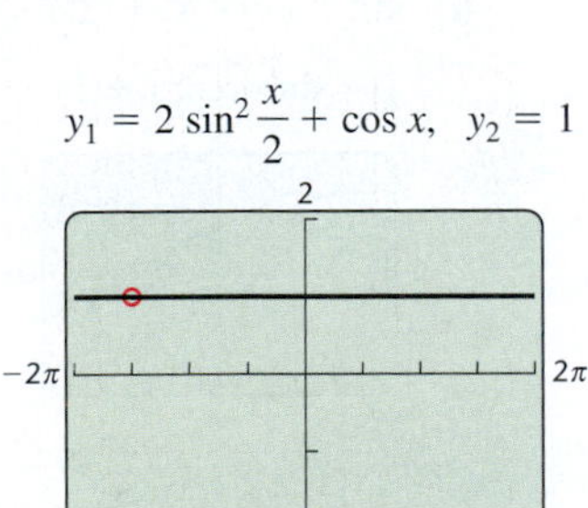

b) We have

$$2\sin^2\frac{x}{2} + \cos x = 2\left(\frac{1 - \cos x}{2}\right) + \cos x$$

$$\text{Using } \sin\frac{x}{2} = \pm\sqrt{\frac{1 - \cos x}{2}}, \text{ or } \sin^2\frac{x}{2} = \frac{1 - \cos x}{2}$$

$$= 1 - \cos x + \cos x$$

$$= 1.$$

X	Y1	Y2
−6.283	1	1
−5.498	1	1
−4.712	1	1
−3.927	1	1
−3.142	1	1
−2.356	1	1
−1.571	1	1

X = −6.28318530718

TblStart = -2π
ΔTbl = $\pi/4$

We can check this result using a graph or a table, as shown at left.

Now Try Exercise 31.

7.2 Exercise Set

1. Given that $\sin(3\pi/10) \approx 0.8090$ and $\cos(3\pi/10) \approx 0.5878$, find each of the following.

a) The other four function values for $3\pi/10$
b) The six function values for $\pi/5$

2. Given that

$$\sin\frac{\pi}{12} = \frac{\sqrt{2-\sqrt{3}}}{2} \quad \text{and} \quad \cos\frac{\pi}{12} = \frac{\sqrt{2+\sqrt{3}}}{2},$$

find exact answers for each of the following.

a) The other four function values for $\pi/12$
b) The six function values for $5\pi/12$

3. Given that $\sin\theta = \frac{1}{3}$ and that the terminal side is in quadrant II, find exact answers for each of the following.

a) The other function values for θ
b) The six function values for $\pi/2 - \theta$
c) The six function values for $\theta - \pi/2$

4. Given that $\cos\phi = \frac{4}{5}$ and that the terminal side is in quadrant IV, find exact answers for each of the following.

a) The other function values for ϕ
b) The six function values for $\pi/2 - \phi$
c) The six function values for $\phi + \pi/2$

Find an equivalent expression for each of the following.

5. $\sec\left(x + \frac{\pi}{2}\right)$

6. $\cot\left(x - \frac{\pi}{2}\right)$

7. $\tan\left(x - \frac{\pi}{2}\right)$

8. $\csc\left(x + \frac{\pi}{2}\right)$

Find the exact value of $\sin 2\theta$, $\cos 2\theta$, $\tan 2\theta$, *and the quadrant in which* 2θ *lies.*

9. $\sin\theta = \frac{4}{5}$, θ in quadrant I

10. $\cos\theta = \frac{5}{13}$, θ in quadrant I

11. $\cos\theta = -\frac{3}{5}$, θ in quadrant III

12. $\tan\theta = -\frac{15}{8}$, θ in quadrant II

13. $\tan\theta = -\frac{5}{12}$, θ in quadrant II

14. $\sin\theta = -\frac{\sqrt{10}}{10}$, θ in quadrant IV

15. Find an equivalent expression for $\cos 4x$ in terms of function values of x.

16. Find an equivalent expression for $\sin^4\theta$ in terms of function values of θ, 2θ, or 4θ, raised only to the first power.

Use the half-angle identities to evaluate exactly.

17. $\cos 15°$

18. $\tan 67.5°$

19. $\sin 112.5°$

20. $\cos\frac{\pi}{8}$

21. $\tan 75°$

22. $\sin\frac{5\pi}{12}$

Given that $\sin\theta = 0.3416$ *and* θ *is in quadrant I, find each of the following using identities.*

23. $\sin 2\theta$

24. $\cos\frac{\theta}{2}$

25. $\sin\frac{\theta}{2}$

26. $\sin 4\theta$

In Exercises 27–30, use a graphing calculator to determine which of the following expressions asserts an identity. Then derive the identity algebraically.

27. $\dfrac{\cos 2x}{\cos x - \sin x} = \cdots$

a) $1 + \cos x$

b) $\cos x - \sin x$

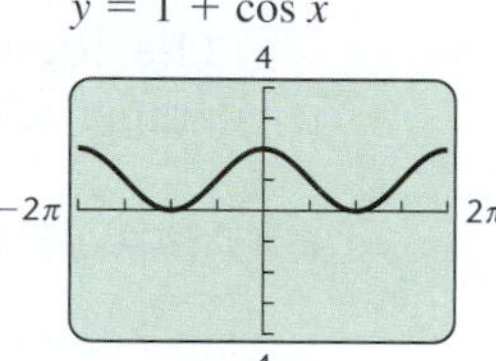

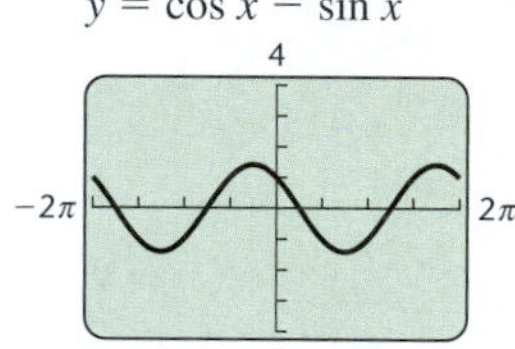

c) $-\cot x$

d) $\sin x(\cot x + 1)$

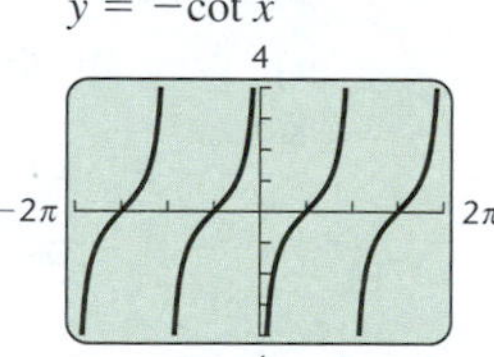

y = sin x (cot x + 1)
4
−2π
2π
−4

28. $2\cos^2\frac{x}{2} = \cdots$

a) $\sin x(\csc x + \tan x)$
b) $\sin x - 2\cos x$
c) $2(\cos^2 x - \sin^2 x)$
d) $1 + \cos x$

29. $\dfrac{\sin 2x}{2\cos x} = \cdots$

a) $\cos x$
b) $\tan x$
c) $\cos x + \sin x$
d) $\sin x$

30. $2 \sin \frac{\theta}{2} \cos \frac{\theta}{2} = \cdots$

a) $\cos^2 \theta$ b) $\sin \frac{\theta}{2}$

c) $\sin \theta$ d) $\sin \theta - \cos \theta$

Simplify. Check your results using a graphing calculator.

31. $2 \cos^2 \frac{x}{2} - 1$

32. $\cos^4 x - \sin^4 x$

33. $\frac{2 - \sec^2 x}{\sec^2 x}$

34. $(\sin x + \cos x)^2$

35. $(\sin x - \cos x)^2 + \sin 2x$

36. $\frac{1 + \sin 2x + \cos 2x}{1 + \sin 2x - \cos 2x}$

37. $(-4 \cos x \sin x + 2 \cos 2x)^2 + (2 \cos 2x + 4 \sin x \cos x)^2$

38. $2 \sin x \cos^3 x - 2 \sin^3 x \cos x$

Skill Maintenance

Complete the identity. [7.1]

39. $1 - \cos^2 x =$

40. $\sec^2 x - \tan^2 x =$

41. $\sin^2 x - 1 =$

42. $1 + \cot^2 x =$

43. $\csc^2 x - \cot^2 x$

44. $1 + \tan^2 x =$

45. $1 - \sin^2 x$

46. $\sec^2 x - 1$

Consider the following functions (a)–(f). Without graphing them, answer questions 47–50 below. [6.5]

a) $f(x) = 2 \sin \left(\frac{1}{2}x - \frac{\pi}{2} \right)$

b) $f(x) = \frac{1}{2} \cos \left(2x - \frac{\pi}{4} \right) + 2$

c) $f(x) = -\sin \left[2\left(x - \frac{\pi}{2} \right) \right] + 2$

d) $f(x) = \sin (x + \pi) - \frac{1}{2}$

e) $f(x) = -2 \cos (4x - \pi)$

f) $f(x) = -\cos \left[2\left(x - \frac{\pi}{8} \right) \right]$

47. Which functions have a graph with an amplitude of 2?

48. Which functions have a graph with a period of π?

49. Which functions have a graph with a period of 2π?

50. Which functions have a graph with a phase shift of $\frac{\pi}{4}$?

Synthesis

51. Given that $\cos 51° \approx 0.6293$, find the six function values for 141°.

52. Find sin 15° first using a difference identity and then using a half-angle identity. Then compare the results.

Simplify. Check your results using a graphing calculator.

53. $\cos (\pi - x) + \cot x \sin \left(x - \frac{\pi}{2} \right)$

54. $\sin \left(\frac{\pi}{2} - x \right) [\sec x - \cos x]$

55. $\frac{\cos^2 y \sin \left(y + \frac{\pi}{2} \right)}{\sin^2 y \sin \left(\frac{\pi}{2} - y \right)}$

56. $\frac{\cos x - \sin \left(\frac{\pi}{2} - x \right) \sin x}{\cos x - \cos (\pi - x) \tan x}$

Find $\sin \theta$, $\cos \theta$, *and* $\tan \theta$ *under the given conditions.*

57. $\tan \frac{\theta}{2} = -\frac{5}{3}, \ \pi < \theta \leq \frac{3\pi}{2}$

58. $\cos 2\theta = \frac{7}{12}, \ \frac{3\pi}{2} \leq 2\theta \leq 2\pi$

59. *Acceleration Due to Gravity.* The acceleration due to gravity is often denoted by g in a formula such as $S = \frac{1}{2}gt^2$, where S is the distance that an object falls in time t. The number g relates to motion near the earth's surface and is generally considered constant. In fact, however, g is not constant, but varies slightly with latitude. *Latitude* is used to measure north–south location on the earth between the equator and the poles. If ϕ stands for latitude, in degrees, g is given with good approximation by the formula

$$g = 9.78049(1 + 0.005288 \sin^2 \phi - 0.000006 \sin^2 2\phi),$$

where g is measured in meters per second per second at sea level.

a) Chicago has latitude 42°N. Find g.
b) Philadelphia has latitude 40°N. Find g.
c) Express g in terms of $\sin \phi$ only. That is, eliminate the double angle.

60. *Nautical Mile.* Sydney, Australia, has latitude 34°S. (See the figure.) In Great Britain, the *nautical mile* is defined as the length of a minute of arc of the earth's radius. Since the earth is flattened slightly at the poles, a British nautical mile varies with latitude. In fact, it is given, in feet, by the function

$$N(\phi) = 6066 - 31 \cos 2\phi,$$

where ϕ is the latitude in degrees.

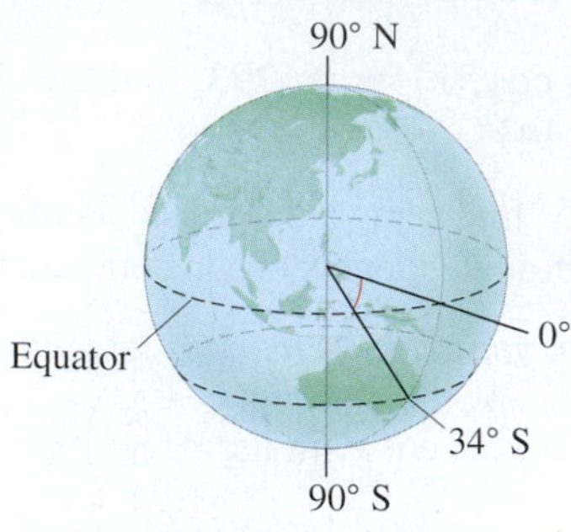

a) What is the length of a British nautical mile at Sydney?

b) What is the length of a British nautical mile at the North Pole?

c) Express $N(\phi)$ in terms of $\cos \phi$ only; that is, do not use the double angle.

7.3 Proving Trigonometric Identities

- Prove identities using other identities.
- Use the product-to-sum identities and the sum-to-product identities to derive other identities.

The Logic of Proving Identities

We outline two algebraic methods for proving identities.

Method 1. Start with either the left side or the right side of the equation and obtain the other side For example, suppose that you are trying to prove that the equation $P = Q$ is an identity. You might try to produce a string of statements ($R_1, R_2, \ldots$ or $T_1, T_2, \ldots$) like the following, which start with P and end with Q or start with Q and end with P:

$$\begin{aligned} P &= R_1 \\ &= R_2 \\ &\vdots \\ &= Q \end{aligned} \qquad \text{or} \qquad \begin{aligned} Q &= T_1 \\ &= T_2 \\ &\vdots \\ &= P. \end{aligned}$$

Method 2. Work with each side separately until you obtain the same expression. For example, suppose that you are trying to prove that $P = Q$ is an identity. You might be able to produce two strings of statements like the following, each ending with the same statement S.

$$\begin{aligned} P &= R_1 \\ &= R_2 \\ &\vdots \\ &= S \end{aligned} \qquad \begin{aligned} Q &= T_1 \\ &= T_2 \\ &\vdots \\ &= S. \end{aligned}$$

The number of steps in each string might be different, but in each case the result is S.

A first step in learning to prove identities is to have at hand a list of the identities that you have already learned. Such a list is on the inside back cover of this text. Ask your instructor which ones you are expected to memorize. The more identities you prove, the easier it will be to prove new ones. A list of helpful hints is shown at left.

Proving Identities

Just in Time 13

EXAMPLE 1 Prove the identity $1 + \sin 2\theta = (\sin\theta + \cos\theta)^2$.

Solution Let's use method 1. We begin with the right side and obtain the left side:

$$(\sin\theta + \cos\theta)^2 = \sin^2\theta + 2\sin\theta\cos\theta + \cos^2\theta \quad \text{Squaring}$$

$$= 1 + 2\sin\theta\cos\theta \quad \text{Recalling the identity } \sin^2 x + \cos^2 x = 1 \text{ and substituting}$$

$$= 1 + \sin 2\theta. \quad \text{Using } \sin 2x = 2\sin x\cos x$$

We could also begin with the left side and obtain the right side:

$$1 + \sin 2\theta = 1 + 2\sin\theta\cos\theta \quad \text{Using } \sin 2x = 2\sin x\cos x$$

$$= \sin^2\theta + 2\sin\theta\cos\theta + \cos^2\theta \quad \text{Replacing 1 with } \sin^2\theta + \cos^2\theta$$

$$= (\sin\theta + \cos\theta)^2. \quad \text{Factoring}$$

Now Try Exercises 13 and 19.

HINTS FOR PROVING IDENTITIES

1. Use either method 1 or method 2.
2. Work with the more complex side first.
3. Carry out any algebraic manipulations, such as adding, subtracting, multiplying, or factoring.
4. When rational expressions are involved, it can be helpful to multiply by forms of 1 to obtain a common denominator.
5. Converting all expressions to sines and cosines is often helpful.
6. Try something! Put your pencil to work and get involved. You will be amazed at how often this leads to success.

EXAMPLE 2 Prove the identity

$$\sin^2 x\tan^2 x = \tan^2 x - \sin^2 x.$$

Solution For this proof, we are going to use method 2, working with each side separately using method 2. We try to obtain the same expression on each side. In actual practice, you might work on one side for a while, then work on the other side, and then go back to the first side. In other words, you work back and forth until you arrive at the same expression. Let's start with the right side.

$$\tan^2 x - \sin^2 x = \frac{\sin^2 x}{\cos^2 x} - \sin^2 x \quad \text{Recalling the identity } \tan x = \frac{\sin x}{\cos x} \text{ and substituting}$$

$$= \frac{\sin^2 x}{\cos^2 x} - \sin^2 x \cdot \frac{\cos^2 x}{\cos^2 x} \quad \text{Multiplying by 1 in order to subtract}$$

$$= \frac{\sin^2 x - \sin^2 x\cos^2 x}{\cos^2 x} \quad \text{Carrying out the subtraction}$$

$$= \frac{\sin^2 x\,(1 - \cos^2 x)}{\cos^2 x} \quad \text{Factoring}$$

$$= \frac{\sin^2 x\sin^2 x}{\cos^2 x} \quad \text{Recalling the identity } 1 - \cos^2 x = \sin^2 x \text{ and substituting}$$

$$= \frac{\sin^4 x}{\cos^2 x}$$

At this point, we stop and work with the left side of the original identity, $\sin^2 x\tan^2 x$, and try to end with the same expression that we ended with on the right side:

$$\sin^2 x\tan^2 x = \sin^2 x\,\frac{\sin^2 x}{\cos^2 x} \quad \text{Recalling the identity } \tan x = \frac{\sin x}{\cos x} \text{ and substituting}$$

$$= \frac{\sin^4 x}{\cos^2 x}.$$

We have obtained the same expression from each side, so the proof is complete.

Now Try Exercise 25.

EXAMPLE 3 Prove the identity

$$\frac{\sin 2x}{\sin x} - \frac{\cos 2x}{\cos x} = \sec x.$$

Solution We will use method 1, starting with the left side since it is the more complex side of the identity.

$$\frac{\sin 2x}{\sin x} - \frac{\cos 2x}{\cos x} = \frac{2 \sin x \cos x}{\sin x} - \frac{\cos^2 x - \sin^2 x}{\cos x} \qquad \text{Using double-angle identities}$$

$$= 2 \cos x - \frac{\cos^2 x - \sin^2 x}{\cos x} \qquad \text{Simplifying}$$

$$= \frac{2 \cos^2 x}{\cos x} - \frac{\cos^2 x - \sin^2 x}{\cos x} \qquad \text{Multiplying } 2 \cos x \text{ by 1, or } \cos x / \cos x$$

$$= \frac{2 \cos^2 x - \cos^2 x + \sin^2 x}{\cos x} \qquad \text{Subtracting}$$

$$= \frac{\cos^2 x + \sin^2 x}{\cos x}$$

$$= \frac{1}{\cos x} \qquad \text{Using } \cos^2 x + \sin^2 x = 1$$

$$= \sec x \qquad \text{Recalling a basic identity}$$

Now Try Exercise 15.

EXAMPLE 4 Prove the identity

$$\frac{\sec t - 1}{t \sec t} = \frac{1 - \cos t}{t}.$$

Solution We use method 1, starting with the left side. Note that the left side involves $\sec t$, whereas the right side involves $\cos t$, so it might be wise to make use of a basic identity that involves these two expressions: $\sec t = 1/\cos t$.

$$\frac{\sec t - 1}{t \sec t} = \frac{\frac{1}{\cos t} - 1}{t \cdot \frac{1}{\cos t}} \qquad \text{Substituting } 1/\cos t \text{ for } \sec t$$

$$= \left(\frac{1}{\cos t} - 1\right) \cdot \frac{\cos t}{t}$$

$$= \frac{1}{t} - \frac{\cos t}{t} \qquad \text{Multiplying}$$

$$= \frac{1 - \cos t}{t}$$

We started with the left side and obtained the right side, so the proof is complete.

Now Try Exercise 5.

EXAMPLE 5 Prove the identity

$$\cot \phi + \csc \phi = \frac{\sin \phi}{1 - \cos \phi}.$$

Solution We are again using method 2, beginning with the left side:

$$\cot \phi + \csc \phi = \frac{\cos \phi}{\sin \phi} + \frac{1}{\sin \phi} \quad \text{Using basic identities}$$

$$= \frac{1 + \cos \phi}{\sin \phi}. \quad \text{Adding}$$

At this point, we stop and work with the right side of the original identity:

$$\frac{\sin \phi}{1 - \cos \phi} = \frac{\sin \phi}{1 - \cos \phi} \cdot \frac{1 + \cos \phi}{1 + \cos \phi} \quad \text{Multiplying by 1}$$

$$= \frac{\sin \phi\,(1 + \cos \phi)}{1 - \cos^2 \phi}$$

$$= \frac{\sin \phi\,(1 + \cos \phi)}{\sin^2 \phi} \quad \text{Using } \sin^2 x = 1 - \cos^2 x$$

$$= \frac{1 + \cos \phi}{\sin \phi}. \quad \text{Simplifying}$$

The proof is complete since we obtained the same expression from each side.

Now Try Exercise 29.

Product-to-Sum and Sum-to-Product Identities

On occasion, it is convenient to convert a product of trigonometric expressions to a sum, or the reverse. The following identities are useful in this connection.

PRODUCT-TO-SUM IDENTITIES

$$\sin x \cdot \sin y = \tfrac{1}{2}[\cos (x - y) - \cos (x + y)] \quad (1)$$

$$\cos x \cdot \cos y = \tfrac{1}{2}[\cos (x - y) + \cos (x + y)] \quad (2)$$

$$\sin x \cdot \cos y = \tfrac{1}{2}[\sin (x + y) + \sin (x - y)] \quad (3)$$

$$\cos x \cdot \sin y = \tfrac{1}{2}[\sin (x + y) - \sin (x - y)] \quad (4)$$

We can derive product-to-sum identities (1) and (2) using the sum and difference identities for the cosine function:

$$\cos (x + y) = \cos x \cos y - \sin x \sin y, \quad \text{Sum identity}$$

$$\cos (x - y) = \cos x \cos y + \sin x \sin y. \quad \text{Difference identity}$$

Subtracting the sum identity from the difference identity, we have

$$\cos (x - y) - \cos (x + y) = 2 \sin x \sin y \quad \text{Subtracting}$$

$$\tfrac{1}{2}[\cos (x - y) - \cos (x + y)] = \sin x \sin y. \quad \text{Multiplying by } \tfrac{1}{2}$$

Thus, $\sin x \sin y = \frac{1}{2}[\cos (x - y) - \cos (x + y)]$.

Adding the cosine sum and difference identities, we have

$$\cos(x - y) + \cos(x + y) = 2\cos x\cos y \qquad \text{Adding}$$
$$\tfrac{1}{2}[\cos(x - y) + \cos(x + y)] = \cos x\cos y. \qquad \text{Multiplying by } \tfrac{1}{2}$$

Thus, $\cos x\cos y = \frac{1}{2}[\cos(x - y) + \cos(x + y)]$.

Identities (3) and (4) can be derived in a similar manner using the sum and difference identities for the sine function.

EXAMPLE 6 Find an identity for $2\sin 3\theta\cos 7\theta$.

Solution We will use the identity

$$\sin x \cdot \cos y = \tfrac{1}{2}[\sin(x + y) + \sin(x - y)].$$

Here $x = 3\theta$ and $y = 7\theta$. Thus,

$$\begin{aligned} 2\sin 3\theta\cos 7\theta &= 2\cdot\tfrac{1}{2}[\sin(3\theta + 7\theta) + \sin(3\theta - 7\theta)] \\ &= \sin 10\theta + \sin(-4\theta) \\ &= \sin 10\theta - \sin 4\theta. \qquad \text{Using } \sin(-x) = -\sin x \end{aligned}$$

Now Try Exercise 37.

SUM-TO-PRODUCT IDENTITIES

$$\sin x + \sin y = 2\sin\frac{x + y}{2}\cos\frac{x - y}{2} \qquad (5)$$

$$\sin x - \sin y = 2\cos\frac{x + y}{2}\sin\frac{x - y}{2} \qquad (6)$$

$$\cos y + \cos x = 2\cos\frac{x + y}{2}\cos\frac{x - y}{2} \qquad (7)$$

$$\cos y - \cos x = 2\sin\frac{x + y}{2}\sin\frac{x - y}{2} \qquad (8)$$

The sum-to-product identities (5)–(8) can be derived using the product-to-sum identities. Proofs are left to the exercises.

EXAMPLE 7 Find an identity for $\cos\theta + \cos 5\theta$.

Solution We will use the identity

$$\cos y + \cos x = 2\cos\frac{x + y}{2}\cos\frac{x - y}{2}.$$

Here $x = 5\theta$ and $y = \theta$. Thus,

$$\begin{aligned} \cos\theta + \cos 5\theta &= 2\cos\frac{5\theta + \theta}{2}\cos\frac{5\theta - \theta}{2} \\ &= 2\cos 3\theta\cos 2\theta. \end{aligned}$$

Now Try Exercise 35.

7.3 Exercise Set

Prove the identity.

1. $\sec x - \sin x \tan x = \cos x$

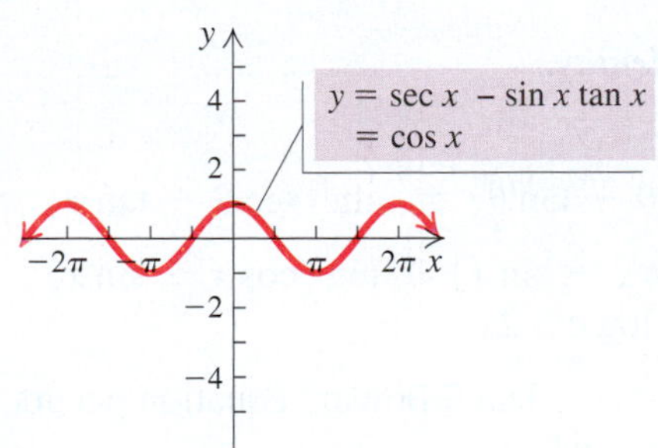

2. $\dfrac{1 + \cos\theta}{\sin\theta} + \dfrac{\sin\theta}{\cos\theta} = \dfrac{\cos\theta + 1}{\sin\theta\cos\theta}$

3. $\dfrac{1 - \cos x}{\sin x} = \dfrac{\sin x}{1 + \cos x}$

4. $\dfrac{1 + \tan y}{1 + \cot y} = \dfrac{\sec y}{\csc y}$

5. $\dfrac{1 + \tan\theta}{1 - \tan\theta} + \dfrac{1 + \cot\theta}{1 - \cot\theta} = 0$

6. $\dfrac{\sin x + \cos x}{\sec x + \csc x} = \dfrac{\sin x}{\sec x}$

7. $\dfrac{\cos^2\alpha + \cot\alpha}{\cos^2\alpha - \cot\alpha} = \dfrac{\cos^2\alpha\tan\alpha + 1}{\cos^2\alpha\tan\alpha - 1}$

8. $\sec 2\theta = \dfrac{\sec^2\theta}{2 - \sec^2\theta}$

9. $\dfrac{2\tan\theta}{1 + \tan^2\theta} = \sin 2\theta$

10. $\dfrac{\cos(u - v)}{\cos u \sin v} = \tan u + \cot v$

11. $1 - \cos 5\theta\cos 3\theta - \sin 5\theta\sin 3\theta = 2\sin^2\theta$

12. $\cos^4 x - \sin^4 x = \cos 2x$

13. $2\sin\theta\cos^3\theta + 2\sin^3\theta\cos\theta = \sin 2\theta$

14. $\dfrac{\tan 3t - \tan t}{1 + \tan 3t\,\tan t} = \dfrac{2\tan t}{1 - \tan^2 t}$

15. $\dfrac{\tan x - \sin x}{2\tan x} = \sin^2\dfrac{x}{2}$

16. $\dfrac{\cos^3\beta - \sin^3\beta}{\cos\beta - \sin\beta} = \dfrac{2 + \sin 2\beta}{2}$

17. $\sin(\alpha + \beta)\sin(\alpha - \beta) = \sin^2\alpha - \sin^2\beta$

18. $\cos^2 x\,(1 - \sec^2 x) = -\sin^2 x$

19. $\tan\theta\,(\tan\theta + \cot\theta) = \sec^2\theta$

20. $\dfrac{\cos\theta + \sin\theta}{\cos\theta} = 1 + \tan\theta$

21. $\dfrac{1 + \cos^2 x}{\sin^2 x} = 2\csc^2 x - 1$

22. $\dfrac{\tan y + \cot y}{\csc y} = \sec y$

23. $\dfrac{1 + \sin x}{1 - \sin x} + \dfrac{\sin x - 1}{1 + \sin x} = 4\sec x\tan x$

24. $\tan\theta - \cot\theta = (\sec\theta - \csc\theta)(\sin\theta + \cos\theta)$

25. $\cos^2\alpha\cot^2\alpha = \cot^2\alpha - \cos^2\alpha$

26. $\dfrac{\tan x + \cot x}{\sec x + \csc x} = \dfrac{1}{\cos x + \sin x}$

27. $2\sin^2\theta\cos^2\theta + \cos^4\theta = 1 - \sin^4\theta$

28. $\dfrac{\cot\theta}{\csc\theta - 1} = \dfrac{\csc\theta + 1}{\cot\theta}$

29. $\dfrac{1 + \sin x}{1 - \sin x} = (\sec x + \tan x)^2$

30. $\sec^4 s - \tan^2 s = \tan^4 s + \sec^2 s$

31. Derive the product-to-sum identities (3) and (4) using the sine sum and difference identities.

32. Derive the sum-to-product identities (5)–(8) using the product-to-sum identities (1)–(4).

Use the product-to-sum identities and the sum-to-product identities to find identities for each of the following.

33. $\sin 3\theta - \sin 5\theta$

34. $\sin 7x - \sin 4x$

35. $\sin 8\theta + \sin 5\theta$

36. $\cos\theta - \cos 7\theta$

37. $\sin 7u\sin 5u$

38. $2\sin 7\theta\cos 3\theta$

39. $7\cos\theta\sin 7\theta$

40. $\cos 2t\sin t$

41. $\cos 55^\circ\sin 25^\circ$

42. $7\cos 5\theta\cos 7\theta$

Use the product-to-sum identities and the sum-to-product identities to prove each of the following.

43. $\sin 4\theta + \sin 6\theta = \cot\theta\,(\cos 4\theta - \cos 6\theta)$

44. $\tan 2x\,(\cos x + \cos 3x) = \sin x + \sin 3x$

45. $\cot 4x\,(\sin x + \sin 4x + \sin 7x)$
$= \cos x + \cos 4x + \cos 7x$

46. $\tan \frac{x+y}{2} = \frac{\sin x + \sin y}{\cos x + \cos y}$

47. $\cot \frac{x+y}{2} = \frac{\sin y - \sin x}{\cos x - \cos y}$

48. $\tan \frac{\theta + \phi}{2} \tan \frac{\phi - \theta}{2} = \frac{\cos \theta - \cos \phi}{\cos \theta + \cos \phi}$

49. $\tan \frac{\theta + \phi}{2}(\sin \theta - \sin \phi)$

$= \tan \frac{\theta - \phi}{2}(\sin \theta + \sin \phi)$

50. $\sin 2\theta + \sin 4\theta + \sin 6\theta = 4 \cos \theta \cos 2\theta \sin 3\theta$

In Exercises 51–56, use a graphing calculator to determine which expression (A)–(F) on the right can be used to complete the identity. Then try to prove that identity algebraically.

51. $\frac{\cos x + \cot x}{1 + \csc x}$

52. $\cot x + \csc x$

53. $\sin x \cos x + 1$

54. $2 \cos^2 x - 1$

55. $\frac{1}{\cot x \sin^2 x}$

56. $(\cos x + \sin x)(1 - \sin x \cos x)$

A. $\frac{\sin^3 x - \cos^3 x}{\sin x - \cos x}$

B. $\cos x$

C. $\tan x + \cot x$

D. $\cos^3 x + \sin^3 x$

E. $\frac{\sin x}{1 - \cos x}$

F. $\cos^4 x - \sin^4 x$

Skill Maintenance

For each function:

a) *Graph the function.* **[5.1]**

b) *Determine whether the function is one-to-one.* **[5.1]**

c) *If the function is one-to-one, find an equation for its inverse.* **[5.1]**

d) *Graph the inverse of the function.* **[5.1]**

57. $f(x) = 3x - 2$

58. $f(x) = x^3 + 1$

59. $f(x) = x^2 - 4,\ x \geq 0$

60. $f(x) = \sqrt{x + 2}$

Solve.

61. $2x^2 = 5x$ **[3.2]**

62. $3x^2 + 5x - 10 = 18$ **[3.2]**

63. $x^4 + 5x^2 - 36 = 0$ **[3.2]**

64. $x^2 - 10x + 1 = 0$ **[3.2]**

65. $\sqrt{x - 2} = 5$ **[3.4]**

66. $x = \sqrt{x + 7} + 5$ **[3.4]**

Synthesis

Prove the identity.

67. $\ln |\tan x| = -\ln |\cot x|$

68. $\ln |\sec \theta + \tan \theta| = -\ln |\sec \theta - \tan \theta|$

69. $\log(\cos x - \sin x) + \log(\cos x + \sin x)$
$= \log \cos 2x$

70. *Mechanics.* The following equation occurs in the study of mechanics:

$$\sin \theta = \frac{I_1 \cos \phi}{\sqrt{(I_1 \cos \phi)^2 + (I_2 \sin \phi)^2}}.$$

It can happen that $I_1 = I_2$. Assuming that this happens, simplify the equation.

71. *Alternating Current.* In the theory of alternating current, the following equation occurs:

$$R = \frac{1}{\omega C(\tan \theta + \tan \phi)}.$$

Show that this equation is equivalent to

$$R = \frac{\cos \theta \cos \phi}{\omega C \sin(\theta + \phi)}.$$

72. *Electrical Theory.* In electrical theory, the following equations occur:

$$E_1 = \sqrt{2}E_t \cos\left(\theta + \frac{\pi}{P}\right)$$

and

$$E_2 = \sqrt{2}E_t \cos\left(\theta - \frac{\pi}{P}\right).$$

Assuming that these equations hold, show that

$$\frac{E_1 + E_2}{2} = \sqrt{2}E_t \cos \theta \cos \frac{\pi}{P}$$

and

$$\frac{E_1 - E_2}{2} = -\sqrt{2}E_t \sin \theta \sin \frac{\pi}{P}.$$

Mid-Chapter Mixed Review

Determine whether the statement is true or false.

1. $\sin x\,(\csc x - \cot x) = 1 - \cos x$ [7.1]

2. $\sin 42^\circ = \sqrt{\dfrac{1 + \cos 84^\circ}{2}}$ [7.2]

3. $\sin \dfrac{\pi}{9} = \cos \dfrac{7\pi}{18}$ [7.2]

4. $\cos^2 x \neq \cos x^2$ [7.1]

For Exercises 5–14, choose one of expressions A–J to complete the identity. [7.1], [7.2]

5. $\cos(-x) =$

6. $\cos(u + v) =$

7. $\tan 2x =$

8. $\tan\left(\dfrac{\pi}{2} - x\right) =$

9. $1 + \cot^2 x =$

10. $\sin \dfrac{x}{2} =$

11. $\sin 2x =$

12. $\sin(u - v) =$

13. $\csc\left(\dfrac{\pi}{2} - x\right) =$

14. $\cos \dfrac{x}{2} =$

A. $2 \sin x \cos x$

B. $\pm\sqrt{\dfrac{1 + \cos x}{2}}$

C. $\csc^2 x$

D. $\dfrac{2 \tan x}{1 - \tan^2 x}$

E. $\pm\sqrt{\dfrac{1 - \cos x}{2}}$

F. $\sec x$

G. $\sin u \cos v - \cos u \sin v$

H. $\cos u \cos v - \sin u \sin v$

I. $\cot x$

J. $\cos x$

Simplify.

15. $\sqrt{\dfrac{\cot x}{\sin x}}$ [7.1]

16. $\dfrac{1}{\sin^2 x} - \left(\dfrac{\cos x}{\sin x}\right)^2$ [7.1]

17. $\dfrac{2\cos^2 x - 5\cos x - 3}{\cos x - 3}$ [7.1]

18. $\dfrac{\sin x}{\tan(-x)}$ [7.1]

19. $(\cos x - \sin x)^2$ [7.2]

20. $1 - 2\sin^2 \dfrac{x}{2}$ [7.2]

21. Rationalize the denominator: $\sqrt{\dfrac{\sec x}{1 - \cos x}}$. [7.1]

22. Write $\cos 41^\circ \cos 29^\circ + \sin 41^\circ \sin 29^\circ$ as a trigonometric function of a single angle and then evaluate. [7.1]

23. Evaluate $\cos \dfrac{3\pi}{8}$ exactly. [7.1]

24. Evaluate $\sin 105^\circ$ exactly. [7.1]

25. Assuming that $\sin \alpha = \frac{5}{13}$ and $\sin \beta = \frac{12}{13}$ and that α and β are between 0 and $\pi/2$, evaluate $\tan(\alpha - \beta)$. [7.1]

26. Find the exact value of $\sin 2\theta$ and the quadrant in which 2θ lies if $\cos \theta = -\frac{4}{5}$ and θ is in quadrant II. [7.2]

Prove the identity. [7.3]

27. $\cos^2 \dfrac{x}{2} = \dfrac{\tan x + \sin x}{2 \tan x}$

28. $\dfrac{1 - \sin x}{\cos x} = \dfrac{\cos x}{1 + \sin x}$

29. $\dfrac{\sin^3 x - \cos^3 x}{\sin x - \cos x} = \dfrac{2 + \sin 2x}{2}$

30. $\sin 6\theta - \sin 2\theta = \tan 2\theta\,(\cos 2\theta + \cos 6\theta)$

COLLABORATIVE DISCUSSION AND WRITING

31. Explain why $\tan(x + 450^\circ)$ cannot be simplified using the tangent sum formula, but can be simplified using the sine and cosine sum formulas. [7.1]

32. Discuss and compare the graphs of $y = \sin x$, $y = \sin 2x$, and $y = \sin(x/2)$. [7.2]

33. What restrictions must be placed on the variable in each of the following identities? Why? [7.3]

a) $\sin 2x = \dfrac{2 \tan x}{1 + \tan^2 x}$ **b)** $\dfrac{1 - \cos x}{\sin x} = \dfrac{\sin x}{1 + \cos x}$

34. Find all errors in the following:

$$\begin{aligned} &2\sin^2 2x + \cos 4x \\ &= 2(2 \sin x \cos x)^2 + 2\cos 2x \\ &= 8 \sin^2 x \cos^2 x + 2(\cos^2 x + \sin^2 x) \\ &= 8 \sin^2 x \cos^2 x + 2. \end{aligned}$$ [7.2]

7.4 Inverses of the Trigonometric Functions

- Find values of the inverse trigonometric functions.
- Simplify expressions involving compositions such as $\sin(\sin^{-1} x)$ and $\sin^{-1}(\sin x)$.
- Simplify expressions involving compositions such as $\sin(\cos^{-1}\frac{1}{2})$ without using a calculator.
- Simplify expressions involving compositions such as $\sin \arctan(a/b)$ by making a drawing and reading off appropriate ratios.

The graphs of the sine, cosine, and tangent functions are shown below. Are the inverses of these functions also functions? We learned earlier that a function has an inverse that is a function if it is one-to-one, which we can check with the horizontal-line test.

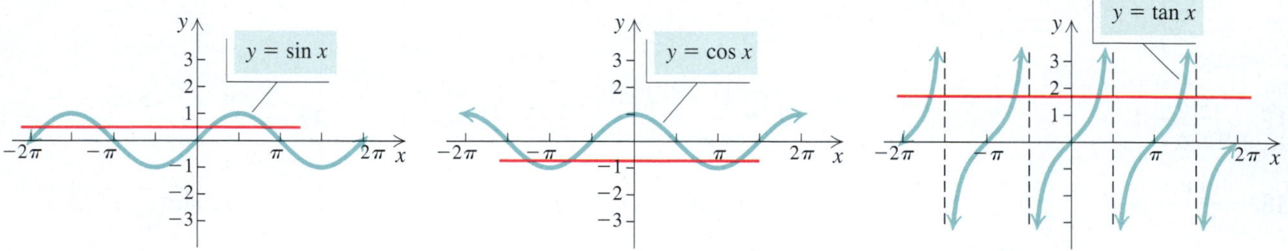

INVERSE FUNCTIONS

REVIEW SECTION 5.1.

Note that for each function, a horizontal line (shown in red) crosses the graph more than once. Therefore, none of them has an inverse that is a function.

The graphs of an equation and its inverse are reflections of each other across the line $y = x$. Let's examine the graphs of the inverses of each of the three functions graphed above.

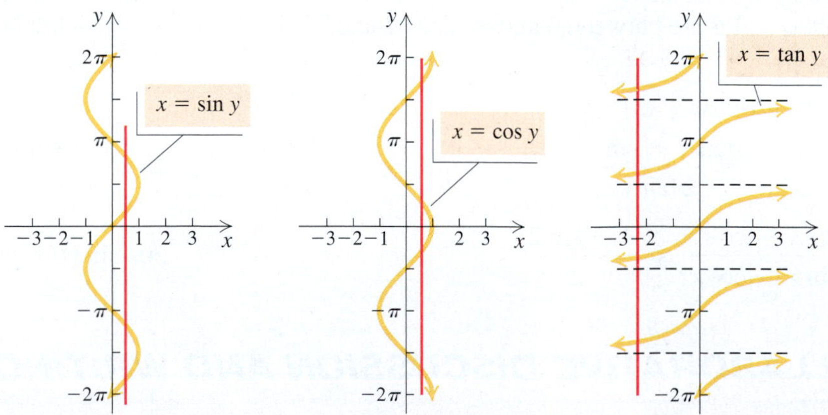

We can check again to see whether these are graphs of functions by using the vertical-line test. In each case, there is a vertical line (shown in red) that crosses the graph more than once, so each *fails* to be a function.

Restricting Ranges to Define Inverse Functions

Recall that a function like $f(x) = x^2$ does not have an inverse that is a function, but by restricting the domain of f to nonnegative numbers, we have a new squaring function, $f(x) = x^2, x \geq 0$, that has an inverse that is a function, $f^{-1}(x) = \sqrt{x}$. This is equivalent to restricting the range of the inverse relation to exclude ordered pairs that contain negative numbers.

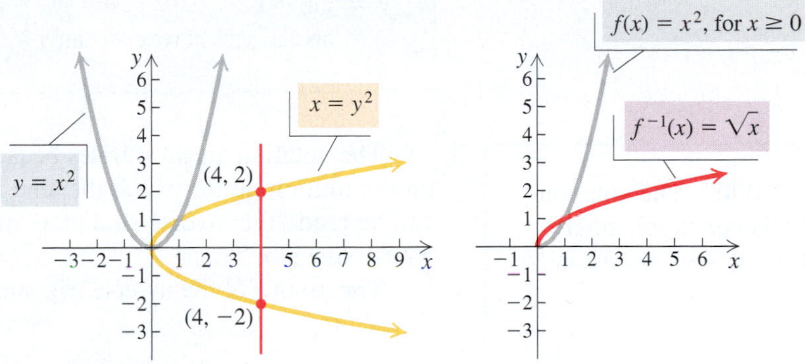

In a similar manner, we can define new trigonometric functions whose inverses are functions. We can do this by restricting either the domains of the basic trigonometric functions or the ranges of their inverse relations. This can be done in many ways, but the restrictions illustrated below with solid red curves are fairly standard in mathematics.

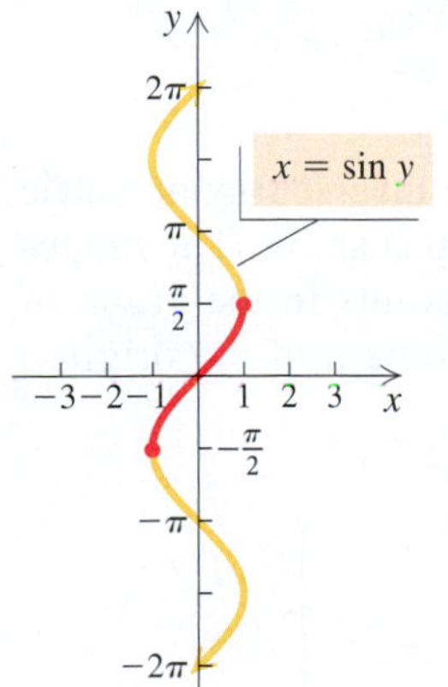

FIGURE 1.

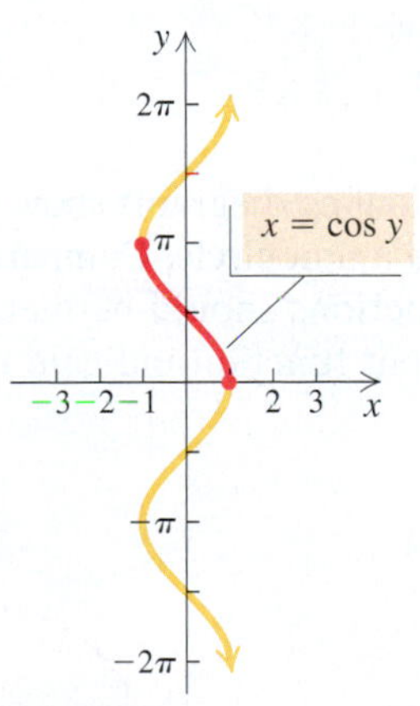

FIGURE 2.

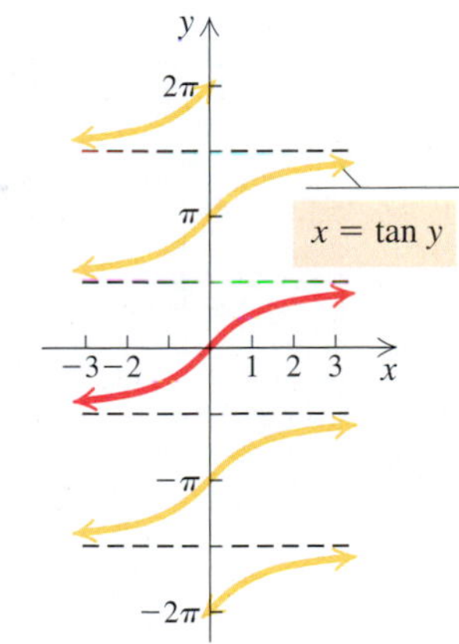

FIGURE 3.

For the inverse sine function, we choose a range close to the origin that allows all inputs on the interval $[-1, 1]$ to have function values. Thus we choose the interval $[-\pi/2, \pi/2]$ for the range (Fig. 1). For the inverse cosine function, we choose a range close to the origin that allows all inputs on the interval $[-1, 1]$ to have function values. We choose the interval $[0, \pi]$ (Fig. 2). For the inverse tangent function, we choose a range close to the origin that allows all real numbers to have function values. The interval $(-\pi/2, \pi/2)$ satisfies this requirement (Fig. 3).

INVERSE TRIGONOMETRIC FUNCTIONS

FUNCTION	DOMAIN	RANGE
$y = \sin^{-1} x$ $= \arcsin x$, where $x = \sin y$	$[-1, 1]$	$[-\pi/2, \pi/2]$
$y = \cos^{-1} x$ $= \arccos x$, where $x = \cos y$	$[-1, 1]$	$[0, \pi]$
$y = \tan^{-1} x$ $= \arctan x$, where $x = \tan y$	$(-\infty, \infty)$	$(-\pi/2, \pi/2)$

CAUTION! The notation $\sin^{-1} x$ is *not* exponential notation. It does *not* mean $\dfrac{1}{\sin x}$!

The notation arcsin x arises because the function value, y, is the length of an arc on the unit circle for which the sine is x. Either of the two kinds of notation above can be read "the inverse sine of x" or "the arc sine of x" or "the number (or angle) whose sine is x."

The graphs of the inverse trigonometric functions are as follows.

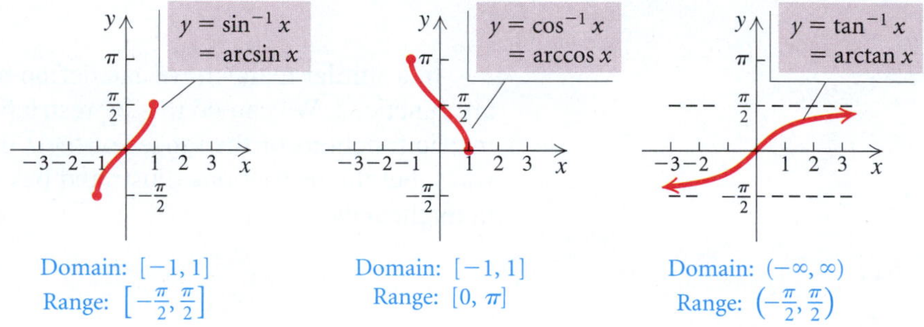

The following diagrams show the restricted ranges for the inverse trigonometric functions on a unit circle. Compare these graphs with the graphs above. The ranges of these functions should be memorized. The missing endpoints in the graph of the arctangent function indicate inputs that are not in the domain of the original function.

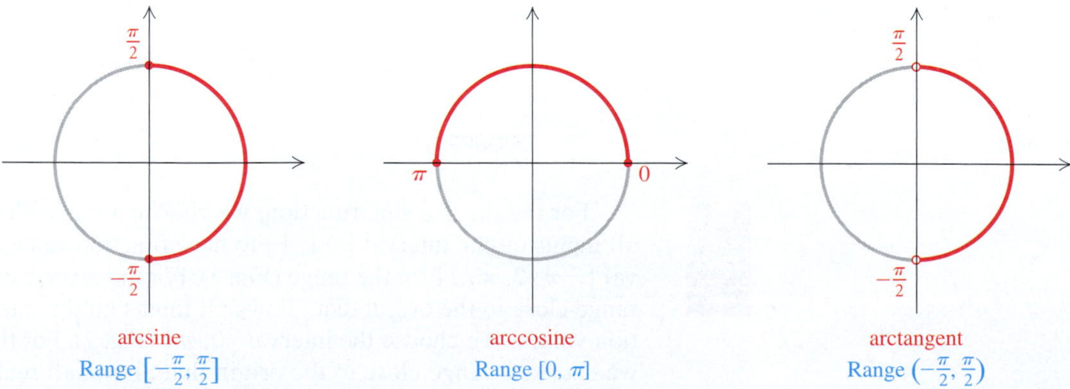

CONNECTING THE CONCEPTS — Domains and Ranges

The following is a summary of the domains and ranges of the trigonometric functions together with a summary of the domains and ranges of the inverse trigonometric functions. For completeness, we have included the arccosecant, the arcsecant, and the arccotangent, though there is a lack of uniformity in their definitions in mathematical literature.

FUNCTION	DOMAIN	RANGE
sin	All reals, $(-\infty, \infty)$	$[-1, 1]$
cos	All reals, $(-\infty, \infty)$	$[-1, 1]$
tan	All reals except $k\pi/2$, k odd	All reals, $(-\infty, \infty)$
csc	All reals except $k\pi$	$(-\infty, -1] \cup [1, \infty)$
sec	All reals except $k\pi/2$, k odd	$(-\infty, -1] \cup [1, \infty)$
cot	All reals except $k\pi$	All reals, $(-\infty, \infty)$

INVERSE FUNCTION	DOMAIN	RANGE
$\sin^{-1}$	$[-1, 1]$	$\left[-\frac{\pi}{2}, \frac{\pi}{2}\right]$
$\cos^{-1}$	$[-1, 1]$	$[0, \pi]$
$\tan^{-1}$	All reals, $(-\infty, \infty)$	$\left(-\frac{\pi}{2}, \frac{\pi}{2}\right)$
$\csc^{-1}$	$(-\infty, -1] \cup [1, \infty)$	$\left[-\frac{\pi}{2}, 0\right) \cup \left(0, \frac{\pi}{2}\right]$
$\sec^{-1}$	$(-\infty, -1] \cup [1, \infty)$	$\left[0, \frac{\pi}{2}\right) \cup \left(\frac{\pi}{2}, \pi\right]$
$\cot^{-1}$	All reals, $(-\infty, \infty)$	$(0, \pi)$

EXAMPLE 1 Find each of the following function values. For parts (d) and (e), see the preceding Connecting the Concepts for the restricted ranges.

a) $\sin^{-1}\frac{\sqrt{2}}{2}$ **b)** $\cos^{-1}\left(-\frac{1}{2}\right)$ **c)** $\tan^{-1}\left(-\frac{\sqrt{3}}{3}\right)$

d) $\cot^{-1}0$ **e)** $\sec^{-1}\left(-\sqrt{2}\right)$

Solution

a) Another way to state "find $\sin^{-1}\sqrt{2}/2$" is to say "find β such that $\sin\beta = \sqrt{2}/2$." In the restricted range $[-\pi/2, \pi/2]$, the only number with a sine of $\sqrt{2}/2$ is $\pi/4$. Thus, $\sin^{-1}\left(\sqrt{2}/2\right) = \pi/4$, or 45°. (See Fig. 4 on the following page.)

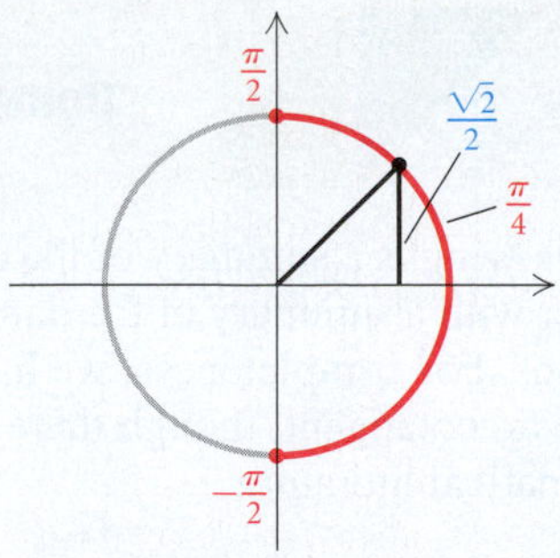

FIGURE 4.

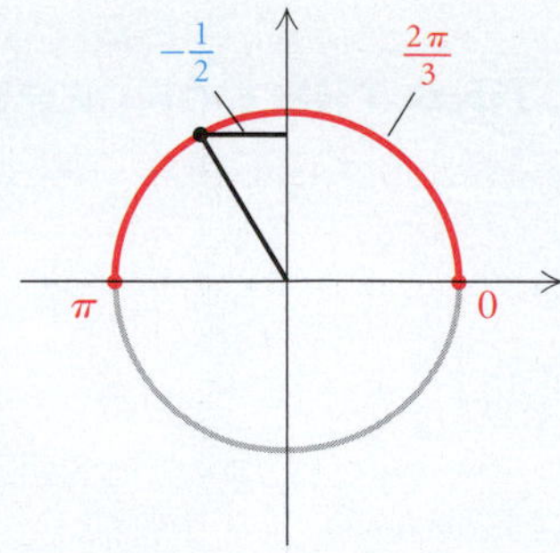

FIGURE 5.

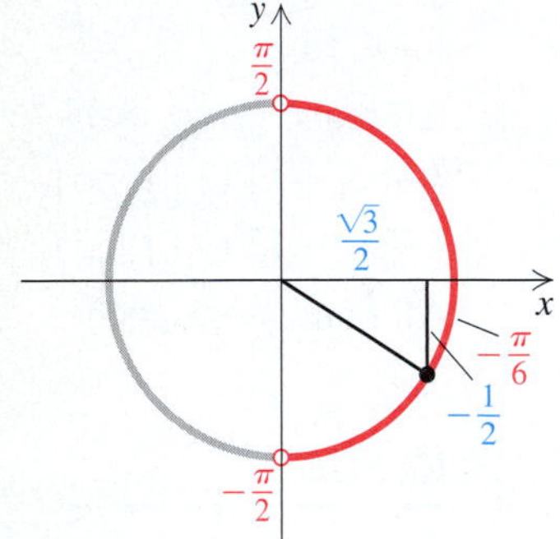

FIGURE 6.

b) The only number with a cosine of $-\frac{1}{2}$ in the restricted range $[0, \pi]$ is $2\pi/3$. Thus, $\cos^{-1}(-\frac{1}{2}) = 2\pi/3$, or 120°. (See Fig. 5 above.)

c) The only number in the restricted range $(-\pi/2, \pi/2)$ with a tangent of $-\sqrt{3}/3$ is $-\pi/6$. Thus, $\tan^{-1}(-\sqrt{3}/3)$ is $-\pi/6$, or $-30°$. (See Fig. 6 at left.)

d) The only number in the restricted range $(0, \pi)$ with a cotangent of 0 is $\pi/2$. Thus, $\cot^{-1} 1 = \pi/2$, or 90°.

e) The only number in the restricted range $[0, \pi/2) \cup (\pi/2, \pi]$ with a secant of $-\sqrt{2}$ is $3\pi/4$. Thus, $\sec^{-1}(-\sqrt{2}) = 3\pi/4$, or 135°.

Now Try Exercises 1 and 11.

We can also use a calculator to find inverse trigonometric function values. The cosecant, the secant, and the cotangent functions are the reciprocals of the sine, the cosine, and the tangent functions, respectively. Thus we have

$$\csc^{-1}(x) = \sin^{-1}\left(\frac{1}{x}\right),$$

$$\sec^{-1}(x) = \cos^{-1}\left(\frac{1}{x}\right), \quad \text{and}$$

$$\cot^{-1}(x) = \tan^{-1}\left(\frac{1}{x}\right).$$

On most graphing calculators, we can find inverse function values in either radians or degrees simply by selecting the appropriate mode. The keystrokes involved in finding inverse function values vary with the calculator. Be sure to read the instructions for the particular calculator that you are using.

EXAMPLE 2 Approximate each of the following function values in both radians and degrees. Round radian measure to four decimal places and degree measure to the nearest tenth of a degree.

a) $\cos^{-1}(-0.2689)$

b) $\tan^{-1}(-0.2623)$

c) $\sin^{-1} 0.20345$

d) $\cos^{-1} 1.318$

e) $\csc^{-1} 8.205$

Solution

	FUNCTION VALUE	MODE	READOUT	ROUNDED
a)	$\cos^{-1}(-0.2689)$	Radian	1.843047111	1.8430
		Degree	105.5988209	105.6°
b)	$\tan^{-1}(-0.2623)$	Radian	-.2565212141	−0.2565
		Degree	-14.69758292	−14.7°
c)	$\sin^{-1}0.20345$	Radian	.2048803359	0.2049
		Degree	11.73877855	11.7°
d)	$\cos^{-1}1.318$	Radian	ERR:DOMAIN	
		Degree	ERR:DOMAIN	

The value 1.318 is not in $[-1, 1]$, the domain of the arccosine function.

e) The cosecant function is the reciprocal of the sine function:

	FUNCTION VALUE	MODE	READOUT	ROUNDED
	$\csc^{-1}8.205 = \sin^{-1}(1/8.205)$	Radian	.1221806653	0.1222
		Degree	7.000436462	7.0°

Now Try Exercises 21, 25, and 27.

Composition of Trigonometric Functions and Their Inverses

Various compositions of trigonometric functions and their inverses often occur in practice. For example, we might want to try to simplify an expression such as

$$\sin(\sin^{-1} x) \quad \text{or} \quad \sin\left(\cot^{-1}\frac{x}{2}\right).$$

COMPOSITION OF FUNCTIONS

REVIEW SECTION 2.3.

In the expression on the left, we are finding "the sine of a number whose sine is x." Recall from Section 5.1 that if a function f has an inverse that is also a function, then

$$f(f^{-1}(x)) = x, \quad \text{for all } x \text{ in the } domain \text{ of } f^{-1},$$

and

$$f^{-1}(f(x)) = x, \quad \text{for all } x \text{ in the } domain \text{ of } f.$$

Thus, if $f(x) = \sin x$ and $f^{-1}(x) = \sin^{-1} x$, then

$$\mathbf{\sin(\sin^{-1}x) = x, \text{ for all } x \text{ in the } \textit{domain} \text{ of } \sin^{-1},}$$

which is any number on the interval $[-1, 1]$. Similar results hold for the other trigonometric functions.

COMPOSITION OF TRIGONOMETRIC FUNCTIONS

$\sin(\sin^{-1} x) = x$, for all x in the domain of $\sin^{-1}$.

$\cos(\cos^{-1} x) = x$, for all x in the domain of $\cos^{-1}$.

$\tan(\tan^{-1} x) = x$, for all x in the domain of $\tan^{-1}$.

EXAMPLE 3 Simplify each of the following.

a) $\cos\left(\cos^{-1}\frac{\sqrt{3}}{2}\right)$ **b)** $\sin\left(\sin^{-1} 1.8\right)$

Solution

a) Since $\sqrt{3}/2$ is in $[-1, 1]$, the domain of $\cos^{-1}$, it follows that

$$\cos\left(\cos^{-1}\frac{\sqrt{3}}{2}\right) = \frac{\sqrt{3}}{2}.$$

b) Since 1.8 is not in $[-1, 1]$, the domain of $\sin^{-1}$, we cannot evaluate this expression. We know that there is no number with a sine of 1.8. Because we cannot find $\sin^{-1} 1.8$, we state that $\sin\left(\sin^{-1} 1.8\right)$ does not exist.

Now Try Exercise 37.

Now let's consider an expression like $\sin^{-1}(\sin x)$. We might also suspect that this is equal to x for any x in the domain of $\sin x$, but this is not true unless x is in the range of the $\sin^{-1}$ function. Note that in order to define $\sin^{-1}$, we had to restrict the domain of the sine function. In doing so, we restricted the range of the inverse sine function. Thus,

$$\sin^{-1}(\sin x) = x, \quad \text{for all } x \text{ in the } range \text{ of } \sin^{-1}.$$

Similar results hold for the other trigonometric functions.

SPECIAL CASES

$\sin^{-1}(\sin x) = x$, for all x in the range of $\sin^{-1}$.
$\cos^{-1}(\cos x) = x$, for all x in the range of $\cos^{-1}$.
$\tan^{-1}(\tan x) = x$, for all x in the range of $\tan^{-1}$.

EXAMPLE 4 Simplify each of the following.

a) $\tan^{-1}\left(\tan\frac{\pi}{6}\right)$ **b)** $\sin^{-1}\left(\sin\frac{3\pi}{4}\right)$

Solution

a) Since $\pi/6$ is in $(-\pi/2, \pi/2)$, the range of the $\tan^{-1}$ function, we can use $\tan^{-1}(\tan x) = x$. Thus,

$$\tan^{-1}\left(\tan\frac{\pi}{6}\right) = \frac{\pi}{6}.$$

b) Note that $3\pi/4$ is not in $[-\pi/2, \pi/2]$, the range of the $\sin^{-1}$ function. Thus we *cannot* apply $\sin^{-1}(\sin x) = x$. Instead we first find $\sin(3\pi/4)$, which is $\sqrt{2}/2$, and substitute:

$$\sin^{-1}\left(\sin\frac{3\pi}{4}\right) = \sin^{-1}\left(\frac{\sqrt{2}}{2}\right) = \frac{\pi}{4}.$$

Now Try Exercises 41 and 43.

Now we find some additional function compositions.

EXAMPLE 5 Simplify each of the following.

a) $\sin\left[\tan^{-1}(-1)\right]$ **b)** $\cos^{-1}\left(\sin\dfrac{\pi}{2}\right)$

Solution

a) We know that $\tan^{-1}(-1)$ is the number (or angle) θ in $(-\pi/2, \pi/2)$ whose tangent is -1. That is, $\tan\theta = -1$. Thus, $\theta = -\pi/4$ and

$$\sin\left[\tan^{-1}(-1)\right] = \sin\left[-\frac{\pi}{4}\right] = -\frac{\sqrt{2}}{2}.$$

b) $\cos^{-1}\left(\sin\dfrac{\pi}{2}\right) = \cos^{-1}(1) = 0$ $\quad \sin\dfrac{\pi}{2} = 1$

Now Try Exercises 47 and 49.

Next, let's consider

$$\cos\left(\sin^{-1}\tfrac{3}{5}\right).$$

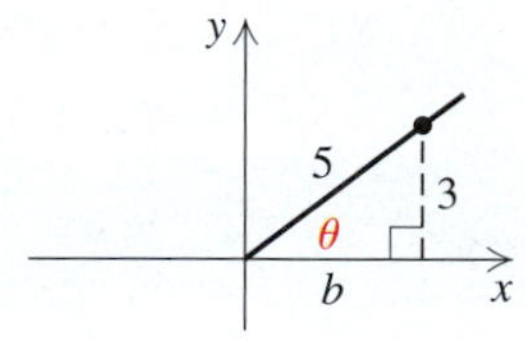

Without using a calculator, we cannot find $\sin^{-1}\frac{3}{5}$. However, we can still evaluate the entire expression by sketching a reference triangle. We are looking for angle θ such that $\sin^{-1}\frac{3}{5} = \theta$, or $\sin\theta = \frac{3}{5}$. Since $\sin^{-1}$ is defined in $[-\pi/2, \pi/2]$ and $\frac{3}{5} > 0$, we know that θ is in quadrant I. We sketch a reference right triangle, as shown at left. The angle θ in this triangle is an angle whose sine is $\frac{3}{5}$. We wish to find the cosine of this angle. Since the triangle is a right triangle, we can find the length of the base, b. It is 4. Thus we know that $\cos\theta = b/5$, or $\frac{4}{5}$. Therefore,

$$\cos\left(\sin^{-1}\tfrac{3}{5}\right) = \tfrac{4}{5}.$$

EXAMPLE 6 Find $\sin\left(\cot^{-1}\dfrac{x}{2}\right)$.

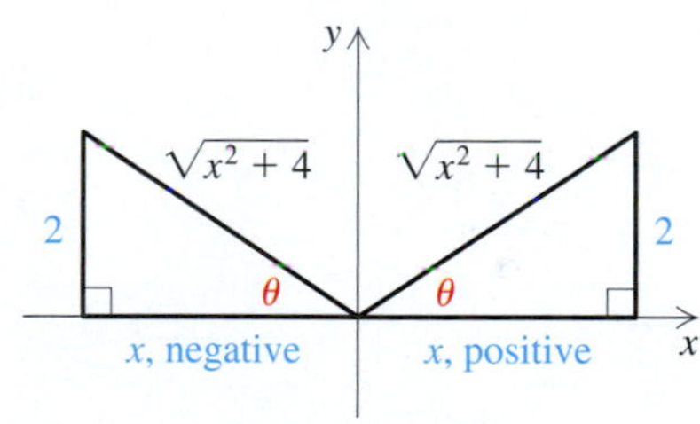

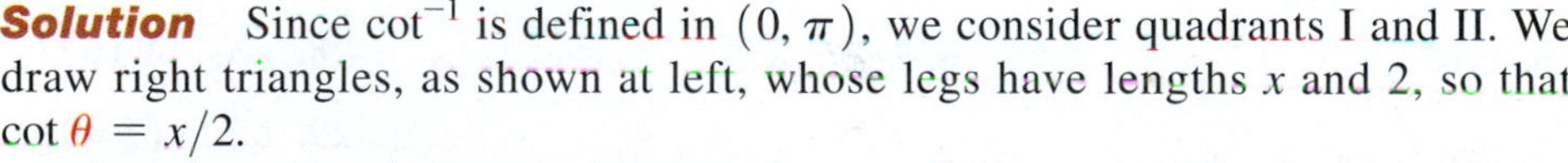

Solution Since $\cot^{-1}$ is defined in $(0, \pi)$, we consider quadrants I and II. We draw right triangles, as shown at left, whose legs have lengths x and 2, so that $\cot\theta = x/2$.

In each, we find the length of the hypotenuse and then read off the sine ratio. We get

$$\sin\left(\cot^{-1}\frac{x}{2}\right) = \frac{2}{\sqrt{x^2+4}}.$$

Now Try Exercise 55.

In the following example, we use a sum identity to evaluate an expression.

EXAMPLE 7 Evaluate:

$$\sin\left(\sin^{-1}\tfrac{1}{2} + \cos^{-1}\tfrac{5}{13}\right).$$

Solution Since $\sin^{-1}\frac{1}{2}$ and $\cos^{-1}\frac{5}{13}$ are both angles, the expression is the sine of a sum of two angles, so we use the identity

$$\sin(u+v) = \sin u\cos v + \cos u\sin v.$$

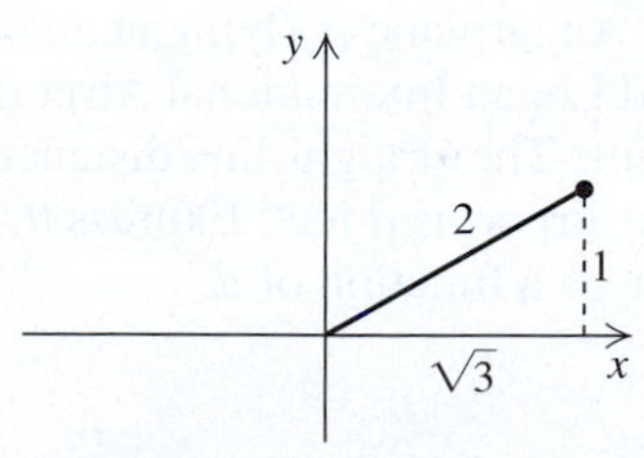

FIGURE 1.

Thus,

$$\begin{aligned}&\sin\left(\sin^{-1}\tfrac{1}{2} + \cos^{-1}\tfrac{5}{13}\right)\\ &\quad= \sin\left(\sin^{-1}\tfrac{1}{2}\right)\cdot\cos\left(\cos^{-1}\tfrac{5}{13}\right) + \cos\left(\sin^{-1}\tfrac{1}{2}\right)\cdot\sin\left(\cos^{-1}\tfrac{5}{13}\right)\\ &\quad= \tfrac{1}{2}\cdot\tfrac{5}{13} + \cos\left(\sin^{-1}\tfrac{1}{2}\right)\cdot\sin\left(\cos^{-1}\tfrac{5}{13}\right). \quad \text{Using composition identities}\end{aligned}$$

Now since $\sin^{-1}\frac{1}{2} = \pi/6$, $\cos\left(\sin^{-1}\frac{1}{2}\right)$ simplifies to $\cos\pi/6$, or $\sqrt{3}/2$. We can illustrate this with a reference triangle in quadrant I. (See Fig. 1.)

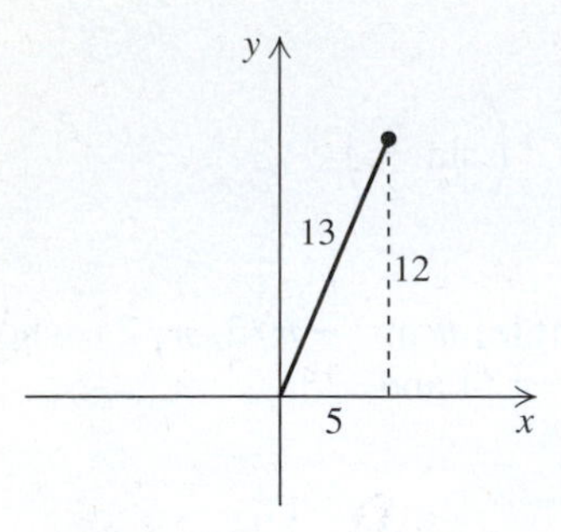

FIGURE 2.

To find $\sin\left(\cos^{-1}\frac{5}{13}\right)$, we use a reference triangle in quadrant I and determine that the sine of the angle whose cosine is $\frac{5}{13}$ is $\frac{12}{13}$. (See Fig. 2.)

Our expression now simplifies to

$$\frac{1}{2}\cdot\frac{5}{13}+\frac{\sqrt{3}}{2}\cdot\frac{12}{13},\quad\text{or}\quad\frac{5+12\sqrt{3}}{26}.$$

Thus,

$$\sin\left(\sin^{-1}\frac{1}{2}+\cos^{-1}\frac{5}{13}\right)=\frac{5+12\sqrt{3}}{26}.$$

Now Try Exercise 63.

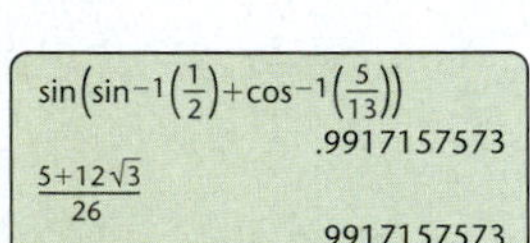

7.4 Exercise Set

Find each of the following exactly in radians and degrees.

1. $\sin^{-1}\left(-\frac{\sqrt{3}}{2}\right)$

2. $\cos^{-1}\frac{1}{2}$

3. $\tan^{-1}1$

4. $\sin^{-1}0$

5. $\cos^{-1}\frac{\sqrt{2}}{2}$

6. $\sec^{-1}\sqrt{2}$

7. $\tan^{-1}0$

8. $\tan^{-1}\frac{\sqrt{3}}{3}$

9. $\cos^{-1}\frac{\sqrt{3}}{2}$

10. $\cot^{-1}\left(-\frac{\sqrt{3}}{3}\right)$

11. $\csc^{-1}2$

12. $\sin^{-1}\frac{1}{2}$

13. $\cot^{-1}(-\sqrt{3})$

14. $\tan^{-1}(-1)$

15. $\sin^{-1}\left(-\frac{1}{2}\right)$

16. $\cos^{-1}\left(-\frac{\sqrt{2}}{2}\right)$

17. $\cos^{-1}0$

18. $\sin^{-1}\frac{\sqrt{3}}{2}$

19. $\sec^{-1}2$

20. $\csc^{-1}(-1)$

Use a calculator to find each of the following in radians, rounded to four decimal places, and in degrees, rounded to the nearest tenth of a degree.

21. $\tan^{-1}0.3673$

22. $\cos^{-1}(-0.2935)$

23. $\sin^{-1}0.9613$

24. $\sin^{-1}(-0.6199)$

25. $\cos^{-1}(-0.9810)$

26. $\tan^{-1}158$

27. $\csc^{-1}(-6.2774)$

28. $\sec^{-1}1.1677$

29. $\tan^{-1}(1.091)$

30. $\cot^{-1}1.265$

31. $\sin^{-1}(-0.8192)$

32. $\cos^{-1}(-0.2716)$

33. State the domains of the inverse sine, inverse cosine, and inverse tangent functions.

34. State the ranges of the inverse sine, inverse cosine, and inverse tangent functions.

35. *Angle of Depression.* An airplane is flying at an altitude of 2000 ft toward Logan International Airport, in Boston, Massachusetts. The straight-line distance from the airplane to the airport is d feet. Express θ, the angle of depression, as a function of d.

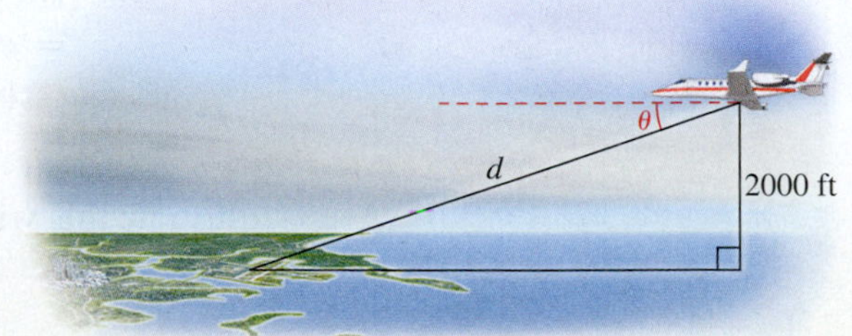

36. *Angle of Inclination.* A guy wire is attached to the top of a 50-ft pole and stretched to a point that is d feet from the bottom of the pole. Express β, the angle of inclination, as a function of d.

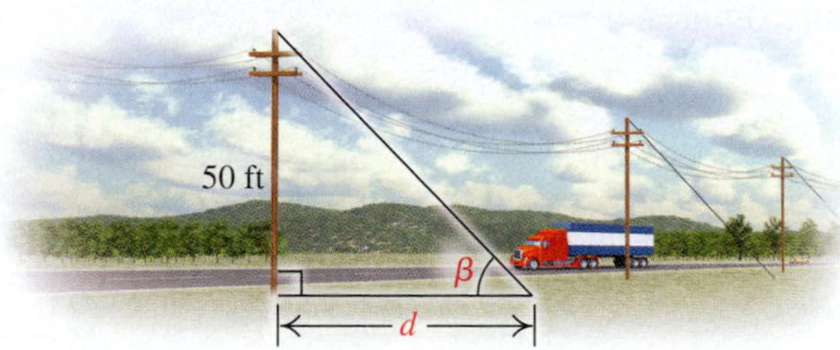

Evaluate.

37. $\sin(\sin^{-1} 0.3)$

38. $\tan[\tan^{-1}(-4.2)]$

39. $\cos^{-1}\left[\cos\left(-\frac{\pi}{4}\right)\right]$

40. $\sin^{-1}\left(\sin\frac{2\pi}{3}\right)$

41. $\sin^{-1}\left(\sin\frac{\pi}{5}\right)$

42. $\cot^{-1}\left(\cot\frac{2\pi}{3}\right)$

43. $\tan^{-1}\left(\tan\frac{2\pi}{3}\right)$

44. $\cos^{-1}\left(\cos\frac{\pi}{7}\right)$

45. $\sin\left(\tan^{-1}\frac{\sqrt{3}}{3}\right)$

46. $\cos\left(\sin^{-1}\frac{\sqrt{3}}{2}\right)$

47. $\tan\left(\cos^{-1}\frac{\sqrt{2}}{2}\right)$

48. $\cos^{-1}(\sin\pi)$

49. $\sin^{-1}\left(\cos\frac{\pi}{6}\right)$

50. $\sin^{-1}\left[\tan\left(-\frac{\pi}{4}\right)\right]$

51. $\tan(\sin^{-1} 0.1)$

52. $\cos\left(\tan^{-1}\frac{\sqrt{3}}{4}\right)$

53. $\sin^{-1}\left(\sin\frac{7\pi}{6}\right)$

54. $\tan^{-1}\left[\tan\left(-\frac{3\pi}{4}\right)\right]$

Find each of the following.

55. $\sin\left(\tan^{-1}\frac{a}{3}\right)$

56. $\tan\left(\cos^{-1}\frac{3}{x}\right)$

57. $\cot\left(\sin^{-1}\frac{p}{q}\right)$

58. $\sin(\cos^{-1} x)$

59. $\tan\left(\sin^{-1}\frac{p}{\sqrt{p^2+9}}\right)$

60. $\tan\left(\frac{1}{2}\sin^{-1}\frac{1}{2}\right)$

61. $\cos\left(\frac{1}{2}\sin^{-1}\frac{\sqrt{3}}{2}\right)$

62. $\sin\left(2\cos^{-1}\frac{3}{5}\right)$

Evaluate.

63. $\cos\left(\sin^{-1}\frac{\sqrt{2}}{2} + \cos^{-1}\frac{3}{5}\right)$

64. $\sin\left(\sin^{-1}\frac{1}{2} + \cos^{-1}\frac{3}{5}\right)$

65. $\sin(\sin^{-1} x + \cos^{-1} y)$

66. $\cos(\sin^{-1} x - \cos^{-1} y)$

67. $\sin(\sin^{-1} 0.6032 + \cos^{-1} 0.4621)$

68. $\cos(\sin^{-1} 0.7325 - \cos^{-1} 0.4838)$

Skill Maintenance

Vocabulary Reinforcement

In each of Exercises 69–76, fill in the blank with the correct term. Some of the given choices will not be used.

linear speed	congruent
angular speed	circular
angle of elevation	periodic
angle of depression	period
complementary	amplitude
supplementary	quadrantal
similar	radian measure

69. A function f is said to be ____________ if there exists a positive constant p such that $f(s + p) = f(s)$ for all s in the domain of f. [6.5]

70. The ____________ of a rotation is the ratio of the distance s traveled by a point at a radius r from the center of rotation to the length of the radius r. [6.4]

71. Triangles are ____________ if their corresponding angles have the same measure. [6.1]

72. The angle between the horizontal and a line of sight below the horizontal is called a(n) ____________. [6.2]

73. ____________ is the amount of rotation per unit of time. [6.4]

74. Two positive angles are ____________ if their sum is 180°. [6.3]

75. The ____________ of a periodic function is one half of the distance between its maximum and minimum function values. [6.5]

76. Trigonometric functions with domains composed of real numbers are called ____________ functions. [6.5]

Synthesis

Prove the identity.

77. $\sin^{-1} x + \cos^{-1} x = \frac{\pi}{2}$

78. $\tan^{-1} x = \sin^{-1}\frac{x}{\sqrt{x^2+1}}$

79. $\sin^{-1} x = \cos^{-1}\sqrt{1-x^2}$, for $x \geq 0$

80. $\cos^{-1} x = \tan^{-1}\frac{\sqrt{1-x^2}}{x}$, for $x > 0$

81. *Height of a Mural.* An art student's eye is at a point A, looking at a mural of height h, with the bottom of the mural y feet above the eye (see the figure). The eye is x feet from the wall. Write an expression for θ in terms of x, y, and h. Then evaluate the expression when $x = 20$ ft, $y = 7$ ft, and $h = 25$ ft.

82. Use a calculator to approximate the following expression:

$$16 \tan^{-1} \tfrac{1}{5} - 4 \tan^{-1} \tfrac{1}{239}.$$

What number does this expression seem to approximate?

7.5 Solving Trigonometric Equations

› Solve trigonometric equations.

When an equation contains a trigonometric expression with a variable, such as $\cos x$, it is called a trigonometric equation. Some trigonometric equations are identities, such as $\sin^2 x + \cos^2 x = 1$. Now we consider equations, such as $2 \cos x = -1$, that are usually not identities. As we have done for other types of equations, we will solve such equations by finding all values for x that make the equation true.

EXAMPLE 1 Solve: $2 \cos x = -1$.

Solution We first solve for $\cos x$:

$$2 \cos x = -1$$
$$\cos x = -\tfrac{1}{2}.$$

The solutions are numbers whose cosine is $-\frac{1}{2}$. To find them, we use the unit circle (see Section 6.5).

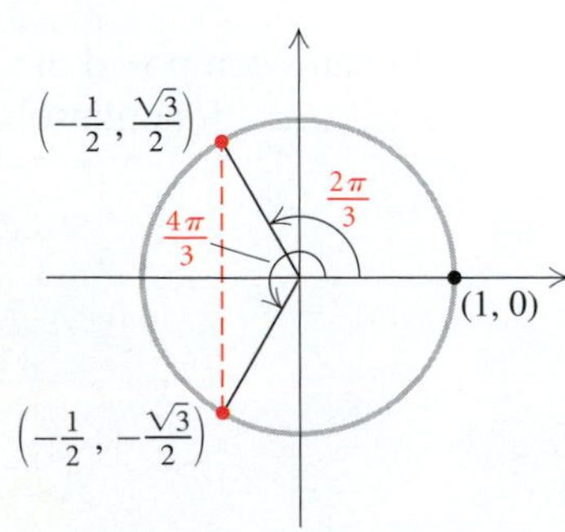

There are just two points on the unit circle for which the cosine is $-\frac{1}{2}$, as shown in the figure at left. They are the points corresponding to $2\pi/3$ and $4\pi/3$. These numbers, plus any multiple of 2π, are the solutions:

$$\frac{2\pi}{3} + 2k\pi \quad \text{and} \quad \frac{4\pi}{3} + 2k\pi,$$

where k is any integer. In degrees, the solutions are

$$120^\circ + k \cdot 360^\circ \quad \text{and} \quad 240^\circ + k \cdot 360^\circ,$$

where k is any integer.

To check the solution to $2\cos x = -1$, we can graph $y_1 = 2\cos x$ and $y_2 = -1$ on the same set of axes and find the *first* coordinates of the points of intersection. Using $\pi/3$ as the Xscl facilitates our reading of the solutions. First, let's graph these equations on the interval from 0 to 2π, as shown in the figure on the left below. The only solutions in $[0, 2\pi)$ are $2\pi/3$ and $4\pi/3$.

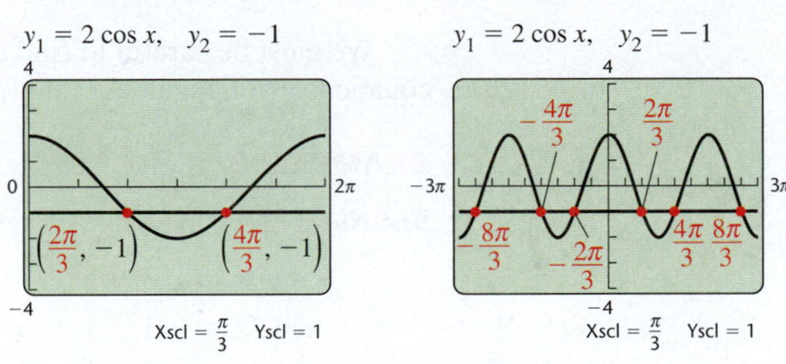

Next, let's change the viewing window to $[-3\pi, 3\pi, -4, 4]$ and graph again. Since the cosine function is periodic, there is an infinite number of solutions. A few of these appear in the graph on the right above. From the graph, we see that the solutions are $2\pi/3 + 2k\pi$ and $4\pi/3 + 2k\pi$, where k is any integer.

Now Try Exercise 1.

EXAMPLE 2 Solve: $4\sin^2 x = 1$.

Solution We begin by solving for $\sin x$:

$$4\sin^2 x = 1$$
$$\sin^2 x = \tfrac{1}{4}$$
$$\sin x = \pm\tfrac{1}{2}.$$

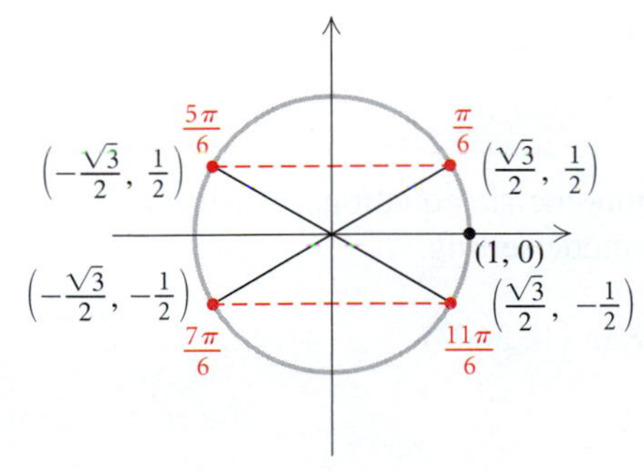

Again, we use the unit circle to find those numbers whose sine is $\frac{1}{2}$ or $-\frac{1}{2}$. The solutions are

$$\frac{\pi}{6} + 2k\pi, \quad \frac{5\pi}{6} + 2k\pi, \quad \frac{7\pi}{6} + 2k\pi, \quad \text{and} \quad \frac{11\pi}{6} + 2k\pi,$$

where k is any integer. In degrees, the solutions are

$$30° + k \cdot 360°, \quad 150° + k \cdot 360°,$$
$$210° + k \cdot 360°, \quad \text{and} \quad 330° + k \cdot 360°,$$

where k is any integer.

The general solutions listed above could be condensed using odd as well as even multiples of π:

$$\frac{\pi}{6} + k\pi \quad \text{and} \quad \frac{5\pi}{6} + k\pi,$$

or, in degrees,

$$30° + k \cdot 180° \quad \text{and} \quad 150° + k \cdot 180°,$$

where k is any integer.

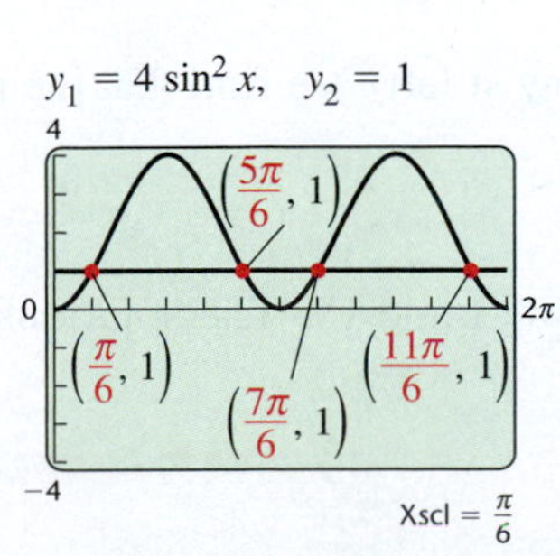

Let's do a partial check using a graphing calculator, checking only the solutions in $[0, 2\pi)$. We graph $y_1 = 4\sin^2 x$ and $y_2 = 1$ and note that the solutions in $[0, 2\pi)$ are $\pi/6$, $5\pi/6$, $7\pi/6$, and $11\pi/6$.

Now Try Exercise 13.

> In most applications, it is sufficient to find just the solutions from 0 to 2π or from $0°$ to $360°$. We then remember that any multiple of 2π, or $360°$, can be added to obtain the rest of the solutions.

We must be careful to find all solutions in $[0, 2\pi)$ when solving trigonometric equations involving double angles.

EXAMPLE 3 Solve $3 \tan 2x = -3$ on the interval $[0, 2\pi)$.

Solution We first solve for $\tan 2x$:

$$3 \tan 2x = -3$$
$$\tan 2x = -1.$$

We are looking for solutions x to the equation for which

$$0 \le x < 2\pi.$$

Multiplying by 2, we get

$$0 \le 2x < 4\pi,$$

which is the interval that we use when solving $\tan 2x = -1$.

Using the unit circle, we find points $2x$ in $[0, 4\pi)$ for which $\tan 2x = -1$. These values of $2x$ are as follows:

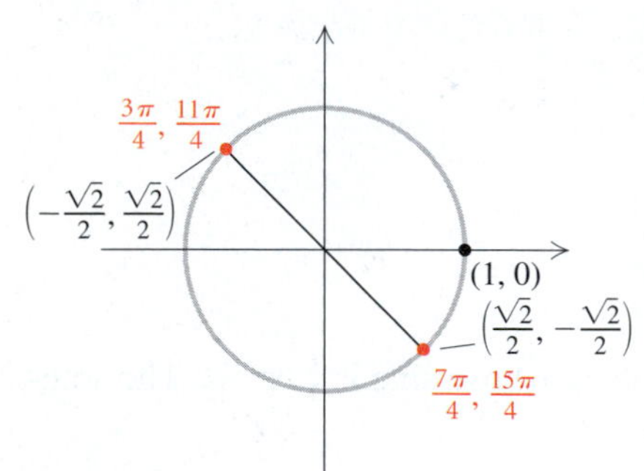
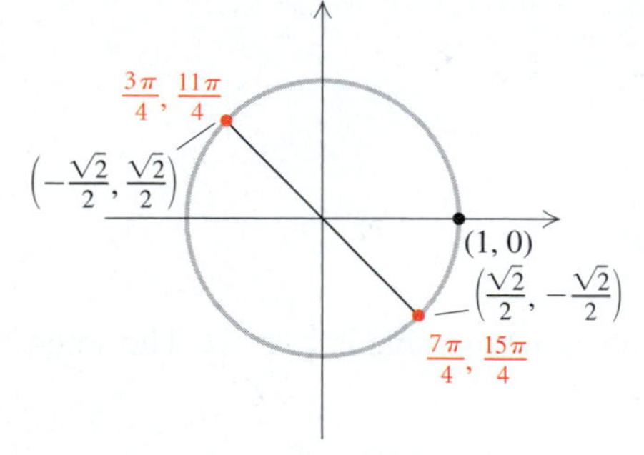

$$2x = \frac{3\pi}{4}, \frac{7\pi}{4}, \frac{11\pi}{4}, \text{ and } \frac{15\pi}{4}.$$

Thus the desired values of x in $[0, 2\pi)$ are each of these values divided by 2—hence,

$$x = \frac{3\pi}{8}, \frac{7\pi}{8}, \frac{11\pi}{8}, \text{ and } \frac{15\pi}{8}.$$

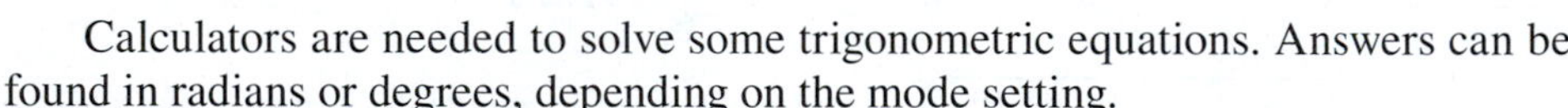

Calculators are needed to solve some trigonometric equations. Answers can be found in radians or degrees, depending on the mode setting.

EXAMPLE 4 Solve $\frac{1}{2}\cos\phi + 1 = 1.2108$ in $[0, 360°)$.

Solution We have

cos⁻¹(0.4216)
65.06435654

$$\tfrac{1}{2}\cos\phi + 1 = 1.2108$$
$$\tfrac{1}{2}\cos\phi = 0.2108$$
$$\cos\phi = 0.4216.$$

Using a calculator set in DEGREE mode (see the window at left), we find that the reference angle, $\cos^{-1} 0.4216$, is

$$\phi \approx 65.06°.$$

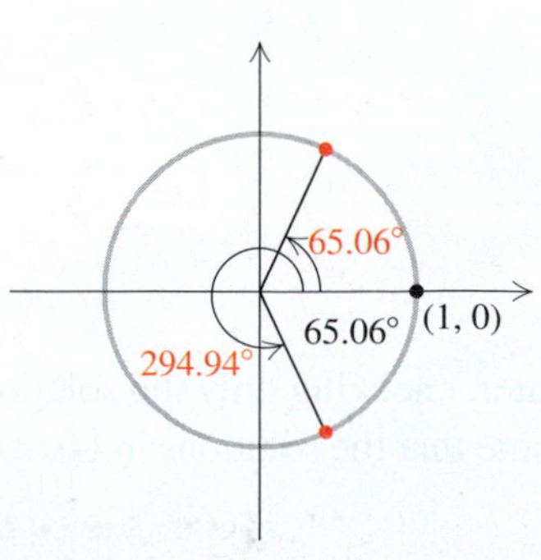

Since $\cos\phi$ is positive, the solutions are in quadrants I and IV. The solutions in $[0, 360°)$ are

$$65.06° \text{ and } 360° - 65.06° = 294.94°.$$

Now Try Exercise 9.

Just in Time 19

EXAMPLE 5 Solve $2\cos^2 u = 1 - \cos u$ in $[0°, 360°)$.

Algebraic Solution

We use the principle of zero products:

$$2\cos^2 u = 1 - \cos u$$
$$2\cos^2 u + \cos u - 1 = 0$$
$$(2\cos u - 1)(\cos u + 1) = 0$$
$$2\cos u - 1 = 0 \quad or \quad \cos u + 1 = 0$$
$$2\cos u = 1 \quad or \quad \cos u = -1$$
$$\cos u = \tfrac{1}{2} \quad or \quad \cos u = -1.$$

Thus,

$$u = 60°, 300° \quad \text{or} \quad u = 180°.$$

The solutions in $[0°, 360°)$ are 60°, 180°, and 300°.

Graphical Solution

We can use either the Intersect method or the Zero method to solve trigonometric equations. Here we illustrate by solving the equation using both methods. We set the calculator in DEGREE mode. With the Intersect method, we graph the equations $y_1 = 2\cos^2 x$ and $y_2 = 1 - \cos x$. With the Zero method, we write the equation in the form $2\cos^2 u + \cos u - 1 = 0$. Then we graph $y = 2\cos^2 x + \cos x - 1$.

Intersect Method

$y_1 = 2\cos^2 x$, $y_2 = 1 - \cos x$

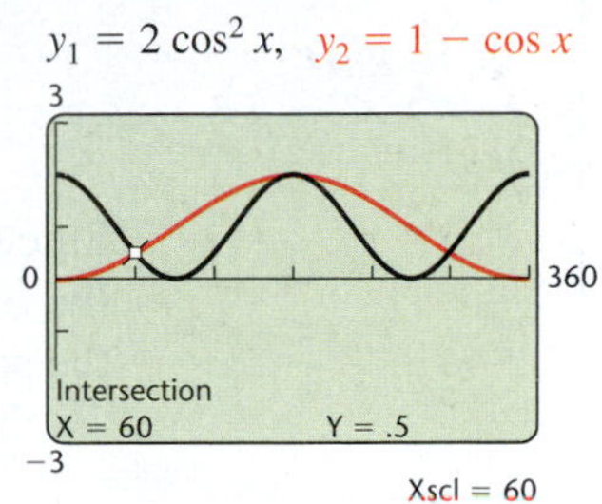

Zero Method

$y = 2\cos^2 x + \cos x - 1$

3
0
360
Zero
X = 60
Y = 0
−3
Xscl = 60

The leftmost solution is 60°. Using the INTERSECT feature or the ZERO feature two more times, we find the other solutions: 180° and 300°. The solutions in $[0°, 360°)$ are 60°, 180°, and 300°.

Now Try Exercise 15.

EXAMPLE 6 Solve $\sin^2 \beta - \sin \beta = 0$ in $[0, 2\pi)$.

Solution We factor and use the principle of zero products:

$$\sin^2 \beta - \sin \beta = 0$$
$$\sin \beta(\sin \beta - 1) = 0 \quad \text{Factoring}$$
$$\sin \beta = 0 \quad or \quad \sin \beta - 1 = 0$$
$$\sin \beta = 0 \quad or \quad \sin \beta = 1$$
$$\beta = 0, \pi \quad or \quad \beta = \pi/2.$$

$y = \sin^2 x - \sin x$

2
0
2π
−1
Xscl = $\frac{\pi}{4}$

The solutions in $[0, 2\pi)$ are 0, $\pi/2$, and π.

Now Try Exercise 17.

QUADRATIC FORMULA

REVIEW SECTION 3.2.

If a trigonometric equation is quadratic but difficult or impossible to factor, we use the quadratic formula.

EXAMPLE 7 Solve $10\sin^2 x - 12\sin x - 7 = 0$ in $[0°, 360°)$.

Solution This equation is quadratic in $\sin x$ with $a = 10$, $b = -12$, and $c = -7$. Substituting into the quadratic formula, we get

$$\sin x = \frac{-b \pm \sqrt{b^2 - 4ac}}{2a} \qquad \text{Using the quadratic formula}$$

$$= \frac{-(-12) \pm \sqrt{(-12)^2 - 4(10)(-7)}}{2 \cdot 10} \qquad \text{Substituting}$$

$$= \frac{12 \pm \sqrt{144 + 280}}{20}$$

$$= \frac{12 \pm \sqrt{424}}{20}$$

$$\approx \frac{12 \pm 20.5913}{20}$$

$$\sin x \approx 1.6296 \quad \textit{or} \quad \sin x \approx -0.4296.$$

$\sin^{-1}(-0.4296)$
-25.44217782

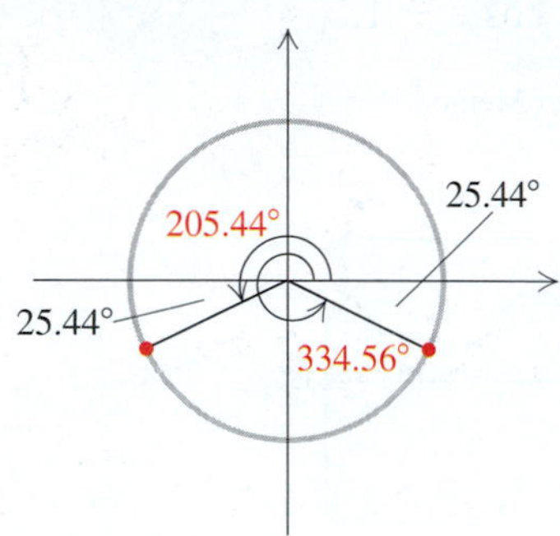

Since sine values are never greater than 1, the first equation has no solution. Using the other equation, we find the reference angle to be 25.44°. Since $\sin x$ is negative, the solutions are in quadrants III and IV.

Thus the solutions in $[0°, 360°)$ are

$$180° + 25.44° = 205.44° \quad \text{and} \quad 360° - 25.44° = 334.56°.$$

Now Try Exercise 23.

Trigonometric equations can involve more than one function.

EXAMPLE 8 Solve $2\cos^2 x \tan x = \tan x$ in $[0, 2\pi)$.

Solution Using a graphing calculator, we can determine that there are six solutions. If we let Xscl $= \pi/4$, the solutions are easily read from the graph. In the figures at left, we show the Intersect method and the Zero method of solving graphically. Each illustrates that the solutions in $[0, 2\pi)$ are

$$0, \quad \frac{\pi}{4}, \quad \frac{3\pi}{4}, \quad \pi, \quad \frac{5\pi}{4}, \quad \text{and} \quad \frac{7\pi}{4}.$$

$y_1 = 2\cos^2 x \tan x, \; y_2 = \tan x$

2, 0, 2π, -2, Xscl $= \frac{\pi}{4}$

We can verify these solutions algebraically, as follows:

$$2\cos^2 x \tan x = \tan x$$

$$2\cos^2 x \tan x - \tan x = 0$$

$$\tan x\,(2\cos^2 x - 1) = 0$$

$$\tan x = 0 \quad \textit{or} \quad 2\cos^2 x - 1 = 0$$

$$\tan x = 0 \quad \textit{or} \quad \cos^2 x = \frac{1}{2}$$

$$\tan x = 0 \quad \textit{or} \quad \cos x = \pm\frac{\sqrt{2}}{2}$$

$$x = 0, \pi \quad \textit{or} \quad x = \frac{\pi}{4}, \frac{3\pi}{4}, \frac{5\pi}{4}, \frac{7\pi}{4}.$$

$y_1 = 2\cos^2 x \tan x - \tan x$

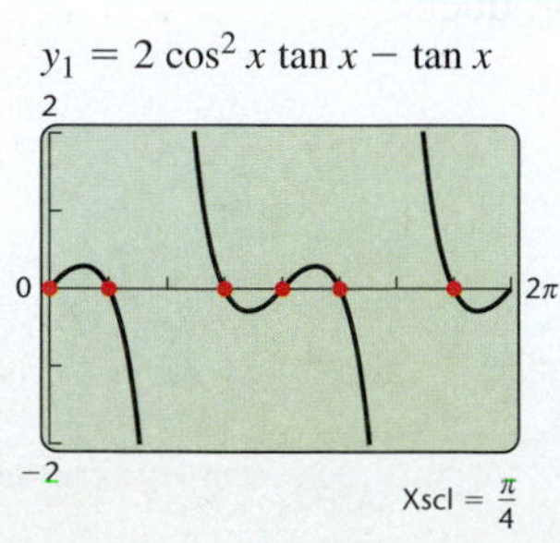

Thus, $x = 0, \pi/4, 3\pi/4, \pi, 5\pi/4$, and $7\pi/4$.

Now Try Exercise 29.

When a trigonometric equation involves more than one function, it is sometimes helpful to use identities to rewrite the equation in terms of a single function.

EXAMPLE 9 Solve $\sin x + \cos x = 1$ in $[0, 2\pi)$.

Algebraic Solution

We have

$$\sin x + \cos x = 1$$
$$(\sin x + \cos x)^2 = 1^2 \quad \text{Squaring both sides}$$
$$\sin^2 x + 2\sin x \cos x + \cos^2 x = 1$$
$$2\sin x \cos x + 1 = 1 \quad \text{Using } \sin^2 x + \cos^2 x = 1$$
$$2\sin x \cos x = 0$$
$$\sin x = 0 \quad or \quad \cos x = 0$$
$$x = 0, \pi \quad or \quad x = \pi/2, 3\pi/2.$$

Now we check these in the original equation $\sin x + \cos x = 1$:

$$\sin 0 + \cos 0 = 0 + 1 = 1,$$
$$\sin \frac{\pi}{2} + \cos \frac{\pi}{2} = 1 + 0 = 1,$$
$$\sin \pi + \cos \pi = 0 + (-1) = -1,$$
$$\sin \frac{3\pi}{2} + \cos \frac{3\pi}{2} = (-1) + 0 = -1.$$

We find that π and $3\pi/2$ do not check, but the other values do. Thus the solutions in $[0, 2\pi)$ are

$$0 \quad \text{and} \quad \frac{\pi}{2}.$$

When the solution process involves squaring both sides, values are sometimes obtained that are not solutions of the original equation. As we saw in this example, it is necessary to check the possible solutions.

Graphical Solution

We can graph the left side and then the right side of the equation as seen in the first window below. Then we look for points of intersection. We could also rewrite the equation as $\sin x + \cos x - 1 = 0$, graph the left side, and look for the zeros of the function, as illustrated in the second window below. In each window, we see that the solutions in $[0, 2\pi)$ are 0 and $\pi/2$.

$y_1 = \sin x + \cos x, \quad y_2 = 1$

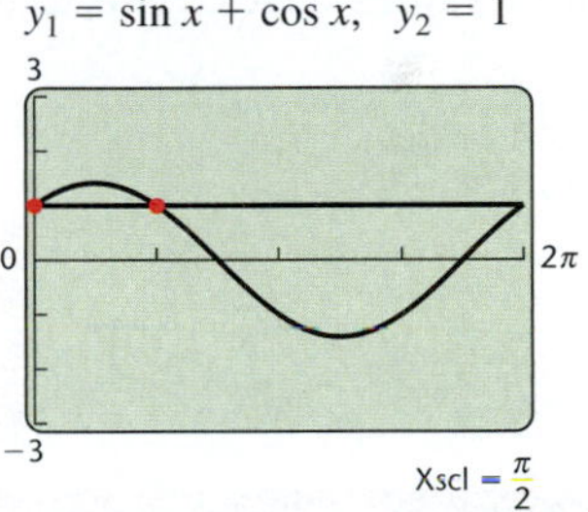

$y = \sin x + \cos x - 1$

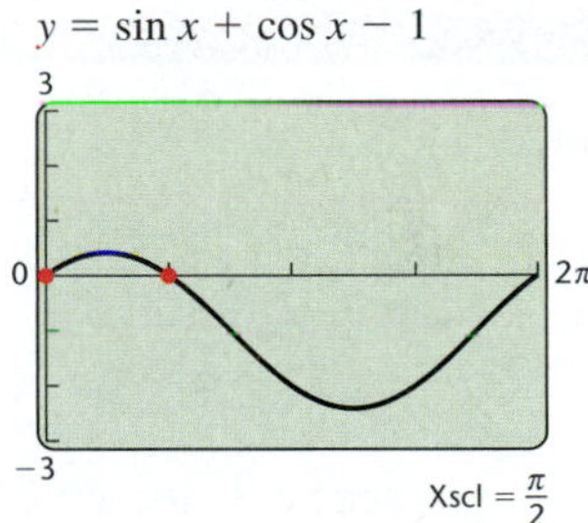

This example illustrates a valuable advantage of the calculator—that is, with a graphing calculator, extraneous solutions do not appear.

Now Try Exercise 39.

EXAMPLE 10 Solve $\cos 2x + \sin x = 1$ in $[0, 2\pi)$.

Algebraic Solution

We have

$$\cos 2x + \sin x = 1$$
$$1 - 2\sin^2 x + \sin x = 1 \quad \text{Using the identity } \cos 2x = 1 - 2\sin^2 x$$
$$-2\sin^2 x + \sin x = 0$$
$$\sin x(-2\sin x + 1) = 0 \quad \text{Factoring}$$
$$\sin x = 0 \quad or \quad -2\sin x + 1 = 0 \quad \text{Principle of zero products}$$
$$\sin x = 0 \quad or \quad \sin x = \frac{1}{2}$$
$$x = 0, \pi \quad or \quad x = \frac{\pi}{6}, \frac{5\pi}{6}.$$

All four values check. The solutions in $[0, 2\pi)$ are 0, $\pi/6$, $5\pi/6$, and π.

Graphical Solution

We graph $y_1 = \cos 2x + \sin x - 1$ and look for the zeros of the function.

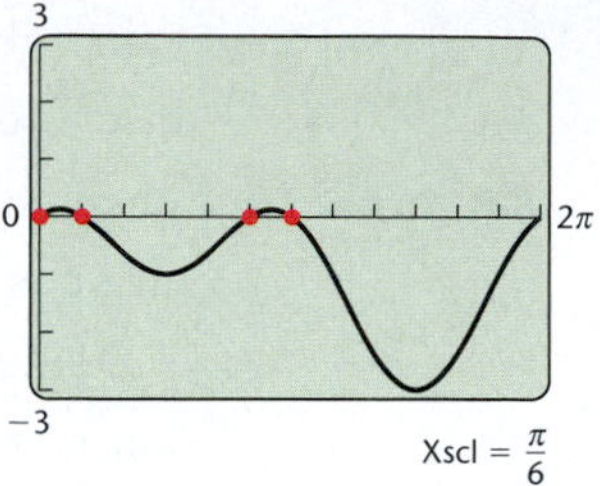

The solutions in $[0, 2\pi)$ are 0, $\pi/6$, $5\pi/6$, and π.

Now Try Exercise 27.

Just in Time 14

EXAMPLE 11 Solve $\tan^2 x + \sec x - 1 = 0$ in $[0, 2\pi)$.

Algebraic Solution

We have

$$\tan^2 x + \sec x - 1 = 0$$
$$\sec^2 x - 1 + \sec x - 1 = 0 \quad \text{Using the identity } 1 + \tan^2 x = \sec^2 x, \text{ or } \tan^2 x = \sec^2 x - 1$$
$$\sec^2 x + \sec x - 2 = 0$$
$$(\sec x + 2)(\sec x - 1) = 0 \quad \text{Factoring}$$
$$\sec x = -2 \quad or \quad \sec x = 1 \quad \text{Principle of zero products}$$
$$\cos x = -\frac{1}{2} \quad or \quad \cos x = 1 \quad \text{Using the identity } \cos x = 1/\sec x$$
$$x = \frac{2\pi}{3}, \frac{4\pi}{3} \quad or \quad x = 0.$$

All these values check. The solutions in $[0, 2\pi)$ are 0, $2\pi/3$, and $4\pi/3$.

Graphical Solution

We graph $y = \tan^2 x + \sec x - 1$, but we enter this equation in the form

$$y_1 = \tan^2 x + \frac{1}{\cos x} - 1.$$

We use the ZERO feature to find zeros of the function.

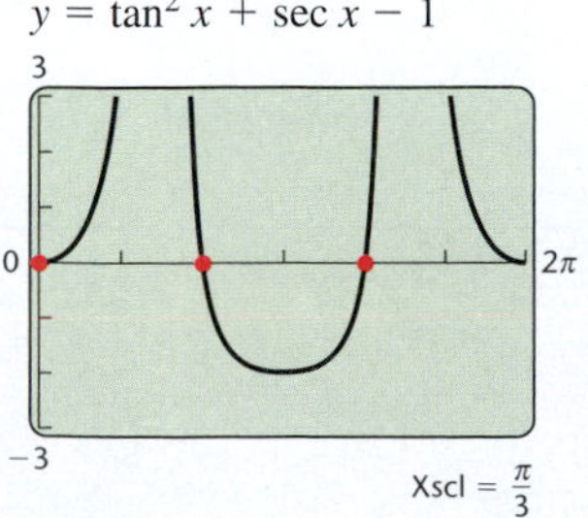

The solutions in $[0, 2\pi)$ are 0, $2\pi/3$, and $4\pi/3$.

Now Try Exercise 37.

Sometimes we cannot find solutions algebraically, but we can approximate them with a graphing calculator.

EXAMPLE 12 Solve $\sin x - \cos x = \cot x$ in $[0, 2\pi)$.

Solution In the screen on the left below, we graph $y_1 = \sin x - \cos x$ and $y_2 = \cot x$ and determine the points of intersection. In the screen on the right, we graph the function $y_1 = \sin x - \cos x - \cot x$ and determine the zeros.

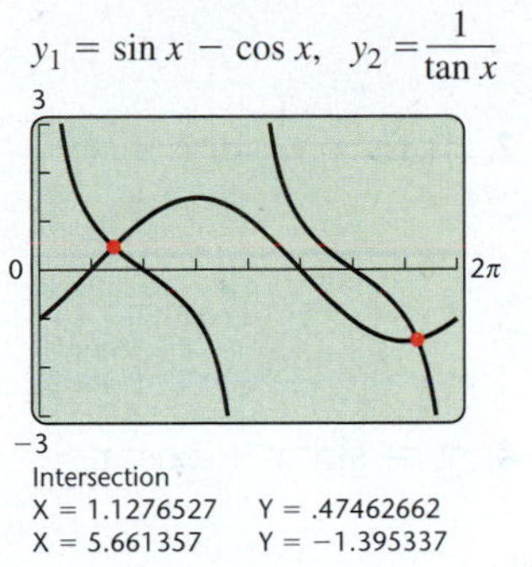

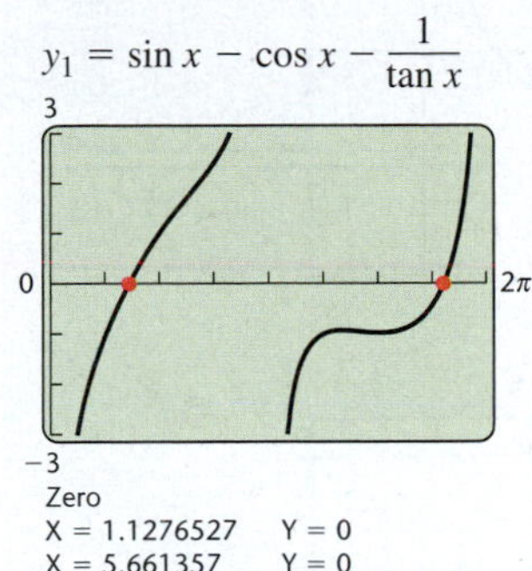

Each method leads to the approximate solutions 1.13 and 5.66 in $[0, 2\pi)$.

Now Try Exercise 47.

Visualizing the Graph

Match the equation with its graph.

1. $f(x) = \dfrac{4}{x^2 - 9}$

2. $f(x) = \dfrac{1}{2}\sin x - 1$

3. $(x - 2)^2 + (y + 3)^2 = 4$

4. $y = \sin^2 x + \cos^2 x$

5. $f(x) = 3 - \log x$

6. $f(x) = 2^{x+3} - 2$

7. $y = 2\cos\left(x - \dfrac{\pi}{2}\right)$

8. $y = -x^3 + 3x^2$

9. $f(x) = (x - 3)^2 + 2$

10. $f(x) = -\cos x$

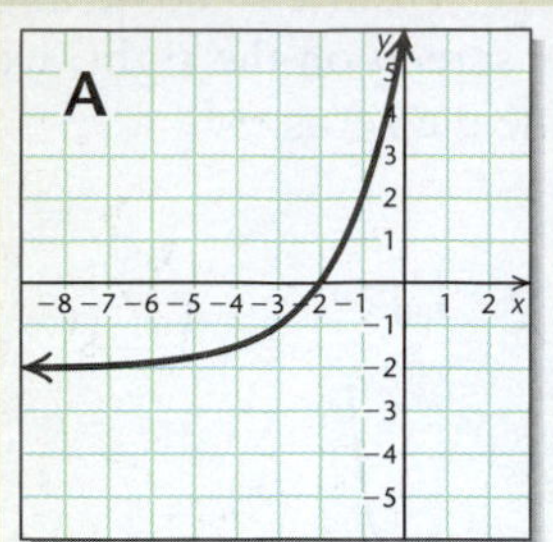

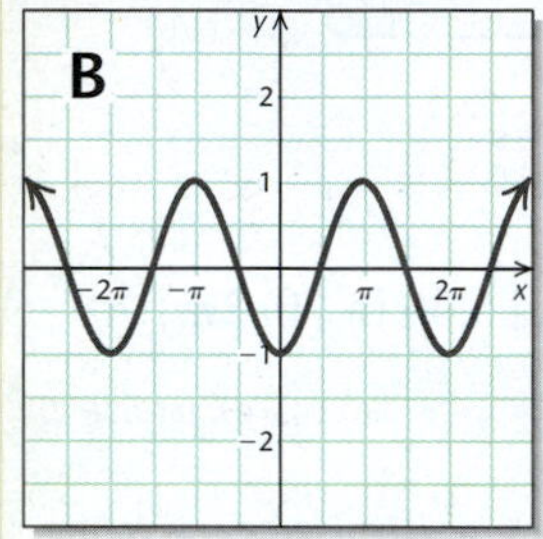

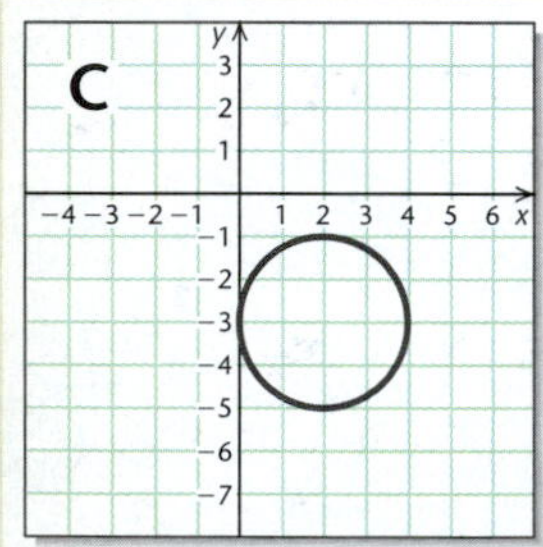

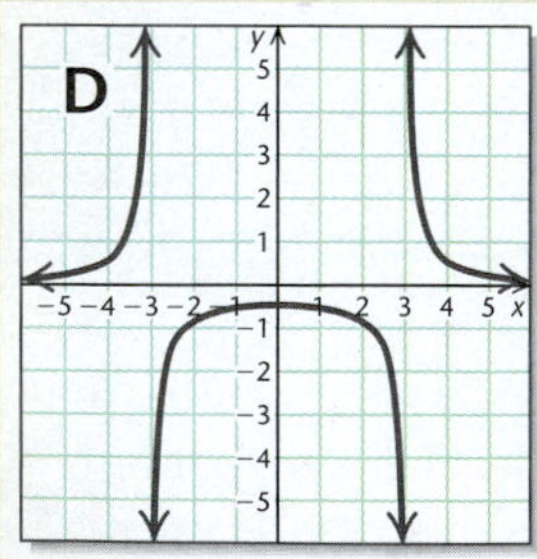

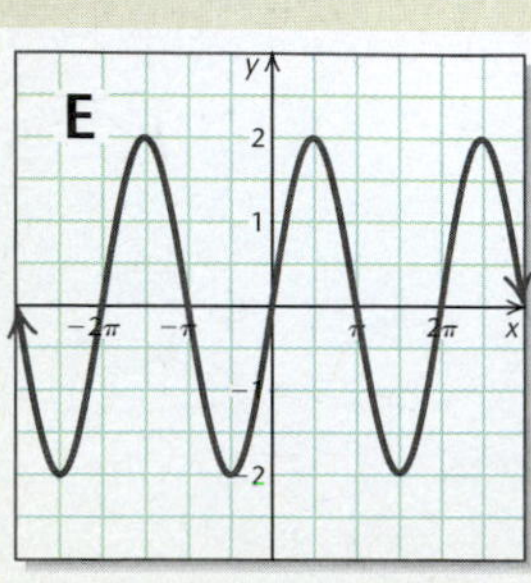

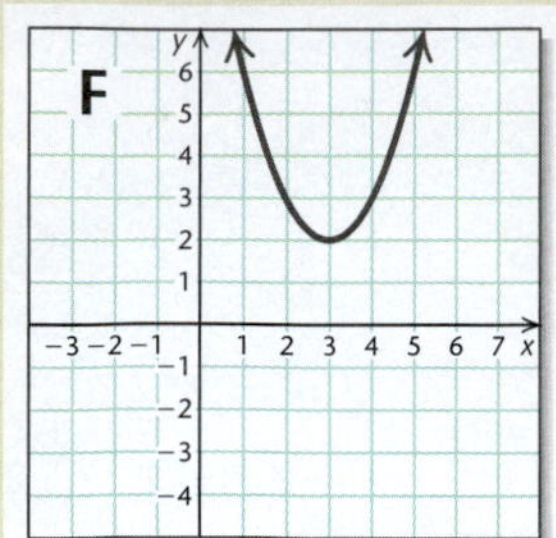

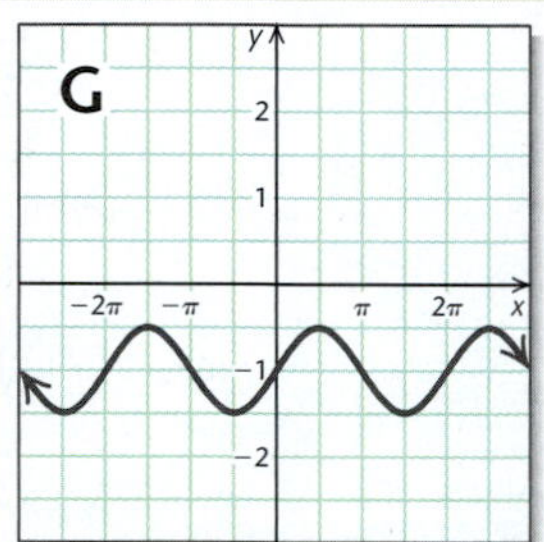

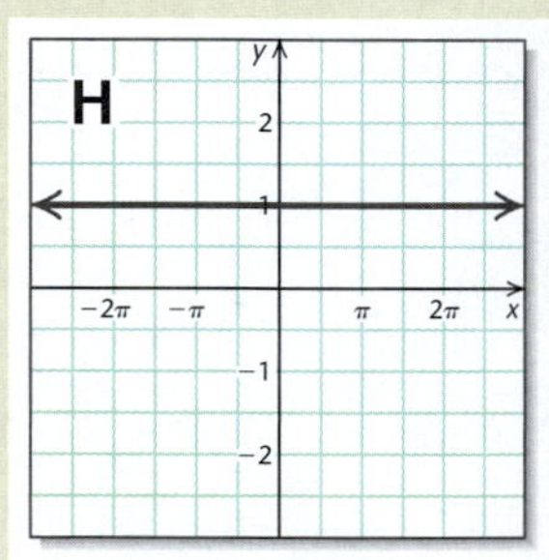

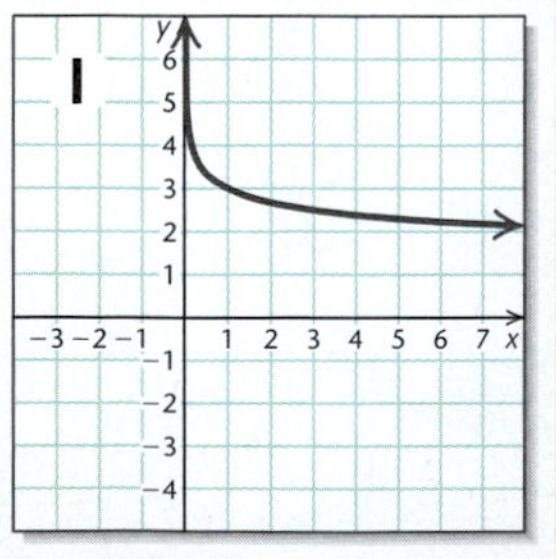

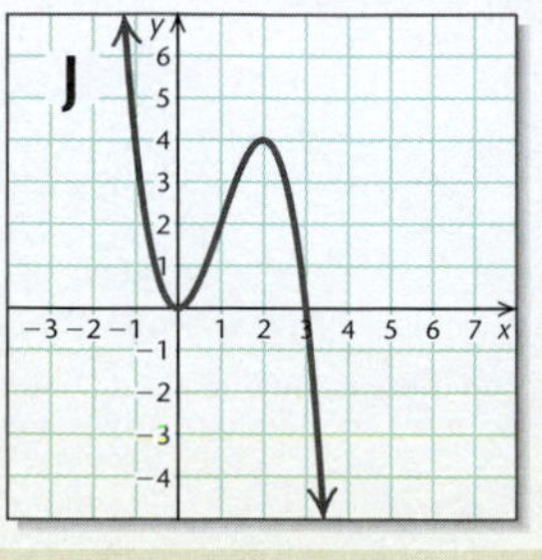

Answers on page A-45

7.5 Exercise Set

Solve, finding all solutions. Express the solutions in both radians and degrees.

1. $\cos x = \frac{\sqrt{3}}{2}$

2. $\sin x = -\frac{\sqrt{2}}{2}$

3. $\tan x = -\sqrt{3}$

4. $\cos x = -\frac{1}{2}$

5. $\sin x = \frac{1}{2}$

6. $\tan x = -1$

7. $\cos x = -\frac{\sqrt{2}}{2}$

8. $\sin x = \frac{\sqrt{3}}{2}$

Solve, finding all solutions in $[0, 2\pi)$ or $[0°, 360°)$. Verify your answer using a graphing calculator.

9. $2 \cos x - 1 = -1.2814$

10. $\sin x + 3 = 2.0816$

11. $2 \sin x + \sqrt{3} = 0$

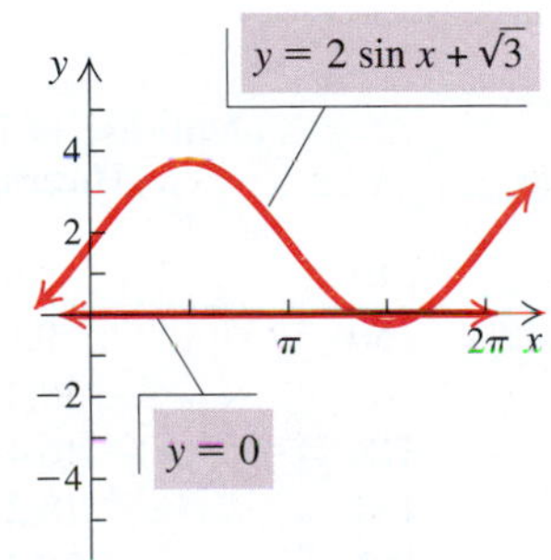

12. $2 \tan x - 4 = 1$

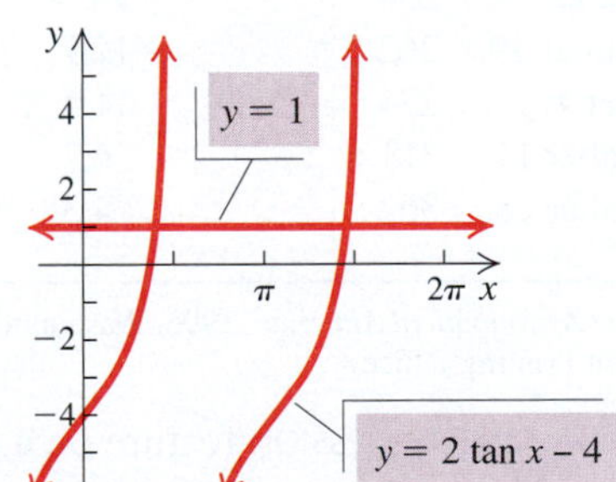

13. $2 \cos^2 x = 1$

14. $\csc^2 x - 4 = 0$

15. $2 \sin^2 x + \sin x = 1$

16. $\cos^2 x + 2 \cos x = 3$

17. $2 \cos^2 x - \sqrt{3} \cos x = 0$

18. $2 \sin^2 \theta + 7 \sin \theta = 4$

19. $6 \cos^2 \phi + 5 \cos \phi + 1 = 0$

20. $2 \sin t \cos t + 2 \sin t - \cos t - 1 = 0$

21. $\sin 2x \cos x - \sin x = 0$

22. $5 \sin^2 x - 8 \sin x = 3$

23. $\cos^2 x + 6 \cos x + 4 = 0$

24. $2 \tan^2 x = 3 \tan x + 7$

25. $7 = \cot^2 x + 4 \cot x$

26. $3 \sin^2 x = 3 \sin x + 2$

Solve, finding all solutions in $[0, 2\pi)$.

27. $\cos 2x - \sin x = 1$

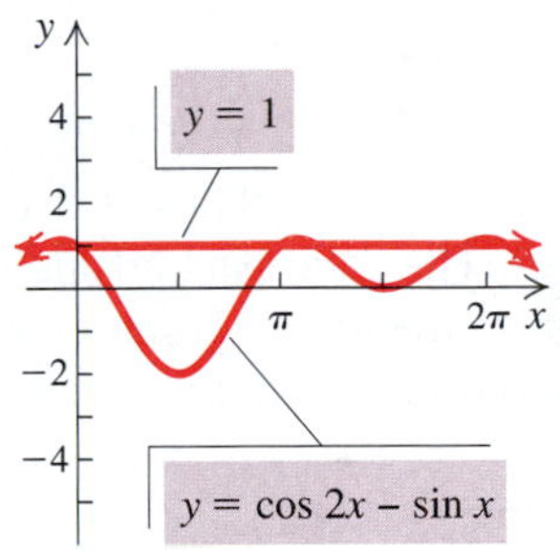

28. $2 \sin x \cos x + \sin x = 0$

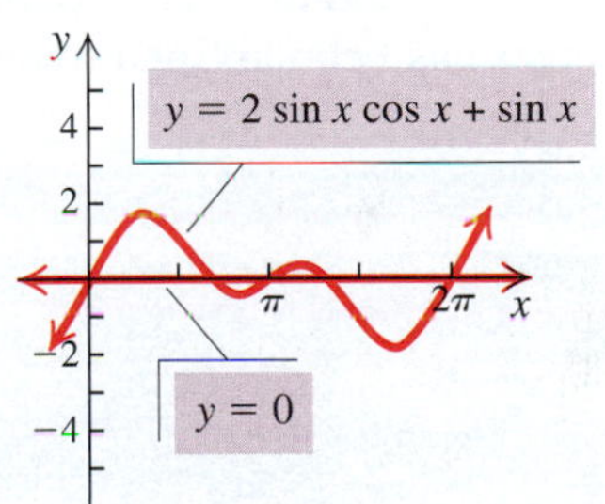

29. $\tan x \sin x - \tan x = 0$

30. $\sin 4x - 2 \sin 2x = 0$

31. $\sin 2x \cos x + \sin x = 0$

32. $\cos 2x \sin x + \sin x = 0$

33. $2 \sec x \tan x + 2 \sec x + \tan x + 1 = 0$

34. $\sin 2x \sin x - \cos 2x \cos x = -\cos x$

35. $\sin 2x + \sin x + 2 \cos x + 1 = 0$

36. $\tan^2 x + 4 = 2 \sec^2 x + \tan x$

37. $\sec^2 x - 2 \tan^2 x = 0$

38. $\cot x = \tan (2x - 3\pi)$

39. $2 \cos x + 2 \sin x = \sqrt{6}$

40. $\sqrt{3} \cos x - \sin x = 1$

41. $\sec^2 x + 2 \tan x = 6$

42. $5\cos 2x + \sin x = 4$

43. $\cos(\pi - x) + \sin\left(x - \frac{\pi}{2}\right) = 1$

44. $\dfrac{\sin^2 x - 1}{\cos\left(\frac{\pi}{2} - x\right) + 1} = \dfrac{\sqrt{2}}{2} - 1$

Solve using a calculator, finding all solutions in $[0, 2\pi)$.

45. $x \sin x = 1$

46. $x^2 + 2 = \sin x$

47. $2\cos^2 x = x + 1$

48. $x \cos x - 2 = 0$

49. $\cos x - 2 = x^2 - 3x$

50. $\sin x = \tan \frac{x}{2}$

51. *Sales of Fishing Boats.* Sales of fishing boats fluctuate in cycles. The following cosine function can be used to estimate the total amount of sales of fishing boats, y, in thousands of dollars, in month x, for a business:

$$y = 15.328 \cos(0.475x - 1.728) + 87.223.$$

Approximate the total amount of sales to the nearest dollar for November and for March. (*Hint*: 1 represents January, 2 represents February, and so on.)

52. *Sales of Skis.* Sales of certain products fluctuate in cycles. The following sine function can be used to estimate the total amount of sales of skis, y, in thousands of dollars, in month x, for a business in a northern climate:

$$y = 9.584 \sin(0.436x + 2.097) + 10.558.$$

Approximate the total amount of sales to the nearest dollar for December and for July. (*Hint*: 1 represents January, 2 represents February, and so on.)

53. *Average High Temperature in Chicago.* The data in the following table give the average high temperature in Chicago for certain months.

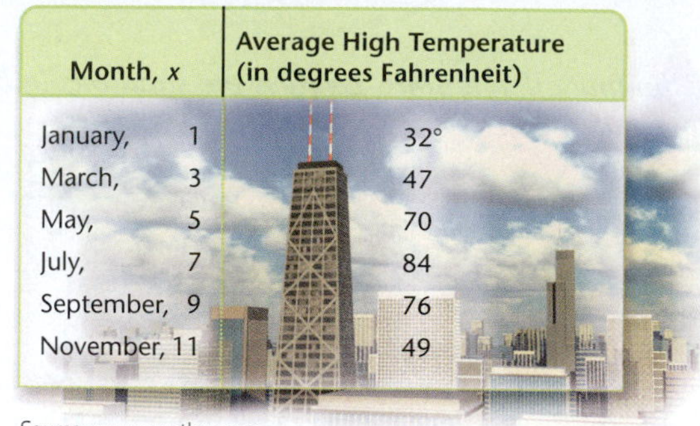

Month, x		Average High Temperature (in degrees Fahrenheit)
January,	1	32°
March,	3	47
May,	5	70
July,	7	84
September,	9	76
November,	11	49

Source: www.weather.com

a) Using the SINE REGRESSION feature on a graphing calculator, fit a sine function of the form $y = A\sin(BX - C) + D$ to this set of data.

b) Approximate the average high temperature in Chicago in April and in December.

c) Determine in which months the average high temperature is about 63°.

54. *Daylight Hours.* The data in the following table give the number of daylight hours for certain days in Kajaani, Finland.

Day, x		Number of Daylight Hours, y
January 10,	10	5.0
February 19,	50	9.1
March 3,	62	10.4
April 28,	118	16.4
May 14,	134	18.2
June 11,	162	20.7
July 17,	198	19.5
August 22,	234	15.7
September 19,	262	12.7
October 1,	274	11.4
November 14,	318	6.7
December 28,	362	4.3

Source: *The Astronomical Almanac*, 1995, Washington: U.S. Government Printing Office

a) Using the SINE REGRESSION feature on a graphing calculator, model these data with an equation of the form $y = A\sin(Bx - C) + D$.

b) Approximate the number of daylight hours in Kajaani for April 22 ($x = 112$), July 4 ($x = 185$), and December 15 ($x = 349$).

c) Determine on which days of the year there will be about 12 hr of daylight.

Skill Maintenance

Solve the right triangle. [6.2]

55.

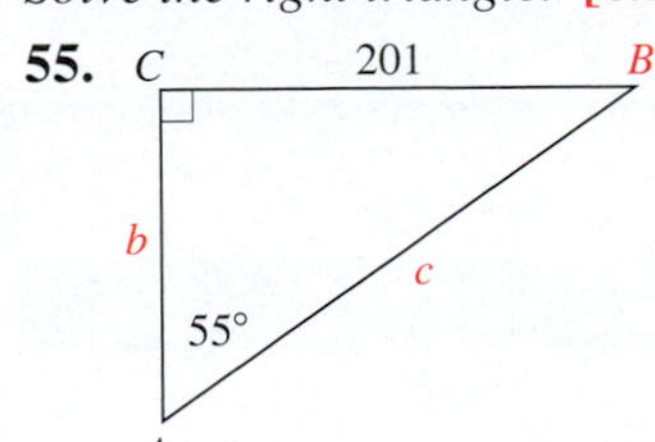

56.

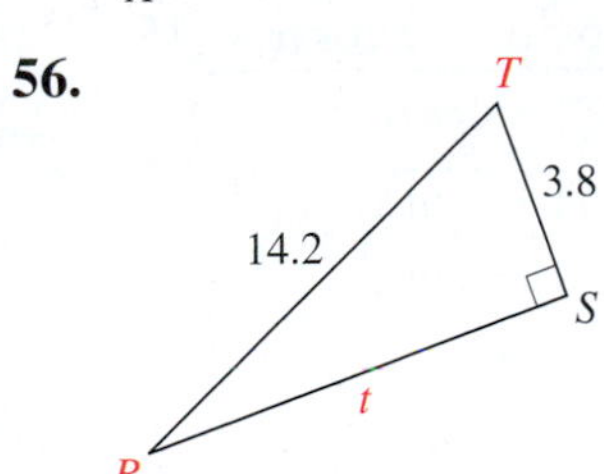

Solve. [1.5]

57. $\frac{x}{27} = \frac{4}{3}$

58. $\frac{0.01}{0.7} = \frac{0.2}{h}$

Synthesis

Solve in $[0, 2\pi)$.

59. $|\sin x| = \frac{\sqrt{3}}{2}$

60. $|\cos x| = \frac{1}{2}$

61. $\sqrt{\tan x} = \sqrt[4]{3}$

62. $e^{\sin x} = 1$

63. $\ln(\cos x) = 0$

64. $e^{\ln(\sin x)} = 1$

65. $\sin(\ln x) = -1$

66. $12 \sin x - 7\sqrt{\sin x} + 1 = 0$

67. *Temperature During an Illness.* The temperature T, in degrees Fahrenheit, of a patient t days into a 12-day illness is given by

$$T(t) = 101.6° + 3° \sin\left(\frac{\pi}{8}t\right).$$

Find the times t during the illness at which the patient's temperature was 103°.

68. *Satellite Location.* A satellite circles the earth in such a manner that it is y miles from the equator (north or south, height from the surface not considered) t minutes after its launch, where

$$Y = 5000\left[\cos \frac{\pi}{45}(t - 10)\right].$$

At what times t on the interval $[0, 240]$, the first 4 hr, is the satellite 3000 mi north of the equator?

69. *Nautical Mile.* (See Exercise 60 in Exercise Set 7.2.) In Great Britain, the *nautical mile* is defined as the length of a minute of arc of the earth's radius. Since the earth is flattened at the poles, a British nautical mile varies with latitude. In fact, it is given, in feet, by the function

$$N(\phi) = 6066 - 31 \cos 2\phi,$$

where ϕ is the latitude in degrees. At what latitude north is the length of a British nautical mile found to be 6040 ft?

70. *Acceleration Due to Gravity.* (See Exercise 59 in Exercise Set 7.2.) The acceleration due to gravity is often denoted by g in a formula such as $S = \frac{1}{2}gt^2$, where S is the distance that an object falls in t seconds. The number g is generally considered constant, but in fact it varies slightly with latitude. If ϕ stands for latitude, in degrees, an excellent approximation of g is given by the formula

$$g = 9.78049(1 + 0.005288 \sin^2\phi - 0.000006 \sin^2 2\phi),$$

where g is measured in meters per second per second at sea level. At what latitude north does $g = 9.8$?

Solve.

71. $\cos^{-1} x = \cos^{-1}\frac{3}{5} - \sin^{-1}\frac{4}{5}$

72. $\sin^{-1} x = \tan^{-1}\frac{1}{3} + \tan^{-1}\frac{1}{2}$

73. Suppose that $\sin x = 5 \cos x$. Find $\sin x \cos x$.

Chapter 7 Summary and Review

STUDY GUIDE

KEY TERMS AND CONCEPTS | **EXAMPLES**

SECTION 7.1: IDENTITIES: PYTHAGOREAN AND SUM AND DIFFERENCE

Identity

An **identity** is an equation that is true for all possible replacements of the variables.

Basic Identities

$\sin x = \dfrac{1}{\csc x}$, $\quad \tan x = \dfrac{\sin x}{\cos x}$,

$\cos x = \dfrac{1}{\sec x}$, $\quad \cot x = \dfrac{\cos x}{\sin x}$,

$\tan x = \dfrac{1}{\cot x}$,

$\sin(-x) = -\sin x$,
$\cos(-x) = \cos x$,
$\tan(-x) = -\tan x$

Pythagorean Identities

$\sin^2 x + \cos^2 x = 1$,
$1 + \cot^2 x = \csc^2 x$,
$1 + \tan^2 x = \sec^2 x$

Simplify.

a) $\dfrac{\tan^2 x \sec x}{\sec^2 x \tan x}$ **b)** $\dfrac{\cos^2\alpha - 2\cos\alpha - 15}{\cos\alpha + 3}$

c) $\dfrac{\cos\theta}{\sin(-\theta)} + \dfrac{1}{\tan\theta}$ **d)** $\sqrt{\dfrac{1 + \sin x}{1 - \sin x}}$

a) $\dfrac{\tan^2 x \sec x}{\sec^2 x \tan x} = \dfrac{\tan x}{\sec x} = \dfrac{\dfrac{\sin x}{\cos x}}{\dfrac{1}{\cos x}} = \dfrac{\sin x}{\cos x}\cdot\dfrac{\cos x}{1} = \sin x$

b) $\dfrac{\cos^2\alpha - 2\cos\alpha - 15}{\cos\alpha + 3} = \dfrac{(\cos\alpha - 5)(\cos\alpha + 3)}{\cos\alpha + 3}$
$= \cos\alpha - 5$

c) $\dfrac{\cos\theta}{\sin(-\theta)} + \dfrac{1}{\tan\theta} = -\dfrac{\cos\theta}{\sin\theta} + \dfrac{\cos\theta}{\sin\theta} = 0$

d) $\sqrt{\dfrac{1 + \sin x}{1 - \sin x}} = \sqrt{\dfrac{1 + \sin x}{1 - \sin x}}\cdot\sqrt{\dfrac{1 - \sin x}{1 - \sin x}}$
$= \dfrac{\sqrt{1 - \sin^2 x}}{1 - \sin x}$
$= \dfrac{\sqrt{\cos^2 x}}{1 - \sin x} = \dfrac{\cos x}{1 - \sin x}$

Sum and Difference Identities

$\sin(u \pm v) = \sin u \cos v \pm \cos u \sin v$,
$\cos(u \pm v) = \cos u \cos v \mp \sin u \sin v$,
$\tan(u \pm v) = \dfrac{\tan u \pm \tan v}{1 \mp \tan u \tan v}$

Evaluate $\cos\dfrac{7\pi}{12}$ exactly.

$\cos\dfrac{7\pi}{12} = \cos\left(\dfrac{3\pi}{4} - \dfrac{\pi}{6}\right) = \cos\dfrac{3\pi}{4}\cos\dfrac{\pi}{6} + \sin\dfrac{3\pi}{4}\sin\dfrac{\pi}{6}$
$= \left(-\dfrac{\sqrt{2}}{2}\right)\left(\dfrac{\sqrt{3}}{2}\right) + \left(\dfrac{\sqrt{2}}{2}\right)\left(\dfrac{1}{2}\right) = \dfrac{\sqrt{2} - \sqrt{6}}{4}$

Write $\sin 51^\circ \cos 24^\circ - \cos 51^\circ \sin 24^\circ$ as a trigonometric function of a single angle. Then evaluate.

We use the identity

$\sin(u - v) = \sin u \cos v - \cos u \sin v$.

Then

$\sin 51^\circ \cos 24^\circ - \cos 51^\circ \sin 24^\circ = \sin(51^\circ - 24^\circ)$
$= \sin 27^\circ \approx 0.4540$.

Assume that $\sin \alpha = \frac{7}{8}$ and $\sin \beta = \frac{9}{10}$ and that α and β are between 0 and $\pi/2$. Then evaluate $\sin(\alpha + \beta)$.

We use the identity

$$\sin(\alpha + \beta) = \sin \alpha \cos \beta + \cos \alpha \sin \beta.$$

Substituting, we have

$$\sin(\alpha + \beta) = \tfrac{7}{8} \cos \beta + \tfrac{9}{10} \cos \alpha.$$

To determine $\cos \alpha$ and $\cos \beta$, we use reference triangles and the Pythagorean theorem.

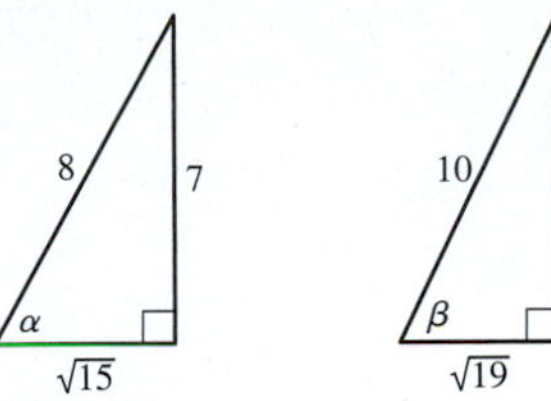

$$\cos \alpha = \frac{\sqrt{15}}{8} \quad \text{and} \quad \cos \beta = \frac{\sqrt{19}}{10}$$

Then

$$\sin(\alpha + \beta) = \frac{7}{8} \cdot \frac{\sqrt{19}}{10} + \frac{9}{10} \cdot \frac{\sqrt{15}}{8} = \frac{7\sqrt{19} + 9\sqrt{15}}{80}.$$

SECTION 7.2: IDENTITIES: COFUNCTION, DOUBLE-ANGLE, AND HALF-ANGLE

Cofunction Identities

$$\sin\left(\frac{\pi}{2} - x\right) = \cos x,$$

$$\cos\left(\frac{\pi}{2} - x\right) = \sin x,$$

$$\tan\left(\frac{\pi}{2} - x\right) = \cot x,$$

$$\cot\left(\frac{\pi}{2} - x\right) = \tan x,$$

$$\sec\left(\frac{\pi}{2} - x\right) = \csc x,$$

$$\csc\left(\frac{\pi}{2} - x\right) = \sec x,$$

$$\sin\left(x \pm \frac{\pi}{2}\right) = \pm \cos x,$$

$$\cos\left(x \pm \frac{\pi}{2}\right) = \mp \sin x$$

Given that $\cos \theta = \frac{3}{5}$ and that the terminal side of θ is in quadrant IV, find the exact function value for $\cos\left(\frac{\pi}{2} - \theta\right)$ and $\cos\left(\theta + \frac{\pi}{2}\right)$.

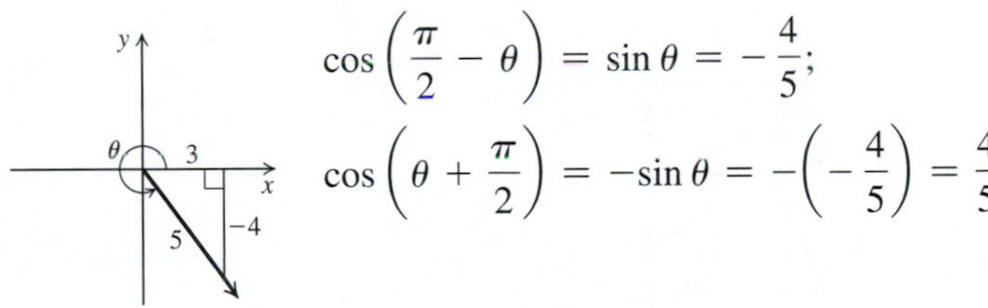

$$\cos\left(\frac{\pi}{2} - \theta\right) = \sin \theta = -\frac{4}{5};$$

$$\cos\left(\theta + \frac{\pi}{2}\right) = -\sin \theta = -\left(-\frac{4}{5}\right) = \frac{4}{5}$$

Double-Angle Identities

$$\sin 2x = 2 \sin x \cos x,$$

$$\cos 2x = \cos^2 x - \sin^2 x = 1 - 2\sin^2 x = 2\cos^2 x - 1,$$

$$\tan 2x = \frac{2 \tan x}{1 - \tan^2 x}$$

Given that $\tan \theta = -3$ and that θ is in quadrant II, find $\sin 2\theta$, $\cos 2\theta$, and the quadrant in which 2θ lies.

$$\sin \theta = \frac{3}{\sqrt{10}};$$

$$\cos \theta = -\frac{1}{\sqrt{10}};$$

$$\sin 2\theta = 2 \sin \theta \cos \theta = 2 \cdot \frac{3}{\sqrt{10}} \cdot \left(-\frac{1}{\sqrt{10}}\right) = -\frac{6}{10} = -\frac{3}{5};$$

$$\cos 2\theta = \cos^2 \theta - \sin^2 \theta = \left(-\frac{1}{\sqrt{10}}\right)^2 - \left(\frac{3}{\sqrt{10}}\right)^2 = \frac{1}{10} - \frac{9}{10} = -\frac{8}{10} = -\frac{4}{5}$$

Since both $\sin 2\theta$ and $\cos 2\theta$ are negative, we have 2θ in quadrant III.

Simplify: $\dfrac{\frac{1}{2}\sin 2x}{1 - \cos^2 x}$.

$$\frac{\frac{1}{2}\sin 2x}{1 - \cos^2 x} = \frac{\frac{1}{2} \cdot 2 \sin x \cos x}{\sin^2 x} = \frac{\cos x}{\sin x} = \cot x$$

Half-Angle Identities

$$\sin \frac{x}{2} = \pm\sqrt{\frac{1 - \cos x}{2}},$$

$$\cos \frac{x}{2} = \pm\sqrt{\frac{1 + \cos x}{2}},$$

$$\tan \frac{x}{2} = \pm\sqrt{\frac{1 - \cos x}{1 + \cos x}} = \frac{\sin x}{1 + \cos x} = \frac{1 - \cos x}{\sin x}$$

Evaluate $\sin \dfrac{3\pi}{8}$ exactly.

Note that $\dfrac{3\pi}{8}$ is in quadrant I. Thus $\sin \dfrac{3\pi}{8}$ is positive.

$$\sin \frac{3\pi}{8} = \frac{\frac{3\pi}{4}}{2} = \sqrt{\frac{1 - \cos \frac{3\pi}{4}}{2}} = \sqrt{\frac{1 - \left(-\frac{\sqrt{2}}{2}\right)}{2}} = \sqrt{\frac{2 + \sqrt{2}}{4}} = \frac{\sqrt{2 + \sqrt{2}}}{2}$$

Simplify: $\sin^2 \dfrac{x}{2} + \cos x$.

$$\sin^2 \frac{x}{2} + \cos x = \frac{1 - \cos x}{2} + \frac{2 \cos x}{2} = \frac{1 + \cos x}{2}$$

SECTION 7.3: PROVING TRIGONOMETRIC IDENTITIES

Proving Identities

Method 1: Start with either the left side or the right side of the equation and obtain the other side.

Method 2: Work with each side separately until you obtain the same expression.

Using method 1, prove $\dfrac{\tan x \sin^2 x}{1 + \cos x} = \tan x - \sin x$.

We begin with the left side and obtain the right side:

$$\begin{aligned}\frac{\tan x \sin^2 x}{1 + \cos x} &= \frac{\tan x\,(1 - \cos^2 x)}{1 + \cos x}\\ &= \frac{\tan x\,(1 - \cos x)(1 + \cos x)}{(1 + \cos x)}\\ &= \tan x\,(1 - \cos x)\\ &= \tan x - \tan x \cos x\\ &= \tan x - \sin x.\end{aligned}$$

Using method 2, prove $\dfrac{\sec \alpha}{1 + \cot^2 \alpha} = \dfrac{1 - \cos^2 \alpha}{\cos \alpha}$.

We begin with the left side:

$$\begin{aligned}\frac{\sec \alpha}{1 + \cot^2 \alpha} &= \frac{\frac{1}{\cos \alpha}}{\csc^2 \alpha} = \frac{1}{\cos \alpha} \cdot \frac{\sin^2 \alpha}{1}\\ &= \sin \alpha \cdot \frac{\sin \alpha}{\cos \alpha} = \sin \alpha \cdot \tan \alpha.\end{aligned}$$

Next, we work with the right side:

$$\frac{1 - \cos^2 \alpha}{\cos \alpha} = \frac{\sin^2 \alpha}{\cos \alpha} = \sin \alpha \cdot \frac{\sin \alpha}{\cos \alpha} = \sin \alpha \cdot \tan \alpha.$$

We have obtained the same expression from each side, so the proof is complete.

Product-to-Sum Identities

$$\sin x \cdot \sin y = \frac{1}{2}[\cos (x - y) - \cos (x + y)],$$

$$\cos x \cdot \cos y = \frac{1}{2}[\cos (x - y) + \cos (x + y)],$$

$$\sin x \cdot \cos y = \frac{1}{2}[\sin (x + y) + \sin (x - y)],$$

$$\cos x \cdot \sin y = \frac{1}{2}[\sin (x + y) - \sin (x - y)]$$

Sum-to-Product Identities

$$\sin x + \sin y = 2 \sin \frac{x + y}{2} \cos \frac{x - y}{2},$$

$$\sin x - \sin y = 2 \cos \frac{x + y}{2} \sin \frac{x - y}{2},$$

$$\cos y + \cos x = 2 \cos \frac{x + y}{2} \cos \frac{x - y}{2},$$

$$\cos y - \cos x = 2 \sin \frac{x + y}{2} \sin \frac{x - y}{2}$$

Use the product-to-sum or the sum-to-product identities to find an identity for $3 \sin 2\beta \sin 5\beta$ and for $\cos 3\alpha - \cos \alpha$.

Using $\sin x \cdot \sin y = \frac{1}{2}[\cos (x - y) - \cos (x + y)]$, we have

$$\begin{aligned}3 \sin 2\beta \sin 5\beta &= 3 \cdot \frac{1}{2}[\cos (-3\beta) - \cos 7\beta]\\ &= \frac{3}{2}[\cos 3\beta - \cos 7\beta].\end{aligned}$$

Using $\cos y - \cos x = 2 \sin \frac{x + y}{2} \sin \frac{x - y}{2}$, we have

$$\begin{aligned}\cos 3\alpha - \cos \alpha &= 2 \sin \frac{4\alpha}{2} \sin \frac{(-2\alpha)}{2}\\ &= 2 \sin 2\alpha \sin (-\alpha)\\ &= -2 \sin 2\alpha \sin \alpha\\ &= -2\,(2 \sin \alpha \cos \alpha)\,(\sin \alpha)\\ &= -4 \sin^2 \alpha \cos \alpha.\end{aligned}$$

SECTION 7.4: INVERSES OF THE TRIGONOMETRIC FUNCTIONS

Inverse Trigonometric Functions

FUNCTION	DOMAIN	RANGE
$y = \sin^{-1}x$	$[-1, 1]$	$\left[-\frac{\pi}{2}, \frac{\pi}{2}\right]$
$y = \cos^{-1}x$	$[-1, 1]$	$[0, \pi]$
$y = \tan^{-1}x$	$(-\infty, \infty)$	$\left(-\frac{\pi}{2}, \frac{\pi}{2}\right)$

The notation $y = \sin^{-1}x$ is equivalent to $y = \arcsin x$. The notation can be read:

- the inverse sine of x,
- the arcsine of x, or
- the number, or the angle, whose sine is x.

Find each of the following function values.

a) $\sin^{-1}\left(-\frac{1}{2}\right)$ **b)** $\cos^{-1}\left(-\frac{\sqrt{2}}{2}\right)$ **c)** $\tan^{-1}\sqrt{3}$

a)

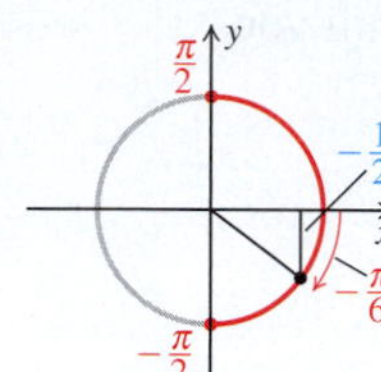

In $[-\pi/2, \pi/2]$, the only number whose sine is $-\frac{1}{2}$ is $-\pi/6$. Thus,

$$\sin^{-1}\left(-\frac{1}{2}\right) = -\pi/6, \text{ or } -30°.$$

b)

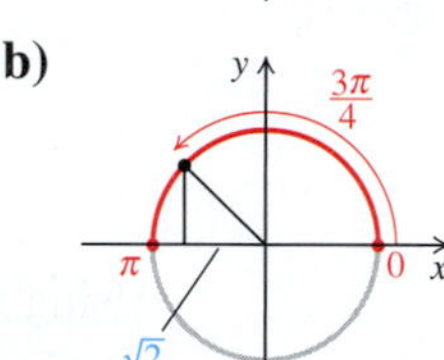

In $[0, \pi]$, the only number whose cosine is $-\frac{\sqrt{2}}{2}$ is $3\pi/4$. Thus,

$$\cos^{-1}\left(-\frac{\sqrt{2}}{2}\right) = 3\pi/4, \text{ or } 135°.$$

c)

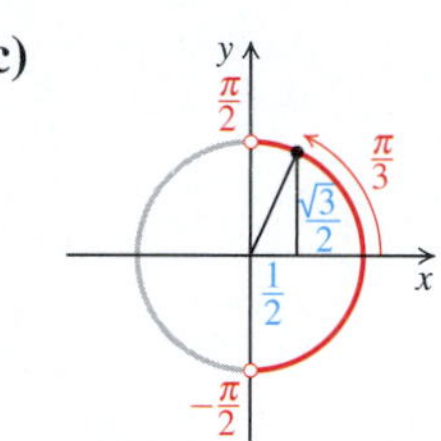

In $(-\pi/2, \pi/2)$, the only number whose tangent is $\sqrt{3}$ is $\pi/3$. Thus, $\tan^{-1}\sqrt{3} = \pi/3$, or $60°$.

Approximate each of the following function values in both radians and degrees. Round radian measure to four decimal places and degree measure to the nearest tenth of a degree.

a) $\cos^{-1}0.3281 \approx 1.2365$, or $70.8°$

b) $\tan^{-1}(-7.1154) \approx -1.4312$, or $-82.0°$

c) $\sin^{-1}(-0.5492) \approx -0.5814$, or $-33.3°$

Composition of Trigonometric Functions

The following are true for any x in the domain of the inverse function:

$$\sin(\sin^{-1}x) = x,$$
$$\cos(\cos^{-1}x) = x,$$
$$\tan(\tan^{-1}x) = x.$$

The following are true for any x in the range of the inverse function:

$$\sin^{-1}(\sin x) = x,$$
$$\cos^{-1}(\cos x) = x,$$
$$\tan^{-1}(\tan x) = x.$$

Simplify each of the following.

a) $\tan(\tan^{-1}\sqrt{3})$

Since $\sqrt{3}$ is in $(-\infty, \infty)$, the domain of $\tan^{-1}$, $\tan(\tan^{-1}\sqrt{3}) = \sqrt{3}$.

b) $\cos\left[\cos^{-1}\left(-\frac{1}{2}\right)\right]$

Since $-\frac{1}{2}$ is in $[-1, 1]$, the domain of $\cos^{-1}$,

$$\cos\left[\cos^{-1}\left(-\frac{1}{2}\right)\right] = -\frac{1}{2}.$$

c) $\sin^{-1}[\sin(\pi/6)]$

Since $\pi/6$ is in $[-\pi/2, \pi/2]$, the range of $\sin^{-1}$, $\sin^{-1}[\sin(\pi/6)] = \pi/6$.

d) $\cos^{-1}[\cos(3\pi/2)]$

Since $3\pi/2$ is not in $[0, \pi]$, the range of $\cos^{-1}$, we cannot apply $\cos^{-1}(\cos x) = x$. Instead, we find $\cos(3\pi/2)$, which is 0, and substitute to get $\cos^{-1} 0 = \pi/2$. Thus, $\cos^{-1}[\cos(3\pi/2)] = \pi/2$.

e) $\sin^{-1}[\cos(-\pi/6)] = \sin^{-1}\frac{\sqrt{3}}{2}$ $\quad$ **$\cos(-\pi/6) = \sqrt{3}/2$**

$= \pi/3$

f) $\tan\left[\cos^{-1}\left(-\frac{1}{2}\right)\right] = \tan\frac{2\pi}{3}$ $\quad$ **$\cos^{-1}\left(-\frac{1}{2}\right) = 2\pi/3$**

$= -\sqrt{3}$

Find: $\sin\left(\cos^{-1}\frac{x}{5}\right)$.

Let θ be the angle whose cosine is $x/5$: $\cos\theta = x/5$. Considering all values of x, we draw right triangles, in which the length of the hypotenuse is 5 and the length of one leg is x. The other leg in each triangle is $\sqrt{25 - x^2}$.

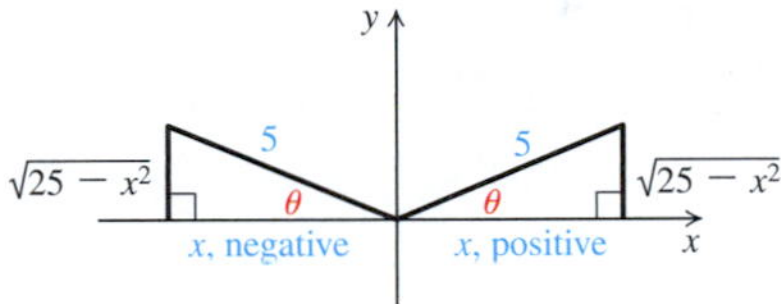

Thus, $\sin\left(\cos^{-1}\frac{x}{5}\right) = \frac{\sqrt{25 - x^2}}{5}$.

SECTION 7.5: SOLVING TRIGONOMETRIC EQUATIONS

Trigonometric Equations

When an equation contains a trigonometric expression with a variable, it is called a trigonometric equation. To solve such equations, we find all values for the variable that make the equation true.

In most applications, it is sufficient to find just the solutions from 0 to 2π, or from 0° to 360°. We then remember that any multiple of 2π, or 360°, can be added to obtain the rest of the solutions.

Solve $3 \tan x = \sqrt{3}$. Find *all* solutions.

$$3 \tan x = \sqrt{3}$$
$$\tan x = \sqrt{3}/3$$

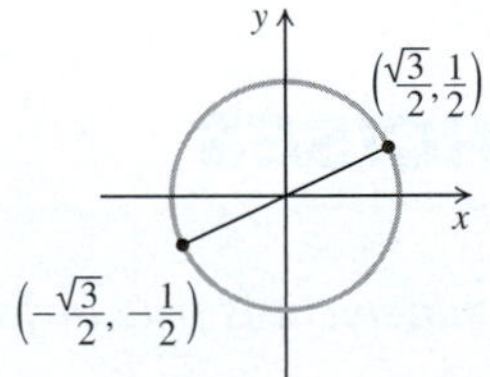

The solutions are numbers whose tangent is $\sqrt{3}/3$. There are only two points on the unit circle for which the tangent is $\sqrt{3}/3$. They are the points corresponding to $\pi/6$ and $7\pi/6$. The solutions are $\pi/6 + k\pi$, where k is any integer.

Solve: $\sin x = 1 - 2\sin^2 x$ in $[0°, 360°)$.

$$\sin x = 1 - 2\sin^2 x$$
$$2\sin^2 x + \sin x - 1 = 0$$
$$(2\sin x - 1)(\sin x + 1) = 0$$
$$2\sin x - 1 = 0 \quad or \quad \sin x + 1 = 0$$
$$\sin x = \tfrac{1}{2} \quad or \quad \sin x = -1$$

Thus, $x = 30°, 150°$ *or* $x = 270°$. All values check. The solutions in $[0°, 360°)$ are 30°, 150°, and 270°.

Solve: $\sin 2x \cos x + \cos x = 0$ in $[0, 2\pi)$.

$$\sin 2x \cos x + \cos x = 0$$
$$\cos x(\sin 2x + 1) = 0$$
$$\cos x = 0 \quad or \quad \sin 2x + 1 = 0$$
$$\cos x = 0 \quad or \quad \sin 2x = -1$$
$$x = \pi/2, 3\pi/2 \quad or \quad 2x = 3\pi/2, 7\pi/2$$
$$x = 3\pi/4, 7\pi/4$$

All values check. The solutions in $[0, 2\pi)$ are $\pi/2$, $3\pi/4$, $3\pi/2$, and $7\pi/4$.

Solve $5\cos^2 x = 2\cos x + 6$ in $[0°, 360°)$.

$$5\cos^2 x - 2\cos x - 6 = 0$$
$$\cos x = \frac{2 \pm \sqrt{124}}{10}$$ **Using the quadratic formula**

$$\cos x \approx 1.3136 \quad or \quad \cos x \approx -0.9136$$

Since cosine values are never greater than 1, $\cos x \approx 1.3136$ has no solution. Using $\cos x \approx -0.9136$, we find that $\cos^{-1}(-0.9136) \approx 156.01°$. Thus the solutions in $[0°, 360°)$ are 156.01° and 360° − 156.01°, or 203.99°.

REVIEW EXERCISES

Determine whether the statement is true or false.

1. $\sin^2 s \neq \sin s^2$. [7.1]

2. Given $0 < \alpha < \pi/2$ and $0 < \beta < \pi/2$ and that $\sin(\alpha + \beta) = 1$ and $\sin(\alpha - \beta) = 0$, then $\alpha = \pi/4$. [7.1]

3. If the terminal side of θ is in quadrant IV, then $\tan\theta < \cos\theta$. [7.1]

4. $\cos 5\pi/12 = \cos 7\pi/12$. [7.2]

5. Given that $\sin\theta = -\frac{2}{5}$, $\tan\theta < \cos\theta$. [7.1]

Complete the Pythagorean identity. [7.1]

6. $1 + \cot^2 x =$

7. $\sin^2 x + \cos^2 x =$

Multiply and simplify. Check using a graphing calculator. [7.1]

8. $(\tan y - \cot y)(\tan y + \cot y)$

9. $(\cos x + \sec x)^2$

Factor and simplify. Check using a graphing calculator. [7.1]

10. $\sec x \csc x - \csc^2 x$

11. $3 \sin^2 y - 7 \sin y - 20$

12. $1000 - \cos^3 u$

Simplify and check using a graphing calculator. [7.1]

13. $\dfrac{\sec^4 x - \tan^4 x}{\sec^2 x + \tan^2 x}$

14. $\dfrac{2 \sin^2 x}{\cos^3 x} \cdot \left(\dfrac{\cos x}{2 \sin x} \right)^2$

15. $\dfrac{3 \sin x}{\cos^2 x} \cdot \dfrac{\cos^2 x + \cos x \sin x}{\sin^2 x - \cos^2 x}$

16. $\dfrac{3}{\cos y - \sin y} - \dfrac{2}{\sin^2 y - \cos^2 y}$

17. $\left(\dfrac{\cot x}{\csc x} \right)^2 + \dfrac{1}{\csc^2 x}$

18. $\dfrac{4 \sin x \cos^2 x}{16 \sin^2 x \cos x}$

In Exercises 19–21, assume that all radicands are nonnegative.

19. Simplify:
$$\sqrt{\sin^2 x + 2 \cos x \sin x + \cos^2 x}.$$ [7.1]

20. Rationalize the denominator: $\sqrt{\dfrac{1 + \sin x}{1 - \sin x}}$. [7.1]

21. Rationalize the numerator: $\sqrt{\dfrac{\cos x}{\tan x}}$. [7.1]

22. Given that $x = 3 \tan \theta$, express $\sqrt{9 + x^2}$ as a trigonometric function without radicals. Assume that $0 < \theta < \pi/2$. [7.1]

Use the sum and difference formulas to write equivalent expressions. You need not simplify. [7.1]

23. $\cos \left(x + \dfrac{3\pi}{2} \right)$

24. $\tan (45^\circ - 30^\circ)$

25. Simplify: $\cos 27^\circ \cos 16^\circ + \sin 27^\circ \sin 16^\circ$. [7.1]

26. Find $\cos 165^\circ$ exactly. [7.1]

27. Given that $\tan \alpha = \sqrt{3}$ and $\sin \beta = \sqrt{2}/2$ and that α and β are between 0 and $\pi/2$, evaluate $\tan (\alpha - \beta)$ exactly. [7.1]

28. Assume that $\sin \theta = 0.5812$ and $\cos \phi = 0.2341$ and that both θ and ϕ are first-quadrant angles. Evaluate $\cos (\theta + \phi)$. [7.1]

Complete the cofunction identity. [7.2]

29. $\cos \left(x + \dfrac{\pi}{2} \right) =$

30. $\cos \left(\dfrac{\pi}{2} - x \right) =$

31. $\sin \left(x - \dfrac{\pi}{2} \right) =$

32. Given that $\cos \alpha = -\frac{3}{5}$ and that the terminal side is in quadrant III:

a) Find the other function values for α. [7.2]
b) Find the six function values for $\pi/2 - \alpha$. [7.2]
c) Find the six function values for $\alpha + \pi/2$. [7.2]

33. Find an equivalent expression for $\csc \left(x - \dfrac{\pi}{2} \right)$. [7.2]

34. Find $\tan 2\theta$, $\cos 2\theta$, and $\sin 2\theta$ and the quadrant in which 2θ lies, where $\cos \theta = -\frac{4}{5}$ and θ is in quadrant III. [7.2]

35. Find $\sin \dfrac{\pi}{8}$ exactly. [7.2]

36. Given that $\sin \beta = 0.2183$ and β is in quadrant I, find $\sin 2\beta$, $\cos \dfrac{\beta}{2}$, and $\cos 4\beta$. [7.2]

Simplify and check using a graphing calculator. [7.2]

37. $1 - 2 \sin^2 \dfrac{x}{2}$

38. $(\sin x + \cos x)^2 - \sin 2x$

39. $2 \sin x \cos^3 x + 2 \sin^3 x \cos x$

40. $\dfrac{2 \cot x}{\cot^2 x - 1}$

Prove the identity. [7.3]

41. $\dfrac{1 - \sin x}{\cos x} = \dfrac{\cos x}{1 + \sin x}$

42. $\dfrac{1 + \cos 2\theta}{\sin 2\theta} = \cot \theta$

43. $\dfrac{\tan y + \sin y}{2 \tan y} = \cos^2 \dfrac{y}{2}$

44. $\dfrac{\sin x - \cos x}{\cos^2 x} = \dfrac{\tan^2 x - 1}{\sin x + \cos x}$

Use the product-to-sum identities and the sum-to-product identities to find identities for each of the following. [7.3]

45. $3 \cos 2\theta \sin \theta$

46. $\sin \theta - \sin 4\theta$

In Exercises 47–50, use a graphing calculator to determine which expression (A)–(D) on the right can be used to complete the identity. Then prove the identity algebraically. [7.3]

47. $\csc x - \cos x \cot x$ **A.** $\dfrac{\csc x}{\sec x}$

48. $\dfrac{1}{\sin x \cos x} - \dfrac{\cos x}{\sin x}$ **B.** $\sin x$

49. $\dfrac{\cot x - 1}{1 - \tan x}$ **C.** $\dfrac{2}{\sin x}$

50. $\dfrac{\cos x + 1}{\sin x} + \dfrac{\sin x}{\cos x + 1}$ **D.** $\dfrac{\sin x \cos x}{1 - \sin^2 x}$

Find each of the following exactly in both radians and degrees. [7.4]

51. $\sin^{-1}\left(-\dfrac{1}{2}\right)$ **52.** $\cos^{-1}\dfrac{\sqrt{3}}{2}$

53. $\tan^{-1} 1$ **54.** $\sin^{-1} 0$

Use a calculator to find each of the following in radians, rounded to four decimal places, and in degrees, rounded to the nearest tenth of a degree. [7.4]

55. $\cos^{-1}(-0.2194)$ **56.** $\cot^{-1} 2.381$

Evaluate. [7.4]

57. $\cos\left(\cos^{-1}\dfrac{1}{2}\right)$ **58.** $\tan^{-1}\left(\tan\dfrac{\sqrt{3}}{3}\right)$

59. $\sin^{-1}\left(\sin\dfrac{\pi}{7}\right)$ **60.** $\cos\left(\sin^{-1}\dfrac{\sqrt{2}}{2}\right)$

Find each of the following. [7.4]

61. $\cos\left(\tan^{-1}\dfrac{b}{3}\right)$ **62.** $\cos\left(2\sin^{-1}\dfrac{4}{5}\right)$

Solve, finding all solutions. Express the solutions in both radians and degrees. [7.5]

63. $\cos x = -\dfrac{\sqrt{2}}{2}$ **64.** $\tan x = \sqrt{3}$

Solve, finding all solutions in $[0, 2\pi)$. [7.5]

65. $4\sin^2 x = 1$

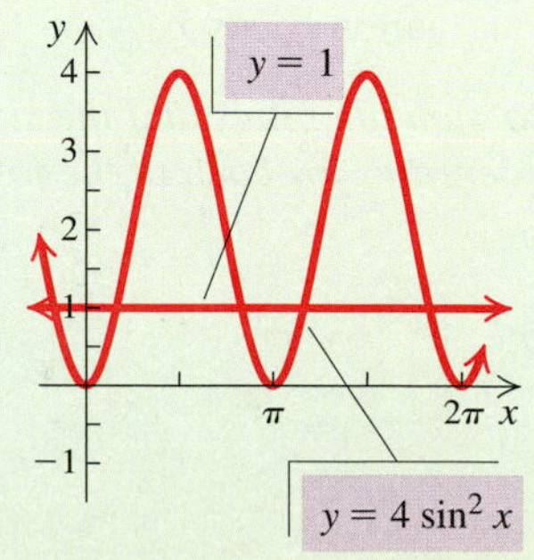

66. $\sin 2x \sin x - \cos x = 0$

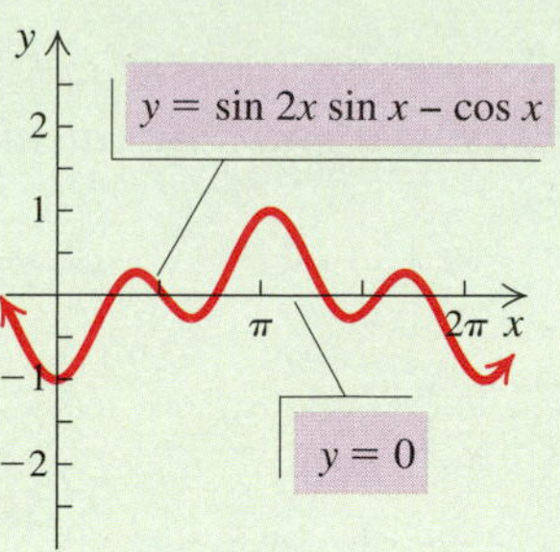

67. $2\cos^2 x + 3\cos x = -1$

68. $\sin^2 x - 7\sin x = 0$

69. $\csc^2 x - 2\cot^2 x = 0$

70. $\sin 4x + 2\sin 2x = 0$

71. $2\cos x + 2\sin x = \sqrt{2}$

72. $6\tan^2 x = 5\tan x + \sec^2 x$

Solve using a graphing calculator, finding all solutions in $[0, 2\pi)$. [7.5]

73. $x\cos x = 1$

74. $2\sin^2 x = x + 1$

75. Which of the following is the domain of the function $\cos^{-1} x$? [7.4]

A. $(0, \pi)$ **B.** $[-1, 1]$
C. $[-\pi/2, \pi/2]$ **D.** $(-\infty, \infty)$

76. Simplify: $\sin^{-1}\left(\sin\dfrac{7\pi}{6}\right)$. [7.4]

A. $-\pi/6$ **B.** $7\pi/6$
C. $-1/2$ **D.** $11\pi/6$

77. The graph of $f(x) = \sin^{-1} x$ is which of the following? [7.4]

A.

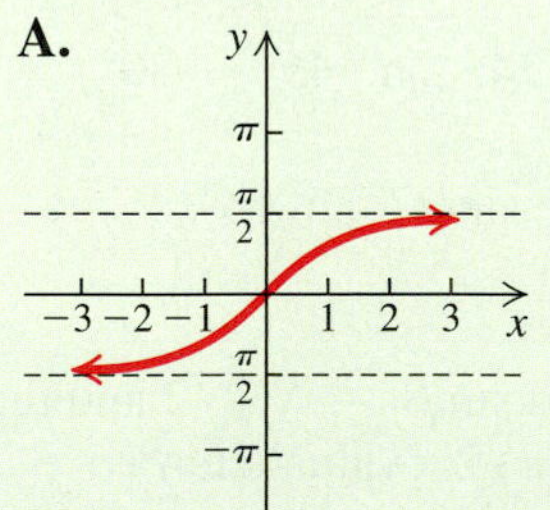

B.

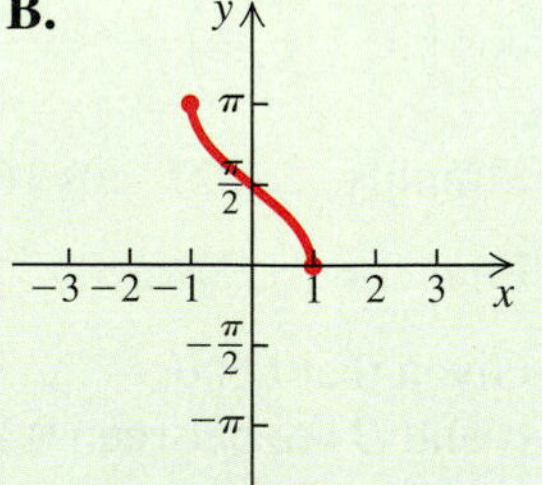

C.

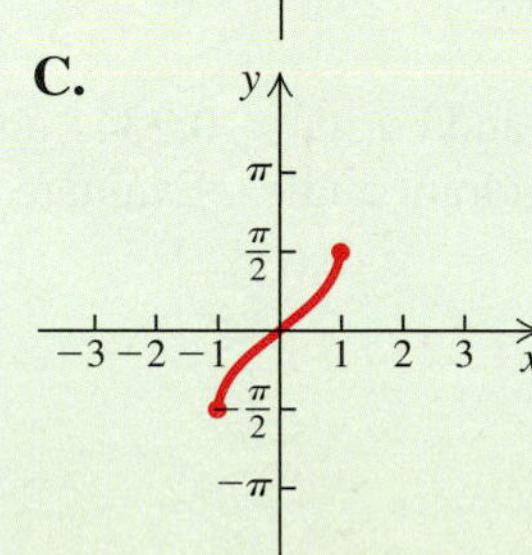

D.

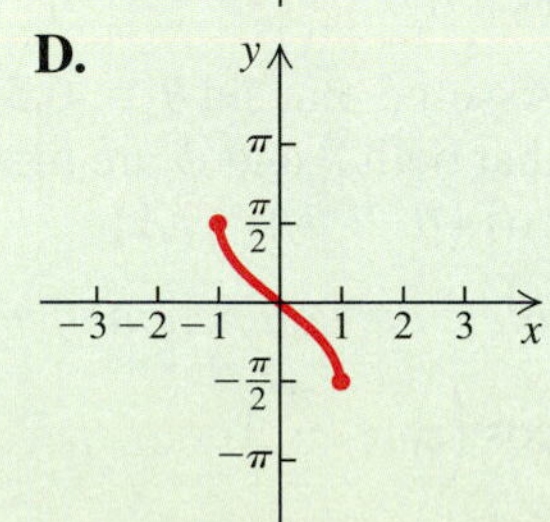

Synthesis

78. Find the measure of the angle from l_1 to l_2:

l_1: $x + y = 3$; l_2: $2x - y = 5$. [7.1]

79. Find an identity for $\cos(u + v)$ involving only cosines. [7.1], [7.2]

80. Simplify: $\cos\left(\frac{\pi}{2} - x\right)[\csc x - \sin x]$. [7.2]

81. Find $\sin\theta$, $\cos\theta$, and $\tan\theta$ under the given conditions:

$\sin 2\theta = \frac{1}{5}$, $\frac{\pi}{2} \leq 2\theta < \pi$. [7.2]

82. Show that

$$\tan^{-1}x = \frac{\sin^{-1}x}{\cos^{-1}x}$$

is *not* an identity. [7.4]

83. Solve $e^{\cos x} = 1$ in $[0, 2\pi)$. [7.5]

Collaborative Discussion and Writing

84. Why are the ranges of the inverse trigonometric functions restricted? [7.4]

85. Miles lists his answer to a problem as $\pi/6 + k\pi$, for any integer k, while Jaylen lists his answer as $\pi/6 + 2k\pi$ and $7\pi/6 + 2k\pi$, for any integer k. Are their answers equivalent? Why or why not? [7.5]

86. How does the graph of $y = \sin^{-1}x$ differ from the graph of $y = \sin x$? [7.4]

87. What is the difference between a trigonometric equation that is an identity and a trigonometric equation that is not an identity? Give an example of each. [7.1], [7.5]

88. Why is it that

$$\sin\frac{5\pi}{6} = \frac{1}{2}, \quad \text{but} \quad \sin^{-1}\left(\frac{1}{2}\right) \neq \frac{5\pi}{6}?$$ [7.4]

7 Chapter Test

Simplify.

1. $\frac{2\cos^2 x - \cos x - 1}{\cos x - 1}$

2. $\left(\frac{\sec x}{\tan x}\right)^2 - \frac{1}{\tan^2 x}$

3. Rationalize the denominator:

$$\sqrt{\frac{1 - \sin\theta}{1 + \sin\theta}}.$$

Assume that the radicand is nonnegative.

4. Given that $x = 2\sin\theta$, express $\sqrt{4 - x^2}$ as a trigonometric function without radicals. Assume that $0 < \theta < \pi/2$.

Use the sum or difference identities to evaluate exactly.

5. $\sin 75°$

6. $\tan\frac{\pi}{12}$

7. Assuming that $\cos u = \frac{5}{13}$ and $\cos v = \frac{12}{13}$ and that u and v are between 0 and $\pi/2$, evaluate $\cos(u - v)$ exactly.

8. Given that $\cos\theta = -\frac{2}{3}$ and that the terminal side is in quadrant II, find $\cos(\pi/2 - \theta)$.

9. Given that $\sin\theta = -\frac{4}{5}$ and θ is in quadrant III, find $\sin 2\theta$ and the quadrant in which 2θ lies.

10. Use a half-angle identity to evaluate $\cos\frac{\pi}{12}$ exactly.

11. Given that $\sin\theta = 0.6820$ and that θ is in quadrant I, find $\cos(\theta/2)$.

12. Simplify: $(\sin x + \cos x)^2 - 1 + 2\sin 2x$.

Prove each of the following identities.

13. $\csc x - \cos x \cot x = \sin x$

14. $(\sin x + \cos x)^2 = 1 + \sin 2x$

15. $(\csc\beta + \cot\beta)^2 = \frac{1 + \cos\beta}{1 - \cos\beta}$

16. $\frac{1 + \sin\alpha}{1 + \csc\alpha} = \frac{\tan\alpha}{\sec\alpha}$

Use the product-to-sum identities and the sum-to-product identities to find identities for each of the following.

17. $\cos 8\alpha - \cos \alpha$

18. $4 \sin \beta \cos 3\beta$

19. Find $\sin^{-1}\left(-\frac{\sqrt{2}}{2}\right)$ exactly in degrees.

20. Find $\tan^{-1}\sqrt{3}$ exactly in radians.

21. Use a calculator to find $\cos^{-1}(-0.6716)$ in radians, rounded to four decimal places.

22. Evaluate $\cos\left(\sin^{-1}\frac{1}{2}\right)$.

23. Find $\tan\left(\sin^{-1}\frac{5}{x}\right)$.

24. Evaluate $\cos\left(\sin^{-1}\frac{1}{2} + \cos^{-1}\frac{1}{2}\right)$.

Solve, finding all solutions in $[0, 2\pi)$.

25. $4\cos^2 x = 3$

26. $2\sin^2 x = \sqrt{2}\sin x$

27. $\sqrt{3}\cos x + \sin x = 1$

28. The graph of $f(x) = \cos^{-1}x$ is which of the following?

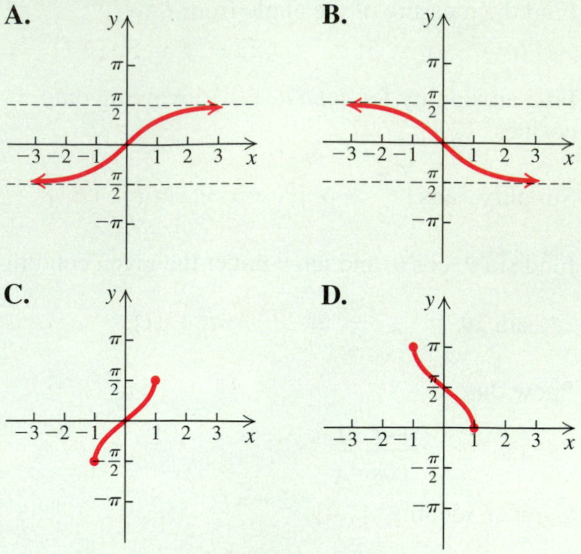

A.

B.

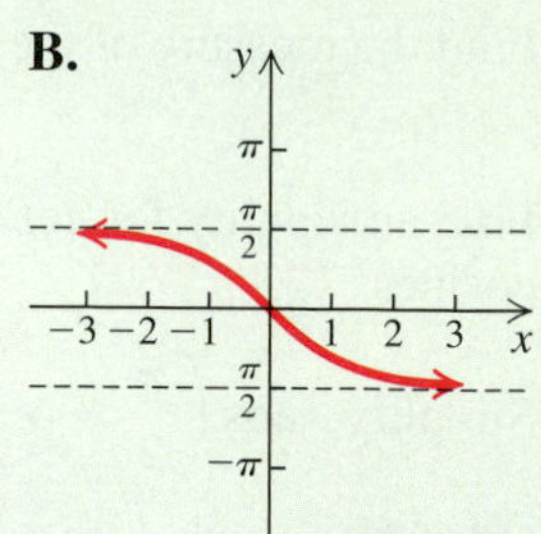

C.

D.

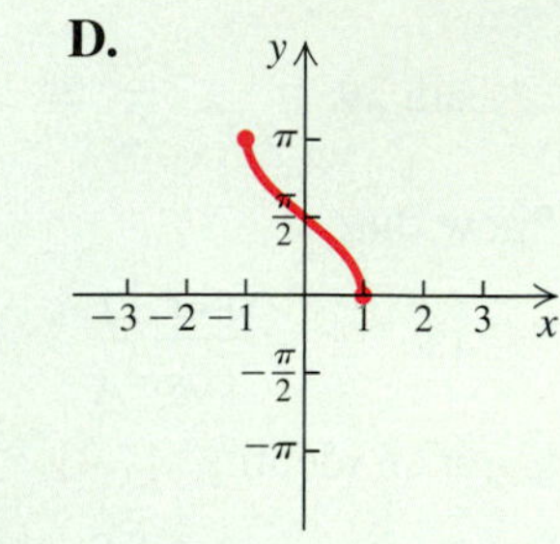

› Synthesis

29. Find $\cos\theta$, given that

$$\cos 2\theta = \frac{5}{6}, \quad \frac{3\pi}{2} < \theta < 2\pi.$$

Applications of Trigonometry

CHAPTER

8

APPLICATION

This problem appears as Exercise 25 in Section 8.2.

A teacup ride for children at an amusement park consists of five teacups equally spaced around a circle. Each cup holds 6 passengers and is at the end of an arm 20 ft long. Find the linear distance between a pair of adjacent cups. Round the length to the nearest tenth of a foot.

8.1 The Law of Sines

- Use the law of sines to solve triangles.
- Find the area of any triangle given the lengths of two sides and the measure of the included angle.

To **solve a triangle** means to find the lengths of all its sides and the measures of all its angles. We solved right triangles in Section 6.2. For review, let's solve the right triangle shown below. We begin by listing the known measures.

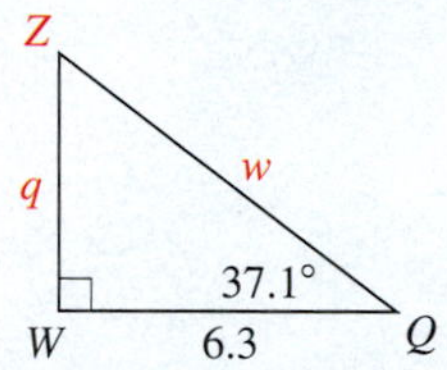

$$\begin{aligned} Q &= 37.1°, & q &= ?, \\ W &= 90°, & w &= ?, \\ Z &= ?, & z &= 6.3 \end{aligned}$$

Since the sum of the three angle measures of any triangle is 180°, we can immediately find the measure of the third angle:

$$\begin{aligned} Z &= 180° - (90° + 37.1°) \\ &= 52.9°. \end{aligned}$$

Then using the tangent and cosine ratios, respectively, we can find q and w:

$$\tan 37.1° = \frac{q}{6.3}, \quad \text{or}$$

$$q = 6.3 \tan 37.1° \approx 4.8,$$

and $$\cos 37.1° = \frac{6.3}{w}, \quad \text{or}$$

$$w = \frac{6.3}{\cos 37.1°} \approx 7.9.$$

Now all six measures are known and we have solved triangle QWZ.

$$\begin{aligned} Q &= 37.1°, & q &\approx 4.8, \\ W &= 90°, & w &\approx 7.9, \\ Z &= 52.9°, & z &= 6.3 \end{aligned}$$

Solving Oblique Triangles

The trigonometric functions can also be used to solve triangles that are not right triangles. Such triangles are called **oblique.** Any triangle, right or oblique, can be solved *if at least one side and any other two measures are known*. The five possible situations are illustrated on the next page.

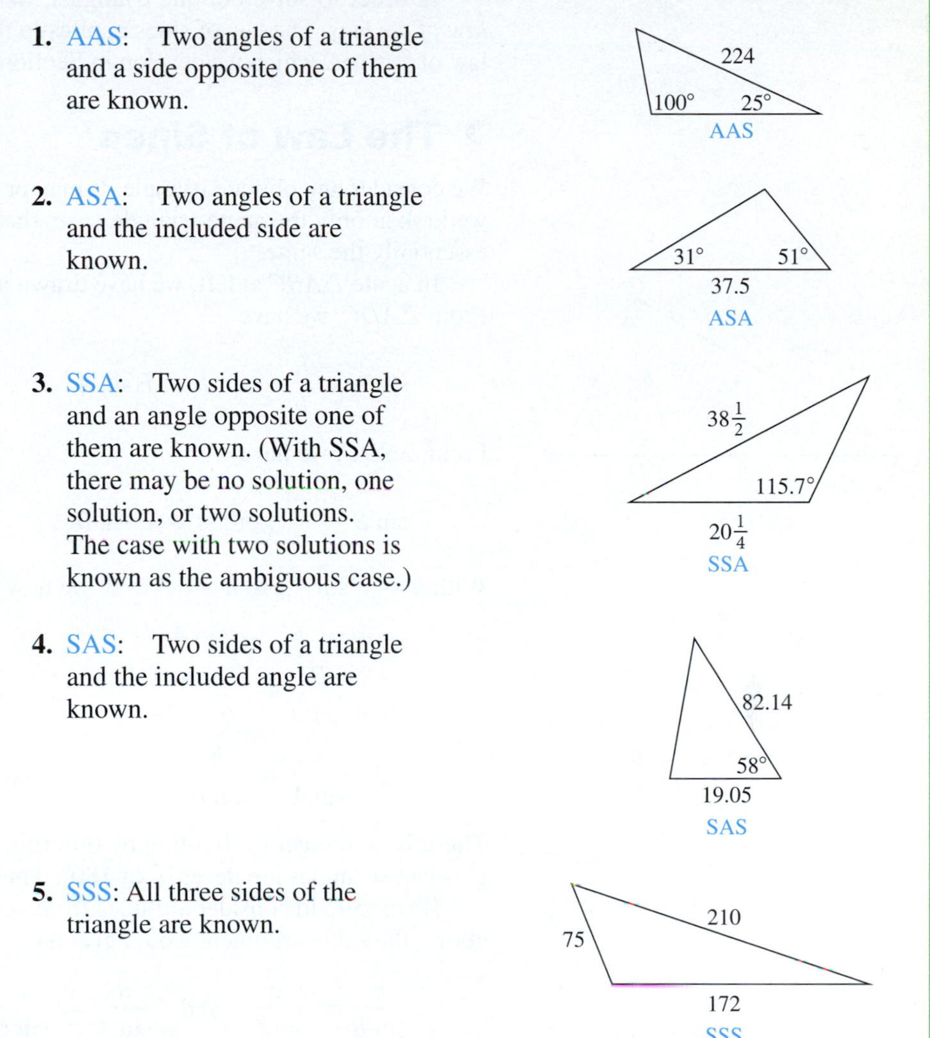

1. AAS: Two angles of a triangle and a side opposite one of them are known.
2. ASA: Two angles of a triangle and the included side are known.
3. SSA: Two sides of a triangle and an angle opposite one of them are known. (With SSA, there may be no solution, one solution, or two solutions. The case with two solutions is known as the ambiguous case.)
4. SAS: Two sides of a triangle and the included angle are known.
5. SSS: All three sides of the triangle are known.

The list above does not include the situation in which only the three angle measures are given. The reason for this lies in the fact that the angle measures determine *only the shape* of the triangle and *not the size*, as shown with the following triangles. Thus we cannot solve a triangle when only the three angle measures are given.

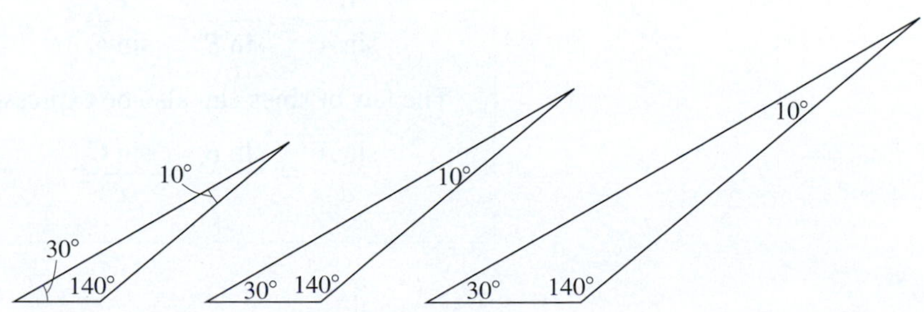

In order to solve oblique triangles, we need to derive the *law of sines* and the *law of cosines*. The law of sines applies to the first three situations listed above. The law of cosines, which we develop in Section 8.2, applies to the last two situations.

The Law of Sines

We consider any oblique triangle. It may or may not have an obtuse angle. Although we look at only the acute-triangle case, the derivation of the obtuse-triangle case is essentially the same.

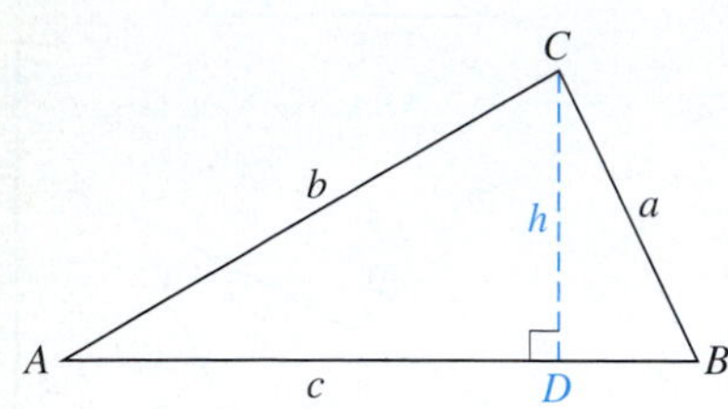

In acute $\triangle ABC$ at left, we have drawn an altitude from vertex C. It has length h. From $\triangle ADC$, we have

$$\sin A = \frac{h}{b}, \quad \text{or} \quad h = b \sin A.$$

From $\triangle BDC$, we have

$$\sin B = \frac{h}{a}, \quad \text{or} \quad h = a \sin B.$$

With $h = b \sin A$ and $h = a \sin B$, we now have

$$a \sin B = b \sin A$$

$$\frac{a \sin B}{\sin A \sin B} = \frac{b \sin A}{\sin A \sin B} \qquad \text{Dividing by } \sin A \sin B$$

$$\frac{a}{\sin A} = \frac{b}{\sin B}. \qquad \text{Simplifying}$$

There is no danger of dividing by 0 in this case because we are dealing with triangles whose angles are never 0° or 180°. Thus the sine value will never be 0.

If we were to consider altitudes from vertex A and vertex B in the triangle shown above, the same argument would give us

$$\frac{b}{\sin B} = \frac{c}{\sin C} \quad \text{and} \quad \frac{a}{\sin A} = \frac{c}{\sin C}.$$

We combine these results to obtain the law of sines.

THE LAW OF SINES

In any triangle ABC,

$$\frac{a}{\sin A} = \frac{b}{\sin B} = \frac{c}{\sin C}.$$

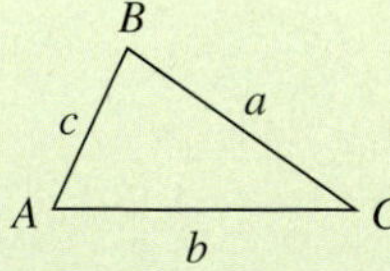

The law of sines can also be expressed as

$$\frac{\sin A}{a} = \frac{\sin B}{b} = \frac{\sin C}{c}.$$

Solving Triangles (AAS and ASA)

When two angles and a side of any triangle are known, the law of sines can be used to solve the triangle.

EXAMPLE 1 In $\triangle EFG$, $e = 4.56$, $E = 43°$, and $G = 57°$. Solve the triangle.

Solution We first make a drawing. We know three of the six measures.

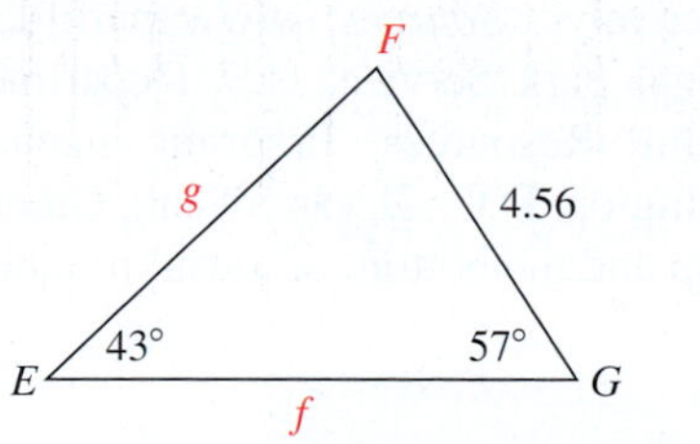

$$E = 43°, \quad e = 4.56,$$
$$F = ?, \quad f = ?,$$
$$G = 57°, \quad g = ?$$

From the figure, we see that we have the AAS situation. We begin by finding F:

$$F = 180° - (43° + 57°) = 80°.$$

We can now find the other two sides, using the law of sines:

$$\frac{f}{\sin F} = \frac{e}{\sin E}$$

$$\frac{f}{\sin 80°} = \frac{4.56}{\sin 43°} \qquad \text{Substituting}$$

$$f = \frac{4.56 \sin 80°}{\sin 43°} \qquad \text{Solving for } f$$

$$f \approx 6.58;$$

$$\frac{g}{\sin G} = \frac{e}{\sin E}$$

$$\frac{g}{\sin 57°} = \frac{4.56}{\sin 43°} \qquad \text{Substituting}$$

$$g = \frac{4.56 \sin 57°}{\sin 43°} \qquad \text{Solving for } g$$

$$g \approx 5.61.$$

Thus we have solved the triangle:

$$E = 43°, \quad e = 4.56,$$
$$F = 80°, \quad f \approx 6.58,$$
$$G = 57°, \quad g \approx 5.61.$$

Now Try Exercise 1.

The law of sines is frequently used in determining distances.

EXAMPLE 2 ***Vietnam Veterans Memorial.*** Designed by Maya Lin, the Vietnam Veterans Memorial, in Washington, D.C., consists of two congruent black granite walls on which 58,307 (as of May 2015) names are inscribed in chronological order of the date of the casualty. Each wall closely approximates a triangle. The height of the memorial at its tallest point is about 120.5 in. The angles formed by the top and the bottom of a wall with the height of the memorial are about 89.4056° and 88.2625°, respectively. (*Sources*: www.tourofdc.org; Maya Lin, Designer, New York, NY; National Park Service, U.S. Department of the Interior; Jennifer Talken-Spaulding, Cultural Resources Program manager, National Mall and Memorial Parks, Washington, D.C.; Bryan Swank, Unique Products, Columbus, IN). Find the lengths of the top and the bottom of a wall rounded to the nearest tenth of an inch.

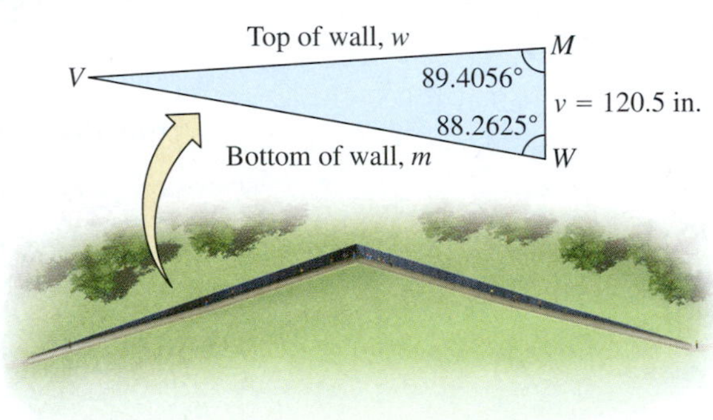

Solution We first find angle V:

$$180° - (89.4056° + 88.2625°) = 2.3319°.$$

Because the application involves the ASA situation, we use the law of sines to determine m and w.

$$\frac{m}{\sin M} = \frac{v}{\sin V}$$

$$\frac{m}{\sin 89.4056°} = \frac{120.5 \text{ in.}}{\sin 2.3319°} \quad \textbf{Substituting}$$

$$m = \frac{120.5 \text{ in.} (\sin 89.4056°)}{\sin 2.3319°} \quad \textbf{Solving for } m$$

$$m \approx 2961.4 \text{ in.}$$

$$\frac{w}{\sin W} = \frac{v}{\sin V}$$

$$\frac{w}{\sin 88.2625°} = \frac{120.5 \text{ in.}}{\sin 2.3319°} \quad \textbf{Substituting}$$

$$w = \frac{120.5 \text{ in.} (\sin 88.2625°)}{\sin 2.3319°} \quad \textbf{Solving for } w$$

$$w \approx 2960.2 \text{ in.}$$

Thus, $m \approx 2961.4$ in. and $w \approx 2960.2$ in.

Now Try Exercise 23.

Solving Triangles (SSA)

When two sides of a triangle and an angle opposite one of them are known, the law of sines can be used to solve the triangle.

Suppose for $\triangle ABC$ that b, c, and B are given. The various possibilities are as shown in the eight cases below: five cases when B is acute and three cases when B is obtuse. Note that $b < c$ in cases 1, 2, 3, and 6; $b = c$ in cases 4 and 7; and $b > c$ in cases 5 and 8.

ANGLE *B* IS ACUTE

Case 1: No solution
$b < c$; side b is too short to reach the base. No triangle is formed.

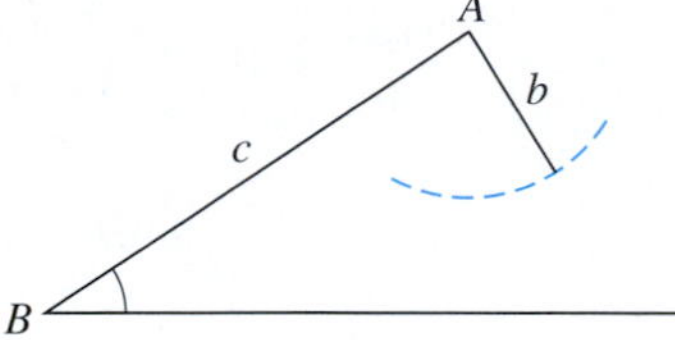

Case 2: One solution
$b < c$; side b just reaches the base and is perpendicular to it.

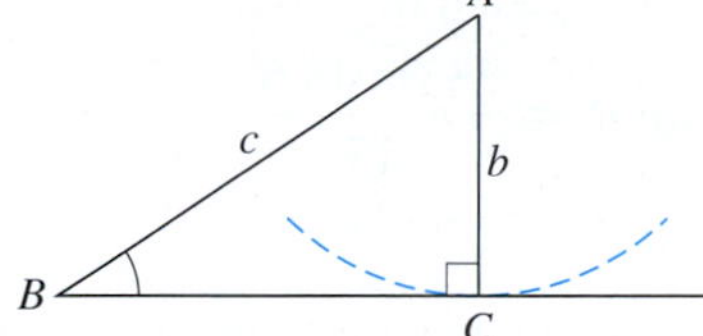

Case 3: Two solutions
$b < c$; an arc of radius b meets the base at two points. (This case is called the **ambiguous case**.)

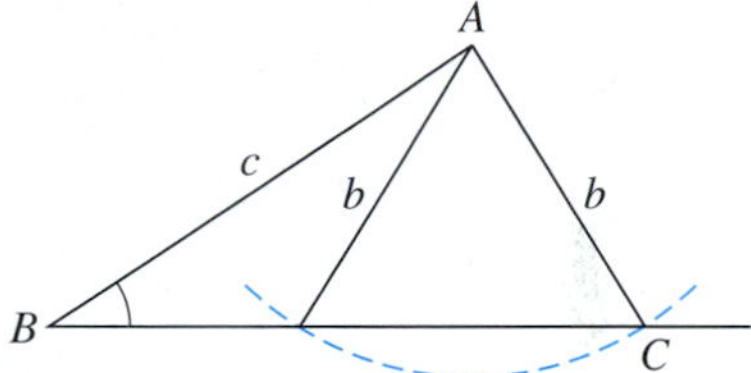

Case 4: One solution
$b = c$; an arc of radius b meets the base at just one point other than B.

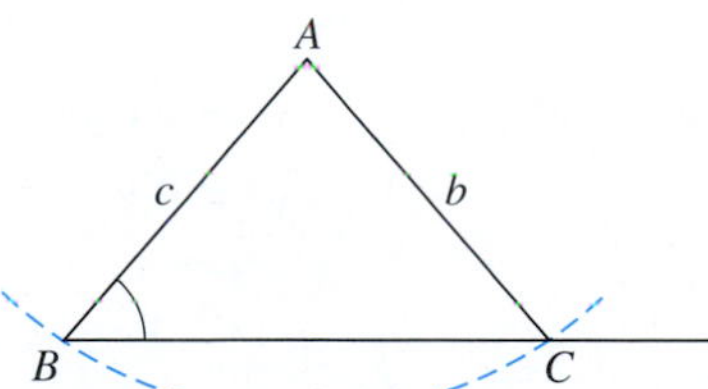

Case 5: One solution
$b > c$; an arc of radius b meets the base at just one point.

ANGLE *B* IS OBTUSE

Case 6: No solution
$b < c$; side b is too short to reach the base. No triangle is formed.

Case 7: No solution
$b = c$; an arc of radius b meets the base only at point B. No triangle is formed.

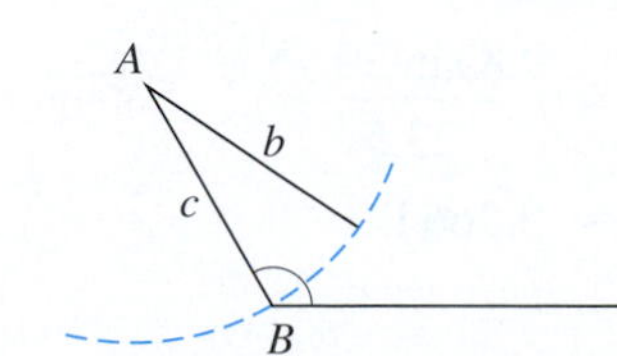

Case 8: No solution
$b > c$; an arc of radius b meets the base at just one point.

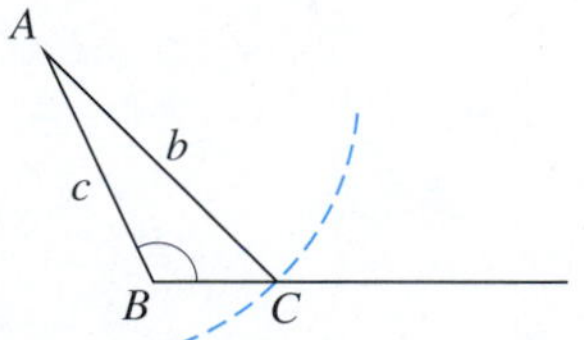

The eight cases illustrated above lead us to three possibilities in the SSA situation: *no* solution, *one* solution, or *two* solutions. Let's investigate these possibilities further, looking for ways to recognize the number of solutions.

EXAMPLE 3 ***No solution.*** In $\triangle QRS$, $q = 15$, $r = 28$, and $Q = 43.6°$. Solve the triangle.

Solution We make a drawing and list the known measures.

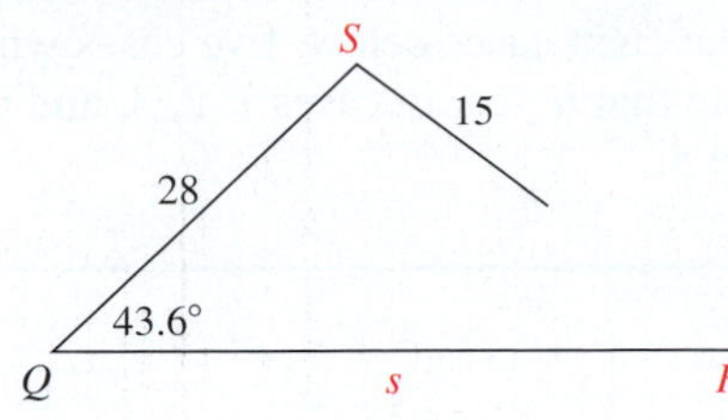

$Q = 43.6°$, $q = 15$,
$R = ?$, $r = 28$,
$S = ?$, $s = ?$

We note the SSA situation and use the law of sines to find R:

$$\frac{q}{\sin Q} = \frac{r}{\sin R}$$

$$\frac{15}{\sin 43.6°} = \frac{28}{\sin R} \quad \text{Substituting}$$

$$\sin R = \frac{28 \sin 43.6°}{15} \quad \text{Solving for } \sin R$$

$$\sin R \approx 1.2873.$$

Since there is no angle with a sine greater than 1, there is *no solution*.

Now Try Exercise 13.

EXAMPLE 4 ***One solution.*** In $\triangle XYZ$, $x = 23.5$, $y = 9.8$, and $X = 39.7°$. Solve the triangle.

Solution We make a drawing and organize the given information.

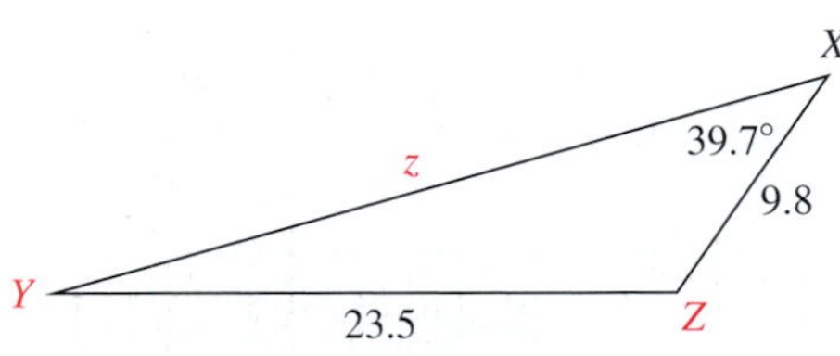

$X = 39.7°$, $x = 23.5$,
$Y = ?$, $y = 9.8$,
$Z = ?$, $z = ?$

We see the SSA situation and begin by finding Y with the law of sines:

$$\frac{x}{\sin X} = \frac{y}{\sin Y}$$

$$\frac{23.5}{\sin 39.7°} = \frac{9.8}{\sin Y} \quad \text{Substituting}$$

$$\sin Y = \frac{9.8 \sin 39.7°}{23.5} \quad \text{Solving for } \sin Y$$

$$\sin Y \approx 0.2664.$$

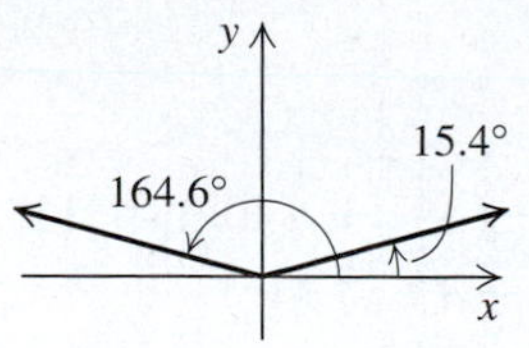

There are two angles less than 180° with a sine of 0.2664. They are 15.4° and 164.6°, to the nearest tenth of a degree. An angle of 164.6° cannot be an angle of this triangle because it already has an angle of 39.7° and these two angles would total more than 180°. Thus, 15.4° is the only possibility for Y. Therefore,

$$Z \approx 180° - (39.7° + 15.4°) \approx 124.9°.$$

We now find z:

$$\frac{z}{\sin Z} = \frac{x}{\sin X}$$

$$\frac{z}{\sin 124.9°} = \frac{23.5}{\sin 39.7°} \qquad \textbf{Substituting}$$

$$z = \frac{23.5 \sin 124.9°}{\sin 39.7°} \qquad \textbf{Solving for } z$$

$$z \approx 30.2.$$

We have now solved the triangle:

$$X = 39.7°, \qquad x = 23.5,$$
$$Y \approx 15.4°, \qquad y = 9.8,$$
$$Z \approx 124.9°, \qquad z \approx 30.2.$$

Now Try Exercise 5.

The next example illustrates the ambiguous case in which there are two possible solutions.

EXAMPLE 5 *Two solutions.* In $\triangle ABC$, $b = 15$, $c = 20$, and $B = 29°$. Solve the triangle.

Solution We make a drawing, list the known measures, and see that we again have the SSA situation.

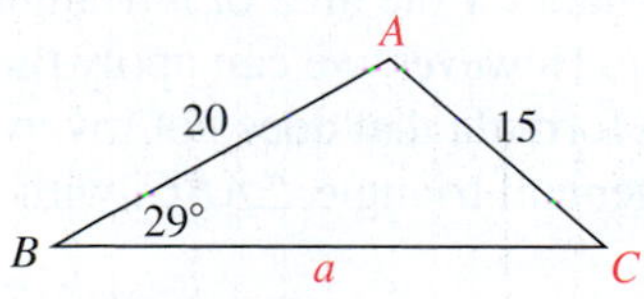

$$A = ?, \qquad a = ?,$$
$$B = 29°, \qquad b = 15,$$
$$C = ?, \qquad c = 20$$

We first find C:

$$\frac{b}{\sin B} = \frac{c}{\sin C}$$

$$\frac{15}{\sin 29°} = \frac{20}{\sin C} \qquad \textbf{Substituting}$$

$$\sin C = \frac{20 \sin 29°}{15} \approx 0.6464. \qquad \textbf{Solving for } \sin C$$

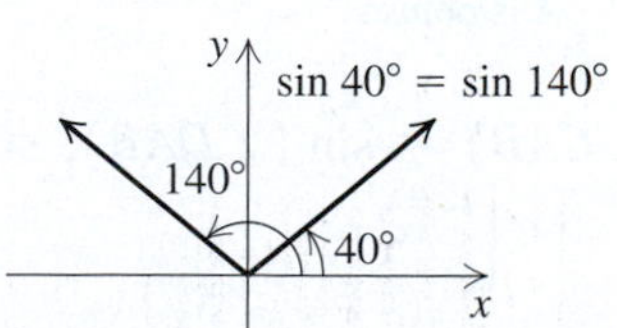

There are two angles less than 180° with a sine of 0.6464. They are 40° and 140°, to the nearest degree. This gives us two possible solutions.

Possible Solution I.

If $C = 40°$, then

$$A = 180° - (29° + 40°) = 111°.$$

Then we find a:

$$\frac{a}{\sin A} = \frac{b}{\sin B}$$

$$\frac{a}{\sin 111°} = \frac{15}{\sin 29°}$$

$$a = \frac{15 \sin 111°}{\sin 29°} \approx 29.$$

These measures make a triangle as shown below; thus we have a solution.

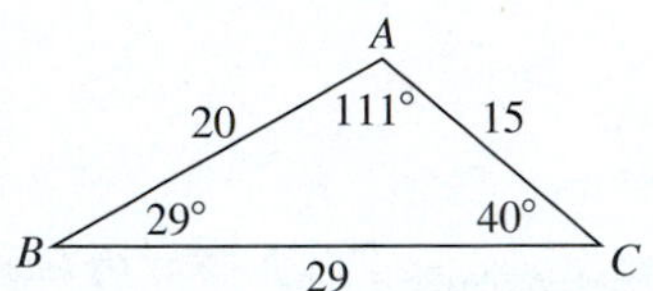

Possible Solution II.

If $C = 140°$, then

$$A = 180° - (29° + 140°) = 11°.$$

Then we find a:

$$\frac{a}{\sin A} = \frac{b}{\sin B}$$

$$\frac{a}{\sin 11°} = \frac{15}{\sin 29°}$$

$$a = \frac{15 \sin 11°}{\sin 29°} \approx 6.$$

These measures make a triangle as shown below; thus we have a second solution.

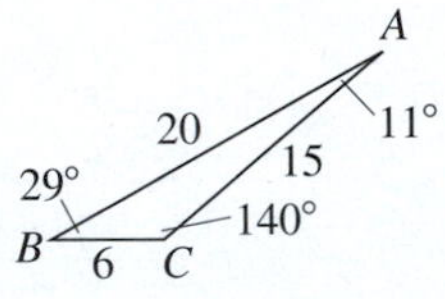

Now Try Exercise 3.

Examples 3–5 illustrate the SSA situation. Note that we need not memorize the relationship between each of the eight cases and the outcome it yields—no solution, one solution, or two solutions. When we are using the law of sines, the sine value leads us directly to the correct solution or solutions.

The Area of a Triangle

The familiar formula for the area of a triangle, $A = \frac{1}{2}bh$, can be used only when the height h is known. However, we can apply the method used to derive the law of sines to derive an area formula that does not involve the height.

Consider a general triangle $\triangle ABC$, with area K, as shown below.

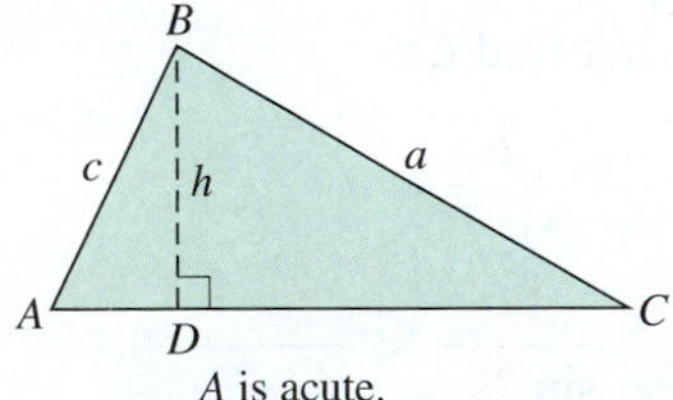

A is acute.

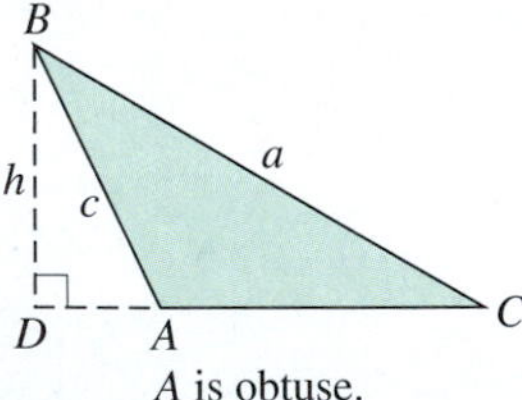

A is obtuse.

Note that in the triangle on the right, $\sin(\angle CAB) = \sin(\angle DAB)$, since $\sin A = \sin(180° - A)$. Then in each $\triangle ADB$,

$$\sin A = \frac{h}{c}, \quad \text{or} \quad h = c \sin A.$$

Substituting into the formula $K = \frac{1}{2}bh$, we get

$$K = \frac{1}{2}bc \sin A.$$

Any pair of sides and the included angle could have been used. Thus we also have

$$K = \frac{1}{2}ab \sin C \quad \text{and} \quad K = \frac{1}{2}ac \sin B.$$

THE AREA OF A TRIANGLE

The area K of any $\triangle ABC$ is one-half of the product of the lengths of two sides and the sine of the included angle:

$$K = \frac{1}{2}bc \sin A = \frac{1}{2}ab \sin C = \frac{1}{2}ac \sin B.$$

EXAMPLE 6 ***Area of a Triangular Garden.*** A university landscaping architecture department is designing a garden for a triangular area in a dormitory complex. Two sides of the garden, formed by the sidewalks in front of buildings A and B, measure 172 ft and 186 ft, respectively, and together form a 53° angle. The third side of the garden, formed by the sidewalk along Crossroads Avenue, measures 160 ft. What is the area of the garden to the nearest square foot?

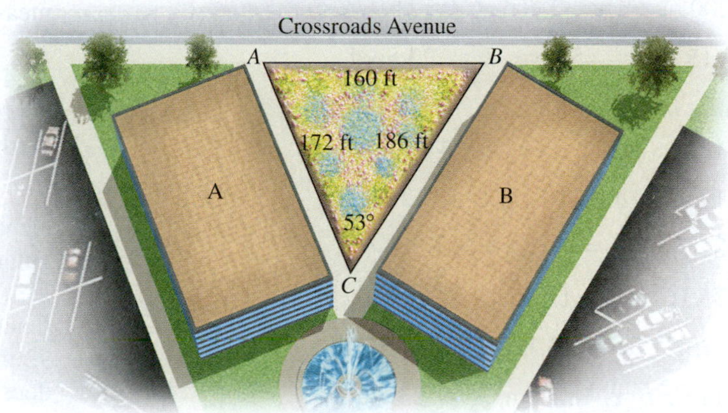

Solution Since we do not know a height of the triangle, we use the area formula:

$$K = \frac{1}{2}ab \sin C$$
$$K = \frac{1}{2} \cdot 186 \text{ ft} \cdot 172 \text{ ft} \cdot \sin 53°$$
$$K \approx 12{,}775 \text{ ft}^2.$$

The area of the garden is approximately 12,775 ft^2.

> **Now Try Exercise 25.**

8.1 Exercise Set

Solve the triangle, if possible.

1.

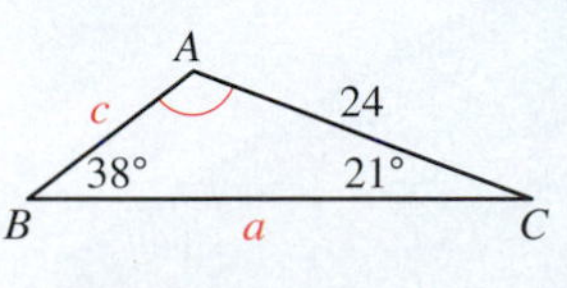

2.

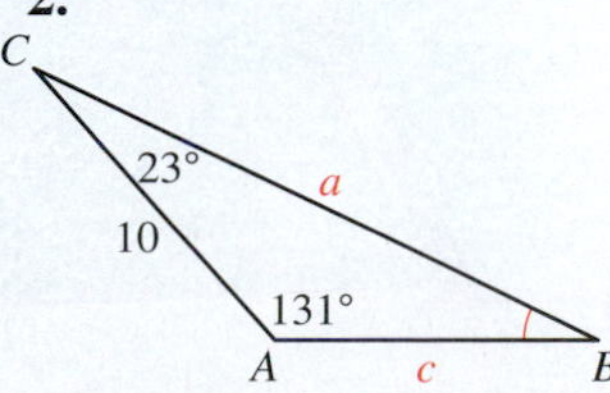

3.

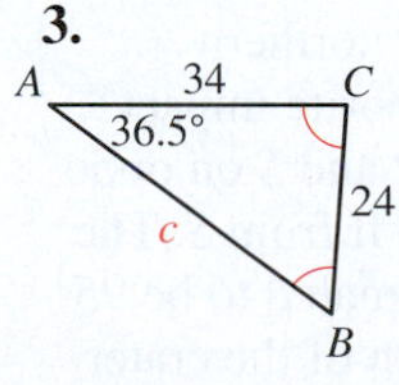

4.

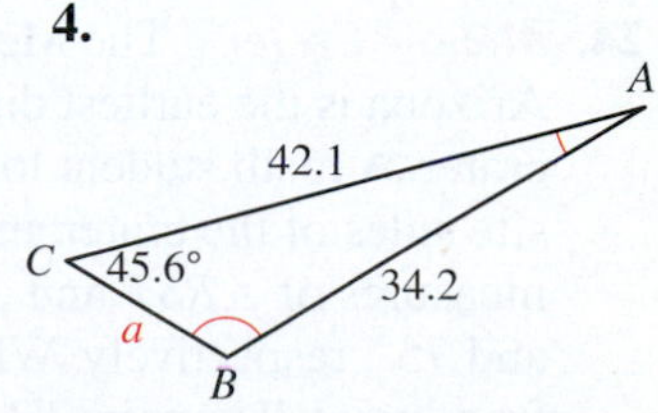

5. $C = 61°10', c = 30.3, b = 24.2$

6. $A = 126.5°, a = 17.2, c = 13.5$

7. $c = 3$ mi, $B = 37.48°, C = 32.16°$

8. $a = 2345$ mi, $b = 2345$ mi, $A = 124.67°$

9. $b = 56.78$ yd, $c = 56.78$ yd, $C = 83.78°$

10. $A = 129°32', C = 18°28', b = 1204$ in.

11. $a = 20.01$ cm, $b = 10.07$ cm, $A = 30.3°$

12. $b = 4.157$ km, $c = 3.446$ km, $C = 51°48'$

13. $A = 89°, a = 15.6$ in., $b = 18.4$ in.

14. $C = 46°32', a = 56.2$ m, $c = 22.1$ m

15. $a = 200$ m, $A = 32.76°, C = 21.97°$

16. $B = 115°, c = 45.6$ yd, $b = 23.8$ yd

Find the area of the triangle.

17. $B = 42°, a = 7.2$ ft, $c = 3.4$ ft

18. $A = 17°12', b = 10$ in., $c = 13$ in.

19. $C = 82°54', a = 4$ yd, $b = 6$ yd

20. $C = 75.16°, a = 1.5$ m, $b = 2.1$ m

21. $B = 135.2°, a = 46.12$ ft, $c = 36.74$ ft

22. $A = 113°, b = 18.2$ cm, $c = 23.7$ cm

23. *Lawn Irrigation.* Sanchez Irrigation is installing a lawn irrigation system with three heads. They determine that the best locations for the heads A, B, and C are such that $\angle CAB$ is 40° and $\angle ACB$ is 45° and that the distance from A to B is 34 ft. Find the distance from B to C.

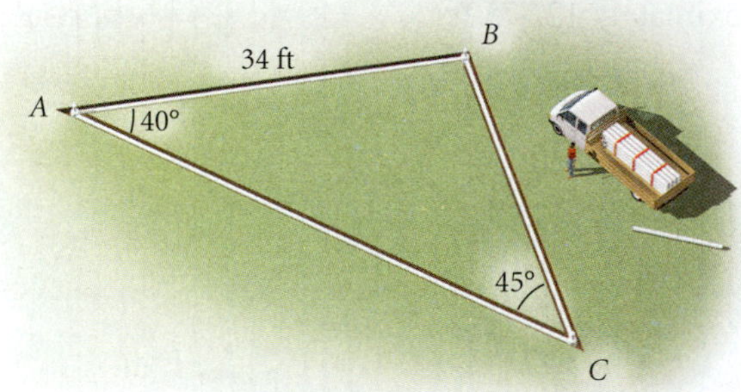

24. *Meteor Crater.* The Meteor Crater in northern Arizona is the earliest discovered meteorite impact crater. A math student locates points R and S on opposite sides of the crater and point T 700 ft from S. The measures of $\angle RST$ and $\angle RTS$ are estimated to be 95° and 75°, respectively. What is the width of the crater from point R to point S?

25. *Area of a Back Yard.* A new homeowner has a triangular-shaped back yard. Two of the three sides measure 53 ft and 42 ft and form an included angle of 135°. To determine the amount of fertilizer and grass seed to be purchased, the owner must know, or at least approximate, the area of the yard. Find the area of the yard to the nearest square foot.

26. *Elephant Zoo Exhibit.* A zoo is expanding its African elephant exhibit and needs to fence an outdoor triangular area, with a barn forming one side of the triangle. The maintenance department has only 168 ft of fencing in stock. If one side of the outdoor area is 92 ft and the angle that it makes with the barn is 78°, does the zoo need to order more fencing?

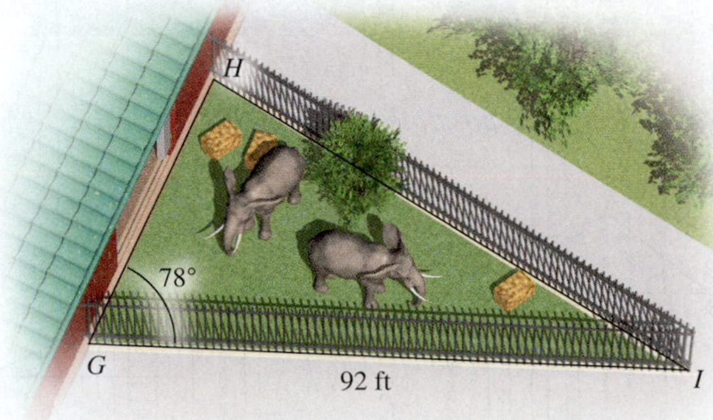

27. *Length of a Pole.* A pole leans away from the sun at an angle of 7° to the vertical. When the angle of elevation of the sun is 51°, the pole casts a shadow 47 ft long on level ground. How long is the pole?

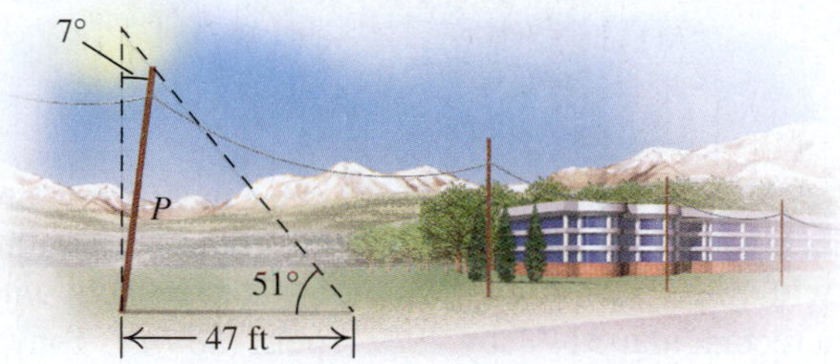

In Exercises 28–31, keep in mind the two types of bearing considered in Sections 6.2 and 6.3.

28. *Reconnaissance Airplane.* A reconnaissance airplane leaves its airport on the east coast of the United States and flies in a direction of 85°. Because of bad weather, it returns to another airport 230 km due north of its home base. To get to the new airport, it flies in a direction of 283°. What is the total distance that the airplane flew?

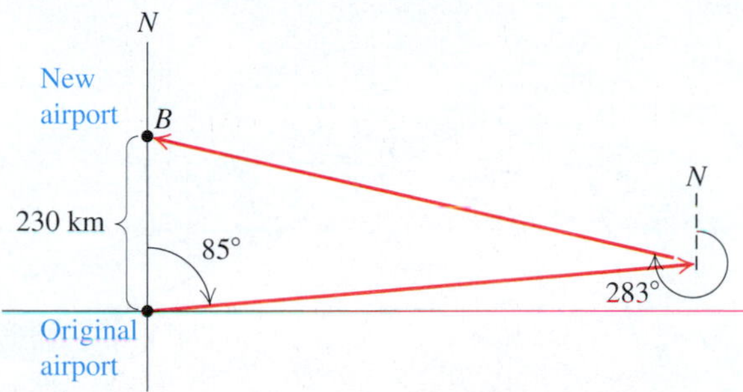

29. *Fire Tower.* A ranger in fire tower A spots a fire at a direction of 295°. A ranger in fire tower B, located 45 mi at a direction of 45° from tower A, spots the same fire at a direction of 255°. How far from tower A is the fire? from tower B?

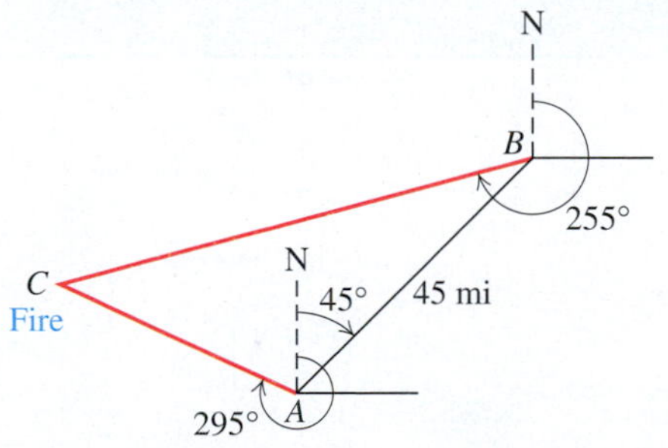

30. *Lighthouse.* A boat leaves lighthouse A and sails 5.1 km. At this time it is sighted from lighthouse B, 7.2 km west of A. The bearing of the boat from B is N65°10′E. How far is the boat from B?

31. *Distance to Nassau.* Miami, Florida, is located 178 mi N73°10′W of Nassau. Because of an approaching hurricane, a cruise ship sailing in the region needs to know how far it is from Nassau. The ship's position is N77°43′E of Miami and N19°35′E of Nassau. How far is the ship from Nassau?

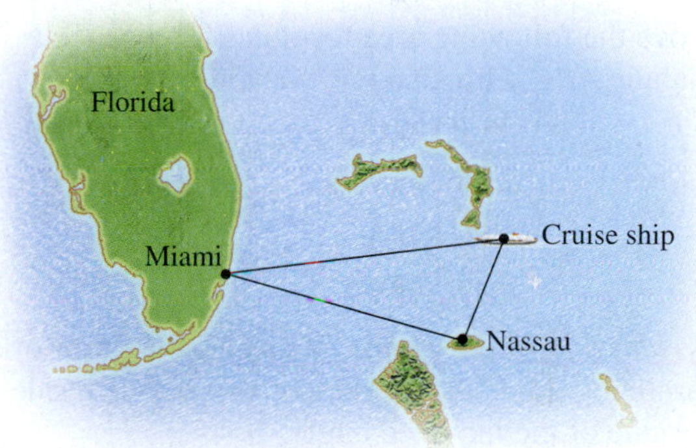

32. *Gears.* Three gears are arranged as shown in the following figure. Find the angle ϕ.

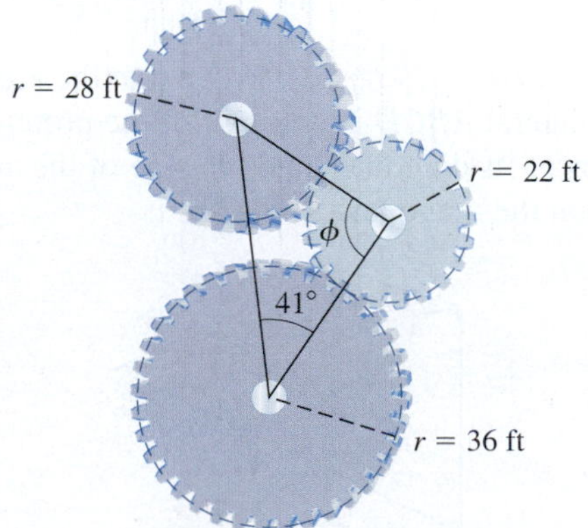

Skill Maintenance

Find the acute angle A, in both radians and degrees, for the given function value. [6.1]

33. $\cos A = 0.2213$

34. $\cos A = 1.5612$

Convert to decimal degree notation. Round to the nearest hundredth. [6.1]

35. $18°14'20''$

36. $125°3'42''$

37. Given that $f(x) = \sqrt{x - 6}$, find $f(31)$. [1.2]

Find the value. [6.3]

38. $\cos \dfrac{\pi}{6}$

39. $\sin 45°$

40. $\sin 300°$

41. $\cos\left(-\dfrac{2\pi}{3}\right)$

42. Multiply: $(1 - i)(1 + i)$. [3.1]

Synthesis

43. Prove the following area formulas for a general triangle ABC with area represented by K.

$$K = \frac{a^2 \sin B \sin C}{2 \sin A}, \quad K = \frac{c^2 \sin A \sin B}{2 \sin C},$$

$$K = \frac{b^2 \sin C \sin A}{2 \sin B}$$

44. *Area of a Parallelogram.* Prove that the area of a parallelogram is the product of two adjacent sides and the sine of the included angle.

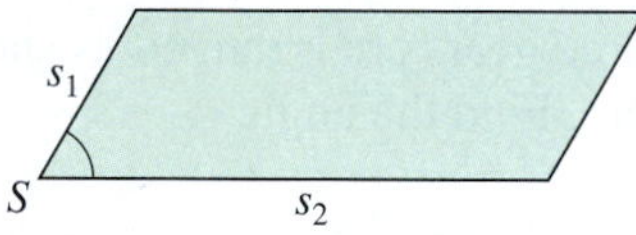

45. *Area of a Quadrilateral.* Prove that the area of a quadrilateral $ABCD$ is one-half of the product of the lengths of its diagonals and the sine of the angle θ between the diagonals.

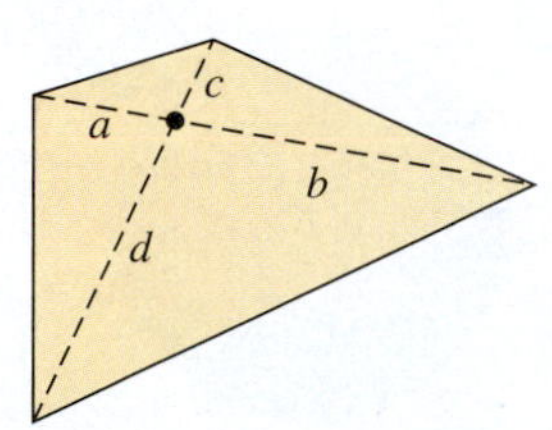

46. Find d.

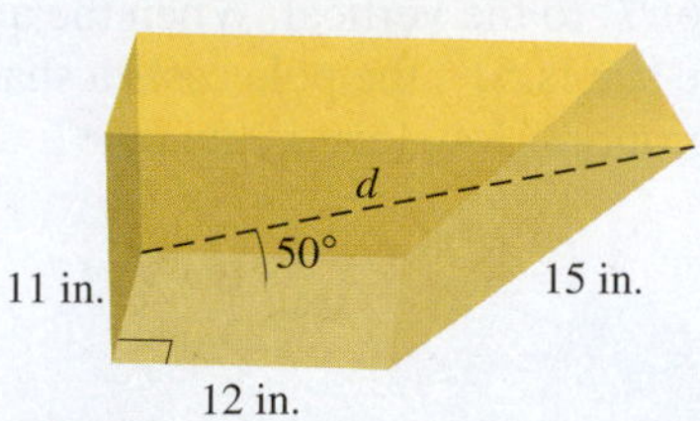

47. *Recording Studio.* A musician is constructing an octagonal recording studio in his home. The studio with dimensions as shown below is to be built within a rectangular $31'9''$ by $29'9''$ room (*Source*: Tony Medeiros, Indianapolis, IN). Point D is $9''$ from wall 2, and points C and B are each $9''$ from wall 1. Using the law of sines and right triangles, determine to the nearest tenth of an inch how far point A is from wall 1 and from wall 4. (For more information on this studio, see Example 2 in Section 8.2.)

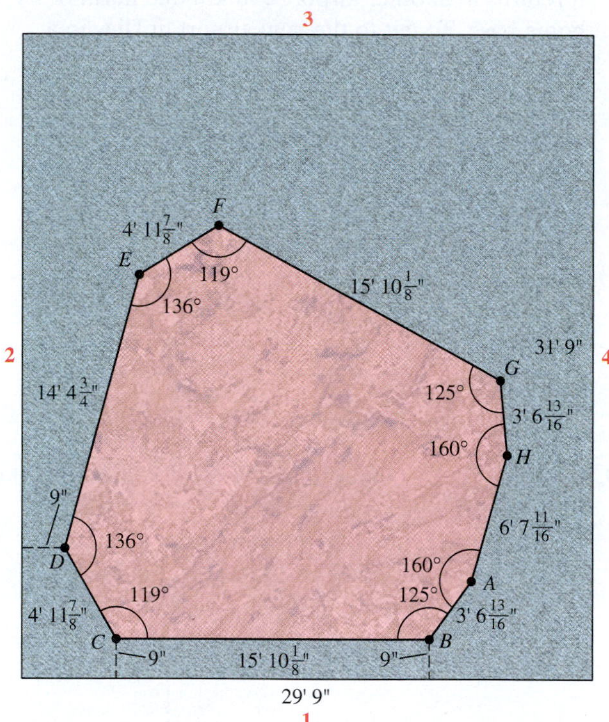

8.2 The Law of Cosines

- Use the law of cosines to solve triangles.
- Determine whether the law of sines or the law of cosines should be applied to solve a triangle.

The law of sines is used to solve triangles given a side and two angles (AAS and ASA) or given two sides and an angle opposite one of them (SSA). A second law, called the *law of cosines*, is needed to solve triangles given two sides and the included angle (SAS) or given three sides (SSS).

The Law of Cosines

To derive this property, we consider any $\triangle ABC$ placed on a coordinate system. We position the origin at one of the vertices—say, C—and the positive half of the x-axis along one of the sides—say, CB. Let (x, y) be the coordinates of vertex A. Point B has coordinates $(a, 0)$ and point C has coordinates $(0, 0)$.

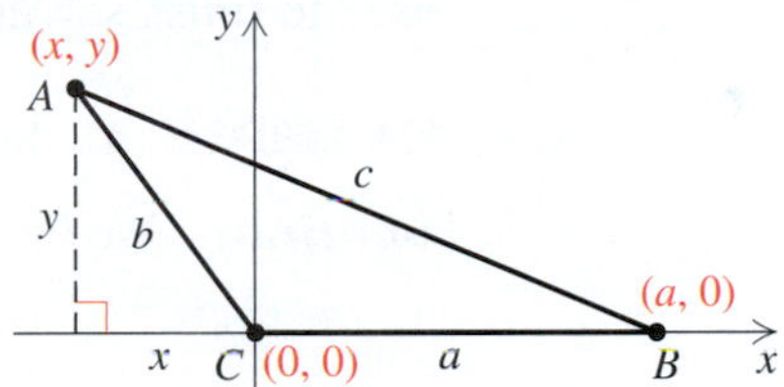

Then $\cos C = \dfrac{x}{b}$, so $x = b\cos C$

and $\sin C = \dfrac{y}{b}$, so $y = b\sin C$.

Thus point A has coordinates $(b\cos C, b\sin C)$.

Next, we use the distance formula to determine c^2:

$$c^2 = (x - a)^2 + (y - 0)^2,$$

or $$c^2 = (b\cos C - a)^2 + (b\sin C - 0)^2.$$

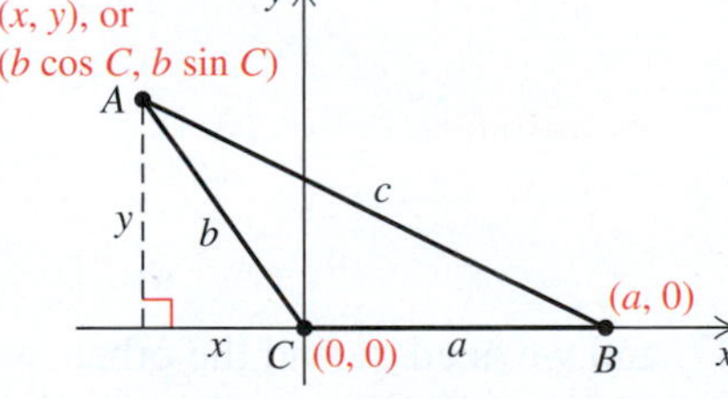

Now we multiply and simplify:

$$\begin{aligned} c^2 &= b^2\cos^2 C - 2ab\cos C + a^2 + b^2\sin^2 C \\ &= a^2 + b^2(\sin^2 C + \cos^2 C) - 2ab\cos C \\ &= a^2 + b^2 - 2ab\cos C. \end{aligned}$$

Using the identity $\sin^2 x + \cos^2 x = 1$

Had we placed the origin at one of the other vertices, we would have obtained

$$a^2 = b^2 + c^2 - 2bc\cos A$$

or $$b^2 = a^2 + c^2 - 2ac\cos B.$$

THE LAW OF COSINES

In any triangle ABC,

$$a^2 = b^2 + c^2 - 2bc \cos A,$$
$$b^2 = a^2 + c^2 - 2ac \cos B,$$
and $$c^2 = a^2 + b^2 - 2ab \cos C.$$

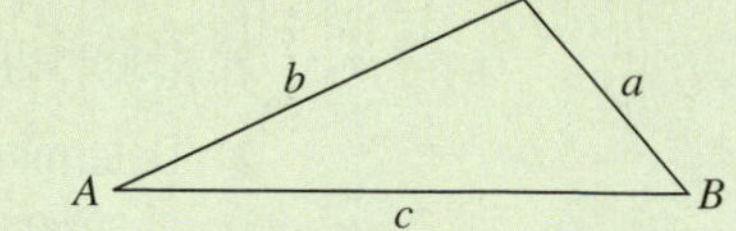

Thus, in any triangle, the square of a side is the sum of the squares of the other two sides, minus twice the product of those sides and the cosine of the included angle. When the included angle is 90°, the law of cosines reduces to the Pythagorean theorem.

The law of cosines can also be expressed as

$$\cos A = \frac{b^2 + c^2 - a^2}{2bc}, \quad \cos B = \frac{a^2 + c^2 - b^2}{2ac}, \quad \text{and} \quad \cos C = \frac{a^2 + b^2 - c^2}{2ab}.$$

Solving Triangles (SAS)

When two sides of a triangle and the included angle are known, we can use the law of cosines to find the third side. The law of cosines or the law of sines can then be used to finish solving the triangle.

EXAMPLE 1 Solve $\triangle ABC$ if $a = 32$, $b = 71$, and $C = 32.8°$.

Solution We first label a triangle with the known and unknown measures:

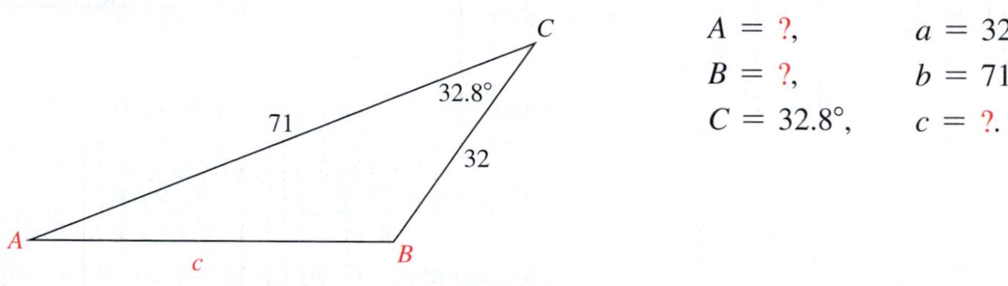

$A = ?$, $\quad a = 32$,
$B = ?$, $\quad b = 71$,
$C = 32.8°$, $\quad c = ?$.

We can find the third side using the law of cosines, as follows:

$$c^2 = a^2 + b^2 - 2ab \cos C$$
$$c^2 = 32^2 + 71^2 - 2 \cdot 32 \cdot 71 \cos 32.8° \qquad \text{Substituting}$$
$$c^2 \approx 2245.5$$
$$c \approx 47.$$

We now have $a = 32$, $b = 71$, and $c \approx 47$, and we need to find the other two angle measures. At this point, we can find them in two ways. One way uses the law of sines. The ambiguous case may arise, however, and we would need to be alert to this possibility. The advantage of using the law of cosines again is that if we solve for the cosine and find that its value is *negative*, then we know that the angle is obtuse. If the value of the cosine is *positive*, then the angle is acute. Thus we use the law of cosines to find a second angle.

Let's find angle A. We select the formula from the law of cosines that contains $\cos A$ and substitute:

$$\begin{aligned} a^2 &= b^2 + c^2 - 2bc \cos A \\ 32^2 &\approx 71^2 + 47^2 - 2 \cdot 71 \cdot 47 \cos A \quad \textbf{Substituting} \\ 1024 &\approx 5041 + 2209 - 6674 \cos A \\ -6226 &\approx -6674 \cos A \\ \cos A &\approx 0.9328738 \\ A &\approx 21.1^\circ. \end{aligned}$$

The third angle is now easy to find:

$$\begin{aligned} B &\approx 180^\circ - (32.8^\circ + 21.1^\circ) \\ &\approx 126.1^\circ. \end{aligned}$$

Thus,

$$\begin{aligned} A &\approx 21.1^\circ, & a &= 32, \\ B &\approx 126.1^\circ, & b &= 71, \\ C &= 32.8^\circ, & c &\approx 47. \end{aligned}$$

Now Try Exercise 1.

Due to differences created by rounding, answers may vary depending on the order in which they are found. Had we found the measure of angle B first in Example 1, the angle measures would have been $B \approx 126.9^\circ$ and $A \approx 20.3^\circ$. Variances in rounding also change the answers. Had we used 47.4 for c in Example 1, the angle measures would have been $A \approx 21.5^\circ$ and $B \approx 125.7^\circ$.

Suppose that we used the law of sines at the outset in Example 1 to find c. We were given only three measures: $a = 32$, $b = 71$, and $C = 32.8^\circ$. When substituting these measures into the proportions, we see that there is not enough information to use the law of sines:

$$\frac{a}{\sin A} = \frac{b}{\sin B} \rightarrow \frac{32}{\sin A} = \frac{71}{\sin B},$$

$$\frac{b}{\sin B} = \frac{c}{\sin C} \rightarrow \frac{71}{\sin B} = \frac{c}{\sin 32.8^\circ},$$

$$\frac{a}{\sin A} = \frac{c}{\sin C} \rightarrow \frac{32}{\sin A} = \frac{c}{\sin 32.8^\circ}.$$

In all three situations, the resulting equation, after the substitutions, still has two unknowns. Thus we cannot use the law of sines to find b.

EXAMPLE 2 ***Recording Studio.*** A musician is constructing an octagonal recording studio in his home and needs to determine two distances for the electrician. The dimensions for the most acoustically perfect studio are shown in the figure on the following page (*Source*: Tony Medeiros, Indianapolis, IN). Determine the distances from D to F and from D to B to the nearest tenth of an inch.

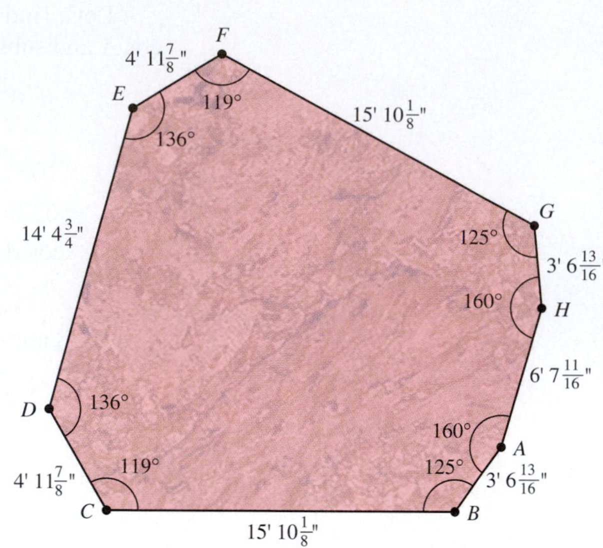

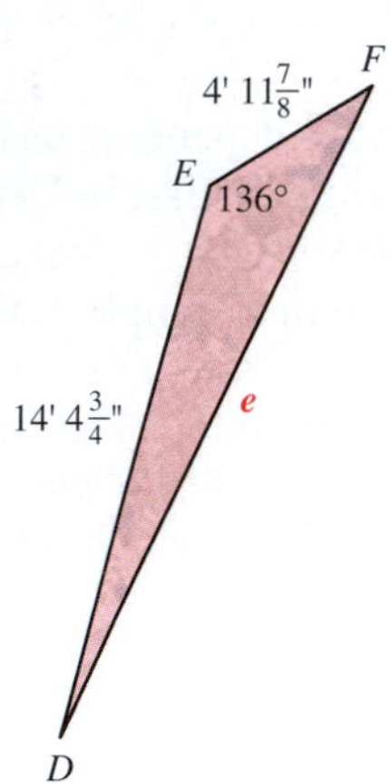

Solution We begin by connecting points D and F and labeling the known measures of $\triangle DEF$. Converting the linear measures to decimal notation in inches, we have

$$d = 4'11\tfrac{7}{8}'' = 59.875 \text{ in.},$$
$$f = 14'4\tfrac{3}{4}'' = 172.75 \text{ in.},$$
$$E = 136°.$$

We can find the measure of the third side, e, using the law of cosines:

$$e^2 = d^2 + f^2 - 2 \cdot d \cdot f \cdot \cos E \qquad \text{Using the law of cosines}$$
$$e^2 = (59.875 \text{ in.})^2 + (172.75 \text{ in.})^2 - 2(59.875 \text{ in.})(172.75 \text{ in.}) \cos 136° \qquad \text{Substituting}$$
$$e^2 \approx 48{,}308.4257 \text{ in}^2$$
$$e \approx 219.8 \text{ in.}$$

Thus it is approximately 219.8 in. from D to F.

We continue by connecting points D and B and labeling the known measures of $\triangle DCB$:

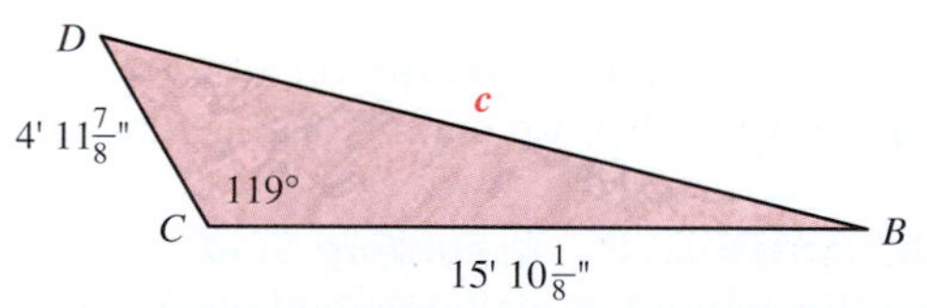

$$b = 4'11\tfrac{7}{8}'' = 59.875 \text{ in.},$$
$$d = 15'10\tfrac{1}{8}'' = 190.125 \text{ in.},$$
$$C = 119°.$$

Using the law of cosines, we can determine c, the length of the third side:

$$c^2 = b^2 + d^2 - 2 \cdot b \cdot d \cdot \cos C \qquad \text{Using the law of cosines}$$
$$c^2 = (59.875 \text{ in.})^2 + (190.125 \text{ in.})^2 - 2(59.875 \text{ in.})(190.125 \text{ in.}) \cos 119° \qquad \text{Substituting}$$
$$c^2 \approx 50{,}770.4191 \text{ in}^2$$
$$c \approx 225.3 \text{ in.}$$

The distance from D to B is approximately 225.3 in.

Now Try Exercise 25.

Solving Triangles (SSS)

When all three sides of a triangle are known, the law of cosines can be used to solve the triangle.

EXAMPLE 3 Solve $\triangle RST$ if $r = 3.5$, $s = 4.7$, and $t = 2.8$.

Solution We sketch a triangle and label it with the given measures.

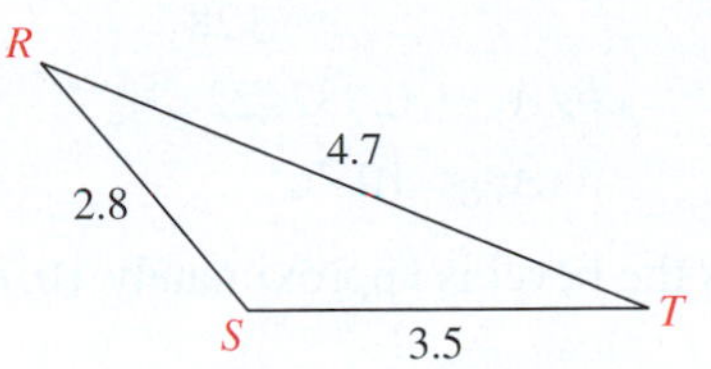

$R = ?,\quad r = 3.5,$
$S = ?,\quad s = 4.7,$
$T = ?,\quad t = 2.8$

Since we do not know any of the angle measures, we cannot use the law of sines. We begin instead by finding an angle with the law of cosines. We choose to find S first and select the formula that contains $\cos S$:

$$s^2 = r^2 + t^2 - 2rt\cos S$$
$$(4.7)^2 = (3.5)^2 + (2.8)^2 - 2(3.5)(2.8)\cos S \qquad \text{Substituting}$$
$$\cos S = \frac{(3.5)^2 + (2.8)^2 - (4.7)^2}{2(3.5)(2.8)}$$
$$\cos S \approx -0.1020408$$
$$S \approx 95.86°.$$

In a similar manner, we find angle R:

$$r^2 = s^2 + t^2 - 2st\cos R$$
$$(3.5)^2 = (4.7)^2 + (2.8)^2 - 2(4.7)(2.8)\cos R$$
$$\cos R = \frac{(4.7)^2 + (2.8)^2 - (3.5)^2}{2(4.7)(2.8)}$$
$$\cos R \approx 0.6717325$$
$$R \approx 47.80°.$$

Then

$$T \approx 180° - (95.86° + 47.80°) \approx 36.34°.$$

Thus,

$R \approx 47.80°,\quad r = 3.5,$
$S \approx 95.86°,\quad s = 4.7,$
$T \approx 36.34°,\quad t = 2.8.$

Now Try Exercise 3.

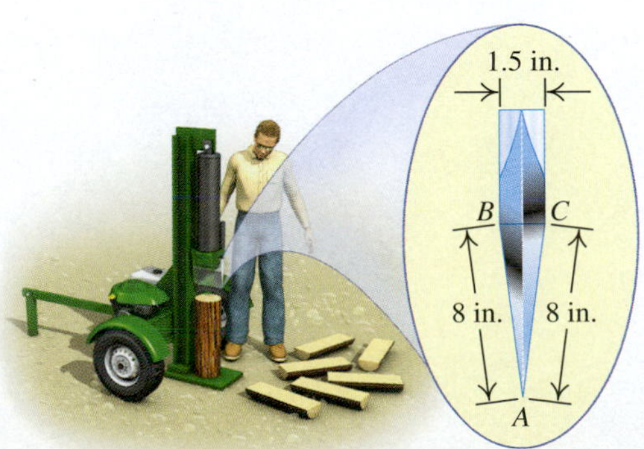

EXAMPLE 4 ***Wedge Bevel.*** The *bevel* of the wedge (the angle formed at the cutting edge of the wedge) of a cutting tool determines the cutting characteristics of the tool. A small bevel like that of a straight razor makes for a keen edge, but is impractical for heavy-duty cutting because the edge dulls quickly and is prone to chipping. A large bevel is suitable for heavy-duty work like chopping wood. The diagram at left illustrates the wedge of a Huskee log splitter (*Source*: Huskee). What is its bevel?

Solution Since we know three sides of a triangle, we can use the law of cosines to find the bevel, angle A:

$$a^2 = b^2 + c^2 - 2bc \cos A$$
$$(1.5)^2 = 8^2 + 8^2 - 2 \cdot 8 \cdot 8 \cdot \cos A$$
$$2.25 = 64 + 64 - 128 \cos A$$
$$\cos A = \frac{64 + 64 - 2.25}{128}$$
$$\cos A \approx 0.982422$$
$$A \approx 10.76°.$$

Thus the bevel is approximately 10.76°.

Now Try Exercise 31.

CONNECTING THE CONCEPTS — Choosing the Appropriate Law

The following summarizes the situations in which to use the law of sines and the law of cosines.

To solve an oblique triangle:

Use the *law of sines* for:	Use the *law of cosines* for:
AAS	SAS
ASA	SSS
SSA	

The law of cosines can also be used for the SSA situation, but since the process involves solving a quadratic equation, we do not include that option here.

EXAMPLE 5 In $\triangle ABC$, three measures are given. Determine which law to use when solving the triangle. You need not solve the triangle.

a) $a = 14,\ b = 23,\ c = 10$

b) $a = 207,\ B = 43.8°,\ C = 57.6°$

c) $A = 112°,\ C = 37°,\ a = 84.7$

d) $B = 101°,\ a = 960,\ c = 1042$

e) $b = 17.26,\ a = 27.29,\ A = 39°$

f) $A = 61°,\ B = 39°,\ C = 80°$

Solution It is helpful to make a drawing of a triangle with the given information. The triangle need not be drawn to scale. The given parts are shown in color.

FIGURE	SITUATION	LAW TO USE
a)	SSS	Law of Cosines
b)	ASA	Law of Sines
c)	AAS	Law of Sines
d)	SAS	Law of Cosines
e)	SSA	Law of Sines
f)	AAA	Cannot be solved

Now Try Exercises 17 and 19.

8.2 Exercise Set

Solve the triangle, if possible.

1.

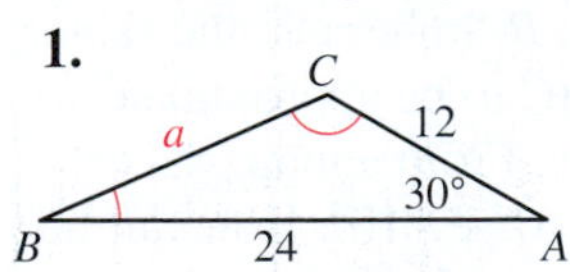

2.

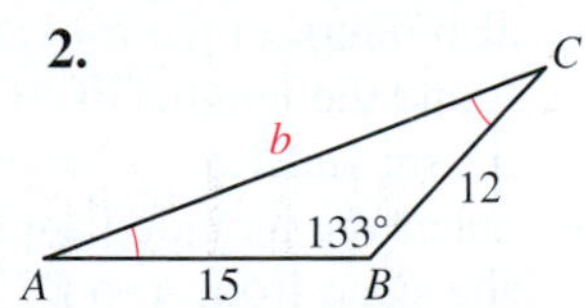

3.

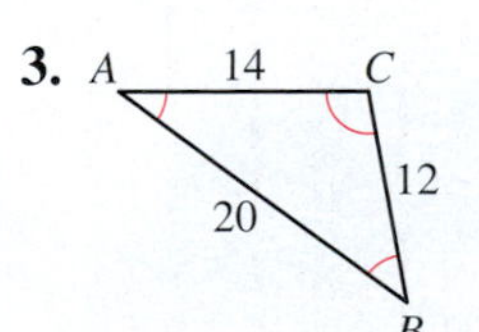

4.

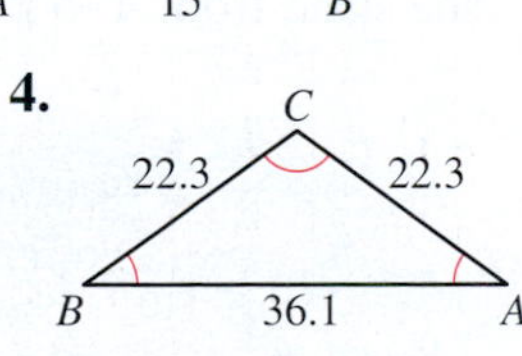

5. $B = 72°40'$, $c = 16$ m, $a = 78$ m

6. $C = 22.28°$, $a = 25.4$ cm, $b = 73.8$ cm

7. $a = 16$ m, $b = 20$ m, $c = 32$ m

8. $B = 72.66°$, $a = 23.78$ km, $c = 25.74$ km

9. $a = 2$ ft, $b = 3$ ft, $c = 8$ ft

10. $A = 96°13'$, $b = 15.8$ yd, $c = 18.4$ yd

11. $a = 26.12$ km, $b = 21.34$ km, $c = 19.25$ km

12. $C = 28°43'$, $a = 6$ mm, $b = 9$ mm

13. $a = 60.12$ mi, $b = 40.23$ mi, $C = 48.7°$

14. $a = 11.2$ cm, $b = 5.4$ cm, $c = 7$ cm

15. $b = 10.2$ in., $c = 17.3$ in., $A = 53.456°$

16. $a = 17$ yd, $b = 15.4$ yd, $c = 1.5$ yd

Determine which law applies. Then solve the triangle.

17. $A = 70°$, $B = 12°$, $b = 21.4$

18. $a = 15$, $c = 7$, $B = 62°$

19. $a = 3.3$, $b = 2.7$, $c = 2.8$

20. $a = 1.5$, $b = 2.5$, $A = 58°$

21. $A = 40.2°$, $B = 39.8°$, $C = 100°$

22. $a = 60$, $b = 40$, $C = 47°$

23. $a = 3.6$, $b = 6.2$, $c = 4.1$

24. $B = 110°30'$, $C = 8°10'$, $c = 0.912$

25. *Amusement Park Ride.* A teacup ride for children at an amusement park consists of five teacups equally spaced around a circle. Each cup holds 6 passengers and is at the end of an arm 20 ft long. Find the linear distance between a pair of adjacent cups. Round the length to the nearest tenth of a foot.

26. *Shark Pool.* Feeding and observing sharks has become a popular attraction at water parks. The floor of a newly constructed shark pool in the shape of a parallelogram has sides that measure 38 ft and 57 ft. To meet the minimum required length for the shortest diagonal, the angles must be 80° and 100°. Find the lengths of the diagonals of the pool. Round the lengths to the nearest foot.

27. *A-Frame Architecture.* William O'Brien, Jr., Assistant Professor of Architecture at the MIT School of Architecture and Planning, designed Allandale House, the asymmetrical A-frame house pictured below (*Source*: William O'Brien, Jr., LLC, 188 Prospect Street, Unit 5, Cambridge, MA 02139). The included angle between sides AB and CB, which measure 32.67 ft and 19.25 ft, respectively, measures 67°. Find the length of side AC.

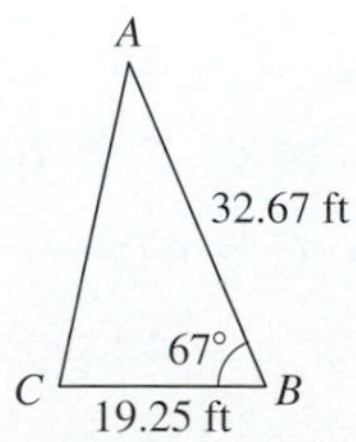

28. *Fish Attractor.* Each year at Cedar Resort, discarded Christmas trees are collected and sunk in the lake to form a fish attractor. Visitors are told that it is 253 ft from the pier to the fish attractor and 415 ft to another pier across the lake. Using a compass, a fisherman finds that the attractor's azimuth (the direction measured as an angle from north) is 340° and that of the other pier is 35°. What is the distance between the fish attractor and the pier across the lake?

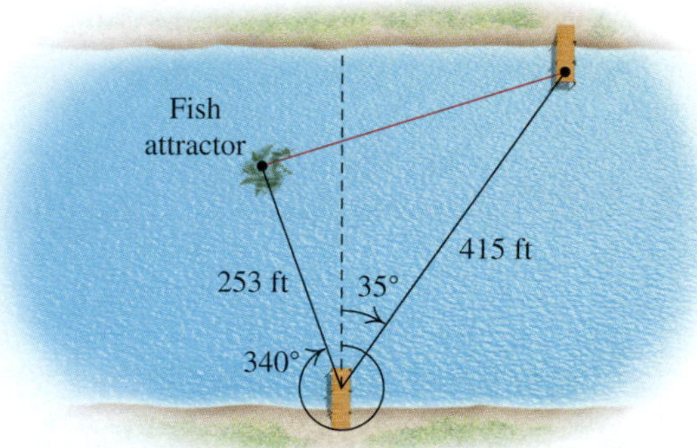

29. *In-line Skater.* An in-line skater skates on a fitness trail along the Pacific Ocean from point A to point B. As shown below, two streets intersecting at point C also intersect the trail at A and B. In her car, the skater found the lengths of AC and BC to be approximately 0.5 mi and 1.3 mi, respectively. From a map, she estimates the included angle at C to be 110°. How far did she skate from A to B?

30. *Baseball Bunt.* A batter in a baseball game drops a bunt down the first-base line. It rolls 34 ft at an angle of 25° with the base path. The pitcher's mound is 60.5 ft from home plate. How far must the pitcher travel to pick up the ball? (*Hint*: A baseball diamond is a square.)

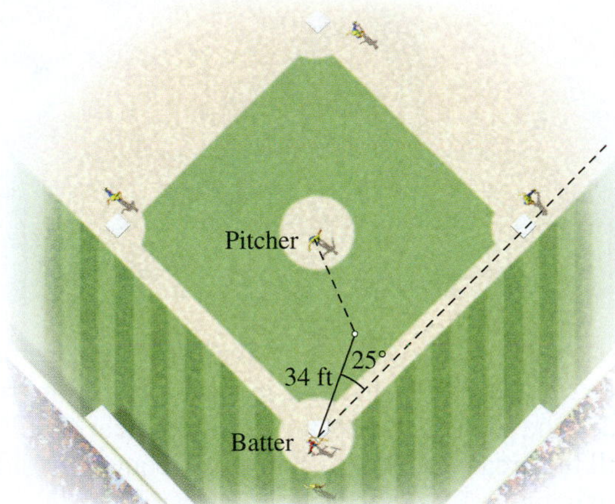

31. *Survival Trip.* A group of college students is learning to navigate for an upcoming survival trip. On a map, they have been given three points at which they are to check in. The map also shows the distances between the points. However, in order to navigate they need to know the angle measurements. Calculate the angles for them.

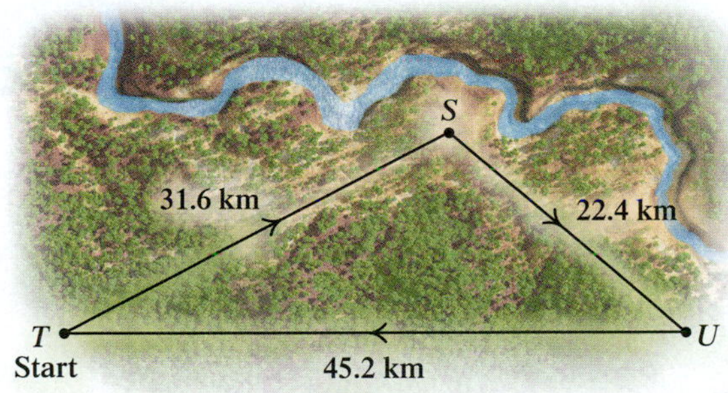

32. *Ships.* Two ships leave harbor at the same time. The first sails N15°W at 25 knots. (A knot is one nautical mile per hour.) The second sails N32°E at 20 knots. After 2 hr, how far apart are the ships?

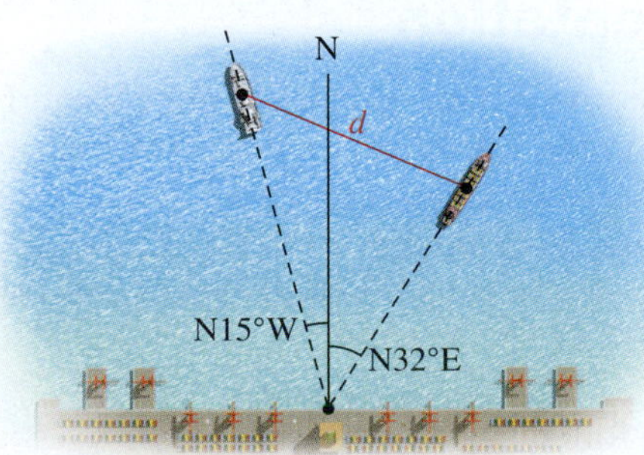

33. *Airplanes.* Two airplanes leave an airport at the same time. The first flies 150 km/h in a direction of 320°. The second flies 200 km/h in a direction of 200°. After 3 hr, how far apart are the planes?

34. *Slow-Pitch Softball.* A slow-pitch softball diamond is a square 65 ft on a side. The pitcher's mound is 46 ft from home plate. How far is it from the pitcher's mound to first base?

35. *Isosceles Trapezoid.* The longer base of an isosceles trapezoid measures 14 ft. The nonparallel sides measure 10 ft, and the base angles measure 80°.

a) Find the length of a diagonal.
b) Find the area.

36. *Dimensions of Sail.* A sail that is in the shape of an isosceles triangle has a vertex angle of 38°. The angle is included by two sides, each measuring 20 ft. Find the length of the other side of the sail.

37. Three circles are arranged as shown in the following figure. Find the length PQ.

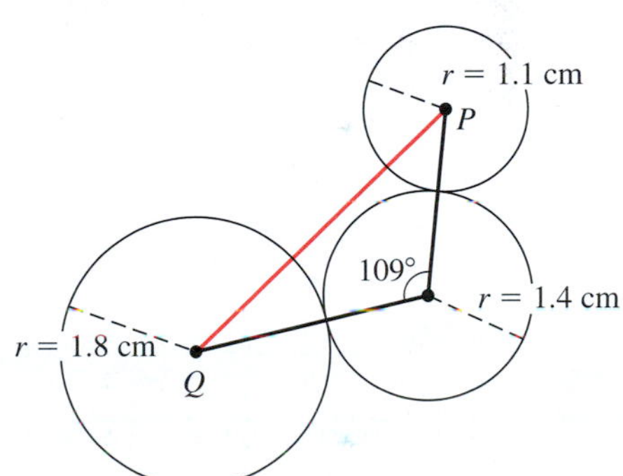

38. *Swimming Pool.* A triangular swimming pool measures 44 ft on one side and 32.8 ft on another side. These sides form an angle that measures 40.8°. How long is the other side?

Skill Maintenance

Classify the function as linear, quadratic, cubic, quartic, rational, exponential, logarithmic, or trigonometric.

39. $f(x) = -\frac{3}{4}x^4$ [4.1]

40. $y - 3 = 17x$ [1.3]

41. $y = \sin^2 x - 3\sin x$ [6.5]

42. $f(x) = 2^{x-1/2}$ [5.2]

43. $f(x) = \dfrac{x^2 - 2x + 3}{x - 1}$ [4.5]

44. $f(x) = 27 - x^3$ [4.1]

45. $y = e^x + e^{-x} - 4$ [5.2]

46. $y = \log_2(x - 2) - \log_2(x + 3)$ **[5.3]**

47. $f(x) = -\cos(\pi x - 3)$ **[6.5]**

48. $y = \frac{1}{2}x^2 - 2x + 2$ **[3.2]**

› Synthesis

49. *Canyon Depth.* A bridge is being built across a canyon. The length of the bridge is 5045 ft. From the deepest point in the canyon, the angles of elevation of the ends of the bridge are 78° and 72°. How deep is the canyon?

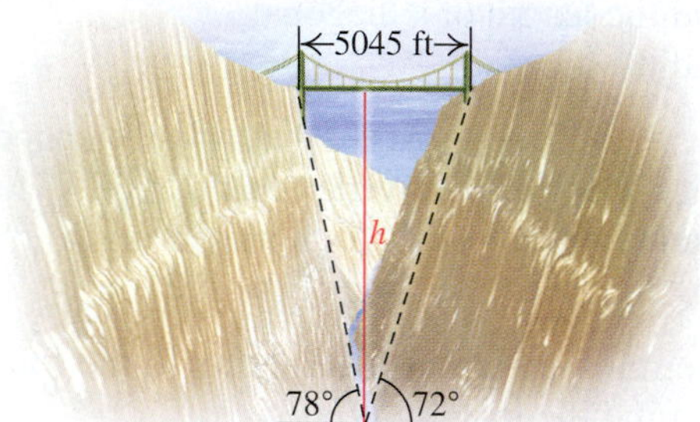

50. *Reconnaissance Plane.* A reconnaissance plane patrolling at 5000 ft sights a submarine at bearing 35° and at an angle of depression of 25°. A carrier is at bearing 105° and at an angle of depression of 60°. How far is the submarine from the carrier?

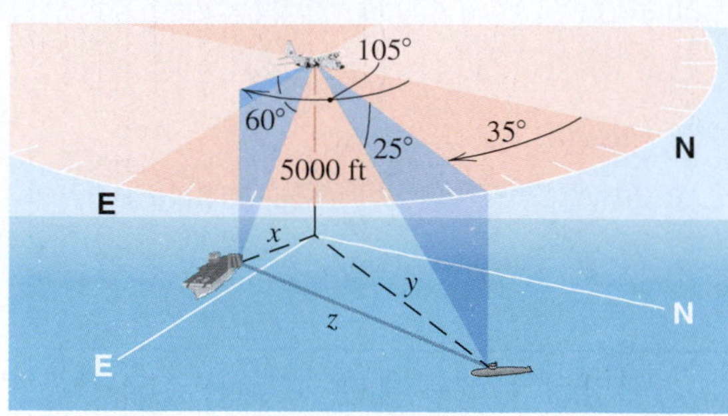

51. *Area of Isosceles Triangle.* Find a formula for the area of an isosceles triangle in terms of the congruent sides and their included angle. Under what conditions will the area of a triangle with fixed congruent sides be maximum?

8.3 Complex Numbers: Trigonometric Notation

- Graph complex numbers.
- Given a complex number in standard form, find trigonometric, or polar, notation; and given a complex number in trigonometric form, find standard notation.
- Use trigonometric notation to multiply and divide complex numbers.
- Use DeMoivre's theorem to raise complex numbers to powers.
- Find the *n*th roots of a complex number.

› Graphical Representation

Just as real numbers can be graphed on a line, complex numbers can be graphed on a plane. We graph a complex number $a + bi$ in the same way that we graph an ordered pair of real numbers (a, b). However, in place of an x-axis, we have a real axis, and in place of a y-axis, we have an imaginary axis. Horizontal distances correspond to the real part of a complex number. Vertical distances correspond to the imaginary part. Recall that $i = \sqrt{-1}$.

COMPLEX NUMBERS

REVIEW SECTION 3.1.

EXAMPLE 1 Graph each of the following complex numbers.

a) $3 + 2i$ **b)** $-4 - 5i$ **c)** $-3i$

d) $-1 + 3i$ **e)** 2

Solution

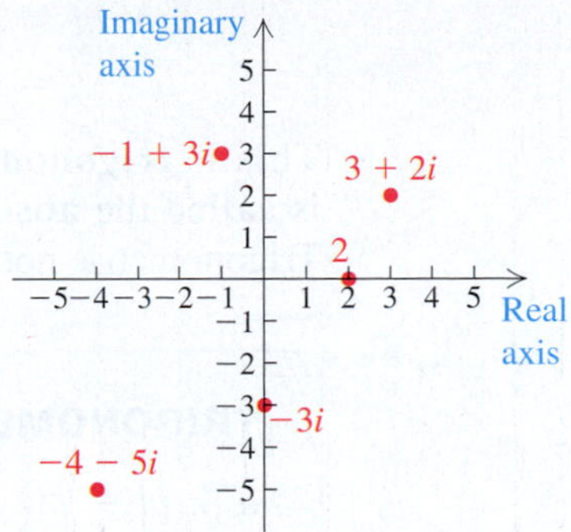

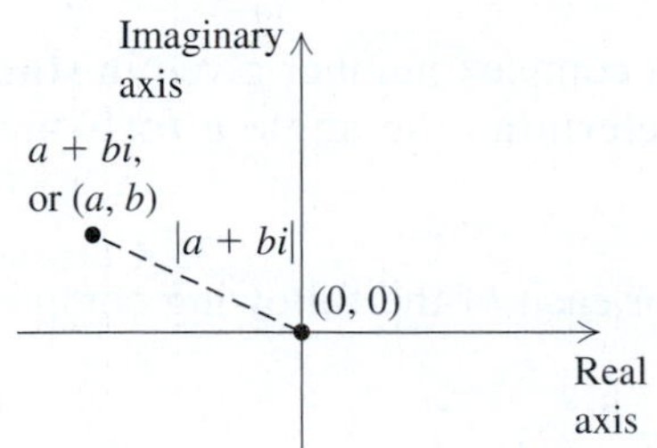

We recall that the absolute value of a real number is its distance from 0 on the number line. The absolute value of a complex number is its distance from the origin in the complex plane. For a point $a + bi$, using the distance formula, we have

$$|a + bi| = \sqrt{(a - 0)^2 + (b - 0)^2} = \sqrt{a^2 + b^2}.$$

ABSOLUTE VALUE OF A COMPLEX NUMBER

The **absolute value of a complex number** $a + bi$ is

$$|a + bi| = \sqrt{a^2 + b^2}.$$

EXAMPLE 2 Find the absolute value of each of the following.

a) $3 + 4i$ **b)** $-2 - i$ **c)** $\frac{4}{5}i$

Solution

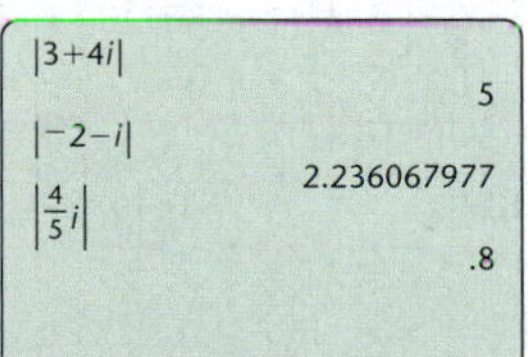

a) $|3 + 4i| = \sqrt{3^2 + 4^2} = \sqrt{9 + 16} = \sqrt{25} = 5$

b) $|-2 - i| = \sqrt{(-2)^2 + (-1)^2} = \sqrt{5}$

c) $\left|\frac{4}{5}i\right| = \left|0 + \frac{4}{5}i\right| = \sqrt{0^2 + \left(\frac{4}{5}\right)^2} = \frac{4}{5}$

We can check these results using a graphing calculator in complex mode as shown at left. Note that $\sqrt{5} \approx 2.236067977$ and $\frac{4}{5} = 0.8$.

Now Try Exercises 3 and 5.

Trigonometric Notation for Complex Numbers

Now let's consider a nonzero complex number $a + bi$. Suppose that its absolute value is r. If we let θ be an angle in standard position whose terminal side passes through the point (a, b), as shown in the figure, then

$$\cos\theta = \frac{a}{r}, \quad \text{or} \quad a = r\cos\theta$$

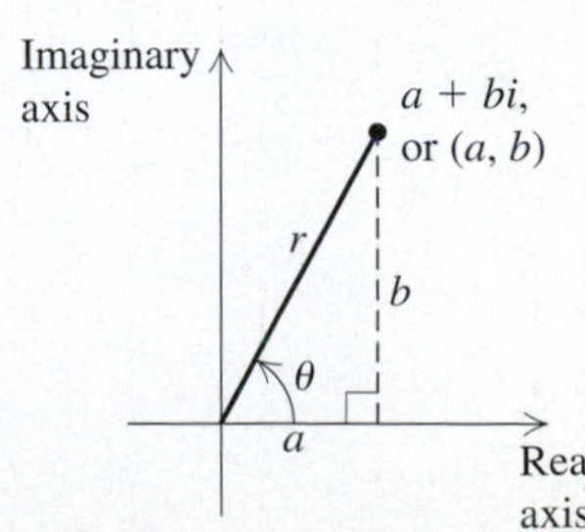

and

$$\sin\theta = \frac{b}{r}, \quad \text{or} \quad b = r\sin\theta.$$

Substituting these values for a and b into the $(a + bi)$ notation, we get

$$\begin{aligned} a + bi &= r\cos\theta + (r\sin\theta)i \\ &= r(\cos\theta + i\sin\theta). \end{aligned}$$

This is **trigonometric notation** for a complex number $a + bi$. The number r is called the **absolute value** of $a + bi$, and θ is called the **argument** of $a + bi$. Trigonometric notation for a complex number is also called **polar notation**.

TRIGONOMETRIC NOTATION FOR COMPLEX NUMBERS

$$a + bi = r(\cos\theta + i\sin\theta)$$

In order to find trigonometric notation for a complex number given in **standard notation**, $a + bi$, we must find r and determine the angle θ for which $\sin\theta = b/r$ and $\cos\theta = a/r$.

EXAMPLE 3 Find trigonometric notation for each of the following complex numbers.

a) $1 + i$ **b)** $\sqrt{3} - i$

Solution

a) We note that $a = 1$ and $b = 1$. Then

$$r = \sqrt{a^2 + b^2} = \sqrt{1^2 + 1^2} = \sqrt{2},$$

$$\sin\theta = \frac{b}{r} = \frac{1}{\sqrt{2}}, \quad \text{or} \quad \frac{\sqrt{2}}{2},$$

and

$$\cos\theta = \frac{a}{r} = \frac{1}{\sqrt{2}}, \quad \text{or} \quad \frac{\sqrt{2}}{2}.$$

Since θ is in quadrant I, $\theta = \pi/4$, or 45°, and we have

$$1 + i = \sqrt{2}\left(\cos\frac{\pi}{4} + i\sin\frac{\pi}{4}\right),$$

or

$$1 + i = \sqrt{2}(\cos 45° + i\sin 45°).$$

b) We see that $a = \sqrt{3}$ and $b = -1$. Then

$$r = \sqrt{(\sqrt{3})^2 + (-1)^2} = 2,$$

$$\sin\theta = \frac{-1}{2} = -\frac{1}{2},$$

and

$$\cos\theta = \frac{\sqrt{3}}{2}.$$

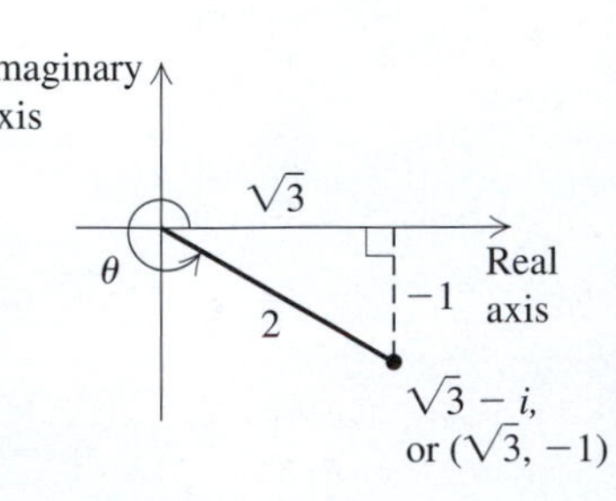

Since θ is in quadrant IV, $\theta = 11\pi/6$, or 330°, and we have

$$\sqrt{3} - i = 2\left(\cos\frac{11\pi}{6} + i\sin\frac{11\pi}{6}\right),$$

or

$$\sqrt{3} - i = 2(\cos 330° + i\sin 330°).$$

angle(1+i)
45
angle($\sqrt{3}$−i)
−30

As shown at left, a graphing calculator (in DEGREE mode) can be used to determine angle values in degrees.

Now Try Exercise 13.

In changing to trigonometric notation, note that there are many angles satisfying the given conditions. We ordinarily choose the *smallest positive* angle.

To change from trigonometric notation to standard notation, $a + bi$, we recall that $a = r\cos\theta$ and $b = r\sin\theta$.

EXAMPLE 4 Find standard notation, $a + bi$, for each of the following complex numbers.

a) $2(\cos 120° + i\sin 120°)$

b) $\sqrt{8}\left(\cos\frac{7\pi}{4} + i\sin\frac{7\pi}{4}\right)$

Solution

a) Rewriting, we have

$$2(\cos 120° + i\sin 120°) = 2\cos 120° + (2\sin 120°)i.$$

Thus,

$$a = 2\cos 120° = 2\cdot\left(-\frac{1}{2}\right) = -1$$

and

$$b = 2\sin 120° = 2\cdot\frac{\sqrt{3}}{2} = \sqrt{3},$$

so

$$2(\cos 120° + i\sin 120°) = -1 + \sqrt{3}i.$$

Degree Mode

2(cos(120)+isin(120))
−1+1.732050808i

b) Rewriting, we have

$$\sqrt{8}\left(\cos\frac{7\pi}{4} + i\sin\frac{7\pi}{4}\right) = \sqrt{8}\cos\frac{7\pi}{4} + \left(\sqrt{8}\sin\frac{7\pi}{4}\right)i.$$

Thus,

$$a = \sqrt{8}\cos\frac{7\pi}{4} = \sqrt{8}\cdot\frac{\sqrt{2}}{2} = \frac{\sqrt{16}}{2} = \frac{4}{2} = 2$$

and

$$b = \sqrt{8}\sin\frac{7\pi}{4} = \sqrt{8}\cdot\left(-\frac{\sqrt{2}}{2}\right) = -\frac{\sqrt{16}}{2} = -\frac{4}{2} = -2,$$

so

$$\sqrt{8}\left(\cos\frac{7\pi}{4} + i\sin\frac{7\pi}{4}\right) = 2 - 2i.$$

Radian Mode

$\sqrt{8}(\cos(\frac{7\pi}{4})+i\sin(\frac{7\pi}{4}))$
2−2i

Now Try Exercises 23 and 27.

Multiplication and Division with Trigonometric Notation

Multiplication of complex numbers is easier to manage with trigonometric notation than with standard notation. We simply multiply the absolute values and add the arguments. Let's state this in a more formal manner.

COMPLEX NUMBERS: MULTIPLICATION

For any complex numbers $r_1(\cos\theta_1 + i\sin\theta_1)$ and $r_2(\cos\theta_2 + i\sin\theta_2)$,

$$r_1(\cos\theta_1 + i\sin\theta_1)\cdot r_2(\cos\theta_2 + i\sin\theta_2)$$
$$= r_1r_2[\cos(\theta_1 + \theta_2) + i\sin(\theta_1 + \theta_2)].$$

Proof. We have

$$r_1(\cos\theta_1 + i\sin\theta_1)\cdot r_2(\cos\theta_2 + i\sin\theta_2)$$
$$= r_1r_2(\cos\theta_1\cos\theta_2 - \sin\theta_1\sin\theta_2) + r_1r_2(\sin\theta_1\cos\theta_2 + \cos\theta_1\sin\theta_2)i.$$

Now, using identities for sums of angles, we simplify, obtaining

$$r_1r_2\cos(\theta_1 + \theta_2) + r_1r_2\sin(\theta_1 + \theta_2)i,$$

or

$$r_1r_2[\cos(\theta_1 + \theta_2) + i\sin(\theta_1 + \theta_2)],$$

which was to be shown. ■

EXAMPLE 5 Multiply and express the answer to each of the following in standard notation.

a) $3(\cos 40° + i\sin 40°)$ and $4(\cos 20° + i\sin 20°)$

b) $2(\cos\pi + i\sin\pi)$ and $3\left[\cos\left(-\frac{\pi}{2}\right) + i\sin\left(-\frac{\pi}{2}\right)\right]$

Solution

a) $3(\cos 40° + i\sin 40°)\cdot 4(\cos 20° + i\sin 20°)$

$$= 3\cdot 4\cdot[\cos(40° + 20°) + i\sin(40° + 20°)]$$
$$= 12(\cos 60° + i\sin 60°)$$
$$= 12\left(\frac{1}{2} + \frac{\sqrt{3}}{2}i\right)$$
$$= 6 + 6\sqrt{3}i$$

Degree Mode

3(cos(40)+isin(40))∗4(cos(20)+isin(20))

6+10.39230485i

b) $2(\cos\pi + i\sin\pi)\cdot 3\left[\cos\left(-\frac{\pi}{2}\right) + i\sin\left(-\frac{\pi}{2}\right)\right]$

$$= 2\cdot 3\cdot\left[\cos\left(\pi + \left(-\frac{\pi}{2}\right)\right) + i\sin\left(\pi + \left(-\frac{\pi}{2}\right)\right)\right]$$
$$= 6\left(\cos\frac{\pi}{2} + i\sin\frac{\pi}{2}\right)$$
$$= 6(0 + i\cdot 1)$$
$$= 6i$$

Radian Mode

2(cos(π)+isin(π))∗3(cos(−π/2)+isin(−π/2))

6i

Now Try Exercise 35.

EXAMPLE 6 Convert to trigonometric notation and multiply:

$$(1 + i)(\sqrt{3} - i).$$

Solution We first find trigonometric notation:

$$1 + i = \sqrt{2}(\cos 45° + i\sin 45°),$$ See Example 3(a).

$$\sqrt{3} - i = 2(\cos 330° + i\sin 330°).$$ See Example 3(b).

Then we multiply:

$$\begin{aligned}\sqrt{2}(\cos 45^\circ + i \sin 45^\circ) \cdot 2(\cos 330^\circ + i \sin 330^\circ) \\ &= 2\sqrt{2}[\cos (45^\circ + 330^\circ) + i \sin (45^\circ + 330^\circ)] \\ &= 2\sqrt{2}(\cos 375^\circ + i \sin 375^\circ) \\ &= 2\sqrt{2}(\cos 15^\circ + i \sin 15^\circ).\end{aligned}$$

375° has the same terminal side as 15°.

Now Try Exercise 37.

To divide complex numbers, we divide the absolute values and subtract the arguments. We state this fact below, but omit the proof.

COMPLEX NUMBERS: DIVISION

For any complex numbers $r_1(\cos \theta_1 + i \sin \theta_1)$ and $r_2(\cos \theta_2 + i \sin \theta_2)$, $r_2 \neq 0$,

$$\frac{r_1(\cos \theta_1 + i \sin \theta_1)}{r_2(\cos \theta_2 + i \sin \theta_2)} = \frac{r_1}{r_2}[\cos (\theta_1 - \theta_2) + i \sin (\theta_1 - \theta_2)].$$

EXAMPLE 7 Divide

$$2\left(\cos \frac{3\pi}{2} + i \sin \frac{3\pi}{2}\right) \quad \text{by} \quad 4\left(\cos \frac{\pi}{2} + i \sin \frac{\pi}{2}\right)$$

and express the solution in standard notation.

Solution We have

$$\begin{aligned}\frac{2\left(\cos \frac{3\pi}{2} + i \sin \frac{3\pi}{2}\right)}{4\left(\cos \frac{\pi}{2} + i \sin \frac{\pi}{2}\right)} &= \frac{2}{4}\left[\cos \left(\frac{3\pi}{2} - \frac{\pi}{2}\right) + i \sin \left(\frac{3\pi}{2} - \frac{\pi}{2}\right)\right] \\ &= \frac{1}{2}(\cos \pi + i \sin \pi) \\ &= \frac{1}{2}(-1 + i \cdot 0) \\ &= -\frac{1}{2}.\end{aligned}$$

EXAMPLE 8 Convert to trigonometric notation and divide:

$$\frac{1 + i}{1 - i}.$$

Solution We first convert to trigonometric notation:

$$1 + i = \sqrt{2}(\cos 45^\circ + i \sin 45^\circ), \quad \text{See Example 3(a).}$$
$$1 - i = \sqrt{2}(\cos 315^\circ + i \sin 315^\circ).$$

We now divide:

$$\frac{\sqrt{2}(\cos 45^\circ + i \sin 45^\circ)}{\sqrt{2}(\cos 315^\circ + i \sin 315^\circ)}$$
$$= 1[\cos (45^\circ - 315^\circ) + i \sin (45^\circ - 315^\circ)]$$
$$= \cos (-270^\circ) + i \sin (-270^\circ)$$
$$= 0 + i \cdot 1$$
$$= i.$$

$\frac{1+i}{1-i}$ i

Now Try Exercise 39.

Powers of Complex Numbers

An important theorem about powers and roots of complex numbers is named for the French mathematician Abraham DeMoivre (1667–1754). Let's consider the square of a complex number $r(\cos \theta + i \sin \theta)$:

$$[r(\cos \theta + i \sin \theta)]^2 = [r(\cos \theta + i \sin \theta)] \cdot [r(\cos \theta + i \sin \theta)]$$
$$= r \cdot r \cdot [\cos (\theta + \theta) + i \sin (\theta + \theta)]$$
$$= r^2(\cos 2\theta + i \sin 2\theta).$$

Similarly, we see that

$$[r(\cos \theta + i \sin \theta)]^3$$
$$= r \cdot r \cdot r \cdot [\cos (\theta + \theta + \theta) + i \sin (\theta + \theta + \theta)]$$
$$= r^3(\cos 3\theta + i \sin 3\theta).$$

DeMoivre's theorem is the generalization of these results.

DEMOIVRE'S THEOREM

For any complex number $r(\cos \theta + i \sin \theta)$ and any natural number n,

$$[r(\cos \theta + i \sin \theta)]^n = r^n(\cos n\theta + i \sin n\theta).$$

EXAMPLE 9 Find each of the following.

a) $(1 + i)^9$ **b)** $(\sqrt{3} - i)^{10}$

Solution

a) We first find trigonometric notation:

$$1 + i = \sqrt{2}(\cos 45^\circ + i \sin 45^\circ).$$ **See Example 3(a).**

Then

$$(1 + i)^9 = [\sqrt{2}(\cos 45^\circ + i \sin 45^\circ)]^9$$
$$= (\sqrt{2})^9[\cos (9 \cdot 45^\circ) + i \sin (9 \cdot 45^\circ)]$$ **DeMoivre's theorem**
$$= 2^{9/2}(\cos 405^\circ + i \sin 405^\circ)$$
$$= 16\sqrt{2}(\cos 45^\circ + i \sin 45^\circ)$$ **405° has the same terminal side as 45°.**
$$= 16\sqrt{2}\left(\frac{\sqrt{2}}{2} + i\frac{\sqrt{2}}{2}\right)$$
$$= 16 + 16i.$$

b) We first convert to trigonometric notation:

$$\sqrt{3} - i = 2(\cos 330° + i \sin 330°).$$ See Example 3(b).

Then

$$\begin{aligned}(\sqrt{3} - i)^{10} &= [2(\cos 330° + i \sin 330°)]^{10} \\ &= 2^{10}(\cos 3300° + i \sin 3300°) \\ &= 1024(\cos 60° + i \sin 60°) \\ &= 1024\left(\frac{1}{2} + i\frac{\sqrt{3}}{2}\right) \\ &= 512 + 512\sqrt{3}i.\end{aligned}$$

3300° has the same terminal side as 60°.

Now Try Exercise 47.

Roots of Complex Numbers

As we will see, every nonzero complex number has two square roots. A nonzero complex number has three cube roots, four fourth roots, and so on. In general, a nonzero complex number has n different nth roots. They can be found using the formula that we now state but do not prove.

Just in Time 27

ROOTS OF COMPLEX NUMBERS

The nth roots of a complex number $r(\cos\theta + i\sin\theta)$, $r \neq 0$, are given by

$$r^{1/n}\left[\cos\left(\frac{\theta}{n} + k\cdot\frac{360°}{n}\right) + i\sin\left(\frac{\theta}{n} + k\cdot\frac{360°}{n}\right)\right],$$

where $k = 0, 1, 2, \ldots, n - 1$.

EXAMPLE 10 Find the square roots of $2 + 2\sqrt{3}i$.

Solution We first find trigonometric notation:

$$2 + 2\sqrt{3}i = 4(\cos 60° + i \sin 60°).$$

Then $n = 2$, $1/n = 1/2$, and $k = 0, 1$. We have

$$\begin{aligned}&[4(\cos 60° + i \sin 60°)]^{1/2} \\ &= 4^{1/2}\left[\cos\left(\frac{60°}{2} + k\cdot\frac{360°}{2}\right) + i\sin\left(\frac{60°}{2} + k\cdot\frac{360°}{2}\right)\right], \quad k = 0, 1 \\ &= 2[\cos(30° + k\cdot 180°) + i\sin(30° + k\cdot 180°)], \quad k = 0, 1.\end{aligned}$$

Thus the roots are

$2(\cos 30° + i \sin 30°)$ for $k = 0$

and $2(\cos 210° + i \sin 210°)$ for $k = 1$,

or $\sqrt{3} + i$ and $-\sqrt{3} - i$.

Now Try Exercise 57.

Imaginary axis; Real axis; $\sqrt{3} + i$; $-\sqrt{3} - i$; −2, −1, 1, 2

In Example 10, we see that the two square roots of the number are opposites of each other. We can illustrate this graphically. We also note that the roots are equally spaced about a circle of radius r—in this case, $r = 2$. The roots are $360°/2$, or $180°$ apart.

EXAMPLE 11 Find the cube roots of 1. Then locate them on a graph.

Solution We begin by finding trigonometric notation:

$$1 = 1(\cos 0° + i \sin 0°).$$

Then $n = 3$, $1/n = 1/3$, and $k = 0, 1, 2$. We have

$$[1(\cos 0° + i \sin 0°)]^{1/3}$$
$$= 1^{1/3}\left[\cos\left(\frac{0°}{3} + k \cdot \frac{360°}{3}\right) + i \sin\left(\frac{0°}{3} + k \cdot \frac{360°}{3}\right)\right], \quad k = 0, 1, 2.$$

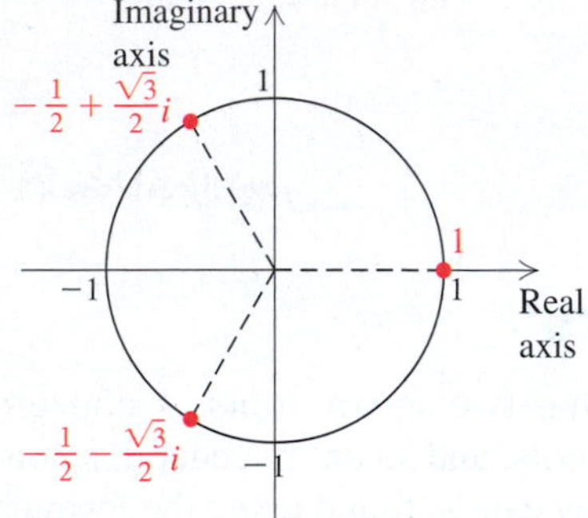

The roots are

$$1(\cos 0° + i \sin 0°), \quad 1(\cos 120° + i \sin 120°),$$

and $\quad 1(\cos 240° + i \sin 240°),$

or $\quad 1, \quad -\frac{1}{2} + \frac{\sqrt{3}}{2}i, \quad$ and $\quad -\frac{1}{2} - \frac{\sqrt{3}}{2}i.$

The graphs of the cube roots lie equally spaced about a circle of radius 1. The roots are $360°/3$, or $120°$ apart.

Now Try Exercise 59.

The nth roots of 1 are often referred to as the ***n*th roots of unity.** In Example 11, we found the cube roots of unity.

EXPLORING WITH TECHNOLOGY

We can approximate the nth roots of a number p with a graphing calculator set in PARAMETRIC mode.

WINDOW

Tmin = 0

Tmax = 360, or 2π

Tstep = $360/n$, or $2\pi/n$

To find the fifth roots of 8, enter

$$X_{1T} = (8 \wedge (1/5)) \cos T$$

and

$$Y_{1T} = (8 \wedge (1/5)) \sin T.$$

Here we use DEGREE mode. Use the TRACE feature to locate the fifth roots. The T, X, and Y values appear on the screen.

$X_{1T} = 8^{1/5} \cos T$
$Y_{1T} = 8^{1/5} \sin T$

2, −3, 3, −2

T = 72
X = .46838218, Y = 1.4415321

One of the fifth roots of 8 is approximately 0.4684. Find the other four.

EXAMPLE 12 Solve: $x^5 + i = 0$.

Solution To find all complex solutions of $x^5 + i = 0$, or $x^5 = -i$, we find the fifth roots of $-i$. We begin by converting $-i$ to trigonometric notation:

$$-i = 1(\cos 270° + i \sin 270°).$$

Then $n = 5$, $1/n = 1/5$, and $k = 0, 1, 2, 3, 4$. We have

$$[1\,(\cos 270° + i \sin 270°)]^{1/5}$$
$$= 1^{1/5}\left[\cos\left(\frac{270°}{5} + k \cdot \frac{360°}{5}\right) + i \sin\left(\frac{270°}{5} + k \cdot \frac{360°}{5}\right)\right], \quad k = 0, 1, 2, 3, 4.$$

The solutions are

$$\cos 54° + i \sin 54°, \quad \cos 126° + i \sin 126°, \quad \cos 198° + i \sin 198°,$$
$$\cos 270° + i \sin 270° \text{ (or } -i\text{)}, \quad \text{and} \quad \cos 342° + i \sin 342°.$$

Now Try Exercise 73.

8.3 Exercise Set

Graph the complex number and find its absolute value.

1. $4 + 3i$
2. $-2 - 3i$
3. i
4. $-5 - 2i$
5. $4 - i$
6. $6 + 3i$
7. 3
8. $-2i$

Express the indicated number in both standard notation and trigonometric notation.

9.

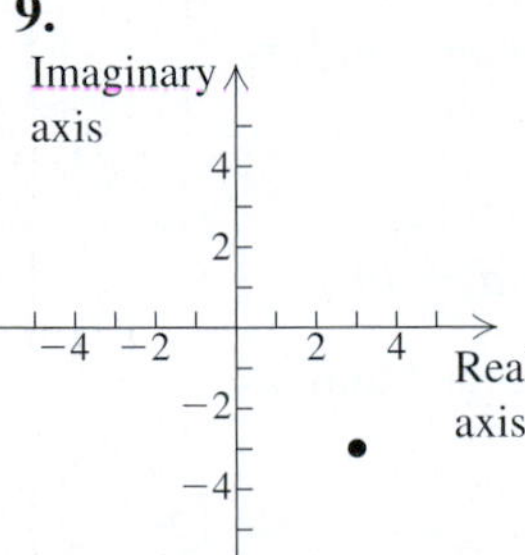

10.

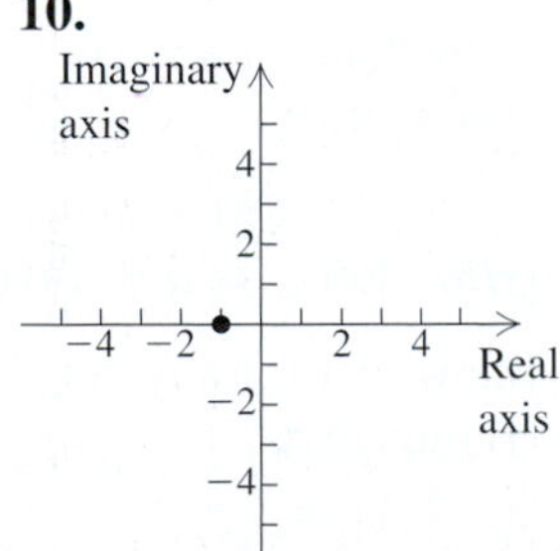

11.

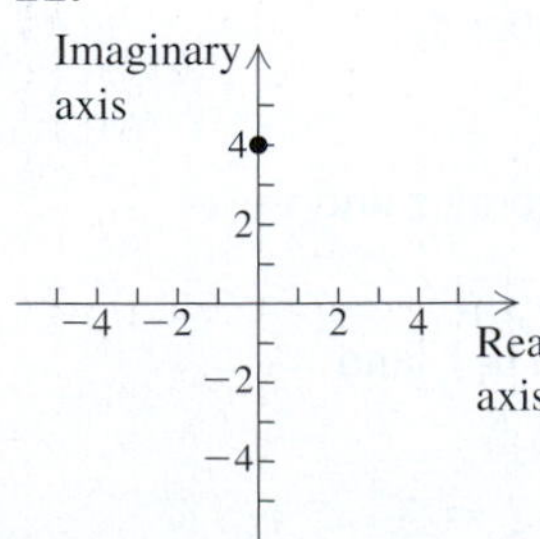

12.

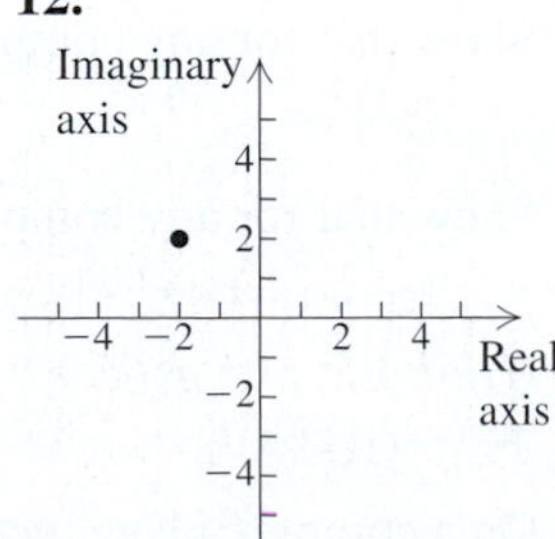

Find trigonometric notation.

13. $1 - i$
14. $-10\sqrt{3} + 10i$
15. $-3i$
16. $-5 + 5i$
17. $\sqrt{3} + i$
18. 4
19. $\frac{2}{5}$
20. $7.5i$
21. $-3\sqrt{2} - 3\sqrt{2}i$
22. $-\frac{9}{2} - \frac{9\sqrt{3}}{2}i$

Find standard notation, $a + bi$.

23. $3(\cos 30° + i \sin 30°)$
24. $6(\cos 120° + i \sin 120°)$
25. $10(\cos 270° + i \sin 270°)$
26. $3(\cos 0° + i \sin 0°)$
27. $\sqrt{8}\left(\cos \frac{\pi}{4} + i \sin \frac{\pi}{4}\right)$
28. $5\left(\cos \frac{\pi}{3} + i \sin \frac{\pi}{3}\right)$
29. $2\left(\cos \frac{\pi}{2} + i \sin \frac{\pi}{2}\right)$
30. $3\left[\cos\left(-\frac{3\pi}{4}\right) + \text{i} \sin\left(-\frac{3\pi}{4}\right)\right]$
31. $\sqrt{2}[\cos(-60°) + i \sin(-60°)]$
32. $4(\cos 135° + i \sin 135°)$

Multiply or divide and leave the answer in trigonometric notation.

33. $\dfrac{12(\cos 48° + i \sin 48°)}{3(\cos 6° + i \sin 6°)}$

34. $5\left(\cos\frac{\pi}{3}+i\sin\frac{\pi}{3}\right)\cdot 2\left(\cos\frac{\pi}{4}+i\sin\frac{\pi}{4}\right)$

35. $2.5(\cos 35^\circ+i\sin 35^\circ)\cdot 4.5(\cos 21^\circ+i\sin 21^\circ)$

36. $\dfrac{\frac{1}{2}\left(\cos\frac{2\pi}{3}+i\sin\frac{2\pi}{3}\right)}{\frac{3}{8}\left(\cos\frac{\pi}{6}+i\sin\frac{\pi}{6}\right)}$

Convert to trigonometric notation and then multiply or divide.

37. $(1-i)(2+2i)$

38. $(1+i\sqrt{3})(1+i)$

39. $\dfrac{1-i}{1+i}$

40. $\dfrac{1-i}{\sqrt{3}-i}$

41. $(3\sqrt{3}-3i)(2i)$

42. $(2\sqrt{3}+2i)(2i)$

43. $\dfrac{2\sqrt{3}-2i}{1+\sqrt{3}i}$

44. $\dfrac{3-3\sqrt{3}i}{\sqrt{3}-i}$

Raise the number to the given power and write trigonometric notation for the answer.

45. $\left[2\left(\cos\frac{\pi}{3}+i\sin\frac{\pi}{3}\right)\right]^3$

46. $[2(\cos 120^\circ+i\sin 120^\circ)]^4$

47. $(1+i)^6$

48. $(-\sqrt{3}+i)^5$

Raise the number to the given power and write standard notation for the answer.

49. $[3(\cos 20^\circ+i\sin 20^\circ)]^3$

50. $[2(\cos 10^\circ+i\sin 10^\circ)]^9$

51. $(1-i)^5$

52. $(2+2i)^4$

53. $\left(\frac{1}{\sqrt{2}}-\frac{1}{\sqrt{2}}i\right)^{12}$

54. $\left(\frac{\sqrt{3}}{2}+\frac{1}{2}i\right)^{10}$

Find the square roots of the number.

55. $-i$

56. $1+i$

57. $2\sqrt{2}-2\sqrt{2}i$

58. $-\sqrt{3}-i$

Find the cube roots of the number.

59. i

60. $-64i$

61. $2\sqrt{3}-2i$

62. $1-\sqrt{3}i$

63. Find and graph the fourth roots of 16.

64. Find and graph the fourth roots of i.

65. Find and graph the fifth roots of -1.

66. Find and graph the sixth roots of 1.

67. Find the tenth roots of 8.

68. Find the ninth roots of -4.

69. Find the sixth roots of -1.

70. Find the fourth roots of 12.

Solve.

71. $x^3=1$

72. $x^5-1=0$

73. $x^4+i=0$

74. $x^4+81=0$

75. $x^6+64=0$

76. $x^5+\sqrt{3}+i=0$

Skill Maintenance

Convert to degree measure. [6.4]

77. $\dfrac{\pi}{12}$

78. 3π

Convert to radian measure. [6.4]

79. 330°

80. -225°

Find the function value using coordinates of points on the unit circle. [6.5]

81. $\sin\dfrac{2\pi}{3}$

82. $\cos\dfrac{\pi}{6}$

83. $\cos\dfrac{\pi}{4}$

84. $\sin\dfrac{5\pi}{6}$

Synthesis

85. Find polar notation for $(\cos\theta+i\sin\theta)^{-1}$.

86. Show that for any complex number z,
$$|z|=|-z|.$$

87. Show that for any complex number z and its conjugate $\bar{z}$,
$$|z|=|\bar{z}|.$$
(*Hint*: Let $z=a+bi$ and $\bar{z}=a-bi$.)

88. Show that for any complex number z and its conjugate $\bar{z}$,
$$|z\bar{z}|=|z^2|.$$
(*Hint*: Let $z=a+bi$ and $\bar{z}=a-bi$.)

89. Show that for any complex number z,
$$|z^2|=|z|^2.$$

90. Show that for any complex numbers z and w,
$$|z\cdot w|=|z|\cdot|w|.$$
(*Hint*: Let $z=r_1(\cos\theta_1+i\sin\theta_1)$ and $w=r_2(\cos\theta_2+i\sin\theta_2)$.)

91. On a complex plane, graph $z+\bar{z}=3$.

92. On a complex plane, graph $|z|=1$.

Mid-Chapter Mixed Review

Determine whether the statement is true or false.

1. Any triangle, right or oblique, can be solved if at least one side and any other two measures are known. [8.1]

2. The absolute value of $-i$ is 1. [8.3]

3. The law of cosines cannot be used to solve a triangle when all three sides are known. [8.2]

4. Since angle measures determine only the shape of a triangle and not the size, we cannot solve a triangle when only the three angle measures are given. [8.1]

Solve $\triangle ABC$, if possible. [8.1], [8.2]

5. $a = 8.3$ in., $A = 52°$, $C = 65°$

6. $A = 27.2°$, $c = 33$ m, $a = 14$ m

7. $a = 17.8$ yd, $b = 13.1$ yd, $c = 25.6$ yd

8. $a = 29.4$ cm, $b = 40.8$ cm, $A = 42.7°$

9. $A = 148°$, $b = 200$ yd, $c = 185$ yd

10. $b = 18$ ft, $c = 27$ ft, $B = 28°$

11. Find the area of the triangle with $C = 54°$, $a = 38$ in., and $b = 29$ in. [8.2]

Graph the complex number and find its absolute value. [8.3]

12. $-5 + 3i$

13. $-i$

14. 4

15. $1 - 5i$

Find trigonometric notation. [8.3]

16. $\frac{\sqrt{2}}{2} + \frac{\sqrt{6}}{2}i$

17. $1 - \sqrt{3}i$

18. $5i$

19. $-2 - 2i$

Find standard notation. [8.3]

20. $2\left(\cos\frac{7\pi}{4} + i\sin\frac{7\pi}{4}\right)$

21. $12(\cos 30° + i\sin 30°)$

22. $\sqrt{5}(\cos 0° + i\sin 0°)$

23. $4\left[\cos\left(-\frac{3\pi}{2}\right) + i\sin\left(-\frac{3\pi}{2}\right)\right]$

Multiply or divide and leave the answer in trigonometric notation. [8.3]

24. $8(\cos 20° + i\sin 20°) \cdot 2(\cos 25° + i\sin 25°)$

25. $3\left(\cos\frac{\pi}{3} + i\sin\frac{\pi}{3}\right) \div \left[\frac{1}{3}\left(\cos\frac{\pi}{4} + i\sin\frac{\pi}{4}\right)\right]$

Convert to trigonometric notation and then multiply or divide. [8.3]

26. $(1 - i)(\sqrt{3} - i)$

27. $\frac{1 - \sqrt{3}i}{1 + i}$

28. Find $(1 - i)^7$ and write trigonometric notation for the answer. [8.3]

29. Find $[2(\cos 15° + i\sin 15°)]^4$ and write standard notation for the answer. [8.3]

30. Find the square roots of $-2 - 2\sqrt{3}i$. [8.3]

31. Find the cube roots of -1. [8.3]

COLLABORATIVE DISCUSSION AND WRITING

32. Try to solve this triangle using the law of cosines. Then explain why it is easier to solve it using the law of sines. [8.2]

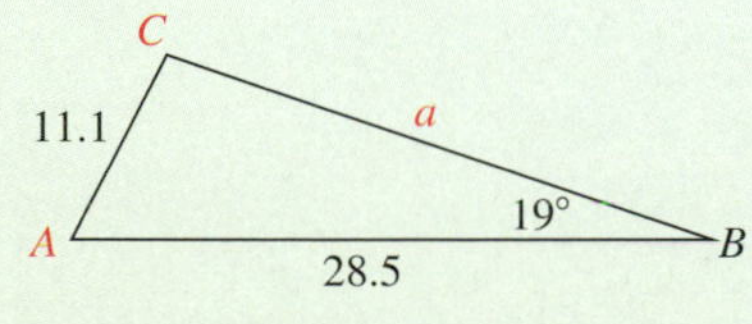

33. Explain why the following statements are not contradictory. [8.3]

The number 1 has one real cube root.
The number 1 has three complex cube roots.

34. Explain why we cannot solve a triangle given SAS with the law of sines. [8.2]

35. Explain why the law of sines cannot be used to find the first angle when solving a triangle given three sides. [8.1]

36. Explain why trigonometric notation for a complex number is not unique, but rectangular, or standard, notation is unique. [8.3]

37. Explain why $x^6 - 2x^3 + 1 = 0$ has three distinct solutions, $x^6 - 2x^3 = 0$ has four distinct solutions, and $x^6 - 2x = 0$ has six distinct solutions. [8.3]

8.4 Polar Coordinates and Graphs

- Graph points given their polar coordinates.
- Convert from rectangular coordinates to polar coordinates and from polar coordinates to rectangular coordinates.
- Convert from rectangular equations to polar equations and from polar equations to rectangular equations.
- Graph polar equations.

Polar Coordinates

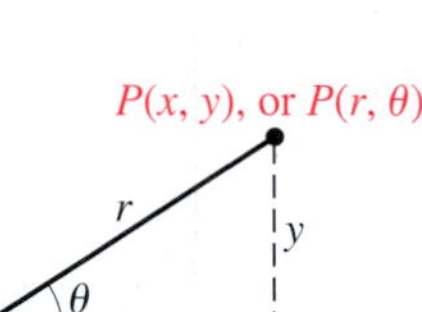

All graphing throughout this text has been done with rectangular coordinates, (x, y), in the Cartesian coordinate system. We now introduce the polar coordinate system. As shown in the diagram at left, any point P has rectangular coordinates (x, y) and polar coordinates (r, θ). On a polar graph, the origin is called the **pole** and the positive half of the x-axis is called the **polar axis.** The point P can be plotted given the directed angle θ from the polar axis to the ray OP and the directed distance r from the pole to the point. The angle θ can be expressed in degrees or radians.

To plot points on a polar graph:

1. Locate the directed angle θ.
2. Move a directed distance r from the pole. If $r > 0$, move along ray OP. If $r < 0$, move in the opposite direction of ray OP.

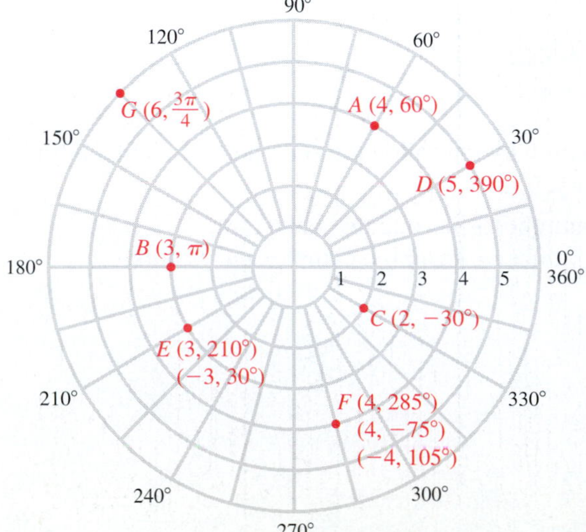

Polar graph paper, shown at left, facilitates plotting. Points B and G illustrate that θ may be in radians. Points E and F illustrate that the polar coordinates of a point are not unique.

EXAMPLE 1 Graph each of the following points.

a) $A(3, 60°)$ **b)** $B(0, 10°)$ **c)** $C(-5, 120°)$

d) $D(1, -60°)$ **e)** $E\left(2, \dfrac{3\pi}{2}\right)$ **f)** $F\left(-4, \dfrac{\pi}{3}\right)$

g) $G\left(-5, -\dfrac{\pi}{4}\right)$ **h)** $H\left(5, \dfrac{3\pi}{4}\right)$

Solution

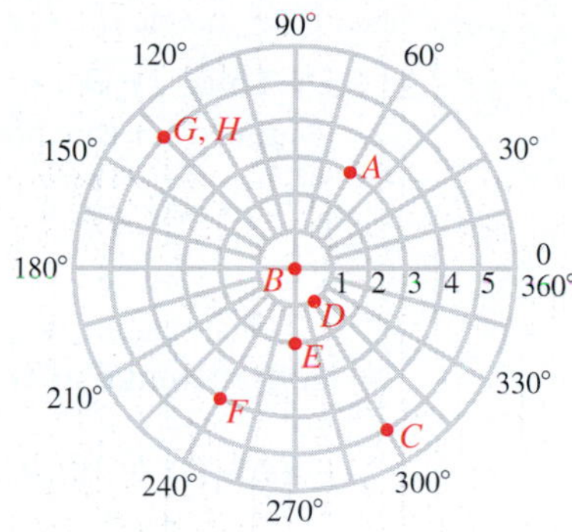

Now Try Exercises 3 and 7.

To convert from rectangular coordinates to polar coordinates and from polar coordinates to rectangular coordinates, we need to recall the following relationships.

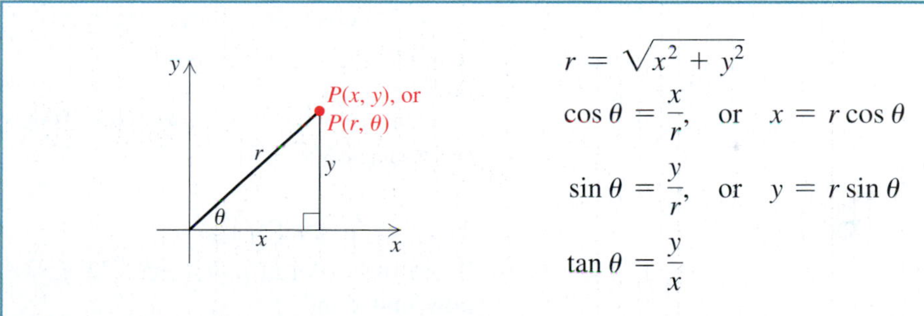

$$r = \sqrt{x^2 + y^2}$$

$$\cos\theta = \frac{x}{r}, \quad \text{or} \quad x = r\cos\theta$$

$$\sin\theta = \frac{y}{r}, \quad \text{or} \quad y = r\sin\theta$$

$$\tan\theta = \frac{y}{x}$$

EXAMPLE 2 Convert each of the following to polar coordinates.

a) $(3, 3)$ **b)** $(2\sqrt{3}, -2)$

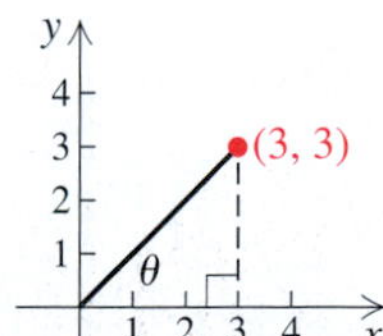

Solution

a) We first find r:

$$r = \sqrt{3^2 + 3^2} = \sqrt{18} = 3\sqrt{2}.$$

Then we determine θ:

$$\tan\theta = \frac{3}{3} = 1; \quad \text{therefore,} \quad \theta = 45°, \text{ or } \frac{\pi}{4}.$$

We know that for $r = 3\sqrt{2}$, $\theta = \pi/4$ and not $5\pi/4$ since $(3, 3)$ is in quadrant I. Thus, $(r, \theta) = (3\sqrt{2}, 45°)$, or $(3\sqrt{2}, \pi/4)$. Other possibilities for polar coordinates include $(3\sqrt{2}, -315°)$ and $(-3\sqrt{2}, 5\pi/4)$.

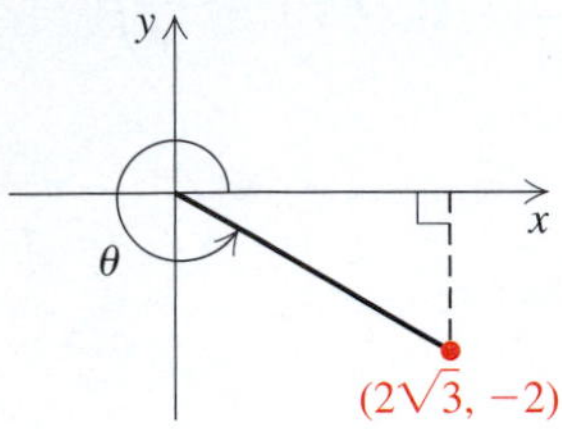

b) We first find r:

$$r = \sqrt{(2\sqrt{3})^2 + (-2)^2} = \sqrt{12 + 4} = \sqrt{16} = 4.$$

Then we determine θ:

$$\tan\theta = \frac{-2}{2\sqrt{3}} = -\frac{1}{\sqrt{3}}; \quad \text{therefore,} \quad \theta = 330°, \text{ or } \frac{11\pi}{6}.$$

Thus, $(r, \theta) = (4, 330°)$, or $(4, 11\pi/6)$. Other possibilities for polar coordinates for this point include $(4, -\pi/6)$ and $(-4, 150°)$.

Now Try Exercise 19.

It is easier to convert from polar coordinates to rectangular coordinates than from rectangular coordinates to polar coordinates.

EXAMPLE 3 Convert each of the following to rectangular coordinates.

a) $\left(10, \frac{\pi}{3}\right)$ **b)** $(-5, 135°)$

Solution

a) The ordered pair $(10, \pi/3)$ gives us $r = 10$ and $\theta = \pi/3$. We now find x and y:

$$x = r\cos\theta = 10\cos\frac{\pi}{3} = 10 \cdot \frac{1}{2} = 5$$

and

$$y = r\sin\theta = 10\sin\frac{\pi}{3} = 10 \cdot \frac{\sqrt{3}}{2} = 5\sqrt{3}.$$

Thus, $(x, y) = (5, 5\sqrt{3})$.

b) From the ordered pair $(-5, 135°)$, we know that $r = -5$ and $\theta = 135°$. We now find x and y:

$$x = -5\cos 135° = -5 \cdot \left(-\frac{\sqrt{2}}{2}\right) = \frac{5\sqrt{2}}{2}$$

and

$$y = -5\sin 135° = -5 \cdot \left(\frac{\sqrt{2}}{2}\right) = -\frac{5\sqrt{2}}{2}.$$

Thus, $(x, y) = \left(\frac{5\sqrt{2}}{2}, -\frac{5\sqrt{2}}{2}\right)$.

Now Try Exercises 31 and 37.

The conversions above can be easily made with some graphing calculators.

Polar Equations and Rectangular Equations

Some curves have simpler equations in polar coordinates than in rectangular coordinates. For others, the reverse is true.

EXAMPLE 4 Convert each of the following to a polar equation.

a) $x^2 + y^2 = 25$ **b)** $2x - y = 5$

Solution

a) We have

$$\begin{aligned} x^2 + y^2 &= 25 \\ (r\cos\theta)^2 + (r\sin\theta)^2 &= 25 && \textbf{Substituting for } x \textbf{ and } y \\ r^2\cos^2\theta + r^2\sin^2\theta &= 25 \\ r^2(\cos^2\theta + \sin^2\theta) &= 25 \\ r^2 &= 25 && \cos^2\theta + \sin^2\theta = 1 \\ r &= 5. \end{aligned}$$

This example illustrates that the polar equation of a circle centered at the origin is much simpler than the rectangular equation.

b) We have

$$\begin{aligned} 2x - y &= 5 \\ 2(r\cos\theta) - (r\sin\theta) &= 5 \\ r(2\cos\theta - \sin\theta) &= 5. \end{aligned}$$

In this example, we see that the rectangular equation is simpler than the polar equation.

Now Try Exercises 47 and 51.

EXAMPLE 5 Convert each of the following to a rectangular equation.

a) $r = 4$

b) $r\cos\theta = 6$

c) $r = 2\cos\theta + 3\sin\theta$

Solution

a) We have

$$\begin{aligned} r &= 4 \\ \sqrt{x^2 + y^2} &= 4 && \textbf{Substituting for } r \\ x^2 + y^2 &= 16. && \textbf{Squaring} \end{aligned}$$

In squaring, we must be careful not to introduce solutions of the equation that are not already present. In this case, we did not, because the graph of either equation is a circle of radius 4 centered at the origin.

b) We have

$$\begin{aligned} r\cos\theta &= 6 \\ x &= 6. && x = r\cos\theta \end{aligned}$$

The graph of $r\cos\theta = 6$, or $x = 6$, is a vertical line.

c) We have

$$r = 2\cos\theta + 3\sin\theta$$

$$r^2 = 2r\cos\theta + 3r\sin\theta \quad \text{Multiplying by } r \text{ on both sides}$$

$$x^2 + y^2 = 2x + 3y. \quad \text{Substituting } x^2 + y^2 \text{ for } r^2, x \text{ for } r\cos\theta, \text{ and } y \text{ for } r\sin\theta$$

Now Try Exercises 59 and 63.

Graphing Polar Equations

To graph a polar equation, we can make a table of values, choosing values of θ and calculating corresponding values of r. We plot the points and complete the graph, as we do when graphing a rectangular equation. A difference occurs in the case of a polar equation, however, because as θ increases sufficiently, points may begin to repeat and the curve will be traced again and again. When this happens, the curve is complete.

EXAMPLE 6 Graph: $r = 1 - \sin\theta$.

Solution We first make a table of values. The TABLE feature on a graphing calculator is the most efficient way to create this list. Note that the points begin to repeat at $\theta = 360°$. We plot these points and draw the curve, as shown below.

θ	r
0°	1
15°	0.7412
30°	0.5
45°	0.2929
60°	0.1340
75°	0.0341
90°	0
105°	0.0341
120°	0.1340
135°	0.2929
150°	0.5
165°	0.7412
180°	1

θ	r
195°	1.2588
210°	1.5
225°	1.7071
240°	1.8660
255°	1.9659
270°	2
285°	1.9659
300°	1.8660
315°	1.7071
330°	1.5
345°	1.2588
360°	1
375°	0.7412
390°	0.5

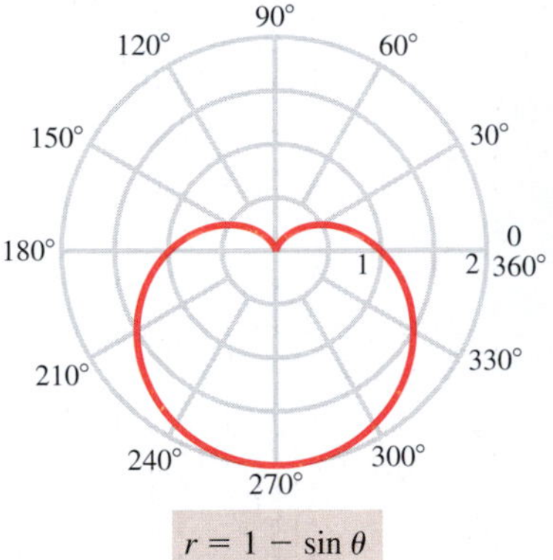

$r = 1 - \sin\theta$

Because of its heart shape, this curve is called a *cardioid*.

Now Try Exercise 77.

We plotted points in Example 6 because we feel that it is important to understand how these curves are developed. We can also graph polar equations using a graphing calculator. The equation usually must be written first in the form $r = f(\theta)$. It is necessary to decide on not only the best window dimensions but also the range of values for θ. Typically, we begin with a range of 0 to 2π for θ in radians and $0°$ to $360°$ for θ in degrees. Because most polar graphs are curved, it is important to *square* the window to minimize distortion.

EXPLORING WITH TECHNOLOGY

Graph $r = 4 \sin 3\theta$. Begin by setting the calculator in POLAR mode, and use either of the following windows. The calculator allows us to view the curve as it is formed.

WINDOW (Radians)	WINDOW (Degrees)
$\theta\text{min} = 0$	$\theta\text{min} = 0$
$\theta\text{max} = 2\pi$	$\theta\text{max} = 360$
$\theta\text{step} = \pi/24$	$\theta\text{step} = 1$

$r = 4 \sin 3\theta$

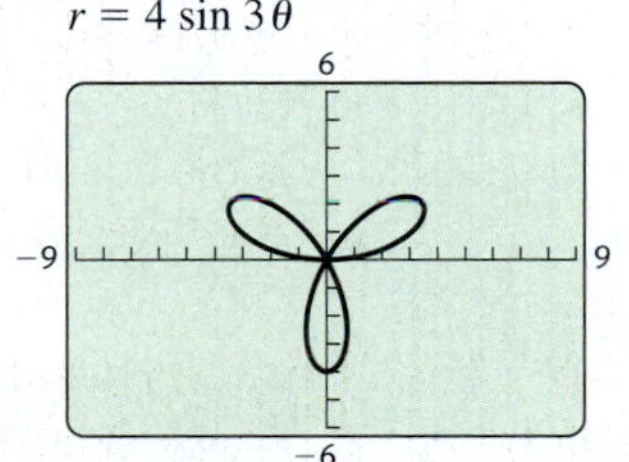

Now graph each of the following equations and note the effect of changing the coefficient of $\sin 3\theta$ and changing the coefficient of θ:

$r = 2 \sin 3\theta$,
$r = 6 \sin 3\theta$,
$r = 4 \sin \theta$, and
$r = 4 \sin 5\theta$.

Polar equations of the form $r = a \cos n\theta$ and $r = a \sin n\theta$ have rose-shaped curves. The number a determines the length of the petals, and the number n determines the number of petals. If n is odd, there are n petals. If n is even, there are $2n$ petals.

EXAMPLE 7 Graph each of the following polar equations. Try to visualize the shape of the curve before graphing it.

a) $r = 3$

b) $r = 5 \sin \theta$

c) $r = 2 \csc \theta$

Solution For each graph, we begin with a table of values. Then we plot points and complete the graph.

a) $r = 3$

For all values of θ, r is 3. Thus the graph of $r = 3$ is a circle of radius 3 centered at the origin.

θ	r
0°	3
60°	3
135°	3
210°	3
300°	3
360°	3

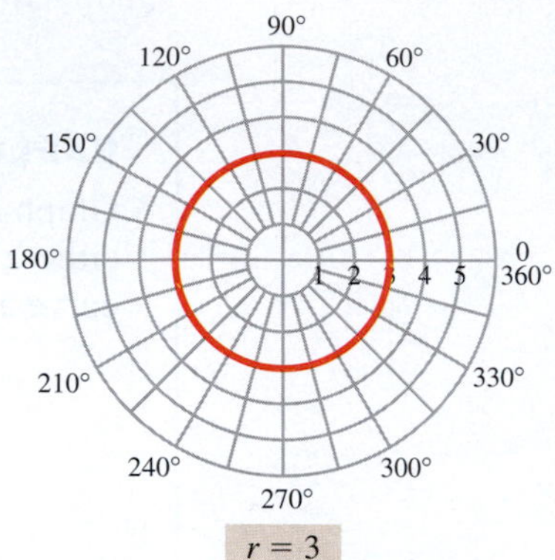

$r = 3$

We can verify our graph by converting to the equivalent rectangular equation. For $r = 3$, we substitute $\sqrt{x^2 + y^2}$ for r and square. The resulting equation,

$$x^2 + y^2 = 3^2,$$

is the equation of a circle with radius 3 centered at the origin.

b) $r = 5 \sin \theta$

$r = 5 \sin \theta$

6, −9, 9, −6

θ	r
0°	0
15°	1.2941
30°	2.5
45°	3.5355
60°	4.3301
75°	4.8296
90°	5
105°	4.8296
120°	4.3301
135°	3.5355
150°	2.5
165°	1.2941
180°	0

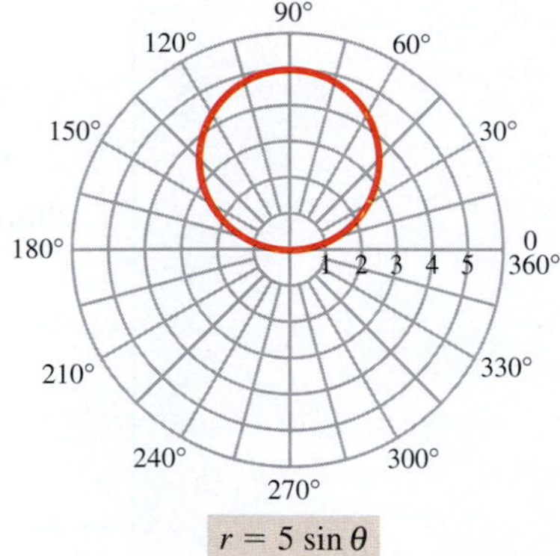

$r = 5 \sin \theta$

c) $r = 2 \csc \theta$

We can rewrite $r = 2 \csc \theta$ as $r = 2/\sin \theta$.

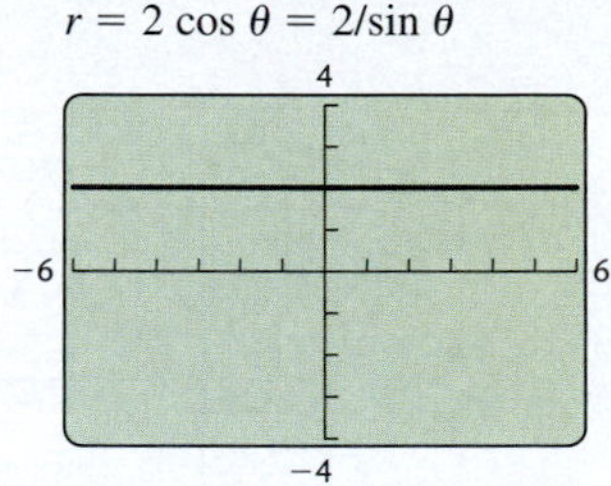

θ	r
0°	Not defined
15°	7.7274
30°	4
45°	2.8284
60°	2.3094
75°	2.0706
90°	2
105°	2.0706
120°	2.3094
135°	2.8284
150°	4
165°	7.7274
180°	Not defined

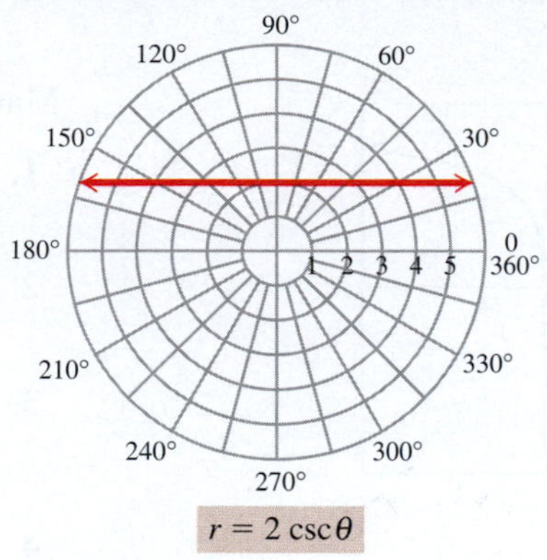

$r = 2 \csc \theta$

Now Try Exercise 71.

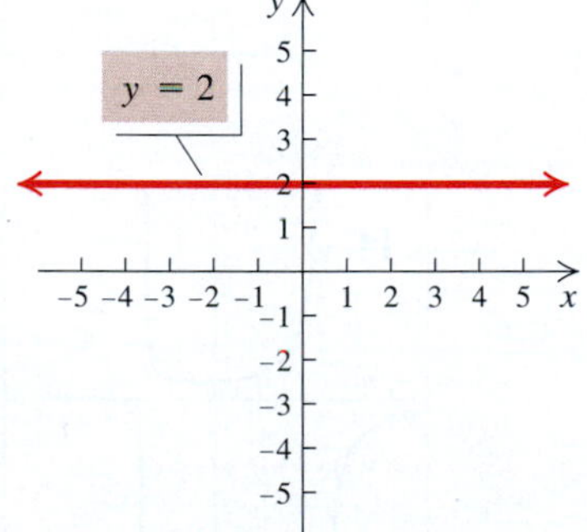

We can check our graph in Example 7(c) by converting the polar equation to the equivalent rectangular equation:

$$r = 2 \csc \theta$$

$$r = \frac{2}{\sin \theta}$$

$$r \sin \theta = 2$$

$$y = 2. \quad \text{Substituting } y \text{ for } r \sin \theta$$

The graph of $y = 2$ is a horizontal line passing through $(0, 2)$ on a rectangular grid, as shown at left.

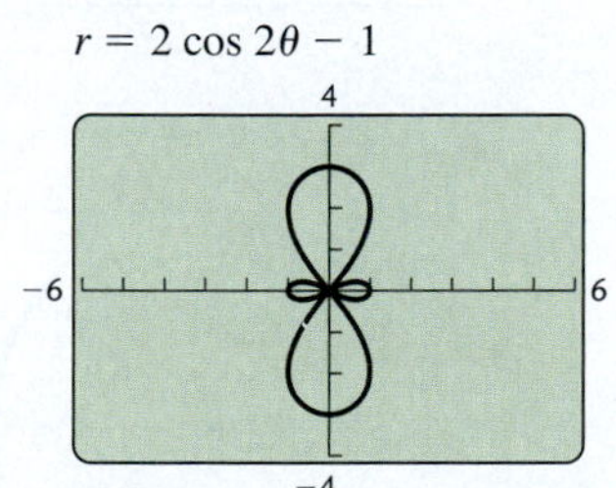

EXAMPLE 8 Graph the equation $r + 1 = 2 \cos 2\theta$ with a graphing calculator.

Solution We first solve for r:

$$r = 2 \cos 2\theta - 1.$$

We then obtain the graph shown at left.

Now Try Exercise 91.

Visualizing the Graph

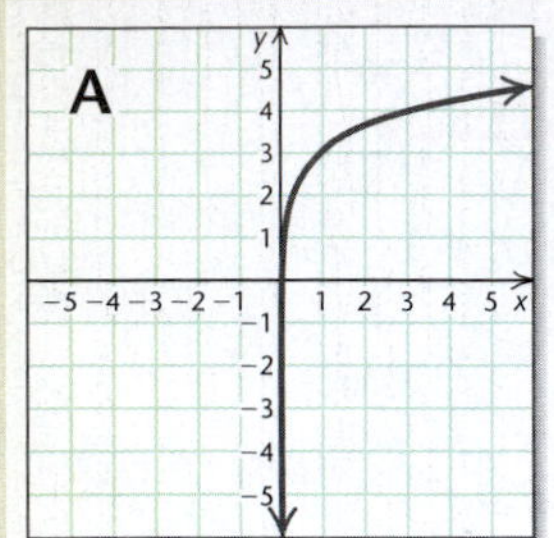

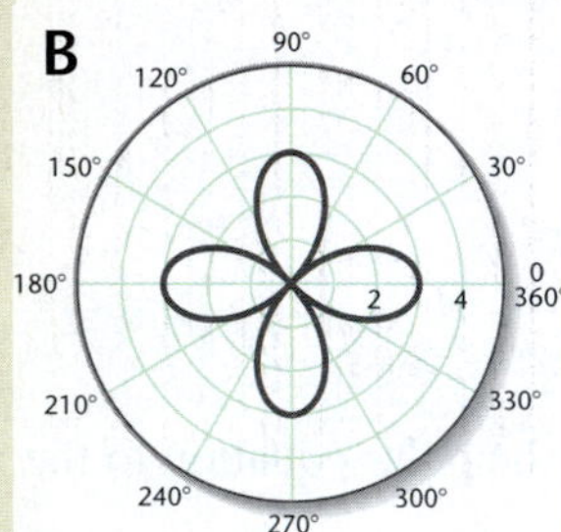

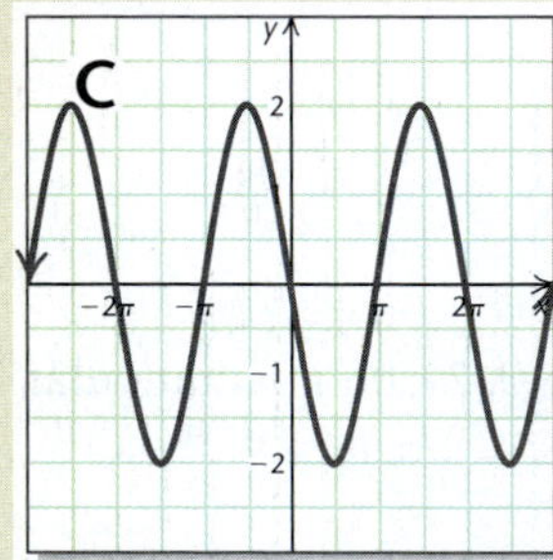

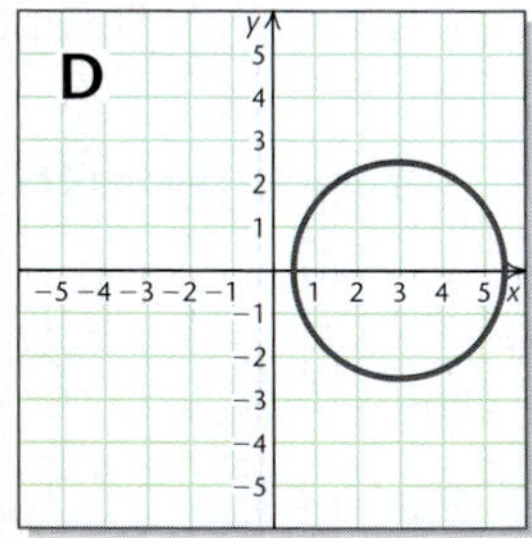

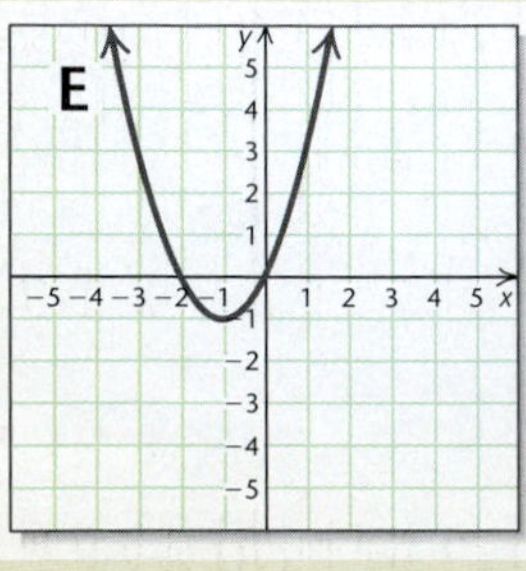

Match the equation with its graph.

1. $f(x) = 2^{(1/2)x}$

2. $y = -2 \sin x$

3. $y = (x + 1)^2 - 1$

4. $f(x) = \dfrac{x - 3}{x^2 + x - 6}$

5. $r = 1 + \sin \theta$

6. $f(x) = 2 \log x + 3$

7. $(x - 3)^2 + y^2 = \dfrac{25}{4}$

8. $y = -\cos\left(x - \dfrac{\pi}{2}\right)$

9. $r = 3 \cos 2\theta$

10. $f(x) = x^4 - x^3 + x^2 - x$

Answers on page A-49

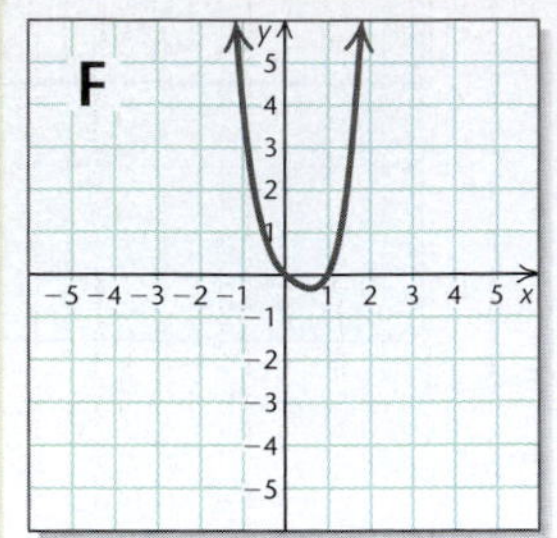

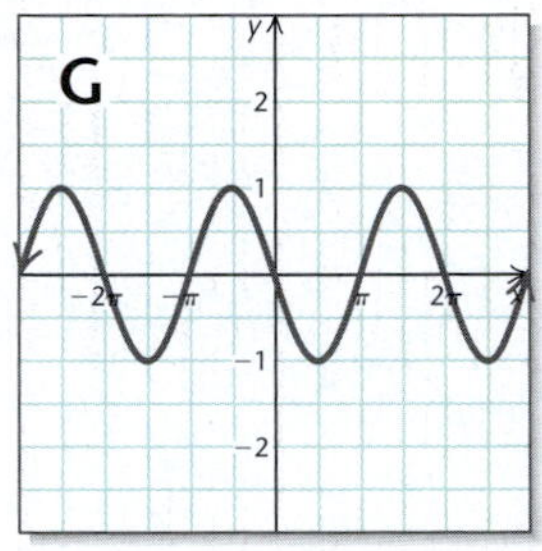

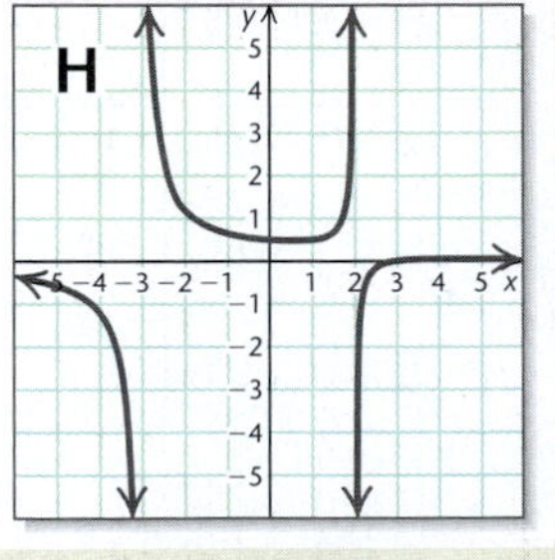

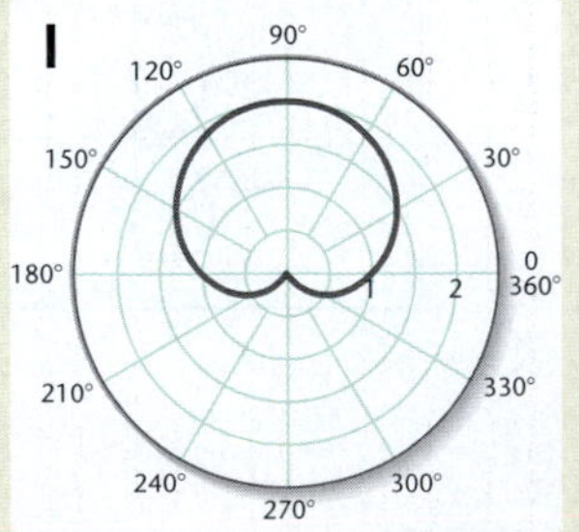

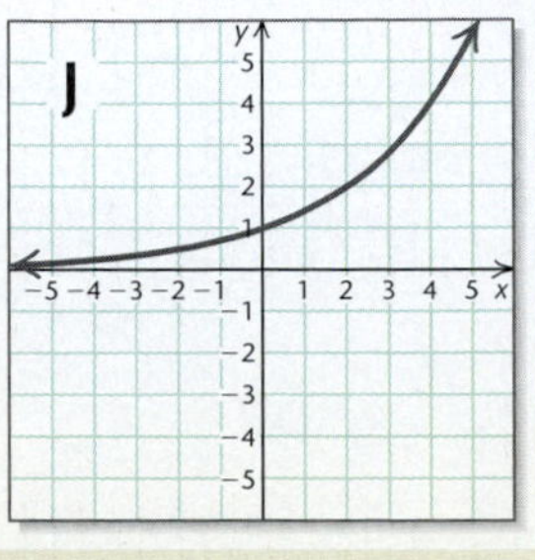

8.4 Exercise Set

Graph the point on a polar grid.

1. $\left(2, \frac{\pi}{4}\right)$

2. $(4, \pi)$

3. $(3.5, 210°)$

4. $(-3, 135°)$

5. $\left(1, \frac{\pi}{6}\right)$

6. $\left(2.75, \frac{5\pi}{6}\right)$

7. $\left(-5, \frac{\pi}{2}\right)$

8. $(0, 15°)$

9. $(3, -315°)$

10. $\left(1.2, -\frac{2\pi}{3}\right)$

11. $(4.3, -60°)$

12. $(3, 405°)$

Find polar coordinates of points A, B, C, and D. Give three answers for each point.

13.

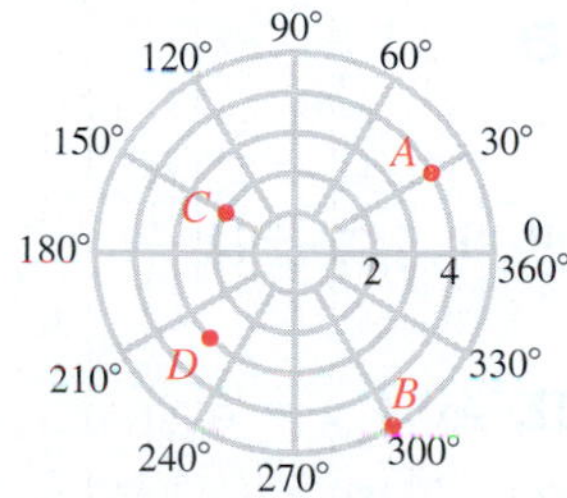

14.

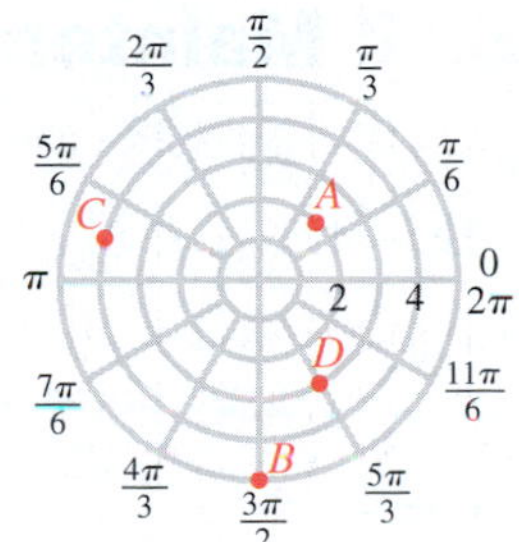

Find the polar coordinates of the point. Express the angle in degrees and then in radians, using the smallest possible positive angle.

15. $(0, -3)$

16. $(-4, 4)$

17. $(3, -3\sqrt{3})$

18. $(-\sqrt{3}, 1)$

19. $(4\sqrt{3}, -4)$

20. $(2\sqrt{3}, 2)$

21. $(-\sqrt{2}, -\sqrt{2})$

22. $(-3, 3\sqrt{3})$

23. $(1, \sqrt{3})$

24. $(0, -1)$

25. $\left(\frac{5\sqrt{2}}{2}, -\frac{5\sqrt{2}}{2}\right)$

26. $\left(-\frac{3}{2}, -\frac{3\sqrt{3}}{2}\right)$

Use a graphing calculator to convert from rectangular coordinates to polar coordinates. Express the answer in both degrees and radians, using the smallest possible positive angle.

27. $(3, 7)$

28. $(-2, -\sqrt{5})$

29. $(-\sqrt{10}, 3.4)$

30. $(0.9, -6)$

Find the rectangular coordinates of the point.

31. $(5, 60°)$

32. $(0, -23°)$

33. $(-3, 45°)$

34. $(6, 30°)$

35. $(3, -120°)$

36. $\left(7, \frac{\pi}{6}\right)$

37. $\left(-2, \frac{5\pi}{3}\right)$

38. $(1.4, 225°)$

39. $(2, 210°)$

40. $\left(1, \frac{7\pi}{4}\right)$

41. $\left(-6, \frac{5\pi}{6}\right)$

42. $(4, 180°)$

Use a graphing calculator to convert from polar coordinates to rectangular coordinates. Round the coordinates to the nearest hundredth.

43. $(3, -43°)$

44. $\left(-5, \frac{\pi}{7}\right)$

45. $\left(-4.2, \frac{3\pi}{5}\right)$

46. $(2.8, 166°)$

Convert to a polar equation.

47. $3x + 4y = 5$

48. $5x + 3y = 4$

49. $x = 5$

50. $y = 4$

51. $x^2 + y^2 = 36$

52. $x^2 - 4y^2 = 4$

53. $x^2 = 25y$

54. $2x - 9y + 3 = 0$

55. $y^2 - 5x - 25 = 0$

56. $x^2 + y^2 = 8y$

57. $x^2 - 2x + y^2 = 0$

58. $3x^2y = 81$

Convert to a rectangular equation.

59. $r = 5$

60. $\theta = \frac{3\pi}{4}$

61. $r \sin \theta = 2$

62. $r = -3 \sin \theta$

63. $r + r \cos \theta = 3$

64. $r = \frac{2}{1 - \sin \theta}$

65. $r - 9 \cos \theta = 7 \sin \theta$

66. $r + 5 \sin \theta = 7 \cos \theta$

67. $r = 5 \sec \theta$

68. $r = 3 \cos \theta$

69. $\theta = \frac{5\pi}{3}$

70. $r = \cos \theta - \sin \theta$

Graph the equation by plotting points. Then check your work using a graphing calculator.

71. $r = \sin \theta$

72. $r = 1 - \cos \theta$

73. $r = 4 \cos 2\theta$

74. $r = 1 - 2 \sin \theta$

75. $r = \cos \theta$

76. $r = 2 \sec \theta$

77. $r = 2 - \cos 3\theta$

78. $r = \dfrac{1}{1 + \cos \theta}$

In Exercises 79–90, use a graphing calculator to match the equation with one of figures (a)–(l) that follow. Try matching the graphs mentally before using a calculator.

a)

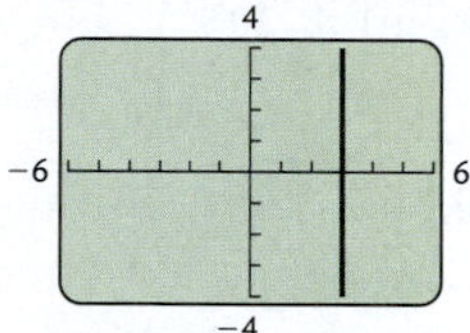

b)

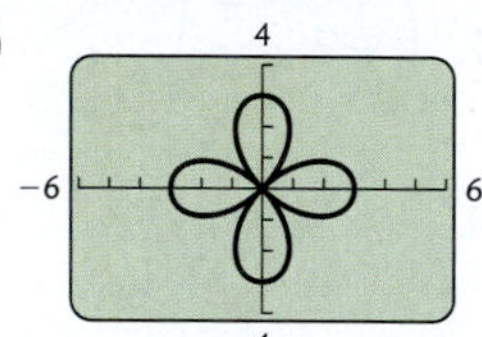

c)

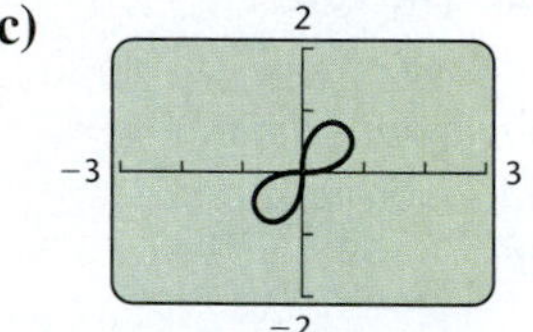

d)

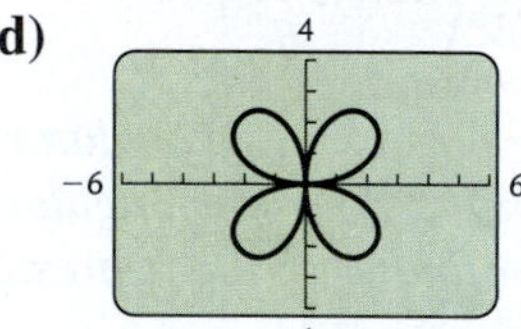

e)

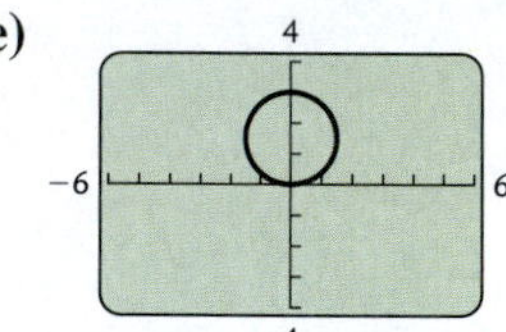

f)

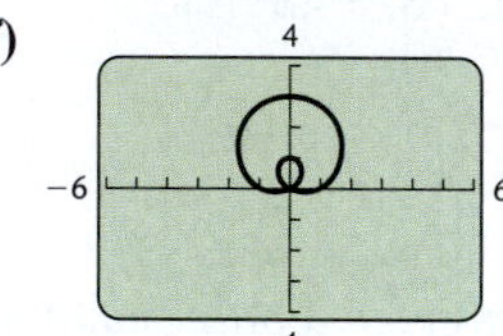

g)

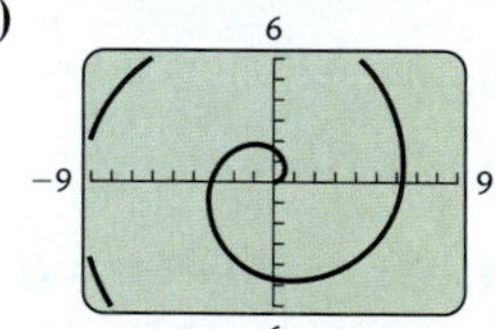

h)

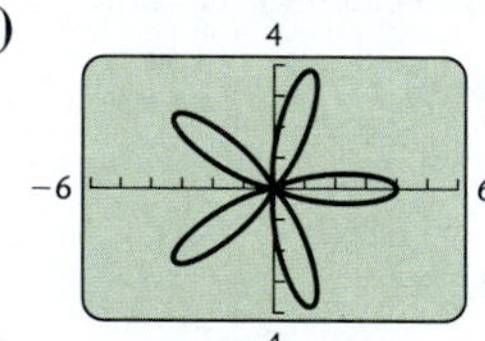

i)

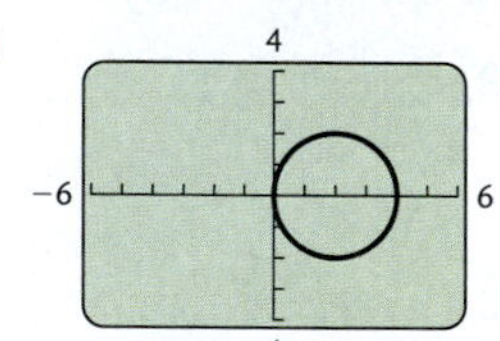

j)

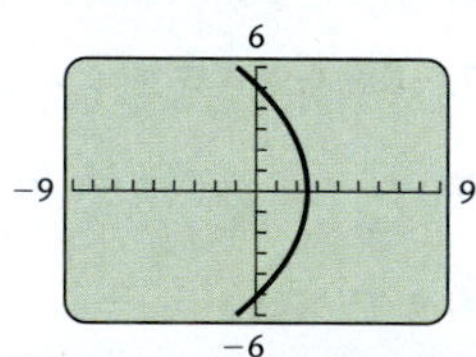

k)

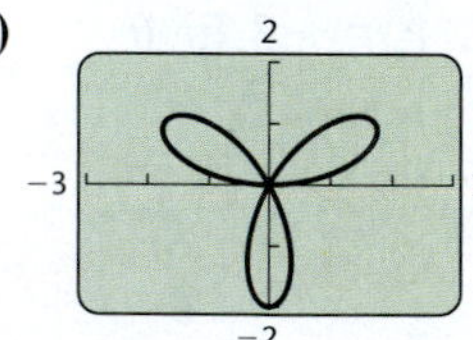

l)

6

−9 9

−6

79. $r = 3 \sin 2\theta$

80. $r = 4 \cos \theta$

81. $r = \theta$

82. $r^2 = \sin 2\theta$

83. $r = \dfrac{5}{1 + \cos \theta}$

84. $r = 1 + 2 \sin \theta$

85. $r = 3 \cos 2\theta$

86. $r = 3 \sec \theta$

87. $r = 3 \sin \theta$

88. $r = 4 \cos 5\theta$

89. $r = 2 \sin 3\theta$

90. $r \sin \theta = 6$

Graph the equation using a graphing calculator.

91. $r = \sin \theta \tan \theta$ (Cissoid)

92. $r = 3\theta$ (Spiral of Archimedes)

93. $r = e^{\theta/10}$ (Logarithmic spiral)

94. $r = 10^{2\theta}$ (Logarithmic spiral)

95. $r = \cos 2\theta \sec \theta$ (Strophoid)

96. $r = \cos 2\theta - 2$ (Peanut)

97. $r = \frac{1}{4} \tan^2 \theta \sec \theta$ (Semicubical parabola)

98. $r = \sin 2\theta + \cos \theta$ (Twisted sister)

Skill Maintenance

Solve. [1.5]

99. $2x - 4 = x + 8$

100. $4 - 5y = 3$

Graph. [1.3]

101. $y = 2x - 5$

102. $4x - y = 6$

103. $x = -3$

104. $y = 0$

Synthesis

105. Convert to a rectangular equation:

$$r = \sec^2 \frac{\theta}{2}.$$

106. The center of a regular hexagon is at the origin, and one vertex is the point $(4, 0°)$. Find the coordinates of the other vertices.

8.5 Vectors and Applications

- Determine whether two vectors are equivalent.
- Find the sum, or resultant, of two vectors.
- Resolve a vector into its horizontal component and its vertical component.
- Solve applied problems involving vectors.

We measure some quantities using only their magnitudes. For example, we describe time, length, and mass using units like seconds, feet, and kilograms, respectively. However, to measure quantities like **displacement**, **velocity**, or **force**, we need to describe a *magnitude* and a *direction*. Together magnitude and direction describe a **vector.** The following are some examples.

Displacement. An object moves a certain distance in a certain direction.

A surveyor steps 20 yd to the northeast.

A hiker follows a trail 5 mi to the west.

A batter hits a ball 100 m along the left-field line.

Velocity. An object travels at a certain speed in a certain direction.

A breeze is blowing 15 mph from the northwest.

An airplane is traveling 450 km/h in a direction of 243°.

Force. A push or pull is exerted on an object in a certain direction.

A force of 200 lb is required to pull a cart up a 30° incline.

A 25-lb force is required to lift a box upward.

A force of 15 newtons is exerted downward on the handle of a jack. (A newton, abbreviated N, is a unit of force used in physics; 1 N $\approx$ 0.22 lb.)

Vectors

Vectors can be graphically represented by directed line segments. The length is chosen, according to some scale, to represent the **magnitude of the vector**, and the direction of the directed line segment represents the **direction of the vector**. For example, if we let 1 cm represent 5 km/h, then a 15-km/h wind from the northwest would be represented by a directed line segment 3 cm long, as shown in the figure at left.

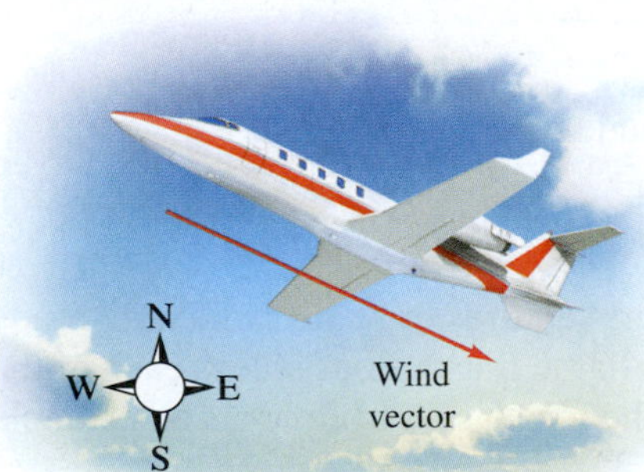

> **VECTOR**
>
> A **vector** in the plane is a directed line segment. Two vectors are **equivalent** if they have the same *magnitude* and the same *direction.*

Consider a vector drawn from point A to point B. Point A is called the **initial point** of the vector, and point B is called the **terminal point.** Symbolic notation for this vector is $\overrightarrow{AB}$ (read "vector AB"). Vectors are also denoted by boldface letters

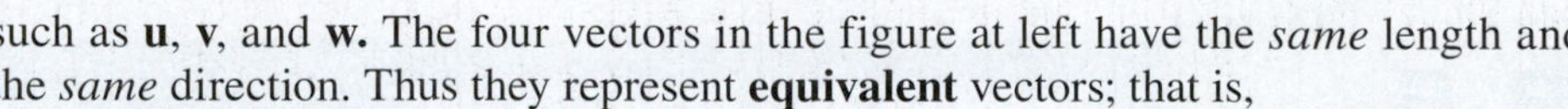

such as **u**, **v**, and **w.** The four vectors in the figure at left have the *same* length and the *same* direction. Thus they represent **equivalent** vectors; that is,

$$\overrightarrow{AB} = \overrightarrow{CD} = \overrightarrow{OP} = \mathbf{v}.$$

In the context of vectors, we use = to mean equivalent.

The length, or **magnitude**, of $\overrightarrow{AB}$ is expressed as $|\overrightarrow{AB}|$. In order to determine whether vectors are equivalent, we find their magnitudes and directions.

EXAMPLE 1 The vectors **u**, $\overrightarrow{OR}$, and **w** are shown in the following figure. Show that $\mathbf{u} = \overrightarrow{OR} = \mathbf{w}$.

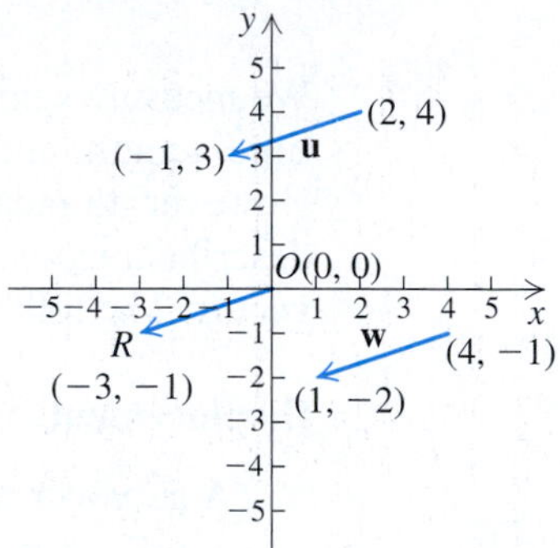

Solution We first find the length of each vector using the distance formula:

$$|\mathbf{u}| = \sqrt{[2 - (-1)]^2 + (4 - 3)^2} = \sqrt{9 + 1} = \sqrt{10},$$
$$|\overrightarrow{OR}| = \sqrt{[0 - (-3)]^2 + [0 - (-1)]^2} = \sqrt{9 + 1} = \sqrt{10},$$
$$|\mathbf{w}| = \sqrt{(4 - 1)^2 + [-1 - (-2)]^2} = \sqrt{9 + 1} = \sqrt{10}.$$

Thus,

$$|\mathbf{u}| = |\overrightarrow{OR}| = |\mathbf{w}|.$$

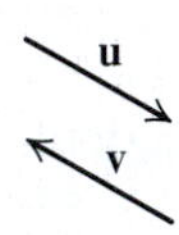

$\mathbf{u} \neq \mathbf{v}$ (not equivalent)
Different magnitudes; different directions

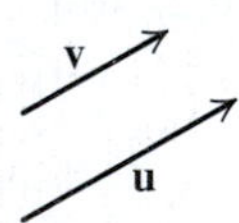

$\mathbf{u} \neq \mathbf{v}$
Same magnitude; different directions

The vectors **u**, $\overrightarrow{OR}$, and **w** appear to go in the same direction so we check their slopes. If the lines that they are on all have the same slope, the vectors have the same direction. We calculate the slopes:

$$\text{Slope} = \overset{\mathbf{u}}{\frac{4 - 3}{2 - (-1)}} = \overset{\overrightarrow{OR}}{\frac{0 - (-1)}{0 - (-3)}} = \overset{\mathbf{w}}{\frac{-1 - (-2)}{4 - 1}} = \frac{1}{3}.$$

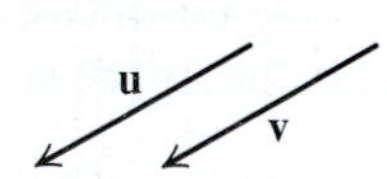

$\mathbf{u} \neq \mathbf{v}$
Different magnitudes; same direction

Since **u**, $\overrightarrow{OR}$, and **w** have the *same* magnitude and the *same* direction,

$$\mathbf{u} = \overrightarrow{OR} = \mathbf{w}.$$

Now Try Exercise 1.

$\mathbf{u} = \mathbf{v}$
Same magnitude; same direction

Keep in mind that the equivalence of vectors requires only the same magnitude and the same direction—not the same location. In the illustrations at left, each of the first three pairs of vectors are not equivalent. The fourth set of vectors is an example of equivalence.

Vector Addition

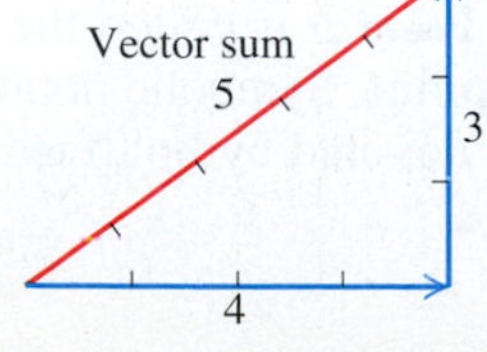

Suppose that a person takes 4 steps east and then 3 steps north. He or she will then be 5 steps from the starting point in the direction shown at left. A vector 4 units long and pointing to the right represents 4 steps east and a vector 3 units long and pointing up represents 3 steps north. The **sum** of the two vectors is the vector 5 steps in magnitude and in the direction shown. The sum is also called the **resultant** of the two vectors.

In general, two nonzero vectors **u** and **v** can be added geometrically by placing the initial point of **v** at the terminal point of **u** and then finding the vector that has the same initial point as **u** and the same terminal point as **v**, as shown in the following figure.

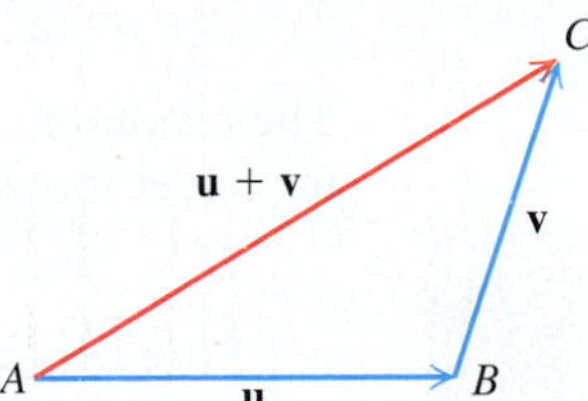

The sum **u** + **v** is the vector represented by the directed line segment from the initial point A of **u** to the terminal point C of **v**. That is, if

$$\mathbf{u} = \overrightarrow{AB} \quad \text{and} \quad \mathbf{v} = \overrightarrow{BC},$$

then

$$\mathbf{u} + \mathbf{v} = \overrightarrow{AB} + \overrightarrow{BC} = \overrightarrow{AC}.$$

We can also describe vector addition by placing the initial points of the vectors together, completing a parallelogram, and finding the diagonal of the parallelogram. (See the figure on the left below.) This description of addition is sometimes called the **parallelogram law** of vector addition. Vector addition is **commutative.** As shown in the figure on the right below, both **u** + **v** and **v** + **u** are represented by the same directed line segment.

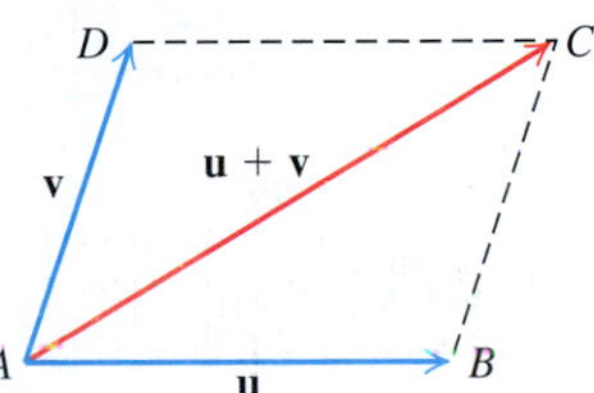

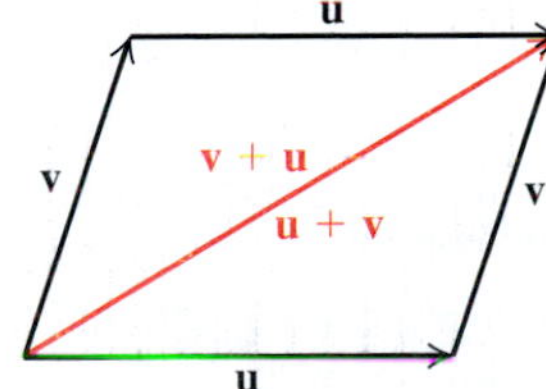

› Applications

If two forces F_1 and F_2 act on an object, the *combined* effect is the sum, or the resultant, $F_1 + F_2$ of the separate forces.

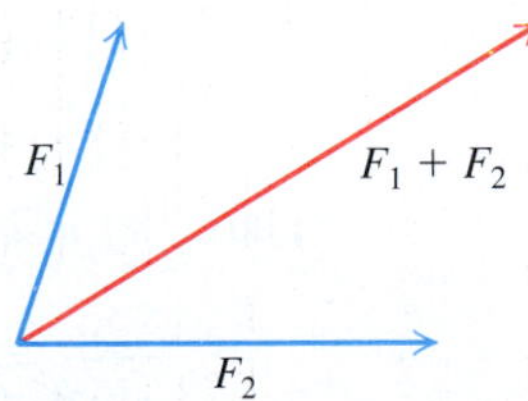

EXAMPLE 2 Forces of 15 newtons and 25 newtons act on an object at right angles to each other. Find their sum, or resultant, giving the magnitude of the resultant and the angle that it makes with the larger force.

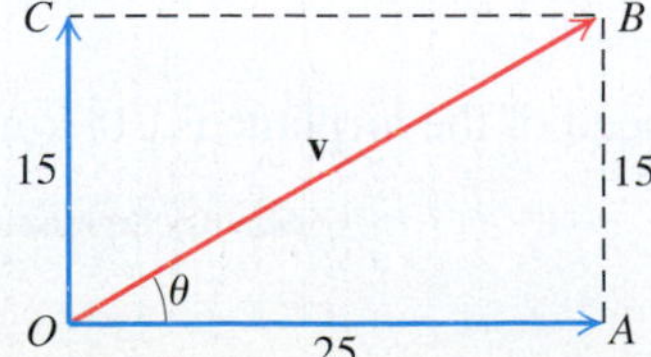

Solution We make a drawing—this time, a rectangle—using **v** or $\overrightarrow{OB}$ to represent the resultant. To find the magnitude, we use the Pythagorean equation:

$$|\mathbf{v}|^2 = 15^2 + 25^2 \qquad \text{|v| denotes the length, or magnitude, of v.}$$
$$|\mathbf{v}| = \sqrt{15^2 + 25^2}$$
$$\approx 29.2.$$

To find the direction, we note that since OAB is a right triangle,

$$\tan\theta = \frac{15}{25} = 0.6.$$

tan⁻¹(.6)
30.96375653

Using a calculator, we find θ, the angle that the resultant makes with the larger force:

$$\theta = \tan^{-1}(0.6) \approx 31°.$$

The resultant $\overrightarrow{OB}$ has a magnitude of 29.2 newtons and makes an angle of 31° with the larger force.

Now Try Exercise 13.

AERIAL BEARINGS

REVIEW SECTION 6.3.

Pilots must adjust the direction of their flight when there is a crosswind. Both the wind and the aircraft velocities can be described by vectors.

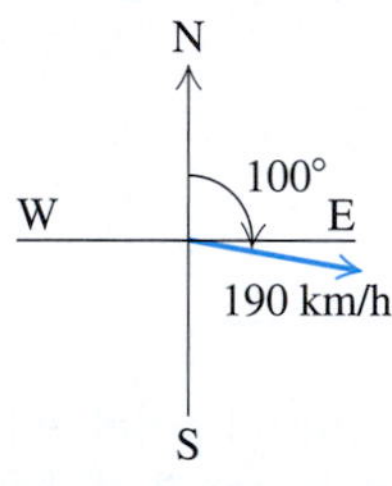

Airplane airspeed

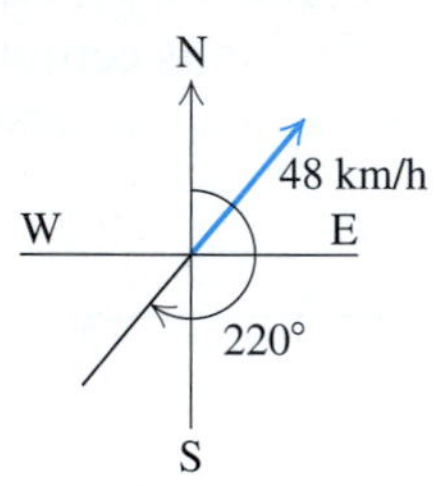

Windspeed

EXAMPLE 3 ***Airplane Speed and Direction.*** An airplane travels on a bearing of 100° at an airspeed of 190 km/h while a wind is blowing 48 km/h from 220°. Find the ground speed of the airplane and the direction of its track, or course, over the ground.

Solution We first make a drawing. The wind is represented by $\overrightarrow{OC}$ and the velocity vector of the airplane by $\overrightarrow{OA}$. The resultant velocity vector is $\mathbf{v}$, the sum of the two vectors. The angle θ between $\mathbf{v}$ and $\overrightarrow{OA}$ is called a **drift angle.**

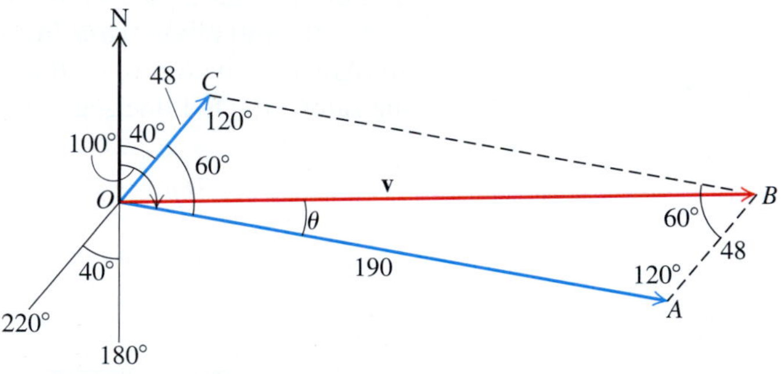

Note that the measure of $\angle COA = 100° - 40° = 60°$. Thus the measure of $\angle CBA$ is also 60° (opposite angles of a parallelogram are equal). Since the sum of all the angles of the parallelogram is 360° and $\angle OCB$ and $\angle OAB$ have the same measure, each must be 120°. By the *law of cosines* in $\triangle OAB$, we have

$$|\mathbf{v}|^2 = 48^2 + 190^2 - 2 \cdot 48 \cdot 190 \cos 120°$$
$$|\mathbf{v}|^2 = 47{,}524$$
$$|\mathbf{v}| = 218.$$

Thus, $|\mathbf{v}|$ is 218 km/h. By the *law of sines* in the same triangle,

$$\frac{48}{\sin\theta} = \frac{218}{\sin 120°},$$

or

$$\sin\theta = \frac{48 \sin 120°}{218} \approx 0.1907$$
$$\theta \approx 11°.$$

sin⁻¹(.1907)
10.99363821

Thus, $\theta = 11°$, to the nearest degree. The ground speed of the airplane is 218 km/h, and its track is in the direction of $100° - 11°$, or 89°.

Now Try Exercise 27.

› Components

Given a vector **w**, we may want to find two other vectors **u** and **v** whose sum is **w.** The vectors **u** and **v** are called **components** of **w** and the process of finding them is called **resolving**, or **representing**, a vector into its vector components.

When we resolve a vector, we generally look for perpendicular components. Most often, one component will be parallel to the x-axis and the other will be parallel to the y-axis. For this reason, they are often called the **horizontal** component and the **vertical** component of a vector. In the figure at left, the vector $\mathbf{w} = \overrightarrow{AC}$ is resolved as the sum of $\mathbf{u} = \overrightarrow{AB}$ and $\mathbf{v} = \overrightarrow{BC}$. The horizontal component of **w** is **u** and the vertical component is **v**.

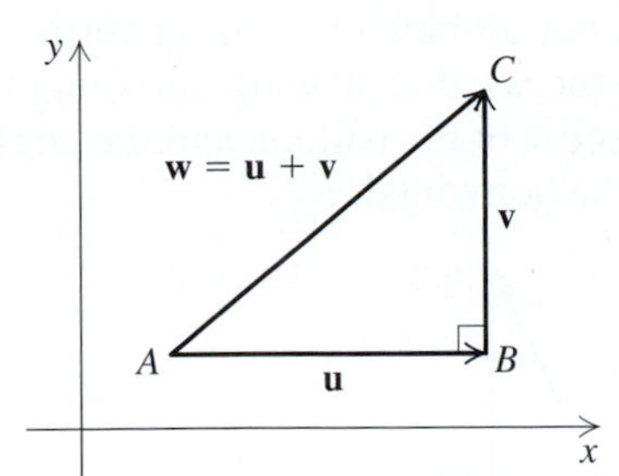

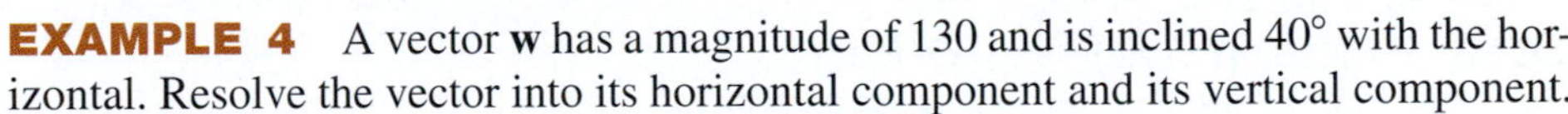

EXAMPLE 4 A vector **w** has a magnitude of 130 and is inclined 40° with the horizontal. Resolve the vector into its horizontal component and its vertical component.

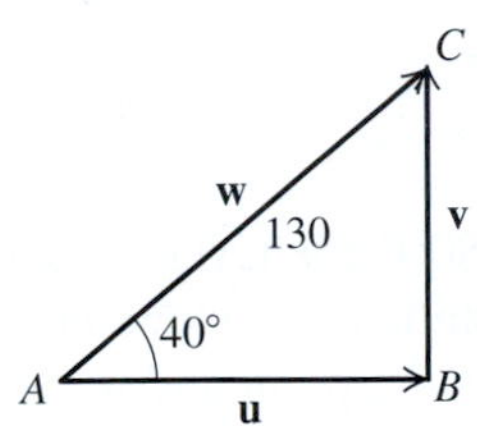

Solution We first make a drawing showing a horizontal vector **u** and a vertical vector **v** whose sum is **w**.

From $\triangle ABC$, we find $|\mathbf{u}|$ and $|\mathbf{v}|$ using the definitions of the cosine and sine functions:

$$\cos 40° = \frac{|\mathbf{u}|}{130}, \quad \text{or} \quad |\mathbf{u}| = 130 \cos 40° \approx 100,$$

$$\sin 40° = \frac{|\mathbf{v}|}{130}, \quad \text{or} \quad |\mathbf{v}| = 130 \sin 40° \approx 84.$$

Thus the horizontal component of **w** is 100 right, and the vertical component of **w** is 84 up.

Now Try Exercise 31.

EXAMPLE 5 ***Shipping Crate.*** A wooden shipping crate that weighs 816 lb is placed on a loading ramp that makes an angle of 25° with the horizontal. To keep the crate from sliding, a chain is hooked to the crate and to a pole at the top of the ramp. Find the magnitude of the components of the crate's weight (disregarding friction) perpendicular to and parallel to the incline.

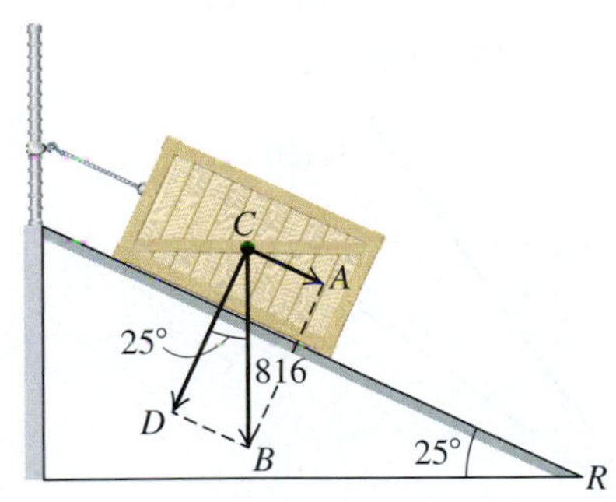

Solution We first make a drawing illustrating the forces with a rectangle. We let

$|\overrightarrow{CB}|$ = the weight of the crate = 816 lb (force of gravity),

$|\overrightarrow{CD}|$ = the magnitude of the component of the crate's weight perpendicular to the incline (force against the ramp), and

$|\overrightarrow{CA}|$ = the magnitude of the component of the crate's weight parallel to the incline (force that pulls the crate down the ramp).

The angle at R is given to be 25° and $\angle BCD = \angle R = 25°$ because the sides of these angles are, respectively, perpendicular. Using the cosine and sine functions, we find that

$$\cos 25° = \frac{|\overrightarrow{CD}|}{816}, \quad \text{or} \quad |\overrightarrow{CD}| = 816 \cos 25° \approx 740 \text{ lb}, \quad \text{and}$$

$$\sin 25° = \frac{|\overrightarrow{DB}|}{816} = \frac{|\overrightarrow{CA}|}{816}, \quad \text{or} \quad |\overrightarrow{CA}| = 816 \sin 25° \approx 345 \text{ lb}.$$

Now Try Exercise 39.

8.5 Exercise Set

Sketch the pair of vectors and determine whether they are equivalent. Use the following ordered pairs for the initial point and the terminal point.

$A(-2, 2)$	$E(-4, 1)$	$I(-6, -3)$
$B(3, 4)$	$F(2, 1)$	$J(3, 1)$
$C(-2, 5)$	$G(-4, 4)$	$K(-3, -3)$
$D(-1, -1)$	$H(1, 2)$	$O(0, 0)$

1. $\overrightarrow{GE}$, $\overrightarrow{BJ}$

2. $\overrightarrow{DJ}$, $\overrightarrow{OF}$

3. $\overrightarrow{DJ}$, $\overrightarrow{AB}$

4. $\overrightarrow{CG}$, $\overrightarrow{FO}$

5. $\overrightarrow{DK}$, $\overrightarrow{BH}$

6. $\overrightarrow{BA}$, $\overrightarrow{DI}$

7. $\overrightarrow{EG}$, $\overrightarrow{BJ}$

8. $\overrightarrow{GC}$, $\overrightarrow{FO}$

9. $\overrightarrow{GA}$, $\overrightarrow{BH}$

10. $\overrightarrow{JD}$, $\overrightarrow{CG}$

11. $\overrightarrow{AB}$, $\overrightarrow{ID}$

12. $\overrightarrow{OF}$, $\overrightarrow{HB}$

13. Two forces of 32 N (newtons) and 45 N act on an object at right angles. Find the magnitude of the resultant and the angle that it makes with the smaller force.

14. Two forces of 50 N and 60 N act on an object at right angles. Find the magnitude of the resultant and the angle that it makes with the larger force.

15. Two forces of 410 N and 600 N act on an object. The angle between the forces is 47°. Find the magnitude of the resultant and the angle that it makes with the larger force.

16. Two forces of 255 N and 325 N act on an object. The angle between the forces is 64°. Find the magnitude of the resultant and the angle that it makes with the smaller force.

In Exercises 17–24, the magnitudes of vectors **u** *and* **v** *and the angle θ between the vectors are given. Find the sum of* $\mathbf{u} + \mathbf{v}$. *Give the magnitude to the nearest tenth and give the direction by specifying to the nearest degree the angle that the resultant makes with* **u**.

17. $|\mathbf{u}| = 45$, $|\mathbf{v}| = 35$, $\theta = 90°$

18. $|\mathbf{u}| = 54$, $|\mathbf{v}| = 43$, $\theta = 150°$

19. $|\mathbf{u}| = 10$, $|\mathbf{v}| = 12$, $\theta = 67°$

20. $|\mathbf{u}| = 25$, $|\mathbf{v}| = 30$, $\theta = 75°$

21. $|\mathbf{u}| = 20$, $|\mathbf{v}| = 20$, $\theta = 117°$

22. $|\mathbf{u}| = 30$, $|\mathbf{v}| = 30$, $\theta = 123°$

23. $|\mathbf{u}| = 23$, $|\mathbf{v}| = 47$, $\theta = 27°$

24. $|\mathbf{u}| = 32$, $|\mathbf{v}| = 74$, $\theta = 72°$

25. *Hot-Air Balloon.* A hot-air balloon is rising vertically 10 ft/sec while the wind is blowing horizontally 5 ft/sec. Find the speed **v** of the balloon and the angle θ that it makes with the horizontal.

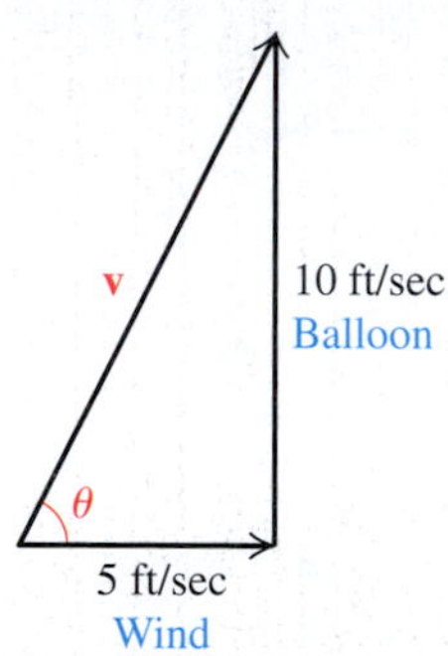

26. *Ship.* A ship first sails N80°E for 120 nautical mi, and then S20°W for 200 nautical mi. How far is the ship, then, from the starting point and in what direction?

27. *Boat.* A boat heads 35°, propelled by a force of 750 lb. A wind from 320° exerts a force of 150 lb on the boat. How large is the resultant force **F**, and in what direction is the boat moving?

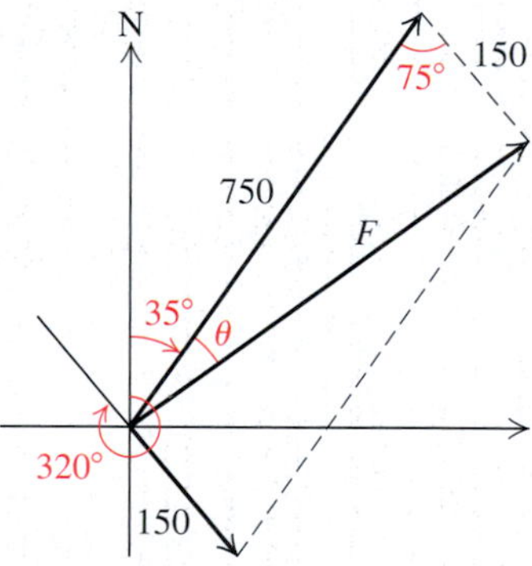

28. *Airplane.* An airplane flies 32° for 210 km, and then 280° for 170 km. How far is the airplane, then, from the starting point and in what direction?

29. *Airplane.* An airplane has an airspeed of 150 km/h. It is to make a flight in a direction of 70° while there is a 25-km/h wind from 340°. What will the airplane's actual heading be?

30. *Wind.* A wind has an easterly component (*from* the east) of 10 km/h and a southerly component (*from* the south) of 16 km/h. Find the magnitude and the direction of the wind.

31. A vector **w** has magnitude 100 and points southeast. Resolve the vector into an easterly component and a southerly component.

32. A vector **u** with a magnitude of 150 lb is inclined to the right and upward 52° from the horizontal. Resolve the vector into components.

33. *Airplane.* An airplane takes off at a speed **S** of 225 mph at an angle of 17° with the horizontal. Resolve the vector **S** into components.

34. *Wheelbarrow.* A wheelbarrow is pushed by applying a 97-lb force **F** that makes a 38° angle with the horizontal. Resolve **F** into its horizontal component and its vertical component. (The horizontal component is the effective force in the direction of motion and the vertical component adds weight to the wheelbarrow.)

35. *Luggage Wagon.* A luggage wagon is being pulled with vector force **V**, which has a magnitude of 780 lb at an angle of elevation of 60°. Resolve the vector **V** into components.

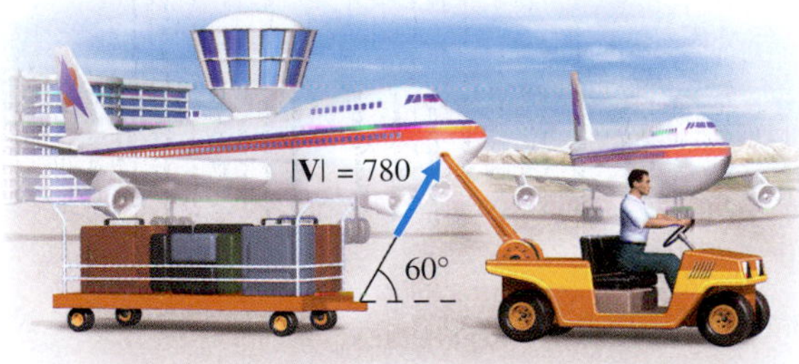

36. *Hot-air Balloon.* A hot-air balloon exerts a 1200-lb pull on a tether line at a 45° angle with the horizontal. Resolve the vector **B** into components.

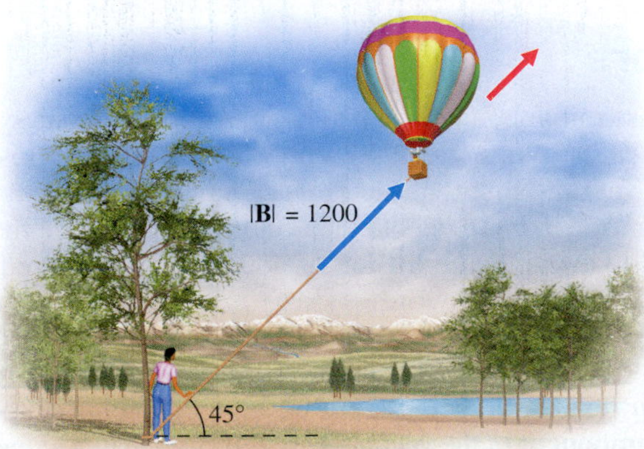

37. *Airplane.* An airplane is flying at 200 km/h in a direction of 305°. Find the westerly component and the northerly component of its velocity.

38. *Baseball.* A baseball player throws a baseball with a speed **S** of 72 mph at an angle of 45° with the horizontal. Resolve the vector **S** into components.

39. A block weighing 100 lb rests on a 25° incline. Find the magnitude of the components of the block's weight perpendicular to and parallel to the incline.

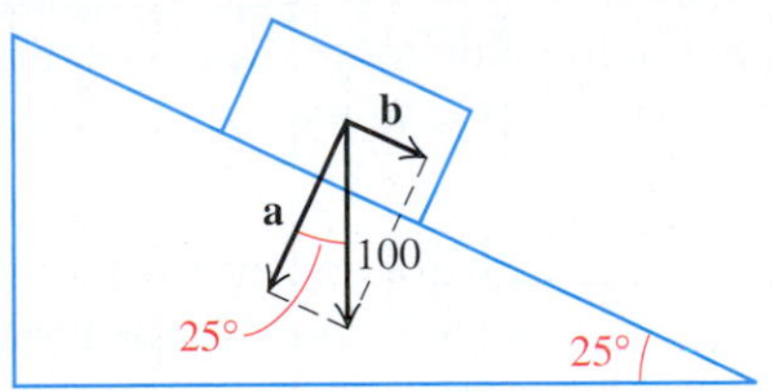

40. A shipping crate that weighs 450 kg is placed on a loading ramp that makes an angle of 30° with the horizontal. Find the magnitude of the components of the crate's weight perpendicular to and parallel to the incline.

41. An 80-lb block of ice rests on a 37° incline. What force parallel to the incline is necessary in order to keep the ice from sliding down?

42. What force is necessary to pull a 3500-lb truck up a 9° incline?

Skill Maintenance

Vocabulary Reinforcement

In each of Exercises 43–52, fill in the blank with the correct term. Some of the given choices will not be used.

angular speed	cosine
linear speed	common
acute	natural
obtuse	horizontal line
secant of θ	vertical line
cotangent of θ	double-angle
identity	half-angle
inverse	coterminal
absolute value	reference angle
sines	

43. Logarithms, base e, are called ____________ logarithms. [5.3]

44. ____________ identities give trigonometric function values of $x/2$ in terms of function values of x. [7.2]

45. ____________ is distance traveled per unit of time. [6.4]

46. The sine of an angle is also the ____________ of the angle's complement. [6.1]

47. A(n) ______________ is an equation that is true for all possible replacements of the variables. [7.1]

48. The ______________ is the length of the side adjacent to θ divided by the length of the side opposite θ. [6.1]

49. If two or more angles have the same terminal side, the angles are said to be ______________. [6.3]

50. In any triangle, the sides are proportional to the ______________ of the opposite angles. [8.1]

51. If it is possible for a(n) ______________ to intersect the graph of a function more than once, then the function is not one-to-one and its ______________ is not a function. [5.1]

52. The ______________ for an angle is the ______________ angle formed by the terminal side of the angle and the x-axis. [6.3]

Synthesis

53. *Eagle's Flight.* An eagle flies from its nest 7 mi in the direction northeast, where it stops to rest on a cliff. It then flies 8 mi in the direction S30°W to land on top of a tree. Place an xy-coordinate system so that the origin is the bird's nest, the x-axis points east, and the y-axis points north.

a) At what point is the cliff located?
b) At what point is the tree located?

8.6 Vector Operations

- Perform calculations with vectors in component form.
- Express a vector as a linear combination of unit vectors.
- Express a vector in terms of its magnitude and its direction.
- Find the angle between two vectors using the dot product.
- Solve applied problems involving forces in equilibrium.

Position Vectors

Let's consider a vector $\mathbf{v}$ whose initial point is the *origin* in an xy-coordinate system and whose terminal point is (a, b). We say that the vector is in **standard position** and refer to it as a **position vector**. Note that the ordered pair (a, b) defines the vector uniquely. Thus we can use (a, b) to denote the vector. To emphasize that we are thinking of a vector and to avoid the confusion of notation with ordered-pair notation and interval notation, we generally write

$$\mathbf{v} = \langle a, b \rangle.$$

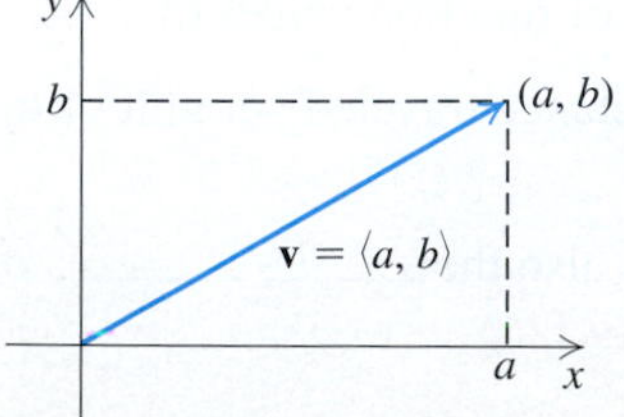

The coordinate a is the *scalar* **horizontal component** of the vector, and the coordinate b is the *scalar* **vertical component** of the vector. By **scalar**, we mean a *numerical* quantity rather than a *vector* quantity. Thus, $\langle a, b \rangle$ is considered to be the *component form* of $\mathbf{v}$. Note that a and b are *not* vectors and should not be confused with the vector component definition given in Section 8.5.

Now consider $\overrightarrow{AC}$ with $A = (x_1, y_1)$ and $C = (x_2, y_2)$. Let's see how to find the position vector equivalent to $\overrightarrow{AC}$. As you can see in the following figure, the initial point A is relocated to the origin $(0, 0)$. The coordinates of P are found by subtracting the coordinates of A from the coordinates of C. Thus, $P = (x_2 - x_1, y_2 - y_1)$ and the position vector is $\overrightarrow{OP}$.

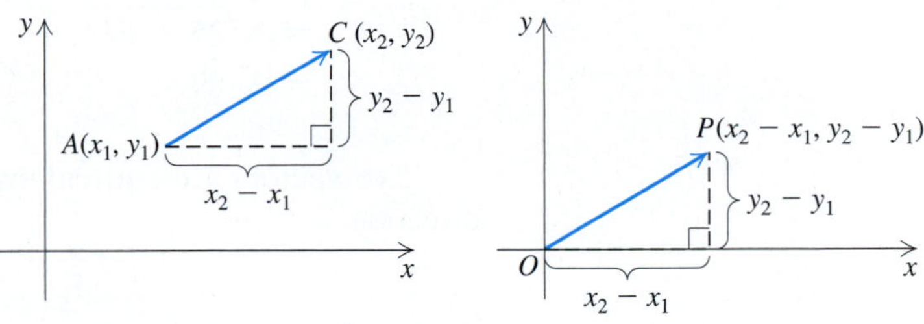

It can be shown that $\overrightarrow{OP}$ and $\overrightarrow{AC}$ have the same magnitude and the same direction and are therefore equivalent. Thus, $\overrightarrow{AC} = \overrightarrow{OP} = \langle x_2 - x_1, y_2 - y_1 \rangle$.

COMPONENT FORM OF A VECTOR

The **component form** of $\overrightarrow{AC}$ with $A = (x_1, y_1)$ and $C = (x_2, y_2)$ is

$$\overrightarrow{AC} = \langle x_2 - x_1, y_2 - y_1 \rangle.$$

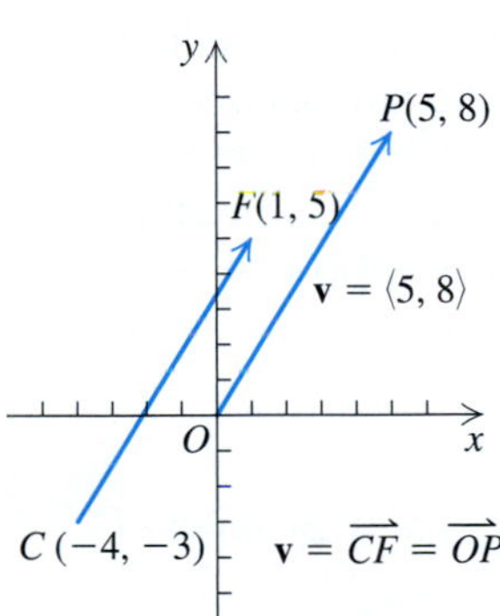

EXAMPLE 1 Find the component form of $\overrightarrow{CF}$ if $C = (-4, -3)$ and $F = (1, 5)$.

Solution We have

$$\overrightarrow{CF} = \langle 1 - (-4), 5 - (-3) \rangle = \langle 5, 8 \rangle.$$

Note that vector $\overrightarrow{CF}$ is equivalent to *position vector* $\overrightarrow{OP}$ with $P = (5, 8)$ as shown in the figure at left.

Now that we know how to write vectors in component form, let's restate some definitions that we first considered in Section 8.5.

The length of a vector $\mathbf{v}$ is easy to determine when the components of the vector are known. For $\mathbf{v} = \langle v_1, v_2 \rangle$, we have

$$|\mathbf{v}|^2 = v_1^2 + v_2^2 \quad \text{Using the Pythagorean equation}$$
$$|\mathbf{v}| = \sqrt{v_1^2 + v_2^2}.$$

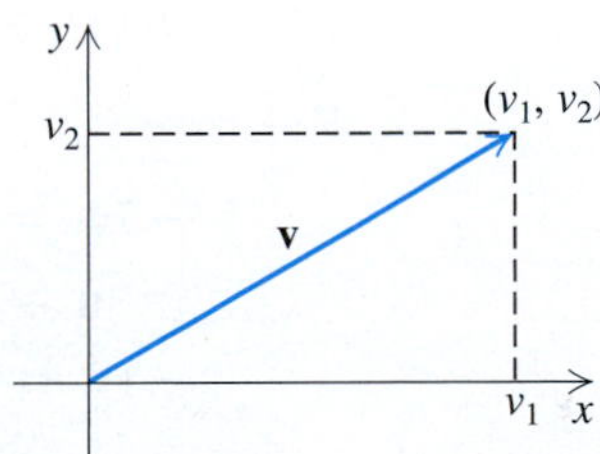

LENGTH OF A VECTOR

The **length**, or **magnitude**, of a vector $\mathbf{v} = \langle v_1, v_2 \rangle$ is given by

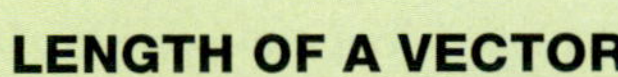

$$|\mathbf{v}| = \sqrt{v_1^2 + v_2^2}.$$

EXAMPLE 2 Find the length, or magnitude, of vector $\mathbf{v} = \langle 5, 8 \rangle$, illustrated in Example 1.

Solution

$$
\begin{aligned}
|\mathbf{v}| &= \sqrt{v_1^2 + v_2^2} && \text{Length of vector } \mathbf{v} = \langle v_1, v_2 \rangle \\
&= \sqrt{5^2 + 8^2} && \text{Substituting 5 for } v_1 \text{ and 8 for } v_2 \\
&= \sqrt{25 + 64} \\
&= \sqrt{89}
\end{aligned}
$$

Now Try Exercises 1 and 7.

Two vectors are **equivalent** if they have the *same* magnitude and the *same* direction.

EQUIVALENT VECTORS

Let $\mathbf{u} = \langle u_1, u_2 \rangle$ and $\mathbf{v} = \langle v_1, v_2 \rangle$. Then

$$\langle u_1, u_2 \rangle = \langle v_1, v_2 \rangle \quad \text{if and only if} \quad u_1 = v_1 \quad \text{and} \quad u_2 = v_2.$$

Operations on Vectors

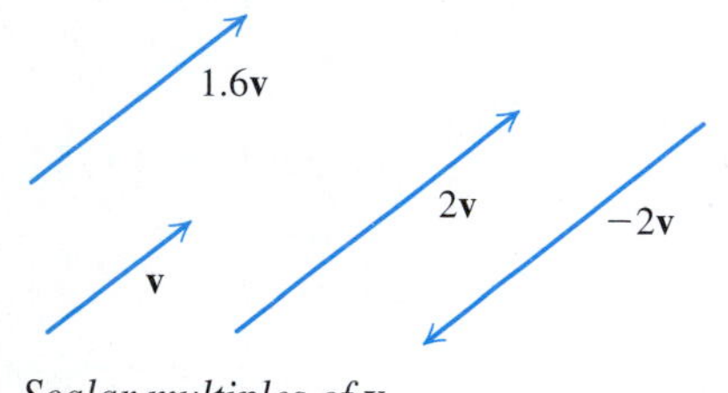

Scalar multiples of **v**

To multiply a vector **v** by a positive real number, we multiply its length by the number. Its direction stays the same. When a vector **v** is multiplied by 2, for instance, its length is doubled and its direction is not changed. When a vector is multiplied by 1.6, its length is increased by 60% and its direction stays the same. To multiply a vector **v** by a negative real number, we multiply its length by the number and reverse its direction. When a vector is multiplied by −2, its length is doubled and its direction is reversed. Since real numbers work like scaling factors in vector multiplication, we call them **scalars** and the products $k\mathbf{v}$ are called **scalar multiples** of **v**.

SCALAR MULTIPLICATION

For a real number k and a vector $\mathbf{v} = \langle v_1, v_2 \rangle$, the **scalar product** of k and **v** is

$$k\mathbf{v} = k\langle v_1, v_2 \rangle = \langle kv_1, kv_2 \rangle.$$

The vector $k\mathbf{v}$ is a **scalar multiple** of the vector **v**.

EXAMPLE 3 Let $\mathbf{u} = \langle -5, 4 \rangle$ and $\mathbf{w} = \langle 1, -1 \rangle$. Find $-7\mathbf{w}$, $3\mathbf{u}$, and $-1\mathbf{w}$.

Solution

$$
\begin{aligned}
-7\mathbf{w} &= -7\langle 1, -1 \rangle = \langle -7, 7 \rangle, \\
3\mathbf{u} &= 3\langle -5, 4 \rangle = \langle -15, 12 \rangle, \\
-1\mathbf{w} &= -1\langle 1, -1 \rangle = \langle -1, 1 \rangle
\end{aligned}
$$

In Section 8.5, we used the parallelogram law to add two vectors, but now we can add two vectors using components. To add two vectors given in component form, we add the corresponding components. Let $\mathbf{u} = \langle u_1, u_2 \rangle$ and $\mathbf{v} = \langle v_1, v_2 \rangle$. Then

$$\mathbf{u} + \mathbf{v} = \langle u_1 + v_1, u_2 + v_2 \rangle.$$

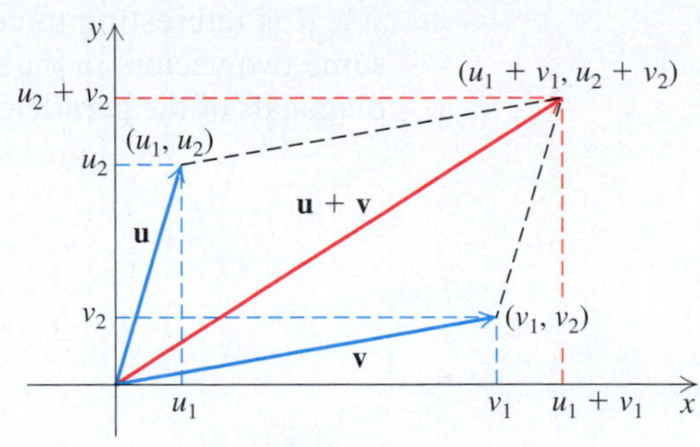

For example, if $\mathbf{v} = \langle -3, 2 \rangle$ and $\mathbf{w} = \langle 5, -9 \rangle$, then

$$\mathbf{v} + \mathbf{w} = \langle -3 + 5, 2 + (-9) \rangle = \langle 2, -7 \rangle.$$

VECTOR ADDITION

If $\mathbf{u} = \langle u_1, u_2 \rangle$ and $\mathbf{v} = \langle v_1, v_2 \rangle$, then

$$\mathbf{u} + \mathbf{v} = \langle u_1 + v_1, u_2 + v_2 \rangle.$$

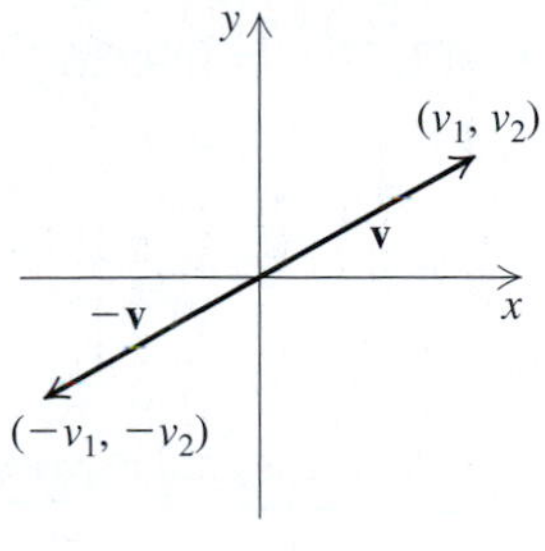

Before we define vector subtraction, we must define $-\mathbf{v}$. The opposite of $\mathbf{v} = \langle v_1, v_2 \rangle$, shown at left, is

$$-\mathbf{v} = (-1)\mathbf{v} = (-1)\langle v_1, v_2 \rangle = \langle -v_1, -v_2 \rangle.$$

Vector subtraction such as $\mathbf{u} - \mathbf{v}$ involves subtracting corresponding components. We show this by rewriting $\mathbf{u} - \mathbf{v}$ as $\mathbf{u} + (-\mathbf{v})$. If $\mathbf{u} = \langle u_1, u_2 \rangle$ and $\mathbf{v} = \langle v_1, v_2 \rangle$, then

$$\begin{aligned} \mathbf{u} - \mathbf{v} = \mathbf{u} + (-\mathbf{v}) &= \langle u_1, u_2 \rangle + \langle -v_1, -v_2 \rangle \\ &= \langle u_1 + (-v_1), u_2 + (-v_2) \rangle \\ &= \langle u_1 - v_1, u_2 - v_2 \rangle. \end{aligned}$$

We can illustrate vector subtraction with parallelograms, just as we did vector addition.

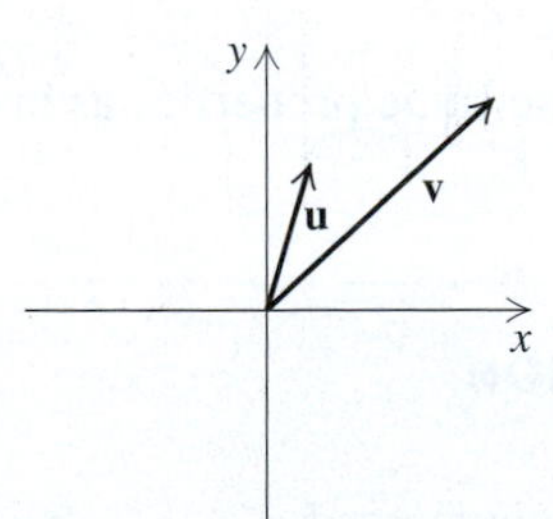

Sketch **u** and **v**.

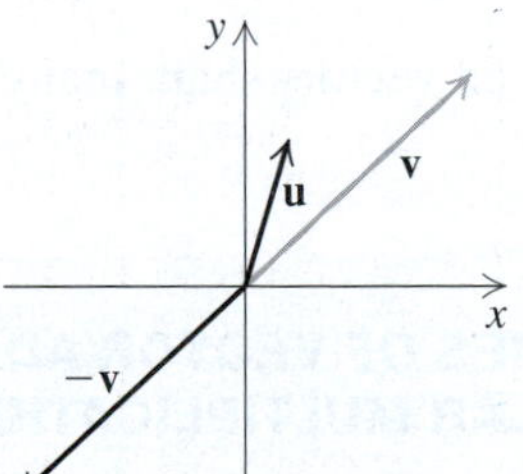

Sketch $-\mathbf{v}$.

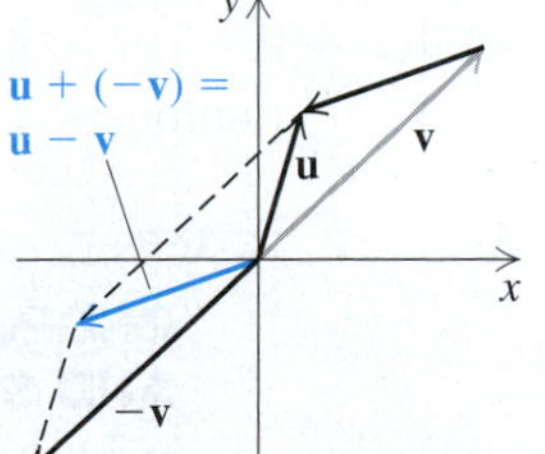

Sketch $\mathbf{u} + (-\mathbf{v})$, or $\mathbf{u} - \mathbf{v}$, using the parallelogram law.

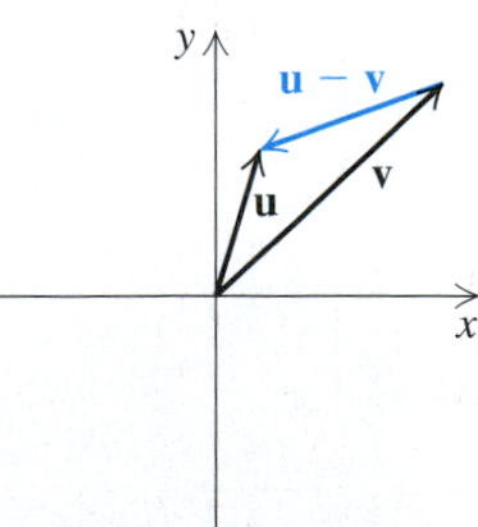

$\mathbf{u} - \mathbf{v}$ is the vector from the terminal point of **v** to the terminal point of **u**.

VECTOR SUBTRACTION

If $\mathbf{u} = \langle u_1, u_2 \rangle$ and $\mathbf{v} = \langle v_1, v_2 \rangle$, then

$$\mathbf{u} - \mathbf{v} = \langle u_1 - v_1, u_2 - v_2 \rangle.$$

It is interesting to compare the sum of two vectors with the difference of the same two vectors in the same parallelogram. The vectors $\mathbf{u} + \mathbf{v}$ and $\mathbf{u} - \mathbf{v}$ are the diagonals of the parallelogram.

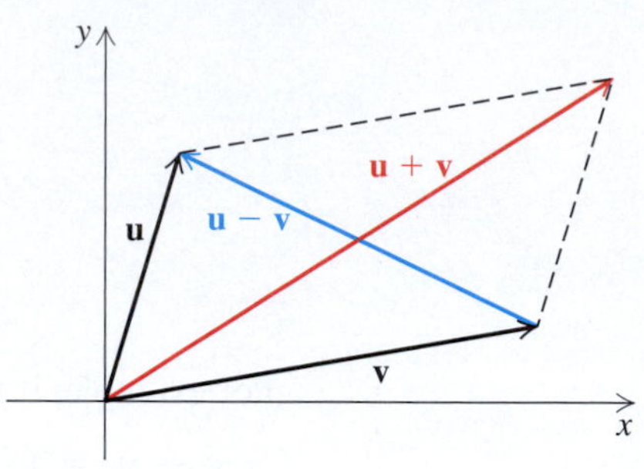

EXAMPLE 4 Do the following calculations, where $\mathbf{u} = \langle 7, 2 \rangle$ and $\mathbf{v} = \langle -3, 5 \rangle$.

a) $\mathbf{u} + \mathbf{v}$ **b)** $\mathbf{u} - 6\mathbf{v}$

c) $3\mathbf{u} + 4\mathbf{v}$ **d)** $|5\mathbf{v} - 2\mathbf{u}|$

Solution

a) $\mathbf{u} + \mathbf{v} = \langle 7, 2 \rangle + \langle -3, 5 \rangle = \langle 7 + (-3), 2 + 5 \rangle = \langle 4, 7 \rangle$

b) $\mathbf{u} - 6\mathbf{v} = \langle 7, 2 \rangle - 6\langle -3, 5 \rangle = \langle 7, 2 \rangle - \langle -18, 30 \rangle = \langle 25, -28 \rangle$

c) $3\mathbf{u} + 4\mathbf{v} = 3\langle 7, 2 \rangle + 4\langle -3, 5 \rangle = \langle 21, 6 \rangle + \langle -12, 20 \rangle = \langle 9, 26 \rangle$

d) $$\begin{aligned} |5\mathbf{v} - 2\mathbf{u}| &= |5\langle -3, 5 \rangle - 2\langle 7, 2 \rangle| = |\langle -15, 25 \rangle - \langle 14, 4 \rangle| \\ &= |\langle -29, 21 \rangle| \\ &= \sqrt{(-29)^2 + 21^2} \\ &= \sqrt{1282} \\ &\approx 35.8 \end{aligned}$$

Now Try Exercises 9 and 11.

Before we state the properties of vector addition and scalar multiplication, we must define another special vector—the zero vector. The vector whose initial and terminal points are both $(0, 0)$ is the **zero vector**, denoted by **O**, or $\langle 0, 0 \rangle$. Its magnitude is 0. In vector addition, the zero vector is the additive identity vector:

$$\mathbf{v} + \mathbf{O} = \mathbf{v}. \qquad \langle v_1, v_2 \rangle + \langle 0, 0 \rangle = \langle v_1, v_2 \rangle$$

Operations on vectors share many of the same properties as operations on real numbers.

PROPERTIES OF VECTOR ADDITION AND SCALAR MULTIPLICATION

For all vectors **u**, **v**, and **w**, and for all scalars b and c:

1. $\mathbf{u} + \mathbf{v} = \mathbf{v} + \mathbf{u}$.
2. $\mathbf{u} + (\mathbf{v} + \mathbf{w}) = (\mathbf{u} + \mathbf{v}) + \mathbf{w}$.
3. $\mathbf{v} + \mathbf{O} = \mathbf{v}$.
4. $1\mathbf{v} = \mathbf{v}$; $0\mathbf{v} = \mathbf{O}$.
5. $\mathbf{v} + (-\mathbf{v}) = \mathbf{O}$.
6. $b(c\mathbf{v}) = (bc)\mathbf{v}$.
7. $(b + c)\mathbf{v} = b\mathbf{v} + c\mathbf{v}$.
8. $b(\mathbf{u} + \mathbf{v}) = b\mathbf{u} + b\mathbf{v}$.

Unit Vectors

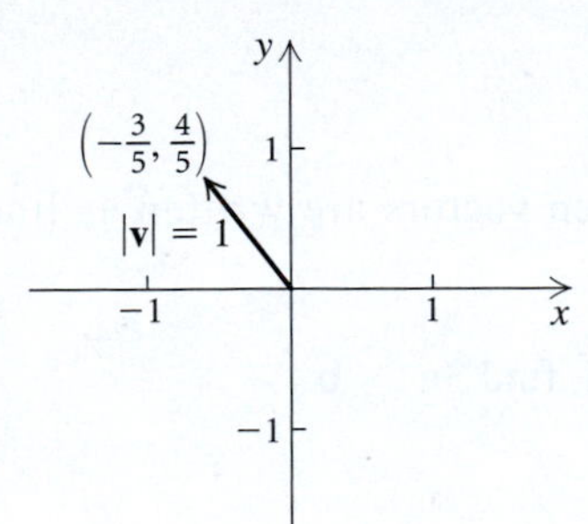

A vector of magnitude, or length, 1 is called a **unit vector**. The vector $\mathbf{v} = \left\langle -\frac{3}{5}, \frac{4}{5} \right\rangle$ is a unit vector because

$$\begin{aligned} |\mathbf{v}| &= \left|\left\langle -\tfrac{3}{5}, \tfrac{4}{5} \right\rangle\right| = \sqrt{\left(-\tfrac{3}{5}\right)^2 + \left(\tfrac{4}{5}\right)^2} \\ &= \sqrt{\tfrac{9}{25} + \tfrac{16}{25}} \\ &= \sqrt{\tfrac{25}{25}} \\ &= \sqrt{1} = 1. \end{aligned}$$

EXAMPLE 5 Find a unit vector that has the same direction as the vector $\mathbf{w} = \langle -3, 5 \rangle$.

Solution We first find the length of $\mathbf{w}$:

$$|\mathbf{w}| = \sqrt{(-3)^2 + 5^2} = \sqrt{34}.$$

Thus we want a vector whose length is $1/\sqrt{34}$ of $\mathbf{w}$ and whose direction is the same as vector $\mathbf{w}$. That vector is

$$\mathbf{u} = \frac{1}{\sqrt{34}}\mathbf{w} = \frac{1}{\sqrt{34}}\langle -3, 5 \rangle = \left\langle \frac{-3}{\sqrt{34}}, \frac{5}{\sqrt{34}} \right\rangle.$$

The vector $\mathbf{u}$ is a *unit vector* because

$$\begin{aligned} |\mathbf{u}| &= \left|\frac{1}{\sqrt{34}}\mathbf{w}\right| = \sqrt{\left(\frac{-3}{\sqrt{34}}\right)^2 + \left(\frac{5}{\sqrt{34}}\right)^2} = \sqrt{\frac{9}{34} + \frac{25}{34}} \\ &= \sqrt{\frac{34}{34}} = \sqrt{1} = 1. \end{aligned}$$

Now Try Exercise 33.

> **UNIT VECTOR**
>
> If $\mathbf{v}$ is a vector and $\mathbf{v} \neq \mathbf{O}$, then
>
> $$\frac{1}{|\mathbf{v}|} \cdot \mathbf{v}, \quad \text{or} \quad \frac{\mathbf{v}}{|\mathbf{v}|},$$
>
> is a **unit vector** in the direction of $\mathbf{v}$.

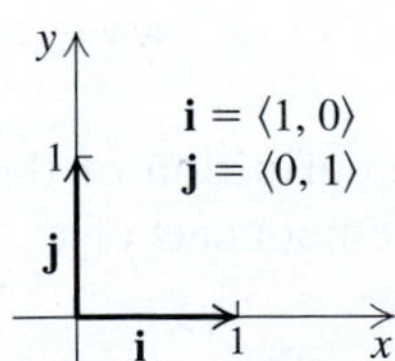

Although unit vectors can have any direction, the unit vectors parallel to the x- and y-axes are particularly useful. They are defined as

$$\mathbf{i} = \langle 1, 0 \rangle \quad \text{and} \quad \mathbf{j} = \langle 0, 1 \rangle.$$

Any vector can be expressed as a **linear combination** of unit vectors $\mathbf{i}$ and $\mathbf{j}$. For example, let $\mathbf{v} = \langle v_1, v_2 \rangle$. Then

$$\begin{aligned} \mathbf{v} = \langle v_1, v_2 \rangle &= \langle v_1, 0 \rangle + \langle 0, v_2 \rangle \\ &= v_1\langle 1, 0 \rangle + v_2\langle 0, 1 \rangle = v_1\mathbf{i} + v_2\mathbf{j}. \end{aligned}$$

EXAMPLE 6 Express the vector $\mathbf{r} = \langle 2, -6 \rangle$ as a linear combination of $\mathbf{i}$ and $\mathbf{j}$.

Solution We have

$$\mathbf{r} = \langle 2, -6 \rangle = 2\mathbf{i} + (-6)\mathbf{j} = 2\mathbf{i} - 6\mathbf{j}.$$

Now Try Exercise 39.

EXAMPLE 7 Write the vector $\mathbf{q} = -\mathbf{i} + 7\mathbf{j}$ in component form.

Solution We have

$$\mathbf{q} = -\mathbf{i} + 7\mathbf{j} = -1\mathbf{i} + 7\mathbf{j} = \langle -1, 7 \rangle.$$

Vector operations can also be performed when vectors are written as linear combinations of **i** and **j**.

EXAMPLE 8 If $\mathbf{a} = 5\mathbf{i} - 2\mathbf{j}$ and $\mathbf{b} = -\mathbf{i} + 8\mathbf{j}$, find $3\mathbf{a} - \mathbf{b}$.

Solution We have

$$\begin{aligned} 3\mathbf{a} - \mathbf{b} &= 3(5\mathbf{i} - 2\mathbf{j}) - (-\mathbf{i} + 8\mathbf{j}) \\ &= 15\mathbf{i} - 6\mathbf{j} + \mathbf{i} - 8\mathbf{j} \\ &= 16\mathbf{i} - 14\mathbf{j}. \end{aligned}$$

Now Try Exercise 45.

Direction Angles

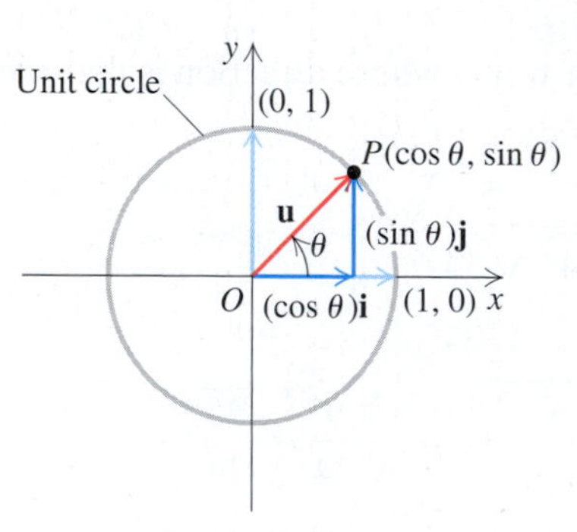

$\mathbf{u} = \langle \cos\theta, \sin\theta \rangle$
$= (\cos\theta)\mathbf{i} + (\sin\theta)\mathbf{j}$

UNIT CIRCLE
REVIEW SECTION 6.5.

The terminal point P of a unit vector in standard position is a point on the unit circle denoted by $(\cos\theta, \sin\theta)$. Thus the unit vector can be expressed in component form,

$$\mathbf{u} = \langle \cos\theta, \sin\theta \rangle,$$

or as a linear combination of the unit vectors **i** and **j**,

$$\mathbf{u} = (\cos\theta)\mathbf{i} + (\sin\theta)\mathbf{j},$$

where the components of **u** are functions of the **direction angle** θ measured counterclockwise from the x-axis to the vector. As θ varies from 0 to 2π, the point P traces the circle $x^2 + y^2 = 1$. This takes in all possible directions for unit vectors so the equation $\mathbf{u} = (\cos\theta)\mathbf{i} + (\sin\theta)\mathbf{j}$ describes every possible unit vector in the plane.

EXAMPLE 9 Calculate and sketch the unit vector

$$\mathbf{u} = (\cos\theta)\mathbf{i} + (\sin\theta)\mathbf{j}$$

for $\theta = 2\pi/3$. Include the unit circle in your sketch.

Solution We have

$$\begin{aligned} \mathbf{u} &= \left(\cos\frac{2\pi}{3}\right)\mathbf{i} + \left(\sin\frac{2\pi}{3}\right)\mathbf{j} \\ &= \left(-\frac{1}{2}\right)\mathbf{i} + \left(\frac{\sqrt{3}}{2}\right)\mathbf{j}. \end{aligned}$$

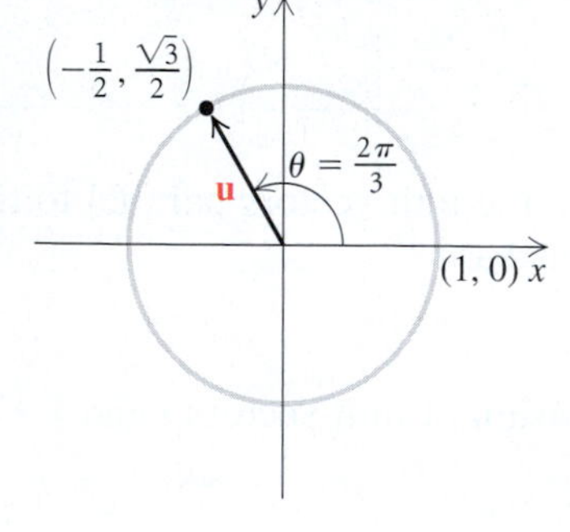

Now Try Exercise 49.

Let $\mathbf{v} = \langle v_1, v_2 \rangle$ with direction angle θ. Using the definition of the tangent function, we can determine the direction angle from the components of **v**:

$\mathbf{v} = \langle v_1, v_2 \rangle$
$= v_1\mathbf{i} + v_2\mathbf{j}$

$$\tan\theta = \frac{v_2}{v_1}$$

$$\theta = \tan^{-1}\frac{v_2}{v_1}.$$

EXAMPLE 10 Determine the direction angle θ of the vector $\mathbf{w} = -4\mathbf{i} - 3\mathbf{j}$.

Solution We know that

$$\mathbf{w} = -4\mathbf{i} - 3\mathbf{j} = \langle -4, -3 \rangle.$$

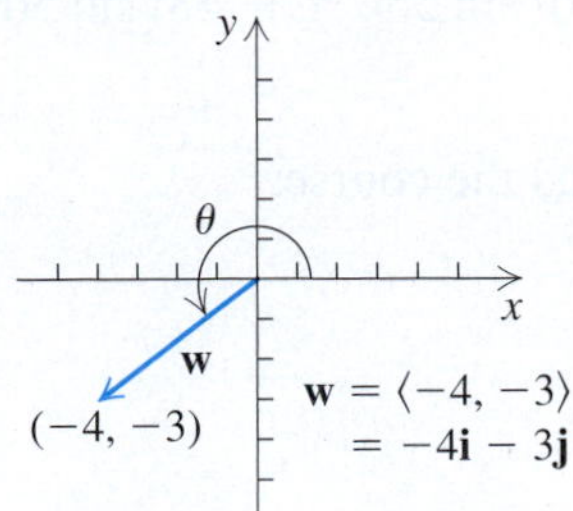

Thus we have

$$\tan\theta = \frac{-3}{-4} = \frac{3}{4} \quad \text{and} \quad \theta = \tan^{-1}\frac{3}{4}.$$

Since $\mathbf{w}$ is in the third quadrant, we know that θ is a third-quadrant angle. The reference angle is

$$\tan^{-1}\frac{3}{4} \approx 37^\circ, \quad \text{and} \quad \theta \approx 180^\circ + 37^\circ, \text{ or } 217^\circ.$$

Now Try Exercise 55.

It is convenient for work with applied problems and in subsequent courses, such as calculus, to have a way to express a vector so that both its magnitude and its direction can be determined, or read, easily. Let $\mathbf{v}$ be a vector. Then $\mathbf{v}/|\mathbf{v}|$ is a unit vector in the same direction as $\mathbf{v}$. Thus we have

$$\frac{\mathbf{v}}{|\mathbf{v}|} = (\cos\theta)\mathbf{i} + (\sin\theta)\mathbf{j}$$

$$\mathbf{v} = |\mathbf{v}|[(\cos\theta)\mathbf{i} + (\sin\theta)\mathbf{j}] \quad \text{Multiplying by } |\mathbf{v}|$$

$$= |\mathbf{v}|(\cos\theta)\mathbf{i} + |\mathbf{v}|(\sin\theta)\mathbf{j}.$$

Let's revisit the applied problem in Example 3 of Section 8.5 and use this new notation.

EXAMPLE 11 ***Airplane Speed and Direction.*** An airplane travels on a bearing of 100° at an airspeed of 190 km/h while a wind is blowing 48 km/h from 220°. Find the ground speed of the airplane and the direction of its track, or course, over the ground.

Solution We first make a drawing. The wind is represented by $\overrightarrow{OC}$ and the velocity vector of the airplane by $\overrightarrow{OA}$. The resultant velocity vector is $\mathbf{v}$, the sum of the two vectors:

$$\mathbf{v} = \overrightarrow{OC} + \overrightarrow{OA}.$$

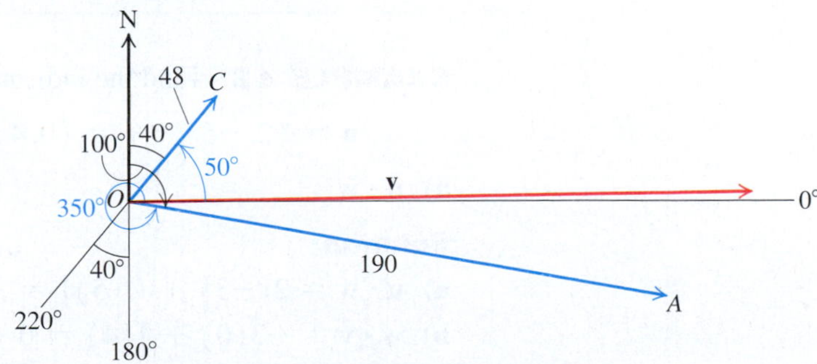

The bearing (measured from north) of the airspeed vector $\overrightarrow{OA}$ is 100°. Its *direction angle* (measured counterclockwise from the positive x-axis) is 350°. The bearing (measured from north) of the wind vector $\overrightarrow{OC}$ is 40°. Its direction angle (measured counterclockwise from the positive x-axis) is 50°. The magnitudes of $\overrightarrow{OA}$ and $\overrightarrow{OC}$ are 190 and 48, respectively. We have

$$\overrightarrow{OA} = 190(\cos 350^\circ)\mathbf{i} + 190(\sin 350^\circ)\mathbf{j}, \quad \text{and}$$

$$\overrightarrow{OC} = 48(\cos 50^\circ)\mathbf{i} + 48(\sin 50^\circ)\mathbf{j}.$$

Thus,

$$\begin{aligned}\mathbf{v} &= \overrightarrow{OA} + \overrightarrow{OC} \\ &= [190(\cos 350°)\mathbf{i} + 190(\sin 350°)\mathbf{j}] + [48(\cos 50°)\mathbf{i} + 48(\sin 50°)\mathbf{j}] \\ &= [190(\cos 350°) + 48(\cos 50°)]\mathbf{i} + [190(\sin 350°) + 48(\sin 50°)]\mathbf{j} \\ &\approx 217.97\mathbf{i} + 3.78\mathbf{j}.\end{aligned}$$

From this form, we can determine the ground speed and the course:

$$\begin{aligned}\text{Ground speed} &\approx \sqrt{(217.97)^2 + (3.78)^2} \\ &\approx 218 \text{ km/h}.\end{aligned}$$

We let α be the direction angle of **v**. Then

$$\tan \alpha = \frac{3.78}{217.97}$$

$$\alpha = \tan^{-1}\frac{3.78}{217.97} \approx 1°.$$

Thus the course of the airplane (the direction from north) is $90° - 1°$, or $89°$.

Now Try Exercise 79.

Angle Between Vectors

When a vector is multiplied by a scalar, the result is a vector. When two vectors are added, the result is also a vector. Thus we might expect the product of two vectors to be a vector as well, but it is not. The *dot product* of two vectors is a real number, or scalar. This product is useful in finding the angle between two vectors and in determining whether two vectors are perpendicular.

DOT PRODUCT

The **dot product** of two vectors $\mathbf{u} = \langle u_1, u_2 \rangle$ and $\mathbf{v} = \langle v_1, v_2 \rangle$ is

$$\mathbf{u} \cdot \mathbf{v} = u_1 v_1 + u_2 v_2.$$

(Note that $u_1 v_1 + u_2 v_2$ is a *scalar*, not a vector.)

EXAMPLE 12 Find the indicated dot product when

$$\mathbf{u} = \langle 2, -5 \rangle, \quad \mathbf{v} = \langle 0, 4 \rangle, \quad \text{and} \quad \mathbf{w} = \langle -3, 1 \rangle.$$

a) $\mathbf{u} \cdot \mathbf{w}$ **b)** $\mathbf{w} \cdot \mathbf{v}$

Solution

a) $\mathbf{u} \cdot \mathbf{w} = 2(-3) + (-5)1 = -6 - 5 = -11$

b) $\mathbf{w} \cdot \mathbf{v} = -3(0) + 1(4) = 0 + 4 = 4$

The dot product can be used to find the angle between two vectors. The angle *between* two vectors is the smallest positive angle formed by the two directed line segments. Thus the angle θ between **u** and **v** is the same angle as that between **v** and **u**, and $0 \le \theta \le \pi$.

ANGLE BETWEEN TWO VECTORS

If θ is the angle between two *nonzero* vectors $\mathbf{u}$ and $\mathbf{v}$, then

$$\cos\theta = \frac{\mathbf{u}\cdot\mathbf{v}}{|\mathbf{u}||\mathbf{v}|}.$$

EXAMPLE 13 Find the angle between $\mathbf{u} = \langle 3, 7\rangle$ and $\mathbf{v} = \langle -4, 2\rangle$.

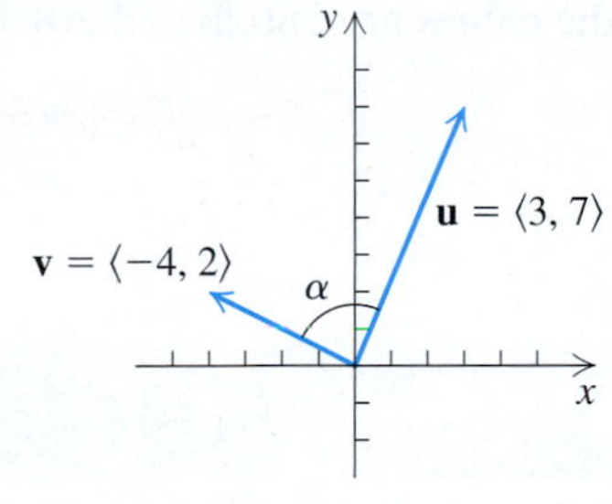

Solution We begin by finding $\mathbf{u}\cdot\mathbf{v}$, $|\mathbf{u}|$, and $|\mathbf{v}|$:

$$\mathbf{u}\cdot\mathbf{v} = 3(-4) + 7(2) = 2,$$
$$|\mathbf{u}| = \sqrt{3^2 + 7^2} = \sqrt{58}, \quad\text{and}$$
$$|\mathbf{v}| = \sqrt{(-4)^2 + 2^2} = \sqrt{20}.$$

Then

$$\cos\alpha = \frac{\mathbf{u}\cdot\mathbf{v}}{|\mathbf{u}||\mathbf{v}|} = \frac{2}{\sqrt{58}\sqrt{20}}$$
$$\alpha = \cos^{-1}\frac{2}{\sqrt{58}\sqrt{20}}$$
$$\approx 86.6^\circ.$$

Now Try Exercise 63.

Forces in Equilibrium

When several forces act through the same point on an object, their vector sum must be $\mathbf{O}$ in order for a balance to occur. When a balance occurs, then the object is either stationary or moving in a straight line without acceleration. The fact that the vector sum must be $\mathbf{O}$ for a balance, and vice versa, allows us to solve many applied problems involving forces.

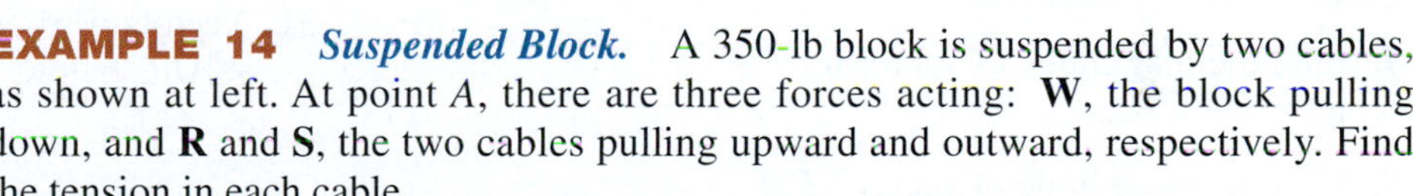

EXAMPLE 14 ***Suspended Block.*** A 350-lb block is suspended by two cables, as shown at left. At point A, there are three forces acting: $\mathbf{W}$, the block pulling down, and $\mathbf{R}$ and $\mathbf{S}$, the two cables pulling upward and outward, respectively. Find the tension in each cable.

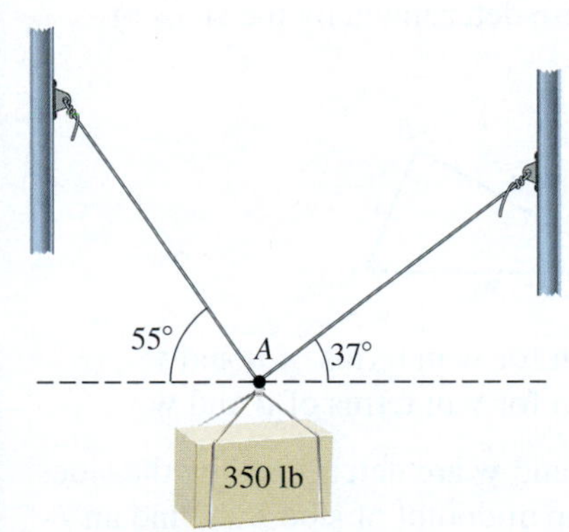

Solution We draw a force diagram with the initial points of each vector at the origin. In order for there to be a balance, the vector sum must be the vector $\mathbf{O}$:

$$\mathbf{R} + \mathbf{S} + \mathbf{W} = \mathbf{O}.$$

We can express each vector in terms of its magnitude and its direction angle:

$$\mathbf{R} = |\mathbf{R}|[(\cos 125^\circ)\mathbf{i} + (\sin 125^\circ)\mathbf{j}],$$
$$\mathbf{S} = |\mathbf{S}|[(\cos 37^\circ)\mathbf{i} + (\sin 37^\circ)\mathbf{j}], \quad\text{and}$$
$$\mathbf{W} = |\mathbf{W}|[(\cos 270^\circ)\mathbf{i} + (\sin 270^\circ)\mathbf{j}]$$
$$= 350(\cos 270^\circ)\mathbf{i} + 350(\sin 270^\circ)\mathbf{j}$$
$$= -350\mathbf{j}. \qquad \cos 270^\circ = 0;\ \sin 270^\circ = -1$$

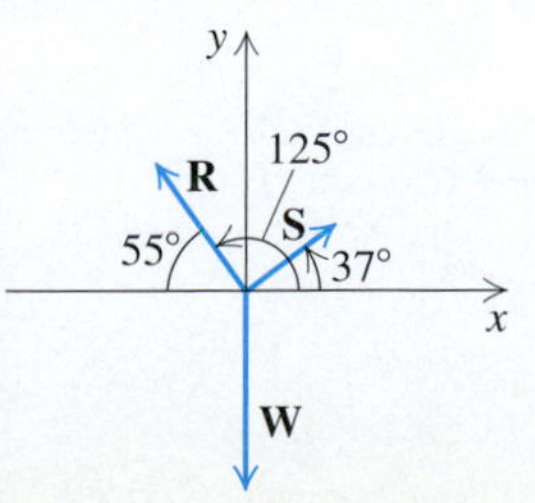

Substituting for $\mathbf{R}$, $\mathbf{S}$, and $\mathbf{W}$ in $\mathbf{R} + \mathbf{S} + \mathbf{W} = \mathbf{O}$, we have

$$[|\mathbf{R}|(\cos 125^\circ) + |\mathbf{S}|(\cos 37^\circ)]\mathbf{i} + [|\mathbf{R}|(\sin 125^\circ) + |\mathbf{S}|(\sin 37^\circ) - 350]\mathbf{j} = 0\mathbf{i} + 0\mathbf{j}.$$

This gives us two equations:

$$|\mathbf{R}|(\cos 125^\circ) + |\mathbf{S}|(\cos 37^\circ) = 0 \quad\text{and} \qquad (1)$$
$$|\mathbf{R}|(\sin 125^\circ) + |\mathbf{S}|(\sin 37^\circ) - 350 = 0. \qquad (2)$$

Solving equation (1) for $|\mathbf{R}|$, we get

$$|\mathbf{R}| = -\frac{|\mathbf{S}|(\cos 37°)}{\cos 125°}. \quad (3)$$

Substituting this expression for $|\mathbf{R}|$ in equation (2) gives us

$$-\frac{|\mathbf{S}|(\cos 37°)}{\cos 125°}(\sin 125°) + |\mathbf{S}|(\sin 37°) - 350 = 0.$$

Then solving this equation for $|\mathbf{S}|$, we get $|\mathbf{S}| \approx 201$, and substituting 201 for $|\mathbf{S}|$ in equation (3), we get $|\mathbf{R}| \approx 280$. The tensions in the cables are 280 lb and 201 lb.

Now Try Exercise 83.

8.6 Exercise Set

Find the component form of the vector given the initial point and the terminal point. Then find the length of the vector.

1. $\overrightarrow{MN}$; $M(6, -7), N(-3, -2)$
2. $\overrightarrow{CD}$; $C(1, 5), D(5, 7)$
3. $\overrightarrow{FE}$; $E(8, 4), F(11, -2)$
4. $\overrightarrow{BA}$; $A(9, 0), B(9, 7)$
5. $\overrightarrow{KL}$; $K(4, -3), L(8, -3)$
6. $\overrightarrow{GH}$; $G(-6, 10), H(-3, 2)$
7. Find the magnitude of vector **u** if $\mathbf{u} = \langle -1, 6 \rangle$.
8. Find the magnitude of vector $\overrightarrow{ST}$ if $\overrightarrow{ST} = \langle -12, 5 \rangle$.

Do the indicated calculations in Exercises 9–26 for the vectors

$\mathbf{u} = \langle 5, -2 \rangle$, $\mathbf{v} = \langle -4, 7 \rangle$, and $\mathbf{w} = \langle -1, -3 \rangle$.

9. $\mathbf{u} + \mathbf{w}$
10. $\mathbf{w} + \mathbf{u}$
11. $|3\mathbf{w} - \mathbf{v}|$
12. $6\mathbf{v} + 5\mathbf{u}$
13. $\mathbf{v} - \mathbf{u}$
14. $|2\mathbf{w}|$
15. $5\mathbf{u} - 4\mathbf{v}$
16. $-5\mathbf{v}$
17. $|3\mathbf{u}| - |\mathbf{v}|$
18. $|\mathbf{v}| + |\mathbf{u}|$
19. $\mathbf{v} + \mathbf{u} + 2\mathbf{w}$
20. $\mathbf{w} - (\mathbf{u} + 4\mathbf{v})$
21. $2\mathbf{v} + \mathbf{O}$
22. $10|7\mathbf{w} - 3\mathbf{u}|$
23. $\mathbf{u} \cdot \mathbf{w}$
24. $\mathbf{w} \cdot \mathbf{u}$
25. $\mathbf{u} \cdot \mathbf{v}$
26. $\mathbf{v} \cdot \mathbf{w}$

The vectors **u**, **v**, *and* **w** *are drawn below. Copy them on a sheet of paper. Then sketch each of the vectors in Exercises 27–30.*

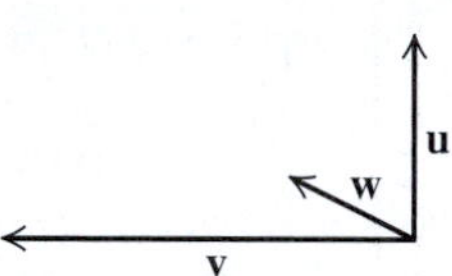

27. $\mathbf{u} + \mathbf{v}$
28. $\mathbf{u} - 2\mathbf{v}$
29. $\mathbf{u} + \mathbf{v} + \mathbf{w}$
30. $\frac{1}{2}\mathbf{u} - \mathbf{w}$
31. Vectors **u**, **v**, and **w** are determined by the sides of $\triangle ABC$ below.

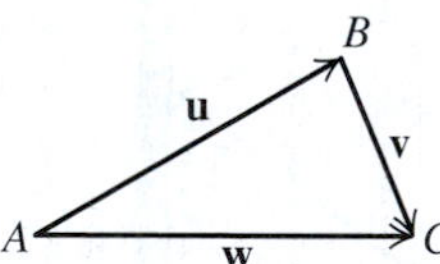

a) Find an expression for **w** in terms of **u** and **v**.
b) Find an expression for **v** in terms of **u** and **w**.

32. In $\triangle ABC$, vectors **u** and **w** are determined by the sides shown, where P is the midpoint of side BC. Find an expression for **v** in terms of **u** and **w**.

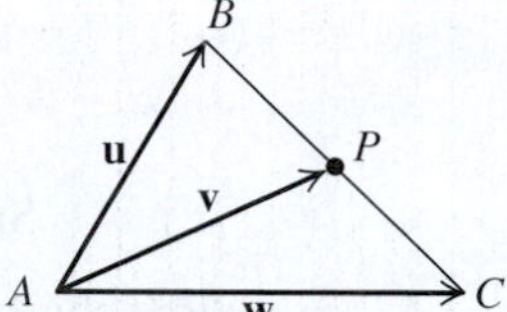

Find a unit vector that has the same direction as the given vector.

33. $\mathbf{v} = \langle -5, 12 \rangle$

34. $\mathbf{u} = \langle 3, 4 \rangle$

35. $\mathbf{w} = \langle 1, -10 \rangle$

36. $\mathbf{a} = \langle 6, -7 \rangle$

37. $\mathbf{r} = \langle -2, -8 \rangle$

38. $\mathbf{t} = \langle -3, -3 \rangle$

Express the vector as a linear combination of the unit vectors **i** *and* **j**.

39. $\mathbf{w} = \langle -4, 6 \rangle$

40. $\mathbf{r} = \langle -15, 9 \rangle$

41. $\mathbf{s} = \langle 2, 5 \rangle$

42. $\mathbf{u} = \langle 2, -1 \rangle$

Express the vector as a linear combination of **i** *and* **j**.

43.

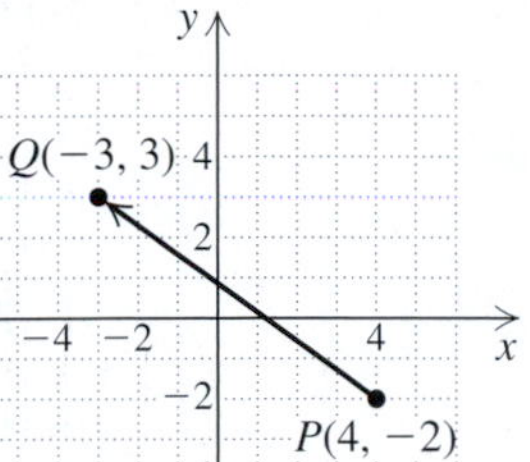

44.

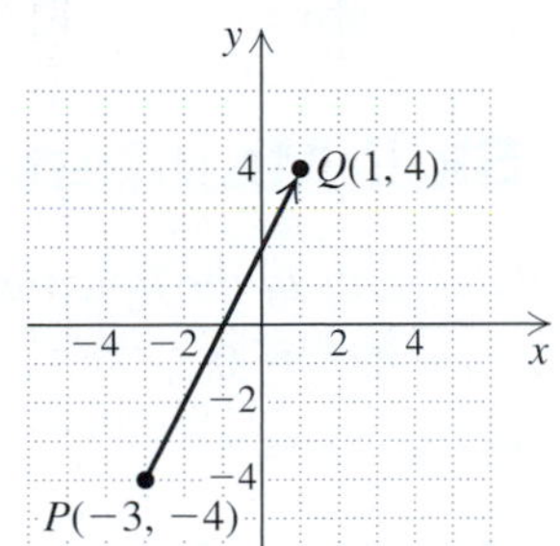

For Exercises 45–48, use the vectors

$$\mathbf{u} = 2\mathbf{i} + \mathbf{j}, \quad \mathbf{v} = -3\mathbf{i} - 10\mathbf{j}, \quad \text{and} \quad \mathbf{w} = \mathbf{i} - 5\mathbf{j}.$$

Perform the indicated vector operations and state the answer in two forms: **(a)** *as a linear combination of* **i** *and* **j** *and* **(b)** *in component form.*

45. $4\mathbf{u} - 5\mathbf{w}$

46. $\mathbf{v} + 3\mathbf{w}$

47. $\mathbf{u} - (\mathbf{v} + \mathbf{w})$

48. $(\mathbf{u} - \mathbf{v}) + \mathbf{w}$

Sketch (include the unit circle) and calculate the unit vector $\mathbf{u} = (\cos\theta)\mathbf{i} + (\sin\theta)\mathbf{j}$ *for the given direction angle.*

49. $\theta = \dfrac{\pi}{2}$

50. $\theta = \dfrac{\pi}{3}$

51. $\theta = \dfrac{4\pi}{3}$

52. $\theta = \dfrac{3\pi}{2}$

Determine the direction angle θ *of the vector, to the nearest degree.*

53. $\mathbf{u} = \langle -2, -5 \rangle$

54. $\mathbf{w} = \langle 4, -3 \rangle$

55. $\mathbf{q} = \mathbf{i} + 2\mathbf{j}$

56. $\mathbf{w} = 5\mathbf{i} - \mathbf{j}$

57. $\mathbf{t} = \langle 5, 6 \rangle$

58. $\mathbf{b} = \langle -8, -4 \rangle$

Find the magnitude and the direction angle θ *of the vector.*

59. $\mathbf{u} = 3[(\cos 45°)\mathbf{i} + (\sin 45°)\mathbf{j}]$

60. $\mathbf{w} = 6[(\cos 150°)\mathbf{i} + (\sin 150°)\mathbf{j}]$

61. $\mathbf{v} = \left\langle -\dfrac{1}{2}, \dfrac{\sqrt{3}}{2} \right\rangle$

62. $\mathbf{u} = -\mathbf{i} - \mathbf{j}$

Find the angle between the given vectors, to the nearest tenth of a degree.

63. $\mathbf{u} = \langle 2, -5 \rangle$, $\mathbf{v} = \langle 1, 4 \rangle$

64. $\mathbf{a} = \langle -3, -3 \rangle$, $\mathbf{b} = \langle -5, 2 \rangle$

65. $\mathbf{w} = \langle 3, 5 \rangle$, $\mathbf{r} = \langle 5, 5 \rangle$

66. $\mathbf{v} = \langle -4, 2 \rangle$, $\mathbf{t} = \langle 1, -4 \rangle$

67. $\mathbf{a} = \mathbf{i} + \mathbf{j}$, $\mathbf{b} = 2\mathbf{i} - 3\mathbf{j}$

68. $\mathbf{u} = 3\mathbf{i} + 2\mathbf{j}$, $\mathbf{v} = -\mathbf{i} + 4\mathbf{j}$

Express each vector in Exercises 69–72 in the form $a\mathbf{i} + b\mathbf{j}$ *and sketch each in the coordinate plane.*

69. The unit vectors $\mathbf{u} = (\cos\theta)\mathbf{i} + (\sin\theta)\mathbf{j}$ for $\theta = \pi/6$ and $\theta = 3\pi/4$. Include the unit circle $x^2 + y^2 = 1$ in your sketch.

70. The unit vectors $\mathbf{u} = (\cos\theta)\mathbf{i} + (\sin\theta)\mathbf{j}$ for $\theta = -\pi/4$ and $\theta = -3\pi/4$. Include the unit circle $x^2 + y^2 = 1$ in your sketch.

71. The unit vector obtained by rotating **j** counterclockwise $3\pi/4$ radians about the origin

72. The unit vector obtained by rotating **j** clockwise $2\pi/3$ radians about the origin

For the vectors in Exercises 73 and 74, find the unit vectors $\mathbf{u} = (\cos\theta)\mathbf{i} + (\sin\theta)\mathbf{j}$ *in the same direction.*

73. $-\mathbf{i} + 3\mathbf{j}$

74. $6\mathbf{i} - 8\mathbf{j}$

For the vectors in Exercises 75 and 76, express each vector in terms of its magnitude and its direction.

75. $2\mathbf{i} - 3\mathbf{j}$

76. $5\mathbf{i} + 12\mathbf{j}$

77. Use a sketch to show that

$$\mathbf{v} = 3\mathbf{i} - 6\mathbf{j} \quad \text{and} \quad \mathbf{u} = -\mathbf{i} + 2\mathbf{j}$$

have opposite directions.

78. Use a sketch to show that

$$\mathbf{v} = 3\mathbf{i} - 6\mathbf{j} \quad \text{and} \quad \mathbf{u} = \tfrac{1}{2}\mathbf{i} - \mathbf{j}$$

have the same direction.

Exercises 79–82 appeared first in Exercise Set 8.5, where we used the law of cosines and the law of sines to solve the applied problems. For this exercise set, solve the problem using the vector form

$$\mathbf{v} = |\mathbf{v}|[(\cos\theta)\mathbf{i} + (\sin\theta)\mathbf{j}].$$

79. *Ship.* A ship first sails N80°E for 120 nautical mi, and then S20°W for 200 nautical mi. How far is the ship, then, from the starting point, and in what direction is the ship moving?

80. *Boat.* A boat heads 35°, propelled by a force of 750 lb. A wind from 320° exerts a force of 150 lb on the boat. How large is the resultant force, and in what direction is the boat moving?

81. *Airplane.* An airplane has an airspeed of 150 km/h. It is to make a flight in a direction of 70° while there is a 25-km/h wind from 340°. What will the airplane's actual heading be?

82. *Airplane.* An airplane flies 032° for 210 mi, and then 280° for 170 mi. How far is the airplane, then, from the starting point, and in what direction is the plane moving?

83. Two cables support a 1000-lb weight, as shown. Find the tension in each cable.

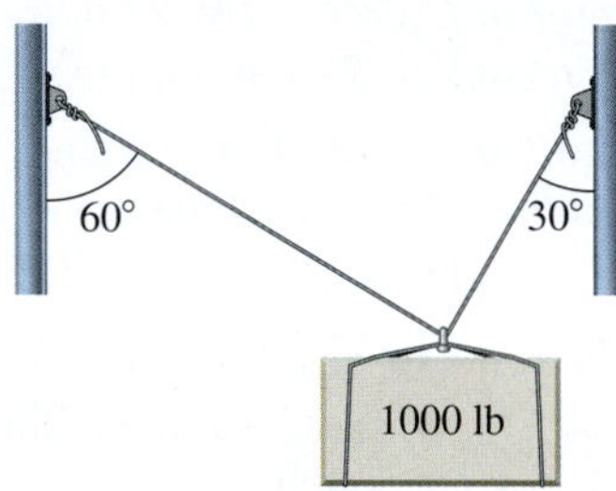

84. A 2500-kg block is suspended by two ropes, as shown. Find the tension in each rope.

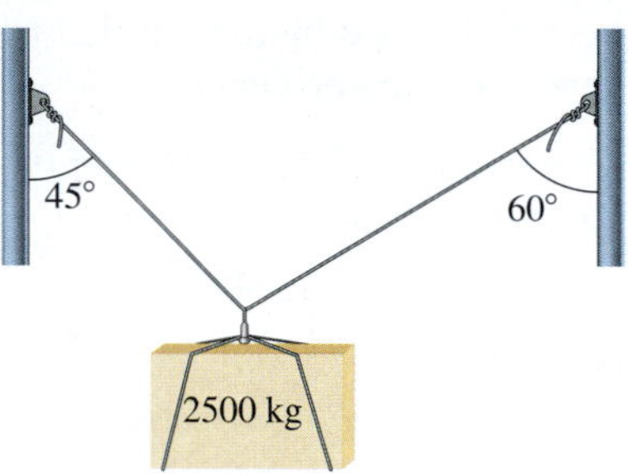

85. A 150-lb sign is hanging from the end of a hinged boom, supported by a cable inclined 42° with the horizontal. Find the tension in the cable and the compression in the boom.

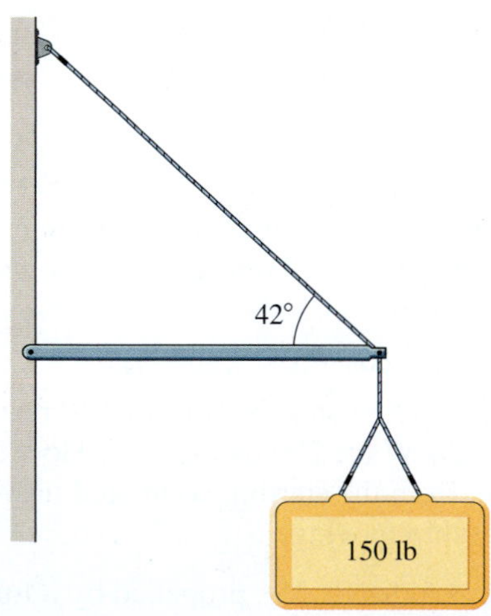

86. A weight of 200 lb is supported by a frame made of two rods and hinged at points A, B, and C. Find the forces exerted by the two rods.

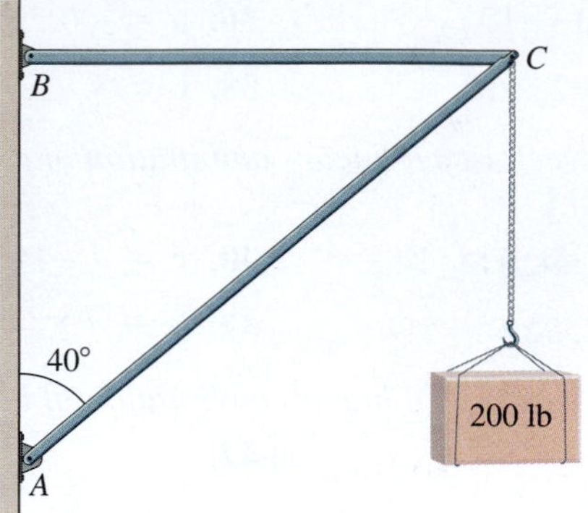

Skill Maintenance

Find the zeros of the function.

87. $x^3 - 4x^2 = 0$ [4.1]

88. $6x^2 + 7x = 55$ [3.2]

Synthesis

89. If the dot product of two nonzero vectors **u** and **v** is 0, then the vectors are perpendicular (**orthogonal**). Let $\mathbf{u} = \langle u_1, u_2 \rangle$ and $\mathbf{v} = \langle v_1, v_2 \rangle$. Prove that if $\mathbf{u} \cdot \mathbf{v} = 0$, then **u** and **v** are perpendicular.

90. If $\overrightarrow{PQ}$ is any vector, what is $\overrightarrow{PQ} + \overrightarrow{QP}$?

91. Find all the unit vectors that are parallel to the vector $\langle 3, -4 \rangle$.

92. Find vector **v** from point A to the origin, where $\overrightarrow{AB} = 4\mathbf{i} - 2\mathbf{j}$ and B is the point $(-2, 5)$.

93. Given the vector $\overrightarrow{AB} = 3\mathbf{i} - \mathbf{j}$ and A is the point $(2, 9)$, find the point B.

Chapter 8 Summary and Review

STUDY GUIDE

KEY TERMS AND CONCEPTS | **EXAMPLES**

SECTION 8.1: THE LAW OF SINES

Solving a Triangle

To **solve a triangle** means to find the lengths of all its sides and the measures of all its angles.

Any triangle, right or oblique, can be solved *if at least one side and any other two measures are known*. We cannot solve a triangle when only the three angle measures are given.

The Law of Sines

In any $\triangle ABC$,

$$\frac{a}{\sin A} = \frac{b}{\sin B} = \frac{c}{\sin C}.$$

The law of sines is used to solve triangles given a side and two angles (AAS and ASA) or given two sides and an angle opposite one of them (SSA). In the SSA situation, there are three possibilities: *no* solution, *one* solution, or *two* solutions.

Solve $\triangle RST$, if $r = 47.6$, $S = 123.5°$, and $T = 31.4°$.

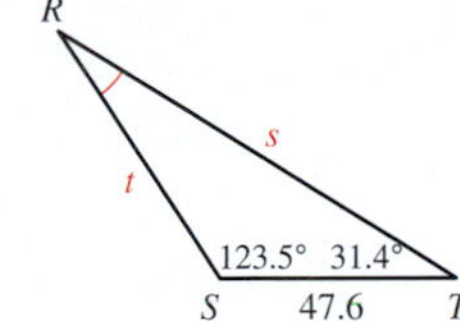

$R = ?,\quad r = 47.6,$
$S = 123.5°,\quad s = ?,$
$T = 31.4°,\quad t = ?$

We have the ASA situation. We first find R:

$$R = 180° - (123.5° + 31.4°) = 25.1°.$$

We then find the other two sides using the law of sines. We have

$$\frac{47.6}{\sin 25.1°} = \frac{s}{\sin 123.5°}. \qquad \text{Using } \frac{r}{\sin R} = \frac{s}{\sin S}$$

Solving for s, we get $s \approx 93.6$. We also have

$$\frac{47.6}{\sin 25.1°} = \frac{t}{\sin 31.4°}. \qquad \text{Using } \frac{r}{\sin R} = \frac{t}{\sin T}$$

Solving for t, we get $t \approx 58.5$.

We have solved the triangle:

$R = 25.1°,\quad r = 47.6,$
$S = 123.5°,\quad s \approx 93.6,$
$T = 31.4°,\quad t \approx 58.5.$

Solve $\triangle DEF$, if $d = 35.6$, $f = 48.1$, and $D = 32.2°$.

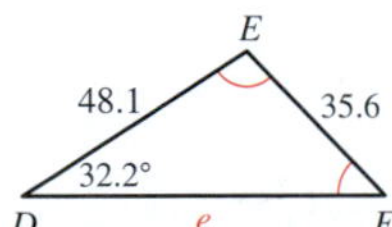

$D = 32.2°,\quad d = 35.6,$
$E = ?,\quad e = ?,$
$F = ?,\quad f = 48.1$

We have the SSA situation. We first find F:

$$\frac{48.1}{\sin F} = \frac{35.6}{\sin 32.2°} \qquad \text{Using } \frac{f}{\sin F} = \frac{d}{\sin D}$$

$$\sin F = \frac{48.1 \sin 32.2°}{35.6} \approx 0.7200.$$

There are two angles less than 180° with a sine of 0.7200. They are 46.1° and 133.9°. This gives us two possible solutions.

(*continued*)

Possible solution I:

If $F \approx 46.1°$, then $E = 180° - (32.2° + 46.1°) = 101.7°$. Then we find e using the law of sines:

$$\frac{e}{\sin 101.7°} = \frac{35.6}{\sin 32.2°}. \quad \text{Using } \frac{e}{\sin E} = \frac{d}{\sin D}$$

Solving for e, we get $e \approx 65.4$. The solution is

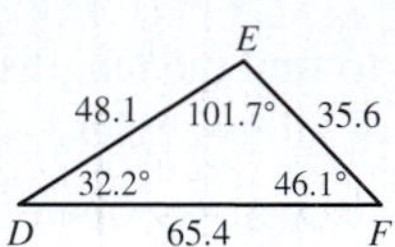

Possible solution II:

If $F \approx 133.9°$, then $E = 180° - (32.2° + 133.9°) = 13.9°$. Then we find e using the law of sines:

$$\frac{e}{\sin 13.9°} = \frac{35.6}{\sin 32.2°}. \quad \text{Using } \frac{e}{\sin E} = \frac{d}{\sin D}$$

Solving for e, we get $e \approx 16.0$. The solution is

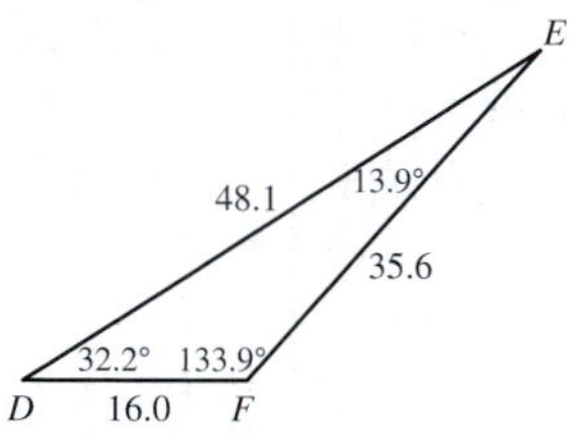

The Area of a Triangle

The area K of any $\triangle ABC$ is one-half of the product of the lengths of two sides and the sine of the included angle:

$$K = \frac{1}{2}bc \sin A = \frac{1}{2}ab \sin C = \frac{1}{2}ac \sin B.$$

Find the area of $\triangle ABC$ if $C = 115°$, $a = 10$ m, and $b = 13$ m.

$$\begin{aligned} K &= \frac{1}{2}ab \sin C \\ &= \frac{1}{2} \cdot 10 \text{ m} \cdot 13 \text{ m} \cdot \sin 115° \\ &\approx 59 \text{ m}^2 \end{aligned}$$

SECTION 8.2: THE LAW OF COSINES

The Law of Cosines

In any $\triangle ABC$,

$$a^2 = b^2 + c^2 - 2bc \cos A,$$
$$b^2 = a^2 + c^2 - 2ac \cos B,$$
$$c^2 = a^2 + b^2 - 2ab \cos C.$$

The law of cosines is used to solve triangles given two sides and the included angle (SAS) or given three sides (SSS).

Solve $\triangle PQR$, if $p = 27$, $r = 39$, and $Q = 110.8°$.

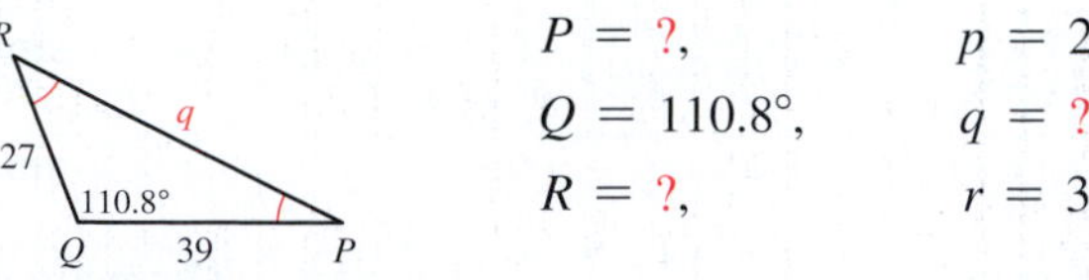

$P = ?$, $p = 27$,
$Q = 110.8°$, $q = ?$,
$R = ?$, $r = 39$

We can find the third side using the law of cosines:

$$\begin{aligned} q^2 &= p^2 + r^2 - 2pr \cos Q \\ q^2 &= 27^2 + 39^2 - 2 \cdot 27 \cdot 39 \cdot \cos 110.8° \\ q^2 &\approx 2997.9 \\ q &\approx 55. \end{aligned}$$

(continued)

We then find angle R using the law of cosines:

$$r^2 = p^2 + q^2 - 2pq \cos R$$
$$39^2 = 27^2 + 55^2 - 2 \cdot 27 \cdot 55 \cdot \cos R$$
$$\cos R = \frac{27^2 + 55^2 - 39^2}{2 \cdot 27 \cdot 55} \approx 0.7519.$$

Since $\cos R$ is positive, R is acute:

$$R \approx 41.2°.$$

Thus angle $P \approx 180° - (41.2° + 110.8°) \approx 28.0°$. We have solved the triangle:

$$P \approx 28.0°, \quad p = 27,$$
$$Q = 110.8°, \quad q \approx 55,$$
$$R \approx 41.2°, \quad r = 39.$$

SECTION 8.3: COMPLEX NUMBERS: TRIGONOMETRIC NOTATION

Complex numbers can be graphed on a plane. We graph $a + bi$ in the same way that we graph an ordered pair of numbers (a, b).

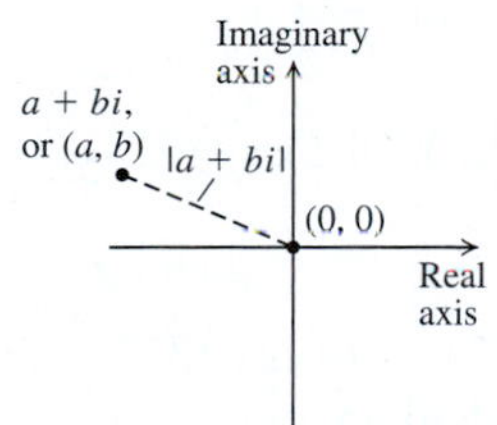

Absolute Value of a Complex Number

$$|a + bi| = \sqrt{a^2 + b^2}$$

Graph the complex numbers $2 - 4i$ and $\frac{7}{3}i$ and find the absolute value of each.

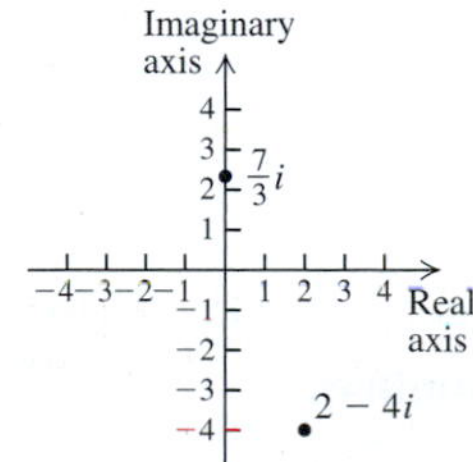

$$|2 - 4i| = \sqrt{2^2 + (-4)^2} = \sqrt{4 + 16} = \sqrt{20} = 2\sqrt{5};$$
$$\left|\frac{7}{3}i\right| = \sqrt{0^2 + \left(\frac{7}{3}\right)^2} = \sqrt{\left(\frac{7}{3}\right)^2} = \frac{7}{3}$$

Trigonometric Notation for Complex Numbers

$$a + bi = r(\cos \theta + i \sin \theta)$$

To find trigonometric notation for a complex number given in *standard notation*, $a + bi$, we find r and determine the angle θ for which $\sin \theta = b/r$ and $\cos \theta = a/r$.

Find trigonometric notation for $-1 + i$.

We have

$a = -1$ and $b = 1$,

$$r = \sqrt{a^2 + b^2} = \sqrt{(-1)^2 + 1^2} = \sqrt{1 + 1} = \sqrt{2},$$
$$\sin \theta = \frac{b}{r} = \frac{1}{\sqrt{2}} = \frac{\sqrt{2}}{2} \quad \text{and} \quad \cos \theta = \frac{a}{r} = \frac{-1}{\sqrt{2}} = -\frac{\sqrt{2}}{2}.$$

Since $\sin \theta$ is positive and $\cos \theta$ is negative, θ is in quadrant II. We have $\theta = \frac{3\pi}{4}$, or 135°, and

$$-1 + i = \sqrt{2}\left(\cos \frac{3\pi}{4} + i \sin \frac{3\pi}{4}\right), \quad \text{or}$$
$$-1 + i = \sqrt{2}(\cos 135° + i \sin 135°).$$

	Find standard notation, $a + bi$, for $2(\cos 210° + i \sin 210°)$. We have $2(\cos 210° + i \sin 210°) = 2 \cos 210° + (2 \sin 210°)i.$ Thus, $a = 2 \cos 210° = 2 \cdot \left(-\frac{\sqrt{3}}{2}\right) = -\sqrt{3},$ and $b = 2 \sin 210° = 2 \cdot \left(-\frac{1}{2}\right) = -1,$ so $2(\cos 210° + i \sin 210°) = -\sqrt{3} - i.$
Multiplication of Complex Numbers $r_1(\cos \theta_1 + i \sin \theta_1) \cdot r_2(\cos \theta_2 + i \sin \theta_2)$ $= r_1 r_2[\cos(\theta_1 + \theta_2) + i \sin(\theta_1 + \theta_2)]$	Multiply $5(\cos \pi + i \sin \pi)$ and $2\left(\cos \frac{\pi}{6} + i \sin \frac{\pi}{6}\right)$ and express the answer in standard notation. $5(\cos \pi + i \sin \pi) \cdot 2\left(\cos \frac{\pi}{6} + i \sin \frac{\pi}{6}\right)$ $= 5 \cdot 2\left[\cos\left(\pi + \frac{\pi}{6}\right) + i \sin\left(\pi + \frac{\pi}{6}\right)\right]$ $= 10\left(\cos \frac{7\pi}{6} + i \sin \frac{7\pi}{6}\right)$ $= 10\left(-\frac{\sqrt{3}}{2} - \frac{1}{2}i\right) = -5\sqrt{3} - 5i$
Division of Complex Numbers $\frac{r_1(\cos \theta_1 + i \sin \theta_1)}{r_2(\cos \theta_2 + i \sin \theta_2)}$ $= \frac{r_1}{r_2}[\cos(\theta_1 - \theta_2) + i \sin(\theta_1 - \theta_2)],$ $r_2 \neq 0$	Divide $3(\cos 315° + i \sin 315°)$ by $6(\cos 135° + i \sin 135°)$ and express the answer in standard notation. $\frac{3(\cos 315° + i \sin 315°)}{6(\cos 135° + i \sin 135°)}$ $= \frac{1}{2}[\cos(315° - 135°) + i \sin(315° - 135°)]$ $= \frac{1}{2}(\cos 180° + i \sin 180°)$ $= \frac{1}{2}(-1 + i \cdot 0) = -\frac{1}{2}$
DeMoivre's Theorem $[r(\cos \theta + i \sin \theta)]^n$ $= r^n(\cos n\theta + i \sin n\theta)$	Find $(-\sqrt{3} - i)^5$. We first find trigonometric notation: $-\sqrt{3} - i = 2(\cos 210° + i \sin 210°).$ Then $(-\sqrt{3} - i)^5 = [2(\cos 210° + i \sin 210°)]^5$ $= 2^5(\cos 1050° + i \sin 1050°)$ $= 32(\cos 330° + i \sin 330°)$ $= 32\left[\frac{\sqrt{3}}{2} - \frac{1}{2}i\right]$ $= 16\sqrt{3} - 16i.$

Roots of Complex Numbers

The nth roots of $r(\cos\theta + i\sin\theta)$ are

$$r^{1/n}\left[\cos\left(\frac{\theta}{n} + k\cdot\frac{360^\circ}{n}\right) + i\sin\left(\frac{\theta}{n} + k\cdot\frac{360^\circ}{n}\right)\right],$$

$r \neq 0, k = 0, 1, 2, \ldots, n - 1.$

Find the cube roots of -8. Then locate them on a graph.

We first find trigonometric notation:

$$-8 = 8(\cos 180^\circ + i\sin 180^\circ).$$

Then $n = 3$, $1/n = 1/3$, and $k = 0, 1, 2$. We have

$$[8(\cos 180^\circ + i\sin 180^\circ)]^{1/3} = 8^{1/3}\left[\cos\left(\frac{180^\circ}{3} + k\cdot\frac{360^\circ}{3}\right) + i\sin\left(\frac{180^\circ}{3} + k\cdot\frac{360^\circ}{3}\right)\right],$$

$k = 0, 1, 2.$

The roots are

$2(\cos 60^\circ + i\sin 60^\circ)$,

$2(\cos 180^\circ + i\sin 180^\circ)$, and

$2(\cos 300^\circ + i\sin 300^\circ)$, or

$1 + \sqrt{3}i$, $\quad -2$, and $1 - \sqrt{3}i$.

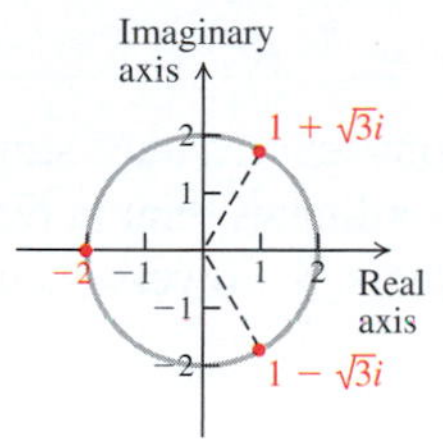

SECTION 8.4: POLAR COORDINATES AND GRAPHS

Plotting Points on a Polar Graph

Any point P has rectangular coordinates (x, y) and polar coordinates (r, θ). To plot points on a polar graph:

1. Locate the direction angle θ.
2. Move a directed distance r from the pole. If $r > 0$, move along ray OP. If $r < 0$, move in the opposite direction of ray OP.

Graph each of the following points:

$A(4, 240^\circ)$,

$B\left(-2, \frac{2\pi}{3}\right)$,

$C(3, -45^\circ)$,

$D(0, \pi)$,

$E\left(-5, \frac{3\pi}{2}\right)$.

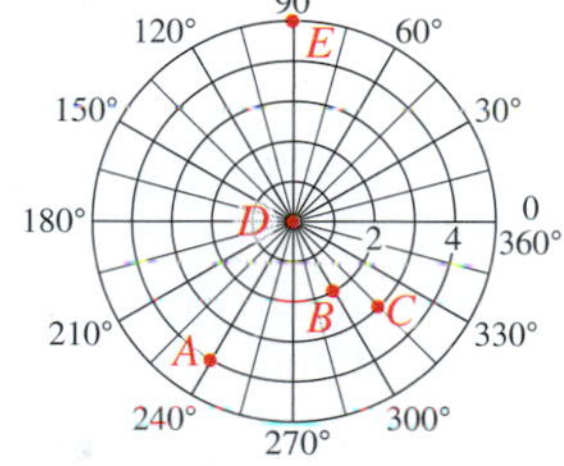

To convert from rectangular coordinates to polar coordinates and from polar coordinates to rectangular coordinates, recall the following relationships:

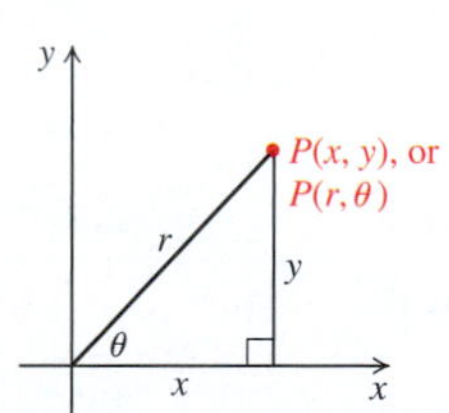

$$r = \sqrt{x^2 + y^2},$$

$$\cos\theta = \frac{x}{r}, \quad \text{or} \quad x = r\cos\theta,$$

$$\sin\theta = \frac{y}{r}, \quad \text{or} \quad y = r\sin\theta,$$

$$\tan\theta = \frac{y}{x}.$$

Convert $(-5, 5\sqrt{3})$ to polar coordinates.

We first find r:

$$r = \sqrt{(-5)^2 + (5\sqrt{3})^2} = \sqrt{25 + 75} = \sqrt{100} = 10.$$

Then we determine θ:

$$\tan\theta = \frac{5\sqrt{3}}{-5} = -\sqrt{3};$$

therefore, $\theta = 120^\circ$, or $\frac{2\pi}{3}$.

Thus, $(r, \theta) = (10, 120^\circ)$, or $(10, 2\pi/3)$.

Other possibilities for polar coordinates for this point include $(10, -4\pi/3)$ and $(-10, 300^\circ)$.

Convert $(4, 210°)$ to rectangular coordinates.

The ordered pair $(4, 210°)$ gives us $r = 4$ and $\theta = 210°$. We now find x and y:

$$x = r\cos\theta = 4\cos 210° = 4\cdot\left(-\frac{\sqrt{3}}{2}\right) = -2\sqrt{3};$$

$$y = r\sin\theta = 4\sin 210° = 4\left(-\frac{1}{2}\right) = -2.$$

Thus, $(x, y) = \left(-2\sqrt{3}, -2\right)$.

Some curves have simpler equations in polar coordinates than in rectangular coordinates. For others, the reverse is true.

Convert each of the following rectangular equations to a polar equation.

a) $x^2 + y^2 = 100$

b) $y - 3x = 11$

a)
$$\begin{aligned} x^2 + y^2 &= 100 \\ (r\cos\theta)^2 + (r\sin\theta)^2 &= 100 \\ r^2\cos^2\theta + r^2\sin^2\theta &= 100 \\ r^2(\cos^2\theta + \sin^2\theta) &= 100 \\ r^2 &= 100 \qquad \cos^2\theta + \sin^2\theta = 1 \\ r &= 10 \end{aligned}$$

b)
$$\begin{aligned} y - 3x &= 11 \\ (r\sin\theta) - 3(r\cos\theta) &= 11 \\ r(\sin\theta - 3\cos\theta) &= 11 \end{aligned}$$

Convert each of the following polar equations to a rectangular equation.

a) $r = 7$

b) $r = 3\sin\theta - 5\cos\theta$

a)
$$\begin{aligned} r &= 7 \\ \sqrt{x^2 + y^2} &= 7 \\ x^2 + y^2 &= 49 \end{aligned}$$

b)
$$\begin{aligned} r &= 3\sin\theta - 5\cos\theta \\ r^2 &= 3r\sin\theta - 5r\cos\theta \qquad \text{Multiplying by } r \text{ on both sides} \\ x^2 + y^2 &= 3y - 5x \end{aligned}$$

To graph a polar equation:

1. Make a table of values, choosing values of θ and calculating corresponding values of r. As θ increases sufficiently, points may begin to repeat.
2. Plot the points and complete the graph.

Graph: $r = 2 - 3 \cos \theta$.

θ	r
0°	−1
15°	−0.8978
30°	−0.5981
45°	−0.1213
60°	0.5
75°	1.2235
90°	2
105°	2.7765
120°	3.5
135°	4.1213
150°	4.5981
165°	4.8978
180°	5

θ	r
195°	4.8978
210°	4.5981
225°	4.1213
240°	3.5
255°	2.7765
270°	2
285°	1.2235
300°	0.5
315°	−0.1213
330°	−0.5981
345°	−0.8978
360°	−1

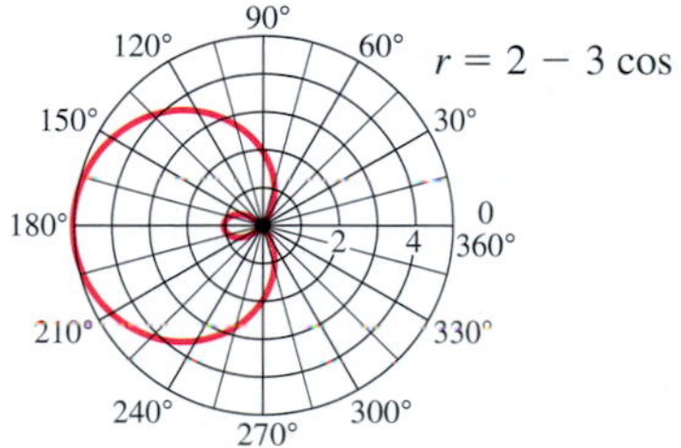

SECTION 8.5: VECTORS AND APPLICATIONS

Vector

A vector in the plane is a directed line segment. Two vectors are equivalent if they have the same *magnitude* and the same *direction*.

The vectors $\mathbf{v}$ and $\overrightarrow{AB}$ are shown in the following figure. Show that $\mathbf{v}$ and $\overrightarrow{AB}$ are equivalent.

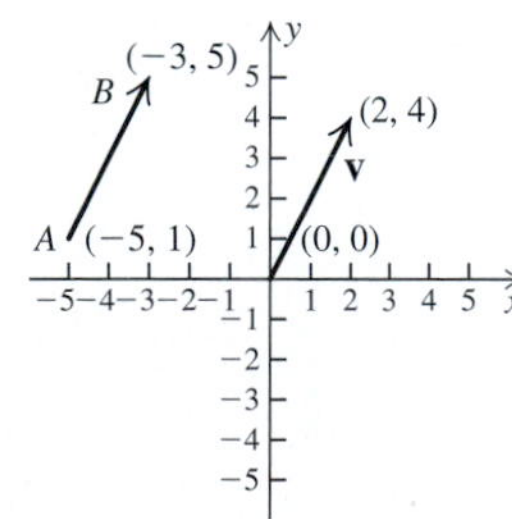

We first find the lengths:

$$|\mathbf{v}| = \sqrt{(2 - 0)^2 + (4 - 0)^2} = \sqrt{2^2 + 4^2} = \sqrt{20};$$

$$|\overrightarrow{AB}| = \sqrt{[-3 - (-5)]^2 + (5 - 1)^2} = \sqrt{2^2 + 4^2} = \sqrt{20}.$$

Thus, $|\mathbf{v}| = |\overrightarrow{AB}|$.

(continued)

The slopes of the lines that the vectors are on are:

$$\text{Slope of } \mathbf{v} = \frac{4-0}{2-0} = \frac{4}{2} = 2;$$

$$\text{Slope of } \overrightarrow{AB} = \frac{5-1}{-3-(-5)} = \frac{4}{2} = 2.$$

Since the slopes are the same, vectors $\mathbf{v}$ and $\overrightarrow{AB}$ have the same direction.

Since $\mathbf{v}$ and $\overrightarrow{AB}$ have the same magnitude and the same direction, $\mathbf{v} = \overrightarrow{AB}$.

If two forces F_1 and F_2 act on an object, the combined effect is the sum, or resultant, $F_1 + F_2$ of the separate forces.

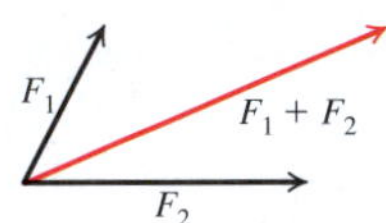

Two forces of 85 N and 120 N act on an object. The angle between the forces is 62°. Find the magnitude of the resultant and the angle that it makes with the larger force.

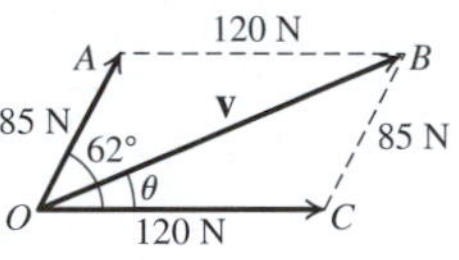

We have

$$\angle A = 180° - 62° = 118°.$$

We use the law of cosines to find the magnitude of the resultant $\mathbf{v}$:

$$|\mathbf{v}|^2 = 120^2 + 85^2 - 2 \cdot 120 \cdot 85 \cdot \cos 118°$$
$$|\mathbf{v}| \approx \sqrt{31{,}202} \approx 177 \text{ N}.$$

We use the law of sines to find θ:

$$\frac{177}{\sin 118°} = \frac{85}{\sin \theta}$$
$$\sin \theta = \frac{85 \sin 118°}{177} \approx 0.4240$$
$$\theta \approx 25°.$$

The magnitude of $\mathbf{v}$ is approximately 177 N, and it makes a 25° angle with the larger force, 120 N.

Horizontal Components and Vertical Components of a Vector

The components of vector $\mathbf{w}$ are vectors $\mathbf{u}$ and $\mathbf{v}$ such that $\mathbf{w} = \mathbf{u} + \mathbf{v}$. We generally look for perpendicular components, with one component parallel to the x-axis (horizontal component) and the other parallel to the y-axis (vertical component).

A vector $\mathbf{w}$ has a magnitude of 200 and is inclined 52° with the horizontal. Resolve the vector into its horizontal component and its vertical component.

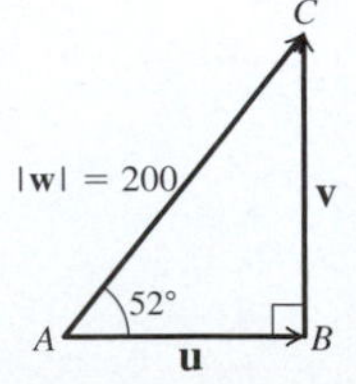

(continued)

From $\triangle ABC$, we find $|\mathbf{u}|$ and $|\mathbf{v}|$:

$$\cos 52^\circ = \frac{|\mathbf{u}|}{200}, \quad \text{or} \quad |\mathbf{u}| = 200 \cos 52^\circ \approx 123;$$

$$\sin 52^\circ = \frac{|\mathbf{v}|}{200}, \quad \text{or} \quad |\mathbf{v}| = 200 \sin 52^\circ \approx 158.$$

Thus the horizontal component of **w** is 123 right, and the vertical component of **w** is 158 up.

Given $|\mathbf{u}| = 16$ and $|\mathbf{v}| = 19$, and that the angle between the vectors is 130°, find the sum $\mathbf{u} + \mathbf{v}$. Give the magnitude to the nearest tenth and give the direction by specifying to the nearest degree the angle that the resultant makes with **u**.

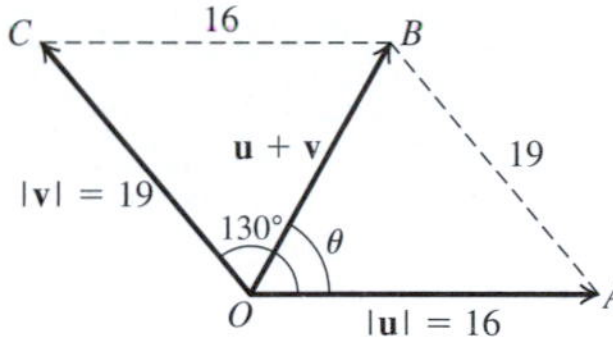

We first find A:

$$A = 180^\circ - 130^\circ = 50^\circ.$$

We use the law of cosines to find $|\mathbf{u} + \mathbf{v}|$:

$$|\mathbf{u} + \mathbf{v}|^2 = 19^2 + 16^2 - 2 \cdot 19 \cdot 16 \cos 50^\circ$$
$$|\mathbf{u} + \mathbf{v}| \approx \sqrt{226.19} \approx 15.0.$$

Next, we use the law of cosines to find θ:

$$19^2 \approx 15^2 + 16^2 - 2 \cdot 15 \cdot 16 \cos \theta$$
$$\cos \theta \approx 0.25$$
$$\theta \approx 76^\circ.$$

The magnitude of $\mathbf{u} + \mathbf{v} \approx 15$, and the angle that $\mathbf{u} + \mathbf{v}$ makes with **u** is about 76°.

SECTION 8.6: VECTOR OPERATIONS

Component Form of a Vector

$$\mathbf{v} = \langle a, b \rangle$$

The coordinate a is the scalar **horizontal component** of the vector, and the coordinate b is the scalar **vertical component**.

The **component form** of $\overrightarrow{AC}$ with $A = (x_1, y_1)$ and $C = (x_2, y_2)$ is

$$\overrightarrow{AC} = \langle x_2 - x_1, y_2 - y_1 \rangle.$$

Find the component form of $\overrightarrow{OR}$ given $O(3, -9)$ and $R(-5, 4)$. Then find the length, or magnitude, of $\overrightarrow{OR}$.

$$\overrightarrow{OR} = \langle -5 - 3, 4 - (-9) \rangle = \langle -8, 13 \rangle;$$
$$|\overrightarrow{OR}| = \sqrt{(-8)^2 + 13^2} = \sqrt{64 + 169} = \sqrt{233} \approx 15.3$$

Vectors

If $\mathbf{u} = \langle u_1, u_2 \rangle$ and $\mathbf{v} = \langle v_1, v_2 \rangle$ and k is a scalar, then:

Length: $|\mathbf{v}| = \sqrt{v_1^2 + v_2^2}$;

Addition: $\mathbf{u} + \mathbf{v} = \langle u_1 + v_1, u_2 + v_2 \rangle$;

Subtraction: $\mathbf{u} - \mathbf{v} = \langle u_1 - v_1, u_2 - v_2 \rangle$;

Scalar Multiplication: $k\mathbf{v} = \langle kv_1, kv_2 \rangle$;

Dot Product: $\mathbf{u} \cdot \mathbf{v} = u_1v_1 + u_2v_2$.

Equivalent Vectors

Let $\mathbf{u} = \langle u_1, u_2 \rangle$ and $\mathbf{v} = \langle v_1, v_2 \rangle$. Then $\langle u_1, u_2 \rangle = \langle v_1, v_2 \rangle$ if and only if $u_1 = v_1$ and $u_2 = v_2$.

Do the indicated calculations, where $\mathbf{u} = \langle -6, 2 \rangle$ and $\mathbf{v} = \langle 8, 3 \rangle$.

a) $\mathbf{v} - 2\mathbf{u}$ **b)** $5\mathbf{u} + 3\mathbf{v}$

c) $|\mathbf{u} - \mathbf{v}|$ **d)** $\mathbf{u} \cdot \mathbf{v}$

a) $$\mathbf{v} - 2\mathbf{u} = \langle 8, 3 \rangle - 2\langle -6, 2 \rangle = \langle 8, 3 \rangle - \langle -12, 4 \rangle = \langle 20, -1 \rangle$$

b) $$5\mathbf{u} + 3\mathbf{v} = 5\langle -6, 2 \rangle + 3\langle 8, 3 \rangle = \langle -30, 10 \rangle + \langle 24, 9 \rangle = \langle -6, 19 \rangle$$

c) $$|\mathbf{u} - \mathbf{v}| = |\langle -6, 2 \rangle - \langle 8, 3 \rangle| = |\langle -14, -1 \rangle| = \sqrt{(-14)^2 + (-1)^2} = \sqrt{197} \approx 14.0$$

d) $$\mathbf{u} \cdot \mathbf{v} = -6 \cdot 8 + 2 \cdot 3 = -42$$

Zero Vector

The vector whose initial and terminal points are both $(0, 0)$ is the **zero vector**, denoted by $\mathbf{O}$, or $\langle 0, 0 \rangle$. Its magnitude is 0. In vector addition, the zero vector is the additive identity vector:

$$\mathbf{v} + \mathbf{O} = \mathbf{v}.$$

Unit Vector

A vector of magnitude, or length, 1 is called a **unit vector**.

If $\mathbf{v}$ is a vector and $\mathbf{v} \neq \mathbf{O}$, then

$$\frac{1}{|\mathbf{v}|} \cdot \mathbf{v}, \quad \text{or} \quad \frac{\mathbf{v}}{|\mathbf{v}|},$$

is a unit vector in the direction of $\mathbf{v}$.

Unit vectors parallel to the x- and y-axes are defined as

$$\mathbf{i} = \langle 1, 0 \rangle \quad \text{and} \quad \mathbf{j} = \langle 0, 1 \rangle.$$

Any vector can be expressed as a linear combination of unit vectors $\mathbf{i}$ and $\mathbf{j}$:

$$\mathbf{v} = \langle v_1, v_2 \rangle = v_1\mathbf{i} + v_2\mathbf{j}.$$

Find a unit vector that has the same direction as the vector $\mathbf{w} = \langle -9, 4 \rangle$.

We first find the length of $\mathbf{w}$:

$$|\mathbf{w}| = \sqrt{(-9)^2 + 4^2} = \sqrt{81 + 16} = \sqrt{97}.$$

We are looking for a vector $\mathbf{u}$ whose length is $1/\sqrt{97}$ of $\mathbf{w}$ and whose direction is the same as vector $\mathbf{w}$:

$$\mathbf{u} = \frac{1}{\sqrt{97}}\mathbf{w} = \frac{1}{\sqrt{97}}\langle -9, 4 \rangle = \left\langle -\frac{9}{\sqrt{97}}, \frac{4}{\sqrt{97}} \right\rangle.$$

Express the vector $\mathbf{q} = \langle 18, -7 \rangle$ as a linear combination of $\mathbf{i}$ and $\mathbf{j}$.

$$\mathbf{q} = \langle 18, -7 \rangle = 18\mathbf{i} + (-7)\mathbf{j} = 18\mathbf{i} - 7\mathbf{j}$$

Write the vector $\mathbf{r} = -4\mathbf{i} + \mathbf{j}$ in component form.

$$\mathbf{r} = -4\mathbf{i} + \mathbf{j} = -4\mathbf{i} + 1\mathbf{j} = \langle -4, 1 \rangle$$

Direction Angles

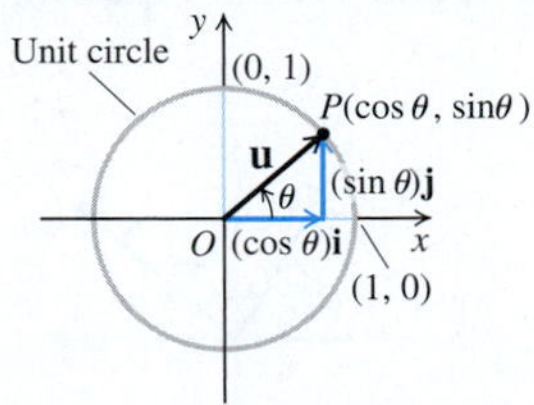

The unit vector can be expressed in component form,

$$\mathbf{u} = \langle \cos\theta, \sin\theta \rangle,$$

or

$$\mathbf{u} = (\cos\theta)\mathbf{i} + (\sin\theta)\mathbf{j},$$

where the components of **u** are functions of the direction angle θ measured counterclockwise from the x-axis to the vector.

Calculate and sketch the unit vector $\mathbf{u} = (\cos\theta)\mathbf{i} + (\sin\theta)\mathbf{j}$ for $\theta = 5\pi/4$.

$$\mathbf{u} = \left(\cos\frac{5\pi}{4}\right)\mathbf{i} + \left(\sin\frac{5\pi}{4}\right)\mathbf{j}$$

$$\mathbf{u} = \left(-\frac{\sqrt{2}}{2}\right)\mathbf{i} + \left(-\frac{\sqrt{2}}{2}\right)\mathbf{j}$$

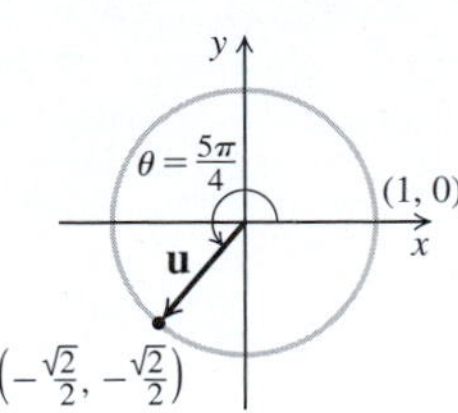

Determine the direction angle θ of the vector to the nearest degree.

a) $\mathbf{t} = \langle -2, 9 \rangle$ **b)** $\mathbf{b} = 5\mathbf{i} + 3\mathbf{j}$

a) $\mathbf{t} = \langle -2, 9 \rangle$

$$\tan\theta = \frac{9}{-2} = -\frac{9}{2} \quad \text{and} \quad \theta = \tan^{-1}\left(-\frac{9}{2}\right) \approx -77°$$

Since **t** is in quadrant II, we know that θ is a second-quadrant angle. The reference angle is 77°. Thus,

$$\theta \approx 180° - 77°, \quad \text{or} \quad 103°.$$

b) $\mathbf{b} = 5\mathbf{i} + 3\mathbf{j} = \langle 5, 3 \rangle$

$$\tan\theta = \frac{3}{5} \quad \text{and} \quad \theta = \tan^{-1}\frac{3}{5} \approx 31°$$

Since **b** is in quadrant I, $\theta \approx 31°$.

Angle Between Two Vectors

If θ is the angle between two nonzero vectors **u** and **v**, then

$$\cos\theta = \frac{\mathbf{u} \cdot \mathbf{v}}{|\mathbf{u}||\mathbf{v}|}.$$

Find the angle between $\mathbf{u} = \langle -2, 1 \rangle$ and $\mathbf{v} = \langle 3, -4 \rangle$.

We first find $\mathbf{u} \cdot \mathbf{v}$, $|\mathbf{u}|$, and $|\mathbf{v}|$:

$$\mathbf{u} \cdot \mathbf{v} = (-2) \cdot 3 + 1 \cdot (-4) = -10,$$
$$|\mathbf{u}| = \sqrt{(-2)^2 + 1^2} = \sqrt{5}, \quad \text{and}$$
$$|\mathbf{v}| = \sqrt{3^2 + (-4)^2} = \sqrt{25} = 5.$$

Then

$$\cos\theta = \frac{\mathbf{u} \cdot \mathbf{v}}{|\mathbf{u}||\mathbf{v}|} = \frac{-10}{\sqrt{5} \cdot 5}$$

$$\theta = \cos^{-1}\left(\frac{-10}{\sqrt{5} \cdot 5}\right) \approx 153.4°.$$

Forces in Equilibrium

When several forces act through the same point on an object, their vector sum must be $\mathbf{O}$ in order for a balance to occur.

A 600-lb block is suspended by two cables as shown. At point B, there are three forces acting: $\mathbf{R}$, $\mathbf{S}$, and $\mathbf{W}$. Find the tension in each cable.

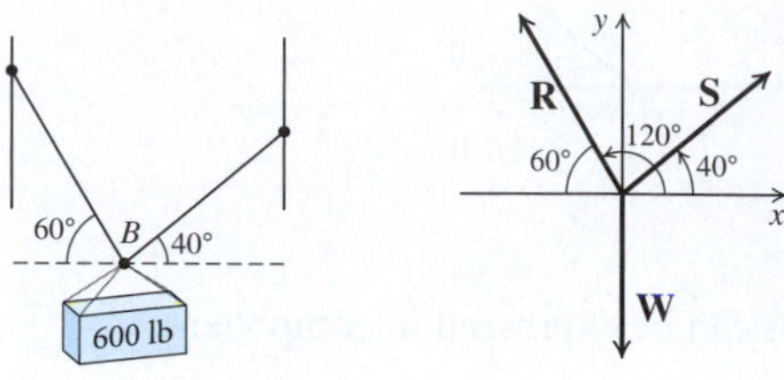

We have

$$\mathbf{R} = |\mathbf{R}|[(\cos 120°)\mathbf{i} + (\sin 120°)\mathbf{j}],$$
$$\mathbf{S} = |\mathbf{S}|[(\cos 40°)\mathbf{i} + (\sin 40°)\mathbf{j}], \quad \text{and}$$
$$\mathbf{W} = 600[(\cos 270°)\mathbf{i} + (\sin 270°)\mathbf{j}] = -600\mathbf{j}.$$

For a balance, the vector sum must be the vector $\mathbf{O}$:

$$\mathbf{R} + \mathbf{S} + \mathbf{W} = \mathbf{O}.$$
$$|\mathbf{R}|[(\cos 120°)\mathbf{i} + (\sin 120°)\mathbf{j}]$$
$$+ |\mathbf{S}|[(\cos 40°)\mathbf{i} + (\sin 40°)\mathbf{j}] - 600\mathbf{j} = 0\mathbf{i} + 0\mathbf{j}$$

This gives us two equations:

$$|\mathbf{R}|(\cos 120°) + |\mathbf{S}|(\cos 40°) = 0,$$
$$|\mathbf{R}|(\sin 120°) + |\mathbf{S}|(\sin 40°) - 600 = 0.$$

Solving this system of equations for $|\mathbf{R}|$ and $|\mathbf{S}|$, we get $|\mathbf{R}| \approx 467$ and $|\mathbf{S}| \approx 305$. The tensions in the cables are 467 lb and 305 lb.

REVIEW EXERCISES

Determine whether the statement is true or false.

1. For any point (x, y) on the unit circle, $\langle x, y \rangle$ is a unit vector. [8.6]

2. The law of sines can be used to solve a triangle when all three sides are known. [8.1]

3. Two vectors are equivalent if they have the same magnitude and the lines that they are on have the same slope. [8.5]

4. Vectors $\langle 8, -2 \rangle$ and $\langle -8, 2 \rangle$ are equivalent. [8.6]

5. Any triangle, right or oblique, can be solved if at least one angle and any other two measures are known. [8.1]

6. When two angles and an included side of a triangle are known, the triangle cannot be solved using the law of cosines. [8.2]

Solve △ABC, if possible. [8.1], [8.2]

7. $a = 23.4$ ft, $b = 15.7$ ft, $c = 8.3$ ft

8. $B = 27°$, $C = 35°$, $b = 19$ in.

9. $A = 133°28'$, $C = 31°42'$, $b = 890$ m

10. $B = 37°$, $b = 4$ yd, $c = 8$ yd

11. Find the area of $\triangle ABC$ if $b = 9.8$ m, $c = 7.3$ m, and $A = 67.3°$. [8.1]

12. A parallelogram has sides of lengths 3.21 ft and 7.85 ft. One of its angles measures 147°. Find the area of the parallelogram. [8.1]

13. *Flower Garden.* A triangular flower garden has sides of lengths 11 m, 9 m, and 6 m. Find the angles of the garden to the nearest degree. [8.2]

14. In an isosceles triangle, the base angles each measure 52.3° and the base is 513 ft long. Find the lengths of the other two sides to the nearest foot. [8.1]

15. *Airplanes.* Two airplanes leave an airport at the same time. The first flies 175 km/h in a direction of 305.6°. The second flies 220 km/h in a direction of 195.5°. After 2 hr, how far apart are the planes? [8.2]

16. *Sandbox.* A child-care center has a triangular-shaped sandbox. Two of the three sides measure 15 ft and 12.5 ft and form an included angle of 42°. To determine the amount of sand that is needed to fill the box, the director must determine the area of the floor of the box. Find the area of the floor of the box to the nearest square foot. [8.1]

Graph the complex number and find its absolute value. [8.3]

17. $2 - 5i$ **18.** 4

19. $2i$ **20.** $-3 + i$

Find trigonometric notation. [8.3]

21. $1 + i$ **22.** $-4i$

23. $-5\sqrt{3} + 5i$ **24.** $\frac{3}{4}$

Find standard notation, $a + bi$. [8.3]

25. $4(\cos 60° + i \sin 60°)$

26. $7(\cos 0° + i \sin 0°)$

27. $5\left(\cos \frac{2\pi}{3} + i \sin \frac{2\pi}{3}\right)$

28. $2\left[\cos\left(-\frac{\pi}{6}\right) + i \sin\left(-\frac{\pi}{6}\right)\right]$

Convert to trigonometric notation and then multiply or divide, expressing the answer in standard notation. [8.3]

29. $(1 + i\sqrt{3})(1 - i)$ **30.** $\frac{2 - 2i}{2 + 2i}$

31. $\frac{2 + 2\sqrt{3}i}{\sqrt{3} - i}$ **32.** $i(3 - 3\sqrt{3}i)$

Raise the number to the given power and write trigonometric notation for the answer. [8.3]

33. $[2(\cos 60° + i \sin 60°)]^3$

34. $(1 - i)^4$

Raise the number to the given power and write standard notation for the answer. [8.3]

35. $(1 + i)^6$ **36.** $\left(\frac{1}{2} + \frac{\sqrt{3}}{2}i\right)^{10}$

37. Find the square roots of $-1 + i$. [8.3]

38. Find the cube roots of $3\sqrt{3} - 3i$. [8.3]

39. Find and graph the fourth roots of 81. [8.3]

40. Find and graph the fifth roots of 1. [8.3]

Find all the complex solutions of the equation. [8.3]

41. $x^4 - i = 0$ **42.** $x^3 + 1 = 0$

43. Find the polar coordinates of each of these points. Give three answers for each point. [8.4]

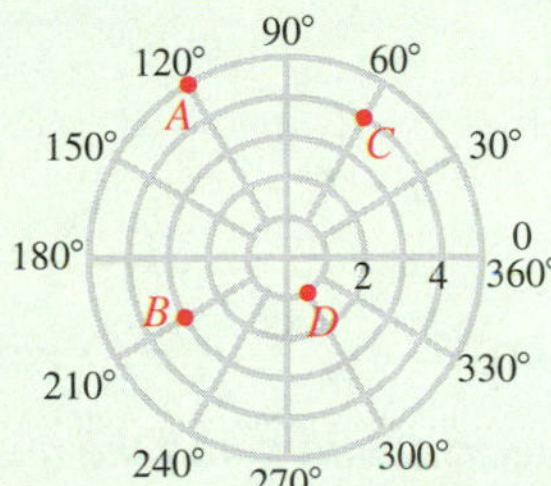

Find the polar coordinates of the point. Express the answer in degrees and then in radians. [8.4]

44. $(-4\sqrt{2}, 4\sqrt{2})$ **45.** $(0, -5)$

Use a graphing calculator to convert from rectangular coordinates to polar coordinates. Express the answer in degrees and then in radians. [8.4]

46. $(-2, 5)$ **47.** $(-4.2, \sqrt{7})$

Find the rectangular coordinates of the point. [8.4]

48. $\left(3, \frac{\pi}{4}\right)$ **49.** $(-6, -120°)$

Use a graphing calculator to convert from polar coordinates to rectangular coordinates. Round the coordinates to the nearest hundredth. [8.4]

50. $(2, -15°)$ **51.** $\left(-2.3, \frac{\pi}{5}\right)$

Convert to a polar equation. [8.4]

52. $5x - 2y = 6$ **53.** $y = 3$

54. $x^2 + y^2 = 9$ **55.** $y^2 - 4x - 16 = 0$

Convert to a rectangular equation. [8.4]

56. $r = 6$

57. $r + r\sin\theta = 1$

58. $r = \dfrac{3}{1 - \cos\theta}$

59. $r - 2\cos\theta = 3\sin\theta$

In Exercises 60–63, match the equation with one of figures (a)–(d) that follow. [8.4]

a)

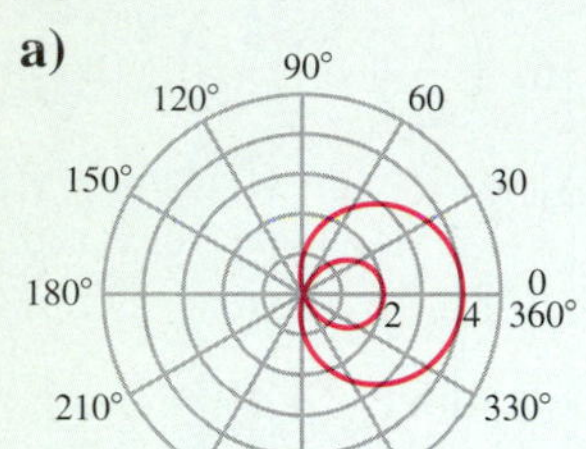

b)

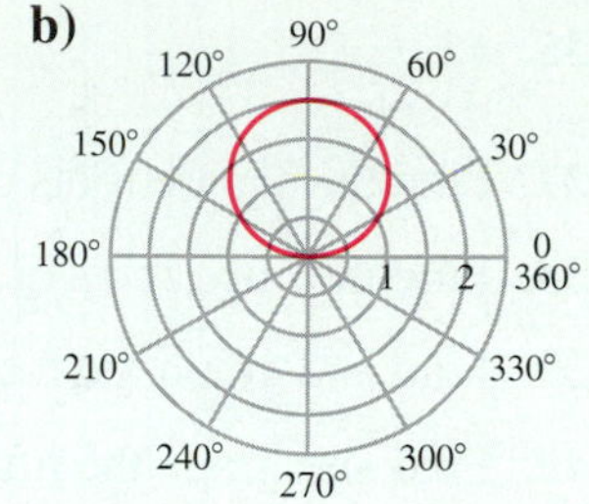

c)

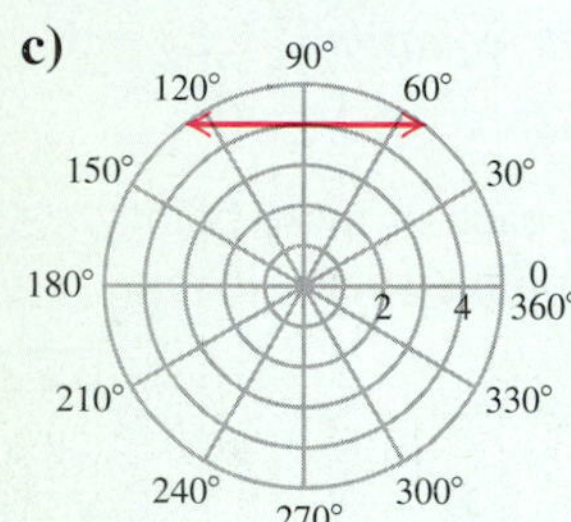

d)

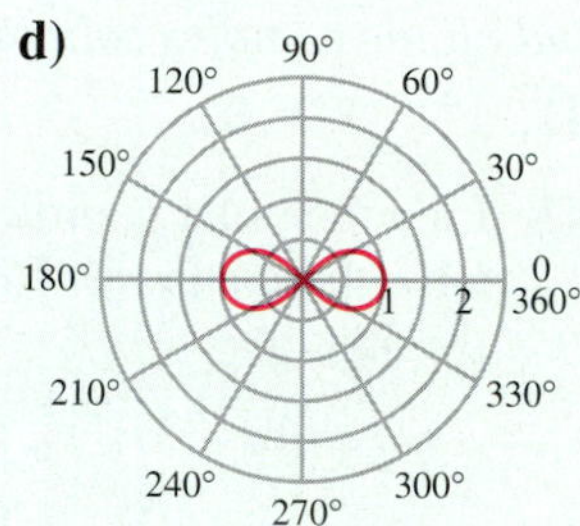

60. $r = 2\sin\theta$

61. $r^2 = \cos 2\theta$

62. $r = 1 + 3\cos\theta$

63. $r\sin\theta = 4$

Magnitudes of vectors **u** *and* **v** *and the angle* θ *between the vectors are given. Find the magnitude of the sum,* **u** + **v**, *to the nearest tenth and give the direction by specifying to the nearest degree the angle that it makes with the vector* **u**. [8.5]

64. $|\mathbf{u}| = 12,\ |\mathbf{v}| = 15,\ \theta = 120°$

65. $|\mathbf{u}| = 41,\ |\mathbf{v}| = 60,\ \theta = 25°$

The vectors **u**, **v**, *and* **w** *are drawn below. Copy them on a sheet of paper. Then sketch each of the vectors in Exercises 66 and 67.* [8.5]

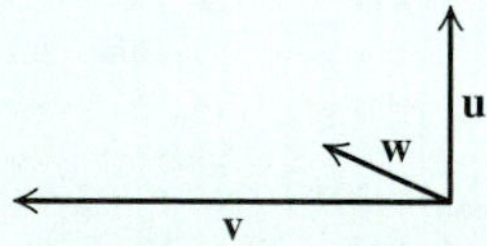

66. $\mathbf{u} - \mathbf{v}$

67. $\mathbf{u} + \frac{1}{2}\mathbf{w}$

68. Forces of 230 N and 500 N act on an object. The angle between the forces is 52°. Find the resultant, giving the angle that it makes with the smaller force. [8.5]

69. *Wind.* A wind has an easterly component of 15 km/h and a southerly component of 25 km/h. Find the magnitude and the direction of the wind. [8.5]

70. *Ship.* A ship first sails N75°E for 90 nautical mi, and then S10°W for 100 nautical mi. How far is the ship, then, from the starting point, and in what direction? [8.5]

Find the component form of the vector given the initial point and the terminal point. [8.6]

71. $\overrightarrow{AB}$; $A(2, -8)$, $B(-2, -5)$

72. $\overrightarrow{TR}$; $R(0, 7)$, $T(-2, 13)$

73. Find the magnitude of vector **u** if $\mathbf{u} = \langle 5, -6 \rangle$. [8.6]

Do the calculations in Exercises 74–77 for the vectors

$\mathbf{u} = \langle 3, -4 \rangle$, $\mathbf{v} = \langle -3, 9 \rangle$, and $\mathbf{w} = \langle -2, -5 \rangle$.

[8.6]

74. $4\mathbf{u} + \mathbf{w}$

75. $2\mathbf{w} - 6\mathbf{v}$

76. $|\mathbf{u}| + |2\mathbf{w}|$

77. $\mathbf{u} \cdot \mathbf{w}$

78. Find a unit vector that has the same direction as $\mathbf{v} = \langle -6, -2 \rangle$. [8.6]

79. Express the vector $\mathbf{t} = \langle -9, 4 \rangle$ as a linear combination of the unit vectors **i** and **j**. [8.6]

80. Determine the direction angle θ of the vector $\mathbf{w} = \langle -4, -1 \rangle$ to the nearest degree. [8.6]

81. Find the magnitude and the direction angle θ of $\mathbf{u} = -5\mathbf{i} - 3\mathbf{j}$. [8.6]

82. Find the angle between $\mathbf{u} = \langle 3, -7 \rangle$ and $\mathbf{v} = \langle 2, 2 \rangle$ to the nearest tenth of a degree. [8.6]

83. *Airplane.* An airplane has an airspeed of 160 mph. It is to make a flight in a direction of 80° while there is a 20-mph wind from 310°. What will the airplane's actual heading be? [8.6]

Do the calculations in Exercises 84–87 for the vectors

$\mathbf{u} = 2\mathbf{i} + 5\mathbf{j}$, $\mathbf{v} = -3\mathbf{i} + 10\mathbf{j}$, and $\mathbf{w} = 4\mathbf{i} + 7\mathbf{j}$.

[8.6]

84. $5\mathbf{u} - 8\mathbf{v}$

85. $\mathbf{u} - (\mathbf{v} + \mathbf{w})$

86. $|\mathbf{u} - \mathbf{v}|$

87. $3|\mathbf{w}| + |\mathbf{v}|$

88. Express the vector $\overrightarrow{PQ}$ in the form $a\mathbf{i} + b\mathbf{j}$, if P is the point $(1, -3)$ and Q is the point $(-4, 2)$. [8.6]

Express each vector in Exercises 89 and 90 in the form $a\mathbf{i} + b\mathbf{j}$ *and sketch each in the coordinate plane.* [8.6]

89. The unit vectors $\mathbf{u} = (\cos\theta)\mathbf{i} + (\sin\theta)\mathbf{j}$ for $\theta = \pi/4$ and $\theta = 5\pi/4$. Include the unit circle $x^2 + y^2 = 1$ in your sketch. [8.6]

90. The unit vector obtained by rotating **j** counterclockwise $2\pi/3$ radians about the origin.

91. Express the vector $3\mathbf{i} - \mathbf{j}$ as a product of its magnitude and its direction. [8.6]

92. Which of the following is the trigonometric notation for $1 - i$? [8.3]

A. $\sqrt{2}\left(\cos\frac{5\pi}{4} + i\sin\frac{5\pi}{4}\right)$

B. $\sqrt{2}\left(\cos\frac{7\pi}{4} - \sin\frac{7\pi}{4}\right)$

C. $\cos\frac{7\pi}{4} + i\sin\frac{7\pi}{4}$

D. $\sqrt{2}\left(\cos\frac{7\pi}{4} + i\sin\frac{7\pi}{4}\right)$

93. Convert the polar equation $r = 100$ to a rectangular equation. [8.4]

A. $x^2 + y^2 = 10{,}000$

B. $x^2 + y^2 = 100$

C. $\sqrt{x^2 + y^2} = 10$

D. $\sqrt{x^2 + y^2} = 1000$

94. The graph of $r = 1 - 2\cos\theta$ is which of the following? [8.4]

A.

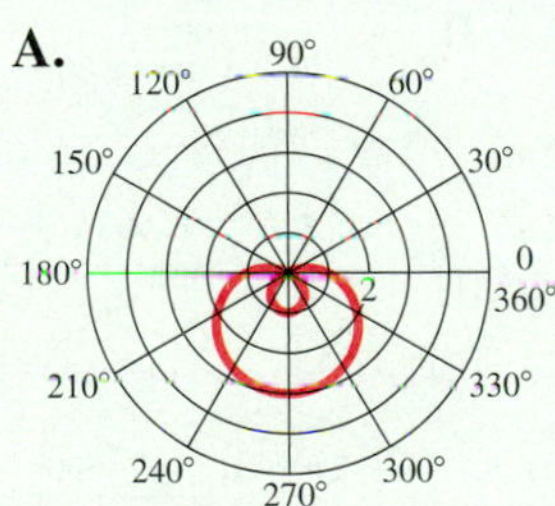

B.

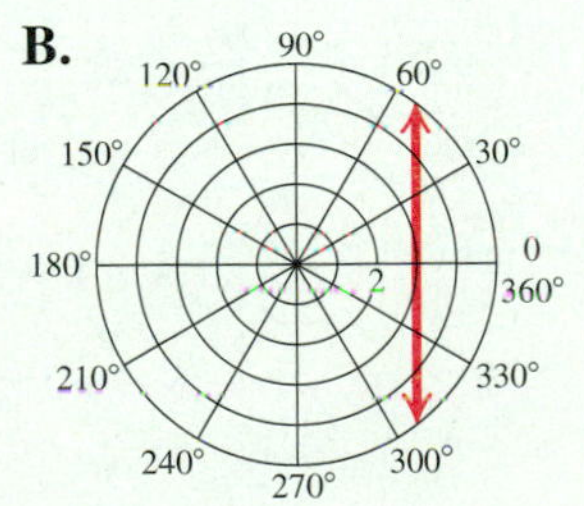

C.

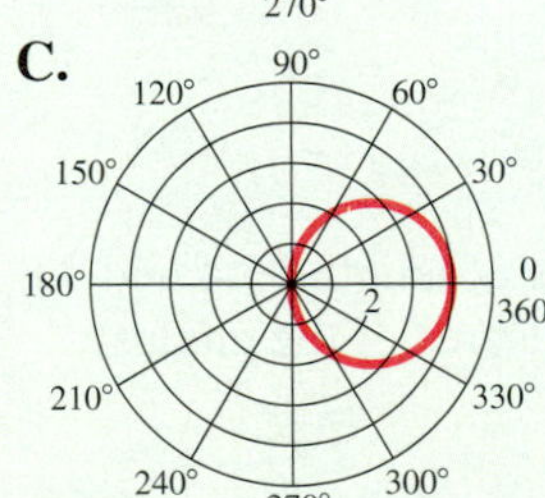

D.

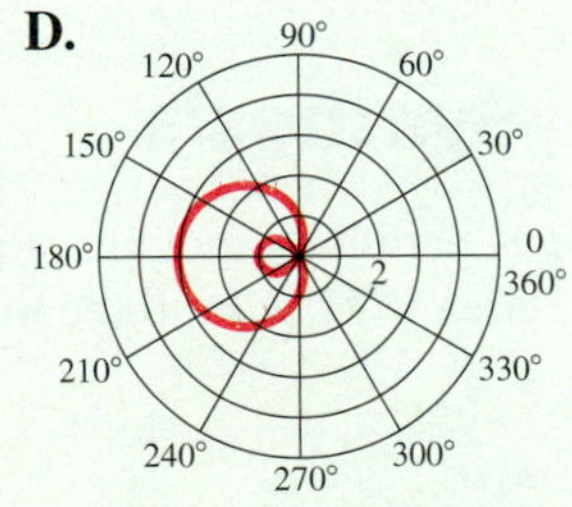

Synthesis

95. Let $\mathbf{u} = 12\mathbf{i} + 5\mathbf{j}$. Find a vector that has the same direction as **u** but has length 3. [8.6]

96. A parallelogram has sides of lengths 3.42 and 6.97. Its area is 18.4. Find the sizes of its angles. [8.1]

Collaborative Discussion and Writing

97. Summarize how you can determine algebraically when solving triangles whether there is no solution, one solution, or two solutions. [8.1], [8.2]

98. Give an example of an equation that is easier to graph in polar notation than in rectangular notation and explain why. [8.4]

99. Explain why the rectangular coordinates of a point are unique and the polar coordinates of a point are not unique. [8.4]

100. Explain why vectors $\overrightarrow{QR}$ and $\overrightarrow{RQ}$ are not equivalent. [8.5]

101. Explain how unit vectors are related to the unit circle. [8.6]

102. Write a vector sum problem for a classmate for which the answer is $\mathbf{v} = 5\mathbf{i} - 8\mathbf{j}$. [8.6]

8 Chapter Test

Solve $\triangle ABC$, if possible.

1. $a = 18$ ft, $B = 54°$, $C = 43°$

2. $b = 8$ m, $c = 5$ m, $C = 36°$

3. $a = 16.1$ in., $b = 9.8$ in., $c = 11.2$ in.

4. Find the area of $\triangle ABC$ if $C = 106.4°$, $a = 7$ cm, and $b = 13$ cm.

5. *Distance Across a Lake.* Points A and B are on opposite sides of a lake. Point C is 52 m from A. The measure of $\angle BAC$ is determined to be 108°, and the measure of $\angle ACB$ is determined to be 44°. What is the distance from A to B?

6. *Location of Airplanes.* Two airplanes leave an airport at the same time. The first flies 210 km/h in a direction of 290°. The second flies 180 km/h in a direction of 185°. After 3 hr, how far apart are the planes?

7. Graph: $-4 + i$.

8. Find the absolute value of $2 - 3i$.

9. Find trigonometric notation for $3 - 3i$.

10. Divide and express the result in standard notation $a + bi$:

$$\frac{2\left(\cos \frac{2\pi}{3} + i \sin \frac{2\pi}{3}\right)}{8\left(\cos \frac{\pi}{6} + i \sin \frac{\pi}{6}\right)}.$$

11. Find $(1 - i)^8$ and write standard notation for the answer.

12. Find the polar coordinates of $(-1, \sqrt{3})$. Express the angle in degrees using the smallest possible positive angle.

13. Convert $\left(-1, \frac{2\pi}{3}\right)$ to rectangular coordinates.

14. Convert to a polar equation: $x^2 + y^2 = 10$.

15. Graph: $r = 1 - \cos \theta$.

16. For vectors **u** and **v**, $|\mathbf{u}| = 8$, $|\mathbf{v}| = 5$, and the angle between the vectors is 63°. Find $\mathbf{u} + \mathbf{v}$. Give the magnitude to the nearest tenth, and give the direction by specifying the angle that the resultant makes with **u**, to the nearest degree.

17. For $\mathbf{u} = 2\mathbf{i} - 7\mathbf{j}$ and $\mathbf{v} = 5\mathbf{i} + \mathbf{j}$, find $2\mathbf{u} - 3\mathbf{v}$.

18. Find a unit vector in the same direction as $-4\mathbf{i} + 3\mathbf{j}$.

19. Which of the following is the graph of $r = 3 \cos \theta$?

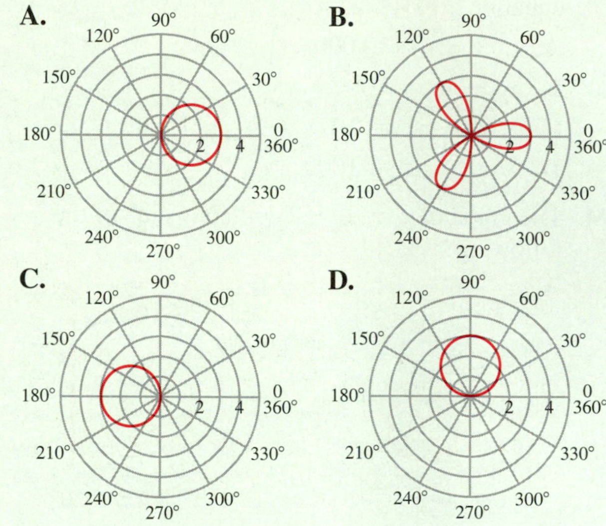

› Synthesis

20. A parallelogram has sides of length 15.4 and 9.8. Its area is 72.9. Find the measures of the angles.

CHAPTER 9

Systems of Equations and Matrices

APPLICATION

This problem appears as Exercise 55 in Section 9.1.

About 791,000 Chinese–Americans live in New York City and Boston. The number of Chinese–Americans in New York City is 95,000 more than five times the number in Boston. (*Source*: U.S. Census Bureau) How many Chinese–Americans live in each city?

9.1 Systems of Equations in Two Variables

- Solve a system of two linear equations in two variables by graphing.
- Solve a system of two linear equations in two variables using the substitution method and the elimination method.
- Use systems of two linear equations to solve applied problems.

A **system of equations** is composed of two or more equations considered simultaneously. For example,

$$x - y = 5,$$
$$2x + y = 1$$

is a **system of two linear equations in two variables**. The solution set of this system consists of all ordered pairs that make *both* equations true. The ordered pair $(2, -3)$ is a solution of the system of equations above. We can verify this by substituting 2 for x and -3 for y in *each* equation.

$x - y = 5$			$2x + y = 1$		
$2 - (-3)$	? 5		$2 \cdot 2 + (-3)$	? 1	
$2 + 3$			$4 - 3$		
5	5	TRUE	1	1	TRUE

GRAPHS OF EQUATIONS

REVIEW SECTION 1.1.

Solving Systems of Equations Graphically

Recall that the graph of a linear equation is a line that contains all the ordered pairs in the solution set of the equation. When we graph a system of linear equations, each point at which the graphs intersect is a solution of *both* equations and therefore a **solution of the system of equations**.

EXAMPLE 1 Solve the following system of equations graphically:

$$x - y = 5,$$
$$2x + y = 1.$$

Solution We graph the equations on the same set of axes, as shown below.

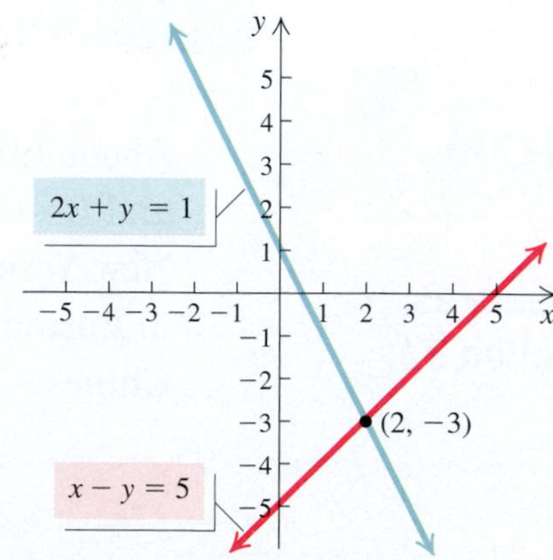

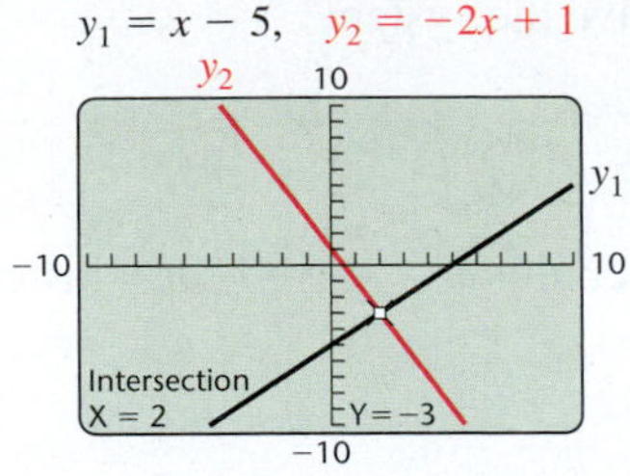

We see that the graphs intersect at a single point, $(2, -3)$, so $(2, -3)$ is the solution of the system of equations. To check this solution, we substitute 2 for x and -3 for y in both equations as we did on the preceding page.

To use a graphing calculator to solve this system of equations, it might be necessary to write each equation in "Y = ⋯" form. If so, we would graph $y_1 = x - 5$ and $y_2 = -2x + 1$ and then use the INTERSECT feature. We see in the window at left that the solution is $(2, -3)$.

Now Try Exercise 7.

The graphs of most of the systems of linear equations that we use to model applications intersect at a single point, like the system above. However, it is possible that the graphs will have no points in common or infinitely many points in common. Each of these possibilities is illustrated below.

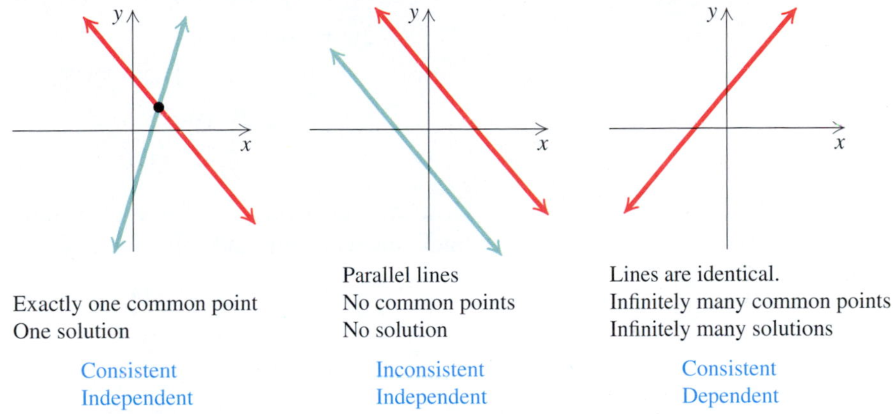

If a system of equations has at least one solution, it is **consistent**. If the system has no solutions, it is **inconsistent**. In addition, for a system of two linear equations in two variables, if one equation can be obtained by multiplying by a constant on both sides of the other equation, the equations are **dependent**. Otherwise, they are **independent**. A system of two dependent linear equations in two variables has an infinite number of solutions.

The Substitution Method

Solving a system of equations graphically is not always accurate when the solutions are not integers. A solution like $\left(\frac{43}{27}, -\frac{19}{27}\right)$, for instance, will be difficult to determine from a hand-drawn graph.

Algebraic methods for solving systems of equations, when used correctly, always give accurate results. One such technique is the **substitution method**. It is used most often when a variable is alone on one side of an equation or when it is easy to solve for a variable. To apply the substitution method, we begin by using one of the equations to express one variable in terms of the other. Then we substitute that expression in the other equation of the system.

EXAMPLE 2 Use the substitution method to solve the system

$$x - y = 5, \quad (1)$$
$$2x + y = 1. \quad (2)$$

Just in Time 17

Solution First, we solve equation (1) for x. (We could have solved for y instead.) We have

$$x - y = 5 \quad (1)$$
$$x = y + 5. \quad \text{Solving for } x$$

Then we substitute $y + 5$ for x in equation (2). This gives an equation in one variable, which we know how to solve:

$$2x + y = 1 \quad (2)$$
$$2(y + 5) + y = 1 \quad \text{The parentheses are necessary.}$$
$$2y + 10 + y = 1 \quad \text{Removing parentheses}$$
$$3y + 10 = 1 \quad \text{Collecting like terms on the left}$$
$$3y = -9 \quad \text{Subtracting 10 on both sides}$$
$$y = -3. \quad \text{Dividing by 3 on both sides}$$

Now we substitute -3 for y in either of the original equations (this is called **back-substitution**) and solve for x. We choose equation (1):

$$x - y = 5 \quad (1)$$
$$x - (-3) = 5 \quad \text{Substituting } -3 \text{ for } y$$
$$x + 3 = 5$$
$$x = 2. \quad \text{Subtracting 3 on both sides}$$

We have previously checked the pair $(2, -3)$ in both equations. The solution of the system of equations is $(2, -3)$. Since there is exactly one solution, the system of equations is consistent, and the equations are independent.

Now Try Exercise 17.

The Elimination Method

Another algebraic technique for solving systems of equations is the **elimination method**. With this method, we eliminate a variable by adding two equations. If the coefficients of a particular variable are opposites, we can eliminate that variable simply by adding the original equations. For example, if the x-coefficient is -3 in one equation and is 3 in the other equation, then the sum of the x-terms will be 0 and thus the variable x will be eliminated when we add the equations.

EXAMPLE 3 Use the elimination method to solve the system

$$2x + y = 2, \quad (1)$$
$$x - y = 7. \quad (2)$$

Algebraic Solution

Since the y-coefficients, 1 and -1, are opposites, we can eliminate y by adding the equations:

$$\begin{aligned} 2x + y &= 2 \quad &(1) \\ x - y &= 7 \quad &(2) \\ \hline 3x \quad &= 9 \quad &\text{Adding} \\ x &= 3. \end{aligned}$$

We then back-substitute 3 for x in either equation and solve for y. We choose equation (1):

$$\begin{aligned} 2x + y &= 2 \quad &(1) \\ 2 \cdot 3 + y &= 2 \quad &\text{Substituting 3 for } x \\ 6 + y &= 2 \\ y &= -4. \end{aligned}$$

We check the solution by substituting the pair $(3, -4)$ in both equations.

$$\begin{array}{r|l} 2x + y & = 2 \\ \hline 2 \cdot 3 + (-4) & ?\ 2 \\ 6 - 4 & \\ 2 & 2 \quad \text{TRUE} \end{array} \qquad \begin{array}{r|l} x - y & = 7 \\ \hline 3 - (-4) & ?\ 7 \\ 3 + 4 & \\ 7 & 7 \quad \text{TRUE} \end{array}$$

The solution is $(3, -4)$. Since there is exactly one solution, the system of equations is consistent, and the equations are independent.

Graphical Solution

We solve each equation for y, getting

$y = -2x + 2$ and $y = x - 7$.

Next, we graph these equations and find the point of intersection of the graphs using the INTERSECT feature.

$y_1 = -2x + 2, \quad y_2 = x - 7$

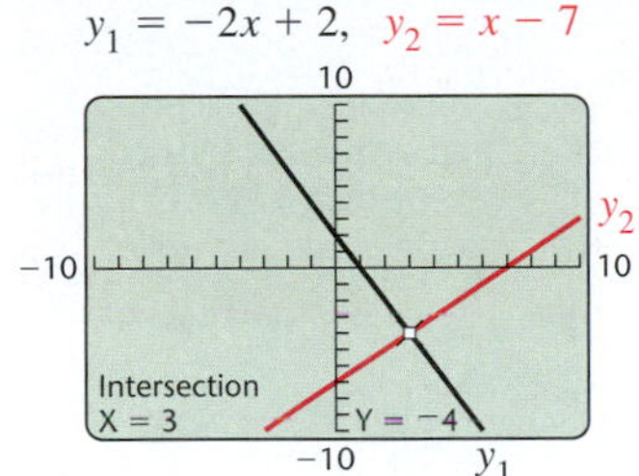

The graphs intersect at the point $(3, -4)$, so the solution of the system of equations is $(3, -4)$.

Now Try Exercise 31.

Before we add, it might be necessary to multiply one or both equations by suitable constants in order to find two equations in which the coefficients of a variable are opposites.

EXAMPLE 4 Use the elimination method to solve the system

$$\begin{aligned} 4x + 3y &= 11, \quad &(1) \\ -5x + 2y &= 15. \quad &(2) \end{aligned}$$

Solution We can obtain x-coefficients that are opposites by multiplying the first equation by 5 and the second equation by 4:

$$\begin{aligned} 20x + 15y &= 55 \quad &\text{Multiplying equation (1) by 5} \\ -20x + 8y &= 60 \quad &\text{Multiplying equation (2) by 4} \\ \hline 23y &= 115 \quad &\text{Adding} \\ y &= 5. \end{aligned}$$

We then back-substitute 5 for y in either equation (1) or (2) and solve for x. We choose equation (1):

$$\begin{aligned} 4x + 3y &= 11 \quad &(1) \\ 4x + 3 \cdot 5 &= 11 \quad &\text{Substituting 5 for } y \\ 4x + 15 &= 11 \\ 4x &= -4 \\ x &= -1. \end{aligned}$$

We can check the pair $(-1, 5)$ by substituting in both equations or by graphing. The solution is $(-1, 5)$. The system of equations is consistent and the equations are independent.

Now Try Exercise 33.

In Example 4, the two systems

$$\begin{aligned} 4x + 3y &= 11, \\ -5x + 2y &= 15 \end{aligned} \quad \text{and} \quad \begin{aligned} 20x + 15y &= 55, \\ -20x + 8y &= 60 \end{aligned}$$

are **equivalent** because they have exactly the same solutions. When we use the elimination method, we often multiply one or both equations by constants to find equivalent equations that allow us to eliminate a variable by adding.

EQUATION-SOLVING SPECIAL CASES

REVIEW SECTION 1.5.

EXAMPLE 5 Solve each of the following systems using the elimination method.

a) $x - 3y = 1$, **(1)**
$-2x + 6y = 5$ **(2)**

b) $2x + 3y = 6$, **(1)**
$4x + 6y = 12$ **(2)**

Solution

a) We multiply equation (1) by 2 and add:

$$\begin{aligned} 2x - 6y &= 2 && \text{Multiplying equation (1) by 2} \\ -2x + 6y &= 5 && (2) \\ \hline 0 &= 7. && \text{Adding} \end{aligned}$$

There are no values of x and y for which $0 = 7$ is true, so the system has *no solution*. The solution set is $\varnothing$. The system of equations is inconsistent, and the equations are independent. The graphs of the equations are parallel lines, as shown in Fig. 1.

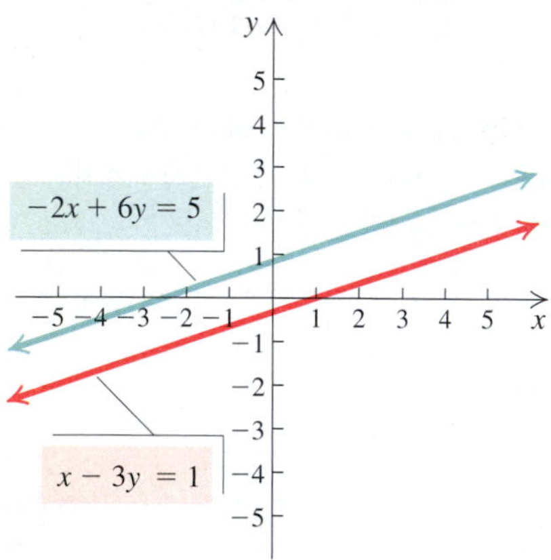

FIGURE 1.

b) We multiply equation (1) by -2 and add:

$$\begin{aligned} -4x - 6y &= -12 && \text{Multiplying equation (1) by } -2 \\ 4x + 6y &= 12 && (2) \\ \hline 0 &= 0. && \text{Adding} \end{aligned}$$

We obtain the equation $0 = 0$, which is true for all values of x and y. This tells us that the equations are dependent, so there are *infinitely many solutions*. That is, any solution of one equation of the system is also a solution of the other. The system of equations is consistent. The graphs of the equations are identical, as shown in Fig. 2.

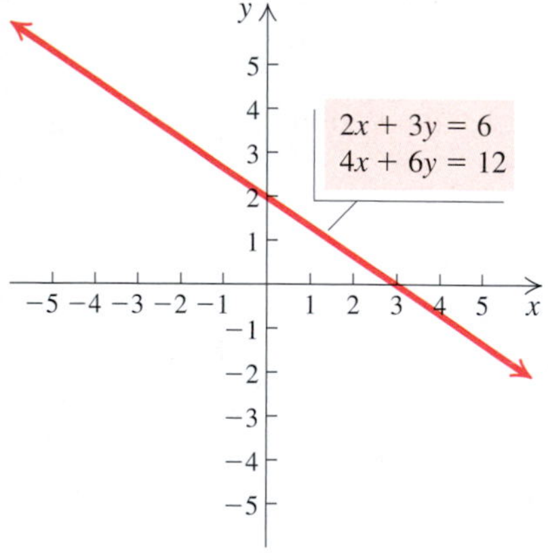

FIGURE 2.

Solving either equation for y, we have $y = -\frac{2}{3}x + 2$, so we can write the solutions of the system as ordered pairs (x, y), where y is expressed as $-\frac{2}{3}x + 2$. Thus the solutions can be written in the form $\left(x, -\frac{2}{3}x + 2\right)$. Any real value that we choose for x then gives us a value for y and thus an ordered pair in the solution set. For example,

$$\begin{aligned} &\text{if } x = -3, && \text{then } -\tfrac{2}{3}x + 2 = -\tfrac{2}{3}(-3) + 2 = 4, \\ &\text{if } x = 0, && \text{then } -\tfrac{2}{3}x + 2 = -\tfrac{2}{3}\cdot 0 + 2 = 2, \quad \text{and} \\ &\text{if } x = 6, && \text{then } -\tfrac{2}{3}x + 2 = -\tfrac{2}{3}\cdot 6 + 2 = -2. \end{aligned}$$

Thus some of the solutions are $(-3, 4)$, $(0, 2)$, and $(6, -2)$.

Similarly, solving either equation for x, we have $x = -\frac{3}{2}y + 3$, so the solutions (x, y) can also be written, expressing x as $-\frac{3}{2}y + 3$, in the form $\left(-\frac{3}{2}y + 3, y\right)$.

Since the two forms of the solutions are equivalent, they yield the same solution set, as illustrated in the table at left. Note, for example, that when $y = 4$, we have the solution $(-3, 4)$; when $y = 2$, we have $(0, 2)$; and when $y = -2$, we have $(6, -2)$.

x	$-\frac{2}{3}x + 2$
$-\frac{3}{2}y + 3$	y
-3	4
0	2
6	-2

Now Try Exercises 35 and 37.

› Applications

Frequently the most challenging and time-consuming step in the problem-solving process is translating a situation to mathematical language. However, in many cases, this task is made easier if we translate to more than one equation in more than one variable.

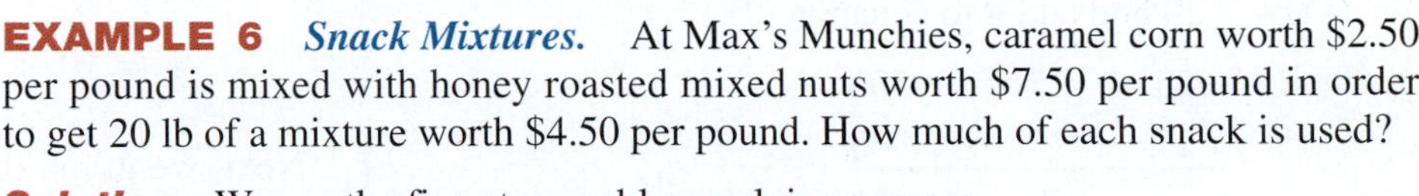

EXAMPLE 6 ***Snack Mixtures.*** At Max's Munchies, caramel corn worth \$2.50 per pound is mixed with honey roasted mixed nuts worth \$7.50 per pound in order to get 20 lb of a mixture worth \$4.50 per pound. How much of each snack is used?

Solution We use the five-step problem-solving process.

1. **Familiarize.** Let's begin by making a guess. Suppose that 16 lb of caramel corn and 4 lb of nuts are used. Then the total weight of the mixture would be 16 lb + 4 lb, or 20 lb, the desired weight. The total values of these amounts of ingredients are found by multiplying the price per pound by the number of pounds used:

   Caramel corn: $\$2.50(16) = \$40$
   Nuts: $\$7.50(4) = \$30$
   Total value: $\$70.$

   The desired value of the mixture is \$4.50 per pound, so the value of 20 lb would be $\$4.50(20)$, or \$90. Thus we see that our guess, which led to a total of \$70, is incorrect. Nevertheless, these calculations will help us to translate.

2. **Translate.** We organize the information in a table. We let $x =$ the number of pounds of caramel corn in the mixture and $y =$ the number of pounds of nuts.

	Caramel Corn	Nuts	Mixture	
Price per Pound	\$2.50	\$7.50	\$4.50	
Number of Pounds	x	y	20	→ $x + y = 20$
Value of Mixture	$2.50x$	$7.50y$	$4.50(20)$, or 90	→ $2.50x + 7.50y = 90$

   From the second row of the table, we get one equation:

   $$x + y = 20.$$

   The last row of the table yields a second equation:

   $$2.50x + 7.50y = 90, \quad \text{or} \quad 2.5x + 7.5y = 90.$$

   We can multiply by 10 on both sides of the second equation to clear the decimals. This gives us the following system of equations:

   $$x + y = 20, \qquad (1)$$
   $$25x + 75y = 900. \qquad (2)$$

3. **Carry out.** We carry out the solution as follows.

Algebraic Solution

Using the elimination method, we multiply equation (1) by -25 and add it to equation (2):

$$\begin{aligned} -25x - 25y &= -500 \\ 25x + 75y &= 900 \\ \hline 50y &= 400 \\ y &= 8. \end{aligned}$$

Then we back-substitute to find x:

$$\begin{aligned} x + y &= 20 \quad &(1) \\ x + 8 &= 20 \quad &\text{Substituting 8 for } y \\ x &= 12. \end{aligned}$$

Graphical Solution

We solve each equation for y, getting

$$y = 20 - x \quad \text{and} \quad y = \frac{900 - 25x}{75}.$$

Next, we graph these equations and find the point of intersection of the graphs. The graphs intersect at the point $(12, 8)$, so the possible solution is $(12, 8)$.

$y_1 = 20 - x, \; y_2 = \dfrac{900 - 25x}{75}$

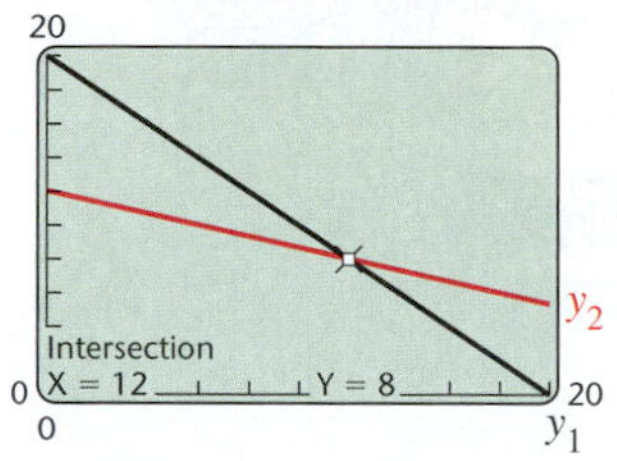

4. **Check.** If 12 lb of caramel corn and 8 lb of nuts are used, the mixture weighs $12 + 8$, or 20 lb. The value of the mixture is $\$2.50(12) + \$7.50(8)$, or $\$30 + \$60$, or \$90. Since the possible solution yields the desired weight and value of the mixture, our result checks.
5. **State.** The mixture should consist of 12 lb of caramel corn and 8 lb of honey roasted mixed nuts.

Now Try Exercise 61.

EXAMPLE 7 ***Airplane Travel.*** An airplane flies the 3000-mi distance from Los Angeles to New York, with a tailwind, in 5 hr. The return trip, against the wind, takes 6 hr. Find the speed of the airplane in still air and the speed of the wind.

Solution

1. **Familiarize.** We first make a drawing, letting p = the speed of the plane in still air, in miles per hour, and w = the speed of the wind, also in miles per hour. When the plane is traveling with a tailwind, the wind increases the speed of the plane, so the speed with the tailwind is $p + w$. On the other hand, the headwind slows the plane down, so the speed with the headwind is $p - w$.

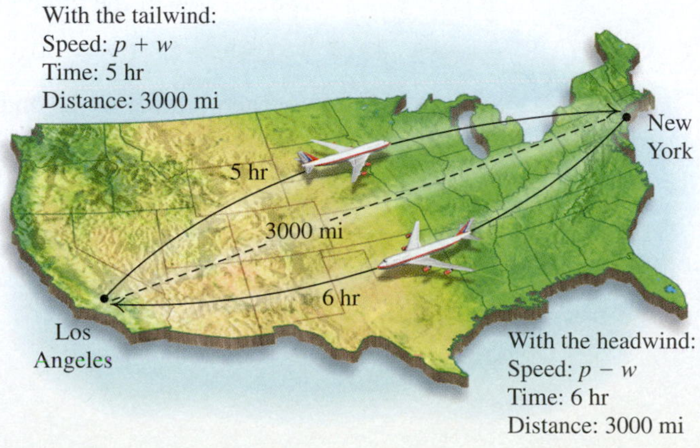

2. **Translate.** We organize the information in a table. Using the formula *Distance* = *Rate* (or *Speed*) · *Time*, we find that each row of the table yields an equation.

	Distance	Rate	Time	
With Tailwind	3000	$p + w$	5	→ $3000 = (p + w)5$
With Headwind	3000	$p - w$	6	→ $3000 = (p - w)6$

We now have a system of equations:

$$3000 = (p + w)5, \quad 600 = p + w, \quad (1) \quad \text{Dividing by 5}$$
or
$$3000 = (p - w)6, \quad 500 = p - w. \quad (2) \quad \text{Dividing by 6}$$

3. **Carry out.** We use the elimination method:

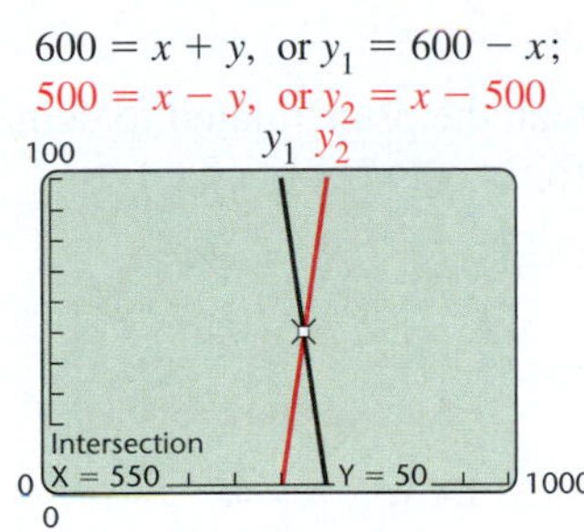

$$600 = p + w \quad (1)$$
$$500 = p - w \quad (2)$$
$$1100 = 2p \quad \text{Adding}$$
$$550 = p. \quad \text{Dividing by 2 on both sides}$$

Now we substitute in one of the equations to find w:

$$600 = p + w \quad (1)$$
$$600 = 550 + w \quad \text{Substituting 550 for } p$$
$$50 = w. \quad \text{Subtracting 550 on both sides}$$

4. **Check.** If $p = 550$ and $w = 50$, then the speed of the plane with the tailwind is $550 + 50$, or 600 mph, and the speed with the headwind is $550 - 50$, or 500 mph. At 600 mph, the time it takes to travel 3000 mi is $3000/600$, or 5 hr. At 500 mph, the time it takes to travel 3000 mi is $3000/500$, or 6 hr. The times check, so the answer is correct.

5. **State.** The speed of the plane in still air is 550 mph, and the speed of the wind is 50 mph.

Now Try Exercise 67.

EXAMPLE 8 *Supply and Demand.* Suppose that the price and the supply of the Star Station satellite radio are related by the equation

$$y = 90 + 30x,$$

where y is the price, in dollars, at which the seller is willing to supply x thousand units. Also suppose that the price and the demand for the same model of satellite radio are related by the equation

$$y = 200 - 25x,$$

where y is the price, in dollars, at which the consumer is willing to buy x thousand units.

The **equilibrium point** for this radio is the pair (x, y) that is a solution of both equations. The **equilibrium price** is the price at which the amount of the product that the seller is willing to supply is the same as the amount demanded by the consumer. Find the equilibrium point for this radio.

Solution

1., 2. Familiarize and Translate. We are given a system of equations in the statement of the problem, so no further translation is necessary.

$$y = 90 + 30x, \qquad \textbf{(1)}$$
$$y = 200 - 25x \qquad \textbf{(2)}$$

We substitute some values for x in each equation to get an idea of the corresponding prices. When $x = 1$,

$$y = 90 + 30 \cdot 1 = 120, \qquad \text{Substituting in equation (1)}$$
$$y = 200 - 25 \cdot 1 = 175. \qquad \text{Substituting in equation (2)}$$

This indicates that the price when 1 thousand units are supplied is lower than the price when 1 thousand units are demanded.

When $x = 4$,

$$y = 90 + 30 \cdot 4 = 210, \qquad \text{Substituting in equation (1)}$$
$$y = 200 - 25 \cdot 4 = 100. \qquad \text{Substituting in equation (2)}$$

In this case, the price related to supply is higher than the price related to demand. It would appear that the x-value we are looking for is between 1 and 4.

3. Carry out. We use the substitution method:

$$y = 90 + 30x \qquad \text{Equation (1)}$$
$$200 - 25x = 90 + 30x \qquad \text{Substituting } 200 - 25x \text{ for } y \text{ in equation (1)}$$
$$110 = 55x \qquad \text{Adding } 25x \text{ and subtracting 90 on both sides}$$
$$2 = x. \qquad \text{Dividing by 55 on both sides}$$

We now back-substitute 2 for x in either equation and find y:

$$y = 200 - 25x \qquad \textbf{(2)}$$
$$= 200 - 25 \cdot 2 \qquad \text{Substituting 2 for } x$$
$$= 200 - 50$$
$$= 150.$$

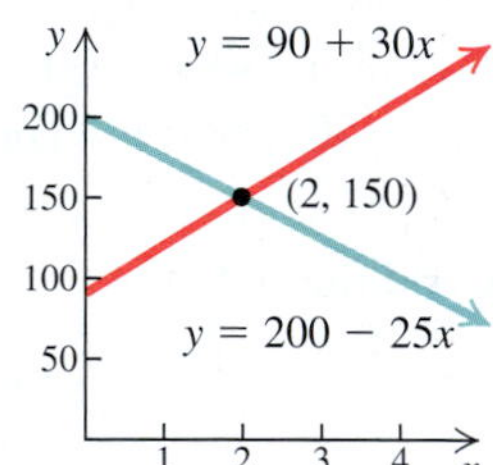

We can visualize the solution as the coordinates of the point of intersection of the graphs of the equations $y = 90 + 30x$ and $y = 200 - 25x$, as shown at left.

4. Check. We can check by substituting 2 for x and 150 for y in both equations. Also note that 2 is between 1 and 4 as expected from the *Familiarize* and *Translate* steps.

5. State. The equilibrium point is $(2, \$150)$. That is, the equilibrium quantity is 2 thousand units and the equilibrium price is \$150.

Now Try Exercise 69.

Visualizing the Graph

Match the equation or system of equations with its graph.

1. $2x - 3y = 6$

2. $f(x) = x^2 - 2x - 3$

3. $f(x) = -x^2 + 4$

4. $(x - 2)^2 + (y + 3)^2 = 9$

5. $f(x) = x^3 - 2$

6. $f(x) = -(x - 1)^2(x + 1)^2$

7. $f(x) = \dfrac{x - 1}{x^2 - 4}$

8. $f(x) = \dfrac{x^2 - x - 6}{x^2 - 1}$

9. $x - y = -1,$
   $2x - y = 2$

10. $3x - y = 3,$
    $2y = 6x - 6$

Answers on page A-52

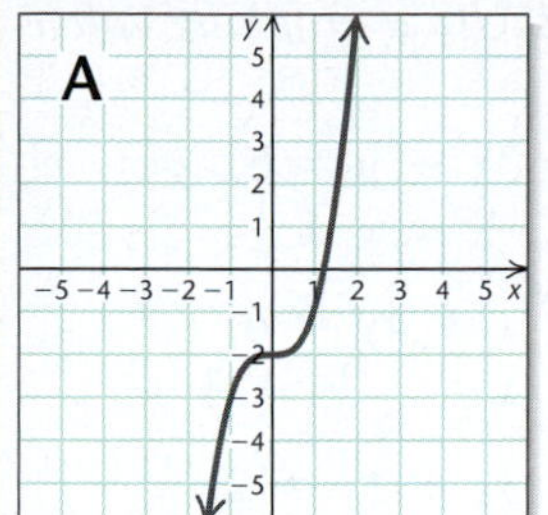

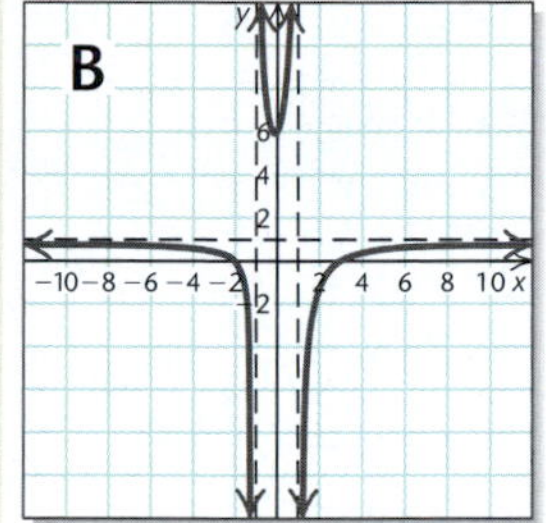

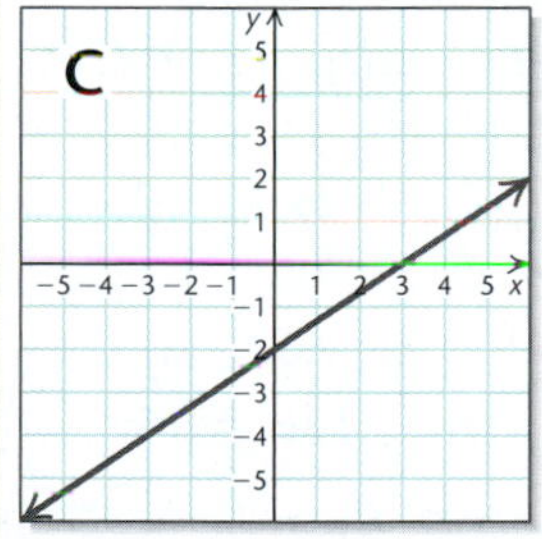

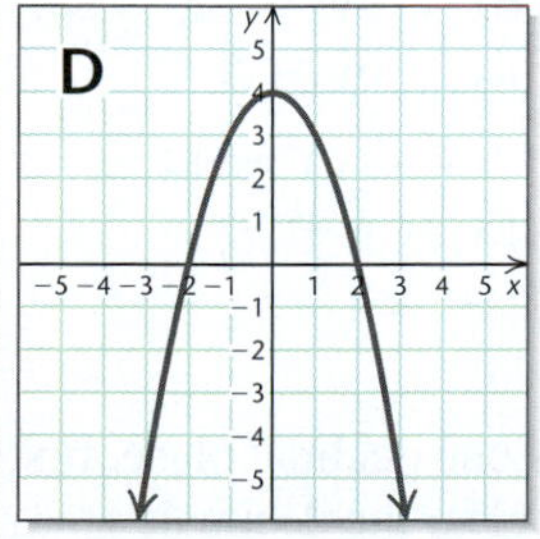

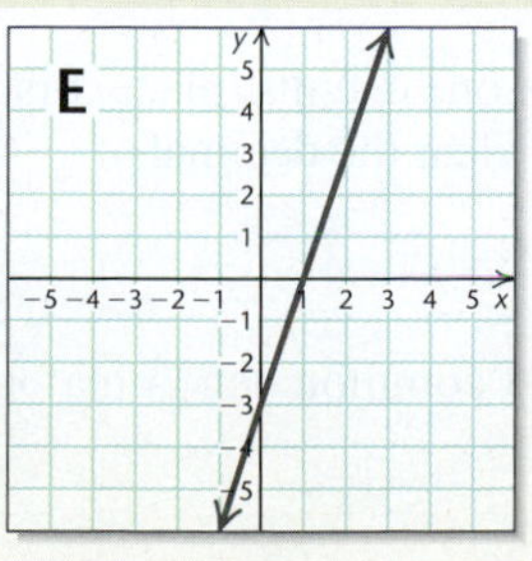

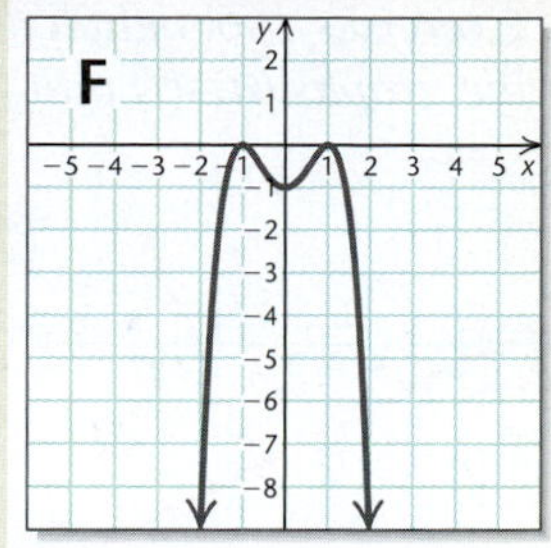

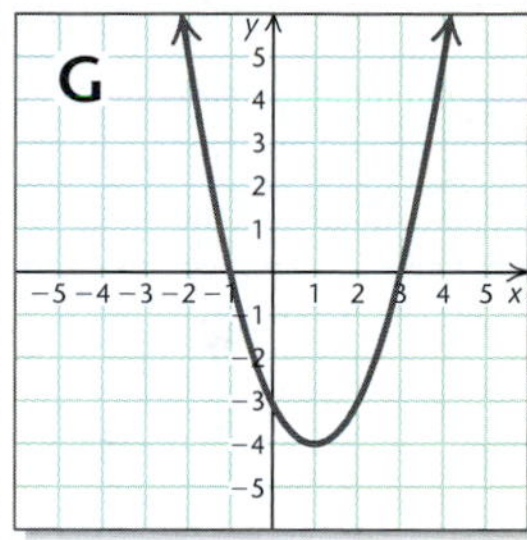

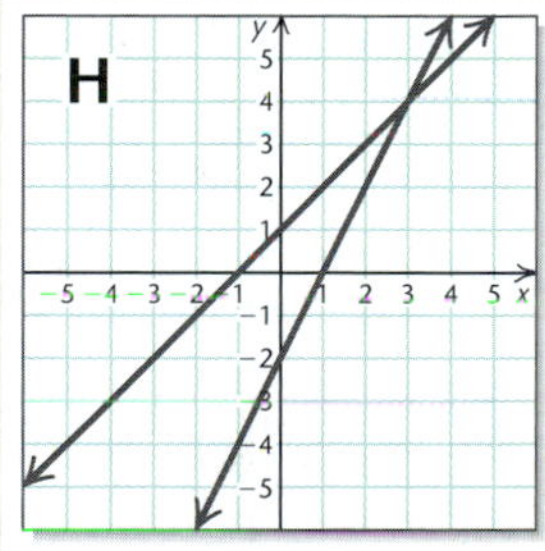

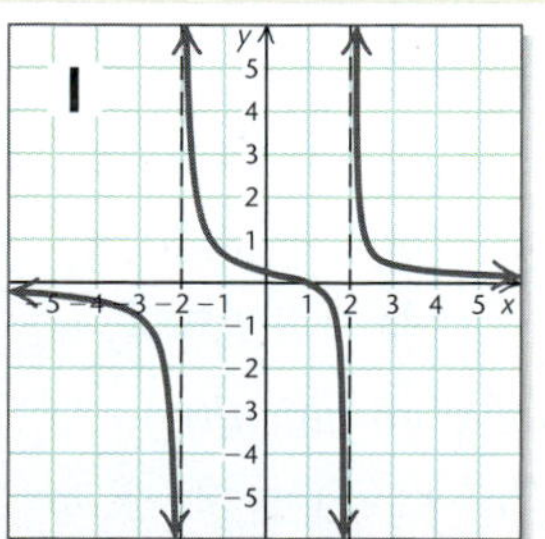

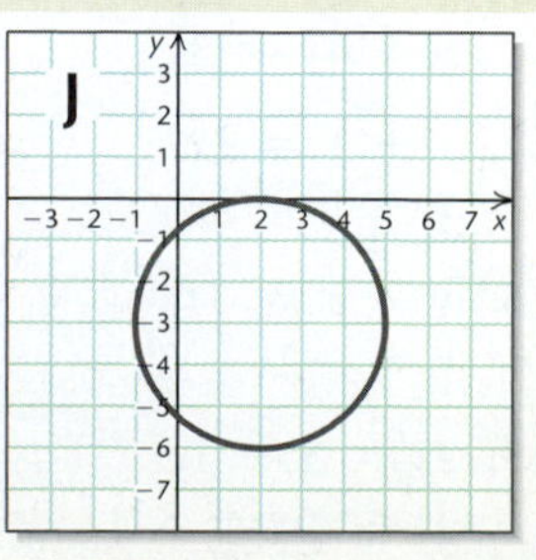

9.1 Exercise Set

In Exercises 1–6, match the system of equations with one of the graphs (a)–(f) that follow.

a)

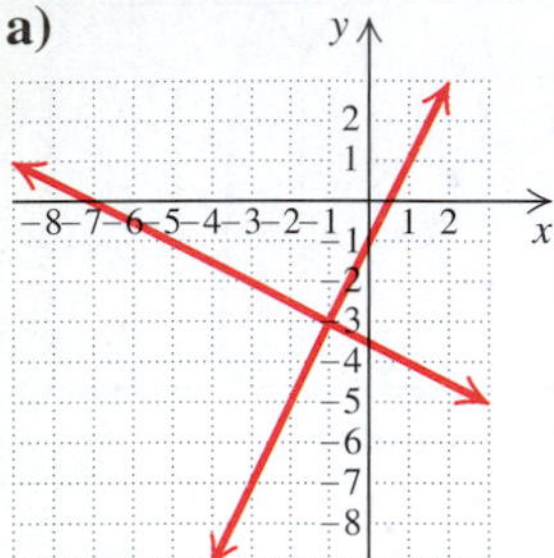

b)

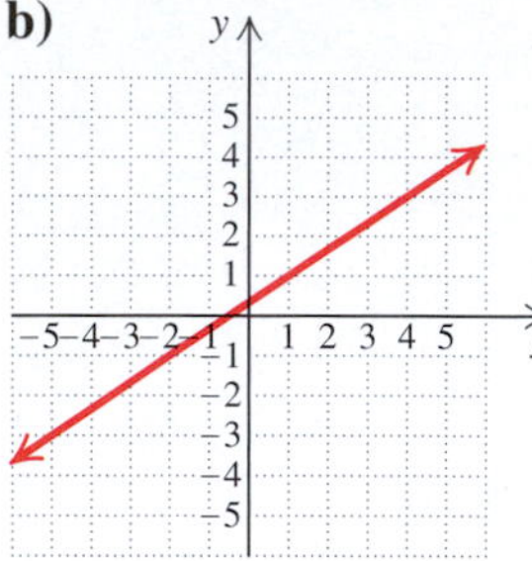

c)

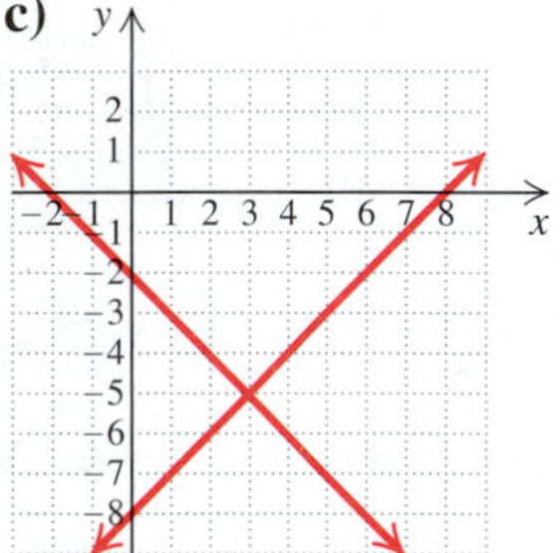

d)

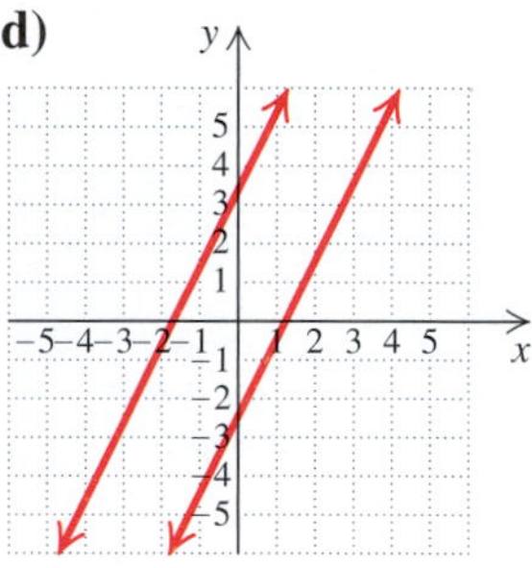

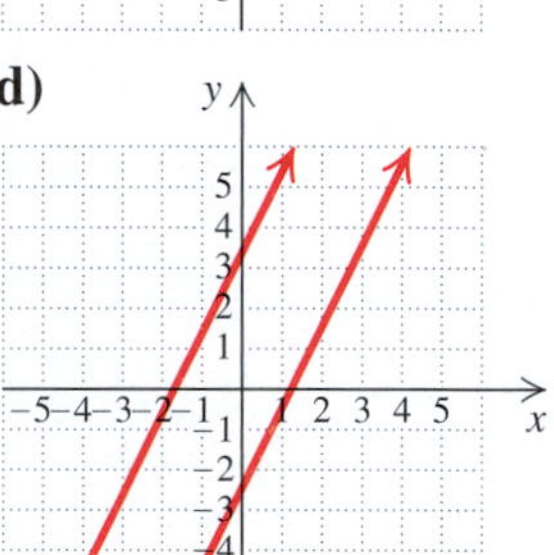

e)

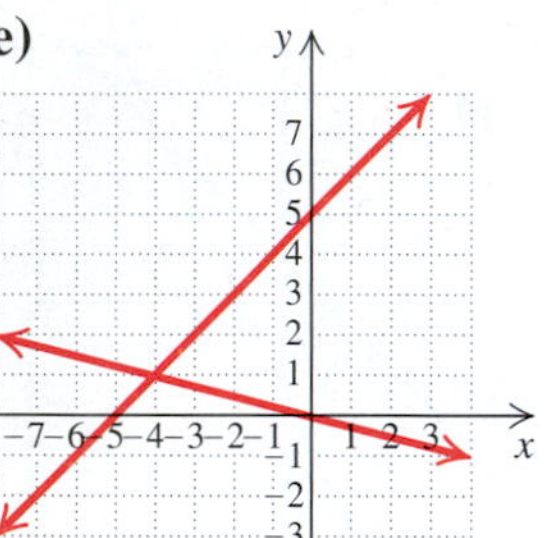

f)

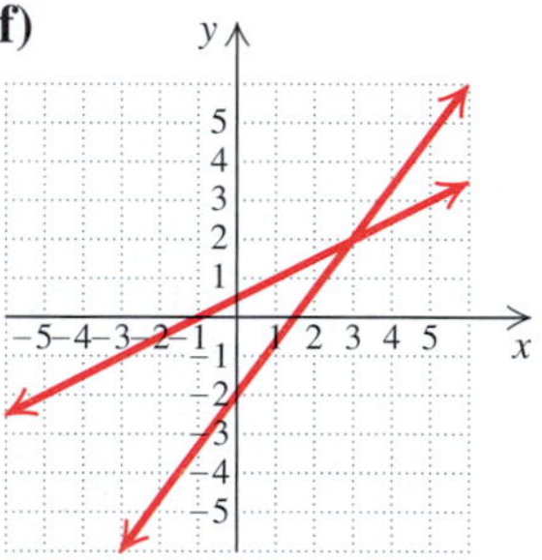

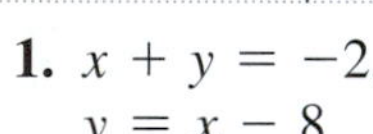

1. $x + y = -2,$
   $y = x - 8$

2. $x - y = -5,$
   $x = -4y$

3. $x - 2y = -1,$
   $4x - 3y = 6$

4. $2x - y = 1,$
   $x + 2y = -7$

5. $2x - 3y = -1,$
   $-4x + 6y = 2$

6. $4x - 2y = 5,$
   $6x - 3y = -10$

Solve graphically.

7. $x + y = 2,$
   $3x + y = 0$

8. $x + y = 1,$
   $3x + y = 7$

9. $x + 2y = 1,$
   $x + 4y = 3$

10. $3x + 4y = 5,$
    $x - 2y = 5$

11. $y + 1 = 2x,$
    $y - 1 = 2x$

12. $2x - y = 1,$
    $3y = 6x - 3$

13. $x - y = -6,$
    $y = -2x$

14. $2x + y = 5,$
    $x = -3y$

15. $2y = x - 1,$
    $3x = 6y + 3$

16. $y = 3x + 2,$
    $3x - y = -3$

Solve using the substitution method. Use a graphing calculator to check your answer.

17. $x + y = 9,$
    $2x - 3y = -2$

18. $3x - y = 5,$
    $x + y = \frac{1}{2}$

19. $x - 2y = 7,$
    $x = y + 4$

20. $x + 4y = 6,$
    $x = -3y + 3$

21. $y = 2x - 6,$
    $5x - 3y = 16$

22. $3x + 5y = 2,$
    $2x - y = -3$

23. $x + y = 3,$
    $y = 4 - x$

24. $x - 2y = 3,$
    $2x = 4y + 6$

25. $x - 5y = 4,$
    $y = 7 - 2x$

26. $5x + 3y = -1,$
    $x + y = 1$

27. $x + 2y = 2,$
    $4x + 4y = 5$

28. $2x - y = 2,$
    $4x + y = 3$

29. $3x - y = 5,$
    $3y = 9x - 15$

30. $2x - y = 7,$
    $y = 2x - 5$

Solve using the elimination method. Also determine whether the system is consistent or inconsistent and whether the equations are dependent or independent. Use a graphing calculator to check your answer.

31. $x + 2y = 7,$
    $x - 2y = -5$

32. $3x + 4y = -2,$
    $-3x - 5y = 1$

33. $x - 3y = 2,$
    $6x + 5y = -34$

34. $x + 3y = 0,$
    $20x - 15y = 75$

35. $3x - 12y = 6,$
    $2x - 8y = 4$

36. $2x + 6y = 7,$
    $3x + 9y = 10$

37. $4x - 2y = 3,$
    $2x - y = 4$

38. $6x + 9y = 12,$
    $4x + 6y = 8$

39. $2x = 5 - 3y,$
    $4x = 11 - 7y$

40. $7(x - y) = 14,$
    $2x = y + 5$

41. $0.3x - 0.2y = -0.9,$
    $0.2x - 0.3y = -0.6$
    (*Hint*: Since each coefficient has one decimal place, first multiply each equation by 10 to clear the decimals.)

42. $0.2x - 0.3y = 0.3,$
    $0.4x + 0.6y = -0.2$
    (*Hint*: Since each coefficient has one decimal place, first multiply each equation by 10 to clear the decimals.)

43. $\frac{1}{5}x + \frac{1}{2}y = 6,$
    $\frac{3}{5}x - \frac{1}{2}y = 2$
    (*Hint*: First multiply by the least common denominator to clear fractions.)

44. $\frac{2}{3}x + \frac{3}{5}y = -17,$
$\frac{1}{2}x - \frac{1}{3}y = -1$
(*Hint*: First multiply by the least common denominator to clear fractions.)

Determine whether the statement is true or false.

45. If the graph of a system of equations is a pair of parallel lines, then the system of equations is inconsistent.

46. If we obtain the equation $0 = 0$ when using the elimination method to solve a system of equations, then the system has no solution.

47. If a system of two linear equations in two variables is consistent, then it has exactly one solution.

48. If a system of two linear equations in two variables is dependent, then it has infinitely many solutions.

49. It is possible for a system of two linear equations in two variables to be consistent and dependent.

50. It is possible for a system of two linear equations in two variables to be inconsistent and dependent.

51. *Cosmetic Surgery.* Liposuction and nose reshaping are two of the five most popular cosmetic surgeries in the United States. Together, these two procedures accounted for 427,676 surgeries in 2014. The number of liposuction surgeries was 6572 fewer than the number of nose-shaping surgeries. (*Source*: American Society of Plastic Surgeons) Find the number of each type of surgery.

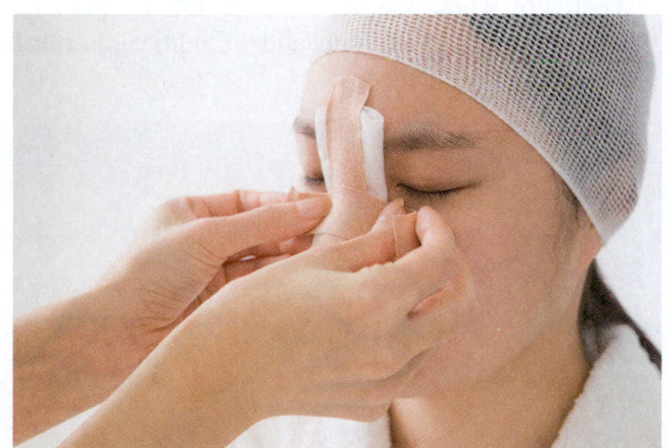

52. *Military Spending.* In 2014, the United States spent \$452 billion more on its military than China spent on its military. Together, China and the United States spent \$710 billion. (*Source*: International Institute for Strategic Studies) Find the amount spent on the military in China and in the United States.

53. *Baggage Fees.* In 2013 and 2014, U.S. airlines collected a total of \$6.85 billion in baggage fees, with the baggage fees in 2014 exceeding those in 2013 by \$0.15 billion (*Source*: U.S. Department of Transportation). Find the amount collected in baggage fees in 2013 and in 2014.

54. *Apartment Rent.* The average apartment rent in the United States is \$1230 per month (*Source*: realtor.com). Jacob has an apartment in Boston and one in San Francisco. The total monthly rent for the two apartments is \$4904. The rent in Boston is \$1142 less than the rent in San Francisco. Find the rent for each apartment.

55. *Chinese–Americans.* About 791,000 Chinese–Americans live in New York City and Boston. The number of Chinese–Americans in New York City is 95,000 more than five times the number in Boston. (*Source*: U.S. Census Bureau) How many Chinese–Americans live in each city?

56. *Calories in Pie.* Using the calorie count of one-eighth of a 9-in. pie, we find that the number of calories in a piece of pecan pie is 221 less than twice the number of calories in a piece of lemon meringue pie. If one eats a piece of each, a total of 865 calories is consumed. (*Source*: *Good Housekeeping*, Good Health, p. 41, November 2007) How many calories are there in each piece of pie?

57. *Mail-Order Business.* A mail-order gardening equipment business shipped 120 packages one day. Customers are charged \$6.50 for each standard-delivery package and \$10.00 for each express-delivery package. Total shipping charges for the day were \$934. How many of each kind of package were shipped?

58. *Concert Ticket Prices.* One evening 1500 concert tickets were sold for the Fairmont Summer Jazz Festival. Tickets cost \$25 for a covered pavilion seat and \$15 for a lawn seat. Total receipts were \$28,500. How many of each type of ticket were sold?

59. *Investment.* Charles inherited \$15,000 and invested it in two municipal bonds that pay 4% and 5% simple interest. The annual interest is \$690. Find the amount invested at each rate.

60. *Sales of Scarves.* During the holiday season, Brianna sold scarves at a kiosk in a shopping mall. Embroidered floral scarves cost \$24 each, and sheer chevron scarves cost \$18 each. One day she sold 39 scarves. Total receipts for the day were \$798. How many of each type of scarf did she sell?

61. *Coffee Mixtures.* The owner of The Daily Grind coffee shop mixes French roast coffee worth \$12.00 per pound with Colombian coffee worth \$9.50 per pound in order to get 20 lb of a mixture worth \$10.50 per pound. How much of each type of coffee was used?

62. *Commissions.* Dyer's Office Solutions offers its sales representatives a choice between being paid a commission of 8% of sales or being paid a monthly salary of \$1500 plus a commission of 1% of sales. For what amount of monthly sales do the two plans pay the same amount?

63. *Nutrition.* A one-cup serving of spaghetti with meatballs contains 260 Cal (calories) and 32 g of carbohydrates. A one-cup serving of chopped iceberg lettuce contains 5 Cal and 1 g of carbohydrates. (*Source*: U.S. Department of Agriculture) How many servings of each would be required in order to obtain 400 Cal and 50 g of carbohydrates?

64. *Nutrition.* One serving of tomato soup contains 100 Cal and 18 g of carbohydrates. One slice of whole wheat bread contains 70 Cal and 13 g of carbohydrates. (*Source*: U.S. Department of Agriculture) How many servings of each would be required in order to obtain 230 Cal and 42 g of carbohydrates?

65. *Motion.* A Leisure Time Cruises riverboat travels 46 km downstream in 2 hr. It travels 51 km upstream in 3 hr. Find the speed of the boat and the speed of the stream.

66. *Motion.* A DC10 airplane travels 3000 km with a tailwind in 3 hr. It travels 3000 km with a headwind in 4 hr. Find the speed of the plane and the speed of the wind.

67. *Motion.* Two private airplanes travel toward each other from cities that are 780 km apart at speeds of 190 km/h and 200 km/h. They leave at the same time. In how many hours will they meet?

68. *Motion.* Aaron's boat travels 45 mi downstream in 3 hr. The return trip upstream takes 5 hr. Find the speed of the boat in still water and the speed of the current.

69. *Supply and Demand.* The supply and demand for an all-terrain skateboard are related to price by the equations

$$y = 140 + 4x,$$
$$y = 275 - 5x,$$

respectively, where y is the price, in dollars, and x is the number of units, in thousands. Find the equilibrium point for this product.

70. *Supply and Demand.* The supply and demand for a particular model of treadmill are related to price by the equations

$$y = 240 + 40x,$$
$$y = 500 - 25x,$$

respectively, where y is the price, in dollars, and x is the number of units, in thousands. Find the equilibrium point for this product.

The point at which a company's costs equal its revenues is the ***break-even point****. In Exercises 71–74, C represents the production cost, in dollars, of x units of a product and R represents the revenue, in dollars, from the sale of x units. Find the number of units that must be produced and sold in order to break even. That is, find the value of x for which* $C = R$.

71. $C = 14x + 350,$
$R = 16.5x$

72. $C = 8.5x + 75,$
$R = 10x$

73. $C = 15x + 12{,}000,$
$R = 18x - 6000$

74. $C = 3x + 400,$
$R = 7x - 600$

75. *Red Meat and Poultry Consumption.* The amount of red meat consumed in the United States has decreased about 7% in recent years while the amount of poultry consumed has increased about 3% during those years, as shown by the data in the following table.

Year	U.S. Red Meat Consumption (in billions of pounds)	U.S. Poultry Consumption (in billions of pounds)
2005	47.385	34.947
2010	45.931	35.201
2014	43.887	36.107

Source: U.S. Department of Agriculture, Economic Research Service

a) Find linear regression functions $r(x)$ and $p(x)$ that represent red meat consumption and poultry consumption, respectively, in billions of pounds x years after 2005.

b) Use the functions found in part (a) to estimate when poultry consumption will equal red meat consumption.

Skill Maintenance

Solve. [1.5]

76. *Paramedic Pay.* In 2012, the annual mean wage of a paramedic in the state of Washington was $50,980. This wage was $8600 less than twice the annual mean wage of a paramedic in Kentucky. (*Source*: U.S. Bureau of Labor Statistics) Find the annual median wage of a paramedic in Kentucky.

77. *Registered Snowmobiles.* There were 251,986 registered snowmobiles in Minnesota in 2013. This was 21,952 more than twice the number of registered snowmobiles in New York. (*Sources*: International Snowmobile Manufacturers Association; Maine Snowmobile Association) Find the number of registered snowmobiles in New York.

78. *International Adoptions.* In 2014, the number of international adoptions in the United States was at its lowest level since 2004. The number of international adoptions in 2014 totaled 6441, a decrease of 71.9% from 2004 (*Source*: U.S. State Department). Find the number of international adoptions in 2004. Round to the nearest ten.

Consider the function

$$f(x) = x^2 - 4x + 3$$

in Exercises 79–82.

79. What are the inputs if the output is 15? [3.2]

80. Given an output of 8, find the corresponding inputs. [3.2]

81. What is the output if the input is -9? [1.2]

82. Find the zeros of the function. [3.2]

Synthesis

83. *Motion.* Nancy jogs and walks to campus each day. She averages 4 km/h walking and 8 km/h jogging. The distance from home to the campus is 6 km, and she makes the trip in 1 hr. How far does she jog on each trip?

84. *e-Commerce.* Shirts.com advertises a limited-time sale, offering 1 turtleneck for $15 and 2 turtlenecks for $25. A total of 1250 turtlenecks are sold and $16,750 is taken in. How many customers ordered 2 turtlenecks?

85. *Motion.* A train leaves Union Station for Central Station, 216 km away, at 9 A.M. One hour later, a train leaves Central Station for Union Station. They meet at noon. If the second train had started at 9 A.M. and the first train at 10:30 A.M., they would still have met at noon. Find the speed of each train.

86. *Antifreeze Mixtures.* An automobile radiator contains 16 L of antifreeze and water. This mixture is 30% antifreeze. How much of this mixture should be drained and replaced with pure antifreeze so that the final mixture will be 50% antifreeze?

87. Two solutions of the equation $Ax + By = 1$ are $(3, -1)$ and $(-4, -2)$. Find A and B.

88. *Ticket Line.* You are in line at a ticket window. There are 2 more people ahead of you in line than there are behind you. In the entire line, there are three times as many people as there are behind you. How many people are ahead of you?

89. *Gas Mileage.* The Jeep Renegade Sport 4×4 vehicle gets 23 miles per gallon (mpg) in city driving and 32 mpg in highway driving (*Source*: *Car and Driver*, May 2015, p. 114). The car is driven 403 mi on 14 gal of gasoline. How many miles were driven in the city and how many were driven on the highway?

9.2 Systems of Equations in Three Variables

- Solve systems of linear equations in three variables.
- Use systems of three equations to solve applied problems.
- Model a situation using a quadratic function.

A **linear equation in three variables** is an equation equivalent to one of the form $Ax + By + Cz = D$, where A, B, C, and D are real numbers and none of A, B, and C is 0. A **solution of a system of three equations in three variables** is an ordered triple that makes all three equations true. For example, the triple $(2, -1, 0)$ is a solution of the system of equations

$$\begin{aligned} 4x + 2y + 5z &= 6, \\ 2x - y + z &= 5, \\ 3x + 2y - z &= 4. \end{aligned}$$

We can verify this by substituting 2 for x, -1 for y, and 0 for z in each equation.

Solving Systems of Equations in Three Variables

We will solve systems of equations in three variables using an algebraic method called **Gaussian elimination**, named for the German mathematician Karl Friedrich Gauss (1777–1855). Our goal is to transform the original system to an equivalent system (one with the same solution set) of the form

$$\begin{aligned} Ax + By + Cz &= D, \\ Ey + Fz &= G, \\ Hz &= K. \end{aligned}$$

Then we solve the third equation for z and back-substitute to find y and then x.

Each of the following operations can be used to transform the original system to an equivalent system in the desired form.

1. Interchange any two equations.
2. Multiply by a nonzero constant on both sides of one of the equations.
3. Add a nonzero multiple of one equation to another equation.

EXAMPLE 1 Solve the following system:

$$\begin{aligned} x - 2y + 3z &= 11, & \textbf{(1)} \\ 4x + 2y - 3z &= 4, & \textbf{(2)} \\ 3x + 3y - z &= 4. & \textbf{(3)} \end{aligned}$$

Solution First, we eliminate x from two pairs of equations. We multiply equation (1) by -4 and add it to equation (2). We also multiply equation (1) by -3 and add it to equation (3).

$$\begin{aligned} -4x + 8y - 12z &= -44 \quad \textbf{Multiplying (1) by } -4 \\ 4x + 2y - 3z &= 4 \quad \textbf{(2)} \\ \hline 10y - 15z &= -40; \quad \textbf{(4)} \end{aligned}$$

$$\begin{aligned} -3x + 6y - 9z &= -33 \quad \textbf{Multiplying (1) by } -3 \\ 3x + 3y - z &= 4 \quad \textbf{(3)} \\ \hline 9y - 10z &= -29. \quad \textbf{(5)} \end{aligned}$$

Now we have

$$\begin{aligned} x - 2y + 3z &= 11, \quad \textbf{(1)} \\ 10y - 15z &= -40, \quad \textbf{(4)} \\ 9y - 10z &= -29. \quad \textbf{(5)} \end{aligned}$$

Next, we multiply equation (5) by 10 to make the y-coefficient a multiple of the y-coefficient in the equation above it:

$$\begin{aligned} x - 2y + 3z &= 11, \quad \textbf{(1)} \\ 10y - 15z &= -40, \quad \textbf{(4)} \\ 90y - 100z &= -290. \quad \textbf{(6)} \end{aligned}$$

Next, we multiply equation (4) by -9 and add it to equation (6):

$$\begin{aligned} -90y + 135z &= 360 \quad \textbf{Multiplying (4) by } -9 \\ 90y - 100z &= -290 \quad \textbf{(6)} \\ \hline 35z &= 70. \quad \textbf{(7)} \end{aligned}$$

We now have the system of equations

$$\begin{aligned} x - 2y + 3z &= 11, \quad \textbf{(1)} \\ 10y - 15z &= -40, \quad \textbf{(4)} \\ 35z &= 70. \quad \textbf{(7)} \end{aligned}$$

Now we solve equation (7) for z:

$$\begin{aligned} 35z &= 70 \\ z &= 2. \end{aligned}$$

Then we back-substitute 2 for z in equation (4) and solve for y:

$$\begin{aligned} 10y - 15 \cdot 2 &= -40 \\ 10y - 30 &= -40 \\ 10y &= -10 \\ y &= -1. \end{aligned}$$

Finally, we back-substitute -1 for y and 2 for z in equation (1) and solve for x:

$$\begin{aligned} x - 2(-1) + 3 \cdot 2 &= 11 \\ x + 2 + 6 &= 11 \\ x &= 3. \end{aligned}$$

We can check the triple $(3, -1, 2)$ in each of the three original equations. Since it makes all three equations true, the solution is $(3, -1, 2)$.

Now Try Exercise 1.

EXAMPLE 2 Solve the following system:

$$x + y + z = 7, \quad (1)$$
$$3x - 2y + z = 3, \quad (2)$$
$$x + 6y + 3z = 25. \quad (3)$$

Solution We multiply equation (1) by -3 and add it to equation (2), getting equation (4). We also multiply equation (1) by -1 and add it to equation (3), getting equation (5).

$$x + y + z = 7, \quad (1)$$
$$-5y - 2z = -18, \quad (4)$$
$$5y + 2z = 18 \quad (5)$$

Next, we add equation (4) to equation (5), getting equation (6).

$$x + y + z = 7, \quad (1)$$
$$-5y - 2z = -18, \quad (4)$$
$$0 = 0. \quad (6)$$

The equation $0 = 0$ tells us that equations (1), (2), and (3) are dependent. This means that the original system of three equations is equivalent to a system of two equations. One way to see this is to note that four times equation (1) minus equation (2) is equation (3). Thus removing equation (3) from the system does not affect the solution of the system. We can say that the original system is equivalent to

$$x + y + z = 7, \quad (1)$$
$$3x - 2y + z = 3. \quad (2)$$

In this particular case, the original system has infinitely many solutions. (In some cases, a system containing dependent equations is inconsistent.) To find an expression for these solutions, we first solve equation (4) for either y or z. We choose to solve for y:

$$-5y - 2z = -18 \quad (4)$$
$$-5y = 2z - 18$$
$$y = -\tfrac{2}{5}z + \tfrac{18}{5}.$$

Then we back-substitute in equation (1) to find an expression for x in terms of z:

$$x - \tfrac{2}{5}z + \tfrac{18}{5} + z = 7 \quad \text{Substituting } -\tfrac{2}{5}z + \tfrac{18}{5} \text{ for } y$$
$$x + \tfrac{3}{5}z + \tfrac{18}{5} = 7$$
$$x + \tfrac{3}{5}z = \tfrac{17}{5}$$
$$x = -\tfrac{3}{5}z + \tfrac{17}{5}.$$

The solutions of the system of equations are ordered triples of the form $\left(-\tfrac{3}{5}z + \tfrac{17}{5}, -\tfrac{2}{5}z + \tfrac{18}{5}, z\right)$, where z can be any real number. Any real number that we use for z then gives us values for x and y and thus an ordered triple in the solution set. For example, if we choose $z = 0$, we have the solution $\left(\tfrac{17}{5}, \tfrac{18}{5}, 0\right)$. If we choose $z = -1$, we have $(4, 4, -1)$.

Now Try Exercise 9.

If we get a false equation, such as $0 = -5$, at some stage of the elimination process, we conclude that the original system is *inconsistent*; that is, it has no solutions.

Although systems of three linear equations in three variables do not lend themselves well to graphical solutions, it is of interest to picture some possible solutions. The graph of a linear equation in three variables is a plane. Thus the solution set of such a system is the intersection of three planes. Some possibilities are shown below.

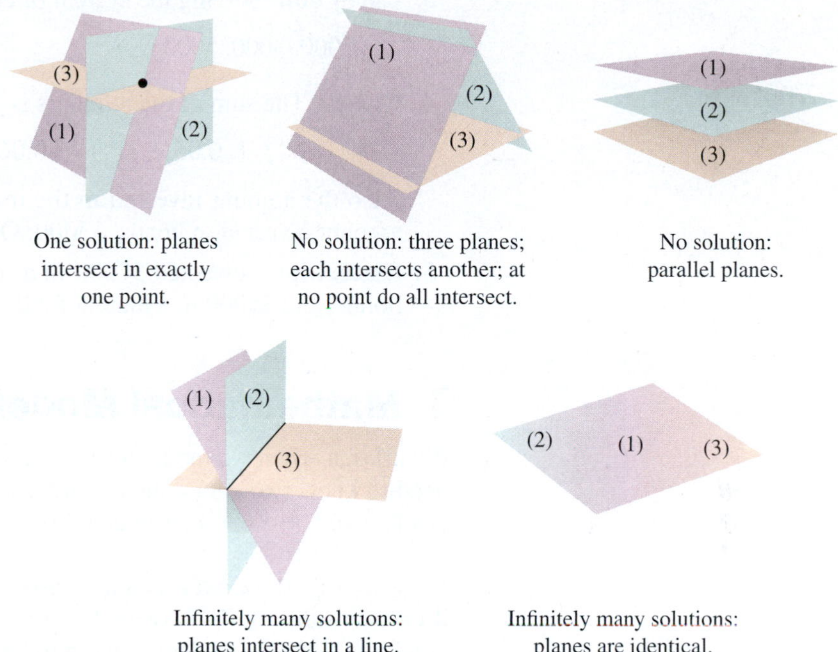

One solution: planes intersect in exactly one point.

No solution: three planes; each intersects another; at no point do all intersect.

No solution: parallel planes.

Infinitely many solutions: planes intersect in a line.

Infinitely many solutions: planes are identical.

Applications

Systems of equations in three or more variables allow us to solve many problems in fields such as business, the social and natural sciences, and engineering.

EXAMPLE 3 ***Investment.*** Moira inherited \$15,000 and invested part of it in a money market account, part in municipal bonds, and part in a mutual fund. After 1 year, she received a total of \$730 in simple interest from the three investments. The money market account paid 4% annually, the bonds paid 5% annually, and the mutual fund paid 6% annually. There was \$2000 more invested in the mutual fund than in bonds. Find the amount that Moira invested in each category.

Solution

1. **Familiarize.** We let x, y, and z represent the amounts invested in the money market account, the bonds, and the mutual fund, respectively. Then the amounts of income produced annually by each investment are given by $4\%x$, $5\%y$, and $6\%z$, or $0.04x$, $0.05y$, and $0.06z$.
2. **Translate.** The fact that a total of \$15,000 is invested gives us one equation:

   $$x + y + z = 15{,}000.$$

   Since the total amount of interest is \$730, we have a second equation:

   $$0.04x + 0.05y + 0.06z = 730.$$

   Another statement in the problem gives us a third equation.

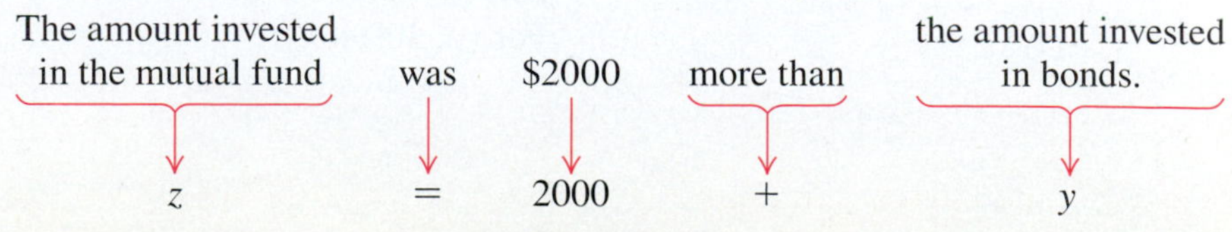

We now have a system of three equations:

$$x + y + z = 15{,}000, \qquad x + y + z = 15{,}000,$$
$$0.04x + 0.05y + 0.06z = 730, \quad \text{or} \quad 4x + 5y + 6z = 73{,}000,$$
$$z = 2000 + y; \qquad -y + z = 2000.$$

3. **Carry out.** Solving the system of equations, we get

$(7000, 3000, 5000)$.

4. **Check.** The sum of the numbers is 15,000. The income produced is

$0.04(7000) + 0.05(3000) + 0.06(5000) = 280 + 150 + 300$, or $730.

Also the amount invested in the mutual fund, $5000, is $2000 more than the amount invested in bonds, $3000. Our solution checks in the original problem.

5. **State.** Moira invested $7000 in a money market account, $3000 in municipal bonds, and $5000 in a mutual fund.

Now Try Exercise 29.

Mathematical Models and Applications

Recall that when we model a situation using a linear function $f(x) = mx + b$, we need to know two data points in order to determine m and b. For a quadratic model, $f(x) = ax^2 + bx + c$, we need three data points in order to determine a, b, and c.

EXAMPLE 4 *Civil Cases in Court.* The following table lists the number of civil cases pending in U.S. federal courts in three recent years. Use the data to find a quadratic function that gives the number of pending civil cases as a function of the number of years after 2009. Then use this function to estimate the number of civil cases pending in U.S. federal courts in 2014.

Year, x	Number of Civil Cases Pending in U.S. Federal Courts
2009, 0	311,353
2011, 2	267,495
2013, 4	300,469

Source: Administrative Office of the U.S. Courts

Solution We let $x =$ the number of years after 2009 and $n(x) =$ the number of pending civil cases. Then $x = 0$ corresponds to 2009, $x = 2$ corresponds to 2011, and $x = 4$ corresponds to 2013. We use three data points $(0, 311{,}353)$, $(2, 267{,}495)$, and $(4, 300{,}469)$ to find a, b, and c in the function $n(x) = ax^2 + bx + c$. First, we substitute:

$$n(x) = ax^2 + bx + c.$$

For $(0, 311{,}353)$: $\quad 311{,}353 = a \cdot 0^2 + b \cdot 0 + c,$

For $(2, 267{,}495)$: $\quad 267{,}495 = a \cdot 2^2 + b \cdot 2 + c,$

For $(4, 300{,}469)$: $\quad 300{,}469 = a \cdot 4^2 + b \cdot 4 + c.$

26. *Spring Cleaning.* In a group of 100 adults, 70 say they are most likely to do spring housecleaning in March, April, or May. Of these 70, the number who clean in April is 14 more than the total number who clean in March and May. The total number who clean in April and May is 2 more than three times the number who clean in March. (*Source*: Zoomerang online survey) Find the number who clean in each month.

27. *Nutrition.* A hospital dietician must plan a lunch menu that provides 485 Cal, 41.5 g of carbohydrates, and 35 mg of calcium. A 3-oz serving of broiled ground beef contains 245 Cal, 0 g of carbohydrates, and 9 mg of calcium. One baked potato contains 145 Cal, 34 g of carbohydrates, and 8 mg of calcium. A one-cup serving of strawberries contains 45 Cal, 10 g of carbohydrates, and 21 mg of calcium. (*Source*: U.S. Department of Agriculture) How many servings of each are required in order to provide the desired nutritional values?

28. *Nutrition.* A diabetic patient wishes to prepare a meal consisting of roasted chicken breast, mashed potatoes, and peas. A 3-oz serving of roasted skinless chicken breast contains 140 Cal, 27 g of protein, and 64 mg of sodium. A one-cup serving of mashed potatoes contains 160 Cal, 4 g of protein, and 636 mg of sodium, and a one-cup serving of peas contains 125 Cal, 8 g of protein, and 139 mg of sodium. (*Source*: U.S. Department of Agriculture) How many servings of each should be used if the meal is to contain 415 Cal, 50.5 g of protein, and 553 mg of sodium?

29. *Investment.* Santiago receives $126 per year in simple interest from three investments. Part is invested at 2%, part at 3%, and part at 4%. There is $500 more invested at 3% than at 2%. The amount invested at 4% is three times the amount invested at 3%. Find the amount invested at each rate.

30. *Investment.* Walter earns a year-end bonus of $5000 and puts it in 3 one-year investments that pay $243 in simple interest. Part is invested at 3%, part at 4%, and part at 6%. There is $1500 more invested at 6% than at 3%. Find the amount invested at each rate.

31. *Price Increases.* In San Francisco, orange juice, a raisin bagel, and a cup of coffee from Katie's Koffee Kart cost a total of $8.15. Katie posts a notice announcing that, effective the following week, the price of orange juice will increase 25% and the price of bagels will increase 20%. After the increase, the same purchase will cost a total of $9.30, and the raisin bagel will cost 30¢ more than coffee. Find the price of each item before the increase.

32. *Cost of Snack Food.* Chad and Brittany pool their loose change to buy snacks on their coffee break. One day, they spent $6.75 on 1 carton of milk, 2 donuts, and 1 cup of coffee. The next day, they spent $8.50 on 3 donuts and 2 cups of coffee. The third day, they bought 1 carton of milk, 1 donut, and 2 cups of coffee and spent $7.25. On the fourth day, they have a total of $6.45 left. Is this enough to buy 2 cartons of milk and 2 donuts?

33. *Hours Spent Studying.* The following table lists the percent of college freshmen who responded that they spent 6 or more hours per week studying during their high school senior year, which is represented in terms of the number of years after 1992.

Year, x	Percent Who Say They Studied 6 or More Hours per Week
1992, 0	43
2002, 10	33
2012, 20	38

Source: 2014 Brown Center Report on Education, compiled from "The American Freshman," UCLA Higher Education Research Institute

a) Use a system of equations to fit a quadratic function $f(x) = ax^2 + bx + c$ to the data.

b) Use the function to estimate the percent of college freshmen who say they spent 6 or more hours per week studying during their high school senior year in 2007 and in 2014.

34. *Existing Home Sales.* The following table lists the number of U.S. existing home sales, in millions, represented in terms of years after 2006.

Year, x	Existing Home Sales (in millions)
2006, 0	6.5
2009, 3	4.3
2012, 6	4.7

Source: National Association of Realtors from Haver Analytics

a) Use a system of equations to fit a quadratic function $f(x) = ax^2 + bx + c$ to the data.

b) Use the function to estimate the number of existing home sales in 2015.

35. *Deportations.* The following table lists the number of foreigners deported, in thousands, represented in terms of the number of years after 2007.

Year, x	U.S. Deportations (in thousands)
2007, 0	291
2010, 3	393
2013, 6	369

Source: U.S. Immigration and Customs Enforcement

a) Use a system of equations to fit a quadratic function $f(x) = ax^2 + bx + c$ to the data.
b) Use the function to estimate the number of deportations in 2014.

36. *Strength of Army.* The following table lists the number of active-duty personnel in the U.S. army, in thousands, represented in terms of the number of years after 2009.

Year, x	Active-Duty Army Personnel (in thousands)
2009, 0	553
2011, 2	565
2013, 4	541

Source: Department of the Army, U.S. Department of Defense

a) Use a system of equations to fit a quadratic function $f(x) = ax^2 + bx + c$ to the data.
b) Use the function to estimate the number of active-duty Army personnel in 2010 and in 2014.

37. *Gasoline Prices.* The following table lists the average U.S. retail price per gallon of gasoline on December 12 for years from 2009 to 2013.

Year, x	U.S. Average Retail Price Per Gallon of Gasoline on December 12
2009, 0	$2.61
2010, 1	2.97
2011, 2	3.27
2012, 3	3.32
2013, 4	3.25

Source: American Automobile Association

a) Use a graphing calculator to fit a quadratic function $f(x) = ax^2 + bx + c$ to the data, where x is the number of years after 2009.
b) Use the function found in part (a) to estimate the average retail price per gallon of gasoline on December 12 in 2014.

38. *Unemployment Rate.* The following table lists the U.S. unemployment rate in October for selected years from 2002 to 2010.

Year	U.S. Unemployment Rate in October
2002, 0	5.7%
2004, 2	5.5
2005, 3	5.0
2008, 6	6.6
2010, 8	9.7

Source: U.S. Bureau of Labor Statistics

a) Use a graphing calculator to fit a quadratic function $f(x) = ax^2 + bx + c$ to the data, where x is the number of years after 2002.
b) Use the function found in part (a) to estimate the unemployment rate in 2003, in 2007, and in 2009.

Skill Maintenance

Vocabulary Reinforcement

In each of Exercises 39–46, fill in the blank with the correct term. Some of the given choices will not be used.

Descartes' rule of signs	constant function
the leading-term test	horizontal asymptote
the intermediate value theorem	vertical asymptote
the fundamental theorem of algebra	oblique asymptote
polynomial function	direct variation
rational function	inverse variation
one-to-one function	horizontal line
	vertical line
	parallel
	perpendicular

39. Two lines with slopes m_1 and m_2 are ______________ if and only if the product of their slopes is -1. [1.4]

40. We can use ______________ to determine the behavior of the graph of a polynomial function as $x \longrightarrow \infty$ or as $x \longrightarrow -\infty$. [4.1]

41. If it is possible for a(n) ______________ to cross a graph more than once, then the graph is not the graph of a function. [1.2]

42. A function is a(n) ______________ if different inputs have different outputs. [5.1]

43. A(n) ______________ is a function that is a quotient of two polynomials. [4.5]

44. If a situation gives rise to a function $f(x) = k/x$, or $y = k/x$, where k is a positive constant, we say that we have ______________. [2.5]

45. A(n) ______________ of a rational function $p(x)/q(x)$, where $p(x)$ and $q(x)$ have no common factors other than constants, occurs at an x-value that makes the denominator 0. [4.5]

46. When the numerator and the denominator of a rational function have the same degree, the graph of the function has a(n) ______________. [4.5]

› Synthesis

47. Let u represent $1/x$, v represent $1/y$, and w represent $1/z$. Solve first for u, v, and w. Then solve the following system of equations:

$$\frac{2}{x} + \frac{2}{y} - \frac{3}{z} = 3,$$
$$\frac{1}{x} - \frac{2}{y} - \frac{3}{z} = 9,$$
$$\frac{7}{x} - \frac{2}{y} + \frac{9}{z} = -39.$$

48. *Transcontinental Railroad.* Use the following facts to find the year in which the first U.S. transcontinental railroad was completed: The sum of the digits in the year is 24. The units digit is 1 more than the hundreds digit. Both the tens and the units digits are multiples of three.

In Exercises 49 and 50, three solutions of an equation are given. Use a system of three equations in three variables to find the constants and write the equation.

49. $Ax + By + Cz = 12$;
$\left(1, \frac{3}{4}, 3\right)$, $\left(\frac{4}{3}, 1, 2\right)$, and $(2, 1, 1)$

50. $y = B - Mx - Nz$;
$(1, 1, 2)$, $(3, 2, -6)$, and $\left(\frac{3}{2}, 1, 1\right)$

In Exercises 51 and 52, four solutions of the equation $y = ax^3 + bx^2 + cx + d$ are given. Use a system of four equations in four variables to find the constants a, b, c, and d and write the equation.

51. $(-2, 59)$, $(-1, 13)$, $(1, -1)$, and $(2, -17)$

52. $(-2, -39)$, $(-1, -12)$, $(1, -6)$, and $(3, 16)$

9.3 Matrices and Systems of Equations

› Solve systems of equations using matrices.

› Matrices and Row-Equivalent Operations

In this section, we consider additional techniques for solving systems of equations. You have probably noted that when we solve a system of equations, we perform computations with the coefficients and the constants and continually rewrite the variables. We can streamline the solution process by omitting the variables until a solution is found. For example, the system

$$2x - 3y = 7,$$
$$x + 4y = -2$$

can be written more simply as

$$\left[\begin{array}{rr|r} 2 & -3 & 7 \\ 1 & 4 & -2 \end{array}\right].$$

The vertical line replaces the equals signs.

A rectangular array of numbers like the one above is called a **matrix** (pl., **matrices**). The matrix above is called an **augmented matrix** for the given system of equations, because it contains not only the coefficients but also the constant terms. The matrix

$$\begin{bmatrix} 2 & -3 \\ 1 & 4 \end{bmatrix}$$

is called the **coefficient matrix** of the system.

The **rows** of a matrix are horizontal, and the **columns** are vertical. The augmented matrix above has 2 rows and 3 columns, and the coefficient matrix has 2 rows and 2 columns. A matrix with m rows and n columns is said to be of **order** $m \times n$. Thus the order of the augmented matrix above is 2×3, and the order of the coefficient matrix is 2×2. When $m = n$, a matrix is said to be **square**. The coefficient matrix above is a square matrix. The numbers 2 and 4 lie on the **main diagonal** of the coefficient matrix. The numbers in a matrix are called **entries** or **elements**.

Gaussian Elimination with Matrices

In Section 9.2, we described a series of operations that can be used to transform a system of equations to an equivalent system. Each of these operations corresponds to one that can be used to produce *row-equivalent matrices.*

ROW-EQUIVALENT OPERATIONS

1. Interchange any two rows.
2. Multiply each entry in a row by the same nonzero constant.
3. Add a nonzero multiple of one row to another row.

We can use these operations on the augmented matrix of a system of equations to solve the system.

EXAMPLE 1 Solve the following system:

$$\begin{aligned} 2x - y + 4z &= -3, \\ x - 2y - 10z &= -6, \\ 3x \qquad + 4z &= 7. \end{aligned}$$

Solution First, we write the augmented matrix, writing 0 for the missing y-term in the last equation:

$$\left[\begin{array}{rrr|r} 2 & -1 & 4 & -3 \\ 1 & -2 & -10 & -6 \\ 3 & 0 & 4 & 7 \end{array}\right].$$

Our goal is to find a row-equivalent matrix of the form shown below.

$$\left[\begin{array}{ccc|c} 1 & a & b & c \\ 0 & 1 & d & e \\ 0 & 0 & 1 & f \end{array}\right]$$

The variables can then be reinserted to form equations from which we can complete the solution. This is done by working from the bottom equation to the top and using back-substitution.

The first step is to multiply and/or interchange rows so that each number in the first column below the first number is a multiple of that number. In this case, we interchange the first and second rows to obtain a 1 in the upper left-hand corner.

$$\left[\begin{array}{rrr|r} 1 & -2 & -10 & -6 \\ 2 & -1 & 4 & -3 \\ 3 & 0 & 4 & 7 \end{array}\right] \quad \begin{array}{l} \text{New row 1} = \text{row 2} \\ \text{New row 2} = \text{row 1} \end{array}$$

Just in Time 5

Next, we multiply the first row by -2 and add it to the second row. We also multiply the first row by -3 and add it to the third row.

$$\left[\begin{array}{rrr|r} 1 & -2 & -10 & -6 \\ 0 & 3 & 24 & 9 \\ 0 & 6 & 34 & 25 \end{array}\right] \quad \begin{array}{l} \textbf{Row 1 is unchanged.} \\ \textbf{New row 2} = -2(\textbf{row 1}) + \textbf{row 2} \\ \textbf{New row 3} = -3(\textbf{row 1}) + \textbf{row 3} \end{array}$$

Now we multiply the second row by $\frac{1}{3}$ to get a 1 in the second row, second column.

$$\left[\begin{array}{rrr|r} 1 & -2 & -10 & -6 \\ 0 & 1 & 8 & 3 \\ 0 & 6 & 34 & 25 \end{array}\right] \quad \textbf{New row 2} = \tfrac{1}{3}(\textbf{row 2})$$

Then we multiply the second row by -6 and add it to the third row.

$$\left[\begin{array}{rrr|r} 1 & -2 & -10 & -6 \\ 0 & 1 & 8 & 3 \\ 0 & 0 & -14 & 7 \end{array}\right] \quad \textbf{New row 3} = -6(\textbf{row 2}) + \textbf{row 3}$$

Finally, we multiply the third row by $-\frac{1}{14}$ to get a 1 in the third row, third column.

$$\left[\begin{array}{rrr|r} 1 & -2 & -10 & -6 \\ 0 & 1 & 8 & 3 \\ 0 & 0 & 1 & -\frac{1}{2} \end{array}\right] \quad \textbf{New row 3} = -\tfrac{1}{14}(\textbf{row 3})$$

Now we can write the system of equations that corresponds to the last matrix above:

$$\begin{aligned} x - 2y - 10z &= -6, &\quad \textbf{(1)} \\ y + 8z &= 3, &\quad \textbf{(2)} \\ z &= -\tfrac{1}{2}. &\quad \textbf{(3)} \end{aligned}$$

We back-substitute $-\frac{1}{2}$ for z in equation (2) and solve for y:

$$\begin{aligned} y + 8\left(-\tfrac{1}{2}\right) &= 3 \\ y - 4 &= 3 \\ y &= 7. \end{aligned}$$

Next, we back-substitute 7 for y and $-\frac{1}{2}$ for z in equation (1) and solve for x:

$$\begin{aligned} x - 2 \cdot 7 - 10\left(-\tfrac{1}{2}\right) &= -6 \\ x - 14 + 5 &= -6 \\ x - 9 &= -6 \\ x &= 3. \end{aligned}$$

The triple $\left(3, 7, -\frac{1}{2}\right)$ checks in the original system of equations, so it is the solution.

Now Try Exercise 27.

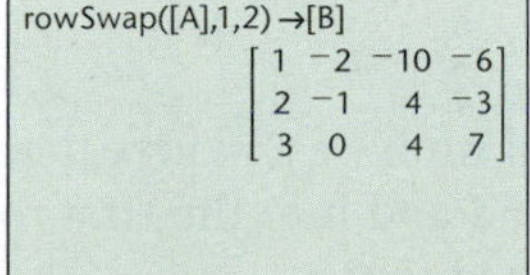

Row-equivalent operations can be performed on a graphing calculator. For example, to interchange the first and second rows of the augmented matrix, as we did in the first step in Example 1, we enter the matrix as matrix **A** and select "rowSwap" from the MATRIX MATH menu. Some graphing calculators will not automatically store the matrix produced using a row-equivalent operation, so when several operations are to be performed in succession, it is helpful to store the result of each operation as it is produced. In the window at left, we see both the matrix produced by the rowSwap operation and the indication that this matrix is stored as matrix **B**.

The procedure followed in Example 1 is called **Gaussian elimination with matrices**. The last matrix in Example 1 is in **row-echelon form**. To be in this form, a matrix must have the following properties.

ROW-ECHELON FORM

1. If a row does not consist entirely of 0's, then the first nonzero element in the row is a 1 (called a **leading 1**).
2. For any two successive nonzero rows, the leading 1 in the lower row is farther to the right than the leading 1 in the higher row.
3. All the rows consisting entirely of 0's are at the bottom of the matrix.

If a fourth property is also satisfied, a matrix is said to be in **reduced row-echelon form**:

4. Each column that contains a leading 1 has 0's everywhere else.

EXAMPLE 2 Which of the following matrices are in row-echelon form? Which, if any, are in reduced row-echelon form?

a) $\left[\begin{array}{ccc|c} 1 & -3 & 5 & -2 \\ 0 & 1 & -4 & 3 \\ 0 & 0 & 1 & 10 \end{array}\right]$ **b)** $\left[\begin{array}{cc|c} 0 & -1 & 2 \\ 0 & 1 & 5 \end{array}\right]$ **c)** $\left[\begin{array}{cccc|c} 1 & -2 & -6 & 4 & 7 \\ 0 & 3 & 5 & -8 & -1 \\ 0 & 0 & 1 & 9 & 2 \end{array}\right]$

d) $\left[\begin{array}{ccc|c} 1 & 0 & 0 & -2.4 \\ 0 & 1 & 0 & 0.8 \\ 0 & 0 & 1 & 5.6 \end{array}\right]$ **e)** $\left[\begin{array}{cccc|c} 1 & 0 & 0 & 0 & \frac{2}{3} \\ 0 & 1 & 0 & 0 & -\frac{1}{4} \\ 0 & 0 & 1 & 0 & \frac{6}{7} \\ 0 & 0 & 0 & 0 & 0 \end{array}\right]$ **f)** $\left[\begin{array}{ccc|c} 1 & -4 & 2 & 5 \\ 0 & 0 & 0 & 0 \\ 0 & 1 & -3 & -8 \end{array}\right]$

Solution The matrices in (a), (d), and (e) satisfy the row-echelon criteria and, thus, are in row-echelon form. In (b) and (c), the first nonzero elements of the first and second rows, respectively, are not 1. In (f), the row consisting entirely of 0's is not at the bottom of the matrix. Thus the matrices in (b), (c), and (f) are not in row-echelon form. In (d) and (e), not only are the row-echelon criteria met but each column that contains a leading 1 also has 0's elsewhere, so these matrices are in reduced row-echelon form.

Gauss–Jordan Elimination

We have seen that with Gaussian elimination we perform row-equivalent operations on a matrix to obtain a row-equivalent matrix in row-echelon form. When we continue to apply these operations until we have a matrix in *reduced* row-echelon form, we are using **Gauss–Jordan elimination**. This method is named for Karl Friedrich Gauss and Wilhelm Jordan (1842–1899).

EXAMPLE 3 Use Gauss–Jordan elimination to solve the system of equations in Example 1.

Solution Using Gaussian elimination in Example 1, we obtained the matrix

$$\left[\begin{array}{ccc|c} 1 & -2 & -10 & -6 \\ 0 & 1 & 8 & 3 \\ 0 & 0 & 1 & -\frac{1}{2} \end{array}\right].$$

We continue to perform row-equivalent operations until we have a matrix in reduced row-echelon form. We multiply the third row by 10 and add it to the first row. We also multiply the third row by -8 and add it to the second row.

$$\left[\begin{array}{ccc|c} 1 & -2 & 0 & -11 \\ 0 & 1 & 0 & 7 \\ 0 & 0 & 1 & -\frac{1}{2} \end{array}\right] \quad \begin{array}{l} \text{New row 1} = 10(\text{row 3}) + \text{row 1} \\ \text{New row 2} = -8(\text{row 3}) + \text{row 2} \\ \end{array}$$

Next, we multiply the second row by 2 and add it to the first row.

$$\left[\begin{array}{ccc|c} 1 & 0 & 0 & 3 \\ 0 & 1 & 0 & 7 \\ 0 & 0 & 1 & -\frac{1}{2} \end{array}\right] \quad \text{New row 1} = 2(\text{row 2}) + \text{row 1}$$

Writing the system of equations that corresponds to this matrix, we have

$$\begin{aligned} x \qquad &= 3, \\ y \quad &= 7, \\ z &= -\tfrac{1}{2}. \end{aligned}$$

We can actually read the solution, $\left(3, 7, -\frac{1}{2}\right)$, directly from the last column of the reduced row-echelon matrix.

rref([A])▸Frac

$$\left[\begin{array}{cccc} 1 & 0 & 0 & 3 \\ 0 & 1 & 0 & 7 \\ 0 & 0 & 1 & -\frac{1}{2} \end{array}\right]$$

We can also use a graphing calculator to solve this system of equations. After the augmented matrix is entered, reduced row-echelon form can be found directly using the "rref" operation from the MATRIX MATH menu.

Now Try Exercise 27.

EXAMPLE 4 Solve the following system:

$$\begin{aligned} 3x - 4y - z &= 6, \\ 2x - y + z &= -1, \\ 4x - 7y - 3z &= 13. \end{aligned}$$

Solution We write the augmented matrix and use Gauss–Jordan elimination.

$$\left[\begin{array}{ccc|c} 3 & -4 & -1 & 6 \\ 2 & -1 & 1 & -1 \\ 4 & -7 & -3 & 13 \end{array}\right]$$

We begin by multiplying the second and third rows by 3 so that each number in the first column below the first number, 3, is a multiple of that number.

$$\left[\begin{array}{ccc|c} 3 & -4 & -1 & 6 \\ 6 & -3 & 3 & -3 \\ 12 & -21 & -9 & 39 \end{array}\right] \quad \begin{array}{l} \text{New row 2} = 3(\text{row 2}) \\ \text{New row 3} = 3(\text{row 3}) \end{array}$$

Next, we multiply the first row by -2 and add it to the second row. We also multiply the first row by -4 and add it to the third row.

$$\left[\begin{array}{ccc|c} 3 & -4 & -1 & 6 \\ 0 & 5 & 5 & -15 \\ 0 & -5 & -5 & 15 \end{array}\right] \quad \begin{array}{l} \text{New row 2} = -2(\text{row 1}) + \text{row 2} \\ \text{New row 3} = -4(\text{row 1}) + \text{row 3} \end{array}$$

Now we add the second row to the third row.

$$\left[\begin{array}{ccc|c} 3 & -4 & -1 & 6 \\ 0 & 5 & 5 & -15 \\ 0 & 0 & 0 & 0 \end{array}\right] \quad \text{New row 3} = \text{row 2} + \text{row 3}$$

We can stop at this stage because we have a row consisting entirely of 0's. The last row of the matrix corresponds to the equation $0 = 0$, which is true for all values of x, y, and z. Therefore, the equations are dependent and the system is equivalent to

$$\begin{aligned} 3x - 4y - z &= 6, \\ 5y + 5z &= -15. \end{aligned}$$

This particular system has infinitely many solutions. (A system containing dependent equations could be inconsistent.)

Solving the second equation for y gives us

$$y = -z - 3.$$

Substituting $-z - 3$ for y in the first equation and solving for x, we get

$$3x - 4(-z - 3) - z = 6$$
$$x = -z - 2.$$

Then the solutions of this system are of the form

$$(-z - 2, -z - 3, z),$$

where z can be any real number.

Now Try Exercise 33.

Similarly, if we obtain a row whose only nonzero entry occurs in the last column, we have an inconsistent system of equations. For example, in the matrix

$$\left[\begin{array}{ccc|c} 1 & 0 & 3 & -2 \\ 0 & 1 & 5 & 4 \\ 0 & 0 & 0 & 6 \end{array}\right],$$

the last row corresponds to the false equation $0 = 6$, so we know the original system of equations has no solution.

9.3 Exercise Set

Determine the order of the matrix.

1. $\begin{bmatrix} 1 & -6 \\ -3 & 2 \\ 0 & 5 \end{bmatrix}$

2. $\begin{bmatrix} 7 \\ -5 \\ -1 \\ 3 \end{bmatrix}$

3. $[2 \quad -4 \quad 0 \quad 9]$

4. $[-8]$

5. $\begin{bmatrix} 1 & -5 & -8 \\ 6 & 4 & -2 \\ -3 & 0 & 7 \end{bmatrix}$

6. $\begin{bmatrix} 13 & 2 & -6 & 4 \\ -1 & 18 & 5 & -12 \end{bmatrix}$

Write the augmented matrix for the system of equations.

7. $2x - y = 7,$
$x + 4y = -5$

8. $3x + 2y = 8,$
$2x - 3y = 15$

9. $x - 2y + 3z = 12,$
$2x - 4z = 8,$
$3y + z = 7$

10. $x + y - z = 7,$
$3y + 2z = 1,$
$-2x - 5y = 6$

Write the system of equations that corresponds to the augmented matrix.

11. $\left[\begin{array}{cc|c} 3 & -5 & 1 \\ 1 & 4 & -2 \end{array}\right]$

12. $\left[\begin{array}{cc|c} 1 & 2 & -6 \\ 4 & 1 & -3 \end{array}\right]$

13. $\left[\begin{array}{ccc|c} 2 & 1 & -4 & 12 \\ 3 & 0 & 5 & -1 \\ 1 & -1 & 1 & 2 \end{array}\right]$

14. $\left[\begin{array}{ccc|c} -1 & -2 & 3 & 6 \\ 0 & 4 & 1 & 2 \\ 2 & -1 & 0 & 9 \end{array}\right]$

Solve the system of equations using Gaussian elimination or Gauss–Jordan elimination. Use a graphing calculator to check your answer.

15. $4x + 2y = 11,$
$3x - y = 2$

16. $2x + y = 1,$
$3x + 2y = -2$

17. $5x - 2y = -3,$
$2x + 5y = -24$

18. $2x + y = 1,$
$3x - 6y = 4$

19. $3x + 4y = 7,$
$-5x + 2y = 10$

20. $5x - 3y = -2,$
$4x + 2y = 5$

21. $3x + 2y = 6,$
$2x - 3y = -9$

22. $x - 4y = 9,$
$2x + 5y = 5$

23. $x - 3y = 8,$
$-2x + 6y = 3$

24. $4x - 8y = 12,$
$-x + 2y = -3$

25. $-2x + 6y = 4,$
$3x - 9y = -6$

26. $6x + 2y = -10,$
$-3x - y = 6$

27. $x + 2y - 3z = 9,$
$2x - y + 2z = -8,$
$3x - y - 4z = 3$

28. $x - y + 2z = 0,$
$x - 2y + 3z = -1,$
$2x - 2y + z = -3$

29. $4x - y - 3z = 1,$
$8x + y - z = 5,$
$2x + y + 2z = 5$

30. $3x + 2y + 2z = 3,$
$x + 2y - z = 5,$
$2x - 4y + z = 0$

31. $x - 2y + 3z = -4,$
$3x + y - z = 0,$
$2x + 3y - 5z = 1$

32. $2x - 3y + 2z = 2,$
$x + 4y - z = 9,$
$-3x + y - 5z = 5$

33. $2x - 4y - 3z = 3,$
$x + 3y + z = -1,$
$5x + y - 2z = 2$

34. $x + y - 3z = 4,$
$4x + 5y + z = 1,$
$2x + 3y + 7z = -7$

35. $p + q + r = 1,$
$p + 2q + 3r = 4,$
$4p + 5q + 6r = 7$

36. $m + n + t = 9,$
$m - n - t = -15,$
$3m + n + t = 2$

37. $a + b - c = 7,$
$a - b + c = 5,$
$3a + b - c = -1$

38. $a - b + c = 3,$
$2a + b - 3c = 5,$
$4a + b - c = 11$

39. $-2w + 2x + 2y - 2z = -10,$
$w + x + y + z = -5,$
$3w + x - y + 4z = -2,$
$w + 3x - 2y + 2z = -6$

40. $-w + 2x - 3y + z = -8,$
$-w + x + y - z = -4,$
$w + x + y + z = 22,$
$-w + x - y - z = -14$

Use Gaussian elimination or Gauss–Jordan elimination in Exercises 41–44.

41. *Borrowing.* Greenfield Manufacturing borrowed \$30,000 to buy a new piece of equipment. Part of the money was borrowed at 8%, part at 10%, and part at 12%. The annual interest was \$3040, and the total amount borrowed at 8% and at 10% was twice the amount borrowed at 12%. How much was borrowed at each rate?

42. *Summer Sports Camps.* One year, a family spent a total of \$2500 on summer sports camps for their three sons. The amount spent on camps for the oldest son was \$200 more than four times what was spent for the youngest son. The total spent for the two younger sons was \$1100. How much was spent on summer sports camps for each son?

43. *Stamp Purchase.* For her business, Olivia spent \$86.80 on both 49¢ and 21¢ stamps. She bought a total of 200 stamps. How many of each type did she buy?

44. *Time of Return.* The Patels pay their babysitter \$11 per hour before 11 P.M. and \$14.50 after 11 P.M. One evening, they went out for 6 hr and paid the sitter \$73. What time did they return?

Skill Maintenance

Classify the function as linear, quadratic, cubic, quartic, rational, exponential, or logarithmic.

45. $f(x) = 3^{x-1}$ [5.2]

46. $f(x) = 3x - 1$ [1.3]

47. $f(x) = \dfrac{3x - 1}{x^2 + 4}$ [4.5]

48. $f(x) = -\frac{3}{4}x^4 + \frac{9}{2}x^3 + 2x^2 - 4$ [4.1]

49. $f(x) = \ln(3x - 1)$ [5.3]

50. $f(x) = \frac{3}{4}x^3 - x$ [4.1]

51. $f(x) = 3$ [1.3]

52. $f(x) = 2 - x - x^2$ [3.2]

Synthesis

In Exercises 53 and 54, three solutions of the equation $y = ax^2 + bx + c$ are given. Use a system of three equations in three variables and Gaussian elimination or Gauss–Jordan elimination to find the constants a, b, and c and write the equation.

53. $(-3, 12)$, $(-1, -7)$, and $(1, -2)$

54. $(-1, 0)$, $(1, -3)$, and $(3, -22)$

55. Find two different row-echelon forms of
$$\begin{bmatrix} 1 & 5 \\ 3 & 2 \end{bmatrix}.$$

56. Consider the system of equations
$$x - y + 3z = -8,$$
$$2x + 3y - z = 5,$$
$$3x + 2y + 2kz = -3k.$$
For what value(s) of k, if any, will the system have

a) no solution?
b) exactly one solution?
c) infinitely many solutions?

Solve using matrices.

57. $y = x + z,$
$3y + 5z = 4,$
$x + 4 = y + 3z$

58. $x + y = 2z,$
$2x - 5z = 4,$
$x - z = y + 8$

59. $x - 4y + 2z = 7,$
$3x + y + 3z = -5$

60. $x - y - 3z = 3,$
$-x + 3y + z = -7$

61. $4x + 5y = 3,$
$-2x + y = 9,$
$3x - 2y = -15$

62. $2x - 3y = -1,$
$-x + 2y = -2,$
$3x - 5y = 1$

9.4 Matrix Operations

- Add, subtract, and multiply matrices when possible.
- Write a matrix equation equivalent to a system of equations.

In addition to being used to solve systems of equations, matrices are useful in many other types of applications. In this section, we study matrices and some of their properties.

An uppercase letter is generally used to name a matrix, and lower-case letters with double subscripts generally denote its entries. For example, a_{47}, read "a sub four seven," indicates the entry in the fourth row and the seventh column. A general term is represented by a_{ij}. The notation a_{ij} indicates the entry in row i and column j. In general, we can write a matrix as

$$\mathbf{A} = [a_{ij}] = \begin{bmatrix} a_{11} & a_{12} & a_{13} & \cdots & a_{1n} \\ a_{21} & a_{22} & a_{23} & \cdots & a_{2n} \\ a_{31} & a_{32} & a_{33} & \cdots & a_{3n} \\ \vdots & \vdots & \vdots & & \vdots \\ a_{m1} & a_{m2} & a_{m3} & \cdots & a_{mn} \end{bmatrix}.$$

The matrix above has m rows and n columns; that is, its order is $m \times n$.

Two matrices are **equal** if they have the same order and corresponding entries are equal.

Matrix Addition and Subtraction

To add or subtract matrices, we add or subtract their corresponding entries. For this to be possible, the matrices must have the same order.

ADDITION AND SUBTRACTION OF MATRICES

Given two $m \times n$ matrices $\mathbf{A} = [a_{ij}]$ and $\mathbf{B} = [b_{ij}]$, their sum is

$$\mathbf{A} + \mathbf{B} = [a_{ij} + b_{ij}]$$

and their difference is

$$\mathbf{A} - \mathbf{B} = [a_{ij} - b_{ij}].$$

Addition of matrices is both commutative and associative.

EXAMPLE 1 Find $\mathbf{A} + \mathbf{B}$ for each of the following.

a) $\mathbf{A} = \begin{bmatrix} -5 & 0 \\ 4 & \frac{1}{2} \end{bmatrix}$, $\mathbf{B} = \begin{bmatrix} 6 & -3 \\ 2 & 3 \end{bmatrix}$

b) $\mathbf{A} = \begin{bmatrix} 1 & 3 \\ -1 & 5 \\ 6 & 0 \end{bmatrix}$, $\mathbf{B} = \begin{bmatrix} -1 & -2 \\ 1 & -2 \\ -3 & 1 \end{bmatrix}$

Solution We have a pair of 2×2 matrices in part (a) and a pair of 3×2 matrices in part (b). Since the matrices within each pair of matrices have the same order, we can add the corresponding entries.

a) $$\mathbf{A} + \mathbf{B} = \begin{bmatrix} -5 & 0 \\ 4 & \frac{1}{2} \end{bmatrix} + \begin{bmatrix} 6 & -3 \\ 2 & 3 \end{bmatrix}$$

$$= \begin{bmatrix} -5 + 6 & 0 + (-3) \\ 4 + 2 & \frac{1}{2} + 3 \end{bmatrix} = \begin{bmatrix} 1 & -3 \\ 6 & 3\frac{1}{2} \end{bmatrix}$$

[A]+[B]
$\begin{bmatrix} 1 & -3 \\ 6 & 3.5 \end{bmatrix}$

We can also enter **A** and **B** in a graphing calculator and then find **A** + **B**.

b) $$\mathbf{A} + \mathbf{B} = \begin{bmatrix} 1 & 3 \\ -1 & 5 \\ 6 & 0 \end{bmatrix} + \begin{bmatrix} -1 & -2 \\ 1 & -2 \\ -3 & 1 \end{bmatrix}$$

$$= \begin{bmatrix} 1 + (-1) & 3 + (-2) \\ -1 + 1 & 5 + (-2) \\ 6 + (-3) & 0 + 1 \end{bmatrix} = \begin{bmatrix} 0 & 1 \\ 0 & 3 \\ 3 & 1 \end{bmatrix}$$

[A]+[B]
$\begin{bmatrix} 0 & 1 \\ 0 & 3 \\ 3 & 1 \end{bmatrix}$

This sum can also be found on a graphing calculator after **A** and **B** have been entered.

Now Try Exercise 5.

EXAMPLE 2 Find **C** − **D** for each of the following.

a) $$\mathbf{C} = \begin{bmatrix} 1 & 2 \\ -2 & 0 \\ -3 & -1 \end{bmatrix}, \quad \mathbf{D} = \begin{bmatrix} 1 & -1 \\ 1 & 3 \\ 2 & 3 \end{bmatrix}$$

b) $$\mathbf{C} = \begin{bmatrix} 5 & -6 \\ -3 & 4 \end{bmatrix}, \quad \mathbf{D} = \begin{bmatrix} -4 \\ 1 \end{bmatrix}$$

Solution

a) Since the order of each matrix is 3×2, we can subtract corresponding entries:

$$\mathbf{C} - \mathbf{D} = \begin{bmatrix} 1 & 2 \\ -2 & 0 \\ -3 & -1 \end{bmatrix} - \begin{bmatrix} 1 & -1 \\ 1 & 3 \\ 2 & 3 \end{bmatrix}$$

$$= \begin{bmatrix} 1 - 1 & 2 - (-1) \\ -2 - 1 & 0 - 3 \\ -3 - 2 & -1 - 3 \end{bmatrix} = \begin{bmatrix} 0 & 3 \\ -3 & -3 \\ -5 & -4 \end{bmatrix}.$$

This subtraction can also be done using a graphing calculator.

b) **C** is a 2×2 matrix and **D** is a 2×1 matrix. Since the matrices do not have the same order, we cannot subtract. If we try to do this subtraction on a graphing calculator, we get an ERROR message indicating that the dimensions are mismatched.

ERR:DIM MISMATCH
1: Quit
2: Goto

Now Try Exercise 13.

The **opposite**, or **additive inverse**, of a matrix is obtained by replacing each entry with its opposite.

EXAMPLE 3 Find −**A** and **A** + (−**A**) for

$$\mathbf{A} = \begin{bmatrix} 1 & 0 & 2 \\ 3 & -1 & 5 \end{bmatrix}.$$

Solution To find $-\mathbf{A}$, we replace each entry of $\mathbf{A}$ with its opposite.

$$-\mathbf{A} = \begin{bmatrix} -1 & 0 & -2 \\ -3 & 1 & -5 \end{bmatrix},$$

$$\mathbf{A} + (-\mathbf{A}) = \begin{bmatrix} 1 & 0 & 2 \\ 3 & -1 & 5 \end{bmatrix} + \begin{bmatrix} -1 & 0 & -2 \\ -3 & 1 & -5 \end{bmatrix}$$

$$= \begin{bmatrix} 0 & 0 & 0 \\ 0 & 0 & 0 \end{bmatrix}$$

[A]+(−[A])
[0 0 0]
[0 0 0]

A matrix having 0's for all its entries is called a **zero matrix**. When a zero matrix is added to a second matrix of the same order, the second matrix is unchanged. Thus a zero matrix is an **additive identity**. For example,

$$\begin{bmatrix} 2 & 3 & -4 \\ 0 & 6 & 5 \end{bmatrix} + \begin{bmatrix} 0 & 0 & 0 \\ 0 & 0 & 0 \end{bmatrix} = \begin{bmatrix} 2 & 3 & -4 \\ 0 & 6 & 5 \end{bmatrix}.$$

The matrix

$$\begin{bmatrix} 0 & 0 & 0 \\ 0 & 0 & 0 \end{bmatrix}$$

is the additive identity for any 2×3 matrix.

Scalar Multiplication

When we find the product of a number and a matrix, we obtain a **scalar product**.

SCALAR PRODUCT

The **scalar product** of a number k and a matrix $\mathbf{A}$ is the matrix denoted $k\mathbf{A}$, obtained by multiplying each entry of $\mathbf{A}$ by the number k. The number k is called a **scalar**.

EXAMPLE 4 Find $3\mathbf{A}$ and $(-1)\mathbf{A}$ for

$$\mathbf{A} = \begin{bmatrix} -3 & 0 \\ 4 & 5 \end{bmatrix}.$$

Solution We have

$$3\mathbf{A} = 3\begin{bmatrix} -3 & 0 \\ 4 & 5 \end{bmatrix} = \begin{bmatrix} 3(-3) & 3\cdot 0 \\ 3\cdot 4 & 3\cdot 5 \end{bmatrix} = \begin{bmatrix} -9 & 0 \\ 12 & 15 \end{bmatrix},$$

$$(-1)\mathbf{A} = -1\begin{bmatrix} -3 & 0 \\ 4 & 5 \end{bmatrix} = \begin{bmatrix} -1(-3) & -1\cdot 0 \\ -1\cdot 4 & -1\cdot 5 \end{bmatrix} = \begin{bmatrix} 3 & 0 \\ -4 & -5 \end{bmatrix}.$$

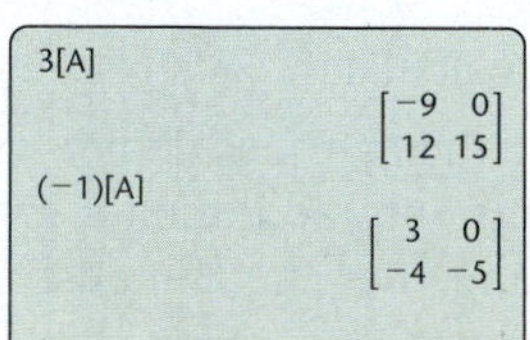

These scalar products can also be found on a graphing calculator after $\mathbf{A}$ has been entered.

Now Try Exercise 9.

The properties of matrix addition and scalar multiplication are similar to the properties of addition and multiplication of real numbers.

PROPERTIES OF MATRIX ADDITION AND SCALAR MULTIPLICATION

For any $m \times n$ matrices **A**, **B**, and **C** and any scalars k and l:

$\mathbf{A} + \mathbf{B} = \mathbf{B} + \mathbf{A}$.	Commutative property of addition
$\mathbf{A} + (\mathbf{B} + \mathbf{C}) = (\mathbf{A} + \mathbf{B}) + \mathbf{C}$.	Associative property of addition
$(kl)\mathbf{A} = k(l\mathbf{A})$.	Associative property of scalar multiplication
$k(\mathbf{A} + \mathbf{B}) = k\mathbf{A} + k\mathbf{B}$.	Distributive property
$(k + l)\mathbf{A} = k\mathbf{A} + l\mathbf{A}$.	Distributive property

There exists a unique matrix **0** such that:

$\mathbf{A} + \mathbf{0} = \mathbf{0} + \mathbf{A} = \mathbf{A}$.	Additive identity property

There exists a unique matrix $-\mathbf{A}$ such that:

$\mathbf{A} + (-\mathbf{A}) = -\mathbf{A} + \mathbf{A} = \mathbf{0}$.	Additive inverse property

EXAMPLE 5 ***Production.*** Waterworks, Inc., manufactures three types of kayaks in its two plants. The following table lists the number of each style produced at each plant in April.

	Whitewater Kayak	Ocean Kayak	Crossover Kayak
Madison Plant	150	120	100
Greensburg Plant	180	90	130

a) Write a 2×3 matrix **A** that represents the information in the table.

b) The manufacturer increased production by 20% in May. Find a matrix **M** that represents the increased production figures.

c) Find the matrix $\mathbf{A} + \mathbf{M}$ and tell what it represents.

Solution

a) We write the entries in the table in a 2×3 matrix **A**.

$$\mathbf{A} = \begin{bmatrix} 150 & 120 & 100 \\ 180 & 90 & 130 \end{bmatrix}$$

b) The production in May will be represented by $\mathbf{A} + 20\%\mathbf{A}$, or $\mathbf{A} + 0.2\mathbf{A}$, or $1.2\mathbf{A}$. Thus,

$$\mathbf{M} = (1.2)\begin{bmatrix} 150 & 120 & 100 \\ 180 & 90 & 130 \end{bmatrix} = \begin{bmatrix} 180 & 144 & 120 \\ 216 & 108 & 156 \end{bmatrix}.$$

c) $$\mathbf{A} + \mathbf{M} = \begin{bmatrix} 150 & 120 & 100 \\ 180 & 90 & 130 \end{bmatrix} + \begin{bmatrix} 180 & 144 & 120 \\ 216 & 108 & 156 \end{bmatrix} = \begin{bmatrix} 330 & 264 & 220 \\ 396 & 198 & 286 \end{bmatrix}$$

The matrix $\mathbf{A} + \mathbf{M}$ represents the total production of each of the three types of kayaks at each plant in April and May.

Now Try Exercise 29.

Products of Matrices

Matrix multiplication is defined in such a way that it can be used in solving systems of equations and in many applications.

MATRIX MULTIPLICATION

For an $m \times n$ matrix $\mathbf{A} = [a_{ij}]$ and an $n \times p$ matrix $\mathbf{B} = [b_{ij}]$, the **product** $\mathbf{AB} = [c_{ij}]$ is an $m \times p$ matrix, where

$$c_{ij} = a_{i1} \cdot b_{1j} + a_{i2} \cdot b_{2j} + a_{i3} \cdot b_{3j} + \cdots + a_{in} \cdot b_{nj}.$$

In other words, the entry c_{ij} in $\mathbf{AB}$ is obtained by multiplying the entries in row i of $\mathbf{A}$ by the corresponding entries in column j of $\mathbf{B}$ and adding the results.

Note that we can multiply two matrices only when the number of columns in the first matrix is equal to the number of rows in the second matrix.

EXAMPLE 6 For

$$\mathbf{A} = \begin{bmatrix} 3 & 1 & -1 \\ 2 & 0 & 3 \end{bmatrix}, \quad \mathbf{B} = \begin{bmatrix} 1 & 6 \\ 3 & -5 \\ -2 & 4 \end{bmatrix}, \quad \text{and} \quad \mathbf{C} = \begin{bmatrix} 4 & -6 \\ 1 & 2 \end{bmatrix},$$

find each of the following.

a) **AB** **b)** **BA**

c) **BC** **d)** **AC**

Solution

a) **A** is a 2×3 matrix and **B** is a 3×2 matrix, so **AB** will be a 2×2 matrix.

$$\mathbf{AB} = \begin{bmatrix} 3 & 1 & -1 \\ 2 & 0 & 3 \end{bmatrix} \begin{bmatrix} 1 & 6 \\ 3 & -5 \\ -2 & 4 \end{bmatrix}$$

$$= \begin{bmatrix} 3 \cdot 1 + 1 \cdot 3 + (-1)(-2) & 3 \cdot 6 + 1(-5) + (-1)(4) \\ 2 \cdot 1 + 0 \cdot 3 + 3(-2) & 2 \cdot 6 + 0(-5) + 3 \cdot 4 \end{bmatrix} = \begin{bmatrix} 8 & 9 \\ -4 & 24 \end{bmatrix}$$

b) **B** is a 3×2 matrix and **A** is a 2×3 matrix, so **BA** will be a 3×3 matrix.

$$\mathbf{BA} = \begin{bmatrix} 1 & 6 \\ 3 & -5 \\ -2 & 4 \end{bmatrix} \begin{bmatrix} 3 & 1 & -1 \\ 2 & 0 & 3 \end{bmatrix}$$

$$= \begin{bmatrix} 1 \cdot 3 + 6 \cdot 2 & 1 \cdot 1 + 6 \cdot 0 & 1(-1) + 6 \cdot 3 \\ 3 \cdot 3 + (-5)(2) & 3 \cdot 1 + (-5)(0) & 3(-1) + (-5)(3) \\ -2 \cdot 3 + 4 \cdot 2 & -2 \cdot 1 + 4 \cdot 0 & -2(-1) + 4 \cdot 3 \end{bmatrix} = \begin{bmatrix} 15 & 1 & 17 \\ -1 & 3 & -18 \\ 2 & -2 & 14 \end{bmatrix}$$

Note in parts (a) and (b) that $\mathbf{AB} \neq \mathbf{BA}$. Multiplication of matrices is generally not commutative.

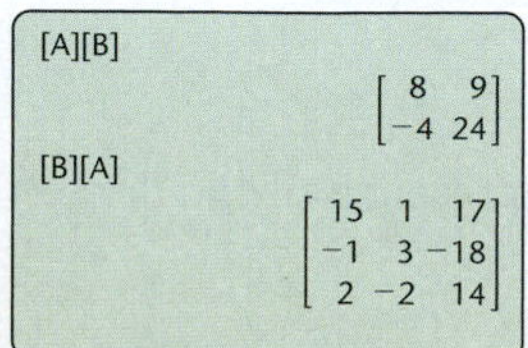

Matrix multiplication can be performed on a graphing calculator. The products in parts (a) and (b) are shown at left.

[A] [C]

ERR:DIM MISMATCH
1: Quit
2: Goto

c) **B** is a 3 × 2 matrix and **C** is a 2 × 2 matrix, so **BC** will be a 3 × 2 matrix.

$$\mathbf{BC} = \begin{bmatrix} 1 & 6 \\ 3 & -5 \\ -2 & 4 \end{bmatrix} \begin{bmatrix} 4 & -6 \\ 1 & 2 \end{bmatrix}$$

$$= \begin{bmatrix} 1 \cdot 4 + 6 \cdot 1 & 1(-6) + 6 \cdot 2 \\ 3 \cdot 4 + (-5)(1) & 3(-6) + (-5)(2) \\ -2 \cdot 4 + 4 \cdot 1 & -2(-6) + 4 \cdot 2 \end{bmatrix} = \begin{bmatrix} 10 & 6 \\ 7 & -28 \\ -4 & 20 \end{bmatrix}$$

d) The product **AC** is not defined because the number of columns of **A**, 3, is not equal to the number of rows of **C**, 2.

When the product **AC** is entered on a graphing calculator, an ERROR message is returned, indicating that the dimensions of the matrices are mismatched.

Now Try Exercises 23 and 25.

EXAMPLE 7 ***Bakery Profit.*** Two of the items sold at Sweet Treats Bakery are gluten-free bagels and gluten-free doughnuts. The following table lists the number of dozens of each product that were sold at the bakery's three stores one week.

	Main Street Store	Avon Road Store	Dalton Avenue Store
Bagels (in dozens)	25	30	20
Doughnuts (in dozens)	40	35	15

The bakery's profit on one dozen bagels is \$5, and its profit on one dozen doughnuts is \$6. Use matrices to find the total profit on these items at each store for the given week.

Solution We can write the table showing the sales of the products as a 2 × 3 matrix:

$$\mathbf{S} = \begin{bmatrix} 25 & 30 & 20 \\ 40 & 35 & 15 \end{bmatrix}.$$

The profit per dozen for each product can also be written as a matrix:

$$\mathbf{P} = [5 \quad 6].$$

Then the total profit at each store is given by the matrix product **PS**:

$$\mathbf{PS} = [5 \quad 6] \begin{bmatrix} 25 & 30 & 20 \\ 40 & 35 & 15 \end{bmatrix}$$

$$= [5 \cdot 25 + 6 \cdot 40 \quad 5 \cdot 30 + 6 \cdot 35 \quad 5 \cdot 20 + 6 \cdot 15]$$

$$= [365 \quad 360 \quad 190].$$

The total profit on gluten-free bagels and gluten-free doughnuts for the given week was \$365 at the Main Street store, \$360 at the Avon Road store, and \$190 at the Dalton Avenue store.

Now Try Exercise 33.

A matrix that consists of a single row, like **P** in Example 7, is called a **row matrix**. Similarly, a matrix that consists of a single column, like

$$\begin{bmatrix} 8 \\ -3 \\ 5 \end{bmatrix},$$

is called a **column matrix**.

We have already seen that matrix multiplication is generally not commutative. Nevertheless, matrix multiplication does have some properties that are similar to those for multiplication of real numbers.

PROPERTIES OF MATRIX MULTIPLICATION

For matrices **A**, **B**, and **C**, assuming that the indicated operations are possible:

$\mathbf{A}(\mathbf{BC}) = (\mathbf{AB})\mathbf{C}$. Associative property of multiplication

$\mathbf{A}(\mathbf{B} + \mathbf{C}) = \mathbf{AB} + \mathbf{AC}$. Distributive property

$(\mathbf{B} + \mathbf{C})\mathbf{A} = \mathbf{BA} + \mathbf{CA}$. Distributive property

Matrix Equations

We can write a matrix equation equivalent to a system of equations.

EXAMPLE 8 Write a matrix equation equivalent to the following system of equations:

$$\begin{aligned} 4x + 2y - z &= 3, \\ 9x \quad\quad + z &= 5, \\ 4x + 5y - 2z &= 1. \end{aligned}$$

Solution We write the coefficients on the left in a matrix. We then write the product of that matrix and the column matrix containing the variables and set the result equal to the column matrix containing the constants on the right:

$$\begin{bmatrix} 4 & 2 & -1 \\ 9 & 0 & 1 \\ 4 & 5 & -2 \end{bmatrix}\begin{bmatrix} x \\ y \\ z \end{bmatrix} = \begin{bmatrix} 3 \\ 5 \\ 1 \end{bmatrix}.$$

If we let

$$\mathbf{A} = \begin{bmatrix} 4 & 2 & -1 \\ 9 & 0 & 1 \\ 4 & 5 & -2 \end{bmatrix}, \quad \mathbf{X} = \begin{bmatrix} x \\ y \\ z \end{bmatrix}, \quad \text{and} \quad \mathbf{B} = \begin{bmatrix} 3 \\ 5 \\ 1 \end{bmatrix},$$

we can write this matrix equation as $\mathbf{AX} = \mathbf{B}$.

Now Try Exercise 41.

9.4 Exercise Set

Find x and y.

1. $[5 \quad x] = [y \quad -3]$

2. $\begin{bmatrix} 6x \\ 25 \end{bmatrix} = \begin{bmatrix} -9 \\ 5y \end{bmatrix}$

3. $\begin{bmatrix} 3 & 2x \\ y & -8 \end{bmatrix} = \begin{bmatrix} 3 & -2 \\ 1 & -8 \end{bmatrix}$

4. $\begin{bmatrix} x-1 & 4 \\ y+3 & -7 \end{bmatrix} = \begin{bmatrix} 0 & 4 \\ -2 & -7 \end{bmatrix}$

For Exercises 5–20, let

$\mathbf{A} = \begin{bmatrix} 1 & 2 \\ 4 & 3 \end{bmatrix}$, $\quad \mathbf{B} = \begin{bmatrix} -3 & 5 \\ 2 & -1 \end{bmatrix}$,

$\mathbf{C} = \begin{bmatrix} 1 & -1 \\ -1 & 1 \end{bmatrix}$, $\quad \mathbf{D} = \begin{bmatrix} 1 & 1 \\ 1 & 1 \end{bmatrix}$,

$\mathbf{E} = \begin{bmatrix} 1 & 3 \\ 2 & 6 \end{bmatrix}$, $\quad \mathbf{F} = \begin{bmatrix} 3 & 3 \\ -1 & -1 \end{bmatrix}$,

$\mathbf{0} = \begin{bmatrix} 0 & 0 \\ 0 & 0 \end{bmatrix}$, $\quad \mathbf{I} = \begin{bmatrix} 1 & 0 \\ 0 & 1 \end{bmatrix}$.

Find each of the following.

5. $\mathbf{A} + \mathbf{B}$ **6.** $\mathbf{B} + \mathbf{A}$ **7.** $\mathbf{E} + \mathbf{0}$

8. $2\mathbf{A}$ **9.** $3\mathbf{F}$ **10.** $(-1)\mathbf{D}$

11. $3\mathbf{F} + 2\mathbf{A}$ **12.** $\mathbf{A} - \mathbf{B}$ **13.** $\mathbf{B} - \mathbf{A}$

14. $\mathbf{AB}$ **15.** $\mathbf{BA}$ **16.** $\mathbf{0F}$

17. $\mathbf{CD}$ **18.** $\mathbf{EF}$ **19.** $\mathbf{AI}$

20. $\mathbf{IA}$

Find the product, if possible.

21. $\begin{bmatrix} -1 & 0 & 7 \\ 3 & -5 & 2 \end{bmatrix}\begin{bmatrix} 6 \\ -4 \\ 1 \end{bmatrix}$

22. $\begin{bmatrix} 6 & -1 & 2 \end{bmatrix}\begin{bmatrix} 1 & 4 \\ -2 & 0 \\ 5 & -3 \end{bmatrix}$

23. $\begin{bmatrix} -2 & 4 \\ 5 & 1 \\ -1 & -3 \end{bmatrix}\begin{bmatrix} 3 & -6 \\ -1 & 4 \end{bmatrix}$

24. $\begin{bmatrix} 2 & -1 & 0 \\ 0 & 5 & 4 \end{bmatrix}\begin{bmatrix} -3 & 1 & 0 \\ 0 & 2 & -1 \\ 5 & 0 & 4 \end{bmatrix}$

25. $\begin{bmatrix} 1 \\ -5 \\ 3 \end{bmatrix}\begin{bmatrix} -6 & 5 & 8 \\ 0 & 4 & -1 \end{bmatrix}$

26. $\begin{bmatrix} 2 & 0 & 0 \\ 0 & -1 & 0 \\ 0 & 0 & 3 \end{bmatrix}\begin{bmatrix} 0 & -4 & 3 \\ 2 & 1 & 0 \\ -1 & 0 & 6 \end{bmatrix}$

27. $\begin{bmatrix} 1 & -4 & 3 \\ 0 & 8 & 0 \\ -2 & -1 & 5 \end{bmatrix}\begin{bmatrix} 3 & 0 & 0 \\ 0 & -4 & 0 \\ 0 & 0 & 1 \end{bmatrix}$

28. $\begin{bmatrix} 4 \\ -5 \end{bmatrix}\begin{bmatrix} 2 & 0 \\ 6 & -7 \\ 0 & -3 \end{bmatrix}$

29. *Produce.* The produce manager at Stan's Market orders 40 lb of tomatoes, 20 lb of zucchini, and 30 lb of onions from a local farmer one week.

a) Write a 1×3 matrix $\mathbf{A}$ that represents the amount of each item ordered.

b) The following week the produce manager increases his order by 10%. Find a matrix $\mathbf{B}$ that represents this order.

c) Find $\mathbf{A} + \mathbf{B}$ and tell what the entries represent.

30. *Budget.* For the month of June, Lexi's budgets \$400 for food, \$160 for clothes, and \$60 for entertainment.

a) Write a 1×3 matrix $\mathbf{B}$ that represents the amounts budgeted for these items.

b) After receiving a raise, Lexi increases the amount budgeted for each item in July by 5%. Find a matrix $\mathbf{R}$ that represents the new amounts.

c) Find $\mathbf{B} + \mathbf{R}$ and tell what the entries represent.

31. *Nutrition.* A 3-oz serving of roasted, skinless chicken breast contains 140 Cal, 27 g of protein, 3 g of fat, 13 mg of calcium, and 64 mg of sodium. One-half cup of potato salad contains 180 Cal, 4 g of protein, 11 g of fat, 24 mg of calcium, and 662 mg of sodium. One broccoli spear contains 50 Cal, 5 g of protein, 1 g of fat, 82 mg of calcium, and 20 mg of sodium. (*Source*: U.S. Department of Agriculture)

a) Write 1×5 matrices $\mathbf{C}$, $\mathbf{P}$, and $\mathbf{B}$ that represent the nutritional values of each food.

b) Find $\mathbf{C} + 2\mathbf{P} + 3\mathbf{B}$ and tell what the entries represent.

32. *Nutrition.* One slice of cheese pizza contains 290 Cal, 15 g of protein, 9 g of fat, and 39 g of carbohydrates. One-half cup of gelatin dessert contains 70 Cal, 2 g of protein, 0 g of fat, and 17 g of carbohydrates. One cup of whole milk contains 150 Cal, 8 g of protein, 8 g of fat, and 11 g of carbohydrates. (*Source*: U.S. Department of Agriculture)

a) Write 1×4 matrices $\mathbf{P}$, $\mathbf{G}$, and $\mathbf{M}$ that represent the nutritional values of each food.

b) Find $3\mathbf{P} + 2\mathbf{G} + 2\mathbf{M}$ and tell what the entries represent.

33. *Food Service Management.* The food service manager at a large hospital is concerned about maintaining reasonable food costs. The following table lists the cost per serving, in dollars, for items on four menus.

Menu	Meat	Potato	Vegetable	Salad	Dessert
1	1.50	0.30	0.36	0.45	0.64
2	1.55	0.28	0.48	0.57	0.75
3	1.62	0.52	0.65	0.38	0.53
4	1.70	0.43	0.40	0.42	0.68

On a particular day, a dietician orders 65 meals from menu 1, 48 from menu 2, 93 from menu 3, and 57 from menu 4.

a) Write the information in the table as a 4×5 matrix **M**.
b) Write a row matrix **N** that represents the number of each menu ordered.
c) Find the product **NM**.
d) State what the entries of **NM** represent.

34. *Food Service Management.* A college food service manager uses a table like the one below to list the number of units of ingredients, by weight, required for various menu items.

	White Cake	Bread	Coffee Cake	Sugar Cookies
Flour	1	2.5	0.75	0.5
Milk	0	0.5	0.25	0
Eggs	0.75	0.25	0.5	0.5
Butter	0.5	0	0.5	1

The cost per unit of each ingredient is 25 cents for flour, 34 cents for milk, 54 cents for eggs, and 83 cents for butter.

a) Write the information in the table as a 4×4 matrix **M**.
b) Write a row matrix **C** that represents the cost per unit of each ingredient.
c) Find the product **CM**.
d) State what the entries of **CM** represent.

35. *Production Cost.* Anja supplies two small campus coffee shops with homemade chocolate chip cookies, oatmeal cookies, and peanut butter cookies. The following table shows the number of each type of cookie, in dozens, that Anja sold in one week.

	Skip's Coffee Shop	The Coffee Club
Chocolate Chip	8	15
Oatmeal	6	10
Peanut Butter	4	3

Anja spends \$4 for the ingredients for one dozen chocolate chip cookies, \$2.50 for the ingredients for one dozen oatmeal cookies, and \$3 for the ingredients for one dozen peanut butter cookies.

a) Write the information in the table as a 3×2 matrix **S**.
b) Write a row matrix **C** that represents the cost, per dozen, of the ingredients for each type of cookie.
c) Find the product **CS**.
d) State what the entries of **CS** represent.

36. *Profit.* A manufacturer produces exterior plywood, interior plywood, and fiberboard, which are shipped to two distributors. The following table lists the number of units of each type of product that are shipped to each warehouse.

	Distributor 1	Distributor 2
Exterior Plywood	900	500
Interior Plywood	450	1000
Fiberboard	600	700

The profits from each unit of exterior plywood, interior plywood, and fiberboard are \$8, \$10, and \$7, respectively.

a) Write the information in the table as a 3×2 matrix **M**.
b) Write a row matrix **P** that represents the profit per unit of each type of product.
c) Find the product **PM**.
d) State what the entries of **PM** represent.

37. *Profit.* In Exercise 35, suppose that Anja's profits on one dozen chocolate chip, oatmeal, and peanut butter cookies are \$7.50, \$4.80, and \$6.25, respectively.

a) Write a row matrix **P** that represents this information.
b) Use the matrices **S** and **P** to find Karin's total profit from each coffee shop.

38. *Production Cost.* In Exercise 36, suppose that the manufacturer's production costs for each unit of exterior plywood, interior plywood, and fiberboard are \$20, \$25, and \$15, respectively.

a) Write a row matrix **C** that represents this information.

b) Use the matrices **M** and **C** to find the total production cost for the products shipped to each distributor.

Write a matrix equation equivalent to the system of equations.

39. $2x - 3y = 7,$
$x + 5y = -6$

40. $-x + y = 3,$
$5x - 4y = 16$

41. $x + y - 2z = 6,$
$3x - y + z = 7,$
$2x + 5y - 3z = 8$

42. $3x - y + z = 1,$
$x + 2y - z = 3,$
$4x + 3y - 2z = 11$

43. $3x - 2y + 4z = 17,$
$2x + y - 5z = 13$

44. $3x + 2y + 5z = 9,$
$4x - 3y + 2z = 10$

45. $-4w + x - y + 2z = 12,$
$w + 2x - y - z = 0,$
$-w + x + 4y - 3z = 1,$
$2w + 3x + 5y - 7z = 9$

46. $12w + 2x + 4y - 5z = 2,$
$-w + 4x - y + 12z = 5,$
$2w - x + 4y = 13,$
$2x + 10y + z = 5$

Skill Maintenance

In Exercises 47–50:

a) *Find the vertex.* [3.3]
b) *Find the axis of symmetry.* [3.3]
c) *Determine whether there is a maximum or a minimum value and find that value.* [3.3]
d) *Graph the function.* [3.3]

47. $f(x) = x^2 - x - 6$

48. $f(x) = 2x^2 - 5x - 3$

49. $f(x) = -x^2 - 3x + 2$

50. $f(x) = -3x^2 + 4x + 4$

Synthesis

For Exercises 51–54, let

$$\mathbf{A} = \begin{bmatrix} -1 & 0 \\ 2 & 1 \end{bmatrix} \quad \text{and} \quad \mathbf{B} = \begin{bmatrix} 1 & -1 \\ 0 & 2 \end{bmatrix}.$$

51. Show that

$$(\mathbf{A} + \mathbf{B})(\mathbf{A} - \mathbf{B}) \neq \mathbf{A}^2 - \mathbf{B}^2,$$

where

$$\mathbf{A}^2 = \mathbf{AA} \quad \text{and} \quad \mathbf{B}^2 = \mathbf{BB}.$$

52. Show that

$$(\mathbf{A} + \mathbf{B})(\mathbf{A} + \mathbf{B}) \neq \mathbf{A}^2 + 2\mathbf{AB} + \mathbf{B}^2.$$

53. Show that

$$(\mathbf{A} + \mathbf{B})(\mathbf{A} - \mathbf{B}) = \mathbf{A}^2 + \mathbf{BA} - \mathbf{AB} - \mathbf{B}^2.$$

54. Show that

$$(\mathbf{A} + \mathbf{B})(\mathbf{A} + \mathbf{B}) = \mathbf{A}^2 + \mathbf{BA} + \mathbf{AB} + \mathbf{B}^2.$$

Mid-Chapter Mixed Review

Determine whether the statement is true or false.

1. For a system of two linear equations in two variables, if the graphs of the equations are parallel lines, then the system of equations has infinitely many solutions. [9.1]

2. One of the properties of a matrix written in row-echelon form is that all the rows consisting entirely of 0's are at the bottom of the matrix. [9.3]

3. We can multiply two matrices only when the number of columns in the first matrix is equal to the number of rows in the second matrix. [9.4]

4. Addition of matrices is not commutative. [9.4]

Solve. [9.1], [9.2]

5. $2x + y = -4,$
$x = y - 5$

6. $x + y = 4,$
$y = 2 - x$

7. $2x - 3y = 8,$
$3x + 2y = -1$

8. $x - 3y = 1,$
$6y = 2x - 2$

9. $x + 2y + 3z = 4,$
$x - 2y + z = 2,$
$2x - 6y + 4z = 7$

10. *e-Commerce.* computerwarehouse.com charges \$8 to ship orders up to 10 lb, \$12 for orders from 10 lb up to 15 lb, and \$15 for orders of 15 lb or more. One day, shipping charges for 150 orders totaled \$1620. The number of orders under 10 lb was three times the number of orders weighing 15 lb or more. Find the number of packages shipped at each rate. [9.2]

Solve the system of equations using Gaussian elimination or Gauss–Jordan elimination. [9.3]

11. $2x + y = 5,$
$3x + 2y = 6$

12. $3x + 2y - 3z = -2,$
$2x + 3y + 2z = -2,$
$x + 4y + 4z = 1$

For Exercises 13–20, let

$$\mathbf{A} = \begin{bmatrix} 3 & -1 \\ 5 & 4 \end{bmatrix}, \quad \mathbf{B} = \begin{bmatrix} -2 & 6 \\ 1 & -3 \end{bmatrix}, \quad \mathbf{C} = \begin{bmatrix} -4 & 1 & -1 \\ 2 & 3 & -2 \end{bmatrix}, \quad \textit{and} \quad \mathbf{D} = \begin{bmatrix} -2 & 3 & 0 \\ 1 & -1 & 2 \\ -3 & 4 & 1 \end{bmatrix}.$$

Find each of the following. [9.4]

13. $\mathbf{A} + \mathbf{B}$

14. $\mathbf{B} - \mathbf{A}$

15. $4\mathbf{D}$

16. $2\mathbf{A} + 3\mathbf{B}$

17. $\mathbf{AB}$

18. $\mathbf{BA}$

19. $\mathbf{BC}$

20. $\mathbf{DC}$

21. Write a matrix equation equivalent to the following system of equations: [9.4]

$2x - y + 3z = 7,$
$x + 2y - z = 3,$
$3x - 4y + 2z = 5.$

COLLABORATIVE DISCUSSION AND WRITING

22. Explain in your own words when using the elimination method for solving a system of equations is preferable to using the substitution method. [9.1]

23. Given two linear equations in three variables,

$$Ax + By + Cz = D \quad \text{and} \quad Ex + Fy + Gz = H,$$

explain how you could find a third equation such that the system contains dependent equations. [9.2]

24. Explain in your own words why the augmented matrix below represents a system of dependent equations. [9.3]

$$\left[\begin{array}{ccc|c} 1 & -3 & 2 & -5 \\ 0 & 1 & -4 & 8 \\ 0 & 0 & 0 & 0 \end{array}\right]$$

25. Is it true that if $\mathbf{AB} = \mathbf{0}$, for matrices $\mathbf{A}$ and $\mathbf{B}$, then $\mathbf{A} = \mathbf{0}$ or $\mathbf{B} = \mathbf{0}$? Why or why not? [9.4]

9.5 Inverses of Matrices

- Find the inverse of a square matrix, if it exists.
- Use inverses of matrices to solve systems of equations.

In this section, we continue our study of matrix algebra, finding the **multiplicative inverse**, or simply **inverse**, of a square matrix, if it exists. Then we use such inverses to solve systems of equations.

The Identity Matrix

Recall that, for real numbers, $a \cdot 1 = 1 \cdot a = a$; 1 is the multiplicative identity. A multiplicative identity matrix is very similar to the number 1.

IDENTITY MATRIX

For any positive integer n, the $n \times n$ **identity matrix** is an $n \times n$ matrix with 1's on the main diagonal and 0's elsewhere and is denoted by

$$\mathbf{I} = \begin{bmatrix} 1 & 0 & 0 & \cdots & 0 \\ 0 & 1 & 0 & \cdots & 0 \\ 0 & 0 & 1 & \cdots & 0 \\ \vdots & \vdots & \vdots & & \vdots \\ 0 & 0 & 0 & \cdots & 1 \end{bmatrix}.$$

Then $\mathbf{AI} = \mathbf{IA} = \mathbf{A}$, for any $n \times n$ matrix $\mathbf{A}$.

EXAMPLE 1 For

$$\mathbf{A} = \begin{bmatrix} 4 & -7 \\ -3 & 2 \end{bmatrix} \quad \text{and} \quad \mathbf{I} = \begin{bmatrix} 1 & 0 \\ 0 & 1 \end{bmatrix},$$

find each of the following.

a) $\mathbf{AI}$ **b)** $\mathbf{IA}$

Solution

a) $$\mathbf{AI} = \begin{bmatrix} 4 & -7 \\ -3 & 2 \end{bmatrix}\begin{bmatrix} 1 & 0 \\ 0 & 1 \end{bmatrix} = \begin{bmatrix} 4 \cdot 1 - 7 \cdot 0 & 4 \cdot 0 - 7 \cdot 1 \\ -3 \cdot 1 + 2 \cdot 0 & -3 \cdot 0 + 2 \cdot 1 \end{bmatrix} = \begin{bmatrix} 4 & -7 \\ -3 & 2 \end{bmatrix} = \mathbf{A}$$

b) $$\mathbf{IA} = \begin{bmatrix} 1 & 0 \\ 0 & 1 \end{bmatrix}\begin{bmatrix} 4 & -7 \\ -3 & 2 \end{bmatrix} = \begin{bmatrix} 1 \cdot 4 + 0(-3) & 1(-7) + 0 \cdot 2 \\ 0 \cdot 4 + 1(-3) & 0(-7) + 1 \cdot 2 \end{bmatrix} = \begin{bmatrix} 4 & -7 \\ -3 & 2 \end{bmatrix} = \mathbf{A}$$

[A] [I] $\begin{bmatrix} 4 & -7 \\ -3 & 2 \end{bmatrix}$
[I] [A] $\begin{bmatrix} 4 & -7 \\ -3 & 2 \end{bmatrix}$

These products can also be found using a graphing calculator after **A** and **I** have been entered.

The Inverse of a Matrix

Recall that for every nonzero real number a, there is a multiplicative inverse $1/a$, or a^{-1}, such that $a \cdot a^{-1} = a^{-1} \cdot a = 1$. The multiplicative inverse of a matrix behaves in a similar manner.

INVERSE OF A MATRIX

For an $n \times n$ matrix $\mathbf{A}$, if there is a matrix $\mathbf{A}^{-1}$ for which $\mathbf{A}^{-1} \cdot \mathbf{A} = \mathbf{I} = \mathbf{A} \cdot \mathbf{A}^{-1}$, then $\mathbf{A}^{-1}$ is the **inverse** of $\mathbf{A}$.

We read $\mathbf{A}^{-1}$ as "$\mathbf{A}$ inverse." Note that not every matrix has an inverse.

EXAMPLE 2 Verify that

$$\mathbf{B} = \begin{bmatrix} 4 & -3 \\ 3 & -2 \end{bmatrix} \text{ is the inverse of } \mathbf{A} = \begin{bmatrix} -2 & 3 \\ -3 & 4 \end{bmatrix}.$$

Solution We show that $\mathbf{BA} = \mathbf{I} = \mathbf{AB}$.

$$\mathbf{BA} = \begin{bmatrix} 4 & -3 \\ 3 & -2 \end{bmatrix}\begin{bmatrix} -2 & 3 \\ -3 & 4 \end{bmatrix} = \begin{bmatrix} 1 & 0 \\ 0 & 1 \end{bmatrix} = \mathbf{I}$$

$$\mathbf{AB} = \begin{bmatrix} -2 & 3 \\ -3 & 4 \end{bmatrix}\begin{bmatrix} 4 & -3 \\ 3 & -2 \end{bmatrix} = \begin{bmatrix} 1 & 0 \\ 0 & 1 \end{bmatrix} = \mathbf{I}$$

Now Try Exercise 1.

We can find the inverse of a square matrix, if it exists, by using row-equivalent operations as in the Gauss–Jordan elimination method. For example, consider the matrix

$$\mathbf{A} = \begin{bmatrix} -2 & 3 \\ -3 & 4 \end{bmatrix}.$$

To find its inverse, we first form an **augmented matrix** consisting of $\mathbf{A}$ on the left side and the 2×2 identity matrix on the right side:

$$\left[\begin{array}{cc|cc} -2 & 3 & 1 & 0 \\ -3 & 4 & 0 & 1 \end{array}\right].$$

The 2×2 matrix $\mathbf{A}$ — The 2×2 identity matrix

Then we attempt to transform the augmented matrix to one of the form

$$\left[\begin{array}{cc|cc} 1 & 0 & a & b \\ 0 & 1 & c & d \end{array}\right].$$

The 2×2 identity matrix — The matrix $\mathbf{A}^{-1}$

If we can do this, the matrix on the right, $\begin{bmatrix} a & b \\ c & d \end{bmatrix}$, is $\mathbf{A}^{-1}$.

EXAMPLE 3 Find $\mathbf{A}^{-1}$, where

$$\mathbf{A} = \begin{bmatrix} -2 & 3 \\ -3 & 4 \end{bmatrix}.$$

Solution First, we write the augmented matrix. Then we transform it to the desired form.

$$\left[\begin{array}{cc|cc} -2 & 3 & 1 & 0 \\ -3 & 4 & 0 & 1 \end{array}\right]$$

$$\left[\begin{array}{cc|cc} 1 & -\frac{3}{2} & -\frac{1}{2} & 0 \\ -3 & 4 & 0 & 1 \end{array}\right] \quad \text{New row 1} = -\tfrac{1}{2}(\text{row 1})$$

$$\left[\begin{array}{cc|cc} 1 & -\frac{3}{2} & -\frac{1}{2} & 0 \\ 0 & -\frac{1}{2} & -\frac{3}{2} & 1 \end{array}\right] \quad \text{New row 2} = 3(\text{row 1}) + \text{row 2}$$

$$\left[\begin{array}{cc|cc} 1 & -\frac{3}{2} & -\frac{1}{2} & 0 \\ 0 & 1 & 3 & -2 \end{array}\right] \quad \text{New row 2} = -2(\text{row 2})$$

$$\left[\begin{array}{cc|cc} 1 & 0 & 4 & -3 \\ 0 & 1 & 3 & -2 \end{array}\right] \quad \text{New row 1} = \tfrac{3}{2}(\text{row 2}) + \text{row 1}$$

Thus,

$$\mathbf{A}^{-1} = \begin{bmatrix} 4 & -3 \\ 3 & -2 \end{bmatrix},$$

which we verified in Example 2.

[A]$^{-1}$
$\begin{bmatrix} 4 & -3 \\ 3 & -2 \end{bmatrix}$

The x^{-1} key on a graphing calculator can also be used to find the inverse of a matrix.

Now Try Exercise 5.

EXAMPLE 4 Find $\mathbf{A}^{-1}$, where

$$\mathbf{A} = \begin{bmatrix} 1 & 2 & -1 \\ 3 & 5 & 3 \\ 2 & 4 & 3 \end{bmatrix}.$$

Solution First, we write the augmented matrix. Then we transform it to the desired form.

$$\left[\begin{array}{ccc|ccc} 1 & 2 & -1 & 1 & 0 & 0 \\ 3 & 5 & 3 & 0 & 1 & 0 \\ 2 & 4 & 3 & 0 & 0 & 1 \end{array}\right]$$

$$\left[\begin{array}{ccc|ccc} 1 & 2 & -1 & 1 & 0 & 0 \\ 0 & -1 & 6 & -3 & 1 & 0 \\ 0 & 0 & 5 & -2 & 0 & 1 \end{array}\right] \quad \begin{array}{l} \text{New row 2} = -3(\text{row 1}) + \text{row 2} \\ \text{New row 3} = -2(\text{row 1}) + \text{row 3} \end{array}$$

$$\left[\begin{array}{ccc|ccc} 1 & 2 & -1 & 1 & 0 & 0 \\ 0 & -1 & 6 & -3 & 1 & 0 \\ 0 & 0 & 1 & -\frac{2}{5} & 0 & \frac{1}{5} \end{array}\right] \quad \text{New row 3} = \tfrac{1}{5}(\text{row 3})$$

$$\left[\begin{array}{ccc|ccc} 1 & 2 & 0 & \frac{3}{5} & 0 & \frac{1}{5} \\ 0 & -1 & 0 & -\frac{3}{5} & 1 & -\frac{6}{5} \\ 0 & 0 & 1 & -\frac{2}{5} & 0 & \frac{1}{5} \end{array}\right] \quad \begin{array}{l} \text{New row 1} = \text{row 3} + \text{row 1} \\ \text{New row 2} = -6(\text{row 3}) + \text{row 2} \end{array}$$

$$\left[\begin{array}{ccc|ccc} 1 & 0 & 0 & -\frac{3}{5} & 2 & -\frac{11}{5} \\ 0 & -1 & 0 & -\frac{3}{5} & 1 & -\frac{6}{5} \\ 0 & 0 & 1 & -\frac{2}{5} & 0 & \frac{1}{5} \end{array}\right] \quad \text{New row 1} = 2(\text{row 2}) + \text{row 1}$$

$$\left[\begin{array}{ccc|ccc} 1 & 0 & 0 & -\frac{3}{5} & 2 & -\frac{11}{5} \\ 0 & 1 & 0 & \frac{3}{5} & -1 & \frac{6}{5} \\ 0 & 0 & 1 & -\frac{2}{5} & 0 & \frac{1}{5} \end{array}\right] \quad \text{New row 2} = -1(\text{row 2})$$

Thus,

$$\mathbf{A}^{-1} = \begin{bmatrix} -\frac{3}{5} & 2 & -\frac{11}{5} \\ \frac{3}{5} & -1 & \frac{6}{5} \\ -\frac{2}{5} & 0 & \frac{1}{5} \end{bmatrix}.$$

Now Try Exercise 9.

If a matrix has an inverse, we say that it is **invertible**, or **nonsingular**. When we cannot obtain the identity matrix on the left using the Gauss–Jordan method, then no inverse exists. This occurs when we obtain a row consisting entirely of 0's in either of the two matrices in the augmented matrix. In this case, we say that **A** is a **singular matrix**.

When we try to find the inverse of a noninvertible, or singular, matrix using a graphing calculator, the calculator returns an error message similar to ERR: SINGULAR MATRIX.

Solving Systems of Equations

MATRIX EQUATIONS

REVIEW SECTION 9.4.

We can write a system of n linear equations in n variables as a matrix equation $\mathbf{AX} = \mathbf{B}$. If **A** has an inverse, then the system of equations has a unique solution that can be found by solving for **X**, as follows:

$$\mathbf{AX} = \mathbf{B}$$

$$\mathbf{A}^{-1}(\mathbf{AX}) = \mathbf{A}^{-1}\mathbf{B} \quad \text{Multiplying by } \mathbf{A}^{-1} \text{ on the left on both sides}$$

$$(\mathbf{A}^{-1}\mathbf{A})\mathbf{X} = \mathbf{A}^{-1}\mathbf{B} \quad \text{Using the associative property of matrix multiplication}$$

$$\mathbf{IX} = \mathbf{A}^{-1}\mathbf{B} \quad \mathbf{A}^{-1}\mathbf{A} = \mathbf{I}$$

$$\mathbf{X} = \mathbf{A}^{-1}\mathbf{B}. \quad \mathbf{IX} = \mathbf{X}$$

MATRIX SOLUTIONS OF SYSTEMS OF EQUATIONS

For a system of n linear equations in n variables, $\mathbf{AX} = \mathbf{B}$, if **A** is an invertible matrix, then the unique solution of the system is given by

$$\mathbf{X} = \mathbf{A}^{-1}\mathbf{B}.$$

Since matrix multiplication is not commutative in general, care must be taken to multiply *on the left* by $\mathbf{A}^{-1}$.

EXAMPLE 5 Use an inverse matrix to solve the following system of equations:

$$-2x + 3y = 4,$$
$$-3x + 4y = 5.$$

Solution We write an equivalent matrix equation, $\mathbf{AX} = \mathbf{B}$:

$$\begin{bmatrix} -2 & 3 \\ -3 & 4 \end{bmatrix} \cdot \begin{bmatrix} x \\ y \end{bmatrix} = \begin{bmatrix} 4 \\ 5 \end{bmatrix}$$
$$\mathbf{A} \quad \cdot \quad \mathbf{X} \quad = \quad \mathbf{B}$$

In Example 3, we found that

$$\mathbf{A}^{-1} = \begin{bmatrix} 4 & -3 \\ 3 & -2 \end{bmatrix}.$$

We also verified this in Example 2. Now we have

$$\mathbf{X} = \mathbf{A}^{-1}\mathbf{B}$$

$$\begin{bmatrix} x \\ y \end{bmatrix} = \begin{bmatrix} 4 & -3 \\ 3 & -2 \end{bmatrix}\begin{bmatrix} 4 \\ 5 \end{bmatrix} = \begin{bmatrix} 1 \\ 2 \end{bmatrix}.$$

The solution of the system of equations is $(1, 2)$.

To use a graphing calculator to solve this system of equations, we enter **A** and **B** and then enter the notation $\mathbf{A}^{-1}\mathbf{B}$ on the home screen.

Now Try Exercises 25 and 35.

9.5 Exercise Set

Determine whether **B** *is the inverse of* **A**.

1. $\mathbf{A} = \begin{bmatrix} 1 & -3 \\ -2 & 7 \end{bmatrix}$, $\mathbf{B} = \begin{bmatrix} 7 & 3 \\ 2 & 1 \end{bmatrix}$

2. $\mathbf{A} = \begin{bmatrix} 3 & 2 \\ 4 & 3 \end{bmatrix}$, $\mathbf{B} = \begin{bmatrix} 3 & -2 \\ -4 & 3 \end{bmatrix}$

3. $\mathbf{A} = \begin{bmatrix} -1 & -1 & 6 \\ 1 & 0 & -2 \\ 1 & 0 & -3 \end{bmatrix}$, $\mathbf{B} = \begin{bmatrix} 2 & 3 & 2 \\ 3 & 3 & 4 \\ 1 & 1 & 1 \end{bmatrix}$

4. $\mathbf{A} = \begin{bmatrix} -2 & 0 & -3 \\ 5 & 1 & 7 \\ -3 & 0 & 4 \end{bmatrix}$, $\mathbf{B} = \begin{bmatrix} 4 & 0 & -3 \\ 1 & 1 & 1 \\ -3 & 0 & 2 \end{bmatrix}$

Use the Gauss–Jordan method to find $\mathbf{A}^{-1}$, *if it exists. Check your answers by using a graphing calculator to find* $\mathbf{A}^{-1}\mathbf{A}$ *and* $\mathbf{A}\mathbf{A}^{-1}$.

5. $\mathbf{A} = \begin{bmatrix} 3 & 2 \\ 5 & 3 \end{bmatrix}$

6. $\mathbf{A} = \begin{bmatrix} 3 & 5 \\ 1 & 2 \end{bmatrix}$

7. $\mathbf{A} = \begin{bmatrix} 6 & 9 \\ 4 & 6 \end{bmatrix}$

8. $\mathbf{A} = \begin{bmatrix} -4 & -6 \\ 2 & 3 \end{bmatrix}$

9. $\mathbf{A} = \begin{bmatrix} 3 & 1 & 0 \\ 1 & 1 & 1 \\ 1 & -1 & 2 \end{bmatrix}$

10. $\mathbf{A} = \begin{bmatrix} 1 & 0 & 1 \\ 2 & 1 & 0 \\ 1 & -1 & 1 \end{bmatrix}$

11. $\mathbf{A} = \begin{bmatrix} 1 & -4 & 8 \\ 1 & -3 & 2 \\ 2 & -7 & 10 \end{bmatrix}$

12. $\mathbf{A} = \begin{bmatrix} -2 & 5 & 3 \\ 4 & -1 & 3 \\ 7 & -2 & 5 \end{bmatrix}$

Use a graphing calculator to find $\mathbf{A}^{-1}$, *if it exists.*

13. $\mathbf{A} = \begin{bmatrix} 4 & -3 \\ 1 & -2 \end{bmatrix}$

14. $\mathbf{A} = \begin{bmatrix} 0 & -1 \\ 1 & 0 \end{bmatrix}$

15. $\mathbf{A} = \begin{bmatrix} 2 & 3 & 2 \\ 3 & 3 & 4 \\ -1 & -1 & -1 \end{bmatrix}$

16. $\mathbf{A} = \begin{bmatrix} 1 & 2 & 3 \\ 2 & -1 & -2 \\ -1 & 3 & 3 \end{bmatrix}$

17. $\mathbf{A} = \begin{bmatrix} 1 & 2 & -1 \\ -2 & 0 & 1 \\ 1 & -1 & 0 \end{bmatrix}$

18. $\mathbf{A} = \begin{bmatrix} 7 & -1 & -9 \\ 2 & 0 & -4 \\ -4 & 0 & 6 \end{bmatrix}$

19. $\mathbf{A} = \begin{bmatrix} 1 & 3 & -1 \\ 0 & 2 & -1 \\ 1 & 1 & 0 \end{bmatrix}$

20. $\mathbf{A} = \begin{bmatrix} -1 & 0 & -1 \\ -1 & 1 & 0 \\ 0 & 1 & 1 \end{bmatrix}$

21. $\mathbf{A} = \begin{bmatrix} 1 & 2 & 3 & 4 \\ 0 & 1 & 3 & -5 \\ 0 & 0 & 1 & -2 \\ 0 & 0 & 0 & -1 \end{bmatrix}$

22. $\mathbf{A} = \begin{bmatrix} -2 & -3 & 4 & 1 \\ 0 & 1 & 1 & 0 \\ 0 & 4 & -6 & 1 \\ -2 & -2 & 5 & 1 \end{bmatrix}$

23. $\mathbf{A} = \begin{bmatrix} 1 & -14 & 7 & 38 \\ -1 & 2 & 1 & -2 \\ 1 & 2 & -1 & -6 \\ 1 & -2 & 3 & 6 \end{bmatrix}$

24. $\mathbf{A} = \begin{bmatrix} 10 & 20 & -30 & 15 \\ 3 & -7 & 14 & -8 \\ -7 & -2 & -1 & 2 \\ 4 & 4 & -3 & 1 \end{bmatrix}$

In Exercises 25–28, a system of equations is given, together with the inverse of the coefficient matrix. Use the inverse of the coefficient matrix to solve the system of equations.

25. $11x + 3y = -4,$ $\quad 7x + 2y = 5;$ $\quad \mathbf{A}^{-1} = \begin{bmatrix} 2 & -3 \\ -7 & 11 \end{bmatrix}$

26. $8x + 5y = -6,$ $\quad 5x + 3y = 2;$ $\quad \mathbf{A}^{-1} = \begin{bmatrix} -3 & 5 \\ 5 & -8 \end{bmatrix}$

27. $3x + y = 2,$ $\quad 2x - y + 2z = -5,$ $\quad x + y + z = 5;$ $\quad \mathbf{A}^{-1} = \frac{1}{9}\begin{bmatrix} 3 & 1 & -2 \\ 0 & -3 & 6 \\ -3 & 2 & 5 \end{bmatrix}$

28. $y - z = -4,$ $\quad 4x + y = -3,$ $\quad 3x - y + 3z = 1;$ $\quad \mathbf{A}^{-1} = \frac{1}{5}\begin{bmatrix} -3 & 2 & -1 \\ 12 & -3 & 4 \\ 7 & -3 & 4 \end{bmatrix}$

Solve the system of equations using the inverse of the coefficient matrix of the equivalent matrix equation.

29. $4x + 3y = 2,$ $\quad x - 2y = 6$

30. $2x - 3y = 7,$ $\quad 4x + y = -7$

31. $5x + y = 2,$ $\quad 3x - 2y = -4$

32. $x - 6y = 5,$ $\quad -x + 4y = -5$

33. $x + z = 1,$ $\quad 2x + y = 3,$ $\quad x - y + z = 4$

34. $x + 2y + 3z = -1,$ $\quad 2x - 3y + 4z = 2,$ $\quad -3x + 5y - 6z = 4$

35. $2x + 3y + 4z = 2,$ $\quad x - 4y + 3z = 2,$ $\quad 5x + y + z = -4$

36. $x + y = 2,$ $\quad 3x + 2z = 5,$ $\quad 2x + 3y - 3z = 9$

37. $2w - 3x + 4y - 5z = 0,$ $\quad 3w - 2x + 7y - 3z = 2,$ $\quad w + x - y + z = 1,$ $\quad -w - 3x - 6y + 4z = 6$

38. $5w - 4x + 3y - 2z = -6,$ $\quad w + 4x - 2y + 3z = -5,$ $\quad 2w - 3x + 6y - 9z = 14,$ $\quad 3w - 5x + 2y - 4z = -3$

39. *Cranberry Production.* In 2012, a total of 660 million lb of cranberries were produced in Wisconsin and Massachusetts. Wisconsin grew 240 million lb more than Massachusetts grew. (*Source*: U.S. Department of Agriculture) How many pounds of cranberries were grown in each state?

40. *Lunch Cost.* Coworkers Jan and Richard purchase lunch from a food truck. Jan buys 1 beef taco and 2 fruit cups for \$5.25. Richard buys 3 beef tacos and 1 fruit cup for \$8.25. Find the price of each item.

41. *Cost of Landscape Materials.* Cheryl's Landscaping bought 15 cubic yards of topsoil, 30 cubic yards of mulch, and 11 cubic yards of decorative rock for \$1693. The next week, the business bought 8 cubic yards of topsoil, 19 cubic yards of mulch, and 7 cubic yards of decorative rock for \$1030. Decorative rock costs \$10 more per cubic yard than mulch. Find the price per cubic yard of each material.

42. *Investment.* Trevor receives \$230 per year in simple interest from three investments totaling \$8500. Part is invested at 2.2%, part at 2.65%, and the rest at 3.05%. There is \$1500 more invested at 3.05% than at 2.2%. Find the amount invested at each rate.

Skill Maintenance

Use synthetic division to find the function values. [4.3]

43. $f(x) = x^3 - 6x^2 + 4x - 8$; find $f(-2)$

44. $f(x) = 2x^4 - x^3 + 5x^2 + 6x - 4$; find $f(3)$

Solve.

45. $2x^2 + x = 7$ [3.2]

46. $\frac{1}{x+1} - \frac{6}{x-1} = 1$ [3.4]

47. $\sqrt{2x+1} - 1 = \sqrt{2x-4}$ [3.2], [3.4]

48. $x - \sqrt{x} - 6 = 0$ [3.2], [3.4]

Factor the polynomial $f(x)$. [4.3]

49. $f(x) = x^3 - 3x^2 - 6x + 8$

50. $f(x) = x^4 + 2x^3 - 16x^2 - 2x + 15$

Synthesis

State the conditions under which $\mathbf{A}^{-1}$ exists. Then find a formula for $\mathbf{A}^{-1}$.

51. $\mathbf{A} = [x]$

52. $\mathbf{A} = \begin{bmatrix} x & 0 \\ 0 & y \end{bmatrix}$

53. $\mathbf{A} = \begin{bmatrix} 0 & 0 & x \\ 0 & y & 0 \\ z & 0 & 0 \end{bmatrix}$

54. $\mathbf{A} = \begin{bmatrix} x & 1 & 1 & 1 \\ 0 & y & 0 & 0 \\ 0 & 0 & z & 0 \\ 0 & 0 & 0 & w \end{bmatrix}$

9.6 Determinants and Cramer's Rule

- Evaluate determinants of square matrices.
- Use Cramer's rule to solve systems of equations.

Determinants of Square Matrices

With every square matrix, we associate a number called its *determinant*.

DETERMINANT OF A 2 × 2 MATRIX

The **determinant** of the matrix $\begin{bmatrix} a & c \\ b & d \end{bmatrix}$ is denoted $\begin{vmatrix} a & c \\ b & d \end{vmatrix}$ and is defined as

$$\begin{vmatrix} a & c \\ b & d \end{vmatrix} = ad - bc.$$

EXAMPLE 1 Evaluate: $\begin{vmatrix} \sqrt{2} & -3 \\ -4 & -\sqrt{2} \end{vmatrix}$.

Solution

$$\begin{vmatrix} \sqrt{2} & -3 \\ -4 & -\sqrt{2} \end{vmatrix}$$ The arrows indicate the products involved.

$$\begin{aligned} &= \sqrt{2}(-\sqrt{2}) - (-4)(-3) \\ &= -2 - 12 \\ &= -14 \end{aligned}$$

Now Try Exercise 5.

We now consider a way to evaluate determinants of square matrices of order 3×3 or higher.

Evaluating Determinants Using Cofactors

Often we first find minors and cofactors of matrices in order to evaluate determinants.

MINOR

For a square matrix $\mathbf{A} = [a_{ij}]$, the **minor** M_{ij} of an entry a_{ij} is the determinant of the matrix formed by deleting the ith row and the jth column of $\mathbf{A}$.

EXAMPLE 2 For the matrix

$$\mathbf{A} = [a_{ij}] = \begin{bmatrix} -8 & 0 & 6 \\ 4 & -6 & 7 \\ -1 & -3 & 5 \end{bmatrix},$$

find each of the following.

a) M_{11}

b) M_{23}

Solution

a) For M_{11}, we delete the first row and the first column and find the determinant of the 2×2 matrix formed by the remaining entries.

$$\begin{bmatrix} -8 & 0 & 6 \\ 4 & -6 & 7 \\ -1 & -3 & 5 \end{bmatrix} \qquad M_{11} = \begin{vmatrix} -6 & 7 \\ -3 & 5 \end{vmatrix}$$

$$= (-6) \cdot 5 - (-3) \cdot 7$$
$$= -30 - (-21)$$
$$= -30 + 21$$
$$= -9$$

b) For M_{23}, we delete the second row and the third column and find the determinant of the 2×2 matrix formed by the remaining entries.

$$\begin{bmatrix} -8 & 0 & 6 \\ 4 & -6 & 7 \\ -1 & -3 & 5 \end{bmatrix} \qquad M_{23} = \begin{vmatrix} -8 & 0 \\ -1 & -3 \end{vmatrix}$$

$$= -8(-3) - (-1)0$$
$$= 24$$

Now Try Exercise 9.

COFACTOR

For a square matrix $\mathbf{A} = [a_{ij}]$, the **cofactor** A_{ij} of an entry a_{ij} is given by

$$A_{ij} = (-1)^{i+j} M_{ij},$$

where M_{ij} is the minor of a_{ij}.

EXAMPLE 3 For the matrix given in Example 2, find each of the following.

a) A_{11} **b)** A_{23}

Solution

a) In Example 2, we found that $M_{11} = -9$. Then

$$A_{11} = (-1)^{1+1} M_{11} = (1)(-9) = -9.$$

b) In Example 2, we found that $M_{23} = 24$. Then

$$A_{23} = (-1)^{2+3} M_{24} = (-1)(24) = -24.$$

Now Try Exercise 11.

Note that minors and cofactors are *numbers.* They are *not matrices.*

Consider the matrix $\mathbf{A}$ given by

$$\mathbf{A} = \begin{bmatrix} a_{11} & a_{12} & a_{13} \\ a_{21} & a_{22} & a_{23} \\ a_{31} & a_{32} & a_{33} \end{bmatrix}.$$

The determinant of the matrix, denoted $|\mathbf{A}|$, can be found by multiplying each element of the first column by its cofactor and adding:

$$|\mathbf{A}| = a_{11}A_{11} + a_{21}A_{21} + a_{31}A_{31}.$$

Because

$$A_{11} = (-1)^{1+1}M_{11} = M_{11},$$
$$A_{21} = (-1)^{2+1}M_{21} = -M_{21},$$

and $A_{31} = (-1)^{3+1}M_{31} = M_{31},$

we can write

$$|\mathbf{A}| = a_{11} \cdot \begin{vmatrix} a_{22} & a_{23} \\ a_{32} & a_{33} \end{vmatrix} - a_{21} \cdot \begin{vmatrix} a_{12} & a_{13} \\ a_{32} & a_{33} \end{vmatrix} + a_{31} \cdot \begin{vmatrix} a_{12} & a_{13} \\ a_{22} & a_{23} \end{vmatrix}.$$

It can be shown that we can find $|\mathbf{A}|$ by choosing *any* row or column, multiplying each element in that row or column by its cofactor, and adding. This is called *expanding* across a row or down a column. We just expanded down the first column. We now define the determinant of a square matrix of any order.

DETERMINANT OF ANY SQUARE MATRIX

For any square matrix $\mathbf{A}$ of order $n \times n$ $(n > 1)$, we define the **determinant** of $\mathbf{A}$, denoted $|\mathbf{A}|$, as follows. Choose any row or column. Multiply each element in that row or column by its cofactor and add the results. The determinant of a 1×1 matrix is simply the element of the matrix. The value of a determinant will be the same no matter which row or column is chosen.

EXAMPLE 4 Evaluate $|\mathbf{A}|$ by expanding across the third row:

$$\mathbf{A} = \begin{bmatrix} -8 & 0 & 6 \\ 4 & -6 & 7 \\ -1 & -3 & 5 \end{bmatrix}.$$

Solution We have

$$\begin{aligned} |\mathbf{A}| &= (-1)A_{31} + (-3)A_{32} + 5A_{33} \\ &= (-1)(-1)^{3+1} \cdot \begin{vmatrix} 0 & 6 \\ -6 & 7 \end{vmatrix} + (-3)(-1)^{3+2} \cdot \begin{vmatrix} -8 & 6 \\ 4 & 7 \end{vmatrix} \\ &\quad + 5(-1)^{3+3} \cdot \begin{vmatrix} -8 & 0 \\ 4 & -6 \end{vmatrix} \\ &= (-1) \cdot 1 \cdot [0 \cdot 7 - (-6)6] + (-3)(-1)[-8 \cdot 7 - 4 \cdot 6] \\ &\quad + 5 \cdot 1 \cdot [-8(-6) - 4 \cdot 0] \\ &= -[36] + 3[-80] + 5[48] \\ &= -36 - 240 + 240 = -36. \end{aligned}$$

The value of this determinant is -36 no matter which row or column we expand on.

Now Try Exercise 13.

Determinants can also be evaluated using a graphing calculator.

EXAMPLE 5 Use a graphing calculator to evaluate $|\mathbf{A}|$:

$$\mathbf{A} = \begin{bmatrix} 1 & 6 & -1 \\ -3 & -5 & 3 \\ 0 & 4 & 2 \end{bmatrix}.$$

det ([A])
26

Solution First, we enter $\mathbf{A}$. Then we select the determinant operation, det, from the MATRIX MATH menu and enter the name of the matrix, $\mathbf{A}$. The calculator will return the value of the determinant of the matrix, 26.

Now Try Exercise 17.

Cramer's Rule

Determinants can be used to solve systems of linear equations. Consider a system of two linear equations:

$$a_1x + b_1y = c_1,$$
$$a_2x + b_2y = c_2.$$

Solving this system using the elimination method, we obtain

$$x = \frac{c_1b_2 - c_2b_1}{a_1b_2 - a_2b_1} \quad \text{and} \quad y = \frac{a_1c_2 - a_2c_1}{a_1b_2 - a_2b_1}.$$

The numerators and the denominators of these expressions can be written as determinants:

$$x = \frac{\begin{vmatrix} c_1 & b_1 \\ c_2 & b_2 \end{vmatrix}}{\begin{vmatrix} a_1 & b_1 \\ a_2 & b_2 \end{vmatrix}} \quad \text{and} \quad y = \frac{\begin{vmatrix} a_1 & c_1 \\ a_2 & c_2 \end{vmatrix}}{\begin{vmatrix} a_1 & b_1 \\ a_2 & b_2 \end{vmatrix}}.$$

If we let

$$D = \begin{vmatrix} a_1 & b_1 \\ a_2 & b_2 \end{vmatrix}, \quad D_x = \begin{vmatrix} c_1 & b_1 \\ c_2 & b_2 \end{vmatrix}, \quad \text{and} \quad D_y = \begin{vmatrix} a_1 & c_1 \\ a_2 & c_2 \end{vmatrix},$$

we have

$$x = \frac{D_x}{D} \quad \text{and} \quad y = \frac{D_y}{D}.$$

This procedure for solving systems of equations is known as *Cramer's rule.*

CRAMER'S RULE FOR 2 × 2 SYSTEMS

The solution of the system of equations

$$a_1x + b_1y = c_1,$$
$$a_2x + b_2y = c_2$$

is given by

$$x = \frac{D_x}{D}, \qquad y = \frac{D_y}{D},$$

where

$$D = \begin{vmatrix} a_1 & b_1 \\ a_2 & b_2 \end{vmatrix}, \quad D_x = \begin{vmatrix} c_1 & b_1 \\ c_2 & b_2 \end{vmatrix},$$

$$D_y = \begin{vmatrix} a_1 & c_1 \\ a_2 & c_2 \end{vmatrix}, \quad \text{and} \quad D \neq 0.$$

Note that the denominator D contains the coefficients of x and y, in the same position as in the original equations. For x, the numerator is obtained by replacing the x-coefficients in D (the a's) with the c's. For y, the numerator is obtained by replacing the y-coefficients in D (the b's) with the c's.

EXAMPLE 6 Solve using Cramer's rule:

$$2x + 5y = 7,$$
$$5x - 2y = -3.$$

Algebraic Solution

We have

$$x = \frac{\begin{vmatrix} 7 & 5 \\ -3 & -2 \end{vmatrix}}{\begin{vmatrix} 2 & 5 \\ 5 & -2 \end{vmatrix}} = \frac{7(-2) - (-3)5}{2(-2) - 5 \cdot 5}$$

$$= \frac{-14 + 15}{-4 - 25}$$

$$= \frac{1}{-29} = -\frac{1}{29};$$

$$y = \frac{\begin{vmatrix} 2 & 7 \\ 5 & -3 \end{vmatrix}}{\begin{vmatrix} 2 & 5 \\ 5 & -2 \end{vmatrix}} = \frac{2(-3) - 5 \cdot 7}{-29}$$

$$= \frac{-6 - 35}{-29} = \frac{-41}{-29} = \frac{41}{29}.$$

The solution is $\left(-\frac{1}{29}, \frac{41}{29}\right)$.

Graphical Solution

To use Cramer's rule to solve this system of equations on a graphing calculator, we first enter the matrices corresponding to D, D_x, and D_y. We enter

$$\mathbf{A} = \begin{bmatrix} 2 & 5 \\ 5 & -2 \end{bmatrix}, \quad \mathbf{B} = \begin{bmatrix} 7 & 5 \\ -3 & -2 \end{bmatrix},$$

and $\mathbf{C} = \begin{bmatrix} 2 & 7 \\ 5 & -3 \end{bmatrix}$.

Then

$$x = \frac{\det(\mathbf{B})}{\det(\mathbf{A})} \quad \text{and} \quad y = \frac{\det(\mathbf{C})}{\det(\mathbf{A})}.$$

det ([B])/det([A])▸Frac $-\frac{1}{29}$
det ([C])/det([A])▸Frac $\frac{41}{29}$

The solution is $\left(-\frac{1}{29}, \frac{41}{29}\right)$.

Now Try Exercise 31.

Cramer's rule works only when a system of equations has a unique solution. This occurs when $D \neq 0$. If $D = 0$ and D_x and D_y are also 0, then the equations are dependent. If $D = 0$ and D_x and/or D_y is not 0, then the system is inconsistent.

Cramer's rule can be extended to a system of n linear equations in n variables. We consider a 3×3 system.

CRAMER'S RULE FOR 3 × 3 SYSTEMS

The solution of the system of equations

$$a_1x + b_1y + c_1z = d_1,$$
$$a_2x + b_2y + c_2z = d_2,$$
$$a_3x + b_3y + c_3z = d_3$$

is given by

$$x = \frac{D_x}{D}, \quad y = \frac{D_y}{D}, \quad z = \frac{D_z}{D},$$

where

$$D = \begin{vmatrix} a_1 & b_1 & c_1 \\ a_2 & b_2 & c_2 \\ a_3 & b_3 & c_3 \end{vmatrix}, \quad D_x = \begin{vmatrix} d_1 & b_1 & c_1 \\ d_2 & b_2 & c_2 \\ d_3 & b_3 & c_3 \end{vmatrix},$$

$$D_y = \begin{vmatrix} a_1 & d_1 & c_1 \\ a_2 & d_2 & c_2 \\ a_3 & d_3 & c_3 \end{vmatrix}, \quad D_z = \begin{vmatrix} a_1 & b_1 & d_1 \\ a_2 & b_2 & d_2 \\ a_3 & b_3 & d_3 \end{vmatrix}, \quad \text{and} \quad D \neq 0.$$

Note that the determinant D_x is obtained from D by replacing the x-coefficients with d_1, d_2, and d_3. D_y and D_z are obtained in a similar manner. As with a system of two equations, Cramer's rule cannot be used if $D = 0$. If $D = 0$ and D_x, D_y, and D_z are 0, then the equations are dependent. If $D = 0$ and at least one of D_x, D_y, or D_z is not 0, then the system is inconsistent.

EXAMPLE 7 Solve using Cramer's rule:

$$\begin{aligned} x - 3y + 7z &= 13, \\ x + y + z &= 1, \\ x - 2y + 3z &= 4. \end{aligned}$$

Solution We have

$$D = \begin{vmatrix} 1 & -3 & 7 \\ 1 & 1 & 1 \\ 1 & -2 & 3 \end{vmatrix} = -10, \quad D_x = \begin{vmatrix} 13 & -3 & 7 \\ 1 & 1 & 1 \\ 4 & -2 & 3 \end{vmatrix} = 20,$$

$$D_y = \begin{vmatrix} 1 & 13 & 7 \\ 1 & 1 & 1 \\ 1 & 4 & 3 \end{vmatrix} = -6, \quad D_z = \begin{vmatrix} 1 & -3 & 13 \\ 1 & 1 & 1 \\ 1 & -2 & 4 \end{vmatrix} = -24.$$

Then

$$x = \frac{D_x}{D} = \frac{20}{-10} = -2, \quad y = \frac{D_y}{D} = \frac{-6}{-10} = \frac{3}{5}, \quad z = \frac{D_z}{D} = \frac{-24}{-10} = \frac{12}{5}.$$

The solution is $\left(-2, \frac{3}{5}, \frac{12}{5}\right)$.

In practice, it is not necessary to evaluate D_z. When we have found values for x and y, we can substitute them into one of the equations to find z.

Now Try Exercise 39.

9.6 Exercise Set

Evaluate the determinant.

1. $\begin{vmatrix} 5 & 3 \\ -2 & -4 \end{vmatrix}$

2. $\begin{vmatrix} -8 & 6 \\ -1 & 2 \end{vmatrix}$

3. $\begin{vmatrix} 4 & -7 \\ -2 & 3 \end{vmatrix}$

4. $\begin{vmatrix} -9 & -6 \\ 5 & 4 \end{vmatrix}$

5. $\begin{vmatrix} -2 & -\sqrt{5} \\ -\sqrt{5} & 3 \end{vmatrix}$

6. $\begin{vmatrix} \sqrt{5} & -3 \\ 4 & 2 \end{vmatrix}$

7. $\begin{vmatrix} x & 4 \\ x & x^2 \end{vmatrix}$

8. $\begin{vmatrix} y^2 & -2 \\ y & 3 \end{vmatrix}$

Use the following matrix for Exercises 9–17:

$$\mathbf{A} = \begin{bmatrix} 7 & -4 & -6 \\ 2 & 0 & -3 \\ 1 & 2 & -5 \end{bmatrix}.$$

9. Find M_{11}, M_{32}, and M_{22}.

10. Find M_{13}, M_{31}, and M_{23}.

11. Find A_{11}, A_{32}, and A_{22}.

12. Find A_{13}, A_{31}, and A_{23}.

13. Evaluate $|\mathbf{A}|$ by expanding across the second row.

14. Evaluate $|\mathbf{A}|$ by expanding down the second column.

15. Evaluate $|\mathbf{A}|$ by expanding down the third column.

16. Evaluate $|\mathbf{A}|$ by expanding across the first row.

17. Use a graphing calculator to evaluate $|\mathbf{A}|$.

Use the following matrix for Exercises 18–24:

$$\mathbf{A} = \begin{bmatrix} 1 & 0 & 0 & -2 \\ 4 & 1 & 0 & 0 \\ 5 & 6 & 7 & 8 \\ -2 & -3 & -1 & 0 \end{bmatrix}.$$

18. Find M_{12} and M_{44}.

19. Find M_{41} and M_{33}.

20. Find A_{22} and A_{34}.

21. Find A_{24} and A_{43}.

22. Evaluate $|\mathbf{A}|$ by expanding down the third column.

23. Evaluate $|\mathbf{A}|$ by expanding across the first row.

24. Use a graphing calculator to evaluate $|\mathbf{A}|$.

Evaluate the determinant.

25. $\begin{vmatrix} 3 & 1 & 2 \\ -2 & 3 & 1 \\ 3 & 4 & -6 \end{vmatrix}$

26. $\begin{vmatrix} 3 & -2 & 1 \\ 2 & 4 & 3 \\ -1 & 5 & 1 \end{vmatrix}$

27. $\begin{vmatrix} x & 0 & -1 \\ 2 & x & x^2 \\ -3 & x & 1 \end{vmatrix}$

28. $\begin{vmatrix} x & 1 & -1 \\ x^2 & x & x \\ 0 & x & 1 \end{vmatrix}$

Solve using Cramer's rule.

29. $-2x + 4y = 3,$
$3x - 7y = 1$

30. $5x - 4y = -3,$
$7x + 2y = 6$

31. $2x - y = 5,$
$x - 2y = 1$

32. $3x + 4y = -2,$
$5x - 7y = 1$

33. $2x + 9y = -2,$
$4x - 3y = 3$

34. $2x + 3y = -1,$
$3x + 6y = -0.5$

35. $2x + 5y = 7,$
$3x - 2y = 1$

36. $3x + 2y = 7,$
$2x + 3y = -2$

37. $3x + 2y - z = 4,$
$3x - 2y + z = 5,$
$4x - 5y - z = -1$

38. $3x - y + 2z = 1,$
$x - y + 2z = 3,$
$-2x + 3y + z = 1$

39. $3x + 5y - z = -2,$
$x - 4y + 2z = 13,$
$2x + 4y + 3z = 1$

40. $3x + 2y + 2z = 1,$
$5x - y - 6z = 3,$
$2x + 3y + 3z = 4$

41. $x - 3y - 7z = 6,$
$2x + 3y + z = 9,$
$4x + y = 7$

42. $x - 2y - 3z = 4,$
$3x - 2z = 8,$
$2x + y + 4z = 13$

43. $6y + 6z = -1,$
$8x + 6z = -1,$
$4x + 9y = 8$

44. $3x + 5y = 2,$
$2x - 3z = 7,$
$4y + 2z = -1$

Skill Maintenance

Determine whether the function is one-to-one, and if it is, find a formula for $f^{-1}(x)$. [5.1]

45. $f(x) = 3x + 2$

46. $f(x) = x^2 - 4$

47. $f(x) = |x| + 3$

48. $f(x) = \sqrt[3]{x} + 1$

Simplify. Write answers in the form $a + bi$, where a and b are real numbers. [3.1]

49. $(3 - 4i) - (-2 - i)$

50. $(5 + 2i) + (1 - 4i)$

51. $(1 - 2i)(6 + 2i)$

52. $\dfrac{3 + i}{4 - 3i}$

Synthesis

Solve.

53. $\begin{vmatrix} y & 2 \\ 3 & y \end{vmatrix} = y$

54. $\begin{vmatrix} x & -3 \\ -1 & x \end{vmatrix} \geq 0$

55. $\begin{vmatrix} 2 & x & 1 \\ 1 & 2 & -1 \\ 3 & 4 & -2 \end{vmatrix} = -6$

56. $\begin{vmatrix} m + 2 & -3 \\ m + 5 & -4 \end{vmatrix} = 3m - 5$

Rewrite the expression using a determinant. Answers may vary.

57. $a^2 + b^2$

58. $\frac{1}{2}h(a + b)$

59. $2\pi r^2 + 2\pi rh$

60. $x^2y^2 - Q^2$

9.7 Systems of Inequalities and Linear Programming

- Graph linear inequalities.
- Graph systems of linear inequalities.
- Solve linear programming problems.

A graph of an inequality is a drawing that represents its solutions. We have already seen that an inequality in one variable can be graphed on the number line. An inequality in two variables can be graphed on a coordinate plane.

SOLVE LINEAR INEQUALITIES
REVIEW SECTION 1.6.

Graphs of Linear Inequalities

A statement like $5x - 4y < 20$ is a linear inequality in two variables.

LINEAR INEQUALITY IN TWO VARIABLES

A **linear inequality in two variables** is an inequality that can be written in the form

$$Ax + By < C,$$

where A, B, and C are real numbers and A and B are not both zero. The symbol $<$ may be replaced with $\leq$, $>$, or $\geq$.

A solution of a linear inequality in two variables is an ordered pair (x, y) for which the inequality is true. For example, $(1, 3)$ is a solution of $5x - 4y < 20$ because $5 \cdot 1 - 4 \cdot 3 < 20$, or $-7 < 20$, is true. On the other hand, $(2, -6)$ is not a solution of $5x - 4y < 20$ because $5 \cdot 2 - 4 \cdot (-6) \not< 20$, or $34 \not< 20$.

The **solution set** of an inequality is the set of all ordered pairs that make it true. The **graph of an inequality** represents its solution set.

EXAMPLE 1 Graph: $y < x + 3$.

Solution We begin by graphing the **related equation** $y = x + 3$. We use a dashed line because the inequality symbol is $<$. This indicates that the line itself is not in the solution set of the inequality.

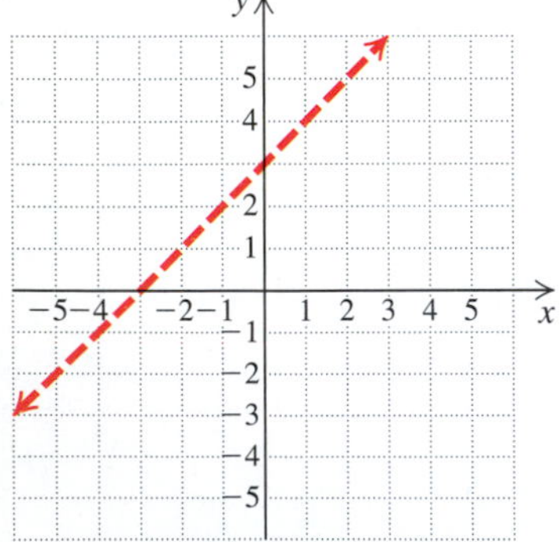

Note that the line divides the coordinate plane into two regions called **half-planes**. One of these half-planes satisfies the inequality. Either *all* points in a half-plane are in the solution set of the inequality or *none* is.

To determine which half-plane satisfies the inequality, we try a test point in either region. The point $(0, 0)$ is usually a convenient choice so long as it does not lie on the line.

$$\begin{array}{c|c} y & x + 3 \\ \hline 0 \ ? & 0 + 3 \\ 0 & 3 \end{array} \quad \text{TRUE} \quad 0 < 3 \text{ is true.}$$

Since $(0, 0)$ satisfies the inequality, so do all points in the half-plane that contains $(0, 0)$. We shade this region to show the solution set of the inequality.

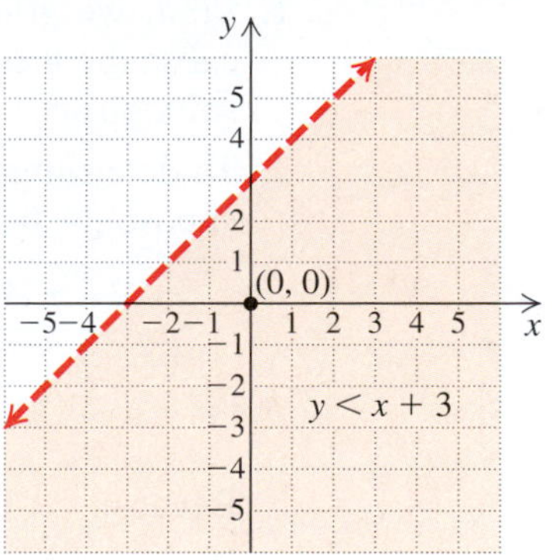

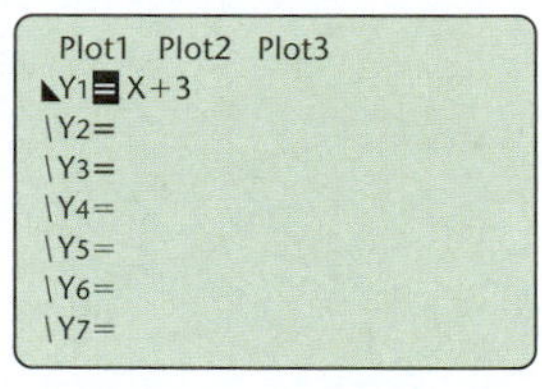

$y = x + 3$

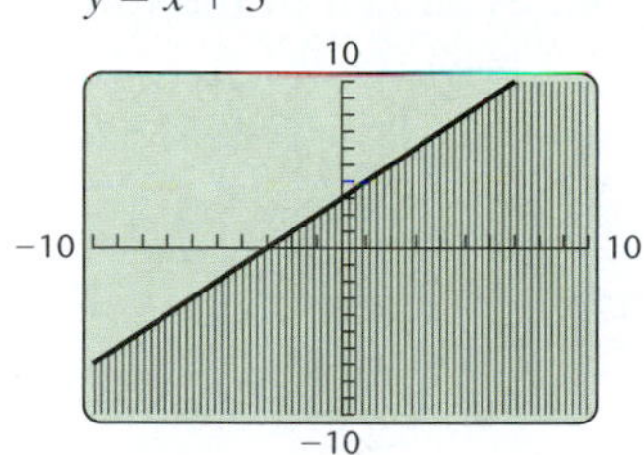

There are several ways to graph this inequality on a graphing calculator. One method is to first enter the related equation, $y = x + 3$, as shown at left. Then, after using a test point to determine which half-plane to shade as described above, we select the "shade below" graph style. Note that we must keep in mind that the line $y = x + 3$ is not included in the solution set. (Even if DOT mode or the dot graph style is selected, the line appears to be solid rather than dashed.)

We could also use the Shade option from the DRAW menu to graph this inequality. Some calculators have a pre-loaded application that can be used to graph an inequality. This application, Inequalz, is found on the APPS menu. Note that when this application is used, the inequality $y < x + 3$ is entered directly and the graph of the related equation appears as a dashed line.

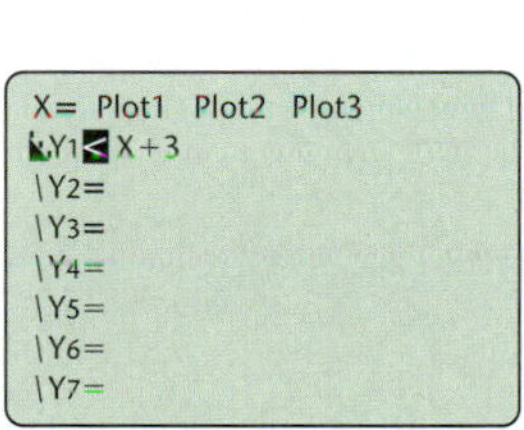

$y < x + 3$

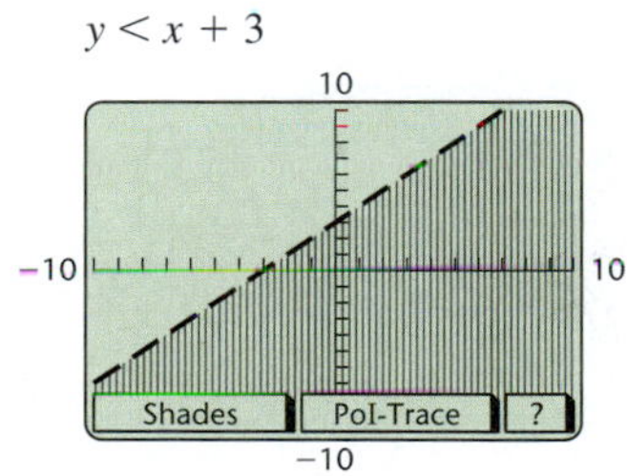

Now Try Exercise 13.

In general, we use the following procedure to graph linear inequalities in two variables by hand.

To graph a linear inequality in two variables:

1. Replace the inequality symbol with an equals sign and graph this related equation. If the inequality symbol is $<$ or $>$, draw the line dashed. If the inequality symbol is $\leq$ or $\geq$, draw the line solid.
2. The graph consists of a half-plane on one side of the line and, if the line is solid, the line as well. To determine which half-plane to shade, test a point not on the line in the original inequality. If that point is a solution, shade the half-plane containing that point. If not, shade the opposite half-plane.

EXAMPLE 2 Graph: $3x + 4y \geq 12$.

Solution

1. First, we graph the related equation $3x + 4y = 12$. We use a solid line because the inequality symbol is $\geq$. This indicates that the line is included in the solution set.
2. To determine which half-plane to shade, we test a point in either region. We choose $(0, 0)$.

$$\begin{array}{c|c} 3x + 4y & \geq 12 \\ \hline 3 \cdot 0 + 4 \cdot 0 & ? \; 12 \\ 0 & 12 \end{array} \quad \text{FALSE} \quad 0 \geq 12 \text{ is false.}$$

Because $(0, 0)$ is *not* a solution, all the points in the half-plane that does *not* contain $(0, 0)$ are solutions. We shade that region, as shown in the following figure.

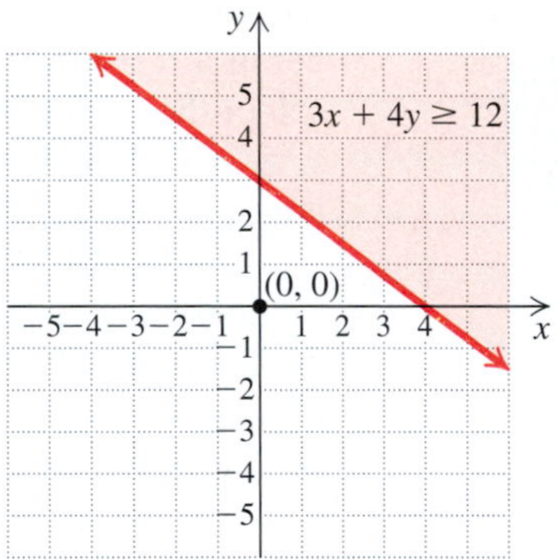

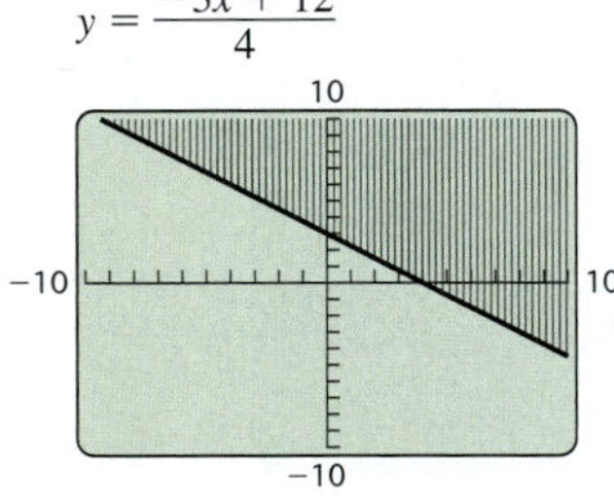

To graph this inequality on a graphing calculator using either the "shade above" graph style or the Shade option from the DRAW menu, we must first solve the related equation for y and enter $y = \frac{-3x + 12}{4}$. Similarly, to use the Inequalz application, we solve the inequality for y and enter $y \geq \frac{-3x + 12}{4}$.

EXAMPLE 3 Graph $x > -3$ on a plane.

Solution

1. First, we graph the related equation $x = -3$. We use a dashed line because the inequality symbol is $>$. This indicates that the line is not included in the solution set.
2. The inequality tells us that all points (x, y) for which $x > -3$ are solutions. These are the points to the right of the line. We can also use a test point to determine the solutions. We choose $(5, 1)$.

$$\begin{array}{c} x > -3 \\ \hline 5 \; ? \; -3 \end{array} \quad \text{TRUE} \quad 5 > -3 \text{ is true.}$$

Because $(5, 1)$ is a solution, we shade the region containing that point—that is, the region to the right of the dashed line.

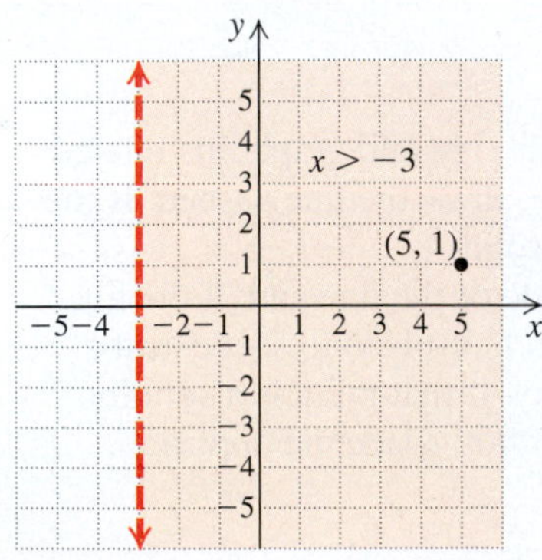

We can also graph this inequality on a graphing calculator that has the Inequalz application on the APPS menu.

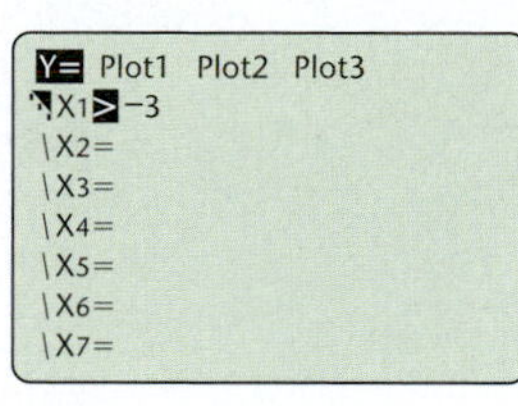

$x > -3$

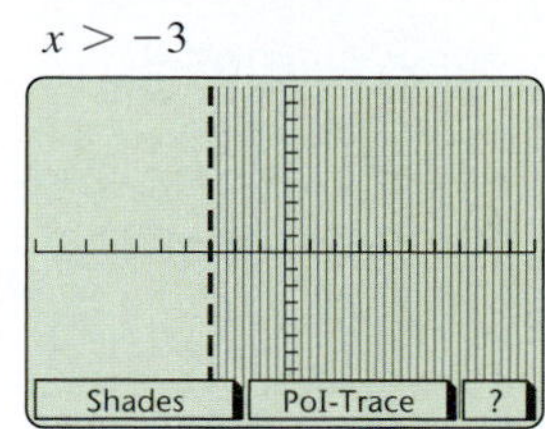

Now Try Exercise 23.

EXAMPLE 4 Graph $y \le 4$ on a plane.

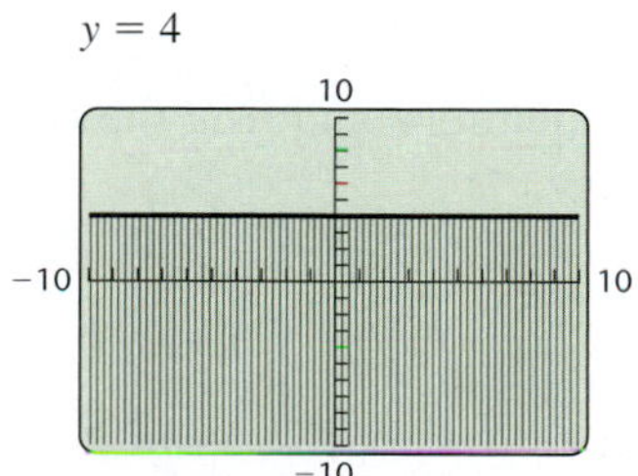

Solution

1. First, we graph the related equation $y = 4$. We use a solid line because the inequality symbol is $\le$.
2. The inequality tells us that all points (x, y) for which $y \le 4$ are solutions of the inequality. These are the points on or below the line. We can also use a test point to determine the solutions. We choose $(-2, 5)$.

$$\frac{y \le 4}{5 \;?\; 4} \quad \text{FALSE} \qquad 5 \le 4 \text{ is false.}$$

Because $(-2, 5)$ is not a solution, we shade the half-plane that does not contain that point.

We can graph the inequality $y \le 4$ using Inequalz or by graphing $y = 4$ and then using the "shade below" graph style, as shown at left.

$y = 4$

10, −10, 10, −10

Now Try Exercise 25.

Systems of Linear Inequalities

A system of inequalities consists of two or more inequalities considered simultaneously. For example,

$$x + y \le 4,$$
$$x - y \ge 2$$

is a system of *two linear inequalities in two variables.*

A solution of a system of inequalities is an ordered pair that is a solution of each inequality in the system. To graph a system of linear inequalities, we graph each inequality and determine the region that is common to *all* the solution sets.

EXAMPLE 5 Graph the solution set of the system

$$x + y \le 4,$$
$$x - y \ge 2.$$

Solution We graph $x + y \le 4$ by first graphing the equation $x + y = 4$ using a solid line. Next, we choose $(0, 0)$ as a test point and find that it is a solution of $x + y \le 4$, so we shade the half-plane containing $(0, 0)$ using red. Next, we graph $x - y = 2$ using a solid line. We find that $(0, 0)$ is not a solution of $x - y \ge 2$, so we shade the half-plane that does not contain $(0, 0)$ using green. The arrows near the ends of each line help to indicate the half-plane that contains each solution set.

The solution set of the system of equations is the region shaded both red and green, or brown, including parts of the lines $x + y = 4$ and $x - y = 2$.

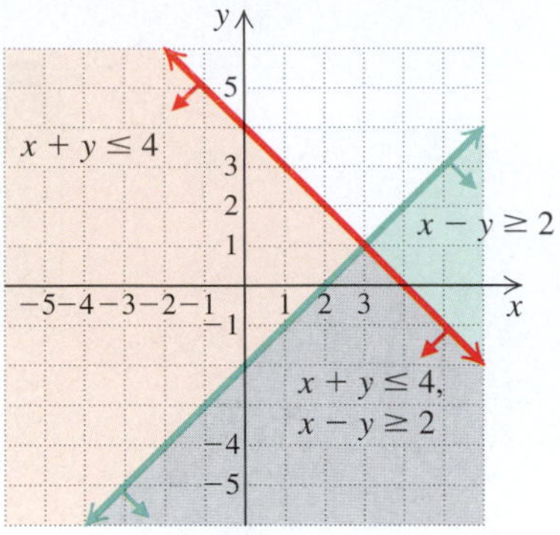

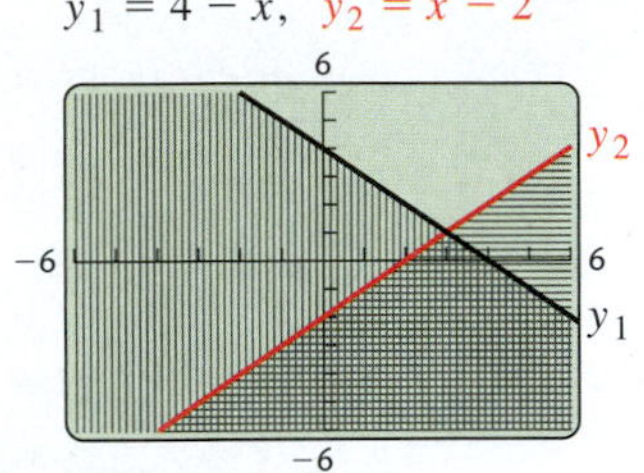

FIGURE 1.

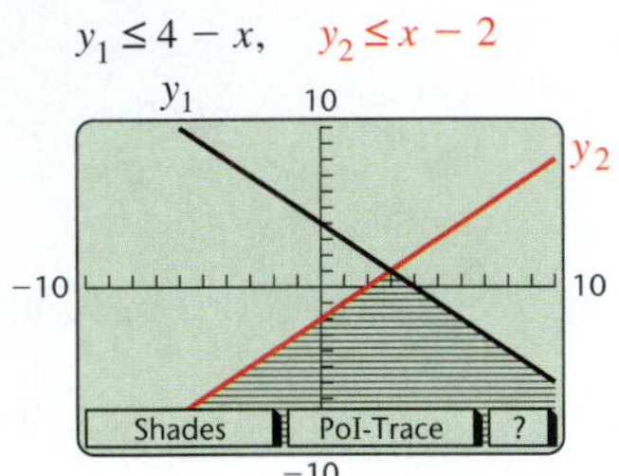

FIGURE 2.

We can use different shading patterns on a graphing calculator to graph this system of inequalities. The solution set is the region shaded using both patterns. See Fig. 1 at left.

We can also use the Inequalz application to graph this system of inequalities. If we choose the Ineq Intersection option from the Shades menu, only the solution set is shaded, as shown in Fig. 2.

A system of inequalities may have a graph that consists of a polygon and its interior. As we will see later in this section, in many applications we will need to know the vertices of such a polygon.

EXAMPLE 6 Graph the following system of inequalities and find the coordinates of any vertices formed:

$$3x - y \leq 6, \quad \textbf{(1)}$$
$$y - 3 \leq 0, \quad \textbf{(2)}$$
$$x + y \geq 0. \quad \textbf{(3)}$$

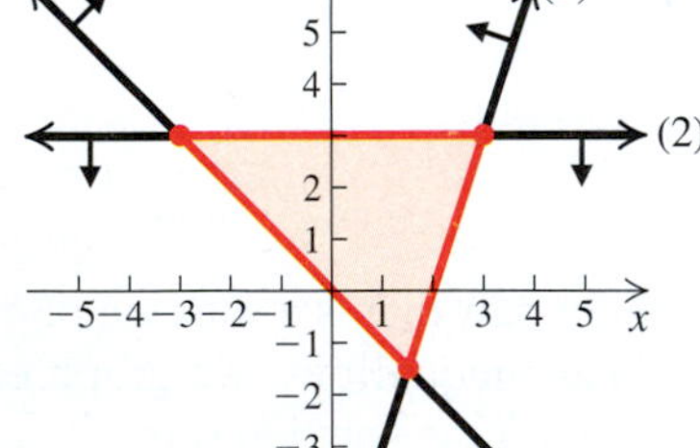

Solution We graph the related equations $3x - y = 6$, $y - 3 = 0$, and $x + y = 0$ using solid lines. The half-plane containing the solution set for each inequality is indicated by the arrows near the ends of each line. We shade the region common to all three solution sets.

To find the vertices, we solve three systems of equations. The system of equations from inequalities (1) and (2) is

$$3x - y = 6,$$
$$y - 3 = 0.$$

Solving, we obtain the vertex $(3, 3)$.

The system of equations from inequalities (1) and (3) is

$$3x - y = 6,$$
$$x + y = 0.$$

Solving, we obtain the vertex $\left(\frac{3}{2}, -\frac{3}{2}\right)$.

The system of equations from inequalities (2) and (3) is

$$y - 3 = 0,$$
$$x + y = 0.$$

Solving, we obtain the vertex $(-3, 3)$.

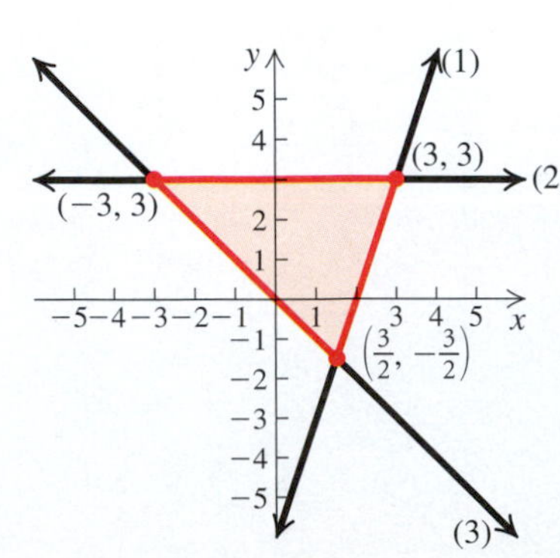

If a system of inequalities is graphed on a graphing calculator using a shading option, the coordinates of the vertices can be found using the INTERSECT feature. If the Inequalz application from the APPS menu is used to graph a system of inequalities, the PoI-Trace feature can be used to find the coordinates of the vertices.

Now Try Exercise 55.

Applications: Linear Programming

In many applications, we want to find a maximum value or a minimum value. In business, for example, we might want to both maximize profit and minimize cost. **Linear programming** can tell us how to do this.

In our study of linear programming, we will consider linear functions of two variables that are to be maximized or minimized subject to several conditions, or **constraints**. These constraints are expressed as inequalities. The solution set of the system of inequalities made up of the constraints contains all the **feasible solutions** of a linear programming problem. The function that we want to maximize or minimize is called the **objective function**.

It can be shown that the maximum and the minimum values of the objective function occur at a vertex of the region of feasible solutions. Thus we have the following procedure.

LINEAR PROGRAMMING PROCEDURE

To find the maximum or the minimum value of a linear objective function subject to a set of constraints:

1. Graph the region of feasible solutions.
2. Determine the coordinates of the vertices of the region.
3. Evaluate the objective function at each vertex. The largest and the smallest of those values are the maximum and the minimum values of the function, respectively.

EXAMPLE 7 *Maximizing Profit.* Aspen Carpentry makes bookcases and desks. Each bookcase requires 5 hr of woodworking and 4 hr of finishing. Each desk requires 10 hr of woodworking and 3 hr of finishing. Each month the shop has 600 hr of labor available for woodworking and 240 hr for finishing. The profit on each bookcase is \$75 and on each desk is \$140. Assume that all that are made are sold. How many of each product should be made each month in order to maximize profit? What is the maximum profit?

Solution We let x = the number of bookcases to be produced and y = the number of desks. Then the profit P is given by the function

$P = 75x + 140y$. To emphasize that P is a function of two variables, we sometimes write $P(x, y) = 75x + 140y$.

We know that x bookcases require $5x$ hr of woodworking and that y desks require $10y$ hr of woodworking. Since there is no more than 600 hr of labor available for woodworking, we have one constraint:

$$5x + 10y \leq 600.$$

Similarly, the bookcases and desks require $4x$ hr and $3y$ hr of finishing, respectively. There is no more than 240 hr of labor available for finishing, so we have a second constraint:

$$4x + 3y \leq 240.$$

We also know that $x \geq 0$ and $y \geq 0$ because the carpentry shop cannot make a negative number of either product.

Thus we want to maximize the objective function $P = 75x + 140y$ subject to the constraints

$$\begin{aligned} 5x + 10y &\leq 600, \\ 4x + 3y &\leq 240, \\ x &\geq 0, \\ y &\geq 0. \end{aligned}$$

We graph the system of inequalities and determine the vertices. Then we evaluate the objective function P at each vertex.

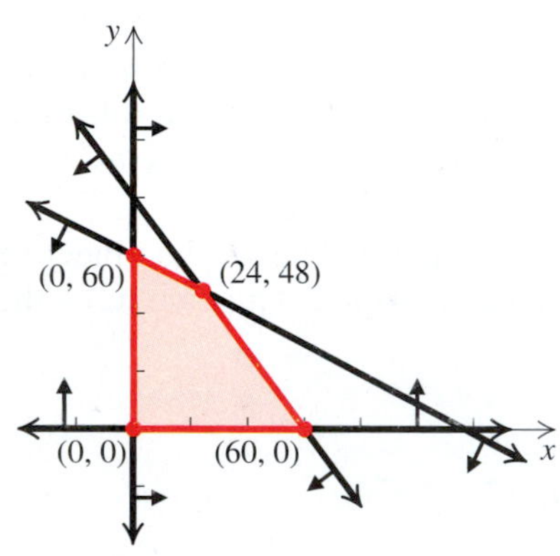

Vertices (x, y)	Profit $P = 75x + 140y$	
$(0, 0)$	$P = 75 \cdot 0 + 140 \cdot 0 = 0$	
$(60, 0)$	$P = 75 \cdot 60 + 140 \cdot 0 = 4500$	
$(24, 48)$	$P = 75 \cdot 24 + 140 \cdot 48 = 8520$	← Maximum
$(0, 60)$	$P = 75 \cdot 0 + 140 \cdot 60 = 8400$	

The carpentry shop will make a maximum profit of \$8520 when 24 bookcases and 48 desks are produced and sold.

We can create a table in which an objective function is evaluated at each vertex of a system of inequalities if the system has been graphed using the Inequalz application in the APPS menu.

Now Try Exercise 65.

9.7 Exercise Set

In Exercises 1–8, match the inequality with one of the graphs (a)–(h) that follow.

a)

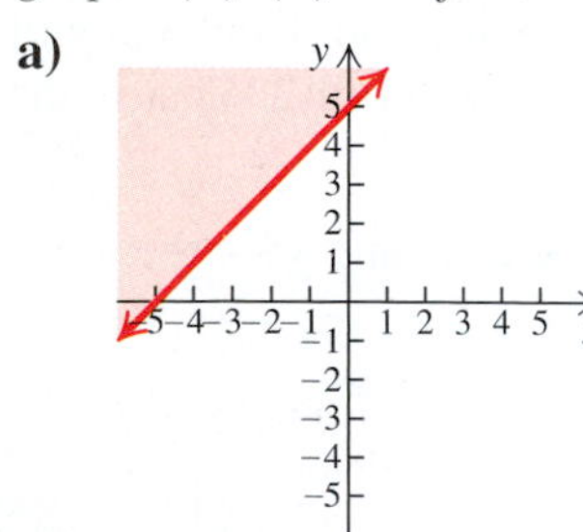

b)

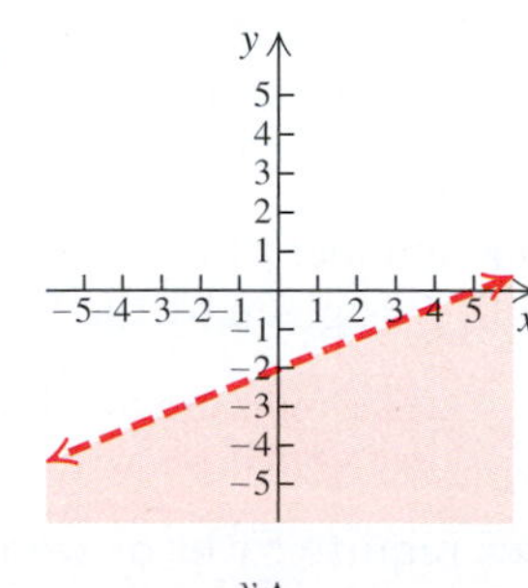

c)

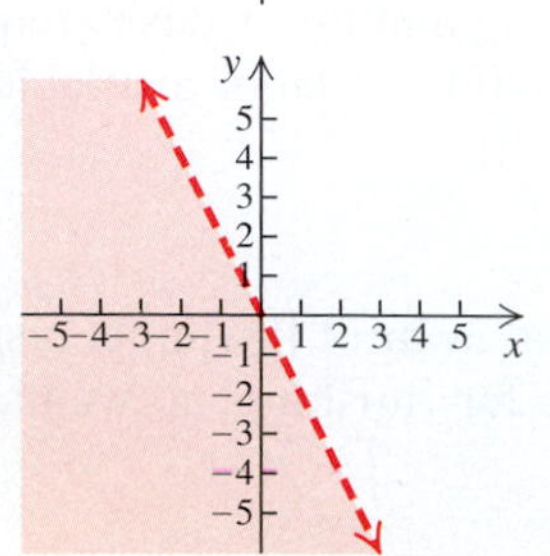

d)

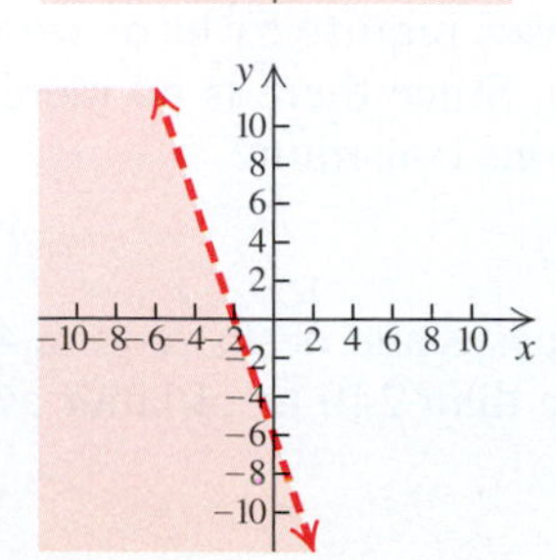

e)

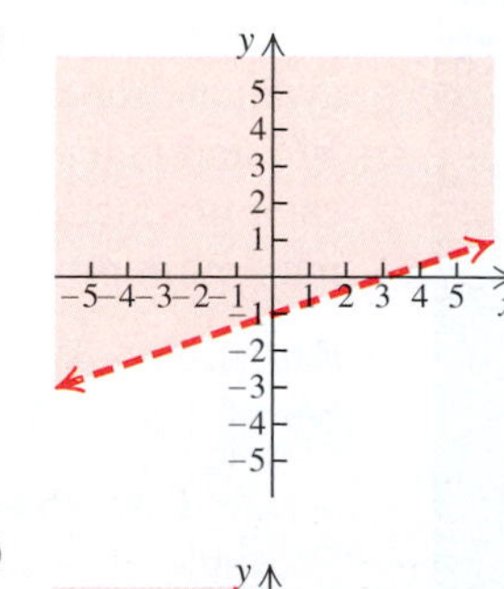

f)

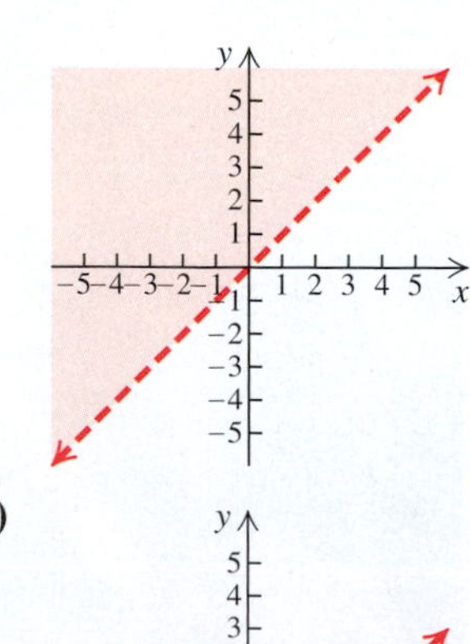

g)

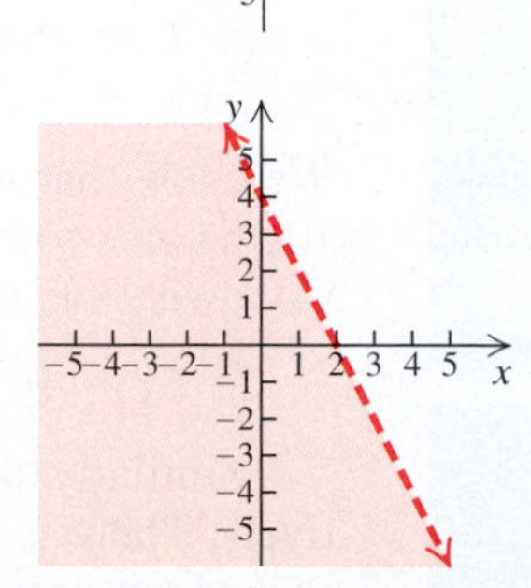

h)

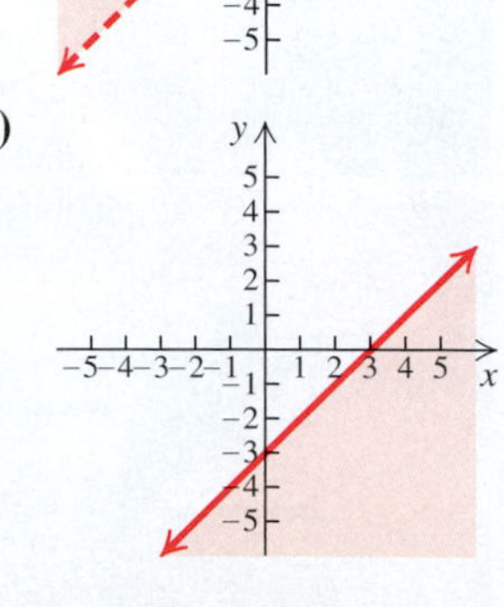

1. $y > x$
2. $y < -2x$
3. $y \leq x - 3$
4. $y \geq x + 5$
5. $2x + y < 4$
6. $3x + y < -6$
7. $2x - 5y > 10$
8. $3x - 9y < 9$

Graph.

9. $y > 2x$
10. $2y < x$
11. $y + x \geq 0$
12. $y - x < 0$
13. $y > x - 3$
14. $y \leq x + 4$
15. $x + y < 4$
16. $x - y \geq 5$
17. $3x - 2y \leq 6$
18. $2x - 5y < 10$
19. $3y + 2x \geq 6$
20. $2y + x \leq 4$
21. $3x - 2 \leq 5x + y$
22. $2x - 6y \geq 8 + 2y$
23. $x < -4$
24. $y > -3$
25. $y \geq 5$
26. $x \leq 5$
27. $-4 < y < -1$
(*Hint*: Think of this as $-4 < y$ *and* $y < -1$.)
28. $-3 \leq x \leq 3$
(*Hint*: Think of this as $-3 \leq x$ *and* $x \leq 3$.)
29. $y \geq |x|$
30. $y \leq |x + 2|$

In Exercises 31–36, match the system of inequalities with one of the graphs (a)–(f) that follow.

a) 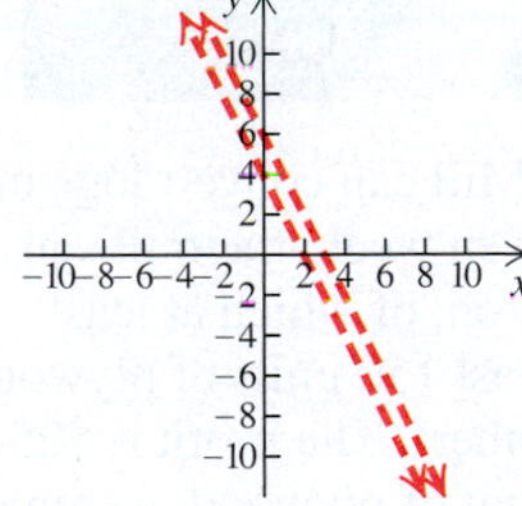

b) 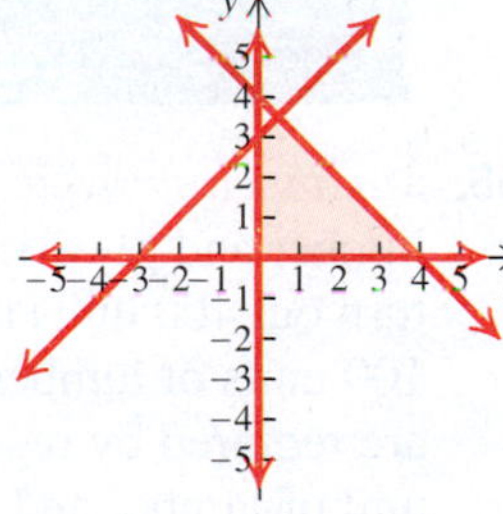

c) 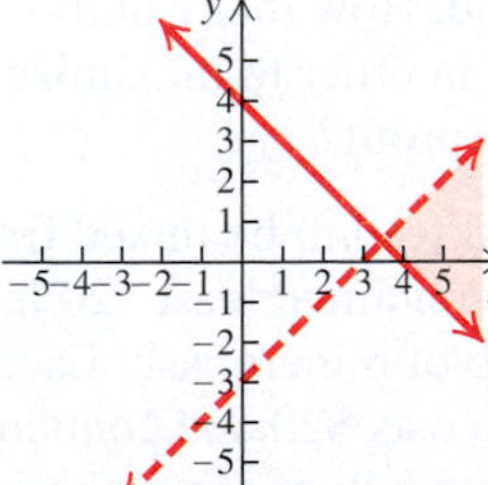

d) 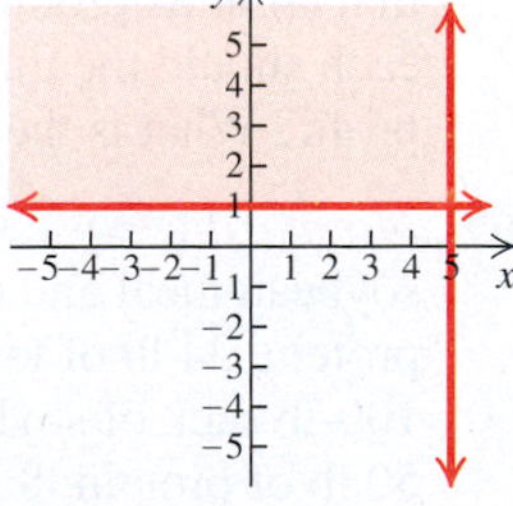

e) 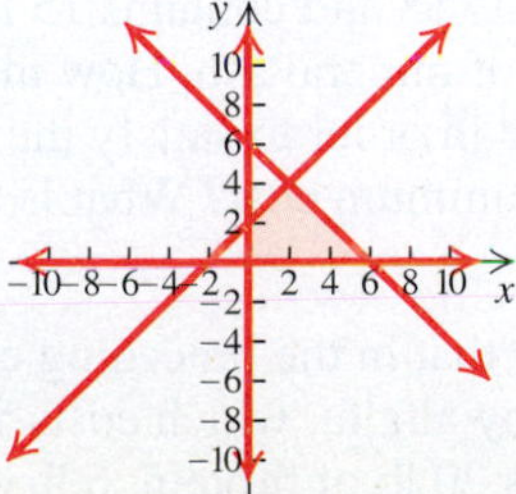

f) 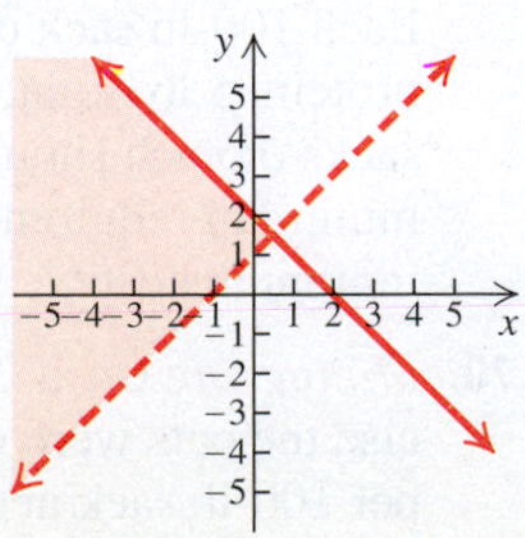

31. $y > x + 1,$
$y \leq 2 - x$
32. $y < x - 3,$
$y \geq 4 - x$
33. $2x + y < 4,$
$4x + 2y > 12$
34. $x \leq 5,$
$y \geq 1$
35. $x + y \leq 4,$
$x - y \geq -3,$
$x \geq 0,$
$y \geq 0$
36. $x - y \geq -2,$
$x + y \leq 6,$
$x \geq 0,$
$y \geq 0$

Find a system of inequalities with the given graph. Answers may vary.

37. 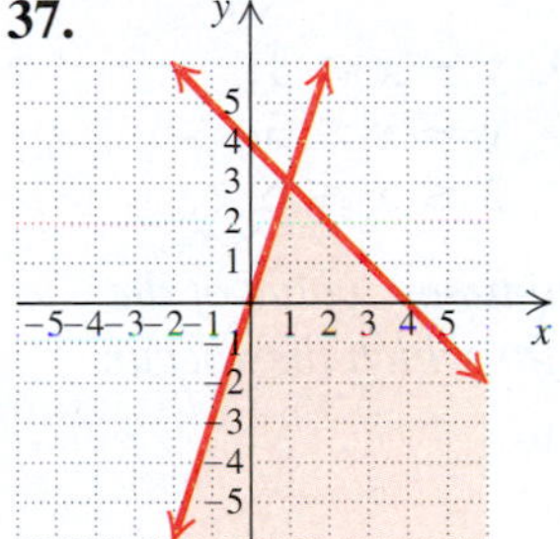

38. 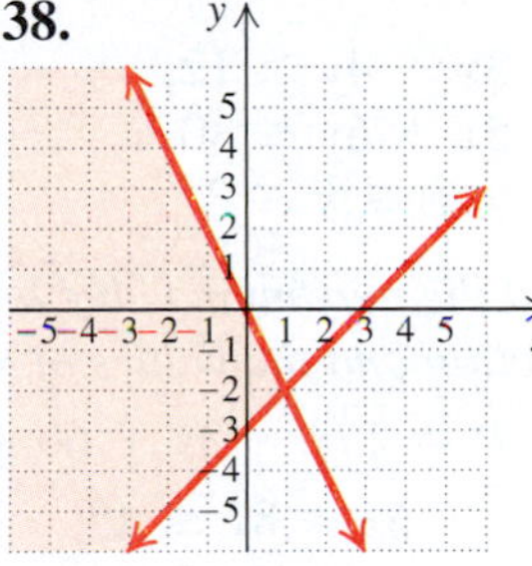

39. 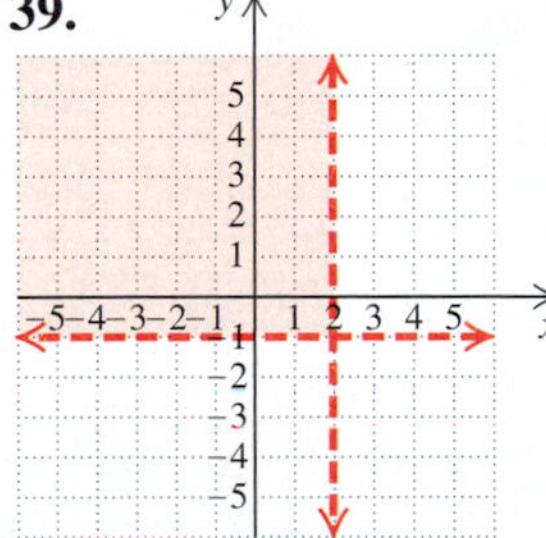

40. 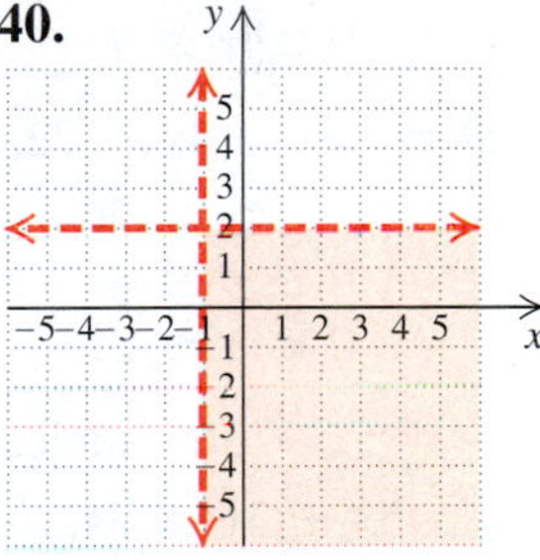

41. 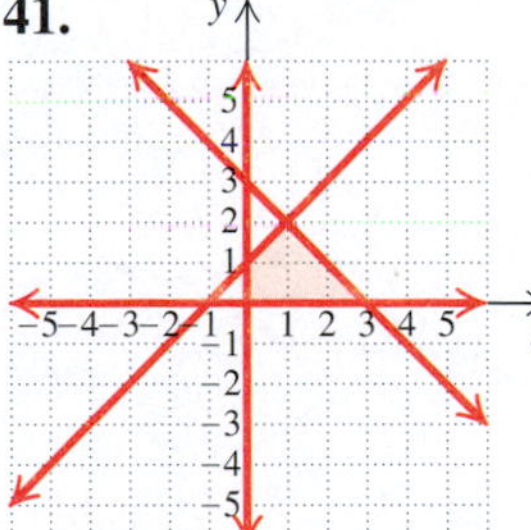

42. 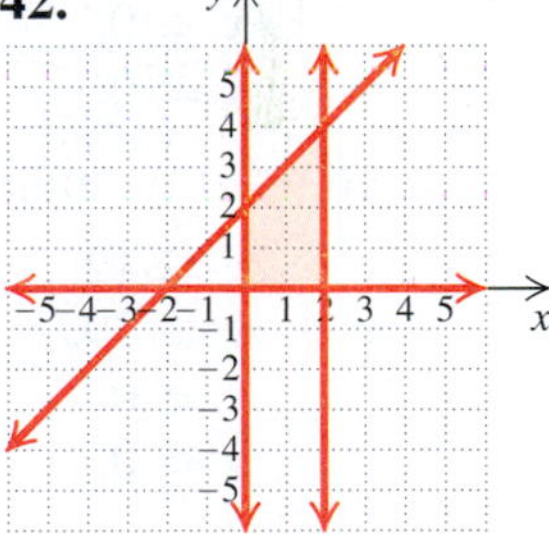

Graph the system of inequalities. Then find the coordinates of the vertices.

43. $y \leq x,$
$y \geq 3 - x$
44. $y \leq x,$
$y \geq 5 - x$
45. $y \geq x,$
$y \leq 4 - x$
46. $y \geq x,$
$y \leq 2 - x$
47. $y \geq -3,$
$x \geq 1$
48. $y \leq -2,$
$x \geq 2$
49. $x \leq 3,$
$y \geq 2 - 3x$
50. $x \geq -2,$
$y \leq 3 - 2x$
51. $x + y \leq 1,$
$x - y \leq 2$
52. $y + 3x \geq 0,$
$y + 3x \leq 2$

53. $2y - x \leq 2,$
$y + 3x \geq -1$

54. $y \leq 2x + 1,$
$y \geq -2x + 1,$
$x - 2 \leq 0$

55. $x - y \leq 2,$
$x + 2y \geq 8,$
$y - 4 \leq 0$

56. $x + 2y \leq 12,$
$2x + y \leq 12,$
$x \geq 0,$
$y \geq 0$

57. $4y - 3x \geq -12,$
$4y + 3x \geq -36,$
$y \leq 0,$
$x \leq 0$

58. $8x + 5y \leq 40,$
$x + 2y \leq 8,$
$x \geq 0,$
$y \geq 0$

59. $3x + 4y \geq 12,$
$5x + 6y \leq 30,$
$1 \leq x \leq 3$

60. $y - x \geq 1,$
$y - x \leq 3,$
$2 \leq x \leq 5$

Find the maximum value and the minimum value of the function and the values of x and y for which they occur.

61. $P = 17x - 3y + 60$, subject to
$6x + 8y \leq 48,$
$0 \leq y \leq 4,$
$0 \leq x \leq 7.$

62. $Q = 28x - 4y + 72$, subject to
$5x + 4y \geq 20,$
$0 \leq y \leq 4,$
$0 \leq x \leq 3.$

63. $F = 5x + 36y$, subject to
$5x + 3y \leq 34,$
$3x + 5y \leq 30,$
$x \geq 0,$
$y \geq 0.$

64. $G = 16x + 14y$, subject to
$3x + 2y \leq 12,$
$7x + 5y \leq 29,$
$x \geq 0,$
$y \geq 0.$

65. *Maximizing Mileage.* Jazmin owns a pickup truck and a moped. He can afford 12 gal of gasoline to be split between the truck and the moped. Jazmin's truck gets 20 mpg and, with the fuel currently in the tank, can hold at most an additional 10 gal of gas. His moped gets 100 mpg and can hold at most 3 gal of gas. How many gallons of gasoline should each vehicle use if Jazmin wants to travel as far as possible on the 12 gal of gas? What is the maximum number of miles that he can travel?

66. *Maximizing Income.* Golden Harvest Foods makes jumbo biscuits and regular biscuits. The oven can cook at most 200 biscuits per day. Each jumbo biscuit requires 2 oz of flour, each regular biscuit requires 1 oz of flour, and there is 300 oz of flour available. The income from each jumbo biscuit is \$0.80 and from each regular biscuit is \$0.50. Assume that all that are made are sold. How many of each size biscuit should be made in order to maximize income? What is the maximum income?

67. *Maximizing Profit.* Waterbrook Farm includes 240 acres of cropland. The farm owner wishes to plant this acreage in both corn and soybeans. The profit per acre in corn production is \$325 and in soybeans is \$180. A total of 320 hr of labor is available. Each acre of corn requires 2 hr of labor, whereas each acre of soybeans requires 1 hr of labor. How should the land be divided between corn and soybeans in order to yield the maximum profit? What is the maximum profit?

68. *Maximizing Profit.* Norris Mill can convert logs into lumber and plywood. In a given week, the mill can turn out 400 units of production, of which at least 100 units of lumber and at least 150 units of plywood are required by regular customers. The profit is \$25 per unit of lumber and \$38 per unit of plywood. Assume that all units produced are sold. How many units of each should the mill produce in order to maximize the profit? What is the maximum profit?

69. *Minimizing Cost.* An animal feed to be mixed from soybean meal and oats must contain at least 120 lb of protein, 24 lb of fat, and 10 lb of mineral ash. Each 100-lb sack of soybean meal costs \$20 and contains 50 lb of protein, 8 lb of fat, and 5 lb of mineral ash. Each 100-lb sack of oats costs \$8 and contains 15 lb of protein, 5 lb of fat, and 1 lb of mineral ash. How many sacks of each should be used in order to satisfy the minimum requirements at minimum cost? What is the minimum cost?

70. *Minimizing Cost.* Suppose that in the preceding exercise the oats were replaced by alfalfa, which costs \$10 per 100-lb sack and contains 20 lb of protein, 6 lb of fat, and 8 lb of mineral ash. How much of each would now be required in order to minimize the cost? What is the minimum cost?

71. *Maximizing Income.* Francisco is planning to invest up to $40,000 in corporate and municipal bonds. The least he is allowed to invest in corporate bonds is $6000, and he does not want to invest more than $22,000 in corporate bonds. He also does not want to invest more than $30,000 in municipal bonds. The interest is 3% on corporate bonds and $4\frac{1}{4}\%$ on municipal bonds. This is simple interest for one year. How much should he invest in each type of bond in order to maximize his income? What is the maximum income?

72. *Maximizing Income.* Mila is planning to invest up to $22,000 in certificates of deposit at City Bank and People's Bank. She wants to invest at least $2000 but no more than $14,000 at City Bank. People's Bank does not insure more than a $15,000 investment, so she will invest no more than that in People's Bank. The interest is $2\frac{1}{2}\%$ at City Bank and $1\frac{3}{4}\%$ at People's Bank. This is simple interest for one year. How much should she invest in each bank in order to maximize her income? What is the maximum income?

73. *Minimizing Transportation Cost.* An airline with two types of airplanes, P_1 and P_2, has contracted with a tour group to provide transportation for a minimum of 2000 first-class, 1500 tourist-class, and 2400 economy-class passengers. For a certain trip, airplane P_1 costs $12 thousand to operate and can accommodate 40 first-class, 40 tourist-class, and 120 economy-class passengers, whereas airplane P_2 costs $10 thousand to operate and can accommodate 80 first-class, 30 tourist-class, and 40 economy-class passengers. How many of each type of airplane should be used in order to minimize the operating cost? What is the minimum operating cost?

74. *Minimizing Transportation Cost.* Suppose that in the preceding problem a new airplane P_3 becomes available, having an operating cost for the same trip of $15 thousand and accommodating 40 first-class, 40 tourist-class, and 80 economy-class passengers. If airplane P_1 were replaced by airplane P_3, how many of P_2 and P_3 should be used in order to minimize the operating cost? What is the minimum operating cost?

75. *Maximizing Profit.* It takes Fena Tailoring 3 hr of cutting and 6 hr of sewing to make a tiered silk organza bridal dress. It takes 6 hr of cutting and 3 hr of sewing to make a lace sheath bridal dress. The shop has at most 27 hr per week available for cutting and at most 36 hr per week for sewing. The profit is $320 on an organza dress and $305 on a lace dress. Assume that all that are made are sold. How many of each kind of bridal dress should be made each week in order to maximize profit? What is the maximum profit?

76. *Maximizing Profit.* Cambridge Metal Works manufactures two sizes of gears. The smaller gear requires 4 hr of machining and 1 hr of polishing and yields a profit of $45. The larger gear requires 1 hr of machining and 1 hr of polishing and yields a profit of $30. The firm has available at most 24 hr per day for machining and 9 hr per day for polishing. Assume that all that are made are sold. How many of each type of gear should be produced each day in order to maximize profit? What is the maximum profit?

77. *Minimizing Nutrition Cost.* Suppose that it takes 12 units of carbohydrates and 6 units of protein to satisfy Jacob's minimum weekly requirements. A particular type of meat contains 2 units of carbohydrates and 2 units of protein per pound. A particular cheese contains 3 units of carbohydrates and 1 unit of protein per pound. The meat costs $3.50 per pound and the cheese costs $4.60 per pound. How many pounds of each are needed in order to minimize the cost and still meet the minimum requirements?

78. *Minimizing Salary Cost.* The Spring Hill school board is analyzing education costs for Hill Top School. It wants to hire teachers and teacher's aides to make up a faculty that satisfies its needs at minimum cost. The average annual salary for a teacher is $53,000 and for a teacher's aide is $23,600. The school building can accommodate a faculty of no more than 50 but needs at least 20 faculty members to function properly. The school must have at least 12 aides, but the number of teachers must be at least twice the number of aides in order to accommodate the expectations of the community. How many teachers and teacher's aides should be hired in order to minimize salary costs? What is the minimum salary cost?

79. *Maximizing Animal Support in a Forest.* A certain area of forest is populated by two species of animal, which scientists refer to as A and B for simplicity. The forest supplies two kinds of food, referred to as F_1 and F_2. For one year, each member of species A requires 1 unit of F_1 and 0.5 unit of F_2. Each member of species B requires 0.2 unit of F_1 and 1 unit of F_2. The forest can normally supply at most 600 units of F_1 and 525 units of F_2 per year. What is the maximum total number of these animals that the forest can support?

80. *Maximizing Animal Support in a Forest.* Refer to Exercise 79. If there is a wet spring, then supplies of food increase to 1080 units of F_1 and 810 units of F_2. In this case, what is the maximum total number of these animals that the forest can support?

Skill Maintenance

Solve.

81. $-5 \le x + 2 < 4$ [1.6]

82. $|x - 3| \ge 2$ [3.5]

83. $x^2 - 2x \le 3$ [4.6]

84. $\dfrac{x-1}{x+2} > 4$ [4.6]

Synthesis

Graph the system of inequalities.

85. $y \ge x^2 - 2,$
$y \le 2 - x^2$

86. $y < x + 1,$
$y \ge x^2$

Graph the inequality.

87. $|x + y| \le 1$

88. $|x| + |y| \le 1$

89. $|x| > |y|$

90. $|x - y| > 0$

91. *Allocation of Resources.* Comfort-by-Design Furniture produces chairs and sofas. Each chair requires 20 ft of wood, 1 lb of foam rubber, and 2 yd^2 of fabric. Each sofa requires 100 ft of wood, 50 lb of foam rubber, and 20 yd^2 of fabric. The manufacturer has in stock 1900 ft of wood, 500 lb of foam rubber, and 240 yd^2 of fabric. The chairs can be sold for \$200 each and the sofas for \$750 each. Assume that all that are made are sold. How many of each should be produced in order to maximize income? What is the maximum income?

9.8 Partial Fractions

- Decompose rational expressions into partial fractions.

There are situations in calculus in which it is useful to write a rational expression as a sum of two or more simpler rational expressions. In the equation

$$\frac{4x - 13}{2x^2 + x - 6} = \frac{3}{x + 2} + \frac{-2}{2x - 3},$$

each fraction on the right side is called a **partial fraction**. The expression on the right side is the **partial fraction decomposition** of the rational expression on the left side. In this section, we learn how such decompositions are created.

Partial Fraction Decompositions

The procedure for finding the partial fraction decomposition of a rational expression involves factoring its denominator into linear factors and quadratic factors.

PROCEDURE FOR DECOMPOSING A RATIONAL EXPRESSION INTO PARTIAL FRACTIONS

Consider any rational expression $P(x)/Q(x)$ such that $P(x)$ and $Q(x)$ have no common factor other than 1 or -1.

1. If the degree of $P(x)$ is greater than or equal to the degree of $Q(x)$, divide to express $P(x)/Q(x)$ as a quotient + remainder$/Q(x)$ and follow steps (2)–(5) to decompose the resulting rational expression.
2. If the degree of $P(x)$ is less than the degree of $Q(x)$, factor $Q(x)$ into linear factors of the form $(px + q)^n$ and/or quadratic factors of the form $(ax^2 + bx + c)^m$. Any quadratic factor $ax^2 + bx + c$ must be *irreducible*, meaning that it cannot be factored into linear factors with rational coefficients.
3. Assign to each linear factor $(px + q)^n$ the sum of n partial fractions:
$$\frac{A_1}{px + q} + \frac{A_2}{(px + q)^2} + \cdots + \frac{A_n}{(px + q)^n}.$$
4. Assign to each quadratic factor $(ax^2 + bx + c)^m$ the sum of m partial fractions:
$$\frac{B_1x + C_1}{ax^2 + bx + c} + \frac{B_2x + C_2}{(ax^2 + bx + c)^2} + \cdots + \frac{B_mx + C_m}{(ax^2 + bx + c)^m}.$$
5. Apply algebraic methods, as illustrated in the following examples, to find the constants in the numerators of the partial fractions.

Just in Time 23

EXAMPLE 1 Decompose into partial fractions:
$$\frac{4x - 13}{2x^2 + x - 6}.$$

Solution The degree of the numerator is less than the degree of the denominator. We begin by factoring the denominator: $(x + 2)(2x - 3)$. We find constants A and B such that
$$\frac{4x - 13}{(x + 2)(2x - 3)} = \frac{A}{x + 2} + \frac{B}{2x - 3}.$$

To determine A and B, we add the expressions on the right:
$$\frac{4x - 13}{(x + 2)(2x - 3)} = \frac{A(2x - 3) + B(x + 2)}{(x + 2)(2x - 3)}.$$

Next, we equate the numerators:
$$4x - 13 = A(2x - 3) + B(x + 2).$$

Since the last equation containing A and B is true for all x, we can substitute any value of x and still have a true equation. If we choose $x = \frac{3}{2}$, then $2x - 3 = 0$ and A will be eliminated when we make the substitution. This gives us
$$4\left(\tfrac{3}{2}\right) - 13 = A\left(2\cdot\tfrac{3}{2} - 3\right) + B\left(\tfrac{3}{2} + 2\right)$$
$$-7 = 0 + \tfrac{7}{2}B.$$

Solving, we obtain $B = -2$.

In order to have $x + 2 = 0$, we let $x = -2$. Then B will be eliminated when we make the substitution. This gives us
$$4(-2) - 13 = A[2(-2) - 3] + B(-2 + 2)$$
$$-21 = -7A + 0.$$

Solving, we obtain $A = 3$.

The decomposition is as follows:

$$\frac{4x - 13}{2x^2 + x - 6} = \frac{3}{x + 2} + \frac{-2}{2x - 3}, \quad \text{or} \quad \frac{3}{x + 2} - \frac{2}{2x - 3}.$$

To check, we can add to see if we get the expression on the left. We can also use the TABLE feature on a graphing calculator, comparing values of

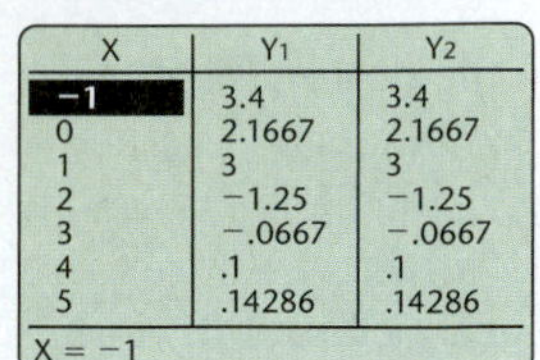

X	Y1	Y2
−1	3.4	3.4
0	2.1667	2.1667
1	3	3
2	−1.25	−1.25
3	−.0667	−.0667
4	.1	.1
5	.14286	.14286

X = −1

$$y_1 = \frac{4x - 13}{2x^2 + x - 6} \quad \text{and} \quad y_2 = \frac{3}{x + 2} - \frac{2}{2x - 3}$$

for the same values of x. Since $y_1 = y_2$ for the given values of x as we scroll through the table, the decomposition appears to be correct.

Now Try Exercise 3.

EXAMPLE 2 Decompose into partial fractions:

$$\frac{7x^2 - 29x + 24}{(2x - 1)(x - 2)^2}.$$

Solution The degree of the numerator is 2 and the degree of the denominator is 3, so the degree of the numerator is less than the degree of the denominator. The denominator is given in factored form. The decomposition has the following form:

$$\frac{7x^2 - 29x + 24}{(2x - 1)(x - 2)^2} = \frac{A}{2x - 1} + \frac{B}{x - 2} + \frac{C}{(x - 2)^2}.$$

As in Example 1, we add the expressions on the right:

$$\frac{7x^2 - 29x + 24}{(2x - 1)(x - 2)^2} = \frac{A(x - 2)^2 + B(2x - 1)(x - 2) + C(2x - 1)}{(2x - 1)(x - 2)^2}.$$

Then we equate the numerators. This gives us

$$7x^2 - 29x + 24 = A(x - 2)^2 + B(2x - 1)(x - 2) + C(2x - 1).$$

Since the equation containing A, B, and C is true for all x, we can substitute any value of x and still have a true equation. In order to have $2x - 1 = 0$, we let $x = \frac{1}{2}$. This gives us

$$7\left(\tfrac{1}{2}\right)^2 - 29 \cdot \tfrac{1}{2} + 24 = A\left(\tfrac{1}{2} - 2\right)^2 + 0 + 0$$
$$\tfrac{45}{4} = \tfrac{9}{4}A.$$

Solving, we obtain $A = 5$.

In order to have $x - 2 = 0$, we let $x = 2$. Substituting gives us

$$7(2)^2 - 29(2) + 24 = 0 + 0 + C(2 \cdot 2 - 1)$$
$$-6 = 3C.$$

Solving, we obtain $C = -2$.

To find B, we choose any value for x except $\frac{1}{2}$ or 2 and replace A with 5 and C with -2. We let $x = 1$:

$$7 \cdot 1^2 - 29 \cdot 1 + 24 = 5(1 - 2)^2 + B(2 \cdot 1 - 1)(1 - 2) + (-2)(2 \cdot 1 - 1)$$
$$2 = 5 - B - 2$$
$$B = 1.$$

The decomposition is as follows:

$$\frac{7x^2 - 29x + 24}{(2x - 1)(x - 2)^2} = \frac{5}{2x - 1} + \frac{1}{x - 2} - \frac{2}{(x - 2)^2}.$$

X	Y1	Y2
−5	−.6382	−.6382
−4	−.7778	−.7778
−3	−.9943	−.9943
−2	−1.375	−1.375
−1	−2.222	−2.222
0	−6	−6
1	2	2
X = −5		

We can check the result using a table of values. We let

$$y_1 = \frac{7x^2 - 29x + 24}{(2x - 1)(x - 2)^2} \quad \text{and} \quad y_2 = \frac{5}{2x - 1} + \frac{1}{x - 2} - \frac{2}{(x - 2)^2}.$$

Since $y_1 = y_2$ for given values of x as we scroll through the table, the decomposition appears to be correct.

Now Try Exercise 7.

POLYNOMIAL DIVISION
REVIEW SECTION 4.3.

EXAMPLE 3 Decompose into partial fractions:

$$\frac{6x^3 + 5x^2 - 7}{3x^2 - 2x - 1}.$$

Solution The degree of the numerator is greater than that of the denominator. Therefore, we divide and find an equivalent expression:

$$\begin{array}{r} 2x + 3 \\ 3x^2 - 2x - 1 \overline{)\, 6x^3 + 5x^2 - 7} \\ \underline{6x^3 - 4x^2 - 2x } \\ 9x^2 + 2x - 7 \\ \underline{9x^2 - 6x - 3} \\ 8x - 4. \end{array}$$

The original expression is thus equivalent to

$$2x + 3 + \frac{8x - 4}{3x^2 - 2x - 1}.$$

We decompose the fraction to get

$$\frac{8x - 4}{(3x + 1)(x - 1)} = \frac{5}{3x + 1} + \frac{1}{x - 1}.$$

The final result is

$$2x + 3 + \frac{5}{3x + 1} + \frac{1}{x - 1}.$$

Now Try Exercise 17.

Systems of equations can be used to decompose rational expressions. Let's reconsider Example 2.

EXAMPLE 4 Decompose into partial fractions:

$$\frac{7x^2 - 29x + 24}{(2x - 1)(x - 2)^2}.$$

Solution The decomposition has the following form:

$$\frac{A}{2x - 1} + \frac{B}{x - 2} + \frac{C}{(x - 2)^2}.$$

We first add as in Example 2:

$$\begin{aligned} \frac{7x^2 - 29x + 24}{(2x - 1)(x - 2)^2} &= \frac{A}{2x - 1} + \frac{B}{x - 2} + \frac{C}{(x - 2)^2} \\ &= \frac{A(x - 2)^2 + B(2x - 1)(x - 2) + C(2x - 1)}{(2x - 1)(x - 2)^2}. \end{aligned}$$

Then we equate numerators:

$$\begin{aligned}7x^2 &- 29x + 24\\ &= A(x-2)^2 + B(2x-1)(x-2) + C(2x-1)\\ &= A(x^2-4x+4) + B(2x^2-5x+2) + C(2x-1)\\ &= Ax^2 - 4Ax + 4A + 2Bx^2 - 5Bx + 2B + 2Cx - C,\end{aligned}$$

or, combining like terms,

$$\begin{aligned}7x^2 &- 29x + 24\\ &= (A+2B)x^2 + (-4A-5B+2C)x + (4A+2B-C).\end{aligned}$$

Next, we equate corresponding coefficients:

$7 = A + 2B,$ **The coefficients of the x^2-terms must be the same.**

$-29 = -4A - 5B + 2C,$ **The coefficients of the x-terms must be the same.**

$24 = 4A + 2B - C.$ **The constant terms must be the same.**

SYSTEMS OF EQUATIONS IN THREE VARIABLES

REVIEW SECTION 9.2, 9.5, OR 9.6.

We now have a system of three equations. You should confirm that the solution of the system is

$$A = 5, \quad B = 1, \quad \text{and} \quad C = -2.$$

The decomposition is as follows:

$$\frac{7x^2 - 29x + 24}{(2x-1)(x-2)^2} = \frac{5}{2x-1} + \frac{1}{x-2} - \frac{2}{(x-2)^2}.$$

Now Try Exercise 15.

EXAMPLE 5 Decompose into partial fractions:

$$\frac{11x^2 - 8x - 7}{(2x^2-1)(x-3)}.$$

Solution The decomposition has the following form:

$$\frac{11x^2 - 8x - 7}{(2x^2-1)(x-3)} = \frac{Ax+B}{2x^2-1} + \frac{C}{x-3}.$$

Adding and equating numerators, we get

$$\begin{aligned}11x^2 - 8x - 7 &= (Ax+B)(x-3) + C(2x^2-1)\\ &= Ax^2 - 3Ax + Bx - 3B + 2Cx^2 - C,\end{aligned}$$

or $\quad 11x^2 - 8x - 7 = (A+2C)x^2 + (-3A+B)x + (-3B-C).$

We then equate corresponding coefficients:

$11 = A + 2C,$ **The coefficients of the x^2-terms**

$-8 = -3A + B,$ **The coefficients of the x-terms**

$-7 = -3B - C.$ **The constant terms**

We solve this system of three equations and obtain

$$A = 3, \quad B = 1, \quad \text{and} \quad C = 4.$$

The decomposition is as follows:

$$\frac{11x^2 - 8x - 7}{(2x^2-1)(x-3)} = \frac{3x+1}{2x^2-1} + \frac{4}{x-3}.$$

Now Try Exercise 13.

9.8 Exercise Set

Decompose into partial fractions. Check your answers using a graphing calculator.

1. $\frac{x+7}{(x-3)(x+2)}$

2. $\frac{2x}{(x+1)(x-1)}$

3. $\frac{7x-1}{6x^2-5x+1}$

4. $\frac{13x+46}{12x^2-11x-15}$

5. $\frac{3x^2-11x-26}{(x^2-4)(x+1)}$

6. $\frac{5x^2+9x-56}{(x-4)(x-2)(x+1)}$

7. $\frac{9}{(x+2)^2(x-1)}$

8. $\frac{x^2-x-4}{(x-2)^3}$

9. $\frac{2x^2+3x+1}{(x^2-1)(2x-1)}$

10. $\frac{x^2-10x+13}{(x^2-5x+6)(x-1)}$

11. $\frac{x^4-3x^3-3x^2+10}{(x+1)^2(x-3)}$

12. $\frac{10x^3-15x^2-35x}{x^2-x-6}$

13. $\frac{-x^2+2x-13}{(x^2+2)(x-1)}$

14. $\frac{26x^2+208x}{(x^2+1)(x+5)}$

15. $\frac{6+26x-x^2}{(2x-1)(x+2)^2}$

16. $\frac{5x^3+6x^2+5x}{(x^2-1)(x+1)^3}$

17. $\frac{6x^3+5x^2+6x-2}{2x^2+x-1}$

18. $\frac{2x^3+3x^2-11x-10}{x^2+2x-3}$

19. $\frac{2x^2-11x+5}{(x-3)(x^2+2x-5)}$

20. $\frac{3x^2-3x-8}{(x-5)(x^2+x-4)}$

21. $\frac{-4x^2-2x+10}{(3x+5)(x+1)^2}$

22. $\frac{26x^2-36x+22}{(x-4)(2x-1)^2}$

23. $\frac{36x+1}{12x^2-7x-10}$

24. $\frac{-17x+61}{6x^2+39x-21}$

25. $\frac{-4x^2-9x+8}{(3x^2+1)(x-2)}$

26. $\frac{11x^2-39x+16}{(x^2+4)(x-8)}$

Skill Maintenance

Find the zeros of the polynomial function.

27. $f(x) = x^3 + x^2 + 9x + 9$ **[4.1], [4.3], [4.4]**

28. $f(x) = x^3 - 3x^2 + x - 3$ **[4.1], [4.3], [4.4]**

29. $f(x) = x^3 + x^2 - 3x - 2$ **[4.4]**

30. $f(x) = x^4 - x^3 - 5x^2 - x - 6$ **[4.3]**

31. $f(x) = x^3 + 5x^2 + 5x - 3$ **[4.1], [4.3], [4.4]**

Synthesis

Decompose into partial fractions.

32. $\frac{9x^3-24x^2+48x}{(x-2)^4(x+1)}$

[*Hint*: Let the expression equal

$$\frac{A}{x+1} + \frac{P(x)}{(x-2)^4}$$

and find $P(x)$].

33. $\frac{x}{x^4-a^4}$

34. $\frac{1}{e^{-x}+3+2e^x}$

35. $\frac{1+\ln x^2}{(\ln x+2)(\ln x-3)^2}$

Chapter 9 Summary and Review

STUDY GUIDE

KEY TERMS AND CONCEPTS | **EXAMPLES**

SECTION 9.1: SYSTEMS OF EQUATIONS IN TWO VARIABLES

A **system of two linear equations in two variables** is composed of two linear equations that are considered simultaneously.

The **solutions** of the system of equations are all ordered pairs that make *both* equations true.

A system of equations is **consistent** if it has at least one solution. A system of equations that has no solution is **inconsistent**.

The equations are **dependent** if one equation can be obtained by multiplying on both sides of the other equation by a constant. Otherwise, the equations are **independent**.

Systems of two equations in two variables can be solved graphically.

Solve: $x + y = 2,$
$y = x - 4.$

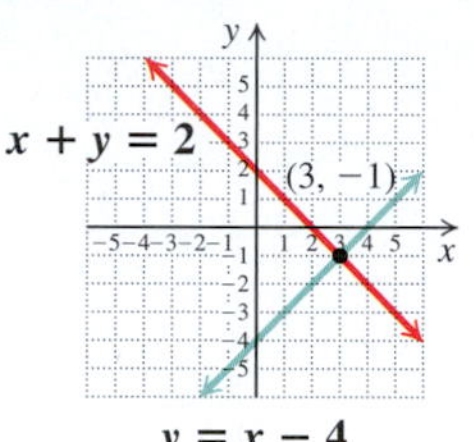

The solution is the point of intersection, $(3, -1)$. The system is consistent. The equations are independent.

Solve: $x + y = 2,$
$x + y = -2.$

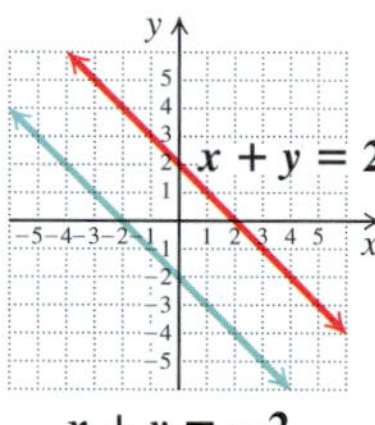

The graphs do not intersect, so there is no solution. The system is inconsistent. The equations are independent.

Solve: $x + y = 2,$
$3x + 3y = 6.$

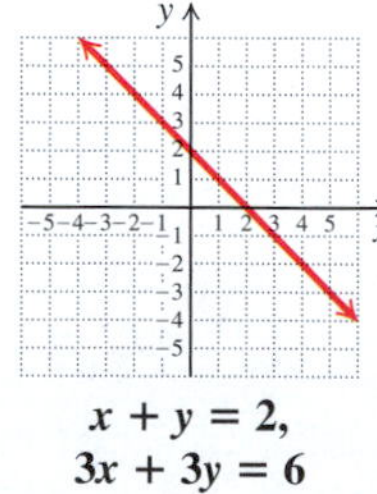

The graphs are the same. There are infinitely many common points, so there are infinitely many solutions. The solutions are of the form $(x, 2 - x)$ or $(2 - y, y)$. The system is consistent. The equations are dependent.

Systems of two equations in two variables can be solved using substitution.	Solve: $x = y - 5,$ $2x + 3y = 5.$ *Substitute and solve for y:* $2(y - 5) + 3y = 5$; $2y - 10 + 3y = 5$; $5y - 10 = 5$; $5y = 15$; $y = 3.$ *Back-substitute and solve for x:* $x = y - 5$; $x = 3 - 5$; $x = -2.$ The solution is $(-2, 3)$.
Systems of two equations in two variables can be solved using elimination.	Solve: $3x + y = -1,$ $x - 3y = 8.$ *Eliminate y and solve for x:* $9x + 3y = -3$; $x - 3y = 8$; $10x = 5$; $x = \frac{1}{2}$ *Back-substitute and solve for y:* $3x + y = -1$; $3 \cdot \frac{1}{2} + y = -1$; $\frac{3}{2} + y = -1$; $y = -\frac{5}{2}.$ The solution is $\left(\frac{1}{2}, -\frac{5}{2}\right)$.
Some applied problems can be solved by translating to a system of two equations in two variables.	See Examples 6–8 on pp. 643–646.

SECTION 9.2: SYSTEMS OF EQUATIONS IN THREE VARIABLES

A **solution** of a system of equations in three variables is an ordered triple that makes *all* three equations true.

We can use **Gaussian elimination** to solve a system of three equations in three variables by using the operations listed on p. 652 to transform the original system to one of the form

$$Ax + By + Cz = D,$$
$$Ey + Fz = G,$$
$$Hz = K.$$

Then we solve the third equation for z and back-substitute to find y and x.

As we see in Example 1 on p. 652, Gaussian elimination can be used to transform the system of equations

$$x - 2y + 3z = 11,$$
$$4x + 2y - 3z = 4,$$
$$3x + 3y - z = 4$$

to the equivalent form

$$x - 2y + 3z = 11,$$
$$10y - 15z = -40,$$
$$35z = 70.$$

Solve for z: $35z = 70$

$$z = 2.$$

Back-substitute to find y and x:

$$10y - 15 \cdot 2 = -40$$
$$10y - 30 = -40$$
$$10y = -10$$
$$y = -1.$$

$$x - 2(-1) + 3 \cdot 2 = 11$$
$$x + 2 + 6 = 11$$
$$x + 8 = 11$$
$$x = 3.$$

The solution is $(3, -1, 2)$.

Some applied problems can be solved by translating to a system of three equations in three variables.	See Example 3 on p. 655.
We can use a system of three equations to model a situation with a quadratic function.	Find a quadratic function that fits the data points $(0, -5)$, $(1, -4)$, and $(2, 1)$. We substitute in the function $f(x) = ax^2 + bx + c$: For $(0, -5)$: $-5 = a \cdot 0^2 + b \cdot 0 + c$, For $(1, -4)$: $-4 = a \cdot 1^2 + b \cdot 1 + c$, For $(2, 1)$: $1 = a \cdot 2^2 + b \cdot 2 + c$. We now have a system of equations: $c = -5$, $a + b + c = -4$, $4a + 2b + c = 1$. Solving this system of equations gives $(2, -1, -5)$. Thus, $f(x) = 2x^2 - x - 5$.

SECTION 9.3: MATRICES AND SYSTEMS OF EQUATIONS

A **matrix** (pl., **matrices**) is a rectangular array of numbers called **entries**, or **elements**, of the matrix.	Row 1 → Row 2 → $\begin{bmatrix} 3 & -2 & 5 \\ -1 & 4 & -3 \end{bmatrix}$ Column 1, Column 2, Column 3 This matrix has 2 rows and 3 columns. Its **order** is 2×3.
We can apply the **row-equivalent operations** on p. 662 to use **Gaussian elimination** with matrices to solve systems of equations.	Solve: $x - 2y = 8$, $2x + y = 1$. We write the augmented matrix and transform it to **row-echelon form** or **reduced row-echelon form:** $\left[\begin{array}{cc\|c} 1 & -2 & 8 \\ 2 & 1 & 1 \end{array}\right] \rightarrow \left[\begin{array}{cc\|c} 1 & -2 & 8 \\ 0 & 1 & -3 \end{array}\right] \rightarrow \left[\begin{array}{cc\|c} 1 & 0 & 2 \\ 0 & 1 & -3 \end{array}\right]$ Row-echelon form; Reduced row-echelon form Thus we have $x = 2$, $y = -3$. The solution is $(2, -3)$.

SECTION 9.4: MATRIX OPERATIONS

Matrices of the same order can be added or subtracted by adding or subtracting their corresponding entries.	Find each of the following. $\begin{bmatrix} 3 & -4 \\ -1 & 2 \end{bmatrix} + \begin{bmatrix} -5 & -1 \\ 3 & 0 \end{bmatrix} = \begin{bmatrix} 3 + (-5) & -4 + (-1) \\ -1 + 3 & 2 + 0 \end{bmatrix} = \begin{bmatrix} -2 & -5 \\ 2 & 2 \end{bmatrix}$ $\begin{bmatrix} 3 & -4 \\ -1 & 2 \end{bmatrix} - \begin{bmatrix} -5 & -1 \\ 3 & 0 \end{bmatrix} = \begin{bmatrix} 3 - (-5) & -4 - (-1) \\ -1 - 3 & 2 - 0 \end{bmatrix} = \begin{bmatrix} 8 & -3 \\ -4 & 2 \end{bmatrix}$

The **scalar product** of a number k and a matrix $\mathbf{A}$ is the matrix $k\mathbf{A}$ obtained by multiplying each entry of $\mathbf{A}$ by k. The number k is called a **scalar**.

The properties of matrix addition and scalar multiplication are given on p. 671.

For $\mathbf{A} = \begin{bmatrix} 2 & 3 & -1 \\ -4 & -2 & 5 \end{bmatrix}$, find $2\mathbf{A}$.

$$\mathbf{2A} = 2\begin{bmatrix} 2 & 3 & -1 \\ -4 & -2 & 5 \end{bmatrix} = \begin{bmatrix} 2\cdot 2 & 2\cdot 3 & 2\cdot(-1) \\ 2\cdot(-4) & 2\cdot(-2) & 2\cdot 5 \end{bmatrix}$$

$$= \begin{bmatrix} 4 & 6 & -2 \\ -8 & -4 & 10 \end{bmatrix}$$

For an $m \times n$ matrix $\mathbf{A} = [a_{ij}]$ and an $n \times p$ matrix $\mathbf{B} = [b_{ij}]$, the **product AB** $= [c_{ij}]$ is an $m \times p$ matrix, where

$$c_{ij} = a_{i1} \cdot b_{1j} + a_{i2} \cdot b_{2j} + a_{i3} \cdot b_{3j} + \cdots + a_{in} \cdot b_{nj}.$$

The properties of matrix multiplication are given on p. 674.

For $\mathbf{A} = \begin{bmatrix} 4 & -1 & 3 \\ 0 & -2 & 1 \end{bmatrix}$ and $\mathbf{B} = \begin{bmatrix} -3 & 1 \\ 3 & 4 \\ 2 & -1 \end{bmatrix}$, find $\mathbf{AB}$.

$$\mathbf{AB} = \begin{bmatrix} 4 & -1 & 3 \\ 0 & -2 & 1 \end{bmatrix}\begin{bmatrix} -3 & 1 \\ 3 & 4 \\ 2 & -1 \end{bmatrix}$$

$$= \begin{bmatrix} 4\cdot(-3)+(-1)\cdot 3 + 3\cdot 2 & 4\cdot 1+(-1)\cdot 4 + 3\cdot(-1) \\ 0\cdot(-3)+(-2)\cdot 3 + 1\cdot 2 & 0\cdot 1+(-2)\cdot 4 + 1\cdot(-1) \end{bmatrix}$$

$$= \begin{bmatrix} -9 & -3 \\ -4 & -9 \end{bmatrix}$$

We can write a matrix equation equivalent to a system of equations.

Write a matrix equation equivalent to the system of equations:

$$2x - 3y = 6,$$
$$x - 4y = 1.$$

This system of equations can be written as

$$\begin{bmatrix} 2 & -3 \\ 1 & -4 \end{bmatrix}\begin{bmatrix} x \\ y \end{bmatrix} = \begin{bmatrix} 6 \\ 1 \end{bmatrix}.$$

SECTION 9.5: INVERSES OF MATRICES

The $n \times n$ **identity matrix I** is an $n \times n$ matrix with 1's on the main diagonal and 0's elsewhere.

For any $n \times n$ matrix $\mathbf{A}$,

$$\mathbf{AI} = \mathbf{IA} = \mathbf{A}.$$

For an $n \times n$ matrix $\mathbf{A}$, if there is a matrix $\mathbf{A}^{-1}$ for which $\mathbf{A}^{-1} \cdot \mathbf{A} = \mathbf{I} = \mathbf{A} \cdot \mathbf{A}^{-1}$, then $\mathbf{A}^{-1}$ is the **inverse** of $\mathbf{A}$.

The inverse of an $n \times n$ matrix $\mathbf{A}$ can be found by first writing an augmented matrix consisting of $\mathbf{A}$ on the left side and the $n \times n$ identity matrix on the right side. Then row-equivalent operations are used to transform the augmented matrix to a matrix with the $n \times n$ identity matrix on the left side and the inverse on the right side.

See Examples 3 and 4 on pp. 680 and 681.

For a system of n linear equations in n variables, $\mathbf{AX} = \mathbf{B}$, if $\mathbf{A}$ has an inverse, then the solution of the system of equations is given by

$$\mathbf{X} = \mathbf{A}^{-1}\mathbf{B}.$$

Since matrix multiplication is not commutative, in general, $\mathbf{B}$ *must* be multiplied *on the left* by $\mathbf{A}^{-1}$.

Use an inverse matrix to solve the following system of equations:

$$x - y = 1,$$
$$x - 2y = -1.$$

First, we write an equivalent matrix equation:

$$\underset{\mathbf{A}}{\begin{bmatrix} 1 & -1 \\ 1 & -2 \end{bmatrix}} \cdot \underset{\mathbf{X}}{\begin{bmatrix} x \\ y \end{bmatrix}} = \underset{\mathbf{B}}{\begin{bmatrix} 1 \\ -1 \end{bmatrix}}.$$

(*continued*)

Then we find $\mathbf{A}^{-1}$ and multiply *on the left* by $\mathbf{A}^{-1}$:

$$\mathbf{X} = \mathbf{A}^{-1} \cdot \mathbf{B}$$

$$\begin{bmatrix} x \\ y \end{bmatrix} = \begin{bmatrix} 2 & -1 \\ 1 & -1 \end{bmatrix}\begin{bmatrix} 1 \\ -1 \end{bmatrix} = \begin{bmatrix} 3 \\ 2 \end{bmatrix}.$$

The solution is $(3, 2)$.

SECTION 9.6: DETERMINANTS AND CRAMER'S RULE

Determinant of a 2×2 Matrix

The determinant of the matrix $\begin{bmatrix} a & c \\ b & d \end{bmatrix}$ is denoted by $\begin{vmatrix} a & c \\ b & d \end{vmatrix}$ and is defined as

$$\begin{vmatrix} a & c \\ b & d \end{vmatrix} = ad - bc.$$

Evaluate: $\begin{vmatrix} 3 & -4 \\ 2 & 1 \end{vmatrix}$.

$$\begin{vmatrix} 3 & -4 \\ 2 & 1 \end{vmatrix} = 3 \cdot 1 - 2(-4) = 3 + 8 = 11$$

The **determinant** of any **square matrix** can be found by *expanding across a row* or *down a column*. See p. 687.

See Example 4 on p. 687.

We can use determinants to solve systems of linear equations.

Cramer's rule for a 2×2 system is given on p. 688. Cramer's rule for a 3×3 system is given on p. 689.

Solve: $2x - 3y = 2,$
$6x + 6y = 1.$

$$x = \frac{\begin{vmatrix} 2 & -3 \\ 1 & 6 \end{vmatrix}}{\begin{vmatrix} 2 & -3 \\ 6 & 6 \end{vmatrix}}, \qquad y = \frac{\begin{vmatrix} 2 & 2 \\ 6 & 1 \end{vmatrix}}{\begin{vmatrix} 2 & -3 \\ 6 & 6 \end{vmatrix}}$$

$$x = \frac{15}{30} = \frac{1}{2}, \qquad y = \frac{-10}{30} = -\frac{1}{3}.$$

The solution is $\left(\frac{1}{2}, -\frac{1}{3}\right)$.

SECTION 9.7: SYSTEMS OF INEQUALITIES AND LINEAR PROGRAMMING

To graph a linear inequality in two variables:

1. Graph the related equation. Draw a dashed line if the inequality symbol is $<$ or $>$. Draw a solid line if the inequality symbol is $\le$ or $\ge$.
2. Use a test point to determine which half-plane to shade.

Graph: $x + y > 2$.

1. Graph $x + y = 2$ using a dashed line.
2. Test a point not on the line. We use $(0, 0)$.

$$\begin{array}{c|c} x + y & > 2 \\ \hline 0 + 0 & ?\ 2 \\ 0 & \end{array} \quad \text{FALSE}$$

Since $0 > 2$ is false, we shade the half-plane that does not contain $(0, 0)$.

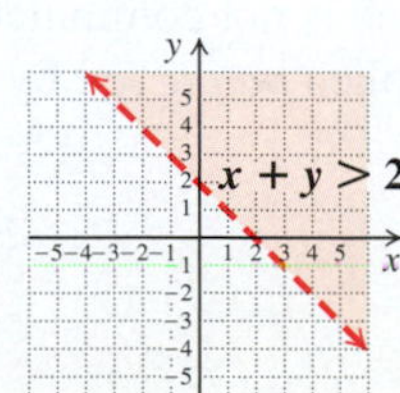

To graph a system of inequalities, graph each inequality and determine the region that is common to all the solution sets.

Graph the solution set of the system

$$x - y \le 3,$$
$$2x + y \ge 4.$$

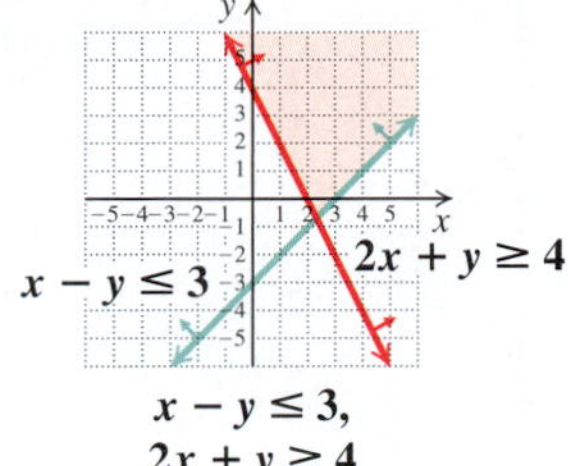

The maximum or the minimum value of an **objective function** over a region of **feasible solutions** is the maximum or the minimum value of the function at a vertex of that region.

Maximize $G = 8x - 5y$ subject to

$$x + y \le 3,$$
$$x \ge 0,$$
$$y \ge 1.$$

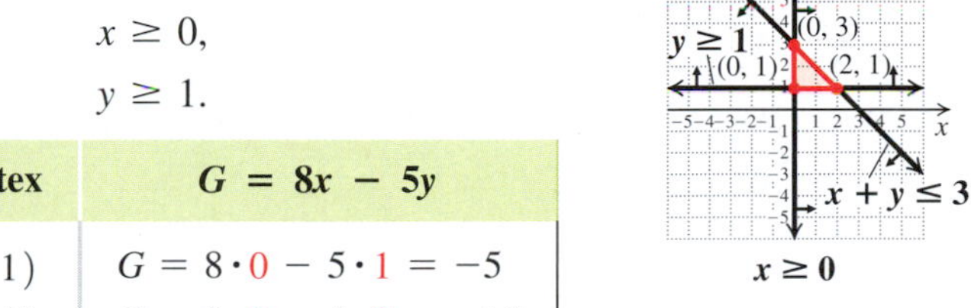

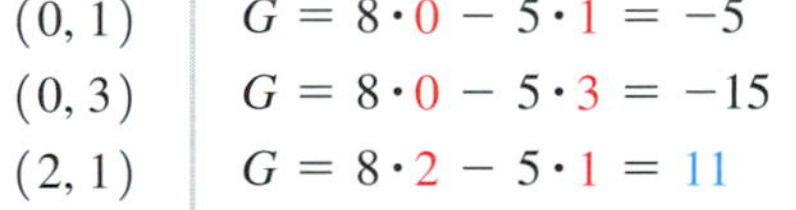

Vertex	$G = 8x - 5y$
$(0, 1)$	$G = 8 \cdot 0 - 5 \cdot 1 = -5$
$(0, 3)$	$G = 8 \cdot 0 - 5 \cdot 3 = -15$
$(2, 1)$	$G = 8 \cdot 2 - 5 \cdot 1 = 11$ ← Maximum

SECTION 9.8: PARTIAL FRACTIONS

The procedure for **decomposing a rational expression into partial fractions** is given on p. 703.

See Examples 1–5 on pp. 703–706.

REVIEW EXERCISES

Determine whether the statement is true or false.

1. A system of equations with exactly one solution is consistent and has independent equations. [9.1]

2. A system of two linear equations in two variables can have exactly two solutions. [9.1]

3. For any $m \times n$ matrices **A** and **B**,

$$\mathbf{A} + \mathbf{B} = \mathbf{B} + \mathbf{A}. \quad [9.4]$$

4. In general, matrix multiplication is commutative. [9.4]

In Exercises 5–12, match the equations or inequalities with one of the graphs (a)–(h) that follow.

a)

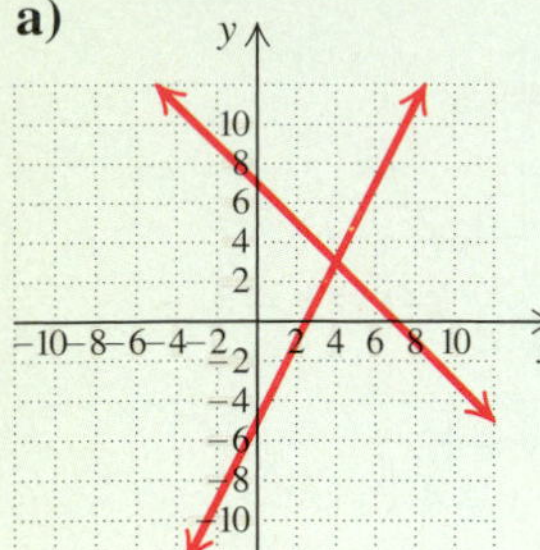

b)

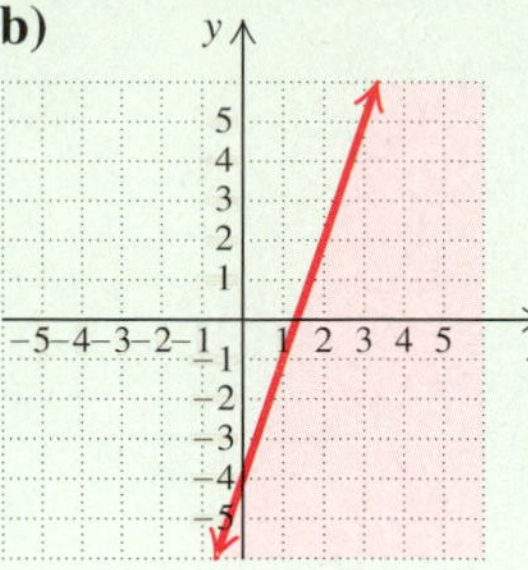

c)

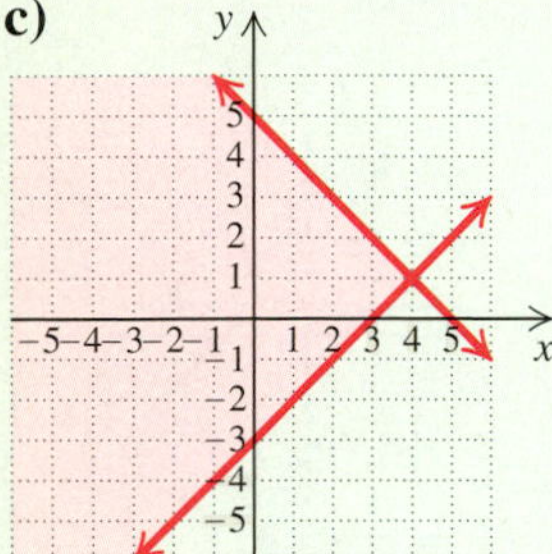

d)

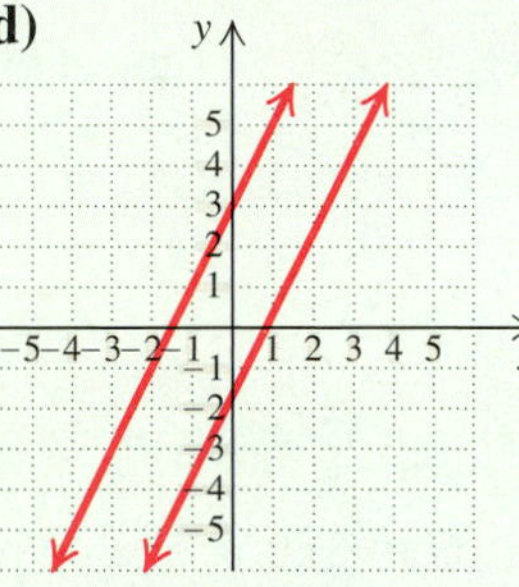

e)

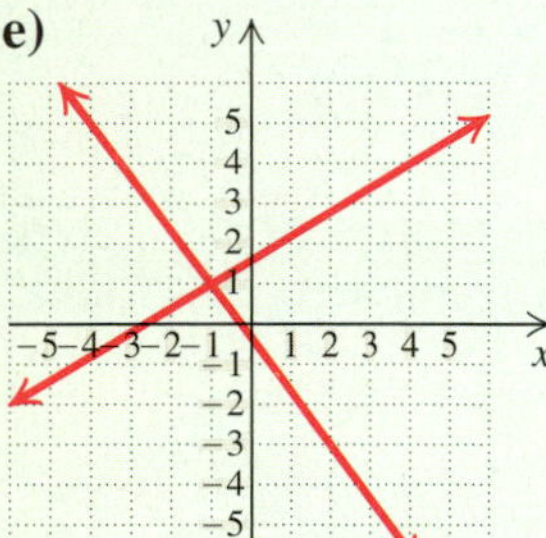

f)

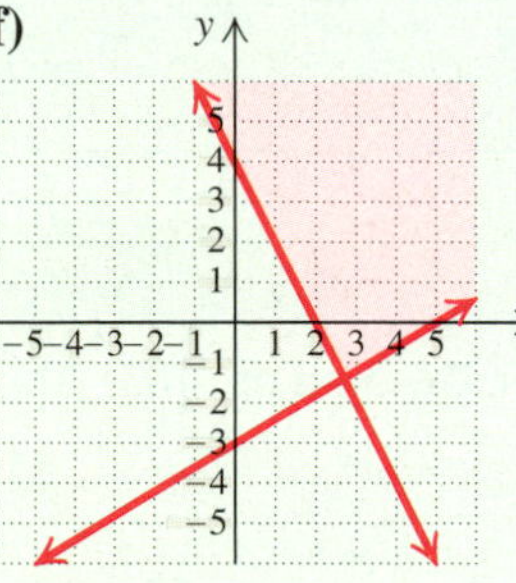

g)

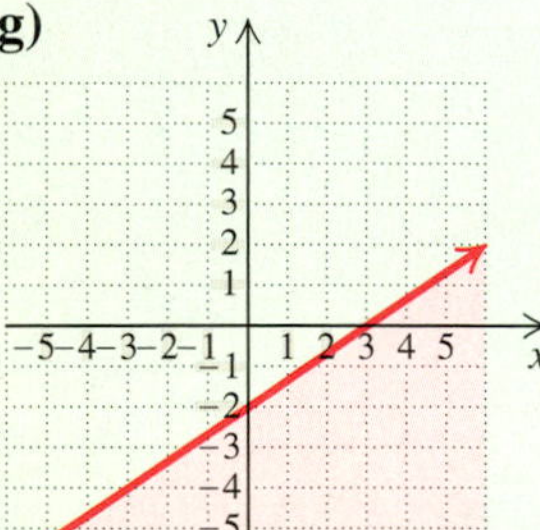

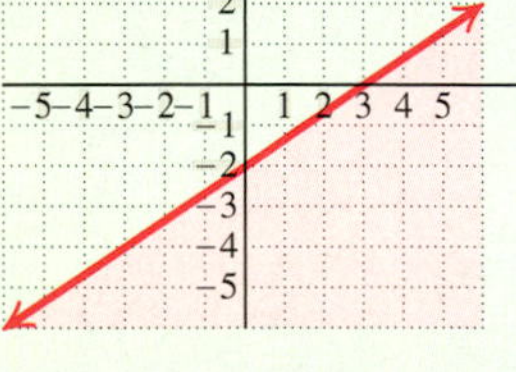

h)

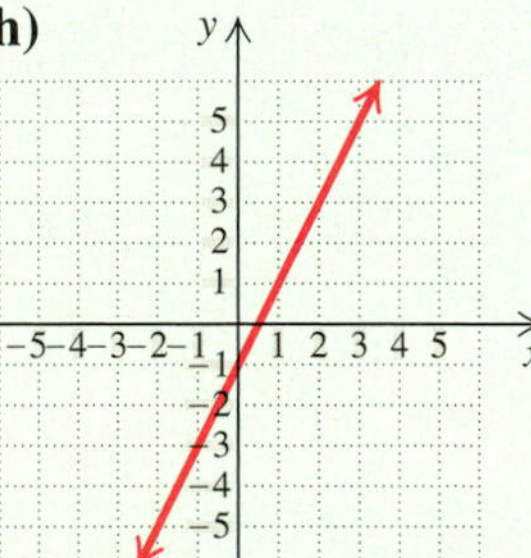

5. $x + y = 7,$
$2x - y = 5$ **[9.1]**

6. $3x - 5y = -8,$
$4x + 3y = -1$ **[9.1]**

7. $y = 2x - 1,$
$4x - 2y = 2$ **[9.1]**

8. $6x - 3y = 5,$
$y = 2x + 3$ **[9.1]**

9. $y \leq 3x - 4$ **[9.7]**

10. $2x - 3y \geq 6$ **[9.7]**

11. $x - y \leq 3,$
$x + y \leq 5$ **[9.7]**

12. $2x + y \geq 4,$
$3x - 5y \leq 15$ **[9.7]**

Solve.

13. $5x - 3y = -4,$
$3x - y = -4$ **[9.1]**

14. $2x + 3y = 2,$
$5x - y = -29$ **[9.1]**

15. $x + 5y = 12,$
$5x + 25y = 12$ **[9.1]**

16. $x + y = -2,$
$-3x - 3y = 6$ **[9.1]**

17. $x + 5y - 3z = 4,$
$3x - 2y + 4z = 3,$
$2x + 3y - z = 5$ **[9.2]**

18. $2x - 4y + 3z = -3,$
$-5x + 2y - z = 7,$
$3x + 2y - 2z = 4$ **[9.2]**

19. $x - y = 5,$
$y - z = 6,$
$-w + z = 7,$
$w + x = 8$ **[9.2]**

20. Classify each of the systems in Exercises 13–19 as consistent or inconsistent. **[9.1], [9.2]**

21. Classify each of the systems in Exercises 13–19 as having dependent or independent equations. **[9.1], [9.2]**

Solve the system of equations using Gaussian elimination or Gauss–Jordan elimination. **[9.3]**

22. $x + 2y = 5,$
$2x - 5y = -8$

23. $3x + 4y + 2z = 3,$
$5x - 2y - 13z = 3,$
$4x + 3y - 3z = 6$

24. $3x + 5y + z = 0,$
$2x - 4y - 3z = 0,$
$x + 3y + z = 0$

25. $w + x + y + z = -2,$
$-3w - 2x + 3y + 2z = 10,$
$2w + 3x + 2y - z = -12,$
$2w + 4x - y + z = 1$

26. *Coins.* The value of 75 coins, consisting of only nickels and dimes, is \$5.95. How many of each kind of coin are there? **[9.1]**

27. *Investment.* The Davidson family invested \$5000, part at 3% and the remainder at 3.5%. The annual income from both investments is \$167. What is the amount invested at each rate? **[9.1]**

28. *Nutrition.* A dietician must plan a breakfast menu that provides 460 Cal, 9 g of fat, and 55 mg of calcium. One plain bagel contains 200 Cal, 2 g of fat, and 29 mg of calcium. A one-tablespoon serving of cream cheese contains 100 Cal, 10 g of fat, and 24 mg of calcium. One banana contains 105 Cal,1 g of fat, and 7 g of calcium. (*Source*: U.S. Department of Agriculture) How many servings of each are required to provide the desired nutritional values? **[9.2]**

29. *Test Scores.* A student has a total of 226 points on three tests. The sum of the scores on the first and second tests exceeds the score on the third test by 62. The first score exceeds the second by 6. Find the three scores. **[9.2]**

30. *Employed Civilians.* The following table lists the number of persons, ages 16 and older, employed in the United States, represented in terms of the number of years after 2008. [9.2]

Year, x	Persons Ages 16 and Older Employed (in millions)
2008, 0	145
2010, 2	139
2012, 4	142

Source: Bureau of Labor Statistics, U.S. Department of Labor

a) Use a system of equations to fit a quadratic function $f(x) = ax^2 + bx + c$ to the data.

b) Use the function to estimate the number of persons employed in 2014.

For Exercises 31–38, let

$$\mathbf{A} = \begin{bmatrix} 1 & -1 & 0 \\ 2 & 3 & -2 \\ -2 & 0 & 1 \end{bmatrix}, \quad \mathbf{B} = \begin{bmatrix} -1 & 0 & 6 \\ 1 & -2 & 0 \\ 0 & 1 & -3 \end{bmatrix},$$

and

$$\mathbf{C} = \begin{bmatrix} -2 & 0 \\ 1 & 3 \end{bmatrix}.$$

Find each of the following, if possible. [9.4]

31. $\mathbf{A} + \mathbf{B}$

32. $-3\mathbf{A}$

33. $-\mathbf{A}$

34. $\mathbf{AB}$

35. $\mathbf{B} + \mathbf{C}$

36. $\mathbf{A} - \mathbf{B}$

37. $\mathbf{BA}$

38. $\mathbf{A} + 3\mathbf{B}$

39. *Food Service Management.* The following table lists the cost per serving, in dollars, for items on four menus that are served at an NFL training camp.

Menu	Meat	Potato	Vegetable	Salad	Dessert
1	2.25	0.38	0.55	0.33	0.85
2	3.09	0.42	0.46	0.48	0.51
3	2.40	0.31	0.59	0.36	0.64
4	1.80	0.29	0.34	0.55	0.52

On a particular day, a dietician orders 41 meals from menu 1, 18 from menu 2, 39 from menu 3, and 36 from menu 4.

a) Write the information in the table as a 4×5 matrix **M**. [9.4]

b) Write a row matrix **N** that represents the number of each menu ordered. [9.4]

c) Find the product **NM**. [9.4]

d) State what the entries of **NM** represent. [9.4]

Find $\mathbf{A}^{-1}$, *if it exists.* [9.5]

40. $\mathbf{A} = \begin{bmatrix} -2 & 0 \\ 1 & 3 \end{bmatrix}$

41. $\mathbf{A} = \begin{bmatrix} 0 & 0 & 3 \\ 0 & -2 & 0 \\ 4 & 0 & 0 \end{bmatrix}$

42. $\mathbf{A} = \begin{bmatrix} 1 & 0 & 0 & 0 \\ 0 & 4 & -5 & 0 \\ 0 & 2 & 2 & 0 \\ 0 & 0 & 0 & 1 \end{bmatrix}$

43. Write a matrix equation equivalent to this system of equations:

$$\begin{aligned} 3x - 2y + 4z &= 13, \\ x + 5y - 3z &= 7, \\ 2x - 3y + 7z &= -8. \end{aligned}$$ [9.4]

Solve the system of equations using the inverse of the coefficient matrix of the equivalent matrix equation. [9.5]

44. $2x + 3y = 5,$
$3x + 5y = 11$

45. $5x - y + 2z = 17,$
$3x + 2y - 3z = -16,$
$4x - 3y - z = 5$

46. $w - x - y + z = -1,$
$2w + 3x - 2y - z = 2,$
$-w + 5x + 4y - 2z = 3,$
$3w - 2x + 5y + 3z = 4$

Evaluate the determinant. [9.6]

47. $\begin{vmatrix} 1 & -2 \\ 3 & 4 \end{vmatrix}$

48. $\begin{vmatrix} \sqrt{3} & -5 \\ -3 & -\sqrt{3} \end{vmatrix}$

49. $\begin{vmatrix} 2 & -1 & 1 \\ 1 & 2 & -1 \\ 3 & 4 & -3 \end{vmatrix}$

50. $\begin{vmatrix} 1 & -1 & 2 \\ -1 & 2 & 0 \\ -1 & 3 & 1 \end{vmatrix}$

Solve using Cramer's rule. [9.6]

51. $5x - 2y = 19,$
$7x + 3y = 15$

52. $x + y = 4,$
$4x + 3y = 11$

53. $3x - 2y + z = 5,$
$4x - 5y - z = -1,$
$3x + 2y - z = 4$

54. $2x - y - z = 2,$
$3x + 2y + 2z = 10,$
$x - 5y - 3z = -2$

Graph. [9.7]

55. $y \leq 3x + 6$

56. $4x - 3y \geq 12$

57. Graph this system of inequalities and find the coordinates of any vertices formed. [9.7]

$2x + y \geq 9,$
$4x + 3y \geq 23,$
$x + 3y \geq 8,$
$x \geq 0,$
$y \geq 0$

58. Find the maximum value and the minimum value of $T = 6x + 10y$ subject to

$x + y \leq 10,$
$5x + 10y \geq 50,$
$x \geq 2,$
$y \geq 0.$ [9.7]

59. *Maximizing a Test Score.* Jackson is taking a test that contains questions in group A worth 7 points each and questions in group B worth 12 points each. The total number of questions answered must be at least 8. If Jackson knows that group A questions take 8 min each and group B questions take 10 min each and the maximum time for the test is 80 min, how many questions from each group must he answer correctly in order to maximize his score? What is the maximum score? [9.7]

Decompose into partial fractions. [9.8]

60. $\dfrac{5}{(x + 2)^2(x + 1)}$

61. $\dfrac{-8x + 23}{2x^2 + 5x - 12}$

62. Solve: $2x + y = 7,$
$x - 2y = 6.$ [9.1]

A. x and y are both positive numbers.
B. x and y are both negative numbers.
C. x is positive and y is negative.
D. x is negative and y is positive.

63. Which of the following is *not* a row-equivalent operation on a matrix? [9.3]

A. Interchange any two columns.
B. Interchange any two rows.
C. Add two rows.
D. Multiply each entry in a row by -3.

64. The graph of the given system of inequalities is which of the following? [9.7]

$x + y \leq 3,$
$x - y \leq 4$

A.

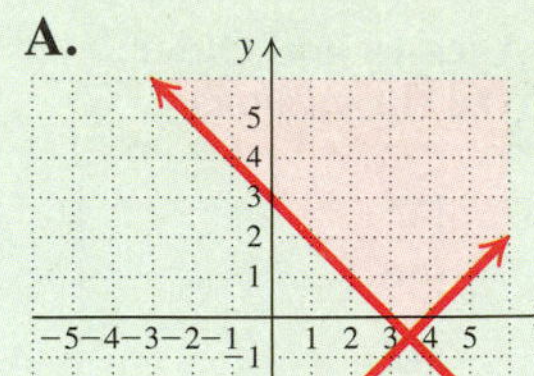

B.

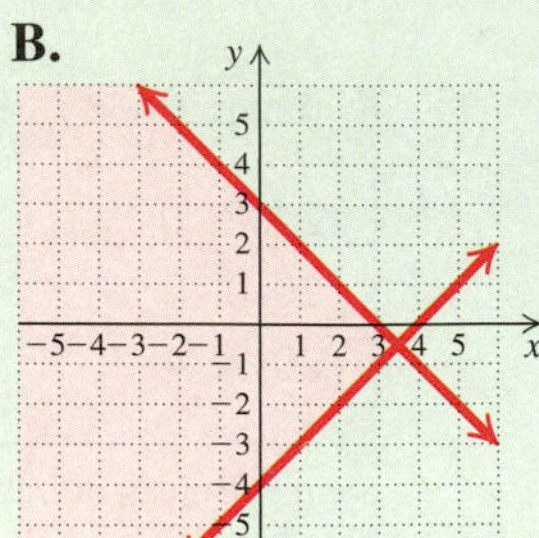

C.

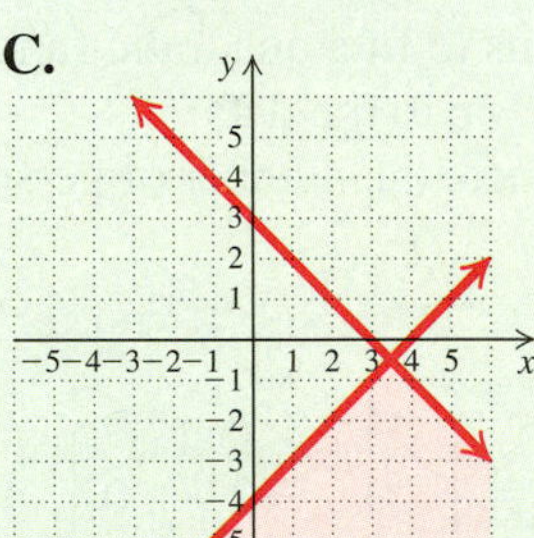

D.

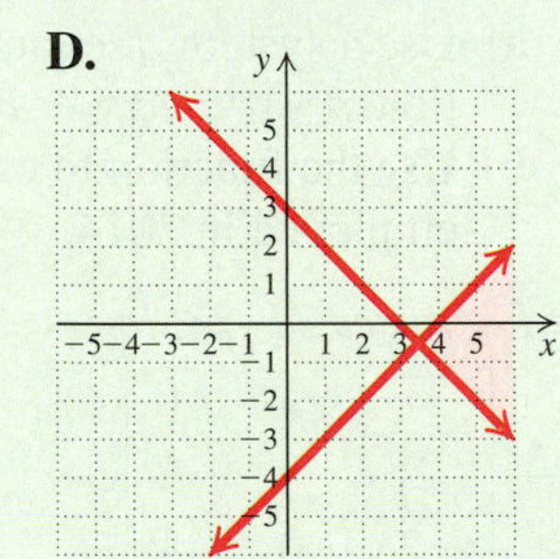

Synthesis

65. One year, Lucia invested a total of \$40,000, part at 4%, part at 5%, and the rest at $5\frac{1}{2}\%$. The total amount of interest received on the investments was \$1990. The interest received on the $5\frac{1}{2}\%$ investment was \$590 more than the interest received on the 4% investment. How much was invested at each rate? [9.2]

Solve.

66. $\dfrac{2}{3x} + \dfrac{4}{5y} = 8,$
$\dfrac{5}{4x} - \dfrac{3}{2y} = -6$ [9.1]

67. $\dfrac{3}{x} - \dfrac{4}{y} + \dfrac{1}{z} = -2,$
$\dfrac{5}{x} + \dfrac{1}{y} - \dfrac{2}{z} = 1,$
$\dfrac{7}{x} + \dfrac{3}{y} + \dfrac{2}{z} = 19$ [9.2]

Graph. [9.7]

68. $|x| - |y| \leq 1$

69. $|xy| > 1$

Collaborative Discussion and Writing

70. Dylon solves the equation $2x + 5 = 3x - 7$ by finding the point of intersection of the graphs of $y_1 = 2x + 5$ and $y_2 = 3x - 7$. She finds the same point when she solves the system of equations

$$y = 2x + 5,$$
$$y = 3x - 7.$$

Explain the difference between the solution of the equation and the solution of the system of equations. [9.1]

71. For square matrices **A** and **B**, is it true, in general, that $(\mathbf{AB})^2 = \mathbf{A}^2\mathbf{B}^2$? Explain. [9.4]

72. Given the system of equations

$$a_1x + b_1y = c_1,$$
$$a_2x + b_2y = c_2,$$

explain why the equations are dependent or the system is inconsistent when

$$\begin{vmatrix} a_1 & b_1 \\ a_2 & b_2 \end{vmatrix} = 0. \text{ [9.6]}$$

73. If the lines $a_1x + b_1y = c_1$ and $a_2x + b_2y = c_2$ are parallel, what can you say about the values of

$$\begin{vmatrix} a_1 & b_1 \\ a_2 & b_2 \end{vmatrix}, \quad \begin{vmatrix} c_1 & b_1 \\ c_2 & b_2 \end{vmatrix}, \quad \text{and} \quad \begin{vmatrix} a_1 & c_1 \\ a_2 & c_2 \end{vmatrix}? \text{ [9.6]}$$

74. Describe how the graph of a linear inequality differs from the graph of a linear equation. [9.7]

75. What would you say to a classmate who tells you that the partial fraction decomposition of

$$\frac{3x^2 - 8x + 9}{(x + 3)(x^2 - 5x + 6)}$$

is

$$\frac{2}{x + 3} + \frac{x - 1}{x^2 - 5x + 6}?$$

Explain. [9.8]

9 Chapter Test

Solve. Use any method. Also determine whether the system is consistent or inconsistent and whether the equations are dependent or independent.

1. $3x + 2y = 1,$
$2x - y = -11$

2. $2x - y = 3,$
$2y = 4x - 6$

3. $x - y = 4,$
$3y = 3x - 8$

4. $2x - 3y = 8,$
$5x - 2y = 9$

Solve.

5. $4x + 2y + z = 4,$
$3x - y + 5z = 4,$
$5x + 3y - 3z = -2$

6. *Ticket Sales.* One evening, 620 tickets were sold for Clearview Community College's talent show. Tickets cost \$8 each for students and \$12 each for nonstudents. Total receipts were \$5592. How many of each type of ticket were sold?

7. Hui, Ashlyn, and Sheriann can process 352 online orders per day. Hui and Ashlyn together can process 224 orders per day while Hui and Sheriann together can process 248 orders per day. How many orders can each of them process alone?

For Exercises 8–13, let

$$\mathbf{A} = \begin{bmatrix} 1 & -1 & 3 \\ -2 & 5 & 2 \end{bmatrix}, \quad \mathbf{B} = \begin{bmatrix} -5 & 1 \\ -2 & 4 \end{bmatrix},$$

and

$$\mathbf{C} = \begin{bmatrix} 3 & -4 \\ -1 & 0 \end{bmatrix}.$$

Find each of the following, if possible.

8. $\mathbf{B} + \mathbf{C}$

9. $\mathbf{A} - \mathbf{C}$

10. $\mathbf{CB}$

11. $\mathbf{AB}$

12. $2\mathbf{A}$

13. $\mathbf{C}^{-1}$

14. *Food Service Management.* The following table lists the cost per serving, in dollars, for items on three lunch menus served at a senior citizens' center.

Menu	Main Dish	Side Dish	Dessert
1	1.55	1.00	0.99
2	1.70	0.95	1.01
3	1.65	0.99	0.96

On a particular day, 26 Menu 1 meals, 18 Menu 2 meals, and 23 Menu 3 meals are served.

a) Write the information in the table as a 3×3 matrix **M**.

b) Write a row matrix **N** that represents the number of each menu served.

c) Find the product **NM**.

d) State what the entries of **NM** represent.

15. Write a matrix equation equivalent to the system of equations

$$\begin{aligned} 3x - 4y + 2z &= -8, \\ 2x + 3y + z &= 7, \\ x - 5y - 3z &= 3. \end{aligned}$$

16. Solve the system of equations using the inverse of the coefficient matrix of the equivalent matrix equation.

$$\begin{aligned} 3x + 2y + 6z &= 2, \\ x + y + 2z &= 1, \\ 2x + 2y + 5z &= 3 \end{aligned}$$

Evaluate the determinant.

17. $\begin{vmatrix} 3 & -5 \\ 8 & 7 \end{vmatrix}$

18. $\begin{vmatrix} 2 & -1 & 4 \\ -3 & 1 & -2 \\ 5 & 3 & -1 \end{vmatrix}$

19. Solve using Cramer's rule. Show your work.

$$\begin{aligned} 5x + 2y &= -1, \\ 7x + 6y &= 1 \end{aligned}$$

20. Graph: $3x + 4y \leq -12$.

21. Find the maximum value and the minimum value of $Q = 2x + 3y$ subject to

$$\begin{aligned} x + y &\leq 6, \\ 2x - 3y &\geq -3, \\ x &\geq 1, \\ y &\geq 0. \end{aligned}$$

22. *Maximizing Profit.* Jane's Cakes prepares pound cakes and carrot cakes. In a given week, at most 100 cakes can be prepared, of which 25 pound cakes and 15 carrot cakes are required by regular customers. The profit from each pound cake is \$6, and the profit from each carrot cake is \$8. How many of each type of cake should be prepared in order to maximize the profit? What is the maximum profit?

23. Decompose into partial fractions:

$$\frac{3x - 11}{x^2 + 2x - 3}.$$

24. The graph of the given system of inequalities is which of the following?

$$\begin{aligned} x + 2y &\geq 4, \\ x - y &\leq 2 \end{aligned}$$

A.

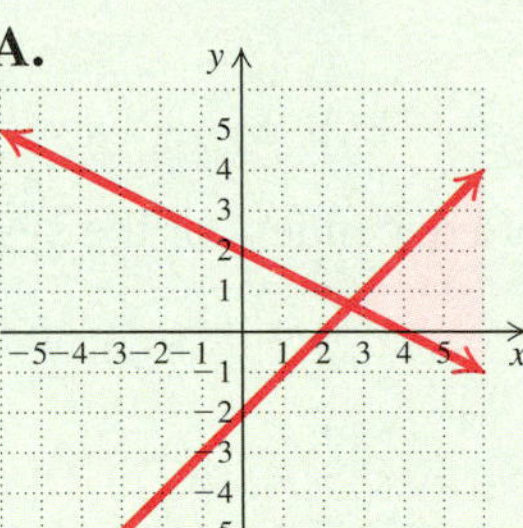

B.

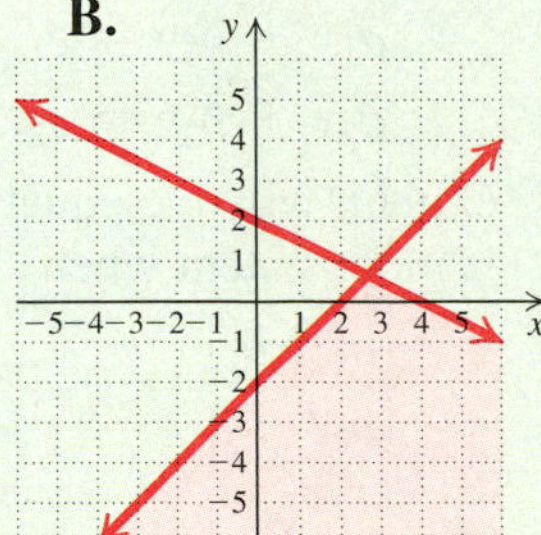

C.

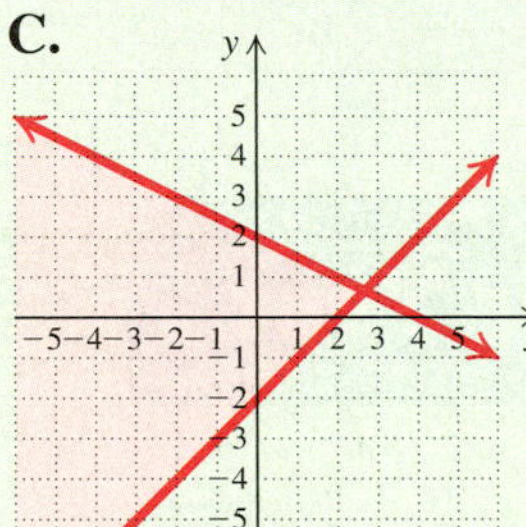

D.

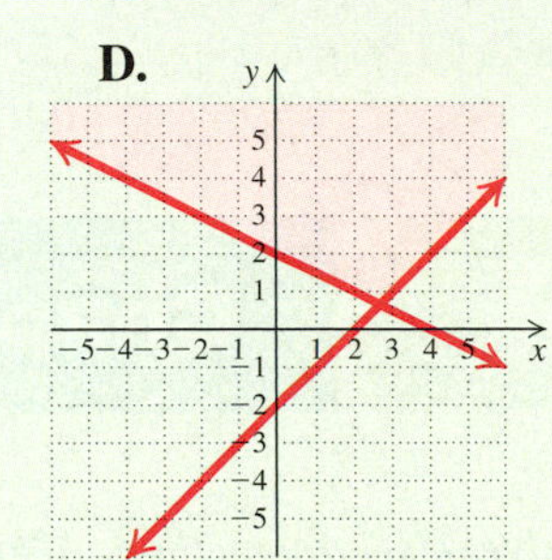

Synthesis

25. Three solutions of the equation $Ax - By = Cz - 8$ are $(2, -2, 2)$, $(-3, -1, 1)$, and $(4, 2, 9)$. Find A, B, and C.

Analytic Geometry Topics

CHAPTER 10

APPLICATION

This problem appears as Exercise 33 in Section 10.1.

An engineer designs a satellite dish with a parabolic cross section. The dish is 15 ft wide at the opening, and the focus is placed 4 ft from the vertex. Find the depth of the satellite dish at the vertex.

10.1 The Parabola

- Given an equation of a parabola, complete the square, if necessary, and then find the vertex, the focus, and the directrix and graph the parabola.

A **conic section** is formed when a right circular cone with two parts, called *nappes*, is intersected by a plane. One of four types of curves can be formed: a parabola, a circle, an ellipse, or a hyperbola.

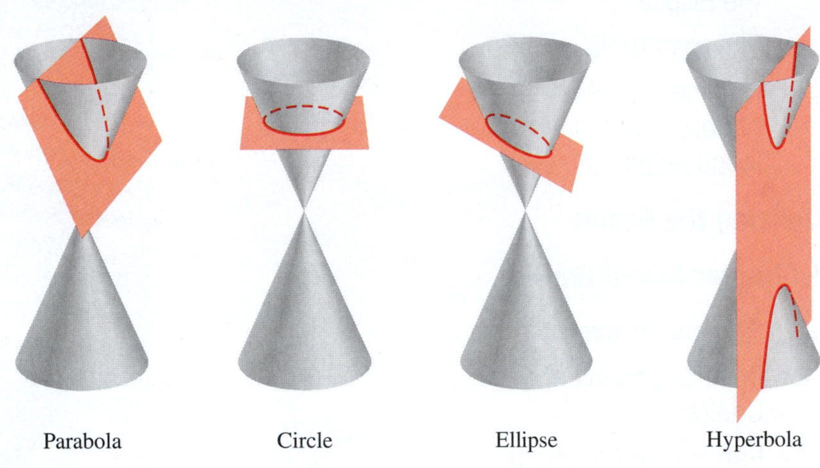

Conic Sections

Conic sections can be defined algebraically using second-degree equations of the form $Ax^2 + Bxy + Cy^2 + Dx + Ey + F = 0$. In addition, they can be defined geometrically as a set of points that satisfy certain conditions.

Parabolas

The graph of the quadratic function $f(x) = ax^2 + bx + c, a \neq 0$, is a parabola. A parabola can be defined geometrically.

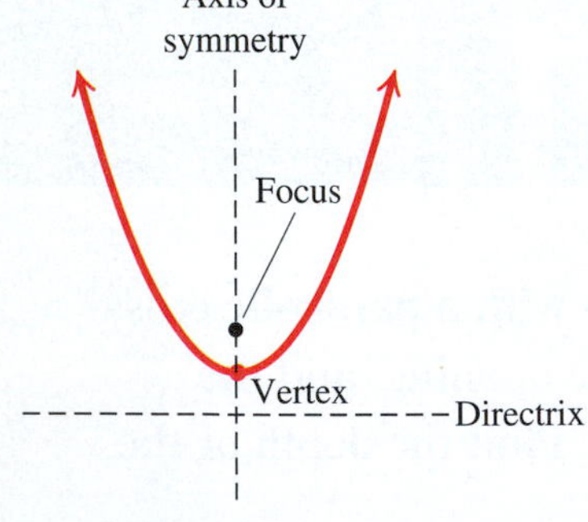

PARABOLA

A **parabola** is the set of all points in a plane equidistant from a fixed line (the **directrix**) and a fixed point not on the line (the **focus**).

The line that is perpendicular to the directrix and contains the focus is the **axis of symmetry**. The **vertex** is the midpoint of the segment between the focus and the directrix. (See the figure at left.)

Let's derive the standard equation of a parabola with vertex $(0, 0)$ and directrix $y = -p$, where $p > 0$. We place the coordinate axes as shown in Fig. 1 on the following page. The y-axis is the axis of symmetry and contains the focus F. The distance from the focus to the vertex is the same as the distance from the vertex to the directrix. Thus the coordinates of F are $(0, p)$.

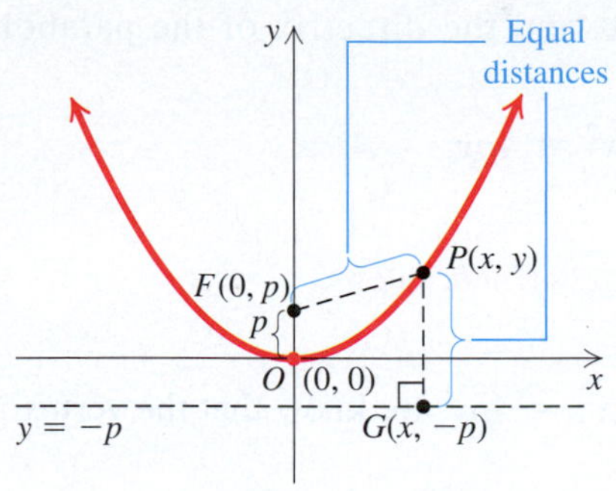

FIGURE 1.

Let $P(x, y)$ be any point on the parabola and consider $\overline{PG}$ perpendicular to the line $y = -p$. The coordinates of G are $(x, -p)$. By the definition of a parabola,

$$PF = PG.$$ **The distance from P to the focus is the same as the distance from P to the directrix.**

Then using the distance formula, we have

$$\sqrt{(x-0)^2 + (y-p)^2} = \sqrt{(x-x)^2 + [y-(-p)]^2}$$

$$x^2 + y^2 - 2py + p^2 = y^2 + 2py + p^2$$ **Squaring both sides and squaring the binomials**

$$x^2 = 4py.$$

We have shown that if $P(x, y)$ is on the parabola shown in Fig. 1 above, then its coordinates satisfy this equation. The converse is also true, but we will not prove it here.

Note that if $p > 0$, as above, the graph opens up. If $p < 0$, the graph opens down.

The equation of a parabola with vertex $(0, 0)$ and directrix $x = -p$ is derived in a similar manner. Such a parabola opens either to the right $(p > 0)$, as shown in Fig. 2 below, or to the left $(p < 0)$.

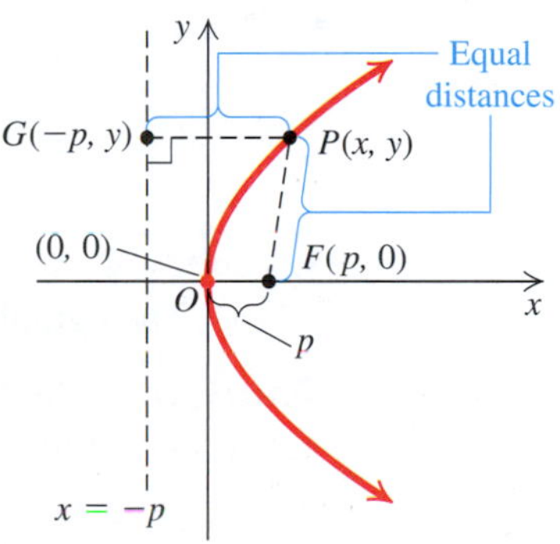

FIGURE 2.

STANDARD EQUATION OF A PARABOLA WITH VERTEX (0, 0) AND VERTICAL AXIS OF SYMMETRY

The standard equation of a parabola with vertex $(0, 0)$ and directrix $y = -p$ is

$$x^2 = 4py.$$

The focus is $(0, p)$ and the y-axis is the axis of symmetry. When $p > 0$, the parabola opens up; when $p < 0$, the parabola opens down.

STANDARD EQUATION OF A PARABOLA WITH VERTEX (0, 0) AND HORIZONTAL AXIS OF SYMMETRY

The standard equation of a parabola with vertex $(0, 0)$ and directrix $x = -p$ is

$$y^2 = 4px.$$

The focus is $(p, 0)$ and the x-axis is the axis of symmetry. When $p > 0$, the parabola opens to the right; when $p < 0$, the parabola opens to the left.

EXAMPLE 1 Find the focus, the vertex, and the directrix of the parabola $y = -\frac{1}{12}x^2$. Then graph the parabola.

Solution We write $y = -\frac{1}{12}x^2$ in the form $x^2 = 4py$:

$$-\tfrac{1}{12}x^2 = y \qquad \text{Given equation}$$
$$x^2 = -12y \qquad \text{Multiplying by } -12 \text{ on both sides}$$
$$x^2 = 4(-3)y. \qquad \text{Standard form}$$

Since the equation can be written in the form $x^2 = 4py$, we know that the vertex is $(0, 0)$.

We have $p = -3$, so the focus is $(0, p)$, or $(0, -3)$. The directrix is $y = -p = -(-3) = 3$.

x	y
0	0
± 1	$-\frac{1}{12}$
± 2	$-\frac{1}{3}$
± 3	$-\frac{3}{4}$
± 4	$-\frac{4}{3}$

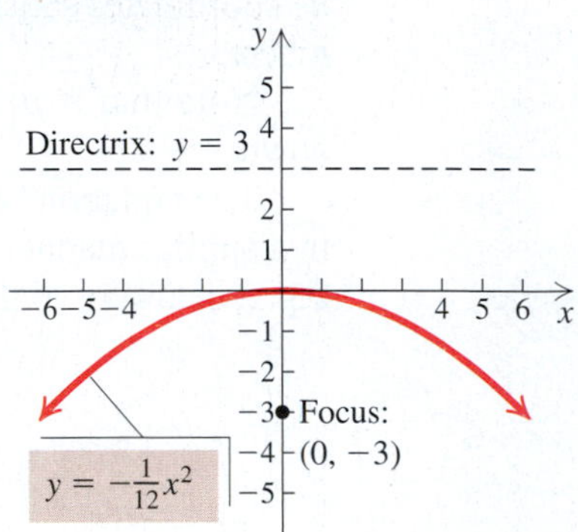

Now Try Exercise 7.

EXAMPLE 2 Find an equation of the parabola with vertex $(0, 0)$ and focus $(5, 0)$. Then graph the parabola.

Solution The focus is on the x-axis so the line of symmetry is the x-axis. The equation is of the type

$$y^2 = 4px.$$

Since the focus $(5, 0)$ is 5 units to the right of the vertex, $p = 5$ and the equation is

$$y^2 = 4(5)x, \quad \text{or} \quad y^2 = 20x.$$

x	y^2	y	(x, y)
0	0	0	$(0, 0)$
1	20	$\pm\sqrt{20}$	$(1, 4.47)$
			$(1, -4.47)$
2	40	$\pm\sqrt{40}$	$(2, 6.32)$
			$(2, -6.32)$
3	60	$\pm\sqrt{60}$	$(3, 7.75)$
			$(3, -7.75)$

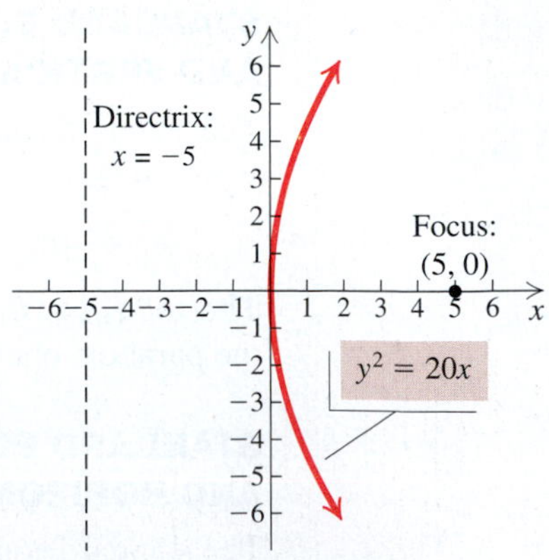

We can also use a graphing calculator to graph parabolas. It might be necessary to solve the equation for y before entering it in the calculator:

$$y^2 = 20x$$
$$y = \pm\sqrt{20x}.$$

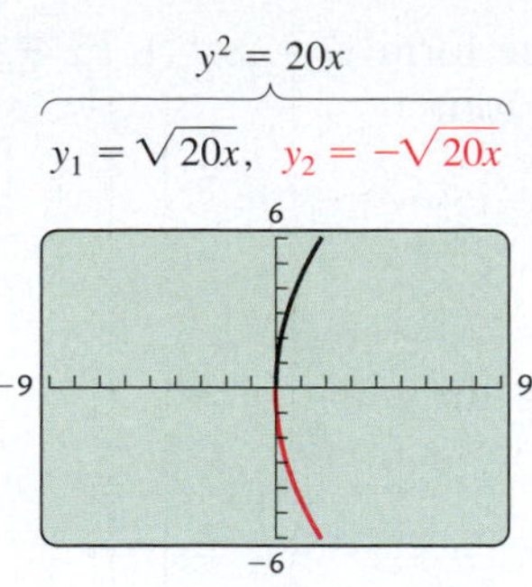

We now graph

$$y_1 = \sqrt{20x} \quad \text{and} \quad y_2 = -\sqrt{20x}, \quad \text{or} \quad y_1 = \sqrt{20x} \quad \text{and} \quad y_2 = -y_1,$$

in a squared viewing window.

On some graphing calculators, the Conics application from the APPS menu can be used to graph parabolas. This method will be discussed in Example 4.

Now Try Exercise 15.

Finding Standard Form by Completing the Square

If a parabola with vertex at the origin is translated horizontally $|h|$ units and vertically $|k|$ units, it has an equation as follows.

STANDARD EQUATION OF A PARABOLA WITH VERTEX (h, k) AND VERTICAL AXIS OF SYMMETRY

The standard equation of a parabola with vertex (h, k) and vertical axis of symmetry is

$$(x - h)^2 = 4p(y - k),$$

where the vertex is (h, k), the focus is $(h, k + p)$, and the directrix is $y = k - p$.

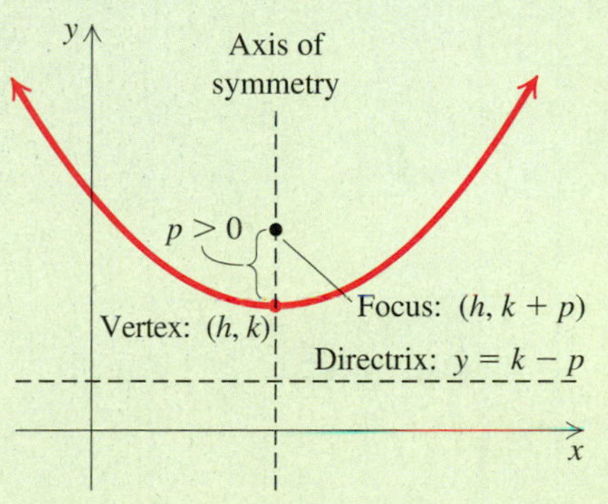

(When $p < 0$, the parabola opens down.)

STANDARD EQUATION OF A PARABOLA WITH VERTEX (h, k) AND HORIZONTAL AXIS OF SYMMETRY

The standard equation of a parabola with vertex (h, k) and horizontal axis of symmetry is

$$(y - k)^2 = 4p(x - h),$$

where the vertex is (h, k), the focus is $(h + p, k)$, and the directrix is $x = h - p$.

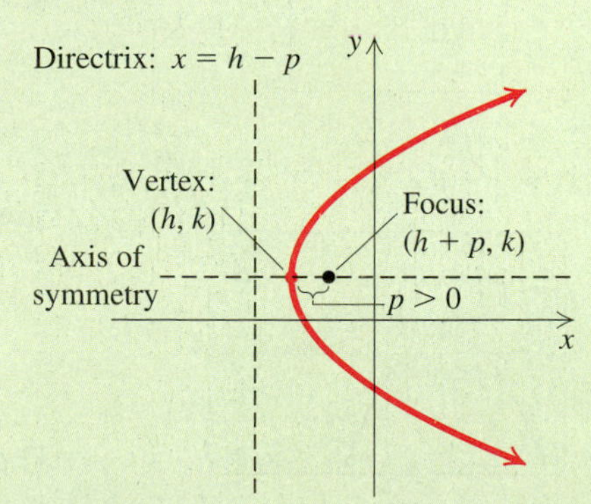

(When $p < 0$, the parabola opens to the left.)

COMPLETING THE SQUARE

REVIEW SECTION 3.2.

We can complete the square on equations of the form $y = ax^2 + bx + c$ or $x = ay^2 + by + c$ in order to write them in standard form.

Just in Time **16**

EXAMPLE 3 For the parabola

$$x^2 + 6x + 4y + 5 = 0,$$

find the vertex, the focus, and the directrix. Then draw the graph.

Solution We first complete the square:

$$x^2 + 6x + 4y + 5 = 0$$

$$x^2 + 6x = -4y - 5 \quad \text{Subtracting } 4y \text{ and } 5 \text{ on both sides}$$

$$x^2 + 6x + 9 = -4y - 5 + 9 \quad \text{Adding 9 on both sides to complete the square on the left side}$$

$$x^2 + 6x + 9 = -4y + 4$$

$$(x + 3)^2 = -4(y - 1) \quad \text{Factoring}$$

$$[x - (-3)]^2 = 4(-1)(y - 1). \quad \text{Writing standard form: } (x - h)^2 = 4p(y - k)$$

We see that $h = -3$, $k = 1$, and $p = -1$, so we have the following:

Vertex (h, k): $(-3, 1)$;
Focus $(h, k + p)$: $(-3, 1 + (-1))$, or $(-3, 0)$;
Directrix $y = k - p$: $y = 1 - (-1)$, or $y = 2$.

x	y
-4	$\frac{3}{4}$
-2	$\frac{3}{4}$
-5	0
-1	0
-6	$-\frac{5}{4}$
0	$-\frac{5}{4}$

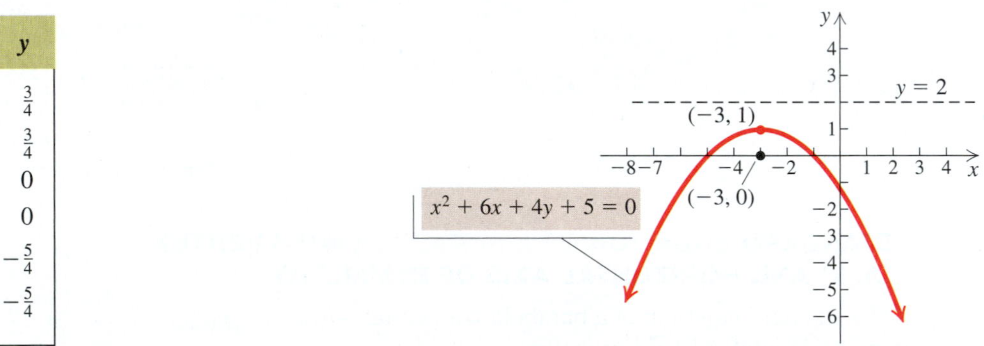

$y = \frac{1}{4}(-x^2 - 6x - 5)$

6, −9, 9, −6

We can check the graph on a graphing calculator using a squared viewing window, as shown at left. It might be necessary to solve for y first:

$$x^2 + 6x + 4y + 5 = 0$$

$$4y = -x^2 - 6x - 5$$

$$y = \tfrac{1}{4}(-x^2 - 6x - 5).$$

The hand-drawn graph appears to be correct.

Now Try Exercise 25.

EXAMPLE 4 For the parabola

$$y^2 - 2y - 8x - 31 = 0,$$

find the vertex, the focus, and the directrix. Then draw the graph.

Solution We first complete the square:

$$y^2 - 2y - 8x - 31 = 0$$

$$y^2 - 2y = 8x + 31 \quad \text{Adding } 8x \text{ and 31 on both sides}$$

$$y^2 - 2y + 1 = 8x + 31 + 1. \quad \text{Adding 1 on both sides to complete the square on the left side}$$

Then

$$\begin{aligned} y^2 - 2y + 1 &= 8x + 32 \\ (y-1)^2 &= 8(x+4) && \text{Factoring} \\ (y-1)^2 &= 4(2)[x-(-4)]. && \text{Writing standard form: } (y-k)^2 = 4p(x-h) \end{aligned}$$

We see that $h = -4$, $k = 1$, and $p = 2$, so we have the following:

Vertex (h, k): $(-4, 1)$;
Focus $(h + p, k)$: $(-4 + 2, 1)$, or $(-2, 1)$;
Directrix $x = h - p$: $x = -4 - 2$, or $x = -6$.

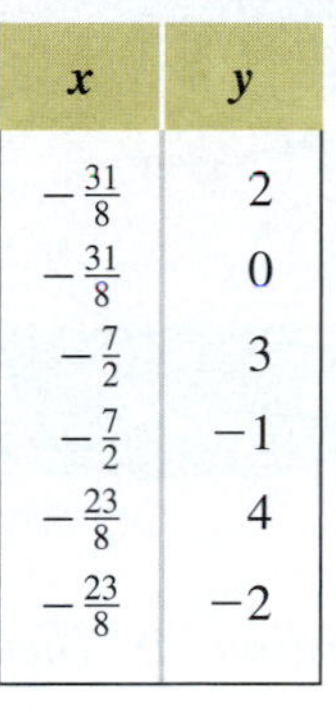

x	y
$-\frac{31}{8}$	2
$-\frac{31}{8}$	0
$-\frac{7}{2}$	3
$-\frac{7}{2}$	-1
$-\frac{23}{8}$	4
$-\frac{23}{8}$	-2

$(-4, 1)$, $(-2, 1)$, $x = -6$

$y^2 - 2y - 8x - 31 = 0$

When the equation of a parabola is written in standard form, we can use the Conics PARABOLA APP to graph it, as shown on the left below.

$(y - 1)^2 = 4(2)[x - (-4)]$

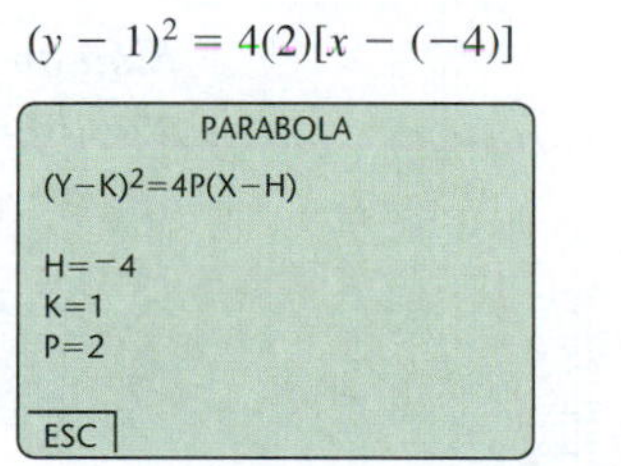

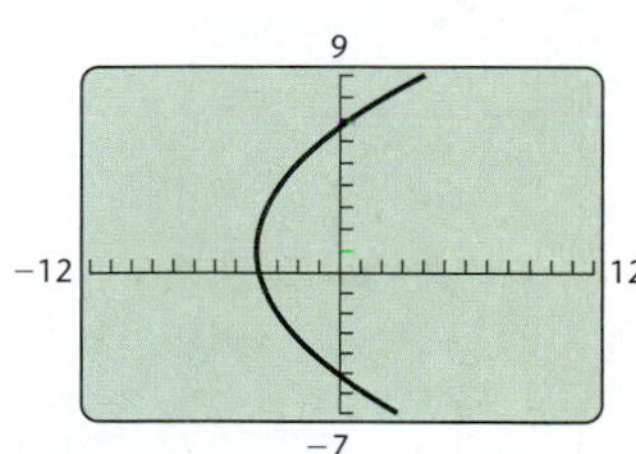

We can also draw the graph or check the hand-drawn graph on a graphing calculator by first solving the original equation for y using the quadratic formula.

Now Try Exercise 31.

Applications

Parabolas have many applications. For example, cross sections of car headlights, flashlights, and searchlights are parabolas. The bulb is located at the focus and light from that point is reflected outward parallel to the axis of symmetry. Satellite dishes and field microphones used at sporting events often have parabolic cross sections. Incoming radio waves or sound waves parallel to the axis are reflected into the focus.

Similarly, in solar cooking, a parabolic mirror is mounted on a rack with a cooking pot hung in the focal area. Incoming sun rays parallel to the axis are reflected into the focus, producing a temperature high enough for cooking.

10.1 Exercise Set

In Exercises 1–6, match the equation with one of the graphs (a)–(f) that follow.

a) 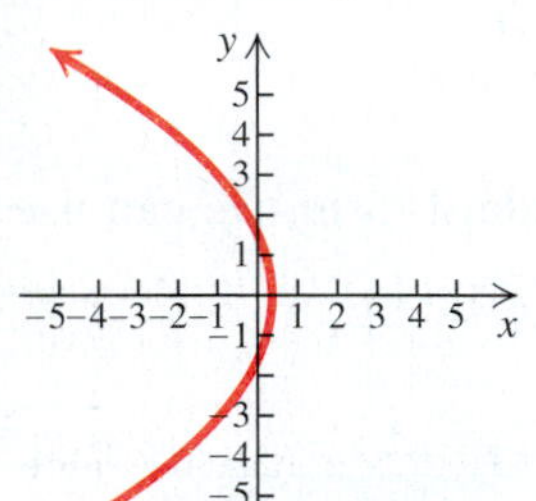

b) 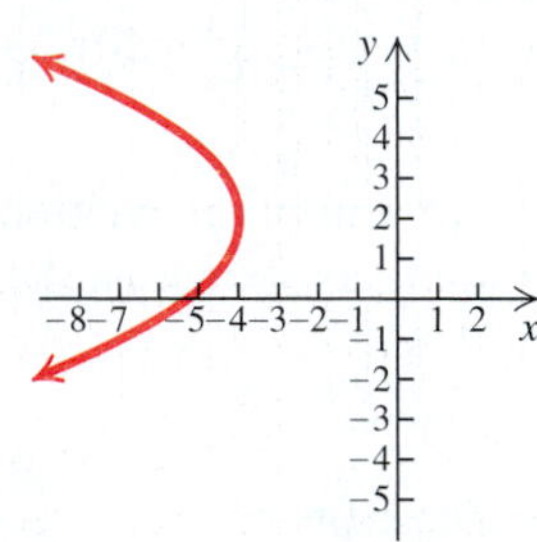

c) 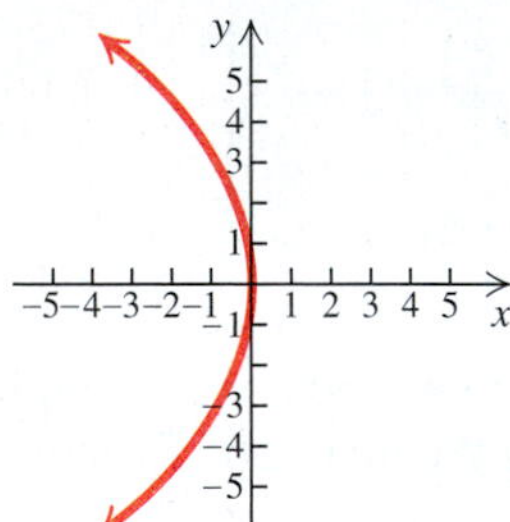

d) 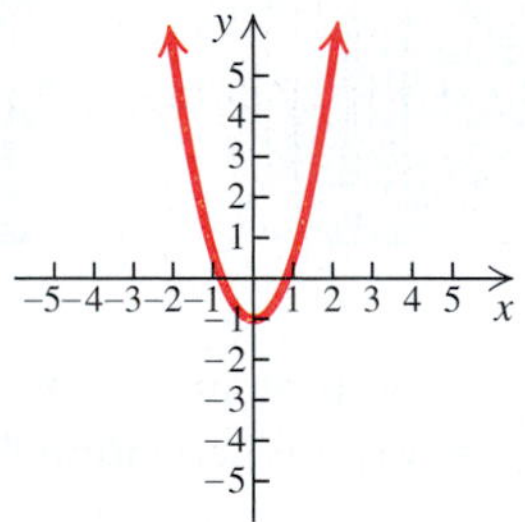

e) 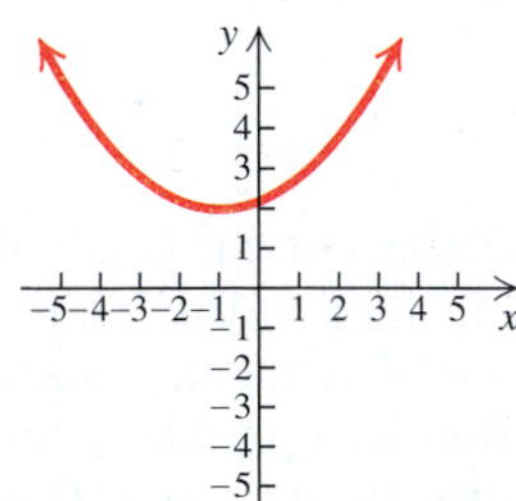

f) 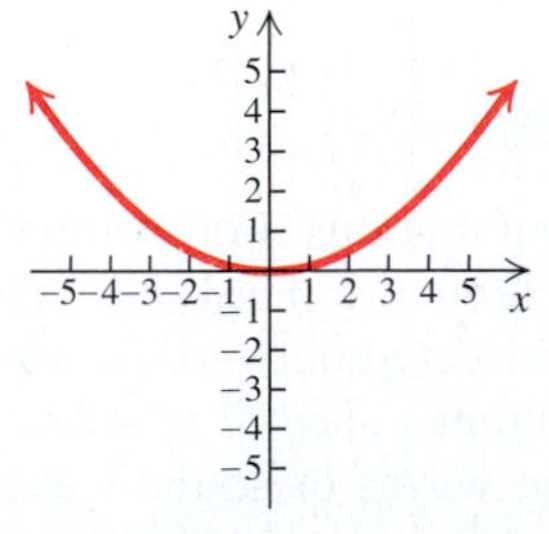

1. $x^2 = 8y$

2. $y^2 = -10x$

3. $(y - 2)^2 = -3(x + 4)$

4. $(x + 1)^2 = 5(y - 2)$

5. $13x^2 - 8y - 9 = 0$

6. $41x + 6y^2 = 12$

Find the vertex, the focus, and the directrix. Then draw the graph.

7. $x^2 = 20y$

8. $x^2 = 16y$

9. $y^2 = -6x$

10. $y^2 = -2x$

11. $x^2 - 4y = 0$

12. $y^2 + 4x = 0$

13. $x = 2y^2$

14. $y = \frac{1}{2}x^2$

Find an equation of a parabola satisfying the given conditions.

15. Vertex $(0, 0)$, focus $(-3, 0)$

16. Vertex $(0, 0)$, focus $(0, 10)$

17. Focus $(7, 0)$, directrix $x = -7$

18. Focus $\left(0, \frac{1}{4}\right)$, directrix $y = -\frac{1}{4}$

19. Focus $(0, -\pi)$, directrix $y = \pi$

20. Focus $(-\sqrt{2}, 0)$, directrix $x = \sqrt{2}$

21. Focus $(3, 2)$, directrix $x = -4$

22. Focus $(-2, 3)$, directrix $y = -3$

Find the vertex, the focus, and the directrix. Then draw the graph.

23. $(x + 2)^2 = -6(y - 1)$

24. $(y - 3)^2 = -20(x + 2)$

25. $x^2 + 2x + 2y + 7 = 0$

26. $y^2 + 6y - x + 16 = 0$

27. $x^2 - y - 2 = 0$

28. $x^2 - 4x - 2y = 0$

29. $y = x^2 + 4x + 3$

30. $y = x^2 + 6x + 10$

31. $y^2 - y - x + 6 = 0$

32. $y^2 + y - x - 4 = 0$

33. *Satellite Dish.* An engineer designs a satellite dish with a parabolic cross section. The dish is 15 ft wide at the opening, and the focus is placed 4 ft from the vertex.

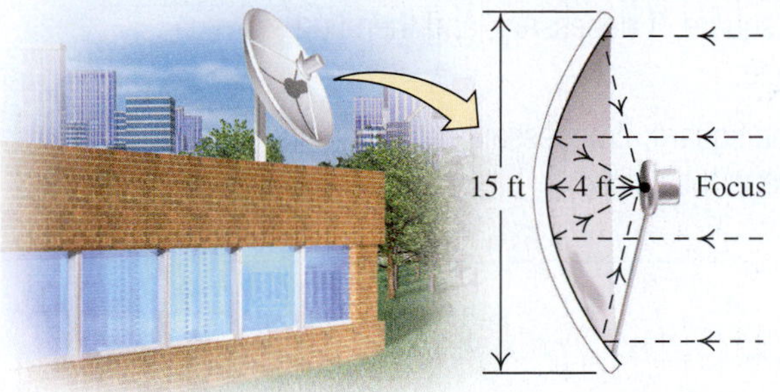

a) Position a coordinate system with the origin at the vertex and the x-axis on the parabola's axis of symmetry and find an equation of the parabola.

b) Find the depth of the satellite dish at the vertex.

34. *Flashlight Mirror.* A heavy-duty flashlight mirror has a parabolic cross section with diameter 6 in. and depth 1 in.

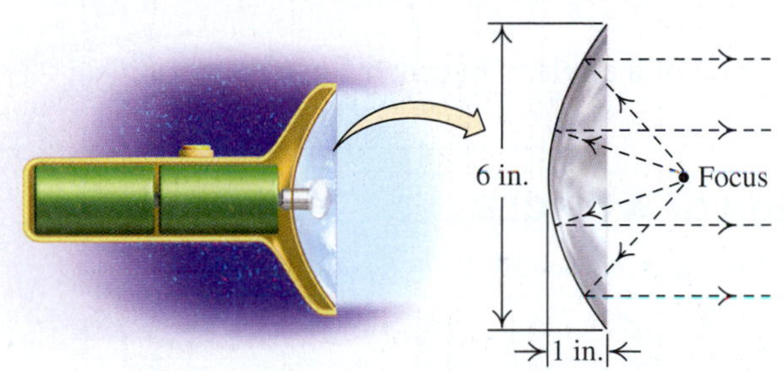

a) Position a coordinate system with the origin at the vertex and the x-axis on the parabola's axis of symmetry and find an equation of the parabola.

b) How far from the vertex should the bulb be positioned if it is to be placed at the focus?

35. *Ultrasound Receiver.* Information Unlimited designed and sells the Ultrasonic Receiver, which detects sounds unable to be heard by the human ear. The HT90P can detect mechanical and electrical sounds such as leaking gases, air, corona, and motor friction noises. It can also be used to hear bats, insects, and even beading water. The receiver has a parabolic cross section and is 2.625 in. deep. The focus is 3.287 in. from the vertex. (*Source*: Information Unlimited, Amherst, NH, Robert Iannini, President) Find the diameter of the outside edge of the receiver.

36. *Spotlight.* A spotlight has a parabolic cross section that is 4 ft wide at the opening and 1.5 ft deep at the vertex. How far from the vertex is the focus?

Skill Maintenance

Consider the following linear equations. Without graphing them, answer the questions below.

a) $y = 2x$ **b)** $y = \frac{1}{3}x + 5$
c) $y = -3x - 2$ **d)** $y = -0.9x + 7$
e) $y = -5x + 3$ **f)** $y = x + 4$
g) $8x - 4y = 7$ **h)** $3x + 6y = 2$

37. Which has/have x-intercept $\left(\frac{2}{3}, 0\right)$? **[1.1]**

38. Which has/have y-intercept $(0, 7)$? **[1.1], [1.4]**

39. Which slant up from left to right? **[1.3]**

40. Which has/have the least steep slant? **[1.3]**

41. Which has/have slope $\frac{1}{3}$? **[1.4]**

42. Which, if any, contain the point $(3, 7)$? **[1.1]**

43. Which, if any, are parallel? **[1.4]**

44. Which, if any, are perpendicular? **[1.4]**

Synthesis

45. Find an equation of the parabola with a vertical axis of symmetry and vertex $(-1, 2)$ and containing the point $(-3, 1)$.

46. Find an equation of a parabola with a horizontal axis of symmetry and vertex $(-2, 1)$ and containing the point $(-3, 5)$.

Use a graphing calculator to find the vertex, the focus, and the directrix of each of the following.

47. $4.5x^2 - 7.8x + 9.7y = 0$

48. $134.1y^2 + 43.4x - 316.6y - 122.4 = 0$

49. *Suspension Bridge.* The cables of a 200-ft portion of the roadbed of a suspension bridge are positioned as shown below. Vertical cables are to be spaced every 20 ft along this portion of the roadbed. Calculate the lengths of these vertical cables.

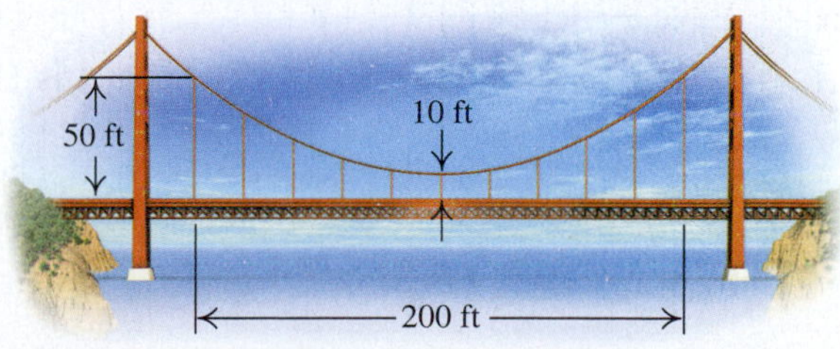

10.2 The Circle and the Ellipse

- Given an equation of a circle, complete the square, if necessary, and then find the center and the radius and graph the circle.
- Given an equation of an ellipse, complete the square, if necessary, and then find the center, the vertices, and the foci and graph the ellipse.

Circles

CIRCLES
REVIEW SECTION 1.1.

We can define a circle geometrically.

CIRCLE

A **circle** is the set of all points in a plane that are at a fixed distance from a fixed point (the **center**) in the plane.

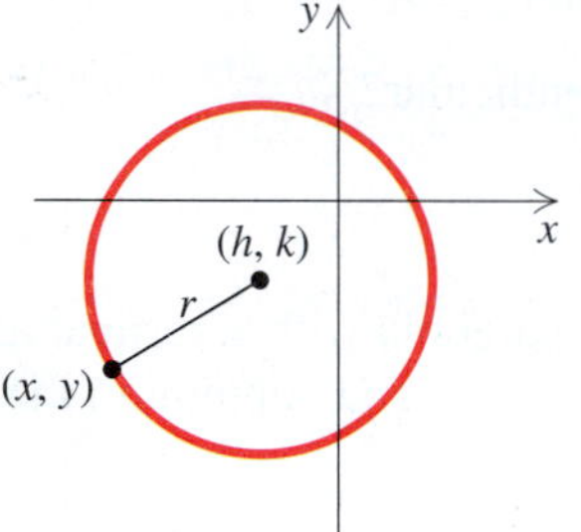

Recall the standard equation of a circle with center (h, k) and radius r.

STANDARD EQUATION OF A CIRCLE

The standard equation of a circle with center (h, k) and radius r is

$$(x - h)^2 + (y - k)^2 = r^2.$$

EXAMPLE 1 For the circle

$$x^2 + y^2 - 16x + 14y + 32 = 0,$$

find the center and the radius. Then graph the circle.

Solution First, we complete the square twice:

$$x^2 + y^2 - 16x + 14y + 32 = 0$$

$$x^2 - 16x \qquad + y^2 + 14y \qquad = -32$$

$$x^2 - 16x + 64 + y^2 + 14y + 49 = -32 + 64 + 49$$

$[\frac{1}{2}(-16)]^2 = (-8)^2 = 64$ and $(\frac{1}{2} \cdot 14)^2 = 7^2 = 49$; adding 64 and 49 on both sides to complete the square twice on the left side

$$(x - 8)^2 + (y + 7)^2 = 81$$

$$(x - 8)^2 + [y - (-7)]^2 = 9^2. \quad \text{Writing standard form}$$

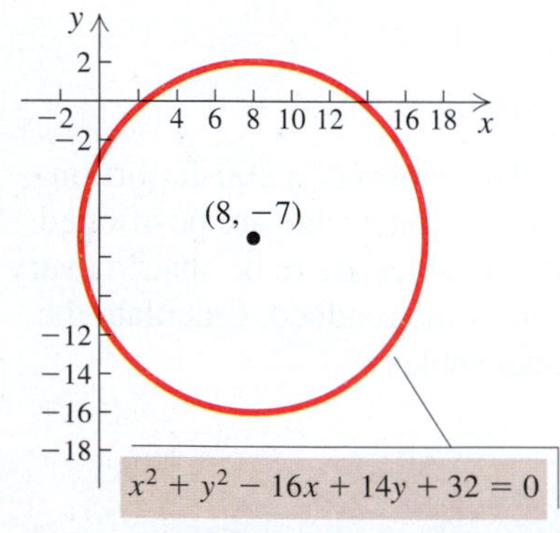

The center is $(8, -7)$ and the radius is 9. We graph the circle as shown at left.

To use a graphing calculator to graph the circle, it might be necessary to solve for y first. The original equation can be solved using the quadratic formula, or the standard form of the equation can be solved using the principle of square roots.

When we use the Conics CIRCLE APP on a graphing calculator to graph a circle, it is not necessary to write the equation in standard form or to solve it for y first. We enter the coefficients of x^2, y^2, x, and y and also the constant term when the equation is written in the form $ax^2 + ay^2 + bx + cy + d = 0$.

$x^2 + y^2 - 16x + 14y + 32 = 0$

CIRCLE
AX2+AY2+BX+CY+D=0

A=1
B=−16
C=14
D=32
ESC

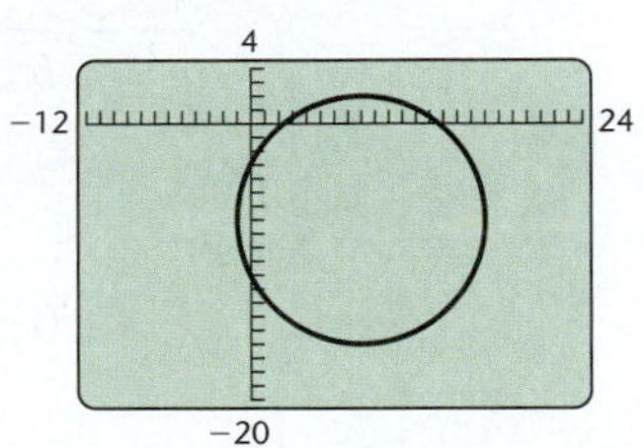

Some graphing calculators have a DRAW feature that provides a quick way to graph a circle when the center and the radius are known. This feature is described on p. 11.

Now Try Exercise 7.

Ellipses

We have studied two conic sections, the parabola and the circle. Now we turn our attention to a third, the *ellipse*.

> **ELLIPSE**
>
> An **ellipse** is the set of all points in a plane, the sum of whose distances from two fixed points (the **foci**) is constant. The **center** of an ellipse is the midpoint of the segment between the foci.

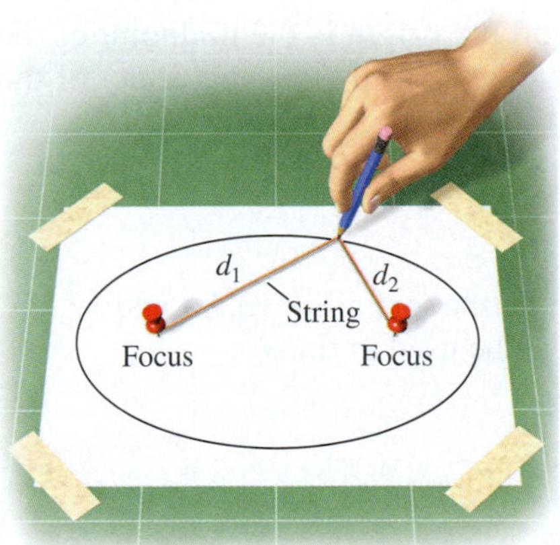

We can draw an ellipse by first placing two thumbtacks in a piece of cardboard, as shown at left. These are the foci (singular, *focus*). We then attach a piece of string to the tacks. Its length is the constant sum of the distances $d_1 + d_2$ from the foci to any point on the ellipse. Next, we trace a curve with a pencil held tight against the string. The figure traced is an ellipse.

Let's first consider the ellipse shown below with center at the origin. The points F_1 and F_2 are the foci. The segment $\overline{A'A}$ is the **major axis**, and the points A' and A are the **vertices**. The segment $\overline{B'B}$ is the **minor axis**, and the points B' and B are the ***y*-intercepts**. Note that the major axis of an ellipse is longer than the minor axis.

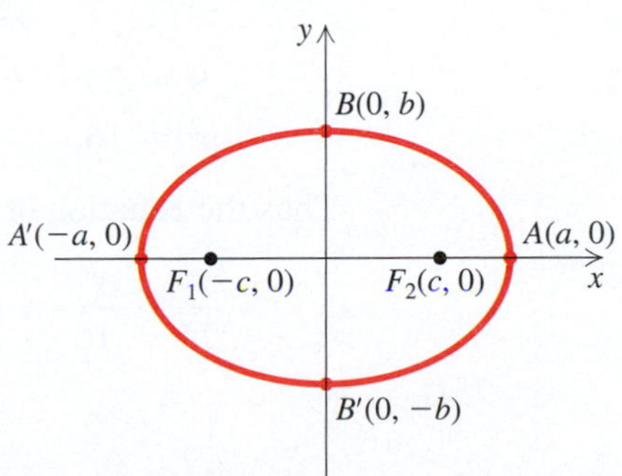

STANDARD EQUATION OF AN ELLIPSE WITH CENTER AT THE ORIGIN

Major Axis Horizontal

$$\frac{x^2}{a^2} + \frac{y^2}{b^2} = 1, \quad a > b > 0$$

Vertices: $(-a, 0)$, $(a, 0)$

y-intercepts: $(0, -b)$, $(0, b)$

Foci: $(-c, 0)$, $(c, 0)$, where $c^2 = a^2 - b^2$

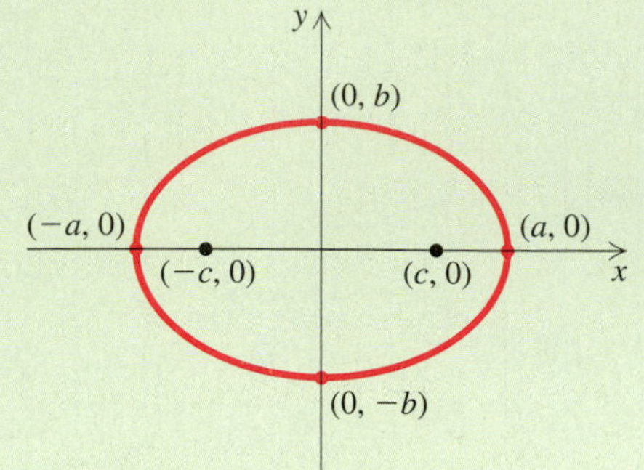

Major Axis Vertical

$$\frac{x^2}{b^2} + \frac{y^2}{a^2} = 1, \quad a > b > 0$$

Vertices: $(0, -a)$, $(0, a)$

x-intercepts: $(-b, 0)$, $(b, 0)$

Foci: $(0, -c)$, $(0, c)$, where $c^2 = a^2 - b^2$

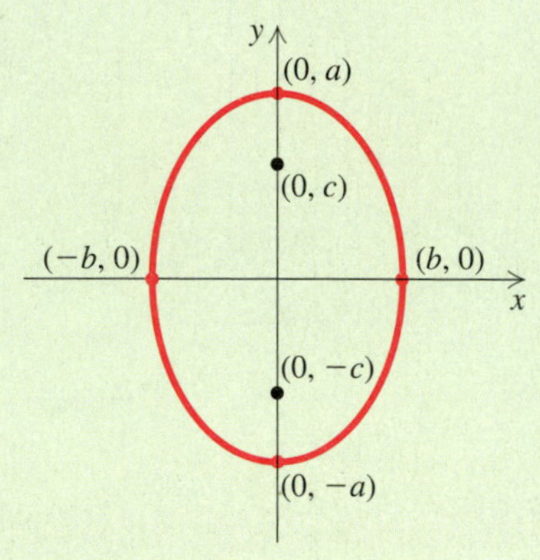

EXAMPLE 2 Find the standard equation of the ellipse with vertices $(-5, 0)$ and $(5, 0)$ and foci $(-3, 0)$ and $(3, 0)$. Then graph the ellipse.

Solution Since the foci are on the x-axis and the origin is the midpoint of the segment between them, the major axis is horizontal and $(0, 0)$ is the center of the ellipse. Thus the equation is of the form

$$\frac{x^2}{a^2} + \frac{y^2}{b^2} = 1.$$

Since the vertices are $(-5, 0)$ and $(5, 0)$ and the foci are $(-3, 0)$ and $(3, 0)$, we know that $a = 5$ and $c = 3$. These values can be used to find b^2:

$$c^2 = a^2 - b^2$$
$$3^2 = 5^2 - b^2$$
$$9 = 25 - b^2$$
$$b^2 = 16.$$

Thus the equation of the ellipse is

$$\frac{x^2}{25} + \frac{y^2}{16} = 1, \quad \text{or} \quad \frac{x^2}{5^2} + \frac{y^2}{4^2} = 1.$$

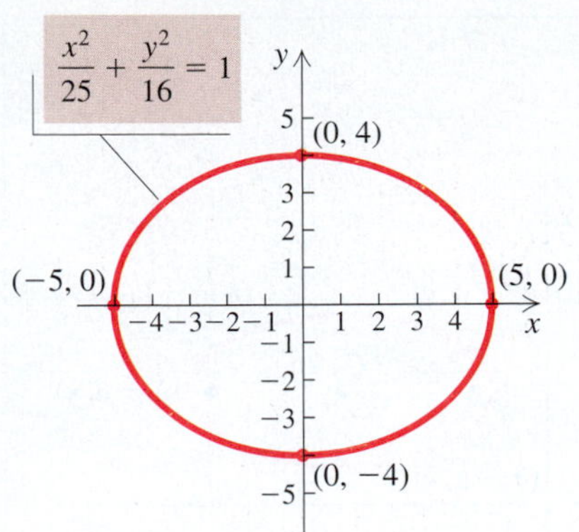

To graph the ellipse, we plot the vertices $(-5, 0)$ and $(5, 0)$. Since $b^2 = 16$, we know that $b = 4$ and the y-intercepts are $(0, -4)$ and $(0, 4)$. We plot these points as well and connect the four points that we have plotted with a smooth curve.

When the equation of an ellipse is written in standard form, we can use the Conics ELLIPSE APP on a graphing calculator to graph it. Note that the center is $(0, 0)$, so $H = 0$ and $K = 0$.

$$\frac{x^2}{5^2} + \frac{y^2}{4^2} = 1$$

ELLIPSE
$\frac{(X-H)^2}{A^2} + \frac{(Y-K)^2}{B^2} = 1$
A=5
B=4
H=0
K=0
ESC

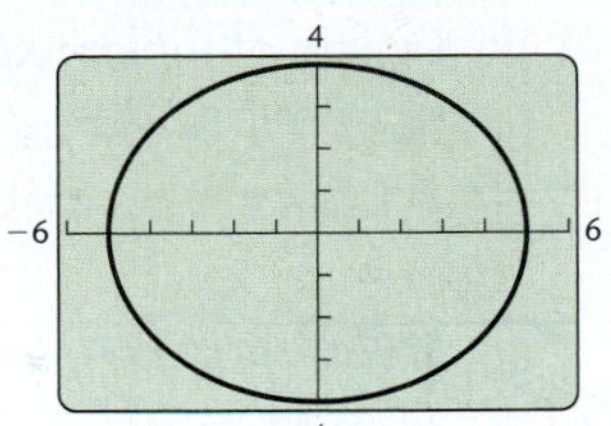

We can also draw the graph or check the hand-drawn graph on a graphing calculator by first solving for y.

Now Try Exercise 31.

EXAMPLE 3 For the ellipse

$$9x^2 + 4y^2 = 36,$$

find the vertices and the foci. Then draw the graph.

Solution We first find standard form:

$$9x^2 + 4y^2 = 36$$

$$\frac{9x^2}{36} + \frac{4y^2}{36} = \frac{36}{36} \quad \text{Dividing by 36 on both sides to get 1 on the right side}$$

$$\frac{x^2}{4} + \frac{y^2}{9} = 1$$

$$\frac{x^2}{2^2} + \frac{y^2}{3^2} = 1. \quad \text{Writing standard form}$$

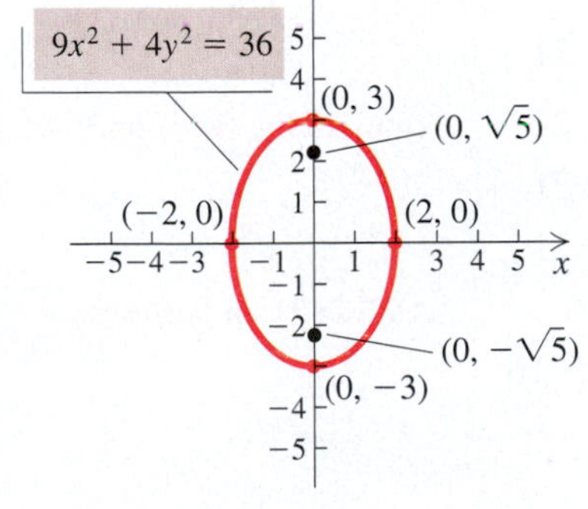

Thus, $a = 3$ and $b = 2$. The major axis is vertical, so the vertices are $(0, -3)$ and $(0, 3)$. Since we know that $c^2 = a^2 - b^2$, we have $c^2 = 3^2 - 2^2 = 9 - 4 = 5$, so $c = \sqrt{5}$ and the foci are $(0, -\sqrt{5})$ and $(0, \sqrt{5})$.

To graph the ellipse, we plot the vertices. Note also that since $b = 2$, the x-intercepts are $(-2, 0)$ and $(2, 0)$. We plot these points as well and connect the four points we have plotted with a smooth curve.

Now Try Exercise 25.

If the center of an ellipse is not at the origin but at some point (h, k), then we can think of an ellipse with center at the origin being translated horizontally $|h|$ units and vertically $|k|$ units.

STANDARD EQUATION OF AN ELLIPSE WITH CENTER AT (*h*, *k*)

Major Axis Horizontal

$$\frac{(x-h)^2}{a^2} + \frac{(y-k)^2}{b^2} = 1, \quad a > b > 0$$

Vertices: $(h-a, k)$, $(h+a, k)$

Length of minor axis: $2b$

Foci: $(h-c, k)$, $(h+c, k)$, where $c^2 = a^2 - b^2$

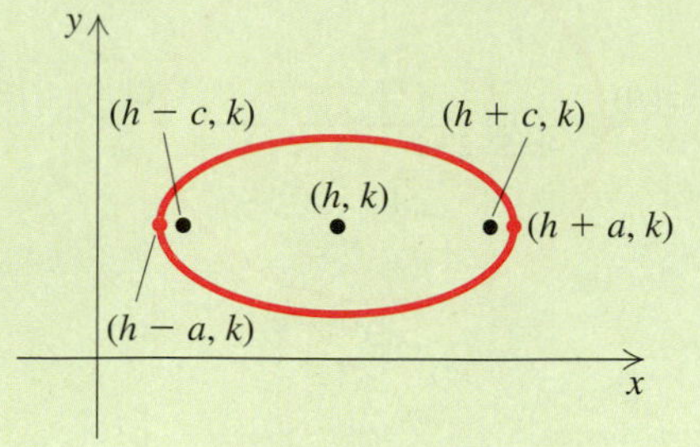

Major Axis Vertical

$$\frac{(x-h)^2}{b^2} + \frac{(y-k)^2}{a^2} = 1, \quad a > b > 0$$

Vertices: $(h, k-a)$, $(h, k+a)$

Length of minor axis: $2b$

Foci: $(h, k-c)$, $(h, k+c)$, where $c^2 = a^2 - b^2$

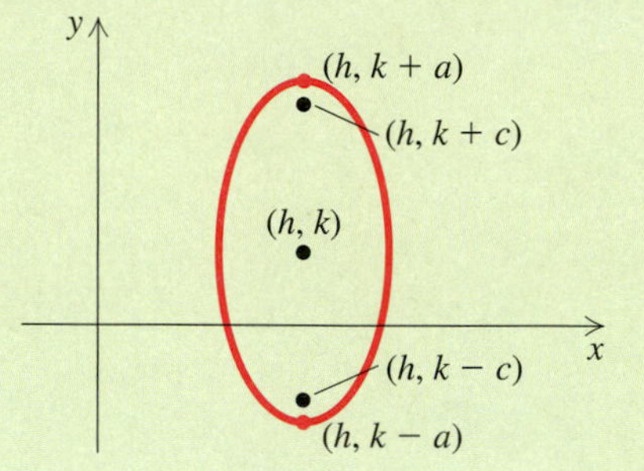

EXAMPLE 4 For the ellipse

$$4x^2 + y^2 + 24x - 2y + 21 = 0,$$

find the center, the vertices, and the foci. Then draw the graph.

Solution First, we complete the square twice to get standard form:

$$\begin{aligned}
4x^2 + y^2 + 24x - 2y + 21 &= 0 \\
4(x^2 + 6x \qquad) + (y^2 - 2y \qquad) &= -21 \\
4(x^2 + 6x + 9 - 9) + (y^2 - 2y + 1 - 1) &= -21 && \text{Completing the square twice} \\
4(x^2 + 6x + 9) + 4(-9) + (y^2 - 2y + 1) + (-1) &= -21 \\
4(x+3)^2 - 36 + (y-1)^2 - 1 &= -21 \\
4(x+3)^2 + (y-1)^2 &= 16 && \text{Adding 37 on both sides} \\
\tfrac{1}{16}[4(x+3)^2 + (y-1)^2] &= \tfrac{1}{16}\cdot 16 \\
\frac{(x+3)^2}{4} + \frac{(y-1)^2}{16} &= 1 \\
\frac{[x-(-3)]^2}{2^2} + \frac{(y-1)^2}{4^2} &= 1. && \text{Writing standard form}
\end{aligned}$$

The center is $(-3, 1)$. Note that $a = 4$ and $b = 2$. The major axis is vertical, so the vertices are 4 units above and below the center:

$$(-3, 1+4) \text{ and } (-3, 1-4), \quad \text{or} \quad (-3, 5) \text{ and } (-3, -3).$$

We know that $c^2 = a^2 - b^2$, so $c^2 = 4^2 - 2^2 = 16 - 4 = 12$ and $c = \sqrt{12}$, or $2\sqrt{3}$. Then the foci are $2\sqrt{3}$ units above and below the center:

$$(-3, 1 + 2\sqrt{3}) \quad \text{and} \quad (-3, 1 - 2\sqrt{3}).$$

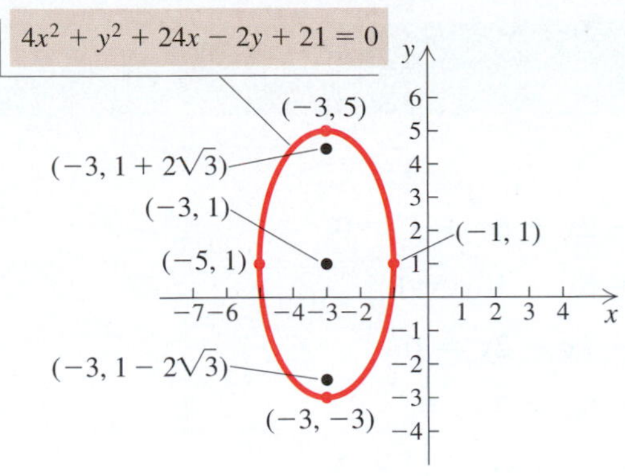

To graph the ellipse, we plot the vertices. Note also that since $b = 2$, two other points on the graph are the endpoints of the minor axis, 2 units right and left of the center:

$$(-3 + 2, 1) \quad \text{and} \quad (-3 - 2, 1),$$

or

$$(-1, 1) \quad \text{and} \quad (-5, 1).$$

We plot these points as well and connect the four points with a smooth curve, as shown at left.

When the equation of an ellipse is written in standard form, we can use the Conics ELLIPSE APP to graph it.

$$\frac{[x - (-3)]^2}{2^2} + \frac{(y - 1)^2}{4^2} = 1$$

ELLIPSE
$\frac{(X-H)^2}{B^2} + \frac{(Y-K)^2}{A^2} = 1$
A=4
B=2
H=−3
K=1
ESC

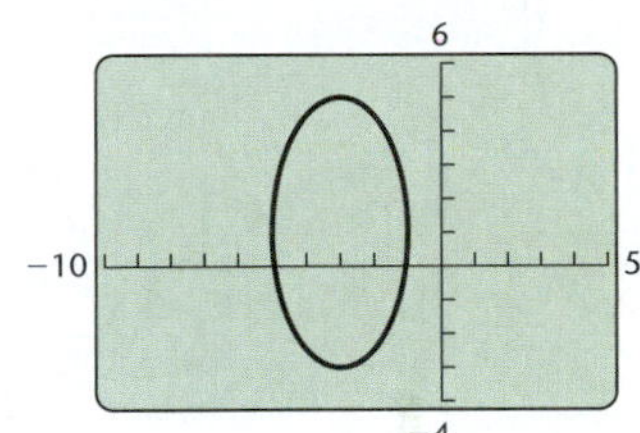

Now Try Exercise 43.

Applications

An exciting medical application of an ellipse is a device called a *lithotripter*. One type of this device uses electromagnetic technology to generate a shock wave to pulverize kidney stones. The wave originates at one focus of an ellipse and is reflected to the kidney stone, which is positioned at the other focus. Recovery time following the use of this technique is much shorter than with conventional surgery.

Ellipses have many other applications. Planets travel around the sun in elliptical orbits with the sun at one focus, for example, and satellites travel around the earth in elliptical orbits as well.

A room with an ellipsoidal ceiling is known as a *whispering gallery*. In such a room, a word whispered at one focus can be clearly heard at the other. Whispering galleries are found in the rotunda of the Capitol Building in Washington, D.C., and in St. Paul's Cathedral in London.

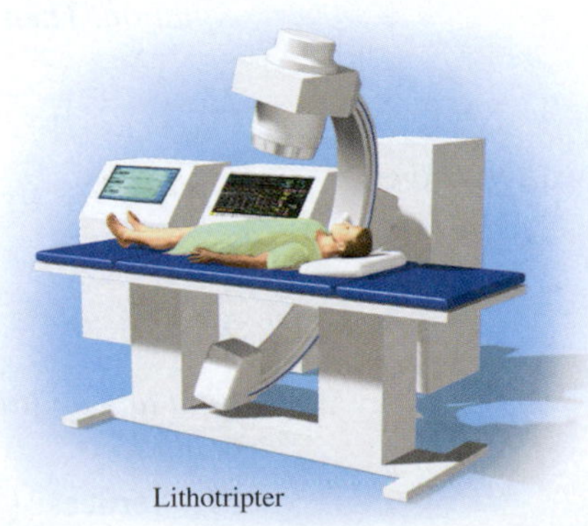
Lithotripter

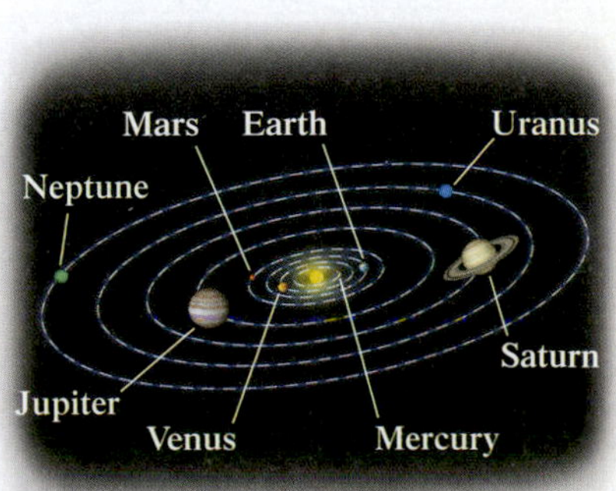

10.2 Exercise Set

In Exercises 1–6, match the equation with one of the graphs (a)–(f) that follow.

a) 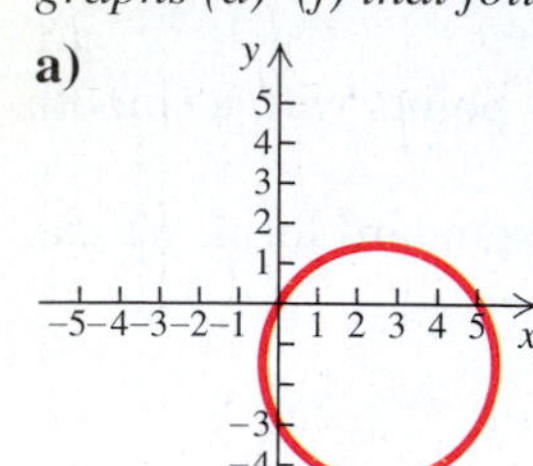

b) 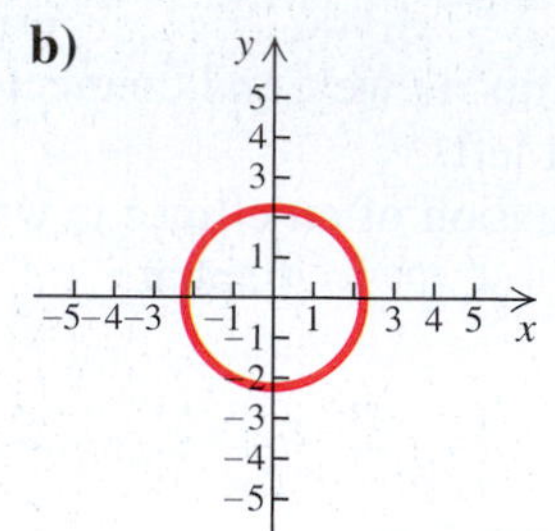

c) 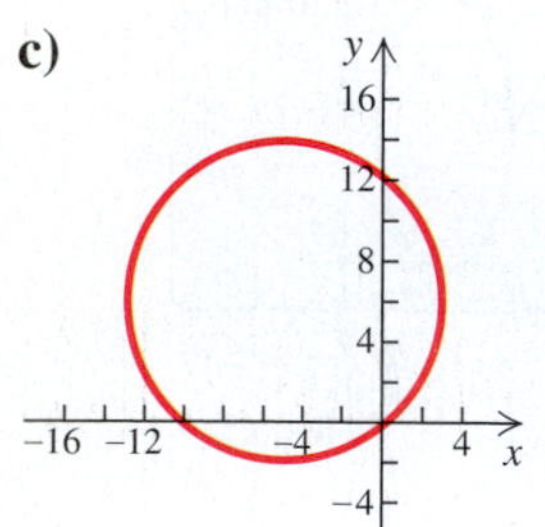

d) 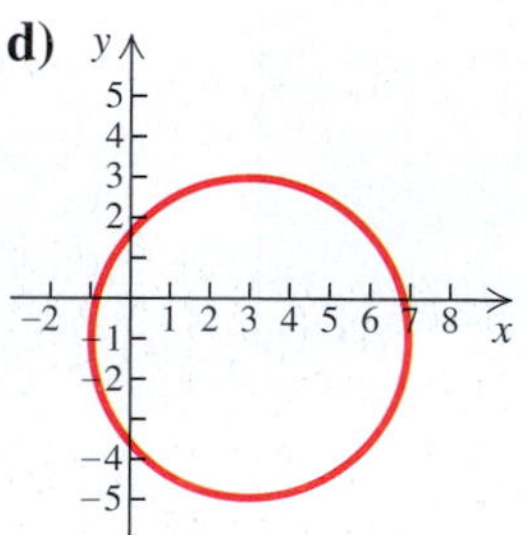

e) 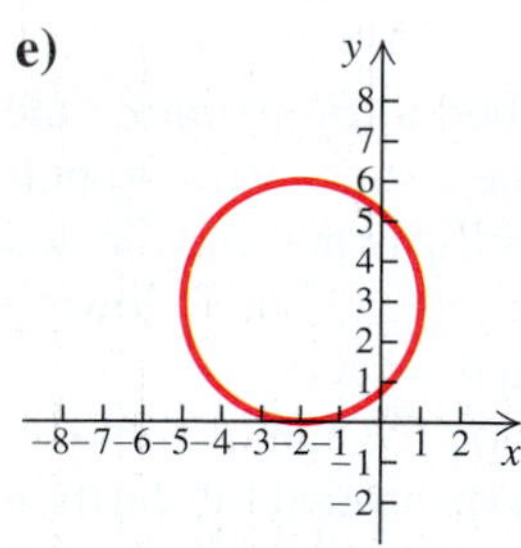

f) 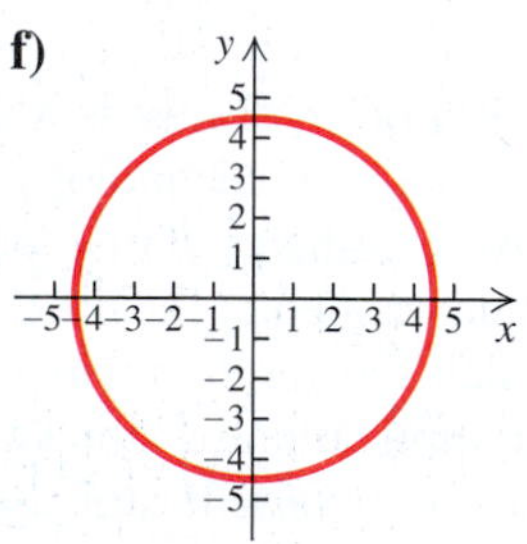

1. $x^2 + y^2 = 5$

2. $y^2 = 20 - x^2$

3. $x^2 + y^2 - 6x + 2y = 6$

4. $x^2 + y^2 + 10x - 12y = 3$

5. $x^2 + y^2 - 5x + 3y = 0$

6. $x^2 + 4x - 2 = 6y - y^2 - 6$

Find the center and the radius of the circle with the given equation. Then draw the graph.

7. $x^2 + y^2 - 14x + 4y = 11$

8. $x^2 + y^2 + 2x - 6y = -6$

9. $x^2 + y^2 + 6x - 2y = 6$

10. $x^2 + y^2 - 4x + 2y = 4$

11. $x^2 + y^2 + 4x - 6y - 12 = 0$

12. $x^2 + y^2 - 8x - 2y - 19 = 0$

13. $x^2 + y^2 - 6x - 8y + 16 = 0$

14. $x^2 + y^2 - 2x + 6y + 1 = 0$

15. $x^2 + y^2 + 6x - 10y = 0$

16. $x^2 + y^2 - 7x - 2y = 0$

17. $x^2 + y^2 - 9x = 7 - 4y$

18. $y^2 - 6y - 1 = 8x - x^2 + 3$

In Exercises 19–22, match the equation with one of the graphs (a)–(d) that follow.

a) 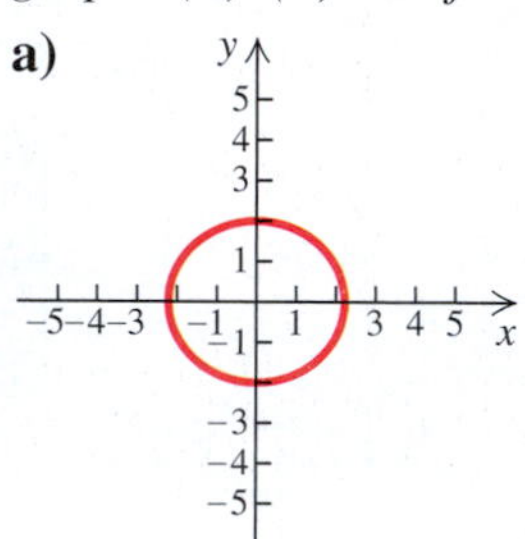

b) 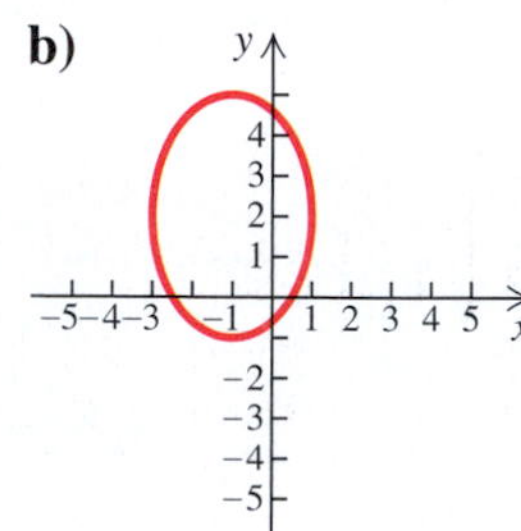

c) 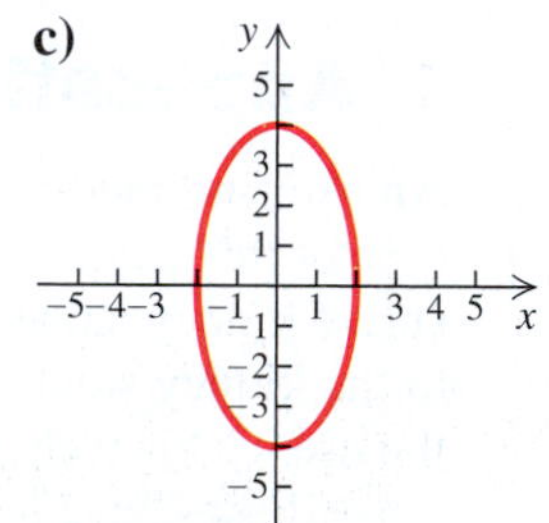

d) 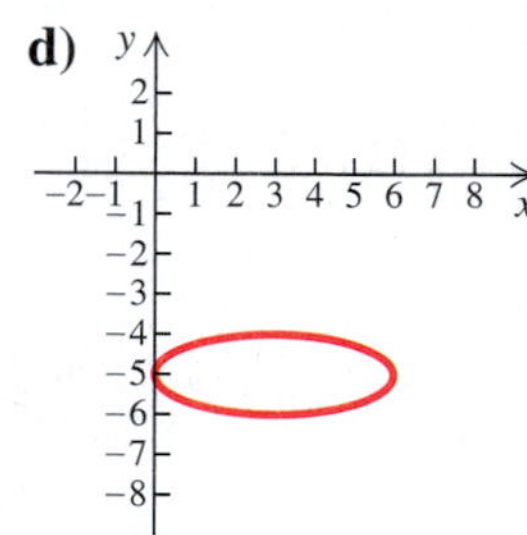

19. $16x^2 + 4y^2 = 64$

20. $4x^2 + 5y^2 = 20$

21. $x^2 + 9y^2 - 6x + 90y = -225$

22. $9x^2 + 4y^2 + 18x - 16y = 11$

Find the vertices and the foci of the ellipse with the given equation. Then draw the graph.

23. $\frac{x^2}{4} + \frac{y^2}{1} = 1$

24. $\frac{x^2}{25} + \frac{y^2}{36} = 1$

25. $16x^2 + 9y^2 = 144$

26. $9x^2 + 4y^2 = 36$

27. $2x^2 + 3y^2 = 6$

28. $5x^2 + 7y^2 = 35$

29. $4x^2 + 9y^2 = 1$

30. $25x^2 + 16y^2 = 1$

Find an equation of an ellipse satisfying the given conditions.

31. Vertices: $(-7, 0)$ and $(7, 0)$;
foci: $(-3, 0)$ and $(3, 0)$

32. Vertices: $(0, -6)$ and $(0, 6)$;
foci: $(0, -4)$ and $(0, 4)$

33. Vertices: $(0, -8)$ and $(0, 8)$; length of minor axis: 10

34. Vertices: $(-5, 0)$ and $(5, 0)$; length of minor axis: 6

35. Foci: $(-2, 0)$ and $(2, 0)$; length of major axis: 6

36. Foci: $(0, -3)$ and $(0, 3)$; length of major axis: 10

Find the center, the vertices, and the foci of the ellipse. Then draw the graph.

37. $\frac{(x-1)^2}{9} + \frac{(y-2)^2}{4} = 1$

38. $\frac{(x-1)^2}{1} + \frac{(y-2)^2}{4} = 1$

39. $\frac{(x+3)^2}{25} + \frac{(y-5)^2}{36} = 1$

40. $\frac{(x-2)^2}{16} + \frac{(y+3)^2}{25} = 1$

41. $3(x+2)^2 + 4(y-1)^2 = 192$

42. $4(x-5)^2 + 3(y-4)^2 = 48$

43. $4x^2 + 9y^2 - 16x + 18y - 11 = 0$

44. $x^2 + 2y^2 - 10x + 8y + 29 = 0$

45. $4x^2 + y^2 - 8x - 2y + 1 = 0$

46. $9x^2 + 4y^2 + 54x - 8y + 49 = 0$

*The **eccentricity** of an ellipse is defined as $e = c/a$. For an ellipse, $0 < c < a$, so $0 < e < 1$. When e is close to 0, an ellipse appears to be nearly circular. When e is close to 1, an ellipse is very flat.*

47. Observe the shapes of the ellipses in Examples 2 and 4. Which ellipse has the smaller eccentricity? Confirm your answer by computing the eccentricity of each ellipse.

48. Which ellipse has the smaller eccentricity? (Assume that the coordinate systems have the same scale.)

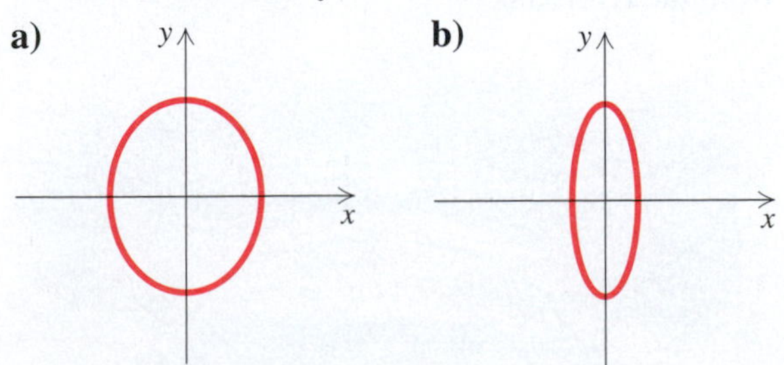

49. Find an equation of an ellipse with vertices $(0, -4)$ and $(0, 4)$ and $e = \frac{1}{4}$.

50. Find an equation of an ellipse with vertices $(-3, 0)$ and $(3, 0)$ and $e = \frac{7}{10}$.

51. *The Ellipse.* The lighting of the National Christmas Tree located on the Ellipse, a large grassy area south of the White House, marks the beginning of the holiday season in Washington, D.C. This area of the lawn is actually an ellipse with major axis of length 1048 ft and minor axis of length 898 ft. Assuming that a coordinate system is superimposed on the area in such a way that the center is at the origin and the major and minor axes are on the x- and y-axes of the coordinate system, respectively, find an equation of the ellipse.

52. *Bridge Supports.* The bridge support shown in the following figure is the top half of an ellipse. Assuming that a coordinate system is super-imposed on the drawing in such a way that point Q, the center of the ellipse, is at the origin, find an equation of the ellipse.

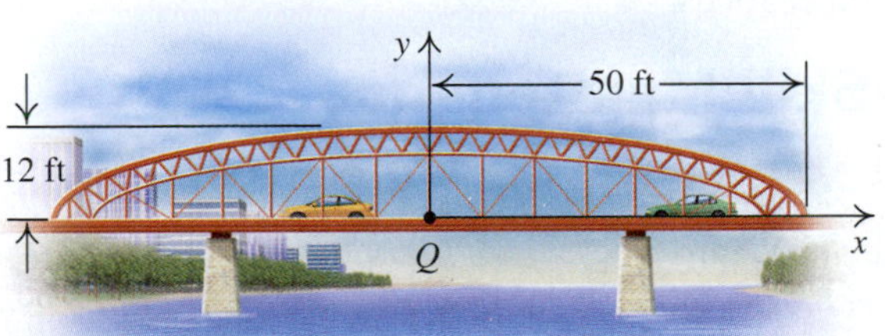

53. *Whispering Gallery.* A whispering gallery, often elliptical in shape, has acoustic properties such that a whisper made at one point can be heard at other distant points. A science museum is designing a new exhibit hall that will illustrate a whispering gallery. The hall will be 90 ft in length with the ceiling 30 ft high at the center. How far are the foci from the center of the ellipse? Round to the nearest tenth of a foot.

54. *Whispering Gallery.* An art museum is adding a new exhibit room in the shape of an ellipse. The director wants to mark the foci so that a tour guide can stand at one focus and without speaking loudly can be clearly heard by a group touring the museum. If the room is 64 ft long and each focus is 5 ft from the outside wall along the major axis, how high is the ceiling? Round to the nearest tenth of a foot.

55. *Carpentry.* A carpenter is cutting a 3-ft by 4-ft elliptical sign from a 3-ft by 4-ft piece of plywood. The ellipse will be drawn using a string attached to the board at the foci of the ellipse.

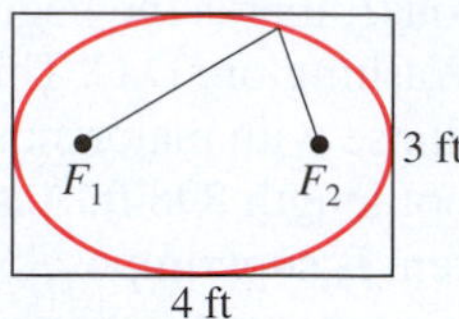

a) How far from the ends of the board should the string be attached?

b) How long should the string be?

56. *The Earth's Orbit.* The maximum distance of the earth from the sun is 9.3×10^7 mi. The minimum distance is 9.1×10^7 mi. The sun is at one focus of the elliptical orbit. Find the distance from the sun to the other focus.

Skill Maintenance

Vocabulary Reinforcement

In each of Exercises 57–64, fill in the blank with the correct term. Some of the given choices will not be used.

piecewise function	ellipse
linear equation	midpoint
factor	distance
remainder	one real-number solution
solution	two different real-number solutions
zero	two different imaginary-number solutions
x-intercept	
y-intercept	
parabola	
circle	

57. The ____________ between two points (x_1, y_1) and (x_2, y_2) is given by $\left(\frac{x_1 + x_2}{2}, \frac{y_1 + y_2}{2}\right)$. [1.1]

58. An input c of a function f is a(n) ____________ of the function if $f(c) = 0$. [1.5]

59. A(n) ____________ of the graph of an equation is a point $(0, b)$. [1.1]

60. For a quadratic equation $ax^2 + bx + c = 0$, if $b^2 - 4ac > 0$, the equation has ____________. [3.2]

61. Given a polynomial $f(x)$, then $f(c)$ is the ____________ that would be obtained by dividing $f(x)$ by $x - c$. [4.3]

62. A(n) ____________ is the set of all points in a plane the sum of whose distances from two fixed points is constant. [10.2]

63. A(n) ____________ is the set of all points in a plane equidistant from a fixed line and a fixed point not on the line. [10.1]

64. A(n) ____________ is the set of all points in a plane that are at a fixed distance from a fixed point in the plane. [10.2]

Synthesis

Find an equation of an ellipse satisfying the given conditions.

65. Vertices: $(3, -4)$, $(3, 6)$;
endpoints of minor axis: $(1, 1)$, $(5, 1)$

66. Vertices: $(-1, -1)$, $(-1, 5)$;
endpoints of minor axis: $(-3, 2)$, $(1, 2)$

67. Vertices: $(-3, 0)$ and $(3, 0)$;
passing through $\left(2, \frac{22}{3}\right)$

68. Center: $(-2, 3)$; major axis vertical;
length of major axis: 4;
length of minor axis: 1

Use a graphing calculator to find the center and the vertices of each of the following.

69. $4x^2 + 9y^2 - 16.025x + 18.0927y - 11.346 = 0$

70. $9x^2 + 4y^2 + 54.063x - 8.016y + 49.872 = 0$

71. *Bridge Arch.* A bridge with a semielliptical arch spans a river as shown here. What is the clearance 6 ft from the riverbank?

10.3 The Hyperbola

› Given an equation of a hyperbola, complete the square, if necessary, and then find the center, the vertices, and the foci and graph the hyperbola.

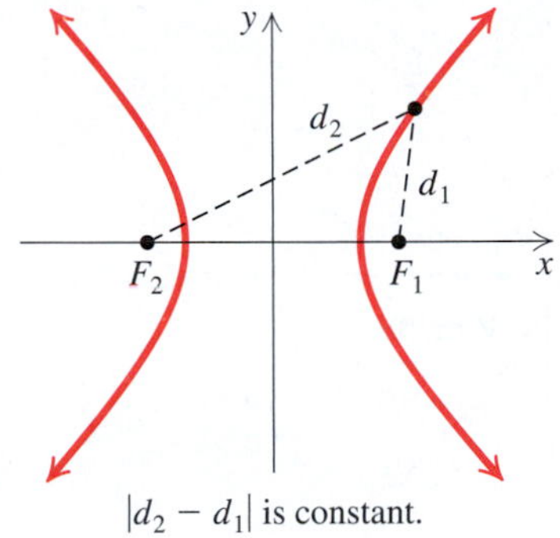

$|d_2 - d_1|$ is constant.

The last type of conic section that we will study is the *hyperbola*.

HYPERBOLA

A **hyperbola** is the set of all points in a plane for which the absolute value of the difference of the distances from two fixed points (the **foci**) is constant. The midpoint of the segment between the foci is the **center** of the hyperbola.

› Standard Equations of Hyperbolas

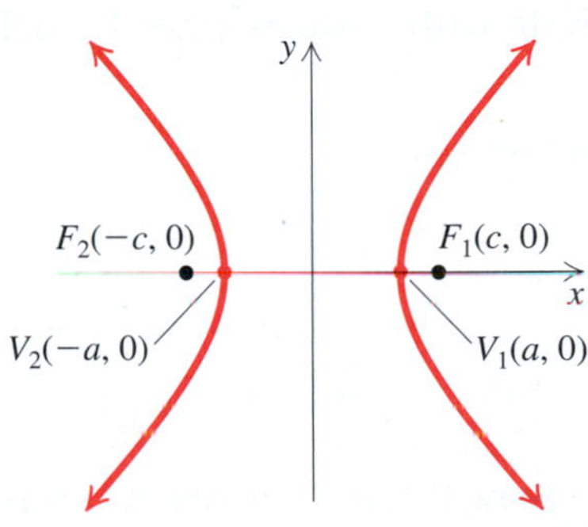

We first consider the equation of a hyperbola with center at the origin. In the figure at left, F_1 and F_2 are the foci. The segment $\overline{V_2V_1}$ is the **transverse axis,** and the points V_2 and V_1 are the **vertices**.

STANDARD EQUATION OF A HYPERBOLA WITH CENTER AT THE ORIGIN

Transverse Axis Horizontal

$$\frac{x^2}{a^2} - \frac{y^2}{b^2} = 1$$

Vertices: $(-a, 0), (a, 0)$

Foci: $(-c, 0), (c, 0)$,
where $c^2 = a^2 + b^2$

Transverse Axis Vertical

$$\frac{y^2}{a^2} - \frac{x^2}{b^2} = 1$$

Vertices: $(0, -a), (0, a)$

Foci: $(0, -c), (0, c)$,
where $c^2 = a^2 + b^2$

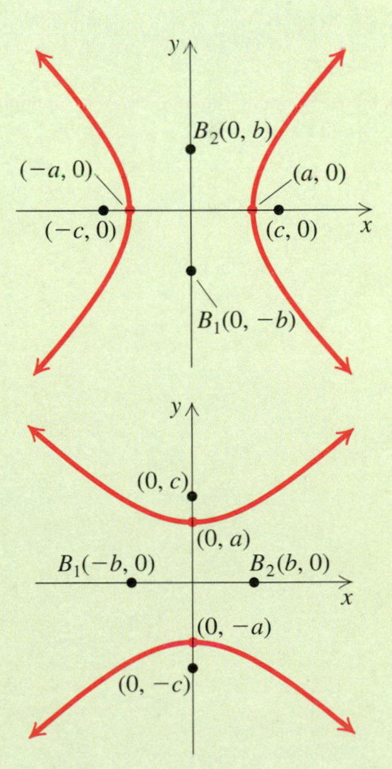

The segment $\overline{B_1B_2}$ is the **conjugate axis** of the hyperbola.

To graph a hyperbola with a horizontal transverse axis, it is helpful to begin by graphing the lines $y = -(b/a)x$ and $y = (b/a)x$. These are the **asymptotes** of the hyperbola. For a hyperbola with a vertical transverse axis, the asymptotes are $y = -(a/b)x$ and $y = (a/b)x$. As $|x|$ gets larger and larger, the graph of the hyperbola gets closer and closer to the asymptotes.

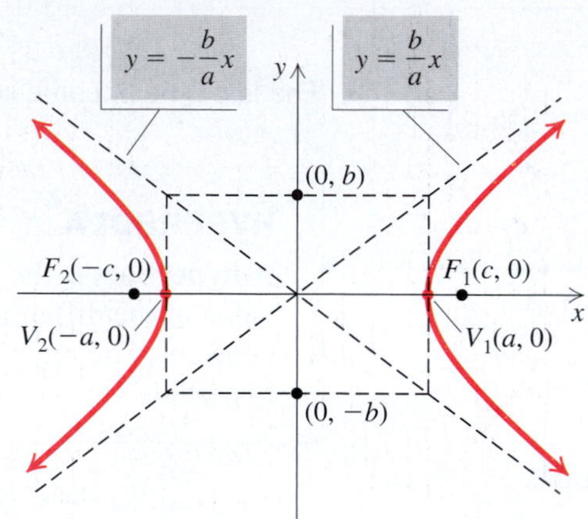

EXAMPLE 1 Find an equation of the hyperbola with vertices $(0, -4)$ and $(0, 4)$ and foci $(0, -6)$ and $(0, 6)$.

Solution We know that $a = 4$ and $c = 6$. We find b^2:

$$c^2 = a^2 + b^2$$
$$6^2 = 4^2 + b^2$$
$$36 = 16 + b^2$$
$$20 = b^2.$$

Since the vertices and the foci are on the y-axis, we know that the transverse axis is vertical. We can now write the equation of the hyperbola:

$$\frac{y^2}{16} - \frac{x^2}{20} = 1. \qquad \frac{y^2}{a^2} - \frac{x^2}{b^2} = 1$$

Now Try Exercise 7.

EXAMPLE 2 For the hyperbola given by

$$9x^2 - 16y^2 = 144,$$

find the vertices, the foci, and the asymptotes. Then graph the hyperbola.

Solution First, we find standard form:

$$9x^2 - 16y^2 = 144$$
$$\frac{1}{144}(9x^2 - 16y^2) = \frac{1}{144} \cdot 144 \qquad \text{Multiplying by } \tfrac{1}{144} \text{ to get 1 on the right side}$$
$$\frac{x^2}{16} - \frac{y^2}{9} = 1$$
$$\frac{x^2}{4^2} - \frac{y^2}{3^2} = 1. \qquad \text{Writing standard form}$$

The hyperbola has a horizontal transverse axis, so the vertices are $(-a, 0)$ and $(a, 0)$, or $(-4, 0)$ and $(4, 0)$. From the standard form of the equation, we know that $a^2 = 4^2$, or 16, and $b^2 = 3^2$, or 9. We find the foci:

$$c^2 = a^2 + b^2$$
$$c^2 = 16 + 9$$
$$c^2 = 25$$
$$c = 5.$$

Thus the foci are $(-5, 0)$ and $(5, 0)$.

Next, we find the asymptotes:

$$y = -\frac{b}{a}x = -\frac{3}{4}x \quad \text{and} \quad y = \frac{b}{a}x = \frac{3}{4}x.$$

To draw the graph, we sketch the asymptotes first. This is easily done by drawing the rectangle with horizontal sides passing through $(0, 3)$ and $(0, -3)$ and vertical sides through $(4, 0)$ and $(-4, 0)$. Then we draw and extend the diagonals of this rectangle. The two extended diagonals are the asymptotes of the hyperbola. Next, we plot the vertices and draw the branches of the hyperbola outward from the vertices toward the asymptotes.

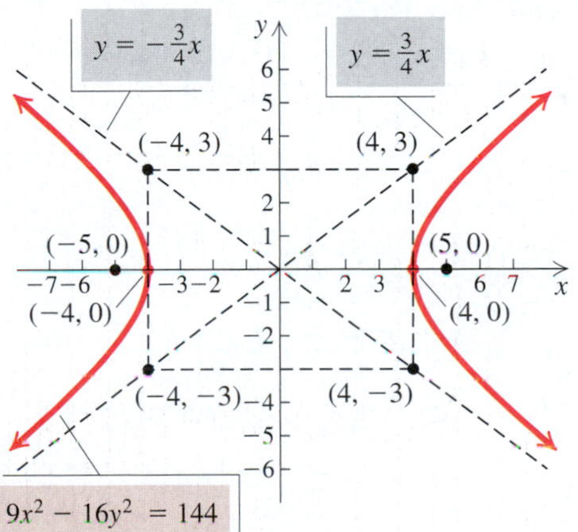

To graph this hyperbola on a graphing calculator, it might be necessary to solve for y first and then graph the top and bottom halves of the hyperbola in the same squared viewing window.

On some graphing calculators, the Conics HYPERBOLA APP can be used to graph hyperbolas, as shown below. Note that the center is $(0, 0)$, so $H = 0$ and $K = 0$.

$$\frac{x^2}{4^2} - \frac{y^2}{3^2} = 1$$

HYPERBOLA
$\frac{(X-H)^2}{A^2} - \frac{(Y-K)^2}{B^2} = 1$
A=4
B=3
H=0
K=0
ESC

$9x^2 - 16y^2 = 144$

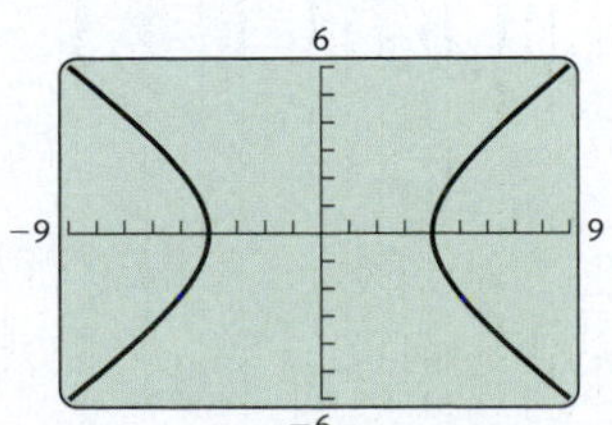

Now Try Exercise 17.

If a hyperbola with center at the origin is translated horizontally $|h|$ units and vertically $|k|$ units, the center is at the point (h, k).

STANDARD EQUATION OF A HYPERBOLA WITH CENTER AT (h, k)

Transverse Axis Horizontal

$$\frac{(x-h)^2}{a^2} - \frac{(y-k)^2}{b^2} = 1$$

Vertices: $(h-a, k), (h+a, k)$

Asymptotes: $y - k = \frac{b}{a}(x-h)$,

$$y - k = -\frac{b}{a}(x-h)$$

Foci: $(h-c, k), (h+c, k)$, where $c^2 = a^2 + b^2$

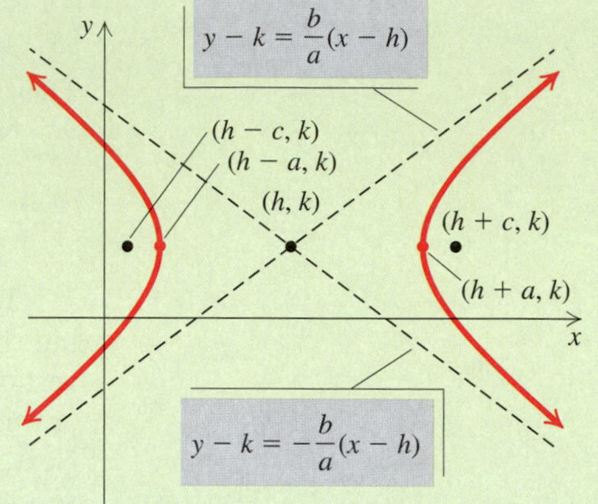

Transverse Axis Vertical

$$\frac{(y-k)^2}{a^2} - \frac{(x-h)^2}{b^2} = 1$$

Vertices: $(h, k-a), (h, k+a)$

Asymptotes: $y - k = \frac{a}{b}(x-h)$,

$$y - k = -\frac{a}{b}(x-h)$$

Foci: $(h, k-c), (h, k+c)$, where $c^2 = a^2 + b^2$

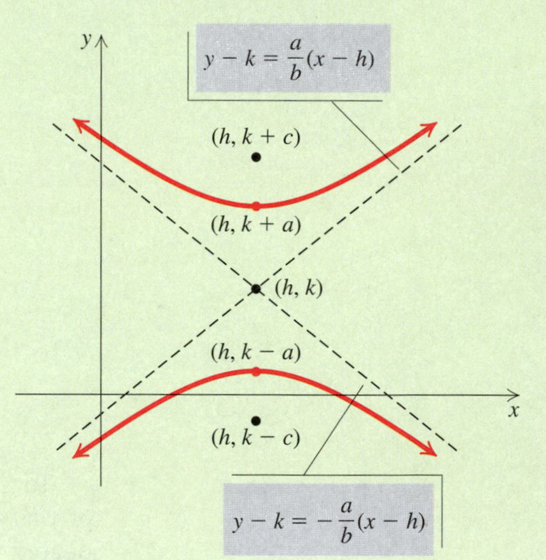

EXAMPLE 3 For the hyperbola given by

$$4y^2 - x^2 + 24y + 4x + 28 = 0,$$

find the center, the vertices, the foci, and the asymptotes. Then draw the graph.

Solution First, we complete the square to get standard form:

$$
\begin{aligned}
4y^2 - x^2 + 24y + 4x + 28 &= 0\\
4(y^2 + 6y \qquad\quad) - (x^2 - 4x \qquad\quad) &= -28\\
4(y^2 + 6y + 9 - 9) - (x^2 - 4x + 4 - 4) &= -28\\
4(y^2 + 6y + 9) + 4(-9) - (x^2 - 4x + 4) - (-4) &= -28\\
4(y^2 + 6y + 9) - 36 - (x^2 - 4x + 4) + 4 &= -28\\
4(y^2 + 6y + 9) - (x^2 - 4x + 4) &= -28 + 36 - 4\\
4(y + 3)^2 - (x - 2)^2 &= 4\\
\frac{(y + 3)^2}{1} - \frac{(x - 2)^2}{4} &= 1 \qquad \text{Dividing by 4}\\
\frac{[y - (-3)]^2}{1^2} - \frac{(x - 2)^2}{2^2} &= 1. \qquad \text{Standard form}
\end{aligned}
$$

The center is $(2, -3)$. Note that $a = 1$ and $b = 2$. The transverse axis is vertical, so the vertices are 1 unit below and above the center:

$$(2, -3 - 1) \text{ and } (2, -3 + 1), \quad \text{or} \quad (2, -4) \text{ and } (2, -2).$$

We know that $c^2 = a^2 + b^2$, so $c^2 = 1^2 + 2^2 = 1 + 4 = 5$ and $c = \sqrt{5}$. Thus the foci are $\sqrt{5}$ units below and above the center:

$$(2, -3 - \sqrt{5}) \quad \text{and} \quad (2, -3 + \sqrt{5}).$$

The asymptotes are

$$y - (-3) = \tfrac{1}{2}(x - 2) \quad \text{and} \quad y - (-3) = -\tfrac{1}{2}(x - 2),$$

or $\quad y + 3 = \tfrac{1}{2}(x - 2) \quad \text{and} \quad y + 3 = -\tfrac{1}{2}(x - 2).$

We sketch the asymptotes, plot the vertices, and draw the graph.

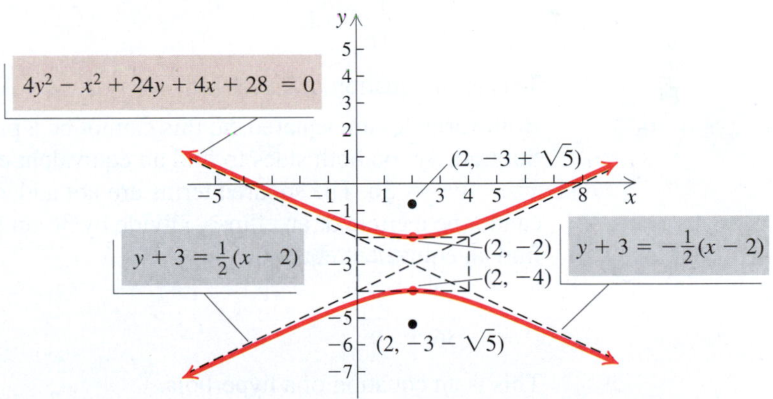

When the equation of a hyperbola is written in standard form, we can use the Conics HYPERBOLA APP to graph it, as shown in the following windows.

$$\frac{[y - (-3)]^2}{1^2} - \frac{(x - 2)^2}{2^2} = 1$$

HYPERBOLA
(Y−K)²/A² − (X−H)²/B² = 1
A=1
B=2
H=2
K=−3
ESC

$$4y^2 - x^2 + 24y + 4x + 28 = 0$$

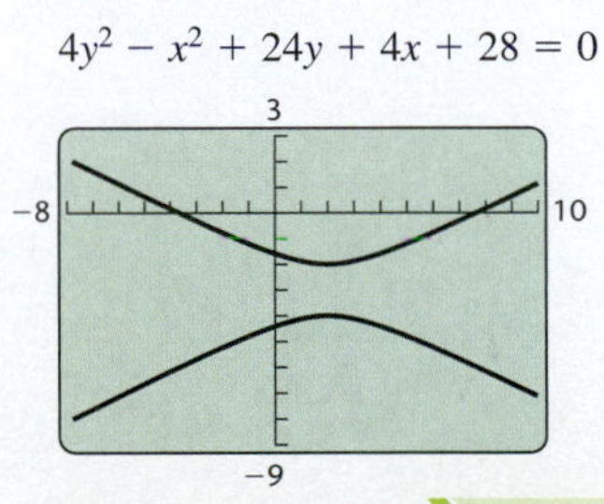

Now Try Exercise 29.

CONNECTING THE CONCEPTS — Classifying Equations of Conic Sections

EQUATION | **TYPE OF CONIC SECTION** | **GRAPH**

$x - 4 + 4y = y^2$

Only one variable is squared, so this cannot be a circle, an ellipse, or a hyperbola. Find an equivalent equation:

$$x = (y - 2)^2.$$

This is an equation of a parabola.

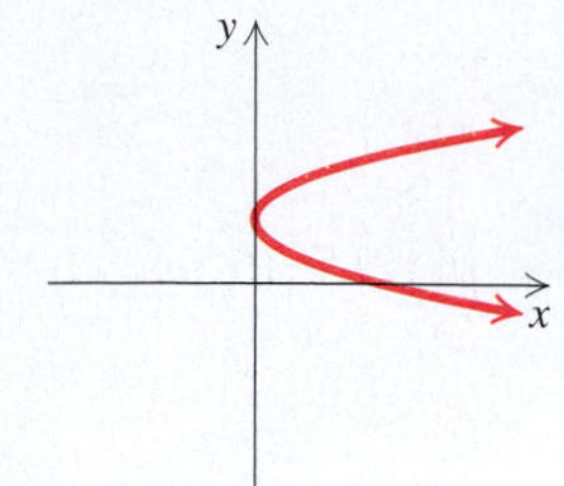

$3x^2 + 3y^2 = 75$

Both variables are squared, so this cannot be a parabola. The squared terms are added, so this cannot be a hyperbola. Divide by 3 on both sides to find an equivalent equation:

$$x^2 + y^2 = 25.$$

This is an equation of a circle.

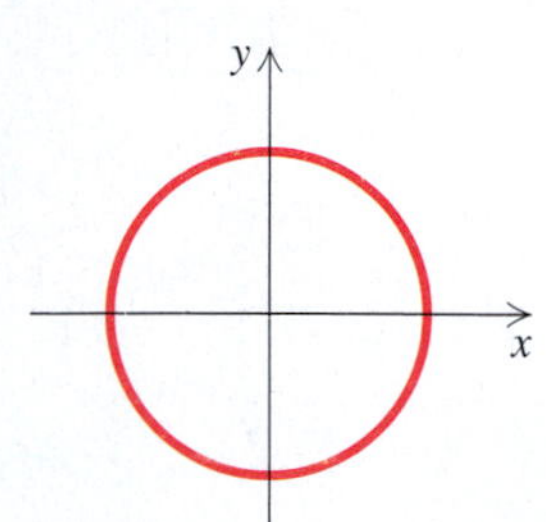

$y^2 = 16 - 4x^2$

Both variables are squared, so this cannot be a parabola. Add $4x^2$ on both sides to find an equivalent equation: $4x^2 + y^2 = 16$. The squared terms are added, so this cannot be a hyperbola. The coefficients of x^2 and y^2 are not the same, so this is not a circle. Divide by 16 on both sides to find an equivalent equation:

$$\frac{x^2}{4} + \frac{y^2}{16} = 1.$$

This is an equation of an ellipse.

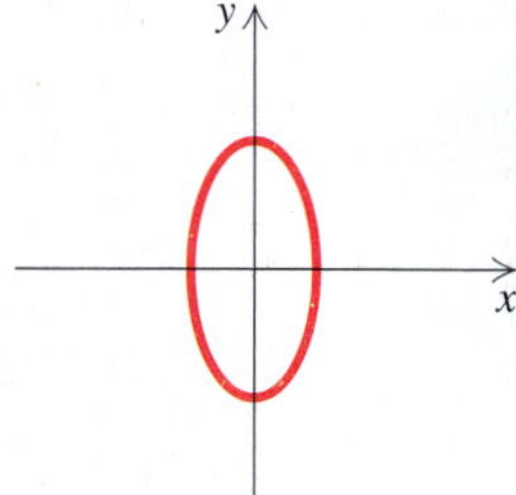

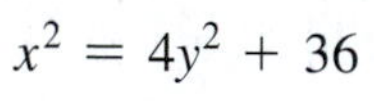

$x^2 = 4y^2 + 36$

Both variables are squared, so this cannot be a parabola. Subtract $4y^2$ on both sides to find an equivalent equation: $x^2 - 4y^2 = 36$. The squared terms are not added, so this cannot be a circle or an ellipse. Divide by 36 on both sides to find an equivalent equation:

$$\frac{x^2}{36} - \frac{y^2}{9} = 1.$$

This is an equation of a hyperbola.

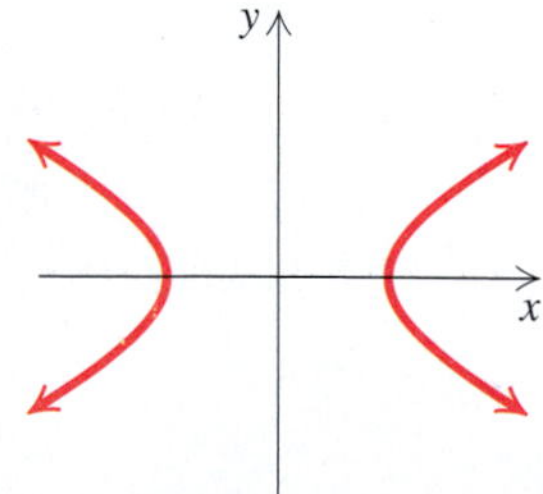

Applications

Some comets travel in hyperbolic paths with the sun at one focus. Such comets pass by the sun only one time, unlike those with elliptical orbits, which reappear at intervals. We also see hyperbolas in architecture, such as in a cross section of a planetarium, an amphitheater, or a cooling tower for a steam or nuclear power plant.

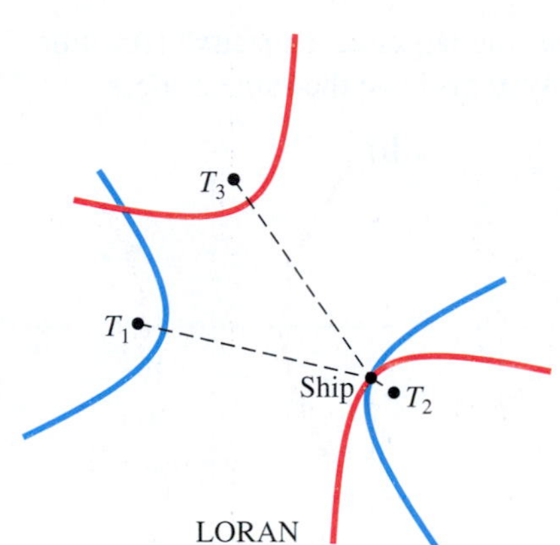

Another application of hyperbolas is in the long-range navigation system LORAN. This system uses transmitting stations in three locations to send out simultaneous signals to a ship or an aircraft. The difference in the arrival times of the signals from one pair of transmitters is recorded on the ship or aircraft. This difference is also recorded for signals from another pair of transmitters. For each pair, a computation is performed to determine the difference in the distances from each member of the pair to the ship or aircraft. If each pair of differences is kept constant, two hyperbolas can be drawn. Each has one of the pairs of transmitters as foci, and the ship or aircraft lies on the intersection of two of their branches.

10.3 Exercise Set

In Exercises 1–6, match the equation with one of the graphs (a)–(f) that follow.

a)

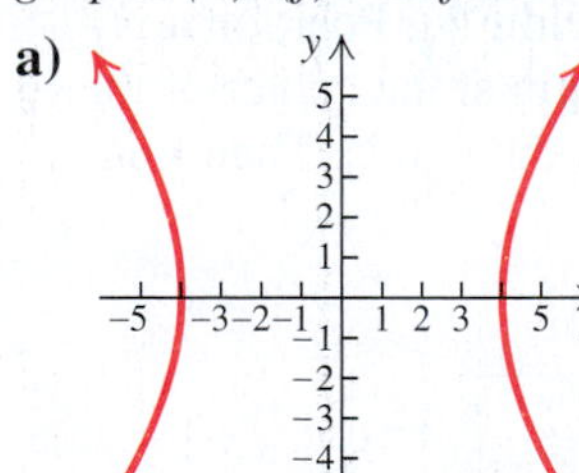

b)

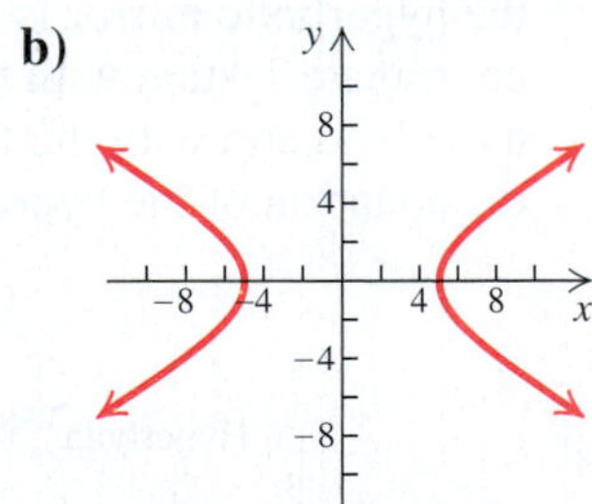

c)

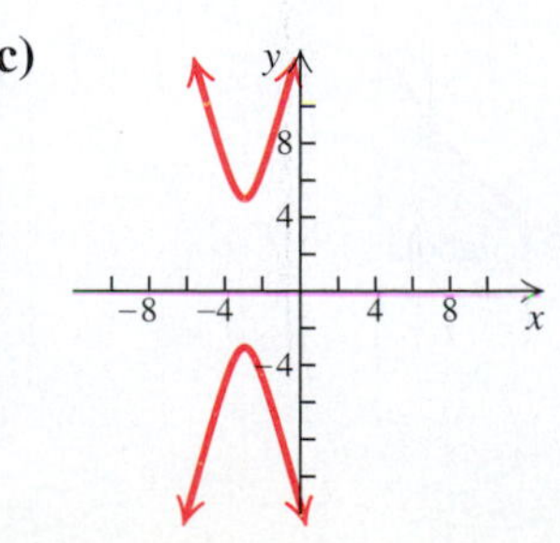

d)

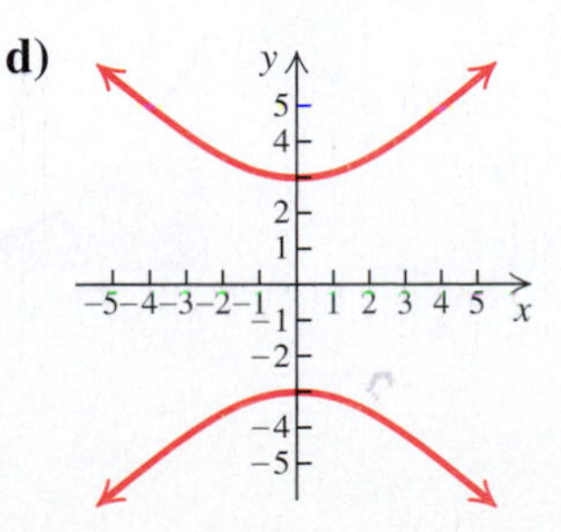

e)

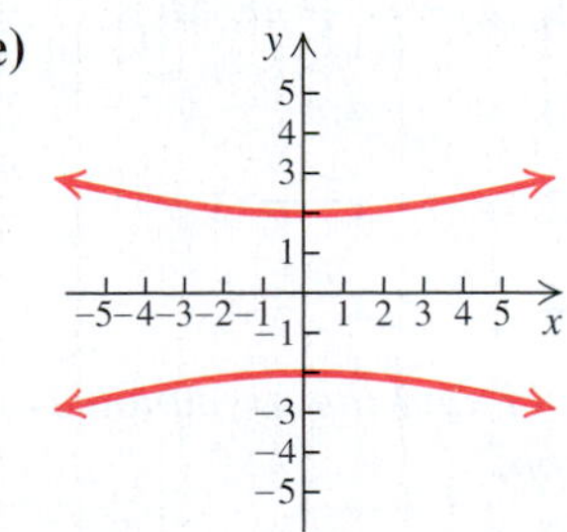

f)

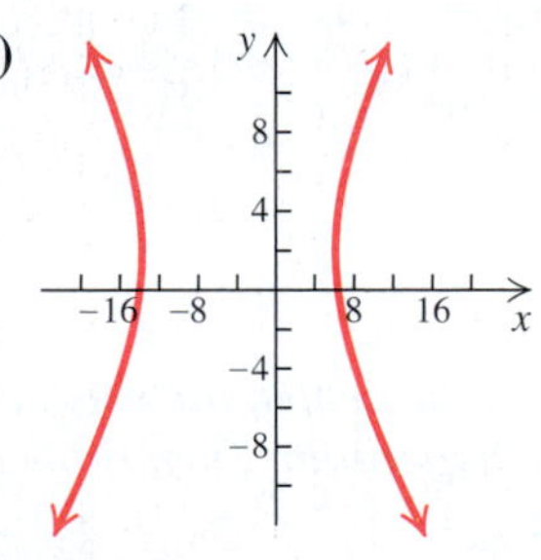

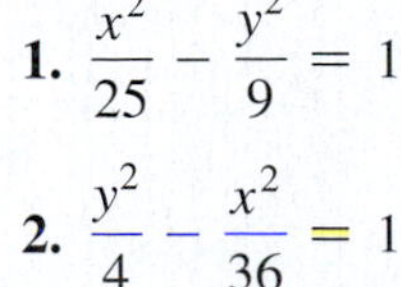

1. $\dfrac{x^2}{25} - \dfrac{y^2}{9} = 1$

2. $\dfrac{y^2}{4} - \dfrac{x^2}{36} = 1$

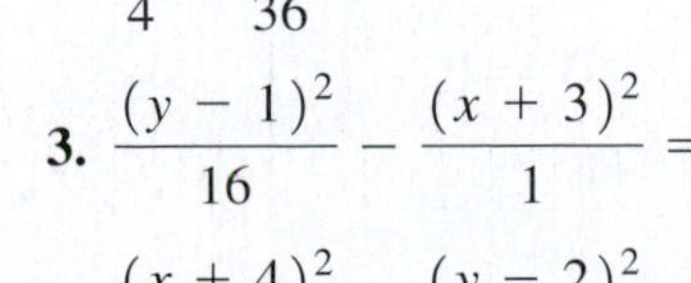

3. $\dfrac{(y-1)^2}{16} - \dfrac{(x+3)^2}{1} = 1$

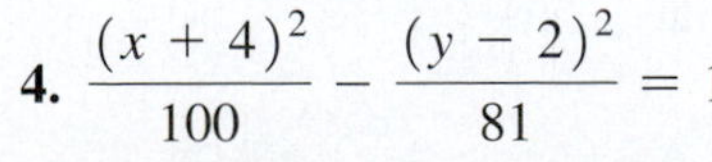

4. $\dfrac{(x+4)^2}{100} - \dfrac{(y-2)^2}{81} = 1$

5. $25x^2 - 16y^2 = 400$

6. $y^2 - x^2 = 9$

Find an equation of a hyperbola satisfying the given conditions.

7. Vertices: $(0, 3)$ and $(0, -3)$; foci: $(0, 5)$ and $(0, -5)$

8. Vertices: $(1, 0)$ and $(-1, 0)$; foci: $(2, 0)$ and $(-2, 0)$

9. Asymptotes: $y = \frac{3}{2}x$, $y = -\frac{3}{2}x$; one vertex: $(2, 0)$

10. Asymptotes: $y = \frac{5}{4}x$, $y = -\frac{5}{4}x$; one vertex: $(0, 3)$

Find the center, the vertices, the foci, and the asymptotes. Then draw the graph.

11. $\dfrac{x^2}{4} - \dfrac{y^2}{4} = 1$

12. $\dfrac{x^2}{1} - \dfrac{y^2}{9} = 1$

13. $\dfrac{(x-2)^2}{9} - \dfrac{(y+5)^2}{1} = 1$

14. $\dfrac{(x-5)^2}{16} - \dfrac{(y+2)^2}{9} = 1$

15. $\dfrac{(y+3)^2}{4} - \dfrac{(x+1)^2}{16} = 1$

16. $\dfrac{(y+4)^2}{25} - \dfrac{(x+2)^2}{16} = 1$

17. $x^2 - 4y^2 = 4$

18. $4x^2 - y^2 = 16$

19. $9y^2 - x^2 = 81$

20. $y^2 - 4x^2 = 4$

21. $x^2 - y^2 = 2$

22. $x^2 - y^2 = 3$

23. $y^2 - x^2 = \frac{1}{4}$

24. $y^2 - x^2 = \frac{1}{9}$

Find the center, the vertices, the foci, and the asymptotes of the hyperbola. Then draw the graph.

25. $x^2 - y^2 - 2x - 4y - 4 = 0$

26. $4x^2 - y^2 + 8x - 4y - 4 = 0$

27. $36x^2 - y^2 - 24x + 6y - 41 = 0$

28. $9x^2 - 4y^2 + 54x + 8y + 41 = 0$

29. $9y^2 - 4x^2 - 18y + 24x - 63 = 0$

30. $x^2 - 25y^2 + 6x - 50y = 41$

31. $x^2 - y^2 - 2x - 4y = 4$

32. $9y^2 - 4x^2 - 54y - 8x + 41 = 0$

33. $y^2 - x^2 - 6x - 8y - 29 = 0$

34. $x^2 - y^2 = 8x - 2y - 13$

*The **eccentricity** of a hyperbola is defined as $e = c/a$. For a hyperbola, $c > a > 0$, so $e > 1$. When e is close to 1, a hyperbola appears to be very narrow. As the eccentricity increases, the hyperbola becomes "wider."*

35. Observe the shapes of the hyperbolas in Examples 2 and 3. Which hyperbola has the larger eccentricity? Confirm your answer by computing the eccentricity of each hyperbola.

36. Which hyperbola has the larger eccentricity? (Assume that the coordinate systems have the same scale.)

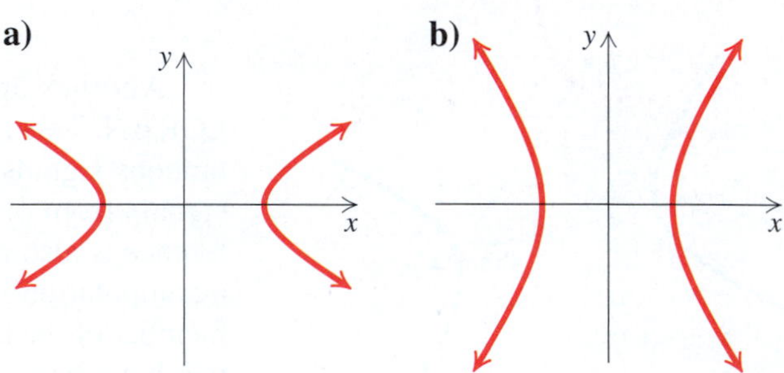

37. Find an equation of a hyperbola with vertices $(3, 7)$ and $(-3, 7)$ and $e = \frac{5}{3}$.

38. Find an equation of a hyperbola with vertices $(-1, 3)$ and $(-1, 7)$ and $e = 4$.

39. *Hyperbolic Mirror.* Certain telescopes contain both a parabolic mirror and a hyperbolic mirror. In the telescope shown in the figure, the parabola and the hyperbola share focus F_1, which is 14 m above the vertex of the parabola. The hyperbola's second focus F_2 is 2 m above the parabola's vertex. The vertex of the hyperbolic mirror is 1 m below F_1. Position a coordinate system with the origin at the center of the hyperbola and with the foci on the y-axis. Then find the equation of the hyperbola.

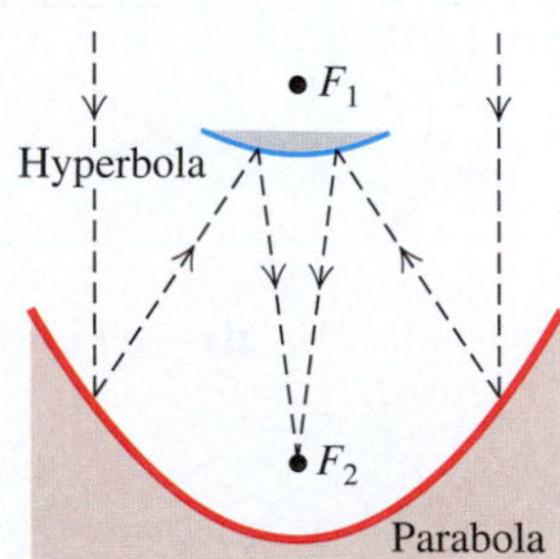

40. *Nuclear Cooling Tower.* A cross section of a nuclear cooling tower is a hyperbola with equation

$$\frac{x^2}{90^2} - \frac{y^2}{130^2} = 1.$$

The tower is 450 ft tall, and the distance from the top of the tower to the center of the hyperbola is half the distance from the base of the tower to the center of the hyperbola. Find the diameter of the top and the base of the tower.

› Skill Maintenance

In Exercises 41–44, given the function:

a) *Determine whether it is one-to-one.* [5.1]

b) *If it is one-to-one, find a formula for the inverse.* [5.1]

41. $f(x) = 2x - 3$

42. $f(x) = x^3 + 2$

43. $f(x) = \dfrac{5}{x - 1}$

44. $f(x) = \sqrt{x + 4}$

Solve. [9.1], [9.3], [9.5], [9.6]

45. $x + y = 5,$
$x - y = 7$

46. $3x - 2y = 5,$
$5x + 2y = 3$

47. $2x - 3y = 7,$
$3x + 5y = 1$

48. $3x + 2y = -1,$
$2x + 3y = 6$

› Synthesis

Find an equation of a hyperbola satisfying the given conditions.

49. Vertices: $(3, -8)$ and $(3, -2)$; asymptotes: $y = 3x - 14$, $y = -3x + 4$

50. Vertices: $(-9, 4)$ and $(-5, 4)$; asymptotes: $y = 3x + 25$, $y = -3x - 17$

Use a graphing calculator to find the center, the vertices, and the asymptotes.

51. $5x^2 - 3.5y^2 + 14.6x - 6.7y + 3.4 = 0$

52. $x^2 - y^2 - 2.046x - 4.088y - 4.228 = 0$

53. *Navigation.* Two radio transmitters positioned 300 mi apart along the shore send simultaneous signals to a ship that is 200 mi offshore, sailing parallel to the shoreline. The signal from transmitter *S* reaches the ship 200 microseconds later than the signal from transmitter *T*. The signals travel at a speed of 186,000 miles per second, or 0.186 mile per microsecond. Find the equation of the hyperbola with foci *S* and *T* on which the ship is located. (*Hint*: For any point on the hyperbola, the absolute value of the difference of its distances from the foci is $2a$.)

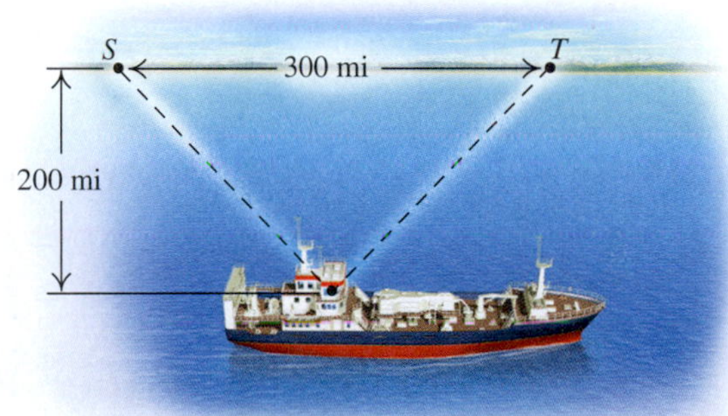

10.4 Nonlinear Systems of Equations and Inequalities

- › Solve a nonlinear system of equations.
- › Use nonlinear systems of equations to solve applied problems.
- › Graph nonlinear systems of inequalities.

The systems of equations that we have studied so far have been composed of linear equations. Now we consider systems of two equations in two variables in which at least one equation is not linear.

Nonlinear Systems of Equations

The graphs of the equations in a nonlinear system of equations can have no point of intersection or one or more points of intersection. The coordinates of each point of intersection represent a solution of the system of equations. When no point of intersection exists, the system of equations has no real-number solution.

Solutions of nonlinear systems of equations can be found using the substitution method or the elimination method. The substitution method is preferable for a system consisting of one linear equation and one nonlinear equation. The elimination method is preferable in most, but not all, cases when both equations are nonlinear.

EXAMPLE 1 Solve the following system of equations:

$x^2 + y^2 = 25,$ **(1)** **The graph is a circle.**

$3x - 4y = 0.$ **(2)** **The graph is a line.**

Algebraic Solution

We use the substitution method. First, we solve equation (2) for x:

$x = \frac{4}{3}y.$ **(3)** **We could have solved for y instead.**

Next, we substitute $\frac{4}{3}y$ for x in equation (1) and solve for y:

$$\left(\tfrac{4}{3}y\right)^2 + y^2 = 25$$
$$\tfrac{16}{9}y^2 + y^2 = 25$$
$$\tfrac{25}{9}y^2 = 25$$
$$y^2 = 9 \quad \textbf{Multiplying by } \tfrac{9}{25}$$
$$y = \pm 3.$$

Now we substitute these numbers for y in equation (3) and solve for x:

$x = \frac{4}{3}(3) = 4,$ **(4, 3) appears to be a solution.**

$x = \frac{4}{3}(-3) = -4.$ **(−4, −3) appears to be a solution.**

Check: For $(4, 3)$:

$x^2 + y^2 = 25$	
$4^2 + 3^2$	? 25
$16 + 9$	
25	25 TRUE

$3x - 4y = 0$	
$3(4) - 4(3)$	? 0
$12 - 12$	
0	0 TRUE

For $(-4, -3)$:

$x^2 + y^2 = 25$	
$(-4)^2 + (-3)^2$	? 25
$16 + 9$	
25	25 TRUE

$3x - 4y = 0$	
$3(-4) - 4(-3)$	? 0
$-12 + 12$	
0	0 TRUE

The pairs $(4, 3)$ and $(-4, -3)$ check, so they are the solutions.

Graphical Solution

We graph both equations in the same viewing window. Note that there are two points of intersection. We can find their coordinates using the INTERSECT feature.

$x^2 + y^2 = 25$: $y_1 = \sqrt{25 - x^2}$, $y_2 = -\sqrt{25 - x^2}$, $y_3 = \frac{3}{4}x$

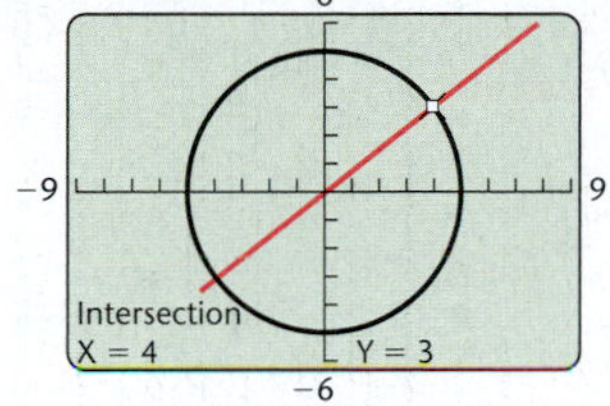

$x^2 + y^2 = 25$: $y_1 = \sqrt{25 - x^2}$, $y_2 = -\sqrt{25 - x^2}$, $y_3 = \frac{3}{4}x$

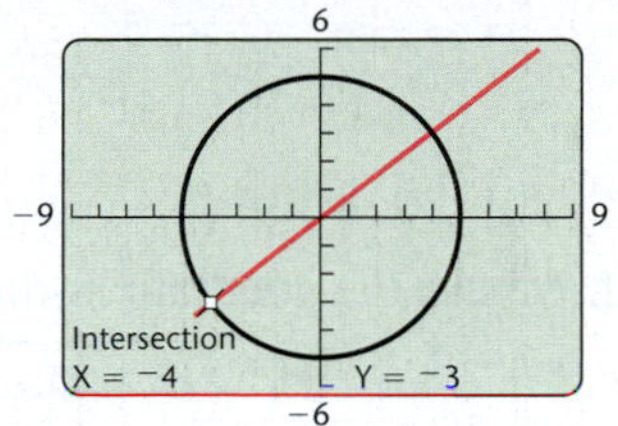

The solutions are $(4, 3)$ and $(-4, -3)$.

Now Try Exercise 7.

In the algebraic solution in Example 1, suppose that to find x we had substituted 3 and -3 in equation (1) rather than equation (3). If $y = 3$, $y^2 = 9$, and if $y = -3$, $y^2 = 9$, so both substitutions can be performed at the same time:

$$
\begin{aligned}
x^2 + y^2 &= 25 \qquad \textbf{(1)} \\
x^2 + (\pm 3)^2 &= 25 \\
x^2 + 9 &= 25 \\
x^2 &= 16 \\
x &= \pm 4.
\end{aligned}
$$

Each y-value produces two values for x. Thus, if $y = 3$, $x = 4$ or $x = -4$, and if $y = -3$, $x = 4$ or $x = -4$. The possible solutions are $(4, 3)$, $(-4, 3)$, $(4, -3)$, and $(-4, -3)$. A check reveals that $(4, -3)$ and $(-4, 3)$ are not solutions of equation (2). Since a circle and a line can intersect in at most two points, it is clear that there can be at most two real-number solutions.

EXAMPLE 2 Solve the following system of equations:

$$x + y = 5, \quad (1) \quad \text{The graph is a line.}$$

$$y = 3 - x^2. \quad (2) \quad \text{The graph is a parabola.}$$

Algebraic Solution

We use the substitution method, substituting $3 - x^2$ for y in equation (1):

$$x + 3 - x^2 = 5$$

$$-x^2 + x - 2 = 0 \quad \text{Subtracting 5 and rearranging}$$

$$x^2 - x + 2 = 0. \quad \text{Multiplying by } -1$$

Next, we use the quadratic formula:

$$x = \frac{-b \pm \sqrt{b^2 - 4ac}}{2a}$$

$$= \frac{-(-1) \pm \sqrt{(-1)^2 - 4(1)(2)}}{2(1)}$$

$$= \frac{1 \pm \sqrt{1 - 8}}{2}$$

$$= \frac{1 \pm \sqrt{-7}}{2}$$

$$= \frac{1 \pm i\sqrt{7}}{2}$$

$$= \frac{1}{2} \pm \frac{\sqrt{7}}{2}i.$$

Now, we substitute these values for x in equation (1) and solve for y:

$$\frac{1}{2} + \frac{\sqrt{7}}{2}i + y = 5$$

$$y = 5 - \frac{1}{2} - \frac{\sqrt{7}}{2}i = \frac{9}{2} - \frac{\sqrt{7}}{2}i$$

and $\frac{1}{2} - \frac{\sqrt{7}}{2}i + y = 5$

$$y = 5 - \frac{1}{2} + \frac{\sqrt{7}}{2}i = \frac{9}{2} + \frac{\sqrt{7}}{2}i.$$

The solutions are

$$\left(\frac{1}{2} + \frac{\sqrt{7}}{2}i, \frac{9}{2} - \frac{\sqrt{7}}{2}i\right) \text{ and } \left(\frac{1}{2} - \frac{\sqrt{7}}{2}i, \frac{9}{2} + \frac{\sqrt{7}}{2}i\right).$$

There are no real-number solutions.

Graphical Solution

We graph both equations in the same viewing window.

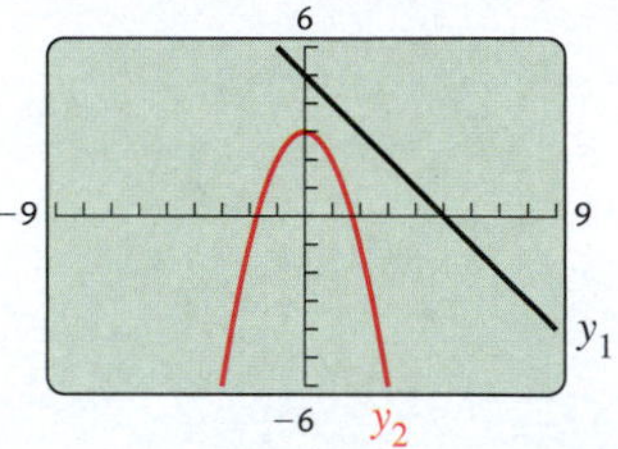

Note that there are no points of intersection. This indicates that there are no real-number solutions. Algebra must be used, as at left, to find the imaginary-number solutions.

Now Try Exercise 17.

EXAMPLE 3 Solve the following system of equations:

$$2x^2 + 5y^2 = 39, \quad (1)$$ **The graph is an ellipse.**

$$3x^2 - y^2 = -1. \quad (2)$$ **The graph is a hyperbola.**

Algebraic Solution

We use the elimination method. First, we multiply equation (2) by 5 and add to eliminate the y^2-term:

$$2x^2 + 5y^2 = 39 \quad (1)$$

$$15x^2 - 5y^2 = -5 \quad \text{Multiplying (2) by 5}$$

$$17x^2 = 34 \quad \text{Adding}$$

$$x^2 = 2$$

$$x = \pm\sqrt{2}.$$

If $x = \sqrt{2}$, $x^2 = 2$, and if $x = -\sqrt{2}$, $x^2 = 2$. Thus substituting $\sqrt{2}$ or $-\sqrt{2}$ for x in equation (2) gives us

$$3(\pm\sqrt{2})^2 - y^2 = -1$$

$$3 \cdot 2 - y^2 = -1$$

$$6 - y^2 = -1$$

$$-y^2 = -7$$

$$y^2 = 7$$

$$y = \pm\sqrt{7}.$$

Each x-value produces two values for y. Thus, for $x = \sqrt{2}$, we have $y = \sqrt{7}$ or $y = -\sqrt{7}$, and for $x = -\sqrt{2}$, we have $y = \sqrt{7}$ or $y = -\sqrt{7}$. The possible solutions are $(\sqrt{2}, \sqrt{7})$, $(\sqrt{2}, -\sqrt{7})$, $(-\sqrt{2}, \sqrt{7})$, and $(-\sqrt{2}, -\sqrt{7})$. All four pairs check, so they are the solutions.

Graphical Solution

We graph both equations in the same viewing window. There are four points of intersection. We can use the INTERSECT feature to find their coordinates.

$y_1 = \sqrt{(39 - 2x^2)/5}$, $y_2 = -\sqrt{(39 - 2x^2)/5}$,
$y_3 = \sqrt{3x^2 + 1}$, $y_4 = -\sqrt{3x^2 + 1}$

(−1.414, 2.646)
(1.414, 2.646)
(−1.414, −2.646)
(1.414, −2.646)
6
−6
−9
9

Note that the algebraic solution yields exact solutions, whereas the graphical solution yields decimal approximations of the solutions on most graphing calculators.

The solutions are approximately $(1.414, 2.646)$, $(1.414, -2.646)$, $(-1.414, 2.646)$, and $(-1.414, -2.646)$.

Now Try Exercise 27.

EXAMPLE 4 Solve the following system of equations:

$$x^2 - 3y^2 = 6, \quad \textbf{(1)}$$
$$xy = 3. \quad \textbf{(2)}$$

Algebraic Solution

We use the substitution method. First, we solve equation (2) for y:

$$xy = 3 \quad \textbf{(2)}$$
$$y = \frac{3}{x}. \quad \textbf{(3)} \quad \textbf{Dividing by } x$$

Next, we substitute $3/x$ for y in equation (1) and solve for x:

$$x^2 - 3\left(\frac{3}{x}\right)^2 = 6$$
$$x^2 - 3 \cdot \frac{9}{x^2} = 6$$
$$x^2 - \frac{27}{x^2} = 6$$
$$x^4 - 27 = 6x^2 \quad \textbf{Multiplying by } x^2$$
$$x^4 - 6x^2 - 27 = 0$$
$$u^2 - 6u - 27 = 0 \quad \textbf{Letting } u = x^2$$
$$(u - 9)(u + 3) = 0 \quad \textbf{Factoring}$$
$$u = 9 \quad or \quad u = -3 \quad \textbf{Principle of zero products}$$
$$x^2 = 9 \quad or \quad x^2 = -3 \quad \textbf{Substituting } x^2 \textbf{ for } u$$
$$x = \pm 3 \quad or \quad x = \pm\sqrt{3}i.$$

Since $y = 3/x$,

when $x = 3$, $\quad y = \frac{3}{3} = 1;$

when $x = -3$, $\quad y = \frac{3}{-3} = -1;$

when $x = \sqrt{3}i$, $\quad y = \frac{3}{\sqrt{3}i} = \frac{3}{\sqrt{3}i} \cdot \frac{-\sqrt{3}i}{-\sqrt{3}i} = -\sqrt{3}i;$

when $x = -\sqrt{3}i$, $\quad y = \frac{3}{-\sqrt{3}i} = \frac{3}{-\sqrt{3}i} \cdot \frac{\sqrt{3}i}{\sqrt{3}i} = \sqrt{3}i.$

The pairs $(3, 1)$, $(-3, -1)$, $(\sqrt{3}i, -\sqrt{3}i)$, and $(-\sqrt{3}i, \sqrt{3}i)$ check, so they are the solutions.

Graphical Solution

We graph both equations in the same viewing window and find the coordinates of their points of intersection.

$x^2 - 3y^2 = 6$

$y_1 = \sqrt{(x^2 - 6)/3}, \quad y_2 = -\sqrt{(x^2 - 6)/3},$

$y_3 = 3/x$

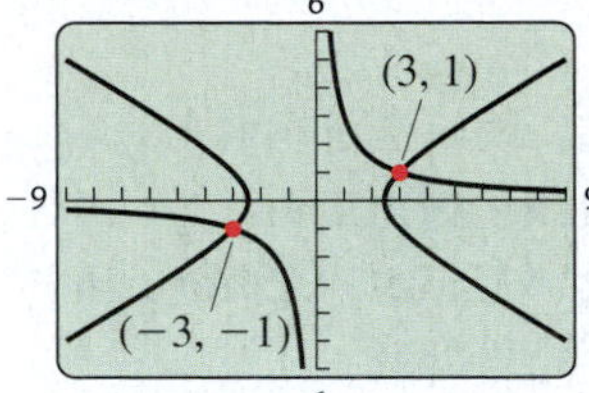

Note that the graphical method yields only the real-number solutions of the system of equations. The algebraic method must be used in order to find *all* the solutions.

Now Try Exercise 19.

Modeling and Problem Solving

EXAMPLE 5 ***Dimensions of a Piece of Land.*** For a student recreation building at Southport Community College, an architect wants to lay out a rectangular piece of land that has a perimeter of 204 m and an area of 2565 m^2. Find the dimensions of the piece of land.

Solution

1. **Familiarize.** We make a drawing and label it, letting l = the length of the piece of land, in meters, and w = the width, in meters.

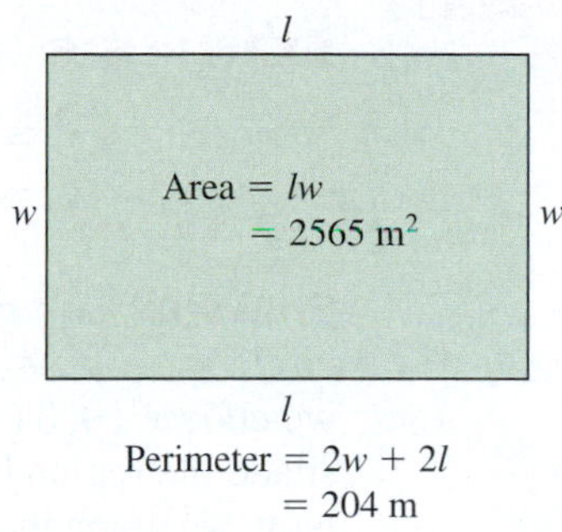

2. **Translate.** We now have the following:

Perimeter: $2w + 2l = 204,$ **(1)**

Area: $lw = 2565.$ **(2)**

3. **Carry out.** We solve the system of equations both algebraically and graphically.

Algebraic Solution

We solve the system of equations

$$2w + 2l = 204, \quad (1)$$
$$lw = 2565. \quad (2)$$

Solving the second equation for l gives us $l = 2565/w$. We then substitute $2565/w$ for l in equation (1) and solve for w:

$$2w + 2\left(\frac{2565}{w}\right) = 204$$

$$2w^2 + 2(2565) = 204w \quad \text{Multiplying by } w$$

$$2w^2 - 204w + 2(2565) = 0$$

$$w^2 - 102w + 2565 = 0 \quad \text{Multiplying by } \tfrac{1}{2}$$

$$(w - 57)(w - 45) = 0$$

$$w = 57 \quad or \quad w = 45. \quad \text{Principle of zero products}$$

If $w = 57$, then $l = 2565/w = 2565/57 = 45$. If $w = 45$, then $l = 2565/w = 2565/45 = 57$. Since length is generally considered to be longer than width, we have the solution $l = 57$ and $w = 45$, or $(57, 45)$.

Graphical Solution

We replace l with x and w with y, graph $y_1 = (204 - 2x)/2$ and $y_2 = 2565/x$, and find the point(s) of intersection of the graphs.

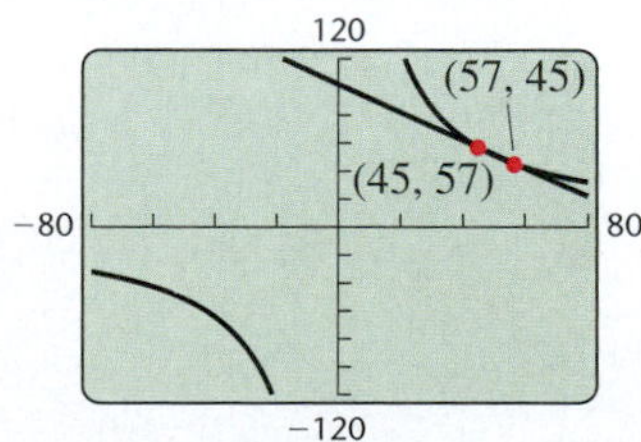

As in the algebraic solution, we have two possible solutions: $(45, 57)$ and $(57, 45)$. Since length, x, is generally considered to be longer than width, y, we have the solution $(57, 45)$.

4. Check. If $l = 57$ and $w = 45$, the perimeter is $2 \cdot 45 + 2 \cdot 57$, or 204. The area is $57 \cdot 45$, or 2565. The numbers check.

5. State. The length of the piece of land is 57 m and the width is 45 m.

Now Try Exercise 61.

Nonlinear Systems of Inequalities

SYSTEMS OF INEQUALITIES

REVIEW SECTION 9.7.

Recall that a solution of a system of inequalities is an ordered pair that is a solution of each inequality in the system. Now we graph a system of nonlinear inequalities.

EXAMPLE 6 Graph the solution set of the system

$$x^2 + y^2 \le 25,$$
$$3x - 4y > 0.$$

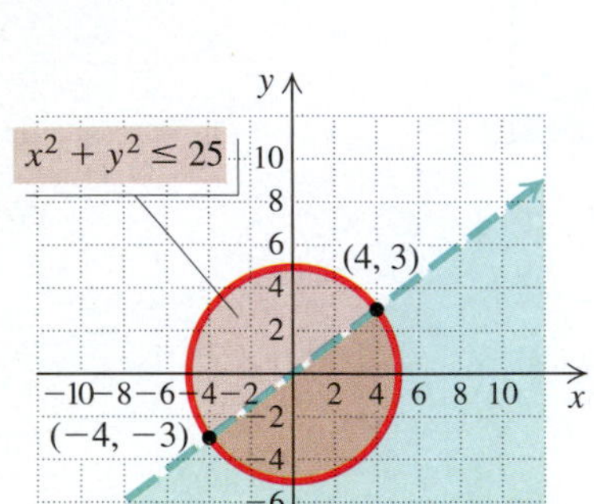

Solution We graph $x^2 + y^2 \le 25$ by first graphing the related equation of the circle $x^2 + y^2 = 25$. We use a solid curve since the inequality symbol is $\le$. Next, we choose $(0, 0)$ as a test point and find that it is a solution of $x^2 + y^2 \le 25$, so we shade the region that contains $(0, 0)$ using red. This is the region inside the circle. Now we graph the line $3x - 4y = 0$ using a dashed line since the inequality symbol is $>$. The point $(0, 0)$ is on the line, so we choose another test point, say, $(0, 2)$. We find that this point is not a solution of $3x - 4y > 0$, so we shade the half-plane that does not contain $(0, 2)$ using green. The solution set of the system of inequalities is the region shaded both red and green, or brown, including part of the circle $x^2 + y^2 = 25$.

To find the points of intersection of the graphs of the related equations, we solve the system composed of those equations:

$$x^2 + y^2 = 25,$$
$$3x - 4y = 0.$$

In Example 1, we found that these points are $(4, 3)$ and $(-4, -3)$.

Now Try Exercise 75.

EXAMPLE 7 Use a graphing calculator to graph the system

$$y \le 4 - x^2,$$
$$x + y \ge 2.$$

Solution We graph $y_1 = 4 - x^2$ and $y_2 = 2 - x$. Using the test point $(0, 0)$ for each inequality, we find that we should shade below y_1 and above y_2. We can find the points of intersection of the graphs of the related equations, $(-1, 3)$ and $(2, 0)$, using the INTERSECT feature.

$y_1 = 4 - x^2, \; y_2 = 2 - x$

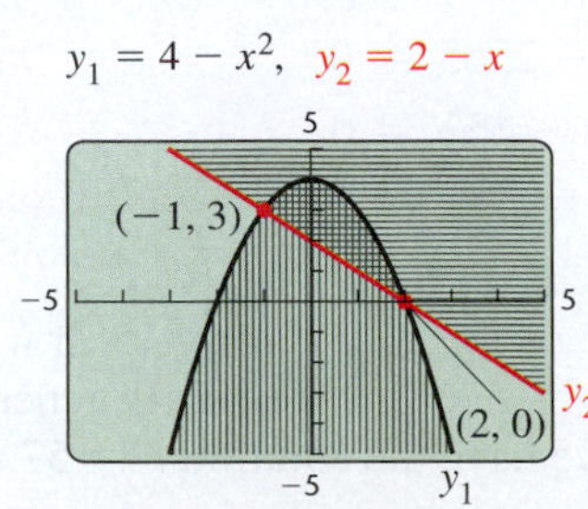

Now Try Exercise 81.

Visualizing the Graph

Match the equation or system of equations with its graph.

1. $y = x^3 - 3x$

2. $y = x^2 + 2x - 3$

3. $y = \dfrac{x - 1}{x^2 - x - 2}$

4. $y = -3x + 2$

5. $x + y = 3,$
   $2x + 5y = 3$

6. $9x^2 - 4y^2 = 36,$
   $x^2 + y^2 = 9$

7. $5x^2 + 5y^2 = 20$

8. $4x^2 + 16y^2 = 64$

9. $y = \log_2 x$

10. $y = 2^x$

Answers on page A-61

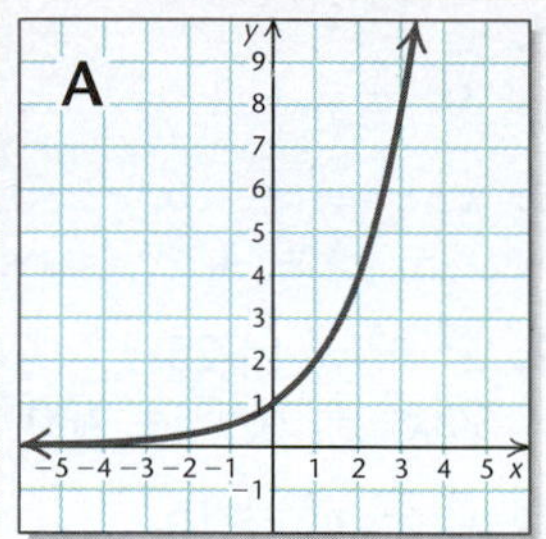

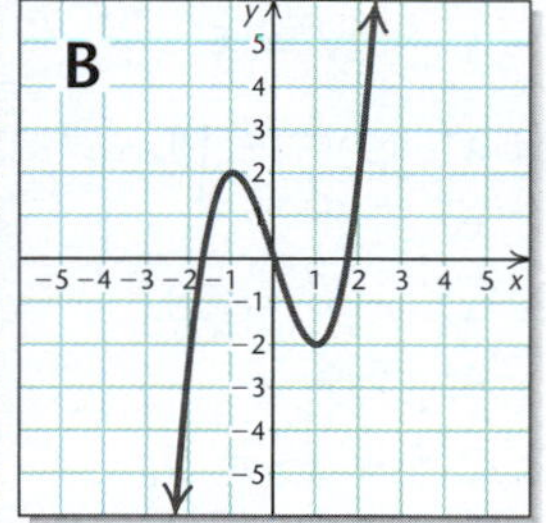

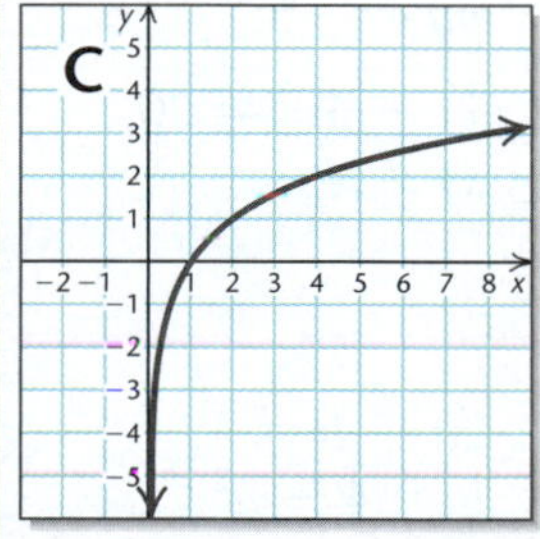

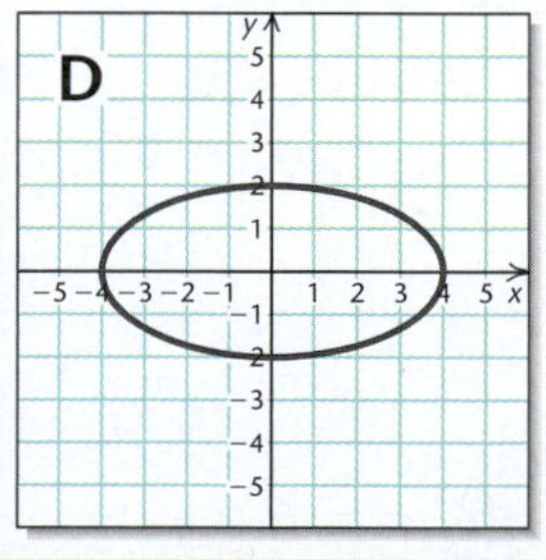

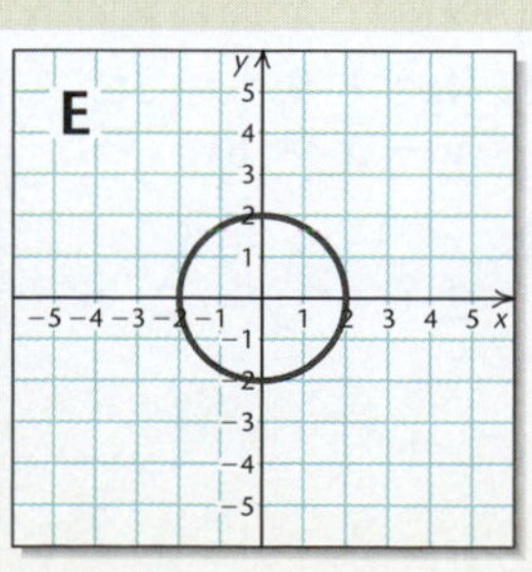

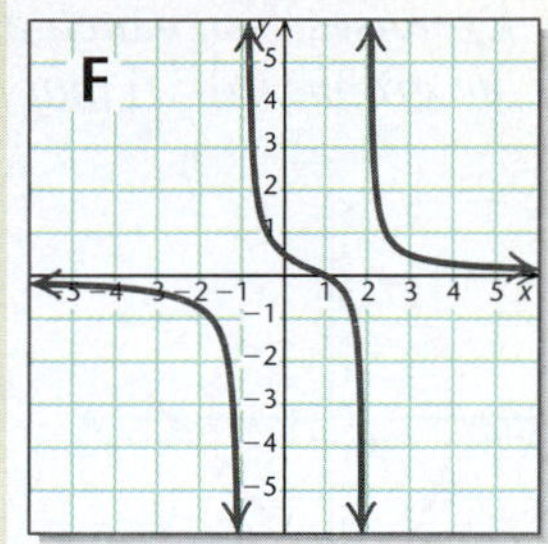

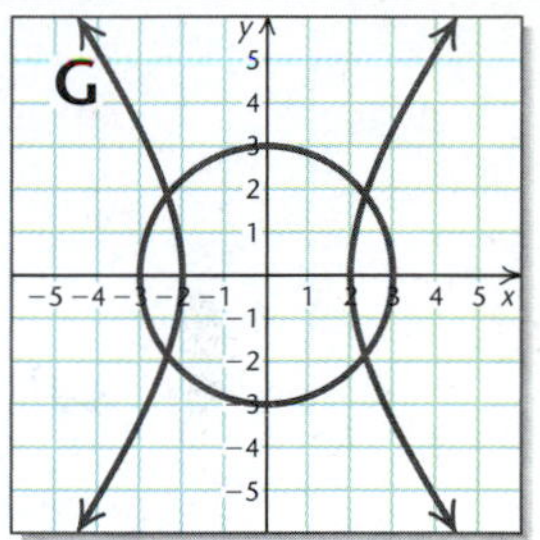

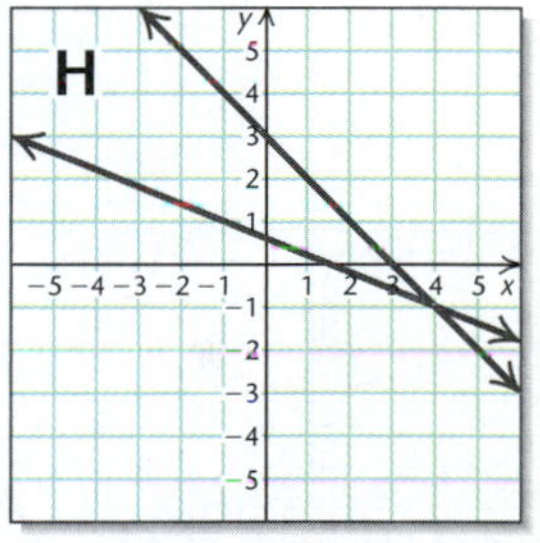

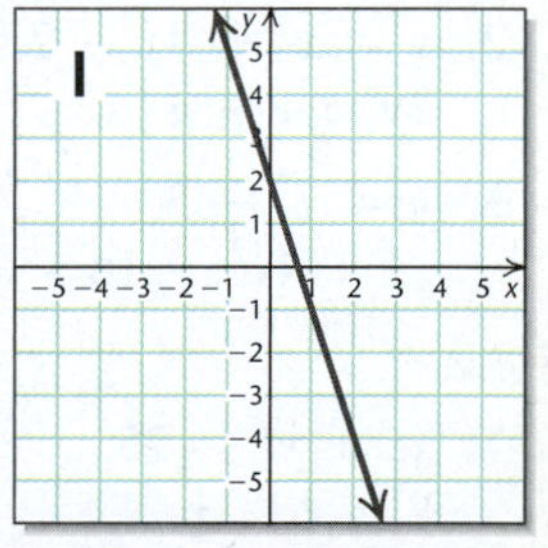

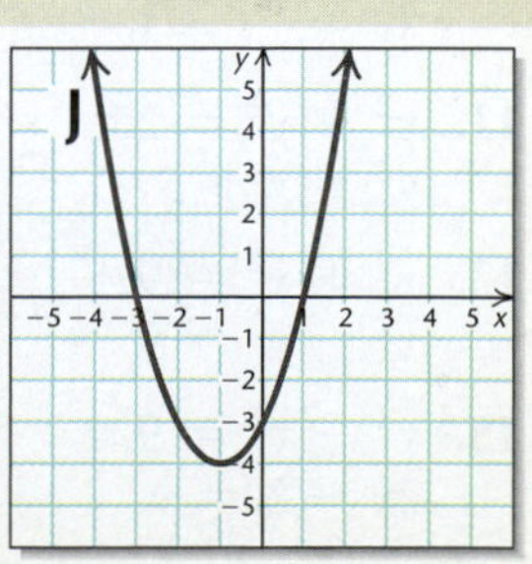

10.4 Exercise Set

In Exercises 1–6, match the system of equations with one of the graphs (a)–(f) that follow.

a)

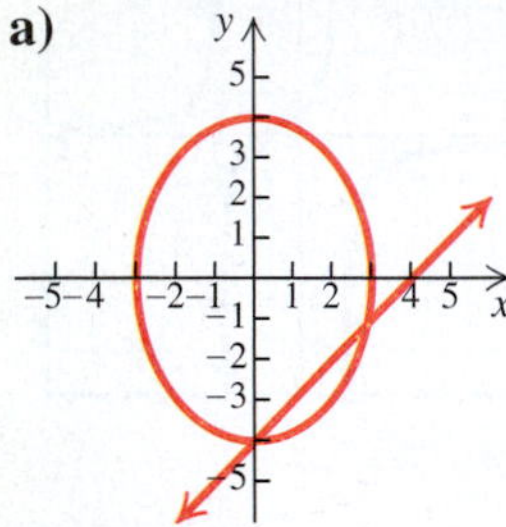

b)

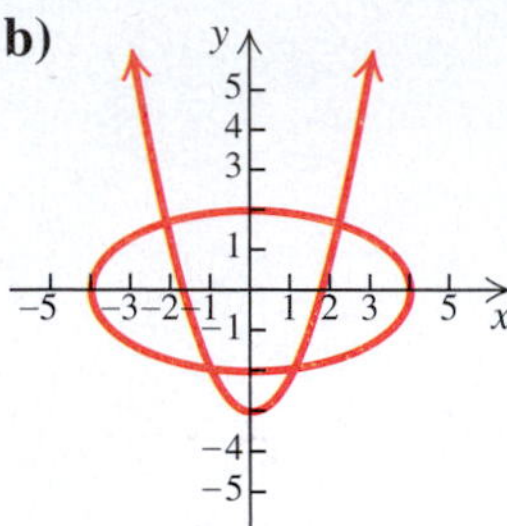

c)

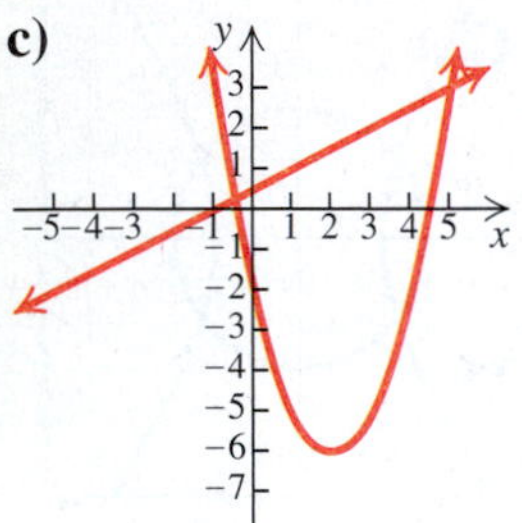

d)

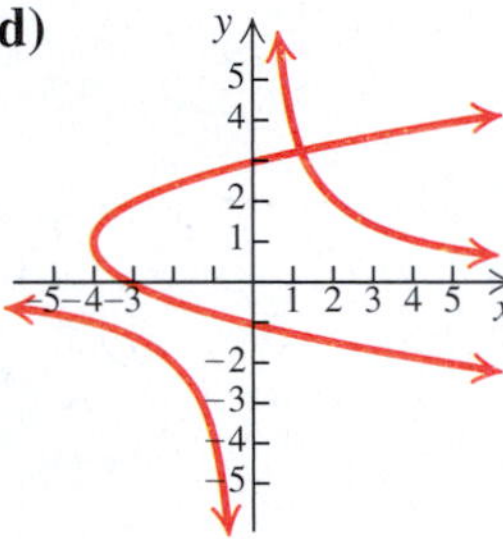

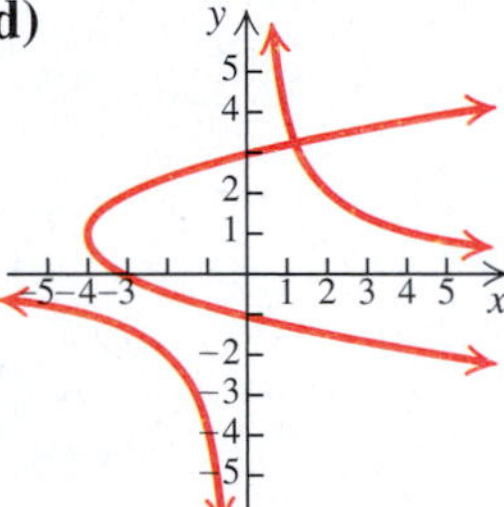

e)

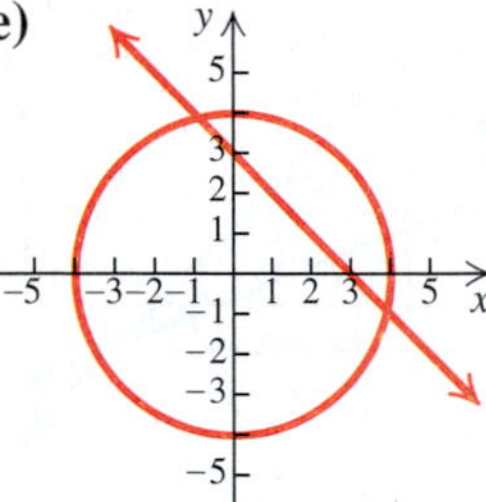

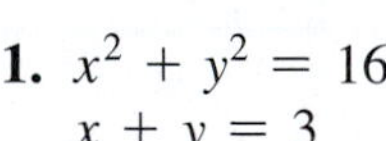

f)

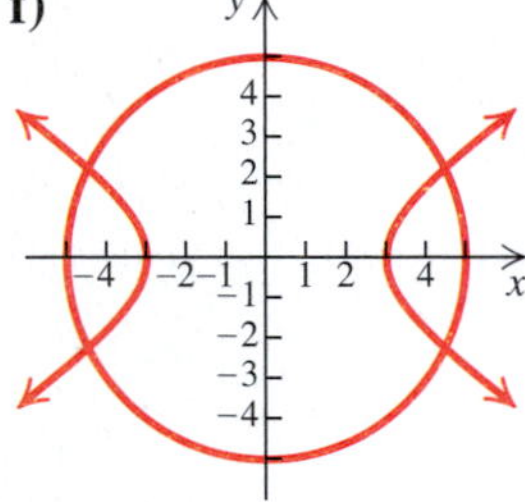

1. $x^2 + y^2 = 16,$
$x + y = 3$

2. $16x^2 + 9y^2 = 144,$
$x - y = 4$

3. $y = x^2 - 4x - 2,$
$2y - x = 1$

4. $4x^2 - 9y^2 = 36,$
$x^2 + y^2 = 25$

5. $y = x^2 - 3,$
$x^2 + 4y^2 = 16$

6. $y^2 - 2y = x + 3,$
$xy = 4$

Solve.

7. $x^2 + y^2 = 25,$
$y - x = 1$

8. $x^2 + y^2 = 100,$
$y - x = 2$

9. $4x^2 + 9y^2 = 36,$
$3y + 2x = 6$

10. $9x^2 + 4y^2 = 36,$
$3x + 2y = 6$

11. $x^2 + y^2 = 25,$
$y^2 = x + 5$

12. $y = x^2,$
$x = y^2$

13. $x^2 + y^2 = 9,$
$x^2 - y^2 = 9$

14. $y^2 - 4x^2 = 4,$
$4x^2 + y^2 = 4$

15. $y^2 - x^2 = 9,$
$2x - 3 = y$

16. $x + y = -6,$
$xy = -7$

17. $y^2 = x + 3,$
$2y = x + 4$

18. $y = x^2,$
$3x = y + 2$

19. $x^2 + y^2 = 25,$
$xy = 12$

20. $x^2 - y^2 = 16,$
$x + y^2 = 4$

21. $x^2 + y^2 = 4,$
$16x^2 + 9y^2 = 144$

22. $x^2 + y^2 = 25,$
$25x^2 + 16y^2 = 400$

23. $x^2 + 4y^2 = 25,$
$x + 2y = 7$

24. $y^2 - x^2 = 16,$
$2x - y = 1$

25. $x^2 - xy + 3y^2 = 27,$
$x - y = 2$

26. $2y^2 + xy + x^2 = 7,$
$x - 2y = 5$

27. $x^2 + y^2 = 16,$
$y^2 - 2x^2 = 10$

28. $x^2 + y^2 = 14,$
$x^2 - y^2 = 4$

29. $x^2 + y^2 = 5,$
$xy = 2$

30. $x^2 + y^2 = 20,$
$xy = 8$

31. $3x + y = 7,$
$4x^2 + 5y = 56$

32. $2y^2 + xy = 5,$
$4y + x = 7$

33. $a + b = 7,$
$ab = 4$

34. $p + q = -4,$
$pq = -5$

35. $x^2 + y^2 = 13,$
$xy = 6$

36. $x^2 + 4y^2 = 20,$
$xy = 4$

37. $x^2 + y^2 + 6y + 5 = 0,$
$x^2 + y^2 - 2x - 8 = 0$

38. $2xy + 3y^2 = 7,$
$3xy - 2y^2 = 4$

39. $2a + b = 1,$
$b = 4 - a^2$

40. $4x^2 + 9y^2 = 36,$
$x + 3y = 3$

41. $a^2 + b^2 = 89,$
$a - b = 3$

42. $xy = 4,$
$x + y = 5$

43. $xy - y^2 = 2,$
$2xy - 3y^2 = 0$

44. $4a^2 - 25b^2 = 0,$
$2a^2 - 10b^2 = 3b + 4$

45. $m^2 - 3mn + n^2 + 1 = 0,$
$3m^2 - mn + 3n^2 = 13$

46. $ab - b^2 = -4,$
$ab - 2b^2 = -6$

47. $x^2 + y^2 = 5,$
$x - y = 8$

48. $4x^2 + 9y^2 = 36,$
$y - x = 8$

49. $a^2 + b^2 = 14,$
$ab = 3\sqrt{5}$

50. $x^2 + xy = 5,$
$2x^2 + xy = 2$

51. $x^2 + y^2 = 25,$
$9x^2 + 4y^2 = 36$

52. $x^2 + y^2 = 1,$
$9x^2 - 16y^2 = 144$

53. $5y^2 - x^2 = 1,$
$xy = 2$

54. $x^2 - 7y^2 = 6,$
$xy = 1$

In Exercises 55–58, determine whether the statement is true or false.

55. A nonlinear system of equations can have both real-number solutions and imaginary-number solutions.

56. If the graph of a nonlinear system of equations consists of a line and a parabola, then the system has two real-number solutions.

57. If the graph of a nonlinear system of equations consists of a line and a circle, then the system has at most two real-number solutions.

58. If the graph of a nonlinear system of equations consists of a line and an ellipse, then it is possible for the system to have exactly one real-number solution.

59. *Photo Dimensions.* Hailey's Frame Shop has been commissioned to frame 5 black-and-white photos for an island resort. Each photo has a perimeter of 68 in. and a diagonal of 26 in. Find the dimensions of the photos.

60. *Sign Dimensions.* Alison's Advertising is building a rectangular sign with an area of 2 yd^2 and a perimeter of 6 yd. Find the dimensions of the sign.

61. *Graphic Design.* Marcia Graham, owner of Graham's Graphics, is designing an advertising brochure for the Art League's spring show. Each page of the brochure is rectangular with an area of 20 in^2 and a perimeter of 18 in. Find the dimensions of the brochure.

62. *Landscaping.* Green Leaf Landscaping is planting a rectangular wildflower garden with a perimeter of 6 m and a diagonal of $\sqrt{5}$ m. Find the dimensions of the garden.

63. *Fencing.* Clark's Country Pet Resort is fencing a new play area for dogs. The manager has purchased 210 yd of fence to enclose a rectangular pen. The area of the pen must be 2250 yd^2. What are the dimensions of the pen?

64. *Carpentry.* Ted Hansen of Hansen Woodworking Designs has been commissioned to make a rectangular tabletop with an area of $\sqrt{2}$ m^2 and a diagonal of $\sqrt{3}$ m for the Decorators' Show House. Find the dimensions of the tabletop.

65. *Banner Design.* A rectangular banner with an area of $\sqrt{3}$ m^2 is being designed to advertise an exhibit at the Davis Gallery. The length of a diagonal is 2 m. Find the dimensions of the banner.

66. *Investment.* Jenna made an investment for 1 year that earned \$7.50 simple interest. If the principal had been \$25 more and the interest rate 1% less, then the interest would have been the same. Find the principal and the interest rate.

67. *Office Dimensions.* The diagonal of the floor of a rectangular office cubicle is 1 ft longer than the length of the cubicle and 3 ft longer than twice the width. Find the dimensions of the cubicle.

68. *Seed Test Plots.* The Burton Seed Company has two square test plots. The sum of their areas is 832 ft^2, and the difference of their areas is 320 ft^2. Find the length of a side of each plot.

In Exercises 69–74, match the system of inequalities with one of the graphs (a)–(f) that follow.

a)
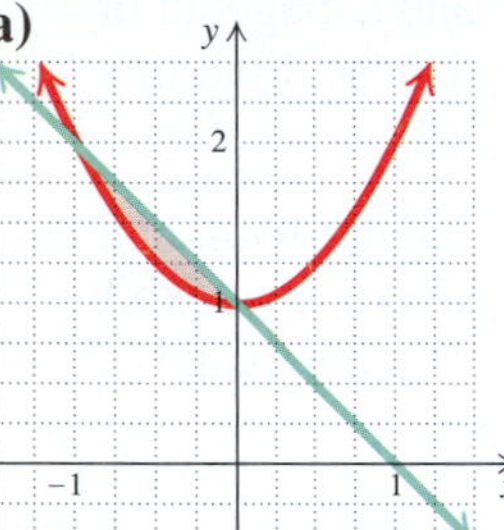

b)
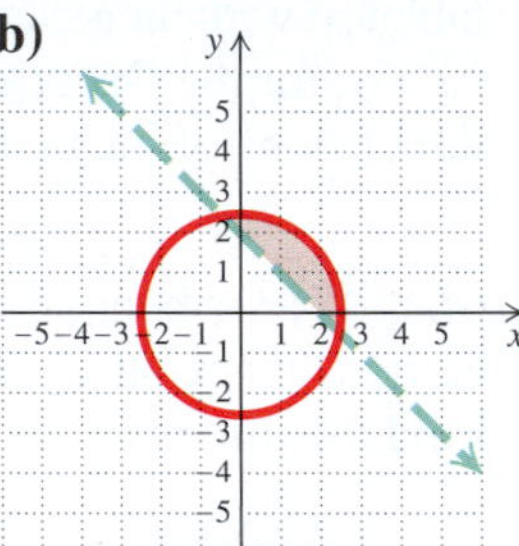

c)
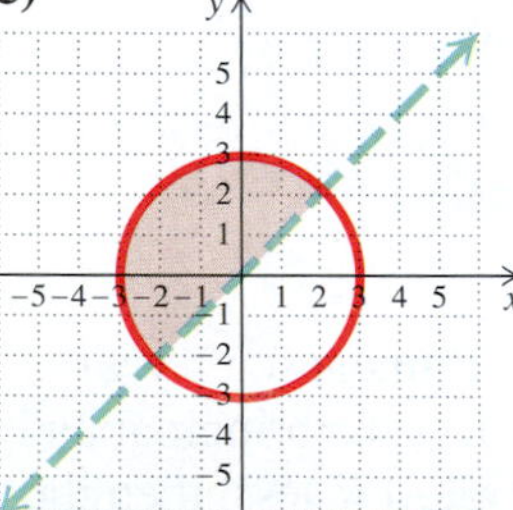

d)
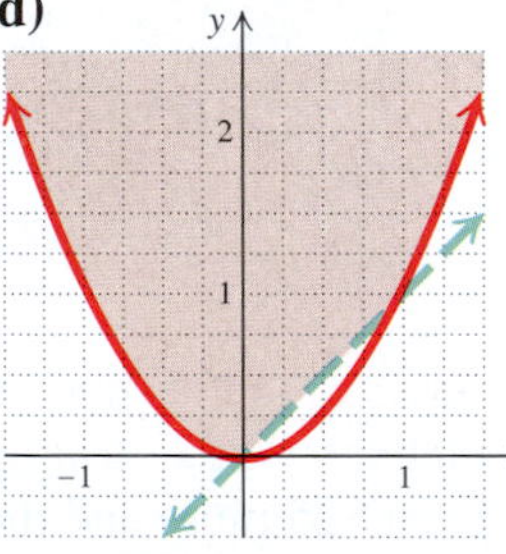

e)
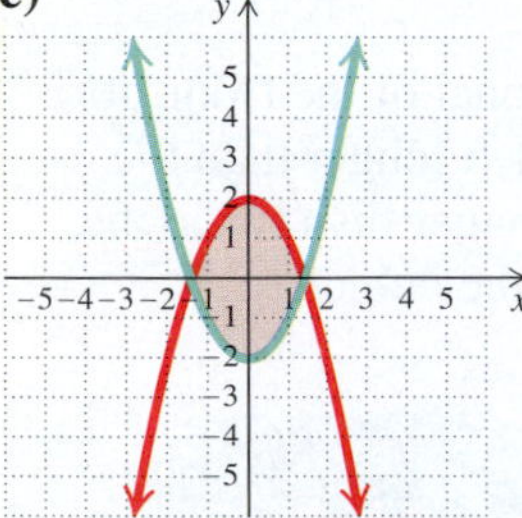

f)
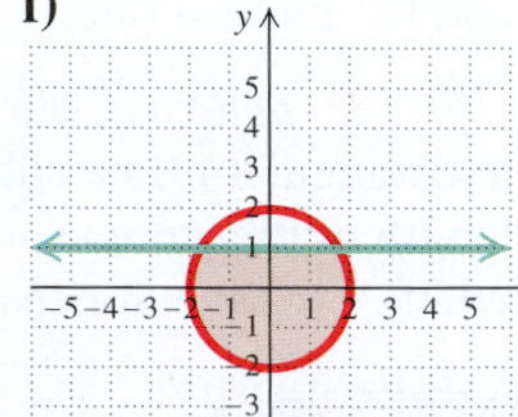

69. $x^2 + y^2 \le 5$,
$x + y > 2$

70. $y \le 2 - x^2$,
$y \ge x^2 - 2$

71. $y \ge x^2$,
$y > x$

72. $x^2 + y^2 \le 4$,
$y \le 1$

73. $y \ge x^2 + 1$,
$x + y \le 1$

74. $x^2 + y^2 \le 9$,
$y > x$

Graph the system of inequalities. Then find the coordinates of the points of intersection of the graphs of the related equations.

75. $x^2 + y^2 \le 16$,
$y < x$

76. $x^2 + y^2 \le 10$,
$y > x$

77. $x^2 \le y$,
$x + y \ge 2$

78. $x \ge y^2$,
$x - y \le 2$

79. $x^2 + y^2 \le 25$,
$x - y > 5$

80. $x^2 + y^2 \ge 9$,
$x - y > 3$

81. $y \ge x^2 - 3$,
$y \le 2x$

82. $y \le 3 - x^2$,
$y \ge x + 1$

83. $y \ge x^2$,
$y < x + 2$

84. $y \le 1 - x^2$,
$y > x - 1$

Skill Maintenance

Solve. **[5.5]**

85. $2^{3x} = 64$

86. $5^x = 27$

87. $\log_3 x = 4$

88. $\log(x - 3) + \log x = 1$

Synthesis

89. Find an equation of the circle that passes through the points $(2, 4)$ and $(3, 3)$ and whose center is on the line $3x - y = 3$.

90. Find an equation of the circle that passes through the points $(2, 3)$, $(4, 5)$, and $(0, -3)$.

91. Find an equation of an ellipse centered at the origin that passes through the points $(1, \sqrt{3}/2)$ and $(\sqrt{3}, 1/2)$.

92. Find an equation of a hyperbola of the type

$$\frac{x^2}{b^2} - \frac{y^2}{a^2} = 1$$

that passes through the points $(-3, -3\sqrt{5}/2)$ and $(-3/2, 0)$.

93. Show that a hyperbola does not intersect its asymptotes. That is, solve the system of equations

$$\frac{x^2}{a^2} - \frac{y^2}{b^2} = 1,$$

$$y = \frac{b}{a}x\left(\text{or } y = -\frac{b}{a}x\right).$$

94. *Numerical Relationship.* Find two numbers whose product is 2 and the sum of whose reciprocals is $\frac{33}{8}$.

95. *Numerical Relationship.* The sum of two numbers is 1, and their product is 1. Find the sum of their cubes. There is a method to solve this problem that is easier than solving a nonlinear system of equations. Can you discover it?

96. *Box Dimensions.* Four squares with sides 5 in. long are cut from the corners of a rectangular metal sheet that has an area of 340 in^2. The edges are bent up to form an open box with a volume of 350 in^3. Find the dimensions of the box.

Solve.

97. $x^3 + y^3 = 72,$
$x + y = 6$

98. $a + b = \dfrac{5}{6},$
$\dfrac{a}{b} + \dfrac{b}{a} = \dfrac{13}{6}$

99. $p^2 + q^2 = 13,$
$\dfrac{1}{pq} = -\dfrac{1}{6}$

100. $e^x - e^{x+y} = 0,$
$e^y - e^{x-y} = 0$

Solve using a graphing calculator. Find all real solutions.

101. $y - \ln x = 2,$
$y = x^2$

102. $y = \ln(x + 4),$
$x^2 + y^2 = 6$

103. $y = e^x,$
$x - y = -2$

104. $y - e^{-x} = 1,$
$y = 2x + 5$

105. $14.5x^2 - 13.5y^2 - 64.5 = 0,$
$5.5x - 6.3y - 12.3 = 0$

106. $2x + 2y = 1660,$
$xy = 35{,}325$

107. $0.319x^2 + 2688.7y^2 = 56{,}548,$
$0.306x^2 - 2688.7y^2 = 43{,}452$

108. $13.5xy + 15.6 = 0,$
$5.6x - 6.7y - 42.3 = 0$

Mid-Chapter Mixed Review

Determine whether the statement is true or false.

1. The graph of $(x + 3)^2 = 8(y - 2)$ is a parabola with vertex $(-3, 2)$. **[10.1]**

2. The graph of $(x - 4)^2 + (y + 1)^2 = 9$ is a circle with radius 9. **[10.2]**

3. The hyperbola $\dfrac{x^2}{5} - \dfrac{y^2}{10} = 1$ has a horizontal transverse axis. **[10.3]**

4. Every nonlinear system of equations has at least one real-number solution. **[10.4]**

In Exercises 5–12, match the equation with one of the graphs (a)–(h) that follow. **[10.1], [10.2], [10.3]**

a)

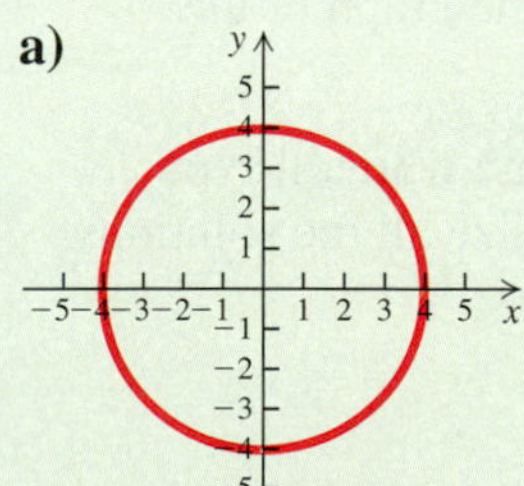

b)

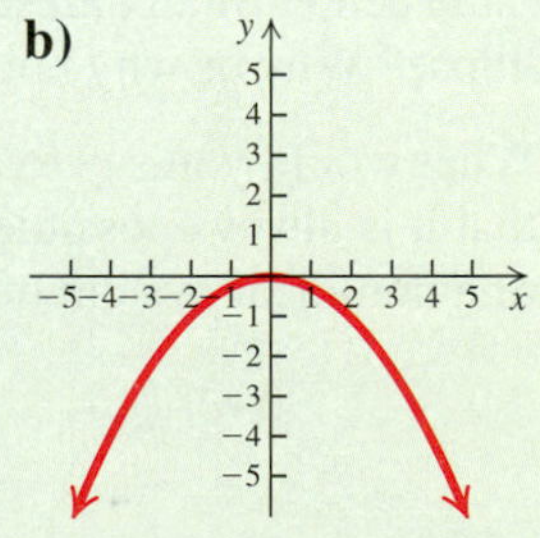

c)

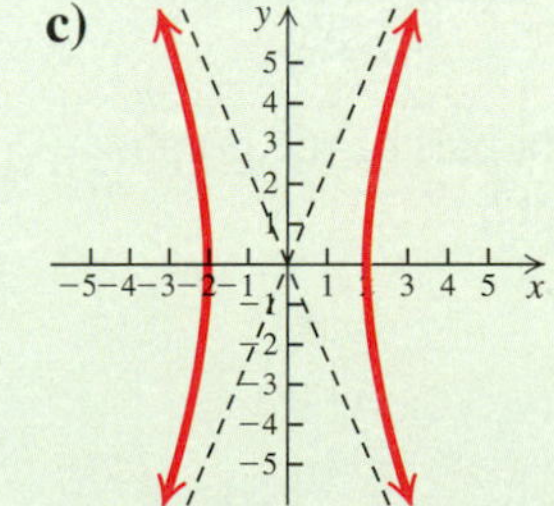

d)

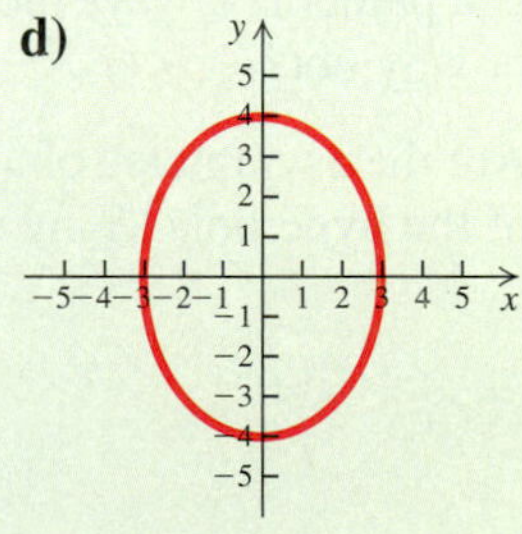

e)

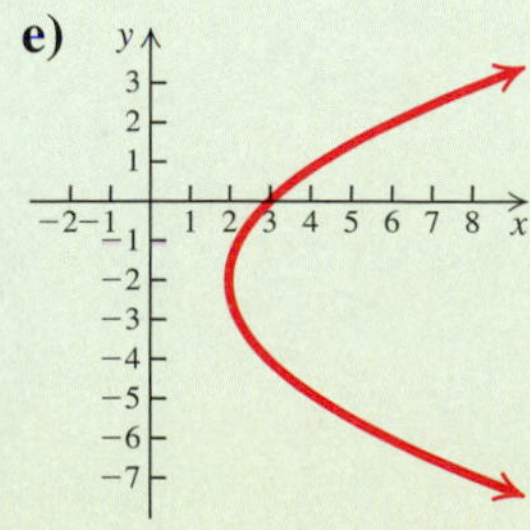

f)

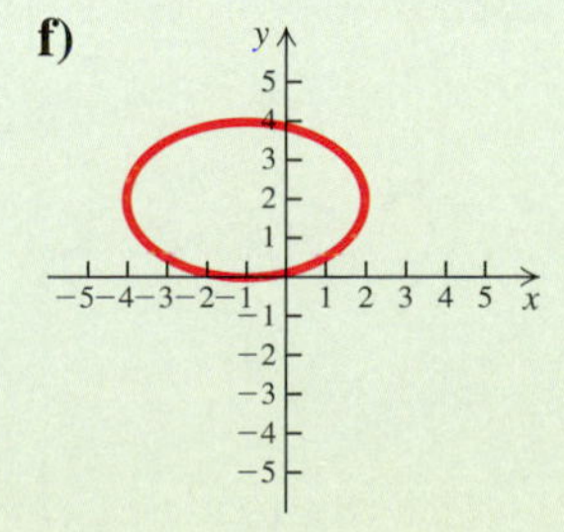

g)

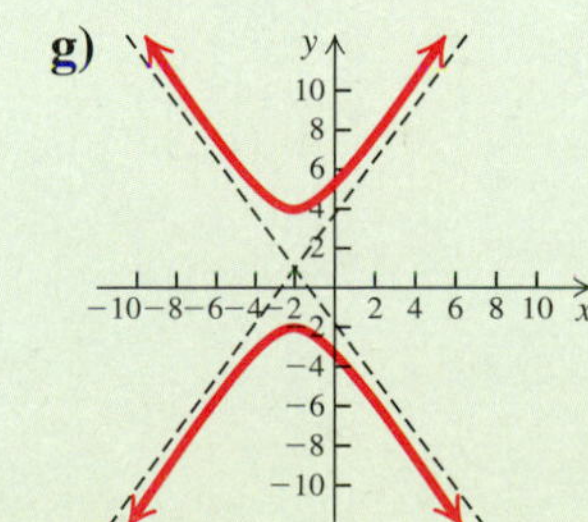

h)

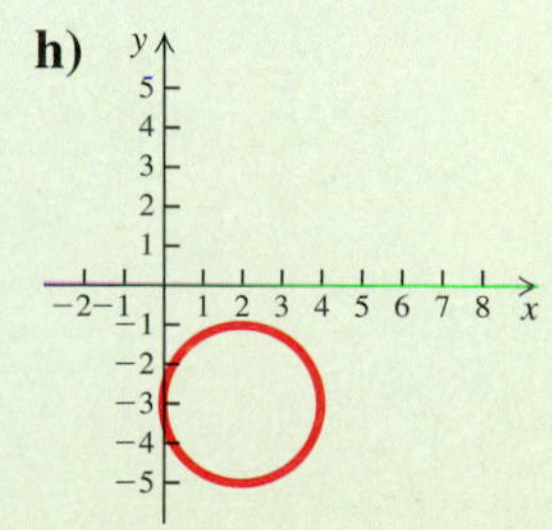

5. $x^2 = -4y$

6. $(y + 2)^2 = 4(x - 2)$

7. $16x^2 + 9y^2 = 144$

8. $x^2 + y^2 = 16$

9. $4(y - 1)^2 - 9(x + 2)^2 = 36$

10. $4(x + 1)^2 + 9(y - 2)^2 = 36$

11. $(x - 2)^2 + (y + 3)^2 = 4$

12. $25x^2 - 4y^2 = 100$

Find the vertex, the focus, and the directrix of the parabola. Then draw the graph. [10.1]

13. $y^2 = 12x$

14. $x^2 - 6x - 4y = -17$

Find the center and the radius of the circle. Then draw the graph. [10.2]

15. $x^2 + y^2 + 4x - 8y = 5$

16. $x^2 + y^2 - 6x + 2y - 6 = 0$

Find the vertices and the foci of the ellipse. Then draw the graph. [10.2]

17. $\frac{x^2}{1} + \frac{y^2}{9} = 1$

18. $25x^2 + 4y^2 - 50x + 8y = 71$

Find the center, the vertices, the foci, and the asymptotes of the hyperbola. Then draw the graph. [10.3]

19. $9y^2 - 16x^2 = 144$

20. $\frac{(x + 3)^2}{1} - \frac{(y - 2)^2}{4} = 1$

Solve. [10.4]

21. $x^2 + y^2 = 29,$
$x - y = 3$

22. $x^2 + y^2 = 8,$
$xy = 4$

23. $x^2 + 2y^2 = 20,$
$y^2 - x^2 = 28$

24. $2x - y = -4,$
$3x^2 + 2y = 7$

25. The sum of two numbers is 1 and the sum of their squares is 13. Find the numbers. [10.4]

Graph the system of inequalities. Then find the coordinates of the points of intersection of the graphs of the related equations. [10.4]

26. $x^2 + y^2 \leq 8,$
$x > y$

27. $y \geq x^2 - 1,$
$y \leq x + 1$

COLLABORATIVE DISCUSSION AND WRITING

28. Is a parabola always the graph of a function? Why or why not? [10.1]

29. Is the center of an ellipse part of the graph of the ellipse? Why or why not? [10.2]

30. Are the asymptotes of a hyperbola part of the graph of the hyperbola? Why or why not? [10.3]

31. What would you say to a classmate who tells you that it is always possible to visualize all the solutions of a nonlinear system of equations? [10.4]

10.5 Rotation of Axes

- Use rotation of axes to graph conic sections.
- Use the discriminant to determine the type of conic represented by a given equation.

CONIC SECTIONS

REVIEW SECTIONS 10.1–10.3.

In Section 10.1, we saw that conic sections can be defined algebraically using a second-degree equation of the form $Ax^2 + Bxy + Cy^2 + Dx + Ey + F = 0$. Up to this point, we have considered only equations of this form for which $B = 0$. Now we turn our attention to equations of conics that contain an xy-term.

Rotation of Axes

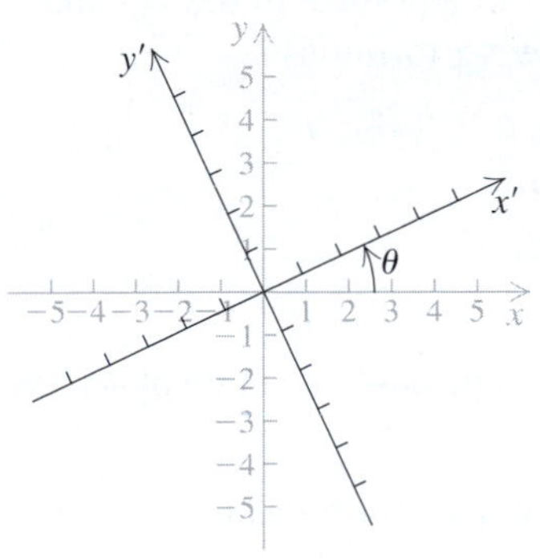

When B is nonzero, the graph of $Ax^2 + Bxy + Cy^2 + Dx + Ey + F = 0$ is a conic section with an axis that is parallel to neither the x-axis nor the y-axis. We use a technique called **rotation of axes** when we graph such an equation. The goal is to rotate the x- and y-axes through a positive angle θ to yield an $x'y'$-coordinate system, as shown at left. For the appropriate choice of θ, the graph of any conic section with an xy-term will have its axis parallel to the x'-axis or the y'-axis.

Algebraically we want to rewrite an equation

$$Ax^2 + Bxy + Cy^2 + Dx + Ey + F = 0$$

in the xy-coordinate system in the form

$$A'(x')^2 + C'(y')^2 + D'x' + E'y' + F' = 0$$

in the $x'y'$-coordinate system. Equations of this second type were graphed in Sections 10.1–10.3.

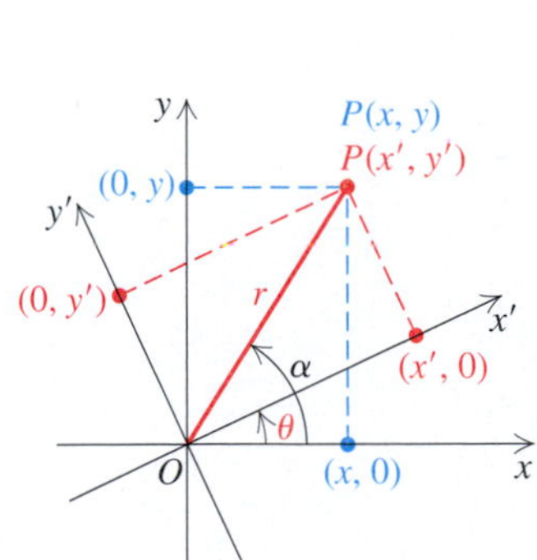

To achieve our goal, we find formulas relating the xy-coordinates of a point and the $x'y'$-coordinates of the same point. We begin by letting P be a point with coordinates (x, y) in the xy-coordinate system and (x', y') in the $x'y'$-coordinate system, as shown at left.

We let r represent the distance OP, and we let α represent the angle from the x-axis to OP. Then

$$\cos \alpha = \frac{x}{r} \quad \text{and} \quad \sin \alpha = \frac{y}{r},$$

so

$$x = r \cos \alpha \quad \text{and} \quad y = r \sin \alpha.$$

We also see from the figure above that

$$\cos(\alpha - \theta) = \frac{x'}{r} \quad \text{and} \quad \sin(\alpha - \theta) = \frac{y'}{r},$$

so

$$x' = r\cos(\alpha - \theta) \quad \text{and} \quad y' = r\sin(\alpha - \theta).$$

Then

$$x' = r\cos\alpha\cos\theta + r\sin\alpha\sin\theta$$

and

$$y' = r\sin\alpha\cos\theta - r\cos\alpha\sin\theta.$$

Substituting x for $r\cos\alpha$ and y for $r\sin\alpha$ gives us

$$x' = x\cos\theta + y\sin\theta \qquad (1)$$

and

$$y' = y\cos\theta - x\sin\theta. \qquad (2)$$

We can use these formulas to find the $x'y'$-coordinates of any point given that point's xy-coordinates and an angle of rotation θ. To express xy-coordinates in terms of $x'y'$-coordinates and an angle of rotation θ, we solve the system composed of equations (1) and (2) above for x and y. (See Exercise 43.) We get

$$x = x'\cos\theta - y'\sin\theta \quad \text{and} \quad y = x'\sin\theta + y'\cos\theta.$$

ROTATION OF AXES FORMULAS

If the x- and y-axes are rotated about the origin through a positive acute angle θ, then the coordinates (x, y) and (x', y') of a point P in the xy- and $x'y'$-coordinate systems are related by the following formulas:

$$x' = x\cos\theta + y\sin\theta, \qquad y' = -x\sin\theta + y\cos\theta;$$
$$x = x'\cos\theta - y'\sin\theta, \qquad y = x'\sin\theta + y'\cos\theta.$$

EXAMPLE 1 Suppose that the xy-axes are rotated through an angle of 45°. Write the equation $xy = 1$ in the $x'y'$-coordinate system.

Solution We substitute 45° for θ in the rotation of axes formulas for x and y:

$$x = x'\cos 45° - y'\sin 45°,$$
$$y = x'\sin 45° + y'\cos 45°.$$

We know that

$$\sin 45° = \frac{\sqrt{2}}{2} \quad \text{and} \quad \cos 45° = \frac{\sqrt{2}}{2},$$

so we have

$$x = x'\left(\frac{\sqrt{2}}{2}\right) - y'\left(\frac{\sqrt{2}}{2}\right) = \frac{\sqrt{2}}{2}(x' - y')$$

and

$$y = x'\left(\frac{\sqrt{2}}{2}\right) + y'\left(\frac{\sqrt{2}}{2}\right) = \frac{\sqrt{2}}{2}(x' + y').$$

Next, we substitute these expressions for x and y in the equation $xy = 1$:

$$\frac{\sqrt{2}}{2}(x' - y') \cdot \frac{\sqrt{2}}{2}(x' + y') = 1$$
$$\frac{1}{2}[(x')^2 - (y')^2] = 1$$
$$\frac{(x')^2}{2} - \frac{(y')^2}{2} = 1, \quad \text{or} \quad \frac{(x')^2}{(\sqrt{2})^2} - \frac{(y')^2}{(\sqrt{2})^2} = 1.$$

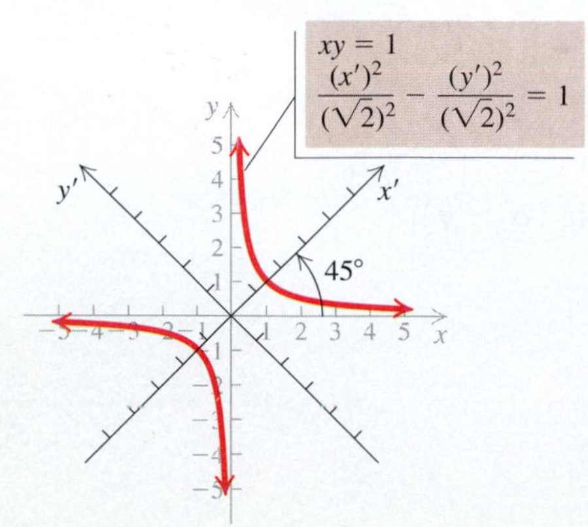

We have the equation of a hyperbola in the $x'y'$-coordinate system with its transverse axis on the x'-axis and with vertices $(-\sqrt{2}, 0)$ and $(\sqrt{2}, 0)$. Its asymptotes are $y' = -x'$ and $y' = x'$. These correspond to the axes of the xy-coordinate system. We now sketch the graph as shown at left.

Now let's substitute the rotation of axes formulas for x and y in the equation

$$Ax^2 + Bxy + Cy^2 + Dx + Ey + F = 0.$$

We have

$$\begin{aligned} A(x' \cos\theta - y' \sin\theta)^2 &+ B(x' \cos\theta - y' \sin\theta)(x' \sin\theta + y' \cos\theta) \\ &+ C(x' \sin\theta + y' \cos\theta)^2 + D(x' \cos\theta - y' \sin\theta) \\ &+ E(x' \sin\theta + y' \cos\theta) + F = 0. \end{aligned}$$

Performing the operations indicated and collecting like terms yields the equation

$$A'(x')^2 + B'x'y' + C'(y')^2 + D'x' + E'y' + F' = 0, \quad \textbf{(3)}$$

where

$$\begin{aligned} A' &= A\cos^2\theta + B\sin\theta\cos\theta + C\sin^2\theta, \\ B' &= 2(C - A)\sin\theta\cos\theta + B(\cos^2\theta - \sin^2\theta), \\ C' &= A\sin^2\theta - B\sin\theta\cos\theta + C\cos^2\theta, \\ D' &= D\cos\theta + E\sin\theta, \\ E' &= -D\sin\theta + E\cos\theta, \quad \text{and} \\ F' &= F. \end{aligned}$$

Recall that our goal is to produce an equation without an $x'y'$-term, or with $B' = 0$. Then we must have

$$\begin{aligned} 2(C - A)\sin\theta\cos\theta + B(\cos^2\theta - \sin^2\theta) &= 0 \\ (C - A)\sin 2\theta + B\cos 2\theta &= 0 \quad \text{Using double-angle formulas} \\ B\cos 2\theta &= (A - C)\sin 2\theta \\ \frac{\cos 2\theta}{\sin 2\theta} &= \frac{A - C}{B} \\ \cot 2\theta &= \frac{A - C}{B}. \end{aligned}$$

Thus, when θ is chosen so that

$$\cot 2\theta = \frac{A - C}{B},$$

equation (3) will have no $x'y'$-term. Although we will not do so here, it can be shown that we can always find θ such that $0° < 2\theta < 180°$, or $0° < \theta < 90°$.

ELIMINATING THE *xy*-TERM

To eliminate the xy-term from the equation

$$Ax^2 + Bxy + Cy^2 + Dx + Ey + F = 0, \quad B \neq 0,$$

select an angle θ such that

$$\cot 2\theta = \frac{A - C}{B}, \quad 0° < 2\theta < 180°,$$

and use the rotation of axes formulas.

EXAMPLE 2 Graph the equation

$$3x^2 - 2\sqrt{3}xy + y^2 + 2x + 2\sqrt{3}y = 0.$$

Solution We have

$$A = 3, \quad B = -2\sqrt{3}, \quad C = 1, \quad D = 2, \quad E = 2\sqrt{3}, \quad \text{and} \quad F = 0.$$

To select the angle of rotation θ, we must have

$$\cot 2\theta = \frac{A - C}{B} = \frac{3 - 1}{-2\sqrt{3}} = \frac{2}{-2\sqrt{3}} = -\frac{1}{\sqrt{3}}.$$

Thus, $2\theta = 120°$, and $\theta = 60°$. We substitute this value for θ in the rotation of axes formulas for x and y:

$$x = x' \cos 60° - y' \sin 60°,$$
$$y = x' \sin 60° + y' \cos 60°.$$

This gives us

$$x = x' \cdot \frac{1}{2} - y' \cdot \frac{\sqrt{3}}{2} = \frac{x'}{2} - \frac{y'\sqrt{3}}{2}$$

and

$$y = x' \cdot \frac{\sqrt{3}}{2} + y' \cdot \frac{1}{2} = \frac{x'\sqrt{3}}{2} + \frac{y'}{2}.$$

Now we substitute these expressions for x and y in the given equation:

$$3\left(\frac{x'}{2} - \frac{y'\sqrt{3}}{2}\right)^2 - 2\sqrt{3}\left(\frac{x'}{2} - \frac{y'\sqrt{3}}{2}\right)\left(\frac{x'\sqrt{3}}{2} + \frac{y'}{2}\right) +$$
$$\left(\frac{x'\sqrt{3}}{2} + \frac{y'}{2}\right)^2 + 2\left(\frac{x'}{2} - \frac{y'\sqrt{3}}{2}\right) + 2\sqrt{3}\left(\frac{x'\sqrt{3}}{2} + \frac{y'}{2}\right) = 0.$$

After simplifying, we get

$$4(y')^2 + 4x' = 0, \quad \text{or}$$
$$(y')^2 = -x'.$$

This is the equation of a parabola with its vertex at $(0, 0)$ of the $x'y'$-coordinate system and axis of symmetry $y' = 0$. We sketch the graph.

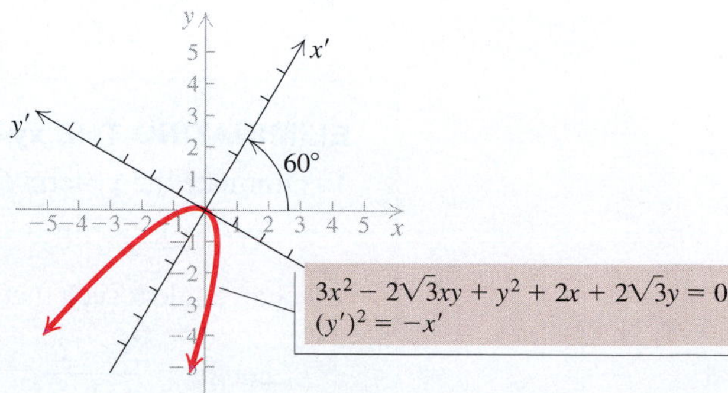

Now Try Exercise 23.

The Discriminant

It is possible to determine the type of conic represented by the equation $Ax^2 + Bxy + Cy^2 + Dx + Ey + F = 0$ before rotating the axes. Using the expressions for A', B', and C' in terms of A, B, C, and θ developed earlier, it can be shown that

$$(B')^2 - 4A'C' = B^2 - 4AC.$$

Now when θ is chosen so that

$$\cot 2\theta = \frac{A - C}{B},$$

rotation of axes gives us an equation

$$A'(x')^2 + C'(y')^2 + D'x' + E'y' + F' = 0.$$

If A' and C' have the same sign, or $A'C' > 0$, then the graph of this equation is an ellipse or a circle. If A' and C' have different signs, or $A'C' < 0$, then the graph is a hyperbola. And, if either $A' = 0$ or $C' = 0$, or $A'C' = 0$, the graph is a parabola.

Because $B' = 0$ and $(B')^2 - 4A'C' = B^2 - 4AC$, it follows that $B^2 - 4AC = -4A'C'$. Then the graph is an ellipse or a circle if $B^2 - 4AC < 0$, a hyperbola if $B^2 - 4AC > 0$, or a parabola if $B^2 - 4AC = 0$. (There are certain special cases, called *degenerate conics*, where these statements do not hold, but we will not concern ourselves with these here.) The expression $B^2 - 4AC$ is the **discriminant** of the equation $Ax^2 + Bxy + Cy^2 + Dx + Ey + F = 0$.

> The graph of the equation
>
> $$Ax^2 + Bxy + Cy^2 + Dx + Ey + F = 0$$
>
> is, except in degenerate cases,
>
> **1.** an ellipse or a circle if $B^2 - 4AC < 0$,
> **2.** a hyperbola if $B^2 - 4AC > 0$, and
> **3.** a parabola if $B^2 - 4AC = 0$.

EXAMPLE 3 Graph the equation $3x^2 + 2xy + 3y^2 = 16$.

2²−4∗3∗3
−32

Solution We have

$$A = 3, \quad B = 2, \quad \text{and} \quad C = 3,$$

so $\quad B^2 - 4AC = 2^2 - 4 \cdot 3 \cdot 3 = 4 - 36 = -32.$

Since the discriminant is negative, the graph is an ellipse or a circle. Now, to rotate the axes, we begin by determining θ:

$$\cot 2\theta = \frac{A - C}{B} = \frac{3 - 3}{2} = \frac{0}{2} = 0.$$

Then $2\theta = 90°$ and $\theta = 45°$, so

$$\sin\theta = \frac{\sqrt{2}}{2} \quad \text{and} \quad \cos\theta = \frac{\sqrt{2}}{2}.$$

As we saw in Example 1, substituting these values for $\sin\theta$ and $\cos\theta$ in the rotation of axes formulas gives

$$x = \frac{\sqrt{2}}{2}(x' - y') \quad \text{and} \quad y = \frac{\sqrt{2}}{2}(x' + y').$$

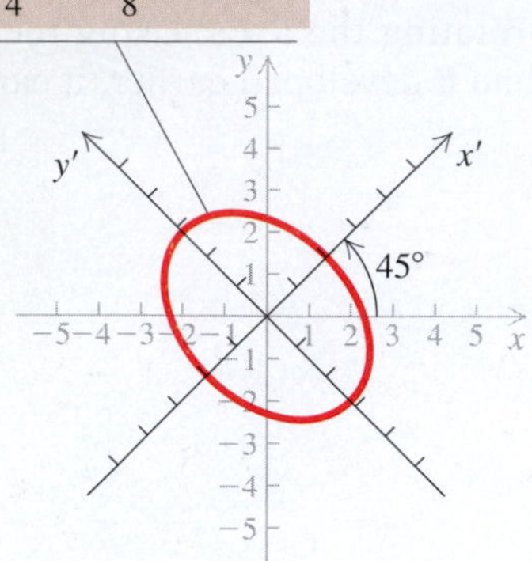

Now we substitute for x and y in the given equation:

$$3\left[\frac{\sqrt{2}}{2}(x' - y')\right]^2 + 2\left[\frac{\sqrt{2}}{2}(x' - y')\right]\left[\frac{\sqrt{2}}{2}(x' + y')\right] + 3\left[\frac{\sqrt{2}}{2}(x' + y')\right]^2 = 16.$$

After simplifying, we have

$$4(x')^2 + 2(y')^2 = 16, \quad \text{or}$$

$$\frac{(x')^2}{4} + \frac{(y')^2}{8} = 1.$$

This is the equation of an ellipse with vertices $(0, -\sqrt{8})$ and $(0, \sqrt{8})$, or $(0, -2\sqrt{2})$ and $(0, 2\sqrt{2})$, on the y'-axis. The x'-intercepts are $(-2, 0)$ and $(2, 0)$. We sketch the graph, shown at left.

Now Try Exercise 19.

EXAMPLE 4 Graph the equation $4x^2 - 24xy - 3y^2 - 156 = 0$.

Solution We have

$$A = 4, \quad B = -24, \quad \text{and} \quad C = -3,$$

so $\quad B^2 - 4AC = (-24)^2 - 4 \cdot 4(-3) = 576 + 48 = 624.$

Since the discriminant is positive, the graph is a hyperbola. To rotate the axes, we begin by determining θ:

$$\cot 2\theta = \frac{A - C}{B} = \frac{4 - (-3)}{-24} = -\frac{7}{24}.$$

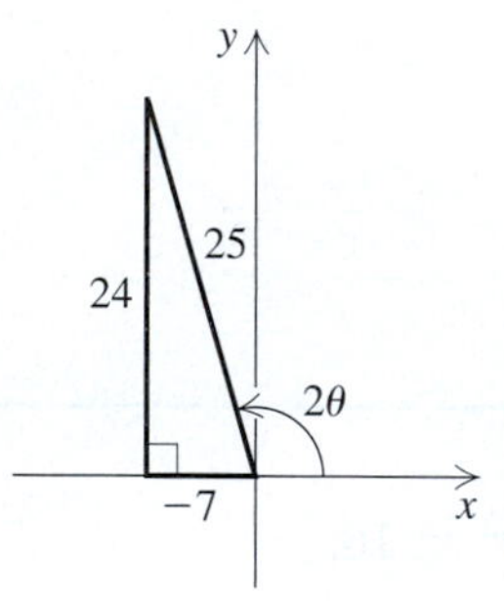

Since $\cot 2\theta < 0$, we have $90° < 2\theta < 180°$. From the triangle at left, we see that $\cos 2\theta = -\frac{7}{25}$.

Using half-angle formulas, we have

$$\sin\theta = \sqrt{\frac{1 - \cos 2\theta}{2}} = \sqrt{\frac{1 - \left(-\frac{7}{25}\right)}{2}} = \frac{4}{5}$$

and

$$\cos\theta = \sqrt{\frac{1 + \cos 2\theta}{2}} = \sqrt{\frac{1 + \left(-\frac{7}{25}\right)}{2}} = \frac{3}{5}.$$

Substituting in the rotation of axes formulas gives us

$$x = x'\cos\theta - y'\sin\theta = \tfrac{3}{5}x' - \tfrac{4}{5}y'$$

and

$$y = x'\sin\theta + y'\cos\theta = \tfrac{4}{5}x' + \tfrac{3}{5}y'.$$

Now we substitute for x and y in the given equation:

$$4\left(\tfrac{3}{5}x' - \tfrac{4}{5}y'\right)^2 - 24\left(\tfrac{3}{5}x' - \tfrac{4}{5}y'\right)\left(\tfrac{4}{5}x' + \tfrac{3}{5}y'\right) - 3\left(\tfrac{4}{5}x' + \tfrac{3}{5}y'\right)^2 - 156 = 0.$$

After simplifying, we have

$$13(y')^2 - 12(x')^2 - 156 = 0$$

$$13(y')^2 - 12(x')^2 = 156$$

$$\frac{(y')^2}{12} - \frac{(x')^2}{13} = 1.$$

The graph of this equation is a hyperbola with vertices $(0, -\sqrt{12})$ and $(0, \sqrt{12})$, or $(0, -2\sqrt{3})$ and $(0, 2\sqrt{3})$, on the y'-axis. Since we know that $\sin\theta = \frac{4}{5}$ and $0° < \theta < 90°$, we can use a calculator to find that $\theta \approx 53.1°$. Thus the xy-axes are rotated through an angle of about 53.1° in order to obtain the $x'y'$-axes. We sketch the graph.

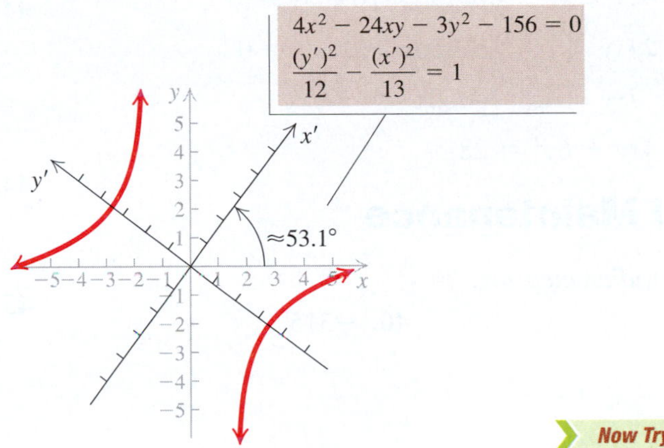

Now Try Exercise 35.

10.5 Exercise Set

For the given angle of rotation and coordinates of a point in the xy-coordinate system, find the coordinates of the point in the x′y′-coordinate system.

1. $\theta = 45°, (\sqrt{2}, -\sqrt{2})$

2. $\theta = 45°, (-1, 3)$

3. $\theta = 30°, (0, 2)$

4. $\theta = 60°, (0, \sqrt{3})$

For the given angle of rotation and coordinates of a point in the x′y′-coordinate system, find the coordinates of the point in the xy-coordinate system.

5. $\theta = 45°, (1, -1)$

6. $\theta = 45°, (-3\sqrt{2}, \sqrt{2})$

7. $\theta = 30°, (2, 0)$

8. $\theta = 60°, (-1, -\sqrt{3})$

Use the discriminant to determine whether the graph of the equation is an ellipse (or a circle), a hyperbola, or a parabola.

9. $3x^2 - 5xy + 3y^2 - 2x + 7y = 0$

10. $5x^2 + 6xy - 4y^2 + x - 3y + 4 = 0$

11. $x^2 - 3xy - 2y^2 + 12 = 0$

12. $4x^2 + 7xy + 2y^2 - 3x + y = 0$

13. $4x^2 - 12xy + 9y^2 - 3x + y = 0$

14. $6x^2 + 5xy + 6y^2 + 15 = 0$

15. $2x^2 - 8xy + 7y^2 + x - 2y + 1 = 0$

16. $x^2 + 6xy + 9y^2 - 3x + 4y = 0$

17. $8x^2 - 7xy + 5y^2 - 17 = 0$

18. $x^2 + xy - y^2 - 4x + 3y - 2 = 0$

Graph the equation.

19. $4x^2 + 2xy + 4y^2 = 15$

20. $3x^2 + 10xy + 3y^2 + 8 = 0$

21. $x^2 - 10xy + y^2 + 36 = 0$

22. $x^2 + 2xy + y^2 + 4\sqrt{2}x - 4\sqrt{2}y = 0$

23. $x^2 - 2\sqrt{3}xy + 3y^2 - 12\sqrt{3}x - 12y = 0$

24. $13x^2 + 6\sqrt{3}xy + 7y^2 - 16 = 0$

25. $7x^2 + 6\sqrt{3}xy + 13y^2 - 32 = 0$

26. $x^2 + 4xy + y^2 - 9 = 0$

27. $11x^2 + 10\sqrt{3}xy + y^2 = 32$

28. $5x^2 - 8xy + 5y^2 = 81$

29. $\sqrt{2}x^2 + 2\sqrt{2}xy + \sqrt{2}y^2 - 8x + 8y = 0$

30. $x^2 + 2\sqrt{3}xy + 3y^2 - 8x + 8\sqrt{3}y = 0$

31. $x^2 + 6\sqrt{3}xy - 5y^2 + 8x - 8\sqrt{3}y - 48 = 0$

32. $3x^2 - 2xy + 3y^2 - 6\sqrt{2}x + 2\sqrt{2}y - 26 = 0$

33. $x^2 + xy + y^2 = 24$

34. $4x^2 + 3\sqrt{3}xy + y^2 = 55$

35. $4x^2 - 4xy + y^2 - 8\sqrt{5}x - 16\sqrt{5}y = 0$

36. $9x^2 - 24xy + 16y^2 - 400x - 300y = 0$

37. $11x^2 + 7xy - 13y^2 = 621$

38. $3x^2 + 4xy + 6y^2 = 28$

Skill Maintenance

Convert to radian measure. [6.4]

39. $120°$

40. $-315°$

Convert to degree measure.

41. $\frac{\pi}{3}$

42. $\frac{3\pi}{4}$

Synthesis

43. Solve this system of equations for x and y:

$$x' = x\cos\theta + y\sin\theta,$$
$$y' = y\cos\theta - x\sin\theta.$$

Show your work.

44. Show that for any angle θ, the equation $x^2 + y^2 = r^2$ becomes $(x')^2 + (y')^2 = r^2$ when the rotation of axes formulas are applied.

45. Show that $A + C = A' + C'$.

10.6 Polar Equations of Conics

- Graph polar equations of conics.
- Convert from polar equations of conics to rectangular equations of conics.
- Find polar equations of conics.

In Sections 10.1–10.3, we saw that the parabola, the ellipse, and the hyperbola have different definitions in rectangular coordinates. When polar coordinates are used, we can give a single definition that applies to all three conics.

CONIC SECTIONS

REVIEW SECTIONS 10.1–10.3.

AN ALTERNATIVE DEFINITION OF CONICS

Let L be a fixed line (the **directrix**), let F be a fixed point (the **focus**) not on L, and let e be a positive constant (the **eccentricity**). A **conic** is the set of all points P in the plane such that

$$\frac{PF}{PL} = e,$$

where PF is the distance from P to F and PL is the distance from P to L. The conic is a parabola if $e = 1$, an ellipse if $0 < e < 1$, and a hyperbola if $e > 1$.

Note that if $e = 1$, then $PF = PL$ and the alternative definition of a parabola is identical to the definition presented in Section 10.1.

Polar Equations of Conics

To derive equations for the conics in polar coordinates, we position the focus F at the pole and position the directrix L either perpendicular to the polar axis or parallel to it. In the figure at left, we place L perpendicular to the polar axis and p units to the right of the focus, or pole.

Note that $PL = p - r\cos\theta$. Then if P is any point on the conic, we have

$$\frac{PF}{PL} = e$$

$$\frac{r}{p - r\cos\theta} = e$$

$$r = ep - er\cos\theta$$

$$r + er\cos\theta = ep$$

$$r(1 + e\cos\theta) = ep$$

$$r = \frac{ep}{1 + e\cos\theta}.$$

Thus we see that the polar equation of a conic with focus at the pole and directrix perpendicular to the polar axis and p units to the right of the pole is

$$r = \frac{ep}{1 + e\cos\theta},$$

where e is the eccentricity of the conic.

For an ellipse and a hyperbola, we can make the following statement regarding eccentricity.

ECCENTRICITY

For an ellipse and a hyperbola, the **eccentricity** e is given by

$$e = \frac{c}{a},$$

where c is the distance from the center to a focus and a is the distance from the center to a vertex.

EXAMPLE 1 Describe and graph the conic $r = \dfrac{18}{6 + 3\cos\theta}$.

Solution We begin by dividing the numerator and the denominator by 6 to obtain a constant term of 1 in the denominator:

$$r = \frac{3}{1 + 0.5\cos\theta}.$$

This equation is in the form

$$r = \frac{ep}{1 + e\cos\theta}$$

with $e = 0.5$. Since $e < 1$, the graph is an ellipse. Also, since $e = 0.5$ and $ep = 0.5p = 3$, we have $p = 6$. Thus the ellipse has a vertical directrix that

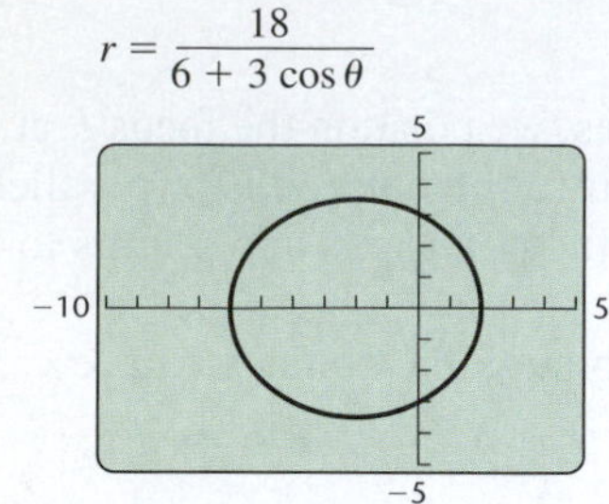

lies 6 units to the right of the pole. We graph the equation in the square window $[-10, 5, -5, 5]$, as shown at left.

It follows that the major axis is horizontal and lies on the polar axis. The vertices are found by letting $\theta = 0$ and $\theta = \pi$. They are $(2, 0)$ and $(6, \pi)$. The center of the ellipse is at the midpoint of the segment connecting the vertices, or at $(2, \pi)$.

The length of the major axis is 8, so we have $2a = 8$, or $a = 4$. From the equation of the conic, we know that $e = 0.5$. Using the equation $e = c/a$, we can find that $c = 2$. Finally, using $a = 4$ and $c = 2$ in $b^2 = a^2 - c^2$ gives us

$$b^2 = 4^2 - 2^2 = 16 - 4 = 12$$
$$b = \sqrt{12}, \text{ or } 2\sqrt{3},$$

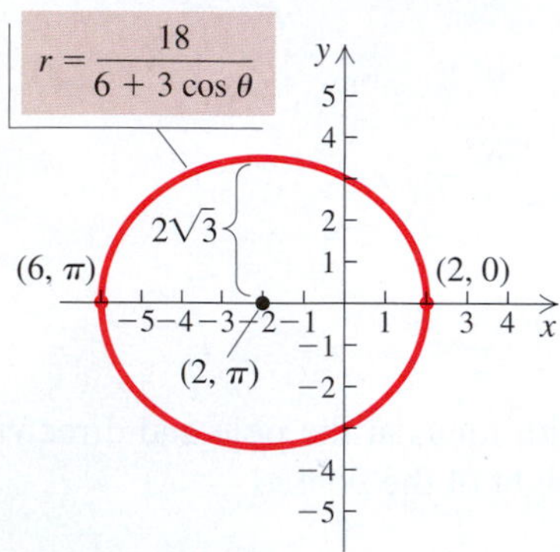

so the length of the minor axis is $2\sqrt{12}$, or $4\sqrt{3}$. This is useful to know when sketching a hand-drawn graph of the conic.

Now Try Exercise 15.

Other derivations similar to the one on p. 767 lead to the following result.

POLAR EQUATIONS OF CONICS

A polar equation of any of the four forms

$$r = \frac{ep}{1 \pm e\cos\theta}, \qquad r = \frac{ep}{1 \pm e\sin\theta}$$

is a conic section. The conic is a parabola if $e = 1$, an ellipse if $0 < e < 1$, and a hyperbola if $e > 1$.

The following table describes the polar equations of conics with a focus at the pole and the directrix either perpendicular to or parallel to the polar axis.

Equation	Description
$r = \dfrac{ep}{1 + e\cos\theta}$	Vertical directrix p units to the right of the pole (or focus)
$r = \dfrac{ep}{1 - e\cos\theta}$	Vertical directrix p units to the left of the pole (or focus)
$r = \dfrac{ep}{1 + e\sin\theta}$	Horizontal directrix p units above the pole (or focus)
$r = \dfrac{ep}{1 - e\sin\theta}$	Horizontal directrix p units below the pole (or focus)

EXAMPLE 2 Describe and graph the conic $r = \dfrac{10}{5 - 5\sin\theta}$.

Solution We first divide the numerator and the denominator by 5:

$$r = \frac{2}{1 - \sin\theta}.$$

This equation is in the form

$$r = \frac{ep}{1 - e\sin\theta}$$

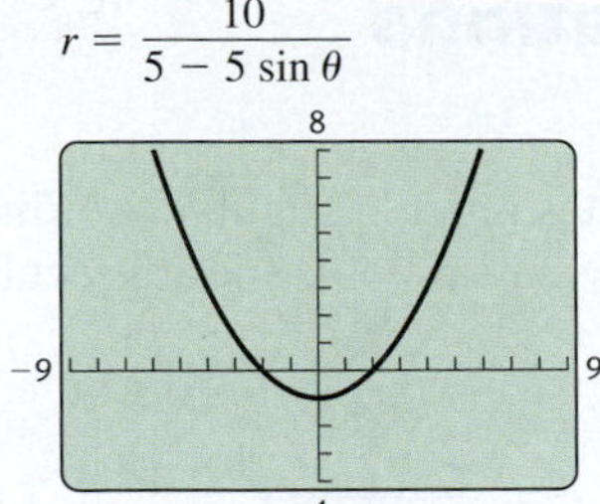

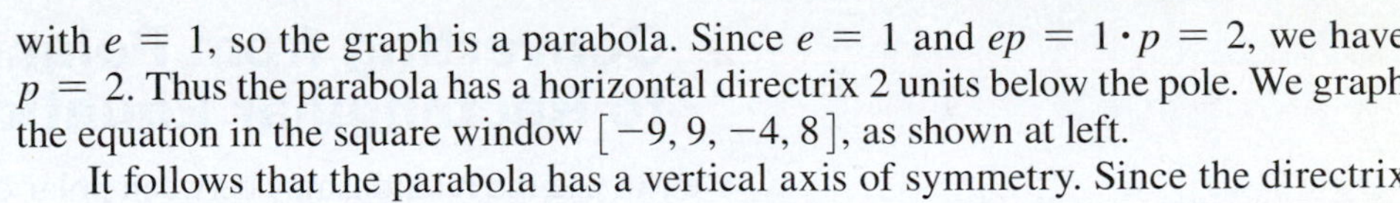

with $e = 1$, so the graph is a parabola. Since $e = 1$ and $ep = 1 \cdot p = 2$, we have $p = 2$. Thus the parabola has a horizontal directrix 2 units below the pole. We graph the equation in the square window $[-9, 9, -4, 8]$, as shown at left.

It follows that the parabola has a vertical axis of symmetry. Since the directrix lies below the focus, or pole, the parabola opens up. The vertex is the midpoint of the segment of the axis of symmetry from the focus to the directrix. We find it by letting $\theta = 3\pi/2$. It is $(1, 3\pi/2)$.

Now Try Exercise 17.

EXAMPLE 3 Describe and graph the conic $r = \dfrac{4}{2 + 6 \sin \theta}$.

Solution We first divide the numerator and the denominator by 2:

$$r = \frac{2}{1 + 3 \sin \theta}.$$

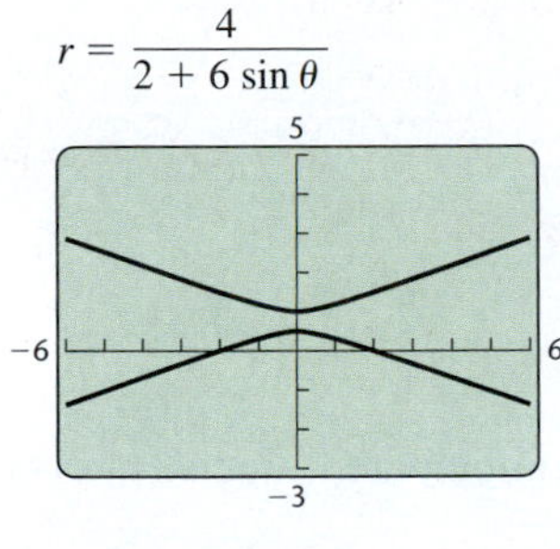

This equation is in the form

$$r = \frac{ep}{1 + e \sin \theta}$$

with $e = 3$. Since $e > 1$, the graph is a hyperbola. We have $e = 3$ and $ep = 3p = 2$, so $p = \frac{2}{3}$. Thus the hyperbola has a horizontal directrix that lies $\frac{2}{3}$ unit above the pole. We graph the equation as shown at left.

It follows that the transverse axis is vertical. To find the vertices, we let $\theta = \pi/2$ and $\theta = 3\pi/2$. The vertices are $(1/2, \pi/2)$ and $(-1, 3\pi/2)$. The center of the hyperbola is the midpoint of the segment connecting the vertices, or $(3/4, \pi/2)$. Thus the distance c from the center to a focus is $3/4$. Using $c = 3/4$, $e = 3$, and $e = c/a$, we have $a = 1/4$. Then since $c^2 = a^2 + b^2$, we have

$$b^2 = \left(\frac{3}{4}\right)^2 - \left(\frac{1}{4}\right)^2 = \frac{9}{16} - \frac{1}{16} = \frac{1}{2}$$

$$b = \frac{1}{\sqrt{2}}, \text{ or } \frac{\sqrt{2}}{2}.$$

Knowing the values of a and b allows us to sketch the asymptotes if we are graphing the hyperbola by hand. We can also easily plot the points $(2, 0)$ and $(2, \pi)$ on the polar axis.

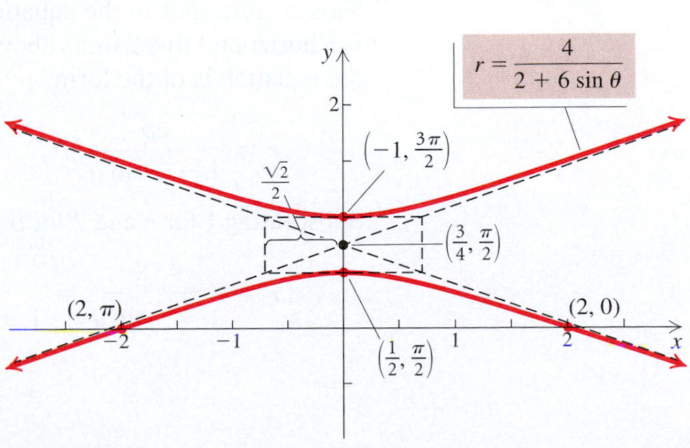

Now Try Exercise 13.

Converting from Polar Equations to Rectangular Equations

We can use the relationships between polar coordinates and rectangular coordinates that were developed in Section 8.4 to convert polar equations of conics to rectangular equations.

EXAMPLE 4 Convert to a rectangular equation: $r = \dfrac{2}{1 - \sin \theta}$.

Solution We have

$$r = \frac{2}{1 - \sin \theta}$$

$$r - r \sin \theta = 2 \qquad \text{Multiplying by } 1 - \sin \theta$$

$$r = r \sin \theta + 2$$

$$\sqrt{x^2 + y^2} = y + 2 \qquad \text{Substituting } \sqrt{x^2 + y^2} \text{ for } r \text{ and } y \text{ for } r \sin \theta$$

$$x^2 + y^2 = y^2 + 4y + 4 \qquad \text{Squaring both sides}$$

$$x^2 = 4y + 4, \quad \text{or}$$

$$x^2 - 4y - 4 = 0.$$

This is the equation of a parabola, as we should have anticipated, since $e = 1$.

Now Try Exercise 23.

Finding Polar Equations of Conics

We can find the polar equation of a conic with a focus at the pole if we know its eccentricity and the equation of the directrix.

EXAMPLE 5 Find a polar equation of the conic with a focus at the pole, eccentricity $\frac{1}{3}$, and directrix $r = 2 \csc \theta$.

Solution The equation of the directrix can be written

$$r = \frac{2}{\sin \theta}, \quad \text{or} \quad r \sin \theta = 2.$$

This corresponds to the equation $y = 2$ in rectangular coordinates, so the directrix is a horizontal line 2 units above the polar axis. Using the table on p. 768, we see that the equation is of the form

$$r = \frac{ep}{1 + e \sin \theta}.$$

Substituting $\frac{1}{3}$ for e and 2 for p gives us

$$r = \frac{\frac{1}{3} \cdot 2}{1 + \frac{1}{3} \sin \theta} = \frac{\frac{2}{3}}{1 + \frac{1}{3} \sin \theta} = \frac{2}{3 + \sin \theta}.$$

Now Try Exercise 39.

10.6 Exercise Set

In Exercises 1–6, match the equation with one of the graphs (a)–(f) that follow.

a)

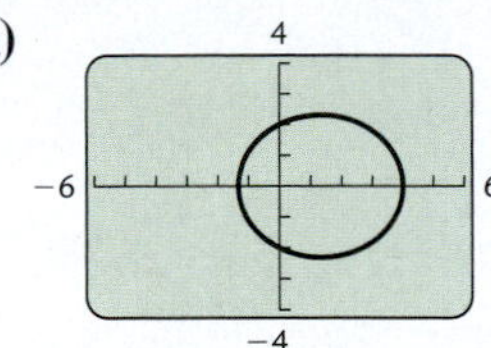

b)

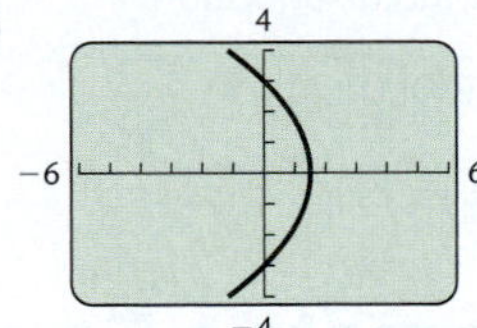

c)

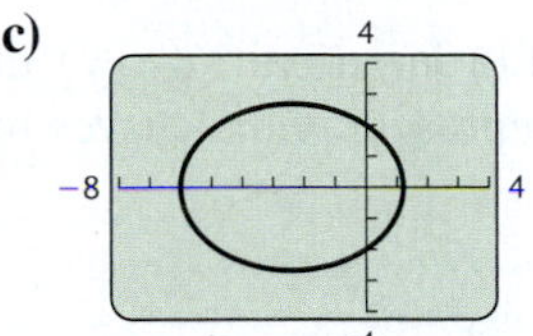

d)

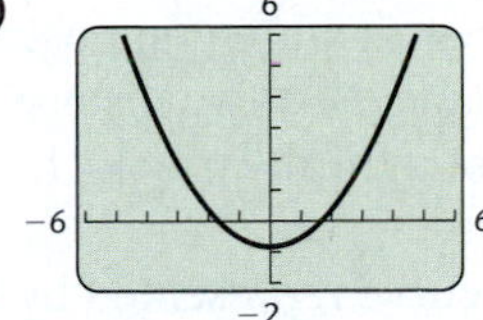

e)

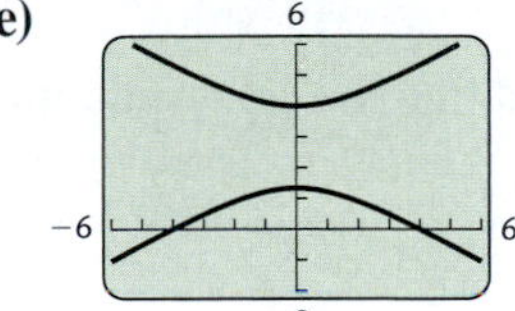

f)

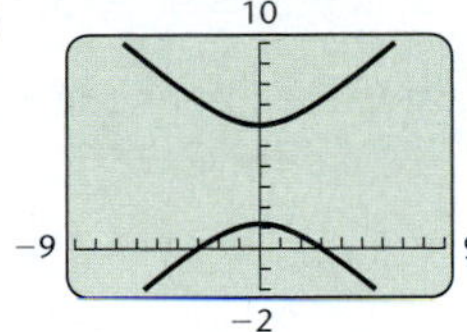

1. $r = \dfrac{3}{1 + \cos\theta}$

2. $r = \dfrac{4}{1 + 2\sin\theta}$

3. $r = \dfrac{8}{4 - 2\cos\theta}$

4. $r = \dfrac{12}{4 + 6\sin\theta}$

5. $r = \dfrac{5}{3 - 3\sin\theta}$

6. $r = \dfrac{6}{3 + 2\cos\theta}$

For each equation:

a) *Tell whether the equation describes a parabola, an ellipse, or a hyperbola.*

b) *State whether the directrix is vertical or horizontal and give its location in relation to the pole.*

c) *Find the vertex or vertices.*

d) *Graph the equation.*

7. $r = \dfrac{1}{1 + \cos\theta}$

8. $r = \dfrac{4}{2 + \cos\theta}$

9. $r = \dfrac{15}{5 - 10\sin\theta}$

10. $r = \dfrac{12}{4 + 8\sin\theta}$

11. $r = \dfrac{8}{6 - 3\cos\theta}$

12. $r = \dfrac{6}{2 + 2\sin\theta}$

13. $r = \dfrac{20}{10 + 15\sin\theta}$

14. $r = \dfrac{10}{8 - 2\cos\theta}$

15. $r = \dfrac{9}{6 + 3\cos\theta}$

16. $r = \dfrac{4}{3 - 9\sin\theta}$

17. $r = \dfrac{3}{2 - 2\sin\theta}$

18. $r = \dfrac{12}{3 + 9\cos\theta}$

19. $r = \dfrac{4}{2 - \cos\theta}$

20. $r = \dfrac{5}{1 - \sin\theta}$

21. $r = \dfrac{7}{2 + 10\sin\theta}$

22. $r = \dfrac{3}{8 - 4\cos\theta}$

23.–38. Convert the equations in Exercises 7–22 to rectangular equations.

Find a polar equation of the conic with a focus at the pole and the given eccentricity and directrix.

39. $e = 2,\ r = 3\csc\theta$

40. $e = \frac{2}{3},\ r = -\sec\theta$

41. $e = 1,\ r = 4\sec\theta$

42. $e = 3,\ r = 2\csc\theta$

43. $e = \frac{1}{2},\ r = -2\sec\theta$

44. $e = 1,\ r = 4\csc\theta$

45. $e = \frac{3}{4},\ r = 5\csc\theta$

46. $e = \frac{4}{5},\ r = 2\sec\theta$

47. $e = 4,\ r = -2\csc\theta$

48. $e = 3,\ r = 3\csc\theta$

Skill Maintenance

For $f(x) = (x - 3)^2 + 4$, find each of the following. [1.2]

49. $f(t)$

50. $f(2t)$

51. $f(t - 1)$

52. $f(t + 2)$

Synthesis

Parabolic Orbit. *Suppose that a comet travels in a parabolic orbit with the sun as its focus. Position a polar coordinate system with the pole at the sun and the axis of the orbit perpendicular to the polar axis. When the comet is the given distance from the sun, the segment from the comet to the sun makes the given angle with the polar axis. Find a polar equation of the orbit, assuming that the directrix lies above the pole.*

53. 100 million miles, $\dfrac{\pi}{6}$

54. 120 million miles, $\dfrac{\pi}{4}$

10.7 Parametric Equations

- Graph parametric equations.
- Determine an equivalent rectangular equation for parametric equations.
- Determine parametric equations for a rectangular equation.
- Solve applied problems involving projectile motion.

Graphing Parametric Equations

We have graphed *plane curves* that are composed of sets of ordered pairs (x, y) in the rectangular coordinate plane. Now we discuss a way to represent plane curves in which x and y are functions of a third variable, t.

EXAMPLE 1 Graph the curve represented by the equations

$$x = \frac{1}{2}t, \quad y = t^2 - 3; \quad -3 \le t \le 3.$$

Solution We can choose values for t between -3 and 3 and find the corresponding values of x and y. When $t = -3$, we have

$$x = \frac{1}{2}(-3) = -\frac{3}{2}, \quad y = (-3)^2 - 3 = 6.$$

The following table lists other ordered pairs. We plot these points and then draw the curve.

t	x	y	(x, y)
-3	$-\frac{3}{2}$	6	$\left(-\frac{3}{2}, 6\right)$
-2	-1	1	$(-1, 1)$
-1	$-\frac{1}{2}$	-2	$\left(-\frac{1}{2}, -2\right)$
0	0	-3	$(0, -3)$
1	$\frac{1}{2}$	-2	$\left(\frac{1}{2}, -2\right)$
2	1	1	$(1, 1)$
3	$\frac{3}{2}$	6	$\left(\frac{3}{2}, 6\right)$

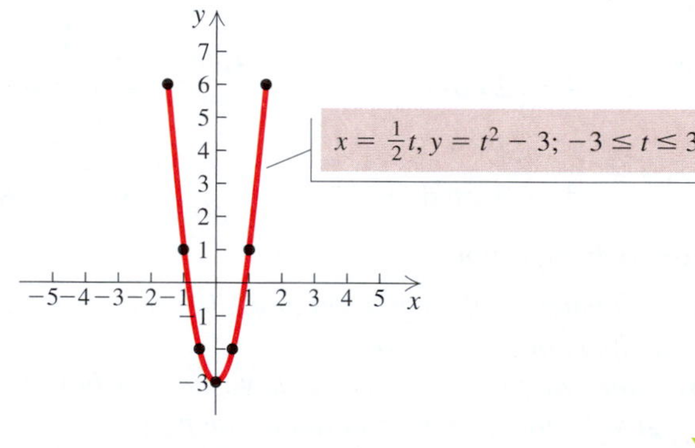

The curve above appears to be part of a parabola. Let's verify this by finding the equivalent rectangular equation. Solving $x = \frac{1}{2}t$ for t, we get $t = 2x$. Substituting $2x$ for t in $y = t^2 - 3$, we have

$$y = (2x)^2 - 3 = 4x^2 - 3.$$

This is a quadratic equation. Hence its graph is a parabola. The curve is part of the parabola $y = 4x^2 - 3$. Since $-3 \le t \le 3$ and $x = \frac{1}{2}t$, we must include the restriction $-\frac{3}{2} \le x \le \frac{3}{2}$ when we write the equivalent rectangular equation:

$$y = 4x^2 - 3, \quad -\frac{3}{2} \le x \le \frac{3}{2}.$$

The equations $x = \frac{1}{2}t$ and $y = t^2 - 3$ are **parametric equations** for the curve. The variable t is the **parameter.** When we write the corresponding rectangular equation, we say that we **eliminate the parameter.**

> **PARAMETRIC EQUATIONS**
>
> If f and g are continuous functions of t on an interval I, then the set of ordered pairs (x, y) such that $x = f(t)$ and $y = g(t)$ is a **plane curve.** The equations $x = f(t)$ and $y = g(t)$ are **parametric equations** for the curve. The variable t is the **parameter**.

Determining a Rectangular Equation for Given Parametric Equations

EXAMPLE 2 Using a graphing calculator, graph each of the following plane curves given their respective parametric equations and the restriction on the parameter. Then find the equivalent rectangular equation.

a) $x = t^2,\ y = t - 1;\ -1 \le t \le 4$

b) $x = \sqrt{t},\ y = 2t + 3;\ 0 \le t \le 3$

Solution

$x = t^2, \quad y = t - 1; \quad -1 \le t \le 4$

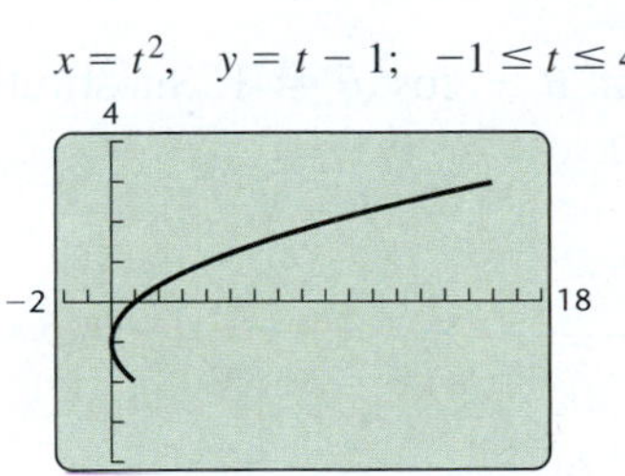

a) To graph the curve (see the window at left), we set the graphing calculator in PARAMETRIC mode, enter the equations, and select minimum and maximum values for x, y, and t.

WINDOW

Tmin = −1	Xmin = −2	Ymin = −4
Tmax = 4	Xmax = 18	Ymax = 4
Tstep = .1	Xscl = 1	Yscl = 1

To find an equivalent rectangular equation, we first solve either equation for t. We choose the equation $y = t - 1$:

$$y = t - 1$$
$$y + 1 = t.$$

We then substitute $y + 1$ for t in $x = t^2$:

$$x = t^2$$
$$x = (y + 1)^2. \qquad \textbf{Substituting}$$

This is an equation of a parabola that opens to the right. Given that $-1 \le t \le 4$, we have the corresponding restrictions on x and y: $0 \le x \le 16$ and $-2 \le y \le 3$. Thus the equivalent rectangular equation is

$$x = (y + 1)^2, \quad -2 \le y \le 3.$$

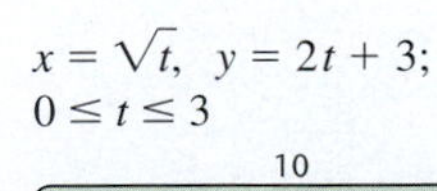

$x = \sqrt{t}, \quad y = 2t + 3;$
$0 \le t \le 3$

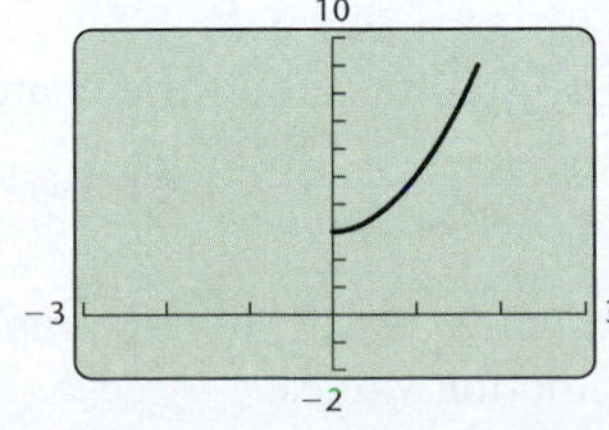

b) To graph the curve (see the window at left), we use PARAMETRIC mode and enter the equations and window settings.

WINDOW

Tmin = 0	Xmin = −3	Ymin = −2
Tmax = 3	Xmax = 3	Ymax = 10
Tstep = .1	Xscl = 1	Yscl = 1

To find an equivalent rectangular equation, we first solve $x = \sqrt{t}$ for t:

$$x = \sqrt{t}$$
$$x^2 = t.$$

Then we substitute x^2 for t in $y = 2t + 3$:

$$y = 2t + 3$$
$$y = 2x^2 + 3. \quad \text{Substituting}$$

When $0 \le t \le 3$, we have $0 \le x \le \sqrt{3}$. The equivalent rectangular equation is

$$y = 2x^2 + 3, \quad 0 \le x \le \sqrt{3}.$$

Now Try Exercise 5.

We first graphed in parametric mode in Section 6.5. There we used an angle measure as the parameter as we do in the next example.

EXAMPLE 3 Graph the plane curve represented by $x = \cos t$ and $y = \sin t$, with t in $[0, 2\pi]$. Then determine an equivalent rectangular equation.

$x = \cos t,\ y = \sin t;\ 0 \le t \le 2\pi$

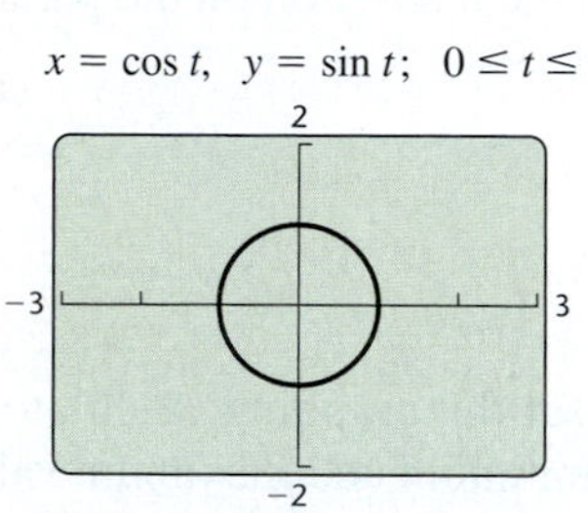

Solution Using a squared window and a Tstep of $\pi/48$, we obtain the graph at left. It appears to be the unit circle.

The equivalent rectangular equation can be obtained by squaring both sides of each parametric equation:

$$x^2 = \cos^2 t \quad \text{and} \quad y^2 = \sin^2 t.$$

This allows us to use the trigonometric identity $\sin^2 \theta + \cos^2 \theta = 1$. Substituting, we get

$$x^2 + y^2 = 1.$$

As expected, this is an equation of the unit circle.

Now Try Exercise 13.

EXAMPLE 4 Graph the plane curve represented by

$$x = 5 \cos t \quad \text{and} \quad y = 3 \sin t; \quad 0 \le t \le 2\pi.$$

Then eliminate the parameter to find the rectangular equation.

$x = 5 \cos t,\ y = 3 \sin t;\ 0 \le t \le 2\pi$

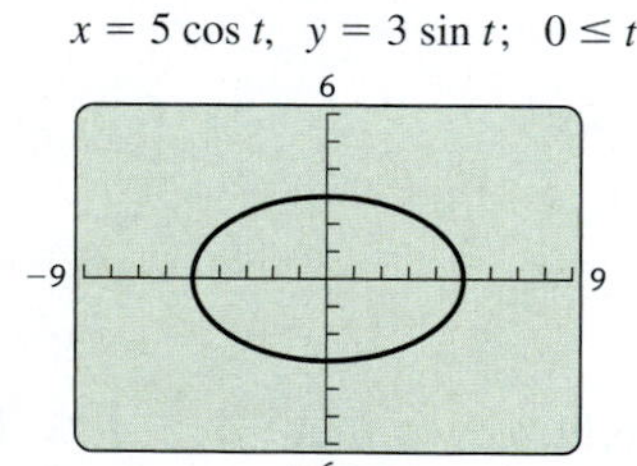

Solution The graph at left appears to be the graph of an ellipse. To find the rectangular equation, we first solve for $\cos t$ and $\sin t$ in the parametric equations:

$$x = 5 \cos t \qquad y = 3 \sin t$$
$$\frac{x}{5} = \cos t, \qquad \frac{y}{3} = \sin t.$$

Using the identity $\sin^2 \theta + \cos^2 \theta = 1$, we can substitute to eliminate the parameter:

$$\sin^2 t + \cos^2 t = 1$$
$$\left(\frac{y}{3}\right)^2 + \left(\frac{x}{5}\right)^2 = 1 \quad \text{Substituting}$$
$$\frac{x^2}{25} + \frac{y^2}{9} = 1. \quad \text{Ellipse}$$

The rectangular form of the equation confirms that the graph is an ellipse centered at the origin with vertices at $(5, 0)$ and $(-5, 0)$.

Now Try Exercise 15.

One advantage of graphing the unit circle parametrically, as we did in Example 3, is that it provides a method of finding trigonometric function values.

EXAMPLE 5 Using the VALUE feature from the CALC menu and the parametric graph of the unit circle, find each of the following function values.

a) $\cos \dfrac{7\pi}{6}$ **b)** $\sin 4.13$

Solution

a) With the calculator set in RADIAN mode and the VALUE feature from the CALC menu, we enter $7\pi/6$ for t. The value of x, which is $\cos t$, appears, as shown at left. Thus,

$$\cos \frac{7\pi}{6} \approx -0.8660.$$

b) With the calculator set in RADIAN mode, we enter 4.13 for t. The value of y, which is $\sin t$, will appear on the screen. The calculator will show that $\sin 4.13 \approx -0.8352$.

Now Try Exercise 21.

Determining Parametric Equations for a Given Rectangular Equation

Many sets of parametric equations can represent the same plane curve. In fact, there are infinitely many such equations.

EXAMPLE 6 Find three sets of parametric equations for the parabola

$$y = 4 - (x + 3)^2.$$

Solution

If $x = t$, then $y = 4 - (t + 3)^2$, or $-t^2 - 6t - 5$.

If $x = t - 3$, then $y = 4 - (t - 3 + 3)^2$, or $4 - t^2$.

If $x = \frac{t}{3}$, then $y = 4 - \left(\frac{t}{3} + 3\right)^2$, or $-\frac{t^2}{9} - 2t - 5$.

Now Try Exercise 29.

Applications

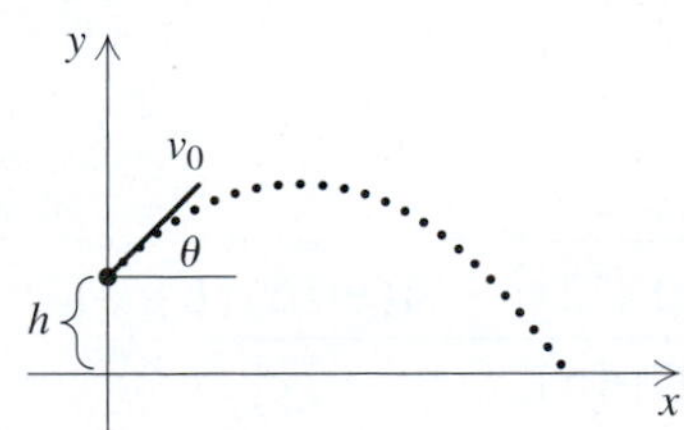

The motion of an object that is propelled upward can be described with parametric equations. Such motion is called **projectile motion.** It can be shown using more advanced mathematics that, neglecting air resistance, the following equations describe the path of a projectile propelled upward at an angle θ with the horizontal from a height h, in feet, at an initial speed v_0, in feet per second:

$$x = (v_0 \cos \theta)t, \qquad y = h + (v_0 \sin \theta)t - 16t^2.$$

We can use these equations to determine the location of the object at time t, in seconds.

EXAMPLE 7 ***Projectile Motion.*** A baseball is thrown from a height of 6 ft with an initial speed of 100 ft/sec at an angle of 45° with the horizontal.

a) Find parametric equations that give the position of the ball at time t, in seconds.

b) Graph the plane curve represented by the equations found in part (a).

c) Find the height of the ball after 1 sec, after 2 sec, and after 3 sec.

d) Determine how long the ball is in the air.

e) Determine the horizontal distance that the ball travels.

f) Find the maximum height of the ball.

Solution

a) We substitute 6 for h, 100 for v_0, and 45° for θ in the equations above:

$$\begin{aligned} x &= (v_0 \cos \theta)t \\ &= (100 \cos 45^\circ)t \\ &= \left(100 \cdot \frac{\sqrt{2}}{2}\right)t = 50\sqrt{2}t; \end{aligned}$$

$$\begin{aligned} y &= h + (v_0 \sin \theta)t - 16t^2 \\ &= 6 + (100 \sin 45^\circ)t - 16t^2 \\ &= 6 + \left(100 \cdot \frac{\sqrt{2}}{2}\right)t - 16t^2 \\ &= 6 + 50\sqrt{2}t - 16t^2. \end{aligned}$$

b) $x = 50\sqrt{2}t,\ y = 6 + 50\sqrt{2}t - 16t^2$

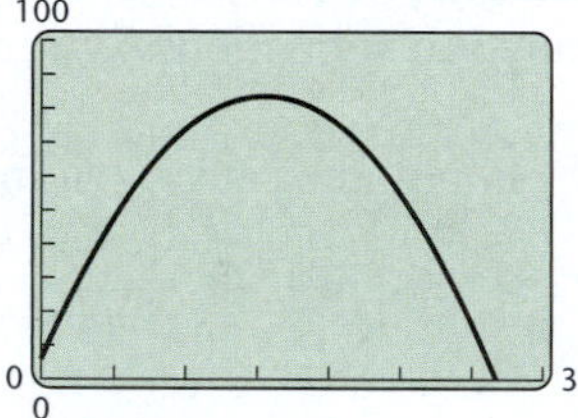

c) The height of the ball at time t is represented by y. We can use a table set in ASK mode to find the desired values of y as shown at left, or we can substitute in the equation for y as shown below.

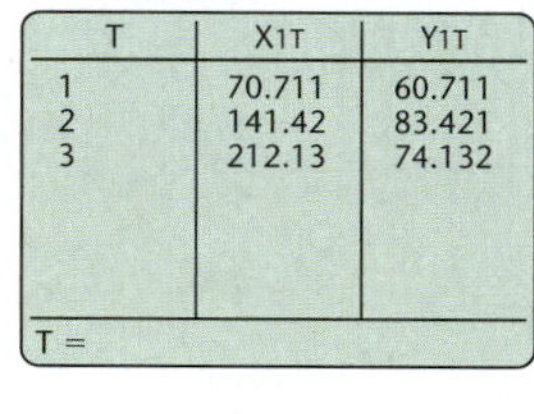

T	X1T	Y1T
1	70.711	60.711
2	141.42	83.421
3	212.13	74.132

T =

When $t = 1,\ y = 6 + 50\sqrt{2}(1) - 16(1)^2 \approx 60.7$ ft.
When $t = 2,\ y = 6 + 50\sqrt{2}(2) - 16(2)^2 \approx 83.4$ ft.
When $t = 3,\ y = 6 + 50\sqrt{2}(3) - 16(3)^2 \approx 74.1$ ft.

d) The ball hits the ground when $y = 0$. Thus, in order to determine how long the ball is in the air, we solve the equation $y = 0$:

QUADRATIC FORMULA
REVIEW SECTION 3.2.

$$6 + 50\sqrt{2}t - 16t^2 = 0$$

$$-16t^2 + 50\sqrt{2}t + 6 = 0 \quad \text{Standard form}$$

$$t = \frac{-50\sqrt{2} \pm \sqrt{(50\sqrt{2})^2 - 4(-16)(6)}}{2(-16)} \quad \text{Using the quadratic formula}$$

$$t \approx -0.1 \quad \text{or} \quad t \approx 4.5.$$

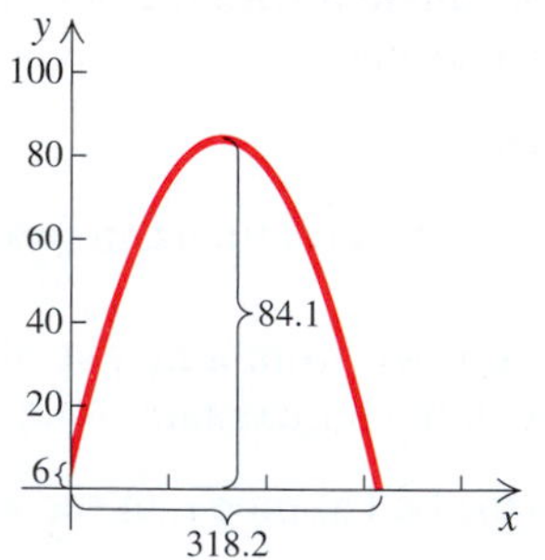

The negative value for t has no meaning in this application. Thus we determine that the ball is in the air for about 4.5 sec.

e) Since the ball is in the air for about 4.5 sec, the horizontal distance that it travels is given by

$$x = 50\sqrt{2}(4.5) \approx 318.2 \text{ ft.}$$

f) To find the maximum height of the ball, we find the maximum value of y. This occurs at the vertex of the quadratic function represented by y. At the vertex, we have

$$t = -\frac{b}{2a} = -\frac{50\sqrt{2}}{2(-16)} \approx 2.2.$$

When $t = 2.2$,

$$y = 6 + 50\sqrt{2}(2.2) - 16(2.2)^2 \approx 84.1 \text{ ft.}$$

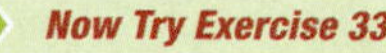
Now Try Exercise 33.

The path of a fixed point on the circumference of a circle as it rolls along a line is called a **cycloid.** For example, a point on the rim of a bicycle wheel traces a cycloid curve.

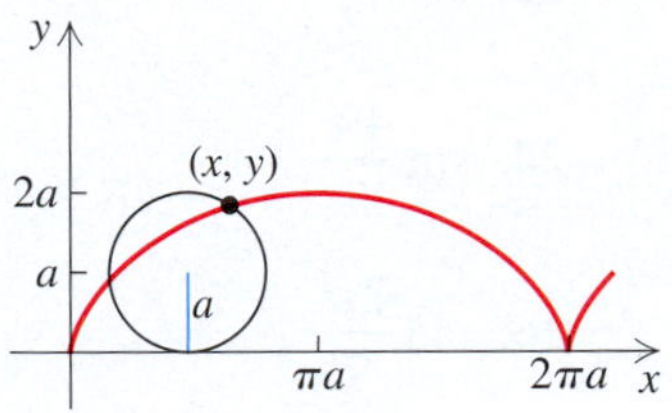

The parametric equations of a cycloid are

$$x = a(t - \sin t), \qquad y = a(1 - \cos t),$$

where a is the radius of the circle that traces the curve and t is in radian measure.

EXAMPLE 8 Graph the cycloid described by the parametric equations

$$x = 3(t - \sin t), \qquad y = 3(1 - \cos t); \quad 0 \le t \le 6\pi.$$

Solution

$x = 3(t - \sin t)$, $y = 3(1 - \cos t)$;
$0 \le t \le 6\pi$

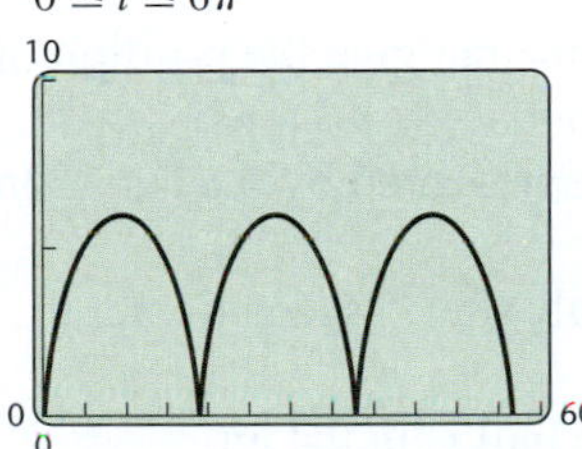

Now Try Exercise 35.

10.7 Exercise Set

Graph the plane curve given by the parametric equations. Then find an equivalent rectangular equation.

1. $x = \frac{1}{2}t,\ y = 6t - 7;\ -1 \le t \le 6$
2. $x = t,\ y = 5 - t;\ -2 \le t \le 3$
3. $x = 4t^2,\ y = 2t;\ -1 \le t \le 1$
4. $x = \sqrt{t},\ y = 2t + 3;\ 0 \le t \le 8$
5. $x = t^2,\ y = \sqrt{t};\ 0 \le t \le 4$
6. $x = t^3 + 1,\ y = t;\ -3 \le t \le 3$
7. $x = t + 3,\ y = \dfrac{1}{t + 3};\ -2 \le t \le 2$
8. $x = 2t^3 + 1,\ y = 2t^3 - 1;\ -4 \le t \le 4$
9. $x = 2t - 1,\ y = t^2;\ -3 \le t \le 3$
10. $x = \frac{1}{3}t,\ y = t;\ -5 \le t \le 5$
11. $x = e^{-t},\ y = e^t;\ -\infty < t < \infty$
12. $x = 2 \ln t,\ y = t^2;\ 0 < t < \infty$
13. $x = 3 \cos t,\ y = 3 \sin t;\ 0 \le t \le 2\pi$
14. $x = 2 \cos t,\ y = 4 \sin t;\ 0 \le t \le 2\pi$
15. $x = \cos t,\ y = 2 \sin t;\ 0 \le t \le 2\pi$
16. $x = 2 \cos t,\ y = 2 \sin t;\ 0 \le t \le 2\pi$
17. $x = \sec t,\ y = \cos t;\ -\dfrac{\pi}{2} < t < \dfrac{\pi}{2}$
18. $x = \sin t,\ y = \csc t;\ 0 < t < \pi$
19. $x = 1 + 2 \cos t,\ y = 2 + 2 \sin t;\ 0 \le t \le 2\pi$
20. $x = 2 + \sec t,\ y = 1 + 3 \tan t;\ 0 < t < \dfrac{\pi}{2}$

Using a parametric graph of the unit circle, find the function value.

21. $\sin \frac{\pi}{4}$

22. $\cos \frac{2\pi}{3}$

23. $\cos \frac{17\pi}{12}$

24. $\sin \frac{4\pi}{5}$

25. $\tan \frac{\pi}{5}$

26. $\tan \frac{2\pi}{7}$

27. $\cos 5.29$

28. $\sin 1.83$

Find two sets of parametric equations for the rectangular equation.

29. $y = 4x - 3$

30. $y = x^2 - 1$

31. $y = (x - 2)^2 - 6x$

32. $y = x^3 + 3$

33. *Projectile Motion.* A ball is thrown from a height of 7 ft with an initial speed of 80 ft/sec at an angle of 30° with the horizontal.

a) Find parametric equations that give the position of the ball at time t, in seconds.
b) Graph the plane curve represented by the equations found in part (a).
c) Find the height of the ball after 1 sec and after 2 sec.
d) Determine how long the ball is in the air.
e) Determine the horizontal distance that the ball travels.
f) Find the maximum height of the ball.

34. *Projectile Motion.* A projectile is launched from the ground with an initial speed of 200 ft/sec at an angle of 60° with the horizontal.

a) Find parametric equations that give the position of the projectile at time t, in seconds.
b) Graph the plane curve represented by the equations found in part (a).
c) Find the height of the projectile after 4 sec and after 8 sec.
d) Determine how long the projectile is in the air.
e) Determine the horizontal distance that the projectile travels.
f) Find the maximum height of the projectile.

Graph the cycloid.

35. $x = 2(t - \sin t),\ y = 2(1 - \cos t);$
$0 \le t \le 4\pi$

36. $x = 4t - 4 \sin t,\ y = 4 - 4 \cos t;$
$0 \le t \le 6\pi$

37. $x = t - \sin t,\ y = 1 - \cos t;$
$-2\pi \le t \le 2\pi$

38. $x = 5(t - \sin t),\ y = 5(1 - \cos t);$
$-4\pi \le t \le 4\pi$

Skill Maintenance

Graph.

39. $y = x^3$ [1.1]

40. $x = y^3$ [1.1]

41. $f(x) = \sqrt{x - 2}$ [1.2]

42. $f(x) = \frac{3}{x^2 - 1}$ [4.5]

Synthesis

43. Graph the curve described by

$$x = 3 \cos t, \qquad y = 3 \sin t; \quad 0 \le t \le 2\pi.$$

As t increases, the path of the curve is generated in the counterclockwise direction. How can this set of equations be changed so that the curve is generated in the clockwise direction?

44. Graph the plane curve described by

$$x = \cos^3 t, \qquad y = \sin^3 t; \quad 0 \le t \le 2\pi.$$

Then find the equivalent rectangular equation.

Chapter 10 Summary and Review

STUDY GUIDE

KEY TERMS AND CONCEPTS	EXAMPLES
SECTION 10.1: THE PARABOLA	
Standard Equation of a Parabola with Vertex (0, 0) and Vertical Axis of Symmetry The standard equation of a parabola with vertex (0, 0) and directrix $y = -p$ is $x^2 = 4py.$ The focus is $(0, p)$ and the y-axis is the axis of symmetry.	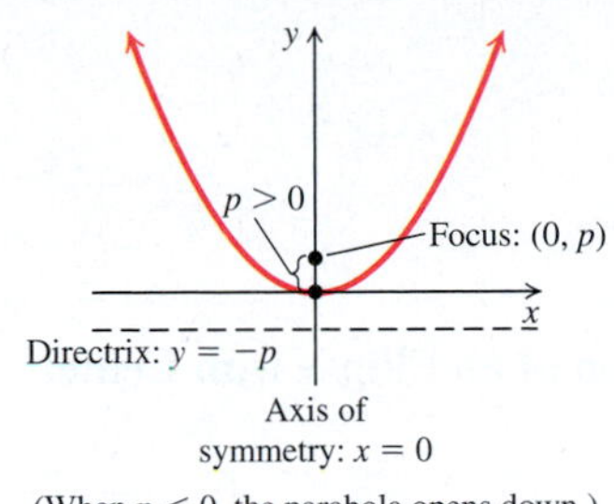 (When $p < 0$, the parabola opens down.) See also Example 1 on p. 722.
Standard Equation of a Parabola with Vertex (0, 0) and Horizontal Axis of Symmetry The standard equation of a parabola with vertex (0, 0) and directrix $x = -p$ is $y^2 = 4px.$ The focus is $(p, 0)$ and the x-axis is the axis of symmetry.	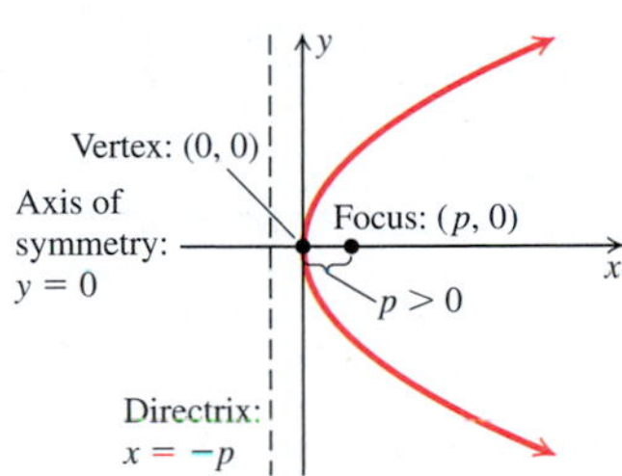 (When $p < 0$, the parabola opens to the left.) See also Example 2 on p. 722.
Standard Equation of a Parabola with Vertex (h, k) and Vertical Axis of Symmetry The standard equation of a parabola with vertex (h, k) and vertical axis of symmetry is $(x - h)^2 = 4p(y - k),$ where the vertex is (h, k), the focus is $(h, k + p)$, and the directrix is $y = k - p$. The parabola opens up if $p > 0$. It opens down if $p < 0$.	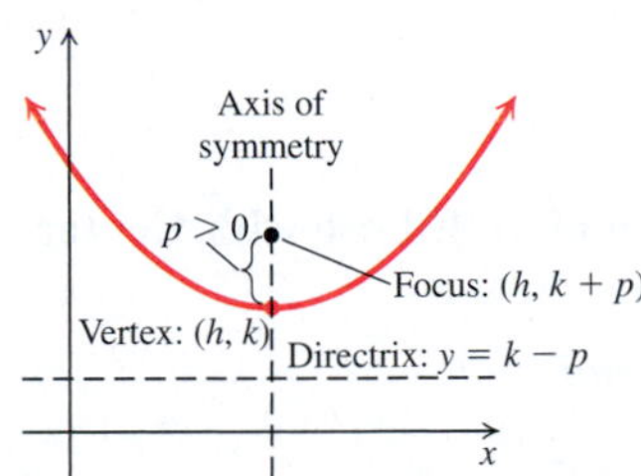 (When $p < 0$, the parabola opens down.) See also Example 3 on p. 724.
Standard Equation of a Parabola with Vertex (h, k) and Horizontal Axis of Symmetry The standard equation of a parabola with vertex (h, k) and horizontal axis of symmetry is $(y - k)^2 = 4p(x - h),$ where the vertex is (h, k), the focus is $(h + p, k)$, and the directrix is $x = h - p$. The parabola opens to the right if $p > 0$. It opens to the left if $p < 0$.	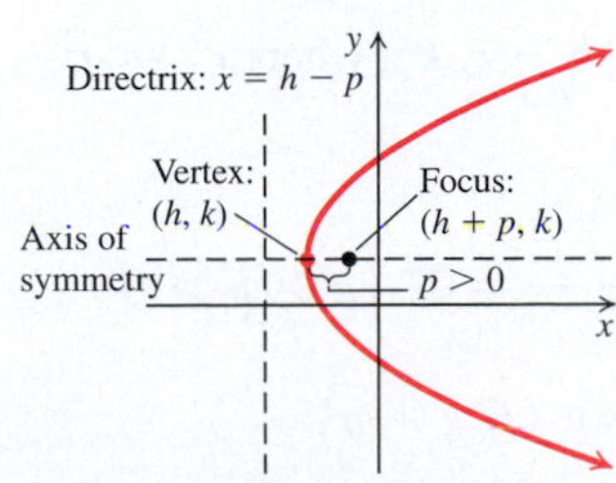 (When $p < 0$, the parabola opens to the left.) See also Example 4 on p. 724.

SECTION 10.2: THE CIRCLE AND THE ELLIPSE

Standard Equation of a Circle The standard equation of a circle with center (h, k) and radius r is $$(x-h)^2 + (y-k)^2 = r^2.$$	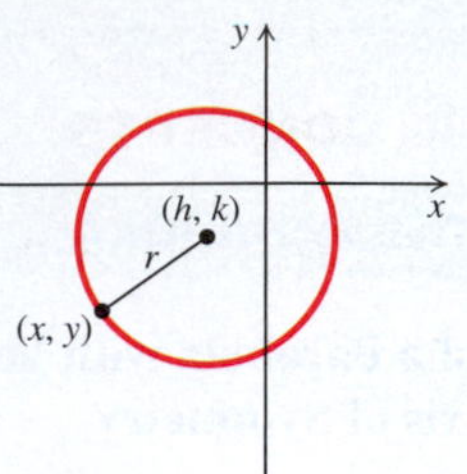 See also Example 1 on p. 728.
Standard Equation of an Ellipse with Center at the Origin *Major Axis Horizontal* $$\frac{x^2}{a^2} + \frac{y^2}{b^2} = 1, \quad a > b > 0$$ Vertices: $(-a, 0), (a, 0)$ y-intercepts: $(0, -b), (0, b)$ Foci: $(-c, 0), (c, 0)$, where $c^2 = a^2 - b^2$ *Major Axis Vertical* $$\frac{x^2}{b^2} + \frac{y^2}{a^2} = 1, \quad a > b > 0$$ Vertices: $(0, -a), (0, a)$ x-intercepts: $(-b, 0), (b, 0)$ Foci: $(0, -c), (0, c)$, where $c^2 = a^2 - b^2$	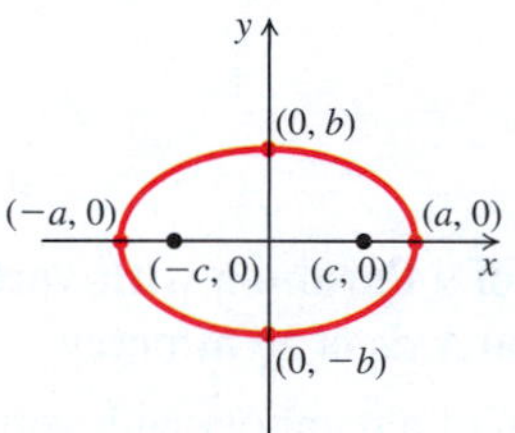 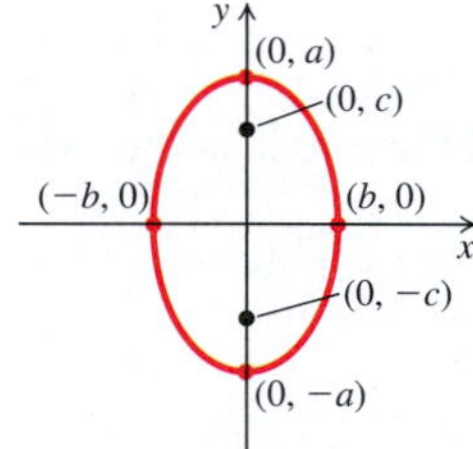 See also Examples 2 and 3 on pp. 730 and 731.
Standard Equation of an Ellipse with Center at (h, k) *Major Axis Horizontal* $$\frac{(x-h)^2}{a^2} + \frac{(y-k)^2}{b^2} = 1, \quad a > b > 0$$ Vertices: $(h-a, k), (h+a, k)$ Length of minor axis: $2b$ Foci: $(h-c, k), (h+c, k)$, where $c^2 = a^2 - b^2$ *Major Axis Vertical* $$\frac{(x-h)^2}{b^2} + \frac{(y-k)^2}{a^2} = 1, \quad a > b > 0$$ Vertices: $(h, k-a), (h, k+a)$ Length of minor axis: $2b$ Foci: $(h, k-c), (h, k+c)$, where $c^2 = a^2 - b^2$	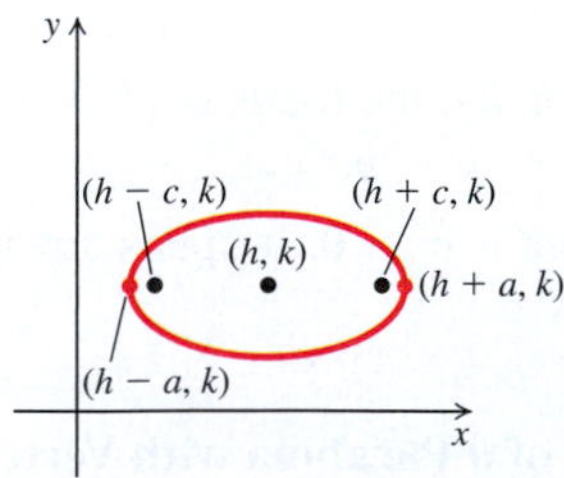 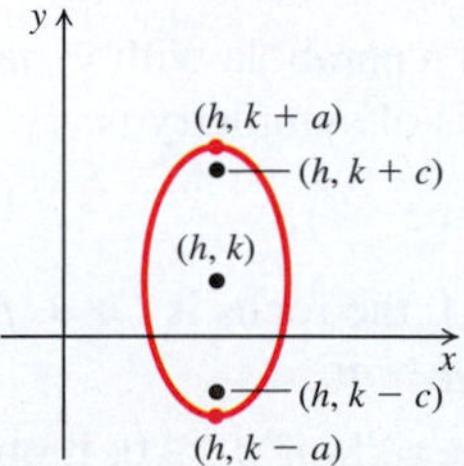 See also Example 4 on p. 732.

SECTION 10.3: THE HYPERBOLA

Standard Equation of a Hyperbola with Center at the Origin

Transverse Axis Horizontal

$$\frac{x^2}{a^2} - \frac{y^2}{b^2} = 1$$

Vertices: $(-a, 0)$, $(a, 0)$

Asymptotes: $y = -\frac{b}{a}x$, $y = \frac{b}{a}x$

Foci: $(-c, 0)$, $(c, 0)$, where $c^2 = a^2 + b^2$

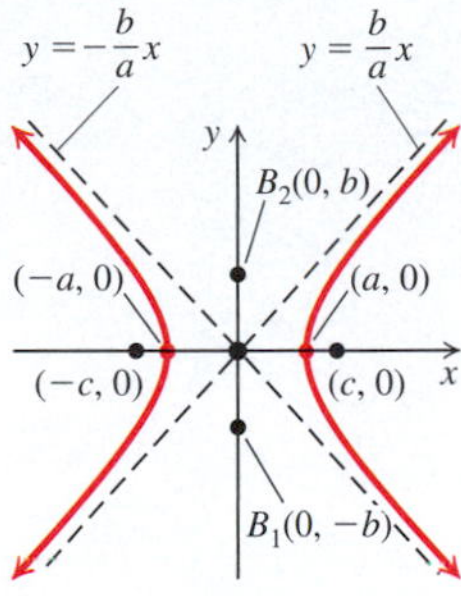

Transverse Axis Vertical

$$\frac{y^2}{a^2} - \frac{x^2}{b^2} = 1$$

Vertices: $(0, -a)$, $(0, a)$

Asymptotes: $y = -\frac{a}{b}x$, $y = \frac{a}{b}x$

Foci: $(0, -c)$, $(0, c)$, where $c^2 = a^2 + b^2$

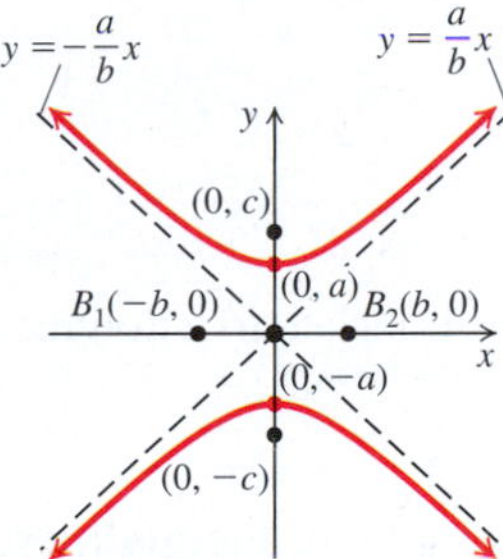

See also Examples 1 and 2 on p. 738.

Standard Equation of a Hyperbola with Center at (h, k)

Transverse Axis Horizontal

$$\frac{(x-h)^2}{a^2} - \frac{(y-k)^2}{b^2} = 1$$

Vertices: $(h - a, k)$, $(h + a, k)$

Asymptotes: $y - k = \frac{b}{a}(x - h)$,

$$y - k = -\frac{b}{a}(x - h)$$

Foci: $(h - c, k)$, $(h + c, k)$, where $c^2 = a^2 + b^2$

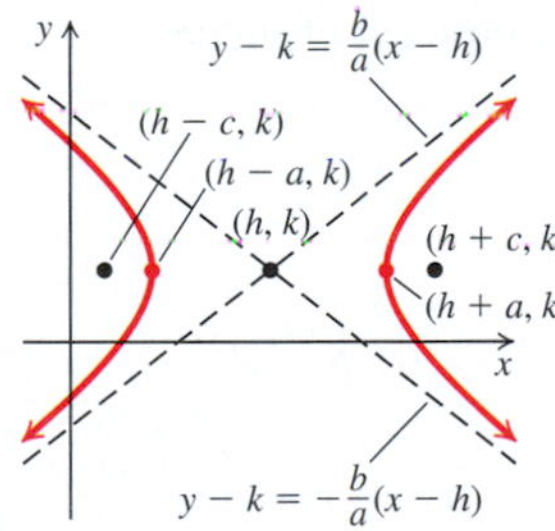

Transverse Axis Vertical

$$\frac{(y-k)^2}{a^2} - \frac{(x-h)^2}{b^2} = 1$$

Vertices: $(h, k - a)$, $(h, k + a)$

Asymptotes: $y - k = \frac{a}{b}(x - h)$,

$$y - k = -\frac{a}{b}(x - h)$$

Foci: $(h, k - c)$, $(h, k + c)$, where $c^2 = a^2 + b^2$

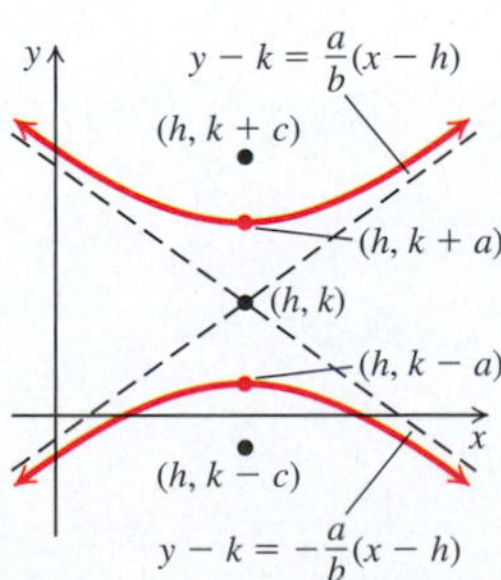

See also Example 3 on p. 740.

SECTION 10.4: NONLINEAR SYSTEMS OF EQUATIONS AND INEQUALITIES

Substitution or elimination can be used to solve **systems of equations containing at least one nonlinear equation**.	Solve: $x^2 - y = 2,$ (1) The graph is a parabola. $x - y = -4.$ (2) The graph is a line. $x = y - 4$ Solving equation (2) for x $(y - 4)^2 - y = 2$ Substituting for x in equation (1) $y^2 - 8y + 16 - y = 2$ $y^2 - 9y + 14 = 0$ $(y - 2)(y - 7) = 0$ $y - 2 = 0$ *or* $y - 7 = 0$ $y = 2$ *or* $y = 7$ If $y = 2$, then $x = 2 - 4 = -2$. If $y = 7$, then $x = 7 - 4 = 3$. The pairs $(-2, 2)$ and $(3, 7)$ check, so they are the solutions.
Some applied problems translate to a nonlinear system of equations.	See Example 5 on p. 751.
To graph a **nonlinear system of inequalities**, graph each inequality in the system and then shade the region where their solution sets overlap. To find the point(s) of intersection of the graphs of the related equations, solve the system of equations composed of those equations.	Graph: $x^2 - y \le 2,$ $x - y > -4.$ 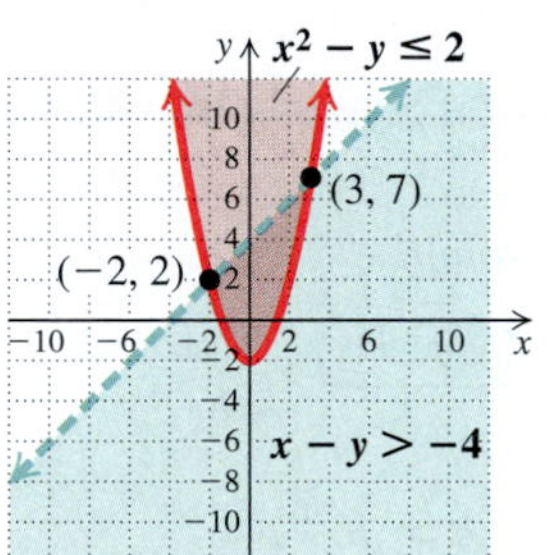 To find the points of intersection of the graphs of the related equations, we solve the system of equations $x^2 - y = 2,$ $x - y = -4.$ We saw in the example above that these points are $(-2, 2)$ and $(3, 7)$.

SECTION 10.5: ROTATION OF AXES

To eliminate the xy-term from the equation

$Ax^2 + Bxy + Cy^2 + Dx + Ey + F = 0,$
$B \neq 0,$

select an angle θ such that

$$\cot 2\theta = \frac{A - C}{B}, \quad 0° < 2\theta < 180°,$$

and use the rotation of axes formulas below.

If the x- and y-axes are rotated about the origin through a positive acute angle θ, then the coordinates (x, y) and (x', y') of a point P in the xy- and $x'y'$-coordinate systems are related by the following formulas:

$$x' = x \cos\theta + y \sin\theta,$$
$$y' = -x \sin\theta + y \cos\theta;$$
$$x = x' \cos\theta - y' \sin\theta,$$
$$y = x' \sin\theta + y' \cos\theta.$$

See Example 2 on p. 762.

The expression $B^2 - 4AC$ is the **discriminant** of the equation

$Ax^2 + Bxy + Cy^2 + Dx + Ey + F = 0.$

The graph of the equation

$Ax^2 + Bxy + Cy^2 + Dx + Ey + F = 0$

is, except in degenerate cases,

1. an ellipse or a circle if $B^2 - 4AC < 0$,
2. a hyperbola if $B^2 - 4AC > 0$, and
3. a parabola if $B^2 - 4AC = 0$.

Use the discriminant to determine whether the equation is an ellipse or a circle, a hyperbola, or a parabola.

a) $x^2 - 2xy - 3y^2 + 5x + 1 = 0$
b) $3x^2 + 6xy + 3y^2 + 4x - 7y + 2 = 0$
c) $5x^2 + 3xy + y^2 - 8 = 0$

a) $A = 1, B = -2$, and $C = -3$, so

$B^2 - 4AC = (-2)^2 - 4 \cdot 1 \cdot (-3) = 4 + 12 = 16 > 0.$

The graph is a hyperbola.

b) $A = 3, B = 6$, and $C = 3$, so

$B^2 - 4AC = 6^2 - 4 \cdot 3 \cdot 3 = 36 - 36 = 0.$

The graph is a parabola.

c) $A = 5, B = 3$, and $C = 1$, so

$B^2 - 4AC = 3^2 - 4 \cdot 5 \cdot 1 = 9 - 20 = -11 < 0.$

The graph is an ellipse or a circle.

SECTION 10.6: POLAR EQUATIONS OF CONICS

A polar equation of any of the four forms

$$r = \frac{ep}{1 \pm e\cos\theta}, \qquad r = \frac{ep}{1 \pm e\sin\theta}$$

is a conic section. The conic is a parabola if $e = 1$, an ellipse if $0 < e < 1$, and a hyperbola if $e > 1$.

Equation	Description
$r = \frac{ep}{1 + e\cos\theta}$	Vertical directrix p units to the right of the pole (or focus)
$r = \frac{ep}{1 - e\cos\theta}$	Vertical directrix p units to the left of the pole (or focus)
$r = \frac{ep}{1 + e\sin\theta}$	Horizontal directrix p units above the pole (or focus)
$r = \frac{ep}{1 - e\sin\theta}$	Horizontal directrix p units below the pole (or focus)

Describe and graph the conic

$$r = \frac{3}{3 + \cos\theta}.$$

We first divide the numerator and the denominator by 3:

$$r = \frac{1}{1 + \frac{1}{3}\cos\theta}.$$

The equation is of the form

$$r = \frac{ep}{1 + e\cos\theta},$$

with $e = \frac{1}{3}$. Since $0 < \frac{1}{3} < 1$, the graph is an ellipse.

Since $e = \frac{1}{3}$ and $ep = \frac{1}{3} \cdot p = 1$, we have $p = 3$. Thus the ellipse has a vertical directrix 3 units to the right of the pole. We find the vertices by letting $\theta = 0$ and $\theta = \pi$. When $\theta = 0$,

$$r = \frac{3}{3 + \cos 0} = \frac{3}{3 + 1} = \frac{3}{4}.$$

When $\theta = \pi$,

$$r = \frac{3}{3 + \cos\pi} = \frac{3}{3 - 1} = \frac{3}{2}.$$

The vertices are $\left(\frac{3}{4}, 0\right)$ and $\left(\frac{3}{2}, \pi\right)$.

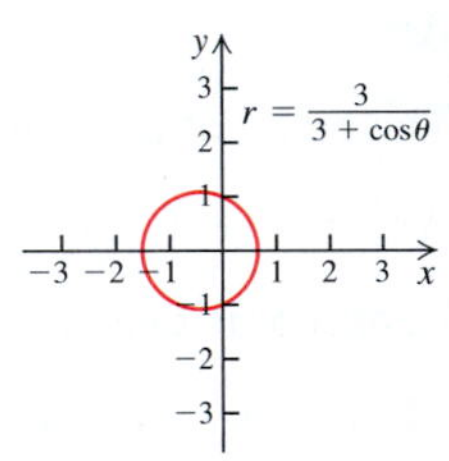

We can use the relationships between polar coordinates and rectangular coordinates to **convert polar equations of conics to rectangular equations.**

Convert to a rectangular equation:

$$r = \frac{4}{3 - \cos\theta}.$$

We have

$$r = \frac{4}{3 - \cos\theta}$$

$$3r - r\cos\theta = 4 \quad \textbf{Multiplying by } 3 - \cos\theta$$

$$3r = r\cos\theta + 4$$

$$3\sqrt{x^2 + y^2} = x + 4 \quad \textbf{Substituting}$$

$$9x^2 + 9y^2 = x^2 + 8x + 16 \quad \textbf{Squaring both sides}$$

$$8x^2 + 9y^2 - 8x - 16 = 0.$$

SECTION 10.7: PARAMETRIC EQUATIONS

Parametric Equations

If f and g are continuous functions of t on an interval I, then the set of ordered pairs (x, y) such that $x = f(t)$ and $y = g(t)$ is a **plane curve.** The equations $x = f(t)$ and $y = g(t)$ are **parametric equations** for the curve. The variable t is the **parameter.**

Graph the plane curve given by the following pair of parametric equations. Then find an equivalent rectangular equation.

$$x = \frac{3}{2}t, \quad y = 2t - 1; \quad -2 \le t \le 3$$

We can choose values of t between -2 and 3 and find the corresponding values of x and y.

t	x	y	(x, y)
-2	-3	-5	$(-3, -5)$
-1	$-\frac{3}{2}$	-3	$(-\frac{3}{2}, -3)$
0	0	-1	$(0, -1)$
1	$\frac{3}{2}$	1	$(\frac{3}{2}, 1)$
2	3	3	$(3, 3)$
3	$\frac{9}{2}$	5	$(\frac{9}{2}, 5)$

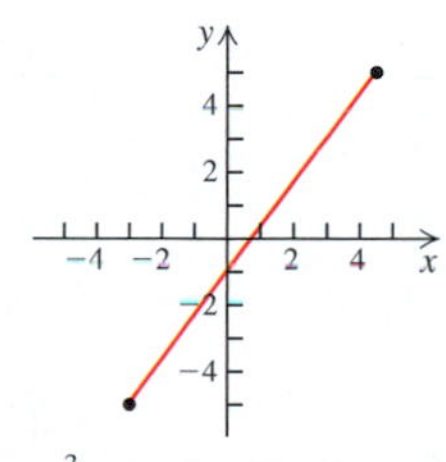

$x = \frac{3}{2}t,\ y = 2t - 1,\ -2 \le t \le 3$

Solving $x = \frac{3}{2}t$ for t, we have $\frac{2}{3}x = t$. Then we have

$$y = 2t - 1 = 2 \cdot \frac{2}{3}x - 1 = \frac{4}{3}x - 1.$$

Since $-2 \le t \le 3$ and $x = \frac{3}{2}t$, we have $-3 \le x \le \frac{9}{2}$. Thus we have the equivalent rectangular equation

$$y = \frac{4}{3}x - 1, \quad -3 \le x \le \frac{9}{2}.$$

REVIEW EXERCISES

Determine whether the statement is true or false.

1. The graph of $x + y^2 = 1$ is a parabola that opens to the left. [10.1]
2. The graph of $\frac{(x-2)^2}{4} + \frac{(y+3)^2}{9} = 1$ is an ellipse with center $(-2, 3)$. [10.2]
3. A parabola must open up or down. [10.1]
4. The major axis of the ellipse $\frac{x^2}{4} + \frac{y^2}{16} = 1$ is vertical. [10.2]
5. The graph of a nonlinear system of equations shows all the solutions of the system of equations. [10.4]

In Exercises 6–13, match the equation with one of the graphs (a)–(h) that follow.

a)

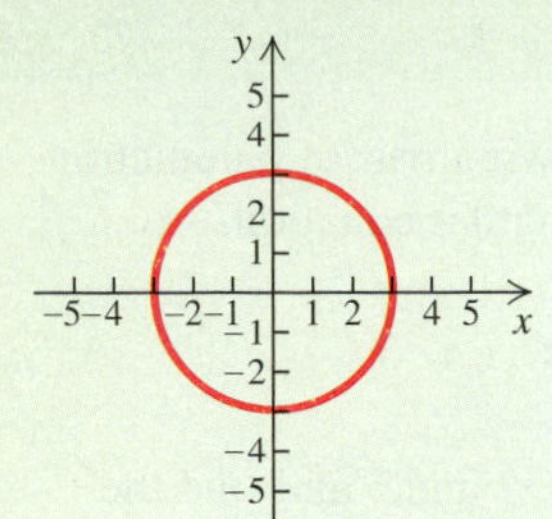

b)

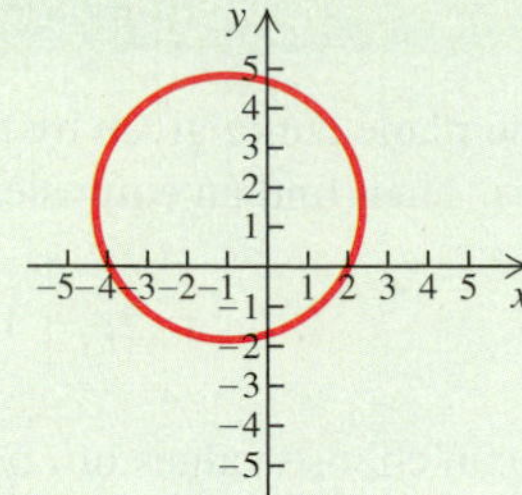

c)

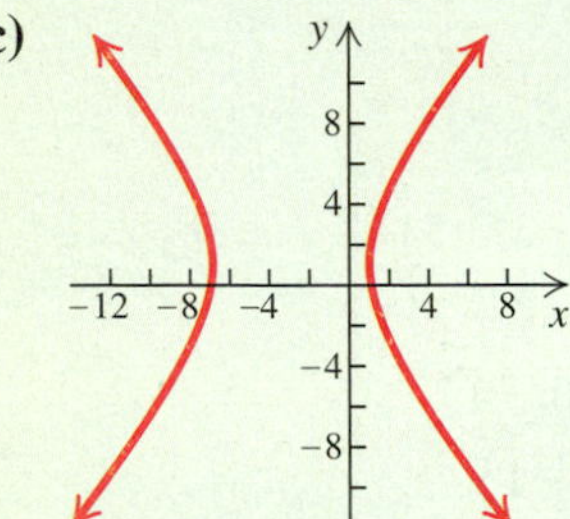

d)

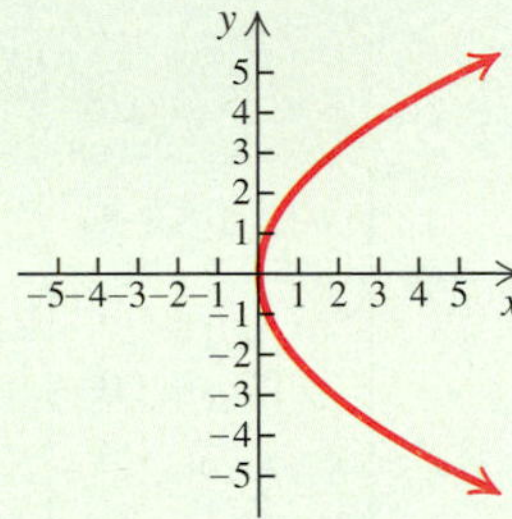

e)

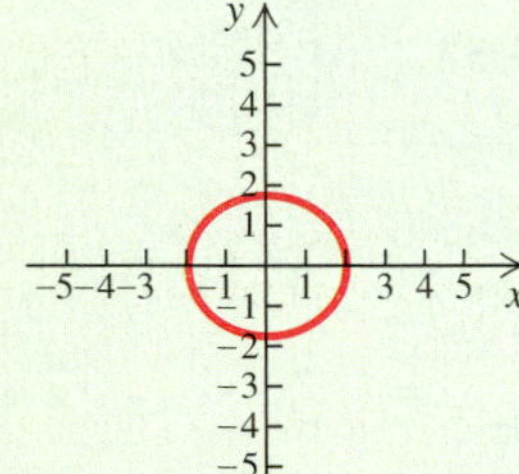

f)

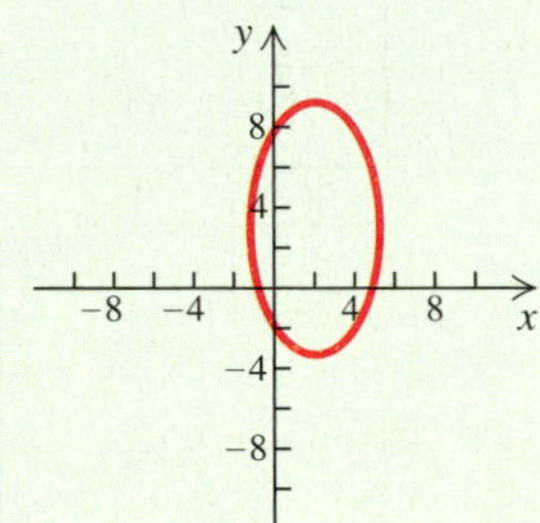

g)

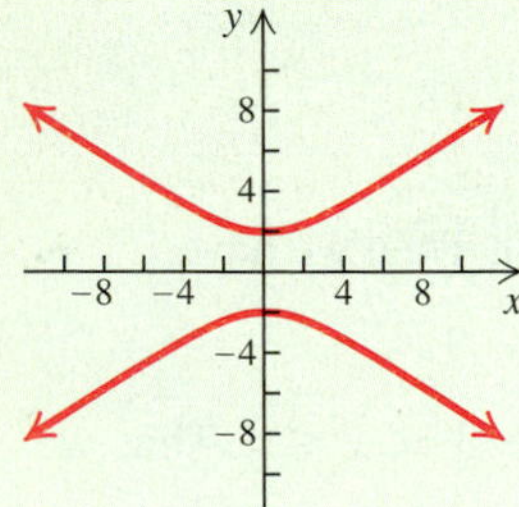

h)

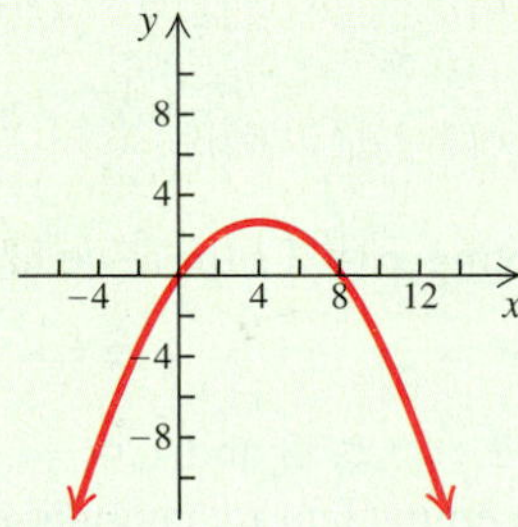

6. $y^2 = 5x$ **[10.1]**

7. $y^2 = 9 - x^2$ **[10.2]**

8. $3x^2 + 4y^2 = 12$ **[10.2]**

9. $9y^2 - 4x^2 = 36$ **[10.3]**

10. $x^2 + y^2 + 2x - 3y = 8$ **[10.2]**

11. $4x^2 + y^2 - 16x - 6y = 15$ **[10.2]**

12. $x^2 - 8x + 6y = 0$ **[10.1]**

13. $\dfrac{(x+3)^2}{16} - \dfrac{(y-1)^2}{25} = 1$ **[10.3]**

14. Find an equation of the parabola with directrix $y = \frac{3}{2}$ and focus $\left(0, -\frac{3}{2}\right)$. **[10.1]**

15. Find the focus, the vertex, and the directrix of the parabola given by $y^2 = -12x$. **[10.1]**

16. Find the vertex, the focus, and the directrix of the parabola given by

$$x^2 + 10x + 2y + 9 = 0.$$ **[10.1]**

17. Find the center, the vertices, and the foci of the ellipse given by

$$16x^2 + 25y^2 - 64x + 50y - 311 = 0.$$

Then draw the graph. **[10.2]**

18. Find an equation of the ellipse having vertices $(0, -4)$ and $(0, 4)$ with minor axis of length 6. **[10.2]**

19. Find the center, the vertices, the foci, and the asymptotes of the hyperbola given by

$$x^2 - 2y^2 + 4x + y - \tfrac{1}{8} = 0.$$ **[10.3]**

20. *Spotlight.* A spotlight has a parabolic cross section that is 2 ft wide at the opening and 1.5 ft deep at the vertex. How far from the vertex is the focus? **[10.1]**

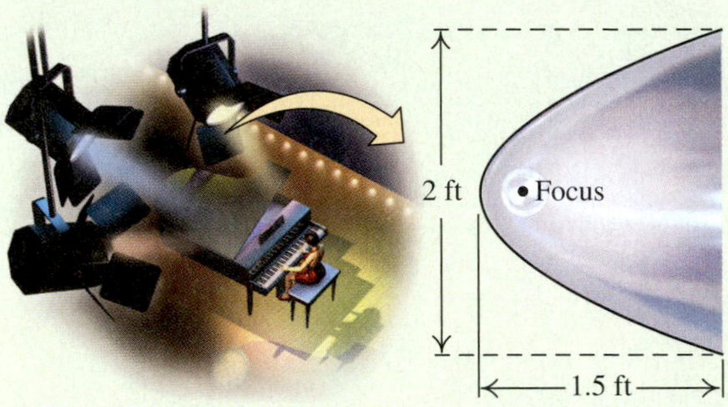

Solve. **[10.4]**

21. $x^2 - 16y = 0,$
$x^2 - y^2 = 64$

22. $4x^2 + 4y^2 = 65,$
$6x^2 - 4y^2 = 25$

23. $x^2 - y^2 = 33,$
$x + y = 11$

24. $x^2 - 2x + 2y^2 = 8,$
$2x + y = 6$

25. $x^2 - y = 3,$
$2x - y = 3$

26. $x^2 + y^2 = 25,$
$x^2 - y^2 = 7$

27. $x^2 - y^2 = 3,$
$y = x^2 - 3$

28. $x^2 + y^2 = 18,$
$2x + y = 3$

29. $x^2 + y^2 = 100,$
$2x^2 - 3y^2 = -120$

30. $x^2 + 2y^2 = 12,$
$xy = 4$

31. *Numerical Relationship.* The sum of two numbers is 11, and the sum of their squares is 65. Find the numbers. **[10.4]**

32. *Dimensions of a Rectangle.* A rectangle has a perimeter of 38 m and an area of 84 m^2. What are the dimensions of the rectangle? **[10.4]**

33. *Numerical Relationship.* Find two positive integers whose sum is 12 and the sum of whose reciprocals is $\frac{3}{8}$. **[10.4]**

34. *Perimeter.* The perimeter of a square is 12 cm more than the perimeter of another square. The area of the first square exceeds the area of the other by 39 cm^2. Find the perimeter of each square. [10.4]

35. *Radius of a Circle.* The sum of the areas of two circles is 130π ft^2. The difference of the areas is 112π ft^2. Find the radius of each circle. [10.4]

Graph the system of inequalities. Then find the coordinates of the points of intersection of the graphs of the related equations. [10.4]

36. $y \le 4 - x^2,$
$x - y \le 2$

37. $x^2 + y^2 \le 16,$
$x + y < 4$

38. $y \ge x^2 - 1,$
$y < 1$

39. $x^2 + y^2 \le 9,$
$x \le -1$

Graph the equation. [10.5]

40. $5x^2 - 2xy + 5y^2 - 24 = 0$

41. $x^2 - 10xy + y^2 + 12 = 0$

42. $5x^2 + 6\sqrt{3}xy - y^2 = 16$

43. $x^2 + 2xy + y^2 - \sqrt{2}x + \sqrt{2}y = 0$

Graph the equation. State whether the directrix is vertical or horizontal, describe its location in relation to the pole, and find the vertex or vertices. [10.6]

44. $r = \dfrac{6}{3 - 3\sin\theta}$

45. $r = \dfrac{8}{2 + 4\cos\theta}$

46. $r = \dfrac{4}{2 - \cos\theta}$

47. $r = \dfrac{18}{9 + 6\sin\theta}$

48.–51. Convert the equations in Exercises 44–47 to rectangular equations. [10.6]

Find a polar equation of the conic with a focus at the pole and the given eccentricity and directrix. [10.6]

52. $e = \frac{1}{2},\ r = 2\sec\theta$

53. $e = 3,\ r = -6\csc\theta$

54. $e = 1,\ r = -4\sec\theta$

55. $e = 2,\ r = 3\csc\theta$

Graph the plane curve given by the set of parametric equations and the restrictions for the parameter. Then find the equivalent rectangular equation. [10.7]

56. $x = t,\ y = 2 + t;\ -3 \le t \le 3$

57. $x = \sqrt{t},\ y = t - 1;\ 0 \le t \le 9$

58. $x = 2\cos t,\ y = 2\sin t;\ 0 \le t \le 2\pi$

59. $x = 3\sin t,\ y = \cos t;\ 0 \le t \le 2\pi$

Find two sets of parametric equations for the given rectangular equation. [10.7]

60. $y = 2x - 3$

61. $y = x^2 + 4$

62. *Projectile Motion.* A projectile is launched from the ground with an initial speed of 150 ft/sec at an angle of 45° with the horizontal. [10.7]

a) Find parametric equations that give the position of the projectile at time t, in seconds.
b) Find the height of the projectile after 3 sec and after 6 sec.
c) Determine how long the projectile is in the air.
d) Determine the horizontal distance that the projectile travels.
e) Find the maximum height of the projectile.

63. The vertex of the parabola $y^2 - 4y - 12x - 8 = 0$ is which of the following? [10.1]

A. $(1, -2)$ **B.** $(-1, 2)$
C. $(2, -1)$ **D.** $(-2, 1)$

64. Which of the following cannot be a number of solutions possible for a system of equations representing an ellipse and a straight line? [10.4]

A. 0 **B.** 1
C. 2 **D.** 4

65. The graph of $x^2 + 4y^2 = 4$ is which of the following? [10.2], [10.3]

A.

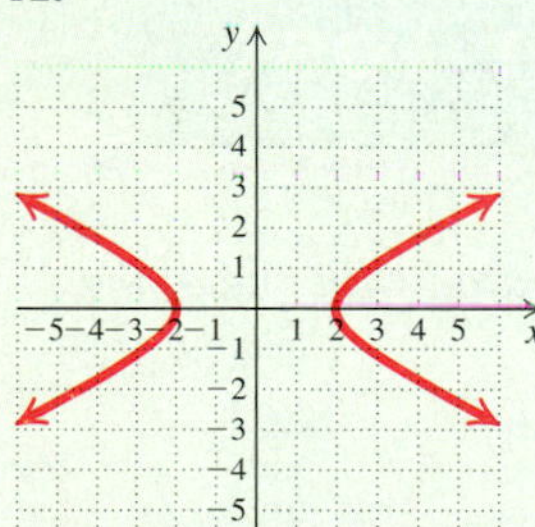

B.

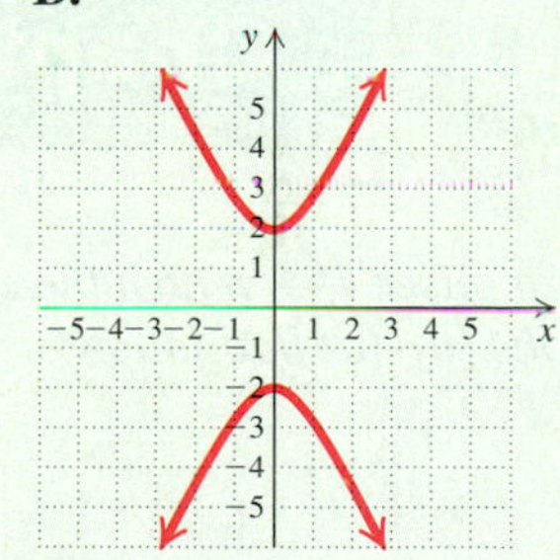

C.

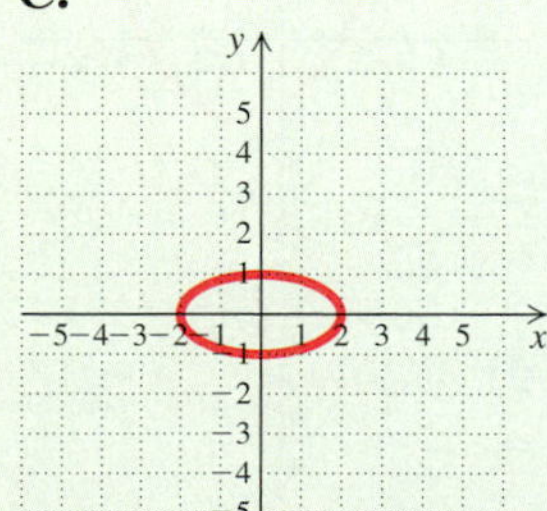

D.

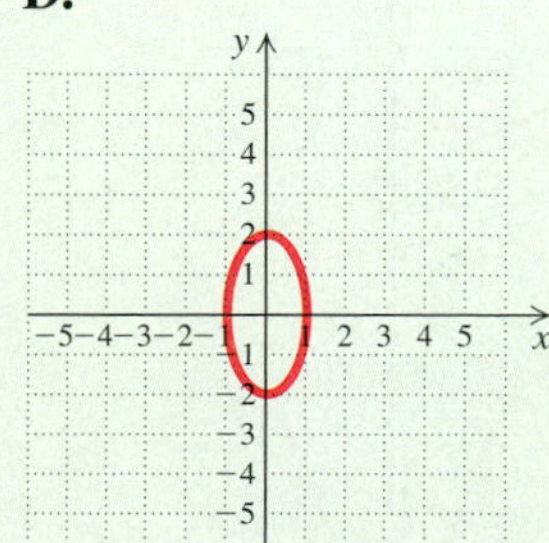

Synthesis

66. Find two numbers whose product is 4 and the sum of whose reciprocals is $\frac{65}{56}$. [10.4]

67. Find an equation of the circle that passes through the points $(10, 7)$, $(-6, 7)$, and $(-8, 1)$. [10.2], [10.4]

68. Find an equation of the ellipse containing the point $(-1/2, 3\sqrt{3}/2)$ and with vertices $(0, -3)$ and $(0, 3)$. **[10.2]**

69. *Navigation.* Two radio transmitters positioned 400 mi apart along the shore send simultaneous signals to a ship that is 250 mi offshore, sailing parallel to the shoreline.

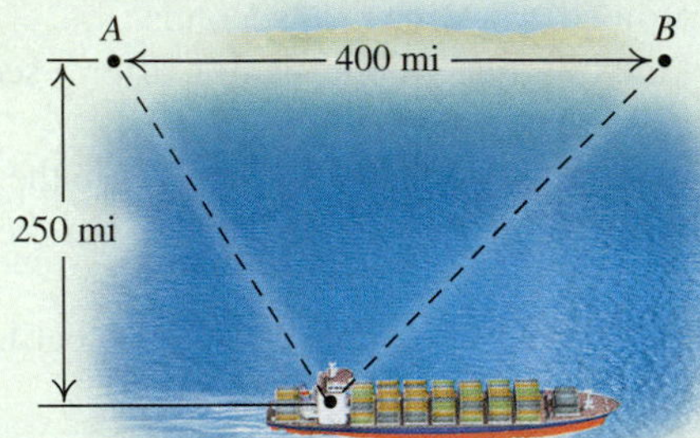

The signal from transmitter A reaches the ship 300 microseconds before the signal from transmitter B. The signals travel at a speed of 186,000 miles per second, or 0.186 mile per microsecond. Find the equation of the hyperbola with foci A and B on which the ship is located. (*Hint*: For any point on the hyperbola, the absolute value of the difference of its distances from the foci is $2a$.) **[10.3]**

Collaborative Discussion and Writing

70. Explain how the distance formula is used to find the standard equation of a parabola. **[10.1]**

71. Explain why function notation is not used in Section 10.2. **[10.2]**

72. Explain how the procedure you would follow for graphing an equation of the form $Ax^2 + Bxy + Cy^2 + Dx + Ey + F = 0$ when $B \neq 0$ differs from the procedure you would follow when $B = 0$. **[10.5]**

73. Consider the graphs of

$$r = \frac{e}{1 - e \sin \theta}$$

for $e = 0.2, 0.4, 0.6$, and 0.8. Explain the effect of the value of e on the graph. **[10.6]**

10 Chapter Test

In Exercises 1–4, match the equation with one of the graphs (a)–(d) that follow.

a)

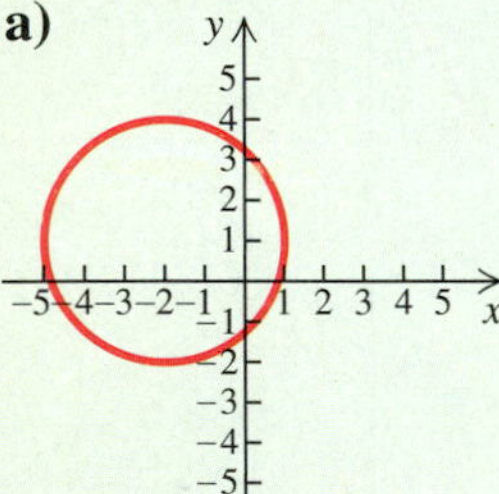

b)

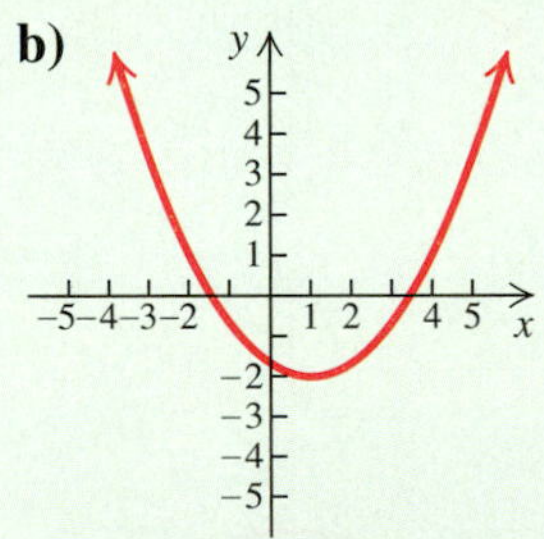

c)

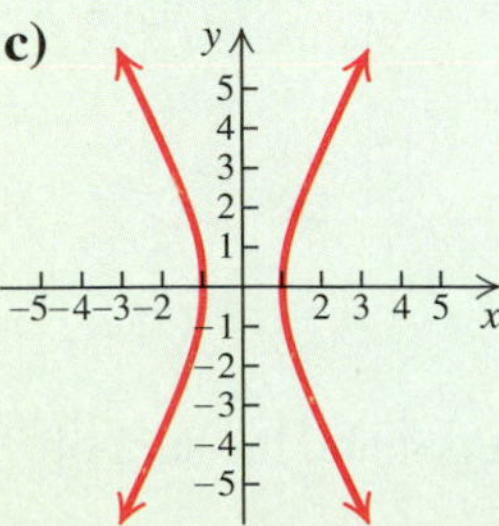

d)

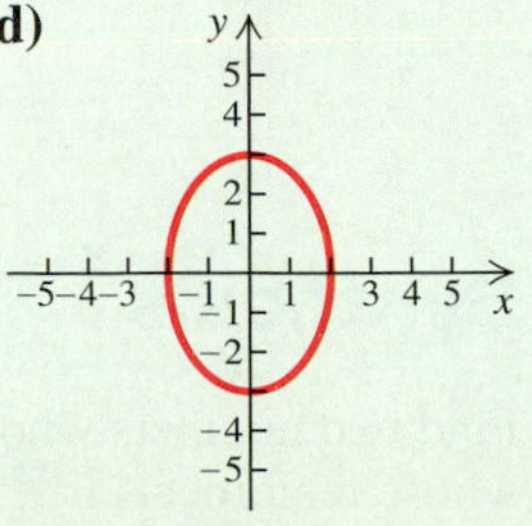

1. $4x^2 - y^2 = 4$

2. $x^2 - 2x - 3y = 5$

3. $x^2 + 4x + y^2 - 2y - 4 = 0$

4. $9x^2 + 4y^2 = 36$

Find the vertex, the focus, and the directrix of the parabola. Then draw the graph.

5. $x^2 = 12y$

6. $y^2 + 2y - 8x - 7 = 0$

7. Find an equation of the parabola with focus $(0, 2)$ and directrix $y = -2$.

8. Find the center and the radius of the circle given by $x^2 + y^2 + 2x - 6y - 15 = 0$. Then draw the graph.

Find the center, the vertices, and the foci of the ellipse. Then draw the graph.

9. $9x^2 + 16y^2 = 144$

10. $\dfrac{(x+1)^2}{4} + \dfrac{(y-2)^2}{9} = 1$

11. Find an equation of the ellipse having vertices $(0, -5)$ and $(0, 5)$ and with minor axis of length 4.

Find the center, the vertices, the foci, and the asymptotes of the hyperbola. Then draw the graph.

12. $4x^2 - y^2 = 4$

13. $\dfrac{(y-2)^2}{4} - \dfrac{(x+1)^2}{9} = 1$

14. Find the asymptotes of the hyperbola given by $2y^2 - x^2 = 18$.

15. *Satellite Dish.* A satellite dish has a parabolic cross section that is 18 in. wide at the opening and 6 in. deep at the vertex. How far from the vertex is the focus?

Solve.

16. $2x^2 - 3y^2 = -10,$
$x^2 + 2y^2 = 9$

17. $x^2 + y^2 = 13,$
$x + y = 1$

18. $x + y = 5,$
$xy = 6$

19. *Landscaping.* Leisurescape is planting a rectangular flower garden with a perimeter of 18 ft and a diagonal of $\sqrt{41}$ ft. Find the dimensions of the garden.

20. *Fencing.* It will take 210 ft of fencing to enclose a rectangular playground with an area of 2700 ft^2. Find the dimensions of the playground.

21. Graph the system of inequalities. Then find the coordinates of the points of intersection of the graphs of the related equations.

$y \geq x^2 - 4,$
$y < 2x - 1$

22. Graph: $5x^2 - 8xy + 5y^2 = 9$.

23. Graph $r = \dfrac{2}{1 - \sin\theta}$. State whether the directrix is vertical or horizontal, describe its location in relation to the pole, and find the vertex or vertices.

24. Find a polar equation of the conic with a focus at the pole, eccentricity 2, and directrix $r = 3\sec\theta$.

25. Graph the plane curve given by the parametric equations $x = \sqrt{t}$, $y = t + 2$; $0 \leq t \leq 16$.

26. Find a rectangular equation equivalent to $x = 3\cos\theta$, $y = 3\sin\theta$; $0 \leq \theta \leq 2\pi$.

27. Find two sets of parametric equations for the rectangular equation $y = x - 5$.

28. *Projectile Motion.* A projectile is launched from a height of 10 ft with an initial speed of 250 ft/sec at an angle of 30° with the horizontal.

a) Find parametric equations that give the position of the projectile at time t, in seconds.
b) Find the height of the projectile after 1 sec and after 3 sec.
c) Determine how long the projectile is in the air.
d) Determine the horizontal distance that the projectile travels.
e) Find the maximum height of the projectile.

29. The graph of $(y-1)^2 = 4(x+1)$ is which of the following?

A. **B.**

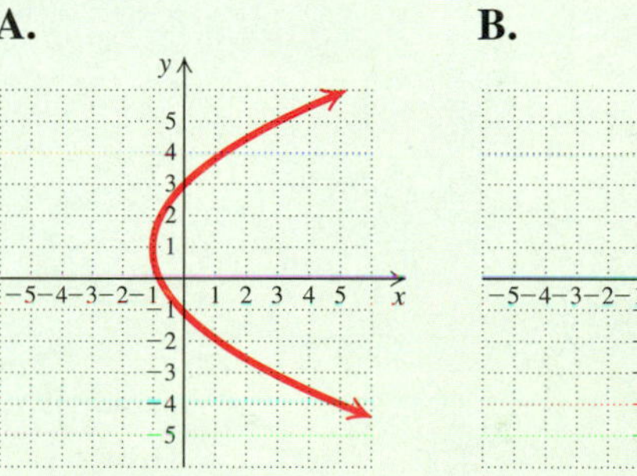

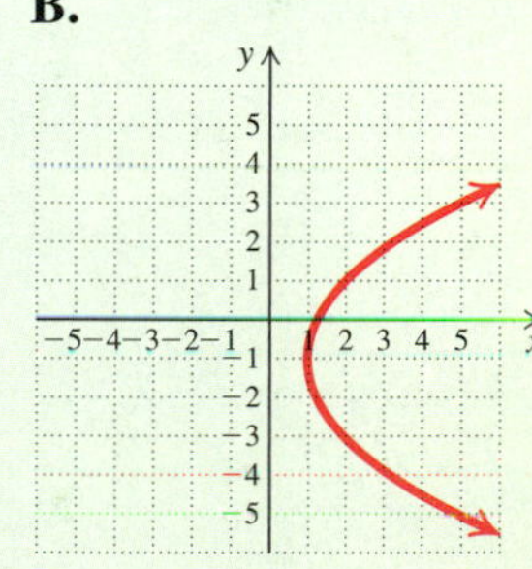

C. **D.**

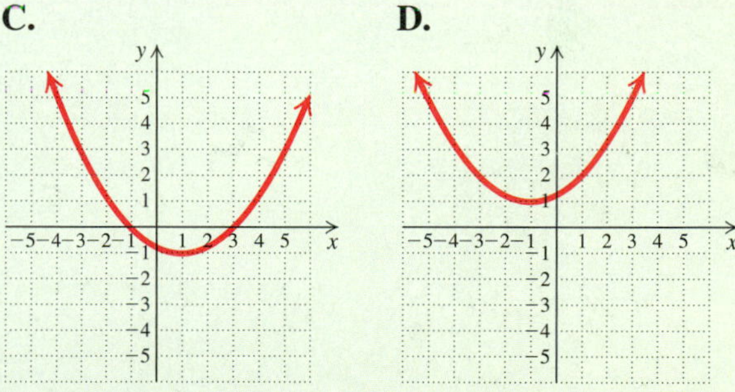

❯ Synthesis

30. Find an equation of the circle for which the endpoints of a diameter are $(1, 1)$ and $(5, -3)$.

Sequences, Series, and Combinatorics

CHAPTER 11

APPLICATION

This problem appears as Exercise 21 in Section 11.8.

An American roulette wheel contains 38 slots numbered 00, 0, 1, 2, 3, . . . , 35, 36. Eighteen of the slots numbered 1–36 are colored red and 18 are colored black. The 00 and 0 slots are considered to be uncolored. The wheel is spun, and a ball is rolled around the rim until it falls into a slot. What is the probability that the ball falls into a black slot?

11.1 Sequences and Series

- Find terms of sequences given the nth term.
- Look for a pattern in a sequence and try to determine a general term.
- Convert between sigma notation and other notation for a series.
- Construct the terms of a recursively defined sequence.

In this section, we discuss sets or lists of numbers, considered in order, and their sums.

Suppose that \$1000 is invested at 4%, compounded annually. The amounts to which the account will grow after 1 year, 2 years, 3 years, 4 years, and so on, form the following sequence of numbers:

(1)	(2)	(3)	(4)
↓	↓	↓	↓
\$1040.00	\$1081.60	\$1124.86	\$1169.86.

We can think of this as a function that pairs 1 with \$1040.00, 2 with \$1081.60, 3 with \$1124.86, and so on. A **sequence** is thus a *function*, where the domain is a set of consecutive positive integers beginning with 1.

If we continue to compute the amounts of money in the account forever, we obtain an **infinite sequence** with function values

\$1040.00, \$1081.60, \$1124.86, \$1169.86, \$1216.65, \$1265.32, ...

The dots ". . . " at the end indicate that the sequence goes on without stopping. If we stop after a certain number of years, we obtain a **finite sequence**:

\$1040.00, \$1081.60, \$1124.86, \$1169.86.

SEQUENCES

An **infinite sequence** is a function having for its domain the set of positive integers, $\{1, 2, 3, 4, 5, \dots\}$.

A **finite sequence** is a function having for its domain a set of positive integers, $\{1, 2, 3, 4, 5, \dots, n\}$, for some positive integer n.

Consider the sequence given by the formula

$$a(n) = 2^n, \quad \text{or} \quad a_n = 2^n.$$

Some of the function values, also known as the **terms** of the sequence, are as follows:

$$a_1 = 2^1 = 2,$$
$$a_2 = 2^2 = 4,$$
$$a_3 = 2^3 = 8,$$
$$a_4 = 2^4 = 16,$$
$$a_5 = 2^5 = 32.$$

The first term of the sequence is denoted as a_1, the fifth term as a_5, and the nth term, or **general term,** as a_n. This sequence can also be denoted as

$$2, 4, 8, \dots, \quad \text{or as} \quad 2, 4, 8, \dots, 2^n, \dots.$$

EXAMPLE 1 Find the first 4 terms and the 23rd term of the sequence whose general term is given by $a_n = (-1)^n n^2$.

Solution We have $a_n = (-1)^n n^2$, so

$$a_1 = (-1)^1 \cdot 1^2 = -1,$$
$$a_2 = (-1)^2 \cdot 2^2 = 4,$$
$$a_3 = (-1)^3 \cdot 3^2 = -9,$$
$$a_4 = (-1)^4 \cdot 4^2 = 16,$$
$$a_{23} = (-1)^{23} \cdot 23^2 = -529.$$

X	Y1	
1	−1	
2	4	
3	−9	
4	16	
23	−529	

X =

We can also use a graphing calculator to find the desired terms of this sequence. We enter $y_1 = (-1)^x x^2$. We then set up a table in ASK mode and enter 1, 2, 3, 4, and 23 as values for x.

Now Try Exercise 1.

Note in Example 1 that the power $(-1)^n$ causes the signs of the terms to alternate between positive and negative, depending on whether n is even or odd. This kind of sequence is called an **alternating sequence.**

EXAMPLE 2 Use a graphing calculator to find the first 5 terms of the sequence whose general term is given by $a_n = n/(n + 1)$.

Solution We can use a table or the SEQ feature, as shown here. We select SEQ from the LIST OPS menu and enter the general term, the variable, and the numbers of the first and last terms desired. The calculator will write the terms horizontally as a list. The list can also be written in fraction notation. The first 5 terms of the sequence are

$$\frac{1}{2}, \quad \frac{2}{3}, \quad \frac{3}{4}, \quad \frac{4}{5}, \quad \text{and} \quad \frac{5}{6}.$$

seq(X/(X+1),X,1,5)▸Frac
$\{\frac{1}{2}\ \frac{2}{3}\ \frac{3}{4}\ \frac{4}{5}\ \frac{5}{6}\}$

We can graph a sequence just as we graph other functions. Consider the function given by $f(x) = x + 1$ and the sequence whose general term is given by $a_n = n + 1$. The graph of $f(x) = x + 1$ is shown on the left below. Since the domain of a sequence is a set of positive integers, the graph of a sequence is a set of points that are not connected. Thus if we use only positive integers for inputs of $f(x) = x + 1$, we have the graph of the sequence $a_n = n + 1$, as shown on the right below.

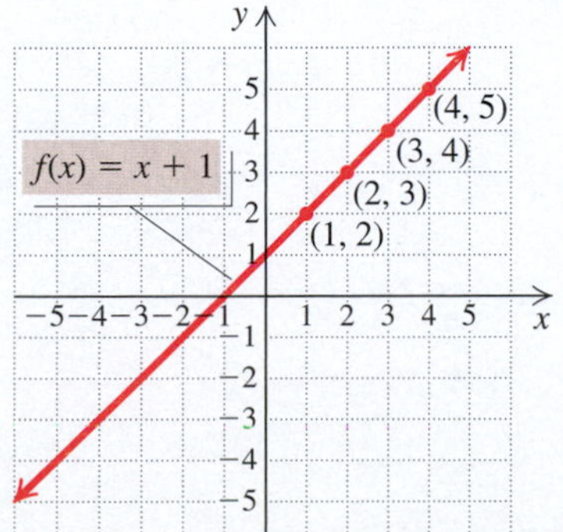

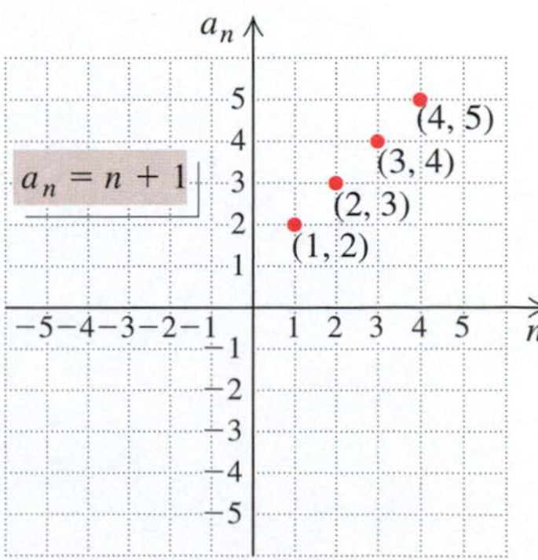

We can also use a graphing calculator to graph a sequence. Since we are graphing a set of unconnected points, we use DOT mode. We also select SEQUENCE mode. In this mode, the variable is n and functions are named $u(n)$, $v(n)$, and $w(n)$ rather than y_1, y_2, and y_3. On many graphing calculators, X,T,Θ,n is used to enter n in SEQUENCE mode.

Plot1 Plot2 Plot3
*n*Min=1
u(*n*)= *n*/(n+1)
u(*n*Min)=
v(*n*)=
v(*n*Min)=
w(*n*)=
w(*n*Min)=

EXAMPLE 3 Use a graphing calculator to graph the sequence whose general term is given by $a_n = n/(n + 1)$.

Solution With the calculator set in DOT mode and SEQUENCE mode, we enter $u(n) = n/(n + 1)$. All the function values will be positive numbers that are less than 1, so we choose the window $[0, 10, 0, 1]$ and we also choose nMin $= 1$, nMax $= 10$, PlotStart $= 1$, and PlotStep $= 1$.

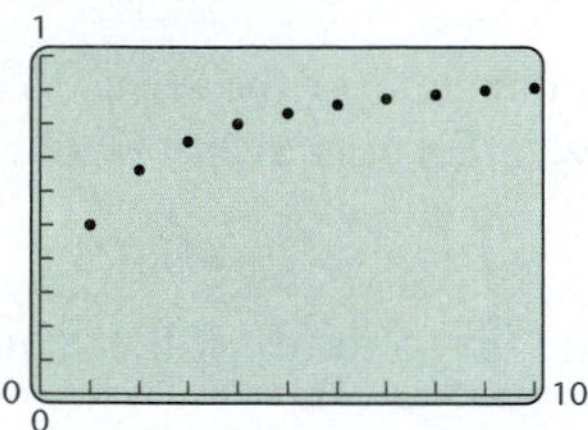

Now Try Exercise 19.

Finding the General Term

When only the first few terms of a sequence are known, we do not know for sure what the general term is, but we might be able to make a prediction by looking for a pattern.

EXAMPLE 4 For each of the following sequences, predict the general term.

a) $1, \sqrt{2}, \sqrt{3}, 2, \ldots$

b) $-1, 3, -9, 27, -81, \ldots$

c) $2, 4, 8, \ldots$

Solution

a) These are square roots of consecutive integers, so the general term might be $\sqrt{n}$.

b) These are powers of 3 with alternating signs, so the general term might be $(-1)^n 3^{n-1}$.

c) If we see the pattern of powers of 2, we will see 16 as the next term and guess 2^n for the general term. Then the sequence could be written with more terms as

$$2, 4, 8, 16, 32, 64, 128, \ldots.$$

If we see that we can get the second term by adding 2, the third term by adding 4, and the next term by adding 6, and so on, we will see 14 as the next term. A general term for the sequence is $n^2 - n + 2$, and the sequence can be written with more terms as

$$2, 4, 8, 14, 22, 32, 44, 58, \ldots.$$

Now Try Exercise 23.

Example 4(c) illustrates that, in fact, you can never be certain about the general term when only a few terms are given. The fewer the given terms, the greater the uncertainty.

Sums and Series

SERIES

Given the infinite sequence

$$a_1, a_2, a_3, a_4, \ldots, a_n, \ldots,$$

the sum of the terms

$$a_1 + a_2 + a_3 + \cdots + a_n + \cdots$$

is called an **infinite series**. A **partial sum** is the sum of the first n terms:

$$a_1 + a_2 + a_3 + \cdots + a_n.$$

A partial sum is also called a **finite series, or *n*th partial sum,** and is denoted S_n.

EXAMPLE 5 For the sequence $-2, 4, -6, 8, -10, 12, -14, \ldots,$ find each of the following.

a) S_1 **b)** S_4 **c)** S_5

Solution

a) $S_1 = -2$

b) $S_4 = -2 + 4 + (-6) + 8 = 4$

c) $S_5 = -2 + 4 + (-6) + 8 + (-10) = -6$

Now Try Exercise 33.

We can also use a graphing calculator to find partial sums of a sequence when a formula for the general term is known.

EXAMPLE 6 Use a graphing calculator to find S_1, S_2, S_3, and S_4 for the sequence whose general term is given by $a_n = n^2 - 3$.

```
cumSum(seq(n²−3,n,1,4))
                {-2 -1 5 18}
```

Solution We can use the CUMSUM feature from the LIST OPS menu. The calculator will write the partial sums as a list. (Note that the calculator can be set in either FUNCTION mode or SEQUENCE mode. Here we show SEQUENCE mode.)

We have $S_1 = -2$, $S_2 = -1$, $S_3 = 5$, and $S_4 = 18$.

Sigma Notation

The Greek letter Σ (sigma) can be used to denote a sum when the general term of a sequence is a formula. For example, the sum of the first four terms of the sequence $3, 5, 7, 9, \ldots, 2k + 1, \ldots$ can be named as follows, using what is called **sigma notation, or summation notation:**

$$\sum_{k=1}^{4} (2k + 1).$$

This is read "the sum as k goes from 1 to 4 of $2k + 1$." The letter k is called the **index of summation.** The index of summation might start at a number other than 1, and letters other than k can be used.

EXAMPLE 7 Find and evaluate each of the following sums.

a) $\sum_{k=1}^{5} k^3$ **b)** $\sum_{k=0}^{4} (-1)^k 5^k$ **c)** $\sum_{i=8}^{11}\left(2 + \frac{1}{i}\right)$

Solution

a) We replace k with 1, 2, 3, 4, and 5. Then we add the results.

$$\sum_{k=1}^{5} k^3 = 1^3 + 2^3 + 3^3 + 4^3 + 5^3$$
$$= 1 + 8 + 27 + 64 + 125 = 225$$

sum(seq(n^3,n,1,5))
225

We can also find this sum on a graphing calculator using the SUM feature from the LIST MATH menu and the SEQ feature from the LIST OPS menu.

b) $$\sum_{k=0}^{4} (-1)^k 5^k = (-1)^0 5^0 + (-1)^1 5^1 + (-1)^2 5^2 + (-1)^3 5^3 + (-1)^4 5^4$$
$$= 1 - 5 + 25 - 125 + 625 = 521$$

c) $$\sum_{i=8}^{11}\left(2 + \frac{1}{i}\right) = \left(2 + \frac{1}{8}\right) + \left(2 + \frac{1}{9}\right) + \left(2 + \frac{1}{10}\right) + \left(2 + \frac{1}{11}\right)$$
$$= 8\frac{1691}{3960}$$

Now Try Exercise 37.

EXAMPLE 8 Write sigma notation for each sum.

a) $1 + 2 + 4 + 8 + 16 + 32 + 64$

b) $-2 + 4 - 6 + 8 - 10$

c) $x + \frac{x^2}{2} + \frac{x^3}{3} + \frac{x^4}{4} + \cdots$

Solution

a) $1 + 2 + 4 + 8 + 16 + 32 + 64$

This is the sum of powers of 2, beginning with 2^0, or 1, and ending with 2^6, or 64. Sigma notation is $\Sigma_{k=0}^{6} 2^k$.

b) $-2 + 4 - 6 + 8 - 10$

Disregarding the alternating signs, we see that this is the sum of the first 5 even integers. Note that $2k$ is a formula for the kth positive even integer, and $(-1)^k = -1$ when k is odd and $(-1)^k = 1$ when k is even. Thus the general term is $(-1)^k(2k)$. The sum begins with $k = 1$ and ends with $k = 5$, so sigma notation is $\Sigma_{k=1}^{5}(-1)^k(2k)$.

c) $x + \frac{x^2}{2} + \frac{x^3}{3} + \frac{x^4}{4} + \cdots$

The general term is x^k/k, beginning with $k = 1$. This is also an infinite series. We use the symbol ∞ for infinity and write the series using sigma notation: $\Sigma_{k=1}^{\infty}(x^k/k)$.

Now Try Exercise 55.

Recursive Definitions

A sequence may be defined **recursively**, or by using a **recursion formula**. Such a definition lists the first term, or the first few terms, and then describes how to determine the remaining terms from the given terms.

EXAMPLE 9 Find the first 5 terms of the sequence defined by

$$a_1 = 5, \qquad a_{n+1} = 2a_n - 3, \quad \text{for } n \geq 1.$$

```
Plot1 Plot2 Plot3
nMin=1
\u(n)=2*u(n-1)-3
 u(nMin)={5}
\v(n)=
 v(nMin)=
\w(n)=
 w(nMin)=
```

Solution We have

$$a_1 = 5,$$

$$a_2 = 2a_1 - 3 = 2 \cdot 5 - 3 = 7,$$

$$a_3 = 2a_2 - 3 = 2 \cdot 7 - 3 = 11,$$

$$a_4 = 2a_3 - 3 = 2 \cdot 11 - 3 = 19,$$

$$a_5 = 2a_4 - 3 = 2 \cdot 19 - 3 = 35.$$

n	u(n)
1	5
2	7
3	11
4	19
5	35
6	67
7	131

n = 1

Many graphing calculators have the capability to work with recursively defined sequences when they are set in SEQUENCE mode. For this sequence, for instance, the function could be entered as $u(n) = 2 * u(n - 1) - 3$ with $u(n\text{Min}) = 5$. We can read the terms of the sequence from a table.

Now Try Exercise 65.

11.1 Exercise Set

In each of the following, the nth term of a sequence is given. Find the first 4 terms, a_{10}, and a_{15}.

1. $a_n = 4n - 1$

2. $a_n = (n - 1)(n - 2)(n - 3)$

3. $a_n = \dfrac{n}{n - 1}, n \geq 2$

4. $a_n = n^2 - 1, n \geq 3$

5. $a_n = \dfrac{n^2 - 1}{n^2 + 1}$

6. $a_n = \left(-\dfrac{1}{2}\right)^{n-1}$

7. $a_n = (-1)^n n^2$

8. $a_n = (-1)^{n-1}(3n - 5)$

9. $a_n = 5 + \dfrac{(-2)^{n+1}}{2^n}$

10. $a_n = \dfrac{2n - 1}{n^2 + 2n}$

Find the indicated term of the given sequence.

11. $a_n = 5n - 6;\ a_8$

12. $a_n = (3n - 4)(2n + 5);\ a_7$

13. $a_n = (2n + 3)^2;\ a_6$

14. $a_n = (-1)^{n-1}(4.6n - 18.3);\ a_{12}$

15. $a_n = 5n^2(4n - 100);\ a_{11}$

16. $a_n = \left(1 + \dfrac{1}{n}\right)^2;\ a_{80}$

17. $a_n = \ln e^n;\ a_{67}$

18. $a_n = 2 - \dfrac{1000}{n};\ a_{100}$

Use a graphing calculator to construct a table of values and a graph for the first 10 terms of the sequence.

19. $a_n = \left(1 + \dfrac{1}{n}\right)^n$

20. $a_n = \sqrt{n + 1} - \sqrt{n}$

21. $a_1 = 2,\ a_{n+1} = \sqrt{1 + \sqrt{a_n}}$

22. $a_1 = 2,\ a_{n+1} = \dfrac{1}{2}\left(a_n + \dfrac{2}{a_n}\right)$

Predict the general term, or nth term, a_n, of the sequence. Answers may vary.

23. $2, 4, 6, 8, 10, \ldots$

24. $3, 9, 27, 81, 243, \ldots$

25. $-2, 6, -18, 54, \ldots$

26. $-2, 3, 8, 13, 18, \ldots$

27. $\frac{2}{3}, \frac{3}{4}, \frac{4}{5}, \frac{5}{6}, \frac{6}{7}, \ldots$

28. $\sqrt{2}, 2, \sqrt{6}, 2\sqrt{2}, \sqrt{10}, \ldots$

29. $1 \cdot 2, 2 \cdot 3, 3 \cdot 4, 4 \cdot 5, \ldots$

30. $-1, -4, -7, -10, -13, \ldots$

31. $0, \log 10, \log 100, \log 1000, \ldots$

32. $\ln e^2, \ln e^3, \ln e^4, \ln e^5, \ldots$

Find the indicated partial sums for the sequence.

33. $1, 2, 3, 4, 5, 6, 7, \ldots$; S_3 and S_7

34. $1, -3, 5, -7, 9, -11, \ldots$; S_2 and S_5

35. $2, 4, 6, 8, \ldots$; S_4 and S_5

36. $1, \frac{1}{4}, \frac{1}{9}, \frac{1}{16}, \frac{1}{25}, \ldots$; S_1 and S_5

Find and evaluate the sum.

37. $\sum_{k=1}^{5} \frac{1}{2k}$

38. $\sum_{i=1}^{6} \frac{1}{2i + 1}$

39. $\sum_{i=0}^{6} 2^i$

40. $\sum_{k=4}^{7} \sqrt{2k - 1}$

41. $\sum_{k=7}^{10} \ln k$

42. $\sum_{k=1}^{4} \pi k$

43. $\sum_{k=1}^{8} \frac{k}{k + 1}$

44. $\sum_{i=1}^{5} \frac{i - 1}{i + 3}$

45. $\sum_{i=1}^{5} (-1)^i$

46. $\sum_{k=0}^{5} (-1)^{k+1}$

47. $\sum_{k=1}^{8} (-1)^{k+1} 3k$

48. $\sum_{k=0}^{7} (-1)^k 4^{k+1}$

49. $\sum_{k=0}^{6} \frac{2}{k^2 + 1}$

50. $\sum_{i=1}^{10} i(i + 1)$

51. $\sum_{k=0}^{5} (k^2 - 2k + 3)$

52. $\sum_{k=1}^{10} \frac{1}{k(k + 1)}$

53. $\sum_{i=0}^{10} \frac{2^i}{2^i + 1}$

54. $\sum_{k=0}^{3} (-2)^{2k}$

Write sigma notation. Answers may vary.

55. $5 + 10 + 15 + 20 + 25 + \cdots$

56. $7 + 14 + 21 + 28 + 35 + \cdots$

57. $2 - 4 + 8 - 16 + 32 - 64$

58. $3 + 6 + 9 + 12 + 15$

59. $-\frac{1}{2} + \frac{2}{3} - \frac{3}{4} + \frac{4}{5} - \frac{5}{6} + \frac{6}{7}$

60. $\frac{1}{1^2} + \frac{1}{2^2} + \frac{1}{3^2} + \frac{1}{4^2} + \frac{1}{5^2}$

61. $4 - 9 + 16 - 25 + \cdots + (-1)^n n^2$

62. $9 - 16 + 25 + \cdots + (-1)^{n+1} n^2$

63. $\frac{1}{1 \cdot 2} + \frac{1}{2 \cdot 3} + \frac{1}{3 \cdot 4} + \frac{1}{4 \cdot 5} + \cdots$

64. $\frac{1}{1 \cdot 2^2} + \frac{1}{2 \cdot 3^2} + \frac{1}{3 \cdot 4^2} + \frac{1}{4 \cdot 5^2} + \cdots$

Find the first 4 terms of the recursively defined sequence.

65. $a_1 = 4, \; a_{n+1} = 1 + \frac{1}{a_n}$

66. $a_1 = 256, \; a_{n+1} = \sqrt{a_n}$

67. $a_1 = 6561, \; a_{n+1} = (-1)^n \sqrt{a_n}$

68. $a_1 = e^Q, \; a_{n+1} = \ln a_n$

69. $a_1 = 2, \; a_2 = 3, \; a_{n+1} = a_n + a_{n-1}$

70. $a_1 = -10, \; a_2 = 8, \; a_{n+1} = a_n - a_{n-1}$

71. *Compound Interest.* Suppose that \$4000 is invested at 3.75%, compounded annually. The value of the investment after n years is given by the sequence

$$a_n = \$4000(1.0375)^n, \quad n = 1, 2, 3, \ldots.$$

a) Find the first 10 terms of the sequence.
b) Find the value of the investment after 20 years.

72. *Salvage Value.* The value of a post-hole digger is \$5200. Its salvage value each year is 75% of its value the year before. Give a sequence that lists the salvage value of the post-hole digger for each year of a 10-year period.

73. *Wage Sequence.* Xavier is paid \$16.20 per hour for working at Red Freight Limited. Each year he receives a \$2.25 hourly raise. Give a sequence that lists Xavier's hourly wage over a 10-year period.

74. *Bacteria Growth.* Suppose that a single cell of bacteria divides into two every 15 min. Suppose that the same rate of division is maintained for 4 hr. Give a sequence that lists the number of cells after successive 15-min periods.

75. *Fibonacci Sequence: Rabbit Population Growth.* One of the most famous recursively defined sequences is the **Fibonacci sequence**. In 1202, the Italian mathematician Leonardo da Pisa, also called Fibonacci, proposed the following model for rabbit population growth. Suppose that every month each mature pair of rabbits in the population produces a new pair that begins reproducing after two months, and also suppose that no rabbits die. Beginning with one pair of newborn rabbits, the population can be modeled by the following recursively defined sequence:

$$a_1 = 1, \; a_2 = 1, \; a_n = a_{n-1} + a_{n-2}, \text{ for } n \geq 3,$$

where a_n is the total number of pairs of rabbits in month n. Find the first 7 terms of the Fibonacci sequence.

76. *Weekly Earnings.* The following table lists the average weekly earnings of U.S. production workers in recent years.

Year, n	Average Weekly Earnings of U.S. Production Workers
2008, 0	\$607.53
2009, 1	616.01
2010, 2	636.25
2011, 3	653.19
2012, 4	665.82
2013, 5	677.67

Sources: U.S. Bureau of Labor Statistics, U.S. Department of Labor

a) Use a graphing calculator to fit a linear sequence regression function

$$a_n = an + b$$

to the data, where n is the number of years since 2008.

b) Estimate the average weekly earnings in 2014, in 2016, and in 2017.

77. *Price of Silver.* The following table lists the price of silver per troy ounce in recent years.

Year, n	Price of Silver per Troy Ounce
2009, 0	\$14.69
2010, 1	20.20
2011, 2	35.26
2012, 3	31.21
2013, 4	23.80

Source: Mineral Commodity Summaries 2014, U.S. Geological Survey, U.S. Department of the Interior

a) Use a graphing calculator to fit a quadratic sequence regression function

$$a_n = an^2 + bn + c$$

to the data, where n is the number of years since 2009.

b) Estimate the price of silver per troy ounce in 2014.

Skill Maintenance

Solve. [9.1], [9.3], [9.5], [9.6]

78. $3x - 2y = 3,$
$2x + 3y = -11$

79. *Harvesting Pumpkins.* A total of 23,400 acres of pumpkins were harvested in Illinois and Ohio in 2012. The number of acres of pumpkins harvested in Ohio was 9000 fewer than the number of acres of pumpkins harvested in Illinois. (*Source*: U. S. Department of Agriculture) Find the number of acres of pumpkins harvested in Illinois and in Ohio in 2012.

Find the center and the radius of the circle with the given equation. [10.2]

80. $x^2 + y^2 - 6x + 4y = 3$

81. $x^2 + y^2 + 5x - 8y = 2$

Synthesis

Find the first 5 terms of the sequence, and then find S_5.

82. $a_n = \frac{1}{2^n}\log 1000^n$

83. $a_n = i^n, i = \sqrt{-1}$

84. $a_n = \ln(1 \cdot 2 \cdot 3 \cdot \cdots \cdot n)$

For each sequence, find a formula for S_n.

85. $a_n = \ln n$

86. $a_n = \frac{1}{n} - \frac{1}{n+1}$

11.2 Arithmetic Sequences and Series

- For any arithmetic sequence, find the nth term when n is given and n when the nth term is given, and given two terms, find the common difference and construct the sequence.
- Find the sum of the first n terms of an arithmetic sequence.

A sequence in which each term after the first term is found by adding the same number to the preceding term is an **arithmetic sequence.**

Arithmetic Sequences

The sequence 2, 5, 8, 11, 14, 17, . . . is arithmetic because adding 3 to any term produces the next term. In other words, the difference between any term and the preceding one is 3. Arithmetic sequences are also called *arithmetic progressions.*

ARITHMETIC SEQUENCE

A sequence is **arithmetic** if there exists a number d, called the **common difference,** such that $a_{n+1} = a_n + d$ for any integer $n \geq 1$.

EXAMPLE 1 For each of the following arithmetic sequences, identify the first term, a_1, and the common difference, d.

a) 4, 9, 14, 19, 24, . . .

b) 34, 27, 20, 13, 6, -1, -8, . . .

c) 2, $2\frac{1}{2}$, 3, $3\frac{1}{2}$, 4, $4\frac{1}{2}$, . . .

Solution The first term, a_1, is the first term listed. To find the common difference, d, we choose any term beyond the first and subtract the preceding term from it.

SEQUENCE	FIRST TERM, a_1	COMMON DIFFERENCE, d
a) 4, 9, 14, 19, 24, . . .	4	5 $(9 - 4 = 5)$
b) 34, 27, 20, 13, 6, -1, -8, . . .	34	-7 $(27 - 34 = -7)$
c) 2, $2\frac{1}{2}$, 3, $3\frac{1}{2}$, 4, $4\frac{1}{2}$, . . .	2	$\frac{1}{2}$ $(2\frac{1}{2} - 2 = \frac{1}{2})$

We obtained the common difference by subtracting a_1 from a_2. Had we subtracted a_2 from a_3 or a_3 from a_4, we would have obtained the same values for d. Thus we can check by adding d to each term in a sequence to see if we progress correctly to the next term.

Check:

a) $4 + 5 = 9$, $9 + 5 = 14$, $14 + 5 = 19$, $19 + 5 = 24$

b) $34 + (-7) = 27$, $27 + (-7) = 20$, $20 + (-7) = 13$,
$13 + (-7) = 6$, $6 + (-7) = -1$, $-1 + (-7) = -8$

c) $2 + \frac{1}{2} = 2\frac{1}{2}$, $2\frac{1}{2} + \frac{1}{2} = 3$, $3 + \frac{1}{2} = 3\frac{1}{2}$, $3\frac{1}{2} + \frac{1}{2} = 4$,
$4 + \frac{1}{2} = 4\frac{1}{2}$

Now Try Exercise 1.

To find a formula for the general, or nth, term of any arithmetic sequence, we denote the common difference by d, write out the first few terms, and look for a pattern:

$$a_1,$$
$$a_2 = a_1 + d,$$
$$a_3 = a_2 + d = (a_1 + d) + d = a_1 + 2d, \qquad \text{Substituting for } a_2$$
$$a_4 = a_3 + d = (a_1 + 2d) + d = a_1 + 3d. \qquad \text{Substituting for } a_3$$

Note that the coefficient of d in each case is 1 less than the subscript.

Generalizing, we obtain the following formula.

nTH TERM OF AN ARITHMETIC SEQUENCE

The ***n*th term** of an arithmetic sequence is given by $a_n = a_1 + (n - 1)d$, for any integer $n \geq 1$.

EXAMPLE 2 Find the 14th term of the arithmetic sequence

4, 7, 10, 13, ...

Solution We first note that $a_1 = 4$, $d = 7 - 4$, or 3, and $n = 14$. Then using the formula for the nth term, we obtain

$$a_n = a_1 + (n - 1)d$$
$$a_{14} = 4 + (14 - 1) \cdot 3 \qquad \text{Substituting}$$
$$= 4 + 13 \cdot 3 = 4 + 39$$
$$= 43.$$

The 14th term is 43.

Now Try Exercise 9.

EXAMPLE 3 In the sequence of Example 2, which term is 301? That is, find n if $a_n = 301$.

Solution We substitute 301 for a_n, 4 for a_1, and 3 for d in the formula for the nth term and solve for n:

$$a_n = a_1 + (n - 1)d$$
$$301 = 4 + (n - 1) \cdot 3 \qquad \text{Substituting}$$
$$\left.\begin{aligned} 301 &= 4 + 3n - 3 \\ 301 &= 3n + 1 \\ 300 &= 3n \\ 100 &= n. \end{aligned}\right\} \text{Solving for } n$$

The term 301 is the 100th term of the sequence.

Now Try Exercise 15.

Given two terms and their places in an arithmetic sequence, we can construct the sequence.

EXAMPLE 4 The 3rd term of an arithmetic sequence is 8, and the 16th term is 47. Find a_1 and d and construct the sequence.

Solution We know that $a_3 = 8$ and $a_{16} = 47$. Thus we would need to add d 13 times to get from 8 to 47. That is,

$$8 + 13d = 47. \qquad a_3 \text{ and } a_{16} \text{ are } 16 - 3, \text{ or } 13, \text{ terms apart.}$$

Solving $8 + 13d = 47$, we obtain

$$13d = 39$$
$$d = 3.$$

Since $a_3 = 8$, we subtract d twice to get a_1. Thus,

$$a_1 = 8 - 2 \cdot 3 = 2.$$ **a_1 and a_3 are $3 - 1$, or 2, terms apart.**

The sequence is 2, 5, 8, 11, ... Note that we could also subtract d 15 times from a_{16} in order to find a_1.

Now Try Exercise 23.

In general, d should be subtracted $n - 1$ times from a_n in order to find a_1.

Sum of the First *n* Terms of an Arithmetic Sequence

Consider the arithmetic sequence

3, 5, 7, 9, ...

When we add the first 4 terms of the sequence, we get S_4, which is

$3 + 5 + 7 + 9$, or 24.

This sum is called an **arithmetic series**. To find a formula for the sum of the first n terms, S_n, of an arithmetic sequence, we first denote an arithmetic sequence, as follows:

This term is two terms back from the last. If you add d to this term, the result is the next-to-last term, $a_n - d$.

$$a_1, \quad (a_1 + d), \quad (a_1 + 2d), \ldots, \quad (a_n - 2d), \quad (a_n - d), \quad a_n.$$

This is the next-to-last term. If you add d to this term, the result is a_n.

Then S_n is given by

$$S_n = a_1 + (a_1 + d) + (a_1 + 2d) + \cdots + (a_n - 2d) + (a_n - d) + a_n. \quad (1)$$

Reversing the order of the addition gives us

$$S_n = a_n + (a_n - d) + (a_n - 2d) + \cdots + (a_1 + 2d) + (a_1 + d) + a_1. \quad (2)$$

If we add corresponding terms of each side of equations (1) and (2), we get

$$2S_n = [a_1 + a_n] + [(a_1 + d) + (a_n - d)] + [(a_1 + 2d) + (a_n - 2d)] + \cdots + [(a_n - 2d) + (a_1 + 2d)] + [(a_n - d) + (a_1 + d)] + [a_n + a_1].$$

In the expression for $2S_n$, there are n expressions in square brackets. Each of these expressions is equivalent to $a_1 + a_n$. Thus the expression for $2S_n$ can be written in simplified form as

$$2S_n = [a_1 + a_n] + [a_1 + a_n] + [a_1 + a_n] + \cdots + [a_n + a_1] + [a_n + a_1] + [a_n + a_1].$$

Since $a_1 + a_n$ is being added n times, it follows that

$$2S_n = n(a_1 + a_n),$$

from which we get the following formula.

SUM OF THE FIRST n TERMS

The sum of the first n terms of an arithmetic sequence is given by

$$S_n = \frac{n}{2}(a_1 + a_n).$$

EXAMPLE 5 Find the sum of the first 100 natural numbers.

Solution The sum is

$$1 + 2 + 3 + \cdots + 99 + 100.$$

This is the sum of the first 100 terms of the arithmetic sequence for which

$$a_1 = 1, \qquad a_n = 100, \quad \text{and} \quad n = 100.$$

Thus substituting into the formula

$$S_n = \frac{n}{2}(a_1 + a_n),$$

we get

$$S_{100} = \frac{100}{2}(1 + 100) = 50(101) = 5050.$$

The sum of the first 100 natural numbers is 5050.

Now Try Exercise 27.

EXAMPLE 6 Find the sum of the first 15 terms of the arithmetic sequence

$$4, 7, 10, 13, \ldots.$$

Solution Note that $a_1 = 4$, $d = 3$, and $n = 15$. Before using the formula

$$S_n = \frac{n}{2}(a_1 + a_n),$$

we find the last term, a_{15}:

$$a_{15} = 4 + (15 - 1)3 \qquad \text{Substituting into the formula } a_n = a_1 + (n - 1)d$$
$$= 4 + 14 \cdot 3 = 46.$$

Thus,

$$S_{15} = \frac{15}{2}(4 + 46) = \frac{15}{2}(50) = 375.$$

The sum of the first 15 terms is 375.

Now Try Exercise 25.

EXAMPLE 7 Find the sum: $\sum_{k=1}^{130}(4k + 5)$.

Solution It is helpful to first write out a few terms:

$$9 + 13 + 17 + \cdots.$$

It appears that this is an arithmetic series coming from an arithmetic sequence with $a_1 = 9$, $d = 4$, and $n = 130$. Before using the formula

$$S_n = \frac{n}{2}(a_1 + a_n),$$

we find the last term, a_{130}:

$$\begin{aligned} a_{130} &= 4 \cdot 130 + 5 \qquad \text{The } k\text{th term is } 4k + 5. \\ &= 520 + 5 \\ &= 525. \end{aligned}$$

Thus,

$$\begin{aligned} S_{130} &= \frac{130}{2}(9 + 525) \qquad \text{Substituting into } S_n = \frac{n}{2}(a_1 + a_n) \\ &= 34{,}710. \end{aligned}$$

```
sum(seq(4X+5,X,1,130))
                 34710
```

This sum can also be found on a graphing calculator. It is not necessary to have the calculator set in SEQUENCE mode in order to do this.

Now Try Exercise 33.

Applications

The translation of some applications and problem-solving situations may involve arithmetic sequences or series. We consider some examples.

EXAMPLE 8 ***Hourly Wages.*** Kendall accepts a job, starting with an hourly wage of \$14.25, and is promised a raise of 15¢ per hour every 2 months for 5 years. At the end of 5 years, what will Kendall's hourly wage be?

Solution It helps to first write down the hourly wage for several 2-month time periods:

Beginning: \$14.25,
After 2 months: \$14.40,
After 4 months: \$14.55,
and so on.

What appears is a sequence of numbers: 14.25, 14.40, 14.55, ... This sequence is arithmetic, because adding 0.15 each time gives us the next term.

We want to find the last term of an arithmetic sequence, so we use the formula $a_n = a_1 + (n - 1)d$. We know that $a_1 = 14.25$ and $d = 0.15$, but what is n? That is, how many terms are in the sequence? Each year there are 12/2, or 6 raises, since Kendall gets a raise every 2 months. There are 5 years, so the total number of raises will be $5 \cdot 6$, or 30. Thus there will be 31 terms: the original wage and 30 increased rates.

Substituting in the formula $a_n = a_1 + (n - 1)d$ gives us

$$a_{31} = 14.25 + (31 - 1) \cdot 0.15 = 18.75.$$

Thus, at the end of 5 years, Kendall's hourly wage will be \$18.75.

Now Try Exercise 43.

The calculations in Example 8 could be done in a number of ways. There is often a variety of ways in which a problem can be solved. In this chapter, we concentrate on the use of sequences and series and their related formulas.

EXAMPLE 9 *Total in a Stack.* A stack of electrical poles has 30 poles in the bottom row. There are 29 poles in the second row, 28 in the next row, and so on. How many poles are in the stack if there are 5 poles in the top row?

Solution A drawing will help in this case. The following figure shows the ends of the poles and the way in which they stack.

Since the number of poles goes from 30 in a row up to 5 in the top row, there must be 26 rows. We want the sum

$$30 + 29 + 28 + \cdots + 5.$$

Thus we have an arithmetic series. We use the formula

$$S_n = \frac{n}{2}(a_1 + a_n),$$

with $n = 26$, $a_1 = 30$, and $a_{26} = 5$.

Substituting, we get

$$S_{26} = \frac{26}{2}(30 + 5) = 455.$$

There are 455 poles in the stack.

Now Try Exercise 39.

11.2 Exercise Set

Find the first term and the common difference.

1. 3, 8, 13, 18, . . .

2. \$1.08, \$1.16, \$1.24, \$1.32, . . .

3. 9, 5, 1, −3, . . .

4. −8, −5, −2, 1, 4, . . .

5. $\frac{3}{2}, \frac{9}{4}, 3, \frac{15}{4}, \ldots$

6. $\frac{3}{5}, \frac{1}{10}, -\frac{2}{5}, \ldots$

7. \$316, \$313, \$310, \$307, . . .

8. Find the 11th term of the arithmetic sequence 0.07, 0.12, 0.17, ...

9. Find the 12th term of the arithmetic sequence 2, 6, 10, ...

10. Find the 17th term of the arithmetic sequence 7, 4, 1, ...

11. Find the 14th term of the arithmetic sequence $3, \frac{7}{3}, \frac{5}{3}, \ldots.$

12. Find the 13th term of the arithmetic sequence \$1200, \$964.32, \$728.64, ...

13. Find the 10th term of the arithmetic sequence \$2345.78, \$2967.54, \$3589.30, ...

14. In the sequence of Exercise 8, what term is the number 1.67?

15. In the sequence of Exercise 9, what term is the number 106?

16. In the sequence of Exercise 10, what term is −296?

17. In the sequence of Exercise 11, what term is -27?

18. Find a_{20} when $a_1 = 14$ and $d = -3$.

19. Find a_1 when $d = 4$ and $a_8 = 33$.

20. Find d when $a_1 = 8$ and $a_{11} = 26$.

21. Find n when $a_1 = 25$, $d = -14$, and $a_n = -507$.

22. In an arithmetic sequence, $a_{17} = -40$ and $a_{28} = -73$. Find a_1 and d. Write the first 5 terms of the sequence.

23. In an arithmetic sequence, $a_{17} = \frac{25}{3}$ and $a_{32} = \frac{95}{6}$. Find a_1 and d. Write the first 5 terms of the sequence.

24. Find the sum of the first 14 terms of the series $11 + 7 + 3 + \cdots$.

25. Find the sum of the first 20 terms of the series $5 + 8 + 11 + 14 + \cdots$.

26. Find the sum of the first 300 natural numbers.

27. Find the sum of the first 400 even natural numbers.

28. Find the sum of the odd numbers 1 to 199, inclusive.

29. Find the sum of the multiples of 7 from 7 to 98, inclusive.

30. Find the sum of all multiples of 4 that are between 14 and 523.

31. If an arithmetic series has $a_1 = 2$, $d = 5$, and $n = 20$, what is S_n?

32. If an arithmetic series has $a_1 = 7$, $d = -3$, and $n = 32$, what is S_n?

Find the sum.

33. $\sum_{k=1}^{40} (2k + 3)$

34. $\sum_{k=5}^{20} 8k$

35. $\sum_{k=0}^{19} \frac{k - 3}{4}$

36. $\sum_{k=2}^{50} (2000 - 3k)$

37. $\sum_{k=12}^{57} \frac{7 - 4k}{13}$

38. $\sum_{k=101}^{200} (1.14k - 2.8) - \sum_{k=1}^{5} \left(\frac{k + 4}{10}\right)$

39. *Total Savings.* If 10¢ is saved on October 1, 20¢ is saved on October 2, 30¢ on October 3, and so on, how much is saved during the 31 days of October?

40. *Stacking Poles.* How many poles will be in a stack of telephone poles if there are 50 in the first layer, 49 in the second, and so on, with 6 in the top layer?

41. *Auditorium Seating.* Auditoriums are often built with more seats per row as the rows move toward the back. Suppose that the first balcony of a theater has 28 seats in the first row, 32 in the second, 36 in the third, and so on, for 20 rows. How many seats are in the first balcony altogether?

42. *Investment Return.* Max, an investment counselor, sets up an investment situation for a client that will return \$5000 the first year, \$6125 the second year, \$7250 the third year, and so on, for 25 years. How much is received from the investment altogether?

43. *Parachutist Free Fall.* When a parachutist jumps from an airplane, the distances, in feet, that the parachutist falls in each successive second before pulling the ripcord to release the parachute are as follows:

$$16, 48, 80, 112, 144, \ldots.$$

Is this sequence arithmetic? What is the common difference? What is the total distance fallen in 10 sec?

44. *Lightning Distance.* The following table lists the distance in miles from lightning d_n when thunder is heard n seconds after lightning is seen. Is this sequence arithmetic? What is the common difference?

n (in seconds)	d_n (in miles)
5	1
6	1.2
7	1.4
8	1.6
9	1.8
10	2

45. *Garden Plantings.* A gardener is making a planting in the shape of a trapezoid. It will have 35 plants in the first row, 31 in the second row, 27 in the third row, and so on. If the pattern is consistent, how many plants will there be in the last row? How many plants are there altogether?

46. *Band Formation.* A formation of a marching band has 10 marchers in the first row, 12 in the second row, 14 in the third row, and so on, for 8 rows. How many marchers are in the last row? How many marchers are there altogether?

47. *Raw Material Production.* In a manufacturing process, it took 3 units of raw materials to produce 1 unit of a product. The raw material needs thus formed the sequence 3, 6, 9, . . . , $3n$, ... Is this sequence arithmetic? What is the common difference?

Skill Maintenance

Solve.

48. $7x - 2y = 4,$
$x + 3y = 17$
[9.1], [9.3], [9.5], [9.6]

49. $2x + y + 3z = 12,$
$x - 3y + 2z = 11,$
$5x + 2y - 4z = -4$
[9.2], [9.3], [9.5], [9.6]

50. Find the vertices and the foci of the ellipse with equation $9x^2 + 16y^2 = 144$. [10.2]

51. Find an equation of the ellipse with vertices $(0, -5)$ and $(0, 5)$ and minor axis of length 4. [10.2]

Synthesis

52. *Straight-Line Depreciation.* An architectural firm buys a large-format copier for \$10,300 on January 1 of a given year. The machine is expected to last for 8 years, at the end of which time its **trade-in value**, or **salvage value**, will be \$2100. If the company's accountant figures the decline in value to be the same each year, then its **book values**, or **salvage values**, after t years, $0 \leq t \leq 8$, form an arithmetic sequence given by

$$a_t = C - t\left(\frac{C - S}{N}\right),$$

where C is the original cost of the item (\$10,300), N is the number of years of expected life (8), and S is the salvage value (\$2100).

a) Find the formula for a_t for the straight-line depreciation of the copier.

b) Find the salvage value after 0 year, 1 year, 2 years, 3 years, 4 years, 7 years, and 8 years.

53. Find a formula for the sum of the first n odd natural numbers:

$$1 + 3 + 5 + \cdots + (2n - 1).$$

54. Find three numbers in an arithmetic sequence such that the sum of the first and third is 10 and the product of the first and second is 15.

55. Find the first term and the common difference for the arithmetic sequence for which

$$a_2 = 40 - 3q \quad \text{and} \quad a_4 = 10p + q.$$

*If p, m, and q form an arithmetic sequence, it can be shown that $m = (p + q)/2$. The number m is the **arithmetic mean, or average**, of p and q. Given two numbers p and q, if we find k other numbers m_1, $m_2, \ldots, m_k$ such that*

$$p, m_1, m_2, \ldots, m_k, q$$

forms an arithmetic sequence, we say that we have "inserted k arithmetic means between p and q."

56. Insert three arithmetic means between −3 and 5.

57. Insert four arithmetic means between 4 and 13.

11.3 Geometric Sequences and Series

- Identify the common ratio of a geometric sequence, and find a given term and the sum of the first n terms.
- Find the sum of an infinite geometric series, if it exists.

A sequence in which each term after the first term is found by multiplying the preceding term by the same number is a **geometric sequence**.

Geometric Sequences

Consider the sequence:

$$2, 6, 18, 54, 162, \ldots.$$

Note that multiplying each term by 3 produces the next term. We call the number 3 the **common ratio** because it can be found by dividing any term by the preceding term. A geometric sequence is also called a *geometric progression.*

> **GEOMETRIC SEQUENCE**
>
> A sequence is **geometric** if there is a number r, called the **common ratio**, such that
>
> $$\frac{a_{n+1}}{a_n} = r, \quad \text{or} \quad a_{n+1} = a_n r, \quad \text{for any integer } n \geq 1.$$

EXAMPLE 1 For each of the following geometric sequences, identify the common ratio.

a) $3, 6, 12, 24, 48, \ldots$

b) $1, -\frac{1}{2}, \frac{1}{4}, -\frac{1}{8}, \ldots$

c) \$5200, \$3900, \$2925, \$2193.75, ...

d) \$1000, \$1060, \$1123.60, ...

Solution

	SEQUENCE	COMMON RATIO	
a)	$3, 6, 12, 24, 48, \ldots$	2	$\left(\frac{6}{3} = 2, \frac{12}{6} = 2, \text{and so on}\right)$
b)	$1, -\frac{1}{2}, \frac{1}{4}, -\frac{1}{8}, \ldots$	$-\frac{1}{2}$	$\left(\frac{-\frac{1}{2}}{1} = -\frac{1}{2}, \frac{\frac{1}{4}}{-\frac{1}{2}} = -\frac{1}{2}, \text{and so on}\right)$
c)	\$5200, \$3900, \$2925, \$2193.75, ...	0.75	$\left(\frac{\$3900}{\$5200} = 0.75, \frac{\$2925}{\$3900} = 0.75, \text{and so on}\right)$
d)	\$1000, \$1060, \$1123.60, ...	1.06	$\left(\frac{\$1060}{\$1000} = 1.06, \frac{\$1123.60}{\$1060} = 1.06, \text{and so on}\right)$

Now Try Exercise 1.

We now find a formula for the general, or nth, term of a geometric sequence. If we let a_1 be the first term and r the common ratio, then the first few terms are as follows:

$$a_1,$$
$$a_2 = a_1 r,$$
$$a_3 = a_2 r = (a_1 r)r = a_1 r^2, \quad \text{Substituting } a_1 r \text{ for } a_2$$
$$a_4 = a_3 r = (a_1 r^2)r = a_1 r^3. \quad \text{Substituting } a_1 r^2 \text{ for } a_3$$

Note that the exponent is 1 less than the subscript.

Generalizing, we obtain the following.

nTH TERM OF A GEOMETRIC SEQUENCE

The ***n*th term** of a geometric sequence is given by

$$a_n = a_1 r^{n-1}, \quad \text{for any integer } n \geq 1.$$

EXAMPLE 2 Find the 7th term of the geometric sequence 4, 20, 100, ...

Solution We first note that

$$a_1 = 4 \quad \text{and} \quad n = 7.$$

To find the common ratio, we can divide any term (other than the first) by the preceding term. Since the second term is 20 and the first is 4, we get

$$r = \frac{20}{4}, \quad \text{or} \quad 5.$$

Then using the formula $a_n = a_1 r^{n-1}$, we have

$$a_7 = 4 \cdot 5^{7-1} = 4 \cdot 5^6 = 4 \cdot 15{,}625 = 62{,}500.$$

Thus the 7th term is 62,500.

Now Try Exercise 11.

EXAMPLE 3 Find the 10th term of the geometric sequence 64, −32, 16, −8, ...

Solution We first note that

$$a_1 = 64, \qquad n = 10, \quad \text{and} \quad r = \frac{-32}{64}, \text{ or } -\frac{1}{2}.$$

Then using the formula $a_n = a_1 r^{n-1}$, we have

$$a_{10} = 64 \cdot \left(-\frac{1}{2}\right)^{10-1} = 64 \cdot \left(-\frac{1}{2}\right)^9 = 2^6 \cdot \left(-\frac{1}{2^9}\right) = -\frac{1}{2^3} = -\frac{1}{8}.$$

Thus the 10th term is $-\frac{1}{8}$.

Now Try Exercise 15.

Sum of the First *n* Terms of a Geometric Sequence

Next, we develop a formula for the sum S_n of the first n terms of a geometric sequence:

$$a_1, a_1 r, a_1 r^2, a_1 r^3, \ldots, a_1 r^{n-1}, \ldots .$$

The associated **geometric series** is given by

$$S_n = a_1 + a_1 r + a_1 r^2 + a_1 r^3 + \cdots + a_1 r^{n-1}. \qquad \textbf{(1)}$$

We want to find a formula for this sum. If we multiply by r on both sides of equation (1), we have

$$rS_n = a_1 r + a_1 r^2 + a_1 r^3 + a_1 r^4 + \cdots + a_1 r^n. \qquad \textbf{(2)}$$

Subtracting equation (2) from equation (1), we see that the differences of the terms shown in red are 0, leaving

$$S_n - rS_n = a_1 - a_1r^n,$$

or

$$S_n(1 - r) = a_1(1 - r^n). \quad \text{Factoring}$$

Dividing by $1 - r$ on both sides gives us the following formula.

> **SUM OF THE FIRST n TERMS**
>
> The sum of the first n terms of a geometric sequence is given by
>
> $$S_n = \frac{a_1(1 - r^n)}{1 - r}, \quad \text{for any } r \neq 1.$$

EXAMPLE 4 Find the sum of the first 7 terms of the geometric sequence 3, 15, 75, 375, ...

Solution We first note that

$$a_1 = 3, \quad n = 7, \quad \text{and} \quad r = \tfrac{15}{3}, \text{ or } 5.$$

Then using the formula

$$S_n = \frac{a_1(1 - r^n)}{1 - r},$$

we have

$$S_7 = \frac{3(1 - 5^7)}{1 - 5} = \frac{3(1 - 78{,}125)}{-4} = 58{,}593.$$

Thus the sum of the first 7 terms is 58,593.

> Now Try Exercise 23.

EXAMPLE 5 Find the sum: $\sum_{k=1}^{11} (0.3)^k$.

Solution This is a geometric series with $a_1 = 0.3$, $r = 0.3$, and $n = 11$. Thus,

$$S_{11} = \frac{0.3(1 - 0.3^{11})}{1 - 0.3} \approx 0.42857.$$

```
sum(seq(.3^X,X,1,11))
                .4285706694
```

We can also find this sum using a graphing calculator set in either FUNCTION mode or SEQUENCE mode.

> Now Try Exercise 41.

Infinite Geometric Series

The sum of the terms of an infinite geometric sequence is an **infinite geometric series.** For some geometric sequences, S_n gets close to a specific number as n gets large. For example, consider the infinite series

$$\frac{1}{2} + \frac{1}{4} + \frac{1}{8} + \frac{1}{16} + \cdots + \frac{1}{2^n} + \cdots.$$

We can visualize S_n by considering the area of a square. For S_1, we shade half the square. For S_2, we shade half the square plus half the remaining half, or $\frac{1}{4}$. For S_3, we shade the parts shaded in S_2 plus half the remaining part. We see that the values of S_n will continue to get close to 1 (shading the complete square).

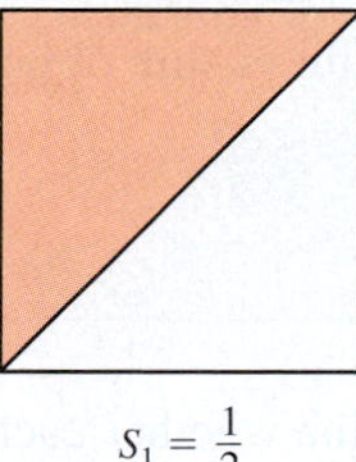
$S_1 = \frac{1}{2}$

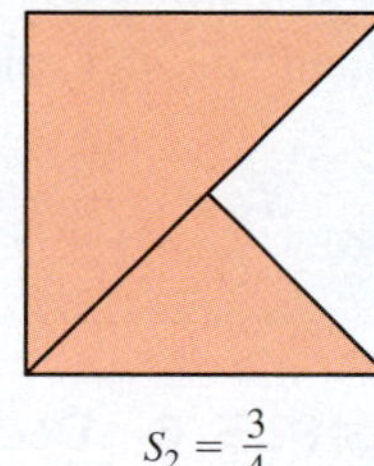
$S_2 = \frac{3}{4}$

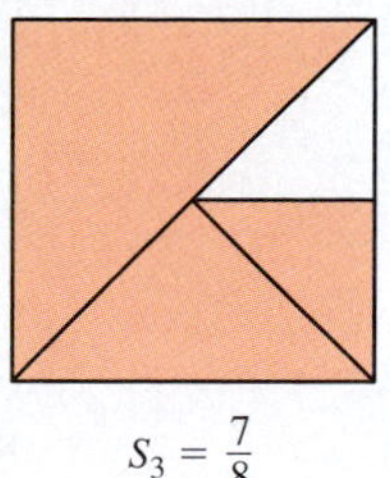
$S_3 = \frac{7}{8}$

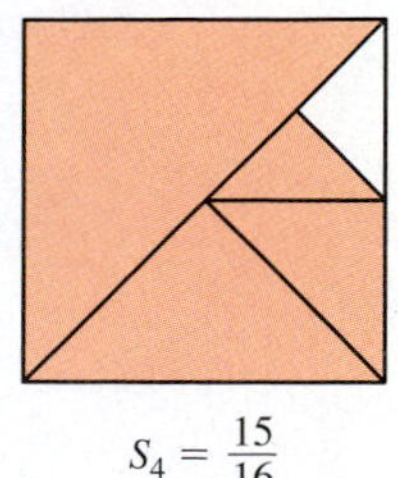
$S_4 = \frac{15}{16}$

We examine some partial sums. Note that each of the partial sums is less than 1, but S_n gets very close to 1 as n gets large.

n	S_n
1	0.5
5	0.96875
10	0.9990234375
20	0.9999990463
30	0.9999999991

```
sum(seq(1/2^X,X,1,20))
                .9999990463
sum(seq(1/2^X,X,1,30))
                .9999999991
```

We say that 1 is the **limit** of S_n and also that 1 is the **sum of the infinite geometric sequence**. The sum of an infinite geometric sequence is denoted S_∞. In this case, $S_\infty = 1$.

Some infinite sequences do not have sums. Consider the infinite geometric series

$$2 + 4 + 8 + 16 + \cdots + 2^n + \cdots.$$

We again examine some partial sums. Note that as n gets large, S_n gets large without bound. This sequence does not have a sum.

n	S_n
1	2
5	62
10	2,046
20	2,097,150
30	2,147,483,646

```
sum(seq(2^X,X,1,20))
                2097150
sum(seq(2^X,X,1,30))
                2147483646
```

It can be shown (but we will not do so here) that the sum of an infinite geometric series exists if and only if $|r| < 1$ (that is, the absolute value of the common ratio is less than 1).

To find a formula for the sum of an infinite geometric series, we first consider the sum of the first n terms:

$$S_n = \frac{a_1(1 - r^n)}{1 - r} = \frac{a_1 - a_1r^n}{1 - r}.$$ Using the distributive law

For $|r| < 1$, values of r^n get close to 0 as n gets large. As r^n gets close to 0, so does $a_1 r^n$. Thus, S_n gets close to $a_1/(1 - r)$.

LIMIT OR SUM OF AN INFINITE GEOMETRIC SERIES

When $|r| < 1$, the limit or sum of an infinite geometric series is given by

$$S_\infty = \frac{a_1}{1 - r}.$$

EXAMPLE 6 Determine whether each of the following infinite geometric series has a limit. If a limit exists, find it.

a) $1 + 3 + 9 + 27 + \cdots$

b) $-2 + 1 - \frac{1}{2} + \frac{1}{4} - \frac{1}{8} + \cdots$

Solution

a) Here $r = 3$, so $|r| = |3| = 3$. Since $|r| > 1$, the series *does not* have a limit.

b) Here $r = -\frac{1}{2}$, so $|r| = |-\frac{1}{2}| = \frac{1}{2}$. Since $|r| < 1$, the series *does* have a limit. We find the limit:

$$S_\infty = \frac{a_1}{1 - r} = \frac{-2}{1 - \left(-\frac{1}{2}\right)} = \frac{-2}{\frac{3}{2}} = -\frac{4}{3}.$$

Now Try Exercises 33 and 37.

EXAMPLE 7 Find fraction notation for 0.78787878. . . , or $0.\overline{78}$.

Solution We can express this as

$$0.78 + 0.0078 + 0.000078 + \cdots.$$

Then we see that this is an infinite geometric series, where $a_1 = 0.78$ and $r = 0.01$. Since $|r| < 1$, this series has a limit:

$$S_\infty = \frac{a_1}{1 - r} = \frac{0.78}{1 - 0.01} = \frac{0.78}{0.99} = \frac{78}{99}, \quad \text{or} \quad \frac{26}{33}.$$

Thus fraction notation for 0.78787878 . . . is $\frac{26}{33}$. You can check this on your calculator.

Now Try Exercise 51.

Applications

The translation of some applications and problem-solving situations may involve geometric sequences or series. Examples 9 and 10, in particular, show applications in business and economics.

EXAMPLE 8 ***A Daily Doubling Salary.*** Suppose that someone offers you a job for the month of September (30 days) under the following conditions. You will be paid \$0.01 for the first day, \$0.02 for the second, \$0.04 for the third, and so on, doubling your previous day's salary each day. How much would you earn? (Would you take the job? Make a conjecture before reading further.)

Solution You earn \$0.01 the first day, \$0.01(2) the second day, \$0.01(2)(2) the third day, and so on. The amount earned is the geometric series

$$\$0.01 + \$0.01(2) + \$0.01(2^2) + \$0.01(2^3) + \cdots + \$0.01(2^{29}),$$

where $a_1 = \$0.01$, $r = 2$, and $n = 30$. Using the formula

$$S_n = \frac{a_1(1 - r^n)}{1 - r},$$

we have

$$S_{30} = \frac{\$0.01(1 - 2^{30})}{1 - 2} = \$10{,}737{,}418.23.$$

The pay exceeds $10.7 million for the month.

Now Try Exercise 57.

EXAMPLE 9 ***The Amount of an Annuity.*** An **annuity** is a sequence of equal payments, made at equal time intervals, that earn interest. Fixed deposits in a savings account are an example of an annuity. Suppose that to save money to buy a car, Janelle deposits $2000 at the *end* of each of 5 years in an account that pays 3% interest, compounded annually. The total amount in the account at the end of 5 years is called the **amount of the annuity**. Find that amount.

Solution The following time diagram can help visualize the problem. Note that no deposit is made until the end of the first year.

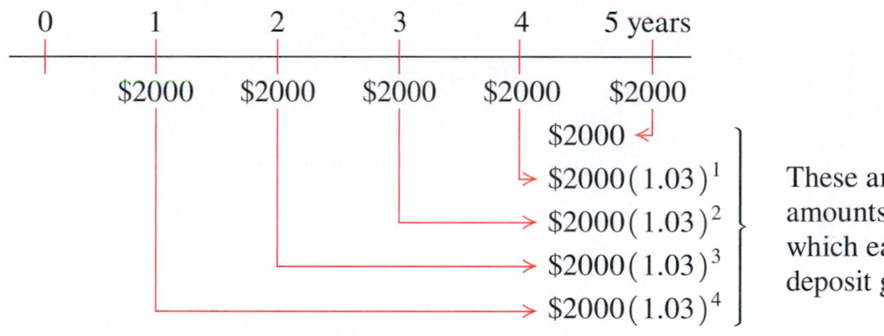

The amount of the annuity is the geometric series

$$\$2000 + \$2000(1.03)^1 + \$2000(1.03)^2 + \$2000(1.03)^3 + \$2000(1.03)^4,$$

where $a_1 = \$2000$, $n = 5$, and $r = 1.03$. Using the formula

$$S_n = \frac{a_1(1 - r^n)}{1 - r},$$

we have

$$S_5 = \frac{\$2000(1 - 1.03^5)}{1 - 1.03} \approx \$10{,}618.27.$$

The amount of the annuity is $10,618.27.

Now Try Exercise 61.

EXAMPLE 10 ***The Economic Multiplier.*** Large sporting events have a significant impact on the economy of the host city. Super Bowl XLVII, hosted by New Orleans, generated a $480-million net impact for the region (*Source*: NewOrleansSaints.com, posted April 18, 2013, Marius M. Mihai, Research Analyst of the Division of Business and Economic Research at the University of New Orleans (DBER)). Assume that 60% of that amount is spent again in the area, and then 60% of that amount is spent again, and so on. This is known as the *economic multiplier effect*. Find the total effect on the economy.

Solution The total economic effect is given by the infinite series

$$\$480{,}000{,}000 + \$480{,}000{,}000(0.6) + \$480{,}000{,}000(0.6)^2 + \cdots.$$

Since $|r| = |0.6| = 0.6 < 1$, the series has a sum. Using the formula for the sum of an infinite geometric series, we have

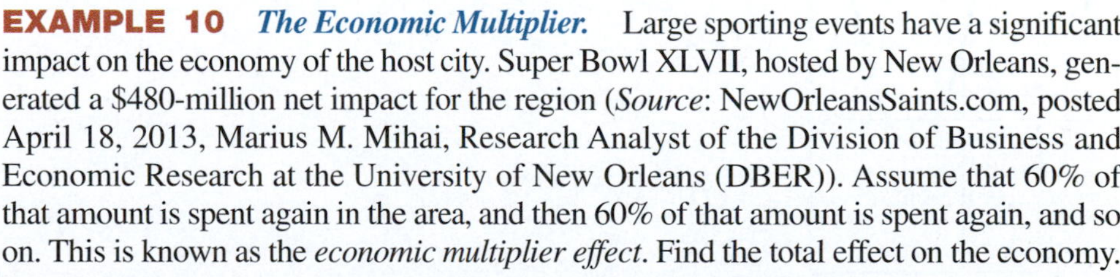

$$S_\infty = \frac{a_1}{1 - r} = \frac{\$480{,}000{,}000}{1 - 0.6} = \$1{,}200{,}000{,}000.$$

The total effect of the spending on the economy is $1,200,000,000.

Now Try Exercise 67.

Visualizing the Graph

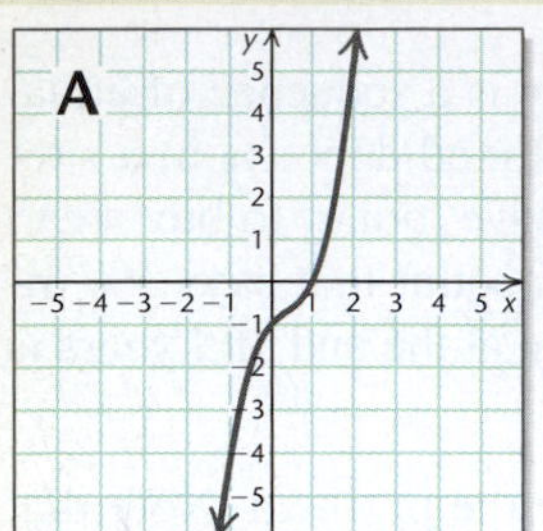

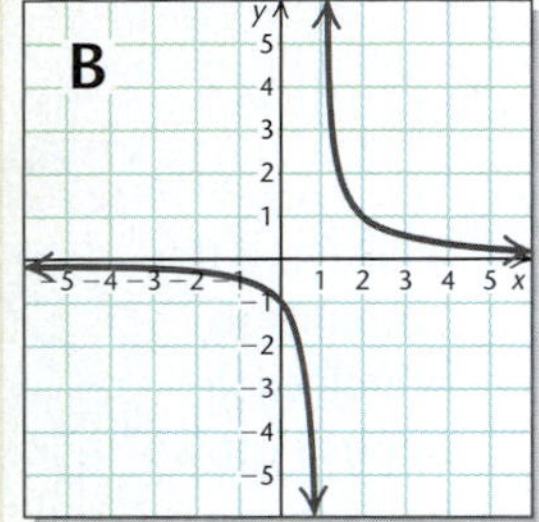

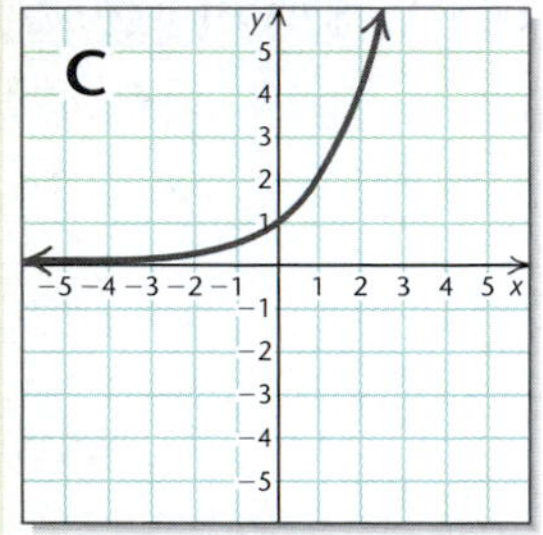

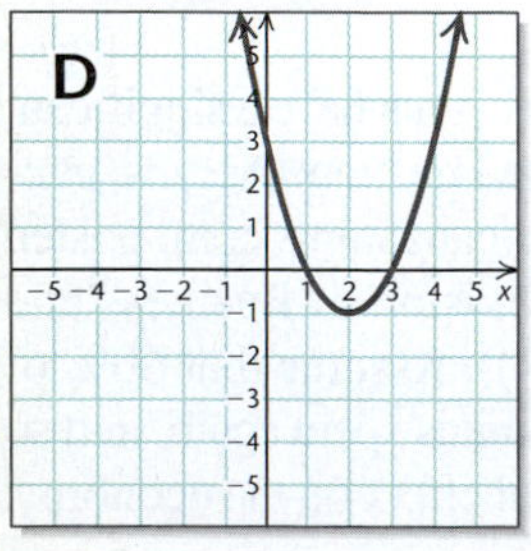

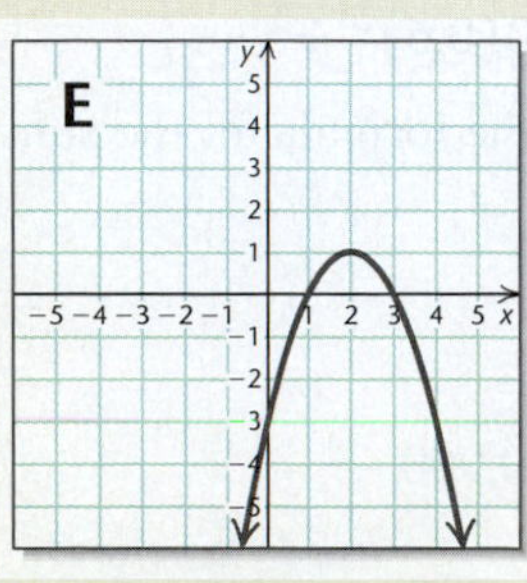

Match the equation with its graph.

1. $(x - 1)^2 + (y + 2)^2 = 9$
2. $y = x^3 - x^2 + x - 1$
3. $f(x) = 2^x$
4. $f(x) = x$
5. $a_n = n$
6. $y = \log(x + 3)$
7. $f(x) = -(x - 2)^2 + 1$
8. $f(x) = (x - 2)^2 - 1$
9. $y = \dfrac{1}{x - 1}$
10. $y = -3x + 4$

Answers on page A-69

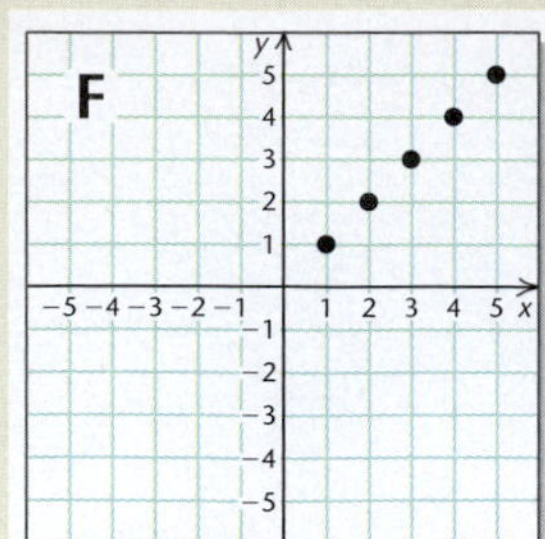

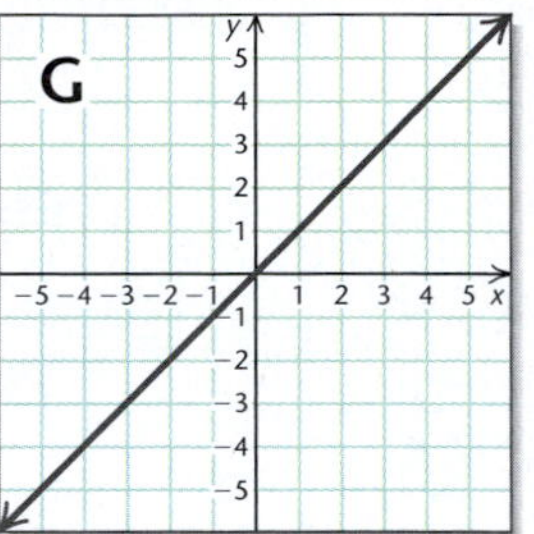

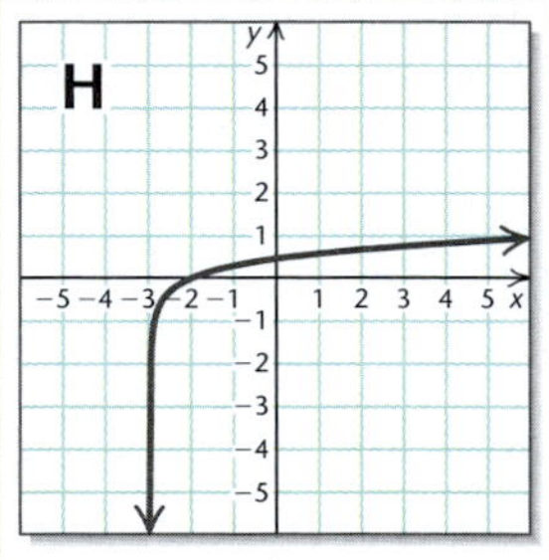

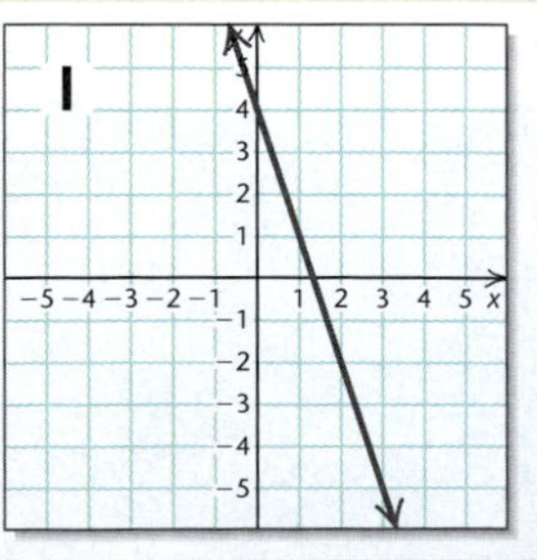

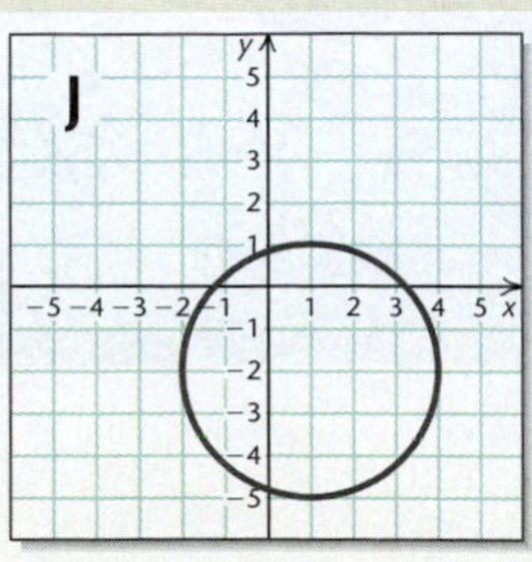

11.3 Exercise Set

Find the common ratio.

1. $2, 4, 8, 16, \ldots$
2. $18, -6, 2, -\frac{2}{3}, \ldots$
3. $-1, 1, -1, 1, \ldots$
4. $-8, -0.8, -0.08, -0.008, \ldots$
5. $\frac{2}{3}, -\frac{4}{3}, \frac{8}{3}, -\frac{16}{3}, \ldots$
6. $75, 15, 3, \frac{3}{5}, \ldots$
7. $6.275, 0.6275, 0.06275, \ldots$
8. $\frac{1}{x}, \frac{1}{x^2}, \frac{1}{x^3}, \ldots$
9. $5, \frac{5a}{2}, \frac{5a^2}{4}, \frac{5a^3}{8}, \ldots$
10. \$780, \$858, \$943.80, \$1038.18, ...

Find the indicated term of the geometric sequence.

11. $2, 4, 8, 16, \ldots$; the 7th term
12. $2, -10, 50, -250, \ldots$; the 9th term
13. $2, 2\sqrt{3}, 6, \ldots$; the 9th term
14. $1, -1, 1, -1, \ldots$; the 57th term
15. $\frac{7}{625}, -\frac{7}{25}, \ldots$; the 23rd term
16. \$1000, \$1060, \$1123.60, ...; the 5th term

Find the nth, or general, term.

17. $1, 3, 9, \ldots$
18. $25, 5, 1, \ldots$
19. $1, -1, 1, -1, \ldots$
20. $-2, 4, -8, \ldots$
21. $\frac{1}{x}, \frac{1}{x^2}, \frac{1}{x^3}, \ldots$
22. $5, \frac{5a}{2}, \frac{5a^2}{4}, \frac{5a^3}{8}, \ldots$

23. Find the sum of the first 7 terms of the geometric series
$$6 + 12 + 24 + \cdots.$$

24. Find the sum of the first 10 terms of the geometric series
$$16 - 8 + 4 - \cdots.$$

25. Find the sum of the first 9 terms of the geometric series
$$\frac{1}{18} - \frac{1}{6} + \frac{1}{2} - \cdots.$$

26. Find the sum of the geometric series
$$-8 + 4 + (-2) + \cdots + \left(-\frac{1}{32}\right).$$

Determine whether the statement is true or false.

27. The sequence $2, -2\sqrt{2}, 4, -4\sqrt{2}, 8, \ldots$ is geometric.
28. The sequence with general term $3n$ is geometric.
29. The sequence with general term 2^n is geometric.
30. Multiplying a term of a geometric sequence by the common ratio produces the next term of the sequence.
31. An infinite geometric series with common ratio -0.75 has a sum.
32. Every infinite geometric series has a limit.

Find the sum, if it exists.

33. $4 + 2 + 1 + \cdots$
34. $7 + 3 + \frac{9}{7} + \cdots$
35. $25 + 20 + 16 + \cdots$
36. $100 - 10 + 1 - \frac{1}{10} + \cdots$
37. $8 + 40 + 200 + \cdots$
38. $-6 + 3 - \frac{3}{2} + \frac{3}{4} - \cdots$
39. $0.6 + 0.06 + 0.006 + \cdots$
40. $\sum_{k=0}^{10} 3^k$
41. $\sum_{k=1}^{11} 15\left(\frac{2}{3}\right)^k$
42. $\sum_{k=0}^{50} 200(1.08)^k$
43. $\sum_{k=1}^{\infty} \left(\frac{1}{2}\right)^{k-1}$
44. $\sum_{k=1}^{\infty} 2^k$
45. $\sum_{k=1}^{\infty} 12.5^k$
46. $\sum_{k=1}^{\infty} 400(1.0625)^k$
47. $\sum_{k=1}^{\infty} \$500(1.11)^{-k}$
48. $\sum_{k=1}^{\infty} \$1000(1.06)^{-k}$
49. $\sum_{k=1}^{\infty} 16(0.1)^{k-1}$
50. $\sum_{k=1}^{\infty} \frac{8}{3}\left(\frac{1}{2}\right)^{k-1}$

Find fraction notation.

51. $0.131313\ldots$, or $0.\overline{13}$
52. $0.2222\ldots$, or $0.\overline{2}$
53. $8.999\overline{9}$
54. $6.1616\overline{16}$
55. $3.4125\overline{125}$
56. $12.7809\overline{809}$

57. *Daily Doubling Salary.* Suppose that someone offers you a job for the month of February (28 days) under the following conditions. You will be paid \$0.01 the 1st day, \$0.02 the 2nd, \$0.04 the 3rd, and so on, doubling your previous day's salary each day. How much would you earn altogether?

58. *Bouncing Ping-Pong Ball.* A ping-pong ball is dropped from a height of 16 ft and always rebounds $\frac{1}{4}$ of the distance fallen.

a) How high does it rebound the 6th time?

b) Find the total sum of the rebound heights of the ball.

59. *Bungee Jumping.* A bungee jumper always rebounds 60% of the distance fallen. A bungee jump is made using a cord that stretches to 200 ft.

a) After jumping and then rebounding 9 times, how far has a bungee jumper traveled upward (the total rebound distance)?

b) About how far will a jumper have traveled upward (bounced) before coming to rest?

60. *Population Growth.* A coastal town has a present population of 32,100, and the population is increasing by 3% each year.

a) What will the population be in 15 years?

b) How long will it take for the population to double?

61. *Amount of an Annuity.* To save for the down payment on a house, the Ramirez family makes a sequence of 10 yearly deposits of \$3200 each in a savings account on which interest is compounded annually at 4.6%. Find the amount of the annuity.

62. *Amount of an Annuity.* To create a college fund, a parent makes a sequence of 18 yearly deposits of \$1000 each in a savings account on which interest is compounded annually at 3.2%. Find the amount of the annuity.

63. *Doubling the Thickness of Paper.* A piece of paper is 0.01 in. thick. It is cut and stacked repeatedly in such a way that its thickness is doubled each time for 20 times. How thick is the result?

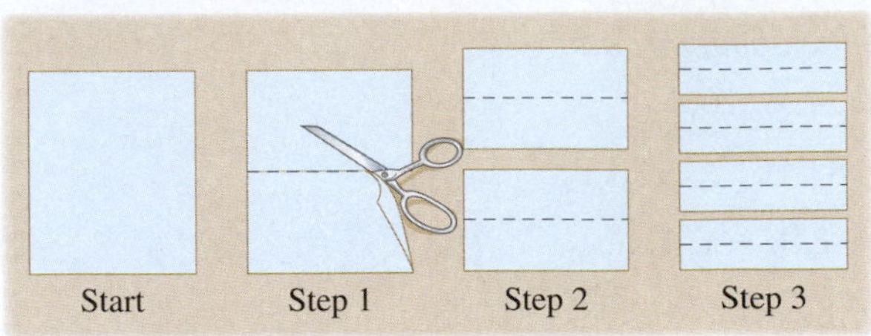

64. *Amount of an Annuity.* A sequence of yearly payments of P dollars is invested at the end of each of N years at interest rate i, compounded annually. The total amount in the account, or the amount of the annuity, is V.

a) Show that

$$V = \frac{P[(1+i)^N - 1]}{i}.$$

b) Suppose that interest is compounded n times per year and deposits are made every compounding period. Show that the formula for V is then given by

$$V = \frac{P\left[\left(1 + \frac{i}{n}\right)^{nN} - 1\right]}{i/n}.$$

65. *Amount of an Annuity.* A sequence of payments of \$300 is invested over 12 years at the end of each quarter at 5.1%, compounded quarterly. Find the amount of the annuity. Use the formula in Exercise 64(b).

66. *Amount of an Annuity.* A sequence of yearly payments of \$750 is invested at the end of each of 10 years at 4.75%, compounded annually. Find the amount of the annuity. Use the formula in Exercise 64(a).

67. *The Economic Multiplier.* Suppose that the government is making a \$13,000,000,000 expenditure for educational improvement. If 85% of this is spent again, and so on, what is the total effect on the economy?

68. *Advertising Effect.* Gigi's Cupcake Truck is about to open for business in a city of 3,000,000 people, traveling to several curbside locations in the city each day to sell cupcakes. The owners plan an advertising campaign that they think will induce 30% of the people to buy their cupcakes. They estimate that if those people like the product, they will induce

30% · 30% · 3,000,000 more to buy the product, and those will induce 30% · 30% · 30% · 3,000,000 and so on. In all, how many people will buy Gigi's cupcakes as a result of the advertising campaign? What percentage of the population is this?

Skill Maintenance

For each pair of functions, find $(f \circ g)(x)$ *and* $(g \circ f)(x)$. [2.3]

69. $f(x) = x^2, g(x) = 4x + 5$

70. $f(x) = x - 1, g(x) = x^2 + x + 3$

Solve. [5.5]

71. $5^x = 35$

72. $\log_2 x = -4$

Synthesis

73. Prove that

$$\sqrt{3} - \sqrt{2}, \quad 4 - \sqrt{6}, \quad \text{and} \quad 6\sqrt{3} - 2\sqrt{2}$$

form a geometric sequence.

74. Assume that $a_1, a_2, a_3, \ldots$ is a geometric sequence. Prove that $\ln a_1, \ln a_2, \ln a_3, \ldots$ is an arithmetic sequence.

75. Consider the sequence

$$x + 3, \quad x + 7, \quad 4x - 2, \ldots .$$

a) If the sequence is arithmetic, find x and then determine each of the 3 terms and the 4th term.

b) If the sequence is geometric, find x and then determine each of the 3 terms and the 4th term.

76. Find the sum of the first n terms of

$$1 + x + x^2 + \cdots .$$

77. Find the sum of the first n terms of

$$x^2 - x^3 + x^4 - x^5 + \cdots .$$

78. The sides of a square are 16 cm long. A second square is inscribed by joining the midpoints of the sides, successively. In the second square, we repeat the process, inscribing a third square. If this process is continued indefinitely, what is the sum of all the areas of all the squares? (*Hint*: Use an infinite geometric series.)

11.4 Mathematical Induction

- Prove infinite sequences of statements using mathematical induction.

In this section, we learn to prove a sequence of mathematical statements using a procedure called *mathematical induction.*

Proving Infinite Sequences of Statements

Infinite sequences of statements occur often in mathematics. In an infinite sequence of statements, there is a statement for each natural number. For example, consider the sequence of statements represented by the following:

"The sum of the first n positive odd integers is n^2," or
$1 + 3 + 5 + \cdots + (2n - 1) = n^2$.

Let's think of this as $S(n)$, or S_n. Substituting natural numbers for n gives a sequence of statements. We list the first four:

S_1: $1 = 1^2$;

S_2: $1 + 3 = 4 = 2^2$;

S_3: $1 + 3 + 5 = 9 = 3^2$;

S_4: $1 + 3 + 5 + 7 = 16 = 4^2$.

The fact that the statement is true for $n = 1, 2, 3$, and 4 might tempt us to conclude that the statement is true for any natural number n, but we cannot be sure that this is the case. We can, however, use the principle of mathematical induction to prove that the statement is true for all natural numbers.

THE PRINCIPLE OF MATHEMATICAL INDUCTION

We can prove an infinite sequence of statements S_n by showing the following.

1. *Basis step*: S_1 is true.
2. *Induction step*: For all natural numbers k, $S_k \rightarrow S_{k+1}$.

Mathematical induction is analogous to lining up a sequence of dominoes. The induction step tells us that if any one domino is knocked over, then the one next to it will be hit and knocked over. The basis step tells us that the first domino can indeed be knocked over. Note that in order for all dominoes to fall, *both* conditions must be satisfied.

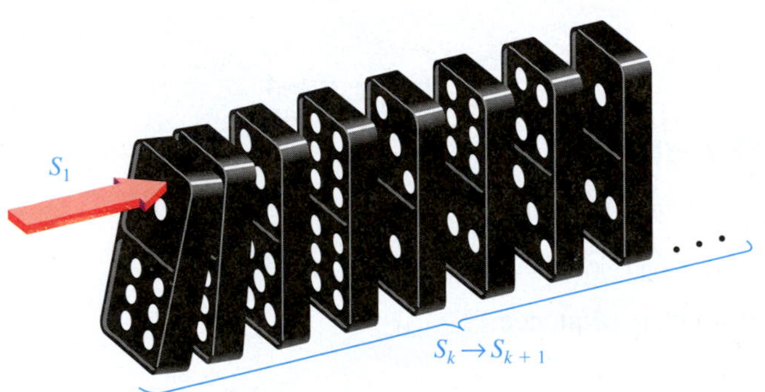

When you are learning to do proofs by mathematical induction, it is helpful to first write out S_n, S_1, S_k, and S_{k+1}. This helps to identify what is to be assumed and what is to be deduced.

EXAMPLE 1 Prove: For every natural number n,

$$1 + 3 + 5 + \cdots + (2n - 1) = n^2.$$

Proof. We first list S_n, S_1, S_k, and S_{k+1}.

S_n: $1 + 3 + 5 + \cdots + (2n - 1) = n^2$

S_1: $1 = 1^2$

S_k: $1 + 3 + 5 + \cdots + (2k - 1) = k^2$

S_{k+1}: $1 + 3 + 5 + \cdots + (2k - 1) + [2(k + 1) - 1] = (k + 1)^2$

1. *Basis step*: S_1, as listed, is true since $1 = 1^2$, or $1 = 1$.
2. *Induction step*: We let k be any natural number. We assume S_k to be true and try to show that it implies that S_{k+1} is true. Now S_k is

$$1 + 3 + 5 + \cdots + (2k - 1) = k^2.$$

Starting with the left side of S_{k+1} and substituting k^2 for $1 + 3 + 5 + \cdots + (2k - 1)$, we have

$$\begin{aligned} &\underbrace{1 + 3 + \cdots + (2k - 1)} + [2(k + 1) - 1] \\ &= k^2 + [2(k + 1) - 1] \quad \text{We assume } S_k \text{ is true.} \\ &= k^2 + 2k + 2 - 1 \\ &= k^2 + 2k + 1 \\ &= (k + 1)^2. \end{aligned}$$

We have shown that for all natural numbers k, $S_k \rightarrow S_{k+1}$. This completes the induction step. It and the basis step tell us that the proof is complete.

Now Try Exercise 5.

EXAMPLE 2 Prove: For every natural number n,

$$\frac{1}{2} + \frac{1}{4} + \frac{1}{8} + \cdots + \frac{1}{2^n} = \frac{2^n - 1}{2^n}.$$

Proof. We first list S_n, S_1, S_k, and S_{k+1}.

$$S_n: \quad \frac{1}{2} + \frac{1}{4} + \frac{1}{8} + \cdots + \frac{1}{2^n} = \frac{2^n - 1}{2^n}$$

$$S_1: \quad \frac{1}{2^1} = \frac{2^1 - 1}{2^1}$$

$$S_k: \quad \frac{1}{2} + \frac{1}{4} + \frac{1}{8} + \cdots + \frac{1}{2^k} = \frac{2^k - 1}{2^k}$$

$$S_{k+1}: \quad \frac{1}{2} + \frac{1}{4} + \frac{1}{8} + \cdots + \frac{1}{2^k} + \frac{1}{2^{k+1}} = \frac{2^{k+1} - 1}{2^{k+1}}$$

1. *Basis step*: We show S_1 to be true as follows:

$$\frac{2^1 - 1}{2^1} = \frac{2 - 1}{2} = \frac{1}{2}.$$

2. *Induction step*: We let k be any natural number. We assume S_k to be true and try to show that it implies that S_{k+1} is true. Now S_k is

$$\frac{1}{2} + \frac{1}{4} + \frac{1}{8} + \cdots + \frac{1}{2^k} = \frac{2^k - 1}{2^k}.$$

We start with the left side of S_{k+1}. Since we assume S_k is true, we can substitute

$$\frac{2^k - 1}{2^k} \quad \text{for} \quad \frac{1}{2} + \frac{1}{4} + \cdots + \frac{1}{2^k}.$$

We have

$$\underbrace{\frac{1}{2} + \frac{1}{4} + \frac{1}{8} + \cdots + \frac{1}{2^k}} + \frac{1}{2^{k+1}}$$

$$= \frac{2^k - 1}{2^k} + \frac{1}{2^{k+1}} = \frac{2^k - 1}{2^k} \cdot \frac{2}{2} + \frac{1}{2^{k+1}}$$

$$= \frac{(2^k - 1) \cdot 2 + 1}{2^{k+1}}$$

$$= \frac{2^{k+1} - 2 + 1}{2^{k+1}}$$

$$= \frac{2^{k+1} - 1}{2^{k+1}}.$$

We have shown that for all natural numbers k, $S_k \rightarrow S_{k+1}$. This completes the induction step. It and the basis step tell us that the proof is complete.

Now Try Exercise 15.

EXAMPLE 3 Prove: For every natural number n, $n < 2^n$.

Proof. We first list S_n, S_1, S_k, and S_{k+1}.

S_n: $n < 2^n$

S_1: $1 < 2^1$

S_k: $k < 2^k$

S_{k+1}: $k + 1 < 2^{k+1}$

1. *Basis step*: S_1, as listed, is true since $2^1 = 2$ and $1 < 2$.
2. *Induction step*: We let k be any natural number. We assume S_k to be true and try to show that it implies that S_{k+1} is true. Now

$k < 2^k$ **This is S_k.**

$2k < 2 \cdot 2^k$ **Multiplying by 2 on both sides**

$2k < 2^{k+1}$ **Adding exponents on the right**

$k + k < 2^{k+1}$. **Rewriting $2k$ as $k + k$**

Since k is any natural number, we know that $1 \leq k$. Thus,

$k + 1 \leq k + k$. **Adding k on both sides of $1 \leq k$**

Putting the results $k + 1 \leq k + k$ and $k + k < 2^{k+1}$ together gives us

$k + 1 < 2^{k+1}$. **This is S_{k+1}.**

We have shown that for all natural numbers k, $S_k \rightarrow S_{k+1}$. This completes the induction step. It and the basis step tell us that the proof is complete.

Now Try Exercise 11.

11.4 Exercise Set

List the first five statements in the sequence that can be obtained from each of the following. Determine whether each of the statements is true or false.

1. $n^2 < n^3$
2. $n^2 - n + 41$ is prime. Find a value for n for which the statement is false.
3. A polygon of n sides has $[n(n - 3)]/2$ diagonals.
4. The sum of the angles of a polygon of n sides is $(n - 2) \cdot 180°$.

Use mathematical induction to prove each of the following.

5. $2 + 4 + 6 + \cdots + 2n = n(n + 1)$
6. $4 + 8 + 12 + \cdots + 4n = 2n(n + 1)$
7. $1 + 5 + 9 + \cdots + (4n - 3) = n(2n - 1)$
8. $3 + 6 + 9 + \cdots + 3n = \dfrac{3n(n + 1)}{2}$
9. $2 + 4 + 8 + \cdots + 2^n = 2(2^n - 1)$
10. $2 \leq 2^n$
11. $n < n + 1$
12. $3^n < 3^{n+1}$
13. $2n \leq 2^n$
14. $\dfrac{1}{1 \cdot 2} + \dfrac{1}{2 \cdot 3} + \cdots + \dfrac{1}{n(n + 1)} = \dfrac{n}{n + 1}$
15. $\dfrac{1}{1 \cdot 2 \cdot 3} + \dfrac{1}{2 \cdot 3 \cdot 4} + \dfrac{1}{3 \cdot 4 \cdot 5} + \cdots + \dfrac{1}{n(n + 1)(n + 2)} = \dfrac{n(n + 3)}{4(n + 1)(n + 2)}$
16. If x is any real number greater than 1, then for any natural number n, $x \leq x^n$.

The following formulas can be used to find sums of powers of natural numbers. Use mathematical induction to prove each formula.

17. $1 + 2 + 3 + \cdots + n = \dfrac{n(n + 1)}{2}$
18. $1^2 + 2^2 + 3^2 + \cdots + n^2 = \dfrac{n(n + 1)(2n + 1)}{6}$
19. $1^3 + 2^3 + 3^3 + \cdots + n^3 = \dfrac{n^2(n + 1)^2}{4}$
20. $1^4 + 2^4 + 3^4 + \cdots + n^4 = \dfrac{n(n + 1)(2n + 1)(3n^2 + 3n - 1)}{30}$

Use mathematical induction to prove each of the following.

21. $\displaystyle\sum_{i=1}^{n} i(i + 1) = \frac{n(n + 1)(n + 2)}{3}$
22. $\left(1 + \dfrac{1}{1}\right)\left(1 + \dfrac{1}{2}\right)\left(1 + \dfrac{1}{3}\right)\cdots\left(1 + \dfrac{1}{n}\right) = n + 1$
23. The sum of n terms of an arithmetic sequence: $a_1 + (a_1 + d) + (a_1 + 2d) + \cdots + [a_1 + (n - 1)d] = \dfrac{n}{2}[2a_1 + (n - 1)d]$

Skill Maintenance

Solve.

24. $2x - 3y = 1,$
$3x - 4y = 3$ **[9.1], [9.3], [9.5], [9.6]**
25. *Investment.* Huiliang received \$150 in simple interest one year from three investments. Part is invested at 1.5%, part at 2%, and part at 3%. The amount invested at 2% is twice the amount invested at 1.5%. There is \$400 more invested at 3% than at 2%. Find the amount invested at each rate. **[9.2], [9.3], [9.5], [9.6]**

Synthesis

Use mathematical induction to prove each of the following.

26. The sum of n terms of a geometric sequence:
$$a_1 + a_1r + a_1r^2 + \cdots + a_1r^{n-1} = \frac{a_1 - a_1r^n}{1 - r}$$
27. $x + y$ is a factor of $x^{2n} - y^{2n}$.

Prove each of the following using mathematical induction. Do the basis step for $n = 2$.

28. For every natural number $n \geq 2$,
$2n + 1 < 3^n$.
29. For every natural number $n \geq 2$,
$\log_a(b_1b_2 \cdots b_n) = \log_a b_1 + \log_a b_2 + \cdots + \log_a b_n$.

Prove each of the following for any complex numbers $z_1, z_2, \ldots, z_n$, where $i^2 = -1$ and $\bar{z}$ is the conjugate of z. (See Section 3.1.)

30. $\overline{z^n} = \bar{z}^n$
31. $\overline{z_1 + z_2 + \cdots + z_n} = \overline{z_1} + \overline{z_2} + \cdots + \overline{z_n}$

32. *The Tower of Hanoi Problem.* There are three pegs on a board. On one peg are n disks, each smaller than the one on which it rests. The problem is to move this pile of disks to another peg. The final order must be the same, but you can move only one disk at a time and can never place a larger disk on a smaller one.

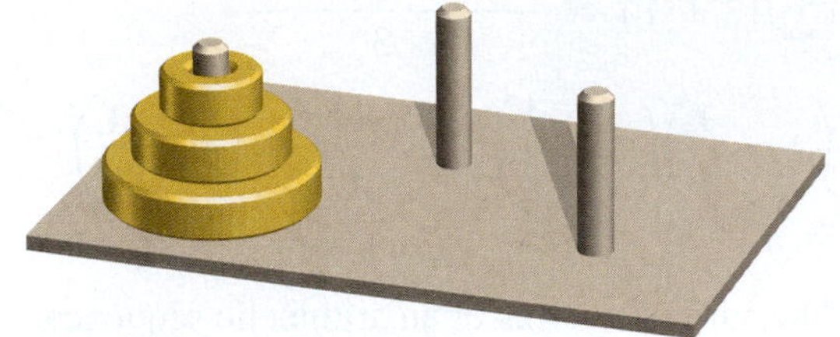

a) What is the *least* number of moves needed to move 3 disks? 4 disks? 2 disks? 1 disk?

b) Conjecture a formula for the *least* number of moves needed to move n disks. Prove it by mathematical induction.

Mid-Chapter Mixed Review

Determine whether the statement is true or false.

1. The general term of the sequence $1, -2, 3, -4, \ldots$ can be expressed as $a_n = n$. [11.1]

2. To find the common difference of an arithmetic sequence, choose any term except the first and then subtract the preceding term from it. [11.2]

3. The sequence $7, 3, -1, -5, \ldots$ is geometric. [11.2], [11.3]

4. If we can show that $S_k \rightarrow S_{k+1}$ for some natural number k, then we know that S_n is true for all natural numbers n. [11.4]

In each of the following, the nth term of a sequence is given. Find the first 4 terms, a_9, and a_{14}.

5. $a_n = 3n + 5$ [11.1]

6. $a_n = (-1)^{n+1}(n - 1)$ [11.1]

Predict the general term, or nth term, a_n, of the sequence. Answers may vary.

7. $3, 6, 9, 12, 15, \ldots$ [11.1]

8. $-1, 4, -9, 16, -25, \ldots$ [11.1]

9. Find the partial sum S_4 for the sequence $1, \frac{1}{2}, \frac{1}{4}, \frac{1}{8}, \frac{1}{16}, \ldots$. [11.1]

10. Find and evaluate the sum $\sum_{k=1}^{5} k(k + 1)$. [11.1]

11. Write sigma notation for the sum $-4 + 8 - 12 + 16 - 20 + \cdots$. [11.1]

12. Find the first 4 terms of the sequence defined by $a_1 = 2, a_{n+1} = 4a_n - 2$. [11.1]

13. Find the common difference of the arithmetic sequence $12, 7, 2, -3, \ldots$. [11.2]

14. Find the 10th term of the arithmetic sequence $4, 6, 8, 10, \ldots$. [11.2]

15. In the sequence in Exercise 14, what term is the number 44? [11.2]

16. Find the sum of the first 16 terms of the arithmetic series $6 + 11 + 16 + 21 + \cdots$. [11.2]

17. Find the common ratio of the geometric sequence $16, -8, 4, -2, 1, \ldots$. [11.3]

18. Find **(a)** the 8th term and **(b)** the sum of the first 10 terms of the geometric sequence $\frac{1}{16}, \frac{1}{8}, \frac{1}{4}, \frac{1}{2}, 1, \ldots$. [11.3]

Find the sum, if it exists.

19. $-8 + 4 - 2 + 1 - \cdots$ [11.3]

20. $\sum_{k=0}^{\infty} 5^k$ [11.3]

21. *Landscaping.* A landscaper is planting a triangular flower bed with 36 plants in the first row, 30 plants in the second row, 24 in the third row, and so on, for a total of 6 rows. How many plants will be planted in all? [11.2]

22. *Amount of an Annuity.* To save money for adding a bedroom to their home, at the end of each of 4 years the Jacobsons deposit \$3200 in an account that pays 4% interest, compounded annually. Find the total amount of the annuity. [11.3]

23. Prove: For every natural number n,

$$1 + 4 + 7 + \cdots + (3n - 2) = \tfrac{1}{2}n(3n - 1).$$

[11.4]

Collaborative Discussion and Writing

24. The sum of the first n terms of an arithmetic sequence can be given by

$$S_n = \frac{n}{2}[2a_1 + (n - 1)d].$$

Compare this formula to

$$S_n = \frac{n}{2}(a_1 + a_n).$$

Discuss the reasons for the use of one formula over the other. [11.2]

25. It is said that as a young child, the mathematician Karl F. Gauss (1777–1855) was able to compute the sum $1 + 2 + 3 + \cdots + 100$ very quickly in his head to the amazement of a teacher. Explain how Gauss might have done this had he possessed some knowledge of arithmetic sequences and series. Then give a formula for the sum of the first n natural numbers. [11.2]

26. Write a problem for a classmate to solve. Devise the problem so that a geometric series is involved and the solution is "The total amount in the bank is $\$900(1.08)^{40}$, or about \$19,552." [11.3]

27. Write an explanation of the idea behind mathematical induction for a fellow student. [11.4]

11.5 Combinatorics: Permutations

› Evaluate factorial notation and permutation notation and solve related applied problems.

In order to study probability, it is first necessary to learn about **combinatorics**, the theory of counting.

› Permutations

In this section, we will consider the part of combinatorics called *permutations.*

> The study of permutations involves *order* and *arrangements*.

EXAMPLE 1 How many 3-letter code symbols can be formed with the letters A, B, C *without* repetition (that is, using each letter only once)?

Solution Consider placing the letters in these boxes.

We can select any of the 3 letters for the first letter in the symbol. Once this letter has been selected, the second must be selected from the 2 remaining letters. After this, the third letter is already determined, since only 1 possibility is left. That is, we can place any of the 3 letters in the first box, either of the remaining 2 letters in the second box, and the only remaining letter in the third box. The possibilities can be determined using a **tree diagram,** as shown below.

TREE DIAGRAM			OUTCOMES
A	B	C	ABC
	C	B	ACB
B	A	C	BAC
	C	A	BCA
C	A	B	CAB
	B	A	CBA

Each outcome represents one permutation of the letters A, B, C.

We see that there are 6 possibilities. The set of all the possibilities is

$$\{ABC, ACB, BAC, BCA, CAB, CBA\}.$$

This is the set of all *permutations* of the letters A, B, C.

Suppose that we perform an experiment such as selecting letters (as in the preceding example), flipping a coin, or drawing a card. The results are called **outcomes.** An **event** is a set of outcomes. The following principle enables us to count actions that are combined to form an event.

THE FUNDAMENTAL COUNTING PRINCIPLE

Given a combined action, or *event,* in which the first action can be performed in n_1 ways, the second action can be performed in n_2 ways, and so on, the total number of ways in which the combined action can be performed is the product

$$n_1 \cdot n_2 \cdot n_3 \cdot \cdots \cdot n_k.$$

Thus, in Example 1, there are 3 choices for the first letter, 2 for the second letter, and 1 for the third letter, making a total of $3 \cdot 2 \cdot 1$, or 6 possibilities.

EXAMPLE 2 How many 3-letter code symbols can be formed with the letters A, B, C, D, E *with* repetition (that is, allowing letters to be repeated)?

Solution Since repetition is allowed, there are 5 choices for the first letter, 5 choices for the second, and 5 for the third. Thus, by the fundamental counting principle, there are $5 \cdot 5 \cdot 5$, or 125 code symbols.

PERMUTATION

A **permutation** of a set of n objects is an ordered arrangement of all n objects.

We can use the fundamental counting principle to count the number of permutations of the objects in a set. Consider, for example, a set of 4 objects

$$\{A, B, C, D\}.$$

To find the number of ordered arrangements of the set, we select a first letter: There are 4 choices. Then we select a second letter: There are 3 choices. Then we select a third letter: There are 2 choices. Finally, there is 1 choice for the last selection. Thus, by the fundamental counting principle, there are $4 \cdot 3 \cdot 2 \cdot 1$, or 24, permutations of a set of 4 objects.

We can find a formula for the total number of permutations of all objects in a set of n objects. We have n choices for the first selection, $n - 1$ choices for the second, $n - 2$ for the third, and so on. For the nth selection, there is only 1 choice.

THE TOTAL NUMBER OF PERMUTATIONS OF *n* OBJECTS

The total number of permutations of n objects, denoted ${}_nP_n$, is given by

$${}_nP_n = n(n-1)(n-2) \cdots 3 \cdot 2 \cdot 1.$$

EXAMPLE 3 Find each of the following.

a) ${}_4P_4$ **b)** ${}_7P_7$

Solution

Start with 4.

a) ${}_4P_4 = \underbrace{4 \cdot 3 \cdot 2 \cdot 1}_{} = 24$

4 factors

b) ${}_7P_7 = 7 \cdot 6 \cdot 5 \cdot 4 \cdot 3 \cdot 2 \cdot 1 = 5040$

We can also find the total number of permutations of n objects using the ${}_nP_r$ operation from the MATH PRB (probability) menu on a graphing calculator.

Now Try Exercise 1.

EXAMPLE 4 In how many ways can 9 packages be placed in 9 mailboxes, one package in a box?

Solution We have

$${}_9P_9 = 9 \cdot 8 \cdot 7 \cdot 6 \cdot 5 \cdot 4 \cdot 3 \cdot 2 \cdot 1 = 362{,}880.$$

Now Try Exercise 23.

Factorial Notation

We will use products such as $7 \cdot 6 \cdot 5 \cdot 4 \cdot 3 \cdot 2 \cdot 1$ so often that it is convenient to adopt a notation for them. For the product

$$7 \cdot 6 \cdot 5 \cdot 4 \cdot 3 \cdot 2 \cdot 1,$$

we write 7!, read "7 factorial."

We now define factorial notation for natural numbers and for 0.

FACTORIAL NOTATION

For any natural number n,

$$n! = n(n-1)(n-2)\cdots 3 \cdot 2 \cdot 1.$$

For the number 0,

$$0! = 1.$$

We define 0! as 1 so that certain formulas can be stated concisely and with a consistent pattern.

Here are some examples of factorial notation.

$$\begin{aligned}
7! &= 7 \cdot 6 \cdot 5 \cdot 4 \cdot 3 \cdot 2 \cdot 1 = 5040\\
6! &= 6 \cdot 5 \cdot 4 \cdot 3 \cdot 2 \cdot 1 = 720\\
5! &= 5 \cdot 4 \cdot 3 \cdot 2 \cdot 1 = 120\\
4! &= 4 \cdot 3 \cdot 2 \cdot 1 = 24\\
3! &= 3 \cdot 2 \cdot 1 = 6\\
2! &= 2 \cdot 1 = 2\\
1! &= 1 = 1\\
0! &= 1 = 1
\end{aligned}$$

Factorial notation can also be evaluated using the **!** operation from the MATH PRB (probability) menu, as shown at left.

We now see that the following statement is true.

$${}_nP_n = n!$$

We will often need to manipulate factorial notation. For example, note that

$$\begin{aligned}
8! &= 8 \cdot 7 \cdot 6 \cdot 5 \cdot 4 \cdot 3 \cdot 2 \cdot 1\\
&= 8 \cdot (7 \cdot 6 \cdot 5 \cdot 4 \cdot 3 \cdot 2 \cdot 1) = 8 \cdot 7!.
\end{aligned}$$

Generalizing, we get the following.

For any natural number n, $n! = n(n-1)!$.

By using this result repeatedly, we can further manipulate factorial notation.

EXAMPLE 5 Rewrite 7! with a factor of 5!.

Solution We have

$$7! = 7 \cdot 6! = 7 \cdot 6 \cdot 5!.$$

In general, we have the following.

> For any natural numbers k and n, with $k < n$,
>
> $$n! = \underbrace{n(n-1)(n-2)\cdots[n-(k-1)]}_{k \text{ factors}} \cdot \underbrace{(n-k)!}_{n-k \text{ factors}}$$

Permutations of n Objects Taken k at a Time

Consider a set of 5 objects

$$\{A, B, C, D, E\}.$$

How many ordered arrangements can be formed using 3 objects from the set without repetition? Examples of such an arrangement are EBA, CAB, and BCD. There are 5 choices for the first object, 4 choices for the second, and 3 choices for the third. By the fundamental counting principle, there are

$$5 \cdot 4 \cdot 3,$$

or 60 *permutations* of a set of 5 objects taken 3 at a time.

Note that

$$5 \cdot 4 \cdot 3 = \frac{5 \cdot 4 \cdot 3 \cdot 2 \cdot 1}{2 \cdot 1}, \quad \text{or} \quad \frac{5!}{2!}.$$

> **PERMUTATION OF n OBJECTS TAKEN k AT A TIME**
>
> A **permutation** of a set of n objects taken k at a time is an ordered arrangement of k objects taken from the set.

Consider a set of n objects and the selection of an ordered arrangement of k of them. There would be n choices for the first object. Then there would remain $n-1$ choices for the second, $n-2$ choices for the third, and so on. We make k choices in all, so there are k factors in the product. By the fundamental counting principle, the total number of permutations is

$$\underbrace{n(n-1)(n-2)\cdots[n-(k-1)]}_{k \text{ factors}}.$$

We can express this in another way by multiplying by 1, as follows:

$$n(n-1)(n-2)\cdots[n-(k-1)] \cdot \frac{(n-k)!}{(n-k)!}$$

$$= \frac{n(n-1)(n-2)\cdots[n-(k-1)](n-k)!}{(n-k)!}$$

$$= \frac{n!}{(n-k)!}.$$

This gives us the following.

THE NUMBER OF PERMUTATIONS OF n OBJECTS TAKEN k AT A TIME

The number of permutations of a set of n objects taken k at a time, denoted ${}_nP_k$, is given by

$$ {}_nP_k = \underbrace{n(n-1)(n-2)\cdots[n-(k-1)]}_{k \text{ factors}} \quad \textbf{(1)} $$

$$ = \frac{n!}{(n-k)!}. \quad \textbf{(2)} $$

EXAMPLE 6 Compute ${}_8P_4$ using both forms of the formula.

Solution Using form (1), we have

$$ {}_8P_4 = \underbrace{8 \cdot 7 \cdot 6 \cdot 5}_{} = 1680. $$

The 8 tells where to start.

The 4 tells how many factors.

Using form (2), we have

$$ \begin{aligned} {}_8P_4 &= \frac{8!}{(8-4)!} \\ &= \frac{8!}{4!} \\ &= \frac{8 \cdot 7 \cdot 6 \cdot 5 \cdot 4!}{4!} = \frac{8 \cdot 7 \cdot 6 \cdot 5 \cdot \cancel{4!}}{\cancel{4!}} \\ &= 8 \cdot 7 \cdot 6 \cdot 5 = 1680. \end{aligned} $$

```
8 nPr 4
                    1680
```

We can also evaluate ${}_8P_4$ using the ${}_nP_r$ operation from the MATH PRB menu on a graphing calculator.

Now Try Exercise 3.

EXAMPLE 7 ***Flags of Nations.*** The flags of many nations consist of three vertical stripes. For example, the flag of Ireland, shown here, has its first stripe green, second white, and third orange.

Suppose that the following 9 colors are available:

{black, yellow, red, blue, white, gold, orange, pink, purple}.

How many different flags of 3 colors can be made without repetition of colors in a flag? This assumes that the order in which the stripes appear is considered.

Solution We are determining the number of permutations of 9 objects taken 3 at a time. There is no repetition of colors. Using form (1), we get

$$ {}_9P_3 = 9 \cdot 8 \cdot 7 = 504. $$

Now Try Exercise 37(a).

EXAMPLE 8 ***Batting Orders.*** A baseball manager arranges the batting order as follows: The 4 infielders will bat first. Then the 3 outfielders, the catcher, and the pitcher will follow, not necessarily in that order. How many different batting orders are possible?

Solution The infielders can bat in ${}_4P_4$ different ways, the rest in ${}_5P_5$ different ways. Then by the fundamental counting principle, we have

$${}_4P_4 \cdot {}_5P_5 = 4! \cdot 5!, \quad \text{or} \quad 2880 \text{ possible batting orders.}$$

Now Try Exercise 31.

If we allow repetition, a situation like the following can occur.

EXAMPLE 9 How many 5-letter code symbols can be formed with the letters A, B, C, and D if we allow a letter to occur more than once?

Solution We can select each of the 5 letters of the code in 4 ways. That is, we can select the first letter in 4 ways, the second in 4 ways, and so on. Thus there are 4^5, or 1024 arrangements.

Now Try Exercise 37(b).

> The number of distinct arrangements of n objects taken k at a time, allowing repetition, is n^k.

Permutations of Sets with Nondistinguishable Objects

Consider a set of 7 marbles, 4 of which are blue and 3 of which are red. When they are lined up, one red marble will look just like any other red marble. In this sense, we say that the red marbles are nondistinguishable and, similarly, the blue marbles are nondistinguishable.

We know that there are 7! permutations of this set. Many of them will look alike, however. We develop a formula for finding the number of distinguishable permutations.

Consider a set of n objects in which n_1 are of one kind, n_2 are of a second kind, . . . , and n_k are of a kth kind. The total number of permutations of the set is $n!$, but this includes many that are nondistinguishable. Let N be the total number of distinguishable permutations. For each of these N permutations, there are $n_1!$ actual, nondistinguishable permutations, obtained by permuting the objects of the first kind. For each of these $N \cdot n_1!$ permutations, there are $n_2!$ nondistinguishable permutations, obtained by permuting the objects of the second kind, and so on. By the fundamental counting principle, the total number of permutations, including those that are nondistinguishable, is

$$N \cdot n_1! \cdot n_2! \cdot \cdots \cdot n_k!.$$

Then we have $N \cdot n_1! \cdot n_2! \cdot \cdots \cdot n_k! = n!$. Solving for N, we obtain

$$N = \frac{n!}{n_1! \cdot n_2! \cdot \cdots \cdot n_k!}.$$

Now, to finish our discussion of the marbles, we have

$$N = \frac{7!}{4!\,3!}$$

$$= \frac{7 \cdot 6 \cdot 5 \cdot 4!}{4! \cdot 3 \cdot 2 \cdot 1} = \frac{7 \cdot \cancel{3} \cdot \cancel{2} \cdot 5 \cdot \cancel{4!}}{\cancel{4!} \cdot \cancel{3} \cdot \cancel{2} \cdot 1}$$

$$= \frac{7 \cdot 5}{1}, \quad \text{or} \quad 35$$

distinguishable permutations of the marbles.

In general, we have the following.

> For a set of n objects in which n_1 are of one kind, n_2 are of another kind, . . . , and n_k are of a kth kind, the number of distinguishable permutations is
>
> $$\frac{n!}{n_1! \cdot n_2! \cdot \cdots \cdot n_k!}.$$

EXAMPLE 10 In how many distinguishable ways can the letters of the word CINCINNATI be arranged?

Solution There are 2 C's, 3 I's, 3 N's, 1 A, and 1 T for a total of 10 letters. Thus,

$$N = \frac{10!}{2! \cdot 3! \cdot 3! \cdot 1! \cdot 1!}, \quad \text{or} \quad 50{,}400.$$

The letters of the word CINCINNATI can be arranged in 50,400 distinguishable ways.

Now Try Exercise 35.

11.5 Exercise Set

Evaluate.

1. $_6P_6$
2. $_4P_3$
3. $_{10}P_7$
4. $_{10}P_3$
5. $5!$
6. $7!$
7. $0!$
8. $1!$
9. $\frac{9!}{5!}$
10. $\frac{9!}{4!}$
11. $(8 - 3)!$
12. $(8 - 5)!$
13. $\frac{10!}{7!3!}$
14. $\frac{7!}{(7 - 2)!}$
15. $_8P_0$
16. $_{13}P_1$
17. $_{52}P_4$
18. $_{52}P_5$
19. $_nP_3$
20. $_nP_2$
21. $_nP_1$
22. $_nP_0$

In each of Exercises 23–41, give your answer using permutation notation, factorial notation, or other operations. Then evaluate.

How many permutations are there of the letters in each of the following words, if all the letters are used without repetition?

23. CREDIT

24. FRUIT

25. EDUCATION

26. TOURISM

27. How many permutations are there of the letters of the word EDUCATION if the letters are taken 4 at a time?

28. How many permutations are there of the letters of the word TOURISM if the letters are taken 5 at a time?

29. How many 5-digit numbers can be formed using the digits 2, 4, 6, 8, and 9 without repetition? with repetition?

30. In how many ways can 7 athletes be arranged in a straight line?

31. *Program Planning.* A program is planned to have 5 musical numbers and 4 speeches. In how many ways can this be done if a musical number and a speech are to alternate and a musical number is to come first?

32. A professor is going to grade her 24 students on a curve. She will give 3 A's, 5 B's, 9 C's, 4 D's, and 3 F's. In how many ways can she do this?

33. *Phone Numbers.* How many 7-digit phone numbers can be formed with the digits 0, 1, 2, 3, 4, 5, 6, 7, 8, and 9, assuming that the first number cannot be 0 or 1? Accordingly, how many telephone numbers can there be within a given area code, before the area needs to be split with a new area code?

34. How many distinguishable code symbols can be formed from the letters of the word BUSINESS? BIOLOGY? MATHEMATICS?

35. Suppose that the expression $a^2b^3c^4$ is rewritten without exponents. In how many distinguishable ways can this be done?

36. *Coin Arrangements.* A penny, a nickel, a dime, and a quarter are arranged in a straight line.

a) Considering just the coins, in how many ways can they be lined up?

b) Considering the coins and heads and tails, in how many ways can they be lined up?

37. How many code symbols can be formed using 5 of the 6 letters A, B, C, D, E, F if the letters:

a) are not repeated?

b) can be repeated?

c) are not repeated but must begin with D?

d) are not repeated but must begin with DE?

38. *License Plates.* A state forms its license plates by first listing a number that corresponds to the county in which the owner of the car resides. (The names of the counties are alphabetized and the number is its location in that order.) Then the plate lists a letter of the alphabet, and this is followed by a number from 1 to 9999. How many such plates are possible if there are 80 counties?

39. *Zip Codes.* A U.S. postal zip code is a five-digit number.

a) How many zip codes are possible if any of the digits 0 to 9 can be used?

b) If each post office has its own zip code, how many possible post offices can there be?

40. *Zip-Plus-4 Codes.* A zip-plus-4 postal code uses a 9-digit number like 75247-5456. How many 9-digit zip-plus-4 postal codes are possible?

41. *Social Security Numbers.* A social security number is a 9-digit number like 243-47-0825.

a) How many different social security numbers can there be?

b) There are about 310 million people in the United States. Can each person have a unique social security number?

Skill Maintenance

Find the zero(s) of the function.

42. $f(x) = 4x - 9$ [1.5]

43. $f(x) = x^2 + x - 6$ [3.2]

44. $f(x) = 2x^2 - 3x - 1$ [3.2]

45. $f(x) = x^3 - 4x^2 - 7x + 10$ [4.4]

Synthesis

Solve for n.

46. ${}_nP_5 = 7 \cdot {}_nP_4$

47. ${}_nP_4 = 8 \cdot {}_{n-1}P_3$

48. ${}_nP_5 = 9 \cdot {}_{n-1}P_4$

49. ${}_nP_4 = 8 \cdot {}_nP_3$

50. Show that $n! = n(n-1)(n-2)(n-3)!$.

51. *Single-Elimination Tournaments.* In a single-elimination sports tournament consisting of n teams, a team is eliminated when it loses one game. How many games are required to complete the tournament?

52. *Double-Elimination Tournaments.* In a double-elimination softball tournament consisting of n teams, a team is eliminated when it loses two games. At most, how many games are required to complete the tournament?

11.6 Combinatorics: Combinations

› Evaluate combination notation and solve related applied problems.

We now consider counting techniques in which order is not considered.

Combinations

We sometimes make a selection from a set *without regard to order.* Such a selection is called a *combination*. If you play cards, for example, you know that in most situations the *order* in which you hold cards is not important. That is,

Each hand contains the same combination of three cards.

EXAMPLE 1 Find all the combinations of 3 letters taken from the set of 5 letters {A, B, C, D, E}.

Solution The combinations are

{A, B, C}, {A, B, D},
{A, B, E}, {A, C, D},
{A, C, E}, {A, D, E},
{B, C, D}, {B, C, E},
{B, D, E}, {C, D, E}.

There are 10 combinations of the 5 letters taken 3 at a time.

When we find all the combinations from a set of 5 objects taken 3 at a time, we are finding all the 3-element subsets. When a set is named, the order of the elements is *not* considered. Thus,

$$\{A, C, B\} \quad \text{names the same set as} \quad \{A, B, C\}.$$

COMBINATION; COMBINATION NOTATION

A **combination** containing k objects chosen from a set of n objects, $k \leq n$, is denoted using **combination notation** ${}_nC_k$.

We want to derive a general formula for ${}_nC_k$ for any $k \leq n$. First, it is true that ${}_nC_n = 1$, because a set with n objects has only 1 subset with n objects, the set itself. Second, ${}_nC_1 = n$, because a set with n objects has n subsets with 1 object each. Finally, ${}_nC_0 = 1$, because a set with n objects has only one subset with 0 objects, namely, the empty set $\varnothing$. To consider other possibilities, let's return to Example 1 and compare the number of combinations with the number of permutations.

COMBINATIONS ($_5C_3$ of these)		PERMUTATIONS ($3! \cdot {}_5C_3$ of these)					
{A, B, C}	→	ABC	BCA	CAB	CBA	BAC	ACB
{A, B, D}	→	ABD	BDA	DAB	DBA	BAD	ADB
{A, B, E}	→	ABE	BEA	EAB	EBA	BAE	AEB
{A, C, D}	→	ACD	CDA	DAC	DCA	CAD	ADC
{A, C, E}	→	ACE	CEA	EAC	ECA	CAE	AEC
{A, D, E}	→	ADE	DEA	EAD	EDA	DAE	AED
{B, C, D}	→	BCD	CDB	DBC	DCB	CBD	BDC
{B, C, E}	→	BCE	CEB	EBC	ECB	CBE	BEC
{B, D, E}	→	BDE	DEB	EBD	EDB	DBE	BED
{C, D, E}	→	CDE	DEC	ECD	EDC	DCE	CED

Note that each combination of 3 objects yields 6, or 3!, permutations.

$$3! \cdot {}_5C_3 = 60 = {}_5P_3 = 5 \cdot 4 \cdot 3,$$

so

$${}_5C_3 = \frac{{}_5P_3}{3!} = \frac{5 \cdot 4 \cdot 3}{3 \cdot 2 \cdot 1} = 10.$$

In general, the number of combinations of n objects taken k at a time, ${}_nC_k$, times the number of permutations of these objects, $k!$, must equal the number of permutations of n objects taken k at a time:

$$k! \cdot {}_nC_k = {}_nP_k$$

$${}_nC_k = \frac{{}_nP_k}{k!} = \frac{1}{k!} \cdot {}_nP_k$$

$$= \frac{1}{k!} \cdot \frac{n!}{(n-k)!} = \frac{n!}{k!(n-k)!}.$$

COMBINATIONS OF *n* OBJECTS TAKEN *k* AT A TIME

The total number of combinations of n objects taken k at a time, denoted ${}_nC_k$, is given by

$$ {}_nC_k = \frac{n!}{k!\,(n-k)!}, \qquad \textbf{(1)} $$

or

$$ {}_nC_k = \frac{{}_nP_k}{k!} = \frac{n(n-1)(n-2)\cdots[n-(k-1)]}{k!}. \qquad \textbf{(2)} $$

Another kind of notation for ${}_nC_k$ is **binomial coefficient notation.** The reason for such terminology will be seen later.

BINOMIAL COEFFICIENT NOTATION

$$ \binom{n}{k} = {}_nC_k $$

You should be able to use either notation and either form of the formula.

EXAMPLE 2 Evaluate $\binom{7}{5}$, using forms (1) and (2).

Solution

a) By form (1),

$$ \binom{7}{5} = \frac{7!}{5!(7-5)!} = \frac{7!}{5!2!} $$
$$ = \frac{7\cdot 6\cdot 5!}{5!\cdot 2!} = \frac{7\cdot 6\cdot \cancel{5!}}{\cancel{5!}\cdot 2!} = \frac{7\cdot 6}{2\cdot 1} = 21. $$

b) By form (2),

The 7 tells where to start.

$$ \binom{7}{5} = \frac{7\cdot 6\cdot 5\cdot 4\cdot 3}{5\cdot 4\cdot 3\cdot 2\cdot 1} = \frac{7\cdot 6}{2\cdot 1} = 21 $$

The 5 tells how many factors there are in both the numerator and the denominator and where to start the denominator.

```
7 nCr 5
                    21
```

We can also find combinations using the ${}_nC_r$ operation from the MATH PRB (probability) menu on a graphing calculator.

Now Try Exercise 11.

Be sure to keep in mind that $\binom{n}{k}$ does not mean $n \div k$, or n/k.

EXAMPLE 3 Evaluate $\binom{n}{0}$ and $\binom{n}{2}$.

Solution We use form (1) for the first expression and form (2) for the second. Then

$$\binom{n}{0} = \frac{n!}{0!(n-0)!} = \frac{n!}{1 \cdot n!} = 1,$$

using form (1), and

$$\binom{n}{2} = \frac{n(n-1)}{2!} = \frac{n(n-1)}{2}, \quad \text{or} \quad \frac{n^2-n}{2},$$

using form (2).

Now Try Exercise 19.

Note that

$$\binom{7}{2} = \frac{7 \cdot 6}{2 \cdot 1} = 21.$$

Using the result of Example 2 gives us

$$\binom{7}{5} = \binom{7}{2}.$$

This says that the number of 5-element subsets of a set of 7 objects is the same as the number of 2-element subsets of a set of 7 objects. When 5 elements are chosen from a set, one also chooses *not* to include 2 elements. To see this, consider the set {A, B, C, D, E, F, G}:

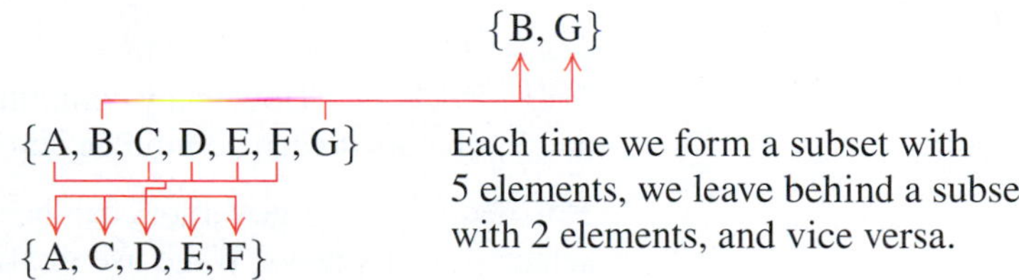

In general, we have the following. This result provides an alternative way to compute combinations.

SUBSETS OF SIZE *k* AND OF SIZE *n* − *k*

$$\binom{n}{k} = \binom{n}{n-k} \quad \text{and} \quad {}_nC_k = {}_nC_{n-k}$$

The number of subsets of size k of a set with n objects is the same as the number of subsets of size $n-k$. The number of combinations of n objects taken k at a time is the same as the number of combinations of n objects taken $n-k$ at a time.

We now solve problems involving combinations.

EXAMPLE 4 ***Indiana Lottery.*** Run by the state of Indiana, Hoosier Lotto is a twice-weekly lottery game with jackpots starting at $1 million. For a wager of $1, a player can choose 6 numbers from 1 through 48. If the numbers match those drawn by the state, the player wins the jackpot. (*Source*: www.hoosierlottery.com)

a) How many 6-number combinations are there?

b) Suppose that it takes you 10 min to pick your numbers and buy a game ticket. How many tickets can you buy in 4 days?

c) How many people would you have to hire for 4 days to buy tickets with all the possible combinations and ensure that you win?

Solution

a) No order is implied here. You pick any 6 different numbers from 1 through 48. Thus the number of combinations is

$$\begin{aligned} {}_{48}C_6 = \binom{48}{6} &= \frac{48!}{6!(48-6)!} = \frac{48!}{6!\,42!} \\ &= \frac{48\cdot 47\cdot 46\cdot 45\cdot 44\cdot 43\cdot \cancel{42!}}{6\cdot 5\cdot 4\cdot 3\cdot 2\cdot 1\cdot \cancel{42!}} \\ &= \frac{48\cdot 47\cdot 46\cdot 45\cdot 44\cdot 43}{6\cdot 5\cdot 4\cdot 3\cdot 2\cdot 1} = 12{,}271{,}512. \end{aligned}$$

b) First, we find the number of minutes in 4 days:

$$4 \text{ days} = 4\,\cancel{\text{days}}\cdot\frac{24\,\cancel{\text{hr}}}{1\,\cancel{\text{day}}}\cdot\frac{60\text{ min}}{1\,\cancel{\text{hr}}} = 5760\text{ min}.$$

Thus you could buy 5760/10, or 576 tickets in 4 days.

c) You would need to hire 12,271,512/576, or about 21,305 people for four days, to buy tickets with all the possible combinations and ensure a win. (This presumes lottery tickets can be bought 24 hours a day.)

Now Try Exercise 23.

EXAMPLE 5 How many committees can be formed from a group of 5 governors and 7 senators if each committee consists of 3 governors and 4 senators?

Solution The 3 governors can be selected in ${}_5C_3$ ways and the 4 senators can be selected in ${}_7C_4$ ways. If we use the fundamental counting principle, it follows that the number of possible committees is

$$\begin{aligned} {}_5C_3\cdot{}_7C_4 &= \frac{5!}{3!\,2!}\cdot\frac{7!}{4!\,3!} = \frac{5\cdot 4\cdot 3!}{3!\cdot 2\cdot 1}\cdot\frac{7\cdot 6\cdot 5\cdot 4!}{4!\cdot 3\cdot 2\cdot 1} \\ &= \frac{5\cdot \cancel{2}\cdot 2\cdot \cancel{3!}}{\cancel{3!}\cdot \cancel{2}\cdot 1}\cdot\frac{7\cdot \cancel{3}\cdot \cancel{2}\cdot 5\cdot \cancel{4!}}{\cancel{4!}\cdot \cancel{3}\cdot \cancel{2}\cdot 1} \\ &= 10\cdot 35 \\ &= 350. \end{aligned}$$

```
5 nCr 3*7 nCr 4
                 350
```

Now Try Exercise 27.

CONNECTING THE CONCEPTS — Permutations and Combinations

PERMUTATIONS	COMBINATIONS
Permutations involve order and arrangements of objects.	Combinations do not involve order or arrangements of objects.
Given 5 books, we can arrange 3 of them on a shelf in ${}_5P_3$, or 60 ways.	Given 5 books, we can select 3 of them in ${}_5C_3$, or 10 ways.
Placing the books in different orders produces different arrangements.	The order in which the books are chosen does not matter.

11.6 Exercise Set

Evaluate.

1. ${}_{13}C_2$
2. ${}_9C_6$
3. $\binom{13}{11}$
4. $\binom{9}{3}$
5. $\binom{7}{1}$
6. $\binom{8}{8}$
7. $\frac{{}_5P_3}{3!}$
8. $\frac{{}_{10}P_5}{5!}$
9. $\binom{6}{0}$
10. $\binom{6}{1}$
11. $\binom{6}{2}$
12. $\binom{6}{3}$
13. $\binom{7}{0} + \binom{7}{1} + \binom{7}{2} + \binom{7}{3} + \binom{7}{4} + \binom{7}{5} + \binom{7}{6} + \binom{7}{7}$
14. $\binom{6}{0} + \binom{6}{1} + \binom{6}{2} + \binom{6}{3} + \binom{6}{4} + \binom{6}{5} + \binom{6}{6}$
15. ${}_{52}C_4$
16. ${}_{52}C_5$
17. $\binom{27}{11}$
18. $\binom{37}{8}$
19. $\binom{n}{1}$
20. $\binom{n}{3}$
21. $\binom{m}{m}$
22. $\binom{t}{4}$

In each of the following exercises, give an expression for the answer using permutation notation, combination notation, factorial notation, or other operations. Then evaluate.

23. *Key Club Officers.* There are 36 students in a high school Key Club, a service organization for teens. How many sets of 4 officers can be selected?

24. *League Games.* How many games can be played in a 9-team sports league if each team plays all other teams once? twice?

25. *Test Options.* On a test, a student is to select 10 out of 13 questions. In how many ways can this be done?

26. *Senate Committees.* Suppose the Senate of the United States consists of 58 Republicans and 42 Democrats. How many committees can be formed consisting of 6 Republicans and 4 Democrats?

27. *Test Options.* Of the first 10 questions on a test, a student must answer 7. Of the second 5 questions, the student must answer 3. In how many ways can this be done?

28. *Lines and Triangles from Points.* How many lines are determined by 8 points, no 3 of which are collinear? How many triangles are determined by the same points?

29. *Poker Hands.* How many 5-card poker hands are possible with a 52-card deck?

30. *Bridge Hands.* How many 13-card bridge hands are possible with a 52-card deck?

31. *Baskin-Robbins Ice Cream.* Burt Baskin and Irv Robbins began making ice cream in 1945. Initially they developed 31 flavors—one for each day of the month. (*Source*: Baskin-Robbins)

a) How many 2-dip cones are possible using the 31 original flavors if order of flavors is to be considered and no flavor is repeated?

b) How many 2-dip cones are possible if order is to be considered and a flavor can be repeated?

c) How many 2-dip cones are possible if order is not considered and no flavor is repeated?

32. *Powerball®.* Powerball® is a biweekly lottery game in which 5 white balls are drawn from a drum of 59 balls numbered 1–59 and 1 red ball is drawn from a drum of 35 balls numbered 1–35. To win the jackpot, a player must select numbers to match in any order the 5 white balls and the 1 red ball. (*Source*: www.powerball.com) How many 6-number combinations are there?

Skill Maintenance

Solve.

33. $3x - 7 = 5x + 10$ [1.5]

34. $2x^2 - x = 3$ [3.2]

35. $x^2 + 5x + 1 = 0$ [3.2]

36. $x^3 + 3x^2 - 10x = 24$ [4.4]

Synthesis

37. *Flush.* A flush in poker consists of a 5-card hand with all cards of the same suit. How many 5-card hands (flushes) are there that consist of all diamonds?

38. *Full House.* A full house in poker consists of three of a kind and a pair (two of a kind). How many full houses are there that consist of 3 aces and 2 queens? (See Section 11.8 for a description of a 52-card deck.)

39. *League Games.* How many games are played in a league with n teams if each team plays each other team once? twice?

40. There are n points on a circle. How many quadrilaterals can be inscribed with these points as vertices?

Solve for n.

41. $\binom{n}{n-2} = 6$

42. $\binom{n+1}{3} = 2 \cdot \binom{n}{2}$

43. $\binom{n+2}{4} = 6 \cdot \binom{n}{2}$

44. $\binom{n}{3} = 2 \cdot \binom{n-1}{2}$

45. Prove that

$$\binom{n}{k-1} + \binom{n}{k} = \binom{n+1}{k}$$

for any natural numbers n and k, $k \leq n$.

11.7 The Binomial Theorem

- Expand a power of a binomial using Pascal's triangle or factorial notation.
- Find a specific term of a binomial expansion.
- Find the total number of subsets of a set of n objects.

In this section, we consider ways of expanding a binomial $(a + b)^n$.

Binomial Expansion Using Pascal's Triangle

Consider the following expanded powers of $(a + b)^n$, where $a + b$ is any binomial and n is a whole number. Look for patterns.

$$\begin{aligned}
(a + b)^0 &= 1\\
(a + b)^1 &= a + b\\
(a + b)^2 &= a^2 + 2ab + b^2\\
(a + b)^3 &= a^3 + 3a^2b + 3ab^2 + b^3\\
(a + b)^4 &= a^4 + 4a^3b + 6a^2b^2 + 4ab^3 + b^4\\
(a + b)^5 &= a^5 + 5a^4b + 10a^3b^2 + 10a^2b^3 + 5ab^4 + b^5
\end{aligned}$$

Each expansion is a polynomial. There are some patterns to be noted.

1. There is one more term than the power of the exponent, n. That is, there are $n + 1$ terms in the expansion of $(a + b)^n$.
2. In each term, the sum of the exponents is n, the power to which the binomial is raised.
3. The exponents of a start with n, the power of the binomial, and decrease to 0. The last term has no factor of a. The first term has no factor of b, so powers of b start with 0 and increase to n.
4. The coefficients start at 1 and increase through certain values about "halfway" and then decrease through these same values back to 1.

Let's explore the coefficients further. Suppose that we want to find an expansion of $(a + b)^6$. The patterns we just noted indicate that there are 7 terms in the expansion:

$$a^6 + c_1a^5b + c_2a^4b^2 + c_3a^3b^3 + c_4a^2b^4 + c_5ab^5 + b^6.$$

How can we determine the value of each coefficient, c_i? We can do so in two ways. The first method involves writing the coefficients in a triangular array, as follows. This is known as **Pascal's triangle**:

$$\begin{array}{lccccccccccc}
(a + b)^0: & & & & & & 1 & & & & & \\
(a + b)^1: & & & & & 1 & & 1 & & & & \\
(a + b)^2: & & & & 1 & & 2 & & 1 & & & \\
(a + b)^3: & & & 1 & & 3 & & 3 & & 1 & & \\
(a + b)^4: & & 1 & & 4 & & 6 & & 4 & & 1 & \\
(a + b)^5: & 1 & & 5 & & 10 & & 10 & & 5 & & 1
\end{array}$$

There are many patterns in the triangle. Find as many as you can.

Perhaps you discovered a way to write the next row of numbers, given the numbers in the row above it. There are always 1's on the outside. Each remaining number is the sum of the two numbers above it. Let's try to find an expansion for $(a + b)^6$ by adding another row using the patterns we have discovered:

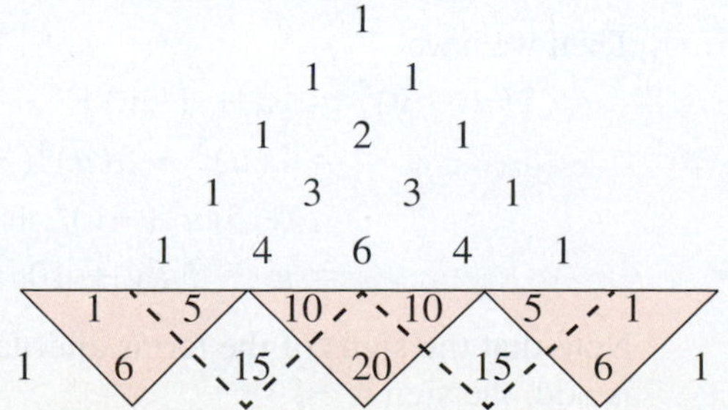

We see that in the last row

the 1st and last numbers are **1**;
the 2nd number is 1 + 5, or **6**;
the 3rd number is 5 + 10, or **15**;
the 4th number is 10 + 10, or **20**;
the 5th number is 10 + 5, or **15**; and
the 6th number is 5 + 1, or **6**.

Thus the expansion for $(a + b)^6$ is

$$(a + b)^6 = 1a^6 + 6a^5b + 15a^4b^2 + 20a^3b^3 + 15a^2b^4 + 6ab^5 + 1b^6.$$

To find an expansion for $(a + b)^8$, we complete two more rows of Pascal's triangle:

```
                        1
                     1     1
                  1     2     1
               1     3     3     1
            1     4     6     4     1
         1     5    10    10     5     1
      1     6    15    20    15     6     1
   1     7    21    35    35    21     7     1
1     8    28    56    70    56    28     8     1
```

Thus the expansion of $(a + b)^8$ is

$$(a + b)^8 = a^8 + 8a^7b + 28a^6b^2 + 56a^5b^3 + 70a^4b^4 + 56a^3b^5 + 28a^2b^6 + 8ab^7 + b^8.$$

We can generalize our results as follows.

THE BINOMIAL THEOREM USING PASCAL'S TRIANGLE

For any binomial $a + b$ and any natural number n,

$$(a + b)^n = c_0a^nb^0 + c_1a^{n-1}b^1 + c_2a^{n-2}b^2 + \cdots + c_{n-1}a^1b^{n-1} + c_na^0b^n,$$

where the numbers $c_0, c_1, c_2, \ldots, c_{n-1}, c_n$ are from the $(n + 1)$st row of Pascal's triangle.

EXAMPLE 1 Expand: $(u - v)^5$.

Solution We have $(a + b)^n$, where $a = u$, $b = -v$, and $n = 5$. We use the 6th row of Pascal's triangle:

1 5 10 10 5 1

Then we have

$$\begin{aligned}(u - v)^5 &= [u + (-v)]^5 \\ &= 1(u)^5 + 5(u)^4(-v)^1 + 10(u)^3(-v)^2 + 10(u)^2(-v)^3 \\ &\quad + 5(u)(-v)^4 + 1(-v)^5 \\ &= u^5 - 5u^4v + 10u^3v^2 - 10u^2v^3 + 5uv^4 - v^5.\end{aligned}$$

Note that the signs of the terms alternate between + and −. When the power of $-v$ is odd, the sign is −.

Now Try Exercise 5.

EXAMPLE 2 Expand: $\left(2t + \frac{3}{t}\right)^4$.

Solution We have $(a + b)^n$, where $a = 2t$, $b = 3/t$, and $n = 4$. We use the 5th row of Pascal's triangle:

1 4 6 4 1

Then we have

$$\left(2t + \frac{3}{t}\right)^4 = 1(2t)^4 + 4(2t)^3\left(\frac{3}{t}\right)^1 + 6(2t)^2\left(\frac{3}{t}\right)^2 + 4(2t)^1\left(\frac{3}{t}\right)^3 + 1\left(\frac{3}{t}\right)^4$$

$$= 1(16t^4) + 4(8t^3)\left(\frac{3}{t}\right) + 6(4t^2)\left(\frac{9}{t^2}\right) + 4(2t)\left(\frac{27}{t^3}\right) + 1\left(\frac{81}{t^4}\right)$$

$$= 16t^4 + 96t^2 + 216 + 216t^{-2} + 81t^{-4}.$$

Now Try Exercise 9.

Binomial Expansion Using Combination Notation

Suppose that we want to find the expansion of $(a + b)^{11}$. The disadvantage in using Pascal's triangle is that we must compute all the preceding rows of the triangle to obtain the row needed for the expansion. The following method avoids this. It also enables us to find a specific term— say, the 8th term—without computing all the other terms of the expansion. This method is useful in such courses as finite mathematics, calculus, and statistics, and it uses the *binomial coefficient notation* $\binom{n}{k}$ developed in Section 11.6.

We can restate the binomial theorem as follows.

THE BINOMIAL THEOREM USING COMBINATION NOTATION

For any binomial $a + b$ and any natural number n,

$$(a + b)^n = \binom{n}{0}a^nb^0 + \binom{n}{1}a^{n-1}b^1 + \binom{n}{2}a^{n-2}b^2 + \cdots + \binom{n}{n-1}a^1b^{n-1} + \binom{n}{n}a^0b^n$$

$$= \sum_{k=0}^{n}\binom{n}{k}a^{n-k}b^k.$$

The binomial theorem can be proved by mathematical induction. (See Exercise 57.) This form shows why $\binom{n}{k}$ is called a *binomial coefficient.*

EXAMPLE 3 Expand: $(x^2 - 2y)^5$.

Solution We have $(a + b)^n$, where $a = x^2$, $b = -2y$, and $n = 5$. Then using the binomial theorem, we have

$$\begin{aligned}(x^2 - 2y)^5 &= \binom{5}{0}(x^2)^5 + \binom{5}{1}(x^2)^4(-2y) + \binom{5}{2}(x^2)^3(-2y)^2 \\ &\quad + \binom{5}{3}(x^2)^2(-2y)^3 + \binom{5}{4}x^2(-2y)^4 + \binom{5}{5}(-2y)^5 \\ &= \frac{5!}{0!\,5!}x^{10} + \frac{5!}{1!\,4!}x^8(-2y) + \frac{5!}{2!\,3!}x^6(4y^2) + \frac{5!}{3!\,2!}x^4(-8y^3) \\ &\quad + \frac{5!}{4!\,1!}x^2(16y^4) + \frac{5!}{5!\,0!}(-32y^5) \\ &= 1 \cdot x^{10} + 5x^8(-2y) + 10x^6(4y^2) + 10x^4(-8y^3) \\ &\quad + 5x^2(16y^4) + 1 \cdot (-32y^5) \\ &= x^{10} - 10x^8y + 40x^6y^2 - 80x^4y^3 + 80x^2y^4 - 32y^5.\end{aligned}$$

Now Try Exercise 11.

EXAMPLE 4 Expand: $\left(\frac{2}{x} + 3\sqrt{x}\right)^4$.

Solution We have $(a + b)^n$, where $a = 2/x$, $b = 3\sqrt{x}$, and $n = 4$. Then using the binomial theorem, we have

$$\begin{aligned}\left(\frac{2}{x} + 3\sqrt{x}\right)^4 &= \binom{4}{0}\left(\frac{2}{x}\right)^4 + \binom{4}{1}\left(\frac{2}{x}\right)^3(3\sqrt{x}) + \binom{4}{2}\left(\frac{2}{x}\right)^2(3\sqrt{x})^2 \\ &\quad + \binom{4}{3}\left(\frac{2}{x}\right)(3\sqrt{x})^3 + \binom{4}{4}(3\sqrt{x})^4 \\ &= \frac{4!}{0!\,4!}\left(\frac{16}{x^4}\right) + \frac{4!}{1!\,3!}\left(\frac{8}{x^3}\right)(3x^{1/2}) \\ &\quad + \frac{4!}{2!\,2!}\left(\frac{4}{x^2}\right)(9x) + \frac{4!}{3!\,1!}\left(\frac{2}{x}\right)(27x^{3/2}) + \frac{4!}{4!\,0!}(81x^2) \\ &= \frac{16}{x^4} + \frac{96}{x^{5/2}} + \frac{216}{x} + 216x^{1/2} + 81x^2.\end{aligned}$$

Now Try Exercise 13.

Finding a Specific Term

Suppose that we want to determine only a particular term of an expansion. The method we have developed will allow us to find such a term without computing all the rows of Pascal's triangle or all the preceding coefficients.

Note that in the binomial theorem, $\binom{n}{0}a^nb^0$ gives us the 1st term, $\binom{n}{1}a^{n-1}b^1$ gives us the 2nd term, $\binom{n}{2}a^{n-2}b^2$ gives us the 3rd term, and so on. This can be generalized as follows.

FINDING THE $(k + 1)$st TERM

The $(k + 1)$st term of $(a + b)^n$ is $\binom{n}{k} a^{n-k} b^k$.

EXAMPLE 5 Find the 5th term in the expansion of $(2x - 5y)^6$.

Solution First, we note that $5 = 4 + 1$. Thus, $k = 4$, $a = 2x$, $b = -5y$, and $n = 6$. Then the 5th term of the expansion is

$$\binom{6}{4}(2x)^{6-4}(-5y)^4, \quad \text{or} \quad \frac{6!}{4!\,2!}(2x)^2(-5y)^4, \quad \text{or} \quad 37{,}500x^2y^4.$$

Now Try Exercise 21.

EXAMPLE 6 Find the 8th term in the expansion of $(3x - 2)^{10}$.

Solution First, we note that $8 = 7 + 1$. Thus, $k = 7$, $a = 3x$, $b = -2$, and $n = 10$. Then the 8th term of the expansion is

$$\binom{10}{7}(3x)^{10-7}(-2)^7, \quad \text{or} \quad \frac{10!}{7!\,3!}(3x)^3(-2)^7, \quad \text{or} \quad -414{,}720x^3.$$

Now Try Exercise 27.

Total Number of Subsets

Suppose that a set has n objects. The number of subsets containing k elements is $\binom{n}{k}$ by a result of Section 11.6. The total number of subsets of a set is the number of subsets with 0 elements, plus the number of subsets with 1 element, plus the number of subsets with 2 elements, and so on. The total number of subsets of a set with n elements is

$$\binom{n}{0} + \binom{n}{1} + \binom{n}{2} + \cdots + \binom{n}{n}.$$

Now consider the expansion of $(1 + 1)^n$:

$$\begin{aligned}(1 + 1)^n &= \binom{n}{0}\cdot 1^n + \binom{n}{1}\cdot 1^{n-1}\cdot 1^1 + \binom{n}{2}\cdot 1^{n-2}\cdot 1^2 \\ &\quad + \cdots + \binom{n}{n}\cdot 1^n \\ &= \binom{n}{0} + \binom{n}{1} + \binom{n}{2} + \cdots + \binom{n}{n}.\end{aligned}$$

Thus the total number of subsets is $(1 + 1)^n$, or 2^n. We have proved the following.

TOTAL NUMBER OF SUBSETS

The total number of subsets of a set with n elements is 2^n.

EXAMPLE 7 The set $\{A, B, C, D, E\}$ has how many subsets?

Solution The set has 5 elements, so the number of subsets is 2^5, or 32.

Now Try Exercise 31.

EXAMPLE 8 Wendy's, a national restaurant chain, offers the following toppings for its hamburgers:

$$\{\text{catsup, mustard, mayonnaise, tomato, lettuce, onions, pickle}\}.$$

How many different kinds of hamburgers can Wendy's serve, excluding size of hamburger or number of patties?

Solution The toppings on each hamburger are the elements of a subset of the set of all possible toppings, the empty set being a plain hamburger. The total number of possible hamburgers is

$$\binom{7}{0} + \binom{7}{1} + \binom{7}{2} + \cdots + \binom{7}{7} = 2^7 = 128.$$

Thus Wendy's serves hamburgers in 128 different ways.

Now Try Exercise 33.

11.7 Exercise Set

Expand.

1. $(x + 5)^4$
2. $(x - 1)^4$
3. $(x - 3)^5$
4. $(x + 2)^9$
5. $(x - y)^5$
6. $(x + y)^8$
7. $(5x + 4y)^6$
8. $(2x - 3y)^5$
9. $\left(2t + \frac{1}{t}\right)^7$
10. $\left(3y - \frac{1}{y}\right)^4$
11. $(x^2 - 1)^5$
12. $(1 + 2q^3)^8$
13. $(\sqrt{5} + t)^6$
14. $(x - \sqrt{2})^6$
15. $\left(a - \frac{2}{a}\right)^9$
16. $(1 + 3)^n$
17. $(\sqrt{2} + 1)^6 - (\sqrt{2} - 1)^6$
18. $(1 - \sqrt{2})^4 + (1 + \sqrt{2})^4$
19. $(x^{-2} + x^2)^4$
20. $\left(\frac{1}{\sqrt{x}} - \sqrt{x}\right)^6$

Find the indicated term of the binomial expansion.

21. 3rd; $(a + b)^7$
22. 6th; $(x + y)^8$
23. 6th; $(x - y)^{10}$
24. 5th; $(p - 2q)^9$
25. 12th; $(a - 2)^{14}$
26. 11th; $(x - 3)^{12}$
27. 5th; $(2x^3 - \sqrt{y})^8$
28. 4th; $\left(\frac{1}{b^2} + \frac{b}{3}\right)^7$
29. Middle; $(2u - 3v^2)^{10}$
30. Middle two; $(\sqrt{x} + \sqrt{3})^5$

Determine the number of subsets of each of the following.

31. A set of 7 elements
32. A set of 6 members
33. The set of letters of the Greek alphabet, which contains 24 letters
34. The set of letters of the English alphabet, which contains 26 letters
35. What is the degree of $(x^5 + 3)^4$?
36. What is the degree of $(2 - 5x^3)^7$?

Expand each of the following, where $i^2 = -1$.

37. $(3 + i)^5$
38. $(1 + i)^6$
39. $(\sqrt{2} - i)^4$
40. $\left(\frac{\sqrt{3}}{2} - \frac{1}{2}i\right)^{11}$
41. Find a formula for $(a - b)^n$. Use sigma notation.
42. Expand and simplify:

$$\frac{(x + h)^{13} - x^{13}}{h}.$$

43. Expand and simplify:

$$\frac{(x+h)^n - x^n}{h}.$$

Use sigma notation.

Skill Maintenance

Given that $f(x) = x^2 + 1$ *and* $g(x) = 2x - 3$, *find each of the following.*

44. $(f+g)(x)$ [2.2]

45. $(fg)(x)$ [2.2]

46. $(f \circ g)(x)$ [2.3]

47. $(g \circ f)(x)$ [2.3]

Synthesis

Solve for x.

48. $\sum_{k=0}^{8} \binom{8}{k} x^{8-k} 3^k = 0$

49. $\sum_{k=0}^{4} \binom{4}{k} (-1)^k x^{4-k} 6^k = 81$

50. Find the ratio of the 4th term of

$$\left(p^2 - \frac{1}{2}\sqrt[3]{q}\right)^5$$

to the 3rd term.

51. Find the term of

$$\left(\sqrt[3]{x} - \frac{1}{\sqrt{x}}\right)^7$$

containing $1/x^{1/6}$.

52. *Money Combinations.* A money clip contains one each of the following bills: \$1, \$2, \$5, \$10, \$20, \$50, and \$100. How many different sums of money can be formed using the bills?

Find the sum.

53. ${}_{100}C_0 + {}_{100}C_1 + \cdots + {}_{100}C_{100}$

54. ${}_nC_0 + {}_nC_1 + \cdots + {}_nC_n$

Simplify.

55. $\sum_{k=0}^{23} \binom{23}{k} (\log_a x)^{23-k} (\log_a t)^k$

56. $\sum_{k=0}^{15} \binom{15}{k} i^{30-2k}$

57. Use mathematical induction and the property

$$\binom{n}{r-1} + \binom{n}{r} = \binom{n+1}{r}$$

to prove the binomial theorem.

11.8 Probability

- Compute the probability of a simple event.

When a coin is tossed, we can reason that the chance, or likelihood, that it will fall heads is 1 out of 2, or the **probability** that it will fall heads is $\frac{1}{2}$. Of course, this does not mean that if a coin is tossed 10 times it will necessarily fall heads 5 times. If the coin is a "fair coin" and it is tossed a great many times, however, it will fall heads very nearly half of the time. Here we give an introduction to two kinds of probability, **experimental** and **theoretical**.

Experimental Probability and Theoretical Probability

If we toss a coin a great number of times—say, 1000—and count the number of times it falls heads, we can determine the probability that it will fall heads. If it falls heads 503 times, we would calculate the probability of its falling heads to be

$$\frac{503}{1000}, \quad \text{or} \quad 0.503.$$

This is an **experimental** determination of probability. Such a determination of probability is discovered by the observation and study of data and is quite common and very useful. Here, for example, are some probabilities that have been determined *experimentally*:

1. 60% of all freshmen entering four-year colleges graduate in 6 years (*Source*: www.satprepct.com, College Planning Partnership's Blog, February 24, 2011, Sam Rosensohn).
2. The probability that a woman will be diagnosed with breast cancer in her lifetime is $\frac{1}{8}$ (*Source*: National Cancer Institute).
3. Anyone who reaches the age of 65 has a 0.4 probability of entering a nursing home during the remaining years of life (*Source*: "Facing the Future," Russ Banham, *Wall Street Journal*).

If we consider a coin and reason that it is just as likely to fall heads as tails, we would calculate the probability that it will fall heads to be $\frac{1}{2}$. This is a **theoretical** determination of probability. Here are some other probabilities that have been determined *theoretically*, using mathematics:

1. If there are 30 people in a room, the probability that two of them have the same birthday (excluding year) is 0.706.
2. While on a trip, you meet someone and, after a period of conversation, discover that you have a common acquaintance. The typical reaction, "It's a small world!", is actually not appropriate, because the probability of such an occurrence is quite high—just over 22%.

In summary, experimental probabilities are determined by making observations and gathering data. Theoretical probabilities are determined by reasoning mathematically. Examples of experimental probability and theoretical probability like those above, especially those we do not expect, lead us to see the value of a study of probability. You might ask, "What is the *true* probability?" In fact, there is none. Experimentally, we can determine probabilities within certain limits. These may or may not agree with the probabilities that we obtain theoretically. There are situations in which it is much easier to determine one of these types of probabilities than the other. For example, it would be quite difficult to arrive at the probability of catching a cold using theoretical probability.

Computing Experimental Probabilities

We first consider experimental determination of probability. The basic principle we use in computing such probabilities is as follows.

PRINCIPLE *P* (EXPERIMENTAL)

Given an experiment in which n observations are made, if a situation, or event, E occurs m times out of n observations, then we say that the **experimental probability** of the event, $P(E)$, is given by

$$P(E) = \frac{m}{n}.$$

EXAMPLE 1 ***Television Ratings.*** There are an estimated 114,200,000 households in the United States that have at least one television. Each week, viewing information is collected and reported. One week, 28,510,000 households tuned in to the 2013 Grammy Awards ceremony on CBS, and 14,204,000 households tuned in to the action series "NCIS" on CBS (*Source*: Nielsen Media Research). What is the probability that a television household tuned in to the Grammy Awards ceremony during the given week? to "NCIS"?

Solution The probability that a television household was tuned in to the Grammy Awards ceremony is P, where

$$P = \frac{28,510,000}{114,200,000} \approx 0.2496 \approx 24.96\%.$$

The probability that a television household was tuned in to "NCIS" is P, where

$$P = \frac{14,204,000}{114,200,000} \approx 0.1244 \approx 12.44\%.$$

Now Try Exercise 1.

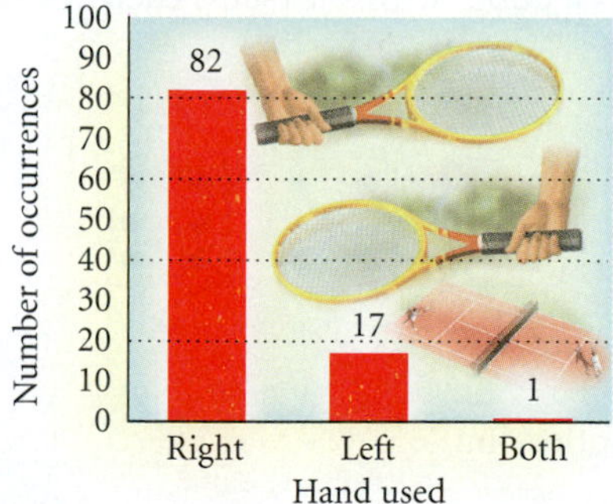

EXAMPLE 2 ***Sociological Survey.*** The authors of this text conducted a survey to determine the number of people who are left-handed, right-handed, or both. The results are shown in the graph at left.

a) Determine the probability that a person is right-handed.

b) Determine the probability that a person is left-handed.

c) Determine the probability that a person is ambidextrous (uses both hands with equal ability).

d) There are 130 students signed up for tennis lessons in a summer program offered by a community school corporation. On the basis of the data in this experiment, how many of the students would you expect to be left-handed?

Solution

a) The number of people who are right-handed is 82, the number who are left-handed is 17, and the number who are ambidextrous is 1. The total number of observations is $82 + 17 + 1$, or 100. Thus the probability that a person is right-handed is P, where

$$P = \frac{82}{100}, \quad \text{or} \quad 0.82, \quad \text{or} \quad 82\%.$$

b) The probability that a person is left-handed is P, where

$$P = \frac{17}{100}, \quad \text{or} \quad 0.17, \quad \text{or} \quad 17\%.$$

c) The probability that a person is ambidextrous is P, where

$$P = \frac{1}{100}, \quad \text{or} \quad 0.01, \quad \text{or} \quad 1\%.$$

d) There are 130 students, and from part (b) we can expect 17% to be left- handed. Since

$$17\% \text{ of } 130 = 0.17 \cdot 130 = 22.1,$$

we can expect that about 22 of the students will be left-handed.

Now Try Exercise 3.

Theoretical Probability

Suppose that we perform an experiment such as flipping a coin, throwing a dart, drawing a card from a deck, or checking an item off an assembly line for quality. Each possible result of such an experiment is called an **outcome**. The set of all possible outcomes is called the **sample space**. An **event** is a set of outcomes, that is, a subset of the sample space.

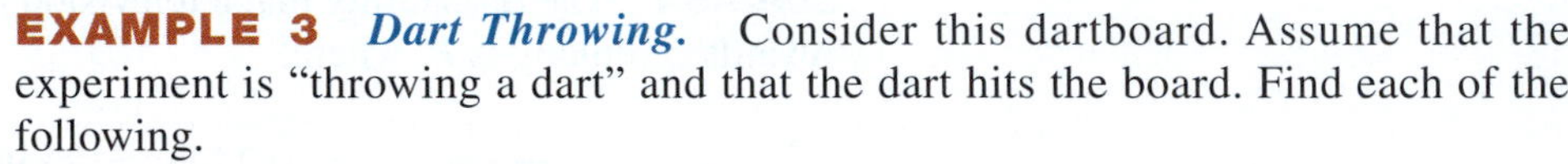

EXAMPLE 3 *Dart Throwing.* Consider this dartboard. Assume that the experiment is "throwing a dart" and that the dart hits the board. Find each of the following.

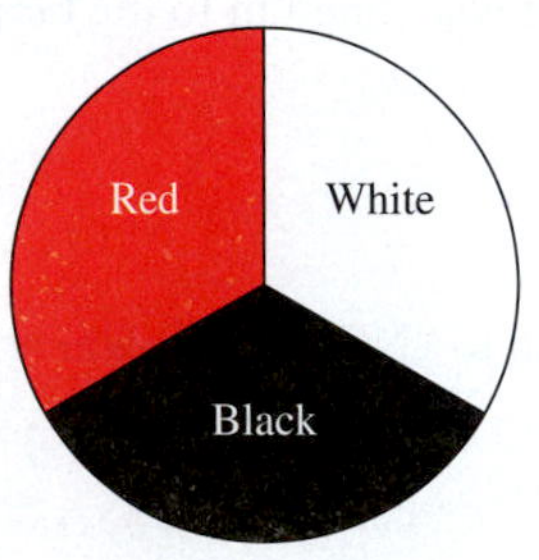

a) The outcomes **b)** The sample space

Solution

a) The outcomes are *hitting black* (B), *hitting red* (R), and *hitting white* (W).

b) The sample space is {*hitting black*, *hitting red*, *hitting white*}, which can be simply stated as {B, R, W}.

EXAMPLE 4 *Die Rolling.* A die (pl., dice) is a cube, with six faces, each containing a number of dots from 1 to 6.

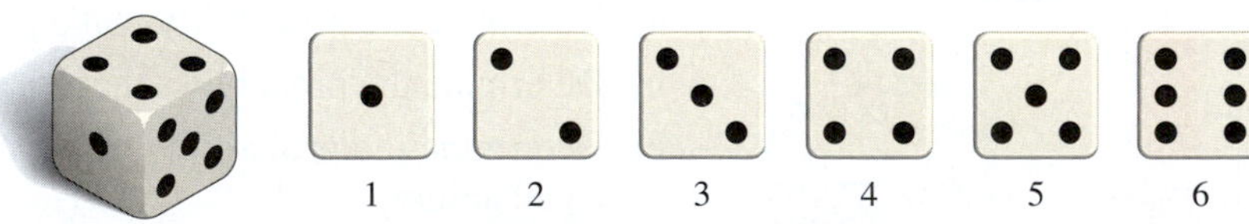

Suppose that a die is rolled. Find each of the following.

a) The outcomes **b)** The sample space

Solution

a) The outcomes are 1, 2, 3, 4, 5, 6.

b) The sample space is {1, 2, 3, 4, 5, 6}.

We denote the probability that an event E occurs as $P(E)$. For example, "a coin falling heads" may be denoted H. Then $P(H)$ represents the probability of the coin falling heads. When all the outcomes of an experiment have the same probability of occurring, we say that they are *equally likely.* To see the distinction between events that are equally likely and those that are not, consider the dartboards shown below.

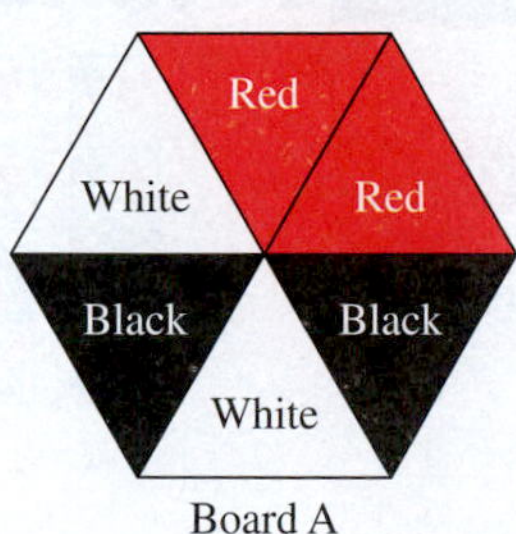

Board A

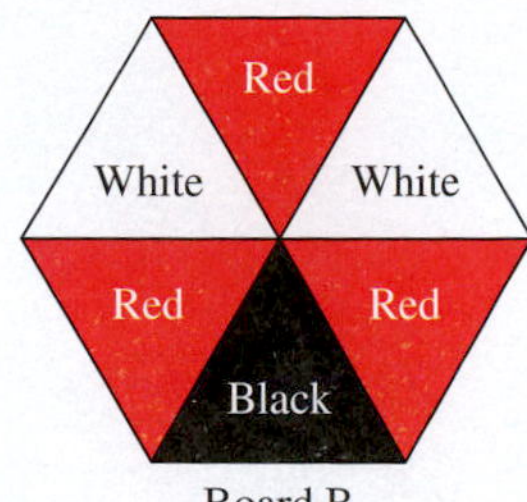

Board B

For board A, the events *hitting black*, *hitting red*, and *hitting white* are equally likely, because the black, red, and white areas are the same. However, for board B the areas are not the same so these events are not equally likely.

PRINCIPLE *P* (THEORETICAL)

If an event E can occur m ways out of n possible equally likely outcomes of a sample space S, then the **theoretical probability** of the event, $P(E)$, is given by

$$P(E) = \frac{m}{n}.$$

EXAMPLE 5 Suppose that we select, without looking, one marble from a bag containing 3 red marbles and 4 green marbles. What is the probability of selecting a red marble?

Solution There are 7 equally likely ways of selecting any marble, and since the number of ways of getting a red marble is 3, we have

$$P(\text{selecting a red marble}) = \frac{3}{7}.$$

Now Try Exercise 5(a).

EXAMPLE 6 What is the probability of rolling an even number on a die?

Solution The event is rolling an *even* number. It can occur 3 ways (rolling 2, 4, or 6). The number of equally likely outcomes is 6. By Principle P, we have

$$P(\text{even}) = \frac{3}{6}, \quad \text{or} \quad \frac{1}{2}.$$

Now Try Exercise 7.

We will use a number of examples related to a standard bridge deck of 52 cards. Such a deck is made up as shown in the following figure.

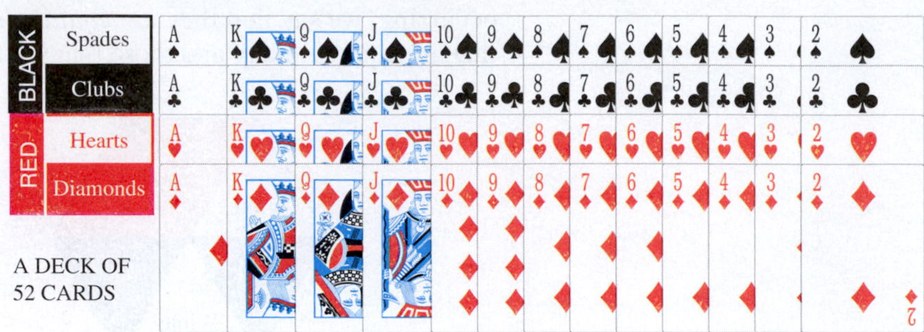

EXAMPLE 7 What is the probability of drawing an ace from a well-shuffled deck of cards?

Solution There are 52 outcomes (the number of cards in the deck), they are equally likely (from a well-shuffled deck), and there are 4 ways to obtain an ace, so by Principle P, we have

$$P(\text{drawing an ace}) = \frac{4}{52}, \quad \text{or} \quad \frac{1}{13}.$$

Now Try Exercise 9(a).

The following are some results that follow from Principle P.

PROBABILITY PROPERTIES

a) If an event E cannot occur, then $P(E) = 0$.
b) If an event E is certain to occur, then $P(E) = 1$.
c) The probability that an event E will occur is a number from 0 to 1: $0 \leq P(E) \leq 1$.

For example, in coin tossing, the event that a coin will land on its edge has probability 0. The event that a coin falls either heads or tails has probability 1.

In the following examples, we use the combinatorics that we studied in Sections 11.5 and 11.6 to calculate theoretical probabilities.

EXAMPLE 8 Suppose that 2 cards are drawn from a well-shuffled deck of 52 cards. What is the probability that both of them are spades?

Solution The number of ways n of drawing 2 cards from a well-shuffled deck of 52 cards is ${}_{52}C_2$. Since 13 of the 52 cards are spades, the number of ways m of drawing 2 spades is ${}_{13}C_2$. Thus,

```
13 nCr 2/52 nCr 2▸Frac
                    1/17
```

$$P(\text{drawing 2 spades}) = \frac{m}{n} = \frac{{}_{13}C_2}{{}_{52}C_2} = \frac{78}{1326} = \frac{1}{17}.$$

Now Try Exercise 11.

EXAMPLE 9 Suppose that 3 people are selected at random from a group that consists of 6 men and 4 women. What is the probability that 1 man and 2 women are selected?

Solution The number of ways of selecting 3 people from a group of 10 is ${}_{10}C_3$. One man can be selected in ${}_6C_1$ ways, and 2 women can be selected in ${}_4C_2$ ways. By the fundamental counting principle, the number of ways of selecting 1 man and 2 women is ${}_6C_1 \cdot {}_4C_2$. Thus the probability that 1 man and 2 women are selected is

6 nCr 1*4 nCr 2/10 nCr 3▸Frac $\frac{3}{10}$

$$P = \frac{{}_6C_1 \cdot {}_4C_2}{{}_{10}C_3} = \frac{3}{10}.$$

Now Try Exercise 13.

EXAMPLE 10 *Rolling Two Dice.* What is the probability of getting a total of 8 on a roll of a pair of dice?

Solution On each die, there are 6 possible outcomes. The outcomes are paired so there are $6 \cdot 6$, or 36, possible ways in which the two can fall. (Assuming that the dice are different—say, one red and one blue—can help in visualizing this.)

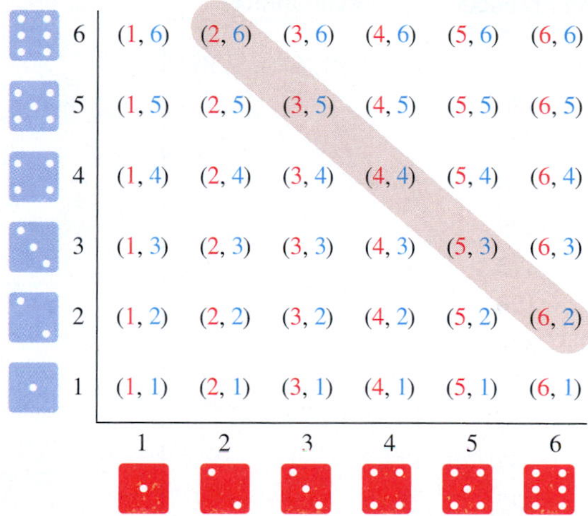

The pairs that total 8 are as shown in the figure above. There are 5 possible ways of getting a total of 8, so the probability is $\frac{5}{36}$.

Now Try Exercise 19.

11.8 Exercise Set

1. *Select a Number.* In a survey conducted by the authors, 100 people were polled and asked to select a number from 1 to 5. The results are shown in the following table.

Number Chosen	1	2	3	4	5
Number Who Chose That Number	18	24	23	23	12

a) What is the probability that the number chosen is 1? 2? 3? 4? 5?

b) What general conclusion might be made from the results of the experiment?

2. *Mason Dots®.* Made by the Tootsie Industries of Chicago, Illinois, Mason Dots® is a gumdrop candy. A box was opened by the authors and was found to contain the following number of gumdrops:

Orange	9
Lemon	8
Strawberry	7
Grape	6
Lime	5
Cherry	4

If we take one gumdrop out of the box, what is the probability of getting lemon? lime? orange? grape? strawberry? licorice?

3. *Marketing via E-mail.* In the second quarter of 2013, the probability that a marketing e-mail would be opened was 28.5% (*Source*: Q2 2013 Email Trends and Benchmarks, Epsilon). A business sent a marketing e-mail to 18,200 subscribers. How many of these e-mails can the business expect will be opened?

4. *Linguistics.* An experiment was conducted by the authors to determine the relative occurrence of various letters of the English alphabet. The front page of a newspaper was considered. In all, there were 9136 letters. The number of occurrences of each letter of the alphabet is listed in the following table.

Letter	Number of Occurrences	Probability
A	853	$853/9136 \approx 9.3\%$
B	136	
C	273	
D	286	
E	1229	
F	173	
G	190	
H	399	
I	539	
J	21	
K	57	
L	417	
M	231	
N	597	
O	705	
P	238	
Q	4	
R	609	
S	745	
T	789	
U	240	
V	113	
W	127	
X	20	
Y	124	
Z	21	$21/9136 \approx 0.2\%$

a) Complete the table of probabilities with the percentage, to the nearest tenth of a percent, of the occurrence of each letter.
b) What is the probability of a vowel occurring?
c) What is the probability of a consonant occurring?

5. *Marbles.* Suppose that we select, without looking, one marble from a bag containing 4 red marbles and 10 green marbles. What is the probability of selecting each of the following?

a) A red marble
b) A green marble
c) A purple marble
d) A red marble or a green marble

6. *Selecting Coins.* Suppose that we select, without looking, one coin from a bag containing 5 pennies, 3 dimes, and 7 quarters. What is the probability of selecting each of the following?

a) A dime
b) A quarter
c) A nickel
d) A penny, a dime, or a quarter

7. *Rolling a Die.* What is the probability of rolling a number less than 4 on a die?

8. *Rolling a Die.* What is the probability of rolling either a 1 or a 6 on a die?

9. *Drawing a Card.* Suppose that a card is drawn from a well-shuffled deck of 52 cards. What is the probability of drawing each of the following?

a) A queen
b) An ace or a 10
c) A heart
d) A black 6

10. *Drawing a Card.* Suppose that a card is drawn from a well-shuffled deck of 52 cards. What is the probability of drawing each of the following?

a) A 7
b) A jack or a king
c) A black ace
d) A red card

11. *Drawing Cards.* Suppose that 3 cards are drawn from a well-shuffled deck of 52 cards. What is the probability that they are all aces?

12. *Drawing Cards.* Suppose that 4 cards are drawn from a well-shuffled deck of 52 cards. What is the probability that they are all red?

13. *Production Unit.* The sales force of a business consists of 10 men and 10 women. A production unit of 4 people is set up at random. What is the probability that 2 men and 2 women are chosen?

14. *Coin Drawing.* A sack contains 7 dimes, 5 nickels, and 10 quarters. Eight coins are drawn at random. What is the probability of getting 4 dimes, 3 nickels, and 1 quarter?

Five-Card Poker Hands. Suppose that 5 cards are drawn from a deck of 52 cards. What is the probability of drawing each of the following?

15. 3 sevens and 2 kings

16. 5 aces

17. 5 spades

18. 4 aces and 1 five

19. *Tossing Three Coins.* Three coins are flipped. An outcome might be HTH.

a) Find the sample space.

What is the probability of getting each of the following?

b) Exactly one head
c) At most two tails
d) At least one head
e) Exactly two tails

Roulette. An American roulette wheel contains 38 slots numbered 00, 0, 1, 2, 3, . . . , 35, 36. Eighteen of the slots numbered 1–36 are colored red and 18 are colored black. The 00 and 0 slots are considered to be uncolored. The wheel is spun, and a ball is rolled around the rim until it falls into a slot. What is the probability that the ball falls in each of the following?

20. A red slot

21. A black slot

22. The 00 slot

23. The 0 slot

24. Either the 00 or the 0 slot

25. A red slot or a black slot

26. The number 24

27. An odd-numbered slot

28. *Dartboard.* The following figure shows a dartboard. A dart is thrown and hits the board. Find the probabilities

$$P(\text{red}), P(\text{green}), P(\text{blue}), P(\text{yellow}).$$

Skill Maintenance

Vocabulary Reinforcement

In each of Exercises 29–36, fill in the blank with the correct term. Some of the given choices will be used more than once. Others will not be used.

range	zero
domain	*y*-intercept
function	one-to-one
inverse function	rational
composite function	permutation
direct variation	combination
inverse variation	arithmetic sequence
factor	geometric sequence
solution	

29. A(n) ______________ of a function is an input for which the output is 0. [1.5]

30. A function is ______________ if different inputs have different outputs. [5.1]

31. A(n) ______________ is a correspondence between a first set, called the ______________, and a second set, called the ______________, such that each member of the ______________ corresponds to exactly one member of the ______________. [1.2]

32. The first coordinate of an x-intercept of a function is a(n) ______________ of the function. [1.5]

33. A selection made from a set without regard to order is a(n) ______________. [11.6]

34. If we have a function $f(x) = k/x$, where k is a positive constant, we have ______________. [2.6]

35. For a polynomial function $f(x)$, if $f(c) = 0$, then $x - c$ is a(n) ______________ of the polynomial. [4.3]

36. We have $\frac{a_{n+1}}{a_n} = r$, for any integer $n \geq 1$, in a(n) ______________. [11.3]

Synthesis

Five-Card Poker Hands. Suppose that 5 cards are drawn from a deck of 52 cards. For the following exercises, give both a reasoned expression and an answer.

37. *Two Pairs.* A hand with *two pairs* is a hand like Q-Q-3-3-A.

a) How many are there?

b) What is the probability of getting two pairs?

38. *Full House.* *A full house* consists of 3 of a kind and a pair such as Q-Q-Q-4-4.

a) How many full houses are there?

b) What is the probability of getting a full house?

39. *Three of a Kind.* A *three-of-a-kind* is a 5-card hand in which exactly 3 of the cards are of the same denomination and the other 2 are not a pair, such as Q-Q-Q-10-7.

a) How many three-of-a-kind hands are there?

b) What is the probability of getting three of a kind?

40. *Four of a Kind.* A *four-of-a-kind* is a 5-card hand in which 4 of the cards are of the same denomination, such as J-J-J-J-6, 7-7-7-7-A, or 2-2-2-2-5.

a) How many four-of-a-kind hands are there?

b) What is the probability of getting four of a kind?

Chapter 11 Summary and Review

STUDY GUIDE

KEY TERMS AND CONCEPTS	EXAMPLES
SECTION 11.1: SEQUENCES AND SERIES	
An **infinite sequence** is a function having for its domain the set of positive integers $\{1, 2, 3, 4, 5, \ldots\}$. A **finite sequence** is a function having for its domain a set of positive integers $\{1, 2, 3, 4, 5, \ldots, n\}$ for some positive integer n.	The first four terms of the sequence whose general term is given by $a_n = 3n + 2$ are $a_1 = 3 \cdot 1 + 2 = 5,$ $a_2 = 3 \cdot 2 + 2 = 8,$ $a_3 = 3 \cdot 3 + 2 = 11,$ and $a_4 = 3 \cdot 4 + 2 = 14.$
The sum of the terms of an infinite sequence is an **infinite series**. A **partial sum** is the sum of the first n terms. It is also called a **finite series** or the ***n*th partial sum** and is denoted S_n.	For the sequence above, $S_4 = 5 + 8 + 11 + 14$, or 38. We can denote this sum using **sigma notation** as $\sum_{k=1}^{4} (3k + 2).$

A sequence can be defined **recursively** by listing the first term, or the first few terms, and then using a **recursion formula** to determine the remaining terms from the given term.

The first four terms of the recursively defined sequence

$$a_1 = 3,\ a_{n+1} = (a_n - 1)^2$$

are

$$a_1 = 3,$$
$$a_2 = (a_1 - 1)^2 = (3 - 1)^2 = 4,$$
$$a_3 = (a_2 - 1)^2 = (4 - 1)^2 = 9, \quad \text{and}$$
$$a_4 = (a_3 - 1)^2 = (9 - 1)^2 = 64.$$

SECTION 11.2: ARITHMETIC SEQUENCES AND SERIES

For an arithmetic sequence:

$a_{n+1} = a_n + d;$ **d is the common difference.**

$a_n = a_1 + (n - 1)d;$ **The nth term**

$S_n = \frac{n}{2}(a_1 + a_n).$ **The sum of the first n terms**

For the arithmetic sequence 5, 8, 11, 14, . . . :

$$a_1 = 5;$$
$$d = 3 \quad (8 - 5 = 3, 11 - 8 = 3, \text{and so on});$$
$$a_6 = 5 + (6 - 1)3 = 5 + 15 = 20;$$
$$S_6 = \frac{6}{2}(5 + 20) = 3(25) = 75.$$

SECTION 11.3: GEOMETRIC SEQUENCES AND SERIES

For a geometric sequence:

$a_{n+1} = a_n r;$ **r is the common ratio.**

$a_n = a_1 r^{n-1};$ **The nth term**

$S_n = \frac{a_1(1 - r^n)}{1 - r};$ **The sum of the first n terms**

$S_\infty = \frac{a_1}{1 - r},\quad |r| < 1.$ **The limit, or sum, of an infinite geometric series**

For the geometric sequence $12, -6, 3, -\frac{3}{2}, \ldots$:

$$a_1 = 12;$$
$$r = -\frac{1}{2} \quad \left(\frac{-6}{12} = -\frac{1}{2}, \frac{3}{-6} = -\frac{1}{2}, \text{and so on}\right);$$
$$a_6 = 12\left(-\frac{1}{2}\right)^{6-1} = 12\left(-\frac{1}{2^5}\right) = -\frac{3}{8};$$
$$S_6 = \frac{12\left[1 - \left(-\frac{1}{2}\right)^6\right]}{1 - \left(-\frac{1}{2}\right)} = \frac{12\left(1 - \frac{1}{64}\right)}{\frac{3}{2}} = \frac{63}{8};$$
$$|r| = \left|-\tfrac{1}{2}\right| = \tfrac{1}{2} < 1, \text{so we have}$$
$$S_\infty = \frac{12}{1 - \left(-\frac{1}{2}\right)} = \frac{12}{\frac{3}{2}} = 8.$$

SECTION 11.4: MATHEMATICAL INDUCTION

The Principle of Mathematical Induction

We can prove an infinite sequence of statements S_n by showing the following.

(1) *Basis step.* S_1 is true.

(2) *Induction step.* For all natural numbers k, $S_k \rightarrow S_{k+1}$.

See Examples 1–3 on pp. 818–820.

SECTION 11.5: COMBINATORICS: PERMUTATIONS

The Fundamental Counting Principle

Given a combined action, or *event*, in which the first action can be performed in n_1 ways, the second action can be performed in n_2 ways, and so on, the total number of ways in which the combined action can be performed is the product

$$n_1 \cdot n_2 \cdot n_3 \cdot \cdots \cdot n_k.$$

The product $n(n-1)(n-2)\cdots 3 \cdot 2 \cdot 1$, for any natural number n, can also be written in **factorial notation** as $n!$. For the number 0, $0! = 1$.

The total number of permutations, or ordered arrangements, of n objects, denoted ${}_nP_n$, is given by

$${}_nP_n = n(n-1)(n-2)\cdots 3 \cdot 2 \cdot 1, \quad \text{or} \quad n!.$$

In how many ways can 7 books be arranged in a straight line?

We have

$${}_7P_7 = 7! = 7 \cdot 6 \cdot 5 \cdot 4 \cdot 3 \cdot 2 \cdot 1 = 5040.$$

The Number of Permutations of *n* Objects Taken *k* at a Time

$${}_nP_k = n(n-1)(n-2)\cdots[n-(k-1)] \quad \textbf{(1)}$$

$$= \frac{n!}{(n-k)!} \quad \textbf{(2)}$$

Compute ${}_7P_4$.

Using form (1), we have

$${}_7P_4 = 7 \cdot 6 \cdot 5 \cdot 4 = 840.$$

Using form (2), we have

$${}_7P_4 = \frac{7!}{(7-4)!} = \frac{7 \cdot 6 \cdot 5 \cdot 4 \cdot 3!}{3!}$$

$$= \frac{7 \cdot 6 \cdot 5 \cdot 4 \cdot \cancel{3!}}{\cancel{3!}} = 840.$$

The number of distinct arrangements of n objects taken k at a time, allowing repetition, is n^k.

The number of 4-number code symbols that can be formed with the numbers 5, 6, 7, 8, and 9, if we allow a number to occur more than once, is 5^4, or 625.

For a set of n objects in which n_1 are of one kind, n_2 are of another kind, . . . , and n_k are of a kth kind, the number of distinguishable permutations is

$$\frac{n!}{n_1! \cdot n_2! \cdot \cdots \cdot n_k!}.$$

Find the number of distinguishable code symbols that can be formed using the letters in the word MISSISSIPPI.

There are 1 M, 4 I's, 4 S's, and 2 P's, for a total of 11 letters, so we have

$$\frac{11!}{1!\,4!\,4!\,2!}, \quad \text{or} \quad 34{,}650.$$

SECTION 11.6: COMBINATORICS: COMBINATIONS

The Number of Combinations of n Objects Taken k at a Time

$$_nC_k = \frac{n!}{k!(n-k)!} \quad \textbf{(1)}$$
$$= \frac{_nP_k}{k!}$$
$$= \frac{n(n-1)(n-2)\cdots[n-(k-1)]}{k!}. \quad \textbf{(2)}$$

We can also use **binomial coefficient notation**:

$$\binom{n}{k} = {_nC_k}.$$

Compute: $_6C_4$, or $\binom{6}{4}$.

Using form (1), we have

$$\binom{6}{4} = \frac{6!}{4!(6-4)!} = \frac{6!}{4!\,2!}$$
$$= \frac{6 \cdot 5 \cdot 4!}{4!\,2!} = \frac{6 \cdot 5 \cdot \cancel{4!}}{\cancel{4!} \cdot 2 \cdot 1} = 15.$$

Using form (2), we have

$$\binom{6}{4} = \frac{_6P_4}{4!} = \frac{6 \cdot 5 \cdot 4 \cdot 3}{4 \cdot 3 \cdot 2 \cdot 1} = 15.$$

SECTION 11.7: THE BINOMIAL THEOREM

The Binomial Theorem Using Pascal's Triangle

For any binomial $a + b$ and any natural number n,

$$(a+b)^n = c_0a^nb^0 + c_1a^{n-1}b^1 + c_2a^{n-2}b^2 + \cdots + c_{n-1}a^1b^{n-1} + c_na^0b^n,$$

where the numbers $c_0, c_1, c_2, \ldots, c_{n-1}, c_n$ are from the $(n+1)$st row of Pascal's triangle. (See Pascal's triangle on p. 839.)

Expand: $(x-2)^4$.

We have $a = x$, $b = -2$, and $n = 4$. We use the fifth row of Pascal's triangle.

$$(x-2)^4 = 1 \cdot x^4 + 4 \cdot x^3(-2)^1 + 6 \cdot x^2(-2)^2 + 4 \cdot x^1(-2)^3 + 1(-2)^4$$
$$= x^4 + 4x^3(-2) + 6x^2 \cdot 4 + 4x(-8) + 16$$
$$= x^4 - 8x^3 + 24x^2 - 32x + 16$$

The Binomial Theorem Using Combination Notation

For any binomial $a + b$ and any natural number n,

$$(a+b)^n = \binom{n}{0}a^nb^0 + \binom{n}{1}a^{n-1}b^1 + \binom{n}{2}a^{n-2}b^2 + \cdots + \binom{n}{n-1}a^1b^{n-1} + \binom{n}{n}a^0b^n$$
$$= \sum_{k=0}^{n}\binom{n}{k}a^{n-k}b^k.$$

Expand: $(x^2+3)^3$.

We have $a = x^2$, $b = 3$, and $n = 3$.

$$(x^2+3)^3 = \binom{3}{0}(x^2)^3 + \binom{3}{1}(x^2)^2(3) + \binom{3}{2}(x^2)3^2 + \binom{3}{3}3^3$$
$$= \frac{3!}{0!\,3!}x^6 + \frac{3!}{1!\,2!}(x^4)(3) + \frac{3!}{2!\,1!}(x^2)(9) + \frac{3!}{3!\,0!}(27)$$
$$= 1 \cdot x^6 + 3 \cdot 3x^4 + 3 \cdot 9x^2 + 1 \cdot 27$$
$$= x^6 + 9x^4 + 27x^2 + 27$$

The $(k+1)$st term of $(a+b)^n$ is

$$\binom{n}{k}a^{n-k}b^k.$$

The third term of $(x^2+3)^3$ is

$$\binom{3}{2}(x^2)^{3-2} \cdot 3^2 = 3 \cdot x^2 \cdot 9 = 27x^2. \qquad (k = 2)$$

The **total number of subsets** of a set with n elements is 2^n.	How many subsets does the set $\{W, X, Y, Z\}$ have? The set has 4 elements, so we have 2^4, or 16.
SECTION 11.8: PROBABILITY	
Principle P (Experimental) Given an experiment in which n observations are made, if a situation, or event, E occurs m times out of n observations, then we say that the **experimental probability** of the event, $P(E)$, is given by $P(E) = \frac{m}{n}$.	From a batch of 1000 gears, 35 were found to be defective. The probability that a defective gear is produced is $\frac{35}{1000} = 0.035$, or 3.5%.
Principle P (Theoretical) If an event E can occur m ways out of n possible equally likely outcomes of a sample space S, then the **theoretical probability** of the event, $P(E)$, is given by $P(E) = \frac{m}{n}$.	What is the probability of drawing 2 red marbles and 1 green marble from a bag containing 5 red marbles, 6 green marbles, and 4 white marbles? Number of ways of drawing 3 marbles from a bag of 15: ${}_{15}C_3$ Number of ways of drawing 2 red marbles from 5 red marbles: ${}_5C_2$ Number of ways of drawing 1 green marble from 6 green marbles: ${}_6C_1$ Probability that 2 red marbles and 1 green marble are drawn: $\frac{{}_5C_2 \cdot {}_6C_1}{{}_{15}C_3} = \frac{10 \cdot 6}{455} = \frac{12}{91}$

REVIEW EXERCISES

Determine whether the statement is true or false.

1. A sequence is a function. [11.1]

2. An infinite geometric series with $r = -1$ has a limit. [11.3]

3. Permutations involve order and arrangements of objects. [11.5]

4. The total number of subsets of a set with n elements is n^2. [11.7]

5. Find the first 4 terms, a_{11}, and a_{23}:
$$a_n = (-1)^n\left(\frac{n^2}{n^4 + 1}\right).$$ [11.1]

6. Predict the general, or nth, term. Answers may vary.
$2, -5, 10, -17, 26, \ldots$ [11.1]

7. Find and evaluate:
$$\sum_{k=1}^{4} \frac{(-1)^{k+1}3^k}{3^k - 1}.$$ [11.1]

8. Use a graphing calculator to construct a table of values and a graph for the first 10 terms of this sequence.
$a_1 = 0.3, \quad a_{k+1} = 5a_k + 1$ [11.1]

9. Write sigma notation. Answers may vary.
$0 + 3 + 8 + 15 + 24 + 35 + 48$ [11.1]

10. Find the 10th term of the arithmetic sequence
$\frac{3}{4}, \frac{13}{12}, \frac{17}{12}, \ldots$. [11.2]

11. Find the 6th term of the arithmetic sequence
$a - b, a, a + b, \ldots$. [11.2]

12. Find the sum of the first 18 terms of the arithmetic sequence
$4, 7, 10, \ldots$. [11.2]

13. Find the sum of the first 200 natural numbers. [11.2]

14. The 1st term in an arithmetic sequence is 5, and the 17th term is 53. Find the 3rd term. [11.2]

15. The common difference in an arithmetic sequence is 3. The 10th term is 23. Find the first term. [11.2]

16. For a geometric sequence, $a_1 = -2$, $r = 2$, and $a_n = -64$. Find n and S_n. [11.3]

17. For a geometric sequence, $r = \frac{1}{2}$ and $S_5 = \frac{31}{2}$. Find a_1 and a_5. [11.3]

Find the sum of each infinite geometric series, if it exists. [11.3]

18. $25 + 27.5 + 30.25 + 33.275 + \cdots$

19. $0.27 + 0.0027 + 0.000027 + \cdots$

20. $\frac{1}{2} - \frac{1}{6} + \frac{1}{18} - \cdots$

21. Find fraction notation for $2.\overline{43}$. [11.3]

22. Insert four arithmetic means between 5 and 9. [11.2]

23. *Bouncing Golfball.* A golfball is dropped from a height of 30 ft to the pavement. It always rebounds three-fourths of the distance that it drops. How far (up and down) will the ball have traveled when it hits the pavement for the 6th time? [11.3]

24. *Amount of an Annuity.* To create a college fund, a parent makes a sequence of 18 yearly deposits of \$2000 each in a savings account on which interest is compounded annually at 2.8%. Find the amount of the annuity. [11.3]

25. *Total Gift.* Suppose that you receive 10¢ on the first day of the year, 12¢ on the 2nd day, 14¢ on the 3rd day, and so on.
   a) How much will you receive on the 365th day? [11.2]
   b) What is the sum of these 365 gifts? [11.2]

26. *The Economic Multiplier.* Suppose that the government is making a \$24,000,000,000 expenditure for travel to Mars. If 73% of this amount is spent again, and so on, what is the total effect on the economy? [11.3]

Use mathematical induction to prove each of the following. [11.4]

27. For every natural number n,
$$1 + 4 + 7 + \cdots + (3n - 2) = \frac{n(3n - 1)}{2}.$$

28. For every natural number n,
$$1 + 3 + 3^2 + \cdots + 3^{n-1} = \frac{3^n - 1}{2}.$$

29. For every natural number $n \geq 2$,
$$\left(1 - \frac{1}{2}\right)\left(1 - \frac{1}{3}\right) \cdots \left(1 - \frac{1}{n}\right) = \frac{1}{n}.$$

30. *Book Arrangements.* In how many ways can 6 books be arranged on a shelf? [11.5]

31. *Flag Displays.* If 9 different signal flags are available, how many different displays are possible using 4 flags in a row? [11.5]

32. *Prize Choices.* The winner of a contest can choose any 8 of 15 prizes. How many different sets of prizes can be chosen? [11.6]

33. *Fraternity–Sorority Names.* The Greek alphabet contains 24 letters. How many fraternity or sorority names can be formed using 3 different letters? [11.5]

34. *Letter Arrangements.* In how many distinguishable ways can the letters of the word TENNESSEE be arranged? [11.5]

35. *Floor Plans.* A manufacturer of houses has 1 floor plan but achieves variety by having 3 different roofs, 4 different ways of attaching the garage, and 3 different types of entrances. Find the number of different houses that can be produced. [11.5]

36. *Code Symbols.* How many code symbols can be formed using 5 out of 6 of the letters of G, H, I, J, K, L if the letters:
   a) cannot be repeated? [11.5]
   b) can be repeated? [11.5]
   c) cannot be repeated but must begin with K? [11.5]
   d) cannot be repeated but must end with IGH? [11.5]

37. Determine the number of subsets of a set containing 8 members. [11.7]

Expand. [11.7]

38. $(m + n)^7$

39. $(x - \sqrt{2})^5$

40. $(x^2 - 3y)^4$

41. $\left(a + \frac{1}{a}\right)^8$

42. $(1 + 5i)^6$, where $i^2 = -1$

43. Find the 4th term of $(a + x)^{12}$. [11.7]

44. Find the 12th term of $(2a - b)^{18}$. Do not multiply out the factorials. [11.7]

45. *Rolling Dice.* What is the probability of getting a 10 on a roll of a pair of dice? on a roll of 1 die? [11.8]

46. *Drawing a Card.* From a deck of 52 cards, 1 card is drawn at random. What is the probability that it is a club? [11.8]

47. *Drawing Three Cards.* From a deck of 52 cards, 3 are drawn at random without replacement. What is the probability that 2 are aces and 1 is a king? [11.8]

48. *Election Poll.* Three people were running for mayor in an election campaign. A poll was conducted to see which candidate was favored. During the polling, 86 favored candidate A, 97 favored B, and 23 favored C. Assuming that the poll is a valid indicator of the election results, what is the probability that the election will be won by A? B? C? [11.8]

49. *Consumption of American Cheese.* The following table lists the number of pounds of American cheese consumed per capita for selected years.

Year, n	American Cheese Consumed per Capita (in pounds)
1930, 0	3.2
1950, 20	5.5
1970, 40	7.0
1990, 60	11.1
2000, 70	12.7
2010, 80	13.3

Sources: Economic Research Service; U.S. Department of Agriculture

a) Find a linear sequence function $a_n = an + b$ that models the data. Let n represent the number of years since 1930. [11.1]

b) Use the sequence found in part (a) to estimate the number of pounds of American cheese consumed per capita in 2016. [11.1]

50. Which of the following is the 25th term of the arithmetic sequence 12, 10, 8, 6, . . . ? [11.2]

A. -38 **B.** -36
C. 32 **D.** 60

51. What is the probability of getting a total of 4 on a roll of a pair of dice? [11.8]

A. $\frac{1}{12}$ **B.** $\frac{1}{9}$
C. $\frac{1}{6}$ **D.** $\frac{5}{36}$

52. The graph of the sequence whose general term is $a_n = n - 1$ is which of the following? [11.1]

A.

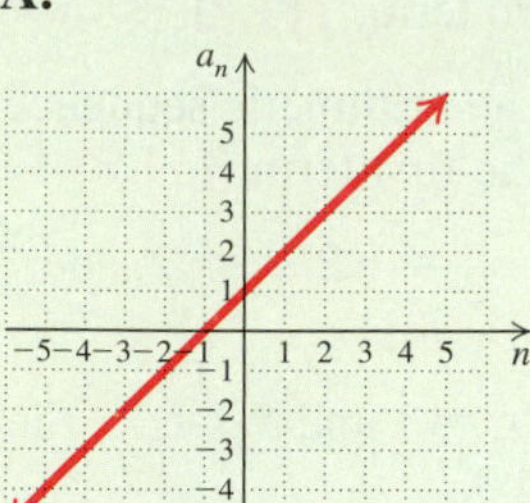

B.

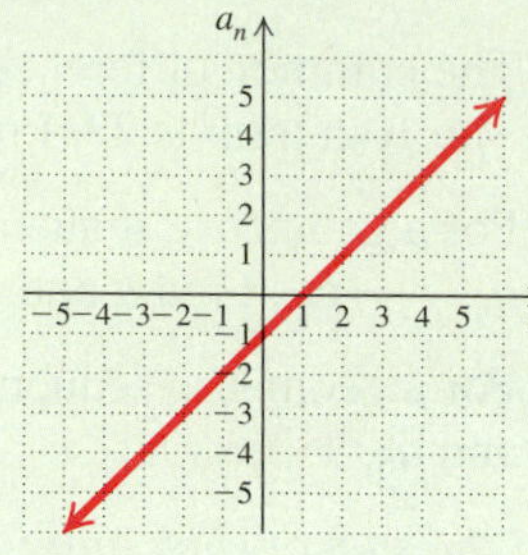

C.

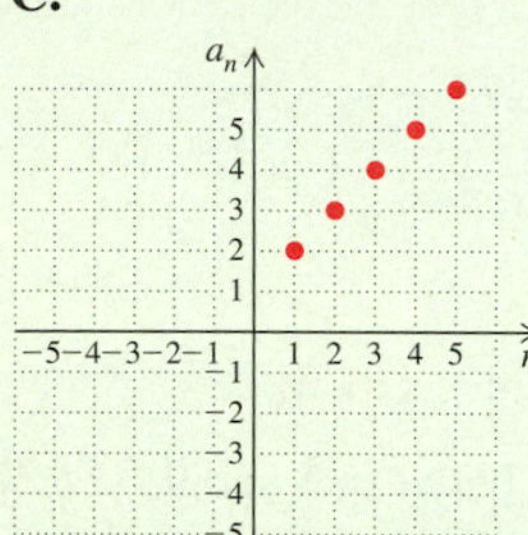

D.

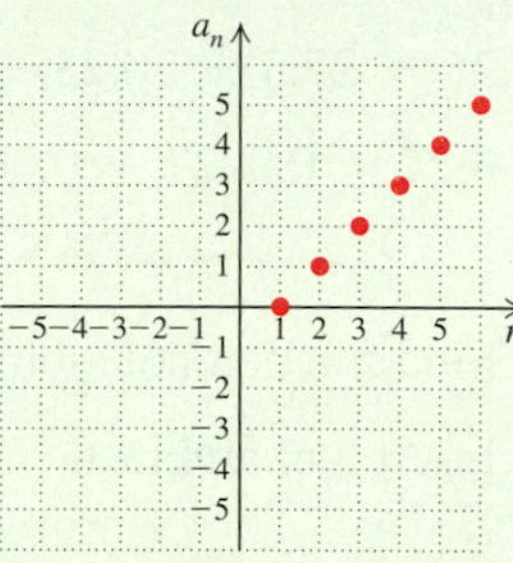

Synthesis

53. Suppose that $a_1, a_2, \ldots, a_n$ is an arithmetic sequence. Is $b_1, b_2, \ldots, b_n$ an arithmetic sequence if:

a) $b_n = |a_n|$? [11.2] b) $b_n = a_n + 8$? [11.2]

c) $b_n = 7a_n$? [11.2] d) $b_n = \frac{1}{a_n}$? [11.2]

e) $b_n = \log a_n$? [11.2] f) $b_n = a_n^3$? [11.2]

54. Suppose that $a_1, a_2, \ldots, a_n$ and $b_1, b_2, \ldots, b_n$ are geometric sequences. Prove that $c_1, c_2, \ldots, c_n$ is a geometric sequence, where $c_n = a_n b_n$. [11.3]

55. Write the first 3 terms of the infinite geometric series with $r = -\frac{1}{3}$ and $S_\infty = \frac{3}{8}$. [11.3]

56. The zeros of this polynomial function form an arithmetic sequence. Find them. [11.2]

$$f(x) = x^4 - 4x^3 - 4x^2 + 16x$$

57. Simplify:

$$\sum_{k=0}^{10} (-1)^k \binom{10}{k} (\log x)^{10-k} (\log y)^k. \text{ [11.6]}$$

Solve for n. [11.6]

58. $\binom{n}{6} = 3 \cdot \binom{n-1}{5}$ **59.** $\binom{n}{n-1} = 36$

60. Solve for a:

$$\sum_{k=0}^{5} \binom{5}{k} 9^{5-k} a^k = 0. \text{ [11.7]}$$

Collaborative Discussion and Writing

61. *Circular Arrangements.* In how many ways can the numbers on a clock face be arranged? See if you can derive a formula for the number of distinct circular arrangements of n objects. Explain your reasoning. [11.5]

62. How "long" is 15!? Suppose that you own 15 books and decide to make up all the possible arrangements of the books on a shelf. About how long, in years, would it take you if you were to make one arrangement per second? Write out the reasoning you used for this problem in the form of a paragraph. [11.5]

63. Explain why a "combination" lock should really be called a "permutation" lock. [11.6]

64. Give an explanation that you might use with a fellow student to explain that

$$\binom{n}{k} = \binom{n}{n-k}. \quad [11.6]$$

11 Chapter Test

1. For the sequence whose nth term is $a_n = (-1)^n(2n + 1)$, find a_{21}.

2. Find the first 5 terms of the sequence with general term

$$a_n = \frac{n+1}{n+2}.$$

3. Find and evaluate:

$$\sum_{k=1}^{4} (k^2 + 1).$$

4. Use a graphing calculator to construct a table of values and a graph for the first 10 terms of the sequence with general term

$$a_n = \frac{n+1}{n+2}.$$

Write sigma notation. Answers may vary.

5. $4 + 8 + 12 + 16 + 20 + 24$

6. $2 + 4 + 8 + 16 + 32 + \cdots$

7. Find the first 4 terms of the recursively defined sequence

$$a_1 = 3, \quad a_{n+1} = 2 + \frac{1}{a_n}.$$

8. Find the 15th term of the arithmetic sequence $2, 5, 8, \ldots$.

9. The 1st term of an arithmetic sequence is 8 and the 21st term is 108. Find the 7th term.

10. Find the sum of the first 20 terms of the series $17 + 13 + 9 + \cdots$.

11. Find the sum: $\sum_{k=1}^{25} (2k + 1)$.

12. Find the 11th term of the geometric sequence $10, -5, \frac{5}{2}, -\frac{5}{4}, \ldots$.

13. For a geometric sequence, $r = 0.2$ and $S_4 = 1248$. Find a_1.

Find the sum, if it exists.

14. $\sum_{k=1}^{8} 2^k$

15. $18 + 6 + 2 + \cdots$

16. Find fraction notation for $0.\overline{56}$.

17. *Salvage Value.* The value of an office machine is \$10,000. Its salvage value each year is 80% of its value the year before. Give a sequence that lists the salvage value of the machine for each year of a 6-year period.

18. *Hourly Wage.* William accepts a job, starting with an hourly wage of \$13.40, and is promised a raise of 30¢ per hour every three months for 4 years. What will William's hourly wage be at the end of the 4-year period?

19. *Amount of an Annuity.* To create a college fund, a parent makes a sequence of 18 equal yearly deposits of \$2500 in a savings account on which interest is compounded annually at 5.6%. Find the amount of the annuity.

20. Use mathematical induction to prove that, for every natural number n,

$$2 + 5 + 8 + \cdots + (3n - 1) = \frac{n(3n+1)}{2}.$$

Evaluate.

21. ${}_{15}P_6$

22. ${}_{21}C_{10}$

23. $\binom{n}{4}$

24. How many 4-digit numbers can be formed using the digits 1, 3, 5, 6, 7, and 9 without repetition?

25. How many code symbols can be formed using 4 of the 6 letters A, B, C, X, Y, Z if the letters:

a) can be repeated?

b) are not repeated and must begin with Z?

26. *Scuba Club Officers.* The Bay Woods Scuba Club has 28 members. How many sets of 4 officers can be selected from this group?

27. *Test Options.* On a test with 20 questions, a student must answer 8 of the first 12 questions and 4 of the last 8. In how many ways can this be done?

28. Expand: $(x + 1)^5$.

29. Find the 5th term of the binomial expansion $(x - y)^7$.

30. Determine the number of subsets of a set containing 9 members.

31. *Marbles.* Suppose that we select, without looking, one marble from a bag containing 6 red marbles and 8 blue marbles. What is the probability of selecting a blue marble?

32. *Drawing Coins.* Ethan has 6 pennies, 5 dimes, and 4 quarters in his pocket. Six coins are drawn at random. What is the probability of getting 1 penny, 2 dimes, and 3 quarters?

33. The graph of the sequence whose general term is $a_n = 2n - 2$ is which of the following?

A.

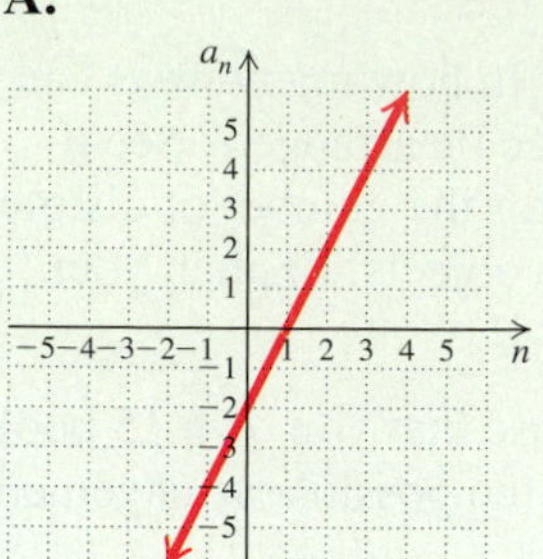

B.

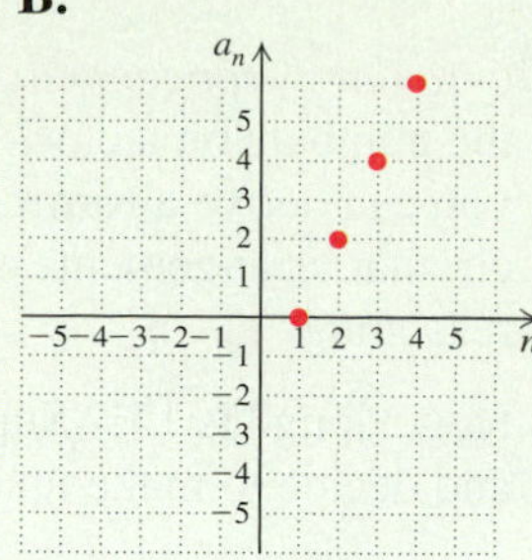

C.

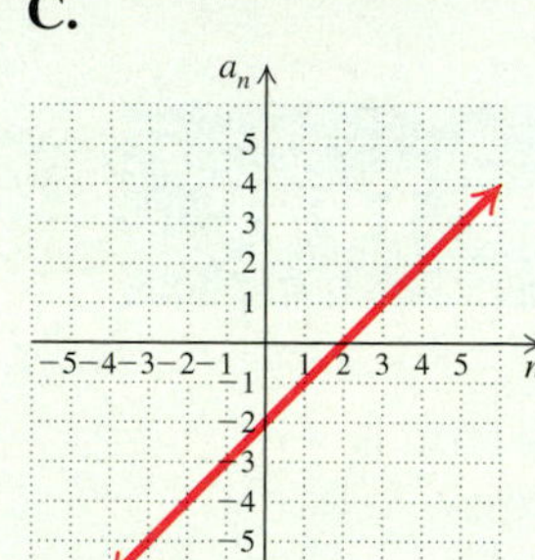

D.

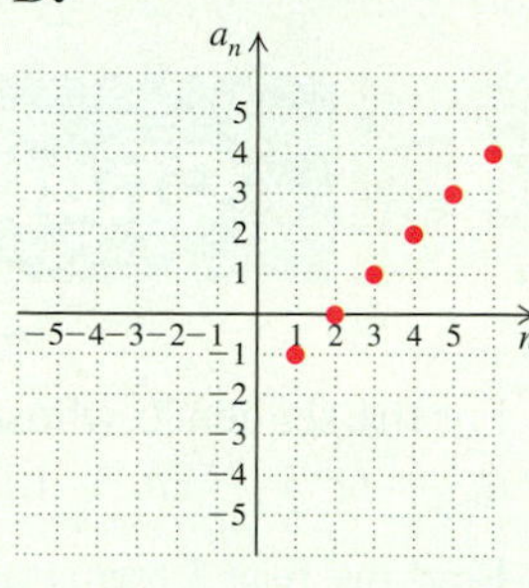

› Synthesis

34. Solve for n: ${}_nP_7 = 9 \cdot {}_nP_6$.

Answers

Chapter 1

Visualizing the Graph

1. H **2.** B **3.** D **4.** A **5.** G **6.** I **7.** C **8.** J
9. F **10.** E

Exercise Set 1.1

1. *A:* $(-5, 4)$; *B:* $(2, -2)$; *C:* $(0, -5)$;
D: $(3, 5)$; *E:* $(-5, -4)$; *F:* $(3, 0)$

3.

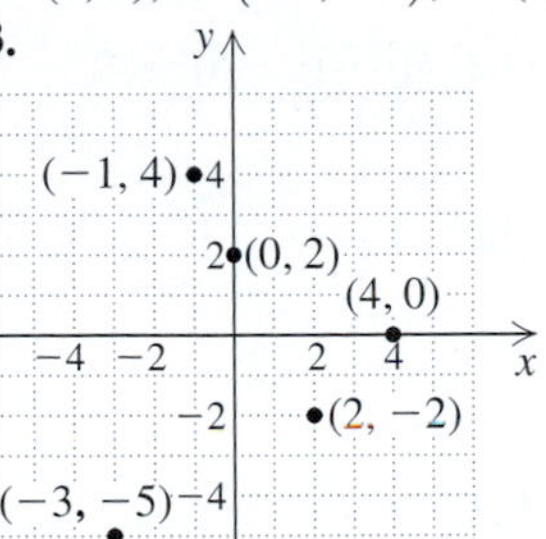

5.

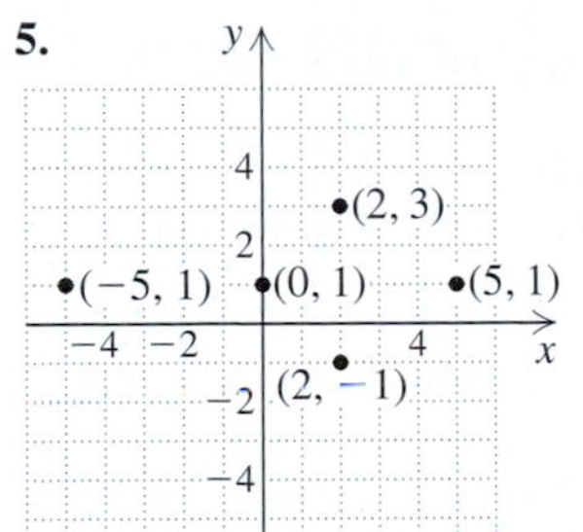

7. $(1971, 3)$, $(1981, 15)$, $(1991, 32)$, $(2001, 59)$, $(2011, 72)$, $(2014, 96)$ **9.** Yes; no **11.** Yes; no **13.** No; yes
15. No; yes
17. x-intercept: $(-3, 0)$; y-intercept: $(0, 5)$;

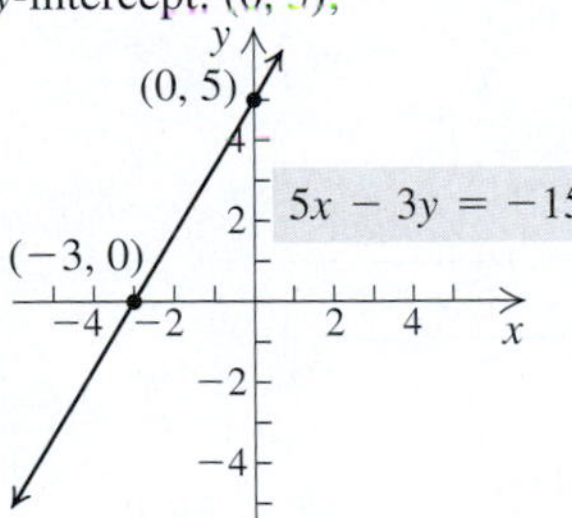

19. x-intercept: $(2, 0)$; y-intercept: $(0, 4)$;

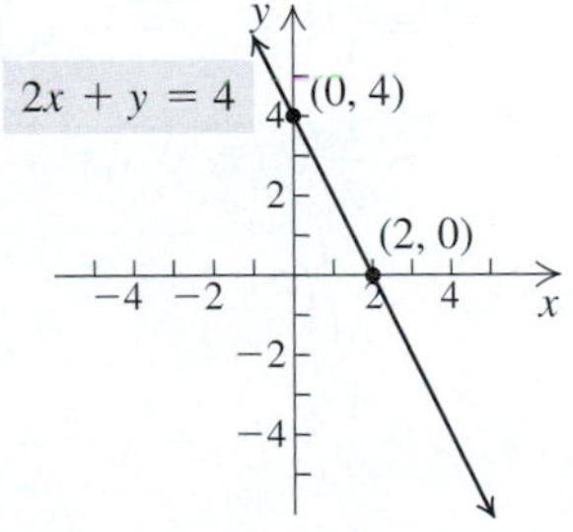

21. x-intercept: $(-4, 0)$; y-intercept: $(0, 3)$;

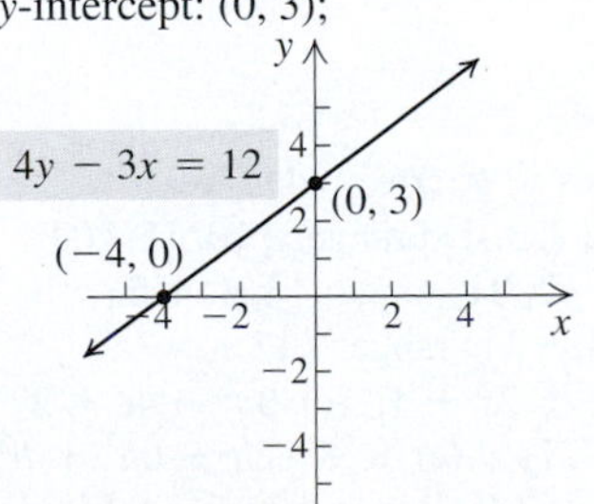

23.

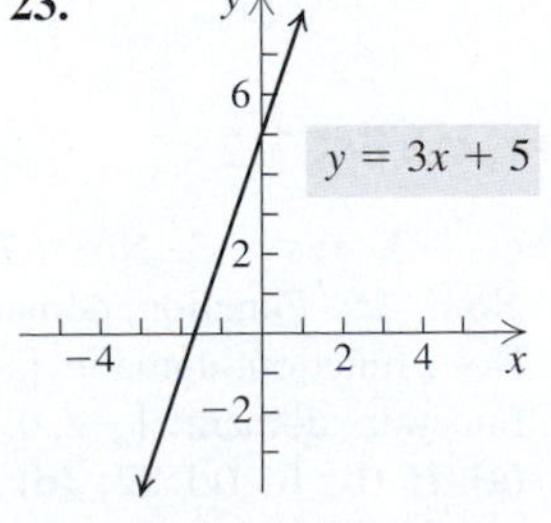

25.

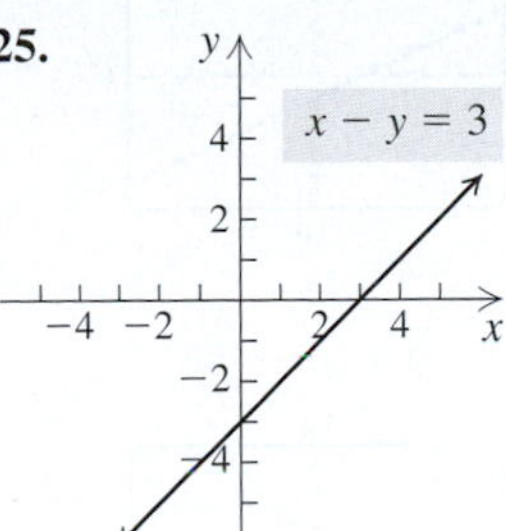

27.

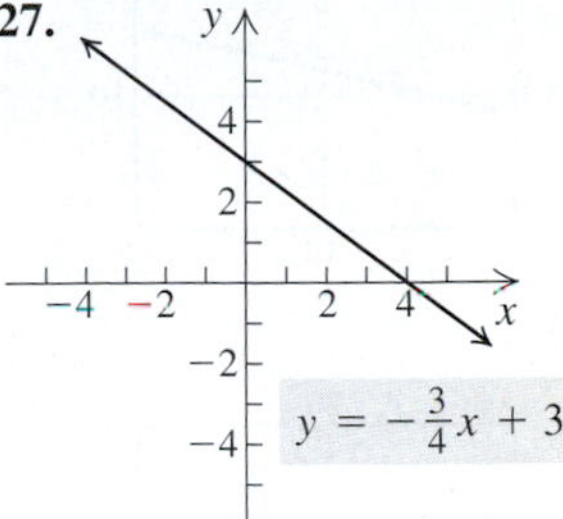

29.

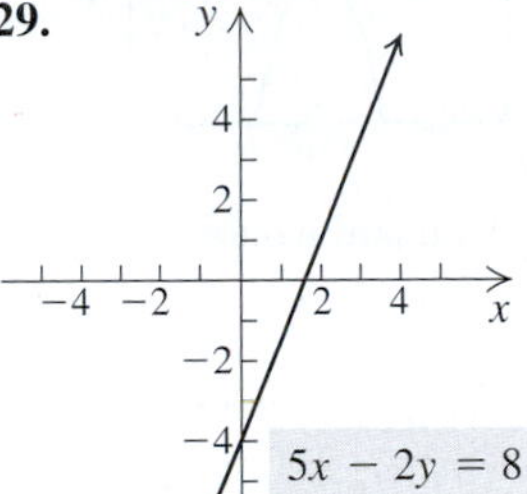

31.

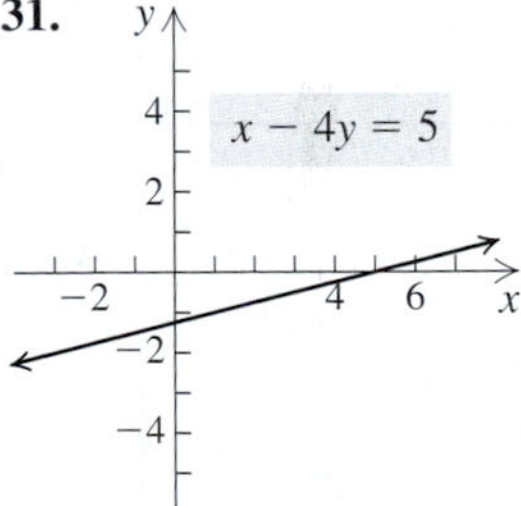

33.

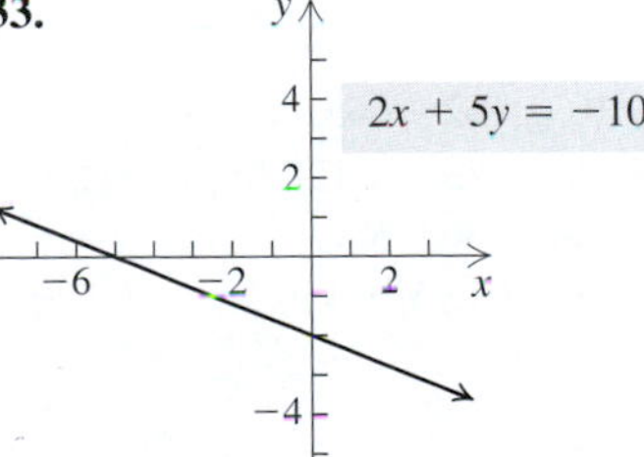

35.

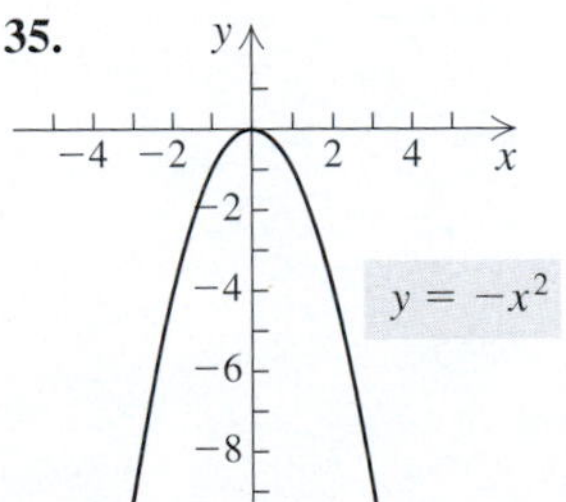

37.

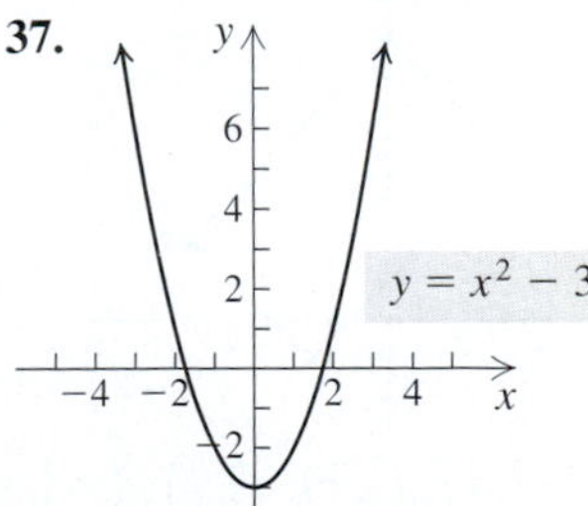

39.

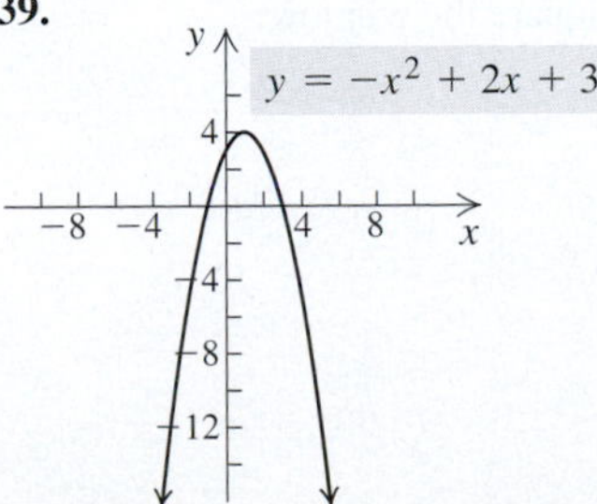

41. (b) **43.** (a)

45. $y = 2x + 1$

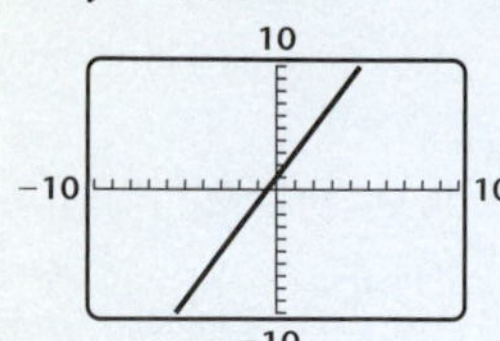

47. $4x + y = 7$

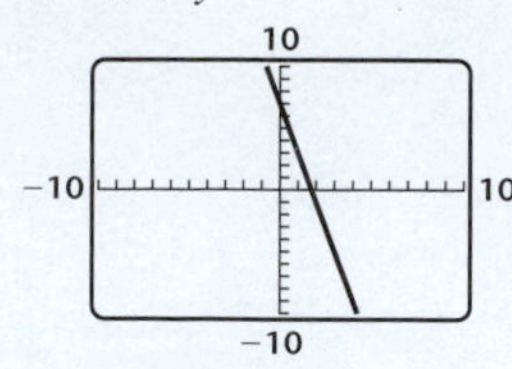

49. $y = \frac{1}{3}x + 2$

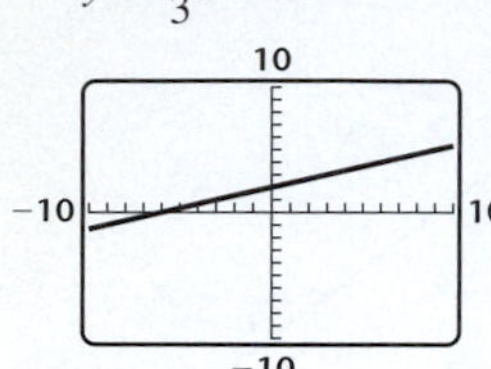

51. $2x + 3y = -5$

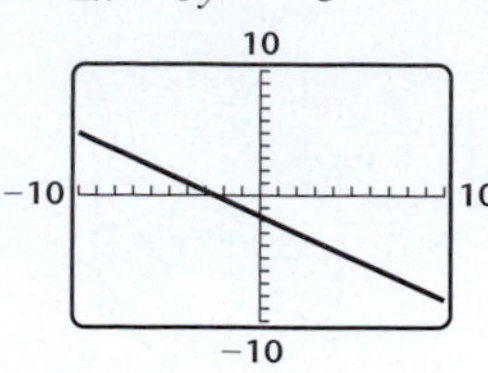

53. $y = x^2 + 6$

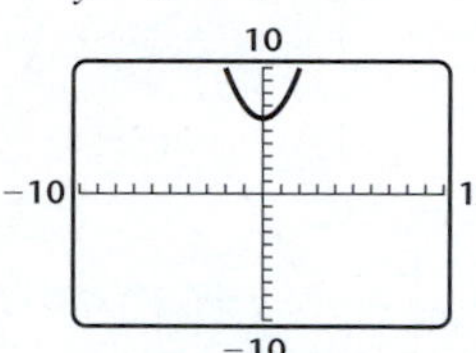

55. $y = 2 - x^2$

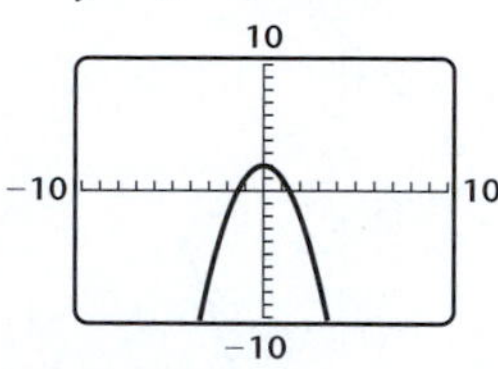

57. $y = x^2 + 4x - 2$

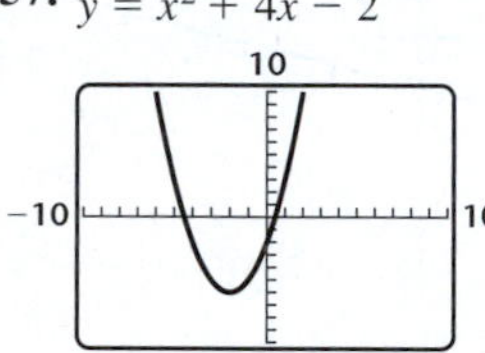
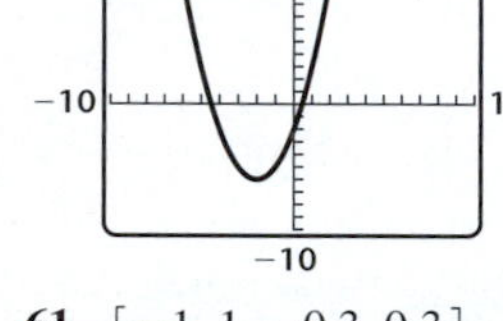

59. Standard window

61. $[-1, 1, -0.3, 0.3]$ **63.** $\sqrt{10}$, 3.162 **65.** 13
67. $\sqrt{45}$, 6.708 **69.** 16 **71.** $\frac{14}{3}$ **73.** $\sqrt{128.05}$, 11.316
75. $\sqrt{a^2 + b^2}$ **77.** 6.5 **79.** Yes **81.** No
83. $(-4, -6)$ **85.** $\left(-\frac{1}{5}, \frac{1}{4}\right)$ **87.** $(4.95, -4.95)$
89. $\left(-6, \frac{13}{2}\right)$ **91.** $\left(-\frac{5}{12}, \frac{13}{40}\right)$
93.

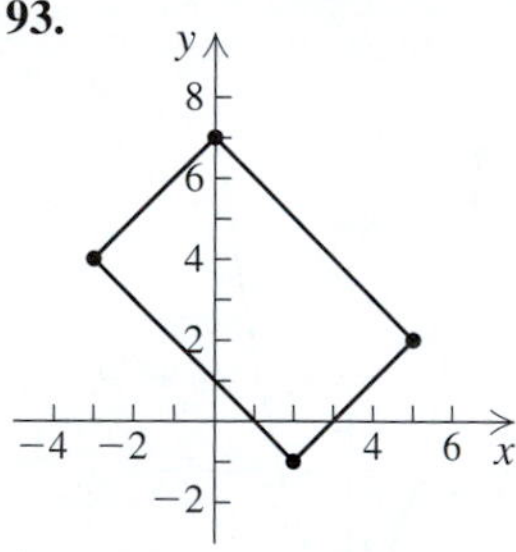

$\left(-\frac{1}{2}, \frac{3}{2}\right), \left(\frac{7}{2}, \frac{1}{2}\right), \left(\frac{5}{2}, \frac{9}{2}\right), \left(-\frac{3}{2}, \frac{11}{2}\right)$; no

95. $\left(\frac{\sqrt{7} + \sqrt{2}}{2}, -\frac{1}{2}\right)$ **97.** Square the window; for example, use $[-12, 9, -4, 10]$.
99. $(x - 2)^2 + (y - 3)^2 = \frac{25}{9}$
101. $(x + 1)^2 + (y - 4)^2 = 25$
103. $(x - 2)^2 + (y - 1)^2 = 169$
105. $(x + 2)^2 + (y - 3)^2 = 4$

107. $(0, 0)$; 2;

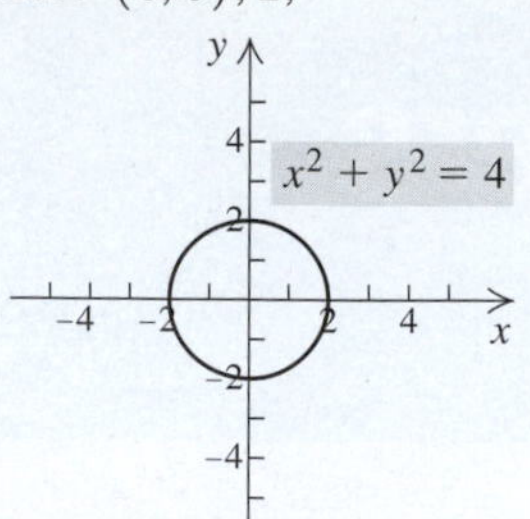

109. $(0, 3)$; 4;

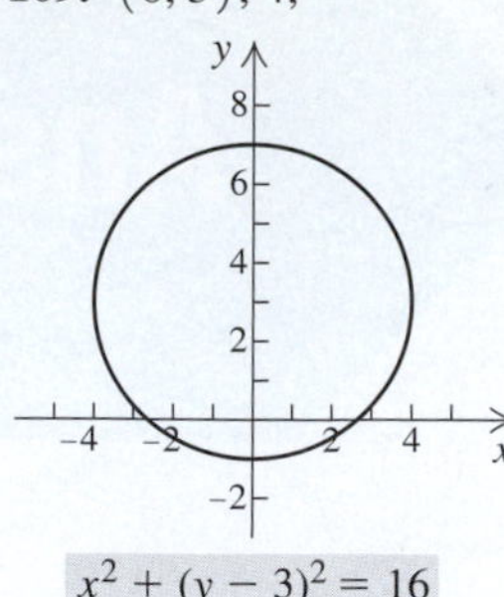

111. $(1, 5)$; 6;

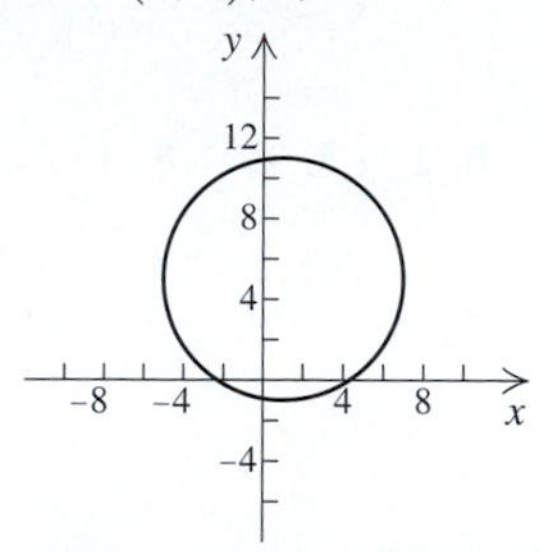

113. $(-4, -5)$; 3;

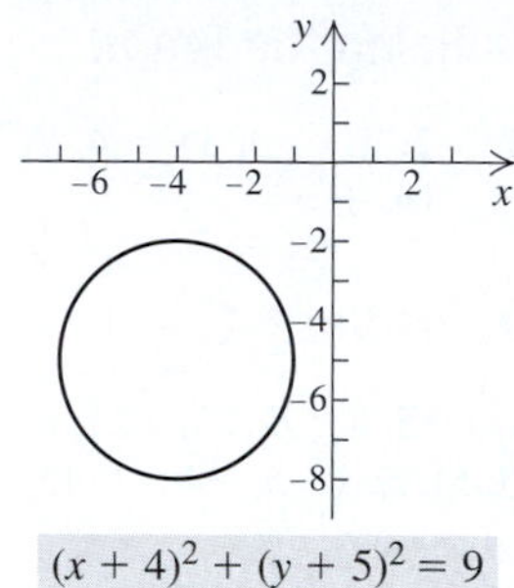

115. $(x + 2)^2 + (y - 1)^2 = 3^2$
117. $(x - 5)^2 + (y + 5)^2 = 15^2$ **119.** Third
121. $\sqrt{h^2 + h + 2a - 2\sqrt{a^2 + ah}}$, $\left(\frac{2a + h}{2}, \frac{\sqrt{a} + \sqrt{a + h}}{2}\right)$
123. $(x - 2)^2 + (y + 7)^2 = 36$ **125.** $(0, 4)$
127. **(a)** $(0, -3)$; **(b)** 5 ft **129.** Yes **131.** Yes
133. Let $P_1 = (x_1, y_1)$, $P_2 = (x_2, y_2)$, and $M = \left(\frac{x_1 + x_2}{2}, \frac{y_1 + y_2}{2}\right)$. Let $d(AB)$ denote the distance from point A to point B.

$$d(P_1M) = \sqrt{\left(\frac{x_1 + x_2}{2} - x_1\right)^2 + \left(\frac{y_1 + y_2}{2} - y_1\right)^2}$$
$$= \frac{1}{2}\sqrt{(x_2 - x_1)^2 + (y_2 - y_1)^2};$$
$$d(P_2M) = \sqrt{\left(\frac{x_1 + x_2}{2} - x_2\right)^2 + \left(\frac{y_1 + y_2}{2} - y_2\right)^2}$$
$$= \frac{1}{2}\sqrt{(x_1 - x_2)^2 + (y_1 - y_2)^2}$$
$$= \frac{1}{2}\sqrt{(x_2 - x_1)^2 + (y_2 - y_1)^2} = d(P_1M).$$

Exercise Set 1.2

1. Yes **3.** Yes **5.** No **7.** Yes **9.** Yes **11.** Yes
13. No **15.** Function; domain: $\{2, 3, 4\}$; range: $\{10, 15, 20\}$
17. Not a function; domain: $\{-7, -2, 0\}$; range: $\{3, 1, 4, 7\}$
19. Function; domain: $\{-2, 0, 2, 4, -3\}$; range: $\{1\}$
21. **(a)** 1; **(b)** 6; **(c)** 22; **(d)** $3x^2 + 2x + 1$; **(e)** $3t^2 - 4t + 2$
23. **(a)** 8; **(b)** -8; **(c)** $-x^3$; **(d)** $27y^3$; **(e)** $8 + 12h + 6h^2 + h^3$
25. **(a)** $\frac{1}{8}$; **(b)** 0; **(c)** does not exist; **(d)** $\frac{81}{53}$, or approximately 1.5283; **(e)** $\frac{x + h - 4}{x + h + 3}$ **27.** 0; does not exist; does not exist as a

real number; $\frac{1}{\sqrt{3}}$, or $\frac{\sqrt{3}}{3}$ **29.** $g(-2.1) \approx -21.8$; $g(5.08) \approx -130.4$; $g(10.003) \approx -468.3$

31.

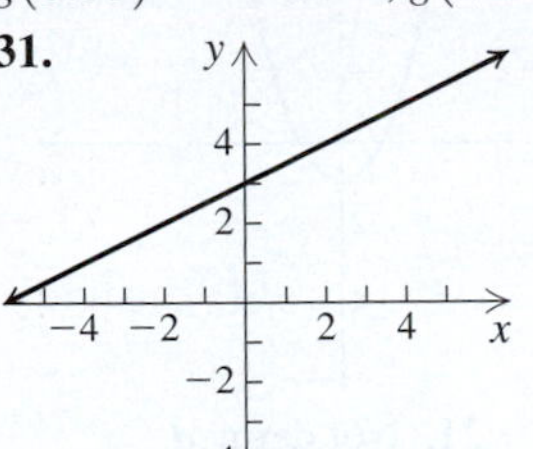

$f(x) = \frac{1}{2}x + 3$

33.

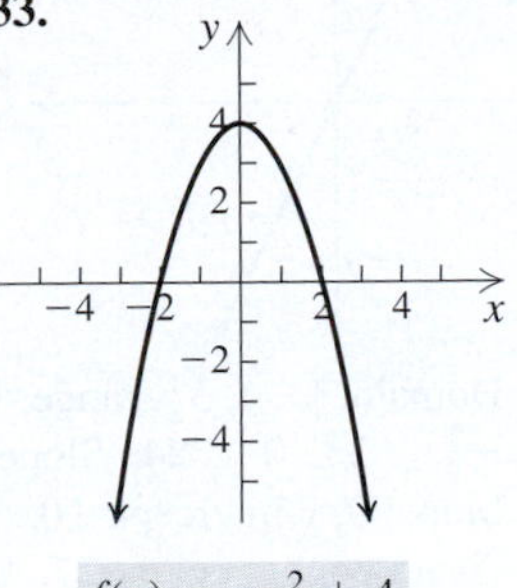

$f(x) = -x^2 + 4$

35.

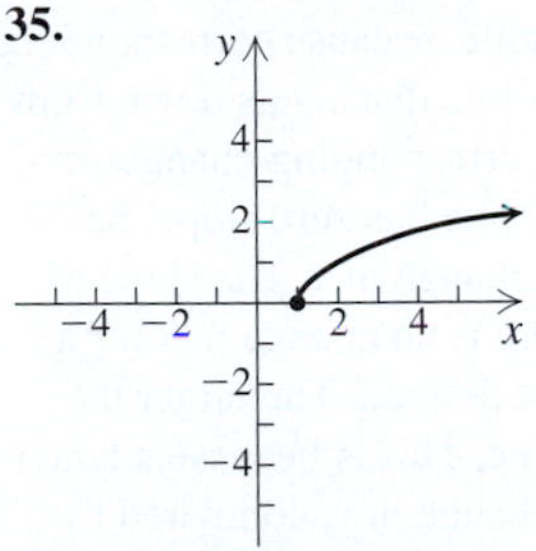

$f(x) = \sqrt{x - 1}$

37. $h(1) = -2$; $h(3) = 2$; $h(4) = 1$
39. $s(-4) = 3$; $s(-2) = 0$; $s(0) = -3$
41. $f(-1) = 2$; $f(0) = 0$; $f(1) = -2$
43. No **45.** Yes **47.** Yes **49.** No
51. All real numbers, or $(-\infty, \infty)$
53. All real numbers, or $(-\infty, \infty)$
55. $\{x|x \neq 0\}$, or $(-\infty, 0) \cup (0, \infty)$
57. $\{x|x \neq 2\}$, or $(-\infty, 2) \cup (2, \infty)$
59. $\{x|x \neq -1 \text{ and } x \neq 5\}$, or $(-\infty, -1) \cup (-1, 5) \cup (5, \infty)$
61. All real numbers, or $(-\infty, \infty)$
63. $\{x|x \neq 0 \text{ and } x \neq 7\}$, or $(-\infty, 0) \cup (0, 7) \cup (7, \infty)$
65. All real numbers, or $(-\infty, \infty)$
67. Domain: $[0, 5]$; range: $[0, 3]$
69. Domain: $[-2\pi, 2\pi]$; range: $[-1, 1]$
71. Domain: $(-\infty, \infty)$; range: $\{-3\}$
73. Domain: $[-5, 3]$; range: $[-2, 2]$
75. Domain: $(-\infty, \infty)$; range: $[0, \infty)$
77. Domain: $(-\infty, \infty)$; range: $(-\infty, \infty)$
79. Domain: $(-\infty, 3) \cup (3, \infty)$; range: $(-\infty, 0) \cup (0, \infty)$
81. Domain: $(-\infty, \infty)$; range: $(-\infty, \infty)$
83. Domain: $(-\infty, 7]$; range: $[0, \infty)$
85. Domain: $(-\infty, \infty)$; range: $(-\infty, 3]$
87. 645 m; 0 m **89.** **(a)** 2018: \$25.21; 2025: \$28.23; **(b)** about 49 years after 1985, or in 2034 **91.** $(-3, -2)$, yes; $(2, -3)$, no **92.** $\left(\frac{4}{5}, -2\right)$, yes; $\left(\frac{11}{5}, \frac{1}{10}\right)$, yes

93.

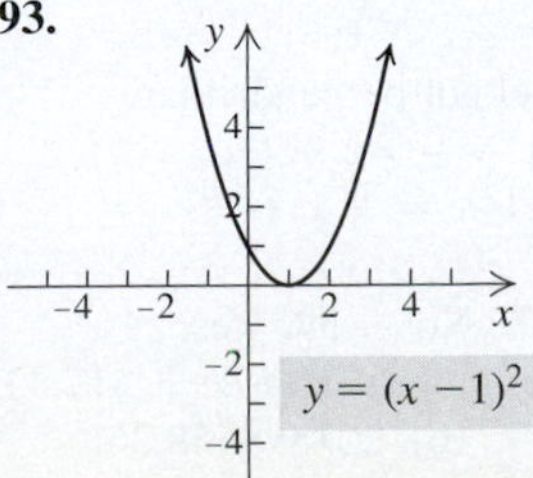

$y = (x - 1)^2$

94.

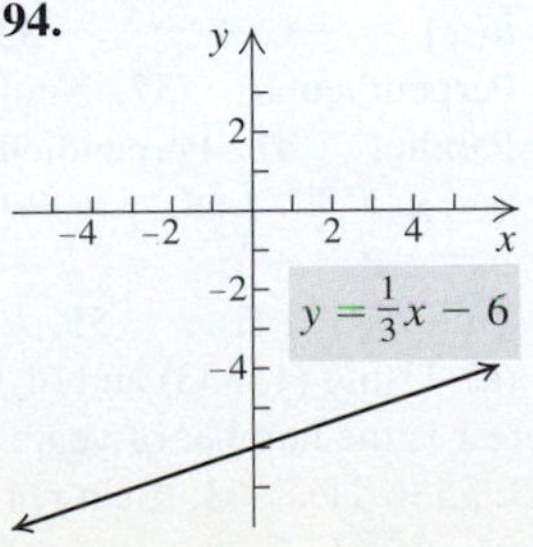

$y = \frac{1}{3}x - 6$

95.

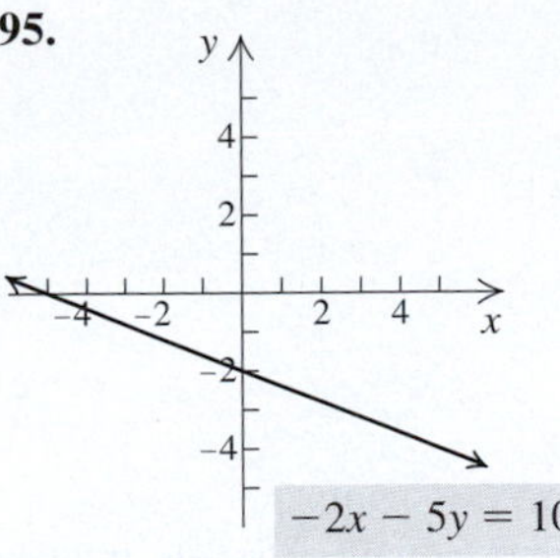

$-2x - 5y = 10$

96.

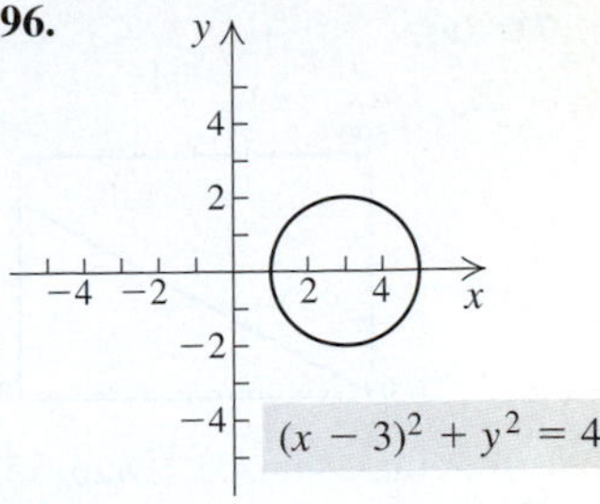

$(x - 3)^2 + y^2 = 4$

97. $\left[-\frac{5}{2}, \infty\right)$ **99.** $[-6, -2) \cup (-2, 3) \cup (3, \infty)$
101. $f(x) = x$, $g(x) = x + 1$ **103.** -7

Visualizing the Graph

1. E **2.** D **3.** A **4.** J **5.** C **6.** F **7.** H **8.** G **9.** B **10.** I

Exercise Set 1.3

1. **(a)** Yes; **(b)** yes; **(c)** yes **3.** **(a)** Yes; **(b)** no; **(c)** no **5.** $\frac{6}{5}$ **7.** $-\frac{3}{5}$ **9.** 0 **11.** $\frac{1}{5}$ **13.** Not defined **15.** 0.3 **17.** 0 **19.** $-\frac{6}{5}$ **21.** $-\frac{1}{3}$ **23.** Not defined **25.** -2 **27.** 5 **29.** 0 **31.** 1.3 **33.** Not defined **35.** $-\frac{1}{2}$ **37.** -1 **39.** 0 **41.** The average rate of change in the lowest price of a World Series ticket from 1946 to 2012 was an increase of about \$1.65 per year. **43.** The average rate of change in the population of Cleveland, Ohio, over the 13-year period was a decrease of about 6792 people per year. **45.** The average rate of change in per-capita consumption of whole milk from 1970 to 2011 was a decrease of about 0.5 gal per year. **47.** The average rate of change in the number of acres used for growing almonds in California from 2003 to 2012 was an increase of about 28,889 acres per year. **49.** $\frac{3}{5}$; $(0, -7)$ **51.** Slope is not defined; there is no y-intercept. **53.** $-\frac{1}{2}$; $(0, 5)$ **55.** $-\frac{3}{2}$; $(0, 5)$ **57.** 0; $(0, -6)$ **59.** $\frac{4}{5}$; $\left(0, \frac{8}{5}\right)$ **61.** $\frac{1}{4}$; $\left(0, -\frac{1}{2}\right)$

63.

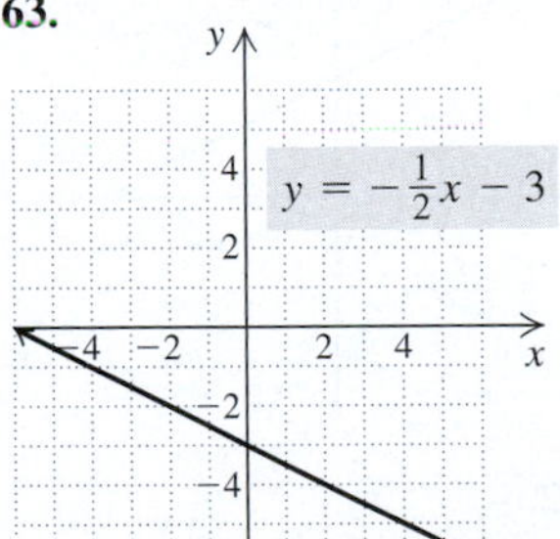

$y = -\frac{1}{2}x - 3$

65.

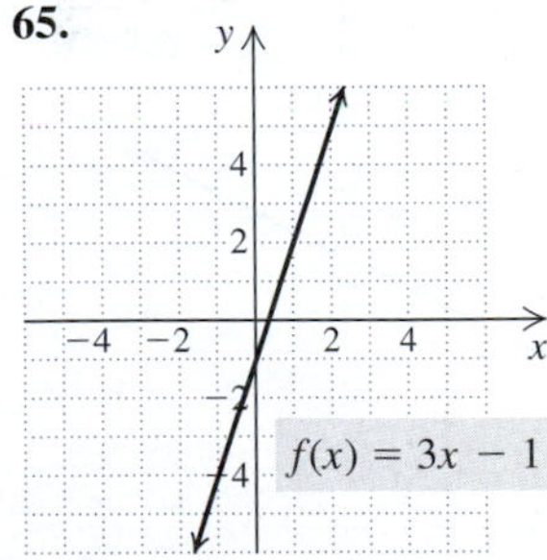

$f(x) = 3x - 1$

67.

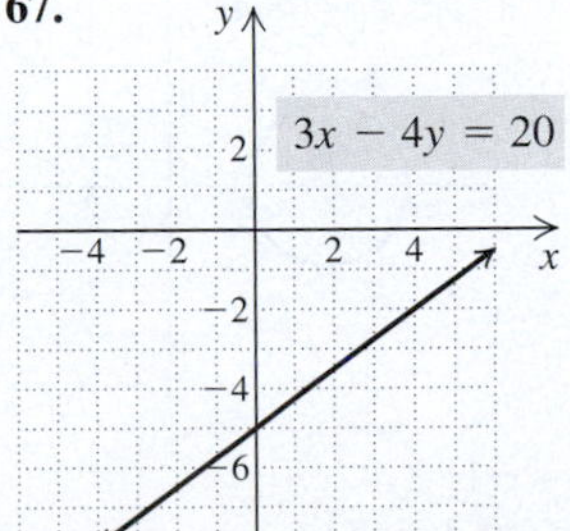

$3x - 4y = 20$

69.

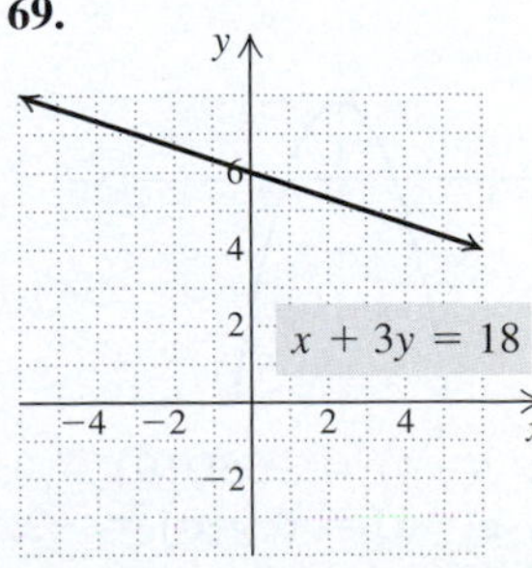

$x + 3y = 18$

71. (a) $y = \frac{1}{33}x + 1$

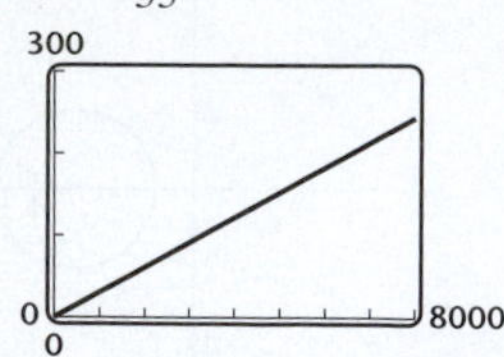

(b) 1 atm, 2 atm, $31\frac{10}{33}$ atm, $152\frac{17}{33}$ atm, $213\frac{4}{33}$ atm
73. (a) $\frac{11}{10}$. For each mile per hour faster that the car travels, it takes $\frac{11}{10}$ ft longer to stop;
(b) $y = \frac{11}{10}x + \frac{1}{2}$

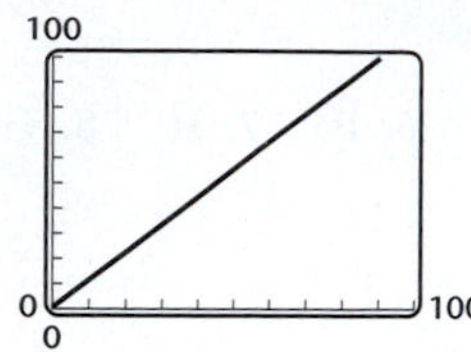

(c) 6 ft, 11.5 ft, 22.5 ft, 55.5 ft, 72 ft; **(d)** $\{r \mid r > 0\}$, or $(0, \infty)$. If r is allowed to be 0, the function says that a stopped car has a reaction distance of $\frac{1}{2}$ ft. **75.** $C(t) = 2250 + 3380t$; $C(20) = \$69{,}850$ **77.** $C(x) = 750 + 15x$; $C(32) = \$1230$ **79.** $-\frac{5}{4}$ **80.** 10 **81.** 40 **82.** $a^2 + 3a$ **83.** $a^2 + 2ah + h^2 - 3a - 3h$ **85.** $2a + h$ **87.** False **89.** $f(x) = x + b$

Mid-Chapter Mixed Review: Chapter 1

1. False **2.** True **3.** False **4.** x-intercept: $(5, 0)$; y-intercept: $(0, -8)$ **5.** $\sqrt{605} = 11\sqrt{5} \approx 24.6$; $\left(-\frac{5}{2}, -4\right)$ **6.** $\sqrt{2} \approx 1.4$; $\left(-\frac{1}{4}, -\frac{3}{10}\right)$ **7.** $(x + 5)^2 + (y - 2)^2 = 169$ **8.** Center: $(3, -1)$; radius: 2

9.

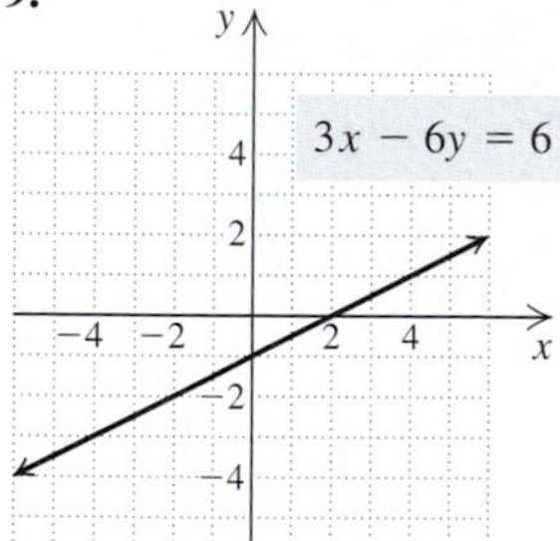

10. $y = -\frac{1}{2}x + 3$

11.

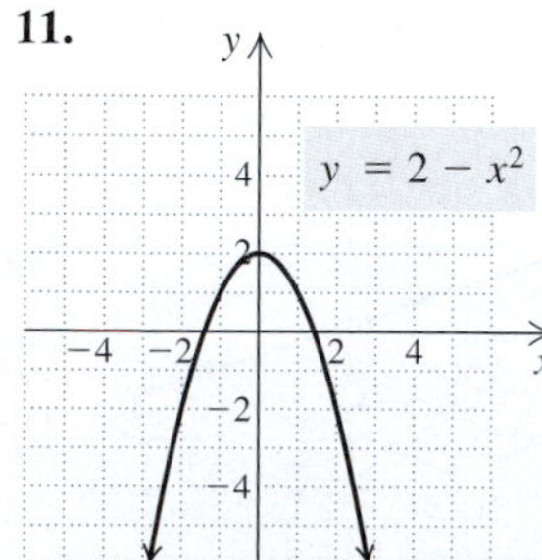

12.

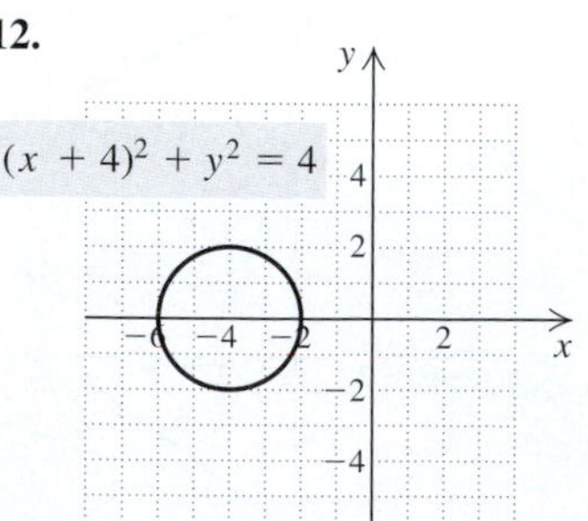

13. $f(-4) = -36$; $f(0) = 0$; $f(1) = -1$
14. $g(-6) = 0$; $g(0) = -2$; $g(3)$ is not defined
15. All real numbers, or $(-\infty, \infty)$
16. $\{x \mid x \neq -5\}$, or $(-\infty, -5) \cup (-5, \infty)$
17. $\{x \mid x \neq -3$ *and* $x \neq 1\}$, or $(-\infty, -3) \cup (-3, 1) \cup (1, \infty)$

18.

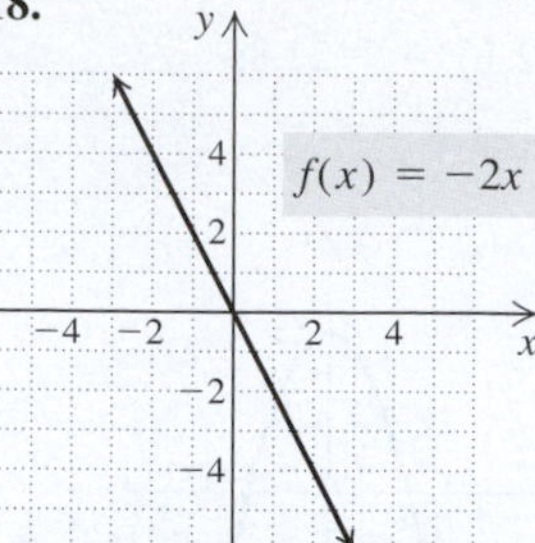

19.

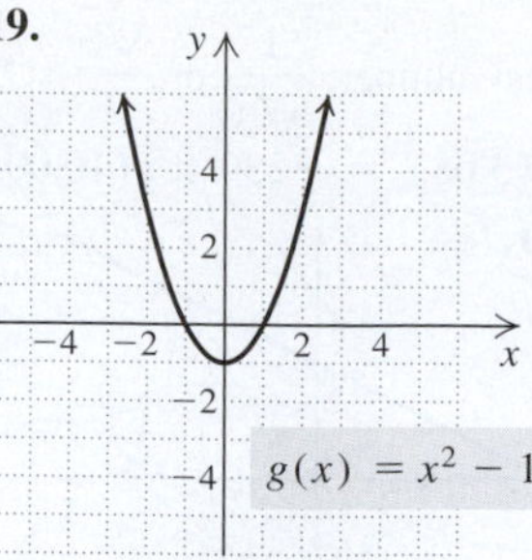

20. Domain: $[-4, 3)$; range: $[-4, 5)$ **21.** Not defined **22.** $-\frac{1}{4}$ **23.** 0 **24.** Slope: $-\frac{1}{9}$; y-intercept: $(0, 12)$ **25.** Slope: 0; y-intercept: $(0, -6)$ **26.** Slope is not defined; there is no y-intercept **27.** Slope: $\frac{3}{16}$; y-intercept: $\left(0, \frac{1}{16}\right)$ **28.** The sign of the slope indicates the slant of a line. A line that slants up from left to right has positive slope, because corresponding changes in x and y have the same sign. A line that slants down from left to right has negative slope, because corresponding changes in x and y have opposite signs. A horizontal line has zero slope, because there is no change in y for a given change in x. The slope of a vertical line is not defined, because there is no change in x for a given change in y and division by 0 is not defined. The larger the absolute value of slope, the steeper the line. This is because a larger absolute value corresponds to a greater change in y, compared to the change in x, than a smaller absolute value. **29.** A vertical line $(x = a)$ crosses the graph more than once. Thus, $x = a$ fails the vertical-line test. **30.** The domain of a function is the set of all inputs of the function. The range is the set of all outputs. The range depends on the domain. **31.** Let $A = (a, b)$ and $B = (c, d)$. The coordinates of a point C one-half of the way from A to B are $\left(\frac{a + c}{2}, \frac{b + d}{2}\right)$. A point D that is one-half of the way from C to B is $\frac{1}{2} + \frac{1}{2} \cdot \frac{1}{2}$, or $\frac{3}{4}$, of the way from A to B. Its coordinates are $\left(\frac{\frac{a + c}{2} + c}{2}, \frac{\frac{b + d}{2} + d}{2}\right)$, or $\left(\frac{a + 3c}{4}, \frac{b + 3d}{4}\right)$. Then a point E that is one-half of the way from D to B is $\frac{3}{4} + \frac{1}{2} \cdot \frac{1}{4}$, or $\frac{7}{8}$, of the way from A to B. Its coordinates are $\left(\frac{\frac{a + 3c}{4} + c}{2}, \frac{\frac{b + 3d}{4} + d}{2}\right)$, or $\left(\frac{a + 7c}{8}, \frac{b + 7d}{8}\right)$.

Exercise Set 1.4

1. 4; $(0, -2)$; $y = 4x - 2$ **3.** -1; $(0, 0)$; $y = -x$ **5.** 0; $(0, -3)$; $y = -3$ **7.** $y = \frac{2}{9}x + 4$ **9.** $y = -4x - 7$ **11.** $y = -4.2x + \frac{3}{4}$ **13.** $y = \frac{2}{9}x + \frac{19}{3}$ **15.** $y = 8$ **17.** $y = -\frac{3}{5}x - \frac{17}{5}$ **19.** $y = -3x + 2$ **21.** $y = -\frac{1}{2}x + \frac{7}{2}$ **23.** $y = \frac{2}{3}x - 6$ **25.** $y = 7.3$ **27.** Horizontal: $y = -3$; vertical: $x = 0$ **29.** Horizontal: $y = -1$; vertical: $x = \frac{2}{11}$ **31.** $h(x) = -3x + 7$; 1 **33.** $f(x) = \frac{2}{5}x - 1$; -1 **35.** Perpendicular **37.** Neither parallel nor perpendicular **39.** Parallel **41.** Perpendicular **43.** $y = \frac{2}{7}x + \frac{29}{7}$; $y = -\frac{7}{2}x + \frac{31}{2}$ **45.** $y = -0.3x - 2.1$; $y = \frac{10}{3}x + \frac{70}{3}$ **47.** $y = -\frac{3}{4}x + \frac{1}{4}$; $y = \frac{4}{3}x - 6$ **49.** $x = 3$; $y = -3$ **51.** True **53.** True **55.** False **57.** No **59.** Yes **61. (a)** Using (1, 333) and (4, 380) gives us $y = 15.67x + 317.33$, where x is the number of years after 2009; **(b)** 2018: \$458.36; 2023: \$536.71 **63.** Using (1, 32.5) and (4, 49.9) gives us

$y = 5.8x + 26.7$, where x is the number of years after 2010 and y is in billions of dollars; 2019: \$78.9 billion **65.** Using (1, 28.3) and (3, 30.8) gives us $y = 1.25x + 27.05$, where x is the number of years after 2009 and y is in gallons; 2017: about 37.1 gal **67. (a)** $y = 20.05714286x + 301.8571429$, where x is the number of years after 2009; **(b)** \$482.37; this value is \$24.01 more than the value found in Exercise 61; **(c)** $r \approx 0.9851$; the line fits the data fairly well. **69. (a)** $y = 6.47x + 23.8$, where x is the number of years after 2010 and y is in billions; **(b)** 2019: \$82.03 billion; this value is \$3.13 billion more than the value found in Exercise 63; **(c)** $r \approx 0.9915$; the line fits the data well. **71. (a)** $M = 0.2H + 156$; **(b)** 164, 169, 171, 173; **(c)** $r = 1$; the regression line fits the data perfectly and should be a good predictor. **73.** -1 **74.** Not defined **75.** $(x + 7)^2 + (y + 1)^2 = \frac{81}{25}$ **76.** $x^2 + (y - 3)^2 = 6.25$ **77.** -7.75 **79.** $y = -\frac{1}{2}x + 7$

Exercise Set 1.5

1. 4 **3.** All real numbers, or $(-\infty, \infty)$ **5.** $-\frac{3}{4}$ **7.** -9 **9.** 6 **11.** No solution **13.** $\frac{11}{5}$ **15.** $\frac{35}{6}$ **17.** 8 **19.** -4 **21.** 6 **23.** -1 **25.** $\frac{4}{5}$ **27.** $-\frac{3}{2}$ **29.** $-\frac{2}{3}$ **31.** $\frac{1}{2}$ **33.** About 51,075 words **35.** \$1300 **37.** 26°, 130°, 24° **39.** \$3.609 billion **41.** 3 hr **43.** 2.1 million tweets **45.** \$5000 **47.** About 287,000 students **49.** Length: 100 yd; width: 65 yd **51.** Length: 93 m; width: 68 m **53.** 2.5 hr **55.** \$2400 at 3%; \$2600 at 4% **57.** IBM: 6809 patents; Samsung: 4676 patents **59.** Michael Jackson: \$140 million; Elvis Presley: \$55 million **61.** 74.25 lb **63.** 4.5 hr **65.** 12 mi **67.** Italy: 30%; Spain: 20%; United States: 8% **69.** -5 **71.** $\frac{11}{2}$ **73.** 16 **75.** -12 **77.** 6 **79.** 20 **81.** 25 **83.** 15 **85. (a)** $(4, 0)$; **(b)** 4 **87. (a)** $(-2, 0)$; **(b)** -2 **89. (a)** $(-4, 0)$; **(b)** -4 **91.** $y = -\frac{3}{4}x + \frac{13}{4}$ **92.** $y = -\frac{3}{4}x + \frac{1}{4}$ **93.** 13 **94.** $\left(-1, \frac{1}{2}\right)$ **95.** $f(-3) = \frac{1}{2}$; $f(0) = 0$; $f(3)$ does not exist. **96.** $m = 7$; y-intercept: $\left(0, -\frac{1}{2}\right)$ **97.** Yes **99.** No **101.** $-\frac{2}{3}$ **103.** No; the 6-oz cup costs about 6.4% more per ounce. **105.** 11.25 mi

Exercise Set 1.6

1. $\{x|x > 5\}$, or $(5, \infty)$;

3. $\{x|x > 3\}$, or $(3, \infty)$;

5. $\{x|x \geq -3\}$, or $[-3, \infty)$;

7. $\left\{y|y \geq \frac{22}{13}\right\}$, or $\left[\frac{22}{13}, \infty\right)$;

9. $\{x|x > 6\}$, or $(6, \infty)$;

11. $\left\{x|x \geq -\frac{5}{12}\right\}$, or $\left[-\frac{5}{12}, \infty\right)$;

13. $\left\{x|x \leq \frac{15}{34}\right\}$, or $\left(-\infty, \frac{15}{34}\right]$;

15. $\{x|x < 1\}$, or $(-\infty, 1)$;

17. $\{x|x \geq 7\}$, or $[7, \infty)$ **19.** $\left\{x|x \leq \frac{1}{5}\right\}$, or $\left(-\infty, \frac{1}{5}\right]$ **21.** $\{x|x > -4\}$, or $(-4, \infty)$

23. $[-3, 3)$;

25. $[8, 10]$;

27. $[-7, -1]$;

29. $\left(-\frac{3}{2}, 2\right)$;

31. $(1, 5]$;

33. $\left(-\frac{11}{3}, \frac{13}{3}\right)$;

35. $(-\infty, -2] \cup (1, \infty)$;

37. $\left(-\infty, -\frac{7}{2}\right] \cup \left[\frac{1}{2}, \infty\right)$;

39. $(-\infty, 9.6) \cup (10.4, \infty)$;

41. $\left(-\infty, -\frac{57}{4}\right] \cup \left[-\frac{55}{4}, \infty\right)$;

43. More than 45 years after 1980 **45.** Less than 10 hr **47.** \$5000 **49.** \$300,000 **51.** Sales greater than \$18,000 **53.** Function; domain; range; domain; exactly one; range **54.** Midpoint formula **55.** x-intercept **56.** Constant; identity **57.** $\left(-\frac{1}{4}, \frac{5}{9}\right]$ **59.** $\left(-\frac{1}{8}, \frac{1}{2}\right)$

Review Exercises: Chapter 1

1. True **2.** True **3.** False **4.** False **5.** True **6.** False **7.** Yes; no **8.** Yes; no

9. x-intercept: (3, 0); y-intercept: (0, −2);

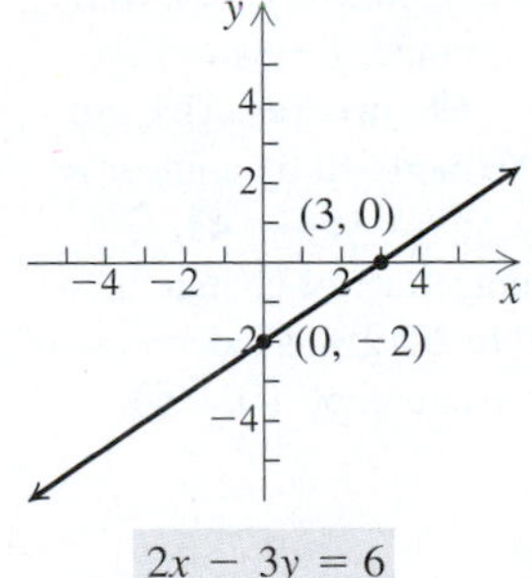

10. x-intercept: (2, 0); y-intercept: (0, 5);

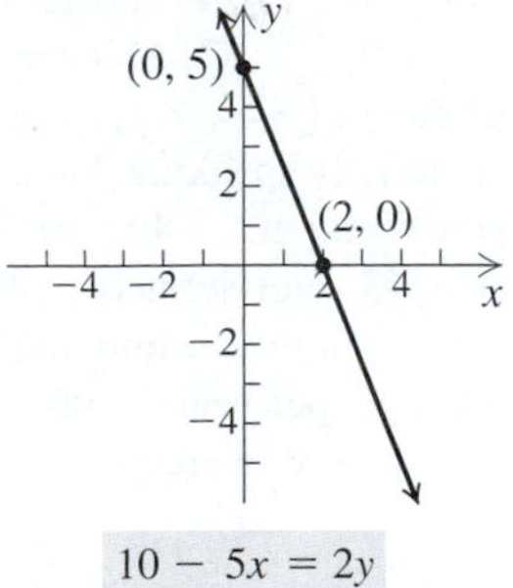

11.

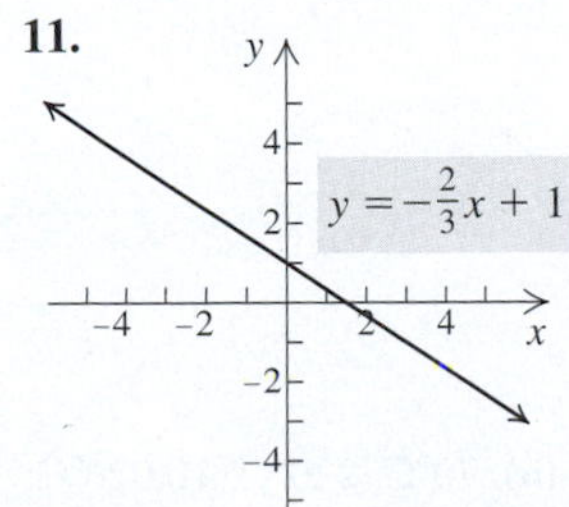

12.

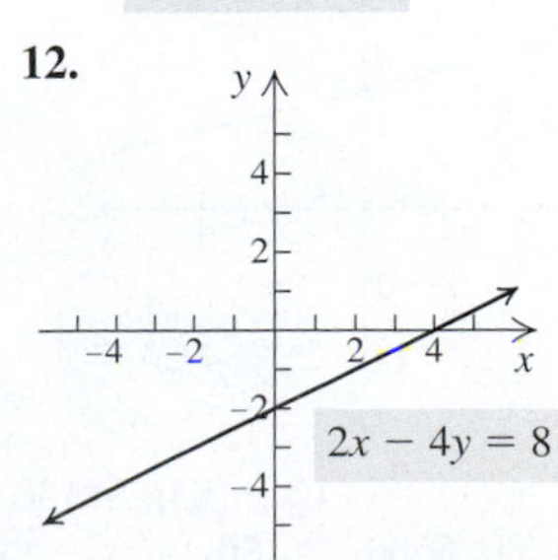

13.

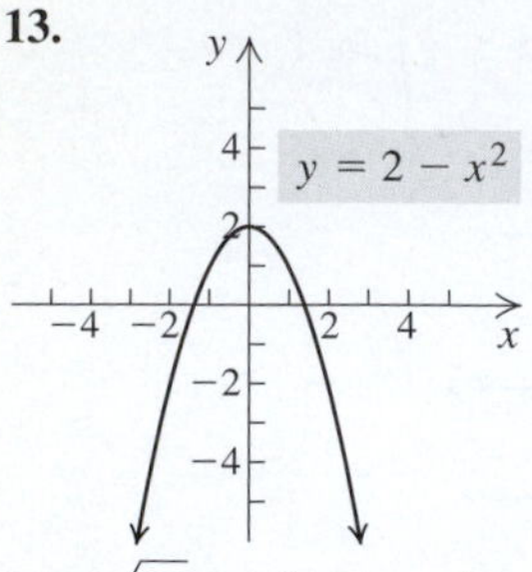

14. $\sqrt{34} \approx 5.831$ **15.** $\left(\frac{1}{2}, \frac{11}{2}\right)$
16. Center: $(-1, 3)$; radius: 3;

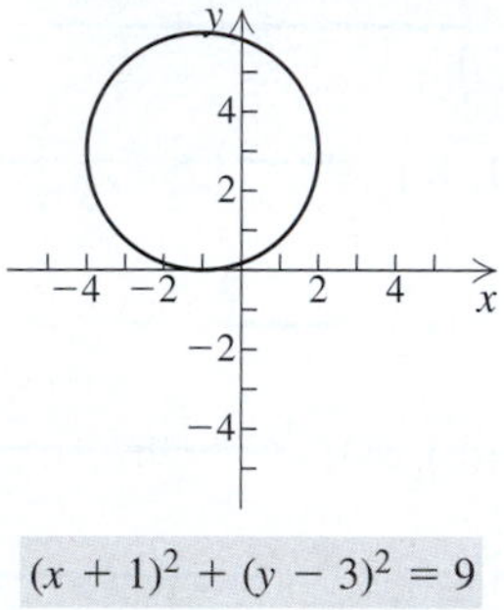

17. $x^2 + (y + 4)^2 = \frac{9}{4}$ **18.** $(x + 2)^2 + (y - 6)^2 = 13$
19. $(x - 2)^2 + (y - 4)^2 = 26$ **20.** No **21.** Yes
22. Not a function; domain: $\{3, 5, 7\}$; range: $\{1, 3, 5, 7\}$
23. Function; domain: $\{-2, 0, 1, 2, 7\}$; range: $\{-7, -4, -2, 2, 7\}$
24. **(a)** -3; **(b)** 9; **(c)** $a^2 - 3a - 1$; **(d)** $x^2 + x - 3$
25. **(a)** 0; **(b)** $\frac{x - 6}{x + 6}$; **(c)** does not exist; **(d)** $-\frac{5}{3}$
26. $f(2) = -1$; $f(-4) = -3$; $f(0) = -1$ **27.** No **28.** Yes
29. No **30.** Yes **31.** All real numbers, or $(-\infty, \infty)$
32. $\{x \mid x \neq 0\}$, or $(-\infty, 0) \cup (0, \infty)$
33. $\{x \mid x \neq 5 \textit{ and } x \neq 1\}$, or $(-\infty, 1) \cup (1, 5) \cup (5, \infty)$
34. $\{x \mid x \neq -4 \textit{ and } x \neq 4\}$, or $(-\infty, -4) \cup (-4, 4) \cup (4, \infty)$
35. Domain: $[-4, 4]$; range: $[0, 4]$ **36.** Domain: $(-\infty, \infty)$; range: $[0, \infty)$ **37.** Domain: $(-\infty, \infty)$; range: $(-\infty, \infty)$
38. Domain: $(-\infty, \infty)$; range: $[0, \infty)$ **39.** **(a)** Yes; **(b)** no; **(c)** no, strictly speaking, but data might be modeled by a linear regression function. **40.** **(a)** Yes; **(b)** yes; **(c)** yes **41.** $\frac{5}{3}$
42. 0 **43.** Not defined **44.** The average rate of change in per-capita coffee consumption from 1990 to 2012 was a decrease of about 0.1 gal per year. **45.** $m = -\frac{7}{11}$; y-intercept: $(0, -6)$
46. $m = -2$; y-intercept: $(0, -7)$
47.

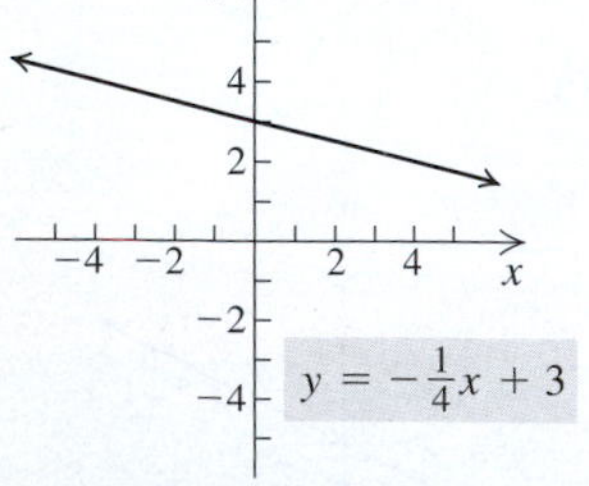

48. $C(t) = 110 + 85t$; \$1130 **49.** **(a)** 70°C, 220°C, 10,020°C; **(b)** $[0, 5600]$ **50.** $y = -\frac{2}{3}x - 4$ **51.** $y = 3x + 5$
52. $y = \frac{1}{3}x - \frac{1}{3}$ **53.** Horizontal: $y = \frac{2}{5}$; vertical: $x = -4$
54. $h(x) = 2x - 5$; -5 **55.** Parallel **56.** Neither
57. Perpendicular **58.** $y = -\frac{2}{3}x - \frac{1}{3}$ **59.** $y = \frac{3}{2}x - \frac{5}{2}$
60. **(a)** Using (2, 7969) and (8, 8576) gives us $W(x) = 101.17x + 7766.64$; 2013: 8475 female graduates; **(b)** $W(x) = 98.9x + 7747.8$, where x is the number of years after 2006; 2013: 8440 female graduates; $r \approx 0.9942$; the line fits the data well. **61.** $\frac{3}{2}$ **62.** -6 **63.** -1 **64.** -21
65. $\frac{95}{24}$ **66.** No solution **67.** All real numbers, or $(-\infty, \infty)$
68. 568 million quarters **69.** \$2300 **70.** 3.4 hr
71. 3 **72.** 4 **73.** 0.2, or $\frac{1}{5}$ **74.** 4
75. $(-\infty, 12)$;

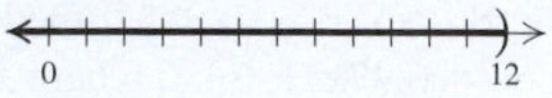

76. $(-\infty, -4]$;

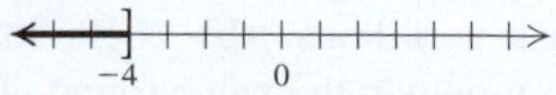

77. $\left[-\frac{4}{3}, \frac{4}{3}\right]$;

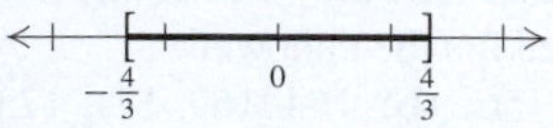

78. $\left(\frac{2}{5}, 2\right]$;

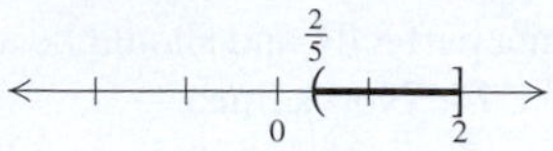

79. $\left(-\infty, -\frac{1}{2}\right) \cup (3, \infty)$;

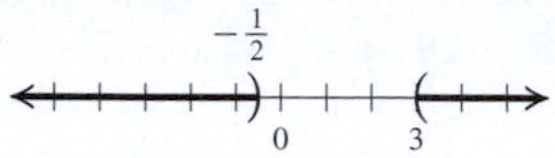

80. $\left(-\infty, -\frac{5}{3}\right] \cup [1, \infty)$;

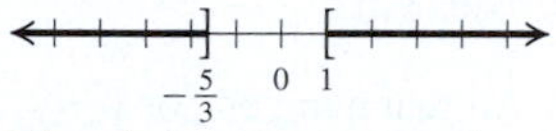

81. Years after 2019 **82.** Fahrenheit temperatures less than 113°
83. B **84.** B **85.** C **86.** $\left(\frac{5}{2}, 0\right)$
87. $\{x \mid x < 0\}$, or $(-\infty, 0)$ **88.** $\{x \mid x \neq -3 \textit{ and } x \neq 0 \textit{ and } x \neq 3\}$, or $(-\infty, -3) \cup (-3, 0) \cup (0, 3) \cup (3, \infty)$ **89.** Think of the slopes as $\frac{-3/5}{1}$ and $\frac{1/2}{1}$. The graph of $f(x)$ changes $\frac{3}{5}$ unit vertically for each unit of horizontal change, whereas the graph of $g(x)$ changes $\frac{1}{2}$ unit vertically for each unit of horizontal change. Since $\frac{3}{5} > \frac{1}{2}$, the graph of $f(x) = -\frac{3}{5}x + 4$ is steeper than the graph of $g(x) = \frac{1}{2}x - 6$.
90. If an equation contains no fractions, using the addition principle before using the multiplication principle eliminates the need to add or subtract fractions. **91.** The solution set of a disjunction is a union of sets, so it is only possible for a disjunction to have no solution when the solution set of each inequality is the empty set.
92. The graph of $f(x) = mx + b$, $m \neq 0$, is a straight line that is not horizontal. The graph of such a line intersects the x-axis exactly once. Thus the function has exactly one zero. **93.** By definition, the notation $3 < x < 4$ indicates that $3 < x$ *and* $x < 4$. The disjunction $x < 3$ *or* $x > 4$ cannot be written $3 > x > 4$, or $4 < x < 3$, because it is not possible for x to be greater than 4 *and* less than 3. **94.** A function is a correspondence between two sets in which each member of the first set corresponds to exactly one member of the second set.

Test: Chapter 1

1. [1.1] Yes **2.** [1.1] x-intercept: $(-2, 0)$; y-intercept: $(0, 5)$;

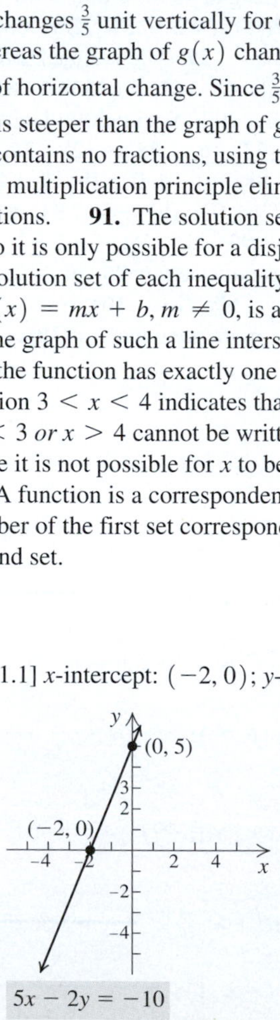

3. [1.1] $\sqrt{45} \approx 6.708$ **4.** [1.1] $\left(-3, \frac{9}{2}\right)$ **5.** [1.1] Center: $(-4, 5)$; radius: 6 **6.** [1.1] $(x + 1)^2 + (y - 2)^2 = 5$
7. [1.2] **(a)** Yes; **(b)** $\{-4, 3, 1, 0\}$; **(c)** $\{7, 0, 5\}$

8. [1.2] **(a)** 8; **(b)** $2a^2 + 7a + 11$ **9.** [1.2] **(a)** Does not exist; **(b)** 0 **10.** [1.2] 0 **11.** [1.2] **(a)** No; **(b)** yes **12.** [1.2] $\{x \mid x \neq 4\}$, or $(-\infty, 4) \cup (4, \infty)$ **13.** [1.2] All real numbers, or $(-\infty, \infty)$ **14.** [1.2] $\{x \mid -5 \leq x \leq 5\}$, or $[-5, 5]$ **15.** [1.2] **(a)**

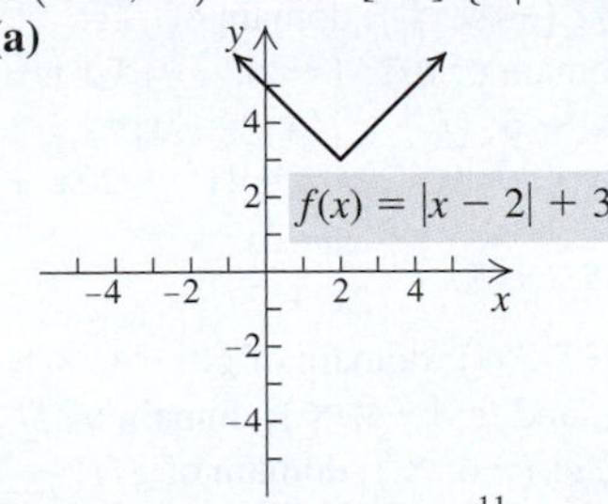

(b) $(-\infty, \infty)$; **(c)** $[3, \infty)$

16. [1.3] Not defined **17.** [1.3] $-\frac{11}{6}$ **18.** [1.3] 0 **19.** [1.3] The average rate of change in the percent of adults who are married for the years from 1960 to 2012 was a decrease of about 0.4% per year. **20.** [1.3] Slope: $\frac{3}{2}$; y-intercept: $\left(0, \frac{5}{2}\right)$ **21.** [1.3] $C(t) = 65 + 48t$; \$173 **22.** [1.4] $y = -\frac{5}{8}x - 5$ **23.** [1.4] $y - 4 = -\frac{3}{4}(x - (-5))$, or $y - (-2) = -\frac{3}{4}(x - 3)$, or $y = -\frac{3}{4}x + \frac{1}{4}$ **24.** [1.4] $x = -\frac{3}{8}$ **25.** [1.4] Perpendicular **26.** [1.4] $y - 3 = -\frac{1}{2}(x + 1)$, or $y = -\frac{1}{2}x + \frac{5}{2}$ **27.** [1.4] $y - 3 = 2(x + 1)$, or $y = 2x + 5$ **28.** [1.4] **(a)** Using (2, 544.05) and (8, 653.19) gives us $y = 18.19x + 507.67$, where x is the number of years after 2003; 2017: \$762.33; **(b)** $y = 16.47728571x + 517.2819048$; 2017: \$747.96; $r \approx 0.9971$ **29.** [1.5] -1 **30.** [1.5] All real numbers, or $(-\infty, \infty)$ **31.** [1.5] -60 **32.** [1.5] $\frac{21}{11}$ **33.** [1.5] Length: 60 m; width: 45 m **34.** [1.5] \$1.80 **35.** [1.5] -3

36. [1.6] $(-\infty, -3]$;

37. [1.6] $(-5, 3)$;

38. [1.6] $(-\infty, 2] \cup [4, \infty)$;

39. [1.6] More than 6 hr **40.** [1.3] B **41.** [1.2] -2

Chapter 2

Exercise Set 2.1

1. **(a)** $(-5, 1)$; **(b)** $(3, 5)$; **(c)** $(1, 3)$ **3.** **(a)** $(-3, -1)$, $(3, 5)$; **(b)** $(1, 3)$; **(c)** $(-5, -3)$ **5.** **(a)** $(-\infty, -8)$, $(-3, -2)$; **(b)** $(-8, -6)$; **(c)** $(-6, -3)$, $(-2, \infty)$ **7.** Domain: $[-5, 5]$; range: $[-3, 3]$ **9.** Domain: $[-5, -1] \cup [1, 5]$; range: $[-4, 6]$ **11.** Domain: $(-\infty, \infty)$; range: $(-\infty, 3]$ **13.** Relative maximum: 3.25 at $x = 2.5$; increasing: $(-\infty, 2.5)$; decreasing: $(2.5, \infty)$ **15.** Relative maximum: 2.370 at $x = -0.667$; relative minimum: 0 at $x = 2$; increasing: $(-\infty, -0.667)$, $(2, \infty)$; decreasing: $(-0.667, 2)$ **17.** Increasing: $(0, \infty)$; decreasing: $(-\infty, 0)$; relative minimum: 0 at $x = 0$ **19.** Increasing: $(-\infty, 0)$; decreasing: $(0, \infty)$; relative maximum: 5 at $x = 0$ **21.** Increasing: $(3, \infty)$; decreasing: $(-\infty, 3)$; relative minimum: 1 at $x = 3$ **23.** Increasing: $(1, 3)$; decreasing: $(-\infty, 1)$, $(3, \infty)$; relative maximum: -4 at $x = 3$; relative minimum: -8 at $x = 1$ **25.** Increasing: $(-1.552, 0)$, $(1.552, \infty)$; decreasing: $(-\infty, -1.552)$, $(0, 1.552)$; relative maximum: 4.07 at $x = 0$; relative minima: -2.314 at $x = -1.552$, -2.314 at $x = 1.552$

27. **(a)** $y = -x^2 + 300x + 6$; **(b)** 22,506 fruit trees

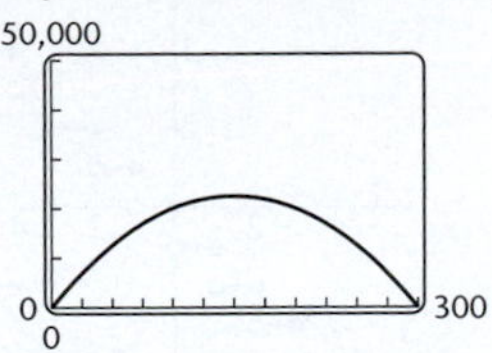

(c) When \$150 thousand is spent on advertising, 22,506 fruit trees will be sold. **29.** Increasing: $(-1, 1)$; decreasing: $(-\infty, -1)$, $(1, \infty)$ **31.** Increasing: $(-1.414, 1.414)$; decreasing: $(-2, -1.414)$, $(1.414, 2)$ **33.** $A(x) = x(240 - x)$, or $240x - x^2$ **35.** $h(d) = \sqrt{d^2 - 3500^2}$

37. $A(w) = 10w - \dfrac{w^2}{2}$ **39.** $d(s) = \dfrac{14}{s}$

41. **(a)** $A(x) = x(240 - 4x)$, or $240x - 4x^2$; **(b)** $\{x \mid 0 < x < 60\}$; **(c)** 120 ft by 30 ft **43.** **(a)** $V(x) = x(12 - 2x)(12 - 2x)$, or $4x(6 - x)^2$; **(b)** $\{x \mid 0 < x < 6\}$; **(c)** $y = 4x(6 - x)^2$;

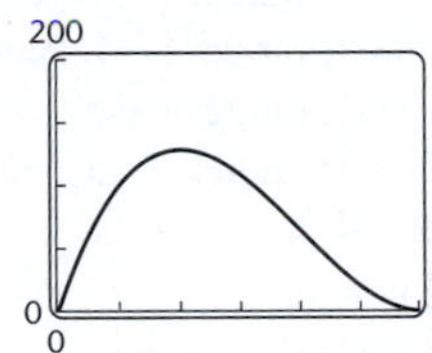

(d) 8 cm by 8 cm by 2 cm **45.** **(a)** $A(x) = x\sqrt{256 - x^2}$; **(b)** $\{x \mid 0 < x < 16\}$; **(c)** $y = x\sqrt{256 - x^2}$

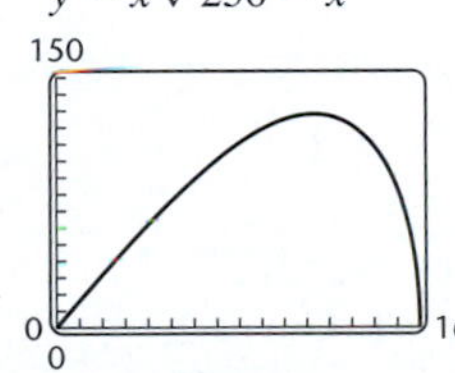

(d) 11.314 ft by 11.314 ft **47.** $g(-4) = 0$; $g(0) = 4$; $g(1) = 5$; $g(3) = 5$ **49.** $h(-6) = 0$; $h(0) = 1$; $h(1) = 3$; $h(4) = 6$

51.

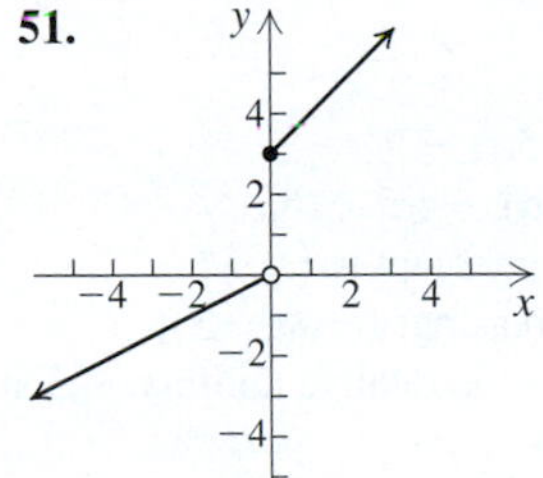

53.

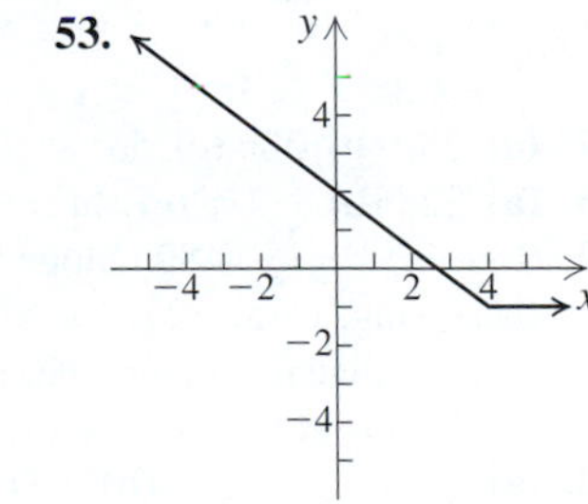

55.

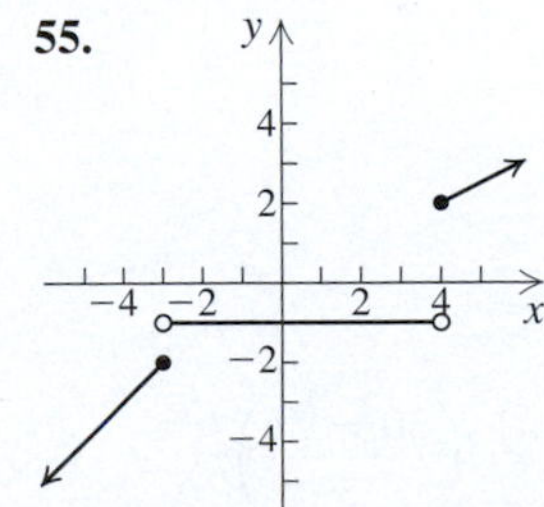

57.

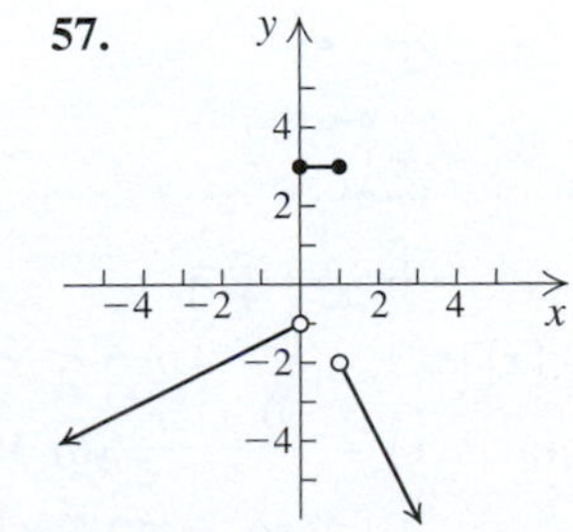

59.

61.

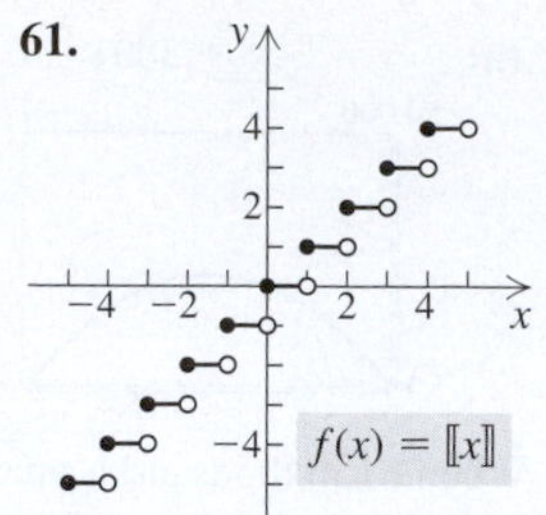

63.

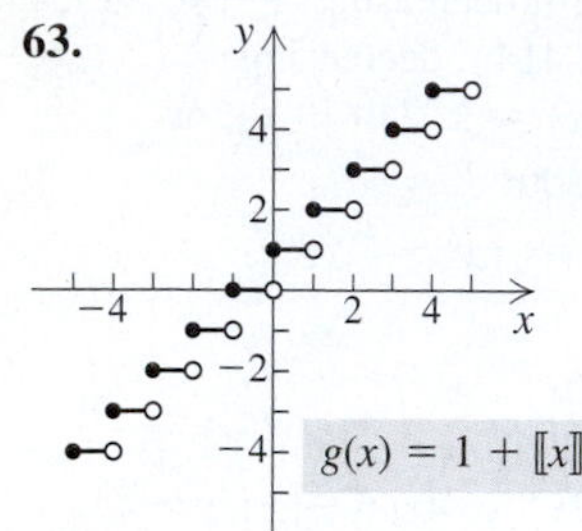

65. Domain: $(-\infty, \infty)$; range: $(-\infty, 0] \cup [3, \infty)$
67. Domain: $(-\infty, \infty)$; range: $[-1, \infty)$
69. Domain: $(-\infty, \infty)$; range: $(-\infty, -2] \cup \{-1\} \cup [2, \infty)$
71. Domain: $(-\infty, \infty)$; range: $\{-5, -2, 4\}$;

$$f(x) = \begin{cases} -2, & \text{for } x < 2, \\ -5, & \text{for } x = 2, \\ 4, & \text{for } x > 2 \end{cases}$$

73. Domain: $(-\infty, \infty)$; range: $(-\infty, -1] \cup [2, \infty)$;

$$g(x) = \begin{cases} x, & \text{for } x \le -1, \\ 2, & \text{for } -1 < x < 2, \\ x, & \text{for } x \ge 2 \end{cases}$$

or

$$g(x) = \begin{cases} x, & \text{for } x \le -1, \\ 2, & \text{for } -1 < x \le 2, \\ x, & \text{for } x > 2 \end{cases}$$

75. Domain: $[-5, 3]$; range: $(-3, 5)$;

$$h(x) = \begin{cases} x + 8, & \text{for } -5 \le x < -3, \\ 3, & \text{for } -3 \le x \le 1, \\ 3x - 6, & \text{for } 1 < x \le 3 \end{cases}$$

77. **(a)** 38; **(b)** 38; **(c)** $5a^2 - 7$; **(d)** $5a^2 - 7$
78. **(a)** 22; **(b)** -22; **(c)** $4a^3 - 5a$; **(d)** $-4a^3 + 5a$
79. $y = -\frac{1}{8}x + \frac{7}{8}$ **80.** Slope is $\frac{2}{9}$; y-intercept is $\left(0, \frac{1}{9}\right)$.
81. Increasing: $(-5, -2)$, $(4, \infty)$; decreasing: $(-\infty, -5)$, $(-2, 4)$; relative maximum: 560 at $x = -2$; relative minima: 425 at $x = -5$, -304 at $x = 4$
83. **(a)**

(b) $C(t) = 3(⟦t⟧ + 1), t > 0$

85. $\{x \mid -5 \le x < -4 \text{ or } 5 \le x < 6\}$
87. **(a)** $h(r) = \dfrac{30 - 5r}{3}$; **(b)** $V(r) = \pi r^2\left(\dfrac{30 - 5r}{3}\right)$;
(c) $V(h) = \pi h\left(\dfrac{30 - 3h}{5}\right)^2$

Exercise Set 2.2

1. 33 **3.** -1 **5.** Does not exist **7.** 0 **9.** 1 **11.** Does not exist **13.** 0 **15.** 5 **17.** **(a)** Domain of f, g, $f + g$, $f - g$, fg, and ff: $(-\infty, \infty)$; domain of f/g: $\left(-\infty, \frac{3}{5}\right) \cup \left(\frac{3}{5}, \infty\right)$; domain of g/f: $\left(-\infty, -\frac{3}{2}\right) \cup \left(-\frac{3}{2}, \infty\right)$; **(b)** $(f + g)(x) = -3x + 6$; $(f - g)(x) = 7x$; $(fg)(x) = -10x^2 - 9x + 9$; $(ff)(x) = 4x^2 + 12x + 9$; $(f/g)(x) = \dfrac{2x + 3}{3 - 5x}$; $(g/f)(x) = \dfrac{3 - 5x}{2x + 3}$
19. **(a)** Domain of f: $(-\infty, \infty)$; domain of g: $[-4, \infty)$; domain of $f + g$, $f - g$, and fg: $[-4, \infty)$; domain of ff: $(-\infty, \infty)$; domain of f/g: $(-4, \infty)$; domain of g/f: $[-4, 3) \cup (3, \infty)$; **(b)** $(f + g)(x) = x - 3 + \sqrt{x + 4}$; $(f - g)(x) = x - 3 - \sqrt{x + 4}$; $(fg)(x) = (x - 3)\sqrt{x + 4}$; $(ff)(x) = x^2 - 6x + 9$; $(f/g)(x) = \dfrac{x - 3}{\sqrt{x + 4}}$; $(g/f)(x) = \dfrac{\sqrt{x + 4}}{x - 3}$ **21.** **(a)** Domain of f, g, $f + g$, $f - g$, fg, and ff: $(-\infty, \infty)$; domain of f/g: $(-\infty, 0) \cup (0, \infty)$; domain of g/f: $\left(-\infty, \frac{1}{2}\right) \cup \left(\frac{1}{2}, \infty\right)$ **(b)** $(f + g)(x) = -2x^2 + 2x - 1$; $(f - g)(x) = 2x^2 + 2x - 1$; $(fg)(x) = -4x^3 + 2x^2$; $(ff)(x) = 4x^2 - 4x + 1$; $(f/g)(x) = \dfrac{2x - 1}{-2x^2}$; $(g/f)(x) = \dfrac{-2x^2}{2x - 1}$ **23.** **(a)** Domain of f: $[3, \infty)$; domain of g: $[-3, \infty)$; domain of $f + g$, $f - g$, fg, and ff: $[3, \infty)$; domain of f/g: $[3, \infty)$; domain of g/f: $(3, \infty)$; **(b)** $(f + g)(x) = \sqrt{x - 3} + \sqrt{x + 3}$; $(f - g)(x) = \sqrt{x - 3} - \sqrt{x + 3}$; $(fg)(x) = \sqrt{x^2 - 9}$; $(ff)(x) = |x - 3|$; $(f/g)(x) = \dfrac{\sqrt{x - 3}}{\sqrt{x + 3}}$; $(g/f)(x) = \dfrac{\sqrt{x + 3}}{\sqrt{x - 3}}$ **25.** **(a)** Domain of f, g, $f + g$, $f - g$, fg, and ff: $(-\infty, \infty)$; domain of f/g: $(-\infty, 0) \cup (0, \infty)$; domain of g/f: $(-\infty, -1) \cup (-1, \infty)$; **(b)** $(f + g)(x) = x + 1 + |x|$; $(f - g)(x) = x + 1 - |x|$; $(fg)(x) = (x + 1)|x|$; $(ff)(x) = x^2 + 2x + 1$; $(f/g)(x) = \dfrac{x + 1}{|x|}$; $(g/f)(x) = \dfrac{|x|}{x + 1}$
27. **(a)** Domain of f, g, $f + g$, $f - g$, fg, and ff: $(-\infty, \infty)$; domain of f/g: $(-\infty, -3) \cup \left(-3, \frac{1}{2}\right) \cup \left(\frac{1}{2}, \infty\right)$; domain of g/f: $(-\infty, 0) \cup (0, \infty)$; **(b)** $(f + g)(x) = x^3 + 2x^2 + 5x - 3$; $(f - g)(x) = x^3 - 2x^2 - 5x + 3$; $(fg)(x) = 2x^5 + 5x^4 - 3x^3$; $(ff)(x) = x^6$; $(f/g)(x) = \dfrac{x^3}{2x^2 + 5x - 3}$; $(g/f)(x) = \dfrac{2x^2 + 5x - 3}{x^3}$ **29.** **(a)** Domain of f: $(-\infty, -1) \cup (-1, \infty)$; domain of g: $(-\infty, 6) \cup (6, \infty)$; domain of $f + g$, $f - g$, and fg: $(-\infty, -1) \cup (-1, 6) \cup (6, \infty)$; domain of ff: $(-\infty, -1) \cup (-1, \infty)$; domain of f/g and g/f: $(-\infty, -1) \cup (-1, 6) \cup (6, \infty)$;
(b) $(f + g)(x) = \dfrac{4}{x + 1} + \dfrac{1}{6 - x}$;

$(f - g)(x) = \frac{4}{x + 1} - \frac{1}{6 - x}$; $(fg)(x) = \frac{4}{(x + 1)(6 - x)}$;
$(ff)(x) = \frac{16}{(x + 1)^2}$; $(f/g)(x) = \frac{4(6 - x)}{x + 1}$;
$(g/f)(x) = \frac{x + 1}{4(6 - x)}$
31. (a) Domain of f: $(-\infty, 0) \cup (0, \infty)$; domain of g: $(-\infty, \infty)$; domain of $f + g$, $f - g$, fg, and ff: $(-\infty, 0) \cup (0, \infty)$; domain of f/g: $(-\infty, 0) \cup (0, 3) \cup (3, \infty)$; domain of g/f: $(-\infty, 0) \cup (0, \infty)$; **(b)** $(f + g)(x) = \frac{1}{x} + x - 3$;
$(f - g)(x) = \frac{1}{x} - x + 3$; $(fg)(x) = 1 - \frac{3}{x}$; $(ff)(x) = \frac{1}{x^2}$;
$(f/g)(x) = \frac{1}{x(x - 3)}$; $(g/f)(x) = x(x - 3)$
33. (a) Domain of f: $(-\infty, 2) \cup (2, \infty)$; domain of g: $[1, \infty)$; domain of $f + g$, $f - g$, and fg: $[1, 2) \cup (2, \infty)$; domain of ff: $(-\infty, 2) \cup (2, \infty)$; domain of f/g: $(1, 2) \cup (2, \infty)$; domain of g/f: $[1, 2) \cup (2, \infty)$; **(b)** $(f + g)(x) = \frac{3}{x - 2} + \sqrt{x - 1}$;
$(f - g)(x) = \frac{3}{x - 2} - \sqrt{x - 1}$;
$(fg)(x) = \frac{3\sqrt{x - 1}}{x - 2}$; $(ff)(x) = \frac{9}{(x - 2)^2}$;
$(f/g)(x) = \frac{3}{(x - 2)\sqrt{x - 1}}$; $(g/f)(x) = \frac{(x - 2)\sqrt{x - 1}}{3}$
35. Domain of F: $[2, 11]$; domain of G: $[1, 9]$; domain of $F + G$: $[2, 9]$ **37.** $[2, 3) \cup (3, 9]$
39.

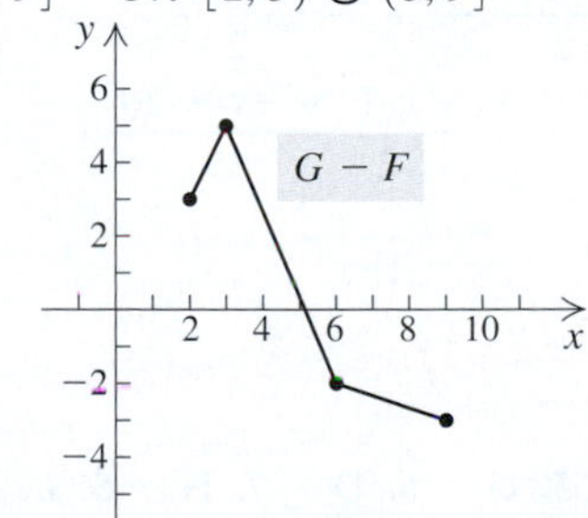

41. Domain of F: $[0, 9]$; domain of G: $[3, 10]$; domain of $F + G$: $[3, 9]$ **43.** $[3, 6) \cup (6, 8) \cup (8, 9]$
45.

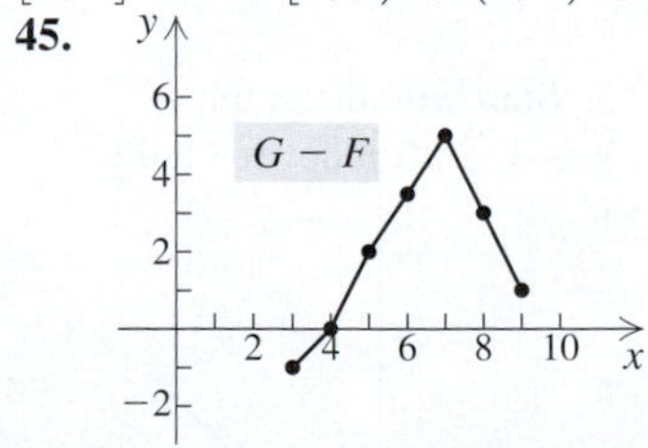

47. (a) $P(x) = -0.4x^2 + 57x - 13$; **(b)** $R(100) = 2000$; $C(100) = 313$; $P(100) = 1687$; **(c)** Left to the student
49. 3 **51.** 6 **53.** $\frac{1}{3}$ **55.** $\frac{-1}{3x(x + h)}$, or $-\frac{1}{3x(x + h)}$
57. $\frac{1}{4x(x + h)}$ **59.** $2x + h$ **61.** $-2x - h$
63. $6x + 3h - 2$ **65.** $\frac{5|x + h| - 5|x|}{h}$
67. $3x^2 + 3xh + h^2$ **69.** $\frac{7}{(x + h + 3)(x + 3)}$

71.

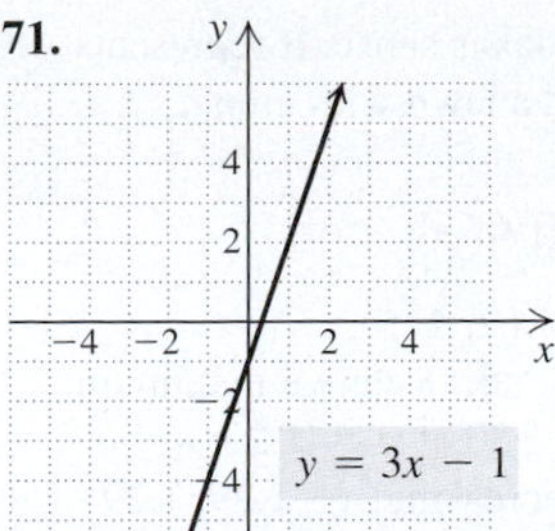

72.

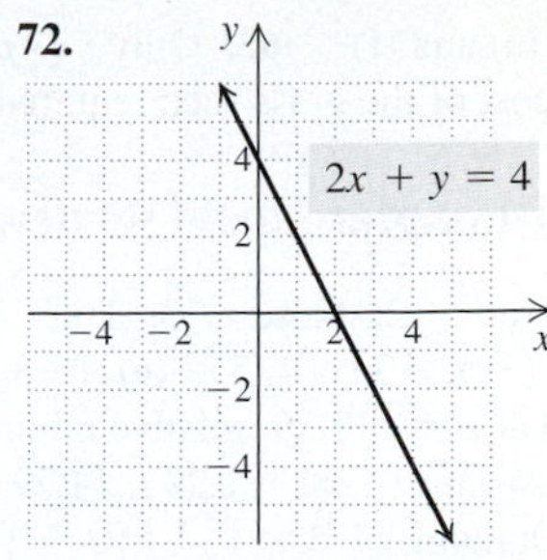

73.

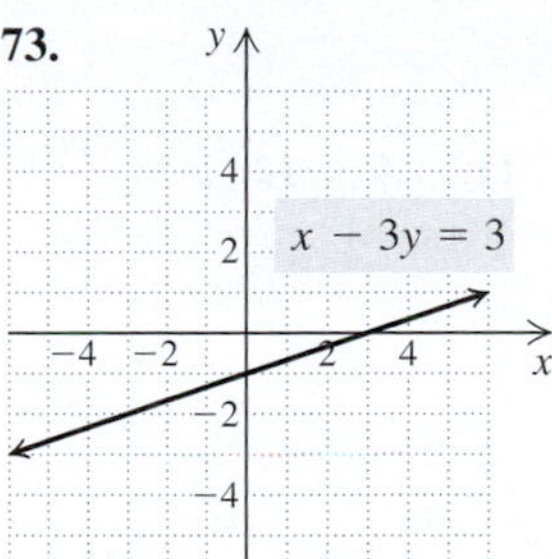

74.

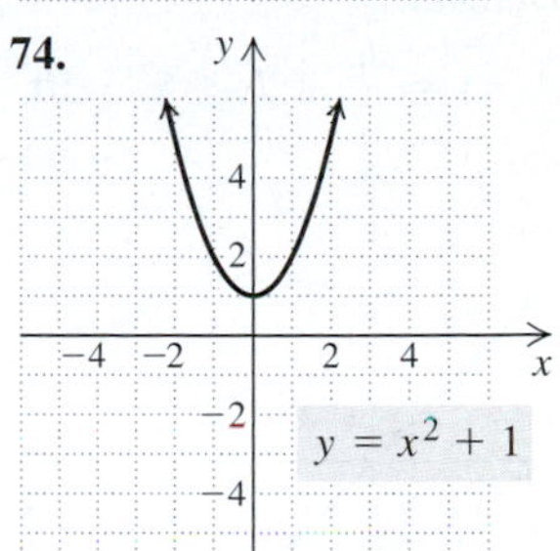

75. $f(x) = \frac{1}{x + 7}$, $g(x) = \frac{1}{x - 3}$; answers may vary
77. $(-\infty, -1) \cup (-1, 1) \cup \left(1, \frac{7}{3}\right) \cup \left(\frac{7}{3}, 3\right) \cup (3, \infty)$

Exercise Set 2.3

1. −8 **3.** 64 **5.** 218 **7.** −80 **9.** −6 **11.** 512
13. −32 **15.** x^9 **17.** $(f \circ g)(x) = (g \circ f)(x) = x$; domain of $f \circ g$ and $g \circ f$: $(-\infty, \infty)$ **19.** $(f \circ g)(x) = 3x^2 - 2x$; $(g \circ f)(x) = 3x^2 + 4x$; domain of $f \circ g$ and $g \circ f$: $(-\infty, \infty)$
21. $(f \circ g)(x) = 16x^2 - 24x + 6$; $(g \circ f)(x) = 4x^2 - 15$; domain of $f \circ g$ and $g \circ f$: $(-\infty, \infty)$
23. $(f \circ g)(x) = \frac{4x}{x - 5}$; $(g \circ f)(x) = \frac{1 - 5x}{4}$; domain of $f \circ g$: $(-\infty, 0) \cup (0, 5) \cup (5, \infty)$; domain of $g \circ f$: $\left(-\infty, \frac{1}{5}\right) \cup \left(\frac{1}{5}, \infty\right)$
25. $(f \circ g)(x) = (g \circ f)(x) = x$; domain of $f \circ g$ and $g \circ f$: $(-\infty, \infty)$ **27.** $(f \circ g)(x) = 2\sqrt{x} + 1$; $(g \circ f)(x) = \sqrt{2x + 1}$; domain of $f \circ g$: $[0, \infty)$; domain of $g \circ f$: $\left[-\frac{1}{2}, \infty\right)$
29. $(f \circ g)(x) = 20$; $(g \circ f)(x) = 0.05$; domain of $f \circ g$ and $g \circ f$: $(-\infty, \infty)$ **31.** $(f \circ g)(x) = |x|$; $(g \circ f)(x) = x$; domain of $f \circ g$: $(-\infty, \infty)$; domain of $g \circ f$: $[-5, \infty)$ **33.** $(f \circ g)(x) = 5 - x$; $(g \circ f)(x) = \sqrt{1 - x^2}$; domain of $f \circ g$: $(-\infty, 3]$; domain of $g \circ f$: $[-1, 1]$ **35.** $(f \circ g)(x) = (g \circ f)(x) = x$; domain of $f \circ g$: $(-\infty, -1) \cup (-1, \infty)$; domain of $g \circ f$: $(-\infty, 0) \cup (0, \infty)$
37. $(f \circ g)(x) = x^3 - 2x^2 - 4x + 6$; $(g \circ f)(x) = x^3 - 5x^2 + 3x + 8$; domain of $f \circ g$ and $g \circ f$: $(-\infty, \infty)$ **39.** $f(x) = x^5$; $g(x) = 4 + 3x$
41. $f(x) = \frac{1}{x}$; $g(x) = (x - 2)^4$ **43.** $f(x) = \frac{x - 1}{x + 1}$; $g(x) = x^3$ **45.** $f(x) = x^6$; $g(x) = \frac{2 + x^3}{2 - x^3}$
47. $f(x) = \sqrt{x}$; $g(x) = \frac{x - 5}{x + 2}$
49. $f(x) = x^3 - 5x^2 + 3x - 1$; $g(x) = x + 2$
51. (a) $r(t) = 3t$; **(b)** $A(r) = \pi r^2$; **(c)** $(A \circ r)(t) = 9\pi t^2$; the function gives the area of the ripple in terms of time t.
53. $f(x) = x + 1$ **55.** (c) **56.** None **57.** (b), (d), (f), and (h) **58.** (b) **59.** (a) **60.** (c) and (g) **61.** (c) and (g)

62. (a) and (f) **63.** Only $(c \circ p)(a)$ makes sense. It represents the cost of the grass seed required to seed a lawn with area a.

Mid-Chapter Mixed Review: Chapter 2

1. True **2.** False **3.** True **4. (a)** $(2, 4)$; **(b)** $(-5, -3), (4, 5)$; **(c)** $(-3, -1)$ **5.** Relative maximum: 6.30 at $x = -1.29$; relative minimum: -2.30 at $x = 1.29$; increasing: $(-\infty, -1.29), (1.29, \infty)$; decreasing: $(-1.29, 1.29)$ **6.** Domain: $[-5, -1] \cup [2, 5]$; range: $[-3, 5]$ **7.** $A(h) = \dfrac{h^2}{2} + h$ **8.** -10; -8; 1; 3

9.

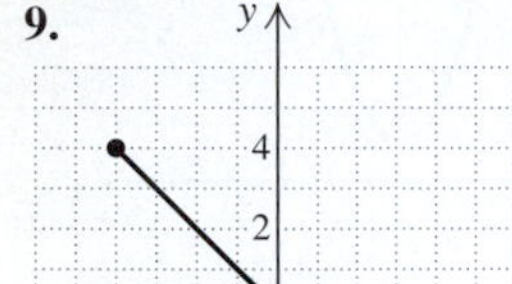

10. 1 **11.** -4 **12.** 5 **13.** Does not exist

14. (a) Domain of f, g, $f + g$, $f - g$, fg, and ff: $(-\infty, \infty)$; domain of f/g: $(-\infty, -4) \cup (-4, \infty)$; domain of g/f: $\left(-\infty, -\frac{5}{2}\right) \cup \left(-\frac{5}{2}, \infty\right)$; **(b)** $(f + g)(x) = x + 1$; $(f - g)(x) = 3x + 9$; $(fg)(x) = -2x^2 - 13x - 20$; $(ff)(x) = 4x^2 + 20x + 25$; $(f/g)(x) = \dfrac{2x + 5}{-x - 4}$; $(g/f)(x) = \dfrac{-x - 4}{2x + 5}$ **15. (a)** Domain of f: $(-\infty, \infty)$; domain of g, $f + g$, $f - g$, and fg: $[-2, \infty)$; domain of ff: $(-\infty, \infty)$; domain of f/g: $(-2, \infty)$; domain of g/f: $[-2, 1) \cup (1, \infty)$; **(b)** $(f + g)(x) = x - 1 + \sqrt{x + 2}$; $(f - g)(x) = x - 1 - \sqrt{x + 2}$; $(fg)(x) = (x - 1)\sqrt{x + 2}$; $(ff)(x) = x^2 - 2x + 1$; $(f/g)(x) = \dfrac{x - 1}{\sqrt{x + 2}}$; $(g/f)(x) = \dfrac{\sqrt{x + 2}}{x - 1}$ **16.** 4 **17.** $-2x - h$ **18.** 6 **19.** 28 **20.** -24 **21.** 102 **22.** $(f \circ g)(x) = 3x + 2$; $(g \circ f)(x) = 3x + 4$; domain of $f \circ g$ and $g \circ f$: $(-\infty, \infty)$ **23.** $(f \circ g)(x) = 3\sqrt{x} + 2$; $(g \circ f)(x) = \sqrt{3x + 2}$; domain of $f \circ g$: $[0, \infty)$; domain of $g \circ f$: $\left[-\frac{2}{3}, \infty\right)$ **24.** The graph of $y = (h - g)(x)$ will be the same as the graph of $y = h(x)$ shifted down b units. **25.** Under the given conditions, $(f + g)(x)$ and $(f/g)(x)$ have different domains if $g(x) = 0$ for one or more real numbers x. **26.** If f and g are linear functions, then any real number can be an input for each function. Thus the domain of $f \circ g$ = the domain of $g \circ f = (-\infty, \infty)$. **27.** This approach is not valid. Consider Exercise 23 in Section 2.3, for example. Since $(f \circ g)(x) = \dfrac{4x}{x - 5}$, an examination of only this composed function would lead to the incorrect conclusion that the domain of $f \circ g$ is $(-\infty, 5) \cup (5, \infty)$. However, we must also exclude from the domain of $f \circ g$ those values of x that are not in the domain of g. Thus the domain of $f \circ g$ is $(-\infty, 0) \cup (0, 5) \cup (5, \infty)$.

Exercise Set 2.4

1. x-axis, no; y-axis, yes; origin, no **3.** x-axis, yes; y-axis, no; origin, no **5.** x-axis, no; y-axis, no; origin, yes **7.** x-axis, no; y-axis, yes; origin, no **9.** x-axis, no; y-axis, no; origin, no **11.** x-axis, no; y-axis, yes; origin, no **13.** x-axis, no; y-axis, no; origin, yes **15.** x-axis, no; y-axis, no; origin, yes **17.** x-axis, yes; y-axis, yes; origin, yes **19.** x-axis, no; y-axis, yes; origin, no **21.** x-axis, yes; y-axis, yes; origin, yes **23.** x-axis, no; y-axis, no; origin, no **25.** x-axis, no; y-axis, no; origin, yes **27.** x-axis, $(-5, -6)$; y-axis, $(5, 6)$; origin, $(5, -6)$ **29.** x-axis, $(-10, 7)$; y-axis, $(10, -7)$; origin, $(10, 7)$ **31.** x-axis, $(0, 4)$; y-axis, $(0, -4)$; origin, $(0, 4)$ **33.** Even **35.** Odd **37.** Neither **39.** Odd **41.** Even **43.** Odd **45.** Neither **47.** Even

49.

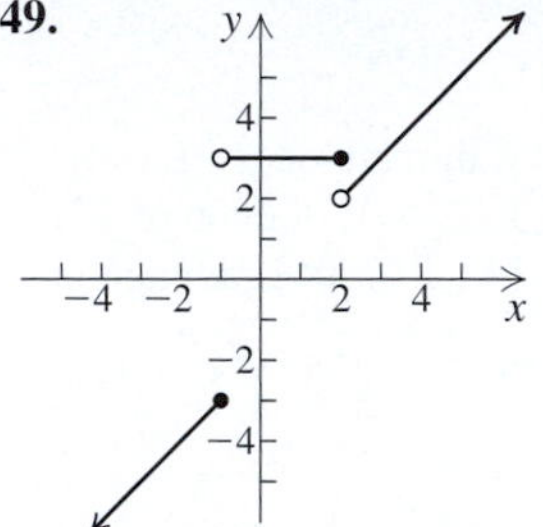

50. University of California–Berkeley: 3576 volunteers; University of Wisconsin–Madison: 3112 volunteers **51.** Odd **53.** x-axis, yes; y-axis, no; origin, no

55. $E(-x) = \dfrac{f(-x) + f(-(-x))}{2} = \dfrac{f(-x) + f(x)}{2} = E(x)$

57. (a) $E(x) + O(x) = \dfrac{f(x) + f(-x)}{2} + \dfrac{f(x) - f(-x)}{2} = \dfrac{2f(x)}{2} = f(x)$; **(b)** $f(x) = \dfrac{-22x^2 + \sqrt{x} + \sqrt{-x} - 20}{2} + \dfrac{8x^3 + \sqrt{x} - \sqrt{-x}}{2}$ **59.** True

Visualizing the Graph

1. C **2.** B **3.** A **4.** E **5.** G **6.** D **7.** H **8.** I **9.** F

Exercise Set 2.5

1. Start with the graph of $y = x^2$. Shift it right 3 units.

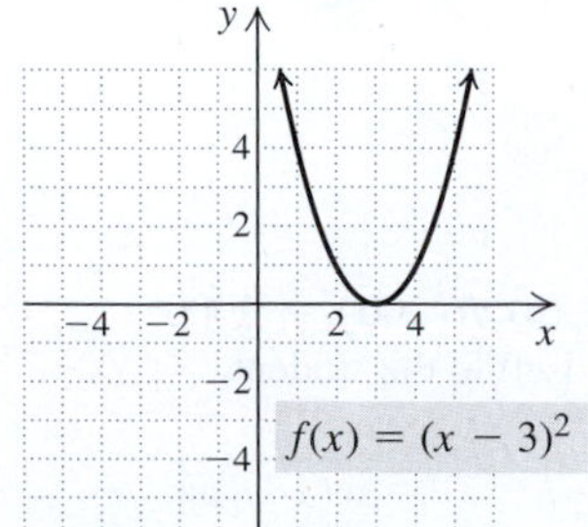

3. Start with the graph of $y = x$. Shift it down 3 units.

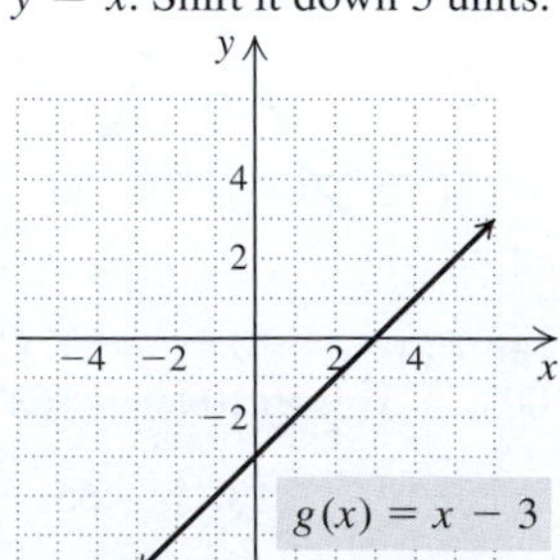

5. Start with the graph of $y = \sqrt{x}$. Reflect it across the x-axis.

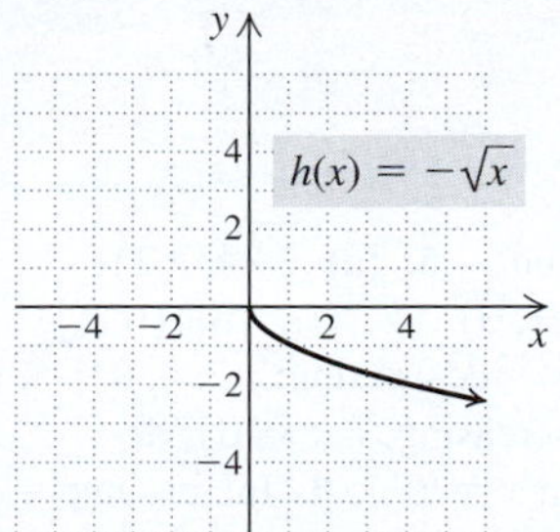

7. Start with the graph of $y = \frac{1}{x}$. Shift it up 4 units.

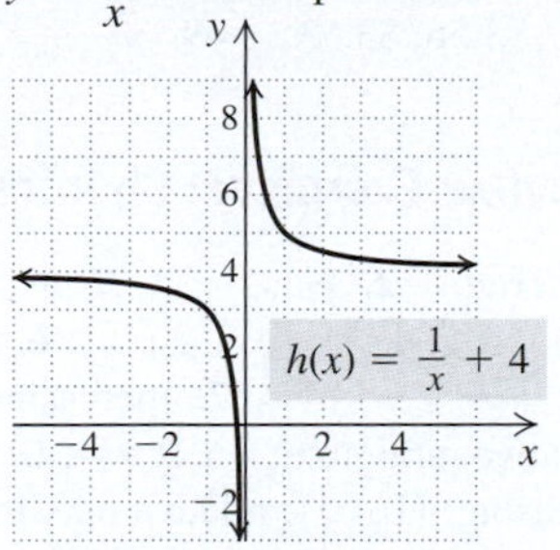

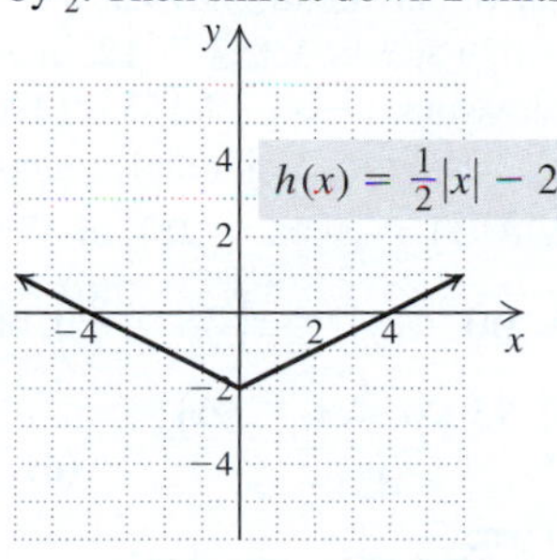

9. Start with the graph of $y = x$. Stretch it vertically by multiplying each y-coordinate by 3. Then reflect it across the x-axis and shift it up 3 units.

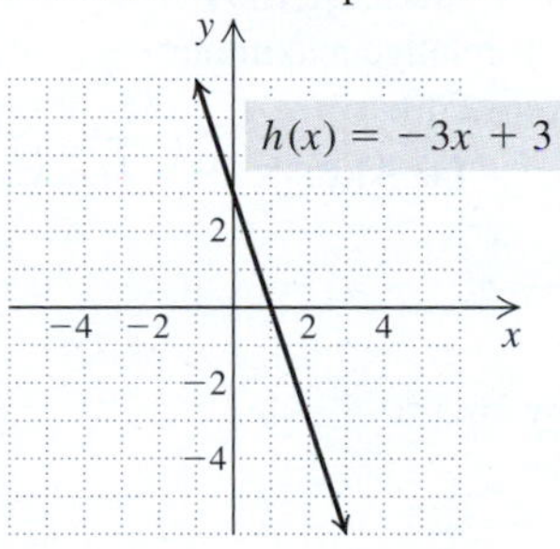

11. Start with the graph of $y = |x|$. Shrink it vertically by multiplying each y-coordinate by $\frac{1}{2}$. Then shift it down 2 units.

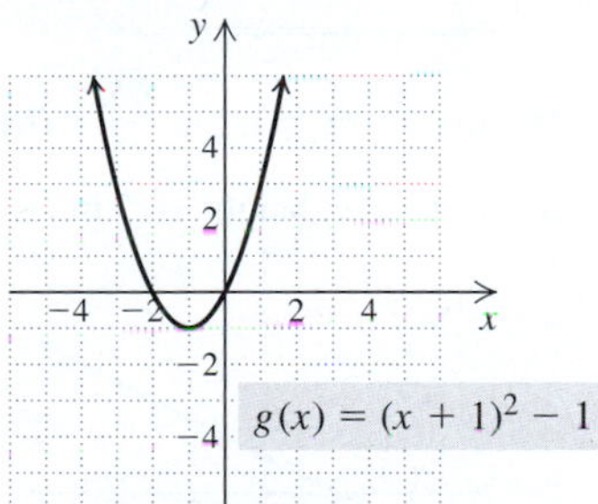

13. Start with the graph of $y = x^3$. Shift it right 2 units. Then reflect it across the x-axis.

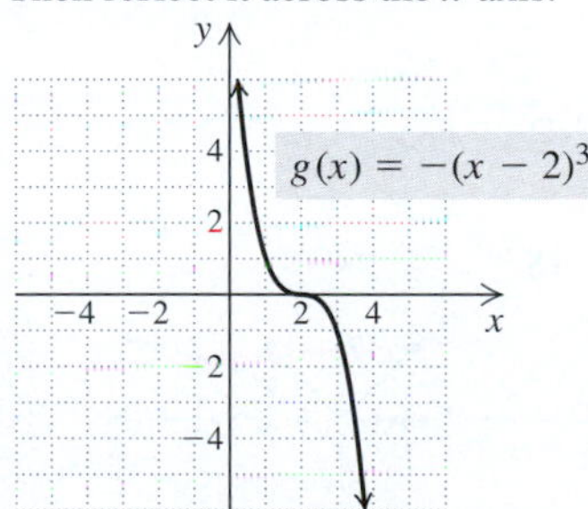

15. Start with the graph of $y = x^2$. Shift it left 1 unit. Then shift it down 1 unit.

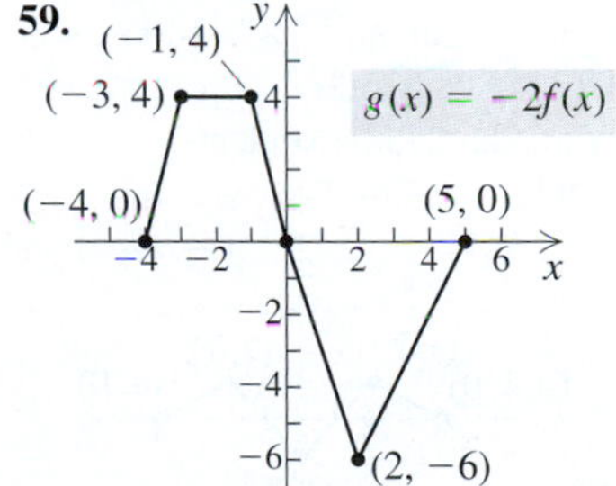

17. Start with the graph of $y = x^3$. Shrink it vertically by multiplying each y-coordinate by $\frac{1}{3}$. Then shift it up 2 units.

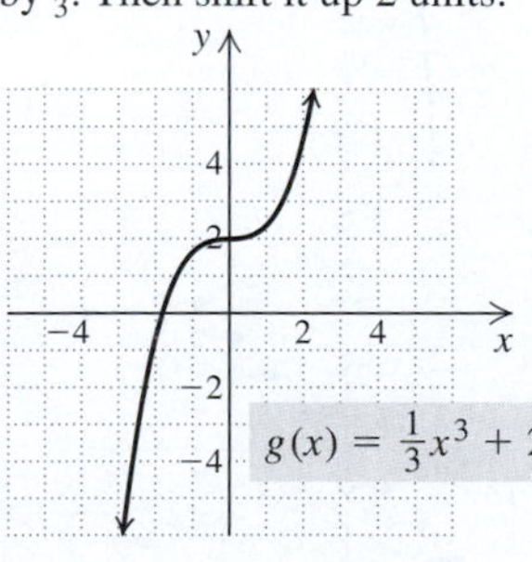

19. Start with the graph of $y = \sqrt{x}$. Shift it left 2 units.

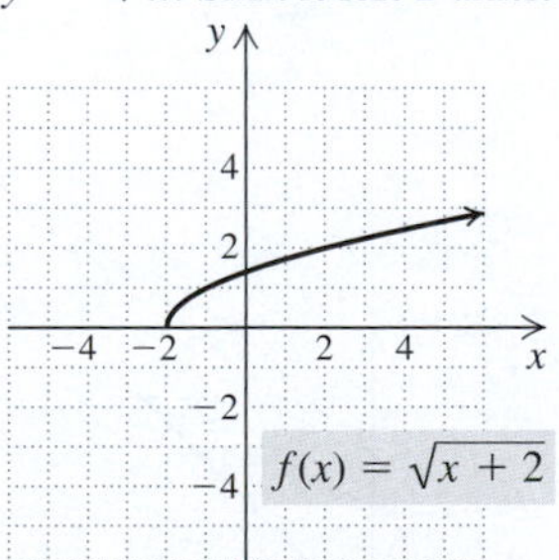

21. Start with the graph of $y = \sqrt[3]{x}$. Shift it down 2 units.

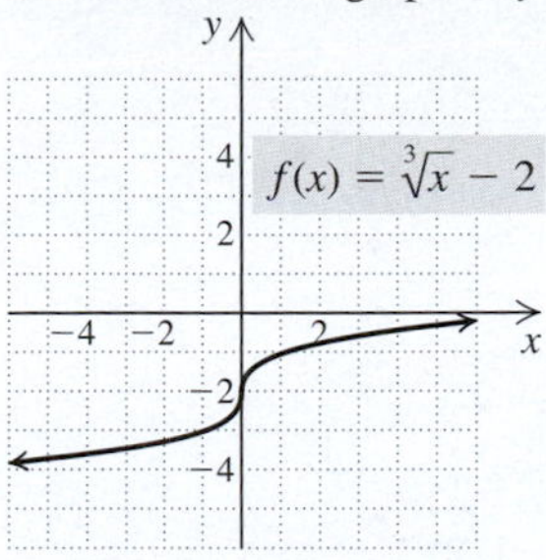

23. Start with the graph of $y = |x|$. Shrink it horizontally by multiplying each x-coordinate by $\frac{1}{3}$ (or dividing each x-coordinate by 3). **25.** Start with the graph of $y = \frac{1}{x}$. Stretch it vertically by multiplying each y-coordinate by 2. **27.** Start with the graph of $y = \sqrt{x}$. Stretch it vertically by multiplying each y-coordinate by 3. Then shift it down 5 units. **29.** Start with the graph of $y = |x|$. Stretch it horizontally by multiplying each x-coordinate by 3. Then shift it down 4 units. **31.** Start with the graph of $y = x^2$. Shift it right 5 units, shrink it vertically by multiplying each y-coordinate by $\frac{1}{4}$, and then reflect it across the x-axis. **33.** Start with the graph of $y = \frac{1}{x}$. Shift it left 3 units, then up 2 units. **35.** Start with the graph of $y = x^2$. Shift it right 3 units. Then reflect it across the x-axis and shift it up 5 units. **37.** $(-12, 2)$ **39.** $(12, 4)$ **41.** $(-12, 2)$ **43.** $(-12, 16)$ **45.** B **47.** A **49.** $f(x) = -(x - 8)^2$ **51.** $f(x) = |x + 7| + 2$ **53.** $f(x) = \frac{1}{2x} - 3$ **55.** $f(x) = -(x - 3)^2 + 4$ **57.** $f(x) = \sqrt{-(x + 2)} - 1$

59.

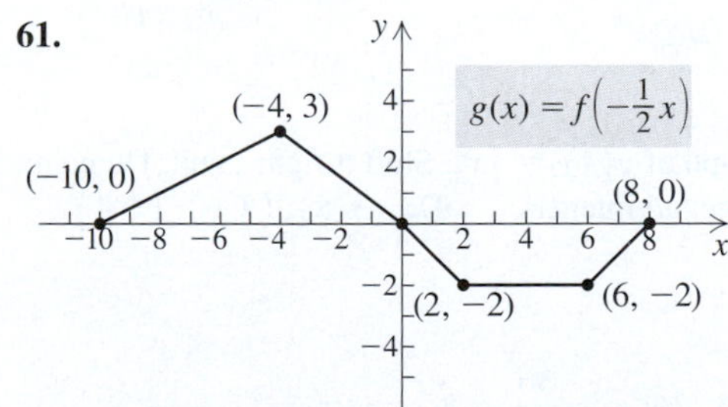

61.

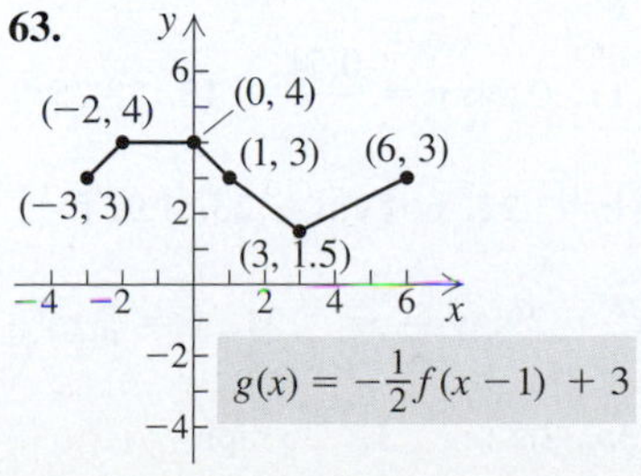

63.

65.

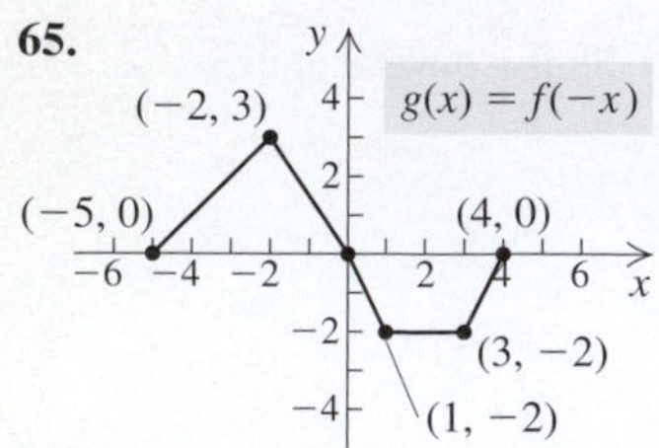

67.

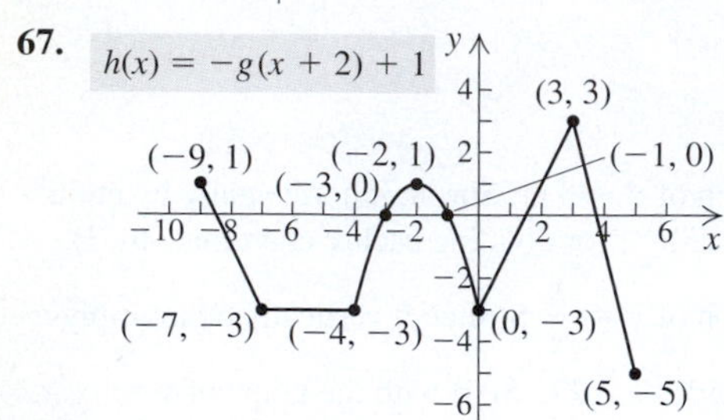

69.

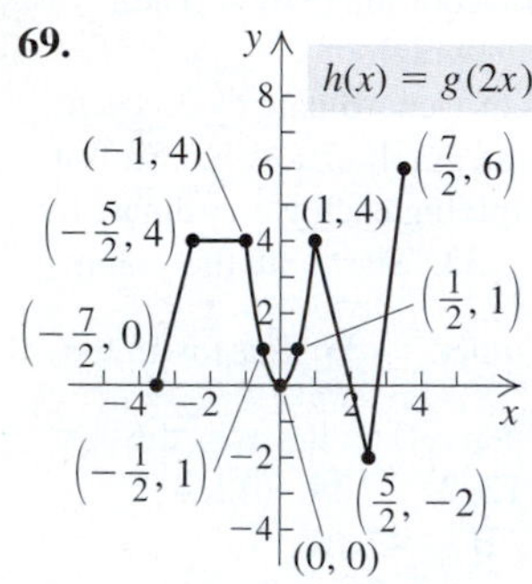

71. (f) **73.** (f) **75.** (d) **77.** (c)
79. $f(-x) = 2(-x)^4 - 35(-x)^3 + 3(-x) - 5 = 2x^4 + 35x^3 - 3x - 5 = g(x)$ **81.** $g(x) = x^3 - 3x^2 + 2$
83. $k(x) = (x + 1)^3 - 3(x + 1)^2$ **85.** x-axis, no; y-axis, yes; origin, no **86.** x-axis, yes; y-axis, no; origin, no **87.** x-axis, no; y-axis, no; origin, yes **88.** \$44,070 **89.** \$7500
90. Saudi Arabia: 53,919 students; Canada: 28,304 students
91. **93.**

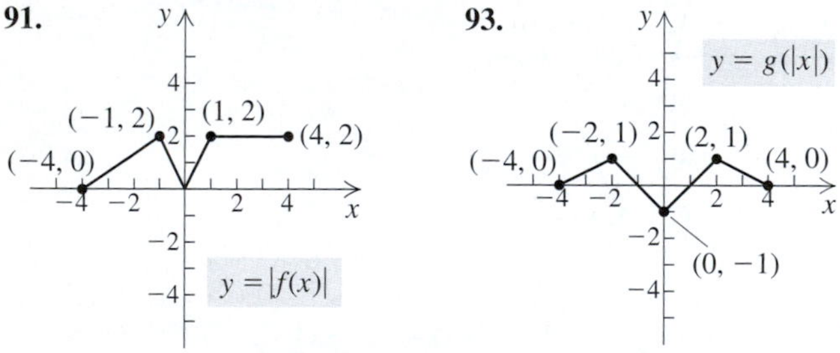

95. Start with the graph of $g(x) = [\![x]\!]$. Shift it right $\frac{1}{2}$ unit. Domain: all real numbers; range: all integers. **97.** $(3, 8)$; $(3, 6)$; $\left(\frac{3}{2}, 4\right)$

Exercise Set 2.6

1. 4.5; $y = 4.5x$ **3.** 36; $y = \frac{36}{x}$ **5.** 4; $y = 4x$

7. 4; $y = \frac{4}{x}$ **9.** $\frac{3}{8}$; $y = \frac{3}{8}x$ **11.** 0.54; $y = \frac{0.54}{x}$ **13.** \$8.25

15. $5\frac{5}{7}$ hr **17.** 90 g **19.** 3.5 hr **21.** $66\frac{2}{3}$ cm **23.** 1.92 ft

25. $y = \frac{0.0015}{x^2}$ **27.** $y = 15x^2$ **29.** $y = xz$ **31.** $y = \frac{3}{10}xz^2$

33. $y = \frac{1}{5} \cdot \frac{xz}{wp}$, or $y = \frac{xz}{5wp}$ **35.** 2.5 m **37.** 36 mph

39. About 43 earned runs **41.** Parallel **42.** Zero
43. Relative minimum **44.** Odd function **45.** Inverse variation
47. \$3.56; \$3.53 **49.** $\frac{\pi}{4}$

Review Exercises: Chapter 2

1. True **2.** False **3.** True **4.** True **5. (a)** $(-4, -2)$; **(b)** $(2, 5)$; **(c)** $(-2, 2)$ **6. (a)** $(-1, 0)$, $(2, \infty)$; **(b)** $(0, 2)$; **(c)** $(-\infty, -1)$ **7.** Increasing: $(0, \infty)$; decreasing: $(-\infty, 0)$; relative minimum: -1 at $x = 0$ **8.** Increasing: $(-\infty, 0)$; decreasing: $(0, \infty)$; relative maximum: 2 at $x = 0$ **9.** Increasing: $(2, \infty)$; decreasing: $(-\infty, 2)$; relative minimum: -1 at $x = 2$
10. Increasing: $(-\infty, 0.5)$; decreasing: $(0.5, \infty)$; relative maximum: 6.25 at $x = 0.5$ **11.** Increasing: $(-\infty, -1.155)$, $(1.155, \infty)$; decreasing: $(-1.155, 1.155)$; relative maximum: 3.079 at $x = -1.155$; relative minimum: -3.079 at $x = 1.155$ **12.** Increasing: $(-1.155, 1.155)$; decreasing: $(-\infty, -1.155)$, $(1.155, \infty)$; relative maximum: 1.540 at $x = 1.155$; relative minimum: -1.540 at $x = -1.155$
13. $A(x) = x(48 - 2x)$, or $48x - 2x^2$ **14.** $A(x) = 2x\sqrt{4 - x^2}$
15. (a) $A(x) = x\left(33 - \frac{x}{2}\right)$, or $33x - \frac{x^2}{2}$;
(b) $\{x \mid 0 < x < 66\}$;
(c) **(d)** 33 ft by 16.5 ft

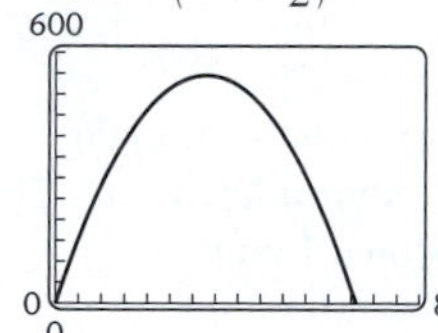

16. (a) $A(x) = x^2 + \frac{432}{x}$; **(b)** $(0, \infty)$;
(c) $x = 6$ in., height $= 3$ in.

17.

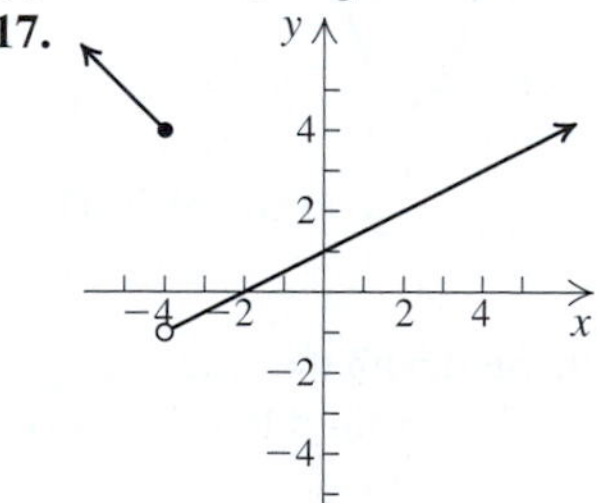

18.

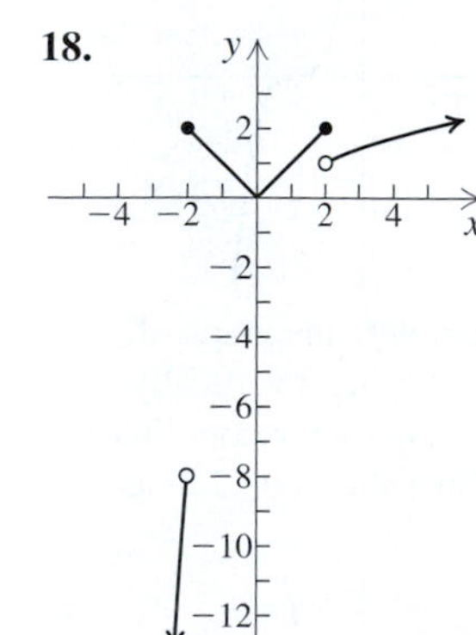

19.

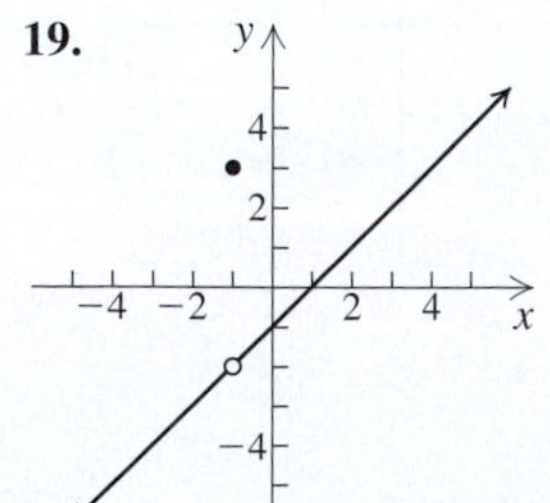

20.

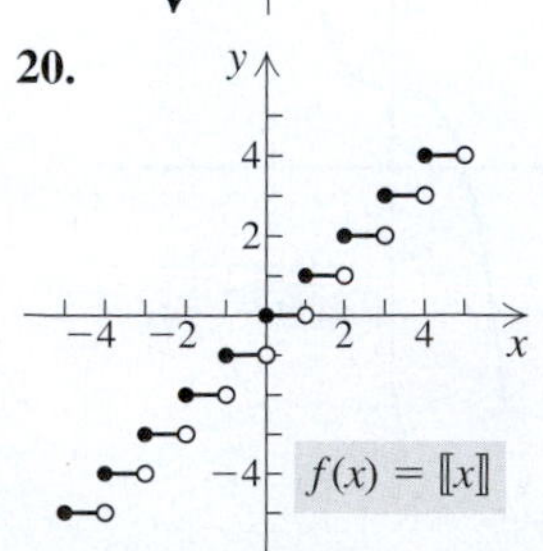

21.

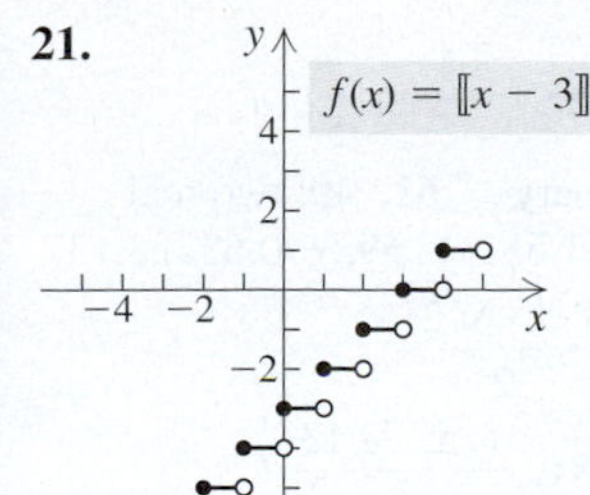

22. $f(-1) = 1; f(5) = 2; f(-2) = 2; f(-3) = -27$
23. $f(-2) = -3; f(-1) = 3; f(0) = -1; f(4) = 3$
24. -33 **25.** 0 **26.** Does not exist
27. (a) Domain of f: $(-\infty, 0) \cup (0, \infty)$; domain of g: $(-\infty, \infty)$; domain of $f + g$, $f - g$, and fg: $(-\infty, 0) \cup (0, \infty)$; domain of f/g: $(-\infty, 0) \cup \left(0, \frac{3}{2}\right) \cup \left(\frac{3}{2}, \infty\right)$ **(b)** $(f + g)(x) = \frac{4}{x^2} + 3 - 2x$; $(f - g)(x) = \frac{4}{x^2} - 3 + 2x$; $(fg)(x) = \frac{12}{x^2} - \frac{8}{x}$; $(f/g)(x) = \frac{4}{x^2(3 - 2x)}$ **28. (a)** Domain of f, g, $f + g$, $f - g$, and fg: $(-\infty, \infty)$; domain of f/g: $\left(-\infty, \frac{1}{2}\right) \cup \left(\frac{1}{2}, \infty\right)$;
(b) $(f + g)(x) = 3x^2 + 6x - 1$; $(f - g)(x) = 3x^2 + 2x + 1$; $(fg)(x) = 6x^3 + 5x^2 - 4x$; $(f/g)(x) = \frac{3x^2 + 4x}{2x - 1}$
29. $P(x) = -0.5x^2 + 105x - 6$ **30.** 2 **31.** $-2x - h$
32. $\frac{-4}{x(x + h)}$, or $-\frac{4}{x(x + h)}$ **33.** 9 **34.** 5 **35.** 128
36. 580 **37.** 7 **38.** -509 **39.** $4x - 3$
40. $-24 + 27x^3 - 9x^6 + x^9$ **41. (a)** $(f \circ g)(x) = \frac{4}{(3 - 2x)^2}$; $(g \circ f)(x) = 3 - \frac{8}{x^2}$; **(b)** domain of $f \circ g$: $\left(-\infty, \frac{3}{2}\right) \cup \left(\frac{3}{2}, \infty\right)$; domain of $g \circ f$: $(-\infty, 0) \cup (0, \infty)$
42. (a) $(f \circ g)(x) = 12x^2 - 4x - 1$; $(g \circ f)(x) = 6x^2 + 8x - 1$; **(b)** domain of $f \circ g$ and $g \circ f$: $(-\infty, \infty)$ **43.** $f(x) = \sqrt{x}$, $g(x) = 5x + 2$; answers may vary. **44.** $f(x) = 4x^2 + 9$, $g(x) = 5x - 1$; answers may vary. **45.** x-axis, yes; y-axis, yes; origin, yes **46.** x-axis, yes; y-axis, yes; origin, yes **47.** x-axis, no; y-axis, no; origin, no
48. x-axis, no; y-axis, yes; origin, no **49.** x-axis, no; y-axis, no; origin, yes **50.** x-axis, no; y-axis, yes; origin, no **51.** Even
52. Even **53.** Odd **54.** Even **55.** Even **56.** Neither
57. Odd **58.** Even **59.** Even **60.** Odd
61. $f(x) = (x + 3)^2$ **62.** $f(x) = -\sqrt{x - 3} + 4$
63. $f(x) = 2|x - 3|$

64.

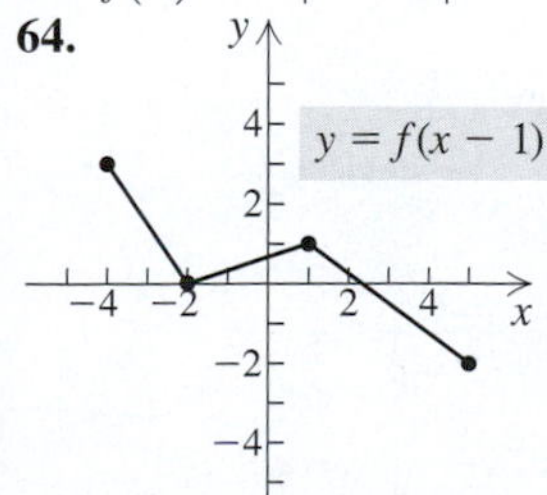

65.

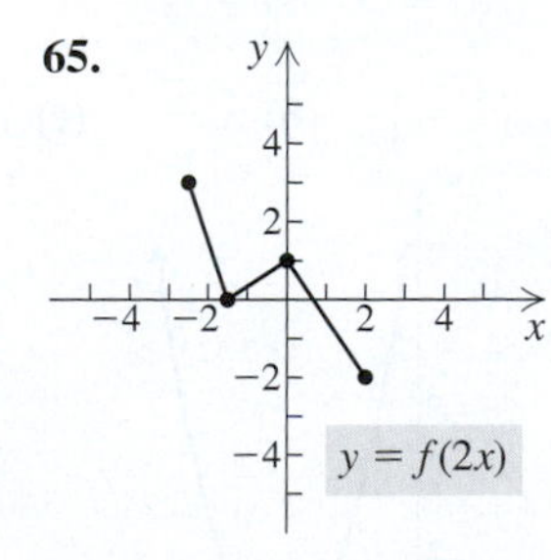

66.

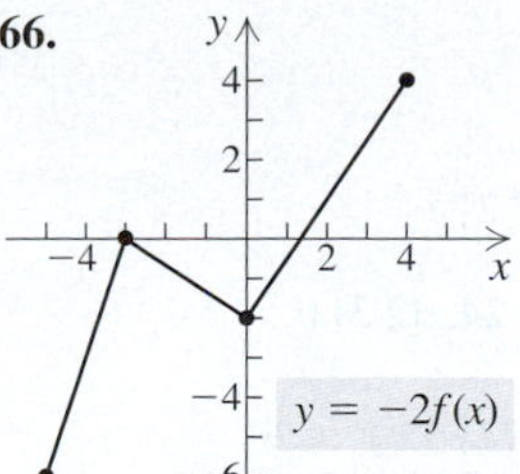

67.

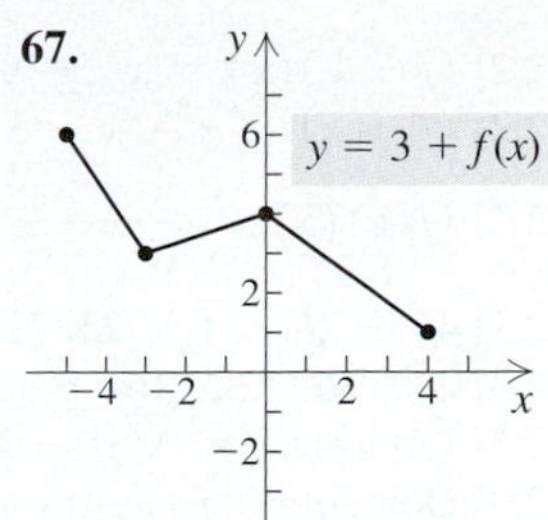

68. $y = 4x$ **69.** $y = \frac{2}{3}x$ **70.** $y = \frac{2500}{x}$ **71.** $y = \frac{54}{x}$
72. $y = \frac{48}{x^2}$ **73.** $y = \frac{1}{10} \cdot \frac{xz^2}{w}$ **74.** 20 min **75.** 75

76. 500 watts **77.** A **78.** C **79.** B **80.** Let $f(x)$ and $g(x)$ be odd functions. Then by definition, $f(-x) = -f(x)$, or $f(x) = -f(-x)$, and $g(-x) = -g(x)$, or $g(x) = -g(-x)$. Thus, $(f + g)(x) = f(x) + g(x) = -f(-x) + [-g(-x)] = -[f(-x) + g(-x)] = -(f + g)(-x)$ and $f + g$ is odd.
81. Reflect the graph of $y = f(x)$ across the x-axis and then across the y-axis. **82. (a)** $4x^3 - 2x + 9$; **(b)** $4x^3 + 24x^2 + 46x + 35$; **(c)** $4x^3 - 2x + 42$. (a) Adds 2 to each function value; (b) adds 2 to each input before finding a function value; (c) adds the output for 2 to the output for x **83.** In the graph of $y = f(cx)$, the constant c stretches or shrinks the graph of $y = f(x)$ horizontally. The constant c in $y = cf(x)$ stretches or shrinks the graph of $y = f(x)$ vertically. For $y = f(cx)$, the x-coordinates of $y = f(x)$ are divided by c; for $y = c\,f(x)$, the y-coordinates of $y = f(x)$ are multiplied by c.
84. The graph of $f(x) = 0$ is symmetric with respect to the x-axis, the y-axis, and the origin. This function is both even and odd.
85. If all the exponents are even numbers, then $f(x)$ is an even function. If $a_0 = 0$ and all the exponents are odd numbers, then $f(x)$ is an odd function. **86.** Let $y(x) = kx^2$. Then $y(2x) = k(2x)^2 = k \cdot 4x^2 = 4 \cdot kx^2 = 4 \cdot y(x)$. Thus doubling x causes y to be quadrupled. **87.** Let $y = k_1 x$ and $x = \frac{k_2}{z}$. Then $y = k_1 \cdot \frac{k_2}{z}$, or $y = \frac{k_1 k_2}{z}$, so y varies inversely as z.

Test: Chapter 2

1. [2.1] **(a)** $(-5, -2)$; **(b)** $(2, 5)$; **(c)** $(-2, 2)$
2. [2.1] Increasing: $(-\infty, 0)$; decreasing: $(0, \infty)$; relative maximum: 2 at $x = 0$ **3.** [2.1] Increasing: $(-\infty, -2.667)$, $(0, \infty)$; decreasing: $(-2.667, 0)$; relative maximum: 9.481 at $x = -2.667$; relative minimum: 0 at $x = 0$
4. [2.1] $A(b) = \frac{1}{2}b(4b - 6)$, or $2b^2 - 3b$
5. [2.1]

6. [2.1] $f\left(-\frac{7}{8}\right) = \frac{7}{8}$; $f(5) = 2$; $f(-4) = 16$ **7.** [2.2] 66
8. [2.2] 6 **9.** [2.2] -1 **10.** [2.2] 0 **11.** [2.1] $(-\infty, \infty)$
12. [2.1] $[3, \infty)$ **13.** [2.2] $[3, \infty)$ **14.** [2.2] $[3, \infty)$
15. [2.2] $[3, \infty)$ **16.** [2.2] $(3, \infty)$
17. [2.2] $(f + g)(x) = x^2 + \sqrt{x - 3}$

18. [2.2] $(f - g)(x) = x^2 - \sqrt{x - 3}$
19. [2.2] $(fg)(x) = x^2\sqrt{x - 3}$
20. [2.2] $(f/g)(x) = \dfrac{x^2}{\sqrt{x - 3}}$ **21.** [2.2] $\frac{1}{2}$
22. [2.2] $4x + 2h - 1$ **23.** [2.3] 83 **24.** [2.3] 0
25. [2.3] 4 **26.** [2.3] $16x + 15$
27. [2.3] $(f \circ g)(x) = \sqrt{x^2 - 4}$; $(g \circ f)(x) = x - 4$
28. [2.3] Domain of $(f \circ g)(x) = (-\infty, -2] \cup [2, \infty)$; domain of $(g \circ f)(x) = [5, \infty)$ **29.** [2.3] $f(x) = x^4$; $g(x) = 2x - 7$; answers may vary **30.** [2.4] x-axis: no; y-axis: yes; origin: no
31. [2.4] Odd **32.** [2.5] $f(x) = (x - 2)^2 - 1$
33. [2.5] $f(x) = -(x + 2)^2 + 3$
34. [2.5]

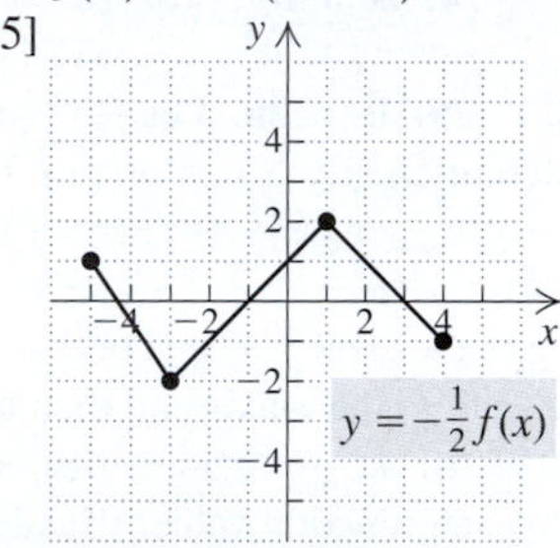

35. [2.6] $y = \dfrac{30}{x}$ **36.** [2.6] $y = 5x$ **37.** [2.6] $y = \dfrac{50xz^2}{w}$
38. [2.6] 50 ft **39.** [2.5] C **40.** [2.5] $(-1, 1)$

Chapter 3

Exercise Set 3.1

1. $\sqrt{3}i$ **3.** $5i$ **5.** $-\sqrt{33}i$ **7.** $-9i$ **9.** $7\sqrt{2}i$
11. $2 + 11i$ **13.** $5 - 12i$ **15.** $4 + 8i$ **17.** $-4 - 2i$
19. $5 + 9i$ **21.** $5 + 4i$ **23.** $5 + 7i$ **25.** $11 - 5i$
27. $-1 + 5i$ **29.** $2 - 12i$ **31.** -12 **33.** -45
35. $35 + 14i$ **37.** $6 + 16i$ **39.** $13 - i$ **41.** $-11 + 16i$
43. $-10 + 11i$ **45.** $-31 - 34i$ **47.** $-14 + 23i$ **49.** 41
51. 13 **53.** 74 **55.** $12 + 16i$ **57.** $-45 - 28i$
59. $-8 - 6i$ **61.** $2i$ **63.** $-7 + 24i$ **65.** $\frac{15}{146} + \frac{33}{146}i$
67. $\frac{10}{13} - \frac{15}{13}i$ **69.** $-\frac{14}{13} + \frac{5}{13}i$ **71.** $\frac{11}{25} - \frac{27}{25}i$
73. $\dfrac{-4\sqrt{3} + 10}{41} + \dfrac{5\sqrt{3} + 8}{41}i$ **75.** $-\frac{1}{2} + \frac{1}{2}i$ **77.** $-\frac{1}{2} - \frac{13}{2}i$
79. $-i$ **81.** $-i$ **83.** 1 **85.** i **87.** 625
89. $y = -2x + 1$ **90.** All real numbers, or $(-\infty, \infty)$
91. $\left(-\infty, -\frac{5}{3}\right) \cup \left(-\frac{5}{3}, \infty\right)$ **92.** $x^2 - 3x - 1$ **93.** $\frac{8}{11}$
94. $2x + h - 3$ **95.** True **97.** True **99.** $a^2 + b^2$
101. $x^2 - 6x + 25$

Exercise Set 3.2

1. $\frac{2}{3}, \frac{3}{2}$ **3.** $-2, 10$ **5.** $-1, \frac{2}{3}$ **7.** $-\sqrt{3}, \sqrt{3}$
9. $-\sqrt{7}, \sqrt{7}$ **11.** $-\sqrt{2}i, \sqrt{2}i$ **13.** $-4i, 4i$ **15.** 0, 3
17. $-\frac{1}{3}, 0, 2$ **19.** $-1, -\frac{1}{7}, 1$ **21.** **(a)** $(-4, 0), (2, 0)$; **(b)** $-4, 2$ **23.** **(a)** $(-1, 0), (3, 0)$; **(b)** $-1, 3$
25. **(a)** $(-2, 0), (2, 0)$; **(b)** $-2, 2$ **27.** **(a)** $(1, 0)$; **(b)** 1
29. $-7, 1$ **31.** $4 \pm \sqrt{7}$ **33.** $-4 \pm 3i$ **35.** $-2, \frac{1}{3}$
37. $-3, 5$ **39.** $-1, \frac{2}{5}$ **41.** $\dfrac{5 \pm \sqrt{7}}{3}$ **43.** $-\dfrac{1}{2} \pm \dfrac{\sqrt{7}}{2}i$
45. $\dfrac{4 \pm \sqrt{31}}{5}$ **47.** $\dfrac{5}{6} \pm \dfrac{\sqrt{23}}{6}i$ **49.** $4 \pm \sqrt{11}$
51. $\dfrac{-1 \pm \sqrt{61}}{6}$ **53.** $\dfrac{5 \pm \sqrt{17}}{4}$ **55.** $-\dfrac{1}{5} \pm \dfrac{3}{5}i$
57. 144; two real **59.** -7; two imaginary **61.** 49; two real
63. 2, 6 **65.** 0.143, 6 **67.** -0.151, 1.651 **69.** -0.637, 3.137
71. $-5, -1$ **73.** $\dfrac{3 \pm \sqrt{21}}{2}$ **75.** $\dfrac{5 \pm \sqrt{21}}{2}$
77. $-1 \pm \sqrt{6}$ **79.** $\dfrac{1}{4} \pm \dfrac{\sqrt{31}}{4}i$ **81.** $\dfrac{1 \pm \sqrt{13}}{6}$
83. $\dfrac{1 \pm \sqrt{6}}{5}$ **85.** $\dfrac{-3 \pm \sqrt{57}}{8}$ **87.** -1.535, 0.869
89. -0.347, 1.181 **91.** $\pm 1, \pm\sqrt{2}$ **93.** $\pm\sqrt{2}, \pm\sqrt{5}i$
95. $\pm 1, \pm\sqrt{5}i$ **97.** 16 **99.** $-8, 64$ **101.** 1, 16
103. $\frac{5}{2}, 3$ **105.** $-\frac{3}{2}, -1, \frac{1}{2}, 1$ **107.** 1995 **109.** 2011
111. About 10.216 sec **113.** Length: 4 ft; width: 3 ft
115. 4 and 9; -9 and -4 **117.** 2 cm **119.** Length: 8 ft; width: 6 ft **121.** Linear **123.** Quadratic **125.** Linear
127. About \$3.95 million **128.** About 12 years after 2004, or in 2016 **129.** x-axis: yes; y-axis: yes; origin: yes **130.** x-axis: no; y-axis: yes; origin: no **131.** Odd **132.** Neither **133.** **(a)** 2; **(b)** $\frac{11}{2}$ **135.** **(a)** 2; **(b)** $1 - i$ **137.** 1
139. $-\sqrt{7}, -\frac{3}{2}, 0, \frac{1}{3}, \sqrt{7}$ **141.** $\dfrac{-1 \pm \sqrt{1 + 4\sqrt{2}}}{2}$
143. $3 \pm \sqrt{5}$ **145.** 19 **147.** $-2 \pm \sqrt{2}, \dfrac{1}{2} \pm \dfrac{\sqrt{7}}{2}i$

Visualizing the Graph

1. C **2.** B **3.** A **4.** J **5.** F **6.** D **7.** I **8.** G
9. H **10.** E

Exercise Set 3.3

1. **(a)** $\left(-\frac{1}{2}, -\frac{9}{4}\right)$; **(b)** $x = -\frac{1}{2}$; **(c)** minimum: $-\frac{9}{4}$
3. **(a)** $(4, -4)$; **(b)** $x = 4$; **(c)** minimum: -4;
(d)

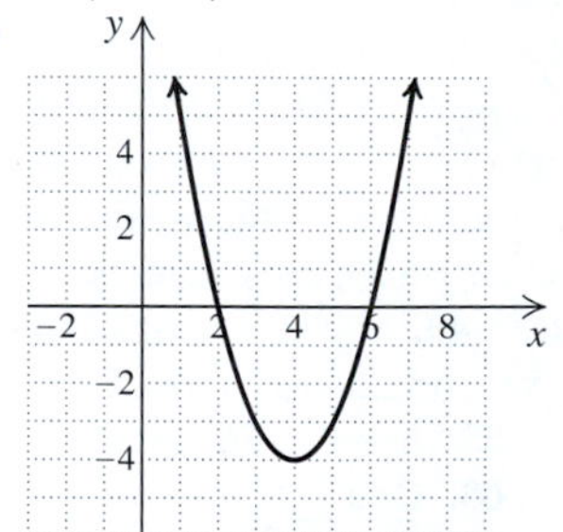

$f(x) = x^2 - 8x + 12$

5. **(a)** $\left(\frac{7}{2}, -\frac{1}{4}\right)$; **(b)** $x = \frac{7}{2}$; **(c)** minimum: $-\frac{1}{4}$;
(d)

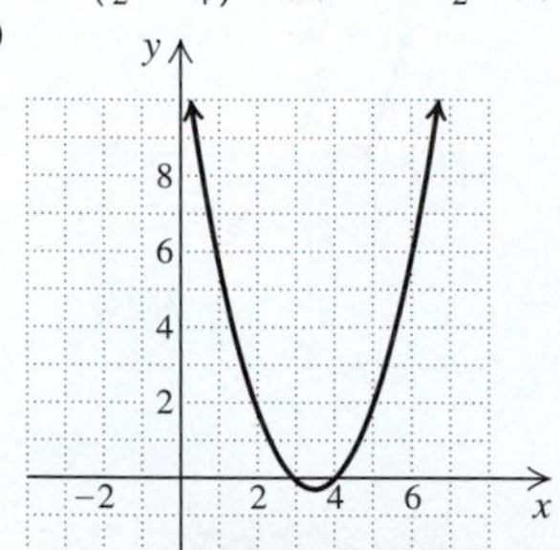

$f(x) = x^2 - 7x + 12$

7. (a) $(-2, 1)$; **(b)** $x = -2$; **(c)** minimum: 1;
(d)

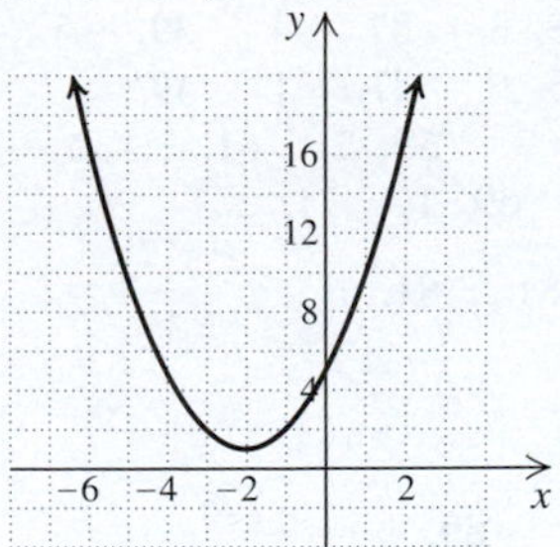

$f(x) = x^2 + 4x + 5$

9. (a) $(-4, -2)$; **(b)** $x = -4$; **(c)** minimum: -2;
(d)

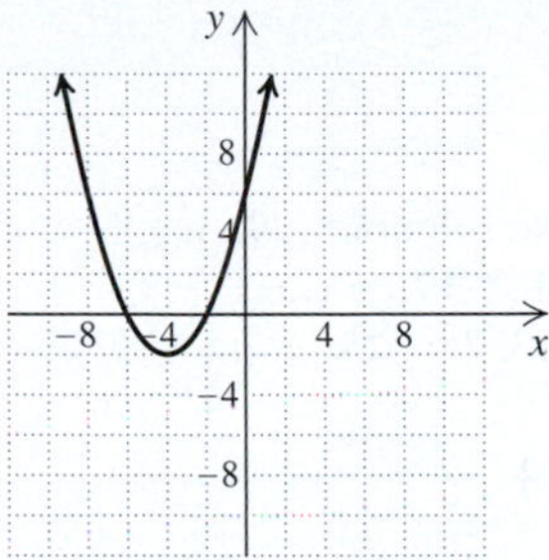

$g(x) = \dfrac{x^2}{2} + 4x + 6$

11. (a) $\left(-\frac{3}{2}, \frac{7}{2}\right)$; **(b)** $x = -\frac{3}{2}$; **(c)** minimum: $\frac{7}{2}$;
(d)

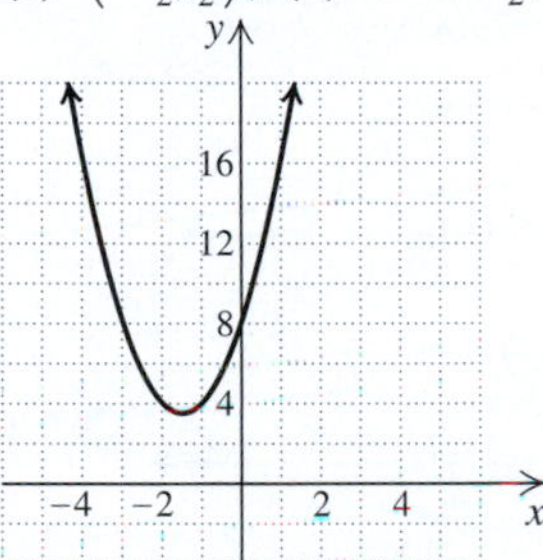

$g(x) = 2x^2 + 6x + 8$

13. (a) $(-3, 12)$; **(b)** $x = -3$; **(c)** maximum: 12;
(d)

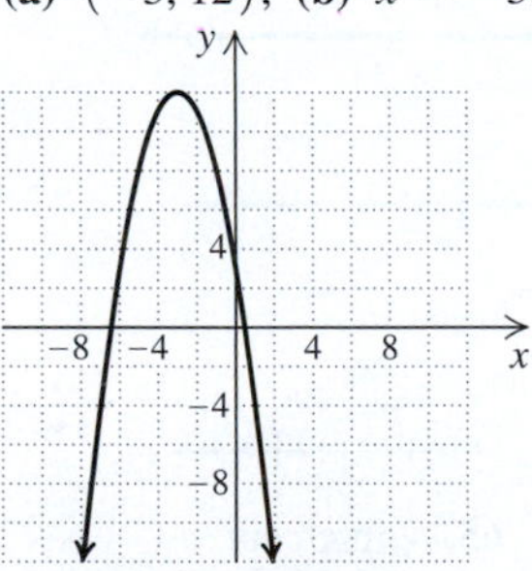

$f(x) = -x^2 - 6x + 3$

15. (a) $\left(\frac{1}{2}, \frac{3}{2}\right)$; **(b)** $x = \frac{1}{2}$; **(c)** maximum: $\frac{3}{2}$;
(d)

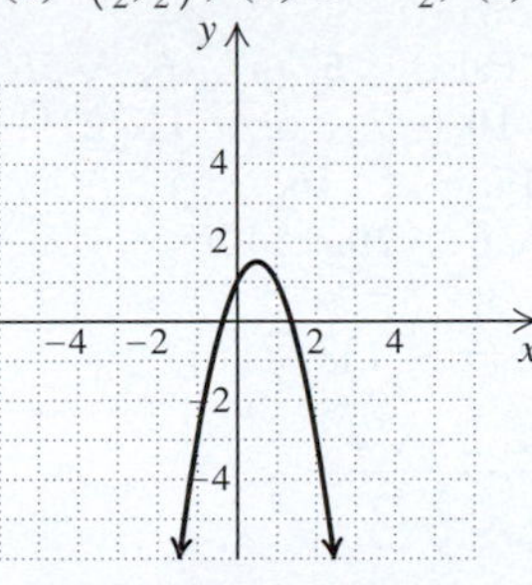

$g(x) = -2x^2 + 2x + 1$

17. (f) **19.** (b) **21.** (h) **23.** (c) **25.** True **27.** False
29. True **31. (a)** $(3, -4)$; **(b)** minimum: -4; **(c)** $[-4, \infty)$;
(d) increasing: $(3, \infty)$; decreasing: $(-\infty, 3)$
33. (a) $(-1, -18)$; **(b)** minimum: -18; **(c)** $[-18, \infty)$;
(d) increasing: $(-1, \infty)$; decreasing: $(-\infty, -1)$
35. (a) $\left(5, \frac{9}{2}\right)$; **(b)** maximum: $\frac{9}{2}$; **(c)** $\left(-\infty, \frac{9}{2}\right]$; **(d)** increasing: $(-\infty, 5)$; decreasing: $(5, \infty)$ **37. (a)** $(-1, 2)$; **(b)** minimum: 2; **(c)** $[2, \infty)$; **(d)** increasing: $(-1, \infty)$; decreasing: $(-\infty, -1)$
39. (a) $\left(-\frac{3}{2}, 18\right)$; **(b)** maximum: 18; **(c)** $(-\infty, 18]$;
(d) increasing: $\left(-\infty, -\frac{3}{2}\right)$; decreasing: $\left(-\frac{3}{2}, \infty\right)$
41. 0.625 sec; 12.25 ft **43.** 3.75 sec; 305 ft **45.** 4.5 in.
47. Base: 10 cm; height: 10 cm **49.** 2100 portable doghouses
51. \$797; 40 units **53.** 4800 yd^2 **55.** About 60.5 ft **57.** 3
58. $4x + 2h - 1$
59.

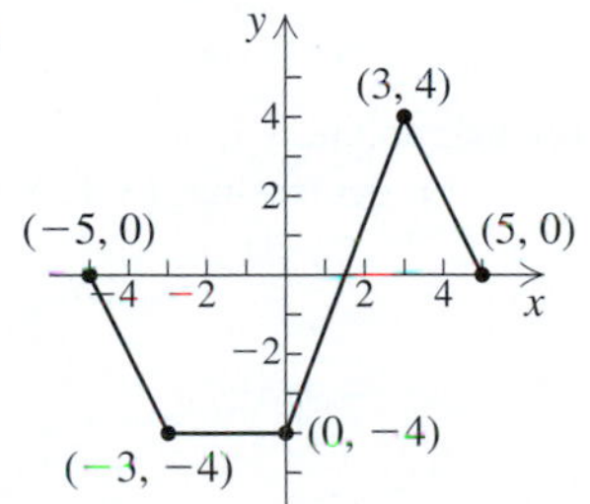

$g(x) = -2f(x)$

60.

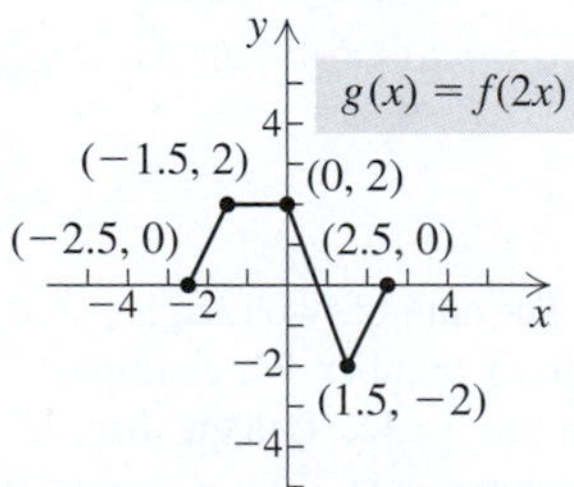

61. -236.25

63. $f(x) = (|x| - 5)^2 - 3$

65. Pieces should be $\dfrac{24\pi}{4 + \pi}$ in. and $\dfrac{96}{4 + \pi}$ in.

Mid-Chapter Mixed Review: Chapter 3

1. True **2.** False **3.** True **4.** False **5.** $6i$ **6.** $\sqrt{5}i$ **7.** $-4i$ **8.** $4\sqrt{2}i$ **9.** $-1 + i$ **10.** $-7 + 5i$ **11.** $23 + 2i$ **12.** $-\frac{1}{29} - \frac{17}{29}i$ **13.** i **14.** 1 **15.** $-i$ **16.** -64 **17.** $-4, 1$ **18.** $-2, -\frac{3}{2}$ **19.** $\pm\sqrt{6}$ **20.** $\pm 10i$ **21.** $4x^2 - 8x - 3 = 0$; $4x^2 - 8x = 3$; $x^2 - 2x = \frac{3}{4}$; $x^2 - 2x + 1 = \frac{3}{4} + 1$; $(x - 1)^2 = \frac{7}{4}$; $x - 1 = \pm\frac{\sqrt{7}}{2}$; $x = 1 \pm \frac{\sqrt{7}}{2} = \frac{2 \pm \sqrt{7}}{2}$

22. **(a)** 29; two real; **(b)** $\frac{3 \pm \sqrt{29}}{2}$; $-1.193, 4.193$ **23.** **(a)** 0; one real; **(b)** $\frac{3}{2}$ **24.** **(a)** -8; two nonreal; **(b)** $-\frac{1}{3} \pm \frac{\sqrt{2}}{3}i$ **25.** $\pm 1, \pm\sqrt{6}i$ **26.** $\frac{1}{4}, 4$ **27.** 5 and 7; -7 and -5 **28.** **(a)** $(3, -2)$; **(b)** $x = 3$; **(c)** minimum: -2; **(d)** $[-2, \infty)$; **(e)** increasing: $(3, \infty)$; decreasing: $(-\infty, 3)$; **(f)**

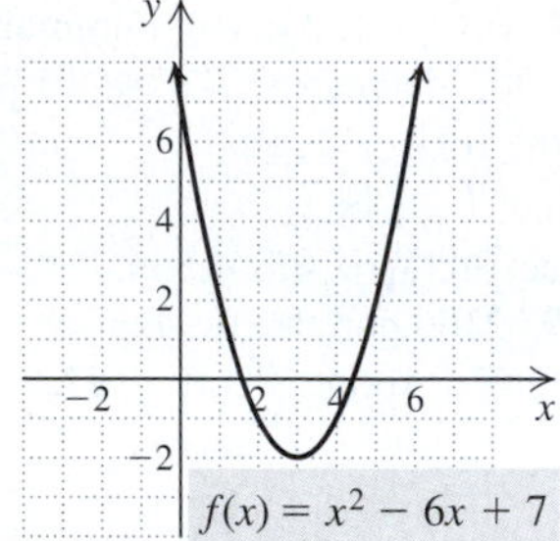

$f(x) = x^2 - 6x + 7$

29. **(a)** $(-1, -3)$; **(b)** $x = -1$; **(c)** maximum: -3; **(d)** $(-\infty, -3]$; **(e)** increasing: $(-\infty, -1)$; decreasing: $(-1, \infty)$; **(f)**

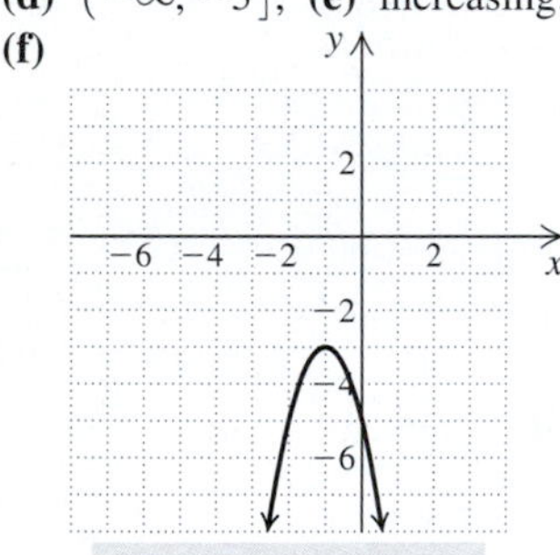

$f(x) = -2x^2 - 4x - 5$

30. Base: 8 in., height: 8 in. **31.** The sum of two imaginary numbers is not always an imaginary number. For example, $(2 + i) + (3 - i) = 5$, a real number. **32.** Use the discriminant. If $b^2 - 4ac < 0$, there are no x-intercepts. If $b^2 - 4ac = 0$, there is one x-intercept. If $b^2 - 4ac > 0$, there are two x-intercepts. **33.** Completing the square was used in Section 3.2 to solve quadratic equations. It was used again in Section 3.3 to write quadratic functions in the form $f(x) = a(x - h)^2 + k$. **34.** The x-intercepts of $g(x)$ are also $(x_1, 0)$ and $(x_2, 0)$. This is true because $f(x)$ and $g(x)$ have the same zeros. Consider $g(x) = 0$, or $-ax^2 - bx - c = 0$. Multiplying by -1 on both sides, we get an equivalent equation $ax^2 + bx + c = 0$, or $f(x) = 0$.

Exercise Set 3.4

1. $\frac{20}{9}$ **3.** 286 **5.** 6 **7.** 6 **9.** 4 **11.** 2, 3 **13.** $-1, 6$ **15.** $\frac{1}{2}, 5$ **17.** 7 **19.** No solution **21.** $-\frac{69}{14}$ **23.** $-\frac{37}{18}$ **25.** 2 **27.** No solution **29.** $\{x \mid x \text{ is a real number } and\ x \neq 0\ and\ x \neq 6\}$ **31.** $\frac{5}{3}$ **33.** $\frac{9}{2}$ **35.** 3 **37.** -4 **39.** -5 **41.** $\pm\sqrt{2}$ **43.** No solution **45.** 6 **47.** -1 **49.** $\frac{35}{2}$ **51.** -98 **53.** -6 **55.** 5 **57.** 7 **59.** 2 **61.** $-1, 2$ **63.** 7 **65.** 7 **67.** No solution **69.** 1 **71.** 3, 7 **73.** 5 **75.** -1 **77.** -8 **79.** 81 **81.** $T_1 = \frac{P_1V_1T_2}{P_2V_2}$ **83.** $C = \frac{1}{LW^2}$ **85.** $R_2 = \frac{RR_1}{R_1 - R}$ **87.** $P = \frac{A}{I^2 + 2I + 1}$, or $\frac{A}{(I + 1)^2}$ **89.** $p = \frac{Fm}{m - F}$ **91.** $\frac{15}{2}$, or 7.5 **92.** 3 **93.** China: 53,800,000 metric tons; United States: 10,508,000 metric tons **94.** About $18,609 **95.** $3 \pm 2\sqrt{2}$ **97.** -1 **99.** 0, 1

Exercise Set 3.5

1. $-7, 7$ **3.** 0 **5.** $-\frac{5}{6}, \frac{5}{6}$ **7.** No solution **9.** $-\frac{1}{3}, \frac{1}{3}$ **11.** $-3, 3$ **13.** $-3, 5$ **15.** $-8, 4$ **17.** $-1, -\frac{1}{3}$ **19.** $-24, 44$ **21.** $-2, 4$ **23.** $-13, 7$ **25.** $-\frac{4}{3}, \frac{2}{3}$ **27.** $-\frac{3}{4}, \frac{9}{4}$ **29.** $-13, 1$ **31.** 0, 1

33. $(-7, 7)$;

35. $[-2, 2]$;

37. $(-\infty, -4.5] \cup [4.5, \infty)$;

39. $(-\infty, -3) \cup (3, \infty)$;

41. $\left(-\frac{1}{3}, \frac{1}{3}\right)$;

43. $(-\infty, -3] \cup [3, \infty)$;

45. $(-17, 1)$;

47. $(-\infty, -17] \cup [1, \infty)$;

49. $\left(-\frac{1}{4}, \frac{3}{4}\right)$;

51. $[-6, 3]$;

53. $(-\infty, 4.9) \cup (5.1, \infty)$;

55. $\left(-\infty, -\frac{1}{2}\right] \cup \left[\frac{7}{2}, \infty\right)$;

57. $\left[-\frac{7}{3}, 1\right]$;

59. $(-\infty, -8) \cup (7, \infty)$;

61. No solution **63.** $(-\infty, \infty)$ **65.** y-intercept **66.** Distance formula **67.** Relation **68.** Function **69.** Horizontal lines **70.** Parallel **71.** Decreasing **72.** Symmetric with respect to the y-axis **73.** $\left(-\infty, \frac{1}{2}\right)$ **75.** No solution **77.** $\left(-\infty, -\frac{8}{3}\right) \cup (-2, \infty)$

Review Exercises: Chapter 3

1. True **2.** True **3.** False **4.** False **5.** $-\frac{5}{2}, \frac{1}{3}$ **6.** $-5, 1$ **7.** $-2, \frac{4}{3}$ **8.** $-\sqrt{3}, \sqrt{3}$ **9.** $-\sqrt{10}i, \sqrt{10}i$

10. 1 **11.** $-5, 3$ **12.** $\frac{1 \pm \sqrt{41}}{4}$ **13.** $-\frac{1}{3} \pm \frac{2\sqrt{2}}{3}i$
14. $\frac{27}{7}$ **15.** $-\frac{1}{2}, \frac{9}{4}$ **16.** 0, 3 **17.** 5 **18.** 1, 7 **19.** $-8, 1$
20. $(-\infty, -3] \cup [3, \infty)$;
21. $\left(-\frac{14}{3}, 2\right)$;
22. $\left(-\frac{2}{3}, 1\right)$;
23. $(-\infty, -6] \cup [-2, \infty)$;
24. $P = \frac{MN}{M + N}$ **25.** $-2\sqrt{10}i$ **26.** $-4\sqrt{15}$ **27.** $-\frac{7}{8}$
28. $2 - i$ **29.** $1 - 4i$ **30.** $-18 - 26i$ **31.** $\frac{11}{10} + \frac{3}{10}i$
32. $-i$ **33.** $x^2 - 3x + \frac{9}{4} = 18 + \frac{9}{4}$; $\left(x - \frac{3}{2}\right)^2 = \frac{81}{4}$; $x - \frac{3}{2} = \pm\frac{9}{2}$; $x = \frac{3}{2} \pm \frac{9}{2}$; $-3, 6$ **34.** $3x^2 - 12x = 6$; $x^2 - 4x = 2$; $x^2 - 4x + 4 = 2 + 4$; $(x - 2)^2 = 6$; $x - 2 = \pm\sqrt{6}$; $x = 2 \pm \sqrt{6}$; $2 - \sqrt{6}, 2 + \sqrt{6}$
35. $-4, \frac{2}{3}$ **36.** $1 - 3i, 1 + 3i$ **37.** $-2, 5$ **38.** 1
39. $\pm\sqrt{\frac{3 \pm \sqrt{5}}{2}}$ **40.** $-\sqrt{3}, 0, \sqrt{3}$ **41.** $-2, -\frac{2}{3}, 3$
42. $-5, -2, 2$ **43.** **(a)** $\left(\frac{3}{8}, -\frac{7}{16}\right)$; **(b)** $x = \frac{3}{8}$; **(c)** maximum: $-\frac{7}{16}$; **(d)** $\left(-\infty, -\frac{7}{16}\right]$;
(e)

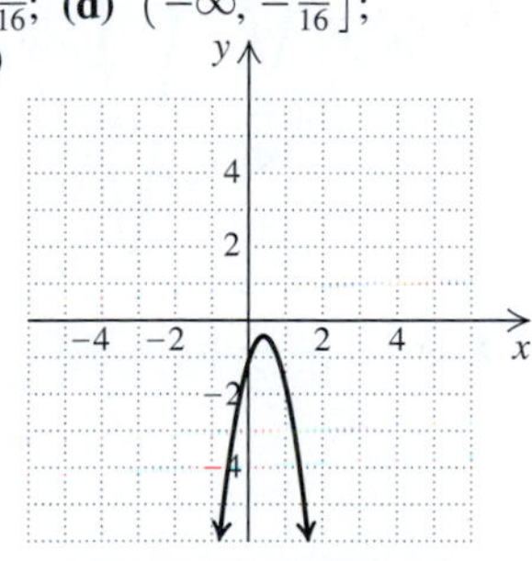

$f(x) = -4x^2 + 3x - 1$

44. **(a)** $(1, -2)$; **(b)** $x = 1$; **(c)** minimum: -2; **(d)** $[-2, \infty)$;
(e)

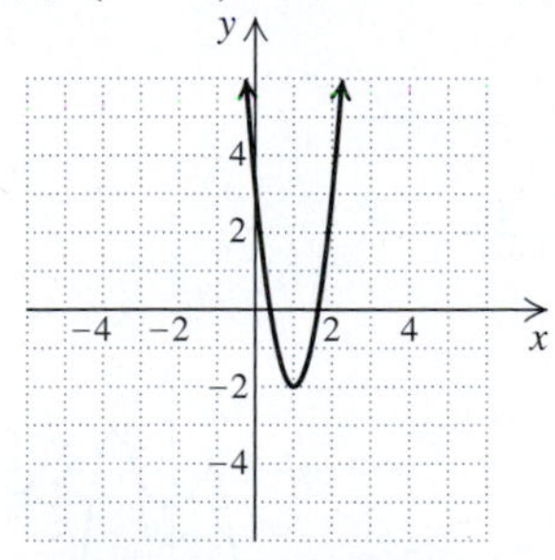

$f(x) = 5x^2 - 10x + 3$

45. (d) **46.** (c) **47.** (b) **48.** (a) **49.** 30 ft, 40 ft
50. Rebecca: 15 km/h; Harry: 8 km/h **51.** $35 - 5\sqrt{33}$ ft, or about 6.3 ft **52.** 6 ft by 6 ft **53.** $\frac{15 - \sqrt{115}}{2}$ cm, or about 2.1 cm **54.** B **55.** B **56.** A **57.** 256
58. $4 \pm 3\sqrt[4]{3}$, or 0.052, 7.948 **59.** $-7, 9$ **60.** $-\frac{1}{4}, 2$
61. -1 **62.** 9% **63.** ± 6 **64.** The product of two imaginary numbers is not always an imaginary number. For example, $i \cdot i = i^2 = -1$, a real number. **65.** No; consider the quadratic formula $x = \frac{-b \pm \sqrt{b^2 - 4ac}}{2a}$. If $b^2 - 4ac = 0$, then $x = \frac{-b}{2a}$, so there is one real zero. If $b^2 - 4ac > 0$, then $\sqrt{b^2 - 4ac}$ is a real number and there are two real zeros. If $b^2 - 4ac < 0$, then $\sqrt{b^2 - 4ac}$ is an imaginary number and there are two imaginary zeros. Thus a quadratic function cannot have one real zero and one imaginary zero. **66.** You can conclude that $|a_1| = |a_2|$ since these constants determine how wide the parabolas are. Nothing can be concluded about the h's and the k's. **67.** When both sides of an equation are multiplied by the LCD, the resulting equation might not be equivalent to the original equation. One or more of the possible solutions of the resulting equation might make a denominator of the original equation 0. **68.** When both sides of an equation are raised to an even power, the resulting equation might not be equivalent to the original equation. For example, the solution set of $x = -2$ is $\{-2\}$, but the solution set of $x^2 = (-2)^2$, or $x^2 = 4$, is $\{-2, 2\}$. **69.** Absolute value is nonnegative.
70. $|x| \geq 0 > p$ for any real number x.

Test: Chapter 3

1. [3.2] $\frac{1}{2}, -5$ **2.** [3.2] $-\sqrt{6}, \sqrt{6}$ **3.** [3.2] $-2i, 2i$
4. [3.2] $-1, 3$ **5.** [3.2] $\frac{5 \pm \sqrt{13}}{2}$ **6.** [3.2] $\frac{3}{4} \pm \frac{\sqrt{23}}{4}i$
7. [3.2] 16 **8.** [3.4] $-1, \frac{13}{6}$ **9.** [3.4] 5 **10.** [3.4] 5
11. [3.5] $-11, 3$ **12.** [3.5] $-\frac{1}{2}, 2$
13. [3.5] $[-7, 1]$;
14. [3.5] $(-2, 3)$;
15. [3.5] $(-\infty, -7) \cup (-3, \infty)$;
16. [3.5] $\left(-\infty, -\frac{2}{3}\right] \cup [4, \infty)$;
17. [3.4] $B = \frac{AC}{A - C}$ **18.** [3.4] $n = \frac{R^2}{3p}$
19. [3.2] $x^2 + 4x = 1$; $x^2 + 4x + 4 = 1 + 4$; $(x + 2)^2 = 5$; $x + 2 = \pm\sqrt{5}$; $x = -2 \pm \sqrt{5}$; $-2 - \sqrt{5}, -2 + \sqrt{5}$
20. [3.2] About 11.4 sec **21.** [3.1] $\sqrt{43}i$ **22.** [3.1] $-5i$
23. [3.1] $3 - 5i$ **24.** [3.1] $10 + 5i$ **25.** [3.1] $\frac{1}{10} - \frac{1}{5}i$
26. [3.1] i **27.** [3.2] $-\frac{1}{4}, 3$ **28.** [3.2] $\frac{1 \pm \sqrt{57}}{4}$
29. [3.3] **(a)** $(1, 9)$; **(b)** $x = 1$; **(c)** maximum: 9; **(d)** $(-\infty, 9]$;
(e)

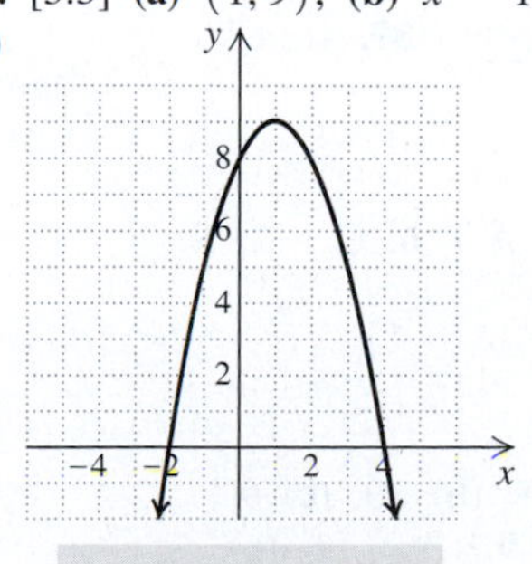

$f(x) = -x^2 + 2x + 8$

30. [3.3] 20 ft by 40 ft **31.** [3.3] C **32.** [3.3], [3.4] $-\frac{4}{9}$

Chapter 4

Exercise Set 4.1

1. $\frac{1}{2}x^3$; $\frac{1}{2}$; 3; cubic **3.** $0.9x$; 0.9; 1; linear **5.** $305x^4$; 305; 4; quartic **7.** x^4; 1; 4; quartic **9.** $4x^3$; 4; 3; cubic **11.** (d) **13.** (b) **15.** (c) **17.** (a) **19.** (c) **21.** (d) **23.** Yes; no; no **25.** No; yes; yes **27.** -3, multiplicity 2; 1, multiplicity 1 **29.** 4, multiplicity 3; -6, multiplicity 1 **31.** ± 3, each has multiplicity 3 **33.** 0, multiplicity 3; 1, multiplicity 2; -4, multiplicity 1 **35.** 3, multiplicity 2; -4, multiplicity 3; 0, multiplicity 4 **37.** $\pm\sqrt{3}$, ± 1, each has multiplicity 1 **39.** -3, ± 1, each has multiplicity 1 **41.** ± 2, $\frac{1}{2}$, each has multiplicity 1 **43.** -1.532, -0.347, 1.879 **45.** -1.414, 0, 1.414 **47.** -1, 0, 1 **49.** -10.153, -1.871, -0.821, -0.303, 0.098, 0.535, 1.219, 3.297 **51.** -1.386; relative maximum: 1.506 at $x = -0.632$, relative minimum: 0.494 at $x = 0.632$; $(-\infty, \infty)$ **53.** -1.249, 1.249; relative minimum: -3.8 at $x = 0$, no relative maxima; $[-3.8, \infty)$ **55.** -1.697, 0, 1.856; relative maximum: 11.012 at $x = 1.258$, relative minimum: -8.183 at $x = -1.116$; $(-\infty, \infty)$ **57.** False **59.** True **61.** 2008: 1.7 million albums; 2012: 4.9 million albums; 2016: 15.6 million albums **63.** 2009: about 771 million print books; 2014: about 635 million print books; 2015: about 795 million print books **65.** 5 sec **67.** 2003: 684,025 admissions; 2006: 739,119 admissions; 2011: 665,806 admissions **69.** 6.3% **71.** (b) **73.** (c) **75.** (a) **77.** **(a)** Quadratic: $y = -83.83100233x^2 - 930.0808858x + 23{,}360.76224$, $R^2 \approx 0.9444$; cubic: $y = 52.94133644x^3 - 877.951049x^2 + 2098.163559x + 21{,}454.87413$, $R^2 \approx 0.9911$; quartic: $y = -2.758741259x^4 + 108.1161616x^3 - 1222.793706x^2 + 2787.848873x + 21{,}256.24476$, $R^2 \approx 0.9919$; the R^2-value for the quartic function, 0.9919, is the closest to 1; thus the quartic function is the best fit; **(b)** 2014: 7476 foreign adoptions **79.** **(a)** Cubic: $y = -0.0018497401x^3 + 0.0678652219x^2 + 0.0636134735x + 2.244194051$, $R^2 \approx 0.8658$; quartic: $y = 0.000024423807x^4 - 0.0037660904x^3 + 0.11514113x^2 - 0.3046875884x + 2.594806309$, $R^2 \approx 0.8686$; the R^2-value for the quartic function, 0.8686, is the closest to 1; thus the quartic function is the better fit. **(b)** 1988: \$10.516 billion; 2002: \$17.158 billion **81.** 5 **82.** $6\sqrt{2}$ **83.** Center: $(3, -5)$; radius: 7 **84.** Center: $(-4, 3)$; radius: $2\sqrt{2}$ **85.** $\{y \mid y \geq 3\}$, or $[3, \infty)$ **86.** $\{x \mid x > \frac{5}{3}\}$, or $(\frac{5}{3}, \infty)$ **87.** $\{x \mid x \leq -13 \text{ or } x \geq 1\}$, or $(-\infty, -13] \cup [1, \infty)$ **88.** $\{x \mid -\frac{11}{12} \leq x \leq \frac{5}{12}\}$, or $[-\frac{11}{12}, \frac{5}{12}]$ **89.** 16; x^{16}

Visualizing the Graph

1. H **2.** D **3.** J **4.** B **5.** A **6.** C **7.** I **8.** E **9.** G **10.** F

Exercise Set 4.2

1. **(a)** 5; **(b)** 5; **(c)** 4 **3.** **(a)** 10; **(b)** 10; **(c)** 9 **5.** **(a)** 3; **(b)** 3; **(c)** 2 **7.** (d) **9.** (f) **11.** (b)

13.

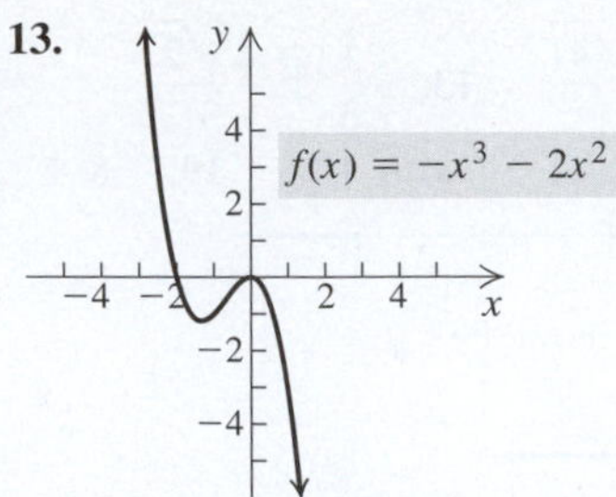

$f(x) = -x^3 - 2x^2$

15.

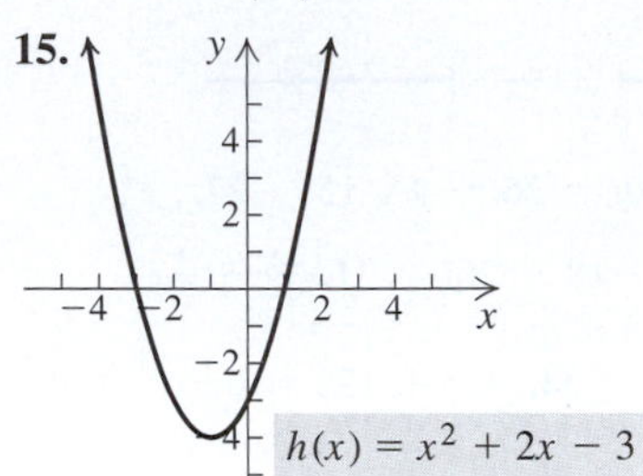

$h(x) = x^2 + 2x - 3$

17.

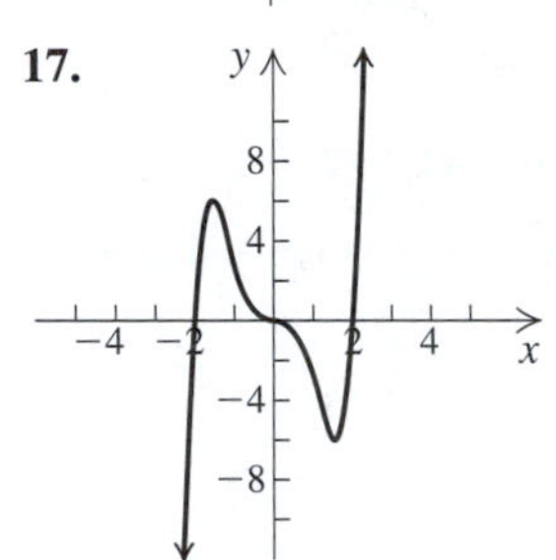

$h(x) = x^5 - 4x^3$

19.

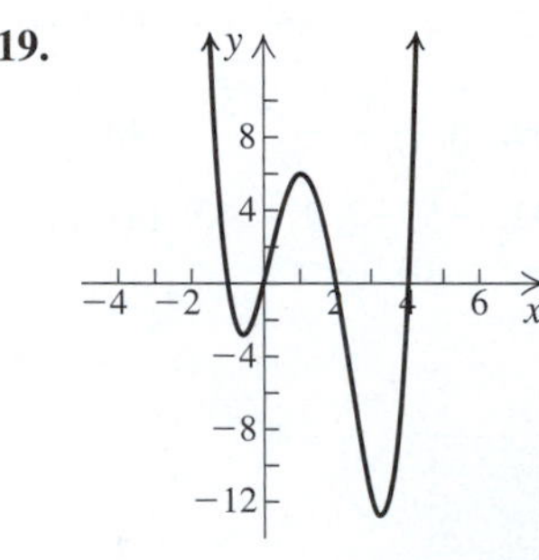

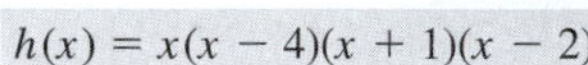

$h(x) = x(x - 4)(x + 1)(x - 2)$

21.

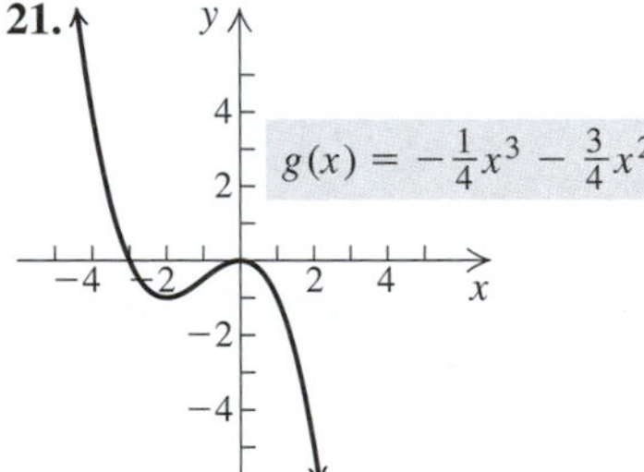

$g(x) = -\frac{1}{4}x^3 - \frac{3}{4}x^2$

23.

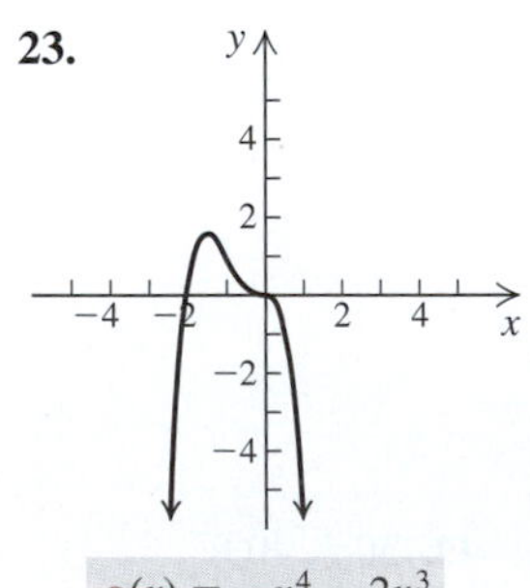

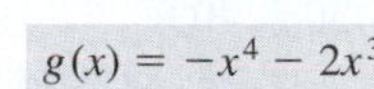

$g(x) = -x^4 - 2x^3$

25.

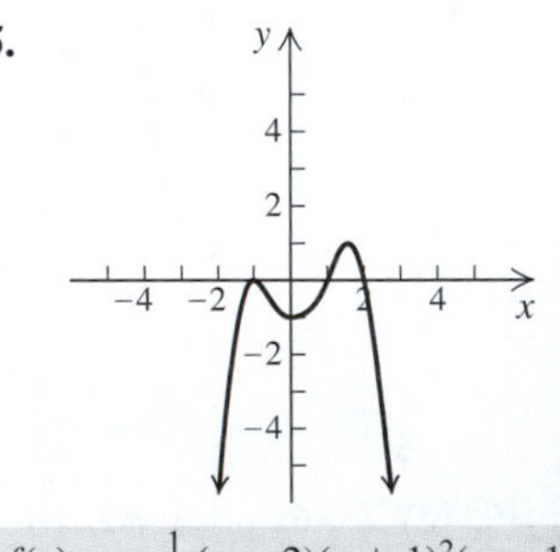

$f(x) = -\frac{1}{2}(x - 2)(x + 1)^2(x - 1)$

27. 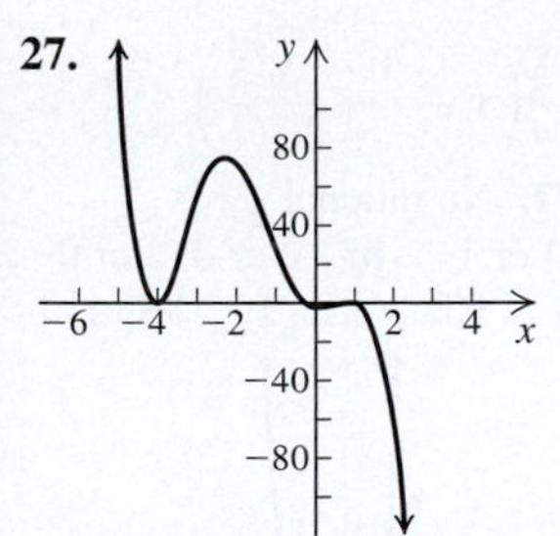

$g(x) = -x(x-1)^2(x+4)^2$

29. 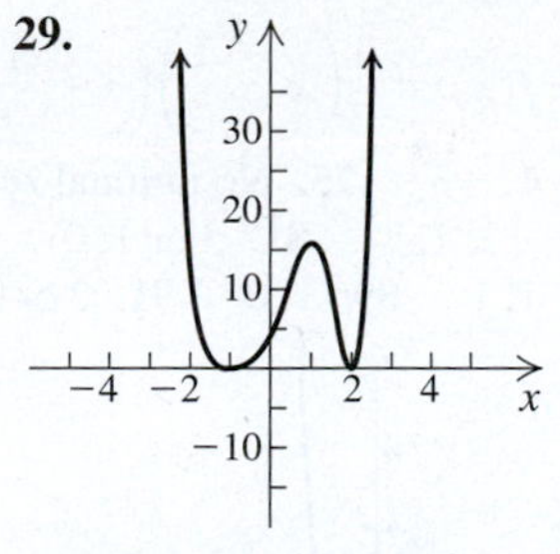

$f(x) = (x-2)^2(x+1)^4$

31. 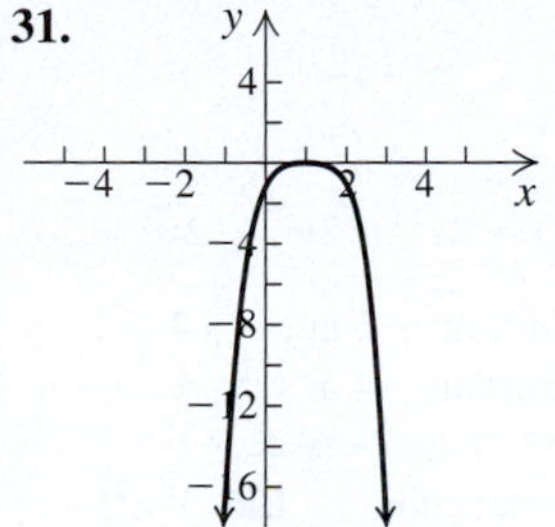

$g(x) = -(x-1)^4$

33. 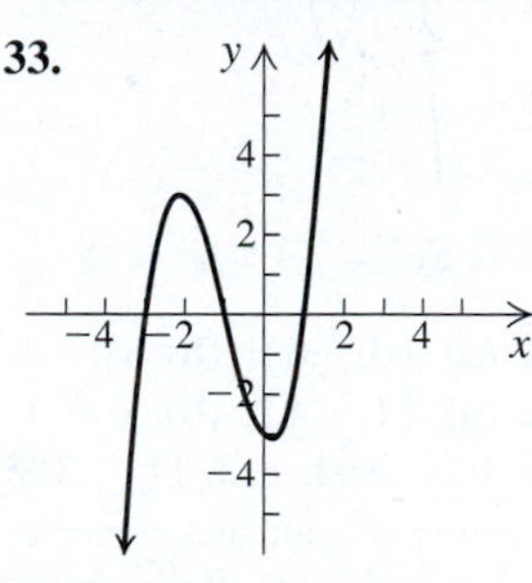

$h(x) = x^3 + 3x^2 - x - 3$

35. 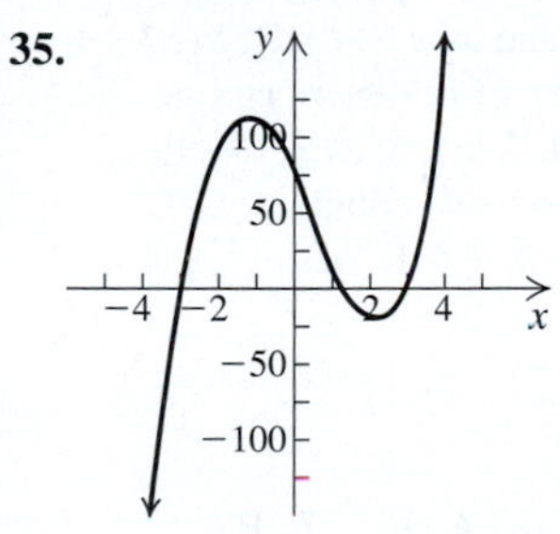

$f(x) = 6x^3 - 8x^2 - 54x + 72$

37. 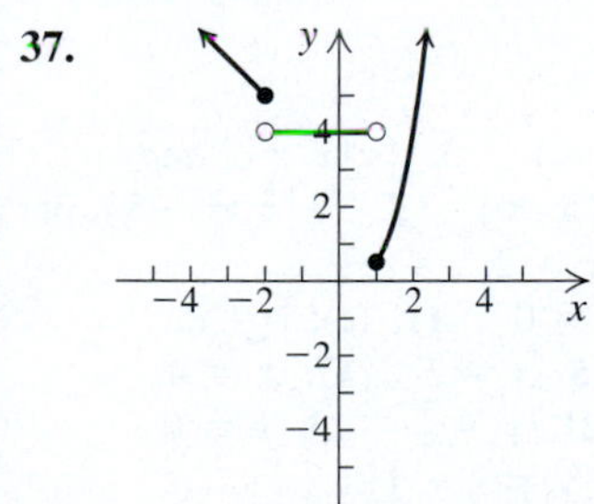

$$g(x) = \begin{cases} -x+3, & \text{for } x \le -2, \\ 4, & \text{for } -2 < x < 1, \\ \frac{1}{2}x^3, & \text{for } x \ge 1 \end{cases}$$

39. $f(-5) = -18$ and $f(-4) = 7$. By the intermediate value theorem, since $f(-5)$ and $f(-4)$ have opposite signs, then $f(x)$ has a zero between -5 and -4. **41.** $f(-3) = 22$ and $f(-2) = 5$. Both $f(-3)$ and $f(-2)$ are positive. We cannot use the intermediate value theorem to determine if there is a zero between -3 and -2. **43.** $f(2) = 2$ and $f(3) = 57$. Both $f(2)$ and $f(3)$ are positive. We cannot use the intermediate value theorem to determine if there is a zero between 2 and 3. **45.** $f(4) = -12$ and $f(5) = 4$. By the intermediate value theorem, since $f(4)$ and $f(5)$ have opposite signs, then $f(x)$ has a zero between 4 and 5. **47.** (d) **48.** (f) **49.** (e) **50.** (a) **51.** (b) **52.** (c) **53.** $\frac{9}{10}$ **54.** $-3, 0, 4$ **55.** $-\frac{5}{3}, \frac{11}{2}$ **56.** $\frac{196}{25}$

Exercise Set 4.3

1. **(a)** No; **(b)** yes; **(c)** no **3.** **(a)** Yes; **(b)** no; **(c)** yes
5. $P(x) = (x+2)(x^2 - 2x + 4) - 16$
7. $P(x) = (x+9)(x^2 - 3x + 2) + 0$
9. $P(x) = (x+2)(x^3 - 2x^2 + 2x - 4) + 11$
11. $Q(x) = 2x^3 + x^2 - 3x + 10, R(x) = -42$
13. $Q(x) = x^2 - 4x + 8, R(x) = -24$
15. $Q(x) = 3x^2 - 4x + 8, R(x) = -18$
17. $Q(x) = x^4 + 3x^3 + 10x^2 + 30x + 89, R(x) = 267$
19. $Q(x) = x^3 + x^2 + x + 1, R(x) = 0$
21. $Q(x) = 2x^3 + x^2 + \frac{7}{2}x + \frac{7}{4}, R(x) = -\frac{1}{8}$ **23.** $0; -60; 0$
25. 10; 80; 998 **27.** 5,935,988; −772 **29.** 0; 0; 65; $1 - 12\sqrt{2}$ **31.** Yes; no **33.** Yes; yes **35.** No; yes
37. No; no **39.** $f(x) = (x-1)(x+2)(x+3); 1, -2, -3$
41. $f(x) = (x-2)(x-5)(x+1); 2, 5, -1$
43. $f(x) = (x-2)(x-3)(x+4); 2, 3, -4$
45. $f(x) = (x-3)^3(x+2); 3, -2$
47. $f(x) = (x-1)(x-2)(x-3)(x+5); 1, 2, 3, -5$

49. 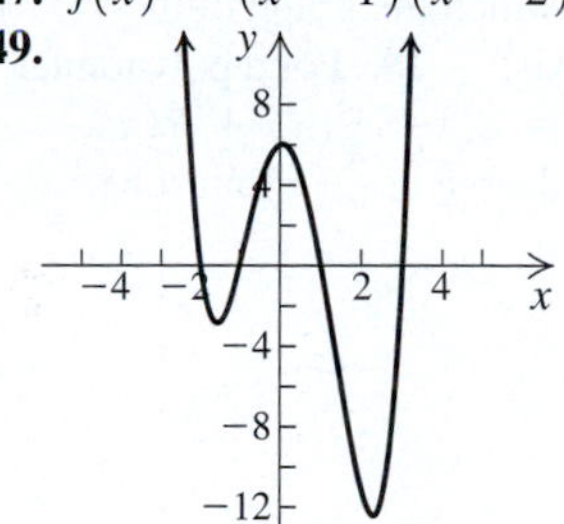

$f(x) = x^4 - x^3 - 7x^2 + x + 6$

51. 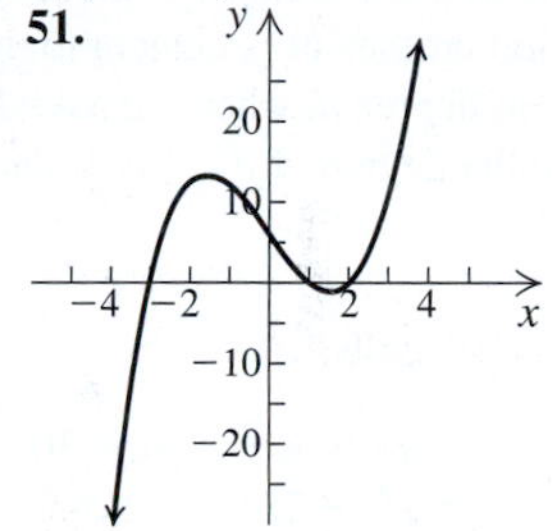

$f(x) = x^3 - 7x + 6$

53. 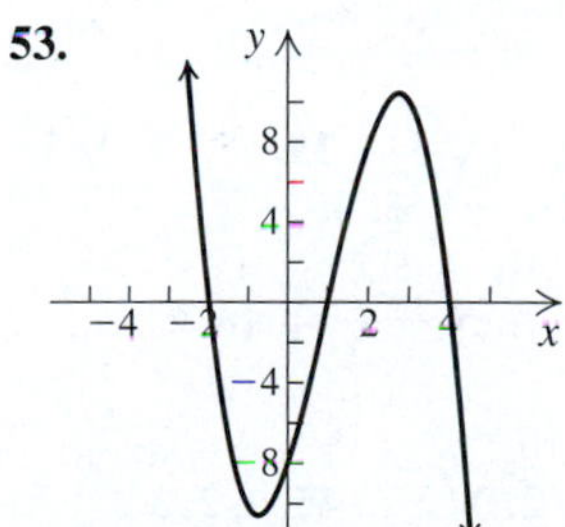

$f(x) = -x^3 + 3x^2 + 6x - 8$

55. $\frac{5}{4} \pm \frac{\sqrt{71}}{4}i$ **56.** $-1, \frac{3}{7}$ **57.** $-5, 0$ **58.** 10
59. $-3, -2$ **60.** $f(x) = 1313x + 339$; 2011: 5591 cases, 2017: 13,469 cases **61.** $b = 15$ in., $h = 15$ in.
63. **(a)** $x + 4, x + 3, x - 2, x - 5$;
(b) $P(x) = (x+4)(x+3)(x-2)(x-5)$; **(c)** yes; two examples are $f(x) = c \cdot P(x)$ for any nonzero constant c and $g(x) = (x-a)P(x)$; **(d)** no
65. $\frac{14}{3}$ **67.** $0, -6$ **69.** Answers may vary. One possibility is $P(x) = x^{15} - x^{14}$. **71.** $x^2 + 2ix + (2 - 4i), R\ -6 - 2i$
73. $x - 3 + i, R\ 6 - 3i$

Mid-Chapter Mixed Review: Chapter 4

1. False **2.** True **3.** True **4.** False **5.** 5; multiplicity 6
6. $-5, -\frac{1}{2}, 5$; each has multiplicity 1 **7.** $\pm 1, \pm\sqrt{2}$; each has multiplicity 1 **8.** 3, multiplicity 2; −4, multiplicity 1
9. (d) **10.** (a) **11.** (b) **12.** (c) **13.** $f(-2) = -13$ and $f(0) = 3$. By the intermediate value theorem, since $f(-2)$ and $f(0)$ have opposite signs, then $f(x)$ has a zero between -2 and 0.

14. $f\left(-\frac{1}{2}\right) = \frac{19}{8}$ and $f(1) = 2$. Both $f\left(-\frac{1}{2}\right)$ and $f(1)$ are positive. We cannot use the intermediate value theorem to determine if there is a zero between $-\frac{1}{2}$ and 1.
15. $P(x) = (x - 1)(x^3 - 5x^2 - 5x - 4) - 6$
16. $Q(x) = 3x^3 + 5x^2 + 12x + 18$, $R(x) = 42$
17. $Q(x) = x^4 - x^3 + x^2 - x + 1$, $R(x) = -6$
18. -380 **19.** -15 **20.** $20 - \sqrt{2}$ **21.** Yes; no
22. Yes; yes **23.** $f(x) = (x - 1)(x - 8)(x + 7)$; 1, 8, -7
24. $f(x) = (x + 1)(x - 2)(x - 4)(x + 3)$; -1, 2, 4, -3
25. The range of a polynomial function with an odd degree is $(-\infty, \infty)$. The range of a polynomial function with an even degree is $[s, \infty)$ for some real number s if $a_n > 0$ and is $(-\infty, s]$ for some real number s if $a_n < 0$. **26.** Since we can find $f(0)$ for any polynomial function $f(x)$, it is not possible for the graph of a polynomial function to have no y-intercept. It is possible for a polynomial function to have no x-intercepts. For instance, a function of the form $f(x) = x^2 + a$, $a > 0$, has no x-intercepts. There are other examples as well. **27.** The zeros of a polynomial function are the first coordinates of the points at which the graph of the function crosses or is tangent to the x-axis. **28.** For a polynomial $P(x)$ of degree n, when we have $P(x) = d(x) \cdot Q(x) + R(x)$, where the degree of $d(x)$ is 1, then the degree of $Q(x)$ must be $n - 1$.

Exercise Set 4.4

1. $f(x) = x^3 - 6x^2 - x + 30$
3. $f(x) = x^3 + 3x^2 + 4x + 12$
5. $f(x) = x^3 - 3x^2 - 2x + 6$ **7.** $f(x) = x^3 - 6x - 4$
9. $f(x) = x^3 + 2x^2 + 29x + 148$
11. $f(x) = x^3 - \frac{5}{3}x^2 - \frac{2}{3}x$
13. $f(x) = x^5 + 2x^4 - 2x^2 - x$
15. $f(x) = x^4 + 3x^3 + 3x^2 + x$ **17.** $-\sqrt{3}$ **19.** i, $2 + \sqrt{5}$
21. $-3i$ **23.** $-4 + 3i$, $2 + \sqrt{3}$ **25.** $-\sqrt{5}$, $4i$
27. $2 + i$ **29.** $-3 - 4i$, $4 + \sqrt{5}$ **31.** $4 + i$
33. $f(x) = x^3 - 4x^2 + 6x - 4$ **35.** $f(x) = x^2 + 16$
37. $f(x) = x^3 - 5x^2 + 16x - 80$
39. $f(x) = x^4 - 2x^3 - 3x^2 + 10x - 10$
41. $f(x) = x^4 + 4x^2 - 45$ **43.** $-\sqrt{2}$, $\sqrt{2}$ **45.** i, 2, 3
47. $1 + 2i$, $1 - 2i$ **49.** ± 1 **51.** ± 1, $\pm\frac{1}{2}$, ± 2, ± 4, ± 8
53. ± 1, ± 2, $\pm\frac{1}{3}$, $\pm\frac{1}{5}$, $\pm\frac{2}{3}$, $\pm\frac{2}{5}$, $\pm\frac{1}{15}$, $\pm\frac{2}{15}$ **55. (a)** Rational: -3; other: $\pm\sqrt{2}$; **(b)** $f(x) = (x + 3)(x + \sqrt{2})(x - \sqrt{2})$
57. (a) Rational: $\frac{1}{3}$; other: $\pm\sqrt{5}$;
(b) $f(x) = 3\left(x - \frac{1}{3}\right)(x + \sqrt{5})(x - \sqrt{5})$, or $(3x - 1)(x + \sqrt{5})(x - \sqrt{5})$ **59. (a)** Rational: -2, 1; other: none; **(b)** $f(x) = (x + 2)(x - 1)^2$ **61. (a)** Rational: $-\frac{3}{2}$; other: $\pm 3i$; **(b)** $f(x) = 2\left(x + \frac{3}{2}\right)(x + 3i)(x - 3i)$, or $(2x + 3)(x + 3i)(x - 3i)$ **63. (a)** Rational: $-\frac{1}{5}$, 1; other: $\pm 2i$; **(b)** $f(x) = 5\left(x + \frac{1}{5}\right)(x - 1)(x + 2i)(x - 2i)$
65. (a) Rational: -2, -1; other: $3 \pm \sqrt{13}$;
(b) $f(x) = (x + 2)(x + 1)(x - 3 - \sqrt{13})(x - 3 + \sqrt{13})$
67. (a) Rational: 2; other: $1 \pm \sqrt{3}$;
(b) $f(x) = (x - 2)(x - 1 - \sqrt{3})(x - 1 + \sqrt{3})$
69. (a) Rational: -2; other: $1 \pm \sqrt{3}i$;
(b) $f(x) = (x + 2)(x - 1 - \sqrt{3}i)(x - 1 + \sqrt{3}i)$
71. (a) Rational: $\frac{1}{2}$; other: $\frac{1 \pm \sqrt{5}}{2}$;
(b) $f(x) = \frac{1}{3}\left(x - \frac{1}{2}\right)\left(x - \frac{1 + \sqrt{5}}{2}\right)\left(x - \frac{1 - \sqrt{5}}{2}\right)$
73. 1, -3 **75.** No rational zeros **77.** No rational zeros
79. -2, 1, 2 **81.** 3 or 1; 0 **83.** 0; 3 or 1 **85.** 2 or 0; 2 or 0
87. 1; 1 **89.** 1; 0 **91.** 2 or 0; 2 or 0 **93.** 3 or 1; 1 **95.** 1; 1
97.

$f(x) = 4x^3 + x^2 - 8x - 2$

99.

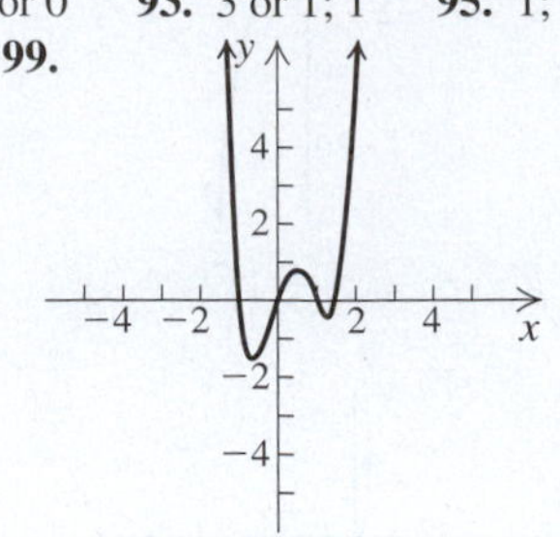

$f(x) = 2x^4 - 3x^3 - 2x^2 + 3x$

101. (a) $(4, -6)$; **(b)** $x = 4$; **(c)** minimum: -6 at $x = 4$
102. (a) $(1, -4)$; **(b)** $x = 1$; **(c)** minimum: -4 at $x = 1$
103. 10 **104.** -3, 11 **105.** $-x^3$; -1; 3; as $x \to \infty$, $g(x) \to -\infty$, and as $x \to -\infty$, $g(x) \to \infty$; cubic **106.** $-x^2$; -1; 2; as $x \to \infty$, $f(x) \to -\infty$, and as $x \to -\infty$, $f(x) \to -\infty$; quadratic **107.** $-\frac{4}{9}$; $-\frac{4}{9}$; 0 degree; for all x, $f(x) = -\frac{4}{9}$; constant **108.** x; 1; 1; as $x \to \infty$, $h(x) \to \infty$, and as $x \to -\infty$, $h(x) \to -\infty$; linear **109.** x^4; 1; 4; as $x \to \infty$, $g(x) \to \infty$, and as $x \to -\infty$, $g(x) \to \infty$; quartic **110.** x^3; 1; 3; as $x \to \infty$, $h(x) \to \infty$, and as $x \to -\infty$, $h(x) \to -\infty$; cubic
111. (a) -1, $\frac{1}{2}$, 3; **(b)** 0, $\frac{3}{2}$, 4; **(c)** -3, $-\frac{3}{2}$, 1; **(d)** $-\frac{1}{2}$, $\frac{1}{4}$, $\frac{3}{2}$
113. -8, $-\frac{3}{2}$, 4, 7, 15

Visualizing the Graph

1. A **2.** C **3.** D **4.** H **5.** G **6.** F **7.** B
8. I **9.** J **10.** E

Exercise Set 4.5

1. $\{x | x \neq 2\}$, or $(-\infty, 2) \cup (2, \infty)$ **3.** $\{x | x \neq 1$ *and* $x \neq 5\}$, or $(-\infty, 1) \cup (1, 5) \cup (5, \infty)$ **5.** $\{x | x \neq -5\}$, or $(-\infty, -5) \cup (-5, \infty)$ **7.** (d); $x = 2$, $x = -2$, $y = 0$ **9.** (e); $x = 2$, $x = -2$, $y = 0$ **11.** (c); $x = 2$, $x = -2$, $y = 8x$ **13.** $x = 0$ **15.** $x = 2$ **17.** $x = 4$, $x = -6$ **19.** $x = \frac{3}{2}$, $x = -1$ **21.** $y = \frac{3}{4}$ **23.** $y = 0$
25. No horizontal asymptote **27.** $y = x + 1$
29. $y = x$ **31.** $y = x - 3$
33. Domain: $(-\infty, 0) \cup (0, \infty)$; no x-intercepts, no y-intercept;

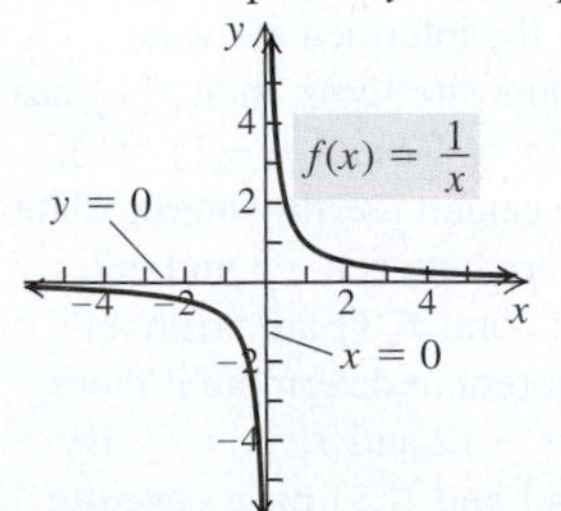

35. Domain: $(-\infty, 0) \cup (0, \infty)$; no x-intercepts, no y-intercept;

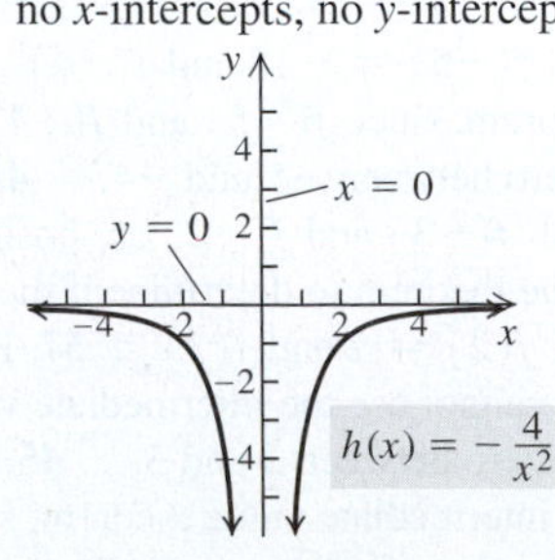

37. Domain: $(-\infty, -1) \cup (-1, \infty)$; x-intercepts: $(1, 0)$ and $(3, 0)$, y-intercept: $(0, 3)$;

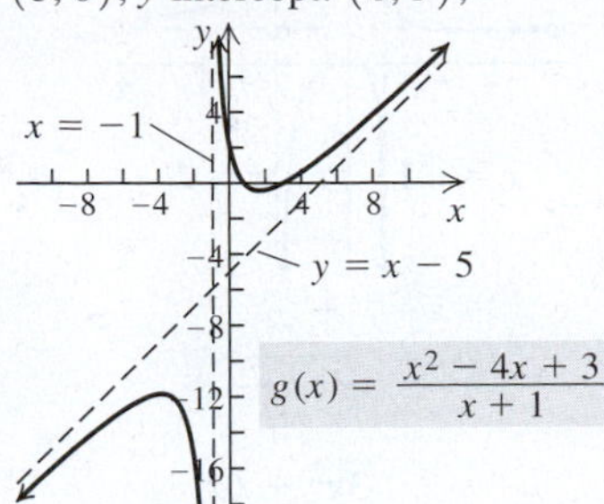

39. Domain: $(-\infty, 5) \cup (5, \infty)$; no x-intercepts, y-intercept: $\left(0, \frac{2}{5}\right)$;

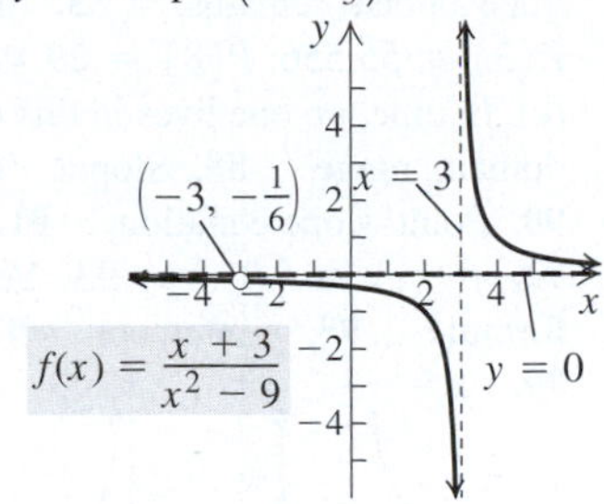

41. Domain: $(-\infty, 0) \cup (0, \infty)$; x-intercept: $\left(-\frac{1}{2}, 0\right)$, no y-intercept;

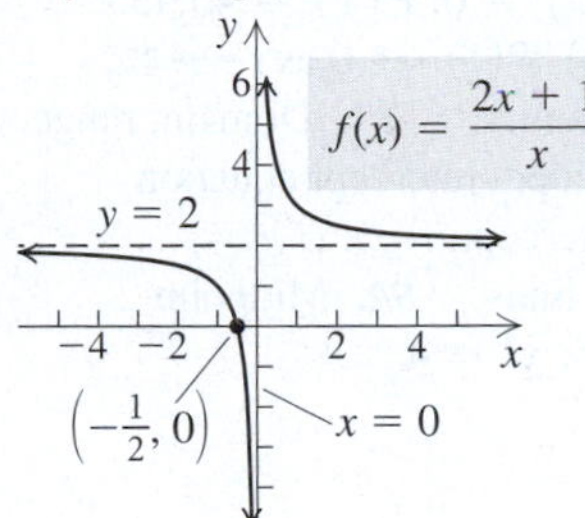

43. Domain: $(-\infty, -3) \cup (-3, 3) \cup (3, \infty)$; no x-intercepts, y-intercept: $\left(0, -\frac{1}{3}\right)$;

45. Domain: $(-\infty, -3) \cup (-3, 0) \cup (0, \infty)$; no x-intercepts, no y-intercept;

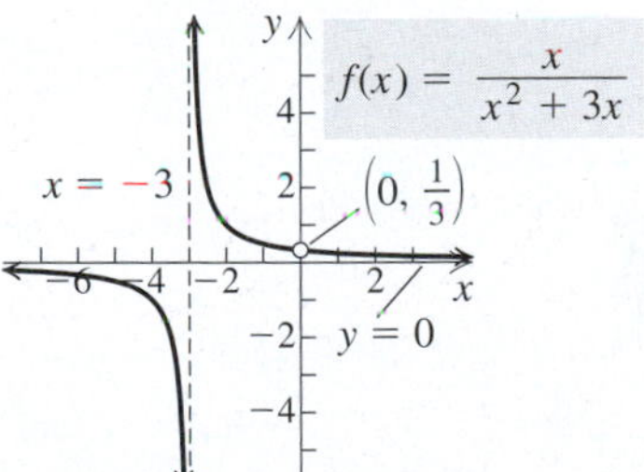

47. Domain: $(-\infty, 2) \cup (2, \infty)$; no x-intercepts, y-intercept: $\left(0, \frac{1}{4}\right)$;

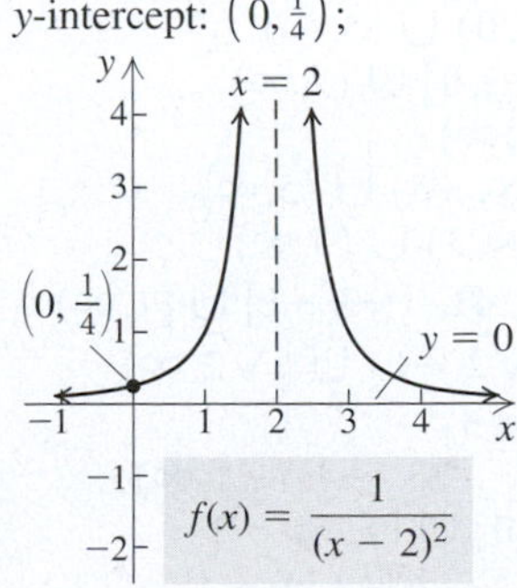

49. Domain: $(-\infty, -3) \cup (-3, -1) \cup (-1, \infty)$; x-intercept: $(1, 0)$, y-intercept: $(0, -1)$;

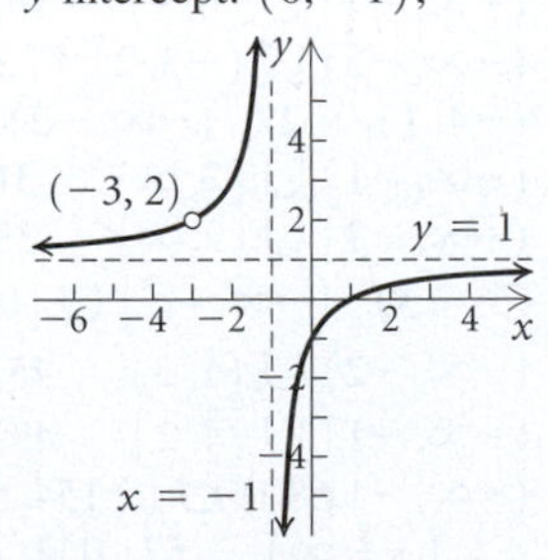

51. Domain: $(-\infty, \infty)$; no x-intercepts, y-intercept: $\left(0, \frac{1}{3}\right)$;

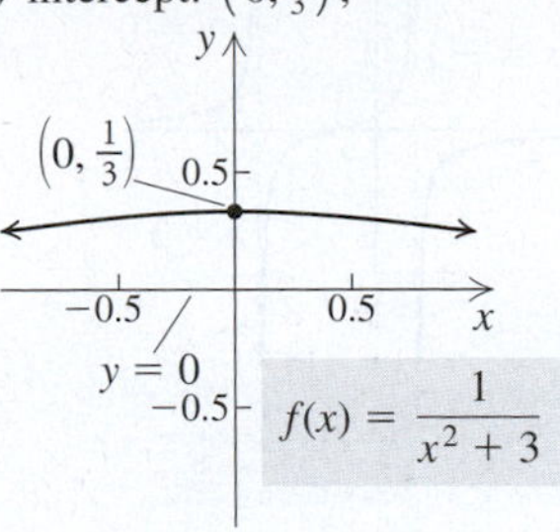

53. Domain: $(-\infty, 2) \cup (2, \infty)$; x-intercept: $(-2, 0)$, y-intercept: $(0, 2)$;

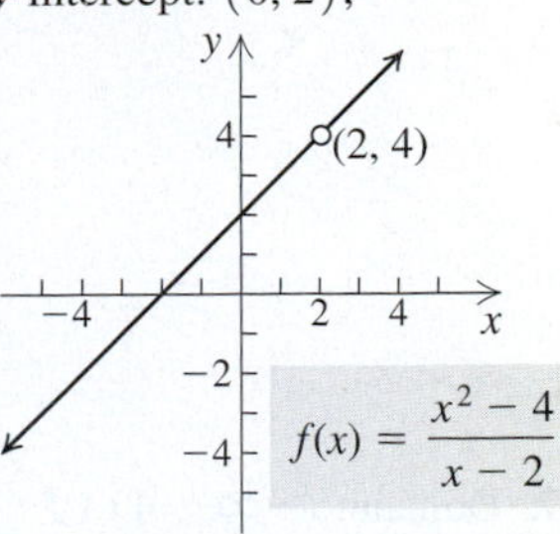

55. Domain: $(-\infty, -2) \cup (-2, \infty)$; x-intercept: $(1, 0)$, y-intercept: $\left(0, -\frac{1}{2}\right)$;

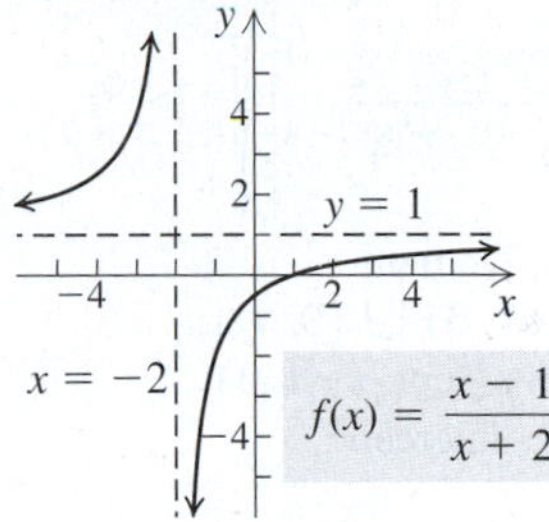

57. Domain: $\left(-\infty, -\frac{1}{2}\right) \cup \left(-\frac{1}{2}, 0\right) \cup (0, 3) \cup (3, \infty)$; x-intercept: $(-3, 0)$, no y-intercept;

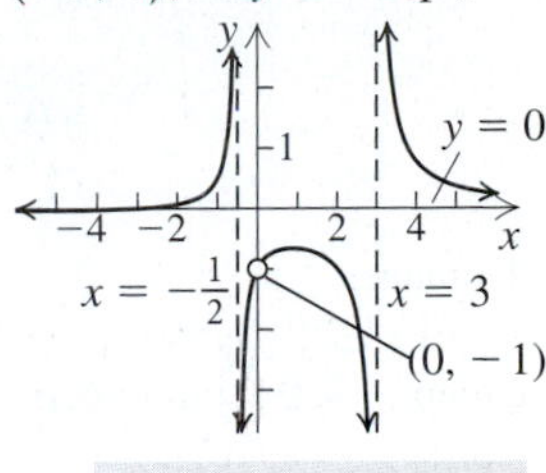

59. Domain: $(-\infty, -1) \cup (-1, \infty)$; x-intercepts: $(-3, 0)$ and $(3, 0)$, y-intercept: $(0, -9)$;

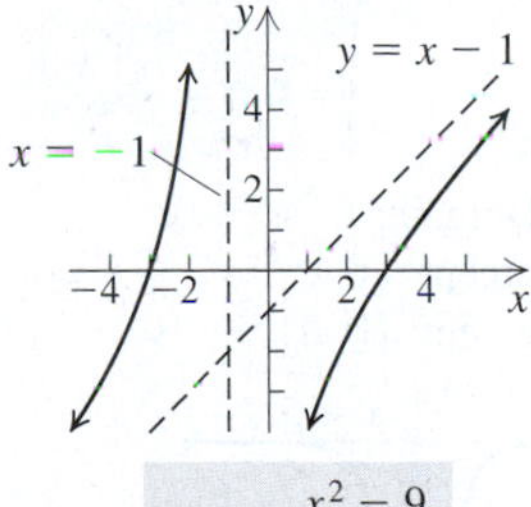

61. Domain: $(-\infty, \infty)$; x-intercepts: $(-2, 0)$ and $(1, 0)$, y-intercept: $(0, -2)$;

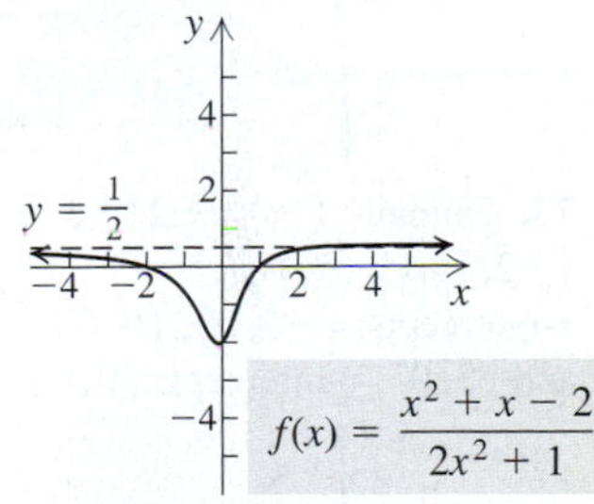

63. Domain: $(-\infty, 1) \cup (1, \infty)$; x-intercept: $\left(-\frac{2}{3}, 0\right)$, y-intercept: $(0, 2)$;

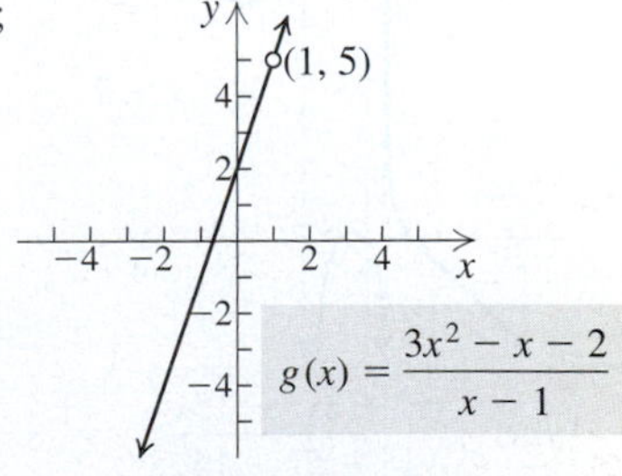

65. Domain: $(-\infty, -1) \cup (-1, 3) \cup (3, \infty)$; x-intercept: $(1, 0)$, y-intercept: $\left(0, \frac{1}{3}\right)$;

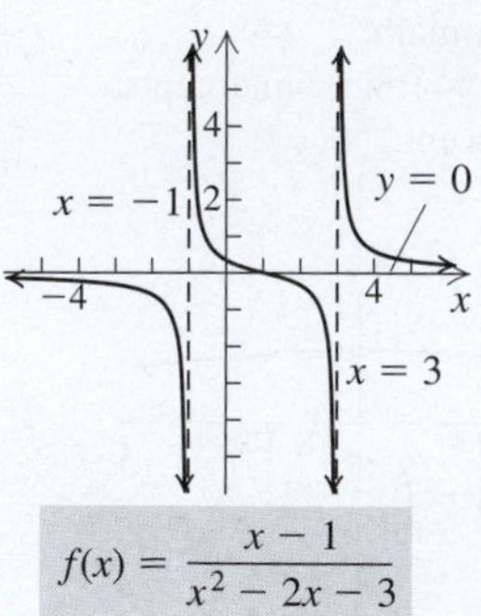

67. Domain: $(-\infty, -4) \cup (-4, 2) \cup (2, \infty)$; x-intercept: $\left(\frac{1}{3}, 0\right)$, y-intercept: $\left(0, \frac{1}{2}\right)$;

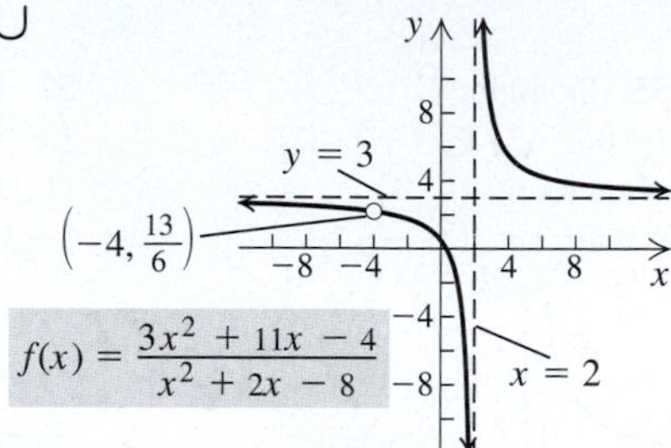

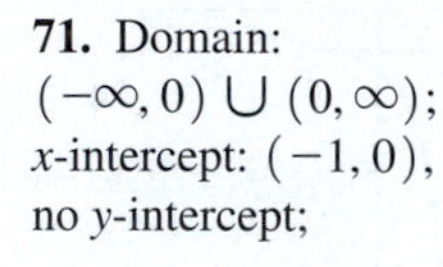

69. Domain: $(-\infty, -1) \cup (-1, \infty)$; x-intercept: $(3, 0)$, y-intercept: $(0, -3)$;

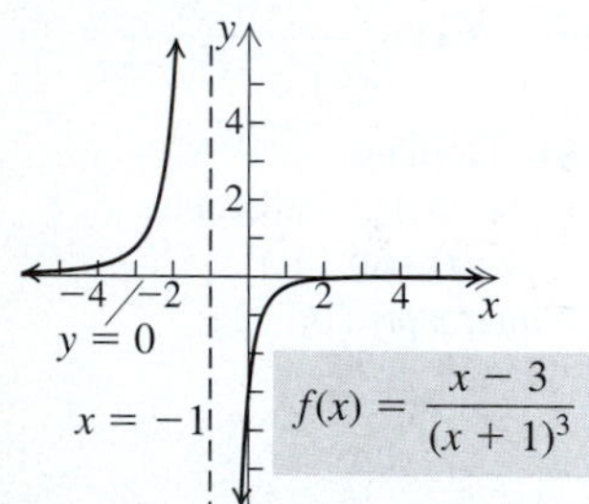

71. Domain: $(-\infty, 0) \cup (0, \infty)$; x-intercept: $(-1, 0)$, no y-intercept;

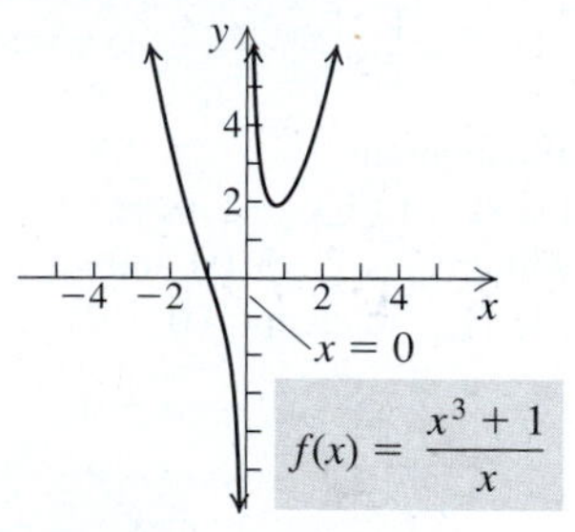

73. Domain: $(-\infty, -2) \cup (-2, 7) \cup (7, \infty)$; x-intercepts: $(-5, 0)$, $(0, 0)$, and $(3, 0)$, y-intercept: $(0, 0)$;

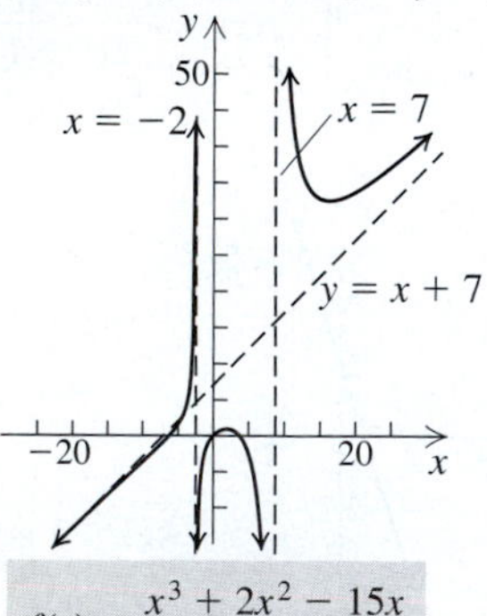

75. Domain: $(-\infty, \infty)$; x-intercept: $(0, 0)$, y-intercept: $(0, 0)$;

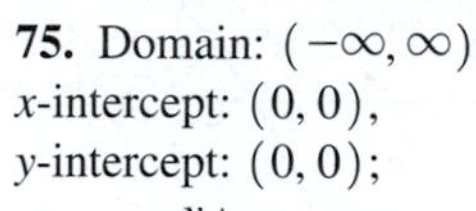

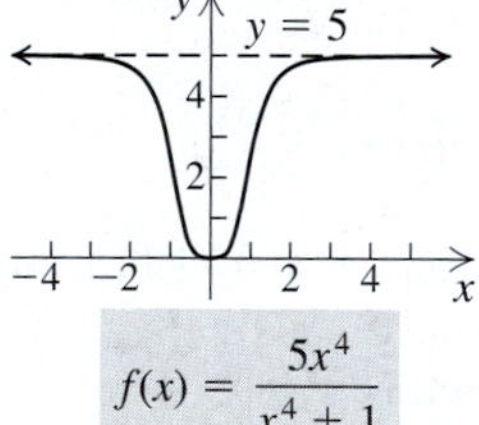

77. Domain: $(-\infty, -1) \cup (-1, 2) \cup (2, \infty)$; x-intercept: $(0, 0)$, y-intercept: $(0, 0)$;

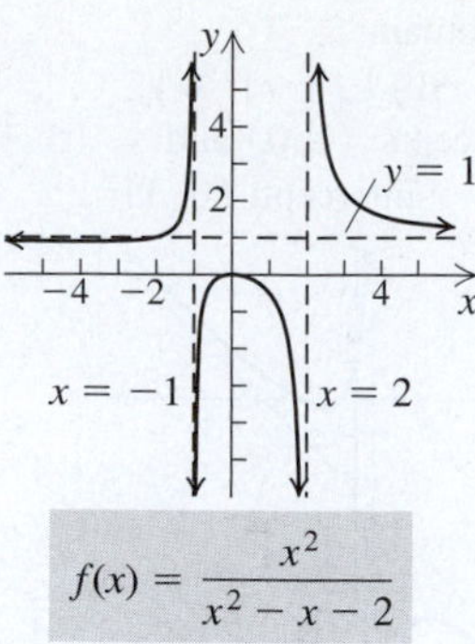

79. $f(x) = \dfrac{1}{x^2 - x - 20}$ **81.** $f(x) = \dfrac{3x^2 + 6x}{2x^2 - 2x - 40}$, or $f(x) = \dfrac{3x^2 + 12x + 12}{2x^2 - 2x - 40}$ **83. (a)** $N(t) \rightarrow 0.16$ as $t \rightarrow \infty$; **(b)** The medication never completely disappears from the body; a trace amount remains. **85. (a)** $P(0) = 0$; $P(1) = 45{,}455$; $P(3) = 55{,}556$; $P(8) = 29{,}197$; **(b)** $P(t) \rightarrow 0$ as $t \rightarrow \infty$; **(c)** In time, no one lives in this community. **87.** Domain, range, domain, range **88.** Slope **89.** Slope–intercept equation **90.** Point–slope equation **91.** x-intercept **92.** $f(-x) = -f(x)$ **93.** Vertical lines **94.** Midpoint formula **95.** y-intercept **97.** $y = x^3 + 4$

99.

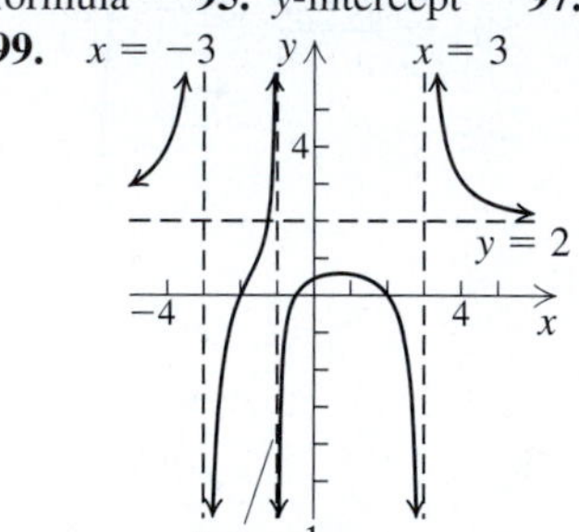

Exercise Set 4.6

1. $-5, 3$ **3.** $[-5, 3]$ **5.** $(-\infty, -5] \cup [3, \infty)$ **7.** $(-\infty, -4) \cup (2, \infty)$ **9.** $(-\infty, -4) \cup [2, \infty)$ **11.** 0 **13.** $(-5, 0] \cup (1, \infty)$ **15.** $(-\infty, -5) \cup (0, 1)$ **17.** $(-\infty, -3) \cup (0, 3)$ **19.** $(-3, 0) \cup (3, \infty)$ **21.** $(-\infty, -5) \cup (-3, 2)$ **23.** $(-2, 0] \cup (2, \infty)$ **25.** $(-4, 1)$ **27.** $(-\infty, -2) \cup (1, \infty)$ **29.** $(-\infty, -1] \cup [3, \infty)$ **31.** $(-\infty, -5) \cup (5, \infty)$ **33.** $(-\infty, -2] \cup [2, \infty)$ **35.** $(-\infty, 3) \cup (3, \infty)$ **37.** $\varnothing$ **39.** $\left(-\infty, -\frac{5}{4}\right] \cup [0, 3]$ **41.** $[-3, -1] \cup [1, \infty)$ **43.** $(-\infty, -2) \cup (1, 3)$ **45.** $\left[-\sqrt{2}, -1\right] \cup \left[\sqrt{2}, \infty\right)$ **47.** $(-\infty, -1] \cup \left[\frac{3}{2}, 2\right]$ **49.** $(-\infty, 5]$ **51.** $(-\infty, -1.680) \cup (2.154, 5.526)$ **53.** -4; $(-4, \infty)$ **55.** $-\frac{5}{2}$; $\left(-\frac{5}{2}, \infty\right)$ **57.** $0, 4$; $(-\infty, 0] \cup (4, \infty)$ **59.** $2, \frac{7}{2}$; $\left(2, \frac{7}{2}\right]$ **61.** $-3, -\frac{1}{5}, 1$; $\left(-3, -\frac{1}{5}\right] \cup (1, \infty)$ **63.** $2, \frac{46}{11}, 5$; $\left(2, \frac{46}{11}\right) \cup (5, \infty)$ **65.** $1 - \sqrt{2}, 0, 1 + \sqrt{2}$; $(1 - \sqrt{2}, 0) \cup (1 + \sqrt{2}, \infty)$ **67.** $-3, 1, 3, \frac{11}{3}$; $(-\infty, -3) \cup (1, 3) \cup \left[\frac{11}{3}, \infty\right)$

69. 0; $(-\infty, \infty)$ **71.** $-3, \frac{1-\sqrt{61}}{6}, -\frac{1}{2}, 0, \frac{1+\sqrt{61}}{6}$; $\left(-3, \frac{1-\sqrt{61}}{6}\right) \cup \left(-\frac{1}{2}, 0\right) \cup \left(\frac{1+\sqrt{61}}{6}, \infty\right)$
73. $-1, 0, \frac{2}{7}, \frac{7}{2}$; $(-1, 0) \cup \left(\frac{2}{7}, \frac{7}{2}\right)$
75. $-6-\sqrt{33}, -5, -6+\sqrt{33}, 1, 5$; $[-6-\sqrt{33}, -5) \cup [-6+\sqrt{33}, 1) \cup (5, \infty)$
77. $(0.408, 2.449)$ **79. (a)** $(10, 200)$; **(b)** $(0, 10) \cup (200, \infty)$ **81.** $\{n \mid 9 \le n \le 23\}$
83. $(x+2)^2 + (y-4)^2 = 9$ **84.** $x^2 + (y+3)^2 = \frac{49}{16}$
85. (a) $\left(\frac{3}{4}, -\frac{55}{8}\right)$; **(b)** maximum: $-\frac{55}{8}$ when $x = \frac{3}{4}$; **(c)** $\left(-\infty, -\frac{55}{8}\right]$ **86. (a)** $(5, -23)$; **(b)** minimum: -23 when $x = 5$; **(c)** $[-23, \infty)$ **87.** $[-\sqrt{5}, \sqrt{5}]$ **89.** $\left[-\frac{3}{2}, \frac{3}{2}\right]$
91. $\left(-\infty, -\frac{1}{4}\right) \cup \left(\frac{1}{2}, \infty\right)$ **93.** $x^2 + x - 12 < 0$; answers may vary **95.** $(-\infty, -3) \cup (7, \infty)$

Review Exercises: Chapter 4

1. True **2.** True **3.** False **4.** False **5.** False
6. (a) $-2.637, 1.137$; **(b)** relative maximum: 7.125 at $x = -0.75$; **(c)** none; **(d)** domain: all real numbers; range: $(-\infty, 7.125]$
7. (a) $-3, -1.414, 1.414$; **(b)** relative maximum: 2.303 at $x = -2.291$; **(c)** relative minimum: -6.303 at $x = 0.291$; **(d)** domain: all real numbers; range: all real numbers
8. (a) 0, 1, 2; **(b)** relative maximum: 0.202 at $x = 0.610$; **(c)** relative minima: 0 at $x = 0$, -0.620 at $x = 1.640$; **(d)** domain: all real numbers; range: $[-0.620, \infty)$
9. $0.45x^4$; 0.45; 4; quartic **10.** -25; -25; 0; constant
11. $-0.5x$; -0.5; 1; linear **12.** $\frac{1}{3}x^3$; $\frac{1}{3}$; 3; cubic
13. As $x \to \infty$, $f(x) \to -\infty$, and as $x \to -\infty$, $f(x) \to -\infty$.
14. As $x \to \infty$, $f(x) \to \infty$, and as $x \to -\infty$, $f(x) \to -\infty$.
15. $\frac{2}{3}$, multiplicity 1; -2, multiplicity 3; 5, multiplicity 2
16. $\pm 1, \pm 5$, each has multiplicity 1 **17.** $\pm 3, -4$, each has multiplicity 1 **18. (a)** 4%; **(b)** 5% **19. (a)** Linear: $f(x) = 0.5408695652x - 30.30434783$; quadratic: $f(x) = 0.0030322581x^2 - 0.5764516129x + 57.53225806$; cubic: $f(x) = 0.0000247619x^3 - 0.0112857143x^2 + 2.002380952x - 82.14285714$; **(b)** the cubic function; **(c)** 298, 498

20.

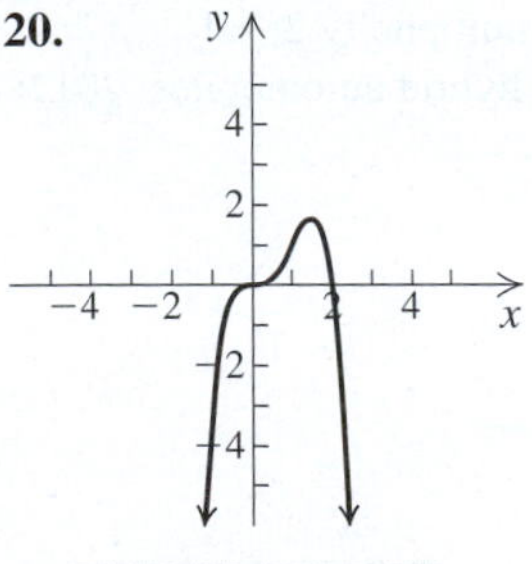

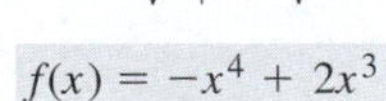
$f(x) = -x^4 + 2x^3$

21.

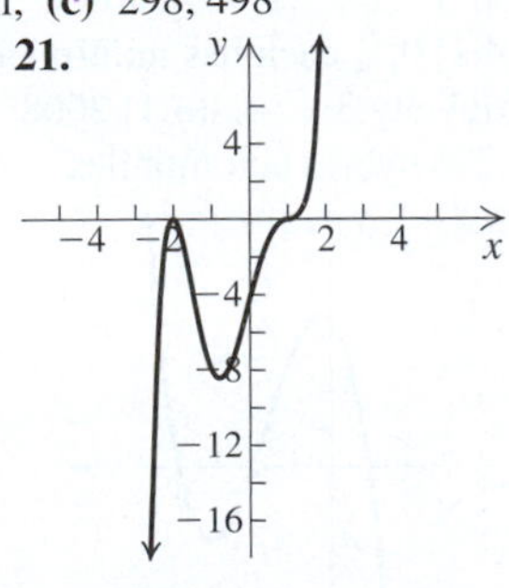
$g(x) = (x-1)^3(x+2)^2$

22.

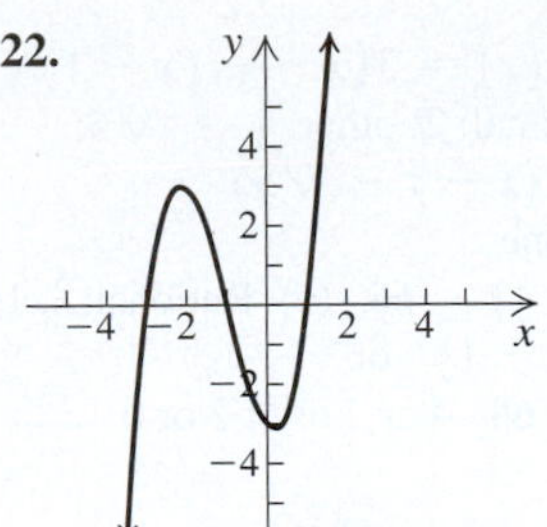
$h(x) = x^3 + 3x^2 - x - 3$

23.

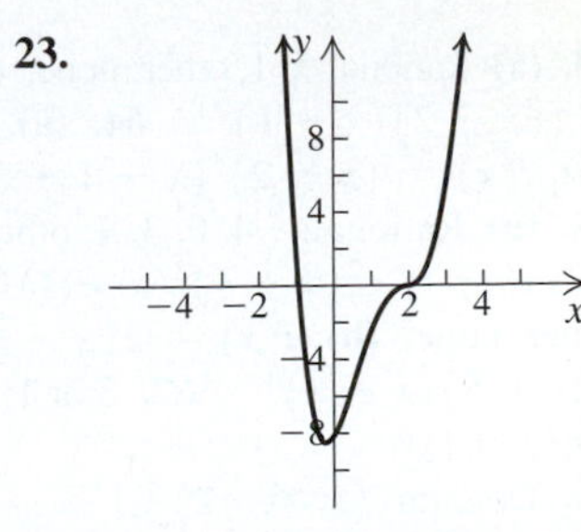
$f(x) = x^4 - 5x^3 + 6x^2 + 4x - 8$

24.

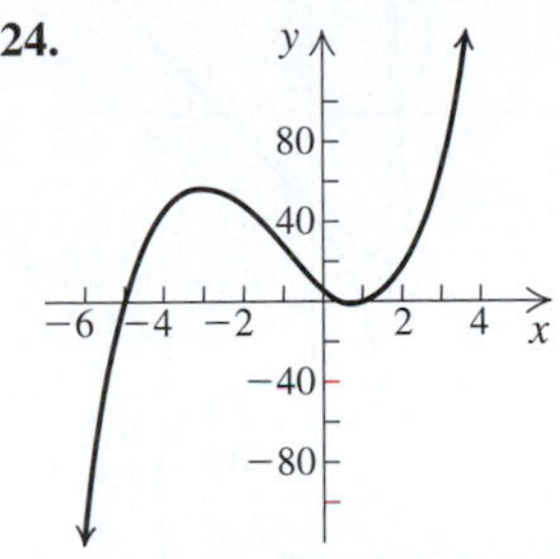
$g(x) = 2x^3 + 7x^2 - 14x + 5$

25. $f(1) = -4$ and $f(2) = 3$. Since $f(1)$ and $f(2)$ have opposite signs, $f(x)$ has a zero between 1 and 2. **26.** $f(-1) = -3.5$ and $f(1) = -0.5$. Since $f(-1)$ and $f(1)$ have the same sign, the intermediate value theorem does not allow us to determine whether there is a zero between -1 and 1. **27.** $Q(x) = 6x^2 + 16x + 52$, $R(x) = 155$; $P(x) = (x-3)(6x^2 + 16x + 52) + 155$
28. $Q(x) = x^3 - 3x^2 + 3x - 2$, $R(x) = 7$; $P(x) = (x+1)(x^3 - 3x^2 + 3x - 2) + 7$
29. $x^2 + 7x + 22$, R 120 **30.** $x^3 + x^2 + x + 1$, R 0
31. $x^4 - x^3 + x^2 - x - 1$, R 1 **32.** 36 **33.** 0
34. $-141{,}220$ **35.** Yes, no **36.** No, yes **37.** Yes, no
38. No, yes **39.** $f(x) = (x-1)^2(x+4)$; $-4, 1$
40. $f(x) = (x-2)(x+3)^2$; $-3, 2$
41. $f(x) = (x-2)^2(x-5)(x+5)$; $-5, 2, 5$
42. $f(x) = (x-1)(x+1)(x-\sqrt{2})(x+\sqrt{2})$; $-\sqrt{2}, -1, 1, \sqrt{2}$ **43.** $f(x) = x^3 + 3x^2 - 6x - 8$
44. $f(x) = x^3 + x^2 - 4x + 6$
45. $f(x) = x^3 - \frac{5}{2}x^2 + \frac{1}{2}$, or $2x^3 - 5x^2 + 1$
46. $f(x) = x^4 + \frac{29}{2}x^3 + \frac{135}{2}x^2 + \frac{175}{2}x - \frac{125}{2}$, or $2x^4 + 29x^3 + 135x^2 + 175x - 125$
47. $f(x) = x^5 + 4x^4 - 3x^3 - 18x^2$ **48.** $-\sqrt{5}, 4 - i$
49. $1 - \sqrt{3}, \sqrt{3}$ **50.** $\sqrt{2}$ **51.** $f(x) = x^2 - 11$
52. $f(x) = x^3 - 6x^2 + x - 6$
53. $f(x) = x^4 - 5x^3 + 4x^2 + 2x - 8$
54. $f(x) = x^4 - x^2 - 20$ **55.** $f(x) = x^3 + \frac{8}{3}x^2 - x$, or $3x^3 + 8x^2 - 3x$ **56.** $\pm\frac{1}{4}, \pm\frac{1}{2}, \pm\frac{3}{4}, \pm 1, \pm\frac{3}{2}, \pm 2, \pm 3, \pm 4, \pm 6, \pm 12$ **57.** $\pm\frac{1}{3}, \pm 1$ **58.** $\pm 1, \pm 2, \pm 3, \pm 4, \pm 6, \pm 8, \pm 12, \pm 24$ **59. (a)** Rational: $0, -2, \frac{1}{3}, 3$; other: none; **(b)** $f(x) = 3x\left(x - \frac{1}{3}\right)(x+2)^2(x-3)$, or $x(3x-1)(x+2)^2(x-3)$ **60. (a)** Rational: 2; other: $\pm\sqrt{3}$; **(b)** $f(x) = (x-2)(x+\sqrt{3})(x-\sqrt{3})$
61. (a) Rational: $-1, 1$; other: $3 \pm i$; **(b)** $f(x) = (x+1)(x-1)(x-3-i)(x-3+i)$
62. (a) Rational: -5; other: $1 \pm \sqrt{2}$; **(b)** $f(x) = (x+5)(x-1-\sqrt{2})(x-1+\sqrt{2})$

63. **(a)** Rational: $\frac{2}{3}$, 1; other: none; **(b)** $f(x) = 3\left(x - \frac{2}{3}\right)(x - 1)^2$, or $(3x - 2)(x - 1)^2$ **64.** **(a)** Rational: 2; other: $1 \pm \sqrt{5}$; **(b)** $f(x) = (x - 2)^3(x - 1 + \sqrt{5})(x - 1 - \sqrt{5})$ **65.** **(a)** Rational: $-4, 0, 3, 4$; other: none; **(b)** $f(x) = x^2(x + 4)^2(x - 3)(x - 4)$ **66.** **(a)** Rational: $\frac{5}{2}$, 1; other: none; **(b)** $f(x) = 2\left(x - \frac{5}{2}\right)(x - 1)^4$, or $(2x - 5)(x - 1)^4$ **67.** 3 or 1; 0 **68.** 4 or 2 or 0; 2 or 0 **69.** 3 or 1; 0

70. Domain: $(-\infty, -2) \cup (-2, \infty)$; x-intercepts: $(-\sqrt{5}, 0)$ and $(\sqrt{5}, 0)$, y-intercept: $\left(0, -\frac{5}{2}\right)$

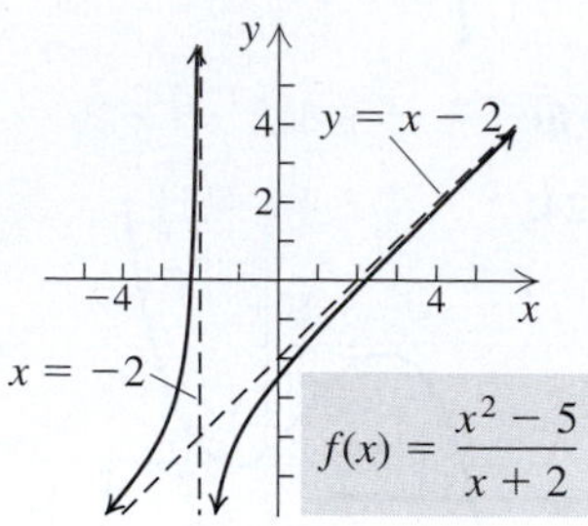

71. Domain: $(-\infty, 2) \cup (2, \infty)$; x-intercepts: none, y-intercept: $\left(0, \frac{5}{4}\right)$

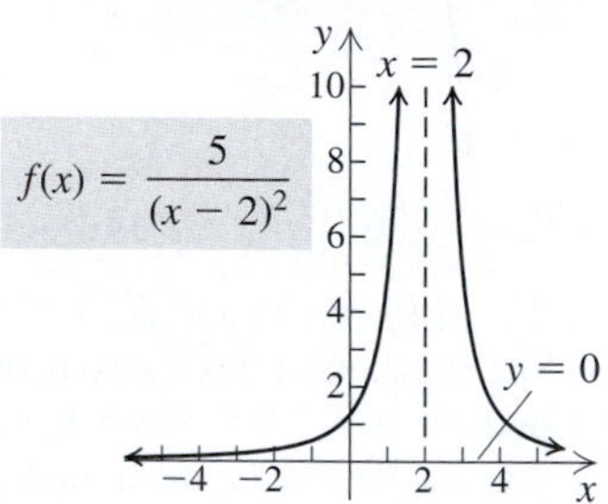

72. Domain: $(-\infty, -4) \cup (-4, 5) \cup (5, \infty)$; x-intercepts: $(-3, 0)$ and $(2, 0)$, y-intercept: $\left(0, \frac{3}{10}\right)$

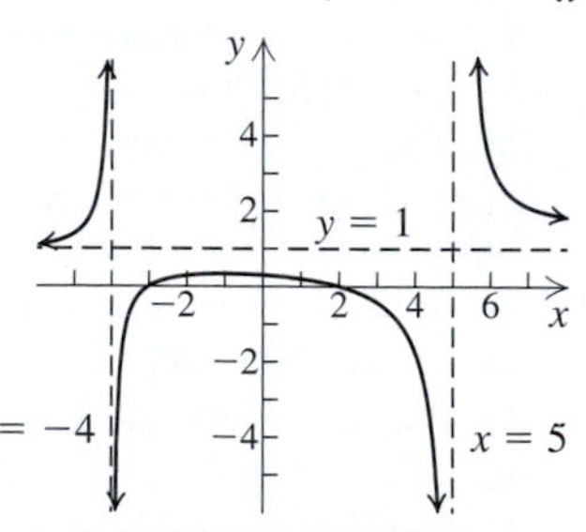

73. Domain: $(-\infty, -3) \cup (-3, 5) \cup (5, \infty)$; x-intercept: $(2, 0)$, y-intercept: $\left(0, \frac{2}{15}\right)$

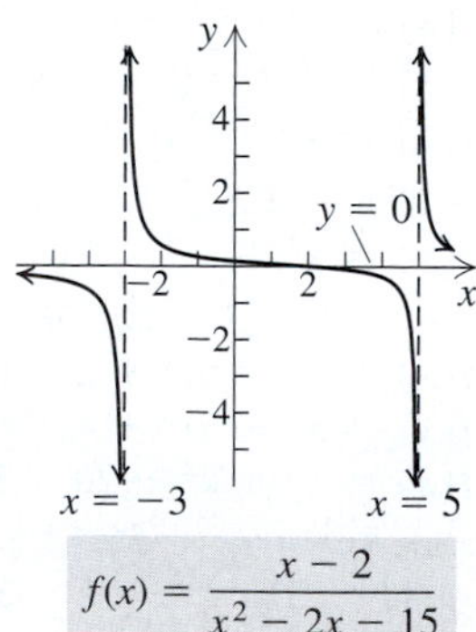

74. $f(x) = \dfrac{1}{x^2 - x - 6}$ **75.** $f(x) = \dfrac{4x^2 + 12x}{x^2 - x - 6}$
76. **(a)** $N(t) \to 0.0875$ as $t \to \infty$; **(b)** The medication never completely disappears from the body; a trace amount remains. **77.** $(-3, 3)$ **78.** $\left(-\infty, -\frac{1}{2}\right) \cup (2, \infty)$ **79.** $[-4, 1] \cup [2, \infty)$ **80.** $\left(-\infty, -\frac{14}{3}\right) \cup (-3, \infty)$

81. **(a)** 7 sec after launch; **(b)** $(2, 3)$ **82.** $\left[\dfrac{5 - \sqrt{15}}{2}, \dfrac{5 + \sqrt{15}}{2}\right]$ **83.** A **84.** C **85.** B **86.** $(-\infty, -1 - \sqrt{6}] \cup [-1 + \sqrt{6}, \infty)$ **87.** $\left(-\infty, -\frac{1}{2}\right) \cup \left(\frac{1}{2}, \infty\right)$ **88.** $1 + i, 1 - i, i, -i$ **89.** $(-\infty, 2)$ **90.** $(x - 1)\left(x + \dfrac{1}{2} - \dfrac{\sqrt{3}}{2}i\right)\left(x + \dfrac{1}{2} + \dfrac{\sqrt{3}}{2}i\right)$ **91.** 7 **92.** -4 **93.** $(-\infty, -5] \cup [2, \infty)$ **94.** $(-\infty, 1.1] \cup [2, \infty)$ **95.** $\left(-1, \frac{3}{7}\right)$
96. A polynomial function is a function that can be defined by a polynomial expression. A rational function is a function that can be defined as a quotient of two polynomials. **97.** No; since imaginary zeros of polynomials with rational coefficients occur in conjugate pairs, a third-degree polynomial with rational coefficients can have at most two imaginary zeros. Thus there must be at least one real zero. **98.** Vertical asymptotes occur at any x-values that make the denominator zero and do not also make the numerator zero. The graph of a rational function does not cross any vertical asymptotes. Horizontal asymptotes occur when the degree of the numerator is less than or equal to the degree of the denominator. Oblique asymptotes occur when the degree of the numerator is 1 greater than the degree of the denominator. Graphs of rational functions may cross horizontal or oblique asymptotes.
99. If $P(x)$ is an even function, then $P(-x) = P(x)$ and thus $P(-x)$ has the same number of sign changes as $P(x)$. Hence, $P(x)$ has one negative real zero also. **100.** A horizontal asymptote occurs when the degree of the numerator of a rational function is less than or equal to the degree of the denominator. An oblique asymptote occurs when the degree of the numerator is 1 greater than the degree of the denominator. Thus a rational function cannot have both a horizontal asymptote and an oblique asymptote. **101.** A quadratic inequality $ax^2 + bx + c \le 0, a > 0$, or $ax^2 + bx + c \ge 0, a < 0$, has a solution set that is a closed interval.

Test: Chapter 4

1. [4.1] $-x^4$; -1; 4; quartic **2.** [4.1] $-4.7x$; -4.7; 1; linear **3.** [4.1] $0, \frac{5}{3}$, each has multiplicity 1; 3, multiplicity 2; -1, multiplicity 3 **4.** [4.1] 2008: 312,848 hybrid automobiles; 2012: 372,750 hybrid automobiles
5. [4.2]

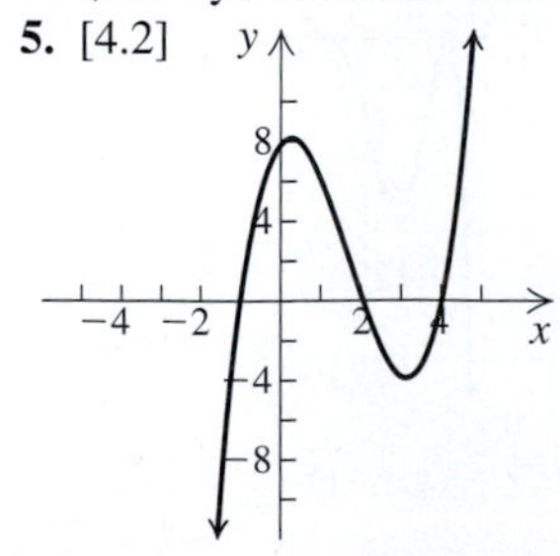

$f(x) = x^3 - 5x^2 + 2x + 8$

6. [4.2]

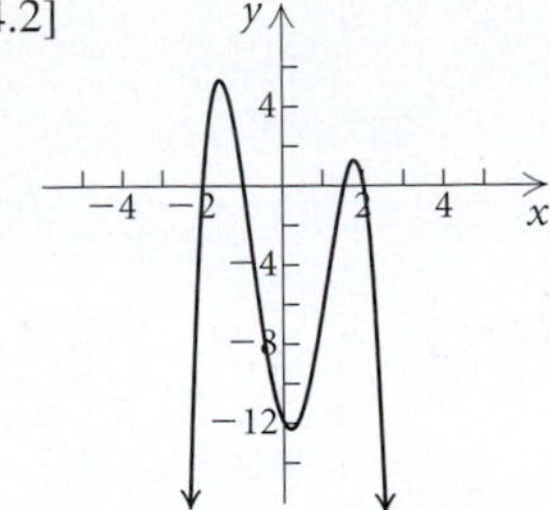

$f(x) = -2x^4 + x^3 + 11x^2 - 4x - 12$

7. [4.2] $f(0) = 3$ and $f(2) = -17$. Since $f(0)$ and $f(2)$ have opposite signs, $f(x)$ has a zero between 0 and 2.
8. [4.2] $g(-2) = 5$ and $g(-1) = 1$. Both $g(-2)$ and $g(-1)$ are positive. We cannot use the intermediate value theorem to determine if there is a zero between −2 and −1.
9. [4.3] $Q(x) = x^3 + 4x^2 + 4x + 6$, $R(x) = 1$; $P(x) = (x - 1)(x^3 + 4x^2 + 4x + 6) + 1$
10. [4.3] $3x^2 + 15x + 63$, R 322 **11.** [4.3] −115
12. [4.3] Yes **13.** [4.4] $f(x) = x^4 - 27x^2 - 54x$
14. [4.4] $-\sqrt{3}, 2 + i$ **15.** [4.4] $f(x) = x^3 + 10x^2 + 9x + 90$
16. [4.4] $f(x) = x^5 - 2x^4 - x^3 + 6x^2 - 6x$
17. [4.4] $\pm 1, \pm 2, \pm 3, \pm 4, \pm 6, \pm 12, \pm\frac{1}{2}, \pm\frac{3}{2}$
18. [4.4] $\pm\frac{1}{10}, \pm\frac{1}{5}, \pm\frac{1}{2}, \pm 1, \pm\frac{5}{2}, \pm 5$
19. [4.4] **(a)** Rational: −1; other: $\pm\sqrt{5}$;
(b) $f(x) = (x + 1)(x - \sqrt{5})(x + \sqrt{5})$ **20.** [4.4]
(a) Rational: $-\frac{1}{2}$, 1, 2, 3; other: none;
(b) $f(x) = 2(x + \frac{1}{2})(x - 1)(x - 2)(x - 3)$, or $(2x + 1)(x - 1)(x - 2)(x - 3)$
21. [4.4] **(a)** Rational: −4; other: $\pm 2i$;
(b) $f(x) = (x - 2i)(x + 2i)(x + 4)$
22. [4.4] **(a)** Rational: $\frac{2}{3}$, 1; other: none;
(b) $f(x) = 3(x - \frac{2}{3})(x - 1)^3$, or $(3x - 2)(x - 1)^3$
23. [4.4] 2 or 0; 2 or 0

24. [4.5] Domain: $(-\infty, 3) \cup (3, \infty)$; x-intercepts: none, y-intercept: $(0, \frac{2}{9})$;

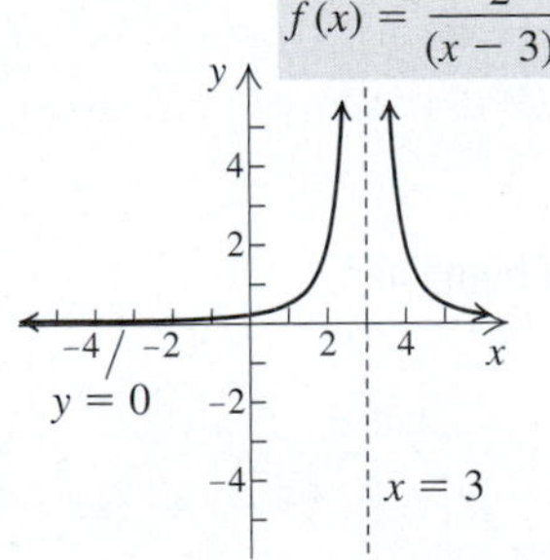

25. [4.5] Domain: $(-\infty, -1) \cup (-1, 4) \cup (4, \infty)$; x-intercept: $(-3, 0)$, y-intercept: $(0, -\frac{3}{4})$;

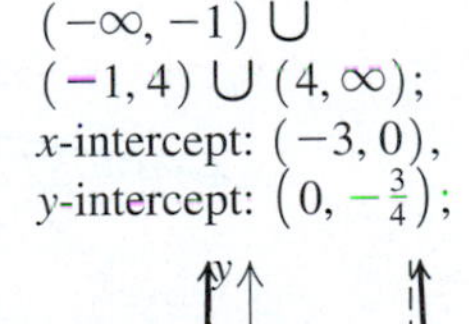

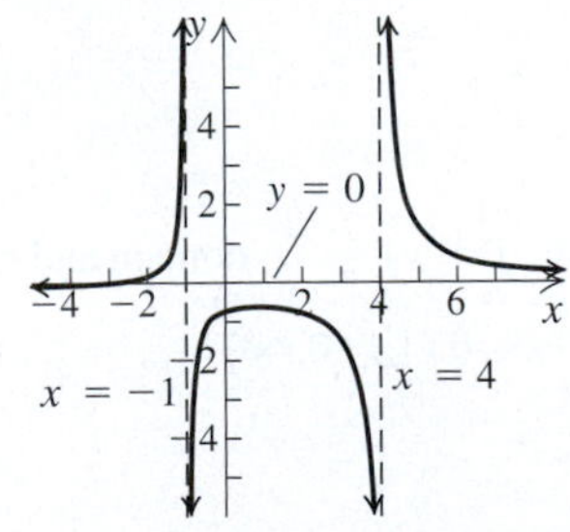

26. [4.5] Answers may vary; $f(x) = \frac{x + 4}{x^2 - x - 2}$
27. [4.6] $(-\infty, -\frac{1}{2}) \cup (3, \infty)$
28. [4.6] $(-\infty, 4) \cup [\frac{13}{2}, \infty)$ **29.** **(a)** [4.1] 6 sec;
(b) [4.1], [4.6] (1, 3) **30.** [4.2] D
31. [4.1], [4.6] $(-\infty, -4] \cup [3, \infty)$

Chapter 5

Exercise Set 5.1

1. $\{(8, 7), (8, -2), (-4, 3), (-8, 8)\}$ **3.** $\{(-1, -1), (4, -3)\}$
5. $x = 4y - 5$ **7.** $y^3x = -5$ **9.** $y = x^2 - 2x$
11.

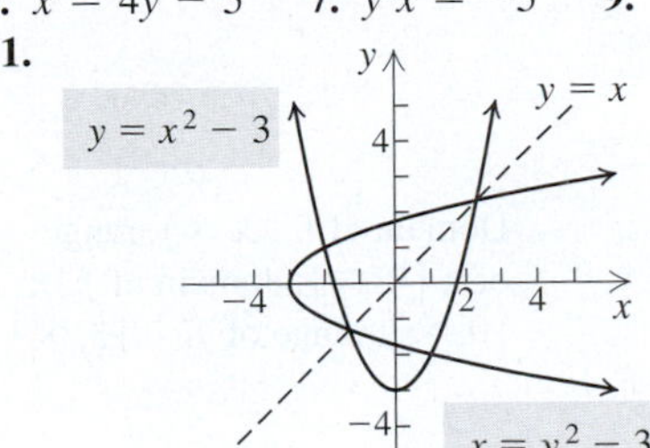

13. **15.**

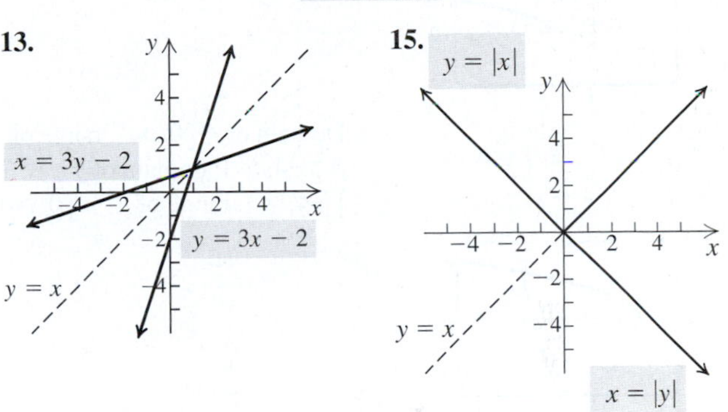

17. Assume that $f(a) = f(b)$ for any numbers a and b in the domain of f. Since $f(a) = \frac{1}{3}a - 6$ and $f(b) = \frac{1}{3}b - 6$, we have

$$\begin{aligned} \tfrac{1}{3}a - 6 &= \tfrac{1}{3}b - 6 \\ \tfrac{1}{3}a &= \tfrac{1}{3}b \quad \text{Adding 6} \\ a &= b. \quad \text{Multiplying by 3} \end{aligned}$$

Thus, if $f(a) = f(b)$, then $a = b$ and f is one-to-one.
19. Assume that $f(a) = f(b)$ for any numbers a and b in the domain of f. Since $f(a) = a^3 + \frac{1}{2}$ and $f(b) = b^3 + \frac{1}{2}$, we have

$$\begin{aligned} a^3 + \tfrac{1}{2} &= b^3 + \tfrac{1}{2} \\ a^3 &= b^3 \quad \text{Subtracting } \tfrac{1}{2} \\ a &= b. \quad \text{Taking the cube root} \end{aligned}$$

Thus, if $f(a) = f(b)$, then $a = b$ and f is one-to-one.
21. Find two numbers a and b for which $a \neq b$ and $g(a) = g(b)$. Two such numbers are −2 and 2, because $g(-2) = g(2) = -3$. Thus, g is not one-to-one.
23. Find two numbers a and b for which $a \neq b$ and $g(a) = g(b)$. Two such numbers are −1 and 1, because $g(-1) = g(1) = 0$. Thus, g is not one-to-one.
25. Yes **27.** No **29.** No **31.** Yes **33.** Yes
35. No **37.** No **39.** Yes **41.** No **43.** No
45. $y_1 = 0.8x + 1.7$, $y_2 = \frac{x - 1.7}{0.8}$ Domain and range of both f and f^{-1}: $(-\infty, \infty)$

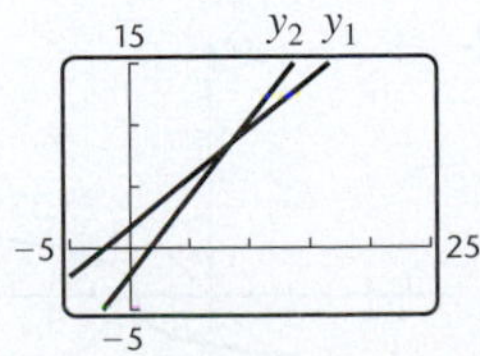

47. $y_1 = \frac{1}{2}x - 4,$
$y_2 = 2x + 8$

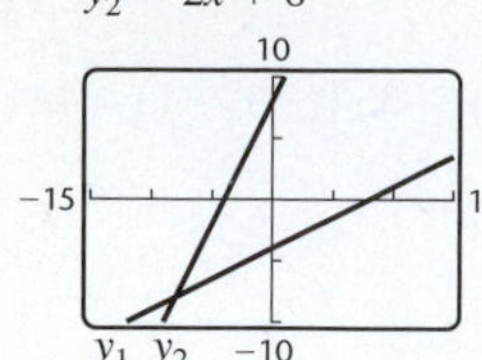

Domain and range of both f and f^{-1}: $(-\infty, \infty)$

49. $y_1 = \sqrt{x - 3},$
$y_2 = x^2 + 3, x \geq 0$

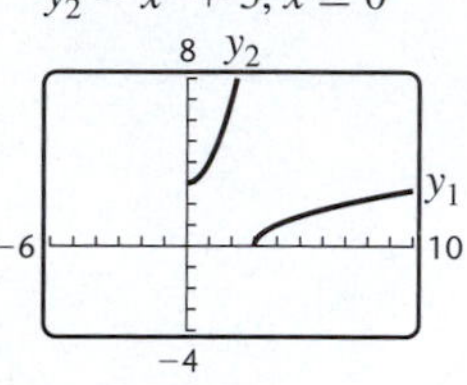

Domain of f: $[3, \infty)$, range of f: $[0, \infty)$; domain of f^{-1}: $[0, \infty)$, range of f^{-1}: $[3, \infty)$

51. $y_1 = x^2 - 4, x \geq 0;$
$y_2 = \sqrt{4 + x}$

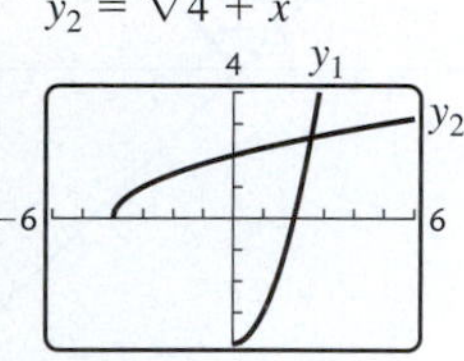

Domain of f: $[0, \infty)$, range of f: $[-4, \infty)$; domain of f^{-1}: $[-4, \infty)$, range of f^{-1}: $[0, \infty)$

53. $y_1 = (3x - 9)^3,$
$y_2 = \dfrac{\sqrt[3]{x} + 9}{3}$

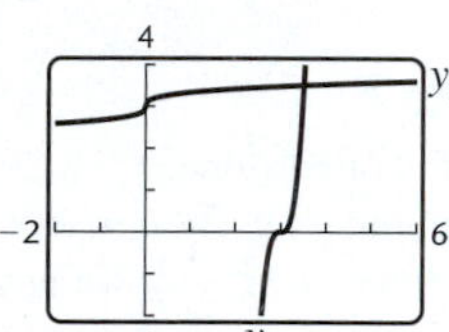

Domain and range of both f and f^{-1}: $(-\infty, \infty)$

55. **(a)** Yes; **(b)** $f^{-1}(x) = x - 4$
57. **(a)** Yes; **(b)** $f^{-1}(x) = \dfrac{x + 1}{2}$
59. **(a)** Yes; **(b)** $f^{-1}(x) = \dfrac{4}{x} - 7$
61. **(a)** Yes; **(b)** $f^{-1}(x) = \dfrac{3x + 4}{x - 1}$
63. **(a)** Yes; **(b)** $f^{-1}(x) = \sqrt[3]{x + 1}$
65. **(a)** No
67. **(a)** Yes; **(b)** $f^{-1}(x) = \sqrt{\dfrac{x + 2}{5}}$
69. **(a)** Yes; **(b)** $f^{-1}(x) = x^2 - 1, x \geq 0$
71. $\frac{1}{3}x$ **73.** $-x$ **75.** $x^3 + 5$
77. **79.**

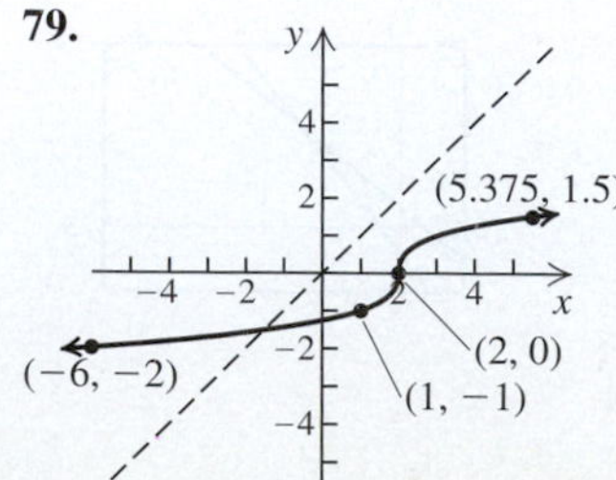

81.

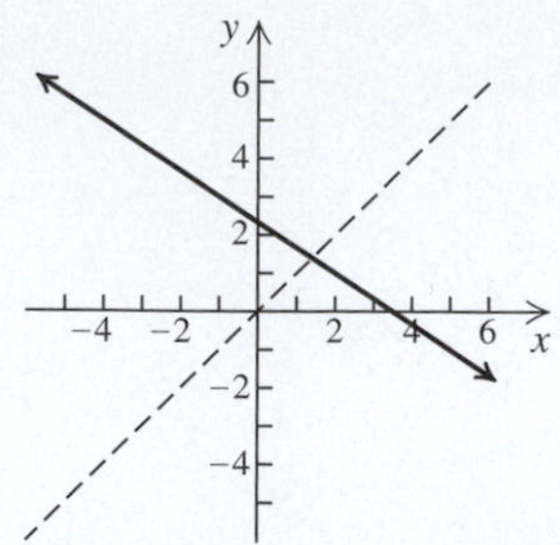

83. $f^{-1}(f(x)) = f^{-1}\left(\frac{7}{8}x\right) = \frac{8}{7} \cdot \frac{7}{8}x = x;$
$f(f^{-1}(x)) = f\left(\frac{8}{7}x\right) = \frac{7}{8} \cdot \frac{8}{7}x = x$

85. $f^{-1}(f(x)) = f^{-1}\left(\dfrac{1 - x}{x}\right) = \dfrac{1}{\dfrac{1 - x}{x} + 1} =$
$\dfrac{1}{\dfrac{1 - x + x}{x}} = \dfrac{1}{\dfrac{1}{x}} = 1 \cdot \dfrac{x}{1} = x; f(f^{-1}(x)) = f\left(\dfrac{1}{x + 1}\right) =$
$\dfrac{1 - \dfrac{1}{x + 1}}{\dfrac{1}{x + 1}} = \dfrac{\dfrac{x + 1 - 1}{x + 1}}{\dfrac{1}{x + 1}} = \dfrac{x}{x + 1} \cdot \dfrac{x + 1}{1} = x$

87. $f^{-1}(f(x)) = f^{-1}\left(\dfrac{2}{5}x + 1\right) = \dfrac{5\left(\dfrac{2}{5}x + 1\right) - 5}{2} =$
$\dfrac{2x + 5 - 5}{2} = \dfrac{2x}{2} = x; f(f^{-1}(x)) = f\left(\dfrac{5x - 5}{2}\right) =$
$\dfrac{2}{5}\left(\dfrac{5x - 5}{2}\right) + 1 = x - 1 + 1 = x$

89. $f^{-1}(x) = \frac{1}{5}x + \frac{3}{5}$; domain and range of both f and f^{-1}: $(-\infty, \infty)$

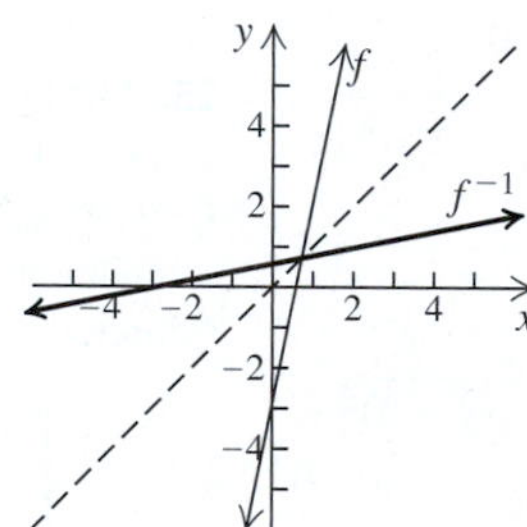

91. $f^{-1}(x) = \dfrac{2}{x}$; domain and range of both f and f^{-1}: $(-\infty, 0) \cup (0, \infty)$

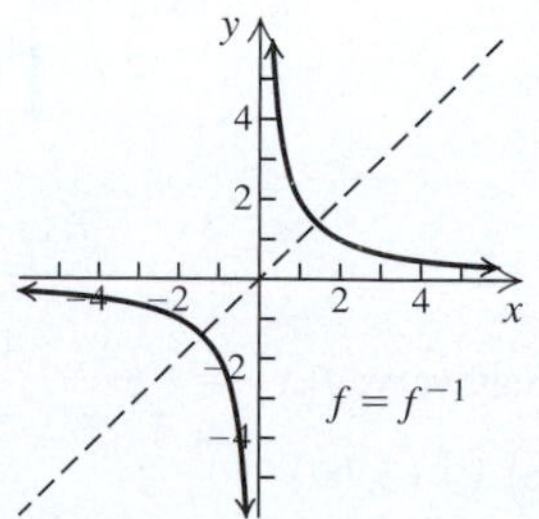

93. $f^{-1}(x) = \sqrt[3]{3x + 6}$; domain of f and f^{-1}: $(-\infty, \infty)$; range of f and f^{-1}: $(-\infty, \infty)$

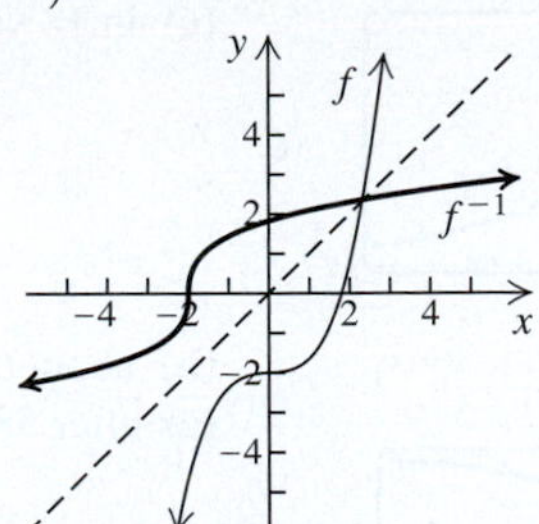

95. $f^{-1}(x) = \dfrac{3x + 1}{x - 1}$; domain of f: $(-\infty, 3) \cup (3, \infty)$;
range of f: $(-\infty, 1) \cup (1, \infty)$;
domain of f^{-1}: $(-\infty, 1) \cup (1, \infty)$;
range of f^{-1}: $(-\infty, 3) \cup (3, \infty)$;

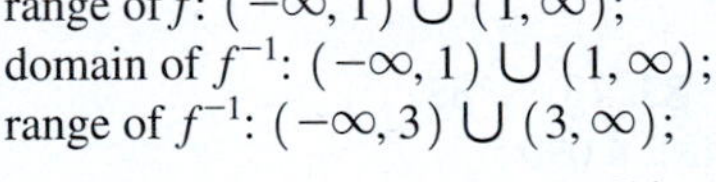

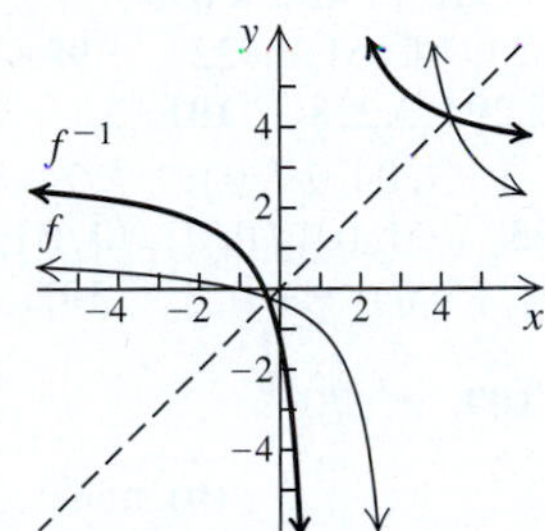

97. 5; a **99.** **(a)** \$38, \$16.40, \$11; **(b)** $C^{-1}(x) = \dfrac{72}{x - 2}$; $C^{-1}(x)$ represents the number of players in the group lesson, where x is the cost per player, in dollars; **(c)** 1 player, 4 players, 8 players **101.** **(a)** 2012: \$42.572 billion; 2016: \$61.764 billion; **(b)** $H^{-1}(x) = \dfrac{x - 28.1778}{4.798}$; $H^{-1}(x)$ represents the number of years after 2009, where x is the amount of e-commerce holiday season sales, in billions of dollars
103. (b), (d), (f), (h) **104.** (a), (c), (e), (g) **105.** (a)
106. (d) **107.** (f) **108.** (a), (b), (c), (d) **109.** Yes
111. $f(x) = x^2 - 3, x \geq 0$; $f^{-1}(x) = \sqrt{x + 3}, x \geq -3$
113. Answers may vary. $f(x) = 3/x$, $f(x) = 1 - x$, $f(x) = x$

Exercise Set 5.2

1. 54.5982 **3.** 0.0856 **5.** (f) **7.** (e) **9.** (a)

11. $f(x) = 3^x$

13. $f(x) = 6^x$

15. $f(x) = \left(\frac{1}{4}\right)^x$

17. $y = -2^x$

19.

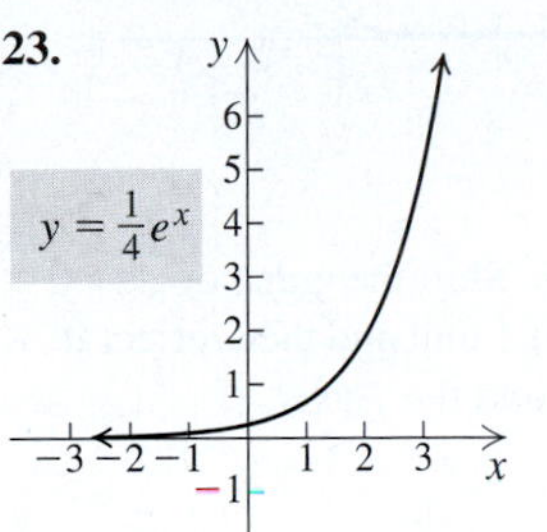

21.

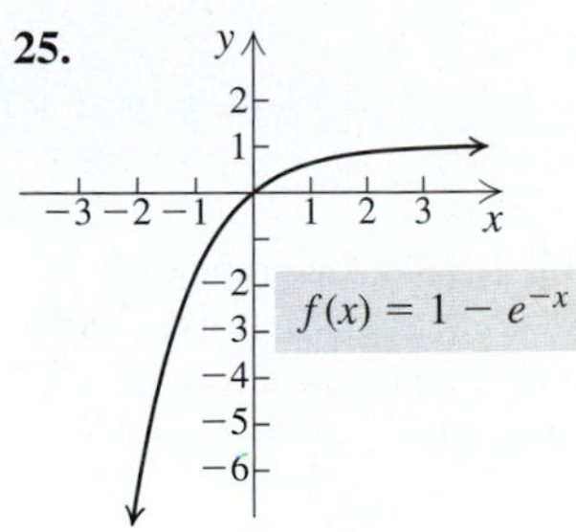

23.

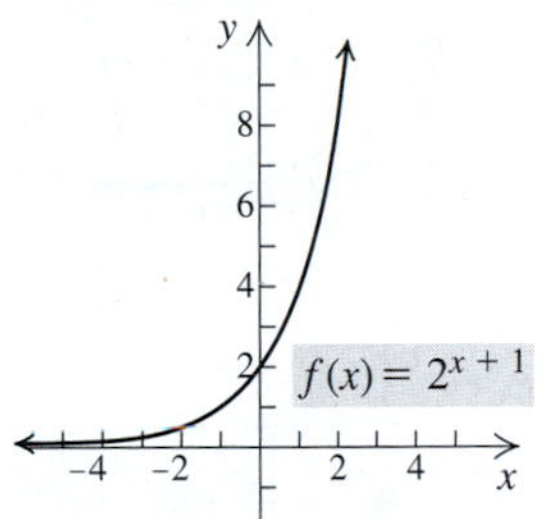

25.

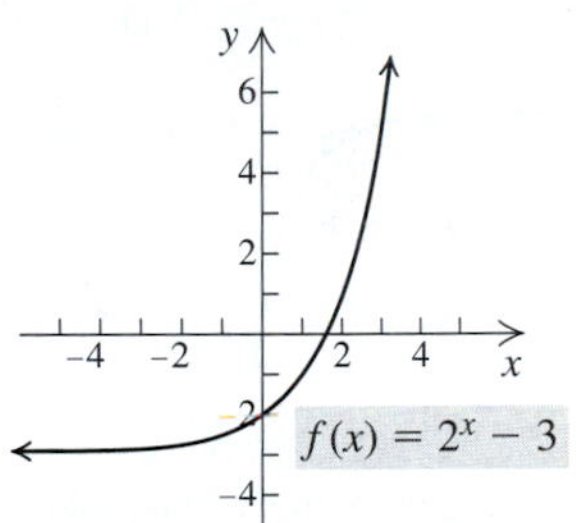

27. Shift the graph of $y = 2^x$ left 1 unit.

$f(x) = 2^{x+1}$

29. Shift the graph of $y = 2^x$ down 3 units.

$f(x) = 2^x - 3$

31. Shift the graph of $y = 2^x$ left 1 unit, reflect it across the y-axis, and shift it up 2 units.

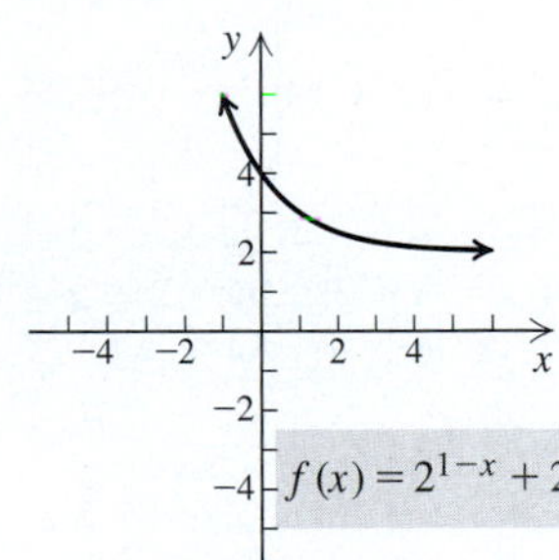

33. Reflect the graph of $y = 3^x$ across the y-axis and then across the x-axis and then shift it up 4 units.

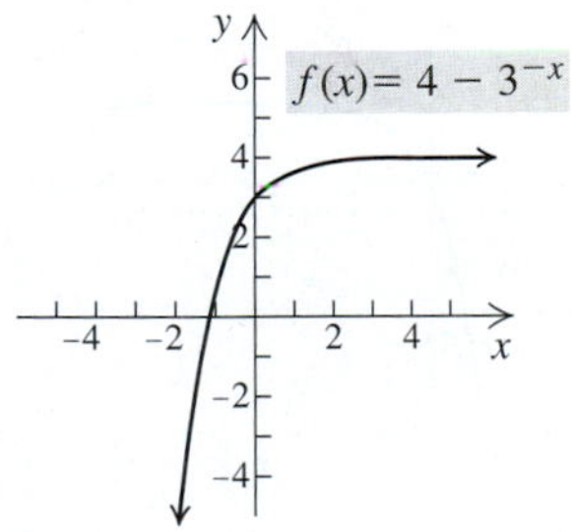

35. Shift the graph of $y = \left(\frac{3}{2}\right)^x$ right 1 unit.

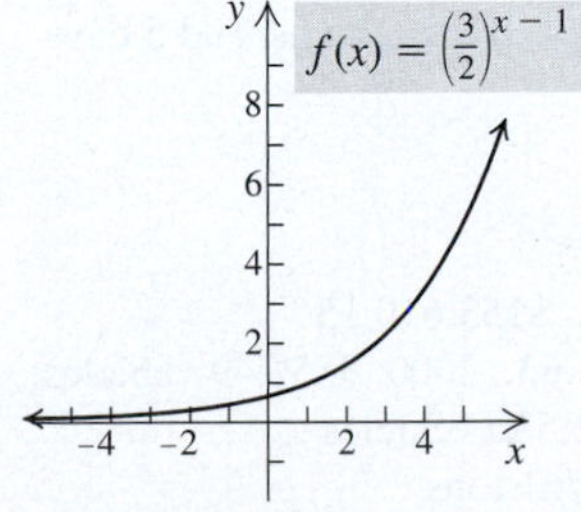

37. Shift the graph of $y = 2^x$ left 3 units and then down 5 units.

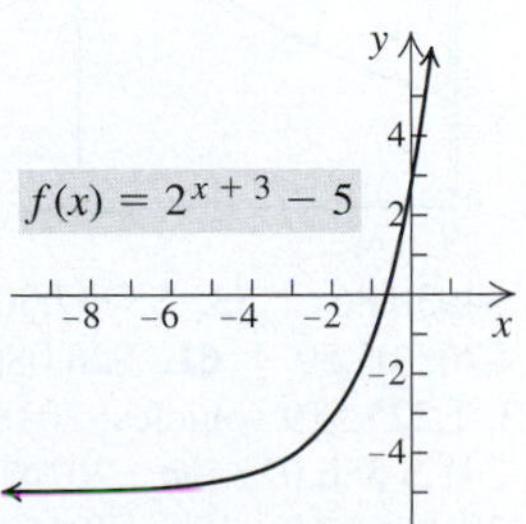

39. Shift the graph of $y = 2^x$ right 1 unit, stretch it vertically, and shift it up 1 unit.

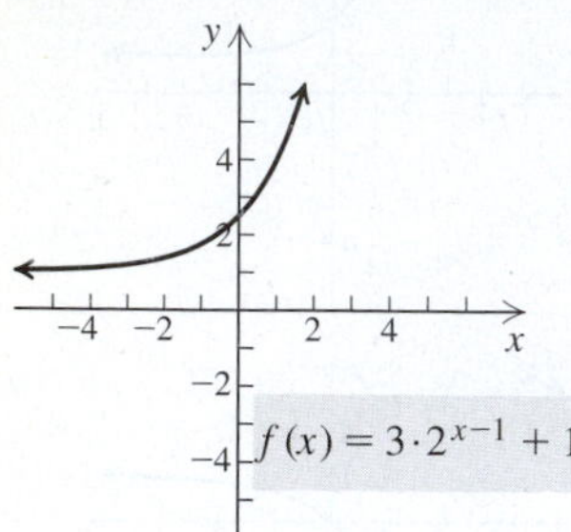

41. Shrink the graph of $y = e^x$ horizontally.

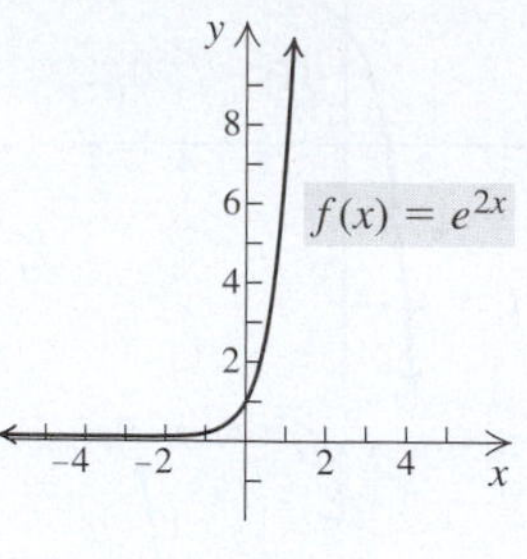

43. Reflect the graph of $y = e^x$ across the x-axis, shift it up 1 unit, and shrink it vertically.

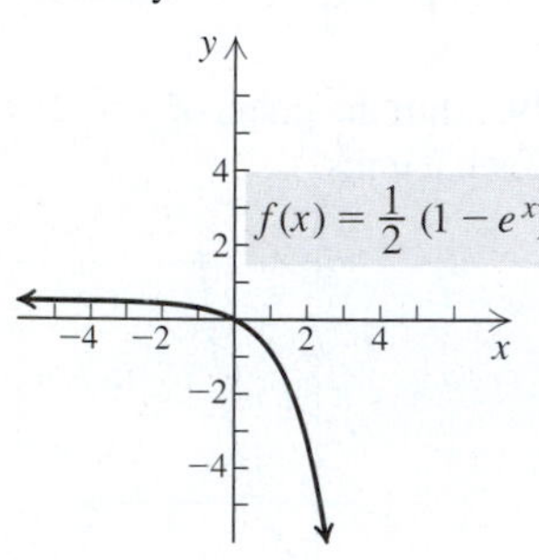

45. Shift the graph of $y = e^x$ left 1 unit and then reflect it across the y-axis.

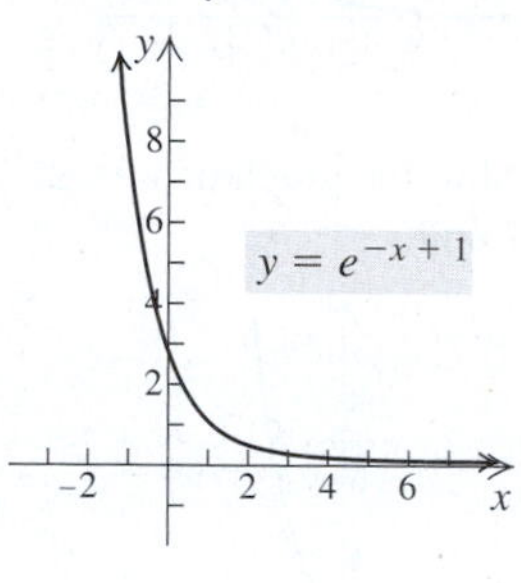

47. Reflect the graph of $y = e^x$ across the y-axis, then across the x-axis, then shift it up 1 unit, and then stretch it vertically.

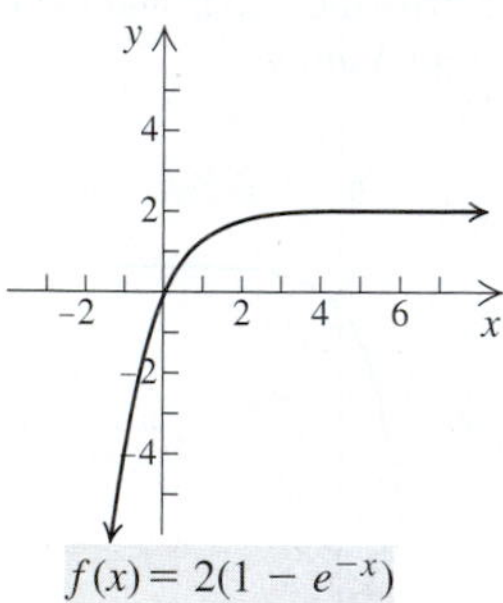

49.

$$f(x) = \begin{cases} e^{-x} - 4, & \text{for } x < -2, \\ x + 3, & \text{for } -2 \le x < 1, \\ x^2, & \text{for } x \ge 1 \end{cases}$$

51. (a) $A(t) = 82{,}000(1.01125)^{4t}$;

(b) $y = 82{,}000(1.01125)^{4x}$;

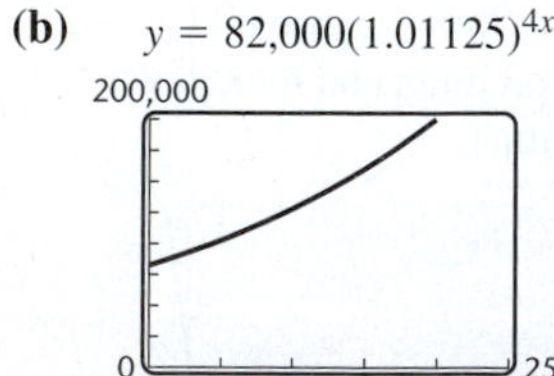

(c) \$82,000; \$89,677.22; \$102,561.54; \$128,278.90; **(d)** about 4.43 years, or about 4 years, 5 months, and 5 days

53. \$4930.86 **55.** \$3247.30 **57.** \$153,610.15 **59.** \$76,305.59 **61.** \$26,086.69 **63.** 2000: 385,249 vehicles; 2013: 1,225,519 vehicles; 2018: 1,912,580 vehicles **65.** 2012: 356,022 tons; 2016: 5,605,761 tons **67.** 2011: \$234 million; 2015: \$5844 million, or \$5.844 billion **69.** 2020: 101,234 centenarians; 2050: 414,387 centenarians **71.** 1982: \$48 billion; 1995: \$109 billion; 2010: \$284 billion; 2017: \$445 billion

73. (a) $y = 6982(0.85)^x$; **(b)** \$6982; \$5935; \$5044; \$3098; \$1903; **(c)** in 12 years

75. (a) $y = 100(1 - e^{-0.04x})$

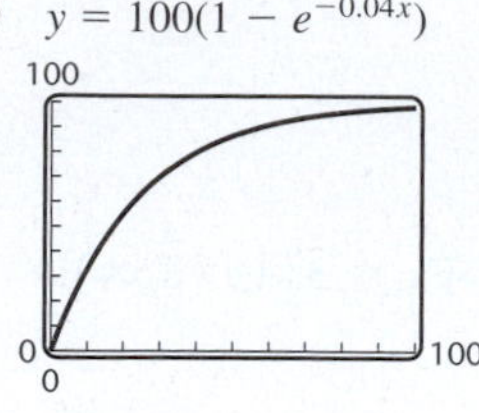
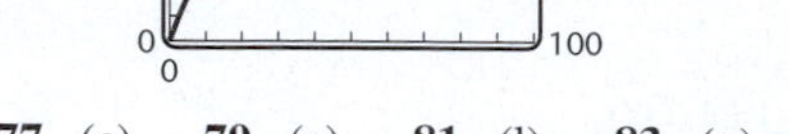

(b) about 63%; **(c)** after 58 days

77. (c) **79.** (a) **81.** (l) **83.** (g) **85.** (i) **87.** (k) **89.** (m) **91.** $(1.481, 4.090)$ **93.** $(-0.402, -1.662)$, $(1.051, 2.722)$ **95.** 4.448 **97.** $(0, \infty)$ **99.** 2.294, 3.228 **101.** $31 - 22i$ **102.** $\frac{1}{2} - \frac{1}{2}i$ **103.** $\left(-\frac{1}{2}, 0\right), (7, 0); -\frac{1}{2}, 7$ **104.** $(1, 0)$; 1 **105.** $(-1, 0), (0, 0), (1, 0); -1, 0, 1$ **106.** $(-4, 0), (0, 0), (3, 0); -4, 0, 3$ **107.** $-8, 0, 2$ **108.** $\dfrac{5 \pm \sqrt{97}}{6}$ **109.** π^7; 70^{80}

111. (a) $y = e^{-x^2}$ **(b)** none; **(c)** relative maximum: 1 at $x = 0$

113. $f(x) = \dfrac{2e^x(e^h - 1)}{h}$

Visualizing the Graph

1. J **2.** F **3.** H **4.** B **5.** E **6.** A **7.** C **8.** I **9.** D **10.** G

Exercise Set 5.3

1. $x = 3^y$

3.

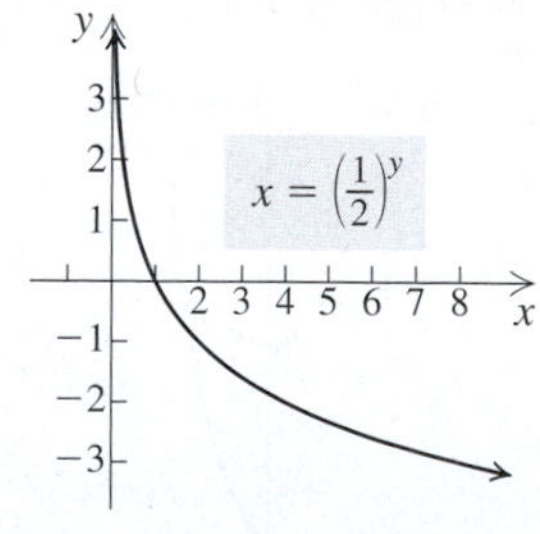

5. $y = \log_3 x$

7. $f(x) = \log x$

9. 4 **11.** 3 **13.** −3 **15.** −2 **17.** 0 **19.** 1
21. 4 **23.** $\frac{1}{4}$ **25.** −7 **27.** $\frac{1}{2}$ **29.** $\frac{3}{4}$ **31.** 0 **33.** $\frac{1}{2}$
35. $\log_{10} 1000 = 3$, or $\log 1000 = 3$ **37.** $\log_8 2 = \frac{1}{3}$
39. $\log_e t = 3$, or $\ln t = 3$
41. $\log_e 7.3891 = 2$, or $\ln 7.3891 = 2$ **43.** $\log_p 3 = k$
45. $5^1 = 5$ **47.** $10^{-2} = 0.01$ **49.** $e^{3.4012} = 30$
51. $a^{-x} = M$ **53.** $a^x = T^3$ **55.** 0.4771 **57.** 2.7259
59. −0.2441 **61.** Does not exist **63.** 0.6931 **65.** 6.6962
67. Does not exist **69.** 3.3219 **71.** −0.2614
73. 0.7384 **75.** 2.2619 **77.** 0.5880

79. $y_1 = 3^x$, $y_2 = \dfrac{\log x}{\log 3}$

81. $y_1 = \log x$, $y_2 = 10^x$

83. Shift the graph of $y = \log_2 x$ left 3 units. Domain: $(-3, \infty)$; vertical asymptote: $x = -3$;

$y = \dfrac{\log (x + 3)}{\log 2}$

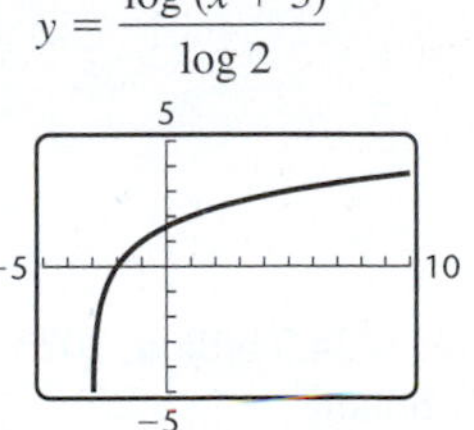

85. Shift the graph of $y = \log_3 x$ down 1 unit. Domain: $(0, \infty)$; vertical asymptote: $x = 0$;

$y = \dfrac{\log x}{\log 3} - 1$

87. Stretch the graph of $y = \ln x$ vertically. Domain: $(0, \infty)$; vertical asymptote: $x = 0$;

$y = 4 \ln x$

89. Reflect the graph of $y = \ln x$ across the x-axis and shift it up 2 units. Domain: $(0, \infty)$; vertical asymptote: $x = 0$;

$y = 2 - \ln x$

91. Shift the graph of $y = \log x$ right 1 unit, shrink it vertically, and shift it down 2 units. Domain: $(1, \infty)$; vertical asymptote: $x = 1$.

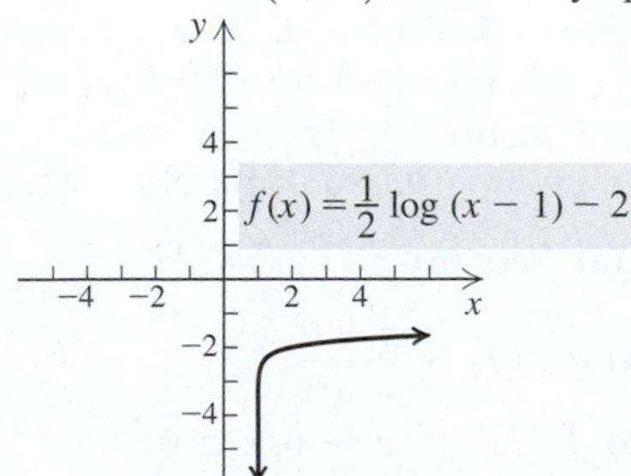

93.

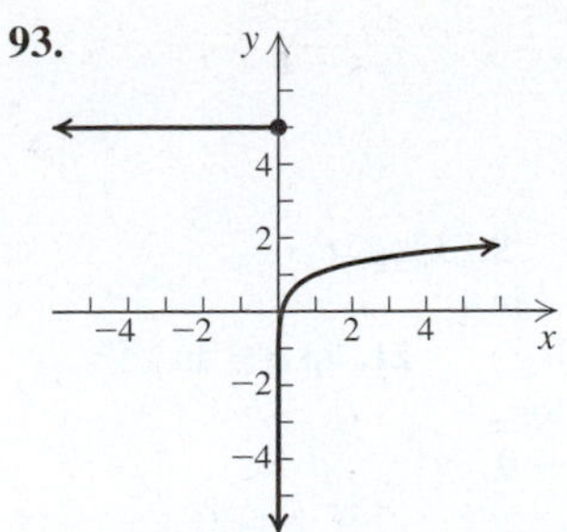

$g(x) = \begin{cases} 5, & \text{for } x \le 0, \\ \log x + 1, & \text{for } x > 0 \end{cases}$

95. **(a)** 2.5 ft/sec; **(b)** 2.8 ft/sec; **(c)** 2.0 ft/sec; **(d)** 2.4 ft/sec; **(e)** 2.2 ft/sec; **(f)** 2.5 ft/sec; **(g)** 2.3 ft/sec; **(h)** 3.1 ft/sec
97. **(a)** 7.7; **(b)** 9.5; **(c)** 6.6; **(d)** 7.6; **(e)** 8.2; **(f)** 7.9; **(g)** 5.1; **(h)** 9.3
99. **(a)** 10^{-7}; **(b)** 4.0×10^{-6}; **(c)** 6.3×10^{-4}; **(d)** 1.6×10^{-5}

101. **(a)** 140 decibels; **(b)** 115 decibels; **(c)** 40 decibels; **(d)** 65 decibels; **(e)** 120 decibels; **(f)** 194 decibels
102. $m = \frac{3}{10}$; y-intercept: $\left(0, -\frac{7}{5}\right)$ **103.** $m = 0$; y-intercept: $(0, 6)$ **104.** Slope is not defined; no y-intercept
105. −280 **106.** −4 **107.** $f(x) = x^3 - 7x$
108. $f(x) = x^3 - x^2 + 16x - 16$ **109.** 3 **111.** $(0, \infty)$
113. $(-\infty, 0) \cup (0, \infty)$ **115.** $\left(-\frac{5}{2}, -2\right)$ **117.** (d) **119.** (b)
121. **(a)** $y = x \ln x$

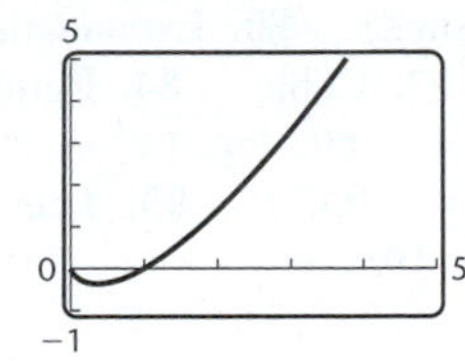

(b) 1; **(c)** relative minimum: −0.368 at $x = 0.368$

123. **(a)** $y = \dfrac{\ln x}{x^2}$

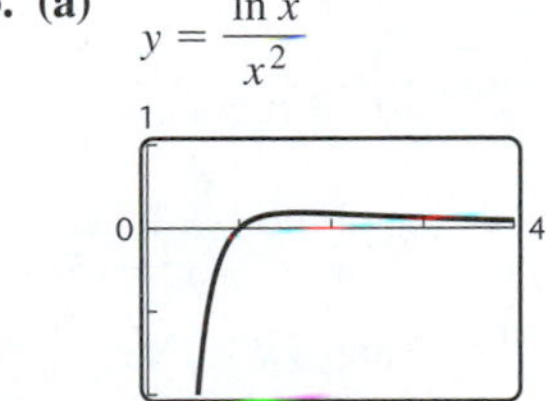

(b) 1; **(c)** relative maximum: 0.184 at $x = 1.649$

125. $(1.250, 0.891)$

Mid-Chapter Mixed Review: Chapter 5

1. False **2.** True **3.** False **4.** Yes; $f^{-1}(x) = -\dfrac{2}{x}$
5. No **6.** Yes; $f^{-1}(x) = \dfrac{5}{x} + 2$
7. $(f^{-1} \circ f)(x) = f^{-1}\left(\sqrt{x - 5}\right) = \left(\sqrt{x - 5}\right)^2 + 5 = x - 5 + 5 = x$; $(f \circ f^{-1})(x) = f(x^2 + 5) = \sqrt{(x^2 + 5) - 5} = \sqrt{x^2} = x$ **8.** $f^{-1}(x) = \sqrt[3]{x - 2}$; domain and range of both f and f^{-1}: $(-\infty, \infty)$ **9.** (d)
10. (h) **11.** (c) **12.** (g) **13.** (b) **14.** (f) **15.** (e)
16. (a) **17.** \$4185.57 **18.** 0 **19.** $-\frac{4}{5}$ **20.** −2
21. 2 **22.** 0 **23.** −4 **24.** 0 **25.** 3 **26.** $\frac{1}{4}$ **27.** 1
28. $\log_e 0.0025 = -6$, or $\ln 0.0025 = -6$
29. $10^r = T$ **30.** 2.7268 **31.** 2.0115 **32.** For an even function f, $f(x) = f(-x)$, so we have $f(x) = f(-x)$ but $x \ne -x$ (for $x \ne 0$). Thus, f is not one-to-one and hence it does not have an inverse. **33.** The most interest will be earned the eighth year, because the principal is greatest during that year.
34. In $f(x) = x^3$, the variable x is the base. The range of f is $(-\infty, \infty)$. In $g(x) = 3^x$, the variable x is the exponent. The range of g is $(0, \infty)$. The graph of f does not have an asymptote. The graph of g has an asymptote $y = 0$.
35. If $\log b < 0$, then $0 < b < 1$.

Exercise Set 5.4

1. $\log_3 81 + \log_3 27 = 4 + 3 = 7$
3. $\log_5 5 + \log_5 125 = 1 + 3 = 4$
5. $\log_t 8 + \log_t Y$ **7.** $\ln x + \ln y$ **9.** $3 \log_b t$
11. $8 \log y$ **13.** $-6 \log_c K$ **15.** $\frac{1}{3} \ln 4$
17. $\log_t M - \log_t 8$ **19.** $\log x - \log y$ **21.** $\ln r - \ln s$
23. $\log_a 6 + \log_a x + 5 \log_a y + 4 \log_a z$
25. $2 \log_b p + 5 \log_b q - 4 \log_b m - 9$
27. $\ln 2 - \ln 3 - 3 \ln x - \ln y$
29. $\frac{3}{2} \log r + \frac{1}{2} \log t$ **31.** $3 \log_a x - \frac{5}{2} \log_a p - 4 \log_a q$
33. $2 \log_a m + 3 \log_a n - \frac{3}{4} - \frac{5}{4} \log_a b$ **35.** $\log_a 150$
37. $\log 100 = 2$ **39.** $\log m^3 \sqrt{n}$
41. $\log_a x^{-5/2} y^4$, or $\log_a \dfrac{y^4}{x^{5/2}}$ **43.** $\ln x$ **45.** $\ln (x - 2)$
47. $\log \dfrac{x-7}{x-2}$ **49.** $\ln \dfrac{x}{(x^2-25)^3}$ **51.** $\ln \dfrac{2^{11/5} x^9}{y^8}$
53. -0.74 **55.** 1.991 **57.** 0.356 **59.** 4.827
61. -1.792 **63.** 0.099 **65.** 3 **67.** $|x - 4|$ **69.** $4x$
71. w **73.** $8t$ **75.** $\frac{1}{2}$ **77.** Quartic
78. Exponential **79.** Linear (constant) **80.** Exponential
81. Rational **82.** Logarithmic **83.** Cubic **84.** Rational
85. Linear **86.** Quadratic **87.** 4 **89.** $\log_a (x^3 - y^3)$
91. $\frac{1}{2} \log_a (x - y) - \frac{1}{2} \log_a (x + y)$ **93.** 7 **95.** True
97. True **99.** True **101.** -2 **103.** 3
105. $e^{-xy} = \dfrac{a}{b}$

107. $$\log_a\left(\frac{x+\sqrt{x^2-5}}{5}\cdot\frac{x-\sqrt{x^2-5}}{x-\sqrt{x^2-5}}\right)$$
$$= \log_a \frac{5}{5(x-\sqrt{x^2-5})} = \log_a \frac{1}{x-\sqrt{x^2-5}}$$
$$= \log_a \left(x-\sqrt{x^2-5}\right)^{-1} = -\log_a\left(x-\sqrt{x^2-5}\right)$$

Exercise Set 5.5

1. 4 **3.** $\frac{3}{2}$ **5.** 5.044 **7.** $\frac{5}{2}$ **9.** $-3, \frac{1}{2}$ **11.** 0.959
13. 0 **15.** 0 **17.** 6.908 **19.** 84.191 **21.** -1.710
23. 2.844 **25.** $-1.567, 1.567$ **27.** 1.869
29. $-1.518, 0.825$ **31.** 625 **33.** 0.0001 **35.** e
37. $-\frac{1}{3}$ **39.** $\frac{22}{3}$ **41.** 10 **43.** 4 **45.** $\frac{1}{63}$ **47.** 2
49. $\frac{2}{5}$ **51.** 5 **53.** $\frac{21}{8}$ **55.** $\frac{8}{7}$ **57.** 6 **59.** 6.192
61. 0 **63.** $-1.911, 4.222$ **65.** 0.621 **67.** $-10, 0.366$
69. 7.062 **71.** 2.444 **73.** $(4.093, 0.786)$
75. $(7.586, 6.684)$ **77.** **(a)** $(0, -6)$; **(b)** $x = 0$;
(c) minimum: -6 when $x = 0$ **78.** **(a)** $(3, 1)$; **(b)** $x = 3$;
(c) maximum: 1 when $x = 3$ **79.** **(a)** $(-1, -5)$; **(b)** $x = -1$;
(c) maximum: -5 when $x = -1$ **80.** **(a)** $(2, 4)$; **(b)** $x = 2$;
(c) minimum: 4 when $x = 2$ **81.** $\dfrac{\ln 2}{2}$, or 0.347 **83.** $1, e^4$ or
1, 54.598 **85.** $\frac{1}{3}, 27$ **87.** $1, e^2$ or 1, 7.389 **89.** $0, \dfrac{\ln 2}{\ln 5}$, or
0, 0.431 **91.** e^{-2}, e^2 or 0.135, 7.389 **93.** $\frac{7}{4}$ **95.** $a = \frac{2}{3}b$

Exercise Set 5.6

1. **(a)** $P(t) = 438{,}000\, e^{0.034t}$, where t is the number of years after 2014; **(b)** 485,034; **(c)** about 4 years after 2014; **(d)** about 20 years
3. **(a)** 0.70%; **(b)** 1.63%; **(c)** 20.9 years; **(d)** 62.4 years; **(e)** 0.18%; **(f)** 29.9 years; **(g)** 54.2 years; **(h)** 0.46%; **(i)** 2.64%; **(j)** 177.7 years **5.** About 750 years after 2015
7. **(a)** $P(t) = 10{,}000e^{0.054t}$; **(b)** \$10,554.85; \$11,140.48; \$13,099.64; \$17,160.07; **(c)** about 12.8 years
9. About 12,320 years **11.** **(a)** 22.4% per minute; **(b)** 3.1% per year; **(c)** 60.3 days; **(d)** 10.7 years; **(e)** 2.4% per year; **(f)** 1.0% per year; **(g)** 0.0029% per year **13.** **(a)** $k \approx 0.0437$; $V(t) = 8000e^{0.0437t}$; **(b)** \$12,938; **(c)** about 15.9 years; **(d)** about 26 years after 2000, or in 2026 **15.** **(a)** $k \approx 0.0681$; $S(t) = 935{,}000e^{0.0681t}$; **(b)** \$3,650,234; **(c)** about 10 years; **(d)** about 37 years after 1980, or in 2017
17. **(a)** $y = \dfrac{3500}{1 + 19.9e^{-0.6x}}$; **(b)** 167; **(c)** 500; 1758; 3007; 3449; 3495; **(d)** as $t \rightarrow \infty$, $N(t) \rightarrow 3500$; the number approaches 3500 but never actually reaches it.
19. 46.7°F **21.** 59.6°F **23.** (d) **25.** (a) **27.** (e)
29. **(a)** $y = 0.1377082721(1.023820625)^x$; $r \approx 0.9824$, the function is a good fit;
(b)

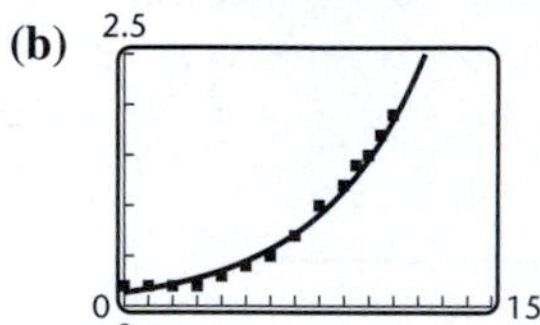

; **(c)** 2007: 1.7%; 2015: 2.1%; 2020: 2.3%
31. **(a)** $y = 173.8943128(1.051418179)^x$; **(b)** ; **(c)** 2016: \$234.9 billion; 2023: \$333.71 billion
33. **(a)** $y = 1454.850619(0.9265336199)^x$; **(b)** 2008: 1348 architects; 2012: 993 architects; **(c)** about 6 years after 2007, or in 2013
35. Multiplication principle for inequalities **36.** Product rule
37. Principle of zero products **38.** Principle of square roots
39. Power rule **40.** Multiplication principle for equations
41. \$166.16 **43.** \$19,609.67 **45.** **(a)** 24.7%; 1.5%; 0.09%; (3.98×10^{-29}); **(b)** 0.00008% **47.** Linear

Review Exercises: Chapter 5

1. True **2.** False **3.** False **4.** True **5.** False
6. True **7.** $\{(-2.7, 1.3), (-3, 8), (3, -5), (-3, 6), (-5, 7)\}$
8. **(a)** $x = -2y + 3$; **(b)** $x = 3y^2 + 2y - 1$;
(c) $0.8y^3 - 5.4x^2 = 3y$ **9.** No **10.** No **11.** Yes
12. Yes **13.** **(a)** Yes; **(b)** $f^{-1}(x) = \dfrac{-x+2}{3}$
14. **(a)** Yes; **(b)** $f^{-1}(x) = \dfrac{x+2}{x-1}$
15. **(a)** Yes; **(b)** $f^{-1}(x) = x^2 + 6, x \geq 0$
16. **(a)** Yes; **(b)** $f^{-1}(x) = \sqrt[3]{x+8}$ **17.** **(a)** No
18. **(a)** Yes; **(b)** $f^{-1}(x) = \ln x$

19. $(f^{-1}\circ f)(x) =$
$f^{-1}(f(x)) = f^{-1}(6x-5) = \dfrac{6x-5+5}{6} = \dfrac{6x}{6} = x;$
$(f\circ f^{-1})(x) =$
$f(f^{-1}(x)) = f\left(\dfrac{x+5}{6}\right) = 6\left(\dfrac{x+5}{6}\right) - 5 = x+5-5 = x$

20. $(f^{-1}\circ f)(x) =$
$f^{-1}(f(x)) = f^{-1}\left(\dfrac{x+1}{x}\right) = \dfrac{1}{\dfrac{x+1}{x} - 1} =$
$\dfrac{1}{\dfrac{x+1-x}{x}} = \dfrac{1}{\dfrac{1}{x}} = x;\ (f\circ f^{-1})(x) =$
$f(f^{-1}(x)) = f\left(\dfrac{1}{x-1}\right) =$
$\dfrac{\dfrac{1}{x-1}+1}{\dfrac{1}{x-1}} = \dfrac{\dfrac{1+x-1}{x-1}}{\dfrac{1}{x-1}} = \dfrac{x}{x-1}\cdot\dfrac{x-1}{1} = x$

21. $f^{-1}(x) = \dfrac{2-x}{5}$; domain and range of both f and f^{-1}: $(-\infty,\infty)$;

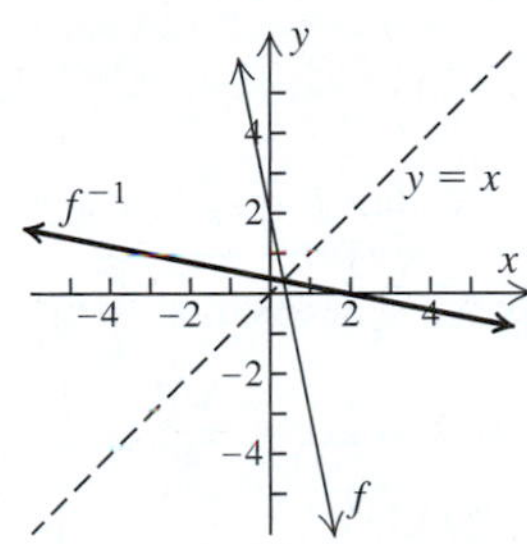

22. $f^{-1}(x) = \dfrac{-2x-3}{x-1}$;
domain of f: $(-\infty,-2)\cup(-2,\infty)$;
range of f: $(-\infty,1)\cup(1,\infty)$;
domain of f^{-1}: $(-\infty,1)\cup(1,\infty)$;
range of f^{-1}: $(-\infty,-2)\cup(-2,\infty)$

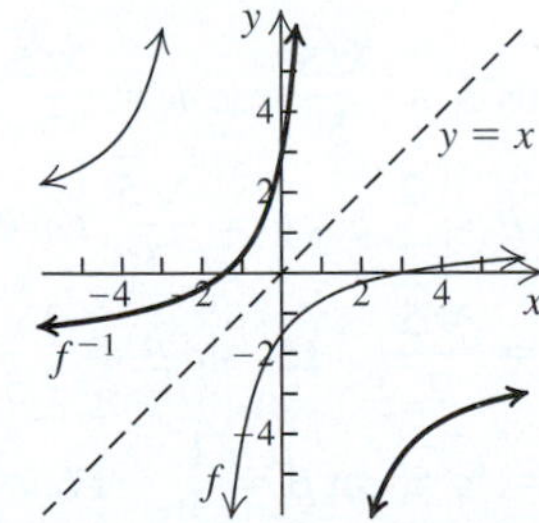

23. 657 **24.** a

25.

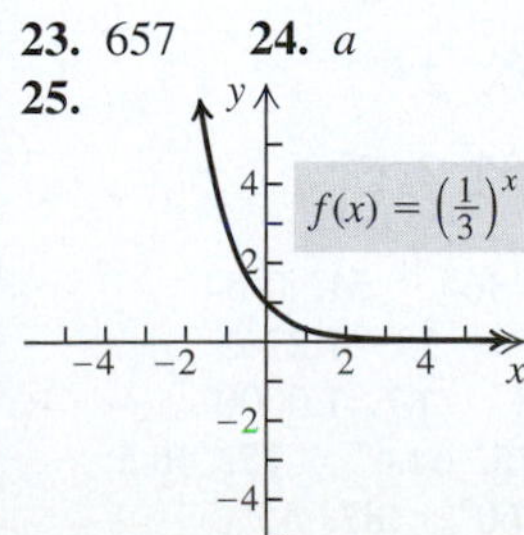

26.

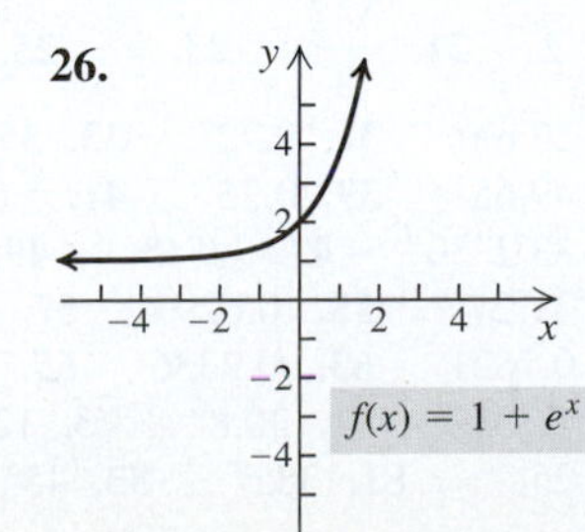

27.

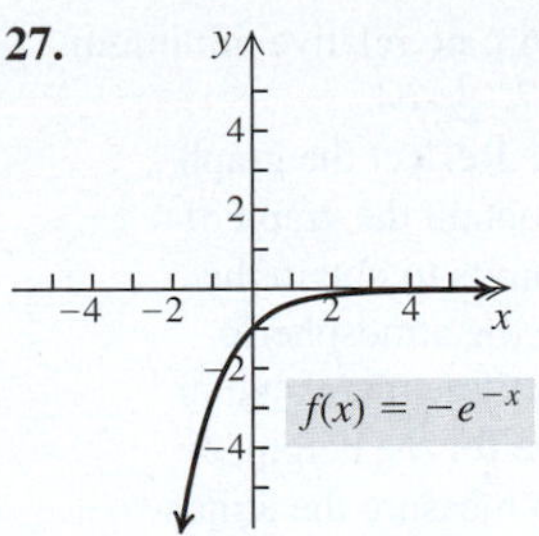

28.

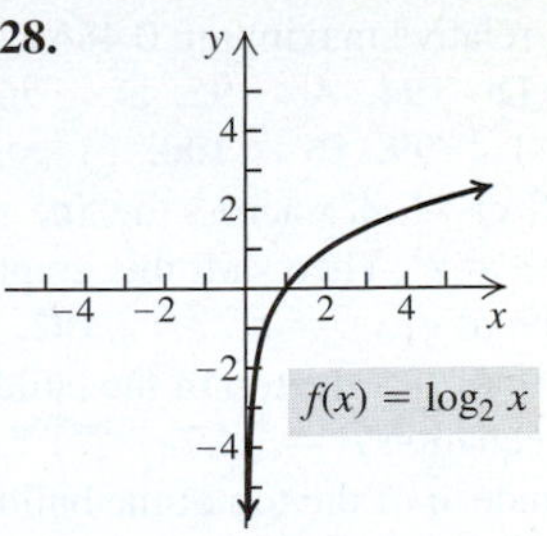

29.

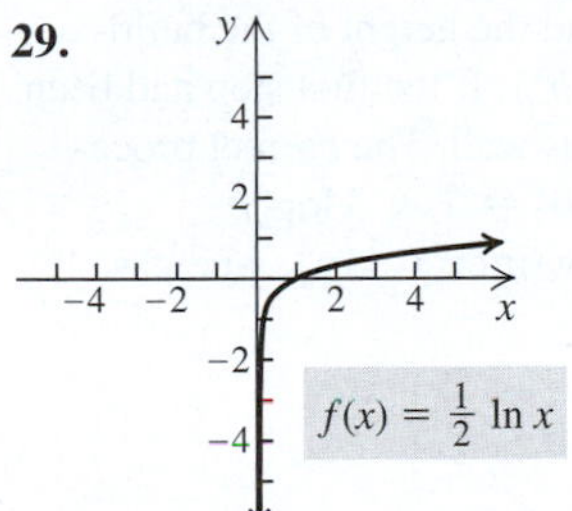

30.

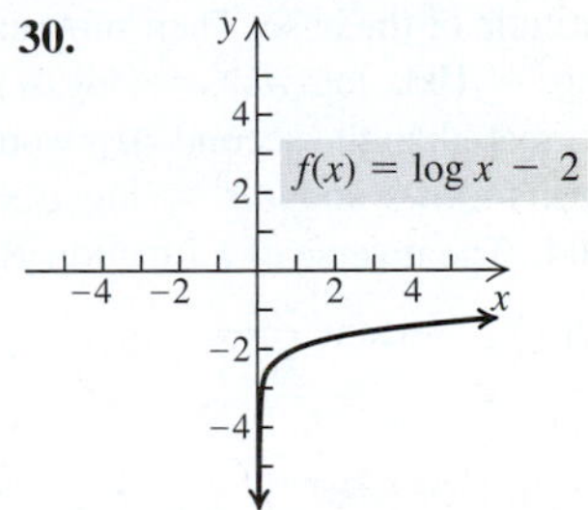

31. (c) **32.** (a) **33.** (b) **34.** (f) **35.** (e) **36.** (d) **37.** 3 **38.** 5 **39.** 1 **40.** 0 **41.** $\frac{1}{4}$ **42.** $\frac{1}{2}$ **43.** 0 **44.** 1 **45.** $\frac{1}{3}$ **46.** -2 **47.** $4^2 = x$ **48.** $a^k = Q$ **49.** $\log_4 \frac{1}{64} = -3$ **50.** $\ln 80 = x$, or $\log_e 80 = x$ **51.** 1.0414 **52.** -0.6308 **53.** 1.0986 **54.** -3.6119 **55.** Does not exist **56.** Does not exist **57.** 1.9746 **58.** 0.5283 **59.** $\log_b \dfrac{x^3\sqrt{z}}{y^4}$, or $\log_b \dfrac{x^3z^{1/2}}{y^4}$ **60.** $\ln(x^2-4)$ **61.** $\frac{1}{4}\ln w + \frac{1}{2}\ln r$ **62.** $\frac{2}{3}\log M - \frac{1}{3}\log N$ **63.** 0.477 **64.** 1.699 **65.** -0.699 **66.** 0.233 **67.** $-5k$ **68.** $-6t$ **69.** 16 **70.** $\frac{1}{5}$ **71.** 4.382 **72.** 2 **73.** $\frac{1}{2}$ **74.** 5 **75.** 4 **76.** 9 **77.** 1 **78.** 3.912 **79.** **(a)** $A(t) = 30{,}000(1.0105)^{4t}$; **(b)** \$30,000; \$38,547.20; \$49,529.56; \$63,640.87 **80.** 2014: 3466 breweries; 2018: 6905 breweries **81.** 15.4 years **82.** 2.7% **83.** About 2623 years **84.** 5.6 **85.** 7.0 **86.** 30 decibels **87.** **(a)** 2.2 ft/sec; **(b)** 8,553,143 **88.** **(a)** $k \approx 0.1392$; **(b)** $S(t) = 0.035e^{0.1392t}$, where t is the number of years after 1940 and S is in billions of dollars; **(c)** 1970: \$2.279 billion; 2000: \$148.353 billion; 2015: \$1197.023 billion, or about \$1.197 trillion; **(d)** in 2019 **89.** **(a)** $P(t) = 15.2e^{0.0167t}$, where t is the number of years after 2013 and P is in millions; **(b)** 2017: 16.3 million; 2020: 17.1 million; **(c)** about 10 years after 2013; **(d)** 41.5 years **90.** **(a)** $y = 99.16310468(1.02186993)^x$;
(b)

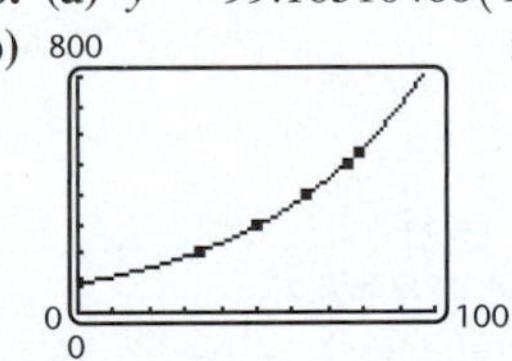

; **(c)** 1970: 221 billion metric tons; 2000: 423 billion metric tons; 2015: 585 billion metric tons; **(d)** about 107 years after 1933, or in 2040

91. No **92.** **(a)** $y = 5e^{-x}\ln x$

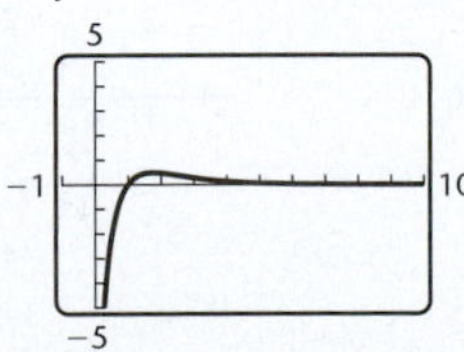

(b) relative maximum: 0.486 at $x = 1.763$; no relative minimum **93.** D **94.** A **95.** D **96.** B **97.** $\frac{1}{64}$, 64 **98.** 1 **99.** 16 **100.** $(1, \infty)$ **101.** Reflect the graph of $f(x) = \ln x$ across the line $y = x$ to obtain the graph of $h(x) = e^x$. Then shift this graph right 2 units to obtain the graph of $g(x) = e^{x-2}$. **102.** Measure the atmospheric pressure P at the top of the building. Substitute that value in the equation $P = 14.7e^{-0.00005a}$, and solve for the height, or altitude, a of the top of the building. Also measure the atmospheric pressure at the base of the building and solve for the altitude of the base. Then subtract to find the height of the building. **103.** $\log_a ab^3 \neq (\log_a a)(\log_a b^3)$. If the first step had been correct, then the second step would be as well. The correct procedure follows: $\log_a ab^3 = \log_a a + \log_a b^3 = 1 + 3\log_a b$. **104.** The inverse of a function $f(x)$ is written $f^{-1}(x)$, whereas $[f(x)]^{-1}$ means $\dfrac{1}{f(x)}$.

Test: Chapter 5

1. [5.1] $\{(5, -2), (3, 4), (-1, 0), (-3, -6)\}$ **2.** [5.1] No **3.** [5.1] Yes **4.** [5.1] **(a)** Yes; **(b)** $f^{-1}(x) = \sqrt[3]{x - 1}$ **5.** [5.1] **(a)** Yes; **(b)** $f^{-1}(x) = 1 - x$ **6.** [5.1] **(a)** Yes; **(b)** $f^{-1}(x) = \dfrac{2x}{1 + x}$ **7.** [5.1] **(a)** No **8.** [5.1] $(f^{-1} \circ f)(x) = f^{-1}(f(x)) = f^{-1}(-4x + 3) = \dfrac{3 - (-4x + 3)}{4} = \dfrac{4x}{4} = x$; $(f \circ f^{-1})(x) = f(f^{-1}(x)) = f\left(\dfrac{3 - x}{4}\right) = -4\left(\dfrac{3 - x}{4}\right) + 3 = -3 + x + 3 = x$ **9.** [5.1] $f^{-1}(x) = \dfrac{4x + 1}{x}$; domain of f: $(-\infty, 4) \cup (4, \infty)$; range of f: $(-\infty, 0) \cup (0, \infty)$; domain of f^{-1}: $(-\infty, 0) \cup (0, \infty)$; range of f^{-1}: $(-\infty, 4) \cup (4, \infty)$;

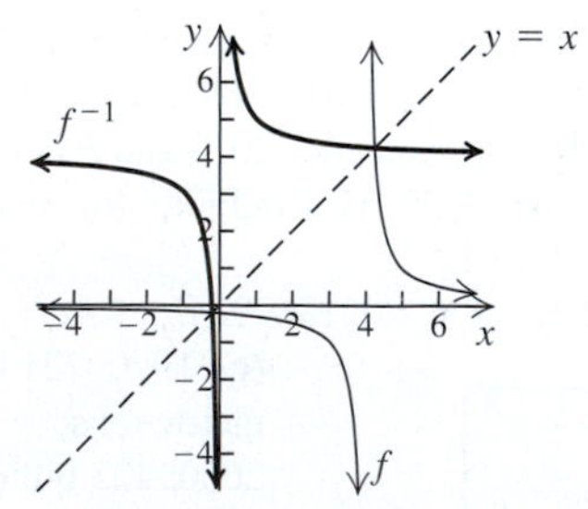

10. [5.2]

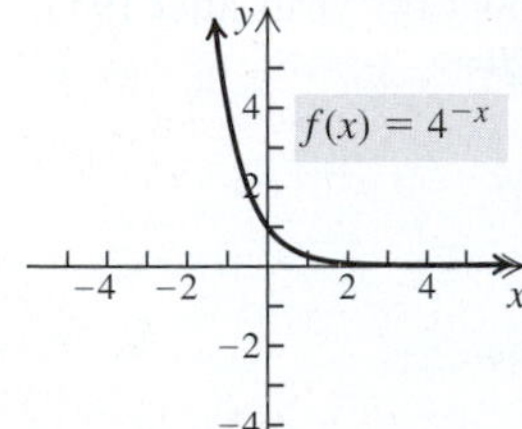

11. [5.3]

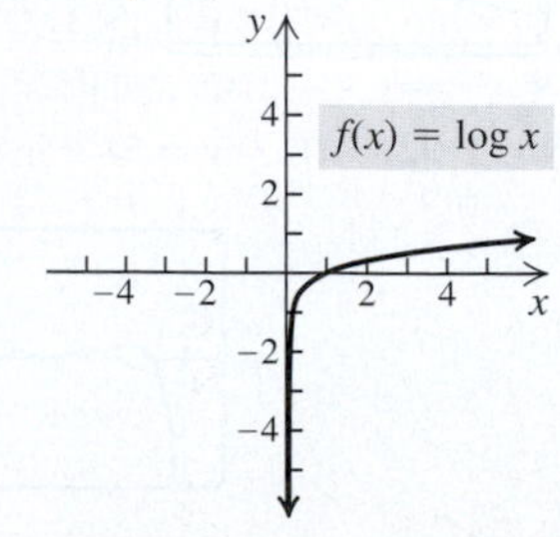

12. [5.2] **13.** [5.3]

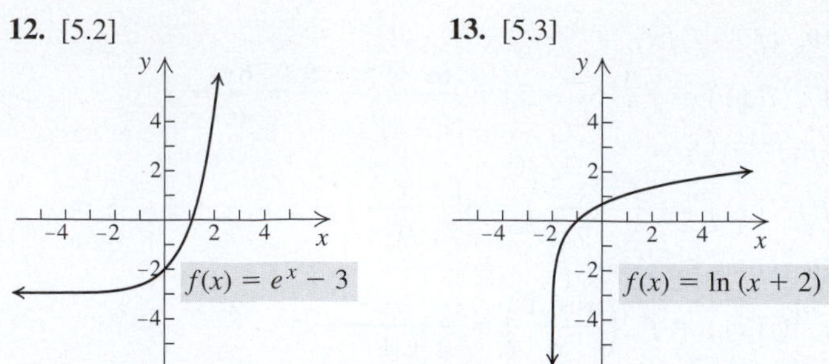

14. [5.3] -5 **15.** [5.3] 1 **16.** [5.3] 0 **17.** [5.3] $\frac{1}{5}$ **18.** [5.3] $x = e^4$ **19.** [5.3] $x = \log_3 5.4$ **20.** [5.3] 2.7726 **21.** [5.3] -0.5331 **22.** [5.3] 1.2851 **23.** [5.4] $\log_a \dfrac{x^2\sqrt{z}}{y}$, or $\log_a \dfrac{x^2z^{1/2}}{y}$ **24.** [5.4] $\frac{2}{5}\ln x + \frac{1}{5}\ln y$ **25.** [5.4] 2.322 **26.** [5.4] $-4t$ **27.** [5.5] $\frac{1}{2}$ **28.** [5.5] 1 **29.** [5.5] 1 **30.** [5.5] 4.174 **31.** [5.3] 9.0 **32.** [5.6] 0.0154 **33.** [5.6] **(a)** 4.5%; **(b)** $P(t) = 1000e^{0.045t}$; **(c)** \$1433.33; **(d)** 15.4 years **34.** [5.2] C **35.** [5.5] $\frac{27}{8}$

Chapter 6

Exercise Set 6.1

1. $\sin\phi = \frac{15}{17}$, $\cos\phi = \frac{8}{17}$, $\tan\phi = \frac{15}{8}$, $\csc\phi = \frac{17}{15}$, $\sec\phi = \frac{17}{8}$, $\cot\phi = \frac{8}{15}$ **3.** $\sin\alpha = \dfrac{\sqrt{3}}{2}$, $\cos\alpha = \dfrac{1}{2}$, $\tan\alpha = \sqrt{3}$, $\csc\alpha = \dfrac{2\sqrt{3}}{3}$, $\sec\alpha = 2$, $\cot\alpha = \dfrac{\sqrt{3}}{3}$ **5.** $\sin\phi = \dfrac{27}{5\sqrt{37}}$, or $\dfrac{27\sqrt{37}}{185}$; $\cos\phi = \dfrac{14}{5\sqrt{37}}$, or $\dfrac{14\sqrt{37}}{185}$; $\tan\phi = \dfrac{27}{14}$; $\csc\phi = \dfrac{5\sqrt{37}}{27}$; $\sec\phi = \dfrac{5\sqrt{37}}{14}$; $\cot\phi = \dfrac{14}{27}$ **7.** $\csc\alpha = \dfrac{3}{\sqrt{5}}$, or $\dfrac{3\sqrt{5}}{5}$; $\sec\alpha = \dfrac{3}{2}$; $\cot\alpha = \dfrac{2}{\sqrt{5}}$, or $\dfrac{2\sqrt{5}}{5}$ **9.** $\cos\theta = \frac{7}{25}$, $\tan\theta = \frac{24}{7}$, $\csc\theta = \frac{25}{24}$, $\sec\theta = \frac{25}{7}$, $\cot\theta = \frac{7}{24}$ **11.** $\sin\phi = \dfrac{2\sqrt{5}}{5}$, $\cos\phi = \dfrac{\sqrt{5}}{5}$, $\csc\phi = \dfrac{\sqrt{5}}{2}$, $\sec\phi = \sqrt{5}$, $\cot\phi = \dfrac{1}{2}$ **13.** $\sin\theta = \dfrac{2}{3}$, $\cos\theta = \dfrac{\sqrt{5}}{3}$, $\tan\theta = \dfrac{2\sqrt{5}}{5}$, $\sec\theta = \dfrac{3\sqrt{5}}{5}$, $\cot\theta = \dfrac{\sqrt{5}}{2}$ **15.** $\sin\beta = \dfrac{2\sqrt{5}}{5}$, $\tan\beta = 2$, $\csc\beta = \dfrac{\sqrt{5}}{2}$, $\sec\beta = \sqrt{5}$, $\cot\beta = \dfrac{1}{2}$ **17.** $\dfrac{\sqrt{2}}{2}$ **19.** 2 **21.** $\dfrac{\sqrt{3}}{3}$ **23.** $\frac{1}{2}$ **25.** 1 **27.** 2 **29.** 22.6 ft **31.** 9.72° **33.** 35.01° **35.** 3.03° **37.** 49.65° **39.** 0.25° **41.** 5.01° **43.** 17°36′ **45.** 83°1′30″ **47.** 11°45′ **49.** 47°49′36″ **51.** 0°54′ **53.** 39°27′ **55.** 0.6293 **57.** 0.0737 **59.** 1.2765 **61.** 0.7621 **63.** 0.9336 **65.** 12.4288 **67.** 1.0000 **69.** 1.7032 **71.** 30.8° **73.** 12.5° **75.** 64.4° **77.** 46.5° **79.** 25.2° **81.** 38.6° **83.** 45° **85.** 60° **87.** 45° **89.** 60° **91.** 30° **93.** $\cos 20° = \sin 70° = \dfrac{1}{\sec 20°}$ **95.** $\tan 52° = \cot 38° = \dfrac{1}{\cot 52°}$ **97.** $\sin 25° \approx 0.4226$,

cos 25° ≈ 0.9063, tan 25° ≈ 0.4663, csc 25° ≈ 2.3662, sec 25° ≈ 1.1034, cot 25° ≈ 2.1445
99. sin 18°49′55″ ≈ 0.3228, cos 18°49′55″ ≈ 0.9465, tan 18°49′55″ ≈ 0.3411, csc 18°49′55″ ≈ 3.0979, sec 18°49′55″ ≈ 1.0565, cot 18°49′55″ ≈ 2.9321
101. sin 8° = q, cos 8° = p, tan 8° = $\frac{1}{r}$, csc 8° = $\frac{1}{q}$, sec 8° = $\frac{1}{p}$, cot 8° = r
102.

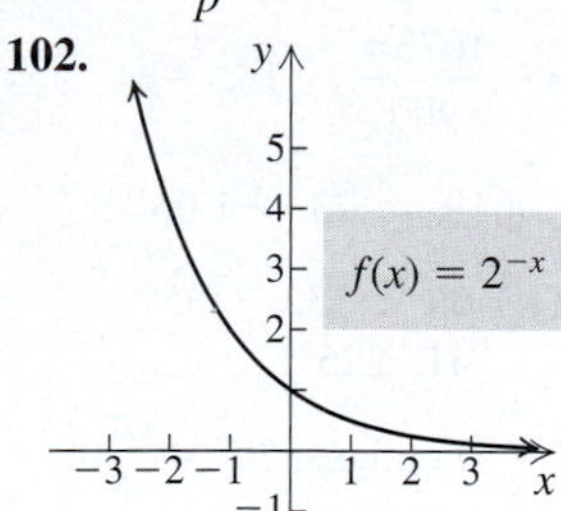

103.

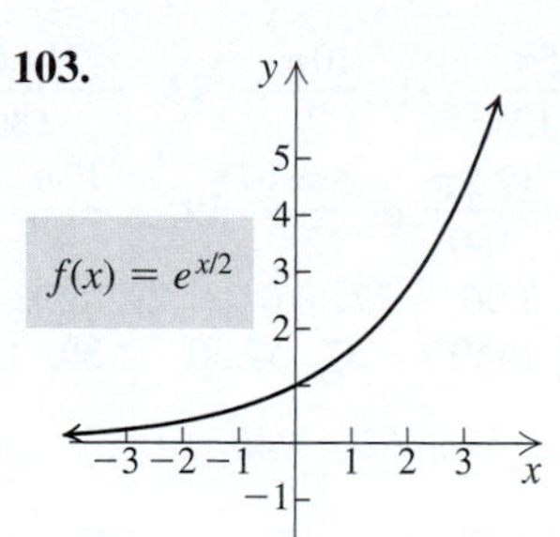

104.

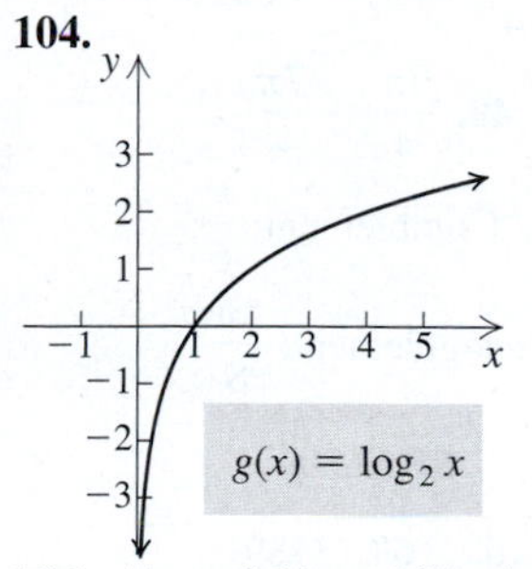

105.

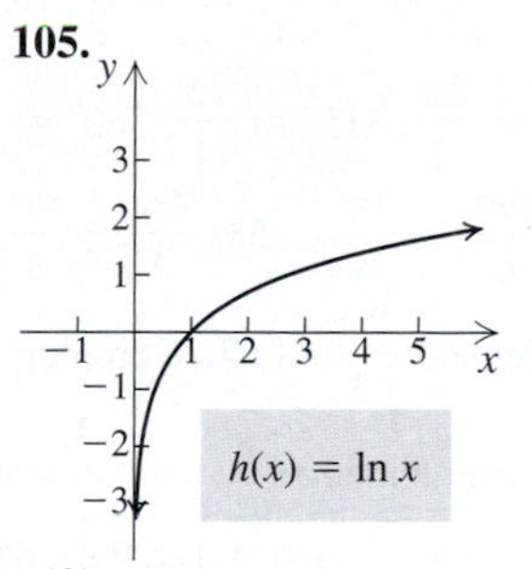

106. About 9.21 **107.** 4 **108.** $\frac{101}{97}$ **109.** 343
111. 0.6534 **113.** Let h = the height of the triangle. Then Area = $\frac{1}{2}bh$, where $\sin\theta = \frac{h}{a}$, or $h = a\sin\theta$, so Area = $\frac{1}{2}ab\sin\theta$.

Exercise Set 6.2

1. $F = 60°, d = 3, f \approx 5.2$ **3.** $A = 22.7°, a \approx 52.7, c \approx 136.6$ **5.** $P = 47°38', n \approx 34.4, p \approx 25.4$
7. $B = 2°17', b \approx 0.39, c \approx 9.74$ **9.** $A \approx 77.2°, B \approx 12.8°, a \approx 439$ **11.** $B = 42.42°, a \approx 35.7, b \approx 32.6$
13. $B = 55°, a \approx 28.0, c \approx 48.8$ **15.** $A \approx 62.4°, B \approx 27.6°, a \approx 3.56$ **17.** Approximately 34° **19.** About 13.9°
21. 154 in., or 12 ft 10 in. **23.** 10.4° **25.** About 424 ft
27. About 92.9 cm **29.** About 45 ft **31.** Radius: 9.15 in.; length: 73.20 in.; width: 54.90 in. **33.** 17.9 ft
35. About 8 km **37.** About 19.5 mi **39.** About 24 km
40. $3\sqrt{10}$, or about 9.487 **41.** $10\sqrt{2}$, or about 14.142
42. $\ln t = 4$ **43.** $10^{-3} = 0.001$ **45.** 3.3

Exercise Set 6.3

1. III **3.** III **5.** I **7.** III **9.** II **11.** II **13.** 434°, 794°, −286°, −646° **15.** 475.3°, 835.3°, −244.7°, −604.7°
17. 180°, 540°, −540°, −900° **19.** 72.89°, 162.89°
21. 77°56′46″, 167°56′46″ **23.** 44.8°, 134.8°
25. $\sin\beta = \frac{5}{13}$, $\cos\beta = -\frac{12}{13}$, $\tan\beta = -\frac{5}{12}$, $\csc\beta = \frac{13}{5}$, $\sec\beta = -\frac{13}{12}$, $\cot\beta = -\frac{12}{5}$
27. $\sin\alpha = -\frac{4\sqrt{97}}{97}$; $\cos\alpha = \frac{9\sqrt{97}}{97}$; $\tan\alpha = -\frac{4}{9}$; $\csc\alpha = -\frac{\sqrt{97}}{4}$; $\sec\alpha = \frac{\sqrt{97}}{9}$; $\cot\alpha = -\frac{9}{4}$
29. $\sin\phi = -\frac{2\sqrt{7}}{7}$, $\cos\phi = -\frac{\sqrt{21}}{7}$, $\tan\phi = \frac{2\sqrt{3}}{3}$, $\csc\phi = -\frac{\sqrt{7}}{2}$, $\sec\phi = -\frac{\sqrt{21}}{3}$, $\cot\phi = \frac{\sqrt{3}}{2}$
31. $\sin\theta = -\frac{2\sqrt{13}}{13}$, $\cos\theta = \frac{3\sqrt{13}}{13}$, $\tan\theta = -\frac{2}{3}$
33. $\sin\theta = \frac{5\sqrt{41}}{41}$, $\cos\theta = \frac{4\sqrt{41}}{41}$, $\tan\theta = \frac{5}{4}$
35. $\cos\theta = -\frac{2\sqrt{2}}{3}$, $\tan\theta = \frac{\sqrt{2}}{4}$, $\csc\theta = -3$, $\sec\theta = -\frac{3\sqrt{2}}{4}$, $\cot\theta = 2\sqrt{2}$ **37.** $\sin\theta = -\frac{\sqrt{5}}{5}$, $\cos\theta = \frac{2\sqrt{5}}{5}$, $\tan\theta = -\frac{1}{2}$, $\csc\theta = -\sqrt{5}$, $\sec\theta = \frac{\sqrt{5}}{2}$ **39.** $\sin\phi = -\frac{4}{5}$, $\tan\phi = -\frac{4}{3}$, $\csc\phi = -\frac{5}{4}$, $\sec\phi = \frac{5}{3}$, $\cot\phi = -\frac{3}{4}$ **41.** 30°; $-\frac{\sqrt{3}}{2}$
43. 45°; 1 **45.** 0 **47.** 45°; $-\frac{\sqrt{2}}{2}$ **49.** 30°; 2
51. 30°; $\sqrt{3}$ **53.** 30°; $-\frac{\sqrt{3}}{3}$ **55.** Not defined **57.** −1
59. 60°; $\sqrt{3}$ **61.** 45°; $\frac{\sqrt{2}}{2}$ **63.** 45°; $-\sqrt{2}$ **65.** 1
67. 0 **69.** 0 **71.** 0 **73.** Positive: cos, sec; negative: sin, csc, tan, cot **75.** Positive: tan, cot; negative: sin, csc, cos, sec
77. Positive: sin, csc; negative: cos, sec, tan, cot **79.** Positive: all
81. sin 319° = −0.6561, cos 319° = 0.7547, tan 319° = −0.8693, csc 319° ≈ −1.5242, sec 319° ≈ 1.3250, cot 319° ≈ −1.1504
83. sin 115° = 0.9063, cos 115° = −0.4226, tan 115° = −2.1445, csc 115° ≈ 1.1034, sec 115° ≈ −2.3663, cot 115° ≈ −0.4663
85. −1.1585 **87.** −1.4910 **89.** 0.8771 **91.** 0.4352
93. 0.9563 **95.** 2.9238 **97.** 275.4° **99.** 200.1°
101. 288.1° **103.** 72.6° **105.** East: about 130 km; south: 75 km
107. About 223 km
109.

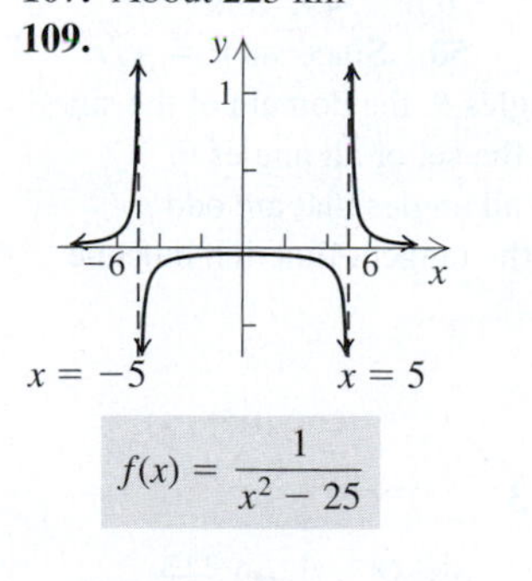

110.

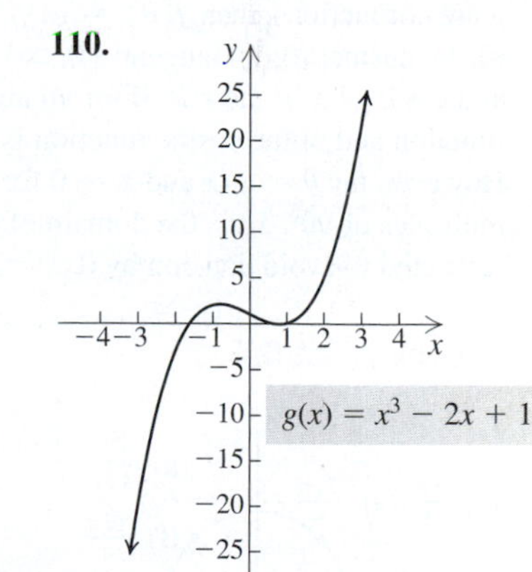

111. Domain: $\{x | x \neq -2\}$; range: $\{x | x \neq 1\}$
112. Domain: $\{x | x \neq -\frac{3}{2} \text{ and } x \neq 5\}$; range: all real numbers
113. 12 **114.** −2, 3 **115.** (12, 0) **116.** (−2, 0), (3, 0)
117. About 70 ft

Mid-Chapter Mixed Review: Chapter 6

1. True **2.** True **3.** True **4.** $S = 47.5°, s \approx 59.9, q \approx 54.9$ **5.** $A \approx 27.8°, B \approx 62.2°, b \approx 27.2$
6. 285°, 645°; −435°, −1155° **7.** 574°30′, 1294°30′; −145°30′, −505°30′ **8.** 71.8°; 161.8° **9.** 2°44′50″; 92°44′50″ **10.** sin 155° = 0.4226, cos 155° = −0.9063,

tan 155° = −0.4663, csc 155° ≈ 2.3663, sec 155° ≈ −1.1034, cot 155° ≈ −2.1445 **11.** $\sin \alpha = -\frac{5}{13}$, $\cos \alpha = -\frac{12}{13}$, $\tan \alpha = \frac{5}{12}$, $\csc \alpha = -\frac{13}{5}$, $\sec \alpha = -\frac{13}{12}$, $\cot \alpha = \frac{12}{5}$

12. $\sin \theta = -\frac{1}{\sqrt{5}}$, or $-\frac{\sqrt{5}}{5}$; $\cos \theta = -\frac{2}{\sqrt{5}}$, or $-\frac{2\sqrt{5}}{5}$; $\tan \theta = \frac{1}{2}$; $\csc \theta = -\sqrt{5}$; $\sec \theta = -\frac{\sqrt{5}}{2}$

13. $\sin \alpha = \frac{\sqrt{77}}{9}$; $\tan \alpha = \frac{\sqrt{77}}{2}$; $\csc \alpha = \frac{9}{\sqrt{77}}$, or $\frac{9\sqrt{77}}{77}$; $\sec \alpha = \frac{9}{2}$; $\cot \alpha = \frac{2}{\sqrt{77}}$, or $\frac{2\sqrt{77}}{77}$

14. 42.1472° **15.** 51°10′48″ **16.** sin 81° ≈ 0.9877, cos 81° ≈ 0.1564, tan 81° ≈ 6.3131, csc 81° ≈ 1.0125, sec 81° ≈ 6.3939, cot 81° ≈ 0.1584 **17.** 67.5°

18. About 290 mi **19.** $\frac{\sqrt{3}}{3}$ **20.** $\frac{\sqrt{2}}{2}$ **21.** $\sqrt{3}$

22. $-\sqrt{2}$ **23.** $\frac{\sqrt{2}}{2}$ **24.** −2 **25.** 1 **26.** 0

27. $\frac{\sqrt{3}}{2}$ **28.** −1 **29.** $-\sqrt{3}$ **30.** 1 **31.** $\frac{1}{2}$

32. $\sqrt{2}$ **33.** $-\frac{\sqrt{3}}{2}$ **34.** Not defined **35.** 2

36. Not defined **37.** 1 **38.** $-\frac{\sqrt{2}}{2}$ **39.** 0.7683

40. 1.5557 **41.** 0.4245 **42.** 0.1817 **43.** −1.0403 **44.** −1.3127 **45.** −0.6441 **46.** 0.0480

47. Given points P and Q on the terminal side of an angle θ, the reference triangles determined by them are similar. Thus corresponding sides are proportional and the trigonometric ratios are the same. See the specific example on p. 423.

48. If f and g are reciprocal functions, then $f(\theta) = \frac{1}{g(\theta)}$. If f and g are cofunctions, then $f(\theta) = g(90° - \theta)$. **49.** Sine: $(0, 1)$; cosine: $(0, 1)$; tangent: $(0, \infty)$ **50.** Since $\sin \theta = y/r$ and $\cos \theta = x/r$ and $r > 0$ for all angles θ, the domain of the sine function and of the cosine function is the set of all angles θ. However, $\tan \theta = y/x$ and $x = 0$ for all angles that are odd multiples of 90°. Thus the domain of the tangent function must be restricted to avoid division by 0.

Exercise Set 6.4

1. **3.**

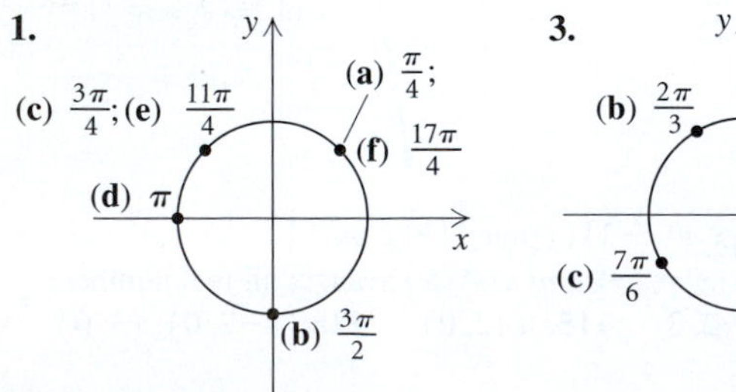

5. M: $\frac{2\pi}{3}, -\frac{4\pi}{3}$; N: $\frac{3\pi}{2}, -\frac{\pi}{2}$; P: $\frac{5\pi}{4}, -\frac{3\pi}{4}$; Q: $\frac{11\pi}{6}, -\frac{\pi}{6}$

7.

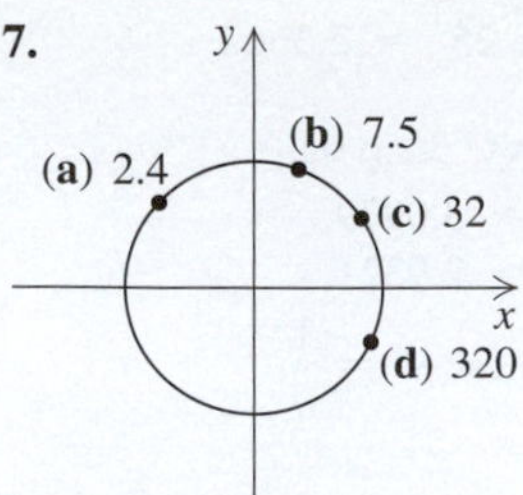

9. $\frac{5\pi}{12}$ **11.** $\frac{10\pi}{9}$ **13.** $-\frac{214.6\pi}{180}$, or $-\frac{1073\pi}{900}$ **15.** $-\pi$

17. $\frac{12.5\pi}{180}$, or $\frac{5\pi}{72}$ **19.** $-\frac{17\pi}{9}$ **21.** 4.19 **23.** −1.05

25. 2.06 **27.** 0.02 **29.** 6.02 **31.** 1.66 **33.** −135° **35.** 1440° **37.** 57.30° **39.** 134.47° **41.** 225°

43. −5156.62° **45.** 51.43° **47.** 0° = 0 radians, $30° = \frac{\pi}{6}$, $45° = \frac{\pi}{4}$, $60° = \frac{\pi}{3}$, $90° = \frac{\pi}{2}$, $135° = \frac{3\pi}{4}$, $180° = \pi$, $225° = \frac{5\pi}{4}$, $270° = \frac{3\pi}{2}$, $315° = \frac{7\pi}{4}$, $360° = 2\pi$ **49.** $\frac{9\pi}{4}, -\frac{7\pi}{4}$

51. $\frac{19\pi}{6}, -\frac{5\pi}{6}$ **53.** $\frac{4\pi}{3}, -\frac{8\pi}{3}$ **55.** Complement: $\frac{\pi}{6}$; supplement: $\frac{2\pi}{3}$ **57.** Complement: $\frac{\pi}{8}$; supplement: $\frac{5\pi}{8}$

59. Complement: $\frac{5\pi}{12}$; supplement: $\frac{11\pi}{12}$ **61.** 2.29

63. 5.50 in. **65.** 1.1; 63° **67.** 3.2 yd **69.** 3489

71. $3150 \frac{\text{cm}}{\text{min}}$ **73.** 0.92 mph **75.** 1047 mph

77. 10 mph **79.** About 202 **81.** 1.676 radians/sec **82.** One-to-one **83.** Cosine of θ **84.** Exponential function **85.** Horizontal asymptote **86.** Odd function **87.** Natural **88.** Horizontal line; inverse **89.** Logarithm **91.** 111.7 km; 69.8 mi **93.** (a) 5°37′30″; (b) 19°41′15″ **95.** 1.46 nautical miles

Exercise Set 6.5

1. (a) $\left(-\frac{3}{4}, -\frac{\sqrt{7}}{4}\right)$; (b) $\left(\frac{3}{4}, \frac{\sqrt{7}}{4}\right)$; (c) $\left(\frac{3}{4}, -\frac{\sqrt{7}}{4}\right)$

3. (a) $\left(\frac{2}{5}, \frac{\sqrt{21}}{5}\right)$; (b) $\left(-\frac{2}{5}, -\frac{\sqrt{21}}{5}\right)$; (c) $\left(-\frac{2}{5}, \frac{\sqrt{21}}{5}\right)$

5. $\left(\frac{\sqrt{2}}{2}, -\frac{\sqrt{2}}{2}\right)$ **7.** 0 **9.** $\sqrt{3}$ **11.** 0 **13.** $-\frac{\sqrt{3}}{2}$

15. Not defined **17.** $\frac{\sqrt{3}}{2}$ **19.** $-\frac{\sqrt{2}}{2}$ **21.** 0 **23.** 0

25. 0.4816 **27.** 1.3065 **29.** −2.1599 **31.** 1 **33.** −1.1747 **35.** −1 **37.** −0.7071 **39.** 0 **41.** 0.8391

43. (a)

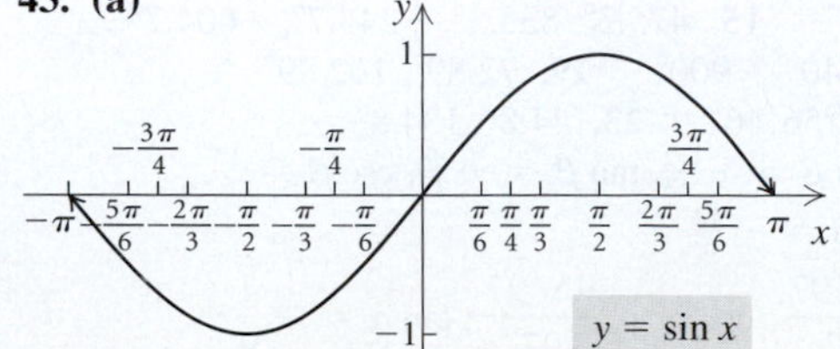

(b)

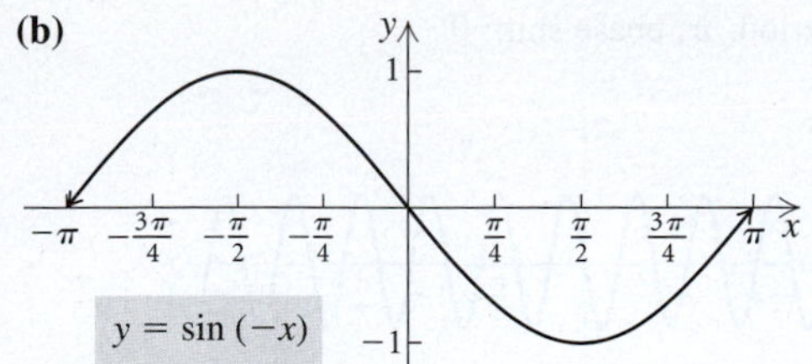

(c) same as (b); **(d)** the same

45. (a)

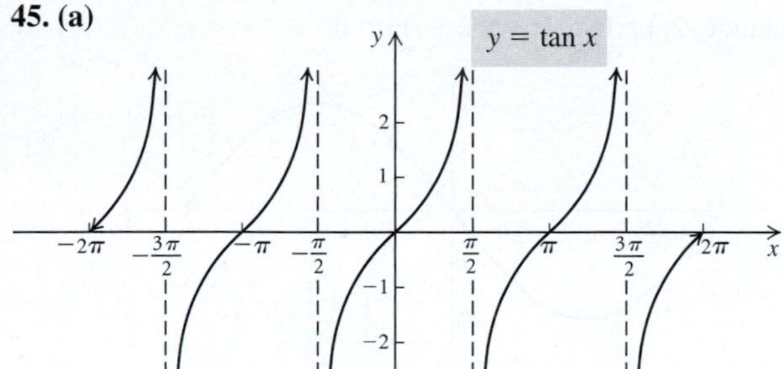

(b)

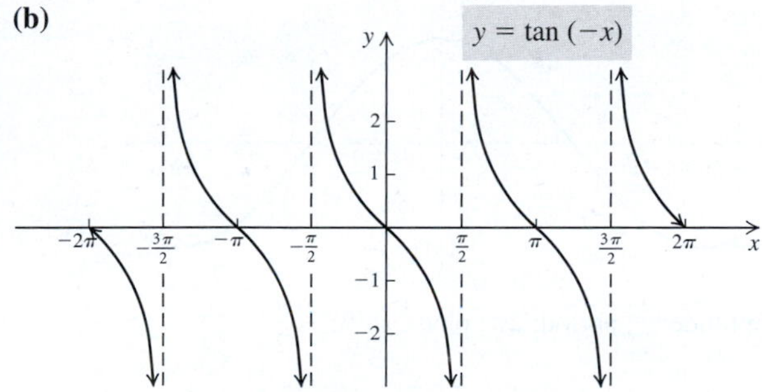

(c) same as (b); **(d)** the same

47. (a) See Exercise 43(a);

(b)

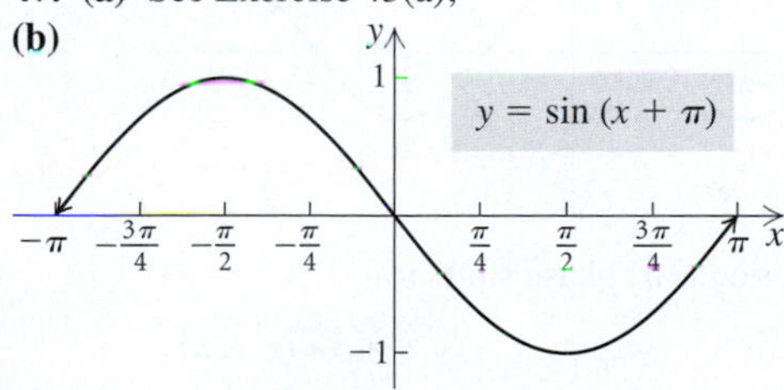

(c) same as (b); **(d)** the same

49. (a)

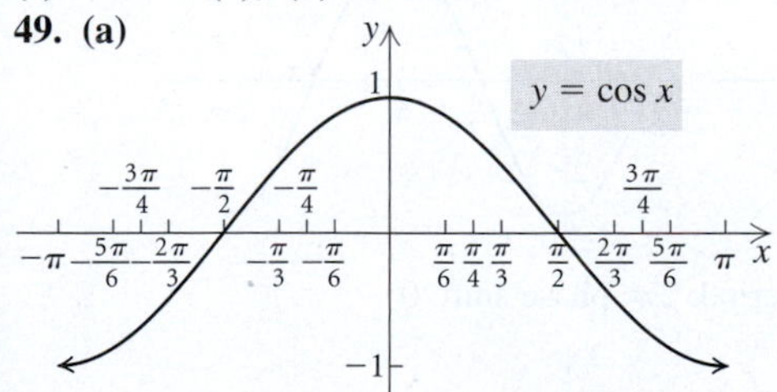

(b)

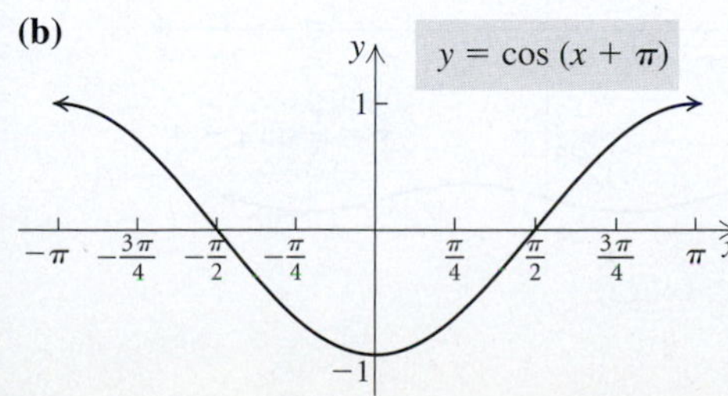

(c) same as (b); **(d)** the same **51.** Even: cosine, secant; odd: sine, tangent, cosecant, cotangent **53.** Positive: I, III; negative: II, IV **55.** Positive: I, IV; negative: II, III

57.

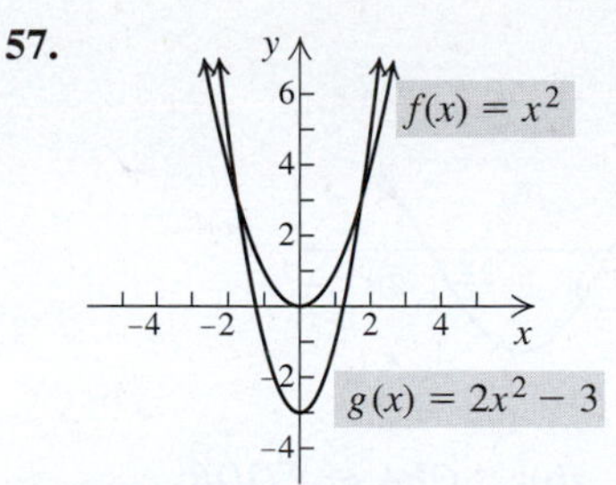

Stretch the graph of *f* vertically, then shift it down 3 units.

58.

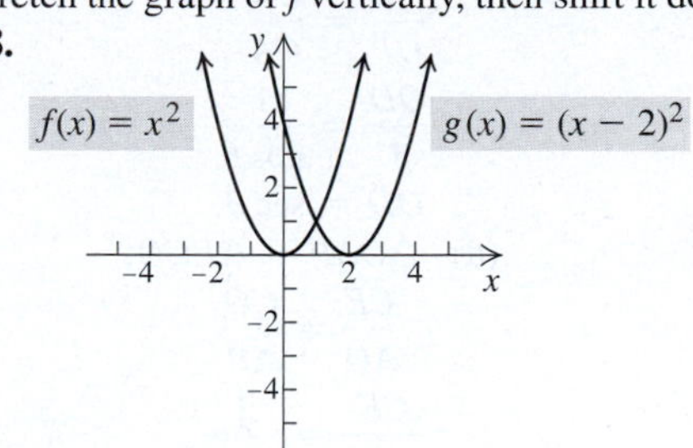

Shift the graph of *f* right 2 units.

59.

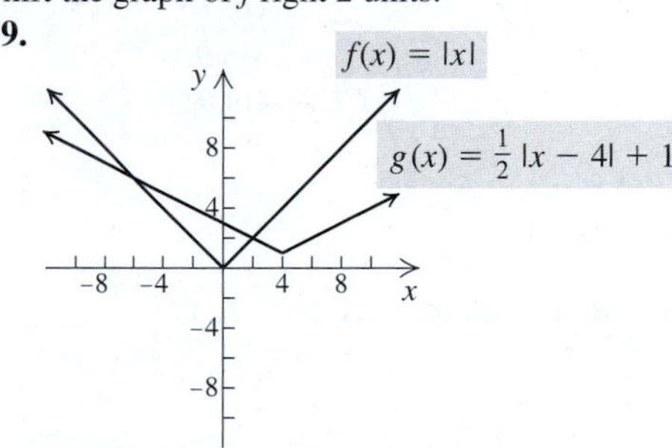

Shift the graph of *f* right 4 units, shrink it vertically, then shift it up 1 unit.

60.

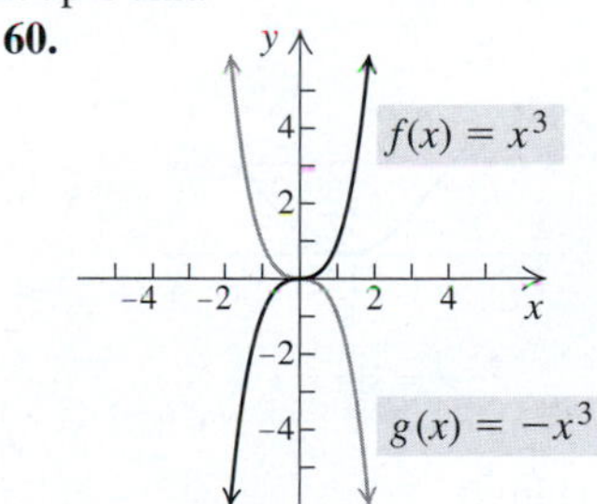

Reflect the graph of *f* across the *x*-axis.

61. $y = -(x - 2)^3 - 1$ **62.** $y = \frac{1}{4x} + 3$ **63.** $\sin x$ **65.** $\sin x$ **67.** $-\cos x$ **69.** $-\sin x$ **71. (a)** $\frac{\pi}{2} + 2k\pi$, k an integer; **(b)** $\pi + 2k\pi$, k an integer; **(c)** $k\pi$, k an integer

73. Domain: $(-\infty, \infty)$; range: $[0, 1]$; period: π; amplitude: $\frac{1}{2}$

75. $\left[-\frac{\pi}{2} + 2k\pi, \frac{\pi}{2} + 2k\pi\right]$, k an integer

77. $\left\{x \,|\, x \neq \frac{\pi}{2} + k\pi, k \text{ an integer}\right\}$

79.

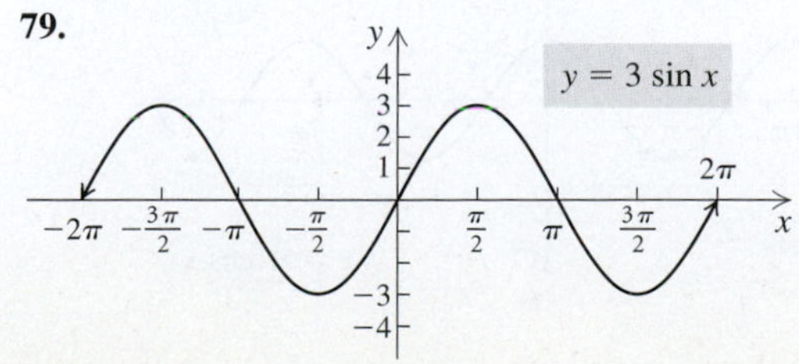

81.

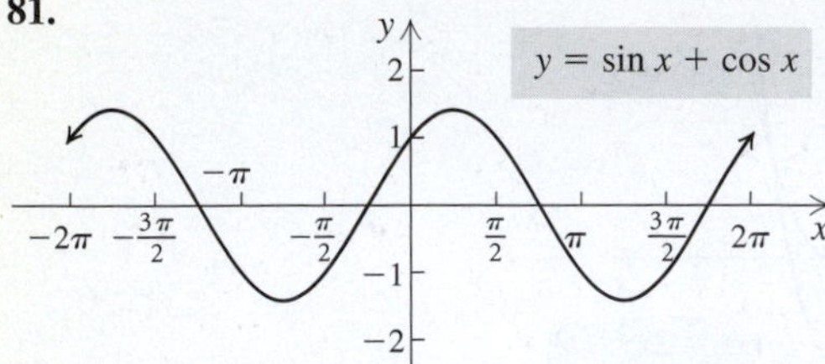

83. (a) $\triangle OPA \sim \triangle ODB$;

Thus, $\dfrac{AP}{OA} = \dfrac{BD}{OB}$

$\dfrac{\sin\theta}{\cos\theta} = \dfrac{BD}{1}$

$\tan\theta = BD$

(b) $\triangle OPA \sim \triangle ODB$;

$\dfrac{OD}{OP} = \dfrac{OB}{OA}$

$\dfrac{OD}{1} = \dfrac{1}{\cos\theta}$

$OD = \sec\theta$

(c) $\triangle OAP \sim \triangle ECO$;

$\dfrac{OE}{PO} = \dfrac{CO}{AP}$

$\dfrac{OE}{1} = \dfrac{1}{\sin\theta}$

$OE = \csc\theta$

(d) $\triangle OAP \sim \triangle ECO$;

$\dfrac{CE}{AO} = \dfrac{CO}{AP}$

$\dfrac{CE}{\cos\theta} = \dfrac{1}{\sin\theta}$

$CE = \dfrac{\cos\theta}{\sin\theta}$

$CE = \cot\theta$

85. 1

Visualizing the Graph

1. J **2.** H **3.** E **4.** F **5.** B **6.** D **7.** G **8.** A **9.** C **10.** I

Exercise Set 6.6

1. Amplitude: 1; period: 2π; phase shift: 0

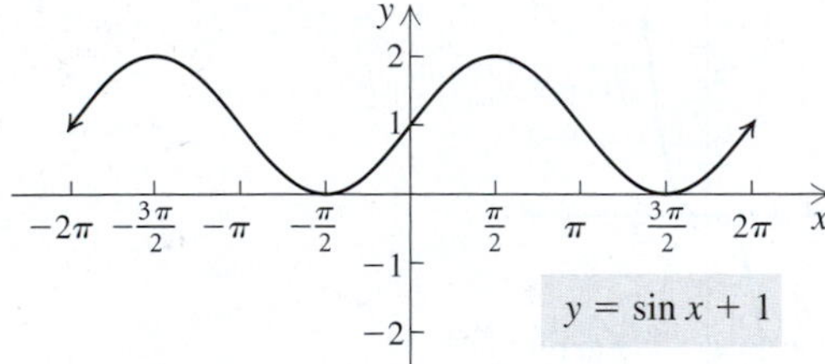

3. Amplitude: 3; period: 2π; phase shift: 0

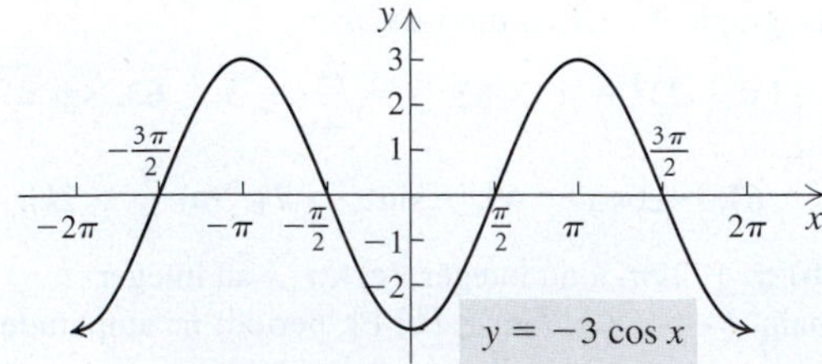

5. Amplitude: $\frac{1}{2}$; period: 2π; phase shift: 0

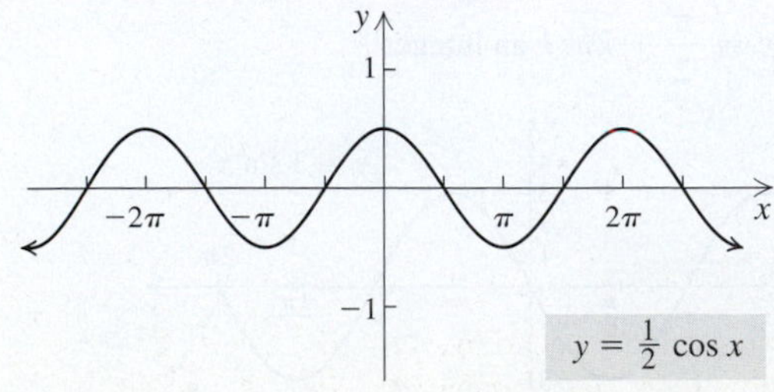

7. Amplitude: 1; period: π; phase shift: 0

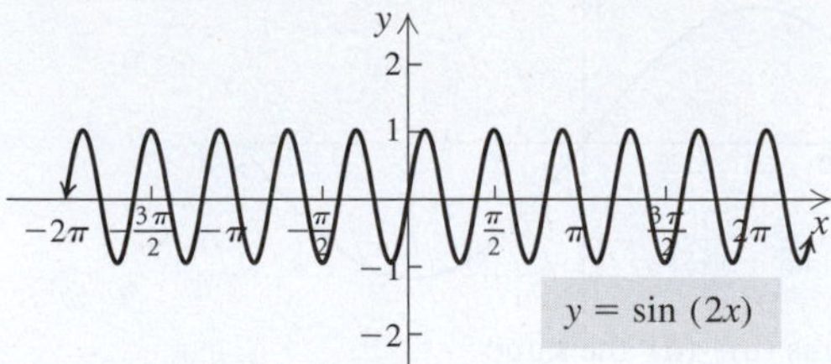

9. Amplitude: 2; period: 4π; phase shift: 0

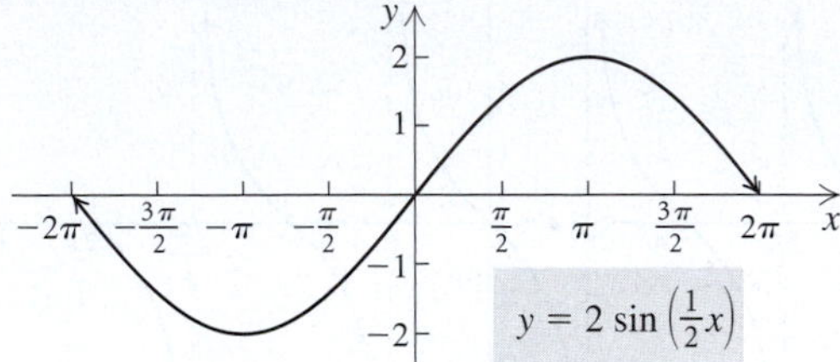

11. Amplitude: 1; period: 4π; phase shift: 0

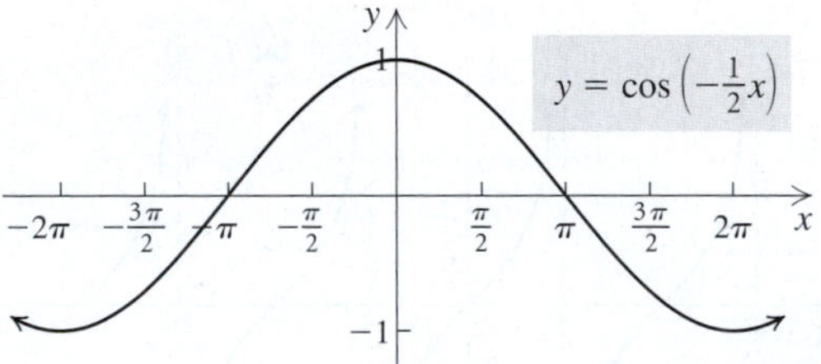

13. Amplitude: $\dfrac{1}{2}$; period: 2π; phase shift: $-\dfrac{\pi}{2}$

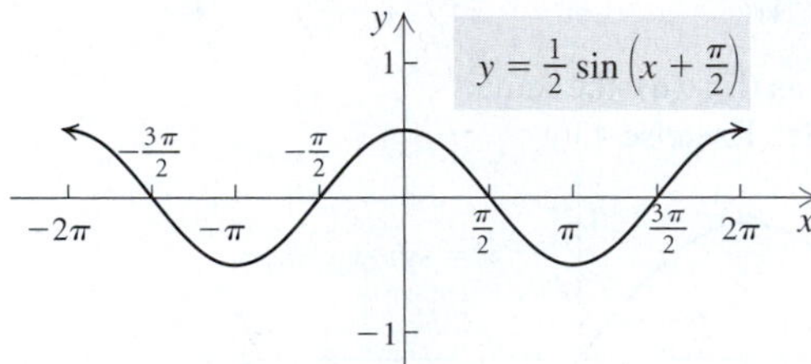

15. Amplitude: 3; period: 2π; phase shift: π

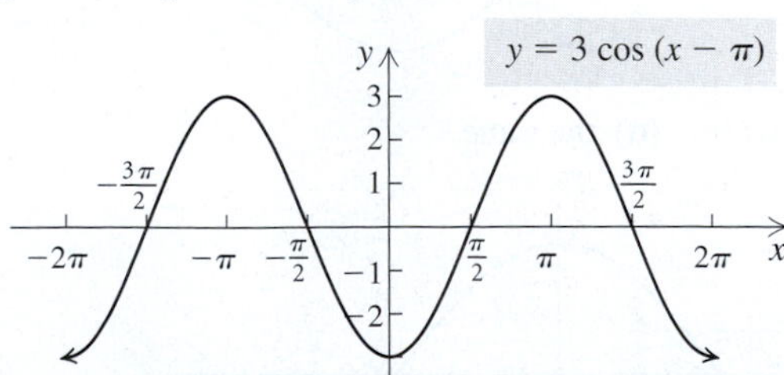

17. Amplitude: $\frac{1}{3}$; period: 2π; phase shift: 0

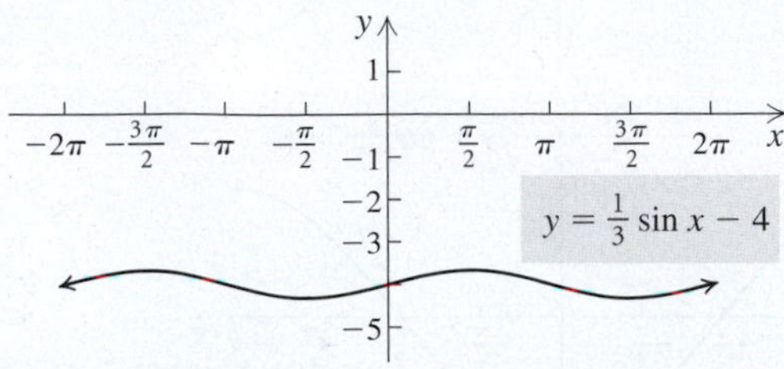

19. Amplitude: 1; period: 2π; phase shift: 0

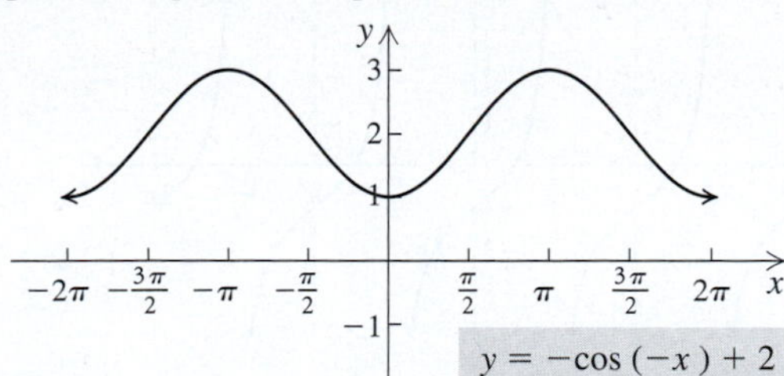

21. Amplitude: 2; period: 4π; phase shift: π

23. Amplitude: $\frac{1}{2}$; period: π; phase shift: $-\frac{\pi}{4}$

25. Amplitude: 3; period: 2; phase shift: $\frac{3}{\pi}$

27. Amplitude: $\frac{1}{2}$; period: 1; phase shift: 0

29. Amplitude: 1; period: 4π; phase shift: π

31. Amplitude: 1; period: 1; phase shift: 0

33. Amplitude: $\frac{1}{4}$; period: 2; phase shift: $\frac{4}{\pi}$

35. (b) **37.** (h) **39.** (a) **41.** (f)

43. $y = \frac{1}{2}\cos x + 1$ **45.** $y = \cos\left(x + \frac{\pi}{2}\right) - 2$

47.

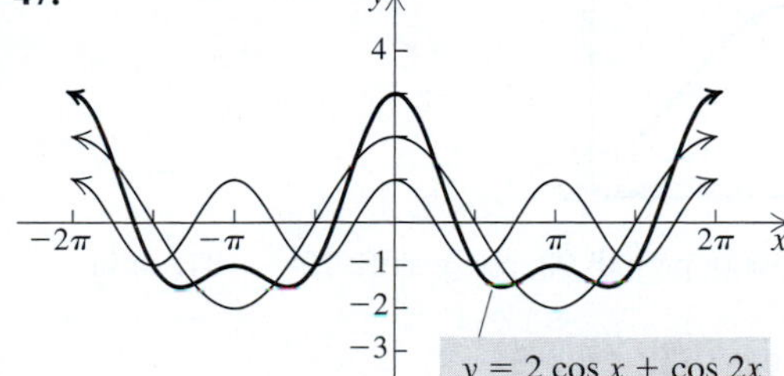

49.

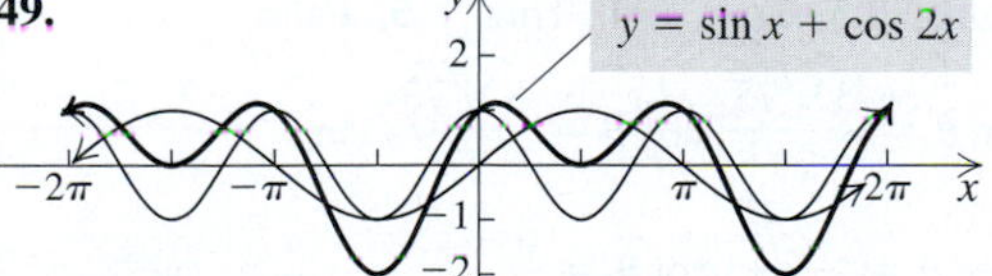

51.

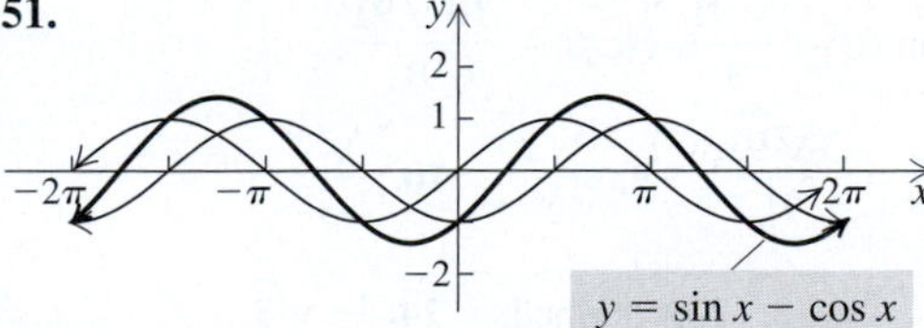

53.

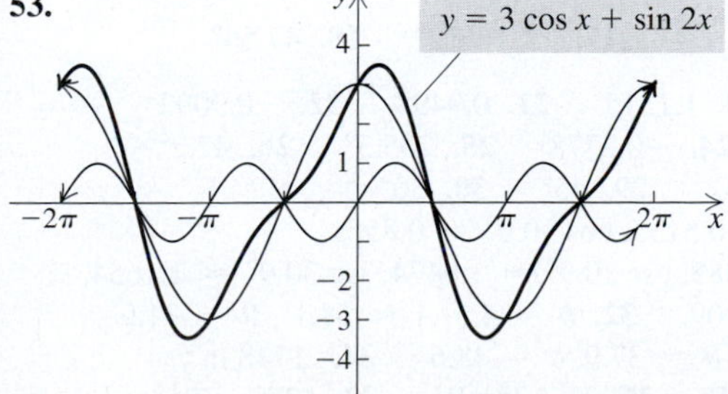

55. $y = x + \sin x$

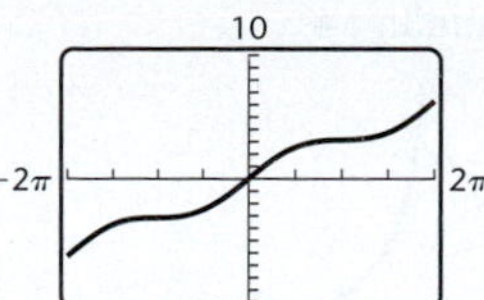

57. $y = \cos x - x$

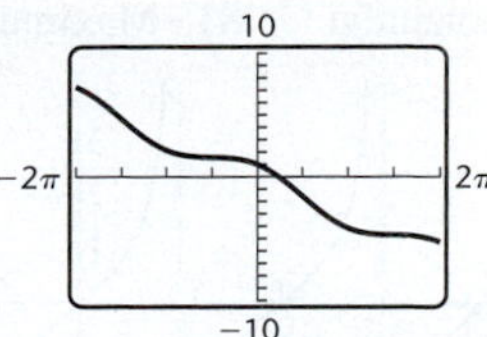

59. $y = \cos 2x + 2x$

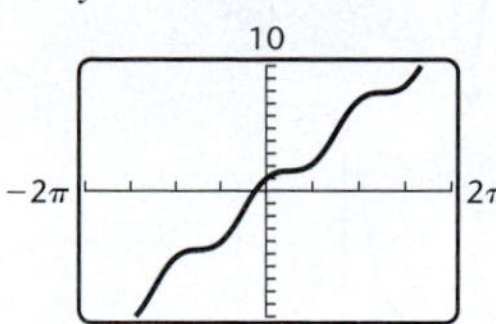

61. $y = 4\cos 2x - 2\sin x$

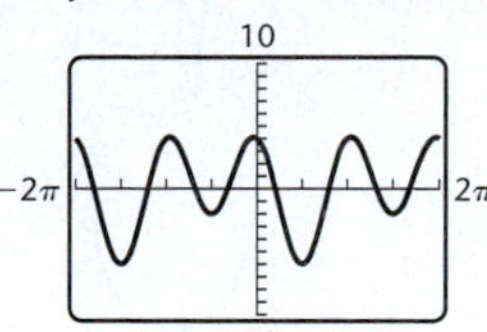

63.

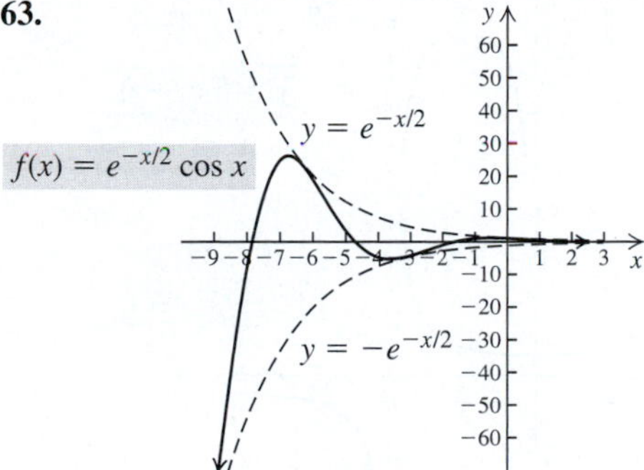

65.

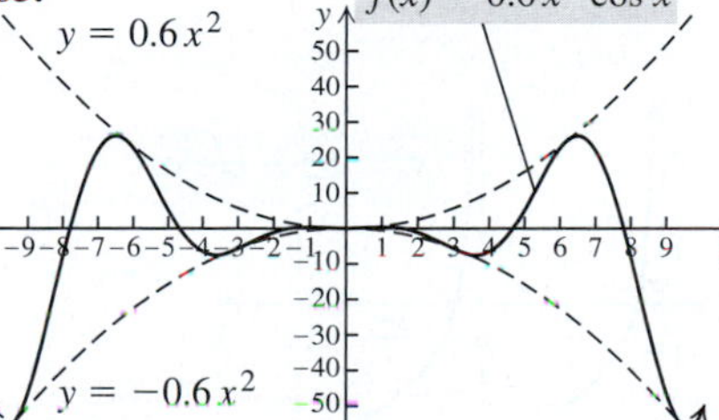

67.

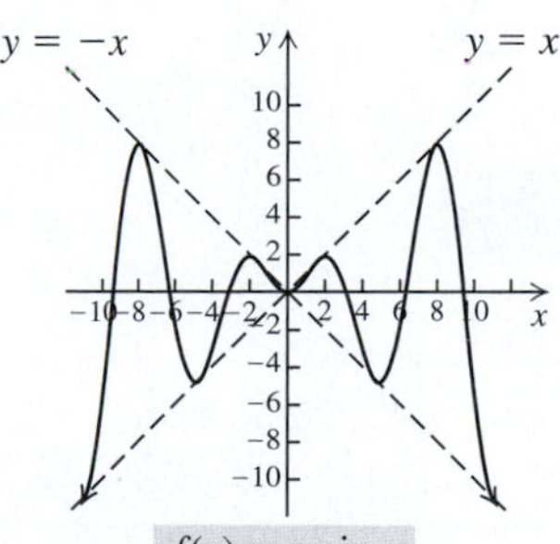

69.

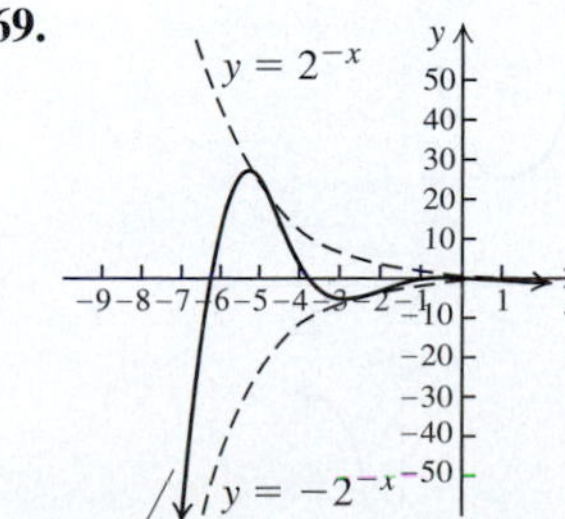

71. Rational **72.** Logarithmic **73.** Quartic
74. Linear **75.** Trigonometric **76.** Exponential

77. Linear **78.** Trigonometric **79.** Cubic
80. Exponential **81.** Maximum: 8; minimum: 4

83.

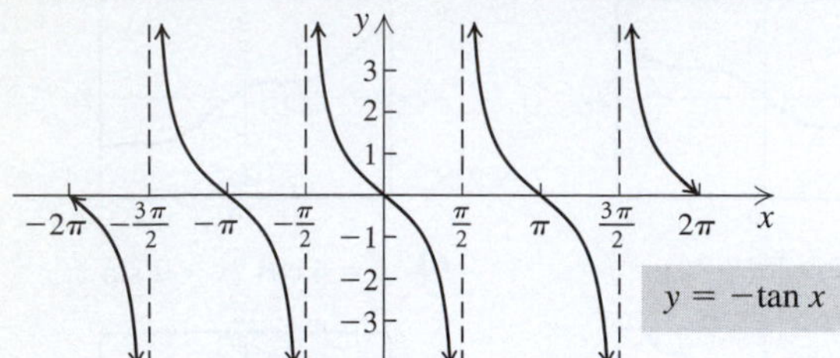

85.

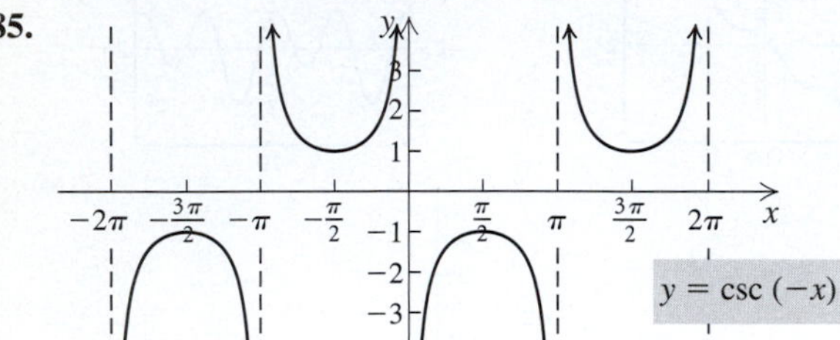

87.

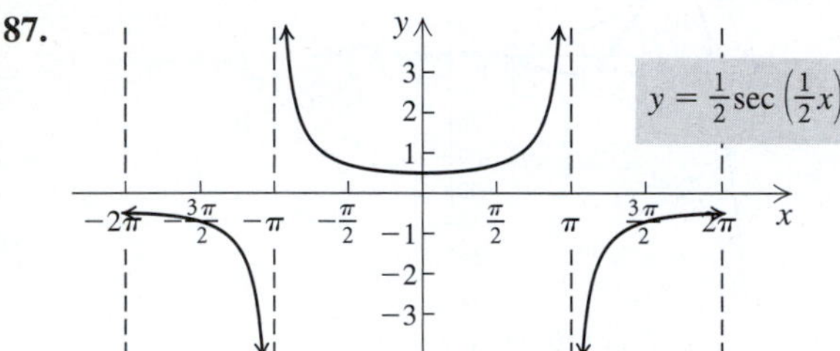

89.

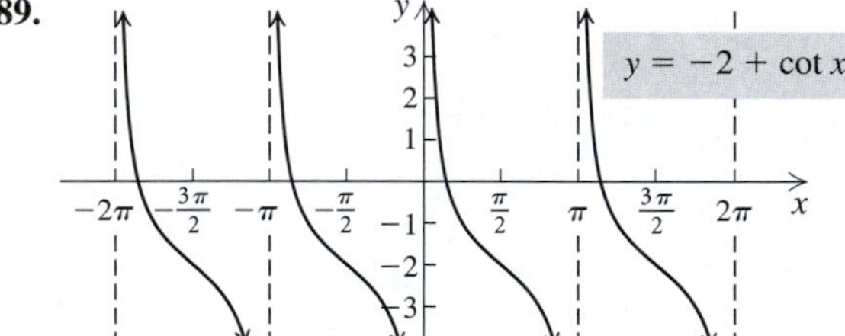

91.

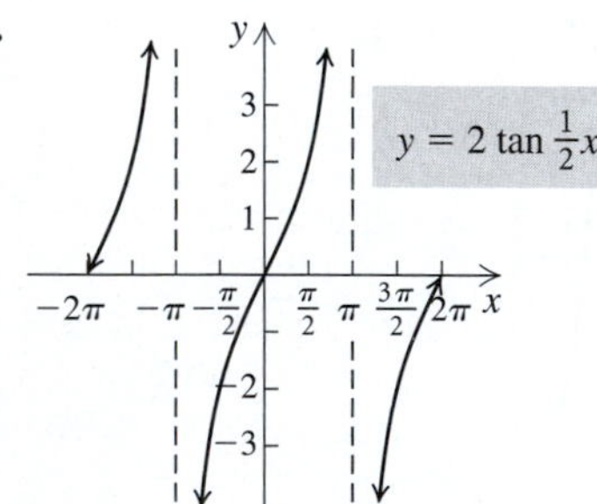

93.

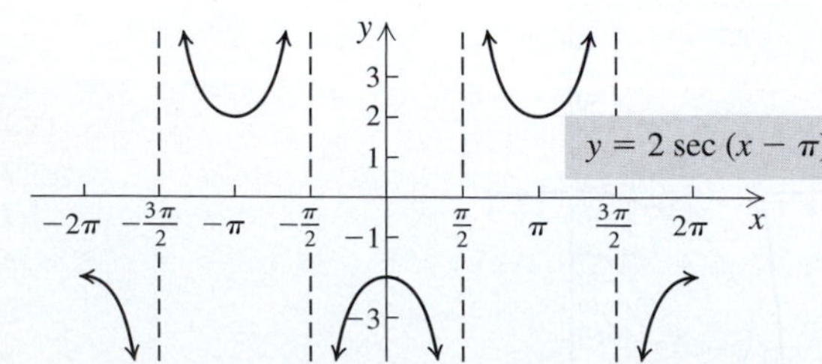

95.

$y = \cot\left(x + \frac{\pi}{2}\right) - 1$

97.

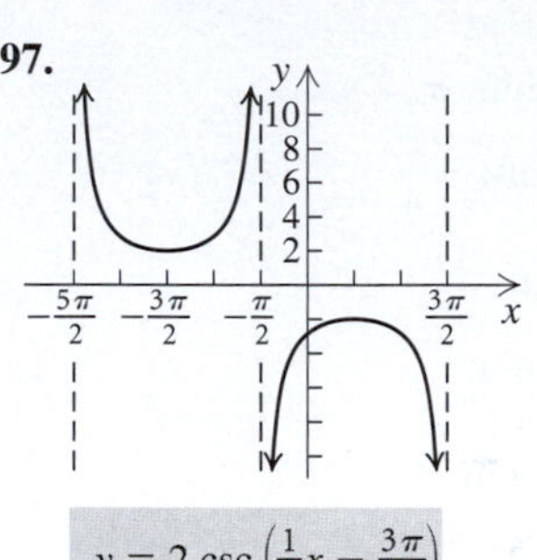

99. $-9.42, -6.28, -3.14, 3.14, 6.28, 9.42$ **101.** $-3.14, 0, 3.14$

103. **(a)** $y = 101.6 + 3 \sin\left(\frac{\pi}{8}x\right)$ **(b)** 104.6°, 98.6°

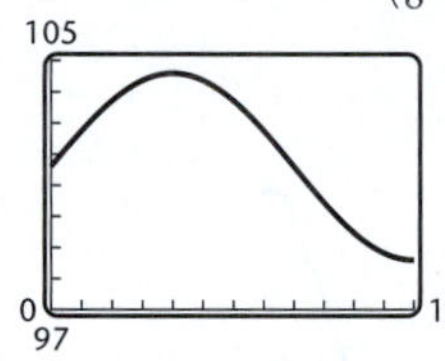

105. Amplitude: 3000; period: 90; phase shift: 10 **107.** 4 in.

Review Exercises: Chapter 6

1. False **2.** False **3.** True **4.** True **5.** False

6. False **7.** $\sin\theta = \frac{3\sqrt{73}}{73}$, $\cos\theta = \frac{8\sqrt{73}}{73}$, $\tan\theta = \frac{3}{8}$, $\csc\theta = \frac{\sqrt{73}}{3}$, $\sec\theta = \frac{\sqrt{73}}{8}$, $\cot\theta = \frac{8}{3}$

8. $\cos\beta = \frac{3}{10}$, $\tan\beta = \frac{\sqrt{91}}{3}$, $\csc\beta = \frac{10\sqrt{91}}{91}$, $\sec\beta = \frac{10}{3}$, $\cot\beta = \frac{3\sqrt{91}}{91}$ **9.** $\frac{\sqrt{2}}{2}$ **10.** $\frac{\sqrt{3}}{3}$

11. $-\frac{\sqrt{2}}{2}$ **12.** $\frac{1}{2}$ **13.** Not defined **14.** $-\sqrt{3}$

15. $\frac{2\sqrt{3}}{3}$ **16.** -1 **17.** 22°16′12″ **18.** 47.56°

19. 0.4452 **20.** 1.1315 **21.** 0.9498 **22.** -0.9092
23. -1.5282 **24.** -0.2778 **25.** 205.3° **26.** 47.2°
27. 60° **28.** 60° **29.** 45° **30.** 30°
31. $\sin 30.9° \approx 0.5135$, $\cos 30.9° \approx 0.8581$, $\tan 30.9° \approx 0.5985$, $\csc 30.9° \approx 1.9474$, $\sec 30.9° \approx 1.1654$, $\cot 30.9° \approx 1.6709$ **32.** $b \approx 4.5$, $A \approx 58.1°$, $B \approx 31.9°$
33. $A = 38.83°$, $b \approx 37.9$, $c \approx 48.6$ **34.** 1748 m
35. 14 ft **36.** II **37.** I **38.** IV **39.** 425°, −295°

40. $\frac{\pi}{3}, -\frac{5\pi}{3}$ **41.** Complement: 76.6°; supplement: 166.6°

42. Complement: $\frac{\pi}{3}$; supplement: $\frac{5\pi}{6}$ **43.** $\sin\theta = \frac{3\sqrt{13}}{13}$, $\cos\theta = \frac{-2\sqrt{13}}{13}$, $\tan\theta = -\frac{3}{2}$, $\csc\theta = \frac{\sqrt{13}}{3}$, $\sec\theta = -\frac{\sqrt{13}}{2}$, $\cot\theta = -\frac{2}{3}$ **44.** $\sin\theta = -\frac{2}{3}$, $\cos\theta = -\frac{\sqrt{5}}{3}$, $\cot\theta = \frac{\sqrt{5}}{2}$, $\sec\theta = -\frac{3\sqrt{5}}{5}$, $\csc\theta = -\frac{3}{2}$ **45.** About 1743 mi

46.

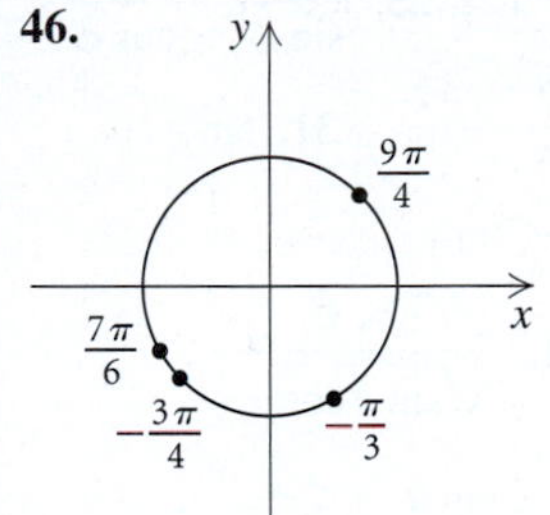

47. $\frac{121}{150}\pi$, 2.53 **48.** $-\frac{\pi}{6}$, −0.52 **49.** 270° **50.** 171.89° **51.** −257.83° **52.** 1980° **53.** $\frac{7\pi}{4}$, or 5.5 cm **54.** 2.25, 129° **55.** About 37.7 ft/min **56.** 497,829 radians/hr **57.** $\left(\frac{3}{5}, \frac{4}{5}\right), \left(-\frac{3}{5}, -\frac{4}{5}\right), \left(-\frac{3}{5}, \frac{4}{5}\right)$ **58.** −1 **59.** 1 **60.** $-\frac{\sqrt{3}}{2}$ **61.** $\frac{1}{2}$ **62.** $\frac{\sqrt{3}}{3}$ **63.** −1 **64.** −0.9056 **65.** 0.9218 **66.** Not defined **67.** 4.3813 **68.** −6.1685 **69.** 0.8090 **70.** $y = \sin x$: see p. 453; $y = \cos x$: see p. 453; $y = \tan x$: see p. 457; $y = \cot x$: see p. 458; $y = \sec x$: see p. 458; $y = \csc x$: see p. 458 **71.** Period of sin, cos, sec, csc: 2π; period of tan, cot: π

72.

Function	Domain	Range
Sine	$(-\infty, \infty)$	$[-1, 1]$
Cosine	$(-\infty, \infty)$	$[-1, 1]$
Tangent	All real numbers except $(\pi/2) + k\pi$, where k is an integer	$(-\infty, \infty)$

73.

Function	I	II	III	IV
Sine	+	+	−	−
Cosine	+	−	−	+
Tangent	+	−	+	−

74. Amplitude: 1; period: 2π; phase shift: $-\frac{\pi}{2}$

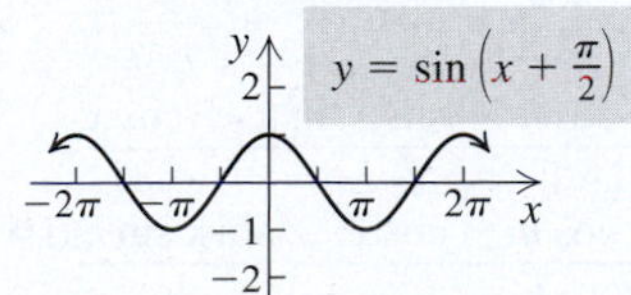

75. Amplitude: $\frac{1}{2}$; period: π; phase shift: $\frac{\pi}{4}$

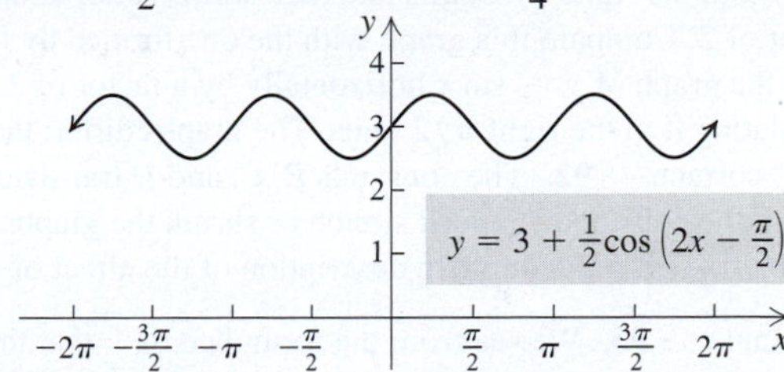

76. (d) **77.** (a) **78.** (c) **79.** (b)

80.

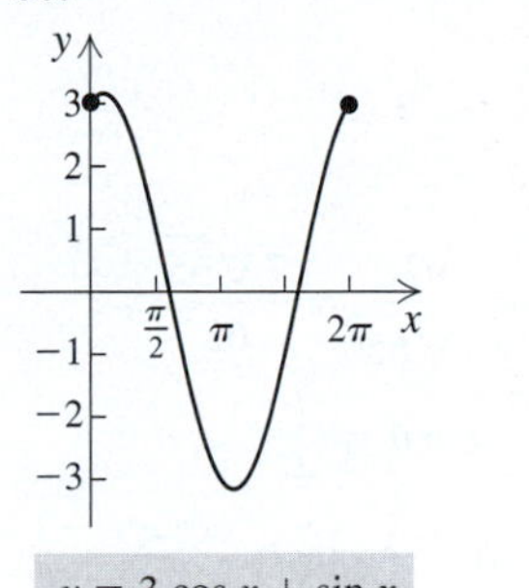

81.

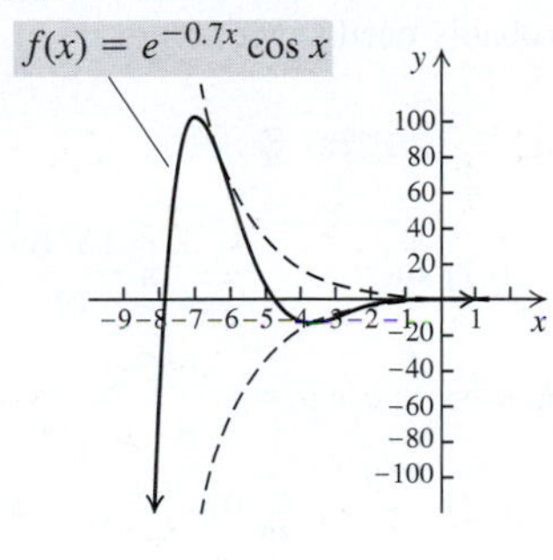

82. C **83.** B **84.** B

85. Domain: $(-\infty, \infty)$; range: $[-3, 3]$; period 4π

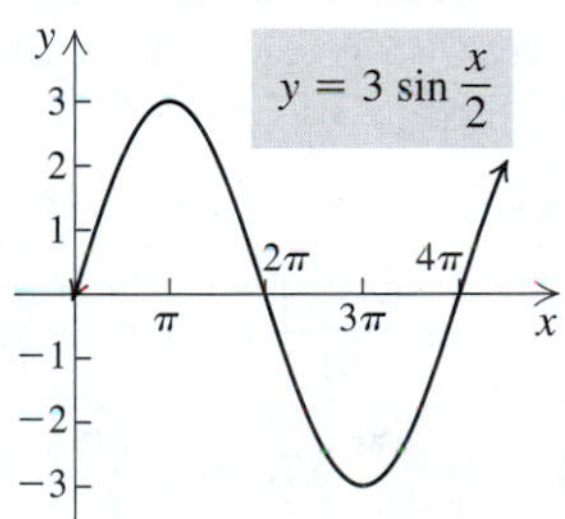

86. $y_2 = 2\sin\left(x + \frac{\pi}{2}\right) - 2$ **87.** The domain consists of the intervals $\left(-\frac{\pi}{2} + 2k\pi, \frac{\pi}{2} + 2k\pi\right)$, k an integer.

88. Both degrees and radians are units of angle measure. Degree notation has been in use since Babylonian times. A degree is defined to be $\frac{1}{360}$ of one complete positive revolution. Radians are defined in terms of intercepted arc length on a circle, with one radian being the measure of the angle for which the arc length equals the radius. There are 2π radians in one complete revolution. **89.** For a point at a distance r from the center of rotation with a fixed angular speed k, the linear speed is given by $v = r \cdot k$, or $r = \frac{1}{k}v$. Thus the length of the radius is directly proportional to the linear speed.

90. The numbers for which the value of the cosine function is 0 are not in the domain of the tangent function. **91.** The denominator B in the phase shift C/B serves to shrink or stretch the translation of C units by the same factor as the horizontal shrinking or stretching of the period. Thus the translation must be done after the horizontal shrinking or stretching. For example, consider $y = \sin(2x - \pi)$.

The phase shift of this function is $\pi/2$. First translate the graph of $y = \sin x$ to the right $\pi/2$ units and then shrink it horizontally by a factor of 2. Compare this graph with the one formed by first shrinking the graph of $y = \sin x$ horizontally by a factor of 2 and then translating it to the right $\pi/2$ units. The graphs differ; the second one is correct. **92.** The constants B, C, and D translate the graphs, and the constants A and B stretch or shrink the graphs. See the chart on p. 470 for a complete description of the effect of each constant. **93.** We see from the formula $\theta = \frac{s}{r}$ that the tire with the 15-in. diameter will rotate through a larger angle than the tire with the 16-in. diameter. Thus the car with the 15-in. tires will probably need new tires first.

Test: Chapter 6

1. [6.1] $\sin\theta = \frac{4}{\sqrt{65}}$, or $\frac{4\sqrt{65}}{65}$; $\cos\theta = \frac{7}{\sqrt{65}}$, or $\frac{7\sqrt{65}}{65}$; $\tan\theta = \frac{4}{7}$; $\csc\theta = \frac{\sqrt{65}}{4}$; $\sec\theta = \frac{\sqrt{65}}{7}$; $\cot\theta = \frac{7}{4}$
2. [6.3] $\frac{\sqrt{3}}{2}$ **3.** [6.3] -1 **4.** [6.4] -1 **5.** [6.4] $-\sqrt{2}$
6. [6.1] 38.47° **7.** [6.3] -0.2419 **8.** [6.3] -0.2079
9. [6.4] -5.7588 **10.** [6.4] 0.7827 **11.** [6.1] 30°
12. [6.1] $\sin 61.6° \approx 0.8796$; $\cos 61.6° \approx 0.4756$; $\tan 61.6° \approx 1.8495$; $\csc 61.6° \approx 1.1369$; $\sec 61.6° \approx 2.1026$; $\cot 61.6° \approx 0.5407$ **13.** [6.2] $B = 54.1°$, $a \approx 32.6$, $c \approx 55.7$
14. [6.3] Answers may vary; 472°, −248° **15.** [6.4] $\frac{\pi}{6}$
16. [6.3] $\cos\theta = \frac{5}{\sqrt{41}}$, or $\frac{5\sqrt{41}}{41}$; $\tan\theta = -\frac{4}{5}$; $\csc\theta = -\frac{\sqrt{41}}{4}$; $\sec\theta = \frac{\sqrt{41}}{5}$; $\cot\theta = -\frac{5}{4}$ **17.** [6.4] $\frac{7\pi}{6}$ **18.** [6.4] 135°
19. [6.4] $\frac{16\pi}{3} \approx 16.755$ cm **20.** [6.6] 1 **21.** [6.6] 2π
22. [6.6] $\frac{\pi}{2}$ **23.** [6.6] (c) **24.** [6.2] 401 ft
25. [6.2] About 272 mi **26.** [6.4] $18\pi \approx 56.55$ m/min
27. [6.6]

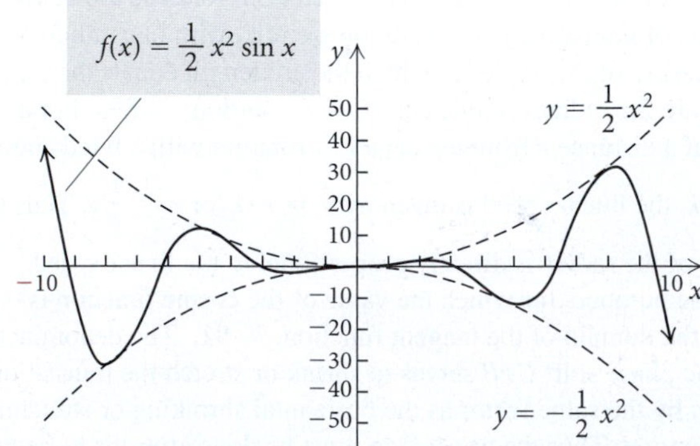

28. [6.6] C
29. [6.5] $\left\{x \middle| -\frac{\pi}{2} + 2k\pi < x < \frac{\pi}{2} + 2k\pi, k \text{ an integer}\right\}$

Chapter 7

Exercise Set 7.1

1. $\sin^2 x - \cos^2 x$ **3.** $\sin y + \cos y$ **5.** $1 - 2\sin\phi\cos\phi$
7. $\sin^3 x + \csc^3 x$ **9.** $\cos x(\sin x + \cos x)$
11. $(\sin x + \cos x)(\sin x - \cos x)$
13. $(2\cos x + 3)(\cos x - 1)$
15. $(\sin x + 3)(\sin^2 x - 3\sin x + 9)$ **17.** $\tan x$
19. $\sin x + 1$ **21.** $\frac{2\tan t + 1}{3\tan t + 1}$ **23.** 1 **25.** $\frac{5\cot\phi}{\sin\phi + \cos\phi}$
27. $\frac{1 + 2\sin s + 2\cos s}{\sin^2 s - \cos^2 s}$ **29.** $\frac{5(\sin\theta - 3)}{3}$ **31.** $\sin x\cos x$
33. $\sqrt{\cos\alpha}(\sin\alpha - \cos\alpha)$ **35.** $1 - \sin y$
37. $\frac{\sqrt{\sin x\cos x}}{\cos x}$ **39.** $\frac{\sqrt{2}\cot y}{2}$ **41.** $\frac{\cos x}{\sqrt{\sin x\cos x}}$
43. $\frac{1 + \sin y}{\cos y}$ **45.** $\cos\theta = \frac{\sqrt{a^2 - x^2}}{a}$, $\tan\theta = \frac{x}{\sqrt{a^2 - x^2}}$
47. $\sin\theta = \frac{\sqrt{x^2 - 9}}{x}$, $\cos\theta = \frac{3}{x}$ **49.** $\sin\theta\tan\theta$
51. $\frac{\sqrt{6} - \sqrt{2}}{4}$ **53.** $\frac{\sqrt{3} + 1}{1 - \sqrt{3}}$, or $-2 - \sqrt{3}$
55. $\frac{\sqrt{6} + \sqrt{2}}{4}$ **57.** $\sin 59° \approx 0.8572$
59. $\cos 24° \approx 0.9135$ **61.** $\tan 52° \approx 1.2799$
63.
$$\tan(\mu + \nu) = \frac{\sin(\mu + \nu)}{\cos(\mu + \nu)} = \frac{\sin\mu\cos\nu + \cos\mu\sin\nu}{\cos\mu\cos\nu - \sin\mu\sin\nu}$$
$$= \frac{\sin\mu\cos\nu + \cos\mu\sin\nu}{\cos\mu\cos\nu - \sin\mu\sin\nu} \cdot \frac{\frac{1}{\cos\mu\cos\nu}}{\frac{1}{\cos\mu\cos\nu}}$$
$$= \frac{\frac{\sin\mu}{\cos\mu} + \frac{\sin\nu}{\cos\nu}}{1 - \frac{\sin\mu\sin\nu}{\cos\mu\cos\nu}} = \frac{\tan\mu + \tan\nu}{1 - \tan\mu\tan\nu}$$
65. 0 **67.** $-\frac{7}{25}$ **69.** $\frac{-24 + 2\sqrt{170}}{63}$ **71.** -1.5789
73. 0.7071 **75.** $2\sin\alpha\cos\beta$ **77.** $\cos u$ **79.** All real numbers **80.** No solution **81.** 1.9417 **82.** 1.6645
83. 0°; the lines are parallel **85.** $\frac{3\pi}{4}$, or 135° **87.** 22.83°
89.
$$\frac{\cos(x + h) - \cos x}{h} = \frac{\cos x\cos h - \sin x\sin h - \cos x}{h}$$
$$= \frac{\cos x\cos h - \cos x}{h} - \frac{\sin x\sin h}{h}$$
$$= \cos x\left(\frac{\cos h - 1}{h}\right) - \sin x\left(\frac{\sin h}{h}\right)$$
91. Let $x = \frac{\pi}{5}$. Then $\frac{\sin 5x}{x} = \frac{\sin\pi}{\pi/5} = 0 \neq \sin 5$. Answers may vary.
93. Let $\alpha = \frac{\pi}{4}$. Then $\cos(2\alpha) = \cos\frac{\pi}{2} = 0$, but $2\cos\alpha = 2\cos\frac{\pi}{4} = \sqrt{2}$. Answers may vary.

95. Let $x = \frac{\pi}{6}$. Then $\frac{\cos 6x}{\cos x} = \frac{\cos \pi}{\cos \frac{\pi}{6}} = \frac{-1}{\sqrt{3}/2} \neq 6.$ Answers may vary. **97.** $\frac{6 - 3\sqrt{3}}{9 + 2\sqrt{3}} \approx 0.0645$

99. 168.7° **101.** $\cos 2\theta = \cos^2 \theta - \sin^2 \theta$, or $1 - 2\sin^2 \theta$, or $2\cos^2 \theta - 1$ **103.** $\tan\left(x + \frac{\pi}{4}\right) = \frac{\tan x + \tan \frac{\pi}{4}}{1 - \tan x \tan \frac{\pi}{4}} = \frac{1 + \tan x}{1 - \tan x}$

105. $\sin(\alpha + \beta) + \sin(\alpha - \beta) = \sin \alpha \cos \beta + \cos \alpha \sin \beta + \sin \alpha \cos \beta - \cos \alpha \sin \beta = 2 \sin \alpha \cos \beta$

Exercise Set 7.2

1. (a) $\tan \frac{3\pi}{10} \approx 1.3763$, $\csc \frac{3\pi}{10} \approx 1.2361$, $\sec \frac{3\pi}{10} \approx 1.7013$, $\cot \frac{3\pi}{10} \approx 0.7266$; **(b)** $\sin \frac{\pi}{5} \approx 0.5878$, $\cos \frac{\pi}{5} \approx 0.8090$, $\tan \frac{\pi}{5} \approx 0.7266$, $\csc \frac{\pi}{5} \approx 1.7013$, $\sec \frac{\pi}{5} \approx 1.2361$, $\cot \frac{\pi}{5} \approx 1.3763$ **3. (a)** $\cos \theta = -\frac{2\sqrt{2}}{3}$, $\tan \theta = -\frac{\sqrt{2}}{4}$, $\csc \theta = 3$, $\sec \theta = -\frac{3\sqrt{2}}{4}$, $\cot \theta = -2\sqrt{2}$;
(b) $\sin\left(\frac{\pi}{2} - \theta\right) = -\frac{2\sqrt{2}}{3}$, $\cos\left(\frac{\pi}{2} - \theta\right) = \frac{1}{3}$, $\tan\left(\frac{\pi}{2} - \theta\right) = -2\sqrt{2}$, $\csc\left(\frac{\pi}{2} - \theta\right) = -\frac{3\sqrt{2}}{4}$, $\sec\left(\frac{\pi}{2} - \theta\right) = 3$, $\cot\left(\frac{\pi}{2} - \theta\right) = -\frac{\sqrt{2}}{4}$;
(c) $\sin\left(\theta - \frac{\pi}{2}\right) = \frac{2\sqrt{2}}{3}$, $\cos\left(\theta - \frac{\pi}{2}\right) = \frac{1}{3}$, $\tan\left(\theta - \frac{\pi}{2}\right) = 2\sqrt{2}$, $\csc\left(\theta - \frac{\pi}{2}\right) = \frac{3\sqrt{2}}{4}$, $\sec\left(\theta - \frac{\pi}{2}\right) = 3$, $\cot\left(\theta - \frac{\pi}{2}\right) = \frac{\sqrt{2}}{4}$
5. $\sec\left(x + \frac{\pi}{2}\right) = -\csc x$ **7.** $\tan\left(x - \frac{\pi}{2}\right) = -\cot x$
9. $\sin 2\theta = \frac{24}{25}$, $\cos 2\theta = -\frac{7}{25}$, $\tan 2\theta = -\frac{24}{7}$; II
11. $\sin 2\theta = \frac{24}{25}$, $\cos 2\theta = -\frac{7}{25}$, $\tan 2\theta = -\frac{24}{7}$; II
13. $\sin 2\theta = -\frac{120}{169}$, $\cos 2\theta = \frac{119}{169}$, $\tan 2\theta = -\frac{120}{119}$; IV
15. $\cos 4x = 1 - 8\sin^2 x \cos^2 x$, or $\cos^4 x - 6\sin^2 x \cos^2 x + \sin^4 x$, or $8\cos^4 x - 8\cos^2 x + 1$
17. $\frac{\sqrt{2 + \sqrt{3}}}{2}$ **19.** $\frac{\sqrt{2 + \sqrt{2}}}{2}$ **21.** $2 + \sqrt{3}$
23. 0.6421 **25.** 0.1735
27. (d);
$$\begin{aligned} \frac{\cos 2x}{\cos x - \sin x} &= \frac{\cos^2 x - \sin^2 x}{\cos x - \sin x} \\ &= \frac{(\cos x + \sin x)(\cos x - \sin x)}{\cos x - \sin x} \\ &= \cos x + \sin x \\ &= \frac{\sin x}{\sin x}(\cos x + \sin x) \\ &= \sin x\left(\frac{\cos x}{\sin x} + \frac{\sin x}{\sin x}\right) \\ &= \sin x(\cot x + 1) \end{aligned}$$

29. (d); $\frac{\sin 2x}{2\cos x} = \frac{2\sin x \cos x}{2\cos x} = \sin x$ **31.** $\cos x$
33. $\cos 2x$ **35.** 1 **37.** 8 **39.** $\sin^2 x$ **40.** 1
41. $-\cos^2 x$ **42.** $\csc^2 x$ **43.** 1 **44.** $\sec^2 x$ **45.** $\cos^2 x$
46. $\tan^2 x$ **47.** (a), (e) **48.** (b), (c), (f) **49.** (d)
50. (e) **51.** $\sin 141° \approx 0.6293$, $\cos 141° \approx -0.7772$, $\tan 141° \approx -0.8097$, $\csc 141° \approx 1.5891$, $\sec 141° \approx -1.2867$, $\cot 141° \approx -1.2350$ **53.** $-\cos x(1 + \cot x)$ **55.** $\cot^2 y$
57. $\sin \theta = -\frac{15}{17}$, $\cos \theta = -\frac{8}{17}$, $\tan \theta = \frac{15}{8}$
59. (a) 9.80359 m/sec²; **(b)** 9.80180 m/sec²;
(c) $g = 9.78049(1 + 0.005264\sin^2 \phi + 0.000024\sin^4 \phi)$

Exercise Set 7.3

1.

$\sec x - \sin x \tan x$	$\cos x$
$\frac{1}{\cos x} - \sin x \cdot \frac{\sin x}{\cos x}$	
$\frac{1 - \sin^2 x}{\cos x}$	
$\frac{\cos^2 x}{\cos x}$	
$\cos x$	

3.

$\frac{1 - \cos x}{\sin x}$	$\frac{\sin x}{1 + \cos x}$
	$\frac{\sin x}{1 + \cos x} \cdot \frac{1 - \cos x}{1 - \cos x}$
	$\frac{\sin x(1 - \cos x)}{1 - \cos^2 x}$
	$\frac{\sin x(1 - \cos x)}{\sin^2 x}$
	$\frac{1 - \cos x}{\sin x}$

5.

$\frac{1 + \tan \theta}{1 - \tan \theta} + \frac{1 + \cot \theta}{1 - \cot \theta}$	0
$\frac{1 + \frac{\sin \theta}{\cos \theta}}{1 - \frac{\sin \theta}{\cos \theta}} + \frac{1 + \frac{\cos \theta}{\sin \theta}}{1 - \frac{\cos \theta}{\sin \theta}}$	
$\frac{\frac{\cos \theta + \sin \theta}{\cos \theta}}{\frac{\cos \theta - \sin \theta}{\cos \theta}} + \frac{\frac{\sin \theta + \cos \theta}{\sin \theta}}{\frac{\sin \theta - \cos \theta}{\sin \theta}}$	
$\frac{\cos \theta + \sin \theta}{\cos \theta} \cdot \frac{\cos \theta}{\cos \theta - \sin \theta} + \frac{\sin \theta + \cos \theta}{\sin \theta} \cdot \frac{\sin \theta}{\sin \theta - \cos \theta}$	
$\frac{\cos \theta + \sin \theta}{\cos \theta - \sin \theta} + \frac{\sin \theta + \cos \theta}{\sin \theta - \cos \theta}$	
$\frac{\cos \theta + \sin \theta}{\cos \theta - \sin \theta} - \frac{\cos \theta + \sin \theta}{\cos \theta - \sin \theta}$	
0	

7.

$$\begin{array}{c|c}
\dfrac{\cos^2\alpha + \cot\alpha}{\cos^2\alpha - \cot\alpha} & \dfrac{\cos^2\alpha\tan\alpha + 1}{\cos^2\alpha\tan\alpha - 1} \\
\dfrac{\cos^2\alpha + \dfrac{\cos\alpha}{\sin\alpha}}{\cos^2\alpha - \dfrac{\cos\alpha}{\sin\alpha}} & \dfrac{\cos^2\alpha\dfrac{\sin\alpha}{\cos\alpha} + 1}{\cos^2\alpha\dfrac{\sin\alpha}{\cos\alpha} - 1} \\
\dfrac{\cos\alpha\left(\cos\alpha + \dfrac{1}{\sin\alpha}\right)}{\cos\alpha\left(\cos\alpha - \dfrac{1}{\sin\alpha}\right)} & \dfrac{\sin\alpha\cos\alpha + 1}{\sin\alpha\cos\alpha - 1} \\
\dfrac{\cos\alpha + \dfrac{1}{\sin\alpha}}{\cos\alpha - \dfrac{1}{\sin\alpha}} & \\
\dfrac{\dfrac{\sin\alpha\cos\alpha + 1}{\sin\alpha}}{\dfrac{\sin\alpha\cos\alpha - 1}{\sin\alpha}} & \\
\dfrac{\sin\alpha\cos\alpha + 1}{\sin\alpha\cos\alpha - 1} &
\end{array}$$

9.

$$\begin{array}{c|c}
\dfrac{2\tan\theta}{1 + \tan^2\theta} & \sin 2\theta \\
\dfrac{2\tan\theta}{\sec^2\theta} & 2\sin\theta\cos\theta \\
\dfrac{2\sin\theta}{\cos\theta}\cdot\dfrac{\cos^2\theta}{1} & \\
2\sin\theta\cos\theta &
\end{array}$$

11.

$$\begin{array}{c|c}
1 - \cos 5\theta\cos 3\theta - \sin 5\theta\sin 3\theta & 2\sin^2\theta \\
1 - [\cos 5\theta\cos 3\theta + \sin 5\theta\sin 3\theta] & 1 - \cos 2\theta \\
1 - \cos(5\theta - 3\theta) & \\
1 - \cos 2\theta &
\end{array}$$

13.

$$\begin{array}{c|c}
2\sin\theta\cos^3\theta + 2\sin^3\theta\cos\theta & \sin 2\theta \\
2\sin\theta\cos\theta(\cos^2\theta + \sin^2\theta) & 2\sin\theta\cos\theta \\
2\sin\theta\cos\theta &
\end{array}$$

15.

$$\begin{array}{c|c}
\dfrac{\tan x - \sin x}{2\tan x} & \sin^2\dfrac{x}{2} \\
\dfrac{1}{2}\left[\dfrac{\dfrac{\sin x}{\cos x} - \sin x}{\dfrac{\sin x}{\cos x}}\right] & \dfrac{1 - \cos x}{2} \\
\dfrac{1}{2}\,\dfrac{\sin x - \sin x\cos x}{\cos x}\cdot\dfrac{\cos x}{\sin x} & \\
\dfrac{1 - \cos x}{2} &
\end{array}$$

17.

$$\begin{array}{c|c}
\sin(\alpha + \beta)\sin(\alpha - \beta) & \sin^2\alpha - \sin^2\beta \\
\left(\sin\alpha\cos\beta + \cos\alpha\sin\beta\right)\left(\sin\alpha\cos\beta - \cos\alpha\sin\beta\right) & 1 - \cos^2\alpha - (1 - \cos^2\beta) \\
\sin^2\alpha\cos^2\beta - \cos^2\alpha\sin^2\beta & \cos^2\beta - \cos^2\alpha \\
\cos^2\beta(1 - \cos^2\alpha) - \cos^2\alpha(1 - \cos^2\beta) & \\
\cos^2\beta - \cos^2\alpha\cos^2\beta - \cos^2\alpha + \cos^2\alpha\cos^2\beta & \\
\cos^2\beta - \cos^2\alpha &
\end{array}$$

19.

$$\begin{array}{c|c}
\tan\theta(\tan\theta + \cot\theta) & \sec^2\theta \\
\tan^2\theta + \tan\theta\cot\theta & \\
\tan^2\theta + 1 & \\
\sec^2\theta &
\end{array}$$

21.

$$\begin{array}{c|c}
\dfrac{1 + \cos^2 x}{\sin^2 x} & 2\csc^2 x - 1 \\
\dfrac{1}{\sin^2 x} + \dfrac{\cos^2 x}{\sin^2 x} & \\
\csc^2 x + \cot^2 x & \\
\csc^2 x + \csc^2 x - 1 & \\
2\csc^2 x - 1 &
\end{array}$$

23.

$$\begin{array}{c|c}
\dfrac{1 + \sin x}{1 - \sin x} + \dfrac{\sin x - 1}{1 + \sin x} & 4\sec x\tan x \\
\dfrac{(1 + \sin x)^2 - (1 - \sin x)^2}{1 - \sin^2 x} & 4\cdot\dfrac{1}{\cos x}\cdot\dfrac{\sin x}{\cos x} \\
\dfrac{(1 + 2\sin x + \sin^2 x) - (1 - 2\sin x + \sin^2 x)}{\cos^2 x} & \dfrac{4\sin x}{\cos^2 x} \\
\dfrac{4\sin x}{\cos^2 x} &
\end{array}$$

25.

$$\begin{array}{c|c}
\cos^2\alpha\cot^2\alpha & \cot^2\alpha - \cos^2\alpha \\
(1 - \sin^2\alpha)\cot^2\alpha & \\
\cot^2\alpha - \sin^2\alpha\cdot\dfrac{\cos^2\alpha}{\sin^2\alpha} & \\
\cot^2\alpha - \cos^2\alpha &
\end{array}$$

27.

$$\begin{array}{c|c}
2\sin^2\theta\cos^2\theta + \cos^4\theta & 1 - \sin^4\theta \\
\cos^2\theta(2\sin^2\theta + \cos^2\theta) & (1 + \sin^2\theta)(1 - \sin^2\theta) \\
\cos^2\theta(\sin^2\theta + \sin^2\theta + \cos^2\theta) & (1 + \sin^2\theta)(\cos^2\theta) \\
\cos^2\theta(\sin^2\theta + 1) &
\end{array}$$

29.

$$\begin{array}{c|c}
\dfrac{1 + \sin x}{1 - \sin x} & (\sec x + \tan x)^2 \\
\dfrac{1 + \sin x}{1 - \sin x}\cdot\dfrac{1 + \sin x}{1 + \sin x} & \left(\dfrac{1}{\cos x} + \dfrac{\sin x}{\cos x}\right)^2 \\
\dfrac{(1 + \sin x)^2}{1 - \sin^2 x} & \dfrac{(1 + \sin x)^2}{\cos^2 x} \\
\dfrac{(1 + \sin x)^2}{\cos^2 x} &
\end{array}$$

31. Sine sum and difference identities:

$$\sin(x + y) = \sin x \cos y + \cos x \sin y,$$
$$\sin(x - y) = \sin x \cos y - \cos x \sin y.$$

Add the sum and difference identities:

$$\sin(x + y) + \sin(x - y) = 2 \sin x \cos y$$
$$\tfrac{1}{2}[\sin(x + y) + \sin(x - y)] = \sin x \cos y. \qquad (3)$$

Subtract the difference identity from the sum identity:

$$\sin(x + y) - \sin(x - y) = 2 \cos x \sin y$$
$$\tfrac{1}{2}[\sin(x + y) - \sin(x - y)] = \cos x \sin y. \qquad (4)$$

33. $\sin 3\theta - \sin 5\theta = 2 \cos \frac{8\theta}{2} \sin \frac{-2\theta}{2} = -2 \cos 4\theta \sin \theta$

35. $\sin 8\theta + \sin 5\theta = 2 \sin \frac{13\theta}{2} \cos \frac{3\theta}{2}$

37. $\sin 7u \sin 5u = \frac{1}{2}(\cos 2u - \cos 12u)$

39. $7 \cos \theta \sin 7\theta = \frac{7}{2}[\sin 8\theta - \sin(-6\theta)]$
$= \frac{7}{2}(\sin 8\theta + \sin 6\theta)$

41. $\cos 55° \sin 25° = \frac{1}{2}(\sin 80° - \sin 30°) = \frac{1}{2} \sin 80° - \frac{1}{4}$

43.

$\sin 4\theta + \sin 6\theta$	$\cot \theta (\cos 4\theta - \cos 6\theta)$
$2 \sin \frac{10\theta}{2} \cos \frac{-2\theta}{2}$	$\frac{\cos \theta}{\sin \theta}\left(2 \sin \frac{10\theta}{2} \sin \frac{2\theta}{2}\right)$
$2 \sin 5\theta \cos(-\theta)$	$\frac{\cos \theta}{\sin \theta}(2 \sin 5\theta \sin \theta)$
$2 \sin 5\theta \cos \theta$	$2 \sin 5\theta \cos \theta$

45.

$\cot 4x (\sin x + \sin 4x + \sin 7x)$	$\cos x + \cos 4x + \cos 7x$
$\frac{\cos 4x}{\sin 4x}\left(\sin 4x + 2 \sin \frac{8x}{2} \cos \frac{-6x}{2}\right)$	$\cos 4x + 2 \cos \frac{8x}{2} \cdot \cos \frac{6x}{2}$
$\frac{\cos 4x}{\sin 4x}(\sin 4x + 2 \sin 4x \cos 3x)$	$\cos 4x + 2 \cos 4x \cdot \cos 3x$
$\cos 4x (1 + 2 \cos 3x)$	$\cos 4x (1 + 2 \cos 3x)$

47.

$\cot \frac{x + y}{2}$	$\frac{\sin y - \sin x}{\cos x - \cos y}$
$\frac{\cos \frac{x + y}{2}}{\sin \frac{x + y}{2}}$	$\frac{2 \cos \frac{x + y}{2} \sin \frac{y - x}{2}}{2 \sin \frac{x + y}{2} \sin \frac{y - x}{2}}$
	$\frac{\cos \frac{x + y}{2}}{\sin \frac{x + y}{2}}$

49.

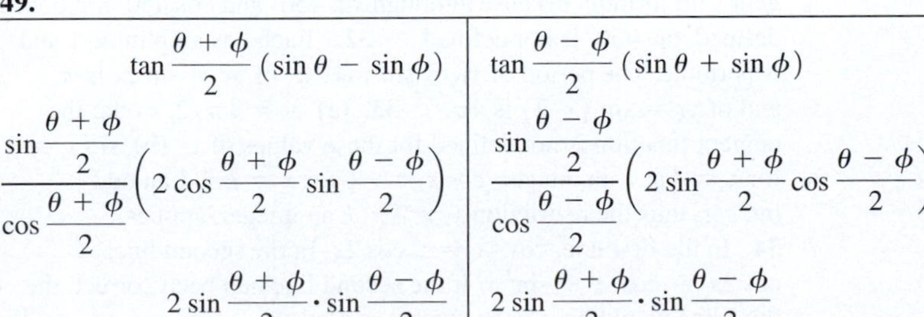

$\tan \frac{\theta + \phi}{2}(\sin \theta - \sin \phi)$	$\tan \frac{\theta - \phi}{2}(\sin \theta + \sin \phi)$
$\frac{\sin \frac{\theta + \phi}{2}}{\cos \frac{\theta + \phi}{2}}\left(2 \cos \frac{\theta + \phi}{2} \sin \frac{\theta - \phi}{2}\right)$	$\frac{\sin \frac{\theta - \phi}{2}}{\cos \frac{\theta - \phi}{2}}\left(2 \sin \frac{\theta + \phi}{2} \cos \frac{\theta - \phi}{2}\right)$
$2 \sin \frac{\theta + \phi}{2} \cdot \sin \frac{\theta - \phi}{2}$	$2 \sin \frac{\theta + \phi}{2} \cdot \sin \frac{\theta - \phi}{2}$

51. B;

$\frac{\cos x + \cot x}{1 + \csc x}$	$\cos x$
$\frac{\frac{\cos x}{1} + \frac{\cos x}{\sin x}}{1 + \frac{1}{\sin x}}$	
$\frac{\sin x \cos x + \cos x}{\sin x} \cdot \frac{\sin x}{\sin x + 1}$	
$\frac{\cos x (\sin x + 1)}{\sin x + 1}$	
$\cos x$	

53. A;

$\sin x \cos x + 1$	$\frac{\sin^3 x - \cos^3 x}{\sin x - \cos x}$
	$\frac{(\sin x - \cos x)(\sin^2 x + \sin x \cos x + \cos^2 x)}{\sin x - \cos x}$
	$\sin^2 x + \sin x \cos x + \cos^2 x$
	$\sin x \cos x + 1$

55. C;

$\frac{1}{\cot x \sin^2 x}$	$\tan x + \cot x$
$\frac{1}{\frac{\cos x}{\sin x} \cdot \sin^2 x}$	$\frac{\sin x}{\cos x} + \frac{\cos x}{\sin x}$
$\frac{1}{\cos x \sin x}$	$\frac{\sin^2 x + \cos^2 x}{\cos x \sin x}$
	$\frac{1}{\cos x \sin x}$

57. **(a), (d)**

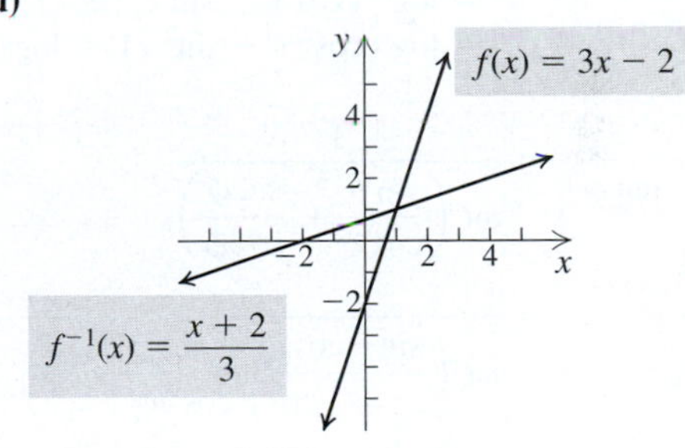

(b) yes; **(c)** $f^{-1}(x) = \frac{x + 2}{3}$

58. **(a), (d)**

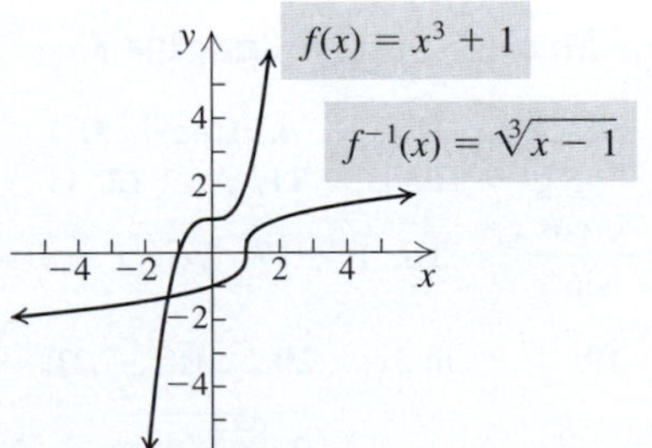

(b) yes; **(c)** $f^{-1}(x) = \sqrt[3]{x - 1}$

59. **(a), (d)**

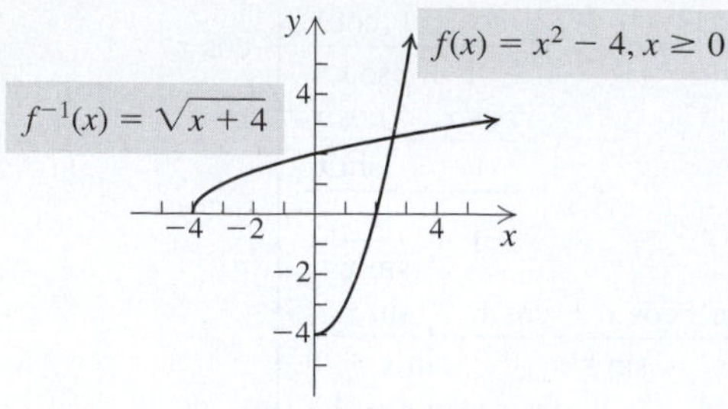

(b) yes; **(c)** $f^{-1}(x) = \sqrt{x+4}$

60. **(a), (d)**

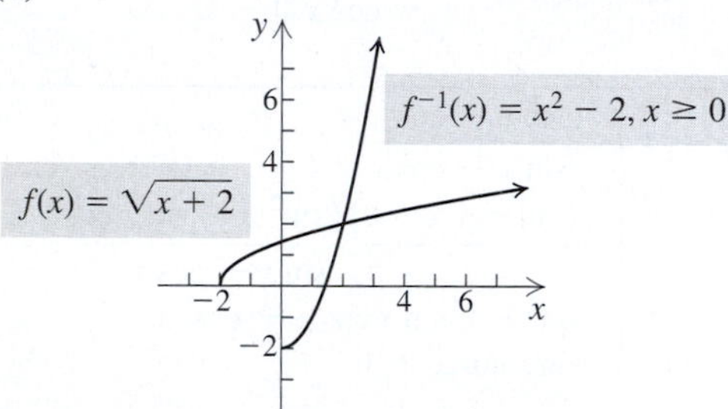

(b) yes; **(c)** $f^{-1}(x) = x^2 - 2, x \geq 0$ **61.** $0, \frac{5}{2}$ **62.** $-4, \frac{7}{3}$
63. $\pm 2, \pm 3i$ **64.** $5 \pm 2\sqrt{6}$ **65.** 27 **66.** 9

67.

$\ln\lvert\tan x\rvert$	$-\ln\lvert\cot x\rvert$
$\ln\left\lvert\dfrac{1}{\cot x}\right\rvert$	
$\ln\lvert 1\rvert - \ln\lvert\cot x\rvert$	
$0 - \ln\lvert\cot x\rvert$	
$-\ln\lvert\cot x\rvert$	

69. $\log(\cos x - \sin x) + \log(\cos x + \sin x)$
$= \log[(\cos x - \sin x)(\cos x + \sin x)]$
$= \log(\cos^2 x - \sin^2 x) = \log\cos 2x$

71.

$$\frac{1}{\omega C(\tan\theta + \tan\phi)} = \frac{1}{\omega C\left(\dfrac{\sin\theta}{\cos\theta} + \dfrac{\sin\phi}{\cos\phi}\right)}$$
$$= \frac{1}{\omega C\left(\dfrac{\sin\theta\cos\phi + \sin\phi\cos\theta}{\cos\theta\cos\phi}\right)}$$
$$= \frac{\cos\theta\cos\phi}{\omega C\sin(\theta + \phi)}$$

Mid-Chapter Mixed Review: Chapter 7

1. True **2.** False **3.** True **4.** True **5.** J **6.** H
7. D **8.** I **9.** C **10.** E **11.** A **12.** G **13.** F
14. B **15.** $\dfrac{\sqrt{\cos x}}{\sin x}$ **16.** 1 **17.** $2\cos x + 1$
18. $-\cos x$ **19.** $1 - \sin 2x$ **20.** $\cos x$ **21.** $\dfrac{\sqrt{\sec x + 1}}{\sin x}$
22. $\cos 12° \approx 0.9781$ **23.** $\dfrac{\sqrt{2 - \sqrt{2}}}{2}$ **24.** $\dfrac{\sqrt{6} + \sqrt{2}}{4}$
25. $-\frac{119}{120}$ **26.** $-\frac{24}{25}$; quadrant IV

27.

$\cos^2\dfrac{x}{2}$	$\dfrac{\tan x + \sin x}{2\tan x}$
$\dfrac{1 + \cos x}{2}$	$\dfrac{1}{2}\left[\dfrac{\dfrac{\sin x}{\cos x} + \sin x}{\dfrac{\sin x}{\cos x}}\right]$
	$\dfrac{1}{2}\dfrac{\sin x + \sin x\cos x}{\cos x} \cdot \dfrac{\cos x}{\sin x}$
	$\dfrac{1}{2}\dfrac{\sin x(1 + \cos x)}{\sin x}$
	$\dfrac{1 + \cos x}{2}$

28.

$\dfrac{1 - \sin x}{\cos x}$	$\dfrac{\cos x}{1 + \sin x}$
$\dfrac{1 - \sin x}{\cos x} \cdot \dfrac{1 + \sin x}{1 + \sin x}$	
$\dfrac{1 - \sin^2 x}{\cos x(1 + \sin x)}$	
$\dfrac{\cos^2 x}{\cos x(1 + \sin x)}$	
$\dfrac{\cos x}{1 + \sin x}$	

29.

$\dfrac{\sin^3 x - \cos^3 x}{\sin x - \cos x}$	$\dfrac{2 + \sin 2x}{2}$
$\dfrac{(\sin x - \cos x)(\sin^2 x + \sin x\cos x + \cos^2 x)}{\sin x - \cos x}$	$\dfrac{2 + 2\sin x\cos x}{2}$
$\sin^2 x + \cos^2 x + \sin x\cos x$	$\dfrac{2(1 + \sin x\cos x)}{2}$
$1 + \sin x\cos x$	$1 + \sin x\cos x$

30.

$\sin 6\theta - \sin 2\theta$	$\tan 2\theta(\cos 2\theta + \cos 6\theta)$
$2\cos\dfrac{8\theta}{2}\sin\dfrac{4\theta}{2}$	$\dfrac{\sin 2\theta}{\cos 2\theta}\left(2\cos\dfrac{8\theta}{2}\cos\dfrac{4\theta}{2}\right)$
$2\cos 4\theta\sin 2\theta$	$\dfrac{\sin 2\theta}{\cos 2\theta}(2\cos 4\theta\cos 2\theta)$
	$2\cos 4\theta\sin 2\theta$

31. The expression $\tan(x + 450°)$ can be simplified using the sine and cosine sum formulas but cannot be simplified using the tangent sum formula because although sin 450° and cos 450° are both defined, tan 450° is not defined. **32.** Each has amplitude 1 and is periodic. The period of $y_1 = \sin x$ is 2π, of $y_2 = \sin 2x$ is π, and of $y_3 = \sin(x/2)$ is 4π. **33.** **(a)** $x \neq k\pi/2$, k odd; the tangent function is not defined for these values of x; **(b)** $\sin x = 0$ for $x = k\pi$, k an integer; $\cos x = -1$ for $x = k\pi$, k an odd integer; thus the restriction $x \neq k\pi$, k an integer, applies.
34. In the first line, $\cos 4x \neq 2\cos 2x$. In the second line, $\cos 2x \neq \cos^2 x + \sin^2 x$. If the second line had been correct, the third line would have been correct as well.

Exercise Set 7.4

1. $-\dfrac{\pi}{3}, -60°$ **3.** $\dfrac{\pi}{4}, 45°$ **5.** $\dfrac{\pi}{4}, 45°$ **7.** $0, 0°$

9. $\dfrac{\pi}{6}, 30°$ **11.** $\dfrac{\pi}{6}, 30°$ **13.** $\dfrac{5\pi}{6}, 150°$ **15.** $-\dfrac{\pi}{6}, -30°$

17. $\frac{\pi}{2}$, 90° **19.** $\frac{\pi}{3}$, 60° **21.** 0.3520, 20.2°
23. 1.2917, 74.0° **25.** 2.9463, 168.8° **27.** −0.1600, −9.2°
29. 0.8289, 47.5° **31.** −0.9600, −55.0° **33.** $\sin^{-1}$: $[-1, 1]$; $\cos^{-1}$: $[-1, 1]$; $\tan^{-1}$: $(-\infty, \infty)$ **35.** $\theta = \sin^{-1}\left(\frac{2000}{d}\right)$
37. 0.3 **39.** $\frac{\pi}{4}$ **41.** $\frac{\pi}{5}$ **43.** $-\frac{\pi}{3}$ **45.** $\frac{1}{2}$ **47.** 1
49. $\frac{\pi}{3}$ **51.** $\frac{\sqrt{11}}{33}$ **53.** $-\frac{\pi}{6}$ **55.** $\frac{a}{\sqrt{a^2+9}}$
57. $\frac{\sqrt{q^2-p^2}}{p}$ **59.** $\frac{p}{3}$ **61.** $\frac{\sqrt{3}}{2}$ **63.** $-\frac{\sqrt{2}}{10}$
65. $xy + \sqrt{(1-x^2)(1-y^2)}$ **67.** 0.9861 **69.** Periodic
70. Radian measure **71.** Similar **72.** Angle of depression
73. Angular speed **74.** Supplementary **75.** Amplitude
76. Circular
77.

$\sin^{-1} x + \cos^{-1} x$	$\frac{\pi}{2}$
$\sin(\sin^{-1} x + \cos^{-1} x)$	$\sin \frac{\pi}{2}$
$[\sin(\sin^{-1} x)][\cos(\cos^{-1} x)] +$ $[\cos(\sin^{-1} x)][\sin(\cos^{-1} x)]$	1
$x \cdot x + \sqrt{1-x^2} \cdot \sqrt{1-x^2}$	
$x^2 + 1 - x^2$	
1	

79.

$\sin^{-1} x$	$\cos^{-1} \sqrt{1-x^2}$
$\sin(\sin^{-1} x)$	$\sin(\cos^{-1} \sqrt{1-x^2})$
x	x

81. $\theta = \tan^{-1} \frac{y+h}{x} - \tan^{-1} \frac{y}{x}$; 38.7°

Visualizing the Graph

1. D **2.** G **3.** C **4.** H **5.** I **6.** A **7.** E
8. J **9.** F **10.** B

Exercise Set 7.5

1. $\frac{\pi}{6} + 2k\pi, \frac{11\pi}{6} + 2k\pi$, or $30° + k \cdot 360°, 330° + k \cdot 360°$
3. $\frac{2\pi}{3} + k\pi$, or $120° + k \cdot 180°$
5. $\frac{\pi}{6} + 2k\pi, \frac{5\pi}{6} + 2k\pi$, or $30° + k \cdot 360°, 150° + k \cdot 360°$
7. $\frac{3\pi}{4} + 2k\pi, \frac{5\pi}{4} + 2k\pi$, or $135° + k \cdot 360°, 225° + k \cdot 360°$
9. 1.7120, 4.5712, or 98.09°, 261.91° **11.** $\frac{4\pi}{3}, \frac{5\pi}{3}$, or 240°, 300° **13.** $\frac{\pi}{4}, \frac{3\pi}{4}, \frac{5\pi}{4}, \frac{7\pi}{4}$, or 45°, 135°, 225°, 315°
15. $\frac{\pi}{6}, \frac{5\pi}{6}, \frac{3\pi}{2}$, or 30°, 150°, 270° **17.** $\frac{\pi}{6}, \frac{\pi}{2}, \frac{3\pi}{2}, \frac{11\pi}{6}$, or 30°, 90°, 270°, 330° **19.** 1.9106, $\frac{2\pi}{3}, \frac{4\pi}{3}$, 4.3726, or 109.47°, 120°, 240°, 250.53° **21.** $0, \frac{\pi}{4}, \frac{3\pi}{4}, \pi, \frac{5\pi}{4}, \frac{7\pi}{4}$, or 0°, 45°, 135°, 180°, 225°, 315° **23.** 2.4402, 3.8430, or 139.81°, 220.19° **25.** 0.6496, 2.9557, 3.7912, 6.0973, or 37.22°, 169.35°, 217.22°, 349.35° **27.** $0, \pi, \frac{7\pi}{6}, \frac{11\pi}{6}$
29. $0, \pi$ **31.** $0, \pi$ **33.** $\frac{3\pi}{4}, \frac{7\pi}{4}$
35. $\frac{2\pi}{3}, \frac{4\pi}{3}, \frac{3\pi}{2}$ **37.** $\frac{\pi}{4}, \frac{3\pi}{4}, \frac{5\pi}{4}, \frac{7\pi}{4}$ **39.** $\frac{\pi}{12}, \frac{5\pi}{12}$
41. 0.967, 1.853, 4.108, 4.994 **43.** $\frac{2\pi}{3}, \frac{4\pi}{3}$ **45.** 1.114, 2.773 **47.** 0.515 **49.** 0.422, 1.756
51. November: $72,853; March: $101,853
53. **(a)** $y = 26.3148 \sin(0.4927x - 1.9612) + 58.2283$; **(b)** April: 58°; December: 39°; **(c)** April and October
55. $B = 35°, b \approx 140.7, c \approx 245.4$
56. $R \approx 15.5°, T \approx 74.5°, t \approx 13.7$ **57.** 36 **58.** 14
59. $\frac{\pi}{3}, \frac{2\pi}{3}, \frac{4\pi}{3}, \frac{5\pi}{3}$ **61.** $\frac{\pi}{3}, \frac{4\pi}{3}$ **63.** 0
65. $e^{3\pi/2 + 2k\pi}$, where k (an integer) ≤ -1
67. 1.24 days, 6.76 days **69.** 16.5°N
71. 1 **73.** $\frac{5}{26}$, or about 0.1923

Review Exercises: Chapter 7

1. True **2.** True **3.** True **4.** False **5.** False
6. $\csc^2 x$ **7.** 1 **8.** $\tan^2 y - \cot^2 y$
9. $\frac{(\cos^2 x + 1)^2}{\cos^2 x}$ **10.** $\csc x(\sec x - \csc x)$
11. $(3 \sin y + 5)(\sin y - 4)$
12. $(10 - \cos u)(100 + 10 \cos u + \cos^2 u)$
13. 1 **14.** $\frac{1}{2} \sec x$ **15.** $\frac{3 \tan x}{\sin x - \cos x}$
16. $\frac{3 \cos y + 3 \sin y + 2}{\cos^2 y - \sin^2 y}$ **17.** 1 **18.** $\frac{1}{4} \cot x$
19. $\sin x + \cos x$ **20.** $\frac{\cos x}{1 - \sin x}$ **21.** $\frac{\cos x}{\sqrt{\sin x}}$
22. $3 \sec \theta$ **23.** $\cos x \cos \frac{3\pi}{2} - \sin x \sin \frac{3\pi}{2}$
24. $\frac{\tan 45° - \tan 30°}{1 + \tan 45° \tan 30°}$ **25.** $\cos(27° - 16°)$, or $\cos 11°$
26. $\frac{-\sqrt{6} - \sqrt{2}}{4}$ **27.** $2 - \sqrt{3}$ **28.** −0.3745
29. $-\sin x$ **30.** $\sin x$ **31.** $-\cos x$
32. **(a)** $\sin \alpha = -\frac{4}{5}$, $\tan \alpha = \frac{4}{3}$, $\cot \alpha = \frac{3}{4}$, $\sec \alpha = -\frac{5}{3}$, $\csc \alpha = -\frac{5}{4}$; **(b)** $\sin\left(\frac{\pi}{2} - \alpha\right) = -\frac{3}{5}$, $\cos\left(\frac{\pi}{2} - \alpha\right) = -\frac{4}{5}$, $\tan\left(\frac{\pi}{2} - \alpha\right) = \frac{3}{4}$, $\cot\left(\frac{\pi}{2} - \alpha\right) = \frac{4}{3}$, $\sec\left(\frac{\pi}{2} - \alpha\right) = -\frac{5}{4}$, $\csc\left(\frac{\pi}{2} - \alpha\right) = -\frac{5}{3}$; **(c)** $\sin\left(\alpha + \frac{\pi}{2}\right) = -\frac{3}{5}$, $\cos\left(\alpha + \frac{\pi}{2}\right) = \frac{4}{5}$, $\tan\left(\alpha + \frac{\pi}{2}\right) = -\frac{3}{4}$, $\cot\left(\alpha + \frac{\pi}{2}\right) = -\frac{4}{3}$, $\sec\left(\alpha + \frac{\pi}{2}\right) = \frac{5}{4}$, $\csc\left(\alpha + \frac{\pi}{2}\right) = -\frac{5}{3}$ **33.** $-\sec x$
34. $\tan 2\theta = \frac{24}{7}$, $\cos 2\theta = \frac{7}{25}$, $\sin 2\theta = \frac{24}{25}$; I

35. $\dfrac{\sqrt{2-\sqrt{2}}}{2}$ **36.** $\sin 2\beta = 0.4261$, $\cos \dfrac{\beta}{2} = 0.9940$, $\cos 4\beta = 0.6369$ **37.** $\cos x$ **38.** 1 **39.** $\sin 2x$ **40.** $\tan 2x$

41.

$\dfrac{1-\sin x}{\cos x}$	$\dfrac{\cos x}{1+\sin x}$
$\dfrac{1-\sin x}{\cos x}\cdot\dfrac{\cos x}{\cos x}$	$\dfrac{\cos x}{1+\sin x}\cdot\dfrac{1-\sin x}{1-\sin x}$
$\dfrac{\cos x-\sin x\cos x}{\cos^2 x}$	$\dfrac{\cos x-\sin x\cos x}{1-\sin^2 x}$
	$\dfrac{\cos x-\sin x\cos x}{\cos^2 x}$

42.

$\dfrac{1+\cos 2\theta}{\sin 2\theta}$	$\cot\theta$
$\dfrac{1+2\cos^2\theta-1}{2\sin\theta\cos\theta}$	$\dfrac{\cos\theta}{\sin\theta}$
$\dfrac{\cos\theta}{\sin\theta}$	

43.

$\dfrac{\tan y+\sin y}{2\tan y}$	$\cos^2\dfrac{y}{2}$
$\dfrac{1}{2}\left[\dfrac{\dfrac{\sin y+\sin y\cos y}{\cos y}}{\dfrac{\sin y}{\cos y}}\right]$	$\dfrac{1+\cos y}{2}$
$\dfrac{1}{2}\left[\dfrac{\sin y\,(1+\cos y)}{\cos y}\cdot\dfrac{\cos y}{\sin y}\right]$	
$\dfrac{1+\cos y}{2}$	

44.

$\dfrac{\sin x-\cos x}{\cos^2 x}$	$\dfrac{\tan^2 x-1}{\sin x+\cos x}$
	$\dfrac{\dfrac{\sin^2 x}{\cos^2 x}-1}{\sin x+\cos x}$
	$\dfrac{\sin^2 x-\cos^2 x}{\cos^2 x}\cdot\dfrac{1}{\sin x+\cos x}$
	$\dfrac{\sin x-\cos x}{\cos^2 x}$

45. $3\cos 2\theta\sin\theta = \frac{3}{2}(\sin 3\theta - \sin\theta)$

46. $\sin\theta - \sin 4\theta = -2\cos\dfrac{5\theta}{2}\sin\dfrac{3\theta}{2}$

47. B;

$\csc x-\cos x\cot x$	$\sin x$
$\dfrac{1}{\sin x}-\cos x\dfrac{\cos x}{\sin x}$	
$\dfrac{1-\cos^2 x}{\sin x}$	
$\dfrac{\sin^2 x}{\sin x}$	
$\sin x$	

48. D;

$\dfrac{1}{\sin x\cos x}-\dfrac{\cos x}{\sin x}$	$\dfrac{\sin x\cos x}{1-\sin^2 x}$
$\dfrac{1}{\sin x\cos x}-\dfrac{\cos^2 x}{\sin x\cos x}$	$\dfrac{\sin x\cos x}{\cos^2 x}$
$\dfrac{1-\cos^2 x}{\sin x\cos x}$	$\dfrac{\sin x}{\cos x}$
$\dfrac{\sin^2 x}{\sin x\cos x}$	
$\dfrac{\sin x}{\cos x}$	

49. A;

$\dfrac{\cot x-1}{1-\tan x}$	$\dfrac{\csc x}{\sec x}$
$\dfrac{\dfrac{\cos x}{\sin x}-\dfrac{\sin x}{\sin x}}{\dfrac{\cos x}{\cos x}-\dfrac{\sin x}{\cos x}}$	$\dfrac{\dfrac{1}{\sin x}}{\dfrac{1}{\cos x}}$
$\dfrac{\cos x-\sin x}{\sin x}\cdot\dfrac{\cos x}{\cos x-\sin x}$	$\dfrac{1}{\sin x}\cdot\dfrac{\cos x}{1}$
$\dfrac{\cos x}{\sin x}$	$\dfrac{\cos x}{\sin x}$

50. C;

$\dfrac{\cos x+1}{\sin x}+\dfrac{\sin x}{\cos x+1}$	$\dfrac{2}{\sin x}$
$\dfrac{(\cos x+1)^2+\sin^2 x}{\sin x\,(\cos x+1)}$	
$\dfrac{\cos^2 x+2\cos x+1+\sin^2 x}{\sin x\,(\cos x+1)}$	
$\dfrac{2\cos x+2}{\sin x\,(\cos x+1)}$	
$\dfrac{2(\cos x+1)}{\sin x\,(\cos x+1)}$	
$\dfrac{2}{\sin x}$	

51. $-\dfrac{\pi}{6}$, $-30°$ **52.** $\dfrac{\pi}{6}$, $30°$ **53.** $\dfrac{\pi}{4}$, $45°$ **54.** 0, 0°

55. 1.7920, 102.7° **56.** 0.3976, 22.8° **57.** $\frac{1}{2}$ **58.** $\dfrac{\sqrt{3}}{3}$

59. $\dfrac{\pi}{7}$ **60.** $\dfrac{\sqrt{2}}{2}$ **61.** $\dfrac{3}{\sqrt{b^2+9}}$ **62.** $-\frac{7}{25}$

63. $\dfrac{3\pi}{4}+2k\pi$, $\dfrac{5\pi}{4}+2k\pi$, or $135° + k\cdot 360°$, $225° + k\cdot 360°$

64. $\dfrac{\pi}{3}+k\pi$, or $60° + k\cdot 180°$ **65.** $\dfrac{\pi}{6}, \dfrac{5\pi}{6}, \dfrac{7\pi}{6}, \dfrac{11\pi}{6}$

66. $\dfrac{\pi}{4}, \dfrac{\pi}{2}, \dfrac{3\pi}{4}, \dfrac{5\pi}{4}, \dfrac{3\pi}{2}, \dfrac{7\pi}{4}$ **67.** $\dfrac{2\pi}{3}, \pi, \dfrac{4\pi}{3}$ **68.** $0, \pi$

69. $\dfrac{\pi}{4}, \dfrac{3\pi}{4}, \dfrac{5\pi}{4}, \dfrac{7\pi}{4}$ **70.** $0, \dfrac{\pi}{2}, \pi, \dfrac{3\pi}{2}$ **71.** $\dfrac{7\pi}{12}, \dfrac{23\pi}{12}$

72. 0.864, 2.972, 4.006, 6.114 **73.** 4.917 **74.** No solution in $[0, 2\pi)$ **75.** B **76.** A **77.** C **78.** 108.4°

79. $\cos(u+v) = \cos u\cos v - \sin u\sin v$

$= \cos u\cos v - \cos\left(\dfrac{\pi}{2}-u\right)\cos\left(\dfrac{\pi}{2}-v\right)$

80. $\cos^2 x$

81. $\sin\theta = \sqrt{\frac{1}{2} + \frac{\sqrt{6}}{5}}$; $\cos\theta = \sqrt{\frac{1}{2} - \frac{\sqrt{6}}{5}}$;
$\tan\theta = \sqrt{\frac{5 + 2\sqrt{6}}{5 - 2\sqrt{6}}}$

82. Let $x = \frac{\sqrt{2}}{2}$. Then $\tan^{-1}\frac{\sqrt{2}}{2} \approx 0.6155$ and

$$\frac{\sin^{-1}\frac{\sqrt{2}}{2}}{\cos^{-1}\frac{\sqrt{2}}{2}} = \frac{\frac{\pi}{4}}{\frac{\pi}{4}} = 1.$$

83. $\frac{\pi}{2}, \frac{3\pi}{2}$

84. The ranges of the inverse trigonometric functions are restricted in order that they might be functions. **85.** Yes; first note that $7\pi/6 = \pi/6 + \pi$. Since $\pi/6 + k\pi$ includes both odd and even multiples of π, it is equivalent to $\pi/6 + 2k\pi$ and $7\pi/6 + 2k\pi$. **86.** The graphs have different domains and ranges. The graph of $y = \sin^{-1} x$ is the reflection of the portion of the graph of $y = \sin x$ for $-\pi/2 \le x \le \pi/2$, across the line $y = x$. **87.** A trigonometric equation that is an identity is true for all possible replacements of the variables. A trigonometric equation that is not true for all possible replacements is not an identity. The equation $\sin^2 x + \cos^2 x = 1$ is an identity whereas $\sin^2 x = 1$ is not. **88.** The range of the arcsine function does not include $5\pi/6$. It is $[-\pi/2, \pi/2]$.

Test: Chapter 7

1. [7.1] $2\cos x + 1$ **2.** [7.1] 1 **3.** [7.1] $\frac{\cos\theta}{1 + \sin\theta}$

4. [7.1] $2\cos\theta$ **5.** [7.1] $\frac{\sqrt{2} + \sqrt{6}}{4}$ **6.** [7.1] $2 - \sqrt{3}$

7. [7.1] $\frac{120}{169}$ **8.** [7.2] $\frac{\sqrt{5}}{3}$ **9.** [7.2] $\frac{24}{25}$, II

10. [7.2] $\frac{\sqrt{2 + \sqrt{3}}}{2}$ **11.** [7.2] 0.9304 **12.** [7.2] $3\sin 2x$

13. [7.3]

$\csc x - \cos x \cot x$	$\sin x$
$\frac{1}{\sin x} - \cos x \cdot \frac{\cos x}{\sin x}$	
$\frac{1 - \cos^2 x}{\sin x}$	
$\frac{\sin^2 x}{\sin x}$	
$\sin x$	

14. [7.3]

$(\sin x + \cos x)^2$	$1 + \sin 2x$
$\sin^2 x + 2\sin x\cos x + \cos^2 x$	
$1 + 2\sin x\cos x$	
$1 + \sin 2x$	

15. [7.3]

$(\csc\beta + \cot\beta)^2$	$\frac{1 + \cos\beta}{1 - \cos\beta}$
$\left(\frac{1}{\sin\beta} + \frac{\cos\beta}{\sin\beta}\right)^2$	$\frac{1 + \cos\beta}{1 - \cos\beta} \cdot \frac{1 + \cos\beta}{1 + \cos\beta}$
$\left(\frac{1 + \cos\beta}{\sin\beta}\right)^2$	$\frac{(1 + \cos\beta)^2}{1 - \cos^2\beta}$
$\frac{(1 + \cos\beta)^2}{\sin^2\beta}$	$\frac{(1 + \cos\beta)^2}{\sin^2\beta}$

16. [7.3]

$\frac{1 + \sin\alpha}{1 + \csc\alpha}$	$\frac{\tan\alpha}{\sec\alpha}$
$\frac{1 + \sin\alpha}{1 + \frac{1}{\sin\alpha}}$	$\frac{\frac{\sin\alpha}{\cos\alpha}}{\frac{1}{\cos\alpha}}$
$\frac{1 + \sin\alpha}{\frac{\sin\alpha + 1}{\sin\alpha}}$	$\sin\alpha$
$\sin\alpha$	

17. [7.4] $\cos 8\alpha - \cos\alpha = -2\sin\frac{9\alpha}{2}\sin\frac{7\alpha}{2}$

18. [7.4] $4\sin\beta\cos 3\beta = 2(\sin 4\beta - \sin 2\beta)$

19. [7.4] $-45°$ **20.** [7.4] $\frac{\pi}{3}$ **21.** [7.4] 2.3072

22. [7.4] $\frac{\sqrt{3}}{2}$ **23.** [7.4] $\frac{5}{\sqrt{x^2 - 25}}$ **24.** [7.4] 0

25. [7.5] $\frac{\pi}{6}, \frac{5\pi}{6}, \frac{7\pi}{6}, \frac{11\pi}{6}$ **26.** [7.5] $0, \frac{\pi}{4}, \frac{3\pi}{4}, \pi$

27. [7.5] $\frac{\pi}{2}, \frac{11\pi}{6}$ **28.** [7.4] D **29.** [7.2] $\sqrt{\frac{11}{12}}$

Chapter 8

Exercise Set 8.1

1. $A = 121°$, $a \approx 33$, $c \approx 14$ **3.** $B \approx 57.4°$, $C \approx 86.1°$, $c \approx 40$ or $B \approx 122.6°$, $C \approx 20.9°$, $c \approx 14$ **5.** $B \approx 44°24'$, $A \approx 74°26'$, $a \approx 33.3$ **7.** $A = 110.36°$, $a \approx 5$ mi, $b \approx 3.4$ mi **9.** $B \approx 83.78°$, $A \approx 12.44°$, $a \approx 12.30$ yd **11.** $B \approx 14.7°$, $C \approx 135.0°$, $c \approx 28.04$ cm **13.** No solution **15.** $B = 125.27°$, $b \approx 302$ m, $c \approx 138$ m **17.** 8.2 ft^2 **19.** 12 yd^2 **21.** 596.98 ft^2 **23.** About 31 ft **25.** 787 ft^2 **27.** About 51 ft **29.** From A: about 35 mi; from B: about 66 mi **31.** About 102 mi **33.** 1.348, 77.2° **34.** No angle **35.** 18.24° **36.** 125.06° **37.** 5 **38.** $\frac{\sqrt{3}}{2}$ **39.** $\frac{\sqrt{2}}{2}$ **40.** $-\frac{\sqrt{3}}{2}$ **41.** $-\frac{1}{2}$ **42.** 2

43. Use the formula for the area of a triangle and the law of sines.

$$K = \frac{1}{2}ab\sin C \quad \text{and} \quad b = \frac{a\sin B}{\sin A},$$
$$\text{so } K = \frac{a^2\sin B\sin C}{2\sin A}.$$
$$K = \frac{1}{2}bc\sin A \quad \text{and} \quad b = \frac{c\sin B}{\sin C},$$
$$\text{so } K = \frac{c^2\sin A\sin B}{2\sin C}.$$
$$K = \frac{1}{2}bc\sin A \quad \text{and} \quad c = \frac{b\sin C}{\sin B},$$
$$\text{so } K = \frac{b^2\sin A\sin C}{2\sin B}.$$

45. For the quadrilateral $ABCD$, we have

$$\begin{aligned}\text{Area} &= \tfrac{1}{2}bd\sin\theta + \tfrac{1}{2}ac\sin\theta \\ &\quad + \tfrac{1}{2}ad(\sin 180^\circ - \theta) + \tfrac{1}{2}bc\sin(180^\circ - \theta) \\ &\quad \textit{Note: } \sin\theta = \sin(180^\circ - \theta). \\ &= \tfrac{1}{2}(bd + ac + ad + bc)\sin\theta \\ &= \tfrac{1}{2}(a+b)(c+d)\sin\theta \\ &= \tfrac{1}{2}d_1d_2\sin\theta,\end{aligned}$$

where $d_1 = a + b$ and $d_2 = c + d$. **45.** $d \approx 18.8$ in.
47. 44.1″ from wall 1 and 104.3″ from wall 4

Exercise Set 8.2

1. $a \approx 15, B \approx 24°, C \approx 126°$ **3.** $A \approx 36.18°$, $B \approx 43.53°, C \approx 100.29°$ **5.** $b \approx 75$ m, $A \approx 94°51'$, $C \approx 12°29'$ **7.** $A \approx 24.15°, B \approx 30.75°, C \approx 125.10°$
9. No solution **11.** $A \approx 79.93°, B \approx 53.55°, C \approx 46.52°$
13. $c \approx 45.17$ mi, $A \approx 89.3°, B \approx 42.0°$ **15.** $a \approx 13.9$ in., $B \approx 36.127°, C \approx 90.417°$ **17.** Law of sines; $C = 98°$, $a \approx 96.7, c \approx 101.9$ **19.** Law of cosines; $A \approx 73.71°$, $B \approx 51.75°, C \approx 54.54°$ **21.** Cannot be solved
23. Law of cosines; $A \approx 33.71°, B \approx 107.08°, C \approx 39.21°$
25. 23.5 ft **27.** 30.76 ft **29.** About 1.5 mi
31. $S \approx 112.5°, T \approx 27.2°, U \approx 40.3°$ **33.** About 912 km
35. **(a)** About 16 ft; **(b)** about 122 ft^2 **37.** About 4.7 cm
39. Quartic **40.** Linear **41.** Trigonometric
42. Exponential **43.** Rational **44.** Cubic
45. Exponential **46.** Logarithmic **47.** Trigonometric
48. Quadratic **49.** About 9386 ft
51. $A = \frac{1}{2}a^2\sin\theta$; when $\theta = 90°$

Exercise Set 8.3

1. 5;

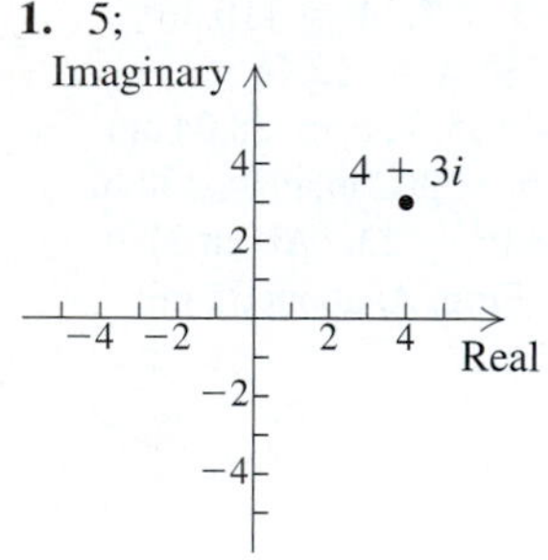

3. 1;

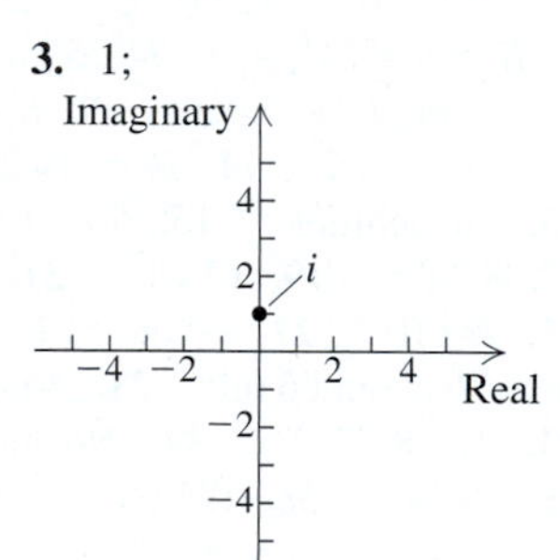

5. $\sqrt{17}$;

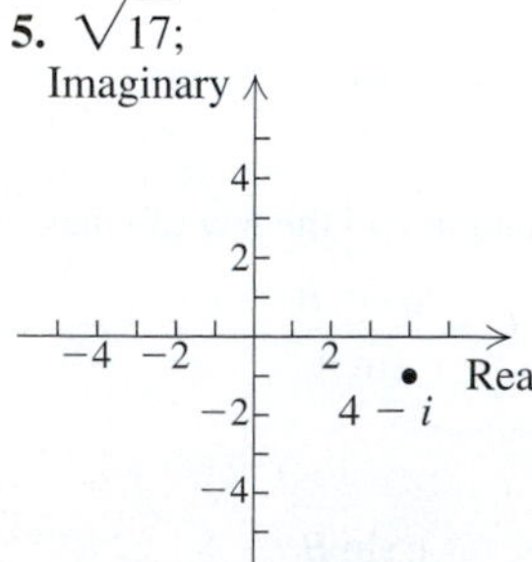

7. 3;

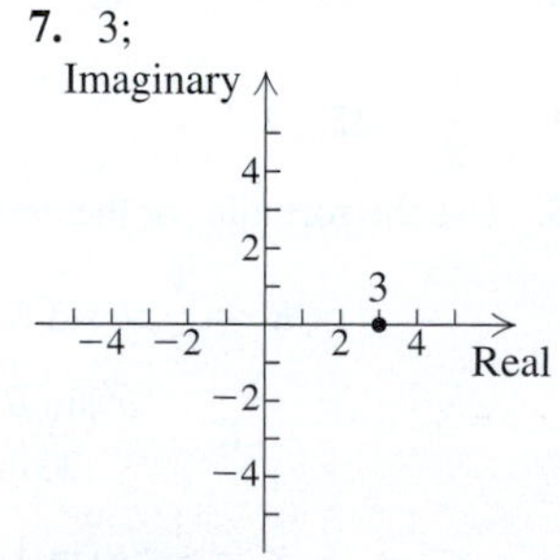

9. $3 - 3i; 3\sqrt{2}\left(\cos\frac{7\pi}{4} + i\sin\frac{7\pi}{4}\right)$, or $3\sqrt{2}(\cos 315° + i\sin 315°)$
11. $4i; 4\left(\cos\frac{\pi}{2} + i\sin\frac{\pi}{2}\right)$, or $4(\cos 90° + i\sin 90°)$
13. $\sqrt{2}\left(\cos\frac{7\pi}{4} + i\sin\frac{7\pi}{4}\right)$, or $\sqrt{2}(\cos 315° + i\sin 315°)$
15. $3\left(\cos\frac{3\pi}{2} + i\sin\frac{3\pi}{2}\right)$, or $3(\cos 270° + i\sin 270°)$
17. $2\left(\cos\frac{\pi}{6} + i\sin\frac{\pi}{6}\right)$, or $2(\cos 30° + i\sin 30°)$
19. $\frac{2}{5}(\cos 0 + i\sin 0)$, or $\frac{2}{5}(\cos 0° + i\sin 0°)$
21. $6\left(\cos\frac{5\pi}{4} + i\sin\frac{5\pi}{4}\right)$, or $6(\cos 225° + i\sin 225°)$
23. $\frac{3\sqrt{3}}{2} + \frac{3}{2}i$ **25.** $-10i$ **27.** $2 + 2i$ **29.** $2i$
31. $\frac{\sqrt{2}}{2} - \frac{\sqrt{6}}{2}i$ **33.** $4(\cos 42° + i\sin 42°)$
35. $11.25(\cos 56° + i\sin 56°)$ **37.** 4 **39.** $-i$
41. $6 + 6\sqrt{3}i$ **43.** $-2i$ **45.** $8(\cos\pi + i\sin\pi)$
47. $8\left(\cos\frac{3\pi}{2} + i\sin\frac{3\pi}{2}\right)$ **49.** $\frac{27}{2} + \frac{27\sqrt{3}}{2}i$
51. $-4 + 4i$ **53.** -1 **55.** $-\frac{\sqrt{2}}{2} + \frac{\sqrt{2}}{2}i, \frac{\sqrt{2}}{2} - \frac{\sqrt{2}}{2}i$
57. $2(\cos 157.5° + i\sin 157.5°), 2(\cos 337.5° + i\sin 337.5°)$
59. $\frac{\sqrt{3}}{2} + \frac{1}{2}i, -\frac{\sqrt{3}}{2} + \frac{1}{2}i, -i$
61. $\sqrt[3]{4}(\cos 110° + i\sin 110°), \sqrt[3]{4}(\cos 230° + i\sin 230°), \sqrt[3]{4}(\cos 350° + i\sin 350°)$
63. $2, 2i, -2, -2i$;

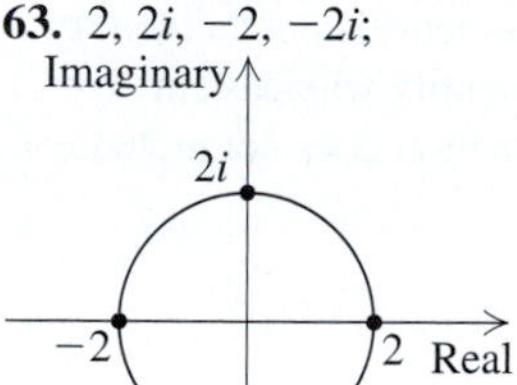

65. $\cos 36° + i\sin 36°$, $\cos 108° + i\sin 108°, -1$, $\cos 252° + i\sin 252°$, $\cos 324° + i\sin 324°$;

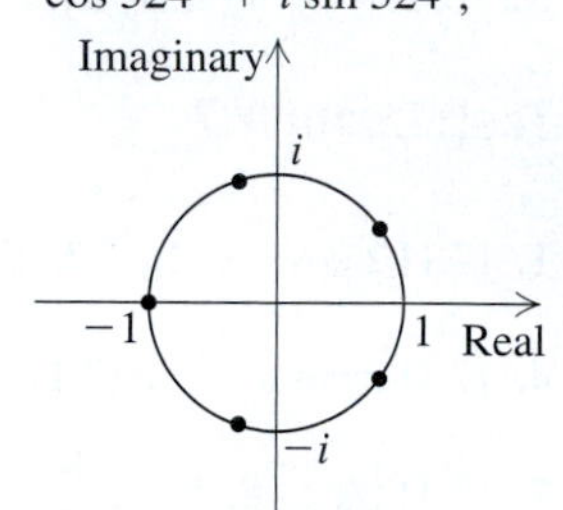

67. $\sqrt[10]{8}, \sqrt[10]{8}(\cos 36° + i\sin 36°)$, $\sqrt[10]{8}(\cos 72° + i\sin 72°), \sqrt[10]{8}(\cos 108° + i\sin 108°)$, $\sqrt[10]{8}(\cos 144° + i\sin 144°), -\sqrt[10]{8}$, $\sqrt[10]{8}(\cos 216° + i\sin 216°), \sqrt[10]{8}(\cos 252° + i\sin 252°)$, $\sqrt[10]{8}(\cos 288° + i\sin 288°), \sqrt[10]{8}(\cos 324° + i\sin 324°)$
69. $\frac{\sqrt{3}}{2} + \frac{1}{2}i, i, -\frac{\sqrt{3}}{2} + \frac{1}{2}i, -\frac{\sqrt{3}}{2} - \frac{1}{2}i, -i, \frac{\sqrt{3}}{2} - \frac{1}{2}i$
71. $1, -\frac{1}{2} + \frac{\sqrt{3}}{2}i, -\frac{1}{2} - \frac{\sqrt{3}}{2}i$
73. $\cos 67.5° + i\sin 67.5°, \cos 157.5° + i\sin 157.5°$, $\cos 247.5° + i\sin 247.5°, \cos 337.5° + i\sin 337.5°$
75. $\sqrt{3} + i, 2i, -\sqrt{3} + i, -\sqrt{3} - i, -2i, \sqrt{3} - i$
77. 15° **78.** 540° **79.** $\frac{11\pi}{6}$ **80.** $-\frac{5\pi}{4}$ **81.** $\frac{\sqrt{3}}{2}$
82. $\frac{\sqrt{3}}{2}$ **83.** $\frac{\sqrt{2}}{2}$ **84.** $\frac{1}{2}$ **85.** $\cos\theta - i\sin\theta$
87. $z = a + bi, |z| = \sqrt{a^2 + b^2}; \bar{z} = a - bi$, $|\bar{z}| = \sqrt{a^2 + (-b)^2} = \sqrt{a^2 + b^2}, \therefore |z| = |\bar{z}|$
89.
$$\begin{aligned}|(a+bi)^2| &= |a^2 - b^2 + 2abi| \\ &= \sqrt{(a^2 - b^2)^2 + 4a^2b^2} \\ &= \sqrt{a^4 + 2a^2b^2 + b^4} = a^2 + b^2, \\ |a + bi|^2 &= (\sqrt{a^2 + b^2})^2 = a^2 + b^2\end{aligned}$$

91. 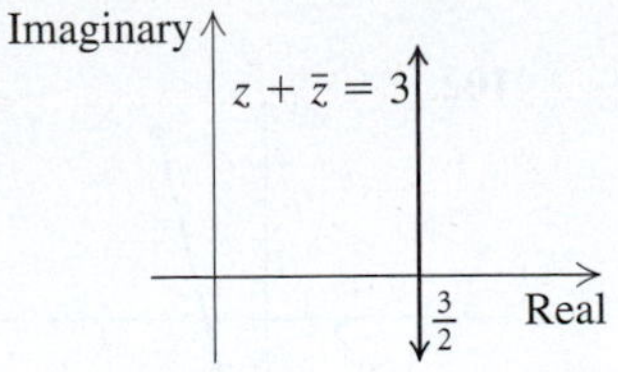

Mid-Chapter Mixed Review: Chapter 8

1. True **2.** True **3.** False **4.** True **5.** $B = 63°$, $b \approx 9.4$ in., $c \approx 9.5$ in. **6.** No solution **7.** $A \approx 40.5°$, $B \approx 28.5°$, $C \approx 111.0°$ **8.** $B \approx 70.2°$, $C \approx 67.1°$, $c \approx 39.9$ cm or $B \approx 109.8°$, $C \approx 27.5°$, $c \approx 20.0$ cm
9. $a \approx 370$ yd, $B \approx 16.6°$, $C \approx 15.4°$ **10.** $C \approx 45°$, $A \approx 107°$, $a \approx 37$ ft, or $C \approx 135°$, $A \approx 17°$, $a \approx 11$ ft
11. About 446 in^2

12. 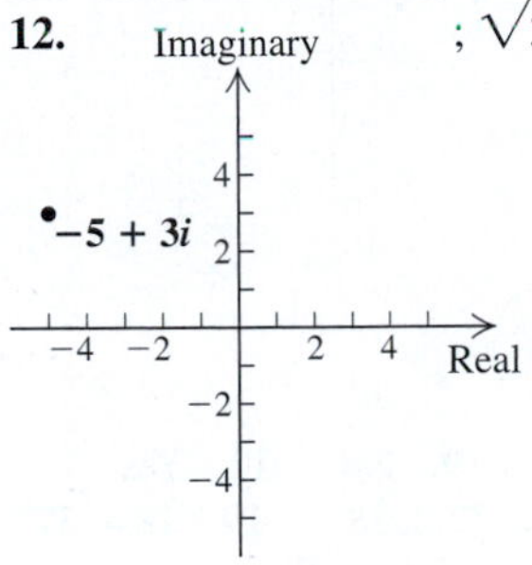

; $\sqrt{34}$

13. 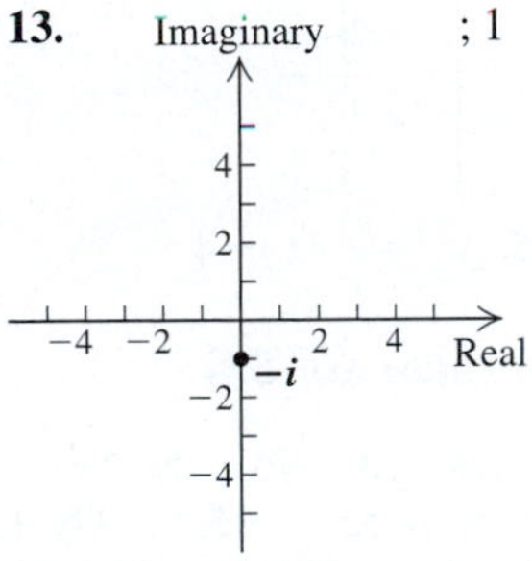

; 1

14. 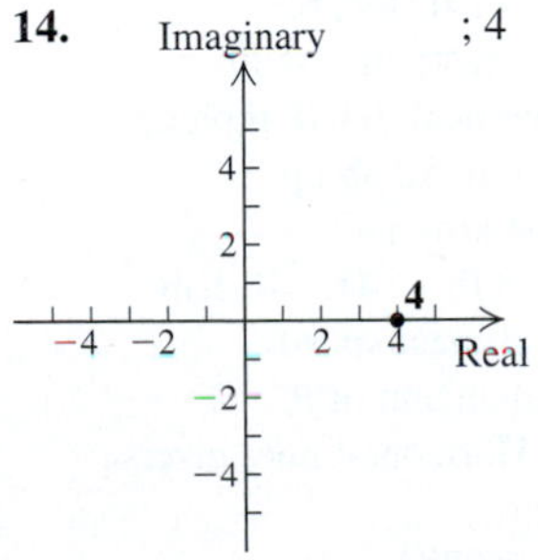

; 4

15. 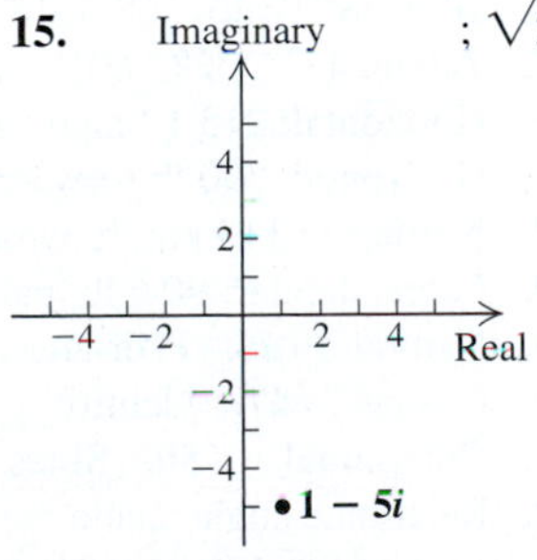

; $\sqrt{26}$

16. $\sqrt{2}\left(\cos\frac{\pi}{3} + i\sin\frac{\pi}{3}\right)$, or $\sqrt{2}(\cos 60° + i\sin 60°)$
17. $2\left(\cos\frac{5\pi}{3} + i\sin\frac{5\pi}{3}\right)$, or $2(\cos 300° + i\sin 300°)$
18. $5\left(\cos\frac{\pi}{2} + i\sin\frac{\pi}{2}\right)$, or $5(\cos 90° + i\sin 90°)$
19. $2\sqrt{2}\left(\cos\frac{5\pi}{4} + i\sin\frac{5\pi}{4}\right)$, or $2\sqrt{2}(\cos 225° + i\sin 225°)$ **20.** $\sqrt{2} - \sqrt{2}i$
21. $6\sqrt{3} + 6i$ **22.** $\sqrt{5}$ **23.** $4i$
24. $16(\cos 45° + i\sin 45°)$
25. $9\left[\cos\left(\frac{\pi}{12}\right) + i\sin\left(\frac{\pi}{12}\right)\right]$
26. $2\sqrt{2}(\cos 285° + i\sin 285°)$
27. $\sqrt{2}(\cos 255° + i\sin 255°)$
28. $8\sqrt{2}(\cos 45° + i\sin 45°)$ **29.** $8 + 8\sqrt{3}i$
30. $2(\cos 120° + i\sin 120°)$ and $2(\cos 300° + i\sin 300°)$, or $-1 + \sqrt{3}i$ and $1 - \sqrt{3}i$ **31.** $1(\cos 60° + i\sin 60°)$, $1(\cos 180° + i\sin 180°)$, and $1(\cos 300° + i\sin 300°)$, or $\frac{1}{2} + \frac{\sqrt{3}}{2}i$, -1, and $\frac{1}{2} - \frac{\sqrt{3}}{2}i$ **32.** Using the law of cosines, it is necessary to solve the quadratic equation:

$$(11.1)^2 = a^2 + (28.5)^2 - 2a(28.5)\cos 19°, \quad \text{or}$$
$$0 = a^2 - [2(28.5)\cos 19°]a + [(28.5)^2 - (11.1)^2].$$

The law of sines requires less complicated computations.
33. A nonzero complex number has n different complex nth roots. Thus, 1 has three different complex cube roots, one of which is the real number 1. The other two roots are complex conjugates. Since the set of real numbers is a subset of the set of complex numbers, the real cube root of 1 is also a complex root of 1. **34.** The law of sines involves two angles of a triangle and the sides opposite them. Three of these four values must be known in order to find the fourth. Given SAS, only two of these four values are known. **35.** The law of sines involves two angles of a triangle and the sides opposite them. Three of these four values must be known in order to find the fourth. Thus we must know the measure of one angle in order to use the law of sines. **36.** Trigonometric notation is not unique because there are infinitely many angles coterminal with a given angle. Standard notation is unique because any point has a unique ordered pair (a, b) associated with it.
37.
$$\begin{aligned} x^6 - 2x^3 + 1 &= 0 \\ (x^3 - 1)^2 &= 0 \\ x^3 - 1 &= 0 \\ x^3 &= 1 \\ x &= 1^{1/3} \end{aligned}$$

This equation has three distinct solutions because there are three distinct cube roots of 1.

$$\begin{aligned} x^6 - 2x^3 &= 0 \\ x^3(x^3 - 2) &= 0 \\ x^3 = 0 \quad &or \quad x^3 - 2 = 0 \\ x^3 = 0 \quad &or \quad x^3 = 2 \\ x = 0 \quad &or \quad x = 2^{1/3} \end{aligned}$$

This equation has four distinct solutions because 0 is one solution and the three distinct cube roots of 2 provide an additional three solutions.

$$\begin{aligned} x^6 - 2x &= 0 \\ x(x^5 - 2) &= 0 \\ x = 0 \quad &or \quad x^5 - 2 = 0 \\ x = 0 \quad &or \quad x^5 = 2 \\ x = 0 \quad &or \quad x = 2^{1/5} \end{aligned}$$

This equation has six distinct solutions because 0 is one solution and the five fifth roots of 2 provide an additional five solutions.

Visualizing the Graph

1. J **2.** C **3.** E **4.** H **5.** I **6.** A
7. D **8.** G **9.** B **10.** F

Exercise Set 8.4

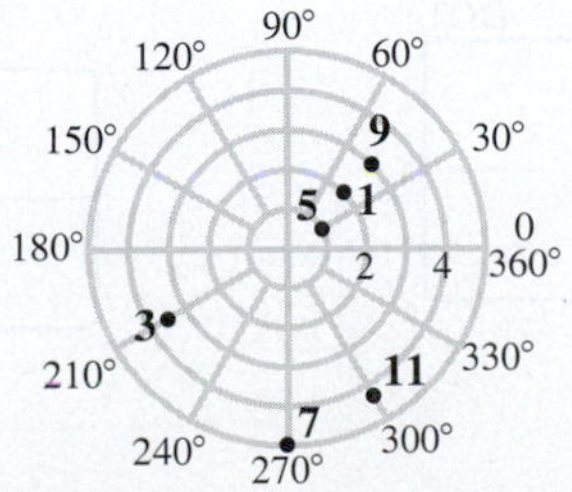

13. A: $(4, 30°)$, $(4, 390°)$, $(-4, 210°)$; B: $(5, 300°)$, $(5, -60°)$, $(-5, 120°)$; C: $(2, 150°)$, $(2, 510°)$, $(-2, 330°)$; D: $(3, 225°)$, $(3, -135°)$, $(-3, 45°)$; answers may vary

15. $(3, 270°), \left(3, \frac{3\pi}{2}\right)$ **17.** $(6, 300°), \left(6, \frac{5\pi}{3}\right)$
19. $(8, 330°), \left(8, \frac{11\pi}{6}\right)$ **21.** $(2, 225°), \left(2, \frac{5\pi}{4}\right)$
23. $(2, 60°), \left(2, \frac{\pi}{3}\right)$ **25.** $(5, 315°), \left(5, \frac{7\pi}{4}\right)$
27. $(7.616, 66.8°), (7.616, 1.166)$
29. $(4.643, 132.9°), (4.643, 2.320)$
31. $\left(\frac{5}{2}, \frac{5\sqrt{3}}{2}\right)$ **33.** $\left(-\frac{3\sqrt{2}}{2}, -\frac{3\sqrt{2}}{2}\right)$
35. $\left(-\frac{3}{2}, -\frac{3\sqrt{3}}{2}\right)$ **37.** $(-1, \sqrt{3})$ **39.** $(-\sqrt{3}, -1)$
41. $(3\sqrt{3}, -3)$ **43.** $(2.19, -2.05)$ **45.** $(1.30, -3.99)$
47. $r(3\cos\theta + 4\sin\theta) = 5$ **49.** $r\cos\theta = 5$
51. $r = 6$ **53.** $r^2\cos^2\theta = 25r\sin\theta$
55. $r^2\sin^2\theta - 5r\cos\theta - 25 = 0$ **57.** $r^2 = 2r\cos\theta$
59. $x^2 + y^2 = 25$ **61.** $y = 2$ **63.** $y^2 = -6x + 9$
65. $x^2 - 9x + y^2 - 7y = 0$ **67.** $x = 5$ **69.** $y = -\sqrt{3}x$

71.

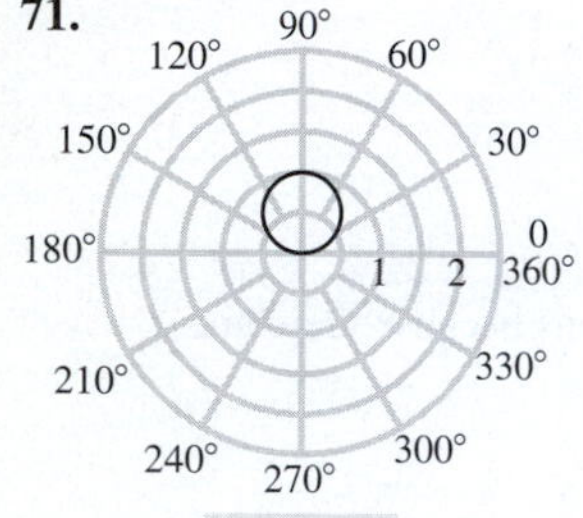

$r = \sin\theta$

73.

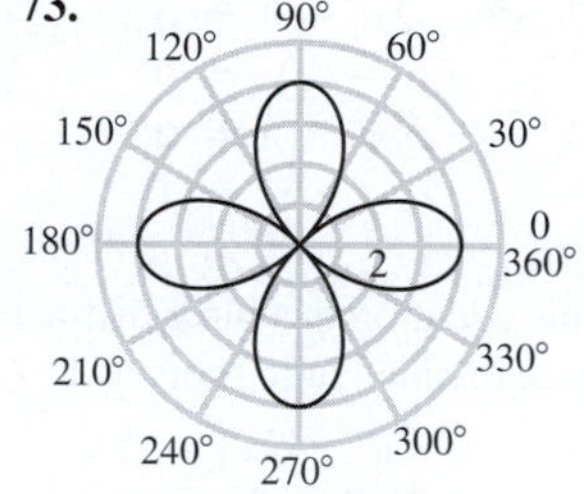

$r = 4\cos 2\theta$

75.

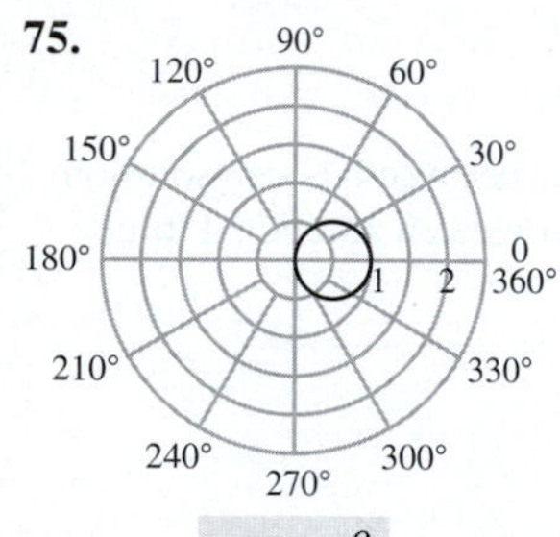

$r = \cos\theta$

77.

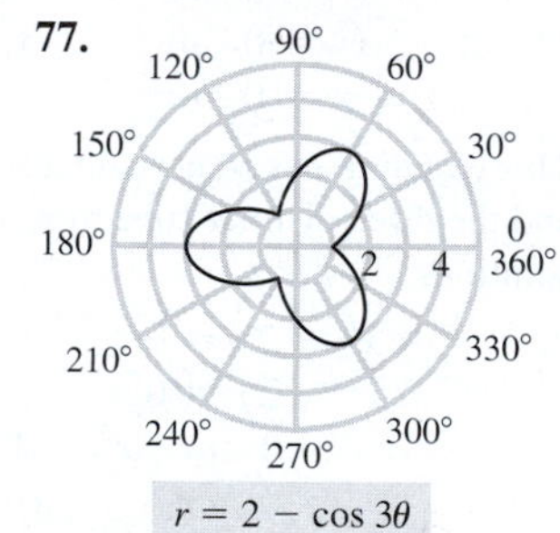

$r = 2 - \cos 3\theta$

79. (d) **81.** (g) **83.** (j) **85.** (b) **87.** (e) **89.** (k)

91. $r = \sin\theta\tan\theta$

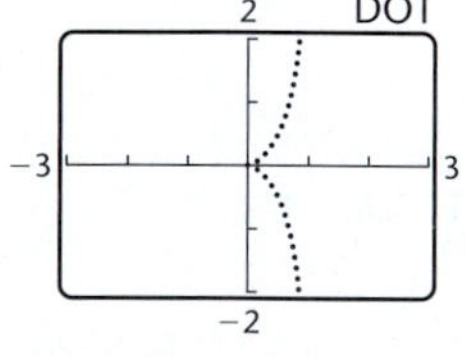

93. $r = e^{\theta/10}$

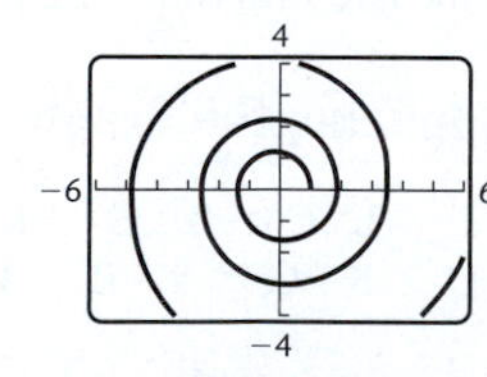

95. $r = \cos 2\theta\sec\theta$

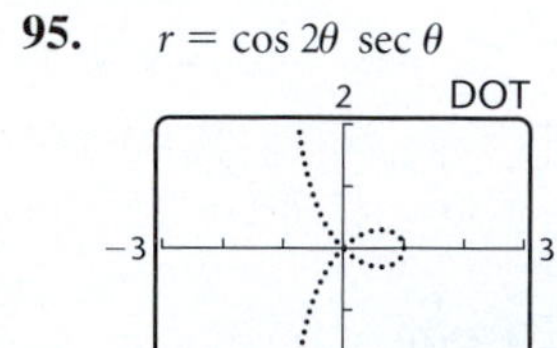

97. $r = \frac{1}{4}\tan^2\theta\sec\theta$

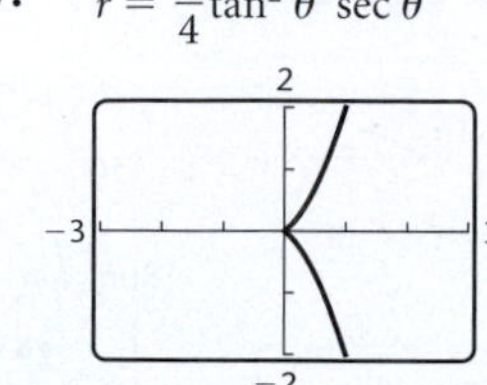

99. 12 **100.** $\frac{1}{5}$

101.

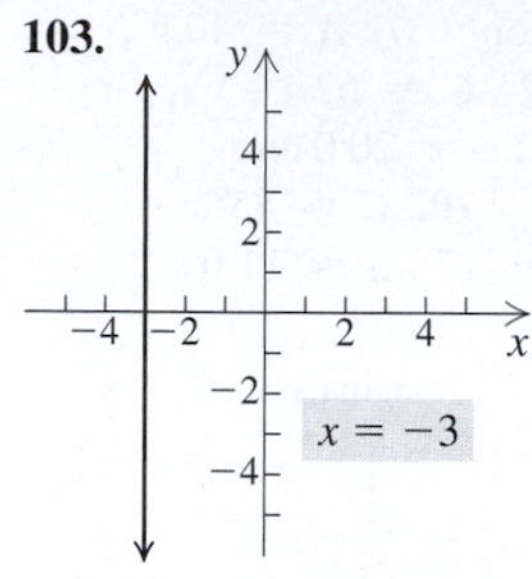

102.

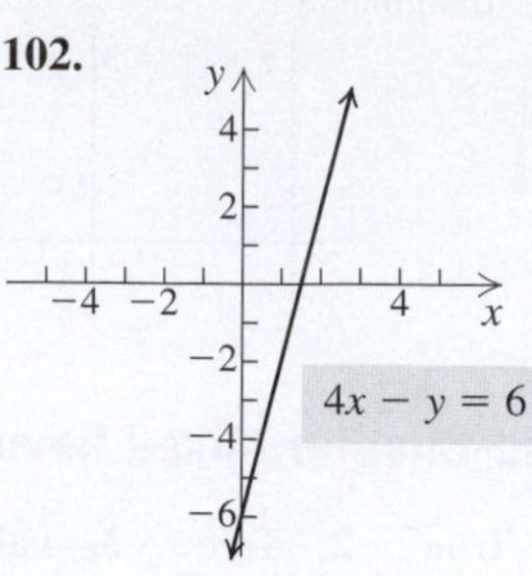

103.

$x = -3$

104.

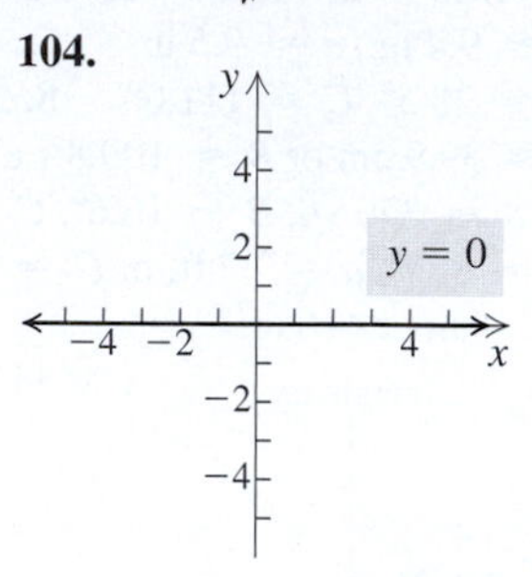

105. $y^2 = -4x + 4$

Exercise Set 8.5

1. Yes **3.** No **5.** Yes **7.** No **9.** No **11.** Yes
13. 55 N, 55° **15.** 929 N, 19° **17.** 57.0, 38° **19.** 18.4, 37°
21. 20.9, 58° **23.** 68.3, 18° **25.** 11 ft/sec, 63°
27. 726 lb, 47° **29.** 60° **31.** 70.7 east; 70.7 south
33. Horizontal: 215.17 mph forward; vertical: 65.78 mph up
35. Horizontal: 390 lb forward; vertical: 675.5 lb up
37. Northerly: 115 km/h; westerly: 164 km/h
39. Perpendicular: 90.6 lb; parallel: 42.3 lb **41.** 48.1 lb
43. Natural **44.** Half-angle **45.** Linear speed
46. Cosine **47.** Identity **48.** Cotangent of θ
49. Coterminal **50.** Sines **51.** Horizontal line; inverse
52. Reference angle; acute
53. **(a)** $(4.950, 4.950)$; **(b)** $(0.950, -1.978)$

Exercise Set 8.6

1. $\langle -9, 5\rangle; \sqrt{106}$ **3.** $\langle -3, 6\rangle; 3\sqrt{5}$ **5.** $\langle 4, 0\rangle; 4$
7. $\sqrt{37}$ **9.** $\langle 4, -5\rangle$ **11.** $\sqrt{257}$ **13.** $\langle -9, 9\rangle$
15. $\langle 41, -38\rangle$ **17.** $\sqrt{261} - \sqrt{65}$ **19.** $\langle -1, -1\rangle$
21. $\langle -8, 14\rangle$ **23.** 1 **25.** -34

27.

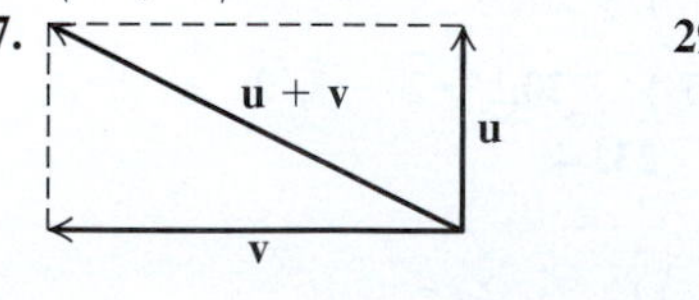

29.

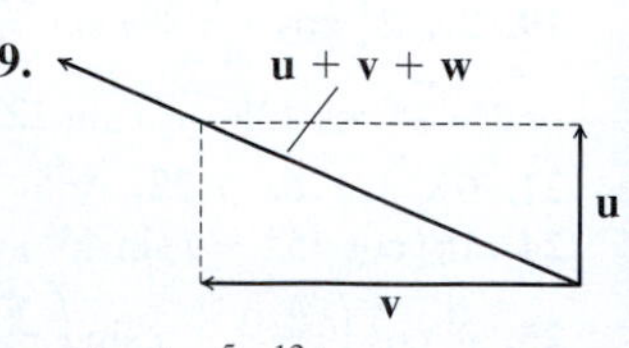

31. **(a)** $\mathbf{w} = \mathbf{u} + \mathbf{v}$; **(b)** $\mathbf{v} = \mathbf{w} - \mathbf{u}$ **33.** $\langle -\frac{5}{13}, \frac{12}{13}\rangle$
35. $\left\langle \frac{1}{\sqrt{101}}, -\frac{10}{\sqrt{101}}\right\rangle$ **37.** $\left\langle -\frac{1}{\sqrt{17}}, -\frac{4}{\sqrt{17}}\right\rangle$
39. $\mathbf{w} = -4\mathbf{i} + 6\mathbf{j}$ **41.** $\mathbf{s} = 2\mathbf{i} + 5\mathbf{j}$ **43.** $-7\mathbf{i} + 5\mathbf{j}$
45. **(a)** $3\mathbf{i} + 29\mathbf{j}$; **(b)** $\langle 3, 29\rangle$ **47.** **(a)** $4\mathbf{i} + 16\mathbf{j}$; **(b)** $\langle 4, 16\rangle$
49. $\mathbf{j}$, or $\langle 0, 1\rangle$ **51.** $-\frac{1}{2}\mathbf{i} - \frac{\sqrt{3}}{2}\mathbf{j}$, or $\left\langle -\frac{1}{2}, -\frac{\sqrt{3}}{2}\right\rangle$
53. 248° **55.** 63° **57.** 50° **59.** $|\mathbf{u}| = 3; \theta = 45°$
61. 1; 120° **63.** 144.2° **65.** 14.0° **67.** 101.3°

69.

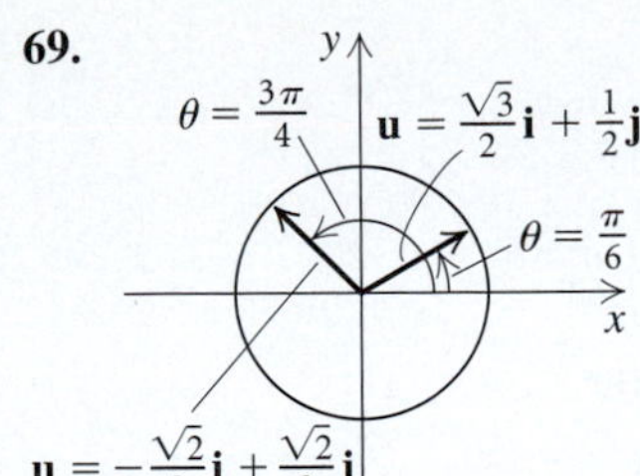

71.

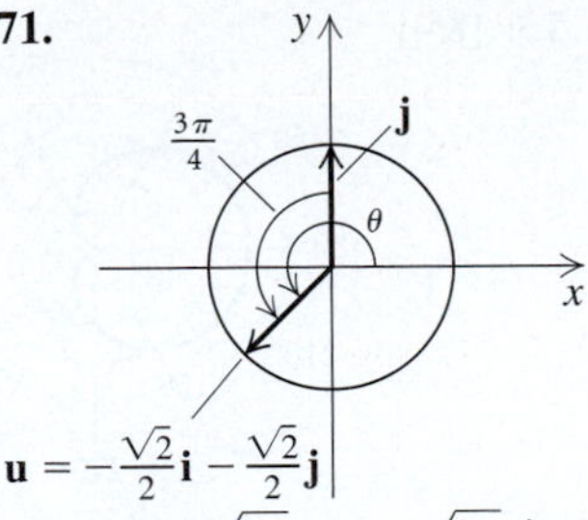

73. $\mathbf{u} = -\frac{\sqrt{10}}{10}\mathbf{i} + \frac{3\sqrt{10}}{10}\mathbf{j}$ **75.** $\sqrt{13}\left(\frac{2\sqrt{13}}{13}\mathbf{i} - \frac{3\sqrt{13}}{13}\mathbf{j}\right)$

77.

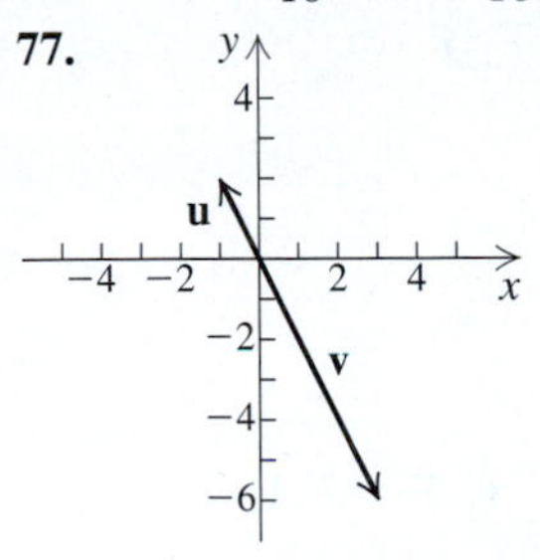

79. 174 nautical mi, S17°E **81.** 60° **83.** 500 lb on left, 866 lb on right **85.** Cable: 224-lb tension; boom: 167-lb compression **87.** 0, 4 **88.** $-\frac{11}{3}, \frac{5}{2}$

89. $\cos\theta = \frac{\mathbf{u}\cdot\mathbf{v}}{|\mathbf{u}||\mathbf{v}|} = \frac{0}{|\mathbf{u}||\mathbf{v}|}$, $\therefore \cos\theta = 0$ and $\theta = 90°$.

91. $\frac{3}{5}\mathbf{i} - \frac{4}{5}\mathbf{j}, -\frac{3}{5}\mathbf{i} + \frac{4}{5}\mathbf{j}$ **93.** (5, 8)

Review Exercises: Chapter 8

1. True **2.** False **3.** False **4.** False **5.** False **6.** True **7.** $A \approx 153°, B \approx 18°, C \approx 9°$ **8.** $A = 118°$, $a \approx 37$ in., $c \approx 24$ in. **9.** $B = 14°50'$, $a \approx 2523$ m, $c \approx 1827$ m **10.** No solution **11.** 33 m^2 **12.** 13.72 ft^2 **13.** 63 ft^2 **14.** 92°, 33°, 55° **15.** 419 ft **16.** About 650 km

17. $\sqrt{29}$;

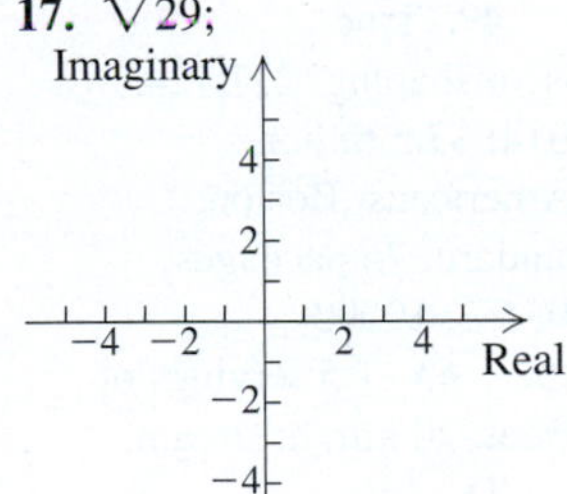

18. 4;

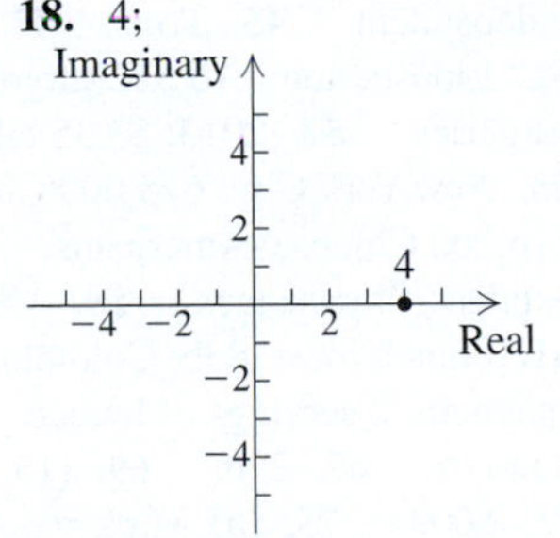

19. 2;

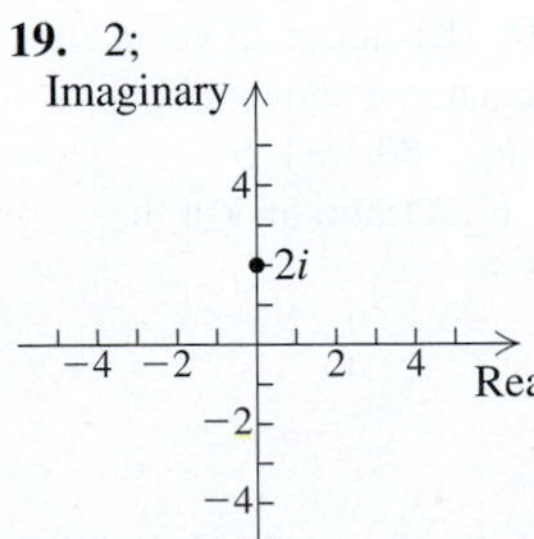

20. $\sqrt{10}$;

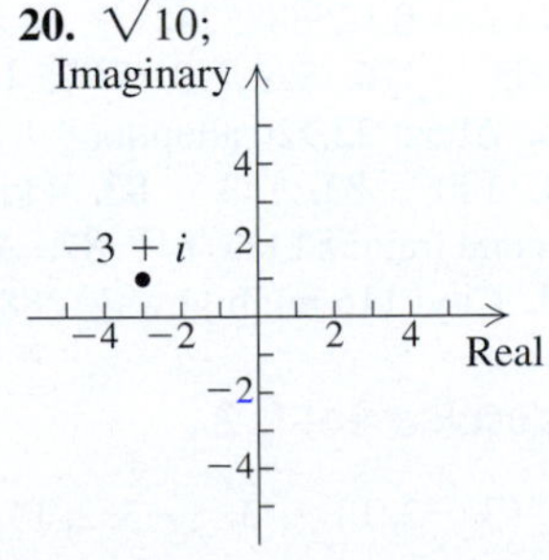

21. $\sqrt{2}\left(\cos\frac{\pi}{4} + i\sin\frac{\pi}{4}\right)$, or $\sqrt{2}(\cos 45° + i\sin 45°)$

22. $4\left(\cos\frac{3\pi}{2} + i\sin\frac{3\pi}{2}\right)$, or $4(\cos 270° + i\sin 270°)$

23. $10\left(\cos\frac{5\pi}{6} + i\sin\frac{5\pi}{6}\right)$, or $10(\cos 150° + i\sin 150°)$

24. $\frac{3}{4}(\cos 0 + i\sin 0)$, or $\frac{3}{4}(\cos 0° + i\sin 0°)$ **25.** $2 + 2\sqrt{3}i$

26. 7 **27.** $-\frac{5}{2} + \frac{5\sqrt{3}}{2}i$ **28.** $\sqrt{3} - i$

29. $1 + \sqrt{3} + (-1 + \sqrt{3})i$

30. $-i$ **31.** $2i$ **32.** $3\sqrt{3} + 3i$

33. $8(\cos 180° + i\sin 180°)$ **34.** $4(\cos 7\pi + i\sin 7\pi)$

35. $-8i$ **36.** $-\frac{1}{2} - \frac{\sqrt{3}}{2}i$

37. $\sqrt[4]{2}\left(\cos\frac{3\pi}{8} + i\sin\frac{3\pi}{8}\right), \sqrt[4]{2}\left(\cos\frac{11\pi}{8} + i\sin\frac{11\pi}{8}\right)$

38. $\sqrt[3]{6}(\cos 110° + i\sin 110°), \sqrt[3]{6}(\cos 230° + i\sin 230°), \sqrt[3]{6}(\cos 350° + i\sin 350°)$

39. $3, 3i, -3, -3i$ **40.** $1, \cos 72° + i\sin 72°, \cos 144° + i\sin 144°, \cos 216° + i\sin 216°, \cos 288° + i\sin 288°$

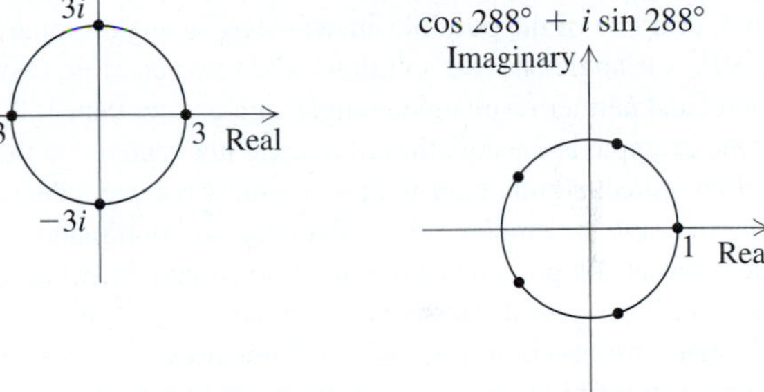

41. $\cos 22.5° + i\sin 22.5°, \cos 112.5° + i\sin 112.5°, \cos 202.5° + i\sin 202.5°, \cos 292.5° + i\sin 292.5°$

42. $\frac{1}{2} + \frac{\sqrt{3}}{2}i, -1, \frac{1}{2} - \frac{\sqrt{3}}{2}i$

43. A: (5, 120°), (5, 480°), (−5, 300°); B: (3, 210°), (−3, 30°), (−3, 390°); C: (4, 60°), (4, 420°), (−4, 240°); D: (1, 300°), (1, −60°), (−1, 120°); answers may vary **44.** (8, 135°), $\left(8, \frac{3\pi}{4}\right)$

45. (5, 270°), $\left(5, \frac{3\pi}{2}\right)$ **46.** (5.385, 111.8°), (5.385, 1.951)

47. (4.964, 147.8°), (4.964, 2.579) **48.** $\left(\frac{3\sqrt{2}}{2}, \frac{3\sqrt{2}}{2}\right)$

49. $(3, 3\sqrt{3})$ **50.** (1.93, −0.52) **51.** (−1.86, −1.35)

52. $r(5\cos\theta - 2\sin\theta) = 6$ **53.** $r\sin\theta = 3$ **54.** $r = 3$

55. $r^2\sin^2\theta - 4r\cos\theta - 16 = 0$ **56.** $x^2 + y^2 = 36$

57. $x^2 + 2y = 1$ **58.** $y^2 - 6x = 9$

59. $x^2 - 2x + y^2 - 3y = 0$ **60.** (b) **61.** (d)

62. (a) **63.** (c) **64.** 13.7, 71° **65.** 98.7, 15°

66.

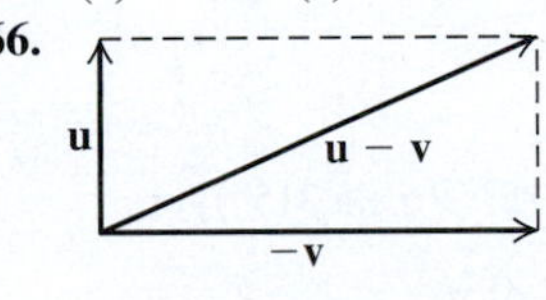

67.

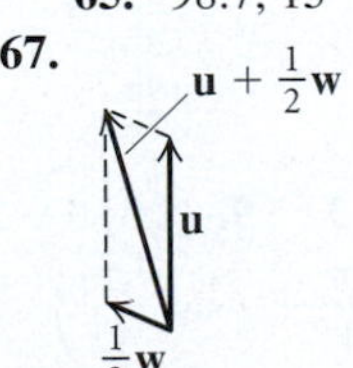

68. 666.7 N, 36° **69.** 29 km/h, 149° **70.** 102.4 nautical mi, S43°E **71.** ⟨−4, 3⟩ **72.** ⟨2, −6⟩ **73.** $\sqrt{61}$

74. ⟨10, −21⟩ **75.** ⟨14, −64⟩ **76.** $5 + \sqrt{116}$

77. 14 **78.** $\left\langle -\frac{3}{\sqrt{10}}, -\frac{1}{\sqrt{10}} \right\rangle$ **79.** $-9\mathbf{i} + 4\mathbf{j}$

80. 194.0° **81.** $\sqrt{34}$; $\theta = 211.0°$ **82.** 111.8° **83.** 85.1°

84. $34\mathbf{i} - 55\mathbf{j}$ **85.** $\mathbf{i} - 12\mathbf{j}$
86. $5\sqrt{2}$ **87.** $3\sqrt{65} + \sqrt{109}$ **88.** $-5\mathbf{i} + 5\mathbf{j}$
89. **90.**

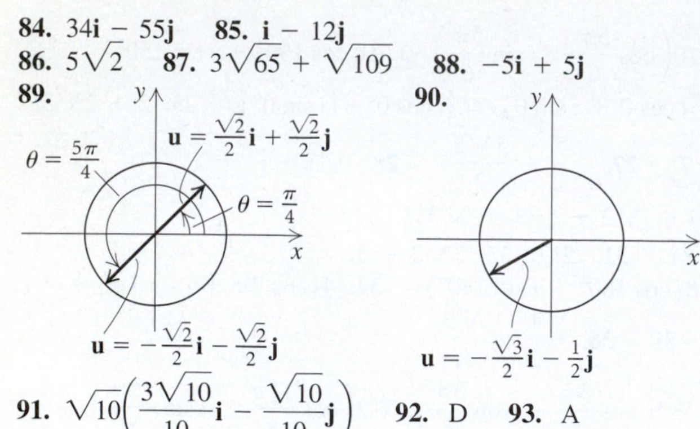

91. $\sqrt{10}\left(\dfrac{3\sqrt{10}}{10}\mathbf{i} - \dfrac{\sqrt{10}}{10}\mathbf{j}\right)$ **92.** D **93.** A
94. D **95.** $\frac{36}{13}\mathbf{i} + \frac{15}{13}\mathbf{j}$ **96.** 50.52°, 129.48°
97. A triangle has no solution when a sine value or a cosine value found is less than -1 or greater than 1. A triangle also has no solution if the sum of the angle measures calculated is greater than 180°. A triangle has only one solution if only one possible answer is found, or if one of the possible answers has an angle sum greater than 180°. A triangle has two solutions when two possible answers are found and neither results in an angle sum greater than 180°.
98. One example is the equation of a circle not centered at the origin. Often, in rectangular coordinates, we must complete the square in order to graph the circle. **99.** Rectangular coordinates are unique because any point has a unique ordered pair (x, y) associated with it. Polar coordinates are not unique because there are infinitely many angles coterminal with a given angle and also because r can be positive or negative depending on the angle used.
100. Vectors $\overrightarrow{QR}$ and $\overrightarrow{RQ}$ have opposite directions, so they are not equivalent. **101.** The terminal point of a unit vector in standard position is a point on the unit circle. **102.** Answers may vary. For $\mathbf{u} = 3\mathbf{i} - 4\mathbf{j}$ and $\mathbf{w} = 2\mathbf{i} - 4\mathbf{j}$, find $\mathbf{v}$, where $\mathbf{v} = \mathbf{u} + \mathbf{w}$.

Test: Chapter 8

1. [8.1] $A = 83°$, $b \approx 14.7$ ft, $c \approx 12.4$ ft
2. [8.1] $A \approx 73.9°$, $B \approx 70.1°$, $a \approx 8.2$ m, or $A \approx 34.1°$, $B \approx 109.9°$, $a \approx 4.8$ m **3.** [8.2] $A \approx 99.9°$, $B \approx 36.8°$, $C \approx 43.3°$ **4.** [8.1] About 43.6 cm^2 **5.** [8.1] About 77 m
6. [8.5] About 930 km
7. [8.3]

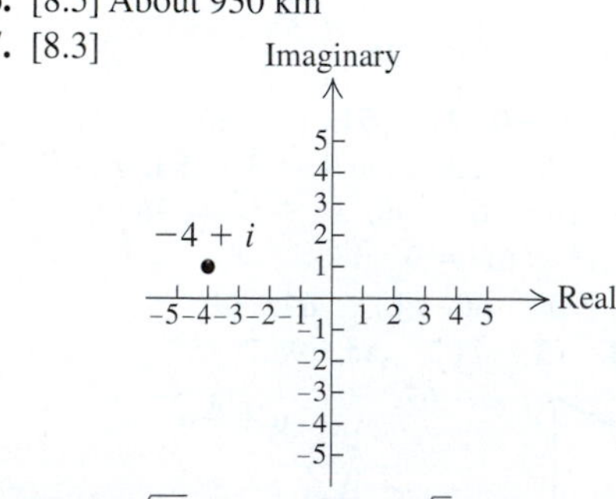

8. [8.3] $\sqrt{13}$ **9.** [8.3] $3\sqrt{2}(\cos 315° + i \sin 315°)$, or $3\sqrt{2}\left(\cos\dfrac{7\pi}{4} + i\sin\dfrac{7\pi}{4}\right)$ **10.** [8.3] $\frac{1}{4}i$
11. [8.3] 16 **12.** [8.4] $2(\cos 120° + i \sin 120°)$
13. [8.4] $\left(\dfrac{1}{2}, -\dfrac{\sqrt{3}}{2}\right)$ **14.** [8.4] $r = \sqrt{10}$
15. [8.4]

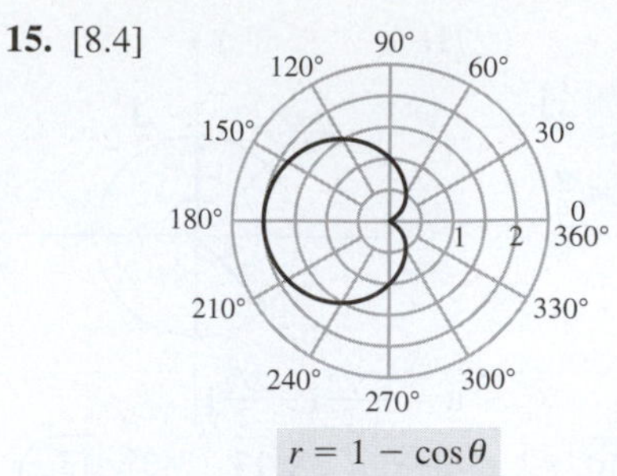

16. [8.5] Magnitude: 11.2; direction: 23.4°
17. [8.6] $-11\mathbf{i} - 17\mathbf{j}$ **18.** [8.6] $-\frac{4}{5}\mathbf{i} + \frac{3}{5}\mathbf{j}$
19. [8.4] A **20.** [8.1] 28.9°, 151.1°

Chapter 9

Visualizing the Graph

1. C **2.** G **3.** D **4.** J **5.** A **6.** F **7.** I
8. B **9.** H **10.** E

Exercise Set 9.1

1. (c) **3.** (f) **5.** (b) **7.** $(-1, 3)$ **9.** $(-1, 1)$
11. No solution **13.** $(-2, 4)$ **15.** Infinitely many solutions; $\left(x, \dfrac{x-1}{2}\right)$, or $(2y + 1, y)$ **17.** $(5, 4)$ **19.** $(1, -3)$
21. $(2, -2)$ **23.** No solution **25.** $\left(\frac{39}{11}, -\frac{1}{11}\right)$ **27.** $\left(\frac{1}{2}, \frac{3}{4}\right)$
29. Infinitely many solutions; $(x, 3x - 5)$ or $\left(\frac{1}{3}y + \frac{5}{3}, y\right)$
31. $(1, 3)$; consistent, independent **33.** $(-4, -2)$; consistent, independent **35.** Infinitely many solutions; $(4y + 2, y)$ or $\left(x, \frac{1}{4}x - \frac{1}{2}\right)$; consistent, dependent **37.** No solution; inconsistent, independent **39.** $(1, 1)$; consistent, independent
41. $(-3, 0)$; consistent, independent **43.** $(10, 8)$; consistent, independent **45.** True **47.** False **49.** True
51. Liposuction: 210,552 surgeries; nose reshaping: 217,124 surgeries **53.** 2013: \$3.35 billion; 2014: \$3.5 billion
55. New York City: 675,000 Chinese–Americans; Boston: 116,000 Chinese–Americans **57.** Standard: 76 packages; express: 44 packages **59.** 4%: \$6000; 5%: \$9000
61. French roast: 8 lb; Colombian: 12 lb **63.** 1.5 servings of spaghetti, 2 servings of lettuce **65.** Boat: 20 km/h; stream: 3 km/h **67.** 2 hr **69.** (15, \$200) **71.** 140
73. 6000 **75. (a)** $r(x) = -0.3846557377x + 47.52939344$; $p(x) = 0.1256885246x + 34.83178689$; **(b)** about 25 years after 2005 **76.** \$29,790 **77.** 115,017 registered snowmobiles
78. About 22,920 adoptions **79.** $-2, 6$ **80.** $-1, 5$
81. 120 **82.** 1, 3 **83.** 4 km **85.** First train: 36 km/h; second train: 54 km/h **87.** $A = \frac{1}{10}$, $B = -\frac{7}{10}$
89. City: 115 mi; highway: 288 mi

Exercise Set 9.2

1. $(3, -2, 1)$ **3.** $(-3, 2, 1)$ **5.** $\left(2, \frac{1}{2}, -2\right)$ **7.** No solution
9. Infinitely many solutions; $\left(\dfrac{11y + 19}{5}, y, \dfrac{9y + 11}{5}\right)$
11. $\left(\frac{1}{2}, \frac{2}{3}, -\frac{5}{6}\right)$ **13.** $(-1, 4, 3)$ **15.** $(1, -2, 4, -1)$

17. Russian Federation: 80 medals; Ukraine: 25 medals; United States: 18 medals **19.** United States: 361 million MT; China: 216 million MT; Brazil: 75 million MT **21.** Bacon: \$142.4 million; Warhol: \$105.4 million; Koons: \$58.4 million **23.** Dogs: 69.9 million; cats: 74.1 million; birds: 8.3 million **25.** United States: \$31.5 billion; Great Britain: \$17.9 billion; Germany: \$14.1 billion **27.** $1\frac{1}{4}$ servings of beef, 1 baked potato, $\frac{3}{4}$ serving of strawberries **29.** 2%: \$300; 3%: \$800; 4%: \$2400 **31.** Orange juice: \$2.40; bagel: \$2.75; coffee: \$3.00 **33.** **(a)** $f(x) = \frac{3}{40}x^2 - \frac{7}{4}x + 43$; **(b)** 2007: 33.625%; 2014: 40.8% **35.** **(a)** $f(x) = -7x^2 + 55x + 291$; **(b)** 333,000 deportations **37.** **(a)** $f(x) = -0.0792857143x^2 + 0.4801428571x + 2.599428571$; **(b)** \$3.02 **39.** Perpendicular **40.** The leading-term test **41.** A vertical line **42.** A one-to-one function **43.** A rational function **44.** Inverse variation **45.** A vertical asymptote **46.** A horizontal asymptote **47.** $\left(-\frac{1}{2}, -1, -\frac{1}{3}\right)$ **49.** $3x + 4y + 2z = 12$ **51.** $y = -4x^3 + 5x^2 - 3x + 1$

Exercise Set 9.3

1. 3×2 **3.** 1×4 **5.** 3×3 **7.** $\left[\begin{array}{rr|r} 2 & -1 & 7 \\ 1 & 4 & -5 \end{array}\right]$

9. $\left[\begin{array}{rrr|r} 1 & -2 & 3 & 12 \\ 2 & 0 & -4 & 8 \\ 0 & 3 & 1 & 7 \end{array}\right]$

11. $3x - 5y = 1,$
$x + 4y = -2$

13. $2x + y - 4z = 12,$
$3x + 5z = -1,$
$x - y + z = 2$

15. $\left(\frac{3}{2}, \frac{5}{2}\right)$ **17.** $\left(-\frac{63}{29}, -\frac{114}{29}\right)$ **19.** $\left(-1, \frac{5}{2}\right)$ **21.** $(0, 3)$ **23.** No solution **25.** Infinitely many solutions; $(3y - 2, y)$ **27.** $(-1, 2, -2)$ **29.** $\left(\frac{3}{2}, -4, 3\right)$ **31.** $(-1, 6, 3)$ **33.** Infinitely many solutions; $\left(\frac{1}{2}z + \frac{1}{2}, -\frac{1}{2}z - \frac{1}{2}, z\right)$ **35.** Infinitely many solutions; $(r - 2, -2r + 3, r)$ **37.** No solution **39.** $(1, -3, -2, -1)$ **41.** 8%: \$8000; 10%: \$12,000; 12%: \$10,000 **43.** 49¢: 160 stamps; 21¢: 40 stamps **45.** Exponential **46.** Linear **47.** Rational **48.** Quartic **49.** Logarithmic **50.** Cubic **51.** Linear **52.** Quadratic

53. $y = 3x^2 + \frac{5}{2}x - \frac{15}{2}$ **55.** $\begin{bmatrix} 1 & 5 \\ 0 & 1 \end{bmatrix}, \begin{bmatrix} 1 & 0 \\ 0 & 1 \end{bmatrix}$

57. $\left(-\frac{4}{3}, -\frac{1}{3}, 1\right)$ **59.** Infinitely many solutions; $\left(-\frac{14}{13}z - 1, \frac{3}{13}z - 2, z\right)$ **61.** $(-3, 3)$

Exercise Set 9.4

1. $x = -3, y = 5$ **3.** $x = -1, y = 1$

5. $\begin{bmatrix} -2 & 7 \\ 6 & 2 \end{bmatrix}$ **7.** $\begin{bmatrix} 1 & 3 \\ 2 & 6 \end{bmatrix}$ **9.** $\begin{bmatrix} 9 & 9 \\ -3 & -3 \end{bmatrix}$

11. $\begin{bmatrix} 11 & 13 \\ 5 & 3 \end{bmatrix}$ **13.** $\begin{bmatrix} -4 & 3 \\ -2 & -4 \end{bmatrix}$ **15.** $\begin{bmatrix} 17 & 9 \\ -2 & 1 \end{bmatrix}$

17. $\begin{bmatrix} 0 & 0 \\ 0 & 0 \end{bmatrix}$ **19.** $\begin{bmatrix} 1 & 2 \\ 4 & 3 \end{bmatrix}$ **21.** $\begin{bmatrix} 1 \\ 40 \end{bmatrix}$

23. $\begin{bmatrix} -10 & 28 \\ 14 & -26 \\ 0 & -6 \end{bmatrix}$ **25.** Not defined **27.** $\begin{bmatrix} 3 & 16 & 3 \\ 0 & -32 & 0 \\ -6 & 4 & 5 \end{bmatrix}$

29. **(a)** $\begin{bmatrix} 40 & 20 & 30 \end{bmatrix}$; **(b)** $\begin{bmatrix} 44 & 22 & 33 \end{bmatrix}$; **(c)** $\begin{bmatrix} 84 & 42 & 63 \end{bmatrix}$; the total amount of each type of produce ordered for both weeks

31. **(a)** $\mathbf{C} = \begin{bmatrix} 140 & 27 & 3 & 13 & 64 \end{bmatrix}$,
$\mathbf{P} = \begin{bmatrix} 180 & 4 & 11 & 24 & 662 \end{bmatrix}$,
$\mathbf{B} = \begin{bmatrix} 50 & 5 & 1 & 82 & 20 \end{bmatrix}$; **(b)** $\begin{bmatrix} 650 & 50 & 28 & 307 & 1448 \end{bmatrix}$, the total nutritional value of a meal of 1 3-oz serving of chicken, 1 cup of potato salad, and 3 broccoli spears

33. **(a)** $\begin{bmatrix} 1.50 & 0.30 & 0.36 & 0.45 & 0.64 \\ 1.55 & 0.28 & 0.48 & 0.57 & 0.75 \\ 1.62 & 0.52 & 0.65 & 0.38 & 0.53 \\ 1.70 & 0.43 & 0.40 & 0.42 & 0.68 \end{bmatrix}$;
(b) $\begin{bmatrix} 65 & 48 & 93 & 57 \end{bmatrix}$;
(c) $\begin{bmatrix} 419.46 & 105.81 & 129.69 & 115.89 & 165.65 \end{bmatrix}$;
(d) the total cost, in dollars, for each item for the day's meals

35. **(a)** $\begin{bmatrix} 8 & 15 \\ 6 & 10 \\ 4 & 3 \end{bmatrix}$; **(b)** $\begin{bmatrix} 4 & 2.50 & 3 \end{bmatrix}$; **(c)** $\begin{bmatrix} 59 & 94 \end{bmatrix}$;
(d) the total cost, in dollars, of ingredients for each coffee shop

37. **(a)** $\begin{bmatrix} 7.50 & 4.80 & 6.25 \end{bmatrix}$; **(b)** $\mathbf{PS} = \begin{bmatrix} 113.80 & 179.25 \end{bmatrix}$

39. $\begin{bmatrix} 2 & -3 \\ 1 & 5 \end{bmatrix}\begin{bmatrix} x \\ y \end{bmatrix} = \begin{bmatrix} 7 \\ -6 \end{bmatrix}$

41. $\begin{bmatrix} 1 & 1 & -2 \\ 3 & -1 & 1 \\ 2 & 5 & -3 \end{bmatrix}\begin{bmatrix} x \\ y \\ z \end{bmatrix} = \begin{bmatrix} 6 \\ 7 \\ 8 \end{bmatrix}$

43. $\begin{bmatrix} 3 & -2 & 4 \\ 2 & 1 & -5 \end{bmatrix}\begin{bmatrix} x \\ y \\ z \end{bmatrix} = \begin{bmatrix} 17 \\ 13 \end{bmatrix}$

45. $\begin{bmatrix} -4 & 1 & -1 & 2 \\ 1 & 2 & -1 & -1 \\ -1 & 1 & 4 & -3 \\ 2 & 3 & 5 & -7 \end{bmatrix}\begin{bmatrix} w \\ x \\ y \\ z \end{bmatrix} = \begin{bmatrix} 12 \\ 0 \\ 1 \\ 9 \end{bmatrix}$

47. **(a)** $\left(\frac{1}{2}, -\frac{25}{4}\right)$;
(b) $x = \frac{1}{2}$;
(c) minimum: $-\frac{25}{4}$;
(d)

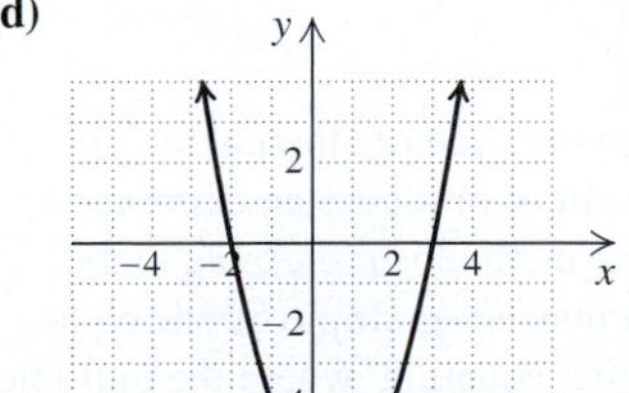

$f(x) = x^2 - x - 6$

48. **(a)** $\left(\frac{5}{4}, -\frac{49}{8}\right)$;
(b) $x = \frac{5}{4}$;
(c) minimum: $-\frac{49}{8}$;
(d)

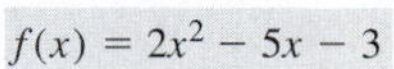

$f(x) = 2x^2 - 5x - 3$

49. **(a)** $\left(-\frac{3}{2}, \frac{17}{4}\right)$;
(b) $x = -\frac{3}{2}$;
(c) maximum: $\frac{17}{4}$;
(d)

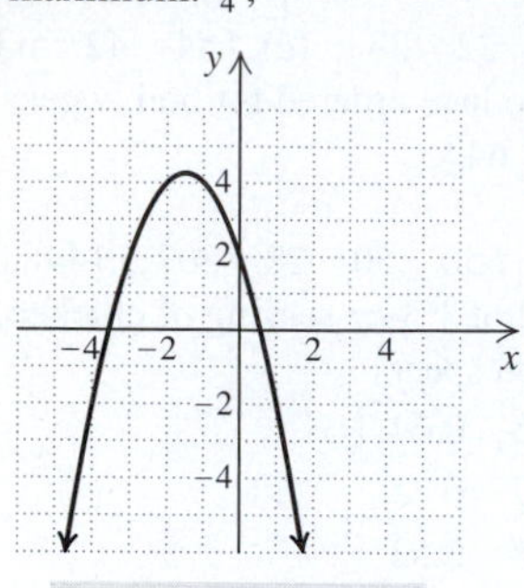

$f(x) = -x^2 - 3x + 2$

50. **(a)** $\left(\frac{2}{3}, \frac{16}{3}\right)$;
(b) $x = \frac{2}{3}$;
(c) maximum: $\frac{16}{3}$;
(d)

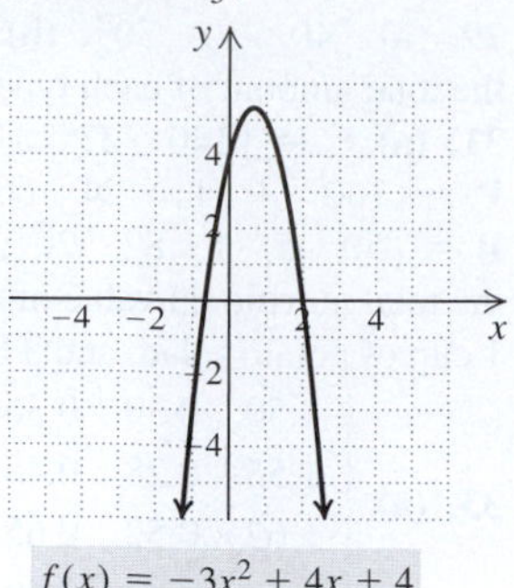

$f(x) = -3x^2 + 4x + 4$

51. $(\mathbf{A} + \mathbf{B})(\mathbf{A} - \mathbf{B}) = \begin{bmatrix} -2 & 1 \\ 2 & -1 \end{bmatrix}$; $\mathbf{A}^2 - \mathbf{B}^2 = \begin{bmatrix} 0 & 3 \\ 0 & -3 \end{bmatrix}$

53. $(\mathbf{A} + \mathbf{B})(\mathbf{A} - \mathbf{B}) = \begin{bmatrix} -2 & 1 \\ 2 & -1 \end{bmatrix}$
$= \mathbf{A}^2 + \mathbf{BA} - \mathbf{AB} - \mathbf{B}^2$

Mid-Chapter Mixed Review: Chapter 9

1. False **2.** True **3.** True **4.** False **5.** $(-3, 2)$
6. No solution **7.** $(1, -2)$ **8.** Infinitely many solutions; $\left(x, \frac{x-1}{3}\right)$ or $(3y + 1, y)$ **9.** $\left(\frac{1}{3}, -\frac{1}{6}, \frac{4}{3}\right)$
10. Under 10 lb: 60 packages; 10 lb up to 15 lb: 70 packages; 15 lb or more: 20 packages **11.** $(4, -3)$ **12.** $(-3, 2, -1)$

13. $\begin{bmatrix} 1 & 5 \\ 6 & 1 \end{bmatrix}$ **14.** $\begin{bmatrix} -5 & 7 \\ -4 & -7 \end{bmatrix}$ **15.** $\begin{bmatrix} -8 & 12 & 0 \\ 4 & -4 & 8 \\ -12 & 16 & 4 \end{bmatrix}$

16. $\begin{bmatrix} 0 & 16 \\ 13 & -1 \end{bmatrix}$ **17.** $\begin{bmatrix} -7 & 21 \\ -6 & 18 \end{bmatrix}$ **18.** $\begin{bmatrix} 24 & 26 \\ -12 & -13 \end{bmatrix}$

19. $\begin{bmatrix} 20 & 16 & -10 \\ -10 & -8 & 5 \end{bmatrix}$ **20.** Not defined

21. $\begin{bmatrix} 2 & -1 & 3 \\ 1 & 2 & -1 \\ 3 & -4 & 2 \end{bmatrix}\begin{bmatrix} x \\ y \\ z \end{bmatrix} = \begin{bmatrix} 7 \\ 3 \\ 5 \end{bmatrix}$

22. When a variable is not alone on one side of an equation or when solving for a variable is difficult or produces an expression containing fractions, the elimination method is preferable to the substitution method. **23.** Add a nonzero multiple of one equation to a nonzero multiple of the other equation, where the multiples are not opposites. **24.** See Example 4 in Section 9.3.
25. No; see Exercise 17 in Section 9.4, for example.

Exercise Set 9.5

1. Yes **3.** No **5.** $\begin{bmatrix} -3 & 2 \\ 5 & -3 \end{bmatrix}$ **7.** Does not exist

9. $\begin{bmatrix} \frac{3}{8} & -\frac{1}{4} & \frac{1}{8} \\ -\frac{1}{8} & \frac{3}{4} & -\frac{3}{8} \\ -\frac{1}{4} & \frac{1}{2} & \frac{1}{4} \end{bmatrix}$ **11.** Does not exist **13.** $\begin{bmatrix} 0.4 & -0.6 \\ 0.2 & -0.8 \end{bmatrix}$

15. $\begin{bmatrix} -1 & -1 & -6 \\ 1 & 0 & 2 \\ 0 & 1 & 3 \end{bmatrix}$ **17.** $\begin{bmatrix} 1 & 1 & 2 \\ 1 & 1 & 1 \\ 2 & 3 & 4 \end{bmatrix}$ **19.** Does not exist

21. $\begin{bmatrix} 1 & -2 & 3 & 8 \\ 0 & 1 & -3 & 1 \\ 0 & 0 & 1 & -2 \\ 0 & 0 & 0 & -1 \end{bmatrix}$

23. $\begin{bmatrix} 0.25 & 0.25 & 1.25 & -0.25 \\ 0.5 & 1.25 & 1.75 & -1 \\ -0.25 & -0.25 & -0.75 & 0.75 \\ 0.25 & 0.5 & 0.75 & -0.5 \end{bmatrix}$

25. $(-23, 83)$ **27.** $(-1, 5, 1)$ **29.** $(2, -2)$ **31.** $(0, 2)$
33. $(3, -3, -2)$ **35.** $(-1, 0, 1)$ **37.** $(1, -1, 0, 1)$
39. Wisconsin: 450 million lb; Massachusetts: 210 million lb
41. Topsoil: \$29 per cubic yard; mulch: \$28 per cubic yard; decorative rock: \$38 per cubic yard **43.** -48 **44.** 194
45. $\frac{-1 \pm \sqrt{57}}{4}$ **46.** $-3, -2$ **47.** 4 **48.** 9
49. $(x + 2)(x - 1)(x - 4)$
50. $(x + 5)(x + 1)(x - 1)(x - 3)$
51. $\mathbf{A}^{-1}$ exists if and only if $x \neq 0$. $\mathbf{A}^{-1} = \begin{bmatrix} \frac{1}{x} \end{bmatrix}$

53. $\mathbf{A}^{-1}$ exists if and only if $xyz \neq 0$. $\mathbf{A}^{-1} = \begin{bmatrix} 0 & 0 & \frac{1}{z} \\ 0 & \frac{1}{y} & 0 \\ \frac{1}{x} & 0 & 0 \end{bmatrix}$

Exercise Set 9.6

1. -14 **3.** -2 **5.** -11 **7.** $x^3 - 4x$ **9.** $M_{11} = 6$, $M_{32} = -9$, $M_{22} = -29$ **11.** $A_{11} = 6$, $A_{32} = 9$, $A_{22} = -29$
13. -10 **15.** -10 **17.** -10 **19.** $M_{41} = -14$, $M_{33} = 20$
21. $A_{24} = 15$, $A_{43} = 30$ **23.** 110 **25.** -109
27. $-x^4 + x^2 - 5x$ **29.** $\left(-\frac{25}{2}, -\frac{11}{2}\right)$ **31.** $(3, 1)$
33. $\left(\frac{1}{2}, -\frac{1}{3}\right)$ **35.** $(1, 1)$ **37.** $\left(\frac{3}{2}, \frac{13}{14}, \frac{33}{14}\right)$ **39.** $(3, -2, 1)$
41. $(1, 3, -2)$ **43.** $\left(\frac{1}{2}, \frac{2}{3}, -\frac{5}{6}\right)$ **45.** $f^{-1}(x) = \frac{x-2}{3}$
46. Not one-to-one **47.** Not one-to-one
48. $f^{-1}(x) = (x - 1)^3$ **49.** $5 - 3i$ **50.** $6 - 2i$
51. $10 - 10i$ **52.** $\frac{9}{25} + \frac{13}{25}i$ **53.** $3, -2$ **55.** 4
57. Answers may vary. $\begin{vmatrix} a & b \\ -b & a \end{vmatrix}$
59. Answers may vary. $\begin{vmatrix} 2\pi r & 2\pi r \\ -h & r \end{vmatrix}$

Exercise Set 9.7

1. (f) **3.** (h) **5.** (g) **7.** (b)

9.

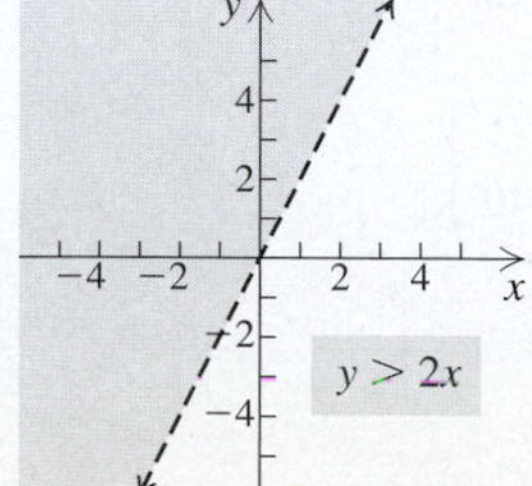

$y > 2x$

11.

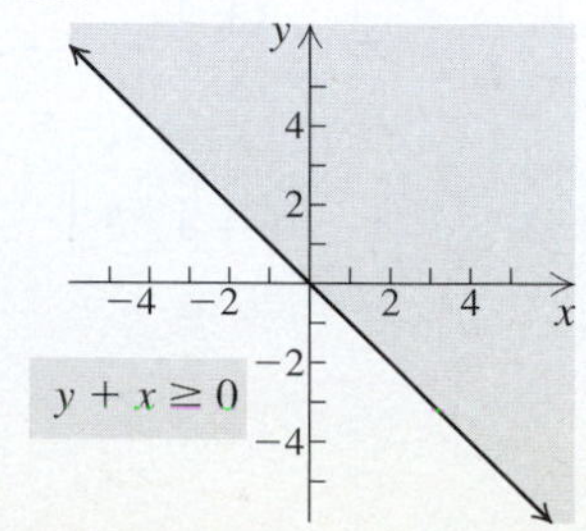

$y + x \geq 0$

13.

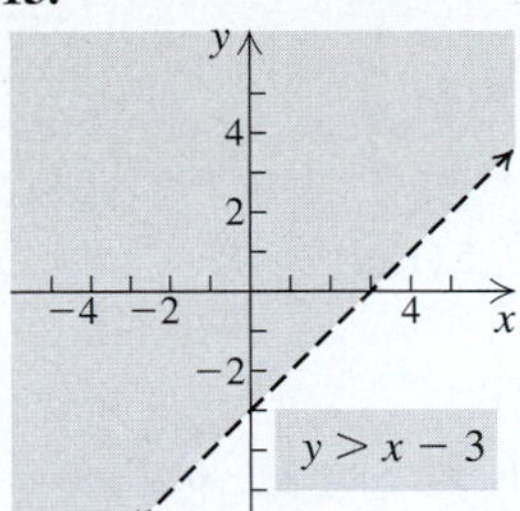

15.

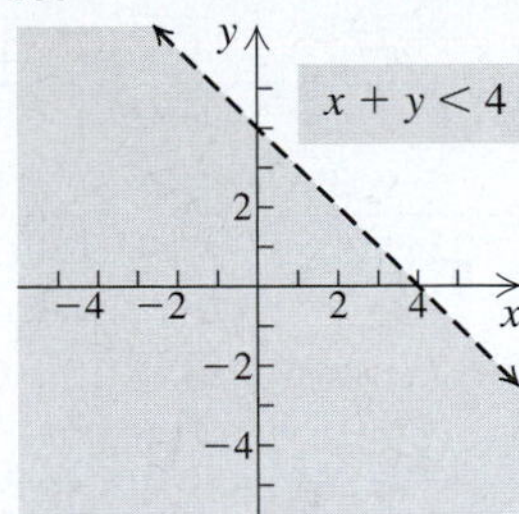

17.

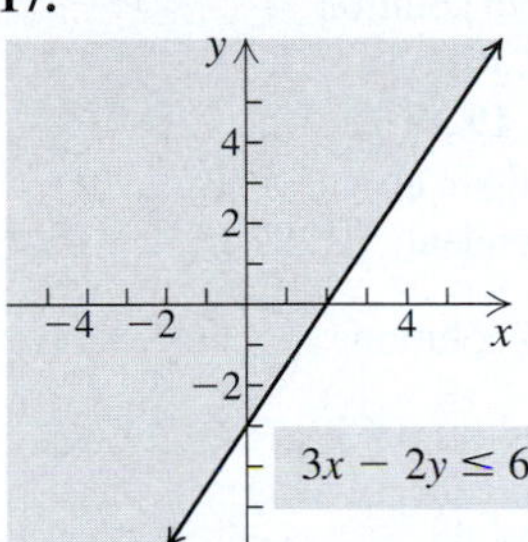

19.

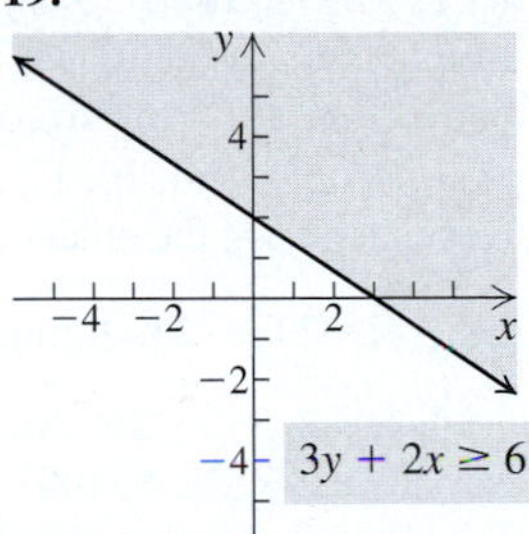

21.

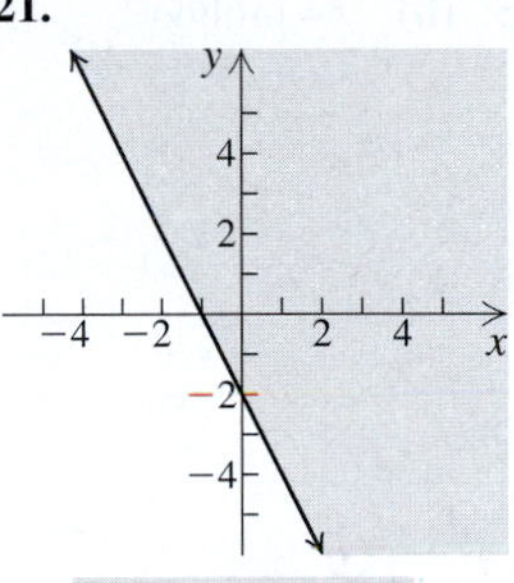

23.

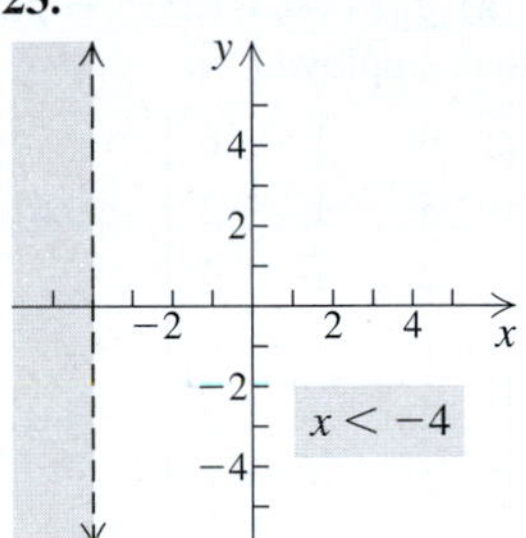

25.

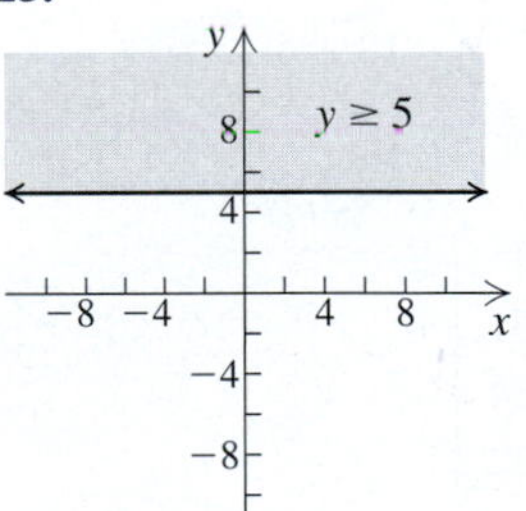

27.

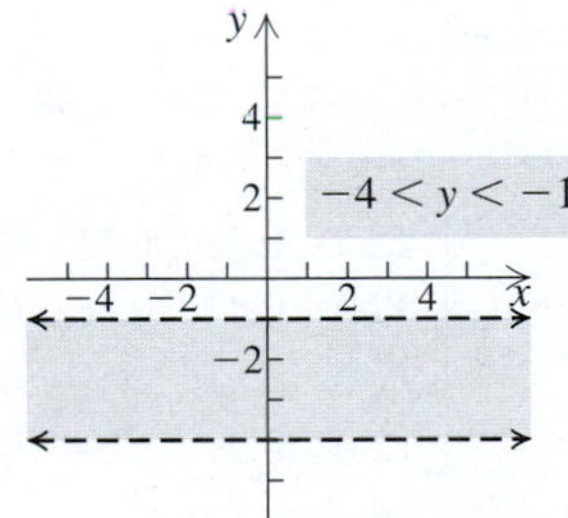

29.

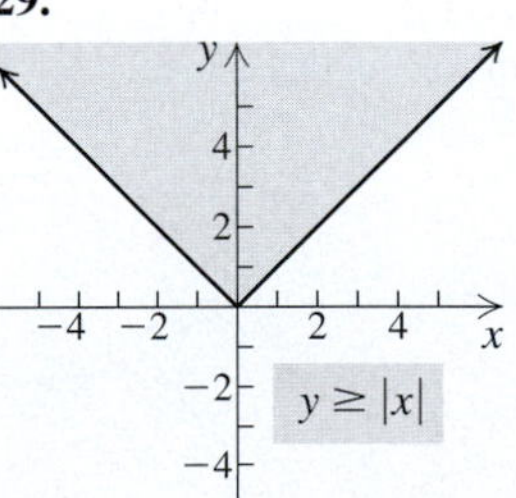

31. (f) **33.** (a) **35.** (b)

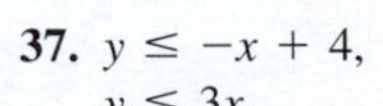

37. $y \leq -x + 4$, $y \leq 3x$ **39.** $x < 2$, $y > -1$ **41.** $y \leq -x + 3$, $y \leq x + 1$, $x \geq 0$, $y \geq 0$

43.

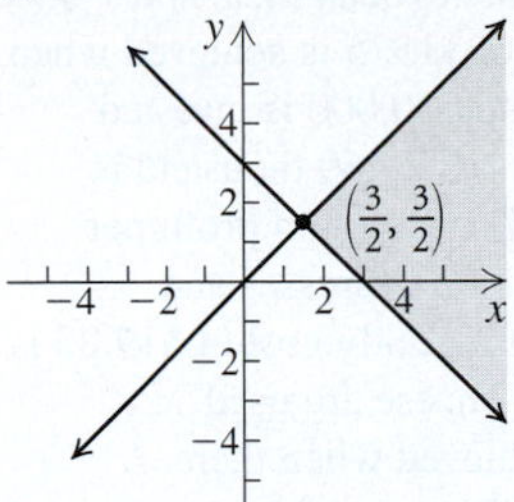

45.

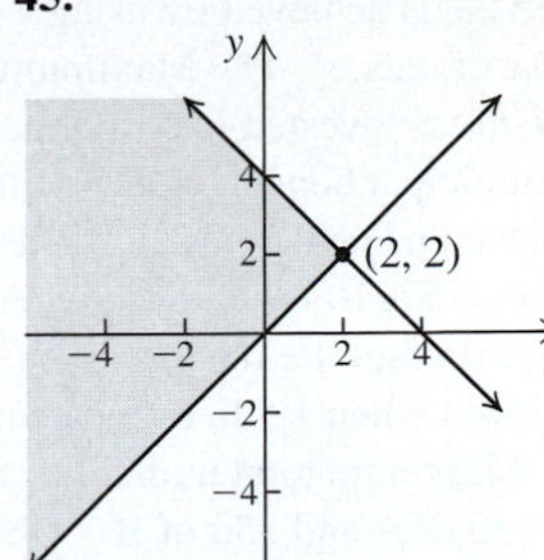

47.

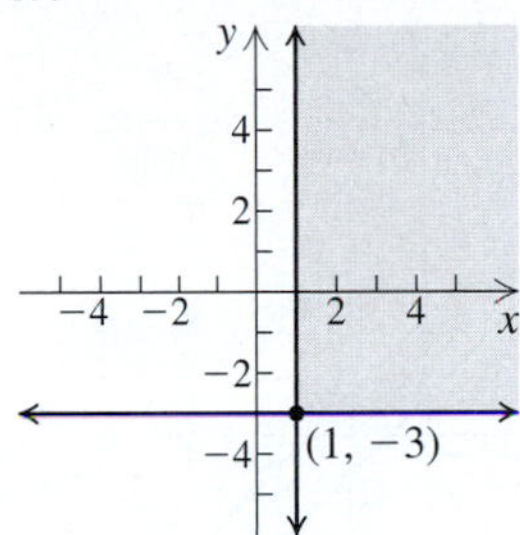

49.

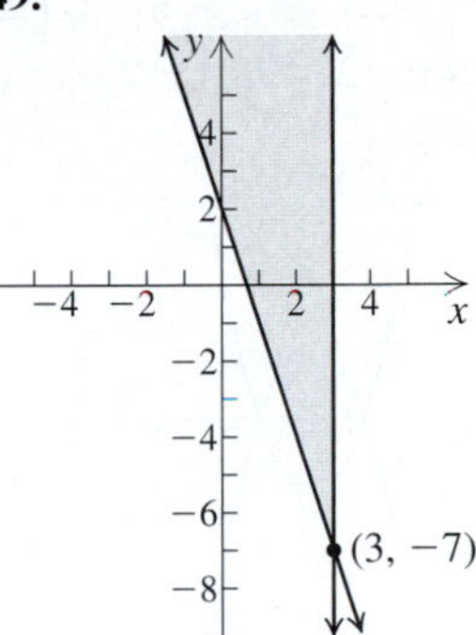

51.

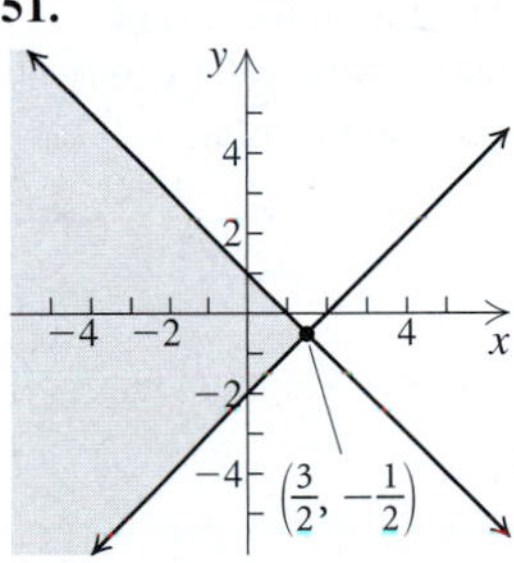

53.

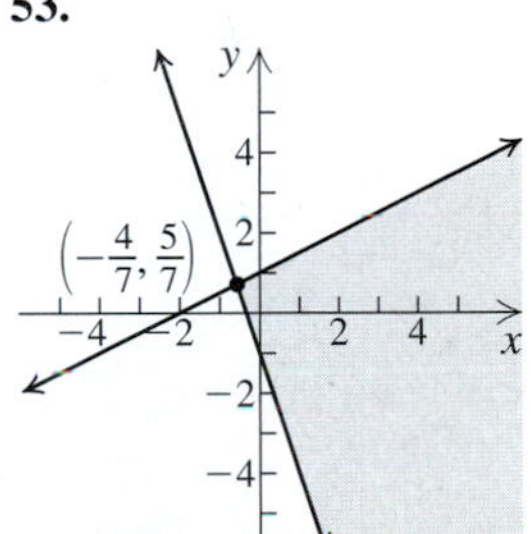

55.

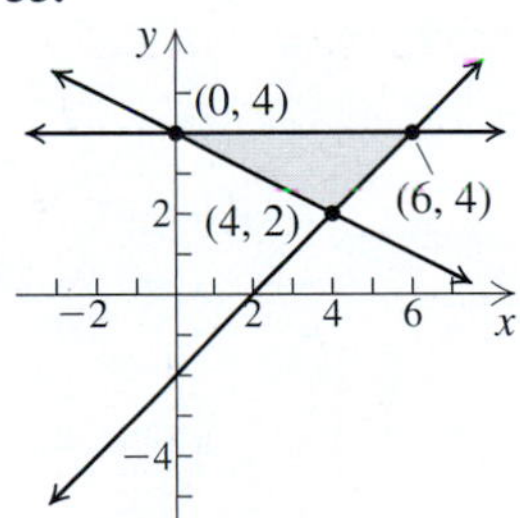

57.

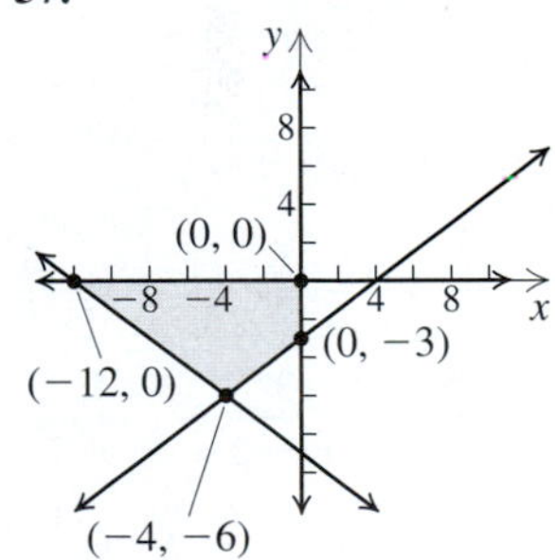

59.

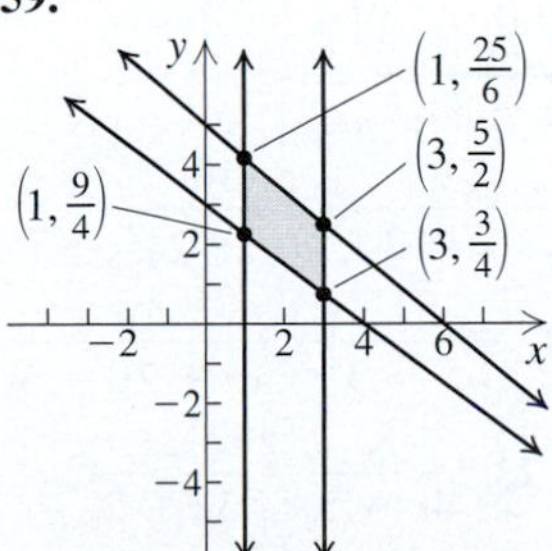

61. Maximum: 179 when $x = 7$ and $y = 0$; minimum: 48 when $x = 0$ and $y = 4$ **63.** Maximum: 216 when $x = 0$ and $y = 6$; minimum: 0 when $x = 0$ and $y = 0$ **65.** Maximum number of miles is 480 when the truck uses 9 gal and the moped uses 3 gal. **67.** Maximum profit of \$54,800 is achieved when 80 acres of corn and 160 acres of soybeans are planted. **69.** Minimum cost

of $51\frac{9}{13}$ is achieved by using $1\frac{11}{13}$ sacks of soybean meal and $1\frac{11}{13}$ sacks of oats. **71.** Maximum income of $1575 is achieved when $10,000 is invested in corporate bonds and $30,000 is invested in municipal bonds. **73.** Minimum cost of $460 thousand is achieved using 30 P_1's and 10 P_2's. **75.** Maximum profit per week of $2210 is achieved when 5 silk organza dresses and 2 lace dresses are made. **77.** Minimum weekly cost of $19.05 is achieved when 1.5 lb of meat and 3 lb of cheese are used. **79.** Maximum total number of 800 is achieved when there are 550 of A and 250 of B. **81.** $\{x \mid -7 \le x < 2\}$, or $[-7, 2)$ **82.** $\{x \mid x \le 1 \text{ or } x \ge 5\}$, or $(-\infty, 1] \cup [5, \infty)$
83. $\{x \mid -1 \le x \le 3\}$, or $[-1, 3]$
84. $\{x \mid -3 < x < -2\}$, or $(-3, -2)$

85.

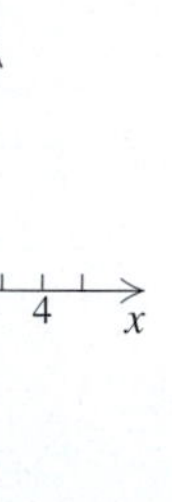

87.

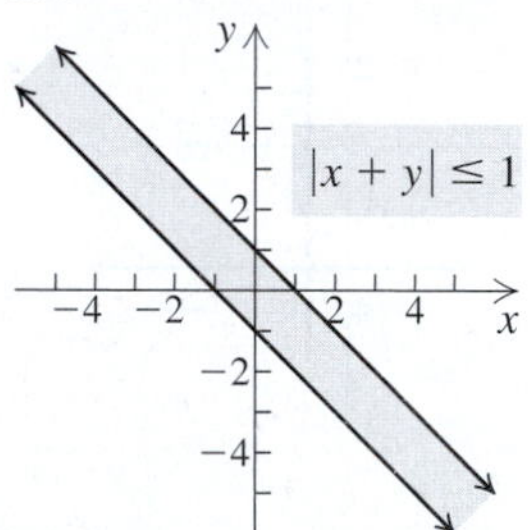

89.

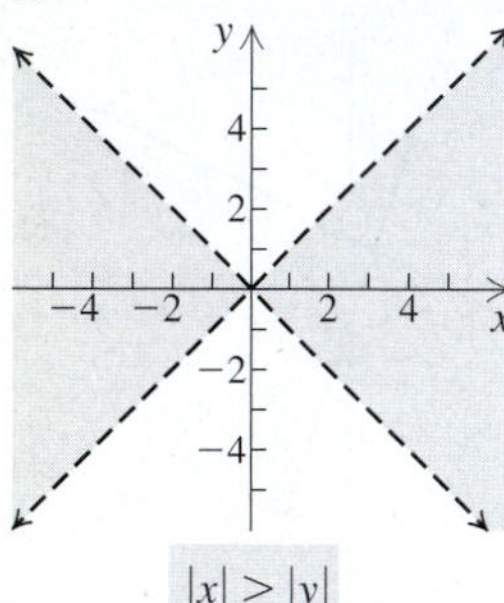

91. Maximum income of $19,000 is achieved by making 95 chairs and 0 sofas.

Exercise Set 9.8

1. $\frac{2}{x-3} - \frac{1}{x+2}$ **3.** $\frac{5}{2x-1} - \frac{4}{3x-1}$

5. $\frac{2}{x+2} - \frac{3}{x-2} + \frac{4}{x+1}$

7. $-\frac{3}{(x+2)^2} - \frac{1}{x+2} + \frac{1}{x-1}$ **9.** $\frac{3}{x-1} - \frac{4}{2x-1}$

11. $x - 2 + \frac{\frac{17}{16}}{x+1} - \frac{\frac{11}{4}}{(x+1)^2} - \frac{\frac{17}{16}}{x-3}$

13. $\frac{3x+5}{x^2+2} - \frac{4}{x-1}$ **15.** $\frac{3}{2x-1} - \frac{2}{x+2} + \frac{10}{(x+2)^2}$

17. $3x + 1 + \frac{2}{2x-1} + \frac{3}{x+1}$ **19.** $-\frac{1}{x-3} + \frac{3x}{x^2+2x-5}$

21. $\frac{5}{3x+5} - \frac{3}{x+1} + \frac{4}{(x+1)^2}$ **23.** $\frac{8}{4x-5} + \frac{3}{3x+2}$

25. $\frac{2x-5}{3x^2+1} - \frac{2}{x-2}$ **27.** $-1, \pm 3i$ **28.** $3, \pm i$

29. $-2, \frac{1 \pm \sqrt{5}}{2}$ **30.** $-2, 3, \pm i$ **31.** $-3, -1 \pm \sqrt{2}$

33. $-\frac{\frac{1}{2a^2}x}{x^2+a^2} + \frac{\frac{1}{4a^2}}{x-a} + \frac{\frac{1}{4a^2}}{x+a}$

35. $-\frac{3}{25(\ln x + 2)} + \frac{3}{25(\ln x - 3)} + \frac{7}{5(\ln x - 3)^2}$

Review Exercises: Chapter 9

1. True **2.** False **3.** True **4.** False **5.** (a) **6.** (e) **7.** (h) **8.** (d) **9.** (b) **10.** (g) **11.** (c) **12.** (f)
13. $(-2, -2)$ **14.** $(-5, 4)$ **15.** No solution
16. Infinitely many solutions; $(-y - 2, y)$, or $(x, -x - 2)$
17. $(3, -1, -2)$ **18.** No solution **19.** $(-5, 13, 8, 2)$
20. Consistent: 13, 14, 16, 17, 19; the others are inconsistent.
21. Dependent: 16; the others are independent. **22.** $(1, 2)$
23. $(-3, 4, -2)$ **24.** Infinitely many solutions; $\left(\frac{z}{2}, -\frac{z}{2}, z\right)$
25. $(-4, 1, -2, 3)$ **26.** Nickels: 31; dimes: 44
27. 3%: $1600; 3.5%: $3400 **28.** 1 bagel, $\frac{1}{2}$ serving of cream cheese, 2 bananas **29.** 75, 69, 82
30. **(a)** $f(x) = 1.125x^2 - 5.25x + 145$; **(b)** 154 million persons employed

31. $\begin{bmatrix} 0 & -1 & 6 \\ 3 & 1 & -2 \\ -2 & 1 & -2 \end{bmatrix}$ **32.** $\begin{bmatrix} -3 & 3 & 0 \\ -6 & -9 & 6 \\ 6 & 0 & -3 \end{bmatrix}$

33. $\begin{bmatrix} -1 & 1 & 0 \\ -2 & -3 & 2 \\ 2 & 0 & -1 \end{bmatrix}$ **34.** $\begin{bmatrix} -2 & 2 & 6 \\ 1 & -8 & 18 \\ 2 & 1 & -15 \end{bmatrix}$

35. Not defined **36.** $\begin{bmatrix} 2 & -1 & -6 \\ 1 & 5 & -2 \\ -2 & -1 & 4 \end{bmatrix}$

37. $\begin{bmatrix} -13 & 1 & 6 \\ -3 & -7 & 4 \\ 8 & 3 & -5 \end{bmatrix}$ **38.** $\begin{bmatrix} -2 & -1 & 18 \\ 5 & -3 & -2 \\ -2 & 3 & -8 \end{bmatrix}$

39. **(a)** $\begin{bmatrix} 2.25 & 0.38 & 0.55 & 0.33 & 0.85 \\ 3.09 & 0.42 & 0.46 & 0.48 & 0.51 \\ 2.40 & 0.31 & 0.59 & 0.36 & 0.64 \\ 1.80 & 0.29 & 0.34 & 0.55 & 0.52 \end{bmatrix}$;
(b) $[41 \quad 18 \quad 39 \quad 36]$;
(c) $[306.27 \quad 45.67 \quad 66.08 \quad 56.01 \quad 87.71]$;
(d) the total cost, in dollars, for each item for the day's meals

40. $\begin{bmatrix} -\frac{1}{2} & 0 \\ \frac{1}{6} & \frac{1}{3} \end{bmatrix}$ **41.** $\begin{bmatrix} 0 & 0 & \frac{1}{4} \\ 0 & -\frac{1}{2} & 0 \\ \frac{1}{3} & 0 & 0 \end{bmatrix}$

42. $\begin{bmatrix} 1 & 0 & 0 & 0 \\ 0 & \frac{1}{9} & \frac{5}{18} & 0 \\ 0 & -\frac{1}{9} & \frac{2}{9} & 0 \\ 0 & 0 & 0 & 1 \end{bmatrix}$

43. $\begin{bmatrix} 3 & -2 & 4 \\ 1 & 5 & -3 \\ 2 & -3 & 7 \end{bmatrix} \begin{bmatrix} x \\ y \\ z \end{bmatrix} = \begin{bmatrix} 13 \\ 7 \\ -8 \end{bmatrix}$

44. $(-8, 7)$ **45.** $(1, -2, 5)$ **46.** $(2, -1, 1, -3)$ **47.** 10
48. -18 **49.** -6 **50.** -1 **51.** $(3, -2)$ **52.** $(-1, 5)$
53. $\left(\frac{3}{2}, \frac{13}{14}, \frac{33}{14}\right)$ **54.** $(2, -1, 3)$

55.

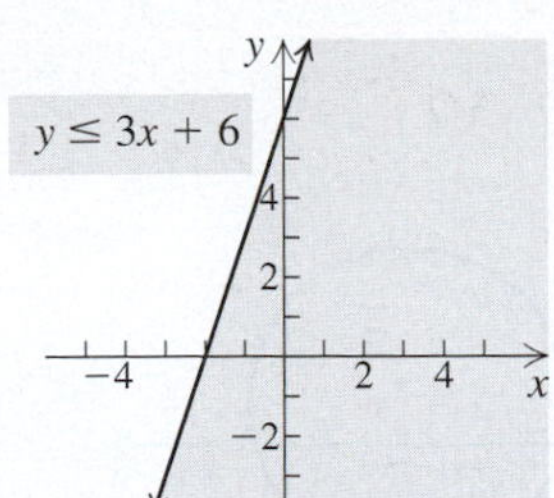

56.

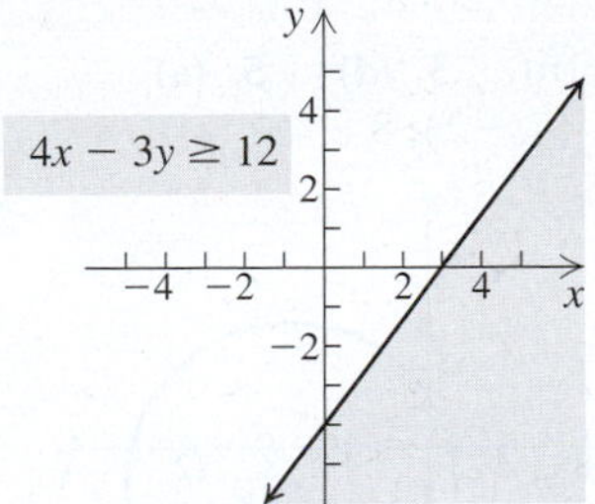

57.

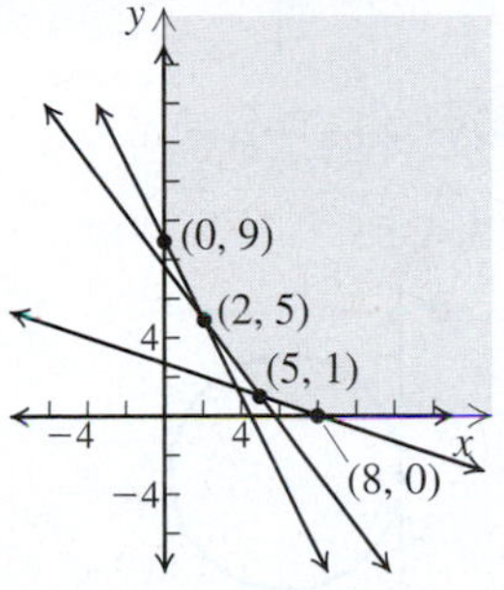

58. Minimum: 52 when $x = 2$ and $y = 4$; maximum: 92 when $x = 2$ and $y = 8$ **59.** Maximum score of 96 is achieved when 0 group A questions and 8 group B questions are answered.

60. $\frac{5}{x+1} - \frac{5}{x+2} - \frac{5}{(x+2)^2}$ **61.** $\frac{2}{2x-3} - \frac{5}{x+4}$

62. C **63.** A **64.** B **65.** 4%: \$10,000; 5%: \$12,000; $5\frac{1}{2}$%: \$18,000 **66.** $\left(\frac{5}{18}, \frac{1}{7}\right)$ **67.** $\left(1, \frac{1}{2}, \frac{1}{3}\right)$

68.

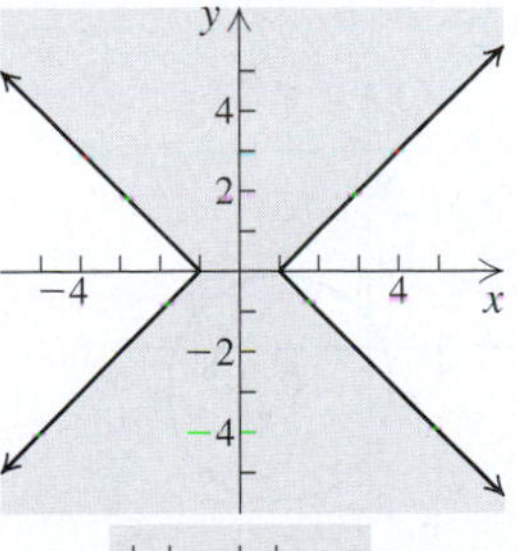

69.

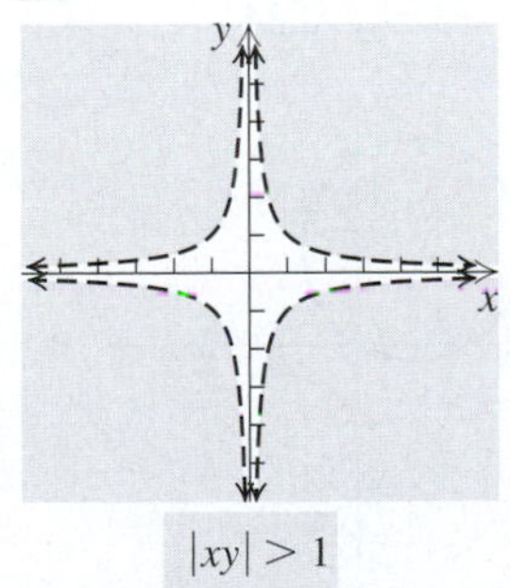

70. The solution of the equation $2x + 5 = 3x - 7$ is the first coordinate of the point of intersection of the graphs of $y_1 = 2x + 5$ and $y_2 = 3x - 7$. The solution of the system of equations $y = 2x + 5$, $y = 3x - 7$ is the ordered pair that is the point of intersection of the two lines. **71.** In general, $(\mathbf{AB})^2 \ne \mathbf{A}^2\mathbf{B}^2$. $(\mathbf{AB})^2 = \mathbf{ABAB}$ and $\mathbf{A}^2\mathbf{B}^2 = \mathbf{AABB}$. Since matrix multiplication is not commutative, $\mathbf{BA} \ne \mathbf{AB}$, so $(\mathbf{AB})^2 \ne \mathbf{A}^2\mathbf{B}^2$. **72.** If

$$\begin{vmatrix} a_1 & b_1 \\ a_2 & b_2 \end{vmatrix} = 0,$$

then $a_1 = ka_2$ and $b_1 = kb_2$ for some number k. This means that the equations $a_1x + b_1y = c_1$ and $a_2x + b_2y = c_2$ are dependent if $c_1 = kc_2$, or the system is inconsistent if $c_1 \ne kc_2$.

73. If $a_1x + b_1y = c_1$ and $a_2x + b_2y = c_2$ are parallel lines, then $a_1 = ka_2$, $b_1 = kb_2$, and $c_1 \ne kc_2$, for some number k. Then

$$\begin{vmatrix} a_1 & b_1 \\ a_2 & b_2 \end{vmatrix} = 0, \begin{vmatrix} c_1 & b_1 \\ c_2 & b_2 \end{vmatrix} \ne 0, \text{ and } \begin{vmatrix} a_1 & c_1 \\ a_2 & c_2 \end{vmatrix} \ne 0.$$

74. The graph of a linear equation consists of a set of points on a line. The graph of a linear inequality consists of the set of points in a half-plane and might also include the points on the line that is the boundary of the half-plane. **75.** The denominator of the second fraction, $x^2 - 5x + 6$, can be factored into linear factors with rational coefficients: $(x-3)(x-2)$. Thus the given expression is not a partial fraction decomposition.

Test: Chapter 9

1. [9.1] $(-3, 5)$; consistent, independent **2.** [9.1] Infinitely many solutions; $(x, 2x - 3)$ or $\left(\frac{y+3}{2}, y\right)$; consistent, dependent

3. [9.1] No solution; inconsistent, independent

4. [9.1] $(1, -2)$; consistent, independent **5.** [9.2] $(-1, 3, 2)$

6. [9.1] Student: 462 tickets; nonstudent: 158 tickets

7. [9.2] Hui: 120 orders; Ashlyn: 104 orders; Sheriann: 128 orders

8. [9.4] $\begin{bmatrix} -2 & -3 \\ -3 & 4 \end{bmatrix}$ **9.** [9.4] Not defined

10. [9.4] $\begin{bmatrix} -7 & -13 \\ 5 & -1 \end{bmatrix}$ **11.** [9.4] Not defined

12. [9.4] $\begin{bmatrix} 2 & -2 & 6 \\ -4 & 10 & 4 \end{bmatrix}$ **13.** [9.5] $\begin{bmatrix} 0 & -1 \\ -\frac{1}{4} & -\frac{3}{4} \end{bmatrix}$

14. [9.4] **(a)** $\begin{bmatrix} 1.55 & 1.00 & 0.99 \\ 1.70 & 0.95 & 1.01 \\ 1.65 & 0.99 & 0.96 \end{bmatrix}$; **(b)** $[26 \quad 18 \quad 23]$;

(c) $[108.85 \quad 65.87 \quad 66.00]$; **(d)** the total cost, in dollars, for each type of menu item served on the given day

15. [9.4] $\begin{bmatrix} 3 & -4 & 2 \\ 2 & 3 & 1 \\ 1 & -5 & -3 \end{bmatrix}\begin{bmatrix} x \\ y \\ z \end{bmatrix} = \begin{bmatrix} -8 \\ 7 \\ 3 \end{bmatrix}$ **16.** [9.5] $(-2, 1, 1)$

17. [9.6] 61 **18.** [9.6] -33 **19.** [9.6] $\left(-\frac{1}{2}, \frac{3}{4}\right)$

20. [9.7]

$3x + 4y \le -12$

21. [9.7] Maximum: 15 when $x = 3$ and $y = 3$; minimum: 2 when $x = 1$ and $y = 0$ **22.** [9.7] Maximum profit of \$750 occurs when 25 pound cakes and 75 carrot cakes are prepared.

23. [9.8] $-\frac{2}{x-1} + \frac{5}{x+3}$ **24.** [9.7] D **25.** [9.2] $A = 1$, $B = -3$, $C = 2$

Chapter 10

Exercise Set 10.1

1. (f) **3.** (b) **5.** (d)

7. V: $(0, 0)$; F: $(0, 5)$; D: $y = -5$

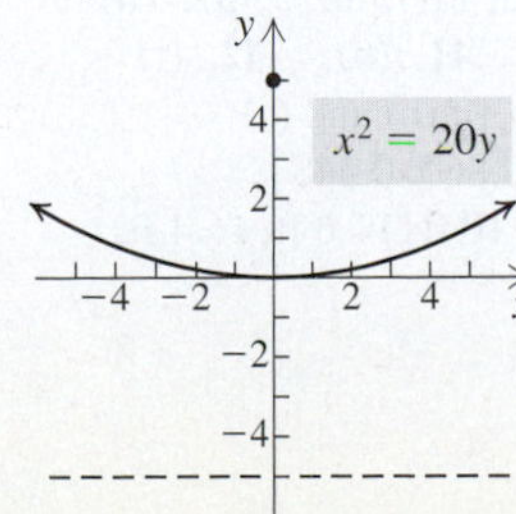

9. V: $(0, 0)$; F: $\left(-\frac{3}{2}, 0\right)$; D: $x = \frac{3}{2}$

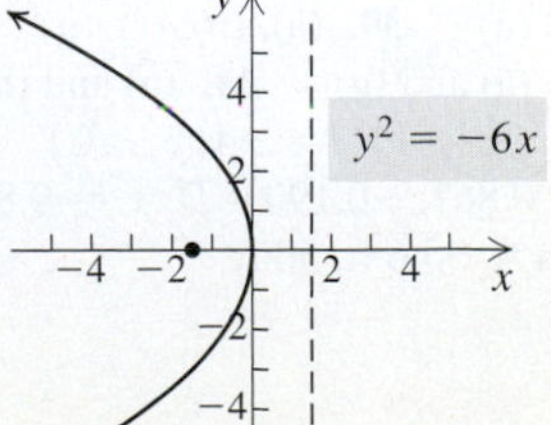

11. V: $(0, 0)$; F: $(0, 1)$; D: $y = -1$

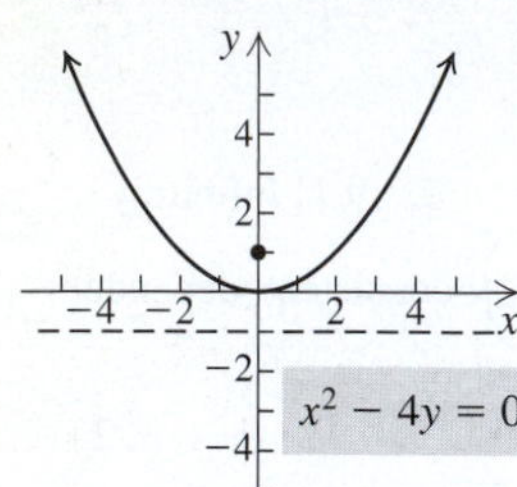

13. V: $(0, 0)$; F: $\left(\frac{1}{8}, 0\right)$; D: $x = -\frac{1}{8}$

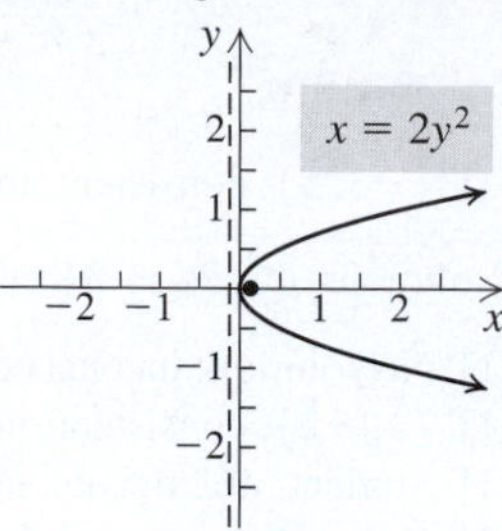

15. $y^2 = -12x$ **17.** $y^2 = 28x$ **19.** $x^2 = -4\pi y$
21. $(y - 2)^2 = 14\left(x + \frac{1}{2}\right)$
23. V: $(-2, 1)$; F: $\left(-2, -\frac{1}{2}\right)$; D: $y = \frac{5}{2}$

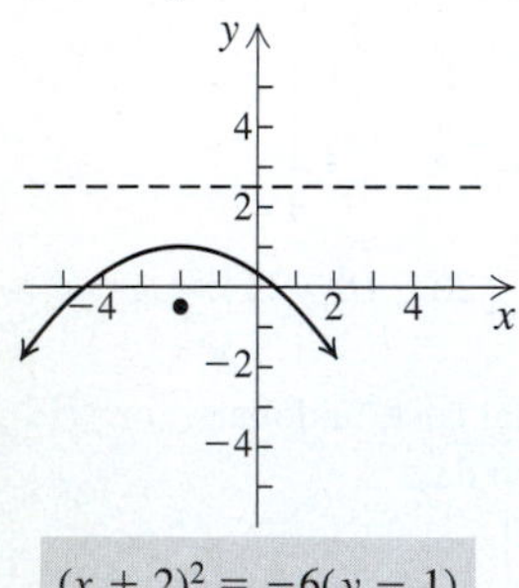

25. V: $(-1, -3)$; F: $\left(-1, -\frac{7}{2}\right)$; D: $y = -\frac{5}{2}$

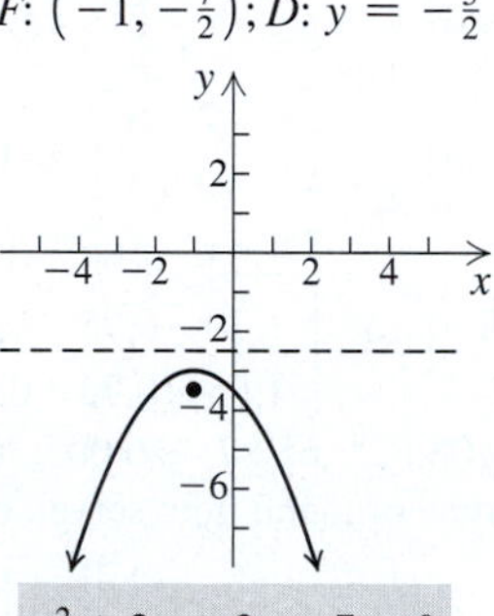

27. V: $(0, -2)$; F: $\left(0, -1\frac{3}{4}\right)$; D: $y = -2\frac{1}{4}$

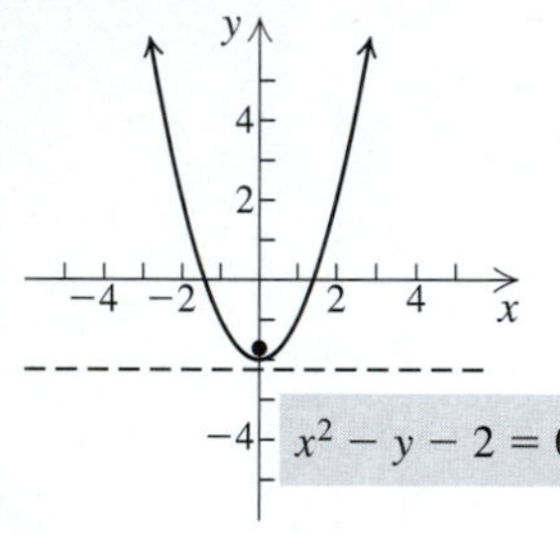

29. V: $(-2, -1)$; F: $\left(-2, -\frac{3}{4}\right)$; D: $y = -1\frac{1}{4}$

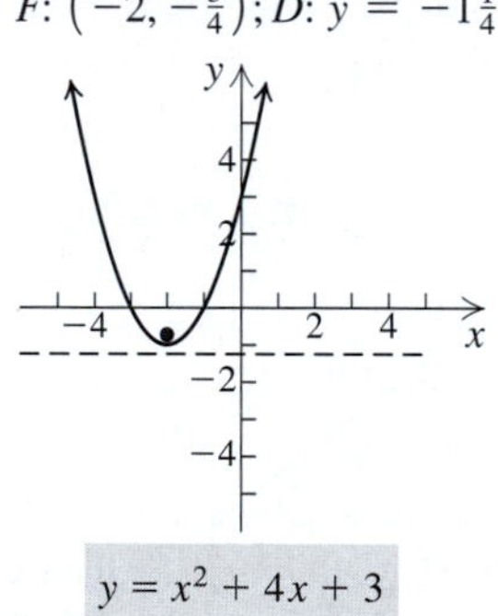
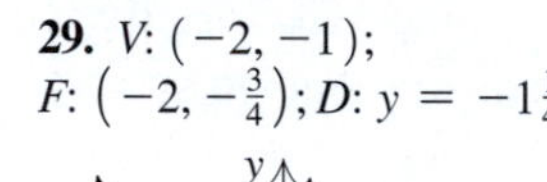

31. V: $\left(5\frac{3}{4}, \frac{1}{2}\right)$; F: $\left(6, \frac{1}{2}\right)$; D: $x = 5\frac{1}{2}$

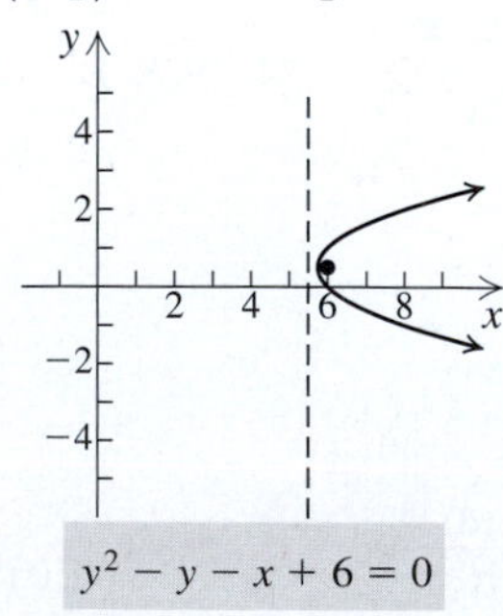

33. (a) $y^2 = 16x$; **(b)** $3\frac{33}{64}$ ft **35.** About 11.75 in. **37.** (h)
38. (d) **39.** (a), (b), (f), (g) **40.** (b) **41.** (b) **42.** (f)
43. (a) and (g) **44.** (a) and (h); (g) and (h); (b) and (c)
45. $(x + 1)^2 = -4(y - 2)$ **47.** V: $(0.867, 0.348)$; F: $(0.867, -0.190)$; D: $y = 0.887$ **49.** 10 ft, 11.6 ft, 16.4 ft, 24.4 ft, 35.6 ft, 50 ft

Exercise Set 10.2

1. (b) **3.** (d) **5.** (a)
7. $(7, -2)$; 8

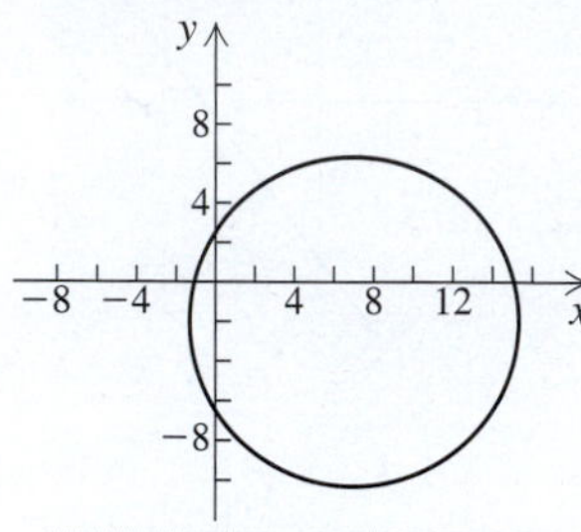

9. $(-3, 1)$; 4

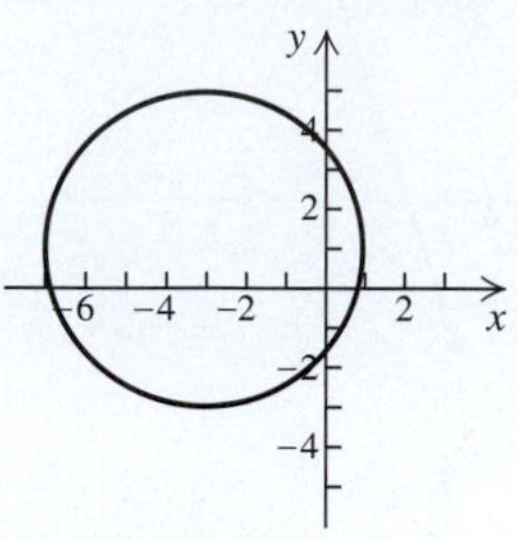
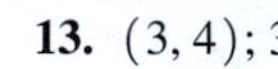

11. $(-2, 3)$; 5

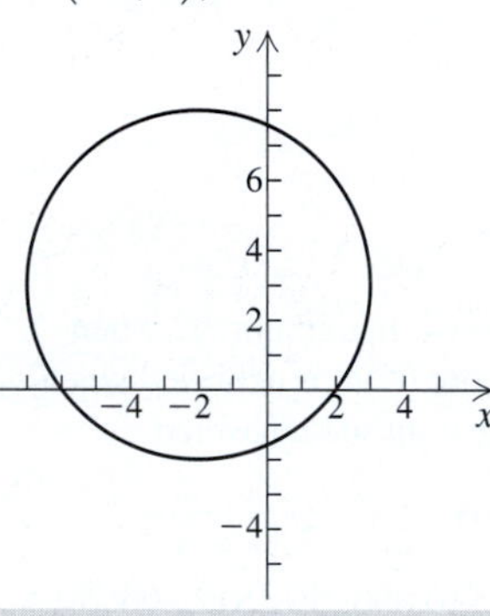

13. $(3, 4)$; 3

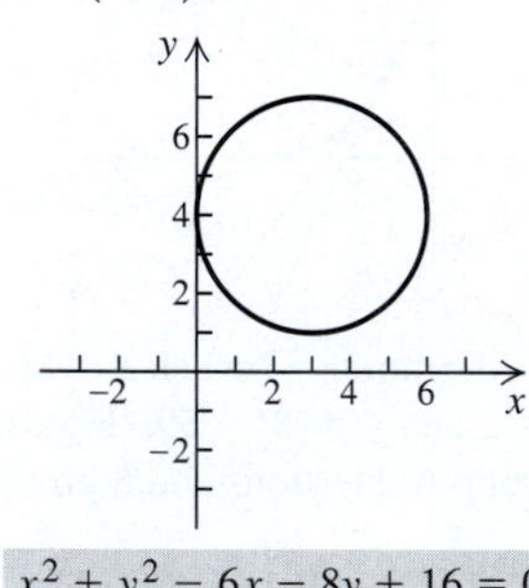
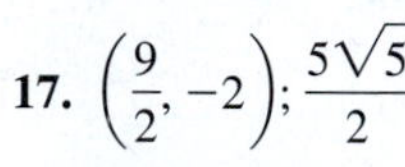

15. $(-3, 5)$; $\sqrt{34}$

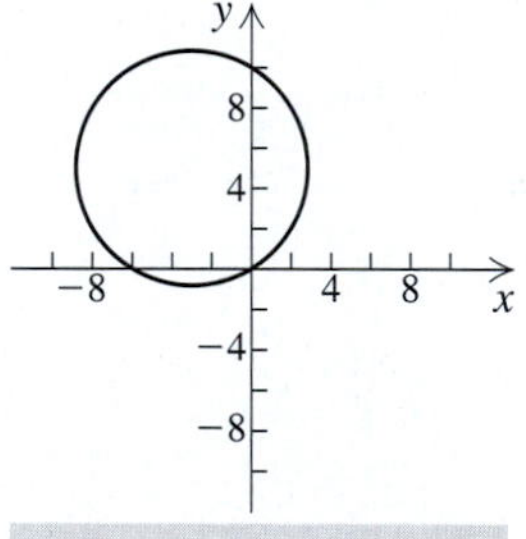

17. $\left(\frac{9}{2}, -2\right)$; $\frac{5\sqrt{5}}{2}$

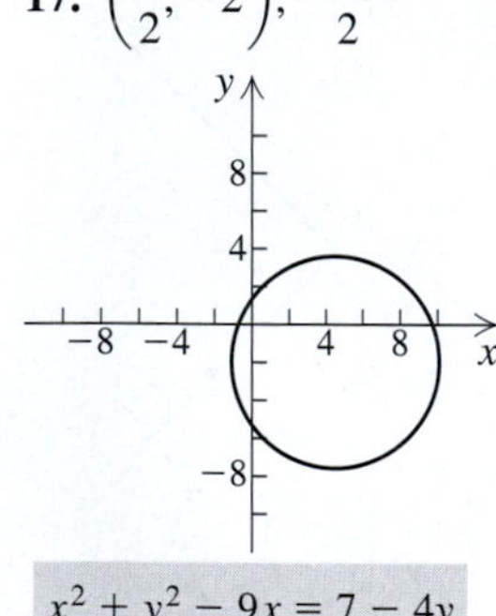

19. (c) **21.** (d)
23. V: $(2, 0)$, $(-2, 0)$; F: $(\sqrt{3}, 0)$, $(-\sqrt{3}, 0)$

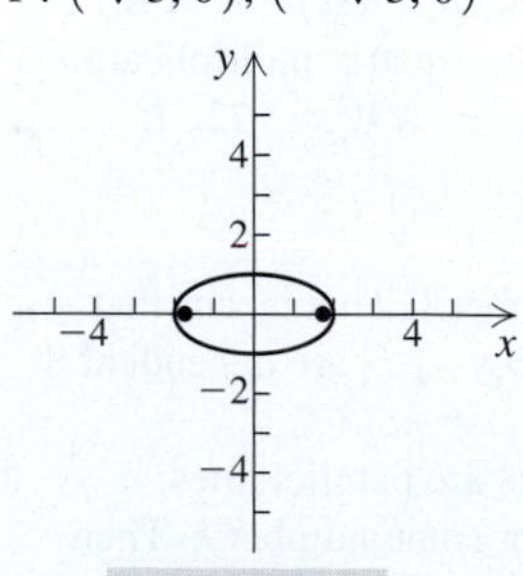

25. V: $(0, 4)$, $(0, -4)$; F: $(0, \sqrt{7})$, $(0, -\sqrt{7})$

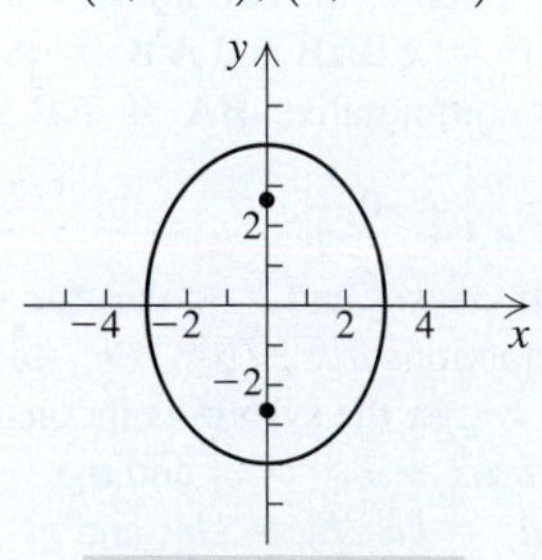
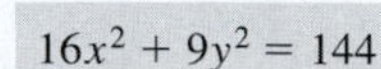

27. V: $(-\sqrt{3}, 0)$, $(\sqrt{3}, 0)$; F: $(-1, 0)$, $(1, 0)$

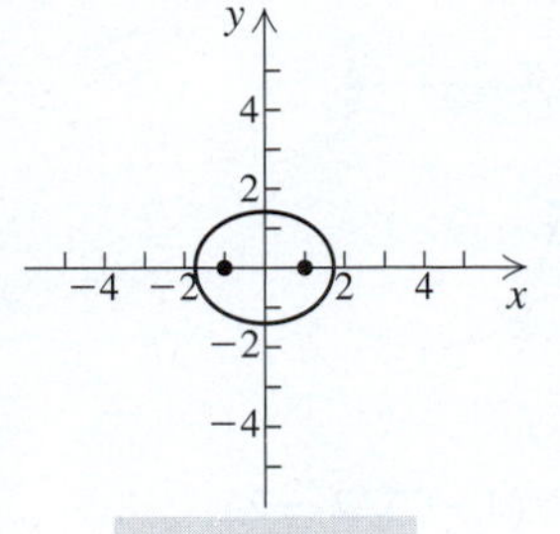

$2x^2 + 3y^2 = 6$

29. V: $\left(-\frac{1}{2}, 0\right)$, $\left(\frac{1}{2}, 0\right)$; F: $\left(-\frac{\sqrt{5}}{6}, 0\right)$, $\left(\frac{\sqrt{5}}{6}, 0\right)$

$4x^2 + 9y^2 = 1$

31. $\frac{x^2}{49} + \frac{y^2}{40} = 1$ **33.** $\frac{x^2}{25} + \frac{y^2}{64} = 1$ **35.** $\frac{x^2}{9} + \frac{y^2}{5} = 1$

37. C: $(1, 2)$; V: $(4, 2)$, $(-2, 2)$; F: $(1 + \sqrt{5}, 2)$, $(1 - \sqrt{5}, 2)$

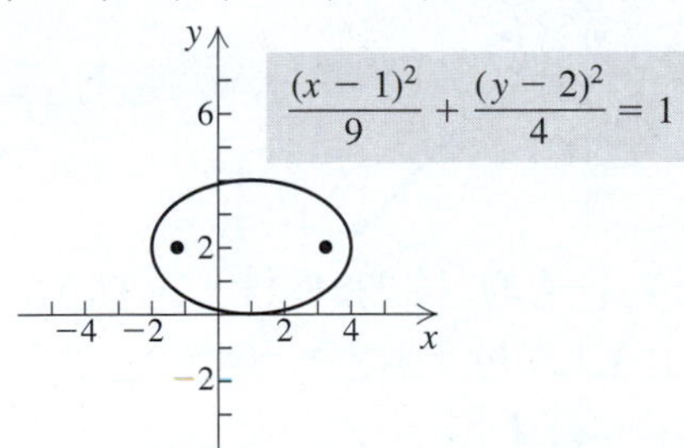

39. C: $(-3, 5)$; V: $(-3, 11)$, $(-3, -1)$; F: $(-3, 5 + \sqrt{11})$, $(-3, 5 - \sqrt{11})$

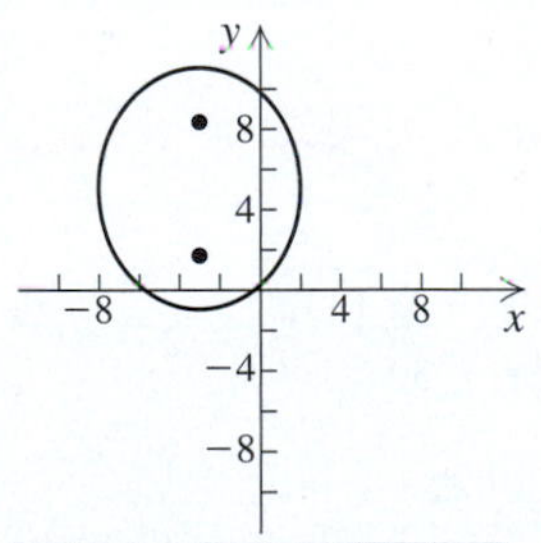

$\frac{(x+3)^2}{25} + \frac{(y-5)^2}{36} = 1$

41. C: $(-2, 1)$; V: $(-10, 1)$, $(6, 1)$; F: $(-6, 1)$, $(2, 1)$

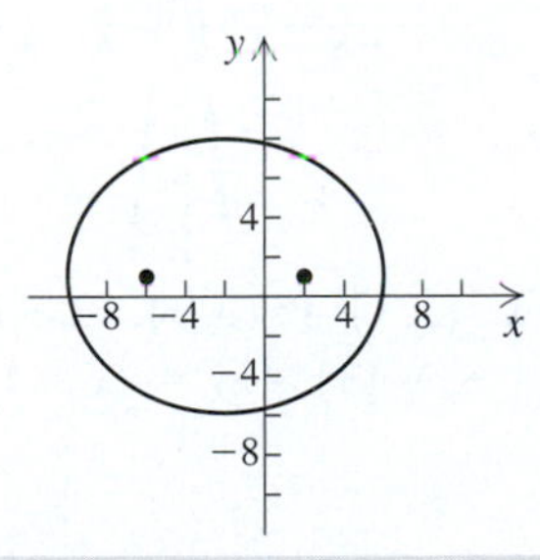

$3(x+2)^2 + 4(y-1)^2 = 192$

43. C: $(2, -1)$; V: $(-1, -1)$, $(5, -1)$; F: $(2 + \sqrt{5}, -1)$, $(2 - \sqrt{5}, -1)$

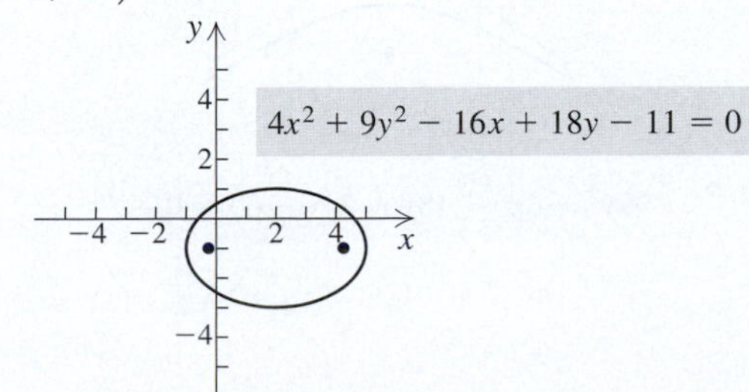

45. C: $(1, 1)$; V: $(1, 3)$, $(1, -1)$; F: $(1, 1 + \sqrt{3})$, $(1, 1 - \sqrt{3})$

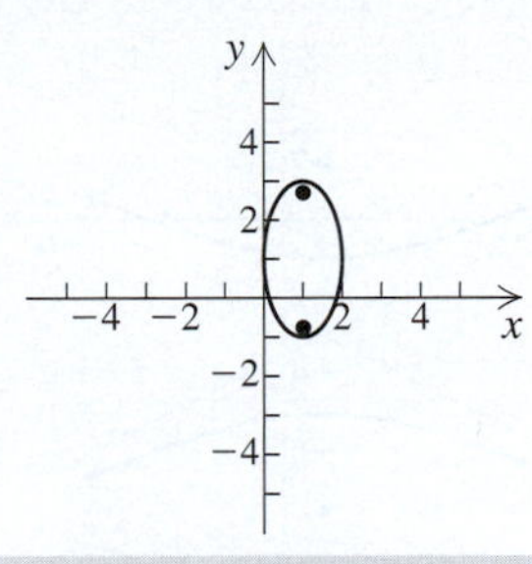

$4x^2 + y^2 - 8x - 2y + 1 = 0$

47. Example 2; $\frac{3}{5} < \frac{\sqrt{12}}{4}$ **49.** $\frac{x^2}{15} + \frac{y^2}{16} = 1$

51. $\frac{x^2}{524^2} + \frac{y^2}{449^2} = 1$, or $\frac{x^2}{274{,}576} + \frac{y^2}{201{,}601} = 1$

53. 33.5 ft **55.** **(a)** 0.7 ft; **(b)** 4 ft **57.** Midpoint **58.** Zero

59. y-intercept **60.** Two different real-number solutions

61. Remainder **62.** Ellipse **63.** Parabola **64.** Circle

65. $\frac{(x-3)^2}{4} + \frac{(y-1)^2}{25} = 1$ **67.** $\frac{x^2}{9} + \frac{y^2}{484/5} = 1$

69. C: $(2.003, -1.005)$; V: $(-1.017, -1.005)$, $(5.023, -1.005)$

71. About 9.1 ft

Exercise Set 10.3

1. (b) **3.** (c) **5.** (a) **7.** $\frac{y^2}{9} - \frac{x^2}{16} = 1$ **9.** $\frac{x^2}{4} - \frac{y^2}{9} = 1$

11. C: $(0, 0)$; V: $(2, 0)$, $(-2, 0)$; F: $(2\sqrt{2}, 0)$, $(-2\sqrt{2}, 0)$; A: $y = x$, $y = -x$

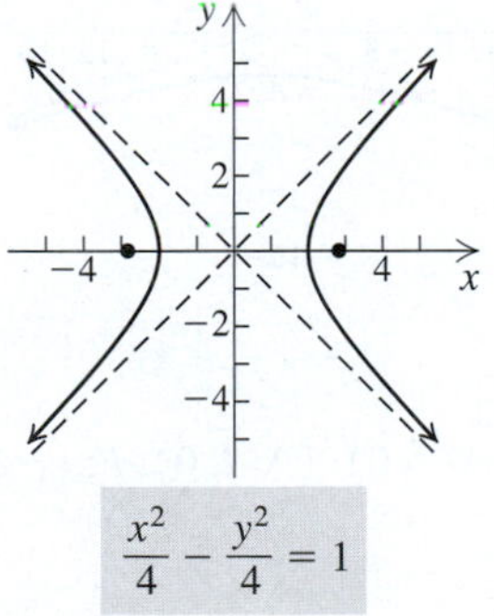

$\frac{x^2}{4} - \frac{y^2}{4} = 1$

13. C: $(2, -5)$; V: $(-1, -5)$, $(5, -5)$; F: $(2 - \sqrt{10}, -5)$, $(2 + \sqrt{10}, -5)$; A: $y = -\frac{x}{3} - \frac{13}{3}$, $y = \frac{x}{3} - \frac{17}{3}$

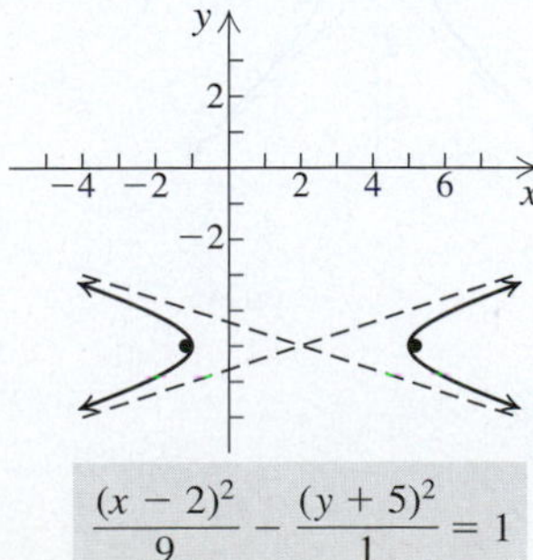

$\frac{(x-2)^2}{9} - \frac{(y+5)^2}{1} = 1$

15. C: $(-1, -3)$; V: $(-1, -1)$, $(-1, -5)$; F: $(-1, -3 + 2\sqrt{5})$, $(-1, -3 - 2\sqrt{5})$; A: $y = \frac{1}{2}x - \frac{5}{2}$, $y = -\frac{1}{2}x - \frac{7}{2}$

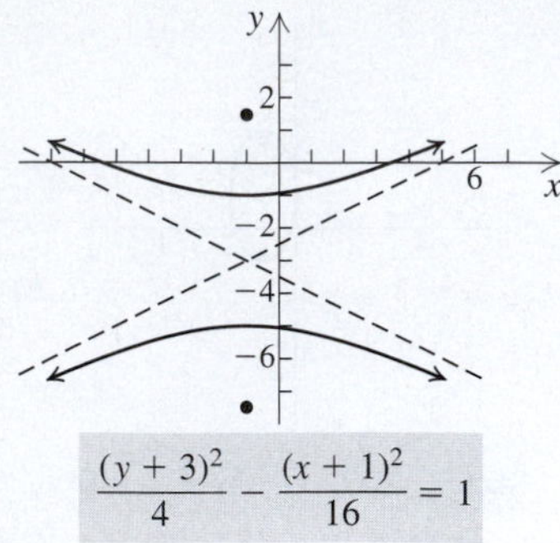

17. C: $(0, 0)$; V: $(-2, 0)$, $(2, 0)$; F: $(-\sqrt{5}, 0)$, $(\sqrt{5}, 0)$; A: $y = -\frac{1}{2}x$, $y = \frac{1}{2}x$

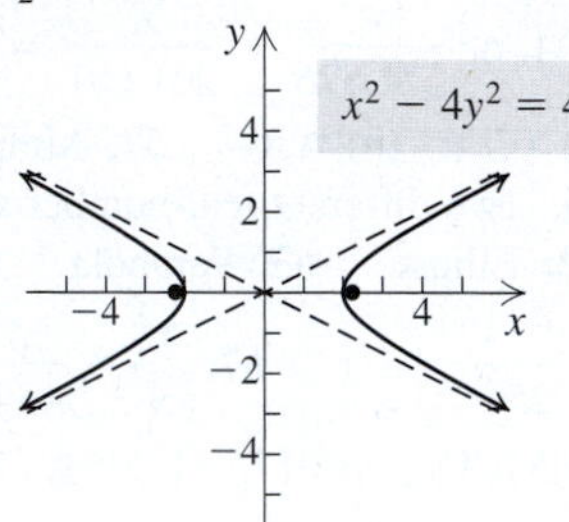

19. C: $(0, 0)$; V: $(0, -3)$, $(0, 3)$; F: $(0, -3\sqrt{10})$, $(0, 3\sqrt{10})$; A: $y = \frac{1}{3}x$, $y = -\frac{1}{3}x$

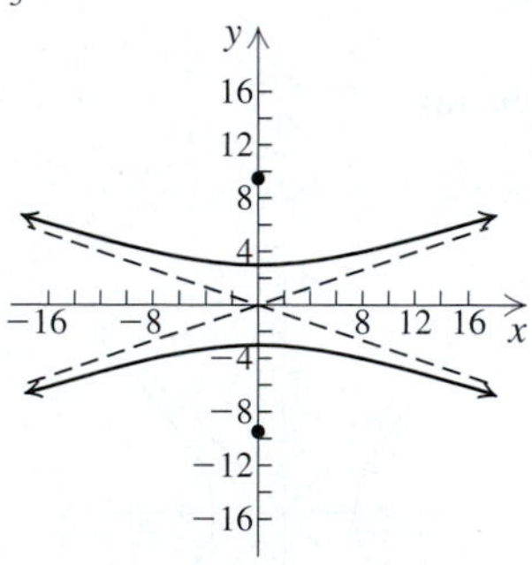

21. C: $(0, 0)$; V: $(-\sqrt{2}, 0)$, $(\sqrt{2}, 0)$; F: $(-2, 0)$, $(2, 0)$; A: $y = x$, $y = -x$

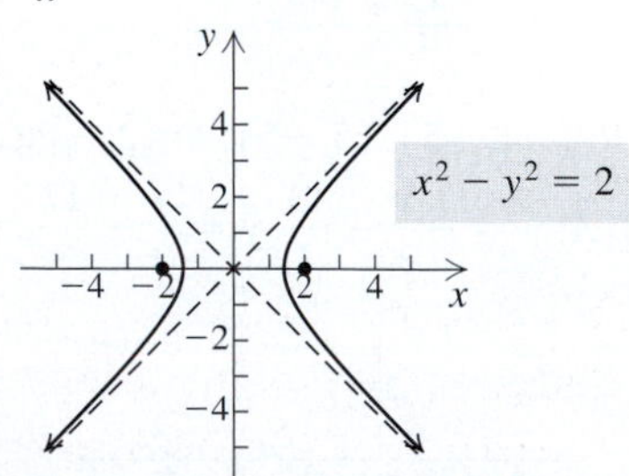

23. C: $(0, 0)$; V: $\left(0, -\frac{1}{2}\right)$, $\left(0, \frac{1}{2}\right)$; F: $\left(0, -\frac{\sqrt{2}}{2}\right)$, $\left(0, \frac{\sqrt{2}}{2}\right)$; A: $y = x$, $y = -x$

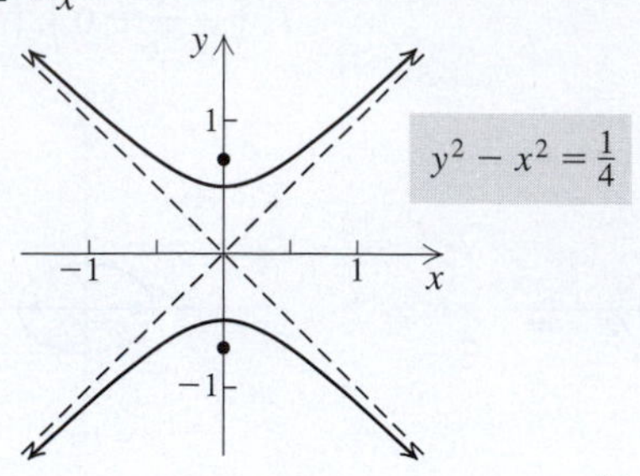

25. C: $(1, -2)$; V: $(0, -2)$, $(2, -2)$; F: $(1 - \sqrt{2}, -2)$, $(1 + \sqrt{2}, -2)$; A: $y = -x - 1$, $y = x - 3$

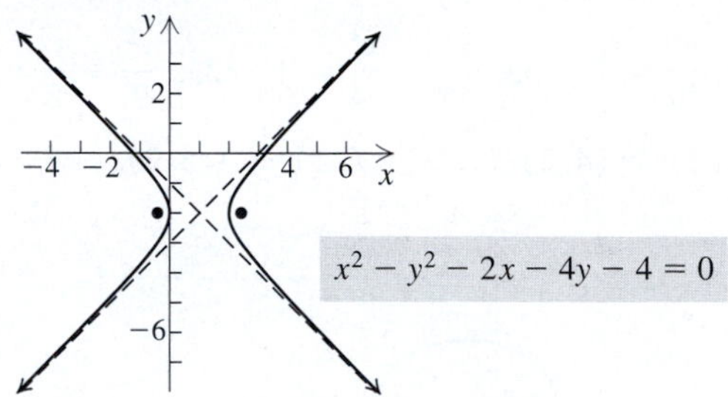

27. C: $\left(\frac{1}{3}, 3\right)$; V: $\left(-\frac{2}{3}, 3\right)$, $\left(\frac{4}{3}, 3\right)$; F: $\left(\frac{1}{3} - \sqrt{37}, 3\right)$, $\left(\frac{1}{3} + \sqrt{37}, 3\right)$; A: $y = 6x + 1$, $y = -6x + 5$

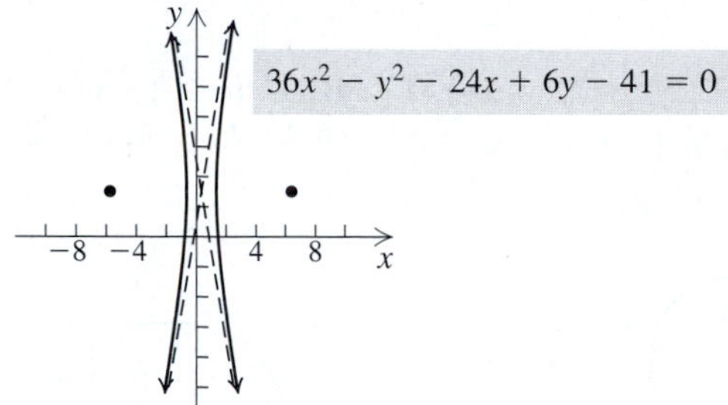

29. C: $(3, 1)$; V: $(3, 3)$, $(3, -1)$; F: $(3, 1 + \sqrt{13})$, $(3, 1 - \sqrt{13})$; A: $y = \frac{2}{3}x - 1$, $y = -\frac{2}{3}x + 3$

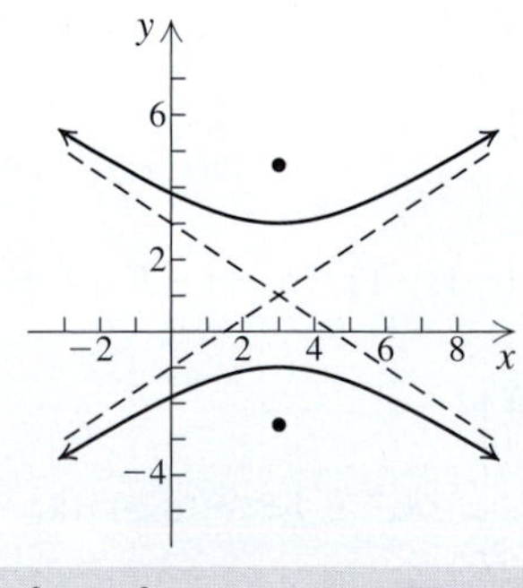

31. C: $(1, -2)$; V: $(2, -2)$, $(0, -2)$; F: $(1 + \sqrt{2}, -2)$, $(1 - \sqrt{2}, -2)$; A: $y = x - 3$, $y = -x - 1$

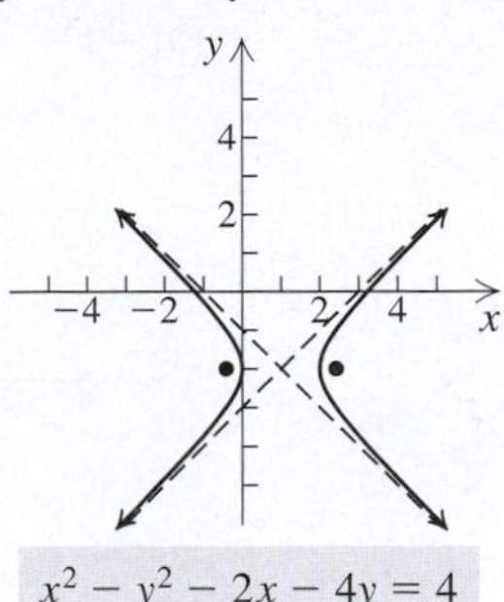

$x^2 - y^2 - 2x - 4y = 4$

33. C: $(-3, 4)$; V: $(-3, 10)$, $(-3, -2)$; F: $(-3, 4 + 6\sqrt{2})$, $(-3, 4 - 6\sqrt{2})$; A: $y = x + 7$, $y = -x + 1$

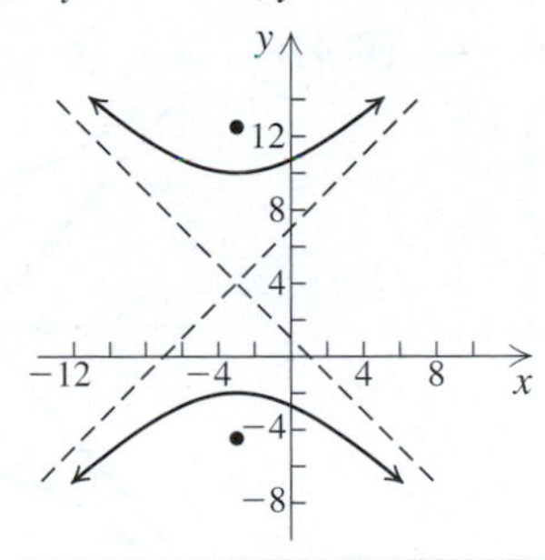

$y^2 - x^2 - 6x - 8y - 29 = 0$

35. Example 3; $\frac{\sqrt{5}}{1} > \frac{5}{4}$ **37.** $\frac{x^2}{9} - \frac{(y - 7)^2}{16} = 1$

39. $\frac{y^2}{25} - \frac{x^2}{11} = 1$ **41. (a)** Yes; **(b)** $f^{-1}(x) = \frac{x + 3}{2}$

42. (a) Yes; **(b)** $f^{-1}(x) = \sqrt[3]{x - 2}$ **43. (a)** Yes;
(b) $f^{-1}(x) = \frac{5}{x} + 1$, or $\frac{5 + x}{x}$ **44. (a)** Yes;
(b) $f^{-1}(x) = x^2 - 4, x \geq 0$ **45.** $(6, -1)$ **46.** $(1, -1)$

47. $(2, -1)$ **48.** $(-3, 4)$ **49.** $\frac{(y + 5)^2}{9} - (x - 3)^2 = 1$

51. C: $(-1.460, -0.957)$; V: $(-2.360, -0.957)$, $(-0.560, -0.957)$; A: $y = -1.20x - 2.70$, $y = 1.20x + 0.79$

53. $\frac{x^2}{345.96} - \frac{y^2}{22{,}154.04} = 1$

Visualizing the Graph

1. B **2.** J **3.** F **4.** I **5.** H **6.** G **7.** E **8.** D **9.** C **10.** A

Exercise Set 10.4

1. (e) **3.** (c) **5.** (b) **7.** $(-4, -3)$, $(3, 4)$
9. $(0, 2)$, $(3, 0)$ **11.** $(-5, 0)$, $(4, 3)$, $(4, -3)$
13. $(3, 0)$, $(-3, 0)$ **15.** $(0, -3)$, $(4, 5)$ **17.** $(-2, 1)$
19. $(3, 4)$, $(-3, -4)$, $(4, 3)$, $(-4, -3)$

21. $\left(\frac{6\sqrt{21}}{7}, \frac{4\sqrt{35}}{7}i\right)$, $\left(\frac{6\sqrt{21}}{7}, -\frac{4\sqrt{35}}{7}i\right)$, $\left(-\frac{6\sqrt{21}}{7}, \frac{4\sqrt{35}}{7}i\right)$, $\left(-\frac{6\sqrt{21}}{7}, -\frac{4\sqrt{35}}{7}i\right)$

23. $(3, 2)$, $\left(4, \frac{3}{2}\right)$ **25.** $\left(\frac{5 + \sqrt{70}}{3}, \frac{-1 + \sqrt{70}}{3}\right)$, $\left(\frac{5 - \sqrt{70}}{3}, \frac{-1 - \sqrt{70}}{3}\right)$ **27.** $(\sqrt{2}, \sqrt{14})$, $(-\sqrt{2}, \sqrt{14})$, $(\sqrt{2}, -\sqrt{14})$, $(-\sqrt{2}, -\sqrt{14})$ **29.** $(1, 2)$, $(-1, -2)$, $(2, 1)$, $(-2, -1)$ **31.** $\left(\frac{15 + \sqrt{561}}{8}, \frac{11 - 3\sqrt{561}}{8}\right)$, $\left(\frac{15 - \sqrt{561}}{8}, \frac{11 + 3\sqrt{561}}{8}\right)$ **33.** $\left(\frac{7 - \sqrt{33}}{2}, \frac{7 + \sqrt{33}}{2}\right)$, $\left(\frac{7 + \sqrt{33}}{2}, \frac{7 - \sqrt{33}}{2}\right)$ **35.** $(3, 2)$, $(-3, -2)$, $(2, 3)$, $(-2, -3)$ **37.** $\left(\frac{5 - 9\sqrt{15}}{20}, \frac{-45 + 3\sqrt{15}}{20}\right)$, $\left(\frac{5 + 9\sqrt{15}}{20}, \frac{-45 - 3\sqrt{15}}{20}\right)$ **39.** $(3, -5)$, $(-1, 3)$

41. $(8, 5)$, $(-5, -8)$ **43.** $(3, 2)$, $(-3, -2)$
45. $(2, 1)$, $(-2, -1)$, $(1, 2)$, $(-1, -2)$

47. $\left(4 + \frac{3\sqrt{6}}{2}i, -4 + \frac{3\sqrt{6}}{2}i\right)$, $\left(4 - \frac{3\sqrt{6}}{2}i, -4 - \frac{3\sqrt{6}}{2}i\right)$

49. $(3, \sqrt{5})$, $(-3, -\sqrt{5})$, $(\sqrt{5}, 3)$, $(-\sqrt{5}, -3)$

51. $\left(\frac{8\sqrt{5}}{5}i, \frac{3\sqrt{105}}{5}\right)$, $\left(\frac{8\sqrt{5}}{5}i, -\frac{3\sqrt{105}}{5}\right)$, $\left(-\frac{8\sqrt{5}}{5}i, \frac{3\sqrt{105}}{5}\right)$, $\left(-\frac{8\sqrt{5}}{5}i, -\frac{3\sqrt{105}}{5}\right)$

53. $(2, 1)$, $(-2, -1)$, $\left(-\sqrt{5}i, \frac{2\sqrt{5}i}{5}\right)$, $\left(\sqrt{5}i, -\frac{2\sqrt{5}i}{5}\right)$

55. True **57.** True **59.** 24 in. by 10 in. **61.** 4 in. by 5 in.
63. 30 yd by 75 yd **65.** Length: $\sqrt{3}$ m; width: 1 m
67. Length: 12 ft; width: 5 ft **69.** (b) **71.** (d) **73.** (a)

75.

77.

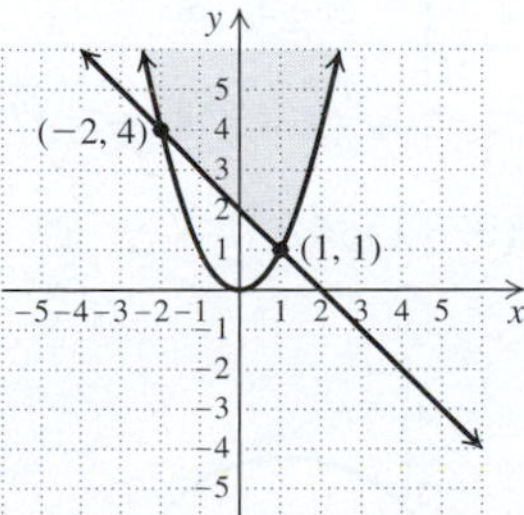

79.

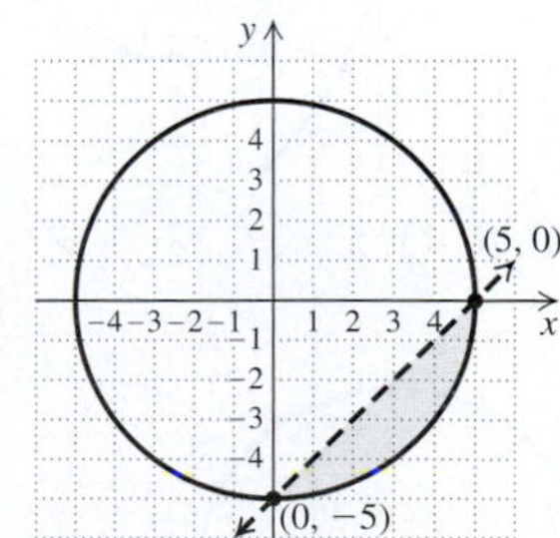

81.

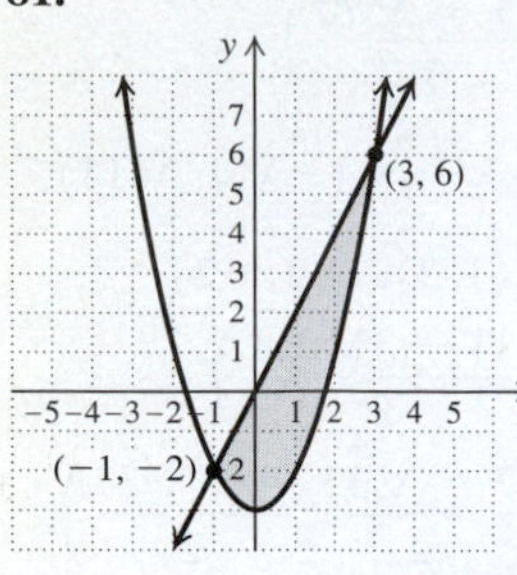

83.

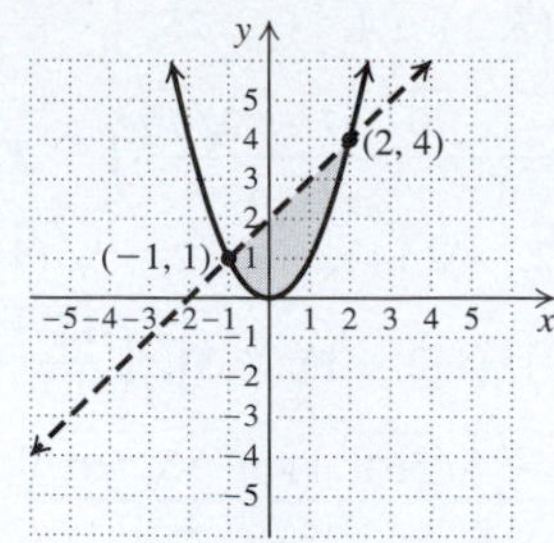

85. 2 **86.** 2.048 **87.** 81 **88.** 5

89. $(x - 2)^2 + (y - 3)^2 = 1$ **91.** $\frac{x^2}{4} + y^2 = 1$

93. There is no number x such that $\frac{x^2}{a^2} - \frac{\left(\frac{b}{a}x\right)^2}{b^2} = 1$, because the left side simplifies to $\frac{x^2}{a^2} - \frac{x^2}{a^2}$, which is 0.

95. Factor: $x^3 + y^3 = (x + y)(x^2 - xy + y^2)$. We know that $x + y = 1$, so $(x + y)^2 = x^2 + 2xy + y^2 = 1$, or $x^2 + y^2 = 1 - 2xy$. We also know that $xy = 1$, so $x^2 + y^2 = 1 - 2 \cdot 1 = -1$. Then $x^3 + y^3 = 1 \cdot (-1 - 1) = -2$. **97.** $(2, 4), (4, 2)$

99. $(3, -2), (-3, 2), (2, -3), (-2, 3)$ **101.** $(1.564, 2.448)$, $(0.138, 0.019)$ **103.** $(1.146, 3.146), (-1.841, 0.159)$

105. $(2.112, -0.109), (-13.041, -13.337)$

107. $(400, 1.431), (-400, 1.431), (400, -1.431), (-400, -1.431)$

Mid-Chapter Mixed Review: Chapter 10

1. True **2.** False **3.** True **4.** False **5.** (b) **6.** (e) **7.** (d) **8.** (a) **9.** (g) **10.** (f) **11.** (h) **12.** (c)

13. $V: (0, 0); F: (3, 0); D: x = -3$

$y^2 = 12x$

14. $V: (3, 2); F: (3, 3); D: y = 1$

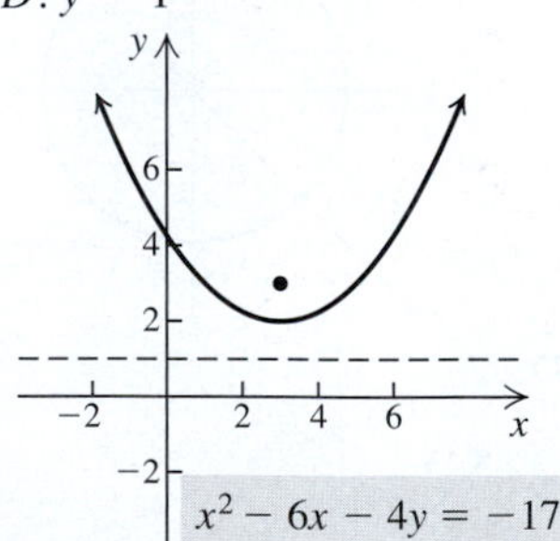

15. $(-2, 4); 5$

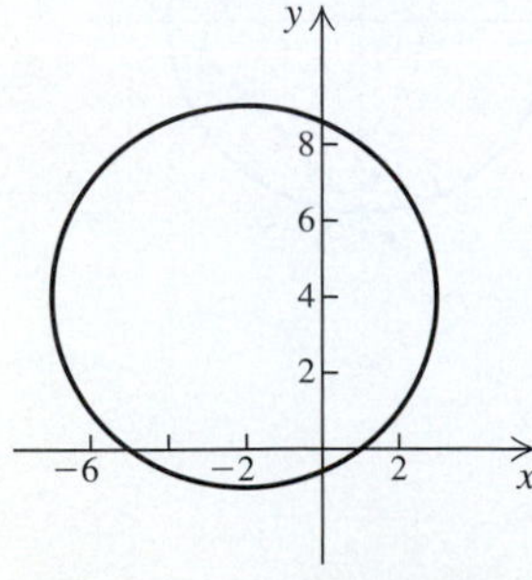

16. $(3, -1); 4$

$x^2 + y^2 - 6x + 2y - 6 = 0$

17. $V: (0, -3), (0, 3)$; $F: (0, -2\sqrt{2}), (0, 2\sqrt{2})$

$\frac{x^2}{1} + \frac{y^2}{9} = 1$

18. $V: (1, -6), (1, 4)$; $F: (1, -1 - \sqrt{21}), (1, -1 + \sqrt{21})$

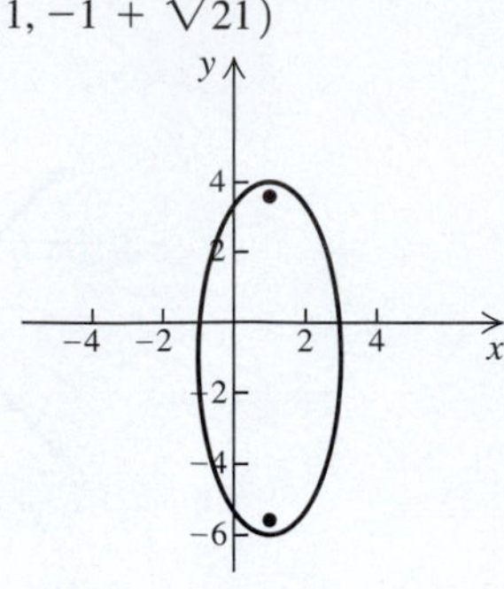

19. $C: (0, 0); V: (0, -4), (0, 4)$; $F: (0, -5), (0, 5)$; $A: y = -\frac{4}{3}x, y = \frac{4}{3}x$

$9y^2 - 16x^2 = 144$

20. $C: (-3, 2); V: (-4, 2), (-2, 2)$; $F: (-3 - \sqrt{5}, 2), (-3 + \sqrt{5}, 2)$; $A: y - 2 = 2(x + 3), y - 2 = -2(x + 3)$

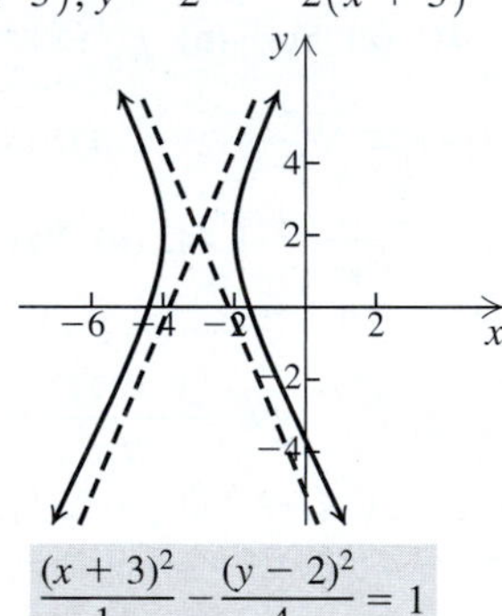

21. $(-2, -5), (5, 2)$ **22.** $(2, 2), (-2, -2)$

23. $(2\sqrt{3}i, 4), (2\sqrt{3}i, -4), (-2\sqrt{3}i, 4), (-2\sqrt{3}i, -4)$

24. $\left(-\frac{1}{3}, \frac{10}{3}\right), (-1, 2)$ **25.** -2 and 3

26.

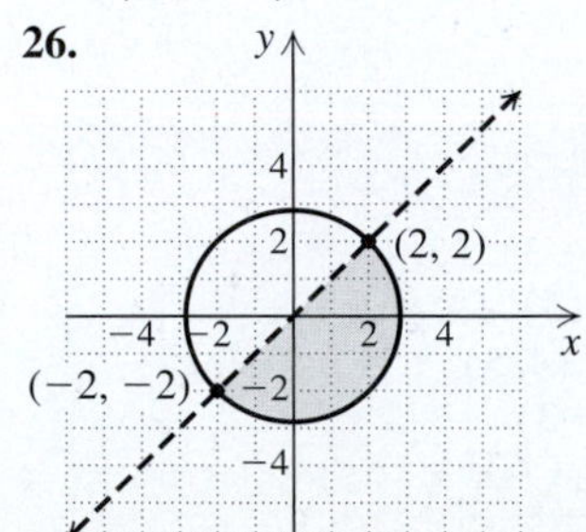

27.

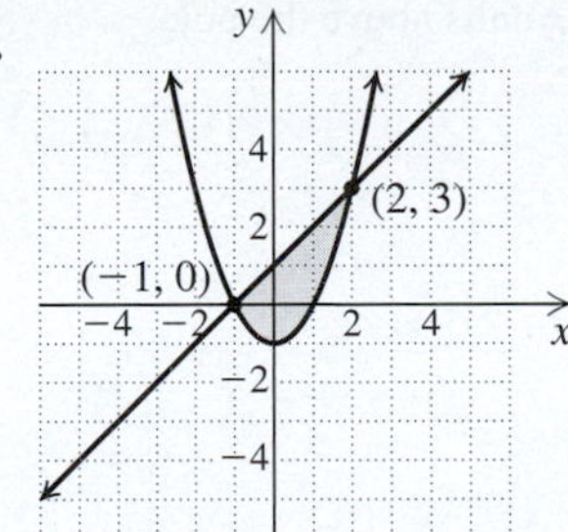

28. No; parabolas with a horizontal axis of symmetry fail the vertical-line test. **29.** No; the center of an ellipse is not part of the graph of the ellipse. Its coordinates do not satisfy the equation of the ellipse. **30.** No; the asymptotes of a hyperbola are not part of the graph of the hyperbola. The coordinates of points on the asymptotes do not satisfy the equation of the hyperbola.
31. Although we can always visualize the real-number solutions, we cannot visualize the imaginary-number solutions.

Exercise Set 10.5

1. $(0, -2)$ **3.** $(1, \sqrt{3})$ **5.** $(\sqrt{2}, 0)$ **7.** $(\sqrt{3}, 1)$
9. Ellipse or circle **11.** Hyperbola **13.** Parabola
15. Hyperbola **17.** Ellipse or circle

19.

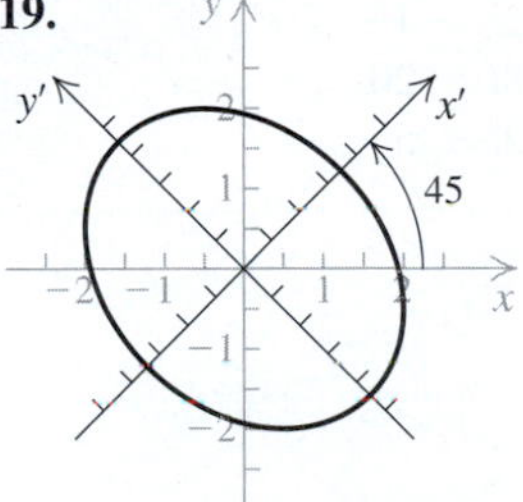

$\frac{(x')^2}{3} + \frac{(y')^2}{5} = 1$

21.

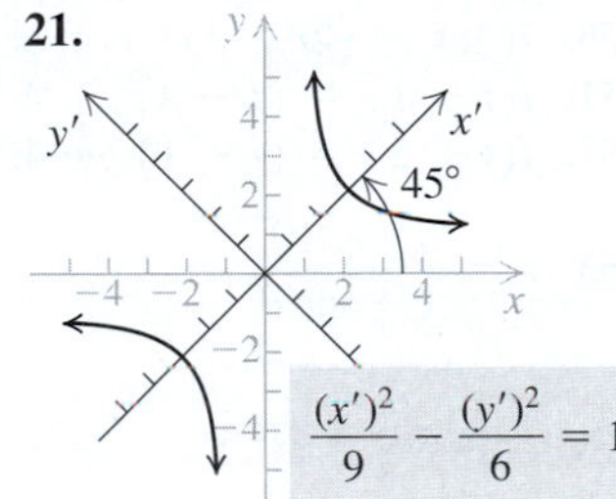

23.

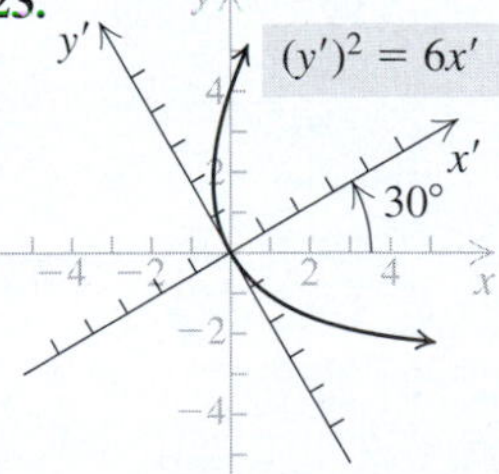

25.

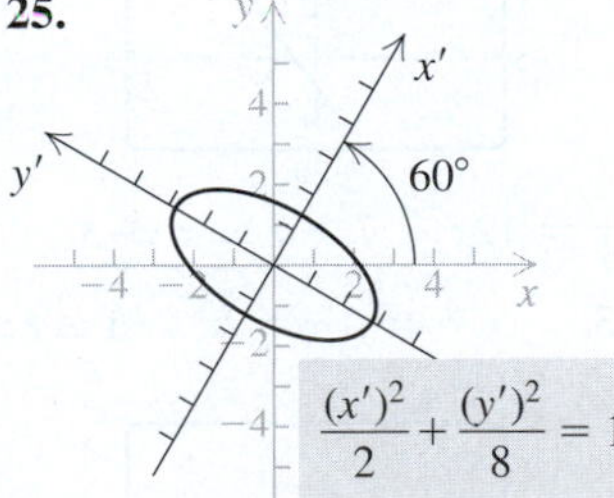

27.

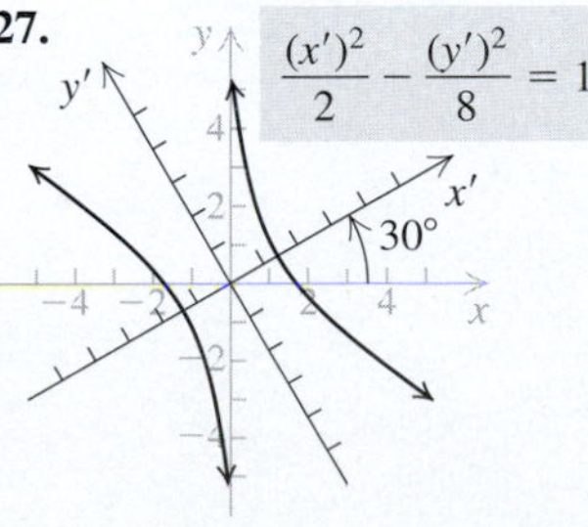

29.

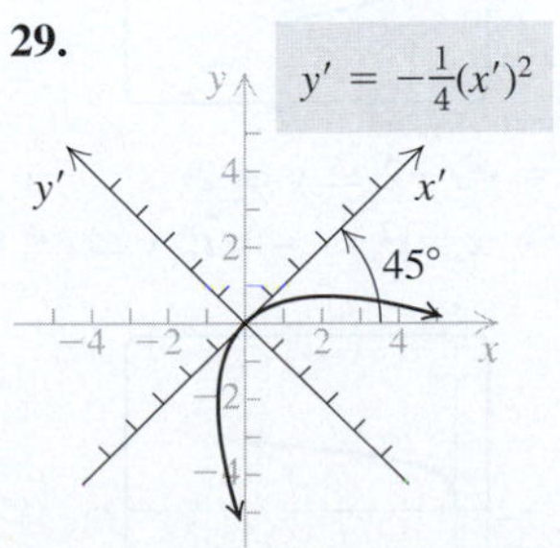

31.

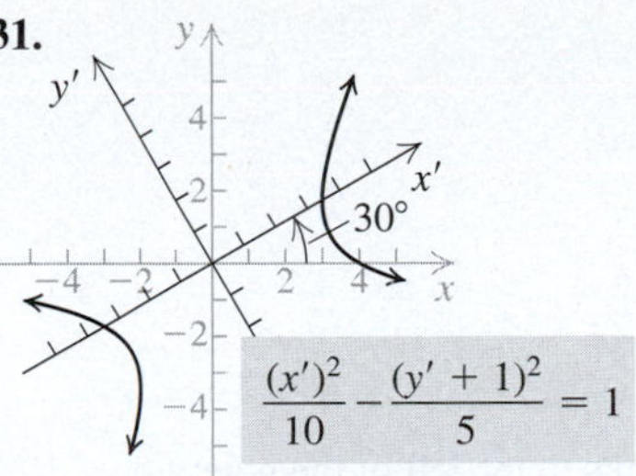

33.

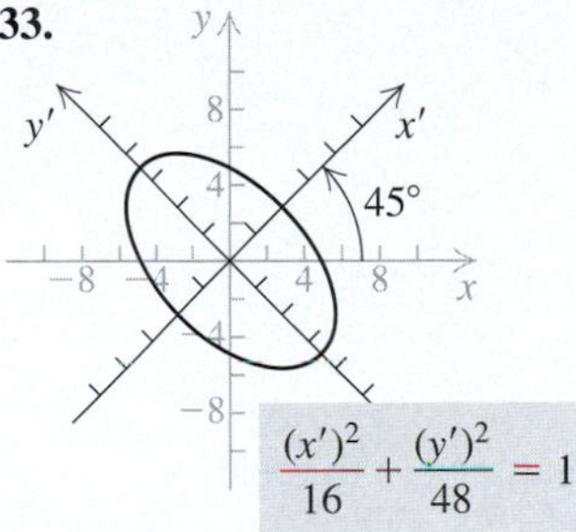

35.

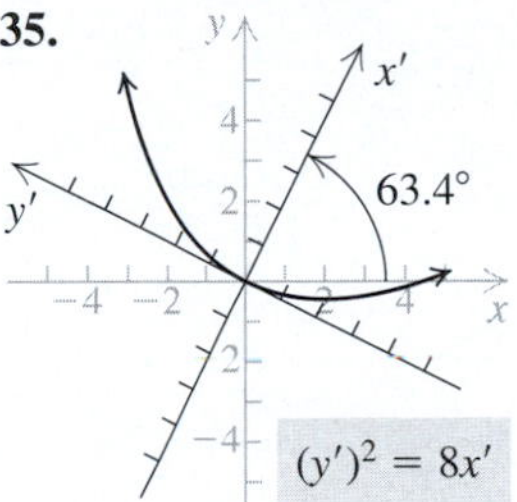

37.

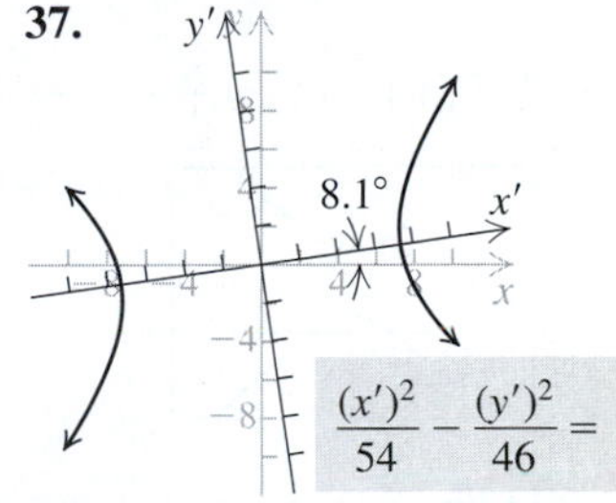

39. $\frac{2\pi}{3}$ **40.** $-\frac{7\pi}{4}$ **41.** $60°$ **42.** $135°$
43. $x = x' \cos\theta - y' \sin\theta,\ y = x' \sin\theta + y' \cos\theta$
45.
$$\begin{aligned} A' + C' &= A\cos^2\theta + B\sin\theta\cos\theta + C\sin^2\theta \\ &\quad + A\sin^2\theta - B\sin\theta\cos\theta + C\cos^2\theta \\ &= A(\sin^2\theta + \cos^2\theta) + C(\sin^2\theta + \cos^2\theta) \\ &= A + C \end{aligned}$$

Exercise Set 10.6

1. (b) **3.** (a) **5.** (d) **7. (a)** Parabola; **(b)** vertical, 1 unit to the right of the pole; **(c)** $(\frac{1}{2}, 0)$;
(d)

$r = \frac{1}{1 + \cos\theta}$

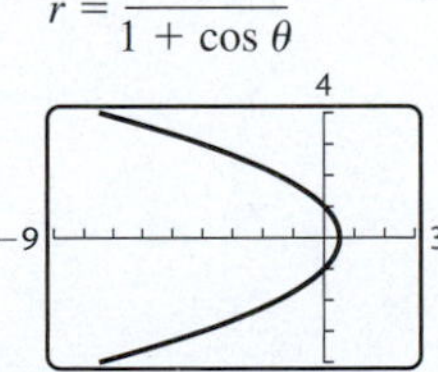

9. (a) Hyperbola; **(b)** horizontal, $\frac{3}{2}$ units below the pole;
(c) $\left(-3, \frac{\pi}{2}\right), \left(1, \frac{3\pi}{2}\right)$;

(d) $r = \dfrac{15}{5 - 10 \sin \theta}$

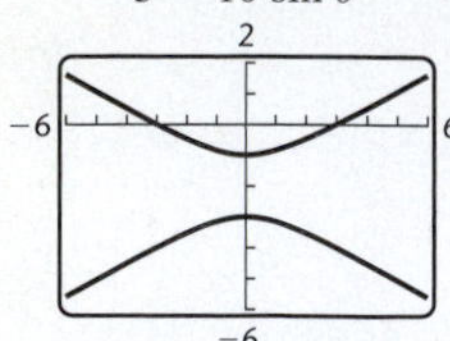

11. **(a)** Ellipse; **(b)** vertical, $\frac{8}{3}$ units to the left of the pole; **(c)** $\left(\frac{8}{3}, 0\right), \left(\frac{8}{9}, \pi\right)$;

(d) $r = \dfrac{8}{6 - 3 \cos \theta}$

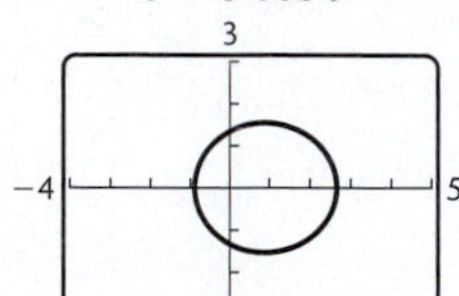

13. **(a)** Hyperbola; **(b)** horizontal, $\frac{4}{3}$ units above the pole; **(c)** $\left(\dfrac{4}{5}, \dfrac{\pi}{2}\right), \left(-4, \dfrac{3\pi}{2}\right)$;

(d) $r = \dfrac{20}{10 + 15 \sin \theta}$

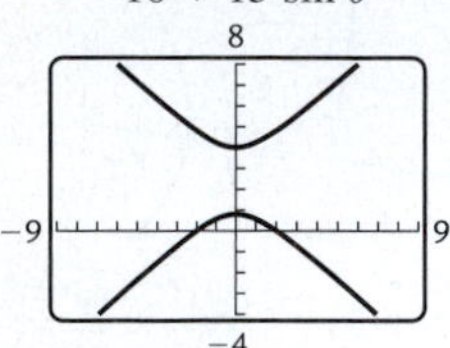

15. **(a)** Ellipse; **(b)** vertical, 3 units to the right of the pole; **(c)** $(1, 0), (3, \pi)$;

(d) $r = \dfrac{9}{6 + 3 \cos \theta}$

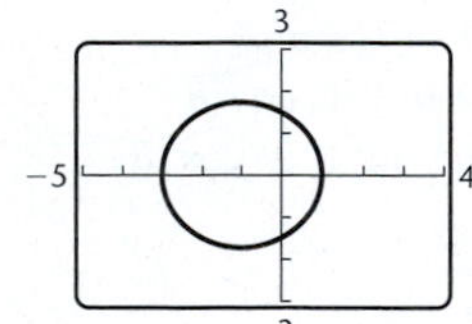

17. **(a)** Parabola; **(b)** horizontal, $\frac{3}{2}$ units below the pole; **(c)** $\left(\dfrac{3}{4}, \dfrac{3\pi}{2}\right)$;

(d) $r = \dfrac{3}{2 - 2 \sin \theta}$

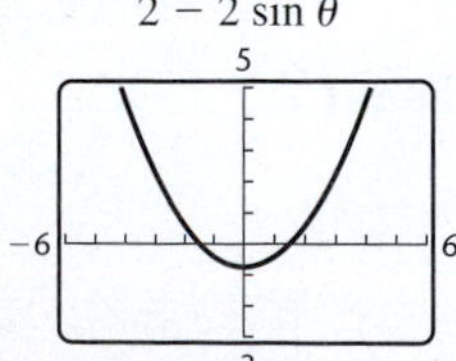

19. **(a)** Ellipse; **(b)** vertical, 4 units to the left of the pole; **(c)** $(4, 0), \left(\frac{4}{3}, \pi\right)$;

(d) $r = \dfrac{4}{2 - \cos \theta}$

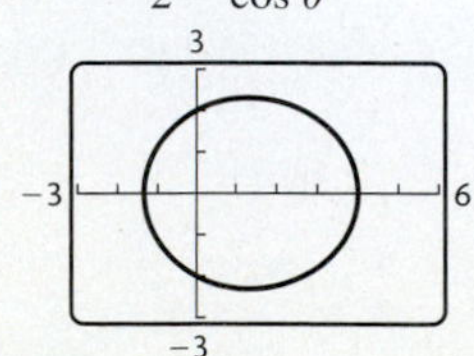

21. **(a)** Hyperbola; **(b)** horizontal, $\frac{7}{10}$ units above the pole; **(c)** $\left(\dfrac{7}{12}, \dfrac{\pi}{2}\right), \left(-\dfrac{7}{8}, \dfrac{3\pi}{2}\right)$;

(d) $r = \dfrac{7}{2 + 10 \sin \theta}$

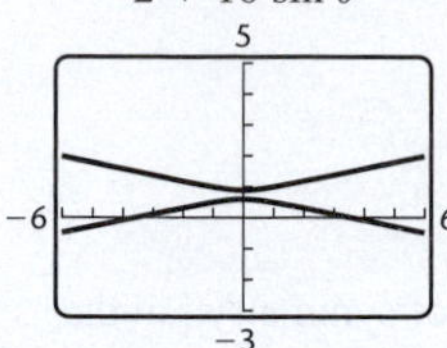

23. $y^2 + 2x - 1 = 0$ **25.** $x^2 - 3y^2 - 12y - 9 = 0$
27. $27x^2 + 36y^2 - 48x - 64 = 0$
29. $4x^2 - 5y^2 + 24y - 16 = 0$
31. $3x^2 + 4y^2 + 6x - 9 = 0$ **33.** $4x^2 - 12y - 9 = 0$
35. $3x^2 + 4y^2 - 8x - 16 = 0$
37. $4x^2 - 96y^2 + 140y - 49 = 0$ **39.** $r = \dfrac{6}{1 + 2 \sin \theta}$
41. $r = \dfrac{4}{1 + \cos \theta}$ **43.** $r = \dfrac{2}{2 - \cos \theta}$
45. $r = \dfrac{15}{4 + 3 \sin \theta}$ **47.** $r = \dfrac{8}{1 - 4 \sin \theta}$
49. $f(t) = (t - 3)^2 + 4$, or $t^2 - 6t + 13$
50. $f(2t) = (2t - 3)^2 + 4$, or $4t^2 - 12t + 13$
51. $f(t - 1) = (t - 4)^2 + 4$, or $t^2 - 8t + 20$
52. $f(t + 2) = (t - 1)^2 + 4$, or $t^2 - 2t + 5$
53. $r = \dfrac{1.5 \times 10^8}{1 + \sin \theta}$

Exercise Set 10.7

1. $x = \frac{1}{2}t,\ y = 6t - 7;\ -1 \le t \le 6$

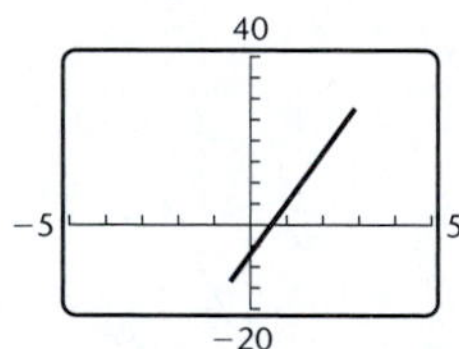

$y = 12x - 7, -\frac{1}{2} \le x \le 3$

3. $x = 4t^2,\ y = 2t;\ -1 \le t \le 1$

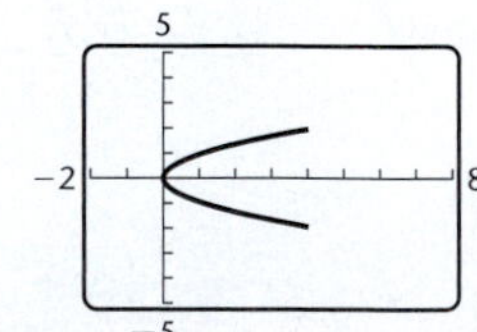

$x = y^2, -2 \le y \le 2$

5. $x = t^2,\ y = \sqrt{t};\ 0 \le t \le 4$

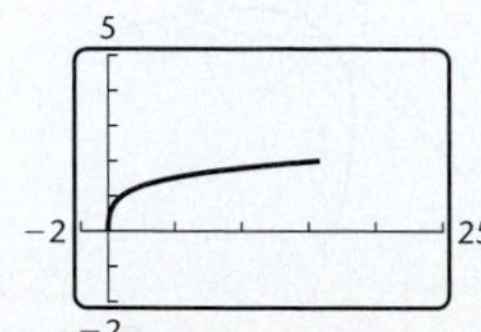

$x = y^4, 0 \le y \le 2$

7. $x = t + 3, \quad y = \dfrac{1}{t+3}; \quad -2 \le t \le 2$

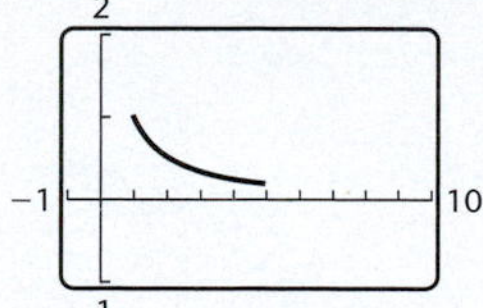

$y = \dfrac{1}{x}, 1 \le x \le 5$

9. $x = 2t - 1, \quad y = t^2; \quad -3 \le t \le 3$

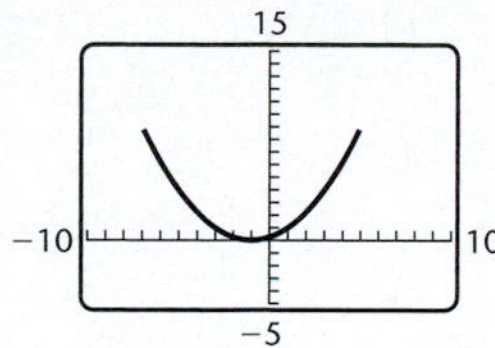

$y = \frac{1}{4}(x + 1)^2, -7 \le x \le 5$

11. $x = e^{-t}, \quad y = e^t; \quad -\infty < t < \infty$

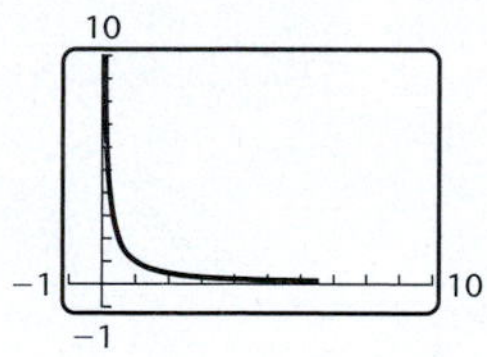

$y = \dfrac{1}{x}, x > 0$

13. $x = 3\cos t, \quad y = 3\sin t; \quad 0 \le t \le 2\pi$

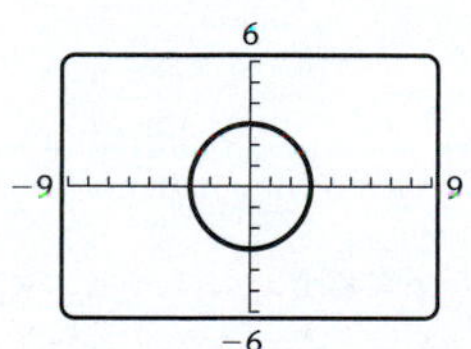

$x^2 + y^2 = 9, -3 \le x \le 3$

15. $x = \cos t, \quad y = 2\sin t; \quad 0 \le t \le 2\pi$

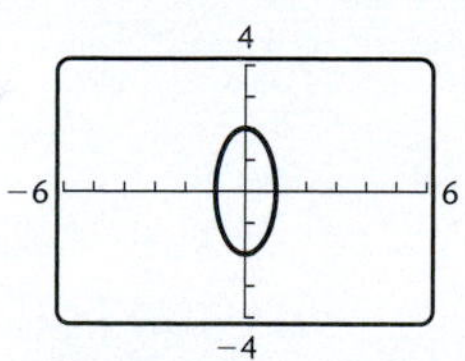

$x^2 + \dfrac{y^2}{4} = 1, -1 \le x \le 1$

17. $x = \sec t, \quad y = \cos t; \quad -\frac{\pi}{2} < t < \frac{\pi}{2}$

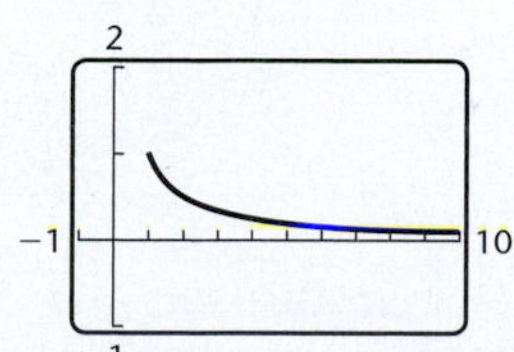

$y = \dfrac{1}{x}, x \ge 1$

19. $x = 1 + 2\cos t, \quad y = 2 + 2\sin t; \quad 0 \le t \le 2\pi$

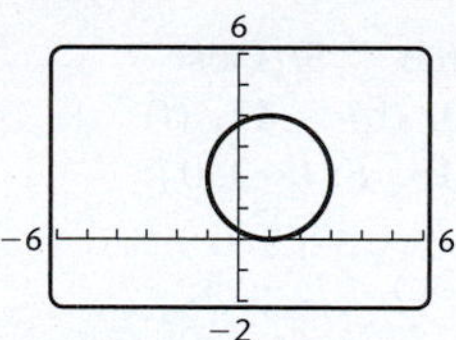

$(x - 1)^2 + (y - 2)^2 = 4, -1 \le x \le 3$

21. 0.7071 **23.** −0.2588 **25.** 0.7265 **27.** 0.5460

29. Answers may vary. $x = t, y = 4t - 3; x = \dfrac{t}{4} + 3, y = t + 9$

31. Answers may vary. $x = t, y = (t - 2)^2 - 6t; x = t + 2, y = t^2 - 6t - 12$ **33.** **(a)** $x = 40\sqrt{3}t, y = 7 + 40t - 16t^2$; **(b)** $x = 40\sqrt{3}t, \quad y = 7 + 40t - 16t^2$

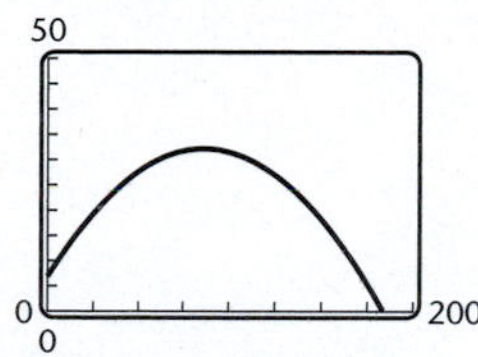

(c) 31 ft, 23 ft; **(d)** about 2.7 sec; **(e)** about 187.1 ft; **(f)** 32 ft

35. $x = 2(t - \sin t), \quad y = 2(1 - \cos t); \quad 0 \le t \le 4\pi$

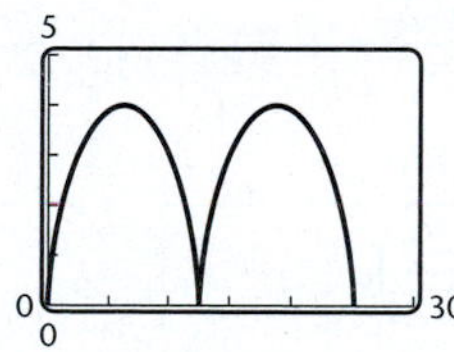

37. $x = t - \sin t, \quad y = 1 - \cos t; \quad -2\pi \le t \le 2\pi$

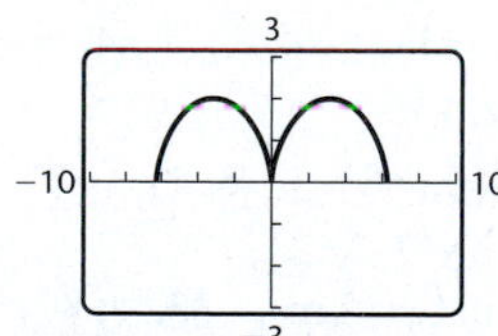

39. $y = x^3$

40. $x = y^3$

41. $f(x) = \sqrt{x - 2}$

42.

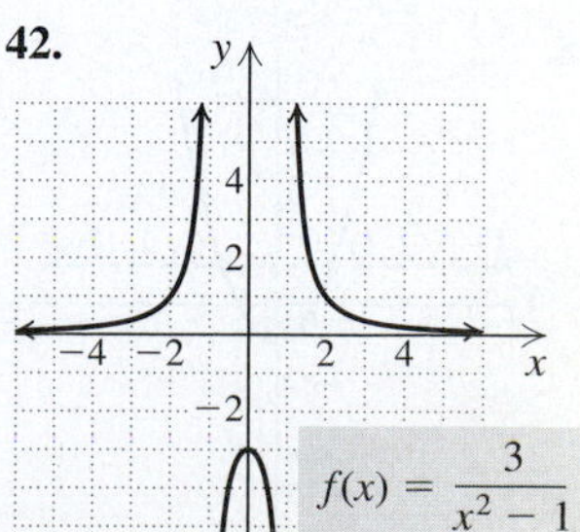

43. $x = 3\cos t, y = -3\sin t$

Review Exercises: Chapter 10

1. True **2.** False **3.** False **4.** True **5.** False **6.** (d) **7.** (a) **8.** (e) **9.** (g) **10.** (b) **11.** (f) **12.** (h) **13.** (c) **14.** $x^2 = -6y$ **15.** F: $(-3, 0)$; V: $(0, 0)$; D: $x = 3$ **16.** V: $(-5, 8)$; F: $\left(-5, \frac{15}{2}\right)$; D: $y = \frac{17}{2}$ **17.** C: $(2, -1)$; V: $(-3, -1)$, $(7, -1)$; F: $(-1, -1)$, $(5, -1)$;

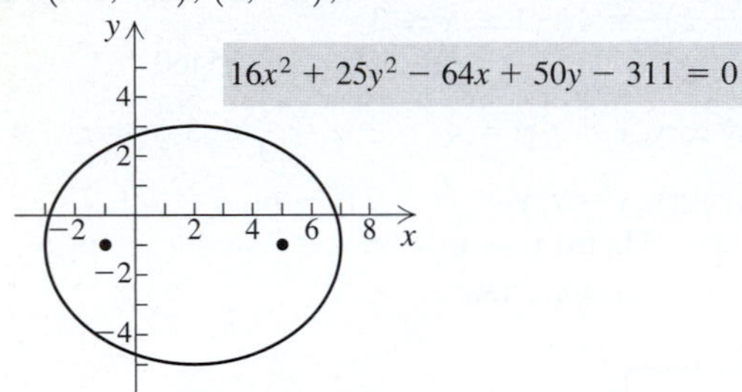

18. $\dfrac{x^2}{9} + \dfrac{y^2}{16} = 1$

19. C: $\left(-2, \frac{1}{4}\right)$; V: $\left(0, \frac{1}{4}\right)$, $\left(-4, \frac{1}{4}\right)$; F: $\left(-2 + \sqrt{6}, \frac{1}{4}\right)$, $\left(-2 - \sqrt{6}, \frac{1}{4}\right)$; A: $y - \frac{1}{4} = \frac{\sqrt{2}}{2}(x + 2)$, $y - \frac{1}{4} = -\frac{\sqrt{2}}{2}(x + 2)$

20. 0.167 ft **21.** $(-8\sqrt{2}, 8)$, $(8\sqrt{2}, 8)$

22. $\left(3, \frac{\sqrt{29}}{2}\right)$, $\left(-3, \frac{\sqrt{29}}{2}\right)$, $\left(3, -\frac{\sqrt{29}}{2}\right)$, $\left(-3, -\frac{\sqrt{29}}{2}\right)$ **23.** $(7, 4)$ **24.** $(2, 2)$, $\left(\frac{32}{9}, -\frac{10}{9}\right)$

25. $(0, -3)$, $(2, 1)$ **26.** $(4, 3)$, $(4, -3)$, $(-4, 3)$, $(-4, -3)$

27. $(-\sqrt{3}, 0)$, $(\sqrt{3}, 0)$, $(-2, 1)$, $(2, 1)$

28. $\left(-\frac{3}{5}, \frac{21}{5}\right)$, $(3, -3)$ **29.** $(6, 8)$, $(6, -8)$, $(-6, 8)$, $(-6, -8)$ **30.** $(2, 2)$, $(-2, -2)$, $(2\sqrt{2}, \sqrt{2})$, $(-2\sqrt{2}, -\sqrt{2})$ **31.** 7, 4 **32.** 7 m by 12 m **33.** 4, 8 **34.** 32 cm, 20 cm **35.** 11 ft, 3 ft

36.

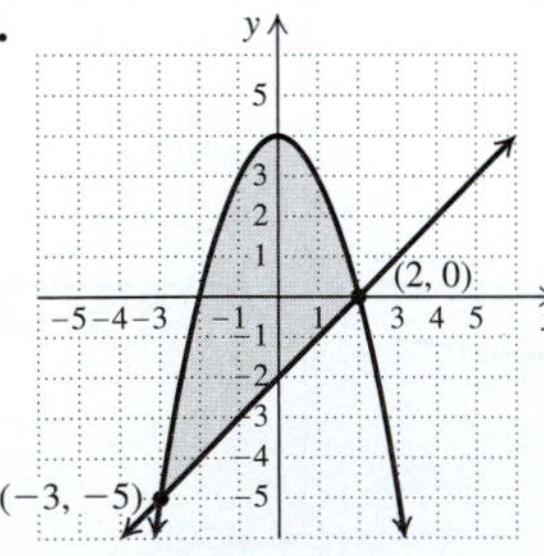

37.

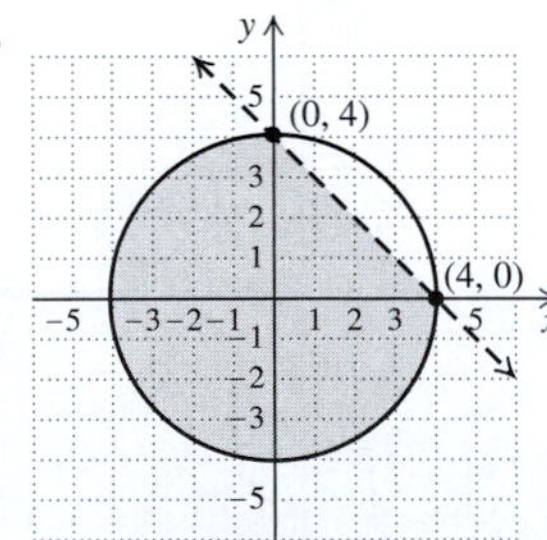

38.

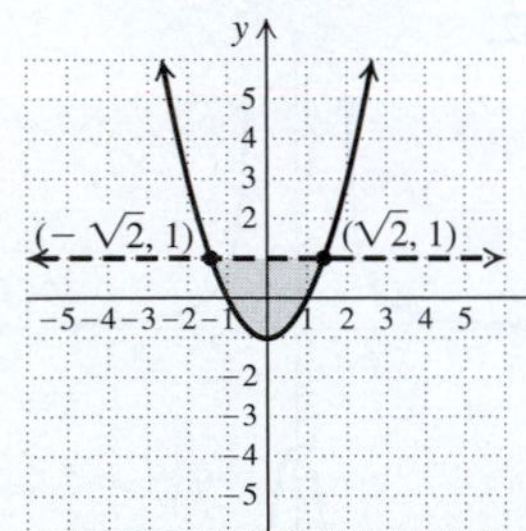

39.

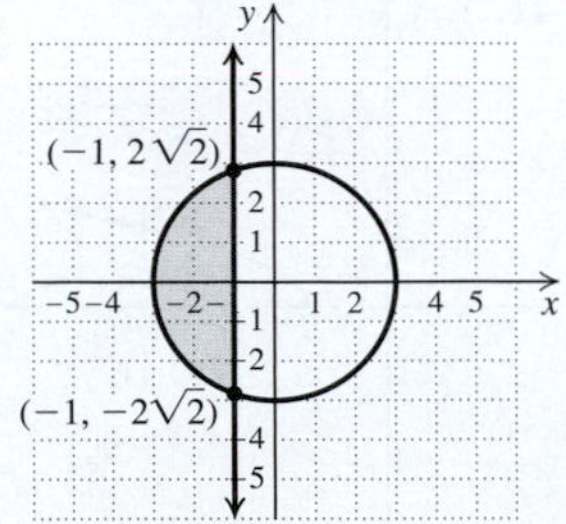

40.

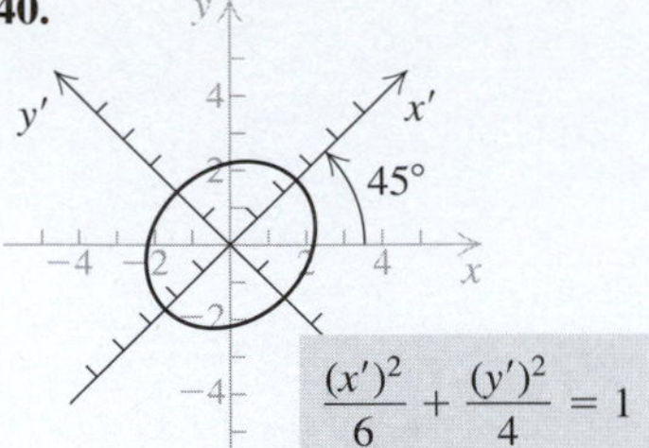

41.

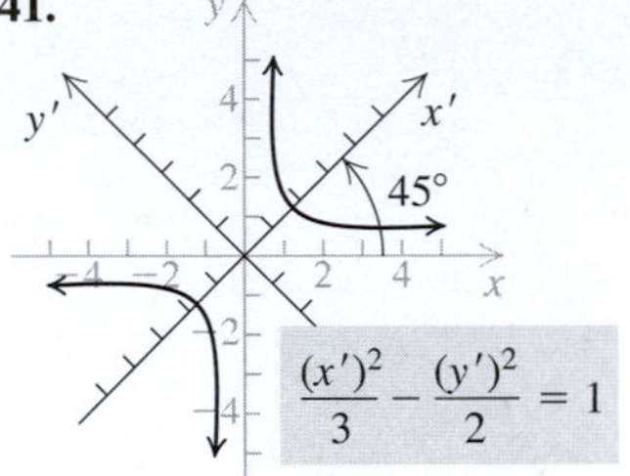

42.

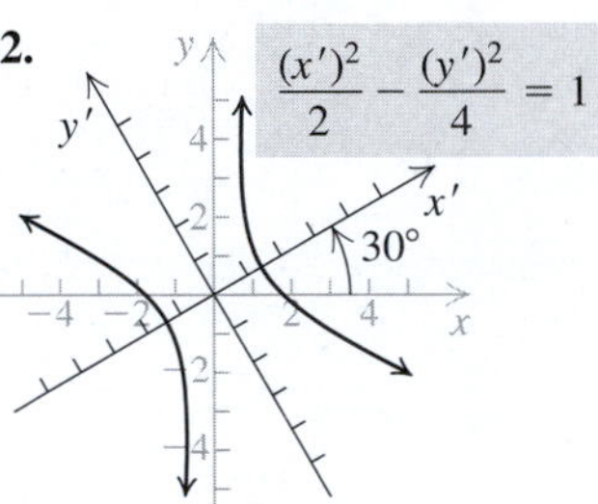

43.

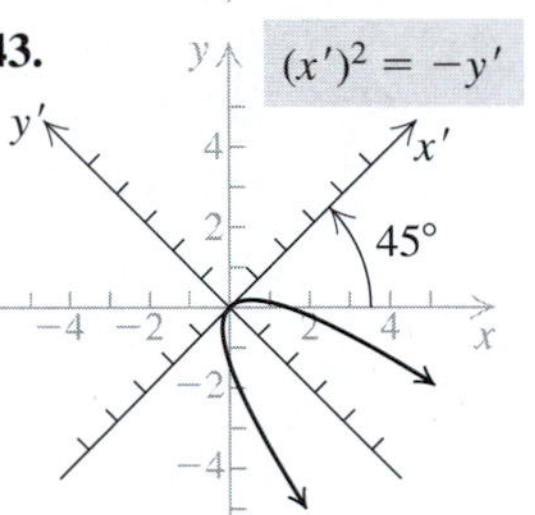

44.

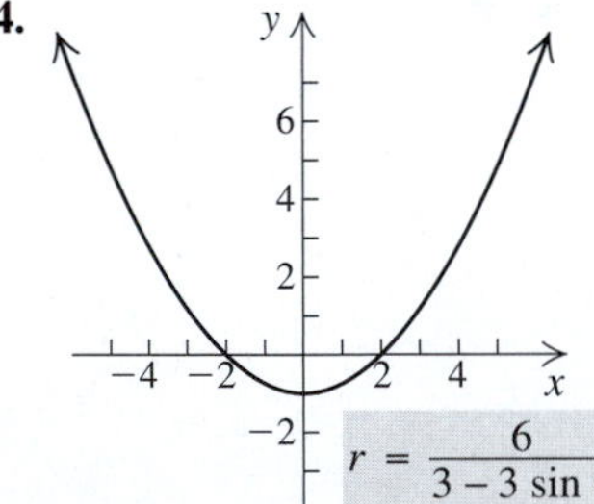

Horizontal directrix 2 units below the pole; vertex: $\left(1, \frac{3\pi}{2}\right)$

45.

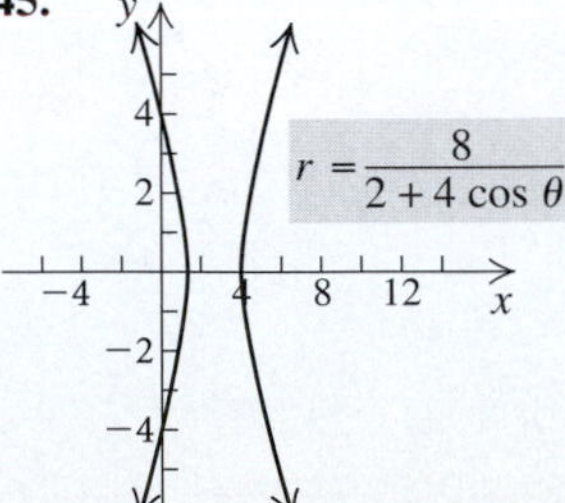

Vertical directrix 2 units to the right of the pole; vertices: $\left(\frac{4}{3}, 0\right)$, $(-4, \pi)$

46.

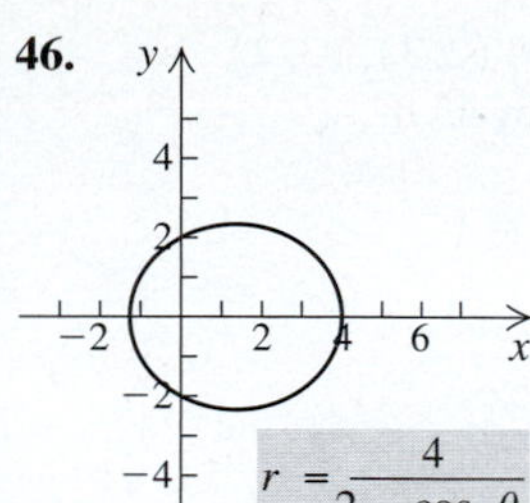

Vertical directrix 4 units to the left of the pole; vertices: $(4, 0)$, $\left(\frac{4}{3}, \pi\right)$

47.

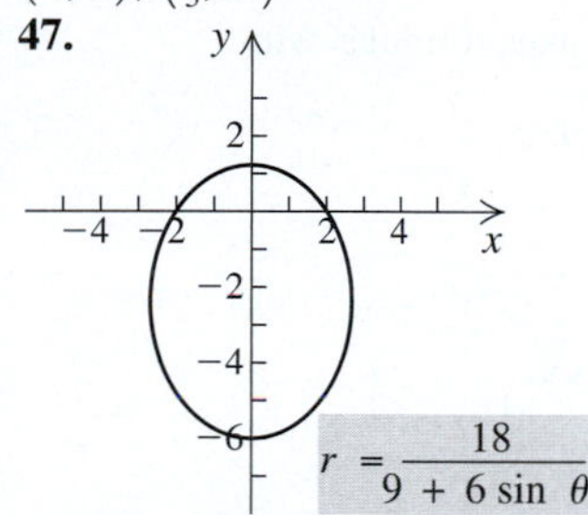

Horizontal directrix 3 units above the pole; vertices: $\left(\dfrac{6}{5}, \dfrac{\pi}{2}\right)$, $\left(6, \dfrac{3\pi}{2}\right)$

48. $x^2 - 4y - 4 = 0$
49. $3x^2 - y^2 - 16x + 16 = 0$
50. $3x^2 + 4y^2 - 8x - 16 = 0$
51. $9x^2 + 5y^2 + 24y - 36 = 0$
52. $r = \dfrac{1}{1 + \frac{1}{2}\cos\theta}$, or $r = \dfrac{2}{2 + \cos\theta}$ **53.** $r = \dfrac{18}{1 - 3\sin\theta}$
54. $r = \dfrac{4}{1 - \cos\theta}$ **55.** $r = \dfrac{6}{1 + 2\sin\theta}$

56.

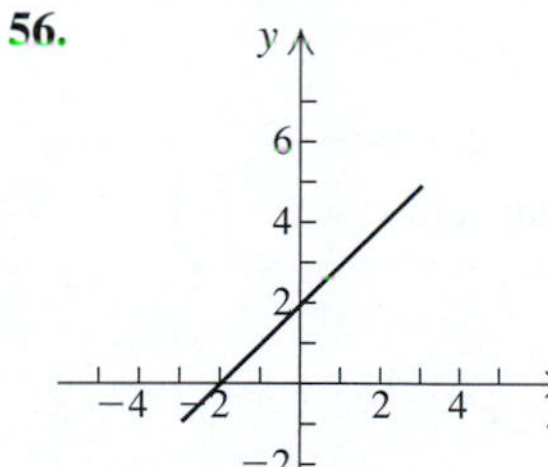

$x = t,\ y = 2 + t; -3 \le t \le 3$

$y = 2 + x, -3 \le x \le 3$

57.

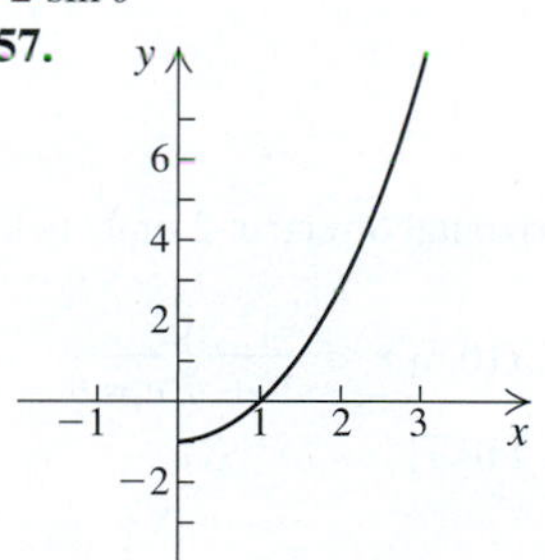

$x = \sqrt{t},\ y = t - 1;\ 0 \le t \le 9$

$y = x^2 - 1, 0 \le x \le 3$

58.

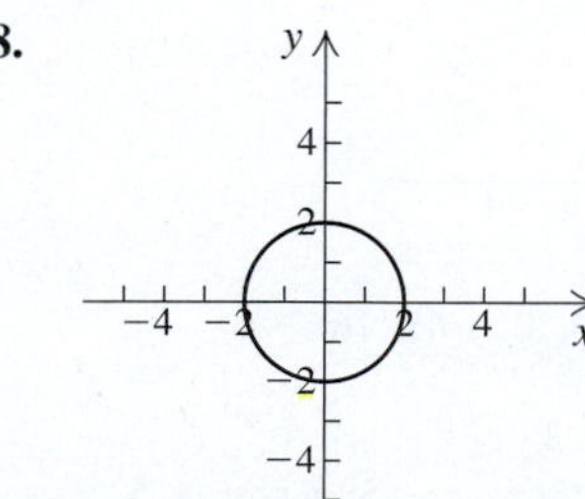

$x = 2\cos t,\ y = 2\sin t;\ 0 \le t \le 2\pi$

$x^2 + y^2 = 4$

59.

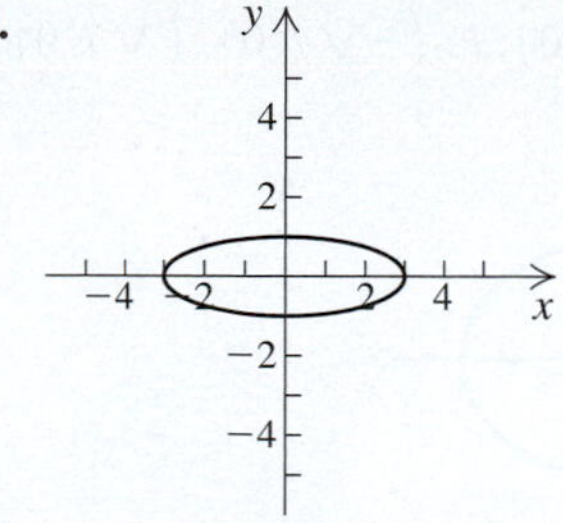

$x = 3\sin t,\ y = \cos t;\ 0 \le t \le 2\pi$

$\dfrac{x^2}{9} + y^2 = 1$

60. Answers may vary. $x = t, y = 2t - 3$; $x = t + 1,\ y = 2t - 1$
61. Answers may vary. $x = t, y = t^2 + 4$; $x = t - 2$, $y = t^2 - 4t + 8$ **62. (a)** $x = 75\sqrt{2}t, y = 75\sqrt{2}t - 16t^2$; **(b)** 174.2 ft, 60.4 ft; **(c)** about 6.6 sec; **(d)** about 700.0 ft; **(e)** about 175.8 ft **63.** B **64.** D **65.** C
66. $\dfrac{8}{7}, \dfrac{7}{2}$ **67.** $(x - 2)^2 + (y - 1)^2 = 100$
68. $x^2 + \dfrac{y^2}{9} = 1$ **69.** $\dfrac{x^2}{778.41} - \dfrac{y^2}{39{,}221.59} = 1$
70. See the development of the formula for the standard form of a parabola that follows Fig. 1 at the beginning of Section 10.1.
71. Circles and ellipses are not functions. **72.** The procedure for rotation of axes would be done first when $B \neq 0$. Then we would proceed as when $B = 0$. **73.** Each graph is an ellipse. The value of e determines the location of the center and the lengths of the major and minor axes. The larger the value of e, the farther the center is from the pole and the longer the axes.

Test: Chapter 10

1. [10.3] (c) **2.** [10.1] (b) **3.** [10.2] (a) **4.** [10.2] (d)
5. [10.1] V: $(0, 0)$; F: $(0, 3)$; D: $y = -3$

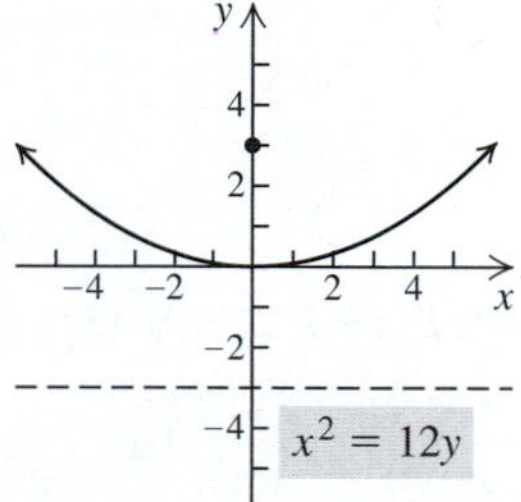

6. [10.1] V: $(-1, -1)$; F: $(1, -1)$; D: $x = -3$

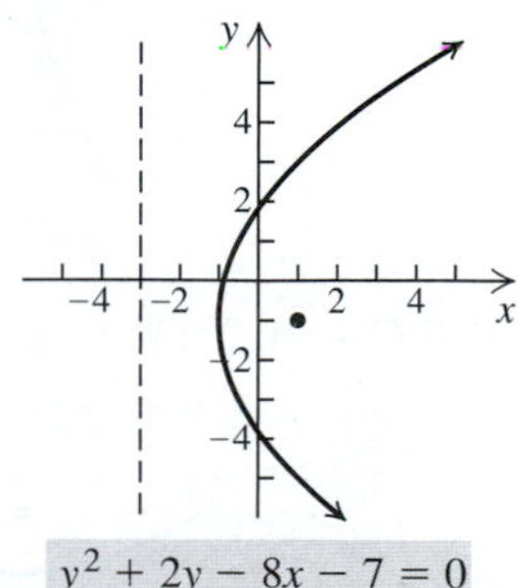

7. [10.1] $x^2 = 8y$
8. [10.2] Center: $(-1, 3)$; radius: 5

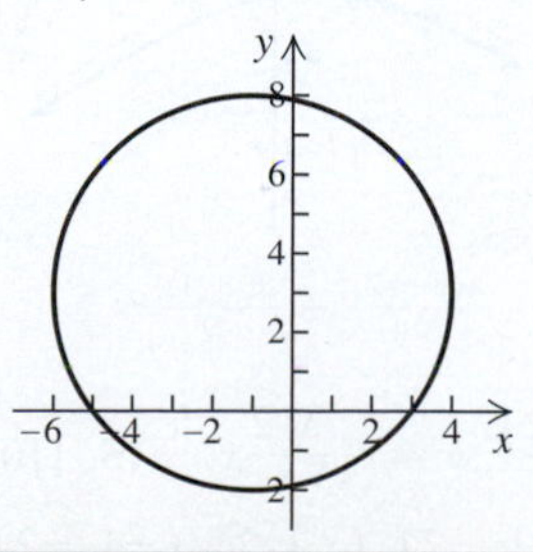

$x^2 + y^2 + 2x - 6y - 15 = 0$

9. [10.2] C: $(0, 0)$; V: $(-4, 0)$, $(4, 0)$; F: $(-\sqrt{7}, 0)$, $(\sqrt{7}, 0)$

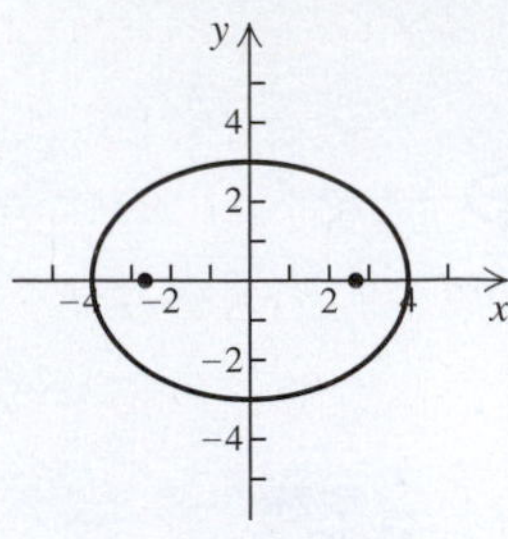

10. [10.2] C: $(-1, 2)$; V: $(-1, -1)$, $(-1, 5)$; F: $(-1, 2 - \sqrt{5})$, $(-1, 2 + \sqrt{5})$

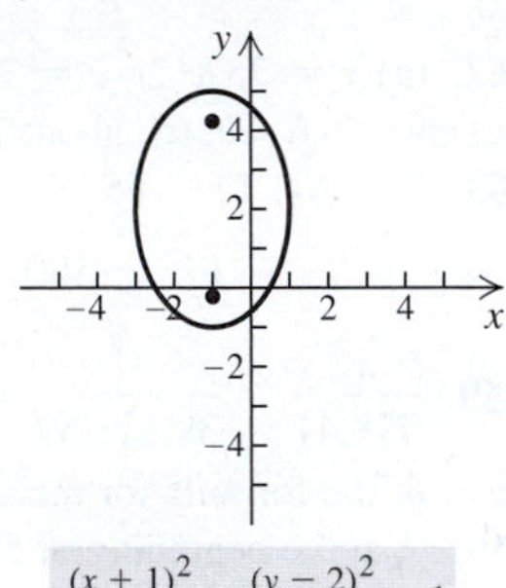

11. [10.2] $\frac{x^2}{4} + \frac{y^2}{25} = 1$

12. [10.3] C: $(0, 0)$; V: $(-1, 0)$, $(1, 0)$; F: $(-\sqrt{5}, 0)$, $(\sqrt{5}, 0)$; A: $y = -2x$, $y = 2x$

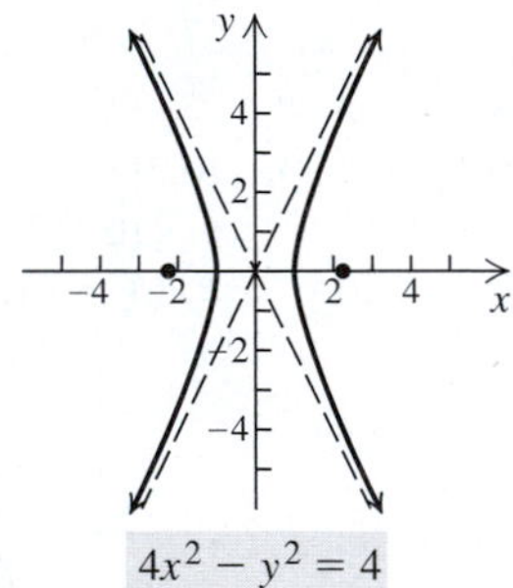

13. [10.3] C: $(-1, 2)$; V: $(-1, 0)$, $(-1, 4)$; F: $(-1, 2 - \sqrt{13})$, $(-1, 2 + \sqrt{13})$; A: $y = -\frac{2}{3}x + \frac{4}{3}$, $y = \frac{2}{3}x + \frac{8}{3}$

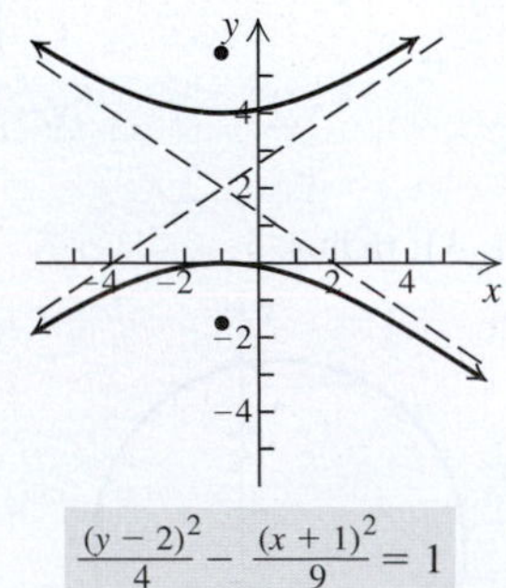

14. [10.3] $y = \frac{\sqrt{2}}{2}x$, $y = -\frac{\sqrt{2}}{2}x$ **15.** [10.1] $\frac{27}{8}$ in.

16. [10.4] $(1, 2)$, $(1, -2)$, $(-1, 2)$, $(-1, -2)$

17. [10.4] $(3, -2)$, $(-2, 3)$ **18.** [10.4] $(2, 3)$, $(3, 2)$

19. [10.4] 5 ft by 4 ft **20.** [10.4] 60 ft by 45 ft

21. [10.4]

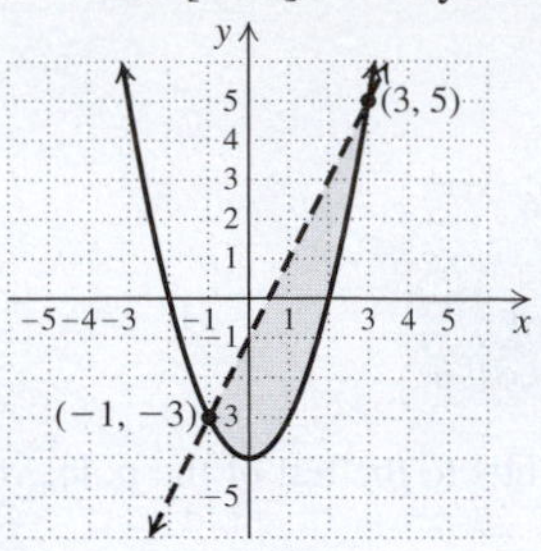

22. [10.5] After using the rotation of axes formulas with $\theta = 45°$, we have $\frac{(x')^2}{9} + (y')^2 = 1$.

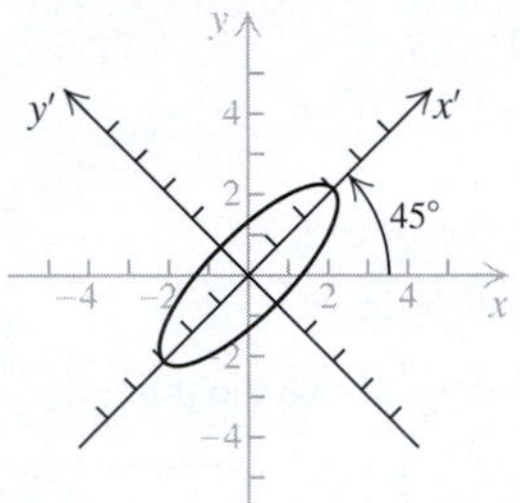

23. [10.6]

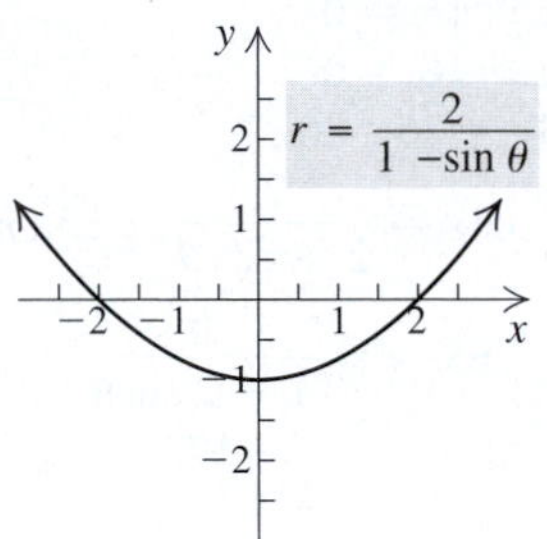

Horizontal directrix 2 units below the pole; vertex: $\left(1, \frac{3\pi}{2}\right)$

24. [10.6] $r = \frac{6}{1 + 2\cos\theta}$

25. [10.7]

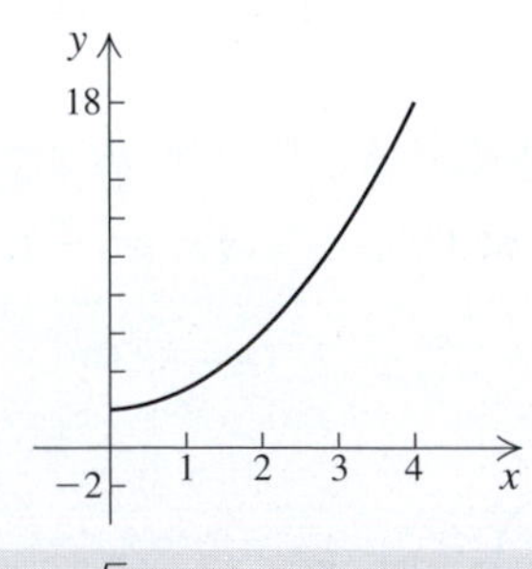

26. [10.7] $x^2 + y^2 = 9$, $-3 \le x \le 3$

27. [10.7] Answers may vary. $x = t$, $y = t - 5$; $x = t + 5$, $y = t$

28. [10.7] **(a)** $x = 125\sqrt{3}t$, $y = 10 + 125t - 16t^2$; **(b)** 119 ft, 241 ft; **(c)** about 7.9 sec; **(d)** about 1710 ft; **(e)** about 254 ft

29. [10.1] A **30.** [10.2] $(x - 3)^2 + (y + 1)^2 = 8$

Chapter 11

Exercise Set 11.1

1. 3, 7, 11, 15; 39; 59 **3.** $2, \frac{3}{2}, \frac{4}{3}, \frac{5}{4}; \frac{10}{9}; \frac{15}{14}$
5. $0, \frac{3}{5}, \frac{4}{5}, \frac{15}{17}; \frac{99}{101}; \frac{112}{113}$ **7.** $-1, 4, -9, 16; 100; -225$
9. 7, 3, 7, 3; 3; 7 **11.** 34 **13.** 225 **15.** $-33{,}880$ **17.** 67

19.

n	u_n
1	2
2	2.25
3	2.3704
4	2.4414
5	2.4883
6	2.5216
7	2.5465
8	2.5658
9	2.5812
10	2.5937

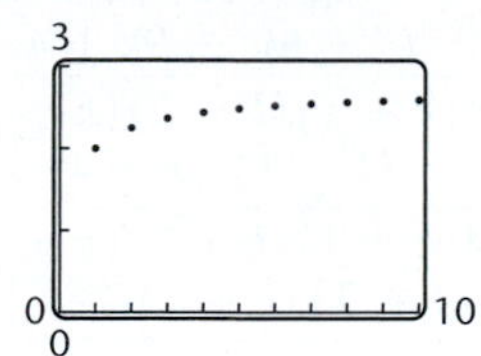

21.

n	u_n
1	2
2	1.5538
3	1.4988
4	1.4914
5	1.4904
6	1.4902
7	1.4902
8	1.4902
9	1.4902
10	1.4902

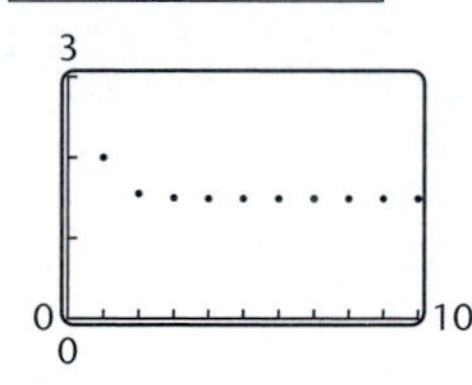

23. $2n$ **25.** $(-1)^n \cdot 2 \cdot 3^{n-1}$ **27.** $\dfrac{n+1}{n+2}$
29. $n(n+1)$ **31.** $\log 10^{n-1}$, or $n - 1$ **33.** 6; 28
35. 20; 30 **37.** $\frac{1}{2} + \frac{1}{4} + \frac{1}{6} + \frac{1}{8} + \frac{1}{10} = \frac{137}{120}$
39. $1 + 2 + 4 + 8 + 16 + 32 + 64 = 127$
41. $\ln 7 + \ln 8 + \ln 9 + \ln 10 = \ln(7 \cdot 8 \cdot 9 \cdot 10) = \ln 5040 \approx 8.5252$
43. $\frac{1}{2} + \frac{2}{3} + \frac{3}{4} + \frac{4}{5} + \frac{5}{6} + \frac{6}{7} + \frac{7}{8} + \frac{8}{9} = \frac{15{,}551}{2520}$
45. $-1 + 1 - 1 + 1 - 1 = -1$
47. $3 - 6 + 9 - 12 + 15 - 18 + 21 - 24 = -12$
49. $2 + 1 + \frac{2}{5} + \frac{1}{5} + \frac{2}{17} + \frac{1}{13} + \frac{2}{37} = \frac{157{,}351}{40{,}885}$
51. $3 + 2 + 3 + 6 + 11 + 18 = 43$
53. $\frac{1}{2} + \frac{2}{3} + \frac{4}{5} + \frac{8}{9} + \frac{16}{17} + \frac{32}{33} + \frac{64}{65} + \frac{128}{129} + \frac{256}{257} + \frac{512}{513} + \frac{1024}{1025} \approx 9.736$ **55.** $\sum_{k=1}^{\infty} 5k$ **57.** $\sum_{k=1}^{6} (-1)^{k+1} 2^k$
59. $\sum_{k=1}^{6} (-1)^k \dfrac{k}{k+1}$ **61.** $\sum_{k=2}^{n} (-1)^k k^2$ **63.** $\sum_{k=1}^{\infty} \dfrac{1}{k(k+1)}$
65. $4, 1\frac{1}{4}, 1\frac{4}{5}, 1\frac{5}{9}$ **67.** $6561, -81, 9i, -3\sqrt{i}$ **69.** 2, 3, 5, 8
71. **(a)** \$4150, \$4305.63, \$4467.09, \$4634.60, \$4808.40, \$4988.71, \$5175.79, \$5369.88, \$5571.25, \$5780.18; **(b)** \$8352.61
73. \$16.20, \$18.45, \$20.70, \$22.95, \$25.20, \$27.45, \$29.70, \$31.95, \$34.20, \$36.45 **75.** 1, 1, 2, 3, 5, 8, 13
77. **(a)** $a_n = -3.210714286n^2 + 15.76585714n + 12.76457143$; **(b)** \$11.33 **78.** $(-1, -3)$ **79.** Illinois: 16,200 acres; Ohio: 7200 acres **80.** $(3, -2)$; 4 **81.** $\left(-\dfrac{5}{2}, 4\right); \dfrac{\sqrt{97}}{2}$
83. $i, -1, -i, 1, i; i$ **85.** $\ln(1 \cdot 2 \cdot 3 \cdot \cdots \cdot n)$

Exercise Set 11.2

1. $a_1 = 3, d = 5$ **3.** $a_1 = 9, d = -4$
5. $a_1 = \frac{3}{2}, d = \frac{3}{4}$ **7.** $a_1 = \$316, d = -\$3$
9. $a_{12} = 46$ **11.** $a_{14} = -\frac{17}{3}$ **13.** $a_{10} = \$7941.62$
15. 27th **17.** 46th **19.** $a_1 = 5$ **21.** $n = 39$
23. $a_1 = \frac{1}{3}; d = \frac{1}{2}; \frac{1}{3}, \frac{5}{6}, \frac{4}{3}, \frac{11}{6}, \frac{7}{3}$ **25.** 670 **27.** 160,400
29. 735 **31.** 990 **33.** 1760 **35.** $\frac{65}{2}$ **37.** $-\frac{6026}{13}$
39. 4960¢, or \$49.60 **41.** 1320 seats **43.** Yes; 32; 1600 ft
45. 3 plants; 171 plants **47.** Yes; 3 **48.** $(2, 5)$
49. $(2, -1, 3)$ **50.** $(-4, 0), (4, 0); (-\sqrt{7}, 0), (\sqrt{7}, 0)$
51. $\dfrac{x^2}{4} + \dfrac{y^2}{25} = 1$ **53.** n^2 **55.** $a_1 = 60 - 5p - 5q$; $d = 5p + 2q - 20$ **57.** $5\frac{4}{5}, 7\frac{3}{5}, 9\frac{2}{5}, 11\frac{1}{5}$

Visualizing the Graph

1. J **2.** A **3.** C **4.** G **5.** F **6.** H **7.** E
8. D **9.** B **10.** I

Exercise Set 11.3

1. 2 **3.** -1 **5.** -2 **7.** 0.1 **9.** $\dfrac{a}{2}$ **11.** 128
13. 162 **15.** $7(5)^{40}$ **17.** 3^{n-1} **19.** $(-1)^{n-1}$ **21.** $\dfrac{1}{x^n}$
23. 762 **25.** $\frac{4921}{18}$ **27.** True **29.** True **31.** True
33. 8 **35.** 125 **37.** Does not exist **39.** $\frac{2}{3}$ **41.** $29\frac{38{,}569}{59{,}049}$
43. 2 **45.** Does not exist **47.** \$4545.45 **49.** $\frac{160}{9}$
51. $\frac{13}{99}$ **53.** 9 **55.** $\frac{34{,}091}{9990}$ **57.** \$2,684,354.55
59. **(a)** About 297 ft; **(b)** 300 ft **61.** \$39,505.71
63. 10,485.76 in. **65.** \$19,694.01 **67.** \$86,666,666,667
69. $(f \circ g)(x) = 16x^2 + 40x + 25$; $(g \circ f)(x) = 4x^2 + 5$
70. $(f \circ g)(x) = x^2 + x + 2$; $(g \circ f)(x) = x^2 - x + 3$
71. 2.209 **72.** $\frac{1}{16}$
73. $(4 - \sqrt{6})/(\sqrt{3} - \sqrt{2}) = 2\sqrt{3} + \sqrt{2}$, $(6\sqrt{3} - 2\sqrt{2})/(4 - \sqrt{6}) = 2\sqrt{3} + \sqrt{2}$; there exists a common ratio, $2\sqrt{3} + \sqrt{2}$; thus the sequence is geometric.
75. **(a)** $\frac{13}{3}; \frac{22}{3}, \frac{34}{3}, \frac{46}{3}, \frac{58}{3}$; **(b)** $-\frac{11}{3}; -\frac{2}{3}, \frac{10}{3}, -\frac{50}{3}, \frac{250}{3}$ or 5; 8, 12, 18, 27
77. $S_n = \dfrac{x^2(1 - (-x)^n)}{x + 1}$

Exercise Set 11.4

1. $1^2 < 1^3$, false; $2^2 < 2^3$, true; $3^2 < 3^3$, true; $4^2 < 4^3$, true; $5^2 < 5^3$, true **3.** A polygon of 3 sides has $\dfrac{3(3-3)}{2}$ diagonals. True; A polygon of 4 sides has $\dfrac{4(4-3)}{2}$ diagonals. True; A polygon of 5 sides has $\dfrac{5(5-3)}{2}$ diagonals. True; A polygon of 6 sides has $\dfrac{6(6-3)}{2}$ diagonals. True; A polygon of 7 sides has $\dfrac{7(7-3)}{2}$ diagonals. True.

5.
S_n: $2 + 4 + 6 + \cdots + 2n = n(n+1)$
S_1: $2 = 1(1+1)$
S_k: $2 + 4 + 6 + \cdots + 2k = k(k+1)$
S_{k+1}: $2 + 4 + 6 + \cdots + 2k + 2(k+1) = (k+1)(k+2)$

(1) *Basis step*: S_1 true by substitution.
(2) *Induction step*: Assume S_k. Deduce S_{k+1}. Starting with the left side of S_{k+1}, we have

$2 + 4 + 6 + \cdots + 2k + 2(k+1)$
$= k(k+1) + 2(k+1)$ **By S_k**
$= (k+1)(k+2).$ **Factoring**

7. S_n: $1 + 5 + 9 + \cdots + (4n - 3) = n(2n - 1)$

S_1: $1 = 1(2 \cdot 1 - 1)$

S_k: $1 + 5 + 9 + \cdots + (4k - 3) = k(2k - 1)$

S_{k+1}: $1 + 5 + 9 + \cdots + (4k - 3) + [4(k + 1) - 3] = (k + 1)[2(k + 1) - 1] = (k + 1)(2k + 1)$

(1) *Basis step: S_1 true by substitution.*

(2) *Induction step*: Assume S_k. Deduce S_{k+1}. Starting with the left side of S_{k+1}, we have

$$\begin{aligned} &1 + 5 + 9 + \cdots + (4k - 3) + [4(k + 1) - 3] \\ &= k(2k - 1) + [4(k + 1) - 3] \quad \textbf{By } S_k \\ &= 2k^2 - k + 4k + 4 - 3 \\ &= 2k^2 + 3k + 1 \\ &= (k + 1)(2k + 1). \end{aligned}$$

9. S_n: $2 + 4 + 8 + \cdots + 2^n = 2(2^n - 1)$

S_1: $2 = 2(2 - 1)$

S_k: $2 + 4 + 8 + \cdots + 2^k = 2(2^k - 1)$

S_{k+1}: $2 + 4 + 8 + \cdots + 2^k + 2^{k+1} = 2(2^{k+1} - 1)$

(1) *Basis step*: S_1 is true by substitution.

(2) *Induction step*: Assume S_k. Deduce S_{k+1}. Starting with the left side of S_{k+1}, we have

$$\begin{aligned} &\underbrace{2 + 4 + 8 + \cdots + 2^k} + 2^{k+1} \\ &= 2(2^k - 1) \quad + 2^{k+1} \quad \textbf{By } S_k \\ &= 2^{k+1} - 2 + 2^{k+1} \\ &= 2 \cdot 2^{k+1} - 2 \\ &= 2(2^{k+1} - 1). \end{aligned}$$

11. S_n: $n < n + 1$

S_1: $1 < 1 + 1$

S_k: $k < k + 1$

S_{k+1}: $k + 1 < (k + 1) + 1$

(1) *Basis step*: Since $1 < 1 + 1$, S_1 is true.

(2) *Induction step*: Assume S_k. Deduce S_{k+1}. Now

$$\begin{aligned} k &< k + 1 && \textbf{By } S_k \\ k + 1 &< k + 1 + 1 && \textbf{Adding 1} \\ k + 1 &< k + 2. && \textbf{Simplifying} \end{aligned}$$

13. S_n: $2n \le 2^n$

S_1: $2 \cdot 1 \le 2^1$

S_k: $2k \le 2^k$

S_{k+1}: $2(k + 1) \le 2^{k+1}$

(1) *Basis step*: Since $2 = 2$, S_1 is true.

(2) *Induction step*: Let k be any natural number. Assume S_k. Deduce S_{k+1}.

$$\begin{aligned} 2k &\le 2^k && \textbf{By } S_k \\ 2 \cdot 2k &\le 2 \cdot 2^k && \textbf{Multiplying by 2} \\ 4k &\le 2^{k+1} \end{aligned}$$

Since $1 \le k$, $k + 1 \le k + k$, or $k + 1 \le 2k$.

Then $2(k + 1) \le 4k$. **Multiplying by 2**

Thus, $2(k + 1) \le 4k \le 2^{k+1}$, so $2(k + 1) \le 2^{k+1}$.

15.

S_n: $\dfrac{1}{1 \cdot 2 \cdot 3} + \dfrac{1}{2 \cdot 3 \cdot 4} + \dfrac{1}{3 \cdot 4 \cdot 5} + \cdots + \dfrac{1}{n(n + 1)(n + 2)} = \dfrac{n(n + 3)}{4(n + 1)(n + 2)}$

S_1: $\dfrac{1}{1 \cdot 2 \cdot 3} = \dfrac{1(1 + 3)}{4(1 + 1)(1 + 2)}$

S_k: $\dfrac{1}{1 \cdot 2 \cdot 3} + \dfrac{1}{2 \cdot 3 \cdot 4} + \cdots + \dfrac{1}{k(k + 1)(k + 2)} = \dfrac{k(k + 3)}{4(k + 1)(k + 2)}$

S_{k+1}: $\dfrac{1}{1 \cdot 2 \cdot 3} + \dfrac{1}{2 \cdot 3 \cdot 4} + \cdots + \dfrac{1}{k(k + 1)(k + 2)} + \dfrac{1}{(k + 1)(k + 2)(k + 3)} = \dfrac{(k + 1)(k + 1 + 3)}{4(k + 1 + 1)(k + 1 + 2)} = \dfrac{(k + 1)(k + 4)}{4(k + 2)(k + 3)}$

(1) *Basis step*: Since $\dfrac{1}{1 \cdot 2 \cdot 3} = \dfrac{1}{6}$ and $\dfrac{1(1 + 3)}{4(1 + 1)(1 + 2)} = \dfrac{1 \cdot 4}{4 \cdot 2 \cdot 3} = \dfrac{1}{6}$, S_1 is true.

(2) *Induction step*: Assume S_k. Deduce S_{k+1}. Add $\dfrac{1}{(k + 1)(k + 2)(k + 3)}$ on both sides of S_k and simplify the right side. Only the right side is shown here.

$$\begin{aligned} &\frac{k(k + 3)}{4(k + 1)(k + 2)} + \frac{1}{(k + 1)(k + 2)(k + 3)} \\ &= \frac{k(k + 3)(k + 3) + 4}{4(k + 1)(k + 2)(k + 3)} \\ &= \frac{k^3 + 6k^2 + 9k + 4}{4(k + 1)(k + 2)(k + 3)} \\ &= \frac{(k + 1)^2(k + 4)}{4(k + 1)(k + 2)(k + 3)} \\ &= \frac{(k + 1)(k + 4)}{4(k + 2)(k + 3)} \end{aligned}$$

17.

S_n: $1 + 2 + 3 + \cdots + n = \dfrac{n(n + 1)}{2}$

S_1: $1 = \dfrac{1(1 + 1)}{2}$

S_k: $1 + 2 + 3 + \cdots + k = \dfrac{k(k + 1)}{2}$

S_{k+1}: $1 + 2 + 3 + \cdots + k + (k + 1) = \dfrac{(k + 1)(k + 2)}{2}$

(1) *Basis step*: S_1 true by substitution.

(2) *Induction step*: Assume S_k. Deduce S_{k+1}. Starting with the left side of S_{k+1}, we have

$$\begin{aligned} &\underbrace{1 + 2 + 3 + \cdots + k} + (k + 1) \\ &= \frac{k(k + 1)}{2} + (k + 1) && \textbf{By } S_k \\ &= \frac{k(k + 1) + 2(k + 1)}{2} && \textbf{Adding} \\ &= \frac{(k + 1)(k + 2)}{2}. && \textbf{Factoring} \end{aligned}$$

19. S_n: $1^3 + 2^3 + 3^3 + \cdots + n^3 = \dfrac{n^2(n + 1)^2}{4}$

S_1: $1^3 = \dfrac{1^2(1 + 1)^2}{4}$

S_k: $1^3 + 2^3 + 3^3 + \cdots + k^3 = \dfrac{k^2(k + 1)^2}{4}$

S_{k+1}: $1^3 + 2^3 + 3^3 + \cdots + k^3 + (k + 1)^3 = \dfrac{(k + 1)^2[(k + 1) + 1]^2}{4}$

(1) *Basis step*: S_1: $1^3 = \frac{1^2(1+1)^2}{4} = 1$. True.

(2) *Induction step*: Assume S_k. Deduce S_{k+1}.

$$1^3 + 2^3 + \cdots + k^3 = \frac{k^2(k+1)^2}{4} \qquad \mathbf{S_k}$$

$$1^3 + 2^3 + \cdots + k^3 + (k+1)^3 = \frac{k^2(k+1)^2}{4} + (k+1)^3 \qquad \textbf{Adding } (k+1)^3$$

$$= \frac{k^2(k+1)^2 + 4(k+1)^3}{4}$$

$$= \frac{(k+1)^2}{4}[k^2 + 4(k+1)]$$

$$= \frac{(k+1)^2}{4}(k^2 + 4k + 4)$$

$$= \frac{(k+1)^2(k+2)^2}{4}$$

21.

S_n: $2 + 6 + 12 + \cdots + n(n+1) = \frac{n(n+1)(n+2)}{3}$

S_1: $1(1+1) = \frac{1(1+1)(1+2)}{3}$

S_k: $2 + 6 + 12 + \cdots + k(k+1) = \frac{k(k+1)(k+2)}{3}$

S_{k+1}:

$2 + 6 + 12 + \cdots + k(k+1) + (k+1)[(k+1)+1]$

$= \frac{(k+1)[(k+1)+1][(k+1)+2]}{3}$

(1) *Basis step*: S_1: $1(1+1) = \frac{1(1+1)(1+2)}{3}$. True.

(2) *Induction step*: Assume S_k:

$2 + 6 + 12 + \cdots + k(k+1) = \frac{k(k+1)(k+2)}{3}$.

Then $2 + 6 + 12 + \cdots + k(k+1) + (k+1)(k+1+1)$

$$= \frac{k(k+1)(k+2)}{3} + (k+1)(k+2)$$

$$= \frac{k(k+1)(k+2) + 3(k+1)(k+2)}{3}$$

$$= \frac{(k+1)(k+2)(k+3)}{3}$$

$$= \frac{(k+1)(k+1+1)(k+1+2)}{3}.$$

23. S_n: $a_1 + (a_1 + d) + (a_1 + 2d) + \cdots +$
$[a_1 + (n-1)d] = \frac{n}{2}[2a_1 + (n-1)d]$

S_1: $a_1 = \frac{1}{2}[2a_1 + (1-1)d]$

S_k: $a_1 + (a_1 + d) + (a_1 + 2d) + \cdots +$
$[a_1 + (k-1)d] = \frac{k}{2}[2a_1 + (k-1)d]$

S_{k+1}: $a_1 + (a_1 + d) + (a_1 + 2d) + \cdots +$
$[a_1 + (k-1)d] + [a_1 + ((k+1)-1)d]$
$= \frac{k+1}{2}[2a_1 + ((k+1)-1)d]$

(1) *Basis step*: Since $\frac{1}{2}[2a_1 + (1-1)d] = \frac{1}{2}\cdot 2a_1 = a_1$, S_1 is true.

(2) *Induction step*: Assume S_k. Deduce S_{k+1}. Starting with the left side of S_{k+1}, we have

$$\underbrace{a_1 + (a_1 + d) + \cdots + [a_1 + (k-1)d]} + [a_1 + kd]$$

$$= \frac{k}{2}[2a_1 + (k-1)d] \qquad + [a_1 + kd] \qquad \mathbf{By\ S_k}$$

$$= \frac{k[2a_1 + (k-1)d]}{2} + \frac{2[a_1 + kd]}{2}$$

$$= \frac{2ka_1 + k(k-1)d + 2a_1 + 2kd}{2}$$

$$= \frac{2a_1(k+1) + k(k-1)d + 2kd}{2}$$

$$= \frac{2a_1(k+1) + (k-1+2)kd}{2}$$

$$= \frac{2a_1(k+1) + (k+1)kd}{2} = \frac{k+1}{2}[2a_1 + kd].$$

24. $(5, 3)$

25. 1.5%: \$1200; 2%: \$2400; 3%: \$2800

27. S_n: $x + y$ is a factor of $x^{2n} - y^{2n}$.
S_1: $x + y$ is a factor of $x^2 - y^2$.
S_k: $x + y$ is a factor of $x^{2k} - y^{2k}$.
S_{k+1}: $x + y$ is a factor of $x^{2(k+1)} - y^{2(k+1)}$.

(1) *Basis step*: S_1: $x + y$ is a factor of $x^2 - y^2$. True.
S_2: $x + y$ is a factor of $x^4 - y^4$. True.

(2) *Induction step*: Assume S_{k-1}: $x + y$ is a factor of $x^{2(k-1)} - y^{2(k-1)}$. Then $x^{2(k-1)} - y^{2(k-1)} = (x+y)Q(x)$ for some polynomial $Q(x)$.

Assume S_k: $x + y$ is a factor of $x^{2k} - y^{2k}$. Then $x^{2k} - y^{2k} = (x+y)P(x)$ for some polynomial $P(x)$.

$x^{2(k+1)} - y^{2(k+1)}$
$= (x^{2k} - y^{2k})(x^2 + y^2) - (x^{2(k-1)} - y^{2(k-1)})(x^2y^2)$
$= (x+y)P(x)(x^2 + y^2) - (x+y)Q(x)(x^2y^2)$
$= (x+y)[P(x)(x^2 + y^2) - Q(x)(x^2y^2)]$

so $x + y$ is a factor of $x^{2(k+1)} - y^{2(k+1)}$.

29. S_2: $\log_a(b_1b_2) = \log_a b_1 + \log_a b_2$
S_k: $\log_a(b_1b_2\cdots b_k) = \log_a b_1 + \log_a b_2 + \cdots + \log_a b_k$
S_{k+1}: $\log_a(b_1b_2\cdots b_{k+1}) = \log_a b_1 + \log_a b_2 + \cdots + \log_a b_{k+1}$

(1) *Basis step*: S_2 is true by the properties of logarithms.

(2) *Induction step*: Let k be a natural number $k \geq 2$. Assume S_k. Deduce S_{k+1}.

$\log_a(b_1b_2\cdots b_{k+1})$ **Left side of S_{k+1}**
$= \log_a(b_1b_2\cdots b_k) + \log_a b_{k+1}$ **By S_2**
$= \log_a b_1 + \log_a b_2 + \cdots + \log_a b_k + \log_a b_{k+1}$ **By S_k**

31. S_2: $\overline{z_1 + z_2} = \overline{z_1} + \overline{z_2}$:
$\overline{(a+bi) + (c+di)} = \overline{(a+c) + (b+d)i}$
$= (a+c) - (b+d)i$
$\overline{(a+bi)} + \overline{(c+di)} = a - bi + c - di$
$= (a+c) - (b+d)i.$

S_k: $\overline{z_1 + z_2 + \cdots + z_k} = \overline{z_1} + \overline{z_2} + \cdots + \overline{z_k}$.
$\overline{(z_1 + z_2 + \cdots + z_k) + z_{k+1}}$
$= \overline{(z_1 + z_2 + \cdots + z_k)} + \overline{z_{k+1}}$ **By S_2**
$= \overline{z_1} + \overline{z_2} + \cdots + \overline{z_k} + \overline{z_{k+1}}$ **By S_k**

Mid-Chapter Mixed Review: Chapter 11

1. False **2.** True **3.** False **4.** False **5.** 8, 11, 14, 17; 32; 47 **6.** 0, −1, 2, −3; 8; −13 **7.** $a_n = 3n$

8. $a_n = (-1)^n n^2$ **9.** $1\frac{7}{8}$, or $\frac{15}{8}$ **10.** $2 + 6 + 12 + 20 + 30 = 70$ **11.** $\sum_{k=1}^{\infty}(-1)^k 4k$ **12.** 2, 6, 22, 86
13. -5 **14.** 22 **15.** 21 **16.** 696 **17.** $-\frac{1}{2}$
18. **(a)** 8; **(b)** $\frac{1023}{16}$, or 63.9375 **19.** $-\frac{16}{3}$ **20.** Does not exist
21. 126 plants **22.** \$13,588.68
23.
$$\begin{aligned}
S_n&: \quad 1 + 4 + 7 + \cdots + (3n - 2) = \tfrac{1}{2}n(3n - 1)\\
S_1&: \quad 3\cdot 1 - 2 = \tfrac{1}{2}\cdot 1(3\cdot 1 - 1)\\
S_k&: \quad 1 + 4 + 7 + \cdots + (3k - 2) = \tfrac{1}{2}k(3k - 1)\\
S_{k+1}&: \quad 1 + 4 + 7 + \cdots + (3k - 2) + [3(k + 1) - 2]\\
&\qquad = \tfrac{1}{2}(k + 1)[3(k + 1) - 1]\\
&\qquad = \tfrac{1}{2}(k + 1)(3k + 2)
\end{aligned}$$
(1) *Basis step*: $S_1: 3\cdot 1 - 2 = \frac{1}{2}\cdot 1(3\cdot 1 - 1)$. **True**
(2) *Induction step*: Assume S_k:
$1 + 4 + 7 + \cdots + (3k - 2) = \frac{1}{2}k(3k - 1)$.
Then $1 + 4 + 7 + \cdots + (3k - 2) + [3(k + 1) - 2]$
$$\begin{aligned}
&= \tfrac{1}{2}k(3k - 1) + [3(k + 1) - 2]\\
&= \tfrac{3}{2}k^2 - \tfrac{1}{2}k + 3k + 1\\
&= \tfrac{3}{2}k^2 + \tfrac{5}{2}k + 1\\
&= \tfrac{1}{2}(3k^2 + 5k + 2)\\
&= \tfrac{1}{2}(k + 1)(3k + 2).
\end{aligned}$$
24. The first formula can be derived from the second by substituting $a_1 + (n - 1)d$ for a_n. When the first and last terms of the sum are known, the second formula is the better one to use. If the last term is not known, the first formula allows us to compute the sum in one step without first finding a_n.
25.
$$\begin{aligned}
&1 + 2 + 3 + \cdots + 100\\
&= (1 + 100) + (2 + 99) + (3 + 98) + \cdots + (50 + 51)\\
&= \underbrace{101 + 101 + 101 + \cdots + 101}_{\text{50 addends of 101}}\\
&= 50\cdot 101\\
&= 5050
\end{aligned}$$
A formula for the first n natural numbers is $\frac{n}{2}(1 + n)$.
26. Answers may vary. One possibility is given. Casey invests \$900 at 8% interest, compounded annually. How much will be in the account at the end of 40 years? **27.** We can prove an infinite sequence of statements S_n by showing that a basis statement S_1 is true and then that for all natural numbers k, if S_k is true, then S_{k+1} is true.

Exercise Set 11.5

1. 720 **3.** 604,800 **5.** 120 **7.** 1 **9.** 3024 **11.** 120
13. 120 **15.** 1 **17.** 6,497,400 **19.** $n(n - 1)(n - 2)$
21. n **23.** $6! = 720$ **25.** $9! = 362{,}880$ **27.** ${}_9P_4 = 3024$
29. ${}_5P_5 = 120$; $5^5 = 3125$ **31.** ${}_5P_5\cdot {}_4P_4 = 2880$
33. $8\cdot 10^6 = 8{,}000{,}000$; 8 million **35.** $\dfrac{9!}{2!\,3!\,4!} = 1260$
37. **(a)** ${}_6P_5 = 720$; **(b)** $6^5 = 7776$; **(c)** $1\cdot {}_5P_4 = 120$; **(d)** $1\cdot 1\cdot {}_4P_3 = 24$ **39.** **(a)** 10^5, or 100,000; **(b)** 100,000
41. **(a)** $10^9 = 1{,}000{,}000{,}000$; **(b)** yes **42.** $\frac{9}{4}$, or 2.25
43. $-3, 2$ **44.** $\dfrac{3 \pm \sqrt{17}}{4}$
45. $-2, 1, 5$ **47.** 8 **49.** 11 **51.** $(n - 1)$ games

Exercise Set 11.6

1. 78 **3.** 78 **5.** 7 **7.** 10 **9.** 1 **11.** 15
13. 128 **15.** 270,725 **17.** 13,037,895 **19.** n
21. 1 **23.** ${}_{36}C_4 = 58{,}905$ **25.** ${}_{13}C_{10} = 286$
27. $\binom{10}{7}\cdot\binom{5}{3} = 1200$ **29.** $\binom{52}{5} = 2{,}598{,}960$
31. **(a)** ${}_{31}P_2 = 930$; **(b)** $31^2 = 961$; **(c)** ${}_{31}C_2 = 465$
33. $-\frac{17}{2}$ **34.** $-1, \frac{3}{2}$ **35.** $\dfrac{-5 \pm \sqrt{21}}{2}$
36. $-4, -2, 3$ **37.** $\binom{13}{5} = 1287$
39. $\binom{n}{2}$; $2\binom{n}{2}$ **41.** 4 **43.** 7
45.
$$\begin{aligned}
&\binom{n}{k-1} + \binom{n}{k}\\
&= \frac{n!}{(k - 1)!(n - k + 1)!}\cdot\frac{l}{k} + \frac{n!}{k!(n - k)!}\cdot\frac{(n - k + 1)}{(n - k + 1)!}\\
&= \frac{n!(k + (n - k + 1))}{k!(n - k + 1)!}\\
&= \frac{(n + 1)!}{k!(n - k + 1)!} = \binom{n+1}{k}
\end{aligned}$$

Exercise Set 11.7

1. $x^4 + 20x^3 + 150x^2 + 500x + 625$
3. $x^5 - 15x^4 + 90x^3 - 270x^2 + 405x - 243$
5. $x^5 - 5x^4y + 10x^3y^2 - 10x^2y^3 + 5xy^4 - y^5$
7. $15{,}625x^6 + 75{,}000x^5y + 150{,}000x^4y^2 + 160{,}000x^3y^3 + 96{,}000x^2y^4 + 30{,}720xy^5 + 4096y^6$
9. $128t^7 + 448t^5 + 672t^3 + 560t + 280t^{-1} + 84t^{-3} + 14t^{-5} + t^{-7}$ **11.** $x^{10} - 5x^8 + 10x^6 - 10x^4 + 5x^2 - 1$
13. $125 + 150\sqrt{5}t + 375t^2 + 100\sqrt{5}t^3 + 75t^4 + 6\sqrt{5}t^5 + t^6$
15. $a^9 - 18a^7 + 144a^5 - 672a^3 + 2016a - 4032a^{-1} + 5376a^{-3} - 4608a^{-5} + 2304a^{-7} - 512a^{-9}$ **17.** $140\sqrt{2}$
19. $x^{-8} + 4x^{-4} + 6 + 4x^4 + x^8$ **21.** $21a^5b^2$
23. $-252x^5y^5$ **25.** $-745{,}472a^3$ **27.** $1120x^{12}y^2$
29. $-1{,}959{,}552u^5v^{10}$ **31.** 2^7, or 128 **33.** 2^{24}, or 16,777,216 **35.** 20 **37.** $-12 + 316i$
39. $-7 - 4\sqrt{2}i$ **41.** $\sum_{k=0}^{n}\binom{n}{k}(-1)^k a^{n-k}b^k$, or $\sum_{k=0}^{n}\binom{n}{k}a^{n-k}(-b)^k$
43. $\sum_{k=1}^{n}\binom{n}{k}x^{n-k}h^{k-1}$ **44.** $x^2 + 2x - 2$
45. $2x^3 - 3x^2 + 2x - 3$ **46.** $4x^2 - 12x + 10$
47. $2x^2 - 1$ **49.** $3, 9, 6 \pm 3i$ **51.** $-\dfrac{35}{x^{1/6}}$
53. 2^{100} **55.** $[\log_a(xt)]^{23}$
57. **(1)** *Basis step*: Since $a + b = (a + b)^1$, S_1 is true.
(2) *Induction step*: Let S_k be the statement of the binomial theorem with n replaced by k and k replaced by r. Multiply both sides of S_k by $(a + b)$ to obtain
$$\begin{aligned}
&(a + b)^{k+1}\\
&= \left[a^k + \cdots + \binom{k}{r-1}a^{k-(r-1)}b^{r-1} + \binom{k}{r}a^{k-r}b^r + \cdots + b^k\right](a + b)\\
&= a^{k+1} + \cdots + \left[\binom{k}{r-1} + \binom{k}{r}\right]a^{(k+1)-r}b^r + \cdots + b^{k+1}
\end{aligned}$$

$$= a^{k+1} + \cdots + \binom{k+1}{r} a^{(k+1)-r} b^r + \cdots + b^{k+1}.$$

This proves S_{k+1}, assuming S_k. Hence S_n is true for $n = 1, 2, 3, \ldots$

Exercise Set 11.8

1. **(a)** 0.18, 0.24, 0.23, 0.23, 0.12; **(b)** Opinions may vary, but it seems that people tend not to pick the first or last numbers.
3. 5187 e-mails **5.** **(a)** $\frac{2}{7}$; **(b)** $\frac{5}{7}$; **(c)** 0; **(d)** 1 **7.** $\frac{1}{2}$
9. **(a)** $\frac{1}{13}$; **(b)** $\frac{2}{13}$; **(c)** $\frac{1}{4}$; **(d)** $\frac{1}{26}$ **11.** $\frac{1}{5525}$ **13.** $\frac{135}{323}$
15. $\frac{1}{108{,}290}$ **17.** $\frac{33}{66{,}640}$ **19.** **(a)** HHH, HHT, HTH, HTT, THH, THT, TTH, TTT; **(b)** $\frac{3}{8}$; **(c)** $\frac{7}{8}$; **(d)** $\frac{7}{8}$; **(e)** $\frac{3}{8}$ **21.** $\frac{9}{19}$
23. $\frac{1}{38}$ **25.** $\frac{18}{19}$ **27.** $\frac{9}{19}$ **29.** Zero
30. One-to-one **31.** Function; domain; range; domain; range
32. Zero **33.** Combination **34.** Inverse variation
35. Factor **36.** Geometric sequence
37. **(a)** $\binom{13}{2}\cdot\binom{4}{2}\cdot\binom{4}{2}\cdot\binom{44}{1}$, or 123,552; **(b)** 0.0475
39. **(a)** $13\cdot\binom{4}{3}\cdot\binom{48}{2} - 3744$, or 54,912; **(b)** $\dfrac{54{,}912}{\binom{52}{5}} \approx 0.0211$

Review Exercises: Chapter 11

1. True **2.** False **3.** True **4.** False
5. $-\frac{1}{2}, \frac{4}{17}, -\frac{9}{82}, \frac{16}{257}; -\frac{121}{14{,}642}; -\frac{529}{279{,}842}$ **6.** $(-1)^{n+1}(n^2+1)$
7. $\frac{3}{2} - \frac{9}{8} + \frac{27}{26} - \frac{81}{80} = \frac{417}{1040}$
8.

n	u_n
1	0.3
2	2.5
3	13.5
4	68.5
5	343.5
6	1718.5
7	8593.5
8	42968.5
9	214843.5
10	1074218.5

1,100,000
0
0
10

9. $\sum_{k=1}^{7}(k^2-1)$ **10.** $\frac{15}{4}$ **11.** $a + 4b$ **12.** 531
13. 20,100 **14.** 11 **15.** −4 **16.** $n = 6$, $S_n = S_6 = -126$
17. $a_1 = 8$, $a_5 = \frac{1}{2}$ **18.** Does not exist **19.** $\frac{3}{11}$
20. $\frac{3}{8}$ **21.** $\frac{241}{99}$ **22.** $5\frac{4}{5}, 6\frac{3}{5}, 7\frac{2}{5}, 8\frac{1}{5}$ **23.** 167.3 ft
24. \$45,993.04 **25.** **(a)** \$7.38; **(b)** \$1365.10
26. \$88,888,888,889

27. S_n: $1 + 4 + 7 + \cdots + (3n-2) = \dfrac{n(3n-1)}{2}$

S_1: $1 = \dfrac{1(3\cdot 1 - 1)}{2}$

S_k: $1 + 4 + 7 + \cdots + (3k-2) = \dfrac{k(3k-1)}{2}$

S_{k+1}: $1 + 4 + 7 + \cdots + (3k-2) + [3(k+1)-2]$
$= 1 + 4 + 7 + \cdots + (3k-2) + (3k+1)$
$= \dfrac{(k+1)(3k+2)}{2}$

(1) *Basis step*: $\dfrac{1(3\cdot 1 - 1)}{2} = \dfrac{2}{2} = 1$ is true.

(2) *Induction step*: Assume S_k. Add $(3k+1)$ on both sides.

$$1 + 4 + 7 + \cdots + (3k-2) + (3k+1)$$
$$= \frac{k(3k-1)}{2} + (3k+1)$$
$$= \frac{k(3k-1)}{2} + \frac{2(3k+1)}{2}$$
$$= \frac{3k^2 - k + 6k + 2}{2}$$
$$= \frac{3k^2 + 5k + 2}{2}$$
$$= \frac{(k+1)(3k+2)}{2}$$

28. S_n: $1 + 3 + 3^2 + \cdots + 3^{n-1} = \dfrac{3^n - 1}{2}$

S_1: $1 = \dfrac{3^1 - 1}{2}$

S_k: $1 + 3 + 3^2 + \cdots + 3^{k-1} = \dfrac{3^k - 1}{2}$

S_{k+1}: $1 + 3 + 3^2 + \cdots + 3^{(k+1)-1} = \dfrac{3^{k+1} - 1}{2}$

(1) *Basis step*: $\dfrac{3^1 - 1}{2} = \dfrac{2}{2} = 1$ is true.

(2) *Induction step*: Assume S_k. Add 3^k on both sides.

$$1 + 3 + \cdots + 3^{k-1} + 3^k$$
$$= \frac{3^k - 1}{2} + 3^k = \frac{3^k - 1}{2} + 3^k\cdot\frac{2}{2}$$
$$= \frac{3\cdot 3^k - 1}{2} = \frac{3^{k+1} - 1}{2}$$

29. S_n: $\left(1 - \frac{1}{2}\right)\left(1 - \frac{1}{3}\right)\cdots\left(1 - \frac{1}{n}\right) = \dfrac{1}{n}$

S_2: $\left(1 - \frac{1}{2}\right) = \dfrac{1}{2}$

S_k: $\left(1 - \frac{1}{2}\right)\left(1 - \frac{1}{3}\right)\cdots\left(1 - \frac{1}{k}\right) = \dfrac{1}{k}$

S_{k+1}: $\left(1 - \frac{1}{2}\right)\left(1 - \frac{1}{3}\right)\cdots\left(1 - \frac{1}{k}\right)\left(1 - \frac{1}{k+1}\right) = \dfrac{1}{k+1}$.

(1) *Basis step*: S_2 is true by substitution.

(2) *Induction step*: Assume S_k. Deduce S_{k+1}. Starting with the left side of S_{k+1}, we have

$$\underbrace{\left(1 - \frac{1}{2}\right)\left(1 - \frac{1}{3}\right)\cdots\left(1 - \frac{1}{k}\right)}\left(1 - \frac{1}{k+1}\right)$$
$$= \frac{1}{k}\cdot\left(1 - \frac{1}{k+1}\right) \quad \textbf{By } S_k$$
$$= \frac{1}{k}\cdot\left(\frac{k+1-1}{k+1}\right)$$
$$= \frac{1}{k}\cdot\frac{k}{k+1}$$
$$= \frac{1}{k+1}. \quad \textbf{Simplifying}$$

30. $6! = 720$ **31.** $9\cdot 8\cdot 7\cdot 6 = 3024$
32. $\binom{15}{8} = 6435$ **33.** $24\cdot 23\cdot 22 = 12{,}144$
34. $\dfrac{9}{1!4!2!2!} = 3780$ **35.** $3\cdot 4\cdot 3 = 36$

36. **(a)** ${}_6P_5 = 720$; **(b)** $6^5 = 7776$; **(c)** $1 \cdot {}_5P_4 = 120$; **(d)** ${}_3P_2 \cdot 1 \cdot 1 \cdot 1 = 6$ **37.** 2^8, or 256
38. $m^7 + 7m^6n + 21m^5n^2 + 35m^4n^3 + 35m^3n^4 + 21m^2n^5 + 7mn^6 + n^7$
39. $x^5 - 5\sqrt{2}x^4 + 20x^3 - 20\sqrt{2}x^2 + 20x - 4\sqrt{2}$
40. $x^8 - 12x^6y + 54x^4y^2 - 108x^2y^3 + 81y^4$
41. $a^8 + 8a^6 + 28a^4 + 56a^2 + 70 + 56a^{-2} + 28a^{-4} + 8a^{-6} + a^{-8}$ **42.** $-6624 + 16{,}280i$ **43.** $220a^9x^3$
44. $-\binom{18}{11}128a^7b^{11} = -4{,}073{,}472a^7b^{11}$ **45.** $\frac{1}{12}$; 0
46. $\frac{1}{4}$ **47.** $\frac{6}{5525}$ **48.** $\frac{86}{206} \approx 0.42$, $\frac{97}{206} \approx 0.47$, $\frac{23}{206} \approx 0.11$
49. **(a)** $a_n = 0.1332631579n + 2.803157895$; **(b)** about 14.3 lb of American cheese **50.** B **51.** A **52.** D
53. **(a)** No (unless a_n is all positive or all negative); **(b)** yes; **(c)** yes; **(d)** no (unless a_n is constant); **(e)** no (unless a_n is constant); **(f)** no (unless a_n is constant)
54. $\frac{a_{k+1}}{a_k} = r_1$, $\frac{b_{k+1}}{b_k} = r_2$, so $\frac{a_{k+1}b_{k+1}}{a_kb_k} = r_1r_2$, a constant.
55. $\frac{1}{2}, -\frac{1}{6}, \frac{1}{18}$ **56.** $-2, 0, 2, 4$ **57.** $\left(\log \frac{x}{y}\right)^{10}$ **58.** 18
59. 36 **60.** -9 **61.** For each circular arrangement of the numbers on a clock face, there are 12 distinguishable ordered arrangements on a line. The number of arrangements of 12 objects on a line is ${}_{12}P_{12}$, or 12!. Thus the number of circular permutations is $\frac{{}_{12}P_{12}}{12} = \frac{12!}{12} = 11! = 39{,}916{,}800$. In general, for each circular arrangement of n objects, there are n distinguishable ordered arrangements on a line. The total number of arrangements of n objects on a line is ${}_nP_n$, or $n!$. Thus the number of circular permutations is $\frac{n!}{n} = \frac{n(n-1)!}{n} = (n-1)!$
62. Put the following in the form of a paragraph.
First find the number of seconds in a year (365 days):
$365 \text{ days} \cdot \frac{24 \text{ hr}}{1 \text{ day}} \cdot \frac{60 \text{ min}}{1 \text{ hr}} \cdot \frac{60 \text{ sec}}{1 \text{ min}} = 31{,}536{,}000 \text{ sec.}$
The number of arrangements possible is 15!.
The time is $\frac{15!}{31{,}536{,}000} \approx 41{,}466$ years.
63. Order is considered in a combination lock. **64.** Choosing k objects from a set of n objects is equivalent to not choosing the other $n - k$ objects.

Test: Chapter 11

1. [11.1] -43 **2.** [11.1] $\frac{2}{3}, \frac{3}{4}, \frac{4}{5}, \frac{5}{6}, \frac{6}{7}$
3. [11.1] $2 + 5 + 10 + 17 = 34$
4. [11.1]

n	u_n
1	0.66667
2	0.75
3	0.8
4	0.83333
5	0.85714
6	0.875
7	0.88889
8	0.9
9	0.90909
10	0.91667

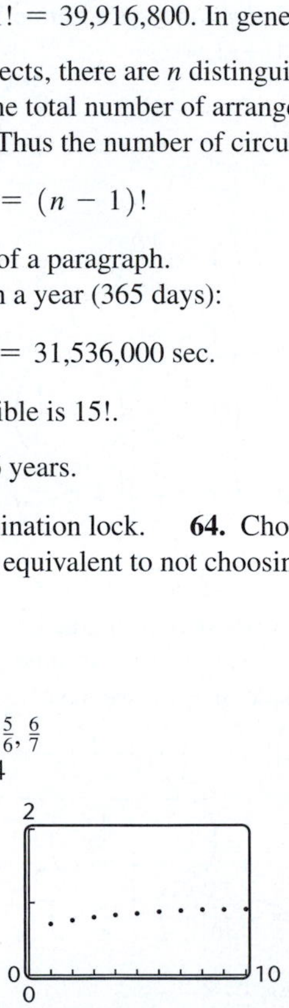

5. [11.1] $\sum_{k=1}^{6} 4k$ **6.** [11.1] $\sum_{k=1}^{\infty} 2^k$ **7.** [11.1] $3, 2\frac{1}{3}, 2\frac{3}{7}, 2\frac{7}{17}$
8. [11.2] 44 **9.** [11.2] 38 **10.** [11.2] -420 **11.** [11.2] 675
12. [11.3] $\frac{5}{512}$ **13.** [11.3] 1000 **14.** [11.3] 510 **15.** [11.3] 27
16. [11.3] $\frac{56}{99}$ **17.** [11.1] \$10,000, \$8000, \$6400, \$5120, \$4096, \$3276.80 **18.** [11.2] \$18.20 **19.** [11.3] \$74,399.77
20. [11.4]

$$S_n:\quad 2 + 5 + 8 + \cdots + (3n - 1) = \frac{n(3n + 1)}{2}$$
$$S_1:\quad 2 = \frac{1(3 \cdot 1 + 1)}{2}$$
$$S_k:\quad 2 + 5 + 8 + \cdots + (3k - 1) = \frac{k(3k + 1)}{2}$$
$$S_{k+1}:\quad 2 + 5 + 8 + \cdots + (3k - 1) + [3(k + 1) - 1] = \frac{(k + 1)[3(k + 1) + 1]}{2}$$

(1) *Basis step*: $\frac{1(3 \cdot 1 + 1)}{2} = \frac{1 \cdot 4}{2} = 2$, so S_1 is true.

(2) *Induction step*:
$$\underbrace{2 + 5 + 8 + \cdots + (3k - 1)} + [3(k + 1) - 1]$$
$$= \frac{k(3k + 1)}{2} + [3k + 3 - 1] \qquad \textbf{By } S_k$$
$$= \frac{3k^2}{2} + \frac{k}{2} + 3k + 2$$
$$= \frac{3k^2}{2} + \frac{7k}{2} + 2$$
$$= \frac{3k^2 + 7k + 4}{2}$$
$$= \frac{(k + 1)(3k + 4)}{2}$$
$$= \frac{(k + 1)[3(k + 1) + 1]}{2}$$

21. [11.5] 3,603,600 **22.** [11.6] 352,716
23. [11.6] $\frac{n(n - 1)(n - 2)(n - 3)}{24}$ **24.** [11.5] ${}_6P_4 = 360$
25. [11.5] **(a)** $6^4 = 1296$; **(b)** $1 \cdot {}_5P_3 = 60$
26. [11.6] ${}_{28}C_4 = 20{,}475$ **27.** [11.6] ${}_{12}C_8 \cdot {}_8C_4 = 34{,}650$
28. [11.7] $x^5 + 5x^4 + 10x^3 + 10x^2 + 5x + 1$
29. [11.7] $35x^3y^4$ **30.** [11.7] $2^9 = 512$ **31.** [11.8] $\frac{4}{7}$
32. [11.8] $\frac{48}{1001}$ **33.** [11.1] B **34.** [11.5] 15

Just-in-Time

1. Real Numbers

1. $\frac{2}{3}, 6, -2.45, 18.\overline{4}, -11, \sqrt[3]{27}, 5\frac{1}{6}, -\frac{8}{7}, 0, \sqrt{16}$
2. $\frac{2}{3}, -2.45, 18.\overline{4}, 5\frac{1}{6}, -\frac{8}{7}$
3. $\sqrt{3}, \sqrt[6]{26}, 7.151551555\ldots, -\sqrt{35}, \sqrt[5]{3}$
4. $6, -11, \sqrt[3]{27}, 0, \sqrt{16}$ **5.** $6, \sqrt[3]{27}, 0, \sqrt{16}$
6. All of them

2. Properties of Real Numbers

1. Additive inverse property **2.** Associative property of multiplication **3.** Distributive property **4.** Commutative property of addition **5.** Multiplicative identity property **6.** Commutative property of multiplication **7.** Additive identity property **8.** Multiplicative inverse property **9.** Associative property of addition **10.** Distributive property

3. Order on the Number Line

1. False **2.** True **3.** True **4.** True **5.** False **6.** True

4. Absolute Value

1. 98 **2.** 0 **3.** 4.7 **4.** $\frac{2}{3}$ **5.** 20 **6.** 12.6 **7.** 11 **8.** $\frac{21}{8}$

5. Operations with Real Numbers

1. 19 **2.** $\frac{1}{10}$ **3.** -5 **4.** -3 **5.** -350 **6.** -5.5 **7.** 24 **8.** 10 **9.** -12.6 **10.** 20 **11.** -15 **12.** $-\frac{1}{6}$ **13.** -8 **14.** -22 **15.** $\frac{4}{5}$

6. Interval Notation

1. $[-5, 5]$ **2.** $(-3, -1]$ **3.** $(-\infty, -2]$ **4.** $(3.8, \infty)$ **5.** $(7, \infty)$ **6.** $(-2, 2)$ **7.** $(-4, 5)$ **8.** $[1.7, \infty)$ **9.** $(-5, -2]$ **10.** $(-\infty, \sqrt{5})$

7. Integers as Exponents

1. $\frac{1}{3^6}$ **2.** $(0.2)^5$ **3.** $\frac{z^9}{w^4}$ **4.** $\frac{z^2}{y^2}$ **5.** 1 **6.** a^8 **7.** $-6x^{-4}y^4$, or $-\frac{6y^4}{x^4}$ **8.** x^{-11}, or $\frac{1}{x^{11}}$ **9.** $m^{-6}n^{-6}$, or $\frac{1}{m^6n^6}$ **10.** t^{-20}, or $\frac{1}{t^{20}}$

8. Scientific Notation

1. 1.85×10^7 **2.** 7.86×10^{-4} **3.** 2.3×10^{-9} **4.** 8.927×10^9 **5.** 0.000000043 **6.** 5,170,000 **7.** 620,300,000,000 **8.** 0.0000294

9. Order of Operations

1. 3 **2.** 103 **3.** -235 **4.** 2048 **5.** 2 **6.** 5 **7.** 32 **8.** 44

10. Introduction to Polynomials

1. 6 **2.** 8 **3.** 4 **4.** 0 **5.** 8 **6.** Binomial **7.** Monomial **8.** Trinomial

11. Add and Subtract Polynomials

1. $9y - 4$ **2.** $-2x^2 + 6x - 2$ **3.** $3x + 2y - 2z - 3$ **4.** $2ab^2 - a^2b + 6ab + 10$ **5.** $-4x^2 + 8xy - 5y^2 + 3$

12. Multiply Polynomials

1. $-21a^6$ **2.** $y^2 + 2y - 15$ **3.** $x^2 + 9x + 18$ **4.** $2a^2 + 13a + 15$ **5.** $4x^2 + 8xy + 3y^2$ **6.** $33t^2 + 41t - 4$

13. Special Products of Binomials

1. $x^2 + 6x + 9$ **2.** $25x^2 - 30x + 9$ **3.** $4x^2 + 12xy + 9y^2$ **4.** $a^2 - 10ab + 25b^2$ **5.** $n^2 - 36$ **6.** $9y^2 - 16$

14. Factor Polynomials; The FOIL Method

1. $3(x + 6)$ **2.** $2z^2(z - 4)$ **3.** $(3x - 1)(x^2 + 6)$ **4.** $(t + 6)(t^2 - 2)$ **5.** $(w - 5)(w - 2)$ **6.** $(t + 3)(t + 5)$ **7.** $2(n - 12)(n + 2)$ **8.** $y^2(y - 2)(y - 7)$ **9.** $(2n - 7)(n + 8)$ **10.** $(2y - 3)(y + 2)$ **11.** $(b - 5t)(b - t)$ **12.** $(x^2 - 10)(x^2 + 3)$

15. Factor Polynomials: The *ac*-Method

1. $(4x + 3)(2x - 3)$ **2.** $2(5t - 3)(t + 1)$ **3.** $3(3a - 1)(2a - 5)$

16. Special Factorizations

1. $(z + 9)(z - 9)$ **2.** $(4x + 3)(4x - 3)$ **3.** $7p(q^2 + y^2)(q + y)(q - y)$ **4.** $(x + 6)^2$ **5.** $(3z - 2)^2$ **6.** $a(a + 12)^2$ **7.** $(x + 4)(x^2 - 4x + 16)$ **8.** $(m - 6)(m^2 + 6m + 36)$ **9.** $3a^2(a - 2)(a^2 + 2a + 4)$ **10.** $(t^2 + 1)(t^4 - t^2 + 1)$

17. Equation-Solving Principles

1. 10 **2.** 12 **3.** -4 **4.** 10 **5.** 2 **6.** -3 **7.** 0 **8.** 5

18. Inequality-Solving Principles

1. $[-125, \infty)$ **2.** $(-\infty, -9)$ **3.** $(-\infty, 2)$ **4.** $(-\infty, -56]$ **5.** $(-\infty, 27)$ **6.** $[11, \infty)$

19. The Principle of Zero Products

1. $0, -21$ **2.** $-7, 1$ **3.** $-\frac{3}{5}, 4$ **4.** $-\frac{5}{3}, \frac{1}{2}$ **5.** $0, 8$ **6.** $-3, 11$ **7.** $-15, 2$ **8.** $-\frac{3}{4}, \frac{4}{3}$

20. The Principle of Square Roots

1. -6 and 6, or ± 6 **2.** $-\sqrt{10}$ and $\sqrt{10}$, or $\pm\sqrt{10}$ **3.** $-\sqrt{3}$ and $\sqrt{3}$, or $\pm\sqrt{3}$ **4.** $-\sqrt{5}$ and $\sqrt{5}$, or $\pm\sqrt{5}$ **5.** -5 and 5, or ± 5 **6.** $-\sqrt{15}$ and $\sqrt{15}$, or $\pm\sqrt{15}$

21. Simplify Rational Expressions

1. The set of all real numbers except 0 and 1 **2.** The set of all real numbers except -7 and 3 **3.** $\frac{x + 2}{x - 2}$ **4.** $\frac{x - 1}{x - 3}$ **5.** $\frac{x - 3}{x}$ **6.** $\frac{2(y + 4)}{y - 1}$

22. Multiply and Divide Rational Expressions

1. 1 **2.** $m + n$ **3.** $\frac{4x+1}{3x-2}$ **4.** $\frac{a+1}{a-3}$ **5.** $\frac{3(x-4)}{2(x+4)}$
6. $\frac{1}{x+y}$

23. Add and Subtract Rational Expressions

1. 2 **2.** $\frac{2(3x^2+2x-7)}{3(3x+1)(x-2)}$ **3.** $\frac{2a}{(a+1)(a-1)}$
4. $\frac{3x-4}{(x-2)(x-1)}$ **5.** $\frac{-y+10}{(y+4)(y-5)}$ **6.** $\frac{y}{(y-2)(y-3)}$

24. Simplify Complex Rational Expressions

1. $x - y$ **2.** $\frac{a}{a+b}$ **3.** $\frac{w^2-2w+4}{w}$ **4.** $\frac{x+y}{x}$
5. $-a - b$

25. Simplify Radical Expressions

1. 21 **2.** $3y$ **3.** $a - 2$ **4.** $-3x$ **5.** $3x^2$ **6.** 2
7. $2xy\sqrt[4]{3x^2}$ **8.** $5\sqrt{21}$ **9.** $\sqrt{5y}$ **10.** $\frac{1}{2x}$ **11.** $x - 2$
12. $2x^2y\sqrt{6}$ **13.** $3x\sqrt[3]{4y}$ **14.** $17\sqrt{2}$ **15.** $12\sqrt{3}$
16. $2\sqrt{2}$ **17.** $4\sqrt{5}$ **18.** $16 + 9\sqrt{3}$ **19.** -12
20. $4 + 2\sqrt{3}$

26. Rationalize Denominators

1. $\frac{4\sqrt{11}}{11}$ **2.** $\frac{\sqrt{21}}{7}$ **3.** $\frac{\sqrt[3]{28}}{2}$ **4.** $\frac{2\sqrt[3]{6}}{3}$
5. $\frac{3\sqrt{30}+12}{14}$ **6.** $\sqrt{7} + \sqrt{3}$ **7.** $\frac{6\sqrt{m}+6\sqrt{n}}{m-n}$
8. $\frac{\sqrt{3}}{3}$

27. Rational Exponents

1. $\sqrt[6]{y^5}$ **2.** $\sqrt[3]{x^2}$ **3.** 8 **4.** 128 **5.** $\frac{1}{5}$ **6.** $\frac{1}{16}$ **7.** $y^{1/3}$
8. $x^{5/2}$ **9.** $x\sqrt[6]{x}$ **10.** $(a-2)^2$ **11.** $n\sqrt[3]{mn^2}$

28. The Pythagorean Theorem

1. 17 **2.** $\sqrt{32} \approx 5.657$ **3.** 12 **4.** 5 **5.** $\sqrt{31} \approx 5.568$

Index

Index of Applications

Consumer

Economics

Education

Engineering

Environment

Finance

General Interest

Geometry

Government

Health/Medicine

Labor

Physics

Social Sciences

Sports/Entertainment

Statistics/Demographics

Transportation

Photo Credits

CHAPTER 1: **1,** Kai Chiang/123rf **7,** fresh_water/Fotolia **17,** WavebreakmediaMicro/Fotolia **18 (top),** Everett Collection Inc./Alamy **18 (bottom),** SimpleVision/Fotolia **25,** Roman Sigaev/Shutterstock **30,** AZP Worldwide/Fotolia **36 (top),** Ruben Sprich/Reuters **36 (middle),** Olesya Shelomova/Fotolia **40,** Rob/Fotolia **43 (left),** Tribune Content Agency LLC/Alamy **43 (top right),** Deyan Georgiev/ Fotolia **43 (bottom right),** happyalex/123rf **44,** Catamando/Fotolia **45,** Jim/Fotolia **53,** Images Distribution Canada/Agence Quebec Presse Canada/Newscom **63 (top),** Olga Serdyuk/123rf **63 (bottom),** Philophoto/Fotolia **66,** Braden Gunem/Alamy **67,** Stuart Monk/ Fotolia **70,** GIPhotoStock X/Alamy **71,** Ponsulak/Shutterstock **72,** Cathy Yeulet/123rf **73,** Alexey Stiop/Alamy **77,** jhk2303/Fotolia **78,** spaceport9/Fotolia **88,** Kai Chiang/123rf **CHAPTER 2:** **93, 103 (left),** Chris Boswell/Fotolia **103 (right),** Taina Sohlman/Fotolia **105,** Gladskikh Tatiana/Shutterstock **114,** pkstock/Fotolia **120,** XYZ/Shutterstock **121,** Chris Willson/Alamy **123 (left),** Aliaksandr Mazurkevich/123rf **123 (top middle),** Tischenko Irina/Shutterstock **123 (bottom middle),** DrHeinrich/Shutterstock **123 (right),** Dimitry Atanasov/Shutterstock **128,** Superstock **132 (left),** gary718/123rf **132 (top middle),** pob_anythink/Fotolia **132 (bottom middle),** Gabriele Maltinti/Fotolia **132 (right),** donvanstaden/Fotolia **141,** Suphatthra China/123rf **143,** Vladimir Salmam/123rf **145,** Eléonore H/ Fotolia **147 (top),** kalichka/Fotolia **147 (bottom),** Denizo71/Shutterstock **148 (left),** David Coleman/Alamy **148 (right),** KazantsevAlexander/ Fotolia **149,** NikoNomad/Shutterstock **CHAPTER 3:** **163,** Big Cheese Photo LLC/Alamy **179,** Rafael Ben-Ari/Fotolia **180,** Big Cheese Photo LLC/Alamy **183,** Cathy Yeulet/123rf **184,** SeanPavonePhoto/Fotolia **193,** Qushe/Shutterstock **206,** Dario Sabljak/ Fotolia **CHAPTER 4:** **219, 228,** MixPix/Alamy **231,** Elena Elisseeva/Shutterstock **232 (left),** selezenj/Fotolia **232 (right),** sakhorn/ Shutterstock **233,** Chuck Berman/Chicago Tribune/MCT via Getty Images **234,** huyangshu/Shutterstock **251,** atosan/123rf **275,** Sylvain Robin/123rf **290,** TATSIANAMA/Shutterstock **CHAPTER 5:** **307,** View Stock/Getty Images **317,** David Madison/Digital Vision/ Getty Images **318,** Yanik Chauvin/Fotolia **327,** viappy/Fotolia **328 (left),** View Stock/Getty Images **328 (right),** Igor Mojzes/Fotolia **329 (left),** Peter Dazeley/Getty Images **329 (right),** Tyler Olson/123rf **344,** Christian Mueller/Shutterstock **345,** Stephen Rees/ Shutterstock **369,** Wiktor Bubniak/Fotolia **371,** Rafael Ben-Ari/Alamy **372,** Tyler Olson/123rf **374,** Andrey_Popov/Shutterstock **375 (left),** Steve Healey **375 (right),** nyiragongo/Fotolia **376 (top left),** Petoo/Fotolia **376 (bottom left),** Rainer W. Schlegelmilch/ Contributor/Getty Images **376 (right),** AFP/Getty Images **377 (left),** Cozyta/Fotolia **377 (right),** Denis Tabler/Shutterstock **379,** Andriy Popov/123rf **CHAPTER 6:** **393,** Lilyana Vynogradova/Shutterstock **399,** Bigfoot/Fotolia **402,** Spiderstock/Getty Images **404,** Michael Nichols/National Geographic/Getty Images **407,** SurangaSL/Sh/Shutterstock **409,** Scott S. Warren/National Geographic/Getty Images **410,** Liunian/Shutterstock **411,** Mrakhr/Fotolia **412,** Jonathan Danie/Getty Images **414,** Icy Straight Point **415,** Lilyana Vynogradova/ Shutterstock **416,** Igor Plotnikov/Shutterstock **430,** Roza/Fotolia **431,** Kobby Dagan/Shutterstock **440,** © 2006 The Children's Museum of Indianapolis **441,** Antonio Gravante/Fotolia **446 (left),** Book printer RR Donnelley **446 (right),** Bryn Lennon/Getty Images **472,** Polder/Shutterstock **478 (left),** Getty Images **478 (right),** ChrisVanLennepPhoto/Fotolia **489,** Marcuspon/Fotolia **CHAPTER 7:** **493,** Amy Walters/Shutterstock **506,** Ikonoklast Fotografie/Shutterstock **514,** Oceanfishing/Shutterstock **542,** Amy Walters/Shutterstock **543,** NASA **CHAPTER 8:** **555,** Alicia Shields/Fotolia **560,** Andre Jenny/Alamy **566 (top),** Oloololo/ Fotolia **566 (bottom),** Four Oaks/Shutterstock **567,** Graphlight/Fotolia **572,** moodboard/Alamy **576 (top left),** Alicia Shields/Fotolia **576 (bottom left),** pio3/Fotolia **576 (right),** William O'Brien Jr., LLC **601,** Andrey Popov/Fotolia **608,** Sean Ne/123rf **633,** Ami Parikh/ Shutterstock **CHAPTER 9:** **637,** George Burba/123rf **649 (left),** yuuuu/Fotolia **649 (top right),** bibiphoto/Shutterstock **649 (bottom right),** George Burba/123rf **650 (left),** alarsonphoto/Fotolia **650 (right),** Adrian Sherratt/Alamy **656,** Ken Wolter/Shutterstock **658 (left),** casadaphoto/Shutterstock **658 (right),** AFP Getty Images **671,** Betty LaRue/Alamy **673,** vanbeets/123rf **675,** GWImages/Shutterstock **676,** Aleph Studio/Shutterstock **684,** Lijuan Guo/Shutterstock **697,** Stanislav Komogorov/123rf **700,** sima/Fotolia **701,** ofoto/123rf **702,** Calin Stan/123rf **CHAPTER 10:** **719,** George Burba/123rf **727,** Information Unlimited **733,** Jason Tench/Shutterstock **735,** Lawrence Jackson/White House/Handout/The White House/Corbis **736,** Les (Ladislav) Palenik/123rf **743,** Frank Oppermann/Shutterstock **755 (left),** Blend Images/Shutterstock **755 (right),** Ivonne Wierink/Fotolia **756,** Denis Nata/Fotolia **CHAPTER 11:** **791,** blotty/123rf **799,** Carola Schubbel/Fotolia **806,** Pavel L Photo and Video/Shutterstock **807,** Shariff Che'Lah/123rf **813,** Jeff Lewis/Icon SMI 910/Jeff Lewis/Icon SMI/Newscom **817,** Terry Mathews/Alamy **836,** ZUMA Press, Inc./Alamy **847,** Thanapol Kuptanisakom/123rf **853,** blotty/123rf **859,** David Coleman/Alamy

Geometry

Plane Geometry

Rectangle
Area: $A = lw$
Perimeter: $P = 2l + 2w$

Square
Area: $A = s^2$
Perimeter: $P = 4s$

Triangle
Area: $A = \frac{1}{2}bh$

Sum of Angle Measures
$A + B + C = 180°$

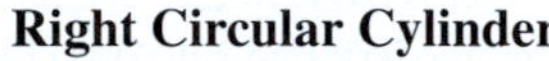

Right Triangle
Pythagorean theorem (equation):
$a^2 + b^2 = c^2$

Parallelogram
Area: $A = bh$

Trapezoid
Area: $A = \frac{1}{2}h(a + b)$

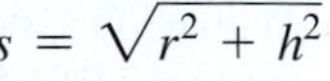

Circle
Area: $A = \pi r^2$
Circumference:
$C = \pi d = 2\pi r$

Solid Geometry

Rectangular Solid
Volume: $V = lwh$

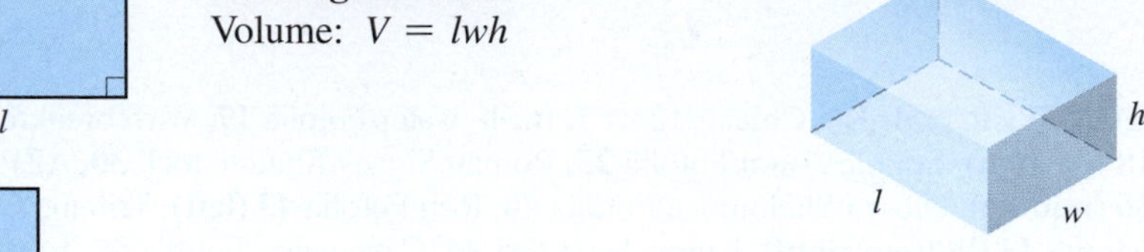

Cube
Volume: $V = s^3$

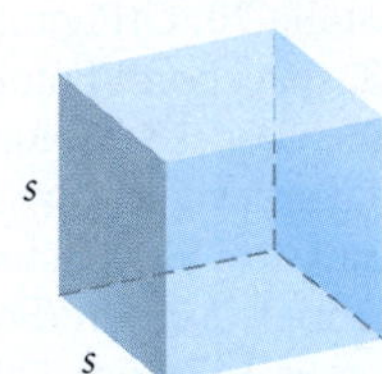

Right Circular Cylinder
Volume: $V = \pi r^2 h$
Lateral surface area:
$L = 2\pi rh$
Total surface area:
$S = 2\pi rh + 2\pi r^2$

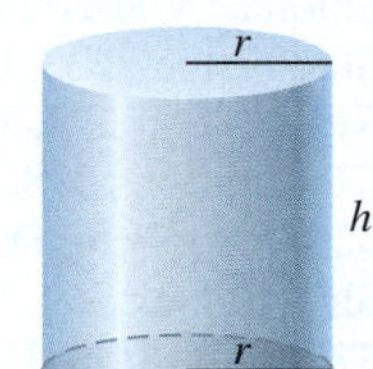

Right Circular Cone
Volume: $V = \frac{1}{3}\pi r^2 h$
Lateral surface area:
$L = \pi rs$
Total surface area:
$S = \pi r^2 + \pi rs$
Slant height:
$s = \sqrt{r^2 + h^2}$

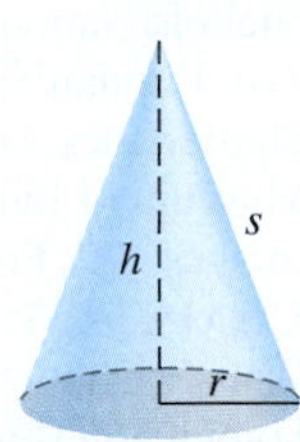

Sphere
Volume: $V = \frac{4}{3}\pi r^3$
Surface area: $S = 4\pi r^2$

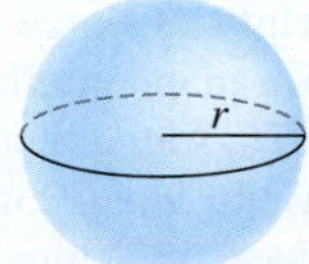

Algebra

Properties of Real Numbers

Commutative: $a + b = b + a; \quad ab = ba$

Associative: $a + (b + c) = (a + b) + c;$
$a(bc) = (ab)c$

Additive Identity: $a + 0 = 0 + a = a$

Additive Inverse: $-a + a = a + (-a) = 0$

Multiplicative Identity: $a \cdot 1 = 1 \cdot a = a$

Multiplicative Inverse: $a \cdot \frac{1}{a} = \frac{1}{a} \cdot a = 1, a \neq 0$

Distributive: $a(b + c) = ab + ac$

Exponents and Radicals

$a^m \cdot a^n = a^{m+n}$ $\qquad \frac{a^m}{a^n} = a^{m-n}$

$(a^m)^n = a^{mn}$ $\qquad (ab)^m = a^m b^m$

$\left(\frac{a}{b}\right)^m = \frac{a^m}{b^m}$ $\qquad a^{-n} = \frac{1}{a^n}$

If n is even, $\sqrt[n]{a^n} = |a|$.

If n is odd, $\sqrt[n]{a^n} = a$.

$\sqrt[n]{a} \cdot \sqrt[n]{b} = \sqrt[n]{ab}, \quad a, b \geq 0$

$\sqrt[n]{\frac{a}{b}} = \frac{\sqrt[n]{a}}{\sqrt[n]{b}}$

$\sqrt[n]{a^m} = (\sqrt[n]{a})^m = a^{m/n}$

Special-Product Formulas

$(a + b)(a - b) = a^2 - b^2$

$(a + b)^2 = a^2 + 2ab + b^2$

$(a - b)^2 = a^2 - 2ab + b^2$

$(a + b)^3 = a^3 + 3a^2b + 3ab^2 + b^3$

$(a - b)^3 = a^3 - 3a^2b + 3ab^2 - b^3$

$(a + b)^n = \sum_{k=0}^{n} \binom{n}{k} a^{n-k} b^k$, where

$$\binom{n}{k} = \frac{n!}{k!\,(n - k)!} = \frac{n(n - 1)(n - 2) \cdots [n - (k - 1)]}{k!}$$

Factoring Formulas

$a^2 - b^2 = (a + b)(a - b)$

$a^2 + 2ab + b^2 = (a + b)^2$

$a^2 - 2ab + b^2 = (a - b)^2$

$a^3 + b^3 = (a + b)(a^2 - ab + b^2)$

$a^3 - b^3 = (a - b)(a^2 + ab + b^2)$

Interval Notation

$(a, b) = \{x | a < x < b\}$

$[a, b] = \{x | a \leq x \leq b\}$

$(a, b] = \{x | a < x \leq b\}$

$[a, b) = \{x | a \leq x < b\}$

$(-\infty, a) = \{x | x < a\}$

$(a, \infty) = \{x | x > a\}$

$(-\infty, a] = \{x | x \leq a\}$

$[a, \infty) = \{x | x \geq a\}$

Absolute Value

$|a| \geq 0$

For $a > 0$,

$|X| = a \rightarrow X = -a \quad or \quad X = a,$

$|X| < a \rightarrow -a < X < a,$

$|X| > a \rightarrow X < -a \quad or \quad X > a.$

Equation-Solving Principles

$a = b \rightarrow a + c = b + c$

$a = b \rightarrow ac = bc$

$a = b \rightarrow a^n = b^n$

$ab = 0 \leftrightarrow a = 0 \quad or \quad b = 0$

$x^2 = k \rightarrow x = \sqrt{k} \quad or \quad x = -\sqrt{k}$

Inequality-Solving Principles

$a < b \rightarrow a + c < b + c$

$a < b$ and $c > 0 \rightarrow ac < bc$

$a < b$ and $c < 0 \rightarrow ac > bc$

(*Algebra continued*)

Algebra *(continued)*

The Distance Formula

The distance from (x_1, y_1) to (x_2, y_2) is given by

$$d = \sqrt{(x_2 - x_1)^2 + (y_2 - y_1)^2}.$$

The Midpoint Formula

The midpoint of the line segment from (x_1, y_1) to (x_2, y_2) is given by

$$\left(\frac{x_1 + x_2}{2}, \frac{y_1 + y_2}{2}\right).$$

Formulas Involving Lines

The slope of the line containing points (x_1, y_1) and (x_2, y_2) is given by

$$m = \frac{y_2 - y_1}{x_2 - x_1}.$$

Slope–intercept equation:	$y = f(x) = mx + b$
Horizontal line:	$y = b$ or $f(x) = b$
Vertical line:	$x = a$
Point–slope equation:	$y - y_1 = m(x - x_1)$

The Quadratic Formula

The solutions of $ax^2 + bx + c = 0, a \neq 0$, are given by

$$x = \frac{-b \pm \sqrt{b^2 - 4ac}}{2a}.$$

Compound Interest Formulas

Compounded n times per year: $A = P\left(1 + \frac{i}{n}\right)^{nt}$

Compounded continuously: $P(t) = P_0 e^{kt}$

Properties of Exponential and Logarithmic Functions

$\log_a x = y \leftrightarrow x = a^y$ $\qquad a^x = a^y \leftrightarrow x = y$

$\log_a MN = \log_a M + \log_a N$ $\qquad \log_a M^p = p \log_a M$

$\log_a \frac{M}{N} = \log_a M - \log_a N$

$\log_b M = \frac{\log_a M}{\log_a b}$

$\log_a a = 1$ $\qquad \log_a 1 = 0$

$\log_a a^x = x$ $\qquad a^{\log_a x} = x$

Conic Sections

Circle: $(x - h)^2 + (y - k)^2 = r^2$

Ellipse: $\frac{(x - h)^2}{a^2} + \frac{(y - k)^2}{b^2} = 1$, $\frac{(x - h)^2}{b^2} + \frac{(y - k)^2}{a^2} = 1$

Parabola: $(x - h)^2 = 4p(y - k)$, $(y - k)^2 = 4p(x - h)$

Hyperbola: $\frac{(x - h)^2}{a^2} - \frac{(y - k)^2}{b^2} = 1$, $\frac{(y - k)^2}{a^2} - \frac{(x - h)^2}{b^2} = 1$

Arithmetic Sequences and Series

$a_1, \quad a_1 + d, \quad a_1 + 2d, \quad a_1 + 3d, \ \ldots$

$a_{n+1} = a_n + d$ $\qquad a_n = a_1 + (n - 1)d$

$S_n = \frac{n}{2}(a_1 + a_n)$

Geometric Sequences and Series

$a_1, \quad a_1 r, \quad a_1 r^2, \quad a_1 r^3, \ \ldots$

$a_{n+1} = a_n r$ $\qquad a_n = a_1 r^{n-1}$

$S_n = \frac{a_1(1 - r^n)}{1 - r}$ $\qquad S_\infty = \frac{a_1}{1 - r}, |r| < 1$

Trigonometry

Trigonometric Functions

Acute Angles

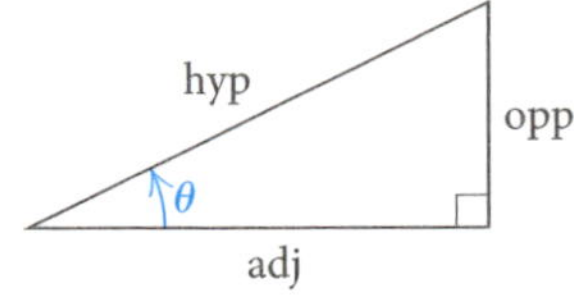

$$\sin\theta = \frac{\text{opp}}{\text{hyp}}, \quad \csc\theta = \frac{\text{hyp}}{\text{opp}},$$
$$\cos\theta = \frac{\text{adj}}{\text{hyp}}, \quad \sec\theta = \frac{\text{hyp}}{\text{adj}},$$
$$\tan\theta = \frac{\text{opp}}{\text{adj}}, \quad \cot\theta = \frac{\text{adj}}{\text{opp}}$$

Any Angle

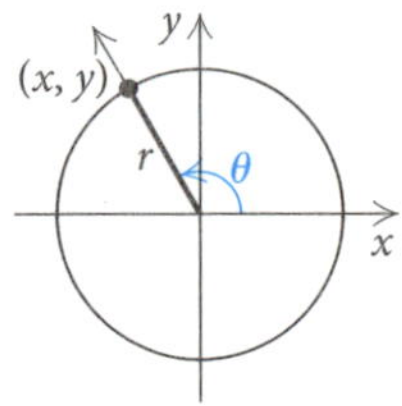

$$\sin\theta = \frac{y}{r}, \quad \csc\theta = \frac{r}{y},$$
$$\cos\theta = \frac{x}{r}, \quad \sec\theta = \frac{r}{x},$$
$$\tan\theta = \frac{y}{x}, \quad \cot\theta = \frac{x}{y}$$

Real Numbers

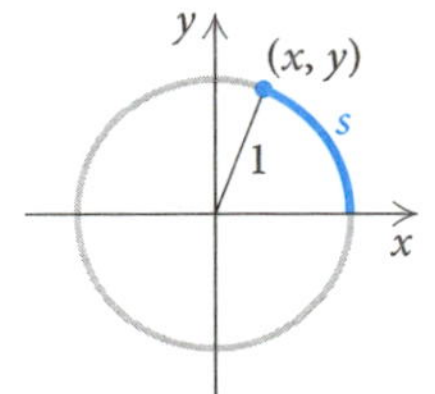

$$\sin s = y, \quad \csc s = \frac{1}{y},$$
$$\cos s = x, \quad \sec s = \frac{1}{x},$$
$$\tan s = \frac{y}{x}, \quad \cot s = \frac{x}{y}$$

Basic Trigonometric Identities

$$\sin(-x) = -\sin x,$$
$$\cos(-x) = \cos x,$$
$$\tan(-x) = -\tan x,$$

$$\tan x = \frac{\sin x}{\cos x},$$
$$\cot x = \frac{\cos x}{\sin x},$$

$$\csc x = \frac{1}{\sin x},$$
$$\sec x = \frac{1}{\cos x},$$
$$\cot x = \frac{1}{\tan x}$$

Pythagorean Identities

$$\sin^2 x + \cos^2 x = 1,$$
$$1 + \cot^2 x = \csc^2 x,$$
$$1 + \tan^2 x = \sec^2 x$$

Identities Involving $\pi/2$

$$\sin(\pi/2 - x) = \cos x,$$
$$\cos(\pi/2 - x) = \sin x, \quad \sin(x \pm \pi/2) = \pm\cos x,$$
$$\tan(\pi/2 - x) = \cot x, \quad \cos(x \pm \pi/2) = \mp\sin x$$

Sum and Difference Identities

$$\sin(u \pm v) = \sin u \cos v \pm \cos u \sin v,$$
$$\cos(u \pm v) = \cos u \cos v \mp \sin u \sin v,$$
$$\tan(u \pm v) = \frac{\tan u \pm \tan v}{1 \mp \tan u \tan v}$$

Double-Angle Identities

$$\sin 2x = 2\sin x \cos x,$$
$$\cos 2x = \cos^2 x - \sin^2 x$$
$$= 1 - 2\sin^2 x$$
$$= 2\cos^2 x - 1,$$
$$\tan 2x = \frac{2\tan x}{1 - \tan^2 x}$$

Half-Angle Identities

$$\sin\frac{x}{2} = \pm\sqrt{\frac{1 - \cos x}{2}}, \quad \cos\frac{x}{2} = \pm\sqrt{\frac{1 + \cos x}{2}},$$
$$\tan\frac{x}{2} = \pm\sqrt{\frac{1 - \cos x}{1 + \cos x}} = \frac{\sin x}{1 + \cos x} = \frac{1 - \cos x}{\sin x}$$

(*Trigonometry continued*)

Trigonometry (continued)

The Law of Sines

In any $\triangle ABC$,

$$\frac{a}{\sin A} = \frac{b}{\sin B} = \frac{c}{\sin C}.$$

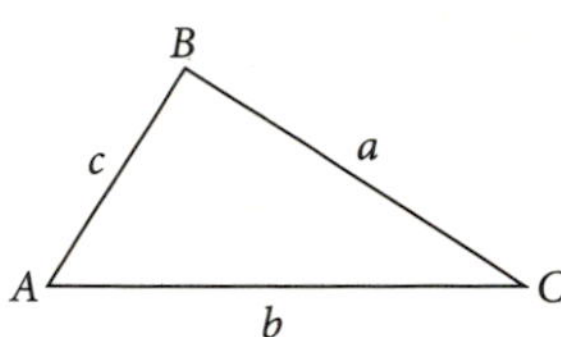

The Law of Cosines

In any $\triangle ABC$,

$$a^2 = b^2 + c^2 - 2bc \cos A,$$
$$b^2 = a^2 + c^2 - 2ac \cos B,$$
$$c^2 = a^2 + b^2 - 2ab \cos C.$$

Trigonometric Function Values of Special Angles

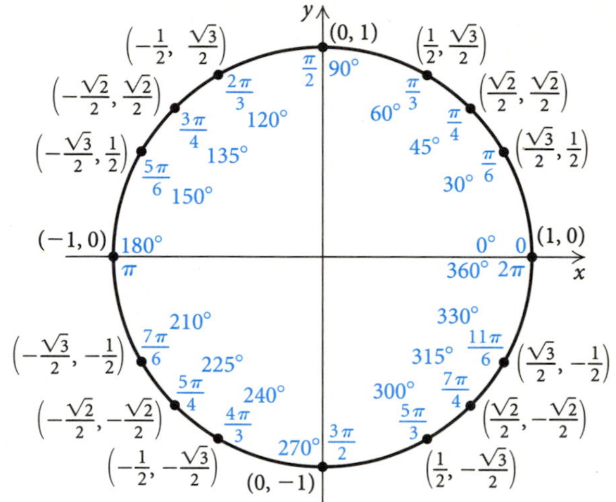

Graphs of Trigonometric Functions

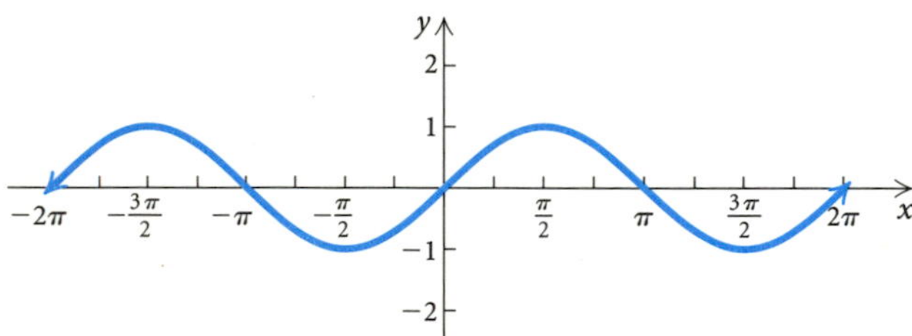

The sine function: $f(x) = \sin x$

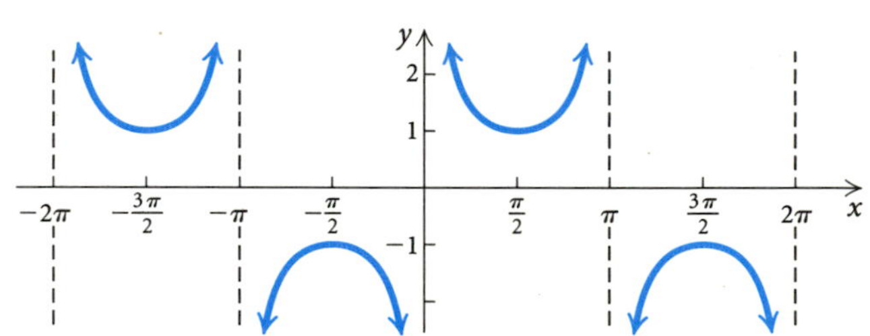

The cosecant function: $f(x) = \csc x$

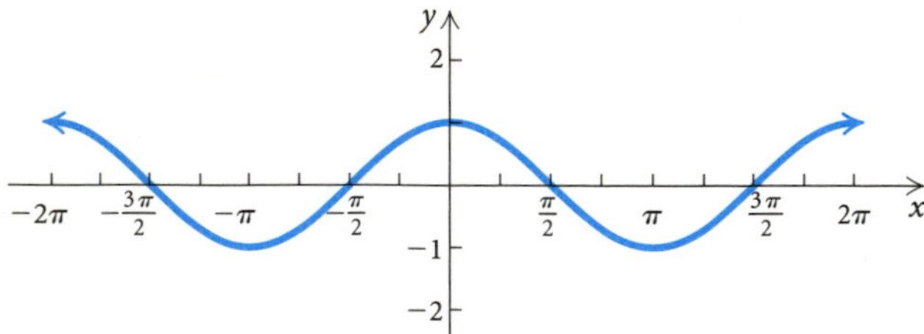

The cosine function: $f(x) = \cos x$

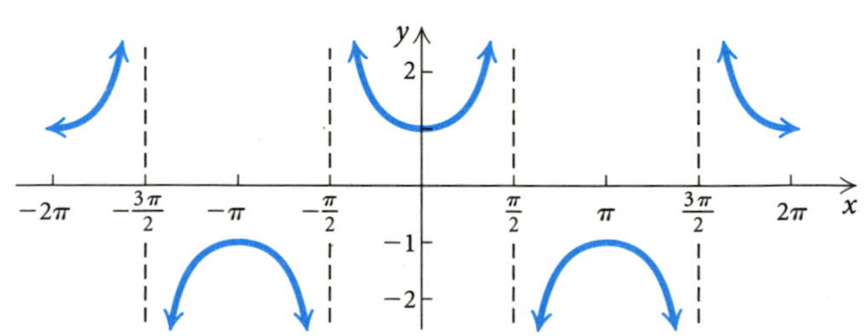

The secant function: $f(x) = \sec x$

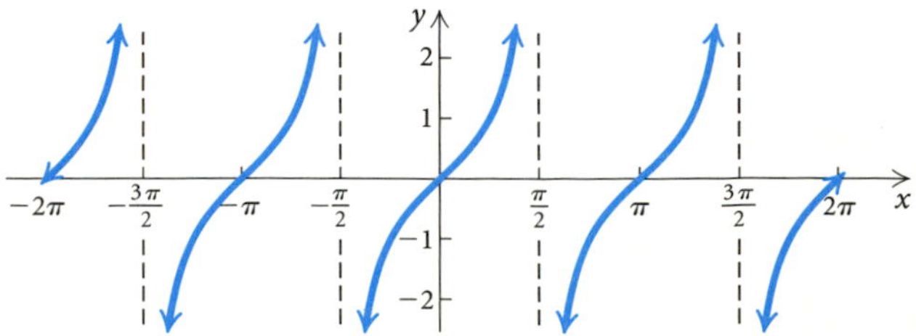

The tangent function: $f(x) = \tan x$

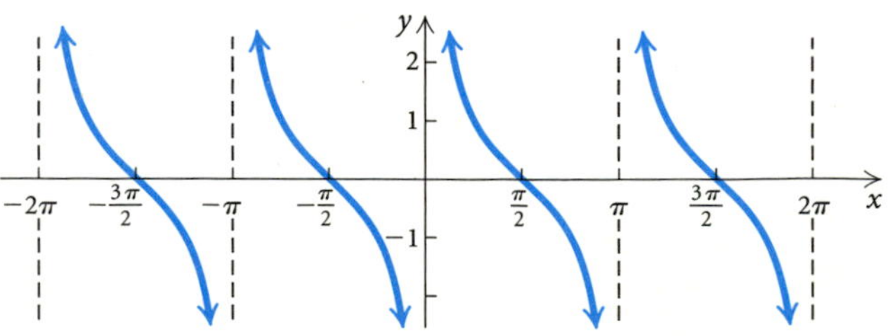

The cotangent function: $f(x) = \cot x$

GENETICS

Principles of GENETICS

EIGHTH EDITION

Eldon John Gardner
Utah State University

Michael J. Simmons
University of Minnesota

D. Peter Snustad
University of Minnesota

JOHN WILEY & SONS, INC.
New York Chichester Brisbane Toronto Singapore

Acquisitions Editor Sally Cheney
Managing Editor Joan Kalkut
Production Manager Joe Ford
Cover and Text Designer Dawn L. Stanley
Production Supervisor Savoula Amanatidis
Manufacturing Manager Lorraine Fumoso
Copy Editor Marjorie Shustak
Photo Researcher John Schultz
Photo Research Assistant Hilary Newman
Photo Research Manager Stella Kupferberg
Illustrations John Balbalis
John Wiley Illustration Dept.
J/B Woolsey Associates
Cover Illustration Steve Jenkins

Library of Congress Cataloging in Publication Data:

Gardner, Eldon John, 1909–1989
Principles of genetics / Eldon John Gardner, D. Peter Snustad, Michael J. Simmons.—8th ed.
p. cm.
Includes bibliographical references and index.
ISBN 0-471-50487-4
1. Genetics. I. Snustad, D. Peter. II. Simmons, Michael J. III. Title.
QH430.G37 1991
575.1—dc20 90-48758
CIP

Recognizing the importance of preserving what has been written, it is a policy of John Wiley & Sons, Inc. to have books of enduring value published in the United States printed on acid-free paper, and we exert our best efforts to that end.

The eighth edition
is dedicated to the memory of

Eldon John Gardner

who skillfully guided this textbook
through seven editions

PREFACE

The eighth edition is a continuing attempt to provide a true "principles of genetics" text that is up-to-date, readable, and challenging. Its major goal is to emphasize the central role that genetics plays in the lives of all organisms—plant, animal, and microbe—on planet Earth. In our complex modern society, with the very survival of our planet at stake, it is imperative that citizens be informed about scientific issues. Thus, our goal has been to write this text for students with only a minimal background in biology and chemistry. It is suitable for one-quarter, one-semester, or two-quarter courses in genetics that are designed for biology majors or for students from other disciplines. We have attempted to provide balanced coverage of many areas in genetics—classical, molecular, and population—but we recognize that each instructor will favor some topics over others. Consequently, this book has been written to allow instructors flexibility in the choice of topics and in the depth of coverage.

Genetics and related sciences have grown explosively in recent years, generating a large body of new and important information. As a result, the eighth edition has expanded considerably when compared with previous editions. A new introductory chapter attempts to convey the excitement of genetics today and to emphasize the central role that it plays in our lives. Two other new chapters cover the major developments in our knowledge of transposable genetic elements and the rapidly advancing methodologies of genetic engineering. Other chapters have been revised and expanded to incorporate new information about the structure and function of genes and their roles in development.

Compared with its predecessor, this edition has been substantially reorganized. The material on the structure and replication of DNA and chromosomes has been divided into two chapters, one focusing on prokaryotes and the other on eukaryotes. The analysis of genetic fine structure is included in two chapters, one dealing with classical studies and the other with molecular approaches, including cloning and sequencing. Also included in this molecular chapter are such important techniques as Southern, northern, and western blotting, PCR amplification of DNA, and oligonucleotide-directed site-specific mutagenesis. Regulation, which was treated in a single chapter in the previous edition, is now treated in four; two of these deal with the basic features of regulation, one focusing on prokaryotes and the other on eukaryotes, and two deal with special eukaryotic phenomena—the immune response and the genetic control of cell division. This last chapter includes an expanded discussion of the structure and function of oncogenes and protooncogenes.

Other changes have been made in the eighth edition to enhance its readability and focus. Multiple alleles are now treated in the same chapter as basic Mendelian genetics. The chapters on quantitative genetics and population and evolutionary genetics have been rewritten to provide more coherence to these somewhat difficult subjects. New material has been added to several other chapters; for example, the chapter on gene expression now contains sections on RNA splicing and the spliceosome, and the chapter on extrachromosomal inheritance contains new sections on the organization of mitochondrial and chloroplast genomes. As in previous editions, each chapter ends with an extensive set of problems and questions of varying difficulty, with the answers provided at the end of the book.

We believe that the eighth edition presents the basic principles of genetics in a clear and logical sequence, but we realize that many instructors will prefer other sequences. As mentioned earlier, this edition begins with a new introductory chapter that attempts to (1) convey some of the excitement that has resulted from recent developments in the field of genetics and (2) emphasize the important role that genetics plays in many aspects of our lives and, indeed, in the very survival of life on our planet. Chapter 2 describes the classic work of Gregor Mendel and the basic principles of Mendelian genetics that resulted from his work. Cells, chromosomes, and the process of meiosis that provide the basis for Mendel's laws are described in Chapter 3. The chromosomes and their behavior during gametogenesis are related to sex determination and sex-linked patterns of inheritance in Chapter 4. The molecular structure of genes and their modes of replication in prokaryotic chromosomes are introduced in Chapter 5. The complex structure and mechanism of replication of eukaryotic chromosomes are the subjects of Chapter 6. Chapter 7 covers the classical genetic topics of linkage, chromosome mapping, tetrad analysis, gene conversion, and the mechanism of crossing over. Recombination mechanisms unique to bacteria are covered in Chapter 8, and the structure and function of transposable genetic elements are the subjects of Chapter 9.

After considering the structure and recombination of genetic elements in Chapters 5 through 9, the focus shifts to the function of genetic elements in Chapter 10 (Gene Expression), which covers transcription, RNA processing ("splicing"), translation, and the genetic code. The mechanisms by which genes change or mutate are covered in Chapter 11. Chapters 12 and 13 dissect genes by both classical (Chapter 12) and molecular (Chapter 13) approaches. Chapter 13 also provides background information on the powerful tools of molecular genetics that are used in these dissetions: cloning vectors, Southern, northern, and western blots, PCR, and site-specific mutagenesis. Chapters 14 through 17 describe our present understanding of the various mechanisms by which gene expression and cell division are regulated. Control of gene expression in prokaryotes is the topic covered in Chapter 14, whereas Chapter 15 focuses on the regulation of gene expression and development in eukaryotes. Although the genetic control of development in higher eukaryotes is much too complex to be covered in detail in an introductory genetics text, we hope the information presented in this chapter will give students an insight into the exciting developments in this field. Although much remains to be learned, molecular pictures of the genetic control of development in *Drosophila melanogaster* and the worm *Caenorhabditis elegans* have rapidly begun to emerge during the last few years.

The genetic control of the vast repertoire of antibodies that can be produced by higher animals was an enigma for decades. In the last few years, many of the mysteries of the immune system have been replaced by fascinating pictures of the genetic control of antibodies and T cell receptors; these new breakthroughs are described in Chapter 16. In Chapter 17, the genetic control of cell division is discussed, with emphases on the roles of protooncogenes and on the perturbations caused by oncogenes. The next two chapters discuss mutation in the broad sense of gross changes in chromosome structure (Chapter 18) and chromosome number (Chapter 19). In Chapter 20, the behavior of genes that are not located on the main nuclear chromosomes of the organism is considered, with an emphasis on the structure of the chromosomes of mitochondria and chloroplasts. Chapter 21 discusses those traits that are controlled by large numbers of genes and thus exhibit quantitative patterns of inheritance. In Chapter 22, the analysis of genes and their effects is extended to the level of the population; major emphases are the distributions of alleles in populations and the changes in these distributions (evolution) over time in response to factors such as mutation, genetic drift, and selection. Chapter 23 provides a brief introduction to the genetic factors that influence behavior.

Chapter 24 introduces the student to some of the present and future applications of genetic engineering. We realize that most introductory genetics classes will cover few, if any, of the topics discussed in Chapter 24. However, we believe that this chapter will be of considerable interest to many students, and we anticipate that those students who find genetics to be fascinating and challenging will read Chapter 24 on their own. Some instructors will probably choose to integrate certain topics from Chapter 24 into their courses. For example, a class with a plant genetics emphasis might appropriately include the material on *Agrobacterium tumefaciens* and plant transformation via the Ti plasmid. Other classes may choose to include the section on somatic-cell gene therapy, and so on. A glossary of important genetic terms, the answers to all the problems in the book, and an index complete the text.

Our thanks to students, teachers, and colleagues who have suggested improvements for the book. We especially thank Rayla Temin, R. H. Whalen, Linda Kosturko, Clint Magill, Peter Weijksnora, Charles Rodell, Hugh Stanley, Larry Puckett, Robert M. Fineman, Irene Uchida, Scott R. Woodward, John R. Simmons, and Franklin D. Enfield, who have all made valuable contributions. We also thank the following reviewers for their valuable suggestions and comments: Judith Van Houten, The University of Vermont; Robert Petters, Pennsylvania State University; Darrell S. English, Northern Arizona University; Romesh C. Mehra, Indiana University at South Bend, Glenn Wolfe, Kansas University; Paul J. Homsher, Old Dominion University; Paul A. Roberts, Oregon State University; Asim Esem, Virginia Polytechnic Institute and State University; Howard Laten, Loyola University; Charles H. Green, Glassboro State College; Alice S. Hunter, College of the Pacific; Martin Bard, Purdue University; John B. Jenkins, Swarthmore College; Kandy D. Baumgardner, Eastern Illinois University; Dwight E. Wilson, Rensselaer Polytechnic Institute; Carter Denniston and William R. Engels, University of Wisconsin, Madison; H. James Price, Texas A & M University; Peter Dawson, Oregon State University; Barry Bean, Lehigh University; Dean Whited, North Dakota State University.

We also gratefully acknowledge the invaluable assistance of Sally Cheney, editor; John Balbalis and John D. Woolsey, illustrators; Dawn L. Stanley, designer; Stella Kupferberg and John Schultz, photo researchers; Joe Ford, production manager; Marjorie Shustak, supervising copy editor; John Thomas, copy editor; Savoula Amanatidis, production supervisor; and Virginia McTavish, manuscript preparator.

Credits for illustrations, tables, and quotations from other individuals or publications are given with the individual items.

Michael J. Simmons
D. Peter Snustad

BRIEF CONTENTS

CONTENTS

Chapter 1

Introduction

(Photo above): View of the Earth from the Apollo 8 spacecraft.

"**GENE Studies Emerging as Key Engine of Science**"—headline, *The New York Times*, September 6, 1988. The significance of this headline is not that it announces any new discovery. Geneticists have known for decades that genes play central roles in science and, in fact, in all aspects of the lives of all living organisms. The significance of this headline is that our society as a whole, or at least the more informed component of our society, now also recognizes the major roles that genes play in shaping our lives and the lives of all living creatures on planet Earth. Of course, such headlines have been commonplace in science journals for decades, but only during the last few years have these headlines appeared frequently in the popular press, particularly the news media. For example, a quick scan of one major newspaper, *The New York Times*, over the 30-day period from September 6, 1988, to October 6, 1988, revealed three major headlines focusing on genetics; one, ***"Panel Backs Gene Transplant For Cancer Treatment Test,"*** was on the front page (Fig. 1.1). In fact, as anyone who makes a significant effort to stay informed about world developments knows, similar headlines have become increasingly prevalent during the last decade in all quality newspapers as well as in the radio and television news media throughout the world.

Gene Studies Emerging As Key Engine of Science

By HAROLD M. SCHMECK Jr.

THE science of genetics is fast becoming what geneticists always knew it was: the central and most provocative science of life.

Toronto, president of the the Toronto meeting wa meetings, which are he Four thousand scientist tries attended, twice t and four times the nur resented three decade:

Panel Backs Gene Transplant For Cancer Treatment Test

By WARREN E. LEARY

Special to The New York Times

WASHINGTON, Oct. 3 — An influential advisory committee recommended today that the Government for the first time approve the transplanting of foreign genes into human patients.

and the Food and Drug Administration.

But the scientist and his principal coworker on the cancer experiment, Dr. Steven Rosenberg, said they had indications from the remaining authorities

DNA Pioneer to Tackle Biggest Gene Project Ever

By HAROLD M. SCHMECK Jr.

DR. James D. Watson, who is about to assume a key role in the largest biological research project ever contemplated, sees that effort as the redemption of a longstanding bargain between scientists and other citizens.

"I think it is imperative to keep our bargain with the people who have supported biomedical research," he said in an interview Friday. The project will immensely expand scientists' knowledge of human genetics. Directly and indirectly, that knowledge will be used to seek new ways of curing or coping with important diseases by teasing out the secrets of the genes.

Dr. Watson was co-discoverer in 1953 of the structure of DNA, the substance of the genes in all living things. In 1962, he shared a Nobel Prize for that work with Dr. Francis H. C. Crick, now at the Salk Institute in La Jolla, Calif., and Dr. Maurice H. F. Wilkins of Cambridge University in England. The discovery showed how DNA functions as the repository of all hereditary information and led to a revolution in biology that continues to accelerate and expand.

The new job that Dr. Watson begins this week involves helping to organize a huge federally supported project to define all of the roughly 50,000 to 100,000 human genes. It is known as the genome project; a genome being the sum of all the genes of any organism. The project, if approved by Congress, is expected to take 15 years and to cost several billion dollars.

The New York Times/Michael Shavel

Dr. James D. Watson, who pioneered research into the structure of DNA and now heads an effort to map the human genome.

Dr. Watson has been appointed associate director of the National Institutes of Health for genome research, a newly created post to oversee the institutes' involvement with the genome project.

In his early years of research, Dr. Watson was known as a brash, witty and often abrasive enfant terrible. The new appointment seemingly completes his metamorphosis into a senior statesman of science.

"People know that I am interested in biology as a whole," he said. "Of course, I am also interested in human beings."

The project is a logical outcome of the discovery made in 1953, but until recently, the idea would have been derided as fantasy because it was so far beyond the technical capabilities of the science. Only a relatively few genes have been completely defined to date but the new project will try to define the chemical compositon of them all. The effort is made possible now by a series of technical and

Continued on Page C16

Figure 1.1 Headlines in the *The New York Times* on September 6 and October 8, 1988, heralding some of the exciting recent developments in genetics. (Copyright 1988 by The New York Times Company. Photograph by Michael Shavel, New York Times Pictures. Reprinted with permission.)

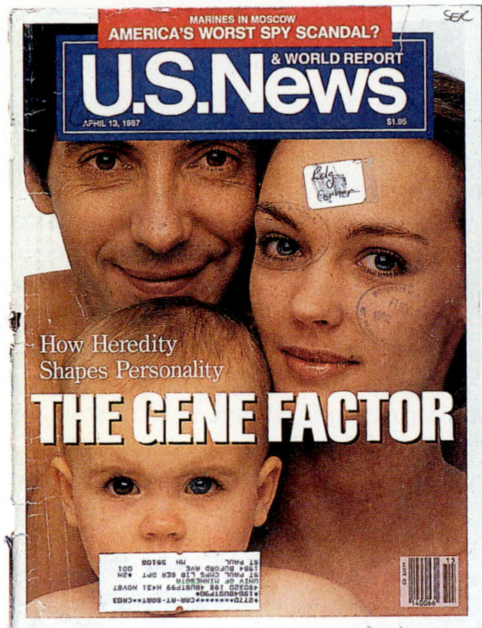

Figure 1.2 Cover of *U.S. News and World Report* on April 13, 1987. This issue contains a discussion of evidence indicating that genetic factors influence personality and behavior in humans. Although environmental factors undoubtedly have the greatest impact on human behavior, at least for most individuals, considerable evidence now indicates that genetic factors also play significant roles. (Reproduced with permission. Copyright 1987 by B. Lipnitsky, New York.)

The summation of the messages highlighted by the hundreds of ***"Genes," "Genetics," "Genetic Engineering,"*** and similar headlines of the last several years is that genes play important roles in essentially all aspects of our lives. They not only control what we look like physically, but also influence our behavior, our personalities (Fig. 1.2), our susceptibility to diseases (be they inherited or infectious), our aging processes, and, indeed, our longevity. Finally, our rapidly expanding understanding of the importance of genetics and the explosive development of new tools and methodologies for genetic engineering applications in agriculture, medicine, microbial fermentations, and other areas now dictate that the genetic factors involved be considered in all areas of our lives from family planning to food production to medical procedures to legal protocols. Clearly, in a democracy where the regulations that govern individual and group activities, some of which will have profound effects on the future of our planet, are established by the citizenry acting through their elected representatives, it is essential that Jane Q. and John Q. Citizen be informed about the issues and be capable of making wise decisions in the voting booth. Given the central role that genetics plays in our lives and in the survival of all living organisms in this world, ***how can Jane Q. and John Q. Citizen make wise, informed decisions without a firm understanding of the basic principles of genetics***—especially in this age of recombinant DNA, gene cloning, and genetic engineering technology?

THE WORLD IS A COLOSSAL GENETICS LABORATORY

Scientists throughout the world are busy identifying and studying genes that control the entire spectrum of phenotypic traits in living organisms, especially humans. These genes are isolated ("cloned"), and their complete nucleotide sequences are determined (they are "sequenced"). The nucleotide sequences are then used to deduce the amino acid sequences of the polypeptide chains (for protein-encoding genes) by using the known nature of the genetic code. Hints as to the function of a particular protein can sometimes be obtained by searching computer data banks maintained by the U.S. National Institutes of Health and by the European Molecular Biology Organization for similar sequences in proteins with known functions. Both genetic and direct molecular approaches are being used to identify pathways of gene expression, that is, the sequences of events by which genes exert their effects on phenotype. At present, major efforts are being focused on the control of gene expression, asking what factors are responsible for turning on and turning off the expression of individual genes. These studies are yielding new, exciting, and sometimes unexpected results. In fact, new information is currently accumulating at an unprecedented rate. No one scientist can possibly keep abreast of all the new developments in the field of genetics.

Two recent events in the genetics arena are attracting large amounts of attention: (1) the launching of the "human genome" project and (2) the imminent green light for the first human somatic-cell "gene therapy" experiment. The goal of the so-called "human genome" project is to map all the estimated 50,000 to 100,000 genes and to sequence the entire 3×10^9 nucleotide-pairs in the genome of the unique species *Homo sapiens*. Obviously, this is a formidable task. It is expected to cost in excess of $3 billion and take 15 to 20 years to complete. This huge project will be carried out jointly by the U.S. National Institutes of Health and the U.S. Department of Energy. The project will be administered by James Watson, who shared the 1962 Nobel Prize in Physiology and Medicine with Francis Crick and Maurice Wilkins for their 1953 discovery of the double-helix structure of DNA. Watson has just accepted the job of Associate Director of the U.S. National Institutes of Health with responsibility for the Office of Human Genome Research. He will be responsible for overseeing and coordinating the activities of researchers in a large number of laboratories that will participate in this huge collaborative effort.

In addition to providing basic knowledge about the organizational and functional aspects of the human genome, this project is expected to have important practical applications. One such application involves identifying and mapping defective genes that cause human diseases. A large number of human diseases are known to be caused by inherited genetic defects, so-called "inborn errors of metabolism." Over 3000 human diseases, fortunately most of them rare, are now believed to result from genetic defects (mutations) in individual genes. Once these genes have been identified and characterized, one can anticipate treating patients with the diseases by somatic-cell "gene therapy," namely, replacing the defective copy of the gene in somatic cells of the patient with a normal, functional copy. Such experiments have proven successful in animal model systems and are likely to be tested in humans in the near future. In fact, the first such experiments are likely to begin within the next year.

In the United States, all experiments involving human subjects require prior approval of the National Institutes of Health. In addition, most of the proposed gene therapy experiments will require prior approval of the U.S. Food and Drug Administration. The Food and Drug Administration must give approval for the utilization of any vector or "carrier DNA" (native, foreign, or engineered *in vitro*) to introduce a gene or DNA sequence of interest into cells that will be implanted in humans.

The National Institutes of Health are responsible for verifying that all the appropriate control experiments have been done with animal systems and that all possible steps are taken to minimize risks to the patients involved. The Institutes rely heavily on their Recombinant DNA Advisory Committee for recommendations when the proposed experiments involve recombinant DNA technologies. This committee has just recommended approval of the first "gene transplant" experiment involving humans. Actually, this first experiment will only involve using a foreign (bacterial) marker gene to trace the survival of white blood cells introduced into about ten volunteer cancer patients. However, if this experiment is successful and there are no serious side effects, actual gene therapy experiments will undoubtedly follow in the not-too-distant future. Several proposals to carry out such gene therapy experiments in attempts to treat specific inherited diseases are currently being reviewed by the National Institutes of Health Recombinant DNA Advisory Committee. However, none of these is likely to be approved until the outcome of the first gene transplant experiment in humans is known.

Scientists are busy making headlines with (1) gene transplant experiments, (2) the human genome project, (3) genetically engineered disease- and insect-resistant plants, (4) the use of recombinant DNA technologies to synthesize medically important gene products such as human insulin and human growth hormone in bacteria, and (5) other laboratory achievements too numerous to mention. However, ***the truly important genetic experiment is continuing outside the research laboratory in the world surrounding us. This colossal genetics experiment is responsible for the continuity of life on planet Earth.*** The survival of every species depends on its ability to transmit its genetic information from parents to offspring generation after generation. This requires each progeny generation of each species to survive to reproductive maturity and, in turn, to produce the next progeny generation. Survival requires that each species be adequately adapted to the existing environmental conditions, that is, that a suitable ecological niche be available.

Unfortunately, humankind has played a very dominant role in changing environmental conditions in many areas of our world, often destroying the environmental niches essential to the survival of species. The result is that many species are now extinct. This trend is continuing today as large regions of rain forests are being burned and cut in South America, rain forests that provided unique habitats for a large number of plant, animal, and microbial species. Many of these species will undoubtedly become extinct in the future. Moreover, the destruction of the tropical rain forests will probably affect the survival of species throughout the world, not just those species native to the rain forests. These forests are major producers of atmospheric O_2, and the burning of these forests results in the production of large amounts of CO_2, thus contributing to the gradual warming of the Earth's atmosphere (the so-called "greenhouse effect"). These effects cause potentially irreversible changes in the environment of our planet, changes that may lead to the extinction of organisms that cannot adapt to the new conditions. When this happens the rich variety of germplasm (the totality of genes in all living organisms) available on our planet will be diminished. Genes that took billions of years to evolve to their present state will be lost forever. Some of the lost genes might have become important to humankind, possibly providing certain plants with resistance to a pathogenic microorganism or an insect pest. Perhaps some of these genes could have been introduced into important food crops by genetic engineering techniques and helped prevent starvation in an overpopulated world of the future. Clearly, every effort must be made to minimize the depletion of the world's germplasm resources and the resulting genetic diversity.

With nuclear bombs and industrial practices that

lead to air and water pollution and the "greenhouse effect," to name a few modern ills, one can easily imagine scenarios in which humankind's dominant manipulation of the environment would render the Earth uninhabitable even for humans. Obviously, the top priority of each generation must be to avoid any of these scenarios. Humankind must learn to live in harmony with the environment and to coexist with the rich variety of life around us. ***If we do not, the great genetic experiment, the continuity of life on Earth, will end.*** Clearly, ***our species ethic must be to protect the environment,*** to forego the temptation to seek short-term gains when these gains come via the development of industrial processes that are potentially destructive to the environment.

Genetics and the Phenotype of Our Biosphere

The phenotype of a living organism is simply what that organism looks like, the summation of its physical characteristics. The phenotype of each organism is controlled by its genotype (all the genes that it carries) interacting with the environment.

genotype + environment ⟶ phenotype

The phenotype of an organism is produced by the coordinated expression of all the genes carried by that organism within the restrictions imposed by the environment. Although our major emphasis in this text will be on the genotypic component, the importance of environmental effects must not be overlooked. A corn plant may have a genotype that will produce a large, vigorous plant when grown under favorable environmental conditions. However, during a drought or when growing in soil without adequate nutrients, this genotype will produce a small, feeble plant.

Although geneticists most often relate the effects of genes to phenotypic traits of individual organisms (e.g., red vs. white flowers in pea plants, or blue vs. brown eyes in humans), these effects are just the first part of a much bigger picture. A population is a group of organisms, and the phenotype of each population is determined by all the genes in all the individual organisms in the population. In turn, an ecosystem includes all the living organisms—all members of all species present—in a particular environment, and its phenotype is determined by the collective action of all the genes in all individuals of all species present. Taken together, all ecosystems on Earth produce the biosphere, the totality of organisms living within the lithosphere, the hydrosphere, and the atmosphere of our planet. The phenotype of the biosphere, what we see in the world around us, is the final product of the action of all the individual genes in all organisms, expressing themselves within the limitations of the environment. When thinking about the phenotype of the entire biosphere, it may be difficult to relate that phenotype to the effects of individual genes. The effect of a single gene may seem infinitesimal when viewed at this level. Yet, the phenotype of our biosphere is indeed the ultimate product of its genotype (the summation of all genes in all living organisms) expressing itself within the restraints of the environment. Thus, the effect of a single gene on the organism carrying this gene is merely the first step in the overall pathway of expression of this gene (Fig. 1.3).

For those of us who live in cities or other ecosystems that include a substantial population of humans, the surrounding environment is controlled almost totally by one species, *Homo sapiens*. The dominant features of our environments are products of human creativity and manipulation: paved highways, streets and sidewalks, houses and skyscrapers, automobiles, trucks, buses, airplanes, and ships. Little space is left for the use of other species; they must survive in any possible ecological niche that is left for them. The few living macroorganisms that remain in our immediate ecosystems are usually positioned in prescribed locations—a tree here, a group of flowers there, all within the restrictions of a carefully manicured bluegrass lawn. Other species are largely restricted to botanical gardens, zoos, parks, and a few wilderness areas. As the human population explosion (Fig. 1.4) continues, will the demand for space by *Homo sapiens* lead to the elimination of the remaining ecosystems suitable for species other than humans?

Clearly, humankind's dominant manipulation of the environment is having a large effect on the evolution of the phenotype of our biosphere. Let's consider one well-documented example of a change in the phenotype of our biosphere brought about by humankind's modification of the environment. "Industrial melanism" is a phenomenon that has been studied in a large number of insect species, especially moths. It occurs when light-colored organisms are gradually replaced by dark-colored ones in populations living in soot-darkened habitats produced by pollution from factories, trucks, tractors, and other large fossil fuel-powered machines in highly industrialized regions of the world. It derives its name from the melanin pigments, dark-colored pigments responsible for coloration in skin, hair, and feathers of animals, and from its direct link to industrialization. Industrial melanism also provides an easily perceived illustration of the effect of a single gene (a few genes in some species) on the phenotype of the biosphere as it changes in response to environmental modifications that result in selection favoring specific genotypes.

Industrial melanism has been extensively studied in the peppered moth, named *Biston betularia* in its light-colored form and *Biston carbonaria* in its dark-

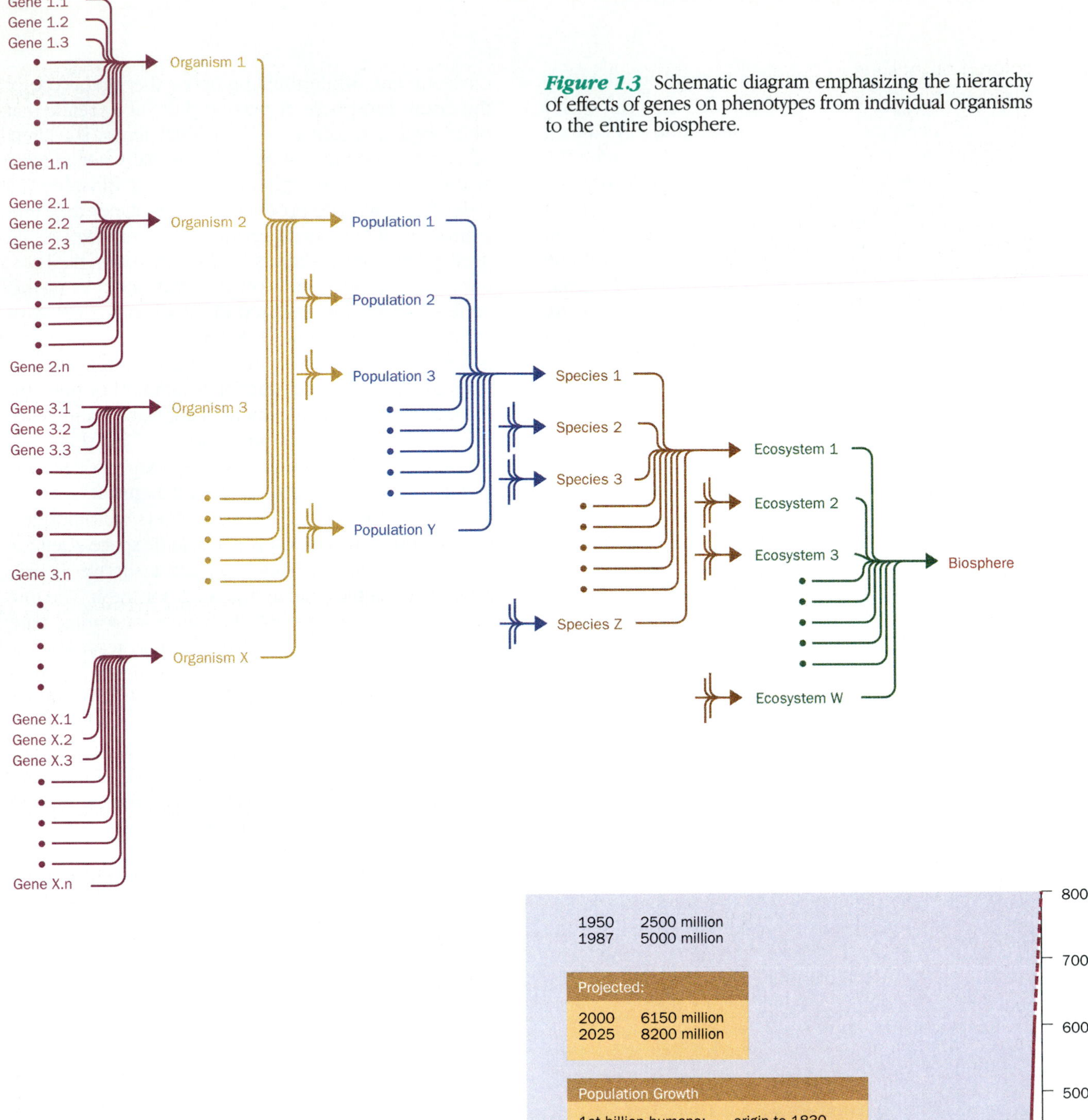

Figure 1.3 Schematic diagram emphasizing the hierarchy of effects of genes on phenotypes from individual organisms to the entire biosphere.

1950	2500 million
1987	5000 million

Projected:

2000	6150 million
2025	8200 million

Population Growth

1st billion humans:	origin to 1830 (>1 million years)
2nd billion humans:	1830 to 1930 (100 years)
3rd billion humans:	1930 to 1960 (30 years)
4th billion humans:	1960 to 1975 (15 years)
5th billion humans:	1975 to 1987 (12 years)

Population in millions: 0, 500, 1000, 2000, 3000, 4000, 5000, 6000, 7000, 8000

8000 B.C. — A.D. 1 — 2025

Figure 1.4 Past and projected worldwide growth in the human population. (Based on data from J.D. Durand, "The Modern Expansion of World Problems," *Proc. Amer. Philos. Soc.*, Vol. III, No. 3, 1967; from C. Haub and M. Mederios Kent, World Population Data Sheet 1988, Population Reference Bureau, Inc., Washington, DC, April 1988; and from P. Hendry, "Food and Population: Beyond Five Billion" Population Bull. 43(2): 1, April 1988.

colored or melanic form, by H.B.D. Kettlewell in England. Since the light-colored form of the peppered moth can give rise to the dark-colored form by a change (mutation) in a single gene, and vice versa, the two forms of the peppered moth clearly should not have been given different species names. Unfortunately, genetic information was either not available or, if available, was not considered when these names were assigned. Today, we should consider these two "species," two forms of the peppered moth, just different genotypes of a single species.

Moths are nocturnal creatures. They are active at night, flying around, feeding, and mating. During the daytime, they sit on tree trunks, rocks, and similar structures. A major factor affecting their survival is how well they are camouflaged from the birds flying overhead while sitting on their daytime perches. In rural areas, tree trunks and boulders tend to be covered by light-colored lichens (composite organisms consisting of fungi and algae living in symbiosis). Light-colored moths are much better camouflaged from predators, primarily birds, on lichen-covered tree trunks (Fig. 1.5*a*). However, in highly industrialized regions of the world, air pollution, especially the carbon particles or soot spewed from smokestacks of fossil fuel-burning factories, kills the lichens and leaves blackened tree trunks. In such environments, the dark-colored moths are better camouflaged than the light-colored moths (Fig. 1.5*b*). The studies of Kettlewell and others have shown that in unpolluted rural areas of the world, light-colored moths are most prevalent, whereas in polluted industrialized regions, the melanic forms are predominant. For example, of 621 moths collected by Kettlewell and colleagues during the summer of 1953 in the forests near Birmingham, England, 559 (90%) were melanic and 62 (10%) were light-colored. Kettlewell and coworkers also carried out release and catch experiments with marked individuals of the two forms of the peppered moth in both unpolluted rural and polluted industrial regions of England. Their results verified exactly what one would predict. The light-colored moths had much higher survival rates in unpolluted rural environments, and the dark-colored moths had greater survival rates in polluted environments. Finally, Kettlewell and colleagues filmed birds in the act of capturing and eating moths as they sat motionless on their daytime perches, to provide further evidence of the role of daytime predation in the survival of these nocturnal species.

The rigorously documented examples of industrial melanism in moths are unlikely to have a major impact on the quality of our lives, in contrast to the "greenhouse effect" and other consequences of humankind's pollution of the environment. However, industrial melanism provides a very clear example of a change in the phenotype of the biosphere brought about by natural selection acting as a consequence of humankind's modification of the environment. Most importantly, it also provides a vivid illustration of the roles that individual genes of living organisms play in the evolution of the phenotype of the biosphere. This total phenotype is the final result, the ultimate product, of the action of the thousands of genes in each of the billions of individual organisms in all the species existing in all the ecosystems on planet Earth.

Figure 1.5 The light- and dark-colored forms of the peppered moth shown (a) (Top) on a lichen-covered tree trunk in an unpolluted rural area, and (b) (Bottom) on the trunk of a tree blackened by soot from factories. Imagine that you are a hungry bird flying overhead and ask yourself which moth you are most likely to spot and eat in each case. (Courtesy of Stephen Dalton/NHPA.)

Clearly, the phenotype of the biosphere is not static, but changes over time in response to changes in the environment, either naturally occurring or imposed by humans. The direction of these changes is determined by natural selection favoring those organisms that are best adapted to their environmental niches. Such well-adapted organisms have the highest probability of surviving to reproductive maturity and transmitting their genes to large numbers of progeny, thus increasing the relative frequencies of these genes in populations. In contrast, less well-adapted individuals leave fewer progeny and, thus, the relative frequencies of their genes will decrease within the respective populations. Although changes occurring from one generation to the next may be relatively small, the cumulative changes occurring over many generations, that is, on the evolutionary scale, can be quite dramatic as in the case of industrial melanism.

Genetics and Food Production

The contributions of genetics to increased food production is without question one of the premier success stories of science in the twentieth century. Moreover, the pinnacle of these achievements was clearly the development of ***hybrid corn***. During the 40-year period from 1940 to 1980, the average yield for corn increased by 251 percent due in large part to the development and introduction of hybrid corn. Impressive increases in yield, though not as striking as those for corn, also have been achieved for most of the other important food crops. These dramatic increases in yields of crop plants, collectively called the "green revolution," have played a major role in our present ability to feed an overpopulated world ***if economic and political systems that would permit adequate worldwide distribution of the food produced were in place.*** Our knowledge of genetics has also been applied to animal improvement with impressive results, such as large increases in meat, milk, and egg production in domestic animals.

Although we will never know for certain exactly when humans carried out the first ***"genetic selection" experiments,*** this probably occurred during the period from 10,000 to 7000 years ago. Fossil records indicate that almost all our present food crops were domesticated during this early Neolithic period coincident with the development of stone tools. Initially, successful selection was done with no knowledge of the genetic basis of the responses that occurred. The largest, most vigorous individuals or those with desired characteristics were simply chosen as parents for subsequent generations. This general approach is still a mainstay of modern plant and animal breeding. However, knowledge about the genetic variability in the population for the trait of interest now permits plant and animal breeders to fine-tune these selection experiments and to predict the changes that will be realized in response to specific selection strategies.

The important role of genetics in the achievements of the "green revolution" is documented by a comparison of the striking increases in agricultural production in the United States with the limited advances made in the Soviet Union during the period from 1937 to 1964. During this period, research in biology and agriculture in the Soviet Union was controlled by one person, ***T. D. Lysenko***. Throughout this period of major developments in genetics in the rest of the world, scientists in the Soviet Union were forced to base their work on a form of Lamarckism now generally called ***"Lysenkoism."*** Lysenkoism was based on the premise that changes in the environmental conditions cause changes in the structure of living organisms, which are then inherited, that is, transmitted to the progeny. Lysenko maintained that if you cut off the tails of mice for enough generations you would eventually produce a strain of tailless mice.

Note that Lysenkoism replaced genetics in the Soviet Union until 1964, eleven years after Watson, Crick, and Wilkins worked out the double-helix structure of DNA. Lysenko's term as czar of Soviet agriculture had disastrous consequences on the development of biology and agriculture in that country. The era of Lysenkoism in the Soviet Union is perhaps described best by I. M. Lerner in the forward to *The Rise and Fall of T. D. Lysenko* by Z. A. Medvedev. The following quotation states the case quite eloquently.

> **The story of Soviet genetics in the period 1937–1964 is, perhaps, the most bizarre chapter in the history of modern science. In a society devoted to the betterment of the lot of peasants and workers, an illiterate and fanatical charlatan was allowed absolute dictatorship and control over both research in biology and practical agriculture. This event not only stifled the development of science, but also had a far reaching and destructive influence on the national economy of the Soviet Union. To the outside world, it was completely incomprehensible that a country capable of developing a nuclear potential rivaling that of the United States, and of establishing itself in the forefront of space exploration, could have entrusted its fundamental agricultural resources to exploitation by an obvious quack.**

Lysenko's ouster in 1964 left Soviet scientists scrambling to rewrite their biology texts, replacing the absurdities of Lysenkoism with the basic principles of genetics that were by that time rigorously documented and very advanced in the rest of the world. Perhaps, above all else, the story of Lysenkoism vividly illustrates the inherent dangers of concentrating power and authority in the hands of one or a few individuals.

Today, Soviet scientists, like scientists throughout the rest of the world, are performing sophisticated plant and animal breeding and selection experiments that are designed on the basis of detailed information about the genetic control of traits such as yield and growth rates in domesticated plants and animals. In addition, scientists are now using ***genetic engineering techniques to design crop plants with desired traits***

such as insect and disease resistance. For example, genes from fungi, other microorganisms, or wild plant species that provide these organisms with resistance to insect pests can now be isolated, tailored for new host organisms by recombinant DNA procedures *in vitro*—for example, by adding new regulatory elements so that they will be expressed in the proper tissues in the new hosts—and introduced into agronomically important domestic plant species by gene transfer technologies. Some of the genetic engineering strategies that are currently being evaluated for potential use in plant and animal breeding and in food microbiology will be discussed in Chapter 24.

In the past, most of the plant improvement research has been directed toward increasing yield, disease resistance, and insect pest resistance in crop plants growing under optimal or near-optimal conditions. However, given the economic and political obstacles to worldwide distribution of food, and given the climatic changes that are predicted to occur in the future as a result of the "greenhouse effect," scientists are now placing greater emphasis on the development of food crop varieties that are more resistant to conditions of environmental stress such as drought, high temperature, and poor soil nutrient conditions. By initiating these strategies now, we may be able to minimize malnourishment in the more severely overpopulated world of the future. ***One key fact must never be forgotten. No matter how successful we are in extending the "green revolution" in the future, we will ultimately fail in our attempt to avoid mass starvation if we do not also manage to stop or at least slow the existing population explosion now rampant in this species*** **Homo sapiens.**

Genetics and Health

The growth and development of a "normal," "healthy" adult human from a single cell, the zygote or ovum after fertilization by a sperm, is one of the most spectacular phenomena in biology. This complex process requires the coordinated action of thousands of genes. ***Each of the thousands of genes involved in the growth and differentiation of the multitude of different cell types in the adult organism must be expressed at the proper time and place in the developmental pathway.*** The striking cellular diversity (Fig. 1.6) that unfolds during development of a higher

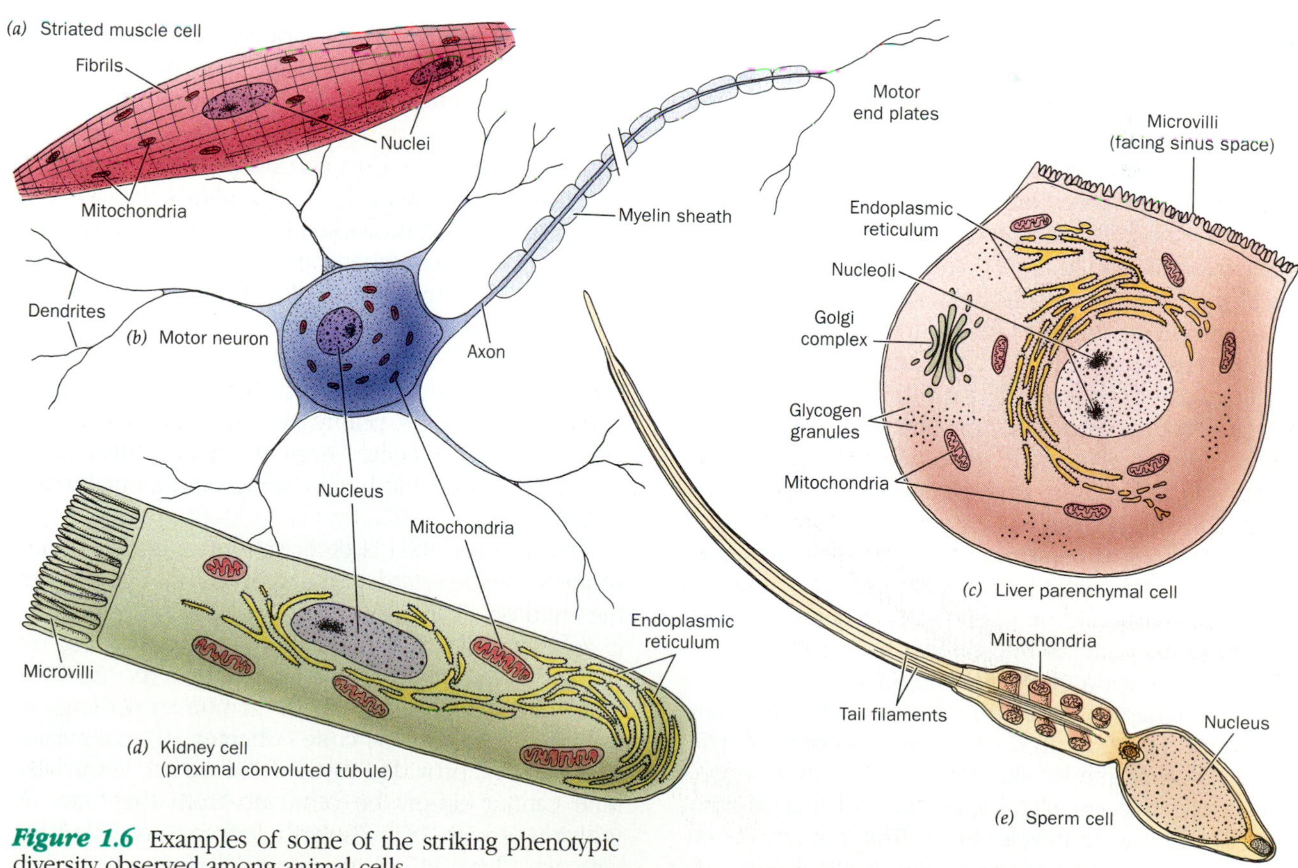

Figure 1.6 Examples of some of the striking phenotypic diversity observed among animal cells.

animal such as a human is known to be encoded in the genome of the zygote and to manifest itself via preprogrammed patterns of gene expression. But what is the nature, the molecular basis, of this genetic preprogramming? How is the programmed gene expression regulated? What types of genetic circuitry assure the orderly unfolding of the cellular diversity observed in the adult organism? These fascinating questions promise to challenge geneticists and stimulate research aimed at providing definitive answers for years to come.

What we do know is that each cell, no matter how divergent its phenotype, contains the same genome, the same set of genes (with rare exceptions such as mature erythrocytes). However, within a given cell type only about 5 to 8 percent of the genes present in the genome are expressed, and the ***divergent cell phenotypes result from the expression of different sets of genes in different cells.*** Thus, the expression of the appropriate sets of genes dictates that bone cell progenitors will develop into bone cells, liver cell progenitors into liver cells, nerve cell progenitors into neurons, and so on. Whereas some genes, generally called "housekeeping genes," are expressed in all cells, other genes exhibit the tissue-specific patterns of expression that are responsible for the differentiation processes observed in complex eukaryotes such as higher animals. But how are these pathways of differentiation regulated? How is this information encoded in the genome of the zygote? Experiments designed to answer these key questions are at the forefront of ongoing research in molecular genetics. Research on the genetic control of development is at the leading edge of investigative genetics today.

One fact is very clear. The genes in the human genome are highly coadapted, and a malfunction in the expression of even a single gene can disrupt the entire developmental process. If the gene involved encodes an essential function, its malfunction will be lethal. In many other cases involving important but not absolutely essential (for survival) genes, such a malfunction will produce an "unhealthy" or "abnormal" phenotype. Often a defect in a single gene will lead to a whole series of phenotypic abnormalities collectively referred to as a ***syndrome.*** One well-known example is ***congenital hyperuricemia*** or ***Lesch–Nyhan syndrome.*** This syndrome is caused by a defect in a single gene. This particular gene encodes an enzyme called hypoxanthine-guanine phosphoribosyltransferase that catalyzes an important reaction in nucleotide metabolism (see Chapter 7). Newborns with this syndrome appear normal but produce excess amounts of uric acid in their urine. By about 10 to 12 months of age, they usually become irritable and uncoordinated, with inadequate muscle development. Within a few years, degeneration of the nervous system is major and self-mutilation (teeth grinding, lip biting, and so on) usually ensues. Death often results from severe neurological and renal damage during adolescence. It is important to understand that ***all of these phenotypic abnormalities result from a defect in a single gene*** that results in the malfunction of a single gene-product. The loss of activity of this single gene-product upsets the normal metabolic balance in the developing infant and leads to the entire array of phenotypic abnormalities. This same general picture is observed for a large number of inherited human syndromes.

As mentioned earlier, over 3000 human diseases are now believed to result from defects in one or more genes. Many of these have only recently been described and will require additional study before we can be sure of their genetic basis. In some cases, an understanding of the genetic basis of the disease has permitted effective treatment. In other cases, gene therapy, replacing the defective gene with a normal, functional copy of the gene, holds promise for future treatment.

Phenylketonuria (PKU) is a classic example of an inherited disease in humans where an understanding of the molecular basis of the disease has led to the development of an effective treatment. This ***treatment does not cure the disease.*** The defective gene is still present and transmitted to the progeny. However, the deleterious effects of the disease can be eliminated by treatment. Individuals with PKU have a defect in the gene encoding the enzyme phenylalanine hydroxylase. This enzyme catalyzes the conversion of the amino acid phenylalanine to the related amino acid tyrosine. This important reaction is the first step that occurs in the catabolism of phenylalanine, which is produced by the hydrolysis of proteins that we eat. When this reaction is blocked because of a mutation in the gene encoding phenylalanine hydroxylase, thus causing no enzyme or a nonfunctional enzyme to be produced, large quantities of phenylalanine will build up in the bloodstream when proteins are consumed. Some of the excess phenylalanine is converted to derivatives such as phenylpyruvic acid, which are highly toxic to the central nervous system. This in turn produces irreversible brain damage and mental retardation.

However, if PKU is detected during infancy, subsequent mental retardation can be avoided by placing the child on a carefully controlled diet. Phenylalanine is an essential amino acid in humans; it cannot be synthesized *de novo* as it is in most microorganisms and plants. Since its presence in human proteins is dictated by the genetic code (Chapter 10), phenylalanine must be provided by the food we eat. Phenylalanine cannot simply be removed from the diets of individuals with PKU. Nevertheless, by providing infants who have PKU with a carefully formulated diet with enough phenylalanine for normal growth and development but not enough to cause a buildup of toxic derivatives, the severe consequences of the disease, specifically mental retardation, can be avoided. The ***hemophilias and erythroblastosis fetalis (Rh factor incompatibility)*** are other well-known examples for which an understanding of the genetic basis and the molecular mechanism underlying the syndrome have led to an effective treatment of the inherited disease.

Other human abnormalities result from disruptions of the normal genetic balance, the expression of the normal coadapted set of genes in the human genome, by the presence of extra chromosomes (Chapter 19). ***Down syndrome*** is the best-known example of this class of genetic abnormalities. The frequency of these chromosome-based syndromes in children is known to increase sharply with increasing age of the mother past 35 years. Clearly, such information is of major importance in family planning.

Genetics impacts on health and medical practices at many other levels far too numerous to describe in any detail in this brief introduction. Examples will be discussed throughout this text; however, a few should be mentioned here. For example, one of the greatest successes of modern medicine has been the development of antibiotics to combat diseases caused by pathogenic bacteria. However, with the widespread use of antibiotics (e.g., they are often added to cattle feeds to prevent dysentery), bacteria have begun to fight back. They have evolved small ***extrachromosomal genetic elements called plasmids*** that carry genes that provide resistance to these antibiotics. Often these genes are present on ***transposable genetic elements ("jumping genes")*** that can transpose themselves ("jump") from one genetic element or chromosome to another. In so doing, they frequently duplicate themselves and leave one copy in the original location while the second copy moves to the new location. Even worse, the elements carrying these antibiotic-resistance genes usually come complete with genetic information that facilitates their replicative transfer from one bacterium to another and their rapid spread through entire populations and even to different bacterial species (Chapters 8 and 9). In this respect, these antibiotic-resistance plasmids behave like computer viruses, spreading rapidly through populations of bacteria and rendering them resistant to our best antibiotics. Should these resistance plasmids become prevalent in populations of pathogenic bacteria, our present success in treating bacterial diseases will be severely impaired.

Cancers and acquired immunodeficiency syndrome (AIDS) are undoubtedly the most dreaded diseases of our day. Cancers are a large group of heterogeneous diseases that all share two common features: (1) a loss of the normal control of cell division such that ***masses of cells or tumors form*** and (2) the ability of the affected cells to ***spread or metastasize to other tissues*** in the body. In animal systems, certain viruses called ***retroviruses*** are capable of transmitting specific types of cancer. These viruses all carry one of a class of genes called ***oncogenes*** that are responsible for the loss of control of host cell division. However, in humans, there has been no demonstration of transmission of any cancer by a retrovirus. But normal human cells are now known to contain genes called ***protooncogenes*** that are closely related to retroviral oncogenes. In fact, it now seems very likely that these retroviral oncogenes have evolved from normal cellular protooncogenes, having been picked up from host cells and adapted to the parasitic viral life cycle. Most importantly, the normal cellular ***protooncogenes have been shown to mutate to cellular oncogenes*** which then play a role in the loss of control of cell division in the cells that carry the mutant gene. The protooncogenes are clearly just normal cellular genes that are directly involved in regulating cell division during the growth and development of normal individuals. The functions of protooncogenes and their derivative cellular oncogenes are being studied intensively in an attempt to understand the genetic control of cell division and, especially, what goes awry with these controls in cancer cells (Chapter 17).

In contrast to cancer in humans, AIDS in humans is known to be caused by retroviruses just as it is in animals. AIDS is caused by a class of retroviruses called the ***human immunodeficiency viruses or HIV***. These viruses infect two types of white blood cells: (1) a subset of lymphocytes called T4 cells, which they kill, and (2) macrophages, within which they reproduce, but which they do not kill. Only these two types of blood cells carry the specific receptor molecule through which the virus particles gain access to host cells. The macrophages are thought to serve as reservoirs for the virus within the body, possibly carrying viruses to the brain, where they subsequently cause trauma to the central nervous system. One factor that probably contributes to the difficulty in diagnosing and treating AIDS is that retroviruses have the ability to incorporate their genetic material, their chromosome, into the chromosome of the host cell (see Chapter 8). In this integrated state, the viral chromosome is called a ***provirus***. As a provirus, the retroviral chromosome replicates each cell generation and is transmitted to all progeny cells just like any other normal segment of the host chromosome. As a result of this alternative provirus state in the viral life cycle, these HIV retroviruses can be present in quiescent form for long periods of time, potentially indefinitely. However, what is most commonly observed is that sporadically the proviruses become active and begin to synthesize viral gene-products and progeny viruses within infected cells. Clearly, scientists would like to know what factors trigger the activation of proviruses and whether or not this activation can somehow be prevented. Investigations of the alternative life strategies of HIV retroviruses are currently an area of intense research activity, and it is hoped that major breakthroughs will occur in the treatment of AIDS before this text appears in print.

One intriguing observation is that at least one retrovirus, the feline leukemia virus, can cause either cancer or immune deficiency depending on minor differences in the genome of the virus. Clearly, these viruses are not static parasites; they mutate and new variants evolve just as in higher forms of life. Indeed, these new viral variants are subject to the forces of natural selection just like all other living organisms. They either find a suitable ecological niche or become extinct. The AIDS viruses are clearly a case where our goal must be the extinction of a genetic element, despite our overall goal of maintaining genetic diversity on our planet. The ability of these viruses to evolve and produce variants with new properties should be noted, for even if we were somehow able to magically cure all presently existing forms of AIDS and cancers today, our job would not be finished. We would need to stay on guard in case new, possibly even more virulent forms of these viruses evolved in the future. This possibility has been underscored by the documented evolution of flu viruses during the last couple of decades. One year we were dealing with the "London" variant, the next year with the "Hong Kong" form, and so on. The variants sometimes caused serious new flu epidemics because they were not sensitive to vaccines prepared to control the previously existing strains of the virus.

Cancers and AIDS cause unbearable pain and suffering, and one has to hope and believe that once we understand the molecular mechanisms responsible for the pathogenesis and the mechanisms by which the replication of retroviral genomes and the expression of proviral genes are regulated, then we should be able to devise effective preventive and treatment strategies. Once we know exactly how the AIDS viruses escape the immune system and persist and reproduce in host cells, we should be able to develop ways to prevent their escape and/or to block their replication or expression in infected cells.

Lastly, what about the hundreds of human diseases that are known to be caused by defects in individual genes? How can these be prevented or treated? It is hoped that somatic-cell gene therapy will provide an effective means of treatment for some of these diseases in the not-too-distant future. However, a major concern is that the number of such diseases that have been identified and studied has increased sharply over the last two decades. This leads to the following question. How many different human inborn errors of metabolism actually exist? Although no precise answer can be given to this question, one informative answer is possible. ***There can be little doubt that the total number of inherited diseases will at least equal the total number of genes that play important roles in the growth and development process.*** All genes have the same basic chemical composition, and all are undoubtedly capable of undergoing mutation to nonfunctional states (Chapter 11). Clearly, then, the number of inherited diseases in humans will be large, probably on the order of 10,000. With this in mind, one of society's goals must be to ***keep mutation rates and the "genetic load" (the accumulation of deleterious mutant genes) in the human gene pool at a tolerable level.*** To do so, we must minimize the pollution of our biosphere with substances that will increase the frequency of mutations, most notably radioactive agents and mutagenic chemicals. How can we accomplish this goal? Above all else, we must elect government officials who place a higher priority on preserving a safe and healthy environment than on achieving short-term economic gains, especially when the latter have potentially deleterious effects on the environment. Our history is one of consistently sacrificing the environment for immediate material gains. This trend must be reversed if we are to have any chance of preserving a quality environment and healthy lives for future generations.

Genetics, Politics, and Law

Genetics impacts on all aspects of our lives, with food production and health being just two of the most important. Recently, genetics has moved front and center in several areas of government and law. The U.S. Food and Drug Administration and the U.S. Department of Agriculture have been involved for years in ***regulating the use of mutagenic chemicals and radioactive substances.*** With recent revelations of horrendous pollution of the environment by toxic chemicals and radioactivity, it is very clear that governments must do a much better job of regulating the use of these dangerous substances. This will only happen if we elect government officials who are knowledgeable of the magnitude of the dangers involved and who will take the actions required to protect the public and all other forms of life as well. In a democracy, all citizens share the responsibility for electing members of government who will vigorously pursue the mandated policies.

A second arena in which genetics now plays an integral role is patent law. The successes of genetic engineering have refocused attention on the question of ***whether living organisms should be patentable***, and, if so, within what limits? In some respects, this is not a new question at all. Plant varieties have been patented for many years with no major controversy. But what about bacteria that have been genetically engineered to degrade chemical pollutants or to make foreign proteins such as human insulin or human growth hormone? Should they be patentable? The courts have ruled that they are indeed subject to patent protection. Or at least the processes that are used to produce these novel bacteria can be protected by patents, and that amounts to the same thing. But to what extremes will the patenting of living organisms be carried? What about animal cells engineered to have useful properties or to produce important products? What about human cells? Do they belong to the person from whom they were originally derived or to the doctor who performed the tissue explant and started the cell line growing in culture? Or to the institution or company that employed the scientist who discovered the economically valuable characteristic of the cell line, or who used genetic engineering technologies to impose the desired properties on the cell line? What about actual organs and tissues, for example, human skin produced by growing epidermal cells under appropriate culture conditions and that may be very valuable in treating burn patients? Should these organs or tissues be patentable? What about genetically modified whole animals such as mice and guinea pigs that may have properties of value for medical research? Should they be subject to patent protection? Two points seem clear. The practice of patent law should be a lucrative profession for years to come, and patent lawyers with a background in genetics and molecular biology should be in particularly strong demand.

Criminal law is another arena in which the spotlight now focuses on genetics (Fig. 1.7). Human ***"DNA prints" or "DNA fingerprints"*** are known to possess greater specificity by several orders of magnitude than human fingerprints. The "DNA print," the specific set of DNA fragments produced by cutting the DNA with a set of sequence-specific enzymes (see Chapter 13), depends on the exact sequence of subunits called nucleotide-pairs in the DNA of the individual. Since the human genome contains about 3×10^9 nucleotide-pairs, it is extremely unlikely that any two individuals other than identical twins will have totally identical "DNA prints."

Thus, "DNA prints" now provide a powerful new tool for establishing identity or nonidentity in paternity, rape, and assault cases, as well as all other identity cases where tissue samples might be available linking a criminal to the crime. The "DNA printing" approach has become particularly powerful with the recent development of techniques by which small amounts of DNA can be extensively amplified *in vitro*. This allows "DNA prints" to be made even when only very minute amounts of tissue are available.

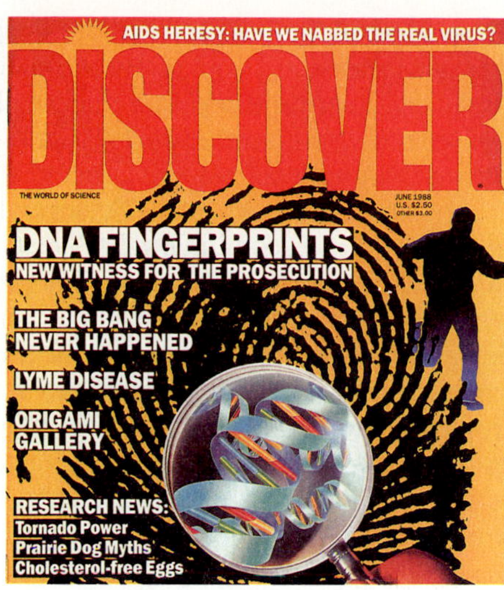

Figure 1.7 Cover of the June 1988 issue of *Discover* magazine with headlines heralding the power of "DNA fingerprints" in criminal identity/nonidentity cases. (Reproduced with permission. Copyright 1988 by *Discover* Publications, New York.)

Genetic Engineering

Today, scientists can isolate essentially any gene of interest from any species, can incorporate that gene into a small self-replicating extrachromosomal genetic element (usually a plasmid or a viral DNA molecule), can introduce that ***"recombinant DNA molecule"*** into host cells such as bacteria or yeasts, can grow large vats or fermentors full of these recombinant organisms, and can subsequently isolate and purify essentially any desired quantity of the gene of interest. They can attach the gene to selected regulatory signals so that the gene will be correctly expressed in bacteria or in plants or in animals. In some cases, its expression can even be restricted to specific tissues or cell types in higher plants and animals. In other cases, the gene can be fitted with controlling sequences such that it will be expressed only after exposure to light, or to heavy metal cations, or to temperature shock. In short, scientists can cut apart and paste together DNA molecules almost at will, regardless of the source of the molecules. All of this has become possible because of the discovery and isolation of (1) a large battery of enzymes called ***restriction endonucleases*** that cut DNA molecules very precisely at known sites and (2) enzymes called ***DNA ligases*** that paste ("ligate") the pieces produced by restriction enzymes back together in the desired configurations. All of these procedures are easily carried out in the test tube (see Chapters 13 and 24) by any reasonably competent technician.

Genetic engineering is a young science. The first restriction enzyme was isolated in 1970, and the first recombinant DNA molecules were generated *in vitro* in 1972. Shortly thereafter, in 1973, several of the scientists involved in recombinant DNA research began to worry about the potential danger of producing pathogenic microorganisms by these procedures. For example, if an oncogene involved in the malignant transformation pathway were cloned into a plasmid and introduced into *Escherichia coli* cells, might these *E. coli* subsequently escape from the laboratory, colonize human intestinal tracts, and play a role in inducing future cancers? In 1974, this group of concerned scientists called for a worldwide moratorium on certain types of recombinant DNA experiments. After extensive discussion and debate about the potential hazards of recombinant DNA research, the first guidelines regulating this work were issued by the National Institutes of Health in 1976. Students who are interested in the details of the recombinant DNA controversy, the voluntary moratorium, and the evolution of the NIH guidelines regulating this work should read *The DNA Story* by J. D. Watson and J. Tooze. This book provides a detailed historical account of these events, complete with photocopies of many of the original documents.

The genetic engineering field is less than two decades old; nevertheless, its impact has already been immense. As would be expected in our highly materialistic society, the popular press has focused largely on the potential commercial applications of genetic engineering and its impact on the stock market (Fig. 1.8). However, the contribution of genetic engineering to the expansion of our basic knowledge of biology has been by far its most important application to date. The development of recombinant DNA technologies allowed scientists to isolate and characterize the structure and function of eukaryotic genes. These studies first demonstrated that eukaryotic genes frequently contain internal noncoding segments called ***intervening sequences*** or ***introns*** that must be removed or ***"spliced out"*** during the expression of these genes (see Chapter 10). Most importantly, genetic engineering methodologies now allow scientists to dissect the genetic control of and molecular mechanisms involved in essentially any biological process of interest, be it photosynthesis in plants or learning and memory in animals.

The commercial applications of genetic engineering are also of importance. There have been several important achievements to date, and there will undoubtedly be many others, some probably rather spectacular, in the future. Some of the ongoing and anticipated applications of genetic engineering will be discussed in Chapter 24.

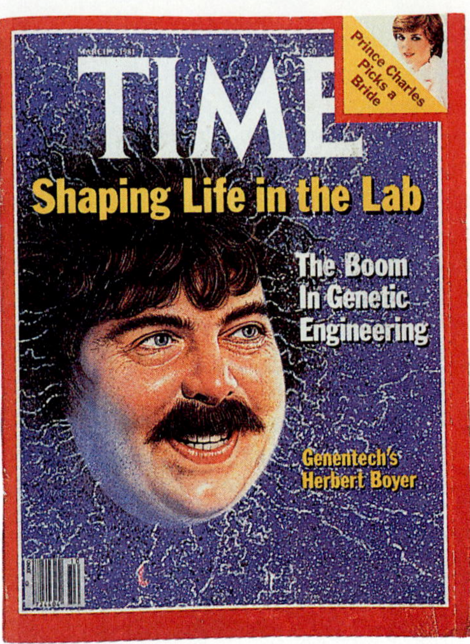

Figure 1.8 Cover of *Time* magazine, March 9, 1981, heralding the "boom" in genetic engineering. (Reproduced with permission. Copyright 1981 by *Time* Inc., New York.)

Numerous products are now produced by microorganisms engineered to synthesize these products via recombinant DNA technologies. Several types of human interferon, human growth hormone, and human insulin are three of the better known products of genetic engineering that are now available. Human insulin produced by recombinant bacteria was first released for sale in 1982. All of these proteins were previously purified from animal tissues; the human forms were available only in very minute quantities. Interferons were not even available in sufficient amounts to permit scientists to carry out basic research on their effects. At one time, scientists hoped that interferons might provide the "magic bullets" needed to treat cancers. Unfortunately, this has proven not to be the case.

Over two million individuals with diabetes mellitus require treatment with insulin in the United States alone. Prior to 1982, all the insulin for diabetics was purified from bovine and porcine pancreas glands that were available as by-products of the meat industry. The insulin supply was thus a function of supply and demand in the meat markets, which fluctuate substantially, and was not particularly responsive to the needs of diabetics for insulin. The commercial production of human insulin in recombinant *E. coli* cells eliminated concerns regarding possible shortages of insulin in the future. In addition, this human insulin was of particular importance to some diabetics who were allergic to bovine and/or porcine insulins.

Finally, the impact of genetic engineering on medicine should be reemphasized here. The possible utility of human gene therapy was discussed earlier; it has tremendous potential as a tool in treating inherited human diseases. Paul Berg shared the 1980 Nobel Prize in Chemistry with F. Sanger and W. Gilbert for key developments in cloning and sequencing DNA. In his acceptance speech in Stockholm, Berg made the following predictions:

> **The development and application of recombinant DNA techniques has opened a new era of scientific discovery, one that promises to influence our future in myriad ways. . . . Just as our present knowledge and practice of medicine relies on a sophisticated knowledge of human anatomy, physiology, and biochemistry, so will dealing with disease in the future demand a detailed understanding of the molecular anatomy, physiology, and biochemistry of the human genome. . . . We shall also have to have physicians who are as conversant with the molecular anatomy and physiology of chromosomes and genes as the cardiac surgeon is with the structure and workings of the heart and circulatory tree. (*Science* 213: 302).**

Indeed, in the future, DNA transplants may be more common than organ transplants are today. Clearly, then, students interested in careers in medicine should obtain the best training available in molecular genetics.

Figure 1.9 Monastery garden at Altbrünn, Austria, where Mendel's experiments on garden peas were conducted. (Courtesy of Professor Jaroslav Křiženecký.)

THE BIRTH OF GENETICS

There can be little doubt that the first humans on Earth pondered the observation that children resembled their parents more than other members of the population. Unfortunately, however, we have no record of their ideas as to why this occurred. The Greek philosophers Hippocrates and Aristotle obviously thought extensively about this fact and developed theories to explain resemblances among relatives. For a synopsis of their ideas, the reader is referred to *History of Genetics* by H. Stubbe. Genetics as we know it today, based on the "gene theory of inheritance," began with the work of ***Gregor Johann Mendel*** (see Chapter 2 frontispiece for photograph); thus, our coverage in this text will begin with Mendel.

Gregor Mendel (1822–1884) is appropriately called the "father of genetics." His precedent-setting experiments with garden peas (*Pisum sativum*), published in 1866, were conducted in the limited space of a monastery garden (Fig. 1.9) while he was also employed as a substitute schoolteacher. The conclusions that he drew from his elegant investigations constitute the foundation of today's science of genetics. Why was Mendel so successful in discovering basic principles of genetics?

Mendel was not the first to perform hybridization experiments, but he was one of the first to consider the results in terms of single traits. Sageret in 1826 had studied the inheritance of contrasting traits. Others of Mendel's predecessors had considered whole organisms, which incorporate a nebulous complex of traits; thus, they could observe only that similarities and differences occurred among parents and progeny, and so missed the significance of individual differences. Employing the scientific method, Mendel designed the necessary experiments, counted and classified the peas resulting from his crosses, compared the proportions with mathematical models, and formulated a hypothesis for these differences. Although Mendel devised a precise mathematical pattern for the transmission of hereditary units, he had no concept of the biological mechanism involved. Nevertheless, on the basis of his preliminary experiments and hypotheses, he predicted precise patterns of transmission of hereditary units and subsequently verified his predictions with the results of later crosses.

In 1900, Mendel's paper was discovered simultaneously by three botanists: Hugo de Vries (Fig. 1.10) in Holland, known for his mutation theory and studies on the evening primrose and maize; Carl Correns in Germany, who investigated maize, peas, and beans; and Eric von Tschermak-Seysenegg in Austria, who worked with several plants, including garden peas. Each of these investigators obtained evidence for Men-

Figure 1.10 Hugo de Vries (1848–1935), Dutch botanist who discovered Mendel's principles in 1900. He was a pioneer in plant genetics. (Portrait by Everett Thorpe.)

Figure 1.11 William Bateson (1861–1926), English biologist who repeated Mendel's experiments and supported his principles, and introduced Mendelian genetics to the English-speaking world. (Portrait by Everett Thorpe.)

del's principles from his own independent studies. They all found Mendel's report while searching the literature for related work and cited it in their own publications. William Bateson (Fig. 1.11), an Englishman, gave this developing science the name ***"genetics"*** in 1905. He coined the term from a Greek word meaning "to generate."

In addition to naming the science, Bateson actively promoted Mendel's view of paired genes. He used the word ***"allelomorph,"*** subsequently shortened to ***"allele,"*** to identify members of pairs that control different alternative traits. Also during the early 1900s, a Frenchman, Lucien Cuenot, showed that genes controlled fur color in the mouse; an American, W. E. Castle, related genes to sex and to fur color and pattern in mammals; and a Dane, W. L. Johannsen, studied the influence of heredity and environment in plants. Johannsen began using the word ***"gene"*** from the last syllable of Darwin's term "pangene." The ***gene concept***, however, ***had been implicit in Mendel's visualization of a physical element or factor (*Anlage*) that acts as the foundation for development of a trait.*** These men and their peers were able to build on the basic principles of cytology, which were established between 1865 (when Mendel's work was completed) and 1900 (when it was discovered). Why were Mendel's important discoveries not recognized for such a long time (35 years) after the studies were completed and reported?

Wilhelm Roux had postulated as early as 1883 that chromosomes within the nucleus of the cell were the bearers of hereditary factors. The only model he was able to devise that would account for his observed genetic results was a row of lined-up objects duplicated exactly. To explain the mechanics of gene transmission from cell to cell, he therefore suggested that nuclei must have invisible structures held in rows or chains that duplicated themselves when the cell divided. Constituents of the nucleus that seemed best designed to carry genes and fill these requirements were ***chromosomes***. Experiments of T. Boveri and W. S. Sutton in 1902 brought confirming evidence that a gene is part of a chromosome. The theory of the gene as a discrete unit of a chromosome was developed by T. H. Morgan (see Fig. 6.1) and his associates from studies on the fruit fly, *Drosophila melanogaster*. H. J. Muller later promoted the merger of the two sciences that had contributed most to the chromosome theory—cytology (the study of cells) and genetics—as ***cytogenetics***.

When Mendel's work was discovered in 1900, the English physician-biochemist Sir Archibald E. Garrod was studying congenital metabolic diseases in humans. One of these was alkaptonuria, which is caused by a block in the catabolism of the amino acids phenylalanine and tyrosine (see Chapter 10). Garrod proposed that alkaptonuria was due to a single defective gene that produced a nonfunctional product resulting in the metabolic block. Garrod's results and his interpretation of these results were described in detail of his book *Inborn Errors of Metabolism*, published in 1909.

Garrod's concept of ***one mutant gene–one metabolic block*** was largely ignored by the scientific community until 1941 when Beadle and Tatum rediscovered this concept during their neoclassical work on *Neurospora* and refined it to ***one gene–one enzyme.*** Subsequently, this concept was revised to the more accurate concept of ***one gene–one polypeptide.*** Thus Garrod's work, like Mendel's, is an example of a major breakthrough that was sufficiently "ahead of its time" that its significance was not recognized by contemporary scientists and, thus, was not appreciated until its rediscovery many years later.

GROWTH OF A SCIENCE: FROM MENDEL TO GENETIC ENGINEERING

Today, genetics is a mature but dynamic science, clearly in its prime and recognized as the very core of modern biology. As such, the science of genetics, built on the foundation put in place by Mendel, owes its present stature to the contributions of a large number of scientists. Indeed, the history of genetics is a subject fully worthy of study in itself. Many of the more important contributions to genetics will be discussed throughout this text. However, the reader who wishes to obtain a more complete perspective of the fascinating history of genetics is encouraged to refer to two excellent books that focus specifically on history. A. H. Sturtevant's *A History of Genetics* covers the history of this science through the early 1960s from the viewpoint of a major participant. Watson and Tooze's *A DNA Story* documents the history of the explosive advances in molecular genetics that occurred in the 1970s.

In an attempt to provide the reader with some appreciation for the chronology of major developments in genetics, some of the most important contributions follow:

1865 ***Mendel read his paper to the Brünn Society for Natural History.***

1866 ***Mendel's paper published*** in the Proceedings of the Brünn Society for Natural History.

1900 ***Mendel's work discovered*** by de Vries, Correns, and von Tschermak-Seysenegg.

1902 Boveri and Sutton demonstrated the presence of ***paired chromosomes*** (homologs) in diploid species.

1905 Bateson named the science ***genetics.***

1908 Hardy and Weinberg formulated the ***"Hardy–Weinberg law"*** relating genotypic frequencies to gene frequencies in randomly mating populations.

1909 Johannsen introduced the word ***gene.***

1909 Garrod's book ***Inborn Errors of Metabolism*** published.

1910 Morgan (Nobel Prize 1933) established the ***sex-linked inheritance*** of white eyes in *Drosophila melanogaster*.

1911 Morgan postulated the chromosomal basis of ***linkage.***

1927 Muller (Nobel Prize 1946) reported the use of the ClB technique to demonstrate that ***X rays are mutagenic.***

1928 Griffith's discovery of ***transformation*** in *Diplococcus pneumoniae*.

1931 Creighton and McClintock's (maize) and Stern's (*Drosophila*) papers appeared, demonstrating that ***genetic recombination is correlated with the exchange of morphological markers*** on chromosomes.

1940 Oliver's demonstration of ***recombination within the*** lozenge ***functional unit*** in *Drosophila*.

1941 Beadle and Tatum's work (Nobel Prize 1958) on *Neurospora* was published, establishing the ***one gene–one enzyme concept.***

1944 Avery, MacLeod, and McCarty's demonstration that the pneumococcal ***"transforming principle" is DNA.***

1946 Lederberg (Nobel Prize 1958) and Tatum's discovery of ***conjugation*** in bacteria.

1950 McClintock's (Nobel Prize 1983) first paper on ***"transposable elements"*** in maize.

1952 Hershey (Nobel Prize 1969) and Chase demonstrated that the ***genetic material of bacteriophage T2 is DNA.***

1952 Zinder and Lederberg's discovery of phage-mediated ***transduction*** in bacteria.

1953 ***Watson and Crick*** (Nobel Prize 1962) ***worked out the double-helix structure of DNA using the X-ray diffraction data of Wilkins*** (Nobel Prize 1962) ***and the base composition data of Chargaff.***

1955 Benzer's first paper on the ***fine structure of the phage T4 rII locus.***

1956 Tjio and Levan establish that the normal diploid ***chromosome number in humans is 46.***

1957 Fraenkel-Conrat and Singer demonstrated that the ***genetic information of tobacco mosaic virus is stored in RNA.***

1958 Meselson and Stahl's demonstration that ***DNA replication is semiconservative.***

1958 Kornberg's (Nobel Prize 1959) isolation of ***DNA polymerase*** I from *E. coli*.

1959 Ochoa's (Nobel Prize 1959) discovery of the first ***RNA polymerase***.

1961 Jacob and Monod (Nobel Prize 1965) proposed the ***"operon model"*** for regulating gene expression.

1964 ***Colinearity*** between genes and polypeptide products established by the work of Yanofsky and colleagues and by Brenner and colleagues.

1964 Temin (Nobel Prize 1975) proposed the ***DNA provirus*** form of RNA tumor viruses.

1965 Holley (Nobel Prize 1968) worked out the first complete ***nucleotide sequence of a tRNA***, a yeast alanine tRNA.

1966 The complete ***genetic code*** was established by the work of Nirenberg and Khorana (Nobel Prize 1968) and coworkers.

1970 Nathans and Smith (Nobel Prize 1978) isolated the first ***restriction endonucleases***.

1970 ***Reverse transcriptase*** of RNA tumor viruses identified by Baltimore (Nobel Prize 1975).

1972 First ***recombinant DNA*** produced *in vitro* in Berg's (Nobel Prize 1980) laboratory.

1976 Bishop and Varmus (Nobel Prize 1989) demonstrate the ***protooncogene to oncogene*** relationship.

1976 Hozumi and Tonigawa's demonstration of ***somatic rearrangements of genes encoding antibodies***.

1977 Demonstration of ***introns*** in eukaryotic genes by Breathnach, Mandel, and Chambon and by Jeffreys and Flavell.

1977 Publication of the ***DNA sequencing techniques*** of Maxam and Gilbert and of Sanger, Nicklen, and Coulson (Sanger and Gilbert, Nobel Prize 1980).

1977 Publication of the ***complete 5387 nucleotide sequence of phage ΦX174*** by Sanger and colleagues.

1978 Discovery of ***"splicing"*** of adenovirus RNAs in three different laboratories.

1982 Publication of the ***complete 48,502-nucleotide-pair sequence of phage lambda*** by Sanger and colleagues.

1983 Cech and Altman (Nobel Prize 1989) establish the existence of ***catalytic RNAs***.

1988 Watson's acceptance of job as coordinator of the ***"human genome project."***

1989 NIH Recombinant DNA Advisory Committee recommends approval of first ***human "gene transplant"*** experiment.

1989 Tsui, Collins, and colleagues clone the ***"cystic fibrosis gene,"*** the gene whose mutant alleles account for the majority of the cases of this dreaded disease that afflicts about one out of every 2000 children in the United States.

The preceding chronicle is necessarily incomplete; many important developments have been omitted. In particular, most of the important contributions that were made in the late 1970s and early 1980s are not included, in part for the sake of brevity and in part because they have not yet passed the "test of time." If the health and vigor of a science can be evaluated by its growth rate, genetics is in great shape. New information is presently accumulating at an unprecedented rate. The next few years promise to be exciting as scientists begin to work out details of the molecular mechanisms responsible for differentiation in complex eukaryotes such as fruit flies, plants, and humans. Perhaps the next challenge beyond differentiation will be to understand the molecular basis of learning and memory, or perhaps aging. Hopefully, humankind will never lose the desire to understand all aspects of the universe of which we are a part, because that desire contributes, probably more than any of us fully realize, to the high quality of life that many of us enjoy.

SUMMARY

Genetic factors play dominant roles in shaping the lives of all living creatures on planet Earth. These genetic factors permeate all aspects of our lives, from food production to health, from individual and family planning to the very survival of our biosphere. The science of genetics was born in a monastery garden in Altbrünn, Austria, where Gregor Mendel performed and carefully analyzed the results of crosses between pea plants that differed in well-defined phenotypic traits. Today, genetics is a mature, but still vigorously growing, science. New developments in genetics are currently accumulating at an unprecedented rate. Genetic engineering is now a reality. One must hope that humans will use the powerful tools of genetic engineering wisely to enhance our quality of life. Few of us would want to forego the use of fire to heat our homes; yet, fire can be extremely dangerous when not used with care. The same will undoubtedly be true of our use of the powerful tools of genetic engineering!

REFERENCES

ABELSON, P.H., editor. 1983. "Biotechnology." *Science* 219: 609–740. (An entire issue of *Science* devoted to biotechnology, with a major focus on applications of recombinant DNA methodologies. Introductory "Overview" by Abelson on pp. 611–613.)

BERG, P. 1981. "Dissections and reconstructions of genes and chromosomes." *Science* 213: 296–303.

DUNN, L.C. 1965. *A short history of genetics.* McGraw–Hill, New York.

GARBER, E.D., editor. 1985. *Genetic perspectives in biology and medicine.* University of Chicago Press, Chicago.

GARDNER, E.J. 1983. *Human heredity.* Wiley, New York.

GROBSTEIN, C. 1979. *A double image of the double helix, the recombinant-DNA debate.* Freeman, San Francisco.

HANSON, E.D., editor. 1983. *Recombinant DNA research and the human prospect.* American Chemical Society, Washington, D.C.

ILTIS, H. 1932. *Life of Mendel.* (Translation by E. and C. Paul.) Norton, New York.

KETTLEWELL, H.B.D. 1959. "Darwin's missing evidence." *Sci. Amer.* 200(3): 48–53. Reprinted in A.M. Srb, R.D. Owen, and R.S. Edgar. 1970. *Facets of genetics.* Freeman, San Francisco.

LANDEGREN, U., R. KAISER, C.T. CASKEY, and L. HOOD. 1988. "DNA diagnostics—Molecular techniques and automation." *Science* 242: 229–237. (A comprehensive discussion of the state-of-the-art DNA technologies, such as automated DNA sequencing and DNA amplification by the DNA polymerase chain reaction technique, that make the "human genome project" feasible today.)

LEWIS, R. 1988. "DNA fingerprints, witness for the prosecution." *Discover* 9(6): 44–52.

MEDVEDEV, Z.A. 1969. *The rise and fall of T.D. Lysenko.* (Translation by I.M. Lerner.) Columbia University Press, New York.

PETERS, J.A., editor. 1959. *Classic papers in genetics.* Prentice–Hall, Englewood Cliffs, NJ.

PORTUGAL, F.H., and J.S. COHEN. 1977. *A century of DNA.* MIT Press, Cambridge, MA.

STUBBE, H. 1972. *History of genetics.* MIT Press, Cambridge, MA.

STURTEVANT, A.H. 1965. *A history of genetics.* Harper & Row, New York.

WATSON, J.D., and J. TOOZE. 1981. *The DNA story.* Freeman, San Francisco.

Chapter 2

Mendelian Genetics

(Photo above): Front page of Mendel's original paper, with a portrait of Mendel. (Courtesy of the New York Public Library and The American Museum of Natural History.)

MENDEL chose the garden pea as his experimental organism because it is an annual plant with well-defined characteristics, and it can be grown and crossed easily. Moreover, garden peas have ***perfect flowers***, flowers that contain ***both female and male (pollen-producing) parts***, and ***they are ordinarily self-fertilized***, that is, the ovule (female gamete) is fertilized by pollen (male gamete) from the same plant (in fact, usually from the same flower). Pollen from another plant can be experimentally introduced to the stigma of a flower, but cross-pollination is rare without human intervention.

Mendel was fortunate in choosing a ***diploid*** plant because diploid organisms contain only ***two sets of chromosomes***. If he had chosen a polyploid organism, an organism with more than two sets of chromosomes, he would not have obtained simple, understandable results. Through many generations of natural self-fertilization, garden peas had developed into pure lines. A single alteration in a trait was therefore demonstrated by a visible difference between varieties. Furthermore, in the seven pairs of contrasting traits Mendel chose to study, the two parental forms exhibited well-defined, contrasting alternative morphologies or visible effects on the plants. Vines were either tall or dwarf; unripe pods were green or yellow and inflated or constricted between the seeds; flowers were either distributed along the stem (axial) or bunched at the top (terminal); nutritive parts of ripe seeds were green or yellow; the outer surface of the seed was smooth or deeply wrinkled; and the seed coats were either white or gray. Flower color was positively correlated with this last trait. Seeds with white seed coats were produced by plants that had white flowers, and those with gray seed coats came from plants that had violet flowers. Much of Mendel's success in his first experiments may be attributed to his good judgment in making crosses, as far as possible, between parents that differed in only one trait. When this was not feasible, he considered only one trait at a time when analyzing the progeny of a cross.

MENDEL'S EXPERIMENTS

Crosses were made with great care when the peas were in blossom. To prevent self-fertilization in "test" flowers, anthers were removed from those chosen to be seed parents before their pollen-receiving parts were fully mature. Pollen from the designated pollen parents was transferred at the appropriate time to the stigma of the seed-parent flower. Seeds were allowed to mature on the vines. With a trait such as seed color, classification could be made immediately; but before traits such as plant size could be classified, the seeds had to be planted in the next season and the plants raised to maturity. Hybridization experiments were carried through several generations, and crosses were made between progeny from the initial crosses and true-breeding parental plants. Mendel ***visualized*** clearly each problem to be solved and ***designed*** his crosses to that end. He observed that weather, soil, and moisture conditions affected the growth characteristics of the peas, but heredity was the main factor under the conditions of his experiments. In a given environment, tall plants were 6 to 7 feet high, whereas dwarfs measured from 9 to 18 inches. No dwarfs ever turned into tall plants, and no tall plants became dwarfs.

PRINCIPLE OF SEGREGATION

In one experiment, Mendel crossed tall and dwarf varieties of garden peas. All offspring in the first (F_1) generation (***F*** symbolized *f*ilial from the Latin, meaning ***"progeny"***; ***1*** for ***first*** progeny generation) were tall. The dwarf trait had disappeared in the F_1 progeny. The progeny produced by crossing organisms that exhibit different inherited traits, such as tall versus dwarf plants, are called ***hybrids***. When tall hybrid plants were self-fertilized and progeny (second, or F_2 generation) were classified, some were tall and some were dwarfs. Careful classification of plants showed that when large numbers were considered, about three-fourths were tall and one-fourth were dwarfs. To be exact, an F_2 of 1064 consisted of 787 tall plants and 277 dwarfs, a nearly perfect 3:1 ***ratio***.

The experiment could have been concluded at this point; but to test his hypothesis that specific entities or "factors" (genes) were responsible for the observed hereditary patterns, Mendel predicted what would occur in the F_3 generation and planted F_2 seeds to test this prediction. On the basis of his hypothesis, he predicted that about one-third of the F_2 tall plants would produce only tall F_3 progeny, whereas two-thirds would produce both tall and short progeny. The F_2 short plants were expected to produce all short F_3 progeny. The predicted results were obtained. Indeed, the tall F_2 plants were of two types; about one-third of the tall plants produced only tall progeny and about two-thirds of them produced both tall and short progeny. In contrast, all the short F_2 plants produced only short progeny.

An important feature of Mendel's results was that the F_1 progeny plants all exhibited the trait expressed in one of the parent plants. The trait expressed in the other parent plant did not appear among the F_1 progeny. The factor (gene) specifying one of the traits had an overriding or dominant controlling effect over the factor specifying the other trait. The other trait disappeared or was masked in the F_1, but reappeared in one-fourth of the F_2 progeny. In genetics, the trait that is expressed in the F_1 generation is referred to as the ***dominant*** trait; the one that is masked in the F_1 hybrids is termed the ***recessive*** trait. Similarly, the factor (now called a gene) that specifies the trait expressed in the F_1 generation is called the dominant factor, and the factor that controls the trait that is masked in the F_1 generation is termed recessive. For the present, a ***gene*** is defined as the genetic factor that controls one trait or one characteristic of the organism. (A more precise definition of the gene will be developed in subsequent chapters.) The different forms of a gene that determine alternate traits or characteristics are called ***alleles***. An allele, then, is one of a pair (or more) of different forms of a given gene.

In Mendel's other crosses, the remaining six of the originally selected seven pairs of contrasting traits were studied. One member of each pair was dominant to the other in the same way as tall was dominant to dwarf. Thus, the dominance of one member of the pair of alleles controlling each trait over the other allele was observed in all of Mendel's crosses.

Mendel's conclusions were based on his concept of ***unit characters***, which was in marked contrast to the prevailing belief in a blending inheritance. On the basis of good experimental evidence, Mendel deduced that traits were controlled by physical entities (the factors or genes) that were present in living organisms, at least in his pea plants, in pairs (as two different alleles in hybrids). In garden peas, for example, a gene for height has two alleles, one for tall and one for dwarf. The allele for tall behaves as a dominant, whereas that for dwarf is recessive. Similarly, the gene for seed coat color has two forms, a dominant one specifying yellow seeds and a recessive one giving rise to green seeds. During sexual reproduction, the members of each pair of alleles separate into different reproductive cells or gametes of the male and female parents that fuse and give rise to the progeny; thus, they occur in different offspring. Mendel called this separating or segregation process the "splitting of hybrids." In these experiments, Mendel had used the scientific method very well indeed.

The significant inference from Mendel's results was that the separation of pairs of determiners resulted in a "purity of gametes." The concept of segregation, identified as Mendel's principle, can be phrased in this way: ***The paired genes (allelic pairs) separate from one another and are distributed to different sex cells.***

SYMBOLS AND TERMINOLOGY

Symbolism is not the important part of genetics and no universal notation has been established. But symbols are useful for providing a language to describe important aspects of genetics. Mendel used letters of the alphabet as symbols for genes. A capital letter signified a dominant and a lowercase letter a recessive member of a pair of alleles. Mendel considered factors as abstract units, any one of which could be symbolized by *A* or *B* or some other letter. Consider that many genes are known and that several may be involved in a single series of experiments. To avoid confusion as to which gene is indicated, appropriate letters are chosen as symbols to represent particular genes. For a given pair of alternate traits, the one that is most frequent in wild or native populations of that species is designated (often quite arbitrarily) as the ***"wild type,"*** and the allele of the gene that produces this trait is called the ***wild-type allele***. Alternate forms of the trait (those less prevalent or rare in natural populations) are referred to as ***mutant*** (from the Latin *mutare*, to change) ***traits***, and the alleles that produce these "nonwild-type" traits are called ***mutant alleles***. The mutant trait that deviates from the wild type is usually chosen as the basis for the genetic symbol. This trait usually is produced by the recessive allele, because most mutations occur as recessives. The dominant allele usually produces a functional product. A mutational change removes all or part of this product and leaves the recessive to come to expression only in the absence of a dominant allele. For example, the mutant vermilion is a recessive for eye color in fruit flies. As compared with the dominant wild type, the color production is diminished and the gene symbol is *v*. A few mutant alleles, such as the one for wrinkled wings in the fruit fly, however, are dominant and are therefore represented by a capital letter (e.g., *W*).

If the mutant is to provide the symbol, the history of the organism under investigation must be known well enough to suggest which member of the contrasting pair (e.g., tall or dwarf in peas) is the mutant trait. World collections of pea species show no dwarfs in natural populations. Dwarf peas occur only in certain cultivated stocks that have been developed. Dwarf is probably the recessive mutant (*d*), and tall is the allele for wild type (*D*). With the aid of these symbols, Mendel's experiment (Fig. 2.1) may be reconstructed in steps. ***Parents (P)***, with either of the two alleles (*DD* or *dd*), are represented as follows:

Tall parent		Dwarf parent
DD	×	*dd*

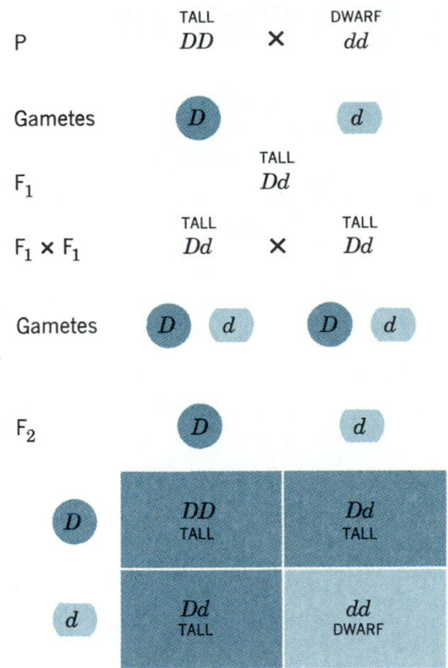

PHENOTYPES	GENOTYPES	GENOTYPIC FREQUENCY	PHENOTYPIC RATIO
Tall	*DD*	1	3
	Dd	2	
Dwarf	*dd*	1	1

Figure 2.1 Mendel's cross between tall and dwarf garden peas and summary of phenotypic and genotypic results.

(Note that in genetic crosses, the female or seed parent is usually written first.)

Segregation, the separation of pairs, occurs during formation of mature reproductive cells or ***gametes***. Each gamete produced by the tall parent carries only one *D* allele, and each gamete from the dwarf parent carries only one *d* allele. Therefore, the fertilized egg (***zygote***), which results from fusion of the male and female gametes, must have one allele of each kind (*Dd*). Because the *D* was always present and dominant, the F_1 plants (first-generation progeny) were all tall. When the F_1 tall (*Dd*) plants were self-fertilized, half the gametes carried the *D* allele and half the *d* allele. Results of self-fertilizing the F_1 indicated to Mendel that alleles were entirely separate from each other.

Gene symbols represented in pairs designate zygotes and individual plants or animals that have arisen from zygotes. Members of pairs of alleles are represented separately to designate mature germ cells or gametes, either eggs or sperm. Circles or brackets placed around gamete symbols indicate mature germ cells, as distinguished from plants or animals. A female and a male gamete combine in fertilization to produce a zygote. Zygotes or individual organisms carrying two units of one allele (*DD* or *dd*) are ***homozygous***, and those with two alleles (*Dd*) are ***heterozygous***. Two other useful terms, ***phenotype*** and ***genotype***, refer to visible expression of trait and actual gene constitution, respectively. Letter symbols are used to represent genotypes.

When the F_1 (*Dd*) plants from Mendel's experiments were crossed back to the dwarf (*dd*) variety, half the progeny were tall and half were dwarf, as illustrated in Fig. 2.2. This demonstrated more conclusively the principle of segregation, but the separation of alleles could be detected only in the parent (*Dd*) that produced two kinds of gametes (*D*) and (*d*). The dwarf parent (*dd*) could produce only one kind of gamete.

As the genetics of a particular species advanced, the 26 letters in the alphabet were soon depleted and more symbols were needed. *Drosophila* (fruit fly) geneticists met this limitation by adding a second letter and a third and fourth, when necessary, taken from the name of the mutant phenotype. Another technical advance from *Drosophila* geneticists is the + to symbolize wild-type alleles of recognized mutant genes. For example, *Cy* is the dominant allele for curly wing, and + is the allele for wild-type wings. The lowercase *b* is assigned to the recessive allele for black body color, and + to the dominant allele for wild-type gray body. If doubt can exist in the meaning of a given +, the symbol is added as a superscript ($+^w$ or w^+). Thus, w^+ symbolizes the wild-type allele for red, and *w* the allele for white eyes in *Drosophila*. Another useful device initiated by *Drosophila* geneticists was the separation of alleles by one or two crossbars or slash marks to indicate chromosomes. A heterozygous pair of alleles at the *w* locus, for example, was symbolized w^+/w. In summary, the various ways of denoting a pair of alleles are: *Aa*, *A/a*, *A//a*, +*a*, +/*a*, $\frac{+}{a}$, and $\frac{+}{\overline{a}}$. Gene symbols are printed in italics.

Hybrids are offspring carrying two nonidentical alleles of at least one gene produced from a cross

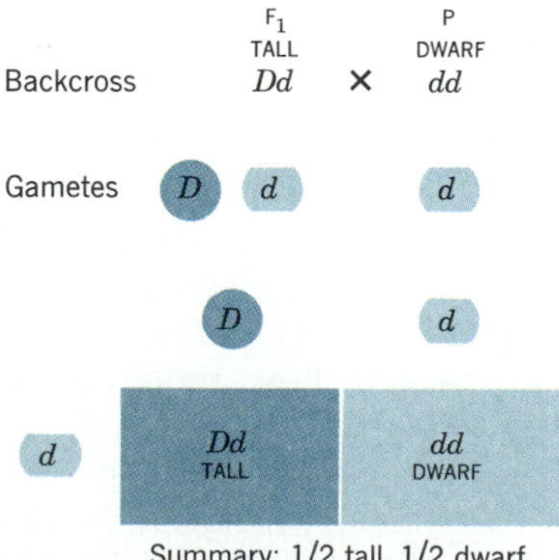

Figure 2.2 Backcross between F_1 tall garden pea and the dwarf parent variety from the cross illustrated in Fig. 2.1.

TABLE 2.1 Crosses Involving One Allelic Pair. Expected Gametes, Genotypic Frequencies of Progeny, and Phenotypic Ratios for Dominance and Intermediate Inheritance Are Given for the Different Combinations

MATING COMBINATIONS	GAMETES: FIRST PARENT	GAMETES: SECOND PARENT	PROGENY: GENOTYPIC FREQUENCIES	PROGENY: PHENOTYPIC RATIO WHEN DOMINANCE IS COMPLETE	PROGENY: PHENOTYPIC RATIO WHEN DOMINANCE IS ABSENT
AA × *aa*	*A*	*a*	all *Aa*	all dom.	all int.
Aa × *Aa*	½*A* ½*a*	½*A* ½*a*	¼*AA*, ½*Aa*, ¼*aa*	3 : 1	1 : 2 : 1
Aa × *AA*	½*A* ½*a*	*A*	½*AA*, ½*Aa*	all dom.	1 : 1
Aa × *aa*	½*A* ½*a*	*a*	½*Aa*, ½*aa*	1 : 1	1 : 1

between two genetically unlike individuals (e.g., *AA* × *aa* ⟶ *Aa*). A monohybrid is heterozygous for one pair of alleles (e.g., *Aa*). By extension, crosses (e.g., *AA* × *aa*) involving parents that differ with respect to one pair of alleles are called "monohybrid crosses."

MONOHYBRID CROSSES

Monohybrid crosses are basic to Mendelian genetics. Pertinent information about genetic segregation as it occurs in monohybrid combination is summarized in Table 2.1 and discussed in the following pages. Such crosses may occur in all major groups of sexually reproducing organisms. Dominance is the interaction between alleles, and the dominant allele is usually the one that produces a functional product while the recessive allele does not. Therefore, the normal or wild-type phenotype is most frequently produced when the dominant allele is present.

Dominance

If the phenotype of allele *A* is dominant, *AA* and *Aa* individuals are alike phenotypically. In the heterozygous (*Aa*) condition, allele *a* is completely masked and the trait is recessive. Phenotypic recognition of dominance may be influenced by factors in the internal and external environment; thus, dominance is not caused by a single allele alone. In practice, however, phenotypes attributable to single allele substitutions are called dominants, and those requiring homozygous combinations for expression are called recessives. Dominants are easier to detect than recessives, because they are expressed when paired with either kind of allele. Criteria for identifying dominant, defect-transmitting alleles from human pedigree studies are summarized as follows: (1) The trait is transmitted by a parent to about half of his or her children. (This assumes that the parent is heterozygous, which is usually the case, because most dominant defective alleles in humans seem to be homozygous lethal.) If each family includes three or four children, the trait usually occurs in every generation. (2) Persons who do not express the trait do not carry the allele and therefore do not transmit it to their children. Each of these criteria is a simple consequence of the definition of dominance. This pattern of dominant inheritance assumes complete expression of dominant alleles in each individual in whom the allele is carried. Some alleles do not have full expression, as discussed later in this chapter in the section entitled Genetic versus Environmental Effects.

Dominant inheritance can be illustrated by results of a study of dentinogenesis imperfecta (opalescent dentin) in a family group (Fig. 2.3). Among the descendants of individual II-1, 16 people have the condition. Fifteen brothers and sisters of those 16 have normal teeth. Thus, about half the children who have one parent with dentinogenesis imperfecta express the trait. This is expected from matings between heterozygous people with dentinogenesis imperfecta and people who are homozygous for the recessive allele for normal white teeth. All the 16 people with the condition had an affected parent. Among the descendants of family members with normal teeth, no opalescent teeth have occurred. Thus, the two criteria for identifying dominant alleles were met. Radiographs of affected teeth (Fig. 2.4) showed that the central pulp cavities of most teeth were filled with dentin. Histological studies confirmed that dentin, which is completely covered in normal white teeth, could be seen through the deficient enamel, and this gave the teeth their opalescent appearance.

From this pattern of dominant inheritance and the probability factors involved, it can be predicted that in future generations those who have dentinogenesis imperfecta (presumably heterozygous) and marry individuals with normal teeth might expect about half of

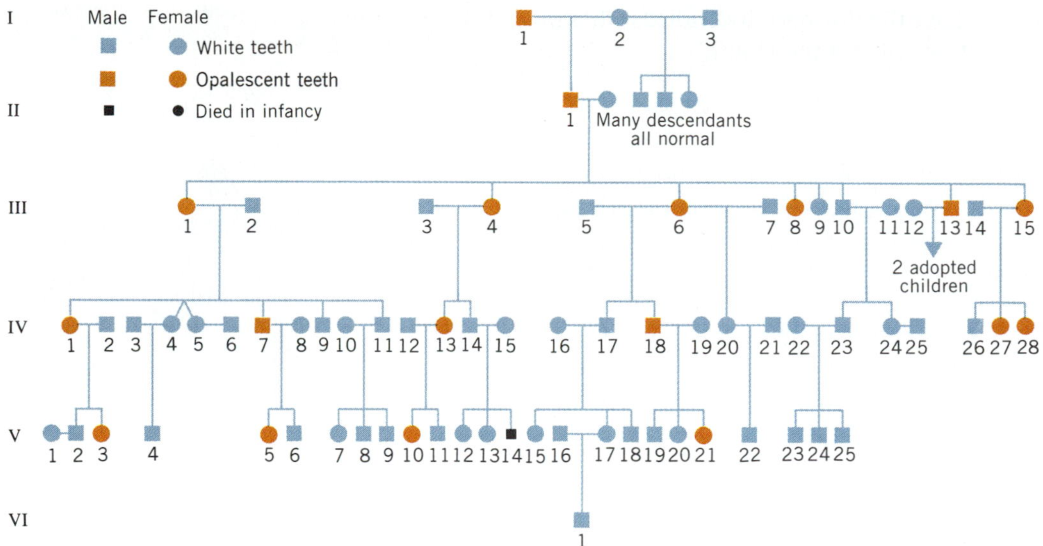

Figure 2.3 Pedigree chart showing the distribution of opalescent dentin in a family group. Generations are identified with Roman numerals in vertical sequence at left. Horizontal lines of squares and circles show individuals in each generation, squares for males and circles for females. A solid line connecting a male and female horizontally represents a marriage. Descending vertical lines illustrate progeny in order of age for each sibship from left to right. Horizontal numbers identify members of generations along with their spouses. Orange squares and circles represent individuals expressing a particular trait (i.e., opalescent dentin, orange symbols; normal teeth, blue symbols). (From E.J. Gardner, *J. Hered.* 42: 289–290, 1951.)

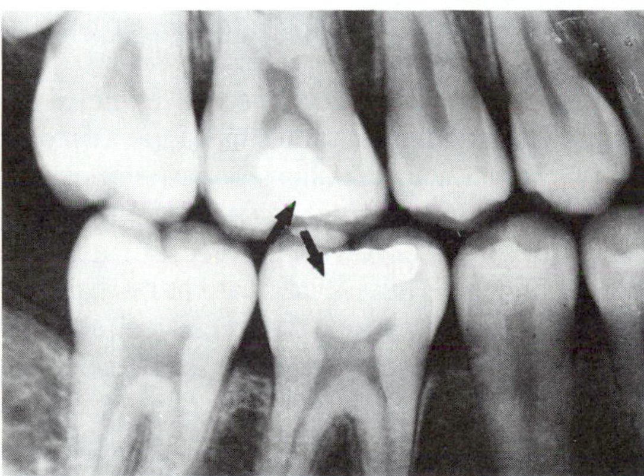

(a)

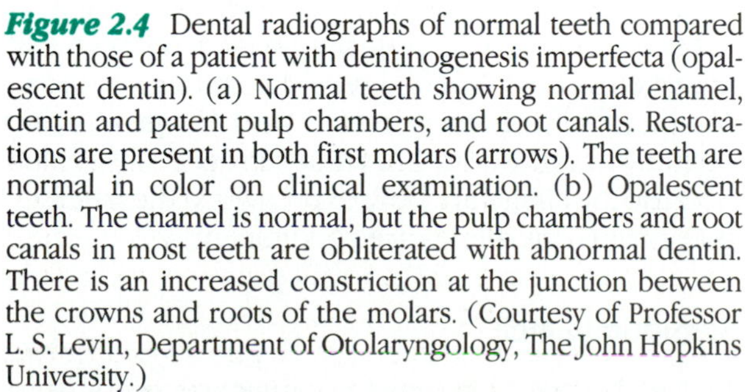

Figure 2.4 Dental radiographs of normal teeth compared with those of a patient with dentinogenesis imperfecta (opalescent dentin). (a) Normal teeth showing normal enamel, dentin and patent pulp chambers, and root canals. Restorations are present in both first molars (arrows). The teeth are normal in color on clinical examination. (b) Opalescent teeth. The enamel is normal, but the pulp chambers and root canals in most teeth are obliterated with abnormal dentin. There is an increased constriction at the junction between the crowns and roots of the molars. (Courtesy of Professor L. S. Levin, Department of Otolaryngology, The John Hopkins University.)

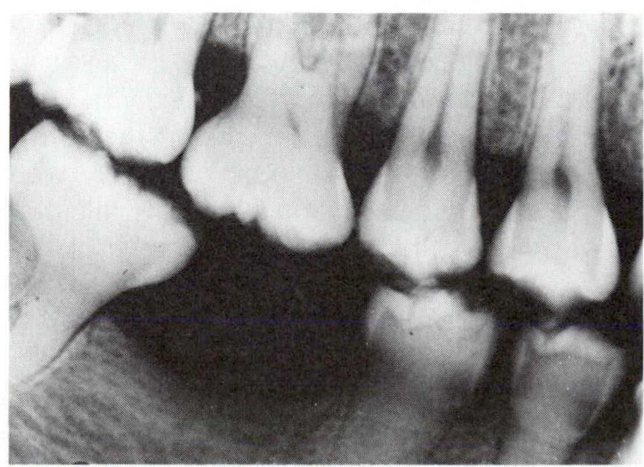

(b)

their children to have the disorder. Individuals who do not express the trait will not transmit it.

Recessiveness

Recessive alleles are expressed only in homozygous (*aa*) individuals. A population of crossbreeding organisms usually includes all three genotypes (*AA*, *Aa*, and *aa*), but it has more heterozygous (*Aa*) ***carriers*** than homozygous (*aa*) individuals who express the trait. Carriers are not detectable phenotypically, but recessive alleles can be identified experimentally by crossing potential carriers to homozygous recessive individuals. Such a cross of an organism with an unknown genotype to a known homozygous recessive organism is called a ***testcross***; such testcrosses are extremely useful in genetics in determining the genotypes of individual organisms.

At the level of function, many recessive alleles are known to encode nonfunctional products, either inactive enzymes or structural proteins. Indeed, the recessive nature of these alleles is the logical consequence of their control of inactive products since a combination of active products and inactive products would, in most cases, be expected to have at least some activity. Thus, an allele encoding an active product would most often be dominant, and an allele encoding an inactive product would most often be recessive.

In humans, the influence of recessive alleles can be detected from pedigree studies using the following criteria: (1) The trait is usually detected among sibs (offspring of the same parents), but not in their parents or other relatives; and (2) on the average, one-fourth of the sibs are affected. These criteria are to be expected on the basis of the definition of a recessive. The allele (*c*) associated with albinism, for example, is a relatively rare recessive. Albino people (*cc*) are characterized by a marked deficiency or complete absence of pigment in the skin, hair, and iris of the eyes. The predicted results of a mating between two normal people, both of whom are carriers of the allele (*c*) for albinism, are diagrammed in Fig. 2.5.

In one Caucasian family of six (Fig. 2.6), two albinos occurred. The parents were within the normal range of pigment for Caucasian people, but they were carriers (*Cc*) and both passed on a recessive allele (*c*) to each of the two albino sons (III-5 and III-13, Fig. 2.7). III-5 married a normally pigmented woman (III-4), who must be a carrier, and they have five albino and two normal children. For an infrequent allele, this occurrence is rare, indeed, if the parents (III-4 and III-5) are not related. This example represents an exception to the first criterion for detecting recessives from pedigrees. The cross is the testcross type (*Cc* × *cc*) and is similar to the one reconstructed in

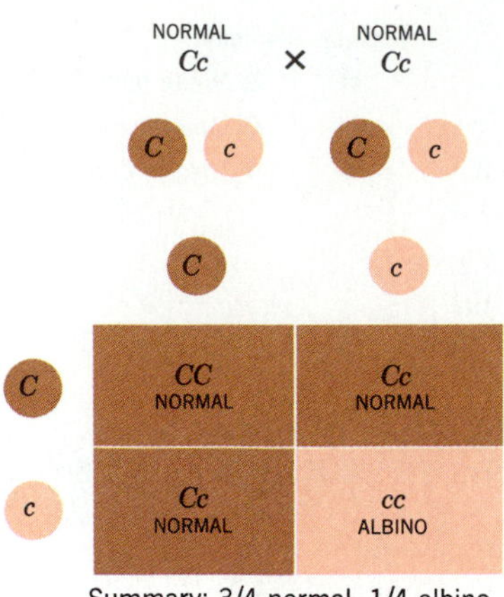

Figure 2.5 Cross between two normally pigmented people, both of whom were carriers of the gene (c) for albinism.

Fig. 2.2, but for a different trait. About half the children of III-4 and III-5 would be expected to express the albino trait, whereas the other half would be normal phenotypically but still carriers (*Cc*).

Codominance

When both alleles of a pair are fully expressed in a heterozygote, they are called ***codominant alleles***. Such alleles exhibit a unique pattern of expression with heterozygotes being phenotypically distinguishable from both of the homozygotes and expressing both alleles equally. In general, the two products are the same with respect to function but different in exact amino acid sequence. In humans, the MN blood group antigens are a good example. Allele L^M for M-type blood is codominant with its allele L^N for N-type blood. The heterozygote (L^ML^N) expresses the characteristics of both M and N antigens (MN-type blood), that is, L^ML^N heterozygotes possess red blood cells that carry both M and N antigens on their surfaces. Since the two alleles encode nonidentical protein products, a mating between a homozygous M-type person (L^ML^M) and a homozygous N-type person (L^NL^N) would result in all heterozygous (L^ML^N) offspring. Mating between heterozygotes ($L^ML^N \times L^ML^N$) would result in a ratio of 1 M-type (L^ML^M) : 2 MN-type (L^ML^N) : 1 N-type (L^NL^N). A phenotypic ratio of 1:2:1 has thus replaced the 3:1 ratio, because the alleles are both expressed in the heterozygote, that is, the alleles are codominant.

Semidominance

In the absence of complete dominance, every genotype has a distinguishable phenotype. ***Semidominant*** alleles may produce the same product but in lesser

Figure 2.6 Family consisting of father, mother (front, center) and six sons, two of whom are albino. (Courtesy of McKay Kunz and Charles Kunz.)

quantity as compared with the dominant allele. Semidominance is sometimes called incomplete dominance. In the heterozygous condition, the total product is intermediate between that of the dominant and recessive alleles. In snapdragons, for example, heterozygotes for color alleles have pink flowers in contrast to red and white for the semidominant and recessive homozygotes, respectively. The phenotypic ratio for the monohybird cross then becomes 1:2:1 instead of 3:1, as occurs for dominant alleles.

Lethals

Genes may affect viability as well as the visible traits of an organism. Appropriate experiments have shown that animals carrying certain genes are disadvantaged through impaired biochemical as well as physical functioning. White-eyed and vestigial-winged *Drosophila*, for example, have lower viabilities than the wild type. Detrimental physiological effects are apparently associated with the genes involved (*w* and *vg*, respectively). Some other genes have no effect on the appearance of the fly, but do influence viability in some way. Other genes have such serious effects that the organism is unable to live. These are called ***lethal genes***. Obviously, if the lethal effect is dominant and immediate in expression, all individuals carrying the gene will die and the gene will be lost. Some dominant lethals, however, have a delayed effect so that the organism lives for a time. Recessive lethals carried in the heterozygous condition have no effect but may come to expression when matings between carriers occur.

The dominant gene (*C*) in chickens is responsible for profound developmental changes that result in aberrant forms called "creepers," and the homozygous genotype (*CC*) is lethal. These birds have short, crooked legs and are of little value except as novelties. When two creepers were mated, a ratio of 2 creepers to 1 normal instead of 3:1 appeared, as illustrated in Fig. 2.8. This is a characteristic ratio for all crosses involving lethals. In this particular case, the *CC* class is missing. All creepers that lived could be shown by testcrosses to be heterozygous (*Cc*). When a creeper was mated with a normal chicken, the expected backcross result of 1 creeper to 1 normal was obtained (Fig. 2.9). How can investigators determine whether they are working with a dominant for a phenotype (e.g., creeper) and a recessive for lethality? If the numbers of progeny from matings between affected individuals (e.g., creepers) are large enough, statistical analysis

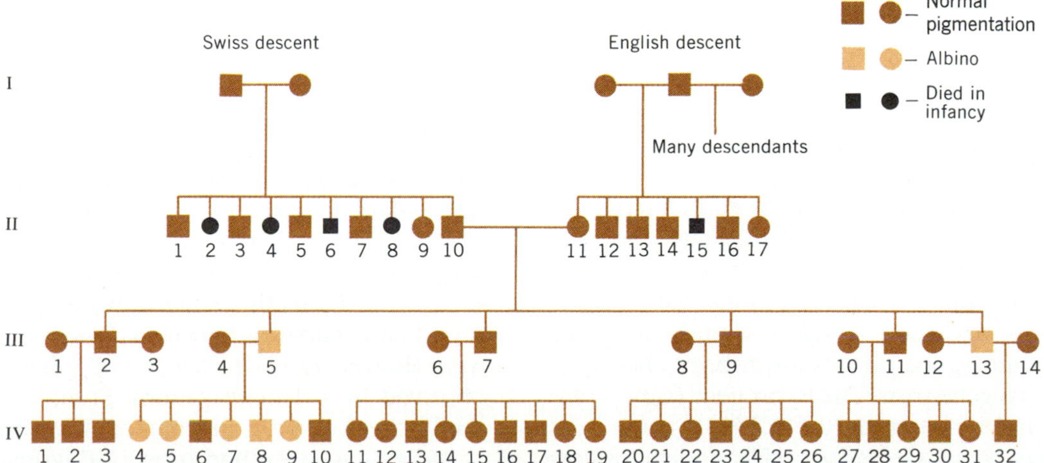

Figure 2.7 Pedigree of a family group in which albinism has occurred.

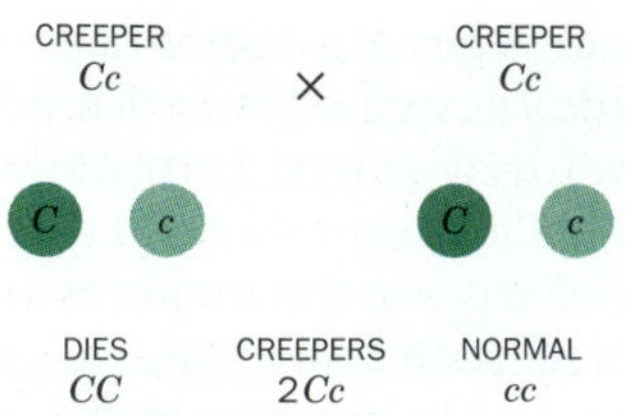

Summary: 2 creepers: 1 normal

Figure 2.8 Cross between two creeper chickens. The 2 : 1 ratio replaces 3 : 1 because the homozygous (*CC*) embryos die.

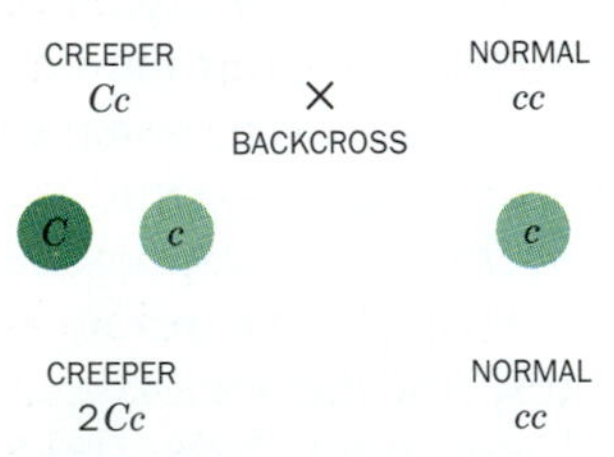

Summary: 1 creeper: 1 normal

Figure 2.9 Cross between a creeper and a normal chicken. Expected ratio is 1 creeper to 1 noncreeper (normal).

might distinguish between a 2:1 and 3:1 ratio. Testcross results could indicate whether progeny from crosses between affected individuals are heterozygous or homozygous. In some animals, such as chickens, it is possible to observe dead or dying embryos at the appropriate time in incubating eggs. The allele *C* is expressed in heterozygous arrangement (*Cc*) in creeper chickens as a dominant. But note that with respect to lethality *C* behaves like a recessive—only *CC* genotypes die. Thus, the *C* allele has a dominant effect on the morphology of the chicken, but a recessive effect with respect to lethality.

PRINCIPLE OF INDEPENDENT ASSORTMENT

Mendel also crossed plants that differed in two pairs of alleles (Fig. 2.10). In this cross, designed to clarify the relation of different pairs of alleles, he crossed plants having round, yellow seeds with plants having wrinkled, green seeds. The F_1 progeny from such a cross between homozygous parents are hybrids (heterozygotes) for two gene pairs. The F_1 progeny (*GgWw*) are dihybrids, and, by extension, the *GGWW* × *ggww* cross is a ***dihybrid cross***. Alleles for both round and yellow were known from previous studies to be dominant over their respective alleles, producing wrinkled and green seeds.

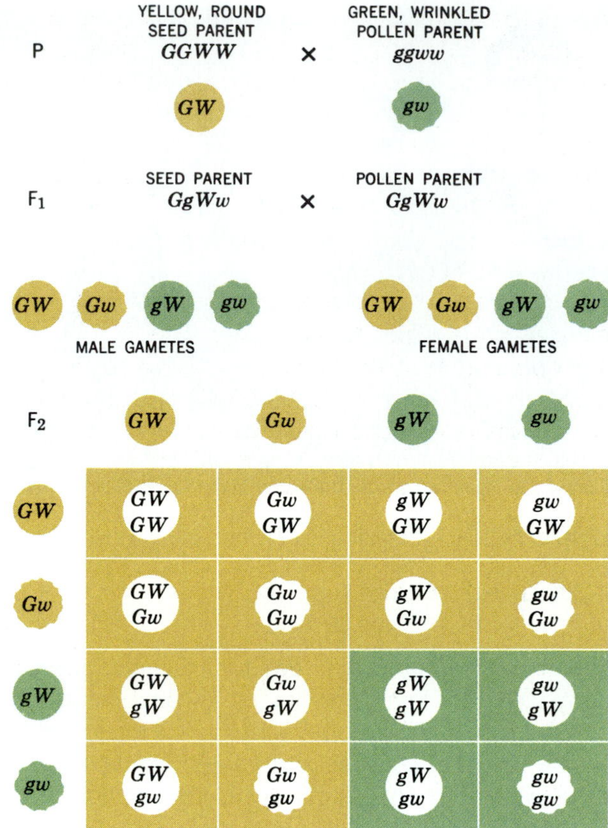

Phenotypes	Genotypes	Genotypic Frequency	Phenotypic Ratio
Yellow, round	*GGWW*	1	9
	GGWw	2	
	GgWW	2	
	GgWw	4	
Yellow, wrinkled	*GGww*	1	3
	Ggww	2	
Green, round	*ggWW*	1	3
	ggWw	2	
Green, wrinkled	*ggww*	1	1

Figure 2.10 Diagram and summary of a cross between a variety of garden peas with yellow, round seeds and a variety with green, wrinkled seeds. The F_1 X F_1 represented illustrates a dihybrid cross.

All the F_1 seeds resulting from the cross were round and yellow, as expected. When the F_1 hybrids were allowed to self-fertilize, four F_2 phenotypes were observed in a definite pattern. From a total of 556 seeds, the following distribution was obtained: 315 round, yellow; 108 round, green; 101 wrinkled, yellow; and 32 wrinkled, green. These results closely fit a ratio of 9:3:3:1 (i.e., 315/556 ≅ 9/16, 108/556 ≅ 3/16, 101/556 ≅ 3/16, and 32/556 ≅ 1/16). Mendel recognized this as the result of ***two monohybrid*** crosses, each expected to result in a 3:1 ratio, operating together. The product of the two monohybrid ratios $(3{:}1)^2$ or $(3+1)^2$ was equal to the ***dihybrid ratio*** $(3+1)^2 = (9+3+3+1)$, thus conforming to the ***law of probability***, called the "product rule," which states: ***The chance of two or more independent events occurring together is the* product *of the chances of their separate occurrences.***

The results were those expected from the assortment of two independent pairs of alleles, each showing dominance of one member. Not only did the members of each pair of alleles segregate, but the allelic pairs of different genes behaved independently with respect to each other. Mendel therefore drew another conclusion: Members of different pairs of alleles ***assort independently into gametes.*** This concept of independent assortment of different pairs of alleles is designated as his second principle. It is a simple corollary of meiosis (Chapter 3). Mendel's two principles were set forth in a paper entitled "Experiments in Plant Hybridization," which was read before the Brünn Natural History Society in 1865 and published in the proceedings of that society in 1866.

An understanding of Mendel's principle of independent assortment is one of the key components considered in the design of many experiments in plant and animal breeding aimed at improving the quantity or quality of agricultural products. Desirable traits carried in different varieties can be combined and maintained in a single type. A variety of barley resistant to rust, for example, was needed in a rust-infested area in the United States. The best available rust-resistant variety, however, like most barley varieties, had hulls on the seeds and did not thresh well. Another variety had no hulls and threshed out clean like wheat, but had poor rust resistance. These two varieties were combined by appropriate crosses, and a valuable new strain with rust resistance and no hulls was obtained.

Dihybrid Ratios

The basic mechanics of genetics were postulated and later established from particular ratios, such as 3:1 and 9:3:3:1. Ratios of this kind merely represent the grouping expected when particular conditions are met. Common patterns such as the 9:3:3:1 ratio may serve as models for analyzing results of experiments. When such a ratio is obtained from a cross in which parental genotypes are not known, the geneticist may postulate that two independent pairs of alleles are involved, and that one member of each pair behaves like a dominant over its allele. Mendel's dihybrid cross between plants with round, yellow seeds and those with wrinkled, green seeds is represented diagrammatically in Fig. 2.10 as a pattern applicable in analyzing other crosses.

When the F_1 plants were selfed (i.e., pollen and eggs from the same plant were united), four kinds of gametes were produced by the male parts and four by the female parts of the F_1. At the top of the checkerboard (Punnett square, Fig. 2.10), the four kinds of gametes from the seed parent are shown. The four possible gametes from the pollen parent are represented in a vertical row at the left. This Punnett square is merely a geometrical device that helps in visualizing all possible combinations of male and female gametes. It is valuable as a learning exercise, but we will shortly turn to methods that are less time-consuming. Letter symbols in the 16 squares represent combinations of independent genes brought together by the fusion of gametes. When these are collected according to the phenotypes represented, the 9:3:3:1 ratio becomes apparent. The completed summary chart (Fig. 2.10) illustrates the F_2 results of the cross in tabular form.

The 1:1:1:1 ratio is expected from a dihybrid backcross to the recessive parent, that is, a cross between an F_1 that carries two heterozygous pairs of alleles and a parent type with the full recessive combination for these two genes. This cross is illustrated in Fig. 2.11. A

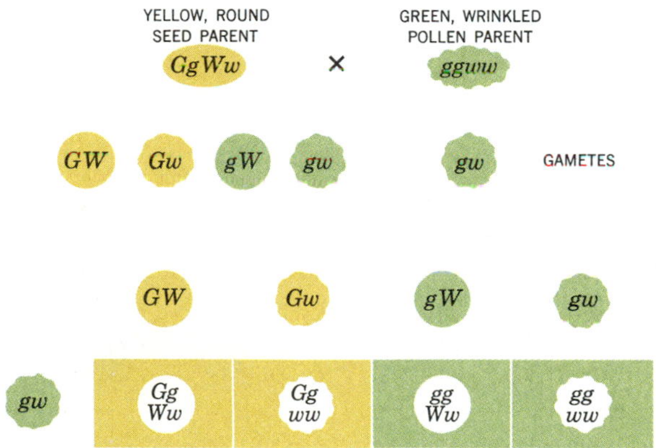

PHENOTYPES	GENOTYPES	GENOTYPIC FREQUENCY	PHENOTYPIC RATIO
Yellow, round	*GgWw*	1	1
Yellow, wrinkled	*Ggww*	1	1
Green, round	*ggWw*	1	1
Green, wrinkled	*ggww*	1	1

Figure 2.11 Diagram and summary illustrating a method for solving a backcross-type problem involving two gene pairs. This cross is between an F_1 garden pea with yellow, round seeds and the fully recessive parent type with green, wrinkled seeds.

cross of this type is used in practical breeding programs to determine the genotype of an individual that may carry a recessive allele, the expression of which could be obscured by a dominant allele. The cross with the double recessive genotype for testing purposes is called a ***testcross***, to distinguish it from a ***backcross***, which is the cross of the F_1 with either of the parent types.

Trihybrid Ratios

Virtually all cross-fertilizing plants or animals differ from other members of their species in more than one or two pairs of alleles. Therefore, matings in natural breeding populations usually produce new combinations of many genes. Genetic analysis of such crosses is complicated. In many cases, however, complex combinations can be simplified by resolving them into monohybrid crosses or by using formulas devised to handle several traits in the same problem. A cross between homozygous parents that differ in three gene pairs (i.e., producing trihybrids) is a combination of three single-pair crosses operating together. Thus, (*AA* × *aa*), (*BB* × *bb*), and (*CC* × *cc*) could be combined in the same cross as *AABBCC* × *aabbcc*.

What results might be expected (1) in the F_1, (2) in the backcross to the fully recessive parent, and (3) in the F_2 from a cross between two varieties of garden peas differing in three traits? A diagram of a cross in which the seed parent is homozygous for the genes producing a tall vine and yellow, round seeds (*DD GGWW*) and the pollen parent is homozygous for a dwarf vine and green, wrinkled seeds (*ddggww*) can answer that question. The three traits represented in the seed parent are known from previous experiments to depend on dominant genes. The first-generation cross may be illustrated as follows:

When the F_1 plants are crossed with the full recessive type, *DdGgWw* × *ddggww*, eight kinds of gametes (*DGW, DGw, DgW, Dgw, dGW, dGw, dgW,* and *dgw*) are produced by the F_1 parent and only one kind, *dgw*, by the full recessive parent. As a result of fertilization, eight kinds of peas are expected in equal proportion. Thus, the trihybrid backcross genotypic and phenotypic ratio of 1:1:1:1:1:1:1:1 is explained by the fertilization of eight different kinds of gametes from the F_1 by the one kind of gamete from the fully recessive parents. The sequence involved in this backcross and the summarized results are illustrated in Fig. 2.12.

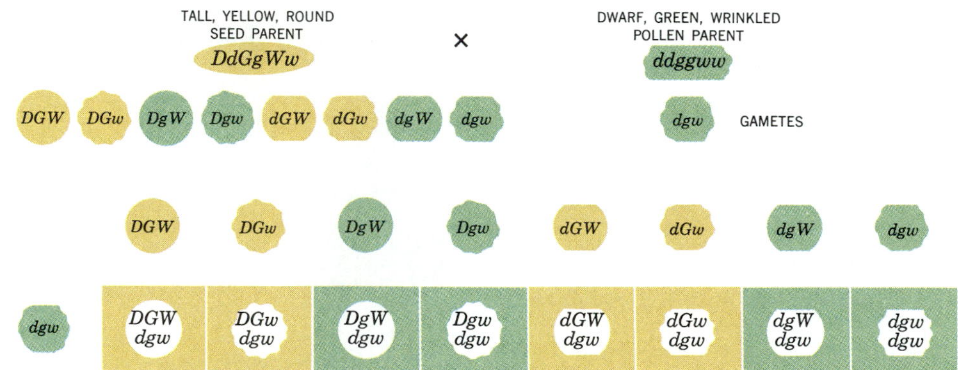

PHENOTYPES	GENOTYPES	GENOTYPIC FREQUENCY	PHENOTYPIC RATIO
Tall, yellow, round	*DdGgWw*	1	1
Tall, yellow, wrinkled	*DdGgww*	1	1
Tall, green, round	*DdggWw*	1	1
Tall, green, wrinkled	*Ddggww*	1	1
Dwarf, yellow round	*ddGgWw*	1	1
Dwarf, yellow wrinkled	*ddGgww*	1	1
Dwarf, green, round	*ddggWw*	1	1
Dwarf, green, wrinkled	*ddggww*	1	1

Figure 2.12 Method for solving backcross-type problems involving three gene pairs. This cross is between an F_1 garden pea with tall vines and yellow, round seeds and the fully recessive parental type with dwarf vines and green, wrinkled seeds.

When the F_1 plants are selfed (i.e., *DdGgWw* × *DdGgWw*), eight kinds of gametes (*DGW, DGw, DgW, Dgw, dGW, dGw, dgW,* and *dgw*) are produced from both the male and female parts. These gametes represent all combinations. If the $F_1 \times F_1$ cross were represented by a Punnett square, 64 (8^2) squares would be required with a phenotypic ratio of 27:9:9:9:3:3:3:1 as a result. In the next section, we look at less time-consuming methods for determining the results of complex combinations.

Forked-Line Method for Genetic Problems

A method for bringing the combinations of a trihybrid cross together may be illustrated as follows: First, visualize the trihybrid cross as three monohybrid crosses—that is, *Dd* × *Dd, Gg* × *Gg,* and *Ww* × *Ww*—operating together. If one member of each pair is dominant, a 3:1 ratio would be predicted from each monohybrid cross. Since the three pairs are independent, each monohybrid segregant may occur with any combination possible from any other pair of alleles. The combinations therefore can be systematically arranged together. The 3:1 ratio from *Dd* × *Dd* may be combined with the 3:1 ratios from each of the other two monohybrid crosses, *Gg* × *Gg* and *Ww* × *Ww*, as shown in Fig. 2.13.

Usually, the genotypes as well as the phenotypes are necessary for the complete solution of such a problem. The same forked-line system may be employed to represent and combine genotypes expected from monohybrid crosses. From each monohybrid cross in the example, a genotypic ratio of 1:2:1 may be predicted. The three monohybrid units may be combined as shown in Fig. 2.14. The forked-line system is merely another device for analyzing crosses.

Mathematical Method for Genetic Problems

This introduction to combinations suggests a third way to anticipate the results from crosses that involve independent pairs of alleles. A mathematical manipulation provides a way to arrive at the product of the combinations without drawing them out mechanically. As an example, consider the crosses made by Toyama between two varieties of the silk moth, *Bombyx mori*. In one variety, the caterpillars were striped and the cocoons were yellow, and in the other variety, the caterpillars were unstriped and the cocoons were white. From previous crosses, striped was known to be dominant over unstriped and yellow over white. What proportions might be expected in the F_2?

If the striped and unstriped alleles are considered separately, three-fourths of the F_2 progeny are expected to be striped and one-fourth unstriped. Similarly, three-fourths are expected to be yellow and one-fourth white. The phenotypes and their proportions are summarized in Table 2.2. Toyama obtained results from actual crosses that satisfied the predictions.

The expected F_2 result from a trihybrid cross that involves independent assortment and dominance of one allele in each pair is the product of three pairs, such as

$$(A{:}a)(B{:}b)(C{:}c)$$

Written algebraically,

$$(3 + 1)(3 + 1)(3 + 1)$$

or $(3 + 1)^3$, which expands to

$$27 + 9 + 9 + 9 + 3 + 3 + 3 + 1$$

or

$$27{:}9{:}9{:}9{:}3{:}3{:}3{:}1$$

A cross with four gene pairs under the same conditions would result in $(3+1)^4$, five gene pairs $(3+1)^5$, and so

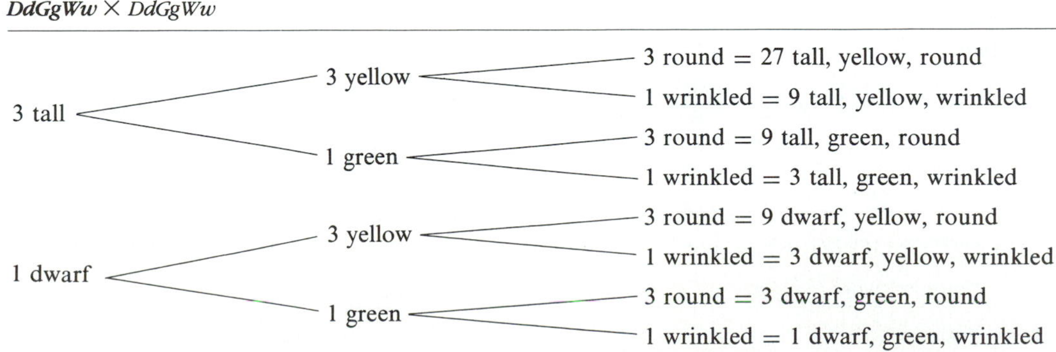

Figure 2.13 Forked-line method for solving genetic problems in which independent assortment is involved. Genotypes of parents are given, but only phenotypes of progeny are listed in this example.

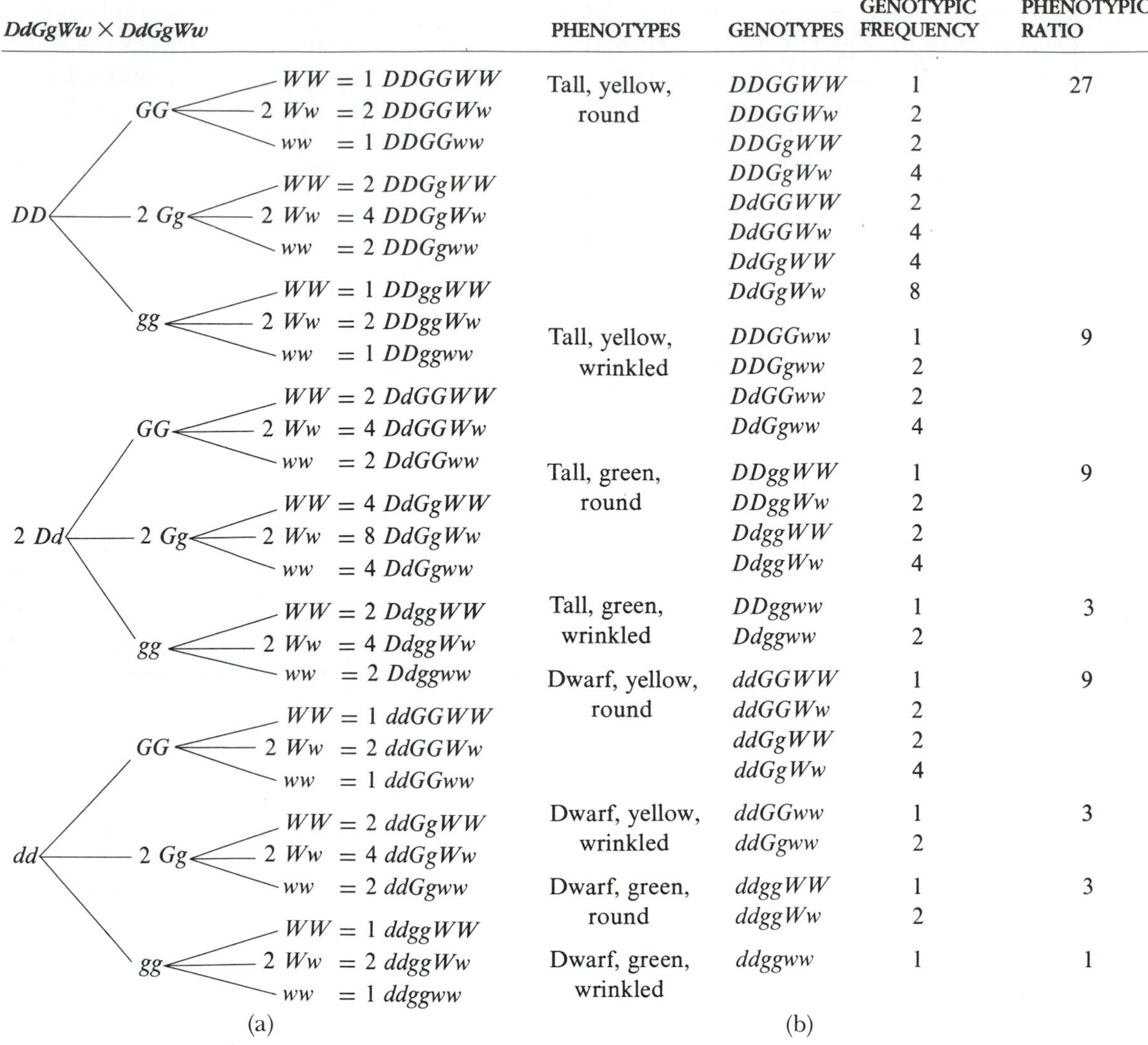

PHENOTYPES	GENOTYPES	GENOTYPIC FREQUENCY	PHENOTYPIC RATIO
Tall, yellow, round	*DDGGWW*	1	27
	DDGGWw	2	
	DDGgWW	2	
	DDGgWw	4	
	DdGGWW	2	
	DdGGWw	4	
	DdGgWW	4	
	DdGgWw	8	
Tall, yellow, wrinkled	*DDGGww*	1	9
	DDGgww	2	
	DdGGww	2	
	DdGgww	4	
Tall, green, round	*DDggWW*	1	9
	DDggWw	2	
	DdggWW	2	
	DdggWw	4	
Tall, green, wrinkled	*DDggww*	1	3
	Ddggww	2	
Dwarf, yellow, round	*ddGGWW*	1	9
	ddGGWw	2	
	ddGgWW	2	
	ddGgWw	4	
Dwarf, yellow, wrinkled	*ddGGww*	1	3
	ddGgww	2	
Dwarf, green, round	*ddggWW*	1	3
	ddggWw	2	
Dwarf, green, wrinkled	*ddggww*	1	1

Figure 2.14 (a) Cross between two F_1 garden peas of the genotype *DdGgWw*. The forked-line method is employed and the genotypes are illustrated. These results represent the F_2 of a cross similar to those obtained from the Punnett square method, which involves 64 squares. (b) Summary of F_2 from trihybrid cross, resulting in a 27 : 9 : 9 : 9 : 3 : 3 : 3 : 1 phenotypic ratio.

TABLE 2.2 Expectations and Results from Crosses between Striped, Yellow and Unstriped, White Moths[a]

PHENOTYPES	GENOTYPES	PROPORTIONS	F_2 EXPECTATIONS	F_2 OBSERVED
Striped, yellow	*S–Y–*	¾ × ¾	$\frac{9}{16}$ × 11,322 = 6,368.6	6,385
Striped, white	*S–yy*	¾ × ¼	$\frac{3}{16}$ × 11,322 = 2,122.9	2,147
Unstriped, yellow	*ssY–*	¼ × ¾	$\frac{3}{16}$ × 11,322 = 2,122.9	2,099
Unstriped, white	*ssyy*	¼ × ¼	$\frac{1}{16}$ × 11,322 = 707.6	691
Total			11,322.0	11,322

[a]Data from Toyama.

TABLE 2.3 Relations among Pairs of Independent Alleles, Gametes, F_2 Genotypes, and F_2 Phenotypes When Dominance Is Present

NUMBER OF HETEROZYGOUS PAIRS	NUMBER OF KINDS OF GAMETES	NUMBER OF F_2 GENOTYPES	NUMBER OF F_2 PHENOTYPES
1	2	3	2
2	4	9	4
3	8	27	8
4	16	81	16
10	1024	59,049	1024
n	2^n	3^n	2^n

on. Numbers of gametes, genotypes, and phenotypes expected from different numbers of heterozygous pairs of genes are summarized in Table 2.3. It will be observed that the number of kinds of ***gametes*** is a multiple of 2, that is 2^n; the number of F_2 ***genotypes*** is a multiple of 3, that is, 3^n; and the number of ***phenotypes is 2^n when dominance is present.*** This pattern forms the basis for predicting results when any number of independent pairs of alleles is involved in the production of hybrids.

Gene Interaction

In the preceding discussion, the emphasis has been placed on the fact that the genes studied by Mendel were segregating independently of each other. What has not yet been mentioned is that these genes must also have been ***functioning*** (exerting their effects on the phenotype) ***independently*** of one another.

If each gene were expressing itself in a separate test tube, it would be reasonable to expect all genes to be functionally independent. But these genes are not in separate test tubes; they are all located in the same nuclei of the same cells. Thus, it should not be surprising that the expression of an allele of one gene will sometimes alter the expression of one or more of the alleles of a second (nonallelic) gene.

A classic example of gene interaction based on the results of crosses between different breeds of chickens was reported in the early part of this century by William Bateson and his associate, R. C. Punnett (after whom the Punnett square was named). Bateson began to confirm and extend Mendel's work immediately after its discovery in 1900 and soon became a pioneer in transmission genetics. He had chickens in his research coops and sweet peas in his garden for immediate use in genetic investigations.

Domestic breeds of chickens have different comb shapes (Fig. 2.15). Wyandottes have a characteristic type of comb called "rose," whereas brahmas have a "pea" comb. Leghorns have "single" combs. The investigators crossed wyandottes and brahmas, and all F_1 chickens had walnut combs, a phenotype not expressed in either parent. When the F_1 chickens were mated among themselves and large F_2 populations were produced, a familiar dihybrid ratio, 9:3:3:1, was recognized, but the phenotypes representing two of the four classes were different from those expressed in the parents. About 9⁄16 of the F_2 birds were walnut, 3⁄16 were rose, 3⁄16 were pea, and 1⁄16 had single combs. Neither single comb nor walnut was expressed in the original parent lines. These two phenotypes were explained as the result of gene product interaction. Results indicated that two different allelic pairs were involved; one pair was introduced by the rose-comb parent and one by the pea-comb parent. A gene for rose and a gene for pea would interact and produce walnut, as in the F_1.

Analysis of the F_2 results and appropriate testcrosses indicated that the 9⁄16 class, with the two dominant genes (*R-P-*), was walnut, like the F_1 chickens. The 1⁄16 class, representing the full recessive combination (*rrpp*), was characterized as single combs. The two 3⁄16 (rose and pea) classes were *R-pp* and *rrP-*. It was then determined that the homozygous genotype of the rose-combed parent (wyandotte) was *RRpp* and that of the pea-combed parent (brahma) was *rrPP*. Although the usual 9:3:3:1 ratio was obtained, the result from this cross was unusual in two important respects: (1) the F_1 progeny differed from those of the parent, that is, none was rose or pea, but all were walnut. (2) Two phenotypes (walnut and single) not expressed in the original parents appeared in the F_2.

Genes R and P were nonallelic, but each was

(a)

(b)

(c)

(d)

Figure 2.15 Comb types characteristic of different breeds of chickens: (a) rose, Wyandottes; (b) pea, Brahmas; (c) walnut, hybrid from cross between chickens with rose and pea combs; and (d) single, Leghorns. (Courtesy of Ralph G. Somes, Jr.)

dominant over its allele (i.e., *R* over *r* and *P* over *p*). When *R* and *P* were together, as in the F_1 (*RrPp*), the two different products interacted to produce a walnut comb. The two nonallelic genes *R* and *P* acted independently in different ways, similar to the ways in which codominant alleles act.

Epistasis

One important type of functional interaction between different genes occurs when an allele or genotype at one locus "masks" or inhibits the expression of a nonallele or genotype at a distinct locus; such an interaction is called ***epistasis*** (Greek for "standing upon"). Any gene that masks the expression of another, nonallelic gene is ***epistatic*** to that gene. ***Epistasis should not be confused with dominance. Epistasis is the interaction between different genes (nonalleles). Dominance is the interaction between different alleles of the same gene.***

Metabolic processes in living organisms take place via sequences of enzyme-catalyzed reactions. Each step of each metabolic pathway thus requires the activity of a specific enzyme. Each enzyme, in turn, is the product of a specific gene. (Some enzymes contain two or more different subunits, each specified by a particular gene. Thus, some enzymes are coded for by two or more different genes. This somewhat more complex picture will be developed further in Chapters 10–12. For now, the assumption that each enzyme is encoded by one specific gene is adequate.)

Consider the following simple two-step pathway:

Precursor
Gene *C* → Enzyme C (Allele *C*) ↓ (Step I)
Intermediate
Gene *P* → Enzyme P (Allele *P*) ↓ (Step II)
Product
(such as an anthocyanin pigment)

If the recessive alleles *c* and *p* of the two genes specify inactive forms of the two enzymes, organisms that are homozygous recessive for either gene (*cc* or *pp*) will not be able to synthesize the product. Organisms of genotype *cc* will be blocked at Step I; those of genotype *pp* will be blocked at Step II.

Suppose that the product of this pathway is an anthocyanin pigment responsible for colored flowers in a plant such as the sweet pea, *Lathyrus odoratus*. If the precursor and intermediate in this pathway are colorless compounds, only plants that carry at least one copy of the dominant allele of each gene (*C-P-*) will have colored flowers.

Clearly, genes controlling the synthesis of enzymes involved at different steps of the same metabolic pathway will ***not*** be functionally independent. Moreover, as a result, such genes will ***not*** yield the classic Mendelian segregation ratios. Instead, they will yield ***modified phenotypic segregation ratios*** such as those listed in Table 2.4.

In sweet peas, the anthocyanin pigment resulting in colored flowers is synthesized by a pathway like that just described. In a classic study, Bateson and Punnett crossed two different white-flowered varieties of sweet peas. The F_1 plants all had purple flowers. When the F_1 plants were intercrossed, they produced F_2 progeny of which 9/16 had purple flowers and 7/16 had white flowers (Fig. 2.16).

In Bateson and Punnett's experiment, one variety of sweet peas was of genotype *ccPP* and was blocked at one of the steps in the synthesis of the anthocyanin pigment. The other variety, of genotype *CCpp*, was blocked at another step in pigment synthesis. The F_1 plants were all of genotype *CcPp*. Since in this case *C* and *P* are dominant to their alleles *c* and *p*, respectively, all of the F_1 plants synthesized the anthocyanin pigment and thus had colored flowers. In the F_2 generation, 9/16 of the progeny would be expected to carry at least one dominant allele of each gene (*C-P-*) and have colored flowers. The other 7/16 of the progeny should be homozygous recessive for one or the other

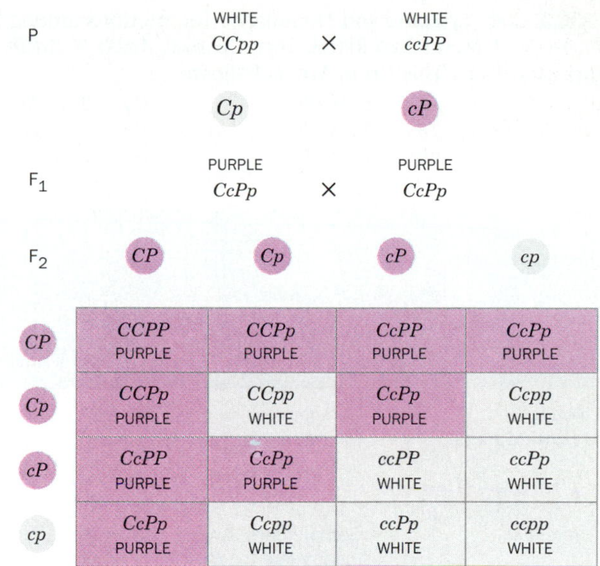

Figure 2.16 Cross between the two white-flowered varieties of sweet peas from which an F_2 ratio of 9 : 7 was obtained.

or both of the genes (3/16 *ccP-* + 3/16 *C-pp* + 1/16 *ccpp*) and have white flowers (Fig. 2.16).

Other modified dihybrid ratios are known to result from different types of epistasis. For example, if two different (nonallelic) genes code for enzymes that are able to catalyze the same reaction, such as by the following duplicate pathways

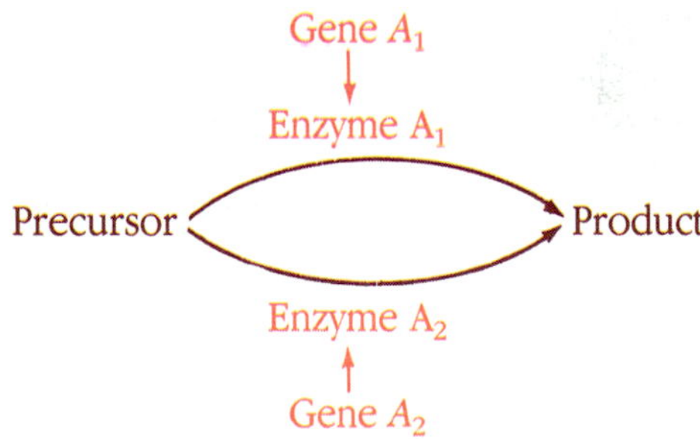

one can easily see how a 15:1 F_2 dihybrid phenotypic segregation ratio might result. If the recessive alleles, a_1 and a_2, of these two genes produce inactive forms of the two enzymes, a cross of $a_1a_1A_2A_2 \times A_1A_1a_2a_2$ would yield a 15 $A_{1 \text{ or } 2}$---: 1 $a_1a_1a_2a_2$ F_2 phenotypic segregation ratio. Such 15:1 ratios have been observed in crosses involving genes controlling kernel color in wheat and seed capsule shape in shepherd's purse.

In other cases of epistasis, the product of one gene may inhibit or otherwise prevent the expression of the product of another gene. If the inhibitor is the product of a dominant allele, a 12:3:1 F_2 dihybrid ratio is produced (Fig. 2.17). If it is the product of a recessive

TABLE 2.4 **Epistatic and Dominance Interactions among Nonalleles and Alleles Resulting in Modifications of the 9 : 3 : 3 : 1 Mendelian Ratios from Crosses, *AaBb* × *AaBb*, Each Pair Assorting Independently. Genotypes Expected from This Cross Are as Follows:**

AA < *BB* → 1 *AA BB*; *2Bb* → 2 *AA Bb*; *bb* → 1 *AA bb*	2*Aa* < *BB* → 2 *Aa BB*; *2Bb* → 4 *Aa Bb*; *bb* → 2 *Aa bb*	*aa* < *BB* → 1 *aa BB*; *2Bb* → 2 *aa Bb*; *bb* → 1 *aa bb*	or for phenotypic classes	9 *A–B–* 3 *A–bb* 3 *aaB–* 1 *aabb*

INTERACTIONS: EPISTASIS	INTERACTIONS: DOMINANCE	GENOTYPES: *AABB*, *AABb*, *AaBB*, *AaBb*	*AAbb*, *Aabb*	*aaBB*, *aaBb*	*aabb*	EXAMPLE	PHENOTYPES
		PHENOTYPIC CLASSES					
None (classical ratio)	Complete for *A*,*B*	9	3	3	1	Garden peas, seed color and surface	9 yellow round, 3 yellow wrinkled, 3 green round, 1 green wrinkled
None	Complete for *A*, incomplete for *B*	× − × − − × − − −	3	○ ● ●	1	Guinea pigs, hair length and color	3 short yellow (×), 6 short cream (−), 3 short white, 1 long yellow (○), 2 long cream (●), 1 long white
aa epistatic to *B*,*b*	Complete for *A*,*B*	9	3	4		Mice, coat pattern and color	9 agouti, 3 colored, 4 white
A epistatic to *B*,*b*	Complete for *A*,*B*	12		3	1	Summer squash, fruit color	12 white, 3 yellow, 1 green
A epistatic to *B*,*b*; *bb* epistatic to *A*,*a*	Complete for *A*,*B*	13		3	(1)	Chickens, color	13 white, 3 colored
aa epistatic to *B*,*b*; *bb* epistatic to *A*,*a*	Complete for *A*,*B*	9	7			Yellow daisy, color	9 purple, 7 yellow center
A epistatic to *B*,*b*; *B* epistatic to *A*,*a*	Complete for *A*,*B*	15			1	Shepherd's purse, seed capsules	15 triangular, 1 ovoid
aa epistatic to *B*,*b*; *bb* epistatic to *A*,*a*; *bb* epistatic to *aa*	Complete for *A*, incomplete for *B*	× − × − − × − − −	4	3	(1)	Flour beetle, color	6 sooty (−), 3 red (×), 3 jet, 4 black
		AABB, *AABb*, *AaBB*, *AaBb*	*AAbb*, *Aabb*	*aaBB*, *aaBb*	*aabb*		

allele, a 9:3:4 F_2 dihybrid ratio will result. Still other kinds of interactions between gene products can produce other modified ratios (Table 2.4).

It should be emphasized that ***genes do not act in isolation***. The final phenotype of an organism is the result of the action of, and the interaction among, a large number of genes. In addition, all of these genes are influenced to a greater or lesser extent by environmental factors. The interaction of many genes and their interaction with the environment in controlling many of the most important components of phenotype are discussed further in Chapter 21.

GENETIC VERSUS ENVIRONMENTAL EFFECTS

The pathway by which a gene produces its effect(s) on the phenotype of an organism is often very complex. (Some of the important steps in this pathway will be described in Chapter 10; for a schematic diagram of the components of such a pathway, see Fig. 10.1.) Given the complexity of pathways of gene expression, it should not be surprising that the phenotypic effects of some genes are variable. Two terms, ***expressivity*** and ***penetrance***, are frequently used in discussing variability of the effects of specific genes on phenotype. When

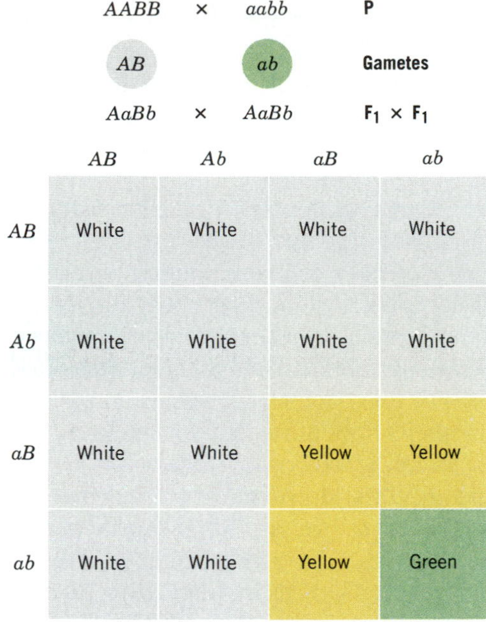

Figure 2.17 Examples of epistasis in fruit color of summer squash. Gene *A* is epistatic to *B* and *b*. Please compare Fig. 2.10 for dominance, where *G* is dominant over allele *g* and *W* is dominant over allele *w*.

the degree of phenotypic expression of a gene varies from individual to individual, the gene is said to have ***variable expressivity***. When the variability in expressivity of a given gene is such that the presence of the gene does not always result in a detectable phenotypic effect, the gene is said to have ***incomplete penetrance***. For example, if a dominant gene is expressed (has a detectable phenotypic effect) in only 70 percent of the individuals known to carry it, its penetrance is said to be 70 percent. Clearly, incomplete penetrance is merely an extension of variable expressivity of genes (where the effects on phenotype are so small that they ***cannot*** be detected). A common type of variable expressivity is the variable "age of onset" observed with certain inherited diseases of humans such as adult forms of diabetes mellitus. It should be apparent that genetic analyses involving genes with incomplete penetrance and/or variable expressivity are much more difficult to interpret than those involving genes with fairly invariant phenotypic effects.

Another potentially confounding factor in genetic analyses, particularly in diagnosing inherited abnormalities in humans, is that environmental factors sometimes induce ***nonhereditary phenotypic changes*** called ***phenocopies*** that are indistinguishable or closely resemble conditions resulting from mutant genes. Epilepsy, for example, may result from either genetic or environmental causes. As a result, the definitive diagnosis of an inherited abnormality requires family pedigree data or other information indicating that the condition is inherited, in addition to the appropriate phenotypic manifestations in an individual.

MULTIPLE ALLELES

An allele is a specific form of a given gene. An allele may be a mutant allele, resulting in an altered phenotype, or a wild-type allele producing an active gene product and a "normal" phenotype. In fact, different wild-type alleles frequently occur in populations. These alleles will all produce phenotypes within the "wild-type" range; however, when examined carefully, they may be found to produce gene products with different levels of activity. Similarly, mutant alleles of a gene may exhibit a wide range of gene-product activity from no activity to almost the wild-type level of activity. Thus, these alleles may produce a whole series of variable phenotypes ranging from an extreme mutant phenotype to a nearly wild-type phenotype. When more than two different forms of a given gene exist in a species, they are referred to as ***multiple alleles***. When the mechanisms by which genes change and the molecular structure of genes are discussed in Chapters 11 and 12, respectively, it will become apparent that multiple alleles probably exist for most, if not all, genes.

A classic example of multiple alleles involves coat color in rabbits. Four alleles of the rabbit coat color (c) gene have been studied: c^+ (wild-type or full color), c^h ("himalayan," white with black tips on the extremities), c^{ch} ("chinchilla," mixed colored and white hairs), and c (albino). These alleles show a gradation in dominance of $c^+ > c^{ch} > c^h > c$. That is, c^+ is dominant to each of the three mutant alleles, while c^h is recessive to c^{ch}, but dominant to c, and so on. Note that the members of the multiple allelic series are conventionally represented by the same letter or symbol (c in the preceding example) with appropriate superscripts to represent the different alleles.

Another example of multiple alleles occurs at a gene controlling eye color in the fruit fly *Drosophila melanogaster*. Wild-type *Drosophila* have red eyes; but a vast array of eye-color mutants have been studied extensively for several decades. Mutant alleles of one gene (*white*, symbolized w) result in flies with eye colors ranging from pure white through a series of intermediate colors up to nearly the wild-type red when present in the homozygous condition. The different alleles of the *white* gene have been given names that reflect the phenotypes that they produce in homozygous flies. For example, the allele that produces the most extreme mutant phenotype was named *white* because the eyes of flies homozygous for this allele are completely white (no coloration). Other alleles of this gene were designated w^a (white apricot), w^e (white eosin), w^{ch} (white cherry), w^{co} (white coral), w^{col} (white colored), w^w (white wine), w^{bl} (white blood), w^{crr} (white carrot), w^{cf} (white coffee), and so on, to indicate that different levels of coloration of the eyes occurred in flies homozygous for each particular allele. As was mentioned earlier, the different alleles of a gene may produce gene-products with different levels of activity. For example, if the quantity of eye pigment present in mutants homozygous for different alleles of the white locus is measured spectrophotometrically, it is found to vary from a very low level in w/w homozygotes to about 25 percent of the "wild-type" level in w^{col}/w^{col} homozygotes (Table 2.5). Actually, flies homozygous for different wild-type alleles vary almost twofold in the amount of eye pigment present (Table 2.5). Thus, the proportion of "wild-type level" of pigment present in various eye mutants depends on which wild-type one selects for comparison. For many genes, alleles probably exist with gene-product activity levels ranging from the complete ***amorph or "null" mutation*** (***no*** activity, as will always be the case for a deletion of the gene) to the wild-type allele producing the optimal level of gene-product activity. In some cases, alleles of a gene produce mutant phenotypes due to the overproduction of the gene-product.

The number of different genotypes possible among diploid organisms is, of course, a function of the number of alleles that exist for any given gene. ***If*** n ***is the number of alleles of a gene, the number of different genotypes possible is*** $n(n+1)/2$. Thus, with 2, 3, 4, or 5 alleles, there are 3, 6, 10, and 15 possible genotypes, respectively. Note that although a large number of different alleles of a given gene may be present in a population or a species, ***only two of those alleles can be present*** (on the two homologous chromosomes carrying that particular gene) ***in any one diploid organism*** (assuming that it is a normal diploid). Note also that although the existence of multiple alleles of a gene results in more different types of crosses than can be done within the population, it does ***not*** change the genetic ratios expected in any given cross between diploid individuals. Since each diploid organism has only two copies of the gene, an individual must be either (1) homozygous for allele a_1, (2) homozygous for allele a_2, or (3) heterozygous for alleles a_1 and a_2. This individual organism can ***not*** harbor allele a_3 or a_4 or any other allele of the multiple allelic series. These other alleles will be present in different individuals in the population and, therefore, will be involved in distinct crosses that will also yield standard Mendelian ratios.

The case of crosses involving multiple alleles should be recognized by the standard monohybrid ratios (i.e., $\frac{1}{4}\, a_1a_1 : \frac{2}{4}\, a_1a_2 : \frac{1}{4}\, a_2a_2$ or $\frac{1}{4}\, a_1a_1 : \frac{2}{4}\, a_1a_3 : \frac{1}{4}\, a_3a_3$, etc.) that result in all cases where a single pair of the alleles are segregating. The ratios observed are standard Mendelian ratios, in distinct contrast to the case of epistasis, which will lead to modified Mendelian ratios (dihybrid, trihybrid, etc., depending on the number of genes segregating in the cross) as discussed in the preceding section.

TABLE 2.5 **Relative Amounts of Eye Pigment Present in Flies Homozygous for Different Alleles of the *white* Locus of *Drosophila melanogaster***

GENOTYPE	RELATIVE CONCENTRATION OF TOTAL PIGMENT[a]
w/w (white)	0.0044
w^a/w^a (white apricot)	0.0197
w^e/w^e (white eosin)	0.0324
w^{ch}/w^{ch} (white cherry)	0.0410
w^{co}/w^{co} (white coral)	0.0798
w^{sat}/w^{sat} (white satsuma)	0.1404
w^{col}/w^{col} (white colored)	0.1636
w^{+s}/w^{+s} (wild-type, Stellenbusch strain)	0.6854
w^{+c}/w^{+c} (wild-type, Canton S strain)	0.9895
w^{+G}/w^{+G} (wild-type, Graff-Reinet strain)	1.2548

[a]Data from D. J. Nolte. *Heredity* 13:219, 1959. For map positions of mutant alleles, see Fig. 12.22.

ABO Blood Type Alleles in Humans

One of the most firmly established series of multiple alleles in humans involves the genetic locus controlling the ***blood types: A, B, AB, and O***. The ABO locus has ***three common alleles:*** I^A, I^B, ***and*** I^O (Table 2.6). I^A and I^B are codominant (I^AI^B heterozygotes have both A and B antigens on their red blood cells) and I^O is recessive (I^OI^O homozygotes have no ABO antigens on their red blood cells; I^AI^O and I^BI^O heterozygotes have A and B antigens, respectively, on their red blood cells).

TABLE 2.6 **Genotypes and the Corresponding Phenotypes (Blood Group Types) for the ABO Locus in Humans**

GENOTYPE	PHENOTYPE
I^AI^A or I^AI^O	A
I^BI^B or I^BI^O	B
I^AI^B	AB
I^OI^O	O

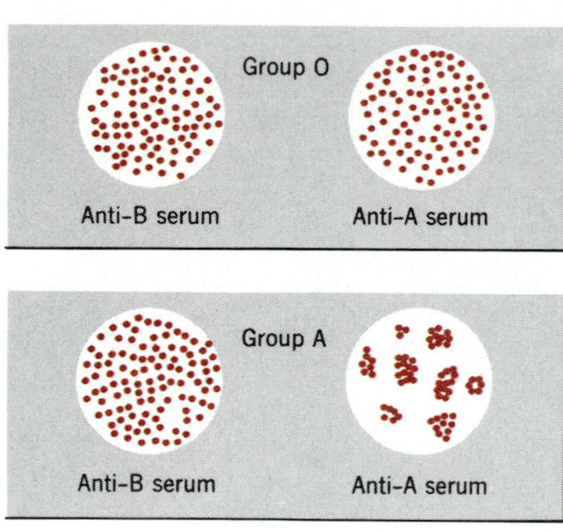

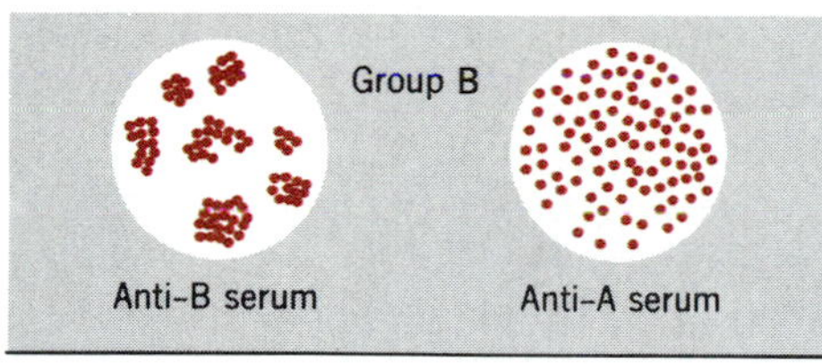

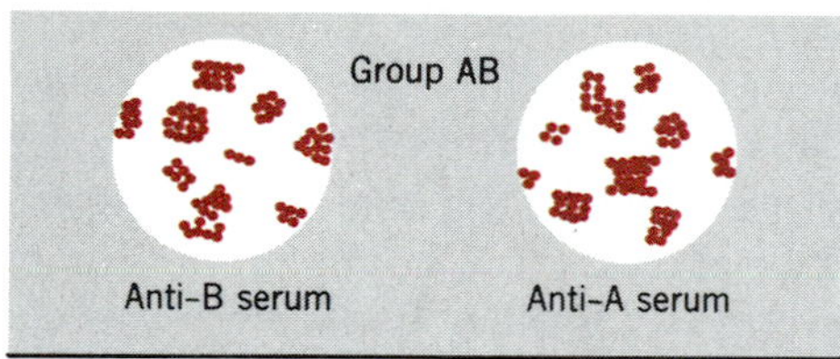

Figure 2.18 Agglutination reactions controlled by the ABO blood-type locus in humans. Red blood cells (erythrocytes) of the type indicated at the top of each slide are mixed with blood serum of the type indicated below each reaction mixture (circle). A clumped pattern of cells within a circle indicates that agglutination occurs.

The ABO locus controls the type of glycolipids found on the surface of erythrocytes, apparently by specifying the type of glycosyl-transferases (enzymes catalyzing the synthesis of polysaccharides) synthesized in the red blood cells. The specific types of glycolipids on the red cell surface provide the antigenic determinants that react with specific antibodies present in the blood serum. Humans, like all other mammals, produce antibodies and circulate them in the blood serum as a defense mechanism against foreign substances (Chapter 16). Fortunately, no antibodies are synthesized (in normal individuals) that react with antigens present on the individual's own cells. However, when type A blood and type B blood are mixed, the anti-A antibodies in the type B blood serum react with the antigens on the type A blood cells, and vice versa, which produces agglutination or clumping of cells (Fig. 2.18). Cross-matching blood types to determine compatibility is thus essential in blood transfusions. In this process, blood donors and recipients are tested for the presence of antigens and antibodies that are incompatible.

Table 2.7 summarizes the cell surface antigenic determinants and the serum antibodies present in the four major ABO blood types. Individuals with blood type AB have both A and B antigens on their erythrocytes, but no anti-A and anti-B antibodies in their blood serum. Type O individuals lack both antigens, but carry both anti-A and anti-B antibodies in their blood serum. ***Type O individuals are*** referred to as ***universal donors***; type O blood can be used in transfusion for individuals of any blood type if the blood is introduced slowly enough to permit sufficient dilution of the anti-A and anti-B antibodies present in the serum of the donor.

Rh Factor Alleles in Humans

The ***Rh factor*** was discovered in 1940 by K. Landsteiner and A. S. Wiener from rabbits immunized with the blood of the monkey *Macaca rhesus*. The resulting antibodies were found to agglutinate not only the red corpuscles of the monkey but those of a high percentage of the Caucasian people of New York. Individuals whose blood cells react with the Rh antibody are termed Rh-positive; those who do not react are Rh-negative. The symbol "Rh" came from the first two letters of the species name of the monkey. A test for Rh incompatibility is accomplished by placing a drop of blood from the subject on a slide and introducing anti-Rh serum. Agglutination of erythrocytes indicates incompatibility, whereas an even distribution of erythrocytes indicates no reaction.

TABLE 2.7 Blood Transfusion Compatibilities for the ABO Blood Groups

BLOOD GROUP	TERMINAL SUGARS OF ANTIGENS PRESENT	ANTIBODIES PRESENT	RED CELL TYPES AGGLUTINATED	TRANSFUSIONS ACCEPTED FROM
A	A (galactosamine)	Anti-B	B, AB	A or O
B	B (galactose)	Anti-A	A, AB	B or O
AB	A (galactosamine) and B (galactose)	None	None	A, B, AB, or O
O	None	Anti-A and anti-B	A, B, and AB	O

The original antigen, now symbolized Rh°, is highly antigenic to humans. Thus, cross-matching of the Rh factor, as well as of the ABO types of donor and recipient blood, is now used to avoid incompatibility agglutination reactions following transfusions. Blood is frequently exchanged between the mother and the fetus during childbirth. Thus, ***Rh-negative mothers are often immunized by blood from Rh-positive fetuses*** (which may result when the fathers are Rh-positive) ***to which they give birth.*** (Rh-positive is dominant; Rh-negative is recessive.) Usually no ill effects are associated with exposure of the mother to the Rh-positive antigen during the first childbirth (unless the mother has been previously exposed to Rh antigen by transfusion). Subsequent Rh-positive children carried by the same mother, however, may be exposed to antibodies produced by the mother against the Rh antigen, which are carried across the placenta in blood serum. Such children may develop symptoms of hemolytic jaundice and anemia, a condition referred to as ***erythroblastosis fetalis.*** The symptoms may be mild or severe, even resulting in the death of the fetus or newborn infant if appropriate steps are not taken by the physician.

Passive immunity for Rh-hemolytic anemia can now be accomplished by use of an incomplete antibody against the Rh° antigen. This antibody does not agglutinate Rh-positive red blood corpuscles. Instead, the antibodies attach to antigen receptors on red cell surfaces and coat the cells. These incomplete antibodies may be injected into an Rh-negative mother immediately after she has given birth to an Rh-positive child. The coating of any Rh antigens from the Rh-positive fetus with incomplete antibodies inhibits the mother's capacity to form Rh antibodies. The injection of incomplete antibodies thus prevents the fetal antigens from eliciting the normal immune response of the mother. Injected antibodies dissipate within a few months and present no danger to the mother or fetuses in subsequent pregnancies.

At first, the genetic control of the Rh system seemed simple. A simple pair of alleles, *R* and *r*, was postulated to account for the difference between Rh-positive and Rh-negative individuals. New antibodies were soon discovered, however, and additional genes were postulated to explain the more complicated situation. Wiener developed a hypothesis based on a series of multiple alleles (Table 2.8). Eight alleles were initially included in the series and more have since been added. Considerable evidence has been presented in support of this hypothesis. On the other hand, R. R. Race, R. A. Fisher, and other investigators explained the same data on the basis of three genes (*C, D,* and *E*) that are located very close together on the same chromosome. Such closely associated genes would be expected to act like alleles in most situations that can be analyzed in human populations. Note that based on the hypotheses presented in Table 2.8, all genotypes carrying either a dominant *R* allele (either R^1, R^2, R^0, or R^z, Wiener hypothesis) or a dominant *D* allele (Fisher and Race hypothesis) will produce an Rh-positive phenotype, and double recessive genotypes *rr* (or *rr'*, *rr"*, rr^y, or any other combination of two copies of the alleles *r, r', r"*, r^y, Wiener hypothesis) or *dd* (Fisher and Race hypothesis) will give rise to an Rh-negative phenotype. At this point, it is still not clear which hypothesis, if either, is correct. However, based on molecular studies of other genes (Chapters 12 and 13), it seems likely that multiple alleles exist at the locus or loci encoding the Rh antigen(s) whether a single gene or several genes are involved.

PROBABILITY IN MENDELIAN INHERITANCE

Probability refers to the chance or likelihood that a specific outcome or form of an event will occur in a particular situation (relative to all possible outcomes or forms of the event). Probability is calculated as the ***ratio*** of (1) the ***number of different ways in which***

TABLE 2.8 Hypotheses for the Genetic Control of the Rh Factors in Human Blood

RH BLOOD TYPE	WIENER HYPOTHESIS (ONE GENE) ALLELES PRESENT	FISHER AND RACE HYPOTHESIS (THREE GENES) ALLELES PRESENT	APPROXIMATE FREQUENCY OF TYPE AMONG CAUCASIANS
Rh_1	R^1	*CDe*	41%
Rh_2	R^2	*cDE*	14%
Rh_0	R^0	*cDe*	3%
Rh_z	R^z	*CDE*	Rare
rh	*r*	*cde*	39%
rh′	*r′*	*Cde*	1%
rh″	*r″*	*cdE*	1%
rh_y	r^y	*CdE*	Very rare

that specific outcome or form of the event can occur to (2) the ***total number of ways in which all possible outcomes or forms of the event can occur.*** The limits of probability are from 0 if an event never occurs to 1 if it always occurs. Mendel had studied probability in his mathematics courses and recognized the 3:1 ratio as a particular mathematical relation that suggested a model for the mechanism of ***segregation.*** If two alleles, one dominant and one recessive, could segregate freely many times in succession, or if several similar pairs were behaving in this way at one time, the expected summation would be about three of the dominant expressions to one recessive. The analysis, therefore, was based on a mathematical relation with which Mendel was familiar, but its application and the concept of segregation were Mendel's own contributions.

The laws of probability apply to genetic mechanisms as well as to other processes in which uncertainty exists. In the F_2 results of Mendel's cross between tall and dwarf garden peas, one-fourth were dwarf, one-fourth homozygous tall, and one-half heterozygous tall. A similar result might be obtained from a simple experiment in tossing coins. A normal coin that is tossed freely is equally likely to fall heads or tails. If one coin is tossed 100 times, it would be expected to fall heads about 50 times and tails about 50 times. When two coins are tossed together, each behaves independently and falls either heads or tails. From 100 trials, about 25 heads–heads, 50 heads–tails in either combination (actually 25 heads–tails and 25 tails–heads), and 25 tails–tails would be expected. Thus, a ratio of 1:2:1 would be expected if one grouped the trials yielding 1 head and 1 tail (either coin heads, either tails), whereas a ratio of 1:1:1:1 would be expected if one classified coin 1 heads–coin 2 tails and coin 1 tails–coin 2 heads separately. These results, which parallel Mendelian segregation, are merely the chance occurrences of sets of independent events. The experimenter would usually not obtain exactly 25 heads–heads, 50 heads–tails (any combination), and 25 tails–tails. It would be surprising if precisely those results were obtained very often. The ratio represents only an average of expected results when independent events occur.

How frequently can the various combinations be expected to occur in succession? The law of probability applied for the solution of this problem is stated as follows: ***If two or more events are independent, the chance that they will occur together is the product of their separate probabilities.*** When a single coin is tossed repeatedly, the chance of heads occurring twice in succession would be $\frac{1}{2} \times \frac{1}{2} = \frac{1}{4}$. The chance of three such occurrences would be $(\frac{1}{2})^3$ or $\frac{1}{8}$, and of four, $(\frac{1}{2})^4$ or $\frac{1}{16}$. When two coins are tossed together, the tails–tails combination is expected in one-fourth of the trials ($\frac{1}{2} \times \frac{1}{2} = \frac{1}{4}$). The chance of occurrence of two such tails–tails combinations for two coins in succession would be $(\frac{1}{4})^2$ or $\frac{1}{16}$.

Fitting Results from Crosses to Hypotheses

The ***"goodness of fit"*** of the numerical result obtained from an actual cross or other experiment, relative to predicted results based on a particular mechanism and a perfect genetic segregation, is a key concept in experimental genetics. The geneticist must know how much the experimental result can differ from the hypothetical or calculated figure and still be regarded as statistically close to expectation. In evaluating the results of crosses and determining which modes of inheritance are involved, how much deviation is permissible without casting some doubt as to whether the data agree with a given hypothesis? Too much deviation would surely make investigators question their hypotheses or discard them entirely. Where should the line be drawn? Unfortunately, there is no precise answer to this question. The best the geneticist can do is to determine the likelihood of the deviation of the results observed in the experiment from the predicted results occurring by chance and use statistical inference to decide whether a particular result supports a given hypothesis. These numerical data are the only means of evaluating goodness of fit of an experimental result as compared with a particular expectation.

Chi-Square

The chi-square (χ^2) test is a valuable tool that aids the investigator in determining goodness of fit. The test takes into account the size of the ***sample*** and the deviations from the expected ratio. It not only can be used for samples of different sizes, but can be adapted to ratios with different numbers of classes, such as those of monohybrid crosses with two classes and those of dihybrid crosses with four classes. Essentially, the chi-square test is a mechanism by which deviations from a hypothetical ratio are reduced to a single value based on the size of the sample. This allows the investigator to determine the probability that a given sum of deviations will occur by chance. Expected values are obtained from the total size of the sample. If the hypothesis is that a 1:1 ratio results from a cross, the total is divided into two equal parts. For any other expected ratio, the total is divided into appropriate proportions.

A formula for χ^2, designed for a sample consisting of two classes (i.e., 1:1 or 3:1 ratios), is symbolized as follows:

$$\chi^2 = \frac{(O_1 - E_1)^2}{E_1} + \frac{(O_2 - E_2)^2}{E_2}$$

where O_1 is the experimentally observed number for the first class, and E_1 is the expected number for the same class derived from the ratio; O_2 is the observed for the second class and E_2 the expected. When each of these deviations is squared $[(O - E)^2]$ and divided by the expected value (E) for that class, the resulting fractions can be added (Σ) to give a single χ^2 value. The formula can be symbolized as follows:

$$\chi^2 = \Sigma \frac{(O - E)^2}{E}$$

where $(O - E)$ is the deviation between each observed and expected class value, E the expected value in the respective class, and the Greek letter Σ the summation sign. If the deviations of expected from observed events are small, χ^2 approaches 0 and the fit is good; if the deviations are large, χ^2 is increased and the fit is poor. As an example, calculate χ^2 for the two arbitrary samples 15:35 and 240:260 on the basis of a 1:1 hypothesis. This example will illustrate how the χ^2 relates the size of the deviation to the size of the sample. For the 15:35 result, with a total of 50, the expected (E) value for each class is $(15 + 35) \div 2$, or 25. The deviations $(O - E)$ on either side of E are 10, that is, $25 - 15 = 10$ and $25 - 35 = -10$. For the larger sample, 240:260, the expected (E) for each class is 250. The deviations $(O - E)$ are $250 - 240 = 10$ and $250 - 260 = -10$.

1.
$$\chi^2 = \Sigma \frac{(O - E)^2}{E} = \frac{10^2}{25} + \frac{(-10)^2}{25} = \frac{200}{25} = 8$$

2.
$$\chi^2 = \Sigma \frac{(O - E)^2}{E} = \frac{10^2}{250} + \frac{(-10)^2}{250} = \frac{200}{250} = 0.8$$

The χ^2 value of 8 for the smaller sample is considerably larger than that of .8 for the larger sample, even though the actual deviations in the two examples are the same. Thus, the χ^2 value is related to sample size as well as to the variability within the sample.

When more than two groups are classified from the sample (e.g., 1:2:1 or 9:3:3:1 ratios), each class is included in the summation, which is χ^2. It should be emphasized that the χ^2 formula is based on actual numerical frequencies and totals, not on percentages that always total 100. When data are reduced to percentages, the factor of sample size is eliminated.

The next step is to interpret the χ^2 value in terms of probability. In any experimental procedure dealing with quantitative data, some variation, called "experimental error," can be attributed to chance alone. It is important to determine whether observed deviations from a hypothesis are significantly different from the experimental error. For interpreting χ^2 values, the number of classes on which a χ^2 is based must be considered. The value for a two-class distribution includes only two squared deviations, whereas that for a distribution with more classes has more squared deviations. It is therefore necessary to consider the number of classes that contribute to a given χ^2 in evaluating the "goodness of fit." The effect of the number of independent classes is included in the mathematical concept as ***degrees of freedom***. As an analogy, a person has two socks, for his or her two feet, but when putting on the socks there is only one degree of freedom. If the individual puts one sock on one foot, the other sock must go on the other foot. When the total number of objects or classes is fixed, and all except one have been placed, the one remaining is not free but must fill a particular vacant niche. In general, therefore, the number of degrees of freedom is one less than the number of classes. Two-class ratios (e.g., 1:1 or 3:1) have one degree of freedom; three-class ratios (1:2:1) have two degrees of freedom; four-class ratios (9:3:3:1) have three degrees of freedom; and so on.

When χ^2 and the degrees of freedom have been determined, Table 2.9 may be consulted for the probability (P) value. Locate the figure representing the number of degrees of freedom at the left, read across horizontally, and find the figures nearest the χ^2 value in the body of the table; then read the P values directly above on the top line. The χ^2 of 8, calculated for the first example, is not on the table. The highest value on the line for one degree of freedom is 6.635, which has a P of 0.01. This indicates that the probability for obtaining, by chance, deviations as great as or greater than those of $\chi^2 = 8$, if the hypothesis is true, would be less than 1 percent. Experimental error is expected, but the difference here is much greater than that expected by chance. Therefore, the fit of these results to a 1:1 ratio is not good. Another hypothesis (other than the 1:1 ratio) might be considered for these data.

In the second example, $\chi^2 = 0.8$ falls between 0.455 and 1.642, or between the P values of 0.50 and 0.20. The probability of obtaining a deviation as great as, or greater than, $\chi^2 = 0.8$ is between 0.20 and 0.50. This probability value indicates that, if numerous independent repetitions of an ideal experiment involving two independent events were conducted, chance de-

TABLE 2.9 Table of Chi-Square (χ^2)[a]

DEGREES OF FREEDOM	P = 0.99	0.95	0.80	0.50	0.20	0.05	0.01
1	0.000157	0.00393	0.0642	0.455	1.642	3.841	6.635
2	0.020	0.103	0.446	1.386	3.219	5.991	9.210
3	0.115	0.352	1.005	2.366	4.642	7.815	11.345
4	0.297	0.711	1.649	3.357	5.989	9.488	13.277
5	0.554	1.145	2.343	4.351	7.289	11.070	15.086
6	0.872	1.635	3.070	5.348	8.558	12.592	16.812
7	1.239	2.167	3.822	6.346	9.803	14.067	18.475
8	1.646	2.733	4.594	7.344	11.030	15.507	20.090
9	2.088	3.325	5.380	8.343	12.242	16.919	21.666
10	2.558	3.940	6.179	9.342	13.442	18.307	23.209
15	5.229	7.261	10.307	14.339	19.311	24.996	30.578
20	8.260	10.851	14.578	19.337	25.038	31.410	37.566
25	11.524	14.611	18.940	24.337	30.675	37.652	44.314
30	14.953	18.493	23.364	29.336	36.250	43.773	50.892

[a]Selected data from R. A. Fisher and F. Yates, *Statistical Tables for Biological, Agricultural and Medical Research.* Oliver & Boyd, London, 1943.

viations as large as, or larger than, those observed here (± 10 corresponding to $\chi^2 = 0.8$) would be expected to occur in 20 to 50 percent of the trials. Such a deviation could be explained readily by chance. The data fit the 1:1 ratio hypothesis very well.

A hypothesis is never proved or disproved by a *P* value. The results of an experiment are evaluated by the investigator as acceptable or unacceptable with respect to the hypothesis. The 5 percent point (0.05) on the table is usually chosen as an arbitrary standard for determining the significance or goodness of fit. Probability at this point is 1 in 20 that a true hypothesis will be rejected. Sometimes the 1 percent point (0.01) is used as a level of significance. At this level, there is a smaller probability (0.01) that a true hypothesis will be rejected, but a correspondingly greater chance that a false hypothesis will be accepted.

It should be emphasized that these are arbitrary points, and judgment is required in making interpretations. In any event, the *P* value represents the probability that a deviation as great as, or greater than, that obtained from the experiment will occur by chance alone. If the *P* is small, it is concluded that the deviations are not due entirely to chance, and the hypothesis is rejected. If the *P* is greater than the predetermined level (e.g., 0.05), the data conform reasonably well with the predictions of the hypothesis, and the hypothesis is accepted.

Independent Assortment and Probability

Probability must be considered in explaining the Mendelian principle of independent combinations or independent assortment as well as that of segregation. It was through Mendel's understanding of the mathematical laws of combinations that he was able to recognize and interpret the dihybrid ratio of 9:3:3:1 as a multiple of the 3:1 monohybrid ratio. If, for example, the 3:1 ratio is changed to an algebraic expression, 3+1, the product of the expected results of two monohybrid crosses is $(3 + 1)^2 = 9+3+3+1$. Because the terms represent separate classes, they are not grouped together, but the product can be converted back to a ratio: 9:3:3:1. This is an example of the ***binomial expansion*** of $(p + q)^n$, in this case $(A + a)^2$, where $A = 3$ and $a = 1$ or as phenotypes 3*A*- + 1*aa*.

Using the F_2 results of the cross between peas with round, yellow seeds and those with wrinkled, green seeds, Mendel tested the mathematical relation between the monohybrid and dihybrid cross. He observed that about 3/4 (423/556) of the F_2 seeds were round and 1/4 (133/556) were wrinkled. Similarly, seeds from about 3/4 (416/556) were yellow, and those from 1/4 (140/556) were green. These observed proportions provided a cross-check for the hypothesis of ***independence***. When the two characters were considered together, the results conformed to the mathematical model expected for two independent events occurring together.

On the basis of the law of probability, Mendel predicted that 9/16 (3/4 × 3/4) of the F_2 would be round, yellow; 3/16 (3/4 × 1/4) round, green; 3/16 (1/4 × 3/4) wrinkled, yellow; and 1/16 (1/4 × 1/4) wrinkled, green. The results that Mendel actually obtained (315:108:101:32) resembled very closely the calculated ratio of 9:3:3:1, based on the hypothesis of complete independence of the genes influencing the shape and color of the seeds. When the χ^2 test for goodness of fit between the actual and the expected result is applied to these figures, the probability (*P*) of finding deviations as great as, or greater than, those obtained by Mendel is between 0.80 and 0.95.

$$\chi^2 = \Sigma \frac{(O-E)^2}{E} = \frac{(2.25)^2}{312.75} + \frac{(3.75)^2}{104.25} + \frac{(-3.25)^2}{104.25} + \frac{(-2.75)^2}{34.75}$$
$$= 0.016 + 0.135 + 0.101 + 0.218$$
$$= 0.470$$
$$P = 0.80\text{–}0.95$$

The actual results fit very closely with those expected on the basis of the hypothesis of independent combinations. In fact, it is so close that the probability is less than 0.20 that further data would fit the hypothesis this well. Investigators are as concerned about results that are very close to a calculated expectation as about those with large deviations. A very close fit may be obtained occasionally by chance; but when the observed data fit "too well," the investigator may have intentionally or unintentionally biased the results.

Mendel presented some experimental results that had poor agreement with expectation. In one experiment from which he expected a 3:1 ratio, he obtained 43 round and only 2 wrinkled seeds. From another experiment with the same expectation, he obtained 32 yellow and 1 green seeds. He explained these results on the basis of fluctuations due to chance, and in the second case there was also some difficulty in distinguishing between yellow and green seeds. This would add a factor of possible classification errors in addition to chance. Fisher has analyzed Mendel's results as a whole and has shown the chance of getting such good agreement to be only 0.0007. Thus, the examples given as 43:2 and 32:1 are atypical.

From his results, Mendel was able to predict the numbers of genotypes to be expected when more than two pairs of alleles were involved in the cross.

Expansion of a Binomial

Various combinations in groups of a given size representing a particular ratio can be calculated by the binomial expansion of $(p+q)^n$, where p and q represent the probabilities of occurrence of two alternative events (e.g., p = the probability of a boy and q = the probability of a girl) and n is the size of the group involved. How many boys and how many girls would be expected in randomly selected families of 2, 3, 4, 5, or more? Combinations for families of a given size may be calculated by the binomial expansion of $(p+q)^n$. Thus, for two-child families,

$$(p+q)^2 = p^2 + 2pq + q^2$$
$$p = q = ½$$
$$p^2 = \text{families of two boys} = ¼$$
$$2pq = \text{families of 1 boy and 1 girl} = ½$$
$$q^2 = \text{families of 2 girls} = ¼$$

Among families of 2 children, ¼ would be expected to be composed of all boys, ½ of one boy and one girl, and ¼ of all girls. For three-child families,

$$(p+q)^3 = p^3 + 3p^2q + 3pq^2 + q^3$$
$$p^3 = \text{families of 3 ♂} = (½)^3 = ⅛$$
$$3p^2q = \text{families of 2 ♂, 1 ♀} = 3(½)^2(½) = ⅜$$
$$3pq^2 = \text{families of 1 ♂, 2 ♀} = 3(½)(½)^2 = ⅜$$
$$q^3 = \text{families of 3 ♀} = (½)^3 = ⅛$$

(♂ is the male symbol, representing the shield and spear of Mars, the Roman war god; ♀ is the female symbol, representing the mirror of Venus, the Roman goddess of love.) The binomial expansion includes all possible combinations of the two events. In three-child families, for example, there are eight combinations.

If the chance of a particular birth order is included in the problem, the probability of each sequence is ⅛. If, on the other hand, only the total number of boys and girls is considered, the probability of all boys is ⅛, 2 boys and 1 girl ⅜, 1 boy and 2 girls ⅜, and all girls ⅛. The expected distributions of males and females in families of one to five children are summarized in Table 2.10.

The probability for each combination can be determined from the binomial coefficient for this combination, as compared with all possible combinations. In sibships of 5, for example, the following proportions of boys and girls would be expected: 1/32, all boys; 5/32, 4 boys and 1 girl; 10/32, 3 boys and 2 girls; 10/32, 2 boys and 3 girls; 5/32, 1 boy and 4 girls; and 1/32, all girls.

When the probability of only a certain combination in a given size group is required, factorials may be employed. These are products of factors derived from functions by successively increasing or decreasing by a constant, usually 1. For example, factorial 4 (4!) is the product of 4 × 3 × 2 × 1, or 24. [Factorial 0 (0!) = 1 and 1! = 1 by definition.] The probability for a particular combination may be calculated from a formula based on the appropriate term in the binomial expansion of

$$(p+q)^n : P = \frac{n!}{x!(n-x)!} p^x q^{(n-x)}$$

where $n!$ is the product of the integers making up the total size of the group (n = the total number in the

TABLE 2.10 Distribution of Boys and Girls in Families

NUMBER OF CHILDREN IN FAMILY	$(p+q)^n$	DISTRIBUTION, $p = ♂$, $q = ♀$
1	$(\frac{1}{2}+\frac{1}{2})^1$	½(1♂) + ½(1♀)
2	$(\frac{1}{2}+\frac{1}{2})^2$	¼(2♂) + ½(1♂:1♀) + ¼(2♀)
3	$(\frac{1}{2}+\frac{1}{2})^3$	⅛(3♂) + ⅜(2♂:1♀) + ⅜(1♂:2♀) + ⅛(3♀)
4	$(\frac{1}{2}+\frac{1}{2})^4$	1/16(4♂) + 4/16(3♂:1♀) + 6/16(2♂:2♀) + 4/16(1♂:3♀) + 1/16(4♀)
5	$(\frac{1}{2}+\frac{1}{2})^5$	1/32(5♂) + 5/32(4♂:1♀) + 10/32(3♂:2♀) + 10/32(2♂:3♀) + 5/32(1♂:4♀) + 1/32(5♀)

group), $x!$ is the product of the integers for the class with probability p, and $(n-x)!$ is the product of the integers for the class with the probability q. The symbol p represents the probability for one occurrence (e.g., a boy), and q is the probability for the other (e.g., a girl). If, for example, 6 babies are born in a given hospital on the same day, what is the probability that 2 will be boys and 4 will be girls? For this problem, assume that $p = q = \frac{1}{2}$. Substituting,

$$\begin{aligned} P &= \frac{n!}{x!(n-x)!} p^x q^{(n-x)} \\ &= \frac{6!}{2!4!}(1/2)^2(1/2)^4 \\ &= \frac{6 \times 5 \times 4 \times 3 \times 2 \times 1}{2 \times 1\,(4 \times 3 \times 2 \times 1)}(1/4)(1/16) \\ &= 15 \times 1/4 \times 1/16 \\ &= 15/64 \end{aligned}$$

The probability of 2 boys and 4 girls in groups of 6 is 15/64.

For examples of boys and girls in families of different sizes, the probability values p and q were equal ($p = q = \frac{1}{2}$). The binomial distribution can be applied for other values of p and q. If, for example, the trait being considered is albinism in a human family, and the parents are known to be heterozygous (*Cc*), the probability for a normally pigmented child (p) would be ¾ and the probability for an albino child (q) would be ¼. In families of 4, what is the probability that 2 will be normally pigmented and 2 will be albino? Substituting,

$$\begin{aligned} P &= \frac{n!}{x!(n-x)!} p^x q^{(n-x)} \\ &= \frac{4!}{2!(2)!}(3/4)^2(1/4)^2 \\ &= \frac{4 \times 3 \times 2 \times 1}{2 \times 1\,(2 \times 1)}(3/4)^2(1/4)^2 \\ &= 54/256 = 27/128. \end{aligned}$$

The probability of 2 normally pigmented and 2 albino children in families of 4 children from heterozygous (*Cc*) parents would be 27/128. Other values can be substituted into the binomial expansion for families of given sizes and the probabilities for various combinations can be calculated.

PROBABILITY IN PEDIGREE ANALYSIS

Practical applications of the Mendelian principles and the laws of probability are made by human geneticists and, in some instances, by animal breeders in analyzing pedigrees. Traits with a simple pattern of inheritance may sometimes be traced accurately enough to justify predictions concerning the likelihood of their expression in further children, if related individuals or those with a family history of such traits marry. The first step in such an analysis is to determine whether the trait in question is behaving as a dominant or a recessive. Although human traits are often affected by many genes, many are associated with the differential action of specific genes and have been identified as dominants or recessives in family groups. Over four thousand human traits and diseases have now been associated with specific genes (V. A. McKusick, *Mendelian Inheritance in Man: Catalogues of Chromosomal Dominant, Chromosomal Recessive, and X-Linked Phenotypes,* Johns Hopkins Univ. Press, 1986; C. DeLisi, "The human genome project," *American Scientist* 76: 488–493, 1988). A confusing feature of this type of analysis is that some phenotypes (for instance, deafness) may behave as dominant in some families and recessive in others. Obviously, several different gene substitutions can result in deafness.

Recessive genes are difficult to follow because they may remain hidden by their dominant alleles generation after generation. Carriers in the population usually cannot be identified until an affected child is born. Traits dependent on recessive genes sometimes appear unexpectedly in families having no visible his-

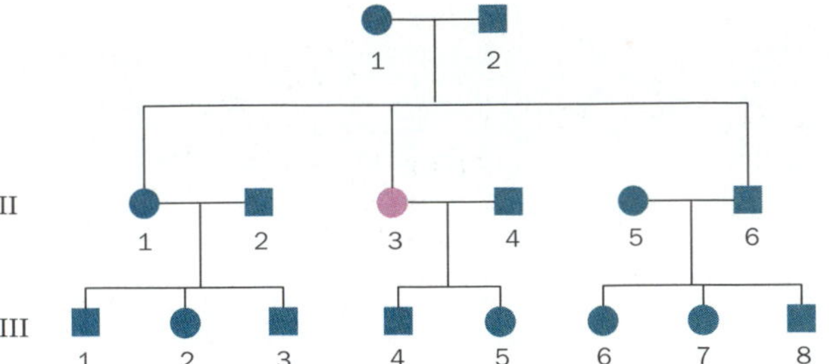

Figure 2.19 Pedigree chart to illustrate probability in pedigree analysis. In this family group, a trait (adherent ear lobes), dependent on a recessive gene, has appeared in one individual (identified by the violet circle). Three basic steps may be followed in all pedigree analysis: (1) What is the chance that one parent is a carrier of the gene in question? (2) What is the chance that the other parent is a carrier for the same gene? (3) What is the chance that a child of these two parents (genotypes) could express the trait involved? The product of the separate probabilities is the chance that a particular future child will express that trait.

tory of such traits. Recessives are expressed more frequently in families in which the father and the mother are more closely related than parents in the general population. The likelihood of similar genes being present is enhanced when the parents have descended from a common ancestor.

In the absence of data to indicate which individuals are carriers, the geneticist may resort to probability as the best available tool for determining the likelihood of expression of a given recessive gene in a certain family. If no expression has occurred in the history of the family, an estimate indicating the frequency of the gene in the general population may be used as a basis of probability. If the trait has appeared in the family, more precise calculations are possible. Probability is then based on the family history, which may be recorded in a pedigree chart.

The use of probability in human pedigree analysis is illustrated in Fig. 2.19. The trait, adherent ear lobes (Fig. 2.20), dependent on a recessive gene, appeared only once in the known history of the family, as indicated by the single darkened circle in Fig. 2.19. No information other than that shown on the chart is available. Unless there is evidence to the contrary, it may be assumed (to avoid dealing with small probabilities) that those individuals who have married into the family are homozygous for the dominant gene and do not carry the recessive gene in question.

The first step is to identify the genotypes of as many individual family members as possible from the information given. The woman (II-3) in whom the trait is expressed must be homozygous (*aa*) for the recessive gene. Each of her parents (I-1 and I-2), who did not express the trait but contributed an *a* gene to their daughter (II-3), must have the heterozygous genotype (*Aa*). The sister (II-1) and brother (II-6) must be *AA* or *Aa*. Obviously, they are not *aa* because they do not express the trait. There is no way to determine whether each of these individuals is *AA* or *Aa*. Therefore, the probability that each individual is a carrier (*Aa*) must be estimated from the available information. From the parent cross (*Aa* × *Aa*), the probability for the occurrence of *Aa* in any child with free earlobes is ⅔ and the probability for the occurrence of *AA* is ⅓. In the absence of more definite information, II-1 and II-6 may be considered *Aa* with ⅔ probability. The children of II-1 and II-6 have a ½ chance of being carriers for the gene (*a*) ***if*** one of their parents is a carrier. Therefore, the probability that III-1, III-2, III-3, III-6, III-7, or III-8 is a carrier is ⅔ × ½ = ⅓. The children of II-3 (III-4 and III-5) must be carriers (probability = 1).

The problem may be carried a step further by calculating the likelihood for the expression (*aa*) of the trait in the first child resulting from a marriage between two of the cousins represented in generation III. The mating III-1 × III-5 will serve as an example.

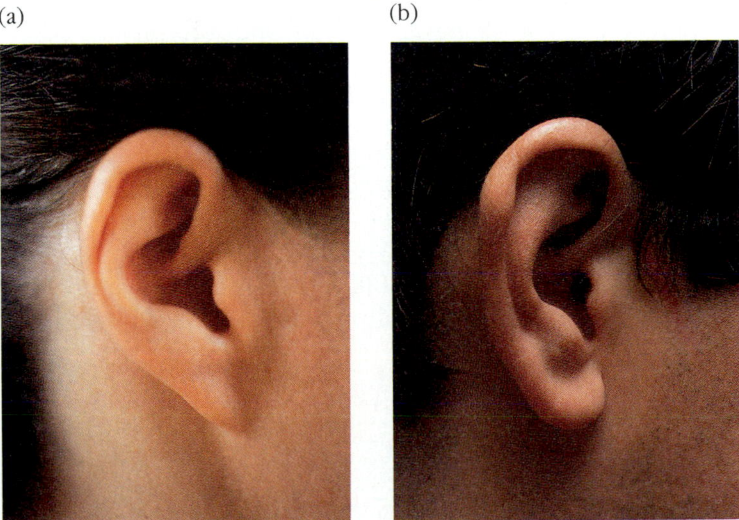

Figure 2.20 Adherent ear lobes (a) compared with free ear lobes (b). (Photographs by Kristine Kirkeby, Biological Sciences Art Services, University of Minnesota.)

The probability of being a carrier (*Aa*) is ⅓ for III-1 and 1 for III-5 (see the preceding paragraph). Both could be carriers and yet avoid an expression of the trait in their family. Therefore, another probability must be included, namely, that of two parents with genotypes *Aa* having an *aa* child ($Aa \times Aa \longrightarrow 1AA, 2Aa, 1aa$), which is ¼. The probability for the expression of the trait in the child of the individuals indicated is

$$(\text{probability III-1 is } Aa) \times (\text{probability III-5 is } Aa) \times (\text{probability child is } aa) = \tfrac{1}{3} \times 1 \times \tfrac{1}{4} = \tfrac{1}{12}$$

If the first child should express the trait, the probability that a second child would also express the trait would be ¼, because evidence would then be available to indicate that the genotypes of III-1 and III-5 were both *Aa*. One element of uncertainty or probability would thus be eliminated. At best, probability is a poor substitute for certainty. It is employed in analyses only when definite information is not available.

MODERN EVALUATIONS OF MENDEL'S CONCLUSIONS

New interpretations are inevitable in scientific disciplines as additional data accumulate. Mendel considered a single gene to be responsible for a single trait. It is now known that many genes are involved in the production of some traits, although single-gene substitutions can influence basic biochemical reactions and thus be responsible for alternative end products. Furthermore, it is the ***genes*** and not the trait that are ***inherited***. Genes behave as separate units, whereas traits may result from complex interactions involving many genes.

Complete dominance was indicated in all seven allelic pairs that Mendel reported. It was natural, therefore, for him to consider dominance an inherent property of genes. When sweet peas and snapdragons were studied, shortly after the discovery of Mendel's paper, intermediate traits were observed in hybrids. Crosses between homozygous snapdragons with red flowers and those with white flowers resulted in F_1 progeny with pink flowers. Heterozygotes could thus be distinguished phenotypically from both parents. Dominance has now been shown to be influenced by factors in the external, internal (hormonal), and genetic environment. Thus, Mendel's view of dominance as a fundamental inherent property of the allele alone is no longer tenable for all cases. Dominance of some genes may eventually be explained on the basis of modifier genes that are present in the genetic environment. In other cases, dominance may depend on the quantity or activity of enzymes that are encoded by the genes involved.

The most important concepts that Mendel inferred from his experiments were (1) ***segregation***, the process through which alleles separate and produce haploid gametes, and (2) ***independent assortment*** of different pairs of alleles. These principles are the basic foundation of Mendelian patterns of inheritance.

SUMMARY

Mendelian genetics is based on the transmission of chemical units (genes) from parents to progeny and thus from generation to generation. The mechanism of transmission includes (1) ***segregation, the separation of pairs of alleles into different gametes,*** and (2) ***independent assortment, the independent segregation of members of different pairs of alleles,*** as demonstrated in dihybrid crosses. Hereditary mechanisms operate in all plants and animals. Probability is involved in genetic mechanisms and must be recognized in predicting the transmission and expression of both ***dominant*** and ***recessive*** alleles. Gene-product interactions such as ***epistasis*** modify phenotypes and Mendelian ratios.

REFERENCES

CARLSON, E.A. 1973. *The gene: A critical history*, 2nd ed. Saunders, Philadelphia.

MENDEL, G. 1866. "Versuche uber pflanzenhybriden." (Available in the original German in *J. Hered.* 42: 1–47. English translation under the title "Experiments in plant hybridization." Harvard University Press, Cambridge, MA.)

MORGAN, T.H. 1926. *The theory of the gene.* Yale University Press, New Haven, CT.

NOVITSKI, E., and S. BLIXT. 1978. "Mendel, linkage, and synteny." *BioScience* 28: 34–35.

PROBLEMS AND QUESTIONS

2.1 On the basis of Mendel's hypothesis and observations, predict the results from the following crosses in garden peas: (a) a tall (dominant and homozygous) variety crossed with a dwarf variety; (b) the progeny of (a) selfed; (c) the progeny from (a) crossed with the original tall parent; (d) the progeny from (a) crossed with the original dwarf-parent variety.

2.2 Mendel crossed pea plants that produced round seeds with those that produced wrinkled seeds. From a total of 7324 F_2 seeds, 5474 were round and 1850 were wrinkled. Using the symbols *W* and *w* for genes, (a) symbolize the original parental cross; (b) the gametes; and (c) F_1 progeny. (d) Represent a cross between two F_1 plants (or one selfed); (e) symbolize the gametes; and (f) summarize the expected F_2 results under the following headings: phenotypes, genotypes, genotypic frequency, and phenotypic ratio.

2.3 The French biologist Cuenot crossed wild, gray-colored mice with white (albino) mice. In the first generation, all were gray. From many litters, he obtained in the F_2 198 gray and 72 white mice. (a) Propose a hypothesis to explain these results. (b) On the basis of the hypothesis, diagram the cross and compare the observed results with those expected.

2.4 A woman has a rare abnormality of the eyelids called ptosis, which makes it impossible for her to open her eyes completely. The condition has been found to depend on a single dominant gene (*P*). The woman's father had ptosis, but her mother had normal eyelids. Her father's mother had normal eyelids. (a) What are the probable genotypes of the woman, her father, and her mother? (b) What proportion of her children will be expected to have ptosis if she marries a man with normal eyelids?

2.5 In pigeons, the checkered pattern is dependent on a dominant gene *C* and a plain exterior on the recessive allele *c*. Red color is controlled by a dominant gene *B* and brown by the recessive allele *b*. Diagram completely a cross between homozygous checkered red birds and plain brown birds. Summarize the expected F_2 results.

2.6 In mice, the gene (*C*) for colored fur is dominant over its allele (*c*) for white. The gene for normal behavior (*V*) is dominant over the one for waltzing (*v*). Give the probable genotypes of the parent mice (each pair was mated repeatedly and produced the following results): (a) colored, normal mice mated with white, normal mice produced 29 colored normal and 10 colored waltzers; (b) colored normal mated with colored normal produced 38 colored normal, 15 colored waltzers, 11 white normal, and 4 white waltzers; (c) colored normal mated with white waltzer produced 8 colored normal, 7 colored waltzers, 9 white normal, and 6 white waltzers.

2.7 In rabbits, black fur is dependent on a dominant gene (*B*) and brown on the recessive allele (*b*). Normal-length fur is determined by a dominant gene (*R*) and short (rex) by the recessive allele (*r*). (a) Diagram and summarize the F_1 and F_2 results of a cross between a homozygous black rabbit with normal-length fur and a brown rex rabbit. (b) What proportion of the normal black F_2 rabbits from the preceding cross may be expected to be homozygous for both gene pairs? (c) Diagram and summarize a testcross between the F_1 and the fully recessive brown rex parent.

2.8 In shorthorn cattle, the gene (*R*) for red coat is not dominant over that for white (*R′*). The heterozygous combination (*RR′*) produces roan. A breeder has white, red, and roan cows and bulls. What phenotypes might be expected from the following matings, and in what proportions?

a. red × red
b. red × roan
c. red × white
d. roan × roan
e. roan × white
f. white × white
g. Would it be easier to establish a true breeding (homogeneous for color) herd of red or a true-breeding herd of roan shorthorns? Explain.

2.9 Albinism in humans is controlled by a recessive gene (*c*). From marriages between normally pigmented people known to be carriers (*Cc*) and albinos (*cc*): (a) What proportion of the children would be expected to be albinos? (b) What is the chance that any pregnancy would result in an albino child? (c) What is the chance in a family of three that one would be normal and two albino?

2.10 If both partners were known to be carriers (*Cc*) for albinism, what is the chance of the following combinations in families of four: (a) all four normal; (b) three normal and one albino; (c) two normal and two albino; (d) one normal and three albino?

2.11 In *Drosophila*, a dominant gene (*D*) for a phenotype called "dichaete" alters the bristles and also makes the wings remain extended from the body while the fly is at rest. It is homozygous lethal. (a) Diagram a cross between two dichaete (*Dd*) flies and summarize the expected results. (b) Diagram a cross between dichaete and wild type and summarize the expected results.

2.12 In humans, two abnormal conditions, cataracts in the eyes and excessive fragility in the bones, seem to depend on separate dominant genes located on different chromosomes. A man with cataracts and normal bones, whose father had normal eyes, married a woman free from cataracts but with fragile bones. Her father had normal bones. What is the probability that their first child will (a) be free from both abnormalities; (b) have cataracts but no fragile bones; (c) have fragile bones but not cataracts; (d) have both cataracts and fragile bones?

2.13 The inheritance pattern represented by violet squares and circles (symbolizing the same trait in different families—see the accompanying figure) may be assumed to depend on a single autosomal dominant or a single autosomal recessive gene. (a) Indicate which is the most likely mode of inheritance for the trait. (b) Based on your answer to (a), symbolize the probable genotype for each individual in each of the four pedigrees.

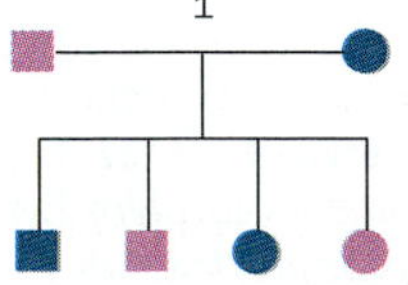

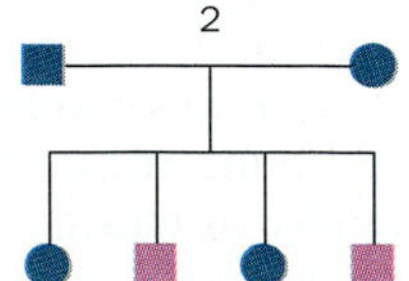

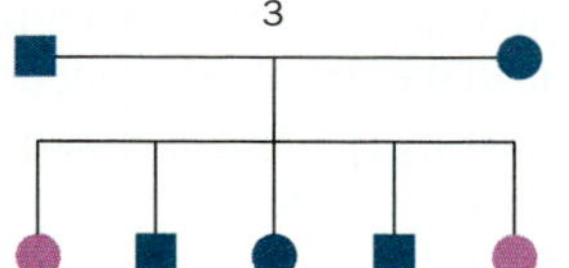

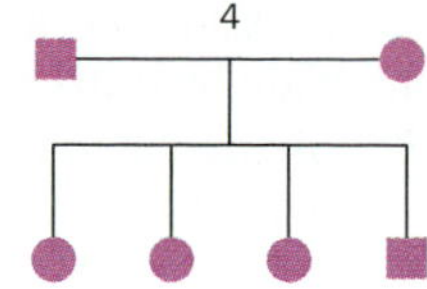

2.14 In garden peas, the genes for tall vine (*D*), yellow seed (*G*), and round seed (*W*) are dominant over their respective alleles for dwarf (*d*), green (*g*), and wrinkled (*w*). (a) Symbolize a cross between a homozygous tall, green, round plant and a dwarf, yellow, wrinkled plant. Represent the gametes possible from each parent and the F_1. (b) Symbolize a cross between two F_1 plants. Complete this cross by making use of the forked-line method and summarize the expected phenotypes. (c) Using the forked-line method, diagram a cross between the F_1 and the dwarf, green, wrinkled parent. Summarize the results for phenotypes, genotypes, genotypic frequency, and phenotypic ratio.

2.15 How many different kinds of F_1 gametes, F_2 genotypes, and F_2 phenotypes would be expected from: (a) *AA* × *aa*; (b) *AABB* × *aabb*; (c) *AABBCC* × *aabbcc?* (d) What general formula can be applied for F_1 gametes, F_2 genotypes, and F_2 phenotypes?

2.16 The shape and the color of radishes are controlled by two independent pairs of alleles that show no dominance; each genotype is distinguishable phenotypically. The color may be red (*RR*), purple (*R'R*), or white (*R'R'*) and the shape may be long (*LL*), oval (*L'L*), or round (*L'L'*). Using the Punnett square method, diagram a cross between red, long (*RRLL*) and white, round (*R'R'L'L'*) radishes and summarize the F_2 results under the headings phenotypes, genotypes, genotypic frequency, and phenotypic ratio.

2.17 In poultry, the genes for rose comb (*R*) and pea comb (*P*) together produce walnut comb. The alternative alleles of both in a homozygous condition (i.e., *rrpp*) produce a single comb. From information concerning interactions of these genes given in the chapter, determine the phenotypes and proportions expected from the following crosses: (a) *RRPp* × *rrPp*; (b) *rrPP* × *RrPp*; (c) *RrPp* × *Rrpp*; and (d) *Rrpp* × *rrpp*.

2.18 Rose-comb chickens mated with walnut-comb chickens produce 15 walnut, 14 rose, 5 pea, and 6 single-comb chicks. Determine the probable genotypes of the parents.

2.19 White fruit color in summer squash is dependent on a dominant allele (*W*), and colored fruit on the recessive allele (*w*). In the presence of *ww* and a dominant gene (*G*), the color is yellow, but when *G* is absent (i.e., *gg*), the color is green. Give the F_2 phenotypes and proportions that are expected from crossing a white-fruited (*WWGG*) plant with a green-fruited (*wwgg*) plant.

2.20 The white leghorn breed of chickens is known to carry in homozygous conditions a color allele (*C*) and a dominant inhibitor (*I*) that prevents the action of *C*. The white wyandotte (*iicc*) has neither the inhibitor nor the color gene. Give the F_2 phenotypes and proportions expected from crossing a white leghorn (*IICC*) with a white wyandotte (*iicc*).

2.21 What phenotypic ratio would be expected from a testcross of F_1 and a full recessive (*AaBb* × *aabb*) if the F_2, resulting from $F_1 \times F_1$ (i.e., *AaBb* × *AaBb*), were as follows: (a) 13:3; (b) 15:1; (c) 9:3:4; (d) 12:3:1; (e) 1:2:1:2:4:2:1:2:1?

2.22 In the F_2 generation of a certain tomato experiment, 3629 fruits were red and 1175 were yellow. A 3:1 ratio was expected. (a) Are the discrepancies between the observed and expected ratios significant? (b) In the same experiment, 671 plants with green foliage and 569 with yellow were counted. This was a backcross and the hypothetical ratio was 1:1. Test with χ^2 and explain.

2.23 The following are some of Mendel's results, with the hypotheses to which they are fitted. Test each for goodness of fit and indicate whether each is significantly different from the hypothesis.

CROSS	RESULTS	HYPOTHESIS
a. Round [X] wrinkled seed (F2)	5474 : 1850	3 : 1
b. Violet [X] white flower (F2)	705 : 224	3 : 1
c. Green [X] yellow pod (F2)	428 : 152	3 : 1
d. Round yellow (F1) [X] wrinkled green	31 : 26 : 27 : 26	1 : 1 : 1 : 1
e. Round yellow (F1) [X] wrinkled green	24 : 25 : 22 : 27	1 : 1 : 1 : 1

2.24 When four coins are tossed together in a series: (a) what proportion of the total results will be in the class of four heads; (b) four tails; (c) three heads and one tail; (d) three tails and one head; (e) two heads and two tails?

2.25 If four babies are born at a given hospital on the same day: (a) What is the chance that two will be boys and two girls? (b) What is the chance that all four will be girls? (c) What combination of boys and girls among four babies is most likely to occur? Why? (d) If a certain family has four girls, what is the chance that the fifth child will be a girl?

2.26 What is the probability in families of six of: (a) one boy and five girls; (b) three boys and three girls; and (c) all six girls?

2.27 The trait represented by violet squares and circles in the following pedigree chart is inherited through a single dominant gene. Calculate the probability of the trait appearing in the offspring if the following cousins should marry: (a) III-1 × III-3; (b) III-2 × III-4.

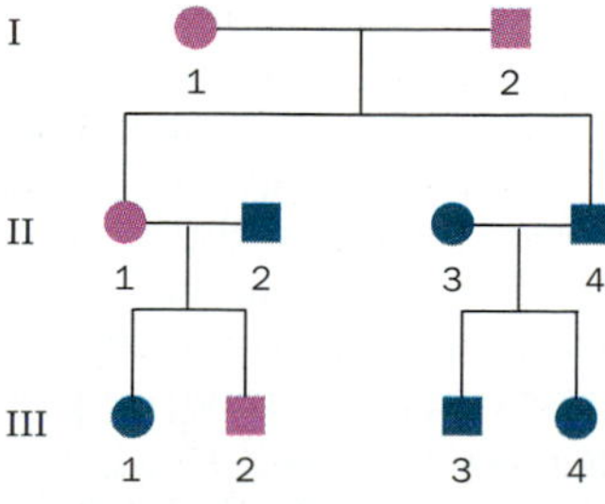

2.28 In the family pedigree shown in the following chart, an abnormal trait is inherited as a simple recessive. Unless there is evidence to the contrary, assume that the individuals who have married into this family do not carry the recessive gene for the trait. Violet squares and circles represent expressions of the trait (see accompanying figure). Calculate the probability of the trait appearing in a given offspring if the following cousins and second cousins in generations III and IV should marry: (a) III-1 × III-12; (b) III-4 × III-14; (c) III-6 × III-13; (d) IV-1 × IV-2.

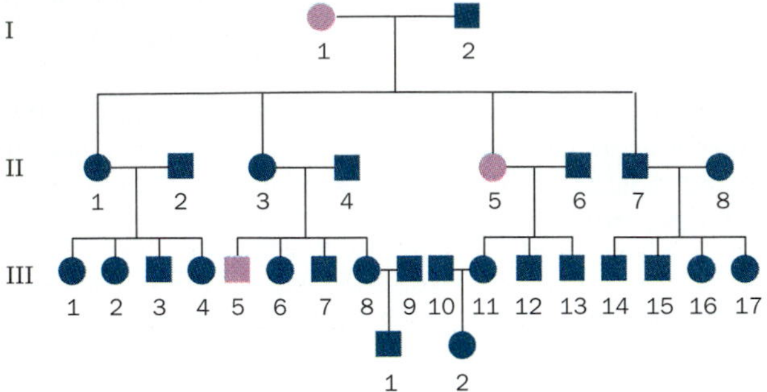

2.29 Phenylketonuria in humans is caused by a recessive allele p. If both partners are known to be carriers (Pp), what is the chance in the following combinations with five children that (a) all are normal; (b) four are normal and one is affected; (c) three are normal and two are affected; (d) two are normal and three are affected; (e) one is normal and four are affected; (f) all are affected?

2.30 (a) How can phenocopies be used to study gene action? (b) What values and limitations do they have for investigations of this kind? (c) How can an investigator determine whether an altered phenotype is due to a mutation or a phenocopy?

2.31 What significance may phenocopies have in medical genetics?

2.32 A series of crosses in a particular controlled environment, from which all progeny were known to carry a dominant autosomal gene A (causing an eye abnormality), resulted in 1400 abnormal and 600 normal flies. What is the penetrance of gene A in this experiment?

2.33 (a) How do multiple alleles originate? (b) How should a series of multiple alleles be symbolized? (c) To what extent do they represent alterations of the same basic phenotype?

2.34 The following, listed in order of dominance, are four alleles in rabbits: c^+, colored; c^{ch}, chinchilla; c^h, himalayan; and c, albino. What phenotypes and ratios would be expected from the following crosses: (a) $c^+c^+ \times cc$; (b) $c^+c \times c^+c$; (c) $c^+c^h \times c^+c^{ch}$; (d) $c^{ch}c \times cc$; (e) $c^+c^h \times c^+c$; and (f) $c^hc \times cc$?

2.35 In mice, a series of five alleles has been associated with fur pattern. These alleles are, in order of dominance, A^Y (homozygous lethal) for yellow fur; A^L, agouti with light belly; A^+, agouti; a^t, black and tan; and a, black. For each of the following crosses, give the coat color of the parents and the phenotypic ratios expected among the progeny: (a) $A^YA^L \times A^YA^+$; (b) $A^Ya \times A^La^t$; (c) $a^ta \times a^Ya$; (d) $A^La^t \times A^LA^L$; (e) $A^LA^L \times A^YA^+$; (f) $A^+a^t \times a^ta$; (g) $a^ta \times aa$; (h) $A^YA^L \times A^+a^t$; and (i) $A^Ya^L \times A^YA^+$.

2.36 If a series of four alleles is known to exist in a given diploid ($2n$) species, how many would be present in: (a) a chromosome; (b) a chromosome pair; and (c) an individual member of the species? (d) How many different combinations might be expected to occur in the entire population?

2.37 Assume that in a certain animal species four alleles (c^+, c^1, c^2, and c) have their locus on chromosome I and another series of two alleles (d^+ and d) have their locus on chromosome II. How many different genotypes, with respect to these two series of alleles, are theoretically possible in the population?

2.38 A series of multiple alleles in a certain species of fish that breeds readily in the laboratory was listed by Myron Gordon as follows: P^{o}, one spot; P^m, moon complete; P^c, crescent; P^{cc}, crescent complete; P^{co}, comet; P^t, twin spot; and P, plain. (a) How many combinations of these alleles might be expected to occur in the population? (b) How could the allelic nature of these genes be indicated by genetic methods?

2.39 In several plants, such as tobacco, primrose, and red clover, combinations of alleles in eggs and pollen have been found to influence the reproductive compatibility of the plants. Homozygous combinations such as S^1S^1 do not develop because S^1 pollen is not effective on S^1 stigmas. S^1 pollen, however, is effective on an S^2S^3 stigma. What progeny might be expected from the following crosses (seed parent always written first): (a) $S^1S^2 \times S^2S^3$; (b) $S^1S^2 \times S^3S^4$; (c) $S^4S^5 \times S^4S^5$; and (d) $S^3S^4 \times S^5S^6$?

2.40 In humans, a series of alleles has been associated with the ABO blood type as follows: I^A, A type; I^B, B type; I^o, O type. I^A and I^B are codominant; I^AI^B heterozygotes have AB type blood; I^O is recessive to both I^A and I^B. What phenotypes and ratios might be expected from the following matings: (a) $I^AI^A \times I^BI^B$; (b) $I^AI^B \times I^OI^O$; (c) $I^AI^O \times I^BI^O$; and (d) $I^AI^O \times I^OI^O$?

2.41 A case was brought before a certain judge in which a woman of blood group O presented a baby of blood group O, which she claimed as her child, and brought suit against a man of group AB whom she claimed was the father of the child. What bearing might the blood-type information have on the case?

2.42 In another case, a woman of blood group AB presented a baby of group O, which she claimed as her baby. What bearing might the blood-type information have on the case?

2.43 A homozygous Rh-positive man (RR) married an Rh-negative (rr) woman. Their first child was normal and their second child had the hemolytic disease of the newborn. (a) What genetic explanation might be offered? (b) What prediction might be made concerning future children by this couple?

2.44 A heterozygous Rh-positive (Rr) man married an Rh-negative (rr) woman. Their first child was normal, and their second child showed the effects of Rh incompatibility. What prediction might be made concerning future children by this couple?

2.45 Consider the following hypothetical scheme of inheritance of coat color in an extinct mammal. Gene A controls the conversion of a white pigment P_0 to a gray pigment P_1; the dominant allele A produces the enzyme necessary for this conversion but the recessive allele a produces an enzyme with no activity. Gene B controls the conversion of the gray pigment P_1 to a black pigment P_2; the dominant allele *B* produces the active enzyme that catalyzes the $P_1 \rightarrow P_2$ reaction. The recessive allele *b* produces a defective enzyme with no activity. The dominant allele *C* of a third gene produces a product that completely inhibits the activity of the enzyme produced by gene *A*, that is, it prevents the reaction $P_0 \rightarrow P_1$. Allele *c* produces a defective product that does ***not*** inhibit the reaction $P_0 \rightarrow P_1$. Genes *A*, *B*, and *C* assort independently, and no other genes are involved. In the F_2 of the cross *AAbbCC* × *aaBBcc*, what is the expected phenotypic segregation ratio?

2.46 What F_2 phenotypic segregation ratio would be expected for the cross described in Problem 2.45 if the dominant allele, *C*, of the third gene produced a product that completely inhibited the activity of the enzyme produced by gene *B*, that is, prevented the reaction $P_1 \rightarrow P_2$, rather than inhibiting the activity of the product of gene *A* ($P_0 \rightarrow P_1$)?

2.47 Consider the following genetically controlled biosynthetic pathway in a hypothetical plant:

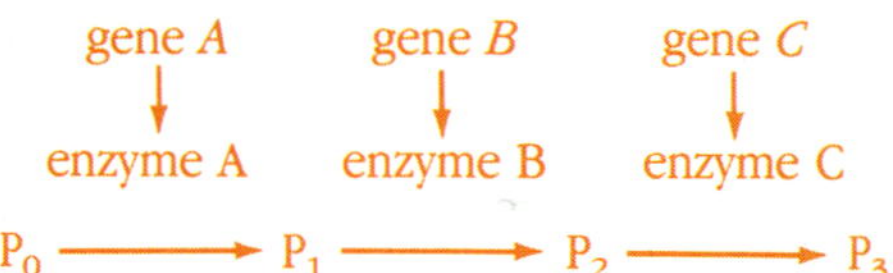

Assume that gene *A* controls the conversion of a white pigment, P_0, to another white pigment, P_1; the dominant allele *A* codes for the enzyme necessary to catalyze this conversion, but the recessive allele *a* codes for a defective enzyme (with no activity). Gene *B* controls the conversion of the white pigment, P_1, to a pink pigment, P_2; again, the dominant allele, *B*, produces the enzyme necessary for the $P_1 \rightarrow P_2$ conversion, but the recessive allele *b* produces an inactive product. The dominant allele, *C*, of a third gene codes for an enzyme that catalyzes the conversion of the pink pigment, P_2, to a red pigment, P_3; its recessive allele, *c*, produces an altered enzyme with no activity.

The dominant allele, *D*, of a fourth gene produces a gene product that completely inhibits the activity of enzyme *C*, that is, it blocks the reaction $P_2 \rightarrow P_3$. Its recessive allele, *d*, produces a defective gene product that does ***not*** block the $P_2 \rightarrow P_3$ reaction.

Assume that flower color is determined solely by these four genes and that they assort independently (i.e., are located on four different chromosomes). In the F_2 of a cross between plants of genotype *AAbbCCDD* and plants of genotype *aaBBccdd*, what proportion of the plants would be expected to have: (a) red flowers? (b) pink flowers? (c) white flowers?

Chapter 3

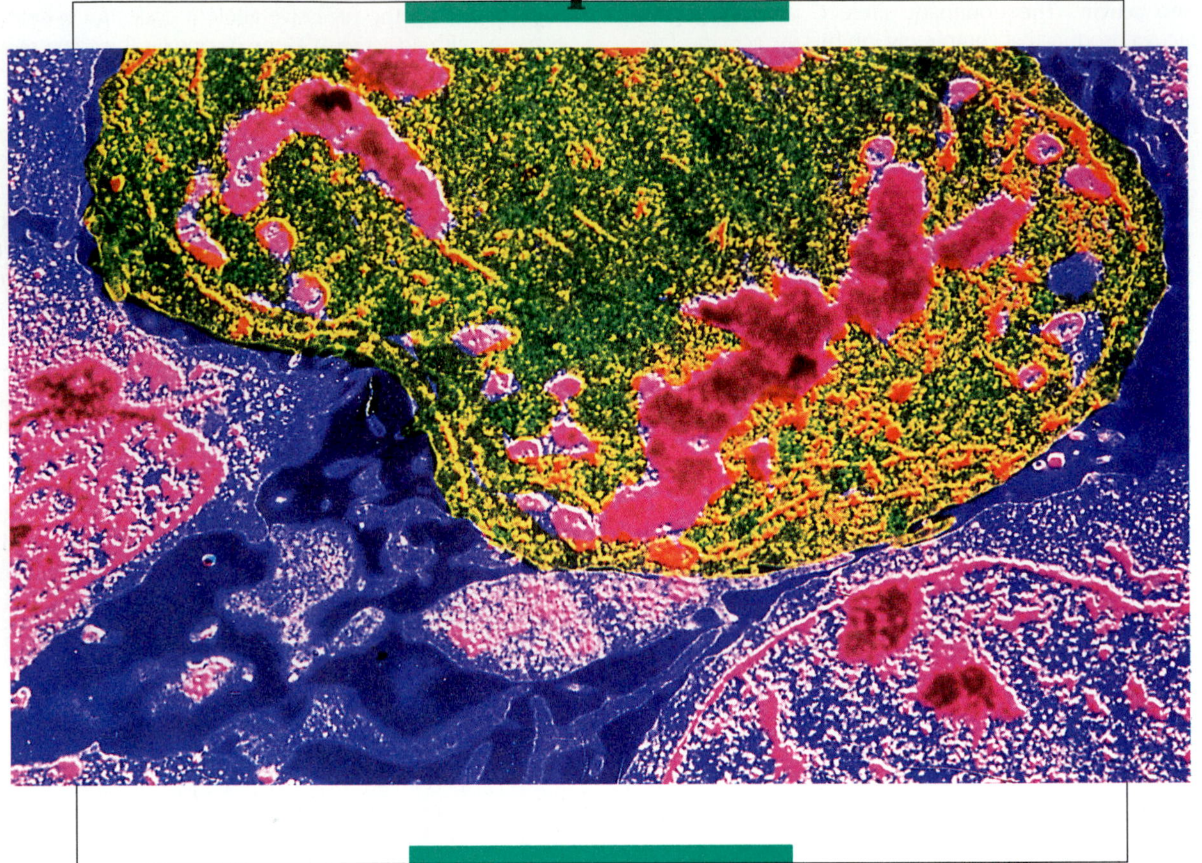

Cell Mechanics

(Photo above): Transmission electron micrograph of a dividing human lymphocyte; X2900. (Courtesy of CNRI/Photo Science Library/Photo Researchers.)

IN the photomicrograph (Fig. 3.1) of a generalized liver cell, structural characteristics are illustrated. Although cells within a single plant or animal vary widely in structure and function, they all represent units of living material and have important properties in common. Both plant and animal cells, for example, are essentially alike in terms of having genes arranged in linear arrays on nuclear structures called ***chromosomes***. Chromosomes are involved in the division cycles of individual cells (Fig. 3.2) as well as in the basic reproduction processes of the entire organism (whether animal or plant).

Mitotic cell division is the process by which single cells reproduce themselves and multicellular organisms grow. When cells divide, each resultant part is a complete, although at first relatively small, cell. Immediately following division, the newly formed progeny cells grow rapidly, soon reaching the size of the original cell. Mitotic cell division is actually ***duplication*** or multiplication rather than just division. In unicellular animals, however, cell duplication equates with reproduction, since two new individuals are formed from the original parent. Each progeny cell receives the genetic information carried by the parent cell.

In higher organisms, growth occurs through mitotic cell division with subsequent enlargement and differentiation of the cells produced. In reaching adulthood, the number of cells increases in all multicellular organisms. Each human being, for example, begins as a single fertilized cell or zygote, which develops eventually into a person with more than a million billion cells. The number of cells then remains approximately

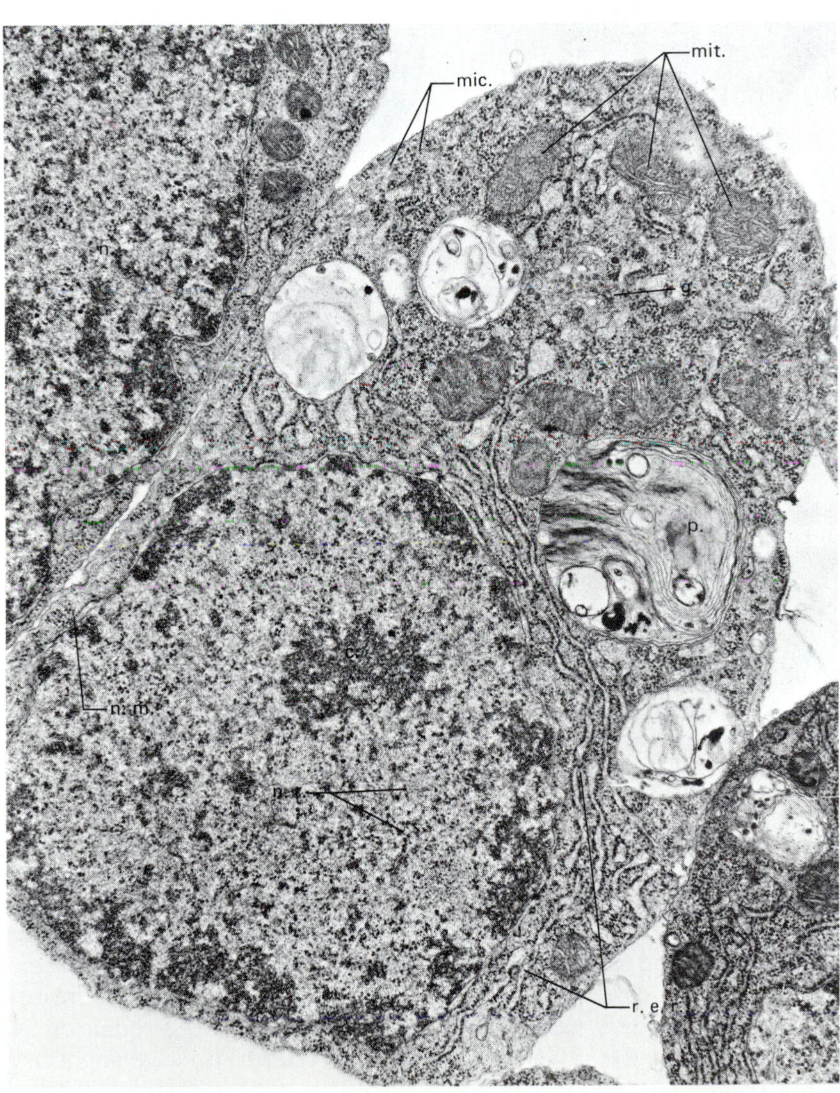

Figure 3.1 Electron photomicrograph of a cell showing structural characteristics: mic., microtubules; mit., mitochondria; g., golgi; p., phagolysosome; r.e.r., rough endoplasmic reticula; n.r., nuclear ribosomes; c., chromatin; n.m., nuclear membrane; n., nucleus of another cell. (Courtesy J.J. Bozzola, Microbiology, S.I.U., Carbondale.)

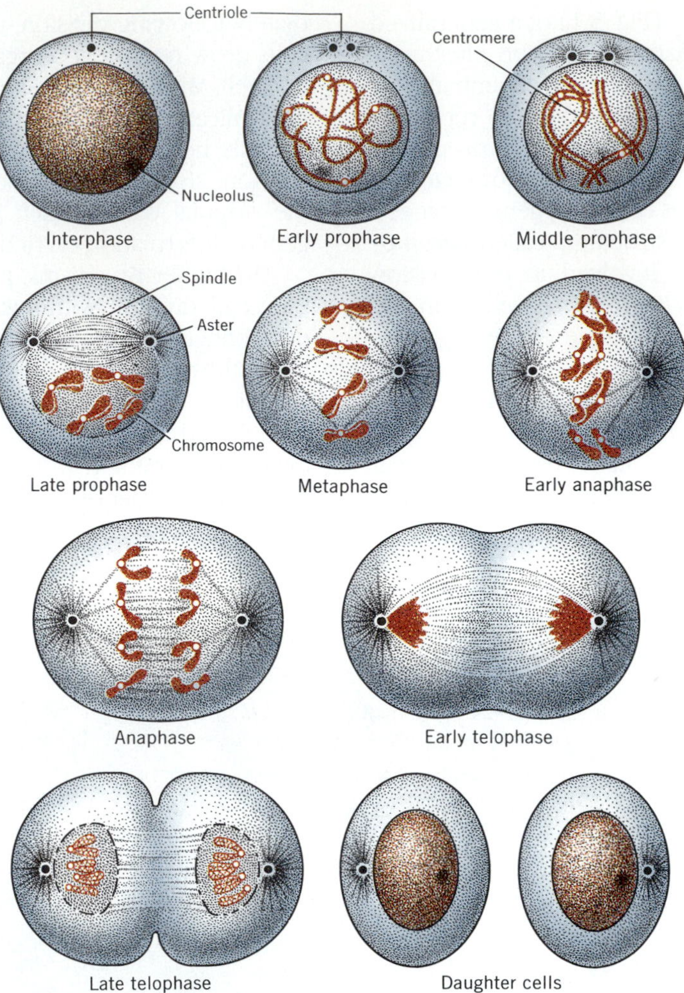

Figure 3.2 Schematic diagram of mitosis in an animal cell with four chromosomes.

constant. Cell division in human skin is rapid throughout life, but this is replacement division and does not increase cell number. In some organisms, such as the adult fruit fly, there is little cell division and little change in cell number.

The complex body of a multicellular organism eventually contains a variety of specialized cells. Epithelial cells have a relatively short life span; replacements must be made continuously. Those in the lining of the respiratory, digestive, and urinary tracts, for example, are replaced within a few days. Some gland cells have a life span of only a few hours. Cells in the human central nervous system, on the other hand, once established do not divide. Loss of cells is not compensated for by mitotic cell division, and the number diminishes through the mature life of the individual.

CELL CYCLE

Growth requires an increase in cell mass, a duplication of the genetic material, and a division assuring that each progeny cell receives an equal complement of the genetic material to ensure perpetuation of the cell line. These steps occur in an ordered progression during the cell life span or cycle (Fig. 3.3). Initially, a diploid cell (containing $2n$ chromosomes) undergoes a period of growth and increase in cell mass. This is called G_1. For a cell requiring 24 hours for the complete cycle, the G_1 stage might take the first 10 hours or so. This stage is devoted to cell growth and chemical preparation for DNA synthesis. At a defined time, duplication of the genetic material commences. During this synthesis (S) phase of about 9 hours, the genetic

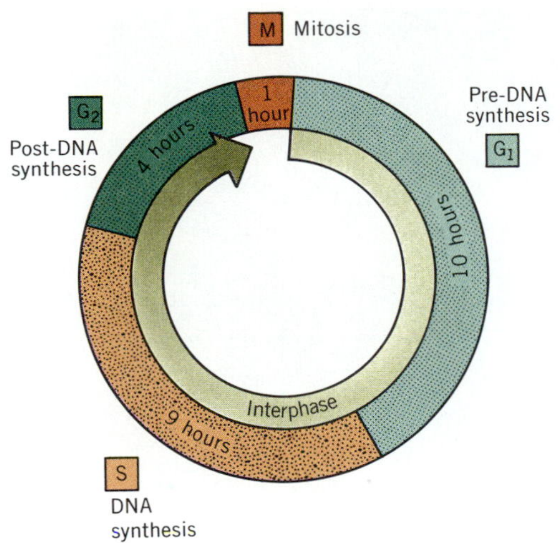

Figure 3.3 Diagrammatic illustration of the stages in the cell cycle of a "typical" mammalian cell growing in tissue culture with a generation time of 24 hours. (After L. E. Hood, J. H. Wilson, and W. B. Wood, *Molecular Biology of Eukaryotic Cells,* Benjamin, New York, 1975.)

material (DNA) of every chromosome is replicated. After the completion of DNA replication, the cell enters a second growth phase called G_2. This post-DNA synthesis stage might typically require about 4 hours, continuing until the onset of mitosis (*M*), which is completed in about 1 hour. During mitosis the sister chromatids of each chromosome separate, one sister chromatid going to each of the two daughter cells (Fig. 3.2).

Mitosis

Details of cell reproduction were elucidated in animal cells in the latter part of the nineteenth century by Walther Flemming, and in plant cells by Edward Strasburger and several other investigators. Two interrelated processes were found to be involved: (1) ***mitosis***, the nuclear division, and (2) ***cytokinesis***, the changes in the cytoplasm that include division of the cell proper. Figure 3.2 illustrates these processes as they occur in animal cells.

The names ***interphase*** (between divisions), ***prophase, metaphase, anaphase***, and ***telophase*** have been associated with different stages of the continuous mitotic cycle for convenience in describing the changes that occur. The prophase and telophase stages of mitosis are usually long and involved, whereas metaphase and anaphase are commonly brief.

The first indications of approaching mitosis in animal cells are observed in the cytoplasm of the interphase cell. A region of differentiated cytoplasm containing the centriole begins to divide. The centriole is a reproducing organell
cells that initiates divisi
apparatus consisting of
collection of microtub
ever, plant cells do
prophase in animal cells, the
apart. Thin, uncoiled, replicated sister
come coiled, shortened, and discrete. The num
coils decreases with a concomitant increase in diameter of each coil (see diagram inside front cover).

By late prophase, the two chromatids of each chromosome are held together at a constricted area called the primary constriction, where the ***centromere*** or spindle fiber attachment region is located. Centrioles have now moved to opposite poles and continuous spindle fibers have connected the poles. The nuclear membrane and nucleolus, which is the site of ribosomal RNA genes, disappear. (Ribosomal RNA genes control the synthesis of the RNA molecules that are structural components of the ribosomes, the macromolecular "workbenches" on which proteins are synthesized. See Chapter 10.) A spindle fiber becomes attached to each chromatid at the centromere. Each of two sister chromatids becomes attached to a different pole of the spindle, but the centromeres remain together. At metaphase, discrete chromatid pairs take places in the center or equatorial plate. Metaphase chromatids are tightly coiled and discrete, thus facilitating chromosome counts and gross structural comparisons. Arms of sister chromatids are extended from the centromeres, but the chromatids are held together by the centromeres until the beginning of anaphase.

After anaphase separation occurs, each chromatid has its own centromere and is now a chromosome. Anaphase chromosomes elongate somewhat by relaxation of the tight metaphase coiling and move to the respective poles of the spindle. Mitosis ensures that each progeny cell has the same genetic information as the mother cell.

During telophase, a nuclear membrane is reconstructed around each daughter nucleus and the nucleolus reappears in a particular place on a chromosome where the ribosomal RNA genes are located. This location corresponds with a structural feature called a secondary constriction in some chromosomes. A secondary constriction is a region of a chromosome where a thin segment joins the distal portion of the chromosome to the main body of the chromosome. The cytoplasmic part of the cell then divides. Animal cells, with their flexible outer layers, accomplish this by a constriction that converges from the two sides and eventually separates the two progeny cells. The surface around the equator pushes in toward the center and pinches the cell into two parts. Plant cells, with their rigid walls, form a partition or cell plate between the

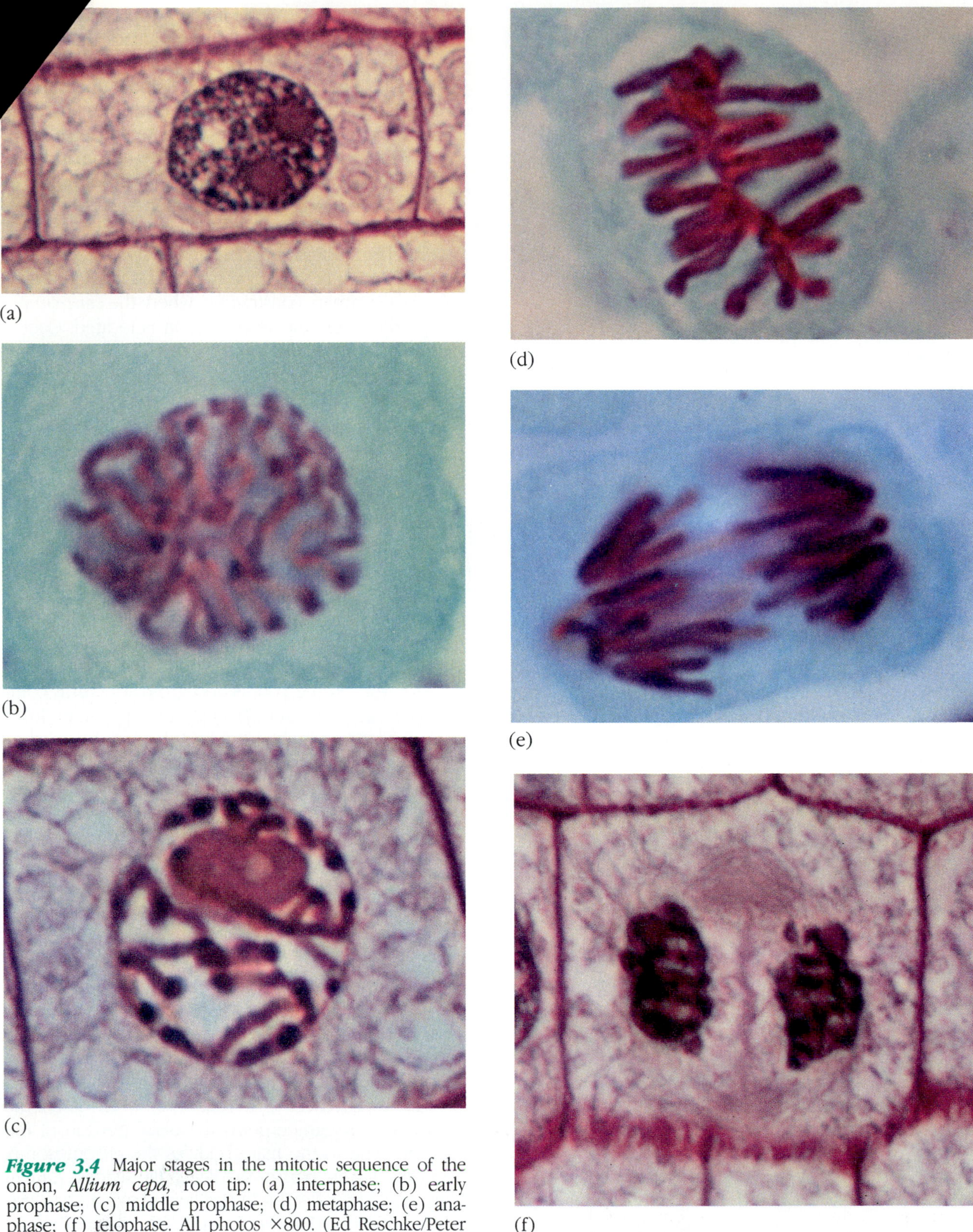

(a)

(b)

(c)

(d)

(e)

(f)

Figure 3.4 Major stages in the mitotic sequence of the onion, *Allium cepa,* root tip: (a) interphase; (b) early prophase; (c) middle prophase; (d) metaphase; (e) anaphase; (f) telophase. All photos ×800. (Ed Reschke/Peter Arnold.)

daughter cells. After the middle lamella (cell plate) is formed, walls of cellulose are deposited on either side.

As indicated earlier, each cell division (nuclear and cytoplasmic phases) is a continuous process from the time a cell first shows evidence of beginning to divide until the two progeny cells are completely formed. Nuclear and cytoplasmic phases are distinct but coordinated processes. The entire procedure ordinarily requires a few hours to several days, with variations dependent on the type of organism and environmental conditions. The actual mitotic sequence is illustrated with photomicrographs of dividing onion cells in Fig. 3.4.

Meiosis

Meiosis is the process in sexual reproduction by which the chromosome number of diploid ($2n$) germ cells is reduced to half (n) in formation of mature reproductive cells or gametes. This reduction is accomplished by two successive cell divisions during which the chromosomes are duplicated only once. Preceding the first of the two divisions, the homologous chromosomes come together in pairs and duplicate. These paired chromosomes are separated, that is, pulled apart in the first meiotic cell division. In the second meiotic cell division, the two chromatids of each chromosome separate, making a total of four separate mature sex cells each capable of fertilization. Through fertilization the diploid ($2n$) chromosome number is restored. A progeny organism characteristic of a species may then be developed largely by mitotic cell divisions.

Yeast cells, *Saccharomyces cerevisiae*, provide a very simple example of meiosis. They are single-celled organisms that reproduce asexually by budding. However, their vegetative cells may also act as sexual cells. The two meiotic divisions occur within a closed sac, the ascus, which holds all four products, called ascospores, of a single meiotic event together (Fig. 3.5).

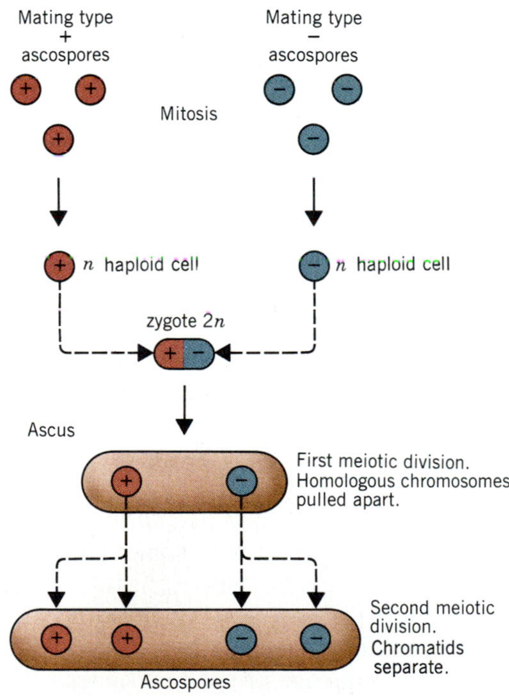

Figure 3.5 Mating types of yeast illustrating mitosis and meiosis in their simplest form. Haploid cells multiply by mitosis (budding). Cells of different mating type fuse to produce a diploid zygote, which may then undergo meiosis. Homologous chromosomes synapse and replicate to produce sister chromatids, and then the homologous chromosomes are pulled into separate nuclei during the first meiotic division. The two progeny cells divide in the second meiotic division giving rise to four haploid cells called ascospores, which are held in the ascus.

GAMETE FORMATION IN ANIMALS

Virtually all normal cells can reproduce themselves. The sex or germs cells, however, can initiate reproduction of the entire organism. A special sequence of events, ***gametogenesis***, the formation of haploid (reduced chromosome number) female and male gametes, results in the development of mature sex cells. Gametogenesis includes meiosis (from the Greek word "to reduce"). During meiosis, the chromosome number is changed from the diploid or $2n$ number, which is characteristic of body cells and premature germ cells, to the haploid or n number, which is characteristic of mature germ cells. Gametogenesis also includes differentiation of eggs and sperm, a process necessary for their functioning. Eggs of animals usually accumulate nutrient materials that sustain the developing embryo for a brief period; sperm of most animal species each develop a flagellum for independent motility.

Spermatogenesis

Sperm originate in the male reproductive organs, or testes, through a sequence called spermatogenesis (Fig. 3.6, left). Considered grossly, the process consists of growth in cell size, two successive cell divisions, and a metamorphosis of the resulting cells from spherical static bodies to elongated motile sperm. Spermatogenesis is initiated in diploid ($2n$), or unreduced, germ

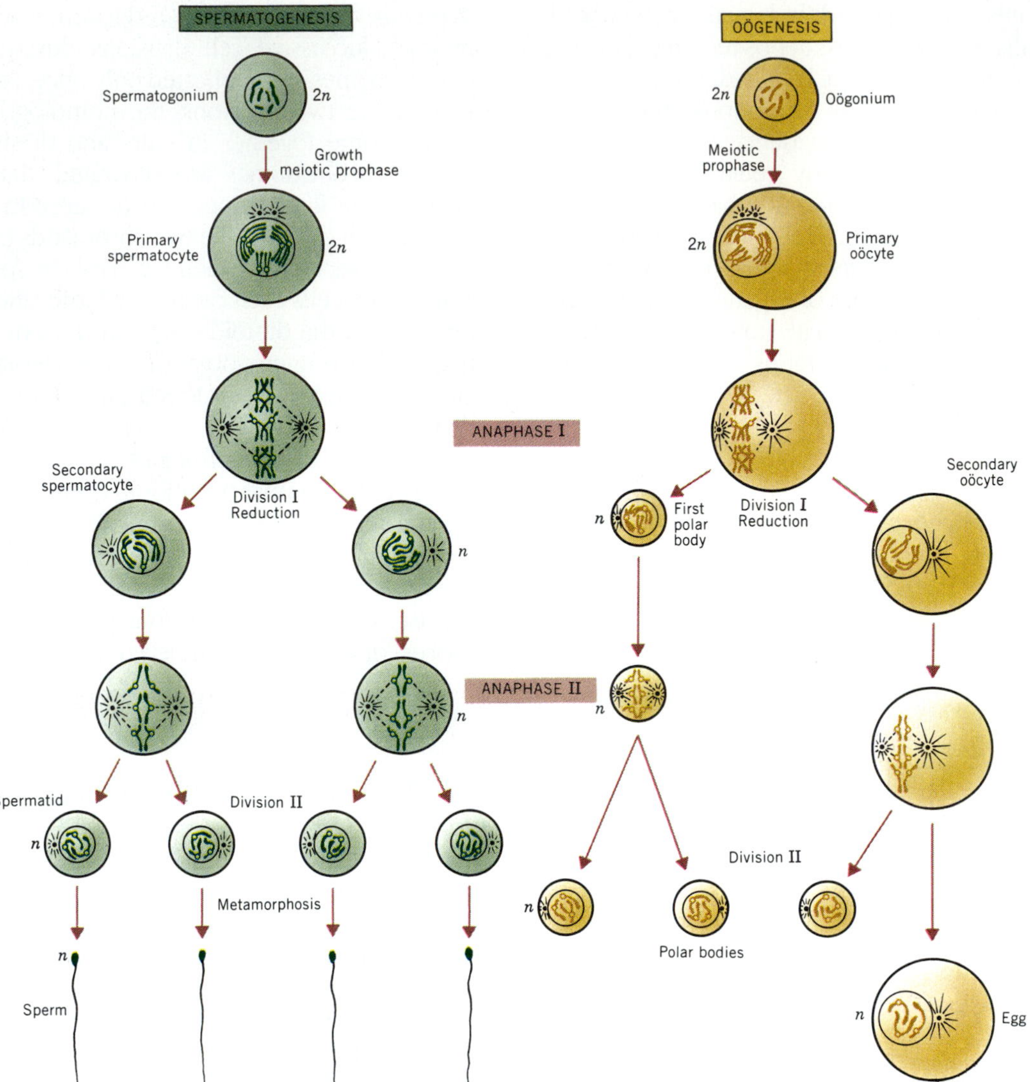

Figure 3.6 Meiotic sequence in a male and female animal. (Left) Process of spermatogenesis resulting in the formation of four sperm; (right) oögenesis resulting in the formation of one egg and three polar bodies.

cells called ***spermatogonia*** that enlarge and become ***primary spermatocytes***. These spermatocytes then undergo the first meiotic division, with each producing two ***secondary spermatocytes***. In turn, each secondary spermatocyte undergoes the second meiotic division to produce two ***spermatids***. Each spermatid then changes shape, develops a flagellum, and becomes a mature sperm. While the cells are dividing ***twice***, the chromosomes duplicate themselves only ***once***. Reduction in chromosome number from **$2n$ to n** is accomplished in the first meiotic division. Units of two chromatids resulting from the reduction division are the sister chromatids whose individual DNA molecules arose from earlier replication. The sister chromatids separate in the second meiotic division, each entering a different mature sperm.

Cells of the salamander *Amphiuma means tridactylum* Cuvier (the three-toed *Amphiuma*) are unusually large, and thus their chromosomes are conducive to studies of mitosis and meiosis. A series of photographs (Fig. 3.7; see also Fig. 3.8) showing chromosomes from this salamander illustrate the actual appearance of stages in the meiotic sequence of spermatogenesis. The sequence of developmental changes in the meiotic prophase is long and involved. The cell in which the ***prophase*** occurs, called the primary spermatocyte, enlarges during the entire prophase. Five major prophase stages are distinguishable in the continuous transition from spermatogonium to metaphase of the first meiotic division. The first stage (Fig. 3.7*a*) is called the ***leptotene*** stage (the noun form is leptonema). Chromosomes ($2n = 28$), which appear as single threads, represent the maternal and paternal chromosomes received by the individual (male) from the

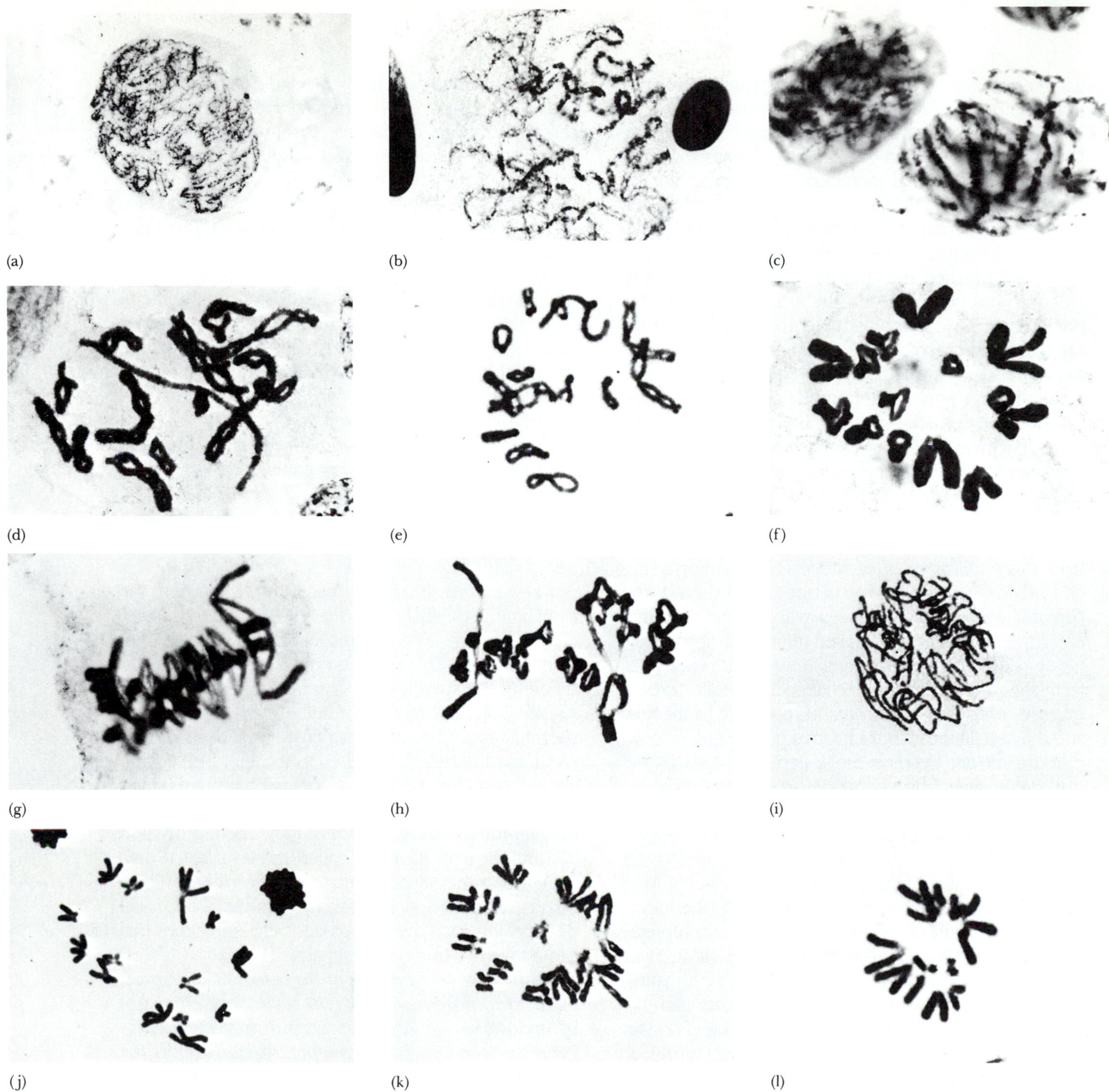

(a) (b) (c) (d) (e) (f) (g) (h) (i) (j) (k) (l)

Figure 3.7 Stages in the meiotic sequence of spermatogenesis in *Amphiuma means tridactylum* Cuvier. (a) Leptonema-zygonema in spermatocyte with single filaments and some pairing visible. Feulgen stained; ×360. (b) Details of zygotene spermatocyte showing paired strands or bivalents; ×680. (c) Early pachytene stage showing some thin filaments that may represent asynaptic regions. Feulgen stained; ×1200. (d) Diplotene spermatocyte showing long arms of chromatids in tetrad (in center) and chiasmata in other tetrads; ×680. (e) Late diplotene with 14 tetrads; ×680. (f) Diakinesis stage showing 14 distinct tetrads; ×680. (g and h) Metaphase I; ×735. (i) Prophase II with identifiable reduced (n) chromosomes, each with 2 chromatids; ×820. (j) Metaphase II; ×740. (k) Anaphase II showing early separation of daughter chromosomes; ×820. (l) The haploid complement of 14 chromosomes after anaphase II separation; ×1120. (Courtesy of G. M. Donnelly, A. H. Sparrow, and R. F. Smith, Brookhaven National Laboratory.)

gametes of his parents. Replication of genetic information occurred in the preceding interphase.

The pairing process, called "synapsis" (from the Greek for "conjunction" or "union"), brings together maternal and paternal members of the same pair of chromosomes (Fig. 3.7*b*). Corresponding segments of particular maternal and paternal chromosomes come together along their length in zipperlike fashion. When synapsis is finished, the number of visible structures is half that in leptonema, and the visible bodies in the nucleus are now homologous chromosome pairs or ***bivalents*** rather than single chromosomes. This is the ***zygotene*** stage (zygonema).

Following zygonema, homologous chromosomes can be observed side by side. This is the ***pachytene*** stage (pachynema; Fig. 3.7*c*). Exchanges between homologs and chiasma formation occur in this stage. The exchange event takes place first and appears visually as a ***chiasma***. At the ends of bivalents, four chromatids can be seen in some preparations. These have resulted from the duplication of each homolog forming sister chromatids. A ribbonlike structure called the ***synaptinemal complex*** can be seen with the electron microscope between the synapsed homologs, apparently binding the chromatids along their length and thus facilitating the chromatid exchange and genetic ***crossing-over*** (Chapter 7). Chromatids continue to shorten and thicken, and the four chromatids in a group, now called a ***tetrad***, become more apparent. This is the ***diplotene*** stage (diplonema). Paired chromatids appear to repel each other, causing the strands to separate longitudinally in some areas and to form loops (Fig. 3.7*d*).

In diplonema (Figs. 3.7*d* and 3.7*e*), the centromere in each chromosome is not split, and the longitudinal separation of the chromosomes is incomplete. Tetrads are held together at various places along their length because of interchanges between chromatids (chiasmata). From one to several chiasmata may be observed in favorable preparations, depending on the length of the homologous chromosomes involved. Each chiasma observable at this stage apparently represents an exchange that occurred between nonsister chromatids during the pachytene stage. In late diplonema (Fig. 3.7*e*) when the homologs repel each other, it is the chiasmata that hold the homologs together in the tetrad. The place where a chiasma appears, however, is not necessarily where the exchange actually occurred, because chiasmata tend to slip toward the ends of the tetrads and thus to become terminalized as the meiotic prophase continues. Genetic implications of chiasmata (crossovers) are discussed in Chapter 7.

Shortening of each tetrad continues through the next stage, ***diakinesis*** (Fig. 3.7*f*), resulting in discrete units (each containing four chromatids) that, in favorable preparations, can be counted and found to represent half the $2n$ chromosome number of the salamander, or 14, as expected following the pairing process. As the meiotic prophase is completed, the tetrads become angular or oval in appearance and take their places in the equatorial plane to form the equatorial plate of meiotic ***metaphase I*** (Figs. 3.7*g* and 3.7*h*).

The first of the two cellular divisions in the meiotic sequence separates the homologous chromosomes (each now containing two sister chromatids) that paired during the zygotene stage. In the anaphase of this so-called reduction division, the original maternal and paternal chromosomes (each composed of one centromere and two chromatids) separate. Thus, the ***number of chromosomes in each resultant cell is reduced from the original* 2n *to the* n *number***. These are shown in ***prophase II*** (Fig. 3.7*i*) and ***metaphase II*** (Fig. 3.7*j*). After the first meiotic division, the cells are called secondary spermatocytes.

During the secondary spermatocyte division, the centromere of each bipartite chromosome divides, providing each new chromosome with its own centromere (Fig. 3.7*k*). Each chromosome then moves to a pole of the spindle. The chromosome number in a spermatid is haploid (n), the same as in the secondary spermatocyte. Chromosomes of spermatids, however, are unipartite (Fig. 3.7*l*), whereas those of the secondary spermatocyte are bipartite being composed of two chromatids. Each ***spermatid*** nucleus has a ***single set of nonhomologous chromosomes***. The second division is a mitotic-type division (called the equational division), because it separates the duplicated (sister) chromatids. The division is different from the reductional division in which the homologous chromosomes that came together in synapsis separate (Fig. 3.8).

A firm understanding of the behavior of homologous chromosomes during the reductional division of meiosis is the key to understanding Mendelian genetics. Mendel's first principle, segregation, is the direct result of the separation of homologous chromosomes during anaphase I of meiosis. Mendel's second principle, independent assortment, results because each pair of homologous chromosomes lines up at the metaphase plate in meiosis I independently of all other pairs of homologous chromosomes. This is illustrated for an organism with two pairs of homologous chromosomes ($2n = 4$) in Fig. 3.9. Thus, meiosis provides the physical basis for segregation and independent assortment.

The terminal process in spermatogenesis is a complicated differentiation called ***spermiogenesis***. Through a progressive sequence of changes, each of the comparatively large, spherical, nonmotile sperma-

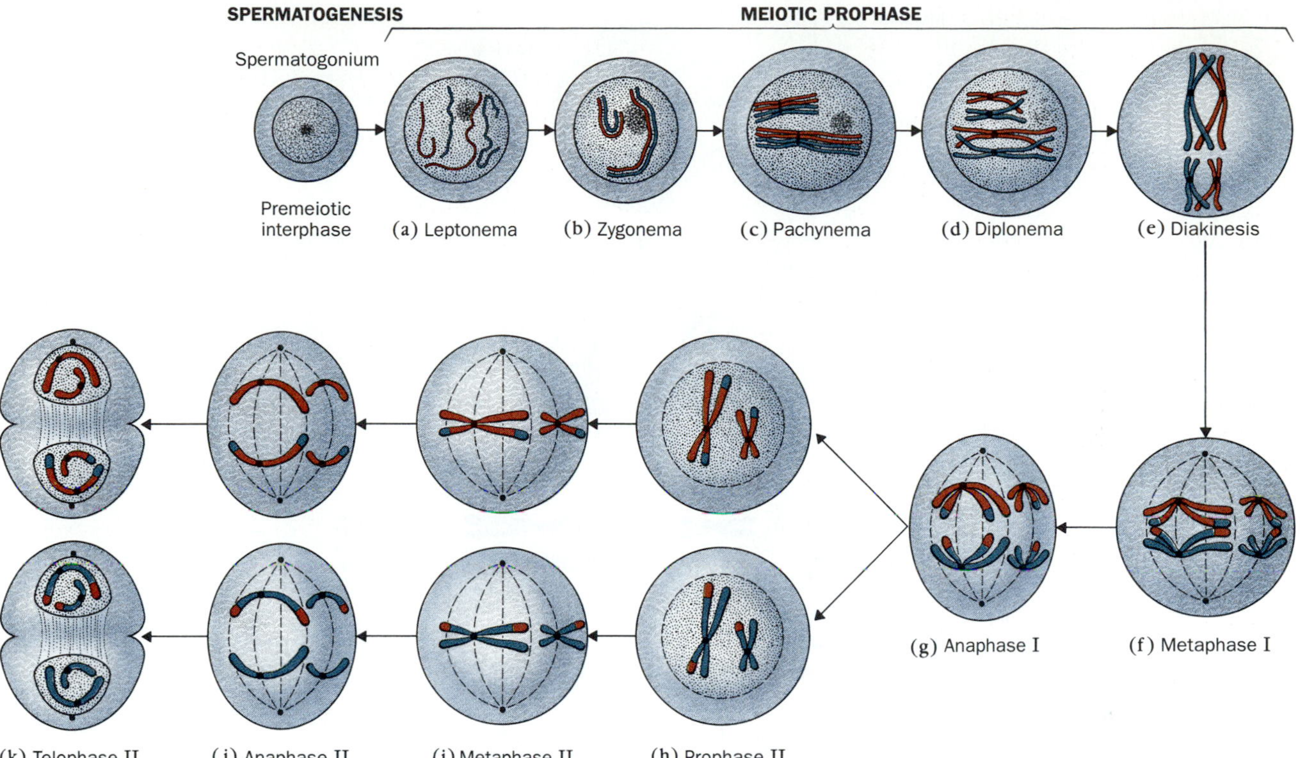

Figure 3.8 Diagrams illustrating chromosome changes that were shown in Fig. 3.7, photographs a to l. Stages in meiosis in an animal with two pairs of chromosomes, one member of each pair having come from the mother (white) and one from the father (colored). (a–e) Meiotic prophase stages parallel with a to e in Fig. 3.7; (f) metaphase I parallels with f in Fig 3.7, showing the results of random assortment; (g–j) second meiotic division stages parallel with g to l in Fig. 3.7, showing the results of segregation of sister chromatids. Two chromatids, one maternal and one paternal, are shown to undergo crossing-over (breakage and exchange of parts) at two sites (colored–white junctions). This important phenomenon is discussed in detail in Chapter 7.

tids is metamorphosed into a small, elongated, motile sperm composed typically of three parts: head, middle piece, and tail. In most animal species, spermatids begin differentiation by secretion of the apical body or acrosome, division of the centriole into two, and production of the flagellum by one centriole. Sloughing off of cytoplasm diminishes the overall size, as the developing sperm changes from a spherical to an oval and finally to an elongate shape. The nucleus moves to one edge of the cell, becoming elongated and increasingly compact. The acrosome, which is produced by the Golgi apparatus, takes its place around the anterior end of the sperm head. It contains lytic enzymes that facilitate sperm entry into ova (Fig. 3.10).

The middle piece contains the centriole, which lies next to the nucleus, and the proximal part of the axial filament, which continues in the tail. The mitochondria of the spermatid become concentrated around the axial filament in the middle piece. The tail of the sperm is composed of two parts: the outer sheath, which is cytoplasmic in origin, and the axial filament inside the sheath, which extends from the base of the head to the posterior end of the tail. Much of the cytoplasm of the spermatid is not used in formation of the sperm and is reabsorbed (taken up by Sertoli cells that presumably have nutritive and endocrine functions).

In most sexually mature male animals, spermatogenesis is constantly or periodically occurring in the testes, and many millions of sperm are produced. Insects generally require only a few days to complete their cycle of spermatogenesis, but in mammals the cycle extends over weeks or months. In mature human males, spermatogenesis occurs in the seminiferous tubules of the testes. Spermatogonia, undergoing mitotic division and continuing the population of stem cells, can be observed in cross sections of tubules at the periphery. The spermatogonia that appear to be in

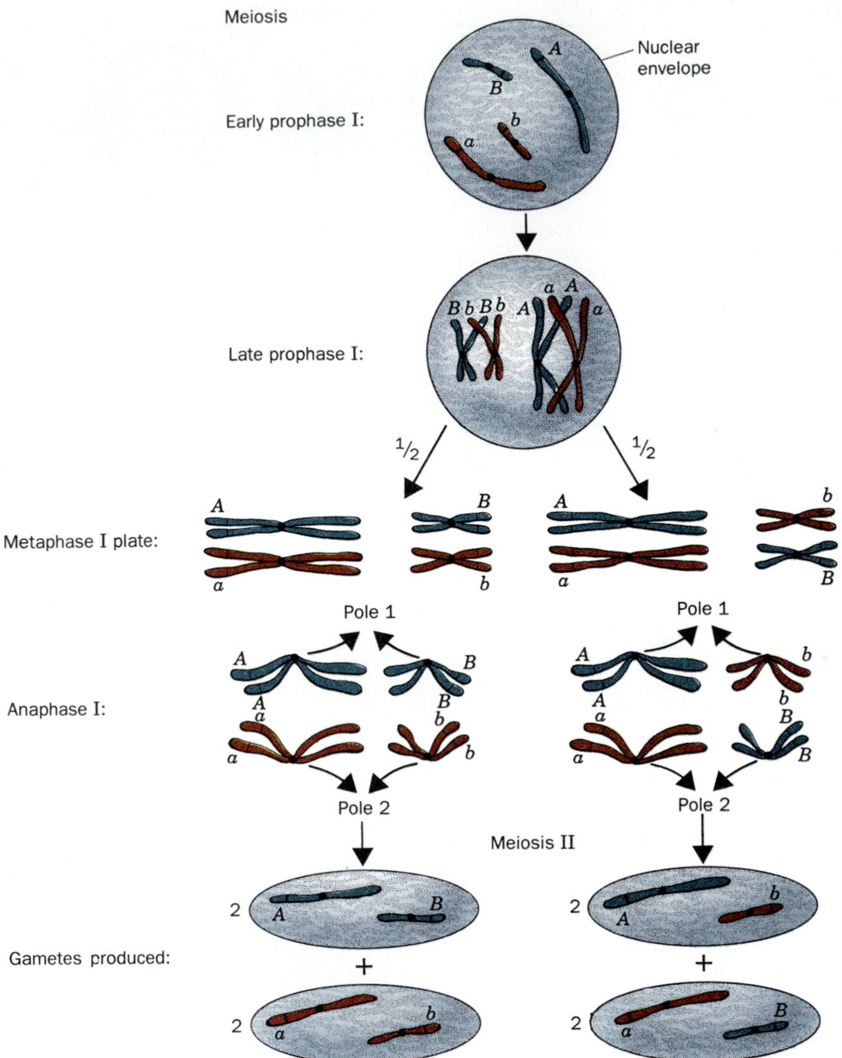

Figure 3.9 Diagram of meiosis in a dihybrid emphasizing the independent alignment of homologous pairs of chromosomes (tetrads) at the metaphase plate of meiosis I. Meiosis I is the reductional division during which segregation and independent assortment occur. Paternal chromosomes carrying genes *A* and *B* are shown in white; maternal chromosomes carrying genes *a* and *b* are shown in color. Only two pairs of homologous chromosomes are included. Note that the two paternal chromosomes align on one side of the metaphase plate, with the two maternal chromosomes on the other side, in one-half of the meiotic events (left). In the other one-half of the meiotic events, one paternal and one maternal chromosome align on each side of the metaphase plate (right). At anaphase I, the two chromosomes (each with two sister chromatids) with the same orientation move to one pole, and the two with the other orientation move to the other pole. The equational division of meiosis II (not shown) separates the two chromatids of each chromosome into different nuclei leading to the production of the four types of gametes, *AB, ab, Ab,* and *aB* (bottom), in a ratio of 1:1:1:1.

the innermost cell layer of the periphery enlarge and form primary spermatocytes with 23 tetrads. These spermatocytes undergo the two mechanically interwoven meiotic divisions in rapid succession and produce spermatids, which develop into sperm with single sets of 23 chromosomes. It should be noted that the sperm cells that develop in the testes of the male descend from original or primordial cells that migrated to the testes during early embryogenesis. The developmental time from primitive spermatogonia to mature human sperm is about 74 days.

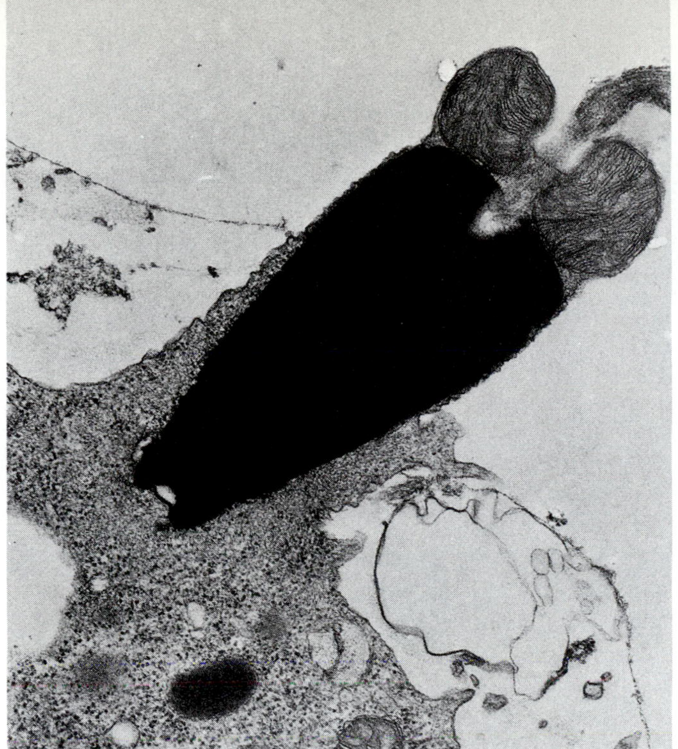

Figure 3.10 Sea urchin sperm entering egg. Magnification ×19,800. The egg cytoplasm bulges up around the sperm head to form a fertilization cone through which the sperm enters. (Photograph courtesy of Everett Anderson, Harvard Medical School, Boston. Reproduced with permission from *J. Cell Biol.* 37: 514, 1968; Copyright © 1968 by The Rockefeller University Press.)

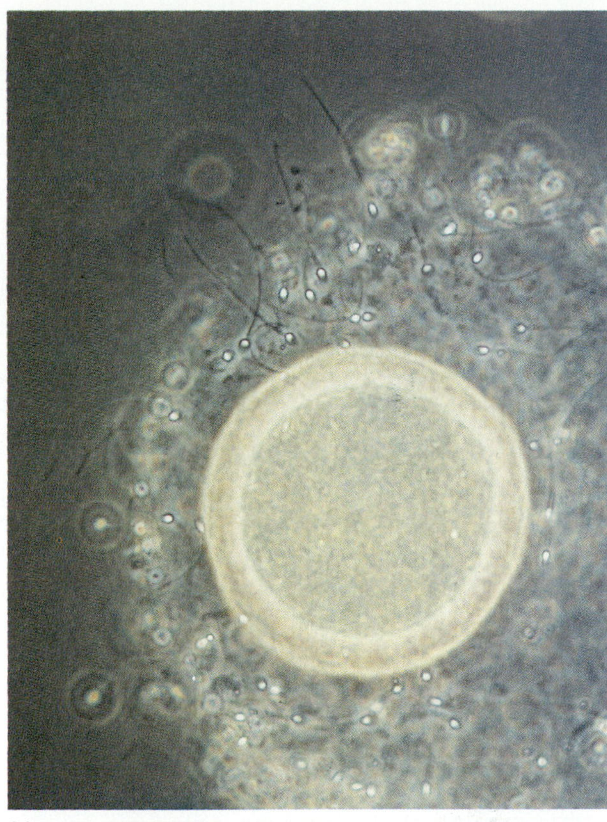

(a)

(b)

Figure 3.11 Human egg and sperm: (a) egg with its surrounding layer of supporting cells as it is being penetrated by sperm; (b) sperm greatly enlarged [(*a*): Lennart Nilsson; (*b*): Manfred Kage/Peter Arnold.)

Oögenesis

The process of gamete formation in the female (oögenesis, the origin of the egg) is also illustrated in Fig. 3.6 (right). Oögenesis is essentially the same as spermatogenesis as far as nuclear division is concerned, but aspects of the cytoplasm are quite different. Much more ***nutrient material*** is accumulated during oögenesis than during spermatogenesis. This is particularly true of oviparous animals—those that lay eggs that hatch outside the body of the mother. These animals must provide yolk material for the nourishment of the developing embryo outside the mother's body. Even in viviparous animals—which retain and nourish the young inside the body of the mother—a considerable amount of nutrient material accumulates. Because of the accumulated nutrient materials, an egg is usually considerably larger than a sperm of the same species (Fig. 3.11).

In addition, the cells that result from divisions in oögenesis are of unequal size. Nutrient material in the primary oöcyte is not divided equally into four cells that result from the meiotic sequence. One large cell (much larger in proportion than that shown in Fig. 3.6, right, for each division) retains essentially all the yolk,

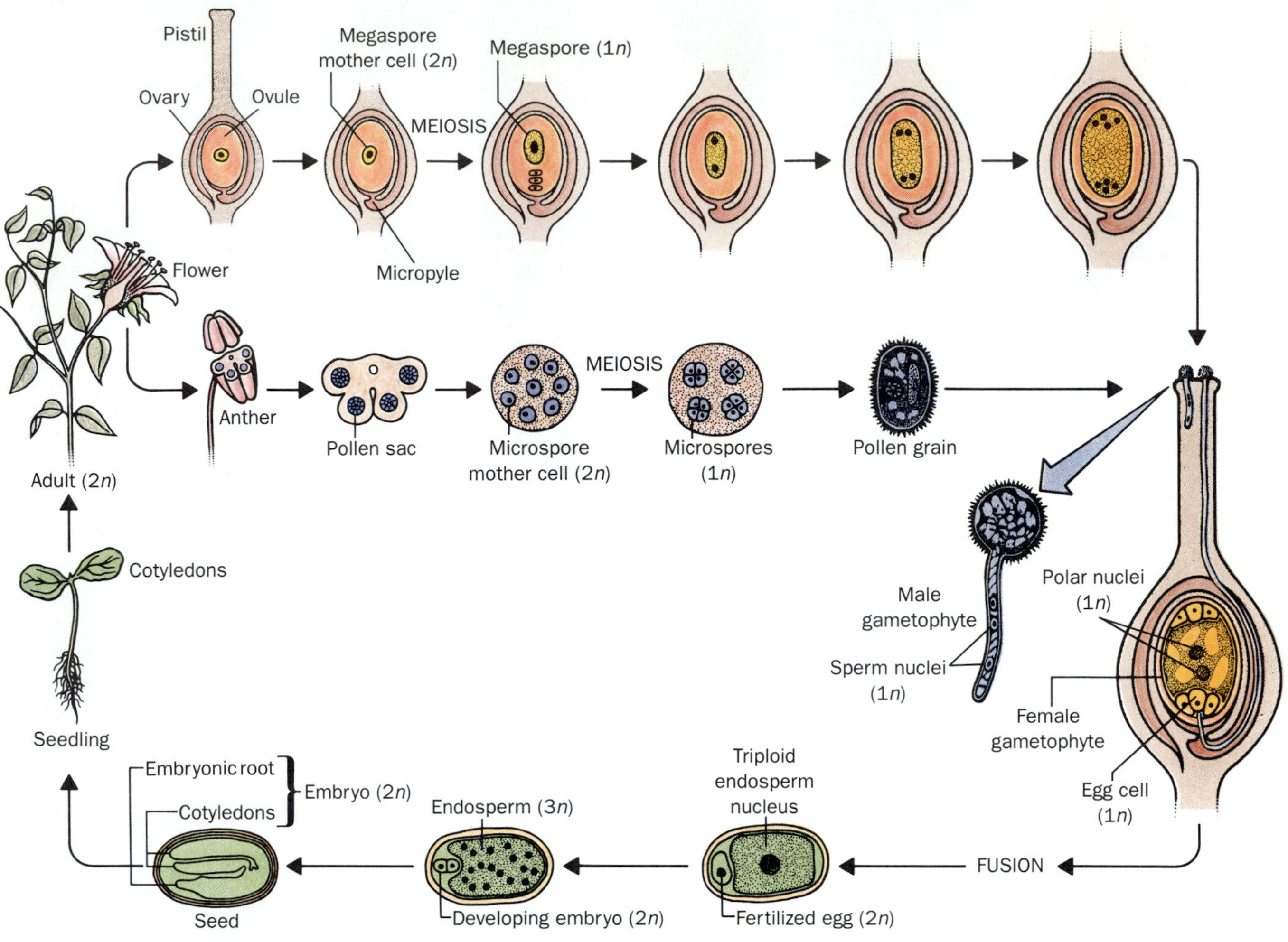

while the other cells, called "polar bodies," get very little. First and second ***polar bodies*** receive the same chromosome complements as the secondary oöcytes and ova from the respective divisions, but they do not become functional sex cells. During differentiation, special egg membranes are formed and the nucleus is generally reduced in size. The cytoplasm contains materials required for cell activity and differentiation.

In some animal species, oögenesis proceeds rapidly and continuously in sexually mature females, and numerous eggs are produced. Usually, these eggs complete the second meiotic division and become mature before encountering sperm. In many other animals, including mammals, the meiotic divisions are not completed until after sperm entry. In the human female, for example, oögenesis begins before birth. Oögonia located in the follicles of cortical tissue in the fetal ovary begin to differentiate into primary oöcytes at about the third month of intrauterine development. At the time of birth of the female infant, all the primary oöcytes are in the prophase of the first meiotic division. They remain in "suspended prophase" for many years until sexual maturity is reached. Then, as the ovarian follicles mature, the meiotic prophase is resumed. The first meiotic division for each developing egg is completed shortly before the time of ovulation for that egg. One cell becomes a secondary oöcyte and the other a polar body.

The second meiotic division is in progress when the developing egg is extruded from the ovary and passes into a Fallopian tube. This division is not completed, however, until after penetration by the sperm, which usually occurs in a Fallopian tube. Sperm entry is a random process in that any available sperm may fuse with any mature egg. If penetration by a sperm is accomplished, the ***secondary oöcyte*** divides and forms a ***mature ovum*** with a pronucleus containing a single set of 23 maternal chromosomes. The other cell

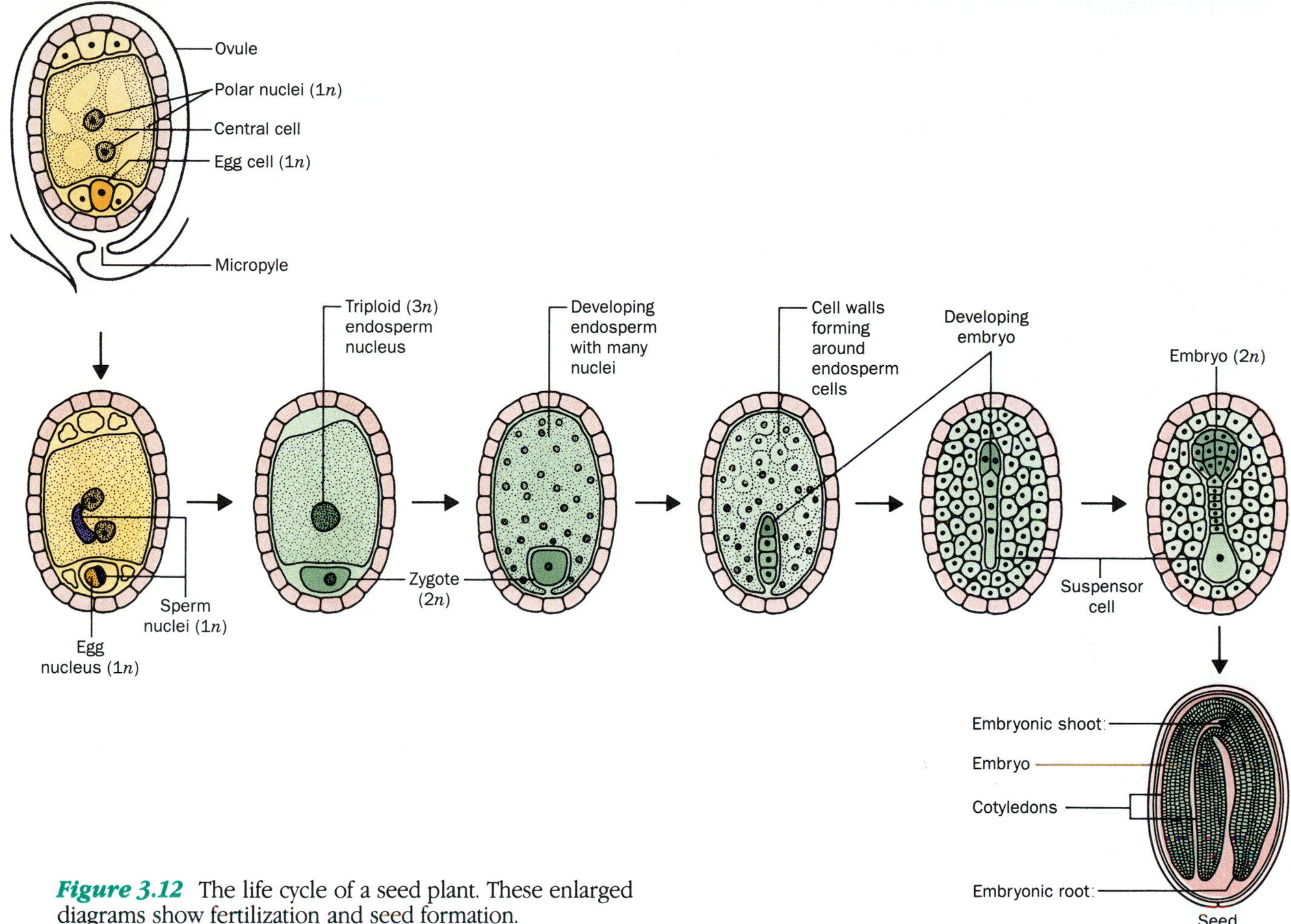

Figure 3.12 The life cycle of a seed plant. These enlarged diagrams show fertilization and seed formation.

resulting from this division is a second polar body not capable of further development. The sperm head forms a pronucleus with 23 paternal chromosomes. After the two pronuclei fuse, the resultant zygote, with ($2n$) 46 chromosomes, begins mitotic division or ***first cleavage***, which produces the two-cell stage of a beginning embryo. If the fertilized egg should undergo a complete mitotic division and the production of two separate fertilized egg cells, identical twins could be produced. If two eggs are produced at the same time and are fertilized by separate sperm, fraternal twins, no more alike genetically than other sibs, will develop together in the same uterus.

SPORE FORMATION IN PLANTS

Gamete formation in plants, like that in animals, requires a reduction in chromosome number from $2n$ to n. The meiotic process itself is similar to that in animals, but the life cycle of plants is somewhat more complicated (Fig. 3.12). Gamete formation usually does not follow meiosis directly. Sporogenesis in plants involves the formation of spores rather than sex cells. Spores produce gametophytes, and gametes come from gametophytes in which the chromosome number is already reduced. An ***alternation of generations*** between a haploid and a diploid phase in virtually all plants can separate the reduction of chromosome number from fertilization in plant reproduction. Diploid plants, called ***sporophytes***, undergo meiosis to produce spores with reduced chromosome number. Spores develop into haploid ***gametophytes***, which ultimately produce gametes capable of fertilization. Zygotes resulting from ***fertilization*** develop into sporophytes, completing the cycle. Sporophyte and gametophyte phases vary in length and importance in different plants.

Plant Fertilization

In the male cycle (Fig. 3.12), three nuclei arise from two mitotic divisions: the tube nucleus and two sperm or generative nuclei. This is the gametophyte phase. The sperm nuclei are carried through the micropyle into the embryo sac and accomplish the ***double fertilization*** process characteristic of the higher plants. One male nucleus fuses with the egg nucleus and gives rise to the $2n$ zygote, which divides repeatedly to form the ***embryo*** of the seed. The second male nucleus unites with the two polar nuclei to form the $3n$ nutrient ***endosperm*** of the seed.

The process of double fertilization introduces genetic material from the pollen parent into the endosperm tissue as well as the embryo. Therefore, it might be expected that both maternal and paternal inheritance would be represented. This hereditary influence of the pollen parent genes on the endosperm is called xenia. When, for example, maize from a variety normally bearing white kernels is pollinated with a yellow-kernel variety, the endosperm of the hybrid kernels is yellow. The dominant gene for yellow from the pollen comes to expression in the endosperm in the same manner as expected for embryonic tissue.

The diploid number of chromosomes is restored in the fertilized cell that gives rise to the plant embryo. Thus, through fertilization, the genetic contributions from each parent are combined. Subsequent continuous mitotic divisions of cells, along with the growth of individual cells, produces a new individual, a representative offspring of the species to which the parent(s) belongs.

MEIOSIS AND MENDEL'S PRINCIPLES

Mendelian segregation and independent assortment are simple ***corollaries of meiosis***. The physical bases of these two principles are inherent in the chromosome mechanism of meiosis. By observing microscopically distinguishable pairs of chromosomes during a reduction division, the maternal and paternal chromosomes can be followed. This was accomplished by E.E. Carothers in studies of grasshoppers. The three distinguishable pairs of maternal and paternal chromosomes are symbolized $AA'BB'CC'$, and the various observed arrangements of the division spindle are illustrated in Fig. 3.13. The independent alignment of the pairs of homologous chromosomes during the reduction division of meiosis explains Mendel's second principle. A chance distribution of maternal and paternal chromosomes occurs in such a way that each has an equal chance of facing one pole or the other on the equatorial plate of the reduction division. When chromosomes become established on the equatorial plate, each goes to the pole it faces and eventually becomes a part of the progeny cell formed around that pole of the spindle.

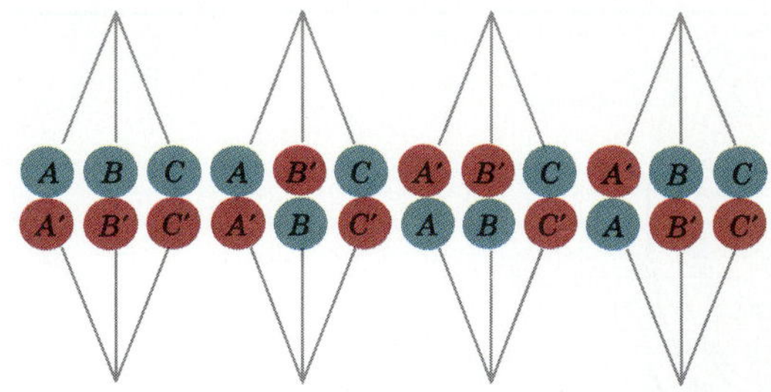

Figure 3.13 Arrangements of maternal and paternal chromosomes that could be identified in Carothers' microscope studies of grasshopper chromosomes.

To illustrate this concept, consider the phenotypes vine height, seed color, and shape that are observed in F_1 peas (Chapter 2). Genes for these three contrasting phenotypes are located on separate homologous chromosome pairs (chromosomes 4, 1, and 7, respectively; E. Novitski and S. Blixt, 1978). Gene *D* in the example is located on the maternal member of one pair (chromosome 4), whereas its allele, *d*, is in the corresponding position of the homologous (paternal) chromosome 4. Figure 3.14*a* shows, diagrammatically, the six chromosomes representing three homologous pairs in unpaired condition before synapsis. Maternal and paternal members of each pair carrying their respective alleles are present in the same nucleus. At synapsis in meiosis, each chromosome finds its mate, and alleles come together in corresponding positions, as illustrated in Fig. 3.14*b*. A particular maternal chromosome is equally likely to face one or the other pole of the spindle.

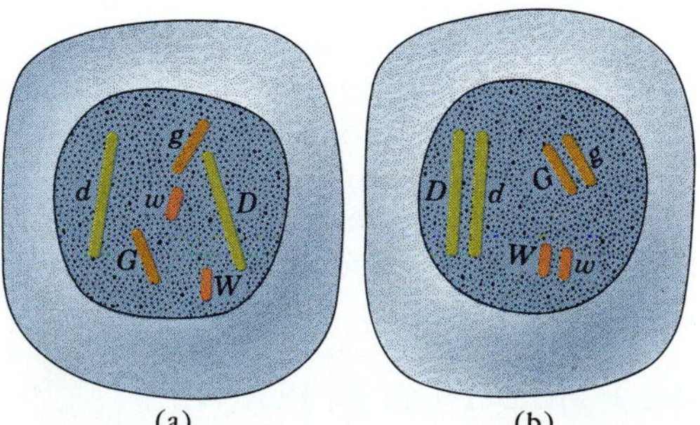

Figure 3.14 Cells showing three chromosome pairs: (a) before synapsis has occurred and (b) after synapsis has occurred.

Zygotes resulting from fusion of haploid (n) gametes are diploid ($2n$) and therefore carry pairs of alleles. As discussed in Chapter 2, alleles are alternate forms of a given gene, located at the same chromosome locus, which may substitute for one another. Body cells of higher organisms originate from the zygote through continued cell duplication and differentiation. Like the zygote, body cells arising from the zygote carry two complete sets of genes present in the two sets of homologous chromosomes. Two pairs of homologous chromosomes illustrating alleles and linked and unlinked nonalleles are diagrammed in Fig. 3.15. A zygote containing these two pairs of homologous chromosomes would be a typical diploid cell and would produce a normal diploid adult by means of a series of mitotic cell divisions.

In animals and plants with a large number of chromosomes, an almost infinite number of possible combinations may be expected. In organisms with 23 pairs of chromosomes, for example, the probability that a gamete produced by an individual in the population will have any specific combination of chromosomes is $(\frac{1}{2})^{23}$, which is on the order of one in eight million. This calculation is an underestimate due to the occurrence of crossing-over, which is another source of variability (see Chapter 7). Further increased numbers of gene combinations are possible in zygotes that result from ***random fertilization***. Much of the variation observed in natural populations can therefore be explained on the basis of the recombination of chromosomes and genes already present in the breeding population.

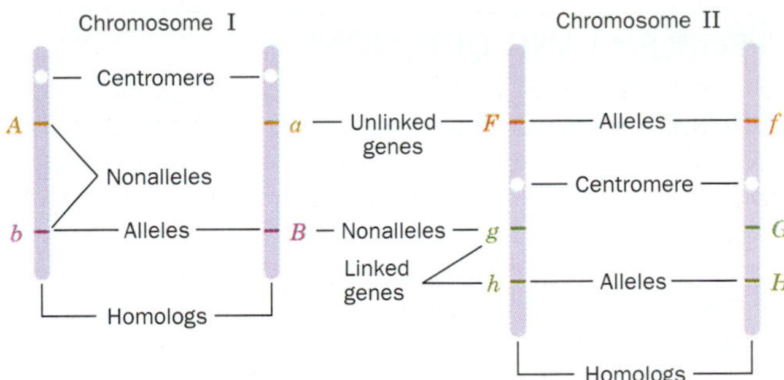

Figure 3.15 Alleles and nonalleles, and linked and unlinked genes.

SUMMARY

Genes are organized into linear arrays in ***chromosomes***. In nondividing cells, the chromosomes are present in intact nuclei, enclosed by the double membrane of the nuclear envelope. Genes replicate in interphase of both ***mitosis*** and ***meiosis***. Each of the two replicated chains becomes a part of a new progeny cell. Most normal living eukaryotic cells can reproduce themselves by mitosis.

Germ cells produce mature sex cells, called gametes, with reduced (n) numbers of chromosomes by the process of meiosis. Through ***fertilization, a male and a female sex cell initiate reproduction*** of the progeny organism. In development of a new organism, the ***2n zygote*** replicates its genes and divides. This process continues and results in the numerous cells that make up the organism. ***Chromosome mechanisms*** occurring in the germ cells ***during meiosis provide the biological basis for the Mendelian principles of segregation and independent assortment.*** Mitosis provides each daughter cell with the same genetic information as the mother cell. Meiosis reduces reproductive cells to haploidy, so that the chromosome number remains constant from generation to generation.

REFERENCES

Alberts, B., D. Bray, J. Lewis, M. Raff, K. Roberts, and J.D. Watson. 1989. *Molecular biology of the cell*, 2nd ed. Garland Publishing, New York.

Dalton, A.J., and F. Haguenau. 1968. *The nucleus*. Academic Press, New York.

De Robertis, E.D.P., and E.M.F. De Robertis, Jr. 1980. *Cell and molecular biology*, 7th ed. Saunders College/Holt, Rinehart & Winston, Philadelphia.

Donnelly, G.M., and A.H. Sparrow. 1965. "Mitotic and meiotic chromosomes of *Amphiuma*." *J. Hered.* 56: 91–98.

Du Praw, E.J. 1968. *Cell and molecular biology*. Academic Press, New York.

Dyson, R.D. 1975. *Essentials of cell biology*. Allyn & Bacon, Boston.

Mittwoch, U. 1973. *Genetics and sex differentiation*. Academic Press, New York.

Novitski, E., and S. Blixt. 1978. "Mendel, linkage, and synteny." *BioScience* 28: 34–35.

Swanson, C.P., and P. Webster. 1977. *The cell*. Prentice-Hall, Englewood Cliffs, NJ.

Voeller, B.R. 1968. *The chromosome theory of inheritance*. Appleton–Century–Crofts, New York.

Zimmerman, A., and A. Forer, eds. 1981. *Mitosis/cytokinesis*. Academic Press, New York.

PROBLEMS AND QUESTIONS

3.1 Mark the true statements with a (+) and the false with a zero (0). (a) Skin cells and gametes of the same animal contain the same number of chromosomes. (b) Any chromosome may pair with any other chromosome in the same cell in meiosis. (c) The gametes of an animal may contain more maternal chromosomes than its body cells contain. (d) Of 10 chromosomes in a mature sperm cell, 5 are always maternal. (e) Of 22 chromosomes in a primary oöcyte, 15 may be paternal. (f) The homologous parts of two chromosomes lie opposite one another in pairing. (g) A sperm has half as many postmitotic chromosomes as a spermatogonium of the same animal.

3.2 In each somatic cell of a particular animal species, there are 46 postmitotic chromosomes. How many should there be in a (a) mature egg, (b) first polar body, (c) sperm, (d) spermatid, (e) primary spermatocyte, (f) brain cell, (g) secondary oöcyte, (h) spermatogonium?

3.3 If spermatogenesis is normal and all cells survive, how many sperm will result from (a) 50 primary spermatocytes and (b) 50 spermatids?

3.4 In humans, a type of myopia (an eye abnormality) is dependent on a dominant gene (*M*). Represent diagrammatically (on the chromosomes) a cross between a woman with myopia who is heterozygous (*Mm*) and a normal man (*mm*). Show the kinds of gametes that each parent could produce and summarize the expected results from the cross.

3.5 Beginning with the myopic woman in Problem 3.4, diagram the oögenesis process that produces the egg involved in the production of a child with myopia. Label all stages.

3.6 In what ways is cell division similar and different in animals and plants?

3.7 How does meiosis differ from mitosis? Consider differences in mechanism as well as end results.

3.8 How does gamete formation in higher plants differ from that in higher animals with reference to (a) gross mechanism and (b) chromosome mechanism?

3.9 How is double fertilization accomplished in plants, and what is the fate of the egg and the endosperm nucleus?

3.10 In humans, an abnormality of the large intestine called intestinal polyposis is dependent on a dominant gene *A*, and a nervous disorder called Huntington's chorea is determined by a dominant gene *H*. A man carrying the gene *A* (*Aahh*) married a woman carrying the gene *H* (*aaHh*). Assume that *A* and *H* are on nonhomologous chromosomes. Diagram the cross and indicate the proportions of the children that might be expected to have each abnormality, neither, or both.

3.11 Beginning with the oögonium in the woman described in Problem 3.10, diagram the steps in the process of oögenesis necessary for formation of the egg that produced an *H* child. Label all stages.

3.12 Diagram completely the process of spermatogenesis involved in the production of the sperm in Problem 3.10 necessary for the production of an *A* child. Label all stages.

3.13 A man produces the following kinds of sperm in equal proportions: *AB*, *Ab*, *aB*, and *ab*. What is his genotype with reference to the genes specified?

3.14 Would greater variability be expected among asexually reproducing organisms, self-fertilizing organisms, or bisexual organisms? Explain.

3.15 If biopsies were taken from follicle tissues of the human ovary at the following developmental periods, what stages in the process of oögenesis might be observed: (a) fifth month of intrauterine development, (b) at birth, (c) 10 years of age, (d) 17 years of age?

Chapter 4

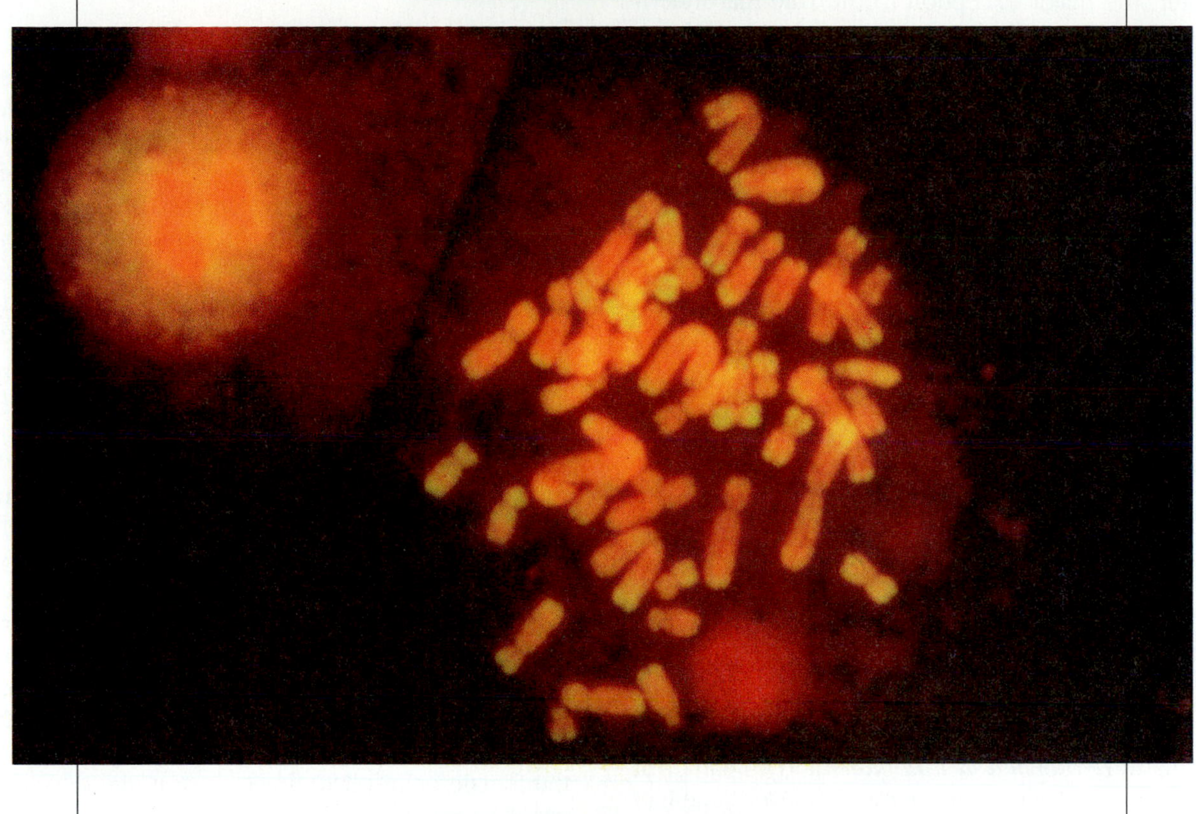

Sex Determination and Sex Linkage

(Photo above): Human chromosomes seen with fluorescent staining. (Courtesy of Jan Hinsch/Science Photo Library/Photo Researchers.)

ONE of the deductions inherent in the results of Mendel's studies of the garden pea was that heredity is equal from both parents (Chapter 2). In this chapter, the major exception to the rule that inheritance is equal from both parents, namely, ***sex linkage,*** is considered. Sex linkage occurs when the gene controlling a trait is located on a chromosome, a so-called ***sex chromosome,*** that is not present in morphologically identical pairs (two homologous copies) in both sexes. Such traits exhibit unique ***sex-linked patterns of inheritance*** that can easily be recognized in pedigree studies.

Before considering sex linkage, one needs to understand the roles of sex chromosomes in controlling sex phenotypes. Thus, this chapter begins with a brief review of the ***mechanisms of sex determination*** and the studies that led to the identification of sex chromosomes in selected species.

MECHANISMS OF SEX DETERMINATION

Nature encompasses a vast array of diverse mechanisms of sex determination in different species. The discussion of sex determination in this chapter is limited to a few species and focuses on two organisms, the fruit fly *Drosophila melanogaster* and humans, that have been especially important in the development of genetic concepts. In these two organisms, and in many other species, especially higher animals, individuals normally exhibit one of two sex phenotypes: ***female*** or ***male.*** In such species, females produce female gametes (eggs, ovules, or macrospores) and males produce male gametes (sperm, pollen, or microspores). Species with separation of the sexes in different individuals are called ***dioecious*** organisms. All higher animals and some higher plants are dioecious. Species in which both male and female gametes are produced by each individual are called ***monoecious*** organisms. In lower animals, the production of both eggs and sperm by the same organisms is more commonly called ***hermaphroditism,*** and individual organisms producing both types of gametes are termed ***hermaphrodites.*** Although the two sex phenotypes are usually quite easily distinguished in humans and fruit flies, this is ***not*** universally the case. In many species, the two sexes are phenotypically indistinguishable except for the reproductive organs. Indeed, in lower eukaryotes, the two genetically distinct types of gametes are sometimes morphologically indistinguishable; this is called ***isogamy*** (***iso*** meaning "same" gametes). Isogamy occurs in several simple eukaryotes such as the green alga *Chlamydomonas reinhardtii* (see the following section). In humans, fruit flies, and most higher eukaryotes, the gametes produced are of two quite different types, namely, eggs and sperm.

Simple Mechanisms: One or a Few Genes

In the green alga *Chlamydomonas reinhardtii*, sexual reproduction involves the fusion of two morphologically identical cells (gametes) that need differ only by the alleles (+ vs. −) at the mating type locus (Fig. 4.1). Similarly, in the yeast *Saccharomyces cerevisiae*, sexual reproduction is very simple and is initiated by fusion of two cells of opposite mating type, again, + vs. − (see Fig. 3.5). Although the process of sexual reproduction is slightly more complex at the morphological level in the common bread mold *Neurospora crassa*, the genetic control of mating type (in this case, *A* vs. *a*) is again monogenic (see Fig. 7.5).

Sex determination in higher eukaryotes is in a few cases known to be controlled or at least strongly influenced by the action of one or a few genes. In fact, the *TDF* gene may prove to be the dominant sex-determining factor in humans (see the section entitled Human Y Chromosome: Dominant Male Determinants, But Few Other Genes). In maize, a monoecious plant with separate male and female flowers (in contrast to the perfect flowers on Mendel's garden peas), the male or staminate flowers occur in the tassel and the female or pistillate flowers are in the developing ear. A substitution of two single gene pairs makes the difference between monoecious and dioecious (separate sexes) maize plants. The allele for barren plant (*ba*), when homozygous, makes the stalk staminate by eliminating silks and ears, whereas the allele for tassel seed (*ts*), when homozygous, transforms the tassel into a pistillate structure that produces no pollen. A plant of the genotype *ba/ba ts/ts* lacks silks on the stalk, but has a transformed tassel and is therefore only pistillate (female). A plant with *ba/ba* ts^+/ts^+ is only staminate (male). These data suggest how monoecious plants could become dioecious, and vice versa, by the alteration (mutation) of just two genes: in this case, ba^+ to *ba* and ts^+ to *ts*.

Identification of Sex Chromosomes

The first investigations relating chromosomes to sex determination were carried out at the turn of the century. H. Henking, a German biologist, discovered in 1891 that a particular nuclear structure could be traced throughout spermatogenesis of certain insects. Half the sperm received this structure and half did not. Henking did not speculate on the significance of this body, but merely identified it as the ***"X body"*** and showed that sperm differed by its presence or absence.

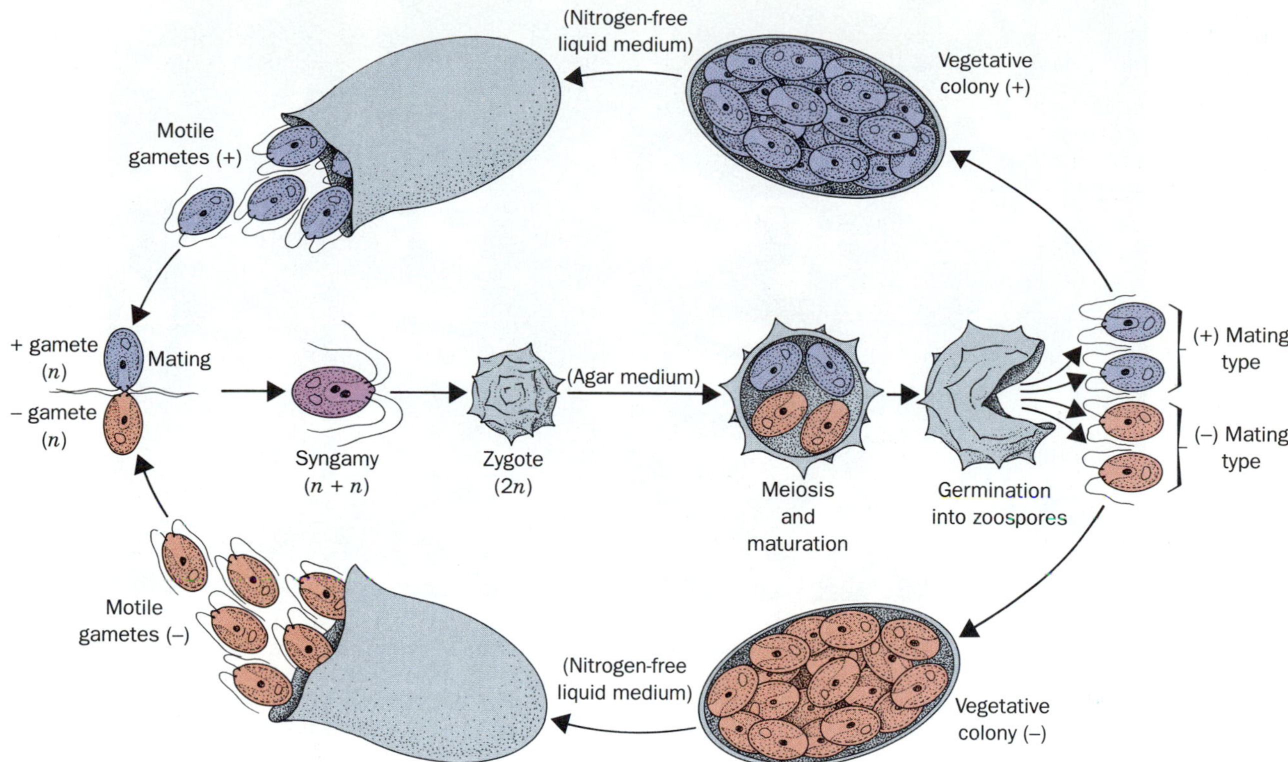

Figure 4.1 Sexual reproduction in the green alga *Chlamydomonas reinhardtii.* (After R. Sager, *Cytoplasmic Genes and Organelles,* Academic Press, New York, 1972.)

In 1902, these observations were verified and extended by C.E. McClung, who made cytological observations on many different species of grasshoppers and demonstrated that the somatic cells in the female grasshopper carry a different chromosome number than do corresponding cells in the male. He followed the X body in spermatogenesis but did not succeed in tracing the oögenesis of the female grasshopper. McClung associated the ***X body with sex determination,*** but erroneously considered it to be peculiar to males. Had he been able to follow oögenesis, his interpretation would undoubtedly have been different.

Contributions to basic knowledge about sex determination were made in the early part of the century by E. B. Wilson and his associates. Wilson reported extensive cytological investigations on several different insects, notably from the genus *Protenor*, an uncommon group of insects closely related to the box elder bug. In these insects, different numbers of chromosomes were observed in germ cells of the two sexes. He succeeded in following oögenesis as well as spermatogenesis and found that unreduced cells of the male carried 13 chromosomes and those of the female carried 14. Some male gametes were found to carry 6 chromosomes, whereas others from the same individual carried 7. The female gametes all had 7 chromosomes. Eggs fertilized with 6-chromosome sperm produced males, and those fertilized with 7-chromosome sperm produced females.

The X body of Henking thus was found to be a chromosome that determined sex. It was identified in several insects and became known as the ***sex or X chromosome***. All eggs of these insects carried an X chromosome, but it was included in only half of the cells forming sperm (Fig. 4.2). All sperm, however, had the usual complement of other chromosomes ***(autosomes)***. Eggs fertilized by sperm containing the X chromosome produced zygotes with ***two X chromosomes,*** which became ***females.*** Eggs receiving sperm without an X chromosome produced zygotes with ***one X,*** which became ***males.*** Males are referred to as ***hemizygous*** ("hemi" or half that present in a zygote) for the X chromosomes or for genes located on the X chromosome. This hemizygosity of males for genes located on the X chromosome is responsible for the altered patterns of inheritance observed for X-linked traits (traits controlled by genes on the X chromosome).

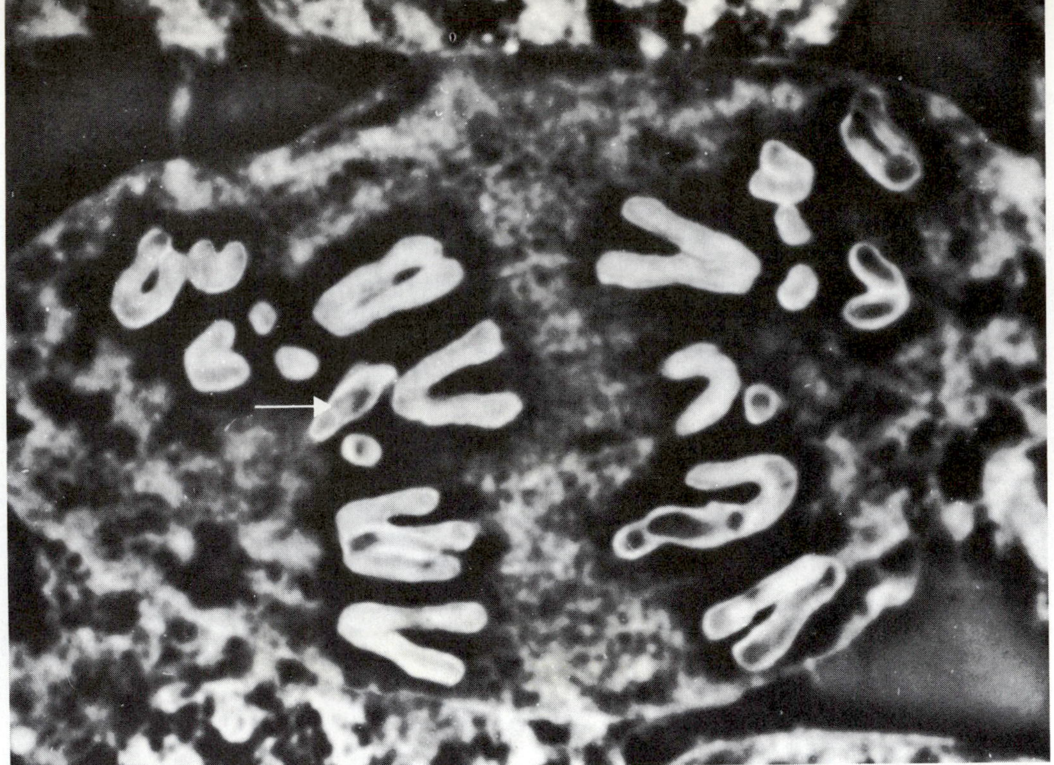

Figure 4.2 Photomicrograph of chromosomes of the grasshopper in anaphase of the first division of spermatogenesis. Twelve chromosomes are at the left of the equatorial plane and 11 are at the right. The difference is the single X chromosome (arrow). Zygotes receiving X chromosomes from the sperm will become females, and those receiving sperm with no X chromosomes will become males. (Courtesy A. M. Winchester, University of Northern Colorado.)

XX–XY Mechanism of Sex Determination

Wilson observed another chromosome arrangement in the milkweed bug, *Lygaeus turcicus*. In this insect, the same number of chromosomes was present in the cells of both sexes. The "homolog" of the X, however, was distinctly smaller and was called the ***Y chromosome***. ***XX zygotes became females, and XY zygotes became males.*** This was called the ***XX–XY type of sex determination***. As evidence was accumulated from a wider variety of animals, the XY mechanism was found to be more prevalent than the XO mechanism. The XY type is now considered characteristic in most higher animals and occurs in some plants (e.g., *Melandrium album*).

The XX–XY pattern of sex chromosomes is also present in the fruit fly *Drosophila melanogaster* and in humans. Both species exhibit the same pattern of transmission of X and Y chromosomes in normal individuals in natural populations. In humans, the X chromosome is considerably longer than the Y chromosome. The total complement of human chromosomes includes 44 autosomes: XX in the female and XY in the male (Fig. 4.3). Eggs produced by the female in oögenesis have the usual complement of autosomes (22) plus an X chromosome. Sperm from the male have the same autosomal number and either an X or a Y. Eggs fertilized with sperm containing a Y chromosome result in zygotes that develop into males; those fertilized with sperm containing an X chromosome develop into females. Segregation of the XY pair and random fertilization thus explain, at least superficially, why some individuals develop into females and some into males, and why about half of the members of each human population are males and half are females.

Historically, the association of the most conspicuous phenotype (i.e., sex) with a particular chromosome greatly strengthened the hypothesis that genes are in chromosomes. This idea originally had been postulated largely because of the parallel observed between the behavior of chromosomes in meiosis and the Mendelian behavior of genes. Control of sex determination by a particular chromosome provided tangible evidence that, indeed, ***genes are in chromosomes***.

Experiments with insects formed a basis for speculation and experimentation concerning the sex-determining mechanisms in higher forms. However, because invertebrate hormones are not functionally comparable with steroid hormones in birds and mammals, animals such as chickens and mice were primarily used in experimental work on secondary sex characteristics (those characteristics that distinguish the two sexes, but have no direct role in reproduction) and hormonal influences on phenotypes.

In species with the XX–XY mechanism of sex determination, the females (XX) produce gametes that all have the same chromosome composition (one X plus one set of autosomes); these females are called the ***homogametic sex*** since all gametes are the same

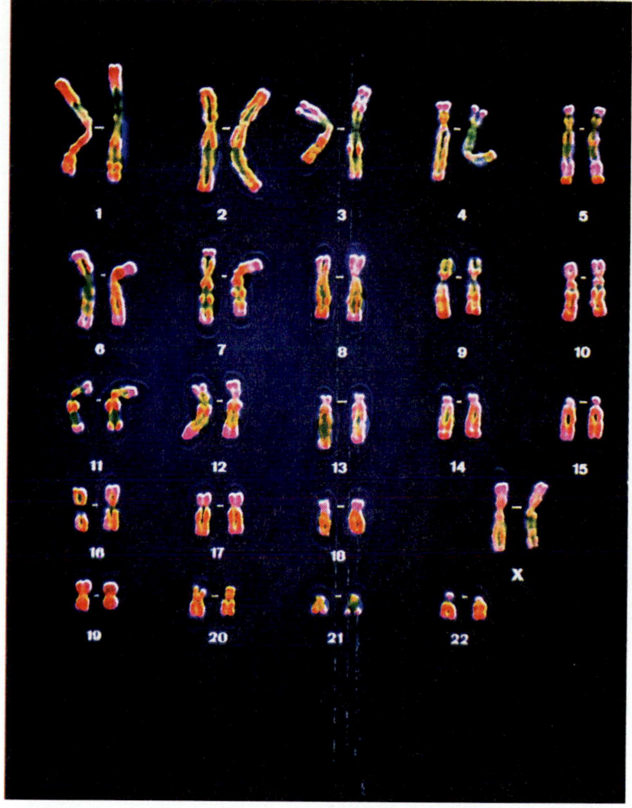

(a)

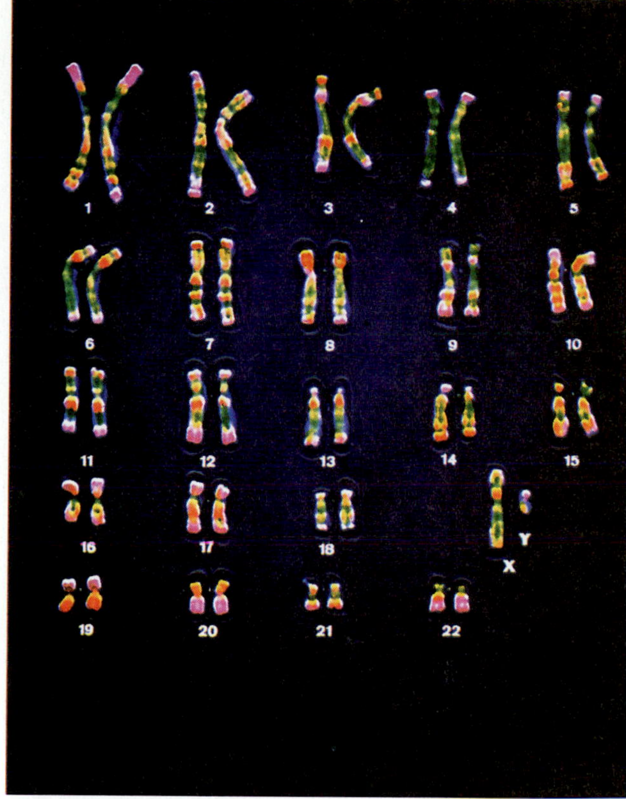

(b)

Figure 4.3 False-color micrographs showing the chromosomes of (a) a woman and (b) a man. The chromosomes were prepared with the normal trypsin/leishman banding technique; X and Y chromosomes as well as the autosomes can be identified by this technique. (CNRI/Science Photo Library/Photo Researchers.)

(***homo***, meaning "same"). The males of these species produce two different types of gametes, one-half contain one X chromosome plus one set of autosomal chromosomes and the other one-half contain one Y chromosome plus one set of autosomes. Males are called ***heterogametic*** (***hetero***, meaning "different") because they produce two different kinds of gametes.

Species with Heterogametic Females

In many species, including most birds, moths, and some fish, a chromosomal mechanism of sex determination occurs that is basically identical to the XX–XY mechanism, but with the ***females*** being ***heterogametic*** (usually designated as ***ZW***) and ***males*** being ***homogametic*** (usually designated ***ZZ***). This mechanism of sex determination is sometimes called ZZ–ZW. However, note that mechanistically this system is identical to the XX–XY system, but with the relationship between sex chromosomes and sex phenotypes reversed. Stated differently, in birds, the chromosome composition of the egg determines the sex of the offspring, whereas in humans and fruit flies, the chromosome composition of the sperm determines the sex of the offspring.

The Y Chromosome and Sex Determination in Mammals

In both *Drosophila* and humans, normal females have an XX sex chromosome composition and normal males have an XY sex chromosome composition. Thus, it might be tempting to assume that in both species the genes for femaleness are on the X chromosomes and the genes for maleness are on the Y chromosomes. Unfortunately for genetics students, this is ***not*** the case. Instead, these two species have very different chromosomal distributions of the genes controlling sex phenotype. In mammals, the presence of a ***Y chromosome is required for the development of a male sex phenotype***. In contrast, as is discussed in the next section, the Y chromosome plays no significant role in sex determination in *Drosophila*.

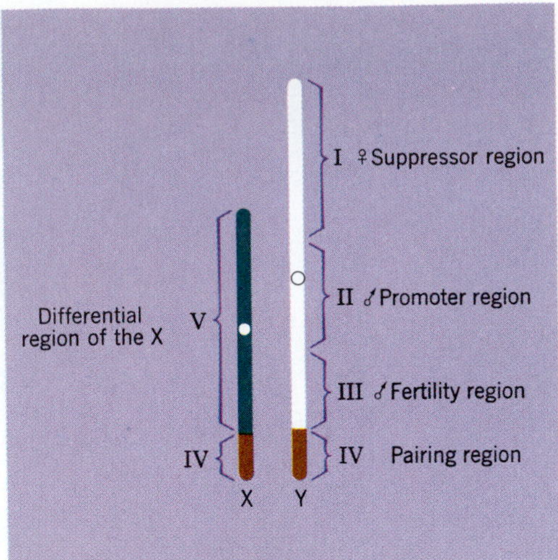

Figure 4.4 Sex chromosomes in *Melandrium.* Regions I, II, and III of the Y chromosome do not have homologous segments in the X, and thus they make up the differential portion of the Y. Regions IV are homologous in the X and Y and are pairing regions at meiosis. Region V is the differential portion of the X chromosome. When I is lost from a Y chromosome, a bisexual plant is produced. When II is lost, a female plant is produced. If region III is absent, male-sterile plants with abortive anthers appear. (After M. Westergaard, *Hereditas* 34: 257–279, 1948.)

Irregular sex chromosome constitutions are fairly common in humans (see Chapter 19). Surprisingly, X chromosomes present in any number (e.g., XXX or XXXX) ***in the absence of a Y chromosome give rise to a female sex phenotype***. A Y chromosome is required for maleness; moreover, the presence of a single Y chromosome is sufficient even in the presence of several X chromosomes (e.g., XXXXY). The Y chromosome induces development of the undifferentiated gonadal medulla into a testis, whereas an XX chromosomal complement induces the undifferentiated gonadal cortex to develop ovaries. The gene on the Y chromosome in humans that is responsible for the development of the testis is called *TDF* (for Testis Determining Factor). The *TDF* gene has been isolated (cloned), characterized structurally (sequenced), and found to encode a protein with features that suggest that it acts by regulating the expression of other genes. If this proves to be true, the *TDF* gene may be the master regulator that triggers the expression of a large number of genes that produce the male sex phenotype. In the absence of the *TDF* gene, the genes that produce the female sex phenotype would be expressed. The *TDF* gene (and possibly other male-determining genes on the Y chromosome) exhibits a very dominant effect on the development of the sex phenotype. Even in the presence of three or more X chromosomes, a single Y chromosome is usually sufficient to produce testes and male characteristics. A Y chromosome with dominant sex-determining properties has also been demonstrated in *Axolotl* (an amphibian genus) and *Melandrium* (a plant genus).

In *Melandrium album*, which follows the XY mode of sex determination, H.E. Warmke, M. Westergaard, and others have shown that sex is determined by a balance between male-determining genes on the Y and female-determining genes on the X and on the autosomes. In this plant, which is normally dioecious, XY individuals are staminate (i.e, pollen bearing) and XX plants are pistillate (egg bearing). The Y chromosome is the largest and most conspicuous member of the complement.

Experimental procedures using spontaneous fragmentations have resulted in mapping of major sections of the *Melandrium* Y chromosome (Fig. 4.4). Three distinct regions influencing sex determination and male fertility have been localized on the differential part of the Y chromosome (which does not have a homologous part on the X). Region I suppresses femaleness. In the absence of this region, plants are bisexual, that is, they express both male and female characteristics. Region II promotes male development. When this region (with or without Region I) is missing, a female plant is produced. Region III carries male fertility genes; loss of this region results in male sterility. A part of the Y chromosome is homologous with a part of the X, but the major part of the X is differential with no structural counterpart on the Y. Although most female-determining genetic material is on the X chromosome, Westergaard found that autosomes were also involved in female determination.

The Balance Concept of Sex Determination in *Drosophila*

Soon after sex chromosomes were identified, it became obvious that sex determination in *Drosophila* was more complicated than preliminary observations had indicated. A more intricate mechanism than segregation of a single pair of chromosomes was in evidence. The investigations on *Drosophila* by C. B. Bridges (Fig. 4.5) showed that female determiners were located on the X chromosomes and male determiners were on the autosomes. No specific loci were identified at the time, but more recent evidence has demonstrated that many chromosome segments are involved. Specifically, female-determining genes were shown to be carried on the X chromosomes, and male-determining genes were shown to be located on the three autosomal chromosomes of *Drosophila*. The

Figure 4.5 Calvin B. Bridges (1889–1938), American cytogeneticist who pioneered in research on chromosomes of *Drosophila melanogaster*. He prepared maps of the giant salivary chromosomes of *Drosophila*. (Courtesy of Everett Thorpe.)

TABLE 4.1 Ratio of X Chromosomes to Autosomes and Corresponding Sex Type in *Drosophila melanogaster*[a]

X CHROMOSOMES (X) AND SETS OF AUTOSOMES (A)	RATIO X/A	SEX
1X 2A	0.5	Male
2X 2A	1.0	Female
3X 2A	1.5	Metafemale
4X 3A	1.33	Metafemale
4X 4A	1.0	Tetraploid female
3X 3A	1.0	Triploid female
3X 4A	0.75	Intersex
2X 3A	0.67	Intersex
2X 4A	0.5	Tetraploid male
1X 3A	0.33	Metamale

[a]After Bridges (1925).

genetic balance theory of sex determination was devised to explain the mechanics of sex determination in *D. melanogaster*.

XO or XY chromosome segregation was interpreted as a means of tipping the balance between maleness and femaleness, whereas more deep-seated processes were involved in the actual process of sex determination. Bridges experimentally produced various combinations of X chromosomes and autosomes in *Drosophila* and deduced from comparisons that one X chromosome (X) and two sets of autosomes (A) produced a normal male. Normal males had a ratio of X chromosomes to sets of autosomes of 0.5. Note that one set of autosomes consists of three chromosomes, one each of chromosomes II, III, and IV. (The X chromosome of *Drosophila* is chromosome I.) This combination of one X and two A's resulted in a normal, diploid male; the combination of two X chromosomes and two sets of autosomes (2X + 2A, ratio of 2:2 = 1) produced a normal diploid female (Table 4.1).

The first irregular chromosome arrangement from Bridges' experiments resulted from ***nondisjunction***, the failure of paired chromosomes to separate in anaphase. X chromosomes, which ordinarily come together in pairs during the meiotic prophase of oögenesis and separate to opposite poles during anaphase, remained together and migrated to the same pole. As a result, some female gametes received two X chromosomes and some received no X chromosome (Fig. 4.6). Following fertilization by sperm from wild-type males (AAXY), all zygotes had $2n$ autosomes (2A), but some received two X's from the mother and an X from the father (3X). The ratio of X chromosomes to sets of autosomes of 3:2 resulted in flies, called metafemales, that were highly inviable. The XXY flies (2X/2A) from the same mating were normal females in appearance; XO (1X/2A) males were sterile, and those with a Y chromosome but no X chromosome did not survive. These results indicated that, in *Drosophila*, the Y chromosome is not involved in sex determination. However, the Y chromosome of *Drosophila* is required for male fertility. In XO *D. melanogaster*, sperm develop but are nonmotile. Interestingly, in a related species of fruit flies, *D. hydei*, sperm do not develop in XO males, indicating that the Y chromosome plays additional roles in spermatogenesis in this species.

Flies produced experimentally with 4X/3A were also metafemales. Those with 4X/4A and also those with 3X/3A, both with an X/A ratio = 1, were females. The combinations 3X/4A = 0.75 and 2X/3A = 0.67 were intermediate in characteristics between males and females and were called "intersexes." Combinations of 2X/4A = 0.5 were males and those of 1X/3A = 0.33 were metamales.

No other animal or plant species has been investigated as thoroughly as *D. melanogaster*, but considerable evidence suggests that similar balance mechanisms of sex determination are probably operating in

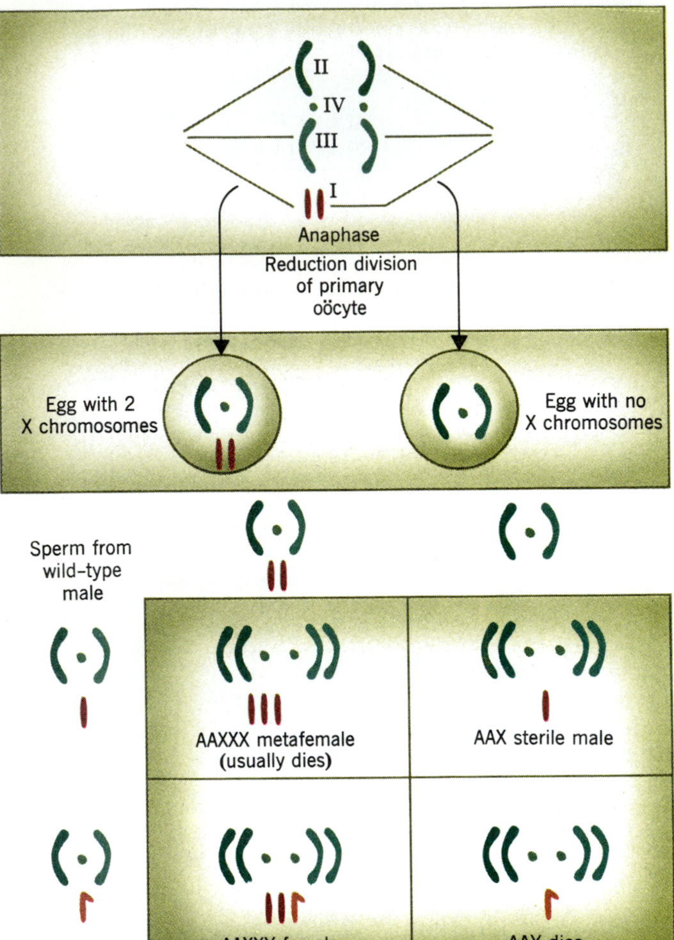

Figure 4.6 Nondisjunction in *Drosophila,* and the zygotes that result from fertilization by wild-type males (AAXY). The AAXXY females and AAX males were exceptional flies in Bridges' experiment. In the primary oöcyte, autosomes II and III are represented by pairs of bent rods; autosomes IV, which are small and take their places in the center of the equatorial plate, are represented by a pair of dots. The X chromosomes (I) are symbolized as short rods; the Y chromosome introduced by the sperm is illustrated by a rod with a hook or short arm. (After Bridges, 1925.)

many other organisms. ***Intersexes*** have been produced experimentally in some animals by upsetting this balance during various stages of development.

Haplodiploidy and Sex Determination in Hymenoptera

More involved mechanisms of sex determination have been described in the insect order Hymenoptera, which includes ants, bees, wasps, and sawflies. In several species of Hymenoptera, ***males develop parthenogenetically*** (from unfertilized eggs) and have a haploid chromosome number (16 in the drone honeybee *Apis mellifera*). The ***queen*** honeybee and the ***workers***, which arise from ***fertilized eggs***, carry the diploid chromosome number (32). In the parasitoid wasp *Bracon hebetor* (formerly *Habrobracon juglandis*), females are diploid with 20 chromosomes, and males are haploid with 10 chromosomes. Females originate from fertilized eggs and males from unfertilized eggs. Because normal males are haploid and normal females are diploid, this mechanism of sex determination is often referred to as ***haplodiploidy***.

Results of experiments by Whiting showed that the homozygous, heterozygous, or hemizygous (gene present in single dose) status of certain chromosome segments controls sex determination. Female determination depends on heterozygosity for part of a chromosome. If different forms of this segment of the chromosome involved are designated X_a, X_b, and X_c, then individuals of chromosome makeup X_aX_b, X_aX_c, or X_bX_c are all females. Hemizygous individuals (X_a, X_b, or X_c) cannot be heterozygous and are therefore male. Whiting showed that sex determination depends on the genetic composition of this region of the chromosome, and not on diploidy versus haploidy per se, by using genetic manipulations to produce homozygous diploid males (X_aX_a, X_bX_b, or X_cX_c).

Mosaics and Gynandromorphs

Abnormal chromosomal behavior in insects can result in the formation of ***gynandromorphs*** or ***sexual mosaics*** in which some parts of the animal are female and other parts are male. When such abnormal chromosome transmission involves autosomes harboring genes that control easily recognized phenotypes, individuals may also be produced that are mosaic for phenotypes unrelated to sex phenotype. Some gynandromorphs in *Drosophila* are bilateral intersexes (Fig. 4.7), with male color pattern, body shape, and sex comb on one half of the body and female characteristics on the other half. Both male and female gonads and genitalia may be present.

Bilateral gynandromorphs have been explained on the basis of an irregularity in mitosis at the first cleavage of the zygote. Infrequently, a chromosome lags in division and does not arrive at the pole in time to be included in the reconstructed nucleus. When one of the ***X chromosomes*** of an XX (female) zygote ***lags*** in the spindle, one daughter nucleus receives only one X chromosome, while the other receives two X chromosomes, as illustrated in Fig. 4.8. A mosaic body pattern is thus established. One nucleus in the two-nuclei stage would be XX (female) and one would be XO (male). If the cleavage plane is oriented so that one

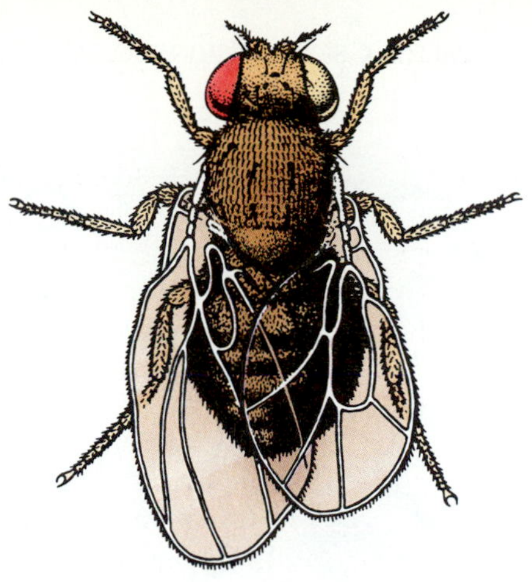

Figure 4.7 Bilateral gynandromorph in *Drosophila,* showing male characteristics including sex comb on the right foreleg. The white eye on the right is the expression of the sex-linked gene *w* when carried in one (hemizygous) X chromosome. (After T. H. Morgan and C. B. Bridges, "Contributions to the genetics of *Drosophila melanogaster.* I. The origin of gynandromorphs." *Carnegie Inst. Wash. Publ.* 278: 1–122, 1919.)

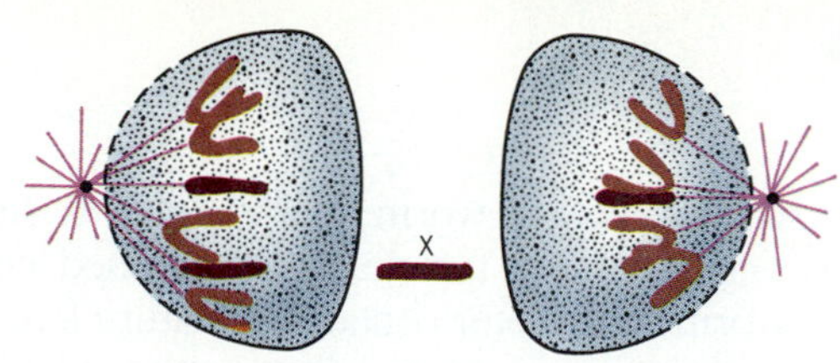

Figure 4.8 A lagging X chromosome in the first cleavage of *Drosophila* illustrating the origin of a bilateral gynandromorph. (After Morgan and Bridges, 1916.)

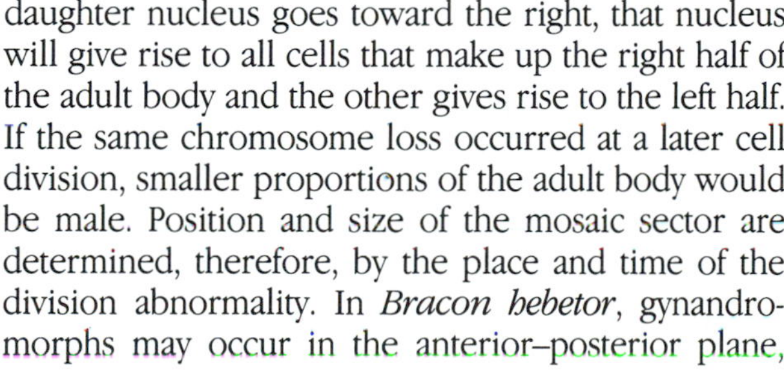

daughter nucleus goes toward the right, that nucleus will give rise to all cells that make up the right half of the adult body and the other gives rise to the left half. If the same chromosome loss occurred at a later cell division, smaller proportions of the adult body would be male. Position and size of the mosaic sector are determined, therefore, by the place and time of the division abnormality. In *Bracon hebetor*, gynandromorphs may occur in the anterior–posterior plane, giving rise to wasps with such peculiar arrangements as male heads with female abdomens and female heads with male abdomens.

Genetically induced mosaics provide one of the most powerful mechanisms for studying pathways of development, especially cell lineages, in multicellular organisms. Such mosaicism may be induced by several genetic methods to facilitate studies of development (Chapter 15).

ENVIRONMENTAL FACTORS AND SEX DETERMINATION

In some lower animals, sex determination is nongenetic and depends on factors in the external environment. Males and females have similar genotypes, but stimuli from environmental sources initiate development toward one sex or the other. Males of the marine worm *Bonellia*, for example, are small and degenerate and live within the reproductive tract of the larger female (Fig. 4.9). All organs of his body are degenerate except those of the reproductive system. F. Baltzer

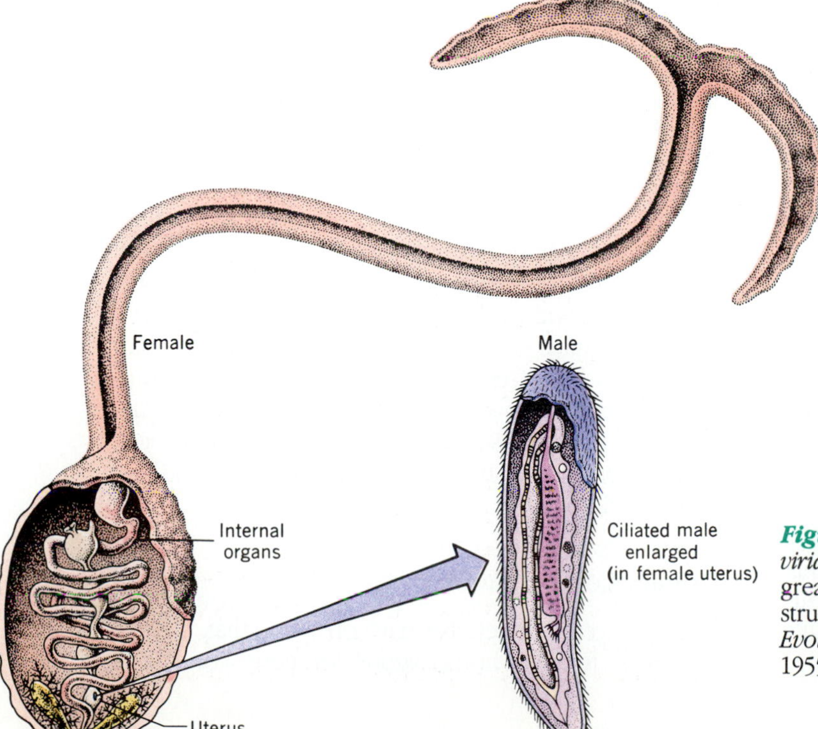

Figure 4.9 Female and male of the marine worm *Bonellia viridis.* The male is shown in the uterus of the female and is greatly enlarged at the right to show details of internal structure. Redrawn with permission from T. Dobzhansky, *Evolution, Genetics and Man,* Wiley, New York, 1955. (© 1955 by John Wiley & Sons, Inc.)

found that any young worm reared from a single isolated egg became a female. If he released newly hatched worms into water containing mature females, however, some young worms were attracted to females and became attached to the female proboscis. These were transformed into males and eventually migrated to the female reproductive tract, where they became parasitic. Genetic determiners for both sexes are obviously present in all young worms. Extracts made from the female proboscis influence young worms toward maleness.

In some reptiles, the temperature at which the fertilized eggs are incubated prior to hatching plays a major role in determining the sex of the offspring. Surprisingly, high temperatures during incubation have opposite effects on sex determination in different species. In the turtle *Chrysema picta*, high egg incubation temperatures result in the production of mostly female progeny, whereas in the lizard *Agama agama*, high incubation temperatures result in mostly male offspring.

These examples are included to illustrate the diversity of sex determination mechanisms that exist in nature. Although the segregation of specific sex-determining genes and chromosomes is responsible for sex phenotype in most species, the genetic potential for both maleness and femaleness is present in every zygote in species such as *Bonellia, Chrysema picta*, and *Agama agama*, and some specific factor in the environment triggers the expression of either the genes producing a male phenotype or the genes producing a female phenotype.

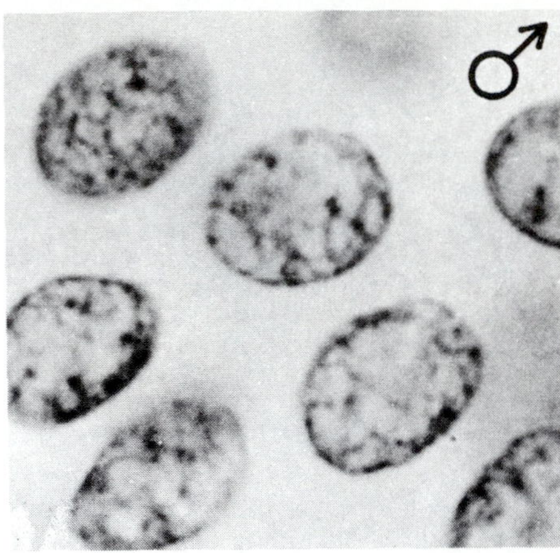

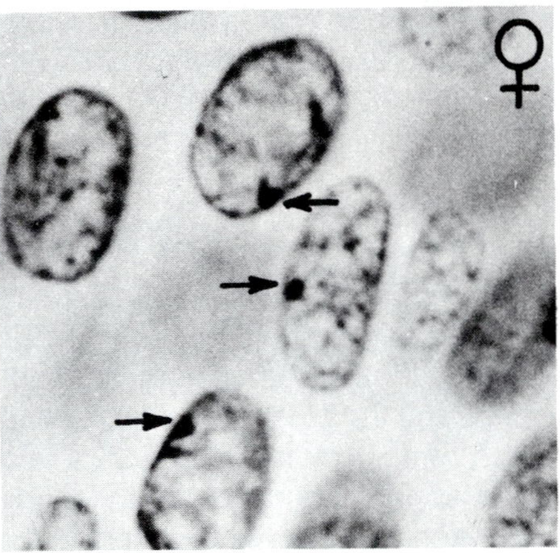

Figure 4.10 Cells of human epidermis illustrating the sex chromatin bodies or Barr bodies. (bottom) Epidermal cells of a female showing sex chromatin bodies (see arrows). (top) Epidermal cells of a male showing no sex chromatin bodies. (Courtesy of M. L. Barr.)

SEX DIFFERENTIATION

Sex Chromatin Bodies

M.L. Barr observed ***chromatin bodies*** in the nerve cells of female cats that were not present in cells of the male. Barr and others reported a consistent cytological difference between the nuclear contents of human male and female cells in several tissues, including epithelium of buccal mucosa (lining inside of the mouth). With appropriate staining techniques, a small chromatin body could be identified in the nucleus of female cells. This ***Barr body*** was related to the number of X chromosomes and was not observed in normal male cells. Nuclei from cells of the two sexes are illustrated in Fig. 4.10. With this technique, the sex of human embryos can be distinguished at early stages of development. More recent studies have shown that most if not all tissues of female mice have some cells with Barr bodies. This cellular characteristic generally seems to apply to all mammals.

Sex chromatin bodies not only distinguish normal female cells from male cells, but are also useful for diagnosing various kinds of sex chromosome abnormalities in humans. Those who have two or more X chromosomes have ***one less chromatin body than the number of X chromosomes present***. Cells of abnormal females with only one X chromosome (Turner's syndrome) have no sex chromatin bodies, and cells of males with two X chromosomes and one Y chromosome (Klinefelter's syndrome) have one sex chromatin body (Chapter 19). Abnormal females with three X chromosomes have two sex chromatin bodies in their cell nuclei.

Dosage Compensation

For many years, geneticists have observed that, in most instances, females homozygous for genes on the X chromosomes do not express a trait more markedly than do hemizygous males. Thus, there must be a mechanism of ***"dosage compensation"*** through which the ***effective dosage of genes of the two sexes is made equal*** or nearly so. Several investigators arrived almost simultaneously at a hypothesis, explaining that the sex chromatin body and dosage compensation in mammals is determined by the inactivation of one X chromosome in the normal female. The hypothesis was named after Mary F. Lyon (Fig. 4.11), who first formulated it in detail from cytological observations and genetic studies on coat color in mice. Female mice, heterozygous for certain coat-color genes, show a mottled effect unlike the homozygotes and very distinct from a uniform intermediate coat color. The fur color pattern (Fig. 4.12) is a ***mosaic*** made up of ***randomly arranged patches*** of the two colors. Normal male mice and abnormal XO female mice never have the mottled phenotype.

The Lyon hypothesis was based on the observation that the number of sex chromatin bodies in interphase cells of adult females is one less than the number of X chromosomes observed in metaphase preparations. It was therefore proposed that the chromatin body is a heterochromatinized X chromosome. If this premise is correct, only one X chromosome is required for normal metabolism in cells of females. Any additional X chromosomes are condensed and become heteropyknotic (i.e., stain more densely than other chromosomes) and are genetically inactive. Thus, the mammalian X chromosome has the capability of being heterochromatic (see Chapter 6) in some cells and euchromatic in others.

Figure 4.12 Female mouse heterozygous for an X-linked gene that controls the coat color tortoiseshell. The mosaic phenotype of such females provided one of the first lines of evidence for the Lyon hypothesis. (Courtesy of The Jackson Laboratory.)

Figure 4.11 Mary F. Lyon, British geneticist who with others formulated the Lyon hypothesis to explain the sex chromatin body and gene dosage compensation in mammals. The hypothesis is based on the inactivation of one X chromosome in the normal female. (Courtesy of Godfrey Argent.)

Which X chromosome becomes inactive is a matter of chance, but once an X has become inactivated, all cells arising from that cell will keep the same inactive X chromosome (a sex chromatin body). In the mouse, the inactivation apparently occurs early in development; in human embryos, sex chromatin bodies have been observed by the sixteenth day of gestation. Some human traits could therefore be influenced by both X chromosomes during the first 16 days. Later (after the sixteenth day), only one X chromosome is functional in a given body cell. Thus, the female is a mosaic with some parts having the alternate allele expressed. X chromosome inactivation occurs only when at least two X chromosomes are present. When several X chromosomes are in the same nucleus, all but one are inactivated. The number of sex chromatin bodies present after inactivation is, therefore, one less than the number of X chromosomes in the original cell.

The Lyon hypothesis explains certain genetic consequences of genes in mammals: (1) the dosage compensation for females with two X chromosomes that regulates enzyme activity to the level of males with only one X chromosome and (2) the variability of expression in heterozygous females because of the random inactivation of one or the other of the two X chromosomes.

Early support for the hypothesis that, in mammalian females, dosage compensation occurs through inactivation of all but one X chromosome, thus producing functional equivalence in the two sexes, came from studies of glucose-6-phosphate dehydrogenase (G-6-PD) activity in cells. ***Enzymatic activity*** was shown to be about equal in the two sexes, that is, activity in females with two X chromosomes was approximately equal to that in males with one X chromosome. Two alleles of the G-6-PD locus, which produce electrophoretically distinct enzymes (F and S), were tested with heterozygous females. Isolated cells from several biopsies of skin from the same heterozygous person were cloned. Each clone contained either the F or S form of the enzyme, but never both. This supports the premise that an X chromosome received from either the mother or the father (of the female being tested) was inactive. After inactivation has occurred, the clone of cells derived from one progenitor cell all express the same allele.

"Fragile X" Chromosome and Mental Retardation in Humans

The inactivation of one of the two X chromosomes in female mammals must, of course, be reversible since females transmit both of their X chromosomes to their progeny in a functional state. This is especially clear in the case of hemizygous male progeny, which receive either of the X chromosomes of the mother with equal probability, because the single X chromosome that each son receives must be fully active given that the X chromosome contains many genes that are vital to growth and development, indeed, to survival. The reactivation of the inactive or heterochromatic X chromosomes of mammalian females occurs in germ cell lineages prior to oögenesis; both X chromosomes of a female are active in oögonial cells. Since normal patterns of transmission of traits controlled by genes on the X chromosome require reactivation of heterochromatic X chromosomes, one is led to wonder whether normal reactivation ever fails to occur. Indeed, considerable evidence suggests that abnormal reactivation of heterochromatic X chromosomes may be partially responsible for the most common form of inherited mental retardation in humans, the so-called ***"fragile X" syndrome***.

The ***"fragile X" chromosome*** of humans contains a ***fragile site*** near the end of the long arm (Fig. 4.13). Fragile sites in chromosomes are defined cytologically (examining the chromosomes with a microscope) as specific regions of chromosomes that exhibit constrictions, gaps, or breaks in cells growing in culture (Figs. 4.13*a* and 4.13*b*). The human "fragile X" chromosome and the syndrome of mental retardation with which it

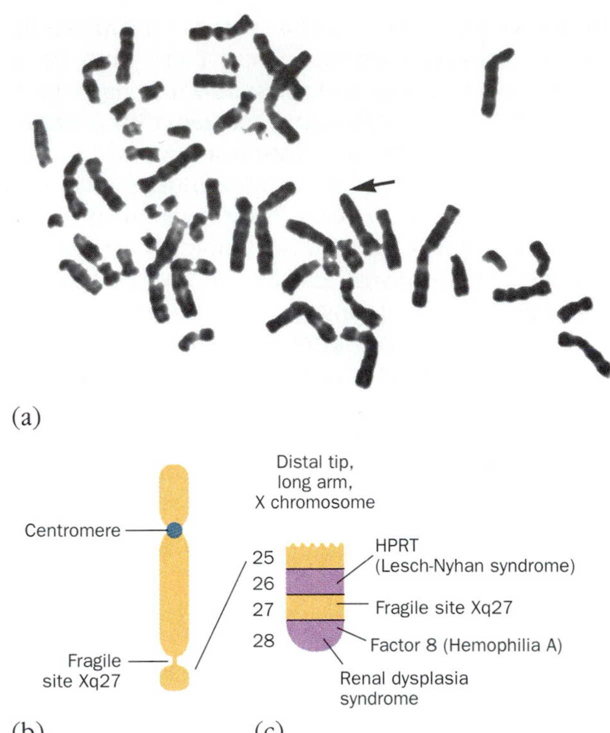

Figure 4.13 Structure of the "fragile X" chromosome that is associated with the most common form of mental retardation in humans. (a) Photomicrograph of the metaphase chromosomes of a patient with Martin–Bell Syndrome showing a "fragile X" chromosome. The arrow points to the fragile site at position Xq27 (X chromosome, q for long arm, 27th "band" or segment from the centromere—see Chapter 18 for a detailed description of the symbolism for human chromosome segments). (Photograph courtesy of Jaroslav Cervenka, Dept. of Oral Sciences, University of Minnesota.) (b) Schematic drawing of the "fragile X" chromosome showing the typical Xq27 fragile site. (c) Partial genetic map of the distal end of the long arm of the X chromosome showing the position of the Xq27 fragile site relative to the positions of genes controlling other well-known traits. The intervals defined by chromosome staining patterns are numbered 25 through 28; numbering always starts with the centromere-proximal band. HPRT is the abbreviation for hypoxanthine-guanine phosphoribosyltransferase, the enzyme that is missing or defective in individuals with Lesch–Nyhan syndrome (see p. 88). Factor 8 is a protein required for normal blood clotting; active factor 8 is lacking in individuals with hemophilia A. [(b) and (c) after G. R. Sutherland, "The enigma of the fragile X chromosome," *Trends in Genetics* 1: 108–112, 1985.]

is associated occur in about one in 2000 to 2500 live births. This particular X chromosome contains a fragile site at position Xq27 (Fig. 4.13*c*). The fragile site in this chromosome is inherited as a codominant with both normal and gapped X chromosomes in cells from heterozygotes. Genetic studies of the "fragile X" syndrome were initially very difficult to interpret because the "fragile X" chromosome exhibits variable expressivity and incomplete penetrance. The affected individuals show a wide range in severity of mental retardation and other phenotypic manifestations of the syndrome (e.g., autistic and hyperactive behavior, prominent jaws and foreheads, or longer than normal ears).

An important finding regarding the "fragile X" chromosome was that the daughters of males who carried the "fragile X" chromosome, but did ***not*** exhibit the phenotypic characteristics of the syndrome, also did ***not*** show any manifestations of the syndrome. This observation indicated that the "fragile X" syndrome depended not only on the presence of the fragile site at Xq27 (and, presumably, some alteration of the X chromosome at this position giving rise to the anomalous "fragile X" chromosome), but also on some event that triggered the manifestation of the presence of this fragile site. Moreover, the event that triggered this manifestation could occur ***only in females***.

One attractive hypothesis that fits all the available data is that the Xq27 alteration (its molecular nature is still unknown) somehow ***interferes with reactivation of the heterochromatic "fragile X" chromosome of females*** that normally occurs in preoögonial cells. This could result in women who carried a "fragile X" chromosome producing offspring with an inactive or not-fully-active X chromosome. Given that many genes on the X chromosome of humans are essential to normal growth and development, the manifestations observed in individuals with "fragile X" syndrome might well be expected. Although the accuracy of this hypothesis is not yet established, it does nicely explain why the "state" of the "fragile X" chromosome ("state" referring to "off" in affected individuals and "on" in unaffected individuals with the "fragile X" chromosome) can only be changed in females, not in males. Fathers always transmit the "fragile X" chromosome to their daughters in the same "state" as they received it from their mothers, that is, it always causes similar manifestations (or lack thereof) in fathers and daughters. Obviously, an important goal of geneticists is to define the molecular basis of the X chromosome inactivation–activation cycle in the near future and, if possible, develop treatments to prevent mental retardation and other manifestations in individuals carrying a "fragile X" chromosome.

Hormones and Sex Differentiation

The hormonal system that regulates the internal or physiological environment of the organism does not directly influence the fundamental process of sex determination. It is important, however, to the development of the ***secondary sex characteristics***. Sex hormones of higher animals are synthesized by the ovaries, testes, and adrenal glands with stimulation from pituitary gland hormones. The adrenals produce steroids that are chemically related to those of the gonads and that also influence the secondary sex characteristics. Ovaries and testes each have a dual function; they are responsible for the production of the gametes, the eggs and sperm, as well as the sex hormones. These hormones influence the development of secondary sexual characteristics such as physiological difference (rate of metabolism, blood pressure, heart beat, and respiration), bone structure, voice, breast development, and hair.

SEX-INFLUENCED DOMINANCE

The dominance of alleles may differ in heterozygotes of the two sexes. This phenomenon is called ***"sex-influenced dominance."*** Gene-products of heterozygotes in the two sexes may be influenced differentially by sex hormones. For example, autosomal genes responsible for horns in some breeds of sheep may behave differently in the presence of the ***male and female sex hormones***. The results are equally well explained by secondary effects due to differences in programming. More than a single pair of genes is involved in the production of horns; but assuming all other genes to be homozygous, the example can be treated as if only a single pair of alleles were involved. Among Dorset sheep, both sexes are horned, and the gene for the horned condition is homozygous (h^+h^+) in all animals of the breed. In Suffolk sheep, neither sex is horned, and the genotype is hh. Among the heterozygous F_1 progeny from crosses between these two breeds, horned males and hornless females are produced. Because both sexes are genotypically alike (h^+h), the gene must behave as a ***dominant in males*** and as a ***recessive in females***, that is, only one allele is required for an expression in the male, but the allele must be homozygous for expression in the female.

When F_1 hybrids are mated, a ratio of 3 horned to 1 hornless is produced among F_2 males, whereas a ratio of 3 hornless to 1 horned is observed among the F_2 females. Genotypes and phenotypes of the two sexes are summarized in Table 4.2. The only departure from the usual pattern is concerned with the heterozygous h^+h genotype. This genotype in the male sheep results in the horned condition, but females with the same genotype are hornless. Dominance of the gene is apparently influenced by sex hormones.

TABLE 4.2 Expression of *b* Alleles in Sex-Influenced Inheritance

GENOTYPES	MALES	FEMALES
b^+b^+	Horned	Horned
b^+b	Horned	Hornless
bb	Hornless	Hornless

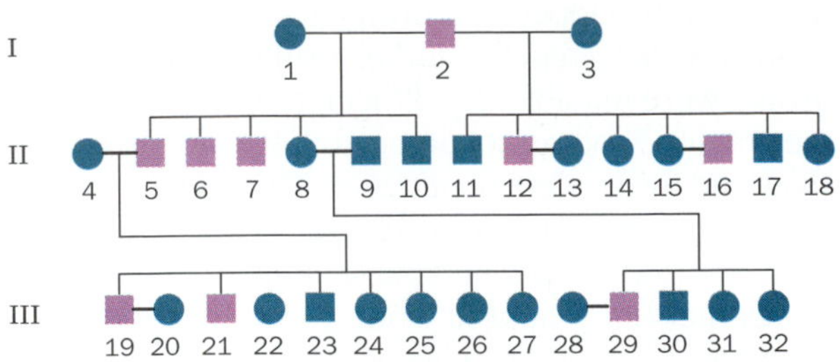

Figure 4.15 Pedigree showing the incidence of premature baldness in a family group. The men represented by the darkened squares were bald-headed before they reached the age of 35. Those symbolized by light squares were over 35 when the study was made and all had thick hair. No women in this family group expressed the trait.

Some human traits, such as a certain type of white forelock, absence of upper lateral incisor teeth, a particular type of enlargement of the terminal joints of the fingers, and premature pattern baldness (Fig. 4.14), have been reported to exhibit a pattern of inheritance characteristic of sex-influenced dominance (Fig. 4.15).

It is important to emphasize that ***traits that show the pattern of inheritance characteristic of sex-influenced dominance*** (Fig. 4.15) ***are controlled by genes located on autosomal chromosomes,*** not on the sex chromosomes. Traits that are governed by genes located on the sex chromosomes exhibit a very distinctive pattern of transmission called sex-linked inheritance (discussed later in this chapter).

SEX-LIMITED GENE EXPRESSION

One sex may be uniform in expression of a particular trait and yet transfer genes that produce a different phenotype in offspring of the other sex. This is called ***"sex-limited gene expression." Sex hormones*** are apparently ***limiting factors*** in the expression of some genes. Other factors may also be involved in controlling the expression of sex-limited characteristics. Milk production in mammals, for example, is limited to females, but certain bulls are in great demand among dairy breeders and artificial insemination associations because their mothers and daughters have increased milk-production records. Another classic example of a sex-limited trait is so-called "cock-feathering" in many different birds. In chickens, roosters generally have long, pointed tail feathers, whereas hens have rather short, blunt tail feathers. Similar examples of "cock-" and "hen-feathering" are seen in many wild and domestic birds.

Figure 4.14 Young man with premature balding beginning on the crown of his head. (Copyright Verjik Martin/Amstock.)

SEX-LINKED INHERITANCE

When parallelism was discovered between the X chromosome cycle and sex determination, it was generally assumed that genes other than sex determiners were also located on the X chromosome. Indeed, many mutant genes such as the $X_g{}^a$ blood alleles in humans have been shown to be located on the X chromosome. Because of their location on the same chromosome as sex determiners, they are said to be ***"sex-linked."***

Morgan's Discovery of Sex Linkage in *Drosophila*

The first extensive experimental evidence for sex linkage in a particular species came in 1910 with the discovery by T. H. Morgan of a white-eyed mutant in *Drosophila*. A gene had undergone a change that resulted in a phenotypic alteration. This change expressed itself as white eyes rather than the normal red eyes. The white-eyed male first discovered was mated with a red-eyed female. The F_1 flies were all red-eyed,

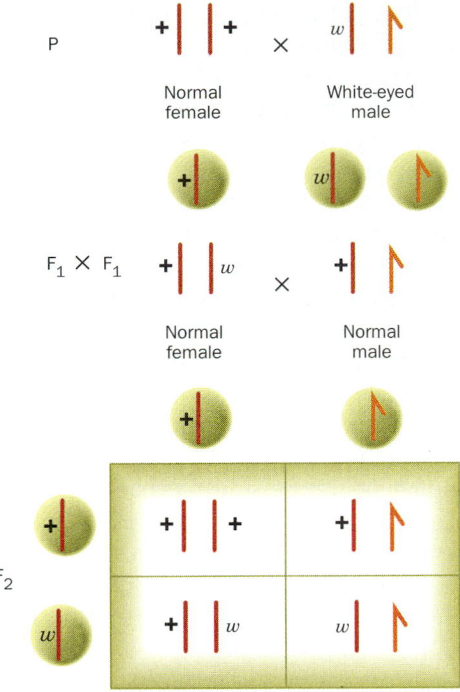

Figure 4.16 A cross illustrating sex linkage in *Drosophila*. This cross is between a red-eyed female and a white-eyed male. Alleles are identified with their chromosomes. The ↾ represents a Y chromosome.

but the F_2 included both red and white in the proportion of about 3 red to 1 white. All white-eyed flies in the F_2 generation, however, were males. About half of the F_2 males had white eyes and half had red eyes, but all females had red eyes. In this experiment, the recessive allele was expressed only in males. Morgan arrived at an explanation by associating this ***gene*** with the ***X chromosome,*** as is illustrated in Fig. 4.16.

Because the male fly had only one X chromosome and a Y chromosome that lacked most genes of the X, it was postulated that the allele for white eyes was hemizygous and thus expressed. Furthermore, the mutant allele present in the X chromosome of the original white-eyed male was passed on to his daughters (he transmitted a Y chromosome to his sons). All the daughters therefore were carriers of the allele. The F_2 ***hemizygous males obtained their X chromosomes from their heterozygous mothers.*** Half received the w^+ allele and developed red eyes, and half received the w allele and developed white eyes. The equal proportion of red-eyed and white-eyed F_2 males was thus explained on the basis of segregation of X chromosomes from the F_1 mothers.

Could white-eyed females occur? On the basis of his hypothesis that the gene was carried in the X chromosome, Morgan predicted that a female of genotype ww could be produced and would have white eyes. This was tested experimentally with crosses between males with white eyes and F_1 (ww^+) females with red eyes. From these crosses, half the females as well as half the males had white eyes, as predicted. Later studies identified many other genes on the X chromosome of *Drosophila*.

Patterns of Inheritance of Sex-Linked Genes

Most sex-linked genes in male heterogametic animals are on the X chromosome. Some animals, however, may carry a few genes on the Y chromosome that produce visible effects on the phenotype of the organism. Such ***"Y-linked"*** or ***holandric*** genes would be transmitted directly from father to son and never appear in females. Y linkage is clearly very rare in higher animals, particularly mammals. X linkage, on the other hand, is very common in all mammals that have been studied; the mammalian X chromosome contains a large number of genes with major effects on phenotype.

Further *Drosophila* studies with attached-X chromosomes (XX) showed that X chromosomes with X-linked genes can be transmitted directly from father to son. Attached-X chromosomes are aberrant compound X chromosomes possessing a fused centromere. When attached-X females are mated with normal XY males, the XX female gametes fertilized with Y male gametes result in XXY females (the Y is not a male sex determiner in *Drosophila*). Male X chromosome gametes that are fertilized with female gametes lacking an X chromosome result in XO-sterile males (see Fig. 4.6) that express the X-linked genes of the father.

In the absence of dominant alleles, a recessive allele, such as the one (w) responsible for white eyes in *Drosophila,* can express itself. The ***crisscross pattern of inheritance,*** which is ***characteristic of sex-linked genes,*** means that traits appearing in males are transmitted (unexpressed) through their daughters to the males in the next generation, where they are expressed. Cytological studies on the structures and behavior of chromosomes have rigorously established the accuracy of Morgan's interpretation of sex linkage. In *D. melanogaster*, the X and Y chromosomes can readily be identified by their appearance. The X is rod-shaped with the centromere near one end, whereas the Y is hooklike, having a long and a short arm.

Although the Y chromosome is composed of heterochromatin and essentially is devoid of genes, one small part of the short arm in *D. melanogaster* has a homologous section on the X chromosome. The allele (*bb*) for bobbed bristles on the X chromosome, or its wild-type allele (bb^+), may occur on the short arm of the Y chromosome. Genes such as *bb* with a locus on the Y chromosome as well as on a homologous part of the X chromosome are said to be incompletely sex-linked. The *Drosophila* Y chromosome also contains at least four fertility factors besides the bobbed region.

Like the Y chromosome in *Drosophila*, supernumerary chromosomes in some species of animals and plants (e.g., mealy bugs and maize) are composed mostly of ***heterochromatin***. Sometimes the heterochromatic regions of different chromosomes in the same cell coalesce and form an amorphous "chromocenter." This occurs in the giant salivary gland chromosomes of *D. melanogaster*, as is shown in Fig. 4.17. The usual metaphase configuration observed in other cells of this species is shown in Fig. 4.17*e* with the heterochromatic regions indicated in black. Giant chromosomes from a cell of a male larva are shown in Fig. 4.17*a*. The heterochromatin near the centromere of each autosome and the X chromosome, along with that making up nearly all the Y chromosome, coalesce to form the chromocenter (Fig. 4.17*c*).

In summary, the Y chromosome of *Drosophila* (and of many other organisms) is mostly heterochromatin and has only a few genes. The X chromosome, with many genes, and the Y, with virtually none, establish the characteristic pattern of inheritance observed for sex-linked traits. The mother, with two X chromosomes, transmits one X to each gamete. Zygotes that receive a Y chromosome from the father develop into males. These hemizygous male progeny express the sex-linked genes received from the mother. Zygotes that receive two X chromosomes, one from each parent, develop into females. The expression of sex-linked traits in females follows the same pattern as autosomal traits, with the recessive phenotype appearing only in homozygotes.

X-Linked Traits in Humans

The inheritance pattern associated with sex linkage is so obvious that it is ideal for genetic studies. This was, in fact, the first pattern of inheritance to be recorded for humans. Early Greek philosophers noticed that some human traits tended to skip a generation. An inherited characteristic was observed in a father but not in any of his children, either male or female, and then it would reappear in males of the next generation. This distinctive ***crisscross pattern, from father through daughter to grandson***, replacing the usual pattern for the F_1 and F_2 generations, is now interpreted as evidence of ***sex linkage***. Since humans are not subjected to experimental procedures, the characteristic inheritance pattern in family groups that can be illustrated in pedigree charts (Fig. 4.18) is the standard means of detecting sex-linked genes.

Sex linkage has been indicated for more than 200 traits in humans, including, in addition to those already mentioned, such important and distinctive traits as optic atrophy (degeneration of the optic nerve), juvenile glaucoma (hardening of the eyeball), myopia (nearsightedness), defective iris, epidermal cysts, distichiasis (double eyelashes), white occipital lock of hair, mitral stenosis (abnormality of the mitral valve in the heart), and some forms of mental retardation (Fig. 4.19). Some of these traits have alternative forms that are dependent on autosomal genes.

Although the pattern now associated with sex linkage was observed in human pedigrees many years ago, the understanding of the genetic mechanism was a direct consequence of Morgan's experimental work with the white-eyed mutant in *Drosophila*. The explanation given for sex-linked inheritance in *Drosophila* applies equally to traits in humans that are associated with genes on the X chromosome.

Several kinds of defective color vision have now been identified, and the genetic mechanisms controlling color perception are more complex than at first suspected. Inherited defects in the ability of individuals to distinguish greens or reds are known to be sex-linked, whereas defects in blue color discrimination are autosomal.

Color perception is controlled by cone-shaped cells in the retina of the eye. Three classes of cone cells, each containing light-absorbing proteins (pigments) that sense a specific portion of the visible spectrum, have been identified. The three types of cone cells are usually referred to as blue-absorbing, green-absorbing, and red-absorbing, and the light sensitivity of these cone cells is known to result from the specific light-absorbing pigment that is present in each type of cone cell (Fig. 4.20).

The genes that encode the three light-absorbing proteins of the retina have been isolated, and their nucleotide sequences determined. The sequences have been used to predict the amino acid sequences of the three light-absorbing proteins. As might be expected for proteins that perform very similar functions, the three light-absorbing proteins have very similar structures. In fact, the red- and the green-receptor proteins differ at only a few of the amino acid subunits that make up these visual pigments (Fig. 4.21). The genes that encode the green- and red-receptor proteins were shown to be located on the X chromosome—thus the sex-linked patterns of inheritance observed for defects in green and red color vision (see the following). The gene encoding the blue-receptor protein was shown to be located on chromosome 7, an autosome.

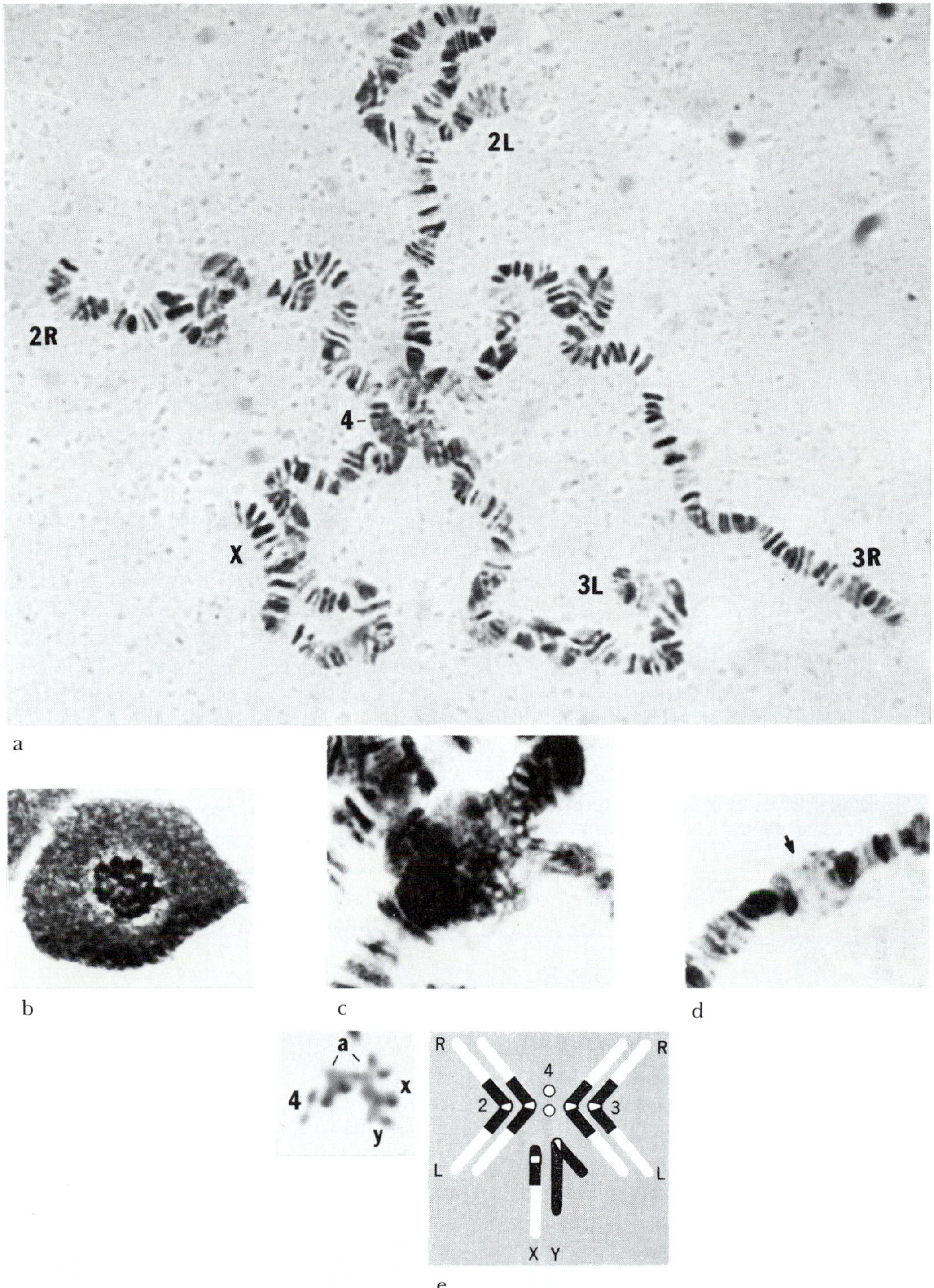

Figure 4.17 Chromosomes of *Drosophila melanogaster*. (a) Polytene chromosomes from salivary gland of a mature larva with chromosomes and chromosome arms identified (× 400). (b) Cell from salivary gland of mature larva. Note polytene chromosomes condensed in nucleus (× 300). (c) Chromocenter of polytene chromosomes: same as center area of chromosome spread above, but stained here to show chromocenter (× 1100). (d) Salivary chromosome puff marked by arrow (× 1100). (e, Left) Photograph of metaphase chromosomes from neuroblast cell of male larval brain: *a* represents the large autosomes, *4* represents the small fourth chromosome dividing, *x* the X chromosome, *y* the Y chromosome. (Right) Diagram of idealized metaphase somatic chromosomes, with chromosomes and chromosome arms identified. (Courtesy of S. W. Rogers.)

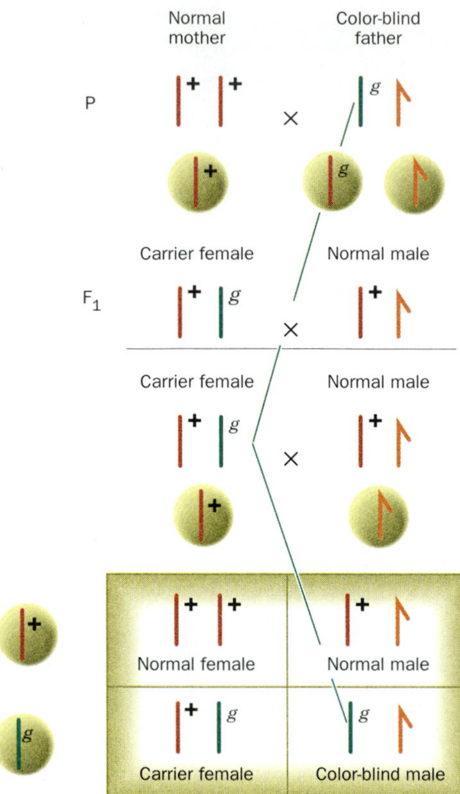

Figure 4.18 Crisscross inheritance from father to daughter to grandson (of original father) that is characteristic of a sex-linked gene. Genes are shown on the chromosomes illustrating a cross between a woman with normal vision and a green (protan) color vision defective man. The symbol g represents the sex-linked recessive gene for green color defective vision.

The most extensively studied color vision anomaly in humans results from an inherited defect in the green-receptor protein (Fig. 4.18). For the sake of simplicity, the gene causing defects in the green-receptor protein will be treated as a single sex-linked recessive allele, without reference to other genes that influence color vision. A man defective in green color vision has a single recessive allele (g) on his X chromosome. Since the Y chromosome carries no color vision locus, the single allele is expressed, causing color blindness. If the wife of this man is homozygous for the dominant allele (g^+) for normal color vision (Fig. 4.18), all of their daughters will receive an X chromosome from the mother carrying g^+ and a g from their father and will be heterozygous. Sons, with only one X chromosome, will have only one allele (g^+) from the mother and will have normal color vision. The Y chromosome, carrying no allele for this trait, will be contributed by the father only to his sons. In the next generation, about half of the sons of the carrier females will be normal and half will have defective color vision because the X chromosome carrying g^+ will segregate to about half of the heterozygous mother's gametes, and the other half will carry g. If the fathers are normal (g^+), half of the daughters of carrier mothers will be carriers, not expressing the trait.

The segregation of X chromosomes and the expression of single recessive alleles in hemizygous males explain the higher incidence of males with defective green color vision than of females with this trait. About 5 percent of Caucasian men in the United States, but less than 1 percent of Caucasian women,

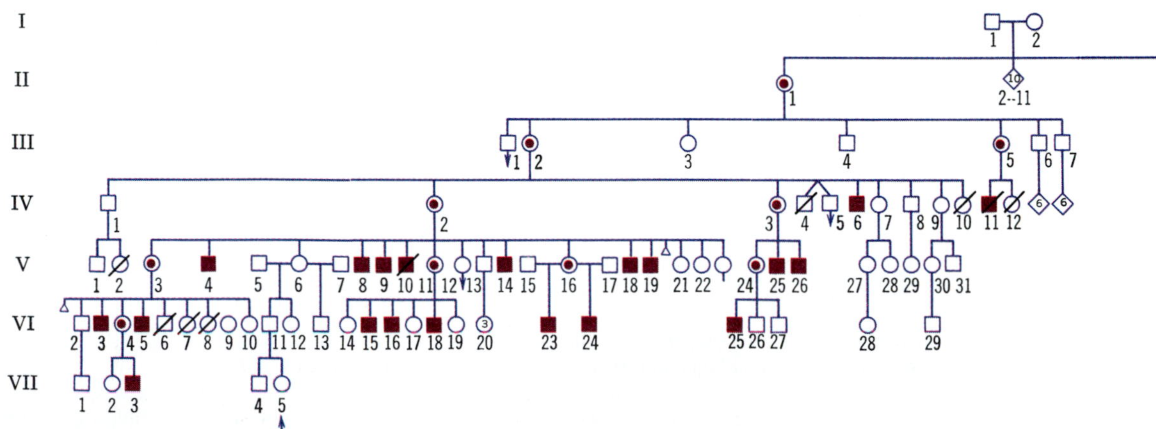

Figure 4.19 Pedigree of a family group in which a particular kind of mental retardation follows the pattern of recessive sex-linked inheritance. Darkened squares and circles represent mentally retarded males and females, respectively. Circles with dark centers represent carrier females, about half of whose sons are mentally retarded and half of whose daughters are carriers of the mental retardation gene. Genotypes of original parents, I-1 and I-2, are unknown, but I-2 is presumed to have been a carrier for mental retardation. (From R. G. Lehrke, "Clinical Studies of Nonspecific X-Linked

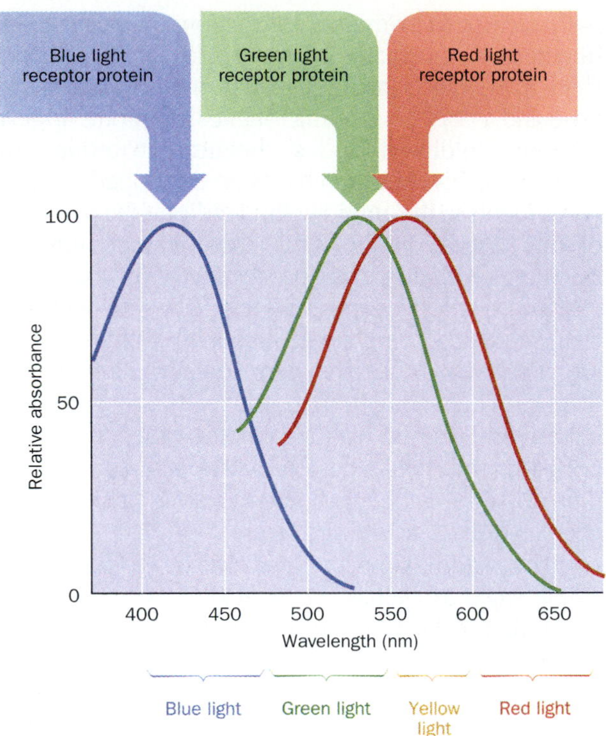

Figure 4.20 Absorption spectra for the three light-sensing proteins present in the cone cells of the retina of the human eye. The ability of humans with normal color vision to distinguish colors throughout the visible spectrum depends on the presence of all three proteins. Defective color vision (see text) results from the absence of, or a defect in, one or more of these pigments (light-absorbing substances are commonly called pigments because of the colors they produce in the reflected light). (After J. Nathans, "The Genes for Color Vision," *Sci. Amer.* 260(2): 42–49, 1989.)

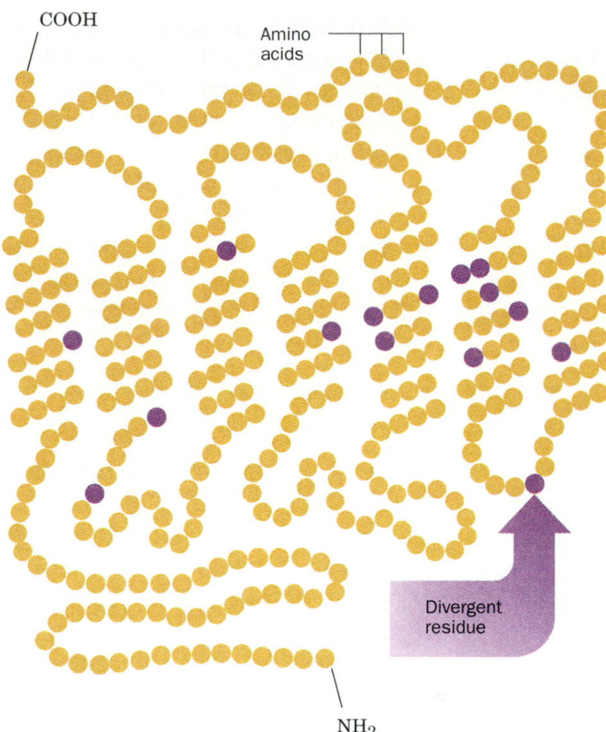

Figure 4.21 Comparison of the structures of the red and the green receptor proteins of humans. The structures of the proteins have been predicted from the known structures (sequences) of the genes encoding these two proteins. Each circle represents one subunit (amino acid) of each of the proteins; the subunits that are different in the two proteins are shown in color. (After J. Nathans, "The Genes for Color Vision," *Sci. Amer.* 260(2): 42–49, 1989.)

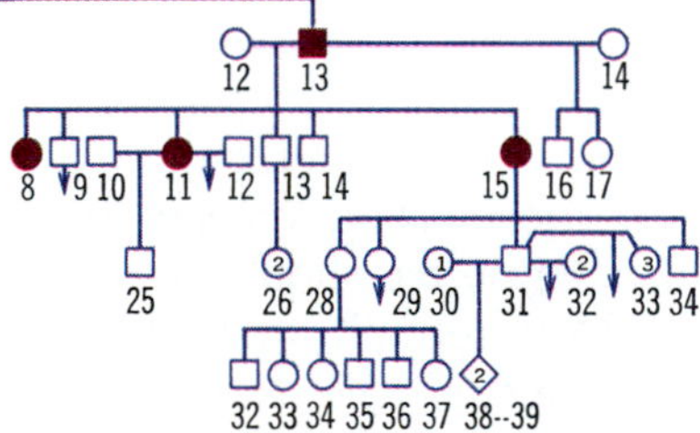

Mental Retardation," in D. Bergsma, (ed.), *X-Linked Mental Retardation and Verbal Disability.* Miami: Symposia Specialists for the National Foundation-March of Dimes *BD: OAS X*(1): 41–70, 1974, with permission.)

have defective green color vision. This difference occurs because females must receive two recessive alleles to exhibit the defect in green color vision. Individuals with defective color vision occur in all human populations. The gene frequency, however, varies among those of different ancestral groups. Only about 4 percent of Negro men in the United States are reported to have inherited defects in green color vision.

The criteria for identifying sex-linked (X-linked) recessive traits from pedigree studies may be summarized as follows: (1) the trait occurs much more frequently in males than in females, (2) traits are transmitted from an affected man through his daughters to half his grandsons, (3) an X-linked allele is never transmitted directly from father to son, and (4) all affected females have an affected father and a carrier or affected mother. Criteria for identifying sex-linked genes are consequences of the definition of sex-linkage.

For dominant X-linked traits, such as the rare blood type X_g^a, males expressing the trait would be expected to transmit it to all of their daughters but none of their sons. Heterozygous females would transmit the trait to half of their children of either sex. If a female expressing the trait is homozygous, all her children will inherit the trait. Sex-linked dominant inheritance cannot be distinguished from autosomal inheritance in the progeny of females expressing the trait, but only in the progeny of affected males.

Deleterious Recessive Sex-Linked Genes in Humans

Congenital hyperuricemia (***Lesch–Nyhan syndrome***), characterized by excess production of uric acid, is inherited through a ***sex-linked recessive*** gene. This means that the mother contributes the X chromosome with the defective gene to a male zygote. Half of the male children of carrier mothers may be expected to inherit the disease. These are deficient for the enzyme hypoxanthine-guanine phosphoribosyltransferase (***HPRT***), an enzyme involved in nucleotide synthesis. Infants who receive the gene appear normal at birth and for several months, but may show symptoms of excessive uric acid in the urine, such as orange "sand" (uric acid crystals). By about 10 months of age, they may become abnormally irritable and lose motor control. Weak and flabby muscles prevent the child from sitting, walking, or speaking normally. By the second year of life, the nervous condition has progressed to a degree that self-mutilation occurs, manifested by lip-biting, finger-chewing, teeth-grinding, and marked swinging of the arms. Death, which is often secondary to severe renal and neurological damage, usually occurs within a few years, but some victims live into their twenties.

Duchene-type muscular dystrophy also depends on a ***sex-linked recessive gene***. If the mother is known to be a carrier of this gene, either from her pedigree or through tests that are available, about half of her male children are expected to be affected. Male fetuses can be identified by a chromosome study. Duchene-type muscular dystrophy afflicts males, usually before they reach their teens, with muscular deterioration that progresses rapidly during the early teen years. Muscles of the legs and shoulders become stiff, and the children usually become paralyzed and crippled during their middle or late teens. Virtually all die before age 21. All female children born to a carrier mother are expected to be normal, since the possibility of their being homozygous for this sex-linked recessive gene is virtually nonexistent.

Another severe disease following the pattern of ***X-linked recessive*** inheritance is ***Hunter syndrome***. It is characterized by mental retardation, coarse features, hirsutism (abnormal hairiness), and a characteristic facial appearance that includes a broad bridge of the nose and a large protruding tongue. Symptoms appear in early childhood, and a chemical procedure for diagnosing this condition has been developed. Certain constituents in the amniotic fluid indicate the presence of this disease in the fetus; these compounds are associated with abnormal processing of mucopolysaccharides synthesized in early pregnancy. Mucopolysaccharides also accumulate in skin cells of persons who are homozygous for the gene causing Hunter syndrome. When amniotic or skin cells are grown in culture and stained with *o*-toluidine blue, any mucopolysaccharide cell inclusions will be stained pink. It is thus possible to identify a heterozygous carrier of the gene as well as an affected fetus.

The pattern of mucopolysaccharide (containing an amino sugar as well as uronic acid units) metabolism by Hunter cells is so strikingly different from the normal that it can be used for prenatal diagnosis of Hunter syndrome in high-risk pregnancies by performing amniocentesis (Chapter 19). Of the many cell types originally present in amniotic fluid, fetal fibroblasts are the only ones that multiply in culture. Like fibroblasts from skin biopsies, fetal fibroblasts show an excessive accumulation of mucopolysaccharide or stainable cell inclusions if the fetus is affected with Hunter syndrome.

Human Y Chromosome: Dominant Male Determinants, But Few Other Genes

Published pedigrees have suggested that the human Y chromosome carries genes controlling holandric traits that are transmitted exclusively through the male line. Pedigree analysis for transmission from father to son provided the first evidence for holandric genes. Although pedigrees have been interpreted to show the holandric pattern for hairy pinna of the ear, other pedigrees clearly show autosomal patterns of inheritance for this trait. A histocompatibility gene (*H-Y*), however, has been mapped on the short arm of the human Y chromosome. More recently, molecular genetic techniques (see Chapter 13) have permitted scientists to isolate individual genes, characterize them, and physically map them to specific regions of individual chromosomes. By using these techniques, researchers have shown that the human Y chromosome contains at least two genes: (1) *H-Y* and (2) *TDF* (encoding a testis determining factor), which appears to be a very important sex-determining gene in humans. Few Y-linked genes have been found in experimental animals. The pedigree pattern is so obvious that if there were many Y-linked genes, more would have been discovered. Thus, it seems quite safe to conclude that the Y chromosome of mammals, including humans, contains very few genes with major effects on phenotype.

SUMMARY

Investigations of ***sex determination*** have shown that the embryo is frequently bipotential, having the ability to develop into either sex. Determination for one sex or the other is usually accomplished by a ***balance between genetic factors for maleness and genetic factors for femaleness.*** Several different factors involving chromosomes, genes, cytoplasm, and hormones have been associated with this balance, particularly in the secondary sex characteristics. Hormones influence the expressions of some genes. Sexual mosaics or gynandromorphs, resulting from irregularities in early cell cleavages of insects, have female characteristics in some parts of the body and male characteristics in other parts. ***Sex chromatin bodies*** result from the ***inactivation of one X chromosome in normal female mammals.*** Cells of normal males have no sex chromatin body. The number of sex chromatin bodies in individuals with more than two X chromosomes is one less than the number of X chromosomes. The "fragile X" syndrome responsible for the most common form of inherited mental retardation in humans may be caused in part by the failure of this anomalous chromosome to undergo normal reactivation of the heterochromatic "fragile X" chromosome prior to oögenesis.

In many species, including humans, many genes that control important traits not related to sex determination are also located on sex chromosomes. Such ***sex-linked traits*** show a very distinctive pattern of inheritance. In humans, recessive sex-linked or ***X-linked*** traits are transmitted from fathers to grandsons and from mothers to half of her sons. This pattern is distinct from autosomal inheritance because males are ***hemizygous*** for the X chromosome (have one copy rather than two). The human Y chromosome is largely devoid of known genes, but contains important male-determining factors; individuals with several X chromosomes and one Y chromosome are phenotypically male.

REFERENCES

BARR, M.L. 1960. "Sexual dimorphism in interphase nuclei." *Amer. J. Hum. Genet.* 12: 118–127.

BRIDGES, C.B. 1925. "Sex in relation to chromosomes and genes." *Amer. Nat.* 59: 127–137.

KOO, G.C., S.S. WACHTEL, K. KRUPEN-BROWN, L.R. MITTL, W.R. BREG, M. GENEL, I.M. ROSENTHAL, D.S. BORGAONKAR, D.A. MILLER, R. TANTRAVAHL, R.R. SCHRECK, B.F. ERLANGER, and O.J. MILLER. 1977. "Mapping the locus of the H-Y gene on the human Y chromosome." *Science* 198: 940–942.

LAIRD, C.D. 1987. "Proposed mechanism of inheritance and expression of the human fragile-X syndrome of mental retardation." *Genetics* 117: 587–599.

LYON, M.F. 1962. "Sex chromatin and gene action in the mammalian X-chromosome." *Amer. J. Hum. Genet.* 14: 135–148.

MITTWOCH, U. 1973. *Genetics and sex differentiation.* Academic Press, New York.

MOORE, K.L. 1966. *The sex chromatin.* Saunders, Philadelphia.

MORGAN, T.H., and C.B. BRIDGES. 1916. "Sex-linked inheritance in Drosophila." *Carnegie Inst. Wash. Publ.* 237: 1–22.

NATHENS, J. 1989. "The genes for color vision." *Sci. Amer.* 260 (2): 42–49.

PAGE, D.C., R. MOSHER, E.M. SIMPSON, E.M.C. FISHER, G. MARDON, J. POLLACK, B. MCGILLIVRAY, A. DE LA CHAPELLE, and L.G. BROWN. 1987. "The sex-determining region of the human Y chromosome encodes a finger protein." *Cell* 51: 1091–1104.

SUTHERLAND, G.R. 1985. "The enigma of the fragile X chromosome." *Trends in Genetics* 1: 108–112.

WACHTEL, S.S. 1977. "H-Y antigen and the genetics of sex determination." *Science* 198: 797–799.

WHITING, P.W. 1945. "The evolution of male haploidy." *Quart. Rev. Biol.* 20: 231–260.

PROBLEMS AND QUESTIONS

4.1 What difference exists between male- and female-determining sperm in animals with heterogametic males?

4.2 In line with Bridges' genetic balance theory for sex determination, what is the expected sex of individuals with each of the following chromosome arrangements: (a) 4X 4A, (b) 3X 4A, (c) 2X 3A, (d) 1X 3A, (e) 2X 2A, and (f) 1X 2A?

4.3 List the expected results in terms of sex and intersex combinations from a cross between a triploid ($3n$) female fly, with two X chromosomes attached and one free, and a normal diploid male. (Assume that the cross is successful and the gametes of the female will carry one or two whole sets of autosomes.)

4.4 In plants of the genus *Melandrium*, which sex will be determined by the following chromosome arrangements: (a) XY, (b) XX, (c) XY with Region I removed, and (d) XY with Region II removed?

4.5 What sex is expected for individuals of the following genotypes in *Bracon hebetor:* (a) X_b, (b) X_aX_b, (c) X_cX_c, and (d) X_bX_c?

4.6 How could maize plants, which are ordinarily monoecious, give rise to plants that are dioecious?

4.7 How many sex chromatin bodies are expected to occur in cell nuclei with each of the following chromosome arrangements: (a) XY, (b) XX, (c) XXY, (d) XXX, (e) XXXX, (f) XYY?

4.8 In sheep, the gene h^+ for the horned condition is dominant in males and recessive in females. If a hornless ram were mated to a horned ewe, what is the chance that (a) an F_2 male sheep will be horned or (b) an F_2 female sheep will be horned?

4.9 In chickens, the gene *h*, which distinguishes hen-feathering from cock-feathering, is sex-limited. Males may be hen-feathered or cock-feathered, but females are always hen-feathered. If a cock-feathered male (*hh*) were mated to a homozygous (h^+h^+) hen-feathered female, what patterns of feathering might be expected among the (a) male F_2 and (b) female F_2 progeny?

4.10 In a particular species of grasshopper, two pairs of autosomes are heteromorphic, that is, they can be distinguished by microscopic observation. In one pair, one homolog is rod-shaped and the other has a small hook at the end. One member of the other pair has a knob on one end. List all distinguishable combinations, with reference to these two pairs of chromosomes, that can be found in the sperm.

4.11 In *Drosophila,* the gene for bobbed bristles (recessive allele *bb,* bobbed bristles; wild-type allele bb^+, normal bristles) is located on the X chromosome and on a homologous segment of the Y chromosome. Give the genotypes and phenotypes of the offspring from the following crosses: (a) $X^{bb}X^{bb} \times X^{bb}Y^{bb+}$, (b) $X^{bb}X^{bb} \times X^{bb+}Y^{bb}$, (c) $X^{bb+}X^{bb} \times X^{bb+}Y^{bb}$, (d) $X^{bb+}X^{bb} \times X^{bb}Y^{bb+}$.

4.12 Consider two different sex-linked traits, one recessive and one dominant, with equal effect on viability and equal frequency in the same population. If males are heterogametic, would the recessive trait or the dominant trait be expressed more frequently in (a) males or (b) females?

4.13 If a white-eyed male fruit fly should occur in a culture of red-eyed flies, how could the investigator obtain evidence to answer the following questions: (a) Is a mutant gene or an environmental change responsible for the new phenotype? (b) If a mutation has occurred, is it sex-linked? (c) Can white-eyed females occur?

4.14 The gene (*w*) for white eyes in *D. melanogaster* is recessive and sex-linked; males are heterogametic. (a) Symbolize on the chromosomes the genotype of a white-eyed male, red-eyed male, red-eyed female (two genotypes), and white-eyed female. (b) Diagram on the chromosomes a cross between a homozygous red-eyed female and a white-eyed male. Carry through the F_2 and summarize the expected sex and eye-color phenotypes. (c) Diagram on the chromosomes and give the expected phenotypes from the cross between F_1 females and (i) a white-eyed male and (ii) a red-eyed male.

4.15 In humans, green defective color vision results from the sex-linked recessive gene (*g*) and normal vision from its allele (g^+). A man (a) and woman (b), both with normal vision, have the following three children, all of whom are married to people with normal vision: a color-defective son (c) who has a daughter of normal vision (f), a daughter of normal vision (d) who has one color-defective son (g) and two normal sons (h), and a daughter of normal vision (e) who has six normal sons (i). Give the probable genotypes of all the individuals in the family (a to i).

4.16 If a mother carried the sex-linked gene for green defective color vision and the father was normal, would their sons or daughters be defective in color vision?

4.17 If a father and son are both defective in green color vision, is it likely that the son inherited the trait from his father?

4.18 Diagram on the chromosomes a cross between a normal woman whose father was defective in green color vision and a green color-defective man. Summarize the expected results for sex and eye condition.

4.19 In humans, the gene (*h*) for hemophilia is sex-linked and recessive to the gene (h^+) for normal clotting. Diagram on the chromosomes the genotypes of the parents of the following crosses and summarize the expected phenotypic ratios resulting from the crosses: (a) hemophiliac woman × normal man, (b) normal (heterozygous) woman × hemophiliac man, and (c) normal (homozygous) woman × hemophiliac man.

4.20 A normal woman, whose father had hemophilia, married a normal man. What is the chance of hemophilia in their children?

4.21 The $X_g^{(a+)}$ allele is dominant and X-linked. If a woman heterozygous for this gene [$X_g^{(a+)}X_g^{(a-)}$] married a man carrying the allele $X_g^{(a-)}$, what is the probability that (a) each daughter and (b) each son will receive the $X_g^{(a+)}$ gene?

Chapter 5

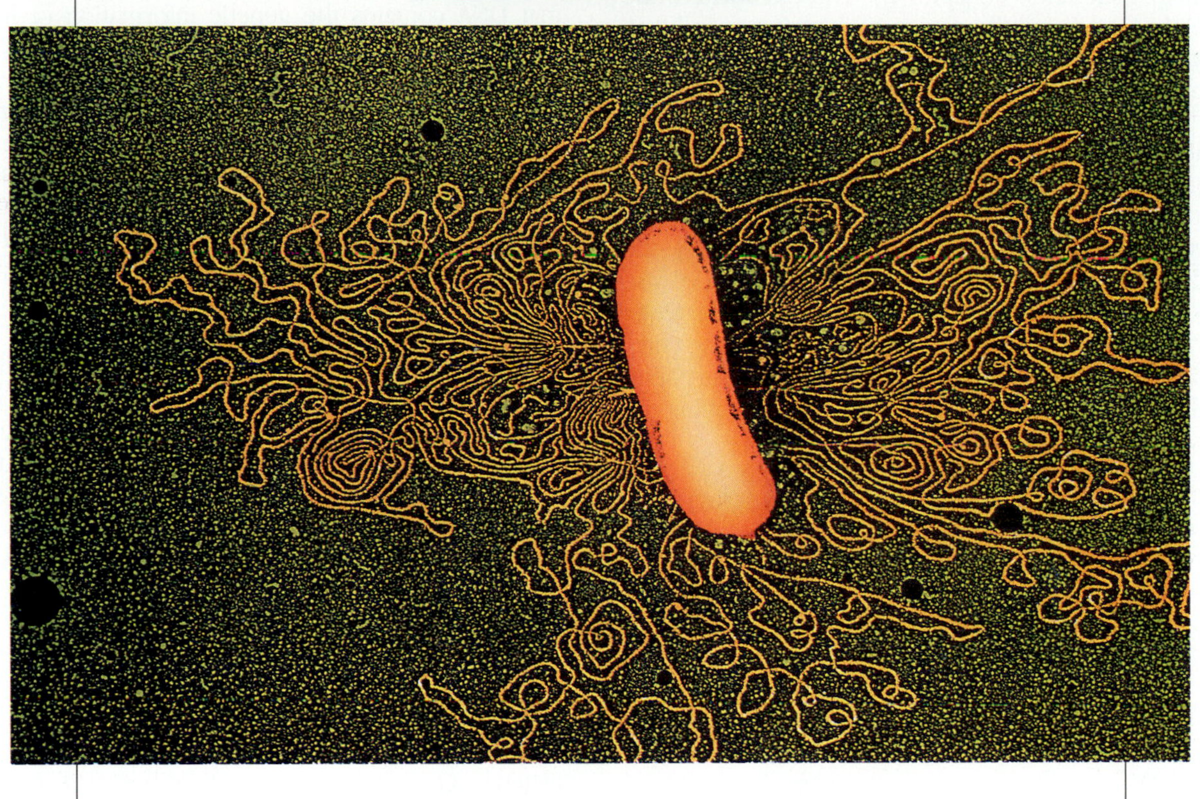

Genetic Material: Properties and Replication

(Photo above): False-color transmission electron micrograph of a ruptured *E. coli* cell from which much of the DNA has been extruded. The DNA is seen as a mass of gold-colored threads surrounding the bacterial cell. (Courtesy of Dr. Gopal Murti/Science Photo Library.)

GENES were first detected and analyzed by Mendel, and subsequently many other scientists, by following their patterns of transmission from generation to generation (Chapter 2). These studies, while greatly elucidating the nature of inheritance in living organisms, provided no insight into the structure or molecular composition of genes.

Subsequent studies established the precise correlation between the patterns of transmission of genes from generation to generation (segregation and independent assortment) and the behavior of chromosomes during sexual reproduction, specifically the reduction division of meiosis and fertilization (Chapter 3). These and related experiments provided strong early evidence that genes are usually located on chromosomes. Thus, in posing questions about the chemical basis of heredity, scientists began by probing the biochemical composition of chromosomes.

Whatever its chemical composition, it was clear even in Mendel's time that the genetic material had to fulfill two key requirements.

1. The ***genotype function*** or ***replication***. The genetic material must be capable of storing genetic information and transmitting this information faithfully from parents to progeny, generation after generation (although, as we will see in Chapter 11, the genetic material does undergo occasional heritable changes called ***mutations***).
2. The ***phenotype function*** or ***gene expression***. The genetic material must control the development of the phenotype of the organism, be it a virus, a bacterium, a plant, or an animal such as a human being. That is, the genetic material must dictate the growth and differentiation of the organism from the single-celled zygote to the mature adult. To control this complex process, the genetic material must not only express itself accurately, but each gene must act at the proper time and place to guarantee that the liver is made up of liver cells, the nervous system of nerve cells, and so on (see Chapters 10 and 15).

Chromosomes are composed of two types of large organic molecules (macromolecules) called ***proteins*** and ***nucleic acids***. The nucleic acids are of two types: ***deoxyribonucleic acid (DNA)*** and ***ribonucleic acid (RNA)***. For many years, there was considerable disagreement among scientists as to which of these three macromolecules carries the genetic information. During the 1940s and early 1950s, several elegant experiments were carried out that clearly established that the genetic information resides in the nucleic acids rather than in proteins. More specifically, these experiments showed that the genetic information resides in DNA. (In a few simple viruses, however, RNA carries the genetic information; these particular viruses contain no DNA.)

DNA, THE GENETIC MATERIAL

Several lines of indirect evidence have long suggested that DNA contains the genetic information of living organisms. Most important, results obtained using several different experimental procedures showed that most of the DNA is located in the chromosomes, whereas RNA and proteins are also abundant in the cytoplasm. Moreover, a precise correlation exists between the amount of DNA per cell and the number of sets of chromosomes per cell. That is, most somatic cells of diploid organisms, for example, contain exactly twice the amount of DNA as the haploid germ cells or gametes of the same species. Finally, the molecular composition of the DNA in all of the different cells of an organism is the same (with rare exceptions), whereas the composition of RNA and proteins varies both qualitatively and quantitatively from one cell type to another. Although these correlations strongly suggest that DNA is the genetic material, they by no means prove it. Fortunately, direct evidence has established that the genetic information is encoded in DNA.

Transformation in Pneumococcus

The first direct evidence showing that the genetic material is DNA rather than protein or RNA was published by O. T. Avery, C. M. Macleod, and M. McCarty in 1944. They demonstrated that the component of the cell responsible for the phenomenon of transformation in the bacterium *Diplococcus pneumoniae* (pneumococcus) is DNA. Transformation is a mode of recombination (exchange or transfer of genetic information between organisms or from one organism to another) occurring in several, but not all, species of bacteria. It does not involve direct contact between the bacterial cells or mediation by any vector such as a virus (see Chapter 8, pp. 206–208).

The phenomenon of transformation was discovered by Frederick Griffith in 1928. It should be emphasized that although Griffith's experiments demonstrated the occurrence of transformation in pneumococcus and thus set the stage for the work of Avery, MacLeod, and McCarty, they provided no evidence that DNA was involved in any way.

Pneumococci, like all other living organisms, exhibit genetic variability that can be recognized by the existence of different phenotypes (Table 5.1). The two phenotypic characteristics of importance in Griffith's demonstration of transformation were (1) the presence or absence of a surrounding polysaccharide (complex sugar polymer) capsule and (2) the type of capsule, that is, the specific molecular composition of the polysaccharides present in the capsules. When grown on appropriate media (such as blood agar) in

TABLE 5.1 **Characteristics of *Diplococcus pneumoniae* Strains When Grown on Blood Agar Medium**

TYPE	COLONY MORPHOLOGY: APPEARANCE	COLONY MORPHOLOGY: SIZE	CAPSULE	VIRULENCE	REACTION WITH ANTISERUM PREPARED AGAINST: TYPE IIS	REACTION WITH ANTISERUM PREPARED AGAINST: TYPE IIIS
IIR[a]	Rough	Small	Absent	Nonvirulent	None	None
IIS	Smooth	Large	Present	Virulent	Agglutination	None
IIIR[a]	Rough	Small	Absent	Nonvirulent	None	None
IIIS	Smooth	Large	Present	Virulent	None	Agglutination

[a]Although Type R cells are nonencapsulated, they carry genes that would direct the synthesis of a specific kind (antigenic Type II or III) of capsule if the block in capsule formation were not present. When Type R cells mutate back to encapsulated Type S cells, the capsule Type (II or III) is determined by these genes. Thus, R cells derived from Type IIS cells are designated Type IIR. When these Type IIR cells mutate back to encapsulated Type S cells, the capsules are of Type II.

petri dishes, pneumococci with a capsule form large, smooth colonies and are thus designated Type S. Such encapsulated pneumococci are quite pathogenic to most mammals (e.g., causing pneumonia in humans). These virulent (disease causing) Type S pneumococci mutate to a nonvirulent (or nonpathogenic) form that has no polysaccharide capsule (at a frequency of about one cell in 10^7). Such nonencapsulated, nonvirulent pneumococci form small, rough-surfaced colonies when grown on blood agar medium and are thus designated Type R (Table 5.1). (The polysaccharide capsule is required for virulence since it protects the bacterial cell against phagocytosis by leukocytes.) When a capsule is present, it may be of several different antigenic types (Type II, III, etc.), depending on the specific molecular composition of the polysaccharides and, of course, ultimately on the genotype of the cell.

The different capsule types can be identified immunologically. If Type II cells are injected into the bloodstream of rabbits, the immune system of the rabbits will produce antibodies (a specific set of large proteins whose function is to protect the organism against foreign substances such as macromolecules, viruses, and bacteria; see Chapter 16) that react specifically with Type II cells. Such Type II antibodies will agglutinate Type II pneumococci but not Type III pneumococci, and vice versa.

Griffith's unexpected discovery was that if he injected heat-killed Type IIIS pneumococci (virulent when alive) plus live Type IIR pneumococci (nonvirulent) into mice, many of the mice succumbed to pneumonia, and live Type IIIS cells were recovered from the carcasses (Fig. 5.1). When mice were injected with heat-killed Type IIIS pneumococci alone (Fig. 5.1, top), none of the mice died. The observed virulence was therefore not due to a few Type IIIS cells that survived the heat treatment. It is critical to note that the live virulent pneumococci recovered from the carcasses were of polysaccharide Type III, since it is known that nonencapsulated Type R cells can mutate back to virulent encapsulated Type S cells. When such a mutation occurs in a Type IIR cell, however, the resulting cell will be Type IIS, not Type IIIS. Thus, the ***"transformation" of nonvirulent Type IIR cells to virulent Type IIIS cells cannot be explained by mutation, rather some component of the dead Type IIIS cells (the "transforming principle") must convert living Type IIR cells to Type IIIS.***

Subsequent experiments showed that the phenomenon described by Griffith, now called ***transformation***, was not mediated in any way by a living host. The same phenomenon occurred in the test tube when live Type IIR cells were grown in the presence of dead Type IIIS cells or extracts of Type IIIS cells. Since it was clearly shown that the new phenotype, Type IIIS, was hereditary, that is, was due to a permanent inherited change in the genotype of the cells, the demonstration of transformation neatly set the stage for determining the chemical basis of heredity in pneumococcus. What remained was to determine what component of the cell extract was responsible for transformation.

Proof That the "Transforming Principle" Is DNA

The "transforming principle" was shown to be DNA in 1944 when Avery, MacLeod, and McCarty published the results of a set of extensive and laborious experiments. They showed that ***if highly purified DNA from Type IIIS pneumococci was present with Type IIR pneumococci, some of the pneumococci were transformed to Type IIIS*** (Fig. 5.2). But how could one be sure that the DNA was really pure? Proving the complete purity of any macromolecular substance is extremely difficult. Maybe the DNA preparation contained a few molecules of protein and these contami-

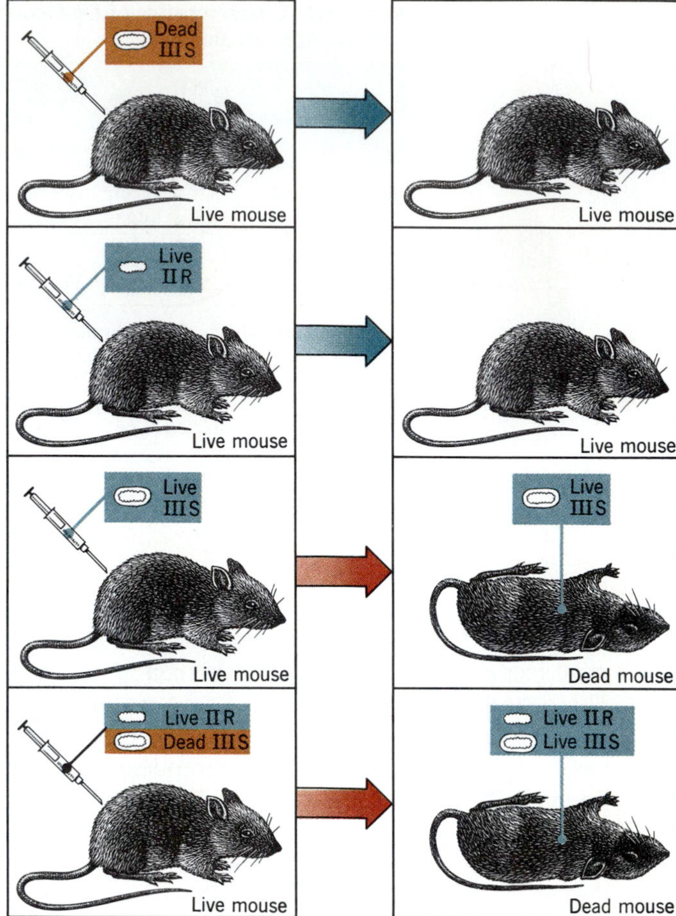

Figure 5.1 Griffith's demonstration of transformation in pneumococcus. When heat-killed encapsulated (designated S for smooth colony formation) Type III pneumococci were injected into mice, the mice did not develop pneumonia. Similarly, when living nonencapsulated (designated R for rough colony formation) Type II cells were injected into mice, the mice showed no ill effects. Injection of living Type IIIS pneumococci resulted in severe pneumonia and the death of many of the mice. Surprisingly, the injection of heat-killed Type IIIS cells (virulent if alive) together with living Type IIR cells (nonvirulent) caused the death of many of the mice.

nating proteins were responsible for the observed transformation. The most definitive experiments in Avery, MacLeod, and McCarty's "proof" that DNA was the transforming principle involved the use of enzymes (proteins that catalyze specific metabolic reactions) that degrade DNA, RNA, or protein. In separate experiments, highly purified DNA from Type IIIS cells was treated with (1) deoxyribonuclease (***"DNase,"*** which degrades DNA), (2) ribonuclease (***"RNase,"*** which degrades RNA), or (3) proteases (which degrade proteins) and then tested for its ability to transform Type IIR cells to Type IIIS. Only DNase had any effect on the transforming activity of the DNA preparation; it totally eliminated all transforming activity (Fig. 5.2).

Although the molecular mechanism by which transformation occurred remained to be worked out in subsequent investigations, the results obtained by Avery and coworkers clearly established that the genetic information in pneumococcus was present in DNA. We now know that the segment of the DNA in the chromosome of pneumococcus that carries the genetic information specifying the synthesis of a Type III capsule is physically integrated into the chromosome of the Type IIR recipient cell by a specific recombination process occurring during transformation (see Chapter 8).

The "Hershey–Chase Experiment"

Additional direct evidence indicating that DNA is the genetic material was published in 1952 by A. D. Hershey (1969 Nobel Prize winner) and M. Chase. These experiments showed that the genetic information of a particular bacterial virus (bacteriophage T2) was present in DNA. Their results, although probably less definitive than the results of Avery, Macleod, and McCarty, had a great impact on the acceptance by scientists of DNA as the genetic material. This large impact undoubtedly was the result of the elegant simplicity of the so-called "Hershey–Chase experiment."

Viruses are the smallest living organisms; they are living at least in the sense that their reproduction is controlled by genetic information stored in nucleic

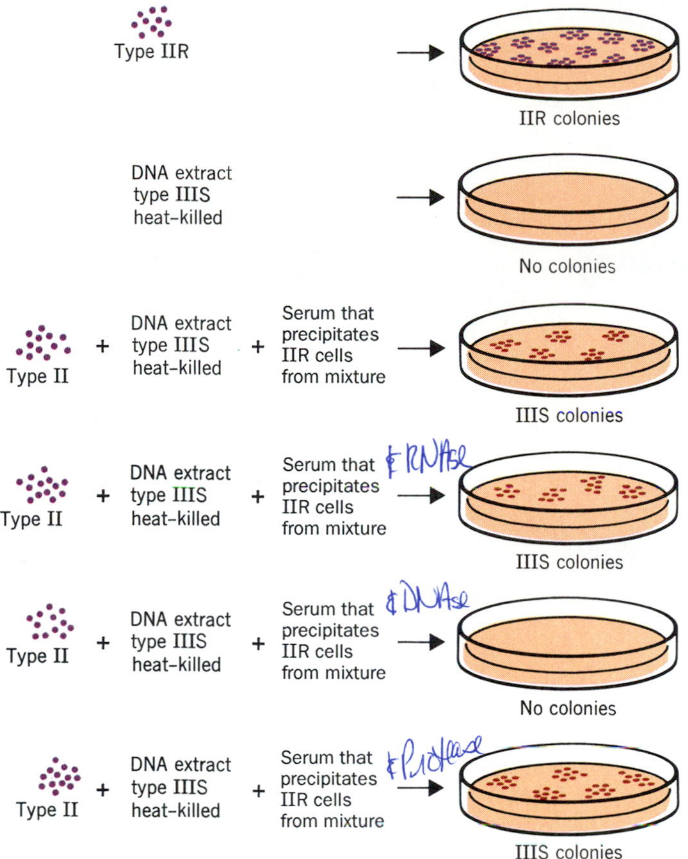

Figure 5.2 Avery, MacLeod, and McCarty's proof that the "transforming principle" is DNA. Transformation of Type IIR pneumococci to Type IIIS could be demonstrated using highly purified DNA from Type IIIS cells as well as using heat-killed Type IIIS cells. Proof that the active component was DNA and not small amounts of contaminating RNA or protein was accomplished by treating the purified DNA with the enzymes DNase, RNase, and trypsin (a protease), which very specifically degrade DNA, RNA, and protein, respectively. Treatment with RNase or protease had no effect on the ability of the purified DNA preparation to transform Type IIR cells to Type IIIS. DNase treatment destroyed the transforming activity of the DNA preparation.

acids via the same processes as in cellular organisms. Viruses, however, are acellular obligate parasites that can reproduce only in appropriate host cells. Their reproduction is totally dependent on the metabolic machinery (ribosomes, energy-generating systems, etc.) of the host. Viruses have been extremely useful in studying many genetic processes because of their simple structure and chemical composition (many contain only proteins and nucleic acids) and their very rapid reproduction (15–20 minutes for some bacterial viruses under optimal conditions).

Bacteriophage T2, which infects the common colon bacillus *Escherichia coli*, is composed of about 50 percent DNA and about 50 percent protein (Fig. 5.3). Experiments prior to 1952 had shown that all bacteriophage T2 reproduction takes place within *E. coli* cells. Therefore, when Hershey and Chase showed that the DNA of the virus particle entered the cell, whereas most of the protein of the virus remained adsorbed to the outside of the cell, this strongly implied that the genetic information necessary for viral reproduction was present in DNA. The basis for the Hershey–Chase experiment is that ***DNA contains phosphorus but no sulfur, whereas proteins contain sulfur but no phosphorus***. Thus, Hershey and Chase were able to specifically label either (1) the phage DNA by growth in a medium containing the radioactive isotope of phosphorus, ^{32}P, in place of the normal isotope, ^{31}P, or (2) the phage protein coats by growth in a medium containing radioactive sulfur, ^{35}S, in place of the normal isotope, ^{32}S (Fig. 5.3). When T2 phage particles labeled with ^{35}S were mixed with *E. coli* cells for a few minutes and were then subjected to shearing forces by placing the infected cells in a Waring blender, it was found that most of the radioactivity (and thus the proteins) could be removed from the cells without affecting progeny phage production. When T2 phage in which the DNA was labeled with ^{32}P were used, however, essentially all the radioactivity was found inside the cells, that is, it was not subject to removal by shearing in a blender. The sheared-off phage coats were separated from the infected cells by low-speed centrifugation which pellets (sediments) cells while leaving phage particles suspended. These results indicated that the DNA of the

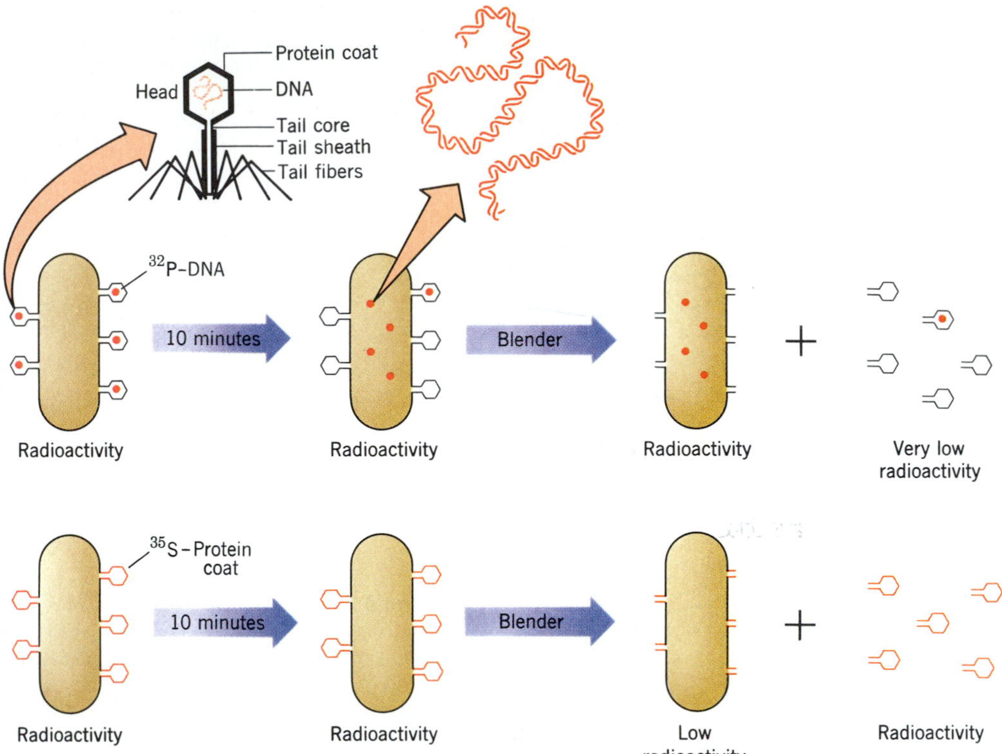

Figure 5.3 The "Hershey–Chase experiment": evidence that DNA is the genetic material in bacteriophage T2. *Escherichia coli* cells were infected with ^{32}P-labeled phage (DNA-labeled), and after being allowed time for infection, they were agitated in a blender, which sheared off the phage coats. The phage coats and the infected cells were then separated by centrifugation. Radioactivity was measured in the cell pellet (the sediment) and in the phage coat suspension. Most of the radioactivity was found in the cells. When the same experiment was done using phage with ^{35}S-labeled proteins, the results were very different. Most of the radioactivity was found in the suspension of phage coats; very little entered the host cells. Since phage reproduction (both DNA synthesis and new-coat protein synthesis) occurs inside the infected cells, and since only the phage DNA enters the host cell, the DNA, not the protein, must carry the genetic information. (Based on R. Sagar and F. J. Ryan, *Cell Heredity,* Wiley, New York, 1961.)

virus enters the host cell, whereas the protein coat remains outside the cell. Since progeny viruses are produced inside the cell, Hershey and Chase's results indicated that the genetic information directing the synthesis of both the DNA molecules and the protein coats of the progeny viruses must be present in the parental DNA. Moreover, the progeny particles were shown to contain some of the ^{32}P, but none of the ^{35}S of the parental phage.

However, the Hershey–Chase experiment did not provide unambiguous proof that the genetic material of phage T2 is DNA. A significant amount of ^{35}S (and thus protein) was found to be injected into the host cells with the DNA. Thus, one could always argue that this small fraction of the phage proteins contained the genetic information. More recently, however, it has been possible to develop conditions in which protoplasts (cells with the walls removed) of *E. coli* can be infected with pure phage DNA. Normal infective progeny phage are produced in these experiments, called ***transfection*** experiments, proving that the genetic material of such bacterial viruses is DNA.

RNA as Genetic Material in Small Viruses

As more and more viruses were identified and studied, it became clear that many of them contain RNA and proteins, but no DNA. In all cases studied to date, it is clear that these ***"RNA viruses"*** store their genetic information in nucleic acids rather than in proteins just like all other organisms, although in these viruses the nucleic acid is RNA. One of the first experiments that established RNA as the genetic material in RNA viruses was the so-called reconstitution experiment of H. Fraenkel-Conrat and B. Singer, published in 1957. Fraenkel-Conrat and Singer's simple, but definitive, experiment was done with tobacco mosaic virus (TMV), a small virus composed of a single molecule of RNA encapsulated in a protein coat. Different strains of TMV can be identified on the basis of differences in the chemical composition of their protein coats.

By using the appropriate chemical treatments, one can separate the protein coats of TMV from the RNA. Moreover, this process is reversible; by mixing the proteins and the RNA under appropriate conditions, "reconstitution" will occur, yielding complete, infective TMV particles. Fraenkel-Conrat and Singer took two different strains of TMV, separated the RNAs from the protein coats, and reconstituted "mixed" viruses by mixing the proteins of one strain with the RNA of the second strain, and vice versa. When these mixed viruses were used to infect tobacco leaves, the progeny viruses produced were always found to be phenotypically and genotypically identical to the parent strain from which the RNA had been obtained (Fig. 5.4). Thus, the genetic information of TMV is stored in RNA, not in protein.

DNA STRUCTURE

The genetic information of all living organisms, except the RNA viruses, is stored in DNA. What, then, is the structure of DNA, and in what form is the genetic information stored? What features of the structure of DNA allow for the transmission of genetic information from generation to generation?

Nucleic acids, first called "nuclein" because they were isolated from cell nuclei by F. Miescher in 1869, are macromolecules composed of repeating subunits called ***nucleotides***. Each nucleotide is composed of (1) a phosphate group, (2) a five-carbon sugar (or pentose), and (3) a cyclic nitrogen-containing compound called a base (Fig. 5.5). In DNA, the sugar is 2-deoxyribose (thus the name ***deoxyribo***nucleic acid); in RNA, the sugar is ribose (thus ***ribo***nucleic acid). There are four different bases commonly found in DNA: ***adenine, guanine, thymine,*** and ***cytosine***. RNA also usually contains adenine, guanine, and cytosine,

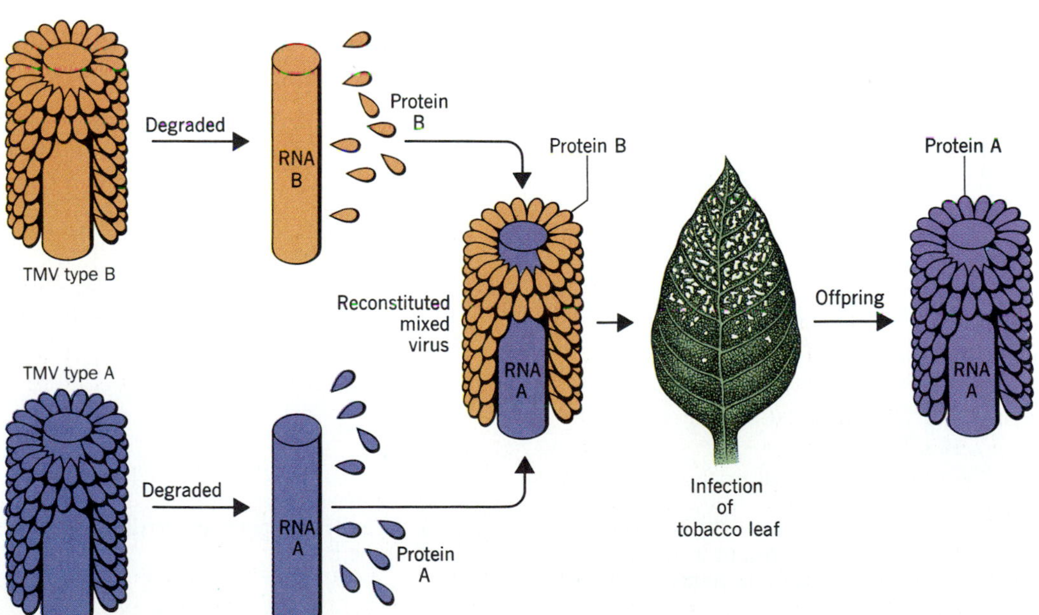

Figure 5.4 Proof that the genetic material of tobacco mosaic virus (TMV) is RNA, not protein. The RNA molecules and the protein coats of two different strains (*A* and *B*) of TMV were separated biochemically. The RNA of strain *A* was then mixed with the protein coats of strain *B* under conditions where complete, infective virus particles are reconstituted. When the reconstituted viruses were rubbed onto live tobacco leaves, the progeny viruses were phenotypically and genotypically identical to strain *A* from which the RNA was obtained and unlike strain *B* from which the protein was obtained. When the reconstituted viruses contained RNA of type *B* and protein of type *A*, the progeny were of type *B*. (After H. Fraenkel-Conrat and B. Singer, *Biochim. Biophys. Acta* 24: 540, 1957.)

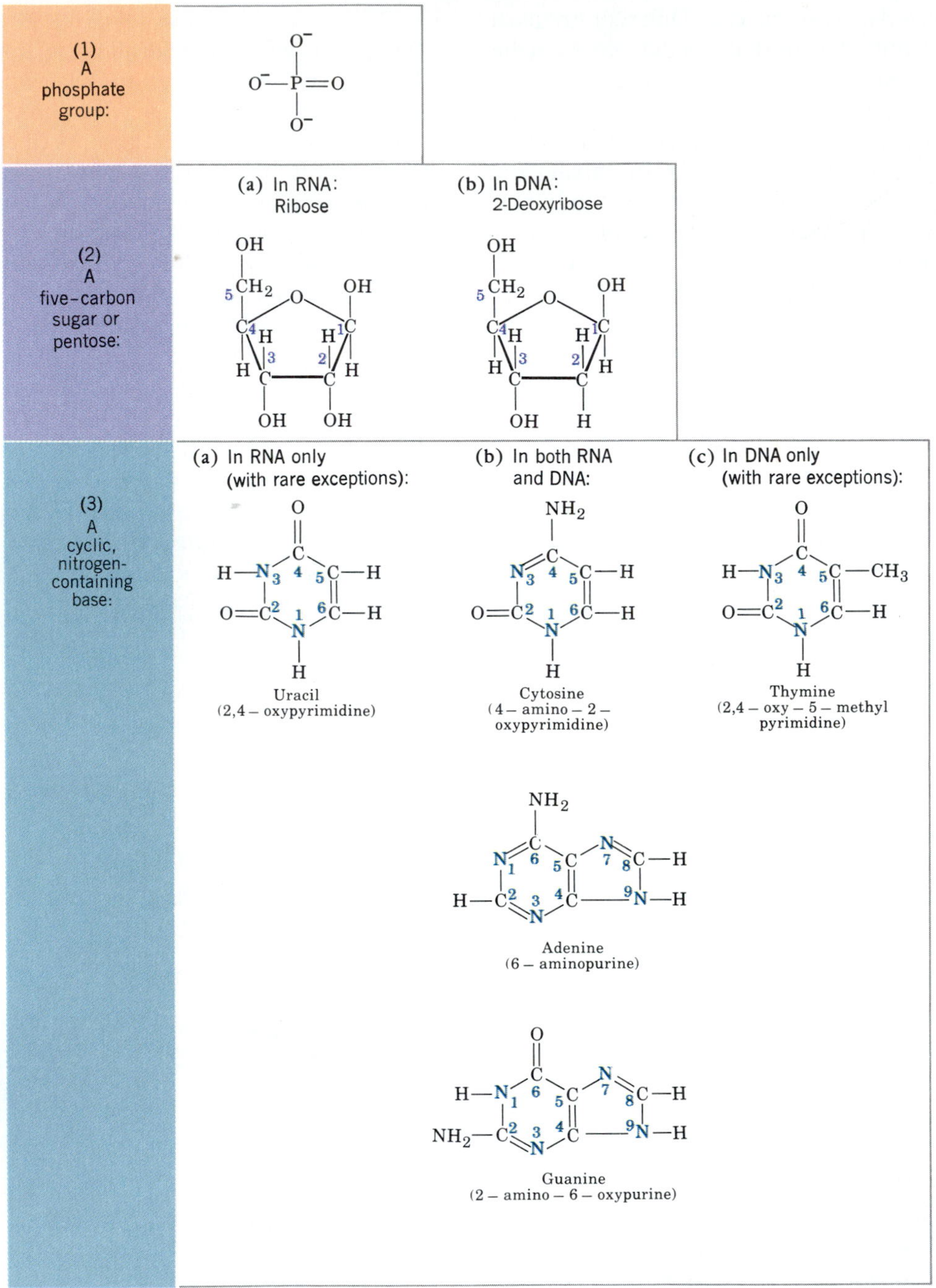

Figure 5.5 Structural formulas for the constituents of nucleic acids. When the pentoses are present in nucleosides, nucleotides, or nucleic acids, the five carbons are numbered 1′, 2′, 3′, 4′, and 5′, respectively, to distinguish them from the carbons of the bases.

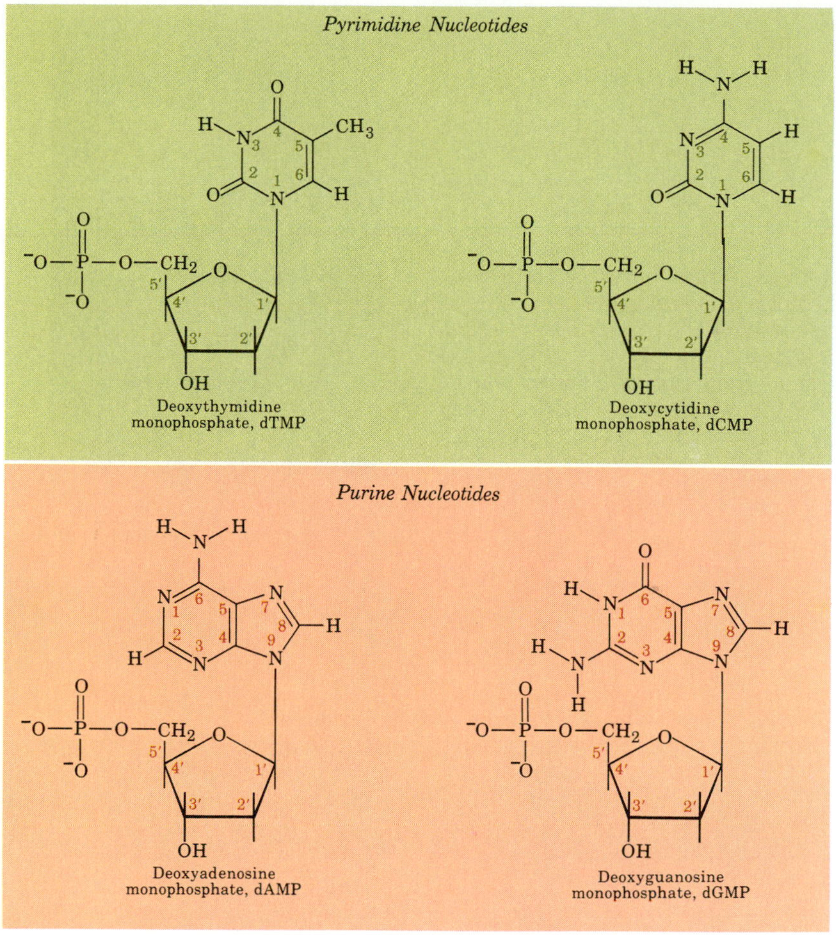

Pyrimidine Nucleotides

Deoxythymidine monophosphate, dTMP

Deoxycytidine monophosphate, dCMP

Purine Nucleotides

Deoxyadenosine monophosphate, dAMP

Deoxyguanosine monophosphate, dGMP

Figure 5.6 The four common deoxyribonucleotides of DNA. RNA contains similar ribonucleotides, which contain the pyrimidines uracil and cytosine and the purines adenine and guanine.

but has a different base, ***uracil,*** in place of thymine. Adenine and guanine are double-ring bases called ***purines***; cytosine, thymine, and uracil are single-ring bases called ***pyrimidines*** (Fig. 5.5). Both DNA and RNA, therefore, contain four different subunits or nucleotides, two purine nucleotides, and two pyrimidine nucleotides (Fig. 5.6). RNA usually exists as a single-stranded polymer that is composed of a long sequence of nucleotides. DNA, however, has one very important additional level of organization; it is usually a double-stranded molecule.

The Watson and Crick DNA Double Helix

The correct structure of DNA was first deduced by J.D. Watson (Fig. 5.7) and F. H. C. Crick in 1953. Their ***double-helix*** model of DNA structure was based on two major kinds of evidence.

1. When the composition of DNA from many different organisms was analyzed by E. Chargaff and colleagues, it was observed that the ***concentration of thymine was always equal to the concentration of adenine and the concentration of cytosine was always equal to the concentration of guanine.*** This strongly suggested that thymine and adenine as well as cytosine and guanine were present in DNA with some fixed interrelationship. Of course, it also necessitated that the total concentration of pyrimidines (thymine plus cytosine) always equal the total concentration of purines (adenine plus guanine; see Table 5.2). However, the (thymine + adenine)/(cytosine + guanine) ratio was found to vary widely in DNAs of different species (Table 5.2).
2. When X rays are focused through isolated macromolecules or crystals of purified molecules, the X rays are deflected by the atoms of the molecules in specific patterns, called diffraction patterns, which provide information about the organization of the components of the molecules. These ***X-ray diffraction patterns*** can be

Figure 5.7 J. D. Watson, American investigator in biochemical genetics. Along with the British investigators F. H. C. Crick and M. H. F. Wilkins, Dr. Watson won the Nobel Prize in physiology and medicine in 1962. The prize was awarded for his contribution to our knowledge and understanding of the chemical nature of the gene. The progress of this team up to 1953 is summarized by the Watson–Crick model, which displays the structure of the DNA molecule. (Courtesy of J. D. Watson.)

TABLE 5.2 Base Composition of DNA from Various Organisms[a]

SPECIES	% ADENINE	% GUANINE	% CYTOSINE	% THYMINE	MOLAR RATIOS $\frac{A+G}{T+C}$	MOLAR RATIOS $\frac{A+T}{G+C}$
I. **Viruses**						
Bacteriophage λ	26.0	23.8	24.3	25.8	0.99	1.08
Bacteriophage T2	32.6	18.1	16.6	32.6	1.03	1.88
Herpes simplex	13.8	37.7	35.6	12.8	1.06	0.36
Pseudorabies	13.2	37.0	36.3	13.5	1.00	0.36
Vaccinia	31.5	18.0	19.0	31.5	0.98	1.70
II. **Bacteria**						
Escherichia coli	26.0	24.9	25.2	23.9	1.04	1.00
Diplococcus pneumoniae	29.8	20.5	18.0	31.6	1.02	1.59
Micrococcus lysodeikticus	14.4	37.3	34.6	13.7	1.07	0.39
Ramibacterium ramosum	35.1	14.9	15.2	34.8	1.00	2.32
III. **Fungi**						
Neurospora crassa	23.0	27.1	26.6	23.3	1.00	0.86
Aspergillus niger	25.0	25.1	25.0	24.9	1.00	1.00
Saccharomyces cerevisiae	31.7	18.3	17.4	32.6	1.00	1.80
IV. **Higher Eukaryotes**						
Zea mays (corn)	25.6	24.5	24.6	25.3	1.00	1.04
Nicotiana tabacum (tobacco)	29.3	23.5	16.5	30.7	1.12	1.50
Arachis hypogaea (peanut)	32.1	17.6	18.0	32.2	0.99	1.80
Drosophila melanogaster	30.7	19.6	20.2	29.4	1.01	1.51
Bombyx mori (silkworm)	30.7	18.9	19.4	31.1	0.98	1.61
Rana pipiens (frog)	26.3	23.5	23.8	26.4	0.99	1.11
Homo sapiens (human)						
liver	30.3	19.5	19.9	30.3	0.99	1.53
thymus	29.8	20.2	18.2	31.8	1.02	1.60
spermatozoa	30.5	19.9	20.6	28.9	1.02	1.47

[a]Selected data from H. Sober (ed.), *Handbook of Biochemistry, Selected Data for Molecular Biology,* 2nd ed. The Chemical Rubber Co., Cleveland, 1970.

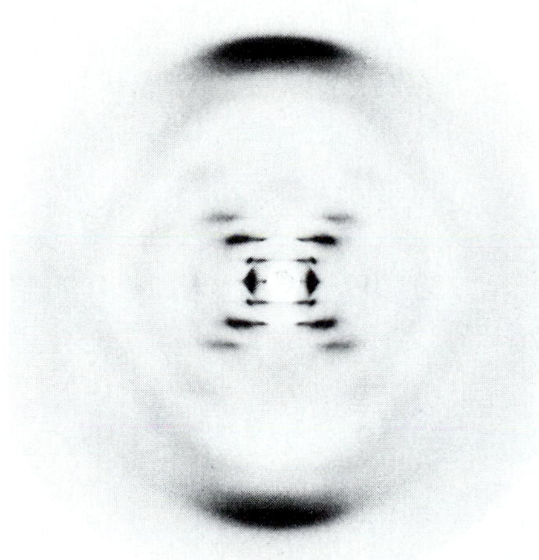

Figure 5.8 One of the X-ray diffraction photographs of DNA that led to the double helix model of DNA structure. An X-ray crystallographer can recognize the central cross-shaped pattern as indicative of a helical structure. The heavy dark patterns (top and bottom) indicate that the bases are stacked perpendicular to the axis of the molecule with a periodicity of 3.4 Å. (Courtesy M. H. F. Wilkins, Biophysics Department, King's College, London.)

recorded on X-ray-sensitive film just as one photographs patterns of light with a camera and light-sensitive film. Watson and Crick had available X-ray crystallographic data on DNA structure from the studies of M. H. F. Wilkins, R. Franklin, and their coworkers (Fig. 5.8). These data indicated that DNA was a highly ordered, multiple-stranded structure with repeating substructures spaced every 3.4 angstroms [1 angstrom (Å) = 10^{-8} cm] along the axis of the molecule.

On the basis of Chargaff's chemical data, Wilkins and Franklin's X-ray diffraction data, and inferences drawn from model building, Watson and Crick proposed that DNA exists as a ***double helix*** in which the two polynucleotide chains are coiled about one another in a spiral (Fig. 5.9). Each polynucleotide chain consists of a sequence of nucleotides linked together by ***phosphodiester bonds***, joining adjacent deoxyribose moieties (Table 5.3 and Fig. 5.10). The two polynucleotide strands are held together in their helical configuration by ***hydrogen bonding*** (Table 5.3) between bases in opposing strands, the resulting base-pairs being stacked between the two chains perpendicular to the axis of the molecule like the steps of a spiral staircase (Fig. 5.9). The base-pairing is specific;

TABLE 5.3 Chemical Bonds Important in DNA Structure

1. *Covalent bonds*
   Strong chemical bonds formed by sharing of electrons between atoms.
   (a) In bases and sugars

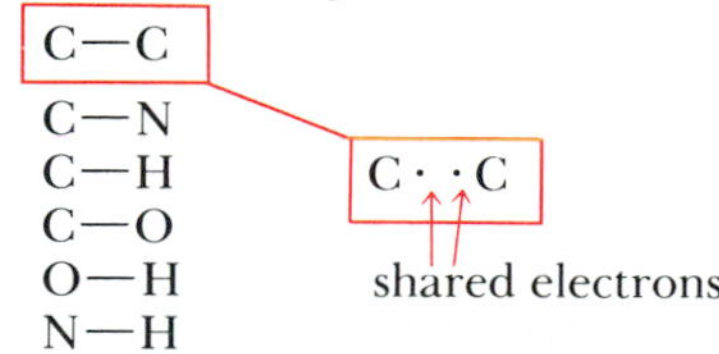

   (b) Phosphodiester linkages

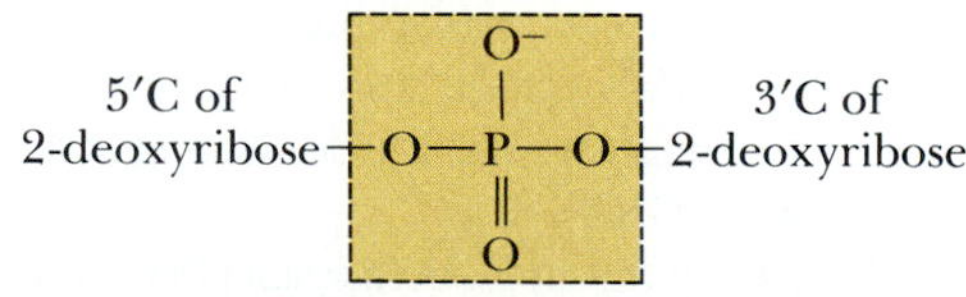

2. *Hydrogen bonds*
   A weak bond between an electronegative atom and a hydrogen atom (electropositive) that is covalently linked to a second electronegative atom.

   δ^+ δ^-
   N—H· · · · ·O—
   N—H· · · · ·N

3. *Hydrophobic "bonds"*
   The association of nonpolar groups with each other when present in aqueous solutions because of their insolubility in water.

   Water molecules are very polar (δ^- O and δ^+ Hs). Compounds which are similarly polar are very soluble in water ("hydrophilic"). Compounds which are nonpolar (no charged groups) are very insoluble in water ("hydrophobic").

   The stacked base-pairs provide a hydrophobic core.

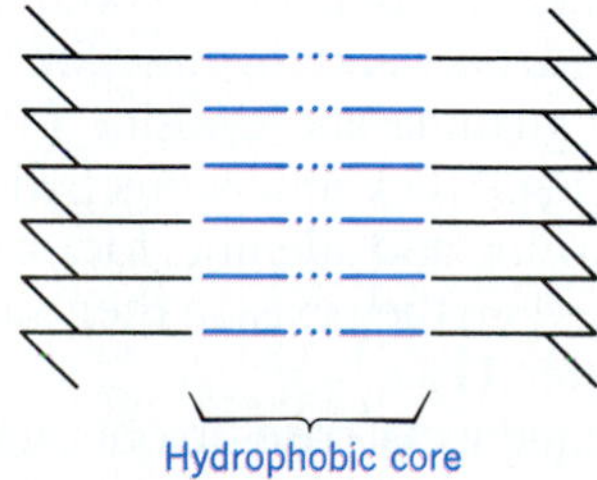

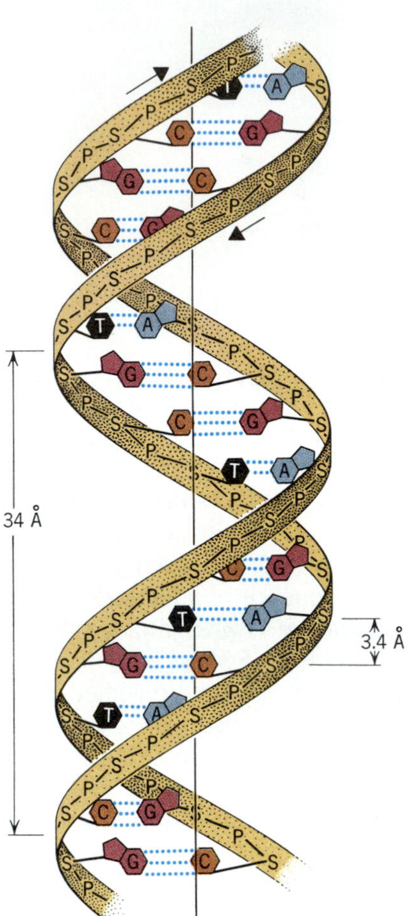

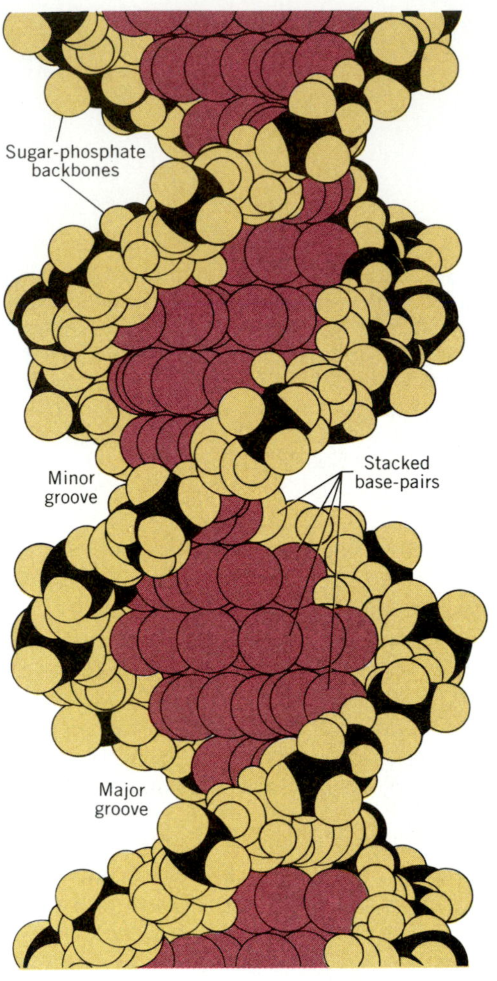

Figure 5.9 Diagram (left) and space-filling model (right) of the Watson–Crick double-helix model of the structure of DNA. A, T, G, and C represent adenine, thymine, guanine, and cytosine, respectively. S and P represent sugar (2-deoxyribose) and phosphate groups. (The space-filling model is based on a diagram by M. Feughelman *et al., Nature* 175: 834, 1955. Reprinted by permission from *Nature*, Vol. 175, p. 834; Copyright © 1955 by Macmillan Magazines Ltd.)

adenine is always paired with thymine, and guanine is always paired with cytosine (Fig. 5.11). Thus, all base-pairs consist of one purine and one pyrimidine. The specificity of base-pairing results from the hydrogen-bonding capacities of the bases in their normal configurations. In their most common structural configurations, adenine and thymine form two hydrogen bonds, and guanine and cytosine form three hydrogen bonds (Fig. 5.11). Analogous hydrogen bonding between cytosine and adenine, for example, is not possible except when they exist in their rare structural states (see Chapter 11).

Once the sequence of bases in one strand of a DNA double helix is known, the sequence of bases in the other strand is also known because of the specific base-pairing. The two strands of a DNA double helix are thus said to be ***complementary (not identical)***; it is this property, complementarity of the two strands, that makes DNA uniquely suited to store and transmit genetic information (see the following section on the replication of DNA).

The base-pairs in DNA are stacked 3.4 Å apart with 10 base-pairs per turn (360°) of the double helix (Fig. 5.9). The sugar–phosphate backbones of the two complementary strands are ***antiparallel***; that is, they have ***opposite chemical polarity*** (Fig. 5.10). As one moves unidirectionally along a DNA double helix, the phosphodiester bonds in one strand go from a 3′ carbon of one nucleotide to a 5′ carbon of the adjacent nucleotide, whereas those in the complementary strand go

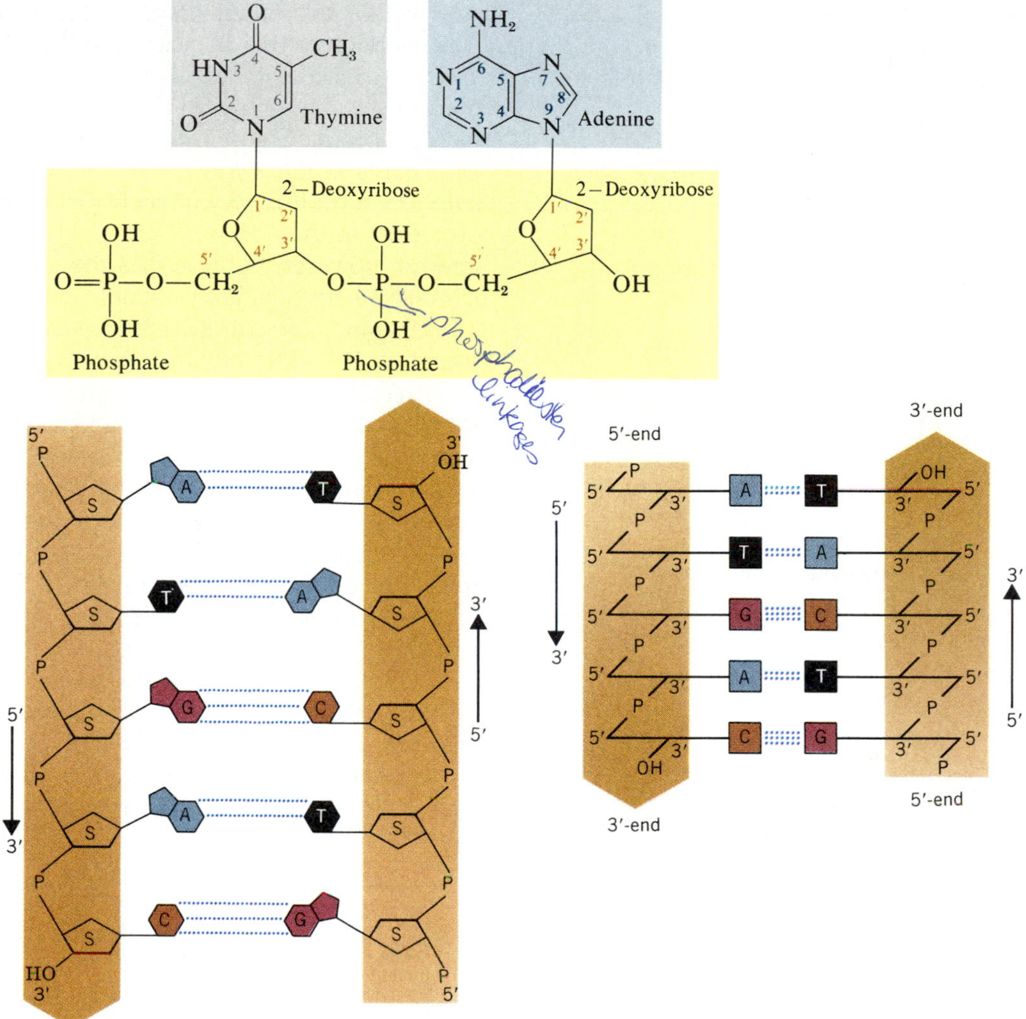

Figure 5.10 Molecular structure of DNA showing the sugar–phosphate backbones of the polynucleotide chains and their antiparallel nature (opposite chemical polarity). (a) A deoxythmidylate–deoxyadenylate dinucleotide, showing the numbering system used for nucleotides and the phosphodiester linkage between the 3′ and 5′ carbons of adjacent nucleotides. Note the 5′ to 3′ chemical polarity going from left to right. (b) Molecular structure and (c) schematic representation of a short segment of a DNA molecule, emphasizing the opposite polarity of the complementary strands.

from a 5′ carbon to a 3′ carbon. This opposite polarity of the complementary strands is very important in considering the mechanism of replication of DNA.

The high degree of stability of DNA double helices results in part from the large number of hydrogen bonds between the base-pairs (even though each hydrogen bond by itself is quite weak, much weaker than a covalent bond) and in part from the hydrophobic bonding (or "stacking forces") between the stacked base-pairs (Table 5.3 and Fig. 5.9). The planar sides of the base-pairs are relatively nonpolar and thus tend to be water insoluble ("hydrophobic"). This hydrophobic core of stacked base-pairs contributes considerable stability to DNA molecules present in the aqueous protoplasms of living cells.

Conformational Flexibility of DNA Molecules

The vast majority of the DNA molecules present in the aqueous protoplasms of living cells almost certainly exists in the Watson–Crick double-helix form just described (Fig. 5.9). This is the ***B*-form** of DNA. The B-form is the conformation that DNA takes under physiological conditions (in aqueous solutions containing low concentrations of salts). However, DNA is not a static, invariant molecule. To the contrary, ***DNA molecules exhibit a considerable amount of conformational flexibility***

The structures of DNA molecules change as a function of their environment. The exact conformation of a given DNA molecule or segment of a DNA mole-

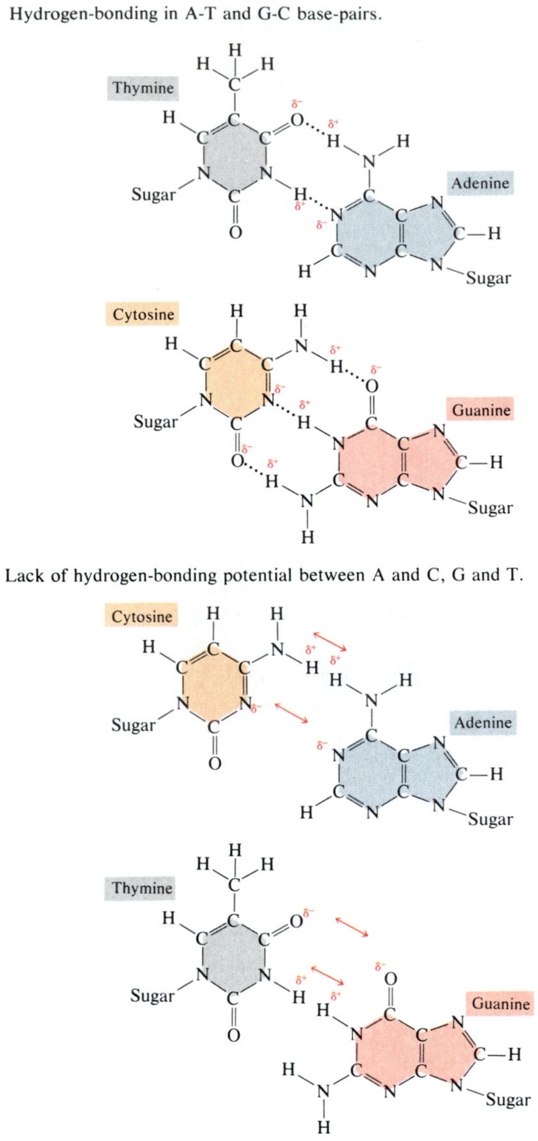

Hydrogen-bonding in A-T and G-C base-pairs.

Thymine Adenine

Cytosine Guanine

Lack of hydrogen-bonding potential between A and C, G and T.

Cytosine Adenine

Thymine Guanine

Figure 5.11 Base-pairing in DNA, adenine with thymine and guanine with cytosine, based on hydrogen bonding between appropriately spaced, negatively charged =O and >N groups and positively charged –Hs. Note that the similar apposition of adenine and cytosine or guanine and thymine results in the juxtaposition of identically charged (+ or −) groups at two of the three sites of potential hydrogen bonding. Thus, adenine is not normally found base-paired with cytosine (nor guanine with thymine) in DNA.

cule will depend on the nature of the molecules with which it is interacting. In fact, intracellular B-form DNA appears to have an average of 10.4 nucleotide-pairs per turn, rather than precisely 10 as shown in Fig. 5.9. In high concentrations of salts or in a dehydrated state, DNA exists in the ***A-form***, which has 11 nucleotide-pairs per turn. It is very unlikely that DNA molecules ever exist in the A-form *in vivo*. This structure is of interest, however, because it is the conformation of DNA–RNA heteroduplexes (double helices containing a DNA strand base-paired with a complementary RNA strand) or RNA–RNA duplexes *in vivo*.

Recently, certain DNA sequences have been shown to exist in a unique left-handed, double helical form called ***Z-DNA*** (Z for the zigzagged path of the sugar–phosphate backbones of the structure). The helices of A- and B-form DNA are wound in a right-handed manner. Moreover, specific segments of DNA molecules can undergo conformational shifts from the B-form to the Z-form and vice versa. In fact, certain regulatory proteins may bind only to the Z-form (or B-form) of a DNA sequence and cause it to shift to the B-form (or Z-form) (see Chapter 15, pp. 432–434). In any case, one must remember that the structure of DNA is not invariant and that structural variations in DNA molecules may play important biological roles.

SEMICONSERVATIVE REPLICATION OF DNA

Living organisms perpetuate their kind through reproduction. This may be simple duplication (cell fission) as in bacteria or complex modes of sexual reproduction as in higher plants and animals. In all cases, however, reproduction entails the faithful transmission of the genetic information of the parents to the progeny. Since the genetic information is stored in DNA, ***the replication of DNA is central to all of biology***

When Watson and Crick proposed the double-helix structure of DNA with its complementary base-pairing, they immediately recognized that the base-pairing specificity could provide the basis for a simple (superficially simple, at least) mechanism for DNA duplication. If the two complementary strands of a double helix separated (by breaking the hydrogen bonds of each base-pair), each parental strand could direct the synthesis of a new complementary strand because of the specific base-pairing requirements (Fig. 5.12). That is, each parental strand could serve as a ***template*** for a new complementary strand. Adenine, for example, in the parent strand would serve as a template via its hydrogen-bonding potential for the incorporation of thymine in the nascent complementary strand. This mechanism of DNA replication is

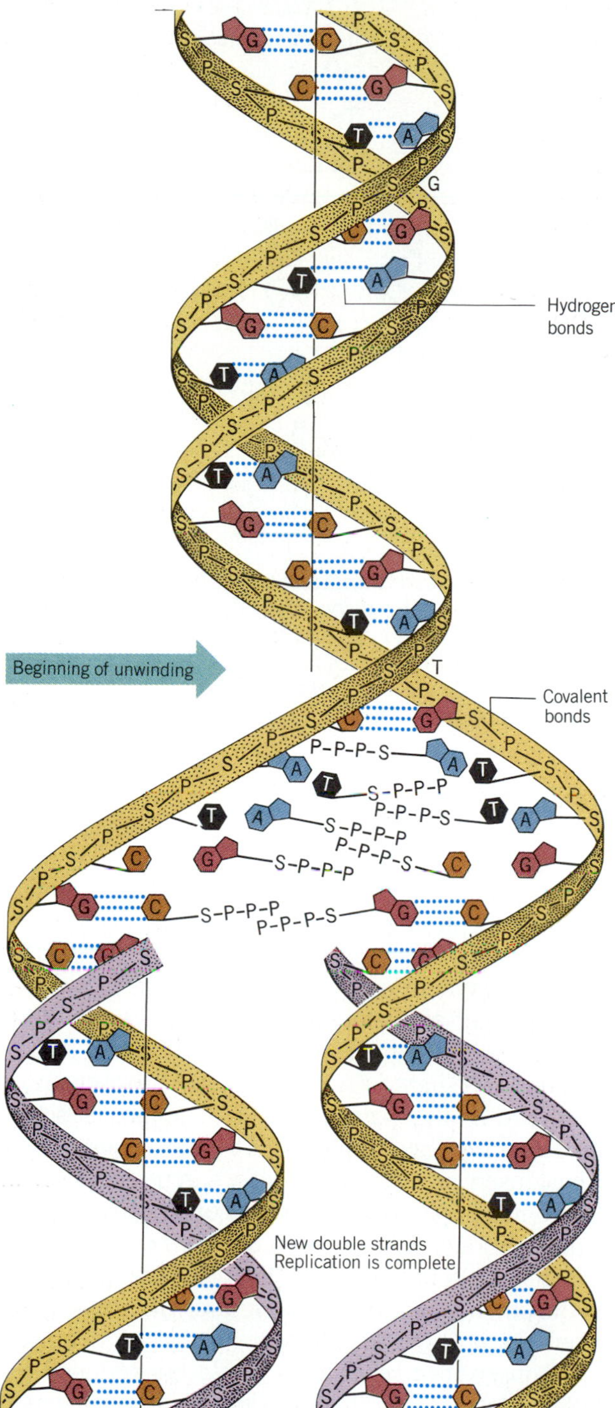

Figure 5.12 Replication of DNA as proposed by Watson and Crick. As the double-stranded parental DNA molecule unwinds, the separated strands serve as templates for the alignment of incoming nucleoside triphosphates, which are then linked by phosphodiester bonds to form new strands complementary to the parental strands. The result is the production of two progeny double helices identical to the parental double helix.

called ***semiconservative*** replication, since each of the complementary strands of the parental double helix is conserved (or the double helix is "half-conserved") during the process.

In considering possible mechanisms of DNA replication, three different hypothetical modes are apparent (Fig. 5.13). In addition to (1) semiconservative replication, one can propose (2) "conservative" replication in which the parental double helix remains intact (is totally conserved) and somehow directs the synthesis of a "progeny" double helix composed of two newly synthesized strands, and (3) "dispersive" replication in which segments of parental strands and progeny or nascent strands become interspersed through some kind of a fragmentation, synthesis, and rejoining process.

The "Meselson–Stahl Experiment"

The results of the first critical test of Watson and Crick's proposal that DNA replicates semiconservatively were published in 1958 by M. S. Meselson and F. W. Stahl. Their results showed that the chromosome (now known to contain a single Watson–Crick double helix of DNA) of the common colon bacillus *Escherichia coli* replicated semiconservatively.

Meselson and Stahl grew *E. coli* cells for many generations in a medium in which the heavy isotope of nitrogen, ^{15}N, had been substituted for the normal, light isotope, ^{14}N. The purine and pyrimidine bases in DNA contain nitrogen; thus, the DNA of cells grown on medium containing ^{15}N will have a greater density (weight per unit volume) than the DNA of cells grown on medium containing ^{14}N. Since molecules of different densities can be separated by a procedure called ***equilibrium density-gradient centrifugation,*** Meselson and Stahl were able to distinguish between the three possible modes of DNA replication by following the changes in the density of DNA of cells grown on ^{15}N medium and then transferred to ^{14}N medium for various periods of time (so-called ***density transfer experiments***).

The density of most DNAs is about the same as the density of concentrated solutions of heavy salts such as cesium chloride (CsCl). For example, the density of 6 *M* CsCl is about 1.7 g/cm^3. *Escherichia coli* DNA containing ^{14}N has a density of 1.710 g/cm^3. Substitution of ^{15}N for ^{14}N increases the density of *E. coli* DNA to 1.724 g/cm^3.

When a heavy salt solution such as 6 *M* CsCl is centrifuged at very high speeds (30,000–50,000 revolutions per minute) for 48–72 hours, an ***equilibrium-density gradient*** is formed (Fig. 5.14). The centrifugal force caused by spinning the solution at high speeds sediments the salt toward the bottom of the tube. Diffusion, on the other hand, results in movement of

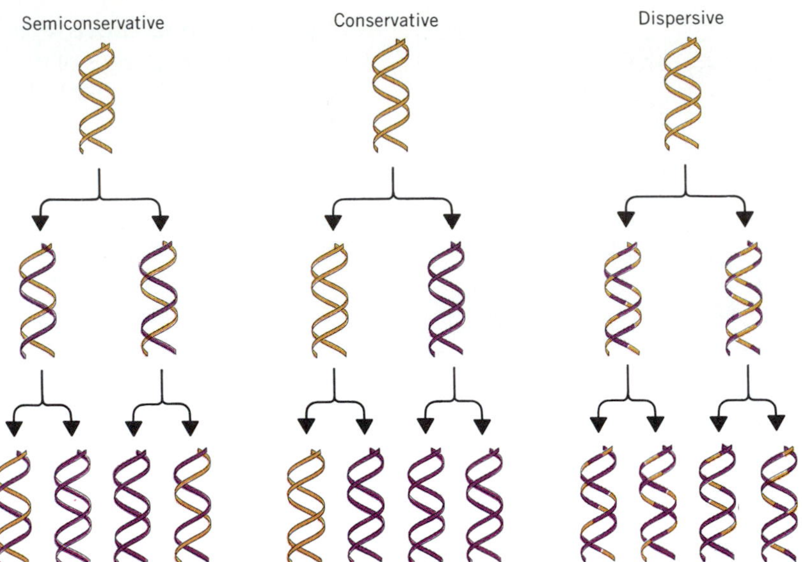

Figure 5.13 Three theoretical modes of DNA replication: semiconservative, conservative, and dispersive. During semiconservative DNA replication (left), each single strand of the parental double helix is conserved, acting as a template for the synthesis of a new complementary strand. During conservative DNA replication (center), the entire parental double helix is conserved and directs the synthesis of a new double helix. During dispersive DNA replication (right), both strands of the parental double helix are fragmented and parental and newly synthesized segments of DNA become interspersed in each of the strands of the two resulting DNA double helices.

salt molecules back toward the top (low salt concentration) of the tube. After a sufficient period of high-speed centrifugation, an equilibrium between sedimentation and diffusion is reached, at which time a linear gradient of increasing density exists from the top of the tube to the bottom of the tube (Fig. 5.14). If DNA is present in such a gradient, it will move to a position where the density of the salt solution is equal to its own density. Thus, if a mixture of *E. coli* DNA containing ^{15}N ("heavy" DNA) and *E. coli* DNA containing ^{14}N ("light" DNA) is subjected to CsCl equilibrium density-gradient centrifugation, the DNA molecules will separate into two "bands," one containing "heavy" DNA and one containing "light" DNA (Fig. 5.14).

Meselson and Stahl took cells that had been growing in medium containing ^{15}N for several generations (and thus contained "heavy" DNA), washed them to remove the ^{15}N-containing medium, and transferred them to medium containing ^{14}N. After allowing the cells to grow in the presence of ^{14}N for varying periods of time, the DNA was extracted and analyzed in CsCl equilibrium-density gradients. The results of their experiment (Fig. 5.15) are only consistent with semiconservative replication, excluding both conservative and dispersive models of DNA synthesis. All the DNA isolated from cells after one generation of growth in medium containing ^{14}N had a density halfway between the densities of "heavy" DNA and "light" DNA. This intermediate density is usually referred to as "hybrid" density. After two generations of growth in medium containing ^{14}N, half of the DNA was of "hybrid" density and half was "light." These results are precisely those predicted by the Watson and Crick semiconservative mode of replication (Fig. 5.15). One generation of semiconservative replication of a parental double helix containing ^{15}N in medium containing only ^{14}N would produce two progeny double helices both of which had ^{15}N in one strand (the "old" strand) and ^{14}N in the other strand (the "new" strand). Such molecules would be of "hybrid" density.

Conservative replication would not produce any DNA molecules with "hybrid" density; after one generation of conservative replication of "heavy" DNA in "light" medium, half of the DNA would still be "heavy" and the other half would be "light." If replication were dispersive, Meselson and Stahl would have observed a shift of the DNA from "heavy" toward "light" in each generation (i.e., "half heavy" or "hybrid" after one generation, "quarter heavy" after two generations, etc.). Meselson and Stahl's results are clearly inconsistent with either of these possibilities.

Subsequent studies have verified Meselson and

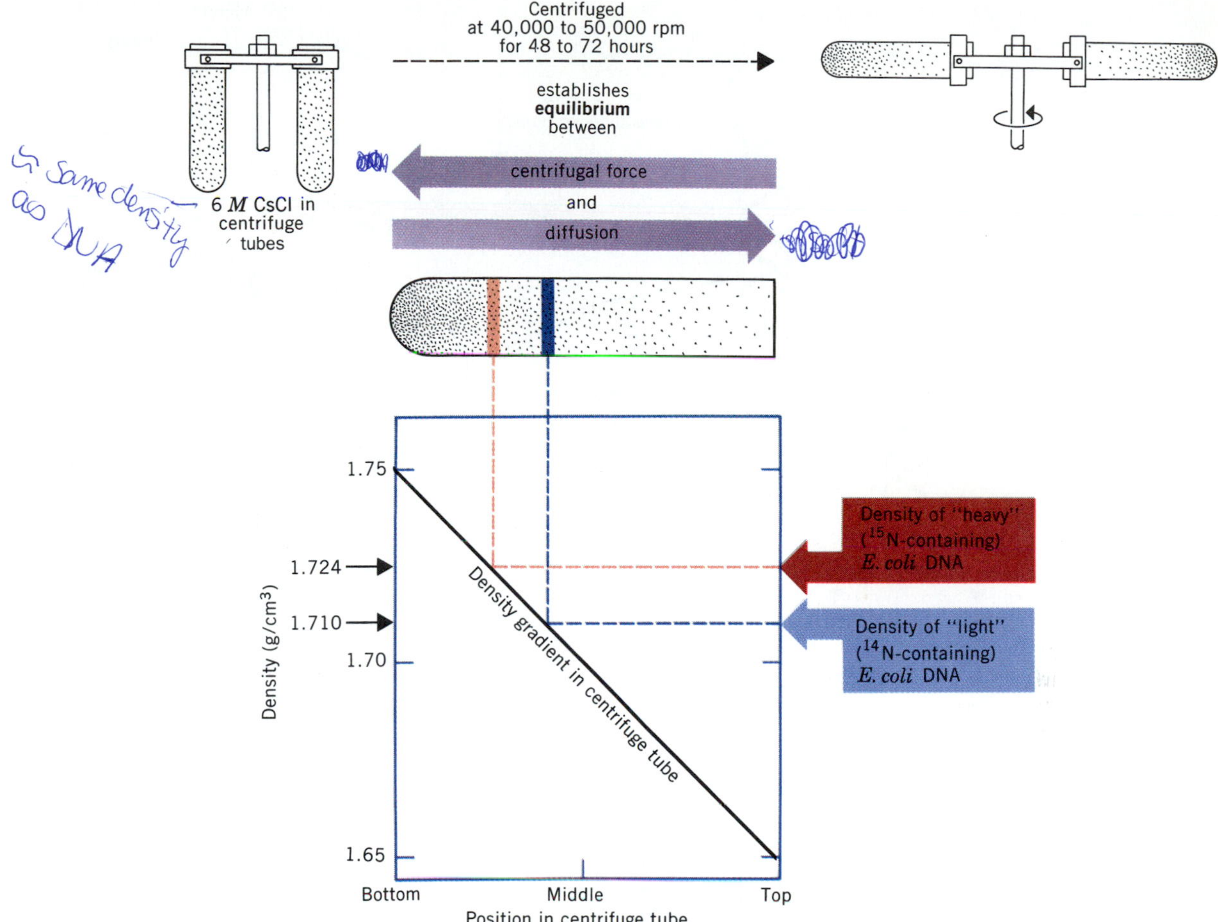

Figure 5.14 The use of cesium chloride (CsCl) in density-gradient centrifugation to separate DNAs of different densities. This procedure was used by Meselson and Stahl to demonstrate that the chromosomes of *E. coli* replicate semiconservatively (see Fig. 5.15). The density of 6 *M* CsCl is about 1.7 g/cm^3. If such a solution is centrifuged at very high speeds for a long enough period of time, a density gradient will be formed because of the equilibrium between (1) sedimentation of the CsCl to the bottom of the centrifuge tube as a result of the centrifugal force and (2) diffusion of the CsCl toward the top of the tube. The densities of most naturally occurring nucleic acids fall within the range covered by such gradients. CsCl density gradients have thus been very useful in the study of nucleic acids.

Stahl's conclusion that DNA replication is semiconservative and have extended it to many other organisms, including higher plants and animals.

Autoradiography of Replicating Bacterial Chromosomes

The visualization of replicating chromosomes was first accomplished by J. Cairns in 1963 using the technique called ***autoradiography***. Autoradiography is a method for detecting and localizing radioactive isotopes in cytological preparations or macromolecules by exposure to a photographic emulsion that is sensitive to low-energy radiation. Autoradiography is particularly useful in studying DNA metabolism because ***DNA can be specifically labeled by growing cells on [^{3}H]thymidine***, the tritiated deoxyribonucleoside of thymine. Thymidine is incorporated exclusively into DNA; it is not present in any other major component of the cell.

Cairns grew *E. coli* cells in medium containing [^{3}H]thymidine for varying periods of time, lysed the cells very gently so as not to break the chromosomes (long DNA molecules are very shear sensitive), and carefully collected the chromosomes on membrane filters. These filters were affixed to glass slides, coated with emulsion sensitive to β-particles (the low-energy electrons emitted during decay of tritium), and stored in the dark for a period of time to allow sufficient

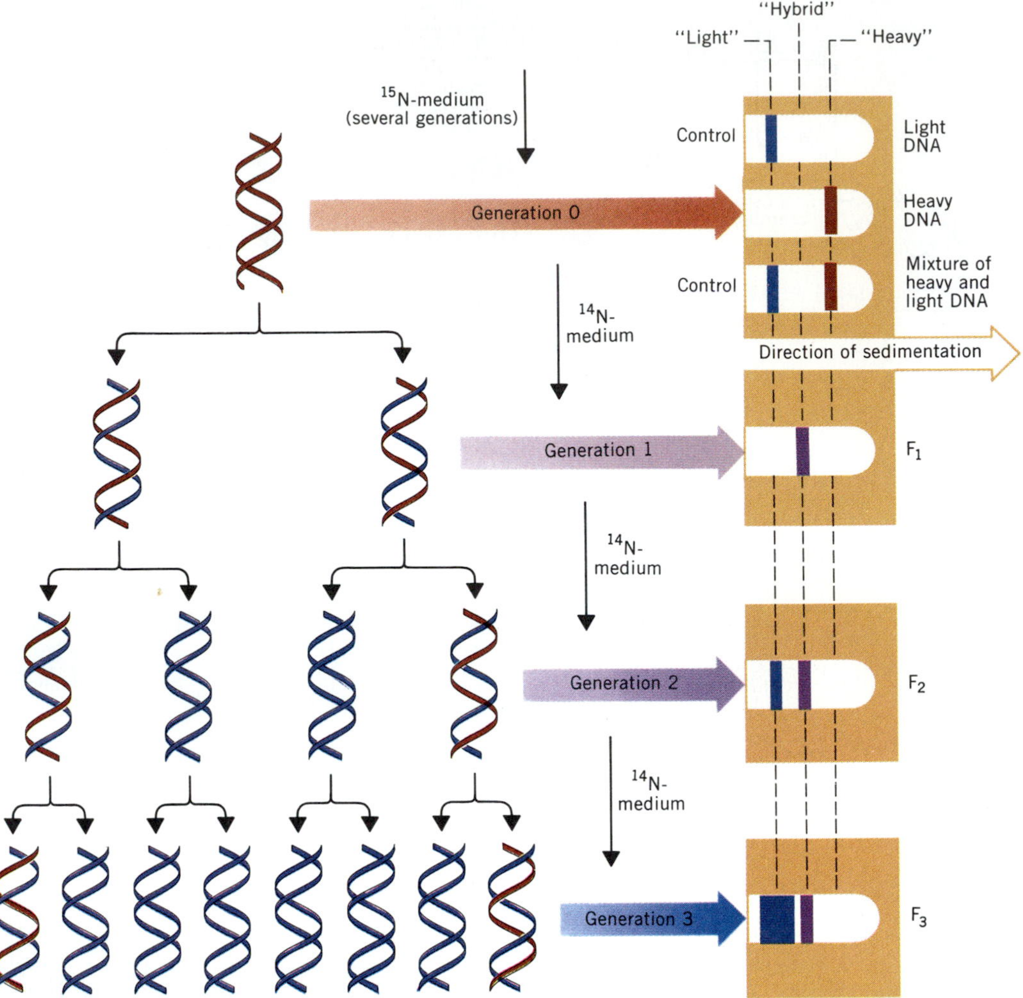

Figure 5.15 Results (right) and interpretation (left) of the Meselson and Stahl experiment that demonstrated that the *E. coli* chromosomes replicate semiconservatively. The DNA of *E. coli* was density-labeled by growing cells for several generations in medium containing the heavy isotope of nitrogen, ^{15}N. The cells were then transferred to medium containing the normal isotope of nitrogen, ^{14}N, and allowed to grow for varying periods of time (one generation, two generations, etc.). The transfer of the density label (^{15}N) from parental DNA molecules to progeny DNA molecules was followed by the extraction of DNA from cells grown for varying periods in ^{14}N and the analysis of the DNA in CsCl density gradients (see Fig. 5.14). DNA that contains ^{15}N in both strands ("heavy" DNA) forms a band in the CsCl density gradient at a higher density position than DNA that contains ^{14}N in both strands ("light" DNA). After one generation of growth in ^{14}N medium, the DNA bands at an intermediate ("hybrid") density. Such "hybrid" DNA contains ^{15}N in one strand and ^{14}N in the other strand. After two generations of growth in ^{14}N medium, half of the DNA bands at the "hybrid" density and half bands at the "light" density. These results are precisely those predicted by semiconservative replication. (After M. S. Meselson and F. W. Stahl, *Proc. Natl. Acad. Sci. U.S.A.* 44: 671, 1958.)

radioactive decays. The autoradiographs observed when the films were developed (Fig. 5.16) showed that the chromosomes of *E. coli* are circular structures that exist as θ-shaped intermediates during replication. These autoradiographs further indicated that the unwinding of the two complementary parental strands (which is necessary for their separation) and their semiconservative replication occur simultaneously or are closely coupled. Since the parental double helix must rotate 360° to unwind each gyre of the helix, this necessitates the existence of some kind of "swivel" in the chromosome. Present evidence suggests that a

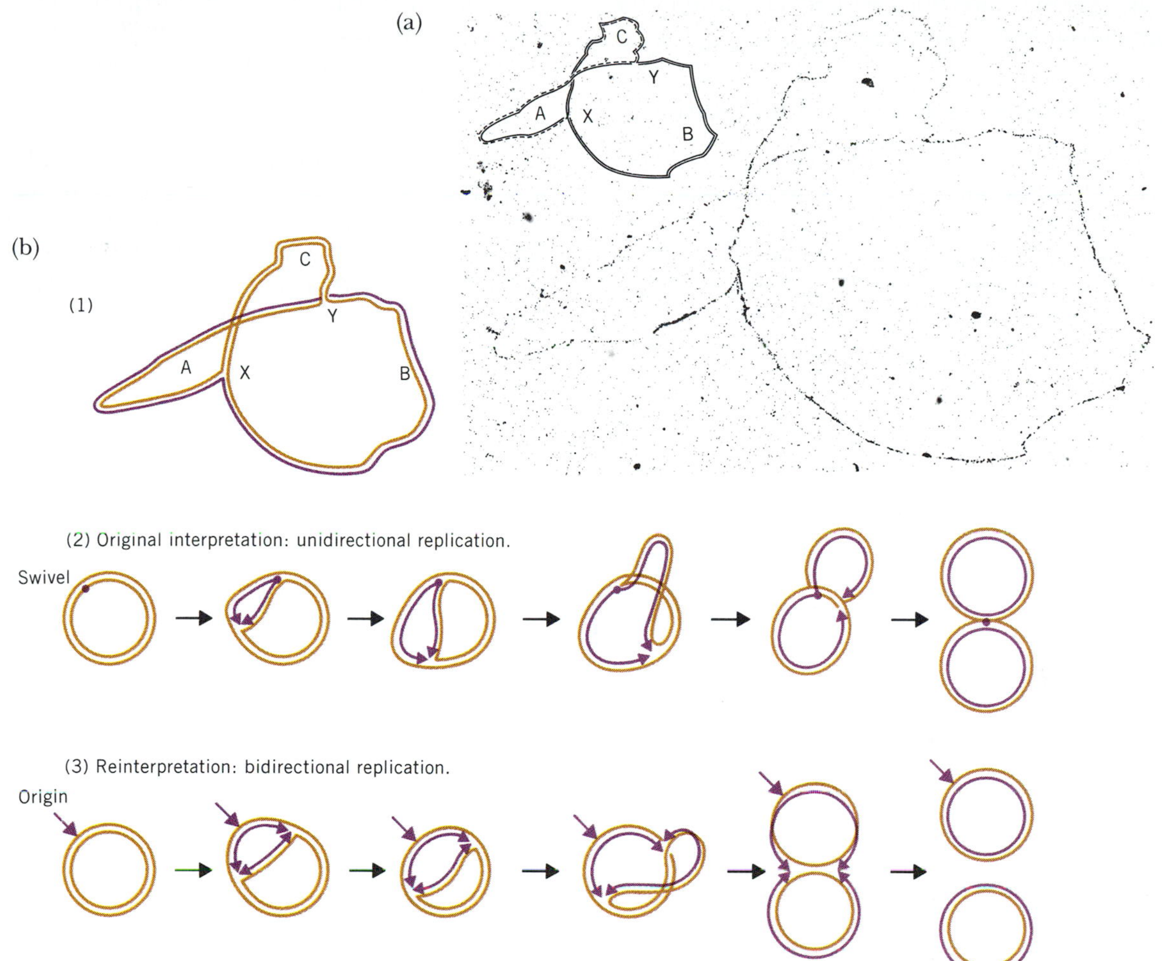

Figure 5.16 Depiction of a replicating chromosome of *E. coli* by autoradiography. (a) Cairns' autoradiograph and an interpretative drawing (upper left inset) of a θ-shaped replicating chromosome from a cell that had been grown for two generations in the presence of ^{3}H-thymidine. Loops A and B have completed a second replication in ^{3}H-thymidine; section C remains to be replicated the second time. Cairns' drawing (upper left inset) shows radioactive strands of DNA as solid lines and nonradioactive strands as dashed lines. Note that the autoradiograph loop B, with two radioactive strands, exhibits about twice the grain density of loop A, with only one radioactive strand. (J. Cairns, Imperial Cancer Research Fund, London, England. Reproduced with permission from *Cold Spring Harbor Sympos. Quant. Biol.* 28:43, 1963. Copyright © 1963 by Cold Spring Harbor Laboratory.) (b) Simplified diagram of Cairns' autoradiograph illustrating only the events of the replication in progress (1) and analogous diagrams illustrating the two possible interpretations of Cairns' results, namely, unidirectional replication (2) and bidirectional replication (3). In these diagrams, the yellow lines represent the two strands of the parental DNA double helix, and the purple lines represent the nascent (or daughter) strands. Originally, one of the two branch points (X and Y) was believed to be a replication fork and the other a terminus containing a "swivel," which served as an axis of rotation for unwinding the double helix (2). Replication has subsequently been shown to be bidirectional in *E. coli.* Thus, both X and Y are replication forks (3).

transient single-strand break (cleavage of one phosphodiester bond in one strand of the double helix) provides an axis of rotation to allow unwinding.

Cairns' interpretation of the autoradiographs was that semiconservative replication started at a site on the chromosome, which he called the "origin," and proceeded sequentially and unidirectionally around the circular structure (Fig. 5.16). Subsequent evidence has shown his original interpretation to be incorrect on one point: replication actually proceeds ***bidirectionally***, not unidirectionally. Each Y-shaped structure is a replication fork, and the two replication forks move in opposite directions sequentially around the circular chromosome (Fig. 5.16).

Unique Origins and Bidirectional Replication

Cairns' results provided no information as to whether the origin of replication (site at which replication is initiated) is a unique site or occurs at random on the chromosome. Moreover, his results did not allow him to differentiate between uni- and bidirectional replication. We now have direct evidence showing that replication in *E. coli* and several other organisms proceeds bidirectionally from a unique origin. These features of DNA replication can be illustrated most simply and convincingly by experiments with some of the small bacterial viruses.

Bacteriophage lambda (phage λ) is, like phage T2, a virus that grows in *E. coli*. It has a small chromosome consisting of a single linear molecule of DNA only 17.5 μm long. The phage λ chromosome is unique in that it has a single-stranded region, 12 nucleotides long, at the 5′ end of each complementary strand (Fig. 5.17).

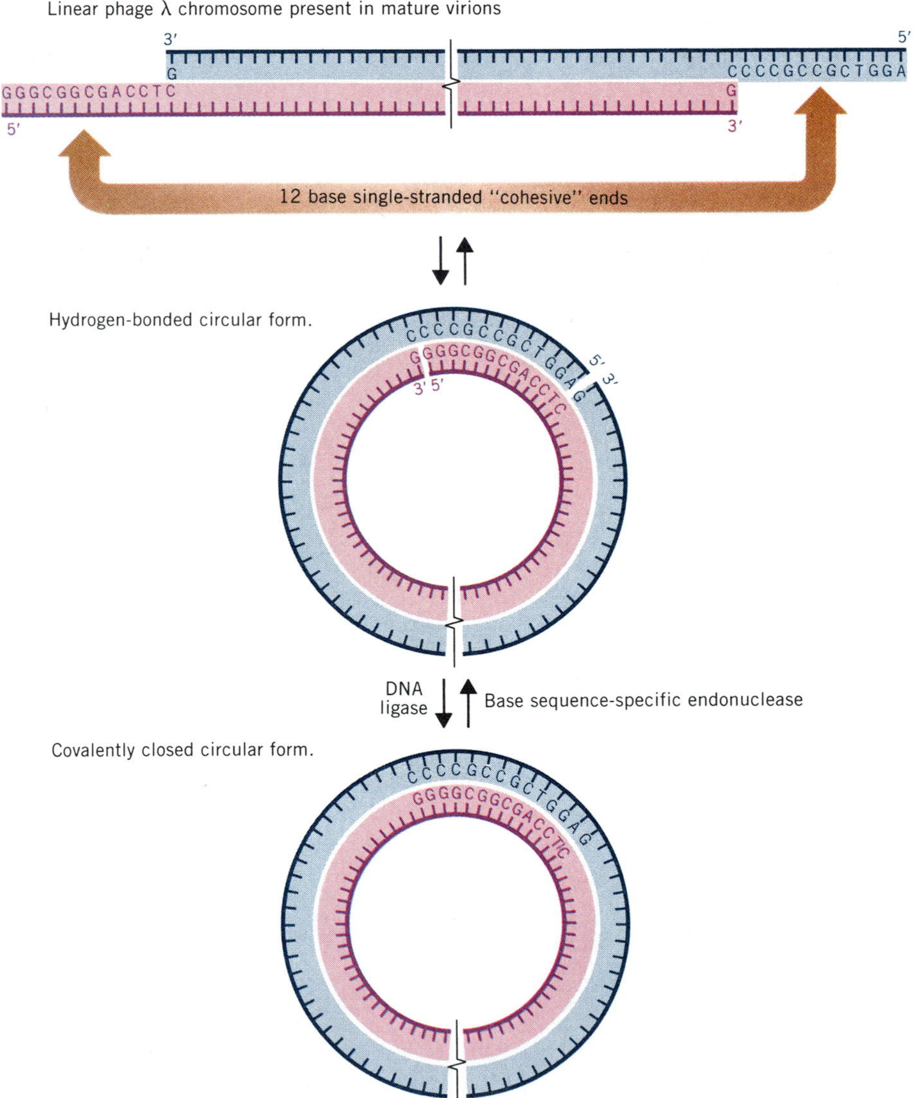

Figure 5.17 Interconversion of the linear lambda chromosome with its complementary "cohesive" ends, the hydrogen-bonded circular lambda chromosome, and the covalently closed circular lambda chromosome. The linear form of the chromosome would appear to be an adaptation to facilitate its injection from the phage head though the small opening in the phage tail into the host cell during infection. Prior to replicating in the host cell, the chromosome is converted to the covalently closed circular form. Only the ends of the chromosome of the mature virion are shown; the jagged vertical line indicates that the central portion of the chromosome is not shown. The entire lambda chromosome is about 4.5×10^4 nucleotide-pairs long.

These single-stranded ends, called "cohesive" or "sticky" ends, are complementary to each other. The cohesive ends of a lambda chromosome can thus base-pair to form a hydrogen-bonded circular structure. One of the first events to occur after a lambda chromosome is injected into a host cell is its conversion to a covalently closed circular molecule (Fig. 5.17). This conversion from the hydrogen-bonded circular form to the covalently closed circular form is catalyzed by ***polynucleotide ligase,*** a very important enzyme that seals single-strand breaks in DNA double helices. (Polynucleotide ligase is required in most, if not all, organisms for DNA replication, DNA repair, and recombination between DNA molecules.) The lambda chromosome replicates in its circular form via θ-shaped intermediates (Fig. 5.18), as does the *E. coli* chromosome.

The feature of the lambda chromosome that facilitated the demonstration of bidirectional replication was its differentiation into regions containing high concentrations of adenine and thymine ("A-T rich" regions) and regions with large amounts of guanine and cytosine ("G-C rich" regions). In particular, it contains a few segments with very high A-T content ("A-T rich clusters"). These A-T rich clusters were used as physical markers by M. Schnös and R. B. Inman to demonstrate, using a technique called "denaturation mapping," that replication of the lambda chromosome is initiated at a unique origin and proceeds bidirectionally rather than unidirectionally.

When DNA molecules are exposed to high temperature (100°C) or high pH (11.4), the hydrogen and hydrophobic bonds that hold the complementary strands together in the double-helix configuration are broken, and the two strands separate. This process is called ***denaturation.*** Because A-T base-pairs are held together by only two hydrogen bonds, compared with three hydrogen bonds in G-C base-pairs, A-T rich molecules denature more easily (at lower pH or temperature) than G-C rich molecules. When lambda chromosomes are exposed to pH 11.05 for 10 minutes under the appropriate conditions, the A-T rich clusters denature to form "denaturation bubbles," which are detectable by electron microscopy, whereas the G-C rich regions remain in the duplex state (Fig. 5.19). These denaturation bubbles can be used as physical markers whether the lambda chromosome is in its mature linear form, its circular form, or its θ-shaped replicative intermediates. By examining the positions of the branch points (Y-shaped structures) relative to the positions of the denaturation bubbles in a large number of θ-shaped replicative intermediates, Schnös and Inman demonstrated that both branch points are replication forks that move in opposite directions around the circular chromosome. A summary of their results is shown in Fig. 5.20. A schematic illustrating

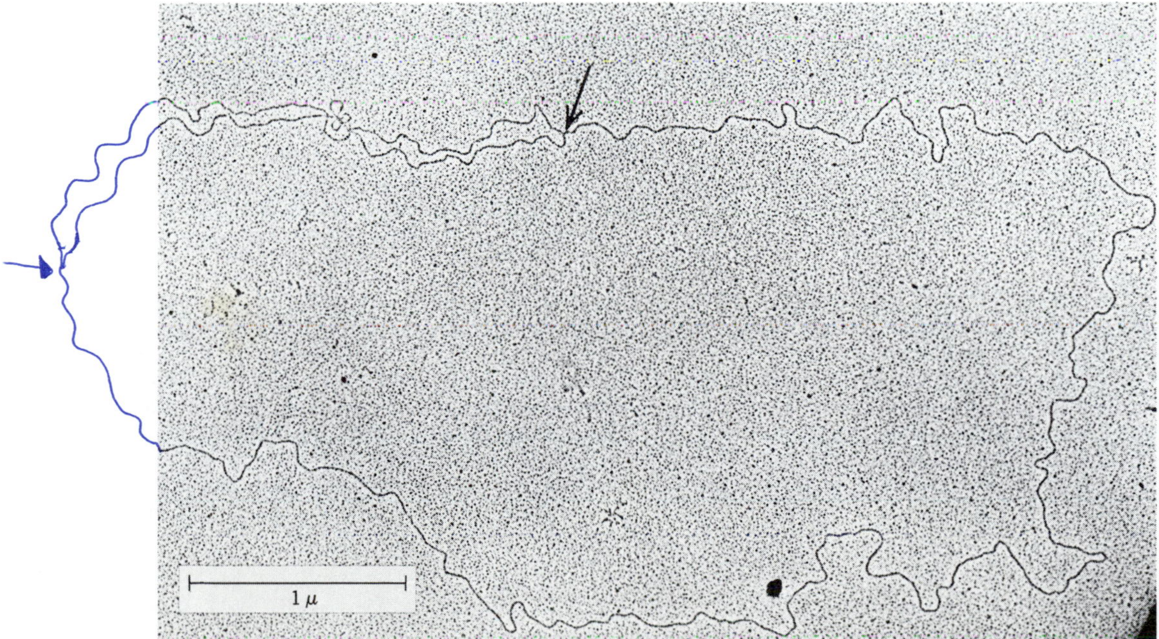

Figure 5.18 Electron micrograph of a θ-form replicative intermediate of the chromosome of bacteriophage lambda. The two replication forks are indicated by the arrow. [Reproduced with permission from M. Schnös and R. B. Inman, *J. Mol. Biol.* 51: 61–73, 1970. Copyright © 1970 by Academic Press, Inc. (London), Ltd.]

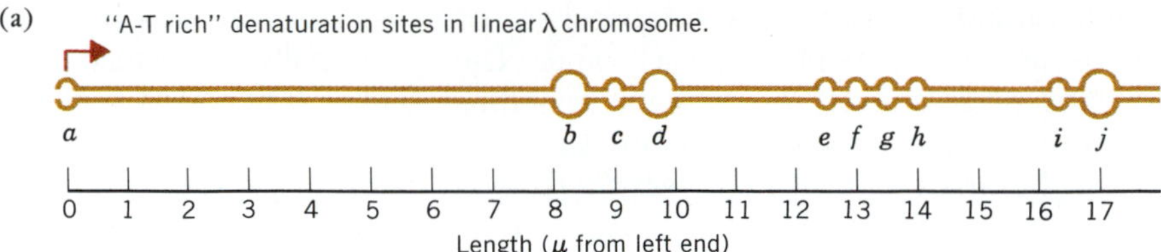

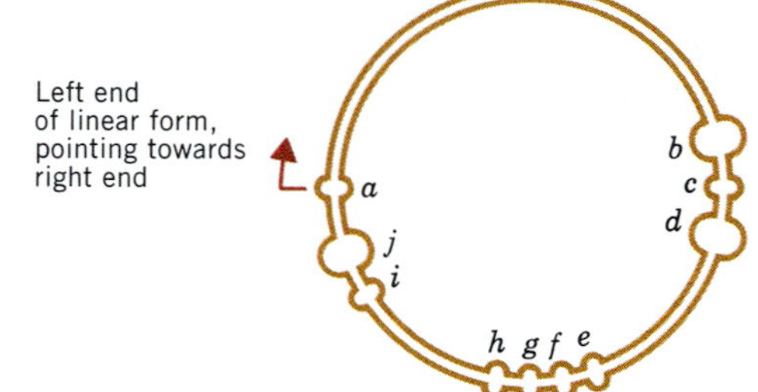

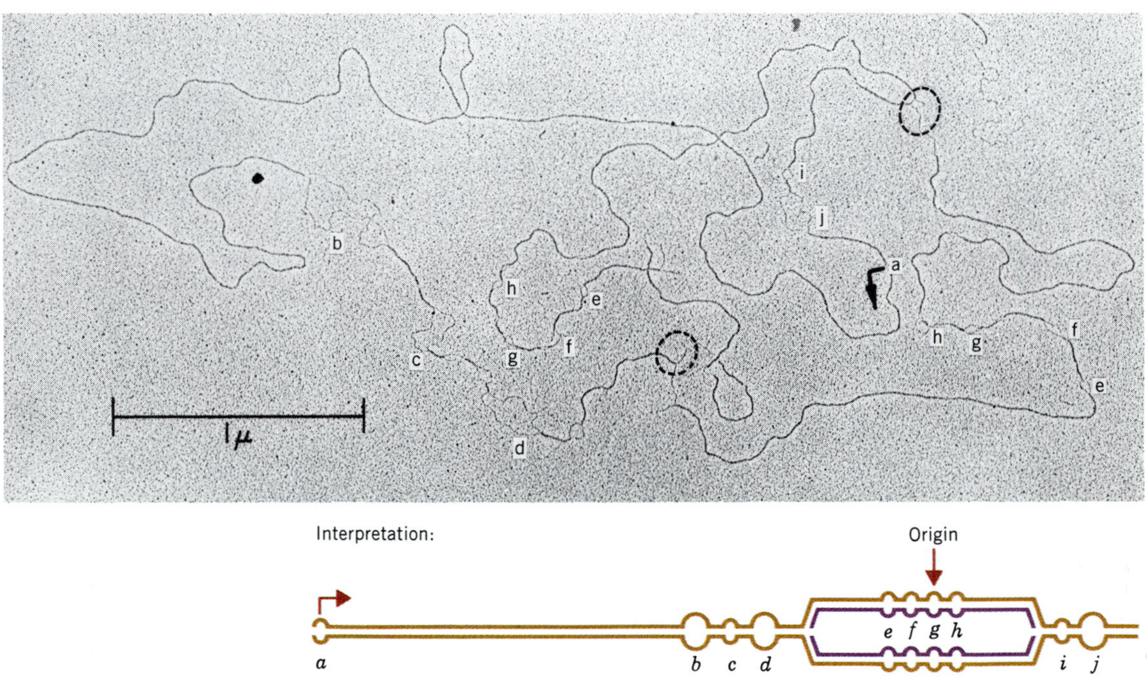

Figure 5.19 The use of "A-T rich" denaturation sites as physical markers on the phage λ chromosome to prove that replication is bidirectional rather than unidirectional. (a) The locations of the denaturation bubbles are seen in the linear form on the 17.5 μ long lambda chromosome after treatment of pH 11.05 for 10 minutes, temperature 25°C. Their use as physical markers on the circular form of the lambda chromosome is shown in (b). An electron micrograph of a partially denatured, θ-shaped replicating λ chromosome is shown in (c), with an interpretation shown in linear form below it. The bent arrow identifies the small denaturation bubble at the left end of the linear chromosome and points toward the right end. Ten of the most obvious denaturation bubbles are labeled *a* to *j*, to aid in orientation. The two replication forks are circled. [The micrograph shown in (*c*) is reproduced with permission from M. Schnös and R. B. Inman, *J. Mol. Biol.* 51: 61–73, 1970. Copyright © 1970 by Academic Press, Inc. (London), Ltd.]

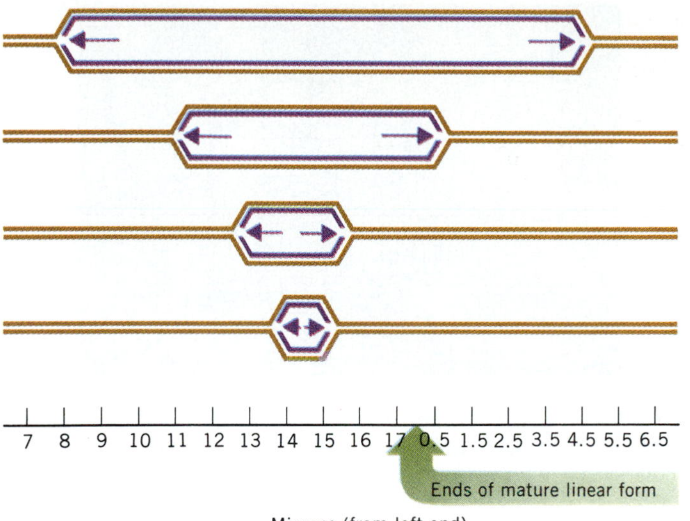

Figure 5.20 Diagrammatic summary of the results of electron microscope denaturation mapping of a population of phage lambda replicative intermediates. The circular replication form of the lambda chromosome is diagrammed in linear form by breaking the circle near the terminus of replication (not at the cohesive ends of the chromosome of the mature virion). The origin of replication is located at about 14.3 μm from the left end of the chromosome (in its mature linear form). Four chromosomes are shown at different stages of replication. Since both branch points occur at variable positions relative to the physical markers (denaturation bubbles; see Figs. 5.19 and 5.21), both are clearly replication forks that move in opposite directions around the circular chromosome during replication. (Based on the data of M. Schnös and R. B. Inman, *J. Mol. Biol.* 51: 61–73, 1970.)

the rationale of this procedure is shown in Fig. 5.21.

Bidirectional replication from a fixed origin has also been demonstrated for several organisms with chromosomes that replicate as linear structures. Replication of the chromosome of phage T7, another small coliphage, begins at a unique site near one end to form a so-called "eye" structure (Fig. 5.22*a*) and then proceeds bidirectionally until one fork reaches the nearest end. Replication of the "Y-shaped" structure (Fig. 5.22*b*) continues until the second fork reaches the other end of the molecule, producing two progeny chromosomes.

Replication of DNA molecules in the chromosomes of eukaryotes is also bidirectional (Chapter 6). However, bidirection replication is not universal. The chromosome of coliphage P2, which like the lambda chromosome is circular during replication, replicates unidirectionally from a unique origin. Whether origins

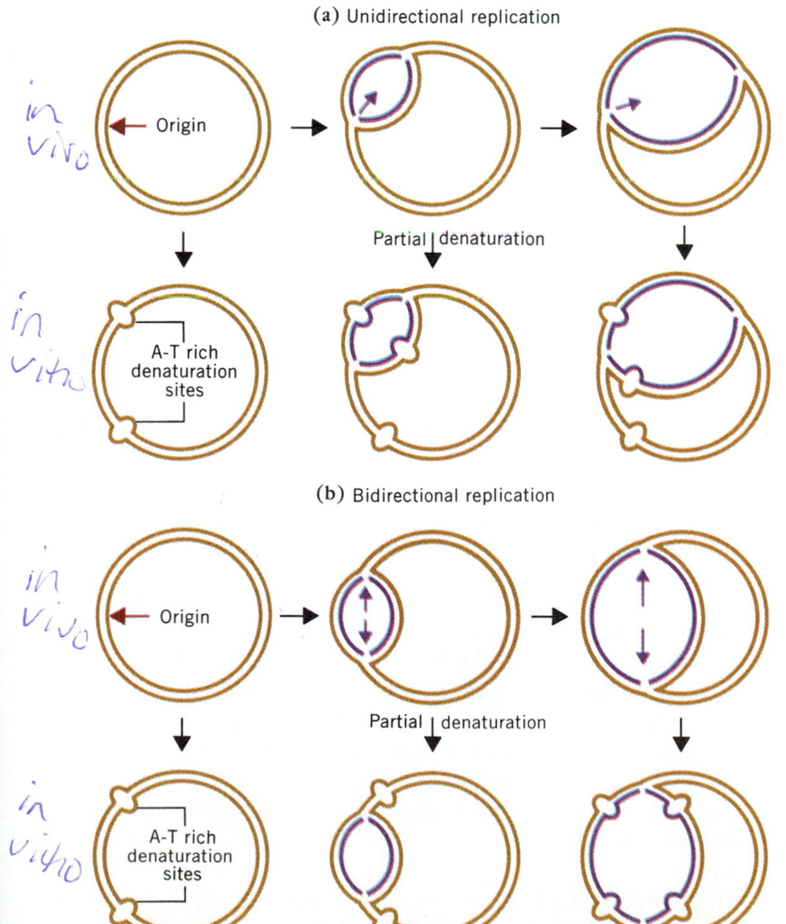

Figure 5.21 The use of electron microscope denaturation mapping to distinguish between (a) unidirectional and (b) bidirectional modes of chromosome replication. Treatment of chromosomes at pH 11.05 for 10 minutes at 25°C under appropriate conditions will denature only regions containing high concentrations of A-T base-pairs. These "denaturation bubbles" can be used as physical markers for specific sites on the chromosome. By examining the positions of the replication forks relative to these markers in a population of replicating chromosomes, one can distinguish between uni- and bidirectional replication.

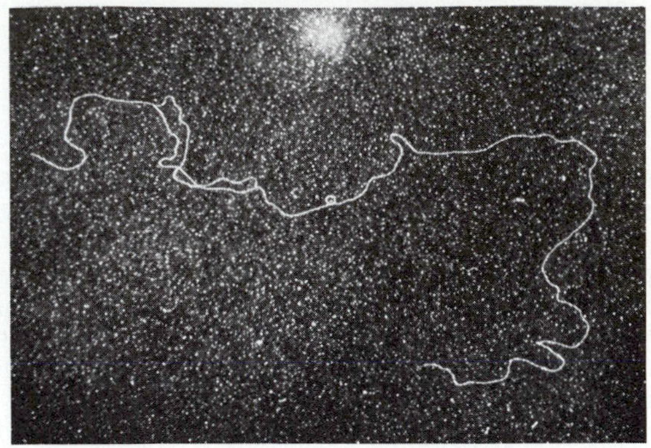

(a)

(b)

Figure 5.22 Electron micrographs of bacteriophage T7 chromosomes in early stages of replication. The phage T7 chromosome, unlike the *E. coli* and phage lambda chromosomes, replicates as a linear structure. Its origin of replication is located 17 percent of the length of the chromosome from one end (the left end of the chromosomes shown). The chromosome in (a) illustrates the "eye" form (-o-) characteristic of early stages in the replication of linear DNA molecules. Replication proceeds bidirectionally until the fork moving in a leftward direction reaches the left end of the molecule, yielding a Y-shaped structure such as that shown in (b). (From J. Wolfson, D. Dressler, and M. Magazin, *Proc. Natl. Acad. Sci., U.S.* 69: 499, 1972. Original micrographs courtesy of D. Dressler.)

of replication in other organisms are always unique sites is also unknown. It is clear that secondary, normally inactive origins exist in some organisms. If the primary origin of replication of phage T7 is deleted (physically removed from the chromosome), a second unique site on the T7 chromosome takes over this function. Why this secondary origin is inactive in the presence of the primary origin is not known.

DNA Polymerases and *In Vitro* DNA Synthesis

Much has been learned about the molecular mechanisms involved in biological processes by fractionating cells into their various organelles, macromolecules, and other components, and then reconstituting systems in the test tube, so-called *in vitro* systems, that are capable of carrying out particular metabolic events. Such *in vitro* systems can be dissected biochemically much more easily than *in vivo* systems. Clearly, the information obtained from studies on *in vitro* systems has been invaluable. One should never assume, however, that a phenomenon demonstrated *in vitro* occurs *in vivo*. Such an extrapolation should be made only when independent evidence from *in vivo* studies validates the *in vitro* studies.

The *in vitro* synthesis of DNA was first accomplished by Arthur Kornberg and his coworkers in 1957. Kornberg, who received the Nobel Prize in 1959 for this work, isolated an enzyme from *E. coli* (initially called ***DNA polymerase*** or ***"Kornberg enzyme,"*** now known as DNA polymerase I) that catalyzes the covalent addition of nucleotides to preexisting DNA chains. The enzyme requires the 5′-triphosphates of each of the four deoxyribonucleosides: deoxyadenosine triphosphate (dATP), deoxythymidine triphosphate (dTTP; often written as TTP in the past because "ribo" derivatives of thymine had not been identified), deoxyguanosine triphosphate (dGTP), and deoxycytidine triphosphate (dCTP). The enzyme is active only in the presence of Mg^{2+} ions and preexisting DNA. This DNA must provide two essential components, one serving a ***primer*** function and the other a ***template*** function (Fig. 5.23). The overall reaction catalyzed by DNA polymerase I is shown in Fig. 5.24; a diagram of the functional sites of the enzyme is shown in Fig. 5.25.

1. **Primer DNA.** DNA polymerase I cannot initiate the synthesis of DNA chains *de novo*. It has an ***absolute requirement for a free 3′-hydroxyl on a preexisting DNA chain.*** DNA polymerase I catalyzes the formation of a phosphodiester bridge between the 3′-OH at the end of the primer DNA chain and the 5′-phosphate of the incoming deoxyribonucleotide. The direction of synthesis is thus always 5′ ⟶ 3′ (Fig. 5.26).
2. **Template DNA.** DNA polymerase I does not contain sequence specificity. That is, the enzyme requires a DNA ***template*** chain (Fig. 5.23) whose base sequence dictates, via the DNA base-pairing requirements, the synthesis of a complementary base sequence in the strand being synthesized.

Since Kornberg's discovery and extensive pioneering work with DNA polymerase I of *E. coli*, a large

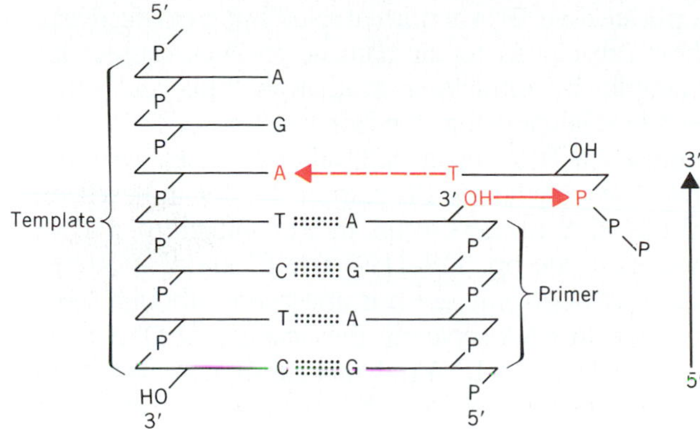

Figure 5.23 Template and primer requirements of DNA polymerases. DNA polymerase requires a primer strand (shown on the right) with a free 3′-hydroxyl. It is the primer strand that is covalently extended by the addition of nucleotides (such as dTMP, from the incoming precursor dTTP shown). In addition, DNA polymerases require a template strand (shown on the left). It is the template strand that determines the base sequence of the strand being synthesized; the new strand will be complementary to the template strand.

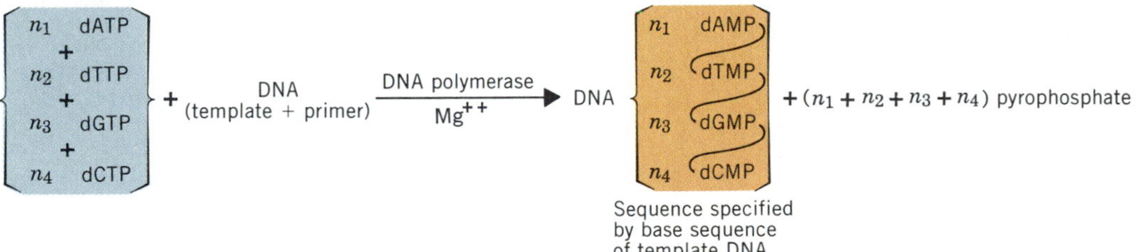

Figure 5.24 Overall reaction for DNA polymerase-catalyzed synthesis of DNA *in vitro*. There are requirements for template and primer (see Fig. 5.23), Mg^{2+} ions, and the four common deoxyribonucleoside triphosphates (dATP, dTTP, dGTP, and dCTP). The values *n*1, *n*2, *n*3, and *n*4 are the moles of dATP, dTTP, dGTP, and dCTP consumed and the moles of dAMP, dTMP, dGMP, and dCMP incorporated during polymerization, respectively. These values will depend on the base composition of the template DNA used in the reaction. The sequence of the nucleotides in the product will also be determined by the base sequence of the template DNA.

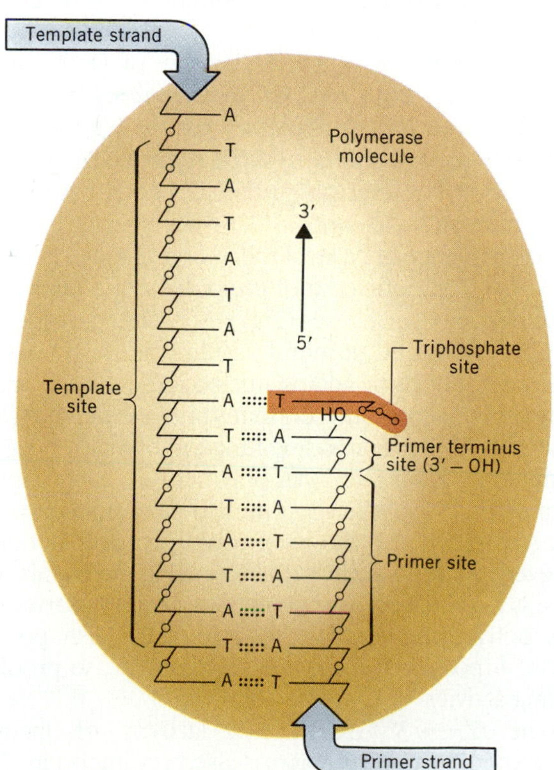

Figure 5.25 The interaction of DNA polymerase I with the template and primer strands of DNA and the incoming nucleoside triphosphate. (From A. Kornberg, *Science* 163: 1410–1418, 1969. Copyright © 1969 by the American Association for the Advancement of Science.)

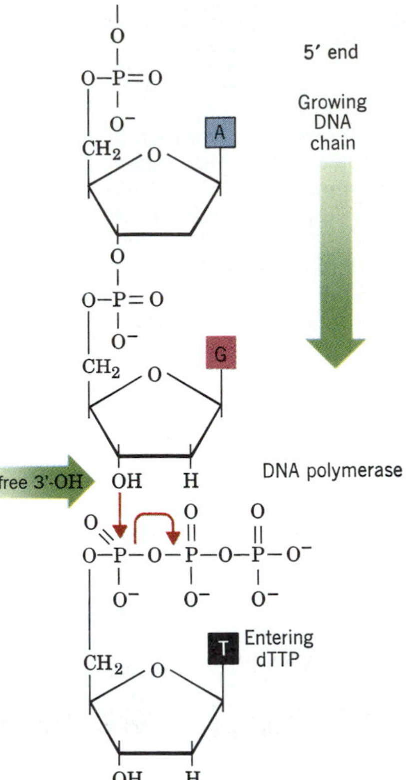

Figure 5.26 Covalent extension of DNA chains in the 5′⟶3′ direction as catalyzed by all known DNA polymerases. The existing chain terminates at the 3′ end with the nucleotide deoxyguanylate (or deoxyguanosine-5′-phosphate). The diagram shows the DNA polymerase-catalyzed addition of deoxythymidine monophosphate (from the precursor deoxythymidine triphosphate, dTTP) to the 3′ end of the chain with the release of pyrophosphate (P_2O_7).

number of DNA polymerases have been isolated and characterized from many different organisms. Three different DNA polymerases (I, II, and III) have been identified and studied in *E. coli* and *B. subtilis*. Similarly, three DNA polymerases, called α, β, and γ, have been identified in several different eukaryotes, and more recently a fourth DNA polymerase, named δ, has been isolated from calf thymus and rabbit bone marrow. Thus, there are at least four different DNA polymerases in eukaryotes.

The precise functions of some of the polymerases are still not clear. However, in *E. coli* and *B. subtilis*, DNA polymerase III, rather than DNA polymerase I as first believed, is the major replicative enzyme. Some of the strongest evidence for this has come from studies of mutants, so-called *polA* mutants, that are deficient in DNA polymerase I. These *polA* mutants of *E. coli* replicate their DNA at normal rates, but are defective in their capacity to repair damage to DNA caused, for example, by ultraviolet irradiation. This and other evidence suggest that a major function of DNA polymerase I is DNA repair. Still other evidence indicates that DNA polymerase I is responsible for the excision of the RNA primers used in the initiation of DNA synthesis (see pp. 118–119). The function(s) carried out by DNA polymerase II is uncertain, although it can function in DNA repair in the absence of DNA polymerases I and III. DNA polymerase III, however, plays an essential role in DNA replication, because in mutant strains growing under conditions where no functional polymerase III is synthesized, DNA synthesis stops.

Most of the prokaryotic DNA polymerases studied so far not only exhibit the 5′⟶3′ polymerase activity discussed earlier, but also have a ***3′⟶5′ exonuclease*** activity. (An ***exonuclease*** is an enzyme that degrades nucleic acids from the ends, as opposed to an ***endonuclease***, which degrades nucleic acids by making internal cuts.) Both activities (polymerase and exonuclease) are present in the same protein macromolecule. The 3′⟶5′ exonuclease activity catalyzes the removal of nucleotides, one by one, from the 3′ ends of polynucleotide chains. Some polymerases, such as DNA polymerase I of *E. coli*, also have ***5′⟶3′ exonuclease*** activity. When present, 5′⟶3′ exonuclease activity is found at the site on the protein molecule distinct from the active side catalyzing the 5′⟶3′ polymerase reaction and the 3′⟶5′ exonuclease reaction. Both of these polymerase-associated exonuclease activities play important roles in DNA metabolism.

The 3′⟶5′ exonuclease activity of DNA polymerases carries out a ***critical "proofreading" or "editing" function that is necessary for the high degree of fidelity characteristic of DNA replication.*** When presented with a template-primer DNA that has a terminal mismatch (an unpaired or incorrectly paired base or sequence of bases at the 3′ end of the primer), the 3′⟶5′ exonuclease of the polymerase clips off the unpaired base or bases (Fig. 5.27). When an appropriately base-paired terminus results, the polymerase begins resynthesis by adding nucleotides to the 3′ end of the primer strand and continues until the template is exhausted. This proofreading function of the 3′⟶5′ exonuclease, built into DNA polymerases, is very important, for DNA replication must be extremely accurate. A tolerable mistake level could probably not be achieved without such a proofreading mechanism. Thus, it is somewhat surprising that of the eukaryotic DNA polymerases studied to date, only DNA polymerase δ possesses the 3′⟶5′ exonuclease proofreading activity.

The 5′⟶3′ exonuclease activity of many prokaryotic DNA polymerases is also very important. It

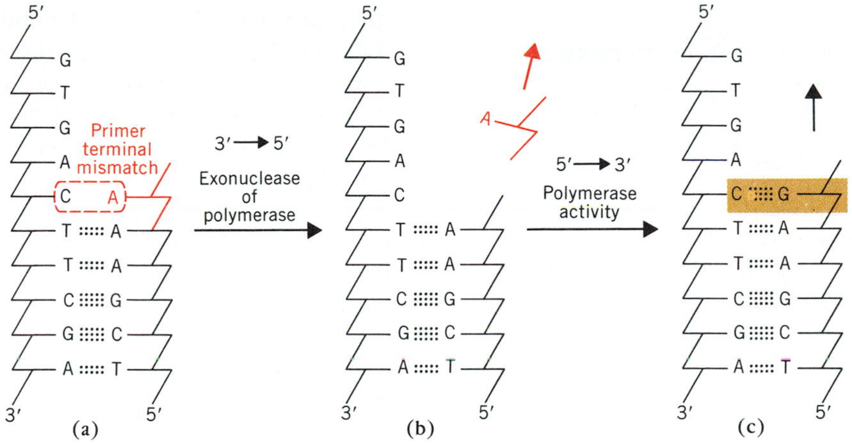

Figure 5.27 "Proofreading" by the $3' \rightarrow 5'$ exonuclease activity of DNA polymerases during DNA replication. If DNA polymerase is presented with a template and primer containing a $3'$ primer terminal mismatch (a), it will not catalyze covalent extension ("polymerization"). Instead, the $3' \rightarrow 5'$ exonuclease activity, an integral part of many DNA polymerases, will cleave off the terminal mismatched nucleotide (b). Then, presented with a correctly base-paired primer terminus, DNA polymerase will catalyze $5' \rightarrow 3'$ covalent extension of the primer strand (c).

functions in the removal of segments of DNA damaged by ultraviolet light irradiation and other agents (see Chapter 11). The $5' \rightarrow 3'$ exonuclease of polymerases, such as *E. coli* DNA polymerase I, also functions in the removal of RNA primers from DNA (see pp. 118–119). This $5' \rightarrow 3'$ exonuclease activity has not been found in any of the eukaryotic DNA polymerases. In eukaryotes, the functions of the $5' \rightarrow 3'$ exonuclease activity associated with DNA polymerases in prokaryotes must be carried out by a separate enzyme.

The "Growing-Point Paradox" and Discontinuous DNA Synthesis

Studies of replicating DNA molecules by autoradiography and electron microscopy indicate that the two progeny strands being synthesized at each replicating fork are being extended in the same overall direction, at least on the macromolecular level. Since the complementary strands of a double helix have opposite polarity, this means that synthesis is occurring at the $5'$ end of one strand (or $3' \rightarrow 5'$) and the $3'$ end of the other strand ($5' \rightarrow 3'$). As discussed in the preceding section, however, all known polymerases have an absolute requirement for a free $3'$-hydroxyl; they only carry out $5' \rightarrow 3'$ synthesis. This paradox existed for many years during which biochemists searched in vain for new polymerases that could carry out $3' \rightarrow 5'$ synthesis. No such polymerase has yet been found. Instead, strong evidence has accumulated indicating that ***all synthesis occurs in the*** $5' \rightarrow 3'$ ***direction***. The resolution of the paradox resulted from the demonstration that the synthesis of one DNA strand is ***discontinuous***.

Autoradiography and electron microscopy show that the two nascent DNA strands being synthesized at each replicating fork are being extended in the ***same direction at the macromolecular level***. Since the complementary strands of a DNA double helix have opposite chemical polarity, one strand is being extended in an overall $5' \rightarrow 3'$ direction and the other strand is being extended in an overall $3' \rightarrow 5'$ direction (Fig. 5.28, top). At the ***molecular level***, however, synthesis is actually occurring in ***opposite directions*** (Fig. 5.28, bottom). At the molecular level both new strands are being synthesized in the $5' \rightarrow 3'$ direction. The strands being extended in the overall $3' \rightarrow 5'$ direction grow by the synthesis of short segments (synthesized $5' \rightarrow 3'$), and the subsequent joining of these short segments by polynucleotide ligase. The evidence for this ***discontinuous*** mode of DNA replication has come from studies in which intermediates in DNA synthesis were radioactively labeled by growth of cells for very short periods of time in medium containing [^{3}H]thymidine ("pulse-labeling"). When *E. coli* cells were pulse-labeled for 15 seconds, for example, all the label was found in small pieces, 1000–2000 nucleotides long. These small pieces or segments of DNA, often called "Okazaki fragments" after R. Okazaki, who first identified them, are smaller, about 100–200 nucleotides long, in eukaryotes. When longer pulse-labeling periods are used, more of the label is recovered in large DNA molecules—probably the size of molecules containing all the DNA present in

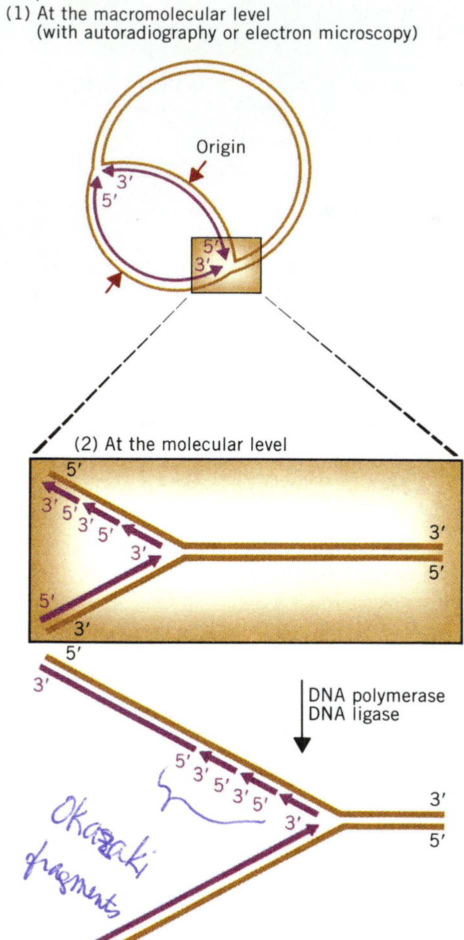

Figure 5.28 Discontinuous DNA synthesis. (1) Relatively low resolution techniques such as autoradiography and electron microscopy show that both nascent DNA chains are extended in the same overall direction at each replication fork. Since the two chains have opposite polarity, the overall or macromolecular direction of extension must be $5' \rightarrow 3'$ on one chain and $3' \rightarrow 5'$ on the other chain. Both are being extended from left to right, for example, at the framed replication fork. (2) Higher-resolution biochemical techniques such as pulse-labeling and density-gradient analysis show that replication is actually discontinuous for the chain being extended in the overall $3' \rightarrow 5'$ direction. Short fragments are synthesized in the $5' \rightarrow 3'$ direction and subsequently are joined by polynucleotide ligase ("DNA ligase").

intact chromosomes. In short pulse-labeling periods, the radioactivity present in short DNA "fragments" becomes incorporated in chromosome-sized DNA molecules during subsequent growth of the cells on medium containing nonradioactive thymidine. This is important because it indicates that the "Okazaki fragments" are true intermediates in DNA synthesis rather than some kind of metabolic by-product. Extensive evidence has shown that DNA synthesis is continuous for the strand growing in the overall $5' \rightarrow 3'$ direction (sometimes called the "leading" strand) and is discontinuous for the strand growing in the overall $3' \rightarrow 5'$ direction (sometimes called the "lagging" strand) as is shown in Fig. 5.28.

Initiation and the "Primer Problem"

As has been emphasized earlier, all known DNA polymerases have an absolute requirement for a free 3′-OH on a DNA primer plus an appropriate DNA template strand for activity. Thus, ***no known DNA polymerase can initiate the synthesis of a new strand of DNA.*** Since the synthesis of each "Okazaki fragment" requires an initiation event, an efficient mechanism of chain initiation is essential for ongoing DNA replication. RNA polymerase, a complex enzyme that catalyzes the synthesis of RNA molecules from DNA templates, has long been known to be capable of initiating the synthesis of new RNA chains at specific sites on the DNA. When this occurs, an RNA–DNA hybrid is formed in which the nascent RNA is hydrogen-bonded to the DNA template. Since DNA polymerases are capable of extending polynucleotide chains containing an RNA primer with a free 3′-OH, scientists in several laboratories began testing the idea that DNA synthesis is initiated by RNA primers. There is now definitive evidence supporting the proposal that ***DNA synthesis is "primed" by short segments of RNA,*** which are later removed by a $5' \rightarrow 3'$ exonuclease and replaced by DNA prior to covalent sealing by polynucleotide ligase (Fig. 5.29). In *E. coli*, the ***RNA primers*** are excised by the $5' \rightarrow 3'$ exonuclease activity of DNA polymerase I. This occurs simultaneously with the synthesis of new DNA strands (replacing the excised RNA primer strands) by the $5' \rightarrow 3'$ polymerase activity of this enzyme (Fig. 5.29).

The synthesis of the RNA primers is catalyzed by enzymes called ***primases,*** which have properties quite distinct from those of the RNA polymerases. The *E. coli* primase is the product of the *dnaG* gene. In prokaryotes, the RNA primers are 10–60 nucleotides in length. In eukaryotes they are quite short, about 10 nucleotides long. The use of RNA primers is almost certainly the most common mechanism used to initiate DNA synthesis. Nevertheless, certain viruses appear to have

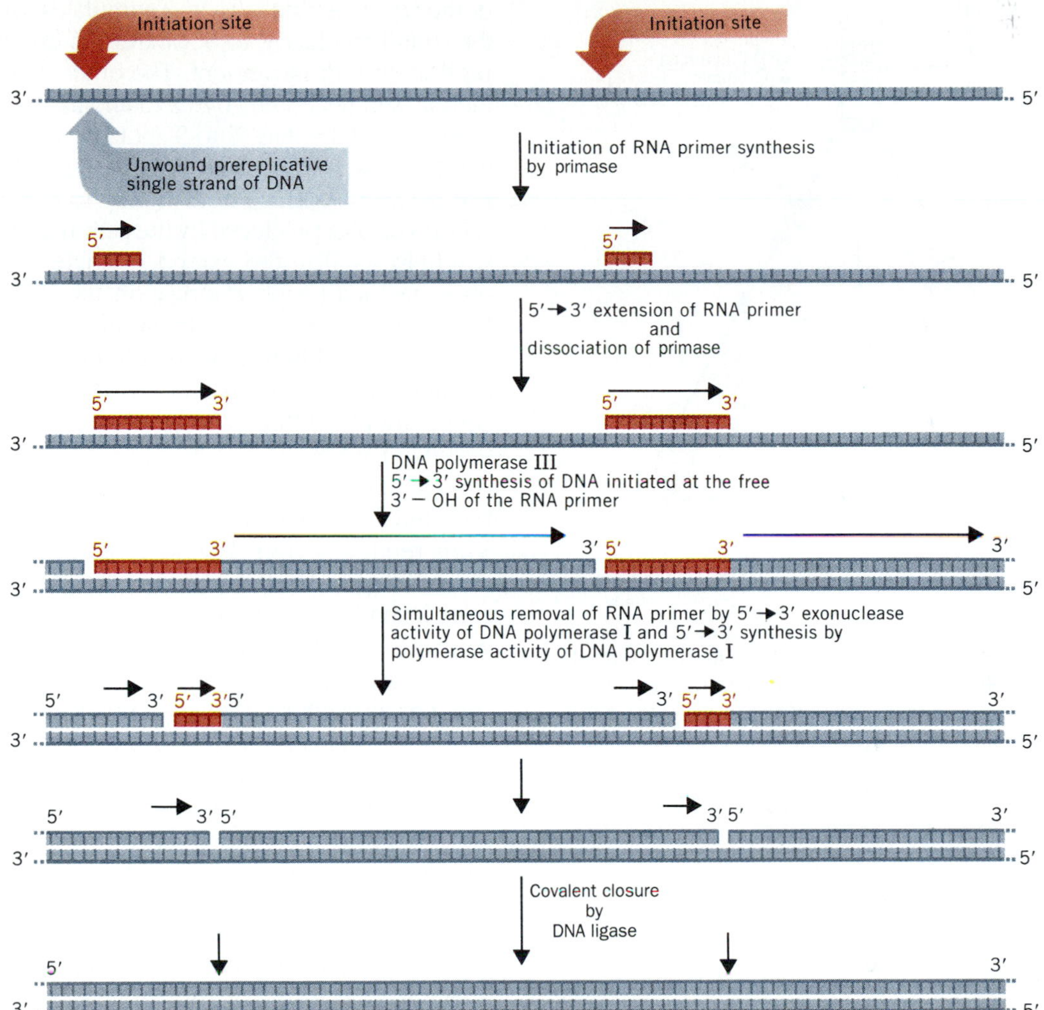

Figure 5.29 Schematic illustration of the initiation of DNA synthesis via RNA primers. A short RNA strand is synthesized to provide a 3′-OH primer for DNA synthesis. This RNA primer is subsequently removed and replaced with DNA by the dual 5′⟶3′ exonuclease and 5′⟶3′ polymerase activities built into DNA polymerase I. DNA ligase then covalently closes the nascent DNA chain, catalyzing the formation of phosphodiester linkages between adjacent 3′-hydroxyls and 5′-phosphates.

evolved quite different mechanisms for the initiation of DNA synthesis (see Kornberg, 1980).

The Complete "Replication Apparatus" Is Complex

When Watson and Crick worked out the double-helix structure of DNA, they immediately recognized that the complementary nature of the two strands provided a simple basis for the faithful duplication of genetic material. Meselson and Stahl's demonstration of the semiconservative replication of the *E. coli* chromosome solidified the concept that the two strands of the double helix unwind and serve as templates for the synthesis of complementary strands. Thus, a parental double helix directs the synthesis of two identical progeny double helices. Kornberg's isolation of an enzyme, DNA polymerase I, capable of synthesizing DNA *in vitro* appeared to provide the final link in what was thought to be an elegantly simple mechanism for the replication of the genetic material—but such was not the case. Twenty years later, scientists are still trying to work out the details of the mechanism of DNA replication.

DNA replication is complex. It is carried out by a ***multienzyme complex,*** often called the ***replication apparatus*** or the ***replisome.*** In eukaryotes, the components of the replication machinery are just begin-

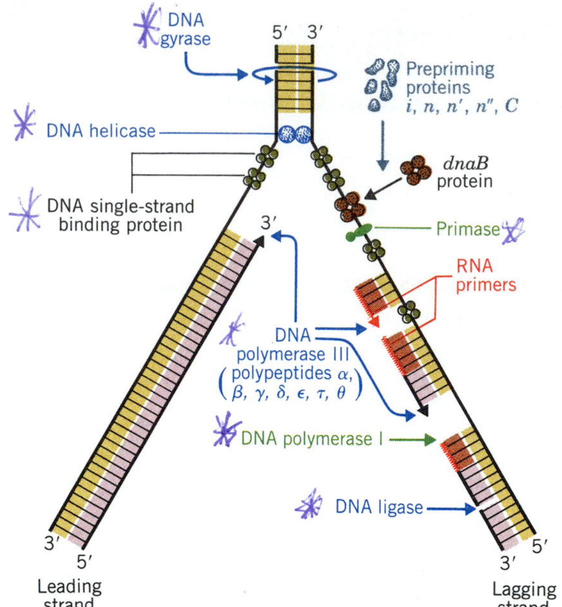

Figure 5.30 Complexity of the *E. coli* replication apparatus. Only those proteins that have been purified (or partially purified) and studied *in vitro* are shown. Other gene-products, such as the products of genes *dnaJ, dnaK, dnaL, dnaP,* and *dnaT,* are known to be required for replication. However, these gene-products have not yet been identified. (After A. Kornberg, *DNA Replication,* Freeman, San Francisco, 1980.)

ning to be identified. Even in prokaryotes, DNA replication requires many different proteins, and the details of how some of these proteins function in DNA replication are still being investigated today. For example, DNA replication in *E. coli* requires at least two dozen different gene-products. Many of these gene-products have been purified and their roles in DNA replication studied *in vitro*. Figure 5.30 shows the involvement of some of these *E. coli* proteins in DNA replication; it is intended to illustrate the complexity of the replication process rather than to illustrate the specific roles of the individual gene-products.

First, the two complementary strands of the parental double helix have to be unwound and separated so that each can serve as a template for the synthesis of a new daughter strand. Unwinding and movement of the replication fork occur ***processively*** with the strands being transiently unwound ahead of the fork as it moves along the chromosome. Three different types of proteins appear to contribute to unwinding the strands of double helices. (1) ***DNA unwinding proteins*** or ***DNA helicases*** are directly involved in catalyzing the unwinding of the double helices. In *E. coli*, two different helicases are involved. One helicase, the product of the *rep* gene, binds to and stimulates separation of the strand that has 3′ to 5′ polarity in the direction of replication fork movement. The other helicase (exact identity still uncertain) binds to and assists unwinding of the strand that has 5′ to 3′ polarity in the direction that the fork is moving. (2) ***DNA single-strand binding proteins (SSBPs)*** bind tightly to single-stranded regions of DNA produced by the action of the helicases and help stabilize the extended single-stranded templates needed for polymerization. The SSBPs bind to DNA as tetramers, and their binding exhibits cooperativity (i.e., the binding of one tetramer stimulates the binding of additional tetramers to adjacent segments of single-stranded DNA). The binding of SSBP to single-stranded DNA tends to hold that DNA in an extended configuration and prevents it from folding back on itself. Single-stranded DNA that is saturated with bound SSBP replicates over 100 times faster than uncomplexed single-stranded DNA *in vitro*. Presumably, uncomplexed single strands of DNA form secondary structures that interfere with the movement of DNA polymerases or other components of the replication complex along the molecule in the normal processive manner. (3) Finally, ***DNA gyrases***, which catalyze the formation of negative supercoils in DNA (see Fig. 5.36), are essential for replication and are believed to play a key role in the unwinding process. Supercoiling has been proposed to help "drive" the unwinding process, however, we still do not know how this works. Very recently, it has been suggested that DNA gyrase may function by removing positive supercoils that accumulate in front of the replication forks as the helicases unwind the double helices. In any case, DNA gyrases are essential for DNA replication and somehow maintain pre- and postreplicative DNAs in the proper topological structures.

Nascent DNA strands are then initiated by the use of ***RNA primers*** by the mechanism discussed earlier (see Fig. 5.29). Synthesis of the RNA primers is catalyzed by a special class of enzymes called ***primases***. Primase activity requires the formation of a complex of primase and at least six other proteins; this complex is called the ***primosome***. In addition to primase, the primosome contains prepriming proteins tentatively designated ***proteins i, n, n′, and n″*** plus the ***products of genes*** dnaB ***and*** dnaC (Table 5.4). The primosome carries out the initial priming reaction for the ***leading strand*** (the strand extended continuously in the overall 5′ to 3′ direction) and the repeating priming of the synthesis of "Okazaki fragments" for the ***lagging strand*** (synthesized discontinuously in the overall 3′ to 5′ direction—but 5′ to 3′ at the molecular level, see Fig. 5.28). The functions of the individual proteins in the primosome are still uncertain.

The covalent extension (see Fig. 5.26) of the primed DNA chains during chromosome replication in

TABLE 5.4 Proteins and Genes Involved in DNA Replication in *E. coli*.[a]

PROTEIN	GENE	FUNCTION
Protein i	Unknown	Primosome component
Protein n	Unknown	Primosome component
Protein n′	Unknown	Primosome component
Protein n″	Unknown	Primosome component
DnaB protein	*dnaB*	Primosome component
DnaC protein	*dnaC*	Primosome component
Primase	*dnaG*	Primosome component; RNA primer synthesis
Rep protein	*rep*	DNA helicase
Helicase II	Unknown	DNA helicase
SSBP	*ssb*	Single-strand binding protein
Gyrase subunit α	*nalA*	DNA gyrase α polypeptide; nalidixate resistance
Gyrase subunit β	*cou*	DNA gyrase β polypeptide; coumermycin resistance
Polymerase III α	*dnaE*	DNA polymerase III α polypeptide
Polymerase III β	*dnaN*	DNA polymerase III β polypeptide
Polymerase III γ	*dnaZ*	DNA polymerase III γ polypeptide
Polymerase III δ	*dnaX*	DNA polymerase III δ polypeptide
Polymerase III ϵ	Unknown	DNA polymerase III ϵ polypeptide
Polymerase III θ	Unknown	DNA polymerase III θ polypeptide
Polymerase III τ	Unknown	DNA polymerase III τ polypeptide
Polymerase I	*polA*	DNA polymerase I; excision of RNA primers
Ligase	*lig*	DNA ligase; covalent closure of "nicks"
Unknown	*dnaA*	Unknown; required to initiate replication
Unknown	*dnaI*	Unknown
Unknown	*dnaJ*	Unknown
Unknown	*dnaK*	Unknown
Unknown	*dnaL*	Unknown
Unknown	*dnaP*	Unknown
Unknown	*dnaT*	Unknown

[a]Selected data from A. Kornberg, *Supplement to DNA Replication*. Freeman, San Francisco, 1982.

E. coli is carried out by DNA polymerase III. Unlike DNA polymerase I of *E. coli* (which is a single polypeptide; see Fig. 5.25), ***DNA polymerase III is a complex enzyme containing seven different polypeptides*** (Fig. 5.31), and all of these polypeptides must be present for proper replicative function. The 5′ to 3′ polymerase activity and the 5′ to 3′ exonuclease activity are both present on the α polypeptide of DNA polymerase III. The 3′ to 5′ proofreading activity (see Fig. 5.27) of polymerase III is present on the ϵ polypeptide. The functions of the other subunits are still uncertain. Subsequent to DNA polymerase III activity at the replication fork, DNA polymerase I catalyzes the removal of the RNA primers by the concerted action of its 5′ to 3′ exonuclease activity and its 5′ to 3′ polymerase activity, and DNA ligase catalyzes covalent closure of the resulting single-stranded "nick" (Fig. 5.30).

Several of the components essential for DNA replication have been identified genetically, that is, *E. coli* strains carrying mutations (heritable changes in the genetic material; see Chapter 11) that result in the inability to replicate DNA under certain conditions (usually high temperature) have been identified. When these mutations were characterized genetically (see Chapters 7 and 8), they were found to identify a set of genes (designated *dnaA*, *dnaB*, etc.) whose products are required for DNA synthesis *in vivo*. The products of some of these genes are known. For example, *dnaE*, *dnaN*, *dnaX*, and *dnaZ* code for four of the seven subunits (polypeptides) of the complete DNA polymerase III enzyme, and *dnaG* specifies the primase. The products and functions of others (see Table 5.4) are still unknown. Other components of the replication enzymes (e.g., some of the subunits of DNA polymerase III) were discovered by biochemical analyses, and the genes that encode these proteins have still not been identified.

It is hoped that the exact functions of the many gene products involved in replication will be worked out during the next few years. Attempts to isolate intact, functional replisomes, however, have been largely unsuccessful. Reconstitution of subcomplexes of replica-

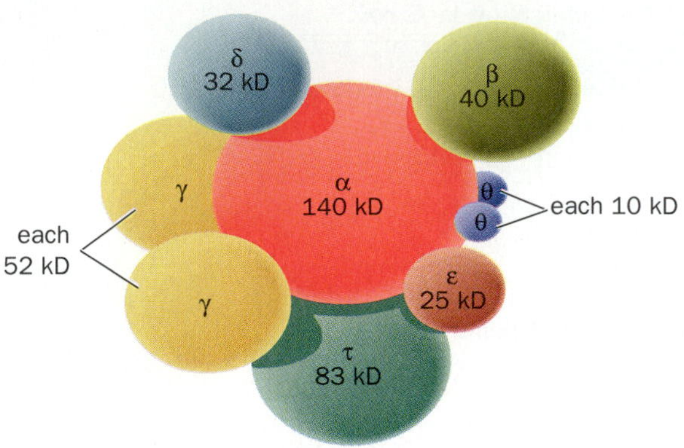

Figure 5.31 Composition of DNA polymerase III holoenzyme of *E. coli.* Size estimates for the polypeptide subunits are given in kilodaltons (kD). (After A. Kornberg, *DNA Replication,* Freeman, San Francisco, 1980.)

tion apparatuses from purified proteins has been more successful. This is undoubtedly a result of the fact that the complexes are held together by relatively weak protein–protein interactions, which are disrupted during the isolation procedures. In addition, replication complexes may be membrane-bound and require membrane structures for their assembly. There is considerable evidence that replication forks are associated with the cell membrane in prokaryotes and with the nuclear envelope in eukaryotes.

For excellent, more detailed accounts of replication and the components of the replication apparatus, the reader is referred to Kornberg's *DNA Replication* and *1982 Supplement to DNA Replication*.

Phage ΦX174 and "Rolling Circle" Replication

Bacteriophage ΦX174 is representative of a group of small viruses, both bacterial and eukaryotic, that store their genetic information in a single-stranded, circular molecule of DNA. When these viruses infect a host cell, *E. coli* in the case of *ΦX174*, the single-stranded viral

Figure 5.32 (Right page) Three stages in the replication of the single-stranded DNA of bacteriophage ΦX174. (Top, a–e) **Stage I: conversion of the single-stranded chromosome (a) to a double-stranded parental replicative (RF) form (e).** (b) Synthesis of the complementary "negative" (−) strand is initiated by the synthesis of a short RNA primer. This reaction is catalyzed by primase and requires the activity of a complex of at least six different priming proteins; this complex is sometimes called the "primosome." (c) DNA polymerase III next catalyzes the covalent addition of deoxyribonucleotides to the 3′ end of the RNA primer. Synthesis of the complementary negative strand then takes place discontinuously (see Figs. 5.28 and 5.29) until the "positive" (+) strand template is exhausted. The primosome appears to travel around the circular template strand, pausing to initiate the synthesis of each new "Okazaki fragment." (d) Excision of the RNA primers and gap filling appear to be catalyzed by DNA polymerase I, as shown in Fig. 5.29. Polynucleotide ligase ("DNA ligase") then catalyzes the formation of a covalent linkage between the adjacent 3′-OH and 5′-PO_4 groups, to produce the closed, double-stranded parental RF (e). (Center e–l) **Stage II: "rolling circle" replication of the parental RF (e) to produce a population of progeny RFs (l).** (f) The positive strand of the parental RF is cut at the origin by the site-specific endonuclease ("nickase") activity of the ΦX174 gene *A* protein. The gene *A* protein nicks the parental RF only at the origin; it will not cut most other DNA molecules at all. (f–h) During the nicking event, the gene *A* protein becomes covalently attached to the 5′-phosphoryl group of the positive strand. It remains linked to the 5′-terminus until a complete progeny positive strand has been synthesized. (g) The 5′ end of the positive strand is displaced from the negative strand and deoxyribonucleotides are added to the free 3′-OH as the circle (maintained by the intact negative strand) rotates about its axis. (g–h) The gene *A* protein remains bound at the replication fork as it travels around the circular negative strand. (h–i) Once a new positive strand origin has been synthesized, the gene *A* protein cleaves the nascent origin and simultaneously ligates the 3′- and 5′-termini to produce a covalently closed circular positive strand. (i–l) Synthesis of the complementary negative strand then takes place discontinuously as in stage I using the nascent positive strand as template. The "Okazaki fragments" are initiated by RNA primers in this stage also; however, these RNA primers are not shown in the diagram. (i–f–l) The parental RF continues to replicate by the rolling circle mode until a population of about 60 progeny RFs are produced. (Bottom, l–q) **Stage III: synthesis of single-stranded progeny chromosomes.** (l–p) Rolling circle replication of progeny RFs occurs just as for parental RFs in stage II, except that negative strands are not synthesized. Instead, the positive strands are packaged in progeny virions. (n–p) The switch from RF synthesis (stage II) to progeny positive strand synthesis (stage III) results from the binding of newly synthesized viral coat proteins to the nascent positive strand, preventing it from serving as a template for negative strand synthesis. (q) Maturation of the progeny virion completes the phage ΦX174 life circle. Approximately 500 progeny virions are produced per infected cell.

DNA (called the *"positive"* (+) strand) is converted to a double helical form (called the *"replicative form,"* RF) by the synthesis of a complementary *"negative"* (−) strand. This double-stranded parental RF then replicates several times to produce a population of progeny RF molecules (double-stranded), which in turn replicate asymmetrically to produce a large population of progeny viral (+) strands. The progeny viral strands are then incorporated into protein coats to complete the reproductive cycle. The replication of the ΦX174 chromosome can thus be divided into three stages: (1) parental (+) strand ⟶ parental RF, (2) parental RF ⟶ progeny RFs, and (3) progeny RFs ⟶ progeny (+) strands (Fig. 5.32). In the last two

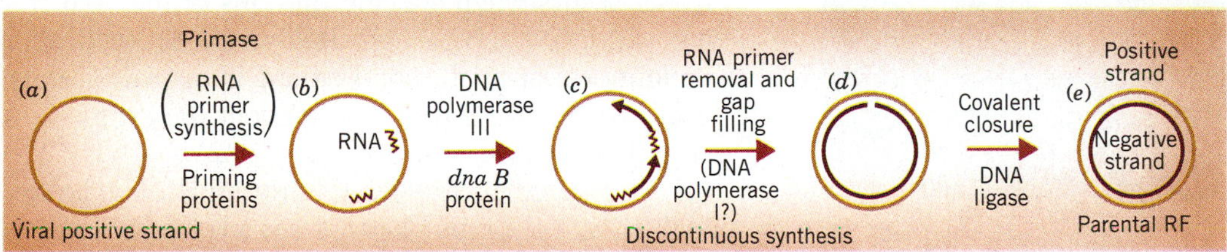

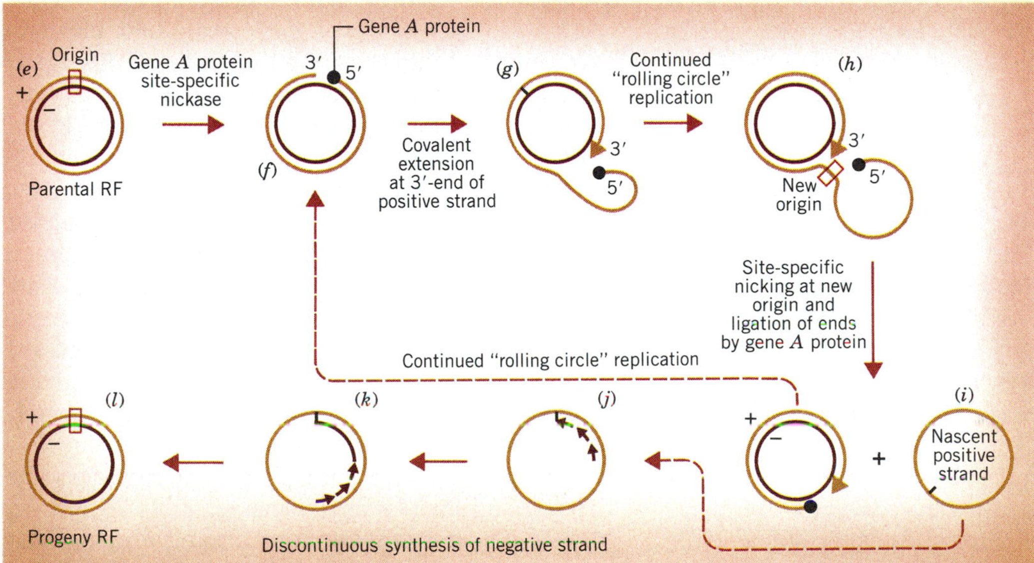

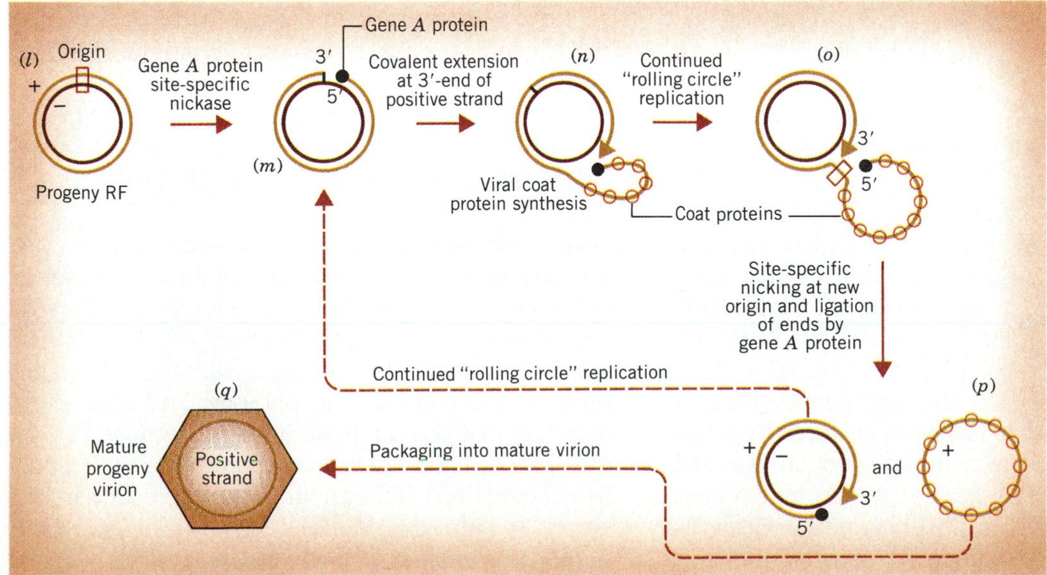

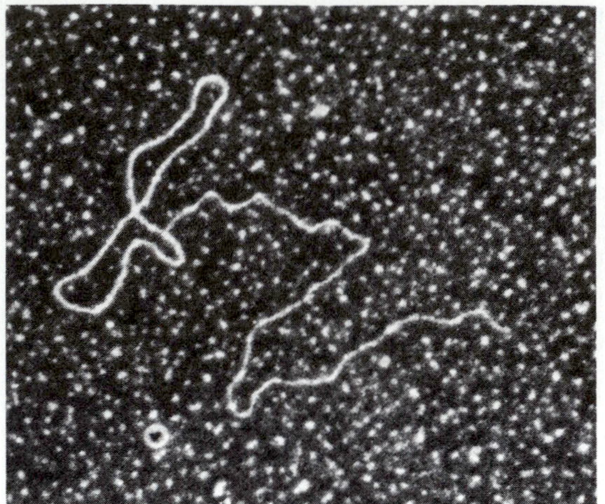

Figure 5.33 Electron micrograph of a "rolling circle" intermediate in the replication of the DNA of bacteriophage ΦX174. A single-stranded tail is seen extending from a double-stranded, circular replication form (RF). (From K. Koths and D. Dressler, *Proc. Natl. Acad. Sci.* 75: 605, 1978.)

stages, DNA synthesis occurs by a different mechanism called ***"rolling circle"*** replication.

Most of the features of "rolling circle" replication are the same as those discussed earlier for replication via the more common θ, "eye," and Y-shaped structures. In this case, however, the replicative structure is a circular DNA molecule with a single-stranded tail (Fig. 5.33).

Rolling circle replication is initiated when the sequence-specific endonuclease activity of the phage ϕX174 gene *A* protein cleaves the positive strand of the parental RF at the origin of replication (Fig. 5.32). This endonuclease activity is site-specific; it ***cuts the ϕX174 chromosome at only one site, the origin of replication.*** It produces 3′-OH and 5′-phosphate termini at the site of the cut in the (+) strand; the (−) strand remains intact. The 5′ end of the (+) strand is unwound and "peeled off" while the (−) strand rotates about its axis (thus the name "rolling circle"). This yields the circle with its tail (Figs. 5.32 and 5.33). As initially proposed by W. Gilbert and D. Dressler, the rolling circle model of DNA replication included a specific enzyme, called a "transferase," which attached the 5′ end of the (+) strand to a specific site on the cell membrane. Although most, if not all, replicating chromosomes are attached to the membrane, little is known about the specific nature of such attachments. In any case, membrane attachment is not an essential feature of rolling circle replication. As the circle rotates and the 5′ end is displaced, DNA polymerase catalyzes covalent extension at the other (3′-OH) end.

During parental RF to progeny RF replication, the nascent positive strands are used as templates for the discontinuous synthesis of complementary negative strands. In some cases, the synthesis of the complementary strand may occur discontinuously on the single-stranded tail before synthesis of the first strand has been completed. In such cases, a double-stranded tail will be produced. The switch from double-stranded RF DNA synthesis to single-stranded viral (+) DNA synthesis occurs when specific proteins of the viral coat are produced in the cell. Rolling circle replication continues, but as the viral strand is displaced, these coat proteins bind to it and prevent the synthesis of complementary (−) strands (Fig. 5.32).

The phage ΦX174 gene *A* protein is a key protein in ΦX174 replication. It possesses a remarkable set of activities. (1) Gene *A* protein possesses a site-specific endonuclease activity that cleaves the positive strand at the origin. (2) Gene *A* protein then maintains the energy of the cleaved phosphodiester linkage by means of a covalent attachment of the 5′-phosphoryl terminus to itself. (3) Gene *A* protein remains bound to the 5′-terminus of the positive strand and to the replication fork while the fork traverses the complete circular minus-strand template. (4) When a complete positive strand has been synthesized, gene *A* protein cleaves the new origin, ligates the 3′-hydroxyl and 5′-phosphoryl termini, and once again becomes covalently linked to the newly generated 5′-positive-strand terminus. This cycle of gene *A* protein activities is repeated until a population of progeny RFs (stage II) or progeny positive strands (stage III) is produced.

To date, evidence for rolling circle replication has been found for (1) single-stranded DNA viruses like ΦX174, (2) the replication associated with chromosome transfer during "mating" (conjugation) in bacteria (see Chapter 8), and (3) the replication of small extrachromosomal DNA molecules carrying clusters of rRNA genes during oögenesis in amphibians (see Chapter 15).

PROKARYOTE NUCLEOID STRUCTURE

Most textbooks in biology have until recently presented a very erroneous picture of the chromosomes of prokaryotes. They have characterized prokaryotic chromosomes as "naked molecules of DNA," in contrast to eukaryotic chromosomes with their associated proteins and complex morphology. This misconception has resulted, at least in part, because (1) the pictures of prokaryotic chromosomes most often published have been autoradiographs and electron micrographs of isolated DNA molecules, ***not metabolically active or functional chromosomes,*** while (2) the most common pictures of eukaryotic chromosomes have been of highly condensed meiotic or mitotic chromosomes—again, ***metabolically inactive chromosomal states***. We now know that functional bacterial chromosomes or "nucleoids" ("nucleoids" rather than "nuclei" since they are not bounded by a nuclear membrane) bear little resemblance to the structures seen in Cairns' autoradiographs just as the metabolically active interphase chromosomes of eukaryotes have little morphological resemblance to mitotic or meiotic metaphase chromosomes.

The contour length of the circular DNA molecule of *E. coli* is about 1100 μm. The *E. coli* cell has a diameter of only 1–2 μm. Clearly, then, the chromosome must exist in a highly folded or coiled configuration within the cell. When the *E. coli* chromosome is isolated by very gentle procedures in the absence of ionic detergents (commonly used to lyse cells) and is kept in the presence of a high concentration of cations such as polyamines (small basic or positively charged proteins) or 1 *M* salt to neutralize the negatively charged phosphate groups of DNA, the chromosome remains in a highly condensed state comparable in size to the nucleoid *in vivo*. This structure, called the ***"folded genome,"*** is apparently the functional state of the *E. coli* chromosome. In its folded genome state, the single DNA molecule of *E. coli* is arranged into about 50 loops or domains, each of which is highly twisted or ***"supercoiled"*** (much like a tightly coiled telephone cord). This structure is dependent on RNA and protein, both of which are components of the folded genome (Fig. 5.34). The folded genome can be relaxed by treatment with either deoxyribonuclease (DNase) or ribonuclease (RNase).

Supercoiling of DNA is an important feature of all chromosomes, from those of the smallest viruses to those of eukaryotes. It occurs whenever the DNA is either underwound ("negative supercoils") or overwound ("positive supercoils"). If one takes a covalently closed, circular double helix of DNA, breaks one strand, and rotates one of the ends that is produced for 360° around the complementary strand, while holding the other end fixed, one supercoil will be introduced into the molecule (Fig. 5.35). If the free end is rotated in the same direction as the DNA double helix is wound (right-handed), a positive supercoil will be produced. If the free end is rotated in the opposite direction (left-handed), a negative supercoil will re-

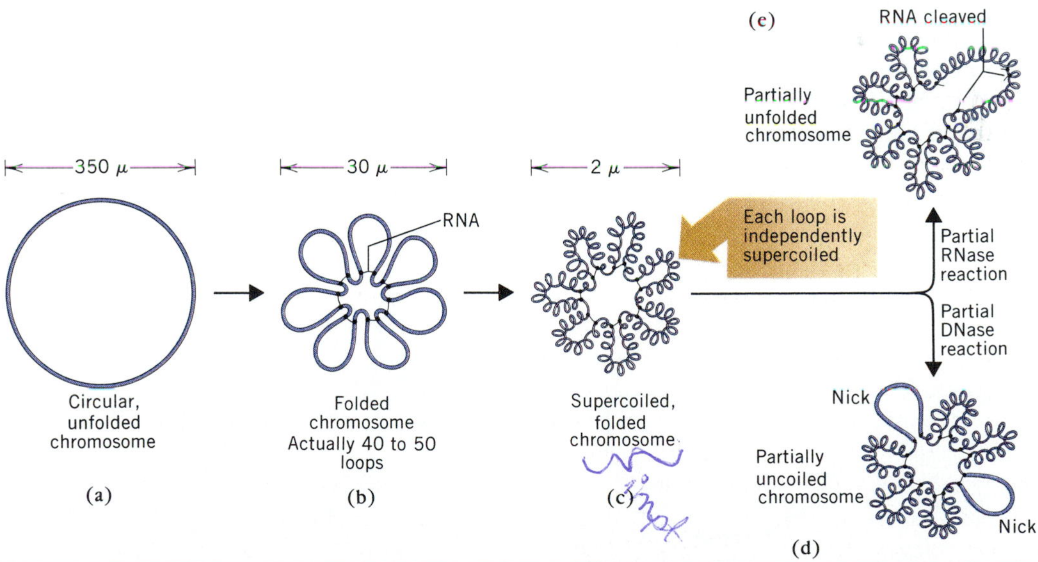

Figure 5.34 (a) Structure of the chromosome of *E. coli* when it is isolated as a completely unfolded, "relaxed" (no supercoils), covalently closed, circular molecule of DNA, much like the structure observed in Cairns' autoradiographs (Fig. 5.16), but here it is nonreplicating. (b) The chromosome segregated into seven loops by RNA "connectors." There are actually 40–50 such loops per chromosome in its *in vivo* folded genome form; only seven are shown in the diagram for the sake of simplicity. (c) Each loop is independently supercoiled. The introduction of single-strand "nicks" (d), by treatment of the folded genomes with an endonuclease, will relax the DNA only in the nicked loops; all other loops will remain supercoiled. Destruction of the RNA "connectors" (e) by ribonuclease (RNase) will unfold the folded genome partially by eliminating the segregation of the DNA molecule into 40–50 loops. However, RNase will not affect supercoiling. (After D. E. Pettijohn and R. Hecht. *Cold Spring Harbor Sympos. Quant. Biol.* 38: 31, 1973.)

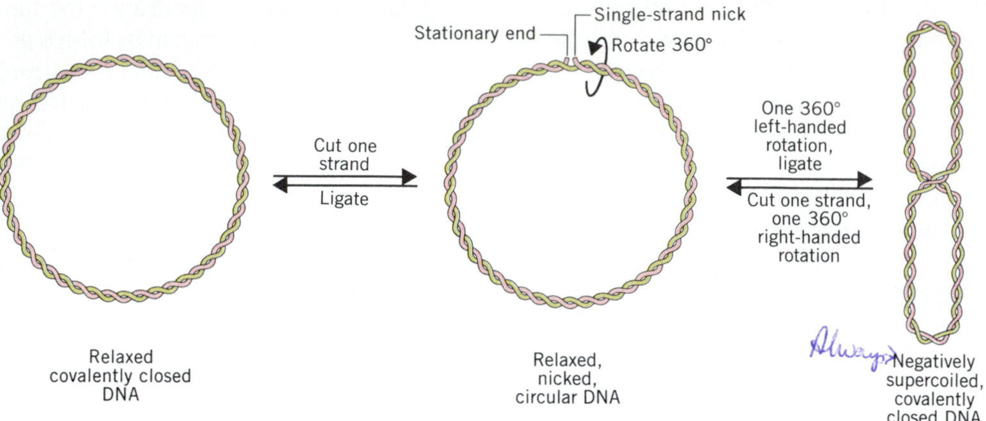

Figure 5.35 Visual definition of negative supercoiling. If one strand of a covalently closed circular molecule of DNA (or a linear molecule with fixed ends) is cleaved and one of the resulting free ends is rotated around the intact complementary strand while the other end remains stationary, supercoils will be produced. One supercoil, as shown on the right, will be produced for each 360° rotation of the free end. If the direction of rotation is the same as the direction in which the DNA double helix is wound (right-handed), a positive supercoil will be produced. If the direction of rotation is left-handed, a negative supercoil will be produced. Note that this diagram is for definition purposes only. This is **NOT** the mechanism used by enzymes such as DNA gyrase to introduce negative supercoils into DNA molecules. That process is illustrated in Fig. 5.36.

sult. Although this is probably the simplest way to visualize the phenomenon of supercoiling in DNA, it is ***not*** the mechanism used by enzymes to produce supercoiling in DNA (see Fig. 5.36).

Many biological functions of chromosomes can be carried out only when the participating DNA molecules are negatively supercoiled. For example, the phage ΦX174 gene *A* protein will nick the ΦX174 replicative form (RF) and initiate replication (see Fig. 5.32) only when the RF is in its negatively supercoiled form. All bacterial chromosomes studied to date appear to be negatively supercoiled in their functional states. When the basic proteins are carefully dissociated from the DNA of the chromosomes of *Drosophila melanogaster*, the DNA is found to contain the same amount of negative supercoiling as is found in the folded genomes of *E. coli*.

It is very likely that negative supercoiling is uni-

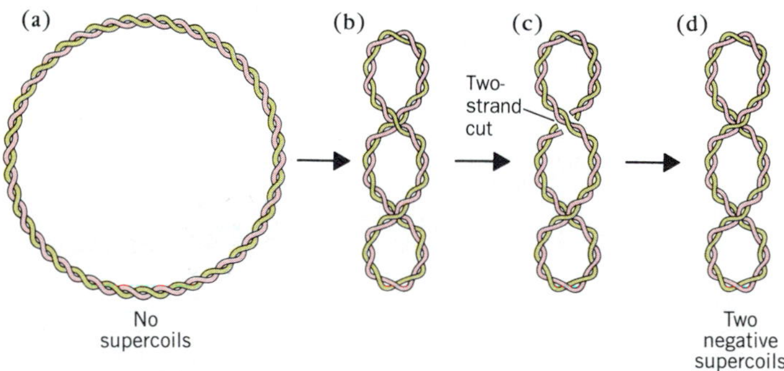

Figure 5.36 Mechanism of DNA gyrase-catalyzed negative supercoiling. A relaxed (no supercoils) DNA molecule (a) is wrapped back on itself in a right-handed manner (b). The underlying strand is then transiently cleaved and the overlying strand is passed through the gap (c). The transient break is resealed to yield two negative supercoils in the molecule (d). The molecule is probably "wrapped back" on itself by winding it around a portion of the enzyme. The energy of the cleaved phosphodiester bond is conserved by means of transient covalent linkages of the ends produced to the enzyme itself. This energy is then apparently used to reseal the cut. (After P. O. Brown and N. R. Cozzarelli, *Science* 206: 1081–1083, 1979.)

versally involved in certain of the biological functions of DNA molecules. Considerable ***evidence suggests that supercoiling is involved in recombination, gene expression, and the regulation of gene expression.*** In addition, ***negative supercoiling is almost certainly required for the replication of all DNA molecules.*** An enzyme called ***DNA gyrase,*** which catalyzes the formation of negative supercoils in DNA, has been isolated from several different organisms, both prokaryotes and eukaryotes. The DNA gyrase from *E. coli* has been the most extensively studied. Its activity is inhibited by the drugs novobiocin and nalidixic acid, two potent inhibitors of DNA synthesis in bacteria. This clearly indicates that DNA gyrase activity is required for DNA replication.

Although the exact role (or roles) of DNA gyrase in DNA replication is not yet established, an obvious possibility is that the introduction of negative supercoils may aid in the unwinding of the strands of the double helix. Mechanistically, DNA gyrase is a most interesting enzyme. *In vitro*, is can tie knots in DNA molecules or join two circular molecules to produce interlocking rings. DNA gyrase is known ***not*** to introduce negative supercoils into DNA molecules by cleaving one strand and rotating one of the ends as originally proposed (see the visual definition of supercoiling in Fig. 5.35). Rather, ***gyrase produces negative supercoils two at a time*** by cleaving both strands of one segment of a DNA molecule, passing another segment of the molecule through the temporary gap, and then rejoining the ends (Fig. 5.36). During each catalytic event, DNA gyrase "holds on" to the cut ends as the other DNA molecule or segment of the DNA molecule is passed through the transient gap, so that it can efficiently rejoin the ends to complete the process. DNA gyrase "holds on" by means of ***covalent linkage of the transient ends of the cleaved DNA molecule to itself,*** in the same manner as the phage ΦX174 gene *A* protein (see the preceding section on phage ΦX174).

SUMMARY

The genetic information of all living organisms, except for certain viruses, is stored in ***DNA. RNA*** is the genetic material of some viruses. DNA usually has a ***double helical structure*** in which the two strands are held together by ***hydrogen bonding*** between the two bases of each nucleotide pair and ***hydrophobic bonding*** in the core of "stacked" base-pairs. The ***base-pairing is specific: adenine always pairs with thymine, and guanine always pairs with cytosine.*** The two strands of a DNA double helix are thus ***complementary***; once the sequence of bases in one strand is established, the sequence of bases in the other (complementary) strands is fixed.

Replication of DNA is semiconservative in both prokaryotes and eukaryotes. During replication the two complementary strands unwind, and each single strand serves as a template directing the synthesis of a new complementary strand. The net result of replication is thus two progeny DNA molecules identical to the parental double helix. Replication often begins at ***unique origins*** and proceeds ***bidirectionally*** from these origins. Prokaryotic chromosomes usually contain a single primary origin. Eukaryotic chromosomes, on the other hand, contain multiple origins of replication.

The two complementary strands of a DNA double helix have ***opposite*** chemical polarity, one 3′ to 5′ and the other 5′ to 3′. All known ***DNA polymerases,*** both prokaryotic and eukaryotic, have an absolute requirement for a free ***3′-hydroxyl primer.*** They only catalyze covalent extension in the ***5′⟶ 3′ direction.*** Thus, synthesis of the strand growing in the overall 3′⟶ 5′ direction is ***discontinuous.*** Short segments are synthesized in the 5′⟶ 3′ direction and then covalently joined by ***polynucleotide ligase.*** No known DNA polymerase can catalyze the initiation of new DNA chains. Instead, nascent DNA chains are initiated by short ***RNA primers.***

Chromosome replication is accomplished by a ***complex replication apparatus*** composed of many different proteins and enzymes. At least three different proteins are involved in unwinding the complementary strands of the DNA double helix, and a DNA single-strand binding protein maintains the DNA in an extended configuration for replication. The enzyme, DNA polymerase III, that catalyzes semiconservative DNA replication in *E. coli* is a multimeric protein containing seven different polypeptides. At least six proteins in addition to primase are present in the ***primosome*** and are involved in initiating the synthesis of new strands. The ***3′⟶ 5′ exonuclease activity built into prokaryotic DNA polymerases provides a proofreading*** function that is essential for accurate DNA replication.

The chromosomes of prokaryotes exist as highly compact ***"folded genomes" in vivo.*** The DNA molecules are separated into many independently ***negatively supercoiled loops by RNA-containing cross-links.*** Supercoils are introduced into and removed from DNA molecules by the enzyme ***DNA gyrase.***

REFERENCES

ALBERTS, B., D. BRAY, J. LEWIS, M. RAFF, K. ROBERTS, and J.D. WATSON. 1989. *Molecular biology of the cell*, 2nd ed. Garland, New York.

AVERY, O.T., C.M. MACLEOD, and M. MCCARTY. 1944. "Studies on the chemical nature of the substance inducing transformation in pneumococcal types." *J. Expl. Med.* 79: 137–158.

CAIRNS, J. 1966. "The bacterial chromosome." *Sci. Amer.* 214(1): 36–44.

CANTOR, C.R. 1981. "DNA choreography." *Cell* 25: 293–295.

COZZARELLI, N.R. 1980. "DNA gyrase and the supercoiling of DNA." *Science* 207: 953–960.

FREIFELDER, D. 1978. *The DNA molecule: Structure and properties.* Freeman, San Francisco.

GILBERT, W., and D. DRESSLER. 1968. "DNA replication: The rolling circle model." *Cold Spring Harbor Sympos. Quant. Biol.* 33: 473–484.

HERSHEY, A.D., and M. CHASE. 1952. "Independent functions of viral protein and nucleic acid in growth of bacteriophage." *J. Gen. Physiol. 36: 39–56.* (Reprinted in G.S. Stent. 1960. *Papers on bacterial viruses,* 2nd ed. Little, Brown, Boston.)

KORNBERG, A. 1980. *DNA replication.* Freeman, San Francisco.

______. 1982. *1982 Supplement to DNA replication.* Freeman, San Francisco.

LEHNINGER, A.L. 1982. *Principles of biochemistry.* Worth, New York.

LEWIN, B. 1974. *Gene expression-1, bacterial genomes.* Wiley, New York.

______. 1977. *Gene expression-3, plasmids and phages.* Wiley, New York.

______. 1990. *Genes IV.* Cell Press, Cambridge, MA./Oxford Univ. Press, Oxford.

LOEB, L.A., and T.A. KUNKEL. 1982. "Fidelity of DNA synthesis." *Ann. Rev. Biochem.* 51: 429–457.

MCHENRY, C.S. 1988. "DNA polymerase III holoenzyme of *Escherichia coli.*" *Ann. Rev. Biochem.* 57: 519–550.

MESELSON, M.S., and F.W. STAHL. 1958. "The replication of DNA in *Escherichia coli.*" *Proc. Natl. Acad. Sci. U.S.A.* 44: 671–682.

SCHNÖS, M., and R.B. INMAN. 1970. "Position of branch points in replicating λ DNA." *J. Mol. Biol.* 51: 61–73.

WANG, J.C. 1982. "DNA topoisomerases." *Sci. Amer.* 247(1): 94–109.

WATSON, J.D., and F.H.C. CRICK. 1953. "A structure for deoxyribose nucleic acid." *Nature* 171: 737–738.

WATSON, J.D., N.H. HOPKINS, J.W. ROBERTS, J.A. STEITZ, and A.M. WEINER. 1987. *Molecular biology of the gene,* 4th ed. Benjamin, Menlo Park. CA

WORCEL, A., and E. BURGI. 1972. "On the structure of the folded chromosome of *Escherichia coli.*" *J. Mol. Biol.* 71: 127–147.

ZIMMERMAN, S.B. 1982. "The three-dimensional structure of DNA." *Ann. Rev. Biochem.* 51: 395–427.

PROBLEMS AND QUESTIONS

5.1 (a) How did the transformation experiments of Griffith differ from those of Avery and his associates? (b) What was the significant contribution of each? (c) Why was Griffith's work not evidence for DNA as the genetic material whereas the experiments of Avery *et al.* provided direct proof that DNA carried the genetic information?

5.2 A cell-free extract is prepared from Type IIIS pneumococcal cells. What effect will treatment of this extract with (a) protease, (b) RNase, and (c) DNase have on its subsequent capacity to transform recipient Type IIR cells to Type IIIS? Why?

5.3 How could it be demonstrated that the mixing of heat-killed Type III pneumococcus with live Type II resulted in a transfer of genetic material from Type III to Type II rather than a restoration of viability to Type III by Type II?

5.4 What is the macromolecular composition of a bacterial virus or bacteriophage such as phage T2?

5.5 What chemical properties do DNA and proteins possess that allow researchers to specifically label one or the other of these macromolecules with a radioactive isotope?

5.6 (a) What was the objective of the experiment carried out by Hershey and Chase? (b) How was the objective accomplished? (c) What is the significance of this experiment?

5.7 (a) What background material did Watson and Crick have available for developing a model of DNA? (b) What was their contribution to the building of the model?

5.8 (a) Why was a double helix chosen for the basic pattern of the molecule? (b) Why were hydrogen bonds placed in the model to connect the bases?

5.9 (a) If a virus particle contains double-stranded DNA with 200,000 base-pairs, how many nucleotides would be present? (b) How many complete spirals would occur on each strand? (c) How many atoms of phosphorus would be present? (d) What would be the length of the DNA configuration in the virus?

5.10 If one strand or helix on the Watson–Crick model has a base sequence of 5′-GTCATGAC-3′, what would be the base sequence of the complementary DNA strand?

5.11 What are the differences between DNA and RNA?

5.12 DNA was extracted from cells of *Staphylococcus afermentans* and analyzed for base composition. It was found that 37 percent of the bases are cytosine. Using this information, is it possible to predict what percentage of the bases are adenine? If so, what percentage? If not, why not?

5.13 RNA was extracted from TMV (tobacco mosaic virus) particles and found to contain 20 percent cytosine (20 percent of the bases were cytosine). Using this information, is it possible to predict what percentage of the bases in TMV are adenine? If so, what percentage? If not, why not?

5.14 Indicate whether each of the following statements about the structure of DNA is true or false. (Each letter is used to refer to the concentration of that base in DNA.) (a)

A + T = G + C. (b) A = G; C = T. (c) A/T = C/G. (d) T/A = C/G. (e) A + G = C + T. (f) G/C = 1. (g) A = T within each single strand. (h) Hydrogen bonding provides stability to the double helix in aqueous cytoplasms. (i) Hydrophobic bonding provides stability to the double helix in aqueous cytoplasms. (j) When separated, the two strands of a double helix are identical. (k) Once the base sequence of one strand of a DNA double helix is known, the base sequence of the second strand can be deduced. (l) The structure of a DNA double helix is fully conserved during its replication. (m) Each nucleotide pair contains two phosphate groups, two deoxyribose molecules, and two bases.

5.15 *Escherichia coli* cells are grown for many generations in a medium in which the only available nitrogen is the heavy isotope ^{15}N. They are then transferred to a medium containing ^{14}N as the only source of nitrogen. (a) What distribution of ^{15}N and ^{14}N would be expected in the DNA molecules of cells that had grown for one generation in the ^{14}N-containing medium ***assuming*** that DNA replication was (i) conservative, (ii) semiconservative, or (iii) dispersive? (b) What distribution would be expected after two generations of growth in the ^{14}N-containing medium ***assuming*** (i) conservative, (ii) semiconservative, or (iii) dispersive replication?

5.16 A culture of bacteria is grown for many generations in a medium in which the only available nitrogen is the heavy isotope (^{15}N). The culture is then switched to a medium containing only ^{14}N for one generation of growth; it is then returned to a ^{15}N-containing medium for one final generation of growth. If the DNA from these bacteria is isolated and centrifuged to equilibrium in a CsCl density gradient, how would you predict the DNA to band in the gradient?

5.17 DNA polymerase I of *E. coli* is a single polypeptide of molecular weight 109,000. (a) What enzymatic activities other than polymerase activity does this polypeptide possess? (b) What are the *in vivo* functions of these activities? (c) Are these activities of major importance to an *E. coli* cell? Why?

5.18 A DNA template plus primer with the structure

3′ Ⓟ — T G C G A A T T A G C G A C A T — Ⓟ 5′
·· ··· ·· ·· ·· ·· ·· ··
5′ Ⓟ — A T C G G T A C G A C G C T T A A C — OH 3′

(where Ⓟ = a phosphate group) is placed in an *in vitro* DNA synthesis system (Mg^{2+}, an excess of the four deoxyribonucleoside triphosphates, etc.) containing a mutant form of *E. coli* DNA polymerase I that lacks 5′⟶3′ exonuclease activity. The 5′⟶3′ polymerase and 3′⟶5′ exonuclease activities of this aberrant enzyme are identical to those of normal *E. coli* DNA polymerase I. It simply has ***no*** 5′⟶3′ exonuclease activity. (a) What will be the structure of the final product? (b) What will be the first step in the reaction sequence?

5.19 How might continuous and discontinuous modes of DNA replication be distinguished experimentally?

5.20 Identify the proteins that are involved in DNA replication in *E. coli* and list their known or putative functions.

Chapter 6

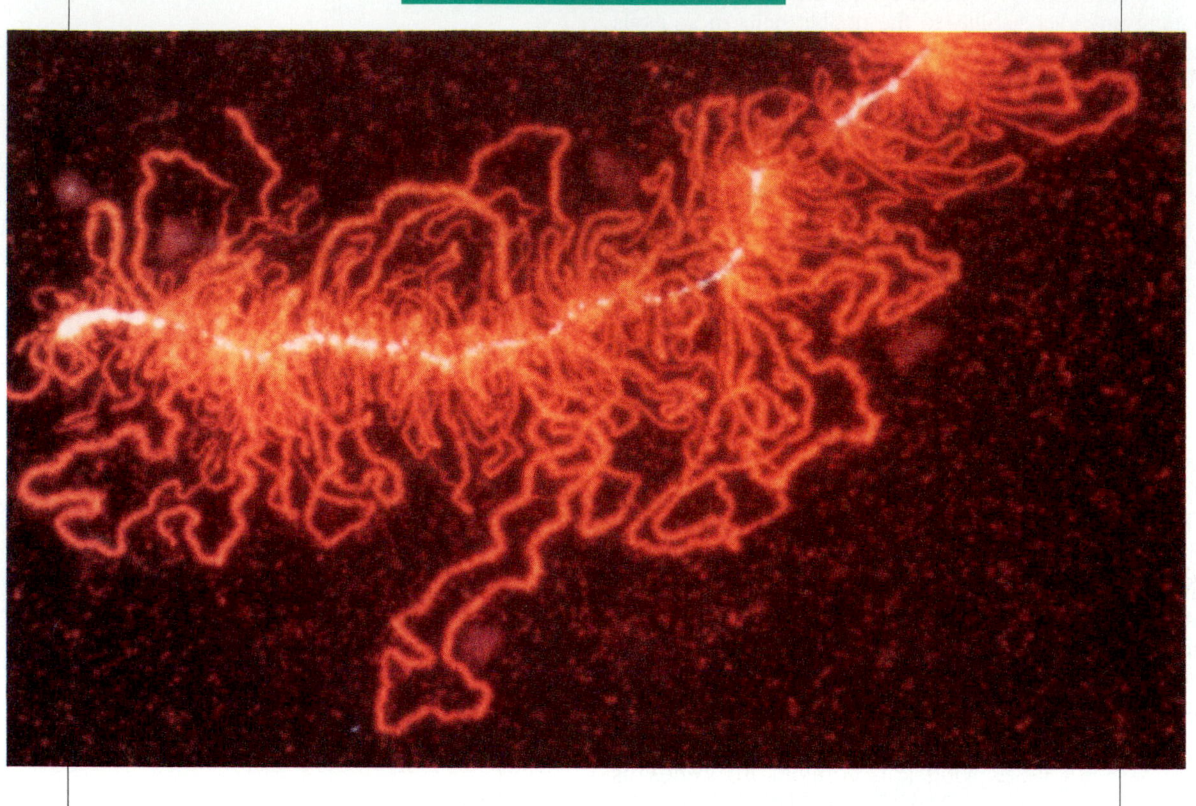

Structure and Replication of Eukaryotic Chromosomes

(Photo above): Scanning electron micrograph of a portion of a giant lampbrush chromosome of the newt *Triturus viridescens*. (Courtesy of Dr. Joseph G. Gail, Carnegie Institution, Baltimore, MD.)

MUCH of our information about the structure and mechanisms of DNA replication has come from studies of prokaryotes. The reason for this is that prokaryotes are less complex, both genetically and biochemically, than eukaryotes. Prokaryotes are monoploid (mono = one; equivalent to the haploid state of diploid organisms); they have only one set of genes (one copy of the genome). Most higher animals and many higher plants, by contrast, are diploid, having two complete sets of genes, one from each parent (see Chapters 2 and 3). Some higher plants are polyploid (poly = many), that is, carry several copies of the genome (see Chapter 19). As discussed in Chapter 5, ***most of the genetic information of most viruses and prokaryotes is stored in a single chromosome, which in turn contains a single molecule of nucleic acid*** (either RNA or DNA).

The smallest known RNA viruses have only three genes. In fact, the complete nucleotide sequences of the genomes of several of these viruses are now known. (The nucleotide sequence of the coat gene of bacteriophage MS2 is shown in Fig. 10.31. The phage MS2 genome is 3569 nucleotides long and contains four genes.) The smallest known DNA viruses have only 9–11 genes. Again, the complete nucleotide sequences are known in a few cases. (Part of the nucleotide sequence of the phage ΦX174 chromosome, which is 5387 nucleotides long and contains 10 genes, is shown in Fig. 12.26.) The largest DNA viruses, like bacteriophage T2 and the animal pox viruses, contain about 150 genes. Bacteria like *E. coli* have 3000–4000 genes, most of which are present in a single molecule of DNA.

GENOME COMPLEXITY

Eukaryotes have from 2 to 10 times as many genes as* E. coli** (see Chapters 12 and 13) ***but have orders of magnitude more DNA (Fig. 6.1). One of the most challenging problems being studied by geneticists today is the question of the function(s) of this "excess" DNA (DNA not required for structural genes that code for proteins; see Chapter 10). This question will be discussed later in this chapter and again in Chapters 10, 12, and 13.

Not only do most eukaryotes contain many times the amount of DNA of prokaryotes, but this DNA is packaged in several chromosomes, and each chromosome is present in two (diploids) or more (polyploids) copies. Recall that the chromosome of *E. coli* has a contour length of 1100 μm or about 1 mm. Now consider that the haploid chromosome complement, or genome, of the human contains about 1000 mm of DNA (or about 2000 mm per diploid cell). This meter of DNA is, of course, subdivided among 23 chromosomes of variable size and shape, each chromosome containing from 15 to 85 mm of DNA. Until recently, we had little information as to how this DNA was arranged in the chromosomes. Is there one molecule of DNA per chromosome as in prokaryotes, or are there many molecules of DNA per chromosome? If many, how are they arranged relative to each other? How does the 85 mm (85,000 μm) of DNA in the largest human chromosome get condensed into a mitotic metaphase structure that is about 0.5 μm in diameter and 10 μm long? What are the structures of the metabolically active interphase chromosomes?

CHEMICAL COMPOSITION OF EUKARYOTIC CHROMOSOMES

Interphase chromosomes are not visible with the light microscope; moreover, electron microscopy of thin sections cut through eukaryotic nuclei has provided essentially no information about their structure. Recently, however, chemical analysis, electron microscopy, and X-ray diffraction studies on isolated ***chromatin*** (the complex of the DNA, chromosomal proteins, and other chromosome constituents isolated from nuclei) have provided a solid framework for a rapidly emerging picture of chromosome structure in eukaryotes.

When chromatin is isolated from interphase nuclei, the individual chromosomes are not recogniz-

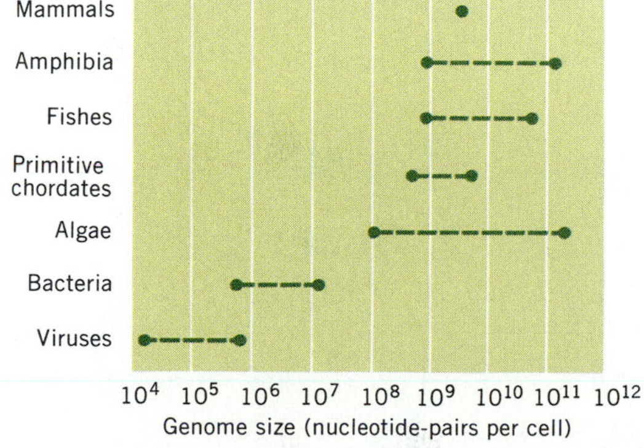

Figure 6.1 Increasing genome size with increasing evolutionary or developmental complexity. The minimum and maximum genome sizes observed in each group are given as nucleotide-pairs of DNA per haploid cell. (From R. J. Britten and D. E. Kohne, "Repeated Segments of DNA," *Sci. Amer.* 222(4): 24–31, 1970. Copyright © 1970 by Scientific American Inc. All rights reserved.)

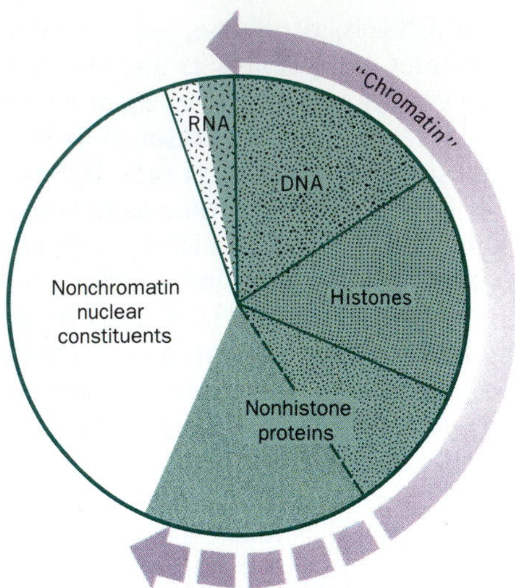

Figure 6.2 Eukaryotic chromatin composition as a fraction of total nuclear content. Estimates of nonhistone protein content are variable depending on the isolation procedures and cell types used. (Reprinted with permission from K. E. Van Holde and I. Isenberg, *Accounts of Chem. Res.* 8: 327–335, 1975. Copyright © 1975 by the American Chemical Society.)

able. Instead, one observes an irregular aggregate of nucleoprotein. Chemical analysis of isolated chromatin shows that it consists primarily of DNA and proteins with lesser amounts of RNA (Fig. 6.2). The proteins are of two major classes: (1) basic proteins (positively charged at neutral pH) called ***histones*** and (2) a heterogeneous, largely acidic (negatively charged at neutral pH) group of proteins collectively referred to as ***nonhistone chromosomal proteins.***

Histones play a major structural role in chromatin. They are present in the chromatin of all higher eukaryotes in amounts equivalent to the amounts of DNA (weight/weight). The histones of all higher plants and animals consist of five different major proteins. These five major histones, called H1, H2a, H2b, H3, and H4 (Fig. 6.3), are present in almost all cell types (a few exceptions exist, most notably some sperm, where the histones are replaced by another class of small basic proteins called ***protamines***).

The five histones are present in molar ratios of approximately 1 H1 : 2 H2a : 2 H2b : 2 H3 : 2 H4. They are specifically complexed with DNA to produce the basic structural subunits of chromatin, small (approximately 110 Å in diameter by 60 Å high) ellipsoidal "beads" called ***nucleosomes*** (see Fig. 6.13). (Nucleosomes are discussed in a subsequent section of this chapter entitled Nucleosome Structure.) The histones have been highly conserved during evolution, four of the five types of histone being very similar in all higher eukaryotes.

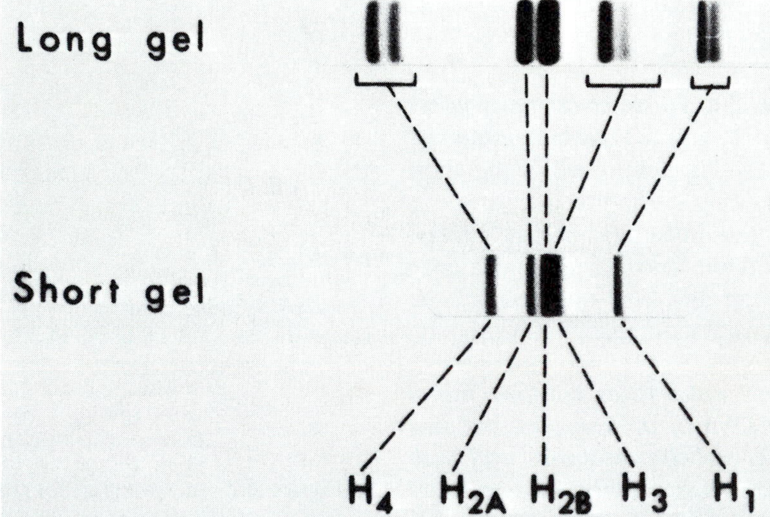

Figure 6.3 Acrylamide gel electropherograms showing the five types of histone found in the chromatin of all higher eukaryotes. Note that the long gel (top) shows heterogeneity of histones H1, H3, and H4. The histones shown are from calf chromatin. (Reprinted with permission from S. Panyim and R. Chalkey, *Biochemistry* 8: 3972–3979, 1969. Copyright © 1969 by the American Chemical Society. Original photograph courtesy of Roger Chalkley.)

```
        H
        |       //O
H3N+ —  C — C
        |       \
H — C — H        O−
        |
H — C — H
        |
H — C — H
        |
       NH
        |
        C = NH
        |
       N+H3
    Arginine

        H
        |       //O
H3N+ —  C — C
        |       \
H — C — H        O−
        |
H — C — H
        |
H — C — H
        |
H — C — H
        |
       N+H3
     Lysine
```

Figure 6.4 Structures of the basic (positively charged at neutral pH) amino acids arginine and lysine. These two amino acids, which make up 20–30 percent of the amino acids of histones, are responsible for the polycationic nature of histones and facilitate their interaction with polyanionic nucleic acids.

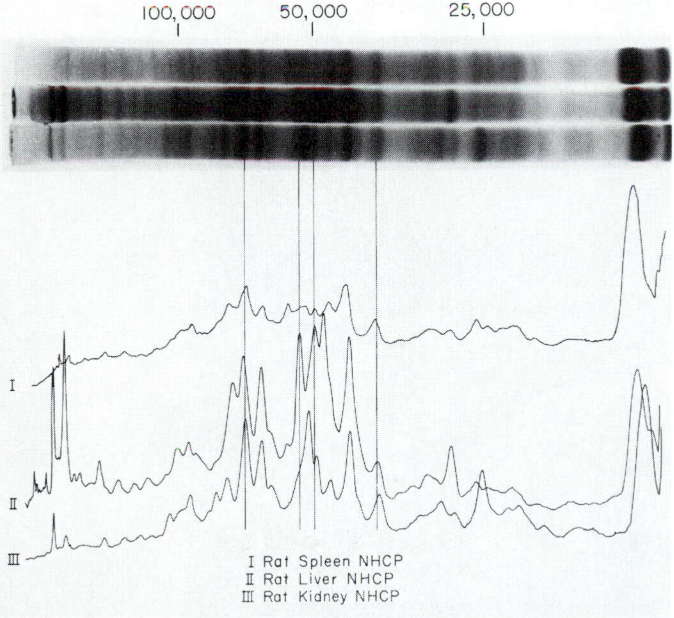

Figure 6.5 Polyacrylamide gel electropherograms (top) and densitometer tracings produced by photometric scans of the gels (bottom) showing the complexity of the nonhistone chromosomal protein fractions from rat spleen, rat liver, and rat kidney. Whereas some nonhistones are present in all three cell types, others are tissue specific. (Reproduced with permission from F. C. Wu, S. C. R. Elgin, and L. E. Hood, *Biochemistry* 12: 2792–2797, 1973. Copyright © 1973 by the American Chemical Society.)

Proteins, like nucleic acids, are large macromolecules composed of a large number of small subunits covalently linked together into long polymers (see Chapter 10). In the case of proteins, the subunits are called amino acids, of which 20 different species make up all proteins. Most of the amino acids are neutral in charge, that is, have no charge at pH 7. However, a few are basic and a few are acidic. The ***histones are basic because they contain 20–30 percent arginine and lysine,*** two positively charged amino acids (Fig. 6.4). The exposed $-NH_3^+$ groups of arginine and lysine allow histones to act as polycations. This is important in their interaction with DNA, which is polyanionic because of the negatively charged phosphate groups.

The remarkable constancy of histones H2a, H2b, H3, and H4 in all cell types of an organism and even between widely divergent species is consistent with the idea that they are important in chromatin structure ("DNA packaging") and are only nonspecifically involved in the regulation of gene expression.

On the other hand, the ***nonhistone protein fraction of chromatin consists of a large number of very heterogeneous proteins*** (Fig. 6.5). Moreover, the composition of the nonhistone chromosomal protein fraction varies widely among different cell types of the same organism. Thus, the nonhistone chromosomal proteins are likely candidates for roles in the regulation of expression of specific genes or sets of genes (see Chapter 15).

ONE GIANT DNA MOLECULE PER CHROMOSOME

How is the 1–20 cm (10^4 to 2×10^5 μm) of DNA, which is present in an average eukaryotic chromosome, arranged in the highly condensed mitotic and meiotic structures that are seen with the light microscope (Chapter 3)? Are there many DNA molecules that run parallel throughout the chromosome (the ***"multineme" or "multistrand" model***), or is there just one DNA double helix extending from one end of the chromosome to the other end (the ***"unineme" or "single-strand" model***)? (Note that ***"strand" here refers to the DNA double helix,*** not the single-strand of DNA.) Are there many DNA molecules joined end-to-end or arranged in some other fashion in the chromosome, or does one giant, continuous molecule of DNA extend from one end to the other in a highly coiled and folded form? The evidence supporting the unineme model of chromosome structure is now overwhelming. In addition, solid evidence presently supports the concept of chromosome-size DNA molecules. That is, ***each chromosome appears to contain a single, giant molecule of DNA that extends from one***

end through the centromere all the way to the other end of the chromosome.

"Lampbrush" Chromosomes in Vertebrate Oöcytes

Some of the strongest evidence supporting the unineme model of chromosome structure has come from studies on the large, so-called ***"lampbrush" chromosomes,*** which are present during prophase I of oögenesis in many vertebrates, particularly amphibians. ***Lampbrush chromosomes are up to 800 μm long;*** they thus provide very favorable material for cytological studies. The homologous chromosomes are paired, and each has duplicated to produce two chromatids at the lampbrush stage. Each lampbrush chromosome contains a central axial region, where the two chromatids are highly condensed, and numerous pairs of lateral loops (Figs. 6.6 and 6.7). The loops are transcriptionally active regions of single chromatids. The integrity of both the central axis and the lateral loops is dependent on DNA. Treatment with DNase fragments both the axis and the loops. Treatment with RNase or proteases removes surrounding matrix material, but does not destroy the continuity of either the axis or the loops. Electron microscopy of RNase- and protease-treated lampbrush chromosomes reveals a central filament of just over 20 Å in diameter in the lateral loops. Since each loop is a segment of one chromatid, and since the diameter of a DNA double helix is 20 Å, these lampbrush chromosomes must be unineme structures (Fig. 6.7). This conclusion is also supported by studies on the kinetics of nuclease digestion of lampbrush chromosomes. That is, the kinetics observed are those expected if a single double helix of DNA is the central filament in the loops. The axial region then contains two DNA molecules, one from each of the two tightly paired chromatids.

Lampbrush chromosomes are meiotic or "germ line" chromosomes. Their structure is thus particularly relevant to an understanding of genetic phenomena. "Nongerm line" or somatic-cell chromosomes may have different structures. Although most are unineme, some are known to be multineme structures (see Chapter 18, pp. 489–490).

Viscoelastometric Evidence for Chromosome-Size DNA Molecules

The question of whether the unineme chromosomes of eukaryotes contain a single large molecule of DNA or many smaller molecules linked end-to-end has

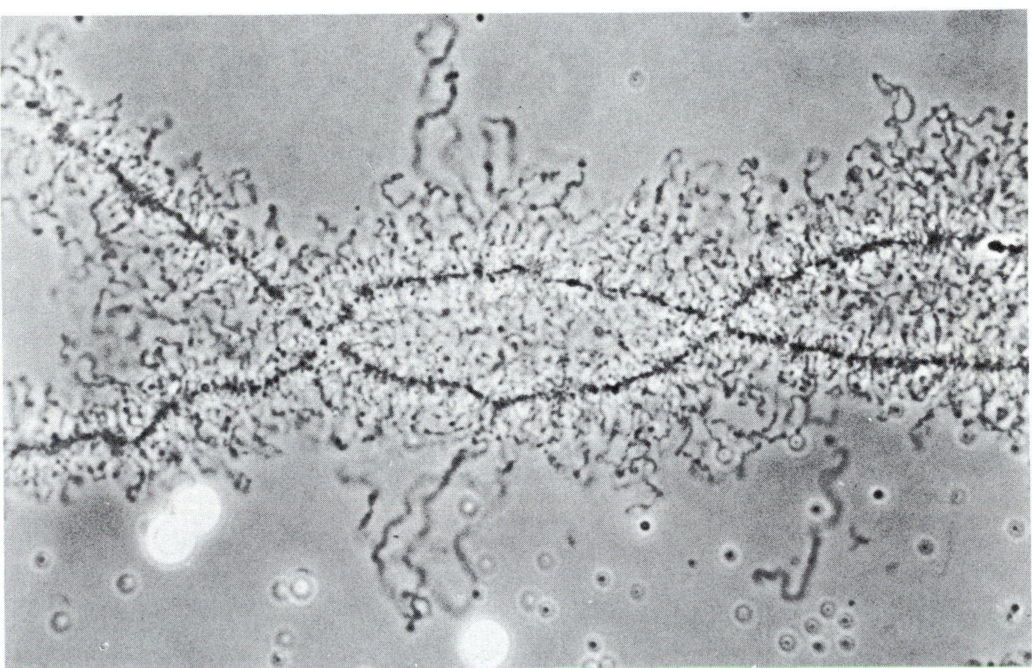

Figure 6.6 Phase contrast micrograph of the giant (400–800 μ long) lampbrush chromosomes in the oöcytes of the newt, *Triturus viridescens.* The homologous chromosomes of the lampbrush bivalent shown are joined at two chiasmata. Recall that each chromosome has already replicated and thus consists of two chromatids. Note the numerous lateral loops of various lengths. (Courtesy of J. G. Gall.)

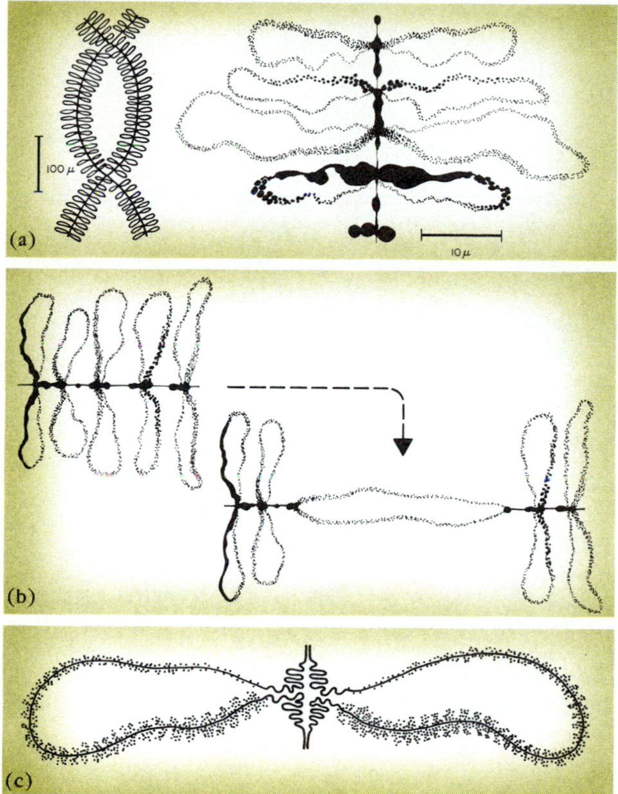

Figure 6.7 Schematic representation of the structure of the giant lampbrush chromosomes of amphibian oöcytes. (Left, a) Diagrammatic view of a lampbrush bivalent (compare Fig. 6.6). (Right, a) Semidiagrammatic view of the central chromosome axis with pairs of morphologically identical loops. Each chromosome has already duplicated in interphase. Each homologous chromosome thus contains two chromatids. (b) A portion of a lampbrush chromosome that has been stretched during preparation, revealing the continuity of the loops and the central axis. (c) A single pair of loops displaying the continuity of the single DNA molecule that extends through the loop and axial regions of each chromatid. The matrix is primarily nascent RNA that is synthesized on the extended DNA in the loop regions. (From J. G. Gall, "Mutation," *Brookhaven Sympos. Biol.* 8: 18, 1955.)

proven very difficult to answer with rigorous experimental evidence. A centimeter-long molecule of DNA has a length-to-width (diameter) ratio of 5 million to 1. Such a structure is very shear sensitive. If such a DNA molecule is in solution in a test tube, the slightest vibration will break the molecule into many fragments. For this and other reasons, accurate estimates of the molecular sizes of eukaryotic DNAs cannot be obtained by conventional biochemical methodology. Recently, biophysicists have used a technique called ***viscoelastometry*** (a procedure for analyzing viscosity parameters of molecules in solution) to estimate the sizes of DNA molecules from eukaryotic chromosomes. When DNA molecules in solution are exposed to a driving force (e.g., by rotating a cylinder in the solution) they are stretched into an extended conformation. When the driving force is removed (e.g., by stopping the rotation of the cylinder), the molecules will return to their lowest energy or relaxed state.

The recoil time is a function of the size of DNA molecules in the solution being analyzed. The recoil decays exponentially over long time intervals, and the time constant, called the retardation time (τ), is a sensitive function of the size of the ***largest DNA molecules in the solution.*** Thus, even if some of the DNA molecules are broken by shearing forces, the viscoelastometric procedure will permit one to estimate the size of the largest (hopefully, intact) DNA molecules present. Moreover, the cells can be lysed right in the chamber of the viscoelastometer to minimize the chance of shearing the molecules when they are released from the cells.

The largest DNA molecules in *Drosophila melanogaster* cells were estimated to have a mass of 4.1×10^{10} daltons by viscoelastometry (Fig. 6.8). (A ***dalton*** is the unit of mass equal to the mass of a single hydrogen atom. It is the most frequently used unit in dealing with the size of macromolecules like DNA.)

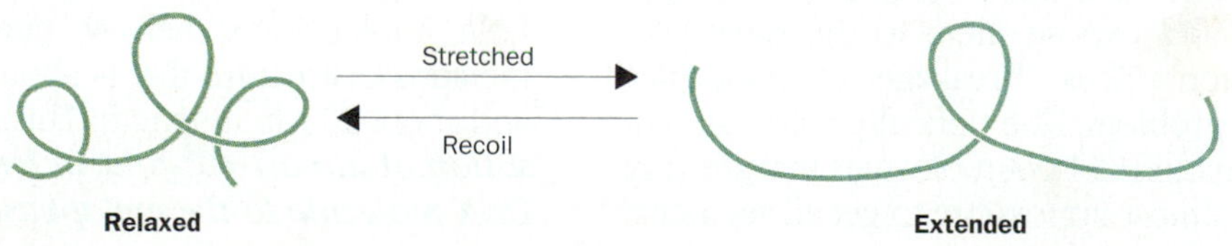

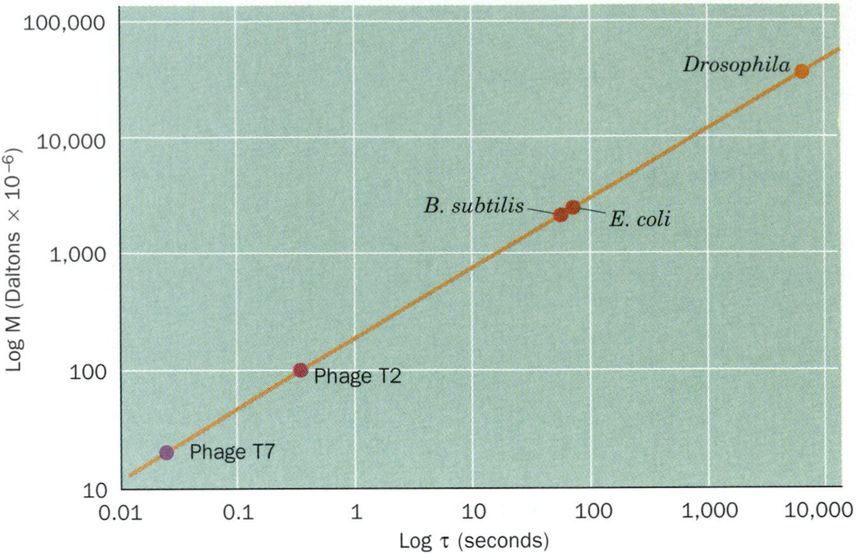

Figure 6.8 Relationship between molecular weight of DNA molecules and the time constant ("retardation time," τ) of the recoil that occurs when the stress is removed in viscoelastometric analysis. See the text for details. The results shown indicate that the largest DNA molecules in wild-type *Drosophila* cells are about 4.1×10^{10} daltons in mass. Since the largest chromosome in *Drosophila* has been estimated to contain about 4.3×10^{10} daltons of DNA by microspectrophotometric methods, the viscoelastometric data support the existence of a single molecule of DNA in the largest *Drosophila* chromosome. (After R. Kavenoff, L. C. Klotz, and B. H. Zimm, *Cold Spring Harbor Sympos. Quant. Biol.* 38: 1–8, 1973.)

The largest chromosomes of *Drosophila* have been shown to contain about 4.3×10^{10} daltons of DNA (total, whether one molecule or many) by direct microspectrophotometric analysis of metaphase chromosomes. Thus, the ***viscoelastometric estimate of the size of the largest DNA molecules in* Drosophila *nuclei correlates almost exactly with the total amount of DNA present in the largest chromosome.***

Autoradiographic Evidence for Chromosome-Size DNA Molecules

Researchers have also used the technique of ***autoradiography*** (see Chapter 5, pp. 107–109) to attempt to detect chromosome-size DNA molecules from eukaryotic cells in the same way that Cairns was able to visualize the intact *E. coli* chromosome (see Fig. 5.16). Like viscoelastometry, autoradiography permits one to lyse cells very gently so as to minimize shearing forces. DNA molecules from the lysed cells are permitted to diffuse onto a membrane that is used to pick up the DNA molecules and expose them to the β-particle-sensitive emulsion. Thus, breakage of molecules should not be a problem. The difficulty with autoradiographic examination of DNA molecules that are very large is that it is almost impossible to get all segments of the molecule sufficiently spread out on the membrane with no tangles or overlaps so that the entire length of the molecule is visible.

Nevertheless, autoradiographic analysis of DNA molecules from *Drosophila* have been successful, and the results of these studies also support the concept of chromosome-size DNA molecules. The ***largest molecules observed have a contour length of 1.2 cm*** (Fig. 6.9). Although these are not as large as the viscoelastometric estimates of the largest molecules, such molecules correspond to a mass in the range of 2.4 to 3.2×10^{10} daltons, two-thirds to three-fourths the size of the largest chromosome-size DNA molecules.

PACKAGING THE GIANT DNA MOLECULES INTO CHROMOSOMES

The largest chromosome in the human genome contains about 85 mm (85,000 μm or 8.5×10^8 Å) of DNA that is believed to exist as one giant molecule. This DNA molecule somehow gets packaged into a metaphase structure that is about 0.5 μm in diameter and about 10 μm in length. This represents a ***condensation of almost 10^4-fold in length from the naked DNA molecule to the metaphase chromosome.*** How

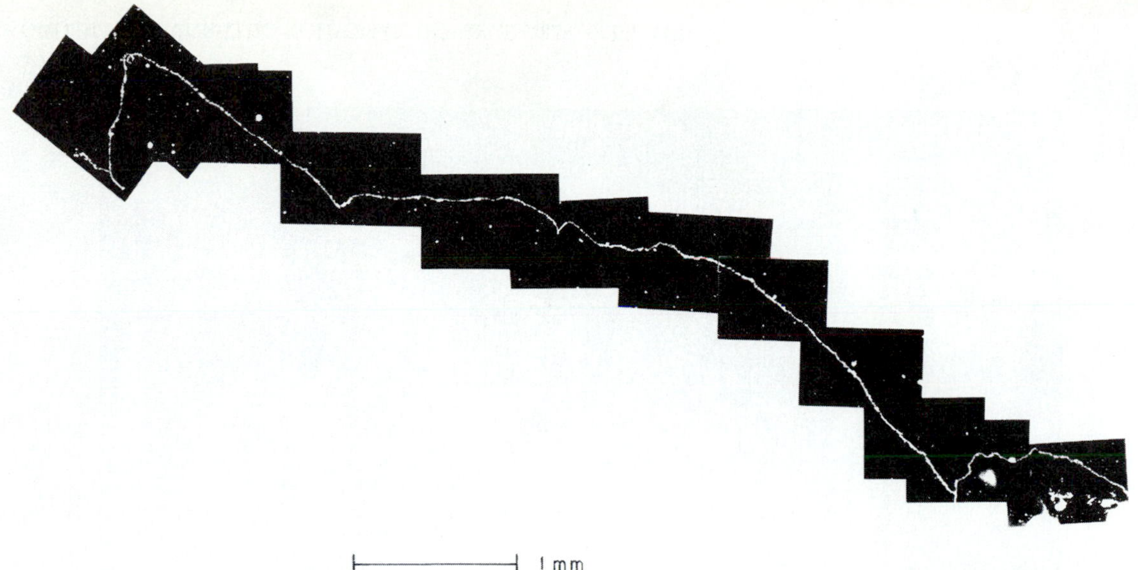

Figure 6.9 Autoradiograph of a giant DNA molecule from *Drosophila melanogaster.* Contour length = 1.2 cm. Such a molecule would have a molecular weight of about 2.8×10^{10}, or about two-thirds of the DNA content of the largest chromosome of *D. melanogaster.* (From R. Kavenoff, L. C. Klotz, and B. H. Zimm, *Cold Spring Harbor Sympos. Quant. Biol.* 38: 1–8, 1973. Copyright © 1973 by Cold Spring Harbor Laboratory. Original photograph courtesy of R. Kavenoff.)

does this condensation occur? What components of the chromosomes are involved in the packaging processes? Are DNA molecules packaged in different chromosomes in different ways or is there a universal packaging scheme? Are there different levels of packaging? Clearly, meiotic and mitotic chromosomes are more extensively condensed than interphase chromosomes. What additional levels of condensation occur in these special structures that are designed to assure the proper segregation of the genetic material during cell divisions? Are DNA sequences of genes that are being expressed packaged differently than those of genes that are not being expressed? In the following sections, some of the evidence that establishes the existence of three different levels of packaging of DNA into chromosomes is presented.

Nucleosome Structure

When isolated chromatin is examined by electron microscopy, it is found to consist of a series of ***ellipsoidal "beads" (about 110 Å in diameter and 60 Å high)*** joined by thin threads (Fig. 6.10). Further evidence for a regular, periodic packaging of DNA has come from studies on the digestion of chromatin with various nucleases. These studies indicated that segments of DNA of 146 nucleotide-pairs in length were somehow protected from degradation by certain nucleases. Moreover, partial digestion of chromatin with these nucleases yielded fragments of DNA in a set of discrete sizes that were integral multiples of the smallest size fragment (Fig. 6.11). These results are neatly explained if chromatin has a repeating structure, supposedly the "bead" seen by electron microscopy, within which the DNA is packaged in a nuclease-resistant form (Fig. 6.12). This "bead" or chromatin subunit is called the ***nucleosome.*** According to this model, the ***"interbead" threads of DNA*** or ***linkers*** are susceptible to nuclease attack.

After partial nuclease digestion of chromatin, an approximately 200-nucleotide-pair length of DNA is found associated with each nucleosome (produced by a cleavage in each linker region). After extensive nuclease digestion, a 146-nucleotide-pair-long segment of DNA remains present in each nucleosome. This nuclease-resistant structure is called the ***nucleosome core.*** Its structure is essentially invariant in all eukaryotes, ***consisting of a 146-nucleotide-pair length of DNA and two molecules each of histones H2a, H2b, H3, and H4.*** Physical studies (X-ray diffraction and similar analyses) of nucleosome-core crystals have shown that the DNA is wound as 1¾ turns of a superhelix around the outside of the histone octamer (Fig. 6.13a).

The complete chromatin subunit consists of the nucleosome core, the linker DNA, an average of one molecule of histone H1, and the associated nonhistone chromosomal proteins. However, note that it has not been firmly established that histone H1 is evenly distributed, one molecule per nucleosome or linker, in chromatin. The size of the linker DNA varies from species to species and from one cell type to another. Linkers as short as 8 nucleotide-pairs and as long as 114 nucleotide-pairs have been reported. Some evi-

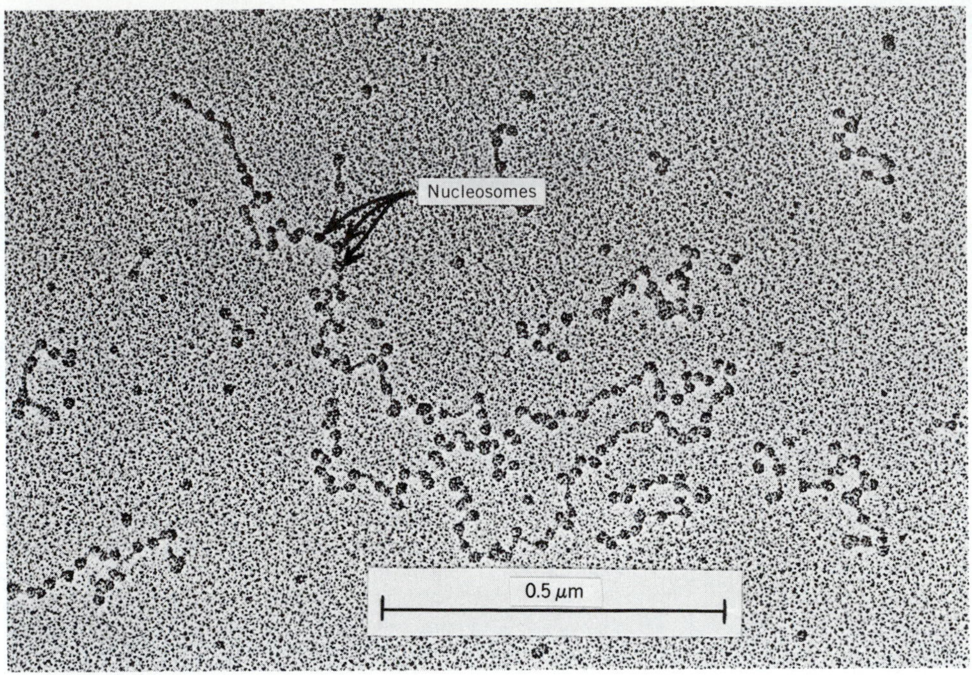

Figure 6.10 Electron micrograph of rat liver chromatin showing the "beads-on-a-string" nucleosome substructure. (From F. Thoma and T. Koller, *Cell* 12: 101–107, 1977. Copyright © 1977 Cell Press; published by Cell Press.)

dence suggests that the "complete nucleosome" (as opposed to the nucleosome core) contains one molecule of histone H1, which stabilizes two full turns of DNA superhelix (a 166-nucleotide-pair length of DNA) on the surface of the histone octamer (Fig. 6.13b). Other evidence indicates that histone H1 is involved in the coiling or folding of the nucleosome fiber to form a 300-Å chromatin fiber (see Fig. 6.16) and may be involved in other higher levels of organization of chromatin.

Clearly, the basic structural component of eukaryotic chromatin is the nucleosome. But are the structures of all nucleosomes the same? How does the replication fork move past the nucleosome during replication? What role(s), if any, does nucleosome structure play in gene expression and the regulation of gene expression? The structure of nucleosomes in genetically active regions of chromatin is known to differ from that of nucleosomes in genetically inactive regions. But what are the details of this structure–function relationship? Present and future studies on the fine structure of nucleosomes will undoubtedly prove very informative with regard to these and other questions.

The 300-Å Chromatin Fiber

Electron micrographs of isolated metaphase chromosomes show masses of tightly coiled or folded lumpy fibers (Fig. 6.14). These ***chromatin fibers have an average diameter of 300 Å.*** When the structures seen by light and electron microscopy during earlier stages of meiosis are compared, it becomes clear that the light microscope simply permits one to see those regions where these 300-Å fibers are tightly packed or condensed (Fig. 6.15).

What is the substructure of the 300-Å fiber seen in mitotic and meiotic chromosomes? Although we do not have a firm answer to this question, we do know that the DNA is wound as a supercoil about a histone octamer to yield the roughly 100 Å in diameter nucleosome. *In vivo,* the nucleosomes are probably in direct juxtaposition with each other without detectable linker regions; if so, they will form a 100-Å nucleosome fiber (see Fig. 6.12, top). If this 100-Å fiber, in turn, is wound in a ***higher-order supercoil (a "super-supercoil" or solenoid),*** a 300-Å fiber can easily be generated (Fig. 6.16). Although we still do not understand all the details of the structure of this 300-Å chromatin

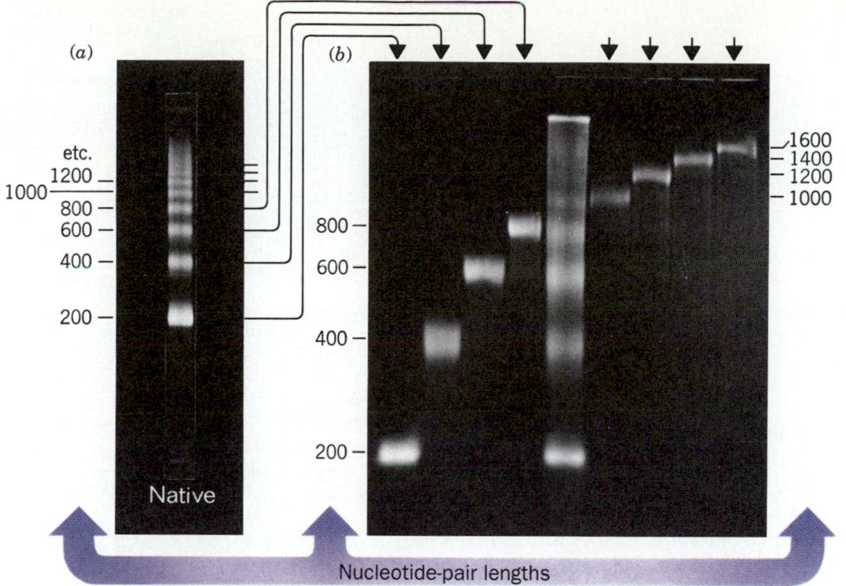

Figure 6.11 Demonstration of the repeating subunit (nucleosome) structure of rat liver chromatin by partial nuclease digestion. Rat liver chromatin was partially digested with micrococcal nuclease (an endonuclease). The DNA fragments produced were then extracted from the chromatin and separated by a technique called polyacrylamide gel electrophoresis. This procedure separates DNA molecules on the basis of size. Each DNA fragment preparation is placed on the top of a tube of polyacrylamide that is permeated with a buffer solution; then a high-voltage field is applied across the tube. The negatively charged DNA fragments move through the polyacrylamide gel toward the anode (the positive electrode). The polyacrylamide acts as a molecular sieve; thus, small fragments move faster than larger fragments with the same charge per unit of mass. (a) Total native (double-stranded) DNA fragments. (b) Each fragment band of the gel shown in (a) was cut out and electrophoresed on a separate gel, this time under denaturing conditions (conditions which cause the complementary strands of the double helices to separate). The central gel shown in (b) contained the same total chromatin, nuclease-digested extract as the gel shown in (a).

Note that the fragments fall into a set of discrete sizes where the larger fragment sizes are integral multiples of the size of the smallest fragments (i.e., 200, 400, 600, 800, etc., nucleotide-pairs). Consider how these discrete fragment sizes might result from endonuclease cuts of the DNA in the "exposed" linker or internucleosome regions of chromatin (Fig. 6.12). (Data from M. Noll, J. O. Thomas, and R. D. Kornberg, *Science* 187:1203–1206, 1975. Copyright © 1975 by the American Association for the Advancement of Science.)

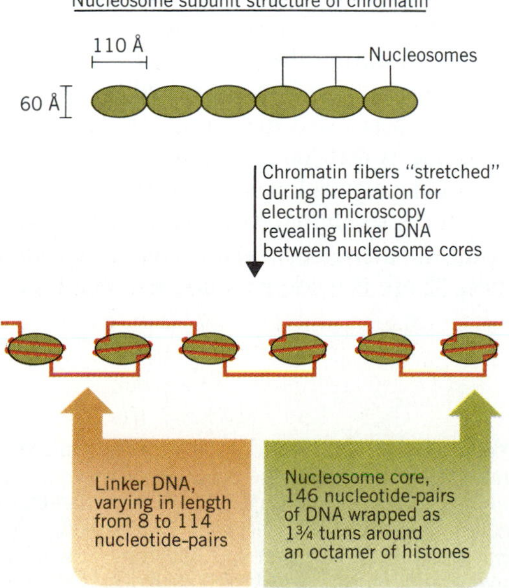

Figure 6.12 Nucleosome structure of chromatin. Chromatin of eukaryotes is composed of repeating subunits called nucleosomes. Each nucleosome core contains 146 nucleotide-pairs of DNA wrapped around an octamer (two molecules each of histones H2a, H2b, H3, and H4) in a relatively nuclease-resistant complex. These nucleosome cores are joined by nuclease-sensitive segments of DNA called linkers. The linkers are of variable size depending on species and cell type. One molecule of histone H1 is somehow associated with each linker or nucleosome. Thus, the complete repeating unit consists of about 200 nucleotide-pairs of DNA and nine histone molecules. *In vivo,* the linker DNA is probably condensed (coiled?) between closely juxtaposed nucleosome cores (top). The "beads-on-a-string" structure of chromatin (bottom) seen in electron micrographs (see Fig. 6.10) probably results when the fibers are stretched during preparation of the chromatin for electron microscopy. Histone H1 is probably not present in this latter structure.

Figure 6.13 (a) Structure of the nucleosome core and (b) the proposed role of histone H1 in stabilizing two complete turns of DNA superhelix around the octamer of histones. The nucleosome core consists of a 146-nucleotide-pair-long segment of DNA wound on the surface of a somewhat cylindrical octamer of histones (two each of H2a, H2b, H3, and H4) to yield a roughly ellipsoidal structure. The exact shapes of the histone octamer and histone H1 are not known. The path of the DNA molecule on the surface of the histone octamer has been firmly established. (After R.D. Kornberg and A. Klug, *Sci. Amer.* 244(2): 52–64, 1981.)

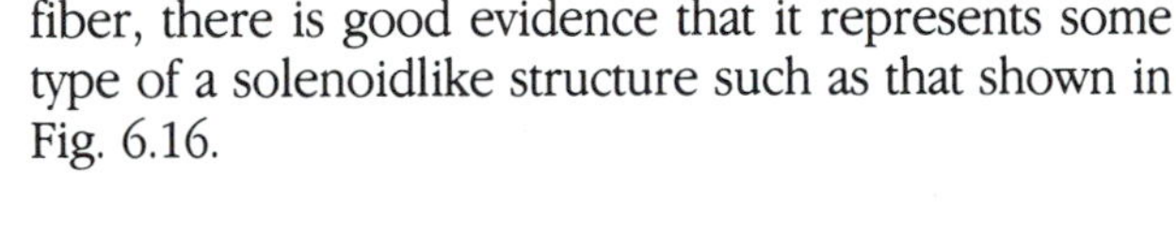

fiber, there is good evidence that it represents some type of a solenoidlike structure such as that shown in Fig. 6.16.

"Scaffolds" Composed of Nonhistone Chromosomal Proteins

Metaphase chromosomes contain the maximum degree of condensation observed in normal eukaryotic chromosomes. Clearly, the role of these highly condensed chromosomes is to organize and package the giant DNA molecules of eukaryotic chromosomes into structures that will facilitate their segregation to daughter nuclei without the DNA molecules of different chromosomes becoming entangled and, as a result, being broken during the anaphase separation of the daughter chromosomes. As was noted in the preceding section, the basic structural unit of the metaphase chromosome is the 300-Å chromatin fiber. However, the next obvious question is how are these 300-Å fibers further condensed into the observed metaphase structure? Unfortunately, there is still no clear answer to this question. There is evidence that the metaphase struc-

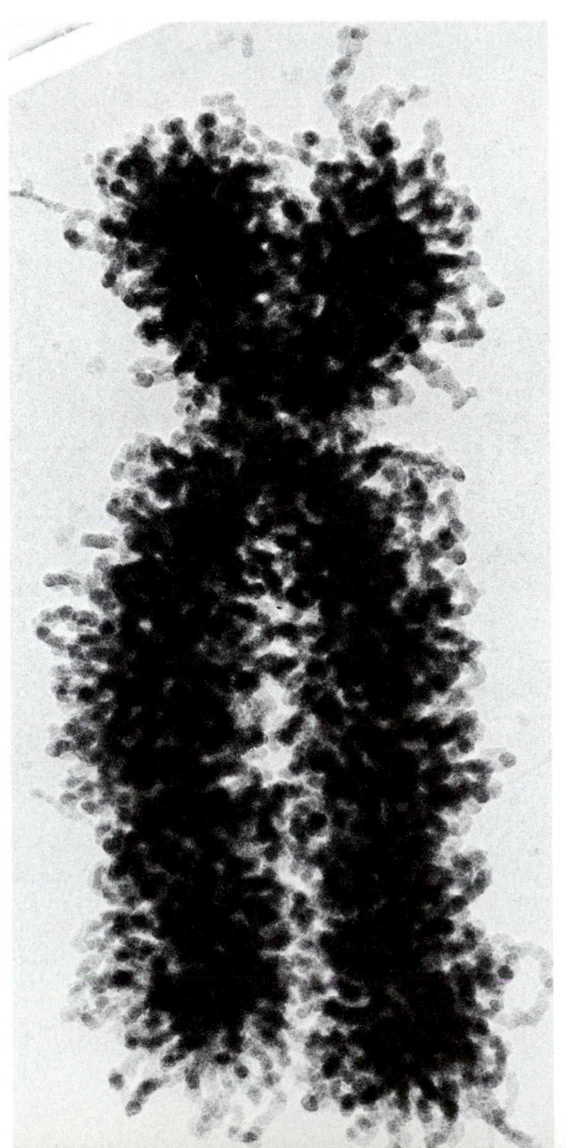

Figure 6.14 Electron micrograph of a human metaphase chromosome. Each of the two chromatids is composed of a densely packed mass of fibers. The average diameter of the individual fibers is 300 Å. (From E. DuPraw, *DNA and Chromosomes,* Holt, Rinehart and Winston, New York, 1970. Original photograph courtesy of E. DuPraw.)

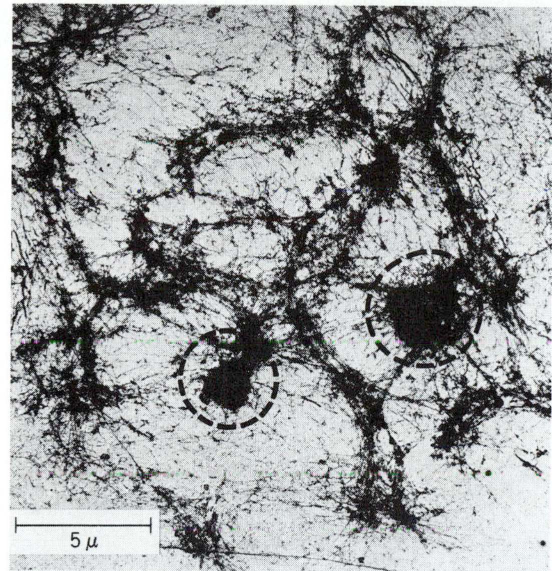

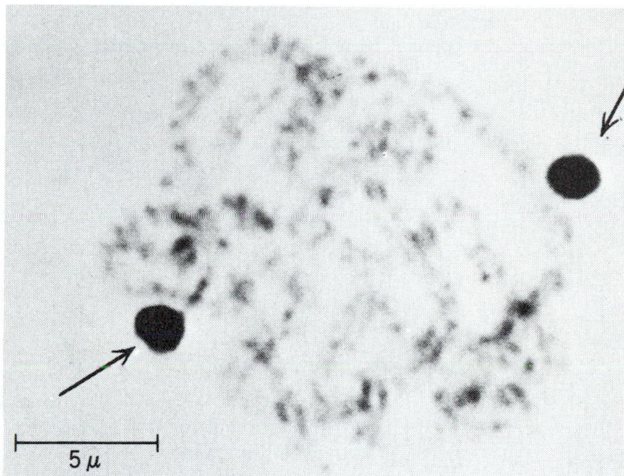

Figure 6.15 Comparison of the electron (left) and light (right) microscope images of meiotic prophase (pachytene) chromosomes of the male milkweed bug, *Oncopeltus fasciatus.* The heterochromatic sex univalents are circled on the electron micrograph and marked by arrows on the light micrograph. Note that the images seen with the light microscope correspond to regions where the 300 Å fibers are present at high density in the electron micrograph. The more dispersed fibers are not visible with the light microscope. (From S. L. Wolfe and B. John, *Chromosoma* 17: 85, 1965.)

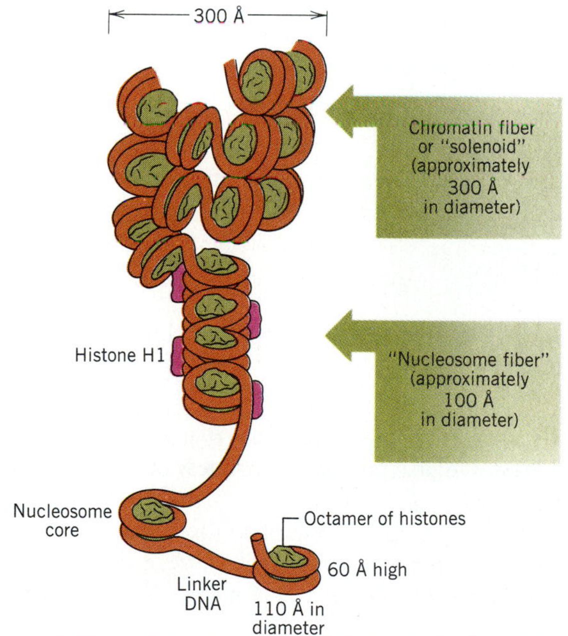

Figure 6.16 "Solenoid" model for the structure of the basic 300-Å fiber of meiotic and mitotic chromosomes of eukaryotes. Two levels of supercoiling are proposed; the first level of supercoiling is firmly established by experimental results. (1) The DNA double helix is wound in a negatively supercoiled form in the nucleosome (bottom, see Fig. 6.13). If the nucleosomes are positioned in a close juxtaposition to each other with the linker regions coiled between them, a 100 to 110-Å "nucleosome fiber" is produced (middle). Coiling of this 100-Å fiber can give rise to a 300-Å fiber or solenoid (top). Histone H1 is believed to be involved in the formation and stabilization of these fibers. (Based in part on the result of F. Thoma *et al., J. Cell Biol.* 83: 403–427, 1979.)

ture is not dependent on histones. Electron micrographs of isolated ***metaphase chromosomes from which the histones have been removed reveal a central core or "scaffold," which is surrounded by a huge pool or "halo" of DNA*** (Fig. 6.17). Note the absence of any apparent ends of DNA molecules in the micrograph, again supporting the concept of one giant DNA molecule per chromosome.

In summary, at least three levels of condensation are required to package the 10^3 to 10^5 μm of DNA in a eukaryotic chromosome into a metaphse structure a few microns long. (1) The first level of condensation involves packaging DNA as a supercoil into ***nucleosomes.*** This produces the ***100-Å-diameter interphase chromatin fiber.*** This clearly involves an ***octamer of histone molecules,*** two each of histones H2a, H2b, H3, and H4. (2) The second level of condensation involves an additional ***folding and/or supercoiling of the 100-Å nucleosome fiber to produce the 300-Å chromatin fiber*** characteristic of mitotic and meiotic chromo-

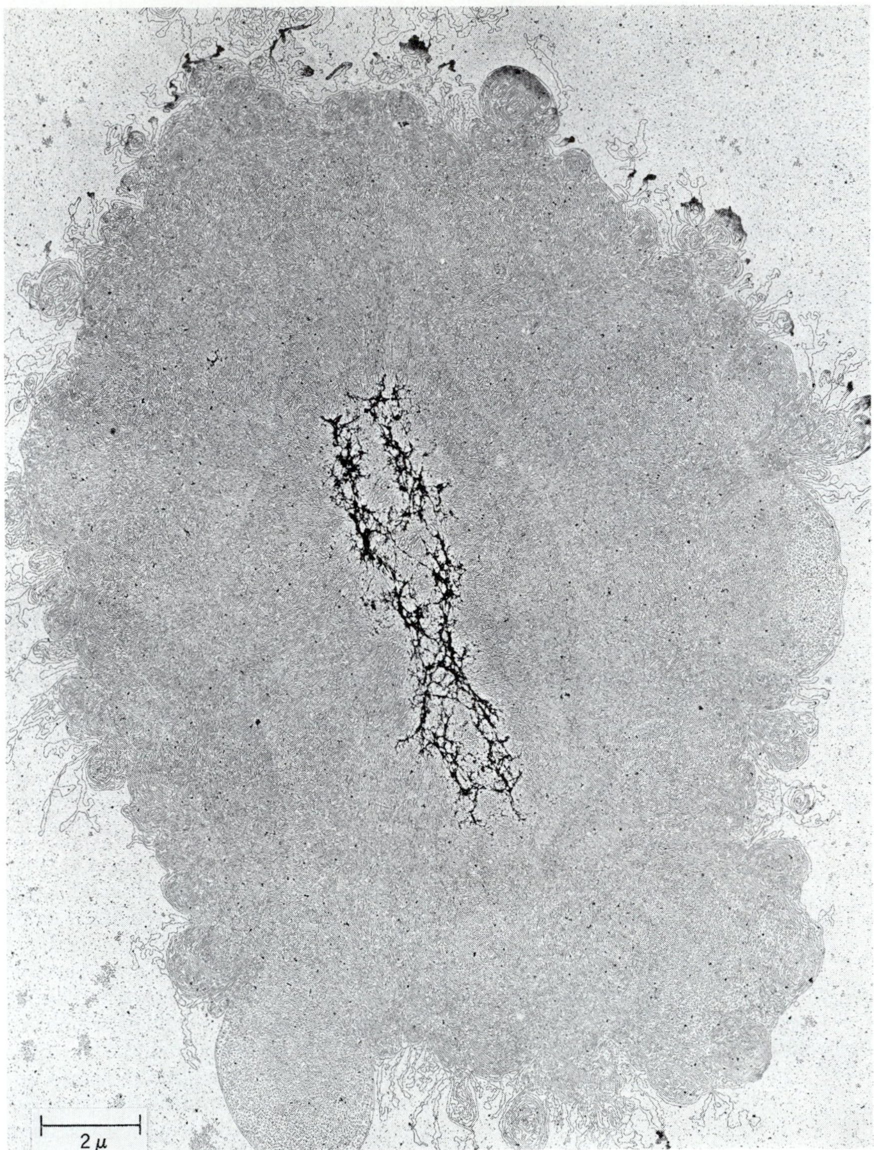

Figure 6.17 Electron micrograph of a human metaphase chromosome from which the histones have been removed. The entire histone-depleted chromosome has been spread on a monolayer of protein (cytochrome C), which is floating on an aqueous surface; the chromosome is then picked up on a grid for electron microscopy. The chromosome consists of a central, dense "scaffold" or core surrounded by a huge halo of DNA. Because of the relatively low magnification, individual DNA molecules can be seen only near the periphery. Note the absence or rarity of ends of DNA molecules in the chromosome. (From J. R. Paulson and U. K. Laemmli, *Cell* 12: 817–828, 1977. Copyright © 1977 Cell Press; published by Cell Press. Original photograph courtesy of U. K. Laemmli.)

somes. Histone H1 is involved in this supercoiling of the 100-Å nucleosome fiber to produce the 300-Å chromatin fiber. (3) Finally, ***nonhistone chromosomal proteins form a "scaffold"*** that is involved in condensing the 300-Å chromatin fiber into the tightly packed metaphase chromosomes. This third level of condensation appears to involve the segregation of segments of the giant DNA molecules present in eukaryotic chromosomes into independently supercoiled domains or loops (see Fig. 6.17). The mechanism by which this third level of condensation occurs is not known.

EUCHROMATIN AND HETEROCHROMATIN

When chromosomes are stained by various procedures, such as the Feulgen reaction, which is specific for DNA, and are examined by light microscopy, some regions of the chromosomes are observed to stain very darkly, whereas other regions stain only lightly. When examined by electron microscopy, the intensely staining chromatin, called ***heterochromatin,*** is seen to consist of densely packed chromatin fibers (300 Å in diameter). The lightly staining chromatin, called ***euchromatin,*** is composed of less tightly packed 300-Å fibers. Heterochromatin can often be shown to remain highly condensed throughout the cell cycle, whereas euchromatin is not visible with the light microscope during interphase.

Genetic analyses indicate that heterochromatin is largely genetically inactive. Most of the genes of eukaryotes that have been extensively characterized are located in euchromatic regions of the chromosomes. A structure–function correlation is thus evident: the highly condensed chromatin tends to be genetically inactive; the less condensed chromatin tends to be genetically active (see Chapters 10 and 15). Furthermore, heterochromatin is often enriched in highly repetitive, tandemly arranged DNA sequences (see the following sections).

REPETITIVE DNA AND SEQUENCE ORGANIZATION

The chromosomes of prokaryotes contain DNA molecules with unique (nonrepeated) base-pair sequences. That is, each gene (consider a gene to be a linear sequence of a few thousand base-pairs; see Chapter 12) is present only once in the genome. (The genes for rRNA molecules are an exception.) If prokaryotic chromosomes are broken into many short fragments, each fragment will contain a different sequence of base-pairs. The chromosomes of eukaryotes are much more complex in this respect. Certain base sequences are repeated many times in the haploid chromosome complement, sometimes as many as a million times. DNA containing such repeated sequences, called ***repetitive DNA,*** often represents a major component (20–50 percent) of the eukaryotic genome.

Satellite DNAs

The first evidence for repetitive DNA came from density-gradient analysis (see Fig. 5.14) of eukaryotic DNA. When the DNA of a prokaryote, such as *E. coli,* is isolated, fragmented, and centrifuged to equilibrium in a cesium chloride (CsCl) density gradient, the DNA usually forms a single band in the gradient. For *E. coli,* this band will form at a position where the CsCl density is equal to the density of DNA containing about 50 percent A-T and 50 percent G-C base-pairs. DNA density increases with increasing G-C content. The extra hydrogen bond in a G-C base-pair is believed to result in a tighter association between the bases and thus a higher density than for A-T base-pairs. On the other hand, CsCl-density-gradient analysis of DNA from eukaryotes usually reveals the presence of one large band of DNA (usually called ***"mainband" DNA***) and one to several small bands. These small bands of DNA are called ***satellite bands,*** and the DNA in these bands is often referred to as ***satellite DNA.*** *Drosophila virilis* DNA, for example, contains three distinct satellite DNAs (Fig. 6.18). When isolated and analyzed, each satellite DNA was found to contain a repeating sequence of seven base-pairs. One satellite repeat se-

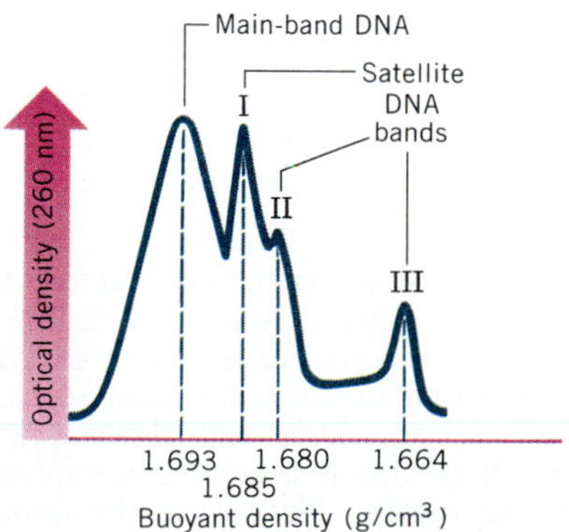

Figure 6.18 The three satellite DNAs of *Drosophila virilis.* DNA from *D. virilis* embryos was centrifuged to equilibrium in 6 *M* CsCl. The DNA in the bands were quantitated by measuring the absorption of ultraviolet light (260 nm). (Courtesy of M. Blumenfeld, Department of Genetics and Cell Biology, University of Minnesota.)

quence is 5′-ACAAACT-3′ (one strand; the other strand will have the complementary sequence). A second satellite DNA has a 5′-ATAAACT-3′ repeat; the third satellite repeat sequence is 5′-ACAAATT-3′. Thus, they differ from each other at only two positions. In three releated species of crabs, a satellite DNA is present that contains 97 percent A-T base-pairs. Some satellite DNAs in other eukaryotes have longer repetitive sequences.

The chromosomal locations of several satellite DNAs have been determined by a technique called **in situ *hybridization.*** The complementary strands of DNA molecules are separated by heat or alkaline denaturation. Conditions can then be reversed by lowering the temperature or lowering the pH, and the separated strands will ***renature*** or ***reanneal*** to reform base-paired double helices, a process called ***DNA renaturation.*** DNA renaturation is a specific type of ***nucleic acid hybridization,*** the formation of hydrogen-bonded double helices by single-stranded molecules containing complementary base sequences. Hybridization will occur between complementary single strands regardless of their source. If both participating strands are DNA, the process is called ***DNA hybridization.*** If one strand is DNA and the complementary strand is RNA, the process is called ***DNA–RNA hybridization.***

In situ hybridization in the case of the satellite DNAs usually involves annealing single strands of isolated radioactive satellite DNA, or complementary RNA sequences synthesized using satellite DNA as template, directly to denatured DNA in chromosome squash preparations. After washing out the nonhybridized radioactive material, the locations of the satellite DNA sequences in chromosomes are determined by autoradiography (Fig. 6.19). All satellite DNAs characterized to date are located in heterochromatic regions of chromosomes. In several cases, the satellite DNA sequences are found in heterochromatin in the centromere regions of chromosomes. In other cases, the satellite sequences are located in heterochromatic telomeres (chromosome termini). Satellite DNA sequences are usually found ***not*** to be transcribed (first step in gene expression; see Chapter 10).

It should be emphasized that a repetitive DNA sequence will be identified as satellite DNA only if the sequence has a base composition sufficiently different from that of main-band DNA; only then is it recognizable as a distinct band in a density gradient. Therefore, many repetitive DNA sequences cannot be identified by this procedure.

DNA Renaturation Kinetics

A much more complete picture of the frequency and complexity of repetitive DNA sequences in eukaryotes has resulted from studies of ***DNA renaturation kinetics.*** Consider a long DNA molecule with no repeated sequences (such as the DNA of a prokaryote). If such a molecule is sheared into fragments of a particular length, say 400 nucleotide-pairs, is denatured, and is then allowed to renature under appropriate conditions, the ***rate of renaturation will depend on (1) the concentration of DNA in solution and (2) the complexity of the DNA,*** that is, the number of different 400 base-pair fragments. Consider a particular fragment composed of two complementary strands, *a* and *a*′.

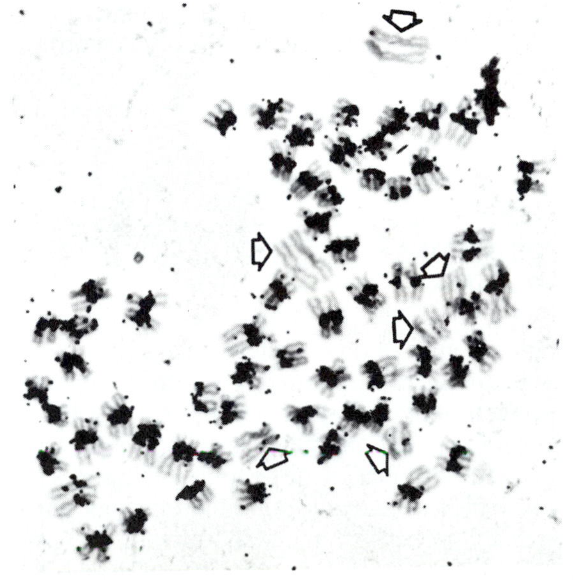

Figure 6.19 Autoradiograph showing the hybridization of ^{3}H-labeled RNA transcripts of the DNA satellite HS-β of the kangaroo rat (*Dipodomys ordii*) to metaphase chromosomes. These chromosomes have undergone endoreduplication, a chromosome duplication without separation of the daughter chromosomes. Thus, each chromosome is represented by a pair of closely associated metaphase chromosomes (four chromatids). Most of the chromosomes show radioactivity localized at or near the centromere. Six endoreduplicated chromosomes (three endoreduplicated pairs of homologs) are unlabeled (arrows). Thus, the HS-β satellite DNA sequence is apparently present in the centromeric regions of all but three of the chromosomes of the kangaroo rat. Magnification × 37,400. (Courtesy of D. M. Prescott and *Chromosoma* 42: 205, 1973.)

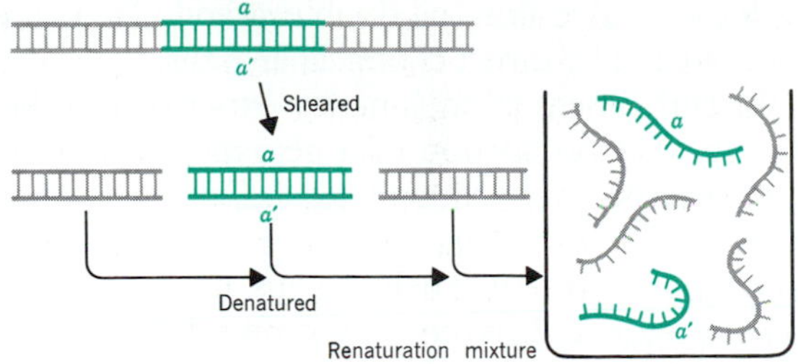

Reassociation of a and a' will require a specific collision between these two single strands. Collisions between a or a' and any other single strand will not lead to hybridization. For a given concentration, the larger the DNA molecule, and thus the more nonidentical 400 base-pair fragments, the slower the reassociation reaction will be, because a smaller proportion of the random collisions will be between complementary single strands such as a and a' (Fig. 6.20).

Note that every reassociation event, like a with a', will require a collision between *two* single strands that are ***present in the renaturation mixture in equal concentration.*** Since the reaction requires the interaction of *two* equally frequent molecules, the rate of renaturation should be a function of the square of the concentration of single strands (so-called ***second-order*** or ***bimolecular reaction kinetics***), or

$$\frac{-dC}{dt} = kC^2$$

$$\left(\text{or } \frac{-dC}{C^2} = k\,dt, \text{ rearranged for integration}\right)$$

where C = the concentration of single-stranded DNA in moles of nucleotide per liter
t = time in seconds
and k = a second-order rate constant in liter per mole seconds

Literally, this equation states that the change (decrease) in concentration of single-stranded DNA ($-dC$)

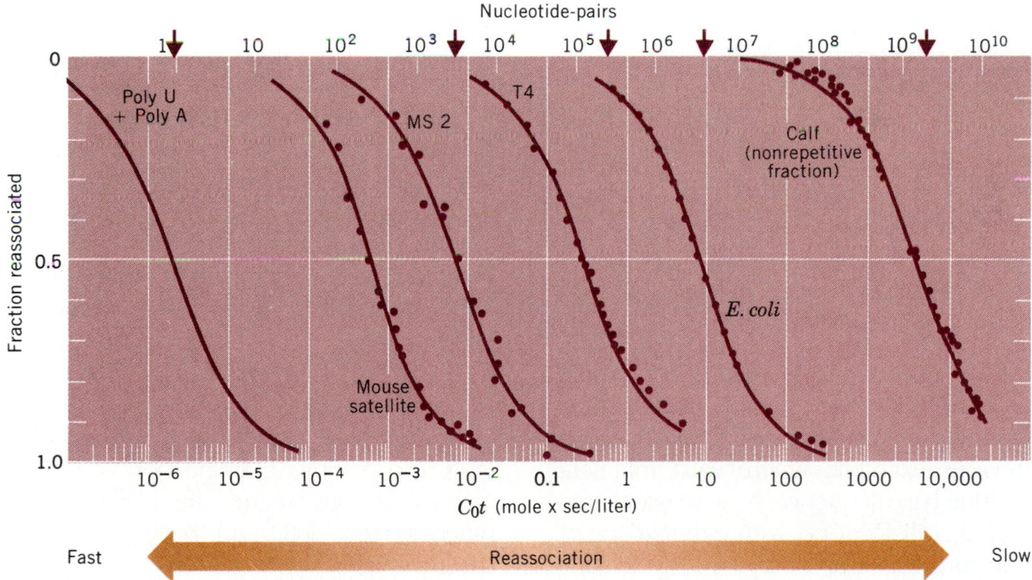

Figure 6.20 Renaturation kinetics as a function of genome size (complexity in this case is equivalent to total genome size). The complexity or size of the DNA in terms of nonrepeating nucleotide-pair sequences, which are plotted along the top, should be read from the midpoints (50 percent reassociation) of the reassociation curves. The fraction reassociated (C/C_0) is plotted as a function of C_0t [the product of the initial concentration (C_0) in moles per liter and time (t) in seconds]. "Poly U + poly A" is a double-helical structure in which one strand contains only U and the other strand contains only A. All base-pairs are thus A : U, and every fragment has the same base sequence. The complexity of such a duplex is one (or the repeating sequence is equivalent to one A : U base-pair). "Poly U + poly A" thus exhibits the fastest bimolecular reassociation possible. The very complex nonrepetitive DNA from calf chromosomes, by contrast, shows very slow reassociation kinetics. DNAs of intermediate complexities show intermediate rates of reassociation. A plot of genome size (excluding repetitive sequences) versus renaturation rate is, in fact, linear. (From R. J. Britten and D. E. Kohne, *Science* 161: 529–540, 1968. Copyright © 1968 by the American Association for the Advancement of Science.)

with time (dt) is equal to the proportionality constant (k) times the square of the concentration of single-stranded DNA.

Integration of the preceding equation from the initial conditions ($t = 0$ seconds and $C = C_0$, where C_0 equals the concentration of single-stranded DNA at $t = 0$) yields

$$\frac{1}{C} = \frac{1}{C_0} = kt$$

or, rearranged,

$$\frac{C}{C_0} = \frac{1}{1 + kC_0t}$$

This equation states that the fraction of input single-stranded DNA remaining in a renaturation reaction mixture (C/C_0) at any given time (t) is a function of the initial concentration (C_0) times elapsed time (t), or C_0t. It is thus convenient to present data on hybridization kinetics in a plot of C/C_0 versus C_0t. These so-called **C_0t** (pronounced "caught") ***curves*** (see Fig. 6.22) have provided a great deal of information about the types of repetitive DNA in eukaryotic genomes.

Consider a DNA molecule containing a 400 base-pair sequence that is repeated (present twice).

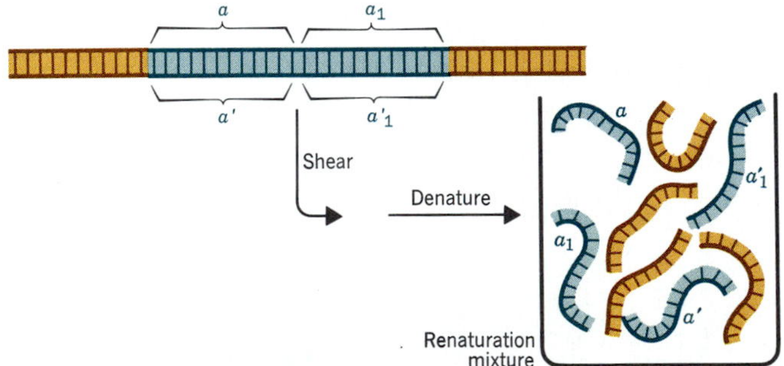

Now a and a_1 are identical single strands, and a' and a'_1 are identical complementary strands. Reassociation of each repeated sequence will take only ***half*** as long as in the previous example, where each fragment contained a unique base sequence. Now reassociation will result from a collision of a with either a' or a'_1 (likewise for a_1 and either a' or a'_1). Thus, the time required for reassociation of a particular DNA sequence is inversely proportional to the number of times that sequence is present in the genome. Clearly, highly repetitive DNA sequences will renature very rapidly.

The proportion of the DNA that has renatured at any time can be quantitated in several ways. One method is simply to treat samples taken at various times with a nuclease that is specific for single-stranded nucleic acids. The DNA remaining in the samples after digestion (all double-stranded) can then be quantitated by direct chemical analysis.

Careful renaturation kinetics experiments show not only what proportion of a genome consists of a particular class of repetitive DNA, but also how many copies of that particular sequence (or class of sequences) are present. Such studies have now been done on many eukaryotes. They reveal a surprising amount of diversity, both in the proportion of the genome consisting of repetitive DNA and in the types of repetitive DNA present (Fig. 6.21). These studies of renaturation kinetics indicate that eukaryotic genomes characteristically contain (1) a fraction (up to 90 percent) of ***unique*** or ***single copy*** DNA sequences present in from one to a few copies, (2) a fraction of ***middle-repetitive*** DNA sequences present in ***10 to 10^5 copies,*** and (3) a fraction of ***highly repetitive*** DNA sequences present in ***greater than 10^5 copies*** (Fig. 6.22). The middle-repetitive class appears to be quite heterogeneous in some eukaryotes; it contains many different sequences with different degrees of reiteration. In *Drosophila melanogaster,* for example, 12 percent of the DNA contains middle-repetitive DNA with an average reiteration frequency of 70. The highly repetitive DNA contains both satellite and nonsatellite DNA sequences.

Interspersion of Single-Copy and Middle-Repetitive DNA Sequences

When the sequence organization of eukaryotic genomes is studied—by combining the techniques of density-gradient analysis, hybridization kinetic analysis, biochemical analyses, and electron microscopy—a pattern emerges. Much of the genome consists of ***middle-repetitive sequences interspersed with single-copy sequences*** (Fig. 6.23). In toads, sea urchins, and humans, the sequences are quite short. The middle-repetitive sequences average 300 nucleotide-pairs in length; the single-copy sequences are about 800–1200 nucleotide-pairs long. *Drosophila melanogaster* DNA also exhibits interspersion of middle-repetitive and single-copy sequences, but the sequences are much longer (5600 and 13,000 nucleotide-pairs, respectively).

What are the functions of the different kinds of DNA sequences in eukaryotes? Most of the structural genes (genes coding for proteins or RNA molecules; see Chapter 10) are single-copy sequences. The genes for histones, rRNA molecules, and ribosomal proteins, all gene-products that are present in the cell in large quantities, are exceptions; these genes are ***redundant (present in middle-repetitive DNA).*** Many geneticists have postulated that most other middle-repetitive sequences may be involved in the regulation of gene

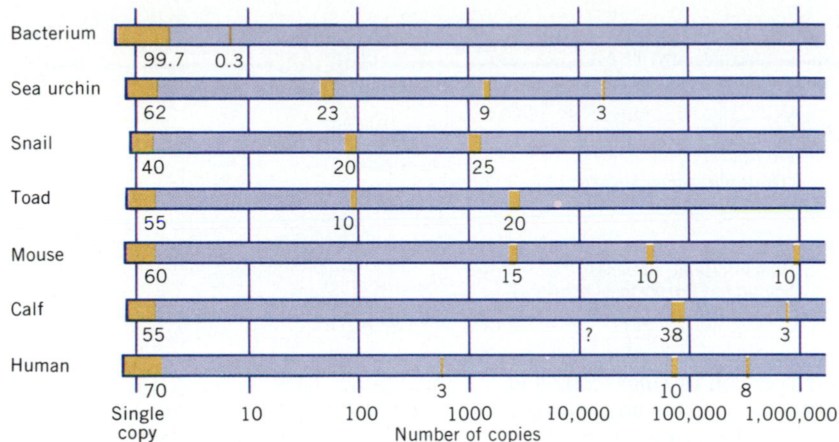

Figure 6.21 Patterns of repetitive DNA sequences in representative organisms. The fractions of the total DNA of the organism that are present in single-copy sequences and in each known repetition class (represented by brown bands) are indicated by the widths of the bands and the numbers below the bands. In certain cases, some of the DNA sequences in the genome are not well characterized; thus the fractions represented do not total 100 percent. The position of the band indicates the number of times that the specific sequence is present in the genome. For example, 25 percent of the DNA of the snail consists of a sequence present in about 1000 copies. (From R. J. Britten and D. E. Kohne, "Repeated Segments of DNA," *Sci. Amer.* 222(4): 24–31, 1970. Copyright © 1970 by Scientific American, Inc. All rights reserved.)

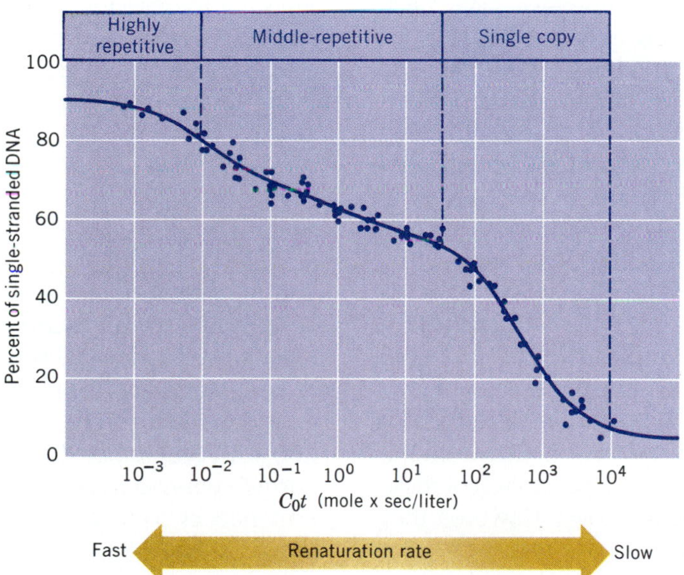

Figure 6.22 The complex renaturation kinetics of human DNA illustrating the presence of highly repetitive sequences and middle-repetitive sequences as well as unique or single-copy sequences. (Data from C. W. Schmid and P. O. Deininger, *Cell* 6: 345–358, 1975. Copyright © 1975 Cell Press; published by Cell Press.)

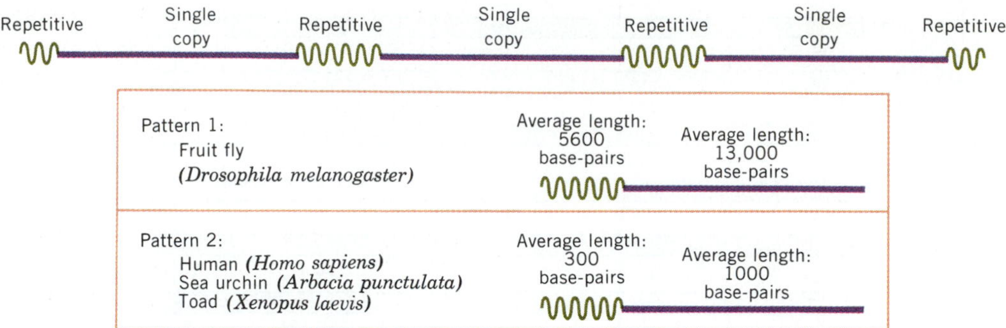

Figure 6.23 Patterns of interspersion of single-copy and middle-repetitive DNA sequences that make up the major components of eukaryotic genomes. The significance of such interspersion patterns is not yet clear. However, most structural genes (genes coding for proteins) are present in a single copy in the genome. Many geneticists believe that middle-repetitive sequences are involved in the regulation of gene expression.

expression (see Chapter 15). Their interspersion with single copy-sequences, and thus their location adjacent to structural genes, is certainly consistent with a regulatory role. Recently, several intermediate-repetitive sequences of *Drosophila melanogaster* have been shown to be capable of moving from one location in a chromosome to another location in the same chromosome or even to a different chromosome. These intermediate-repetitive DNA sequences have been called ***"nomadic" sequences*** ***transposable elements,*** because of their ability to migrate from one position to another in the genome. These "nomadic" middle-repetitive DNA sequences are clearly analogous to the well-characterized transposable elements of prokaryotes (see Chapter 9, pp. 232–236). The ***transposability*** of these intermediate repetitive sequences has led to much speculation of their possible involvement in the regulation of gene expression during development and their almost certain involvement in the evolution of eukaryotic genomes. The actual function(s) of these "nomadic" sequences is still unknown. However, they are known to be responsible for a surprisingly large number of the mutations of yeast and *Drosophila* (see Chapter 9).

Function of Highly Repetitive DNA Sequences?

The function(s) of highly repetitive DNA—most, if not all, of which is located in genetically inactive heterochromatic regions of chromosomes—is also completely unknown. Postulated functions for highly repetitive DNA include (1) structural or organizational roles in chromosomes; (2) involvement in chromosome pairing during meiosis; (3) involvement in crossing-over or recombination (see Chapter 7); (4) "protection" of important structural genes, like histone, rRNA, or ribosomal protein genes; (5) a repository of unessential DNA sequences for use in the future evolution of the species; and (6) no function at all—just "junk" DNA that is carried along by the processes of replication and segregation of chromosomes. The validity of any of these postulated roles remains to be established.

Telomere Structure

The ***telomeres*** or ends of eukaryotic chromosomes have been known to have unique properties for several decades. In a neoclassical study of maize chromosomes, Barbara McClintock demonstrated that new ends of broken chromosomes are "sticky" and tend to fuse with each other. In contrast, the natural ends of normal (unbroken) chromosomes are stable and show no tendency to fuse with other ends (native or broken). McClintock's results indicated that telomeres must have special structures different from the ends produced by breakage of chromosomes.

Another reason for postulating that telomeres have unique structures is that the known DNA polymerases should ***not*** be able to replicate the terminal DNA segment of the lagging strand (the strand being synthesized discontinuously; see Chapter 5) of a linear chromosome. At the end of the DNA molecule being replicated discontinuously, there would be no DNA strand to provide a free 3′-OH (primer) for polymerization of deoxyribonucleotides as the RNA primer of the terminal "Okazaki fragment" is excised, that is,

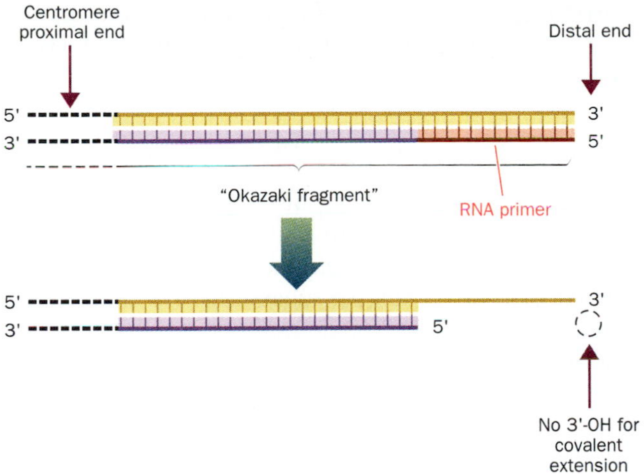

Clearly, the telomere must have a unique structure that facilitates its replication, or there must be some special replication enzyme that resolves this enigma of replicating the terminus of the lagging strand.

Until recently, little was known about the molecular structure of telomeres. However, during the last few years, telomeres have been isolated and characterized from several species. In all cases, the telomeres have been shown to have unique structures that include short nucleotide sequences present as tandemly repeated units. Although these sequences are somewhat variable in different species, the ***basic repeat unit*** in all species studied to date has the pattern $5'\text{-}T_{1-4}A_{0-1}G_{1-8}\text{-}3'$. For example, the repeat sequence in humans is ***TTAGGG,*** that of *Tetrahymena thermophila* is ***TTGGGG,*** and that of the plant *Arabidopsis thaliana* is ***TTTAGGG.*** The number of copies of this basic repeat unit in telomeres varies from species to species, from chromosome to chromosome within a species, and even on the same chromosome at different stages of the life cycle. The significance of variation in telomere length and the factors that control this variation are still unknown.

At least in some species, the telomeres terminate with a single-stranded region of the DNA strand with the 3′ end (a so-called 3′ overhang). Terminal bases of this single-stranded end exhibit unique patterns of methylation (covalently attached methyl groups) that probably contribute to the formation of a unique "hairpin" or "folded" structure at the very tip of the telomeric DNA. Additional repetitive DNA sequences are present adjacent to the telomere; these are referred to as telomere-associated sequences. Although the actual three-dimensional structures of telomeres and the mode of replication of the telomere sequences are still uncertain, it is clear that the telomeres do possess special structures that give them their unique properties. In addition, it is known that telomere sequences can be added to chromosomes by a special enzyme called a ***telomere transferase*** or a ***telomerase.***

These unique telomere structures of eukaryotic chromosomes must provide at least three important functions. They must (1) prevent exonucleases from degrading the ends of the linear DNA molecules, (2) prevent fusion of the ends with other DNA molecules, and (3) facilitate replication of the ends of the linear DNA molecules without loss of the termini. If telomeres did not provide the third function, chromosomes would become progressively shorter during each round of replication.

REPLICATION OF EUKARYOTIC CHROMOSOMES

The duplication of the chromosomes of eukaryotes involves not only the replication of their giant DNA molecules, but also the synthesis of the associated histones and nonhistone chromosomal proteins. The packaging of DNA and histones into nucleosomes supposedly is the final stage in chromosome duplication. Thus, some features of the replication of eukaryotic chromosomes are clearly unique to eukaryotes. However, at the molecular level, the replication of DNA in eukaryotes appears to involve the same enzymes and mechanisms as in prokaryotes. The eukaryotic DNA polymerases have the same absolute requirements for template and primer as prokaryote polymerases. DNA replication is semiconservative and is discontinuous, at least for the strand synthesized in the overall 3′⟶ 5′ direction. However, the "Okazaki fragments" produced by discontinuous synthesis are only 100–200 nucleotides long in eukaryotes, whereas they are 1000 to 2000 nucleotides long in prokaryotes. This may be related to the fact that DNA synthesis is slower (about 1 μm of DNA per minute) in eukaryotes than in prokaryotes (about 30 μm per minute). The slower rate of replication, in turn, may result from the need to disassemble or conformationally rearrange the nucleosomes ahead of the advancing replication forks.

The Cell Cycle

DNA synthesis occurs continuously in most prokaryotic organisms growing under optimal conditions, that is, DNA synthesis occurs from the time a new cell is formed by cell fission until the time that cell divides again. Eukaryotic organisms, on the other hand, exhibit a more complex cell cycle. The important chromosome condensation and segregation events of mitosis (**M** phase; see Fig. 3.3) occur within a short period (often about one hour) of the total cell cycle. Most DNA synthesis occurs during interphase over a period (designated **S** for synthesis) of several hours. In many eukaryotic cells, there is an interval called G_1 (for "first gap") after the completion of mitosis and prior to the initiation of DNA synthesis. Another interval, called G_2, follows the completion of DNA synthesis and precedes mitosis. Interphase can thus be divided into three stages, G_1, S, and G_2. These are periods of intense metabolic activity and growth of cells; most cellular metabolism stops during mitosis. The sequence $G_1 \rightarrow S \rightarrow G_2 \rightarrow M \rightarrow G_1$ is known as the ***cell cycle.*** The durations of the various stages, particularly G_1 and G_2, are highly variable, depending on the cell type and species involved.

Semiconservative Replication of Eukaryotic Chromosomes

Eukaryotic chromosomes, like their simpler prokaryotic counterparts, replicate in a semiconservative manner. This was first demonstrated in 1957 by J. H. Taylor, P. Woods, and W. Hughes for root-tip cells of the broad bean, *Vicia faba.* Taylor and his colleagues labeled *V. faba* chromosomes by growing root tips for 8 hours (less than one cell cycle) in medium containing radioactive [^{3}H] thymidine. The root tips were then removed from the radioactive medium, washed, and transferred to nonradioactive medium containing ***colchicine.*** Colchicine binds to microtubules and prevents the formation of functional spindle fibers. As a result, daughter chromosomes do not undergo their normal anaphase separation. Thus, the number of chromosomes per nucleus will double once per cell cycle in the presence of colchicine. This allowed Taylor and his colleagues to determine how many DNA duplications each cell had undergone ***subsequent*** to the incorporation of radioactive thymidine. At the first metaphase in colchicine (***c-metaphase***), nuclei will contain 12 pairs of chromatids (still joined at the centromeres). At the second c-metaphase, nuclei will contain 24 pairs, and so on. The distributions of radioactive DNA at the first and second c-metaphases were determined by autoradiography (Fig. 6.24*a* and *b*, respectively). Both chromatids of each pair were similarly labeled at the first c-metaphase. However, at the second c-metaphase, only one of the chromatids of each pair was radioactive. These results are precisely what is predicted by the semiconservative replication of DNA, given one DNA molecule per chromosome (Fig. 6.24*c*). However, at the time, Taylor and his colleagues could only conclude that the chromosomal DNA segregated in a semiconservative manner during each cell cycle because it had not yet been established that each chromosome contained a single molecule of DNA. Analogous experiments have subsequently been carried out with several other eukaryotes, and, in all cases, the results indicate that replication is semiconservative.

Multiple "Replicons" per Chromosome

When eukaryotic cells are pulse-labeled with [^{3}H]thymidine and the DNAs are extracted from the chromosomes and autoradiographed, tandem arrays of radioactivity are observed, suggesting that single macromolecules of DNA contain multiple origins of replication (Fig. 6.25*a*.) Moreover, when the pulse-labeling period is followed by a short period of growth in nonradioactive medium ("pulse-chase" experiments), the tandem arrays contain central regions of high grain density with "tails" of decreasing grain density ***at each end*** (Fig. 6.25*b*). This result indicates that replication in eukaryotes is bidirectional just as it is in most prokaryotes. The tails of decreasing grain density result from the gradual dilution of the intracellular pools of [^{3}H]thymidine by cold [^{1}H]thymidine as the replication forks move bidirectionally from central origins toward replication termini (Fig. 6.25*c*). A segment of the chromosome whose replication is under the control of an origin and two termini has been called a ***replicon.*** In prokaryotes, the entire chromosome is usually one replicon.

The existence of many units of replication (or replicons) per eukaryotic chromosome, each with its own origin, has been verified by electron microscopy (Fig. 6.26). Some evidence indicates that replication may be initiated at more sites during the very rapid cell divisions of embryogenesis than it is during later stages of development. If so, it would be very interesting to know what determines which origins are operational at a given time or in a particular type of cell! One point that is clear is that the newly replicated DNA of eukaryotes is packaged into nucleosomes very rapidly. Electron microscopy of replicating chromatin reveals the presence of nucleosomes with apparently normal spacing on both sides of the replication forks (Fig. 6.27).

The multiple origins of replication in eukaryotic chromosomes are clearly required to allow the very large DNA molecules in these chromosomes to complete replication within the observed cell division times. In *Drosophila melanogaster,* for example, the giant DNA molecules of the largest chromosomes contain about 6.5×10^7 nucleotide-pairs (see pp. 134–136). The rate of DNA replication in *Drosophila* is about 2600 nucleotide-pairs per minute at 25°C. A single replication fork would therefore take about 17½ days to traverse one such giant DNA molecule. With two replication forks moving bidirectionally from a central origin, such a DNA molecule could be repli-

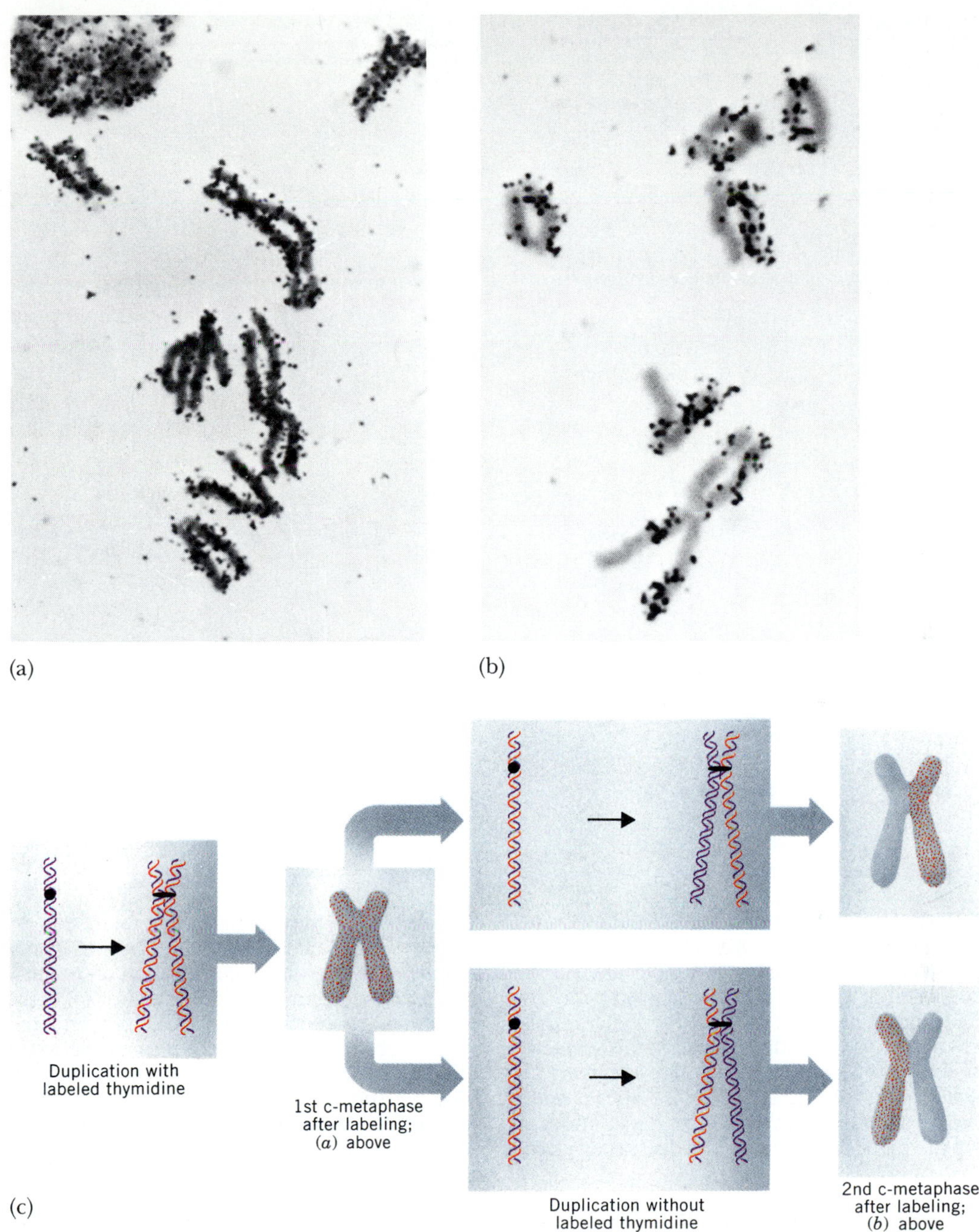

Figure 6.24 (Top) Autoradiographs of *Vicia faba* chromosomes at the first metaphase after replication in the presence of ^{3}H-thymidine (a) and at the second metaphase after an additional replication in nonradioactive medium (b). (c) Diagrammatic interpretation of the results shown in a and b. Radioactive single strands of DNA are shown in orange. Radioactive chromatids at metaphase are indicated by orange dots. Colchicine has been used to inhibit spindle fiber formation and thus the anaphase separation of sister chromatids. Under these "c-metaphase" conditions, separation of sister centromeres is delayed. (From J. H. Taylor, "The Replication and Organization of DNA in Chromosomes," *Molecular Genetics,* Part I, J. H. Taylor (ed.), Academic Press, New York, 1963.)

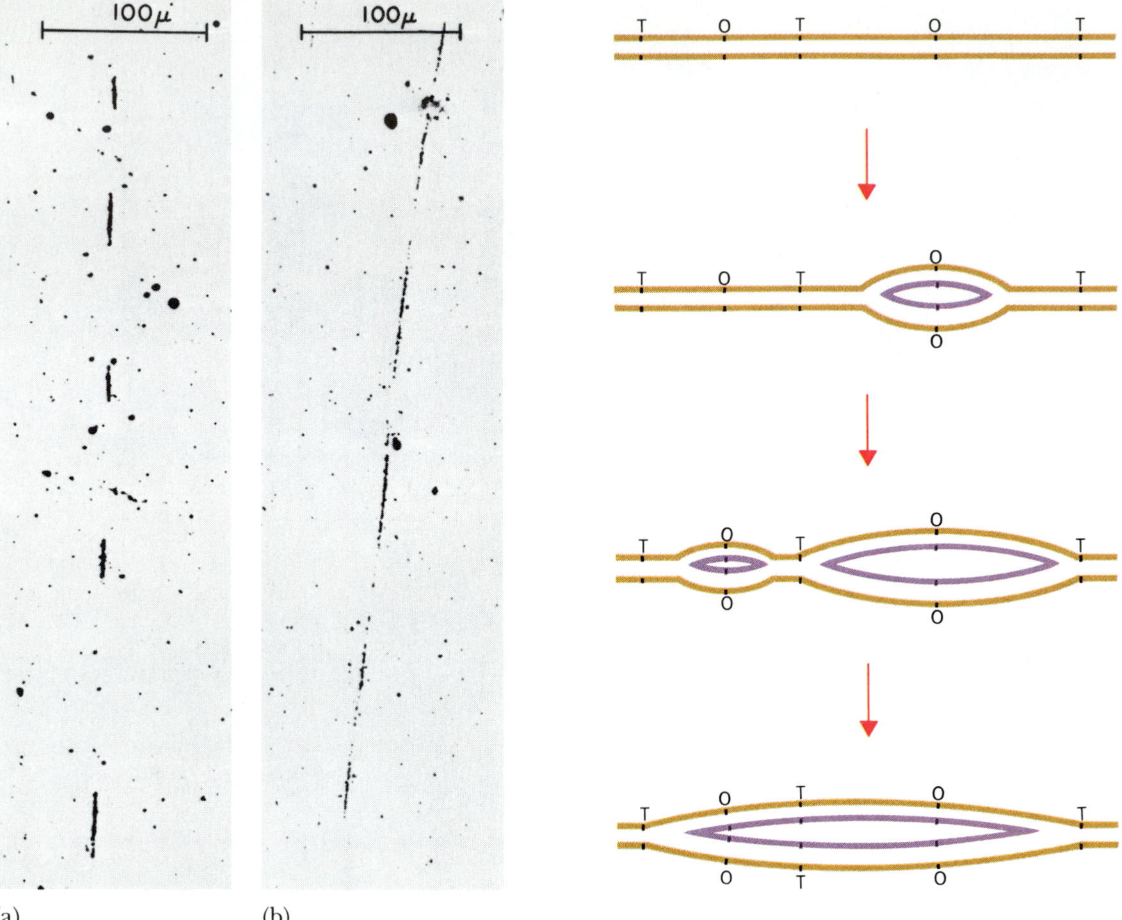

Figure 6.25 Bidirectional replication of the multiple replicons of DNA molecules of mammalian chromosomes. (a) An autoradiograph of a segment of a DNA molecule from a Chinese hamster cell that has been pulse-labeled with ^{3}H-thymidine. The tandem arrays of radioactivity indicate that replication was occurring at five distinct sites in this segment of the DNA molecule. (b) An autoradiograph of a segment of a DNA molecule from a cell that was pulse-labeled with tritiated thymidine and then allowed to grow for a period in nonradioactive medium (pulse-chase). The decreasing grain densities from the centers to both ends of the tracks indicate that two replication forks were moving in opposite directions. (c) A schematic interpretation of the autoradiographs. Replication begins at multiple origins (*O*) of the DNA molecules and proceeds bidirectionally toward termini (*T*). Initiation does not necessarily occur simultaneously at all origins. [Reproduced with permission from J. A. Huberman and A. D. Riggs, *J. Mol. Biol.* 32: 327–341, 1968. Copyright © 1968 by Academic Press, Inc. (London) Ltd. Original photographs courtesy of J. A. Huberman.]

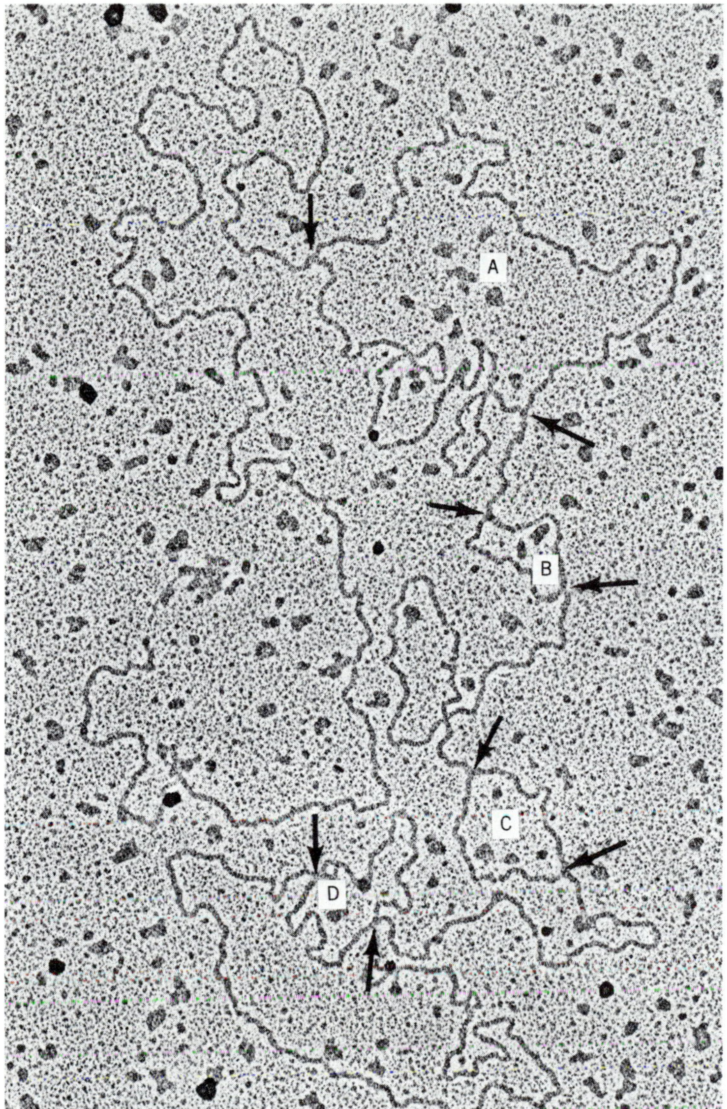

Figure 6.26 Electron micrograph of a DNA molecule in *Drosophila melanogaster* showing multiple sites of replication. Four eye-shaped (-o-) replication structures (labeled A–D) are present in the segment of the DNA molecule shown. (From D. R. Wolstenholme, *Chromosoma* 43: 1, 1973. Original photograph courtesy of D. R. Wolstenholme.)

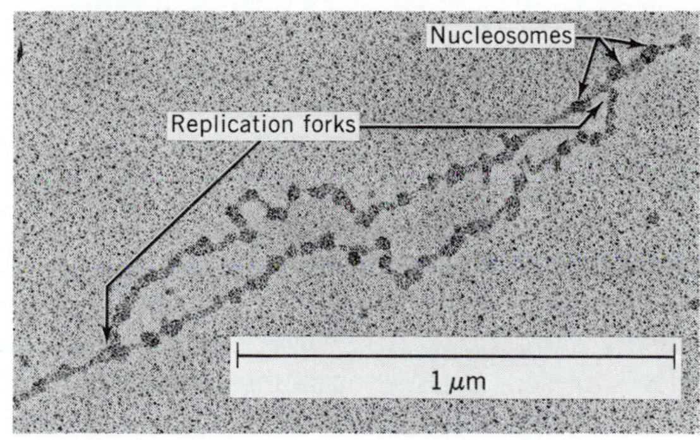

Figure 6.27 Electron micrograph of replicating chromatin from blastoderm embryos of *Drosophila melanogaster*. Note the organization of the chromatin into nucleosomes both ahead of the replication forks ("prereplicative chromatin") and behind the replication forks ("postreplicative chromatin"). If the nucleosomes are disassembled to allow the replication forks to pass, as would seem necessary, they are clearly reassembled quite rapidly after DNA replication has occurred. Recall that replication is bidirectional; thus each branch point is a replication fork. (From S. L. McKnight and O. L. Miller, Jr., *Cell* 12: 795–804, 1977. Copyright © 1977 Cell Press; published by Cell Press. Original photograph courtesy of O. L. Miller, Jr.)

cated in just over 8½ days. The chromosomes of *Drosophila* embryos replicate within 3 to 4 minutes, and the nuclei divide once every 9 to 10 minutes during the early cleavage divisions. Complete replication of the DNA of the largest chromosomes within 3½ minutes would require over 7000 replication forks distributed at equal intervals along the molecules.

The number of active origins varies at different stages of development and in different cell types. In *Drosophila,* there are about 10 times as many active origins per nuclei during early cleavage divisions as in dividing cells of adults. Unfortunately, we still do not know what determines how many or which origins are active in different cells and at different stages of development. Whether all origins of replication in eukaryotes are unique sites is also still unknown. It is likely that secondary, normally inactive origins exist in some eukaryotes, and that these "origins" may become active under certain circumstances.

Components of the Eukaryotic "Replication Apparatus"

Given the complexity of the "replication apparatus" of the simple bacterium *Escherichia coli* (Chapter 5), it seems safe to conclude that the replication machinery of eukaryotes will be equally complex, if not more complex. Although our knowledge of the structure of the "replication apparatus" in eukaryotes is still quite limited, some differences between eukaryotes and prokaryotes are already apparent. ***Four distinct DNA polymerases,*** named α, β, δ, ***and*** γ have been identified in some eukaryotes. Polymerases α, β, and δ are located in nuclei; polymerase γ is located in organelles (mitochondria and chloroplasts). Thus, DNA polymerase γ is probably responsible for the replication of the DNA present in these organelles (see Chapter 20). Of course, one should keep in mind that these organelles may also contain other, yet-to-be-identified DNA polymerases. For many years, DNA polymerase α was thought to be solely responsible for the replication of chromosomal DNA in nuclei. However, recent evidence has shown that DNA polymerases α and δ are both required for DNA replication in nuclei (Fig. 6.28). The role of polymerase β is still unclear.

As in the case of prokaryotes, much of our information about DNA synthesis in eukaryotes has come from the development and dissection of *in vitro* DNA replication systems. Studies of the replication of DNA viruses of eukaryotes have proven very informative, and, of these viruses, Simian virus 40 (SV40) has proven particularly useful. The replication of SV40 is carried out almost entirely by the host cell's "replication apparatus." Only one viral protein, the so-called T antigen, is required for the replication of the SV40 chromosome. Two of the host cell's DNA polymerases, α and δ, are required to replicate the SV40 DNA molecule *in vitro* (Fig. 6.28). Polymerase α appears to be the most similar to the true replicase, DNA polymerase III, of *E. coli.* Polymerase α is multimeric in structure, and two of its subunits apparently interact to provide DNA primase activity. Polymerase δ lacks primase activity, and evidence suggests that it is involved in the polymerization of the leading strand (the strand being extended by continuous synthesis). A helicase is required to unwind the double helix ahead of the replication fork, just as in prokaryotes. Several other proteins have been identified as "replication factors" (e.g., RF-A, RF-B, RF-C, etc.); their functions remain to be elucidated. A protein called PCNA (proliferating cell nuclear antigen) appears to be a cofactor of polymerase δ; its presence results in a much increased rate of replication. Other accessory proteins will undoubtedly be identified in the future. One enigma is that to date no eukaryotic replication protein has been identified that has proofreading activity. Since the fidelity of DNA replication is comparable in eukaryotes and prokaryotes, one must assume that such an activity is present in eukaryotes, but simply has not yet been detected.

Clearly, a picture of the "replication apparatus" of eukaryotic organisms is beginning to emerge, and, given the utility of the *in vitro* systems that are now

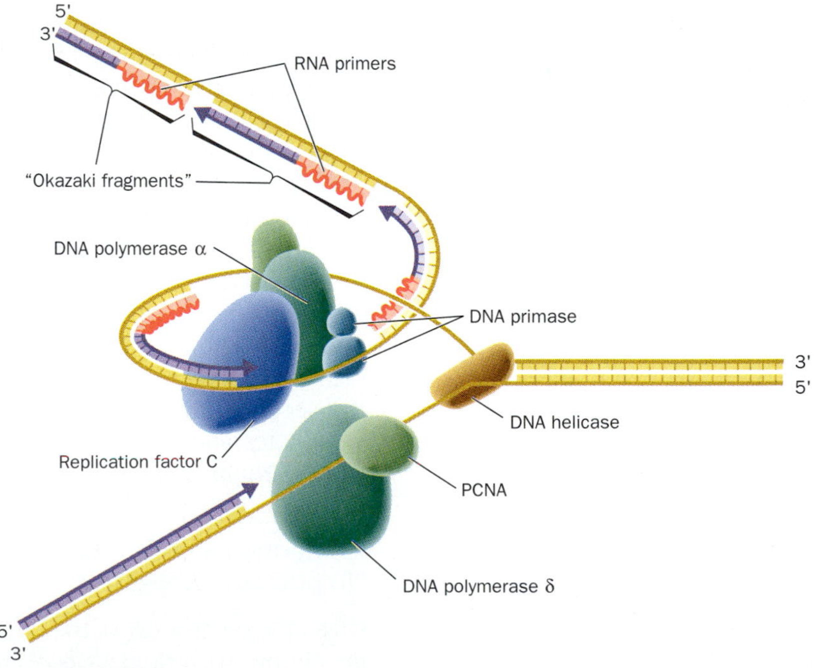

Figure 6.28 Schematic drawing of a eukaryotic *in vitro* DNA replication system showing some of the known components. PCNA = proliferating cell nuclear antigen. For details, see text. (After B. Stillman, *Bio Essays* 9: 56, 1988.)

available, one can anticipate a complete understanding of DNA replication in eukaryotes in the near future. A greater challenge, however, is to identify the signals that regulate DNA replication during growth and differentiation in multicellular organisms such as higher plants and animals (see Chapter 17).

SUMMARY

Each eukaryotic chromosome apparently contains a single giant molecule of DNA that extends from one end of the chromosome through the centromere to the other end of the chromosome. In interphase chromatin with its characteristic 100-Å-diameter fibers, this DNA is held in a ***negatively supercoiled*** configuration by ***histone*** packaging. Relaxed or swollen interphase chromatin has a "beads-on-a-string" structure. Each "bead" or ***nucleosome*** is an ellipsoidlike structure about 110 Å in diameter and 60 Å high. Each nucleosome contains a highly conserved ***core*** consisting of a ***146 nucleotide-pair-long segment of DNA wrapped around an octamer of histones.*** Each octamer contains two molecules each of the histones H2a, H2b, H3, and H4. During mitosis and meiosis, the chromatin is further condensed, possibly by a second level of supercoiling, into ***300-Å-diameter chromatin fibers.*** At metaphase, the coiled or folded 300-Å chromatin fibers are in turn coiled or folded into the highly condensed structures that can be seen with the light microscope; this third level of condensation involves a chromosomal ***"scaffold" that is composed of nonhistone chromosomal proteins.***

Eukaryotic chromosomes contain some DNA sequences that are repeated many times in the genome. DNA containing repeated sequences is called ***repetitive DNA.*** DNA containing sequences that are repeated more than 10^5 times is called ***highly repetitive DNA.*** Highly repetitive sequences are usually located in ***heterochromatin,*** chromatin that remains condensed during interphase and is genetically inactive. Heterochromatin is found at specific sites in chromosomes, often in centromeric or telomeric regions of chromosomes. ***Euchromatin*** is not visible during interphase with the light microscope; it contains most, if not all, of the structural genes. Euchromatin consists of both ***unique or single-copy sequences*** (sequences present once or a few times) and base sequences repeated 10 to 10^5 times (***middle-repetitive sequences***). In all eukaryotes studied to date, a major fraction of the ***genome contains single-copy sequences of DNA interspersed with middle-repetitive sequences.***

Most structural genes are presumed to correspond to the single-copy DNA sequences; many are known to be single-copy sequences. A few structural genes are redundant (middle-repetitive sequences). There is much speculation that many middle-repetitive sequences may have regulatory functions.

REFERENCES

BRITTEN, R. J., and D. E. KOHNE. 1970. "Repeated segments of DNA." *Sci. Amer.* 222(4): 24–31.

HUBERMAN, J. A., and A. D. RIGGS. 1968. "On the mechanism of DNA replication in mammalian chromosomes." *J. Mol. Biol.* 32: 327–341.

IGO-KEMENES, T., W. HORZ, and H. G. ZACHAU. 1982. "Chromatin." *Annu. Rev. Biochem.* 51: 89–121.

JELINEK, W. R., and C. W. SCHMID. 1982. "Repetitive sequences in eukaryotic DNA and their expression." *Annu. Rev. Biochem.* 51: 813–844.

KORNBERG, A. 1980. *DNA replication.* Freeman, San Francisco.

______. 1982. *1982 Supplement to DNA replication.* Freeman, San Francisco.

KORNBERG, R. D., and A. KLUG. 1981. "The nucleosome." *Sci. Amer.* 244(2): 52–64.

LEWIN, B. 1980. *Gene expression-2, eucaryotic chromosomes,* 2nd ed. Wiley, New York.

______. 1990. *Genes IV.* Cell Press, Cambridge, MA/Oxford Univ. Press, Oxford.

TAYLOR, J. H., P. S. WOODS, and W. L. HUGHES. 1957. "The organization and duplication of chromosomes as revealed by autoradiographic studies using tritium-labeled thymidine." *Proc. Natl. Acad. Sci. U.S.A.* 43: 122–128.

WATSON, J. D., N. H. HOPKINS, J. W. ROBERTS, J. A. STEITZ, and A. M. WEINER. 1987. *Molecular biology of the gene,* 4th ed. Benjamin, Menlo Park, CA.

PROBLEMS AND QUESTIONS

6.1 The available evidence indicates that each eukaryotic chromosome (excluding polytene chromosomes) contains a single giant molecule of DNA. What different levels of organization of this DNA molecule are apparent in chromosomes of eukaryotes at various times during the cell cycle?

6.2 A diploid nucleus of *Drosophila melanogaster* contains about 2×10^8 nucleotide-pairs. Assume (1) that all the nuclear DNA is packaged in nucleosomes and (2) that an average internucleosome linker size is 60 nucleotide-pairs. How many nucleosomes would be present in a diploid nucleus of *D. melanogaster?* How many molecules of histone H2a, H2b, H3, and H4 would be required?

6.3 The satellite DNAs of *Drosophila virilis* can be isolated, essentially free of main-band DNA, by density-gradient centrifugation. If these satellite DNAs are sheared into approximately 40-nucleotide-pair-long fragments and are analyzed in denaturation–renaturation experiments, how would you expect their hybridization kinetics to compare with the renaturation kinetics observed using similarly sheared main-band DNA under the same conditions? Why?

6.4 It has been demonstrated experimentally that most highly repetitive DNA sequences in the chromosomes of eukaryotes are **not** transcribed. What does this indicate about the function of highly repetitive DNA?

6.5 The Boston straggler is an imaginary plant with a diploid chromosome number of 4. Boston straggler cells are easily grown in suspended cell cultures. [^{3}H]Thymidine was added to the culture medium in which a G_1-stage cell of this plant was growing. After one cell generation of growth in [^{3}H]thymidine-containing medium, colchicine was added to the culture medium. The medium now contained both [^{3}H]thymidine and colchicine. After two "generations" of growth in [^{3}H] thymidine-containing medium (the second "generation" occurring in the presence of colchicine as well), the two progeny cells (each now containing eight chromosomes) were transferred to culture medium containing nonradioactive thymidine ([^{1}H]thymidine) ***plus colchicine.*** Note that a "generation" in the presence of colchicine consists of a normal cell cycle's chromosomal duplication, but no cell division. The two progeny cells were allowed to continue to grow, proceeding through the "cell cycle," until ***each cell*** contained a set of metaphase chromosomes that looked like the following.

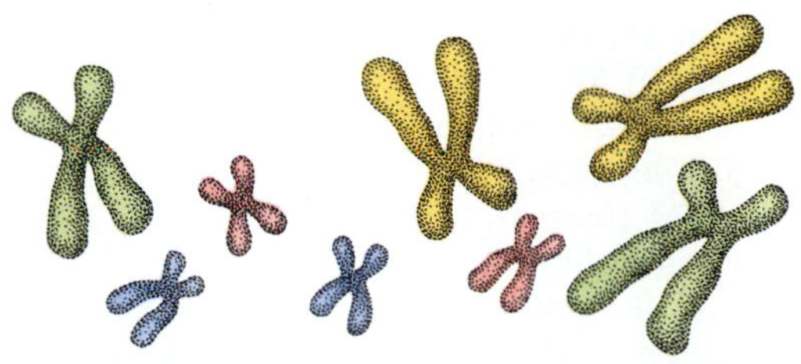

If autoradiography were carried out on these metaphase chromosomes (four large plus four small), what pattern of radioactivity (as indicated by silver grains on the autoradiograph) would be expected? (Assume no recombination between DNA molecules.)

6.6 Suppose that the experiment described in Problem 6.5 were carried out again, except this time replacing the [^{3}H]thymidine with nonradioactive thymidine at the same time that the colchicine was added (after one cell generation of growth in [^{3}H]thymidine-containing medium). The cells were then maintained in colchicine plus nonradioactive thymidine until the metaphase shown in Problem 6.5 occurred. What would the autoradiographs of these chromosomes look like?

6.7 Suppose that the DNA of cells (growing in a cell culture) in an unknown eukaryotic species was labeled for a short period of time by the addition of [^{3}H]thymidine to the medium. Next assume that the label was removed and the cells were resuspended in nonradioactive medium. After a short period of growth in nonradioactive medium, the DNA was extracted from these cells, diluted, gently layered on filters, and autoradiographed. If autoradiographs of the type

———.. . .   ———... . .   ———.. . . .

were observed, what would this indicate about the nature of DNA replication in these cells? Why?

6.8 (a) Are (1) single-copy DNA sequences, (2) middle-repetitive DNA sequences, (3) highly repetitive DNA sequences more prevalent in euchromatin or heterochromatin? (b) Are most (1) single-copy DNA sequences, (2) highly repetitive DNA sequences transcribed? (c) What are the presumed functions of most (1) single-copy DNA sequences, (2) highly repetitive DNA sequences? (d) What roles are the middle-repetitive DNA sequences believed to play in higher eukaryotes?

6.9 Are eukaryotic chromosomes metabolically most active during prophase, metaphase, anaphase, telophase, or interphase?

6.10 (a) What functions do telomeres provide? (b) Do telomeres have any unique structural features? (c) What is the function of telomere transferase? (d) When chromosomes are broken by exposure to high energy radiation such as X-rays, the resulting broken ends exhibit a pronounced tendency to stick to each other and fuse. Why might this occur?

6.11 Of what special interest is the biophysical technique of viscoelastometry to geneticists?

6.12 How many DNA molecules are present in (a) the axial regions, (b) the lateral loops of "lampbrush" chromosomes?

6.13 Are the "scaffolds" of eukaryotic chromosomes composed of histone or nonhistone chromosomal proteins? How has this been determined experimentally?

6.14 Assuming that each pre-replication human chromosome contains a single giant molecule of DNA, how many DNA molecules are present in the chromosomes of the nucleus of (a) a human egg, (b) a human sperm, (c) a human diploid somatic cell in stage G_1, (d) a human diploid somatic cell in stage G_2, (e) a human primary oöcyte?

6.15 (a) Which class of chromosomal proteins, histones or nonhistones, is the more highly conserved in different eukaryotic species? Why might this be expected? (b) If one compares the histone and nonhistone chromosomal proteins of chromatin isolated from different tissues or cell types of a given eukaryotic organism, which class of proteins will exhibit the greater heterogeneity? Why are both classes of proteins not expected to be equally homogeneous in chromosomes from different tissues or cell types?

6.16 During DNA renaturation experiments, why is the rate of renaturation ($-dC/dt$) proportional to the square of the concentration of single strands (kC^2) rather than simply proportional to the concentration of single strands (kC)?

6.17 Studies of the renaturation kinetics of DNA from the human genome reveal that at $t = 0$ (the earliest time point at which the proportion of double-stranded DNA can be measured in standard renaturation experiments) approximately 6% of the DNA is already double stranded (see Figure 6.22). Assuming that all of the DNA was initially denatured in these experiments, how can this result (6% renatured at $t = 0$) be explained?

6.18 Four distinct DNA polymerases: α, β, γ, and δ, have been characterized in several higher eukaryotes. What are the intracellular locations and functions of these polymerases?

6.19 The *E. coli* chromosome contains approximately 4×10^6 nucleotide-pairs and replicates as a single bidirectional replicon in approximately 40 minutes under a wide variety of growth conditions. The largest chromosome of *D. melanogaster* contains about 6×10^7 nucleotide-pairs. (a) If this chromosome contains one giant molecule of DNA that replicates bidirectionally from a single origin located precisely in the middle of the DNA molecule, how long would it take to replicate the entire chromosome if replication in *Drosophila* occurred at the same rate as replication in *E. coli?* (b) Actually, replication rates are slower in eukaryotes than in prokaryotes. If each replication fork moves at a rate of 5,000 nucleotide-pairs per minute in *Drosophila* and 100,000 nucleotide-pairs per minute in *E. coli,* how long will it take to replicate the largest *Drosophila* chromosome if it contains a single bidirectional replicon as described in (a) above? (c) During the early cleavage divisions in *Drosophila* embryos, the nuclei divide every 9 to 10 minutes. Based on your calculations in (a) and (b), what do these rapid nuclear divisions indicate about the number of replicons per chromosome in *Drosophila?*

6.20 (a) If the haploid human genome contains 3×10^9 nucleotide-pairs and the average molecular weight of a nucleotide-pair is 660, how many copies of the human genome are present, on average, in 1 μg of human DNA? (b) What is the mass of one copy of the human genome? (c) If the haploid genome of the small plant *Arabidopsis thaliana* contains 7.7×10^7 nucleotide-pairs, how many copies of the *A. thaliana* genome are present, on average, in 1 μg of *A. thaliana* DNA? (d) What is the mass of one copy of the *A. thaliana* genome? (e) Of what importance are calculations of the type above to geneticists?

Chapter 7

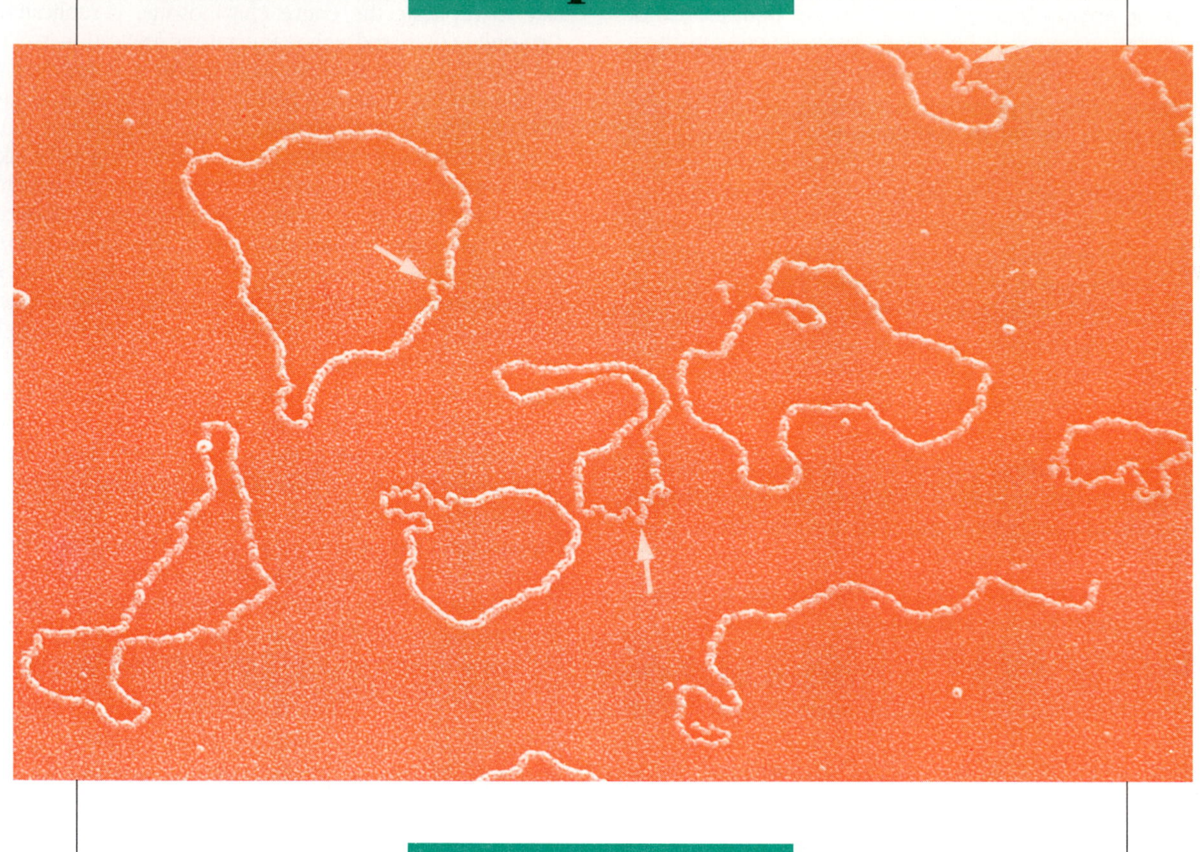

Linkage, Crossing-over, and Chromosome Mapping

(Photo above): Phage M13 single-stranded DNA molecules complexed with single-stranded DNA-binding protein and a protein (*rec A* protein, arrows) from *Escherichia coli* that catalyzes a key step in recombination between homologous DNA molecules. (Reproduced with permission from R. J. Thresher, G. Christiansen, and J. D. Griffith, *J. Mol. Biol.* 201: 101–113, 1988. Original photograph courtesy of J. D. Griffith, Department of Microbiology and Immunology, University of North Carolina Medical School, Chapel Hill.)

IN his classic paper, Mendel reported the results of crosses involving alleles of seven genes that controlled seven different phenotypic characteristics of the garden pea (Chapter 2). The observed segregation of the two alleles of each gene was the basis of Mendel's first law—the law of segregation. When Mendel crossed pea plants differing in two phenotypic characteristics, so-called dihybrid crosses, the segregation of the alleles of the second gene occurred independently of the segregation of the alleles of the first gene. This provided the basis of Mendel's second law—the law of independent assortment.

The garden pea, *Pisum sativum*, has seven pairs of chromosomes. Like all other organisms, it has many more genes than chromosomes. In fact, in all organisms except the simplest viruses, there are orders of magnitude more genes than chromosomes. We don't know exactly how many genes the garden pea has, but it seems safe to conclude that the correct number is over 10,000. Indeed, the correct number might be considerably higher. Clearly, then, there must be hundreds to thousands of genes per chromosome.

Mendel's paper detailed the results of his dihybrid cross between pea plants with yellow, round seeds and pea plants with green, wrinkled seeds. The genes controlling these two traits, yellow versus green seed color and round versus wrinkled seed shape, are now known to be located on chromosomes 1 and 7, respectively, of the garden pea. Had these traits been controlled by genes located near one another on the same chromosome, Mendel would not have observed the independent assortment of the alleles of these genes, and, thus, would not have been able to deduce his second law from the result of this cross. Combinations of ***genetic markers*** (alleles of each of the segregating genes) that are present on a particular chromosome of the parents tend to remain together on that chromosome in the progeny. That is, such genetic markers behave as though they are partially ***linked***.

Clearly, ***each chromosome must contain many genes, and these genes would not be expected to assort independently since the basis of independent assortment is (usually) the independent segregation of the different pairs of homologous chromosomes during the reductional division of meiosis*** (Chapter 3). Genes that are located far apart on the same chromosome will also assort independently for reasons that will be explained later in this chapter.

As mentioned in the foregoing, Mendel studied the segregation of the alleles of seven different genes. On the basis of Mendel's descriptions of the seven pairs of alternative phenotypic traits and of present knowledge of the chromosomal locations of known genes, S. Blixt has concluded that the genes studied by Mendel are located on four of the seven pairs of chromosomes in the garden pea. Two of the genes are on chromosome 1 and three are on chromosome 4. Why, then, did Mendel not detect the linked segregation of the alleles of pairs of genes in his dihybrid and trihybrid crosses?

For the two genes on chromosome 1 and for two of the three combinations of the three genes on chromosome 4, the answer is that the genes are located so far apart on the chromosome that linked segregation is not observed or expected. Only in the case of the genes controlling plant height (tall versus dwarf) and pod shape (inflated versus constricted) is linked segregation of the alleles expected. Mendel states in his classic paper, after presenting the data for dihybrid and trihybrid crosses involving the segregation of the alleles of the genes controlling seed shape, seed endosperm color, and flower color, that "further experiments were made with a smaller number of experimental plants in which the remaining characters by twos and threes were united as hybrids: all yielded approximately the same results" (from W. Bateson's translation of Mendel's paper, reprinted in *Classic Papers in Genetics*, J. A. Peters, ed.). This statement clearly implies that Mendel carried out all possible di- and trihybrid crosses involving the seven genes that he studied. However, H. Lamprecht has investigated this question and has concluded that Mendel probably did not study the simultaneous segregation of the alleles of the genes controlling plant height and pod shape.

LINKAGE

Consider a dihybrid cross, a cross in which two pairs of alleles are segregating. For example, let us look at a cross in plants and, for simplicity, assume (1) that both genes are autosomal, (2) that one allele of each gene is completely dominant to the other allele, and (3) that no epistasis (gene interaction) is involved. In Chapter 2, such a cross might have been diagrammed as follows:

Parental genotypes:	*AABB* × *aabb*
Parental gametes:	*AB* — *ab*
Progeny or F_1 genotype:	*AaBb*

Next, suppose that the F_1 plants are testcrossed, that is, crossed to fully homozygous recessive plants, as follows:

	F_1 Plant		Testcross Plant
Genotypes:	*AaBb*	×	*aabb*
	↓		↓
Gamete types:	?		*ab*

What types and frequencies of offspring would we expect from such a testcross? Clearly, all the gametes produced by the doubly homozygous recessive testcross parent will be *ab* in genotype. But what types of gametes will the F_1 parent (*AaBb*) produce, and in what frequencies? We find that this question does not have a simple answer!

Up to this point, we have considered only crosses where the two genes involved were located on different (nonhomologous) chromosomes and thus showed independent assortment. Such crosses might be more specifically diagrammed as

A B / A B × a b / a b

to indicate that the two loci involved are on different chromosomes. The F_1 plants would then be

A B / a b

From Chapter 2, we know that such an F_1 plant would produce gametes of type

A B   A b   a B   a b

in equal proportions (1:1:1:1). Thus, the expected frequencies of testcross progeny will be

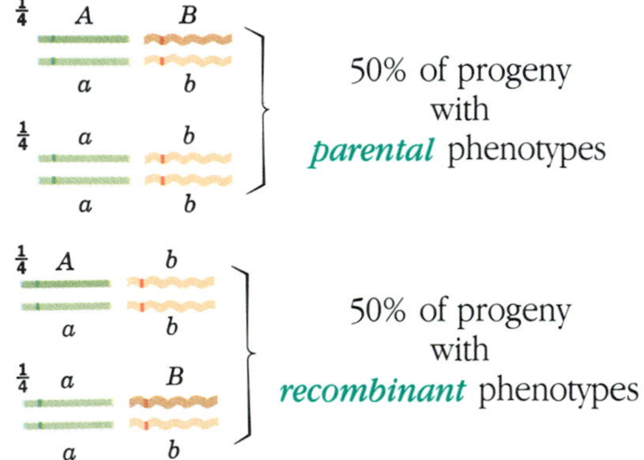

(Note that "recombinant phenotypes" include all new combinations of the inherited phenotypic traits, i.e., all combinations except those of the parents.)

Now, suppose that gene *C* is located on the same chromosome as gene *A*. A dihybrid cross in which alleles of genes *A* and *C* are segregating might be diagrammed as

Parental genotypes: A C / A C × a c / a c

Parental gametes: A C   a c

Progeny or F_1 genotype: A C / a c

A testcross of the F_1 plants would be

A C / a c × a c / a c

What progeny types and frequencies might we expect from this testcross? One's first guess might be that the F_1 plants would produce gametes of only two types:

A C   a c

If so, genes *A* and *C* would be ***completely linked***, and all gametes would carry the ***parental combinations*** of markers. ***No recombinant gametes*** would be produced, in sharp contrast to the ***50 percent recombinant gametes characteristic of independent assortment***. However, complete linkage seldom if ever occurs. The actual results of such testcrosses involving two genes located on the same chromosome are usually somewhere between the results expected for complete linkage and the results expected for independent assortment. ***At some frequency less than 50 percent, recombinant gametes are formed.*** Again, an examination of the testcross progeny will show this directly:

A C / a c
a c / a c
Greater than 50% of progeny with ***parental*** phenotypes

A c / a c
a C / a c
Less than 50% of progeny with ***recombinant*** phenotypes

Gametes that contain recombinant combinations of genetic markers located on the same chromosome are produced by a phenomenon called ***crossing-over***, which will be discussed in detail in the following section.

Genetic markers located on the same chromosome thus tend to remain together during sexual reproduction. That is, they do not exhibit independent assortment. Such genetic markers are said to be ***linked***, and the phenomenon, or transmission pattern, of linked genes is called ***linkage***.

Strictly speaking, genetic markers are said to be ***linked whenever over 50 percent of the gametes produced contain parental combinations*** of the markers and ***less than 50 percent of the gametes contain recombinant combinations*** of the markers. There are rare cases when genes located on different chromosomes exhibit linkage and fairly common cases when genes located on the same chromosome assort independently. Usually, however, when linkage is observed, it results because the genes involved are located close together on the same chromosome.

The effects of linkage were first evident in the results of a dihybrid cross in sweet peas that were reported by W. Bateson and R. C. Punnett in 1906. However, Bateson and Punnett did not interpret their results in terms of the behavior of genes located on the same chromosome. T. H. Morgan (Fig. 7.1) was the first to relate linkage to the segregation of homologous chromosomes and the occurrence of crossing-over between homologous chromosomes during meiosis. Morgan's interpretation of linkage was published in 1911 in a paper in which he reported the results of crosses involving linked genes in the fruit fly *Drosophila melanogaster*. Many of our current concepts about linkage, crossing-over, and chromosome mapping have evolved from the work of Morgan and his students, C. B. Bridges, H. J. Muller, and A. H. Sturtevant.

The effects of linkage can be illustrated by considering the results of two crosses involving pairs of alleles of two genes located on the second chromosome of *D. melanogaster*. One gene affects body color. Flies homozygous for its recessive mutant allele b have black bodies. The presence of its dominant wild-type allele b^+ results in gray bodies. The second gene affects the phenotype of the wing. In the homozygous recessive mutant form, it results in "vestigial" or short, underdeveloped wings. The symbol for this mutant allele is *vg*; however, we will take the liberty of substituting the symbol s for short wings, and use s^+ for its dominant wild-type allele, which results in long wings. First, consider a cross between homozygous flies with long wings and gray bodies (wild-type flies) and homozygous flies with short wings and black bodies. This cross will produce heterozygous F_1 flies with long wings and gray bodies, as diagrammed in Cross I. Of the progeny of this testcross, 82 percent exhibited one or the other (41 percent each) of the parental combinations of traits. The other 18 percent of the progeny had new or recombinant combinations.

Next, consider a different cross, one between homozygous flies with long wings and black bodies and homozygous flies with short wings and gray bodies, as shown in Cross II. Again, in Cross II, 82 percent of the testcross progeny have parental phenotypes (phenotypes identical to one or the other of the original parents) and 18 percent have new or recombinant phenotypes.

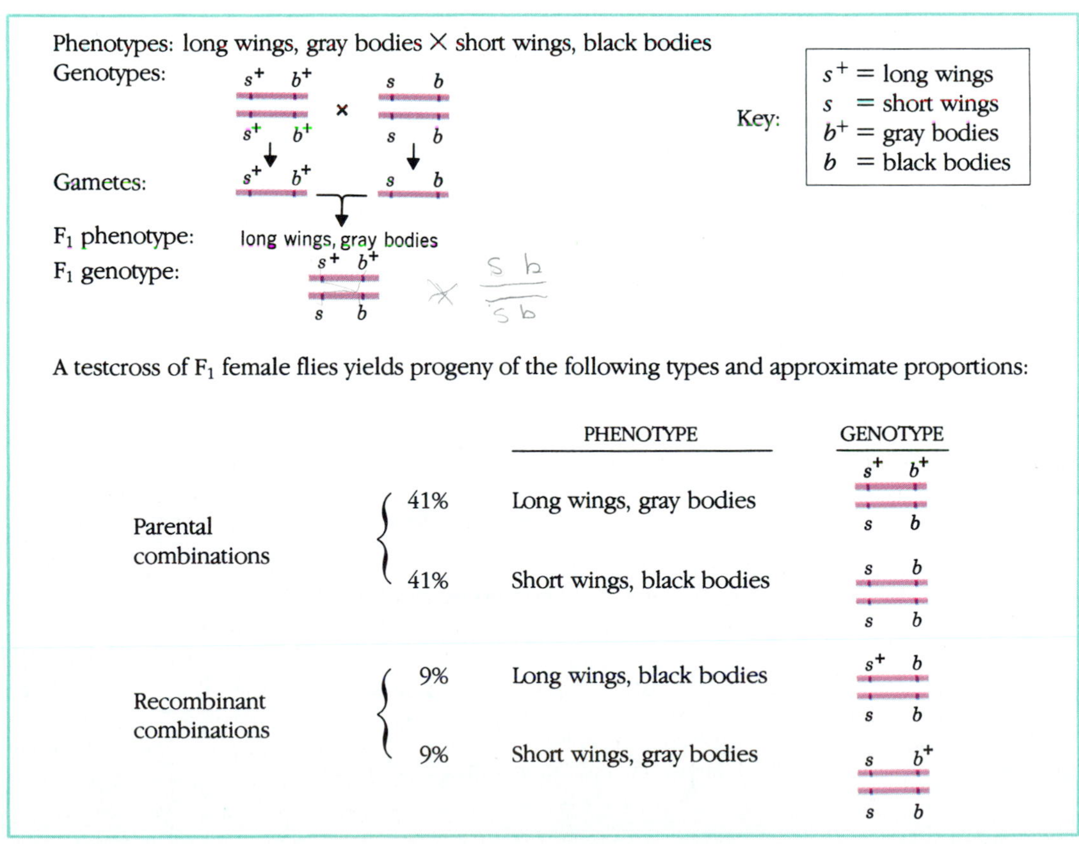

CROSS I

Figure 7.1 Thomas Hunt Morgan, American geneticist and embryologist. Dr. Morgan was Nobel Laureate in biology and medicine in 1933. (Culver Pictures.)

Although the F_1 flies have the same phenotype (long wings, gray bodies) in both crosses, the testcross progeny of the F_1 female flies contain very different frequencies of the four phenotypic classes in the two cases. (The reason for using F_1 females and not F_1 males in the testcross is that crossing-over occurs infrequently in wild-type male *Drosophila*; see p. 193.) For example, 41 percent of the testcross progeny in Cross I are wild type (have long wings and gray bodies); in Cross II, only 9 percent are wild type. Clearly, the allelic forms of the two genes that are present together on the homologous chromosomes of the parents tend to remain together on the chromosomes of the progeny.

In Cross I, the F_1 flies carried the wild-type forms ($s^+\ b^+$) of the two genes on one homolog and the mutant forms ($s\ b$) on the other homolog. The genotype of a heterozygote of this type is frequently written as

$$\underline{s^+\ \ b^+} \,/\, \underline{s\ \ \ b}$$

or simply $s^+\ b^+/s\ b$. This arrangement of mutant and wild-type forms of two genes in a heterozygote is called the ***coupling*** state or **cis-*configuration***. The alternative arrangement, illustrated in Cross II [where each homolog contains one mutant gene and one wild-type gene ($s^+\ b/s\ b^+$)], is called the ***repulsion*** state or **trans-*configuration***. (If you have taken or-

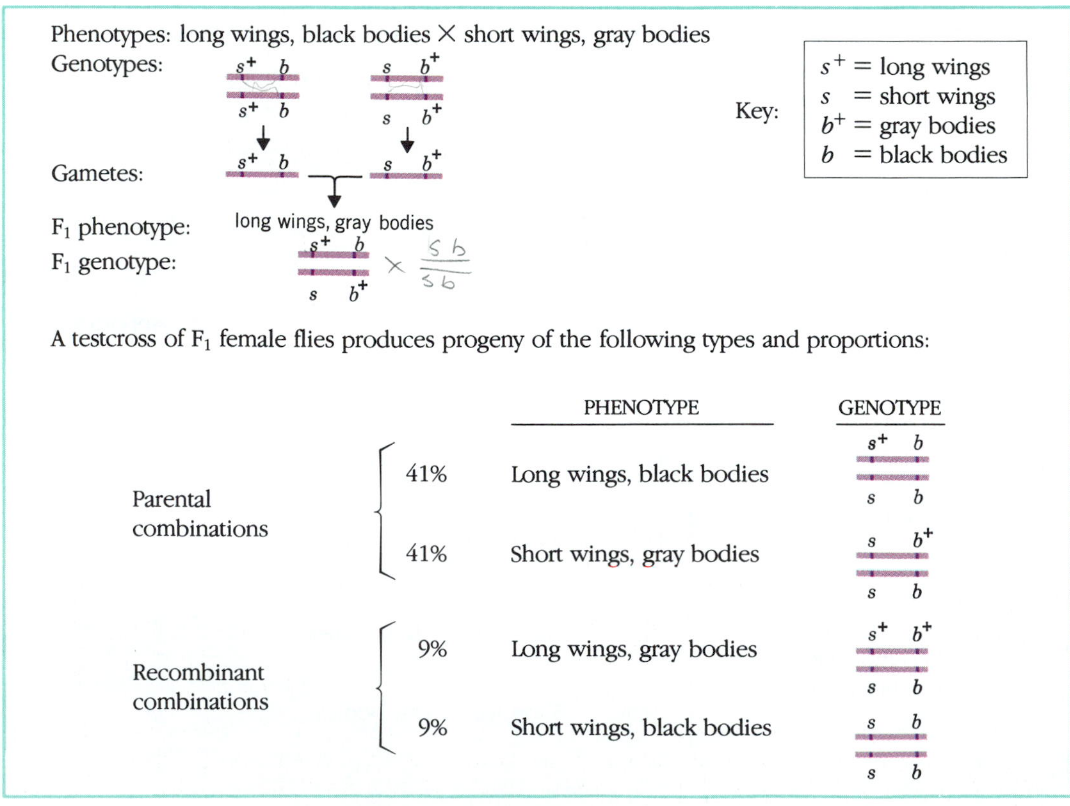

CROSS II

ganic chemistry, you may recall that *cis* and *trans* represent substitutions or side groups on the ***same*** (*cis*) and on ***opposite*** (*trans*) sides of a carbon–carbon double bond; these usages may help in remembering the analogous usages of *cis* and *trans* in genetics.)

An important feature of the two crosses just described is that—regardless of whether the cross was done in *cis* (coupling; Cross I) or in *trans* (repulsion; Cross II)—the F_1 female flies produced gametes, 18 percent of which carried recombinant combinations of the genetic markers. How are the recombinant chromosomes in these gametes formed, and why is their frequency the same in the two crosses?

CROSSING-OVER

Recombination of genes on the same chromosome is accomplished by ***crossing-over,*** a process by which parts of homologous chromosomes are interchanged. Many scientists have proposed that crossing-over is important in evolution. Crossing-over and independent assortment are mechanisms that produce new combinations of genes. Natural selection can then act to preserve those combinations that produce organisms with maximum fitness, that is, maximum probability of perpetuation of the genotype.

We can summarize the most important features of the concept of crossing-over as follows.

1. The location of a gene on a chromosome is called a ***locus*** (plural ***loci***). The loci of the genes on a chromosome are arranged in a linear sequence.
2. The two alleles of a gene in a heterozygote occupy corresponding positions in the homologous chromosomes, that is, allele *A* occupies the same position in homolog 1 that allele *a* occupies in homolog 2.
3. Crossing-over involves the breakage of each of two homologous chromosomes (actually chromatids) and the exchange of parts.
4. Crossing-over occurs at pachytene after the ***synapsis of the homologous chromosomes has occurred in prophase I*** of meiosis. Since chromosome replication occurs during interphase, meiotic crossing-over occurs in the ***postreplication tetrad stage,*** that is, after each chromosome has doubled such that ***four chromatids are present for each pair of homologous chromosomes.*** Crossing-over that involves sister chromatids (the two chromatids of one homolog) occurs, but it is seldom detectable genetically since the sister chromatids are usually identical.
5. Chromosomes with recombinant combinations of linked genes are formed by the occurrence of crossing-over in the region between the two loci.
6. The probability that crossing-over will occur between two loci increases with increasing distance between the two loci on the chromosome.

Some of these features of crossing-over are illustrated in Fig. 7.2. The molecular mechanism by which crossing-over occurs is discussed on pp. 186–196 of this chapter.

Cytological Basis of Crossing-Over

Morgan first proposed crossing-over to explain the formation of recombinant combinations of genes that were shown to be linked by genetic data. He hypothesized that this linkage was the result of the location of these genes on the same chromosome. If crossing-over occurs as diagrammed in Fig. 7.2, one might expect to be able to observe it (or the consequences of it) cytologically. In fact, cross-shaped structures, in which two of the four chromatids of homologous chromosome pairs appear to exchange pairing partners, are readily detected in cytological studies of meiosis in many organisms (Fig. 7.3). These cross-shaped structures were first detected in amphibians and called ***chiasmata*** (singular ***chiasma***) by F. Janssens (see Chapter 3). Studies in which chiasma frequencies were correlated with recombination frequencies have

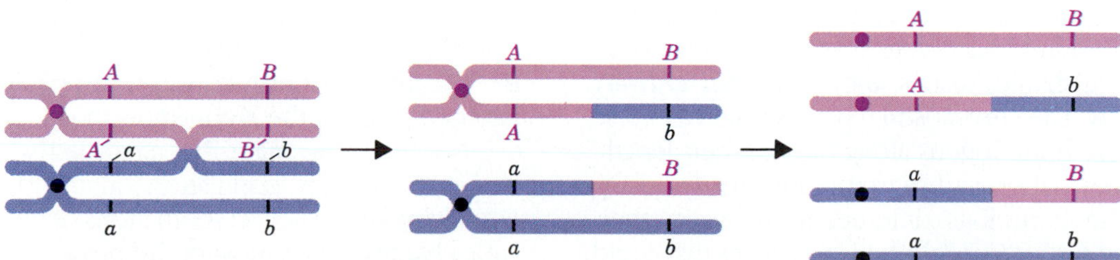

Figure 7.2 Diagram illustrating the occurrence of crossing-over between two loci. Note that the crossover involves only two of the four chromatids of the pair of homologous chromosomes. These two chromatids interchange corresponding segments by a breakage and exchange mechanism. Also notice that of the four products of this meiotic event only two contain recombinant arrangements of the alleles of the two genes. The other two daughter chromosomes (top and bottom, right) carry parental arrangements of the alleles of the genes.

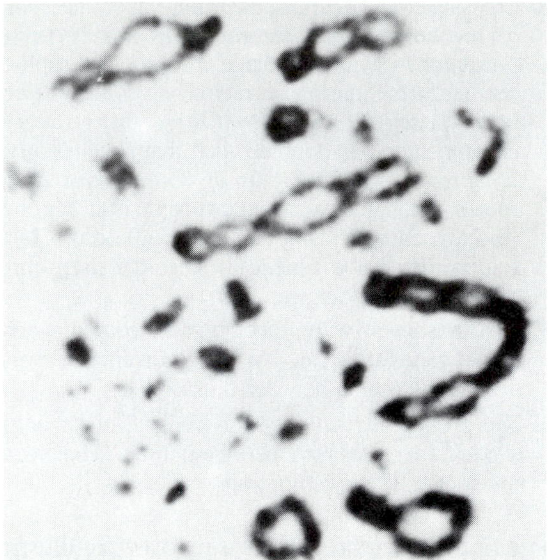

Figure 7.3 Meiotic prophase I chromosomes of the fowl (*Gallus domesticus*), showing multiple chiasmata in the large bivalents. The preparation is from the testis of a White Leghorn male. All domestic fowl and the ancestral Red Jungle Fowl contain 76 pairs of autosomes plus a pair of sex chromosomes (ZZ homogametic males and ZW heterogametic females). The chromosomes of fowl exhibit a vast variation in size, from a few pairs of very large chromosomes to many pairs of very small chromosomes. (Photograph courtesy of R. N. Shoffner, Cytogenetics Laboratory, Department of Animal Science, University of Minnesota.)

shown a direct relationship between crossing-over and chiasmata.

Direct cytological evidence that homologous chromosomes exchange parts during crossing-over was first obtained in 1931 by C. Stern (working with *Drosophila*) and by H. B. Creighton and B. McClintock (working with maize). Normally, the two chromosomes of any homologous pair are morphologically indistinguishable. Stern, Creighton, and McClintock, however, identified homologs that were morphologically distinguishable, that is, they were ***not entirely homologous***. The chromosome pairs studied by these workers were homologous along most of their length, such that they paired and segregated normally during meiosis. The homologs differed, however, at their ends, having distinct morphological features that could be recognized by microscopy.

Stern studied two X chromosomes that differed from the normal X chromosomes of *Drosophila*. One X chromosome had part of a Y chromosome (that had broken off its normal location) attached to one end. (Such chromosomal rearrangements are well known and are called translocations; see Chapter 18.) The second X chromosome was shorter than normal; a piece had been broken off and translocated to another chromosome (in this case, to chromosome IV). Crosses were done to produce female flies heterozygous for these two morphologically distinguishable X chromosomes. These female flies were also heterozygous for alleles of two genes that are located on the X chromosome. One gene affects eye shape, and the other eye color. The partially dominant mutant allele *B* results in bar-shaped eyes; its wild-type allele B^+ produces round eyes when homozygous. The mutant allele *car* of the second gene results in carnation-colored eyes; its dominant wild-type allele car^+ yields red eyes. The females used in Stern's study carried the allelic pairs in the *cis*-configuration, as shown in the following figure.

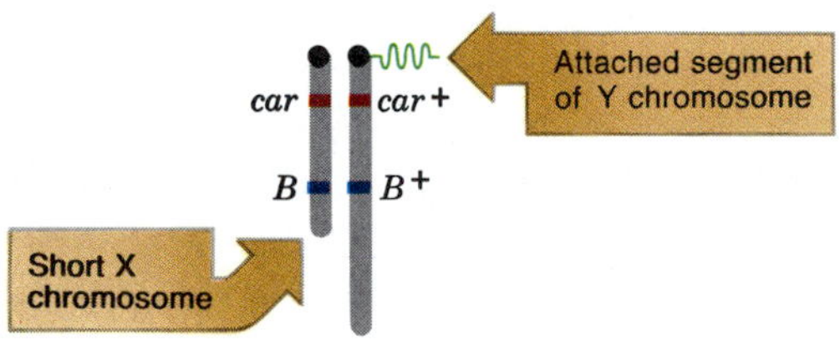

Stern crossed such heterozygous females with males having carnation-colored, normal-shaped eyes (*car* B^+/Y males). This cross and the results Stern obtained are diagrammed in Fig. 7.4. Stern determined the genotypes of the progeny and asked whether the progeny with recombinant genotypes carried X chromosomes with recombinant combinations of the morphological markers. He observed that the combination of morphological markers present on the X chromosomes of each recombinant progeny were precisely those predicted (from the arrangement of the markers in the heterozygous female parent) if crossing-over involved the breakage and exchange of parts of homologous chromosomes. Recombinant male progeny with bar-shaped red eyes (car^+B/Y), for example, were found to carry the short X chromosome, but now with the translocated piece of the Y chromosome at one end. Male flies with normal-shaped carnation eyes (*car* B^+/Y) contained long X chromosomes, but without the attached piece of the Y chromosome.

Note that the male flies analyzed were selected from the progeny population on the basis of their phenotypes—the new combinations of the two phenotypic characteristics. Most of the progeny had parental combinations of these traits. Some progeny flies would be expected to carry X chromosomes that were produced by a crossover event between the *car* and *B* loci as shown in Fig. 7.4 and another crossover event occurring either between the *car* locus and the centromere or between the *B* locus and the centromere–distal end of the chromosome. The presence of X chromosomes produced by two crossover events of this type might be expected to make Stern's results difficult to interpret. Indeed, such chromosomes are produced. The reason that they did not interfere with the interpretation of Stern's results is that such chromosomes are rare (see pp. 174–178). Progeny flies carrying such chromosomes are not frequent enough to confound the observed correlation between crossing-over and genetic recombination.

Creighton and McClintock also demonstrated the correlation between crossing-over and the exchange of parts of homologous chromosomes. They analyzed crosses involving two loci on chromosome 9 of maize. One gene controlled kernel color (*C*, colored; *c*, colorless) and the other controlled the type of carbohydrates in the kernel (*Wx*, starchy; *wx*, waxy). They made use of a chromosome with a densely staining "knob" at one end and an extra (translocated) piece of chromosome at the other end:

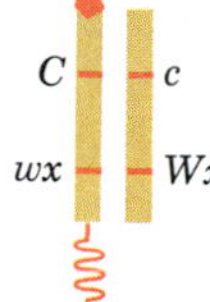

When a testcross of the trans-heterozygote was made, the *C Wx* and *c wx* recombinant progeny were found to carry a chromosome with recombinant combinations of the morphological markers, as shown in the following diagram (plus, of course, a normal chromosome from the testcross parent).

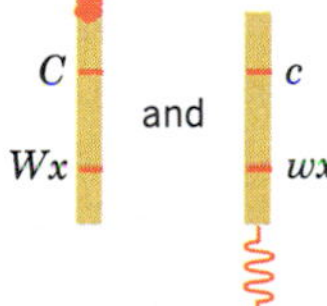

These two experiments, by Stern and by Creighton and McClintock, are true classics in genetics. They provided confirmation of Morgan's hypothesis that crossing-over involves the interchange of parts of homologous chromosomes; they also provided strong evidence that indicated that genes are indeed located on chromosomes.

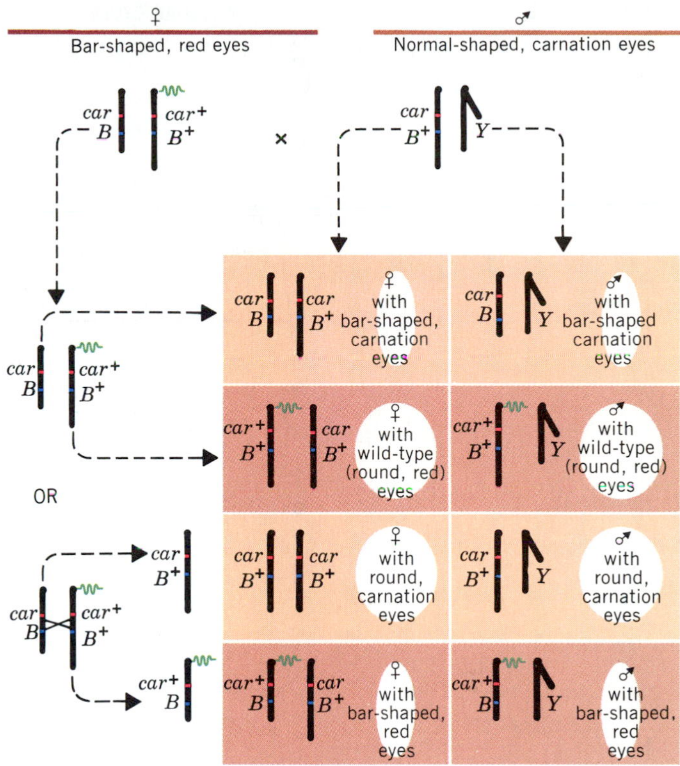

Figure 7.4 C. Stern's classic experiment demonstrating that crossing-over involves the interchange of parts of homologous chromosomes. Details of the experiment are described in the text. The × (bottom left) indicates the occurrence of crossing-over between the *car* and *B* loci. The centromeres on the chromosomes are represented by circles. The two X chromosomes in the female parent (with bar-shaped, red eyes) are morphologically distinguishable. The short one (left) is missing the distal end; its homolog has a piece of a Y chromosome attached to the centromere end (shown as a horizontal zigzagged line).

Crossing-Over Occurs in the Postreplication Tetrad Stage

An important feature of crossing-over as illustrated in Fig. 7.2 is that crossing-over occurs after chromosome duplication, when each homologous chromosome is represented by two chromatids. Each pair of synapsed homologs is called a ***tetrad***, because it consists of ***four chromatids***. A genetically detectable crossover occurs between any two nonsister chromatids of a homologous pair of chromosomes. Note that ***crossing-over*** refers to the ***exchange of segments of chromosomes*** (the process), whereas ***recombination*** implies the ***formation of new combinations of genes*** (one possible outcome of the process). Thus ***crossing-over occurs in completely homozygous organisms, but new combinations can be formed only in organisms that are heterozygous at two or more loci.***

Proof that crossing-over occurs after duplication of the chromosomes rather than prior to their duplication is most easily obtained by studying fungi of the class known as Ascomycetes. Of the Ascomycetes, the bread mold *Neurospora crassa* has been of particular importance in genetic studies. Five features of *N. crassa* make it highly suited for certain types of genetic analysis. (1) After two haploid nuclei from cells of two

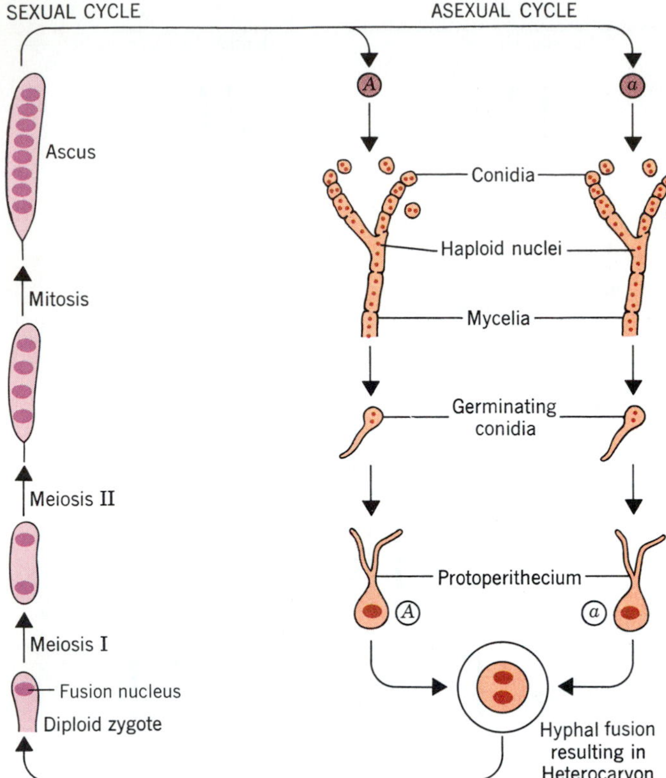

Figure 7.5 Sexual and asexual reproduction in the bread mold *Neurospora crassa.* During sexual reproduction, the products of meiosis are maintained in a tubelike structure called an ascus. Each ascus contains four pairs of ascospores, with the members of each pair being identical twins, as a result of the mitotic (equational) division following meiosis. Asexual reproduction occurs by mitotic divisions of haploid cells to form spores called conidia. ***Hyphal fusion*** can also occur between mycelia. If the fusion occurs between cells with nuclei that are genetically identical, the resulting cells are called ***homokaryons*** ("same nuclei"). If the nuclei are of two different genotypes, the resulting cells are called ***heterokaryons*** ("different nuclei"). Note that during the formation of homokaryons and heterokaryons, the nuclei do ***not*** fuse.

different mating types fuse to form a diploid fusion nucleus (much like fertilization in higher organisms), meiosis occurs, just as it does in higher plants and animals. (2) The haploid products of meiosis, called ***ascospores,*** are maintained in linear order within an elongate, tubelike structure called an ***ascus*** (plural ***asci***). Thus each ascus contains ***all four products of a single meiotic event.*** Moreover, all of the ascospores in each ascus can usually be recovered and analyzed genetically. (3) These haploid ascospores germinate and grow to produce multicellular ***mycelia*** (singular ***mycelium***), and all of these cells are also haploid. The genotype of each product of meiosis can thus be determined without carrying out testcrosses or other genetic manipulations. Because of the haploid state of the mycelium, the presence of recessive markers is never masked by dominant alleles. (4) *Neurospora crassa* can be grown on a simple synthetic medium containing only inorganic salts, a carbohydrate (such as sucrose), and one other organic compound (the vitamin biotin). (5) *Neurospora crassa* reproduces asexually as well as sexually, facilitating the maintenance of strains of particular genotypes. The sexual and asexual life cycles of *N. crassa* are illustrated in Fig. 7.5.

In most organisms, such as *Drosophila* and maize, one ***cannot*** recover and analyze the genotypes of all four haploid products of a ***single meiotic event.*** Instead, one is forced to perform testcrosses, for example, and to examine a ***population of randomly selected products of meiosis.*** The genotypes of the progeny in these random samples are then used to deduce what has occurred during meiosis. In the case of *N. crassa*, however, one can isolate and determine the genotypes of all four products of a single meiotic event. The ability to analyze all four products of individual meiotic events has provided geneticists with several kinds of useful information (e.g., see pp. 178–183). Data from crosses in which the genotypes of all the products of meiosis have been determined are called ***tetrad data.***

An analysis of tetrad data from crosses in which pairs of alleles of two genes located on the same chromosome were segregating readily shows that crossing-over occurs after replication, in the "four-strand" or tetrad stage, rather than before replication, in the "two-strand" (per homologous chromosome pair) stage. If crossing-over occurred prior to replication (or chromosome duplication), all the products of a meiotic event in which a crossover occurred between the two loci would have recombinant combinations of the genetic markers (Fig. 7.6*a*). On the other hand, if crossing-over occurs after replication, in the tetrad stage, only two of the four products of each such meiotic event will be recombinant (Fig. 7.6*b*). The other two products will have parental combinations of the segregating genetic markers. Tetrad data clearly show that the second alternative is correct. Tetrads produced by meiotic events in which a single crossover has occurred between the loci involved give rise to asci containing 50 percent genotypically recombinant ascospores and 50 percent genotypically parental ascospores (Fig. 7.6*b*). Tetrads containing all recombinant ascospores are rare, indicating that they are formed by meiotic events in which two crossovers have occurred (see pp. 178–183).

(a) If crossing-over occurs before chromosome replication:

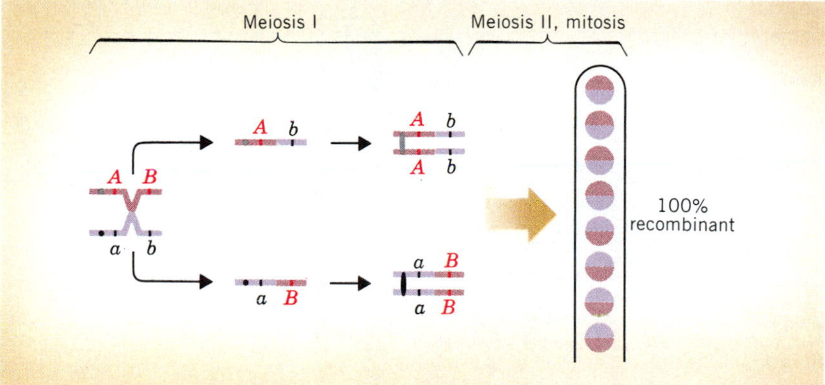

(b) If crossing-over occurs after chromosome replication:

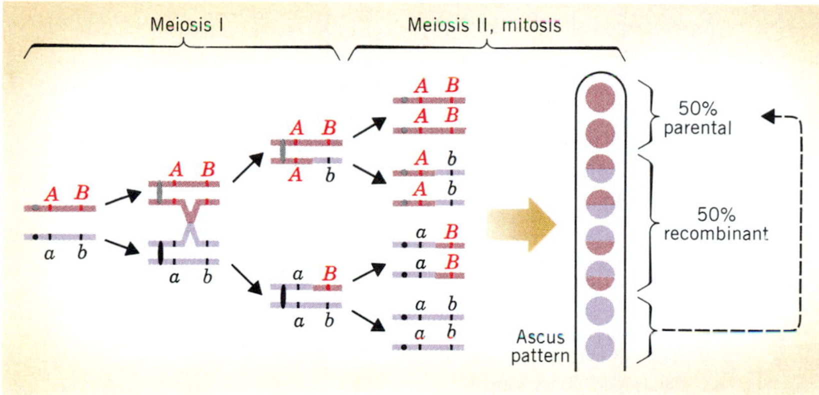

Figure 7.6 The *Neurospora* asci patterns predicted to result from a single crossover between two loci (*A*, a and *B*, b) if crossing-over occurs (a) prior to replication (when only the two homologous chromosomes are present) or (b) after replication (in the four-chromatid or tetrad stage). Recall (Fig. 7.5) that in *Neurospora* that the four haploid nuclei produced by meiosis immediately undergo one mitotic division to produce eight ascospores (or four pairs of ascospore "twins") per ascus as shown. The chromosomes, chromatids, and ascospores containing the chromosomes from the two parents are shown in two different colors. The centromeres are represented by circles. Analyses of tetrad data from many crosses carried out in many different laboratories have shown conclusively that crossing-over occurs in the postreplication tetrad stage as diagrammed in b.

CHROMOSOME MAPPING

Earlier in this chapter, we considered the results of crosses in which pairs of alleles of two genes located on the same chromosome were segregating. Such genes were shown to be ***linked*** in that ***over 50 percent of the gametes*** produced by F_1 double heterozygotes contained ***parental combinations*** of the genetic markers and ***less than 50 percent of the gametes*** contained ***recombinant combinations*** of the markers. This result was shown to contrast with the predictions of independent assortment, namely, 50 percent parental and 50 percent recombinant combinations. Subsequently, we showed that recombinant combinations of linked genes are produced by crossing-over, a process during which the breakage and exchange of parts of homologous chromosomes occur.

When the preceding kinds of analysis have been used to investigate the linkage relationships of a large number of genes of any species, the genes have always been found to occur in a distinct set of ***linkage groups***, with each linkage group corresponding to one of the pairs of homologous chromosomes in the genome of that species. In *D. melanogaster*, for example, there are 4 linkage groups, corresponding to the 4 pairs of chromosomes (Fig. 7.7). Maize has 10 linkage groups and 10 pairs of chromosomes (Fig. 7.8). The mouse has

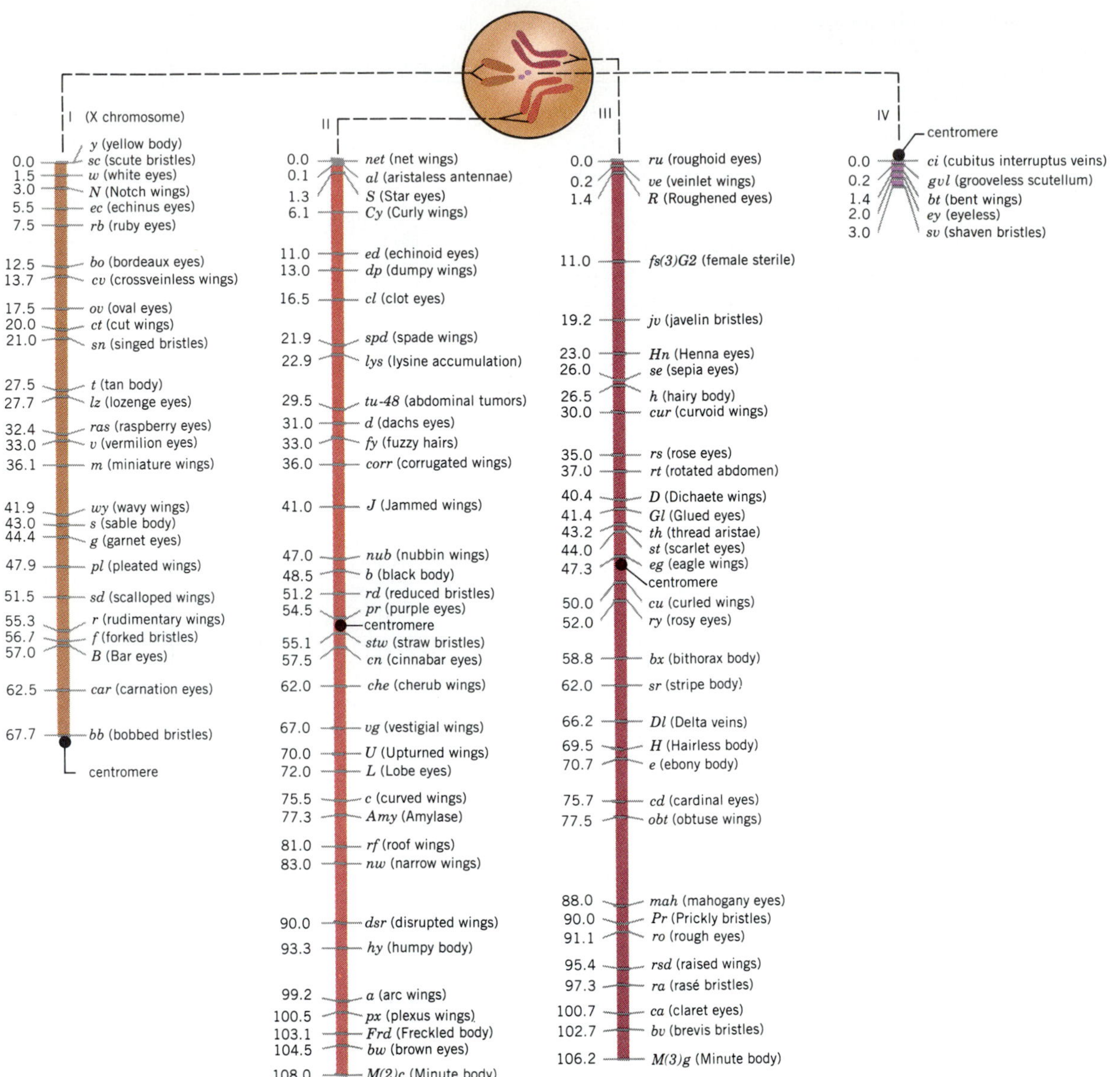

Figure 7.7 Abbreviated linkage map of *Drosophila melanogaster*. The correlation of each linkage group with one of the pairs of chromosomes (circular inset) is shown by the dashed lines. Map positions are given on the left of each linkage map; the genetic symbols used to represent the mutant allele(s) of each gene are shown on the right of each linkage map. The phenotypic description commonly used for each mutant allele is given in parentheses after the gene symbol. The map positions are those given by D. L. Lindsley and E. H. Grell, *Genetic Variations of Drosophila melanogaster,* Publication 627, Carnegie Institution of Washington, Washington, D.C., 1968, except where updated by R. C. King, pp. 625–652, in *Handbook of Genetics,* Vol. 3, *Invertebrates of Genetic Interest,* Plenum Press, New York, 1975.

20 pairs of chromosomes and 20 linkage groups, and so on. The linkage groups of humans can***not*** be investigated by the designed crosses used with other species. However, new methodologies, especially somatic-cell hybridization (see pp. 183–186) and chromosome banding techniques, have facilitated the construction of linkage maps for each of the 23 pairs of human chromosomes.

An important feature of all linkage maps is their ***linearity***; all genes in a given linkage group can be shown to map in a linear array. [A circle or ring exhibits linearity; it is simply a closed line (no ends). Most importantly, it has no branches. The linkage maps of viruses and prokaryotes (see Chapters 8 and 12) and of eukaryotic organelles (mitochondria and chloroplasts, see Chapter 20) are usually circular.] In Chapter

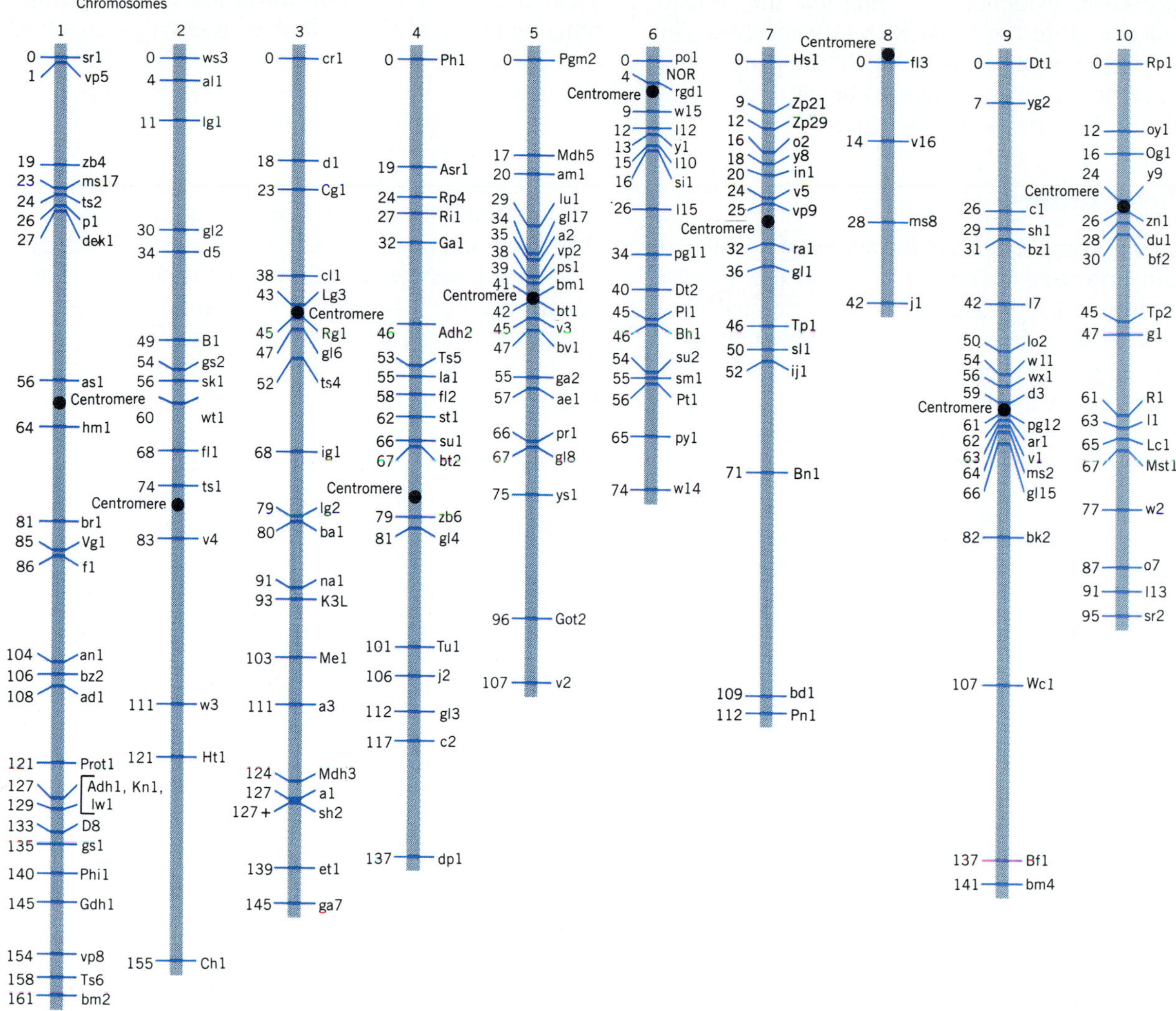

Figure 7.8 Linkage map of maize. The following are descriptions of phenotypes caused by mutant alleles of a few genes. Designations are in the form: chromosome no.-map position, gene symbol, name of mutant allele, mutant phenotype. 1-0, *sr1*, striate-1, white striations on leaves; 1-56, *as1*, asynaptic-1, failure of the meiotic prophase chromosomes to synapse; 2-30, *g12*, glossy-2, cuticle wax altered such that leaf surface appears glossy; 2-121, *Ht1, Helminthosporium turcicum* resistance, resistance to the northern leaf blight caused by this fungus; 3-0, *cr1*, crinkly-1, leaves broad and crinkled; 3-139, *et1*, etched-1, pitted or scarred endosperm; 4-24, *Rp4*, rust resistance-4, resistance to the fungus *Puccinia sorghi;* 4-101, *Tu1*, tunicate-1, kernels enclosed in long glumes; 5-20, *am1*, ameiotic-1, meiosis fails; 6-3, *NOR*, nucleolus organizer, tandem repeats of genes coding for rRNA; 6-74, *w14*, white-14, white seedling; 7-16, *o2*, opaque-2, endosperm starch soft and opaque, has high lysine content; 8-28, *ms8*, male sterile-8, anthers not exserted; 9-26, *c1*, colorless-1, nonpigmented aleurone; 9-56, *wx1*, waxy-1, no amylose in endosperm or pollen; 10-16, *Og1*, old gold-1, variable bright yellow stripes on leaf blades. (Based on the map of Coe, Hoisington, and Neuffer, 1982. For additional details and descriptions of the mutant phenotypes for all the genes shown, see E. H. Coe, Jr., D. A. Hoisington, and M. G. Neuffer, "Linkage Map of Corn (Maize) (*Zea mays* L.)," pp. 377–393 in *Genetic Maps,* Vol. 2, S. J. O'Brien, editor, National Cancer Institute, Frederick, MD, 1982.)

5, we presented evidence indicating that the genetic information is stored in DNA (RNA in some cases) in the form of the specific sequences of base-pairs. Since a sequence of base-pairs in DNA is linear, it is appropriate that linkage maps are also linear.

Two-Factor Crosses

Recombinant combinations of the alleles of two linked genes are produced by crossing-over in the interval between the two segregating loci. The rationale behind genetic mapping is that the probability of a crossover occurring between two loci is a function of the length of the interval separating the loci. Intuitively, this seems very reasonable. Consider, for example, the following two crosses:

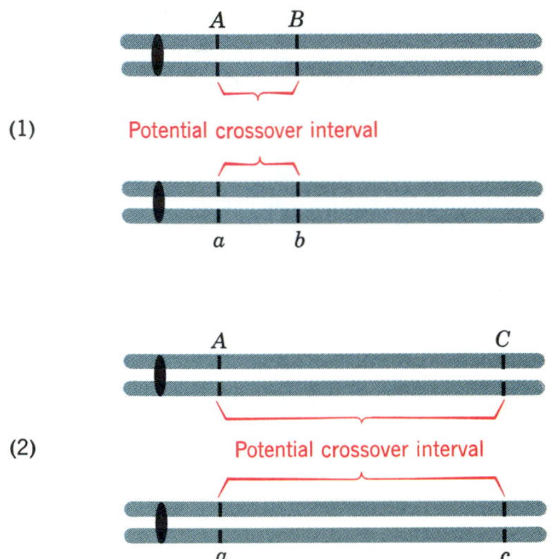

All three genes, *A*, *B*, and *C*, are on the same chromosome. The *A* and *B* loci are close together, whereas *A* and *C* are quite far apart. A crossover occurring anywhere within the long interval between the *A* locus and the *C* locus will produce recombinant combinations (*A c* and *a C*) of the two pairs of alleles segregating in Cross 2. Recombinants (*A b* and *a B*) will be produced in Cross 1 only when a crossover occurs within the short interval between the *A* locus and the *B* locus. It seems very reasonable, therefore, to expect more recombinants to be produced in Cross 2 than in Cross 1.

On the basis of the foregoing type of logic, one of Morgan's students, A. H. Sturtevant (while still an undergraduate), suggested that the frequency of recombinant gametes produced be used as an index of the distance between two loci on a chromosome. From Sturtevant's suggestion, ***linkage maps*** of the type shown in Figs. 7.7 and 7.8 have evolved. ***Linkage maps*** are made ***quantitative*** by ***defining one map unit as the distance that yields, on average, 1 percent recombinant chromosomes or gametes.*** Thus, the genes *b* (black body) and *vg* (short or vestigial wings) that yielded ***18 percent recombinant*** testcross progeny—whether the F_1 dihybrid (double heterozygote) carried the markers in the *cis* (coupling)-configuration or in the *trans* (repulsion)-configuration (see pp. 161–163)—are about ***18 map units*** apart on linkage Group II of the genetic map of *D. melanogaster* (Fig. 7.7).

As is discussed in the next section of this chapter, the relationship between recombination frequencies in crosses and map distances becomes more complex over long distances. A more precise definition of map distance is that it is the average number of crossover events that a progeny chromosome has undergone in the interval between the genetic markers. Recombination frequency is the percentage of progeny chromosomes that have undergone an odd number of crossover events in the interval between the markers. Hopefully, the reason for this distinction will be made clear in the next section, which discusses the relationship between map distance and recombination frequency in detail. For short distances, the two are essentially identical. Thus, for the sake of simplicity in introducing the concept of linkage distances, assume that recombination frequency and map distance are equivalent in the following discussion with 1 map unit being the linkage distance that yields 1 percent recombinant progeny chromosomes.

It is very important not to confuse the frequency of ***crossing-over, the event occurring in meiotic tetrads,*** with the frequency of ***crossover*** or ***recombinant chromosomes, the products of crossing-over. Linkage map distances are determined by the frequencies of crossover*** or ***recombinant chromosomes.*** As is illustrated in Fig. 7.2, ***each meiotic crossing-over event yields two crossover chromosomes*** (two recombinant chromosomes if the interval within which crossing-over occurred is flanked by heterozygous loci). Thus, if a single crossover occurs between two loci in 100 percent of the tetrads, only 50 percent of the progeny chromosomes will be recombinant (i.e., the recombination frequency will be 50 percent).

If one ***assumes*** that the average number of crossovers per chromatid occurring between two loci is directly proportional to the distance between the two loci, that is,

$$\text{Number of crossovers} = K\,(\text{distance})$$

where *K* is a proportionality constant, then one would ***predict*** that ***map distances would be additive.*** This property of ***additivity*** can be illustrated by the following example. ***If*** loci P and Q ***are linked*** and are ***8 map units apart*** (yield 8 percent recombinant gametes), and loci P and R ***are linked*** and are ***3 map units apart,*** then ***loci*** Q ***and*** R ***are also linked*** and are ***either (1) 5 map units apart or (2) 11 map units apart.*** That is,

additivity can be achieved only by the following two linkage arrangements:

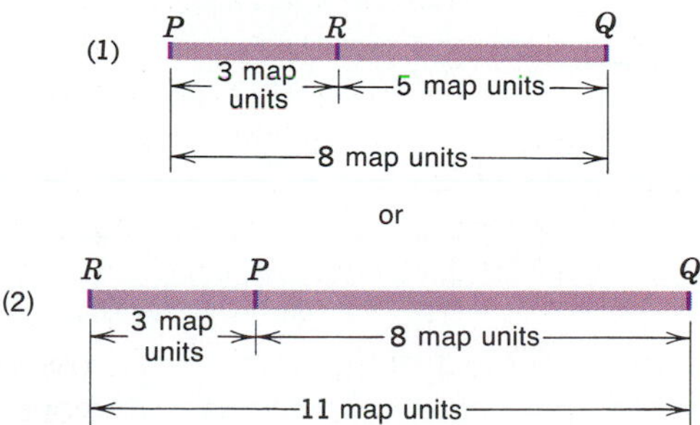

However, one can ***not*** assume that the frequency of recombination is directly proportional to distance. This assumption is reasonably accurate, at least as a first approximation, for linkage distances up to 10–20 map units. As distances become large, however, significant deviations from additivity are observed. These deviations are of the following nature.

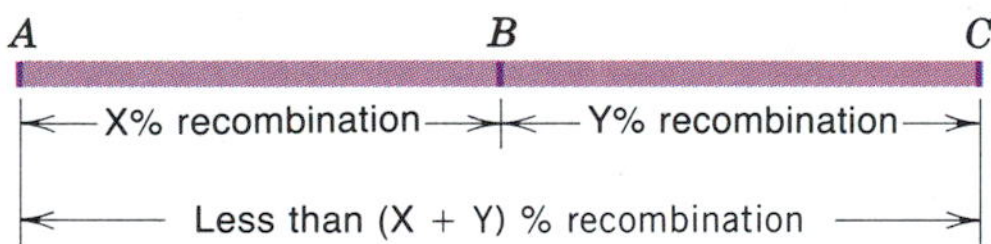

Consider, for example, that X and Y are both 15, that is, two-factor crosses, in which alleles of either genes *A* and *B* or genes *B* and *C* are segregating, would produce about 15 percent recombinant gametes. Additivity would predict that 30 percent recombinant gametes would be produced in two-factor crosses involving genes *A* and *C*. In actual crosses analogous to the gene *A*–gene *C* cross, recombinant gametes are produced at frequencies significantly lower than 30 percent.

The reason for this systematic deviation from additivity as distances become large is that, when loci are far apart, double crossovers occur at significant frequencies. Consider any two homologous, nonsister chromatids that have recombinant combinations of genetic markers as a result of a crossover occurring between the two marked loci. A second crossover involving these two chromatids in the interval between the two loci will, in affect, cancel out the first crossover.

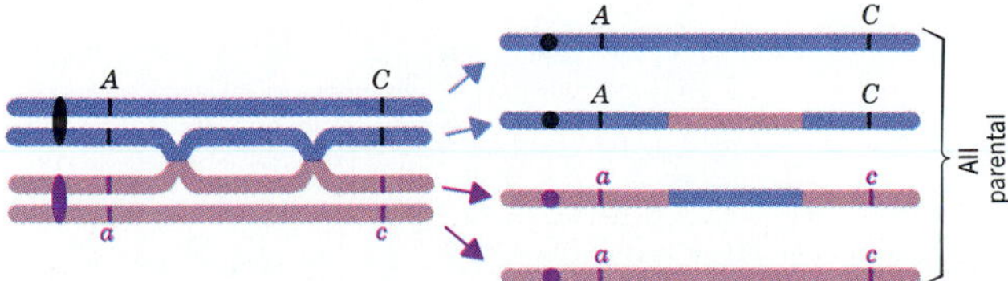

Although two crossovers have occurred, they will not be counted when the data for such a two-factor cross are recorded. For this reason, two-factor crosses involving loci that are far apart (10–20 map units or more) will underestimate true linkage distances.

Maximum Frequency of Recombination Is 50 Percent

The ***maximum frequency of recombination*** that can result from crossing-over between linked genes is ***50 percent***. Fifty percent is also, of course, the frequency of recombination observed for genes on different chromosomes that are assorting independently. Thus, recombination frequencies never exceed 50 percent. For two-factor crosses, recombination frequencies are observed to asymptotically approach 50 percent as the ***additive map distance*** between the loci involved increases (Fig. 7.9). The additive map distance is the summation of the map distance between genetic markers spanning the shortest marked intervals. The additive map distance between genes *A* and *Z* in Fig. 7.9 is 130 map units, the sum of the distances between *A* and *B*, *B* and *C*, *C* and *D*, . . ., *X* and *Y*, and *Y* and *Z*. Yet, in a two-factor cross between *A Z* and *a z*, only 50 percent or less of the progeny chromosomes will be recombinant. Clearly, then, the observed recombination frequencies are not additive. The deviation from additivity results from the increasing number of multiple crossovers as distances become large. All multiple crossovers with an ***even number of exchanges*** between two segregating loci will yield parental combinations of the genetic markers and thus go undetected. When intervals between loci are small, such multiple crossovers will be rare and have a negligible effect on additivity relationships.

The maximum frequency of recombination for genes located on the same chromosome is 50 percent—the same frequency as observed for genes located on different chromosomes and thus assorting independently. The reason for the 50 percent maximum frequency can be illustrated as follows. Assume that a single crossover occurs between two loci in every meiotic event. In reality, that will not happen, since crossovers are random events such that no crossovers will occur in some meiotic events, double crossovers will occur in others, and so on. But assume that it does happen. The resulting recombination frequency will be 50 percent because each crossover involves only two of the four chromatids, and the gametes carrying the two noncrossover chromatids will have parental combinations of the genetic markers (as shown in Fig. 7.2).

Now consider an analogous hypothetical situation. Assume that two crossovers always occur between two loci. Again, the recombination frequency will be 50 percent. As is illustrated in Fig. 7.10, double crossovers can occur in three different ways. (1) ***Two-strand double crossovers*** occur when both crossovers involve the same two chromatids. (2) ***Three-strand double crossovers*** are those in which the second crossover involves one of the same two chromatids as the first

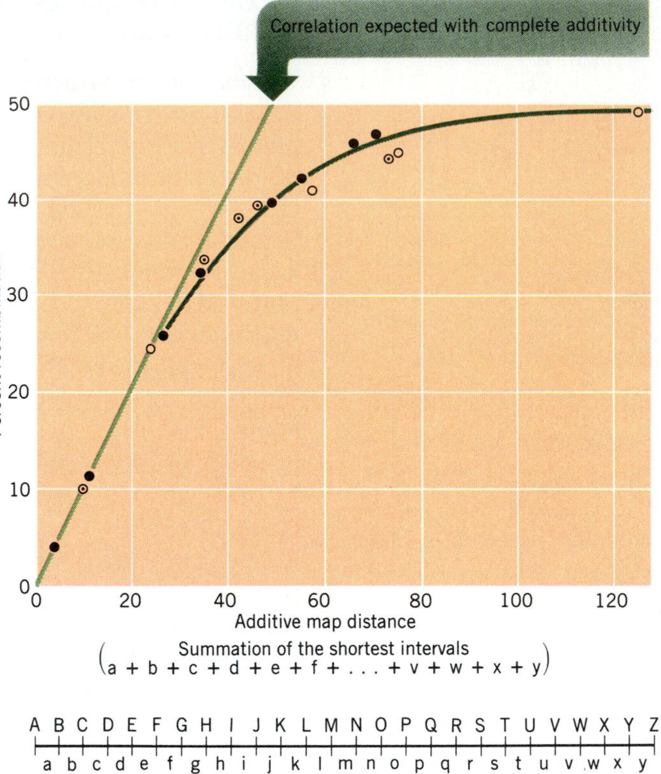

Figure 7.9 Frequency of recombination in two-factor crosses as a function of the additive map distance separating two loci. The data shown are those summarized by Perkins (*Genetics* 47: 1253–1274, 1962) for *Drosophila* (●), *Neurospora* (⊙), and maize (○). The frequency of recombination asymptotically approaches 50 percent, the value characteristic of unlinked genes, as the additive distance between two loci increases. At map distances in excess of about 20 map units, the deviation from linearity (predicted if recombination frequencies and thus map distances were completely additive) becomes quite large; this results from the inability to detect double crossovers (or any even number of multiple exchanges) in two-factor crosses. The additive map distance is the summation of the map distances obtained from two-factor crosses in which pairs of alleles of loci spanning the shortest intervals are segregating. Thus, the additive map distance from *A* to *Z* is 130 map units, the summation of the distance from *A* to *B* (*a* map units), the distance from *B* to *C* (*b* map units), and so on. Even though genes *A* and *Z* are 130 map units apart, less than 50 percent of the gametes produced by an *A Z*/*a z* heterozygote will be recombinant (*A z* and *a Z*). Many linkage maps are over 100 map units long. In Fig. 7.7 the linkage map corresponding to chromosome II of *D. melanogaster* is 108 map units long. The linkage map corresponding to chromosome 1 of maize (Fig. 7.8) is over 160 map units long.

This relationship between recombination frequencies and additive map distances can be described mathematically by so-called ***mapping functions***. The first mapping function was developed by J. B. S. Haldane in 1919; it is often referred to as the "Haldane function." Since then, many different mapping functions have been devised, some of which are applicable to only one or a few related species. These mapping functions are useful in allowing one to convert recombination frequencies to physical distances (usually in terms of nucleotide-pairs) with fair accuracy. The reader is referred to Stahl, *Genetic Recombination*, 1979, for an excellent discussion of mapping functions.

crossover plus one different chromatid. Three-strand double crossovers can occur in twice as many ways as two-strand or four-strand double crossovers (see Fig. 7.10). (3) ***Four-strand double crossovers*** occur when the second crossover involves the two chromatids not involved in the first crossover. If the second crossover in a tetrad involves two chromatids at random, irrespective of which chromatids were involved in the first crossover, then the ratio of two-strand to three-strand to four-strand double crossovers will be 1:2:1. Devia-

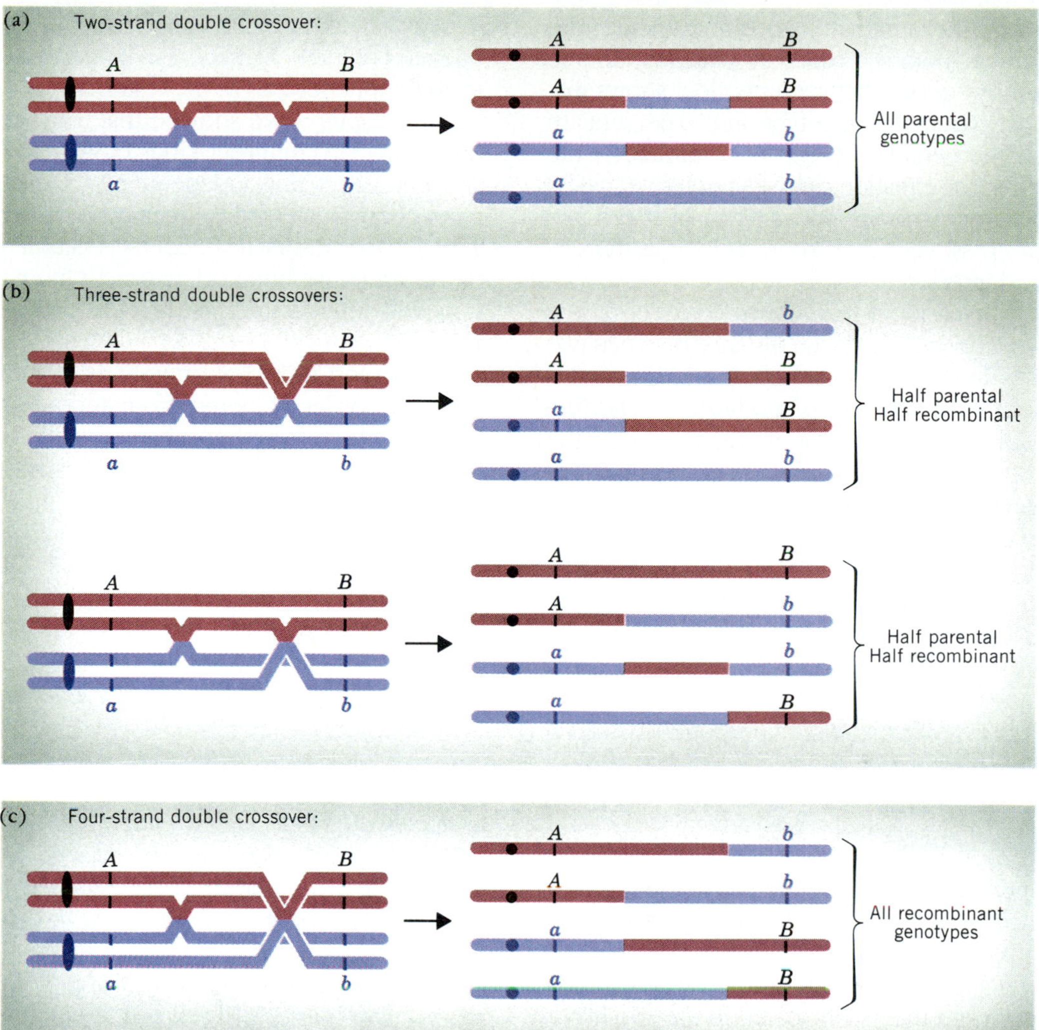

Figure 7.10 Diagram illustrating the consequences of (a) two-strand, (b) three-strand, and (c) four-strand double crossovers between two loci. All heterozygotes are in the *cis* (coupling)-configuration. The chromatids of the two homologs are shown in two different colors. The circles represent the centromeres. All the gametes resulting from meiotic events in which two-strand double crossovers have occurred between the two loci have parental combinations of the segregating pairs of alleles. Those resulting from meiotic events in which three-strand doubles have occurred are half parental and half recombinant in genotype. Only meiotic events in which four-strand double crossovers have occurred yield all gametes with recombinant genotypes. Three-strand double crossovers can occur in twice as many ways as two-strand or four-strand double crossovers. Given the 1:2:1 ratio of two-strand to three-strand to four-strand double crossovers, the total double crossover gamete population will be 50 percent parental and 50 percent recombinant.

tions from this 1:2:1 ratio are said to result from ***chromatid interference:*** the effect of the involvement of two chromatids in a crossover on which chromatids will be involved in additional crossovers in nearby regions. Ratios of two- to three- to four-strand double crossovers usually approximate 1:2:1, indicating that little, if any, chromatid interference is occurring.

All the gametes produced by meiotic events in which two-strand double crossovers have occurred between two loci will contain parental combinations of the genetic markers. Half of the gametes produced by meiotic events in which three-strand double crossovers have occurred between two loci will contain recombinant combinations of the genetic markers; the other half will contain parental combinations. All the gametes resulting from four-strand double crossovers between the loci will be recombinant. Given the 1:2:1 ratio of two-strand to three-strand to four-strand double crossovers, the summation of the gamete types produced from meiotic events in which a double crossover always occurs between two loci shows that 50 percent are parental in genotype and 50 percent are recombinant in genotype (Fig. 7.10). Similarly, the frequency of recombination can be shown to equal 50 percent if three crossovers always occur between two loci, or four crossovers, and so on, as long as there is no chromatid interference.

This may be easier to visualize by considering any two nonsister chromatids. As the distance between two loci becomes large, the frequency of multiple crossovers with an ***even number of exchanges (yielding parental combinations)*** will approximately equal the frequency of multiple crossovers with an ***odd number of exchanges (yielding recombinant combinations).*** If chromosomes were infinitely long, the two (even number of exchanges and odd number exchanges) would occur with exactly equal frequency, yielding 50 percent recombinant and 50 percent parental gamete types. Only as a result of very unlikely events, such as a preponderance of four-strand double crossovers, would the observed recombination frequency exceed 50 percent.

Three-Factor Crosses

How can one detect and take into account the double crossovers that go unrecognized in a two-factor cross? This can be accomplished (and then only in part) only by identifying a third gene with distinguishable alleles that maps between the loci involved in the two-factor cross:

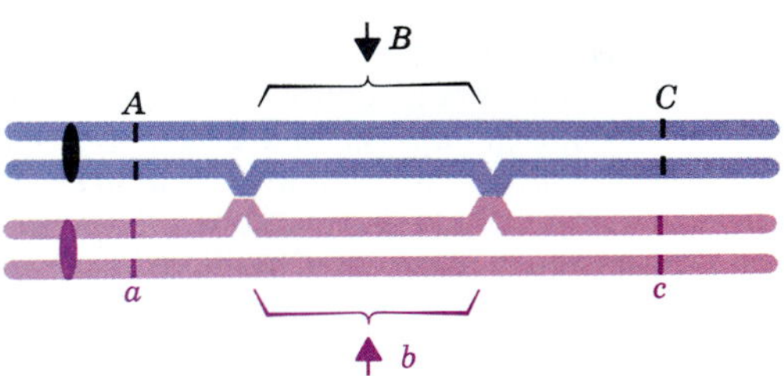

Given a third genetic marker on a chromosome, one can carry out a ***three-factor cross***, a cross in which three pairs of alleles are segregating. In addition to allowing the detection of some of the double crossovers not recognizable in two-factor crosses, ***three-factor crosses*** allow one to ***order the markers*** involved. The three-factor cross is undoubtedly the most important tool used in chromosome mapping. In diploids, the three-factor cross is usually used in a manner analogous to the use of the two-factor cross in detecting linkage. That is, homozygotes are crossed to produce triple heterozygotes or trihybrids (such as *A B C*/*a b c*), and the trihybrids are then testcrossed (to *a b c*/*a b c* homozygotes), so that the kinds and frequencies of the F_1 gametes can be determined directly from the phenotypes of the testcross progeny.

In Chapter 2 we showed that trihybrids can produce eight (2^3) different kinds of gametes. If all three genes are assorting independently, the eight gamete types occur with equal frequency. Deviations from this 1:1:1:1:1:1:1:1 ratio will, of course, occur if two or more of the genes are linked. The exact ratio will be determined by the degree of linkage between the loci involved.

First, and most importantly, how can one determine the order of genes from three-factor crosses? Consider three hypothetical pairs of alleles (*X*, *x*; *Y*, *y*; and *Z*, *z*) of linked genes. Assume that the order of the three genes is known to be *X*-*Y*-*Z*. Now, ask yourself the following questions. (1) What kinds of gametes would a triple heterozygote of genotype *X Y Z*/*x y z* be expected to produce? (2) In what relative frequencies would these gametes be produced? (3) What different patterns of meiotic crossing-over can be detected by examining the testcross progeny of this triple heterozygote?

The eight gamete types produced by the *X Y Z*/*x y z* triple heterozygote will result from four different patterns with respect to crossing-over, as shown in Fig. 7.11. ***Although crossing-over occurs in the tetrad stage of meiosis, for simplicity the diagrams show only the two chromatids involved in each crossover type.***

In three-factor crosses of the type diagrammed, the ***parental gamete types*** (as revealed by testcrosses) ***will necessarily be present in the highest frequencies.*** In the present case, the markers exist in the complete *cis* (coupling)-configuration. Of course, they may be present on the parental chromosomes in any combination. The parental combinations can always be determined, however, by identifying the two classes of testcross progeny that occur at the highest frequencies. These two parental progeny classes are expected to occur in approximately equal frequencies (the deviations representing simply random sampling variations).

The ability to identify the parental progeny classes allows one to determine the linkage relationships of the genetic markers involved, that is, whether the alleles of each pair of genes are present in the *cis*- or *trans*-configuration on the homologous chromosomes of the F_1 progeny. In the preceding example, the markers are all in the *cis*-configuration (*X Y Z* and *x y z*).

The two reciprocal double crossover progeny classes can also be identified. They will always be the two classes present in the lowest frequencies among the eight possible combinations. If the two crossovers involved in a double crossover are independent events (we will discuss whether they are later), then the probability of a double crossover occurring should equal the product of the probabilities of the two single crossovers occurring. Stated differently, if crossovers in Interval I occur with a frequency of p and crossovers in Interval II occur with a frequency of q, then double crossovers with one exchange in Interval I and a second exchange in Interval II should occur with a frequency of pq. Since p and q are fractions of less than ½ (usually much less than ½), pq is necessarily smaller than p or q. Thus, the reciprocal double crossover progeny classes will always be less frequent than either of the pairs of reciprocal single crossover classes. In actuality, double crossover types usually occur even less frequently than expected if the two crossovers represent independent events. (This will be discussed later.)

The ability to ***identify the two reciprocal parental progeny classes*** and the ***two double crossover progeny classes*** of a three-factor cross ***allows one to determine the order of the three loci***. A double crossover always results in the center marker on each parental chromosome being placed in a recombinant combination with the outside markers from the other parental chromosome. Stated differently, the two outside markers will maintain the linkage relationship in the double crossover chromosomes that were present in the parental chromosomes. In our hypothetical *X-Y-Z* example, the double crossover gametes carried *X-y-Z* and *x-Y-z*. The middle marker *Y* of one parent is now present on a chromosome with the outside markers *x* and *z* from the other parent, and vice versa. ***Only one of the three possible orders of the markers in a three-factor cross will allow the observed double crossover types to be produced from the observed parental types by a double crossover.*** Assume, for the moment, that the order of the genes is unknown in our hypothetical *X-Y-Z* example, but that we know that the parental classes are *X Y Z* and *x y z* and the double crossover classes are *X y Z* and *x Y z*. The three possible orders of the loci and the double crossover gametes that would be produced from each order are diagrammed in Fig. 7.12. ***Again, for clarity, only two of the four chromatids of each tetrad are shown.*** Clearly, the known double crossover gamete types (*X y Z* and *x Y z*) can be produced from the parental chromosomes by a double crossover only in the case of Order I (*X-Y-Z*).

Once the order of the markers is determined by the procedure just detailed, the linkage distances for each interval can be calculated from the observed frequency of recombination (in percent) for each adjacent pair of markers. This can best be illustrated by an example. Suppose that the data given in Table 7.1

(1) No crossover within the marked region,

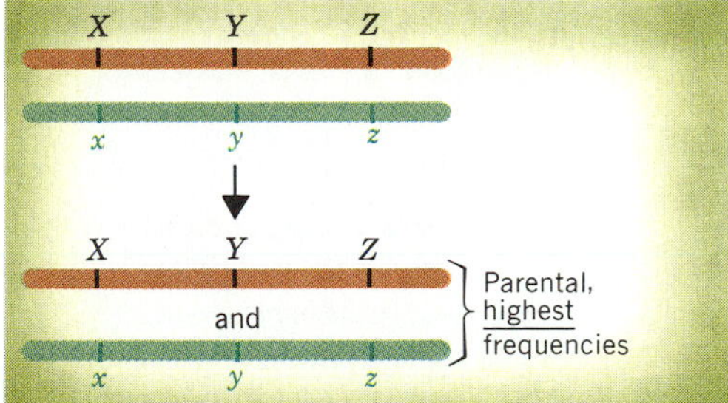

(2) Single crossover between loci *X* and *Y*,

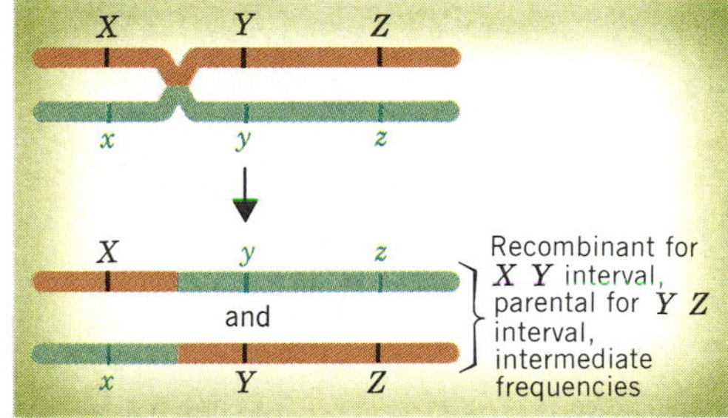

(3) Single crossover between loci *Y* and *Z*,

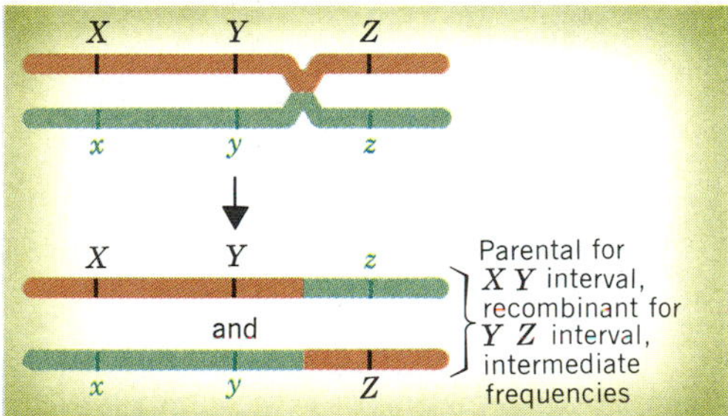

(4) Double crossover, one between *X* and *Y*, the other between *Y* and *Z*.

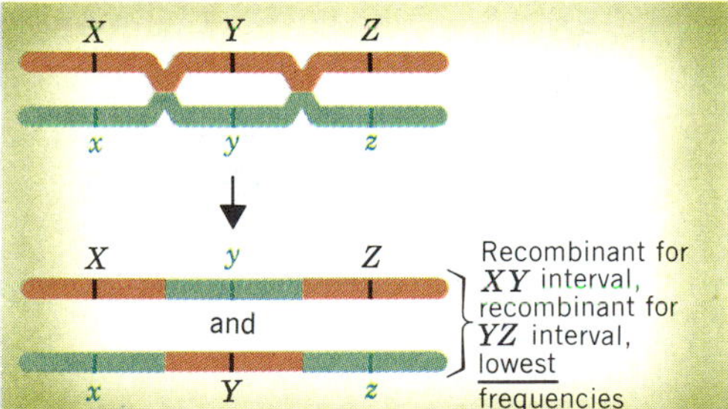

Figure 7.11 Gametes produced by the *X Y Z*/*x y z* triple heterozygote. Note that for clarity only two of the four chromatids of the tetrad are shown.

resulted from a testcross of an F_1 triple heterozygote. Note that we do not know before analyzing the data whether the markers are in *cis* (coupling)- or *trans* (repulsion)-configuration, nor do we know the order of the loci. They are arbitrarily written in alphabetical order.

Note that all eight possible genotypes are present, but in very different frequencies, clearly showing that the three genes are linked. The reciprocal types, such as *A b C* and *a B c*, are present in about equal fre-

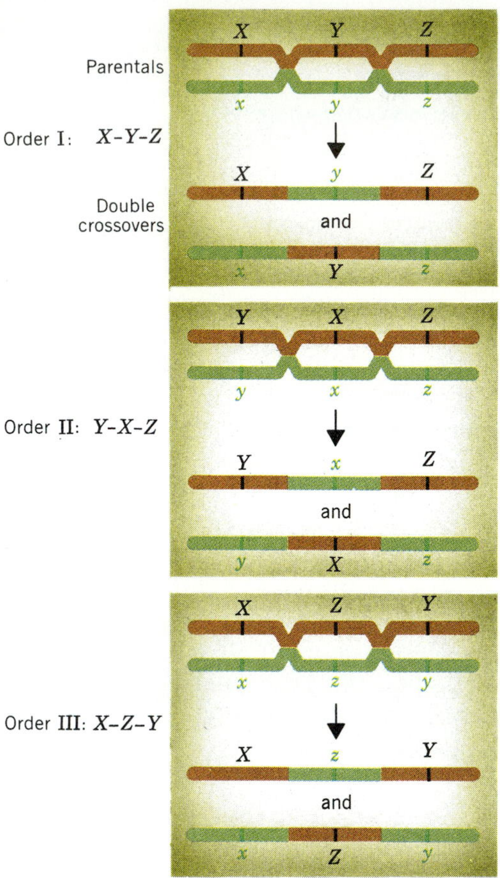

Figure 7.12 Double crossover chromosomes produced for the three ***possible*** orders of the linked markers *X*, *Y*, and *Z*. Note that for clarity only two of the four chromatids of the tetrad are shown.

TABLE 7.1 Three-Factor Testcross Data (Note That the Alleles Are Listed Arbitrarily in Alphabetical Order)

GENOTYPE	NUMBER OF PROGENY	
ABC/abc	370	Parental
abc/abc	385	
Abc/abc	45	Single crossover in Interval I
aBC/abc	50	
ABc/abc	2	Double crossover: one in Interval I and one in Interval II
abC/abc	3	
AbC/abc	75	Single crossover in Interval II
aBc/abc	70	
Total =	1000	

quency, within the range of random sampling variation. Since the triple recessive (*a b c*) chromosome was contributed by the gamete from the testcross parent, we need only concern ourselves with the other chromosome (carrying the markers shown on the left). Based on relative frequencies, the ***parental chromosomes must have been of genotypes* A B C *and* a b c** (highest frequencies), and the ***double crossover progeny chromosomes are of genotype* A B c *and* a b C** (lowest frequencies). This tells us that ***gene* C *must be in the center. Gene* C *was present with* A *and* B *on one parental chromosome; it is present with* a *and* b *on the double crossover chromosome*** (and vice versa for *c*). If you are uncertain about the order, draw out the predicted double crossover genotypes for all three possible orders as we did with the *X-Y-Z* example. Only the order *A-C-B* (*A C B*/*a c b*, F_1) will yield the observed *A c B* and *a C b* (written *A B c* and *a b C* in the foregoing) double crossover classes.

Now that the order of the markers is known, the genetic map can be completed by calculating the linkage distances between loci *A* and *C* and between loci *C* and *B*. The parental marker combinations for the *A–C* interval are *A C* and *a c*; the recombinants are *A c* and *a C*. The frequency of recombination for the *A–C* interval is thus (45 + 50 + 2 + 3)/1000 or 0.10. Note that the progeny with genotypes resulting from double crossovers (*A B c* and *a b C*) are included in the numerator. One of the two crossovers occurred within the *A–C* interval. Thus, these progeny carry recombinant (*A c* and *a C*) arrangements of the alleles of these two loci.

Recall that one map unit equals the distance that yields 1 percent recombination. Thus, *A* and *C* are 10 map units apart on the linkage map. Similarly, *C* and *B* are 15 map units (calculated from [2 + 3 + 75 + 70] × 100/1000) apart. Again, note that the progeny with genotypes resulting from double crossovers are included in the numerator. One of the crossovers occurred within the *C–B* interval and produced chromosomes with recombinant (*C b* and *c B*) arrangements of the alleles of the *B* and *C* loci. Thus, the progeny with genotypes produced by double crossovers are included in the numerator (recombinants) in calculating both linkage distances.

The linkage map defined by the data in Table 7.1 is, therefore,

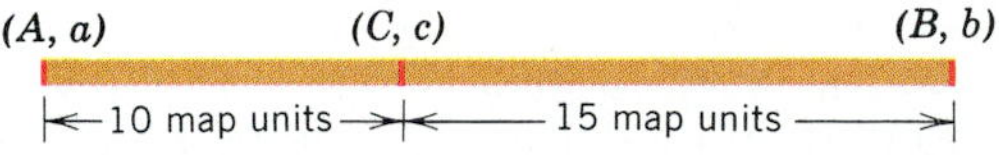

In this example, all the markers were present on the parental chromosomes in the *cis*-arrangement. One can visualize and remember the products of the

various crossover events most easily when the markers are in this arrangement, an important aid in working through a three-factor cross for the first time. However, the markers could just as likely have been present on the chromosomes of the F_1 trihybrid in any of the other three possible arrangements: (1) *A C b/a c B*, (2) *A c b/a C B*, or (3) *A c B/a C b*. The same procedures are used to order the loci and to determine linkage distances in all cases.

Consider, for example, the results (Table 7.2) of a testcross in *Drosophila* between female F_1 flies heterozygous at three loci on the third chromosome, cu/cu^+ (curled versus straight wings), e/e^+ (ebony versus gray bodies), and st/st^+ (scarlet versus red eyes), and completely homozygous recessive (*cu e st/cu e st*) male flies.

The data in Table 7.2 show that one of the parental chromosomes carried markers *cu*, *e*, and st^+, whereas the other carried cu^+, e^+ and *st*. Thus the alleles at the *curled* and *ebony* loci were in the *cis* (coupling)-configuration, whereas the alleles at the *scarlet* locus were in the *trans* (repulsion)-configuration relative to those of *curled* and *ebony*. Since the parental (most frequent) combinations are $cu\ e\ st^+$ and $cu^+\ e^+\ st$, and the double crossover (least frequent) combinations are $cu\ e^+\ st$ and $cu^+\ e\ st^+$, the *cu* locus must lie between the *st* and *e* loci. (Note that loci are usually identified using the symbol for their mutant alleles, rather than those for their wild-type alleles).

Observe that a double crossover moves the middle marker onto a chromosome with the two outside markers from the other parent. By contrast, the two outside markers on each parental chromosome remain together. In our example, *e* and st^+ are together on one of the parental chromosomes ***and*** on one of the double crossover chromosomes. The same is true for e^+ and *st*. Thus, these must be the outside markers. The alleles at the *curled* locus, on the other hand, are present on the double crossover chromosomes with the alleles of the *ebony* and *scarlet* loci inherited from the other parent. Thus, the *cu* locus must be in the middle.

With a little practice, you will be able to order the three loci by simply looking at the parental and double crossover genotypes. Until you reach this level of proficiency, however, diagram the three possible orders of the parental markers and determine which order will give rise to the ***observed*** double crossover genotypes by means of a double crossover.

Having determined the order of the loci to be *st-cu-e*, we simply have to calculate the distances from *st* to *cu* and from *cu* to *e* to complete the linkage map. For the *st–cu* interval, the recombinant genotypes are *st cu* and $st^+\ cu^+$. The distance in map units is thus $(24 + 30 + 2 + 4)/1000 \times 100 = 6$ map units. For the

TABLE 7.2 Results of a Three-Factor Testcross in *Drosophila melanogaster*

GENOTYPE	NUMBER OF PROGENY	
$cu\ e\ st^+/cu\ e\ st$	366	Parental
$cu^+\ e^+\ st\ /cu\ e\ st$	380	
$cu\ e\ st\ /cu\ e\ st$	24	Single crossover in Interval I
$cu^+\ e^+\ st^+/cu\ e\ st$	30	
$cu^+\ e\ st\ /cu\ e\ st$	89	Single crossover in Interval II
$cu\ e^+\ st^+/cu\ e\ st$	105	
$cu\ e^+\ st\ /cu\ e\ st$	2	Double crossover; one in each interval
$cu^+\ e\ st^+/cu\ e\ st$	4	
Total = 1000		

cu–e interval, the recombinant genotypes are $cu\ e^+$ and $cu^+\ e$. The linkage distance, based on the results of this cross, is $(89 + 105 + 2 + 4)/1000 \times 100 = 20$ map units. The complete linkage map defined by the results of this one cross is, therefore,

which agrees reasonably well with the known map positions of these three loci (Fig. 7.7). If additional linkage data are available for this region of the chromosome, a more accurate estimate of distance may result if one uses the sum of the shortest intervals (see Fig. 7.9).

By combining linkage data from many three factor crosses that have at least one marker in common, detailed linkage maps, such as those shown in Figs. 7.7 and 7.8, can be derived.

Interference

In the three-factor testcross data given in Table 7.1, five progeny resulting from double crossovers in meiosis in the F_1 trihybrid were observed. The ***observed frequency of double crossover gametes is thus 0.005.*** If crossovers occur completely at random, such that double crossovers are the result of two independent events as was discussed earlier, then the ***expected frequency of double crossover gametes would be 0.10 × 0.15 or 0.015.*** Fewer double crossovers are usually observed than would be expected if crossovers were independent events. This phenomenon, which was first observed by H. J. Muller in 1916, is called ***chromosome interference*** or ***chiasma interference.*** It is observed in most three-factor crosses in eukaryotes.

Although its molecular basis is unknown, the observed levels of interference indicate that the occurrence of one crossover decreases the likelihood of (or "interferes" with) another crossover occurring nearby. The degree of interference is measured by ***the coefficient of coincidence***, which is simply the ratio of the observed to the excepted frequency of double crossover gametes, that is,

$$\text{Coefficient of coincidence} = \frac{\text{observed double crossover frequency}}{\text{expected double crossover frequency}}$$

where the expected double crossover frequency is that predicted if the two crossovers are independent events. In the absence of interference, the coefficient of coincidence will equal 1. Sometimes levels of interference are given as the ***coefficient of interference***, which equals 1 minus the coefficient of coincidence. The coefficient of interference will thus be 0 in the absence of interference.

Interference values between 0 and 1 indicate that ***positive interference*** has occurred. For example, in our three-factor testcross example, the coefficient of coincidence was 0.005/0.015 or 0.33. The coefficient of interference for these data is therefore $1 - 0.33 = 0.67$. Positive interference indicates that one crossover interferes with the occurrence of a second crossover nearby.

In certain microorganisms, particularly bacteriophages, coefficients of coincidence greater than 1 often occur. Coefficients of coincidence larger than 1 indicate that the occurrence of one crossover increases the likelihood of additional crossovers occurring nearby (so-called "clustering" of crossovers). This second type of interference is called ***negative interference***

Ordered Tetrad Data

As was discussed earlier (see pp. 165–167), in certain fungi all the products of a single meiotic event can be recovered and analyzed genetically. In some species, the individual haploid products of meiosis are maintained in a linear cylindrical structure (called an ***ascus***) in the order in which they are formed during meiosis. In *Neurospora crassa*, a single mitotic division follows meiosis, producing eight haploid ascospores (four pairs of identical "ascospore twins") that are maintained in the tubelike ascus (see Fig. 7.5). The individual ascospores can be removed from the ascus and separated to discrete cells by micromanipulation, germinated, and then analyzed genetically. Their order within the ascus can be recorded during dissection, providing the geneticist with ***ordered tetrad data***

Ordered tetrad data are unique in allowing the experimenter (1) to determine whether recombination is reciprocal (as it has been assumed to be in all the examples discussed so far in this chapter) or nonreciprocal (as in the case of ***gene conversion***; see pp. 196–199) and (2) to map the centromere.

Consider the segregation of two alleles (say *A* and *a*) of a gene in *N. crassa*. The arrangement of ascospore genotypes in an ascus (called the ***ascus pattern***) produced when a crossover occurs in the interval between that gene and the centromere on the chromosome will be very different from the ascus pattern produced when no crossover occurs in that interval (compare Figs. 7.13 and 7.14; see also Fig. 7.15). If no crossover occurs between the *A* locus and the centromere, the four ascospores at one end of the ascus will all be of one genotype (either *A* or *a*), and the four ascospores at the other end will be of the other genotype (Figs. 7.13 and 7.15). This 4:4 pattern is called the ***first-division segregation pattern*** because the ***segregation*** of the two alleles ***occurs during the first division of meiosis***. There are actually two first division segregation patterns: (1) the four *A* spores at the tip of the ascus and the four *a* spores at the base, and (2) the four *a* spores at the tip and the four *A* spores at the base. Which pattern forms in a given ascus depends on which way the tetrad is orientated at the metaphase I plate.

If a crossover occurs between the *A* locus and the centromere, a ***second-division segregation pattern*** (e.g., 2*A* : 2*a* : 2*A* : 2*a*) will result (Figs. 7.14 and 7.15). In this case, the alleles are not segregated into different nuclei until the second division of meiosis. Actually, four different second-division segregation patterns occur, depending on the orientation of the recombinant chromatids during anaphase II of meiosis. Thus, the four second-division segregation patterns shown in Fig. 7.16 occur with approximately equal frequency. Three of these four patterns are represented among the asci shown in the photomicrograph in Fig. 7.15. All four of these patterns provide equivalent information, namely, the occurrence of second-division segregation. They may thus be considered one ascus pattern. In presenting ordered tetrad data, only the genotypes of the four identical pairs of ascospores (rather than all eight ascospores) in each ascus are usually given. The preceding second-division segregation pattern is often abbreviated as *A* : *a* : *A* : *a*, as shown (top to bottom) in Fig. 7.16.

The recombination frequency for the interval from a gene to the centromere is ***one-half the frequency of asci that exhibit second-division segregation patterns for the alleles of the gene.*** The ***one-half*** corrects for the fact that crossing-over involves only two of the four chromatids in a tetrad (Fig. 7.6*b*). Half of the ascospores in each ascus with a second-division

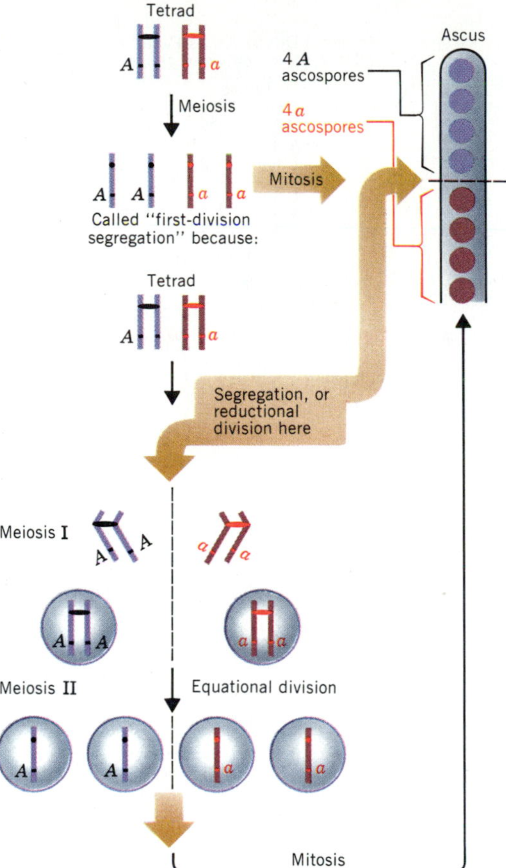

Figure 7.13 Diagram showing the first-division segregation ascus pattern that is produced from meiotic events in which no crossover occurs between the segregating locus and the centromere. Note that the alleles (*A* and *a*) have segregated into different nuclei at the completion of the first division of meiosis. The second division of meiosis is equational, just like the subsequent mitotic division. The result is an ascus with four ascospores of genotype *A* at one end and four spores of genotype *a* at the other end, the so-called first-division segregation pattern. Two first-division segregation patterns occur with equal frequency: (1) *A* alleles at the tip and *a* alleles at the base of the ascus and (2) *a* alleles at the tip and *A* alleles at the base of the ascus.

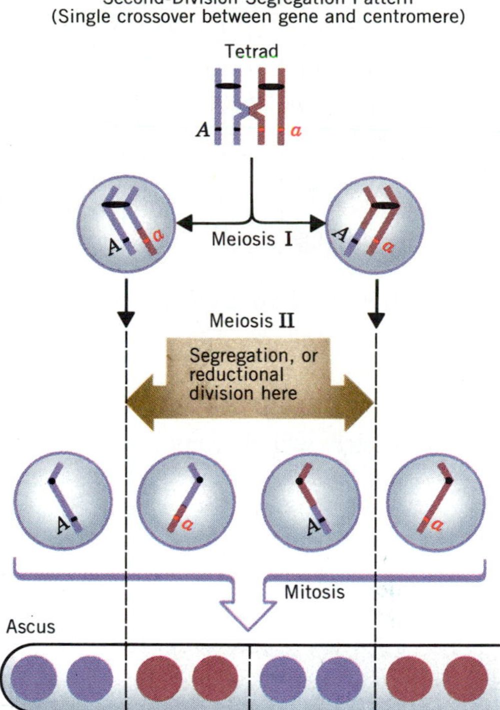

Figure 7.14 Diagram showing the second-division segregation ascus pattern that is produced from meiotic events in which a single crossover occurs between the segregating locus and the centromere. Note that the two alleles (*A* and *a*) do not segregate into different nuclei until the second division of meiosis. The result is a 2:2:2:2 ascus pattern with alternating pairs of ascospores of genotype *A* and pairs of ascospores of genotype *a*. Compare this ascus pattern with that shown in Fig. 7.13, which resulted when no crossover occurred.

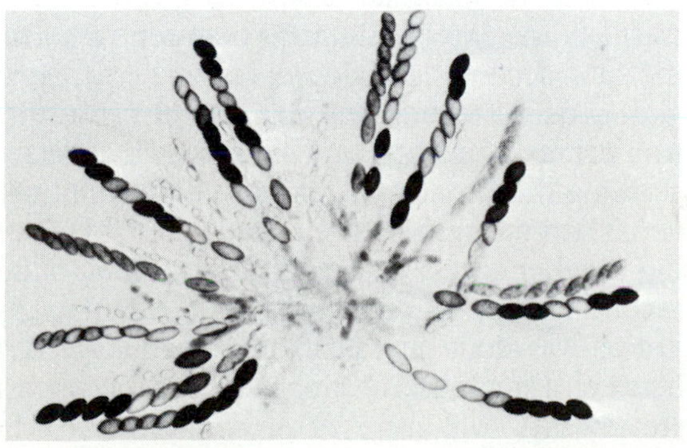

Figure 7.15 Photomicrograph of the asci from one perithecium of *Neurospora,* showing both first-division segregation patterns and three of the four second-division segregation patterns expected for the segregation of a single pair of alleles. The dark ascospores carry the wild-type allele ws^+; the white or colorless ascospores carry the mutant allele *ws* (white spore). (Reproduced with permission from R. L. Phillips and A. M. Srb, *Canad. J. Genet. Cytol.* 9: 766–775, 1967. Original photograph courtesy of R. L. Phillips.)

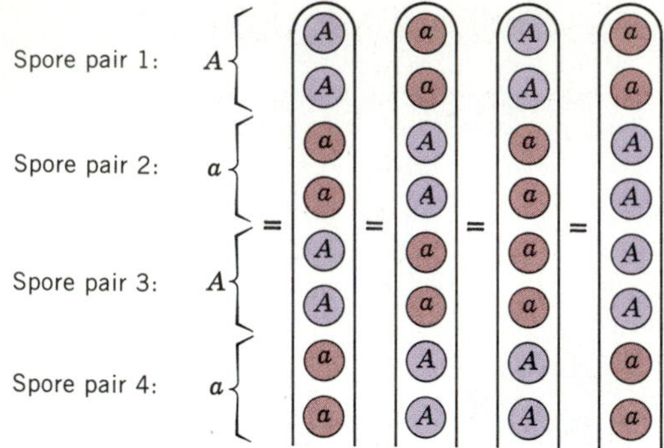

Figure 7.16 The four second-division segregation ascus patterns.

segregation pattern will therefore contain chromosomes that were not involved in crossing-over. Thus, the map distance (assume no other data are available) from gene *A* to the centromere equals

$$\frac{\frac{1}{2}(\text{number of asci with second-division segregation patterns})}{\text{total number of asci}} \times 100$$

The equivalent of a three-factor cross can be done in *N. crassa* by using alleles of just two linked genes, since the centromere can be used as a third marker. That is, ordered tetrad data can be used to order two loci with respect to the centromere. However, before considering how two markers can be ordered relative to the centromere, we should examine the ascus patterns that can occur with two segregating pairs of alleles, for example, *A*, *a* and *B*, *b*. Three different ascus patterns occur; these patterns and their formation are illustrated in Figs. 7.17, 7.18, and 7.19.

Meiotic events in which ***no*** crossover occurs between the two loci will yield asci with four spores of one parental genotype at one end and four spores of the other parental genotype at the other end (Fig. 7.17). A two-strand double crossover between the two loci will yield the same pattern. An ascus with this pattern is called a ***parental ditype,*** because it contains two types of spores, and both are parental in genotype. When a single crossover or a three-strand double crossover occurs between the two loci, a 2 : 2 : 2 : 2 ascus pattern will result, with two spores being of one parental genotype (e.g., *A B*), two spores being of the second parental genotype (*a b*), and two spores being of each reciprocal recombinant type (two *A b* and two *a B*) (Fig. 7.18). Such an ascus is called a ***tetratype*** (four different ascospore genotypes). The third ascus pattern results only from a four-strand double crossover between the two loci (Fig. 7.19). This pattern is called the ***nonparental ditype*** pattern, because such

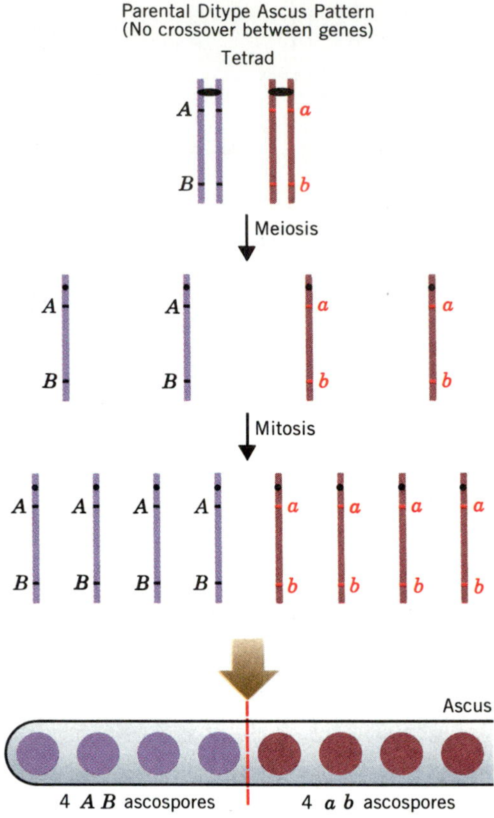

Figure 7.17 Diagram showing the ***parental ditype*** ascus pattern, which results from meiotic events in which no crossover occurs in the interval between the segregating genes. The four ascospores at one end of the ascus are of one parental genotype; the four at the other end are of the second parental genotype. The same pattern will occur as a result of a two-strand double crossover in the interval between the two genes.

asci contain only two types of spores and both have recombinant genotypes (four *A b* and four *a B*).

The relative frequencies of parental ditype, tetratype, and nonparental ditype asci can be used to calculate the linkage distance between two loci (see the next section, Unordered Tetrad Data). However, with ordered tetrad data, one can also use the centromere as a marker and construct a linkage map showing the order and relative positions of the two genes and the centromere. This is done by determining whether each ascus pattern is the result of first- or second-division segregation of the alleles of each gene, just as was done in calculating the linkage distance between a single segregating gene and the centromere (review Figs. 7.13 and 7.14).

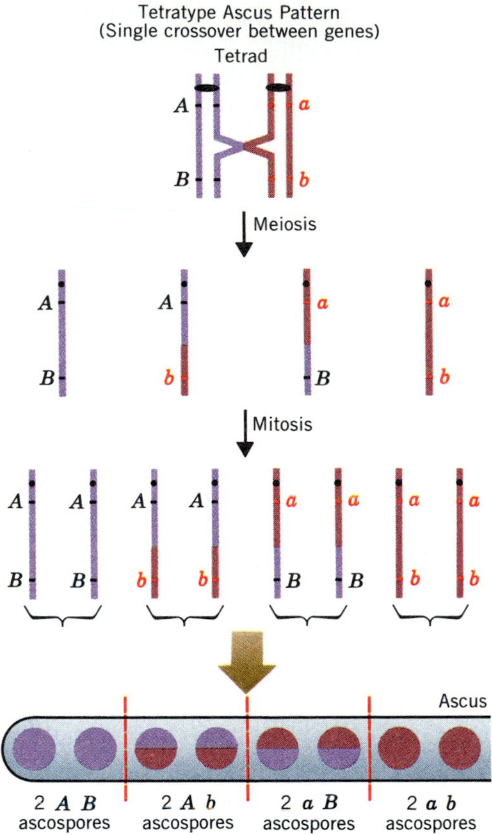

Figure 7.18 Diagram showing the ***tetratype*** ascus pattern, which results from meiotic events in which a single crossover occurs in the interval between the segregating genes. Four types of ascospores are present in the ascus. Half the spores are parental in genotype and half are recombinant. The actual sequences of spore pairs within each ascus will depend on the orientation of the homologous chromosomes at metaphase I of meiosis and the orientation of the two chromatids of each chromosome at metaphase II of meiosis. The tetratype pattern will also occur as a result of a three-strand double crossover in the interval between the two genes.

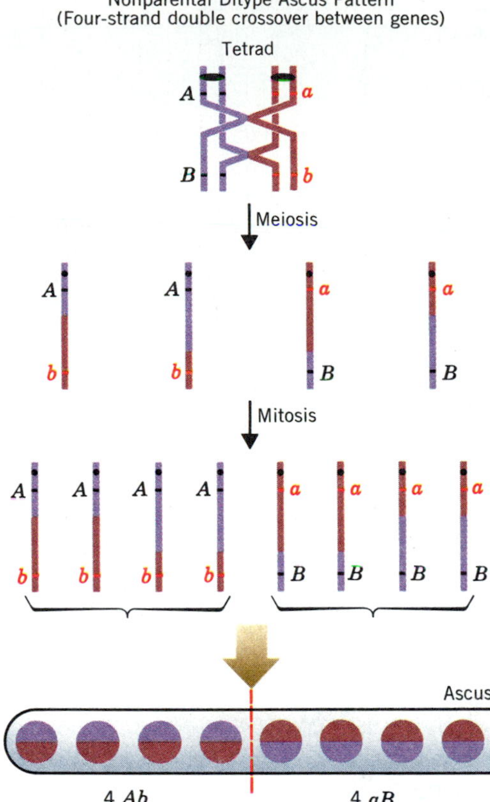

Figure 7.19 Diagram showing the ***nonparental ditype*** ascus pattern, which results from meiotic events in which a four-strand double crossover occurs in the interval between the segregating genes. Only two types of spores, the two reciprocal recombinant genotypes, are present in the ascus. Note that this type of ascus will be rare in the case of linked genes, but will be just as frequent as the parental ditype class (Fig. 7.17) for independently assorting genes.

Consider a hypothetical cross in *Neurospora* of the type *AB* × *ab*. If the asci resulting from this cross are dissected so that ordered tetrad data are available, the results of this cross will define two intervals (using the centromere as a third marker). Ordered tetrad data will allow one to determine which of the following three possible orders is correct.

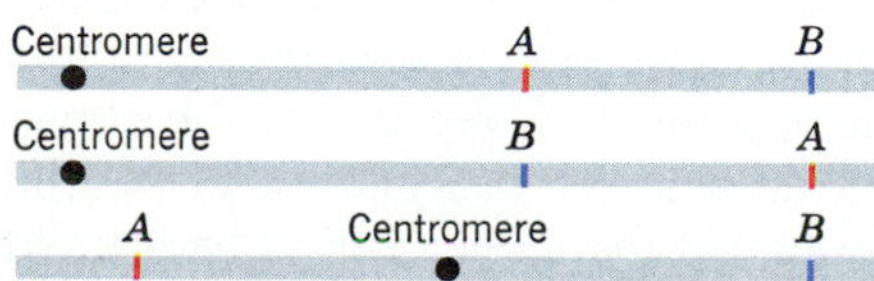

For example, assume that the ordered tetrad data for this cross were as given in Table 7.3. Note that only the genotypes of the four spore pairs in each ascus are given and that all ascus patterns of the same type are pooled. That is, spore pair 1 may be either the top or the bottom spore pair in an ascus, and all second-division segregation patterns for a locus are pooled under one genotype sequence ($A : a : A : a$ includes $a : A : a : A$, $A : a : a : A$, and $a : A : A : a$).

In crosses involving unlinked genes, the frequency of nonparental ditype asci (Fig. 7.19) is expected to equal the frequency of parental ditype asci (Fig. 7.17). In our example, the frequency of nonparental ditype asci is zero. Thus, the two genes are clearly linked. Of the four types of ascus patterns observed, the parental ditype pattern (ascus type 1) is the most frequent, as expected. Type 4 asci are least

TABLE 7.3 Ordered Tetrad Data

	TYPE OF ASCUS PATTERN			
SPORE PAIR	(1)	(2)	(3)	(4)
1	*A B*	*A B*	*A B*	*A B*
2	*A B*	*A b*	*a b*	*a B*
3	*a b*	*a B*	*A B*	*A b*
4	*a b*	*a b*	*a b*	*a b*
Total number of asci:	112	48	38	2

frequent (2/200); the type 4 pattern clearly must result from meiotic events in which some type of double crossover has occurred. Ascus pattern types 2 and 3 occur at intermediate frequencies. They must result from meiotic events in which a single crossover has occurred in one or the other of the two marked intervals.

If one recalls that first-division segregation occurs when no crossover occurs between a locus and the centromere whereas second-division segregation occurs when a single crossover occurs between a locus and the centromere, it becomes apparent that ascus pattern types 2 and 3 can only be produced by a single crossover for one of the three possible orders. This conveniently allows one to order the two markers relative to the position of the centromere.

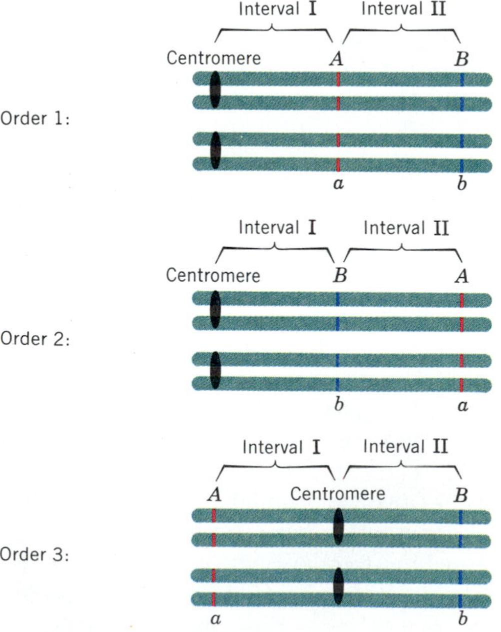

Figure 7.20 The three ***possible*** orders of locus *A*, locus *B*, and the centromere.

The meiotic tetrad that will be involved in our cross can be diagrammed as shown in Fig. 7.20 for each of the three possible orders.

Ascus pattern type 2 shows first-division segregation for the alleles at the *A* locus and second-division segregation for the alleles at the *B* locus. This requires that the single crossover ***(remember, based on frequency, we know that only one crossover is involved)*** occurring in the meiotic events giving rise to these asci be located between the *B* locus and the centromere, but *not* between the *A* locus and the centromere. That, of course, is ***not possible for order 2***. Similarly, ascus pattern type 3 exhibits second-division segregation for both loci, *A* and *B*. Thus, the single crossover occurring in the meiotic events giving rise to these asci must be located between both loci and the centromere. That is ***not possible if the loci are located on opposite sides of the centromere, order 3.*** Ascus pattern types 2 and 3 can both result from single crossovers only in the case of order 1. Given the order centromere-*A*-*B*, ascus pattern type 2 will be produced when a single crossover occurs in interval II, the interval between loci *A* and *B*. Ascus pattern type 3 will result when a single crossover occurs in interval I, between the *A* locus and the centromere. By using the preceding procedure, one can always eliminate two of the three possible orders and thus deduce the correct order. By a simple diagram, one can easily show that the rare ascus pattern type 4 results from a two-strand double crossover, with one crossover in interval I and the second in interval II.

All that is needed to allow one to draw the linkage map defined by the tetrad data given is the calculation of the linkage distances. We have deduced (1) that ascus pattern type 2 results from a single crossover in interval II, (2) that ascus pattern type 3 results from a single crossover in interval I, and (3) that ascus pattern type 4 results from a double crossover, with one crossover in each interval. The map distance (assume that no other data are available) from the *A* locus to the *B* locus (interval II) will thus be [½ (number of asci of type 2 + number of asci of type 4)/(total number of asci)] × 100, or

$$\frac{½(48 + 2)}{200} \times 100 = 12.5 \text{ map units}$$

These 50 asci all resulted from meiotic events in which there was one crossover in the interval between the *A* and *B* loci. Thus 12.5 percent of the progeny have recombinant arrangements of the alleles of the *A* and *B* genes. Note that the 2 (number of two-strand double crossover asci) is multiplied by ½. Even though there were two crossovers in the meiotic events that gave rise to these asci, these crossovers involved ***only two of the four chromatids in the tetrads.***

Similarly, the distance (again, assume no other data are available) from the *A* locus to the centromere (interval I) will be

$$\frac{\frac{1}{2}(38 + 2)}{200} \times 100 = 10 \text{ map units}$$

Again, these 40 asci all resulted from meiotic events in which there was one crossover in the interval between the *A* locus and the centromere. ***Don't forget the factor one-half that corrects for the fact that each crossover involves only two of the four chromatids in a tetrad and thus only one-half of the ascospores in each ascus.*** Only in the case of nonparental ditype asci (resulting from four-strand double crossovers; see Fig. 7.19) is the factor one-half not used. Thus the linkage map defined by our hypothetical tetrad data is

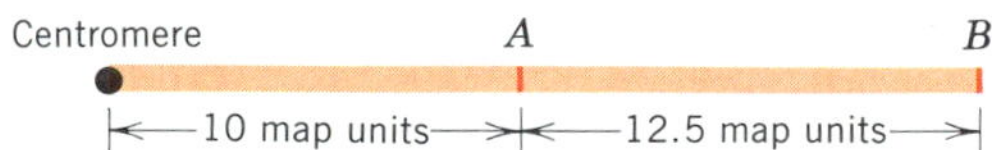

Unordered Tetrad Data

The dissection of a large number of asci is a tedious task; it is much easier to allow the asci to mature and burst open themselves. By this procedure, all the ascospores in individual asci can be analyzed, but the order of the ascospores in the asci cannot be determined. Tetrad data obtained in this way are called ***unordered tetrad data.*** In addition, some ascomycetes such as the yeasts do not produce ordered tetrads. One can still determine whether recombination events are reciprocal using unordered tetrad data. One can***not,*** however, determine whether alleles segregate at the first or the second meiotic division, and thus one can ***not*** map the centromere by the procedure described in the preceding section.

Linkage distances between loci can be calculated from unordered tetrad data using the frequencies of parental ditype, tetratype, and nonparental ditype asci (Figs. 7.17, 7.18, and 7.19). The map distance between two loci is

$$\frac{\frac{1}{2}\text{ (number of tetratype asci)} + \text{(number of nonparental ditype asci)}}{\text{total number of asci}} \times 100$$

This formula gives the frequency of ascospores with recombinant arrangements of the alleles of the genes involved. It is equivalent to the formula used to calculate the linkage distance between the *A* and *B* loci in our earlier example of ordered tetrad data. It can***not*** be used with ordered tetrad data to calculate linkage distances from genes to centromeres. Unordered tetrad data have been of great value in the study of nonreciprocal recombination or "gene conversion" (see pp. 196–199).

Somatic-Cell Hybridization

Obviously, human genes cannot be mapped by designed crosses of the types described in earlier sections of this chapter. Instead, one has to study family histories or pedigrees and identify families in which traits of interest are segregating. The pedigrees of these families are then studied to identify linked genes and to extract data on linkage relationships. Prior to about 1965, the only data available for assigning the location of genes to individual linkage groups in humans were pedigree data. During the last two decades, however, much progress has been made in mapping genes in humans. This progress is largely the result of the development of (1) a valuable new technique called somatic-cell hybridization, (2) improved cytological techniques for identifying individual human chromosomes and specific segments of individual chromosomes, and (3) recombinant DNA techniques used to isolate and identify the locations of individual genes within DNA molecules and chromosomes.

Somatic-cell hybridization, first demonstrated with mouse cells by G. Barski and colleagues in 1960, involves the fusion of ***somatic cells growing in vitro*** (Fig. 7.21). (Somatic cells include all diploid body cells or nongerm-line cells of an organism.) Cell fusion produces ***binucleate hybrid cells*** or ***heterokaryons.*** Cell fusion is followed by nuclear fusion to produce ***uninucleate hybrid cells*** or ***synkaryons.*** The frequency of spontaneous cell fusion is very low, about one cell fusion per million cells. Fortunately, the frequency can be greatly increased by the addition of ultraviolet-inactivated ***Sendai virus*** or the chemical ***polyethylene glycol.*** These agents stimulate cell fusion by increasing cell contact and altering cell membranes. Sendai viruses are very small relative to the sizes of cells. Thus, when the viruses bind to two cells, they tend to hold them in close juxtaposition to each other. Moreover, the processes involved in normal viral penetration of cells probably facilitate fusion of the membranes of the juxtaposed cells.

Somatic-cell hybrids can be produced between two cells of the same species ***(monospecific hybrids)*** or two cells from different species ***(interspecific hybrids). Mouse–human somatic-cell hybrids*** have been particularly useful for mapping human genes.

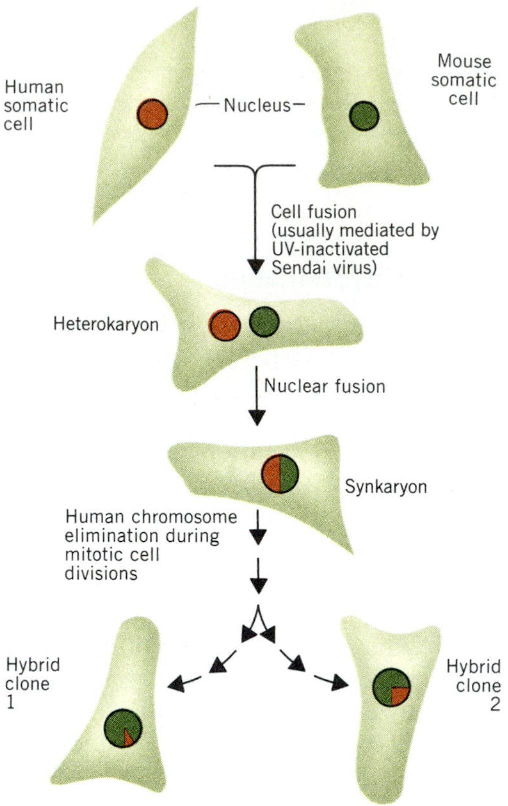

Figure 7.21 Schematic illustration of the formation of mouse–human somatic-cell hybrids and the ***elimination of human chromosomes during subsequent mitotic divisions.*** The established mouse cell culture lines used in these experiments rarely contain the normal diploid number of 40 chromosomes. Usually, they have much higher numbers of chromosomes. The primary human culture cells usually do contain the normal diploid number of 46 chromosomes. After about 30 cell generations, these mouse–human hybrids produce clones that usually contain all of the mouse chromosomes and from 1 to 20 of the original 46 human chromosomes.

They are usually made using established mouse cell culture lines and human fibrocytes or leukocytes.

Mouse–human somatic-cell hybrids have three important features that make them especially useful for genetic analyses. (1) ***The mouse and human chromosomes are easily distinguished,*** making the identification of the human chromosomes present in hybrid cells a fairly simple task. (2) All the mouse chromosomes are usually retained in the hybrid cells, but ***only 1–20 of the human chromosomes are retained*** in hybrid cell clones. Thus, most of the 46 human chromosomes are eliminated during subsequent mitotic divisions of the hybrid cells, and, most important, they are ***eliminated at random or nearly at random.*** (3) ***Both sets of chromosomes, human and mouse, are expressed in the hybrid cells.*** The mouse and human genomes contain many closely related genes that code for gene-products with identical, or nearly identical, functions. Nevertheless, these gene-products are usually different enough to be distinguished biochemically.

Genes are shown to be located on specific chromosomes by correlating the phenotypes (e.g., enzyme activities) controlled by these genes with the presence of individual human chromosomes in hybrid cell clones. For example, suppose that human cells that make the enzyme thymidine kinase (TK^+ cells) were hybridized with mouse cells that can ***not*** make this enzyme (TK^- cells). Assume that the hybrid cells were cultured for about 30 cell generations, and then five different cell clones were examined to determine which human chromosomes were present. If the results shown in Table 7.4 were obtained, one could deduce that the gene controlling the synthesis of thymidine kinase (gene *TK1*) is located on chromosome 17, because it is the only human chromosome present in all four TK^+ hybrid cell clones.

Each of the 24 different human chromosomes can now be accurately identified using improved staining procedures such as G-banding (see Chapter 4, Fig. 4.3).

Most somatic-cell hybridization experiments are currently done using selection procedures that pre-

TABLE 7.4 Somatic-Cell Hybridization Data

HYBRID CLONE NO.	THYMIDINE KINASE ACTIVITY	HUMAN CHROMOSOMES (+ = PRESENT, – = ABSENT)											
		X	Y	1	2	4	7	9	10	15	17	18	21
1	+	+	+	–	+	–	–	+	+	+	+	–	+
2	+	+	–	+	+	–	+	+	–	+	+	+	–
3	+	–	–	+	+	+	+	–	–	–	+	–	–
4	+	–	–	+	–	–	–	–	+	–	+	–	+
5	–	+	–	–	+	–	–	+	–	+	–	–	–

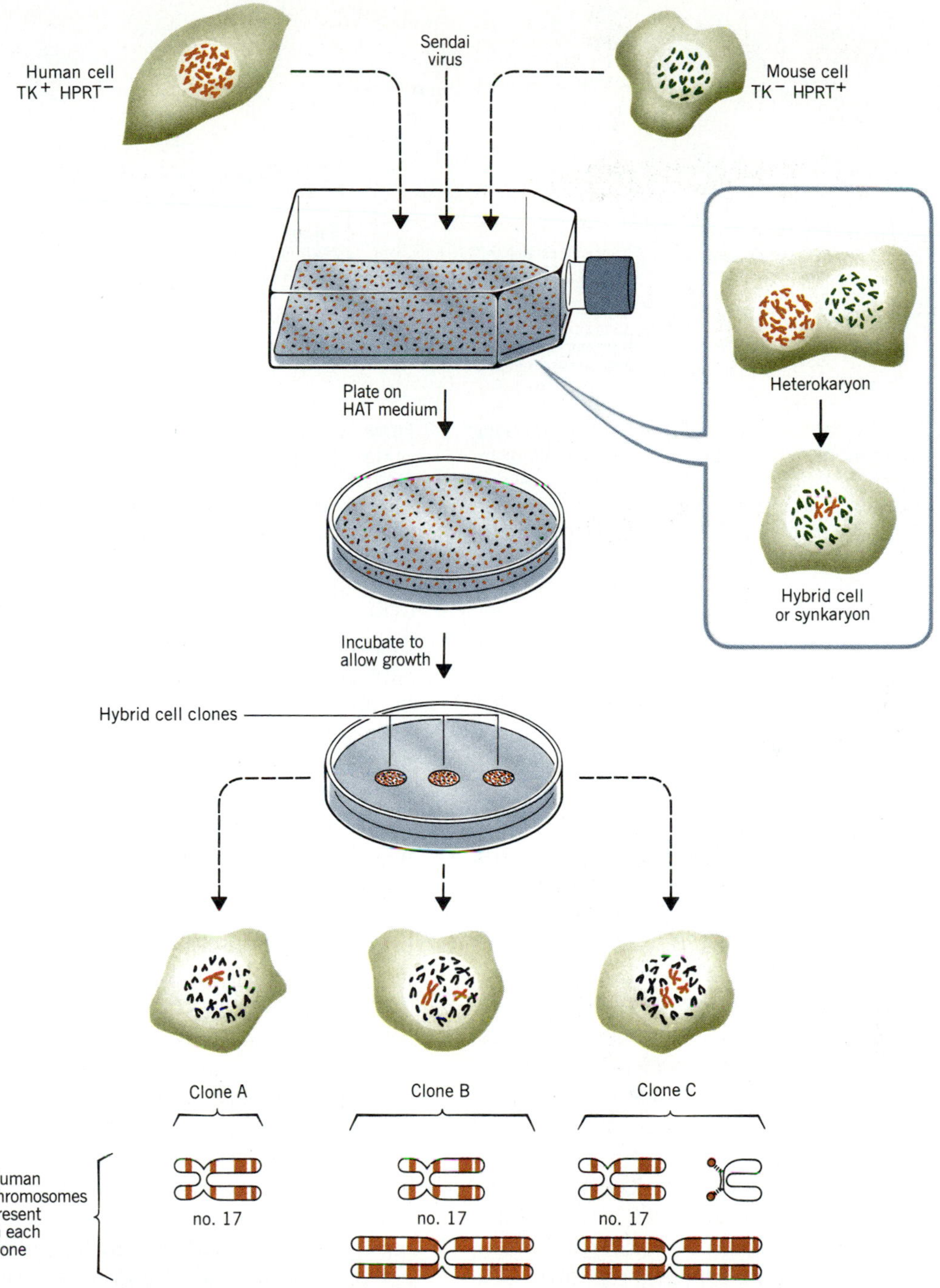

Figure 7.22 Identification of the chromosomal location of human genes by mouse–human somatic-cell hybridization using the HAT selection procedure. (After F. H. Ruddle and R. S. Kucherlapati, "Hybrid Cells and Human Genes," *Sci. Amer.* 231(1): 36–44, 1974.)

vent the growth of the parental cells while allowing growth of hybrid cells produced by cell fusion. One widely used method of selection (illustrated in Fig. 7.22) is to fuse mouse cells that cannot make the enzyme thymidine kinase (TK) with human cells that cannot make the enzyme hypoxanthine-guanine phosphoribosyltransferase (HPRT). (HPRT is deficient in cells of individuals with Lesch–Nyhan syndrome, which is caused by an X-linked recessive allele; see p. 88.)

In the appropriate selective medium, both of these enzymes are required for growth. Thus, neither parent

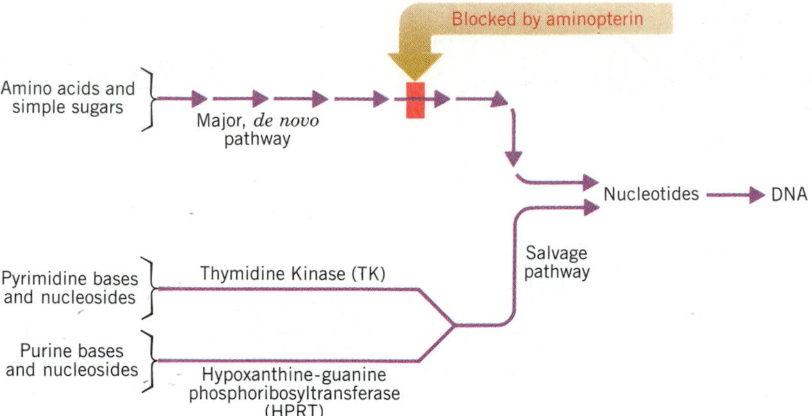

Figure 7.23 Pathways of nucleotide biosynthesis, showing only those components important to the ***HAT*** (***h***ypoxanthine, ***a***minopterin, ***t***hymidine) selection procedure used in somatic-cell hybridization experiments. Aminopterin blocks the major, *de novo* pathway of nucleotide biosynthesis. However, in HAT medium, pyrimidine and purine nucleotides can be synthesized by the salvage pathway from the hypoxanthine and thymidine in this medium if the cells make thymidine kinase and hypoxanthine-guanine phosphoribosyltransferase (TK^+ $HPRT^+$ cells).

cell can grow, but hybrids produced by cell fusion can grow. The TK^- mouse cells are $HPRT^+$ (they make enzyme HPRT), and the $HPRT^-$ human cells are TK^+ (they make enzyme TK). Since both the mouse and the human chromosomes are expressed in the hybrid cells, they have both enzymes required for growth in the selective medium.

The biochemical basis of this selection procedure involves two important pathways for the biosynthesis of the nucleotides required for DNA synthesis (Fig. 7.23). One is the major pathway by which nucleotides are synthesized *de novo* from simple sugars and amino acids. This pathway is ***blocked by the antimetabolite aminopterin***. In the presence of aminopterin, however, cells can still synthesize the nucleotides essential for growth ***if certain precursors are added*** to the medium and ***if*** the second pathway, usually called the ***salvage pathway***, is ***operative***.

Enzymes TK and HPRT catalyze key steps in the salvage pathway. As the name suggests, hypoxanthine-guanine phosphoribosyltransferase catalyzes the conversion of hypoxanthine to guanine, and thymidine kinase ("kinases" catalyze phosphate group transfers) catalyzes the phosphorylation of thymidine. Thus, ***TK^+ $HGPRT^+$ cells can grow in the presence of aminopterin if hypoxanthine and thymidine are added to the medium.*** Medium containing these three compounds, ***h***ypoxanthine, ***a***minopterin, and ***t***hymidine, is called ***HAT medium***. HAT medium selects for TK^+ $HGPRT^+$ hybrid cells, since neither the TK^- $HGPRT^+$ parental mouse cells nor the TK^+ $HGPRT^-$ parental human cells can grow in it.

If the selection procedure described is used, human chromosome elimination may occur at random, but those cells that have lost the human chromosome carrying the gene coding for thymidine kinase will ***not*** be able to grow. Only those hybrid cell clones that have retained this chromosome will grow on HAT medium. The analysis of a set of such clones should reveal that they all have one human chromosome in common. In fact, clones of hybrid cells selected in this way always contain human chromosome 17, indicating that it carries the gene coding for thymidine kinase.

Several selection procedures similar to the HAT procedure just described have been developed for use in somatic-cell hybridization experiments. As a result, the localization of genes on the 24 human chromosomes has progressed very rapidly. The 24 human "linkage groups" and the locations of a few selected "anchor" or reference genes are shown in Fig. 7.24 (pp. 188–192). For a complete listing of the 4344 loci for which the mode of inheritance had been quite firmly established by March 1, 1988, see V.A. McKusick, *Mendelian Inheritance in Man*, 8th edition. Remember that genes located on the same chromosome are not necessarily linked. Thus, the term ***syntenic*** has been introduced to refer to genes known to be located on the same chromosome, but for which no information is available regarding their linkage relationship(s).

MOLECULAR MECHANISM OF CROSSING-OVER

Despite a wealth of genetic data that describes certain aspects of crossing-over and seemingly innumerable models proposed to explain these data, the molecular details of the mechanism of crossing-over are still only partly known.

The models that have been proposed to account for crossing-over have been of two general types: (1) ***breakage and reunion*** and (2) ***copy choice***. Breakage and reunion involves the breakage of two homologous chromatids and the reunion of the parts in new (recombinant) arrangements (as diagrammed in Figs. 7.2 and 7.6). Extensive data now document the occurrence of breakage and reunion during the crossing-over process. Copy choice models were based on the assumption that molecules of DNA in the process of being synthesized could switch from using the DNA of one homolog as template in one region to using the DNA of the other homolog as template in another region, as in the following diagram.

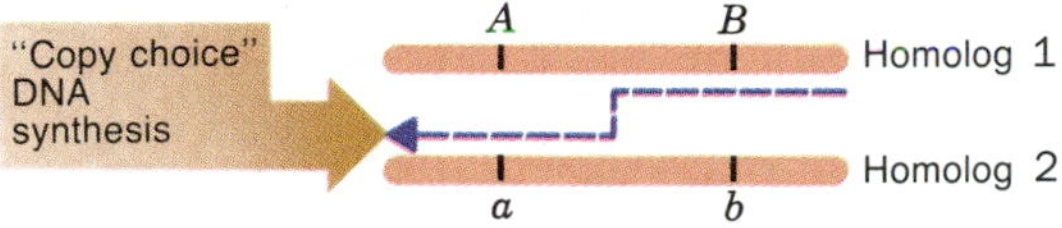

Most such copy choice models of crossing-over were based on the assumption that DNA synthesis was conservative (see Chapter 5, pp. 104–107). Copy choice models rapidly lost support once DNA replication was proven to be semiconservative. [Recalling that a newly synthesized strand of DNA is wound around its template strand in double-helix form, draw the structure that would be produced by semiconservative DNA replication and copy choice crossing-over, and try to visualize the segregation of the resulting DNA molecules. Segregation will be possible only by (1) completely unwinding the newly synthesized strand from one of the template strands or (2) breaking the newly synthesized strand at the position of the switch in template (breakage and reunion?).] Pure copy choice (with no associated breakage and reunion) is simply not mechanistically compatible with semiconservative DNA replication. Moreover, pure copy choice requires that crossing-over occur during replication, rather than in the tetrad stage of meiosis after replication (see Fig. 7.6). The copy choice mechanism warrants discussion, however, because it now seems clear that a small amount of copy choice DNA repair synthesis is associated with crossing-over by breakage and reunion. This copy choice repair synthesis may be responsible for the occurrence of ***nonreciprocal recombination*** or ***gene conversion,*** which is frequently observed for genetic markers that are located very close to the sites of crossing-over (see pp. 196–199). Crossing-over, therefore, probably occurs by a complex mechanism that includes some aspects of both breakage and reunion and copy choice models.

The classic experiments of Stern and of Creighton and McClintock (pp. 163–165) were consistent with, but did not prove, the hypothesis that crossing-over occurred by breakage and reunion. If the morphology of chromosomes is determined by the genetic information they carry, as seems likely, at least in part, then copy choice models of crossing-over would also have predicted Stern's and Creighton and McClintock's results. Their experiments did not demonstrate that the recombinant chromosomes contained parts physically derived from (i.e., segments of DNA actually present in) the parental chromosomes.

Direct evidence for breakage and reunion (or possibly breakage and resynthesis) has been obtained in eukaryotes by labeling chromosomes with radioactive [^{3}H]thymidine and following the distribution of radioactive chromatids by autoradiography. Recall that in Taylor's classic experiment, which showed semiconservative chromosome replication in mitosis of *Vicia faba*, one chromatid of each pair was labeled and the other was unlabeled at the second metaphase after labeling (see Chapter 6, Fig. 6.24). In subsequent experiments by Taylor involving meiosis in the grasshopper *Romalea*, crossover patterns of labeling were frequently observed, that is, labeled and unlabeled segments of chromatids of homologous chromosome pairs were often observed to be interchanged (Fig. 7.25). Taylor's experiments thus provided direct evidence for breakage and reunion. They did not exclude, however, the possibility that breakage and ***resynthesis*** is occurring, that is, that the unlabeled segment is the result of new synthesis rather than the reunion of an unlabeled segment of a parental chromosome.

Definitive evidence for ***breakage and reunion***, excluding the possibility of breakage and resynthesis, has been obtained only for crossing-over in bacteriophage lambda. M. Meselson crossed $^{13}C^{15}N$-density-labeled lambda phage of genotype *h c* (*host* range and *clear* plaque mutant alleles, respectively) with $^{13}C^{15}N$-density-labeled lambda phage of genotype $h^+ c^+$. The cross was done in medium containing the common "light" isotopes (^{12}C, ^{14}N) of carbon and nitrogen. Meselson then analyzed the progeny phage in CsCl density gradients (see Chapter 5, pp. 105–108). He identified $h^+ c$ recombinant phage with the same (or nearly the same) density as the fully "heavy" parental phage. These phage contained no more than 1–3 percent newly synthesized "light" DNA, probably resulting from a small amount of repair synthesis. ***The "heavy" recombinants could have been produced only by breakage and reunion,*** since breakage and resynthesis would have produced phage of intermediate density (Fig. 7.26).

Although the evidence is convincing that crossing-over involves breakage and reunion, probably with a small amount of DNA repair synthesis, the mechanism (or mechanisms) by which the breakage and reunion

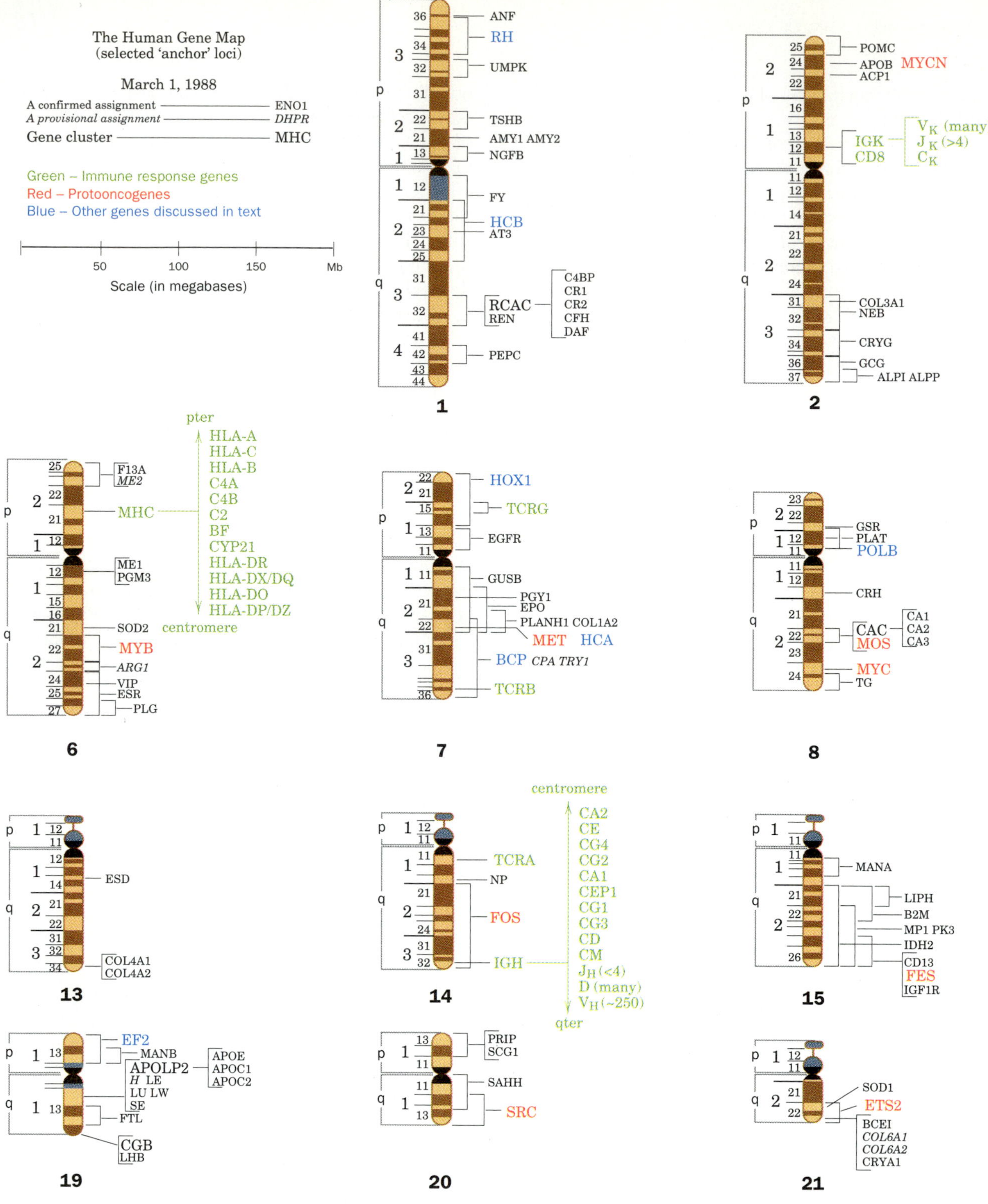

Figure 7.24 Abbreviated map of the 24 human chromosomes showing the locations of a few selected "anchor" or reference genes or loci. Loci specifying products involved in the immune response (Chapter 16) are labeled in green; protooncogenes or cellular oncogenes (Chapter 17) are labeled in red; other genes or loci that are discussed elsewhere in this text are labeled in blue. The short and long arms of each chromosome are designated ***p*** and ***q***, respectively; each arm, in turn, is divided into segments (numbered intervals) based on chromosome banding patterns. The symbols for some of the genes shown on the map and their products or the phenotypic traits that they control are listed below by chromosome. For a complete listing of the 4,344 human loci for which the mode of inheritance was reasonably well established by March 1, 1988, see McKusick, *Mendelian Inheritance in Man,* 8th edition.

CHROMOSOME 1

RH	Rhesus blood group (erythroblastosis fetalis)
UMPK	Uridine monophosphate kinase
TSHB	Thyroid stimulating hormone, beta subunit
AMY 1	Amylase, salivary
AMY 2	Amylase, pancreatic

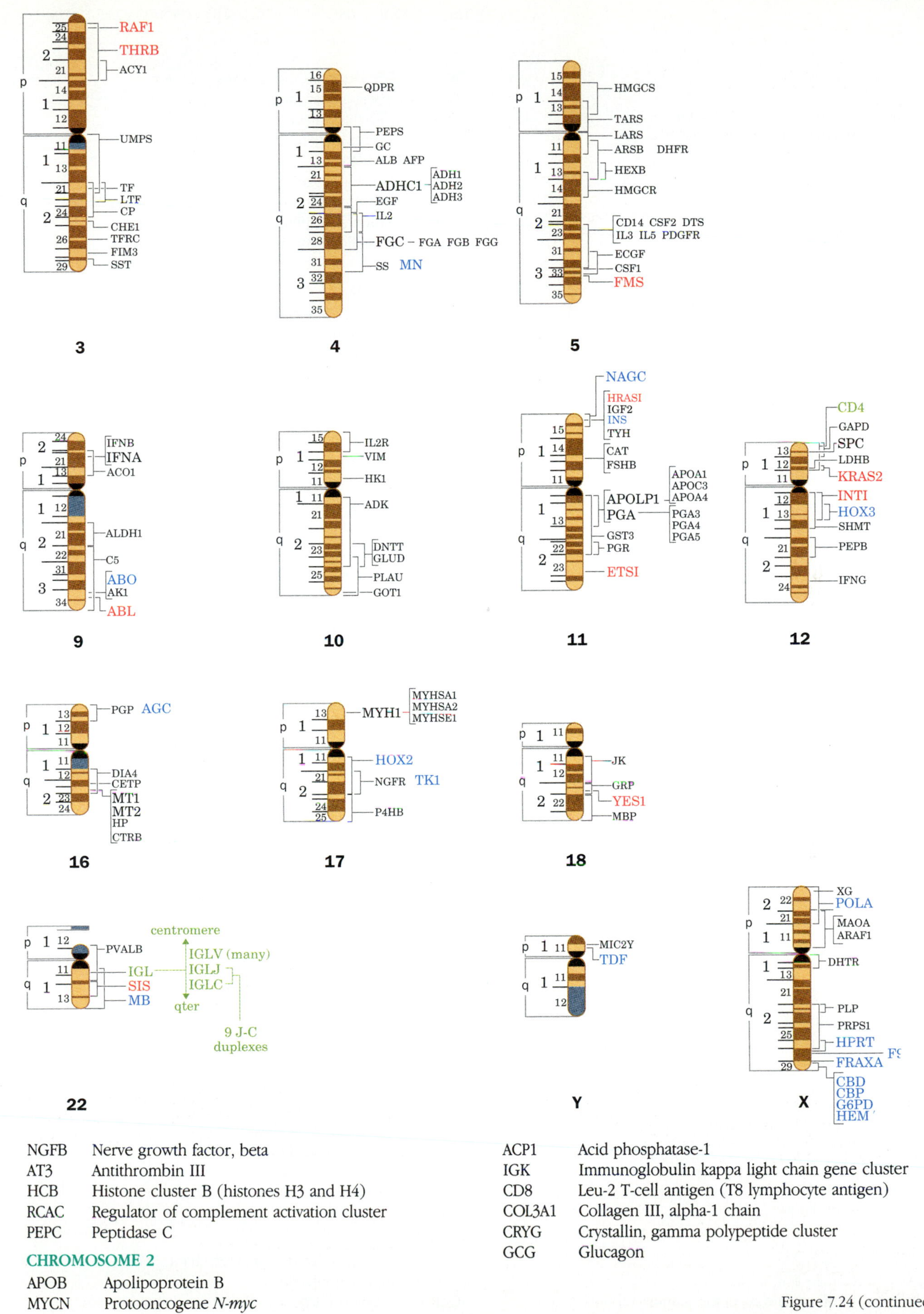

NGFB Nerve growth factor, beta
AT3 Antithrombin III
HCB Histone cluster B (histones H3 and H4)
RCAC Regulator of complement activation cluster
PEPC Peptidase C

CHROMOSOME 2

APOB Apolipoprotein B
MYCN Protooncogene *N-myc*

ACP1 Acid phosphatase-1
IGK Immunoglobulin kappa light chain gene cluster
CD8 Leu-2 T-cell antigen (T8 lymphocyte antigen)
COL3A1 Collagen III, alpha-1 chain
CRYG Crystallin, gamma polypeptide cluster
GCG Glucagon

Figure 7.24 (continued)

Figure 7.24 (continued)

ALPI	Alkaline phosphatase, adult intestinal
ALPP	Alkaline phosphatase, placental

CHROMOSOME 3

RAF1	Protooncogene *raf1*
THRB	Thyroid hormone receptor, beta (Protooncogene *erbA2*)
ACY1	Aminoacylase-1
UMPS	UMP synthase
TF	Transferrin
LTF	Lactotransferrin
CP	Ceruloplasmin
CHE1	Pseudocholinesterase-1
TFRC	Transferrin receptor
SST	Somatostatin

CHROMOSOME 4

PEPS	Peptidase S
ALB	Albumin
ADHC1	Alcohol dehydrogenase, class I, cluster
EGF	Epidermal growth factor
IL2	Interleukin-2 (T-cell growth factor)
FGC	Fibrinogen gene cluster
SS	Ss blood group (glycophorin B)
MN	MN blood group (glycophorin A)

CHROMOSOME 5

HMGCS	3-Hydroxy-3-methylglutaryl coenzyme A synthase
TARS	Threonyl-tRNA synthetase
LARS	Leucyl-tRNA synthetase
DHFR	Dihydrofolate reductase
HMGCR	3-Hydroxy-3-methylglutaryl coenzyme A reductase
CSF2	Granulocyte-macrophage colony-stimulating factor
DTS	Diphtheria toxin sensitivity
IL3	Interleukin-3
IL5	Interleukin-5
PDGFR	Platelet-derived growth factor receptor
ECGF	Endothelial cell growth factor
CSF1	Macrophage colony stimulating factor
FMS	Protooncogene *fms*

CHROMOSOME 6

F13A	Clotting factor XIII, A component
ME2	Malic enzyme, mitochondrial
MHC	Major histocompatibility complex
ME1	Malic enzyme, cytoplasmic
PGM3	Phosphoglucomutase-3
SOD2	Superoxide dismutase-2, mitochondrial
MYB	Protooncogene *myb*
ARG1	Arginase, liver
ESR	Estrogen receptor
PLG	Plasminogen

CHROMOSOME 7

HOX1	Homeo box-1
TCRG	T-cell antigen receptor, gamma subunit
EGFR	Epidermal growth factor receptor
GUSB	Beta-glucuronidase
EPO	Erythropoietin
PLANH1	Plasminogen activator inhibitor-1
COLIA2	Collagen I, alpha-2 chain
MET	Protooncogene *met*
HCA	Histone cluster A (histones H1, H2A, H2B)
BCP	Blue cone pigment (blue color vision)
TCRB	T-cell antigen receptor, beta subunit

CHROMOSOME 8

GSR	Glutathione reductase
PLAT	Plasminogen activator, tissue type
POLB	DNA polymerase, beta subunit
CRH	Corticotropin releasing hormone
CAC	Carbonic anhydrase cluster
MOS	Protooncogene *mos*
MYC	Protooncogene *myc*
TG	Thyroglobulin

CHROMOSOME 9

IFNB	Fibroblast interferon (beta-interferon)
IFNA	Leukocyte interferon gene cluster (alpha-interferon)
ACO1	Aconitase, soluble
ALDH1	Aldehyde dehydrogenase-1
CS	Citrate synthase, mitochondrial
ABO	ABO blood group
AK1	Adenylate kinase-1, soluble
ABL	Protooncogene *abl*

CHROMOSOME 10

IL2R	Interleukin-2 receptor
VIM	Vimentin
HK1	Hexokinase-1
ADK	Adenosine kinase
DNTT	Terminal deoxynucleotidyltransferase
GLUD	Glutamate dehydrogenase
PLAU	Plasminogen activator, urinary (urokinase)
GOT1	Glutamate oxaloacetate transaminase, soluble

CHROMOSOME 11

NAGC	Non-alpha (beta) globin cluster
HRAS1	Protooncogene *H-ras1*
IGF2	Insulin-like growth factor II (somatomedin A)
INS	Insulin
TYH	Tyrosine hydroxylase
CAT	Catalase
FSHB	Follicle stimulating hormone, beta polypeptide
APOLP1	Apolipoprotein cluster I
PGA	Pepsinogen A cluster
GST3	Glutathione S-transferase-3
PGR	Progesterone receptor
ETS1	Protooncogene *ets1*

CHROMOSOME 12

CD4	T-cell antigen CD4 (molecule to which HIV binds during infection)
GAPD	Glyceraldehyde-3-phosphate dehydrogenase

SPC	Salivary proline-rich protein complex
LDHB	Lactate dehydrogenase B
KRAS2	Protooncogene *K-ras2*
INT1	Protooncogene *int1*
HOX3	Homeo box 3
SHMT	Serine hydroxymethyltransferase
PEPB	Peptidase B
IFNG	Interferon, gamma or immune type

CHROMOSOME 13

ESD	Esterase D
COL4A1	Collagen IV, alpha-1 chain
COL4A2	Collagen IV, alpha-2 chain

CHROMOSOME 14

TCRA	T-cell antigen receptor, alpha subunit
NP	Nucleoside phosphorylase
FOS	Protooncogene *fos*
IGH	Immunoglobulin heavy chain gene cluster

CHROMOSOME 15

MANA	Alpha-mannosidase-A, cytoplasmic
LIPH	Hepatic lipase
B2M	Beta-2-microglobulin
MPI	Mannosephosphate isomerase
PK3	Pyruvate kinase-3
IDH2	Isocitrate dehydrogenase, mitochondrial
CD13	GP150,95 myeloid membrane antigen, alpha subunit
FES	Protooncogene *fes*
IGF1R	Insulin-like growth factor-1 receptor

CHROMOSOME 16

PGP	Phosphoglycolate phosphatase
AGC	Alpha globin cluster
DIA4	Disphorase-4
CETP	Cholesterol ester transfer protein, plasma
MT1	Metallothionein I cluster
MT2	Metallothionein II cluster
HP	Haptoglobin
CTRB	Chymotrypsinogen B

CHROMOSOME 17

MYH1	Myosin, heavy chain cluster
HOX2	Homeo box-2
NGFR	Nerve growth factor receptor
TK1	Thymidine kinase-1
P4HB	Prolyl-4-hydroxylase, beta subunit

CHROMOSOME 18

JK	Kidd blood group
GRP	Gastrin releasing peptide
YES1	Protooncogene *yes1*
MBP	Myelin basic protein

CHROMOSOME 19

EF2	Elongation factor-2
MANB	Lysosomal alpha-D-mannosidase-B
APOLP2	Apolipoprotein cluster II
H	Bombay phenotype
LE	Lewis blood group
LU	Lutheran blood group
LW	Landsteiner-Weiner blood group
SE	Secretor
FTL	Ferritin, light chain
CGB	Chorionic gonadotropin, beta chain
LBH	Luteinizing hormone, beta chain

CHROMOSOME 20

PRIP	Prion protein
SCG1	Chromogranin B (secretogranin B)
SAHH	S-adenosylhomocysteine hydrolase
SRC	Protooncogene *src*

CHROMOSOME 21

SOD1	Superoxide dismutase-1, soluble
ETS2	Protooncogene *ets2*
BCEI	Breast cancer estrogen-inducible sequence
COL6A1	Collagen VI, alpha-1 chain
COL6A2	Collagen VI, alpha-2 chain
CRYA1	Crystallin, alpha A

CHROMOSOME 22

PVALB	Parvalbumin
IGL	Immunoglobulin lambda light chain gene cluster
SIS	Protooncogene *sis*
MB	Myoglobin

CHROMOSOME Y

MIC2Y	Antigen determined by monoclonal antibody 12E7, Y homolog
TDF	Testis determining factor

CHROMOSOME X

XG	Xg blood group
POLA	DNA polymerase, alpha subunit
MAOA	Monoamine oxidase A
ARAF1	Protooncogene *araF1*
DHTR	Testicular feminization (androgen receptor)
PLP	Myelin proteolipid protein
PRPS1	Phosphoribosylpyrophosphate synthetase
HPRT	Hypoxanthine-guanine phosphoribosyltransferase
F9	Hemophilia B (clotting factor IX)
FRAXA	Fragile site Xq27.3 (Martin-Bell syndrome)
CBD	Deutan colorblindness (green cone pigment)
CBP	Protan colorblindness (red cone pigment)
G6PD	Glucose-6-phosphate dehydrogenase
HEMA	Hemophilia A (clotting factor VIII)

This human gene map has been reproduced with permission from V. A. McKusick, *Mendelian Inheritance in Man,* 8th edition, copyright © 1988, The Johns Hopkins University Press. A few selected genes that are discussed in this textbook have been added to McKusick's map. For a complete presentation of the human gene map, see McKusick's text plus recent updates in the *Journal of Human Genetics.*

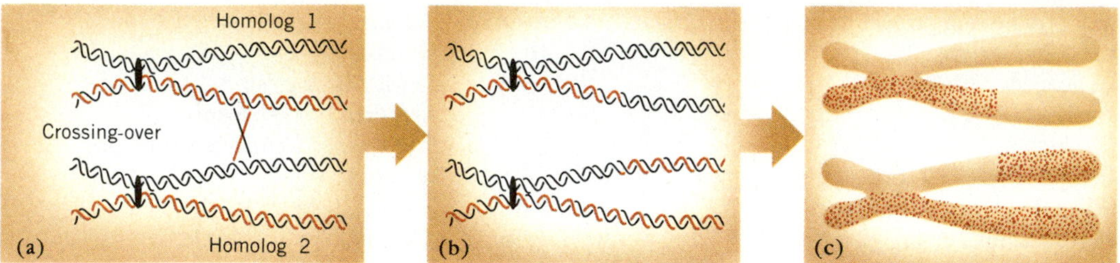

Figure 7.25 Schematic diagram of Taylor's autoradiographic evidence for breakage and reunion of homologous chromatids during meiotic crossing-over in the grasshopper (*Romalea*). Note that for the sake of simplicity each chromatid is shown with a single DNA double helix extending linearly from one end to the other. However, remember that the giant DNA molecule in each chromatid actually exists in a highly condensed (coiled and/or folded) configuration (see Figs. 6.14, 6.16, and 6.17). Crossing-over is shown between a chromatid that is radioactive due to the presence of one strand of DNA containing [^{3}H]thymidine (see Fig. 6.24) and a nonradioactive chromatid. Radioactive strands of DNA are shown in orange; nonradioactive strands are shown in black. The centromeres are represented by filled ovals. The tetrad of four chromatids is shown (a) before and (b) after crossing-over. The orange dots on the metaphase chromosomes shown in (c) indicate the locations of radioactivity as indicated by autoradiography. (Based on the results of J. H. Taylor, *J. Cell Biol.* 25: 57, 1965.)

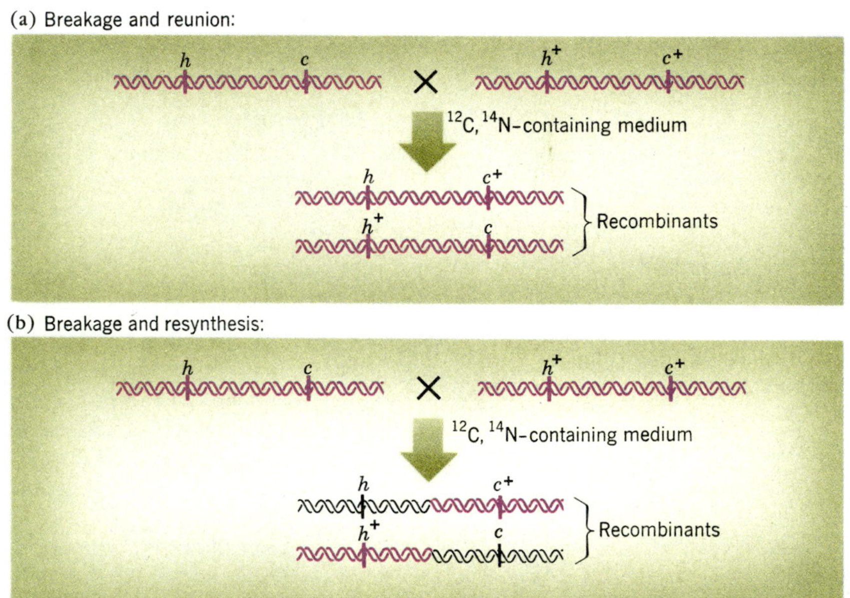

Figure 7.26 Meselson's proof of crossing-over by breakage and reunion in phage lambda. Both parental phage (*h c* and h^+ c^+) contained "heavy" DNA due to the presence of the heavy isotopes of carbon and nitrogen—^{13}C and ^{15}N. The cross was carried out in host cells that were growing in medium containing the normal ("light") isotopes ^{12}C and ^{14}N. Progeny phage were separated on the basis of density in a CsCl density gradient and then analyzed for genotype. Meselson demonstrated that rare h^+ *c* recombinant progeny containing fully heavy DNA (DNA in which both strands contain ^{13}C and ^{15}N throughout) were produced. Under these conditions, recombinants containing fully heavy DNA can be produced by (a) breakage and reunion, but not by (b) breakage and resynthesis. Any newly synthesized DNA will be "light" since the cross is being done in medium containing ^{12}C and ^{14}N. (These "heavy" h^+ *c* recombinant progeny are rare since most of the progeny, recombinant and parental, will be "light" or "hybrid" because of several rounds of semiconservative DNA replication in "light" medium. The rare h^+ *c* recombinants were detected by a selection technique not applicable to the reciprocal *h* c^+ recombinant class.)

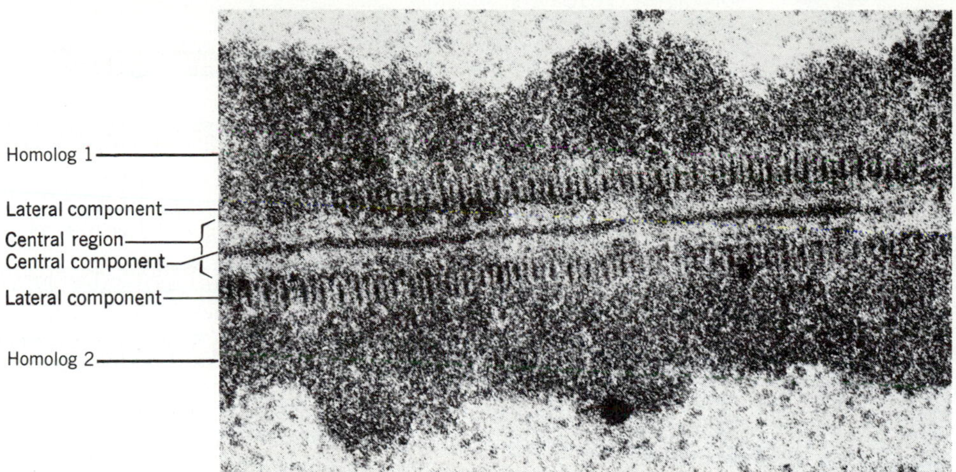

Figure 7.27 Structure of the synaptinemal complex of the ascomycete *Neottiella rutilans* (magnified ×61,200). An electron-dense "central component" is surrounded by a less electron-dense space; together these make up the so-called central region (about 1000 Å in width). The "lateral components" (each about 500 Å in width) are present on each side of the central region. The two homologous chromosomes are juxtaposed to the two lateral elements. (From M. Westergaard and D. von Wettstein, *Annu. Rev. Genet.* 6: 71–110, 1972. Reproduced with permission, from the *Annual Review of Genetics,* Vol. 6; copyright © 1972 by Annual Reviews Inc. Original photograph courtesy of D. von Wettstein.)

events occur is still not completely established. Clearly the process is complex, especially in eukaryotes. In eukaryotes, crossing-over has been associated with the formation of a structure (or set of structures) called the ***synaptinemal complex*** (Fig. 7.27), which forms during prophase of the first meiotic division. This structure, which is composed primarily of proteins and RNA, has been identified in a large number of eukaryotic species. For unknown reasons, crossing-over only rarely occurs, and synaptinemal complexes have not been observed in *Drosophila* males. (Crossing-over does occur in both sexes of most species; the near absence of crossing-over in the heterogametic sex is unique to *Drosophila* and a few other species.) A mutation is also known in *Drosophila melanogaster* that suppresses or prevents crossing-over from occurring in females homozygous for this mutation. Such females fail to form synaptinemal complexes during oögenesis. Unfortunately, essentially nothing is known about the functions of the various components of the synaptinemal complex. It is known that a small amount of DNA synthesis (an amount equivalent to less than 1 percent of the total DNA in the genome) occurs at about the time that the synaptinemal complex forms. It seems likely that this DNA synthesis is somehow involved in synapsis and/or crossing-over; however, that remains largely speculation at present.

Despite the paucity of information about the mechanism of recombination, evidence has accumulated, particularly in prokaryotic systems, that indicates that crossing-over is an enzymatic process, like most other metabolic processes. In *E. coli*, phage lambda, and phage T4, ***recombination-deficient mutants*** have been identified and shown to result from mutations in genes coding for nucleases and other enzymes involved in DNA metabolism. However, the complete pathway of enzyme-catalyzed steps has not been worked out for any organism.

Most of the currently popular models of crossing-over are derived from the Holliday model and the Whitehouse model, named after the two scientists who proposed them in 1964. These models were the first that took into account all the various types of genetic data that must be consistent with the mechanism of crossing-over in terms of breakage and reunion with associated repair synthesis. Many modifications of

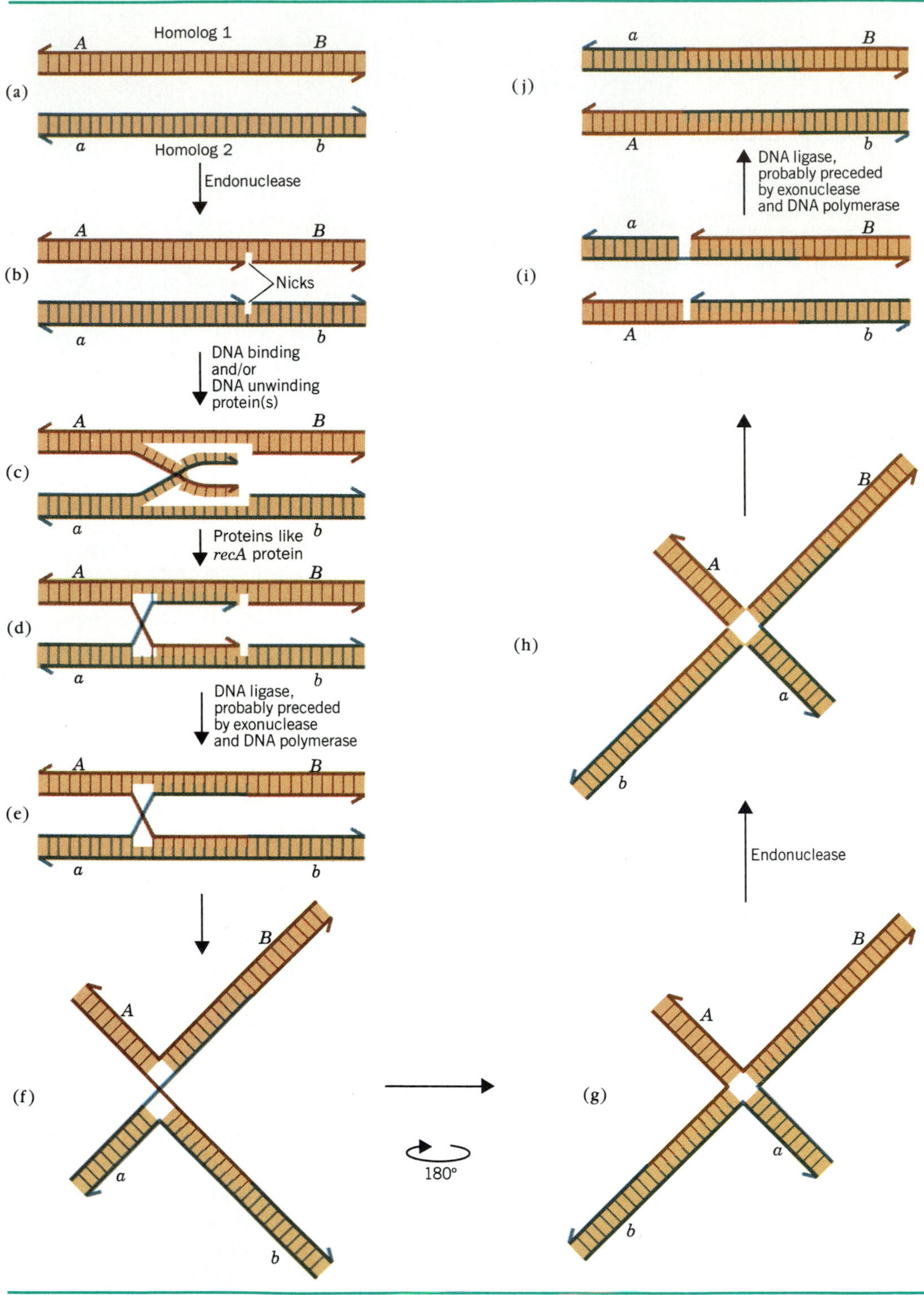

these two models have subsequently been proposed. One possible pathway for the occurrence of crossing-over, based largely on the Holliday model, is illustrated in Fig. 7.28. This pathway, like many others that have been invoked, begins when an ***endonuclease*** cleaves single strands of each of the two parental DNA molecules (***breakage***). Segments of the single strands on one side of each cut are displaced from their complementary strands, probably with the aid of ***"DNA-binding" proteins, "helix-destabilization" proteins,*** and ***"DNA-unwinding" proteins*** (also called ***"DNA helicases"***).

The displaced strands then exchange pairing partners, base-pairing with the intact complementary

Figure 7.28 A putative pathway for the occurrence of crossing-over by breakage and reunion showing some of the enzymes believed to be involved. This pathway is based on the model of R. Holliday (*Genet. Res.* 5: 282, 1964) with modifications as described by H. Potter and D. Dressler (*Proc. Natl. Acad. Sci. U.S.A.* 72: 3000, 1976). It involves (a) the synapsis of homolog 1 (shown in brown), carrying markers *A* and *B*, with homolog 2 (shown in aqua), carrying markers *a* and *b*. The opposite polarities of the complementary strands of each double helix are indicated by the arrowheads at the ends of each single strand. (Base-pairing must always involve two strands of opposite polarity!) (b) Crossing-over is initiated by endonuclease-induced nicks (single-strand cuts) in each DNA molecule. In the Holliday model shown, these cuts occur in strands with the ***same polarity.*** (c) The free ends on one side of the cuts are displaced from their complementary strands, probably with the aid of "DNA-binding," "helix-destabilization," and/or "DNA-unwinding" proteins. (d) The displaced strands then base-pair with the complementary intact strands of the other homolog, that is, the cleaved single strand of homolog 1 base-pairs with the complementary intact strand of homolog 2, and vice versa. The protein product of the *recA* gene of *E. coli* has been shown to promote this reaction *in vitro.* (e) The cleaved strands are then covalently rejoined in recombinant combinations by DNA ligase. If the original cuts did not occur between precisely the same two base-pairs, excision by an exonuclease and resynthesis by a DNA polymerase will be required to provide the proper ***adjacent*** "unlinked-bases" substrate for DNA ligase. At this point, the two homologs are joined by a single-stranded exchange or bridge. The three-dimensional structure of this intermediate in crossing-over may be more easily visualized when drawn as in (f) or (g), equivalent structures shown from different planar views. Such x-shaped intermediates (called "Holliday intermediates" or "chi forms," after the Greek letter chi, written χ) have been observed in prokaryotes (Fig. 7.29). (h) If the previously intact single strands are then cleaved at the "bridge" or "intersection," the structures shown in (i) will be formed. Covalent closure of the single-stranded interruptions will yield the intact ***recombinant chromosomes*** shown in (j).

strands of the homologous chromosomes. This process is also aided by certain proteins. In *E. coli*, the product of the *rec A* gene, referred to simply as the **rec A *protein***, has been shown to promote just such reciprocal exchanges of DNA single strands between two DNA double helices. The *rec A* protein-mediated reaction proceeds in two stages, at least *in vitro*. As a single strand of one double helix "invades" (displaces the identical or homologous strand and base-pairs with the complementary strand) a second, homologous double helix, the displaced single strand of the second double helix, in turn, similarly invades the first double helix. The *rec A* protein mediates such a reaction by binding to the unpaired strand of DNA, somehow "searching" for a homologous DNA sequence, and, once a homologous double helix is found, promoting the displacement of a segment of one strand of the double helix by the unpaired strand.

If complementary sequences already exist as unpaired strands, as in Fig. 7.28*c*, the presence of the **rec A *protein will increase the rate of renaturation by over 50-fold.***

The cleaved strands are subsequently covalently joined in recombinant arrangements ***(reunion)*** by DNA ligase. If the original breaks in the two strands do not occur at exactly the same site in the two homologs, some "tailoring" will be required before DNA ligase can catalyze the reunion step; this tailoring involves excision of a limited number of bases by an ***exonuclease*** and repair synthesis by a ***DNA polymerase.*** The sequence of events described so far will produce an X-shaped recombination intermediate (called a "chi" form); such intermediates have been observed by electron microscopy in several prokaryotic systems (Fig. 7.29). A similar sequence of enzyme-catalyzed breakage and reunion events, involving the other two single strands, occurs to complete the process of crossing-over.

Actually, a substantial body of evidence has accumulated that indicates that homologous recombination occurs by more than one mechanism—very possibly by several different mechanisms. Evidence from studies of the yeast *Saccharomyces cerevisiae* shows that ends of molecules produced by double-strand cuts or breaks are very recombinogenic. This and a variety of other evidence have led to the conclusion that recombination in yeast often involves an intermediate with a double-strand break in at least one of the parental double helices. Thus, a ***double-strand-break model*** of crossing-over was proposed by Szostak *et al.* in 1983. An updated version of this model is diagrammed in Fig. 7.30. A major difference between this model and the Holliday model (see Fig. 7.28) is that recombination is mediated by a double-strand break in one of the parental double helices, not by just single-strand breaks. The initial breaks in the two strands are then enlarged to gaps in both strands. The two single-stranded termini produced at the double-stranded gap of the broken double helix invade the intact double helix, displacing the homologous strand in this region. Repair DNA synthesis next fills in the gaps using the complementary single strands of the other chromo-

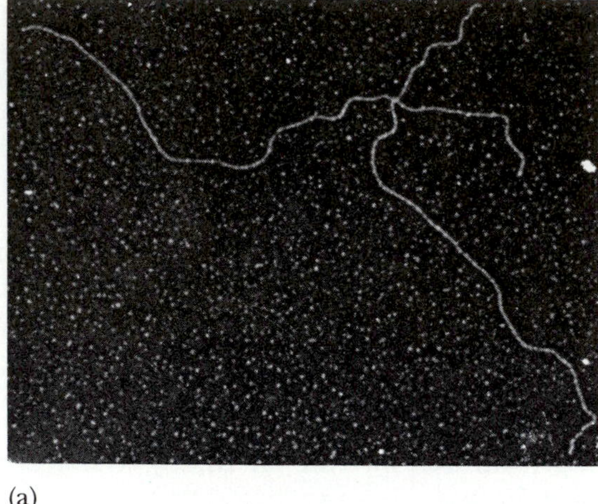

(a)

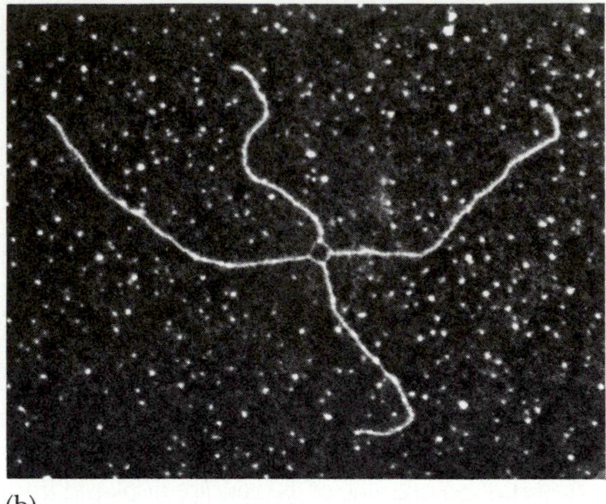

(b)

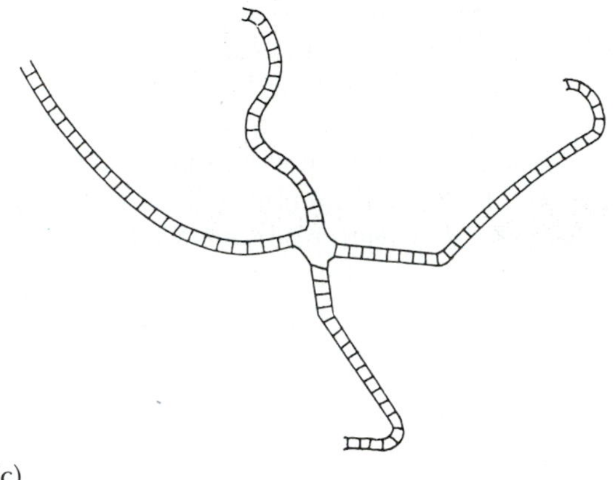

(c)

Figure 7.29 Electron micrographs of χ-shaped recombination intermediates ("chi forms" or "Holliday intermediates"). (a) A typical chi form equivalent to that diagrammed in Fig. 7.28g. (b) A chi form in which the covalent single-stranded connections at the intersection of the four arms are particularly clear (apparently due to stretching forces during preparation for electron microscopy). (c) Diagram of the structure seen in b. The DNA molecules are those of the *E. coli* "minichromosome" or plasmid Col E1. These DNA molecules are circular *in vivo,* but have been cleaved to produce the linear forms shown using a specific endonuclease. (From H. Potter and D. Dressler, *Proc. Natl. Acad. Sci. U.S.A.* 73: 3000, 1976.)

some as templates. This process yields a recombination intermediate with the two double helices joined by two single-stranded bridges (chi structures). The bridges are resolved by endonuclease cleavages by the same process as for chi structures formed by the Holliday mechanism (see Figs. 7.28*g–j*). Thus, the double-strand-break model, like the Holliday model, nicely explains the production of chromosomes that are recombinant for genetic markers flanking the region in which the crossover occurs.

Although the details of the mechanism of crossing-over remain uncertain, enzyme-catalyzed events similar to these are clearly involved. Proteins such as the *rec A* protein play very important roles in the process of recombination between ***homologous chromosomes or homologous DNA molecules.*** This type of recombination is called ***general recombination.*** It should ***not*** be confused with ***site-specific recombination,*** recombination that always occurs at specific sites or sequences of DNA molecules, or with ***"illegitimate" recombination,*** recombination between nonhomologous DNA molecules. The latter processes, which will be discussed in subsequent chapters, occur by quite different mechanisms.

GENE CONVERSION

Up to this point, we have only discussed recombination events that can be explained by breakage of homologous chromatids and the ***reciprocal exchange*** of parts. When crosses are done in Ascomycetes (fungi that produce ascospores) using genetic markers that are very closely linked (such as two mutations at different sites within the same gene; see Chapter 12) and tetrad data are collected and analyzed, recombination is frequently found to be ***nonreciprocal.*** For example, if crosses are done between two very closely linked mutations in *Neurospora*, and asci containing wild-type recombinants are analyzed, these asci frequently do ***not contain the reciprocal double mutant recombinant.*** Consider a cross involving two very closely linked mutations m_1 and m_2. In a cross of type

$m_1\ m_2^+ \times m_1^+\ m_2$, asci of the following type are frequently observed:

Spore pair 1: $m_1^+\ m_2$
Spore pair 2: $m_1^+\ m_2^+$
Spore pair 3: $m_1\ \ m_2^+$
Spore pair 4: $m_1\ \ m_2^+$

The wild-type $m_1^+\ m_2^+$ spores are present, but the reciprocal recombinant $m_1\ m_2$ double mutant spores are not present in the ascus. The $m_2^+ : m_2$ ratio is 3 : 1 rather than 2 : 2 as expected. It is as though one of the m_2 alleles is "converted" to the m_2^+ allelic form. Thus this type of ***nonreciprocal recombination*** ***conversion,*** and, despite its somewhat misleading connotation, the term has been extensively used for over two decades. One might assume that gene conversion is a result of mutation (see Chapter 11), except that it occurs at a higher frequency than the corresponding mutation events, always produces the allele present on the homologous chromosome, and is correlated about 50 percent of the time with reciprocal recombination of outside markers. The last observation strongly suggests that gene conversion is a result of processes involved in recombination. In fact, it now seems likely that gene conversion is a direct consequence of classical recombination that is observed when very closely linked genetic markers are used, such that the markers are within the region in which the breakage, excision, repair, and rejoining events involved in crossing-over are occurring.

With closely linked markers, gene conversion or nonreciprocal recombination is more frequent than reciprocal recombination. In a study of tetrad data from crosses involving mutations at different sites within the histidine-1 gene of yeast, S. Fogel and D. Hurst found that 980 out of 1081 asci containing wild-type *his*$^+$ recombinants exhibited gene conversion of one or the other of the segregating mutations, whereas only 101 showed classical reciprocal recombination.

The most striking feature of gene conversion is that the input 1:1 allele ratio expected from Mendel's law of segregation (2:2 in yeast, 4:4 in *Neurospora*) is not maintained. This can most easily be explained if, during the recombination process, short segments of strands of parental DNA are degraded and then resynthesized using a strand of DNA containing the other parental allele as template (a so-called excision–repair reaction). Several DNA repair mechanisms are known in different organisms (see Chapter 11, pp. 305–307). Given mechanisms for the repair of mismatched segments of DNA (segments in which one or more base-pairs cannot properly form hydrogen bonds), the mechanism presented in Fig. 7.28 to explain classical reciprocal recombination would be expected to give

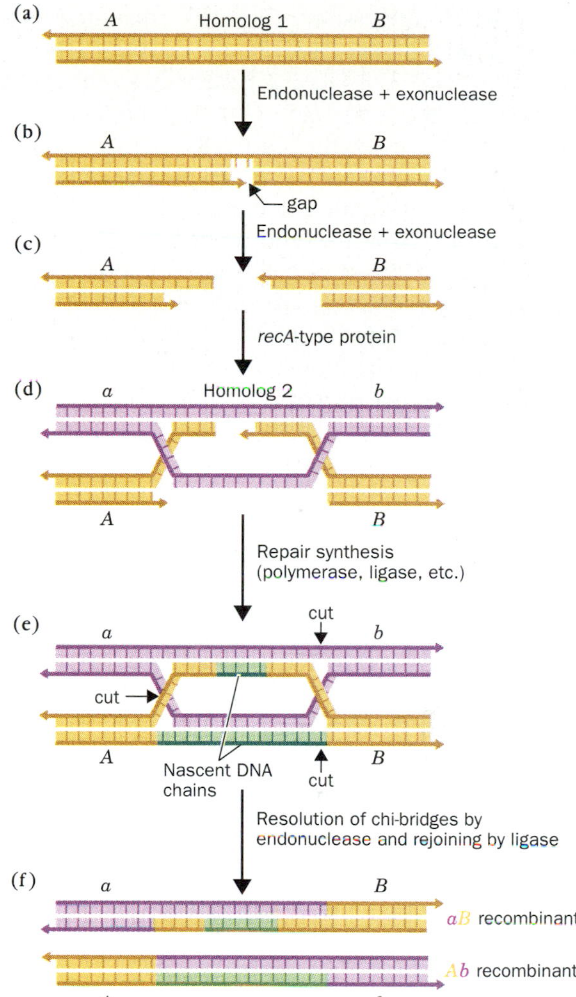

Figure 7.30 Double-strand-break model of recombination. (a–c) A double-strand gap is produced in the DNA of one chromosome (homolog 1) by sequential endonuclease and exonuclease activities. (d) The free single-stranded ends produced at the gap invade the second DNA molecule (homolog 2) with the aid of proteins such as *recA*-type proteins, DNA single-strand binding protein, helicase, etc., displacing the homologous strand of the intact double helix. (e) Repair synthesis fills in the gap by using the complementary strands of the intact, but locally unwound, double helix as templates. Ligation produces two double helices joined at two sites by single-stranded bridges (chi structures). (f) The bridges are resolved by the action of endonucleases, with subsequent rejoining catalyzed by DNA ligase (see g–j of Fig. 7.28). The recombinant chromosomes produced by this sequence of reactions are shown at the bottom. Their exact structures will depend on which strands are cut during resolution of the single-stranded bridges. (Based on Version II of D. S. Thaler and F. W. Stahl, "DNA Double-Chain Breaks in Recombination of Phage λ and of Yeast," *Annu. Rev. Genet.* 22: 169–197, 1988.)

rise to gene conversion if genetic markers were present in the immediate vicinity of the crossover. Note that in Fig. 7.28*d* to 7.28*j* there is a segment of DNA between the *A* and *B* loci where complementary strands of DNA from the two different parents are base-paired. If a third pair of alleles located within this segment were segregating in the cross, mismatches in the two double helices would be present. DNA molecules containing such mismatches, or different alleles in the two complementary strands of a single double helix, are called ***heteroduplexes***. Such heteroduplex molecules are now known to occur frequently as intermediates in the process of recombination.

If Fig. 7.28*e* is modified to include a third pair of alleles, and the other two chromatids are added to show the expected tetrad composition, it would appear as follows (again, note that for simplicity the DNA molecule in each chromatid is shown as an extended linear structure):

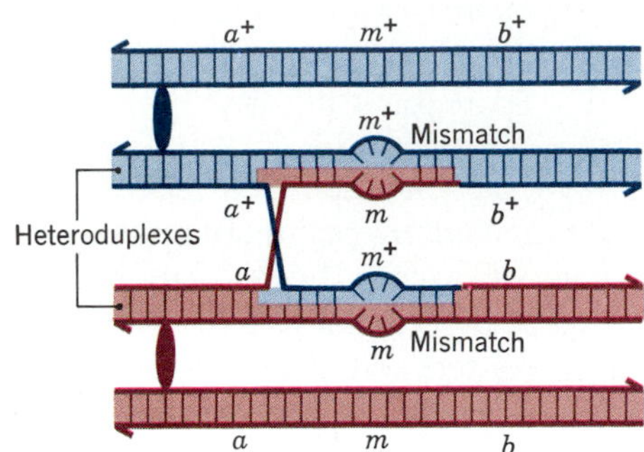

Figure 7.31 Diagram showing the formation of either the recombinant combinations (bottom left) or the parental combinations (bottom right) of outside markers in association with gene conversion. The recombination intermediate at the top is the equivalent to Fig. 7.28g for the mismatch repaired chromatids of the tetrad shown in the text (giving rise to the 3 m^+ to 1 m gene conversion ascus pattern). Depending on which two of the four strands are cleaved to resolve the bridge and allow segregation, either the parental (a^+b^+ and $a\ b$) or the recombinant ($a^+\ b$ and $a\ b^+$) combinations of outside markers may result. Cleavage in a vertical plane, as shown on the left, yields recombinant combinations of outside markers; cleavage in a horizontal plane, as shown on the right, yields parental combinations of outside markers.

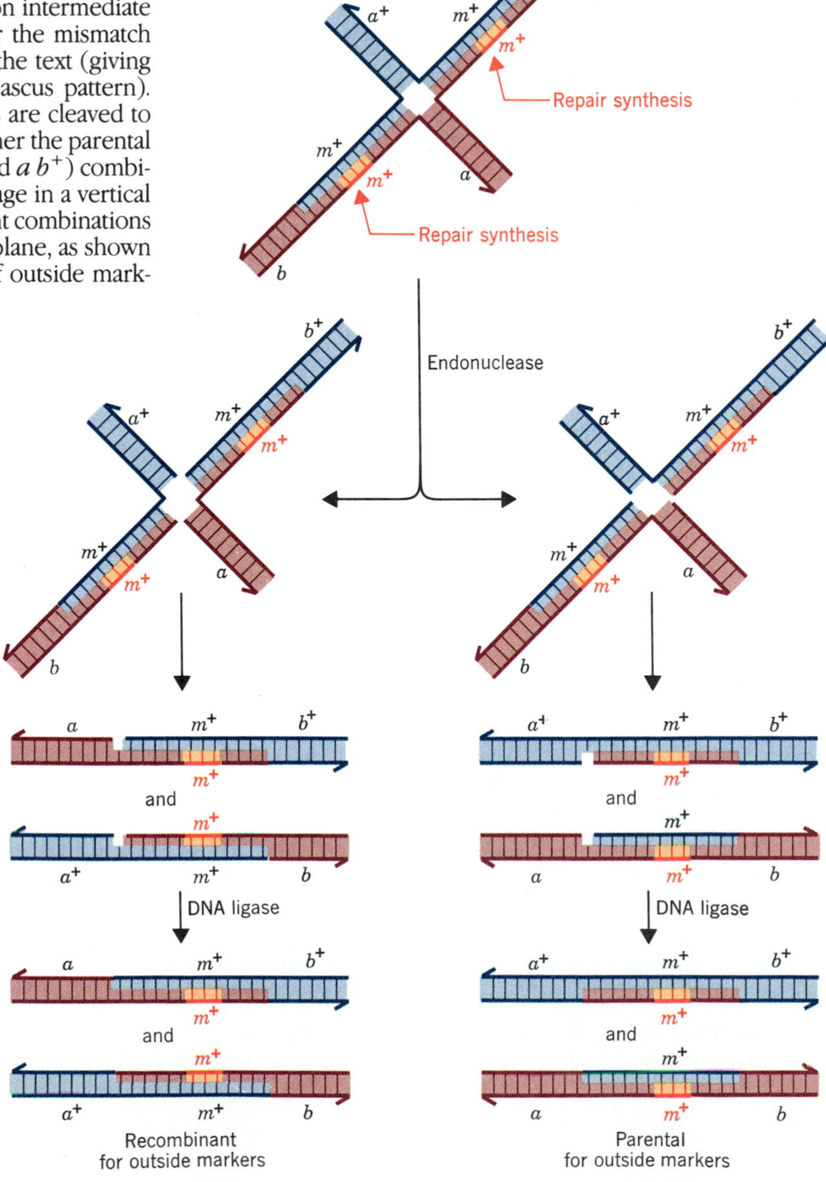

If the mismatches are resolved by an excision–repair pathway, such as that diagrammed in Fig. 11.22 (Chapter 11), with the m strands being excised and resynthesized using the complementary m^+ strands as templates, the following tetrad will result:

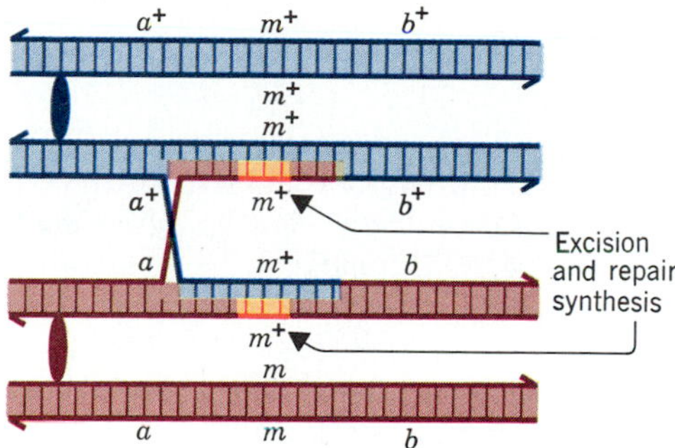

As a result of semiconservative DNA replication during the subsequent mitotic division, this tetrad will yield an ascus containing six m^+ ascospores and two m ascospores, or the 3 : 1 gene conversion ratio.

Suppose that only one of the two mismatches in the tetrad just described was repaired prior to the mitotic division. In that case, the semiconservative replication of the remaining heteroduplex would yield one m^+ homoduplex and one m homoduplex, and the resulting ascus would contain a 5 m^+ : 3 m ratio of ascospores. Such 5 : 3 gene conversion ratios are in fact observed. They represent postmeiotic or mitotic segregations of unrepaired heteroduplexes.

Another important aspect of gene conversion is the approximately 50 percent correlation with reciprocal recombination of outside markers; this is also nicely explained by the mechanism of recombination presented in Fig. 7.28. If the two recombining chromatids of the preceding diagrammed tetrad (the two center chromatids connected by a single-stranded bridge) are drawn in a planar form equivalent to Fig. 7.28g, one can easily see how gene conversion could be associated with either a parental arrangement of outside markers or a recombinant arrangement of outside markers. This is illustrated in Fig. 7.31. The analysis of asci produced from tetrads in which gene conversion has occurred has thus provided an important approach by which to study the mechanism of recombination.

SUMMARY

Genes that are located on the same chromosome do not assort independently during meiosis. Instead, they tend to segregate together. Such genes are said to be ***linked***. By definition, two genes are linked whenever a dihybrid produces over 50 percent gametes with parental arrangements of the segregating pairs of alleles and less than 50 percent gametes with recombinant arrangements.

Recombinant arrangements of genes located on the same chromosome are produced by ***crossing-over***, which involves the breakage of individual chromatids and the exchange of parts. This process of ***breakage and reunion*** is usually associated with a small amount of DNA repair synthesis. Crossing-over occurs after chromosome duplication, in the ***tetrad*** or ***four-chromatid*** stage of meiosis (early prophase I). A given crossover involves any two of the four chromatids.

The relative locations of genes on chromosomes can be ***mapped*** by two- and three-factor crosses. ***One map unit is defined as the linkage distance that yields 1 percent recombination.*** When a large number of genes are mapped in any given species, the genes are observed to occur in ***linkage groups, with one linkage group corresponding to each pair of chromosomes*** (diploids). All linkage maps exhibit ***linearity*** (unbranched lines or circles). The maximum frequency of recombination that can occur in any two-factor cross is 50 percent (the value observed for independently assorting genes). In Ascomycetes, such as *Neurospora*, all the products of a single meiotic event are maintained in a linear, tubelike structure called an ***ascus***. By dissecting the haploid products of meiosis, or ***ascospores***, out of an ascus, one can determine ***whether alleles segregate during the first meiotic division or the second meiotic division***. The alleles will ***segregate*** during the ***second meiotic division only when a crossover occurs between the locus involved and the centromere on the chromosome.*** This allows one to map the centromere and to use it as a marker in genetic crosses. Data in which the genotypes of all the products of a single meiotic event are available are called ***tetrad data***. Tetrad data permit one to determine whether recombination is reciprocal or nonreciprocal.

When crosses are performed using ***genetic markers that are far apart, recombination is*** almost always observed to be ***reciprocal***. When ***very closely linked markers***, such as two mutations at different sites within the same gene, ***are used, recombination is frequently nonreciprocal***. Such nonreciprocal recombination is called ***gene conversion***. Gene conversion apparently results from DNA excision and DNA repair synthesis that occur in heteroduplex regions during the process of breakage and reunion.

Genetic mapping in humans has been facilitated by the development of techniques for the hybridization of somatic cells. ***Somatic-cell hybridization*** involves the fusion of cells (and their nuclei) that are growing *in vitro* to form hybrids. Both monospecific and interspecific cell hybrids can be produced. In rodent–human somatic-cell hybrids, most of the human chromosomes are eliminated (for unknown reasons) in the cell divisions following fusion. This elimination of human chromosomes in rodent–human somatic-cell hybrids has proven to be extremely advantageous in the determination of linkage relationships in humans.

REFERENCES

CREIGHTON, H.B., and B. MCCLINTOCK. 1931. "A correlation of cytological and genetical crossing over in *Zea mays*." *Proc. Natl. Acad. Sci. U.S.A.* 17: 492–497.

DRESSLER, D., and H. POTTER. 1982. "Molecular mechanisms in genetic recombination." *Annu. Rev. Biochem.* 51: 727–761.

HSU, T.C. 1973. "Longitudinal differentiation of chromosomes." *Annu. Rev. Genet.* 7: 153–176.

KING, R.C., editor. 1974. *Handbook of genetics,* Vol. 2, *Plants, plant viruses, and protists.* Plenum, New York.

______. 1975. *Handbook of genetics,* Vol. 3, *Invertebrates of genetic interest.* Plenum, New York.

______. 1975. *Handbook of genetics,* Vol. 4, *Vertebrates of genetic interest.* Plenum, New York.

LEWIN, B. 1990. *Genes IV,* Chap. 32, "Recombination and other topological manipulations of DNA." Cell Press, Cambridge, MA/Oxford Univ. Press, Oxford.

LINDSLEY, D.L., and E.H. GRELL. 1968. *Genetic variations of Drosophila melanogaster.* Publication 627, Carnegie Institution of Washington, Washington, D.C.

MCKUSICK, V.A. 1980. "The anatomy of the human genome." *J. Hered.* 71: 370–391.

______. 1988. *Mendelian inheritance in man,* 8th ed. Johns Hopkins Univ. Press, Baltimore.

MESELSON, M. 1964. "On the mechanism of genetic recombination between DNA molecules." *J. Mol. Biol.* 9: 734–745.

MIZUUCHI, K., B. KEMPER, J. HAYS, and R.A. WEISBERG. 1982. "T4 endonuclease VII cleaves Holliday structures." *Cell* 29: 357–365.

MORGAN, T.H., C.B. BRIDGES, and A.H. STURTEVANT. 1925. "The genetics of *Drosophila*." *Bibliographia Genetica* 2: 1–262 (The Hague, Netherlands).

MORROW, J. 1983. *Eukaryotic cell genetics.* Academic Press, New York.

O'BRIEN, S.J., editor. 1982. *Genetic maps,* Vol. 2. National Cancer Institute, Frederick, MD.

PETERS, J.A., editor. 1959. *Classic papers in genetics.* Prentice-Hall, Englewood Cliffs, N.J.

RADDING, C.M. 1978. "Genetic recombination: Strand transfer and mismatch repair." *Annu. Rev. Biochem.* 47: 847–880.

RUDDLE, F.H., and R.S. KUCHERLAPATI. 1974. "Hybrid cells and human genes." *Sci. Amer.* 231(1): 36–44.

SHERIDAN, W.F., editor. 1982. *Maize for biological research.* Plant Molecular Biology Association, Charlottesville, VA.

STAHL, F.W. 1979. *Genetic recombination.* Freeman, San Francisco.

STURTEVANT, A.H., and G.W. BEADLE. 1939. *An introduction to genetics.* Saunders, Philadelphia. (Republished in 1962 by Dover, New York.)

THALER, D.S., and F.W. STAHL. 1988. "DNA double-chain breaks in recombination of phage λ and of yeast." *Annu. Rev. Genet.* 22: 169–197.

VASIL, I.K., W.R. SCOWCROFT, and K.J. FREY, editors. 1982. *Plant improvement and somatic cell genetics.* Academic Press, New York.

PROBLEMS AND QUESTIONS

7.1 Suggest experiments on some organism to determine genetically (a) whether two genes are located on the same chromosome pair and (b) whether they are in the coupling or repulsion phase. (c) What are the advantages and disadvantages of the testcross method for determining linkage relationships?

7.2 From a cross between individuals with the genotypes *Cc Dd Ee* × *cc dd ee*, 1000 offspring were produced. The class appearing *C-D-ee* included 351 individuals. Are the genes *c*, *d*, and *e* on the same or different chromosome pairs? Explain.

7.3 If an animal with the genotype *Rr Ss Tt* produced 1020 eggs, of which 127 are *r S t*, 121 *r S T*, and 130 *R S T*, are the three pairs of alleles on the same chromosome, on two different chromosomes, or on three different chromosomes? Explain.

7.4 If two loci are 10 map units apart, what proportion of the meiotic events will contain a single crossover in the region between these two loci, assuming that no multiple crossovers occur?

7.5 Genes *a* and *b* are linked and are 20 map units apart. An $a^+ b^+/a^+ b^+$ individual was mated with an $a\ b/a\ b$ individual. (a) Represent the cross on the chromosomes, illustrate the gametes produced by each parent, and illustrate the F_1. (b) What gametes can the F_1 produce and in what proportions? (c) If the F_1 was crossed with the double recessive, what offspring would be expected and in what proportions? (d) Is this an example of coupling or repulsion?

7.6 If the original cross in Problem 7.5 was $a^+ b/a^+ b$ × $a\ b^+/a\ b^+$, diagram (a) the arrangement of the genetic markers on the chromosomes in the F_1; (b) the gametes produced by the F_1 and proportions; and (c) expected testcross results. (d) Is this coupling or repulsion?

7.7 If the recombination frequency in Problems 7.5 and 7.6 were 40 percent instead of 20 percent, what difference would it make in the proportions of gametes and testcross progeny?

7.8 If Problems 7.5 (a) and 7.6 (b) with 20 percent recombination were carried to the F_2 ($F_1 \times F_1$), and a^+ and b^+ were dominant over their alleles, what phenotypic classes would be produced and in what proportions?

7.9 A fully heterozygous F_1 corn plant was red with normal seed. This plant was crossed with a green plant (*b*) that had tassel seed (*ts*). The following results were obtained: red, normal 124; red, tassel 126; green, normal 125; and green, tassel 123. (a) Does this indicate linkage? (b) If so, what is the percentage of crossing-over? (c) If not, show that the recombination frequency is 50 percent. (d) Diagram the cross showing the arrangement of the genetic markers on the chromosomes.

7.10 A fully heterozygous gray-bodied (b^+), normal-winged (vg^+) female F_1 fruit fly crossed with a black-bodied (*b*), vestigial-winged (*vg*) male gave the following results: gray, normal 126; gray, vestigial 24; black, normal 26; and black, vestigial 124. (a) Does this indicate linkage? (b) If so, what is the percentage of crossing-over? (c) If not, show that the recombination frequency is 50 percent. (d) Diagram the cross showing the arrangement of the genetic markers on the chromosomes.

7.11 Another fully heterozygous gray-bodied, normal-winged female F_1 fruit fly crossed with a black-bodied, vestigial-winged male gave the following results: gray, normal 23; gray, vestigial 127; black, normal 124; and black, vestigial 26. (a) Does this indicate linkage? (b) If so, what is the percentage of crossing-over? (c) Diagram the cross showing the arrangement of the genetic markers on the chromosomes.

7.12 In rabbits, color results from a dominant gene (c^+) and albinism from its recessive allele (*c*). Black is the result of a dominant gene (b^+) and brown of its recessive allele (*b*). Fully homozygous brown rabbits were crossed with albinos, which carried the gene for black in the homozygous state. The F_1 rabbits were then crossed to double recessive rabbits ($\frac{c\ b}{c\ b}$ or $\frac{cb}{cb}$). From many such crosses the results were: black 34; brown 66; and albino 100. (a) Are these genes linked? (b) If so, what is the percentage of crossing-over? (c) Diagram the original cross showing the arrangement of the genetic markers on the chromosomes.

7.13 In tomatoes, tall vine (d^+) is dominant over dwarf (*d*), and spherical fruit shape (p^+) is dominant over pear (*p*). Vine height and fruit shape are linked with 20 percent crossing-over. A certain tall, spherical tomato plant (a) crossed with a dwarf, pear-fruited plant produced 81 tall, spherical; 79 dwarf, pear; 22 tall, pear; and 17 dwarf, spherical. Another tall, spherical plant (b) that was crossed with a dwarf, pear produced 21 tall, pear; 18 dwarf, spherical; 5 tall, spherical; and 4 dwarf, pear. Represent on the chromosomes the arrangements of the genes in these two tall, spherical plants. (c) If these two plants were crossed with each other, what phenotypic classes would be expected and in what proportions?

7.14 Genes *a* and *b* are located on chromosome II with a crossover frequency of 20 percent. Genes *c* and *d* are located on chromosome III with a crossover frequency of 40 percent. An individual homozygous for a^+ b^+ c^+ d^+ was crossed with a fully recessive individual. The F_1 was backcrossed to the full recessive. Diagram (a) the original cross showing the arrangement of genetic markers on the chromosomes, (b) the F_1, and (c) the gametes that the F_1 is expected to produce with their proportions.

7.15 A student has two dominant traits dependent on single genes: cataract (an eye abnormality), which he inherited from his mother, and polydactyly (an extra finger), which he inherited from his father. His wife has neither trait. If the genes for these two traits are closely linked, would the student's child be more apt to have (a) either cataract or polydactyly, (b) cataract and polydactyly, or (c) neither trait? Explain.

7.16 In *Drosophila*, the recessive genes *sr* (stripe) and *e* (ebony body) are located at 62 and 70 map units, respectively, from the left end of the third chromosome. A striped female (homozygous for e^+) was mated with a male having an ebony body (homozygous for sr^+). (a) What kind of gametes will be produced by the F_1 females and in what proportions? (b) If F_1 females are mated with striped, ebony males, what phenotypes would be expected and in what proportions?

7.17 In *Drosophila*, the gene (*vg*) for vestigial wing is recessive and is located at 67 units from the left end of the second chromosome. Another gene (*cn*) for cinnabar eye color is also recessive and is located at 57 units from the left end of the second chromosome. A fully homozygous female with vestigial wings was crossed with a fully homozygous cinnabar male. (a) How many different kinds of gametes could the F_1 females produce and in what proportions? (b) If the females are mated with cinnabar, vestigial males, what phenotypes would be expected and in what proportions?

7.18 In *Drosophila*, the recessive genes *st* (scarlet eye), *ss* (spineless bristles), and *e* (ebony body) are located on the same (third) chromosome in the following positions (map distances), starting from the left end of the chromosome: *st* 44, *ss* 58, *e* 70. Fully heterozygous females with the genotype *st ss* e^+/st^+ ss^+ *e* are mated with fully recessive males *st ss e*/*st ss e*. If many flies are produced and no interference occurs, what phenotypes will be expected and in what percentages?

7.19 In maize, genes *Pl* for purple (dominant over Pl^+ for green), *sm* for salmon silk (recessive to sm^+ for yellow silk), and *py* for pigmy (recessive to py^+ for normal size) are on chromosome 6 (Fig. 7.8). From the cross

$$\frac{Pl\ sm\ py}{Pl\ sm\ py} \times \frac{Pl^+\ sm^+\ py^+}{Pl^+\ sm^+\ py^+}$$

and the testcross between the F_1 and fully recessive plants, what phenotypes would be expected and in what proportions? Assume no interference and refer to Fig. 7.8 for map distances.

7.20 In maize, the genes *Tu*, j_2 and gl_3 are on chromosome 4 (Fig. 7.8). If plants carrying these three genes in the homozygous recessive condition are crossed with plants homozygous for the three dominant alleles, and F_1 plants are testcrossed to fully recessive plants, what genotypes would be expected and in what proportions? Assume no interference and refer to Fig. 7.8 for map distances.

7.21 A cross was made between yellow-bodied (*y*), echinus (*ec*), white-eyed (*w*) female (*y ec w*/*y ec w*) flies and wild males. The F_1 females were mated with *y ec w* males. The following genotypes were present in a sample of 1000 progeny flies:

FEMALES	MALES	NUMBER OF PROGENY
+ + +/*y ec w*	+ + +/*Y*	475
y ec w/*y ec w*	*y ec w*/*Y*	469
y + +/*y ec w*	*y* + +/*Y*	8
+ *ec w*/*y ec w*	+ *ec w*/*Y*	7
y + *w*/*y ec w*	*y* + *w*/*Y*	18
+ *ec* +/*y ec w*	+ *ec* +/*Y*	23
+ + *w*/*y ec w*	+ + *w*/*Y*	0
y ec +/*y ec w*	*y ec* +/*Y*	0

Determine the order in which the three loci *y*, *ec*, and *w* occur on the chromosome and prepare a linkage map.

7.22 A cross involving X-linked genes was made between yellow, bar, vermilion female flies and wild males, and the F_1 females were crossed with $y\,B^+\,v$ males. The following phenotypes were obtained when 1000 progeny were examined:

y B v	and	+ + +	546
y + +	and	+ *B v*	244
y + *v*	and	+ *B* +	160
y B +	and	+ + *v*	50

Determine the order in which the three loci occur on the chromosome and prepare a linkage map.

7.23 Female *Drosophila* heterozygous for ebony (e^+/e), scarlet (st^+/st), and spineless (ss^+/ss) were testcrossed, and the following progeny were obtained:

PROGENCY PHENOTYPES	NUMBER
Wild type	67
Ebony	8
Ebony, scarlet	68
Ebony, spineless	347
Ebony, scarlet, spineless	78
Scarlet	368
Scarlet, spineless	10
Spineless	54

(a) Are these genes linked? Justify your answer. (b) Write the genes given on a chromosome symbol with the genes in correct order. (c) Write the genotypes of the flies involved in the parental cross and testcross. (d) What is the map distance between the loci for ebony and scarlet? (e) What is the map distance between the loci for ebony and spineless? (f) Calculate the coefficient of coincidence.

7.24 Consider a female *Drosophila* having the following X chromosome constitution:

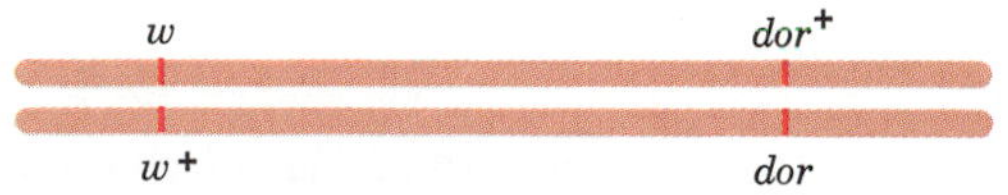

The genes *w* and *dor* are recessive mutations resulting in mutant eye color (white and deep orange, respectively). If these loci (white and deep orange) exhibit 40 percent recombination in two-factor crosses, what proportion of the male progeny of this female will be expected to show ***a mutant*** phenotype?

7.25 Assume that in *Drosophila* there are three pairs of alleles, +/*x*, +/*y*, and +/*z*. As shown by the symbols, each mutant gene is recessive to its wild-type allele. A cross between females heterozygous at these three loci and wild-type males yielded the following progeny:

Females:	+ + +	1010
Males:	+ + +	39
	+ + *z*	430
	+ *y z*	32
	x + +	27
	x y +	441
	x y z	31
		Total = 2010

Draw the appropriate linkage map for these data showing the order of the three markers and the map distances for each marked interval. Calculate the coefficient of coincidence for these data.

7.26 In the nematode *Caenorhabditis elegans, dpy* (dumpy) and *unc* (uncoordinated) are recessive linked genes with a recombination frequency of ***P***. A heterozygote is made with genes in repulsion, for example,

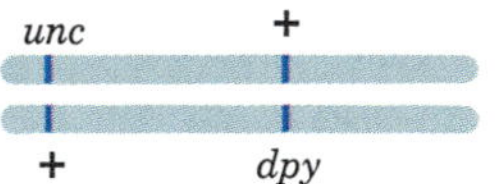

The heterozygote is allowed to self-fertilize. If crossing-over in sperm production is independent of crossing-over in oöcytes, what fraction of progeny would be expected to be ***both*** dumpy and uncoordinated?

7.27 In the following testcross, genes *a* and *b* are 20 map units apart and genes *b* and *c* are 10 map units apart:

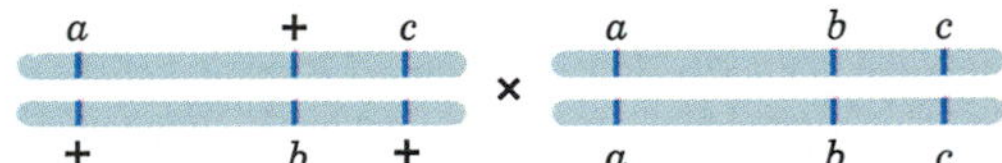

If the coefficient of coincidence is 0.5, how many triple homozygous recessive individuals would you expect to find among 1000 progeny from the cross?

7.28 Consider *a*, *b*, and *c* to be three recessive mutations in *Drosophila*. The following data are the results of a testcross in which F_1 females heterozygous at all three loci were crossed to males homozygous for all three recessive mutants.

+ + +	75	*a* + +	110
+ + *c*	348	*a b* +	306
+ *b c*	96	*a b c*	65
	Total = 1000		

Construct a linkage map showing the correct order of the three genes and the map distances between adjacent genes.

7.29 Singed bristles (*sn*), crossveinless wings (*cv*), and vermilion eye color (*v*) are due to recessive mutant alleles of three sex-linked genes in *Drosophila melanogaster*. When a female heterozygous for each of the three genes was testcrossed with a singed, crossveinless, vermilion male, the following progeny were obtained:

Singed, crossveinless, vermilion	3
Crossveinless, vermilion	392
Vermilion	34
Crossveinless	61
Singed, crossveinless	32
Singed, vermilion	65
Singed	410
Wild type	3
	Total = 1000

What is the correct order of the three genes on the chromosome? What are the genetic map distances between *sn* and *cv*, *sn* and *v*, and *cv* and *v*? What is the coefficient of coincidence?

7.30 What will the progeny phenotype ratio be if a plant of the genotype $l_1 A l_2^+/l_1^+ a l_2$ (where *A* and *a* are alleles with *A* dominant to *a*; l_1 and *A* as well as l_1^+ and *a* are completely linked; the *A* and l_2 loci are unlinked; and both l_1 and l_2 are lethal in the homozygous condition but have no detectable effect in the heterozygous condition) is self-fertilized?

7.31 Assume that a cross is made between a plant having the genotype *AAbb* and a second plant having the genotype *aaBB*. The F_1 plants were intercrossed (i.e., *AaBb* × *AaBb*) to produce the following F_2 progeny:

PHENOTYPE	NUMBER OF PLANTS
A–B–	102
A–bb	46
aaB–	50
aabb	2

How can one make a quick rough estimate of the linkage distance between the two loci? Based on these data, what would this estimate be for the map distance between the *A* locus and the *B* locus?

7.32 On the basis of the following ordered tetrad data from *Neurospora*:

Cross	Asci Patterns				
(*A*) × (*a*)	SPORE PAIRS				
	1–2	3–4	5–6	7–8	NUMBER OF ASCI
	(*A*)	(*A*)	(*a*)	(*a*)	61
	(*a*)	(*a*)	(*A*)	(*A*)	55
	(*a*)	(*A*)	(*a*)	(*A*)	40
	(*A*)	(*a*)	(*A*)	(*a*)	44
					Total = 200

what is the linkage distance in map units between the *A*, *a* locus and the centromere?

7.33 Of what genetic significance is the relatively frequent occurrence of tetratype asci in *Neurospora*?

7.34 The following tetrad data were obtained from the cross *AB* × *ab* in *Neurospora*:

Asci Patterns

SPORE PAIRS				
1–2	3–4	5–6	7–8	NUMBER OF ASCI
(*AB*)	(*AB*)	(*ab*)	(*ab*)	1766
(*AB*)	(*aB*)	(*Ab*)	(*ab*)	220
(*AB*)	(*Ab*)	(*aB*)	(*ab*)	14
				Total = 2000

On the basis of these data, draw a linkage map showing the relative positions of the two genes and the centromere.

7.35 Three mutations (*arg*, *thi*, and *leu*) in *Neurospora* are known to result in blocks in the synthesis of arginine, thiamine, and leucine, respectively. You are given the following data from asci dissections from crosses of *arg* with *thi* and *arg* with *leu*:

Asci Patterns

	SPORE PAIRS				
CROSS	1–2 or 7–8	3–4 or 5–6	5–6 or 3–4	7–8 or 1–2	NUMBER OF ASCI
(*arg* +) × (+ *thi*)	(*arg* +)	(*arg* +)	(+ *thi*)	(+ *thi*)	46
	(*arg thi*)	(*arg thi*)	(+ +)	(+ +)	56
					Total = 100
(*arg* +) × (+ *leu*)	(*arg* +)	(*arg* +)	(+ *leu*)	(+ *leu*)	155
	(*arg* +)	(*arg leu*)	(+ +)	(+ *leu*)	44
	(*arg* +)	(+ *leu*)	(*arg* +)	(+ *leu*)	1
					Total = 200

[Note that the three mutant alleles are represented by *arg*, *thi*, and *leu*, and the three wild-type alleles simply by plus signs (+) in each case.] Designate each chromosome involved by a single straight line and each centromere by an open circle, and draw the linkage map(s) dictated by these data, that is, show order and each map distance that can be calculated from the data.

7.36 Given the following ***unordered tetrad data*** from *Neurospora*:

Cross (+ + +) × (*x y z*)

CLASS	UNORDERED SPORE PAIRS				NUMBER OF ASCI
1	*x y z*	*x y z*	+ + +	+ + +	24
2	*x y z*	+ *y z*	*x* + +	+ + +	2
3	*x y* +	*x y* +	+ + *z*	+ + *z*	20
4	*x y z*	*x y* +	+ + +	+ + *z*	8
5	*x* + +	*x* + +	+ *y z*	+ *y z*	20
6	*x* + *z*	*x* + *z*	+ *y* +	+ *y* +	26

(a) What are the linkage relationships between the genes, that is, which genes are linked and what are the map distances between the linked genes? (b) Which gene is the greatest distance from a centromere? Explain the basis for your answer.

7.37 Two mutant genes (*a* and *t*) in *Neurospora* are known to interfere, respectively, with the synthesis of the amino acid arginine and the vitamin thiamine. After a cross in which these genes were segregating, the following ordered spore arrangements were found in the frequencies noted. (Only one member of each pair of spores is indicated.)

PAIR 1	PAIR 2	PAIR 3	PAIR 4	NUMBER OF ASCI
at	*at*	+ +	+ +	42
+ *t*	+ *t*	*a* +	*a* +	40
+ +	+ +	*a t*	*a t*	39
a +	*a* +	+ *t*	+ *t*	42

How are the *a* and *t* genes located on the chromosomes with respect to their centromere(s) and with respect to each other?

7.38 In *Neurospora* the following is the order of three genes and a centromere:

A cross between + + + and *x y z* gave one ascus with the following ordered ascospore composition:

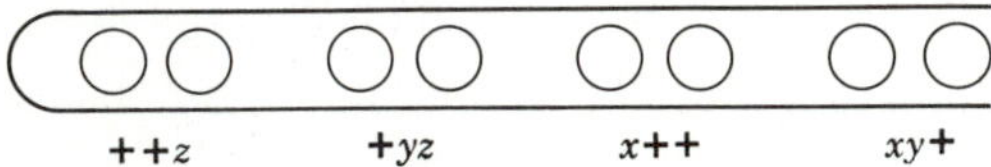

(a) Is this ascus most likely the result of a meiotic event in which 0, 1, 2, or 3 crossovers occurred? (b) In what interval(s) did the crossover(s) most likely occur? (c) If double or triple crossovers were involved, were they 2-strand, 3-strand, or 4-strand multiple crossovers?

7.39 The rosy (*ry*) locus of *Drosophila melanogaster* is located at map position 52.0 on chromosome 3. It controls the structure of the enzyme xanthine dehydrogenase. The rosy locus is particularly susceptible to genetic fine structure analysis because zygotes possessing little or no xanthine dehydrogenase activity are unable to complete development and die before eclosion on standard *Drosophila* culture medium supplemented with purines. The *kar* (karmoisin eyes) locus of *D. melanogaster* is located at map position 53.0 on chromosome 3; it provides an appropriate outside marker in fine structure mapping of the *ry* locus. A. Chovnick and colleagues performed crosses of the type

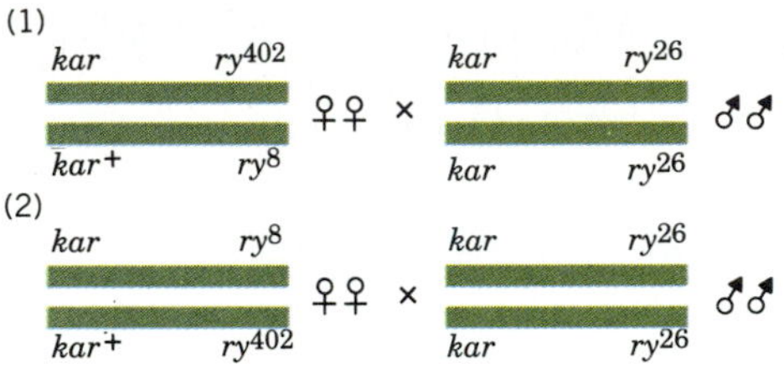

and scored the frequency of the outside marker *kar* among ry^+ progeny (progeny flies surviving on purine-enriched medium; note that ry^{402}, ry^{26}, and ry^8 are all completely recessive alleles of the *rosy* locus).

In cross (1), the majority of the ry^+ progeny were kar^+, whereas in cross (2) the majority of the ry^+ progeny were *kar* (had karmoisin eyes). What is the relative order of *kar*, ry^8, and ry^{402} on chromosome 3 of *D. melanogaster*?

7.40 A patient with Lesch–Nyhan disease was found to be heterozygous for alleles of the *gal K* gene, which codes for the enzyme galactokinase. Fibroblasts of this patient produced two different forms of this enzyme, which could be separated by polyacrylamide gel electrophoresis and identified using an enzyme-specific stain. Somatic-cell hybrids were produced between fibroblasts of this patient and thymidine kinase-deficient mouse cells. The hybrid cells were selected on HAT medium (see p. 186) and cultured for about 30 generations to allow human chromosome elimination to occur. Each of 100 independent clones was then examined by electrophoresis to determine its galactokinase phenotype. All the hybrid clones produced one or the other of the two forms of galactokinase; a few produced both forms. On which human chromosome is the *gal K* gene probably located?

Chapter 8

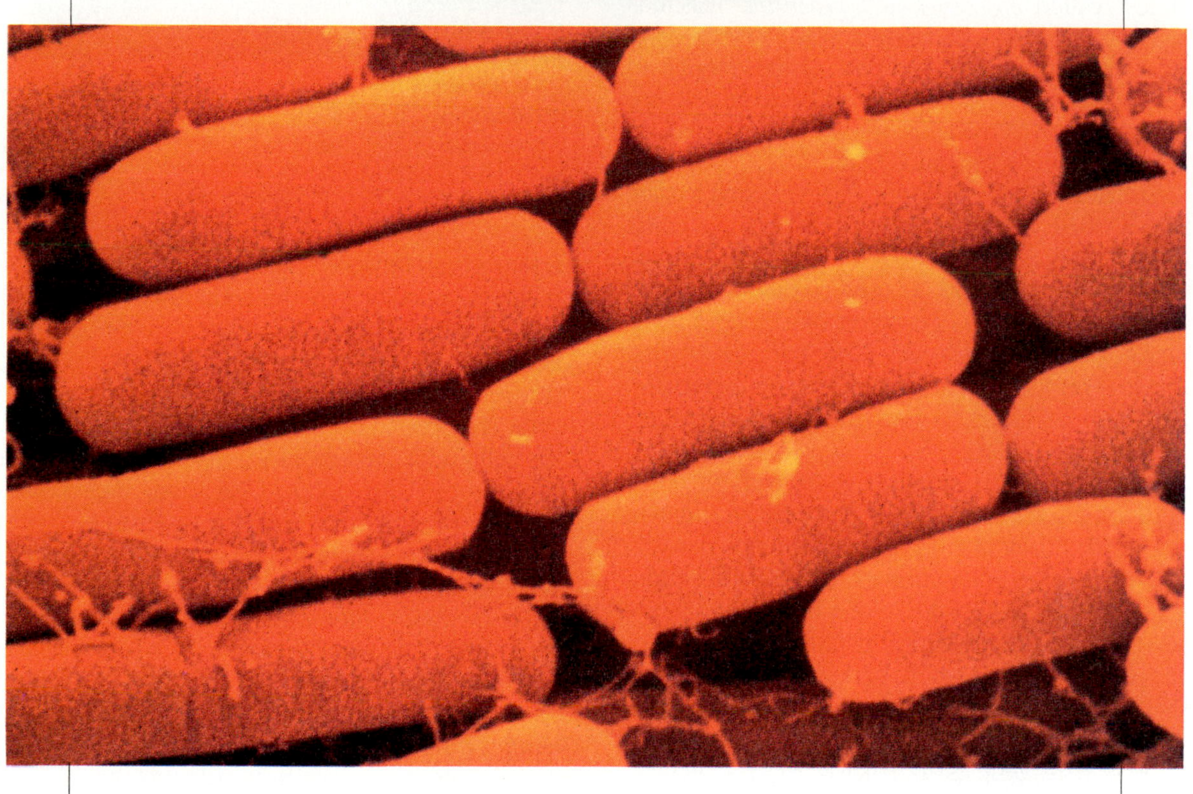

Recombination in Bacteria

(Photo above): False-color scanning electron micrograph of a group of *E. coli* cells. (Courtesy of Dr. Tony Brain/Science Photo Library/Photo Researchers.)

THE genetic information of bacteria is stored in (1) a ***single main chromosome,*** carrying a few thousand genes, plus (2) from zero to several "minichromosomes" called ***episomes*** and ***plasmids.*** Plasmids are highly variable in size, ranging from those carrying no more than three genes to those large enough to carry several hundred genes. Certain bacterial cells are known to harbor as many as 11 different plasmids in addition to the main chromosome.

Each bacterial chromosome characterized to date contains a single, circular molecule of DNA that exists in a highly condensed ("folded," negatively supercoiled; see Chapter 5, pp. 124–127) conformation *in vivo.* The chromosomes of bacteria, like those of other prokaryotes, are ***not*** enclosed in nuclear membranes. The nuclear regions of bacteria are thus called ***nucleoids,*** rather than nuclei, to distinguish them from their eukaryotic counterparts.

Bacteria divide by simple fission, with an equational distribution of their genetic material to the two progeny cells. They are usually haploid (or partially diploid, during recombination processes), but "multinucleate." That is, the nucleoid of a bacterium usually contains two or more identical copies of the chromosome (plus plasmids, when present). The chromosomes of bacteria do not go through the mitotic and meiotic condensation cycles associated with cell division and gametogenesis in eukaryotes. The genetic recombination events associated with sexual reproduction in higher organisms (segregation, independent assortment, and meiotic crossing-over) thus do not constitute integral parts of the life cycles of bacteria.

Recombination, however, is undoubtedly important in the evolution of bacteria just as it is in the evolution of eukaryotes. Three different processes have evolved that mediate transfer of genetic material from one bacterium to another, making possible the subsequent recombination events. The most obvious difference between these three processes is the mode of transfer of DNA from one cell to another. (1) ***Transformation*** involves the uptake of naked DNA molecules from one bacterium (the ***donor*** cell) by another bacterium (the ***recipient*** cell) (see Chapter 5). (2) ***Transduction*** occurs when bacterial genes are carried from a donor cell to a recipient cell by a bacteriophage. (3) ***Conjugation*** is the process during which DNA from a donor or male cell is transferred to a recipient or female cell through a specialized sex pilus or "conjugation tube."

The three modes of recombination in bacteria can be distinguished by two simple criteria: (1) sensitivity to the presence of deoxyribonuclease (DNase) and (2) dependence on cell contact (Table 8.1). These two criteria can be easily tested experimentally. The first criterion can be tested by simply adding DNase to the medium containing the bacteria involved in recombination. If recombination occurs in the absence of DNase, but not in its presence, then the DNase-sensitive recombination process, transformation, is occurring. The protein coat of the bacteriophage vector and the cell wall and membrane enclosing the conjugation tube protect the donor DNA from degradation by DNase in the medium during transfer to the recipient cell during transduction and conjugation, respectively. Whether cell contact is required is tested by carrying out a so-called U-tube experiment, in which bacteria of two different genotypes are placed in opposite arms of a U-shaped culture tube (Fig. 8.1). The two cultures of bacteria are separated by a semipermeable filter containing pores large enough to allow DNA molecules and viruses, but ***not cells,*** to pass through it. Conjugation cannot occur between cells separated in this way. If recombination occurs in the presence of DNase and in the absence of cell contact, it must be due to transduction or an unknown transductionlike process.

TABLE 8.1 Criteria for Determining the Mode of Recombination in Bacteria

RECOMBINATION PROCESS	CRITERION: CELL CONTACT REQUIRED?	CRITERION: SENSITIVE TO DNASE?
Transformation	no	yes
Transduction	no	no
Conjugation	yes	no

TRANSFORMATION

As was discussed in Chapter 5 (pp. 92–93), transformation was first discovered in pathogenic strains of *Diplococcus pneumoniae* by Griffith in 1928. The details of Avery, MacLeod, and McCarty's (1944) proof that the "transforming principle" (the cellular component mediating transformation) is DNA were also described in Chapter 5 (pp. 93–95). Some of the details of the mechanism by which transformation occurs are still unknown, 60 years after Griffith's discovery. However, a reasonably complete picture of the overall process of transformation has been established.

The uptake of DNA molecules by recipient bacteria is an active, energy-requiring process. It does not involve passive entry of DNA molecules through permeable cell walls and membranes (although this type of uptake of DNA molecules may be induced by experimental manipulations of bacteria in the laboratory, such as in the *Escherichia coli* recombinant DNA-

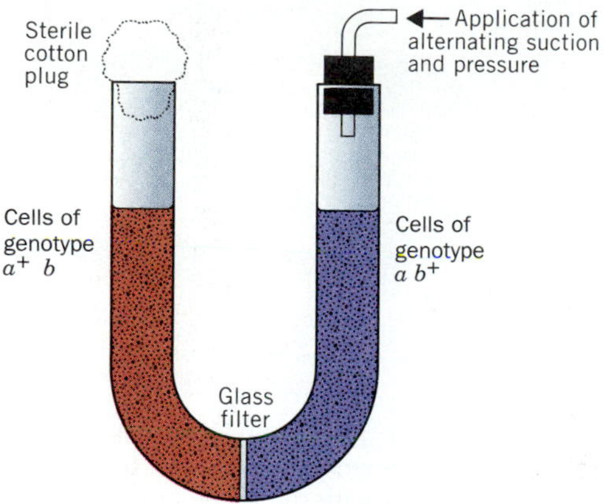

Figure 8.1 Diagram of the U-tube experiment used to determine whether cell contact is required for recombination in bacteria. The two arms of the Davis U-tube are separated by a glass filter containing pores of a size that permit the passage of DNA molecules and viruses, but not cells. Bacteria of one genotype (e.g., a^+ b) are placed in one arm of the Davis U-tube; those of another genotype (e.g., a b^+) are placed in the other arm. Alternating suction and pressure are applied to one arm of the tube during incubation to mix the medium and any small particles (such as phage) suspended in it. However, the glass filter prevents cells from passing from one arm of the tube to the other. If recombinants (e.g., a^+ b^+) are formed during the U-tube experiment, cell contact is not required for the recombination process to occur; thus this excludes conjugation as a possible mechanism.

cloning experiments; see Chapter 13, pp. 349–356). Thus, transformation does not occur "naturally" in all species of bacteria, only in those species possessing the enzymatic machinery involved in the active uptake and recombination processes. Most of the studies on transformation have been done with three species, *D. pneumoniae, Bacillus subtilis,* and *Hemophilus influenzae.* Even in these species, all cells in a given population are not capable of active uptake of DNA. Only ***competent*** cells, which possess a so-called ***"competence factor"*** (probably a cell-surface protein or enzyme involved in binding or in taking up DNA), are capable of serving as recipients in transformation. The proportion of the bacteria in a culture that are in the physiologically competent state varies with the growth conditions and the stage of the growth curve (becoming maximal in late log-phase).

The process of transformation can be divided into several stages: (1) ***reversible binding of double-stranded DNA*** molecules to ***receptor sites*** on the cell surface; (2) ***irreversible uptake*** of the donor DNA (at which time the donor DNA becomes resistant to DNase in the medium); (3) ***conversion*** of the double-stranded donor DNA molecules ***to single-stranded molecules*** by nucleolytic degradation of one strand; (4) ***integration (covalent insertion)*** of all or part of the single strand of donor DNA into the chromosome of the recipient; and (5) the ***segregation*** and ***phenotypic expression*** of the integrated donor gene or genes in the recombinant ("transformed") cell. Steps (2) and (3) may well be coincident effects of a single process. One attractive model, for which there is supporting evidence in the case of pneumococcus, proposes that a specific exonuclease (or ***DNA "translocase"***) pulls one strand of donor DNA into the cell using energy derived from the degradation of the complementary strand (Fig. 8.2). Whether degradation of the complementary strand of DNA actually occurs during uptake or immediately after uptake is uncertain. Moreover, considerable evidence suggests that these processes may vary in different species.

The first three steps in transformation—binding, uptake, and degradation of one strand of the double-stranded DNA—are not specific for homologous DNA. In fact, competent bacteria will carry out these three processes equally well given calf thymus DNA or other foreign DNAs. However, the ***integration,*** or ***DNA recombination*** step, ***is specific*** for ***homologous DNA.*** That is not to say that the integration of segments of heterologous (foreign) DNA never occurs. If it does occur, however, it does so at frequencies very much lower than the frequencies observed using homologous DNA. Although very small fragments of DNA are taken up by competent cells, a minimum length of about 500 nucleotide-pairs appears to be required for integration to occur. During integration, a single strand (either strand, the previous degradation of one strand is at random) of donor DNA is physically inserted into the recipient chromosome, replacing a segment of one strand of the recipient chromosome (Fig. 8.2 and pp. 223–225). In most transformation experiments, donor DNA fragments are about 20,000 nucleotide-pairs (or about 1/200 of the total chromosome) in length. This means that mapping experiments can be done using transformation only if the genetic markers employed are located close together on the host chromosome.

If two genes are far apart on the chromosome, they will never be carried on the same molecule of transforming DNA. Thus, double transformants for the two genes (say a to a^+ and b to b^+, using an a^+ b^+ donor and an a b recipient) will require two independent transformation events (uptake and integration of one DNA molecule carrying a^+ and another molecule carrying b^+). The probability of two such independent events occurring together will equal the product of the

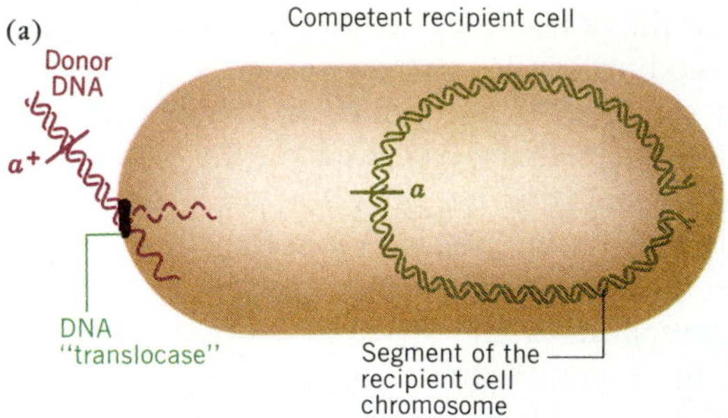

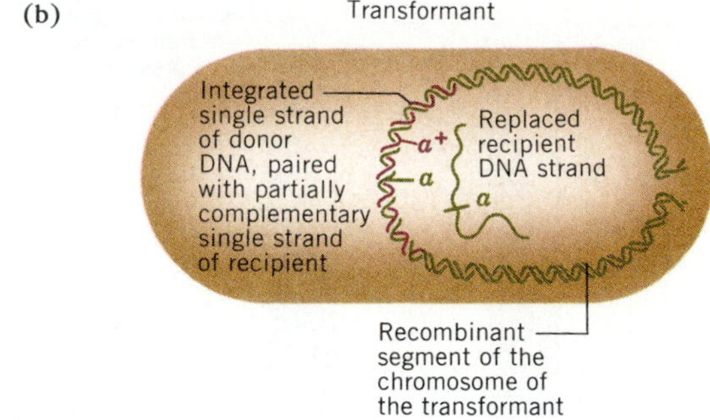

Figure 8.2 Diagrammatic illustration of the two key steps, uptake and integration, in transformation of pneumococcus. (a) This model for the uptake of donor DNA was proposed by S. Lacks in 1962. A membrane-bound exonuclease or DNA "translocase" pulls one strand of donor DNA into the cell, using energy derived from degradation of the complementary strand of DNA. The donor DNA fragment (shown in reddish color) carries the genetic marker a^+; the recipient chromosome (greenish color, only a segment of which is shown) carries the allelic marker a. (b) The single strand of donor DNA (reddish color) is shown covalently inserted (integrated) into the chromosome of the recipient (now called a transformant). The donor strand base-pairs over most of its length with the complementary (except at the a site) single strand of the recipient chromosome. A "mismatch" occurs at the a^+/a site where the donor strand carries the a^+ allele, whereas the recipient strand carries the a allele. Such DNA molecules, carrying different alleles in the complementary (partially) strands, are called DNA heteroduplexes. DNA heteroduplexes are important intermediates in mutation, recombination, and DNA repair processes. They segregate during subsequent semiconservative replication.

probability of each occurring alone. Since transformation of any single marker occurs with a low frequency, double independent transformation events of this type will be extremely rare. If, on the other hand, two genes are closely linked, they may be carried on a single molecule of transforming DNA. In this case, double transformants can be formed by the uptake and integration of one molecule of donor DNA carrying both genes. Thus, if two genes or genetic markers are very closely linked, double transformants may be formed at a frequency approaching the frequency of single transformants in comparable single-marker experiments. The frequency with which two genetic markers are cotransformed can thus be used as a crude estimate of the linkage distance between them. Genetic markers can also be ordered by means of three-factor transformation experiments using the same rationale as in three-factor transduction, conjugation, or sexduction experiments (see pp. 220–223).

TRANSDUCTION

Transduction, discovered by N. Zinder and J. Lederberg in 1952, occurs when a ***bacteriophage particle carries a segment of the chromosome*** from one bacterium (the ***donor***) to another bacterium (the ***recipient***), facilitating subsequent recombination of the genetic markers of the two cells. Two very different types of transduction are known. (1) In ***generalized transduction,*** a random or nearly random segment of bacterial DNA is "wrapped up" during phage maturation in place of, or along with, the phage chromosome in a few "progeny" particles, called ***transducing particles. Generalized transducing phages*** can therefore transport any gene of the donor cell to the recipient cell. Since all the genes of the donor are represented in a population of these transducing particles (although any one transducing particle contains only one segment of host DNA, representing 1/100 to 1/50 of the total donor chromosome), this type of transduction was named "generalized" transduction. In some cases, generalized transducing particles contain only bacterial DNA. In other cases, they contain both phage and bacterial DNA. (2) In ***specialized transduction*** (also called ***restricted transduction***), a recombination event involving the host chromosome and the phage chromosome occurs, producing a phage chromosome containing a segment of bacterial DNA. ***Specialized transducing particles*** thus always ***contain both phage and bacterial DNA.*** Specialized transduction is so named because a given virus transduces only genetic markers of the host that are located in one small region of the bacterial chromosome. (The reason for this is discussed on pp. 210–213). Bacteriophage lambda, the best-known specialized transducing phage, for example, usually mediates transduction of only the *gal* and *bio* genes of *E. coli.*

Generalized Transduction

Bacteriophages have been classified into two types on the basis of their interactions with the bacterial cell. ***Virulent phages*** always multiply and lyse the host cell after infection. ***Temperate phages*** have a choice between two life-styles after infection (Fig. 8.3). They can either (1) enter the ***lytic cycle,*** during which they reproduce and lyse their host cells just like virulent phages, or, alternatively, they can (2) enter the ***lysogenic pathway,*** during which their chromosomes are integrated into the chromosomes of the host and replicate like any other segments of the host chromosomes. ***Generalized transduction*** is mediated by some virulent bacteriophages and by certain temperate bacteriophages whose chromosomes are not integrated at specified attachment sites on the host chromosome. Generalized transducing particles are produced during the lytic cycles of these phages.

Not all virulent phages mediate transduction. The T-even bacteriophages (T2, T4, and T6), for example, degrade the host DNA and reutilize the mononucle-

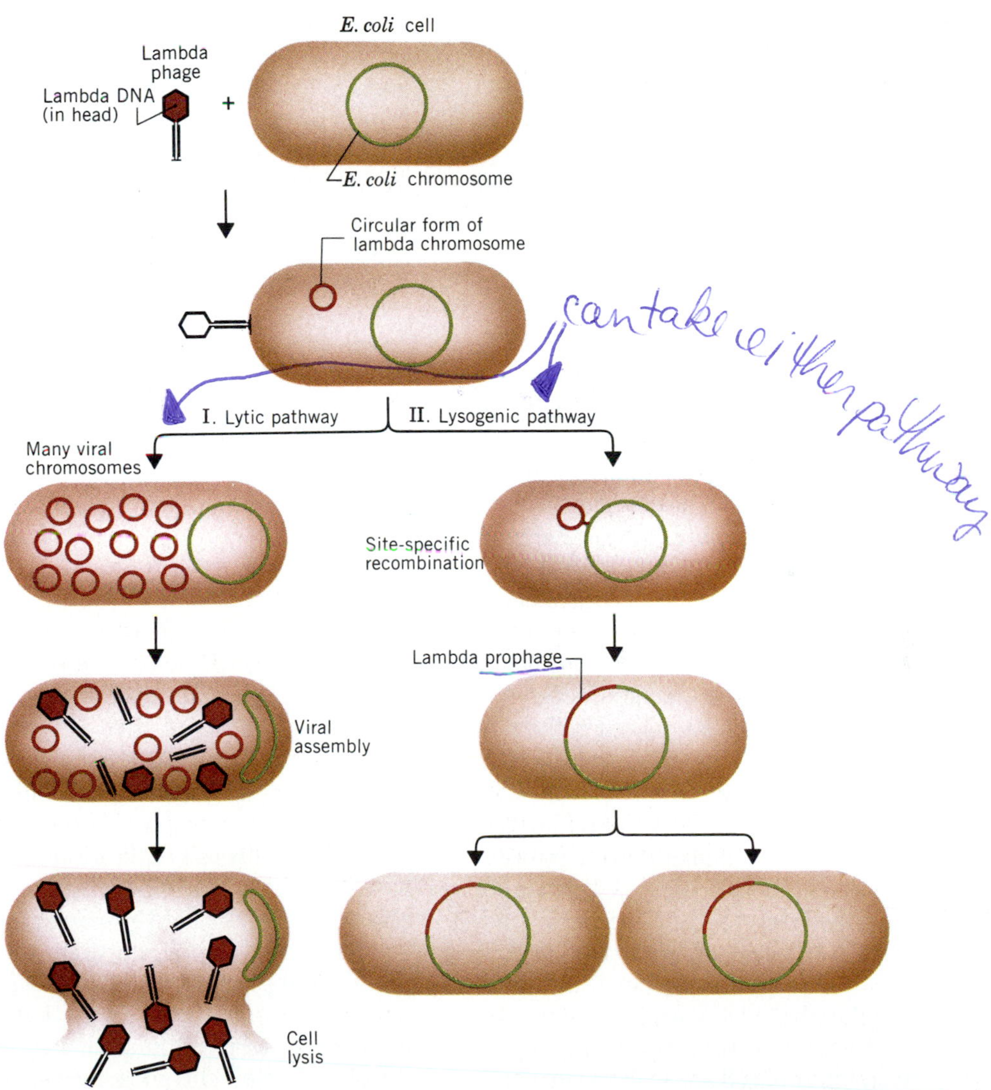

Figure 8.3 Schematic diagram illustrating the two alternative life-styles of temperate bacteriophages such as the *E. coli* phage lambda (λ). After the injection of the linear lambda chromosome into an *E. coli* cell, the chromosome is converted to its circular form. It then either (I) commences replicating autonomously and lyses the host cell, releasing 100 or more progeny viruses (the ***lytic pathway***), or (II) integrates via a site-specific recombination event into the *E. coli* chromosome (the ***lysogenic pathway***). In the integrated ***prophage*** state, the lambda chromosome replicates as part of the host chromosome and is transmitted to both progeny cells at each cell division. With a low probability, the prophage may be spontaneously excised from the host chromosome and enter the lytic pathway. It can be induced to enter the lytic pathway with a high frequency by irradiation with ultraviolet light.

otides produced for the synthesis of phage DNA. The host DNA is thus not available for packaging in progeny particles. Other viruses may not degrade the host DNA at all, and, since the host chromosome is too large to be packaged intact, they would not be able to form transducing particles. In still other viruses, the maturation process may be highly specific for phage DNA, excluding the packaging of fragments of host DNA. In any case, only a limited number of the virulent phages known mediate transduction. Of the generalized transducing phages, *E. coli* phage P1, *Salmonella* phage P22, and *Bacillus subtilis* phages PBS1 and SP10 have been extensively used for genetic fine structure mapping (mapping mutant sites within individual genes or short segments of the chromosome; see Chapter 12).

After a transducing phage injects a fragment of host DNA into a recipient cell, that DNA may either (1) be integrated into the host chromosome in a manner similar to the integration of transforming DNA, except that the integrated segment is double-stranded, or (2) remain free in the cytoplasm. If it is not integrated, it will not replicate and will be transmitted to only one progeny cell during each cell division. The genes located on the transduced chromosome fragments may be expressed, even if they are not integrated. Cells carrying nonintegrated transducing fragments are called ***abortive transductants.*** They are partially diploid and can be used to carry out complementation tests. The complementation test provides the operational definition of the gene. It is used to determine whether different mutations are in the same gene or in different genes (see Chapter 12).

The conclusion that the generalized transducing particles produced by certain phages carry only bacterial DNA (no phage DNA) is based on the results of density transfer experiments (see Chapter 5, pp. 105–108). If bacteria are grown for several generations in medium containing a precursor of DNA that is heavier than normal (such as ^{15}N in place of ^{14}N or 5-bromouracil in place of thymine), their DNA will be heavier than normal and will band at a higher density position in a CsCl density gradient. If such bacteria (containing "heavy" DNA) are transferred to "light" medium (medium with normal DNA precursors such as ^{14}N) containing a radioactive DNA precursor (such as ^{32}P or [^{3}H]thymidine), all the phage DNA synthesized will be "light" and radioactive. If the "progeny" particles produced from such an infected cell are analyzed in a CsCl density gradient, any transducing particles (particles containing bacterial DNA) present should band at a heavier than normal position. This was indeed found to be the case when this experiment was done using coliphage P1. Moreover, no radioactive "phage DNA" (DNA synthesized after infection) was found associated with the band of transducing particles. All the radioactivity in the gradient was present at the "light" density position, in the band containing the infectious progeny phage. Thus, generalized transducing particles formed during phage P1 infections do not contain any phage DNA. (If this experiment is done using a specialized transducing phage, however, the results are very different.)

Transducing particles are produced at a low frequency. Only one out of 10^5–10^7 of the "progeny" particles present in a lysate contains bacterial DNA. Thus, the probability of a cell being doubly transduced for genetic markers carried in two different transducing particles is negligible. (If cells are simultaneously infected with 100 or more phage particles, they are rapidly killed by a process called "lysis-from-without," which apparently results from simply punching too many holes in the cell membrane.) The cotransduction of two or more genetic markers therefore indicates that the markers are relatively closely linked, and the frequency of cotransduction of any two markers is indicative of the degree of linkage between them. Occasionally, genetic markers can be ordered by cotransduction patterns. If (1) markers a^+ and b^+ are cotransduced, (2) markers b^+ and c^+ are cotransduced, but (3) markers a^+ and c^+ are ***not*** cotransduced, then the order of the three markers must be a^+-b^+-c^+. More frequently, however, three-factor transduction experiments must be used to unambiguously order genetic markers (see pp. 220–223 for the rationale involved).

Specialized Transduction

Specialized transduction is mediated by temperate bacteriophages whose chromosomes are able to ***integrate at one*** or a ***few specified attachment sites*** on the host chromosome. The chromosomes of temperate phages of this type are thus capable of both (1) ***autonomous replication*** (replication independent of the replication of the host chromosome) and (2) ***integrated replication*** (replication as a segment of the host chromosome). As such, they are examples of genetic elements called ***episomes*** (see pp. 224–226).

Integration of the chromosome of a specialized transducing phage, such as the coliphage lambda, involves a recombination event between the circular intracellular form of the phage chromosome and the circular bacterial chromosome at ***specific attachment sites*** on the two chromosomes. This ***site-specific*** recombination event results in the covalent linear insertion of the phage chromosome into the chromosome of the bacterium (Fig. 8.4). In its integrated state, the phage chromosomes is called a ***prophage.*** The lytic genes of the virus, those involved in viral reproduction and lysis of the host, are ***repressed*** (turned off) when

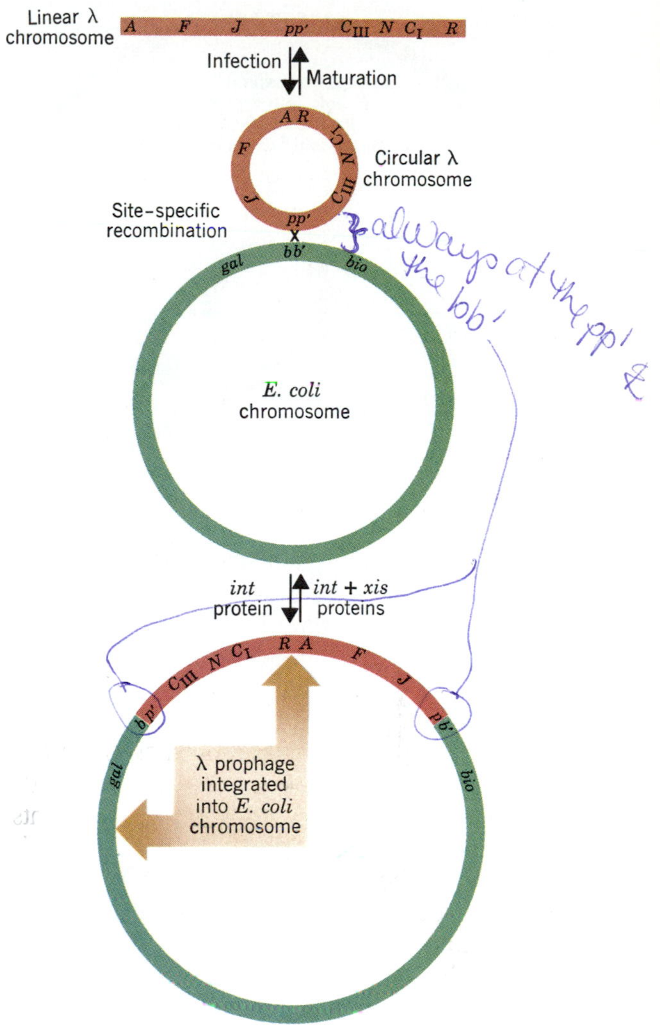

Figure 8.4 Integration and excision of the phage lambda chromosome. When the lambda chromosome is packaged inside the phage head, it exists as a linear molecule with gene sequence *A* through *R*, as shown at the top (red). After infection, it is converted to its circular intracellular form by the mechanism described in Fig. 5.17. When lambda enters the lysogenic pathway (Fig. 8.3), it undergoes a site-specific recombination event between the *pp'* site on the phage chromosome and the lambda attachment site, *bb'*, on the *E. coli* chromosome (yellow-green). This integrative recombination step is catalyzed by the lambda *int* gene-product. In the prophage state, the lambda gene sequence is *p'-CIII-R-A-J-p*. This sequence is a circular permutation of the gene sequence in the linear form packaged in the lambda head (top), which results from a specific endonuclease cut between genes *A* and *R* during phage maturation.

During spontaneous or UV-induced excision, essentially the reverse of integration occurs. We can visualize this as the lambda prophage "looping out" to form a "figure-8" structure; this step is followed by a site-specific recombination event between the *bp'* sequence and *pb'* sequence. The site-specific excision step requires the product of the *int* gene plus the product of another lambda gene called *xis*. Normal excision produces an intact circular lambda chromosome plus an intact circular *E. coli* chromosome. Landy and Ross have sequenced both the lambda chromosome attachment site (*pp'*) and the *E. coli* chromosome lambda attachment site (*bb'*), as well as the lambda prophage-*E. coli* DNA junctions (*bp'* and *pb'*), and found that all four contain the identical 15 nucleotide-pair sequence:

—GCTTTTTTATACTAA→
←CGAAAAAATATGATT—

For this to be true, the site-specific recombination events involved in integration and excision must occur within or at one end of these 15 nucleotide-pair sequences.

the chromosome is in the prophage state. (The mechanism by which the prophage genes are repressed is discussed in Chapter 14, pp. 402–403.) A bacterium harboring a prophage is said to be ***lysogenic;*** the prophage–host relationship is called ***lysogeny.*** A lysogenic cell is ***immune*** to secondary infections by the same virus (homologous to the prophage), because the lytic genes of the infecting virus will be repressed just as those of the prophage are repressed.

Temperate phages undergo rare (about one in 10^5 cell divisions) spontaneous transitions from the lysogenic or prophage state to the lytic state. Such transitions can also be ***induced,*** for example, by irradiation with ultraviolet light. During the switch from the lysogenic state to lytic growth, the prophage is ***excised*** from the host chromosome (Fig. 8.4) and commences replicating autonomously. The ***excision*** process is ***site specific,*** like the integration process. The site-specific integration and excision processes are catalyzed by enzymes encoded by phage genes.

The excision process is usually very precise in cutting out the phage chromosome in exactly the form in which it existed prior to its integration. Occasionally, however, the excision event occurs at a site other than the original attachment site. When this happens, a portion of the phage chromosome is left in the host chromosome and a portion of the bacterial chromosome is excised with the phage DNA (Fig. 8.5). Such

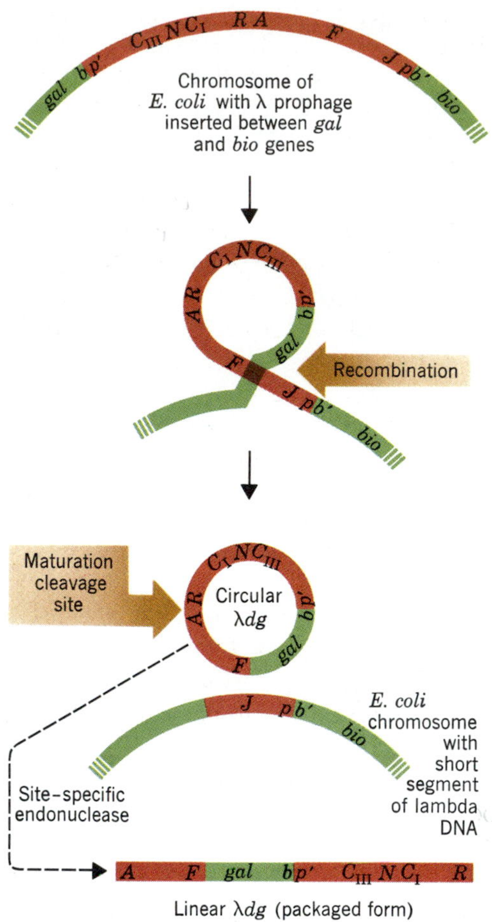

Figure 8.5 Diagram illustrating the formation of a λ*dg* specialized transducing particle. The sequence of events is essentially the same as in normal prophage excision (see Fig. 8.4), except that the recombination event does ***not*** occur between the *bp'* and *pb'* homology sites. Instead, the recombination event occurs at a position within the *gal* region of the host chromosome and a position in the gene *F* to gene *J* region of the lambda chromosome. As a result, the excised lambda chromosome includes a segment of host DNA containing part of the *gal* region, and at a portion of the lambda prophage, containing gene *J*, is left in the host chromosome. Transducing particles carrying *bio* markers are similarly produced when a segment of host DNA containing part of the *bio* region is excised. (Keep the circle the same size and roll it to the right before excision.) The circular λ*dg* chromosome is converted to the linear form (bottom) that is packaged within the head of the phage particle by a site-specific endonuclease that cleaves between genes *A* and *R*. This endonuclease makes staggered cuts in the complementary strands of DNA, producing the 12-nucleotide-long cohesive (complementary single-stranded) ends of the mature lambda chromosome.

"mistakes" during prophage excision are responsible for the formation of specialized transducing particles. Only host genes located close to the site of prophage insertion can be excised with the phage DNA and packaged in "phage" particles. Thus specialized transduction is restricted to the transfer of genes located within a short distance on each side of the prophage attachment site. Phage lambda integrates between the *gal* genes (required for the utilization of galactose as an energy source) and the *bio* genes (essential for the synthesis of biotin) on the *E. coli* chromosome (Figs. 8.4 and 8.5); lambda thus usually only transduces *gal* and *bio* markers. Specialized transducing phage Φ80, on the other hand, integrates near the *E. coli trp* genes (required for the synthesis of the amino acid tryptophan) and transduces *trp* markers.

If specialized transducing particles are formed during prophage excision, as shown in Fig. 8.5, then only phage lysates produced by ***induction*** of lysogenic cells should have transducing activity. This is indeed the case. If bacteria are infected by specialized transducing phages under conditions where only lytic infections occur, no transducing particles are present in the phage lysates. The frequency of transducing particles in lysates produced by induction of lysogenic cells is about one in 10^6 progeny particles.

The chromosome composition of transductants produced by specialized transduction is quite different from the chromosome composition of transductants resulting from generalized transduction or transformation. In the latter cases, recombination ***replaces*** a segment of the recipient's chromosome with a segment of the donor's chromosome. In specialized transduction, the segment of donor DNA and the phage chromosome in which it is covalently inserted are ***added*** to the recipient's chromosome, producing a ***partially diploid transductant*** (Fig. 8.6*a*). This partial diploidy has several important consequences, which are most easily illustrated by considering a specific transducing phage. Since coliphage lambda is by far the best known of the specialized transducing phages, it is the logical choice for discussion. Specifically, consider a lambda chromosome carrying *gal* genes, like the one illustrated in Fig. 8.5. The transducing phage produced is called a λ*dg* (for lambda defective *gal*). It is ***defective*** because genes required for growth

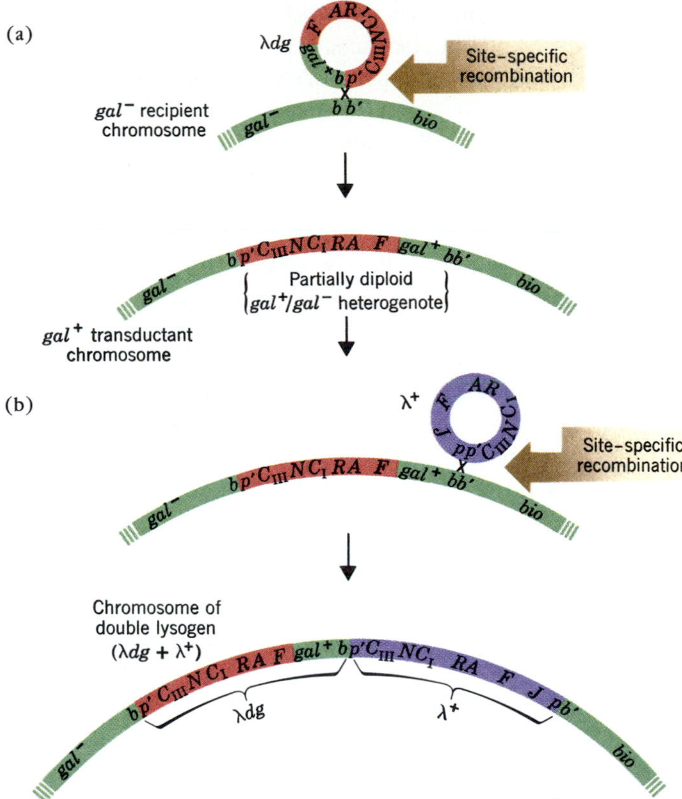

Figure 8.6 (a) The integration of a λ*dg* chromosome, which carries gal^+, into a gal^- recipient chromosome to form a partially diploid gal^+/gal^- transductant (heterogenote). (b) The subsequent integration of a λ^+ chromosome ("helper" phage) to form a $\lambda^+/\lambda dg$ double lysogen. Note that gal^+ and gal^- refer to wild-type and mutant alleles, respectively, at the *gal* locus. Cells that are gal^+ can utilize galactose as a carbon source; cells that are gal^- cannot. Induction of a double lysogen of this type with UV will produce an Hft (high-frequency transduction) lysate containing 50 percent λ*dg* particles and 50 percent λ^+ phage. If a recipient cell is infected with a single λ*dg* particle (low ratio of phage/bacteria), then the gal^+ transductant formed will have the genotype shown in a. If a recipient is simultaneously infected with both a λ*dg* particle and one or more λ^+ phage (high ratio of phage/bacteria), then the transductant will usually carry both λ*dg* and λ^+ as shown in b. Integration of the λ*dg* and the λ^+ chromosomes can occur in either order.

and maturation under lytic conditions have been replaced by bacterial DNA. Thus λ*dg* transducing particles can reproduce only in the presence of a wild-type lambda ***"helper"*** phage. The *g* indicates that the bacterial genes present are *gal* genes rather than *bio* genes (as in λ*db* or λ*bio* transducing particles).

When gal^+ λ lysogens (cells lysogenic for λ) are induced by ultraviolet light, rare λ*dg* particles are formed, carrying the donor gal^+ gene or genes (depending on the size of the segment of bacterial DNA that is carried along). When these λ*dg* particles infect gal^- recipient cells, the λ*dg* integrates by a crossover event within the *gal* region or at the λ attachment site (Fig. 8.6*a*). (Note that microbiologists use a minus superscript to designate a mutant allele of a gene. Thus gal^- cells are unable to utilize galactose as a carbon source.) If the ratio of phage/bacteria is high enough, such that the recipient cells are also infected with wild-type λ phage, a wild-type genome usually integrates by recombination within the normal λ attachment site, producing transductants that are double lysogens (carrying one λ^+ prophage and one λ*dg*; see Fig. 8.6*b*). The transductants are thus gal^+/gal^- partial diploids. They are called **gal^+/gal^- *heterogenotes*** and contain a **gal^+ *exogenote*** (donor DNA fragment) and a **gal^- *endogenote*** (recipient chromosome).

The primary transductants, or gal^+/gal^- heterogenotes, are unstable; they segregate gal^- cells at a frequency of about one in 1000 cell divisions. These gal^- segregants can be explained by the excision of the λ*dg* chromosome. Since λ*dg* cannot reproduce in the absence of a helper phage, it may simply be lost or diluted out during cell division. Recombination may also occur between the gal^+ exogenote and the gal^- endogenote, transferring the gal^+ marker to the endogenote and producing stable gal^+ transductants.

Because of the presence of the λ genes controlling immunity in λ*dg* chromosomes, the partially diploid transductants are immune to subsequent lambda infections. If the transductants are λ*dg*-λ^+ double lysogens (Fig. 8.6*b*) and are induced with ultraviolet light, they will produce lysates containing 50 percent λ*dg* particles and 50 percent λ^+ particles. Both prophages will be excised and will replicate with equal efficiency using the gene-products coded for by the λ^+ genome. Such lysates are called *Hft* (high-frequency transduction) lysates. *Hft* lysates greatly facilitate genetic analyses using specialized transduction by dramatically increasing the frequency of transduction events. *Hft* lysates can also be obtained from gal^+/gal^- heterogenotes that are single λ*dg* lysogens by simultaneously infecting them with wild-type λ and inducing with ultraviolet light.

CONJUGATION

Conjugation was discovered in 1946 by J. Lederberg (Fig. 8.7) and E. L. Tatum (1958 Nobel Prize corecipients). During conjugation, DNA is transferred from a ***donor cell to a recipient cell*** through a specialized intercellular connection, or ***conjugation tube,*** that forms between them (Fig. 8.8). (The donor and recip-

Figure 8.7 J. Lederberg, 1958 Nobel Laureate, who with colleagues discovered transduction and conjugation in bacteria. (Courtesy of The Rockefeller University.)

ient cells are sometimes referred to as male and female cells, respectively.) The transfer of genetic information is thus a ***one-way transfer*** during conjugation, just as in transformation and transduction, rather than a reciprocal exchange of genetic material. Cells that have the capacity to serve as donors during conjugation are differentiated by the presence of specialized cell-surface appendages called ***F pili*** (Fig. 8.8). The synthesis of these F pili is controlled by several (nine, based on current data) genes that are carried by a small circular molecule of DNA or "minichromosome" (about 94,500 nucleotide-pairs long) called an ***F factor*** (for fertility factor; also called "sex factor" and "F plasmid"). Cells carrying an F factor (donor cells) form conjugation tubes and initiate DNA transfer after making contact with ***cells not carrying an F factor,*** called ***F^- cells*** (recipient cells).

The F factor can exist in two different states: (1) the ***autonomous state,*** in which it replicates independently of the host chromosome, and (2) the ***integrated state,*** in which it is covalently inserted into the host chromosome and replicates along with the host chromosome like any other set of chromosomal genes (Fig. 8.9). The F factor is thus, like the chromosomes of specialized transducing phages, an example of a class of genetic elements called ***episomes*** (see pp. 224–226).

A donor cell containing the F factor in the autonomous state is called an ***F^+ cell. When an F^+ donor cell conjugates with an F^- recipient cell, only the autonomous F factor is transferred.*** Both exconjugants (cells that have been involved in conjugation) become F^+ because the F factor replicates during transfer. Thus, mixing a population of F^+ cells with a

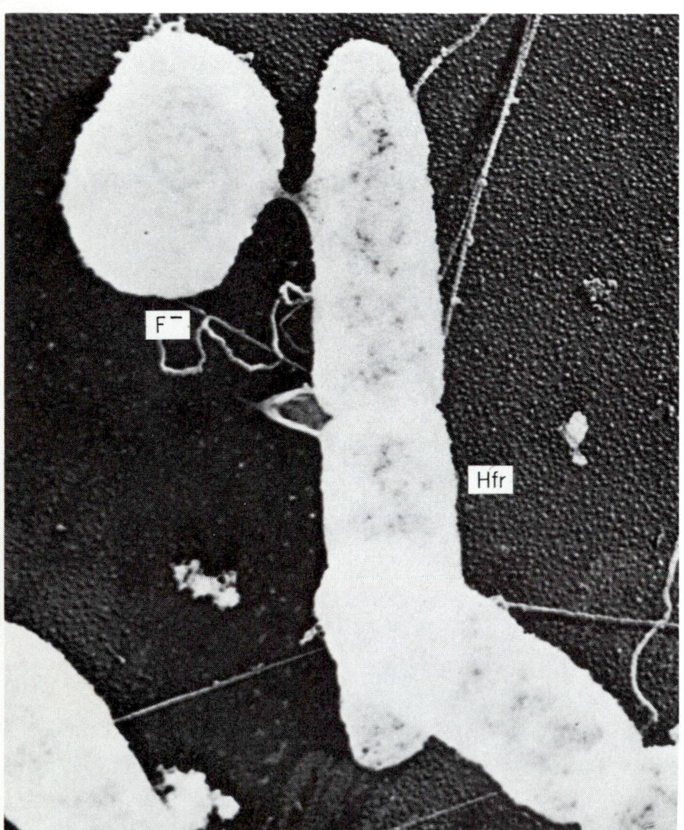

Figure 8.8 An electron micrograph of conjugating *E. coli* cells. The elongated donor (Hfr) cell (on the right) is undergoing cell division. It is joined to the short, plump recipient (F^-) cell (top left) by a conjugation tube, through which the Hfr chromosome passes. These morphological differences are ***not*** characteristic differences between Hfr cells and F^- cells. Cells with these different morphologies were used in this experiment to aid in distinguishing donor and recipient cells. The only characteristic morphological difference between donor (Hfr and F^+) cells and recipient (F^-) cells is the presence of F pili on donor cells. The ropelike fibers on the Hfr cell are F pili; their synthesis is controlled by genes located on the F factor. Magnification is ×32,700. (Photograph by T. F. Anderson, E. Wollman, and F. Jacob, from *Ann. Inst. Pasteur* 93: 450–455, 1957.)

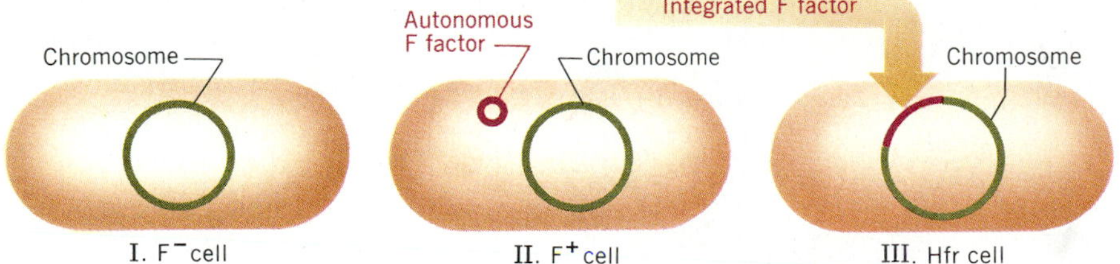

Figure 8.9 The three states of an *E. coli* cell with respect to the F factor: (I) F$^-$ cell, no F factor present; (II) F$^+$ cell, containing an autonomously replicating F factor (reddish color); (III) Hfr cell, containing an F factor integrated into the *E. coli* chromosome (greenish color). Conjugation occurs when an Hfr or an F$^+$ cell contacts an F$^-$ cell and forms a conjugation tube. In F$^+$ by F$^-$ matings, only the F factor is transferred. In Hfr by F$^-$ matings, the Hfr chromosome is nicked within the integrated F factor, and a sequential transfer of chromosomal genes ensues.

population of F$^-$ cells results in virtually all the cells in the new population becoming F$^+$.

The F factor can integrate into the host chromosome at any one of many sites by a mechanism that appears analogous to the integration of the chromosome of a specialized transducing phage, namely, a site-specific recombination event (Fig. 8.10). The integration of the F factor is believed to be mediated by short DNA sequences called ***IS elements*** (see pp. 224–226). A cell carrying an integrated F factor is called an ***Hfr*** (for ***high-frequency recombination***). ***In the integrated state, the F factor mediates the transfer of a chromosome of the Hfr cell to a recipient (F$^-$) cell.*** Usually, only a portion of the Hfr chromosome is transferred before the cells separate, thus breaking the chromosome. Only rarely will an entire Hfr chromosome be transferred.

The mechanism of transfer of DNA from a donor cell to a recipient cell during conjugation appears to be the same whether just the F factor is being transferred, as in F$^+$ by F$^-$ matings, or the chromosome is being transferred, as in Hfr by F$^-$ matings. Transfer is believed to be initiated by an endonucleolytic nick in one strand at a specific site (the "origin" of transfer) on the F factor. The 5′ end of the nicked strand is then transferred through the conjugation tube into the recipient cell. Transfer is believed to be coupled to ***rolling circle replication*** (see Chapter 5, pp. 122–

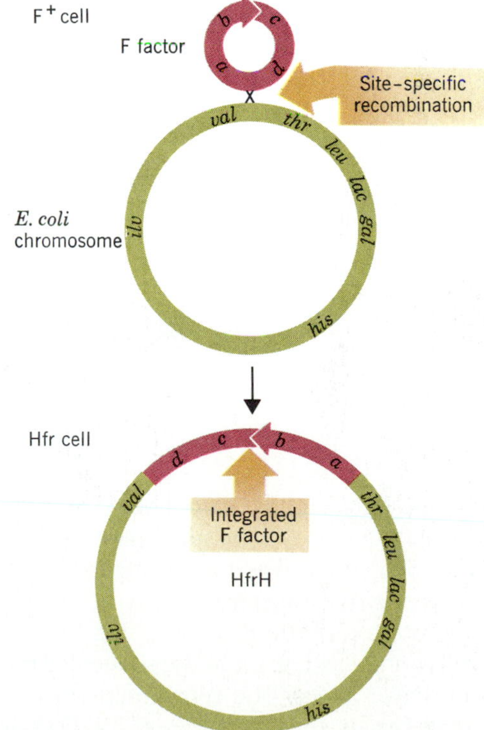

Figure 8.10 Conversion of an F$^+$ cell to an Hfr cell by the integration of the autonomous F factor into the host chromosome. The F factor is covalently inserted into the host chromosome by a site-specific recombination event like that occurring during the integration of the phage lambda chromosome. Integration of the F factor appears to be mediated by IS elements (see Chapter 9). The markers *a, b, c,* and *d* shown on the F factor are hypothetical; they were arbitrarily positioned to illustrate the linear insertion occurring during integration. The arrowhead indicates the site of origin and direction of the sequential transfer that occurs during conjugation. The site of integration illustrated is that occurring during the formation of the classic Hfr strain, HfrH (or Hfr Hayes). A few *E. coli* genes are shown for the sake of orientation. Genes *thr, leu, his, ilv,* and *val* are involved in the biosynthesis of the amino acids threonine, leucine, histidine, isoleucine plus valine, and valine, respectively. Genes *lac* and *gal* are required for the utilization of lactose and galactose, respectively, as energy sources.

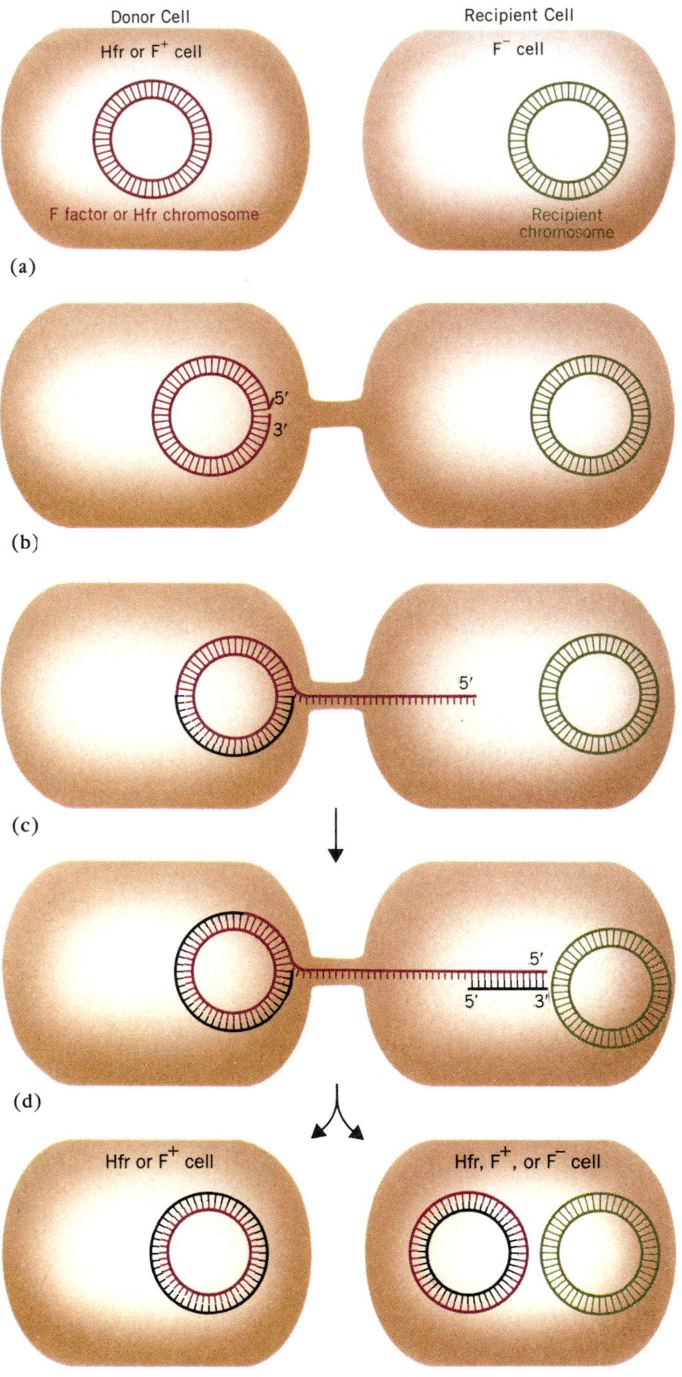

Figure 8.11 Mechanism of DNA transfer during conjugation. (a) The recipient chromosome is shown in greenish color. The chromosome shown (reddish color) in the donor is either carrying the integrated F factor (Hfr cell) or is the F factor (F^+ cell). (b) After cell contact occurs, a conjugation tube forms between the two cells. An endonuclease cleaves one strand of DNA at a unique site (the "origin" of transfer) on the F factor (either integrated in the case of an Hfr or autonomous in the case of an F^+ cell). (c) The 5′ end of the cleaved strand is then displaced, as in normal rolling circle replication, except that during conjugation the 5′ end moves through the conjugation tube into the recipient cell. (d) Transfer occurs concurrently with, possibly driven by, rolling circle replication, with the intact circular strand serving as a template for the synthesis of a complementary strand in the donor cell and the transferred linear strand being replicated discontinuously (by the synthesis of short 5′→3′ Okazaki fragments that are then joined by DNA ligase in an overall 3′→5′ reaction) in the recipient cell immediately after transfer. (DNA strands synthesized during the conjugation process are shown in black; parental DNA strands are shown in reddish color.) (e) In F^+ by F^- matings, both exconjugants will be F^+, since both will have a complete copy of the F factor. In Hfr by F^- matings, the donor cell will remain an Hfr and the recipient cell will usually remain F^-, since a portion of the integrated F factor is the last segment of DNA to be transferred. Although the diagram shows complete chromosome transfer to be comparable to F factor transfer in F^+ by F^- matings, the conjugation tube and chromosome usually break spontaneously in Hfr by F^- matings before the entire chromosome is transferred. In the rare cases where the entire Hfr chromosome is transferred, the recipient cell becomes an Hfr after conjugation.

124), with the intact circular strand being replicated in the donor cell and the displaced strand being replicated in the recipient cell as it is transferred (Fig. 8.11). Because the origin of transfer is within the integrated F factor, one portion of the F factor is transferred from an Hfr cell to an F^- cell prior to the sequential transfer of chromosomal genes. The remaining part of the F factor, however, is the last segment of DNA to be transferred. Thus, in Hfr by F^- matings, the recipient F^- cell acquires a complete F factor (thus becoming an Hfr donor) only in those rare cases when an entire Hfr chromosome, with its integrated F factor, is transferred.

Several of the details of conjugation were worked out using one particular Hfr strain called ***HfrH*** or ***Hfr Hayes*** (for W. Hayes, who isolated it). In this strain, the

F factor is integrated near the *thr* (threonine) and *leu* (leucine) loci, as shown in Fig. 8.10. The most famous and most informative conjugation experiment was the ***interrupted mating experiment*** of E. Wollman and F. Jacob. They crossed HfrH cells of genotype thr^+ leu^+ *azi-s* *T1-s* lac^+ gal^+ *str-s* with F^- cells of genotype thr^- leu^- *azi-r* *T1-r* lac^- gal^- *str-r*. [The *thr* gene and the *leu* gene are responsible for the syntheses of the amino acids threonine and leucine, respectively. Allele pairs, *azi-s*/*azi-r*, *T1-s*/*T1-r*, and *str-s*/*str-r* control sensitivity (*s*) or resistance (*r*) to sodium azide, bacteriophage T1, and streptomycin, respectively. Alleles lac^+ and lac^- and alleles gal^+ and gal^- govern the ability (+) or inability (−) to utilize lactose and galactose, respectively, as energy sources.] At varying times after the HfrH and F^- cells were mixed to initiate matings, samples were removed and agitated vigorously in a blender to break the conjugation tubes and separate the conjugating cells. The cells were then plated on medium containing the antibiotic streptomycin, but lacking the amino acids threonine and leucine. On this ***selective medium, only recombinant cells*** carrying the thr^+ and leu^+ genes of the HfrH parent and the *str-r* gene of the F^- parent ***can grow.*** The HfrH donor cells are killed by the streptomycin; the F^- recipient cells cannot grow in the absence of threonine and leucine. Colonies produced by thr^+ leu^+ *str-r* recombinants were then ***replica-plated*** (see Chapter 11, Fig. 11.1) to a series of plates containing different selective media to determine which of the other donor markers were present. The series included medium containing (1) sodium azide, to score cells as *azi-r* or *azi-s*, (2) bacteriophage T1, to score cells as *T1-r* or *T1-s*, (3) lactose as the sole carbon source, to score recombinants as lac^+ or lac^-, and (4) galactose as the sole carbon source, to score recombinant cells as gal^+ or gal^-.

When the conjugating cells were interrupted (separated by agitating in a blender) at any time prior to 8 minutes after mixing the HfrH cells and the F^- cells, **no** thr^+ leu^+ *str-r* recombinants were detected. Recombinants (thr^+ leu^+ *str-r*) first appeared about 8½ minutes after mixing the HfrH and F^- cells and accumulated with linear kinetics up to a maximum frequency within a few minutes. Of most importance, when the presence of the other donor (HfrH) markers was scored among thr^+ leu^+ *str-r* recombinants from subcultures interrupted at varying times after the initiation of mating, the donor markers were observed to be transferred in a specific temporal sequence (Fig. 8.12). The HfrH *azi-s* gene first appeared among recombinants formed by conjugating cells that were separated by blending at about 9 minutes after the matings were initiated. The *T1-s*, lac^+, and gal^+ markers first appeared after 11, 18, and 25 minutes of mating, respectively. These results indicated that the HfrH genes were being transferred to the F^- cells in a specific linear sequence (Fig. 8.13).

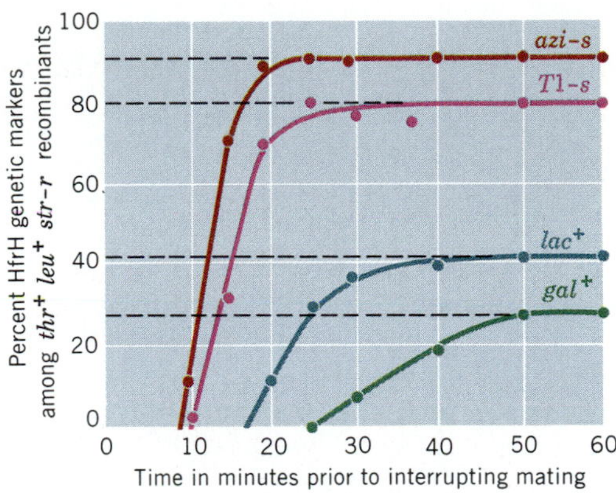

Figure 8.12 Kinetics of appearance of the unselected HfrH markers *azi-s, T1-s,* lac^+, and gal^+ in thr^+ leu^+ *str-r* recombinants produced during conjugation between an HfrH strain of genotype thr^+ leu^+ *azi-s T1-s* lac^+ gal^+ *str-s* and an F^- strain of genotype thr^- leu^- *azi-r T1-r* lac^- gal^- *str-r*. Matings were interrupted by agitating in a blender at the times shown on the abscissa. The frequencies of the unselected donor markers (in percentages), as determined by replica-plating to various indicator media, are shown on the ordinate. The dashed lines indicate the plateau frequencies observed for the various donor markers; they correspond to the frequencies observed in comparable uninterrupted mating experiments. (Based on the data of F. Jacob and E. L. Wollman, *Sexuality and the Genetics of Bacteria,* Academic Press, New York, 1961.)

Subsequent studies with different Hfr strains revealed similar fixed transfer sequences, although different Hfr's initiated transfer from different sites on the chromosome. It is now clear that the F factor can integrate at many different sites in the circular *E. coli* chromosome (Fig. 8.14), and the site of integration determines the origin of transfer characteristic of a given Hfr. Moreover, the ***orientation*** of F integration—either *d c b a* reading clockwise (as shown for HfrH in Fig. 8.10), or *a b c d* reading clockwise—determines whether the sequence of chromosome markers transferred is clockwise relative to the *E. coli* linkage map (Fig. 8.14) or counterclockwise. The inner circle in Fig. 8.14 shows the sites of F integration and the direction of chromosome transfer for a few Hfr strains.

The transfer of a complete chromosome from an Hfr to an F^- cell takes from 90 to 100 minutes, depending on the strain. Chromosome transfer appears to proceed at a fairly constant rate. Thus, the time interval between the transfer of any two markers (easily deter-

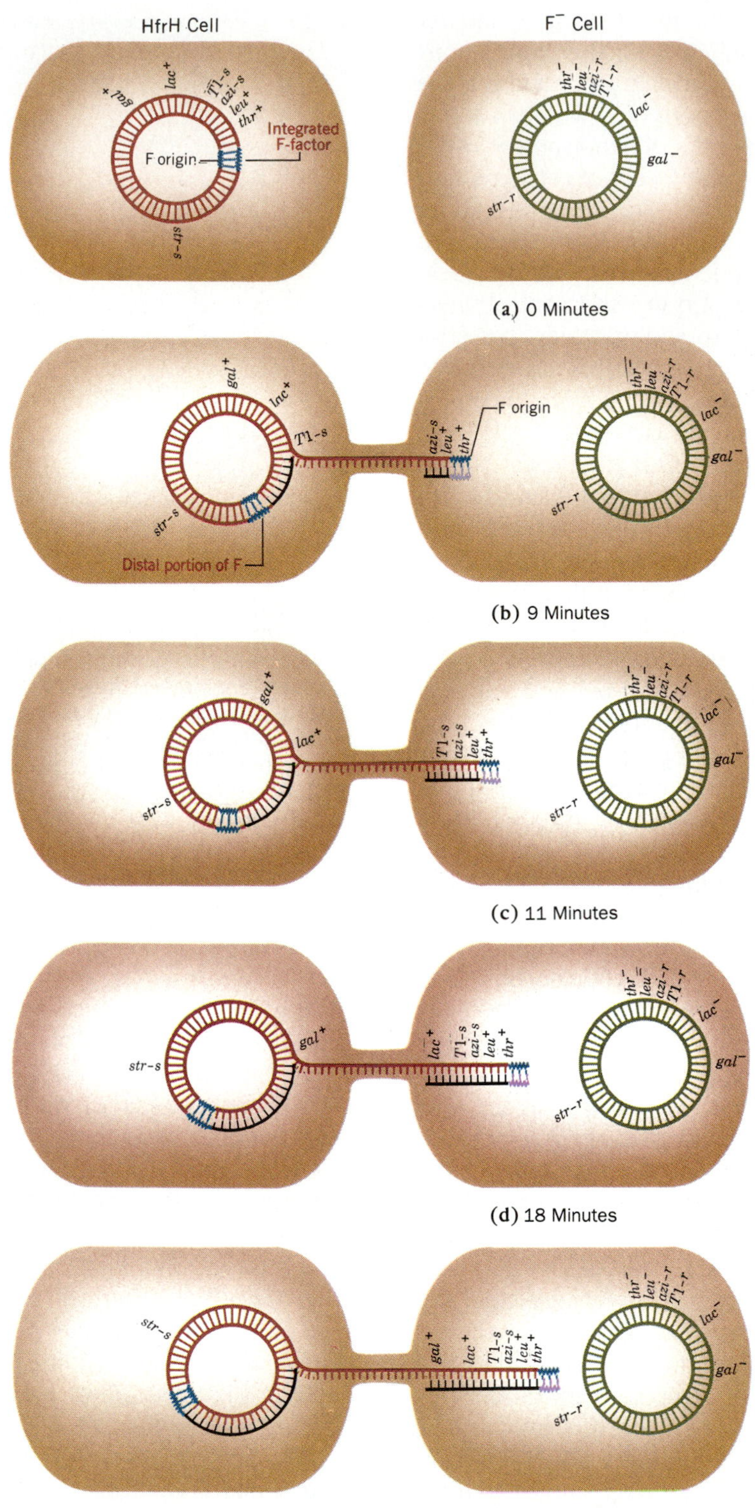

Figure 8.13 Interpretation of the interrupted mating data of Jacob and Wollman (Fig. 8.12) in terms of the linear transfer of the HfrH chromosomes to the F^- cells. The genetic markers carried by the HfrH strain are all dominant or codominant (both alleles expressed) to the alleles carried by the F^- strain. The times given indicate the number of minutes since conjugation was initiated. The diagram is not drawn to scale; distances between markers are only approximate.

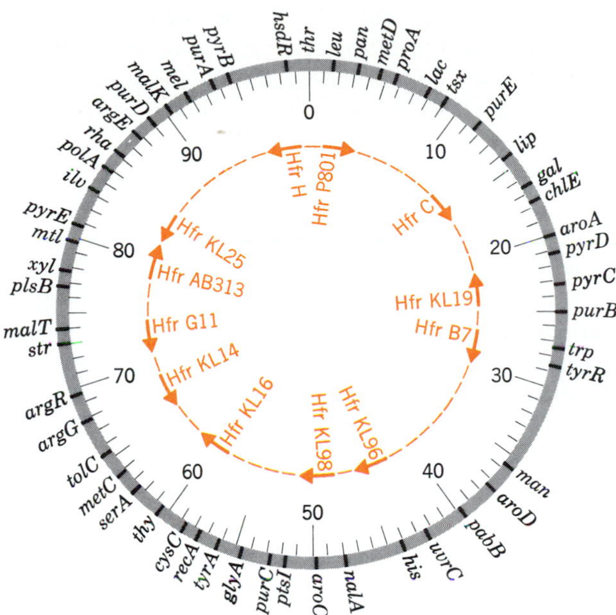

Figure 8.14 The circular linkage map of *E. coli* strain K12. The map is divided into 100 1-minute intervals based on conjugation chromosome transfer times. Only 52 of the 1027 loci that had been mapped in *E. coli* K12 in 1983 are shown. The loci shown were selected by B. J. Bachmann, K. B. Low, and A. L. Taylor on the basis of greatest accuracy of map position, utility as reference points in further mapping experiments, and/or familiarity as well-known landmarks on the K12 linkage map. The arrows on the dashed inner circle indicate the origin and direction of transfer for a few well-known Hfr strains. The origin and direction of transfer (clockwise or counterclockwise) are determined by the site and the orientation, respectively, of integration of the F factor. Genetic symbols, descriptions of the loci, and references to the mapping studies on which the map positions are based are given in Bachmann, Low, and Taylor (1976). (After B. J. Bachmann, K. B. Low, and A. L. Taylor, *Bact. Rev.* 40: 116, 1976. For a complete map of *E. coli* and a description of all mapped loci, see B. J. Bachmann, "Linkage Map of *Escherichia coli* K12, Edition 7," pp. 807–876 in *Escherichia coli and Salmonella typhimurium, Cellular and Molecular Biology*, F. C. Neidhardt, editor in chief, American Society for Microbiology, Washington, DC, 1987.)

mined by interrupted mating experiments) is a good estimate of the physical distance separating the markers on the chromosome. It has therefore proven convenient to use the ***minute,*** representing the time interval between the transfer of markers in interrupted mating experiments, as the standard unit for measuring linkage in *E. coli.* A map distance of 1 minute corresponds to the length of the segment of the chromosome transferred in 1 minute during conjugation. The standard *E. coli* linkage map (Fig. 8.14) is thus divided into minute intervals from 0 (arbitrarily set at the *thrA* gene) to 100 minutes on the basis of interrupted mating experiments.

Linkage relationships can also be determined from uninterrupted mating experiments. Consider, for example, the HfrH by F^- cross discussed earlier, except now with the matings being allowed to proceed uninterrupted for 1–2 hours. If thr^+ leu^+ *str-r* recombinants are selected and scored for the presence of the other segregating markers by replicating as before, what will the frequencies of the donor *azi-s*, *T*1*-s*, lac^+, and gal^+ markers be among the recombinants? Will these donor markers all be present with the same frequency? The frequencies of these donor markers are observed to vary, with the frequency of a marker decreasing as a function of its distance from the selected (thr^+ and leu^+) donor markers. The frequencies will, in fact, be identical to the plateau frequencies observed in the interrupted mating experiment (Fig. 8.12). Donor markers *azi-s*, *T*1*-s*, lac^+, and gal^+ will occur among thr^+ leu^+ *str-r* recombinants with percentage frequencies of 90, 80, 40, and 25, respectively. The farther a marker is from the selected donor marker (in the HfrH experiment, thr^+ and leu^+), the lower its frequency among the recombinants. The marker frequency gradient is caused by two major factors: (1) the approximately constant probability per unit time of spontaneous rupture of the conjugation tube and the chromosome and (2) the decreasing probability that any two donor markers will be incorporated into the recipient chromosome by a single pair of recombination events (incorporation of a donor fragment into a recipient chromosome always requires ***two*** recombination events; see pp. 220–223) as the distance separating the two markers increases. Although uninterrupted conjugation experiments of this type can be used to determine linkage relationships, interrupted mating experiments are simpler and more direct. Thus, when a new mutation is identified, its approximate location is usually first determined by interrupted conjugation mapping. Its exact location is then usually determined by transduction mapping.

F-MEDIATED SEXDUCTION

Specialized transducing particles are occasionally formed during the excision of prophages (Fig. 8.5). The Hfr strains are formed when the F factor integrates into the chromosome (Fig. 8.10) by a mechanism similar to prophage formation. Moreover, rare F^+ cells are present in Hfr cultures, indicating that excision of the F factor also occurs, presumably by a mechanism analogous to the excision of a prophage (Fig. 8.4). Therefore, one might expect that an occasional anomalous excision event would produce an F factor carrying bacterial genes (Fig. 8.15). E. A. Adelberg and S. N. Burns first identified such modified F factors, called F′ ("F-prime") factors or F′ elements in 1959. Since F′ factors do not require packaging in a phage head like specialized transducing elements (e.g., λ*dg*'s), the size of the segment of the bacterial chromosome carried by the F′ is not restricted. The F′ factors range in size from those carrying a single bacterial marker to those carrying up to half of the bacterial chromosome (Fig. 8.16).

Transfer of F′ factors to recipient (F^-) cells apparently occurs by the same mechanism as F factor transfer in F^+ by F^- matings and chromosome transfer in Hfr by F^- matings (Fig. 8.11). The F′ factors have some interesting and useful properties, however. Consider an F′ *thr*$^+$ *leu*$^+$ factor generated by anomalous excision of the F factor from HfrH, as shown in Fig. 8.15. Matings between F′ *thr*$^+$ *leu*$^+$ donor cells and *thr*$^-$ *leu*$^-$ recipient (F^-) cells result in essentially all the recipient cells being converted to *thr*$^-$ *leu*$^-$/F′ *thr*$^+$ *leu*$^+$ partial diploids (or heterogenotes). These partial diploids are unstable like the partial diploids formed by specialized transduction. The F′ factor may be lost, producing *thr*$^-$ *leu*$^-$ haploids, or recombination may occur between the chromosome and the F′, producing stable *thr*$^+$ *leu*$^+$ recombinants. Recombination of this type, mediated by F′ factors, is called ***sexduction*** or ***F-duction***. Because of the partial diploidy resulting from sexduction, it provides an important method for determining dominance relationships between alleles and defining genes by complementation tests in bacteria (see Chapter 12). Since F′ factors are available carrying almost any segment of the *E. coli* chromosome (Fig. 8.16), sexduction analysis can be carried out with almost any mutation of interest.

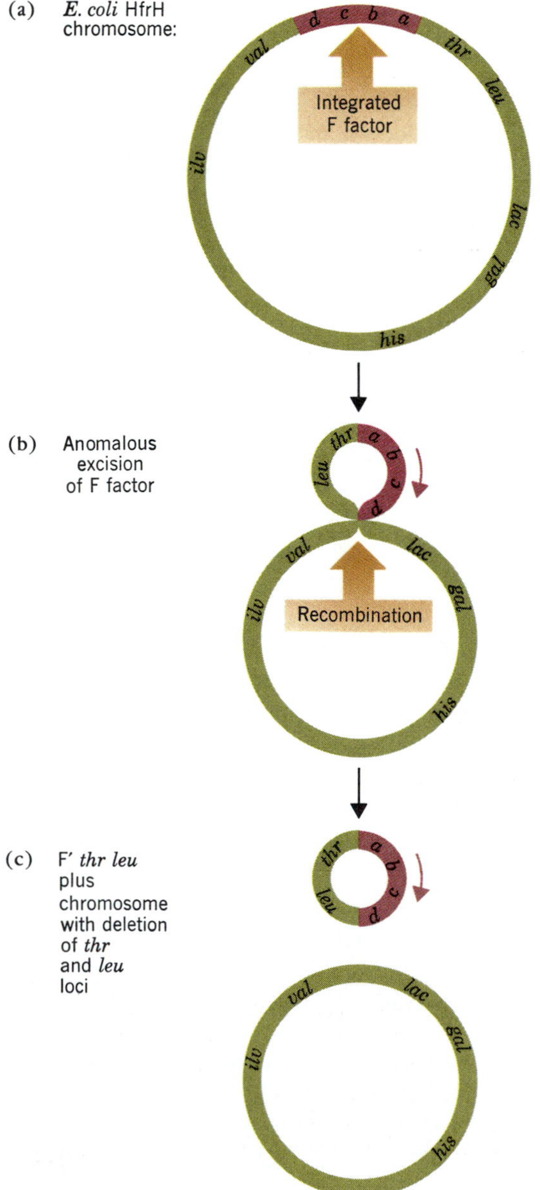

Figure 8.15 Anomalous excision of the F factor from an HfrH chromosome producing an F′ factor carrying the *thr* and *leu* loci. (a) The HfrH chromosome is shown with the F factor integrated between the *val* and *thr* loci. The arrowhead indicates the origin and direction of chromosome transfer during conjugation; *a, b, c,* and *d* are hypothetical genetic markers on the F factor. (b) The F factor "loops out" in a "figure-8" structure as in normal F excision, but in this case the "loop" containing the F factor includes the segment of the chromosome carrying the *thr* and *leu* loci. Recombination produces (c) an F′ factor carrying *thr* and *leu* plus a chromosome with a deletion of the *thr–leu* segment. An *E. coli* cell carrying the F′ *thr leu* and the deletion chromosome remains haploid for all genetic loci, like the Hfr cell from which it was derived; the genes are just present in a new arrangement. When such an F′ cell conjugates with a normal (nondeletion) F^- cell, however, the recipient cell becomes partially diploid, with two copies of the *thr–leu* region.

RATIONALE OF FINE STRUCTURE MAPPING IN MEROZYGOTES

Definitive fine structure mapping in bacteria, particularly the ordering of closely linked markers, is usually done using three-factor crosses, just as it is in eukaryotes (see Chapter 7). The rationale behind such mapping experiments with bacteria is essentially the same as for eukaryotes. The key differences between mapping in eukaryotes and in prokaryotes will be briefly discussed in this section. Those mapping techniques unique to bacteria, such as interrupted mating, cotransformation, cotransduction, and sexduction, have been described in the preceding sections of this chapter and will not be discussed further here.

The recombination events occurring in bacteria usually take place (except for rare occurrences of complete chromosome transfer during conjugation) in ***partial zygotes*** called ***merozygotes*** rather than in true zygotes or diploid cells as in eukaryotes. With the aforementioned exception, the recipient cells in transformation, transduction, conjugation, and sexduction are converted to merozygotes, containing a fragment

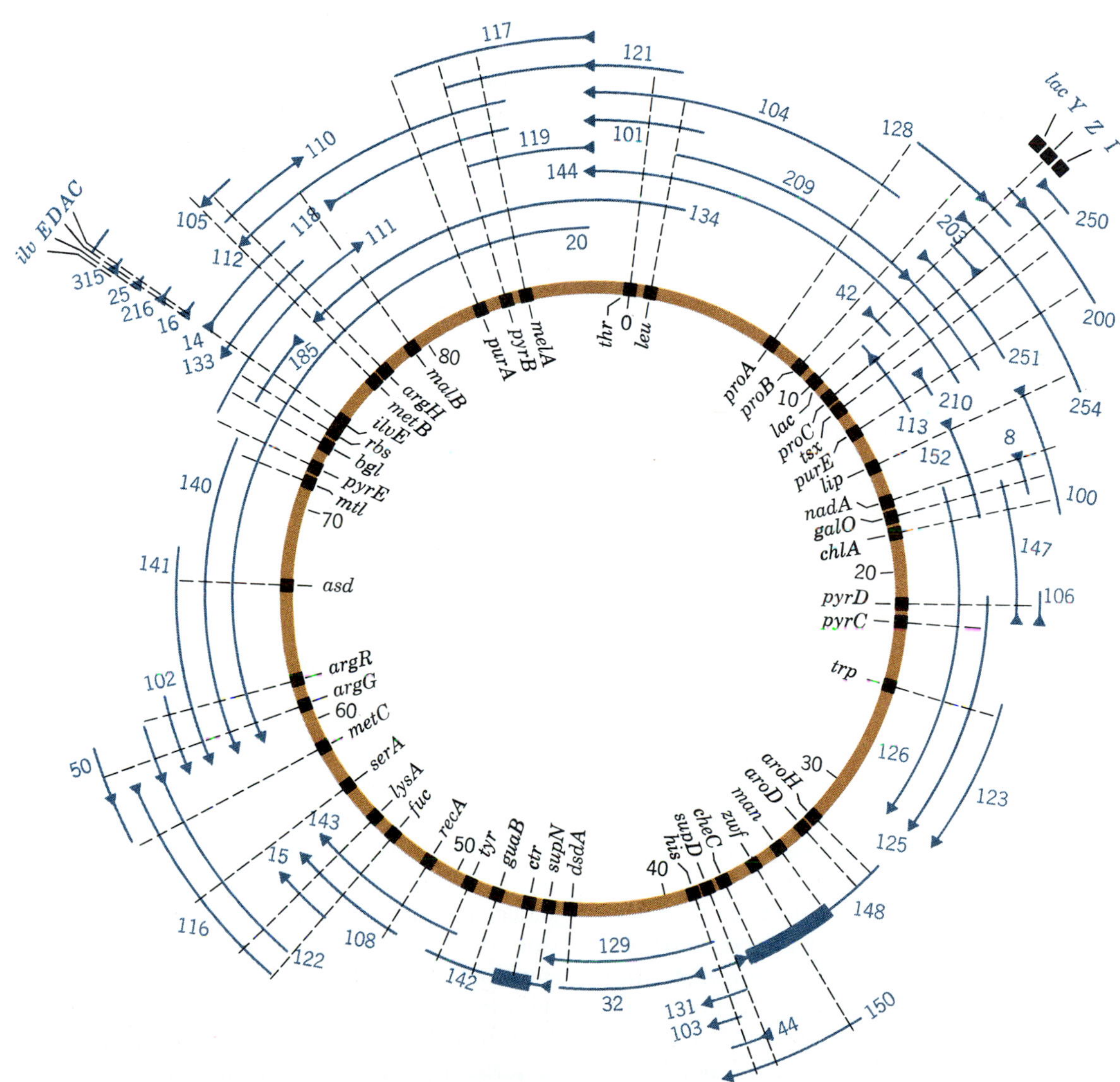

Figure 8.16 Summary of some of the F′ factors of *E. coli* K12. The chromosomal genes carried by each F′ factor are indicated by the lines paralleling the circular genetic map (center), which shows the locations of a few selected loci. Note, however, that all the characterized F's exist as circular DNA molecules; the true structures of these F′ factors thus require joining of the ends of the line representing each numbered F′ factor. The arrowheads on each F′ factor indicate the origin and direction of transfer during conjugation. F′ 142 and F′ 148 carry deletions, as indicated by the boxes. This map was prepared before the revision of the *E. coli* linkage map (Fig. 8.14); it is thus divided into 90 1-minute intervals instead of the currently accepted 100 1-minute intervals. (From B. Low and J. O. Falkinham, *Handbook of Microbiology,* Vol. IV, p. 593, A. I. Laskin and H. Lechevalier, editors, The Chemical Rubber Co., Copyright © 1974, CRC Press, Inc., Boca Raton, Florida.)

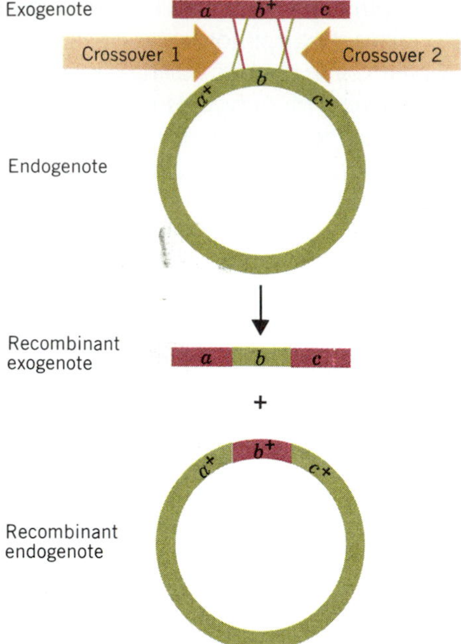

Figure 8.17 Incorporation of a genetic marker (b^+) on an exogenote (donor chromosome fragment, shown in reddish color) into an endogenote (recipient chromosome, shown in greenish color) in a merozygote. The exogenote is only a fragment of the donor chromosome and is not capable of autonomous replication. Two crossovers are therefore required to incorporate a segment of the exogenote into the recipient chromosome, if the latter is to remain structurally intact and capable of replication.

of the donor chromosome, the ***exogenote,*** and a complete recipient chromosome, the ***endogenote.*** Recombination in merozygotes does not take place between two complete chromosomes. ***In merozygotes, the recombination process must incorporate a segment of the exogenote into the endogenote*** (the complete chromosome of the recipient cell) ***to yield an intact recombinant chromosome. Such an incorporation requires two crossovers or recombination events*** (Fig. 8.17). Single crossovers between exogenotes and endogenotes will ***not*** yield structurally intact chromosomes. Since the incorporation of each donor DNA segment requires two crossovers, only recipient chromosomes involved in an ***even number of crossovers with exogenotes will remain structurally intact.***

This restriction to paired recombination events in the case of merozygotes has little effect on the overall rationale of fine structure mapping. Consider, for example, a standard three-factor cross. In diploids, genetic markers are ordered by using three-factor cross data and by comparing the frequencies of the various recombinant progeny genotypes (Chapter 7). The rationale for diploids (or other complete zygotes) is that a genotype that requires two crossover events to be produced should be less frequent than a genotype that requires just one crossover. Specifically, if each single crossover type occurs with a frequency of x, then each double crossover type should occur with a frequency of about x^2. (Recall that the maximum frequency of each of the reciprocal recombinants produced by a crossover and thus the ***maximum value of x is $1/4$;*** therefore x^2 is always much smaller than x.) Exactly the same rationale holds for merozygotes, except that the frequencies of genotypes formed by two crossovers (say, frequencies of y) are compared with the frequencies of genotypes formed by four crossovers (frequencies of about y^2; see Fig. 8.18).

In merozygotes, each three-factor cross can be performed in two ways: (1) double mutant donor × single mutant recipient and (2) single mutant donor × double mutant recipient. The progeny genotype frequencies of these reciprocal crosses are not necessarily the same. In fact, the results of such reciprocal crosses can be used to order the markers involved. Suppose that one wishes to order two markers (b_1 and b_2) in a particular gene (gene b) relative to a marker (a) in a nearby gene. The two reciprocal crosses are

(1) $a^+\ b_1^+\ b_2$ donor × $a\ b_1\ b_2^+$ recipient

(2) $a\ b_1\ b_2^+$ donor × $a^+\ b_1^+\ b_2$ recipient

[One may also consider crosses (3) $a^+\ b_1\ b_2^+$ donor × $a\ b_1^+\ b_2$ recipient and (4) $a\ b_1^+\ b_2$ donor × $a^+\ b_1\ b_2^+$ recipient; but only crosses (1) and (2) need be considered to illustrate the rationale involved. Note that the order $a\ b_1\ b_2$ is arbitrary; the crosses are being done to determine the real order of the b_1 and b_2 markers relative to the known outside marker a.] Usually, such crosses are done with conditional lethal mutants so that one can use selective media or conditions under which only the $a^+\ b_1^+\ b_2^+$ recombinants can grow. Such mutants carry mutations that are lethal to the organism when grown in one set of environmental conditions. In another set of environmental conditions, the mutations are not lethal, allowing the mutant organisms to grow and reproduce. Such mutations are thus called ***conditional lethal mutations;*** they are extremely useful in many kinds of genetic analysis (see Chapters 11 and 12).

Two orders are possible for the two gene b mutations relative to the outside marker in gene a: (1) a-b_1-b_2 and (2) a-b_2-b_1. If the correct order is a-b_1-b_2, $a^+\ b_1^+\ b_2^+$ recombinants will occur with about the same frequency in both crosses. They will be formed by two crossovers in both crosses. If, on the other hand, the order is a-b_2-b_1, $a^+\ b_1^+\ b_2^+$ recombinants will occur much more frequently in cross (2) than in cross

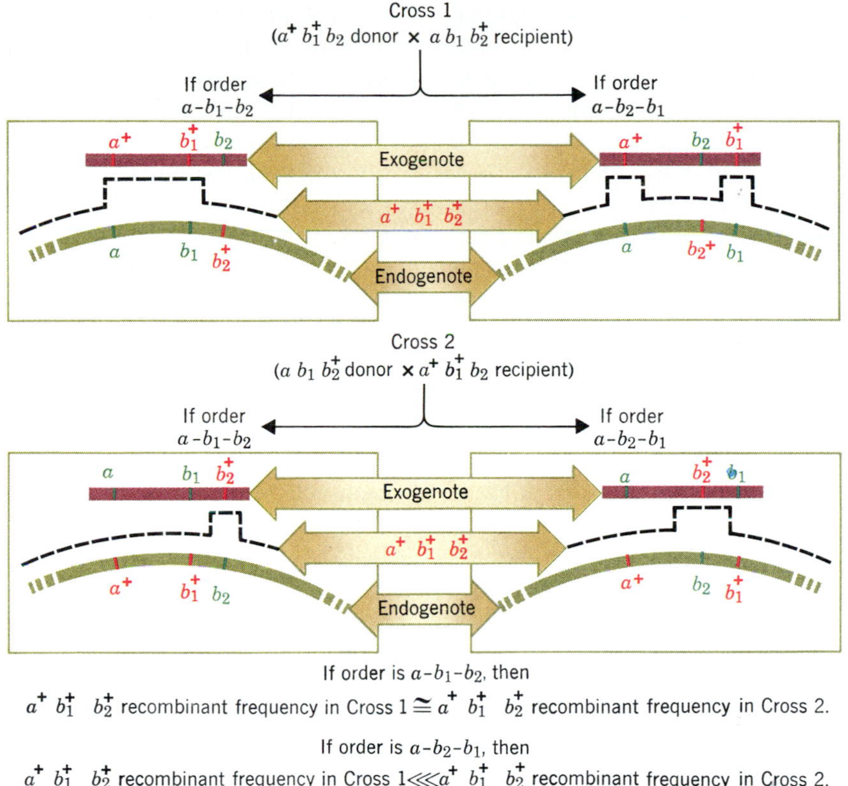

Figure 8.18 Diagram illustrating the rationale used in ordering markers by three-factor crosses involving merozygotes. The rationale is essentially the same as that for true zygotes, except that recombination events must always occur in pairs so as to incorporate markers present on the exogenote (donor fragment, shown in reddish color) into the endogenote (recipient chromosome, shown in greenish color). In the example illustrated, conditional lethal mutations are used as genetic markers, and the mapping is based on the frequencies of selected wild-type recombinants (a^+ b_1^+ b_2^+ genotypes) formed in the two reciprocal crosses. Mutant organisms carrying conditional lethal mutations are viable when grown in one environment, but are nonviable in another environment. Mutations b_1 and b_2 are ordered relative to the known outside marker a. ***The dashed lines indicate the segments of the exogenote and the endogenote that must be present in a recombinant of the selected*** a^+ b_1^+ b_2^+ ***genotype.*** If the order of the markers is a-b_1-b_2, as is shown on the left, then a^+ b_1^+ b_2^+ recombinants will be formed by two crossovers in both crosses. If the order of the markers is a-b_2-b_1 as shown on the right, then a^+ b_2^+ b_1^+ recombinants will be formed by two crossovers in Cross 2, but will require four crossovers in Cross 1. Thus, if the frequencies of wild-type recombinants are about the same in the two reciprocal crosses, the order is a-b_1-b_2. If, on the other hand, the frequency of wild-type recombinants is much higher in Cross 2 than in Cross 1, the correct order is a-b_2-b_1.

(1). If the order is a-b_2-b_1, four crossovers will be required to produce a^+ b_1^+ b_2^+ recombinants in cross (1) whereas only two crossovers will be required in cross (2). This is diagrammed in Fig. 8.18.

The same rationale can be applied to three-factor merozygote crosses in which the markers are not conditional lethal mutations. Such mapping experiments are much more laborious, however, because the genotypes of all the progeny must be individually determined by direct visualization of colony phenotypes or other screening techniques. Most of the detailed fine structure maps in bacteria are thus based on data obtained using conditional lethal mutations.

BREAKAGE AND REUNION AS THE MECHANISM OF RECOMBINATION IN BACTERIA

During recombination in bacteria, be it during transformation, transduction, conjugation, or sexduction, a gene or a set of closely linked genes is transferred from a donor cell to a recipient cell. If the appropriate recombination events occur (see the preceding section), the donor gene or set of genes becomes a stable part of the genetic material of the recipient cell and is transmitted to its descendants, generation after generation, just like any of its other genes. This indicates that the incorporated genetic material of the exogenote becomes associated with the endogenote in a stable way. In fact, there is extensive evidence indicating that the donor DNA is physically inserted, by covalent linkages, into the chromosome of the recipient.

In Chapter 7, evidence was presented that demonstrated that crossing-over occurred by the actual breakage of parental chromosomes (or chromatids) and the reunion of their parts, usually, if not always, with the occurrence of a small amount of repair synthesis. Recombinant chromosomes are composed, therefore, of parts derived from two different parental chromosomes. Recombination in bacteria also occurs by breakage and reunion, with limited amounts of DNA repair synthesis. This has been experimentally demonstrated in several systems by density transfer experiments (see Chapter 5, pp. 105–108). These experiments are done using donor cells containing DNA that is both radioactive and "heavy," that is, has a higher than normal density, and recipient cells containing nonradioactive DNA of normal density. Donor cells with heavy, radioactive DNA are prepared by growing them on medium containing one or more heavy isotopes such as ^{2}H and ^{15}N and a radioactive isotope such as ^{3}H or ^{32}P. After recombination (transformation, transduction, or conjugation) has occurred, the state of the donor DNA, which can be followed because of its radioactivity, is examined in the recombinant by CsCl density-gradient analysis combined with various biochemical techniques. The details of one such study, the pneumococcus transformation experiment of M. S. Fox and M. K. Allen, are described in Fig. 8.19. Similar conjugation and transduction studies have shown that the donor DNA is covalently inserted into the chromosome of the recipient by a breakage and reunion mechanism, probably involving a small amount of DNA repair synthesis, during these processes.

PLASMIDS AND EPISOMES

In the introductory section of this chapter, we stated that the genetic material of bacteria is carried in one main chromosome plus, in many cases, from one to several extrachromosomal DNA molecules or "minichromosomes" called ***plasmids.*** By definition, ***a plasmid is a replicon*** (unit of genetic material capable of independent replication) that is ***stably inherited*** (maintained without specific selection) ***in an extrachromosomal state. Most,*** but not all, ***plasmids are dispensable,*** that is, they are not required for survival of the cell in which they reside. In many cases, however, they are essential under certain environment conditions, such as in the presence of an antibiotic.

The importance of plasmids has become increasingly recognized during the last two decades. Plasmids have been identified in almost all strains of bacteria tested. They are known to have major practical significance in two areas: (1) the spread of multiple antibiotic and drug resistance in pathogenic bacteria and (2) the instability of industrially important microorganisms. Multiple antibiotic and drug resistance will be discussed in some detail in Chapter 9. In *Streptococcus lactis* and related bacteria used in cheese processing, multiple plasmids have been identified and shown to carry genes coding for enzymes important in the fermentation processes involved in making cheeses. These observations explain in part why the cheese "starter cultures" of these bacteria are unstable and frequently must be discarded, at considerable expense to the cheese-making industry (see Chapter 24).

Three major types of bacterial plasmids have been extensively studied: (1) ***F and F′ plasmids,*** the conjugation fertility factors previously discussed; (2) ***R plasmids*** (previously called RTF, or resistance transfer factors), plasmids carrying genes for resistance to antibiotics or other antibacterial drugs; and (3) ***Col plasmids*** (previously called colicinogenic factors), plasmids coding for ***colicins,*** which are proteins that kill sensitive *E. coli* cells. Plasmids are also known in bacteria that encode bacteriocins other than colicins. For example, plasmids are known that code for ***vibriocins;*** these are proteins that kill sensitive *Vibrio cholerae* cells. They appear to be analogous to Col plasmids.

In some respects, the chromosomes of mitochondria and chloroplasts in eukaryotes also fit the definition of plasmids. They will be discussed in Chapter 20, along with other examples of extranuclear inheritance in eukaryotes.

Plasmids may be divided into two groups on the basis of whether or not they mediate conjugative self-transfer. ***Conjugative or transmissible plasmids mediate transfer of DNA by conjugation*** (as in F^+ by F^- matings; see pp. 213–219). All F and F′ plasmids, many R plasmids, and some Col plasmids are conjugative. The conjugative nature of many R plasmids has major significance in the rapid spread of antibiotic and drug resistance genes through populations of pathogenic bacteria. ***Nonconjugative or nontransmissible plasmids are those that do not mediate transfer of DNA by conjugation.*** Many R and Col plasmids are nonconjugative.

Some plasmids, such as F factors, also fit the definition of genetic elements called episomes. ***Episomes are genetic elements that can replicate in either of two alternative states: (1) as an integrated (covalently inserted) part of the main host chromosome or (2) as an autonomous genetic element,***

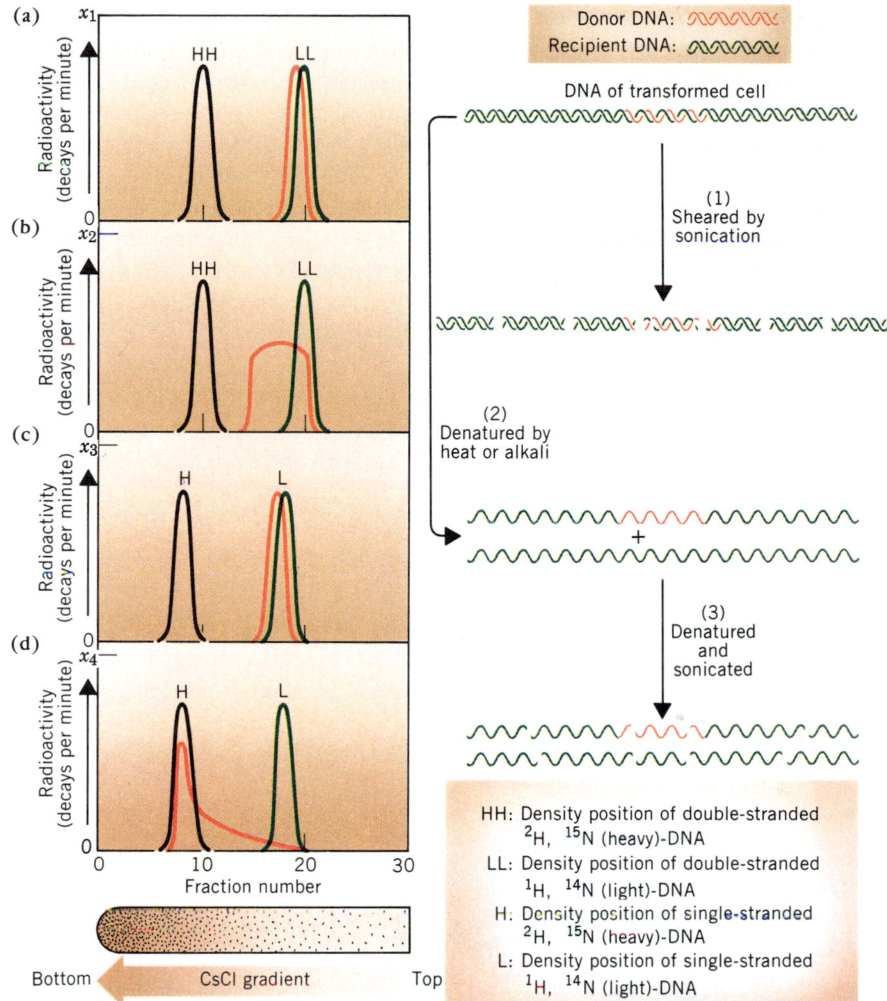

Figure 8.19 Diagram illustrating the results (left) and interpretation (right) of the experiment by M. S. Fox and M. K. Allen, demonstrating the covalent insertion of a single strand of donor DNA into the recipient chromosome during transformation in *Diplococcus pneumoniae.* Donor cells were grown for several generations in medium containing the radioactive isotope of phosphorus, ^{32}P, and the heavy isotopes of hydrogen and nitrogen, ^2H and ^{15}N, respectively. The DNA of donor cells was thus both heavy and radioactive (shown in orange). Recipient cells were grown in "normal" (^1H, ^{14}N, nonradioactive) medium; their DNA was therefore light and nonradioactive (shown in green). Recipient cells were transformed with heavy, radioactive donor DNA, and the state of the donor DNA fragments in the transferred cells was analyzed using cesium chloride density gradients. (a) When the DNA of transformants was simply extracted and analyzed in CsCl density gradients, the donor DNA (followed by its radioactivity) banded (peak shown in orange) at a density position only slightly heavier than that of light recipient DNA (shown in green). This indicates that short segments of donor DNA were associated with long segments of light recipient DNA (top right). (b) If the DNA of transformants was first sheared into short double-stranded fragments by sonication [exposure to sound waves (1)], the donor DNA exhibited densities ranging from "hybrid" density (one strand heavy and one strand light) to near-light density (both strands light). No fully heavy donor DNA was observed after shearing. (c) Denaturation (2) of the DNA from transformed cells had no significant effect on the density position of the donor DNA, except the denatured (single-stranded) DNA had a slightly higher density than native (double-stranded) DNA. The donor DNA still banded at the near-light density. This result demonstrates that the donor DNA is covalently inserted into the recipient chromosome. (d) Only after the DNA of transformants had been both sonicated and denatured (3) was donor DNA recovered in the gradient at the heavy density position. These results can be explained only if a single strand of donor DNA is covalently inserted into the chromosome of the recipient cell during transformation in pneumococcus. (Based on the results of M. S. Fox and M. K. Allen, *Proc. Natl. Acad. Sci. U.S.A.,* 52:412, 1964.)

independent of the main host chromosomes. The terms plasmid and episome are ***not*** synonyms. Many plasmids do not exist in integrated states and are thus not episomes. Similarly, many temperate phage chromosomes, such as the phage λ genome, are episomes but are not plasmids.

Spectacular progress has been made in our understanding of the structures and properties of plasmids and episomes during the last two decades. Many of their properties are now known to depend on the presence of short DNA sequences called ***IS elements*** or insertion sequences. The IS elements are also present in the main host chromosomes. These short sequences (from about 800 to about 1400 nucleotide-pairs in length) are ***transposable,*** that is, they can move from one position to another within a chromosome, or move from one chromosome to a different chromosome. In addition, IS elements mediate recombination between otherwise nonhomologous genetic elements within which they reside. Considerable evidence indicates that IS elements mediate the integration of episomes into host chromosomes. This is particularly clear in the case of the integration of the *E. coli* K12 F plasmid (F factor) during the formation of Hfr's (Fig. 8.20).

The first four IS elements to be extensively characterized and sequenced—IS1, IS2, IS3, and IS4—are 768, 1327, 1300, and 1426 nucleotide-pairs in length, respectively. Several other IS elements have subsequently been identified, characterized, and, in some cases, sequenced.

The *E. coli* K12 chromosome apparently contains eight copies of IS1 and five copies of IS2, plus one or more copies of IS3 and IS4. The *E. coli* K12 F factor contains one copy of IS2 and two copies of IS3 (Fig. 8.20). The positions of the IS elements in the various F factors and in the chromosomes of various *E. coli* strains are believed to determine the sites of integration of the F factor during the formation of Hfr strains (Fig. 8.20).

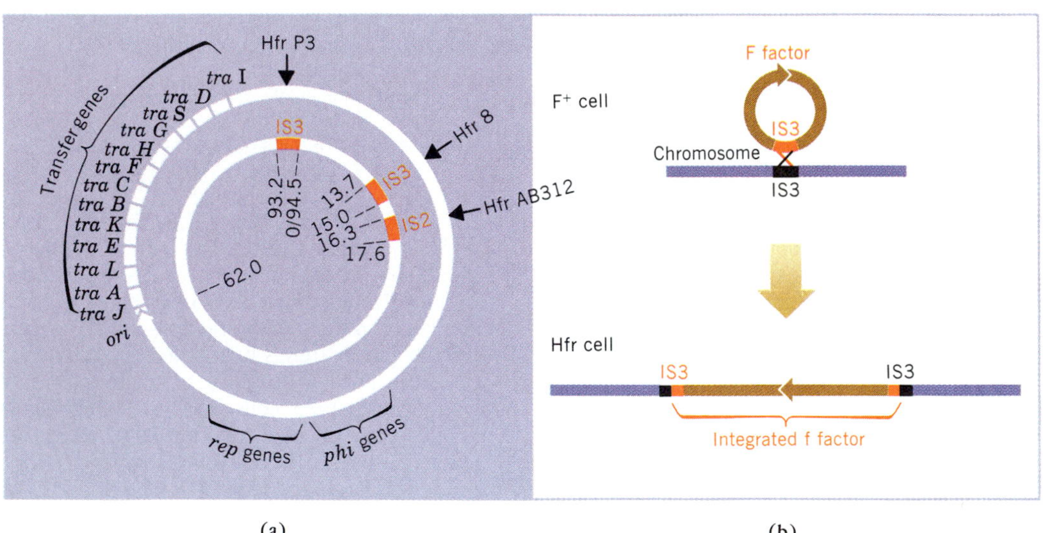

Figure 8.20 (a) Abbreviated genetic map of the *E. coli* K12 F factor (F plasmid) and (b) the postulated IS element-mediated integration of the F factor during Hfr formation. (a) The inner circle shows the location of one IS2 element and two IS3 elements. The distances given within the inner circle are in kilobases (1000 nucleotide-pair units); the *E. coli* K12 F factor is 94,500 nucleotide-pairs in length. The approximate locations of 13 genes involved in chromosome transfer (*tra* genes), the origin of transfer (*ori*), the genes required for replication (*rep* genes), and the genes involved in the inhibition of growth of F^--specific bacteriophages such as phage T7 (*phi*, for phage inhibition, genes) are shown on the outer circle. The sites of recombination with the host chromosome during integration are shown by the arrows for three Hfr strains. Note that these sites coincide with the known locations of the three IS elements. (b) Proposed mechanism of integration of the F factor mediated by the homology of an IS element in the chromosome and in the F factor. Such integration results in the F factor being flanked by identical IS elements when covalently inserted into the chromosome of an Hfr cell. (Based on the data summarized by J. A. Shapiro, p. 671, in *DNA Insertion Elements, Plasmids, and Episomes,* A. I. Bukhari, J. A. Shapiro, and S. L. Adhya, editors, Cold Spring Harbor Laboratory Press, Cold Spring Harbor, NY, 1977.)

SUMMARY

In bacteria, there are three different mechanisms for the transfer of genetic material from one cell, the ***donor*** cell, to a second cell, the ***recipient*** cell. During (1) ***transformation,*** naked donor DNA molecules are taken up enzymatically by ***competent*** recipient cells. (2) ***Transduction*** occurs when a fragment of the donor chromosome is carried to and injected into the recipient cell by a bacterial virus (bacteriophage). (3) ***Conjugation*** requires direct cell contact and involves the transfer of donor DNA to recipient cells through a ***conjugation tube*** that forms between the two cells.

Transduction is of two types: (1) ***generalized transduction,*** in which all genetic markers of the donor cell are represented in a population of transducing phage, and (2) ***restricted transduction,*** in which only genetic markers near the ***prophage*** (integrated phage chromosome) site are transduced.

A ***plasmid*** is an extrachromosomal DNA molecule or "minichromosome" that can replicate independently of the main cellular chromosome. The three main types of plasmids are (1) ***F plasmids,*** the F factors responsible for DNA transfer during conjugation; (2) ***R plasmids,*** DNA molecules carrying genes for resistance to various antibiotics and antibacterial drugs; and (3) ***Col plasmids,*** plasmids that code for proteins called ***colicins,*** which kill sensitive *E. coli* cells. All known plasmids are circular molecules of DNA.

Episomes are genetic elements that can replicate (1) in an ***integrated state,*** covalently inserted in the host chromosome, and (2) in an ***autonomous*** or ***extrachromosomal state.*** The *E. coli* K12 F factor and the phage λ chromosome are the best-known episomes. A cell carrying the F factor in the autonomous state is called an ***F^+ donor.*** During conjugation between an F^+ donor and an ***F^- recipient,*** only the F factor is transferred. A cell carrying the F factor in the integrated state is called an ***Hfr*** (for high-frequency recombination). During conjugation between an Hfr cell and an F^- cell, the Hfr chromosome undergoes linear transfer to the F^- cell. Usually, only part of the Hfr chromosome is transferred before the cells separate. The origin and direction of transfer are determined by the site and orientation of the F factor in the chromosome.

Occasionally, anomalous excisions of the F factor from Hfr chromosome occur, producing recombinant F factors, called ***F′ factors,*** that carry chromosomal genes. The conjugative transfer of donor chromosomal genes carried by F′ factors to recipient cells is called ***sexduction.***

Transformation, transduction, and conjugation almost always produce cells that are partial zygotes or partial diploids called ***merozygotes.*** Merozygotes contain only a part of the donor chromosome, the ***exogenote,*** plus the intact recipient chromosome, the ***endogenote.*** Crossovers in merozygotes must therefore always occur ***in pairs,*** to yield intact chromosomes. Recombination in bacteria occurs by ***breakage and reunion*** of parental chromosomes.

The integration of episomes and the evolution of plasmids, particularly R plasmids, are mediated by short (about 800–1400 nucleotide-pairs long) DNA sequences called ***insertion sequences*** or ***IS elements.*** These IS elements are ***transposable,*** that is, they can move from one position to another in the genome of a cell. The IS elements can also mediate recombination between genetic elements in which they are inserted.

REFERENCES

ADELBERG, E. A., editor. 1966. *Papers on bacterial genetics,* 2nd ed. Little, Brown, Boston.

BACHMANN, B. J., 1990. "Linkage map of *Escherichia coli* K-12, Edition 8." *Microbiol. Rev.* 54: 130–197.

BUKHARI, A. I., J. A. SHAPIRO, and S. L. ADHYA, editors. 1977. *DNA insertion elements, plasmids, and episomes.* Cold Spring Harbor Laboratory Press, Cold Spring Harbor, NY.

DAVIS, R. W., D. BOTSTEIN, and J. R. ROTH. 1980. *Advanced bacterial genetics.* Cold Spring Harbor Laboratory Press, Cold Spring Harbor, NY.

FOX, M. S., and M. K. ALLEN. 1964. "On the mechanism of deoxyribonucleate integration in pneumococcal transformation." *Proc. Natl. Acad. Sci. U.S.A.* 52: 412–419.

FREIFELDER, D. 1983. *Molecular biology, a comprehensive introduction to prokaryotes and eukaryotes.* Science Books International, Boston.

GOODGAL, S. H. 1982. "DNA uptake in *Haemophilus* transformation." *Annu. Rev. Genet.* 16: 169–192.

HAYES, W. 1968. *The genetics of bacteria and their viruses,* 2nd ed. Wiley, New York.

HERSHEY, A. D., editor. 1971. *The bacteriophage lambda.* Cold Spring Harbor Laboratory Press, Cold Spring Harbor, NY.

KLECKNER, N. 1981. "Transposable elements in prokaryotes." *Annu. Rev. Genet.* 15: 341–404.

LEWIN, B. 1974. *Gene expression-1. Bacterial genomes.* Wiley, New York.

______. 1977. *Gene expression-3. Plasmids and phages.* Wiley, New York.

______. 1990. *Genes IV.* Cell Press, Cambridge, MA/Oxford Univ. Press, Oxford.

MATTHEWS, C., E. M. KUTTER, G. MOSIG, and P. BERGET, editors. 1983. *The bacteriophage T4.* American Society for Microbiology, Washington, DC.

MEYNELL, G. G. 1973. *Bacterial plasmids.* MIT Press, Massachusetts Institute of Technology, Cambridge, MA.

MILLER, J. H. 1972. *Experiments in molecular genetics.* Cold Spring Harbor Laboratory Press, Cold Spring Harbor, NY.

NASH, H. A. 1981. "Integration and excision of bacteriophage λ: The mechanism of conservative site-specific recombination." *Annu. Rev. Genet.* 15: 143–167.

NEIDHARDT, F. C., editor in chief, J. L. INGRAHAM, K. B. LOW, B. MAGASANIK, M. SCHAECHTER, and H. E. UMBARGER, editors. 1987. *Escherichia coli and Salmonella typhimurium, cellular and molecular biology.* American Society for Microbiology, Washington, DC.

NOVICK, R. P. 1980. "Plasmids." *Sci. Amer.* 243(6): 102–127.

STARLINGER, P., and H. SAEDLER. 1976. "IS-elements in microorganisms." *Current Topics Microbiol. Immunol.* 75: 111–152.

WATSON, J. D., N. H. HOPKINS, J. W. ROBERTS, J. ARGETSINGER STEITZ, and A. M. WEINER. 1987. *Molecular biology of the gene,* 4th ed. Benjamin/Cummings, Menlo Park, CA.

PROBLEMS AND QUESTIONS

8.1 A nutritionally defective *E. coli* strain grows only on a medium containing thymine, whereas another nutritionally defective strain grows only on medium containing leucine. When these two strains were grown together, a few progeny were able to grow on a minimal medium with neither thymine or leucine. How can this result be explained?

8.2 Assume that you have just demonstrated genetic recombination (e.g., when a strain of genotype $a\ b^+$ is present with a strain of genotype $a^+\ b$, some recombinant genotypes, $a^+\ b^+$ and $a\ b$, are formed) in a previously undescribed species of bacteria. Describe how you would go about (operationally) determining whether the observed recombination was the result of a process similar to transformation, a process similar to transduction, or a process similar to conjugation.

8.3 Compare, in table form, similarities and differences of the mechanisms through which (a) transduction, (b) sexduction, and (c) transformation may occur.

8.4 (a) What are the genotypic differences between F^- cells, F^+ cells, and Hfr cells? (b) What are the phenotypic differences? (c) By what mechanism are F^- cells converted to F^+ cells? F^+ cells to Hfr cells?

8.5 (a) Of what use are F′ factors in genetic analysis? (b) How are F′ factors formed? (c) By what mechanism does sexduction occur?

8.6 What are the basic differences between generalized transduction and specialized transduction?

8.7 (a) What is the difference between a phage and a prophage? (b) Are the linkage maps of the mature phage lambda chromosome and the lambda prophage identical? (c) If not, in what way(s) do they differ?

8.8 (a) What are IS elements? (b) What role(s) do IS elements play in recombination? (c) What role(s) do they play in the evolution of genetic elements?

8.9 How can genes be mapped by interrupted mating conjugation experiments?

8.10 What does the term cotransduction mean? How can cotransduction frequencies be used to map genetic markers?

8.11 An F^+ strain, marked at 10 loci, gives rise spontaneously to Hfr progeny whenever the F factor becomes incorporated into the chromosome of the F^+ strain. The F factor can integrate into the circular chromosome at many points, so that the various Hfr segregants transfer the genetic markers in different sequences. For any Hfr strain, the order of markers entering early can be determined by interrupted mating experiments using a Waring blender. From the following data for several Hfr strains derived from the same F^+, determine the order of markers in the F^+ strain.

HFR STRAIN	MARKERS DONATED IN ORDER
1	— *Z-H-E-R* →
2	— *O-K-S-R* →
3	— *K-O-W-I* →
4	— *Z-T-I-W* →
5	— *H-Z-T-I* →

8.12 In *E. coli,* the ability to utilize lactose as a carbon source requires the presence of the enzymes β-galactosidase and β-galactoside permease. These enzymes are coded for by two closely linked genes, *lacZ* and *lacY*, respectively. Another gene, *proC*, controls, in part, the ability of *E. coli* cells to synthesize the amino acid proline. The alleles str^r and str^s control resistance and sensitivity, respectively, to streptomycin. HfrH is known to transfer the two *lac* genes, *proC*, and *str*, in that order, during conjugation.

A cross was made between HfrH of genotype $lacZ^-\ lacY^+\ proC^+\ str^s$ and an F^- of genotype $lacZ^+\ lacY^-\ proC^-\ str^r$. After about 2 hours, the mixture was diluted and plated out on medium containing streptomycin, but no proline. When the resulting $proC^+\ str^r$ recombinant colonies were checked for their ability to grow on medium containing lactose as the sole carbon source, very few of them were capable of fermenting lactose. When the reciprocal cross (HfrH $lacZ^+\ lacY^-\ proC^+\ str^s \times F^-\ lacZ^-\ lacY^+\ proC^-\ str^r$) was done, many of the $proC^+\ str^r$ recombinants were able to grow on medium containing lactose as the sole carbon source. What is the order of the *lacZ* and *lacY* genes relative to *proC*?

8.13 The data in the following table were obtained from three-point transduction tests made to determine the order of mutant sites in the *A* gene for tryptophan synthetase in *E. coli. Anth* is a linked, unselected marker. In each cross, trp^+

recombinants were selected and then scored for the *anth* marker (*anth*$^+$ or *anth*$^-$). What is the linear order of *anth* and the three mutant alleles of the *A* gene indicated by the data in the table?

CROSS	DONOR MARKERS	RECIPIENT MARKERS	*anth* ALLELE IN RECOMBINANTS	% *anth*$^+$
1	*anth*$^+$-*A34*	*anth*$^-$-*A223*	72 *anth*$^+$: 332 *anth*$^-$	18
2	*anth*$^+$-*A46*	*anth*$^-$-*A223*	196 *anth*$^+$: 180 *anth*$^-$	52
3	*anth*$^+$-*A223*	*anth*$^-$-*A34*	380 *anth*$^+$: 379 *anth*$^-$	50
4	*anth*$^+$-*A223*	*anth*$^-$-*A46*	60 *anth*$^+$: 280 *anth*$^-$	20

8.14 Two additional mutations in the *trp A* gene of *E. coli, trp A58* and *trp A487*, were ordered relative to *trp A223* and the outside marker *anth* by three-factor transduction crosses as described in Problem 8.13. The results of these crosses are summarized in the following table. What is the linear order of *anth* and the three mutant sites in the *trp A* gene?

CROSS	DONOR MARKERS	RECIPIENT MARKERS	*anth* ALLELE IN *trp*$^+$ RECOMBINANTS	% *anth*$^-$
1	*anth*$^+$-*A487*	*anth*$^-$-*A223*	72 *anth*$^+$: 332 *anth*$^-$	82
2	*anth*$^+$-*A58*	*anth*$^-$-*A223*	196 *anth*$^+$: 180 *anth*$^-$	48
3	*anth*$^+$-*A223*	*anth*$^-$-*A487*	380 *anth*$^+$: 379 *anth*$^-$	50
4	*anth*$^+$-*A223*	*anth*$^-$-*A58*	60 *anth*$^+$: 280 *anth*$^-$	80

8.15 Mutations *nrd* 11 (gene *nrd B*, coding for the B submit of the enzyme ribonucleotide reductase), *am* M69 (gene 63, coding for a protein that aids tail-fiber attachment), and *nd* 28 (*denA*, coding for the enzyme endonuclease II) are known to be located between gene 31 and gene 32 on the bacteriophage T4 chromosome. Mutations *am* N54 and *am* A453 are located in genes 31 and 32, respectively. Given the three-factor cross data in the following table, what is the linear order of the five mutant sites?

Three-Factor Cross Data

CROSS	% RECOMBINATION[a]
1. *am A453—am M69* × *nrd* 11	2.6
2. *am A453—nrd* 11 × *am M69*	4.2
3. *am A453—am M69* × *nd* 28	2.5
4. *am A453—nd* 28 × *am M69*	3.5
5. *am A453—nrd* 11 × *nd* 28	2.9
6. *am A453—nd* 28 × *nrd* 11	2.1
7. *am N54—am M69* × *nrd* 11	3.5
8. *am N54—nrd* 11 × *am M69*	1.9
9. *am N54—nd* 28 × *am M69*	1.7
10. *am N54—am M69* × *nd* 28	2.7
11. *am N54—nd* 28 × *nrd* 11	2.9
12. *am N54—nrd* 11 × *nd* 28	1.9

[a]All recommbination frequencies are given as $\frac{2\text{ (wild-type progeny)}}{\text{total progeny}} \times 100$.

Chapter 9

Transposable Genetic Elements

(Photo above): Color variation among kernels of maize. Studies of the genetic basis of this variation led to the discovery of transposable elements. Credit: J. Beckner.

A great part of classical genetic analysis has been devoted to the localization of genes on chromosomes. As discussed in previous chapters, genetic mapping depends on the assumption that genes do not move from one position to another. To a great extent, this assumption has been satisfied. Most genes occupy fixed sites on the chromosomes, and the overall structure of the genetic map is practically invariant. However, beginning in the 1940s, researchers have found that some DNA sequences can actually change position.

These mobile sequences are called ***transposable genetic elements***, or simply ***transposons***. Typically they are quite small, ranging from 500 to 10,000 nucleotide-pairs, but some are larger. Studies with diverse organisms, including bacteria, fungi, nematodes, insects, plants, and mammals, suggest that transposable elements are widespread among both prokaryotes and eukaryotes. At the molecular level, these elements exhibit considerable variation in structure and function.

GENETIC INSTABILITY AND THE DISCOVERY OF TRANSPOSABLE ELEMENTS

Transposable elements were discovered by B. McClintock through an analysis of genetic instability in maize. The instability involved chromosome breakage and was found to occur at sites where transposable elements were located. In McClintock's analysis, breakage events were detected by following the loss of certain genetic markers. In some experiments, McClintock used a marker that controlled the deposition of pigmentation in the ***aleurone***, the outermost layer of the endosperm of maize kernels. Recall that the endosperm is triploid, being produced by the union of two maternal nuclei and one paternal nucleus. McClintock's marker was an allele of the C locus on the short arm of chromosome 9. Since this allele, called C^I, is a dominant inhibitor of aleurone coloration, any kernel possessing it should be colorless. McClintock fertilized CC ears with pollen from C^IC^I tassels, producing kernels in which the endosperm was C^ICC. Although many of these kernels were colorless, as expected, some showed patches of brownish-purple pigment (Fig. 9.1). McClintock guessed that in such mosaics, the inhibitory C^I allele had been lost sometime during endosperm development, leading to a clone of tissue that was capable of producing pigment. The genotype in such a clone would be $-CC$, where the dash indicates loss of the C^I allele. Further analysis demonstrated that this allele had been lost through chromosome breakage.

The situation is diagrammed in Fig. 9.2. A break at the site labeled by the arrow would detach a segment of the chromosome from its centromere, creating what cytologists call an ***acentric fragment***. Such a fragment tends to be lost during cell division, so all the descendants of this cell would lack part of the paternally derived chromosome. Since the lost fragment carried the C^I allele, none of the cells in this clone would be inhibited from forming pigment, and if any of them produced a part of the aleurone, a patch of color would appear. This is the result that is seen in the kernel shown in Fig. 9.1.

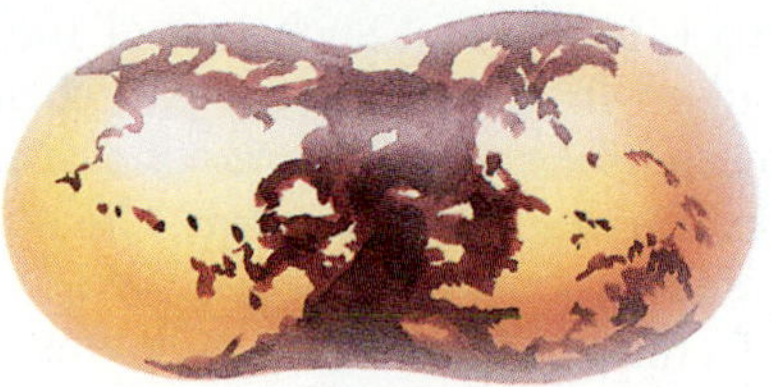

Figure 9.1 Maize kernel showing loss of the C^I allele for the inhibition of pigmentation in the aleurone. The pigmented patches are $-CC$, whereas the colorless patches are C^ICC. (After B. McClintock, *Cold Spring Harbor Sympos. Quant. Biol.* 16: 13–47, 1951.)

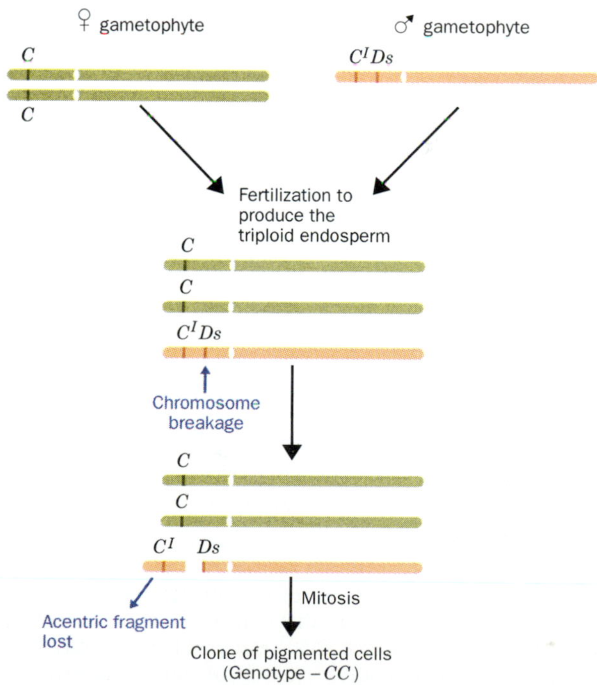

Figure 9.2 Chromosome breakage caused by the transposable element *Ds* in maize. The allele C on the short arm of chromosome 9 produces normal pigmentation in the aleurone; the allele C^I inhibits this pigmentation.

McClintock found that such kernel mosaics frequently resulted from breaks at a particular site on chromosome 9. She named the factor that produced these breaks **Ds**, for ***"Dissociation."*** In her experiments, the chromosome that carried the C^I allele also carried the *Ds* factor. However, by itself, this factor was not capable of inducing chromosome breakage. Through careful work, McClintock found that *Ds* had to be activated by another factor, called **Ac**, for ***"Activator."*** The *Ac* factor was present in some maize stocks, but absent in others. By crossing different stocks, *Ac* could be combined with *Ds*, creating the condition that led to chromosome breakage.

This two-factor system provided an explanation for the genetic instability that McClintock had observed on chromosome 9. However, additional experiments demonstrated that this was only one of many instabilities present in the maize genome. McClintock found other instances of breakage at different sites on chromosome 9, and also on other chromosomes. Since breakage at these sites depended on activation by *Ac*, McClintock hypothesized that *Ds* factors were also involved. A simple explanation was that *Ds* could exist at many different sites in the genome, and that it was possible for *Ds* to change its position.

This explanation has been borne out by subsequent analysis. ***Both* Ac *and* Ds *are members of a family of transposable elements.*** These elements are structurally related to each other and can insert at many different sites on the chromosomes. In fact, there often are multiple copies of the *Ac* and *Ds* elements present in the maize genome. Through genetic analysis, McClintock demonstrated that both *Ac* and *Ds* could move. When one of these elements was inserted in or near a gene, McClintock sometimes found that the gene's function was altered. In extreme cases, the function was completely abolished. Because of this influence on gene expression, McClintock referred to *Ac* and *Ds* as ***controlling elements.***

Sometimes mutations that are due to controlling element insertions are unstable (Fig. 9.3). For instance, one of the mutations of the *bronze* locus, *bz-m2*, reverts spontaneously at a very high rate. This mutation contains an insertion of the *Ac* element and reverts when the *Ac* element is excised. Another mutation, *bz-m1*, contains a *Ds* insertion; however, in this case, reversions occur only if an *Ac* element is present elsewhere in the genome. The difference between these two mutable alleles demonstrates an important feature of the *Ac/Ds* system. *Ac* elements can activate themselves, but *Ds* elements cannot. ***Whenever a transposon is self-activating, it is said to be functionally autonomous; whenever it is not self-activating, it is considered to be nonautonomous.***

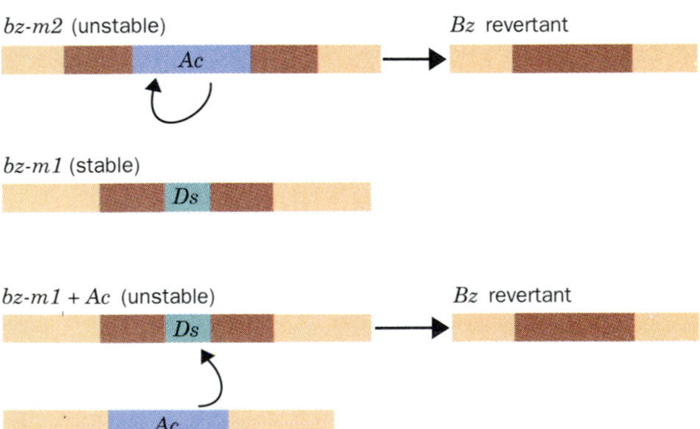

Figure 9.3 Mutable alleles of the *bronze* locus in maize. *Bronze* is one of many loci that control pigmentation in the endosperm. Mutations in this locus reduce the amount of pigment. The mutant allele *bz-m2* contains an insertion of the *Ac* element; the mutant allele *bz-m1* contains an insertion of a *Ds* element. Since *Ac* is self-activating, *bz-m2* is spontaneously unstable, reverting to the wild-type *Bz* allele; however, the *Ds* insertion mutation *bz-m1* reverts to *Bz* only if an *Ac* element is present elsewhere in the genome. This indicates the *trans*-acting nature of the *Ac* element.

TRANSPOSABLE ELEMENTS IN BACTERIA

Genetic instabilities have also been found in bacteria, and in many cases these have led to the identification of transposable elements. These bacterial transposons were the first to be studied at the molecular level and therefore provided important clues about the organization and behavior of eukaryotic transposons. The simplest bacterial transposons are the ***insertion sequences,*** or ***IS elements,*** which were introduced in Chapter 8. These are typically less than 1500 nucleotide-pairs long and contain only genes involved in promoting or regulating transposition. Sometimes two homologous IS elements combine with other genes to form a composite transposon, denoted by the symbol Tn. This symbol is also used to denote transposons that do not contain IS elements, such as the element known as Tn*3*. Like the composite transposons, this element contains a gene that is not necessary for transposition. The integrating bacteriophage λ (see Chapter 8, pp. 210–213) is also considered to be a transposable element because it can insert inself into a bacterial chromosome. However, this phage represents the upper limit of transposon size and obviously contains many genes that are not necessary for the insertion behavior.

IS Elements

IS elements are compactly organized. Typically there is a single coding sequence with short, identical, or

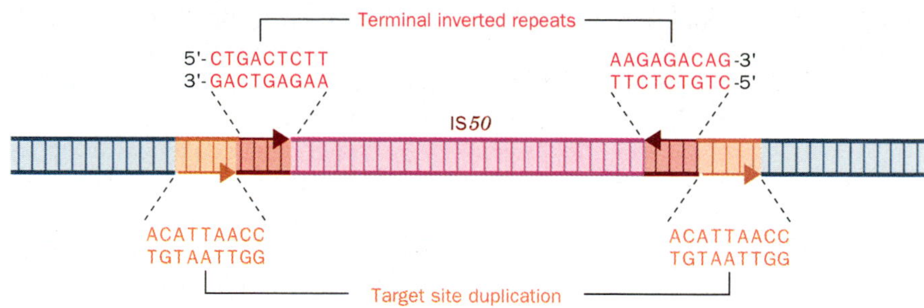

Figure 9.4 Structure of IS elements. The terminal inverted repeats of the element IS*50* and a target site duplication caused by the insertion of this element are indicated. Notice that the terminal inverted repeats are imperfect; the fourth nucleotide-pair in from each end is different. (After D. E. Berg *et al., Proc. Natl. Acad. Sci. U.S.A.* 79: 2632–2635, 1982.)

nearly identical sequences at both ends (Fig. 9.4). These terminal sequences are always in inverted orientation with respect to each other, so they are called ***inverted terminal repeats***. Their lengths range from 9 to 40 nucleotide-pairs.

When IS elements insert into chromosomes or plasmids, they create a duplication of the DNA sequence at the site of the insertion. One copy of the duplication is located on each side of the element. These short (3–12 nucleotide-pairs), directly repeated sequences are therefore called ***target site duplications*** and are thought to arise from staggered breaks in double-stranded DNA (Fig. 9.5). As discussed in Chapter 8, IS elements mediate the integration of episomes into bacterial chromosomes. This process involves homologous recombination between IS elements located in the episome and in the chromosome.

Composite transposons are created when two IS elements insert near each other. The sequence between them can then be transposed by the joint action of the flanking elements. Figure 9.6 gives three examples. In Tn*9*, the flanking IS elements are in direct orientation with respect to each other, whereas in Tn*5* and Tn*10*, the orientation is inverted. Each of these composite transposons carries a gene for antibiotic resistance—chloramphenicol resistance in Tn*9*, kanamycin resistance in Tn*5*, and tetracycline resistance in Tn*10*. It should be noted that sometimes the flanking IS elements in a composite transposon are not quite identical. For instance, in Tn*5*, the element on the left, called IS*50*L, is incapable of stimulating transposition, but the element on the right, called IS*50*R, is capable. This difference is due to a single nucleotide-pair change that prevents IS*50*L from synthesizing a necessary transposition factor. However, since this factor, a protein called the ***transposase***, is synthesized by IS*50*R, the entire composite transposon can be mobilized.

Tn*5* also illustrates another feature of the IS class of element: transpositional activity is regulated (Fig. 9.7). This can be seen when a bacterial cell is infected with a nonlytic bacteriophage that carries Tn*5* on its chromosome. In such infections, the frequency of Tn*5* transposition is dramatically reduced whenever the infected cell already carries a copy of Tn*5*. This reduction implies that the resident transposon inhibits the transposition of an incoming transposon, possibly by synthesizing a repressor. Detailed analyses have shown that this hypothesis is correct. The IS*50*R element of Tn*5* actually produces two proteins. One, the transposase, catalyzes transposition, whereas the other, a shortened version of the transposase, prevents transposition. Since the shorter protein is the more abundant, Tn*5* transposition tends to be repressed.

The Tn*3* Family

The elements in this group of transposons have inverted terminal repeats that are 38–40 nucleotide-pairs long and produce target site duplications of 5 nucleotide-pairs upon insertion. They are larger than the IS

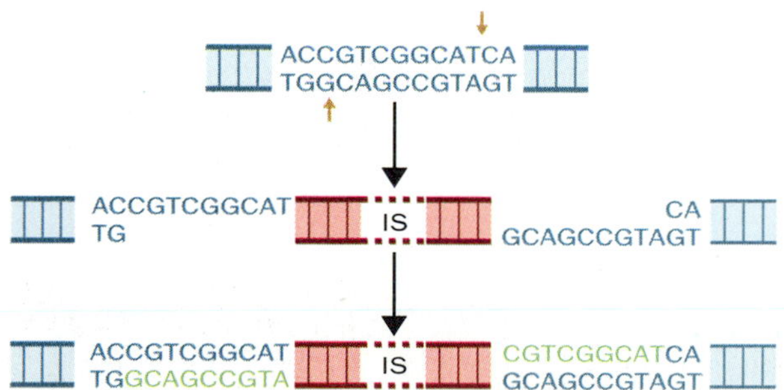

Figure 9.5 Production of target site duplications by the insertion of an IS element. The element causes breaks in the target DNA at the sites indicated by the arrows. After insertion, DNA synthesis (green) fills in the gaps left by the staggered breaks, producing direct repeats on each side of the element. (After S. N. Cohen and J. A. Shapiro, *Sci. Amer.* 242: 40–49, 1980.)

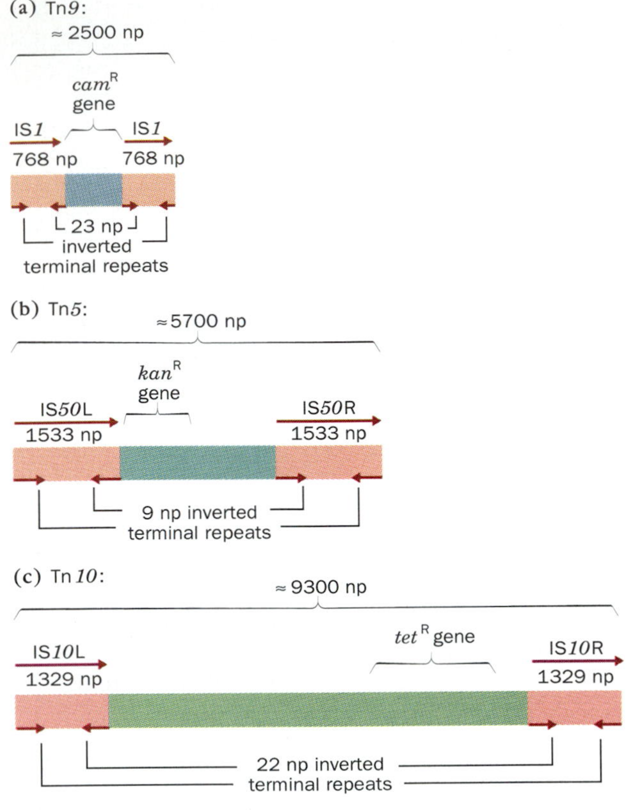

elements (typically 5000 nucleotide-pairs long, or longer) and usually contain accessory genes as well as the genes needed for transposition. Tn*3* is the most thoroughly studied example.

The genetic organization of Tn*3* is shown in Fig. 9.8. There are three genes, *tnpA*, *tnpR*, and *bla*, encoding, respectively, a ***transposase***, a ***resolvase/repressor***, and an enzyme called ***beta lactamase***. The beta lactamase confers resistance to the antibiotic ampicillin, whereas the other two proteins play important roles in transposition.

The transposition of Tn*3* occurs in two stages (Fig. 9.9). First, the transposase mediates the fusion of two

Figure 9.6 Genetic organization of composite transposons. The orientation and length of constituent nucleotide sequences are indicated; sequence lengths are given in nucleotide-pairs (np). (a) Tn*9*. This transposon consists of two IS*1* elements flanking a gene for chloramphenicol resistance. (b) Tn*5*. This transposon consists of two IS*50* elements flanking a gene for kanamycin resistance. (c) Tn*10*. This transposon consists of two IS*10* elements flanking a gene for tetracycline resistance.

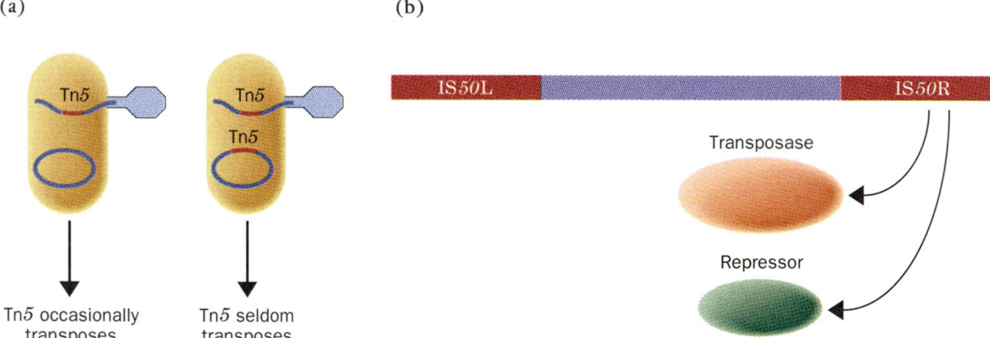

Figure 9.7 Regulation of Tn*5*. (a) Infection of *E. coli* cells with bacteriophages carrying Tn*5*. Cells that already possess a copy of Tn*5* repress transposition. (b) Genetic basis of Tn*5* regulation. The IS*50*R element produces two proteins. One, a transposase, catalyzes transposition, but the other, a repressor, inhibits transposition. The repressor protein is more abundant than the transposase, so its effect usually prevails.

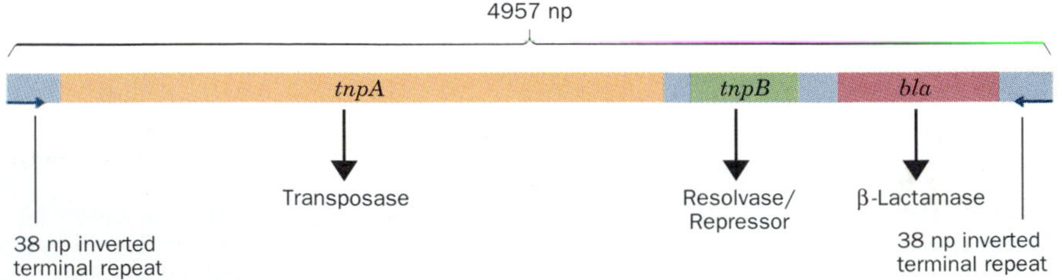

Figure 9.8 Genetic organization of Tn*3*. This transposon contains three genes that produce the proteins indicated. Lengths of DNA sequences are given in nucleotide-pairs (np).

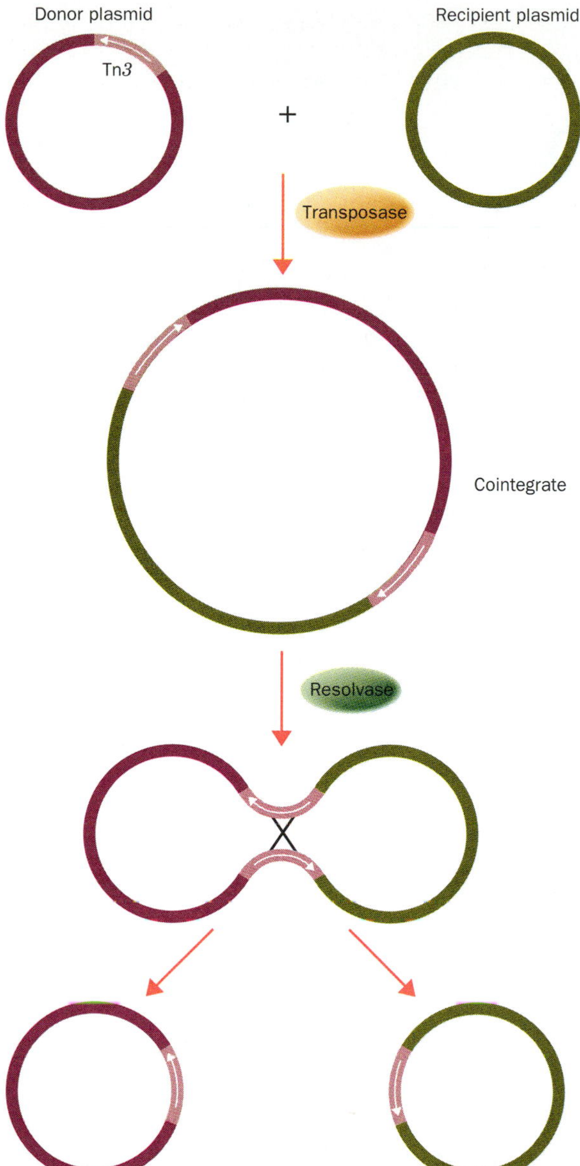

Figure 9.9 Transposition of Tn*3*. The donor plasmid carries a single copy of Tn*3* and the recipient plasmid lacks Tn*3* altogether. A transposase produced by the *tnpA* gene of Tn*3* catalyzes the formation of a cointegrate in which the two plasmids are fused. During this process, Tn*3* is also replicated so there is a copy of the element at each junction in the cointegrate. A resolvase produced by the *tnpR* gene resolves the cointegrate by mediating a recombination event between the two Tn*3* elements. Donor and recipient plasmids are thereby separated, each with a copy of Tn*3*.

molecules, forming a structure called a ***cointegrate***. During this process, the transposon is replicated, and one copy is inserted at each junction in the cointegrate. Notice that the two Tn*3* elements are oriented in the same direction. In the second stage of transposition, the *tnpR*-encoded resolvase mediates a site-specific recombination event between the two Tn*3* elements. This event occurs at a sequence in Tn*3* called *res*, the ***resolution site***, and generates two molecules, each with a copy of the transposon.

The *tnpR* gene-product also has another function, namely, to repress the synthesis of both the transposase and resolvase proteins. This repression occurs because the *res* site is located in between the *tnpA* and *tnpR* genes. By binding to this site, the *tnpR* protein interferes with the synthesis of both gene-products, leaving them in chronic short supply. Consequently, the Tn*3* element tends to remain immobile.

Medical Significance of Bacterial Transposons

Bacterial transposons are clearly responsible for the transposition of genes controlling resistance to antibiotics (and other drugs) from one molecule to another. They are believed to play a role in the observed rapid evolution of R plasmids (Chapter 8). All conjugative R plasmids have at least two components, one segment carrying a set of genes involved in conjugative DNA transfer (probably analogous to the *tra* genes of F plasmids) and a second segment carrying the antibiotic and/or drug resistance gene or genes (Fig. 9.10). The segment carrying the transfer genes is called the ***RTF*** (resistance transfer factor) ***component***; the segment carrying the resistance gene or genes is called the ***R-determinant***. The RTF components of several different conjugative R plasmids appear to have a large amount of homology, based on DNA–DNA cross-hybridization experiments. The R-determinant components exhibit more divergence. In several R plasmids, the R-determinant is flanked by homologous IS elements. In some cases, these are present in the same orientation, and in others they are inserted with opposite orientations. In either case, they can mediate the transposition of an R-determinant from one R plasmid to another. Several compound R plasmids have been characterized as carrying two or more R-determinants, each flanked by IS elements. These IS elements are almost certainly responsible for the rapid evolution of bacterial plasmids that carry multiple antibiotic and drug resistance factors.

The transmissibility of R plasmids, the transposability of the R-determinants, and the rapid evolution of compound R plasmids, which carry genes for resistance to a whole battery of our most effective antibiotics and drugs, are of great concern to medical practitioners. Not only are these plasmids rapidly dispersed within a bacterial species, but they are also transmitted across species and even generic lines. For example, *E. coli* R plasmids are known to be transferred to several genera, including *Proteus*, *Salmonella*, *Hemophilus*, *Pasturella*, and *Shigella*, all of which include

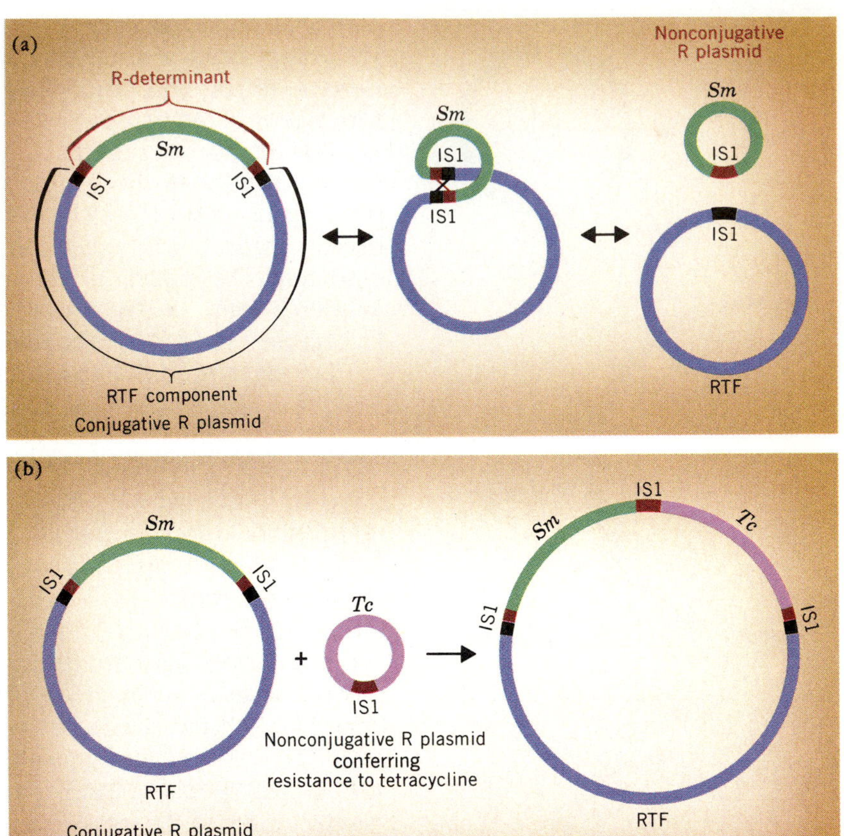

Figure 9.10 Structure of conjugative and nonconjugative R plasmids (a) and a proposed mechanism for the evolution of compound R plasmids, which provide multiple-antibiotic resistance and drug resistance to bacterial cells (b). (a) Simple conjugative R plasmids contain two major components: (1) the RTF component (shown blue), which contains the *tra* genes responsible for conjugative transfer of the plasmid, and (2) the R-determinant component (shown in yellow green), which carries the gene or genes responsible for antibiotic or drug resistance. The R-determinant in several characterized conjugative R plasmids is flanked by identical IS elements. These IS elements are believed to mediate the transposition of R-determinants from one R plasmid to another (or to other genetic elements). Nonconjugative R plasmids (top right) do not carry the RTF (*tra* genes) component. (b) IS element-mediated formation of a compound R plasmid, carrying genes that provide resistance to both streptomycin and tetracycline. The mechanism is as shown in (a), from right to left. This process may continue with R-determinants carrying genes for resistance to other antibiotics and drugs until conjugative R plasmids have evolved, which provide the host cell with resistance to a whole battery of antibiotics and drugs. *Sm* and *Tc* symbolize the plasmid-carried genes providing streptomycin and tetracycline resistance, respectively.

pathogenic species. The increased frequency of bacteria carrying plasmids with R-determinants, which result in resistance to antibiotics such as penicillin, tetracycline, streptomycin, and kanamycin, in hospital populations (which are continuously exposed to these antibiotics) has been extensively documented. Of even greater concern are the results of studies in Japan that show that, in less than 10 years, natural populations of bacteria (in sewers and in polluted lakes and streams) have evolved from very low frequencies (less than 1 percent) of R plasmid-mediated antibiotic resistance to relatively high frequencies (50–80 percent).

The results of these studies make it eminently clear that we should restrict our use of antibiotics to serious bacterial infections and not use them for every minor infection that comes along. If we do not restrict use, the antibiotics and drugs that are so effective today may have little, if any, utility in the future.

TRANSPOSABLE ELEMENTS IN EUKARYOTES

Although some of the most detailed studies of transposable elements have been done with bacteria, there has also been extensive research on transposons in eukaryotes, beginning with the classic work of McClintock. Some of the more recent discoveries with yeast, maize, and *Drosophila* transposons are discussed next.

Yeast *TY* Elements

The yeast *Saccharomyces cerevisiae* carries about 35 copies of a transposable element called **Ty** in its haploid genome. These transposons are about 5900 nucleotide-pairs long and are bounded at each end by a DNA segment called the ***δ sequence***, which is approximately 340 base-pairs long (Fig. 9.11). Each δ sequence is oriented in the same direction, forming what are known as direct ***long terminal repeats***, or ***LTRs***. Sometimes an LTR becomes detached from a *Ty* element, creating a so-called solo δ. It is thought that these solo δ's are generated by recombination between the LTRs of a complete *Ty* element, as shown in Fig. 9.12. The fate of the circular molecule that is formed as a by-product of this event is not known, but such molecules have been detected in yeast cells, giving plausibility to the model. *Ty* elements are flanked by five nucleotide-pair direct repeats created by the duplication of DNA at the site of the *Ty* insertion. These target site duplications do not have a standard sequence, but they tend to contain AT base pairs. This may indicate that *Ty* elements preferentially insert into A-T-rich regions of the genome.

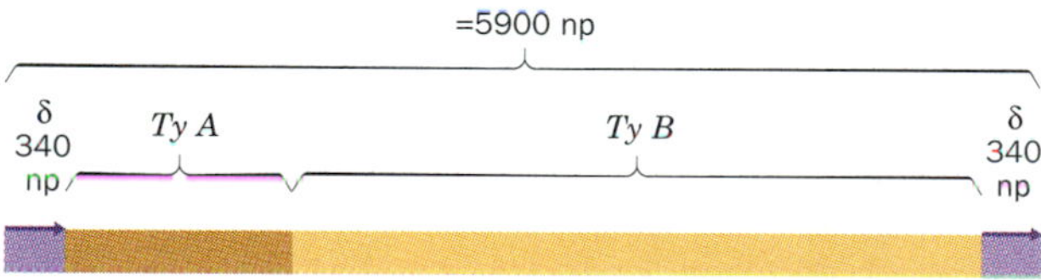

Figure 9.11 Genetic organization of yeast *Ty* elements. The long terminal repeat (LTR) sequences are denoted by the Greek letter δ. *Ty* elements apparently contain two genes, *TyA* and *TyB*, as shown. Sequence lengths are given in nucleotide-pairs (np).

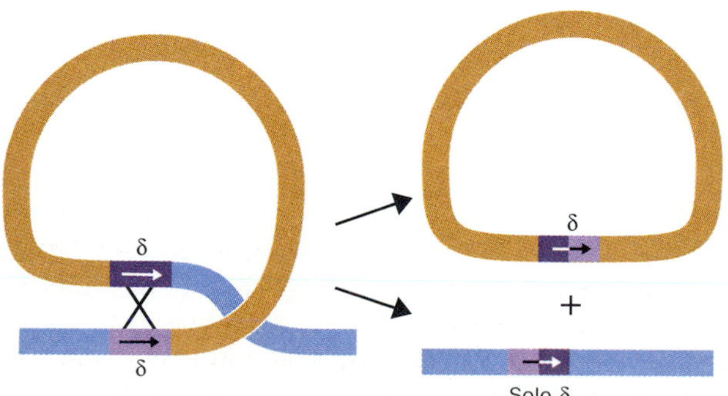

Figure 9.12 Formation of a solor δ sequence by the excision of a yeast *Ty* element. The excision event is thought to involve homologous recombination between the δ sequences at the ends of the element.

The genetic organization of the* Ty *elements resembles that of the eukaryotic retroviruses (Fig. 9.13*a*). These single-stranded RNA viruses synthesize DNA from their RNA after entering a cell. The DNA then inserts itself into a site in the genome, creating a target site duplication. This inserted material has the same overall structure as a yeast *Ty* element—a DNA sequence bounded by LTRs—and is called a ***provirus***. The simplest of the proviruses possess three genes, *gag*, *pol*, and *env*, which encode structural, catalytic, and membrane proteins, respectively. *Ty* elements have only two genes, *A* and *B*, which are analogous to the *gag* and *pol* genes of the retroviruses. Biochemical studies have shown that the products of these two genes can form viruslike particles inside yeast cells. However, it is not known whether these particles are genuinely infectious. One hypothesis is that yeast *Ty* elements are primitive retroviruses, capable of moving from one site to another inside a cell, but not capable of moving between cells. In this regard, it has been shown that the transposition of *Ty* elements involves an RNA intermediate (Fig. 9.13*b*). After the RNA is synthesized from *Ty* DNA, a product of the *TyB* gene uses the RNA to make double-stranded DNA. The process is called reverse transcription. Then the newly synthesized DNA is inserted somewhere in the genome, creating a new *Ty* element. Because of their overall similarity to the retroviruses, yeast *Ty* elements are sometimes called ***retrotransposons***.

Maize Transposons

Transposable elements have been found in several plants, most notably maize (*Zea mays*) and snapdragons (*Antirrhinum majus*). The most extensive investigations have involved maize, in which several transposon families have been identified.

Ac and Ds Elements

The *Ac/Ds* family of maize, initially discovered by McClintock, comprises numerous elements scattered throughout the genome. Molecular studies have shown that the functionally autonomous element, *Ac*, consists of 4563 nucleotide-pairs bounded by an 8-nucleotide-pair direct repeat (Fig. 9.14*a*). This direct repeat is created at the time that the element inserts

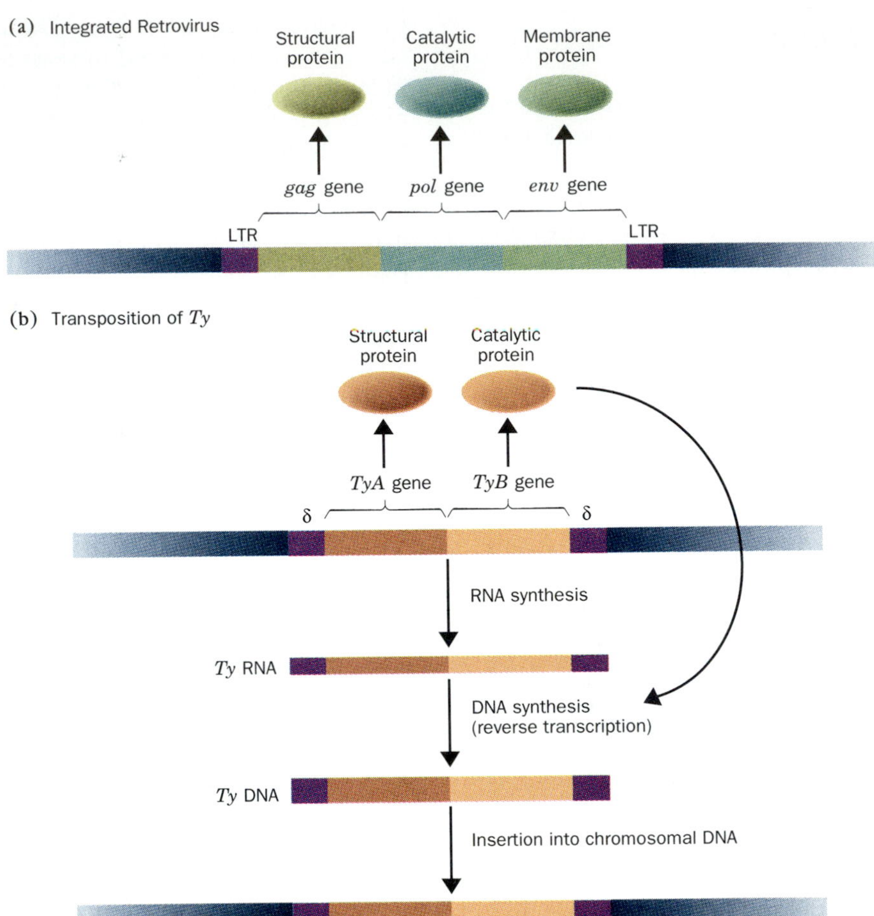

Figure 9.13 The structure of integrated retroviruses and the mechanism of *Ty* element transposition. (a) A retrovirus has inserted into a site on a chromosome (wavy line). The viral DNA (rectangle) contains LTR sequences and three genes that synthesize the proteins indicated. (b) A yeast *Ty* element inserted in chromosomal DNA (wavy line) has the same basic organization as an integrated retrovirus, however, it lacks one of the retroviral genes. The *Ty* element synthesizes RNA, which is then used to synthesize DNA via a process called reverse transcription. A product of the *TyB* gene probably catalyzes this process. Newly formed *Ty* DNA then inserts into yeast chromosomal DNA, creating a new *Ty* element.

into a site on a chromosome. Other repeat sequences are found within the element itself. The most conspicuous of these are at the ends, where an 11-nucleotide-pair sequence at one end is repeated in the opposite orientation at the other end. These inverted terminal repeats are thought to play an important role in transposition.

All the *Ac* elements in the maize genome appear to be structurally similar, if not identical. This is not the case with the *Ds* elements, in which considerable structural heterogeneity has been observed. One class of *Ds* elements is derived from *Ac* elements by deletions of internal sequences. Figure 9.14*b* gives some examples. Another class possesses the characteristic inverted terminal repeat sequences of *Ac*, as well as some of the subterminal sequences, but the remainder of the DNA is different (Fig. 9.14*c*). These unusual members of the *Ac*/*Ds* family are called ***aberrant* Ds** elements. A third class of *Ds* elements is characterized by a peculiar piggybacking arrangement (Fig. 9.14*d*). One *Ds* element is inserted into another, but in an inverted orientation. It has been shown that these ***double* Ds** elements are responsible for chromosome breakage.

The activating function of *Ac* elements is associated with a protein that they synthesize. Since this protein is involved in transposition, it is sometimes called the ***transposase*** of the *Ac*/*Ds* family. Deletions or mutations in the gene that encodes this protein abolish the activating signal and explain why *Ds* ele-

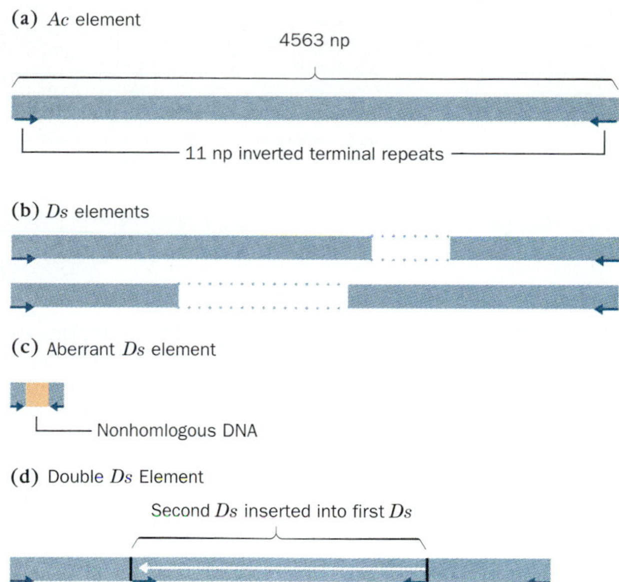

Figure 9.14 Structural organization of autonomous and nonautonomous members of the *Ac/Ds* family of transposable elements in maize. The terminal inverted repeats are indicated by the short arrows under each element. (a) *Ac,* the autonomous element. The lengths of DNA sequences are given in nucleotide-pairs (np). (b) Nonautonomous *Ds* elements derived from *Ac* by the deletion of internal sequences (gaps). (c) An aberrant *Ds* element with terminal and subterminal sequences from *Ac* flanking non-*Ac* DNA. (d) A double *Ds* element, in which one *Ds* element has inserted in the middle of another *Ds* element, but in inverted orientation. (After N. V. Federoff, *Cell* 56: 181–191, 1989.)

ments, which have such lesions, cannot activate themselves. However, since this transposase is diffusable, a single *Ac* element can provide it to all the *Ac* and *Ds* elements in the genome (see Fig. 9.3). We therefore say that the *Ac/Ds* transposase is ***trans-acting***.

Genetic analysis has provided some information about the mechanism whereby *Ac* (and presumably *Ds*) elements transpose. After an *Ac* element has been replicated as part of the DNA in a chromosome, it can excise from its position and move to a new one, usually nearby. The situation is diagrammed in Fig. 9.15. The line represents a chromosome in the process of replication. Once the replication fork has passed over the *Ac* element, a copy of that element can transpose to a site ahead of the replication fork. When the replication process is finished, there will be two sister chromatids, one with a single copy of *Ac* (in the new location only) and one with two copies (one in the new location and one in the old). Notice that in this process, the *Ac* element does not replicate itself during transposition; rather, it is copied by the normal replication machinery before and after movement. For this reason, the actual transposition of an *Ac* element is considered to be nonreplicative.

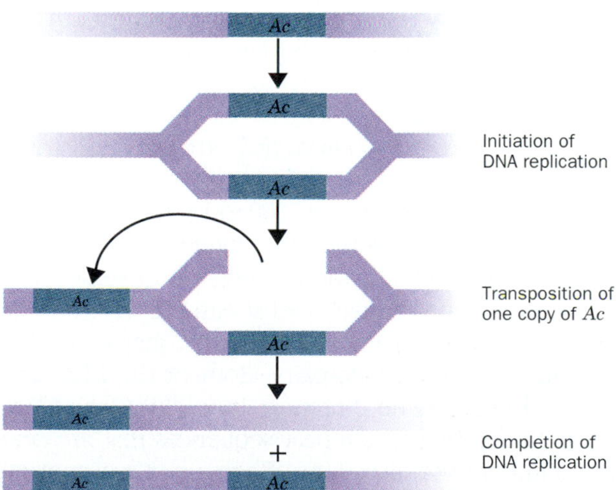

Figure 9.15 Transposition of an *Ac* element in maize. Initially, the element is replicated by the normal cellular machinery. Then one copy of *Ac* transposes ahead of the replication fork. After transposition, chromosome replication is completed, producing two sister chromatids as shown at the bottom. One of these has an *Ac* element in the new position, whereas the other has an *Ac* element at the old position as well. (After N. V. Federoff, *Cell* 56: 181–191, 1989.)

Spm and dSpm Elements

Another maize transposon family discovered by McClintock is the ***Suppressor-mutator*** family (Fig. 9.16). In this family, the autonomous element is called *Spm* and the nonautonomous elements are called *dSpm* ("*d*" stands for deleted or defective). *Spm* elements are 8287 nucleotide-pairs long, including 13-nucleotide-pair inverted terminal repeats. When they insert into a chromosome, they create a 3-nucleotide-pair target site duplication. The *dSpm* elements are smaller than the *Spm* elements because part of the DNA sequence has been deleted. These deletions disrupt the function of a gene carried by complete *Spm* elements and therefore prevent the synthesis of the gene's product. Since this product is necessary for transposition, deleted *dSpm* elements are unable to stimulate their own movement.

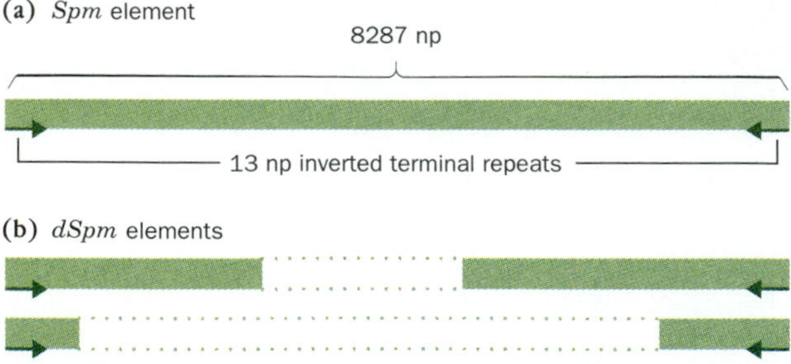

Figure 9.16 Structure of autonomous and nonautonomous members of the *Spm*/*dSpm* family of transposable elements in maize. Terminal inverted repeats are indicated by the short arrows at the ends of each element. (a) *Spm*, the autonomous element. The lengths of DNA sequences are given in nucleotide-pairs (np). (b) Nonautonomous *dSpm* elements derived from an *Spm* element by the deletion of internal sequences (gaps). (After N. V. Federoff, *Cell* 56: 181–191, 1989.)

The *Spm* family was named because its elements can suppress the function of genes into which they have transposed. This occurs when an inserted *dSpm* element interacts with an *Spm* element located elsewhere in the genome. Figure 9.17 shows an example in which a *dSpm* element has inserted into one of the genes controlling pigmentation in the kernel. Notice that although the *dSpm* insertion reduces the expression of this gene, it does not abolish it completely. However, when an autonomous *Spm* element is introduced into the genome, the expression of the pigmentation gene is completely inhibited in most of the kernel. This indicates the "suppressor" action of the *Spm* element. In addition, this element stimulates the excision of the *dSpm* element in some of the cells, leading to clones in which gene function is partially restored. These clones, which are recognized by their moderate to heavy pigmentation, demonstrate the trans-acting, "mutator" function of the *Spm* element.

Recently, biochemical analysis has indicated that the activity of *Ac* and *Spm* elements is controlled by the methylation of selected nucleotides in the DNA sequence. Research on this phenomenon is currently under way and may lead to a deeper understanding of the mechanisms that regulate the behavior of these transposable element families.

Drosophila Transposons

Transposable elements have been discovered in many animals, but some of the best information comes from studies with *Drosophila*, in which as much as 15 percent of the DNA is mobile. Several classes of *Drosophila* transposons have been identified.

Full pigmentation | Uniformly reduced pigmentation | Patches of pigmentation on colorless background

Wild-type | *dSpm* insertion | *dSpm* insertion + *Spm* elsewhere

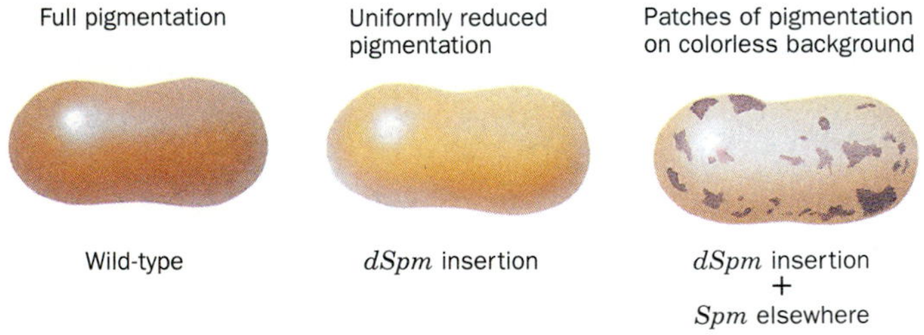

Figure 9.17 Suppressor-mutator behavior in maize kernels resulting from interactions between a *dSpm* element inserted in the *A* locus, which controls pigmentation in the endosperm, and an *Spm* element located elsewhere in the genome. The wild-type kernel at the left shows full pigmentation. The kernel in the middle shows uniformly reduced pigmentation due to the insertion of a *dSpm* element in the *A* locus. The kernel on the right shows the suppressor-mutator action of an *Spm* element on this *dSpm* insertion. For the most part, this kernel is colorless, indicating that the function of the mutant *A* locus is suppressed by a product of the *Spm* element. However, patches of colored tissue indicate that this element is also capable of inducing the excision of the *dSpm* element responsible for the mutation, restoring *A* locus function. (After N. V. Federoff, *Cell* 56: 181–191, 1989.)

Retrotransposons

The largest group of *Drosophila* transposons comprises the retroviruslike elements, or ***retrotransposons***. These elements are 5000 to 15,000 nucleotide-pairs long and resemble the integrated forms of retroviruses, much like the *TY* elements of yeast. Each retrotransposon is demarcated at either end by a long terminal repeat sequence, or LTR, that may contain a few hundred nucleotide-pairs. Both of the LTRs are oriented in the same direction. In addition, the LTRs are bounded by short repeat sequences that are oriented in opposite directions. When a retrotransposon inserts into a chromosome, it creates a target site duplication, with one copy on each side of the transposon. The size of this duplication is characteristic of each retrotransposon family. For instance, the members of the *copia* family (Fig. 9.18*a*) produce a 5-nucleotide-pair duplication, whereas the members of the *gypsy* family (Fig. 9.18*b*) produce a 4-nucleotide-pair duplication. Target site duplications are always oriented in the same direction.

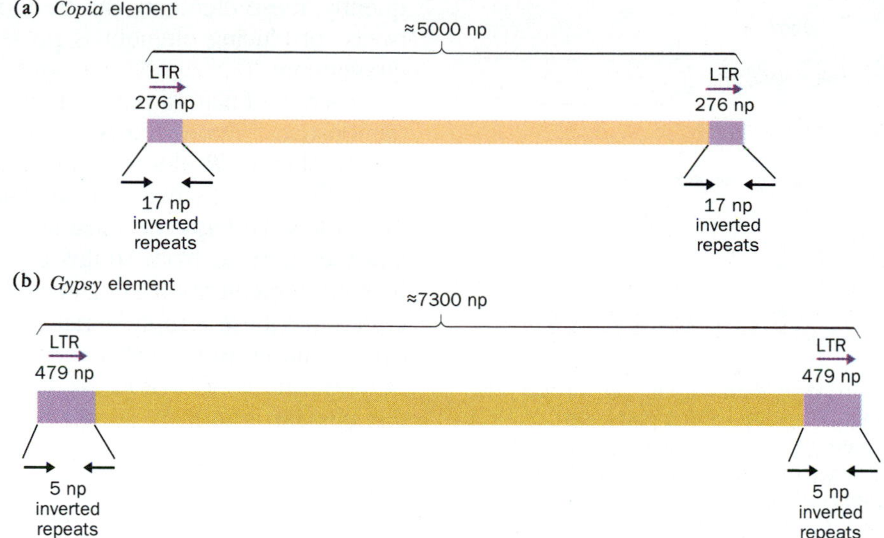

Figure 9.18 Structures of retrotransposons in *Drosophila*. The orientations and lengths (in nucleotide-pairs, np) of DNA sequences are shown. (a) the *copia* element. (b) The *gypsy* element.

It is not clear how many different retrotransposon families exist in *Drosophila*, but probably there are not more than 30. Studies with different strains indicate that the size of each family varies. For example, some *Drosophila* strains have only a few *gypsy* elements, whereas others have more than 100. Moreover, these elements may be scattered all over the genome, occupying different positions in different strains. Much of this variation is probably random, but some researchers have speculated that the number of elements in a retrotransposon family might be regulated.

Retrotransposons are responsible for many of the mutations of classic *Drosophila* genetics. Figure 9.19 shows four alleles of the X-linked *white* locus that are due to retrotransposon insertions. In one case, the expression of the locus is essentially abolished, whereas in the others it is merely reduced from wild-type levels. Although most of the retrotransposon in-

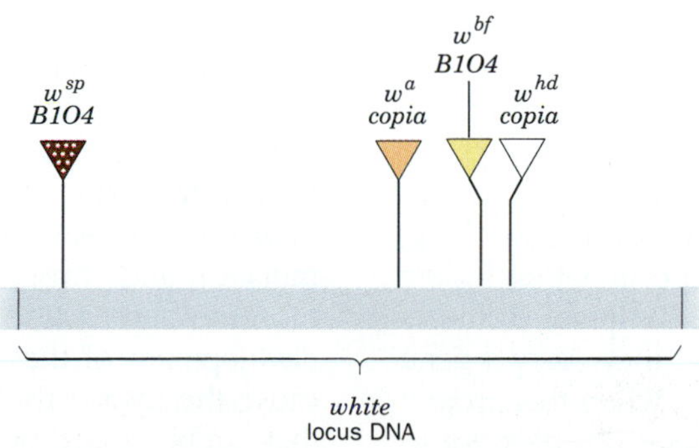

Figure 9.19 Retrotransposon insertion mutations of the *white* locus in *Drosophila*. The position and nature of each insertion is shown. *B104* and *copia* are different retrotransposons found in the *Drosophila* genome. The pigmentation in the eyes of each mutant is indicated by the color of the insertion. The identities of the mutant alleles are also given: w^{hd} = *white-hybrid dysgenesis* gives a bleach white phenotype; w^{bf} = *white-buff* gives a light yellow phenotype; w^{a} = *white-apricot* gives a yellow-orange phenotype; w^{sp} = *white-spotted* gives a dark, speckled phenotype. (After K. O'Hare, R. Levis, and G. M. Rubin, *Proc. Natl. Acad. Sci. U.S.A.* 80:6917–6921, 1983.)

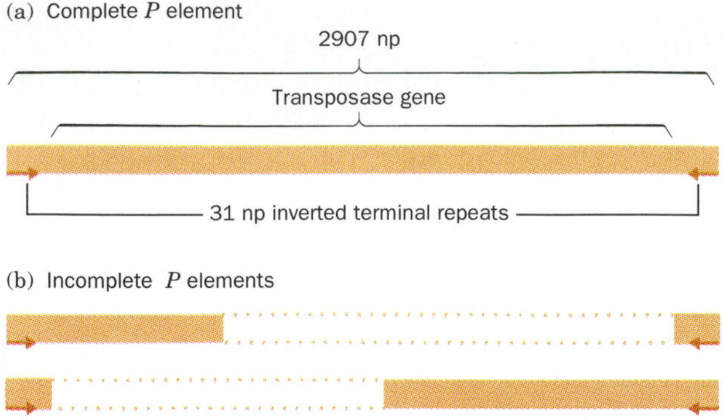

Figure 9.20 Structure of *P* elements in *Drosophila*. The orientations and lengths (in nucleotide-pairs, np) of DNA sequences are indicated. (a) A complete, or autonomous, *P* element. (b) Incomplete, or nonautonomous, *P* elements that were derived from complete elements by the deletion of internal sequences (gaps). (After K. O'Hare and G. M. Rubin, *Cell* 34: 25–35, 1983.)

sertion mutations in *Drosophila* are stable, a few revertants have been observed. For instance, a *gypsy* insertion mutation of the *cut* wings locus sometimes reverts to wild-type. Reversions of this allele are apparently associated with the excision of the *gypsy* element.

As with the *Ty* elements of yeast, transposition of the *Drosophila* retrotransposons seems to involve an RNA intermediate. The detailed mechanism is not known, but it is thought to resemble the process of retroviral infection. In some cases, there is evidence for the formation of viruslike particles, but to date no one has demonstrated that these are able to leave one cell and enter another one.

P Elements and Hybrid Dysgenesis

Some of the most extensive research on *Drosophila* transposons has focused on the members of the *P* element family (Fig. 9.20). These small transposons terminate in 31-nucleotide-pair inverted repeats and are flanked by 8-nucleotide-pair target site duplications.

The members of the *P* element family vary in size. The largest elements are 2907 nucleotide-pairs long, including the terminal inverted repeats but excluding the target site duplications. These complete elements are autonomously mobile because they carry a gene that encodes a transposase protein. When this protein attaches to the element, it can move the element to another position in the genome. Other *P* elements are structurally incomplete (Fig. 9.20*b*). Although such elements lack the ability to produce the transposase, they do possess the terminal and subterminal sequences that are needed for transposition. Consequently, these elements can be mobilized if a transposase-producing element is present somewhere in the genome.

Surveys of natural populations of *Drosophila* have demonstrated that there is considerable variation in the number of *P* elements present in the genome. Some flies have as many as 50, whereas others have only a few. Perhaps the most surprising discovery is that flies derived from strains captured before 1950 have no *P* elements at all. Detailed studies have suggested that these "empty" strains represent the primitive condition, and that *P* elements have invaded natural populations of *Drosophila* during recent times. In this context, it is interesting to note that the closest relatives of *D. melanogaster* have preserved the "empty" condition, whereas other, more distantly related species have acquired *P* elements. At the present time, it is not possible to say how these elements were introduced into so many different *Drosophila* species; however, some researchers have speculated that they gained entry by hitchhiking on viruses that naturally infect *Drosophila*. Such a process would be analogous to the transduction of *E. coli* cells by a bacteriophage that carried an IS element.

Populations of *Drosophila* that possess *P* elements have evolved mechanisms to regulate their movement. In some strains, this regulation depends on a maternally inherited property called the P cytotype. *Drosophila* with this condition repress *P* element movement more or less completely. This can be seen by crossing P cytotype flies with flies that have neither *P* elements nor the ability to regulate *P* element movement (Fig. 9.21). This absence of regulatory ability is called the M cytotype. Hybrids derived from a cross between P cytotype females and M cytotype males inherit *P* elements from their mothers; however, since they also inherit the P cytotype through the maternal cytoplasm, the movement of these elements is repressed. This is not the case with hybrids from the reciprocal cross, P cytotype ♂♂ × M cytotype ♀♀. Such hybrids do not inherit the P cytotype even though they do inherit *P* elements from their fathers. Consequently, the *P* elements present in these hybrids transpose freely, leading to a syndrome of genetic abnormalities called ***P–M hybrid dysgenesis***. This includes high frequencies of mutation and chromosome breakage, aberrant chromosome segregation, and, in extreme cases, faulty development of the gonads. This last condition may cause the hybrid fly to be sterile. Given the damage that can be caused by extensive *P* element movement, it may seem surprising that P ♂ × M ♀ crosses produce any viable progeny at all. However, the progeny of these crosses are reasonably healthy because *P* element movement is confined to the cells of the germ line. In the somatic tissues, where the mobilization of

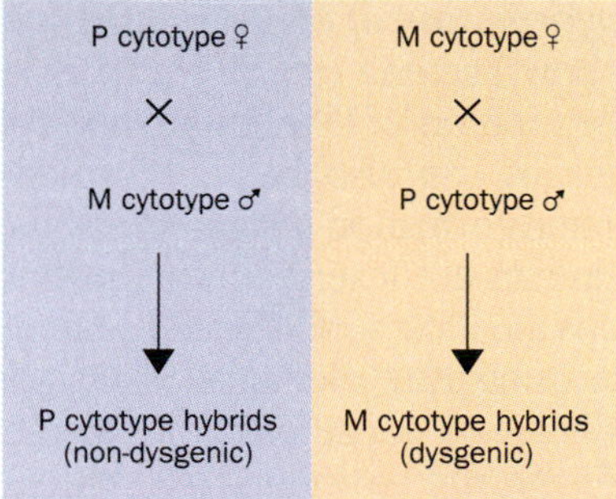

Figure 9.21 *P* element-mediated hybrid dysgenesis in *Drosophila*. *P* element activity is repressed by a cellular state called the P cytotype. Since this state is maternally inherited, the hybrid progeny of crosses between P cytotype females and M cytotype males do not experience much *P* element movement. However, the progeny of P cytotype males and M cytotype females do not inherit the P cytotype so they experience considerable *P* element activity. This activity leads to a syndrome of genetic abnormalities called hybrid dysgenesis.

P elements would certainly cause very serious problems, there is little, if any, transposition. This selective inhibition of transposition occurs because the transposase gene that is carried by complete *P* elements cannot be expressed in somatic tissues. Current research in several laboratories is attempting to determine why this is so.

THE GENETIC AND EVOLUTIONARY SIGNIFICANCE OF TRANSPOSABLE ELEMENTS

Mutation and Chromosome Breakage

There is little doubt that transposable elements are responsible for mutations in a wide variety of organisms. The best evidence for this comes from *Drosophila*, in which many mutant alleles have been shown to involve transposon insertions. However, experimental work with different kinds of transposable elements suggests that the occurrence of an insertion mutation is still a rather rare event, possibly because many transposon families are stringently regulated. When this regulation is upset, a burst of transposition may occur, causing many mutations simultaneously. This is apparently what happens when *P* elements are mobilized in dysgenic hybrids of *Drosophila*.

Transposable elements also produce chromosome breakage. This is demonstrated by the behavior of the double *Ds* elements in maize and by the *P* elements in *Drosophila*. In both cases, breaks can lead to the loss or rearrangement of chromosomal material. A full discussion of these structural aberrations is presented in Chapters 18 and 19.

Sometimes transposable elements mediate recombination events between DNA molecules. One example is the IS-mediated insertion of F plasmids into the *E. coli* chromosome. Another is the structural rearrangement of X chromosomes in *Drosophila* following recombination between homologous transposons that are located in different positions. J. K. Lim has found that one family of transposable elements (called *hobo*) apparently mediates these events, leading to the deletion or inversion of large segments of the chromosome. This and other findings suggest that transposons may play an important role in the evolution of chromosome structure.

Use in Genetic Analysis

The natural ability of transposable elements to cause mutations has been harnessed in the laboratory. ***In several organisms, it is now feasible to stimulate the transposition of a particular family of elements, thereby increasing the natural mutation rate.*** This procedure has an advantage over traditional methods of inducing mutations (see Chapter 11) because a transposable element that has caused a mutation by inserting into a gene can serve as a landmark for more detailed studies. This feature is best seen in *Drosophila*, in which the technique of *in situ* hybridization can be used to locate the site of a transposon insertion. In this technique, radioactively labeled transposon sequences are made single-stranded and then hybridized to single-stranded DNA in the giant chromosomes of the salivary glands. The hybridization reaction takes place on the surface of a microscope slide, where the chromosomes have been spread by squashing dissected glands. When the hybridization reaction is completed, the location of the radioactive sequences can be determined by autoradiography. Figure 9.22 shows a typical result. The chromosomes in this squash were hybridized with a *gypsy* element sequence, autoradiographed, and stained. Each of the dark spots in Fig. 9.22 indicates where the radioactive *gypsy* sequences have hybridized with the chromosomal DNA. These spots therefore identify the chromosomal sites that contain *gypsy* elements.

Genes that have been mutated by the insertion of a transposable element are said to have been "tagged." This word is used deliberately to convey the sense that the gene can be readily identified. When used together with gene cloning and colony or plaque hybridization procedures (Chapter 13), transposon tagging provides an extraordinarily useful way of identifying gene sequences in large, heterogeneous mixtures of DNA. It is therefore a standard technique in genetic engineering.

Transposable elements are also useful in the

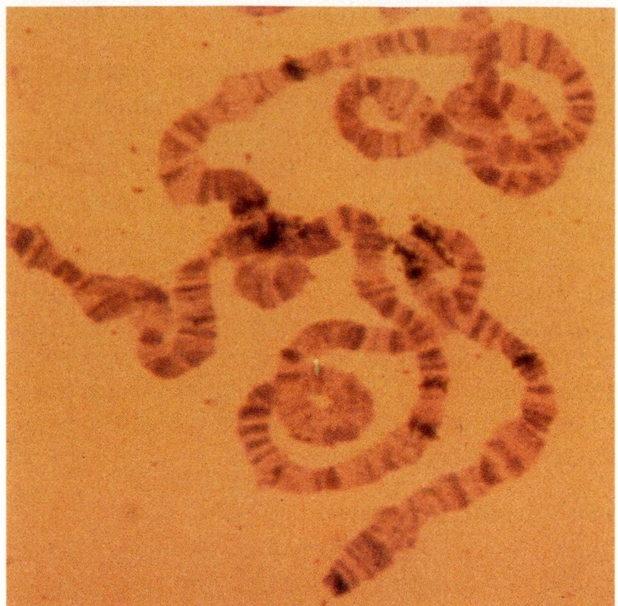

Figure 9.22 *In situ* hybridization with *gypsy* transposon sequences. Radioactively labeled *gypsy* DNA has been hybridized with DNA from the giant salivary gland chromosomes to produce this autoradiogram. The darkened regions over the chromosomes indicate where the *gypsy* sequences have bound to chromosomal DNA. Each darkened region therefore identifies the location of *gypsy* elements in the chromosomes. (Photo courtesy of Fang-miin Sheen, University of Minnesota.)

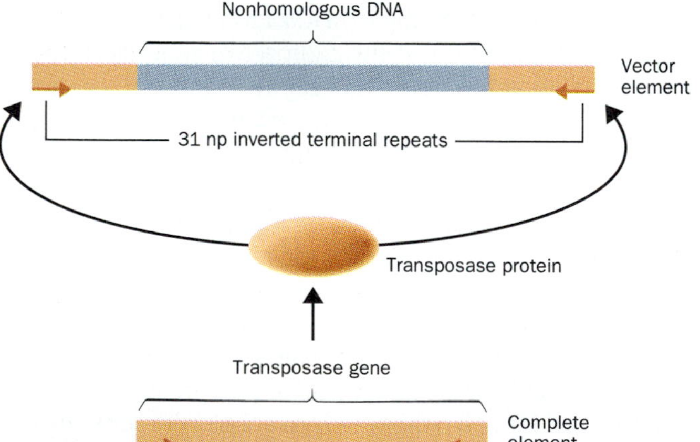

Figure 9.23 Genetic transformation of *Drosophila* by means of *P* element vectors. *P* element sequences (yellow) are combined with a sequence of non-*P* DNA to produce a vector *P* element. This element is then mixed with a complete *P* element and the mixture is injected into *Drosophila* embryos to produce genetic transformants. The complete element furnishes the transposase that is needed to insert the vector element into the chromosomes of an embryo.

genetic transformation of higher organisms. Chapter 8 discussed how bacteria can be transformed by the physical incorporation of DNA fragments. The cells of higher organisms can also be transformed, but the frequency of transformation is significantly increased if the DNA fragments are inserted into transposable elements. Perhaps the most sophisticated system has been developed by using the *P* elements of *Drosophila* (Fig. 9.23). In this system, a nonautonomous element serves as the ***transformation vector*** and a complete element serves as the source of the transposase that is needed to insert the vector into the chromosomes of a *Drosophila* cell. Practically any DNA sequence can be placed into the vector element (the term vector comes from the Latin word for "carrier" and is used because the vector element carries a nonhomologous sequence of DNA). The only concern in constructing the vector is to preserve the terminal and subterminal sequences of the *P* element that are needed for transposition. In a typical experiment, a mixture of the vector and complete elements is injected into very young *Drosophila* embryos. If the injection is not too traumatic, the embryos will survive and develop into healthy, fertile adults. During development, there is a chance that transposase from the complete element will catalyze the insertion of the vector element into one of the *Drosophila* chromosomes. If this event occurs in a cell that eventually gives rise to part of the adult germ line, the vector element may be passed on to the next generation, producing genetically transformed progeny. Using this technique, hundreds of *Drosophila* have been transformed, some carrying DNA from other organisms. Other aspects of the process of *P* element-mediated transformation are discussed in Chapter 24.

Evolutionary Issues

The widespread distribution of transposable elements suggests that they have played a role in evolution. One hypothesis is that these elements are nature's tools for genetic engineering. Their ability to copy, transpose, and rearrange other DNA sequences, such as genes for antibiotic resistance, can be construed as a benefit for the organisms that carry them. In this view, transposable elements have spread because they have given a selective advantage to their carriers. Another hypothesis is that transposable elements have spread simply because they have an ability to multiply independently of the normal replication machinery. This may not be true for all transposons, but it probably applies to some of them, such as the bacterial elements that transpose by a replicative mechanism. In this view, transposable elements are little more than genomic parasites—segments of DNA that replicate selfishly, possibly even to the detriment of their hosts.

How might the first transposons have evolved? N. Kleckner has suggested that a primordial transposon might arise by the modification of a gene encoding an enzyme for the creation and repair of DNA breaks. All that would be needed is for the enzyme to develop a modest degree of specificity, perhaps by recognizing a particular DNA sequence of six or eight nucleotide-pairs. Such a sequence might occur by chance in inverted orientation on either side of the gene, creating a situation in which the gene's product could interact with each of these flanking sequences. By "cutting and pasting" the DNA, this modified enzyme could then transpose the entire unit to a new position in the genome. Such a unit would therefore behave as a primordial transposon.

Other questions concern the evolutionary relationship between retrotransposons, such as the *Ty* elements of yeast, and full-fledged retroviruses. Collectively, these entities have been referred to as ***retroelements***. A. J. Kingsman and S.M. Kingsman have proposed that retroviruses have developed from the simpler retrotransposons by the addition of a gene (called *env*) that synthesizes a membrane protein. With this addition, the retroelement could produce a particle capable of escaping from one cell and entering another one. Such a particle would be infectious and would therefore provide the retroelement with the opportunity to transpose between genomes as well as within them.

SUMMARY

Transposable elements, or ***transposons***, are present in the genomes of a wide variety of organisms. These elements move from one position to another in the genome and are responsible for an assortment of ***genetic instabilities***, including ***mutation*** and ***chromosome breakage***. In bacteria, some transposons carry genes for antibiotic resistance. Eukaryotic transposons exhibit considerable diversity in structure and behavior. Some, such as the **Ty** elements of yeast, resemble the integrated forms of retroviruses and transpose by means of an RNA intermediate. These elements are therefore called ***retrotransposons***. Other eukaryotic transposons are organized into structurally heterogeneous families in which some elements are ***autonomously transposable*** but others are not. Examples are the **Ac/Ds** and **Spm/dSpm** families of maize and the **P** family of *Drosophila*. Transposable elements have been used in genetic analysis to ***induce mutations*** and to ***tag*** genes with a readily identifiable DNA sequence. They have also been used as vectors to ***transform*** experimental organisms genetically.

REFERENCES

BERG, D.E., L. JOHNSRUD, L. MCDIVITT, R. RAMABHADRAN, and B.J. HIRSCHEL. 1982. "Inverted repeats of Tn5 are transposable elements." *Proc. Natl. Acad. Sci. U.S.A.* 79: 2632–2635.

COHEN, S.N., and J.A. SHAPIRO. 1980. "Transposable genetic elements." *Sci. Amer.* 242(2): 40–49.

ENGELS, W.R. 1989. "*P* elements in *Drosophila*." In D.E. Berg and M.M. Howe, editors, *Mobile DNA*. American Society for Microbiology, Washington, DC.

FARABAUGH, P.J., and G.R. FINK. 1980. "Insertion of the eukaryotic transposable element Ty1 creates a 5-base pair duplication." *Nature* 286: 352–356.

FEDEROFF, N.V. 1989. "About maize transposable elements and development." *Cell* 56: 181–191.

GRINDLEY, N.D.F. 1983. "Transposition of Tn3 and related transposons. *Cell* 32: 3–5.

KINGSMAN, A.J., and S.M. KINGSMAN. 1988. "Ty: A retroelement moving forward." *Cell* 53: 333–335.

KLECKNER, N. 1981. "Transposable elements in prokaryotes." *Annu. Rev. Genet.* 15: 341–404.

LIM, J.K. 1988. "Intrachromosomal rearrangements mediated by hobo transposons in *Drosophila melanogaster.*" *Proc. Natl. Acad. Sci. U.S.A.* 85: 9153–9157.

MCCLINTOCK, B. 1950. "The origin and behavior of mutable loci in maize." *Proc. Natl. Acad. Sci. U.S.A.* 36: 344–355.

______. 1951. "Chromosome organization and genic expression." *Cold Spring Harbor Sympos. Quant. Biol.* 16: 13–47.

______. 1956. "Controlling elements and the gene." *Cold Spring Harbor Sympos. Quant. Biol.* 21: 197–216.

O'HARE, K., R. LEVIS, and G.M. RUBIN. 1983. "Transcription of the *white* locus in *Drosophila melanogaster.*" *Proc. Natl. Acad. Sci. U.S.A.* 80: 6917–6921.

O'HARE, K., and G.M. RUBIN. 1983. "Structure of *P* transposable elements and their sites of insertion and excision in the *Drosophila melanogaster* genome." *Cell* 34: 25–35.

PEREIRA, A., H. CUYPERS, A. GIERL, Z. SCHWARZ-SOMMER, and H. SAEDLER. 1986. "Molecular analysis of the *En/Spm* transposable element system of *Zea mays.*" *EMBO J.* 5: 835–841.

REZNIKOFF, W.S. 1982. "Tn5 transposition and its regulation." *Cell* 31: 307–308.

ROEDER, G.S., P.J. FARABAUGH, D.T. CHALEFF, and G.R. FINK. 1980. "The origin of gene instability in yeast." *Science* 209; 1375–1380.

SHAPIRO, J.A., editor. 1983. *Mobile genetic elements.* Academic Press, New York.

PROBLEMS AND QUESTIONS

9.1 It has been proposed that the *hobo* transposable elements in *Drosophila* mediate intrachromosomal recombination, that is, two *hobo* elements on the same chromosome pair and recombine with each other. What would such a recombination event produce if the *hobo* elements were oriented in the same direction on the chromosome? What if they were oriented in opposite directions?

9.2 The X-linked *singed* locus is one of several in *Drosophila* that controls the formation of bristles on the adult cuticle. Males that are hemizygous for a mutant *singed* allele have bent, twisted bristles that are often much reduced in size. Several *P* element insertion mutations of the *singed* locus have been characterized and some have been shown to revert to the wild-type allele by excision of the inserted element. What conditions must be present to allow such reversions to occur?

9.3 In maize, the recessive allele *bz* produces a lighter color in the aleurone than does the dominant allele, *Bz*. Ears on a homozygous *bz/bz* plant were fertilized by pollen from a homozygous *Bz/Bz* plant. The resulting cobs contained kernels that were uniformly dark except for a few on which light spots occurred. Suggest an explanation.

9.4 Which of the following pairs of DNA sequences could qualify as the terminal repeats of a bacterial IS element? (a) 5′-GAATCCGCA-3′ and 5′-ACGCCTAAG-3′, (b) 5-GAATCCGCA-3′ and 5′-CTTAGGCGT-3′, (c) 5′-GAATCCGCA-3′ and 5′-GAATCCGCA-3′, (d) 5′-GAATCCGCA-3′ and 5′-TGCGGATTC-3′. Explain.

9.5 Sometimes solo copies of the LTR of a retrotransposon, such as *gypsy*, are found in *Drosophila* chromosomes. How might these solo LTRs originate?

9.6 *Escherichia coli* cells that are resistant to kanamycin are infected with bacteriophage lambda. Phage particles that were purified from the lysed cells were used to infect a kanamycin-sensitive strain called C107 and cells that were lysogenized were picked from turbid plaques. These cells were subsequently plated on a medium containing kanamycin and several kanamycin-resistant colonies appeared. Suggest an explanation for the transfer of kanamycin resistance to the C107 strain.

9.7 Which of the following pairs of DNA sequences could qualify as target site duplications at the point of an IS*50* insertion? (a) 5′-AATTCGCGT-3′ and 5′-AATTCGCGT-3′, (b) 5′-AATTCGCGT-3′ and 5′-TGCGCTTAA-3′, (c) 5′-AATTCGCGT-3′ and 5′-TTAAGCGCA-3′, (d) 5′-AATTCGCGT-3′ and 5′-ACGCGAATT-3′. Explain.

Chapter 10

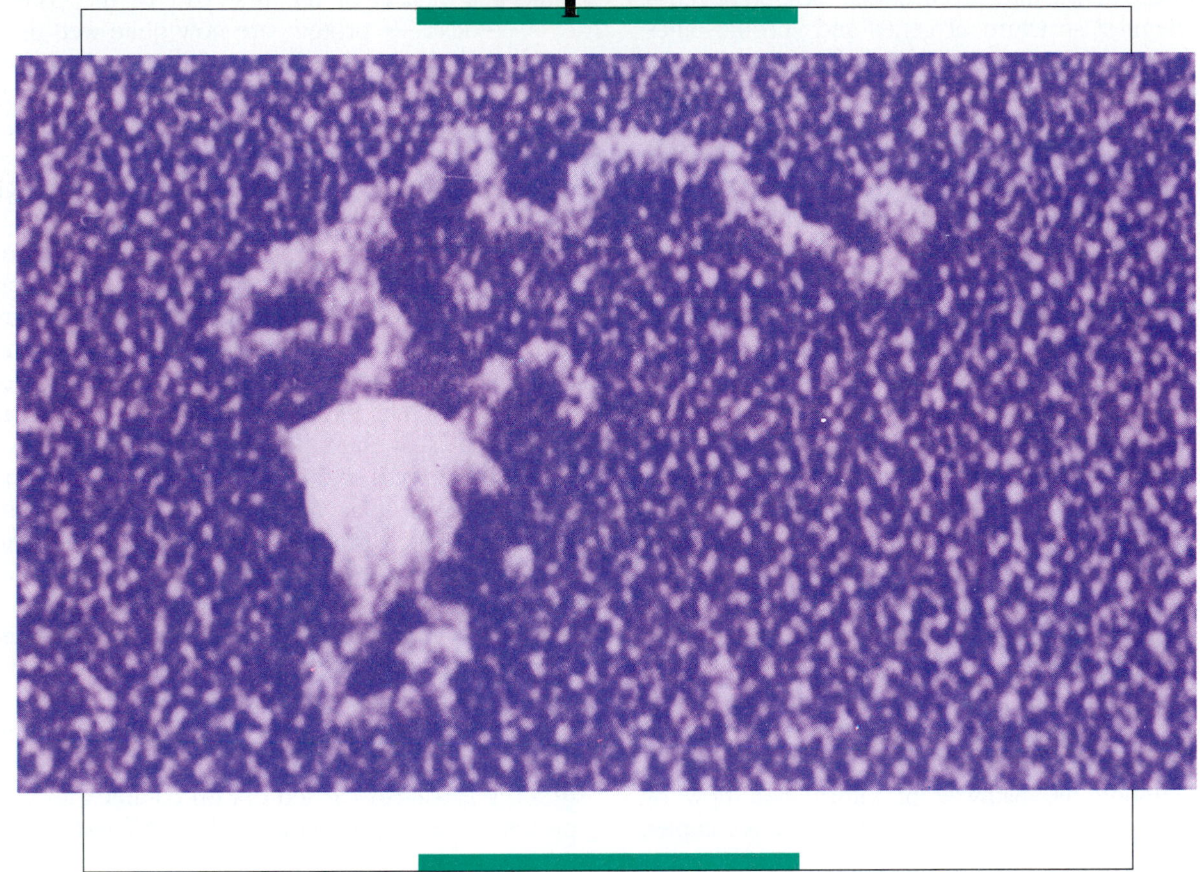

Gene Expression

(Photo above): Electron micrograph of a purified spliceosome processing a pre-mRNA molecule. Bar, 100 nm. (Reproduced with permission from R. Reed, J. Griffith, and T. Maniatis, *Cell* 53: 949–961, 1988. Copyright © 1988 by Cell Press; published by Cell Press. Original micrograph courtesy of J. D. Griffith, Department of Microbiology and Immunology, University of North Carolina Medical School, Chapel Hill.)

IN the preceding chapters, we have considered the patterns of transmission of independently assorting and linked genes, the chromosomal location of genes, the chemical structure of genes and chromosomes, and the mechanism of replication of the genetic material. In this chapter, we will discuss how genes perform their phenotypic functions. That is, how do genes exert their effects on the phenotype of a virus, a cell, or an organism? This question does not have a simple answer. Different genes clearly exert their effects in different ways. Moreover, all the genes of an organism are located in the same cells and in the same nuclei. They do not function independently. The ultimate phenotype of an organism is the product of the action of all the genes (including many gene-product interactions) and their interaction with the environment. The pathway of transfer or flow of information—from the gene to the final effect of the gene on the phenotype—is thus frequently very complex (Fig. 10.1).

Each gene also has an effect on the population of which the organism carrying the gene is a part; these effects will be discussed in Chapter 22. Ultimately, each gene has a potential effect, small though it may be, on the cumulative phenotype of our biosphere, as each gene may affect the ability of the organism, or the population, or the species to compete for an ecological niche in our biosphere (see Chapters 21 and 22).

Although the pathway of information flow by which a gene exerts its effect on phenotype is complex (Fig. 10.1), rather spectacular progress has been made in the last few decades in elucidating this pathway. The molecular details of the first part of the pathway, DNA ⟶ RNA ⟶ protein, are now quite well understood, and we are currently extending our knowledge of this pathway to the cellular level. However, this extension to the cellular level involves a large increase in organizational complexity (protein ⟶ cell) and promises to challenge several generations of biologists before all the details are clearly understood.

We know that the information stored in the form of the nucleotide-pair sequence in a gene is transferred by a process called ***transcription*** to an intermediary, a single-stranded ***messenger RNA (mRNA)*** molecule, which carries that information from the genes in the chromosomes to the ***ribosomes,*** the cytoplasmic sites of protein synthesis. Many nuclear genes of higher eukaryotes and some genes of lower eukaryotes and of cellular organelles contain ***noncoding sequences*** called ***introns*** or ***intervening sequences*** that separate ***coding sequences*** called ***exons*** of these genes. The entire sequences of these split genes are transcribed into ***primary transcripts*** called ***pre-messenger RNAs*** (***pre-mRNAs***) or sometimes ***heterogeneous nuclear RNAs*** (***hnRNAs***) (for nuclear transcripts). The noncoding intron sequences of these pre-mRNAs are subsequently removed by two-step ***splicing processes*** carried out on complex RNA- and protein-containing structures called ***spliceosomes.*** In-

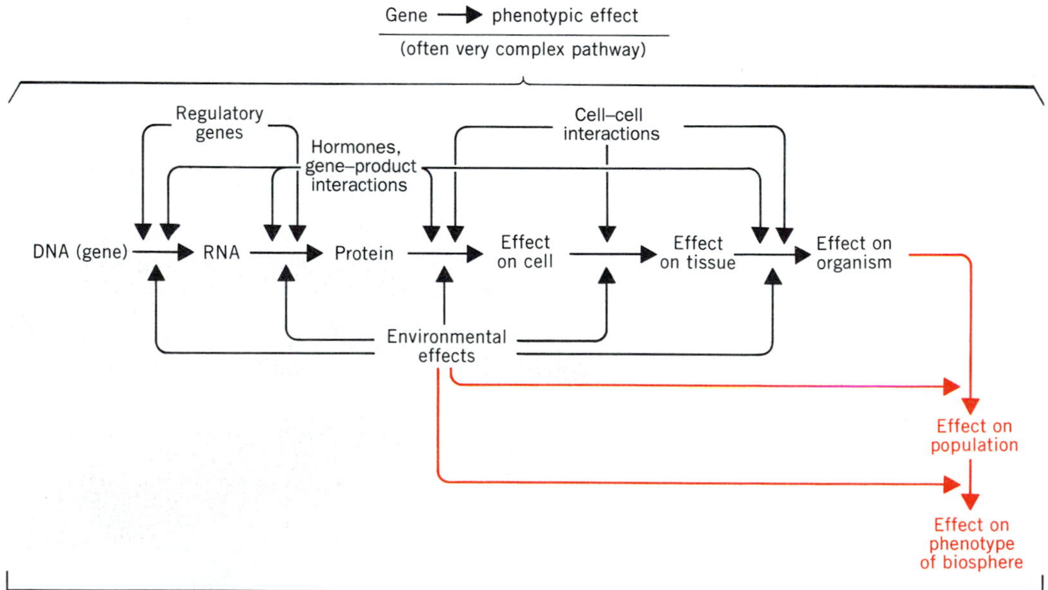

Figure 10.1 The complex pathway from the gene to its effect on the phenotype of a higher plant or animal and, ultimately, on the phenotype of the biosphere. Some of the known interactions involved in the frequently very complex pathway of gene expression in eukaryotes are indicated. The population and biopshere levels of gene effects are included for completeness.

tron sequences also occur in certain genes that produce important RNA molecules. These introns are removed from the primary transcripts of the genes by different splicing mechanisms; in some cases, the excision of the intron sequences is ***autocatalytic*** or ***"self-catalyzed."***

We also understand the most important features of the mechanism by which the information stored in the sequences of nucleotides in the mRNA molecules is converted, in a process called ***translation,*** into the sequences of amino acids in the protein gene-products. The ***genetic code*** that governs this translation process has been worked out in full. Finally, these protein gene-products (enzymes and structural proteins) are known to control the metabolic processes that occur in cells. Enzymes catalyze the numerous metabolic reactions that occur in living systems (see the section of this chapter entitled *Genetic Control of Metabolism*). Other proteins are important structural constituents of various cellular components, such as membranes, ribosomes, and chromosomes. In this chapter, we will focus on these two key processes, transcription and translation, on the properties of the genetic code, and on the overall genetic control of metabolism.

EVOLUTION OF THE ONE GENE–ONE POLYPEPTIDE CONCEPT

At the time of the discovery of Mendel's work in 1900, the English physician-biochemist Sir Archibald E. Garrod was studying several congenital metabolic diseases in humans. One of these was the inherited disease alkaptonuria, which is easily detected because of the blackening of the urine upon exposure to air. The substance responsible for this blackening is alkapton or homogentisic acid, an intermediate in the degradation of the aromatic amino acids tyrosine and phenylalanine (see Fig. 10.4). Garrod believed that the presence of alkapton or homogentisic acid in the urine was due to a block in the normal pathway of metabolism of this compound. Moreover, on the basis of the family pedigree studies, Garrod proposed that alkaptonuria was inherited as a single recessive gene. The results of Garrod's studies of alkaptonuria and a few other congenital diseases in humans, such as albinism, were presented in detail in the first edition of Garrod's book, *Inborn Errors of Metabolism,* published in 1909. The title of his book is clear evidence of Garrod's insight into the genetic control of metabolism. Although the details of the biochemical pathway affected by the recessive mutation responsible for alkaptonuria were not worked out until many years later, Garrod clearly understood the gene–metabolic reaction relationship. His concept might be best stated as ***one mutant gene–one metabolic block.*** Garrod's concept was the forerunner to the one gene–one enzyme concept and our present one gene–one polypeptide concept. Like Mendel's work, Garrod's results were largely unknown or ignored prior to their rediscovery many years later (in Garrod's case, by Beadle and Tatum in 1941).

As was discussed in Chapter 7, the fungus *Neurospora crassa* can grow on medium containing only (1) inorganic salts, (2) a simple sugar, and (3) one vitamin, biotin. Medium containing only these components is called "minimal medium." G. W. Beadle (Fig. 10.2) and E. L. Tatum reasoned that *Neurospora* must be capable of synthesizing all the other essential metabolites, such as the purines, pyrimidines, amino acids, and other vitamins, *de novo.* Furthermore, they reasoned that the biosynthesis of these growth factors must be under genetic control. If so, mutations in genes whose products are involved in the biosynthesis of essential metabolites would be expected to produce mutant strains with additional growth-factor requirements.

Beadle and Tatum tested this prediction by irradiating asexual spores (conidia) of wild-type *Neurospora,* with X rays or ultraviolet light, and screening the clones produced by the mutagenized spores for new

Figure 10.2 George Beadle, distinguished researcher, teacher of genetics, and university administrator. Dr. Beadle, along with E. L. Tatum and J. Lederberg, was Nobel Laureate in medicine and physiology in 1958. (Photo courtesy G. W. Beadle.)

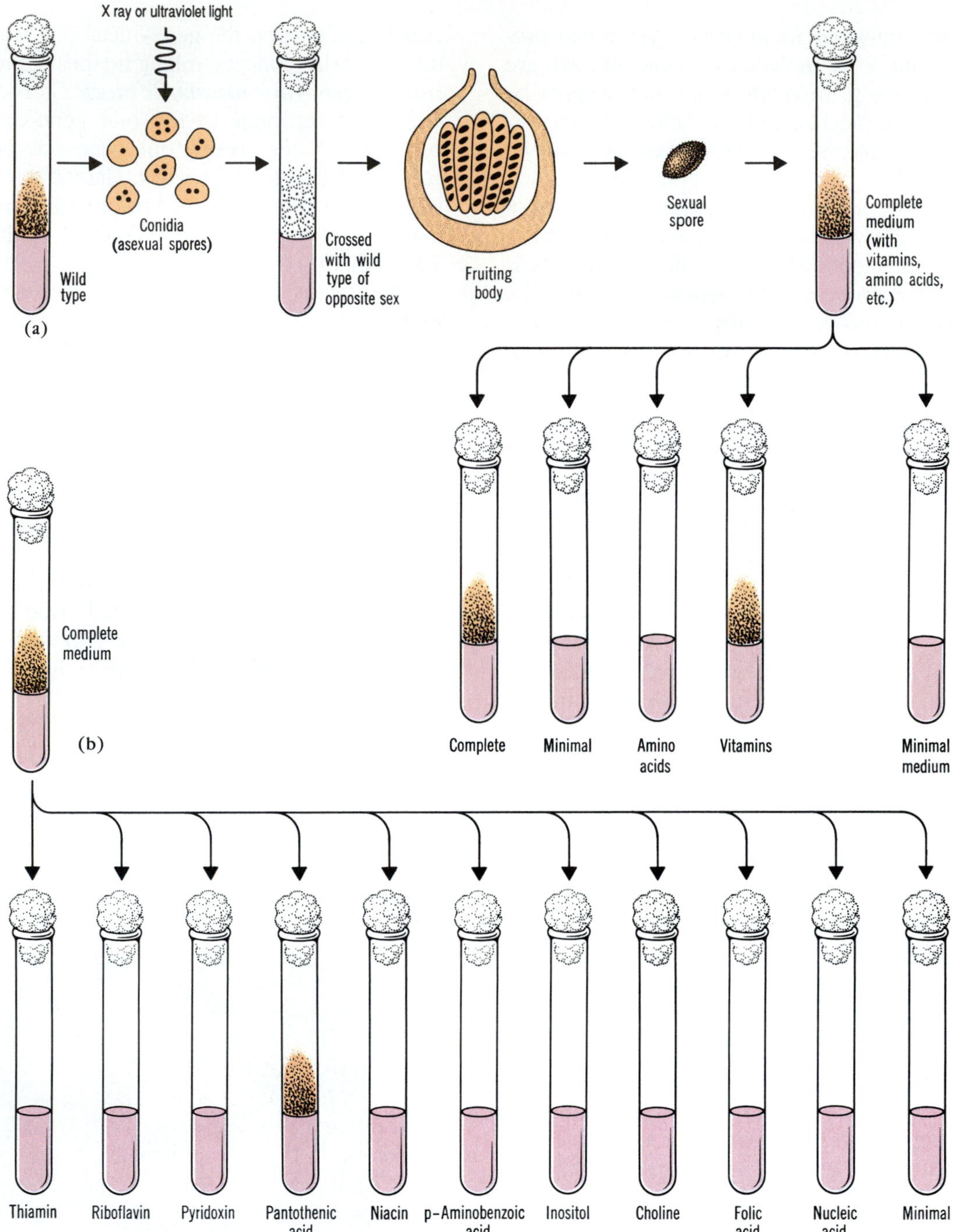

Figure 10.3 Diagram of the classic experiment of Beadle and Tatum, showing that one mutation results in a new requirement for one growth factor. *Neurospora* conidia were mutagenized with X rays or ultraviolet light and then screened for their ability to grow on "minimal medium" (containing only inorganic salts, a simple sugar, and biotin) and "complete medium" (supplemented with all the vitamins, purines, pyrimidines, and amino acids). Wild-type *Neurospora* will grow on either minimal or complete medium. Mutant strains that could grow on complete medium, but not minimal medium, were analyzed further to determine their exact growth-factor requirements (a and b). A backcross to wild type (shown at top) was done with each mutant clone to select mutant strains carrying only one mutation. Single mutants should yield a 1 : 1 mutant to wild type segregation pattern. Correlated biochemical analyses demonstrated that each mutation resulted in the loss or inactivation of one enzyme. (From G. W. Beadle, "Genes and the Chemistry of the Organism," *Amer. Scientist* 34: 31–53, 1946. Reprinted by permission of *American Scientist,* Journal of Sigma Xi, "The Scientific Research Society.")

growth-factor requirements (Fig. 10.3). Only mutant strains that yielded a 1 : 1 mutant to wild-type ratio when crossed to wild type were studied further, so as to select strains with a mutation in only one gene. Mutant clones that could grow on medium supplemented with all the amino acids, purines, pyrimidines, and vitamins (called "complete medium"), but could not grow on minimal medium, were further analyzed for their ability to grow on medium supplemented with just amino acids, or just vitamins, and so on (Fig. 10.3*a*). Strains that grew when just vitamins were added, for example, were in turn analyzed for their ability to grow on media supplemented with each of the vitamins separately (Fig. 10.3*b*). In this way, Beadle and Tatum demonstrated that each mutation resulted in a requirement for ***one*** growth factor. By correlating their genetic analyses with biochemical studies of the mutant strains, Beadle and Tatum demonstrated in several cases that one mutation resulted in the loss of one enzyme activity. This work, for which Beadle and Tatum received a Nobel Prize in 1958, was soon verified by similar studies of many other organisms in many laboratories. The ***one gene–one enzyme*** concept thus became a central tenet of molecular genetics.

Appropriately, in his Noble Prize acceptance speech, Beadle stated:

> In this long, roundabout way, we had discovered what Garrod had seen so clearly so many years before. By now we knew of his work and were aware that we had added little if anything new in principle . . . . Thus we were able to demonstrate that what Garrod had shown for a few genes and a few chemical reactions in man was true for many genes and many reactions in *Neurospora*. (Beadle, 1958)

Subsequently, many enzymes, as well as the hemoglobins (see Chapter 11, pp. 295–297), were shown to consist of two or more different polypeptide chains, and each polypeptide was found to be the product of a separate gene. Tryptophan synthetase of *E. coli,* for example (see Chapter 12, pp. 336–337), contains an α-polypeptide, the product of the *trpA* gene, and a β-polypeptide, the product of the *trpB* gene. It was necessary, therefore, to change the concept of one gene–one enzyme to ***one gene–one polypeptide.***

GENETIC CONTROL OF METABOLISM

Even if the one gene–one polypeptide concept requires modification or refinement in the future, it is clear that most genes exert their effects on phenotype via the polypeptides that they encode. Metabolism occurs by sequences of chemical reactions, each step of which is catalyzed by a specific enzyme. Each enzyme is, in turn, specified by one or more genes. The genetic control of a metabolic pathway may thus be diagrammed as shown at the bottom of this page. The number of steps in a pathway may vary from 2 to 10 or more.

Enzymes are very specific. They usually catalyze only one or a few very similar chemical reactions. This specificity depends on their unique, complex, three-dimensional structures, which in turn are determined by their primary structures (amino acid sequences). Most importantly, their amino acid sequences are controlled by the base-pair sequences in the structural genes (the genes coding for the particular polypeptide; see the section in this chapter entitled *Protein Synthesis*).

When mutations occur in genes that result in the synthesis of inactive or otherwise defective enzymes (see Chapter 11), they produce what Garrod so aptly called "inborn errors of metabolism." Hundreds of "inborn errors" or congenital diseases are now known in humans (a few of these are described in Chapter 4). We can illustrate the genetic control of metabolism and the consequences of gene defects in organisms by considering the metabolism of phenylalanine and tyrosine in humans (Fig. 10.4).

Alkaptonuria (Fig. 10.4*a*) is caused by a recessive mutation that results in the loss of activity of the enzyme homogentisic acid oxidase. In the absence of this enzyme, which catalyzes cleavage of the benzene ring of homogentisic acid, homogentisic acid accumulates and is excreted in the urine.

Another defect in phenylalanine–tyrosine metabolism occurs in individuals with the inherited disorder phenylketonuria (PKU; Fig. 10.4*b*). PKU results from a recessive mutation that causes a loss of phenylalanine hydroxylase activity. Phenylalanine hydroxylase converts phenylalanine to tyrosine, the first step in the catabolism of phenylalanine. As protein is consumed, phenylalanine accumulates in the blood of individuals

Precursor
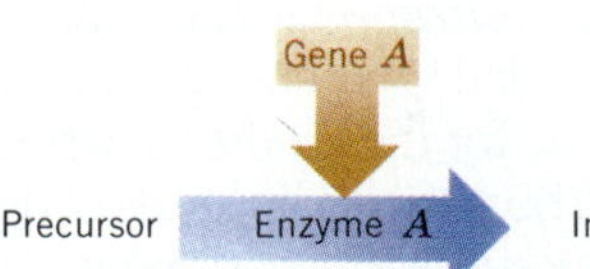

Intermediate 1
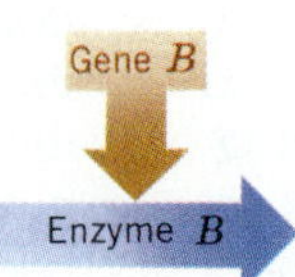

Intermediate 2
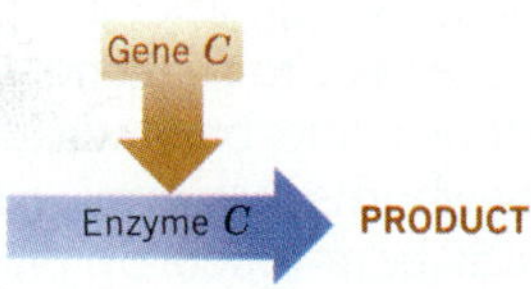

PRODUCT

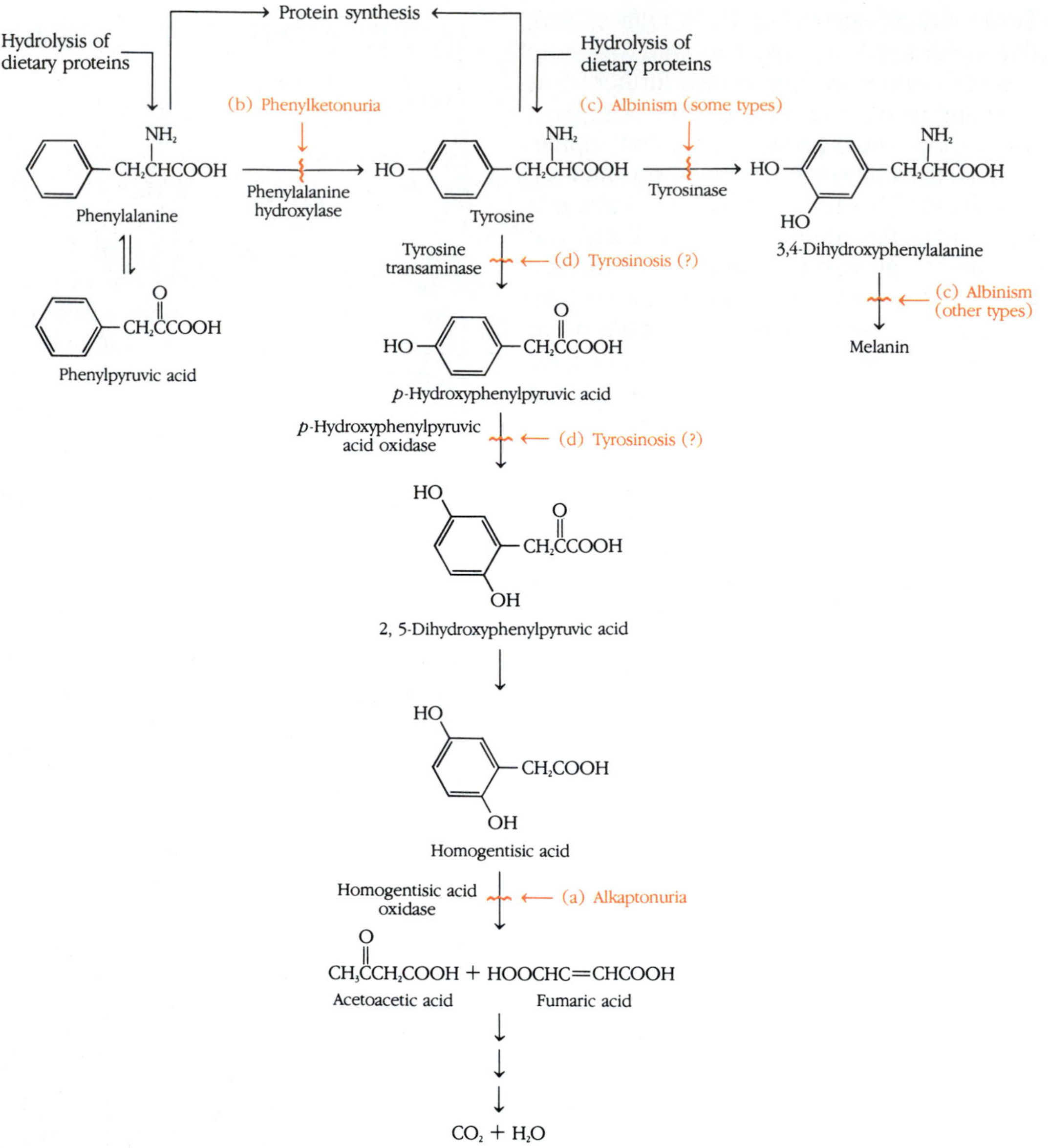

Figure 10.4 Inherited human diseases ("inborn errors") with defects in phenylalanine–tyrosine metabolism. (a) Alkaptonuria, extensively studied by Garrod at the turn of the century, which results in blackening and hardening of cartilage and blackening of urine upon exposure to air; (b) phenylketonuria, which results in severe mental retardation if untreated; (c) albinism, the inability to synthesize melanin pigments; and (d) tyrosinosis, the inability to convert tyrosine to 2,5-dihydroxyphenylpyruvic acid, an early stage in the tyrosine degradative pathway. At least two different types of albinism occur in humans. One type is caused by a mutation resulting in the loss of tyrosinase activity. A second type of albinism apparently results from a mutational block in a subsequent step in melanin synthesis or in the ability to deposit melanin pigment in melanocytes.

with PKU, sometimes up to 100 times the normal level. As a result, metabolic derivatives of phenylalanine, such as phenylpyruvic acid, are formed. Some of the derivatives are toxic to the central nervous system and produce irreversible brain damage. If PKU is diagnosed in newborns, however, subsequent mental retardation can be avoided by placing the child on a carefully controlled diet. Phenylalanine is an essential amino acid in humans; it can ***not*** be synthesized *de novo* as it can in *Neurospora.* Thus, individuals with PKU must be provided with enough phenylalanine in their diet to provide for the synthesis of body proteins, but not enough to build up toxic levels of derivatives of phenylalanine. The successful treatment of infants with

PKU to avoid mental retardation is one of the more dramatic achievements of the application of genetics and biochemistry in modern medicine. It should be emphasized, however, that successful treatment of PKU is not a "cure" of the disease. The defective gene is still present and is transmitted to the progeny.

Albinism (Fig. 10.4*c*) is another recessive condition caused by a defect in phenylalanine–tyrosine metabolism. In certain types of albinism, the enzyme tyrosinase is inactive or lacking, resulting in a block in the pathway of conversion of tyrosine to the dark-colored pigment melanin. Last, tyrosinosis (Fig. 10.4*d*) appears to result from a mutation in another gene coding for an enzyme involved in the degradation of phenylalanine and tyrosine.

Although we have focused here on the genetic control of the metabolism of phenylalanine and tyrosine, and the "inborn errors" resulting in blocks in this metabolic pathway, this picture of the genetic control of metabolism and the consequences of mutations is valid for all metabolic pathways in humans and in all other organisms. Moreover, given this picture, it should be obvious that epistatic interactions will be important components of the genetic control of most phenotypic traits.

PROTEIN SYNTHESIS

Most genes exert their effects on phenotype through the proteins (enzymes and structural proteins) whose structures they specify. Proteins are complex macromolecules that exhibit a high degree of functional specificity. A given enzyme, for example, will usually only catalyze one specific reaction. This explains why a given gene usually has a specific effect on the phenotype of an organism. The one primary effect of each gene may, however, lead to many secondary effects.

Proteins are composed of from one to several polypeptides, with each species of polypeptide being encoded by one gene (or, in a few cases, multiple copies of a redundant gene). Each polypeptide consists of a long sequence of amino acids linked together by ***peptide bonds.*** Twenty different amino acids are commonly found in natural proteins. Their structures (Fig. 10.5), with one exception, proline, include a free amino group and a free carboxyl group:

Amino group — H, H—N — C(H)(R) — C(=O)—OH — Carboxyl group

R ⟵ Side group

The amino acids differ from each other by the side groups that are present (Fig. 10.5). During protein synthesis, the amino acids become covalently linked by peptide bonds formed by hydrolysis from the amino and carboxyl groups (see bottom of this page).

A ***peptide*** is a compound composed of two or more amino acids. ***Polypeptides*** are long sequences of amino acids, ranging in length from 51 amino acids in insulin to over 1000 amino acids in the silk protein fibroin. Given the 20 different amino acids commonly found in polypeptides, the number of different polypeptides that are possible is enormous. Calculate, for example, the number of different sequences that can occur in a polypeptide that is 100 amino acids long (20^{100})!

The sequence of amino acids in a polypeptide is called its ***primary structure.*** Proteins, however, have highly varied and very complex three-dimensional structures, a few of which have been determined by X-ray diffraction studies. The ***secondary structure*** of a polypeptide refers to the spatial interrelationships of

$$H_2N-C(H)(R_1)-C(=O)-OH + H-N(H)-C(H)(R_2)-C(=O)-OH \rightarrow H_2N-C(H)(R_1)-C(=O)-N(H)-C(H)(R_2)-C(=O)-OH + H_2O$$

Peptide bond

Dipeptide

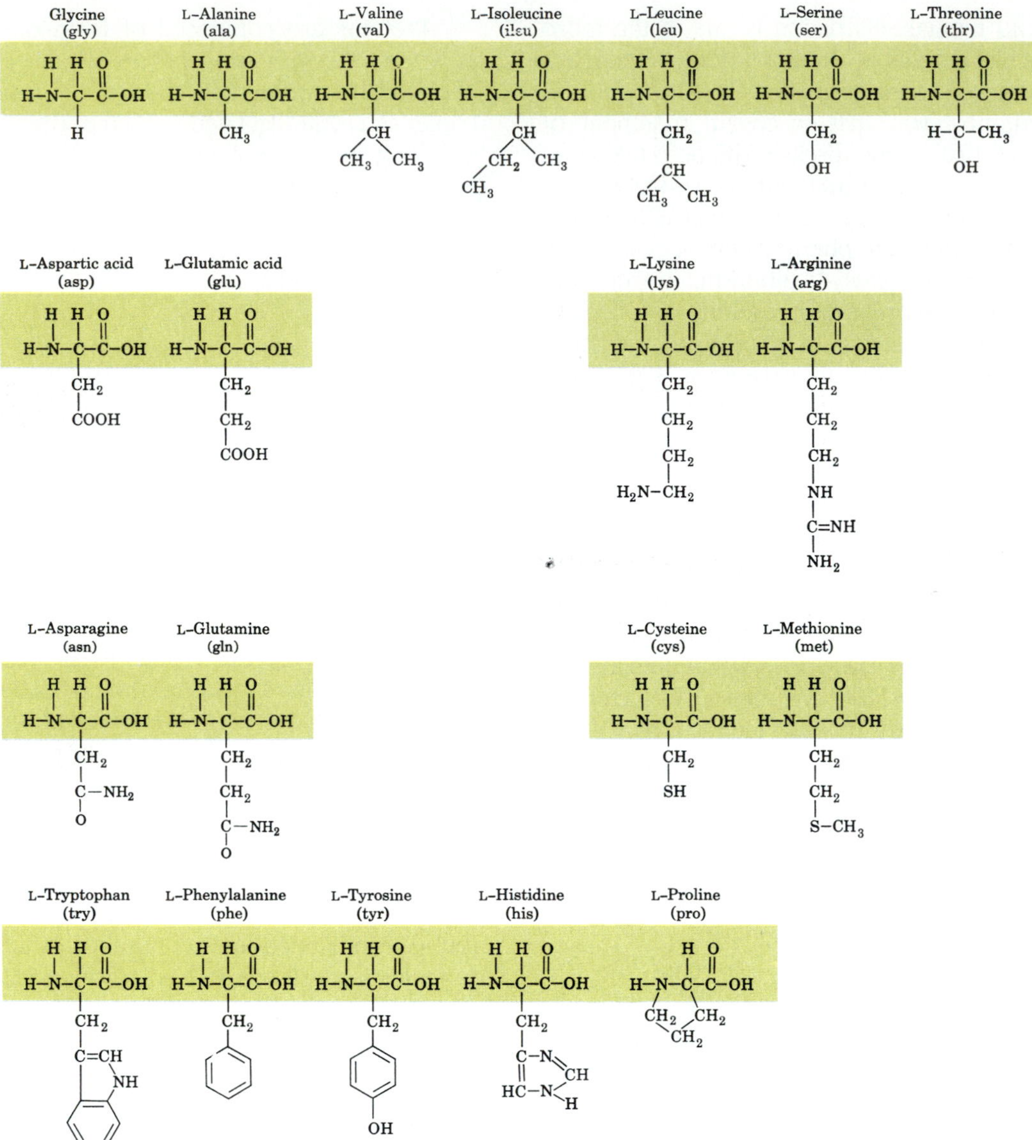

Figure 10.5 Structures of the 20 amino acids commonly found in proteins, arranged in groups with similar chemical properties of the side groups (shown below the olive areas). The amino and carboxyl groups, which participate in peptide bond formation during protein synthesis, are shown in the olive areas. The commonly used abbreviation is shown below the name of each amino acid.

the amino acids in segments of the polypeptide. For example, a segment of a polypeptide may exhibit a helical arrangement referred to as an α-helix. The ***tertiary structure*** of a protein refers to the folding of the polypeptide in three-dimensional space. Last, ***quaternary structure*** refers to the association of two or more polypeptides in a multimeric protein.

Hemoglobin provides an excellent example of the complexity of proteins, exhibiting all four levels of structural organization (Fig. 10.6). It consists of four polypeptides, two identical α-chains and two identical β-chains, all of which are highly folded. In most if not all proteins, the secondary, tertiary, and quaternary structures are determined by the primary structure(s) of the polypeptide(s) involved. Therefore, for the rest of this chapter, we will concern ourselves with the mechanism by which genes control the primary structure of polypeptides.

The ***central dogma*** of molecular genetics is that genetic information is transferred (1) from DNA to DNA (replication; see Chapter 5) during its transmission from generation to generation and (2) from DNA to RNA to protein during its phenotypic expression in an organism (Fig. 10.7). (Recently, genetic information has also been shown to be transferred from RNA to DNA in provirus formation by RNA tumor viruses.) The

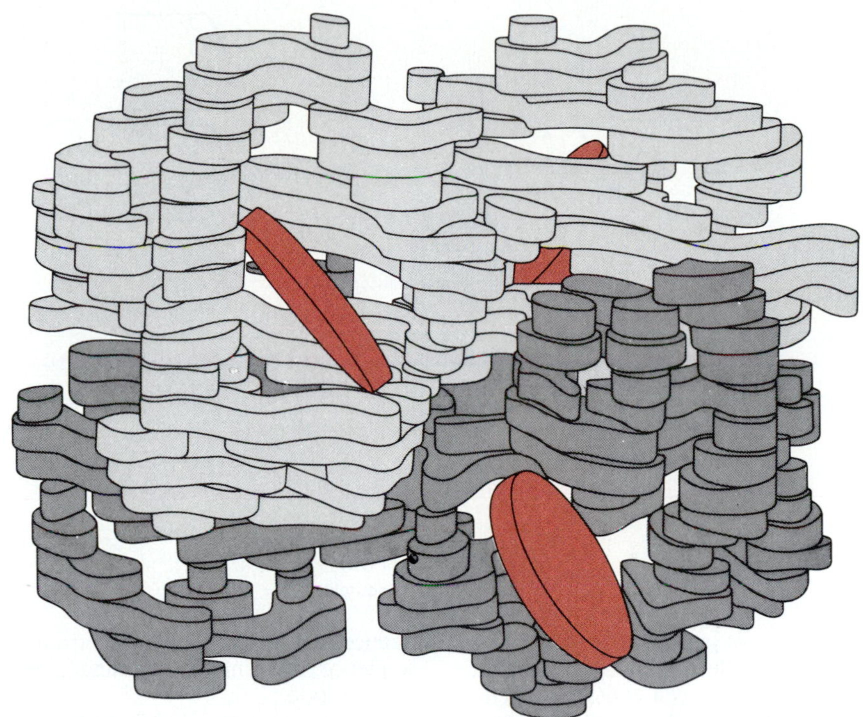

Figure 10.6 The three-dimensional structure of hemoglobin as deduced from X-ray diffraction studies. The hemoglobin molecule consists of two α-polypeptides (shown in light gray), two β-polypeptides (shown in dark gray), and four heme groups (shown as flat disks; only three are visible from this side view). (Reproduced with permission from M. F. Perutz, "The Hemoglobin Molecule," *Sci. Amer.* 211(5): 64–67. Copyright © 1964 by Scientific American, Inc. All rights reserved.)

transfer of genetic information from DNA to RNA to protein, or protein synthesis, involves (1) ***transcription,*** the transfer of the genetic information from DNA to RNA, and (2) ***translation,*** the transfer of information from RNA to protein.

Transcription

In eukaryotic organisms, the chromosomal genes, consisting of DNA, remain in the nuclei of cells, whereas proteins are synthesized in the cytoplasm. DNA ***cannot,*** therefore, serve directly as a template for protein synthesis. Instead, one strand of DNA, called the ***sense strand,*** is used as a template for the synthesis of a complementary strand of RNA, called ***messenger RNA*** (or ***mRNA***) or ***pre-messenger RNA*** (if it requires processing prior to translation), in a process called ***transcription*** (Fig. 10.8). The transcribed (sense) strand of two different genes, even adjacent genes, is not always the same strand. However, for a given gene only one strand is usually transcribed. The mRNA then carries the genetic information from its site of synthesis in the nucleus to the sites of protein synthesis, the ribosomes in the cytoplasm. The synthesis of mRNA (or pre-mRNA) in the nucleus and its subsequent transport to the cytoplasm can be documented by pulse-labeling experiments, pulse-chase labeling experiments, and autoradiography (Fig. 10.9). If a cell is exposed to a radioactive RNA precursor (such as [^{3}H]uridine or [^{3}H]cytidine) for a few minutes, and the intracellular location of the incorporated radioactivity is determined by autoradiography, almost all the nascent, labeled RNA is found in the nucleus (Fig. 10.9*a*). If, on the other hand, the short exposure ("pulse") to the labeled RNA precursor is followed by a period of growth in nonradioactive medium (a "pulse-chase experiment") before doing the autoradiography, most of the incorporated radioactivity is found transported to the cytoplasm (Fig. 10.9*b*). [Pulse-chase experiments are done by either (1) sedimenting the cells from the radioactive medium or (2) adding a vast excess of nonradioactive precursor so as to dilute the radioactive precursor to negligible concentrations.]

Strong evidence for an RNA intermediary in protein synthesis also resulted from studies on T2 phage-infected *E. coli* cells. The phage proteins were shown to be synthesized on ribosomes present in the cell prior to infection. The specificity determining the amino acid sequences of polypeptides was not therefore an integral part of ribosome structure. E. Volkin

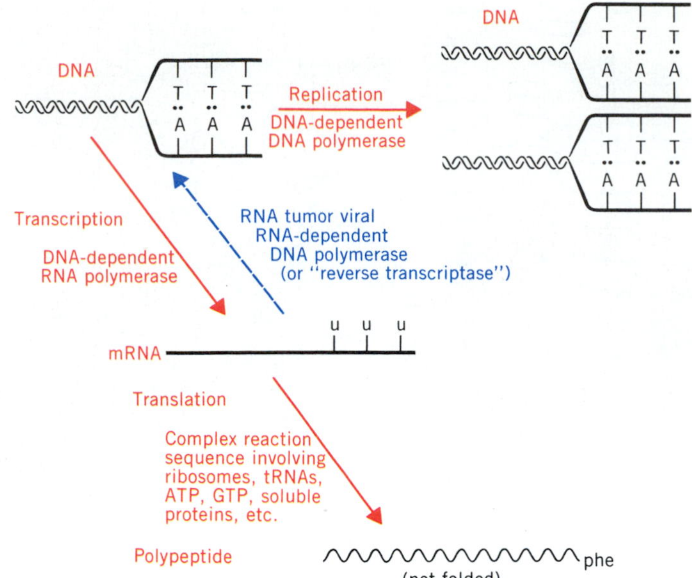

Figure 10.7 Routes of transfer of genetic information according to the central dogma of molecular biology: (1) DNA to DNA during transmission from one cell and/or one generation to the next, and (2) DNA→ mRNA→ protein during gene expression. The segment of three base-pairs on the right end of each DNA molecule is enlarged to show the base-pair sequence. During transcription, the bottom (AAA-containing) strand of DNA is used as a template for the synthesis of a complementary (UUU-containing) molecule of mRNA. Transcription is catalyzed by DNA-dependent RNA polymerase. During translation, the sequence of nucleotides in mRNA is decoded in a complex sequence of reactions to produce the polypeptide gene-product. The amino acid sequence of a polypeptide is determined by the nucleotide sequence of the mRNA according to the specifications of the genetic code.

More recently, genetic information has also been shown to be transferred from RNA to DNA (dashed line) in cells infected with RNA tumor viruses. The genetic information stored in the RNA of these tumor virus genomes is converted to DNA form before it can be stably integrated into the chromosomes (DNA) of the host cells as a provirus. These RNA tumor viruses thus direct the synthesis of an RNA-dependent DNA polymerase (also called a reverse transcriptase, since it catalyzes the reaction that is the reverse of transcription) to carry out the RNA to DNA conversion.

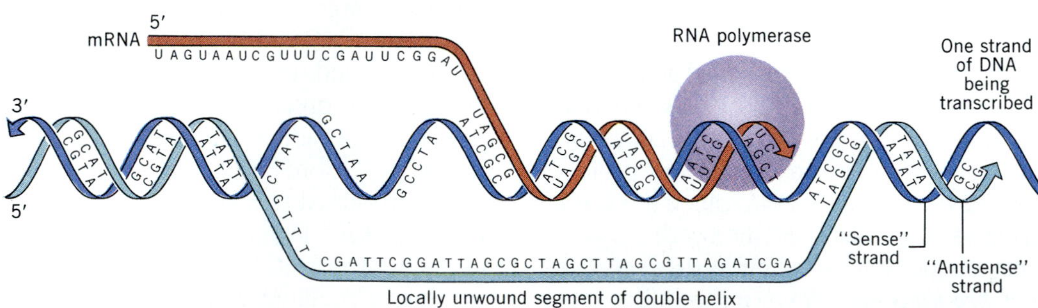

Figure 10.8 Transcription, or messenger RNA synthesis. The "sense" strand of DNA functions as a template for the synthesis of a complementary molecule of RNA (messenger RNA or mRNA). The process is catalyzed by the enzyme DNA-dependent RNA polymerase, usually shortened to just RNA polymerase. RNA synthesis, like DNA synthesis, always occurs in a 5′ to 3′ direction. Base sequences are shown along the DNA and mRNA strands to illustrate their complementary nature. (Recall that uracil is present in RNA at positions where thymine would occur in DNA). Because of the complementarity of the two strands of DNA, their base sequences cannot evolve independently. It is not surprising, therefore, that only one of the two strands, the sense strand, of any given gene specifies the amino acid sequence of the polypeptide gene-product. The other, nontranscribed, strand is usually referred to as the "antisense" strand. The sense and antisense strands of DNA play equally important roles in replication; however, only the sense strand is involved in transcription.

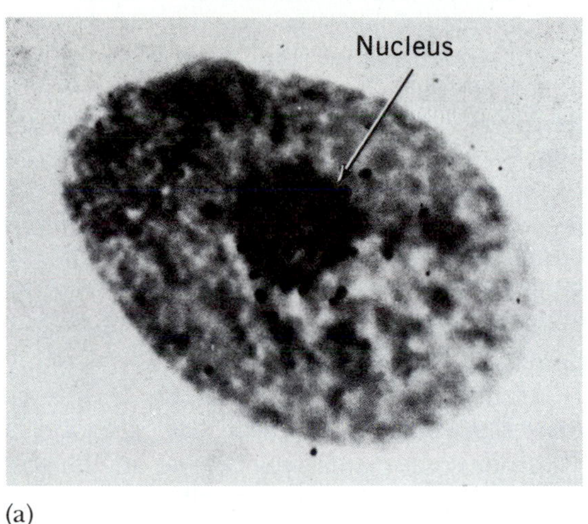

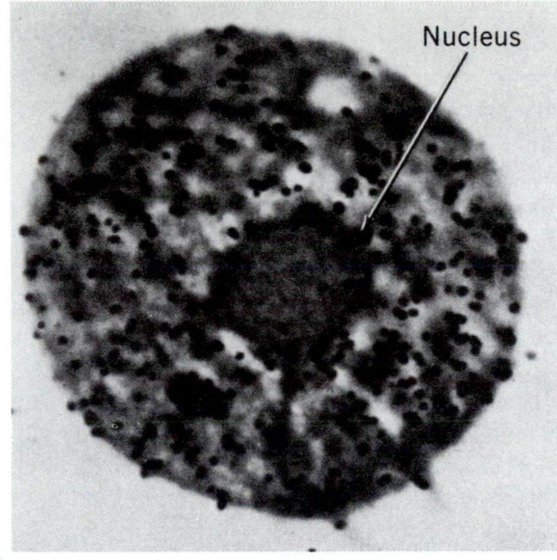

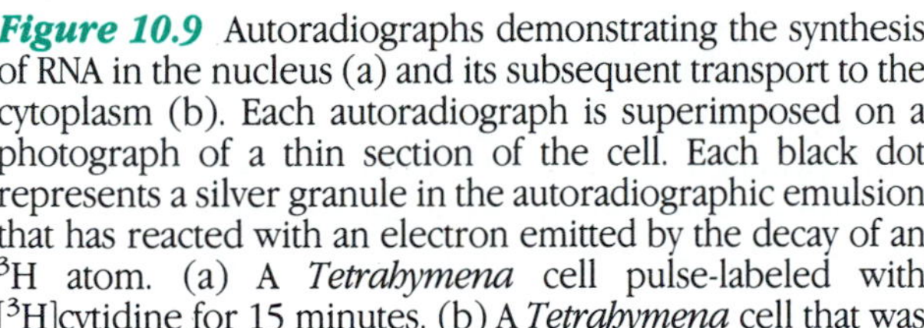
Figure 10.9 Autoradiographs demonstrating the synthesis of RNA in the nucleus (a) and its subsequent transport to the cytoplasm (b). Each autoradiograph is superimposed on a photograph of a thin section of the cell. Each black dot represents a silver granule in the autoradiographic emulsion that has reacted with an electron emitted by the decay of an ^3H atom. (a) A *Tetrahymena* cell pulse-labeled with [^3H]cytidine for 15 minutes. (b) A *Tetrahymena* cell that was grown on nonradioactive medium for 88 minutes after exposure to [^3H]cytidine for 12 minutes ("pulse-chase" experiment). Note that most of the label is present in the nucleus of the cell shown in a, whereas the nucleus of the cell shown in b is almost free of radioactivity, all the labeled RNA molecules having been transported to the cytoplasm. (From D. M. Prescott, "Cellular Sites of RNA Synthesis," *Prog. Nucleic Acid Res. Mol. Biol.* 3: 33–57, 1964.)

and L. Astrachan demonstrated that there is a large burst of RNA synthesis shortly after T2 phage infection, and these short-lived (half-lives of only a few minutes) RNA molecules have nucleotide compositions like the T2 phage DNA and not like the host DNA. Shortly thereafter, these unstable RNA molecules were shown to be complementary to segments of strands of DNA in the phage chromosome. Many different phage mRNA molecules have now been isolated and shown to direct the synthesis of specific phage proteins by *in vitro* protein synthesis (see the following section entitled *Translation*).

The DNA-dependent RNA polymerases that catalyze transcription are usually complex, multimeric proteins. The *E. coli* RNA polymerase, the most extensively studied of all the RNA polymerases, has a molecular weight of about 490,000 and consists of six polypeptides (Fig. 10.10). Two of these are identical; thus

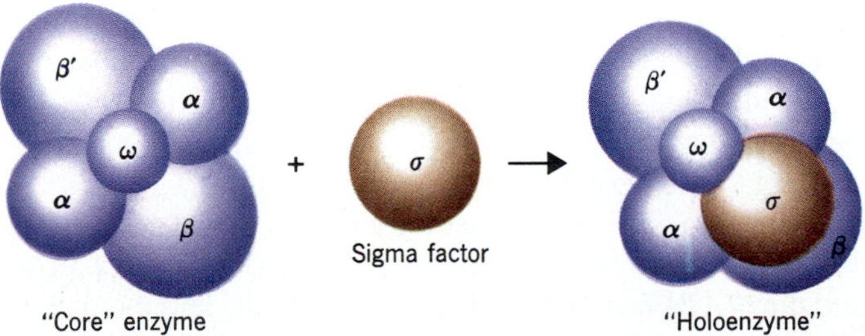

Figure 10.10 Schematic illustration of the complex structure of the RNA polymerase of *E. coli.* The "core" enzyme, which catalyzes covalent chain extension, consists of two α-polypeptides (each of molecular weight about 41,000), one β-polypeptide (molecular weight about 155,000), one β′-polypeptide (molecular weight about 165,000), and one ω-polypeptide (molecular weight about 12,000). The "holoenzyme" contains, in addition, a σ-polypeptide (molecular weight about 95,000). The presence of the sigma factor is required for initiation at the proper transcription initiation (or promotor) sites. After each RNA chain-initiation event, sigma is released, and the core enzyme catalyzes chain elongation.

the enzyme contains five distinct polypeptides. One of these subunits, the sigma (σ) factor, is only involved in the initiation of transcription; it does not have a catalytic function. The complete RNA polymerase molecule, called the "holoenzyme," contains two α-polypeptides and one polypeptide of each of the following types: β, β', ω, and σ. After initiation of the synthesis of an RNA chain, the σ-factor is released and chain elongation is catalyzed by the so-called core enzyme, which has the composition 2α, β, β', ω. The function of sigma is to recognize and bind RNA polymerase to the correct initiation sites, the ***promoter*** sites (see Chapter 14), on the DNA. RNA core polymerase (sigma absent) will catalyze RNA synthesis from DNA templates *in vitro,* but, in so doing, it initiates at random sites on both strands of DNA. The holoenzyme (sigma present), on the other hand, initiates *in vitro* only at sites used *in vivo.*

Hundreds of promoters have now been sequenced in *E. coli,* and these sequences have surprisingly little in common. Two short sequences within these promoters are sufficiently conserved to be recognized, but even these are seldom identical in two different promoters. The midpoints of the two conserved sequences occur at about 10 and 35 nucleotide-pairs, respectively, before the transcription–initiation site. Thus, they are often called the ***"–10 sequence"*** and the ***"–35 sequence,"*** respectively. The most common or ***consensus*** –10 sequence (also called the ***"Pribnow box"***) is ***TATAAT.*** The consensus –35 sequence (also called the ***"recognition sequence"***) is ***TTGACA.***

In eukaryotes, there are three different RNA polymerases, I, II, and III. RNA polymerase I is located in the nucleolus and catalyzes the synthesis of rRNA. RNA polymerases II and III are present in the nucleoplasm (outside the nucleolus). RNA polymerase III transcribes the genes for small nuclear RNAs and tRNAs. Its binding sites are located within these genes, rather than on the 5′ or "upstream" sides of the genes.

RNA polymerase II transcribes the majority of the nuclear structural genes; it is responsible for pre-mRNA synthesis. Comparisons of the sequences of the promoters or RNA polymerase II-binding sites of over 200 different eukaryotic genes reveal consensus sequences located at about 25 and 75 nucleotide-pairs, respectively, before the transcription–start site. The consensus sequences for the ***"–25 sequence"*** (also called the ***"Hogness box"*** and the ***"TATA box"***) and the ***"–75 sequence"*** (also called the ***"CAAT box"***) are ***TATAAAA*** and ***GGCCAATCT,*** respectively.

The mechanism of RNA synthesis is analogous to DNA synthesis (see Chapter 5) except that (1) the precursors are ribonucleoside triphosphates, (2) only limited segments of single strands are copied, and (3) the complementary RNA is released from the template as it is synthesized. Covalent extension occurs, as with DNA synthesis (see Chapter 5), by the addition of mononucleotides to the 3′ end of the chain, with the release of pyrophosphate.

The termination of transcription occurs at specific ***terminator sequences*** in DNA. Most prokaryotic mRNAs terminate with the sequence 5′UUUUUUA-3′, suggesting that the sequence 3′-AAAAAAT-5′ in the sense strand of DNA is at least part of the transcription terminator sequence (see Chapter 14, Fig. 14.11*a*). Some transcription–termination signals require the presence of a protein called ***rho,*** whereas others do not.

Although only one of the two DNA strands is transcribed in any given region, both strands of DNA in a chromosome usually participate in transcription, with some mRNAs being transcribed from one strand and other mRNAs (from different genes) being transcribed from the other strand. Because the two strands of a DNA double helix have opposite polarity, transcription events using opposite strands as templates will proceed in opposite direction along the DNA molecule. Even in simple viruses such as λ and T4 (see Chapter 12, Fig. 12.11), transcription occurs off both DNA strands (but only rarely in the same region).

Translation

The ***process during which the genetic information*** (which is stored in the sequence of nucleotides in an mRNA molecule) ***is translated, following the dictations of the genetic code, into the sequence of amino acids in the polypeptide gene-product is complex,*** requiring the functions of a large number of macromolecules. These include (1) over 50 polypeptides and from 3 to 5 RNA molecules present in each ribosome (the exact composition varies from species to species), (2) at least 20 amino acid-activating enzymes (aminoacyl-tRNA synthetases), (3) from 40 to 60 different tRNA molecules, and (4) at least 9 soluble proteins involved in polypeptide chain initiation, elongation, and termination. Since many of these macromolecules, particularly the components of the ribosome, are present in large quantities in each cell, the translation system makes up a major portion of the metabolic machinery of each cell.

An overview of protein synthesis, illustrating its complexity and the major macromolecules involved, is presented in Fig. 10.11. The translation process occurs on ***ribosomes,*** complex macromolecular structures located in the cytoplasm. Translation involves three types of RNA, all of which are transcribed from DNA templates (chromosomal genes). In addition to mRNA (see the preceding section entitled *Transcription*), 3 to 5 RNA molecules (***rRNA molecules***) are present as part

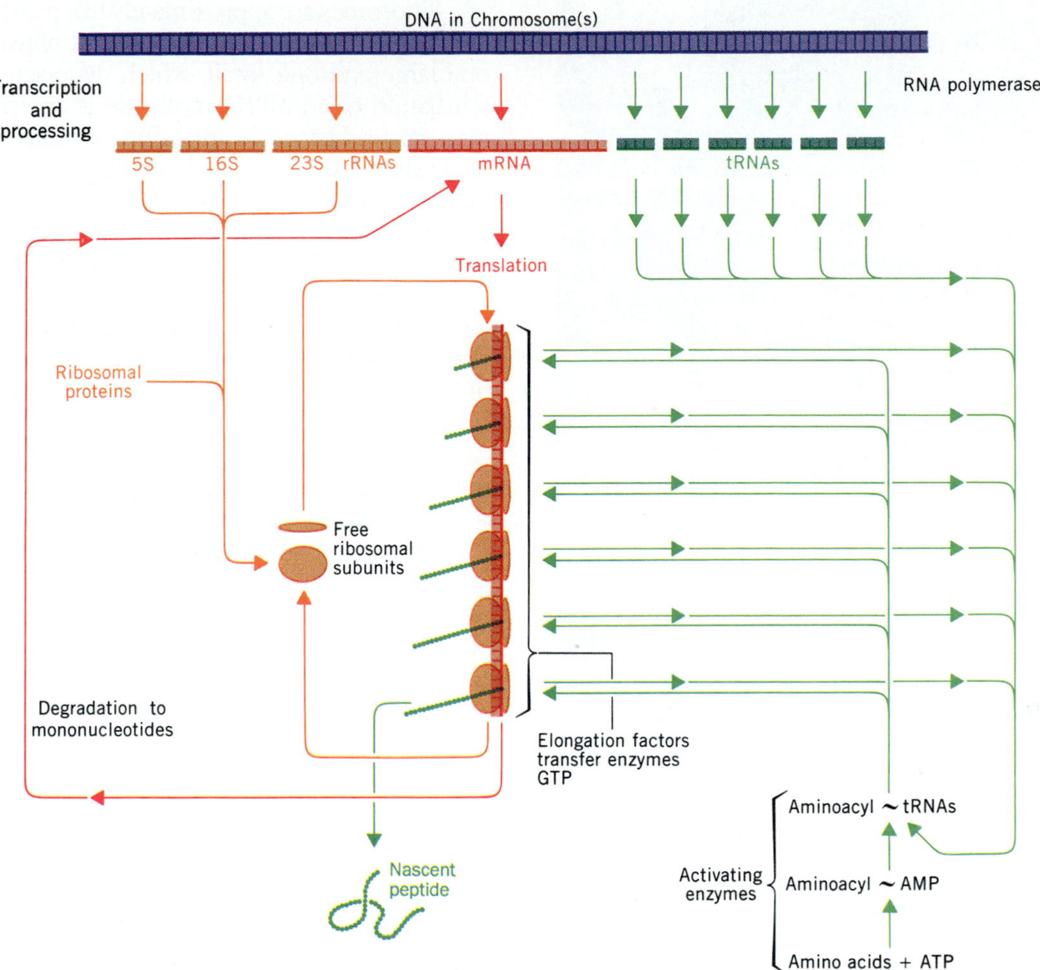

Figure 10.11 Diagrammatic overview of protein synthesis showing (1) the origin of the three types of RNA (rRNA, mRNA, and tRNA) by transcription from a DNA template, (2) translation of the mRNA on polyribosomes (several ribosomes simultaneously translating a single mRNA molecule), (3) the formation of ribosomal subunits from ribosomal RNAs and proteins, and the ribosome cycle (shown in orange), (4) the synthesis and degradation of mRNA molecules (cycle shown in red), (5) the formation of aminoacyl-tRNAs (catalyzed by activating enzymes) and the release of the tRNA after each peptide bond is formed (cycle shown in green), and (6) the covalent extension and eventual release of the nascent polypeptide chain. For simplicity, all the different species of RNA are shown being transcribed from contiguous segments of a single DNA molecule. In different organisms, the various RNA molecules are actually transcribed from genes located at different positions on from one to many chromosomes.

of the structure of each ribosome, and 40–60 small (70–80 nucleotides) RNA molecules ***tRNA molecules***) function as adaptors mediating the incorporation of the proper amino acids in response to specific codons in mRNA. The ribosomes may be thought of as workbenches, complete with machines and tools needed to make a polypeptide. They are nonspecific in the sense that they can synthesize any polypeptide (any amino acid sequence) specified by a particular mRNA molecule. Given this superficial overview of protein synthesis, we will now examine more closely some of the more important components and steps involved in the translation process.

Early studies using pulse-labeling (with radioactive amino acids) and autoradiography showed that proteins are synthesized largely in the cytoplasm on small, but complex, macromolecular structures called ribosomes. In prokaryotes, the ribosomes are distributed throughout the cells; in eukaryotes, they are located in the cytoplasm, frequently on an extensive intracellular membrane network called the ***endoplasmic reticulum*** (Fig. 10.12).

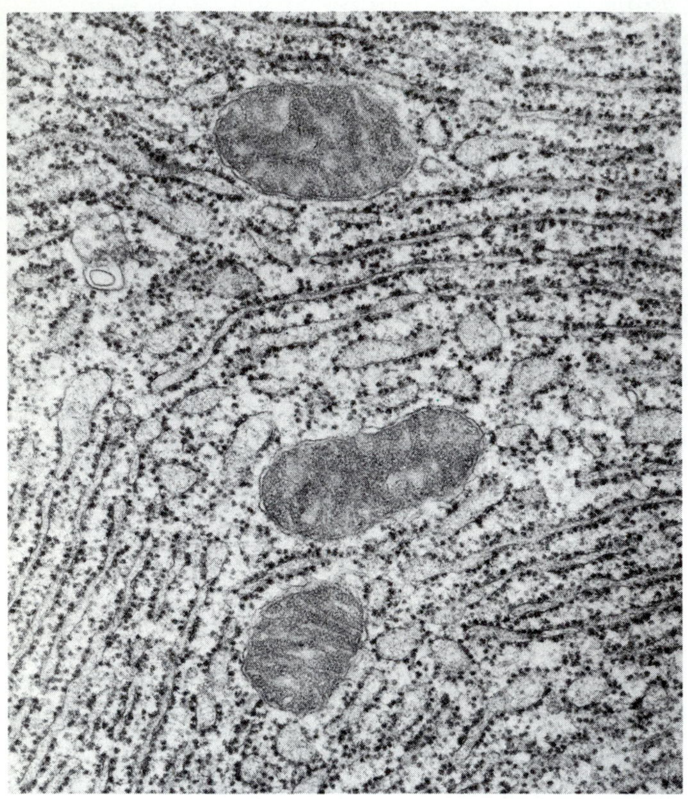

Figure 10.12 Electron micrograph of a thin section of a secretory cell from rat pancreas. Ribosomes (the small, densely stained structures) are seen associated with membranous sheets of endoplasmic reticulum (sectioned in a crosswise manner). Three mitchochondria are also present. Magnification ×28,900. (Courtesy of Ross Johnson, Department of Genetics and Cell Biology, University of Minnesota.)

Ribosomes are approximately half protein and half RNA (Fig. 10.13). They are composed of two subunits, one large and one small, which dissociate when the translation of an mRNA molecule is completed; they reassociate during the initiation of translation. Ribosome sizes are most frequently expressed in terms of their rates of sedimentation during centrifugation, in units called ***S*** or ***Svedberg units.*** The *E. coli* ribosome, like those of most prokaryotes, has a molecular weight of 2.7×10^6 and a "size" of "70S." The ribosomes of eukaryotes are larger (usually about 80S); however, size varies from species to species. The ribosomes present in organelles (mitochondria and chloroplasts) of eukaryotic cells are smaller (usually about 60S).

In all cases, each ribosome consists of two subunits. In the case of *E. coli,* the small (30S) ribosomal subunit contains a 16S (mol. wt. about 6×10^5) RNA molecule plus 21 different polypeptides, and the large (50S) subunit contains two RNA molecules (5S, mol. wt. about 4×10^4, and 23S, mol. wt. about 1.2×10^6) plus 31 polypeptides. In eukaryotic ribosomes, the small subunit contains an 18S (average size) RNA molecule and the large subunit contains a 5S, a 5.8S, and a 28S (average size) RNA molecule. *Drosophila* ribosomes, but not those of several other eukaryotes examined, also appear to contain a small 2S RNA molecule. In organelles, the corresponding rRNA sizes are 5S, 13S, and 21S.

In the case of *E. coli,* M. Nomura and his colleagues have been able to completely dissociate the 30S ribosomal subunit into the individual macromolecules and then reconstitute functional 30S subunits from the components. This has allowed them to study the function(s) of individual RNA and protein molecules.

The ribosomal RNA molecules are transcribed from a DNA template, just like mRNA molecules. In eukaryotes, however, rRNA synthesis occurs in the nucleolus and is catalyzed by a special RNA polymerase

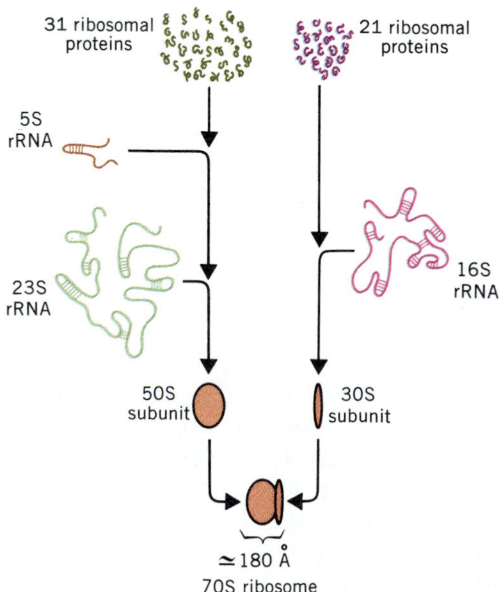

Figure 10.13 Macromolecular composition of the *E. coli* ribosome. All other ribosomes, prokaryotic and eukaryotic, have a similar structure and composition, although their sizes, and the sizes and number of individual components, vary. The ribosomal RNA molecules exhibit considerable secondary structure because of intrastrand hydrogen bonding. Their actual structures within the ribosome are unknown. The molecular weights of the 5S, 16S, and 23S rRNA molecules are 4×10^4, 6×10^5, and 1.2×10^6, respectively. The complete 70S ribosome has a molecular weight of about 2.7 million.

present only in the nucleolus. Moreover, transcription of the rRNA genes produces rRNA precursors that are larger than the RNA molecules found in ribosomes. These rRNA precursors undergo posttranscriptional processing to produce the rRNA molecules involved in translation. In *E. coli,* the rRNA gene transcript is a 30S precursor, which undergoes cleavage by endoribonucleases to produce the 5S, 16S, and 23S rRNAs plus a 4S transfer RNA molecule (Fig. 10.14). In eukaryotes, the 2S (when present), 5.8S, 18S, and 28S rRNAs are cut from a 40S to 45S (depending on the species) precursor, whereas the 5S rRNA is produced by posttranscriptional processing of a separate gene transcript.

In addition to the posttranscriptional cleavages of rRNA precursors, posttranscriptional methylation of many of the nucleotides in rRNA occurs. Presumably, the methylation protects the rRNA molecules from intracellular ribonucleases that are involved in mRNA degradation; its exact function is not yet clear, however.

Multiple copies of the genes for rRNA are present in the genomes of all organisms studied to date. This is not surprising considering the large number of ribosomes present per cell. In *E. coli,* there are estimated to be 5–10 copies of the rRNA (*rrn*) gene, with at least one copy at each of three distinct sites on the chromosome. In eukaryotes, the rRNA genes are present in hundreds to thousands of copies. The 5.8S–18S–28S rRNA genes of eukaryotes are present in tandem duplication in the ***nucleolar organizer*** regions of the chromosomes. In some eukaryotes, such as maize, there is a single pair of nucleolar organizers (on chromosome pair 6 in maize). In *Drosophila* and the extensively studied South African clawed toad, *Xenopus laevis,* the sex chromosomes carry the nucleolar organizers. Humans, on the other hand, have five pairs of nucleolar organizers, located on the short arms of chromosomes 13, 14, 15, 21, and 22. Careful studies indicate that there are about 500 copies of the 5.8S–18S–28S rRNA gene per nucleolar organizer in *Xenopus laevis.* Similar levels of redundancy have been estimated to occur in several other animals. Plants exhibit a greater variation in rRNA gene redundancy, with several thousand copies present in some genomes. Intraspecies variation in the amount of rRNA gene redundancy has also been documented in several species.

The 5S rRNA genes in eukaryotes are not located in the nucleolar organizer regions. Instead, they are usually distributed over several chromosomes. They are, however, highly redundant, like the 5.8S–18S–28S rRNA genes.

Although the ribosomes provide the workbenches and much of the machinery required for protein synthesis, and the specifications for each polypeptide are encoded in an mRNA molecule, the translation of a coded mRNA message into a sequence of amino acids in a polypeptide requires one additional class of RNA molecules, the ***transfer RNA (tRNA)*** molecules. Chemical considerations suggest that a direct interaction between the amino acids and the nucleotide sequences or ***codons*** in mRNA is unlikely. (A ***codon*** is a nucleotide sequence in mRNA that specifies the incorporation of one amino acid.) In 1958, Crick therefore proposed that some kind of an "adapter" molecule mediates amino acid codon recognition during protein synthesis. The "adapter" molecules were soon identified and found to be small (4S, 70–80 nucleotides long) RNA molecules. These molecules, first called "soluble RNA" (sRNA) molecules and subsequently transfer RNA (tRNA) molecules, contain a triplet base sequence, called the ***anticodon*** sequence, which is complementary to and recognizes the ***codon*** sequence in mRNA during translation. There are from one to four known tRNAs for each amino acid.

The amino acids are attached to the tRNAs by high-energy (very reactive) bonds between the carboxyl groups of the amino acids and the 3′-hydroxyl termini of the tRNAs. These reactive ***aminoacyl tRNAs***

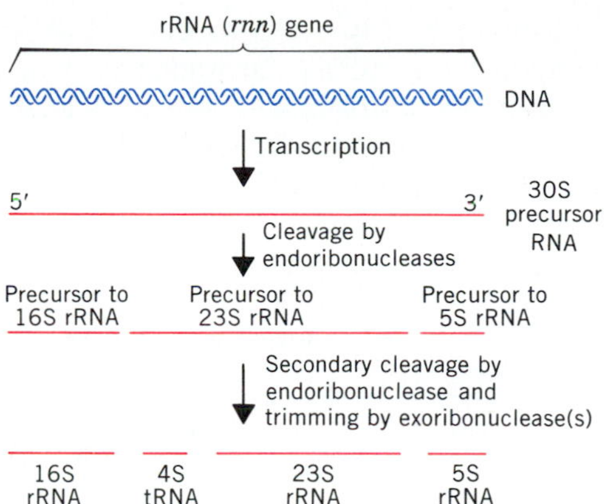

Figure 10.14 Transcription of the *E. coli* rRNA (*rrn*) genes to produce the 30S rRNA precursor and its subsequent processing (cleavage and trimming) to form the 5S, 16S, and 23S rRNA molecules, plus one 4S tRNA molecule. The 30S gene transcript is rapidly cleaved (in fact, intact 30S rRNA precursors can only be isolated from ribonuclease III-deficient mutant strains of *E. coli*) to three smaller molecules, one containing the 5S rRNA sequence, another containing the 16S rRNA sequence, and the third containing the 23S rRNA sequence. These three intermediates are then further processed to produce the mature RNA molecules present in the ribosome. During the second stage of rRNA maturation, a 4S tRNA sequence is cleaved from the intermediate precursor to either the 16S or the 23S rRNA.

are formed in a two-step process, both steps being catalyzed by a specific "activating enzyme" or ***aminoacyl-tRNA synthetase.*** There is at least one aminoacyl-tRNA synthetase for each of the 20 amino acids. The first step in aminoacyl-tRNA synthesis involves the ***activation*** of the amino acid using energy from adenosine triphosphate (ATP):

amino acid + ATP

aminoacyl-tRNA synthetase ↓

amino acid ~ AMP + Ⓟ ~ Ⓟ

The amino acid ~ AMP intermediate is not normally released from the enzyme before undergoing the second step in aminoacyl-tRNA synthesis, namely, the reaction with the appropriate tRNA:

amino acid ~ AMP + tRNA

aminoacyl-tRNA synthetase ↓

amino acid ~ tRNA + AMP

The aminoacyl~ tRNAs (amino acid ~ tRNAs) are the immediate precursors of polypeptide synthesis on ribosomes, with each activated tRNA recognizing the correct mRNA codon and presenting the amino acid in a steric configuration (three-dimensional structure) that facilitates peptide bond formation.

The tRNAs are transcribed from chromosomal genes. As in the case of rRNAs, the tRNAs are transcribed in the form of larger precursor molecules that undergo posttranscriptional processing (cleavage, trimming, methylation, etc.). The mature tRNA molecules contain several nucleosides not present in mRNA or in the primary tRNA gene transcripts. These unusual nucleosides, such as inosine, pseudouridine, dihydrouridine, 1-methylguanosine, and several others, are produced by posttranscriptional, enzyme-catalyzed modifications of the four nucleosides incorporated into RNA during transcription.

Because of their small size (70–80 nucleotides long), tRNAs have been more amenable to structural analysis than the other, larger molecules of RNA involved in protein synthesis. The complete nucleotide sequence and proposed "cloverleaf" structure of the alanine tRNA of yeast (Fig. 10.15) was published by R. W. Holley and colleagues in 1965; Holley shared the 1968 Nobel Prize in physiology and medicine for this work. Since then, many tRNAs have been sequenced, and the yeast alanine tRNA gene has even been synthesized *in vitro* from mononucleotides by H. G. Khorana (another 1968 Nobel Prize winner; in Khorana's case, for work on the nature of the genetic code) and coworkers. The three-dimensional structure of the

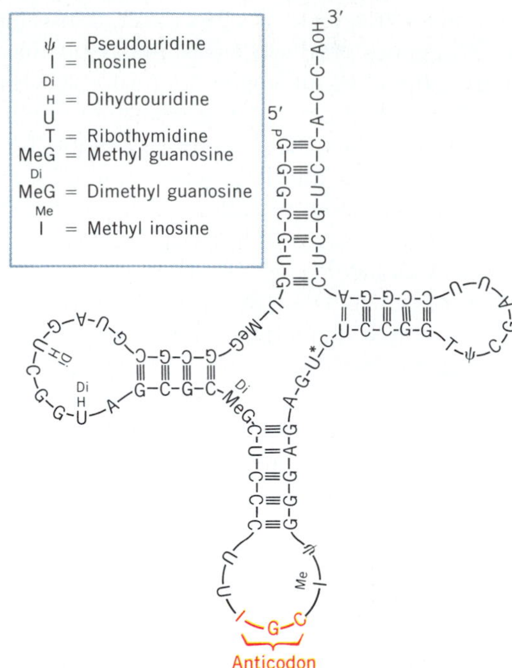

Figure 10.15 Nucleotide sequence and cloverleaf configuration of the yeast alanine tRNA. The secondary structure results from hydrogen bonding between bases in different segments of the molecule as shown. The structure contains three loops, within which no base-pairing is possible. One of these loops contains the anticodon sequence; in the case of the yeast alanine tRNA, the sequence is 3′-CGI-5′. The names of the modified nucleosides present in the yeast alanine tRNA are shown in the inset. (Adapted from R. W. Holley *et al., Science* 147: 1462–1465, 1965. Copyright © 1965 by the American Association for the Advancement of Science.)

phenylalanine tRNA of yeast has been determined by X-ray diffraction studies (Fig. 10.16). The anticodons of the alanine (Fig. 10.15) and phenylalanine (Fig. 10.16) tRNAs of yeast occur within a loop (nonhydrogen-bonded region) near the center of the molecule. In fact, the anticodons of all the tRNAs sequenced to date (over 70 from all organisms) have been found within comparably located anticodon loops.

Each ribosome has two tRNA binding sites (Fig. 10.17). The **A** or ***aminoacyl site*** binds the incoming aminoacyl-tRNA, the tRNA carrying the amino acid that is next to be added to the growing polypeptide chain. The **P** or ***peptidyl site*** binds the tRNA to which the growing polypeptide is attached. The specificity for aminoacyl-tRNA binding in these sites is provided by the mRNA codons that make up part of the *A* and *P* binding sites. As the ribosome moves along an mRNA (or as the mRNA is shuttled across the ribosome), the specificity for the aminoacyl-tRNA binding in the *A* and *P* sites changes as different mRNA codons move into

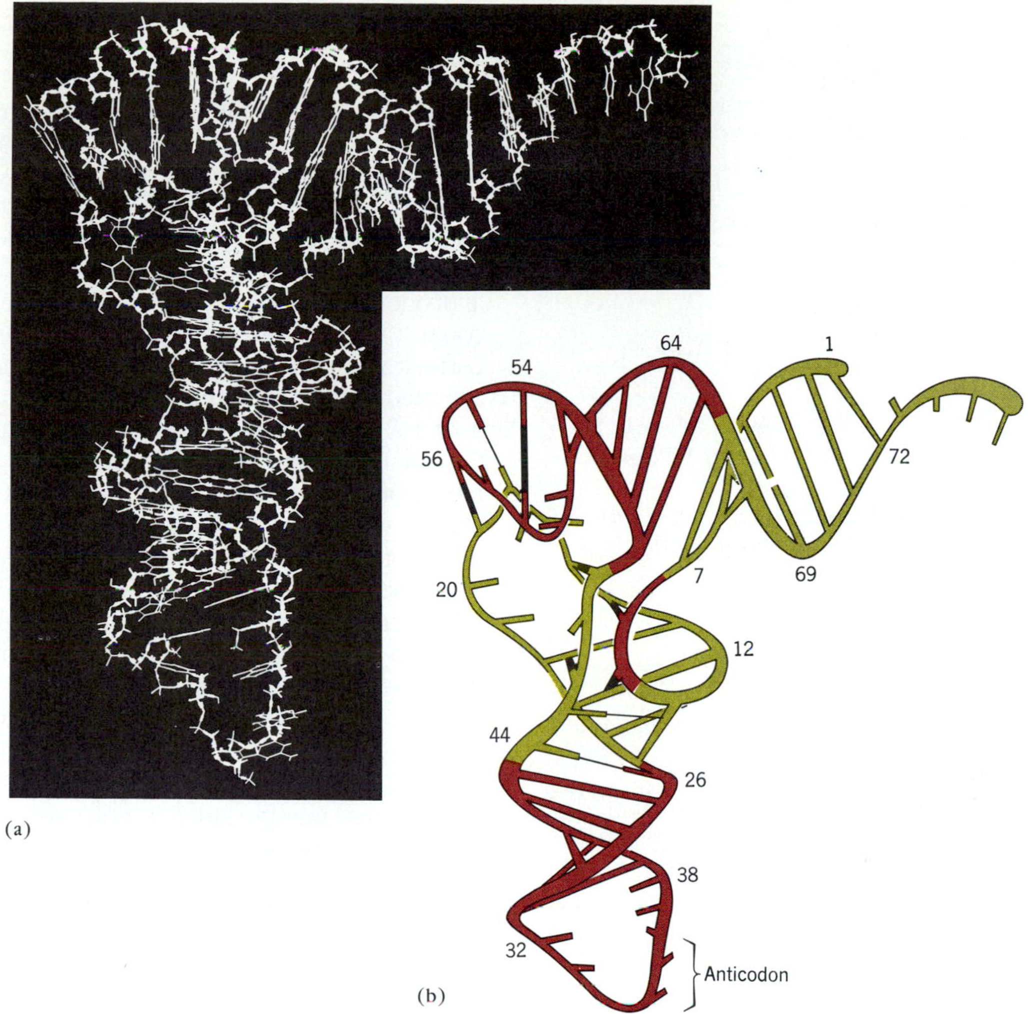

Figure 10.16 Photograph (a) and interpretative drawing (b) of a molecular model of the yeast phenylalanine tRNA based on X-ray diffraction data. The anticodon loop is at the bottom. The ribose–phosphate backbone of the tRNA is drawn as a continuous cylinder. The crossbars indicate hydrogen-bonded base-pairs. Unpaired bases are indicated by shortened rods. (From S. H. Kim, F. L. Suddath, G. J. Quigley, A. McPherson, J. L. Sussman, A. H. J. Wang, N. C. Seeman, and A. Rich, *Science* 185: 435–440, 1974. Copyright © 1974 by the American Association for the Advancement of Science.)

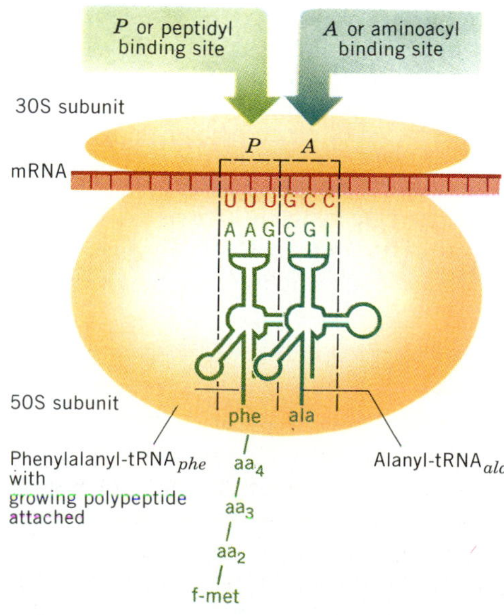

Figure 10.17 The aminoacyl-tRNA binding sites formed by each ribosome–mRNA complex. The *A* or aminoacyl-tRNA site is shown occupied by an alanyl-tRNA$_{ala}$ complex. The *P* or peptidyl site is shown occupied by a phenylalanyl-tRNA$_{phe}$ complex, with the growing polypeptide chain covalently linked to the phenylalanine tRNA. The next step in protein synthesis will involve the formation of a peptide bond between phenylalanine and alanine, followed by the translocation of the alanine tRNA (with the growing polypeptide now attached) from the *A* site to the *P* site.

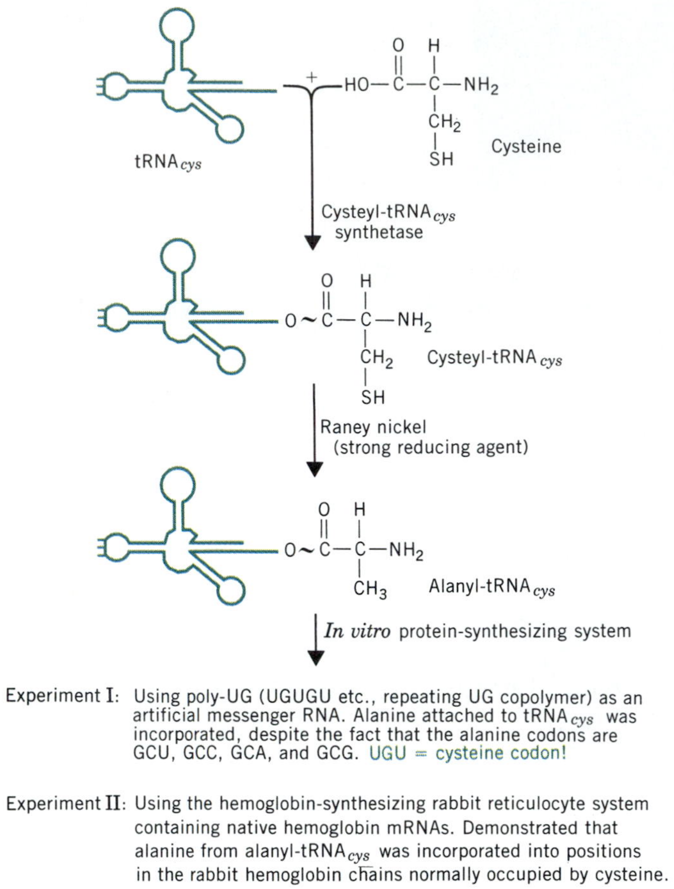

$tRNA_{cys}$ + Cysteine

Cysteyl-$tRNA_{cys}$ synthetase

Cysteyl-$tRNA_{cys}$

Raney nickel (strong reducing agent)

Alanyl-$tRNA_{cys}$

In vitro protein-synthesizing system

Experiment I: Using poly-UG (UGUGU etc., repeating UG copolymer) as an artificial messenger RNA. Alanine attached to $tRNA_{cys}$ was incorporated, despite the fact that the alanine codons are GCU, GCC, GCA, and GCG. UGU = cysteine codon!

Experiment II: Using the hemoglobin-synthesizing rabbit reticulocyte system containing native hemoglobin mRNAs. Demonstrated that alanine from alanyl-$tRNA_{cys}$ was incorporated into positions in the rabbit hemoglobin chains normally occupied by cysteine.

Figure 10.18 Proof that the codon-recognizing specificity of an aminoacyl-tRNA complex resides in the tRNA rather than in the amino acid. The tRNA specific for cysteine was activated with cysteine using cysteyl-$tRNA_{cys}$ synthetase. The cysteine, attached to $tRNA_{cys}$, was then reduced to alanine with Raney nickel (fine powdered nickel). The "hybrid" alanyl-$tRNA_{cys}$ was then placed in an *in vitro* protein-synthesizing system. In one experiment, protein synthesis was directed by a synthetic poly UG mRNA. Poly UG usually stimulates the incorporation of cysteine (codon UGU), but not alanine (codon GCX, where X equals any one of the four common bases in RNA). When the "hybrid" alanyl-$tRNA_{cys}$ was present, poly UG stimulated the incorporation of the alanine into polypeptides. In a second experiment, protein synthesis was directed by native rabbit hemoglobin mRNAs. When the alanyl-$tRNA_{cys}$ "hybrid" was added, alanine was found incorporated into hemoglobin chains at positions normally occupied by cysteine. (Based on the results of Chapeville *et al., Proc. Natl. Acad. Sci. U.S.A.* 48: 1086–1092, 1962, and of von Ehrenstein *et al., Proc. Natl. Acad. Sci. U.S.A.* 49: 669–675, 1963.)

register in the binding sites. The ribosomal binding sites by themselves (minus mRNA) are thus capable of binding any aminoacyl-tRNA.

It should be quite apparent that the tRNA molecules contain a great deal of specificity despite their small size. Not only must they (1) have the correct anticodon sequences, so as to respond to the right codons, but they must also (2) be recognized by the correct aminoacyl-tRNA synthetases, so that they are activated with the right amino acids, and (3) bind to the *A* and *P* sites on the ribosomes. F. Chapeville and G. von Ehrenstein and colleagues have proven, by means of a simple and direct experiment (Fig. 10.18), that the specificity for codon recognition resides in the tRNA portion of an aminoacyl-tRNA, rather than in the amino acid. They treated cysteyl-$tRNA_{cys}$ (the cysteine tRNA activated with cysteine) with a strongly reducing nickel powder (Raney nickel), which converted (reduced) the cysteine to alanine—still attached, however, to the cysteine tRNA. When this "hybrid" aminoacyl-tRNA, ***alanyl-$tRNA_{cys}$***, was used in *in vitro* protein-synthesizing systems, alanine was found to be incorporated into positions in polypeptides normally occupied by cysteine.

Protein synthesis is initiated by a special ***initiator tRNA,*** designated ***$tRNA_f^{Met}$***. This means that all polypeptides begin with methionine during synthesis. The amino-terminal methionine is subsequently cleaved from many polypeptides. Thus, functional proteins need not have an amino-terminal methionine. In prokaryotes and in eukaryote organelles, the ***methionine*** on the initiator $tRNA_f^{Met}$ has the ***amino group blocked with a formyl*** $\left(\begin{array}{c} O \\ \| \\ -C-H \end{array}\right)$ ***group.*** In eukaryotic cytoplasmic systems, a special initiator $tRNA_i^{Met}$ also exists, but the amino group is not formylated. A distinct methionine tRNA, $tRNA^{Met}$, which responds to internal methionine codons, exists in both prokaryotic and eukaryotic systems. Both methionine tRNAs have the same anticodon and both respond to the same codon (AUG) for methionine. In prokaryotes, the formylated amino group on methionyl-$tRNA_f^{Met}$ prevents the formation of a peptide bond between the amino group and the carboxyl group of the amino acid at the end of the growing polypeptide chain. In eukaryotes, however, the amino group of methionyl-$tRNA_i^{Met}$ is not blocked. What, then, prevents methionyl-$tRNA_i^{Met}$ (*i* for initiator) from responding to internal AUG codons and methionyl-$tRNA^{Met}$ from responding to initiator AUG codons in eukaryotic mRNAs? Apparently, only methionyl-$tRNA_i^{Met}$ will react with the protein initiation factors, IF-1, IF-2, and IF-3, and only methionyl-$tRNA^{Met}$ will interact with the protein elongation factors, Ef-T_s and Ef-T_u. In any case, only methionyl-$tRNA_i^{Met}$ responds to AUG initiation codons and only methionyl-$tRNA^{Met}$ responds to AUG internal codons. Methionyl-$tRNA_f^{Met}$ also responds to an alternate initiator codon, GUG (a valine codon when present at internal positions), known to be present in certain natural mRNAs.

In prokaryotes, polypeptide chain initiation occurs with the formation of a complex between mRNA, methionyl-$tRNA_f^{Met}$, and the 30S ribosomal subunit (Figs. 10.19*a* and *b*). The formation of this initiation complex requires the activity of three protein initiation factors, designated ***IF-1, IF-2,*** and ***IF-3,*** plus guanosine triphosphate (GTP). It may be facilitated by a base-pairing interaction between a base sequence near the 3′ end of the 16S rRNA and a base sequence in the ***"leader sequence"*** of the mRNA. (***Leader sequences*** of mRNA molecules are the nontranslated sequences from the 5′ end to the first AUG or GUG initiation codon. These nontranslated leader sequences vary in length from a few nucleotides to several hundred nucleotides. Little is known about their biological significance.)

The initiation complex then combines with the 50S ribosomal subunit, and the methionyl-$tRNA_f^{Met}$ becomes bound to the peptidyl site (Figs. 10.19*b* and *c*). This requires the hydrolysis of one molecule of GTP. The alignment of the AUG initiation codon with the anticodon of $tRNA_f^{Met}$ (in the *P* site) fixes the codon present at the A site, thus establishing the specificity for aminoacyl-tRNA binding at the *A* site (for alanyl-$tRNA_{ala}$ in the example diagrammed in Fig. 10.19*d*). The binding of alanyl-$tRNA_{ala}$ in the *A* site (and all subsequent aminoacyl-tRNA binding) requires the hydrolysis of one molecule of GTP and the protein elongation factors designated ***Ef-T_s*** and ***Ef-T_u*** (Figs. 10.19*c* and *d*). Peptide bond formation between the carboxyl group of f-methionine bound to the $tRNA_f^{Met}$ in the *P* site and the amino group of the alanine molecule bound to $tRNA_{ala}$ in the *A* site is then catalyzed by ***peptidyl transferase,*** an enzyme bound to the 50S ribosomal subunit (Figs. 10.19*d* and *e*). This reaction leaves the f-met-ala dipeptide attached to $tRNA_{ala}$ bound to the *A* site of the ribosome (Fig. 10.19*e*).

The next step in translation, called ***translocation,*** involves (1) movement of f-met-ala-$tRNA_{ala}$ from the *A* site to the *P* site and (2) movement of the mRNA molecule exactly three nucleotides, relative to the position of the ribosome, so that the codon previously in register with the *A* site moves into register with the *P* site (Figs. 10.19*e* and *f*). In Figs. 10.19*e* and *f*, the alanine codon GCC moves from its position at the *A* site into register with the *P* site, and the subsequent codon, the serine codon UCC, moves into register with the *A* site. Translocation requires the activity of the elongation factor designated ***Ef-G*** and the hydrolysis of one molecule of GTP.

The next aminoacyl-tRNA specified by the mRNA codon at the *A* site (seryl-$tRNA_{ser}$ in Figs. 10.19*f* and *g*) then binds at the *A* site, and the peptide bond formation and translocation steps are repeated. The just-described sequence is repeated for each codon of the mRNA (about 300 codons on average) until a ***chain-termination codon*** is reached (Figs. 10.19*g* and *h*). The formyl group on the amino-terminal methionine is usually removed by a deformylase (Figs. 10.19*g* and *h*) before synthesis of the polypeptide is completed. When one of the three polypeptide chain-termination codons (UAG, UAA, or UGA; see pp. 280–281) comes into register with the *A* site (Fig. 10.19*h*), the nascent polypeptide, the tRNA in the *P* site, and the mRNA are released, and the ribosomal subunits dissociate (Figs. 10.19*h* and *i*). Termination requires the activity of one of two protein ***release factors,*** designated RF-1 and RF-2. The dissociated ribosomal subunits are then free to initiate the translation of another mRNA molecule (Fig. 10.11).

Rather than each mRNA molecule being translated by a single ribosome, most mRNAs are simultaneously translated by several ribosomes, spaced about 90 nucleotides apart along the mRNA molecule. The size of these translation complexes, called ***polyribosomes*** or ***polysomes,*** is highly variable but is correlated with the size of the polypeptide being synthesized. Hemoglobin chains (about 150 amino acids), for example, are synthesized on pentaribosome (average size) complexes.

Coupled Transcription and Translation in Prokaryotes

In prokaryotes, the translation of an mRNA molecule frequently begins before its synthesis (transcription) is complete. This is possible because mRNA molecules are both synthesized and translated in the 5′ to 3′ direction, and because there is no nuclear membrane separating transcription from translation as in eukaryotes. This coupling between transcription and translation facilitates the very rapid and efficient "turn-on" and "turn-off" of gene expression that is observed in prokaryotes (see Chapter 14).

O. L. Miller, B. A. Hamkalo, and colleagues have developed techniques by which transcription and translation can be visualized directly using electron microscopy. Figure 10.20, for example, shows the coupled transcription and translation of a gene in *E. coli.*

Transcription, RNA Processing and Transport, and Translation in Eukaryotes

In eukaryotes, transcription and translation cannot be coupled, since transcription occurs in the nucleus and translation occurs in the cytoplasm. This poses the

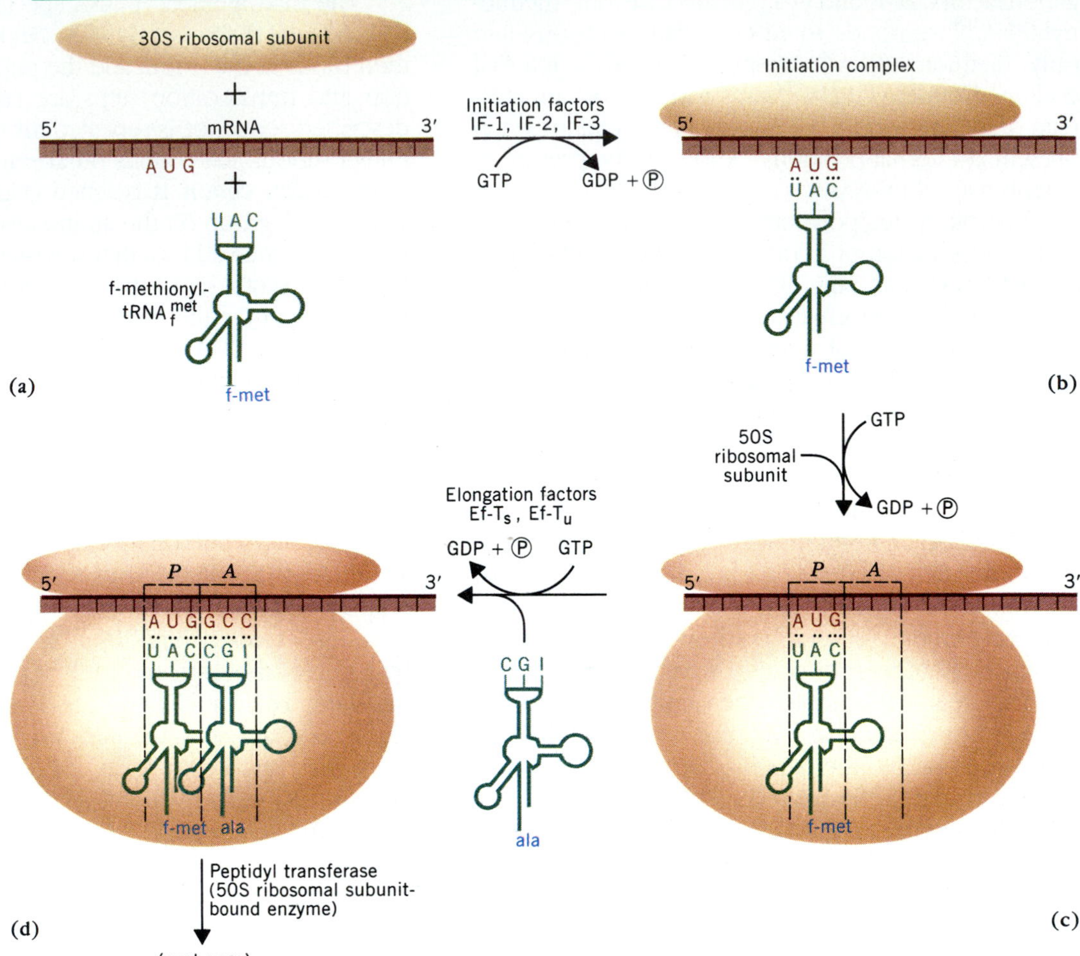

Figure 10.19 (Above and right page): Schematic drawing of the steps involved in translation.

questions of how gene transcripts are transported from the nucleus to the cytoplasm and what determines the time and place of mRNA translation. Unfortunately, we do not yet have answers to these questions. We do know that the transcription and translation processes in eukaryotes are more complex than those in prokaryotes, involving several intermediate mRNA processing steps, as well as transport from nucleus to cytoplasm.

The mRNAs of eukaryotes are derived from the primary gene transcript by several types of processing (Fig. 10.21). These include (1) cleavages of ***large mRNA precursors (pre-***mRNAs) to smaller mRNA molecules, (2) the addition of ***7-methyl guanosine groups (mRNA "caps")*** to the 5′ ends of the molecules, (3) the addition of approximately ***200-nucleotide-long sequences of adenylate nucleotides ("poly-A tails")*** to the 3′ ends of the molecules, and (4) formation of complexes with specific proteins. The cleavages involved in the conversion of pre-mRNAs to mRNAs frequently involve the removal of ***leader sequences,*** sequences from the 5′ end to the translation initiation

Figure 10.20 (Right page): Electron micrograph showing the coupled transcription and translation of a gene in *E. coli.* The nascent polypeptide chains being synthesized on the ribosomes are not visible as they fold into their three-dimensional configuration during synthesis. However, the silk protein fibroin, synthesized in the silk gland of the silkworm, remains rodlike on the ribosome and can be seen during synthesis by electron microscopy (see Fig. 10.22). (From O. L. Miller, Jr., B. A. Hamkalo, and C. A. Thomas, Jr., *Science* 169: 392–395, 1970. Copyright © 1970 by the American Association for the Advancement of Science. Original micrograph courtesy of O. L. Miller, Jr.)

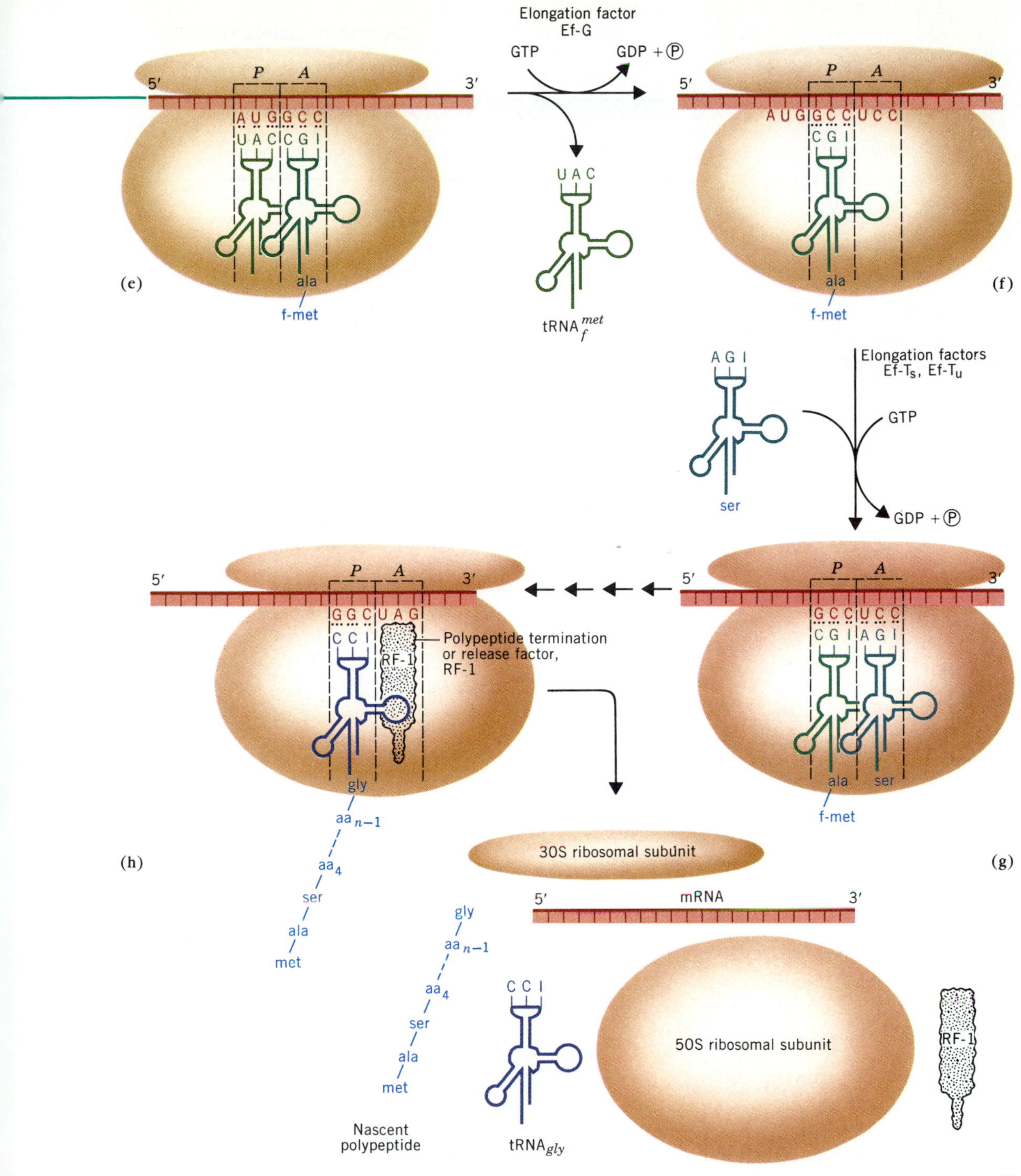

Elongation factor Ef-G

GTP → GDP + Ⓟ

(e) 5′ P A 3′ — AUG GCC / UAC CGI — ala, f-met

tRNA$_f^{met}$ (UAC)

(f) 5′ P A 3′ — AUG GCC UCC / CGI — ala, f-met

AGI — ser

Elongation factors Ef-T_s, Ef-T_u

GTP → GDP + Ⓟ

(g) 5′ P A 3′ — GCC UCC / CGI AGI — ala, ser, f-met

(h) 5′ P A 3′ — GGC UAG / CCI — RF-1

Polypeptide termination or release factor, RF-1

gly / aa $_{n-1}$ / aa $_4$ / ser / ala / met

(i) 30S ribosomal subunit

5′ mRNA 3′

gly / aa $_{n-1}$ / aa $_4$ / ser / ala / met

Nascent polypeptide

CCI tRNA$_{gly}$

50S ribosomal subunit

RF-1

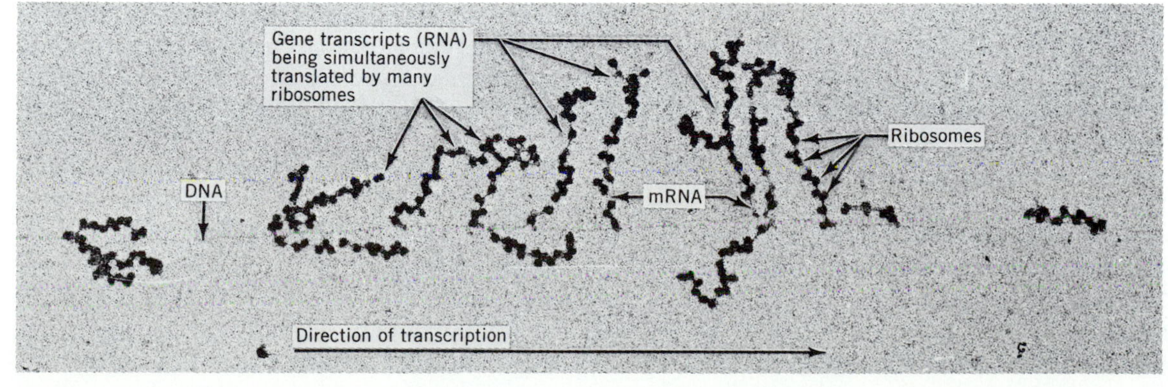

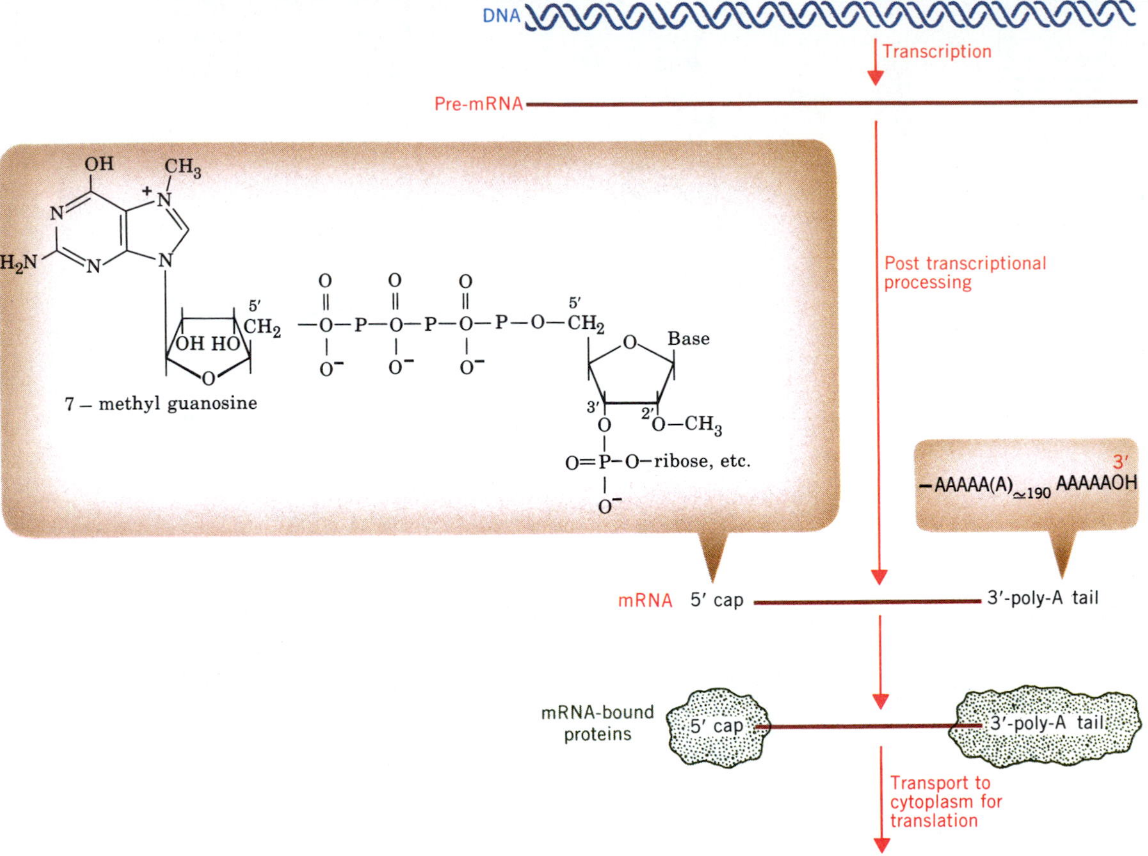

Figure 10.21 Processing of primary gene transcripts (pre-mRNA) in eukaryotes. Processing involves: (1) the cleavage of the pre-mRNA, with the removal of ends and/or the excision of noncoding introns, (2) the enzymatic addition of a 5′ cap, a 7-methyl guanosine group linked to a 5′ subterminal, 2′-*O*-methyl nucleotide by an unusual 5′–5′ triphosphate linkage, (3) the addition of a 200-nucleotide-long poly-A tail (average size) to the 3′ end, and (4) the association with two specific proteins. The addition of the 3′-poly-A tail is catalyzed by an enzyme called poly-A polymerase.

codon, and ***noncoding sequences*** (called ***"intervening sequences"*** or ***"introns"***) that are located between coding sequences (see the following section and Chapter 13, pp. 356–359). Individual gene transcripts may undergo some or all of these four types of processing. Not all mRNAs contain the 5′ cap, nor do all of them have poly-A tails. This has made it difficult to determine the functions of these posttranscriptional modifications.

Most of the nonribosomal RNAs synthesized in the nuclei of eukaryotic cells consist of very large molecules that are highly variable in size (10S–200S, or about 1000–50,000 nucleotides in length). This RNA has been called ***heterogeneous nuclear RNA,*** abbreviated ***hnRNA.*** It now seems clear that much, if not most, of this hnRNA is really pre-mRNA. Rapid processing of these giant hnRNA or pre-mRNA molecules in the nucleus soon after transcription apparently results in (1) the bulk of the nonribosomal RNA synthesized in the nucleus (probably large segments of each primary transcript) being rapidly degraded (average half-life of about 30 minutes) and (2) the formation of the smaller mRNA molecules that are transported to the cytoplasm. However, it is not yet clear whether all, most, or only part of the hnRNA molecules synthesized in the nuclei of eukaryotic cells are, in fact, pre-mRNA molecules.

Definitive evidence for pre-mRNA processing in the formation of eukaryotic mRNAs has been obtained in the case of the β-globin gene transcript of the mouse. In this case, a 15S hnRNA (or pre-mRNA; 1200–1500 nucleotides long) is processed to a 9S (about 600 nucleotides long) β-globin mRNA.

Similar evidence for hnRNA or pre-mRNA processing in the formation of mature mRNA molecules is now available for many other eukaryotic gene transcripts. This processing often involves the excision of noncoding intervening sequences or introns located between coding sequences (called ***exons*** for ***ex***pression).

In addition, the mRNAs of some eukaryotic viruses are known to contain leader sequences (the sequence from the 5′ ends to the translation–initiation codons of mRNAs) that are transcribed from DNA sequences that are ***not*** contiguous with the structural genes. Several different mRNAs may, in fact, contain identical leader sequences. These leader sequences are apparently spliced onto the 5′ ends of gene transcripts during processing. It is believed that these leader sequences must be involved in the regulation of translation. However, their exact function(s) is not yet known.

Translation in eukaryotes appears to be analogous to translation in prokaryotes except that (1) the amino group of methionyl-tRNA$_i^{Met}$ (the initiation tRNA) is not formylated and (2) most eukaryotic mRNAs studied to date appear to be monogenic, such that only one polypeptide species is translated from each mRNA. In prokaryotes, many mRNAs are ***polygenic,*** that is, two or more different polypeptides are synthesized from nonoverlapping segments of a single mRNA (see Chapter 14, pp. 393–397).

The synthesis of one eukaryotic protein, the silk protein fibroin, can be visualized by electron microscopy using the techniques developed by O. L. Miller, B. A. Hamkalo, and colleagues. Fibroin does not fold up on the surface of the ribosome as other polypeptides do under the conditions used. As a result, nascent polypeptide chains of increasing length can be seen attached to the ribosomes as one scans from one end (the mRNA 5′ end) of the giant polysomes (containing 50–80 ribosomes; fibroin has a molecular weight of over 200,000) to the other end (Fig. 10.22).

Removal of Intron Sequences by RNA Splicing

Most, but not all, genes of higher eukaryotes contain ***noncoding intervening sequences*** or ***introns separating the coding sequences or exons.*** (See Chapter 13 for a discussion of the fine structure of eukaryotic genes.) Fewer, but still many, of the genes of lower eukaryotes such as the yeasts and *Neurospora* contain noncoding introns. Rare genes of a few viruses of prokaryotes, for example, *E. coli* bacteriophage T4 and *B. subtilis* bacteriophage SP01, and of an archebacterium (a primitive bacterium) have also been found to contain introns. In the case of these "split" or "mosaic" genes (coding sequences separated by noncoding sequences), the primary transcript contains the entire sequence of the gene and the ***noncoding sequences are "spliced" out during processing*** (Fig. 10.23). (One

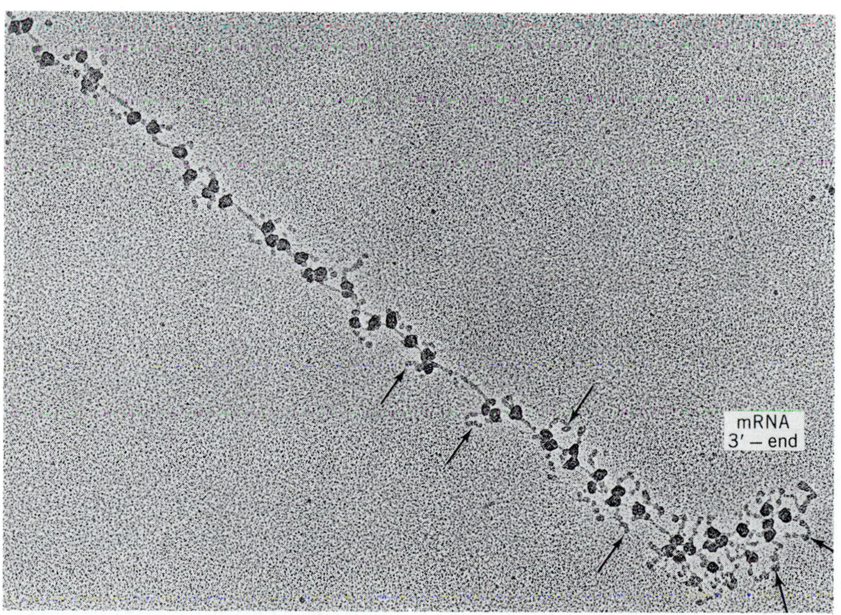

Figure 10.22 Electron micrograph showing the translation of the fibroin mRNA in cells of the posterior silk gland of the silkworm *Bombyx mori.* The arrows point to putative nascent fibroin polypeptides. Note their increasing length as one approaches the 3′ end of the mRNA molecule. Each fibroin polysome consists of 50–80 ribosomes that simultaneously translate each giant fibroin mRNA; only a portion of one polyribosome is shown. (See S. L. McKnight, N. L. Sullivan, and O. L. Miller, Jr., *Prog. Nucleic Acid Res. Mol. Biol.* 19: 313–318, 1976. Micrograph courtesy of S. L. McKnight and O. L. Miller, Jr., University of Virginia.)

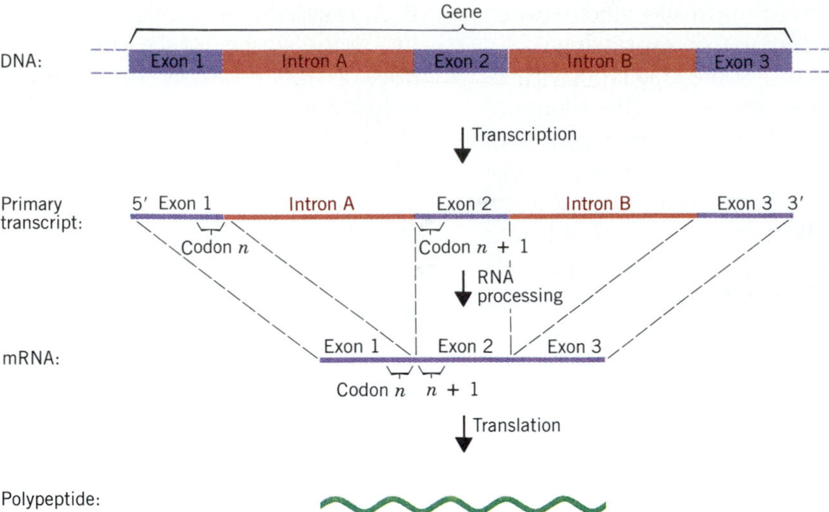

Figure 10.23 Splicing intron sequences out of primary transcripts during RNA processing. The splicing process must be accurate to the single nucleotide so that codons in exons distal to introns are in the proper reading frame. (See Chapter 13, p. 357, for evidence that intron sequences are removed from primary transcripts during processing.)

phage T4 intron appears to be an exception; evidence suggests that this intron is somehow "bypassed" during the transcription process.)

Clearly, the splicing mechanism must be very precise; it must join exon sequences with accuracy to the single nucleotide to assure that codons in exons distal to introns are read in the correct reading frame. Accuracy to this degree would seem to require very precise splicing signals, presumably nucleotide sequences within introns and at the exon–intron junctions. However, in primary transcripts of nuclear genes of higher animals, the only completely conserved sequences of different introns are the dinucleotide sequences at the ends of introns, namely,

$$\text{exon-}\overbrace{GT \ldots\ldots\ldots\ldots AG}^{\text{intron}}\text{-exon}$$

The sequences shown here are for the DNA strand that is equivalent to the RNA transcript (T in place of U). In addition, there are short consensus sequences at the exon–intron junctions. For nuclear genes, the consensus junctions are

$$\overbrace{A_{64}\,G_{73}}^{\text{exon}}\,\overbrace{G_{100}\,T_{100}\,A_{68}\,A_{68}\,G_{84}\,T_{63} \ldots\ldots\ldots\ldots 6Py_{74\text{–}87}\,N\,C_{65}\,A_{100}\,G_{100}}^{\text{intron}}\,\overbrace{N}^{\text{exon}}$$

The numerical subscripts indicate the percentage frequencies of the consensus bases at each position; thus a 100 subscript indicates that a base is always present at that position. *N* indicates that any of the four standard nucleotides may be present at the indicated position. The exon–intron junctions are different in the case of tRNA genes and structural genes in mitochondria and chloroplasts, which utilize different RNA splicing mechanisms (see the following section). There is only one short conserved sequence within the introns of nuclear genes, the so-called ***"TACTAAC box,"*** and it is rather poorly conserved. The "TACTAAC box" does exhibit a strong preference for either a purine or a pyrimidine at each site as follows:

$$Py_{80}\,N\,Py_{80}\,Py_{87}\,Pu_{75}\,A_{100}\,Py_{95}$$

The adenine residue at position six in the "TACTAAC box" is completely conserved and is known to play a key role in the splicing reaction (see the section *Pre-mRNA Splicing: snRNAs, snRNPs, and the Spliceosome* later in this chapter). The remaining sequences of introns (often very large) of nuclear genes are highly divergent and appear to be entirely random sequences. The introns of genes of mitochondria and chloroplasts contain conserved sequences that are different from those of nuclear genes.

Three Distinct Types of RNA Splicing

The discovery of noncoding introns in genes stimulated intense interest in the mechanism(s) by which intron sequences are removed during gene expression. The early demonstration that the intron sequences in eukaryotic genes were transcribed along with the exon sequences focused research on the processing of primary gene transcripts. Just as *in vitro* transcription and translation systems were instrumental in elucidating those processes, the key to deciphering RNA splicing events was the development of *in vitro* splicing systems. As it turned out, there are three totally distinct types of intron excision from RNA transcripts. We present them in the order of increasing complexity, not in order of importance. Class 3 introns are the most prevalent and the most important in terms of overall eukaryotic gene expression.

1. The introns of tRNA precursors are excised by precise endonucleolytic cleavage and ligation reactions catalyzed by special splicing endonuclease and ligase activities.
2. The introns of *Tetrahymena* rRNA precursors are removed autocatalytically in a unique reaction mediated by the RNA molecule itself (no protein enzymatic activity is involved).
3. The introns of nuclear pre-mRNA (hnRNA) transcripts are spliced out in two-step reactions carried out by complex ribonucleoprotein particles called "spliceosomes."

RNA splicing mechanisms are the focus of considerable research at present, and it is likely that new variations of these splicing mechanisms will be discovered in the future. Many genes contain large numbers of introns (e.g., the chicken α2 collagen gene contains over 50 introns), which leads to the question of the order in which multiple introns are removed. For certain genes that have been studied, the introns are excised in a preferred, but not fixed, order. Other introns have been found to undergo alternative pathways of splicing leading to mRNAs that produce different proteins (i.e., with some common and some distinct amino acid sequences). Finally, one intron in the cytochrome *b* gene of yeast mitochondria includes part of the coding sequence for a protein, an "RNA maturase," that is responsible for excising the second intron from the transcript of that gene. This suggests interesting mechanisms for regulating the expression of genes at the level of intron processing. Clearly, then, interesting variations in the use, structure, and excision of intron sequences exist in nature, and novel intron structures are almost certain to be discovered in the future. The three major established types of intron excision are described briefly in the following three sections.

Splicing tRNA Precursors: Unique Nuclease and Ligase

The tRNA precursor splicing reaction has been worked out in detail in the yeast *Saccharomyces cerevisiae*. Both *in vitro* splicing systems and temperature-sensitive splicing mutants have been used in dissecting the tRNA splicing mechanism in *S. cerevisiae*. The excision of introns from yeast tRNA precursors occurs in two stages (Fig. 10.24). First, ***a nuclear membrane-bound splicing endonuclease makes two cuts precisely at the ends of the intron.*** Then, in a fairly complex set of

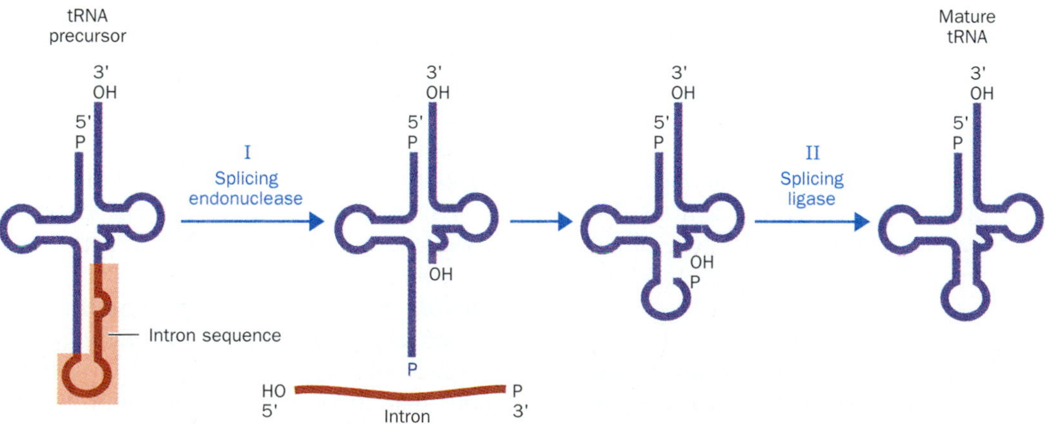

Figure 10.24 Diagram illustrating the two stages in the excision of introns from tRNA precursors. The first stage involves the introduction of ***two cuts precisely at the ends of the intron;*** these cleavages are ***catalyzed by a nuclear envelope-bound splicing endonuclease.*** Intramolecular base-pairing dictates the formation of the mature three-dimensional form after the intron sequence has been cut out in stage I. The second stage involves the ***ligation of the newly formed exon ends*** to produce the mature tRNA molecule. The ligation process actually involves three distinct reactions, all catalyzed by the 90,000-dalton ***splicing ligase.*** This enzyme apparently contains phosphodiesterase, kinase, and ligase catalytic activities (see text).

reactions, ***a splicing ligase joins the two halves of the tRNA to produce the mature form of the tRNA molecule.*** The specificity for these reactions resides in the conserved three-dimensional structural features of the tRNA precursors, not in the nucleotide sequences per se.

Cleavage of the tRNA precursor yields 5′-OH termini and 2′–3′ cyclic phosphate groups at the 3′ termini. The stage II ligation process actually involves four separate reactions. (1) The first reaction is the addition of a phsophate group to the 5′-OH terminus; this reaction requires kinase activity and a phosphate donor (ATP). (2) Then, the 5′ phosphate group is activated by the transfer of an AMP group to the terminus from an AMP–ligase intermediate (the AMP originally having been derived from ATP also). (3) The 2′–3′ cyclic phosphate is opened by a cyclic phosphodiesterase activity that produces a 2′ phosphate and a free 3′ hydroxyl. (4) The final ligation reaction occurs via a nucleophilic attack of the free 3′-OH on the interior 5′ phosphate with the release of AMP. All four of these reactions are catalyzed by the splicing ligase. Finally, the 2′ phosphate group (remaining from the 2′–3′ cyclic phosphate produced by the original cleavage reaction) is removed by a phosphatase activity to yield the mature tRNA molecule.

The overall two-stage mode of tRNA intron excision appears to occur in other organisms as well. In fact, the mechanism may involve the same reactions in plants. However, in mammals, the reactions are not the same. Splicing still occurs in two stages, but the ligation reaction appears to directly join the 2′–3′ cyclic phosphate terminus to the 5′-OH terminus. The details of this process of tRNA precursor splicing in mammalian cells are not yet as clearly established as they are in yeast.

Autocatalytic Splicing of *Tetrahymena* rRNA Precursor

A general theme in biology is that metabolism occurs via sequences of enzyme-catalyzed reactions. Moreover, these all-important enzymes are generally proteins, albeit sometimes single polypeptides and sometimes complex heteromultimers, that occasionally require nonprotein cofactors to perform their functions. Therefore, when covalent bonds are being altered (removed, transferred, or formed), we expect that the reaction is being catalyzed by an enzyme. Thus, the discovery that the intron in the rRNA precursor of *Tetrahymena thermophila* was excised without the involvement of any protein was quite surprising to most biologists. However, it is now clearly established that ***the splicing activity that excises the intron from this rRNA precursor is intrinsic to the RNA molecule itself.*** Moreover, such ***self-splicing*** or ***autocatalytic activity*** has been shown to occur in rRNA precursors of several lower eukaryotes and in a large number of rRNA, tRNA, and mRNA precursors in mitochondria and chloroplasts of many different species. In the case of many of these introns (the so-called group I introns) in RNA precursor molecules, the self-splicing mechanism is the same as or very similar to that for *Tetrahymena* rRNA precursors described next. For others (called group II introns), the self-splicing mechanism is similar to the splicing mechanism observed with nuclear mRNA precursors (see the following section) except that it requires no "spliceosome" activity.

The autocatalytic excision of the intron in the *Tetrahymena* rRNA precursor (and other group I introns) requires no external energy source (no ATP, etc.) and no protein. Instead, it involves a series of ***phosphoester bond transfers,*** with no bonds lost or gained in the process. The ***reaction requires a guanine nucleoside or nucleotide with a free 3′-OH group (GTP, GDP, GMP, or guanosine all work) as a cofactor*** plus a monovalent cation and a divalent cation. The requirement for the G-3′-OH is absolute; no other base can be substituted in the nucleoside or nucleotide cofactor. ***The intron is excised by means of two phosphoester bond transfers,*** and the excised intron can subsequently circularize by means of another phosphoester bond transfer. These reactions are diagrammed in Fig. 10.25. The autocatalytic circularization of the excised intron suggests that the self-splicing of these rRNA precursors resides primarily, if not entirely, within the intron structure itself. Presumably, the autocatalytic activity is dependent on the secondary structure of the intron or at least the secondary structure of the RNA precursor molecule. The secondary structures of these self-splicing RNAs must bring the reactive groups into close juxtaposition to allow the phosphoester bond transfers to occur. Since the self-splicing phosphoester bond transfers are potentially reversible reactions, rapid degradation of the excised introns or export of the spliced rRNAs to the cytoplasm may drive splicing in the forward direction.

A key point is that the ***autocatalytic splicing reactions are intramolecular*** in nature and, thus, are not concentration dependent. Moreover, the ***RNA precursors are capable of forming an active center in which the guanosine-3′-OH cofactor binds.*** Thus, catalytic sites are not restricted to proteins; but also note that there is no *trans* catalytic activity as for enzymes, only *cis* catalytic activity.

Pre-mRNA Splicing: snRNAs, snRNPs, and the Spliceosome

The introns in nuclear mRNA precursors (nuclear pre-mRNAs) are excised in two steps like the introns in yeast pre-tRNAs and *Tetrahymena* pre-rRNAs discussed

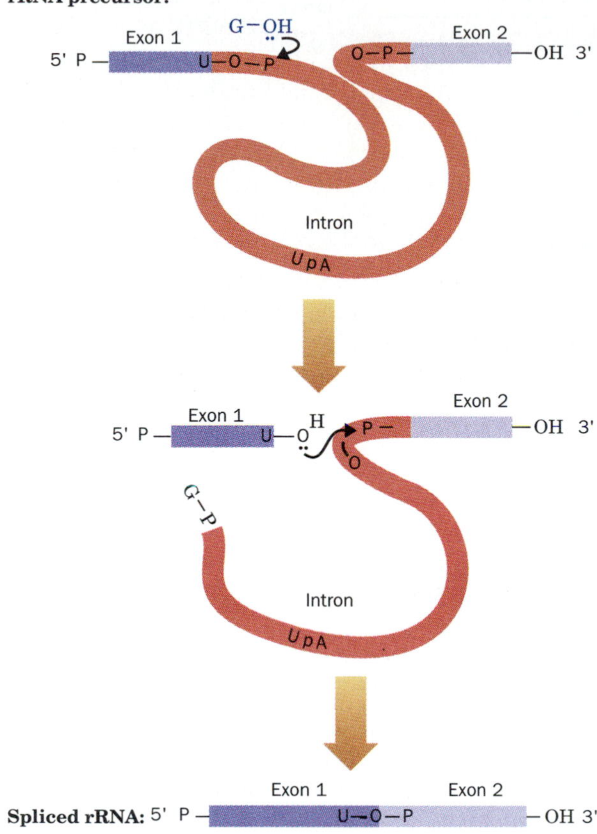

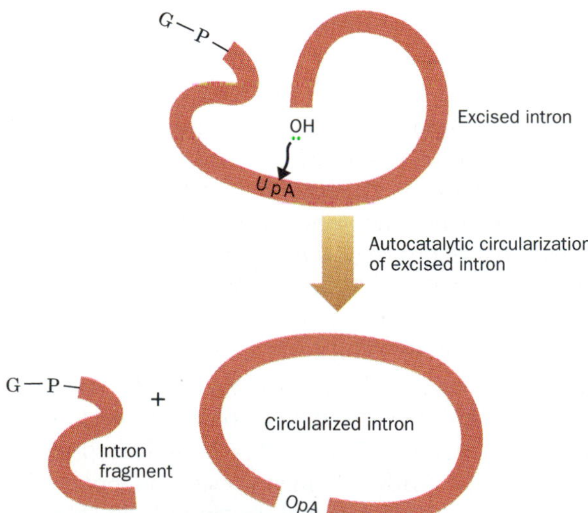

Figure 10.25 Diagram of the mechanism of self-splicing of the *Tetrahymena thermophila* rRNA precursor and the subsequent circularization of the excised intron. The reactions are described in the text. The guanosine cofactor and exon sequence are shown in blue; and the intron in reddish brown. (After A. J. Zaug, P. J. Grabowski, and T. R. Cech, *Nature* 301: 578–583, 1983.)

in the preceding two sections. However, the introns are not excised by simple splicing nucleases and ligases or autocatalytically. Instead, nuclear pre-mRNA splicing is carried out by complex RNA/protein structures called ***spliceosomes***. These spliceosomes (Fig. 10.26) are in many ways like small ribosomes. They contain a set of small RNA molecules called ***snRNAs*** (***s***mall ***n***uclear ***RNA***s) and a set of proteins that are still not completely defined. The two steps in nuclear pre-mRNA splicing are known (Fig. 10.27); however, some of the details of the splicing process are still uncertain.

Five snRNAs called ***U1, U2, U4, U5,*** and ***U6*** are involved in nuclear pre-mRNA splicing as components of the spliceosome. (snRNA U3 is localized in the nucleolus and probably is involved in the formation of ribosomes.) In mammals, these snRNAs range in size from 100 nucleotides (U6) to 215 nucleotides (U3). Some of the snRNAs in the yeast *S. cerevisiae* are much larger. These snRNAs do not exist as free RNA molecules. Instead, they are present in small nuclear RNA–protein complexes called ***snRNP***s (***s***mall ***n***uclear ***r***ibo***n***ucleo***p***roteins). Characterization of snRNPs has been facilitated by the discovery that some patients with a disease called systemic lupus erythematosus produce antibodies that react with snRNP proteins. These antibodies are called autoantibodies because they react

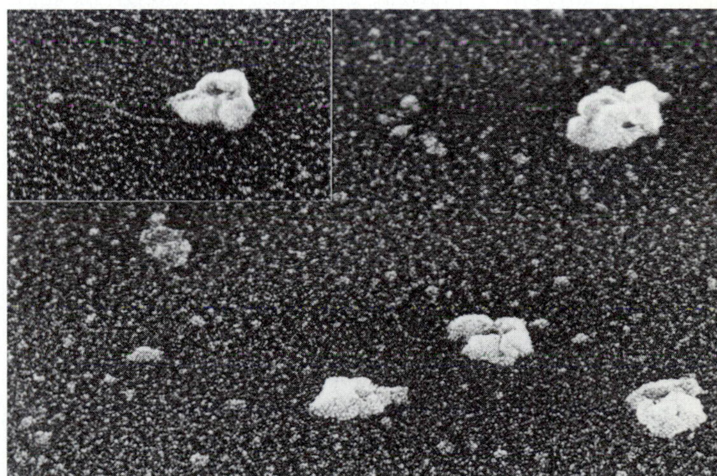

Figure 10.26 Electron microscopic visualization of purified spliceosomes. Note the striking substructure of the particles, which have dimensions of 40 to 60 nanometers (1 nanometer = 10^{-9} meter or 10^{-3} micron). The inset (top left) shows a particle with a thin filament and a smaller particle at the end of the filament. (Reproduced with permission from R. Reed, J. Griffith, and T. Maniatis, "Purification and Visualization of Native Spliceosomes," *Cell* 53: 949–961, 1988. Copyright © 1988 Cell Press; published by Cell Press. Original photograph courtesy of J. Griffith.)

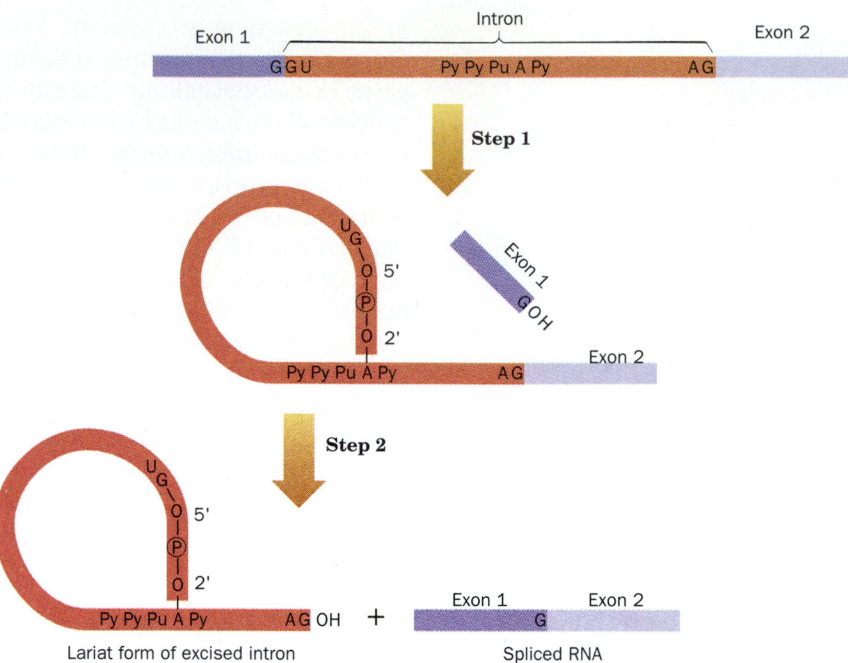

Figure 10.27 Diagram of the two-step pathway of excision of introns from nuclear pre-mRNAs (hnRNAs). This process occurs on spliceosomes (see Fig. 10.26). In the first step, a precise cleavage occurs at the 5′ end of the intron, and a 2′–5′ phosphodiester linkage is formed between the 5′ position of the G at the intron cleavage site and a conserved A residue located near the 3′ end of the intron. Exon 1 is held in place by components of the spliceosome that are not shown here (see Fig. 10.28). In the second step, the two exons are joined by a normal 3′–5′ phosphodiester bond, and the lariat-shaped intron is released.

with the patient's own proteins ("self" proteins); normally, only antibodies that react with foreign proteins will be produced by the immune system. These antibodies can be used to precipitate snRNPs; thus, they greatly facilitate the purification of snRNPs for structural and functional studies.

The snRNAs U1, U2, and U5 are present in three different snRNP particles, each of which contains a single snRNA. snRNAs U4 and U6 are present together in a fourth snRNP; U4 and U6 snRNAs contain two regions of intermolecular complementarity that probably are base-paired in the U4/U6 snRNP. Each of the four types of snRNP particles contains a subset of seven well-characterized snRNP proteins plus one or more proteins unique to the particular type of snRNP particle. All four snRNP complexes are present in the isolated spliceosomes shown in Fig. 10.26. However, the exact protein composition of intact spliceosomes is still not established.

The first step in nuclear pre-mRNA splicing involves cleavage at the 5′ intron splice site (↓GT-intron) and the formation of an intramolecular phosphodiester linkage between the 5′ carbon of the G at the cleavage site and the 2′carbon of a conserved A residue near the 3′ end of the intron (Fig. 10.27). This step occurs on complete spliceosomes (Fig. 10.28) and requires the hydrolysis of ATP. Evidence indicates that the U1 snRNP must bind at the 5′ splice site prior to the initial cleavage reaction. The recognition of the cleavage site at the 5′ end of the intron probably involves base-pairing between the consensus sequence at this site and a complementary sequence near the 5′ terminus of snRNA U1 (Fig. 10.29). However, the specificity of the binding of at least some of the snRNPs to intron consensus sequences involves both the snRNAs and specific snRNP proteins; thus, the base-pairing between the intron 5′ consensus sequence and the complementary sequence in the snRNA U1 may provide only part of the specificity for the functional binding of the U1 snRNP to the pre-mRNA molecule.

The second snRNP to be added to the splicing complex appears to be the U2 snRNP; it binds at the consensus sequence that contains the 100 percent conserved A residue that forms the branch point in the lariat structure of the spliced intron (Fig. 10.27). Thereafter, the U5 snRNP binds at the 3′ splice site, and the U4/U6 snRNP is added to the complex to yield the complete spliceosome (see Figs. 10.26 and 10.28). When the 5′ intron splice site is cleaved in step 1, the

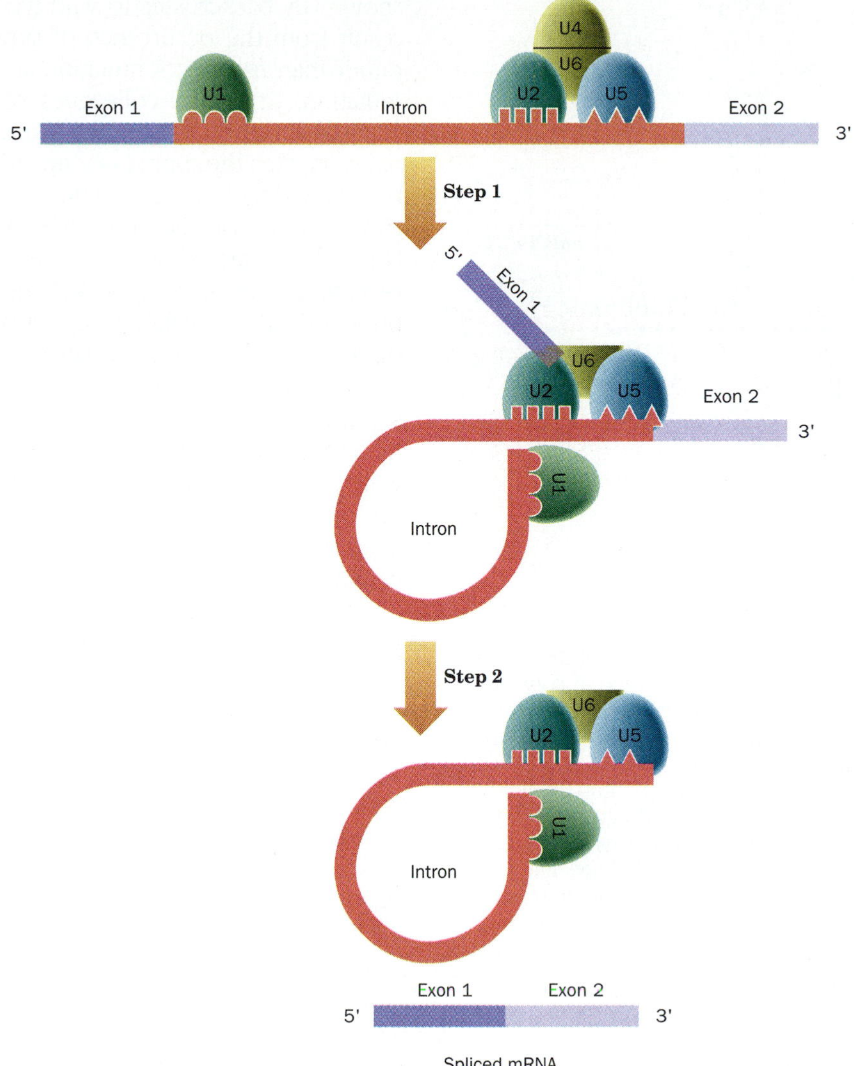

Figure 10.28 Diagram illustrating the postulated roles of the snRNA-containing snRNPs in nuclear pre-mRNA splicing. The two steps in the splicing pathway and the binding of the U1 snRNP at the 5′ splice site of the intron are well established. The specific interactions of the U2, U5, and U4/U6 snRNP complexes are less rigorously established. However, the interactions shown are consistent with the available data. See the text and Fig. 10.27 for further details on the splicing process. (After D. A. Brow and C. Guthrie, "Splicing a Spliceosomal RNA," *Nature* 337: 14–15, 1989.)

U4 snRNA is released from the spliceosome (at least *in vitro*). ***In step 2 of the splicing reaction, the 3′ splice site of the intron is cleaved, and the two exons are joined by a normal 5′ to 3′ phosphodiester linkage*** (Fig. 10.28). The spliced mRNA is now ready for export to the cytoplasm and translation on ribosomes.

THE GENETIC CODE

As it became evident that genes controlled the structure of polypeptides, attention focused on how the sequence of the four base-pairs in DNA could control the sequence of the 20 amino acids found in proteins. With the discovery of the mRNA intermediary, the question became one of how the sequence of the four bases present in mRNA molecules could specify the amino acid sequence of a polypeptide. What is the nature of the ***genetic code*** relating mRNA base sequences (or DNA base-pair sequences) to amino acid sequences? Clearly, the symbols or "letters" used in the code must be the bases; but what comprises a ***codon,*** the unit or "word" specifying one amino acid (or, actually, one aminoacyl-tRNA complex)?

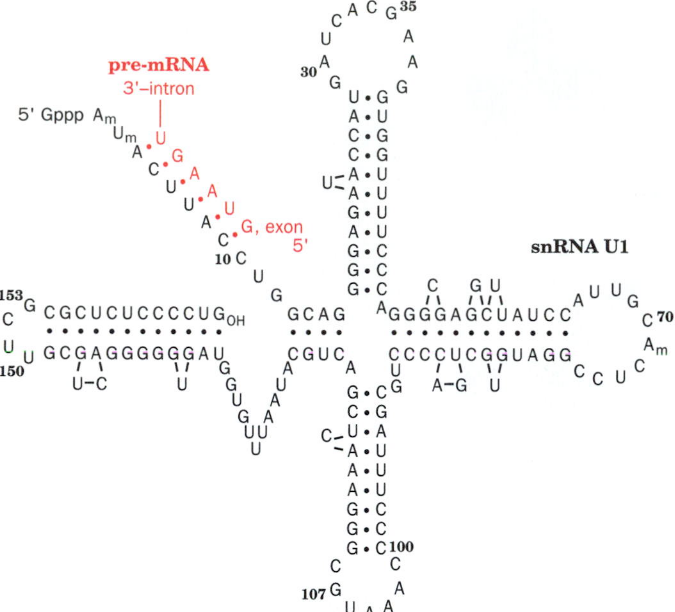

Figure 10.29 Probable secondary structure of the human snRNA U1 and its potential base-pairing to the 5′ consensus sequence of introns in nuclear pre-mRNAs. The pre-mRNA sequence is shown in red with the 5′ consensus sequence (5′-GUAAGU-3′) of the intron written 3′ to 5′ to align it with the complementary sequence near the 5′ end of the U1 snRNA. The m symbols refer to methyl side groups. The U1 snRNA has a trimethyl G cap at the 5′ terminus linked to the subterminal A_m by a 5′–5′ triphosphate linkage like most eukaryotic mRNAs (see Fig. 10.21). (After P. Epstein, R. Reddy, and H. Busch, "Site-Specific Cleavage by T1 RNase of U-1 RNA in U-1 Ribonucleoprotein Particles," *Proc. Natl. Acad. Sci. U.S.A.* 78:1562–1566, 1981.)

Three Nucleotides per Codon

Twenty different amino acids are incorporated during translation. Thus, at least 20 different codons must be formed using the four symbols (bases) available in the "message" (mRNA). Two bases per codon would yield only 4^2 or 16 possible codons—clearly not enough. Three bases per codon yield 4^3 or 64 possible codons—an apparent excess.

The first strong evidence that the genetic code was in fact a ***triplet code*** (three nucleotides per codon) resulted from a genetic analysis of proflavin-induced mutations in the *r*II locus of phage T4 carried out by F. H. C. Crick and colleagues in 1961. Crick and colleagues isolated proflavin-induced revertants of a proflavin-induced mutation. (Proflavin, an acridine dye, induces single base-pair additions and deletions; see Chapter 11, pp. 310–311.) These revertants were shown (by backcrosses to wild type; see Fig. 11.9) to result from the occurrence of suppressor mutations rather than from back-mutation at the original site of mutation. Crick and colleagues reasoned that if the original mutation was a single base-pair addition or deletion, then the suppressor mutations must be single base-pair deletions or additions, respectively, occurring at a site or sites near the original mutation. A single base-pair addition or deletion will alter the ***reading frame*** of the gene and mRNA (the codons in phase during translation) for that portion of the gene distal to the mutation (relative to the direction of translation). This is illustrated in Fig. 10.30*a*. When the suppressor mutations were isolated as single mutants by screening progeny of backcrosses to wild type, they were found to produce mutant phenotypes, just like the original mutation. Crick and colleagues next isolated proflavin-induced suppressor mutations of the original suppressor mutations, and so on.

All the isolated mutations were then classified into two groups, plus (+) and minus (−) (for additions and deletions, although Crick *et al.* had no idea which group was which) using the reasoning that a (+) mutation would suppress a (−) mutation, but not another (+) mutation, and vice versa (see Fig. 10.30 and legend for additional details). Next, Crick *et al.* constructed recombinants that carried various combinations of the (+) and the (−) mutations. Recombinants with two (+) mutations or two (−) mutations always had mutant phenotypes, just like the single

Figure 10.30 (Right page): Schematic illustration of Crick and coworker's proof that the genetic code is a triplet code (three bases per codon). Crick and colleagues studied a series of suppressor mutations of a mutation at the *r*II locus of phage T4. The original *r*II mutation had been induced with the acridine dye proflavin; it was most likely, therefore, the result of a single base-pair addition or deletion. In (a), the original mutation is shown arbitrarily as a single base-pair addition, specifically as an AT base-pair insertion (wild-type allele → mutant allele). The nucleotide-pair sequence shown for the wild-type allele (and thus the mRNA base sequence and amino acid sequence of the polypeptide) is hypothetical. Crick and coworkers selected phenotypic revertants of this mutant and demonstrated by backcrosses that these revertants resulted from suppressor mutations, not back-mutations at the original mutant site. If the original mutation is an addition, gene-product activity might be restored by a deletion (single base-pair) mutation in a nearby region, for example, the deletion of a CG base-pair as shown (mutant allele → revertant allele). The original addition mu-

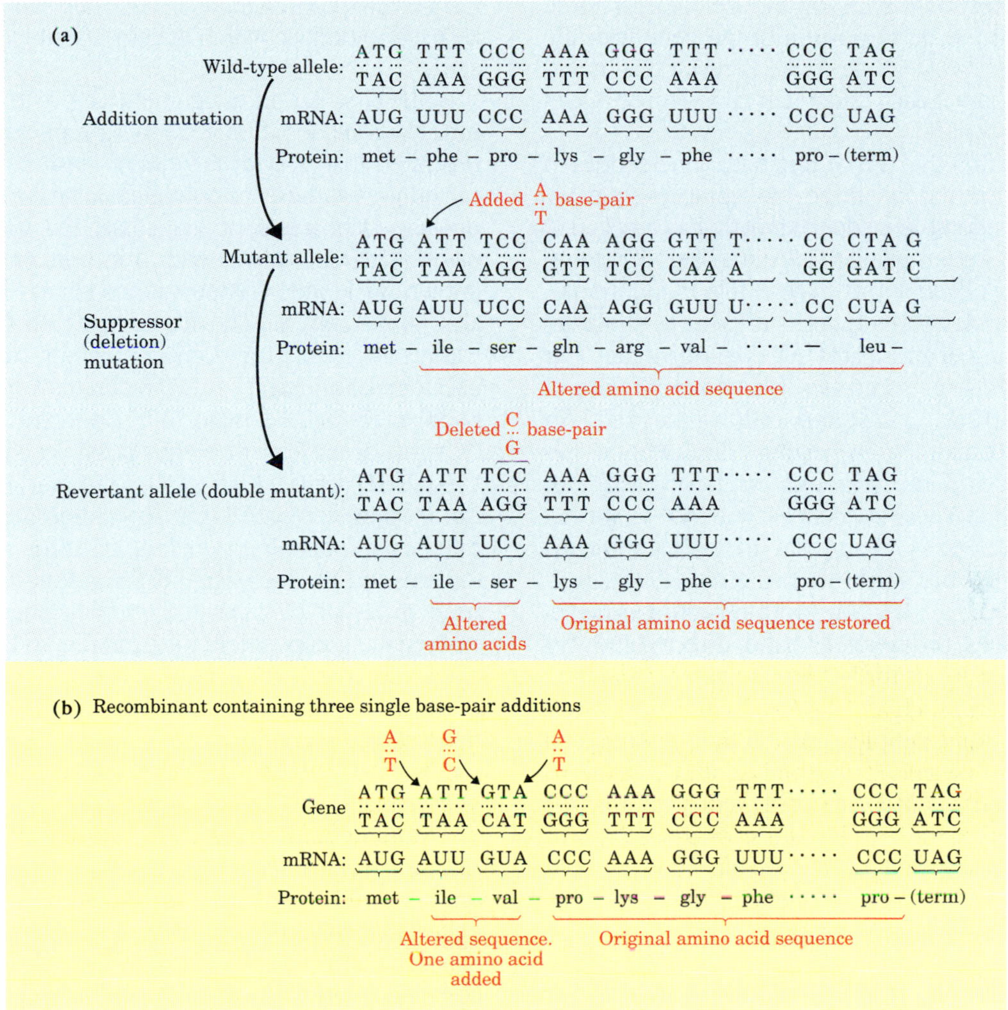

tation will change the ***reading frame*** (determining codons in phase for translation) for all codons distal (relative to the direction of translation) to the site of the mutation. The subsequent deletion (suppressor mutation) will restore the reading frame for the distal portion of the gene. If the altered amino acid sequence is not critical to function, the protein produced by the doubly mutant gene will be active. When the suppressor mutations were isolated in single-mutant strains by backcrosses to wild type, all were found to yield mutant phenotypes, just like the original mutations that they suppressed. Crick and colleagues then isolated proflavin-induced suppressor mutations of the previously isolated suppressor mutations (present in single mutants recovered from backcrosses). After repeating this process for several cycles, all the mutations were classified as plus (for single base-pair addition) or minus (for single base-pair deletion) on the basis that a plus mutation would suppress a minus mutation, but not another plus mutation, and vice versa. Actually, Crick and colleagues had no idea whether the plus group of mutations represented additions or not; they could just as likely have been deletions. The only important point was that all deletions ended up in one group (be it the plus group or the minus group), and all additions ended up in the other group.

Having so classified the mutations, Crick and coworkers next performed the critical experiment. They isolated recombinants carrying various combinations of plus and minus mutations. When two plus mutations were present in a recombinant, its phenotype was always mutant. The same was true in the case of recombinants with two minus mutations. However, recombinants containing ***three plus mutations*** (b) or ***three minus mutations*** frequently had wild-type phenotypes. Thus, the wild-type reading frame for the distal portion of the gene was restored by three single base-pair additions or three single base-pair deletions, but was altered by either one or two base-pair additions or deletions. These results are most easily explained if each codon contains three bases.

mutants. Recombinants carrying ***three (+) mutations*** (Fig. 10.30*b*) or ***three (−) mutations,*** however, often had wild-type phenotypes. This indicated that the additions of three base-pairs or the deletion of three base-pairs left the distal portion of the gene with the correct (wild-type) reading frame, a result that would be expected only if each codon contained three nucleotides.

Confirmation that the coding ratio (nucleotides to amino acids) is indeed three has come from many sources. Considerable evidence favoring a triplet code evolved from studies using *in vitro* translation systems. The following observations were of major importance. (1) Trinucleotides were found sufficient to stimulate specific binding of aminoacyl-tRNAs to ribosomes. For example, 5′-UUC-3′ stimulates ribosomal binding of phenylalanyl-$tRNA_{phe}$. (2) Chemically synthesized RNA molecules, containing repeating dinucleotide sequences, directed the synthesis of copolymers with alternating amino acid sequences. Poly $(UG)_n$, for example, when used as an artificial mRNA in an *in vitro* system, directed the synthesis of the repeating copolymer $(cys\text{-}val)_n$. (3) Molecules with repeating trinucleotide sequences, on the other hand, directed the synthesis of a mixture of three homopolymers (initiation being random on such an mRNA in an *in vitro* system). Poly $(UGG)_n$, for example, directed the synthesis of a mixture of polyserine, polyarginine, and polyvaline. Again, these results are only consistent with a triplet code. Ultimately, the triplet nature of the code was definitively established by the results of correlated nucleic acid and protein sequencing (e.g., see Fig. 10.31 and Chapter 12, Fig. 12.26).

Deciphering the Code

The deciphering of the genetic code—that is, determining (1) which codons specify which amino acids, (2) how many of the 64 possible codons are used, (3) how the code is punctuated, and (4) whether different species use the same or different codons—took place during the early 1960s and was one of the most exciting periods in the history of science. The "cracking" of the genetic code had an effect on the life sciences like the splitting of the atom did on the physical sciences. It opened up a vast new field of study of gene expression.

The first major breakthrough came in 1961 when M. W. Nirenberg (1968 Nobel Prize recipient) and J. H. Matthaei and then S. Ochoa (1959 Nobel Prize recipient) and coworkers demonstrated that synthetic RNA molecules could be used as artificial mRNAs to direct *in vitro* protein synthesis. That is, when ribosomes, aminoacyl-tRNAs, and the soluble protein factors required for translation are purified free of natural mRNAs, these components can be combined *in vitro* and stimulated to synthesize polypeptides by the addition of chemically synthesized RNA molecules. If these synthetic mRNA molecules are of known composition, the composition of the polypeptides synthesized can be used to deduce which codons specify which amino acids.

The first codon assignment (UUU for phenylalanine) was made when Nirenberg and Matthaei demonstrated that ***polyuridylic acid*** [***poly U*** = $(U)_n$] directed the synthesis of polyphenylalanine [$(\text{phenylalanine})_n$]. Ochoa and others continued this approach using synthetic RNAs with random sequences of known nucleotide composition, such as 50 percent U and 50 percent G. The frequencies of the different triplets in such a random copolymer can be easily calculated. For example, the 50 percent U/50 percent G copolymer will contain 12.5 percent ($\frac{1}{2} \times \frac{1}{2} \times \frac{1}{2} = \frac{1}{8}$) of each of the eight possible codons: UUU, UUG, UGU, GUU, UGG, GUG, GGU, and GGG. These can then be compared with the amino acids incorporated (phenylalanine, leucine, cysteine, valine, tryptophan, and glycine) when this random copolymer is used in an *in vitro* protein-synthesizing system. By varying the composition, for example, to 75 percent U and 25 percent G, one can vary the relative frequencies of the eight codons and correlate them with the relative frequencies of the amino acids in the polypeptides synthesized. Such experiments provided a great deal of information about the nature of the code.

More definitive data were later obtained by H. G. Khorana using *in vitro* systems that were activated by synthetic mRNAs of known nucleotide sequences. Khorana's experiments permitted direct comparisons between nucleotide sequences and the amino acids incorporated in response to these sequences. The ultimate "cracking" of the code occurred when trinucleotides were found to function as "mini-mRNAs" in directing the specific binding of aminoacyl-tRNAs to ribosomes. By using all the 64 possible trinucleotide sequences in such aminoacyl-tRNA binding experiments, it was possible to verify the codon assignments made from data of earlier experiments.

On the basis of extensive data accumulated over several years, the codon assignments shown in Table 10.1 became firmly established. Two important questions remained to be answered. (1) Are the assignments based on *in vitro* experiments valid *in vivo?* (2) Is the code ***universal,*** that is, do the codons specify the same amino acids in all organisms? Several lines of evidence now indicate that these codon assignments are correct for protein synthesis *in vivo* for most, if not all, species. When the amino acid substitutions that result from mutations induced with chemical mutagens with specific mutagenic effects (see Chapter 11) are determined by amino acid sequencing, the substi-

Coat Protein initiation codon

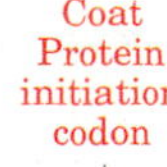

. . . (G) – AUA – GAG – CCC – UCA – ACC – GGA – GUU – UGA – AGC – AUG –

GCU –	UCU –	AAC –	UUU –	ACU –	CAG –	UUC –	GUU –	CUC –	GUC –	GAC –	AAU –	GGC –	GGA –	ACU –	GGC –	GAC –	GUG –	ACU –	GUC –	GCC –	CCA –	AGC –	AAC –	UUC –
Ala	Ser	Asn	Phe	Thr	Gln	Phe	Val	Leu	Val	Asp	Asn	Gly	Gly	Thr	Gly	Asp	Val	Thr	Val	Ala	Pro	Ser	Asn	Phe
1				5					10					15					20					25

GCU –	AAC –	GGG –	GUC –	GCU –	GAA –	UGG –	AUC –	AGC –	UCU –	AAC –	UCG –	CGU –	UCA –	CAG –	GCU –	UAC –	AAA –	GUA –	ACC –	UGU –	AGC –	GUU –	CGU –	CAG –
Ala	Asn	Gly	Val	Ala	Glu	Trp	Ile	Ser	Ser	Asn	Ser	Arg	Ser	Gln	Ala	Tyr	Lys	Val	Thr	Cys	Ser	Val	Arg	Gln
				30					35					40					45					50

AGC –	UCU –	GCG –	CAG –	AAU –	CGC –	AAA –	UAC –	ACC –	AUC –	AAA –	GUC –	GAG –	GUG –	CCU –	AAA –	GUG –	GCA –	ACC –	CAG –	ACU –	GUU –	GGU –	GGU –	GUA –
Ser	Ser	Ala	Gln	Asn	Arg	Lys	Tyr	Thr	Ile	Lys	Val	Glu	Val	Pro	Lys	Val	Ala	Thr	Gln	Thr	Val	Gly	Gly	Val
				55					60					65					70					75

GAG –	CUU –	CCU –	GUA –	GCC –	GCA –	UGG –	CGU –	UCG –	UAC –	UUA –	AAU –	AUG –	GAA –	CUA –	ACC –	AUU –	CCA –	AUU –	UUC –	GCU –	ACG –	AAU –	UCC –	GAC –
Glu	Leu	Pro	Val	Ala	Ala	Trp	Arg	Ser	Tyr	Leu	Asn	Met	Glu	Leu	Thr	Ile	Pro	Ile	Phe	Ala	Thr	Asn	Ser	Asp
				80					85					90					95					100

UGC –	GAG –	CUU –	AUU –	GUU –	AAG –	GCA –	AUG –	CAA –	GGU –	CUC –	CUA –	AAA –	GAU –	GGA –	AAC –	CCG –	AUU –	CCC –	UCA –	GCA –	AUC –	GCA –	GCA –	AAC –
Cys	Glu	Leu	Ile	Val	Lys	Ala	Met	Gln	Gly	Leu	Leu	Lys	Asp	Gly	Asn	Pro	Ile	Pro	Ser	Ala	Ile	Ala	Ala	Asn
				105					110					115					120					125

UCC –	GGC –	AUC –	UAC –	UAA –	UAG –	ACG –	CCG –	GCC –	AUU –	CAA –	ACA –	UGA –	GGA –	UUA –	CCC	AUG	UCG	AAG	ACA	ACA	AAG	AAG	(U)
Ser	Gly	Ile	Tyr														Ser	Lys	Thr	Thr	Lys	Lys	
			129														1				5		

Tandem *ochre* and *amber* termination codons (UAA – UAG)

Opal termination codon (UGA)

Replicase initiation codon (AUG)

Figure 10.31 Correlated nucleotide sequence of the coat protein gene of the RNA bacteriophage MS2 and the amino acid sequence of the polypeptide (coat protein) that it specifies. The initial sequence of the MS2 replicase (RNA polymerase) gene and the correlated six amino-terminal amino acids are also shown. An untranslated intergenic sequence separates the genes. Translation proceeds from left to right as drawn, and from the coat gene to the replicase gene. Both polypeptides are initiated by f-methionine at the AUG codons indicated. The methionine is cleaved off following (or during) translation, yielding the alanine terminus on the coat protein and the serine terminus on the replicase. The coat protein gene has two tandem "periods" (two termination codons) at the end, as though to make absolutely certain that translation terminates at this point. In addition, a third termination codon is located, in proper reading frame, seven base triplets from the second tandem termination. Each of the three termination codons is present once between the translated sequence of the coat gene and the translated sequence of replicase gene. Note that the amino acid sequence of this protein, synthesized *in vivo,* is precisely that predicted from the nucleotide sequence using the codon assignment presented in Table 10.1. (Data from W. Min Jou, G. Haegeman, Y. Ysebaert, and W. Fiers, *Nature* 237: 82–88, 1972.)

tutions are almost always consistent with the codon assignments given in Table 10.1 and the known effect of the mutagen. More convincingly, when the nucleotide sequences of genes or of mRNAs are determined and compared with the amino acid sequences of the polypeptides encoded by those genes or mRNAs, the observed correlations are always found to be those predicted from the accepted codon assignments (Table 10.1). This can be illustrated by comparing the nucleotide sequence of the gene coding for the protein coat or capsid of bacteriophage MS2 with the amino acid sequence of the capsid polypeptide (Fig. 10.31). Phage MS2 stores its genetic information in RNA (like TMV virus; see Chapter 5, pp. 96–97). Its chromosome is equivalent to an mRNA molecule in organisms with DNA genomes. (Also see Chapter 12, Fig. 12.26.)

Degeneracy and Wobble

All the amino acids except methionine and tryptophan are specified by more than one codon (Table 10.1). Three amino acids, leucine, serine, and arginine, are each specified by six different codons. Isoleucine has three codons. The other amino acids each have either two or four codons. The occurrence of more than one codon per amino acid is called ***degeneracy*** (though the usual connotations of the term are hardly appropriate). The degeneracy in the genetic code is not at random; instead, it is highly ordered. Usually, the

TABLE 10.1 The Genetic Code[a]

FIRST (5′) LETTER	SECOND LETTER				THIRD (3′) LETTER
	U	C	A	G	
U	UUU, UUC } Phe UUA, UUG } Leu	UCU, UCC, UCA, UCG } Ser	UAU, UAC } Tyr UAA *Ochre* (terminator) UAG *Amber* (terminator)	UGU, UGC } Cys UGA *Opal* (terminator) UGG Trp	U C A G
C	CUU, CUC, CUA, CUG } Leu	CCU, CCC, CCA, CCG } Pro	CAU, CAC } His CAA, CAG } Gln	CGU, CGC, CGA, CGG } Arg	U C A G
A	AUU, AUC, AUA } Ileu AUG Met (initiator)	ACU, ACC, ACA, ACG } Thr	AAU, AAC } Asn AAA, AAG } Lys	AGU, AGC } Ser AGA, AGG } Arg	U C A G
G	GUU, GUC, GUA, GUG } Val GUG (initiator)	GCU, GCC, GCA, GCG } Ala	GAU, GAC } Asp GAA, GAG } Glu	GGU, GGC, GGA, GGG } Gly	U C A G

[a]Each triplet nucleotide sequence or codon refers to the nucleotide sequence in **mRNA** (not DNA) that specifies the incorporation of the indicated amino acid or polypeptide chain termination.

multiple codons specifying an amino acid differ by only one base, the third or 3′ base of the codon. The degeneracy is primarily of two types. (1) Partial degeneracy occurs when the third base may be either one of the two pyrimidines (U and C) or, alternatively, either one of the two purines (A and G). With partial degeneracy, changing the third base from a purine to a pyrimidine, or vice versa, will change the amino acid specified by the codon. (2) In the case of complete degeneracy, any of the four bases may be present at the third position in the codon, and the codon will still specify the same amino acid. For example, valine is specified by GUU, GUC, GUA, and GUG (Table 10.1).

It has been speculated that the order in the genetic code has evolved as a way of minimizing mutational lethality. Many base substitutions at the third position of codons do not change the amino acid specified by the codon. Moreover, amino acids with similar chemical properties (such as leucine, isoleucine, and valine) have codons that differ from each other by only one base. Thus, many single base-pair substitutions will result in the substitution of one amino acid for another amino acid with very similar chemical properties (e.g., valine for isoleucine). In most cases, such substitutions will not result in inactive gene-products; again, this minimizes the effects of mutations.

Because of the degeneracy of the genetic code, there must either be several different tRNAs that recognize the different codons specifying a given amino acid or the anticodon of a given tRNA must be able to base-pair with several different codons. Actually, both of these occur. Several tRNAs exist for certain amino acids, and some tRNAs recognize more than one codon. The hydrogen bonding between the bases in the anticodon of tRNA and the codon of mRNA appears to follow strict base-pairing rules (i.e., be "tight") only for the first two bases of the codon. The base-pairing involving the third base of the codon is apparently less stringent, allowing what Crick has called wobble at this site.

On the basis of molecular distances and steric (three-dimensional structure) considerations, Crick proposed that wobble would allow several types, but not all types, of base-pairing at the third codon base in the codon–anticodon interaction. His proposal has since been strongly supported by experimental data. Table 10.2 shows the base-pairing predicted by the wobble hypothesis. It necessitates that there be at least two tRNAs for each amino acid whose codons exhibit complete degeneracy at the third position. This has indeed been found to be true. The wobble hypothesis predicted the occurrence of three tRNAs for the six serine codons. Three serine tRNAs have been characterized: (1) $tRNA_{ser1}$ (anticodon AGG) binds to codons UCU and UCC, (2) $tRNA_{ser2}$ (anticodon AGU) binds to codons UCA and UCG, and (3) $tRNA_{ser3}$ (anticodon UCG) binds to codons AGU and AGC. These specificities were verified by the trinucleotide-stimulated binding of purified aminoacyl-tRNAs to ribosomes *in vitro*.

Finally, several tRNAs contain the base inosine (produced by posttranscriptional enzymatic modification). Crick's wobble hypothesis predicted that inosine could pair (at the wobble position) with adenine, uracil, or cytosine (in the codon). In fact, purified alanyl-tRNA containing inosine (I) at the 5′ position of the anticodon (Fig. 10.15) binds to ribosomes activated with GCU, GCC, or GCA trinucleotides. The same result has been obtained with other purified tRNAs with inosine at the 5′ position of the anticodon. The wobble hypothesis thus fits several observations; whether it is entirely accurate remains unknown.

TABLE 10.2 Base-pairing between the 5′ Base of the Anticodon of tRNAs and the 3′ Base of Codons of mRNAs According to the Wobble Hypothesis

BASE IN ANTICODON	BASE IN CODON
G	U or C
C	G
A	U
U	A or G
I	A, U, or C

Initiation and Termination Codons

The genetic code also provides for punctuation of genetic information at the level of translation. Three codons, UAA, UAG, and UGA, specify polypeptide chain termination. These codons are recognized by protein release factors, rather than by tRNAs. One of these proteins, designated RF-1, is apparently specific for UAA and UAG. The other, RF-2, causes termination at UAA and UGA codons. Two codons, AUG and GUG, are recognized by the initiator tRNA, $tRNA_i^{Met}$, but apparently only when they follow an appropriate nucleotide sequence in the leader segment of an mRNA molecule. At internal positions, AUG is recognized by $tRNA^{Met}$, and GUG is recognized by a valine tRNA. In the case of the initiation codons AUG and GUG and $tRNA_i^{Met}$, the wobble base appears to be the first or 5′ base of the codon. Since wobble at the first base is unique to initiation, it may be related to base-pairing at the *P* site rather than at the *A* site on the ribosome.

Universality of the Code

A vast amount of data is now available from *in vitro* studies, from amino acid replacements due to mutations, and from correlated nucleic acid and polypeptide sequencing—all suggesting that the genetic code is the same or very nearly the same in all organisms. These data (e.g., see the human hemoglobin substitutions, Chapter 11, Fig. 11.8; the correlated nucleotide and amino acid sequences in the overlapping genes of the DNA bacteriophage ΦX174, Chapter 12, Fig. 12.26; and of the RNA bacteriophage MS2, Fig. 10.31) all indicate that the genetic code is largely ***universal***.

The major exception to the universality of the code occurs in mitochondria of humans, yeast, and several other species, where UGA is a tryptophan codon. UGA is a termination codon in nonmitochondrial systems. Also, in yeast mitochondria, CUA specifies threonine instead of the usual leucine, and, in mammalian mitochondria, AUA specifies methionine instead of the usual isoleucine. Excluding these and a few related exceptions, the code appears to be universal.

Suppressor Mutations Producing tRNAs with Altered Codon Recognition

The nonmitochondrial genetic code is not absolutely universal, at least not in the most restrictive sense of the word, because minor variations in codon recognition and translation are well documented in a few cases. In *E. coli* and yeast, for example, mutations occur in tRNA genes, resulting in altered codon recognition by the anticodons of tRNAs. These mutations were initially detected as suppressor mutations, mutations that suppressed the effects of other mutations. They were subsequently shown to be located in tRNA genes and to change the specificity of codon recognition by the tRNAs.

The best-known examples of suppressor mutations that alter tRNA specificity are the suppressors of mutations that produce UAG chain-termination triplets within the coding sequences of structural genes. Such mutations, called *amber* mutations, result in the synthesis of amino-terminal fragments of the polypeptides specified by the affected genes. Mutations that produce chain-termination triplets within genes have come to be known as ***nonsense*** mutations, in contrast to ***missense*** mutations, which change a triplet to another triplet specifying a different amino acid. Missense mutations thus result in complete polypeptides, but with an amino acid substitution (such as the mutant hemoglobins described in Fig. 11.8). Nonsense mutations result in fragments of the polypeptides, whose lengths depend on the position of the mutation within the gene (see Fig. 10.33). Nonsense mutations frequently result from single base-pair substitutions (as shown in Fig. 10.32). The polypeptide fragments produced from genes containing nonsense mutations are almost always totally nonfunctional.

Suppression of nonsense mutations has been shown to result from mutations in tRNA genes that cause the mutant tRNAs to recognize the nonsense (UAG, UAA, or UGA) codons, albeit with varying efficiencies. These mutant tRNAs are referred to as suppressor tRNAs. When one of the **amber** (UAG) suppressor tRNAs, the one resulting from the **amber su3** ***mutation,*** was sequenced, it was found to have an altered anticodon. This particular *amber* suppressor mutation occurs in the $tRNA_{tyr2}$ gene (specifying one of the two tyrosine tRNAs in *E. coli*). The anticodon of the wild-type (nonsuppressor) $tRNA_{tyr2}$ was shown to be 5′-G′UA-3′ (where G′ is a derivative of guanine). The anticodon of the mutant (suppressor) $tRNA_{tyr2}$ is 5′-CUA-3′. This mutant anticodon is thus able to base-pair with the 5′-UAG-3′ *amber* codon (pairing with opposite polarity, of course), that is

3′-AUC-5′ (anticodon)
5′-UAG-3′ (codon)

The suppressor tRNAs thus permit complete polypeptides to be synthesized from mRNAs containing nonsense codons. These polypeptides will be functional as long as the amino acid inserted by the suppressor tRNA is acceptable at that position, yielding a functional gene-product (Fig. 10.32*c*).

Escherichia coli strains containing *amber* suppressor genes (and tRNAs) exhibit growth rates comparable to strains without suppressor genes. This might not be expected if the UAG codon is frequently used for the normal termination of translation at the ends of coding sequences of mRNAs. In suppressor strains, one would expect translation occasionally (how frequently depends on the efficiency of the particular suppressor) to continue through into normally untranslated intergenic spacer sequences and even into other coding sequences located translationally distal on the mRNAs. This would generate long "poly-proteins," which one would not expect to be very functional. Why suppressor strains do not show deleterious effects of such "read-through" is not understood. Perhaps most genes are like the phage MS2 coat gene (Fig. 10.31) and terminate with two or more different translation-termination codons. The nucleotide sequences of several genes, however, reveal the presence of single termination codons. The apparent lack of deleterious effects of suppressor mutations thus remains an enigma.

COLINEARITY OF GENE AND POLYPEPTIDE

The genetic information is stored in linear sequences of nucleotide-pairs in DNA (or nucleotides in RNA, in some cases). Transcription and translation convert this genetic information into polypeptides (linear sequences of amino acids), which function as the key intermediaries in the genetic control of phenotype.

We now know that the amino acid sequences of the polypeptides and the nucleotide-pair sequences of the genes coding for these polypeptides are ***colinear.*** That is, the first three base-pairs of a gene specify the first amino acid of the polypeptide, the next three base-pairs (four to six) specify the second amino acid, and so on, in a colinear fashion. The first strong evidence for colinearity of genes and polypeptides resulted from studies on one of the two polypeptides in tryptophan synthetase of *E. coli.* C. Yanofsky and colleagues demonstrated that there was a perfect correlation between the map positions of mutations in the tryptophan synthetase *A* gene and the positions of the resultant amino acid substitutions in the tryptophan synthetase α-polypeptide. Yanofsky's results are summarized in Chapter 12, Fig. 12.19.

At about the same time, A. Sarabhai and associates demonstrated a similar colinearity between the positions of mutations in the gene of bacteriophage T4 that codes for the major structural protein of the phage head and the positions in the polypeptide affected by the mutations. Sarabhai and colleagues studied *amber* (UAG chain-termination) mutations and demonstrated a direct correlation between the length of the polypeptide fragment produced and the position of the mutation within the gene (Fig. 10.33).

Definitive evidence for colinearity has been provided by correlated nucleic acid sequencing and polypeptide sequencing studies. These studies (see Fig. 10.31 and Chapter 12, Fig. 12.26) show a direct colinearity between the nucleotide sequences of the genes and the amino acid sequences of the polypeptide gene-products.

In eukaryotes, the data available to date also support colinearity, ***but the linear sequence of nucleotide-pairs in the gene that specifies a colinear polypeptide may not always consist of contiguous nucleotide-pairs.*** Instead, there frequently are ***noncoding intron sequences intervening between coding sequences*** (see Chapter 13, pp. 356–359). This does not violate the concept of colinearity, but only

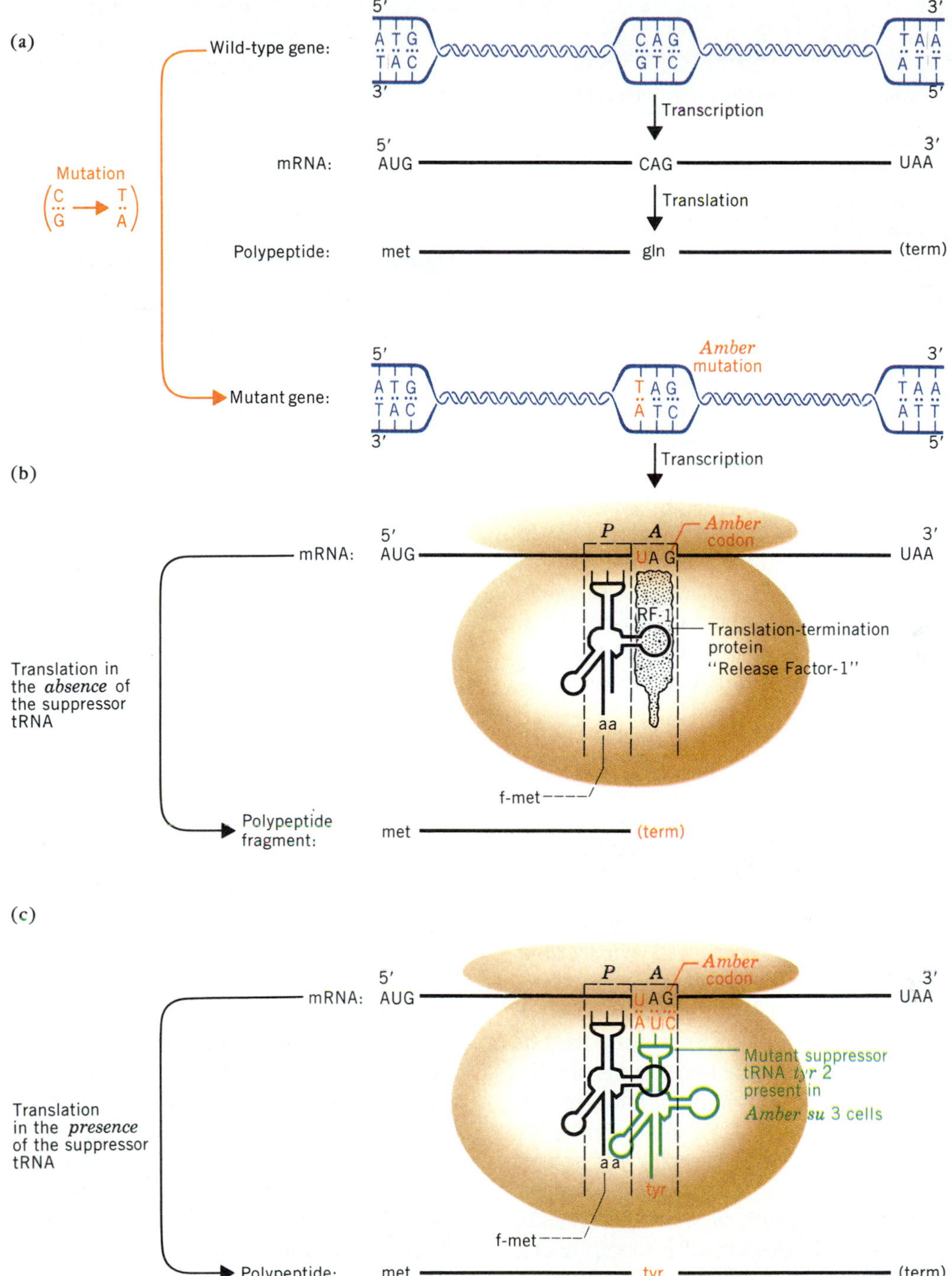

Figure 10.32 Schematic illustration showing the formation of an *amber* (UAG) chain-termination mutation (a to b) and its effect in (b) the absence of a suppressor tRNA and (c) the presence of a suppressor tRNA. The *amber* mutation shown changes a CAG mRNA codon that specifies glutamine to a UAG chain-termination codon. In the absence of a suppressor mutation resulting in a suppressor tRNA in the cell, the *amber* mutation causes the synthesis of only a fragment of the polypeptide gene-product. When the presence of the suppressor mutation yields a suppressor tRNA (c), a modified tyrosine tRNA mediates the insertion of tyrosine into the polypeptide at the site specified by the UAG codon. Other *amber* suppressors insert serine, lysine, leucine, or glutamine in response to UAG codons. *Amber* suppressors will yield a functional product of a mutant gene containing a UAG codon only if the amino acid inserted by the suppressor is an acceptable substitute for the original ("wild-type") amino acid present at that position. For example, the tyrosine inserted in c replaces a glutamine at that position in the wild-type polypeptide (a). The polypeptide containing tyrosine may or may not be functional. Suppression will, of course, be observed only if the tyrosine-containing polypeptide is active.

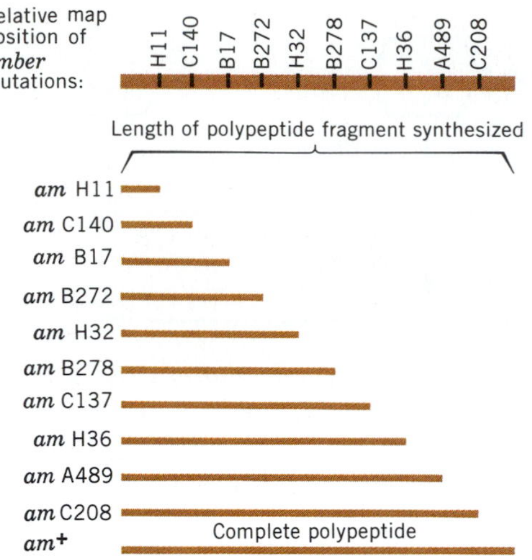

Figure 10.33 Colinearity between the map positions of *amber* (UAG chain-termination) mutations in gene 23 (the gene coding for the major structural protein of the head) of bacteriophage T4 and the positions of the resulting polypeptide chain terminations in the mutant gene-products. The wild-type (am^+) polypeptide is represented at the bottom for comparison. (Based on the data of A. S. Sarabhai, A. O. W. Stretton, and S. Brenner, *Nature* 201: 13–17, 1964.)

demonstrates that the sequences of base-pair triplets in the gene specifying mRNA codons and amino acids in the colinear polypeptide are ***not*** always ***uninterrupted***. In one case, F. Sherman and colleagues have demonstrated an uninterrupted colinear relationship between the map positions of mutations in the *CYCl* gene of yeast and the amino acid substitutions in the mutant forms of iso-1-cytochrome *c*, the polypeptide specified by the *CYCl* gene. Whether colinearity is a universal feature of gene–polypeptide relationships is still an open question.

SUMMARY

The pathway of information flow by which a gene exerts its effect on the phenotype of an organism is often very complex. The main features of the first two steps in this pathway, ***transcription*** and ***translation***, are now quite well understood. Transcription involves the synthesis of a ***messenger RNA*** or ***pre-messenger RNA*** molecule, which functions as an intermediary in protein synthesis, using one of the two strands of DNA in the gene as a template. Transcription is catalyzed by a complex, multimeric enzyme, ***DNA-dependent RNA polymerase***. In eukaryotes, the primary transcripts usually undergo several types of processing. Processing frequently includes excision of specific segments of the primary transcript, addition of a 5′ cap and a 3′-poly-A tail, and association with specific proteins.

Many eukaryotic genes and a few prokaryotic genes contain ***noncoding sequences*** called ***introns*** or ***intervening sequences*** that separate the ***coding sequences*** called ***exons***. In these cases, the entire nucleotide-pair sequence of each split gene, including both intron and exon sequences, is transcribed to produce a ***pre-mRNA molecule***. The intron sequences are then removed from the primary transcript or pre-mRNA by ***splicing processes*** to produce the mature, functional RNA molecule. ***Three distinct splicing processes are known***. (1) The introns in many ***pre-tRNA molecules*** are excised by simple ***splicing nucleases*** and ***ligases***. (2) The introns in certain ***rRNA precursors*** and some primary transcripts of genes in mitochondria and chloroplasts are excised ***autocatalytically*** (self-catalyzed with no protein involved). (3) Finally, the introns of ***nuclear pre-mRNAs*** are excised by ***complex ribonucleoprotein structures*** called ***spliceosomes***.

The messenger RNA produced is then ***translated***, according to the specifications of the ***genetic code***, into the sequence of amino acids in the ***polypeptide*** gene-product. Translation takes place on complex, macromolecular structures called ***ribosomes*** that are located in the cytoplasm of cells. Each ribosome consists of two subunits, which together contain 3–5 different RNA molecules and 50–60 different proteins. The exact composition varies from species to species. Translation requires the participation of 40–60 small RNA molecules called ***transfer RNAs***. Each transfer RNA is activated for protein synthesis by the covalent attachment of a specific amino acid. This attachment is catalyzed by an enzyme called ***aminoacyl-tRNA synthetase***. There is at least one specific tRNA and one specific aminoacyl-tRNA synthetase for each of the 20 amino acids commonly found in natural proteins. Translation also requires several soluble proteins called ***initiation, elongation, and termination factors***. The result of the transcription of a gene and the subsequent translation of the mRNA produced during transcription is the synthesis of a polypeptide whose amino acid sequence is ***colinear*** with the sequence of nucleotide-pairs in the gene. In eukaryotes, this sequence of nucleotide-pairs in the gene is often interrupted by ***noncoding intron sequences*** that separate the coding sequences.

The ***genetic code*** is a ***triplet code,*** a sequence of three nucleotides making up each ***codon.*** The code is ***degenerate*** in the sense that all 64 possible triplet nucleotide sequences are used and several different codons frequently code for the same amino acid. Each of the 61 amino acid-specifying codons in mRNA is capable of base-pairing with a three-nucleotide sequence present at the ***anticodon*** site of one or more tRNAs. The base-pairing between the third (3′) base of the codon and the 5′ base of the anticodon does not follow the normal strict base-pairing rules; instead, there is ***"wobble"*** at this site, permitting base-pairs to form other than the usual four base-pairs. Thus, the anticodon of a single tRNA may recognize one, two, or three different codons. Two codons are used for polypeptide chain initiation, and three codons specify polypeptide chain termination. The genetic code is nearly ***universal;*** the codons appear to have almost the same meaning in all species.

The polypeptides produced by transcription and translation make up the structural proteins and enzymes that control metabolic processes in cells.

Blocks in metabolic processes resulting from genetic defects frequently have severe consequences. When the molecular bases of such congenital diseases are known, the symptoms of the disease can sometimes be relieved by appropriate treatment.

REFERENCES

Breathnach, R., and P. Chambon. 1981. "Organization and expression of eucaryotic split genes coding for proteins." *Annu. Rev. Biochem.* 50: 349–383.

Brow, D. A., and C. Guthrie. 1989. "Splicing a spliceosomal RNA." *Nature* 337: 14–15.

Brown, D. D. 1981. "Gene expression in eukaryotes." *Science* 211: 667–674.

Busch, H., R. Reddy, L. Rothblum, and Y. C. Choi. 1982. "SnRNAs, SnRNPs, and RNA processing." *Annu. Rev. Biochem.* 51: 617–654.

Cech, T. R. 1986. "The generality of self-splicing RNA: Relationship to nuclear mRNA splicing." *Cell* 44: 207–210.

Crick, F. H. C. 1966. "Codon–anticodon pairing; The wobble hypothesis." *J. Mol. Biol.* 19: 548–555.

______. 1966. "The genetic code: III." *Sci. Amer.* 215(4): 55–62.

Fiers, W., R. Contreras, F. Duerinck, G. Haegeman, D. Iserentant, J. Meerregaert, W. Min Jou, F. Molemans, A. Raeymekers, A. Van den Berghe, G. Volckaert, and M. Ysebaert. 1976. "Complete nucleotide sequence of bacteriophage MS2 RNA: Primary and secondary structure of the replicase gene." *Nature* 260: 500–507.

Friefelder, D. 1983. *Molecular biology, a comprehensive introduction to prokaryotes and eukaryotes.* Science Books International, Boston.

Garen, A. 1968. "Sense and nonsense in the genetic code." *Science* 160: 149–159.

Gold, L., D. Pribnow, T. Schneider, S. Shinedling, B. S. Singer, and G. Stormo. 1981. "Translational initiation in prokaryotes." *Annu. Rev. Microbiol.* 35: 365–403.

Goodman, H. M., J. Abelson, A. Dandry, S. Brenner, and J. D. Smith. 1968. "*Amber* suppression: A nucleotide change in the anticodon of a tyrosine transfer RNA." *Nature* 217: 1019–1024.

Guthrie, C., and B. Patterson. 1988. "Spliceosomal snRNAs." *Annu. Rev. Genet.* 22: 387–419.

Hall, B. D., S. G. Clarkson, and G. Tocchini-Valentini. 1982. "Transcription initiation of eukaryotic transfer RNA genes." *Cell* 29: 3–5.

Holley, R. W. 1966. "The nucleotide sequence of a nucleic acid." *Sci. Amer.* 214(2): 30–39.

Kruger, K., P. J. Grabowski, A. J. Zaug, J. Sands, D. E. Gottschiling, and T. R. Cech. 1982. "Self-splicing RNA: Autoexcision and autocyclization of the ribosomal RNA intervening sequence of *Tetrahymena.*" *Cell* 31: 147–157.

Lewin, B. 1974. *Gene expression-1. Bacterial genomes.* Wiley, New York.

______. 1977. *Gene expression-3. Plasmids and phages.* Wiley, New York.

______. 1980. *Gene expression-2. Eucaryotic genomes,* 2nd ed. Wiley, New York.

______. 1990. *Genes IV.* Cell Press, Cambridge, MA/Oxford Univ. Press, New York.

Maitra, U., E. A. Stringer, and A. Chaudhuri. 1982. "Initiation factors in protein synthesis." *Annu. Rev. Biochem.* 51: 869–900.

Nirenberg, M. W. 1963. "The genetic code: II." *Sci. Amer.* 208(3): 80–94.

Nirenberg, M. W., and J. H. Matthaei. 1961. "The dependence of cell-free protein synthesis in *E. coli* upon naturally occurring or synthetic polyribonucleotides." *Proc. Natl. Acad. Sci. U.S.A.* 47: 1588–1602.

Nomura, M., editor. 1974. *Ribosomes.* Cold Spring Harbor Laboratory Press, Cold Spring Harbor, NY.

Padgett, R. A., P. J. Grabowski, M. M. Konarska, S. Seiler, and P. A. Sharp. 1986. "Splicing of messenger RNA precursors." *Annu. Rev. Biochem.* 55: 1119–1150.

Pederson, T. 1981. "Messenger RNA biosynthesis and nuclear structure." *Amer. Scientist* 69: 76–84.

Reed, R., J. Griffith, and T. Maniatis. 1988. "Purification and visualization of native spliceosomes." *Cell* 53: 949–961.

Sarabhai, A. S., A. O. W. Stretton, and S. Brenner. 1964. "Co-linearity of the gene with the polypeptide chain." *Nature* 201: 13–17.

Sharp, P. A. 1981. "Speculations on RNA splicing." *Cell* 23: 643–646.

The genetic code. 1966. *Cold Spring Harbor Sympos. Quant. Biol.,* Vol. 31. Cold Spring Harbor Laboratory Press, Cold Spring Harbor, NY.

The mechanism of protein synthesis. 1970. *Cold Spring Harbor Sympos. Quant. Biol.,* Vol. 34. Cold Spring Harbor Laboratory Press, Cold Spring Harbor, NY.

Watson, J. D., N. H. Hopkins, J. W. Roberts, J. A. Steitz, and A. M. Weiner. 1987. *Molecular biology of the gene,* 4th ed. Benjamin/Cummings, Menlo Park, CA.

Weissbach, H., and S. Ochoa. 1976. "Soluble factors required for eukaryotic protein synthesis." *Annu. Rev. Biochem.* 45: 191–216.

Woese, C. R. 1967. *The genetic code, the molecular basis for genetic expression.* Harper & Row, New York.

Yanofsky, C. 1967. "Gene structure and protein structure." *Sci. Amer.* 216(5): 80–94.

Zang, A. J., P. J. Grabowski, and T. R. Cech. 1983. "Autocatalytic cyclization of an excised intervening sequence RNA is a cleavage–ligation reaction." *Nature* 301: 578–583.

PROBLEMS AND QUESTIONS

10.1 Distinguish between DNA and RNA (a) chemically, (b) functionally, and (c) by location in the cell.

10.2 What bases on the mRNA transcript would represent the following DNA sequence: 5′-TGCAGACA-3′?

10.3 What bases in the transcribed strand of DNA would give rise to the following mRNA base sequence: 5′-CUGAU-3′?

10.4 From what evidence was the messenger RNA hypothesis established?

10.5 In a general way, describe the molecular organization of proteins and distinguish proteins from DNA, chemically and functionally. Why is the synthesis of proteins of particular interest to geneticists?

10.6 At what different locations in the cell may protein synthesis occur?

10.7 Characterize ribosomes in general as to size, location, function, and macromolecular composition.

10.8 (a) Where in the cells of higher organisms do ribosomes originate? (b) Where in the cells are ribosomes most active in protein synthesis?

10.9 Identify three different major kinds of RNA and give the principal locations, characteristics, and functions of each in the living cell.

10.10 (a) How is messenger RNA related to polysome formation? (b) How does rRNA differ from mRNA and tRNA in specificity? (c) How does the tRNA molecule differ from that of DNA and mRNA in size and helical arrangement?

10.11 Outline the process of activation of amino acids.

10.12 (a) How was the genetic code first decoded? (b) What refinements have since been incorporated in the technique?

10.13 In what sense and to what extent is the genetic code (a) degenerate and (b) universal?

10.14 What evidence supports the hypothesis of colinearity between the nucleotide sequence in a gene and the amino acid sequence in a polypeptide?

10.15 Why is colinearity between codons and polypeptides significant?

10.16 Draw an analogy between the processes of transcription and translation and the process of building a house.

10.17 The thymine analog 5-bromouracil is a chemical mutagen that induces single base-pair substitutions in DNA called transitions (substitutions of one purine for another purine and one pyrimidine for another pyrimidine; see Chapter 11). Using the known nature of the genetic code (Table 10.1), which of the following amino acid substitutions should you expect to be induced by 5-bromouracil with the highest frequency: (a) Met → Val; (b) Met → Leu; (c) Lys → Thr; (d) Lys → Gln; (e) Pro → Arg; or (f) Pro → Gln? Why?

10.18 Using the information referred to in Problem 10.17, would you expect 5-bromouracil to induce a higher frequency of His → Arg or His → Pro substitutions? Why?

10.19 How is transcription (a) initiated and (b) terminated? How is translation (c) initiated and (d) terminated?

10.20 If the average molecular mass of an amino acid is assumed to be 100 daltons, about how many nucleotides will be present in an mRNA coding sequence specifying a single polypeptide with a molecular mass of 27,000 daltons?

10.21 How can a mutation in a tRNA gene lead to the incorrect translation of a specific codon? Could mutations in any genes other than tRNA genes lead to incorrect or ambiguous translation of a particular codon? If so, in genes coding for what?

10.22 What base-pair sequence in a segment of a gene in *Drosophila melanogaster* would code for the dipeptide sequence (NH_2) Met-Trp (COOH)?

10.23 If you were to (1) purify cysteine transfer RNA and charge it with labeled cysteine (i.e., activate it by attaching ^{3}H-labeled cysteine), (2) use Raney nickel (a highly reducing nickel powder) to convert the cysteine to alanine still attached to the cysteyl-specific transfer RNA, and (3) place the alanine-charged cysteyl transfer RNA into an *in vitro* protein-synthesizing system activated with poly UG templates that normally stimulate the incorporation of cysteine ***but not alanine*** into polypeptide chains (i.e., when you use the normal alanine-charged alanyl and cysteine-charged cysteyl transfer RNAs), what result would you expect?

10.24 A. Garen extensively studied a particular nonsense (chain-termination) mutation in the alkaline phosphatase gene of *E. coli.* This mutation resulted in the termination of the alkaline phosphatase polypeptide chain at a position where the amino acid tryptophan occurred in the wild-type polypeptide. Garen induced revertants (in this case, mutations altering the same codon) of this mutant with chemical mutagens that induced single base-pair substitutions (see

Chapter 11) and sequenced the mutant polypeptides. Seven different types of revertants were found, each with a different amino acid at the tryptophan position of the wild-type polypeptide (termination position of the mutant polypeptide fragment). The amino acids present at this position in the various revertants included tryptophan, serine, tyrosine, leucine, glutamic acid, glutamine, and lysine. Was the nonsense mutation studied by Garen an *amber* (UAG), an *ochre* (UAA), or an *opal* (UGA) nonsense mutation? Explain the basis of your deduction.

10.25 The goal of many genetic engineering experiments is to place eukaryotic genes into bacteria and to have the bacteria synthesize the eukaryotic gene-products, preferably in large amounts. Based on what is known about gene expression in bacteria and eukaryotes, what obstacles might have to be overcome to achieve this goal?

10.26 Many eukaryotic genes contain ***noncoding introns*** that separate the ***coding sequences*** or ***exons*** of these genes. At what stage during the expression of these mosaic genes are the noncoding intron sequences removed?

10.27 For several decades, the dogma in biology has been that molecular reactions occurring in living cells are catalyzed by enzymes composed of polypeptides. We now know that the introns of some precursor RNA molecules such as the rRNA precursors in *Tetrahymena* are removed autocatalytically ("self-spliced") with no involvement of any protein. What does the demonstration of autocatalytic splicing indicate about the dogma that biological reactions are always catalyzed by proteinaceous enzymes?

10.28 In what ways are ribosomes and spliceosomes similar? Different?

10.29 What components of the introns of nuclear genes that encode proteins in higher eukaryotes are conserved and required for the correct excision of intron sequences from primary transcripts by spliceosomes?

10.30 (a) Which of the following nuclear pre-mRNA nucleotide sequences potentially contains an intron?

(1) 5′—UGACCAUGGCGCUAACACUGCCAAUUGGCCAAU-ACUGACCUGAUAGCAUCAGCCAA—3′

(2) 5′—UAGUCUCAUCUGUCCAUUGACUUCGAAACUGAAU-CGUAACUCCUACGUCUAUGGA—3′

(3) 5′—UAGCUGUUUGUCAUGACUGACUGGUCACUAUCG-UACUAACCUGUCAUGCAAUGUC—3′

(4) 5′—UAGCAGUUCUGUCGCCUCGUGGUGCUGGCUGG-CCCUUCGUCGCUCGGGCUUAGCUA—3′

(5) 5′—UAGGUUCGCAUUGACGUACUUCUGAGACUACUA-ACUACUAACGCAUCGAGUCUCAA—3′

(b) One of the five pre-mRNAs shown above is a likely candidate for splicing out an intron sequence. What mRNA nucleotide sequence would one expect to result from this splicing event?

(Also see Chapter 11, Problems 11.28–11.35.)

Chapter 11

Mutation

Scanning electron micrograph of *Tetraptera,* a mutant in *Drosophila* that has four wings instead of two. (Photograph by David Scharf/Peter Arnold, Inc.)

INHERITANCE is based on genes that are faithfully transmitted from parents to offspring during reproduction (see Chapter 2). In higher organisms, these genes are located in chromosomes, which are duplicated and passed on to the progeny through the gametes during sexual reproduction (see Chapter 3). Structurally, these genes are almost always DNA and encode their information in sequences of base-pairs that are accurately duplicated during semiconservative replication. The DNA polymerases that catalyze the replication of DNA contain built-in $3' \rightarrow 5'$ exonuclease activities to proofread the progeny DNA molecules and correct mistakes made during the initial polymerization reaction (see Chapter 5). Mechanisms have thus evolved to facilitate the faithful transmission of genetic information from generation to generation. Nevertheless, "mistakes" or changes in the genetic material do occur. ***Such sudden, heritable changes in the genetic material are called mutations.***

The term ***mutation*** refers both to the ***change in the genetic material*** and to the ***process by which the change occurs.*** An organism exhibiting a novel phenotype as a result of the presence of a mutation is referred to as a ***mutant.*** Used in its broad historical sense, mutation refers to any sudden, heritable change in the genotype of an organism not explainable by recombination of preexisting genetic variability (see Chapters 7 and 8). Such genotypic changes include changes in chromosome number (***euploidy*** and ***aneuploidy;*** see Chapter 19), gross changes in the structure of chromosomes (***chromosome aberrations;*** see Chapter 18), and changes in individual genes. The term mutation is also frequently used in a narrow sense to refer only to changes in individual genes (excluding changes in chromosome number or structure). This chapter will deal with mutation in this narrower sense.

Many mutations involve changes in single base-pairs, the substitution of one base-pair for another, or the duplication or deletion of single base-pairs. Such mutations are referred to as ***point mutations.***

Mutation is the ultimate source of all genetic variation; it provides the raw material for evolution. Recombination (independent assortment plus recombination of genetic variability present in individual chromosomes; see Chapters 7, 8, and 12) merely rearranges this genetic variability into new combinations, and natural (or artificial) selection simply preserves the combinations best adapted to the existing (or desired) environmental conditions. ***Without mutation, all genes would exist in only one form. Alleles would not exist, and thus genetic analysis would not be possible.*** Most important, ***organisms would not be able to evolve and adapt to environmental changes.*** Mutation, then, is an important phenomenon. Some level of mutation is essential to provide new genetic variability to allow organisms to adapt to new environments. At the same time, if mutations occurred too frequently, they would totally disrupt the transmission of genetic information from generation to generation. What, then, is an appropriate level of mutation? This question will be returned to later in this chapter.

SPONTANEOUS VERSUS INDUCED MUTATION

Spontaneous mutations are those that occur without a known cause. They may be truly spontaneous, resulting from an inherent low level of metabolic errors, that is, mistakes during DNA replication, or they may actually be caused by mutagenic agents present in the environment. ***Induced mutations*** are those resulting from exposure of organisms to mutagenic agents such as ionizing irradiation, ultraviolet light, or various chemicals that react with DNA (or RNA, in RNA viruses). Operationally, it is impossible to prove that a particular mutation occurred spontaneously or that a specific mutation was induced by a mutagenic agent. The potential mutagenicity of both human-made and naturally occurring chemicals is only beginning to be assessed. Solar radiations are one cause of mutation, one that life can hardly do without. Fortunately, the ultraviolet rays in sunlight are of very low energy and do not penetrate tissue far enough to expose the gonads or germ cells of higher organisms to possible mutagenic effects. Ultraviolet light is mutagenic only to the skin cells of humans. This is probably of little comfort, however, to individuals with the inherited disease xeroderma pigmentosum, who often develop severe skin cancers following exposure to sunlight (see pp. 307–308). Thus, many, if not most, of the apparently "spontaneous" mutations may actually be induced by physical and chemical agents.

On the basis of chemical considerations, the intrinsic mistake frequency in nucleotide selection during DNA polymerization has been estimated to be about 10^{-5}. One cycle of proofreading by the $3' \rightarrow 5'$ exonuclease activity of DNA polymerases, assuming the same degree of accuracy in replacing the mismatched bases, would reduce the intrinsic mistake frequency to about 10^{-10} ($10^{-5} \times 10^{-5}$) errors per incorporated nucleotide. Additional repair mechanisms may reduce the intrinsic error frequency even further. Nevertheless, it seems clear that some minimal intrinsic error frequency does exist, and that this inherent frequency is subject to both genetic and environmental influences.

One cannot, therefore, distinguish between "spontaneous" and "induced" mutations when considering individual mutations. Such distinctions must be restricted to the population level. If the mutation rate is increased a hundredfold by treatment of a population with a mutagen, then an average of 99 out of every 100 mutations present in the population will have been induced by the mutagen. Valid comparisons between spontaneous and induced mutations can thus be made statistically by comparing populations exposed to a mutagenic agent with control populations that have not been exposed to the mutagenic agent.

Spontaneous mutations occur infrequently, although the observed frequencies vary from gene to gene and from organism to organism. Measurements of spontaneous forward mutation frequencies for various genes of phage and bacteria range from about 10^{-8} to 10^{-10} detectable mutations per nucleotide pair per generation. For eukaryotes, estimates of forward mutation rates range from about 10^{-7} to 10^{-9} detectable mutations per nucleotide pair per generation (considering only those genes for which extensive data are available). (In comparing mutation rates ***per nucleotide*** with rates ***per gene,*** the average gene is usually assumed to be 1000 nucleotide-pairs in length.)

Treatment with mutagenic agents can increase mutation frequencies by orders of magnitude. The mutation frequency ***per gene*** in bacteria and viruses, for example, can easily be increased to over 1 percent by treatment with potent chemical mutagens. That is, over 1 percent of the genes of the treated organisms will contain a mutation, or stated differently, over 1 percent of the individual phage or bacteria in the population will have a mutation in a given gene.

MUTATION: RANDOM RATHER THAN DIRECTED BY THE ENVIRONMENT

Mouse populations in many cities are no longer affected by the anticoagulants that have traditionally been used as rodent poisons. Many cockroach populations are insensitive to chlordane, the poison used to control them in the 1950s. Housefly populations often exhibit high levels of resistance to insecticides like DDT. More and more pathogenic microorganisms are becoming resistant to antibiotics such as penicillin and streptomycin developed by modern medicine to control them. The introduction of these pesticides and antibiotics by humans produced a new environment for the organisms involved. These organisms responded to the imposed environmental changes by evolving to forms that are resistant to these chemicals. Mutations producing resistance to these pesticides and antibiotics occurred, and the mutant organisms were at a large selective advantage in environments where these agents were present. The sensitive organisms were killed, and the mutants multiplied to produce new resistant populations. Many such cases of evolution via mutation and natural selection are now well documented.

These examples raise a basic question about the nature of mutation. Is mutation a purely random event with the environmental stress merely preserving preexisting mutations? Or is mutation directed by the environmental stress? As a specific example, consider a population of bacteria such as *E. coli* that is isolated from an environment where streptomycin has not been present. Upon exposure of the population of bacteria to streptomycin, the vast majority of them will be killed. If the population is large enough, however, it will soon be repopulated with bacteria, all of which are resistant to streptomycin. Does streptomycin simply select rare, randomly occurring mutants preexisting in the large population, or do all the cells have some low probability of developing resistance in response to the environmental stress—in this case streptomycin? Distinguishing between these two possibilities is not a trivial task. Evidence for the existence of resistance to streptomycin is usually ascertained only by treatment of the culture with the antibiotic. How can one determine whether resistant mutants are present ***prior*** to exposure to streptomycin?

This was first accomplished by J. and E. M. Lederberg using a technique called ***replica-plating*** (Fig. 11.1). Replica-plating experiments are carried out as follows. Bacterial cultures are diluted, and the cells are spread on the surface of semisolid nutrient agar medium in petri dishes. After a sufficient period of growth, each bacterium will produce a visible colony on the surface of the agar. Each plate is then inverted and pressed onto sterile velvet, which is covering a wood block. Some of the cells from each colony stick to the velvet. A sterile plate of nutrient agar medium containing streptomycin (or any other antibiotic) is then pressed onto the velvet. This replica-plating procedure is repeated with a large number of bacterial colonies. After incubation of the selective plates (containing streptomycin), rare streptomycin-resistant colonies will have formed. When the colonies on the nonselective plates (***not*** containing streptomycin) are subsequently tested for streptomycin resistance, those that gave rise to colonies on the selective replica plates are always found to contain resistant cells, whereas those that did not grow on the selective medium seldom contain any streptomycin-resistant cells (Fig. 11.1). The existence of streptomycin-resistant mutants in a population of bacteria prior to their exposure to the antibiotic can thus be demonstrated by replica-

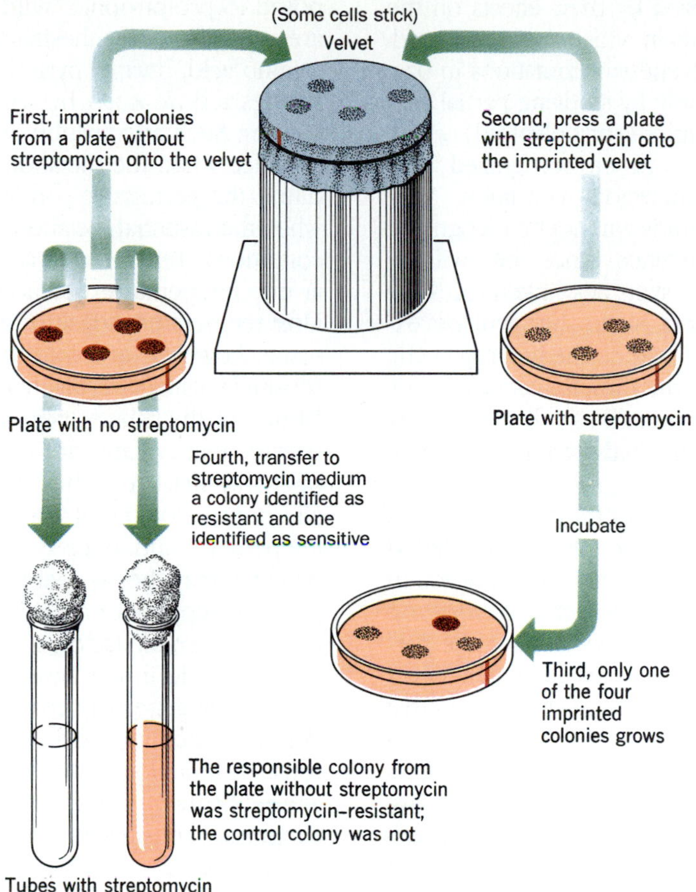

Figure 11.1 Diagram illustrating the use of the replica-plating technique to demonstrate the random or nondirected nature of mutation. For simplicity, only four colonies are shown on each plate, and only two are tested for streptomycin resistance in step 4. Actually, each plate will contain 50–100 colonies, and a large number of plates may be used to find one mutant colony. Also, many colonies are tested in step 4. The definitive result is that all the colonies from the nonselective plates that give rise to colonies on the selective plates are found to contain streptomycin-resistant mutants, and all the colonies on the nonselective plates that do not grow on the selective plates contain only streptomycin-sensitive cells. (After R. Sager and F. J. Ryan, *Cell Heredity,* Wiley, New York, 1961.)

plating experiments. The environmental stress does not direct or cause genetic changes; it simply selects rare preexisting mutations that result in phenotypes that are better adapted to the new environment.

PHENOTYPIC EFFECTS OF MUTATIONS

Mutations must normally cause some detectable ***phenotypic change*** for their presence to be recognized. The effects of mutations on phenotype range from alterations so minor that they can be detected only by special genetic or biochemical techniques to gross modifications of morphology to lethals. A gene is a specific sequence of nucleotide-pairs coding for a particular polypeptide. Any mutation occurring within a given gene will thus produce a new form or ***new allele*** of that gene. Because of the degeneracy of the genetic code (see Chapter 10), some base-pair changes do not change the protein products encoded by the genes in any way. Genes containing mutations with small effects that can be recognized only by special techniques are called ***"isoalleles."*** Other mutations result in total loss of gene-product activity. If mutations of the latter type occur in essential genes (genes required for viability), they will, of course, be lethal.

Mutations may be either recessive or dominant. In haploid (or, more accurately, monoploid) organisms like viruses and bacteria, both recessive and dominant mutations can be recognized by their effects on the phenotype of the organisms in which they originated. The dominance or recessiveness of mutations in bacteria can be determined only by studying partial diploids (see Chapter 8). In diploid (or polyploid) organisms, recessive mutations will be recognized only when present in the homozygous condition. Most recessive mutations in diploids will not be recognized at the time of their occurrence, since they will be present in the heterozygous state. Sex-linked recessive mutations are an exception, since they will be expressed in the hemizygous state in the heterogametic sex (males in humans and fruit flies; females in birds). Sex-linked, recessive lethal mutations will alter the sex ratio, since hemizygous individuals carrying the lethal will not survive (Fig. 11.2).

The most useful mutations for the genetic analyses of many biological processes are ***conditional lethal mutations***. These are mutations that are (1) ***lethal*** in one environment, the so-called ***restrictive conditions***, but are (2) ***viable*** in a second environment, the ***permissive conditions***. Such mutations allow geneticists to identify and study mutations in essential genes that result in complete loss of gene-product activity even in haploid organisms. Mutants carrying conditional lethals can be propagated under permissive conditions, and information about the functions of the gene-products can be deduced by studying the consequences of their absence under the restrictive conditions. The use of conditional lethal mutations to dissect biological processes is illustrated on pp. 315–316. Conditional lethal mutations also provide valuable selective sieves for genetic fine structure analysis (see Chapter 12).

The three major classes of mutants with conditional lethal mutations are (1) ***auxotrophic mutants***, (2) ***temperature-sensitive mutants***, and (3) ***suppressor-sensitive mutants***. Auxotrophic mutants (as opposed to prototrophic "wild types") are mutants that are unable to synthesize an essential metabolite (amino acid, purine, pyrimidine, vitamin, etc.) that is synthesized *de novo* by wild-type individuals of the species. Such auxotrophic mutants will grow and reproduce when the metabolite is supplied in the medium (the permissive conditions); they will not grow when the essential metabolite is absent (the restrictive condition). Temperature-sensitive mutants will grow at one temperature but not at another temperature. Most temperature-sensitive mutants are heat-sensitive; some, however, are cold-sensitive. The temperature sensitivity usually results from increasing heat or cold lability of the mutant gene-product, for example, an enzyme that is active at low temperature but partially or totally inactive at higher temperatures. Occasionally, only the synthesis of a gene-product is sensitive to temperature and, once synthesized, the mutant gene-product may be as stable as the wild-type gene-product. Suppressor-sensitive mutants are viable when a second genetic factor, a ***suppressor***, is present, but are nonviable in the absence of the suppressor. The suppressor gene may correct or compensate for the defect in phenotype that is caused by the suppressor-sensitive mutations, or it may render the gene-product, altered by the mutation, nonessential. Special types of suppressor-sensitive mutations are discussed in Chapters 10 and 12.

Most of the thousands of mutations that have been identified and studied by geneticists have been found to be deleterious and recessive. This is to be expected, considering what we know about the genetic control of metabolism and the techniques available for identifying mutations. Metabolism occurs by sequences of chemical reactions, each step of which is catalyzed by a specific enzyme encoded by one or more genes (see Chapter 10). Mutations in these genes frequently produce blocks in metabolic pathways (Fig. 11.3). These blocks occur because changes in the base-pair sequences of genes often (but not always) cause changes in the amino acid sequences of polypeptides (Fig. 11.4), which may, in turn, result in the loss of function (Fig. 11.3). This, in fact, has been the most commonly observed effect of easily detected mutations. Given a wild-type allele encoding an active enzyme and mutant alleles encoding less active or totally inactive enzymes, it is apparent why most of the observed mutations might be recessive, as is observed. If the enzyme catalyzes an essential reaction, the mutations causing total loss of activity will be recessive lethals.

But why should most mutations with phenotypically recognizable effects result in decreased gene-product activity or no gene-product activity? This result can be predicted if one accepts the effectiveness of natural selection and if one assumes the existence of a

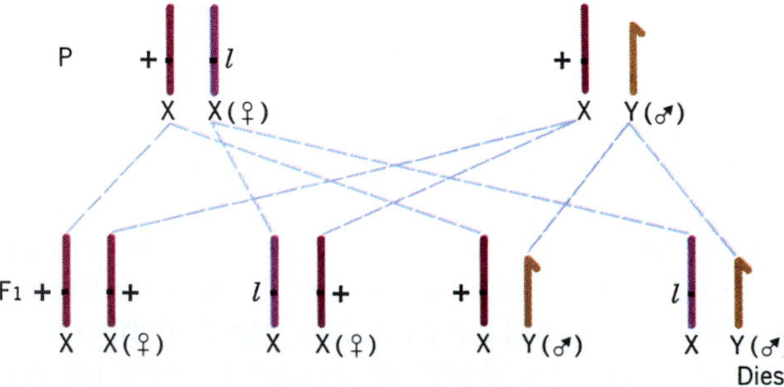

Figure 11.2 Altered sex ratio that results from a sex-linked recessive lethal. Heterozygous individuals of the homogametic sex will produce progeny with a 2 : 1 (homogametic sex to heterogametic sex) sex ratio.

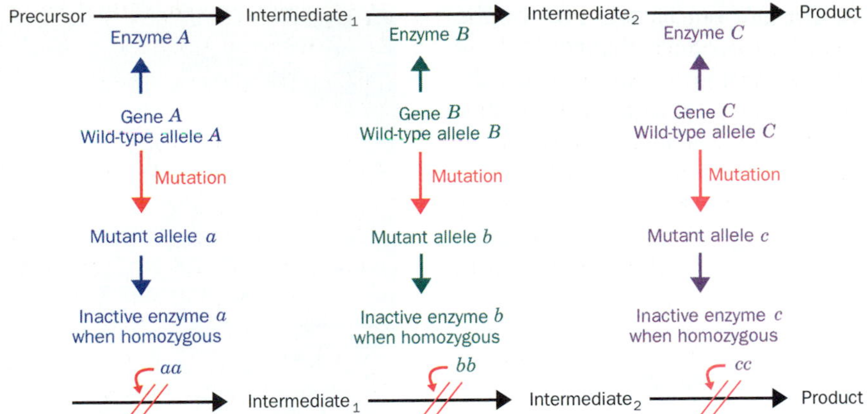

Figure 11.3 Blocks in metabolic pathways caused by recessive mutant alleles. Metabolic pathways may be only a few steps long, as diagrammed here, or many steps in length. The wild-type form of each gene (or wild-type allele) codes for an active enzyme that efficiently catalyzes the appropriate reaction. Most mutations occurring in a wild-type gene result in the synthesis of an altered form of the enzyme with reduced activity or no activity. Such mutant alleles are usually recessive to their wild-type alleles. In the homozygous state, such mutant alleles often result in metabolic blocks (⇸) because of the inability of the mutant form of the enzyme to catalyze the requisite reaction. This picture of the genetic control of metabolism is valid for both anabolic (biosynthetic) and catabolic (degradative) pathways. Consider, for example, that the product is an anthocyanin pigment, yielding colored flowers in plants. Homozygosity for the recessive mutant allele of any one of the three genes (*aa, bb,* or *cc*) will result in the inability to synthesize the pigment.

semiconstant environment during the recent (on the evolutionary scale) evolution of life-forms on earth. A "wild-type" allele of a gene encoding a "wild-type" enzyme or structural protein will have been selected for optimal activity for many generations. Mutations resulting in amino acid changes that increased the efficiencies of enzymes in carrying out particular functions will have been preserved by natural selection, and, as the most "fit," they will have become the new "wild types." Given a sufficient period of time in a semiconstant environment, most, if not all, sequences of amino acids (at least all sequences of amino acids that differ from the "wild type" by a single mutation) will have been tried, and natural selection will have

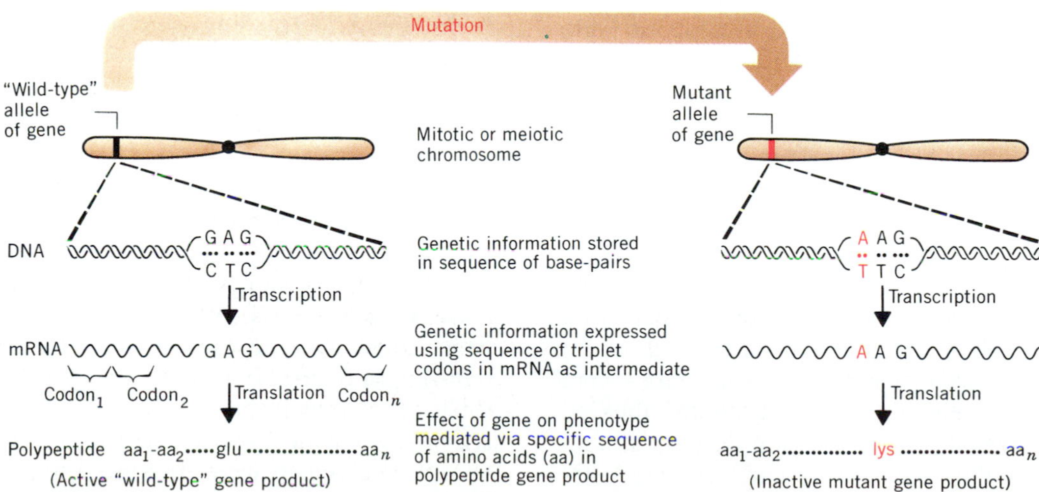

Figure 11.4 Mutation and expression of wild-type and mutant alleles. Mutations occurring in genes alter the base sequences and thus usually change the amino acid sequences of the gene-products. In the example shown, a GC base-pair (top, left) has mutated to an AT base-pair (top, right), changing one codon in the mRNA from GAG to AAG and one amino acid in the polypeptide gene-product from glutamic acid (glu) to lysine (lys) (see Table 10.1). Such changes frequently cause the gene-products to be nonfunctional.

preserved the most efficient one. It will now be the wild type. Mutations, which produce changes in these very specific sequences of amino acids, will usually result in less activity or no activity at all. As such, they will most frequently be recessive and deleterious.

An analogy can be made with any complex, carefully engineered machine. If you randomly modify any one essential component (e.g., of a watch or an automobile), it seldom performs as well as it did prior to the random change.

Somatic and Germinal Mutations

Mutations may occur in any cell and at any state in the cell cycle. The immediate effect of the mutation and its ability to produce a phenotypic change are determined by its dominance, the type of cell in which it occurs, and when it happens relative to the life cycle of the organism.

If the mutation occurs in a somatic cell (all cells other than reproductive cells), which can produce cells like itself but not the whole organism, the mutant change will be perpetuated only in ***somatic cells*** that descend from the original cell in which the mutation occurred. The Delicious apple and the navel orange, for example, originally were mosaics in somatic tissues. Changes that give these two fruits their desirable qualities apparently followed spontaneous mutation in single cells. In each case, the cell carrying the mutant gene reproduced, eventually producing an entire branch that had the characteristics of the mutant type. Fortunately, ***vegetative propagation*** was feasible for both the Delicious apple and the navel orange, and today numerous progeny from grafts and buds have perpetuated the original mutation. Descendants of the mutant types are now widespread in apple orchards and orange groves.

If dominant mutations occur in ***germ*** cells, their effects may be expressed immediately in progeny. If the mutations are ***recessive,*** their effects are often obscured in diploids. Germinal mutations, like somatic mutations, may occur at any stage in the reproductive cycle of the organism, but they are more common during some stages than others. If the mutation arises in a ***gamete,*** only a single member of the progeny is likely to have the mutant gene. If, on the other hand, a mutation occurs in gonial cells, several gametes may receive the mutant gene and thus enhance its potential for perpetuation. In any case, the dominance of the mutant allele and the stage in the reproductive cycle when the mutation occurs are major factors in determining the likelihood of the mutant allele being manifest in an organism and a population.

The earliest recorded dominant germinal mutation in domestic animals was that observed by Seth Wright in 1791 on his farm by the Charles River in Dover, Massachusetts. Wright noticed a peculiar male lamb with unusually short legs in his flock of sheep. It occurred to him that it would be an advantage to have a whole flock of these short-legged sheep (Fig. 11.5), which could not jump over the low stone fences in his New England neighborhood. Wright used the new short-legged ram for breeding his 15 ewes in the next season. Two of the 15 lambs produced had short legs. Short-legged sheep were then bred together, and a line was developed in which the new trait was expressed in all individuals. The mutation that gave rise to the short-legged sheep was obviously of the germinal type, because the cell carrying the mutation had the capacity to reproduce the entire organism. Germinal

(a)

(b)

Figure 11.5 Effect of a dominant mutation. (a) Short-legged sheep of the Ancon breed; (b) sheep with normal-length legs. (Courtesy of: (a) Frank P. Rossotto, The Stock Market; (b) Kenneth W. Fink, Bruce Colman, Inc.).

mutations have since been described in a variety of animals and plants.

Mutant Hemoglobins

Hemoglobin is the oxygen-transporting macromolecule present in the red corpuscles of chordate animals. It serves the essential function of transporting oxygen from the lungs to all the various tissues of the body. Two major types of hemoglobin occur in humans at different stages of development: fetal hemoglobin (hemoglobin F) and adult hemoglobin (hemoglobin A). (Other forms of hemoglobin—embryonic and minor adult forms—exist, but are not discussed here.) Each molecule of hemoglobin is composed of four polypeptide chains plus a heme group. Hemoglobin A contains ***two identical alpha chains*** and ***two identical beta chains.*** Hemoglobin F contains two alpha chains and two ***identical gamma chains.*** Each polypeptide chain—alpha, beta, or gamma—is encoded by a specific gene. Hemoglobin F is present in the developing fetus, but is normally replaced by hemoglobin A within the first six months after birth. Of course, this raises the interesting question of how the expression of genes is regulated. How are genes turned on and off? This intriguing question is discussed in Chapters 14 and 15.

Each alpha polypeptide consists of a specific sequence of 141 amino acids. The beta and gamma chains are 146 amino acids long. Because of similarities in their amino acid sequences, all the hemoglobin chains (and thus their structural genes) are believed to have evolved from a common progenitor (see Chapter 22, pp. 585–587).

Many different variants of the hemoglobins have been identified in human populations, most being variants of hemoglobin A. Some of the human hemoglobin variants (or mutant forms) have severe phenotypic effects; others have very minor phenotypic effects. Many of the variants were initially detected by their altered electrophoretic behavior (movement in an electric field due to charge differences). The hemoglobin variants provide an excellent illustration of the effects of mutation on the structures and functions of gene-products and, ultimately, on the phenotypes of the affected individuals.

Sickle cell hemoglobin (hemoglobin S) is one such variant. Individuals homozygous for the hemoglobin S allele (Hb^S_β/HB^S_β; the β subscript is used because hemoglobin S is a variant of the beta chain) develop severe hemolytic anemia. Hemoglobin S molecules precipitate when deoxygenated, forming crystalloid aggregates that distort the morphology of red blood corpuscles. The erythrocytes (red blood corpuscles) elongate and form sickle-shaped cells (Fig. 11.6). These sickle-shaped cells clog small blood vessels and cut off oxygen transport to various tissues. In many, but not all cases, sickle-cell anemia results in death during childhood.

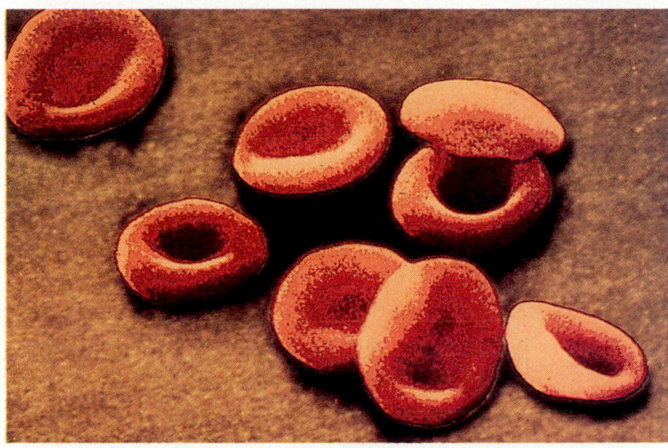

(a)

(b)

Figure 11.6 Scanning electron micrographs of human erythrocytes. (a) Normal cells; (b) sickle-shaped cells. (Courtesy of Bill Longcore.)

When the amino acid sequences of the beta chains of hemoglobin A and hemoglobin S were determined and compared, the hemoglobin S beta chain was found to differ from that of hemoglobin A by only one amino acid (Fig. 11.7). The sixth amino acid from the amino (NH_2-)-terminal end of the beta chain of hemoglobin A is glutamic acid (a negatively charged amino acid). The beta chain of hemoglobin S contains valine (no charge at neutral pH) at that position. The alpha chains of hemoglobin A and hemoglobin S are identical. Thus, changing a single amino acid in one polypeptide can have severe effects on phenotype. A large number

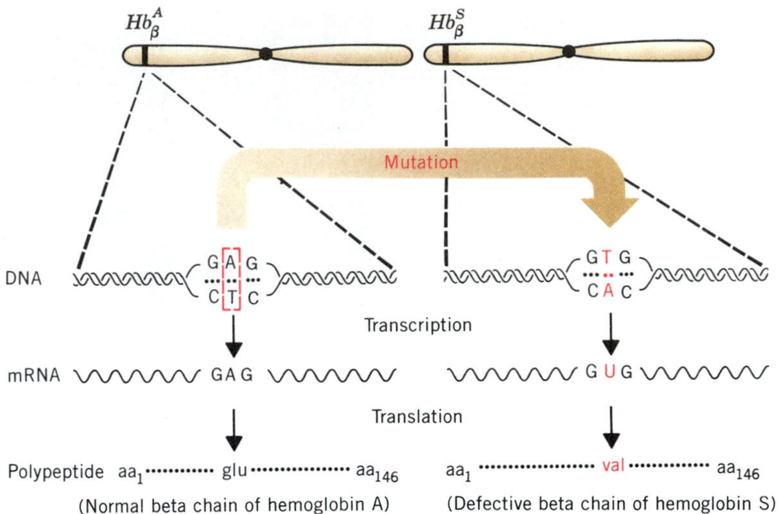

Figure 11.7 Mutational origin of sickle-cell hemoglobin (hemoglobin S). The amino acid sequences of the "normal" (hemoglobin A) beta polypeptide and the sickle-cell beta polypeptide have been determined by direct chemical analysis (amino acid sequencing). From our knowledge of the nature of the genetic code (which triplet codons specify which amino acids; see Table 10.1) and the observed substitution of valine at amino acid position 6 in the beta chain of hemoglobin S, one can deduce that the mutation giving rise to hemoglobin S (or, more accurately, to the Hb_{β}^{S} gene) is the substitution of an adenine for a thymine in the transcribed strand of DNA, or stated in base-pair terms, the substitution of a $\begin{smallmatrix}\leftarrow T - \\ - A \rightarrow\end{smallmatrix}$ base-pair for an $\begin{smallmatrix}\leftarrow A - \\ - T \rightarrow\end{smallmatrix}$ base-pair, where the bottom strand is the one transcribed (complementary to mRNA).

of similar effects of mutation on protein structure, and ultimately on phenotype, are now well documented.

In the case of hemoglobin S, it is hypothesized that the substitution of valine for glutamic acid at the sixth position in the beta chain allows a new bond to form that changes its conformation and leads to molecular stacking or aggregation of hemoglobin molecules. In any case, the change results in the gross abnormality of the morphology of the red blood cells. The mutational change in the Hb_{β}^{A} gene that gave rise to Hb_{β}^{S} can be deduced from our knowledge about the genetic code (the triplet base sequences or codons in mRNA that specify the amino acids in the polypeptide gene-product; see Chapter 10). The glutamic acid to valine change is nicely explained by a substitution of an AT base-pair for a TA base-pair (where the T is in the transcribed strand in the first case and the A is in the transcribed strand in the second case). When the Hb_{β}^{A} and Hb_{β}^{S} genes were sequenced, they were indeed found to differ by a single TA⟶AT base-pair substitution (see Fig. 11.7). Over 100 hemoglobin variants with amino acid changes in the beta chain are known. Most of them differ from the normal beta chain (of hemoglobin A) by a single amino acid substitution. A few examples are illustrated in Fig. 11.8. Numerous variants of the alpha and gamma polypeptides have also been identified. Mutation, then, is a process in which changes in gene structure, often changes in one or a few base-pairs, cause changes in the amino acid sequences of the polypeptide gene-products. These, in turn, cause the changes in phenotype that are recognized as mutant. The pathways by which the polypeptides, the primary gene-products, exert their effects on the phenotype are often very complex, particularly in higher plants and animals (see Chapter 10).

Pleiotropy

The term ***pleiotropy*** refers to the situation in which a gene influences more than one trait. Many such instances have been discovered. In fact, all genes (whether mutant or wild-type allelic forms) may be pleiotropic, with their various effects simply not yet recognized. Even though a structural gene may have many end effects, it usually has only one primary function, that of producing one polypeptide (in some cases, one RNA molecule). This polypeptide may give rise to different expressions at the phenotypic level.

The Hb_{β}^{S} allele provides a classic example of pleiotropy. It not only causes hemolytic anemia (in the homozygous state) but also results in increased resistance to one type of malaria, that caused by the parasite *Plasmodium falciparum*. Because the increased resistance to falciparum malaria occurs in $Hb_{\beta}^{A}/Hb_{\beta}^{S}$ heterozygotes, such heterozygous individuals have a selective advantage in geographic regions where this

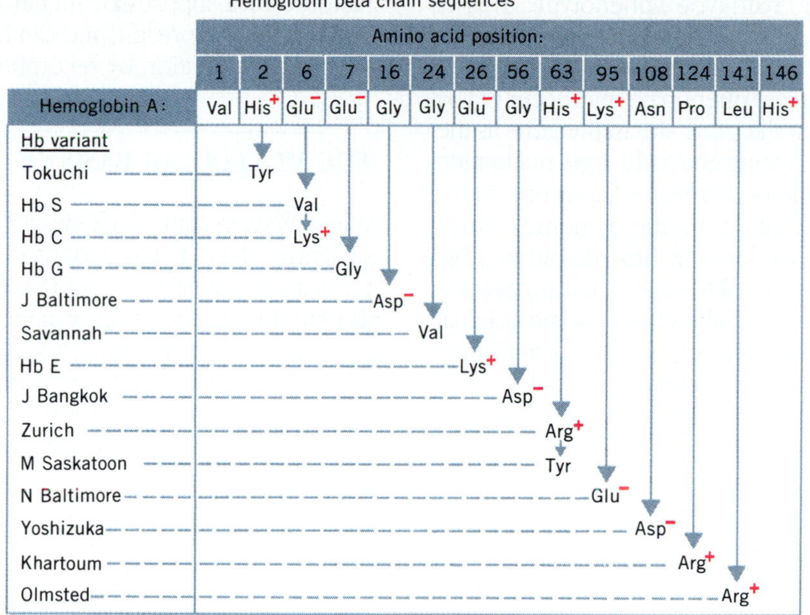

Figure 11.8 A few of the large number of hemoglobin beta chain variants that have been determined by amino acid sequencing. All the examples shown differ from the sequence of the beta chain of hemoglobin A by a single amino acid substitution as indicated. They all can be explained by a single base-pair substitution as the mutational event in the Hb^A_β gene giving rise to the variant (as illustrated for hemoglobin S in Fig. 11.7). In fact, almost all of the over 100 hemoglobin variants known in the alpha, beta, and gamma polypeptides can be explained by single base-pair substitutions. This not only provides insight into the mechanism of mutation in humans, but also indicates that the codon assignments (see Table 10.1) determined experimentally with lower organisms are in large part valid for humans. Note that many of the amino acid substitutions involve changes in charge, even though only 5 of the 20 common amino acids have a net charge at neutral pH. This may be explained in part because of the ease of detecting hemoglobins with altered charges by electrophoresis. In addition, however, amino acid substitutions involving charge changes are probably more likely to alter protein structure and function than substitutions not involving charge changes.

type of malaria is prevalent (see Chapter 22, pp. 580–581). The sickle-cell allele also has pleiotropic effects on the development of many tissues and organs such as bones, the lungs, the kidneys, the spleen, and the heart.

Cystic fibrosis is a hereditary, metabolic disorder in children that is controlled by a single autosomal recessive gene. The gene specifies an enzyme that produces a unique glycoprotein. This glycoprotein results in the production of mucus with abnormally high viscosity. Overly viscous mucus interferes with the normal functioning of several exocrine glands, including those in the skin (sweat), lungs (mucus), liver, and pancreas. The ***syndrome (a group of symptoms that characterizes the disease)*** is related directly or indirectly to the abnormal mucus. Abnormally high levels of sodium chloride occur in the sweat, and mucus stagnates in tubules of the lungs, which frequently become infected, giving rise to bronchitis. Secreting cells in the liver and the pancreas are impaired, curtailing production of fat-emulsifying agents and digestive enzymes and thus interfering with digestion and absorption of food. Several different phenotypic effects may thus result from the action of a single pleiotropic gene.

Back-Mutations and Suppressor Mutations

The mutation of a "wild-type" gene to a form that results in a mutant phenotype is usually referred to as "forward mutation." Sometimes, however, the designation of the "wild type" and the "mutant" is quite arbitrary. They may simply represent two ***different*** phenotypes. For example, we consider the two alleles controlling brown and blue eye color in humans both to be the "wild type." However, in a population composed almost entirely of brown-eyed individuals, the allele for blue eyes might be thought of as a mutant allele. In any case, the mutation events are often reversible. That is, a subsequent mutation may occur that restores the original "wild-type" phenotype. This is referred to as ***back-mutation, reverse mutation,*** or ***reversion.*** Reversion may occur in two different ways. Restoration of the original phenotype may occur (1) by a true back-mutation at the same site in the gene as the original mutation, restoring the wild-type nucleotide sequence, or (2) by the occurrence of a second mutation at a different location in the genome, which somehow compensates for the first mutation (Fig. 11.9). Mutations of the latter kind are called ***suppressor mutations:*** they "suppress" the effects of the original mutations. Suppressor mutations may occur at distinct sites in the same gene as the original mutation or in different genes, even in different chromosomes. Suppressor mutations can be distinguished from true back-mutations by backcrossing the revertants to the original wild type. If the wild-type phenotype was restored by a suppressor mutation, the original mutation will still be present and can be separated from the suppressor mutation by recombination (Fig. 11.9).

THE MOLECULAR BASIS OF MUTATION

When Watson and Crick described the double-helix structure of DNA and proposed its semiconservative replication based on specific base-pairing to explain the faithful transmission of genetic information from generation to generation, they also proposed a mechanism to explain spontaneous mutations. Watson and Crick pointed out that the structures of the bases in DNA are not static. Hydrogen atoms can move from one position in a purine or pyrimidine to another position, for example, from an amino group to a ring

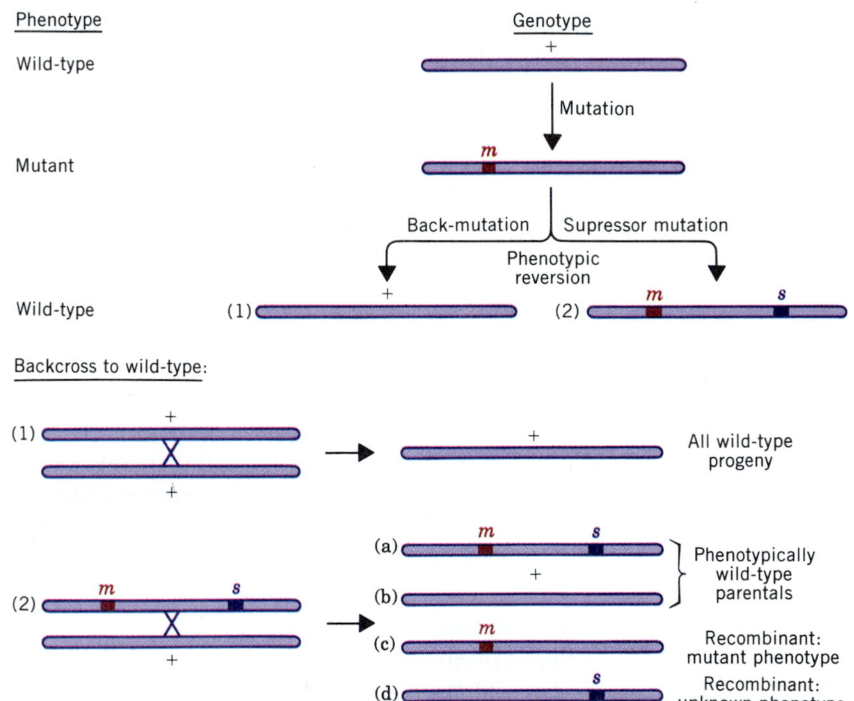

Figure 11.9 Restoration of the original "wild-type" phenotype by (1) true back-mutation and (2) suppressor mutation. Many mutants can revert to the wild-type phenotype by both mechanisms. Revertants of the two types can be distinguished by backcrosses to the original wild type. All progeny with parental genotypes (a and b) will have wild-type phenotypes. Revertants containing suppressor mutations will produce some recombinant progeny with mutant phenotypes (c). The frequency will depend on whether the initial mutation and the suppressor mutation are linked (as shown) and, if linked, how far apart they are on the chromosome. In higher eukaryotes, it may be difficult to analyze enough progeny to assure detection of the recombinants with mutant phenotypes if the suppressor mutation and the original mutation are closely linked. The recombinant progeny carrying the suppressor mutation but not the initial mutation (d) may have either the mutant or the wild-type phenotype, depending on the nature of the mutation. Note that only single chromosomes carrying the various genetic markers being discussed are shown. For diploids, homologous chromosomes (not shown) will also be present. In diploids, the backcross will involve crossing a homozygous phenotypic revertant with a homozygous wild-type organism and examining testcross progeny of the F_1 heterozygotes. At the chromosome level, however, the scheme is applicable to both haploid and diploid organisms.

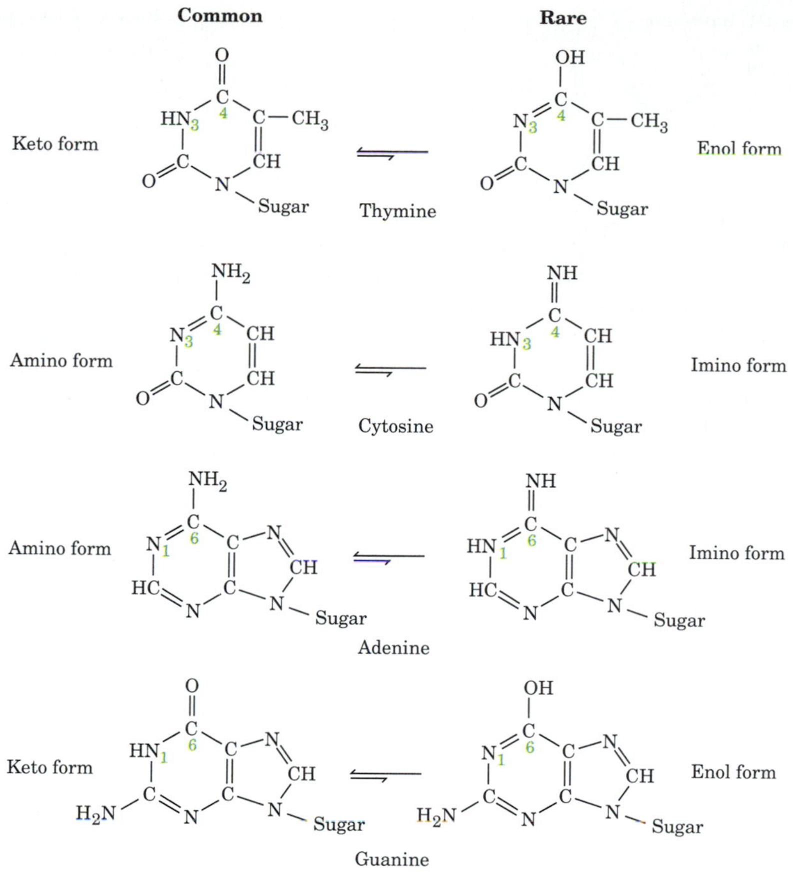

Common | Rare

Keto form — Thymine — Enol form

Amino form — Cytosine — Imino form

Amino form — Adenine — Imino form

Keto form — Guanine — Enol form

Figure 11.10 Tautomeric forms of the four common bases in DNA. The shifts of hydrogen atoms between the number 3 and number 4 positions of the pyrimidines and between the number 1 and number 6 positions of the purines change the base-pairing potential of the bases.

nitrogen. Such chemical fluctuations are called ***tautomeric shifts.*** Although tautomeric shifts are rare, they may be of considerable importance in DNA metabolism since some can alter the pairing potential of the bases. The structures of the bases shown in Fig. 5.5 are the common, more stable forms, in which adenine always pairs with thymine and guanine always pairs with cytosine (Fig. 5.11). The ***more stable keto forms*** of thymine and guanine and ***amino forms*** of adenine and cytosine may ***infrequently*** undergo tautomeric shifts to ***less stable enol and imino forms,*** respectively (Fig. 11.10). The bases would be expected to exist in their less stable tautomeric forms for only very short periods of time. However, if a base existed in the rare form at the moment that it was being replicated or being incorporated into a nascent DNA chain, a mutation might result. When the bases are present in their rare imino or enol states, they can form adenine–cytosine and guanine–thymine base-pairs (Fig. 11.11). The net effect of such an event, and the subsequent replication required to segregate the "mismatched" base-pair, is an AT to GC or a GC to AT base-pair substitution (Fig. 11.12).

Mutations resulting from tautomeric shifts in the bases of DNA involve the replacement of a purine in one strand of DNA with the other purine and the replacement of a pyrimidine in the complementary strand with the other pyrimidine. Such base-pair substitutions are called ***transitions.*** Base-pair substitutions involving the substitution of a purine for a pyrimidine and vice versa are called ***transversions.*** Four different transitions and eight different transversions are possible (Fig. 11.13). A third type of point mutation involves the addition or deletion of one or a few base-pairs. Base-pair additions and deletions are collectively referred to as ***frameshift mutations*** because they alter the reading frame of all base-pair triplets (specifying codons in mRNA and amino acids in the polypeptide gene-product) in the gene distal to the mutation (Fig. 11.14).

All three types of point mutations—transitions, transversions, and frameshift mutations—are present

Rare A–C base-pair

Cytosine
(rare imino form)

Adenine
(common amino form)

Rare G–T base-pair

Thymine
(common keto form)

Guanine
(rare enol form)

Figure 11.11 Examples of mismatched AC and GT base-pairs that can form when one of the purines or pyrimidines exists in a rare tautomeric form. Similar mismatched base-pairs can form when thymine exists in its rare enol form or adenine exists in its rare imino form.

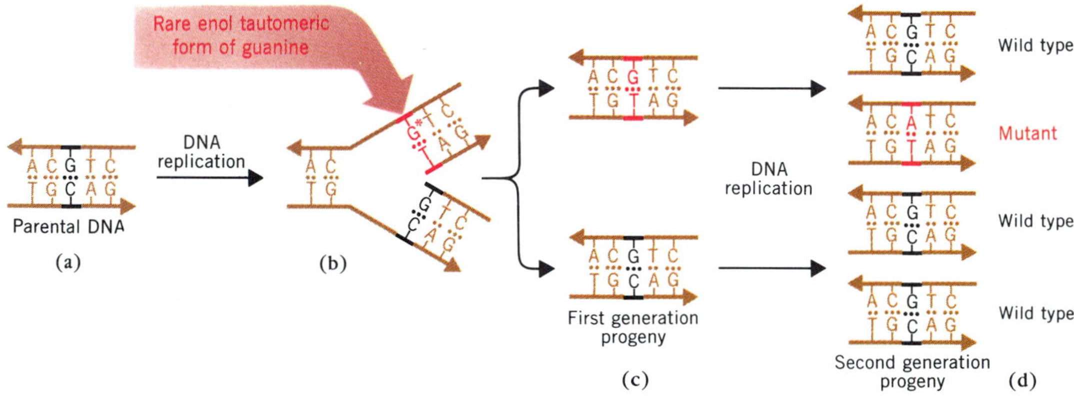

Figure 11.12 Mutation via tautomeric shifts in the bases of DNA. In the example diagrammed, a guanine (a) undergoes a tautomeric shift to its rare enol form (G*) at the time of replication (b). In its enol form, it pairs with thymine (b). During the next replication (c to d), the guanine shifts back to its more stable keto form. The thymine incorporated opposite the enol form of guanine (b) directs the incorporation of adenine during the subsequent replication (c to d). The net result is a GC to AT mutation. If a guanine undergoes a tautomeric shift from the common keto form to the rare enol form at the time of incorporation (as a nucleoside triphosphate, rather than in a template strand as diagrammed), it will be incorporated opposite thymine in the template strand and cause an AT to GC mutation.

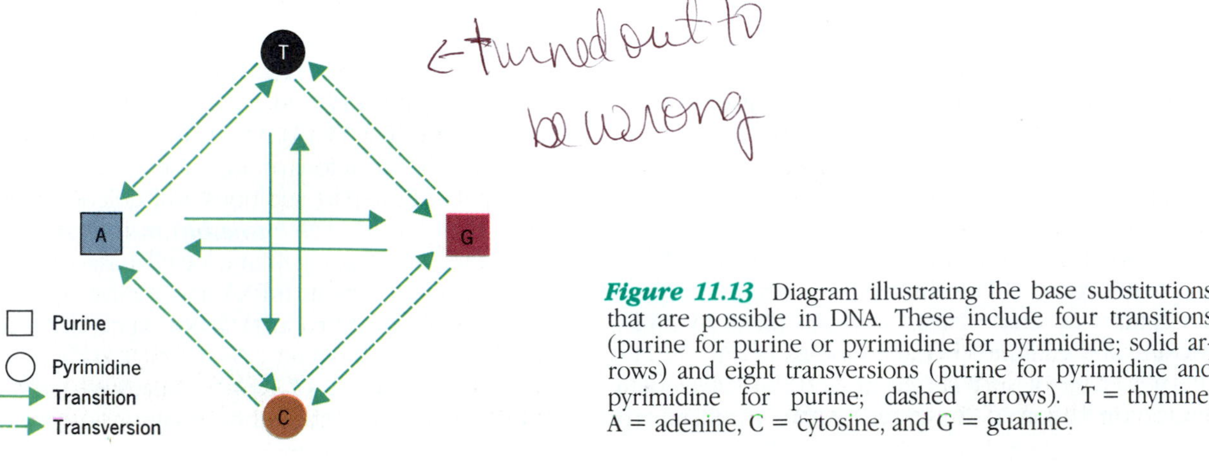

Figure 11.13 Diagram illustrating the base substitutions that are possible in DNA. These include four transitions (purine for purine or pyrimidine for pyrimidine; solid arrows) and eight transversions (purine for pyrimidine and pyrimidine for purine; dashed arrows). T = thymine, A = adenine, C = cytosine, and G = guanine.

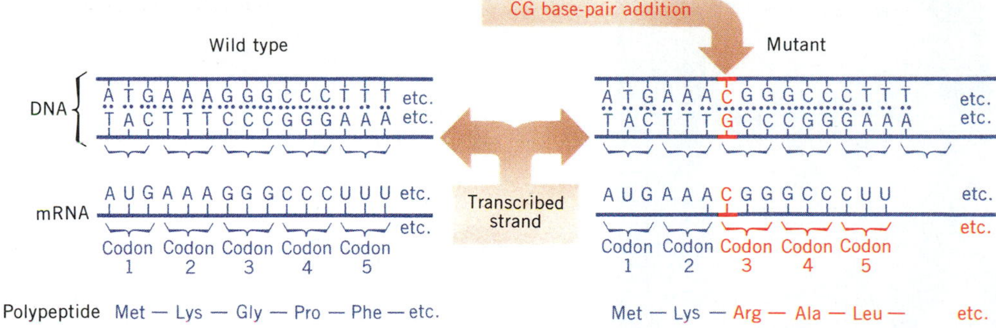

Figure 11.14 Diagram of a frameshift mutation that results from the addition of a single base-pair to a structural gene. The mutant gene (top, right) was produced by the insertion of a CG base-pair between the sixth and seventh base-pairs of the wild-type gene (top, left). This alters the "reading frame" of that portion of the gene distal, relative to the direction of transcription and translation (left to right, as diagrammed), to the mutation. As a result, all the codons of the mRNA and all the amino acids of the polypeptide that correspond to base-pair triplets distal to the mutation are altered.

among spontaneously occurring mutations. A surprisingly large proportion of the spontaneous mutations studied in prokaryotes are found to be single base-pair additions and deletions rather than base-pair substitutions.

Although there is still a great deal to be learned about the causes, molecular mechanisms, and frequencies of various types of spontaneous mutations, the three key factors appear to be (1) the ***accuracy of the DNA replication machinery,*** (2) the ***efficiency*** of the numerous mechanisms that have evolved for the ***repair of damaged DNA*** (see pp. 305–307), and (3) the degree of ***exposure to mutagenic agents present in the environment.*** Any perturbations of the DNA replication apparatus or DNA repair systems clearly increase mutation rates.

RADIATION-INDUCED MUTATION

That portion of the electromagnetic spectrum containing wavelengths that are shorter and of higher energy than visible light (wavelengths below about 0.1 μm) can be subdivided into ionizing radiation (X rays, gamma rays, and cosmic rays) and nonionizing radiation (ultraviolet light). Ionizing radiations such as X rays (about 0.1 to 1 nm) are of high energy and thus are useful for medical diagnosis because they can penetrate living tissues. In the process of penetrating matter, these high-energy rays collide with atoms and cause the release of electrons, leaving positively charged free radicals or ions. These ions, in turn, collide with other molecules, causing the release of further electrons. The net result is that a "core" of ions is formed along the track of each high-energy ray as it passes through matter or living tissues. This process of ionization (thus the name *ionizing radiation*) is induced by machine-produced X rays, protons, and neutrons, as well as by the alpha, beta, and gamma rays released by radioactive isotopes of the elements (e.g., ^{32}P, ^{35}S, radium, cobalt-90, etc.). Ultraviolet rays, having lower energy, penetrate only the surface layer of cells in higher plants and animals and do not induce ionizations. Ultraviolet rays dissipate their energy to atoms that they encounter, raising the electrons in the outer orbitals to higher energy levels, a state referred to as ***excitation.*** Molecules containing atoms in either ***ionic*** forms or ***excited*** states are chemically more reactive than those containing atoms in their normal stable states. The increased reactivity of atoms present in DNA molecules is the basis of the mutagenic effects of ultraviolet light and ionizing radiation.

Ionizing Radiation

In 1927, H. J. Muller (Fig. 11.15) first demonstrated that mutation could be induced by an external factor. Muller demonstrated that X-ray treatment markedly increased the frequency of sex-linked recessive lethal mutations in *D. melanogaster*. Muller's unambiguous demonstration of the mutagenicity of X rays became possible by his development of a technique facilitating the simple and accurate identification of lethal mutations in the X chromosome of *Drosophila*. This technique, called the **ClB** ***method,*** involves the use of females heterozygous for a normal X chromosome and an X chromosome (the *ClB* chromosome) specifically constructed for Muller's experiment.

The *ClB* chromosome has three essential components. (1) The **C** (for ***crossover*** "suppressor") refers to the presence of a long inversion (see Chapter 18) that

Figure 11.15 H. J. Muller, who first demonstrated the mutagenic effects of X rays in 1927. Professor Muller made many contributions to our understanding of basic genetics through his extensive research. He was awarded the Nobel Prize in 1949. (National Archives.)

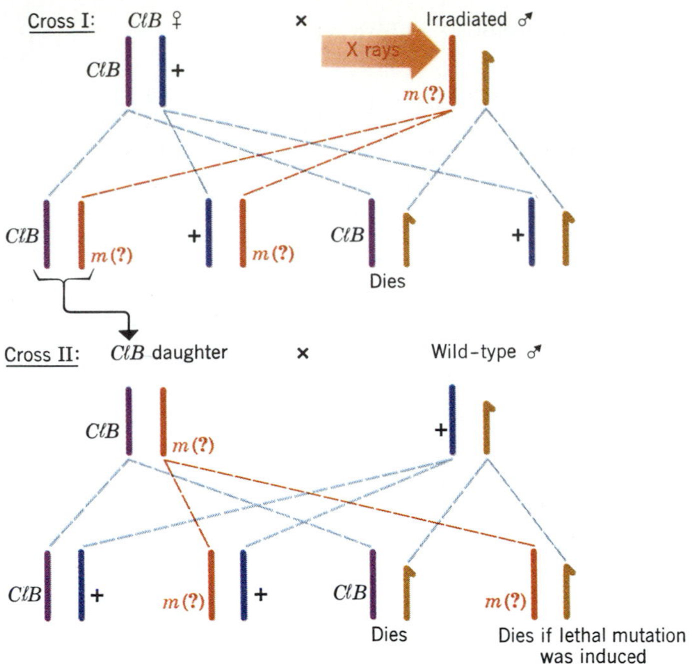

Figure 11.16 Muller's *ClB* method for detecting sex-linked recessive lethal mutations in *Drosophila.* Females heterozygous for the *ClB* X chromosome are mated with irradiated males (Cross I). *ClB* daughters, which will always carry the *ClB* chromosome of the female parent and the irradiated X chromosome of the male parent, are then mated with wild-type males (Cross II). This second mating will produce only female progeny if an X-linked recessive lethal has occurred on that particular X chromosome. It will produce both female and male progeny (in a ratio of 2:1) if no X-linked lethal was induced on the irradiated X chromosome that the *ClB* daughter received. Scoring for the presence or absence of sex-linked recessive lethal mutations thus involves simply scoring for the presence or absence of male progeny from matings of Cross II.

prevents recombination of genetic markers on the *ClB* chromosome and alleles on the normal X chromosome. The inversion does not actually prevent crossing-over, but causes gametes containing X chromosomes produced by crossing-over between the *ClB* chromosomes and the normal X chromosome to be unbalanced, and therefore unable to support development. Chromosomes resulting from crossing-over between a chromosome containing an inversion (an inverted segment of the chromosome) and a normal chromosome will contain duplications and deficiencies (repeated and missing sets of genes; see Chapter 18). The inversion is required in Muller's experiment to assure that the markers on the *ClB* chromosome stay together through meiosis. (2) The ***l*** refers to a recessive ***lethal*** in the *ClB* chromosome. (3) The ***B*** refers to the presence of the ***partially dominant*** mutation that causes the ***bar*** eye phenotype, which is a narrow, slit-shaped eye. Because it is partially dominant, it allows females heterozygous for the *ClB* chromosome to be identified readily. Both the recessive lethal (*l*) and the bar eye mutation (*B*) are located within the inverted segment of the *ClB* chromosome.

Given females heterozygous for the *ClB* chromosome, Muller's experiment was operationally quite simple (Fig. 11.16). Male flies were irradiated and mated with *ClB* females. The bar-eyed daughters of this mating will all carry the *ClB* chromosome of the female parent and the irradiated X chromosome of the male parent. Since the entire population of reproductive cells of the males was irradiated, each bar-eyed daughter carries a potentially mutated X chromosome. That is, each male is likely to produce some sperm that contain X chromosomes carrying new lethal mutations and some sperm without X-linked lethal mutations. The bar-eyed daughters were then mated individually (in separate bottles) with wild-type males. If the irradiated X chromosome carried by a bar-eyed daughter contains a sex-linked lethal, all the progeny of the mating will be female. Since males are hemizygous for the X chromosome, those receiving the *ClB* chromosome will die because of the recessive lethal (*l*) that it carries. Those receiving the irradiated X chromosome will also die if a recessive lethal has been induced in it (Fig. 11.16). Matings of bar-eyed daughters carrying an irradiated X chromosome, in which no lethal mutation has been induced, with wild-type males will produce female and male progeny in a ratio of 2 : 1 (only the males with the *ClB* chromosome will die). Scoring for the presence of recessive sex-linked lethals is thus unambiguous and error free using the *ClB* technique—simply scoring for the presence or absence of male progeny. By using this technique, Muller was able to demonstrate an increase in mutation rate of up to 150-fold after X-ray treatment.

Another technique that facilitates the detection of mutations in *Drosophila,* in this case sex-linked mutations with visible effects on phenotype (often called "visible mutations"), involves the use of ***attached-X chromosomes.*** Attached-X chromosomes undergo compulsory nondisjunction (failure of homologous chromosomes or chromatids to disjoin or separate during anaphase) since the two X chromosomes are joined to a single centromere. If females with two attached-X chromosomes plus a Y chromosome ($\widehat{XX}Y$) are mated to normal males, any mutation that occurs on the X chromosome of the male will be expressed in the surviving male progeny (Fig. 11.17). In such attached-X matings, the male progeny receive their X chromosome from their male parent rather than from their female parent as in a normal mating. If the male parent is treated with a mutagenic agent such as X rays, the increased frequency of recessive visible mutations can be easily assessed by screening the male progeny of the attached-X mating.

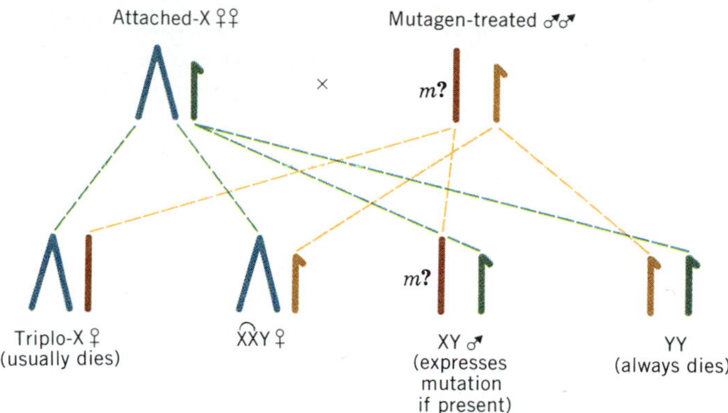

Figure 11.17 Attached-X method for detecting sex-linked visible mutations. The mutagen-treated males are crossed with $\widehat{XX}Y$ females (the two X chromosomes are attached to a single centromere). In such a mating, the viable male progeny receive their X chromosomes from their male parents. If a mutation with a visible effect on phenotype has been induced in the X chromosome of a mutagenized male, it should be expressed in all viable male progeny.

X rays and most other forms of ionizing radiation are quantitated in ***roentgen units*** (***r units,*** pronounced "rŭntgen"), which are measured in terms of the number of ionizations per unit volume under a standard set of conditions. More specifically, one roentgen unit is a quantity of ionizing radiation that produces one electrostatic unit of charge in a 1-cm^3 volume. Note that the dosage of irradiation in roentgen units does ***not*** involve a time scale. The ***same dosage*** may be obtained by a ***low intensity of irradiation over a long period of time*** or ***a high intensity of irradiation for a short period of time.*** This is very important because in most studies the ***frequency of induced point mutations is directly proportional to the dosage of irradiation*** (Fig. 11.18). In *Drosophila* sperm, for example, there is an increase of approximately 3 percent in the mutation rate for each 1000-r increase in irradiation dosage. This linear relationship between mutation frequency and dosage is indicative of so-called "single-hit kinetics." That is, only one event (ionization?) or one "hit" is required to cause a mutation. Or, stated differently, every ionization has some fixed (under a specific set of conditions) probability of inducing a mutation. If the cumulative effects of many ionizations (or other radiation-induced events) were required to induce a mutation, the relationship shown in Fig. 11.18 would plot as a curve that was concave upward.

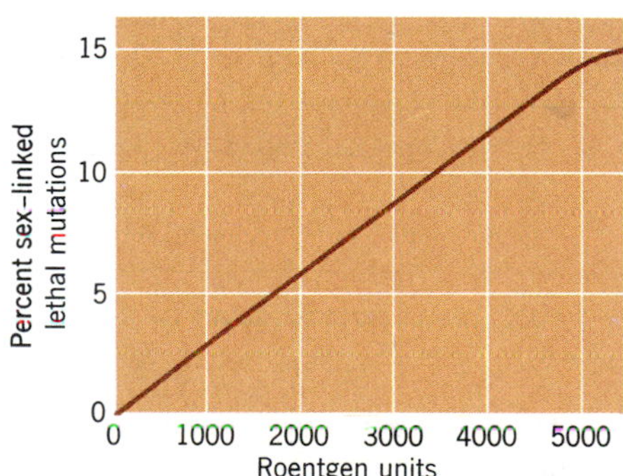

Figure 11.18 Relationship between the frequency of sex-linked lethal mutations induced in *Drosophila* sperm and ionizing irradiation dosage.

The linear relationship between mutation rate and radiation dosage is important because it speaks directly to the frequently asked question: ***What is a safe level of irradiation?*** Even very low levels of irradiation have certain low, but very real, probabilities of inducing mutations. The question is thus meaningless. There is no such thing as a safe level. In *Drosophila* sperm, for example, very low levels of irradiation over long periods of time (chronic irradiation) are as effective in inducing mutations as the same total dosage of irradiation administered at high intensity for short periods of time (acute irradiation). This clearly has major practical significance in evaluating the effects of the increased exposure of living organisms to radiation that results from the testing and use of nuclear weapons and nuclear reactors in generators, spaceships, and so on.

In mice, chronic irradiation has been found to induce somewhat fewer mutations than the same dosage of acute irradiation. Moreover, when mice were treated with intermittent doses of irradiation, the mutation frequency was slightly lower than when they were treated with the same total amount of irradiation in a continuous dose. It should be emphasized that ***all*** of these irradiation treatments were mutagenic, albeit to different degrees, to both *Drosophila* and mice. The different responses of fruit flies and mice to chronic irradiation may result from differences in their ability to repair damaged DNA. Repair mechanisms (see pp. 305–307) may exist in the spermatogonia and oöcytes of mice that do not exist in *Drosophila* sperm.

The single-hit theory implies that one ionization can produce one mutation, but it does not imply anything about the ***efficiency*** with which this happens. Several factors have been shown to affect the efficiency with which irradiation induces mutation. The receptivity of a cell at different stages of its metabolic cycle is an important factor in determining rates for induced mutations. A. H. Sparrow has shown marked variations in numbers of chromosome fragments, assumed to be directly related to mutational changes, at different stages in meiosis and cleavage (Fig. 11.19) in the plant genus *Trillium*. Chromosome aberrations were induced about 60 times more frequently at metaphase than at interphase. Nondividing cells in *Trillium* showed little radiation damage, whereas rapidly dividing cells were very sensitive.

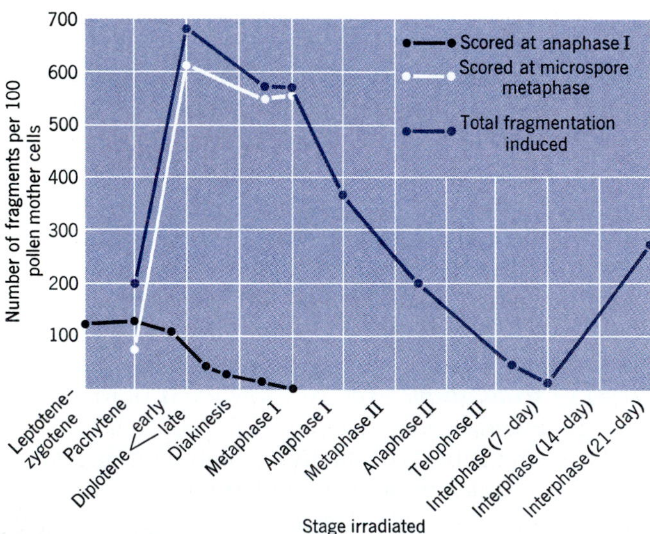

Figure 11.19 Relationship between the number of chromosomal fragments induced in meiotic cells of *Trillium erectum* at a specific dosage of radiation (50 r) and the stage at which the irradiation occurred. (From A. H. Sparrow, *New York. Acad. Sci. Ann.* 51: 1508–1540, 1951.)

Oxygen tension and temperature change, when associated with irradiation, also may significantly alter the frequency of mutations. Low oxygen tension decreases mutations. Oxygen can magnify the effect of radiation, but only if it is present during the irradiation. Oxygen has less effect with intense conditions than with moderate conditions of ionization. Environmental agents that protect germ cells from radiation damage often do so by lowering the oxygen concentration of tissue, and those that enhance the effectiveness of radiation add oxygen.

Ionizing radiation also induces various kinds of gross changes in chromosome structure (chromosome aberrations), such as deletions, duplications, inversions, and translocations (see Chapter 18). These changes in chromosome structure result from breaks in chromosomes caused by ionizing radiation. Since they require two breaks, the kinetics of induction are "two hit" as expected, rather than "single hit."

Ultraviolet Radiation

Ultraviolet rays do not possess sufficient energy to induce ionizations. They are, however, readily ***absorbed*** by certain substances such as ***purines*** and ***pyrimidines,*** which then enter a more reactive or ***excited*** state. Because of their lower energy, they penetrate tissues only slightly, usually only the surface layer of cells in multicellular organisms. Nevertheless, ultraviolet light (UV) is a potent mutagen for unicellular organisms. The maximum absorption of UV by DNA is at a wavelength of 254 nm. Maximum mutagenicity also occurs at 254 nm, suggesting that the UV-induced mutation process is mediated directly by the absorption of UV by purines and pyrmidines. *In vitro* studies show that the pyrimidines (especially thymine) absorb strongly at 254 nm and, as a result, become very reactive. The two major products of UV absorption by pyrimidines appear to be pyrimidine hydrates and pyrimidine dimers (Fig. 11.20). Several lines of evidence indicate that thymine dimerization is probably the major mutagenic effect of UV. Thymine dimers appear to cause mutations indirectly in two ways. (1) Dimers apparently perturb the DNA double helix and interfere with accurate DNA replication. (2) Occasional errors are made during the cell processes for the repair of "damaged" DNA, such as DNA containing thymine dimers (see the following section).

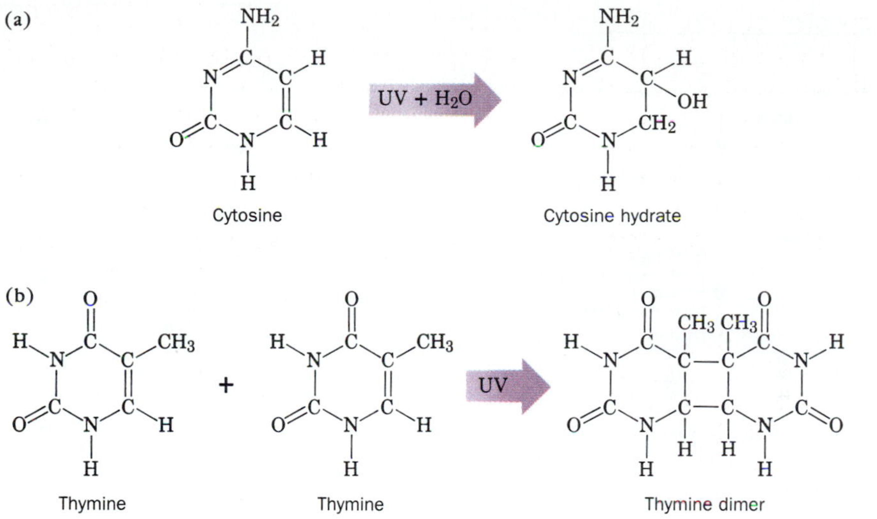

Figure 11.20 Pyrimidine photoproducts of UV irradiation. (a) Hydrolysis of cytosine to a hydrate form that may cause mispairing of bases during replication. (b) Cross-linking of adjacent thymine molecules to form thymine dimers, which block replication.

The relationship between mutation rate and UV dosage is highly variable, depending on the type of mutation, the organism, and the conditions employed. "Single-hit kinetics" are only occasionally observed, in contrast to ionizing radiation.

DNA Repair Mechanisms

An indication of the importance of keeping mutation, both somatic and germ line, at a tolerable level is the multiplicity of repair mechanisms in organisms ranging from the simple bacterial viruses to humans. Mechanisms for the repair of damaged DNA are probably universal. *Escherichia coli,* for example, possesses at least three different mechanisms for the repair of DNA containing thymine dimers: (1) ***photoreactivation,*** (2) ***excision repair,*** and (3) ***postreplication recombination repair.***

Photoreactivation involves an enzyme that splits thymine dimers directly without the removal of any nucleotides (Fig. 11.21). This enzyme will bind to thymine dimers in DNA in the dark, but it cannot catalyze cleavage of the bonds joining the thymine molecules without energy derived from visible light, specifically light within the blue region of the spectrum. The enzyme is also active on cytosine dimers and cytosine–thymine dimers. Thus, when ultraviolet light is used as an experimental mutagen, the treatment is usually carried out in the dark to maximize the mutation frequency.

Excision repair involves a sequence of enzyme-catalyzed steps in which the thymine dimers are removed from the DNA molecule and a new segment of DNA is synthesized (Fig. 11.22). Excision repair occurs as efficiently in the dark as in the presence of blue light. The first step in excision repair is catalyzed by an endonuclease that recognizes thymine dimers, or the distortion in the double helix that they cause, and cleaves the phosphodiester backbone of the DNA strand containing the dimer at a site near the damage. An exonuclease, probably the $5' \longrightarrow 3'$ exonuclease activity of DNA polymerase I (at least it catalyzes the removal of thymine dimers *in vitro*), then removes a segment of the strand adjacent to the endonuclease cut, including the dimer. A DNA polymerase, probably DNA polymerase I, then fills in the gap using the complementary strand as template. DNA ligase then catalyzes covalent closure or formation of the final phosphodiester linkage between adjacent nucleotides. Although the exact mechanism of excision repair is not yet known, the general features of the pathway just described are well documented. DNA polymerase I appears to be involved because mutants lacking this enzyme are defective in excision repair. In addition, the products of three genes of *E. coli*—*uvrA, uvrB,* and *uvrC*—are required for this process.

The second type of dark repair of UV damage that occurs in *E. coli* involves both replication and recombination and has thus been called postreplication recombination repair. The exact details of this repair process are still uncertain; however, the process clearly involves replication of the damaged chromosomes followed by recombination. A simplified model of postreplication recombination repair is shown in

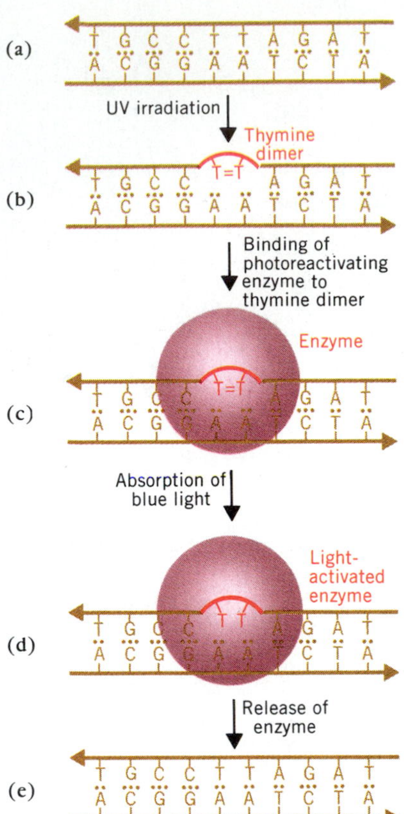

Figure 11.21 Cleavage of thymine dimer cross-links by photoreactivation. (a) Segment of DNA containing two adjacent thymine bases. (b) Thymine dimer formation as a result of UV irradiation. (c) Binding of the photoreactivating enzyme to the thymine dimer-containing segment of the DNA molecule. (d) Cleavage of the dimer cross-links by the photoreactivating enzyme using energy from the absorption of blue light. (e) Release of the photoreactivating enzyme, leaving a repaired, structurally normal DNA molecule. The arrows indicate the opposite polarity of the complementary strands.

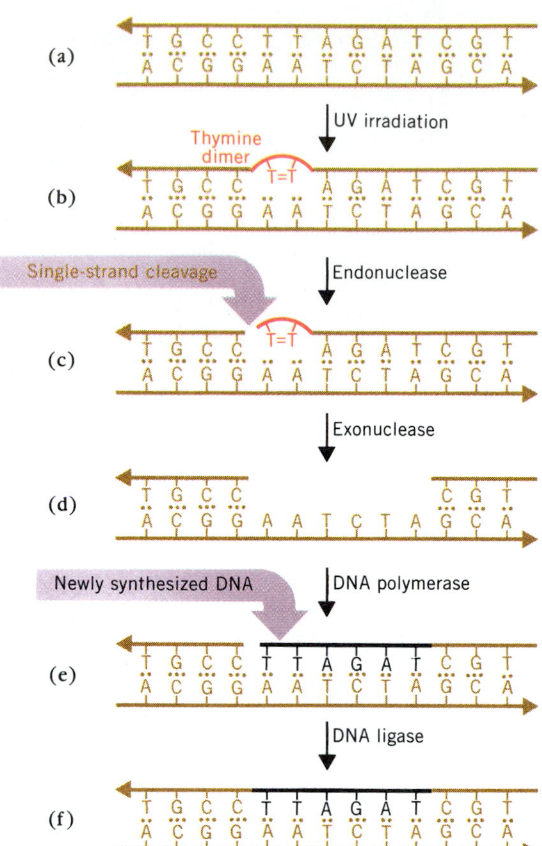

Figure 11.22 Diagram of the excision repair pathway for the removal of thymine dimers from DNA. (a) Segment of DNA containing two adjacent thymine bases. (b) Thymine dimer formation as a result of UV radiation. (c) An endonuclease-induced single-strand cut adjacent to the thymine dimer. (d) An exonuclease-catalyzed removal of six nucleotides, including the thymine dimer. (e) Resynthesis of the excised segment by a DNA polymerase. (f) Formation of the final phosphodiester bond by DNA ligase. The arrows indicate the opposite polarity of the two complementary strands.

Fig. 11.23. When DNA molecules containing thymine dimers are replicated, gaps are formed in the nascent complementary strands opposite the dimers because DNA polymerase cannot use the distorted strands as templates. After a lag due to the thymine dimer blockage, replication is reinitiated at secondary initiation sites beyond the dimer. This results in progeny double helices with thymine dimers in one strand and gaps in the complementary strand. If these two "sister" chromosomes recombine such that the dimers and gaps end up in one chromosome and the intact, undamaged segments end up in the other chromosome, the latter will be functional and produce a viable cell.

Another DNA repair enzyme, ***uracil DNA glycosylase,*** has been characterized in *E. coli.* This enzyme efficiently removes uracil from DNA (thus its name). It is called a glycosylase rather than a nuclease because it removes uracil by cleaving the glycosidic bond between the base and the sugar rather than phosphodiester bonds in the DNA chain. The enzyme thus produces apyrimidinic sites in the DNA. Such apyrimidinic sites have been shown to undergo an excision repair mechanism much like that described in Fig. 11.22. Uracil is produced in DNA by deamination of cytosine (e.g., by nitrous acid; see Fig. 11.27). *Escherichia coli* mutants deficient in uracil-DNA glycosylase exhibit increased sensitivity to deaminating agents such as nitrous acid, indicating that the enzyme is important in the repair of this type of damage to DNA.

Mechanisms for the efficient repair of DNA dam-

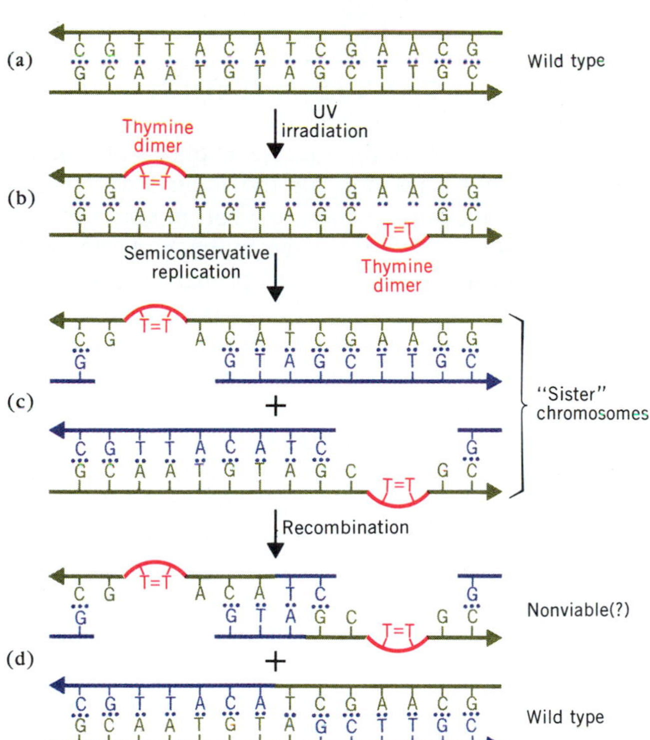

Figure 11.23 Simplified version of postreplication recombination repair of UV-induced damage in DNA. (a) Segment of a DNA molecule containing adjacent thymine bases. (b) Formation of thymine dimers during UV irradiation. (c) Replication of dimer-containing DNA. Gaps are present in the nascent strands (blue) because of the inability of the distorted dimer-containing regions of the parental strands to function as templates for DNA polymerase. After a lag period, synthesis is reinitiated past the dimers, leaving the indicated gaps. (d) Recombination between the "sister chromosomes," producing one undamaged DNA molecule and one doubly damaged DNA molecule. The net effect is the production of one undamaged chromosome from two damaged ones or, quite possibly, one viable product from two inviable ones. Details of the recombination process have been purposely omitted since they are not yet clear. Nucleases, a DNA polymerase, and DNA ligase are undoubtedly involved.

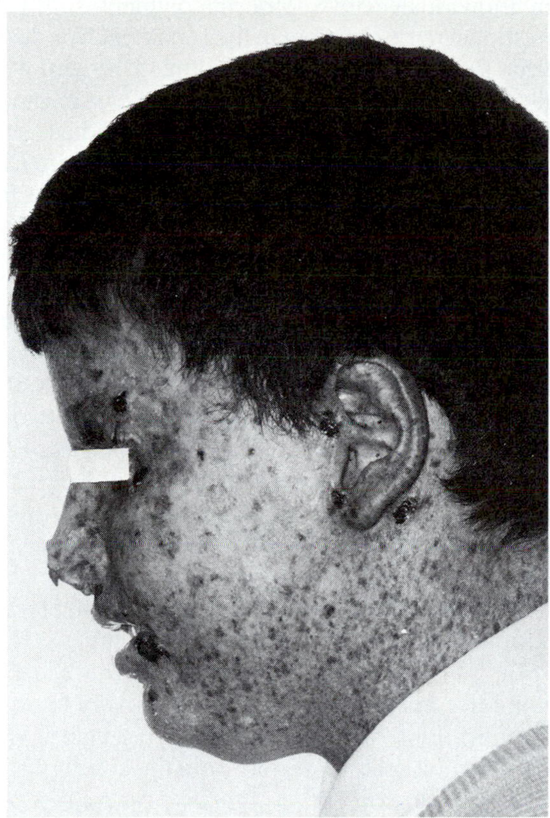

Figure 11.24 Phenotypic effects of the inherited disease xeroderma pigmentosum. Individuals with this malignant disease develop extensive skin tumors after exposure to sunlight. Homozygotes for the autosomal recessive mutation responsible for xeroderma pigmentosum are less efficient in the repair of DNA damaged by exposure to ultraviolet light. (Photograph by P. E. Polani, Guy's Hospital, London. Reprinted with permission of Macmillan Publishing Co., Inc., from E. Novitski, *Human Genetics,* Copyright © 1977 by Edward Novitski.)

aged by mutagenic agents are clearly important even in bacteria and viruses. One can argue that they should be increasingly important with the increasing biological complexity and the correlated increase in genome size of higher organisms. The ***larger the number of essential genes,*** the more sites at which a lethal mutation can occur, and, given equivalent repair processes, the ***greater the probability that a given dose of mutagen*** such as irradiation ***will cause a lethal mutation.***

Xeroderma Pigmentosum

DNA repair mechanisms similar to those described for *E. coli* have also been demonstrated in humans in studies done with cell cultures. In addition, their existence is indicated by genetic studies on individuals with inherited sensitivity to mutagenic agents such as ultraviolet light and/or ionizing radiation. The best-known example of an inherited deficiency in the repair of radiation damage in humans is the autosomal recessive disease called ***xeroderma pigmentosum.*** The skin of individuals with xeroderma pigmentosum exhibits extreme sensitivity to sunlight, resulting in a high frequency of tumors in the epidermal cells of exposed areas such as the face (Fig. 11.24). The cells of many individuals with xeroderma pigmentosum have been shown to be deficient in the repair of UV-induced damage to DNA, such as thymine dimers. It seems very likely, therefore, that the high incidence of skin cancer results directly from the inability of individuals with xeroderma pigmentosum to repair damage in DNA induced by UV in sunlight. In some cases, the defective enzyme appears to be the endonuclease that recognizes the thymine dimer and catalyzes the first step in excision repair (see Fig. 11.22). Genetic analyses of cells from individuals with xeroderma pigmentosum suggest that mutations in as many as six different genes can cause the disease. This is not unexpected since (1) many enzymes are known to be composed of two or more different polypeptides (gene-products) and (2) a mutation in any of the genes coding for polypeptides involved in the multiple-step repair processes might be expected to cause a block (see Fig. 11.3) in a repair pathway.

Some individuals with xeroderma pigmentosum develop neurological abnormalities that appear to result from premature death of nerve cells. This effect on the long-lived nerve cells has potentially very interesting implications for the mechanism of aging. One theory of aging is that it results from the accumulation of somatic mutations. If so, a defective repair system would be expected to speed up the aging process; this appears to be the case with the nerve cells of some xeroderma pigmentosum patients. F. M. Burnet has emphasized that a low but significant level of spontaneous mutation must be maintained in living cells to provide sufficient flexibility for evolution to occur, and that aging may be an unavoidable consequence of this low level of mutation. Certainly, somatic mutations would be expected to contribute to the aging process. At present, however, essentially nothing is known about the relative importance of somatic mutation in senescence.

Fanconi's anemia, ataxia-telangiectasia, and Bloom's syndrome are three other inherited (autosomal recessive) diseases in humans in which the primary defect may be in a DNA repair pathway. All three conditions result in a high risk of malignancy, particularly leukemia in the case of Fanconi's anemia and ataxia-telangiectasia. Cells of patients with ataxia-telangiectasia are abnormally sensitive to ionizing radiation, suggesting a defect in the repair of radiation-induced DNA damage. Cells of individuals with Fanconi's anemia are impaired in the removal of DNA interstrand cross-links, such as those formed by the antibiotic mitomycin C. Individuals with Bloom's syndrome exhibit a high frequency of chromosome breakage that results in aberrations (see Chapter 18) and sister chromatid exchanges. Although all three of these malignant-prone, inherited diseases are probably caused by defects in DNA metabolism, possibly the repair processes, the primary lesions are still unknown.

CHEMICALLY INDUCED MUTATION

Muller's discovery of the mutagenic effects of irradiation provided a method for the induction of large numbers of mutations with a minimal amount of effort. However, because of the large number of effects of radiation on living tissues, both primary and secondary, studies on radiation-induced mutation provided little information about the molecular details of mutation processes. Subsequent discoveries of the mutagenic effect of a large variety of chemicals, many with very specific effects on DNA, provided the tools needed to work out many of the details of mutation processes at the molecular level. Hundreds of chemicals are now known to have from slight to very large mutagenic effects. Only a few of those with very strong mutagenic effects will be discussed here. Some of the more potent chemical mutagens are listed in Table 11.1.

The first chemical mutagen discovered was mustard gas (sulfur mustard). When C. Auerbach and her associates first discovered the mutagenic effects of mustard gas and related compounds during World War II, their data were classified. These compounds are examples of a large class of chemical mutagens that transfer alkyl (CH_3-, CH_3CH_2-, etc.) groups to the bases in DNA; thus they are called ***alkylating agents.***

Chemical mutagens can be divided into two classes: (1) those that are mutagenic to both replicating and nonreplicating DNA, such as the alkylating agents and ***nitrous acid,*** and (2) those that are mutagenic only to replicating DNA. The latter class includes the ***acridine dyes,*** which bind to DNA and increase the probability of mistakes during replication, and ***base analogs,*** purines and pyrimdines with structures similar to the normal bases of DNA. The base analogs must be incorporated into DNA chains in the place of normal bases during replication to exert their mutagenic effects.

Base Analogs

The base analogs that are mutagenic have structures sufficiently similar to the normal bases so that they are metabolized and incorporated into DNA during replication, but sufficiently different such that they increase the frequency of mispairing and thus mutation. The two most commonly used base analogs are ***5-bromouracil*** and ***2-aminopurine.*** The pyrimidine 5-bromouracil is a thymine analog, the bromine at the 5 position being similar in several respects to the methyl ($-CH_3$) group at the 5 position in thymine. The presence of the bromine, however, changes the charge distribution and increases the probability of tautomeric shifts (see Fig. 11.10). In its more stable keto form, 5-bromouracil pairs with adenine. After a tautomeric shift to its enol form, 5-bromouracil pairs with guanine (Fig. 11.25). The mutagenic effect of 5-bromouracil should be the same as that predicated for tautomeric shifts in normal bases (see Fig. 11.12), namely, transitions. If 5-bromouracil is present in its less frequent enol form as a nucleoside triphosphate at the time of its incorporation into a nascent strand of DNA, it will be incorporated opposite guanine in the template strand and cause a GC⟶AT transition (Fig. 11.26*a*). If, on the other hand, 5-bromouracil is incorporated in its more frequent keto form opposite adenine (in place of thymine) and undergoes a tautomeric shift to its enol form during a subsequent replication, it will cause an AT⟶GC transition (Fig. 11.26*b*). Thus, 5-bromouracil induces transitions in both directions, AT⟷GC. An important consequence of the bidirectionality of 5-bromouracil-induced transitions is that mutations originally induced

TABLE 11.1 Some of the More Potent Chemical Mutagens

CHEMICAL NAME	COMMON NAME OR ABBREVIATION	STRUCTURE
I. Alkylating agents		
Di-(2-chloroethyl) sulfide	Mustard gas or sulfur mustard	Cl—CH_2—CH_2—S—CH_2—CH_2—Cl
Di-(2-chloroethyl) methylamine	Nitrogen mustard	CH_3 | Cl—CH_2—CH_2—N—CH_2—CH_2—Cl
Ethylmethane sulfonate	EMS	CH_3—CH_2—O—SO_2—CH_3
Ethylethane sulfonate	EES	CH_3—CH_2—O—SO_2—CH_2—CH_3
N-Methyl-*N'*-nitro-*N*-nitrosoguanidine	NTG	HN=C—NH—NO_2 | O=N—N—CH_3
II. Base analogs		
5-Bromouracil	5-BU	O ‖ H C Br N C C C O N H | H
2-Aminopurine	2-AP	H | C N C N C—H C C N H_2N N | H
III. Acridines		
2,8-Diamino acridine	Proflavin	H H H H C C C H C C C C C C C H_2N C C $\overset{+}{N}$ C C $\overset{+}{NH_3}$ H H H
IV. Deaminating agents		
Nitrous acid	—	HNO_2
V. Miscellaneous		
Hydroxylamine	HA	NH_2OH

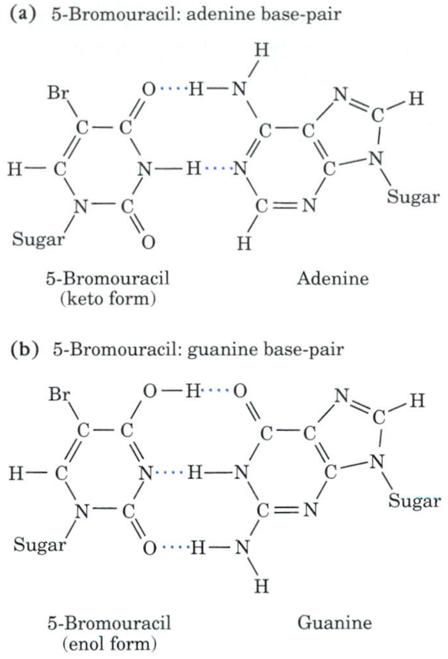

(a) 5-Bromouracil: adenine base-pair

5-Bromouracil (keto form) Adenine

(b) 5-Bromouracil: guanine base-pair

5-Bromouracil (enol form) Guanine

Figure 11.25 Base-pairing of 5-bromouracil with (a) adenine and (b) guanine in its two tautomeric forms.

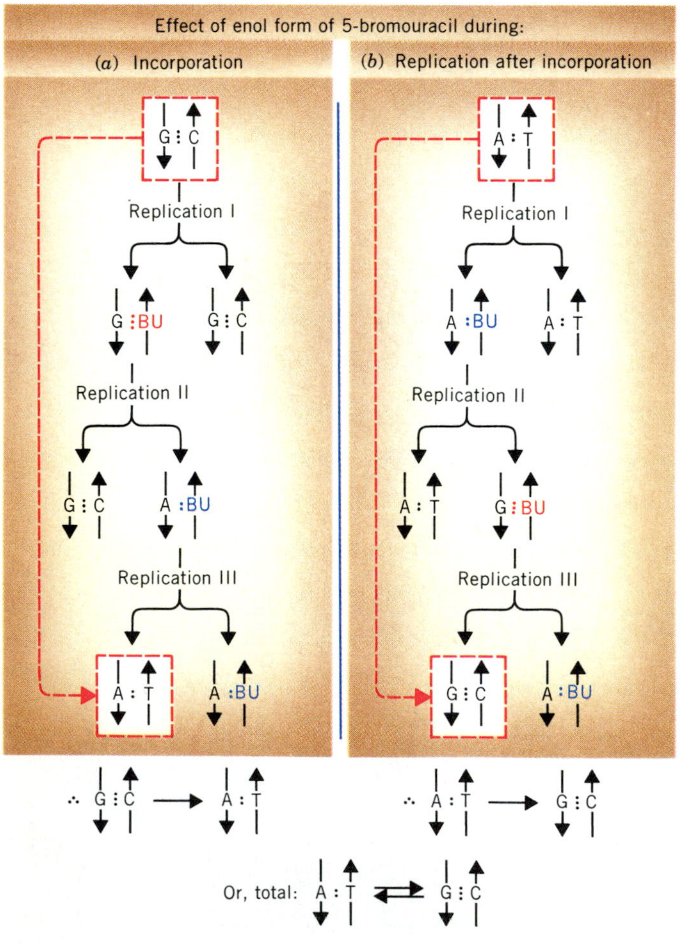

Effect of enol form of 5-bromouracil during:

(a) Incorporation

(b) Replication after incorporation

Replication I

Replication II

Replication III

Or, total: A : T ⇄ G : C

Figure 11.26 Summary of the mutagenic effects of 5-bromouracil. When 5-bromouracil (abbreviated BU in the diagram) is in its less frequent enol form (abbreviated in red) at the time of incorporation into DNA, it induces GC→AT transitions (a). When it is incorporated into DNA in its more common keto form (shown in blue) and shifts to its enol form during a subsequent replication, it induces AT→GC transitions (b). The summation of these two effects is that 5-bromouracil can induce transitions in both directions, AT↔GC.

with this thymine analog can also be induced to revert with it. The purine base analog 2-aminopurine is believed to induce mutations in a similar manner.

Nitrous Acid

Nitrous acid (HNO_2) is a very potent mutagen that acts directly on either replicating or nonreplicating DNA by oxidative deamination of the bases that contain amino groups—adenine, guanine, and cytosine. Conversion of the amino groups to keto groups changes the hydrogen-bonding potential of the bases (Fig. 11.27). Adenine is deaminated to hypoxanthine, which base-pairs with cytosine rather than thymine. Cytosine is converted to uracil, which base-pairs with adenine instead of guanine. Deamination of guanine produces xanthine, but xanthine base-pairs with cytosine just like guanine. Thus, the deamination of guanine is not directly mutagenic like that of adenine and cytosine. Since the deamination of adenine leads to AT→GC transitions, and the deamination of cytosine results in GC→AT transitions, nitrous acid induces transitions in both directions, AT↔GC. Nitrous acid-induced mutations can thus also be induced to revert with nitrous acid.

Acridines

The acridine dyes such as proflavin (see Table 11.1), acridine orange, and a whole series of compounds called ICR-170, ICR-191, and so on, are very powerful mutagens that induce frameshift mutations (see Fig. 11.14). The ICR compounds have acridine moieties with various side chains, often alkylating agents. The positively charged acridines intercalate or "sandwich themselves" between the stacked base-pairs in DNA (Fig. 11.28). In so doing, they increase the rigidity and alter the conformation of the double helix, possibly causing slight "kinks" in the molecule. In any case, when DNA molecules containing intercalated acridines replicate, additions and deletions of from one to a few base-pairs occur. As would be expected, these small additions and deletions, usually of a single base-

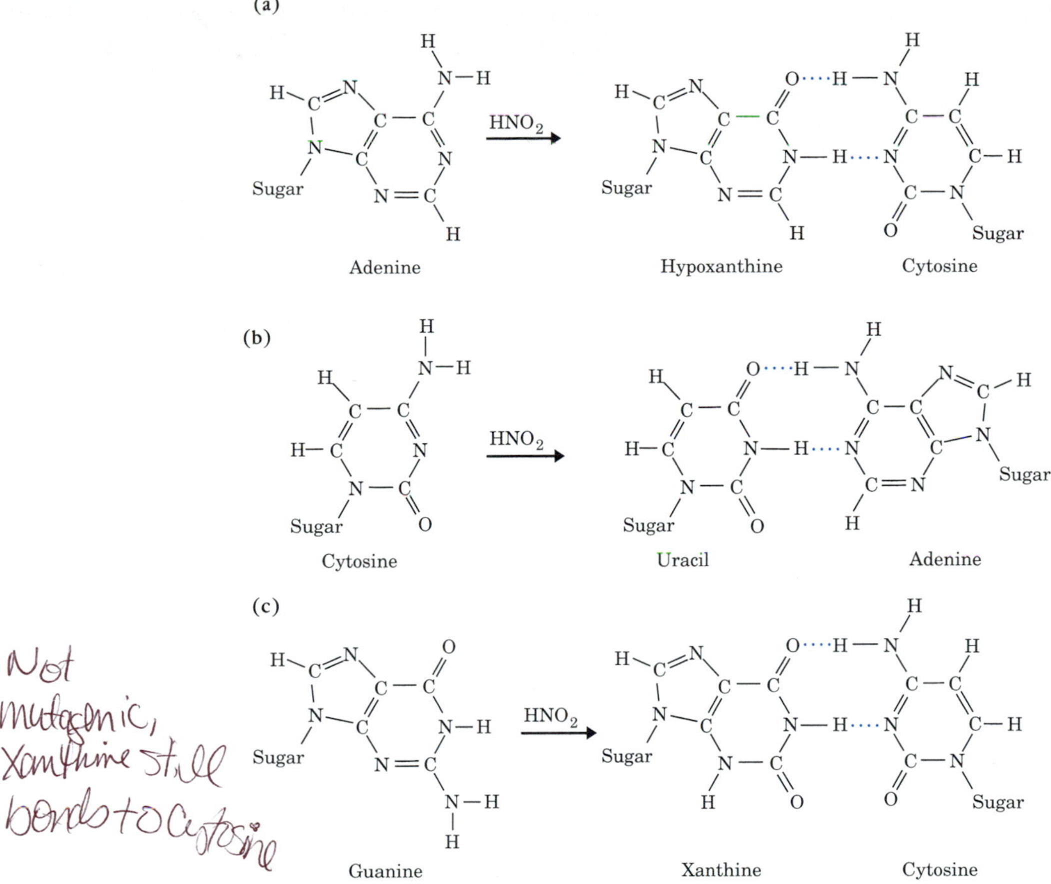

Figure 11.27 Mutagenic action of nitrous acid. Oxidative deamination converts (a) adenine to hypoxanthine, causing AT⟶GC transitions; (b) cytosine to uracil, causing GC⟶AT transitions; and (c) guanine to xanthine, which is not directly mutagenic. The combined effects of nitrous acid on adenine and cytosine explain its ability to induce transitions in both directions, AT⟷GC.

pair, result in reading frameshifts for the portion of the gene distal to the mutation (see Fig. 11.14).

Alkylating and Hydroxylating Agents

Alkylating agents such as nitrogen and sulfur mustards, methyl and ethyl methanesulfonate (MMS and EMS), nitrosoguanidine (NTG; see Table 11.1), and many others have several effects on DNA. One major mechanism of mutagenesis by alkylating agents involves the transfer of methyl or ethyl groups to the bases such that their base-pairing potentials are altered and transitions result. For example, ethylation at the 7-*N* position and at the 6-*O* position are believed to be two effects of EMS. 7-Ethylguanine is then believed to base-pair with thymine (Fig. 11.29*a*). Other base alkylation products are believed to somehow "activate" repair processes in much the same way as thymine

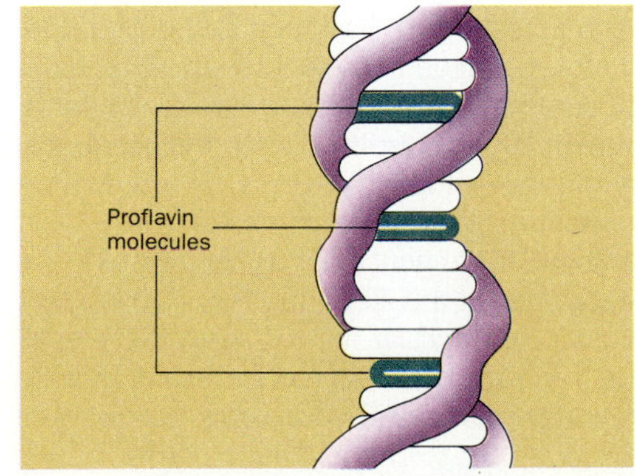

Figure 11.28 Intercalation of the acridine proflavin into the DNA double helix. X-ray diffraction studies have shown that the positively charged acridines become "sandwiched" between the stacked base-pairs. Genetic studies have shown that acridines such as proflavin induce frameshift mutations by the addition or deletion of one or a few nucleotide-pairs during the replication of DNA containing such intercalated molecules. (After L. S. Lerman, *J. Cellular Comparative Physiol.* 64 (Suppl. 1): 1, 1964.)

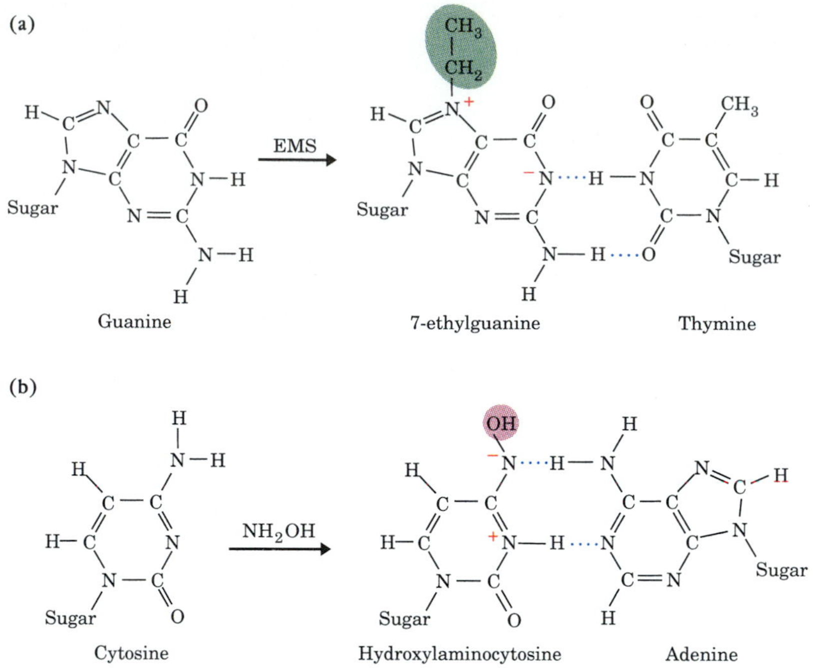

Figure 11.29 One proposed mutagenic effect of (a) the alkylating agent ethyl methanesulfonate (EMS) and (b) the hydroxylating agent hydroxylamine (NH_2OH). Alkylating agents such as EMS clearly induce mutations by other mechanisms as well (see text). The mutagenic action of hydroxylamine is known to involve cytosine, but its exact effect on cytosine is still somewhat disputed.

dimers do (see Figs. 11.21, 11.22, and 11.23). The occasional errors occurring in these repair processes may lead to transversions and frameshift mutations in addition to transitions. Finally, some alkylating agents, particularly difunctional alkylating agents (those with two reactive alkyl groups), cross-link DNA strands and/or molecules and induce chromosome breaks and the various kinds of chromosomal aberrations (see Chapter 18) found correlated with breaks. Alkylating agents as a class therefore exhibit less specificity in their mutagenicity than base analogs, nitrous acid, or acridines. They induce all types of mutations, including transitions, transversions, frameshifts, and even chromosome aberrations, with various relative frequencies, depending on the specific alkylating agent employed.

Nitrosoguanidine (NTG), one of the most potent chemical nutagens known, has been found to induce clusters of closely linked mutations in the segment of the chromosome that is replicating during the mutagenic treatment. Mutants isolated after NTG treatment often carry multiple, closely linked mutations, making them less useful for most genetic studies.

The hydroxylating agent ***hydroxylamine,*** NH_2OH, in contrast to many of the alkylating agents, has a very specific mutagenic effect. It induces ***only GC ⟶ AT transitions.*** Although the exact mechanism of hydroxylamine mutagenesis is still somewhat uncertain, it apparently acts by hydroxylating the amino group of cytosine (Fig. 11.29*b*). The resulting hydroxylaminocytosine can base-pair with adenine to produce the observed GC ⟶ AT transitions. Because of its specificity, hydroxylamine has been very useful in classifying transition mutations. Mutants that are revertible by nitrous acid or base analogs, and therefore resulting from transitions, can be divided into two classes on the basis of their revertibility with hydroxylamine. (1) Those with an AT base-pair at the mutated site will not be induced to revert by hydroxylamine. (2) Those with a GC base-pair at the mutated site will be induced to revert by hydroxylamine. Thus, hydroxylamine can be used to determine whether a particular mutation was an AT ⟶ GC or a GC ⟶ AT transition.

CORRELATION BETWEEN MUTAGENICITY AND CARCINOGENICITY

For many years it has been recognized that most of the strongly mutagenic agents, such as ionizing radiations, ultraviolet light, and chemicals like those discussed in the preceding sections, are also ***carcinogenic,*** that is, ***induce cancers.*** In recent years, sensitive techniques have been developed to test chemicals and other agents for mutagenicity and carcinogenicity. Carcinogenicity tests are usually done with rodents, most frequently newborn mice. These studies involve feeding or injecting the substance being tested and subsequent examination of the animal for tumor formation. The mutagenicity tests have often been done in similar fashion. However, because mutation is a low-frequency event and because maintaining large populations of animals such as mice is an expensive undertaking, such animal tests have usually been quite insensitive, that is, low levels of mutagenicity would seldom have been detected.

B. Ames and his associates have developed very sensitive techniques by which large numbers of chemicals and other agents can be screened for mutagenicity. Their procedures are rapid and inexpensive. Ames' procedures involve the use of specially constructed strains of *Salmonella typhimurium* (a bacterium related to *E. coli*) that carry auxotrophic mutations of various types: transitions, transversions, and frameshifts. The reversion of these auxotrophic mutants to prototrophy, either spontaneous or induced with various agents, can easily be monitored by placing a known number of mutant cells in a petri plate containing medium supplemented with a trace of the growth factor required by the auxotroph (enough to support a few cell divisions, but not enough to allow the formation of visible colonies) and counting the number of colonies produced by prototrophic revertants after an appropriate period of incubation. (Some growth is required to detect the effects of mutagens that are only mutagenic to replicating DNA.) The frequency of reversion induced by a particular substance can be compared directly with the spontaneous reversion frequency to obtain an estimate of its mutagenicity. Its ability to induce different types of mutations can be assessed by using a set of test strains that carry different types of mutations—one strain with a transition, one with a frameshift mutation, and so forth. The procedures developed by Ames and his colleagues are very sensitive (Fig. 11.30) and facilitate the detection of even weak mutagens.

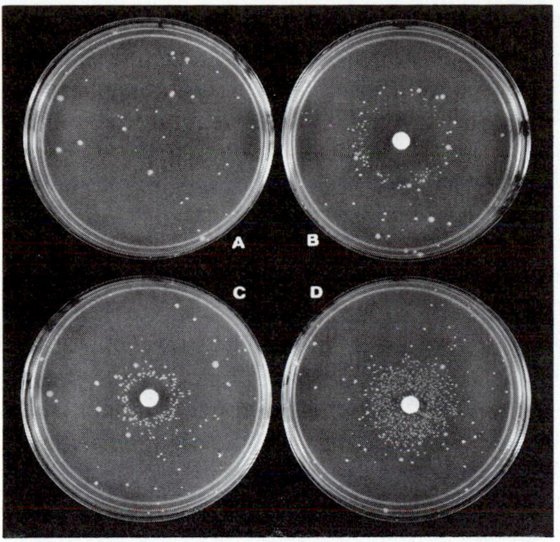

Figure 11.30 *Salmonella* mutagenicity tests. Each petri dish contains a layer of agar medium containing only a trace of histidine and a known number of *Salmonella* "tester strain" cells, which are carrying, in this case, a frameshift mutation that results in histidine auxotrophy. Plates C and D also contain the rat-liver microsomal activation system. The potential mutagens are applied to 6-mm-diameter filter paper disks, which are placed on the surface of the agar in the centers of plates B, C, and D. (A) Control plate with no addition, showing the background level of spontaneous reversion to prototrophy. (B) Plate to which the carcinogen furylfuramide (a food additive) has been added. (C) Plate to which the carcinogen aflatoxin B (a fungal toxin) has been added. (D) Plate to which the carcinogen 2-aminofluorene has been added. (From B. N. Ames, J. McCann, and E. Yamasaki, *Mutat. Res.* 31: 347, 1975. Photograph courtesy of B. N. Ames.)

Over a period of several years during which hundreds of different chemicals were tested, Ames and his colleagues have observed a greater than 90 percent correlation between the mutagenicity and the carcinogenicity of the substances tested. Initially, several potent carcinogens were found to be nonmutagenic to the tester strains. Subsequently, many of these carcinogens, although nonmutagenic per se, were found to be metabolized to strongly mutagenic derivatives in eukaryotic cells. Ames and his associates therefore added a rat liver microsomal fraction to their assay systems in an attempt to detect the mutagenicity of metabolic derivatives of substances being tested. Coupling of the rat liver microsomal activation system to the microbial mutagenicity tests expanded the utility of the system considerably. Nitrates, for example, are not mutagenic or carcinogenic themselves. *In vivo,* however, in eukaryotic cells, nitrates are converted via a series of enzyme-catalyzed reactions to nitrosamines, which are highly mutagenic and carcinogenic.

Ames' *Salmonella* mutagenicity tests demonstrated the presence of frameshift mutagens in several different components of chemically fractionated cigarette smoke condensates. In some cases, activation by the liver microsomal preparation was required for mutagenicity; in other cases, activation was not required.

The observed correlation between mutagenicity and carcinogenicity is consistent with the theory that cancer is caused by somatic mutations. Most geneticists would agree that somatic mutations ***can*** cause cancer. This has received strong support from the discovery of cellular oncogenes (cancer-causing genes) and the demonstration that the oncogene responsible for human bladder carcinoma resulted from a single base-pair change in its normal cellular counterpart (see Chapter 17, pp. 476–478). The common feature of all of the many different types of cancer is that the malignant cells, whatever their type, continue to divide when their normal counterparts would no longer be undergoing cell division. That is, all cancers exhibit a loss of the normal control of cell division, with the resultant formation of tumors. Cell division is undoubtedly, at least in part, under genetic control. Thus, mutation in a gene involved in the control of cell division, like a mutation in any other gene, can cause a loss of function and thus a loss of the normal control of cell division. One can also easily envision epigenetic causes of cancer, that is, changes in states of differentiation, rather than permanent changes in the genetic material per se. The key question would appear to be, therefore, not whether somatic mutations can cause cancer, but what proportion of human cancers are, in fact, caused by somatic mutations.

MUTATION FREQUENCY

Although mutation is necessary to provide the genetic variability required for the evolutionary adaptation of species to environmental changes, most mutations are deleterious. Thus, if mutation were to become too frequent in a species, it would create a sizable "genetic load" of deleterious effects. Clearly, if this "genetic load" became too large, the species would face extinction. Human technology has already contaminated the Earth with increased levels of radiation and chemicals that are known to be mutagenic. Although the consequences of the present increased level of mutagens in the environment cannot yet be accurately assessed, most scientists agree that further increase in the levels of mutagens in the environment should be avoided. Yet significant quantities of hundreds of new chemicals are introduced into the enviroment each year, most of them with insufficient, if any, mutagenicity tests. Whether they will cause harmful increases in mutation rate (and/or cancer incidence, see pp. 312–314) is a question of utmost concern to everyone.

Each gene probably has its own characteristic mutational behavior (e.g., see Fig. 12.16). Some genes undergo mutations more frequently than others in the same organism. Those with unusually high mutation rates are called unstable or mutable, but a wide range of mutation rates exists among genes that are considered stable. The mutation rate per gene in bacteria is on the order of 1 in 100,000 to 1 in 10 million (10^{-5} to 10^{-7}) per cell generation. For fruit flies, the average for mutation in a particular gene is on the order of 1 in 100,000, with a range from 1 in 20,000 to 1 in 200,000 gametes. Since most of the data on mutation rates in fruit flies have been obtained from experiments with males, questions have arisen concerning a possible sex difference in overall mutation rates. B. Wallace has shown through extensive experiments that mutation rates are not significantly different in the two sexes of *D. melanogaster*. However, mutation rates do differ in different strains.

Estimates of mutation rates for humans indicate a somewhat greater frequency than those cited for stable genes in most other organisms. Samples collected thus far have been small, and the methods used were indirect and subject to large errors. Genes associated with human traits such as intestinal polyposis and muscular dystrophy have been estimated to mutate once in 10^4 to 10^5 people. A human generation is equal to about 50 cell generations. By expressing any mutation rate as a probability of mutation per cell per generation, a mutation rate is defined independently of exact physiological conditions and stage in the life cycle. This definition is based on a time unit proportional to a cell's division time. When expressed in terms of cell generations, rates for fruit flies and humans are generally comparable with those for bacteria.

PRACTICAL APPLICATIONS OF MUTATIONS

As was discussed earlier, mutations are invaluable to the process of evolution since they provide the raw material required for its occurrence. Moreover, mutations provide the alleles required for the various types of genetic analysis, from Mendel's two-factor crosses (Chapter 2) to chromosome mapping (Chapters 7, 8, and 12) to studies of genetic structures of populations (Chapter 22). Without mutation, genetic analyses of the types described in this textbook would be impossible. Mutations, and the ability to induce them with mutagenic agents, have thus proven useful in many ways.

Beneficial Mutations

Even though most mutations make the organism less efficient and are thus disadvantageous, the possibility of developing new desirable traits through induced mutations has intrigued many plant breeders. Plant breeders have reported induced mutants in barley, wheat, oats, soybeans, tomatoes, and fruit trees that may improve presently cultivated strains. Barley mutants, for example, have been obtained that provide increased yield, resistance to smut (Fig. 11.31), stiff straw, increased protein content, and hull-less seeds.

One application of induced mutations came from concentrated efforts to improve the yield of penicillin by the mold *Penicillium*. When penicillin was first discovered, the yield was low and production was seriously limited. Then millions of spores were irradiated and a few of the surviving colonies produced considerably more penicillin than the average. Such mutant overproducers of penicillin have proven invaluable in the commercial production of this important antibiotic.

Figure 11.31 Heads of barley demonstrating the effects of resistance to loose smut. The brown heads are affected with the smut, whereas the yellow-green heads are resistant. (Courtesy of Grant Heilman.)

High-Resolution Dissection of Biological Processes via Mutation Analysis

Mutations have been used extensively to elucidate the pathways by which biological processes occur. Metabolism occurs via sequences of enzyme-catalyzed reactions. By isolating and studying mutations in the genes coding for the enzymes involved, the sequence of steps in a pathway can often be determined. Morphogenesis frequently involves the sequential addition of proteins in the formation of specific three-dimensional structures. Again, the sequence of protein additions can often be determined by isolating and studying mutant organisms with mutations in the genes coding for the proteins involved. Because an appropriate mutation will eliminate the activity of a single polypeptide, mutations provide an extremely powerful probe with which to dissect biological processes.

Consider, for example, the following simple pathway:

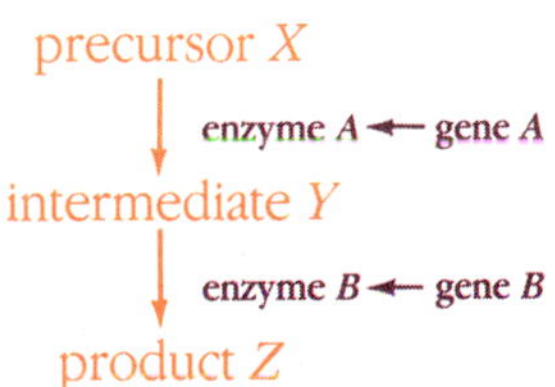

Intermediate *Y* is produced from precursor *X* by the action of enzyme *A*, the product of gene *A*. Intermediate *Y*, however, may be rapidly converted to product *Z* by enzyme *B*, the product of gene *B*. If so, intermediate *Y* may be present in very low quantities and be very difficult to identify biochemically. However, in a mutant organism that has a mutation in gene *B*, resulting in the synthesis of either an inactive form of enzyme *B* or no enzyme *B*, intermediate *Y* will often accumulate to much higher concentrations, facilitating its isolation and identification. Similarly, a mutation in gene *A* may aid in the identification of precursor *X*. Furthermore, the functions and mechanisms of action of individual gene-products can often be deduced by comparative biochemical and biophysical analyses of mutant and wild-type organisms.

The resolving power of mutational dissection in biological processes has been emphatically demonstrated by the elucidation of the pathway of morphogenesis in bacteriophage T4. This complex process involves the products of about 50 genes. Each gene encodes a structural protein of the virus or an enzyme that catalyzes one or more steps in the morphogenetic pathway. By (1) isolating mutant strains of phage T4 with temperature-sensitive and suppressor-sensitive conditional lethal mutations (see p. 292) in each of the approximately 50 genes, and (2) analyzing the structures that accumulate when these mutant strains are grown under the restrictive conditions by electron microscopy and biochemical techniques, R. S. Edgar, W. B. Wood, J. King, and colleagues have been able to work out almost the entire pathway of phage T4 morphogenesis (Figs. 11.32 and 11.33).

Other biological processes that have been successfully dissected by mutational analysis include the photosynthetic electron transport chains in *Chlamydomonas reinhardi* and maize and nitrogen fixation in bacteria. This approach is currently being used to dissect differentiation and development in higher plants and animals. S. Benzer and colleagues are using mutations to dissect behavior and learning in *Drosophila*. In theory, mutational dissection should be applicable to any process that is under genetic control. The existence of a gene that does not mutate to a nonfunctional state seems extremely unlikely. Thus, mutational dissection of biological processes is limited only by the ingenuity of scientists in identifying mutations of the desired types.

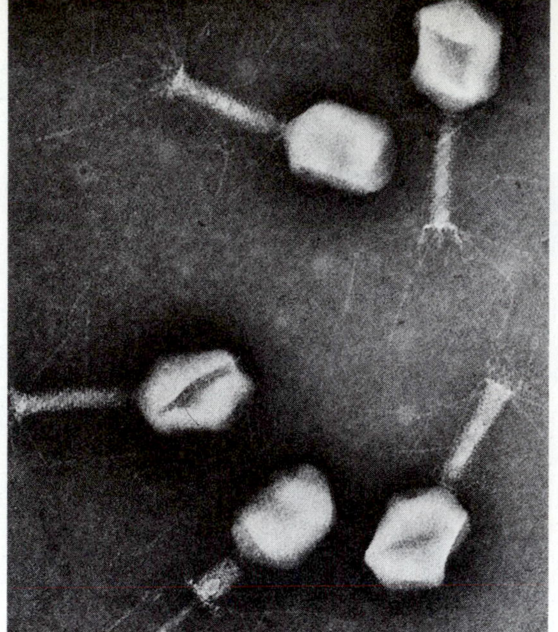

Figure 11.32 Electron micrograph of negatively stained T4 bacteriophages, showing their intricate morphology. The DNA-filled heads, the tails, and the six-pointed baseplates with their attached tail fibers are easily recognized. On some of the phage (particularly the one at the lower right), the collar between the head and the tail is visible. The tail sheath of the phage at the lower left is contracted, revealing the distal portion of the inner core or tube of the tail through which the DNA is ejected during infection. The tail sheath functions like a little muscle. After the tail fibers and spikes on the baseplate have become firmly attached to receptor sites on the bacterial cell, the tail sheath contracts, pushing the tail core through the cell envelope, much like an injection needle. Magnification is ×37,000. (Courtesy E. Boy de la Tour and E. Kellenberger.)

MUTATIONS AND HUMANS

Purposeful artificial selection is not practiced in humans (except possibly in the broadest sense, involving selective mating and personal decisions to limit reproduction) and, therefore, the possible advantages cited for domestic animals and plants do not apply to humankind. Variations do exist in populations, however, and presumably they originated through past mutations. Since mutations, on the average, are detrimental, it would seem advantageous, from the standpoint of short-term effects, for humans to avoid excessive exposure to mutagenic agents.

In the case of acute irradiation, two types of danger should be considered: (1) the immediate damage to the exposed person and (2) the more insidious damage to the DNA in his or her reproductive cells, which would affect future generations. The immediate damage is indicated by burns and other direct or secondary effects on body tissues. When doses are on the order of 50 mr (milliroentgens) or lower, no immediate damage can be detected, although some unseen harmful effects such as induction of leukemia and general shortening of the life span may occur. Doses that exceed safety standards may be prescribed by physicians as therapeutic measures, such as for cancer treatment. In these cases, possible benefits must be carefully weighed against possible damage.

Effects of the second type of damage will be observed only in future generations. There is reason to

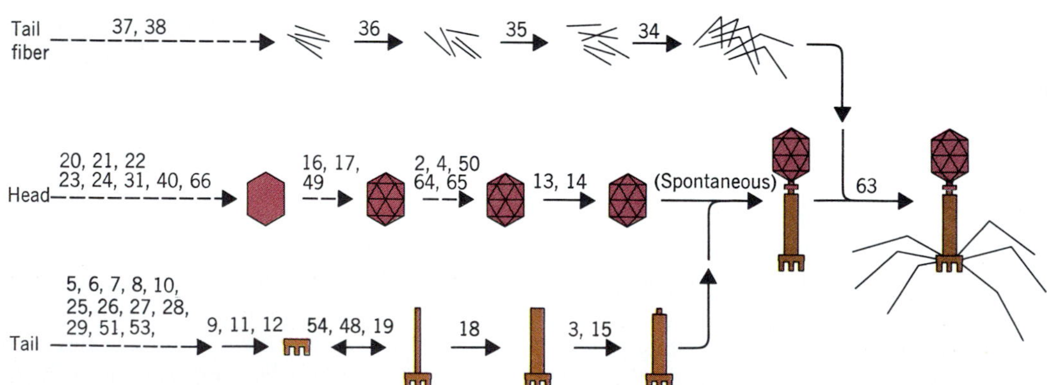

Figure 11.33 Summary of the pathway of morphogenesis in bacteriophage T4, as revealed by mutational dissection and electron microscopic and biochemical analyses. The head, the tail, and the tail fibers are synthesized by means of separate branches of the pathway, which are then joined in terminal steps of the pathway. The numbers beside the arrows indicate the T4 genes whose products are required at that particular step in the pathway. The pathway is obligatorily sequential in all but a few steps, that is, the morphogenesis must proceed in the sequence indicated. The tail fibers will not attach to headless tails, for example, even though the structure of the baseplate appears complete prior to the joining of the tail to the head. (After W. B. Wood, R. S. Edgar, J. King, I. Lielausis, and M. Henninger, *Fed. Proc.* 27: 1160–1166, 1968.)

believe, however, that exposure to high-energy irradiations of any kind, at any dosage level, is potentially harmful. In the few clear experimental results to date, mutations have generally been proportional to the dosage and the effects have been cumulative.

The relation between dosage and effects cannot be accurately measured in humans at present because of the complexity of the subject and the special difficulties of dealing with the genetics of humans. Problems have been recognized in investigations concerning the survivors of the Hiroshima and Nagasaki bombings. The Atomic Bomb Casualty Commission is investigating the bomb's effect on the people exposed to atomic bomb irradiation and their descendants. Preliminary reports, including data on children born to parents who survived the bombing, have revealed a significant increase in the incidence of leukemia. The normal sex ratio has been altered, and the changes have been interpreted as resulting from induced sex-linked lethals. Most of the data available on mutation rates and the nature of mutations have come from other organisms and only by inference are they applied to humans. However, the genetic material in humans is DNA, just as it is in most other organisms. It should not be surprising, therefore, that the effects of irradiation seem fairly comparable for most organisms. In the absence of specific data, the facts learned from other organisms should be considered relevant to humans. All such facts indicate that increased exposure to irradiation will be detrimental to future generations.

SUMMARY

Sudden, heritable changes in the genetic material are called ***mutations.*** Mutation also refers to the process by which such changes are produced. Mutations may occur "spontaneously" (for unknown reasons) or may be induced by agents that interact with DNA and RNA. Various kinds of irradiation and many chemicals that react with DNA and RNA are very potent ***mutagenic agents.***

New mutations provide the genetic variability that fuels evolution. Some level of mutation is thus required to provide the raw material for evolution. However, most mutations are detrimental. High frequencies of mutation would thus be disadvantageous to a species, except possibly in a rapidly changing environment.

The potential benefits of the use of irradiation (solar irradiation, X rays, nuclear reactors) must be carefully weighed against the known and estimated potential risks. Similar precautions must be taken to prevent the continued pollution of our environment with mutagenic (and/or carcinogenic) chemicals. These risk estimates and precautions must take into account the potential harm to future generations of living organisms, including humans, keeping in mind the increased frequencies of deleterious recessive mutations that may result.

REFERENCES

Ames, B.N., J. McCann, and E. Yamasaki. 1975. "Methods for detecting carcinogens and mutagens with the *Salmonella*/mammalian-microsome mutagenicity test." *Mutation Res.* 31: 347–364.

Auerbach, C. 1962. *Mutation, an introduction to research on mutagenesis.* Oliver & Boyd, Edinburgh, Great Britain.

Burnet, F.M. 1974. *Intrinsic mutagenesis: A genetic approach to aging.* Wiley, New York.

Cleaver, J.E. 1967. "Defective repair replication of DNA in xeroderma pigmentosum." *Nature* 218: 652–656.

Cox, E.C. 1976. "Bacterial mutator genes and the control of spontaneous mutation." *Annu. Rev. Genet.* 10: 135–156.

Denniston, C. 1982. "Low level radiation and genetic risk estimation in man." *Annu. Rev. Genet.* 16: 329–355.

Drake, J.W. 1970. *The molecular basis of mutation.* Holden–Day, San Francisco.

______, editor. 1973. "The genetic control of mutation." *Genetics* 73, supplement: 1–205.

Drake, J.W., and R.H. Baltz. 1976. "The biochemistry of mutagenesis." *Annu. Rev. Biochem.* 45: 11–37.

Howard-Flanders, P. 1981. "Inducible repair of DNA." *Sci. Amer.* 245(5): 72–80.

Kikuchi, Y., and J. King. 1975. "Genetic control of bacteriophage T4 baseplate morphogenesis, I, II, and III." *J. Molec. Biol.* 99: 645–672, 673–694, 695–716.

King, J.L. 1976. "Progress in the neutral mutation–random drift controversy." *Fed. Proc.* 35: 2087–2091.

Krieg, D.R. 1963. "Specifity of chemical mutagenesis." *Prog. Nucleic Acid Res.* 3: 125–168.

Lederberg, J., and E.M. Lederberg. 1952. "Replica plating and indirect selection of bacterial mutants." *J. Bacteriol.* 63: 399–406. (Reprinted in E. A. Adelberg, editor, 1966. *Papers on bacterial genetics,* 2nd ed. Little, Brown, Boston.)

Lindahl, T. 1982. "DNA repair enzymes." *Annu. Rev. Biochem.* 51: 61–87.

Little, J.W., and D.W. Mount. 1982. "The SOS-regulatory system of *Escherichia coli.*" *Cell* 29: 11–22.

Muller, H.J. 1927. "Artificial transmutation of the gene." *Science* 66: 84–87. (Reprinted in J. A. Peters, editor. 1959.

Classical papers in genetics. Prentice–Hall, Englewood Cliffs, NJ.)

Schull, W.J., M. Otake, and J.V. Neel. 1981. "Genetic effects of the atomic bombs: A reappraisal." *Science* 213: 1220–1227.

Shortle, D., D. DiMaio, and D. Nathans. 1981. "Directed mutagenesis." *Annu. Rev. Genet.* 15: 265–294.

Singer, B., and J.T. Kusmierek. 1982. "Chemical mutagenesis." *Annu. Rev. Biochem.* 51: 655–693.

Vogel, F. 1970. *Chemical mutagenesis in mammals and man.* Springer-Verlag, New York.

Witkin, E.M. 1969. "Ultraviolet-induced mutation and DNA repair." *Annu. Rev. Genet.* 3: 525–552.

Wood, W.B., R.S. Edgar, J. King, I. Lielausis, and M. Henninger. 1968. "Bacteriophage assembly." *Fed. Proc.* 27: 1160–1166.

PROBLEMS AND QUESTIONS

11.1 Identify the following point mutations represented in DNA and in RNA as (1) transitions, (2) transversions, or (3) reading frameshifts. (a) A to G; (b) C to T; (c) C to G; (d) T to A; (e) UAU ACC UAU to UAU AAC CUA; (f) UUG CUA AUA to UUG CUG AUA.

11.2 Both lethal and visible mutations are expected to occur in fruit flies that are subjected to irradiation. Outline a method for detecting (a) sex-linked lethals and (b) visible mutations in irradiated *Drosophila.*

11.3 How can mutations in bacteria causing resistance to a particular drug be detected? How can it be determined whether a particular drug causes mutations or merely identifies mutations already present in the organisms under investigation?

11.4 Published spontaneous mutation rates for humans are generally higher than those for bacteria. Does this indicate that individual genes of humans mutate more frequently than those of bacteria? Explain.

11.5 A precancerous condition (intestinal polyposis) in a particular human family group is determined by a single dominant gene. Among the descendants of one woman who died with cancer of the colon, 10 people have died with the same type of cancer and 6 now have intestinal polyposis. All other branches of the large kindred have been carefully examined and no cases have been found. Suggest an explanation for the origin of the defective gene.

11.6 Juvenile muscular dystrophy in humans is dependent on a sex-linked recessive gene. In an intensive study, 33 cases were found in a population of some 800,000 people. The investigators were confident that they had found all cases that were well enough advanced to be detected at the time the study was made. The symptoms of the disease were expressed only in males. Most of those who had it died at an early age and none lived beyond 21 years of age. Usually only one case was detected in a family, but sometimes two or three cases occurred in the same family. Suggest an explanation for the sporadic occurrence of the disease and the tendency for the gene to persist in the population.

11.7 Products of somatic mutation, such as the navel orange and the Delicious apple, have become widespread in citrus groves and apple orchards, but they are uncommon in animals. Why?

11.8 If a single short-legged sheep should occur in a flock, suggest experiments to determine whether it is the result of mutation or an environmental modification, and if it is a mutation, is it dominant or recessive?

11.9 How might enzymes such as DNA polymerase be involved in the rate of mutation?

11.10 How could spontaneous mutation rates be optimized by natural selection?

11.11 A mutator gene *Dt* in maize increases the rate at which the gene for colorless aleurone (*a*) mutates to the dominant allele (*A*) for colored aleurone. When reciprocal crosses were made (i.e., seed parent *dt*/*dt*, *a*/*a* × *Dt*/*Dt*, *a*/*a* and seed parent *Dt*/*Dt*, *a*/*a* × *dt*/*dt*, *a*/*a*), the cross with *Dt*/*Dt* seed parents produced three times as many dots per kernel as the reciprocal cross. Explain these results.

11.12 A single gene change blocks the normal conversion of phenylalanine to tyrosine. (a) Is the mutant gene expected to be pleiotropic? (b) Explain.

11.13 How can normal hemoglobin (hemoglobin A) and hemoglobin S be distinguished?

11.14 If CTT is a DNA triplet (transcribed strand of DNA) specifying glutamic acid, what DNA and mRNA base triplet alterations could account for valine and lysine in position 6 of the beta hemoglobin chain?

11.15 Why is sickle-cell anemia called a molecular disease?

11.16 Assuming that the beta hemoglobin chain originated in evolution from the alpha chain, what mechanisms might explain the differences that now exist in these two chains? What changes in DNA and mRNA codons would account for the differences that have resulted in unlike amino acids in corresponding positions?

11.17 In a strain of bacteria, all organisms are usually killed when a given amount of streptomycin is introduced into the medium. Mutations sometimes occur that make the bacteria resistant to streptomycin. Resistant mutants are of two types: some can live with or without streptomycin; others cannot live unless this drug is present in the medium. Given a nonresistant strain, outline an experimental procedure by which resistant strains of the two types might be established.

11.18 One sample of fruit flies was treated with X rays at 1000 roentgens (r). The mutation rate of a particular gene was found to be increased by 2 percent. What percentage increases would be expected at 1500 r, 2000 r, and 3000 r?

11.19 Why does the frequency of chromosome breaks induced by X rays vary with the total dosage and not with the rate at which it is delivered?

11.20 One person was in an accident and received 50 roentgens (r) of X rays at one time. Another person received 5 r in each of 20 treatments. Assuming no intensity effect, what proportionate number of mutations would be expected in each person?

11.21 How does ultraviolet light produce revertible mutations?

11.22 How does nitrous acid induce mutations? What specific end results might be expected on DNA and mRNA from treatment of viruses with nitrous acid?

11.23 Are mutational changes induced by nitrous acid more likely to be transitions or transversions?

11.24 How does the action and mutagenic effect of 5-bromouracil differ from that of nitrous acid?

11.25 How has induced mutation been put to practical use in improving crops of economic value?

11.26 Evaluate the effects, immediate and potential, that might come from intense, mass irradiation of people.

11.27 How do acridine-induced changes in DNA result in inactive proteins?

Use the known codon–amino acid assignments given in Chapter 10 on p. 280 to work the following problems.

11.28 Bacteriophage MS2 carries its genetic information in RNA. Its chromosome is analogous to a polygenic molecule of mRNA in organisms that store their genetic information in DNA. The MS2 "minichromosome" encodes 4 polypeptides (i.e., it has 4 genes). One of these 4 genes codes for its coat protein, a polypeptide 129 amino acids long. The entire nucleotide sequence in the RNA of MS2 is known. Codon 112 of the coat protein gene is CUA, which specifies the amino acid leucine. If you were to treat a replicating population of bacteriophage MS2 with the mutagen 5-bromouracil, what amino acid substitutions would you expect to be induced at position 112 of the MS2 coat protein (i.e., Leu → other amino acid)? (*Note:* Bacteriophage MS2 RNA replicates using a complementary strand of RNA and base-pairing like DNA.)

11.29 Would the different amino acid substitutions induced by 5-bromouracil at position 112 of the coat polypeptide that you indicated in Problem 11.28 be expected to occur with equal frequency? If so, why? If not, why not? Which one(s), if any, would occur more frequently?

11.30 Would such mutations occur if a nonreplicating suspension of MS2 phage was treated with 5-bromouracil?

11.31 Recalling that nitrous acid deaminates adenine, cytosine, and guanine (adenine → hypoxanthine, which base-pairs with adenine; cytosine → uracil, which base-pairs with adenine; and guanine → xanthine, which base-pairs with cytosine), would you expect nitrous acid to induce any mutations that result in the substitution of another amino acid for a glycine residue in a wild-type polypeptide (i.e., glycine → another amino acid) if the mutagenesis was carried out on a suspension of a mature (nonreplicating) T4 bacteriophages? (*Note:* After the mutagenic treatment of the phage suspension, the nitrous acid is removed. The treated phage are then allowed to infect *E. coli* cells to express any induced mutations.) If so, by what mechanism? If not, why not?

11.32 Keeping in mind the known nature of the genetic code, the information given about phage MS2 in Problem 11.28, and the information we have learned about nitrous acid in Problem 11.31, would you expect nitrous acid to induce any mutations that would result in amino acid substitutions of the type ***glycine → another amino acid*** if the mutagenesis were carried out on a suspension of mature (nonreplicating) MS2 bacteriophage? If so, by what mechanism? If not, why not?

11.33 Would you expect nitrous acid to induce a higher frequency of Tyr → Ser or Tyr → Cys substitutions? Why?

11.34 Which of the following amino acid substitutions should you expect to be induced by 5-bromouracil with the highest frequency? (a) Met → Leu; (b) Met → Thr; (c) Lys → Thr; (d) Lys → Gln; (e) Pro → Arg; or (f) Pro → Gln. Why?

11.35 Acridine dyes such as proflavin are known to induce primarily single base-pair additions and deletions. Suppose that the wild-type nucleotide sequence in the mRNA of a gene is

(5′ end) AUGCCCUUUGGGAAAGGGUUUCCCUAA (3′ end)

Also, assume that a mutation is induced within this gene by proflavin and, subsequently, a revertant of this mutation is likewise induced with proflavin and shown to be a second-site suppressor mutation within the same gene. If the amino acid sequence of the polypeptide coded for by this gene in the revertant (double mutant) strain is

(amino end)
Met-Pro-Phe-Gly-Glu-Arg-Phe-Pro
(carboxyl end)

what would be the most likely nucleotide sequence in the ***mRNA*** of this gene in the revertant (double mutant)?

Chapter 12

Genetic Fine Structure: Classical Genetics

PRE-1940 studies of the gene were carried out primarily with higher plants and *Drosophila*. These studies focused on the gene as the unit of inheritance controlling one phenotypic trait. The genes were found to undergo rare changes or mutations to new forms or new alleles. New combinations of the wild-type and mutant forms of genes located at different positions on a chromosome were shown to result from crossing-over or recombination between homologous chromosomes. These studies provided no information about the structure of the gene itself, however.

Prior to 1940, recombination was not believed to occur within genes. The genes in a chromosome were considered analogous to beads on a string. Recombination was believed to occur only between the beads or genes. The gene was not considered subdivisible.

Since recombination frequencies are correlated with the physical distances separating genetic markers on chromosomes, the failure to detect recombination between markers located at different positions in the same gene is not surprising. Such recombinants occur very infrequently. Their detection involves screening thousands to hundreds of thousands of progeny. With higher plants, and even with *Drosophila*, examining enough progeny to detect such rare recombinants is a formidable task.

Subsequent studies with microorganisms using selection techniques that permit millions and even billions of progeny to be screened within one to a few days have provided a detailed picture of the fine structure of the gene. In particular, the use of conditional lethal mutations (see pp. 291–293) has greatly facilitated the detection of rare recombination events occurring within genes, or ***intragenic*** recombination.

CLASSICAL VERSUS MOLECULAR CONCEPT OF THE GENE

According to the pre-1940 "beads-on-a-string" concept, the gene was the basic unit of inheritance defined by three criteria: (1) function, (2) recombination, and (3) mutation. More specifically, the gene was:

1. **The unit of function**: the unit of genetic material that controlled the inheritance of one "unit character" or one attribute of phenotype.
2. **The unit of structure**: this could be operationally defined in two ways:
   a. **By recombination**: as the unit of inheritance not subdivisible by recombination.
   b. **By mutation**: as the smallest unit of genetic material capable of independent mutation.

(Photo Left page): False-color transmission electron micrograph of T4 bacteriophage. (Courtesy of M. Wurtz/Biozentrum, University of Basel/Science Photo Library/Photo Researchers.)

The classical view was that all three criteria defined the same basic unit of inheritance, namely, the gene.

We now know that these criteria define two very different units of inheritance. According to our current molecular concept, the gene is:

The unit of function: the unit of inheritance

1. coding for one polypeptide chain and
2. operationally defined by the *cis–trans* or complementation test (see pp. 322–324).

The other two criteria, recombination and mutation, define the ***unit of structure,*** which is ***equivalent to the single nucleotide-pair*** (see pp. 336–337). Since it clearly does not make sense to call each nucleotide-pair a gene, emphasis has been shifted to the original definition of the gene as the unit of function. Defining the gene as the unit of function, while discarding the recombination and mutation criteria, clearly seems to be the most appropriate solution to the dilemma posed by the demonstration that the original three criteria define two very different units. The emphasis in Mendel's work was on the "anlage" (factor, or gene, as we now call it) controlling one phenotypic characteristic. Thus, Mendel's unit of inheritance correlates more directly with the gene as a unit of function than with the unit of structure or single nucleotide-pair.

EARLY EVIDENCE THAT THE GENE IS SUBDIVISIBLE

The first evidence that the gene was subdivisible by mutation and recombination came from studies of the X-linked lozenge locus of *Drosophila melanogaster* by C. P. Oliver in 1940. The two mutations lz^s ("spectacle" eye) and lz^g ("glassy" eye) were considered to be alleles, that is, different forms of the same gene. The data available prior to 1940 indicated that they mapped at the same position on the X chromosome. They had similar effects on the phenotype of the eye, and, most important, lz^s/lz^g heterozygous females had lozenge rather than wild-type eyes. When lz^s/lz^g females were crossed with either lz^s or lz^g males and large numbers of progeny were examined, however, wild-type progeny occurred with a frequency of about 0.2 percent.

These rare wild-type progeny could be explained by reversion of either the lz^s or the lz^g mutation, but there were two arguments against the reversion hypothesis. (1) The frequency of reversion of lz^s or lz^g to wild type in hemizygous lozenge males was much lower than 0.2 percent. (2) When the lz^s/lz^g heterozygotes carried outside genetic markers bracketing the lozenge locus, the rare progeny with wild-type eyes always carried an X chromosome with lz^+ that was flanked by a ***recombinant arrangement of outside markers.*** Moreover, the same combination of outside markers always occurred, as though the sites of lz^s and lz^g were fixed relative to each other and reciprocal recombination was occurring. Different sets of outside markers were used and yielded the same result. If the lz^s/lz^g heterozygous female carried X chromosomes of the type

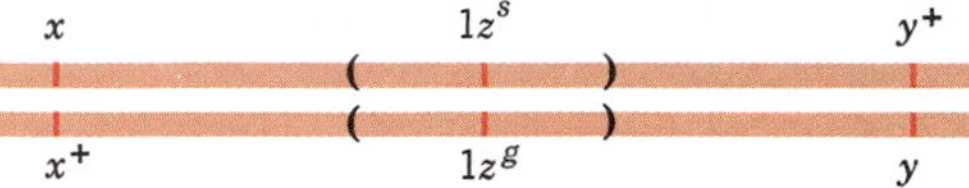

the rare progeny with wild-type eyes all (with one exception) contained an X chromosome with the composition

x 1z$^+$ y

Among progeny of these matings, the reciprocal combination of outside markers (x^+-y^+) was never found together with lz^+. This strongly suggested that recombination was involved. Definitive evidence of the involvement of recombination required the recovery and identification of the $lz^s lz^g$ double mutant with the reciprocal combination of outside markers, that is,

x$^+$ 1z^g 1z^s y$^+$

Oliver was not able to identify this double-mutant, reciprocal recombinant because of the inability to distinguish it from the parental single mutant phenotypes. The identification of both products, the wild type and the double mutant, of recombination between two functionally allelic mutations was first accomplished by two of Oliver's students, E. B. Lewis and M. M. Green, in studies of the *star-asteroid* (small rough eye) locus and the *white* (eye color) locus, respectively, of *D. melanogaster*.

These pioneering studies first indicated that the gene was in fact more complex than a "bead-on-a-string." They were the first steps toward our present concept of the gene as a long sequence of nucleotide pairs that is capable of mutating and recombining at many different sites along its length.

THE *CIS–TRANS* OR COMPLEMENTATION TEST FOR FUNCTIONAL ALLELISM

The functional allelism of any two recessive mutations is determined experimentally by the **cis–trans** or ***complementation*** test. (To be completely accurate, we should state that the complementation test is only part, the *trans*-component, of the *cis–trans* test.) The **cis–trans *test provides an operational definition of the gene as the unit of function—the unit controlling the synthesis of one polypeptide.***

The **cis-*test*** involves putting the two mutations being examined ***in a common protoplasm in the* cis- *or coupling configuration*** (Fig. 12.1*a*). The organism or cell containing the two mutations in the *cis*- or coupling configuration (on the same chromosome) is called a **cis-*heterozygote.*** The *cis*-heterozygote must have the ***wild-type phenotype*** (Fig. 12.2, right) for the results of the *trans*-test with the same pair of mutations to be valid. If the *cis*-heterozygote has the mutant phenotype, the *trans*-test ***cannot*** be used to determine whether the two mutations are in the same gene. Thus the *trans*-test can never be used with dominant mutations. The *cis*-test is an important control, which establishes the validity of the correlated *trans*- or complementation test.

The **trans-*test,*** more commonly called the complementation test, involves putting the ***two mutations*** to be tested ***in a common protoplasm in the* trans- *or repulsion configuration,*** that is, on separate chromosomes (Fig. 12.1*b*). This is accomplished in different ways in different organisms. With diploids like fruit flies and corn plants, it simply involves crossing two homozygotes: one homozygous for one mutation, the other homozygous for the second mutation. In the case of bacteriophage, it involves simultaneously infecting host cells with two different mutants. The organism or cell containing the two mutations in the *trans*- or repulsion configuration (on different chromosomes) is called a **trans-*heterozygote.***

1. ***If the* trans-*heterozygote has the mutant phenotype*** (the phenotype of organisms or cells homozygous for either one of the two mutations), ***then the two muta-***

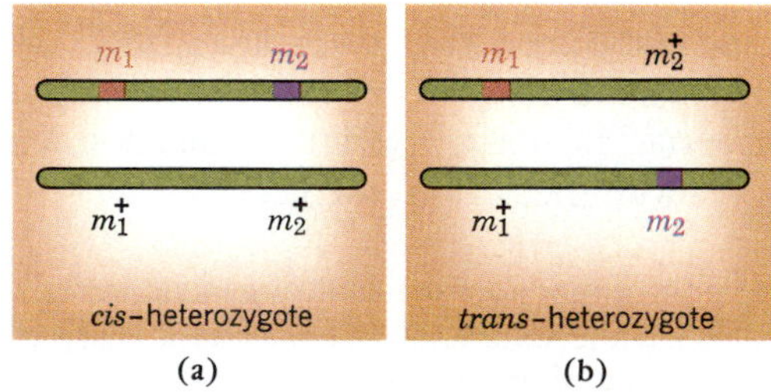

Figure 12.1 Composition of *cis*- and *trans*-heterozygotes for two mutations m_1 and m_2. The corresponding wild-type sequences indicated by m_1^+ and m_2^+ will represent one sequence in (a) and will not exist in (b) if m_1 and m_2 overlap a common site, that is, have mutational defects that overlap by one nucleotide-pair or more.

tions are in the same unit of function, the same gene (Fig. 12.2, top left).

2. ***If the* trans*-heterozygote has the wild-type phenotype, then the two mutations are in two different units of function, two different genes*** (Fig. 12.2, bottom left). In this case the two mutations are said to ***complement*** each other.

Because construction of the *cis*-heterozygotes requires the often difficult identification of chromosomes carrying both mutations (the "double mutants"), the *cis*-controls are often omitted in routine complementation analyses. Omission of the *cis*-tests may occasionally result in erroneous conclusions. In most cases, however, the *trans*-tests can be interpreted correctly without the *cis*-controls. Because of the time and expense that would usually be required to construct all the *cis*-heterozygotes, their omission is usually justified.

The rationale behind the *cis–trans* or complementation test is illustrated in Fig. 12.2. The production of a wild-type phenotype depends on whether functional copies of all the relevant gene-products (usually polypeptides) are produced in the common protoplasm

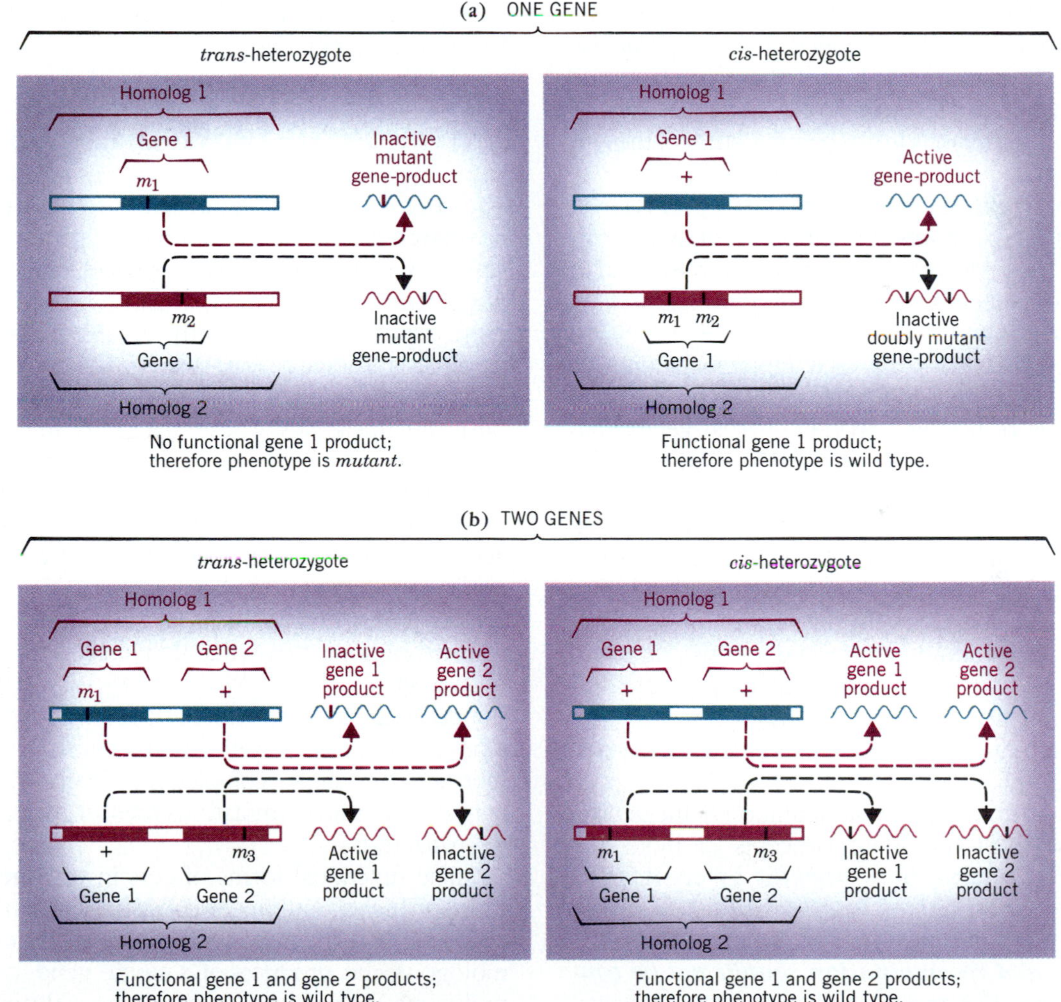

Figure 12.2 Rationale of the *cis–trans* or complementation test for functional allelism. The two mutations m_1 and m_2 are both in the same gene (a); mutation m_3 is in a second gene (b). The phenotype of the *cis*-heterozygote (right) must be wild type or the *trans*-test (left) will not be valid. If the two mutations are located in the same functional unit (the same gene), the *trans*-heterozygote will have the mutant phenotype, because no functional gene-product will be synthesized (a, left). If the two mutations are in two different genes, the phenotype of the *trans*-heterozygote will be wild type, since both gene-products will be synthesized in the common protoplasm (b, left). In the case shown, mutations m_1 and m_2 are both in gene 1; they do not complement each other. Mutations m_1 and m_3, on the other hand, are in two different genes and do complement each other. Homologous chromosome 1 and its gene-products are shown in red; homologous chromosome 2 and its gene-products are shown in aqua.

(cell or organism). If no functional product is produced from one or more of the relevant genes, then the cell or organism will exhibit a mutant phenotype. The *cis*-heterozygotes (Fig. 12.2, right) carry one chromosome with wild-type copies of all the relevant genes. These should produce functional gene-products and thus a wild-type phenotype, unless their activity is somehow interfered with or inhibited by the mutant gene-products produced from the double-mutant chromosome. In the latter case, the result of the correlated *trans*-tests will not be informative. The *trans*-heterozygotes (Fig. 12.2, left), on the other hand, will produce active forms of all the gene-products only when the two mutations are in two different genes (Fig. 12.2, bottom left). If the two mutations are in the same gene, the *trans*-heterozygote will contain only mutant (nonfunctional) products of that particular gene (Fig. 12.2, top left). It will therefore have the mutant phenotype, the same phenotype as cells or organisms homozygous for one or the other of the two mutations.

S. Benzer introduced the term ***cistron*** to refer to the unit of function operationally defined by the *cis–trans* test. The concept of the gene has now evolved to where most geneticists use the terms ***gene*** and ***cistron as exact synonyms.***

The results of *cis–trans* or complementation tests are usually unambiguous when mutations that result in the synthesis of no gene-product, partial gene-products, or totally defective gene-products are used, for example, deletions of segments of genes, frameshift mutations, or polypeptide chain-terminating mutations. The mutations must, of course, be recessive. When mutations causing single-amino acid substitutions are used, and when the active form of the gene-product is a multimer composed of two or more polypeptides, the results of *cis–trans* tests are frequently ambiguous because of the occurrence of a phenomenon called ***intragenic*** complementation (see pp. 324–327).

The *cis–trans* operational definition of the gene is one of the most important concepts of molecular genetics. ***The information provided by* cis–trans *or complementation tests is totally distinct from that obtained from recombination analyses. Complementation and recombination should not be confused.*** Figures 12.3 and 12.4 illustrate complementation and recombination tests, respectively, with the same mutant strains of phage T4.

Complementation is the result of the interaction of the ***gene-products*** produced by chromosomes carrying two different mutations when they are present in a ***common protoplasm.*** It does not depend on recombination of the two chromosomes or involve any direct interaction between the chromosomes. ***Complementation, or the lack of it, is determined by the phenotype (wild-type or mutant) of each* trans*-heterozygote.***

Recombination, on the other hand, involves ***direct interaction between the chromosomes*** carrying the mutations, the actual ***breakage of chromosomes and reunion of parts*** in such a way that wild-type and double-mutant chromosomes are produced (see Chapter 7). ***Complementation should occur in every* trans*-heterozygote*** containing mutations in two different genes. ***Recombination is normally detected by examining the progeny of heterozygotes,*** and a *trans*-heterozygote will never produce more than 25 percent gametes (or progeny for haploids) with wild-type chromosomes. If the mutations are closely linked, for example, as in Fig. 12.4*b*, this frequency will be much lower than 25 percent.

LIMITATIONS OF THE *CIS–TRANS* TEST

The *cis–trans* or complementation test has proven to be extremely useful in delimiting genes. In many cases, two or more mutations producing the same phenotype can be unambiguously assigned to one or more genes on the basis of the results of complementation tests. There are four cases in which the results of complementation tests ***cannot*** be used to unambiguously delimit genes. These are (1) dominant or codominant mutations, (2) genes in which mutations occur that exhibit ***intra***genic complementation (see the following), (3) "polar mutations" (mutations that affect the expression of adjacent genes) within "operons" (coordinately regulated polygenic units of expression; see Chapter 14), and (4) *cis*-acting genes (most commonly genes that do not code for diffusible products).

cis-Heterozygotes will ***not*** have wild-type phenotypes (as required for valid *cis–trans* tests) unless both mutations present are completely recessive. Thus, dominant or codominant mutations ***cannot*** be analyzed by *cis–trans* tests.

The functional forms of certain enzymes are dimers or higher multimers consisting of two or more polypeptides. These polypeptides may be either homologous, the products of a single gene, or nonhomologous, the products of two or more distinct genes. When the active form of the enzyme contains two or more homologous polypeptides (it may or may not also contain nonhomologous polypeptides), ***intragenic complementation*** may occur. ***Intra***genic complementation is a phenomenon totally distinct from ***inter***genic complementation, on which the *cis–trans* test is based. In organisms that are homozygous for the wild-type allele of a given gene, all the protein dimers

Complementation between mutations *am*B17 and *am*E18.

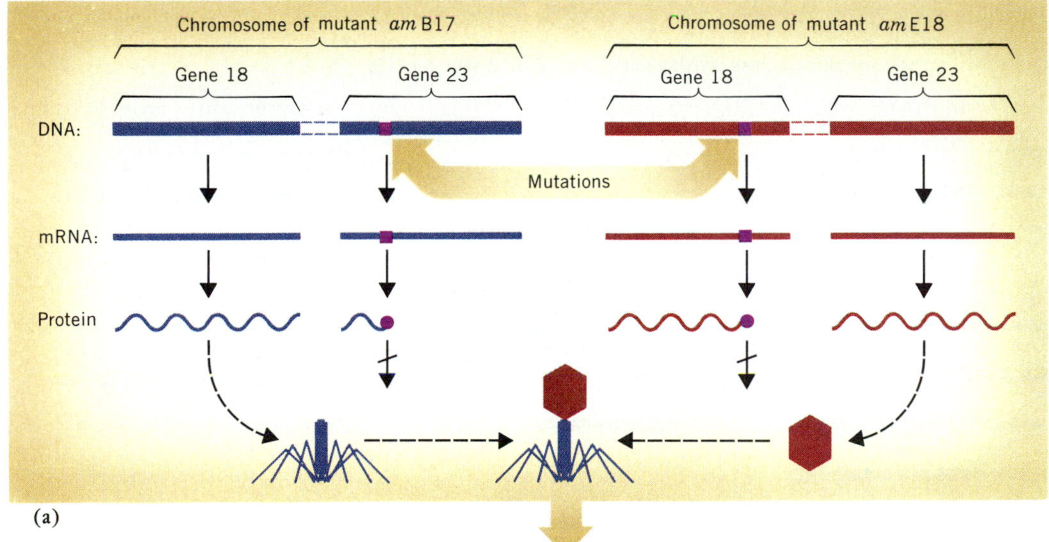

Yields infective progeny phage upon lysis.

Lack of complementation between mutations *am*B17 and *am*H32.

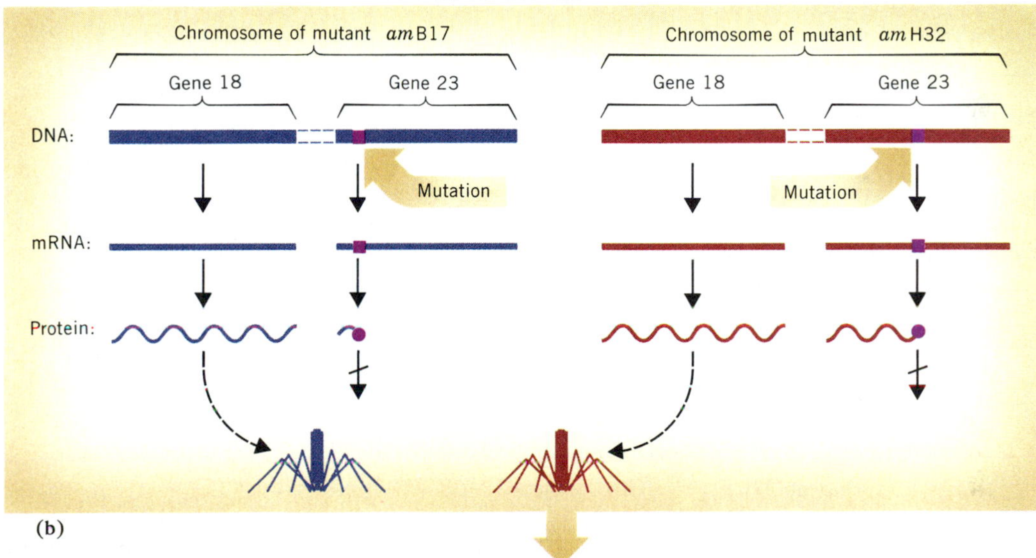

No functional heads; therefore, no infective progeny phage.

Figure 12.3 Diagram illustrating (a) complementation or (b) lack of complementation in *trans*-heterozygotes. (a) Complementation between a mutation (*am* B17) in the geme (gene 23) coding for the major structural protein of the phage T4 head and a mutation (*am* E18) in the gene (gene 18) coding for a major structural protein of the phage tail. Both heads and tails are synthesized in the common protoplasm with the result that infective progeny phage are produced. (b) When both mutations (*am* B17 and *am* H32) are in the gene (gene 23) coding for the major head protein, no heads are synthesized. Thus, no infective progeny phage can be assembled. In both cases, the *trans*-heterozygotes are produced by simultaneously infecting *E. coli* strain B with T4 phage strains that carry the mutations being analyzed. In this host, *amber* mutations cause only a fragment of the polypeptide gene–product to be synthesized, since these mutations yield mRNA codons that are chain-terminating codons. (After D. P. Snustad and D. S. Dean, *Genetics Experiments with Bacterial Viruses,* Freeman, San Francisco, 1971.)

or higher multimers will contain identical wild-type polypeptides. Similarly, organisms that are homozygous for any mutation in that gene will contain dimers or higher multimers, all of which contain identical copies of the mutant polypeptide (Fig. 12.5). An organism that is ***heterozygous for two different mutations in the gene*** will usually produce ***some dimers or higher multimers that contain one or more of each of the two different mutant polypeptides*** (Fig. 12.5). These "heteromultimers" (protein multimers composed of the polypeptide products of two different alleles of a gene) ***may*** have partial or complete (wild-

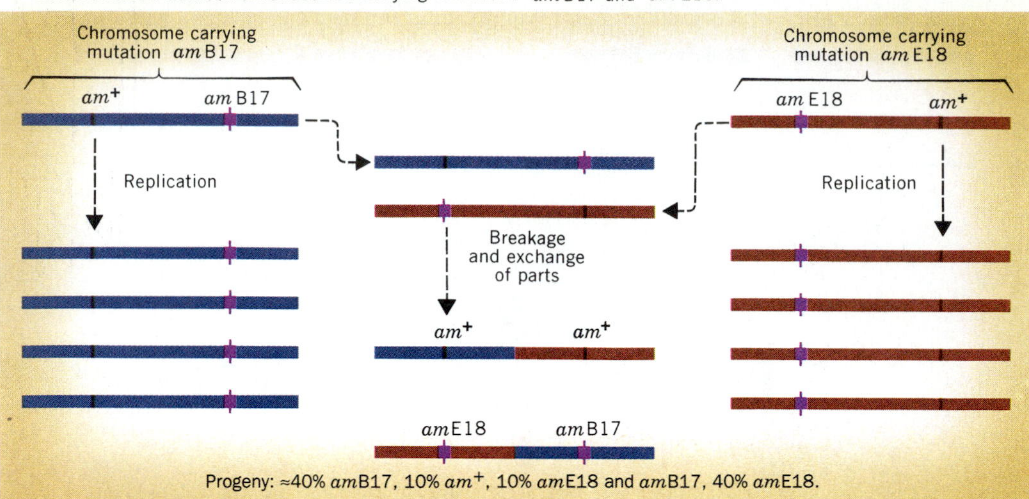

(a)

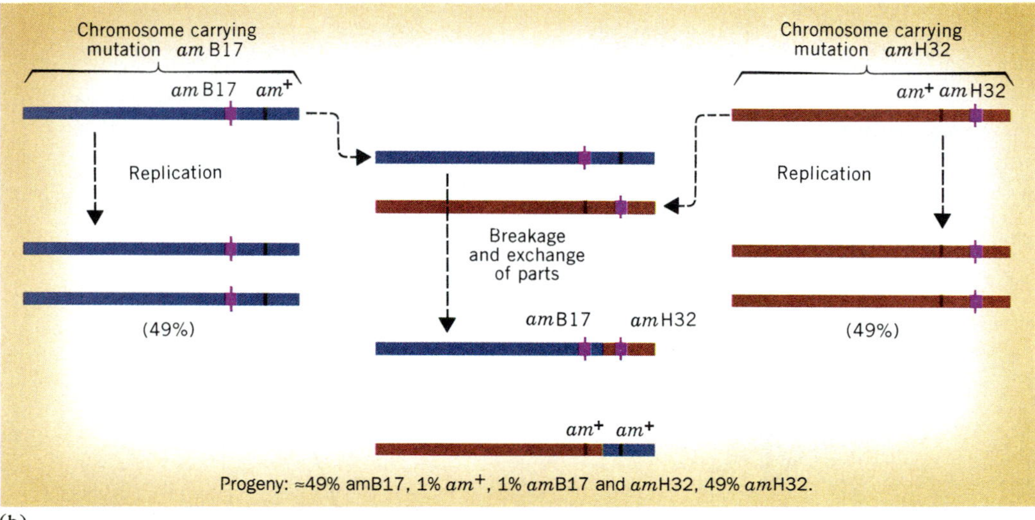

(b)

Figure 12.4 Diagram illustrating ***recombination*** between (a) the complementing mutations *am* B17 (gene 23) and *am* E18 (gene 18) and (b) the noncomplementing mutations *am* B17 and *am* H32 (both in gene 23). In both cases, recombination is analyzed by simultaneously infecting *E. coli* strain CR63 (*amber* permissive) with the two mutant T4 phages. In this host, the *amber* mutations are suppressed, allowing the mutants to grow. The genotypes of the progeny are determined by their ability or inability to grow on restrictive (suppressor negative) hosts, such as *E. coli* strain B. (After D. P. Snustad and D. S. Dean, *Genetic Experiments with Bacterial Viruses,* Freeman, San Francisco, 1971.)

type) activity. In the case of ***noncomplementing*** mutations in a gene coding for a multimeric enzyme, the "heteromultimers" are ***inactive*** just like the mutant "homomultimers" (protein multimers composed of two or more identical mutant polypeptides).

In several known cases of ***intra***genic complementation, the active form of the enzyme in the heterozygote has been purified and shown to be a heterodimer or heterotetramer containing the two distinct mutant polypeptides. Why such heteromultimers should be active whereas the two corresponding homomultimers are inactive is not clear. Apparently, the sequence of amino acids in the "nonmutant" segment of one mutant polypeptide stabilizes or somehow compensates for the mutant segment of the polypeptide encoded by the second mutant allele, and vice versa (Fig. 12.5). Most proteins, however, have very complex three-dimensional structures (see Chapter 10, pp. 253–255), and until the exact structures of a wild-type homomultimer, two mutant homomultimers, and an active heteromultimer composed of the two mutant polypeptides have been determined, the molecular

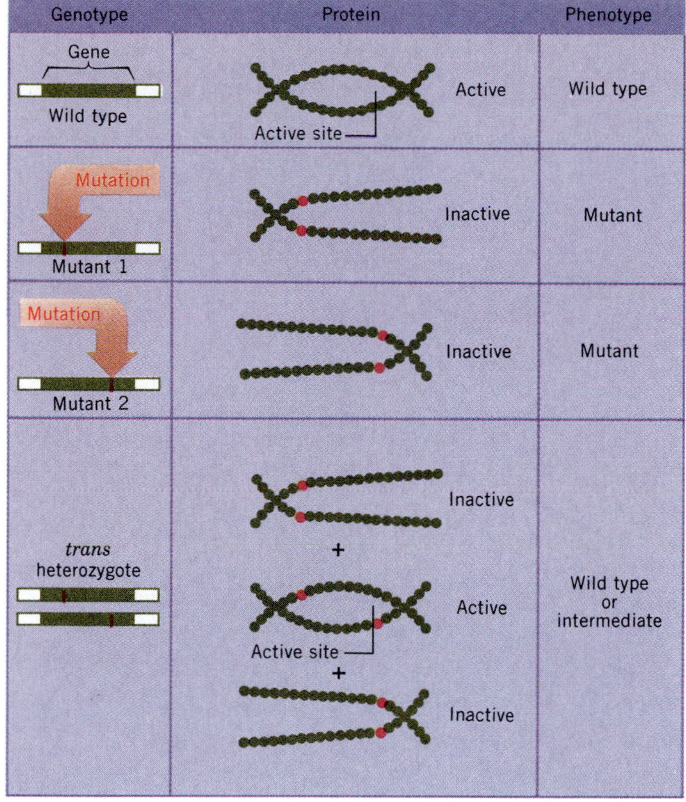

Figure 12.5 Schematic illustration of the mechanism of intragenic complementation. Intragenic complementation may occur whenever the active form of the enzyme (or protein) is a multimer that ***contains at least two copies of any one gene-product (polypeptide).*** In the case illustrated, the functional form of the enzyme is a dimer composed of two polypeptide products of a single gene. In mutant organisms carrying either mutation 1 or mutation 2 (either homozygous diploids or haploids), only nonfunctional dimers are produced (thus the mutant phenotype). In a *trans*-heterozygote carrying both mutant alleles (1 and 2), part of the dimers may contain one copy of each of the mutant polypeptides. Such "heterodimers" (one mutant 1 polypeptide plus one mutant 2 polypeptide) (bottom, center) are sometimes functional. These heterodimers may be totally active (wild-type level of activity), partially active (some activity, but less than the wild-type level), or inactive. As a result, such *trans*-heterozygotes may have a wild-type phenotype or a phenotype intermediate between the mutant and wild type. The positions of the two mutations within the gene are indicated in red. The amino acids altered by the mutations are shown as red circles in the polypeptide chains.

basis of ***intra***genic complementation will continue to be a subject of speculation.

A ***"polar mutation"*** is a mutation that not only results in a defective product of the gene in which it is located, but also interferes with the expression of one or more adjacent genes. The latter are always located on one side of the gene carrying the mutation (thus the term "polar mutation"). Such polar mutations are frequently observed in coordinately regulated sets of genes called ***operons*** in prokaryotes (see Chapter 14). Most commonly, they are mutations resulting in polypeptide chain-termination signals (nucleotide-pair triplets yielding UAA, UAG, and UGA codons in mRNA; see Table 10.1) within structural genes (genes coding for polypeptide products). One would not expect a polar mutation to complement a mutation in another gene if the latter cannot be expressed because of the epistatic effect of the polar mutation. *Cis–trans* analyses of polar mutations thus frequently yield ambiguous results.

In addition, **"cis-*acting" genes*** or segments of genetic material such as "operator regions" (binding sites for regulator proteins) or "promoter regions" (binding sites for RNA polymerase) cannot be analyzed by the *cis–trans* test. Clearly, the *cis–trans* test depends on the gene-products being diffusible. Regulatory binding sites on chromosomes like operators and promoters do not code for diffusible products. They only affect the expression of genes located *cis* to them, that is, on the same chromosome (see Chapter 14). Such "*cis*-acting" genes or sequences are not tractable by *cis–trans* analysis.

FINE STRUCTURE OF THE PHAGE T4 *r*II LOCUS

The most extensive fine structure map constructed to date is that of the phage T4 *r*II locus. S. Benzer isolated over 3000 independent mutant strains carrying mutations that mapped within the *r*II locus. His detailed fine structure analysis of these *r*II mutants showed that the mutations mapped at over 300 distinct sites, all located with the two genes that make up the *r*II locus.

Phage T4, like phage T2 (see Chapter 5, pp. 94–96), is an obligate parasite that grows in the common colon bacillus *Escherichia coli*. It consists of a single linear molecule of DNA about 200,000 nucleotide-pairs long, which is packed within the head of the protein phage coat (Figs. 12.6 and 12.7; also see Chapter 11, pp. 315–316). When a wild-type T4 phage particle infects an *E. coli* (Fig. 12.8) under optimal conditions, it will produce 200–400 progeny viruses

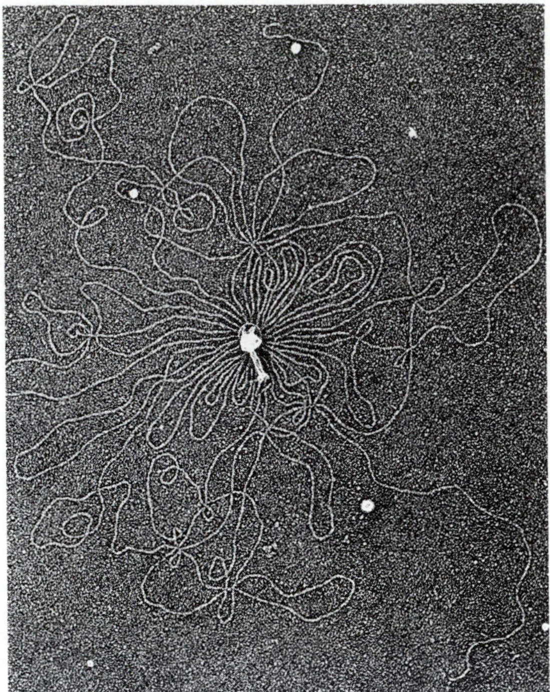

Figure 12.6 Electron micrograph of a bacteriophage T2 particle (center) from which the DNA (a single linear molecule 20 Å in diameter and about 500,000 Å long) has been released by osmotic shock. The DNA molecule has been shadowed with platinum to make it visible. The diameter of the visible DNA–platinum thread is thus much greater than 20 Å, making it appear impossible to pack all of the DNA inside the phage head. Note the two ends of the linear molecule—one at the top center and the other at the bottom right. (Courtesy of Dr. A. K. Kleinschmidt. Reproduced with permission from *Biochem. Biophys. Acta* 61, p. 861 (Fig. 1), 1962.)

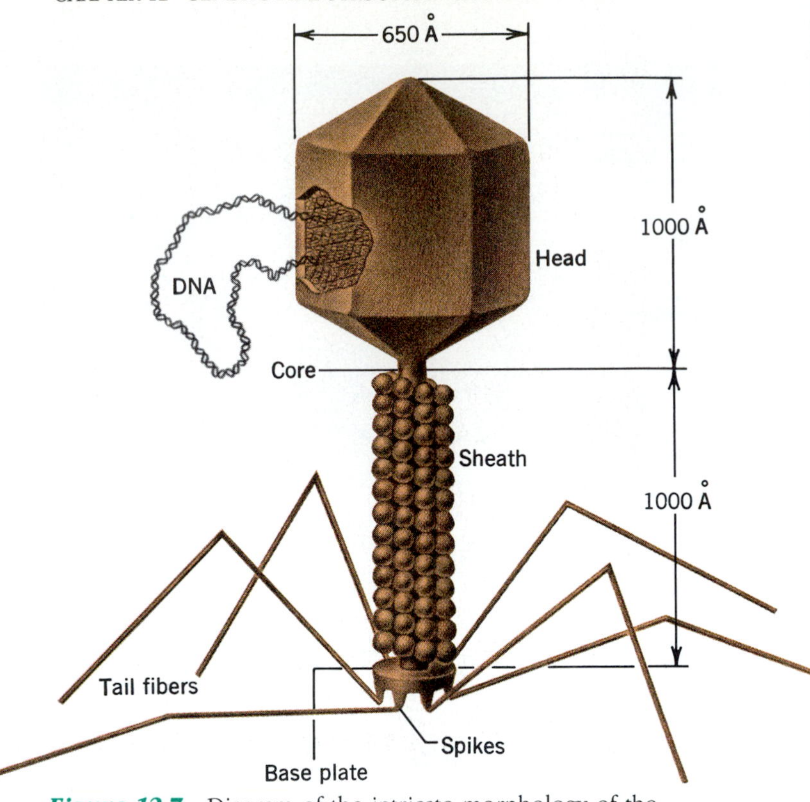

Figure 12.7 Diagram of the intricate morphology of the T-even (T2, T4, and T6) bacteriophages. (After Nason and DeHaaun, *The Biological World,* 1973.)

and lyse the host cell in about 20–25 minutes (Fig. 12.9). If a T4 phage particle is placed on the surface of semisolid agar nutrient medium seeded with a confluent growth ("lawn") of *E. coli* cells in a petri dish, the phage will infect one cell, lyse it, and release about 300 progeny phage particles. Each of these progeny phage will, in turn, infect another *E. coli* cell, lyse it, and release about 300 progeny viruses. After a few hours and several cycles of infection, all the cells in the immediate vicinity of the original phage particle will have been lysed, leaving a clear zone or ***plaque*** in the lawn of bacteria (Fig. 12.10). (Each plaque will ultimately contain up to 5×10^8 phage particles.)

Plaques produced by wild-type T4 phage have fuzzy or turbid margins or halos (Fig. 12.10). When a wild-type T4-infected *E. coli* is superinfected with a second T4 phage prior to lysis, synthesis of new cell wall material is triggered, delaying lysis for up to 2 hours. This phenomenon, called "lysis inhibition," is responsible for the turbid halos or margins around plaques produced by wild-type T4 phage. Progeny phage, released by cells lysing early, superinfect surrounding cells and delay their lysis. The margins of wild-type T4 plaques thus contain a mixture of lysed and lysis-inhibited cells, producing a partially clear zone (Fig. 12.10).

Certain mutants of phage T4, designated *r* mutants for ***rapid lysis,*** do not exhibit lysis inhibition. As a result, *r* mutants produce easily recognized plaques with clear margins (Fig. 12.10). The *r* mutant plaques are also larger under certain conditions. The mutations in these *r* mutants map primarily at three different locations on the T4 chromosome (Fig. 12.11).

Benzer discovered that mutants carrying mutations located at one of these three loci, the *r*II locus, were unable to grow on strains of *E. coli* K12 carrying an integrated (covalently inserted, see pp. 208–213) chromosome of another bacteriophage, phage lambda

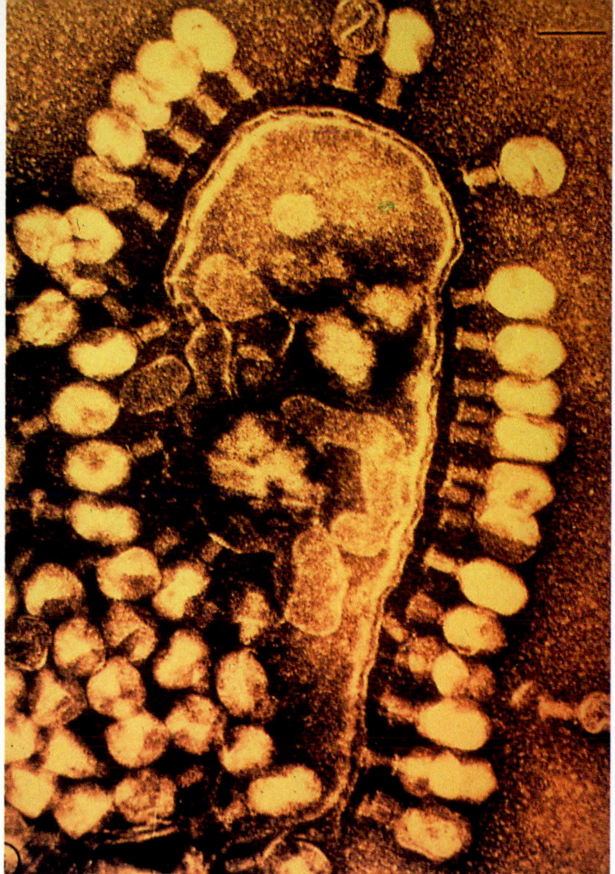

Figure 12.8 False-color transmission electron micrograph showing an *E. coli* cell that has been infected by a large number of T4 bacteriophages. The tail sheaths have condensed like small muscles, and the tail cores protrude into the host cell. The DNA molecules that were packaged inside the phage heads have been injected into the cell through openings in the tail cores. Magnification approximately ×50,000. (Micrograph by Dr. Lee D. Simon/Science Photo Library.)

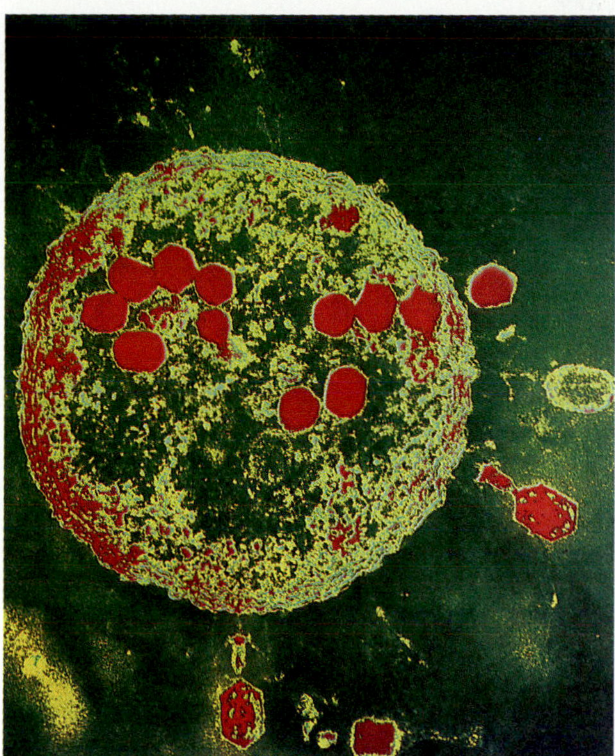

Figure 12.9 False-color transmission electron micrograph of newly assembled T2 bacteriophages, with some phage particles being released from the lysed *E. coli* cell. (Micrograph courtesy of Dr. Lee D. Simon.)

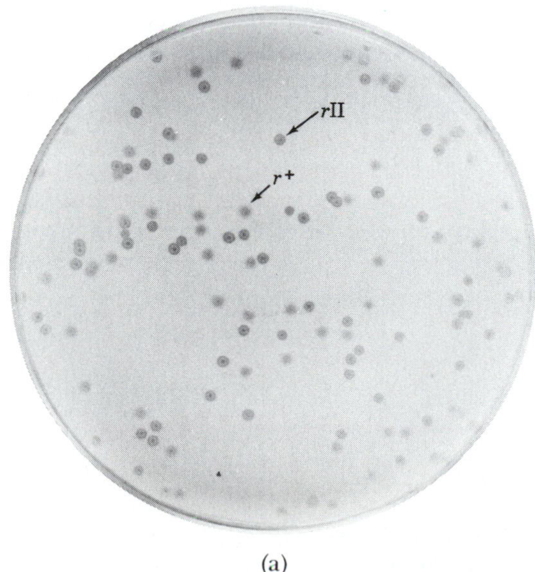

(a)

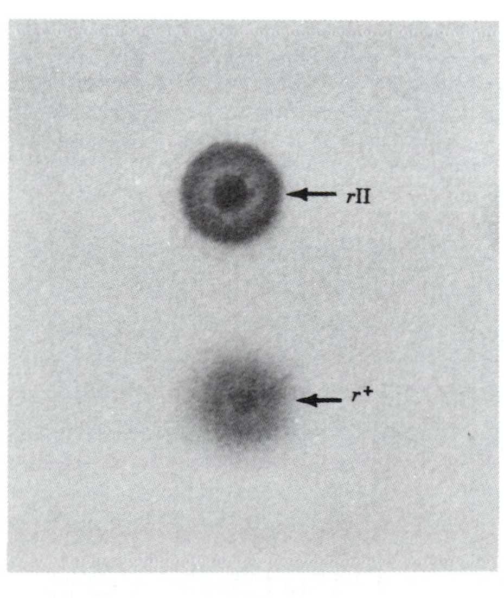

(b)

Figure 12.10 (a) Photograph of a mixture of phage T4 *r*II mutant and r^+ plaques on a confluent lawn of *E. coli* strain B cells. (b) Higher magnification of individual *r*II and r^+ plaques, showing the sharp margin on the *r*II plaque and the fuzzy margin on the r^+ plaque. (Photographs by R. Wagner and D. P. Snustad.)

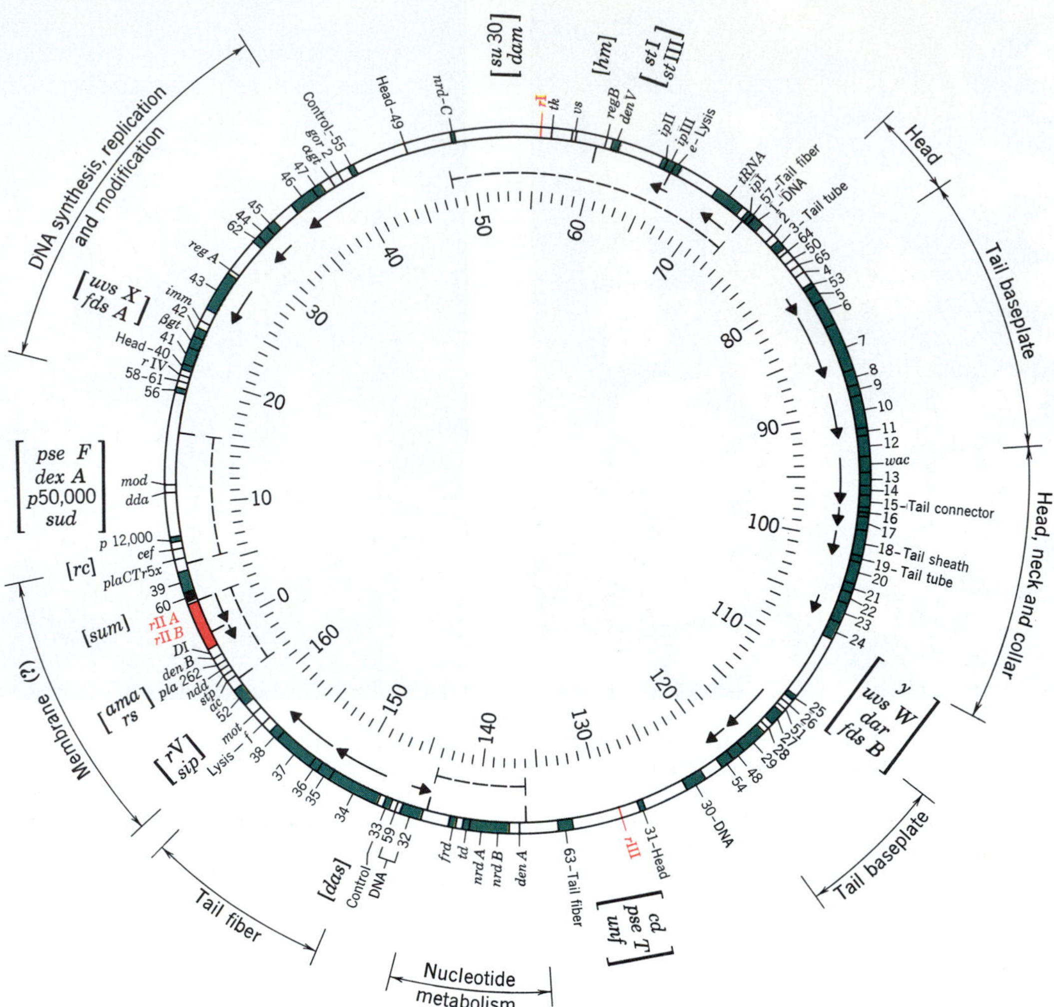

Figure 12.11 The circular linkage map of phage T4, showing the location of the *r*II locus (genes *A* and *B*, orange) studied by Benzer as well as the *r*I and *r*III loci (orange). Most *r* mutations map at these three loci. The interior numerical scale gives physical distances in 1000 base-pair ("kilobase") units, starting arbitrarily at the junction between the *r*II*A* and *r*II*B* genes. The map includes most of the known genes of phage T4. Genes whose exact order is not known are enclosed in parentheses in arbitrary order. The arrows indicate the direction of transcription where known. Functional descriptions of clusters of genes are included around the periphery of the map. (From W. B. Wood and H. R. Revel, *Bacteriol. Rev.* 40: 847, 1976. Copyright © 1976 by the American Society for Microbiology.)

[designated *E. coli* K12(λ)]. They grew rapidly, however, on other strains of *E. coli*, like strain B or K12 strains not carrying the integrated lambda chromosome. Although neither Benzer nor anyone else has been able to explain the basis of the lethality or *r*II mutants in *E. coli* K12(λ), Benzer recognized that the *r*II mutants provided him with a conditional lethal system (see pp. 291–293) that could be exploited to study genetic fine structure.

Phage carrying *r*II mutations can be easily isolated by serially transferring inocula with sterile toothpicks from *r*-type plaques on *E. coli* B to lawns of *E. coli* K12(λ) and *E. coli* B. All *r*II mutants will produce *r*-type plaques on *E. coli* B, but will not grow on *E. coli* K12(λ) (Fig. 12.12). The *r*I and *r*III mutants will grow on both B and K12(λ). As mentioned earlier, Benzer isolated over 3000 independent *r*II mutants. The first question that one might ask, as Benzer did, is how many genes do these mutations identify—one, two, three, . . ., or 3000? When analyzed by *cis–trans* or complementation tests (see Fig. 12.2), all the *r*II mutations were found to be located in one of two genes or cistrons. Benzer designated these two genes *r*II*A* and *r*II*B*. Fortunately, not all pairwise combinations of the over 3000 mutants needed to be tested. Once a mutation in each gene had been identified, it was used as a reference for that gene. Each new mutation that was isolated only had to be tested for complementation with the *A* gene and *B* gene reference mutations. Additional complementation tests were unnecessary since all the *r*II mutations identified failed to complement with one of the two reference mutations. That is, one group of mutations complemented all gene *A* mutations and failed to complement all other gene *B* mutations. The second group of mutations complemented all gene *B* mutations and failed to complement all other gene *A* mutations. (A few mutations, "deletion" mutations, failed to complement all other *r*II mutations; see pp. 332–335.)

The *r*II and r^+ phenotypes are the most distinct in *E. coli* K12(λ), with *r*II mutants being lethal and r^+ phage exhibiting normal growth. Thus, complementation tests between *r*II mutants are done by simultaneous pairwise infections of *E. coli* K12(λ) cells. After allowing a period of time for phage growth and lysis of the host cells (usually 90–120 minutes), the infected K12(λ) cells, or lysates, are spread on lawns of *E. coli* B cells to see if any progeny phage were produced. If complementation has occurred, 200–400 progeny phage will have been produced per infected cell. In the absence of complementation, no progeny phage will have been produced.

Benzer also developed a simplified procedure, the so-called ***complementation spot test,*** for examining complementation qualitatively. This procedure involves placing a small drop of a mixed suspension of

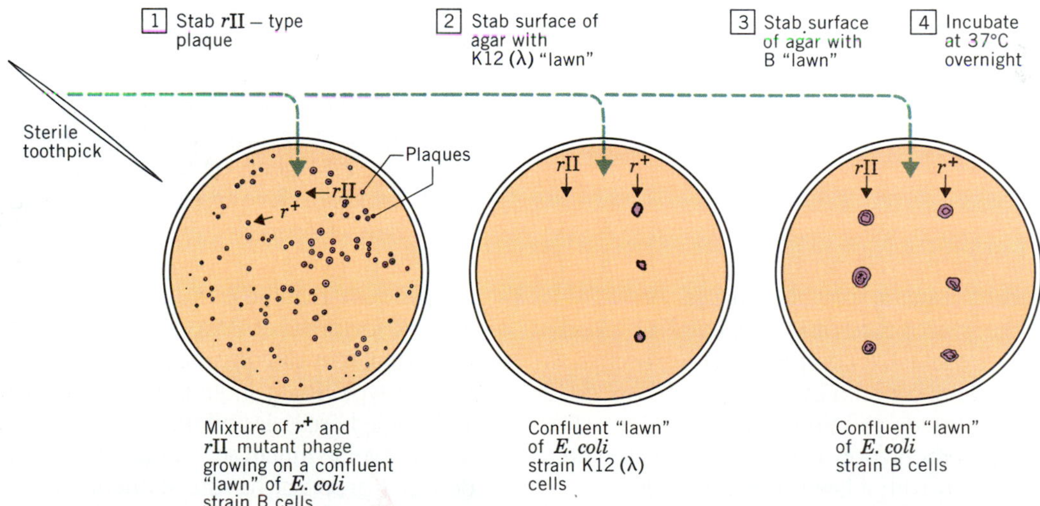

Figure 12.12 Identification of *r*II mutants by sterile toothpick transfers of phage from individual plaques growing on *E. coli* strain B (*r*II-permissive) "lawns" to lawns of *E. coli* strain K12(λ) (*r*II-restrictive) and lawns of *E. coli* strain B. Each plaque to be tested (left) is stabbed with a sterile toothpick, which is subsequently touched to a marked area in a petri dish with a K12(λ) lawn (center) and then to an identically marked area in a dish with an *E. coli* B lawn (right). Mutants that fail to grow (are lethal) on K12(λ) (center plate, left side) can be recovered from the plaques on the *E. coli* B plates (right). Mutants at the *r*I and *r*III loci as well as r^+ phage (center plate, right side) will grow on both K12(λ) and B. (After D. P. Snustad and D. S. Dean, *Genetics Experiments with Bacterial Viruses,* Freeman, San Francisco, 1971.)

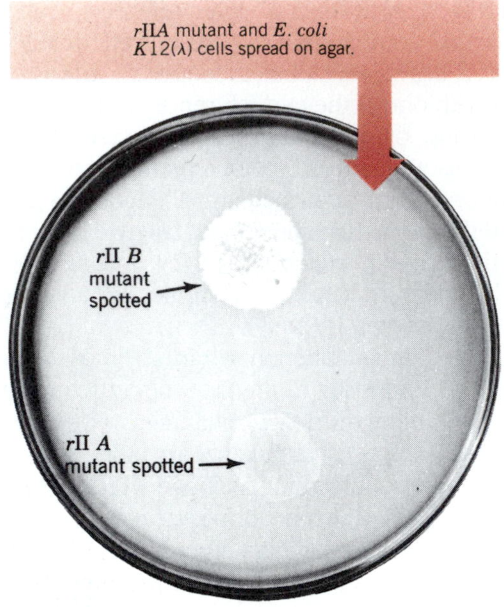

Figure 12.13 Complementation spot tests with phage T4 *r*II mutants. T4 phage particles with a mutation in the *A* gene of the *r*II locus and enough *E. coli* K12(λ) cells to yield a confluent growth of cells after incubation were spread on the surface of nutrient agar medium. A drop of solution that contained a suspension of phage particles (about 10^6–10^7 phage) with a mutation in the *B* gene was then applied to a spot at the top of the plate. Similarly, a drop containing phage particles with a mutation in the *A* gene was applied to a spot at the bottom of the plate. The large number of plaques (clear zones) formed in the top "spot" after incubation is evidence of complementation. The absence of plaques in the bottom "spot" indicates that complementation did not occur between the two *rIIA* mutants. (Photograph by R. Wagner and D. P. Snustad.)

each pair of mutants being tested on a lawn of *E. coli* K12(λ) cells. Alternatively, small drops, each containing a different mutant, may be placed on a lawn containing the appropriate ratio of *E. coli* K12(λ) cells and a reference mutant. If complementation occurs, all the cells in the area covered by the drop of suspended phage will be lysed, yielding a large plaque (Fig. 12.13). If complementation does not occur, the area will be overgrown with a lawn of *E. coli* cells.

Although complementation tests showed that all the *r*II mutations were located in two genes, recombination tests demonstrated that these mutations were located at many distinct sites within these two genes. Recombination between *r*II mutations is analyzed by simultaneously infecting *E. coli* B cells (**permissive** host cells) with the two *r*II mutants in question and "plating" (spreading on the surface of agar medium in a petri dish) the progeny phage on a lawn of *E. coli* K12(λ) to see if any wild-type recombinants have been produced. The total number of progeny phage produced is determined by plating the lysate on an *E. coli* B lawn. As many as 10^8 progeny phage may be examined in a single petri dish. If only one r^+ recombinant is present, only one plaque will be formed on K12(λ). Thus, recombination events as rare as 1 in 100 million can easily be detected. Such a conditional lethal system thus provides an extremely powerful sieve for selecting rare recombinants.

Only the r^+ recombinants form plaques on K12(λ). The equally frequent double-mutant recombinants do not produce plaques on K12(λ). Thus, the

$$\text{Recombination frequency} = \frac{2[\text{number of plaques on K12}(\lambda)]}{\text{number of plaques on B}}$$

However, to map over 3000 independent mutations by standard two- and three-factor crosses (see Chapter 7), even with a phage system, is a formidable task. All possible two-factor crosses of 3000 mutants total about 4½ million [(3000)(2999)/2] crosses. Benzer was able to avoid such a laborious undertaking by developing a shortcut method of mapping that used overlapping deletion mutations.

Deletion Mapping

Many of the spontaneous *r*II mutants isolated by Benzer failed to recombine with two or more mutants that recombined with each other. They behaved as though they contained "multisite" mutations or defects that extended over a segment of the *r*II locus. Some of them appeared to involve only short segments within either the *rIIA* gene or the *rIIB* gene. Others were much longer; a few appeared to span the entire *r*II locus. Benzer proposed that these multisite mutations resulted from the deletion or loss of segments of DNA. His proposal that these multisite mutations were deletions of segments of DNA was supported by the observation that they did not undergo reverse mutation back to wild type. Point mutations (base-pair substitutions and single base-pair additions and deletions) are capable of reverse mutation, albeit at highly variable rates. Benzer's proposal has subsequently been verified directly by analysis of several of these putative deletion mutations by electron microscopy (see the following section).

The extents of the deleted segments can be determined by crossing the deletion mutants to a set of reference point mutants previously mapped by two- and three-factor crosses (Fig. 12.14). Once a set of overlapping deletions has been mapped, their end

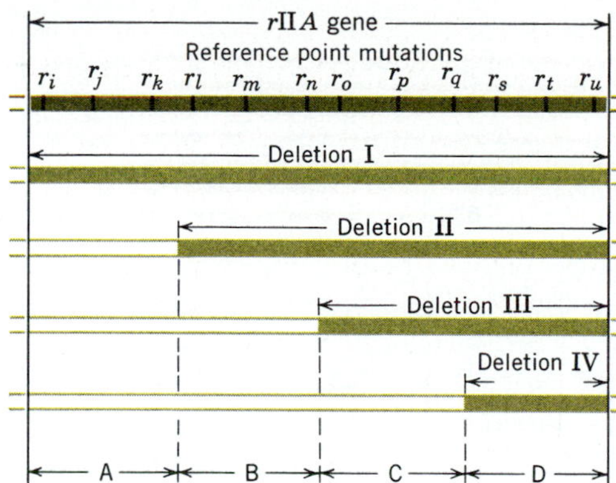

Figure 12.14 Illustration of the use of overlapping deletions in genetic fine structure mapping. Four hypothetical deletion mutants are shown; the olive shaded areas indicate the extents of the deleted segments. Their end points have been determined by crossing them to a set of *r*II mutant strains that carry point mutations (top) previously mapped by standard two- and three-factor crosses. These deletions divide the *r*IIA gene into four intervals. The interval A, B, C, or D in which an unknown point mutation is located can be determined by crossing a mutant that carries the unknown mutation with each of the four deletion mutants. A mutation in interval D will not produce any wild-type recombinant progeny in any of the four crosses. A mutation in interval C will recombine with deletion IV, but not with the other three deletions, and so on.

points will divide the region spanned by the longest deletion into a set of intervals (such as A, B, C, and D in Fig. 12.14). When a new mutant carrying a point mutation is isolated, the mutation can quickly be mapped to a defined interval by crossing the mutant strain with each of the overlapping deletion mutants. For example, suppose that five point mutants r_1, r_2, r_3, r_4, and r_5 are crossed with the four deletion mutants I, II, III, and IV, shown in Fig. 12.14. Furthermore, suppose that the results of these two-factor crosses are:

POINT MUTANTS	DELETIONS I	II	III	IV
r_1	0	+	+	+
r_2	0	0	+	+
r_3	0	0	0	+
r_4	0	0	0	0
r_5	0	0	0	0

where + indicates that recombination has occurred and 0 indicates that no recombinants have been formed. These results would immediately allow one to localize the mutation present in mutant r_1 to interval A. Similarly, the mutations in r_2 and r_3 could be assigned to intervals B and C, respectively. Finally, the defects in both r_4 and r_5 must be in interval D. The r_4 and r_5 mutations cannot be ordered with respect to each other from these data. If a set of smaller overlapping deletions in interval D were available, positions of the mutations present in mutants r_4 and r_5 might be further localized by crossing each of these point mutants with each mutant of a set of smaller overlapping deletion mutants. Benzer characterized a large number of deletion mutants that ***divided the rII locus into 47 small segments*** (Fig. 12.15). By using these to localize the *r*II point mutants in newly isolated mutant strains, Benzer was able to greatly reduce the number of crosses needed to establish the detailed fine structure map of the *r*II locus. After the shortcut deletion mapping was completed, the mutations located within each of the 47 intervals had to be mapped relative to each other by standard two- and three-factor crosses. Since most of the intervals contained a relatively small number of mutations, this was a manageable task, in contrast to the 4½ million two-factor crosses required without the deletion mapping step.

Benzer and colleagues identified ***over 300 sites of mutation that were separable by recombination.*** Figure 12.16 shows the 250 sites at which spontaneous mutations occurred. Note that the ***frequency of mutation at different sites is highly variable.*** In particular, two sites, one in segment A6c and one in segment B4, are highly mutable (so-called "hot spots" for mutation). It would be very interesting to know why these two sites mutate so frequently.

Electron Microscope Heteroduplex Mapping

The existence of genetically well-defined deletion mutations at the *r*II locus has permitted E. K. F. Bautz and colleagues to determine the physical size of the *r*II locus. This was done using a technique called ***heteroduplex mapping.*** A ***DNA heteroduplex is a DNA molecule in which the two strands are not entirely complementary.*** One strand of a DNA double helix may contain one allele of a gene and the other strand (largely, but not totally, complementary) may carry a different allele of the gene. The noncomplementary regions of DNA heteroduplexes vary in size from one mismatched base-pair to large segments of the molecule. DNA heteroduplexes occur *in vivo* as intermediates in certain processes of mutation (see Chapter 11) and recombination (see Chapter 7).

Heteroduplex mapping involves the *in vitro* preparation of DNA heteroduplexes and their analysis by electron microscopy. If, for example, a preparation of DNA from wild-type T4 phage and a preparation of

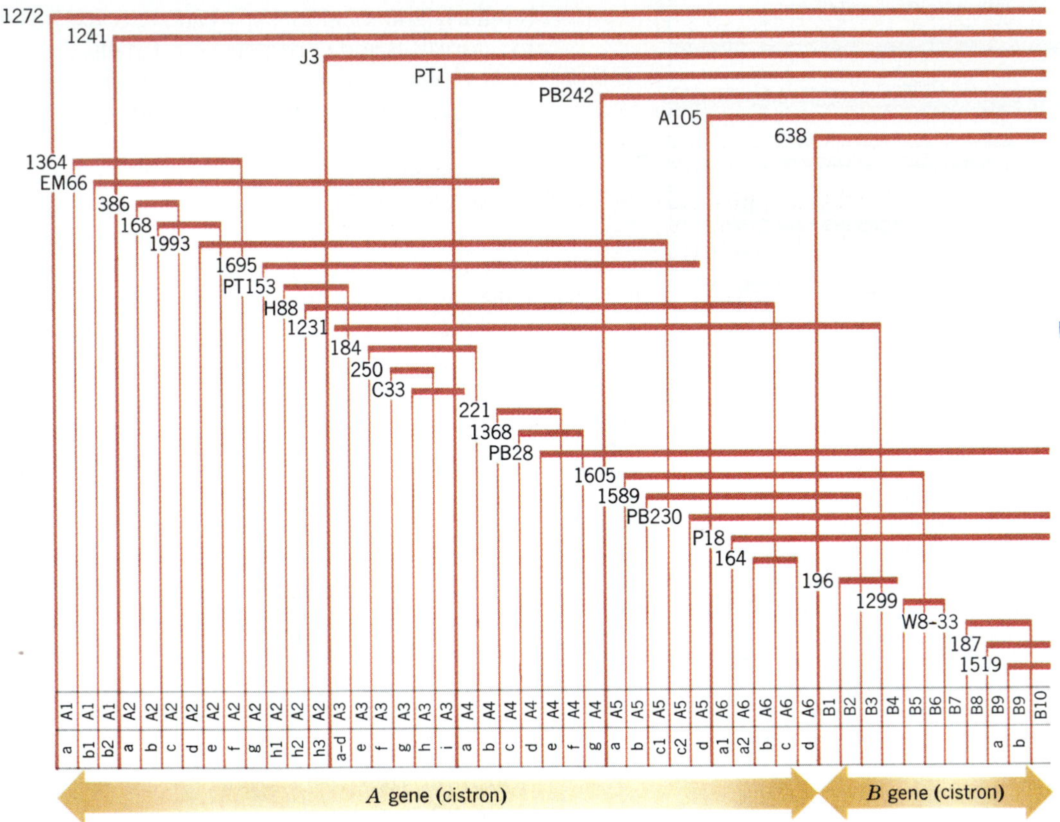

Figure 12.15 The overlapping deletions used by Benzer and coworkers to divide the *r*II locus into 47 smaller intervals. Benzer's designations for each of the 47 segments are given in the boxes near the bottom; they correspond to the segments shown on the map in Fig. 12.16. The limits of the *A* and *B* genes, as defined by complementation tests, are shown by the arrows at the bottom. (After S. Benzer, *Proc. Natl. Acad. Sci. U.S.A.* 47: 410, 1961.)

DNA from an *r*II mutant with a deletion of the *r*IIA gene are denatured (such as by heating to 100°C or by alkali at pH 11.4), mixed together, and then allowed to renature (e.g., 60°C for 8–12 hours in the appropriate solution), some of the renatured DNA double helices (half of them if equal amounts of the two DNA types are mixed) will be heteroduplexes. These heteroduplexes (Fig. 12.17) will contain one strand of DNA from the wild-type T4 and one strand from the *r*IIA deletion mutant. The wild-type strands of these heteroduplexes will not have complementary base sequences with which to pair in the segment of the molecule corresponding to the *r*IIA gene. The wild-type strands will thus form single-stranded loops, the length of which can be measured on electron micrographs.

Bautz and coworkers prepared heteroduplexes between DNA from T4 r^+ phage and DNA from each of several genetically well-characterized *r*II deletion mutants; they then analyzed them by electron microscopy. Their results yielded estimates of ***1800 ± 70 nucleotide-pairs and 845 ± 50 nucleotide-pairs for the sizes of the rIIA and rIIB genes, respectively.*** These results, combined with the extensive genetic data of Benzer and colleagues, provide a fairly clear picture of the fine structure of the *r*II locus (Fig. 12.18).

Since the mutable site is clearly equivalent to the single nucleotide-pair (see Chapter 11 and the following section), the over 300 mutable sites identified by Benzer represent only a small fraction of the total number of "sites" or nucleotide-pairs (about 2645) in the *r*II locus. A mutable site can be identified, of course, only when a base-pair substitution at that site causes a mutant phenotype, usually as a result of the production of an inactive or partially inactive gene-product. Because of the "degeneracy" in the genetic code (more than one triplet codon per amino acid; see Chapter 10), many base-pair substitutions will not change the amino acid sequences of the polypeptide

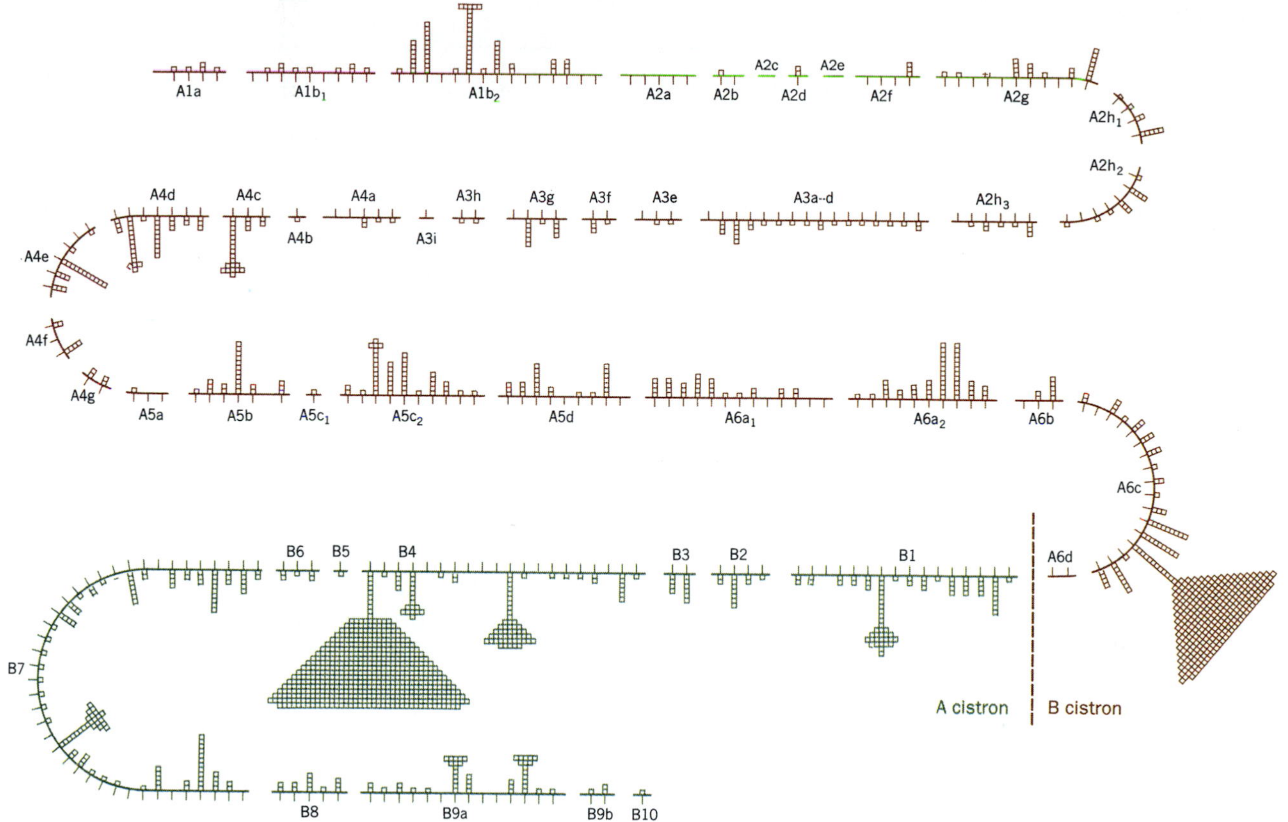

Figure 12.16 A genetic map of the spontaneous mutations in the *r*II locus studied by Benzer and colleagues. Each small square symbolizes one mutation that occurred at a particular mutable site. Mutable sites that are separable by recombination are indicated by short vertical lines. Mutable sites at which no squares occur were identified by mutations induced with mutagenic agents; no spontaneous mutations were detected at these sites. Each short segment of the map (defined by the discontinuities) represents one of the 47 intervals defined by the overlapping deletions shown in Fig. 12.15. No spontaneous mutations were detected in intervals A2a, A2c, A2e, or A6d; in fact, no mutations, spontaneous or induced, were identified in intervals A2c and A2e. Spontaneous mutation occurred very frequently at two sites ("hot spots"), one in interval A6c and one in interval B4. The boundary between the *A* and *B* genes is indicated by the dashed line. (After S. Benzer, *Proc. Natl. Acad. Sci. U.S.A.* 47: 410, 1961.)

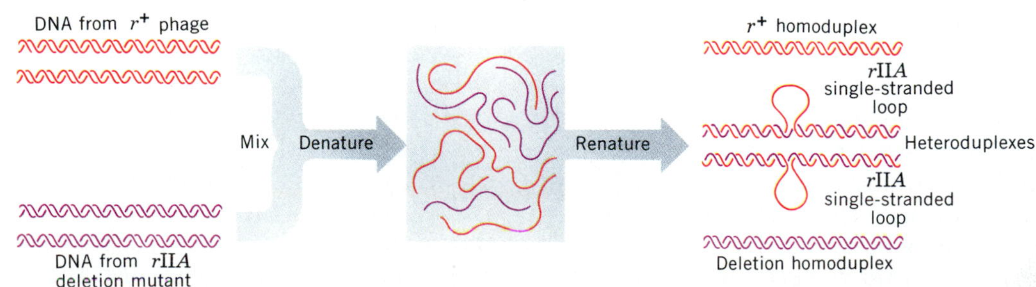

Figure 12.17 Diagram illustrating the use of deletion heteroduplex mapping of genes or other segments of DNA. The example shown illustrates the formation of heteroduplex molecules in which one strand of DNA is r^+ and the other strand contains a deletion of the entire *r*IIA gene. Such DNA heteroduplexes will contain single-stranded loops, the length of which corresponds to the length of the *r*IIA gene. If equal numbers of the two kinds of DNA molecules are mixed, the renatured double helices should contain, on average, 50 percent heteroduplexes.

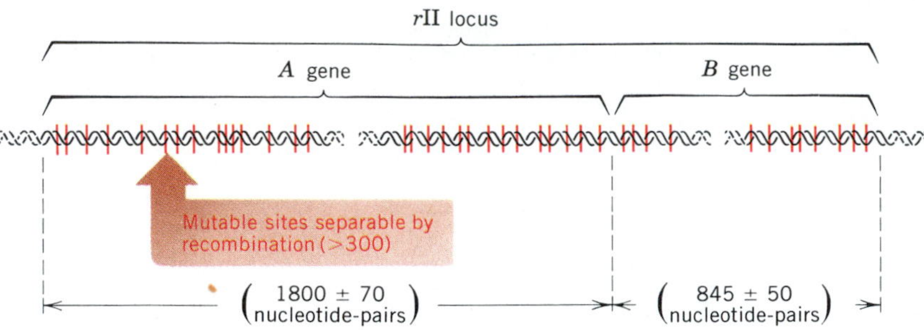

Figure 12.18 Summary of the fine structure of the phage T4 *r*II locus, based largely on the results of Benzer and coworkers and Bautz and coworkers. The total length of the *r*II locus is, according to Bautz's data, 2645 ± 100 nucleotide-pairs of the DNA double helix. For simplicity, only the distal portions of each gene are shown. Benzer identified over 300 distinct (by recombination) sites of mutation within these two genes.

gene-products at all. Many other base-pair substitutions will result in the substitution of one amino acid for another, where both amino acids have very similar chemical properties. Such substitutions usually do not significantly alter the activity of the protein. Thus, although it is clear that many mutable sites in the *r*II locus were not detected in Benzer's study, it is also clear that ***many nucleotide-pairs are not potential sites for changes that lead to a mutant phenotype.***

THE UNIT OF STRUCTURE IS THE SINGLE NUCLEOTIDE-PAIR

Benzer's extensive genetic analysis of the *r*II locus of phage T4 demonstrated that many mutable sites that are separable by recombination can exist within a single gene (defined by the *cis–trans* test as the unit of function coding for one polypeptide). This picture of the fine structure of the gene was soon verified by the results of studies of many different genes in many different organisms. Based on the known structure of DNA and the nature of base-pair substitution in mutagenesis (Chapter 11), it seemed likely that the unit of structure not divisible by recombination was a single nucleotide-pair. This was first demonstrated experimentally by C. Yanofsky and coworkers in studies of mutations in the *trpA* gene of *E. coli.*

The enzyme tryptophan synthetase catalyzes the final step in the biosynthesis of the amino acid tryptophan. In *E. coli*, this enzyme is composed of two different polypeptides, designated A and B (the products of genes *trpA* and *trpB*, respectively). Yanofsky and colleagues isolated and characterized a large number of tryptophan auxotrophic mutants with mutations in the *trpA* gene. The *trpA* gene codes for a polypeptide (the tryptophan synthetase A polypeptide) that is 268 amino acids long, and Yanofsky and associates used the techniques of protein sequencing to work out the complete amino acid sequence of this polypeptide. They also determined the amino acid sequences of several mutant forms of the tryptophan synthetase A polypeptide. When the positions of the amino acid changes in the polypeptide were compared with the locations of the mutations on the genetic map, a precise ***colinear*** relationship was observed (Fig. 12.19).

Convincing evidence for ***colinearity*** between the sequence of nucleotide-pairs in the gene and the sequence of amino acids in the polypeptide gene-product has also been obtained in several other studies, in both prokaryotes and eukaryotes (see Chapter 10, pp. 282–284). The recent demonstrations of "noncoding segments" or "introns" within certain eukaryotic genes (see pp. 356–359), however, indicate that the concept of an ***uninterrupted*** sequence of triplet base-pairs in the gene specifying a colinear sequence of amino acids in the polypeptide gene-product may have to be modified in many cases.

Yanofsky's correlated genetic and biochemical data for the *trpA* gene and the tryptophan synthetase A polypeptide (Fig. 12.19) showed that independent mutations that alter the same amino acid may either (1) affect the same mutable site and not recombine (such as *A*3 and *A*33) or (2) occur at different mutable sites (such as *A*23 and *A*46 or *A*78 and *A*58). Moreover, the data show that ***independent mutations at the same mutable site*** (as shown by their failure to give rise to *trp*$^+$ recombinants in crosses) ***are not always identical.*** Mutation *trpA*3 changes glutamic acid at position 49 to valine. Mutation *trpA*33 changes the glutamic acid at position 49 to methionine. No recombinants are produced in crosses between mutant strains carrying mutations *A*3 and *A*33. Thus, trpA3 and trpA33 ***are structurally allelic (as determined by recombination tests) and functionally allelic (as determined by*** cis–trans ***tests). Both mutations have occurred at***

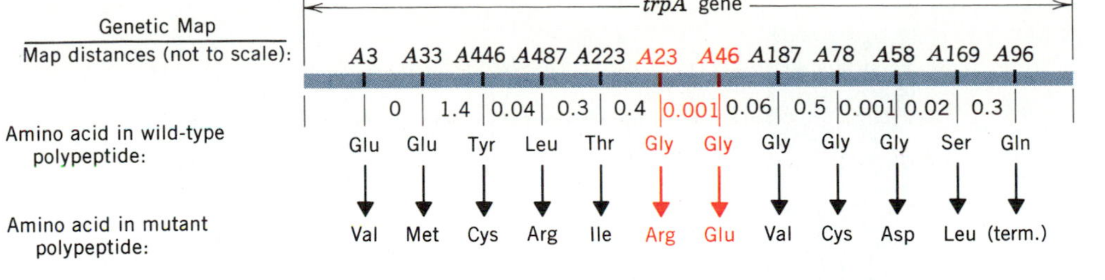

Figure 12.19 Genetic map of the *trpA* gene of *E. coli,* and the amino acid replacements that have been shown to occur in the tryptophan synthetase A polypeptide as a result of the indicated mutations (*A3*, *A33*, etc.). C. Yanofsky and colleagues have worked out the entire sequence of this 268-amino-acid-long polypeptide. The positions (numbered from the amino terminus) of the amino acid changes that occurred due to the mutations shown are given at the bottom. Note that mutations *A23* and *A46* (shown in red) do recombine despite the fact that they both alter the same amino acid, namely, the glycine at position number 211 of the wild-type polypeptide. The tryptophan synthetase A polypeptide of mutant *A23* has arginine at position 211; that of mutant *A46* has glutamic acid at position 211. Thus both mutations must have occurred within the sequence of three base-pairs of the *trpA* gene that specifies (via the mRNA triplet codon) the amino acid present at position 211 of the tryptophan synthetase A polypeptide. (After C. Yanofsky and V. Horn, *J. Biol. Chem.* 247: 4494–4498, 1972.)

the same unit of structure, the same nucleotide-pair. H. Roman has proposed that such mutations be called ***"homoalleles"*** ("homo" = same, or ***alleles at the same site***), to distinguish them from ***"heteroalleles"*** ("hetero" = different, or ***alleles at different sites***). ***Heteroalleles are functionally allelic (in the same gene or unit of function, as operationally defined by the cis–trans test), but are structurally nonallelic (in different units of structure or at different sites, as operationally defined by the recombination test).***

Mutations *trpA23* and *trpA46* both result in the substitution of another amino acid (arginine in the case of *A23* and glutamic acid in the case of *A46*) for the glycine present at position 211 of the wild-type tryptophan synthetase A polypeptide (Fig. 12.19). However, these two mutations occur at different mutable sites. That is, the *A23* and *A46* sites are separable by recombination. By determining the amino acids present at position 211 of the tryptophan synthetase A polypeptide in other mutants as well as in revertants and partial revertants of the *trpA23* and *trpA46* mutants, Yanofsky and coworkers were able to determine which of the glycine, arginine, and glutamic acid codons (see p. 338) were present in the messenger RNA (mRNA) at the position corresponding to amino acid number 211 of the polypeptide gene-product in trp^+, *trpA23*, and *trpA46* cells, respectively (Fig. 12.20). (A partial revertant is the result of a mutation that occurs in a mutant strain and that restores some, but not wild-type, levels of gene-product activity. The partial revertants referred to here are capable of slow growth on medium lacking tryptophan. These partial revertants can further mutate to "full revertants" that grow at wild-type rates in the absence of tryptophan.) Once the specific codons (triplet base sequences in mRNA) are known, the corresponding base-pair sequences in the structural gene from which the mRNA is transcribed are also known. One strand of DNA will be complementary to the mRNA, and the second strand of DNA will, of course, be complementary to the first strand. Thus, Yanofsky's data demonstrated that the mutation events producing the *trpA23* and *trpA46* mutants were GC to AT transitions occurring at adjacent nucleotide-pairs. The trp^+ cells produced by recombination between chromosomes carrying mutations *A23* and *A46* (Fig. 12.21) demonstrate that recombination can occur between adjacent nucleotide-pairs. This shows that the unit of genetic material not divisible by recombination is the single nucleotide-pair. Clearly, then, the unit of structure of genetic material, as defined by the operational criteria of mutation and recombination, is the nucleotide-pair (or single nucleotide in genomes consisting of single-stranded DNA or RNA).

FINE STRUCTURE OF GENES AND "COMPLEX LOCI" IN EUKARYOTES

Examining enough progeny of a cross to detect rare ***intra***genic recombination in eukaryotes is a laborious project. Nevertheless, intragenic recombination was first detected in studies with *Drosophila* (p. 322). Genetic fine structure maps have now been constructed for many genes of *Drosophila* and maps have also been worked out for several other higher animals and higher plants. In almost every case where an

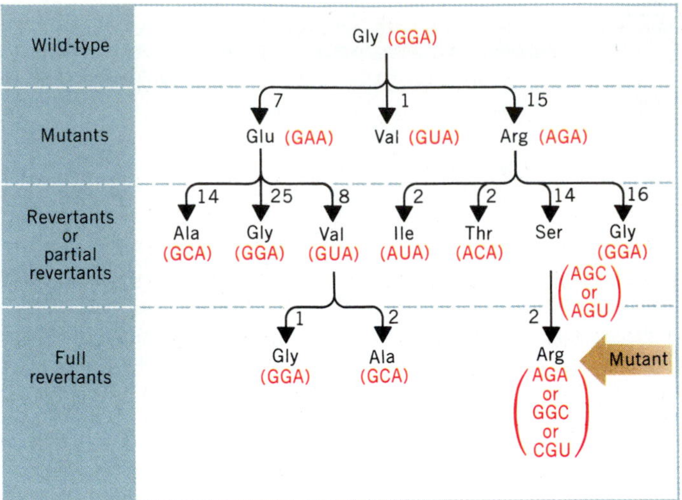

Figure 12.20 Pedigree of the tryptophan synthetase A polypeptide position number 211 (211th amino acid from the NH_2 terminus) — the position of the *trpA*23 and *trpA*46 mutation-induced amino acid substitutions (see Fig. 12.19). The triplet codons shown in parentheses (in red) are the only codons that are specific for these amino acids and that will also permit all the observed amino acid replacements at position 211 to occur as a result of single base-pair substitutions. The number beside each arrow indicates the number of times that particular substitution has been observed to occur. These data strongly indicate that the arginine and glutamic acid codons used to specify amino acid number 211 of the tryptophan synthetase A polypeptide in *trpA*23 and *trpA*46 are AGA and GAA, respectively. The corresponding base-pair sequences in the *trpA* gene would be —A G A→ / ←T C T— and —G A A→ / ←C T T— for *A*23 and *A*46, respectively, indicating that these two mutations alter adjacent base-pairs in this gene. (After C. Yanofsky, J. Ito, and V. Horn, *Cold Spring Harbor Sympos. Quant. Biol.* 31: 151, 1966.)

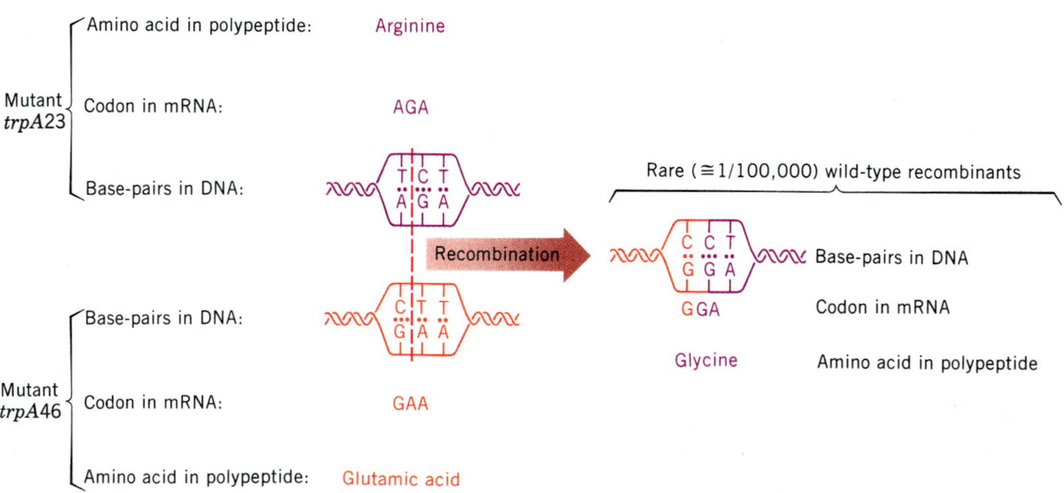

Figure 12.21 Recombination between mutations at adjacent base-pairs. Mutations *A*23 and *A*46 both result in an amino acid substitution at position number 211 of the tryptophan synthetase A polypeptide. *A*23 causes a glycine to arginine substitution; *A*46 causes a glycine to glutamic acid substitution. Analysis of the pedigree of amino acid substitutions that have been observed at position 211 as a result of mutations in the *trpA* gene has allowed Yanofsky and colleagues to deduce the exact sequence of base-pairs at the position in the *trpA* gene corresponding to amino acid number 211 in the A polypeptide of both mutants (*A*23 and *A*46) as well as wild type. (Based on the data of C. Yanofsky, J. Ito, and V. Horn, *Cold Spring Harbor Sympos. Quant. Biol.* 31: 151, 1966.)

extensive analysis of a gene has been carried out, mutable sites that are separable by recombination have been detected. Clearly, then, the genes of eukaryotes are sequences of nucleotide-pairs specifying the amino acid sequences of polypeptides. In eukaryotes, however, some genes have interesting structural features not observed in most prokaryotic genes (see pp. 356–359).

In many cases, determining how many genes are present at a locus has proven difficult in eukaryotes. Attempts to delimit genes by complementation tests have often yielded ambiguous results. In some cases, the ambiguities resulted from ***intra***genic complementation (see pp. 324–327), as at the *rosy* locus (discussed later in this section) in *Drosophila*. In other cases, it is still not clear whether a particular locus (a location on the chromosome at which mutations causing similar mutant phenotypes map) contains one or more than one gene. For these cases, where two or more mutable sites that are separable by recombination are known, but where the number of genes involved is not known (because of complex or ambiguous complementation data), the term **complex locus** has been introduced. This term is appropriate because ***locus*** just implies a region on the chromosome and does not have a precise operational definition such as that for ***gene*** (coding for one polypeptide and defined by the *cis–trans* test).

Sometimes the ambiguities that have arisen in attempts to operationally define functional units may be the result of differential effects of different mutations in a gene on pleiotropic interactions of the gene-product. It seems reasonable to expect more pleiotropic effects in eukaryotic organisms with their greater developmental complexity than in prokaryotic organisms. Finally, some of the ambiguities may result from the structural organization of genes in eukaryotes. What effect(s) the noncoding "introns" within the coding sequences of structural genes in eukaryotes might have on *cis–trans* tests is still not entirely clear.

As an example of a complex locus, consider the *white* (white eyes) locus in *Drosophila*. A large number of mutations that have similar, but phenotypically distinguishable, effects on eye color are known to map at this locus (Fig. 12.22). Recombination analyses have demonstrated that these mutations map at seven or more distinct sites within this locus. Complementation tests suggest that two units of function exist. Some of the mutations at the four sites on the left (identified by w^{Bwx}, w^{bf}, w^{a}, w^{cb}) complement mutations at the site on the right (the *sp-w* site). This could mean that two genes are present within the *white* locus. A different functional subdivision of the locus is suggested by the ability of the mutations to act as dominant suppressors of mutations at another locus, *zeste* (a nearby, but distinct, eye color locus on the X chromosome). Mutations at the two sites on the right of the *white* locus (identified by w^{cb} and *sp-w*) are dominant suppressors of *zeste* mutations. Mutations at the three sites on the left (w^{Bwx}, w^{bf}, and w^{a}) do not suppress *zeste* mutations.

Does the *white* locus contain one, two, three, or seven genes? The definitive evidence needed to answer this question—the nucleotide sequence of the *white* locus—is now available. The *white* locus is one gene, albeit a complex gene with an interrupted coding sequence (see Chapter 13). Many of the classical mutations at the locus, like w^{a}, are known to be due to

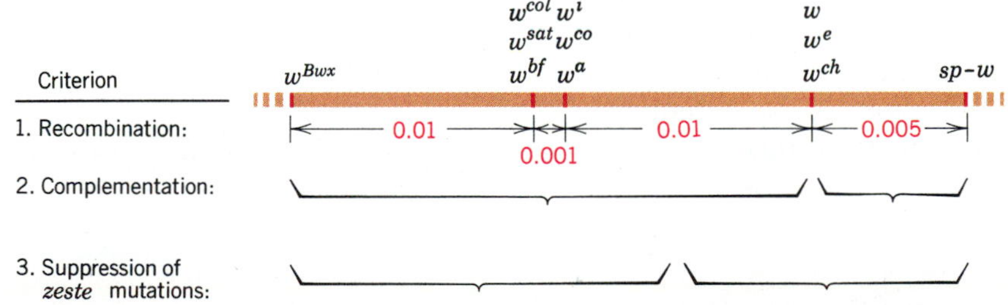

Figure 12.22 An early map of the *white* locus of *D. melanogaster,* showing a few of the phenotypically distinguishable mutations. Those shown are w^{Bwx} (white Brownex), w^{bf} (white buff), w^{sat} (white satsuma), w^{col} (white colored), w^{a} (white apricot), w^{co} (white coral), w^{i} (white ivory), w^{cb} (white cherry), w^{e} (white eosin), *w* (white), and *sp-w* (spotted-white). Others mapping at the *white* locus and not shown include w^{ti} (white tinged), w^{bl} (white blood), w^{w} (white wine), w^{cf} (white coffee), and w^{crr} (white carrot). Different mutations occurring at the *white* locus thus produce a large number of similar, but distinguishable, effects on phenotype. The subdivision of the locus based on recombination analysis (1) as well as the functional subdivisions suggested by complementation analysis (2) and the ability of the various mutations to suppress mutations at a different (*zeste*) locus (3) are also indicated.

the insertion of transposable elements. The observed complementation is clearly ***intra***genic complementation.

Early support for the preceding interpretation of the structure of the *white* locus was provided by W. J. Welshons' study of the *Notch* locus in *Drosophila*. The *Notch* (*N*) locus corresponds to a single salivary chromosome band (see pp. 341–342) located near the centromere–distal end of the X chromosome. *Notch* mutants have "notched" wings, thickened wing veins, and minor bristle abnormalities. *N* mutations are classified as ***dominants***, since they exert these phenotypic effects in *N*/+ heterozygous females. These *N* mutations are also ***recessive lethals*** because *N*/*N* females and *N*/*Y* males are inviable. Other mutations at the *Notch* locus exhibit recessive visible effects on phenotype and are nonlethal. The recessive visible mutations, such as *fa* (facet), *spl* (split), fa^{no} (facet-notchoid), and *nd* (notchoid), map interspersed among the various recessive lethal and dominant visible *N* mutations (Fig. 12.23).

Welshons has shown that when the recessive lethal *N* mutations are analyzed by *cis–trans* tests, they define a single gene, that is, they do not complement each other. He refers to these as ***amorphic mutations—mutations that result in a complete loss of gene-product function.*** Welshons presented evidence that indicates that the nonlethal, recessive visible mutations (*fa*, *spl*, etc.) are ***hypomorphic mutations—mutations that result in reduced, but still some, gene-product activity.*** (The terms amorphic and hypomorphic were initially introduced by H. J. Muller.) These hypomorphic mutations do complement each other in certain combinations in *trans*-heterozygotes. Taken at face value, such complementation might suggest that the *Notch* locus contains more than one gene. Based on his study of the ***Notch*** locus, Welshons makes a very important point, namely, that ***complementation data obtained using hypomorphic mutations often are not useful in delimiting genes.*** This has become eminently clear in studies with prokaryotes. Attempts to delimit genes using temperature-sensitive mutations, which are often hypomorphic, have frequently yielded ambiguous results due to ***intra***genic complementation. When *amber* mutations (UAG polypeptide chain-termination mutations resulting in the absence of the carboxyl terminus), which are usually amorphic, are used, clear-cut results are almost always obtained.

If the *white* locus of *Drosophila* is considered, keeping in mind Welshons' data for the *Notch* mutants, it is apparent that most of the *white* locus mutations, such as w^{ch}, w^{a}, and so on, are hypomorphs, since they have phenotypes intermediate between the phenotype of the most extreme mutation, *w* (white), and wild type. As such, they might be expected to exhibit ***intra***genic complementation in certain combinations. However, whether this explanation of the complementation observed with mutations at the *white* locus is correct is not yet firmly established.

The most complete picture available for a locus in eukaryotes is of the *rosy* (*ry*) locus in *Drosophila*. *Rosy* mutants have brownish-colored eyes; they are deficient in the normal red pigment drosopterin, which results from the absence of a single enzyme, xanthine dehydrogenase. Studies in the laboratories of A. Chovnick and E. Glassman have shown that xanthine dehydrogenase is a dimer composed of two identical polypeptides. As might be expected, some of the *ry* mutant alleles (Fig. 12.24) exhibit ***intra***genic complementation, just as was observed for certain mutations at the *white* and *Notch* loci. In the case of *rosy*, however, definitive data are available, indicating that the locus contains a single structural gene.

Drosophila embryos deficient in xanthine dehydrogenase activity (e.g., *ry*/*ry* homozygotes) are inviable on culture medium supplemented with purine. This has provided Chovnick and colleagues with a very powerful selective sieve with which to carry out fine structure analysis. Only those progeny of *trans*-heterozygotes (e.g., ry^1/ry^2) that carry a rare wild-type recombinant chromosome will be viable on purine-supplemented medium. *Rosy* mutants thus provide a specialized (applicable only to this one locus) conditional lethal system, which is ideal for genetic fine structure analysis.

Similar fine structure maps have been obtained in higher plants. ***Intra***genic recombination in plants was first detected at the *waxy* (*wx*) locus in *Zea mays*

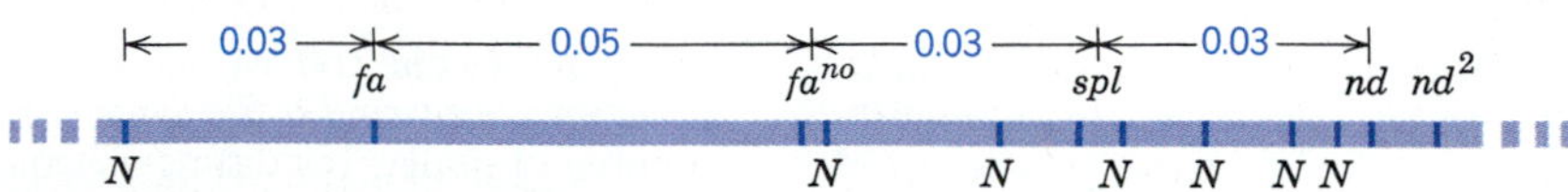

Figure 12.23 Map of the *Notch* locus of *Drosophila melanogaster*. The map positions of several dominant visible and recessive lethal *N* mutations are given below the line. The locations of recessive visible mutations are shown above the line. Distances are in map units; total length of the defined interval is about 0.14 map unit. (Data are from W. J. Wel shons, *Science* 150: 1122, 1965.)

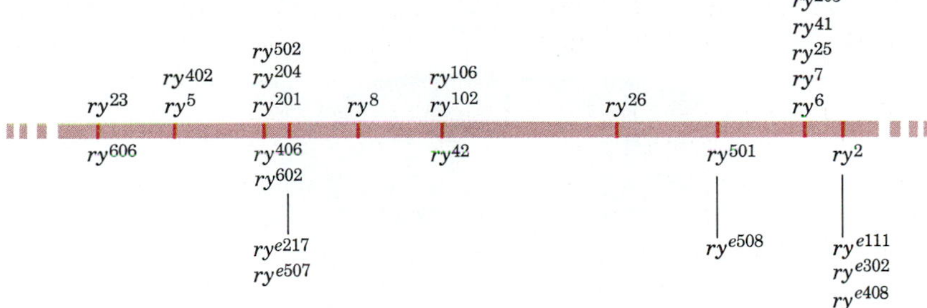

Figure 12.24 Map of the *rosy* locus of *Drosophila melanogaster,* showing the map positions of mutations in the structural gene for the enzyme xanthine dehydrogenase. Noncomplementing xanthine dehydrogenase-deficient *ry* mutations are shown above the line. Mutations that exhibit intragenic complementation are shown directly below the line. Alterations in the *ry* gene that result in active xanthine dehydrogenase molecules, but with altered electrophoretic mobilities (changes in the charge of the polypeptide gene-product), are shown at the bottom (superscripts with *e* prefixes). Chovnick and colleagues have also identified and mapped mutations that alter the amount of xanthine dehydrogenase synthesized. These mutations are in regulatory regions of the *rosy* locus; they map outside the structural gene in the region just to the left of ry^{23} as shown here. (Based on the data of A. Chovnick, W. Gelbart, and M. McCarron, *Cell* 11: 1, 1977.)

(corn) by O. Nelson. Wild-type (*Wx*, "starchy") corn plants contain two types of starch, called amylose and amylopectin, deposited in the endosperm of the kernels and in the pollen. *Waxy* (*wx*) mutants contain amylopectin, but no amylose; they lack a single enzyme, ADPG transferase, required for the synthesis of amylose. Fine structure mapping of *wx* mutations was facilitated by the fact that this phenotype can be scored in the male gametes or pollen grains. *Wx* pollen grains turn blue whereas *wx* pollen grains turn light red after staining with iodine. Since each corn plant produces millions of pollen grains, rare wild-type recombinants can be detected by screening pollen produced by a wx^1/wx^2 *trans*-heterozygote. Thus, Nelson was able to demonstrate that the gene in higher plants, like the gene in prokaryotes and in *Drosophila*, contains mutable sites that are separable by recombination.

ONE GENE–ONE BAND IN *DROSOPHILA* SALIVARY GLAND CHROMOSOMES?

The somatic chromosomes of the salivary glands of dipteran flies such as *Drosophila* are giant structures with a banded appearance resulting from the alternation of regions of highly condensed DNA (densely staining regions) with regions of less condensed DNA (lightly staining regions; see Chapter 18, Figs. 18.2 and 18.3). Each giant salivary gland chromosome contains over 1000 times as much DNA as the homologous meiotic germ line chromosome. These giant chromosomes are produced by replication of the single molecule of DNA in each germ line chromosome some 10 times without separation of the replication products into different chromosomes. Each chromosome thus contains about 2^{10} or 1024 DNA molecules, which are associated in a lateral array. Such "multistranded" (containing many DNA molecules running longitudinally throughout the chromosome) chromosomes are called ***polytene*** chromosomes. (Details of the structure of the polytene salivary gland chromosomes of *Drosophila melanogaster* are discussed in Chapter 18.)

Extensive genetic and cytological analyses of specific regions of the *D. melanogaster* salivary gland chromosomes have provided strong evidence that there is a 1 : 1 correlation between genes and bands (or ***chromomeres***, as they are often called). In a very detailed study, B. H. Judd and colleagues identified and characterized a large number of mutations in one region of the X chromosome. They identified 16 genes ("complementation groups") in a segment of the chromosome containing 15 distinct bands (Fig. 12.25). Similarly, the small chromosome 4 of *Drosophila* has been shown to contain 43 genes (units of function) to date. Chromosome 4 has somewhere between 33 and 50 bands, again indicating a good fit to the one gene–one band hypothesis.

In total, the salivary gland chromosomes contain about 5000–6000 bands. If the one gene–one band hypothesis is correct, *Drosophila* has 5000–6000 genes. This estimate of the number of genes is consistent with the ***independent*** estimate of the number of genes in *Drosophila* based on mutation frequencies. However, estimates of the number of different mRNA molecules indicate that there are about 17,000 structural genes—or about three times as many genes as polytene chromosome bands—in *Drosophila*.

The haploid chromosome complement of *D. mel-*

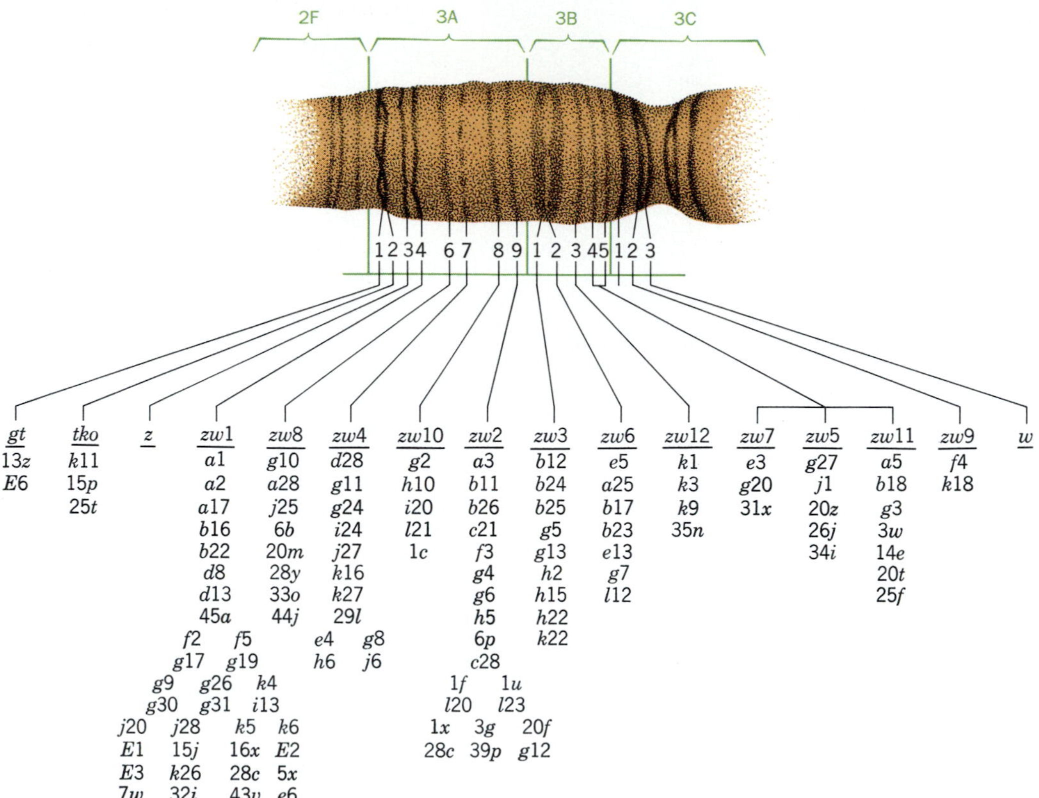

Figure 12.25 Correlation between the number of genes (complementation groups) and the number of bands in one segment (region 3A1 to 3C2) of the polytene X chromosome in the salivary glands of *Drosophila melanogaster*. The mutations identified and localized in the 16 complementation groups are shown below the schematic drawing of the 3A1 to 3C2 segment of the X chromosome. (Data are from B. H. Judd, M. W. Shen, and T. C. Kaufman, *Genetics* 71: 139, 1972.)

anogaster contains about 10^8 nucleotide-pairs. If all of this DNA represented structural genes of an average size of 1000 nucleotide-pairs, the *Drosophila* genome would contain 100,000 genes, and each band (containing about 30,000 nucleotide-pairs) would contain about 30 genes. What is (are) the function(s) of this apparent excess of DNA in *Drosophila* and all other eukaryotes? We know that part of this DNA resides in noncoding introns within eukaryotic genes (pp. 356–359), but what is (are) the role(s) of the rest of this DNA? This is one of the most intriguing questions challenging geneticists today (see Chapter 6, pp. 146–148, for further discussion).

GENES-WITHIN-GENES IN PHAGE ΦX174

Bacteriophage ΦX174 is a small virus that stores its genetic information in a single-stranded, circular molecule of DNA. Phage ΦX174 replicates its DNA using a double-stranded replicative intermediate (see pp. 122–124). Until recently, the ΦX174 genome was believed to contain nine genes coding for nine different polypeptides. All nine of these gene-products had been characterized and their molecular weights determined. By using the known molecular weights of the nine polypeptides, and knowing that the genetic code is a triplet code (three nucleotides per amino acid), it was possible to calculate the length of DNA required to code for all nine polypeptides. Assuming that the nine structural genes represented nonoverlapping nucleotide sequences, the minimum length of DNA required to code for the nine polypeptides was shown to be 6078 nucleotides. However, the ΦX174 chromosome is only 5387 nucleotides long, not long enough to carry the coding sequences for all nine polypeptides in nonoverlapping genes. For several years (until 1976), this presented a puzzling dilemma. This dilemma was not resolved until nucleotide sequence data became available.

The nucleotide sequence data for ΦX174 resolved the dilemma of not enough DNA by showing that the coding sequences of two genes are located within the coding sequences of two other genes with the reading frames (the triplet sequences read during translation;

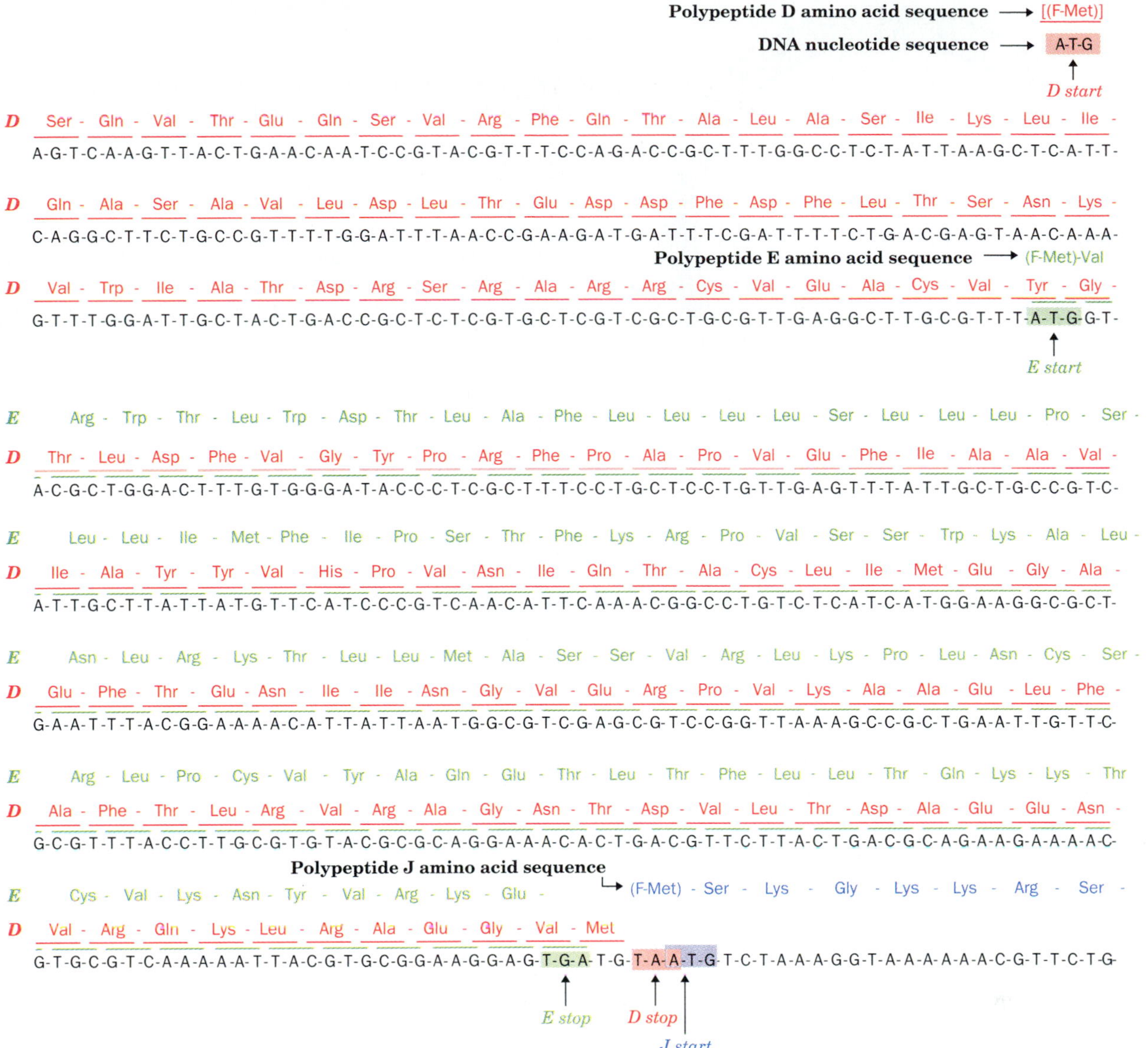

Figure 12.26 The nucleotide sequence for the segment of the phage ΦX174 chromosome containing the coding sequence of gene *E*, which is located within the coding sequence of gene *D*. The reading frame of gene *E* is offset to the right by one nucleotide from the reading frame of gene *D*. The reading frame of gene *D* and the correlated amino acid sequence of the D polypeptide are shown (in red) immediately above the nucleotide sequence. The reading frame of gene *E* and the amino acid sequence of the E polypeptide predicted from the nucleotide sequence are shown (in green) above the reading frame of the *D* gene and the amino acid sequence of the D polypeptide, respectively. The amino-terminal segment of the J polypeptide and the corresponding nucleotide sequence of the *J* gene are also shown. Note that the initiation triplet (ATG) of the *J* gene also overlaps the termination triplet (TAA) of the *D* gene by one nucleotide. (Based on the data of B. G. Barrell, G. M. Air, and C. A. Hutchison, III, *Nature* 264: 34, 1976.)

see Chapter 10) offset by one nucleotide in each case (Figs. 12.26 and 12.27). This surprising result has important genetic implications. Clearly, the amino acid sequences of genes with overlapping coding sequences cannot evolve totally independently of one another. (They have some independence due to the degeneracy of the genetic code; see Chapter 10.) Moreover, a single mutation (e.g., a single base addition or deletion) can result in the loss of two gene-product activities (pleiotropy of the most direct type).

How common overlapping genes or genes-within-genes are in various organisms remains unknown. Recently, a fourth gene has been discovered in phage MS2 that overlaps two of the three previously charac-

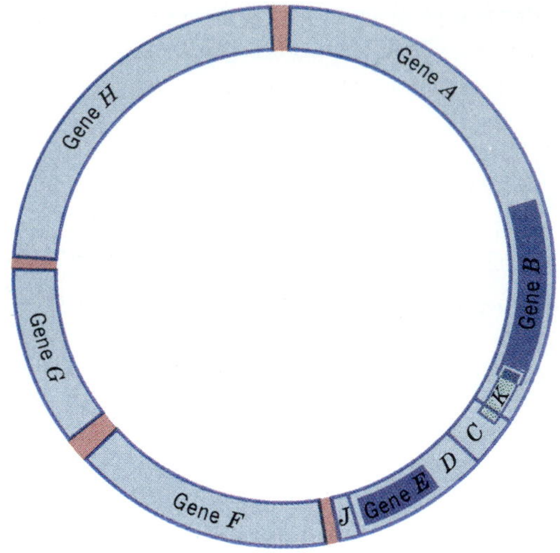

Figure 12.27 Organization of the 10 genes present in the circular chromosome of phage ΦX174 showing the two cases of a gene-within-a-gene. The coding sequences of genes *B* and *E* (shown in blue) are located entirely within the coding sequences of genes *A* and *D*, respectively. The coding sequence of gene *K* (shown in green, stippled) overlaps the coding sequences of the *A*, *B*, and *C*. It only overlaps with the termination codon of gene *B*. In all cases, the reading frames of the two overlapping genes are offset from each other (see Fig. 12.26). The four short noncoding segments of the chromosomes are shown in gray. Relative distances are only approximate. (Based on the data of F. Sanger, G. M. Air, B. G. Barrell, N. L. Brown, A. R. Coulson, J. C. Fiddes, C. A. Hutchison, III, P. M. Slocombe, and M. Smith, *Nature* 265: 687, 1977, and E. S. Tessman, I. Tessman, and T. J. Pollock, *J. Virol.* 33: 557, 1980.)

terized genes of MS2. Moreover, partial-coding-sequence overlaps do occur in the Simian virus (SV40) chromosome. Whether such overlaps exist in the chromosomes of organisms other than viruses remains a question to challenge present and future generations of geneticists.

Of course, the ultimate fine structure map is the nucleotide sequence of a gene complete with the nucleotide changes occurring during mutation events that result in the loss of specific functions. Such information allows one to correlate particular functions with specific signals coded in nucleotide sequences. We now know the complete nucleotide sequence of (1) the single-stranded RNA chromosome of phage MS2 (3569 nucleotides long), (2) the double-stranded DNA chromosome of SV40 (5226 nucleotide-pairs long), (3) the single-stranded DNA chromosome of phage ΦX174 (5387 nucleotides long), (4) the double-stranded DNA chromosome of phage λ (48,502 nucleotide-pairs long), and (5) the double-stranded DNA chromosome of the chloroplast of tobacco (155,844 nucleotide-pairs long).

SUMMARY

Traditionally, the ***gene*** has been defined as the unit of genetic material controlling the inheritance of ***one phenotypic characteristic*** or ***one trait***. Today, the ***gene*** is more precisely defined as the unit of genetic material ***coding for one polypeptide*** (or one RNA molecule in the case of genes not specifying proteins). Prior to 1941, the ***gene*** was also believed ***not to be subdivisible by mutation or recombination***. The unit of genetic material ***not subdivisible*** by mutation or recombination is now known to be the ***single nucleotide-pair*** (or single nucleotide in the case of single-stranded nucleic acid genomes). Many ***mutable sites separable by recombination thus exist within each gene***.

The gene is operationally defined by the* cis–*trans or complementation test. This test is used to determine whether organisms heterozygous for two mutations have mutant or wild-type phenotypes. cis-***Heterozygotes*** (organisms in which the two mutations are on one chromosome and the corresponding wild-type alleles are on another chromosome) must exhibit the wild-type phenotype for the *cis–trans* test to be informative. If ***both*** the cis-***heterozygote and*** the trans-***heterozygote*** (a cell or organism in which the two mutations are on different chromosomes) containing a given pair of mutations have ***wild-type phenotypes***, then the ***two mutations are in two different genes***. If the cis-***heterozygote has*** the ***wild-type phenotype*** and the trans-***heterozygote has*** the ***mutant phenotype***, then the ***two mutations are in the same gene***.

The utility of the *cis–trans* test is limited in some cases by (1) ***intragenic complementation***, (2) ***polar effects*** of chain-termination mutations in ***operons***, and (3) cis-***acting genes or gene-products***. ***Intragenic complementation*** frequently occurs when the active form of the protein gene-product is a ***multimer*** consisting of ***two or more polypeptide products of a given gene***. Delimiting genes by complementation tests should be done whenever possible using ***amorphic*** or ***null mutations*** (mutations resulting in no gene-product, a partial gene-product, or an otherwise totally nonfunctional gene-product) to minimize the possibility of confounding effects of ***intra***genic complementation.

Detailed genetic and cytological analyses of segments of the giant salivary chromosomes in *D. melanogaster* support the idea that each band of the chromosome contains one gene. If so, *Drosophila* would have only 5000–6000 genes, rather than 100,000 as estimated from the total amount of DNA in the genome, and an average of 1000 base-pairs per gene. However, estimates of the total number of mRNA molecules suggest that *Drosophila* has about 17,500 genes.

The complete nucleotide sequence of the bacteriophage ΦX174 chromosome is known. Surprisingly, of the 10 genes of ΦX174, ***two are located entirely within the coding sequences of two different genes.*** A third overlaps the sequences of three different genes.

REFERENCES

BENZER, S. 1959. "On the topology of the genetic fine structure." *Proc. Natl. Acad. Sci. U.S.A.* 45: 1607–1620.

______. 1961. "On the topography of the genetic fine structure. *Proc. Natl. Acad. Sci. U.S.A.* 47: 403–415.

BRINK, R.A., editor. 1967. *Heritage from Mendel.* Univ. of Wisconsin Press, Madison.

CARLSON, E.A. 1966. *The gene: A critical history.* Saunders, Philadelphia.

CHOVNICK, A., W. GELBART, and M. MCCARRON. 1977. "Organization of the *rosy* locus in *Drosophila melanogaster.*" *Cell* 11: 1–10.

FINCHAM, J.R.S. 1966. *Genetic complementation.* Benjamin, Menlo Park, CA.

JUDD, B.H., M.W. SHEN, and T.C. KAUFMAN. 1972. "The anatomy and function of a segment of the X chromosome of *Drosophila melanogaster.*" *Genetics* 71: 139–156.

STENT, G.S. and R. CALENDAR. 1978. *Molecular genetics, an introductory narrative,* 2nd ed. Freeman, San Francisco.

WELSHONS, W.J. 1965. "Analysis of a gene in *Drosophila.*" *Science* 150: 1122–1129.

YANOFSKY, C., and V. HORN. 1972. "Tryptophan synthetase α chain positions affected by mutations near the ends of the genetic map of *trpA* of *Escherichia coli.*" *J. Biol. Chem.* 247: 4494–4498.

ZACHAR, Z., and P.M. BINGHAM. 1982. "Regulation of *white* locus expression: The structure of mutant alleles at the *white* locus of *Drosophila melanogaster.*" *Cell* 30: 529–541.

PROBLEMS AND QUESTIONS

12.1 In what ways does our present concept of the gene differ from the pre-1940 or classical concept of the gene?

12.2 What was the first evidence that indicated that the unit of function and the unit of structure of genetic material were not the same?

12.3 What is the currently accepted operational definition of the gene?

12.4 Of what value are conditional lethal mutations for genetic fine structure analysis?

12.5 You are given five female *E. coli* strains, each carrying a different deletion in the *z* gene (for β-galactosidase). ***The light-colored boxes illustrated here represent the relative positions and extents of the five deletions.***

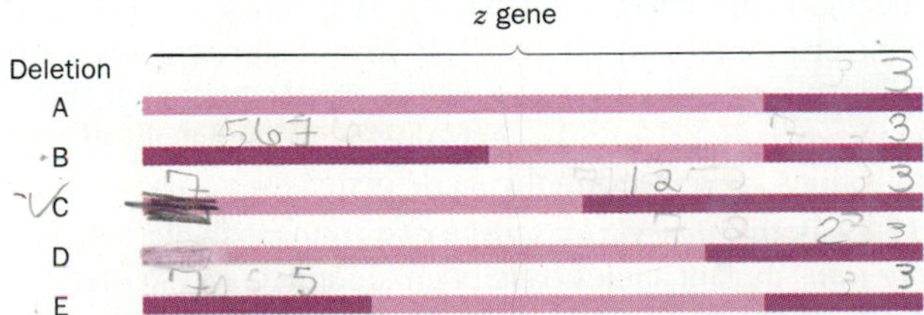

Seven independent z^- point mutations were introduced by 5-bromouracil in an Hfr (male) strain. When each of the seven mutants is mated with each of the five deletion strains, the following results are obtained (+ indicates the occurrence of recombination with the formation of z^+ recombinants, and 0 indicates that no recombination is detected).

	POINT MUTANTS						
DELETION:	1	2	3	4	5	6	7
A	0	0	+	0	0	0	0
B	0	0	+	0	+	+	+
C	+	+	+	0	0	0	+
D	0	+	+	0	0	0	0
E	0	0	+	0	+	0	+

What is the linear order of the seven point mutations?

12.6 Eight independently isolated mutants of *E. coli*, all of which are unable to grow in the absence of histidine (his⁻), were examined in all possible *cis*- and *trans*-heterozygotes (partial diploids). All the *cis*-heterozygotes were able to grow in the absence of histidine. The *trans*-heterozygotes yielded two different responses: some of them grew in the absence of histidine and some did not. The experimental results, using + to indicate growth and 0 to indicate no growth, are given in the following table. How many genes are defined by these eight mutations? Which mutant strains carry mutations in the same gene(s)?

	GROWTH OF *TRANS*-HETEROZYGOTES (−HISTIDINE)							
MUTANT:	1	2	3	4	5	6	7	8
8	0	0	0	0	0	0	+	0
7	+	+	+	+	+	+	0	
6	0	0	0	0	0	0		
5	0	0	0	0	0			
4	0	0	0	0				
3	0	0	0					
2	0	0						
1	0							

12.7 Assume that the mutants described in Problem 12.6 yielded the following results. How many genes would they have defined? Which mutations would have been in the same gene(s)?

	GROWTH OF *TRANS*-HETEROZYGOTES (−HISTIDINE)							
MUTANT:	1	2	3	4	5	6	7	8
8	+	+	+	+	+	+	0	0
7	+	+	+	+	+	+	0	
6	+	+	+	+	0	0		
5	+	+	+	+	0			
4	+	+	0	0				
3	+	+	0					
2	0	0						
1	0							

12.8 What determines the maximum number of different alleles that can exist for a given gene?

12.9 What is the difference between a pair of homoalleles and a pair of heteroalleles?

12.10 Two different inbred varieties of a particular plant species have white flowers. All other varieties of this species have red flowers. What experiments might be done to obtain evidence to determine whether the difference in flower color in these varieties is the result of different alleles of a single gene or the result of genetic variation in two or more genes?

12.11 Why are mutant (alternative) alleles essential for identifying wild-type alleles and locating the positions of gene loci on chromosomes?

12.12 (a) What are some of the genetic implications of overlapping genes or "genes-within-genes"? (b) What is the maximum number of different amino acid sequences that can be produced from the same segment of one strand of DNA? (c) From both strands of a DNA double helix?

12.13 The *amber* mutants of phage T4 are conditional lethal mutants. They grow on *E. coli* strain CR63, but are lethal on *E. coli* strain B. An *amber* mutant almost never exhibits ***intra***genic complementation with any other *amber* mutant; for this problem, assume that no ***intra***genic complementation occurs between any of the mutants involved. The following results were obtained when eight *amber* mutants were analyzed for complementation by infecting the restrictive host (*E. coli* strain B) with each possible pair of mutants. The results of mixed infections by pairs of mutants are shown as 0 if no progeny are produced, and as + if progeny phage resulted from the infection with that particular pair of mutants.

MUTANT:	1	2	3	4	5	6	7	8
8	+	+	+	+	+	+	0	0
7	+	+	+	+	+	+	0	
6	+	+	+	+	+	0		
5	0	+	0	+	0			
4	+	+	+	0				
3	0	+	0					
2	+	0						
1	0							

a. These data indicate that the eight amber mutations are located in how many different genes?
b. Which mutations are located in the same gene or genes?

12.14 In *Drosophila*, *white*, *cherry*, and *vermilion* are all sex-linked mutations affecting eye color. A white-eyed female crossed with a vermilion-eyed male produces white-eyed male offspring and red-eyed (wild-type) female offspring. A white-eyed female crossed with a cherry-eyed male produces white-eyed sons and light cherry-eyed daughters. Do these results indicate whether or not any of these mutations affecting eye color are functionally allelic? If so, which mutations are functional alleles?

12.15 Are the following statements concerning the genetic element referred to as the ***gene*** true or false?
a. The classical (pre-1940) conception of the gene was that it was (1) a unit of physiological function or expression, (2) the smallest unit that could undergo mutation, and (3) a unit ***not*** subdivisible by recombination.
b. The *cis–trans* test provides an operational definition by which we can often identify a gene as the unit which specifies one mRNA molecule.
c. Our present knowledge of the structure of the gene indicates that the units defined by criteria (2) and (3) in statement (a) above are both equivalent to a single nucleotide-pair.
d. Studies in the 1940s demonstrated the existence of heteroalleles (pseudoalleles), clearly indicating that many mutations which were allelic by the functional criterion could be separated by recombination, thereby indicating that the units of function, mutation, and recombination are ***not*** equivalent.
e. Homoalleles are functionally and structurally allelic; heteroalleles are functionally allelic, but structurally nonallelic.

12.16 The rosy (*ry*) gene of *Drosophila* encodes the enzyme xanthine dehydrogenase; the active form of xanthine dehydrogenase is a dimer containing two copies of the *rosy* gene-product. Mutations ry^2 and ry^{42} are both located within the region of the *rosy* gene that encodes the *rosy* polypeptide gene-product. However, ry^2/ry^{42} *trans*-heterozygotes have wild-type eye color. How can the observed complementation between ry^2 and ry^{42} be explained given that these two mutations are located in the same gene?

12.17 Both *temperature-sensitive* (*ts*) mutant alleles and *amber* (*am*) mutant alleles have been identified and studied for many of the genes of bacteriophage T4. Different *ts* mutations within the same gene are frequently found to complement each other, whereas different *am* mutations within the same gene practically never complement one another. Why is this difference to be expected?

12.18 Suppressor-sensitive (*sus*) mutants of bacteriophage lambda can grow on *E. coli* strain C600, but cannot grow (i.e., are lethal) on *E. coli* strain W3350. They are thus conditional-lethal mutants like the *r*II mutants of bacteriophage T4. Seven *sus* mutants were analyzed for complementation by simultaneously infecting the restrictive host (*E. coli* strain W3350) with each possible pair of mutants. Single infections with each mutant and with wild-type lambda were also done as controls. The results of these complementation or *trans* tests and the controls are given as progeny per infected cell in the table below. Several infections with wild-type lambda yielded 120 to 150 progeny phage per infected cell. The results of the *cis*-heterozygote controls are not given, but assume that all of the *cis*-heterozygotes yielded >100 progeny phage per infected cell. Also, assume that ***no intra***genic complementation occurred between any of these *sus* mutants.

	LAMBDA PROGENY PER INFECTED CELL						
MUTANT:	1	2	3	4	5	6	7
7	0.01	133	0.01	146	134	128	0.01
6	131	142	161	0.06	0.1	0.1	
5	120	126	134	0.05	<0.01		
4	147	129	134	0.06			
3	<0.01	147	<0.01				
2	170	<0.01					
1	0.02						

(a) Based on the above data, how many genes are defined by the seven *sus* mutants? (b) Which *sus* mutations are located in the same gene(s)?

12.19 The first five nucleotide-pairs of a particular gene are present in the following sequence:

A T G C A
·· ·· ··· ··· ··
T A C G T
↑

How many distinct homoalleles of the above gene are there that differ from the above sequence by a single nucleotide substitution occurring at position no. 5 (nucleotide-pair number 5 in this gene—see arrow)?

12.20 ***Assume*** that the gene *mut A* and its polypeptide product, the enzyme trinucleotide mutagenase, have been intensively studied during December 1990 in the previously unknown organism *Escherichia mutaphilium*. This organism has been shown to use the established, nearly universal genetic code and to behave in all other respects relevant to molecular genetics like *Escherichia coli*.

The sixth amino acid from the amino terminus of the wild-type trinucleotide mutagenase is histidine, and the wild-type *mut A* gene has the triplet nucleotide-pair sequence

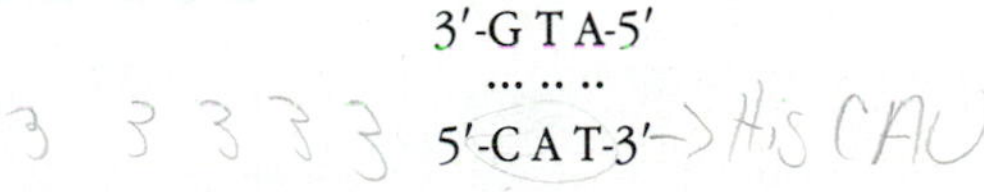

3′-G T A-5′
··· ·· ··
5′-C A T-3′

at the position corresponding to the sixth amino acid of the gene-product. Seven independently isolated mutants with single nucleotide-pair substitutions within the triplet nucleotide sequence of the *mut A* gene given have also been characterized. Moreover, the mutant mutagenases have all been purified and sequenced. All seven are different; they contain glutamine, tyrosine, asparagine, aspartic acid, arginine, proline, and leucine as the sixth amino acid from the amino terminus.

Mutants *mut A*1, *mut A*2, and *mut A*3 will not recombine with each other, but each will recombine with each of the other four mutants (*mut A*4, *mut A*5, *mut A*6, and *mut A*7) to yield true wild-type recombinants. Similarly, mutants *A*4, *A*5, and *A*6 will not recombine with each other, but will each yield true wild-type recombinants in crosses with each of the other four mutants. Finally, crosses between *mut A*1 and *mut A*7 yield about twice as many true wild-type recombinants as do crosses between *mut A*6 and *mut A*7.

Mutants *A*1 and *A*6 can be induced to revert to true wild type by mutagenesis with 5-bromouracil; mutants *A*2, *A*3, *A*4, *A*5, and *A*7 ***cannot*** be induced to revert with 5-bromouracil. Mutants *A*2 and *A*4 are capable of slow growth on minimal medium, whereas mutants *A*3 and *A*5 carry null mutations (producing completely inactive gene-products) and are incapable of growth on minimal medium. This has been used to select for mutation events from genotypes *mut A*3 and *mut A*5 to genotypes *mut A*2 and *mut A*4. Mutants *A*3 and *A*5 can be induced to mutate to *A*2 and *A*4, respectively, by treatment with 5-bromouracil or hydroxylamine. However, mutant *A*3 ***cannot*** be induced to mutate to *A*4, nor *A*5 to *A*2, by treatment with either mutagen.

Given that (a) 5-bromouracil induces both GC ⟶ AT and AT ⟶ GC transitions and (b) hydroxylamine induces only GC ⟶ AT transitions, and using the nature of the genetic code (Chapter 10, p. 280), deduce which mutant allele specifies the mutant polypeptide with each of the seven different amino acid substitutions at position 6 of trinucleotide mutagenase, and describe the rationale behind each of your deductions.

Chapter 13

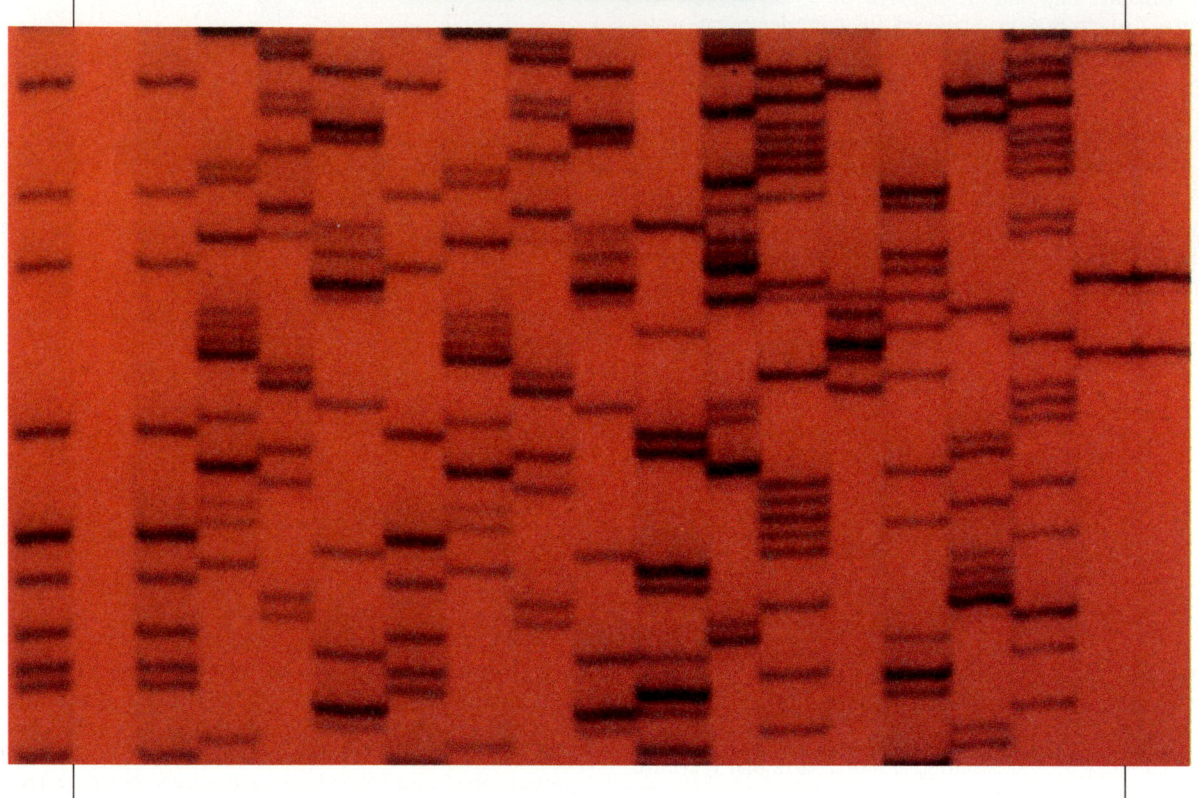

Genetic Fine Structure: Gene Cloning and Sequencing

THE results discussed in Chapter 12 demonstrated the great resolving power of classical genetic techniques in analyzing the fine structure of genes and chromosomes. Indeed, these classical genetic studies established the unit of structure of genetic elements to be the single nucleotide-pair. Recombination was shown to occur between adjacent nucleotide-pairs. However, this level of resolution has only been achieved for genes of bacteria and bacteriophages. Similar high levels of resolution are possible with lower eukaryotes such as the yeast *Saccharomyces cerevisiae.* However, in the case of higher plants and animals, it is very difficult to examine large enough progeny populations to detect rare recombination events within short segments of a gene. Moreover, in humans, where controlled crosses and large progeny populations are not feasible, classical genetic approaches have quite limited utility in analyzing genetic fine structure. Thus, almost all of our information about the structure of genes of humans has been obtained by ***"cloning"*** these genes (putting them into small plasmids or viral chromosomes) and analyzing their structures by direct ***DNA sequencing*** (determining their nucleotide-pair sequences). The results of these cloning and sequencing studies yielded a major surprise. ***Most of the genes of higher eukaryotes such as humans contain interrupted coding sequences with noncoding DNA sequences intervening between coding sequences.*** In this chapter, we will discuss (1) the techniques that are used to clone and sequence genes, (2) the surprising structure of most genes of higher eukaryotes, and (3) some of the methodologies that are used in molecular analyses of the structure and function of genes.

"RECOMBINANT DNA" AND "GENE CLONING" AS A TOOL FOR FINE STRUCTURE ANALYSIS

Nucleic acid sequencing is a very powerful technique for studying the structure and function of genes and chromosomes (e.g., see the section entitled Genes-within-Genes in Phage ΦX174 in Chapter 12). However, sequencing is presently practicable only with relatively short nucleic acid sequences, such as those of the chromosomes of small viruses. Sequencing the entire chromosome of even a bacterium such as *E. coli,* about 4 million nucleotide-pairs in length, let alone the giant DNA molecules in chromosomes of eukaryotes, would be an almost inconceivable task. One might suggest isolating a particular segment that contained only one to a few thousand nucleotide-pairs and sequencing just that segment. The haploid genome of mammals, for example, contains about 3×10^9 nucleotide-pairs. Any one gene or sequence of one to a few thousand nucleotide-pairs represents only about one unit out of a total of approximately a million such units in the genome. Thus, ***isolating any one gene is like searching for a needle in a haystack.*** Nucleic acid sequencing techniques require that the sequence be available in significant quantities in pure or essentially pure form. How, then, can one hope to extend the nucleic acid sequencing techniques to the more complex chromosomes of eukaryotes?

The *In Vitro* Synthesis of Recombinant DNA and Gene Cloning Techniques

"Recombinant DNA" and ***"gene cloning"*** technologies now provide molecular biologists with a method by which genes or other segments of large chromosomes can be isolated, replicated, and studied by nucleic acid sequencing techniques, electron microscopy, and other analytical techniques. Various procedures for gene cloning are now in use. In all cases, however, the overall procedure involves (1) ***the* in vitro *incorporation of the gene or segment of DNA of interest into a small self-replicating chromosome*** such as that of phage lambda or a plasmid (see Chapter 8, pp. 209–213 and 224–225) and (2) ***the introduction of the recombinant "minichromosome" into a host cell where it will replicate.*** Step one involves the *in vitro* synthesis of ***recombinant DNA,*** for example, an *E. coli* plasmid containing one gene of *Drosophila.* Step two is the ***gene cloning*** step, in which the recombinant DNA (or recombinant minichromosome) is replicated or "cloned" to produce many identical copies for subsequent biochemical analysis. Thus, although the entire procedure is often referred to as the "recombinant DNA" or "gene cloning" technique (particularly by the popular press), these terms actually refer to two separate steps in the overall process, both of which are essential to its utility.

Restriction Endonucleases

The original cloning experiments by A. Chang and S. Cohen in 1973, and many subsequent cloning experi-

(Left page): Photo of the audiograph of a 2′,3′-dideoxynucleotide chain-terminator sequencing gel that was used to separate the products of the four standard reactions (ddGTP, ddATP, ddTTP, and ddCTP) for each of five template DNAs (cloned segments of the *ndd* gene of phage T4). The outside lanes are repeated in a pattern that permits the autoradiograph to be oriented relative to the gel; this eliminates the need to label or otherwise keep track of the orientation of the X-ray film during processing. (Photograph courtesy of D. G. Oppenheimer and D. P. Snustad, Department of Genetics and Cell Biology, University of Minnesota.)

ments, have been done using a special class of endonucleases (nucleases that make internal cuts in nucleic acids) called ***restriction endonucleases.*** Many endonucleases make random cuts in DNA. The ***restriction endonucleases are site specific,*** however, and cleave DNA molecules only at ***specific nucleotide sequences called restriction sites.*** Different restriction endonucleases, present in different bacteria, recognize different nucleotide sequences (Table 13.1).

A function served by the restriction nucleases is to protect the genetic material of the bacterium from "invasion" by foreign DNA. All restriction sites in the host chromosome of a bacterium must be protected from its own restriction nucleases to prevent suicidal self-degradation. This is accomplished by methylation of one or more nucleotides in each nucleotide sequence that is recognized by the organism's own restriction nuclease(s). Methylation occurs rapidly after replication and is catalyzed by site-specific methylases encoded by the host chromosome. Each restriction nuclease will cleave a foreign DNA molecule (a DNA molecule from another species) into a fixed number of fragments, the number depending on the number of restriction sites in the particular DNA molecule.

An interesting feature of restriction endonucleases is that they commonly recognize DNA sequences that are ***palindromes,*** that is, nucleotide-pair sequences that are the same reading forward or backward from a central axis of symmetry, like the nonsense phrase

AND MADAM DNA

In addition, a very useful feature of many restriction nucleases is that they make "staggered" cuts, that is, they cleave the two strands of a double helix at different points. Most important, because of the palindromic nature of the restriction sites, the staggered cuts produce segments of DNA with ***complementary single-stranded ends.*** For example, cleaving a DNA molecule of the type

with the restriction endonuclease *Eco*RI (an *E. coli* enzyme) will yield

TABLE 13.1 Recognition Sequences and Cleavage Sites of Representative Restriction Endonucleases

ENZYME	SOURCE	RECOGNITION SEQUENCE[a] AND CLEAVAGE SITES[b]	NUMBER OF RECOGNITION SEQUENCES PER CHROMOSOME OF: PHAGE λ	SV40 VIRUS
*Eco*R1	*Escherichia coli*	↓ GAA TTC ● CTT AAG ↑	5	1
*Hind*II	*Hemophilus influenzae*	↓ GTPy PuAC[c] ● CAPu PyTG ↑	34	7
*Hind*III	*Hemophilus influenzae*	↓ AAG CTT ● TTC GAA ↑	6	6
*Hpa*I	*Hemophilus parainfluenzae*	↓ GTT AAC ● CAA TTG ↑	11	4
*Hpa*II	*Hemophilus parainfluenzae*	↓ CC GG ● GG CC ↑	750	1

[a]The axis of symmetry in each palindromic recognition sequence is indicated by the colored dot.
[b]The position of each bond cleaved is indicated by a colored arrow. Note that the cuts are staggered (at different positions in the two complementary strands) with some restriction nucleases.
[c]Pu indicates that either purine (adenine or guanine) may be present at this position; Py indicates that either pyrimidine (thymine or cytosine) may be present.

Because all the fragments produced will have complementary single-stranded 5′ ends, they can be rejoined under appropriate renaturation conditions using the enzyme DNA ligase (see Chapter 5, pp. 110–111).

Cloning DNAs with Complementary Single-Stranded Ends

The single-stranded ends of "*Eco*RI fragments" (as fragments produced by digestion of DNA with *Eco*RI restriction enzyme are called) will be complementary regardless of the source of DNA. Thus, *Eco*RI fragments cut from mouse chromosomes, chicken chromosomes, and so on can be incorporated into phage lambda chromosomes or plasmids that have been similarly cleaved with *Eco*RI restriction enzyme (Fig. 13.1). Once a gene or restriction enzyme fragment is inserted into a plasmid, the recombinant plasmid is introduced into a host cell, such as *E. coli,* by transformation (uptake of naked DNA by cells; see Chapter 8, pp. 206–208). When *E. coli* is used, the cells must first be made permeable to DNA by treatment with a calcium salt. An important advantage of this procedure over some other cloning procedures (see Fig. 13.2) is that after cloning, the cloned sequence can be precisely cut out of the recombinant chromosome by using the same restriction enzyme that was used in the original cloning process (the restriction sites will remain at the same positions).

Cloning DNAs with Blunt Ends by Adding Complementary Tails

A second procedure for the *in vitro* construction of recombinant DNA molecules involves the synthesis of complementary single-stranded tails on the 3′ ends of strands of DNA by using the enzyme ***terminal transferase*** (Fig. 13.2). This procedure has an advantage over the restriction enzyme procedure (see Fig. 13.1) in that it can be used to join DNA molecules that terminate with blunt ends produced by cleavage with restriction enzymes that cut both strands of DNA at the same point. Once recombinant DNA molecules are produced using the terminal transferase procedure, they are cloned in the same way as those produced using complementary ends generated by restriction enzymes that make staggered cuts.

The two procedures described above are two of the original techniques used to construct recombinant DNA molecules. Several modifications of these procedures as well as several new techniques for synthesizing recombinant DNA molecules are currently in use. A few of these techniques are described in a subsequent section of this chapter entitled Advanced Cloning Vectors.

Cloning cDNAs Produced from Purified mRNAs

Cloning experiments can be divided into two types: (1) those where the cloned DNA segment is of known composition and function and (2) those where the cloned DNA segment is of uknown composition and function, such as uncharacterized restriction fragments.

When pure mRNA or pre-mRNA transcripts of a gene can be obtained (such as for hemoglobin genes using reticulocytes and the ovalbumin gene using chicken oviduct cells), they can be used to carry out cloning experiments of the first type. The purified mRNA is used as a template for the *in vitro* synthesis of a complementary strand of DNA (called ***cDNA***) by the enzyme ***reverse transcriptase*** (because it catalyzes the reaction that is the reverse of transcription, that is, RNA ⟶ DNA rather than DNA ⟶ RNA as in transcription). The complementary DNA strand in turn serves as a template for DNA polymerase-catalyzed synthesis of a second strand of DNA (complementary to the first). Thus, ***reverse transcriptase mediates the conversion of the genetic information present in a single-stranded molecule of RNA into a double-stranded molecule of DNA*** (Fig. 13.3). It is therefore more accurately called an ***RNA-dependent DNA polymerase.*** The double-stranded DNA molecule synthesized with reverse transcriptase and DNA polymerase can then be inserted into a phage chromosome or plasmid by the terminal transferase procedure (Fig. 13.2).

Reverse transcriptases are encoded by the genomes of RNA tumor viruses. Their normal function is to convert the genetic information of the virus from the RNA form in which it is stored in the mature virion into a double-stranded DNA form after infection of a host cell. The DNA form is called a ***provirus,*** and it is this DNA provirus form that integrates by covalent linear insertion into the chromosome of the host (see Chapter 17, pp. 468–469). In its integrated provirus form, the viral chromosome replicates along with the host chromosome and segregates in a Mendelian fashion during sexual reproduction (just like any other gene or set of genes of the host). The *in vitro* synthesis of cDNA from mRNA and other RNA molecules using reverse transcriptases has become a very important tool in molecular genetics.

Screening a Genomic DNA Library by *In Situ* Colony Hybridization

The second approach to cloning genes has been to prepare clonable-sized fragments of DNA by digestion of the entire genome of an organism with one or more restriction enzymes or by shearing forces, and to use these to clone random fragments of the genome. Such

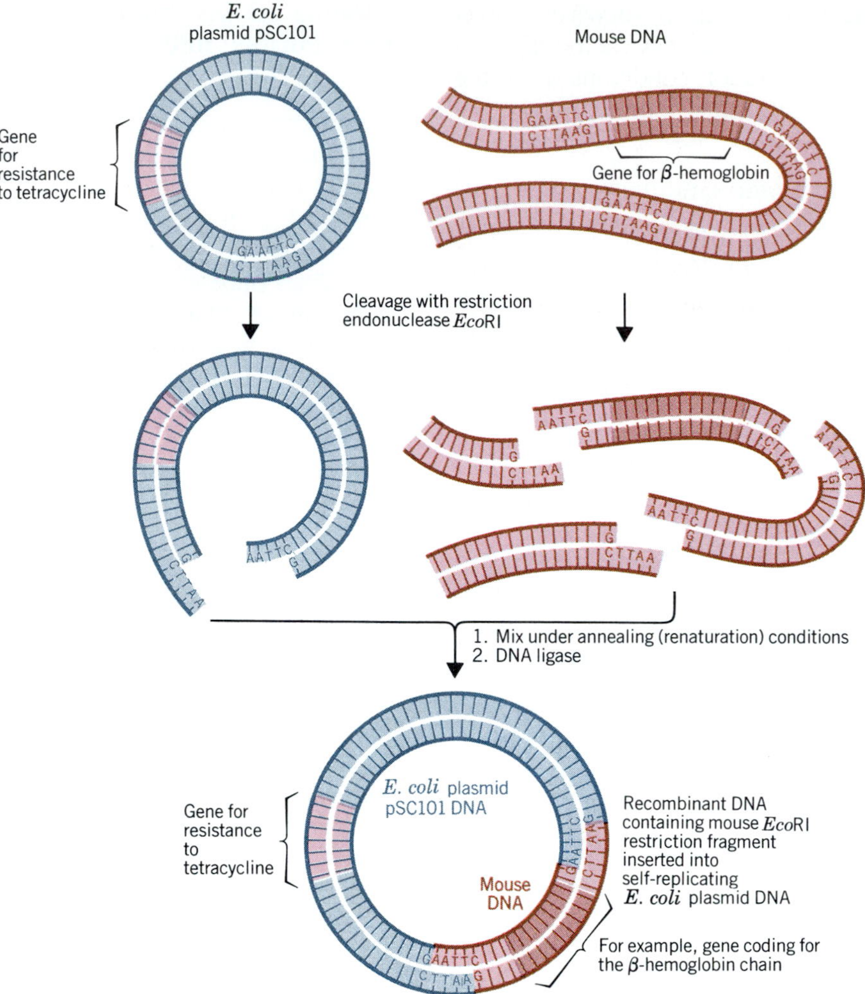

Figure 13.1 Schematic illustration of the *Eco*RI restriction nuclease technique for the *in vitro* construction of recombinant DNA molecules for subsequent cloning. In the example illustrated, the autonomously replicating *E. coli* plasmid pSC101 is used as the cloning vector. It has two important features: (1) it carries a gene that provides tetracycline resistance to the host cell and (2) it has a single *Eco*RI restriction site (see Table 13.1) facilitating the insertion of a fragment of foreign DNA after cleavage by *Eco*RI endonuclease. *Eco*RI restriction fragments from any DNA (mouse DNA is used for illustrative purposes) may be inserted into the cleaved pSC101 plasmid by mixing the two DNA preparations under annealing conditions and sealing with the enzyme DNA ligase. The recombinant DNA molecules are then used to transform *E. coli* cells. If tetracycline-sensitive *E. coli* cells are used and the cells are subsequently grown in medium containing tetracycline, only cells that were "transformed" and thus contain a pSC101 plasmid will be able to grow. Once an *E. coli* cell that contains a recombinant plasmid carrying the desired gene or nucleotide sequence has been identified (the most difficult step in the gene cloning procedure), that cell can be used to grow a population of identical cells (or "clone"), all containing the desired gene or sequence.

experiments are commonly referred to as ***"shotgun" cloning experiments***. The difficulty with the shotgun approach is in identifying a cell or clone containing the gene or sequence of interest (which can be expected to occur at maximum frequencies of about 10^{-6} in the case of higher eukaryotes, see pp. 349–350). Where selection can be applied, this is no problem. For example, cloning the *E. coli* gene that confers resistance to penicillin could easily be accomplished by "shotgun" cloning the entire genome of a pen^r strain using a tetracycline-resistant plasmid (like pSC101; see Fig. 13.1) and selecting for penicillin-resistant, tetracycline-resistant transformants (where untransformed cells were pen^s). In other cases, complementation can be used as a basis of selection. *His* genes of yeast, for example, were cloned in *E. coli* histidine

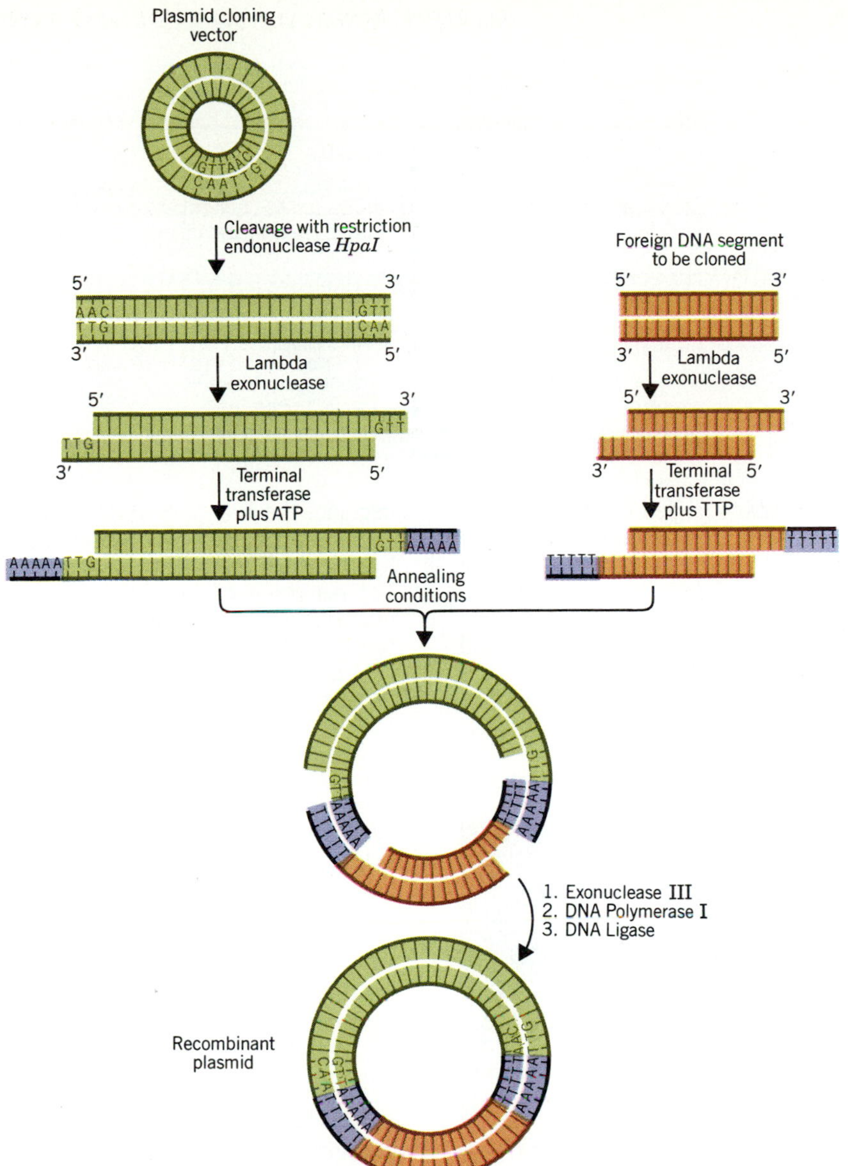

Figure 13.2 Diagram illustrating the terminal transferase procedure for the construction of recombinant DNA molecules. A segment of foreign DNA (brown) is cloned in a plasmid (olive) containing a single endonuclease *Hpa*I restriction site (see Table 13.1). After cleavage of the plasmid with *Hpa*I, the linear plasmid DNA molecules and the segments of foreign DNA are treated briefly with phage lambda exonuclease to prepare appropriate substrates for the terminal transferase reaction. (Lambda exonuclease cleaves off nucleotides from the 5′ ends of each chain, leaving double helices with single-stranded 3′ tails.) Poly-A tails are then added to the 3′ ends of each plasmid DNA molecule, using terminal transferase in the presence of ATP (with no other nucleoside triphosphate present). Poly-T tails are similarly added to the 3′ ends of the foreign DNA molecules, using TTP as substrate. The two DNA preparations are then mixed under conditions facilitating the annealing of the complementary poly-A and poly-T tails. Phosphates are removed from the 3′ ends with the enzyme exonuclease III, and the gaps are filled by using the four nucleoside triphosphates and DNA polymerase I. Covalent closure is catalyzed by DNA ligase. The recombinant, self-replicating plasmids are then cloned following introduction into *E. coli* cells by transformation.

auxotrophs (containing nonreverting deletion mutations), and the desired clones were identified by their ability to grow on medium containing no histidine. Indeed, maize glutamine synthetase cDNA clones have recently been identified on the basis of their ability to rescue *E. coli* glutamine synthetase-deficient mutants growing on medium lacking glutamine (Fig. 13.4). Of course, this complementation selection procedure depends on the correct transcription and translation of the cloned gene or genes in the new host, which may not always be the case. The signals that regulate gene expression are different in *E. coli* and eukaryotes.

Another approach to identifying the desired clones after "shotgun" cloning, which does not depend on the correct expression of the cloned genes, involves ***replica plating the colonies formed by transformed***

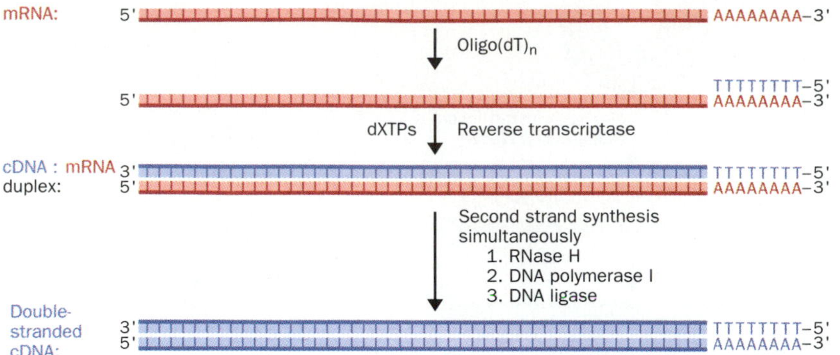

Figure 13.3 Synthesis of double-stranded cDNA from mRNA. First-strand cDNA synthesis is primed by oligo(dT)$_n$, which forms a duplex with the poly-(A) tail at the 3′ terminus of the mRNA molecule. The synthesis of the DNA strand complementary to the mRNA is catalyzed by the enzyme reverse transcriptase (see text). The replacement of the mRNA strand in the mRNA:cDNA duplex with a second strand of DNA is accomplished by the concerted actions of RNase H (which degrades the RNA strand), DNA polymerase I (which catalyzes the synthesis of the second strand of DNA), and DNA ligase (which seals the breaks in the nascent DNA strand). The net result of these reactions is that the sequence of nucleotides in the mRNA molecule is converted to a homologous sequence of nucleotide-pairs in a double-stranded DNA molecule.

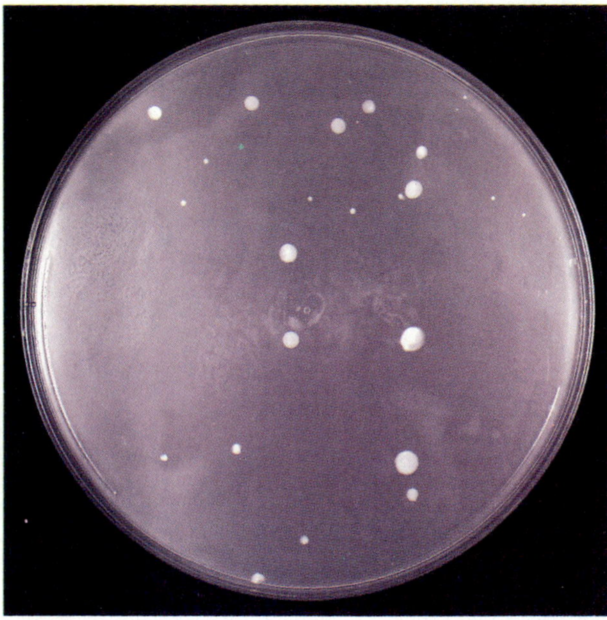

Figure 13.4 Illustration of the use of genetic selection to isolate cDNA clones. The enzyme glutamine synthetase catalyzes the last step in the biosynthesis of the amino acid glutamine. The photograph shows colonies of glutamine synthetase-deficient mutants of *E. coli* that are able to grow on medium lacking glutamine as a result of functional rescue by plasmids expressing maize glutamine synthetase cDNAs. Because of the great resolving power of genetic selection, large numbers of independent cDNA clones can be isolated on a single plate. (From D. P. Snustad, J. P. Hunsperger, B. M. Chereskin, and J. Messing, *Genetics* 120: 1111–1124, 1988. Reproduced with permission. Copyright © 1988 by the Genetics Society of America.)

cells onto nitrocellulose filters **in situ** ***hybridization with a radioactive probe*** (labeled DNA or RNA molecule), and ***autoradiography*** (Fig. 13.5). If purified mRNA containing the gene or sequence of interest can be obtained, it can be used as a template for the *in vitro* synthesis of radioactive cDNA (using labeled nucleoside triphosphate precursors) by reverse transcriptase. The labeled cDNA is then used as a probe for *in situ* hybridization to denatured DNA of colonies that are lysed right on nitrocellulose filters (Fig. 13.5). After allowing a sufficient period of time for hybridization to occur, the filters are washed to remove nonhybridized cDNA (the DNA from the lysed cells is tightly bound to the nitrocellulose filter before hybridization so that it will not wash off) and then autoradiographed (see Chapter 5, pp. 107–109). Only colonies containing DNA sequences complementary to the radioactive cDNA will give rise to radioactive spots (Fig. 13.5). These are used in turn to identify colonies that contain the desired sequence on the original replicated plates.

Applications and Potential Hazards of Recombinant DNA and Gene Cloning Technology

As the technology for constructing recombinant DNA molecules and cloning was developing, many molecular biologists recognized that this technology would provide a powerful tool with which to study the structure and function of genetic material. They also wondered whether it might introduce potential hazards. For example, if the gene or genes responsible for transformation to malignancy carried by a tumor virus

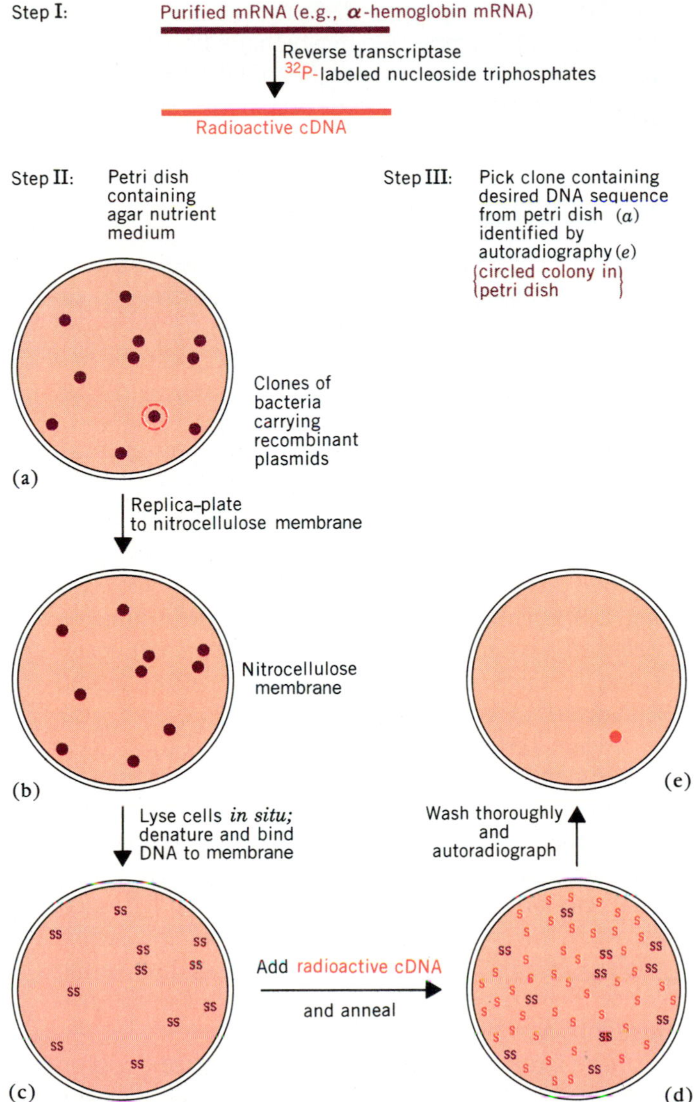

Figure 13.5 Diagrammatic illustration of the autoradiographic technique for identifying an *E. coli* clone carrying a recombinant plasmid with a DNA sequence homologous to a purified mRNA. Step I: A labeled DNA sequence complementary to the mRNA is synthesized using reverse transcriptase. Step II: (*a*) Cells transformed with recombinant DNA molecules (such as recombinant plasmids) are plated on agar medium and incubated until visible colonies form. (*b*) The colonies are replica-plated onto nitrocellulose membranes. (*c*) The cells on the nitrocellulose replica are lysed *in situ*, and their DNA is denatured and bound to the membrane (e.g., 0.1 *N* NaOH for a few seconds, followed by heat and vacuum). (*d*) Then the membrane is placed in a solution containing the radioactive cDNA under annealing conditions for several hours. (*e*) The membrane is washed thoroughly to remove nonannealed cDNA and is autoradiographed. The autoradiograph will show the exposure to radioactive decays only at the locations of colonies formed by cells that contain DNA sequences complementary to the cDNA probe (only one such colony is shown). Step III: The colony at the corresponding position in the petri dish is picked for subsequent cloning and analysis.

were cloned in an *E. coli* plasmid, might the *E. coli* strain carrying the viral genes serve as a vector for the transmission of cancer-causing genes? Thus, several of the leading molecular biologists called for a voluntary moratorium on all research of this type. Extensive conferences and discussions on the pros and cons of doing research involving recombinant DNA and cloning were held. A complete discussion of the pros and cons of recombinant DNA and cloning research is beyond the scope of this book. The reader is referred to the published proceedings of a forum entitled *Research with Recombinant DNA,* which was organized by the National Academy of Sciences (U.S.) for a discussion of the arguments presented by proponents and opponents of recombinant DNA research.

Clearly, a major concern of opponents of recombinant DNA work is that a pathogenic or otherwise harmful new organism might be produced and introduced into the ecosystem. Proponents argue against the likelihood of this occurring, and emphasize the potential benefits of recombinant DNA research. Proponents emphasize the information that can be obtained about the structure, mode of expression, and regulation of expression of the genetic material of eukaryotes. The importance of potential practical applications, particularly in medicine and agriculture, is also emphasized.

After extensive hearings and discussions, the National Institutes of Health (NIH) of the U.S. Department of Health, Education, and Welfare (now Health and Human Services) established specific guidelines under which recombinant DNA research of various types could be continued with ***prior*** approval from NIH. These guidelines emphasized both ***physical*** and ***biological containment*** of the recombinant molecules constructed. Physical containment includes the use of sterile techniques, containment hoods, and specially designed laboratories to prevent vectors containing recombinant DNA molecules from being transferred or "escaping" from laboratories to natural ecosystems. Biological containment involves the use of organisms with specially constructed, "weakened" genotypes as vectors in cloning experiments. Ideally, these organisms should be unable to survive under conditions existing in any natural ecosystem.

As more work with recombinant DNA was done, it became evident that bacteria and viruses that carry foreign genes are simply not very "healthy." They have been found not to survive in competition with wild-type organisms under natural conditions. Thus, the NIH guidelines have been gradually relaxed for the more routine types of gene cloning experiments. Other kinds of experiments still require approval by an NIH panel and by a local institutional biosafety committee ***prior*** to their initiation.

The potential of recombinant DNA and gene cloning methodology has been rapidly realized. New knowledge about the structures of eukaryotic genes and chromosomes has accumulated at an unprecedented rate since the development of these techniques. [For example, see (1) the following section of this chapter entitled Noncoding Intervening Sequences or Introns within Eukaryotic Genes, (2) Chapter 16, Genetic Control of the Immune Response, and (3) Chapter 17, Genetic Control of Cell Division: Oncogenes and Protooncogenes.] As for practical applications, human insulin and human growth hormone produced in bacteria engineered by recombinant DNA and gene cloning techniques are both now commercially available and other products are on the horizon (see Chapter 24). For diabetics who are allergic to animal insulins or for individuals with defects in the synthesis of human growth hormone, these early achievements of scientists using recombinant DNA technologies are of major significance.

NONCODING INTERVENING SEQUENCES OR INTRONS WITH EUKARYOTIC GENES

The application of the recombinant DNA and cloning technology to the organization of the genetic material in eukaryotes has already yielded important and surprising results. In prokaryotes, we know that all the well-characterized genes consist of continuous sequences of nucleotide-pairs, which specify colinear sequences of amino acids in the polypeptide gene-products (see Chapters 10 and 12). However, analyses of a large number of cloned eukaryotic genes—including genes of animals, plants, and eukaryotic microorganisms—have shown that the ***coding sequences*** (called ***"exons,"*** for ***ex***pressed sequences) ***of these genes*** (sequences specifying the functional gene-products, RNA or protein) ***are often, but not always, interrupted by noncoding "intervening sequences" or "introns."*** Several kinds of evidence, the most important being via electron microscopy, the location of restriction enzyme cleavage sites, and correlated nucleotide and amino acid sequences, document the existence of these intervening sequences within the structural genes.

R-Loop Mapping

When purified RNA transcripts are annealed with double-stranded DNA containing the gene from which the RNA was transcribed using conditions that favor DNA–RNA duplex formation (e.g., high concentrations of formamide), the RNA strands will displace the homologous DNA strands. As a result, single-stranded loops of DNA, called ***R-loops,*** will form spanning that segment of the DNA molecule in which the RNA is hybridized to the complementary strand of DNA (Fig. 13.6*a*). R-loops can be characterized directly by electron microscopy.

Mammalian Hemoglobin Genes

When P. Leder and coworkers hybridized purified mouse β-globin mRNA to a recombinant DNA molecule that contained the mouse β-globin gene, cloned in a phage lambda chromosome, they observed two R-loops separated by a loop of double-stranded DNA (Fig. 13.6*b*). Their result demonstrated the presence of a sequence of nucleotide-pairs in the middle of the β-globin gene that is not present in β-globin mRNA and, therefore, does not code for amino acids in the β-globin polypeptide (Fig. 13.7). Leder and colleagues verified this interpretation of their results by sequencing a segment of the gene spanning one junction between a coding sequence and an intron and correlating the nucleotide sequence with the known amino acid sequence of mouse β-hemoglobin.

When Leder and coworkers repeated the R-loop experiments using purified β-globin gene transcripts ***isolated from nuclei*** and, thus, believed to be ***primary gene transcripts or pre-mRNA molecules*** (see Chapter 10), in place of cytoplasmic β-globin mRNA, only one R-loop was observed (Fig. 13.6*a*). This indicates that the intron sequence is cut out and the coding sequences are spliced together during processing events that convert the primary transcript to the mature mRNA.

The large intron sequence in the mouse β-globin gene is 653 nucleotide-pairs long. Shortly after the discovery of the large intron, Leder and colleagues found that the mouse β-globin gene contains a second, smaller intron that is 116 nucleotide-pairs in length (Fig. 13.7).

Studies by T. Maniatis, R. Flavell, and many others have shown that the structure of the human β-globin gene is very similar to that of the mouse β-globin gene, with two introns separating three exons. Each of the two human α-globin genes also contains two introns. In fact, the human genes coding for the embryonic α-like (ζ) and β-like (ε) hemoglobin chains, the fetal β-like (γ) hemoglobin chain, and the adult minor β-like (δ) hemoglobin chain, all contain two introns and three exons. Moreover, all the human genes coding for α-like globins have the introns at the same positions (separating codon 31 from 32 and codon 99 from 100), as do all the human genes coding for β-like globins (separating codon 30 from 31 and codon 104 from 105).

The human genes coding for the α-like globins and the β-like globins are located in clusters on chro-

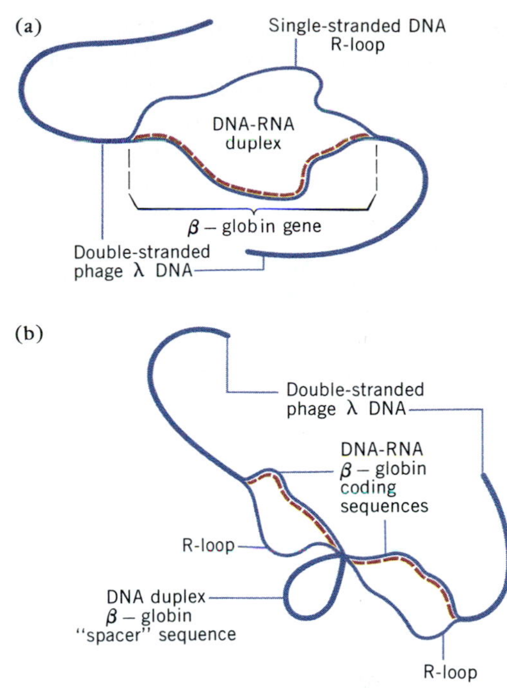

Figure 13.6 Diagram illustrating the R-loop DNA–RNA hybridization technique used by P. Leder and colleagues to demonstrate the 653-nucleotide-long "intron" (noncoding sequence) in the mouse β-globin gene. (a) When the hnRNA (heterogeneous nuclear RNA) precursor to β-globin mRNA was hybridized to a phage lambda chromosome carrying a segment of mouse DNA encompassing the β-globin gene under conditions favoring DNA–RNA hybridization, a single R-loop was observed. (The RNA strand in DNA–RNA hybrids is shown by the dashed red lines.) This R-loop corresponds to the nontranscribed strand of the segment of double-stranded DNA from which the mouse β-globin hnRNA was transcribed. (b) When β-globin mRNA was used in the experiment, two R-loops were observed that were separated by a double-stranded DNA loop. This result indicates that the mouse β-globin gene contains a noncoding sequence or intron for which no complementary nucleotide sequence exists in the β-globin mRNA. The intron nucleotide sequence is excised during the conversion of hnRNA to mRNA. This excision must occur by a very precise mechanism to produce the correct codon reading frame in the distal portion of the β-globin mRNA.

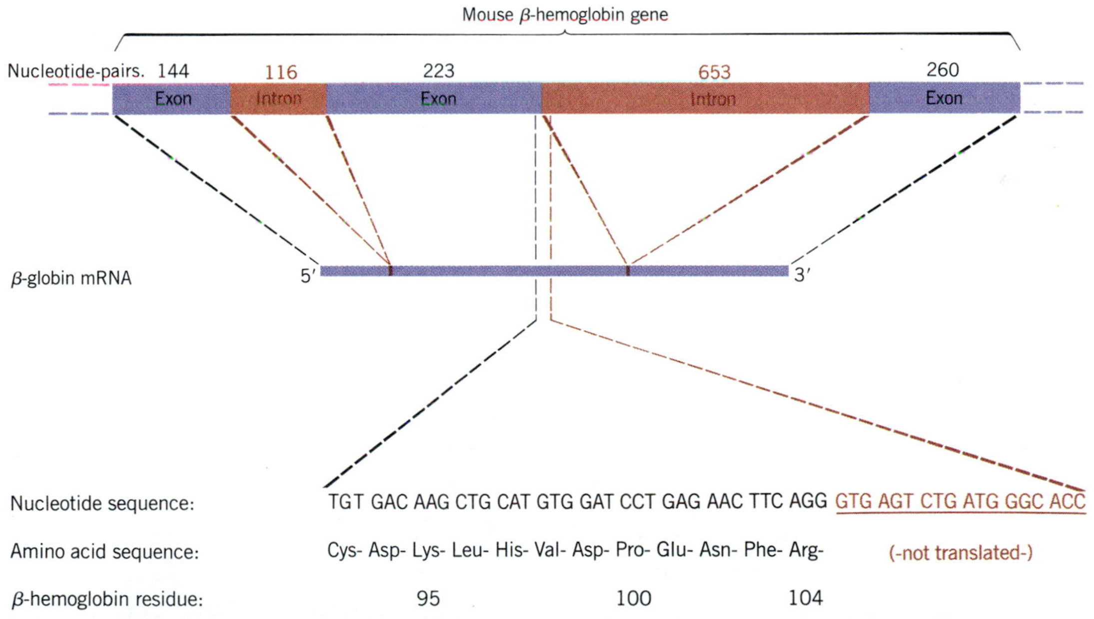

Figure 13.7 Organization of the mouse β-globin gene, showing the locations of noncoding intron sequences (red) and coding ("exon") sequences (blue). The correspondence between segments of the gene and segments of the mRNA is indicated by dashed diagonal lines. The nucleotide sequence of the "antisense" (nontranscribed) strand of DNA for the indicated juncture (dashed lines) between the second coding sequence and the large intron sequence is shown at the bottom, along with the amino acid sequence of the segment of the β-globin chain that it encodes. Amino acid number 104 (numbering from the amino terminus) is arginine (Arg). Amino acids 105 through 110 are Leu-Leu-Gly-Asn-Met-Ile. If the contiguous (intron) nucleotide sequence (underlined in red on the right) was transcribed and translated, amino acids 105 through 110 should be Val-Ser-Leu-Met-Gly-Thr. (Based on the data of S. M. Tilghman, D. C. Tiemeier, J. G. Seidman, B. M. Peterlin, M. Sullivan, J. V. Maizel, and P. Leder, *Proc. Natl. Acad. Sci. U.S.A.* 75: 725, 1978, and the data summarized in T. Pederson, *Amer. Sci.* 69: 76, 1981.)

mosomes 16 and 11, respectively. The globin genes in each cluster are lined up on their respective chromosomes in the order in which they are expressed during development. We do not yet know how this relates to the regulation of their expression, if at all, but this represents an interesting correlation.

Eukaryotic Genes with Many Introns and with No Introns

While Leder and coworkers were demonstrating the presence of two introns in the mouse β-globin gene, work in two other laboratories, B. O'Malley's and P. Chambon's, showed that the chicken ovalbumin gene contains seven intervening sequences or introns separating eight exons (Fig. 13.8).

Subsequent studies have demonstrated the presence of noncoding introns in numerous other eukaryotic genes. The rat serum albumin gene contains 13 introns, and the *Xenopus laevis* vitellogenin (which ends up as egg yolk protein) A2 gene contains 33 introns. The record for introns per gene is presently held by the chicken α2 collagen gene, which has over 50 introns. This gene spans 38,000 nucleotide-pairs, but gives rise to an mRNA molecule only about 5000 nucleotides long. The first evidence for introns in plant genes was the demonstration that the phaseolin (major storage protein) gene of the French bean contains three introns. Subsequent studies have shown that most of the genes of higher plants contain multiple introns and exons.

The only structural features that seem to be shared by different introns are the dinucleotide sequences at their ends. The primary transcripts of genes almost always begin introns with the sequence GU (5′) and end them with the sequence AG (3′); these consensus sequences at intron–exon junctions must be important in the mechanism by which introns are spliced out of primary transcripts (Chapter 10, pp. 269–275).

Finally, introns are not essential components of

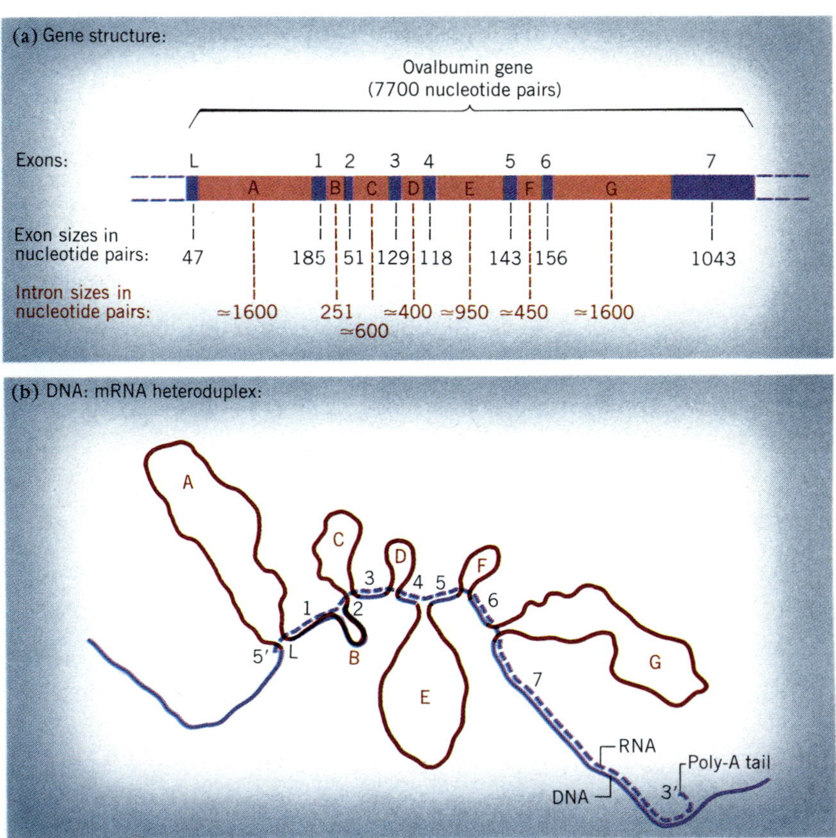

Figure 13.8 Structure of the chicken ovalbumin gene. (a) Diagram showing the seven introns (A–G, shown in red) and eight exons (L and 1–7) of the ovalbumin gene. Exon L specifies the 5′ leader sequence of the ovalbumin mRNA; exons 1–7 encode amino acid sequences of ovalbumin. (b) Tracing of a DNA: mRNA heteroduplex containing the ovalbumin mRNA and a single strand of DNA carrying the transcribed strand of the chicken ovalbumin gene, as observed by electron microscopy. The mRNA and DNA strands are shown as dashed and solid lines, respectively. The unpaired intron segments of the DNA strand are shown in red; they correspond to the segments of the gene shown in red in (a). (After P. Chambon, *Sci. Amer.* 244(5): 60–71, 1981.)

eukaryotic genes since not all eukaryotic genes contain introns. For example, the sea urchin histone genes and four *Drosophila* heat shock genes were among the first to be shown ***not*** to contain introns. Subsequently, numerous genes of both higher animals and higher plants have been found not to contain introns, and there is no apparent pattern regarding which genes contain introns and which do not contain introns.

Biological Significance of Introns?

We do not yet understand much about the biological significance of the exon–intron structure of eukaryotic genes. Introns are highly variable in size, ranging from a few nucleotide-pairs to thousands of nucleotide-pairs in length. In several genes the introns have been shown to accumulate new mutations much more rapidly than the exons. This suggests that their specific nucleotide-pair sequences, excluding the ends, are not very important.

In some cases, the different exons of genes code for different functional domains of the protein gene-products. This is most apparent in the case of the genes coding for antibodies (Chapter 16). In the case of the globin genes, the middle exon codes for the domain of the protein containing the heme-binding site. This has led to speculation that the exon–intron structure of eukaryotic genes results from the evolution of new genes by bringing together exons that were ancestral genes. If this is correct, introns may be merely relics of the evolutionary process.

Alternatively, introns may provide a selective advantage by increasing the rate at which coding sequences in different exons of a gene can reassort by recombination, thus speeding up the process of evolution. In the case of the mitochondrial gene of yeast coding for cytochrome *b*, the introns are themselves part of the exons for genes coding for enzymes ("maturases") involved in processing of the primary transcript of the gene. (For further details, see Lewin, *Genes IV,* Chap. 25). Thus, different introns may indeed play different roles, and some introns may have no biologically important function. Finally, we should emphasize that since many eukaryotic genes contain no introns, these noncoding regions cannot be essential for any function required for the expression of genes.

MAPPING RESTRICTION ENZYME CLEAVAGE SITES OF CHROMOSOMES

Most restriction endonucleases cleave DNA molecules in a site-specific manner (see Table 13.1). As a result, they can be used to generate ***physical maps of chromosomes*** that are of tremendous value in assisting researchers in isolating small segments of chromosomes carrying genes or other DNA sequences of interest. These maps are invaluable in DNA sequencing experiments because the sequencing techniques can be used only with small DNA molecules, usually sequences of a few hundred to a few thousand nucleotide-pairs in length.

The sizes of the restriction fragments can be determined by ***polyacrylamide or agarose gel electrophoresis*** (see Figs. 13.9, 13.11, and 13.23). Because of the nucleotide subunit structure of DNA, with one phosphate group per nucleotide, DNA has an essentially constant charge per unit of mass. Thus, the rates of migration of DNA fragments during electrophoresis provide accurate estimates of their lengths, with the rate of migration being inversely proportional to length. The polyacrylamide or agarose gel simply acts as a molecular sieve, with small fragments able to move through the sieve faster than large molecules. Agarose works somewhat better than polyacrylamide for large fragments as it has a slightly larger pore size; polyacrylamide gels yield better resolution of short fragments.

A simple diagram illustrating the procedure that is used to map the restriction enzyme cleavage sites on a DNA molecule is shown in Fig. 13.9. A given DNA molecule (Fig. 13.9*a*) is shown by agarose gel electrophoresis (see Fig. 6.11) to be approximately 6000 nucleotide-pairs in length (Fig. 13.9*e*, lane *a*). Sizes of the DNA molecule and restriction fragments are estimated by using a set of DNA markers of known size; in this example, a size-marker ladder (Fig. 13.9*e*, left lane) that consists of a set of DNA molecules that differ in length by 1000 nucleotide-pairs is used (such marker DNA ladders are commercially available). When this DNA molecule is cut with *Eco*RI, two fragments 4000 and 2000 nucleotide-pairs long are produced (Fig. 13.9*e*, lane *b*). The possible positions of the single *Eco*RI cleavage site in the molecule are shown in Fig. 13.9*b*. When the same DNA molecule is cleaved with *Hin*dIII, two fragments 5000 and 1000 nucleotide-pairs long result (Fig. 13.9*e*, lane *c*). The possible positions of the single *Hin*dIII cleavage site are shown in Fig. 13.9*c*. Note that at this stage of our analysis we cannot make any deductions about the relative positions of the *Eco*RI and *Hin*dIII cleavage sites. The *Hin*dIII cleavage site may be located on either of the two *Eco*RI restriction fragments. The molecule is then simultaneously digested with both *Eco*RI and *Hin*dIII, and three fragments of sizes 3000, 2000, and 1000 nucleotide-pairs are produced (Fig. 13.9*e*, lane *d*). This result establishes the positions of the two cleavage sites relative to one another on the molecule. Since the 2000-nucleotide-pair *Eco*RI restriction fragment is still present (not cut by *Hin*dIII), the *Hin*dIII cleavage site

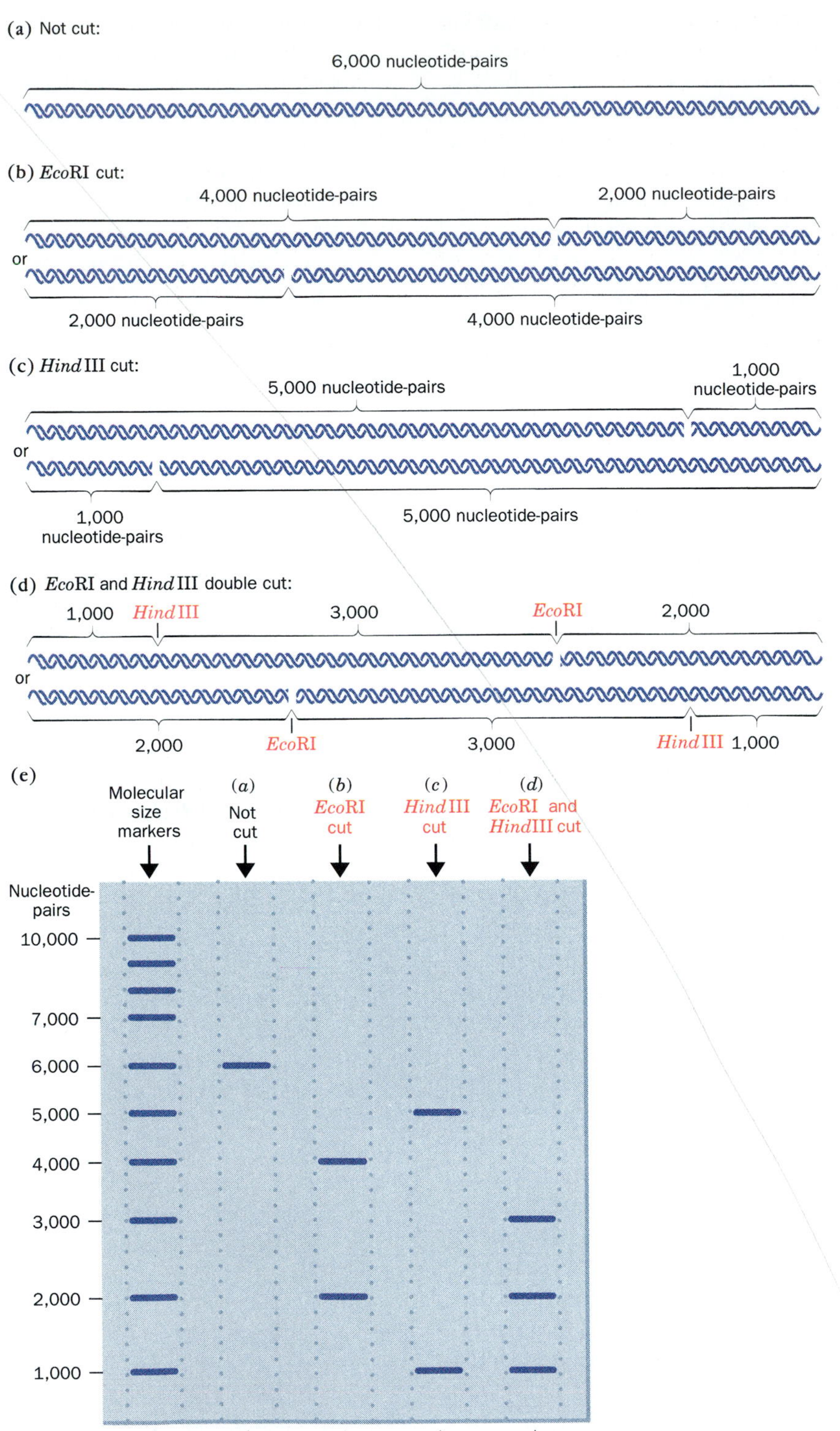

Figure 13.9 Schematic diagram illustrating the procedure used to map restriction enzyme cleavage sites in DNA molecules. (a–d) Structures of the DNA molecule or of restriction fragments of the molecule either (a) uncut or cut with (b) *Eco*RI, (c) *Hind*III, or (d) *Eco*RI and *Hind*III. (e) Diagram of the separation of these DNA molecules and fragments by agarose gel electrophoresis (see Figure 13.23 for a description of this technique). The left lane on the gel contains a set of molecular size markers—in this case, a set of DNA molecules 1000 nucleotide-pairs long and multiples thereof. See the text for a discussion of the results and the deductions that can be made from them. See Fig. 13.10 for another illustration of the procedures used to map the positions of restriction enzyme cleavage sites on chromosomes.

must be at the opposite end of the molecule from the *Eco*RI cleavage site (Fig. 13.9*d*). By extending this type of analysis to include the use of several different restriction enzymes, more extensive restriction maps can be constructed (see Figs. 13.10 and 13.11 for a slightly more complex example). When large numbers of restriction enzymes are employed, detailed maps of entire chromosomes can be constructed (see Fig. 13.13).

During their early work on the characterization of restriction endonucleases, H. Smith and D. Nathans used the DNA chromosome of ***simian virus 40 (SV40)*** for their studies. (They shared the 1978 Nobel Prize in Physiology and Medicine with W. Arber for the discovery of restriction enzymes.) SV40 is an animal virus that can transform cells to the cancerous state. Its circular chromosome (now completely sequenced) is only 5226 nucleotide-pairs in length, making it well suited for restriction enzyme studies. As a result of Smith and Nathans' work, ***restriction enzyme cleavage site maps***, now commonly called simply ***restriction maps***, of the SV40 chromosome were among the first such maps constructed.

Restriction enzyme *Eco*RI cleaves the SV40 chromosome at only one site (Table 13.1 and Fig. 13.10*a*). This site has been arbitrarily set as position 0 on the SV40 chromosome and restriction map. Restriction enzymes *Hpa*I and *Hin*dIII cleave the SV40 chromosome at four and six sites, respectively (Figs. 13.10*b* and *c*). When the SV40 chromosome is digested with all three of these enzymes, 11 distinct restriction fragments are produced (Fig. 13.10*d*).

There are three major approaches to ordering restriction fragments—all of which depend on the use of polyacrylamide or agarose gel electrophoresis to obtain accurate estimates of fragment size. These are (1) ***sequential digestion of chromosomes*** with two or more different restriction enzymes, (2) ***partial digestion of chromosomes*** after labeling the ends with a radioactive isotope, and (3) determining whether the strands of ***different restriction fragments can hybridize with each other*** after denaturation.

If the SV40 chromosome is cleaved with *Hin*dIII (Fig. 13.10*c*), a large fragment *E* is produced (Fig. 13.11*c*). This same fragment is still present after subsequent digestion with *Hpa*I (Fig. 13.11*d*), indicating that there are no *Hpa*I cleavage sites within the *Hin*dIII *E* fragment. By contrast, when *Hpa*I restriction fragment *B* + *C* (Figs. 13.10*b* and 13.11*b*) is subsequently digested with *Hin*dIII, it is cleaved into two fragments *B* and *C* (Figs. 13.10*d* and 13.11*d*). Similarly, subsequent digestion of *Hin*dIII fragment *C* + *D* (Figs. 13.10*c* and 13.11*c*) with *Hpa*I yields fragments *C* and *D* (Figs. 13.10*d* and 13.11*d*). These data, taken together, show that the common sequence (or overlap sequence) in the *Hpa*I fragment *B* + *C* and *Hin*dIII fragment *C* + *D* is fragment *C* and establish the order *B-C-D* (or *D-C-B*). By continuing to analyze the fragments produced by sequential digestions with these and other restriction enzymes, the order of the fragments can usually be unambiguously established.

Partial restriction digests of radioactively end-labeled DNA molecules provide an even simpler mechanism for ordering restriction sites. The radioactive isotope ^{32}P is added to the ends of a linear DNA molecule (such as an *Eco*RI-cleaved SV40 chromosome; Fig. 13.10*a*) using ^{32}P-ATP and the enzyme polynucleotide kinase. The molecule is then usually cut into two unequal halves by using a restriction

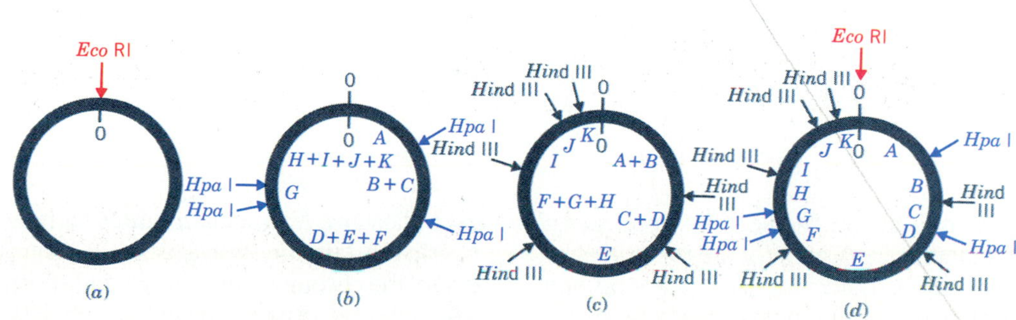

Figure 13.10 Restriction enzyme cleavage site maps of the simian virus 40 (SV40) chromsome. The restriction enzymes (a) *Eco*RI, (b) *Hpa*I, and (c) *Hin*dIII cleave the SV40 chromosome at 1, 4, and 6 sites, respectively. The 11 fragments produced when the SV40 chromosome is cut with all three restriction enzymes are shown in (d); they are arbitrarily labeled alphabetically starting at the unique *Eco*RI cleavage site and progressing in the clockwise direction. The sizes of the various fragments can be determined by polyacrylamide or agarose gel electrophoresis (see Fig. 13.23) as shown in Figure 13.11. (Based on the data summarized by D. Nathans and H. O. Smith, *Annu. Rev. Biochem.* 44: 273–293, 1975.)

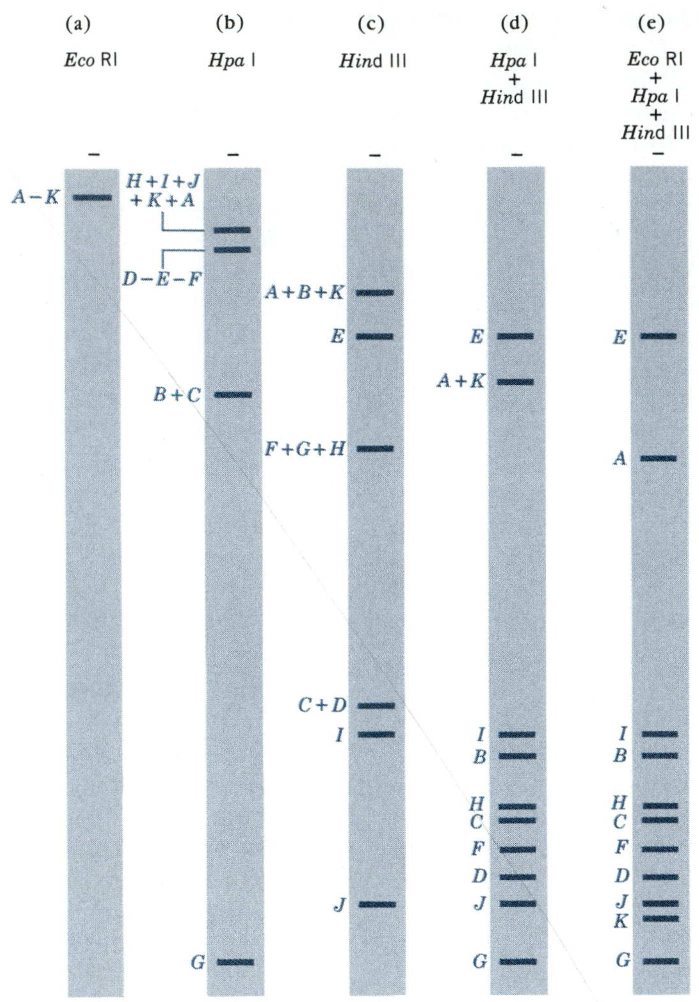

Figure 13.11 Separation of SV40 restriction fragments by polyacrylamide gel electrophoresis. See Fig. 13.23 for a description of the procedure. The fragments are separated on the basis of size—the smaller the fragment, the greater the distance that it migrates in the gel. SV40 chromosomes were cleaved with (a) *Eco*RI, (b) *Hpa*I, (c) *Hind*III, (d) *Hpa*I + *Hind*III, or (e) *Eco*RI + *Hpa*I + *Hind*III. The fragments produced by digestion with all three restriction enzymes have been arbitrarily labeled A through K, in clockwise order, starting at the *Eco*RI cleavage site. The restriction fragment bands are visualized under ultraviolet light after staining with ethidium bromide. Ethidium bromide binds to DNA and in the bound state exhibits an increased fluorescence under UV light, making DNA bands easy to detect. The positions of the various restriction fragment bands shown above are only approximate. (The relative sizes of the restriction fragments are based on the data summarized by D. Nathans and H. O. Smith, *Annu. Rev. Biochem.* 44: 273–293, 1975.)

enzyme that cleaves the molecule only once, in an asymmetrical manner. Separation of the two halves by electrophoresis yields molecules with only one end labeled; the two halves are then analyzed separately.

The fragments produced by ***partial digestion*** of the isolated half-chromosomes are separated by gel electrophoresis, and the radioactive fragments are identified by autoradiography (see Chapter 5, p. 107). Only those fragments that contain an original radioactive end will be detected by this procedure. The shortest radioactive fragment will indicate how far the first restriction enzyme cleavage site is from the labeled end of the chromosome. The second shortest fragment will give the distance from the labeled end to the second restriction enzyme cleavage site, and so on. This is illustrated in Fig. 13.12.

Finally, overlapping sequences of different restriction fragments can be detected by their ability to hybridize with each other ("cross-hybridization"). Two fragments that cross-hybridize must overlap a com-

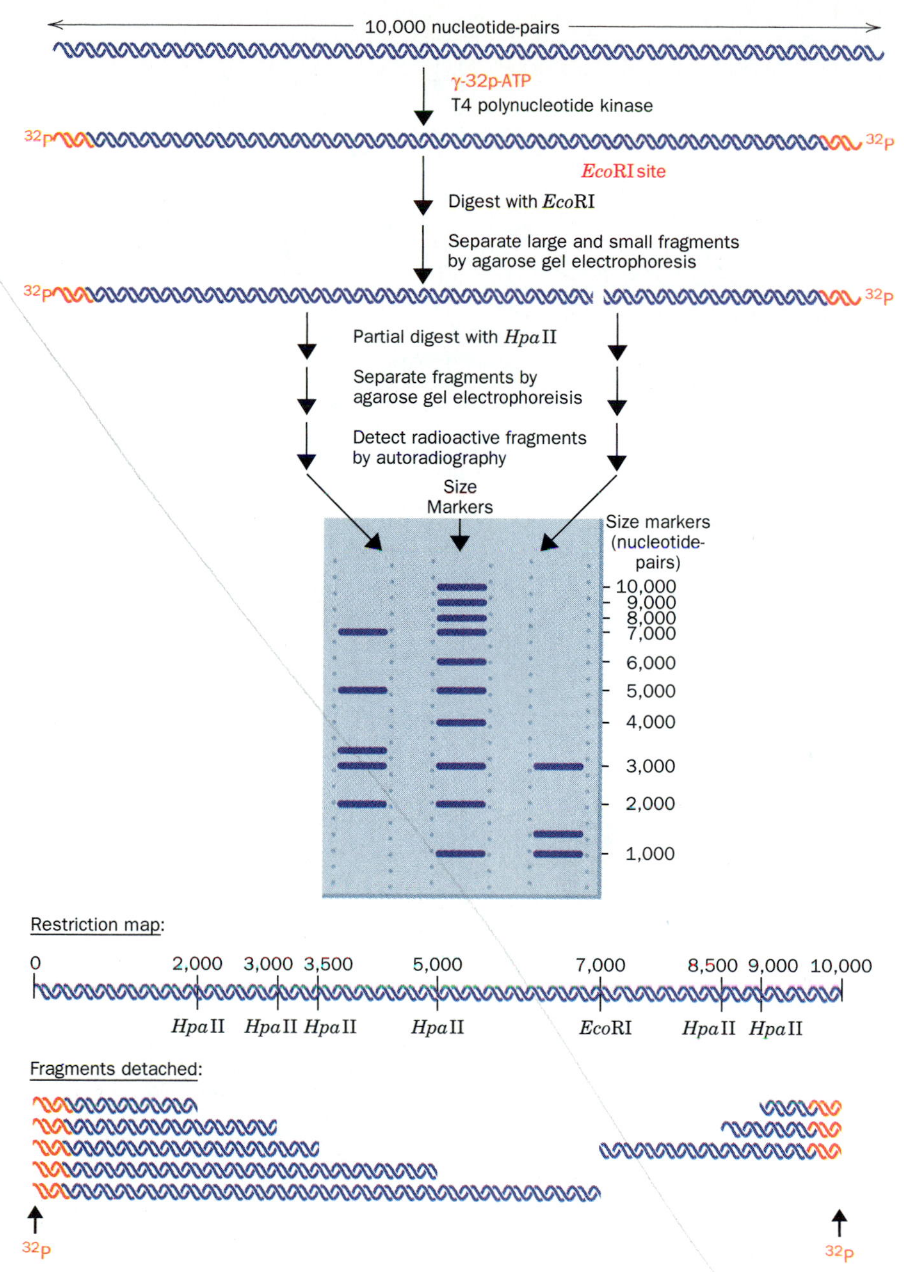

Figure 13.12 Diagram illustrating the construction of restriction maps by analyzing the products of partial digestion of end-labeled DNA molecules. The molecules are first end-labeled with ^{32}P by using phage T4 polynucleotide kinase to transfer radioactive phosphate groups (from γ-^{32}P-ATP) to the 5′-OH ends of the DNA strands. At this point, the molecules are all labeled at both ends with ^{32}P. The end-labeled molecules are then cleaved with a restriction enzyme (*Eco*RI in the example shown) that cuts them only once asymmetrically to yield "half" molecules of two different sizes. The short and long "halves" are then separated by agarose gel electrophoresis, and the two DNA fragments are extracted from the excised slices of the gel. The two end-labeled "halves" (now in separate tubes) are next subjected to partial digestion with another restriction enzyme (*Hpa*II in the example shown), and the fragments that result are separated by agarose gel electrophoresis. The positions of the end-labeled fragments in the gel are determined by autoradiography. Note that internal restriction fragments will not be labeled, and thus will not be detected by autoradiography. The only bands on the autoradiogram will be those corresponding to the positions of fragments that are ^{32}P end-labeled. The sizes of the fragments are determined by using a set of size markers (also end-labeled with ^{32}P) as described in the legend to Fig. 13.9. The restriction fragments that are detected by autoradiography (center) and the restriction map that they define are shown at the bottom.

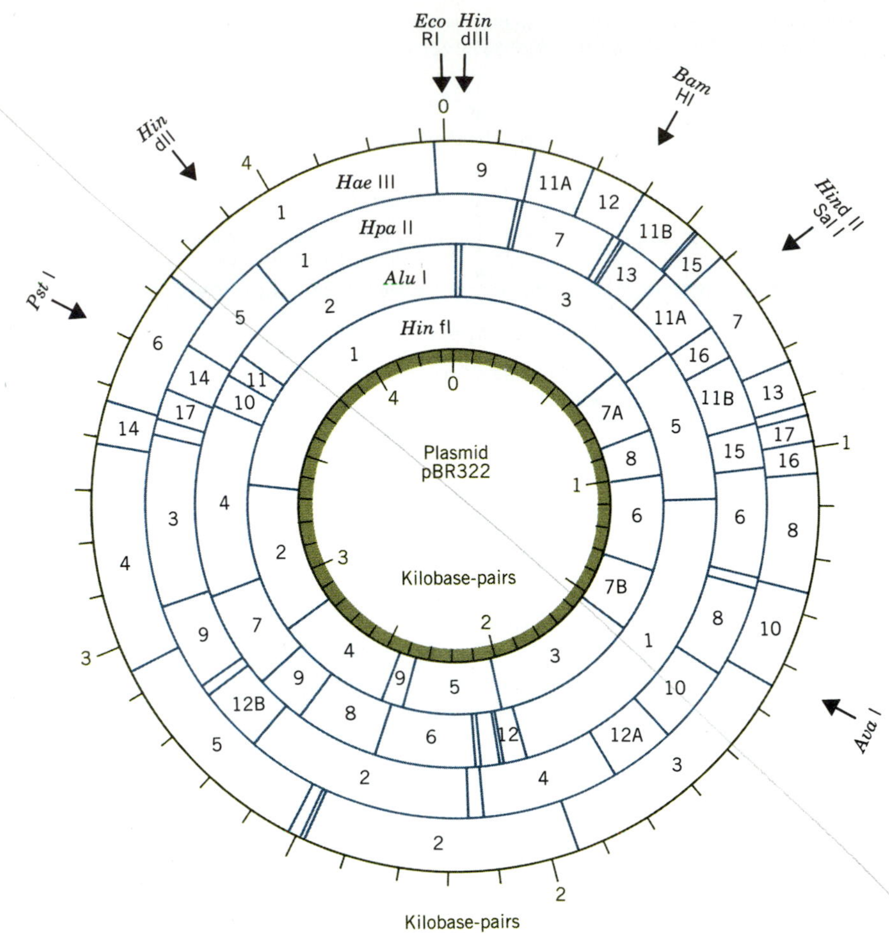

Figure 13.13 Restriction enzyme cleavage map of plasmid pBR322, one of the more popular cloning vehicles for recombinant DNA research. The cleavage sites for enzymes *Hin*fI, *Alu*I, *Hpa*II, and *Hae*III are shown in the concentric rings going from the inside to the outside, respectively. The restriction fragments produced by each enzyme are numbered 1 through *n* in order of decreasing size. The cleavage sites for enzymes that cut the pBR322 DNA molecule only once or twice are shown around the perimeter. The scale given on the inner and outer rings shows distances in kilobase-pairs (1000-nucleotide-pair units). The 4362-nucleotide-pair sequence of plasmid pBR322 has been determined by J. G. Sutcliffe. (After J. G. Sutcliffe, *Nucleic Acid Res.* 5: 2721–2728, 1978.)

mon sequence, whereas two fragments that do not cross-hybridize cannot overlap by more than a few nucleotide-pairs.

By combining these approaches, detailed restriction maps (Fig. 13.13) have been worked out for many different chromosomes. In the future, restriction maps of this kind will be available for entire genomes, even those of the most complex eukaryotes.

NUCLEOTIDE SEQUENCES: THE ULTIMATE FINE STRUCTURE MAPS

The ***ultimate fine structure map of a gene or a chromosome is its nucleotide-pair sequence,*** complete with a chart of all nucleotide-pair changes that alter the function of that gene or chromosome. Prior to 1975, the thought of trying to sequence entire chromosomes was barely conceivable—at best, a laborious task requiring years of work. By late 1976, the entire 5387-nucleotide-long chromosome of phage ΦX174 had been sequenced. Today, the entire chromosomes of several viruses, including the complete 48,502-nucleotide-pair sequence of the circular form of the phage λ chromosome, and segments of eukaryotic chromosomes tens of thousands of nucleotide-pairs in length have been sequenced. Within the next few years, the nucleotide sequences of hundreds of genes and even entire eukaryotic chromosomes will be determined and stored in computer data banks for subsequent reference. As was discussed in Chapter 1, the goal of the "Human Genome Project" is to sequence the entire human genome in the course of the next decade or two.

Our present ability to sequence essentially any DNA molecule is the result of four major developments. (1) The most important breakthrough was the discovery of restriction enzymes and their use in preparing homogeneous samples of specific segments of chromosomes. (2) A second major advance was the improvement of gel electrophoresis procedures to the point where DNA fragments that differed in length by a single nucleotide could be resolved. (3) The development of gene cloning techniques greatly facilitated the preparation of large quantities of a particular gene or DNA sequence of interest for use in sequencing studies. (4) Finally, two different approaches to the rapid sequencing of DNA molecules were developed in the period from 1974 to 1977.

Both approaches depend on the generation of a ***population of DNA fragments*** that all have ***one end in common*** (all end at exactly the same nucleotide) and ***terminate at all possible positions*** (every consecutive nucleotide) ***at the other end.*** These fragments are then separated on the basis of chain length by polyacrylamide gel electrophoresis. In both cases, ***four separate parallel reactions are carried out, each of which generates a set of fragments terminating at one of the four bases (A, G, C, or T) in DNA.***

The ***Maxam and Gilbert procedure,*** named after A. Maxam and W. Gilbert, who developed it, uses four different ***chemical reactions*** to cleave DNA chains specifically at As, Gs, Cs, or Cs + Ts. The second approach, developed by F. Sanger and colleagues, uses an ***enzymatic procedure*** and ***specific chain-terminators*** to generate four populations of fragments that terminate at As, Gs, Cs, and Ts, respectively.

2′,3′-Dideoxyribonucleoside triphosphates (Fig. 13.14) are the chain-terminators most commonly used in DNA sequencing studies. Recall that DNA polymerases have an absolute requirement for a free 3′-OH on the DNA primer strand (see Fig. 5.26). If a 2′,3′-dideoxynucleotide is added to the end of a chain, it will block subsequent extension of that chain since the ***2′,3′-dideoxynucleotides have no 3′-OH.*** By using (1) 2′,3′-dideoxythymidine triphosphate (ddTTP), (2) 2′,3′-dideoxycytidine triphosphate (ddCTP), (3) 2′,3′-dideoxyadenosine triphosphate (ddATP), and (4) 2′,3′-dideoxyguanosine triphosphate (ddGTP) as chain-terminators in four separate DNA biosynthetic reactions, one can generate four populations of fragments, each population containing chains that all terminate with the same base (T, C, A, or G) (Fig. 13.15).

Adenine

The normal DNA precursor
2′-deoxyadenosine triphosphate (dATP)

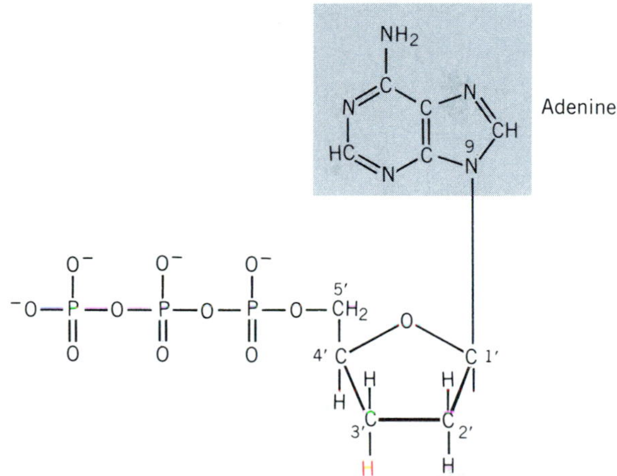

The chain-termination precursor
2′,3′-dideoxyadenosine triphosphate (ddATP)

Figure 13.14 Comparison of the structures of the normal DNA precursor 2′-deoxyadenosine triphosphate (dATP) and the chain-terminator 2′,3′-dideoxyadenosine triphosphate (ddATP) used in DNA sequencing reactions (see Fig. 13.15). Note the absence of the 3′-OH on ddATP. Once a 2′,3′-dideoxynucleotide is added to the 3′ terminus of a nascent DNA chain, no further covalent extension of the chain is possible.

In a given reaction, the ratio of dXTP : ddXTP (where X can be any one of the four bases) is kept at approximately 100 : 1 so that the probability of termination at a given "X" in the nascent chain is about 1/100. This yields a population of fragments terminating at all potential ("X") termination sites within a distance of more than 100 nucleotides from the original primer terminus (Figs. 13.15 and 13.16).

After the nascent DNA fragments generated in the four parallel reactions are released from the template strands by denaturation, they can be separated by polyacrylamide gel electrophoresis and their positions

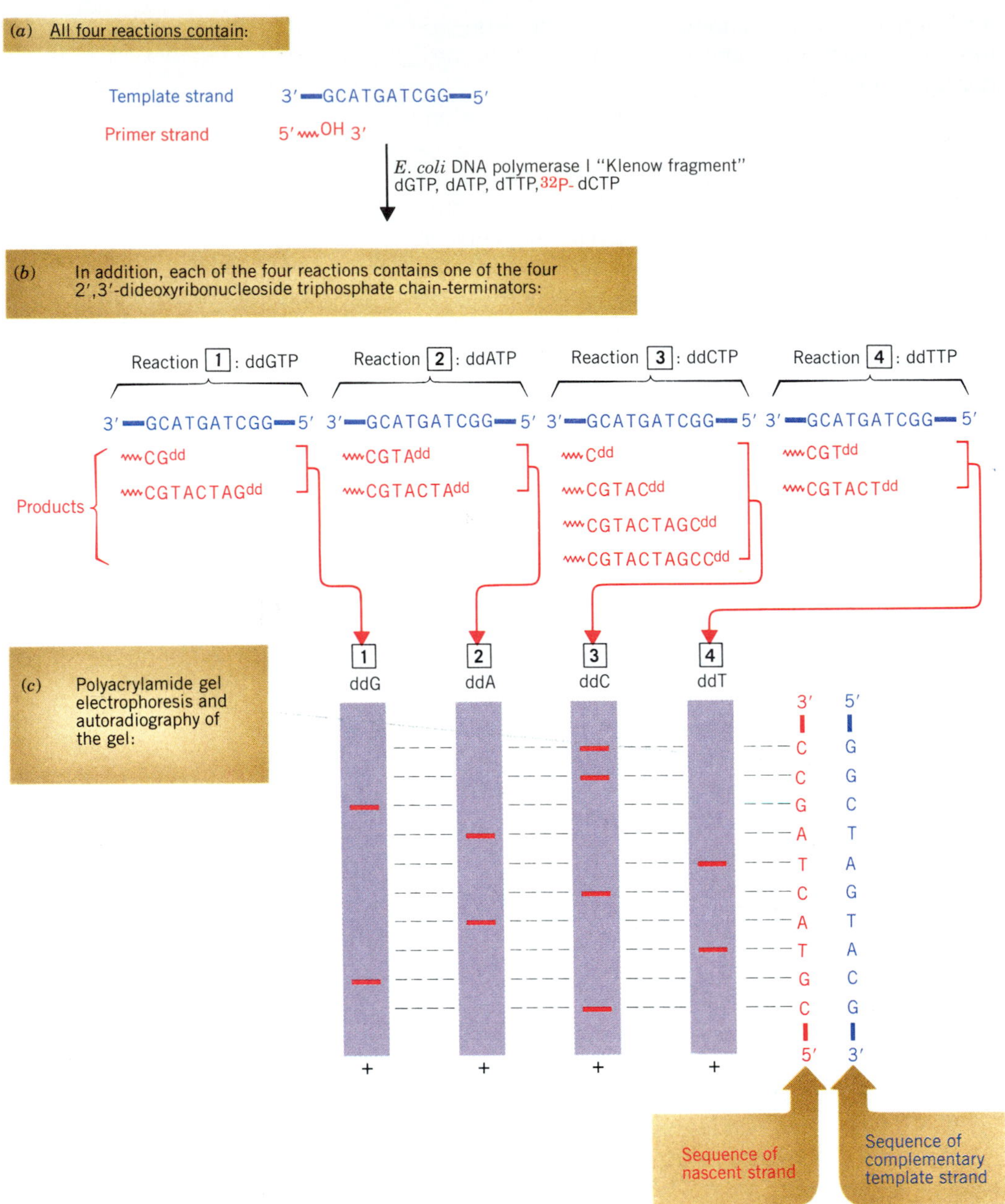

Figure 13.15 Diagram illustrating the technique of DNA sequencing using 2′,3′-dideoxynucleoside triphosphate chain-terminators (see Fig. 13.14). Four reactions are carried out in parallel, each of which contains one of the four 2′,3′-dideoxy chain-terminators: ddGTP, ddATP, ddCTP, and ddTTP. (a) All the reaction mixtures contain (1) a template strand, the nucleotide sequence of which is to be determined; (2) a primer strand with a free 3′-hydroxyl (usually obtained from a restriction fragment or by chemical synthesis); and (3) the four DNA precursors, dGTP, dATP, dTTP, and dCTP, at least one of which is radioactive (^{32}P-dCTP in the scheme shown), and the "Klenow fragment" of *E. coli* DNA polymerase I. The "Klenow fragment" is the "large half" of *E. coli* DNA polymerase I produced by cleavage with the proteolytic enzyme trypsin. The "Klenow fragment" retains the 5′→3′ polymerase and 3′→5′ exonuclease activities of DNA polymerase I, but has ***no 5′→3′ exonuclease activity*** (see Chapter 5, pp. 116–117). (The 5′→3′ exonuclease activity is on the "small half" of the molecule cleaved off with trypsin.) The absence of 5′→3′ exonuclease activity is critical to the technique; if this activity is present, it will cut back the primer strand from the 5′ end. ***The 5′ ends must remain fixed at the same nucleotide position on all nascent chains so that their lengths will be determined by the position of the dideoxy chain-termination events at the 3′ ends.*** (b) The key to the sequencing technique is to use the correct ratio of the normal deoxyribonucleoside triphosphate (e.g., dGTP in reaction 1) and the terminator dideoxyribonucleoside triphosphate (e.g., ddGTP in reaction 1), so as to obtain a population of nascent chains terminating at all possible nucleotide positions (e.g., all template-strand C's in reaction 1). The ratio of 100 deoxyribonucleoside triphos-

in the gel detected by autoradiography. The bands on the autoradiograms will correspond to radioactive chains of different lengths; they will produce a "ladder" defining the nucleotide sequence of the longest nascent chain synthesized (Figs. 13.15 and 13.16).

The shortest fragment will migrate the greatest distance, giving rise to the band nearest the anode (the positive electrode). Each successive band will contain chains that are one nucleotide longer than the chains in the preceding band of the "ladder." The 3′-terminal nucleotide of the chain in each band will be the dideoxynucleotide chain-terminator present in the reaction mixture (1, 2, 3, or 4) in which that specific chain was produced (see Fig. 13.15). By "reading" the "ladder" produced by autoradiography of the polyacrylamide gels used to separate the fragments generated in each of the four parallel reactions, the complete nucleotide sequence of a DNA chain can be determined. This is illustrated for a hypothetical nucleotide sequence in Fig. 13.15. A photograph of an autoradiogram of an actual dideoxynucleotide chain-terminator sequencing gel is shown in Fig. 13.16. Under optimal conditions, long sequences of several hundred nucleotides can be "read" from a single sequencing gel.

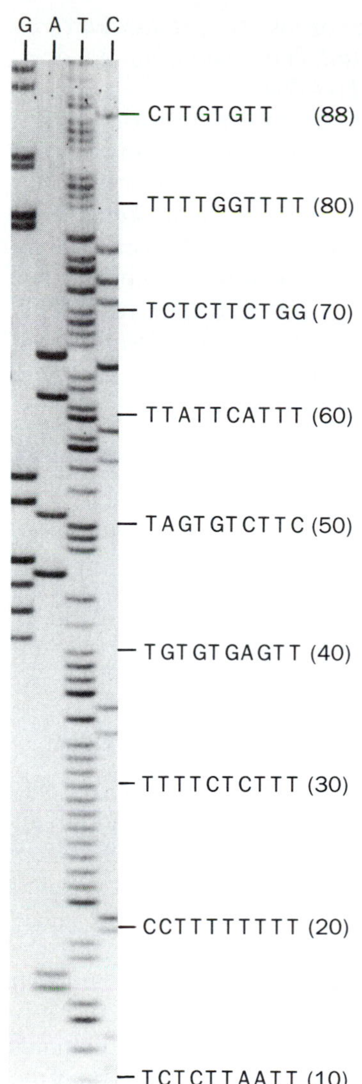

Figure 13.16 Photograph of a section of a 2′,3′-dideoxyribonucleotide chain-terminator sequencing gel. The sequence shown is that of a segment of one strand of a glutamine synthetase cDNA of maize. (Courtesy of D. P. Snustad, Department of Genetics and Cell Biology, University of Minnesota.)

ADVANCED CLONING VECTORS

Since the first recombinant DNA and gene cloning experiments were carried out in the early 1970s, technological advances in this field have been rapid and continuous. Essentially all the early gene cloning experiments were done using just a few ***vectors*** (the self-replicating DNA molecules into which foreign DNA molecules are inserted). For the most part, the vectors used in these initial experiments were derivatives of either (1) the phage λ chromosome or (2) small plasmids of *E. coli.* However, specialized cloning vectors have subsequently been developed for use in many different species of bacteria and several simple eukaryotes, especially the yeast *Saccharomyces cerevisiae* (see Chapter 24), and for use in the delivery of foreign genes into the cells and chromosomes of higher plants and animals (see Chapter 24). A discussion of the different vectors and their uses is far beyond the scope of this text. For excellent discussions of various cloning vectors and their applications, the

phates to 1 dideoxyribonucleoside triphosphate usually gives the desired frequency of termination at each potential termination site. (c) The products of the four reactions are separated in parallel by polyacrylamide gel electrophoresis and the positions of the radioactive reaction products (nascent DNA chains) are determined by autoradiography. Since the shortest chain migrates the greatest distance in the gel, the nucleotide sequence defined by the nascent chains is given for the 5′→3′ direction by "reading" from the bottom (anode) to the top (cathode) of the gel. This sequence is shown in red to the immediate right of the autoradiogram. The sequence of the template strand (shown in blue) will be complementary to that of the longest nascent chain. (For additional details, see F. Sanger, S. Nicklen, and A. R. Coulson, *Proc. Natl. Acad. Sci. U.S.A.* 74: 5463–5467, 1977.)

reader is referred to Sambrook *et al., Molecular Cloning, A Laboratory Manual,* 2nd edition, and Wu *et al., Recombinant DNA Methodology*.

Two widely used *E. coli* plasmid vectors are discussed in the following two sections of this chapter. Actually, each vector system consists of a pair of closely related vectors that facilitate insertion of foreign DNAs into the vector in each of the two possible orientations. This makes it very easy to isolate each of the two complementary strands of DNA of the foreign gene in large quantities. The first vector system described has many features that make it easy to clone a variety of foreign DNAs and to isolate single strands of these cloned DNAs for sequencing and site-specific mutagenesis. The second vector system has additional components that permit the cloned gene, actually each strand of the cloned gene, to be transcribed and translated *in vitro.* The latter vectors are particularly useful in identifying and characterizing the products of cloned genes.

Plasmids That Provide an Option to Package Single DNA Strands in Viral Coats

Single-stranded DNA provides the optimal substrate for DNA sequencing experiments and is required for certain *in vitro* mutagenesis protocols. Several vectors have been constructed that provide a simple biological mechanism for purifying large amounts of single strands of cloned genes. The most useful of these are "hybrid" vectors containing components from both plasmids and phage chromosomes. These vectors replicate in *E. coli* as normal double-stranded plasmids until a "helper" phage is provided. After addition of the "helper" phage, they switch to the phage mode of replication (rolling circle replication like that observed for phage ΦX174; see Fig. 5.32) and package single strands of DNA in phage particles. The "helper" phage is a mutant that replicates its own DNA inefficiently, but provides viral replication enzymes and structural proteins for the production of plasmid DNA molecules that are packaged in phage coats.

To understand these phage–plasmid hybrid vectors, one must first have some background on the single-stranded DNA-containing filamentous phages. The best known of the filamentous single-stranded DNA phages are M13, f1, and fd. Their genetic material is single-stranded DNA just as for phage ΦX174 (see Chapter 5). Moreover, the genomes of these phages replicate by the rolling circle mechanism (see Figs. 5.32 and 5.33) just like the DNA of ΦX174. The virions (phage particles) of the M13, f1, and fd phage group are long, threadlike filaments, rather than icosahedrons (20-face polyhedrons that are almost spherical) like phage ΦX174. Furthermore, the filamentous single-stranded DNA phages infect cells by adsorbing to and entering through F pili; thus, these phages only infect F^+ or Hfr cells, not F^- cells. Lastly, these phages do not lyse the host cells (releasing progeny phage plus bacterial DNA from the ruptured cells) like phage λ during the lytic cycle (see Chapter 8) or phage T4 (see Chapter 12). Instead, the progeny viruses are extruded through the layers of the cell membrane and cell wall without major interference with cell growth. Infected cells will continue to grow and extrude thousands of progeny virus particles, each containing a single-stranded genome, to the medium. Because the virus particles are very small relative to the size of the host bacteria, the host cells can be removed by low-speed centrifugation. The virus particles can then be collected from the supernatant suspension by high-speed centrifugation, and their single-stranded DNA molecules can be isolated by simple phenol–chloroform extractions. The same DNA strand of the virus is always packaged; it is called the "+" strand (its complement is the "−" strand). The packaged "+" strand has the same "sense" as the mRNA; its nucleotide triplets correspond to the mRNA codons, but with T in place of U.

Figure 13.17 illustrates the major features of the M13 life cycle. Note that the packaging of single strands of phage DNA in progeny phage provides a neat biological purification of single-stranded DNA. Importantly, this will be true for a foreign gene cloned in the viral chromosome just as for the phage genes themselves. The phage–plasmid "hybrid" cloning vectors exploit this aspect of M13 reproduction.

The phage–plasmid "hybrid" vectors pUC 118 and pUC 119 are a pair of vectors that are essentially identical except that the regions into which foreign DNAs are inserted are present in opposite orientations (turned end-for-end) relative to the rest of the genes of the vector (Fig. 13.18). Thus, if a foreign DNA is inserted into a specific restriction site in both vectors, one vector will package one strand of the gene, and the other vector will package the complementary strand of the gene. Therefore, both strands of the gene can be isolated, sequenced, subjected to site-specific mutagenesis, and so on. The vectors were designated pUC for ***p**lasmid*, ***U**niversity* of ***C**alifornia* (where the first pUC vectors in the series were constructed by J. Messing and J. Vieira), and 118, 119 to distinguish them from earlier members (lower numbers) of the series. Vectors pUC 118 and pUC 119 differ from earlier vectors in the pUC series by the addition of the origin of replication from phage M13. This permits pUC 118 and pUC 119 to replicate either (1) as a double-stranded plasmid in the absence of "helper" phage or (2) as a single-stranded DNA that is packaged in M13 phage coats and extruded from the cell in the presence of "helper" phage. In the absence of "helper" phage, replication is controlled by the plasmid origin of replication. The pUC vectors were derived from an early plasmid cloning vector called pBR322 by a series of directed modifications. The pUC vectors, like pBR322, contain an origin of replication initially present in plasmid ColE1. In the presence of "helper" phage, replication of pUC 118 and pUC 119 is directed by the M13 phage origin of replication that has been added to these plasmids.

The important features of the pUC 118 and pUC 119 cloning vectors (Fig. 13.18) are the following:

1. ***small, supercoiled, covalently closed circular DNA***—thus, easily isolated and manipulated *in vitro;*
2. ***carry the amp^r gene as a selectable marker***—thus, only bacteria harboring a plasmid will grow on medium containing ampicillin;
3. ***high copy number, up to 500 copies per bacterium***—thus, large yields of DNA from small cell cultures;
4. ***carry a polycloning region containing a variety of restriction enzyme cleavage sites***—thus, many different types of restriction fragments can be inserted without modification;
5. ***the polycloning region interrupts the coding region of the 5′ end of the E. coli* lac z *gene***—thus, colonies harboring plasmids with foreign DNA inserts can be distinguished from those carrying plasmids with no insert by a simple color test (see the following and Fig. 13.19);
6. ***the* lac z *gene is under the control of the* lac *promoter***—thus, genes inserted in frame (codons in proper reading frame) can be expressed to produce β-galactosidase-foreign protein fusion products;
7. ***carry a plasmid origin of replication***—thus, replication of the vector DNA produces large numbers of double-stranded plasmid DNAs in the absence of "helper" phage;
8. ***carry a phage M13 origin of replication***—thus, production of single-stranded DNA and packaging of this DNA in phage coats in the presence of "helper" phage; and
9. ***the polycloning regions are present in pUC 118 and pUC 119 in opposite orientations***—thus, if pUC 118 packages one strand of a cloned gene, pUC 119 will package the complementary strand (see Fig. 13.20).

Different members of the pUC vector series contain different, but related, sets of restriction enzyme cleavage sites. The polycloning regions of pUC 118 and pUC 119 contain 10 clustered restriction enzyme cleavage sites (Fig. 13.18). Some of these sites are substrates for two or more different restriction enzymes.

The utility of the pUC vectors is greatly enhanced by a simple color test that allows one to distinguish cells harboring plasmids with foreign DNA inserts from those harboring plasmids with no insert. The basis of this color indicator test is the functional inactivation of the 5′ segment of the *lac z* gene present in the vector (Fig. 13.18) by the insertion of foreign DNA into the polycloning region.

The *E. coli lac z* gene encodes β-galactosidase, the enzyme that cleaves lactose into glucose and galactose. This is the first step in the catabolism of lactose in *E.*

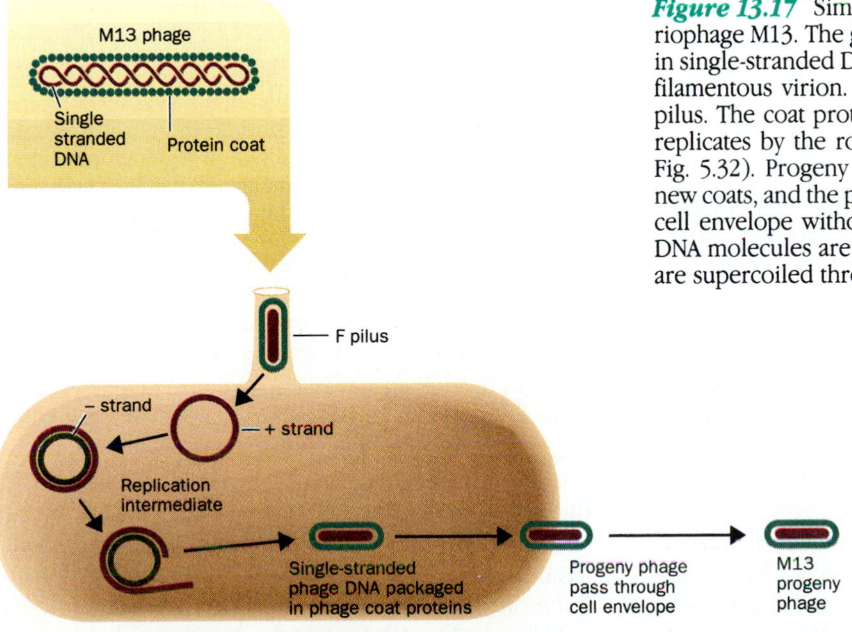

Figure 13.17 Simplified diagram of the life cycle of bacteriophage M13. The genetic information of the virus is stored in single-stranded DNA packaged within a long threadlike or filamentous virion. The virus enters the cell through an F pilus. The coat proteins are removed from the DNA, and it replicates by the rolling circle mechanism (see Chapter 5, Fig. 5.32). Progeny single strands of DNA are packaged in new coats, and the progeny virions are extruded through the cell envelope without killing the host cell. The replicating DNA molecules are shown as relaxed circles; in reality, they are supercoiled throughout the duplication process.

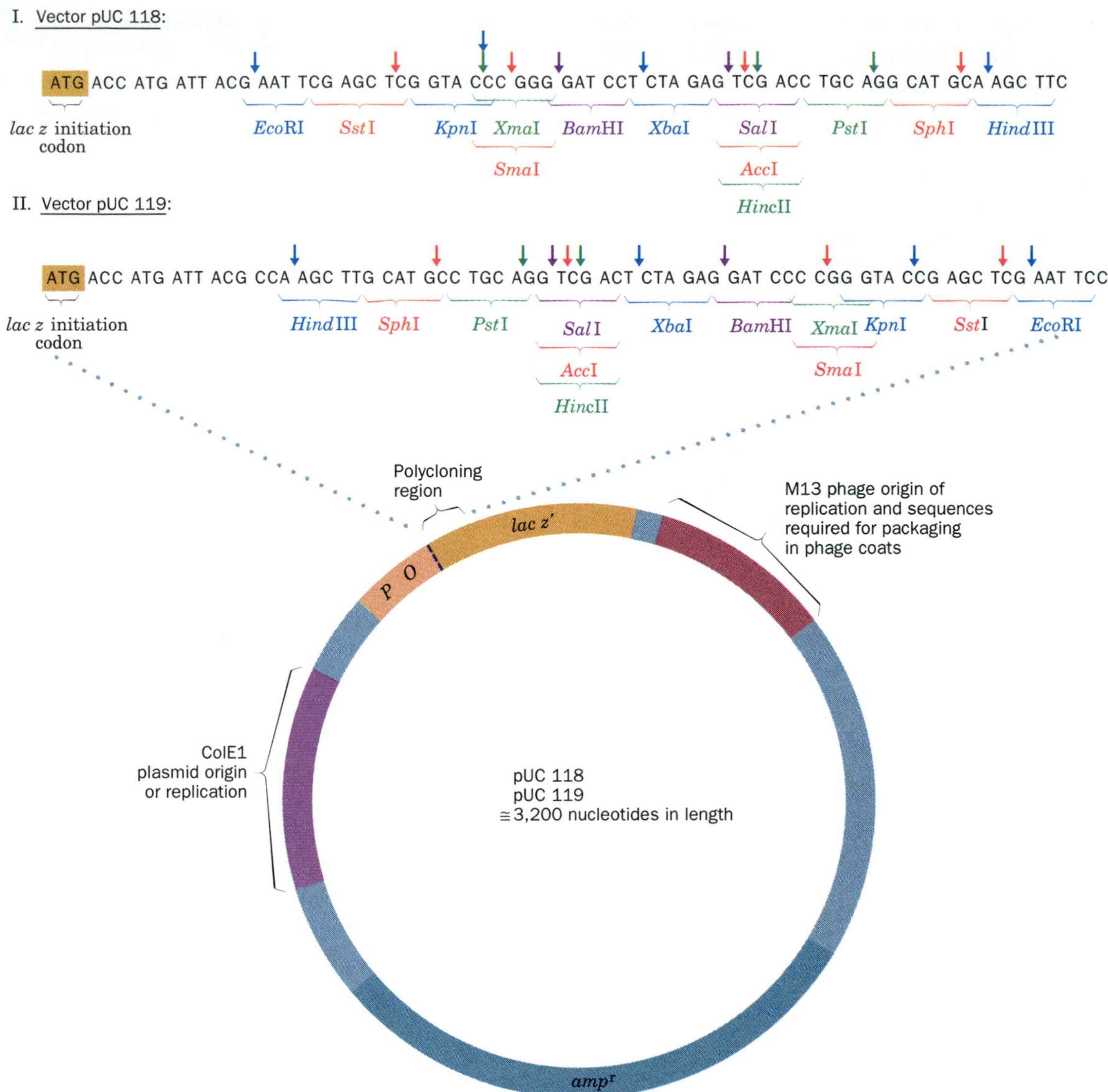

Figure 13.18 Diagram showing the important components of pUC 118 and pUC 119 vectors. *P* and *O* are the promoter (RNA polymerase binding site) and operator (repressor binding site) regions of the *lac* operon (regulated unit of transcription, see Chapter 14). Gene segment *lac z'* encodes the amino-terminal 146 amino acids of the β-galactosidase polypeptide (see text). The polycloning site lies within *lac z'* and contains a cluster of restriction enzyme cleavage sites (top). Note that the polycloning site is present in pUC 118 and pUC 119 in opposite orientations. The ATG triplet shown (top) is the *lac z* translation-initiation codon. The *ColE1* plasmid origin controls replication in the absence of "helper" phage. The M13 origin controls replication in the presence of "helper" phage; the phage mode of replication produces single-stranded progeny DNA molecules that are packaged in protein coats (provided by the "helper" phage) and extruded through the cell envelope (see Fig. 13.17). The *amp^r* gene encodes the enzyme β-lactamase, which inactivates the antibiotic ampicillin and renders the host cell ampicillin resistant.

coli (see Chapter 14). The presence of β-galactosidase in cells can be monitored on the basis of its ability to cleave the substrate 5-bromo-4-chloro-3-indolyl-β-D-galactoside (usually called Xgal) to galactose and 5-bromo-4-chloroindigo. ***Xgal is colorless; 5-bromo-4-chloroindigo is blue.*** Thus, ***cells containing active β-galactosidase produce blue colonies on agar medium containing Xgal. Cells lacking β-galactosidase activity produce white colonies on Xgal plates*** (Fig. 13.19).

The molecular basis of the β-galactosidase activity that provides the color indicator test for pUC vectors is somewhat more complex. The *lac z* gene of *E. coli* is over 3000 nucleotide-pairs long, and placing the entire gene on the plasmid would make the vector undesirably large. The pUC vectors contain only a small part of the 5′ end of the *lac z* gene. This *lac z* gene segment encodes the amino-terminal portion of β-galactosidase. However, the presence of a functional copy of the *lac z* gene segment can be detected because of a unique type of intragenic complementation. When a functional copy of the *lac z* gene segment on the pUC plasmid is present in a cell that contains a particular *lac z* mutant allele on the chromosome or on an F′ plasmid, the two defective *lac z* sequences yield polypeptide products that complement each other and produce an active β-galactosidase. The mutant allele, designated *lac z*Δ*M15,* synthesizes a *lac* protein that lacks amino acids 11 through 14 from the amino terminus. The absence of these amino acids prevents the mutant polypeptides from forming the active tetramer (4 subunits) form of the enzyme. By some mechanism that is still not completely understood, the presence of the amino-terminal fragment (first 146 amino acids) of the *lac z* polypeptide encoded by the *lac z* gene fragment on pUC plasmids facilitates tetramer formation by the ΔM15 deletion polypeptides. This yields active β-galactosidase that can be detected by the Xgal color test. This unique type of intragenic complementation observed between the amino-terminal fragment of β-galactosidase and the complete, but defective polypeptide is called α ***complementation.*** It conveniently permits one to utilize the Xgal color test without placing the entire *lac z* gene on the pUC vectors. However, for this color test to work with these vectors, the host cell must contain a mutant *lac z*Δ*M15* gene on the chromosome or, more commonly, on an F′ plasmid in a cell that has a large deletion in the chromosomal copy of the *lac z* gene.

The standard protocol for cloning foreign DNAs into pUC vectors involves (1) ***ligation*** of the foreign DNA of interest into one of the restriction enzyme cleavage sites in the polycloning region of the vector, (2) ***transformation*** of cells of an appropriate *amp*^{s} *E. coli* strain carrying the *lac z*Δ*M15* gene with the ligation products, and (3) ***plating the transformed cells*** in plates with a ***nutrient agar medium containing Xgal and ampicillin.*** Only transformed cells that contain a pUC plasmid will be able to grow in the presence of ampicillin. Colonies produced by cells harboring pUC plasmids without foreign DNA inserts will be blue (Fig. 13.19). Colonies produced by bacteria harboring pUC vectors with foreign DNA inserts will be white (due to functional inactivation of the *lac z* gene segment product by the inserted DNA). Thus, white colonies are picked and used for further characterization of the foreign DNAs.

Another important advantage of the pUC 118–119 vector system is that foreign DNAs can often be ***force cloned*** into the polycloning region in a specific, predetermined orientation. Consider a DNA sequence that has a *Sst*I site at one end and a *Pst*I site at the other end. If this DNA is cleaved with both enzymes, the internal *Sst*I–*Pst*I fragment can be inserted into pUC 118 double-cut with these enzymes in only one orientation (see Fig. 13.18). Importantly, this *Sst*I–*Pst*I fragment can also be inserted in pUC 119 in only one orientation, and this will be the reverse of its orientation in pUC 118. Thus, if a gene or other DNA fragment is force cloned in both pUC 118 and pUC 119, one vector will package one strand and the other the complementary strand after infection of the host cells with "helper" phage (Fig. 13.20). By using both vectors, then, each of the two complementary single strands of a cloned gene or other DNA sequence of interest can easily be isolated in large quantities.

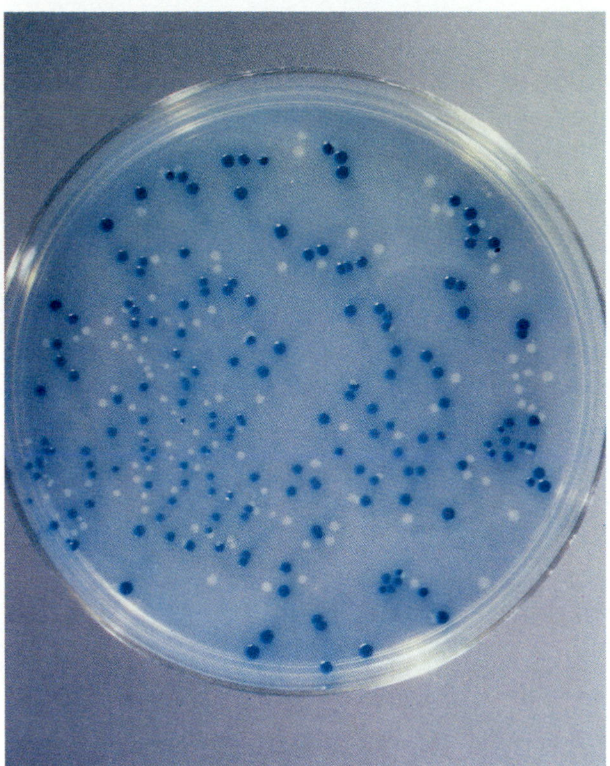

Figure 13.19 Photograph showing the use of the X-gal visual selection procedure to identify *E. coli* cells that harbor pUC 119 plasmids carrying insertions of foreign DNA in the polycloning site. The blue colonies contains pUC 119 plasmids without inserts; the white colonies contain pUC 119 plasmids with inserts (in this case, an *Arabidopsis thaliana* tubulin gene). See the text and Fig. 13.18 for details. The petri dish shown contains medium supplemented with (1) the antibiotic ampicillin to select for transformed cells that harbor plasmid pUC 119 and (2) the compound X-gal that results in *lac*$^+$ colonies becoming bright blue in color, whereas *lac*$^-$ colonies remain white (see text). (Photograph courtesy of S. Kopczak and D. P. Snustad, Department of Genetics and Cell Biology, University of Minnesota.)

(a) Replication of pUC 118 or pUC 119 without "helper" phage:

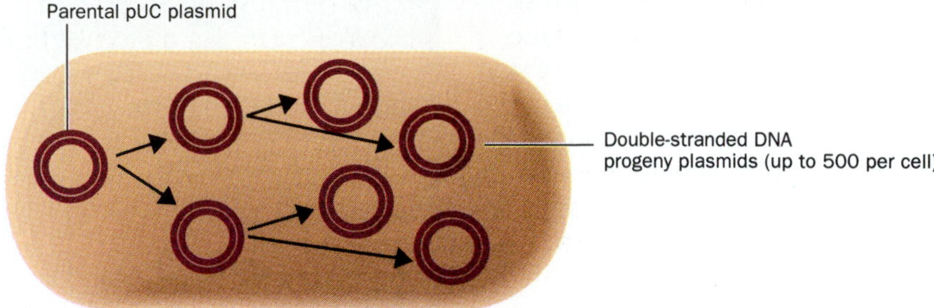

(b) Replication of pUC 118 or pUC 119 with "helper" phage:

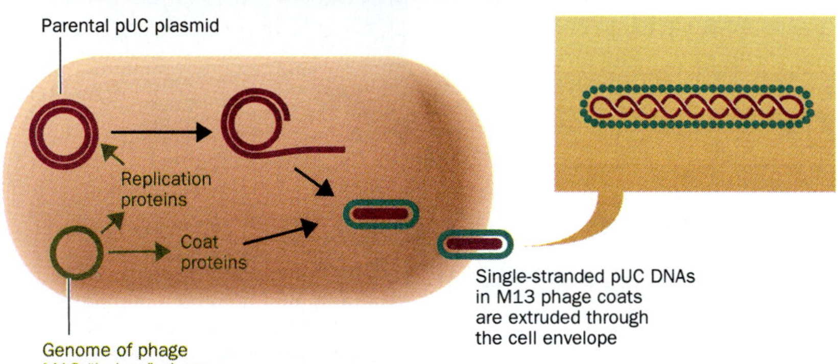

Figure 13.20 Schematic diagram showing the two modes of replication of the pUC 118 and pUC 119 plasmid–phage "hybrid" cloning vectors. (a) Replication occurs in the plasmid mode when M13 "helper" phage is not present in the host cell; this replication is controlled by an origin of replication derived from plasmid ColE1. The plasmid mode of replication produces double-stranded vector molecules that remain within the host bacterium. (b) Replication switches to the M13 phage mode of replication (rolling circle replication) when pUC 118- or pUC 119-containing cells are infected with an M13 "helper" phage. This phage mode of replication is controlled by the M13 phage origin present in these pUC vectors and produces single-stranded progeny DNA molecules that are packaged in phage coats and extruded through the cell envelope into the surrounding medium. The phage coat proteins are encoded by genes of the "helper" phage. The "helper" phage (most commonly an M13 derivative called K07) contains a modified origin of replication that causes it to replicate inefficiently. Thus, most of the extruded phage particles contain single-stranded pUC DNA molecules. However, the diagram has been oversimplified somewhat for clarity. Some M13 "helper" phage DNA replication does occur, and some "helper" phage DNA molecules are packaged. (For further details, see J. Vieira and J. Messing, "Production of Single-Stranded Plasmid DNA," *Methods in Enzymol.* 153: 3–11, 1987.)

In Vitro Transcription Vectors

Several bacteriophages, such as the *E. coli* phages T3 and T7 and *B. subtilis* phage SP6, encode their own RNA polymerases. The phage gene specifying the RNA polymerase is transcribed by the host cell's RNA polymerase, but then the viral "late proteins" or structural components of the progeny viruses are produced from genes that are transcribed by the respective phage RNA polymerase. These phage RNA polymerases only transcribe genes with unique phage-specific promoter sequences. They do not transcribe any genes of the host bacterium. Moreover, each phage RNA polymerase is highly specific for its own promoter sequence, in particular for a conserved sequence of 18 to 22 nucleotide-pairs just upstream from the 5′ GTP transcription start site (see Chapter 10).

By adding T3, T7, or SP6 promoter sequences adjacent to the polycloning regions of vectors like pUC 118 and pUC 119, one can construct vectors such that foreign DNAs inserted into the polycloning region can be transcribed *in vitro* by using the appropriate viral RNA polymerase. Because of the specificity of the viral RNA polymerase for its own promoter sequence, only the inserted gene or DNA of interest will be transcribed *in vitro*. No other plasmid genes will be adjacent to that particular phage promoter, thus none of the plasmid genes will be transcribed.

A variety of *in vitro* transcription vectors have been constructed. Most commonly, they contain promoters for two different viral RNA polymerases on opposite sides of the polycloning region. These promoters are oriented so as to direct transcription of foreign DNAs located within the polycloning region. One set of transcription vectors, called pBS (+) and pBS (−)—one packages the "+" strands of DNA and the other the "−" strands of DNA—is essentially identical to pUC 118 and pUC 119 except for the addition of two phage promoter sequences. The general structure of a transcription vector such as pBS (+) is shown in Fig. 13.21. Vector pBS (+) contains a T7 promoter on one side of the polycloning region and a T3 promoter in the opposite orientation on the other side of the polycloning region. These promoters direct transcription through the polycloning region and any foreign DNA that is inserted in it. One promoter directs transcription with one strand of the DNA as template, and the other directs transcription using the complementary strand as template. Since the foreign DNA insert may not contain a transcription termination signal, the insert-vector DNA is usually linearized by cleavage with a restriction enzyme that cuts the polycloning region just distal to the insert from the promoter being used (Figs. 3.21*b* and *c*). This prevents transcription from continuing beyond the insert into vector sequences.

Transcription vectors of this type are extremely useful. They are used to prepare radioactive DNA hybridization probes to screen genomic and cDNA clones, for Southern blot analysis of genome structure, and for northern blot studies of gene expression (see subsequent sections of this chapter). The transcripts produced *in vitro* can also be used to synthesize protein gene-products in an *in vitro* translation system. These proteins are usually ***fusion proteins*** containing an amino-terminal β-galactosidase peptide joined to a product of the insert gene. Of course, proper translation requires that the coding sequence of the inserted gene be fused to the *lac z′* gene segment in the correct reading frame. For a randomly inserted DNA sequence, this will be the case one-third of the time. In the other cases, one will have to modify the reading frame by adding or deleting bases or inserting linkers (see the next section). The present recombinant DNA technology is so advanced that one can manipulate DNA sequences and reading frames at will by a variety of procedures.

Use of Linkers in Manipulating DNAs

The powerful tools of recombinant DNA and gene cloning technologies are based almost exclusively on utilizing the sequence specificity of restriction endonucleases. In dissecting genes, chromosomes, and genomes of organisms, molecular biologists rely on their ability to cut DNA molecules at desired positions with restriction enzymes. Sometimes the cleavage sites of choice are not present at the desired locations. To circumvent this problem, scientists have taken two different approaches. One approach is to change the native DNA sequence at this position to the desired sequence, that is, to add the restriction enzyme cleavage site of choice. This is done by the procedure of site-specific mutagenesis (which is discussed in the last section of this chapter). The other approach is to add synthetic oligonucleotides containing the restriction sites of choice to the ends of DNA molecules produced by cleavage with a different enzyme.

Suppose that a gene of interest is flanked by two *Nco*I cleavage sites, and that one wants to clone this gene into pUC 119. Vector pUC 119 does not contain any *Nco*I sites. This presents no major obstacle to cloning the gene into pUC 119, in fact, into any one of several sites in the pUC 119 polycloning region. Suppose that for some reason one wants to clone this gene into the *Bam*HI site of pUC. This can be done by simply (1) excising the gene from its native site in the chromosome with *Nco*I, (2) isolating the *Nco*I restriction fragment, (3) filling in the 3′ ends *in vitro* by using the Klenow fragment of *E. coli* DNA polymerase to produce blunt-ended molecules, (4) ligating on ***BamHI linkers,*** (5) cleaving the *Bam*HI sites in the linkers by digestion with *Bam*HI, and (6) inserting the *Bam*HI fragment that has been produced into *Bam*HI-cut pUC 119. This procedure is illustrated in Fig. 13.22.

Linkers are short, double-stranded DNAs that contain restriction enzyme cleavage sites; they are synthesized from mononucleotides by simple protocols of organic chemistry, and any desired sequence can be synthesized. In today's world of biotechnology, linkers containing a wide variety of restriction sites are simply purchased from any one of several companies that supply chemical reagents to research laboratories.

Other manipulations of DNA ends are done by using so-called ***adaptors,*** synthetic oligonucleotides that contain two restriction enzyme cleavage sites. By adding adaptors to the ends of a DNA sequence of interest, one can convert the ends from one restriction enzyme cleavage sequence to another restriction enzyme cleavage sequence. For example, *Eco*RI–*Hin*dIII adaptors will allow one to convert an *Eco*RI restriction fragment (a DNA fragment with *Eco*RI sites at its ends) to a *Hin*dIII restriction fragment (one with *Hin*dIII sites at the ends). Indeed, the present tools of molecular biology allow one to modify DNA molecules at will—to change sequences, to add sequences, or to delete sequences.

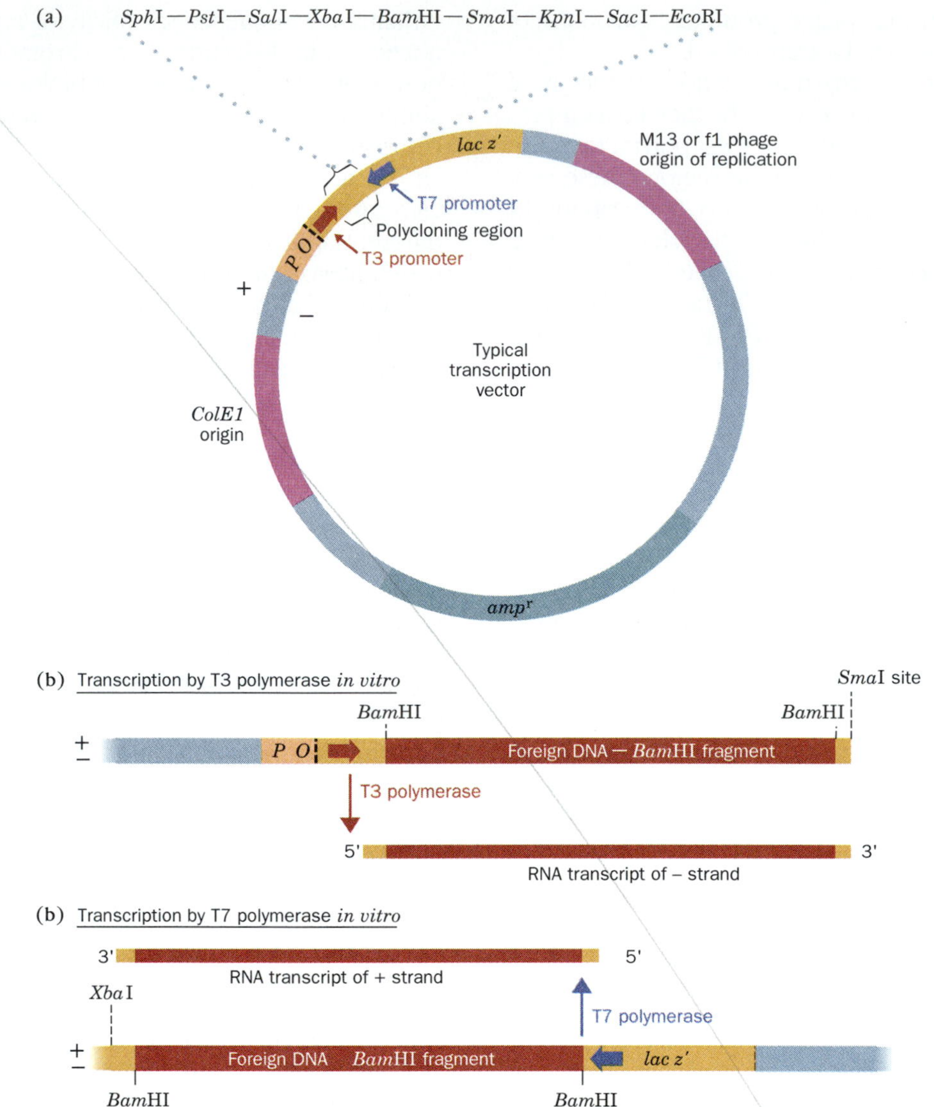

Figure 13.21 (a) Structure of a typical transcription vector and *in vitro* transcription of (b) the "−" strand of a foreign DNA by T3 RNA polymerase or (c) the "+" strand by T7 RNA polymerase. Components of the vector, for example vector pBS(+), are essentially as described for the pUC 118 and pUC 119 vectors in Fig. 13.18 except for the addition of T3 and T7 promoter sequences flanking the polycloning region. The promoter sequences are shown as red and blue arrows. In the vector shown, T3 RNA polymerase will synthesize a transcript of the "−" strand after the vector has been linearized by cleavage with *Sma*I. T7 RNA polymerase will synthesize a transcript of the "+" strand after the vector has been linearized by cleavage with *Xba*I. The use of linearized vector molecules as shown in (b) and (c) prevents transcription from continuing beyond the DNA insert into vector sequences when no transcription-termination sequence is present in the foreign DNA insert.

MOLECULAR ANALYSIS OF GENES AND GENE-PRODUCTS

The development of the recombinant DNA and gene cloning technologies discussed in the preceding sections of this chapter have spawned a whole array of new approaches to analyzing the structure and expression of genes, chromosomes, and, indeed, entire genomes. Many questions that were totally unapproachable just fifteen years ago can now be investigated with ease. One can isolate and characterize essentially any component of any genetic element. In addition, one can study the expression of any gene or set of genes in even the most complex organisms such as humans.

Is a particular gene of interest expressed in the kidney, the liver, bone cells, hair follicles, erythrocyes, or lymphocytes? Is this gene expressed throughout the

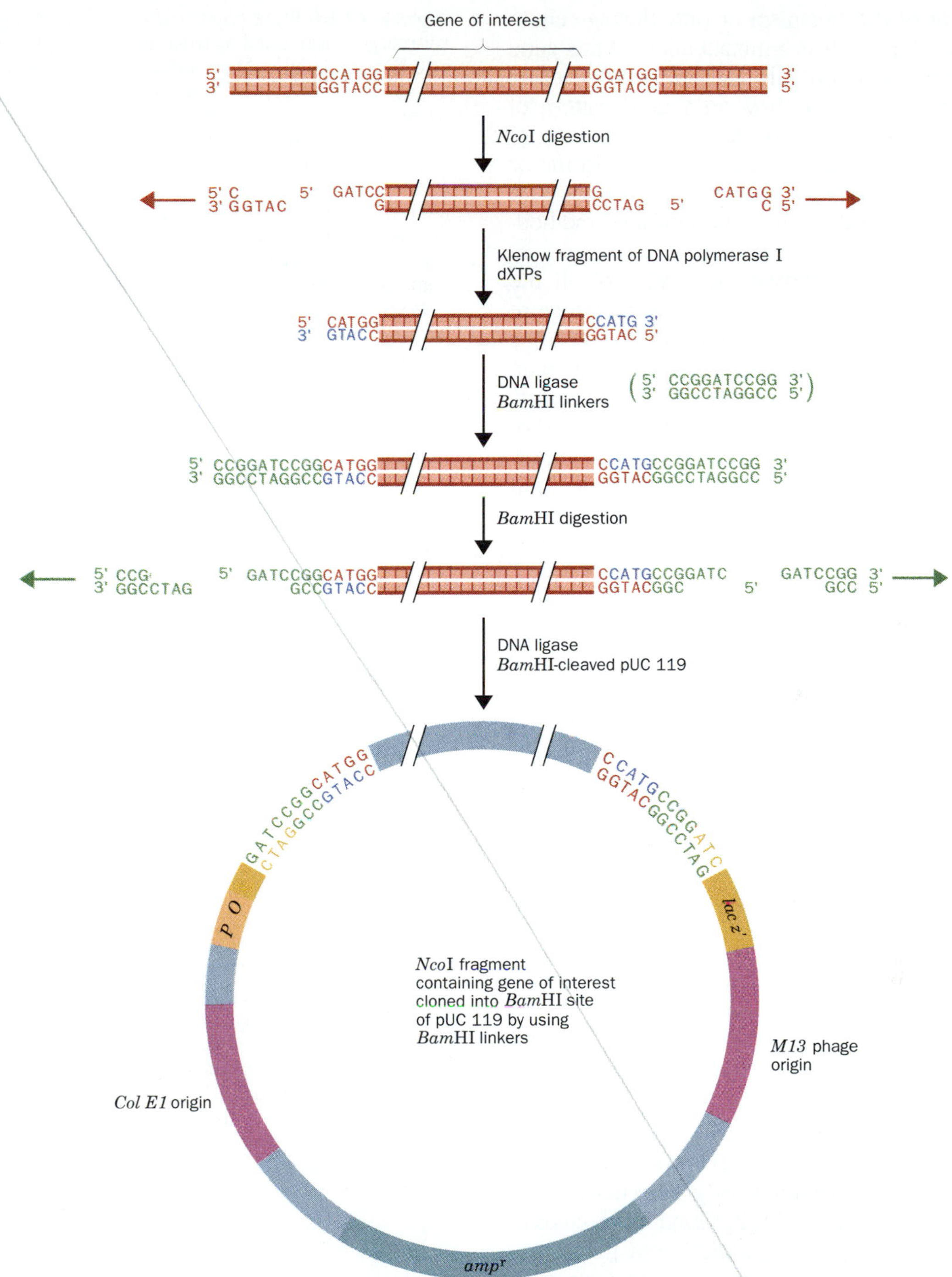

Figure 13.22 Diagram illustrating the use of *Bam*HI linkers to clone a gene on an *Nco*I restriction fragment into the *Bam*HI site in the polycloning region of pUC 119. Recall that the Klenow fragment of *E. coli* DNA polymerase I contains the 3′→5′ proofreading activity and the 5′→3′ polymerase activity of this enzyme; the 5′→3′ exonuclease activity is present on the small fragment of DNA polymerase I that is removed during the purification of the Klenow fragment. Thus, 5′ ends of DNA strands remain intact, and the 3′ ends are filled in by the polymerase until the template is exhausted. This yields blunt-ended molecules that can be ligated to synthetic linkers.

development of the organism or only during certain stages of development? Is a mutant allele of this gene similarly expressed, spatially and temporally during development, or does it show an altered pattern of expression? If the latter, is this altered pattern of expression responsible for an inherited syndrome or disease? These questions and many others can now be routinely investigated using well-established methodologies.

Clearly, a comprehensive discussion of all the techniques that are used to investigate gene structure and function is far beyond the scope of this text. For additional information, the reader is referred to Sambrook *et al., Molecular Cloning, A Laboratory Manual,* 2nd edition, and Ausubel *et al.,* editors, *Current Protocols in Molecular Biology.* A few of the most important methods used to investigate the structure and function of genetic elements are discussed briefly in the following sections of this chapter.

Analysis of DNAs by Southern Blot Hybridizations

Gel electrophoresis provides a powerful tool for the separation of macromolecules with different sizes and charges. DNA molecules have an essentially constant charge per unit mass, thus they separate in agarose and acrylamide gels almost entirely on the basis of size or conformation. Agarose or acrylamide gels act as molecular sieves by retarding the passage of large molecules more than small molecules. Agarose gels provide better sieves for large molecules (larger than a few hundred nucleotides); acrylamide gels are better for separating small DNA molecules. As is shown in Figs. 13.15 and 13.16, an essential feature of DNA sequencing is that DNA molecules that differ in length by even a single nucleotide can be separated by acrylamide gel electrophoresis.

In this and the following two sections of this chapter, techniques are discussed by which macromolecules that have been separated by gel electrophoresis can be analzyed further. As background for discussion of these techniques, Fig. 13.23 outlines one procedure for the separation of DNA molecules or restriction fragments of large DNA molecules by agarose gel electrophoresis. The procedures used to separate RNA and protein molecules are largely the same in principle, but involve slightly different protocols because of the unique properties of each class of macromolecule.

In 1975, E. M. Southern published an important new procedure that allowed investigators to identify the locations of genes and other DNA sequences on restriction fragments separated by gel electrophoresis. The essential feature of this technique is the transfer of the DNA molecules separated by gel electrophoresis to a nitrocellulose or nylon membrane (Fig. 13.24). The DNA is denatured either prior to or during transfer by placing the gel in an alkaline solution. After transfer is complete, the DNA is immobilized on the membrane by drying or UV-induced cross-linking to the filter. A radioactive DNA ("probe") containing the sequence of interest is then hybridized or annealed (Chapter 5) with the immobilized DNA on the membrane. The

Figure 13.23 Right page: Diagram illustrating the use of submerged horizontal agarose gel electrophoresis to separate DNA molecules. Many variations of the protocol shown are used, however, the basic principles remain the same irrespective of the apparatus employed. (a) Agarose gels may be poured directly on a glass plate. However, thick gels will require using tape around the plate so that the molten agarose solution does not run off the plate. A comb with teeth of the desired size is used to form the sample wells in the gel. (b) The comb is positioned about 0.5 mm above the glass plate so that agarose can run under the comb. If the comb is inserted too deeply, the DNA sample will run out the bottom and diffuse between the plate and the agarose. (c) After the agarose gel has cooled and solidified, it is placed in an electrophoresis chamber, an appropriate running buffer solution is added, and the DNA samples are loaded into the sample wells with a micropipet. Gel-loading buffer is added to the DNA solution prior to loading the samples; this loading buffer contains sucrose, glycerol, or Ficoll so that its density is greater than that of the running buffer. This causes the DNA samples to settle to the bottom of the wells and not to diffuse into the running buffer. The loading buffer also usually contains at least one dye such as bromophenol blue that is used to monitor the rate of migration of molecules through the gel during electrophoresis. (d) The cover is positioned on the electrophoresis buffer chamber, and current is applied by attaching the electrodes to a power supply and turning it to the desired amperage or voltage. Care must be used in attaching the electrodes. DNA is negatively charged and will migrate toward the positive electrode. If the electrodes are reversed, the DNA will quickly run off the end of the gel near the sample wells. After the indicator dye has migrated the desired distance (for bromophenol blue, usually to the end of the gel), the gel is removed from the chamber and placed in a solution of ethidium bromide. The ethidum bromide binds to DNA in the gel and fluoresces when illuminated with ultraviolet light. (e) The stained gel is photographed with an appropriate UV light source and filter to provide a permanent record of the results. The mini-gel shown contained eight sample wells; each well was loaded with an *Eco*RI-digested plasmid DNA preparation. Lanes 1–8, left to right. Lane 3 was loaded with *Eco*RI-cleaved pUC 119 (no insert). Lanes 1, 6, 7, and 8 were loaded with *Eco*RI-cut pMGS2, a pUC 119 derivative containing a maize glutamine synthetase cDNA flanked by *Eco*RI sites. Lanes 2, 4, and 5 were loaded with *Eco*RI-cut pMGS1, a pUC 119 derivative similar to pMGS2, but containing a slightly smaller maize glutamine synthetase cDNA insert. (Photograph by D. P. Snustad.)

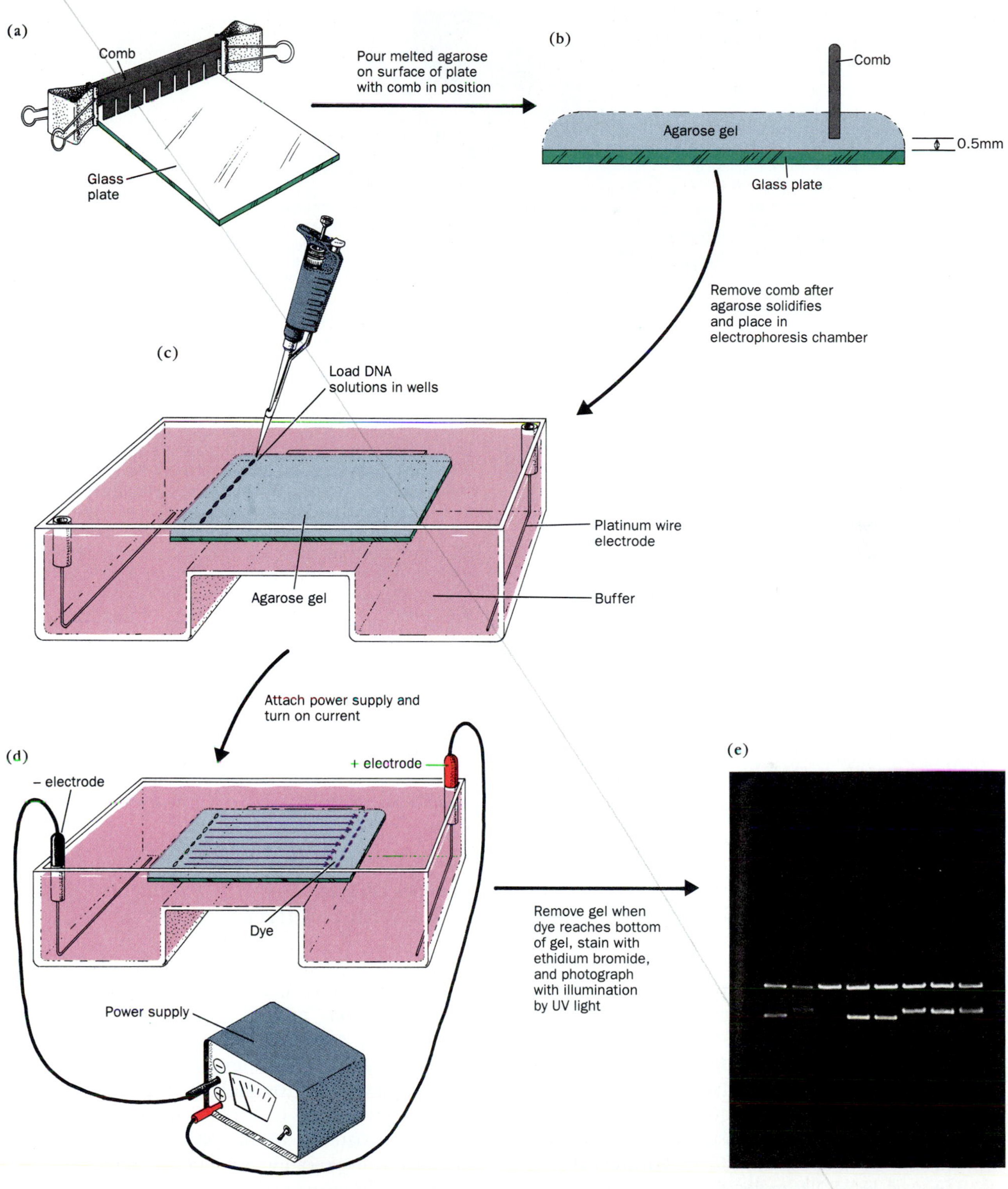

probe will anneal (form a double helix) only with DNA molecules on the membrane that contain a nucleotide sequence complementary to the sequence of the probe. Nonannealed probe is then washed off the membrane, and the washed membrane is exposed to X-ray film that detects the presence of the radioactivity in the bound probe. After the autoradiogram is developed, the dark bands show the position(s) of DNA sequences that have hybridized with the probe (Fig. 13.25).

The ability to transfer DNA restriction fragments or other DNA molecules that have been separated by gel electrophoresis to nitrocellulose or nylon membranes for hybridization studies and other types of

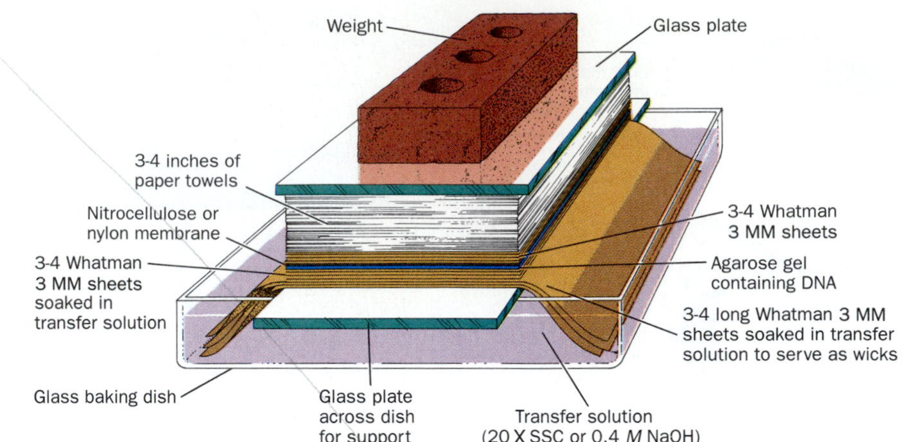

Figure 13.24 Diagram illustrating the Southern transfer procedure. If the DNA fragments are relatively small (up to a few kilobases), they can be transferred directly after denaturation in NaOH. Some protocols use 0.4 *M* NaOH as the transfer solution; in this case denaturation occurs during transfer. If the DNA molecules are large, they should first be fragmented to increase the efficiency of transfer. This is usually done by brief submersion of the gel in 0.25 *M* HCl. This causes depurinations, and when the gel is transferred to 0.4–0.5 *M* NaOH for denaturation, the alkaline solution hydrolyzes the DNA molecules at the sites of depurination. The gel is then neutralized in a buffer solution, and transfer of the DNA from the gel to a nitrocellulose or nylon membrane placed on top of it is performed as shown. The transfer salt solution carries the DNA with it from the gel to the membrane above it as the dry paper towels on top draw the transfer solution from the reservoir through the gel. The DNA sticks to the membrane on contact and does not pass through it. Transfer is usually carried out for anywhere from two hours to overnight. The membrane with the DNA bound to it is dried and baked for 30 minutes to an hour in a vacuum oven to affix the DNA firmly prior to hybridization. Some recently developed nylon membranes just require drying at room temperature to immobilize the transferred DNA. The original transfer solution employed by Southern was 20X SSC, which is a concentrated solution of sodium chloride and sodium citrate with the pH adjusted to 7.0.

analyses has proven to be extremely useful. Such transfers of DNA to membranes are called ***Southern blots*** after E. M. Southern, who developed the technique. Radioactive probes for use in Southern blot hybridization experiments are prepared by a variety of procedures including (1) producing radioactive RNA probes by *in vitro* transcription of cloned sequences in transcription vectors (see Fig. 13.21) with at least one radioactive ribonucleoside triphosphate precursor, (2) transferring ^{32}P groups to the ends of DNA molecules from labeled nucleotide triphosphates by the action of polynucleotide kinases, and (3) ***nick-translation*** of double-stranded DNA molecules (Fig. 13.26). Nick-translation is done by making single-strand cuts ("nicks") in double helices with endonucleases, and then "translating" the nick along the DNA molecule by the concerted action of the $5' \longrightarrow 3'$ exonuclease activity and the $5' \longrightarrow 3'$ polymerase activity of *E. coli* DNA polymerase I (Chapter 5) in the presence of a labeled deoxyribonucleotide triphosphate precursor.

Analysis of RNAs by Northern Blot Hybridizations

If DNA molecules can be transferred from agarose gels to nitrocellulose or nylon membranes for hybridization studies, one might expect that RNA molecules separated by agarose gel electrophoresis could be similarly transferred and analyzed. Indeed, such RNA transfers are used routinely in molecular genetics laboratories. These RNA blots are called ***northern blots*** in recognition of the fact that the procedure is the mirror image of the Southern blotting technique. (As will be discussed in the next section, this somewhat unusual terminology has been extended to the transfer of proteins from gels to membranes—so-called ***western blots***.)

The northern blot procedure is essentially identical to that used for Southern blot transfers (see Fig. 13.24), however, RNA molecules are more sensitive to degradation by RNases. Thus, gloves must be worn at

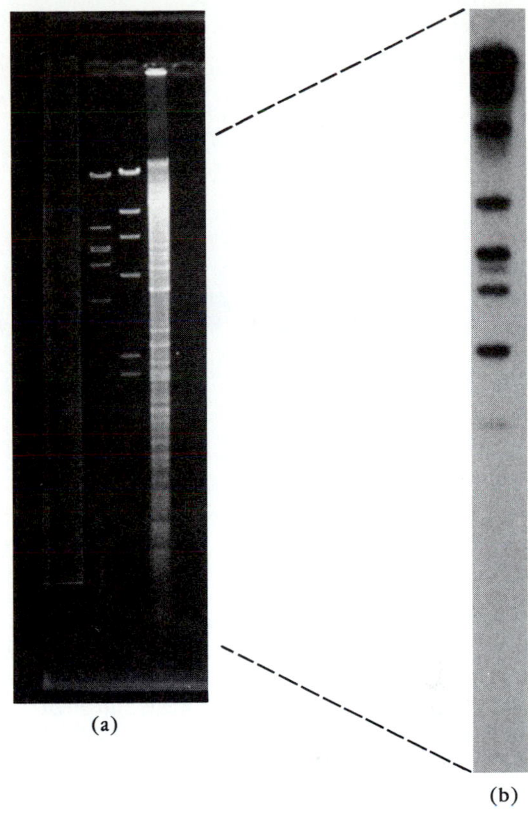

Figure 13.25 Illustration of the use of the Southern blot hybridization procedure to identify genomic restriction fragments that contain specific genes or DNA sequences of interest. (a) Photograph of an ethidim bromide-stained agarose gel containing (left lane) phage λ DNA digested with *Eco*RI, (center lane) phage λ DNA digested with *Hin*dIII, and (right lane) *Arabidopsis thaliana* (a small plant) total DNA digested with *Eco*RI. The λ DNA digests provide size markers; the sizes of all the *Eco*RI and *Hin*dIII fragments of λ DNA are known. The distinct bands present in the *A. thaliana* digest are sequences present in DNA of organelles (mitochondria and chloroplasts) and perhaps a few repetitive nuclear DNA sequences. Unique or low-copy DNA sequences are not detected as bands in the photograph; they contribute to the background fluorescence observed in the *A. thaliana* DNA lane. The *A. thaliana* total DNA digest was transferred to a nylon membrane by the Southern blotting procedure and hybridized to a short segment of a cloned gene encoding the protein β-tubulin. The β-tubulin probe was labeled with ^{32}P by nick-translation (see Fig. 13.26). The autoradiogram of the Southern blot is shown in (b) at slightly higher magnification. Note that nine different restriction fragments hybridize with the β-tubulin coding sequence probe. Other results have established that the *A. thaliana* genome contains nine different β-tubulin genes. (Photographs courtesy of D. G. Oppenheimer and D. P. Snustad, Department of Genetics and Cell Biology, University of Minnesota.)

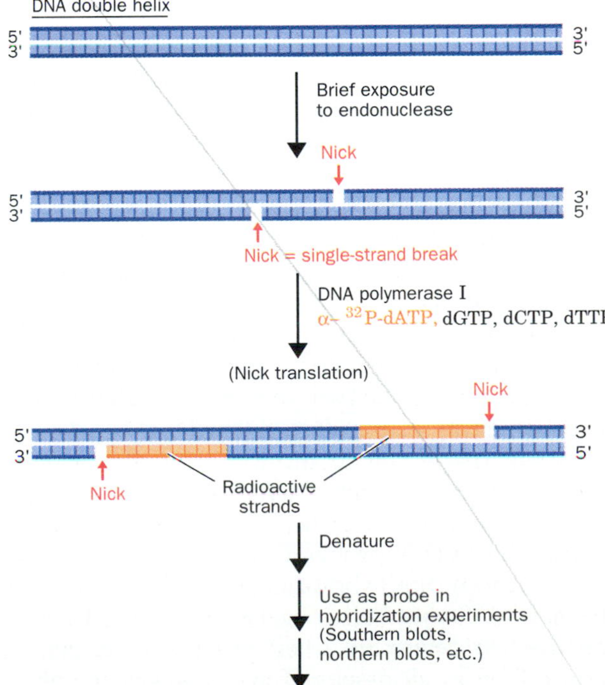

Figure 13.26 Preparation of radioactive probes by nick-translation. Single-strand breaks ("nicks") are first introduced into the double-stranded DNA molecule by brief exposure to a dilute solution of an endonuclease (usually deoxyribonuclease I of *E. coli*). DNA polymerase I is then used in the presence of at least one radioactive precursor (e.g., α-^{32}P-dATP) to "translate" the nick along the molecule in the 5′ to 3′ direction. This occurs by the concerted action of the 5′ → 3′ polymerase and the 5′ → 3′ exonuclease activities built into DNA polymerase I. The 5′ → 3′ polymerase adds nucleotides to the 3′-OH terminus at each nick, while the 5′ → 3′ exonuclease removes mononucleotides from the 5′-phosphate terminus at the nick. The net result is that a nonradioactive strand of DNA is replaced by a radioactive strand. The DNA is then denatured and used as a radioactive probe in hybridization experiments (Southern blots, northern blots, etc.).

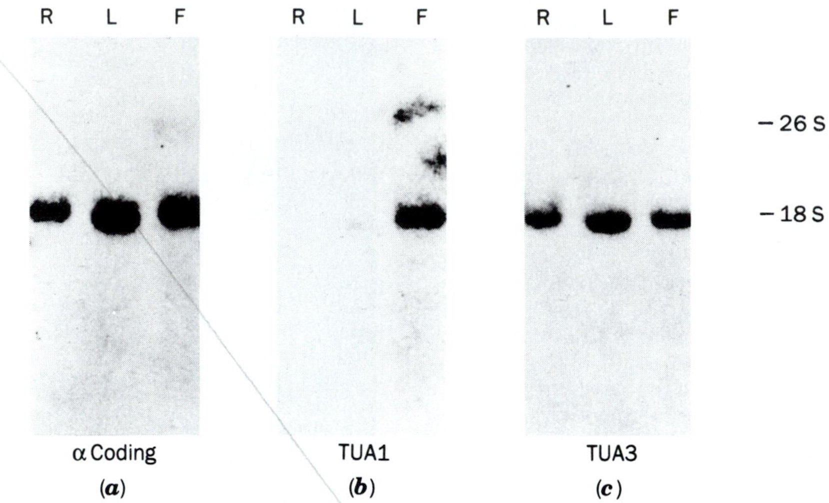

Figure 13.27 Typical northern blot hybridization data. Total RNAs were isolated from whole roots (R), leaves plus petioles (L), and whole flowers (F) of *Arabidopsis thaliana* plants, were separated by electrophoresis through an agarose gel under denaturing conditions (formaldehyde present), and were transferred to a nylon membrane by the northern blotting procedure. The autoradiogram shown in (a) is of a blot that was hybridized to a radioactive probe containing an α-tubulin coding sequence. *Arabidopsis thaliana* contains five α-tubulin genes with highly conserved coding sequences. Thus, this probe hybridizes to transcripts of all five genes that all happen to be approximately the same size. Note that α-tubulin transcripts are present in all three RNA samples (R, L, and F). The hybridized probe is then removed from the nylon membrane by denaturation and washing under conditions that leave the RNA immobilized on the membrane. The RNA blot is then hybridized sequentially with gene-specific probes that anneal (b) only with the transcript of α1-tubulin gene (*TUA*1) or (c) only with the transcript of the α3-tubulin gene (*TUA*3) of *Arabidopsis*. These gene-specific probes are 3′ noncoding DNA sequences (sequences from the translation-termination codon to the end of the transcript), which are different in each tubulin gene. Note that the α3-tubulin transcript is present in all tissues analyzed, but the α1-tubulin transcript is present only in flowers. The 18S and 26S labels show the positions of the 18S and 26S ribosomal RNAs that were visible in the ethidium bromide-stained gel photographed prior to transfer. (Photograph courtesy of S. R. Ludwig and D. P. Snustad, Department of Genetics and Cell Biology, University of Minnesota.)

all times to prevent contamination of solutions with RNases on one's fingers. In addition, all glassware must be baked overnight to inactivate RNases, and all chemical reagents must be baked or be highly purified to ensure that they are RNase-free. RNase molecules are extremely stable enzymes; they are not inactivated by boiling or other treatments that would totally destroy most enzymes. Most RNA molecules contain considerable secondary structure. Therefore, they must be kept denatured during electrophoresis if one wants to separate them on the basis of size. Denaturation is accomplished by adding formaldehyde or some other chemical denaturant to the loading buffer and the electrophoresis running buffer. After transfer to an appropriate membrane, the RNA blot is hybridized to either RNA or DNA probes just as with Southern blots.

Northern blot hybridizations (Fig. 13.27) are extremely useful in studies of gene expression. They can be used to determine whether a particular gene is transcribed in all tissues of an organism or only in certain tissues. They can also be used to study the temporal pattern of expression of individual genes during growth and development. However, one must remember that northern blot hybridizations only measure the accumulation of RNA transcripts. They say nothing about why the observed accumulation occurs. Changes in transcript levels may be due to changes in the rate of transcription or changes in the rate of transcript turnover. More sophisticated procedures must be used to distinguish between these possibilities; northern blot data only provide information on transcript levels present in a given organism or tissue at the stage or stages of development analyzed.

Analysis of Proteins by Western Blot Techniques

Polyacrylamide gel electrophoresis has been used as an important tool for the separation and characterization of proteins. Since many functional proteins are

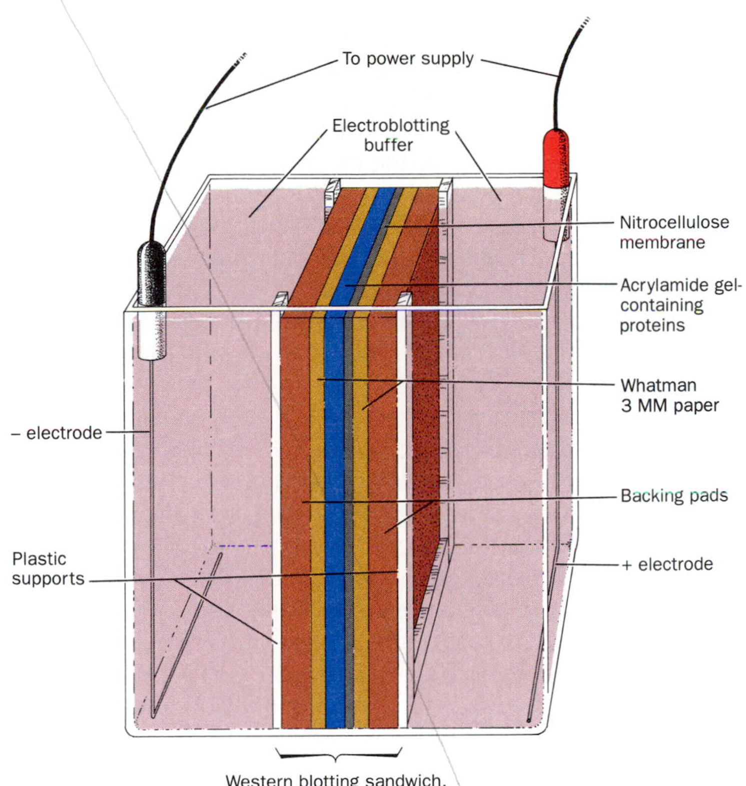

Figure 13.28 Diagram of a typical western blotting or electroblotting apparatus. An electric current is used to transfer the proteins from the gel to the nitrocellulose membrane placed next to it in the blotting sandwich. All other components of the sandwich function to provide gentle but firm support; tight contact between the gel and the membrane is essential for good transfer. All bubbles must be carefully removed from the sandwich to assure uniform flow of current and complete transfer of all proteins.

composed of two or more subunits, individual polypeptides are separated by electrophoresis in the presence of the detergent sodium dodecyl sulfate (SDS), which denatures the proteins. After electrophoresis, the proteins are commonly detected by treating with Coomassie blue or silver stain. However, the separated polypeptides in the gel can also be transferred to a nitrocellulose membrane, and individual proteins can be detected by using specific antibodies. This transfer of proteins from acrylamide gels to nitrocellulose membranes is called ***western blotting*** and is performed by using an electric current—so-called "electroblotting" (Fig. 13.28). After transfer, a specific protein of interest is identified by placing the membrane with the immobilized proteins in a solution containing an antibody to the protein. Nonbound antibodies are then washed off the membrane, and the presence of the initial ("primary") antibody is detected by placing the membrane in a solution containing a "secondary" antibody. This "secondary" antibody reacts with immunoglobulins (the group of proteins containing all antibodies) in general (see Chapter 16). The secondary antibody is conjugated to either a radioactive isotope (permitting autoradiography) or an enzyme that produces a visible product when the proper substrate is added. Figure 13.29 shows a western blot illustrating the use of an antibody to detect a single protein in an acrylamide gel containing total cellular proteins from maize roots and leaves. Obviously, the western blot procedure is a very powerful tool for identifying and characterizing specific gene-products.

Amplification of DNAs by the Polymerase Chain Reaction

The ***polymerase chain reaction,*** usually called just ***PCR,*** is an extremely powerful new procedure that allows one to amplify a selected DNA sequence in a genome a millionfold or more. PCR can be used to clone a given DNA sequence *in vitro*—without using living cells during the cloning process. However, the procedure can only be applied when the nucleotide sequence of at least one short DNA segment on each side of the region of interest is known. The PCR procedure involves using synthetic oligonucleotides complementary to the known sequences spanning the region of interest to prime ***enzymatic amplification of this segment of DNA in the test tube.*** The PCR procedure involves three steps, each repeated many times to produce cycles of amplification (Fig. 13.30). Step 1: The genomic DNA containing the sequence to be amplified is denatured by heating. Step 2: The denatured DNA is annealed to an excess of the synthetic oligonucleotide primers. Step 3: DNA polymerase is used to replicate the DNA segment between the sites complementary to the oligonucleotide primers. The primer provides the free 3′-OH required for covalent extension; the denatured genomic DNA provides the required template function (see Chapter 5). The products of the first cycle of replication are then denatured, annealed to oligonucleotide primers, and replicated again with DNA polymerase. The cycle is repeated many times (Fig. 13.30) until the desired level of amplification is achieved. Note that amplification occurs exponentially. One DNA double helix will yield 2 double helices after one cycle of replication, 4 after 2 cycles, 8 after 3 cycles, 16 after 4 cycles, 1024 after 10 cycles, and so on.

Initially, PCR was performed with the Klenow fragment of DNA polymerase I of *E. coli* as the replicase. Because this enzyme is heat-inactivated during the denaturation step, new enzyme had to be added at Step 3 of each cycle. A major improvement in PCR amplification of DNA came with the discovery of a

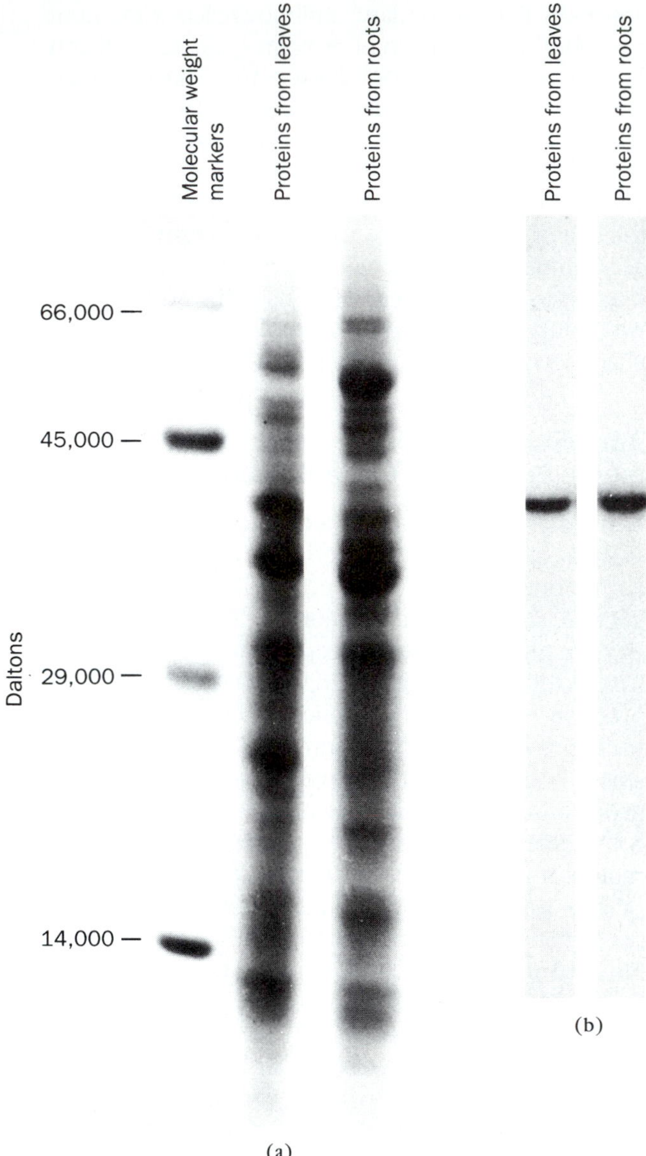

Figure 13.29 Illustration of the use of western blots to identify specific proteins after separation by acrylamide gel electrophoresis. (a) Total proteins in maize roots and leaves, respectively, were separated by electrophoresis through a 10% acrylamide gel, and proteins in the gel were stained with Coomassie blue. (b) Detection of the maize chloroplastic glutamine synthetase polypeptide by western blot analysis of the gel shown in (a). The probe was a monoclonal antibody specific for maize chloroplastic glutamine synthetase. The primary antibody probe was visualized by using a secondary antibody (see text) coupled to the enzyme horseradish peroxidase. In the presence of hydrogen peroxide, horseradish peroxidase converts the substrate diaminobenzidine to an insoluble dark product that yields the band observed on the western blot. (Photograph courtesy of M.-g. Li and D.P. Snustad, Department of Genetics and Cell Biology, University of Minnesota.)

heat-stable DNA polymerase in the thermophilic ("heat-loving") bacterium *Thermus aquaticus.* This polymerase, called ***Taq polymerase*** (after the genus and species name of the bacterium from which it was isolated), remains active during the heat denaturation step in each cycle of amplification. Thus, excess Taq polymerase and oligonucleotide primers can be added at the start of the PCR process, and amplification cycles can be carried out by cyclic alterations in temperature. PCR machines are now available that do this automatically, making the PCR amplification of a specific DNA sequence a rather simple task.

PCR technologies thus provide shortcuts for many cloning and sequencing applications. These procedures permit scientists to obtain definitive structural data on genes and DNA sequences when very small amounts of DNA are available. One important application involves the diagnosis of inherited human diseases, especially in cases of prenatal diagnosis (Chapter 19), where very limited amounts of fetal DNA are available. A second major application occurs in forensic cases involving individual identities. No criteria can provide more definitive evidence of identity than DNA sequences; after all, other phenotypic traits are controlled by the expression of these DNA sequences. By using PCR amplification, DNA sequences can be obtained from very small amounts of DNA isolated from small blood samples, sperm, or even individual human hairs. Thus, PCR DNA printing experiments are likely to play a major role in legal cases involving disputed identity.

TAILORING GENES TO SUIT ONE'S NEEDS: *IN VITRO* SITE-SPECIFIC MUTAGENESIS

Recombinant DNA and gene cloning technologies not only allow scientists to isolate and characterize any given gene from any living organism, but, in addition, these methodologies permit researchers to modify these genes *in vitro* to produce new restriction enzyme cleavage sites or to change a particular codon to a different codon or even a whole set of different codons. In short, scientists can now dissect genes at the nucleotide level by changing one nucleotide at a time and examining the effect of each change on the function of the gene or its product. Such nucleotide changes are accomplished by a procedure called ***site-specific mutagenesis*** or, more precisely, ***oligonucleotide-directed site-specific mutagenesis.***

Site-specific mutagenesis (Fig. 13.31) can be carried out quite easily on genes cloned in vectors that provide for the isolation of DNA single strands, for example, pUC 118 and pUC 119 (see Figs. 13.18 and 13.20). An oligonucleotide of 12 to 15 nucleotides in

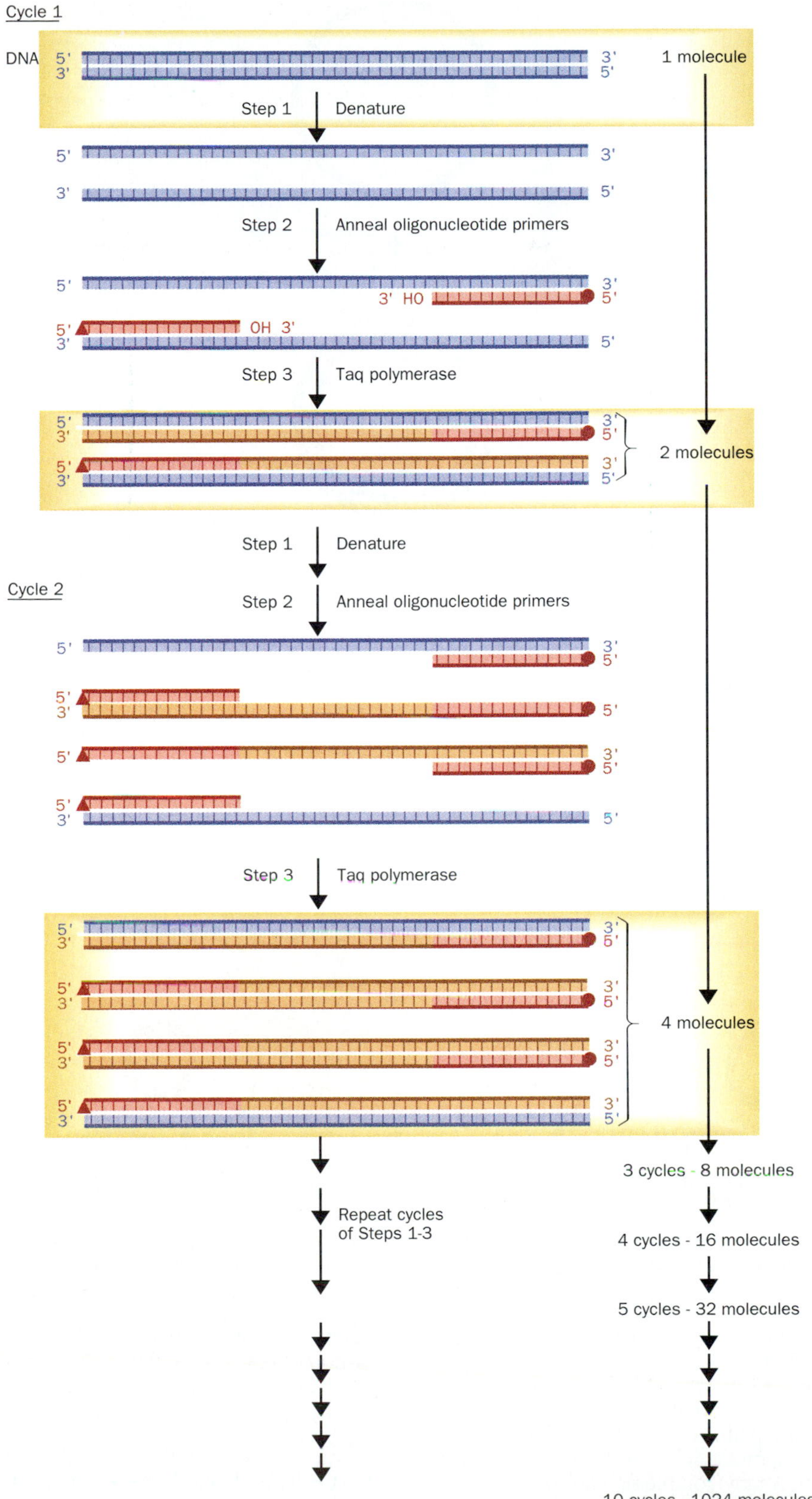

Figure 13.30 Diagram illustrating the use of PCR to amplify DNA molecules *in vitro*. Each cycle of amplification involves three steps: (1) denaturation of the genomic DNA being analyzed, (2) annealing of the denatured DNA to chemically synthesized oligonucleotide primers with sequences complementary to sites on opposite sides of the DNA region of interest, and (3) enzymatic replication of the region of interest by Taq polymerase. For details, see text.

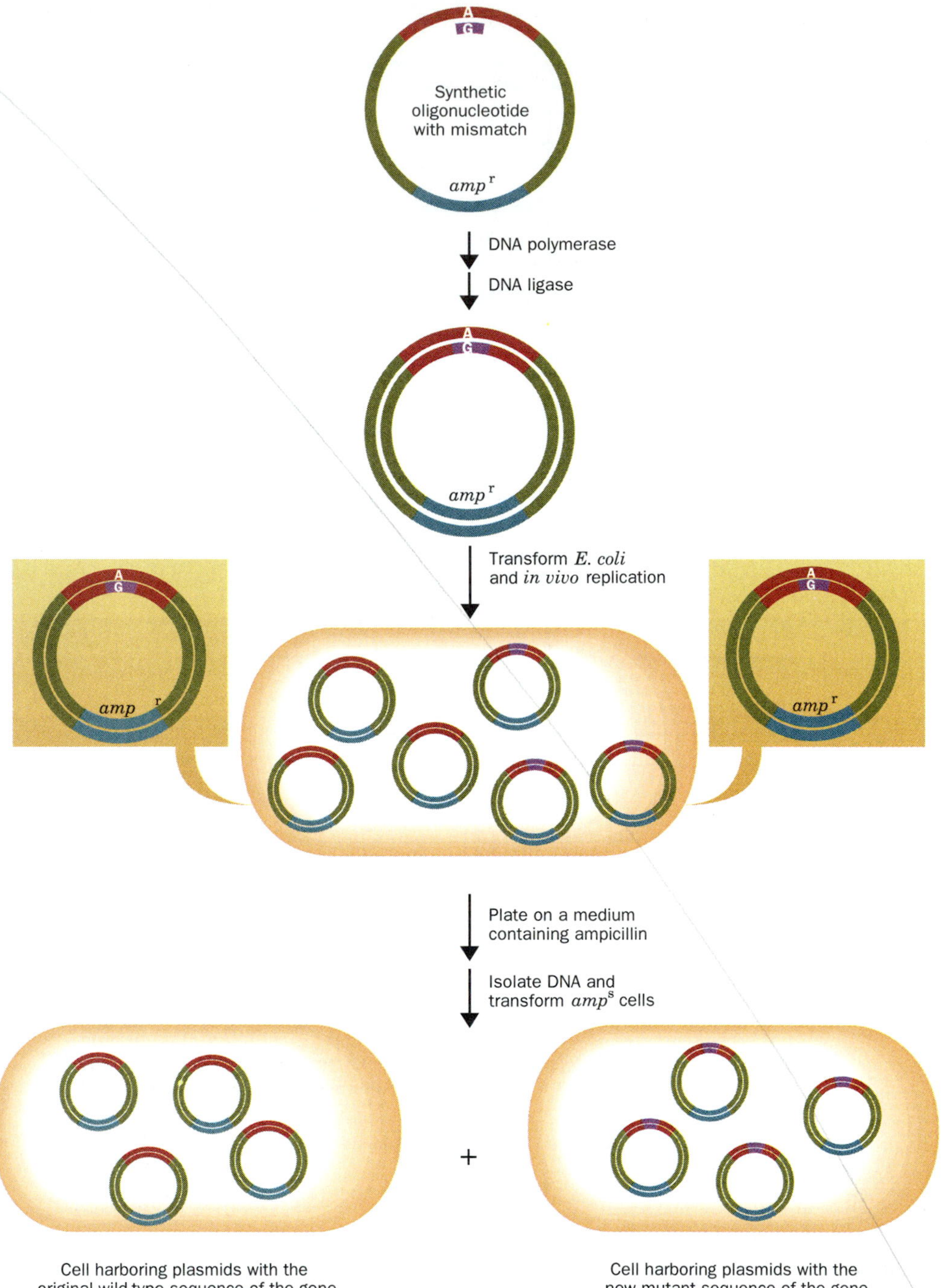

Figure 13.31 Oligonucleotide-directed site-specific mutagenesis performed with pUC 119 or related vectors. Single-stranded pUC 119 DNA carrying the gene of interest is purified and annealed to a synthetic oligonucleotide that contains the desired change in sequence. In the example diagrammed, an AT base-pair is converted to a CG base-pair. The annealed primer–template complex contains a single-base (A/G) mismatch. A new complementary strand is synthesized *in vitro* with DNA polymerase, and the resulting double helix is covalently sealed with DNA ligase. The heteroduplex DNA molecules are introduced into ampicillin-sensitive (*amp*^{s}) *E. coli* cells by transformation, and the cells are plated on medium containing ampicillin. Only transformed cells will be able to form colonies on this medium. Semiconservative replication *in vivo* will produce two types of homoduplexes: (1) half will contain the original, unmutated DNA sequence, and (2) half will contain the new, mutant DNA sequence. The two types of molecules can be separated by isolation of plasmid DNA from the original transformants and retransformation of *amp*^{s} cells using low DNA/cell ratios. The desired mutant DNA can be identified by hybridization to the original synthetic oligonucleotide (^{32}P end-labeled by using polynucleotide kinase) under conditions of high stringency, which require perfect complementarity for duplex formation.

length is synthesized that is largely complementary to the purified single-stranded copy of the gene in the region of interest, but that contains one or more noncomplementary or "mismatched" bases. The mismatched bases will provide the desired mutant sequences. The synthetic oligonucleotide is annealed to the single strand of the gene in the vector, and a complementary strand is synthesized *in vitro* with DNA polymerase (either the Klenow fragment of DNA polymerase I of *E. coli* or a modified form of the phage T7 polymerase called Sequenase or Taq polymerase). DNA ligase is then used to seal the single-strand break left by polymerase, and the double-stranded circular DNA products are introduced into *E. coli* cells by transformation. The semiconservative replication of the DNA molecules in *E. coli* will produce a population of progeny DNA molecules, half of which carry the original DNA sequence and half of which carry the new or mutant DNA sequence. The two types of molecules are separated by a second cycle of transformation, and the desired mutant genes are identified by hybridization to radioactively labeled synthetic oligonucleotide under conditions of high stringency (conditions that require perfect complementarity of the strands for duplexes to form) or by sequencing the modified region of the cloned gene.

One can enrich for the desired mutant sequence among progeny DNA molecules produced in an oligonucleotide-directed mutagenesis experiment by using single-stranded template DNA that contains some uracil residues incorporated at positions normally occupied by thymine bases. Such uracil-containing DNA can be produced in *E. coli dut$^-$ ung$^-$* mutant strains that lack two enzymes, dUTPase and uracil *N*-glycosylase. The enzyme dUTPase removes dUTP from the intracellular precursor pool and prevents uracil from being incorporated into DNA. In the absence of dUTPase, dUTP accumulates and deoxyuridine monophosphate groups are incorporated into DNA at positions normally occupied by thymidine monophosphate moieties. DNA polymerases will utilize both dUTPs and dTTPs as substrates for polymerization. The enzyme uracil *N*-glycosylase excises uracil bases from DNA leaving apyrimidinic sites (phosphodeoxyribose backbones with no base attached). These apyrimidinic sites block the replication of template strands in which they occur.

Single-stranded foreign DNA insert-vector DNA molecules synthesized in *dut$^-$ ung$^-$ E. coli* will contain some uracil (Fig. 13.32). This uracil-containing DNA can be annealed to synthetic oligonucleotides with mismatches, and complementary strands that contain only thymine (no uracil) bases can then be synthesized *in vitro*. In the test tube, DNA polymerases replicate uracil-containing templates just as efficiently

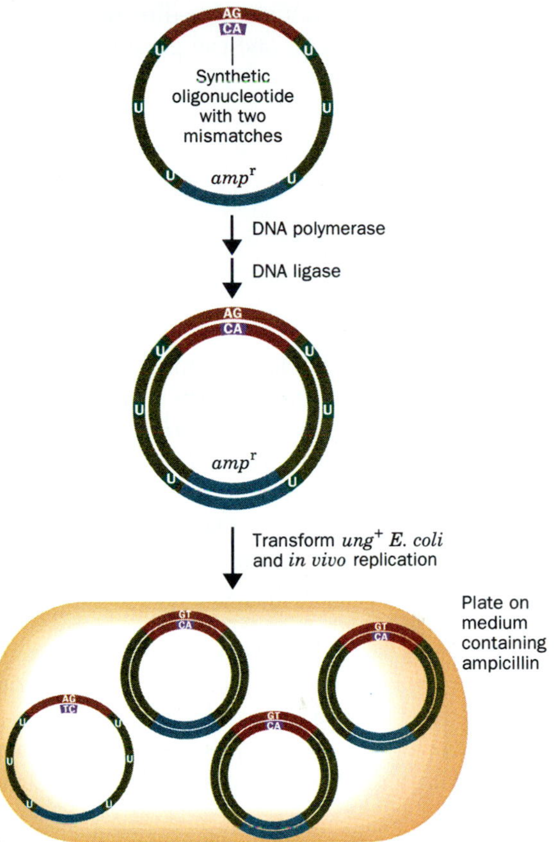

Figure 13.32 Enrichment for DNA molecules carrying the desired site-specific mutant by the use of uracil-containing template strands. The procedure is the same as that described in Fig. 13.31 with two exceptions. (1) The original template strands were synthesized in a *dut$^-$ ung$^-$ E. coli* double mutant and thus contain some uracil residues at positions normally occupied by thymine bases (see text). (2) The DNA heteroduplexes resulting from *in vitro* synthesis are introduced into *ung$^+$ E. coli* cells by transformation with the result that replication of the uracil-containing strand is largely blocked by the formation of apyrimidinic sites produced by uracil *N*-glycosylase, the *ung$^+$* gene-product (see text). Thus, most of the progeny DNA molecules produced in transformed cells will carry the desired mutant sequence.

as thymine-containing templates. After ligation, the double-stranded products ("old" strand containing U, "new" strand containing T) are introduced into *ung$^+$ E. coli* by transformation. The *ung$^+$* gene-product, uracil *N*-glycosylase, removes the uracil residues, leaving apyrimidinic sites that interfere with replication of the U-containing template strand. The strand synthesized *in vitro* by covalent extension of the oligonucleotide primer contains only thymine (no uracil). Thus, it will replicate efficiently and produce a large population of progeny molecules that contain the desired mutant sequence (Fig. 13.32). By using this simple modification of the oligonucleotide-directed site-specific mutagenesis protocol, one can greatly enrich for the desired mutant product and thus reduce the effort required to screen progeny molecules in order to identify a clone with the desired sequence.

SUMMARY

Technology is now available for constructing ***recombinant DNA molecules that contain sequences from totally unrelated species and cloning these molecules in appropriate host cells.*** The application of recombinant DNA technology to the question of the structure of genes has shown that ***eukaryotic genes*** frequently ***contain noncoding "introns"*** or ***"intervening sequences" located between coding sequences*** or ***"exons."*** The ***different exons*** of eukaryotic genes often code for ***different functional domains*** of the protein gene-products. In some cases, the introns of one gene may contain exons of a different gene.

Restriction maps of chromosomes are prepared by using various restriction enzymes to cleave DNA molecules and agarose and acrylamide gel electrophoresis to estimate the sizes of the fragments produced. Such restriction maps have proven invaluable in dissecting the molecular structure of genes and even of entire chromosomes.

DNA sequencing technologies permit scientists to determine the nucleotide-pair sequences of DNA molecules. Indeed, scientists are currently making plans to sequence the entire human genome within the next few decades. ***The nucleotide sequences of genes and chromosomes provide the ultimate fine structure maps.*** When complete with catalogs that list the nucleotide-pair changes that result in the loss of specific functions, such nucleotide sequence maps will provide a detailed picture of the structure and function of the gene or chromosome.

Cloning vectors designed for a wide variety of applications make the isolation and characterization of genes and other DNA sequences quite routine. ***Plasmid–phage "hybrid" vectors*** provide for the isolation of double-stranded or single-stranded DNA copies of any cloned gene. They also provide a simple color indicator screen for vectors that contain foreign DNA inserts. ***Transcription vectors*** permit one to transcribe either strand of a cloned gene *in vitro*. ***Synthetic oligonucleotide linkers*** can be used to add new restriction sites to DNA molecules thus facilitating their manipulation in the test tube.

Agarose and acrylamide gel electrophoresis are powerful tools for the separation of DNA and RNA molecules of different sizes. ***SDS polyacrylamide gel electrophoresis*** is a similarly powerful procedure for separating different polypeptides. DNA, RNA, or protein macromolecules that have been separated by gel electrophoresis can be transferred to nitrocellulose or nylon membranes for further analysis. The nitrocellulose or nylon membranes containing the immobilized DNA, RNA, or protein macromolecules after transfer from an agarose or acrylamide gel are called ***Southern blots, northern blots,*** and ***western blots,*** respectively. Southern blots and northern blots can be hybridized with radioactive DNA or RNA probes carrying sequences of interest, and the DNA or RNA bands containing complementary sequences can be detected by autoradiography. Individual proteins of interest can be detected on western blots by the use of antibody probes that have been conjugated to radioactive isotopes or to enzymes that provide a visible detection system.

The ***polymerase chain reaction (PCR)*** can be used to amplify a specific DNA sequence a millionfold or more by enzymatic replication of the sequence *in vitro*. PCR provides a powerful tool for studying the structure and function of DNA and RNA sequences. In addition, PCR has important applications in the diagnosis of human diseases and in forensic cases involving questions of identity.

Oligonucleotide-directed site-specific mutagenesis permits scientists to change nucleotide sequences in a predetermined manner. Site-specific mutagenesis can be used to dissect gene structure and function with resolution at the nucleotide level. In addition, it can be used to engineer modified genes and to introduce new restriction enzyme cleavage sites into DNA molecules at desired locations. Recombinant DNA and gene cloning technologies have advanced at an explosive pace during the last two decades. Genetic engineering is now a well-established discipline that has applications not only throughout biology, but also in plant and animal breeding, in medicine, and even in law.

REFERENCES

Anderson, W. F., and E. G. Diacumakos. 1981. "Genetic engineering in mammalian cells." *Sci. Amer.* 245(1): 60–93.

Ausubel, F. M., R. Brent, R. E. Kingston, D. D. Moore, J. D. Seidman, J. A. Smith, and K. Struhl. 1988. *Current protocols in molecular biology.* Green Publishing Associates/Wiley, New York.

Breathnach, R., and P. Chambon. 1981. "Organization and expression of eucaryotic split genes coding for proteins." *Annu. Rev. Biochem.* 50: 349–383.

Chambon, P. 1981. "Split genes." *Sci. Amer.* 244(5): 60–71.

Cohen, S. N. 1975. "The manipulation of genes." *Sci. Amer.* 233(1): 24–33.

Erlich, H. A., editor. 1989. *PCR technology.* Stockton Press, New York.

Hackett, P. B., J. A. Fuchs, and J. W. Messing. 1984. *An introduction to recombinant DNA techniques.* Benjamin/Cummings, Menlo Park, CA.

Innis, M. A., D. H. Gelfand, J. J. Sninsky, and T. J. White. 1989. *PCR protocols, a guide to methods and applications.* Academic Press, San Diego.

Lewin, B. 1990. *Genes IV.* Cell Press, Cambridge, MA. Oxford Univ. Press, New York.

Maxam, A. M., and W. Gilbert. 1977. "A new method for sequencing DNA." *Proc. Natl. Acad. Sci. U.S.A.* 74: 560–564.

Sambrook, J., E. F. Fritsch, and T. Maniatis. 1989. *Molecular cloning, a laboratory manual,* 2nd ed. Cold Spring Harbor Laboratory Press, Cold Spring Harbor, NY.

Sanger, F., A. R. Coulson, G. F. Hong, D. F. Hill, and G. B. Petersen. 1982. "Nucleotide sequence of bacteriophage λ DNA." *J. Molec. Biol.* 162: 729–773.

Sanger, F., S. Nicklen, and A. R. Coulson. 1977. "DNA sequencing with chain-terminating inhibitors." *Proc. Natl. Acad. Sci. U.S.A.* 74: 5463–5467.

Smith, H. O. 1979. "Nucleotide sequence specificity of restriction endonucleases." *Science* 205: 455–462.

Southern, E. M. 1975. "Detection of specific sequences among DNA fragments separated by gel electrophoresis." *J. Molec. Biol.* 98: 503–517.

Tilghman, S. M., D. C. Tiemeier, J. G. Seidman, B. M. Peterlin, M. Sullivan, J. V. Maizel, and P. Leder. 1978. "Intervening sequence of DNA identified in the structural portion of a mouse β-globin gene." *Proc. Natl. Acad. Sci. U.S.A.* 75:725–729.

Vieira, J., and J. Messing. 1987. "Production of single-stranded plasmid DNA." *Methods in Enzymol.* 153: 3–11.

Watson, J. D., and J. Tooze. 1981. *The DNA story.* Freeman, San Francisco.

White, T. J., N. Arnheim, and H. J. Erlich. 1989. "The polymerase chain reaction." *Trends in Genet.* 5: 185–189.

Wu, R., L. Grossman, and K. Moldave, editors. 1989. *Recombinant DNA methodology.* Academic Press, San Diego.

PROBLEMS AND QUESTIONS

13.1 (a) In what ways is the introduction of recombinant DNA molecules into host cells similar to mutation? (b) In what ways is it different?

13.2 What is the function of the "introns" or "intervening sequences" (noncoding) located between coding sequences within genes in eukaryotic organisms?

13.3 Of what value are recombinant DNA and gene cloning technologies to geneticists?

13.4 What determines the sites at which DNA molecules will be cleaved by a restriction endonuclease?

13.5 Restriction endonucleases are invaluable tools for biologists. However, genes encoding restriction enzymes obviously did not evolve to provide tools for scientists. Of what possible value are restriction endonucleases to the microorganisms that produce them?

13.6 Why is the DNA of a microrganism not degraded by a restriction endonuclease that it produces even though its DNA contains recognition sequences normally cleaved by the endonuclease?

13.7 One of the procedures for cloning foreign DNA segments takes advantage of restriction endonucleases like *Hin*dIII (see Table 13.1) that produce complementary single-stranded ends. These enzymes produce identical complementary ends on cleaved foreign DNAs and on the vector DNAs into which the foreign DNAs are inserted. What major advantage does this cloning strategy have over procedures that use terminal transferase to synthesize complementary single-stranded ends on foreign DNAs and vector DNAs *in vitro?*

13.8 The maize gene (*gln2*) that encodes the chloroplastic form of the enzyme glutamine synthetase is known to contain a single cleavage site for *Hin*dIII, but no cleavage site for *Eco*RI. You are given an *E. coli* plasmid cloning vector that contains a unique *Hin*dIII cleavage site within the gene (amp^r) that confers resistance to the antibiotic ampicillin on the host cell and a unique *Eco*RI cleavage site within a second gene (tet^r) that makes the host cell resistant to the antibiotic tetracycline. You are also given an *E. coli* strain that is sensitive to both ampicillin and tetracycline (amp^s tet^s). How would you go about constructing a maize genomic library (set of clones containing various fragments of genomic DNA) that would include clones carrying a complete *gln2* gene?

13.9 When one compares the nucleotide-pair sequences of genomic DNA clones and cDNA clones of specific genes of higher plants and animals, what is the most common difference observed?

13.10 Most of the genes of higher plants and animals that were cloned soon after the development of recombinant DNA and gene cloning technologies were genes encoding products that are synthesized in very large quantities in specialized cells or tissues. For example, about 90% of the protein synthesized in mature red blood cells of mammals consists of α- and β-hemoglobin chains, and the globin genes were among the first mammalian genes cloned. Why were genes of this type so prevalent among the first higher eukaryotic genes cloned?

13.11 Genomic clones of the chloroplastic glutamine synthetase gene (*gln2*) of maize are cleaved into two fragments by digestion with restriction endonuclease *Hin*dIII, whereas full-length maize *gln2* cDNA clones are not cut by *Hin*dIII. How can these results be explained?

13.12 You are studying a gene in *E. coli* that is expressed at 45°C, but not at 37°C. You have shown that the regulation of this gene occurs at the level of transcription. In addition, you have isolated a protein required for induction of transcription of the gene at high temperature and have demonstrated that it binds to a specific octameric nucleotide-pair sequence upstream (5′) from the gene. You have cloned the complete gene plus the upstream regulatory sequence in a plasmid, and you have available an *E. coli* mutant carrying a deletion of the entire gene. You now want to determine which of the nucleotide pairs in the 5′ octameric protein binding site are involved in the interaction with (binding of) the regulatory protein. How could this be determined experimentally?

13.13 What major advantage does the polymerase chain reaction (PCR) technique have over all other methods for analyzing nucleic acid structure and function?

13.14 Certain types of molecular analyses are facilitated by the availability of large quantities of pure single-stranded DNA from a normally double-stranded DNA molecule. However, the separation and purification of the two single strands of a given double helix is usually very difficult to accomplish by standard biochemical techniques. How have molecular biologists taken advantage of a natural biological mechanism for the purification of DNA single strands?

13.15 Almost all of the sophisticated cloning vectors—phage, plasmid, or phage–plasmid hybrid—in use today have one component (in addition to some type of origin of replication) in common. What is this component, and what is its function?

13.16 What advantages do transcription vectors have over other types of cloning vectors?

13.17 Suppose that a gene that you wish to clone is located on a *Hpa*I restriction fragment (i.e., the gene is flanked by restriction endonuclease *Hpa*I cleavage sites). Further suppose that you want to clone this *Hpa*I fragment containing the gene of interest into the *Hin*dIII site of pUC119. What is the simplest way to obtain the desired clone? (The *Hpa*I cleavage site is shown in Table 13.1.)

13.18 An important tool of molecular biologists involves the transfer of proteins that have been separated by gel electrophoresis to nitrocellulose membranes and the detection of specific proteins on the membranes by using antibodies and coupled radioactive labels or coupled enzymatic reactions. When this procedure is used, the resulting display of the visualized protein bands is called a "western blot." What is the significance of the name "western blot"?

13.19 Cystic fibrosis (CF) is an autosomal recessive disorder of humans characterized by unusually viscous mucus that interferes with the normal functioning of several exocrine glands including those in the skin (sweat glands), lungs (mucus glands), liver, and pancreas. The available evidence suggests that the CF gene-product is a transmembrane protein involved in the regulation of ion transport across epithelial cell membranes, but the exact function of the CF gene-product remains unknown. The CF gene (location: chromosome 7, region q31) has been cloned and sequenced, and studies of CF patients have shown that about ***70% of them are homozygous for a mutant CF allele that has a specific three nucleotide-pair deletion*** (equivalent to one codon). This deletion results in the loss of a phenylalanine residue at position 508 in the predicted CF gene-product. Assume that you are a genetic counselor responsible for advising families with CF in their pedigrees regarding the risk of CF among their offspring. How might you screen putative CF patients and their parents and relatives for the presence of the CFΔ508 mutant gene? What would the detection of this mutant gene in a family allow you to say about the chances of CF occurring again in the family?

13.20 Cereal grains are major food sources for humans and other animals in many regions of the world. However, most cereal grains contain inadequate supplies of certain of the amino acids that are essential for monogastric animals. For example, corn contains insufficient amounts of lysine, tryptophan, and threonine. Thus, a major goal of plant geneticists has been, and is, to produce corn varieties with increased kernel lysine content. As a prerequisite to the engineering of high-lysine corn, molecular biologists need more basic information about the regulation of the biosynthesis and the activity of the enzymes involved in the synthesis of lysine. The first step in the anabolic pathway unique to the biosynthesis of lysine is catalyzed by the enzyme dihydrodipicolinate synthase. Assume that (1) you have recently been hired by a major U.S. plant research institute, and (2) the job you have been assigned is to isolate a clone of the nucleic acid sequence encoding dihydrodipicolinate synthase in maize. Briefly describe four different approaches you might take in attempting to isolate such a clone and include at least one genetic approach.

13.21 (a) What common experimental procedure is carried out in Southern, northern, and western blot analyses? (b) What is the major difference between Southern, northern, and western blot analyses?

13.22 You have isolated a cDNA clone encoding a protein of interest in a higher eukaryote. This cDNA clone is ***not*** cleaved by restriction endonuclease *Eco*RI. When this cDNA is used as a radioactive probe for blot hybridization analysis of *Eco*RI-digested genomic DNA, three radioactive bands are seen on the resulting Southern blot. Does this result indicate that the genome of the eukaryote in question contains three copies of the gene encoding the protein of interest?

13.23 You are studying a circular plasmid DNA molecule of size 10.5 kilobase pairs (kb). When you digest this plasmid with restriction endonucleases *Bam*HI, *Eco*RI, and *Hin*dIII, singly and in all possible combinations, you obtain linear restriction fragments of the following sizes.

ENZYME(S)	FRAGMENT SIZE(S) IN KB
*Bam*HI	7.3, 3.2
*Eco*RI	10.5
*Hin*dIII	5.1, 3.4, 2.0
*Bam*HI + *Eco*RI	6.7, 3.2, 0.6
*Bam*HI + *Hin*dIII	4.6, 2.7, 2.0, 0.7, 0.5
*Eco*RI + *Hin*dIII	4.0, 3.4, 2.0, 1.1
*Bam*HI + *Eco*RI + *Hin*dIII	4.0, 2.7, 2.0, 0.7, 0.6, 0.5

Draw a restriction map (map of the restriction enzyme cleavage sites) for the plasmid that fits your data.

13.24 A particular gene is cloned using phage lambda as the cloning vector and is shown to contain two introns. (a) The primary transcript of this gene is purified from isolated nuclei. When this primary transcript is hybridized under R-loop conditions with the recombinant lambda chromosome carrying the gene, what will the R-loop structure(s) look like? Label your diagram. (b) The mRNA produced from the primary transcript of this gene is then isolated from cytoplasmic polyribosomes and similarly examined by the R-loop hybridization procedure using the recombinant lambda chromosome carrying the gene. Diagram what the R-loop structure(s) will look like when the cytoplasmic mRNA is used. Again, label the components of your diagram.

13.25 Ten micrograms of a decanucleotide-pair *Hpa*I restriction fragment were isolated from the double-stranded DNA chromosome of a small virus. Octanucleotide poly-A tails were then added to the 3′ ends of both strands using terminal transferase and dATP, that is,

5′-X X X X X X X X X X-3′
3′-X′X′X′X′X′X′X′X′X′X′-5′

↓ terminal transferase, dATP

5′-X X X X X X X X X X A A A A A A A A-3′
3′-A A A A A A A A X′X′X′X′X′X′X′X′X′X′-5′

where X and X′ can be any of the four standard nucleotides, but X′ is always complementary to X.

The two complementary strands (Watson strand and Crick strand) were then separated and sequenced by the 2′,3′-dideoxyribonucleoside triphosphate chain-termination method. The reactions were all primed using a radioactive (^{32}P-labeled) synthetic poly-T octamer, that is,

Watson strand

3′-A A A A A A A A X′X′X′X′X′X′X′X′X′X′-5′
5′-^{32}P-T T T T T T T T OH

Crick strand

5′-X X X X X X X X X X A A A A A A A A-3′
HO T T T T T T T T-^{32}P-5′

The usual four parallel reactions, (1) ddTTP, (2) ddCTP, (3) ddATP, and (4) ddGTP (plus the Klenow fragment of DNA polymerase I and all other substrates and required components), were carried out for both strands. Each reaction mixture was applied to a lane in a polyacrylamide slab gel, electrophoresed, and autoradiographed. The autoradiogram of the sequencing gel for one of the strands is shown on the left in the following diagram.

Draw the banding pattern that would be expected on the autoradiogram of the gel of reaction mixtures for the complementary strand on the diagram shown on the right.

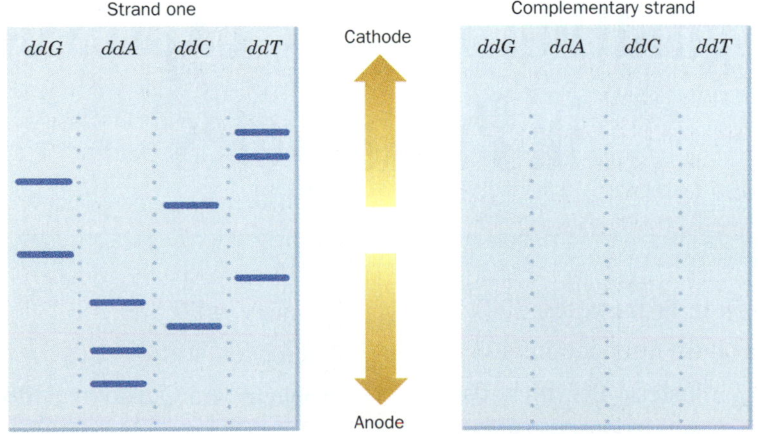

Chapter 14

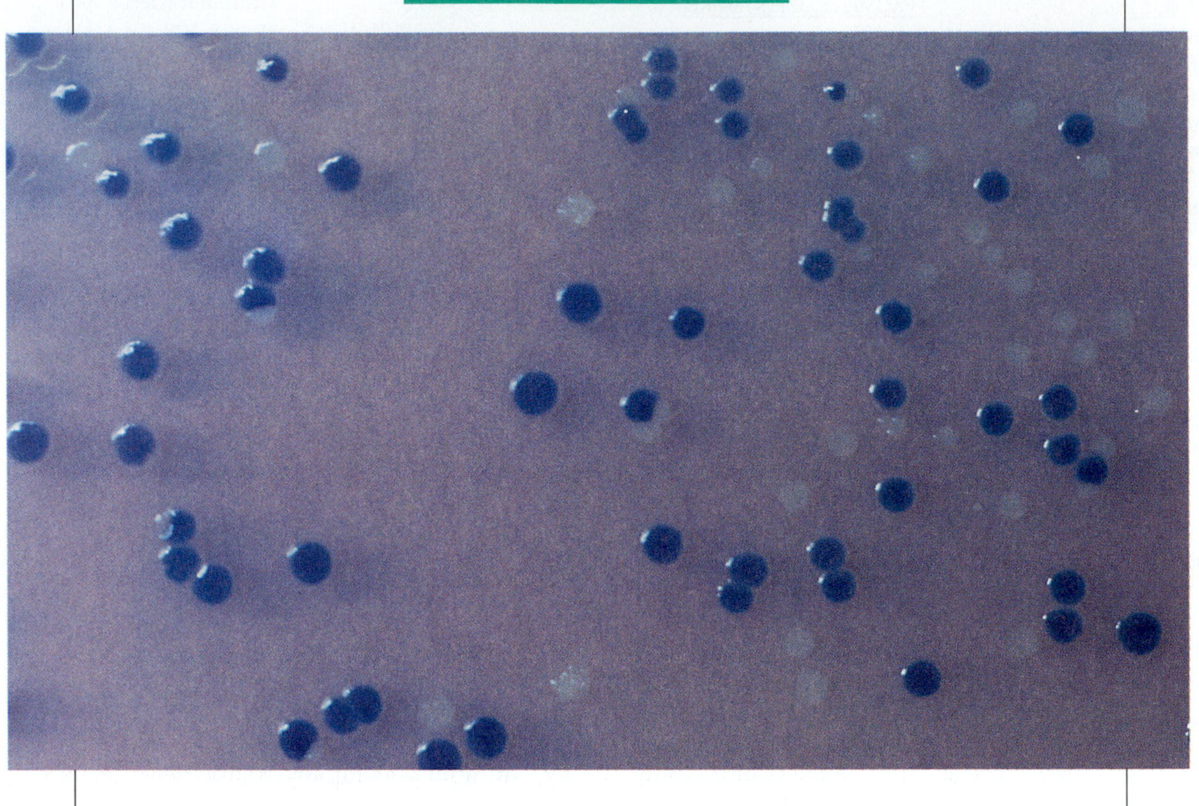

Regulation of Gene Expression in Prokaryotes

(Photo above): Colonies of lac^+ *E. coli* cells (blue) and lac^- *E. coli* cells (white) growing on X-gal medium. (Photo by Kristine Kirkeby, Biological Sciences Art Services, University of Minnesota.)

BACTERIA such as *Escheriscia coli* are exposed to a wide variety of environmental conditions. For example, *E. coli* cells may encounter rapidly changing growth conditions as they pass from mammalian intestinal tracts to sewer systems to polluted rivers, lakes, ponds, and so on. Each of these ecological niches will provide different organic molecules for use as energy sources. Thus, it seems quite reasonable to assume that natural selection will have preserved those organisms that have evolved ways of adapting to the wide range of environmental conditions encountered during their evolution. Indeed, the available information indicates that most prokaryotes such as *E. coli* exhibit remarkable capacities to adapt to diverse environmental conditions.

To a considerable degree, the adaptability of bacteria and other prokaryotes depends on their ability to "turn on" and "turn off" the expression of specific sets of genes in response to the specific demands of the environmental milieu. Stated differently, these organisms exhibit a striking ability to ***regulate the expression of specific genes in response to environmental signals***. The expression of particular genes is "turned on" when the products of these genes are needed for growth in a given environment. Their expression is "turned off" when their products are no longer needed for growth in the existing milieu. Clearly, the ability of an organism to regulate gene expression in this way will increase its overall "fitness" (its ability to grow and leave progeny under a variety of environmental conditions). The synthesis of gene transcripts and translation products requires the expenditure of considerable energy. By "turning off" the expression of genes when their products are not needed, an organism can avoid wasting energy and can utilize the conserved energy to synthesize products that maximize the growth rate in the existing milieu. What, then, are the mechanisms by which these organisms regulate gene expression in response to changes in the environment? Is there a single mechanism by which the expression of different genes or sets of genes are regulated? Or are different genes controlled by different mechanisms?

Certain genes, for example, the genes specifying ribosomal RNAs, ribosomal proteins, and transfer RNAs, are undoubtedly expressed at some time in virtually all cells regardless of the environmental conditions. The products of these genes are required for growth of all cells in all environments. However, the products of many other genes are required for growth only in certain environments, and the expression of these genes is regulated such that the products are synthesized only when they are needed. As a result, the expression of these genes is continually being "turned on" and "turned off" in response to changes in the environment.

As it turns out, ***gene expression can be (and is) regulated at several different levels***—for example, transcription, mRNA processing, mRNA turnover, translation, and enzyme function. However, extensive data indicate that the ***regulation of transcription is the most important mode of the control of gene expression***, at least in prokaryotes. That is not to say that regulation does not occur at other levels. Regulatory fine-tuning at translational levels is clearly important in the overall control of metabolic processes in living organisms. The regulatory mechanisms with the largest effects on phenotype, however, have been shown to act at the level of transcription.

Based on what is presently known about the regulation of transcription in both prokaryotes and eukaryotes, the various regulatory mechanisms seem to fit into two general categories. The first, and best understood, category includes mechanisms involved in the ***rapid turn-on and turn-off of gene expression in response to environmental changes***. Regulatory mechanisms of this type are very important in microorganisms because of the frequent exposure of these organisms to sudden changes in environment. They provide microorganisms with a great deal of "plasticity," an ability to rapidly adjust their metabolic processes in order to achieve maximal growth and reproduction under highly variable environmental conditions. These quick responding on–off switches seem to be less important in higher eukaryotes. This might be expected since the circulatory systems of higher eukaryotes buffer their cells against many sudden environmental changes.

The second major category of regulatory mechanisms includes what might be called ***preprogrammed circuits of gene expression***. In these cases, some event (such as infection by a virus) triggers the expression of one set of genes. The product (or products) of one (or more) of these genes functions by turning off the transcription of the first set of genes and/or turning on the transcription of a second set of genes. In turn, one or more of the products of the second set acts by turning on a third set, and so on. In these cases, the sequential expression of genes is genetically preprogrammed, and the genes usually ***cannot*** be turned on out of sequence. Such preprogrammed sequences of gene expression in viral infections are well documented. In most of these preprogrammed sequences, it seems the circuitry is cyclical. For example, in viral infections some event associated with the packaging of the viral DNA or RNA inside the protein coat somehow seems to reset the program so that the first set of genes will again be expressed when a progeny virus subsequently infects another host cell.

INDUCTION AND REPRESSION IN PROKARYOTES

Certain gene-products, such as tRNA molecules, rRNA molecules, ribosomal proteins, RNA polymerase components (polypeptides), and other enzymes catalyzing metabolic processes that are frequently referred to as cellular "housekeeping" functions, are essential components of almost all living cells. Genes that specify products of this type are ***continually being expressed*** in most cells. Such genes are said to be expressed ***constitutively*** and are frequently referred to as ***constitutive genes.***

Other gene-products are needed for cell growth only under certain environmental conditions. Constitutive synthesis of such gene-products would clearly be wasteful, using energy that could otherwise be utilized for more rapid growth and reproduction under the existing environmental conditions. The evolution of regulatory mechanisms that would provide for the synthesis of such gene-products only when and where they were needed would clearly provide organ isms possessing these regulatory mechanisms with a selective advantage over organisms lacking these mechanisms. This undoubtedly explains why presently existing organisms, including the "primitive" bacteria and viruses, exhibit highly developed and very efficient mechanisms for the control of gene expression.

Escherichia coli and most other bacteria are capable of growth using any one of several carbohydrates (e.g., glucose, sucrose, galactose, arabinose, lactose) as an energy source. If glucose is present in the environment, it will be preferentially metabolized by *E. coli* cells. In the absence of glucose, however, *E. coli* cells can grow very well on other carbohydrates. Cells growing in medium containing the sugar lactose, for example, as the sole carbon source synthesize two enzymes, β-galactosidase and β-galactoside permease, that are uniquely required for the catabolism of lactose. (A third enzyme, β-galactoside transacetylase, is also synthesized. It has no ***known*** metabolic function, however.) β-galactosidase cleaves lactose into glucose and galactose, and β-galactoside permease pumps β-galactosides into the cell. Neither of these enzymes is of any use to *E. coli* cells when present in environments not containing lactose. The synthesis of these two enzymes, of course, requires the utilization of considerable energy (in the form of ATP and GTP; see Chapter 10). Thus, *E. coli* cells have evolved a regulatory mechanism by which the synthesis of these lactose-catabolizing enzymes is turned on in the presence of lactose and turned off in its absence.

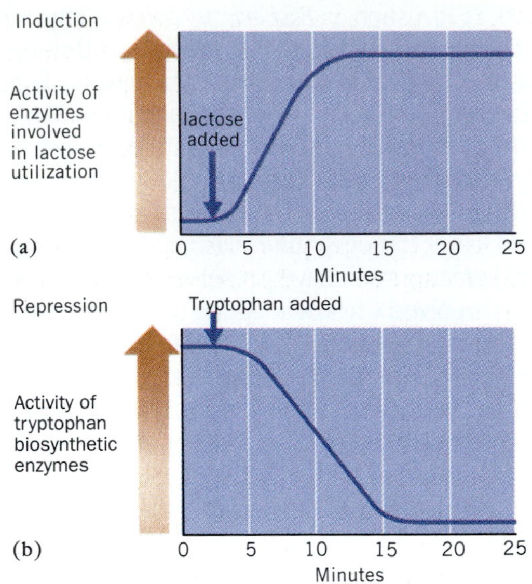

Figure 14.1 ***Induction*** (a) and ***repression*** (b) of enzyme synthesis in bacteria. Induction is characteristic of catabolic (degradative) pathways; repression is characteristic of anabolic (biosynthetic) pathways. (a) The induction of the synthesis of enzymes required for the utilization of the sugar lactose as an energy source in *E. coli* is illustrated. In the absence of lactose in the environment, *E. coli* cells synthesize only very minute amounts of the lactose-utilizing enzymes. When such cells are transferred to an environment containing lactose as the sole carbon source (occurring at the time indicated by the arrow labeled ***lactose added***), the synthesis of the enzymes required for lactose catabolism is rapidly ***induced*** (turned on). (b) The repression of the synthesis of the enzymes required for the biosynthesis of tryptophan in *E. coli* is illustrated. When tryptophan is not present in the environment, the *E. coli* cells synthesize the enzymes required for tryptophan biosynthesis. If tryptophan is added to the environment of such cells (e.g., at the time indicated by the arrow labeled ***tryptophan added***), the synthesis of the tryptophan biosynthetic enzymes is rapidly ***repressed*** (turned off). The kinetics shown are only approximate.

In natural environments (intestinal tracts and sewers), *E. coli* cells probably encounter an absence of glucose and the presence of lactose relatively infrequently. Most of the time, therefore, the *E. coli* genes coding for the enzymes involved in lactose utilization are not being expressed. If cells growing on a carbohydrate other than lactose are transferred to medium containing lactose as the only carbon source, they rapidly begin synthesizing the enzymes required for lactose utilization (Fig. 14.1*a*). This process, by which the expression of genes is turned on in response to a substance in the environment, is called ***induction***. Genes whose expression are so regulated are called ***inducible genes***; their products, if enzymes, are called ***inducible enzymes***. The substances or molecules responsible for induction are known as ***inducers***.

Enzymes that are involved in ***catabolic (degradative) pathways***, such as in lactose, galactose, or arabinose utilization, are characteristically inducible. As will become apparent in the following sections of this chapter, ***induction occurs at the level of transcription***. It alters the rate of synthesis of enzymes, not the activity of existing enzyme molecules. Induction should not be confused with ***enzyme activation***, in which the binding of a small molecule to an enzyme increases the ***activity*** of the enzyme (but does not affect its rate of synthesis).

Bacteria possess the metabolic capacity to synthesize most of the organic molecules (such as amino acids, purines, and vitamins) required for their growth. For example, *E. coli* has five genes coding for enzymes that are required in the synthesis of tryptophan. These five genes must be expressed in *E. coli* cells growing in an environment devoid of tryptophan in order to provide adequate amounts of this amino acid for ongoing protein synthesis.

When *E. coli* cells are present in an environment containing concentrations of tryptophan sufficient to support optimal growth, the continued synthesis of the tryptophan biosynthetic enzymes would be a waste of energy, because these bacteria have the capacity to take in external tryptophan. Thus, a regulatory mechanism has evolved in *E. coli* by which the synthesis of the tryptophan biosynthetic enzymes is turned off when tryptophan is present in the external milieu (Fig. 14.1*b*). This process of "turning off" the expression of sets of genes is called ***repression***. A gene whose expression has been turned off in this way is said to be ***repressed***; when its expression is turned on, a gene of this type is said to be ***derepressed***.

Enzymes that are components of ***anabolic (biosynthetic) pathways*** are frequently subject to ***repression*** (are ***repressible***). ***Repression***, like induction, ***occurs at the level of transcription***. Repression should not be confused with ***feedback inhibition***, in which the binding of an end product to the first enzyme in a biosynthetic pathway ***inhibits*** the ***activity*** of the enzyme (but does not affect its synthesis).

The Operon Model

Induction and repression of gene expression can be accomplished by essentially the same mechanism. This mechanism was first accurately described in 1961 when F. Jacob and J. Monod, both 1965 Nobel Prize recipients, proposed the ***operon model*** to explain the regulation of genes encoding the enzymes required for lactose utilization in *E. coli*. Jacob and Monod proposed that the transcription of one or a set of contiguous structural genes (genes coding for polypeptides) is regulated by two controlling elements (Fig. 14.2*a*). One of these elements, called the ***regulator gene*** (or ***repressor gene***), codes for a protein called the ***repressor***; under the appropriate conditions, the repressor binds to the second element, the ***operator*** (or ***operator sequence***). The operator is always located contiguous to the structural gene or genes whose expression it regulates. ***When the repressor is bound to the operator, transcription of the structural genes cannot occur***. We now know that this results because the binding of the repressor to the operator sterically prevents RNA polymerase from binding at the ***promoter site*** (the RNA polymerase binding site; see Chapter 10), which is always located contiguous with (or even overlapping) the operator sequence. The operator is usually located between the promoter and the structural genes (Fig. 14.2*a*). (The promoter was not recognized at the time of Jacob and Monod's proposal, but has since been shown to be an essential component of an operon.) The complete contiguous unit, including the structural gene or genes, the operator, and the promoter, is called an ***operon***.

Whether the repressor will bind to the operator and turn off the transcription of the structural genes in an operon is determined by the presence or absence of ***effector molecules*** (small molecules such as amino acids and sugars) in the environment. In the case of inducible operons, these effector molecules are called ***inducers***. Those active on repressible operons are called ***co-repressors***. These effector molecules (inducers and co-repressors) act by binding to (or forming a complex with) the repressors.

The only essential difference between inducible operons and repressible operons is whether the naked repressor or the repressor–effector molecule complex is active in binding to the operator. (1) In the case of an inducible operon, the ***free repressor binds to the operator***, turning off transcription (Fig. 14.2*b*). When the effector molecule (the inducer) is present, it binds to the repressor, releasing the repressor from the operator, that is, the ***repressor–inducer complex cannot bind to the operator***. Thus, the addition of inducer turns on or induces the transcription of the structural genes in the operon (Fig. 14.2*b*). (2) In the case of a repressible operon, the situation is just reversed. ***The free repressor cannot bind to the operator. Only the repressor–effector molecule (co-repressor) complex is active in binding to the operator*** (Fig. 14.2*c*). Thus, transcription of the structural genes in a repressible operon is turned on in the absence of and turned off in the presence of the effector molecule (co-repressor).

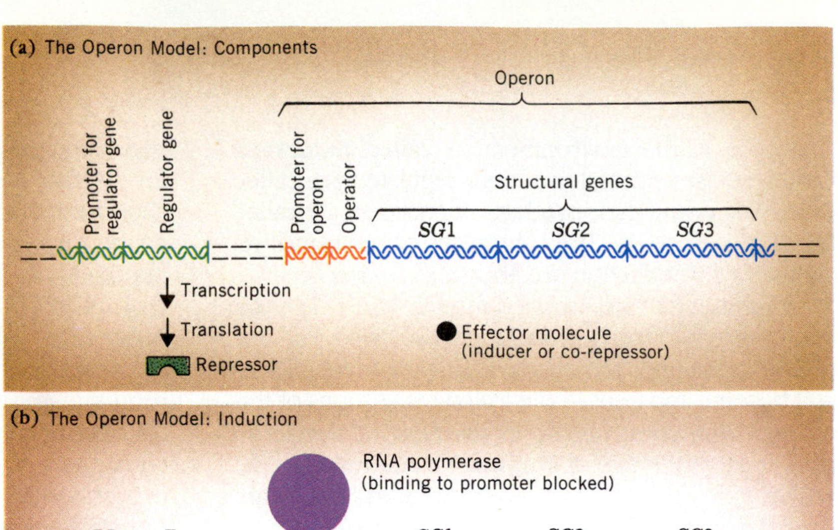

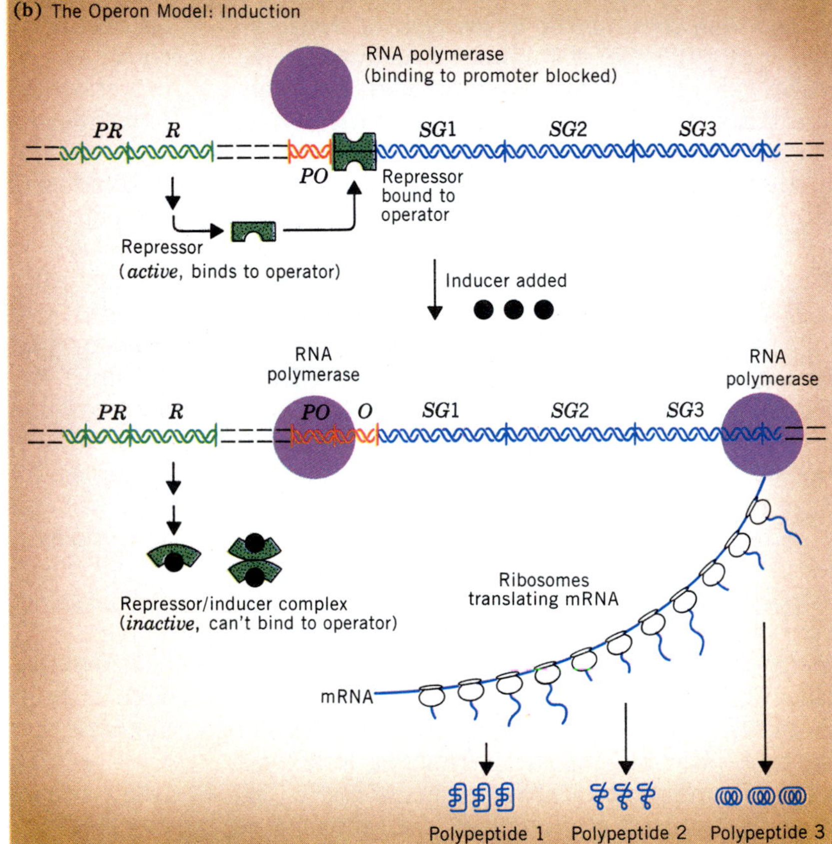

Figure 14.2 The operon model for regulation of gene expression. (a) Diagram showing the essential components of regulation specified by the operon model. The ***operon*** consists of one or more ***structural genes*** (three—*SG1, SG2,* and *SG3*—are arbitrarily shown) and the adjoining ***operator*** and ***promoter*** sequences. The promoter for the operon (*PO*) is the site at which RNA polymerase must bind to initiate transcription of the structural genes. The operator (*O*) is the site at which the protein ***repressor***—the product of the ***regulator gene*** (or ***repressor gene***)—binds. The regulator gene need not be closely linked to the operon; in fact, it can be located at any position in the genome. The transcription of the regulator gene is initiated by RNA polymerase, which binds to its promoter (labeled *PR*, for promoter for regulator gene). When the repressor is bound to the operator, it sterically prevents RNA polymerase from binding to the adjoining promoter (*PO*) and from initiating transcription of the structural genes. Whether the repressor binds to the operator or not depends on the presence or absence of a metabolite called an ***effector molecule.*** (b) Mode of regulation of gene expression for an ***inducible operon.*** The product of the regulator gene (*R*), the repressor, in the ***absence*** of

Except for this difference in the operator-binding behavior of the repressor, inducible and repressible operons are comparable.

A single mRNA transcript carries the coding information of an entire operon. Thus, the mRNAs of operons consisting of more than one structural gene are polygenic. For example, the tryptophan operon mRNA of *E. coli* is a huge macromolecule carrying the coding sequences that specify five different polypeptides (see Fig. 14.5). Because of their cotranscription, all the structural genes in an operon are ***coordinately expressed.***

Because the product of the regulator gene, the repressor, acts by shutting off the transcription of structural genes, the operon model, as originally proposed by Jacob and Monod, is referred to as a ***negative control*** system. In ***positive control*** systems, the products of regulator genes are required to turn on tran-

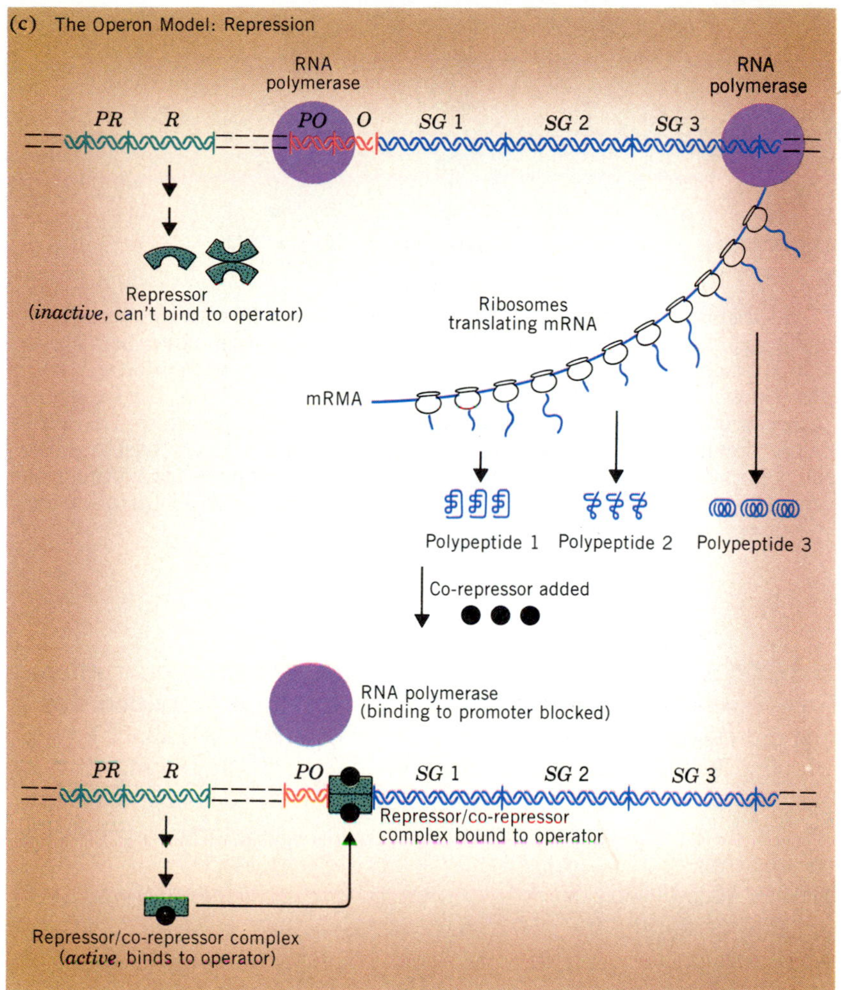

the effector molecule (called an ***inducer,*** for inducible operons), binds to the operator, preventing RNA polymerase from binding to the promoter for the operon (*PO*). Thus, transcription of the structural genes ***cannot*** occur. When inducer is added, it binds to the repressor, causing it to be released from the operator (*O*). This, in turn, allows RNA polymerase to bind to the promoter (*PO*) and initiate transcription of the structural genes. The resulting multigenic mRNA is rapidly translated by ribosomes, producing the three polypeptide products of the structural genes. (c) Mode of regulation of gene expression for a ***repressible operon.*** In this case, the repressor can ***only bind to the operator*** in the ***presence*** of the ***effector molecule*** (called a ***co-repressor,*** for repressible operons). In its absence, the operator is free, permitting RNA polymerase to bind at the adjoining promoter (*PO*) and to initiate transcription of the structural genes. When co-repressor is added, it forms a complex with the repressor. This repressor–co-repressor complex then binds to the operator (*O*). This, in turn, prevents RNA polymerase from binding at *PO* and transcribing the three structural genes.

scription. We will discuss examples of positive control mechanisms later in this chapter.

lac, an Inducible Operon

Jacob and Monod proposed the operon model largely as a result of their studies of the *lac* operon of *E. coli*. More is known about the *lac* operon than any other operon. The *lac* operon contains a promoter, an operator, and three structural genes, *z*, *y*, and *a*, coding for the enzymes β-galactosidase, β-galactoside permease, and β-galactoside transacetylase, respectively (Fig. 14.3). β-galactoside permease "pumps" lactose into the cell, where β-galactosidase cleaves it into glucose and galactose (Fig. 14.4). The function of the transacetylase is still not clear.

The *lac* regulator gene, designated the *i* gene, codes for a repressor that is 360 amino acids in length.

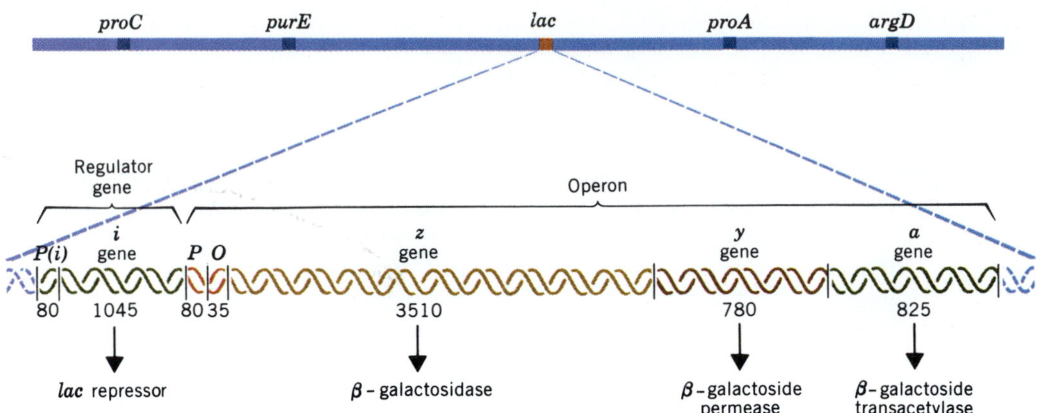

Figure 14.3 The *lac* operon of *E. coli,* an inducible operon. The *lac* operon consists of three structural genes, *z*, *y*, and *a*, plus the promoter (*P*) and operator (*O*) regions adjoining the *z* gene. The regulator gene (*i*) is contiguous with the operon in the case of *lac*. Regulator genes of other operons are frequently located at a considerable distance from the operon. (Note that the regulator gene is normally not considered to be part of the operon proper, which consists of the structural genes plus their promoter and operator.) The regulator gene has its own promoter (*P(i)*). The numbers below the various genes indicate their approximate lengths in nucleotide-pairs. The nucleotide sequences for *P(i)*, *P*, and *O* are known (see Fig. 14.7). The *lac* operon functions as shown in Fig. 14.2*b*, with allolactose, a derivative formed from lactose in the presence of β-galactosidase, acting as an inducer. Thus, the *lac* inducer is produced by the small amounts of β-galactosidase and β-galactoside permease that are present in uninduced cells.

The active form of the *lac* repressor, however, is a tetramer that contains four copies of the *i* gene-product. In the absence of the inducer, the repressor binds to the *lac* operator sequence, preventing RNA polymerase from binding to the promoter and transcribing the structural genes (see Fig. 14.2*b*). A few molecules of the *z*, *y*, and *a* gene-products are synthesized in the uninduced state, providing a low background level of enzyme activity. However, this background activity is essential for induction of the *lac* operon, because the ***inducer*** of the operon, ***allolactose***, is derived from lactose in a reaction that is catalyzed by β-galactosidase (Fig. 14.4). Once formed, allolactose binds to the repressor, causing it to be released from the operator; in so doing, it ***induces*** transcription of the *z*, *y*, and *a* structural genes (see Fig. 14.2*b*).

Formation of *lac* inducer

Lactose + H_2O β-galactosidase

Allolactose

Catabolism of lactose

Glucose + Galactose

Figure 14.4 The two physiologically important reactions that are catalyzed by β-galactosidase: (1) conversion of lactose to the *lac* operon inducer allolactose and (2) cleavage of lactose to produce the monosaccharides glucose and galactose, the first step in the degradation of lactose.

The *lac i* gene, operator, and promoter were all initially identified genetically by the isolation of mutations within these genetic units that rendered them nonfunctional. Mutations within the *i* gene and the operator frequently result in the constitutive synthesis of the lactose-utilizing enzymes. These mutations are designated i^- and o^c, respectively. The i^- and o^c constitutive mutations can be distinguished not only by map position, but also by their behavior in F′ merozygotes (see Chapter 8, p. 220) in which they are located in *cis*- and *trans*-configurations relative to mutations in *lac* structural genes.

Merozygotes (partial diploids) of genotype F′ $i^+ o^+ z^+ y^+ a^+/i^+ o^+ z^- y^- a^-$ or of genotype F′ $i^+ o^+ z^- y^- a^-/i^+ o^+ z^+ y^+ a^+$ are inducible for the utilization of lactose as a carbon source, just like haploid wild-type $(i^+ o^+ z^+ y^+ a^+)$ cells. The wild-type alleles (z^+, y^+, and a^+) of the three structural genes are dominant to their mutant alleles (z^-, y^-, and a^-). This is to be expected because the wild-type alleles produce functional enzymes, whereas the mutant alleles produce no enzymes or defective (inactive) enzymes. F′ merozygotes that have the genotype $i^+ o^+ z^+ y^+ a^+/i^- o^+ z^+ y^+ a^+$ are also inducible for the synthesis of the three enzymes specified by the *lac* operon. Thus i^+ is dominant to i^- as expected, since i^+ specifies an active protein (repressor molecule) and i^- yields an inactive protein.

Merozygotes that have the genotype F′ $i^+ o^+ z^+ y^+ a^+/i^- o^+ z^- y^- a^-$ or the genotype F′ $i^+ o^+ z^- y^- a^-/i^- o^+ z^+ y^+ a^+$ are ***inducible*** for β-galactosidase, β-galactoside permease, and β-galactoside transacetylase, like wild-type cells. This indicates that the *i* gene codes for a diffusible product, since it affects the expression of structural genes located in either *cis*- or *trans*-configuration in relation to itself. On the other hand, the operator constitutive o^c mutations ***act only in* cis**, that is, o^c mutations cause only the constitutive expression of structural genes located on the same chromosome. Of course, this is to be expected if the operator is the repressor binding site; as such, it does not code for any product, diffusible or otherwise. A merozygote of genotype F′ $i^+ o^c z^- y^- a^-/i^+ o^+ z^+ y^+ a^+$ is thus inducible for the three enzymes specified by the structural genes of the *lac* operon, whereas a merozygote of genotype F′ $i^+ o^c z^+ y^+ a^+/i^+ o^+ z^- y^- a^-$ synthesizes these enzymes constitutively.

Some of the *i* gene mutations, those designated i^{-d}, are dominant to the wild-type allele (i^+). This dominance apparently results from the inability of heteromultimers (recall that the *lac* repressor functions as a tetramer), which contain both wild-type and mutant polypeptides, to bind to the operator sequence. Other *i* gene mutations, those designated i^{-s} (for "superrepressed"), cause the *lac* operon to be ***uninducible***. (Strains carrying these i^{-s} mutations can usually be induced to some degree by using extremely high concentrations of inducer. They are not induced at normal inducer concentrations.) When studied *in vitro*, the mutant i^{-s} polypeptides form tetramers that bind to *lac* operator DNA. They either do not bind inducer, however, or exhibit a very low affinity for inducer. The i^{-s} mutations therefore modify the inducer binding site of the *lac* repressor.

Promoter mutations do not alter the inducibility of the *lac* operon, instead they modify the levels of gene expression in the induced and uninduced state by changing the frequency of initiation of *lac* operon transcription (i.e., the efficiency of RNA polymerase binding).

The *lac* promoter actually contains two functionally distinct components: (1) the RNA polymerase binding site and (2) a binding site for another protein, called catabolite activator protein (abbreviated CAP), that functions such that the *lac* operon is not transcribed in the presence of glucose at concentrations sufficient to support optimal growth. This second control circuit assures the preferential utilization of glucose as an energy source when it is available.

trp, a Repressible Operon

The *trp* (tryptophan) operon of *E. coli* is probably the best-known repressible operon. The organization of the five structural genes and the adjacent regulatory sequences of the *trp* operon (Fig. 14.5) has been analyzed in detail by Charles Yanofsky and colleagues. The regulation of transcription of the *trp* operon occurs as diagrammed in Fig. 14.2*c*. The *trp* operon repressor is the product of gene *trpR*, which is not closely linked to the *trp* operon (Fig. 14.5).

In the absence of tryptophan (the co-repressor), RNA polymerase binds to the promoter region and transcribes the structural genes of the operon. In the presence of tryptophan, the co-repressor/repressor complex binds to the operator region and prevents the binding of RNA polymerase to the promoter. The operator sequence of the *trp* operon lies entirely within the promoter region (Fig. 14.5).

The rate of transcription of the *trp* operon in the derepressed state (absence of tryptophan) is 70 times the rate that occurs in the repressed state (presence of tryptophan). In *trpR* mutants that cannot make repressor, the rate of synthesis of the tryptophan biosynthetic enzymes (the products of the structural genes of the *trp* operon) is still reduced about 10-fold by the addition of tryptophan to the medium. This reduction is due to a second level of regulation of *trp* operon expression called attenuation. Attenuation occurs by

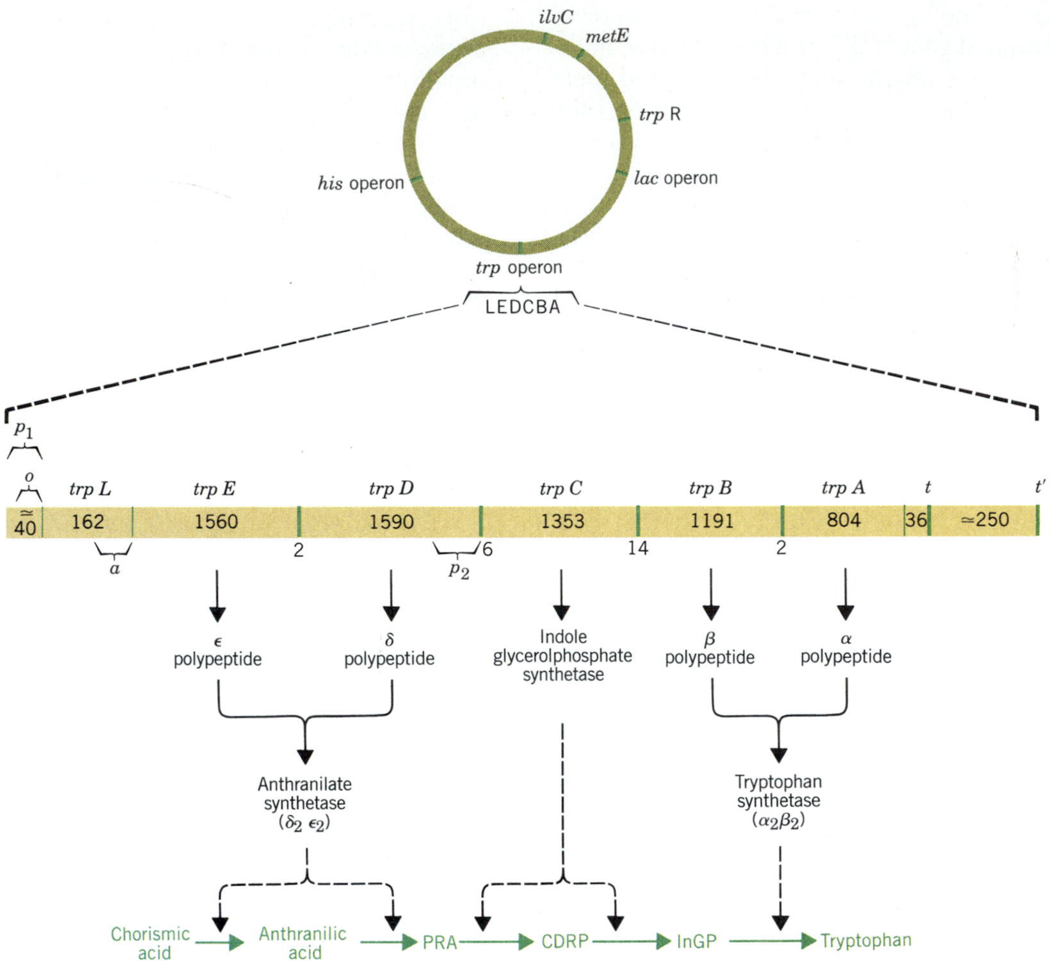

Figure 14.5 Organization of the tryptophan (*trp*) operon of *E. coli.* (The *Salmonella typhimurium trp* operon is essentially identical.) The operon contains five structural genes that encode the enzymes involved in the biosynthesis of tryptophan as shown at the bottom. The *trpR* gene, which encodes the *trp* repressor, is not closely linked to the *trp* operon (top). The operator (*o*) region of the *trp* operon lies entirely within the primary promoter (p_1) region. There is also a weak promoter (p_2), at the operator–distal end of the *trpD* gene, that results in a somewhat increased basal level of constitutive transcription of the *trpC, B* and *A* genes. There are two transcription termination sequences (*t* and *t'*) downstream from *trpA*. The *trpL* region specifies a 162-nucleotide-long mRNA leader sequence; it contains the attenuator (*a*) region that provides a second level of control of *trp* operon expression (see pp. 403–406). The p_1 promoter actually extends about 18 nucleotide-pairs into *trpL*. The length of each gene or region is given in nucleotide-pairs; the intergenic distances are given in nucleotide-pairs below the gene sequence. The abbreviations used are: PRA = phosphoribosyl anthranilate; CDRP = carboxyphenylamino-deoxyribulose phosphate; InGP = indole-glycerol phosphate. (Based on the data summarized by C. Yanofsky, *Nature* 289: 751–758, 1981, and by T. Platt, *Cell* 24: 10–23, 1981.)

tryptophan-mediated termination of transcription in the *trpL* (mRNA leader) region of the operon (see pp. 403–406).

Positive Control of the *lac* Operon by CAP and Cyclic AMP

The operon model was proposed by Jacob and Monod to explain the induction of the biosynthesis of the enzymes involved in lactose utilization when this sugar is added to the medium in which *E. coli* cells are growing. The presence of glucose, however, has long been known to prevent the induction of the *lac* operon, as well as other operons controlling enzymes involved in carbohydrate catabolism (e.g., the arabinose and galactose operons). This phenomenon, called ***catabolite repression*** (or the ***glucose effect***), has apparently evolved to assure that glucose is metabolized when present, in preference to other, less efficient, energy sources.

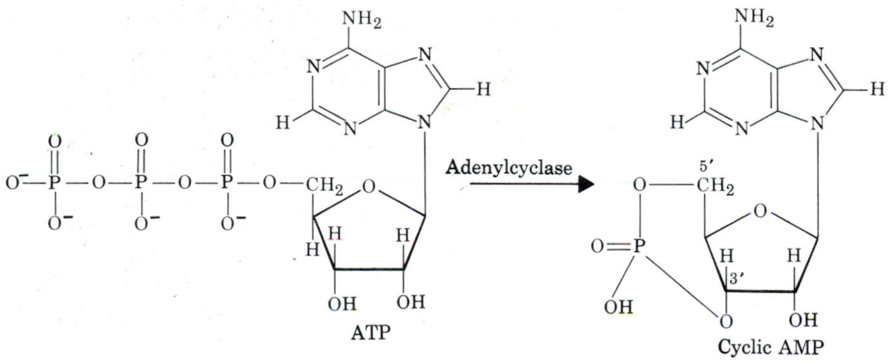

ATP

Adenylcyclase

Cyclic AMP

Figure 14.6 Formation of the regulatory molecule cyclic AMP (adenosine-3′,5′-phosphate) from ATP by the action of the enzyme adenylcyclase. Cyclic AMP binds to the CAP protein; the CAP–cAMP complex then binds to the CAP site of the *lac* promoter, stimulating the binding of RNA polymerase to its binding site in the *lac* promoter. In the absence of sufficient quantities of cAMP (which results for unknown reasons in the presence of high concentrations of glucose), CAP ***cannot*** bind to the *lac* promoter and the *lac* operon ***cannot*** be induced.

Catabolite repression of the *lac* operon is now known to be mediated via positive control of transcription by a regulatory protein called ***CAP*** (for Catabolite Activator Protein) and a small effector molecule called ***cyclic AMP*** (adenosine-3′,5′-phosphate; Fig. 14.6). (CAP is also sometimes called ***cyclic AMP receptor protein.***) As mentioned earlier, the *lac* promoter contains two separate binding sites: (1) one for RNA polymerase and (2) one for the CAP–cAMP complex (Fig. 14.7; cyclic AMP is abbreviated cAMP).

The CAP–cAMP complex must be bound to its binding site in the *lac* promoter in order for the operon to be induced. The CAP–cAMP complex thus exerts positive control over the transcription of the *lac* operon. It has an effect exactly opposite to that of repressor binding to an operator. Although the precise mechanism by which CAP–cAMP stimulates RNA polymerase binding to the promoter is still uncertain, its positive control of *lac* operon transcription is firmly established by both *in vivo* and *in vitro* experiments.

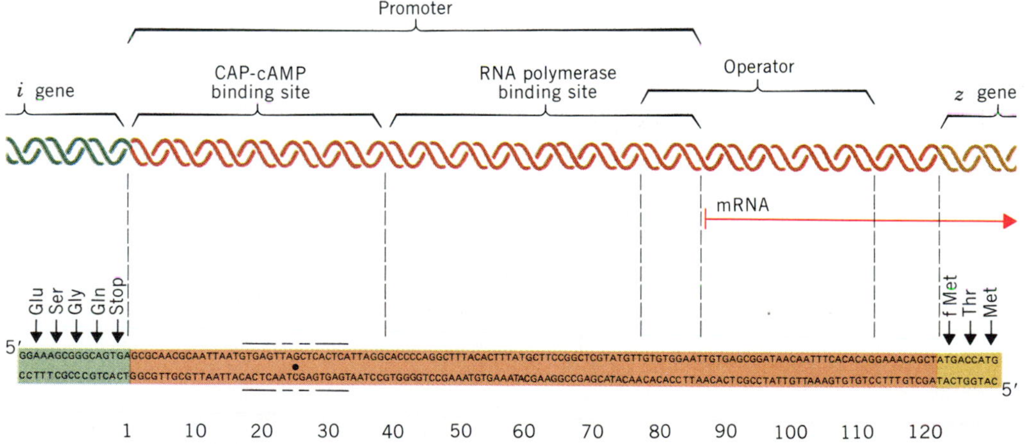

Figure 14.7 Organization and nucleotide sequence of the promoter–operator region of the *lac* operon. The promoter consists of two components: (1) the site that binds the catabolite activator protein (abbreviated CAP)–cyclic AMP (or cAMP) complex and (2) the RNA polymerase binding site. Note that the promoter and the operator (repressor binding site) overlap slightly. The adjacent segments of the *i* (repressor) and *z* (β-galactosidase) structural genes are also shown. The horizontal line labeled mRNA indicates the position at which transcription of the operon begins (5′ end of mRNA). The numbers at the bottom give distances in nucleotide-pairs from the end of the *i* gene (beginning of the promoter). The dot between the nucleotide chain sequences (in the CAP–cAMP binding site) indicates the center of symmetry of an imperfect palindrome [a sequence where the nucleotide-pair sequences read almost the same in both directions (indicated by the parallel bars); see Chapter 13, p. 350]. This is a potential site of primary interaction with the CAP–cAMP complex. (Based on the data of R. C. Dickson, J. Abelson, W. M. Barnes, and W. S. Reznikoff, *Science* 187: 27–35, 1975.)

CAP is known to function as a dimer; thus, like the *lac* repressor, it is multimeric in its functional state.

Only the CAP–cAMP complex binds to the *lac* promoter; in the absence of cAMP, CAP does not bind. Thus cAMP acts as the effector molecule, determining the effect of CAP on *lac* operon transcription. The intracellular cAMP concentration is sensitive to the presence or absence of glucose. High concentrations of glucose cause sharp decreases in the intracellular concentration of cAMP. How glucose controls the cAMP concentration is not clear. Perhaps glucose, or some metabolite that forms in the presence of sufficient concentrations of glucose, inhibits the activity of ***adenylcyclase,*** the enzyme that catalyzes the formation of cAMP from ATP. Whatever the mechanism, the presence of glucose results in a decrease in the intracellular concentration of cAMP. In the absence of (or in the presence of a low concentration of) cAMP, CAP ***cannot*** bind to the *lac* operon promoter. In turn, RNA polymerase cannot bind efficiently to the *lac* promoter in the absence of bound CAP. The overall result of the positive control of transcription of the *lac* operon by the CAP–cAMP complex is that in the presence of glucose, *lac* operon transcription never exceeds 2 percent of the induced rates observed in the absence of glucose.

The complete nucleotide-pair sequence of the *lac* operon regulatory region (the promoter and operator sequences) is now known (Fig. 14.7). Comparative nucleotide sequence studies of mutant and wild-type promoters and operators (plus *in vitro* CAP–cAMP, RNA polymerase, and repressor binding studies) are beginning to provide detailed information about the nature of these important ***sequence-specific protein–nucleic acid interactions.*** These studies provide an excellent example of the advantages of integrating genetic and biochemical approaches in attempting to understand biological phenomena.

Complex Regulation of the *ara* Operon

Almost all the details about the mechanisms by which the *lac* and *trp* operons are regulated are known and supported by an extensive body of experimental data. However, other operons like the ***arabinose (*ara*) operon*** of *E. coli* exhibit much more complex patterns of regulation that are still not completely understood. In the *lac* and *trp* operons, the product of the regulator gene, the repressor, functions in a negative manner, turning off transcription of the operon. On the other hand, the catabolite activator protein (CAP) exerts a positive control over the *lac* operon by stimulating transcription of the operon. The major regulatory protein of the *ara* operon exhibits both negative and positive regulatory effects on the transcription of the structural genes of the operon depending on the environmental conditions. Moreover, the regulatory components that control transcription of the *ara* operon include one element that acts from a distance of over 200 nucleotide-pairs away from the promoter that it helps to control. Although all the details of the *ara* operon regulatory circuitry are not yet established unambiguously, the current working model is presented here to give the student an appreciation for the complexity of the regulation of some bacterial operons.

The ***arabinose (*ara*) operon*** of *E. coli* contains three structural genes (*araB*, *araA*, and *araD*) that encode the three enzymes involved in the catabolism of arabinose (Fig. 14.8*a*). These three genes are cotranscribed on a single mRNA that is initiated at a promoter called P_{BAD}. (Active transport of arabinose into cells is carried out by the products of genes *araE*, *araF*, and *araG*. These genes are located at sites quite distant from the *araBAD* operon of interest here and will not be discussed further.) The major regulatory protein of the *ara* operon (the *araC* protein) is produced from a transcript that is initiated at a promoter called P_C. The P_C promoter is only slightly over 100 nucleotide-pairs away from P_{BAD}, but the two promoters initiate transcription in opposite directions (Figs. 14.8*a* and *b*). The *araC* protein acts as a negative regulator (a repressor) of transcription of the *araB*, *araA*, and *araD* structural genes from the P_{BAD} promoter in the absence of arabinose and cyclic AMP (cAMP). It acts as a positive regulator (an activator) of transcription of these genes from the P_{BAD} promoter when arabinose and cAMP are present. Thus, depending on the presence or absence of the effector molecules arabinose and cAMP, the *araC* regulatory gene-product may exert either a positive or a negative effect on transcription of the *araB*, *araA*, and *araD* structural genes. Since the *ara* operon is subject to catabolite repression like the *lac* operon (see the preceding section), and thus to positive control by CAP and cAMP, induction of the *ara* operon depends on the positive

Figure 14.8 (Right page): Structure and complex regulation of the arabinose (*ara*) operon of *E. coli* (a) The *araB*, *araA*, and *araD* structural genes encode the three enzymes that catalyze the catabolism of arabinose to D-xylulose-5-phosphate. Cotranscription of these three structural genes is initiated at a promoter called P_{BAD}. The *araC* gene encodes a protein that regulates the transcription of the *araBAD* operon. The *araC* protein acts as a repressor of the operon in the absence of arabinose and cAMP; however, it functions as an activator of transcription of the operon in the presence of arabinose and cAMP. Transcription of the *araC* gene is

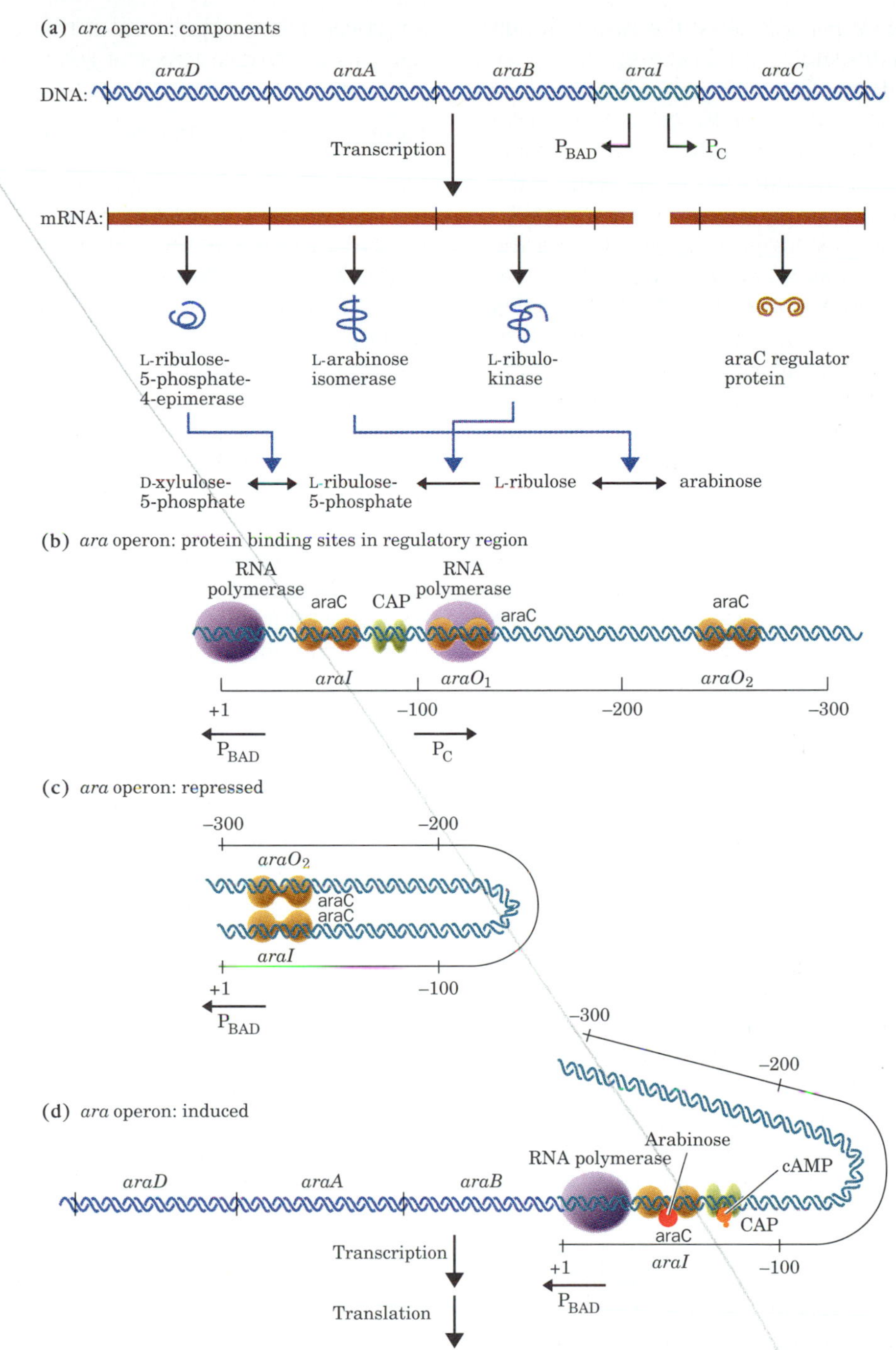

initiated at a promoter called P_C. Transcript extension occurs in opposite directions from the P_{BAD} and P_C promoters. The *araI* region contains the binding sites for the *araC* regulator protein and the cAMP–CAP complex. (b) Enlargement of the *araI* region showing the positions of the binding sites for *araC* protein, cAMP–CAP, and RNA polymerase, and the location of the *araC* protein binding site at *araO*$_2$. (c) Model for the repression of the *ara* operon. *AraC* protein dimers bind at the *araI* and *araO*$_2$ sites, and then bind to each other to form a DNA loop. RNA polymerase cannot bind at P_{BAD} to initiate transcription when the DNA is in this looped conformation. (d) Induction of the *ara* operon occurs when the arabinose–*araC* protein and cAMP–CAP complexes both bind at their *araI* sites. When these complexes are bound at their *araI* sites, the DNA loop structure cannot form, and RNA polymerase is able to bind at P_{BAD} and initiate transcription of the *araB, araA,* and *araD* structural genes. Additional details are given in the text. (After D. H. Lee and R. F. Schleif, "*In Vivo* DNA Loops in *araCBAD:* Size limits and Helical Repeat," *Proc. Natl. Acad. Sci. U.S.A.* 86: 476–480, 1989.)

regulatory effects of two proteins, the *araC* protein and CAP (the cAMP-binding catabolite activator protein). The binding sites for these two proteins and for RNA polymerase all appear to be located in a region of the *ara* operon historically called *araI* (*I* for *i*nduction), located between the three structural genes of the operon and the regulator gene (*araC*) (Fig. 14.8*a*).

Initially, scientists studying the regulation of the *ara* operon thought that all the binding sites for the *araC* regulatory protein and the cAMP–CAP complex were in the *araI* region. The surprising discovery was that repression of the *ara* operon depended on the binding of *araC* protein at a site called *araO*$_2$ (*O* for operator, *2* because it was the ***second*** *ara* operator identified) located 211 nucleotide-pairs upstream (relative to the direction of transcription from P_{BAD}) from the *araC* protein-binding site in *araI* (Fig. 14.8*b*). (Operator *araO*$_1$—the first *ara* operator to be identified—controls the transcription of the *araC* regulator gene initiated at P_C.) The currently accepted model for the repression of the *ara* operon is that the *araC* protein must bind (as a dimer) at both the *araI* site and the *araO*$_2$ site, and that these proteins then bind to each other to form a DNA loop (Fig. 14.8*c*). In fact, there is now considerable evidence in support of this model. For example, if five nucleotide-pairs are inserted or deleted in the region between *araI* and *araO*$_2$, normal repression of the operon ***cannot*** occur. Such an insertion or deletion will rotate one *araC* protein-binding site halfway around the double helix (to the opposite face) relative to the other *araC* protein-binding site. This presumably would make it difficult or impossible for the *araC* dimers bound at *araI* and *araO*$_2$ to interact and form the loop required for repression.

When the loop structure (Fig. 14.8*c*) is formed, it must prevent or interfere with the binding of RNA polymerase at the adjacent promoter (P_{BAD}) of the operon (Fig. 14.8*b*). In the presence of arabinose and cAMP, the *ara* operon is induced. Moreover, under these conditions, the *araC* protein has been shown to become an activator of transcription of the operon. The details of the mechanism by which arabinose causes the *araC* protein to become a positive regulator of transcription of the operon are not clear. Somehow, the arabinose–*araC* protein complex and the cAMP–CAP complex must open the loop by binding at their *araI* sites. This, in turn, must permit RNA polymerase to bind at the P_{BAD} site and initiate transcription of the *ara* structural genes (Fig. 14.8*d*).

Clearly, the regulation of transcription of the *ara* operon of *E. coli* is considerably more complex than the regulation of transcription of the *lac* operon of this bacterium. It will be interesting to determine whether or not this loop formation mechanism is commonly employed in the regulation of transcription of other operons in prokaryotes or of genes in eukaryotes.

Lambda Prophage Repression During Lysogeny

When a temperate bacteriophage such as lambda exists in the prophage state in a lysogenic cell (see Chapter 8, pp. 210–213), the genes coding for products involved in the lytic pathway—namely, the genes controlling phage DNA replication, phage morphogenesis, and lysis of the host cell—must ***not*** be expressed. This is accomplished by a repressor–operator–promoter circuit, much like that involved in bacterial operons. Specifically, the C_I gene of phage lambda codes for repressor, a well-characterized protein with a molecular weight of 27,000, which in the dimer or tetramer state binds to two operator regions that control transcription of the lambda genes involved in lytic growth (Fig. 14.9). These two operator regions, termed O_L (for transcription in a leftward direction) and O_R (for transcription in the rightward direction), overlap with promoter sequences at which RNA polymerase binds and initiates transcription of the genes controlling lytic development. Thus, with the repressor bound to the two operators, RNA polymerase ***cannot*** bind to the two promoters and ***cannot,*** therefore, initiate transcription. In this way, the phage genes are kept repressed, allowing the "dormant" prophage to be transmitted from parental host cells to progeny cells generation after generation.

In experiments in which the operator and promoter regions of phage lambda were sequenced, each operator was unexpectedly found to contain three repressor binding sites with similar, but not identical, sequences of 17 nucleotide-pairs. Each repressor binding site has partial twofold symmetry around the central base-pair (i.e., is partially palindromic; see Chapter 13, pp. 349–351). It has been speculated that this partial symmetry may facilitate interaction with repressor dimers, which might also have twofold symmetry. Although this is an attractive possibility, it is nothing more than that at present.

The interaction of lambda repressor with DNA sequences O_LP_L and O_RP_R nicely explains how the lambda prophage genes are maintained in the repressed state. The mechanism responsible for the decision between lytic development and lysogenic development after infection of an *E. coli* cell with a lambda phage is considerably more complex, involving interactions among several other lambda regulatory genes. The reader is referred to one of the papers by Ptashne and coworkers (see the References for this chapter) for a discussion of the mechanism by which the choice between the lytic pathway and the lysogenic pathway is made.

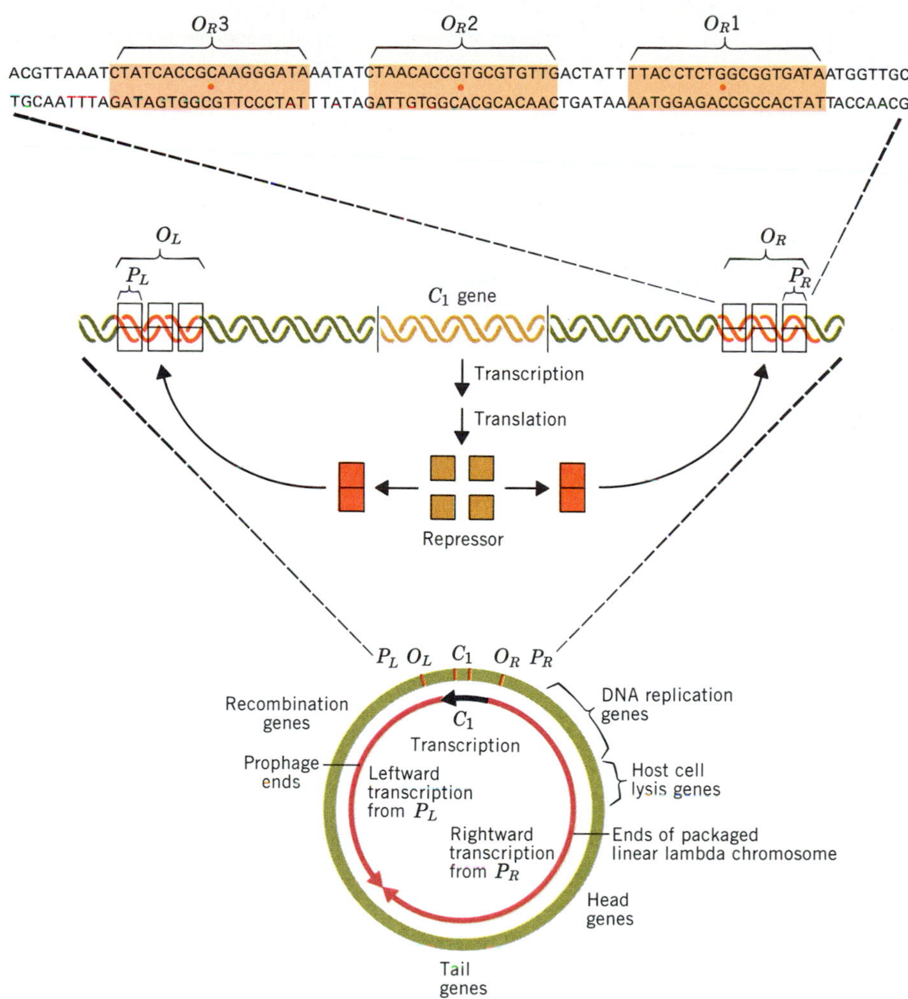

Figure 14.9 Repression of the phage lambda genes that control lytic development by the binding of the C_1 gene-product (repressor) to the two operator sequences (O_L and O_R) that control leftward and rightward transcription of the lambda chromosome. Transcription of the C_1 gene itself is in a leftward direction (solid black arrow), beginning at the C_1 promoter (not shown), which is located between the C_1 gene and O_R. The circular intracellular form of the lambda chromosome (see Chapter 5, pp. 110–113) is shown at the bottom, with the approximate location of the C_1 gene, P_LO_L (promoter leftward, operator leftward), O_RP_R (operator rightward, promoter rightward), the ends of the linear forms of the lambda chromosome (prophage and mature forms), and clusters of some of the genes controlling lytic development. The orange arrows indicate the transcription of genes from P_L and P_R, respectively. The repression of the genes controlling lytic development by the binding of repressor molecules to O_L and O_R is illustrated in the center enlargement. Note that each operator sequence has three repressor binding sites and that the operator and promoter (RNA polymerase binding site) overlap. The nucleotide-pair sequence of the O_R region is shown at the top, with the 17-nucleotide-pair sequence of each repressor binding site in brackets. The orange dot between the two strands of DNA within each repressor binding site indicates the axis of partial symmetry. (The nucleotide-pair sequence data are from M. Ptashne, K. Backman, M. Z. Humayun, A. Jeffrey, R. Maurer, B. Meyer, and R. T. Sauer, *Science* 194: 156, 1976.)

CONTROL OF THE *trp* OPERON BY ATTENUATION

Repression and derepression can change the level of expression of the structural genes of the *trp* operon by about 70-fold. There is a second level of regulation of *trp* operon expression, however. In *trpR* mutants that cannot make repressor, the addition of tryptophan to a culture of cells growing in the absence of tryptophan will cause an 8- to 10-fold decrease in the rates of synthesis of the tryptophan biosynthetic enzymes.

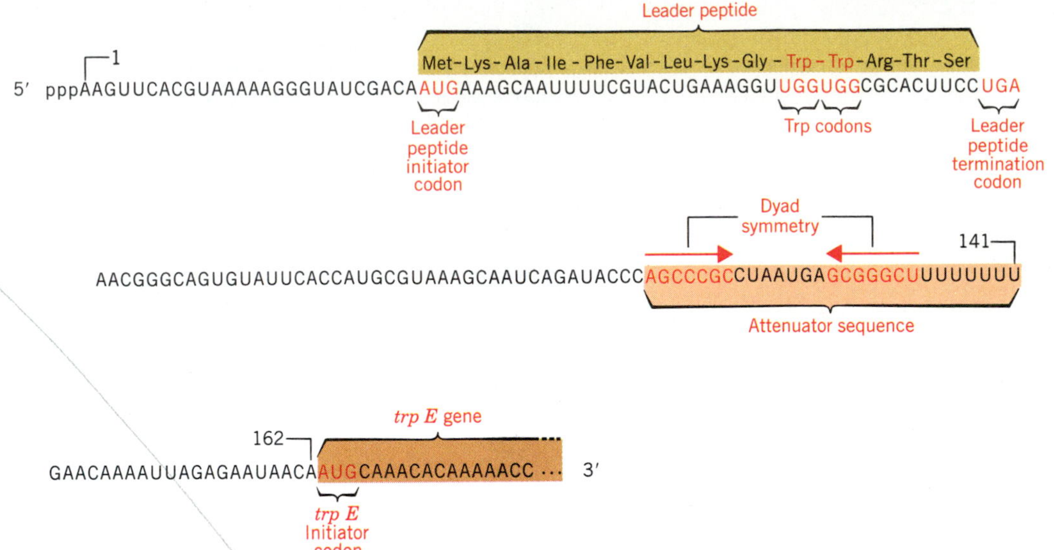

Figure 14.10 Nucleotide sequence of the leader of the *trp* operon mRNA. The region of dyad symmetry in the attenuator forms the transcription-termination "hairpin" shown in Fig. 14.11c. Note the two tandem tryptophan codons in the sequence coding for the putative leader peptide. These Trp codons are believed to be responsible for the control of attenuation by tryptophan (see Figs. 14.11b and c). (Based on the data summarized by C. Yanofsky, *Nature* 289: 751–758, 1981.)

Moreover, deletions that remove part of the *trpL* region (Fig. 14.5) result in increased rates of expression of the *trp* operon. The effects of these deletions are independent of repression; the increase occurs in both the repressed and the derepressed state.

This second level of regulation of the *trp* operon is called ***attenuation***, and the sequence within *trpL* that controls this phenomenon is called the ***attenuator*** (Fig. 14.10). Attenuation occurs by control of the termination of transcription at a site near the end of the mRNA leader sequence. This "premature" termination of *trp* operon transcription occurs only in the presence of tryptophan-charged $tRNA_{trp}$ and yields a 140-nucleotide-long leader-sequence transcript (Fig. 14.10).

The attenuator region has a nucleotide-pair sequence essentially identical to the transcription-termination signals found at the ends of most bacterial operons (including the *trp* operon; see Fig. 14.11*a*). These termination signals contain a GC-rich palindrome followed by several AT base-pairs. Transcription of these termination signals yields a nascent RNA with the potential to form a ***hydrogen-bonded "hairpin" structure followed by several U's*** (Fig. 14.11*a*). When a nascent transcript forms this hairpin structure, it is believed to cause a conformational change in the associated RNA polymerase, resulting in termination of transcription within the following, more weakly hydrogen-bonded $[(A\colon U)_n]$ region of

Figure 14.11 (Right page): Control of the *trp* operon by attenuation. Attenuation occurs by tryptophan-regulated termination of transcription within *trpL* (controlling the *trp* operon mRNA leader sequence; see Fig. 14.5). The normal transcription-termination signal in bacteria is a region of dyad symmetry (two sequences that read the same in opposite directions) followed by 4–8 AT base-pairs. The region of dyad symmetry results in mRNA sequences that can base-pair to form "hairpin" structures as shown in (a) and (c). The sequence shown in (a) is, in fact, the base-pair sequence for the first transcription-termination signal (*t* in Fig. 14.5) at the end of the *trp* operon. Note the similarity of the mRNA hairpin structures formed by the normal *t* terminator (a) and by the attenuator (c). (b) In the absence of tryptophan, the ribosome translating the mRNA to form the leader peptide becomes stalled at one or the other of the two Trp codons. This allows the base-pairing formed between leader sequences 74–85 and 108–119 to remain intact, which in turn prevents the transcription-termination hairpin from forming. Thus, transcription continues through the rest of the *trp* operon. (c) If tryptophan is present, a ribosome can translate on past the Trp codons to the UGA leader-peptide translation-termination codon. In so doing, it disrupts the base-pairing between leader sequences 74–85 and 108–119. This, in turn, leaves the latter sequence free to base-pair with sequence 126–134, forming the transcription-termination hairpin. Thus, transcription is terminated at the attenuator sequence and does not continue through the rest of the *trp* operon. Remember that transcription starts at the 5′ end of the mRNA. For this reason, the 74–85 and 108–119 sequences will be produced and be able to base-pair before the 126–134 sequence has even been synthesized. (After C. Yanofsky, *Nature* 289: 751–758, 1981, and T. N. Platt, *Cell* 24: 10–23, 1981.)

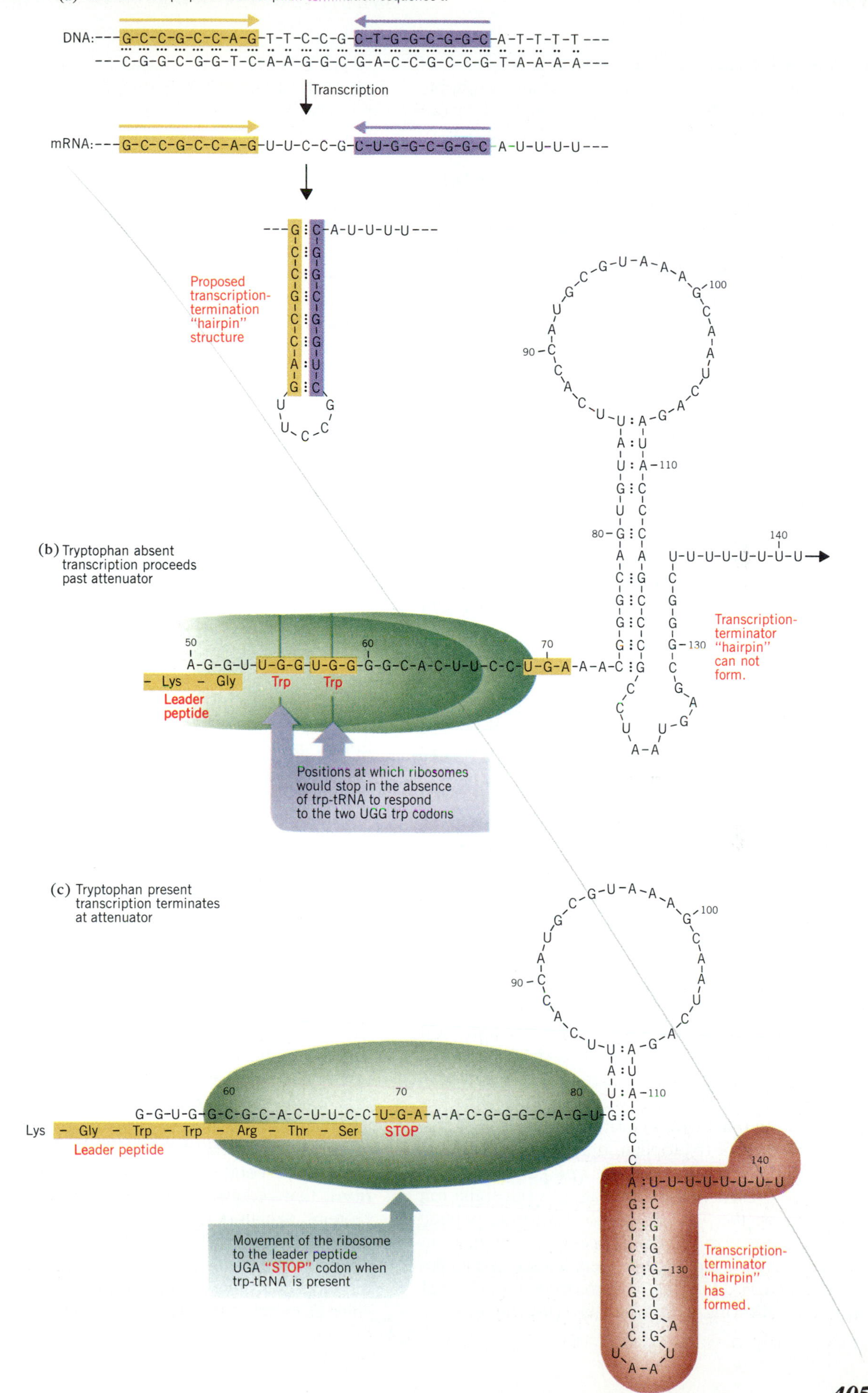

(a) Structure of trp operon transcription-termination sequence t.
DNA:---G-C-C-G-C-C-A-G-T-T-C-C-G-C-T-G-G-C-G-G-C-A-T-T-T-T---
---C-G-G-C-G-G-T-C-A-A-G-G-C-G-A-C-C-G-C-C-G-T-A-A-A-A---
Transcription
mRNA:---G-C-C-G-C-C-A-G-U-U-C-C-G-C-U-G-G-C-G-G-C-A-U-U-U-U---
Proposed transcription-termination "hairpin" structure
(b) Tryptophan absent transcription proceeds past attenuator
A-G-G-U-U-G-G-U-G-G-G-G-C-A-C-U-U-C-C-U-G-A-A-A-C
- Lys - Gly
Trp
Trp
Leader peptide
Positions at which ribosomes would stop in the absence of trp-tRNA to respond to the two UGG trp codons
Transcription-terminator "hairpin" can not form.
(c) Tryptophan present transcription terminates at attenuator
G-G-U-G-G-C-G-C-A-C-U-U-C-C-U-G-A-A-A-C-G-G-G-C-A-G-U-G
Lys - Gly - Trp - Trp - Arg - Thr - Ser STOP
Leader peptide
Movement of the ribosome to the leader peptide UGA "STOP" codon when trp-tRNA is present
Transcription-terminator "hairpin" has formed.

DNA–RNA base-pairing. The nucleotide sequence of the attenuator therefore explains its ability to prematurely terminate *trp* operon transcription. But how can this be regulated by the presence or absence of tryptophan?

First, recall that ***transcription and translation are coupled in prokaryotes,*** that is, ribosomes begin translating mRNAs while they are still being produced by transcription. Thus, events occurring during translation may also affect transcription.

Second, note that the 162-nucleotide-long leader sequence of the *trp* operon mRNA (Fig. 14.10) contains sequences that can base-pair to form alternate secondary structures. Two of these sequences form the previously mentioned transcription-termination hairpin (Fig. 14.11*c*). This hairpin is formed by base-pairing between nucleotide sequences 114–121 and 126–134 (nucleotide 1 is at the 5′ terminus). An alternate secondary structure results from base-pairing between leader sequences 74–85 and 108–119 (Fig. 14.11*b*). Obviously, only one of these structures can exist at one time, since nucleotides 114–119 are part of both. Thus, ***if sequences 74–85 and 108–119 are base-paired, the attenuator transcription-termination hairpin cannot form.***

Third, note that the leader sequence contains an AUG translation-initiation codon, followed by 13 codons for amino acids, followed in turn by a UGA translation-termination codon (Fig. 14.10). Moreover, the *trp* leader sequence has been shown to contain an efficient ribosome-binding site located in the appropriate position for the initiation of translation at the leader AUG initiation codon. It seems very likely that a 14-amino-acid-long "leader peptide" is synthesized as diagrammed in Fig. 14.10. This putative leader peptide has not yet been detected *in vivo*, but short peptides of this type are very rapidly degraded in *E. coli*, so failure to detect it is not unexpected.

Note that the leader peptide contains two contiguous tryptophan residues. The two Trp codons are positioned such that in the absence of tryptophan (and thus the absence of Trp–$tRNA_{trp}$), the ribosome will become stalled before it encounters the base-paired structure formed by leader sequences 74–85 and 108–119 (Fig. 14.11*b*). This base-pairing precludes the formation of the transcription-termination hairpin. Thus, in the absence of tryptophan, transcription will continue past the attenuator into the *trpE* gene.

In the presence of tryptophan, the ribosome can translate past the Trp codons to the leader-peptide termination codon. In the process, it will have to disrupt the base-pairing between leader sequences 74–85 and 108–119. This, in turn, frees the 114–121 sequence, allowing it to base-pair with the 126–134 sequence and form the transcription-termination hairpin (Fig. 14.11*c*). Thus, in the presence of tryptophan, transcription frequently terminates at the attenuator, reducing the amount of mRNA for the *trp* structural genes.

The transcription of the *trp* operon can be regulated over a range of almost 700-fold by the combined effects of repression (up to 70-fold) and attenuation (up to 10-fold).

Regulation of transcription by attenuation is not unique to the *trp* operon. Six operons (*trp*, *thr*, *ilv*, *leu*, *phe*, and *his*) are known to be regulated by attenuation. Of these, *trp* and possibly *phe* are also regulated by repression. The *his* operon, which has long been thought to be repressible, is now believed to be regulated entirely by attenuation. Although minor details vary from operon to operon, the main features of attenuation are the same for all six operons.

FEEDBACK INHIBITION AND ALLOSTERIC ENZYMES

Earlier in this chapter, we described the mechanism by which the transcription of bacterial genes coding for enzymes in a biosynthetic pathway is repressed when the end product of the pathway is present in the medium in which the cells are growing. A second, and more rapid, regulatory fine-tuning of metabolism often occurs at the level of enzyme activity. The presence of sufficient concentrations of an end product (such as histidine or tryptophan) of a biosynthetic pathway will frequently result in the inhibition of the first enzyme in the pathway. This phenomenon is called ***feedback inhibition*** or ***end product inhibition***; it should ***not be confused with repression (inhibition of enzyme synthesis).*** Feedback inhibition results in an almost instantaneous arrest of the synthesis of an end product when it is added to the medium.

Feedback inhibition-sensitive enzymes have been shown to have an end product binding site (or sites) in addition to the substrate binding site (or sites). In the case of some multimeric enzymes, the ***end product or regulatory binding site*** is on a different subunit (polypeptide) than the substrate site. Upon binding the end product, such enzymes are believed to undergo changes in conformation, called ***allosteric transitions,*** that reduce their affinity for their substrates. Proteins that undergo such conformational changes are usually referred to as ***allosteric proteins.*** Many examples are known, including numerous feedback inhibition-sensitive enzymes and the repressor molecules discussed in the preceding sections.

TEMPORAL SEQUENCES OF GENE EXPRESSION DURING PHAGE INFECTION

Regulation of gene expression during the lytic life cycles of bacteriophages is quite different from the reversible on–off switches characteristic of bacterial operons. Instead, viral genes are expressed in ***genetically preprogrammed sequences***, possibly analogous to the preprogrammed sequences of gene expression putatively involved in differentiation in higher organisms. Although different bacterial viruses exhibit variations of the specific mechanisms involved, a common picture emerges. One set of phage genes, usually called "early" genes, is expressed immediately after infection. The product(s) of one or more of the "early" genes is responsible for turning off the expression of the "early" genes and turning on the expression of the next set of genes, and so on. Two to four sets of genes, depending on the virus, are characteristically involved. In all cases studied so far, the regulation of ***sequential gene expression*** during phage infection ***occurs primarily at the level of transcription***.

In three of the most extensively studied bacterial viruses—*E. coli* phages T4 and T7 and *Bacillus subtilis* phage SP01—the sequential gene expression is controlled by modifying the promoter specificity of RNA polymerase, either by the synthesis of a new RNA polymerase (T7) or by phage-induced alterations of the host cell's RNA polymerase (T4 and SP01).

In phage T7-infected cells, the "early" genes are transcribed by the *E. coli* RNA polymerase. One of the "early" genes codes for T7 RNA polymerase, which then transcribes all the "late" genes (coding for T7 structural proteins, lysozyme, etc.). *Bacillus subtilis* phage SP01 exhibits a slightly more complex pathway of sequential gene expression, involving three sets of genes. These three sets of genes are called "early," "middle," and "late" genes in reference to their time of expression during the phage reproductive cycle. The SP01 "early" genes are transcribed by the *B. subtilis* RNA polymerase. One of the "early" gene-products is a polypeptide that binds to the host cell's RNA polymerase, changing its specificity such that it transcribes the "middle" genes of SP01. Two of the products of "middle" genes are, in turn, polypeptides that associate with the *B. subtilis* RNA polymerase, further changing its specificity so that it then transcribes the "late" genes of SP01.

Phage T4 exhibits an even more complex pattern of sequential gene expression, involving several different modifications of the host cell's RNA polymerase. Thus, in the case of these bacterial viruses, the control of the observed sequential gene expression occurs primarily at the level of transcription and is mediated by specific RNA polymerase–promoter sequence interactions.

SUMMARY

Gene expression is frequently under the control of ***regulator genes***. These regulator genes act ***primarily at the level of transcription***, often affecting the ability of ***RNA polymerase to bind to promoter sequences***. The effects of regulator gene-products are, in turn, controlled by the presence or absence of specific ***effector molecules*** in the environment.

In prokaryotes, genes with related functions are frequently present in coordinately regulated units called ***operons***. Each operon is one unit of transcription, that is, a single mRNA transcript carries the coding sequences for all the genes in the operon.

Certain regulator genes code for proteins called ***repressors***, which function by means of their ***sequence-specific binding to DNA***. Some repressors (those for ***repressible operons***) bind to DNA only in the presence of specific effector molecules called ***co-repressors***; others (those for ***inducible operons***) bind to DNA only in the absence of specific effector molecules called ***inducers***.

Repressors are ***allosteric proteins***, that is, proteins that undergo conformational shifts and correlated changes in activity in response to the binding of specific effector molecules. Repressors bind to their specific DNA sequences in one conformation, but not in the other conformation.

Repressors act by binding to DNA sequences called ***operators***, which are located adjacent to the structural genes whose transcription they control. When a repressor, or the complex of repressor and co-repressor, is bound to the operator sequence of an operon, it prevents RNA polymerase from binding to the contiguous promoter sequence and initiating the transcription of the operon.

Operons containing genes that code for enzymes involved in catabolic pathways are also often under ***positive control*** by the ***catabolite activator protein (CAP)*** and ***cyclic AMP (cAMP)***. The binding of the CAP–cAMP complex to a binding site within the promoter region of such an operon is required for the efficient binding of RNA polymerase to the RNA polymerase binding site of the promoter region. High concentrations of glucose result in low intracellular concentrations of cAMP. Thus, such operons cannot be induced in the presence of high concentrations of glucose, a phenomenon known as ***catabolite repression.***

Regulation of the expression of the ***arabinose operon (*araBAD*) is complex*** with the ***product of the* araC *regulator gene*** acting (1) as a ***negative regulator*** (repressor) of transcription ***in the absence of arabinose and cAMP*** and (2) as a ***positive regulator*** (activator) of transcription ***in the presence of arabinose and cAMP. Repression*** of the *araBAD* operon involves the ***binding of* araC *protein dimers at two sites,*** one near the promoter and one located about 200 nucleotide-pairs upstream from the promoter. The bound *araC* dimers then bind to each other to form a ***DNA loop,*** which prevents RNA polymerase from binding at the promoter to initiate transcription.

Operons controlling enzymes involved in amino acid biosynthetic pathways are frequently controlled by ***attenuation.*** Attenuation occurs by premature termination of transcription at an ***attenuator*** site located within the mRNA leader sequence. The attenuators contain a transcription-termination sequence; however, the ability of the attenuator RNA sequence to form the secondary structure that leads to termination of transcription depends on the presence of the amino acid that is the end product of the pathway controlled by the operon in question.

Preprogrammed sequences of viral gene expression occur in bacteriophage-infected cells where the products of genes expressed early after infection interact with RNA polymerase to change its promoter specificity and switch transcription to a second set of genes. In the case of some of the more complex bacteriophages, this switching process may be repeated several times during the life cycle of the virus.

REFERENCES

BECKWITH, J.R., and D. ZIPSER. 1970. *The lactose operon.* Cold Spring Harbor Laboratory Press, Cold Spring Harbor, NY.

DICKSON, R.C., J. ABELSON, W.M. BARNES, and W.S. REZNIKOFF. 1975. "Genetic regulation: The *lac* control region." *Science* 187: 27–35.

FREIFELDER, D. 1983. *Molecular biology, a comprehensive introduction to prokaryotes and eukaryotes.* Science Books International, Boston.

GILBERT, W., and B. MULLER-HILL. 1967. "The *lac* operator is DNA." *Proc. Natl. Acad Sci. U.S.A.* 58: 2415–2421.

JACOB, F., and J. MONOD. 1961. "Genetic regulatory mechanisms in the synthesis of proteins." *J. Molec. Biol.* 3: 318–356.

KOLTER, R., and C. YANOFSKY. 1982. "Attenuation in amino acid biosynthetic operons." *Annu. Rev. Genet.* 16: 113–134.

KOSHLAND, JR., D.E. 1973. "Protein shape and biological control." *Sci. Amer.* 229(4): 52–64.

LEE, D.-H., and R.F. SCHLEIF. 1989. "*In vivo* DNA loops in *araCBAD*: Size limits and helical repeat. *Proc. Natl. Acad. Sci. U.S.A.* 86: 476–480.

LEWIN, B. 1990. *Genes IV.* Cell Press, Cambridge, MA/Oxford Univ. Press, New York.

MILLER, J.H., and W.S. REZNIKOFF. 1978. *The operon.* Cold Spring Harbor Laboratory Press, Cold Spring Harbor, NY.

MANIATIS, T., and M. PTASHNE. 1976. "A DNA operator–repressor system." *Sci. Amer.* 234(1): 64–76.

NEIDHARDT, F.C., editor in chief, J.L. INGRAHAM, K.B. LOW, B. MAGASANIK, M. SCHAECHTER, and H.E. UMBARGER, editors. 1987. *Escherichia coli and Salmonella typhimurium, cellular and molecular biology.* American Society for Microbiology, Washington, DC. (See in particular Vol. 2, Part IV, entitled "Regulation of Gene Expression," pp. 1231–1526.)

PABO, C.O., and M. LEWIS. 1982. The operator-binding domain of λ repressor: Structure and DNA recognition." *Nature* 298: 443–447.

PLATT, T. 1981. "Termination of transcription and its regulation in the tryptophan operon of *E. coli.*" *Cell* 24: 10–23.

PTASHNE, M. 1987. *A genetic switch: Gene control and phage* λ. Cell Press, Cambridge, MA.

______. 1989. "How gene activators work." *Sci. Amer.* 260(1): 40–47.

PTASHNE, M., A.D. JOHNSON, and C.O. PABO. 1982. "A genetic switch in a bacterial virus." *Sci. Amer.* 247(5): 128–140.

WATSON, J.D., N.H. HOPKINS, J.W. ROBERTS, J.A. STEITZ, and A.M. WEINER. 1987. *Molecular biology of the gene,* 4th ed. Benjamin, Menlo Park, CA.

YANOFSKY, C. 1981. "Attenuation in the control of expression of bacterial operons." *Nature* 289: 751–758.

PROBLEMS AND QUESTIONS

14.1 How can inducible and repressible enzymes of microorganisms be distinguished?

14.2 Distinguish between (a) repression and (b) feedback inhibition caused by the end product of a biosynthetic pathway. How do these two regulatory phenomena complement each other to provide for the efficient regulation of metabolism?

14.3 In the lactose operon of *E. coli* (Fig. 14.3), what is the function of each of the following genes or sites: (a) regulator, (b) operator, (c) promoter, (d) structural gene *z*, and (e) structural gene *y*?

14.4 What would be the result of inactivation by mutation of the following genes or sites in the *E. coli* lactose operon: (a) regulator, (b) operator, (c) promoter, (d) structural gene *z*, and (e) structural gene *y*?

14.5 Groups of alleles associated with the lactose operon (Fig. 14.3) are as follows (in order of dominance for each allelic series): repressor, i^s (superrepressor), i^+ (inducible), and i^- (constitutive); operator, o^c (constitutive, *cis*-dominant) and o^+ (inducible, *cis*-dominant); structural, z^+ and z^-. (a) Which of the following genotypes will produce β-galactosidase and permease if lactose is present: (1) $i^+ o^+ z^+ y^+$, (2) $i^- o^c z^+ y^+$, (3) $i^s o^c z^+ y^+$, (4) $i^s o^+ z^+ y^+$, (5) $i^- o^+ z^+ y^+$? (b) Which of these genotypes will produce β-galactosidase and permease if lactose is absent?

14.6 For each of the following partial diploids indicate whether enzyme formation is constitutive or inducible (see Problem 14.5 for dominance relationships): (a) $i^+ o^+ z^+ y^+/i^+ o^+ z^+ y^+$, (b) $i^+ o^+ z^+ y^+/i^+ o^c z^+ y^+$, (c) $i^+ o^c z^+ y^+/i^+ o^c z^+ y^+$, (d) $i^+ o^+ z^+ y^+/i^- o^+ z^+ y^+$, (e) $i^- o^+ z^+ y^+/i^- o^+ z^+ y^+$.

14.7 Write the partial diploid genotype for a strain that will (a) produce β-galactosidase constitutively and permease inducibly and (b) produce β-galactosidase constitutively but not permease either constitutively or inducibly, even though a y^+ gene is known to be present.

14.8 Constitutive mutations produce elevated enzyme levels at all times; they may be of two types: o^c or i^-. Assume that all other DNA present is wild type. Outline the way by which the two constitutive mutants can be distinguished with respect to: (a) map position, (b) regulation of enzyme levels in o^c/o^+ versus i^-/i^+ partial diploids, and (c) the position of the structural genes affected by an o^c mutation versus the genes affected by an i^- mutation in a partial diploid.

14.9 How could the tryptophan operon in *S. typhimurium* have developed and been maintained by evolution?

14.10 Of what biological significance is the phenomenon of catabolite repression?

14.11 How might the concentration of glucose in the medium in which an *E. coli* cell is growing regulate the intracellular level of cyclic AMP?

14.12 Is the CAP–cAMP effect on the transcription of the *lac* operon an example of positive or negative regulation? Why?

14.13 Would it be possible to isolate *E. coli* mutants in which the transcription of the *lac* operon is not sensitive to catabolite repression? If so, in what genes might the mutations be located?

14.14 Using examples, distinguish between regulation based on negative mechanisms and that based on positive regulatory mechanisms.

14.15 A deletion of the regulator (*i*) gene of the *lac* operon in *E. coli* would be expected to result in the constitutive synthesis of the *lac* operon enzymes. What would the expected phenotype be for a mutant strain of *E. coli* with a deletion of the regulator gene (*araC*) of the *ara* operon? Why?

14.16 The following table gives the relative activities of the enzymes β-galactosidase and β-galactoside permease in cells with different genotypes at the *lac* locus in *E. coli*. The level of activity of each enzyme in wild-type *E. coli* not carrying F′s was arbitrarily set at 100; all other values are relative to the observed levels of activity in these wild-type bacteria. Based on the data given in the table for the first four genotypes, fill in the levels of enzyme activity that would be expected for the fifth genotype.

	β-GALACTOSIDASE		β-GALACTOSIDE PERMEASE	
GENOTYPE	− INDUCER	+ INDUCER	− INDUCER	+ INDUCER
$i^+ o^+ z^+ y^+$	0.1	100	0.1	100
$i^- o^+ z^+ y^+$	100	100	100	100
$i^+ o^c z^+ y^+$	25	100	25	100
$i^- o^+ z^+ y^-/F'$ $i^- o^+ z^+ y^+$	200	200	100	100
$i^- o^c z^- y^+/F'$ $i^+ o^+ z^+ y^+$	—	—	—	—

14.17 What effect, if any, would deletion of the $araO_2$ site have on the expression of the *araBAD* operon?

14.18 The rate of transcription of the *trp* operon in *E. coli* is controlled by both (1) repression/derepression and (2) attenuation. By what mechanisms do these two regulatory processes modulate *trp* operon transcript levels?

14.19 What effect would deletion of the *trpL* region of the *trp* operon have on the rates of synthesis of the enzymes encoded by the five genes in the *trp* operon in *E. coli* cells growing in the presence of tryptophan?

14.20 By what mechanism does the presence of tryptophan in the medium in which *E. coli* cells are growing result in "premature" termination or attenuation of transcription of the *trp* operon?

14.21 Suppose that you used site-specific mutagenesis to modify the *trpL* sequence such that the two UGG Trp codons at positions 54–56 and 57–60 (see Fig. 14.11*b*) in the mRNA leader sequence were changed to GGG Gly codons. Would attenuation of the *trp* operon still be regulated by the presence or absence of tryptophan in the medium in which the *E. coli* cells were growing?

14.22 Would attenuation of the type that regulates the level of *trp* transcripts in *E. coli* be likely to occur in eukaryotic organisms?

Chapter 15

Regulation of Gene Expression and Development in Eukaryotes

AN amoeba is not likely to be confused with an *E. coli* cell. The phenotype of a corn plant is very different from the phenotype of an oak tree. Different human beings, excluding identical twins, can usually be readily distinguished. These different phenotypes, we know, result from different genes and different forms of genes (different alleles) in the genomes of various organisms and individuals.

Similarly, if we microscopically examine the nerve cells, kidney cells, liver cells, bone cells, blood cells, hair cells, skin cells, and so forth of any one organism or any one individual, we find vast differences in the phenotypes of these cells (see Chapter 1, Fig. 1.6). Some are short and fat; some are long and thin; some have appendages; and others are roughly spherical. Yet all of these cells (with a few notable exceptions) contain the same set of genes. All of them are produced from a single cell, the zygote in sexually reproducing species, by purely equational mitotic divisions. How, then, can their highly diverse phenotypes be explained?

At present, this question cannot be answered in detail. The superficial answer is that not all the genes present in a nucleus are ever expressed in a given cell at the same time. Furthermore, in differentiated cells of higher eukaryotes, only a small proportion (less than 10 percent) of the genes are ever expressed in any one tissue at any given point in time.

Nevertheless, detailed answers to several aspects of this question are emerging rapidly in the case of the fruit fly *Drosophila melanogaster* and the tiny worm *Caenorhabditis elegans*. Many of the ***key genes that regulate specific patterns of morphogenesis, both spatial and temporal, have been identified*** in these two model systems, and an elegant picture of the genetic control of development is beginning to emerge. This picture already shows that ***development of a fly or a worm unfolds under the control of a cascade of regulatory genes*** each in turn adding new pattern information to the blueprints that were in place at the time of the expression of that particular factor. By sequential rounds of regulatory gene fine-tuning, each group of cells produces the appropriate tissue or organ of each individual. In the end, each fly has the appropriate organs in the proper places: antennae and eyes on the head, legs and wings on the thorax, and genitals on the abdomen. However, mutations in key regulatory genes can destroy the normal pattern of development and produce flies with leglike structures where antennae should occur and so on. By studying the function of such mutant genes and their wild-type alleles, geneticists have discovered that many of these genes encode *trans*-acting regulatory proteins that bind to *cis*-acting regulatory regions of sets of genes and control their expression—in some cases turning them on and in other cases turning them off. Frequently, one or more of the genes that are activated by the original regulatory gene-product will in turn produce a *trans*-acting regulatory protein that turns on another set of genes. Development and cell differentiation thus result from the action of a cascade of key regulatory genes. Recently, the products of many of these key regulatory genes have been shown to be transcriptional activators that bind to the promoters or other regulatory sequences of the structural genes that they regulate. A molecular map of the process of development in higher animal model systems is thus rapidly emerging. Indeed, it seems likely that the important features of this map will be clearly established within the next few years.

As was discussed in Chapter 14, certain "housekeeping" genes such as those encoding ribosomal RNAs, ribosomal proteins, and transfer RNAs are necessarily expressed at some time in all cells. However, many other genes are apparently expressed for only a short period of time in one or a few cell types at a specific stage in development. In other words, the ***expression of genes is regulated, and the coordinated regulation of sequential pathways of gene expression is primarily responsible for the diversity of cell phenotypes that unfold during the development of a higher plant or animal.***

(Photo at left): Expression of the *Escherichia coli* β-glucuronidase coding sequence in flowers of transgenic *Arabidopsis* plants when regulated by (left) no promoter, (center) the promoter of the α1-tubulin gene of *Arabidopsis*, and (right) the 35S promoter of cauliflower mosaic virus. In the histochemical assay employed, β-glucuronidase converts the colorless substrate 5-bromo-4-chloro-3-indolyl glucuronide (X-gluc) to a product that undergoes oxidative dimerization to form an insoluble blue indigo dye or pigment. (Photographs courtesy of Jeffrey L. Carpenter, D. Peter Snustad, and Carolyn D. Silflow, Department of Genetics and Cell Biology, Univ. of Minnesota, St. Paul.)

CELLULAR DIFFERENTIATION IN HIGHER EUKARYOTES

During the development of a higher eukaryote, a single cell—the zygote—gives rise by mitotic cell divisions to a vast array of cell types (in animals, skin cells, nerve cells, bone cells, blood cells, etc.) with highly divergent morphologies (Fig. 15.1) and macromolecular compositions. These different cell types are often highly specialized, carrying out only a few specific metabolic functions. For example, red blood cells are highly specialized for the synthesis and storage of hemoglobin. Over 90 percent of the protein molecules synthesized in red blood cells during their period of maximal biosynthetic activity are hemoglobin chains. Nerve cells are apparently the only cells capable of synthesizing neurotransmitters. What, then, is the mechanism by which these various cells differentiate

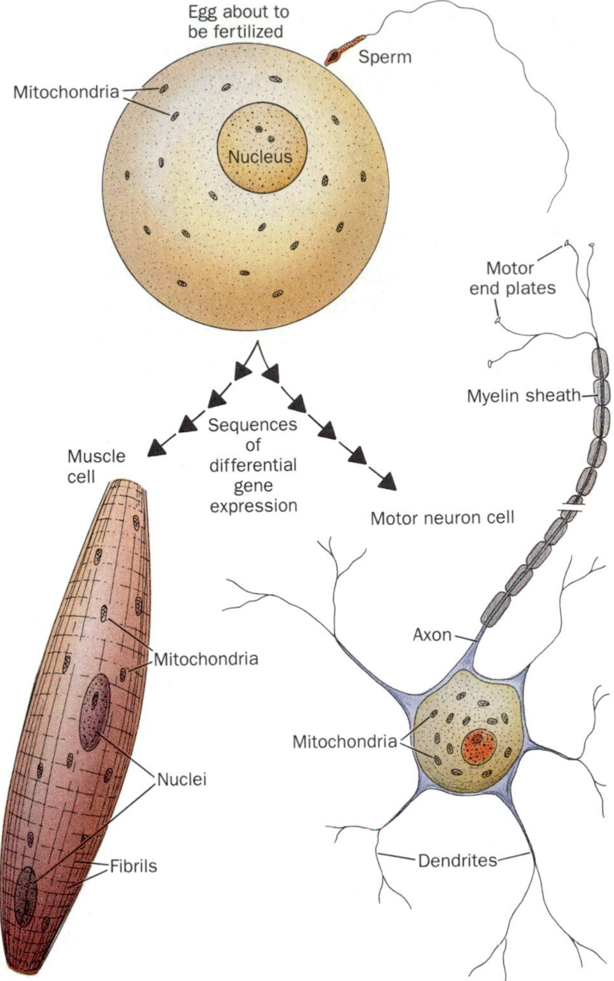

Figure 15.1 Schematic drawing illustrating the dramatic effects of differential gene expression during development in higher animals. A single cell, the fertilized egg or zygote, gives rise to the vast array of distinct cell phenotypes in the adult organism. The striated muscle and motor neuron cells shown are only two of the many cell types that exhibit highly divergent phenotypes in a higher animal such as a human being.

from one another during the growth and development of an organism? By what mechanism is the expression of the hemoglobin genes in red blood cells or the neurotransmitter genes in neurons, and the lack of expression of these genes in other cell types, brought about?

The only definitive answer that can be given to these questions at present is that the expression of these genes is controlled, at least in part, at the levels of transcription and transcript processing. Hemoglobin mRNA molecules are present in red blood cells, but are absent in other types of cells not synthesizing hemoglobin.

Differentiation occurs by the regulation of gene expression, rather than by changes in genome composition. This has been demonstrated by various techniques in many different organisms. In amphibians, for example, nuclei from differentiated cells can be transplanted into enucleated eggs (eggs from which the original nuclei have been removed) and shown to direct the development of normal embryos. Thus none of the genetic information required for the normal development of entire embryos is lost during the differentiation of the amphibian nuclei-donor cells. In addition, biochemical analyses of the DNA from the nuclei of various differentiated cells have in almost all cases shown that the genomes contain the same set of nucleotide-pair sequences. Rare exceptions are known. Mammalian red blood cells, for example, extrude their nuclei during the last stages of differentiation. By this time, large intracellular pools of stable hemoglobin mRNA have already been synthesized, so that the nuclei are no longer needed.

During the development of a complex plant or animal, gene expression has been shown to be regulated in different instances at essentially all possible levels—transcription, pre-mRNA processing, mRNA transport, mRNA stability, translation, posttranslational protein processing, protein stability, and enzyme function. However, extensive data now indicate that gene expression is regulated primarily at the levels of transcription and pre-mRNA processing. Clearly, regulation does occur at other levels as well. Regulatory fine-tuning at translational levels is obviously important in the overall control of metabolic processes in living organisms. However, the regulatory mechanisms with the largest effects on phenotype have been shown to act at the levels of transcription and RNA processing.

Many of the developmental processes in higher eukaryotes seem to be controlled, at least in part, by ***preprogrammed circuits of gene expression.*** In these cases, some event (such as release of a hormone in the bloodstream or fertilization of an egg) triggers the expression of one set of genes. The product (or products) of one (or more) of these genes functions by turning off the transcription of the first set of genes and/or turning on the transcription of a second set of genes. In turn, one or more of the products of the second set acts by turning on a third set, and so on. In these cases, the sequential expression of genes is genetically preprogrammed, and the genes ***cannot*** usually be turned on out of sequence.

In eukaryotes, we know that hormones can trigger the sequential expression of sets of genes. In addition, we know that regulatory genes are involved in the control of patterns of differentiation. In some cases, we know that regulatory elements called ***enhancers*** and ***silencers*** modulate levels of gene expression from nearby promoters. However, the question of how these enhancers and silencers work in controlling gene expression remains as a challenge to present and future generations of geneticists. In the future, will we understand the pathway of morphogenesis of a fruit fly, a worm, a plant, or a human being as we now understand the pathway of morphogenesis of bacteriophage T4 (see Chapter 11, Fig. 11.33)? The answer to this question is simple—YES!—and sooner than you think!

GENETIC CONTROL OF DEVELOPMENT IN *DROSOPHILA*

The mechanism by which information stored in a linear format—the sequences of nucleotide-pairs in genes—can control the ***development of three-dimensional form*** or ***morphogenesis*** has intrigued scientists since the birth of genetics. Of course, we now know that morphogenesis is merely the result of intricately regulated pathways of gene expression. The appropriate genes are expressed at the proper time in the correct sequence to produce the proteins that interact to produce a virus (see Chapter 11, Fig. 11.33), a cell, a tissue, an organ, or even an entire living organism such as a fruit fly. However, the programs of gene expression and the pathways of gene-product action that are required to produce an intricate living creature like a fly are very complex, and working out all the programs and pathways is a major challenge. Nevertheless, rather spectacular progress has been made and continues to be realized in understanding the key features of morphogenesis in higher eukaryotes, especially *Drosophila.* This rapid progress is the result of the combination of the unique advantages that *Drosophila* offers for genetic analysis and the application of the powerful tools of recombinant DNA and gene cloning to the study of key genes that regulate morphogenesis. In the following sections of this chapter, we briefly examine some of the progress that has been made to date.

Indeed, the picture that emerges is quite clear and quite elegant. Development in *Drosophila* is controlled by a cascade of regulatory genes that act in a specific sequence to impose different patterns of gene expression on different regions of the fertilized egg, then the developing embryo, then different cells of the blastula, and so on, until the complex morphology of the adult fly is produced. Although thousands of genes may be expressed in a neuron that are not expressed in cells of the eye, the leg, or the wing, these large differences in the patterns of gene expression in the different cells do not happen all at once in response to a single triggering event. Instead, the different cell lineages that will give rise to neurons, cells of the eye, and so on gradually diverge, with different programs of gene expression being turned on by a cascade of different regulator genes acting in preprogrammed sequences throughout development. In *Drosophila,* many of these key regulator genes have been identified, and about half of them are now known to encode *trans*-acting proteins that regulate the transcription of other genes, often including the next key regulator gene(s) in the cascade of morphogenetic regulators.

The stages in the development of *Drosophila* are summarized in Fig. 15.2. Development begins even before fertilization of the egg with molecular gradients being established in the mature egg. Development continues after fertilization with specific patterns of gene expression in the various stages of embryonic development (Fig. 15.3). Early embryogenesis occurs very rapidly without cell growth. The large egg initially contains two haploid nuclei, the male and female pronuclei. They then fuse to produce the diploid nucleus of the zygote, which in turn divides about nine times to yield a large single cell called a syncytium containing about 512 nuclei. The nuclei migrate to the surface of the egg at this stage and undergo four additional divisions. Cell membranes then form around the nuclei to produce the ***blastula*** containing about 4000 cells. Separate germ line cells, called ***pole cells,*** form at this stage.

In some ways, development in insects is more complex than development in many other higher eukaryotes. Genetic programs for the development of both the immature, wormlike larvae and pupae, as well as genetic programs for the adult insects, are required. Metamorphoses from larva to pupa and from pupa to adult require sudden changes in programs of gene expression. These changes in gene expression are triggered, at least in part, by the steroid hormone ecdysone (see p. 439). Interestingly, ecdysone has been shown to stimulate the transcription of different sets of genes in different tissues. The genes activated by ecdysone in cells destined to form adult wings or adult legs, for example, are different from those activated by ecdysone in early larval salivary glands.

Imaginal Discs and Determination

Preparations for insect metamorphosis are made during early embryogenesis. *Drosophila* blastoderm cells differentiate into two kinds of cells: (1) those that will give rise to larval tissues and (2) those that will develop into the various tissues and organs of the adult fly. Specifically, certain groups of undifferentiated cells form structures called ***imaginal discs.*** In all, 17 imaginal discs form, each of which is destined to give rise to a specific organ of the adult fly (Fig. 15.4). These include pairs of labial discs, antenna discs, eye discs,

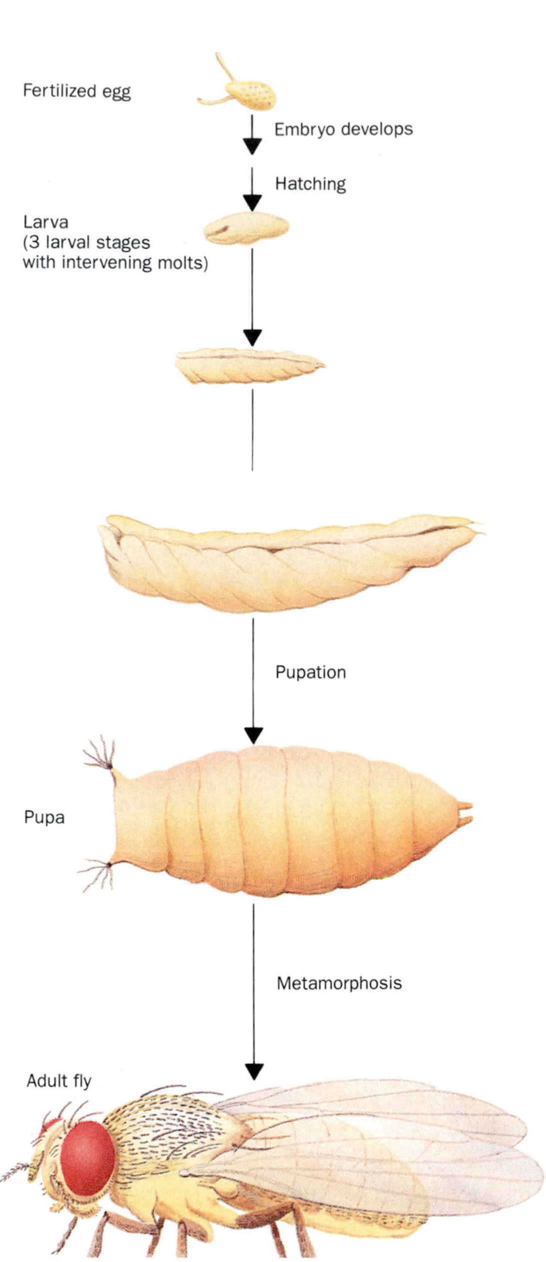

Figure 15.2 Stages of development in *Drosophila*. Embryonic development (see Fig. 15.3) of the fertilized egg occurs very rapidly. The first larval stage hatches within about one day after fertilization. Pupation occurs at about five days postfertilization, and metamorphosis of the adult fly occurs at about nine days postfertilization.

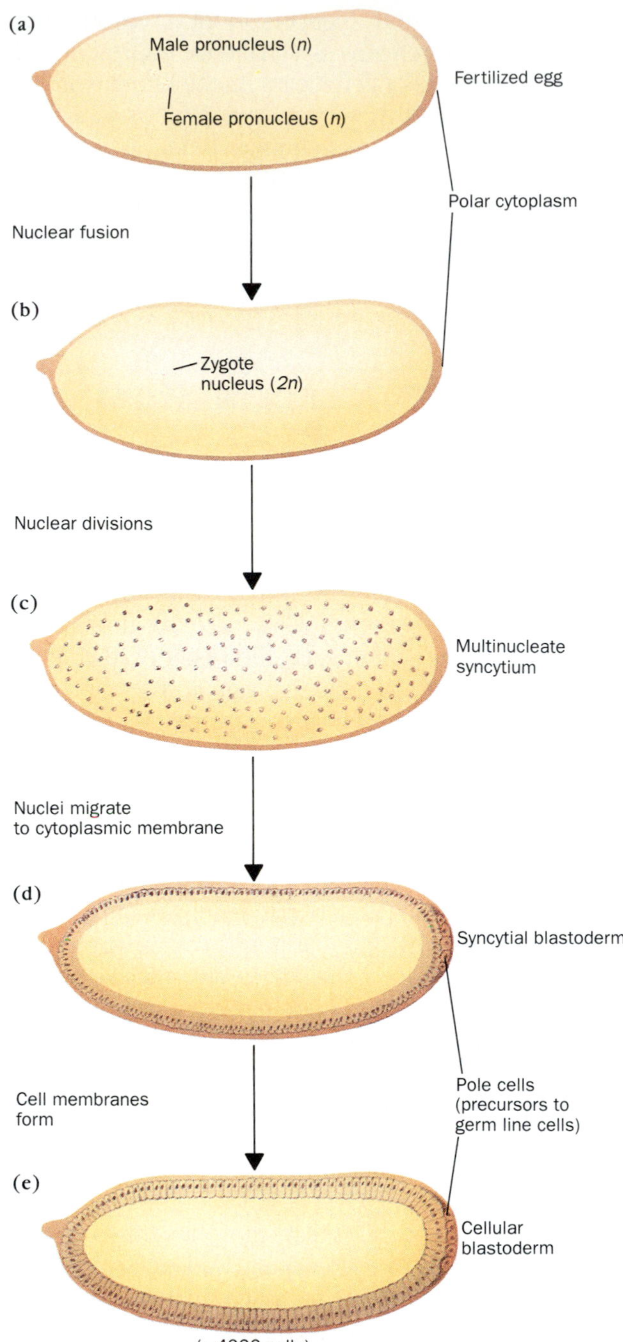

Figure 15.3 Embryonic development in *Drosophila*. (a) The fertilized egg initially contains two haploid (n) nuclei, one from the egg and one from the sperm. (b) The two haploid nuclei then fuse to form the diploid nucleus ($2n$) of the zygote. (c) The zygotic nucleus divides very rapidly to produce a single cell (syncytium) with many nuclei. (d) After about 9 nuclear divisions have produced about 512 nuclei, the nuclei migrate to the periphery of the syncytium and continue to divide (about 4 more times each). A few nuclei migrate into the polar cytoplasm and form pole cells, which are the progenitors of all the germ line cells of the fly. (e) Cell membranes then form around the nuclei to produce the cellular blastoderm, composed of about 4000 cells. (After V. E. Foe and B. M. Alberts, *J. Cell Sci.* 61: 31–70, 1983.)

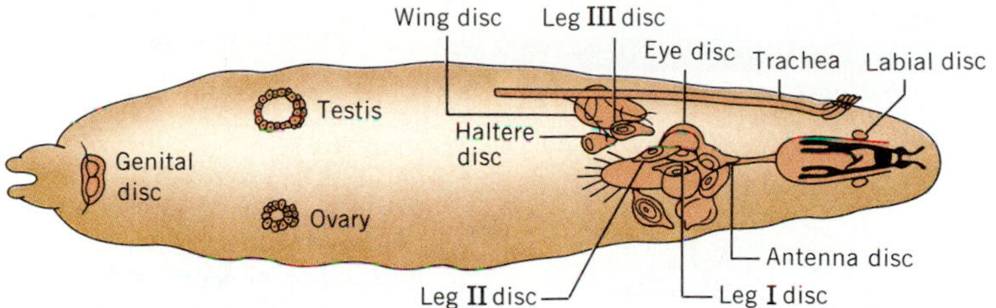

Figure 15.4 The location of the imaginal discs of *Drosophila* larvae. Each disc gives rise to the indicated organ in the adult fly. (After D. Bodenstein, *Biology of Drosophila,* M. Demerec, editor, Wiley, New York, 1950.)

wing discs, haltere discs, three different pairs of leg discs, and a genital disc. In preparation for the bilateral symmetry of the adult fly, each disc, except the genital disc, is present in duplicate.

Although the cells of imaginal discs are (for the most part) undifferentiated, their destinies have already been decided. This process by which cells (actually groups of cells, and usually groups of from 5 to 50 cells in *Drosophila*) make irreversible commitments to specific patterns of differentiation is called ***determination. Determination*** thus occurs at the time that the ***destiny of a cell lineage becomes fixed.*** Imaginal disc cells are already determined prior to disc formation; in fact, ***imaginal disc determination occurs in the cellular blastoderm stage.*** The time at which cell determination occurs can be ascertained by experiments in which genetic and embryological approaches are combined. Such experiments allow one to work out ***fate maps***—diagrams relating the fates (or destinies) of various cell lineages.

The determined state of imaginal disc cells can be demonstrated by dissecting them from larvae and transplanting them into abnormal positions in other larvae. Regardless of their new locations, transplanted imaginal discs develop into the previously ***determined*** organ. If an eye disc, for example, is transplanted into another larvae, the fly that emerges following metamorphosis will have three eyes, one of which may occur in a totally bizarre location.

Homeotic Mutants

A key step in understanding differentiation is understanding the mechanism(s) by which cell determination occurs. Determination is clearly under genetic control; this is demonstrated by the existence of ***homeotic mutations,*** that is, mutations that cause altered states of determination in specific imaginal discs. The mutation *Antennapedia,* for example, transforms the determination of the antenna disc to that of a leg disc, which results in a fly with a leglike appendage extending from its head (Fig. 15.5). *Ophthalmoptera* results in the development of wing structures from eye discs, and so on.

A large number of different homeotic mutants have been identified and studied in *Drosophila* and the majority of the mutations map within one of two clusters of homeotic genes on chromosome 3. These two clusters are referred to as the **Antennapedia *complex (ANT-C)*** and the **bithorax *complex (BX-C).*** Both gene complexes are very large: ANT-C spans over 100 kb and BX-C covers at least 300 kb of DNA. Different homeotic mutations within each complex give rise to distinct, but often related changes in phenotype. Homeotic mutations in ANT-C affect the development of the head and thorax regions of the fly, whereas those in BX-C alter the development of the thorax and the abdominal segments of the body (Fig. 15.6). However,

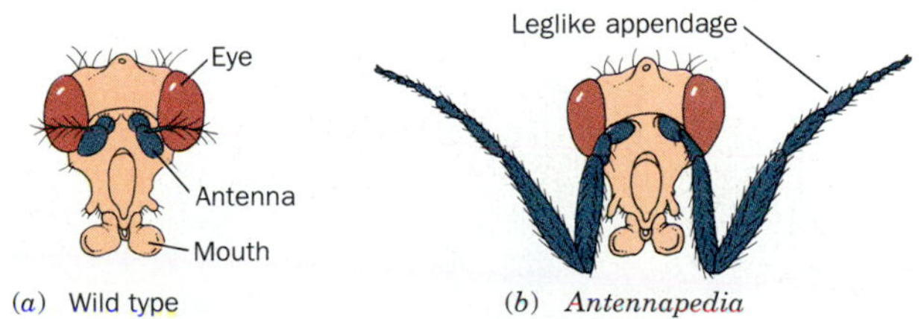

Figure 15.5 Diagram illustrating the effects of the homeotic mutation *Antennapedia* on the phenotype of *Drosophila.* (a) Wild-type fly showing the antennae on the head. (b) *Antennapedia* mutant showing the replacement of the distal portion of the antennae with leg structures. (For photographs of various *Antennapedia* mutants, see J. H. Postlewait and H. A. Schneiderman, *Develop. Biol.* 25: 606–640, 1971.)

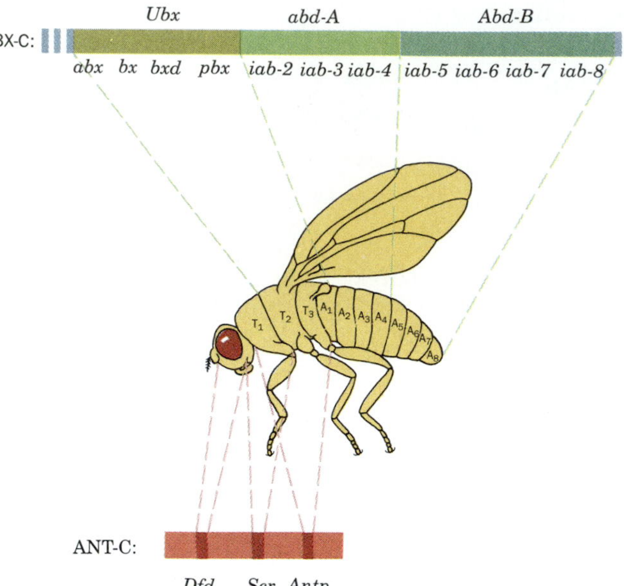

Figure 15.6 Diagram showing the body segments of *Drosophila* primarily affected by homeotic mutations in the two major clusters of homeotic genes on chromosome 3. The *bithorax* complex (BX-C) contains genes that control the development of thorax segments 2 and 3 (T2 and T3) and the segments of the abdomen (A1–A8). The *Antennapedia* complex (ANT-C) contains three genes (*Dfd, Scr,* and *Antp*) that are essential for development of the head and thorax segments. The dotted lines indicate the zones of influence of the various genes of BX-C and ANT-C. Each of the homeotic mutants shown converts a particular body segment to a phenotype more like an adjacent segment. For example, *pbx (postbithorax)* converts thorax segment 3 to the normal phenotype of thorax segment 2 (see Fig. 15.7). Mutant *iab-8 (infraabdominal 8)* converts abdomen segment 8 to the phenotype of abdomen segment 7, and so on. (After K. Harding *et al., Science* 229: 1236–1242, 1985.)

we should hasten to emphasize that homeotic mutations usually have such extreme effects on the phenotypes of the mutant fly that these mutants die before reaching the adult stage. Other mutations with less extreme effects on phenotype produce flies that survive, but exhibit abnormal phenotypes such as flies with four wings (Fig. 15.7). Actually, the four-winged fly shown in Fig. 15.7 is a triple mutant harboring three defects in the BX-C gene cluster. The three mutations (*abx, bx^3,* and *pbx*) convert the halteres (organs used to maintain balance) on the third thorax segment to nearly normal-looking wings.

Figure 15.7 Photograph of a fruit fly with four wings produced by the presence of three homeotic mutations (*abx, bx^3,* and *pbx*) in the *Ultrabithorax* region of the *bithorax* complex. For details, see W. Bender *et al.* "Molecular Genetics of the Bithorax Complex in *Drosophila melanogaster.*" *Science* 221: 23–29, 1983. (Courtesy Edward B. Lewis.)

The ANT-C region contains at least four genes: *Deformed (Dfd), Sex combs reduced (Scr), fushi tarazu (ftz),* and *Antennapedia (Antp).* The BX-C region contains three complementation groups: *Ultrabithorax (Ubx), abdominal-A (abd-A),* and *Abdominal-B (Abd-B).* Note that both dominant and recessive homeotic mutations occur within each gene complex.

The early discovery of homeotic mutations was of major importance in establishing that different pathways of morphogenesis were under the control of key "switch" or "selector" genes that could turn off one pattern of gene expression and turn on a different pattern. The *bithorax* complex has been studied intensively by E. B. Lewis and coworkers since the 1940s. This complex locus has now been cloned and much of it sequenced. Clearly, all of Lewis' work is now yielding a wealth of information about the genetic control of development in *Drosophila.* We will examine the *bithorax* complex in further detail on pp. 422–423.

The homeotic mutations such as *Antennapedia* (see Fig. 15.5) cause changes in the state of determination of certain imaginal discs. However, only a subset of all possible transformations in imaginal disc determination are observed among the known homeotic mutations. Other transformations have never been observed. This suggests that some determined states are more closely related than others. A single mutation (alteration of one "selector" gene?) can convert cells from antenna determination to leg determination, but not from antenna determination to eye determination. Does this tell us anything about the programs of gene expression controlling antenna, leg, and eye determination and morphogenesis?

Transdetermination

Interestingly, the transformations observed in homeotic mutants are the same as the changes occasionally observed following repeated transplantations of larval imaginal discs in adult flies. E. Hadorn observed that after many serial transplantations of genital discs to adult abdomens, some of the genital discs would develop into leg structures when subsequently implanted in metamorphosing larvae. These events were called ***transdeterminations.*** They did not occur at random; only specific transdeterminations were observed (Fig. 15.8). A genital disc occasionally underwent transdetermination to a leg structure or an antenna structure, but never to an eye or a wing structure.

The same transformations in determination are observed in homeotic mutations and in transdetermination events. This suggests that certain preprogrammed circuits of gene expression have more closely related controlling elements. For example, mutations in a single ***"program selector gene"*** may divert the cells of a leg imaginal disc from the expression of the genes in the "leg program" to the expression of the genes in the "wing program," but ***not*** to the expression of the genes in the "eye program." The transformation from the "leg program" to the "eye program" would supposedly require changes in two or more "program selector genes."

Regardless of the molecular details, the existence of homeotic mutations provides clear evidence of the role of preprogrammed patterns of gene expression in determination and differentiation.

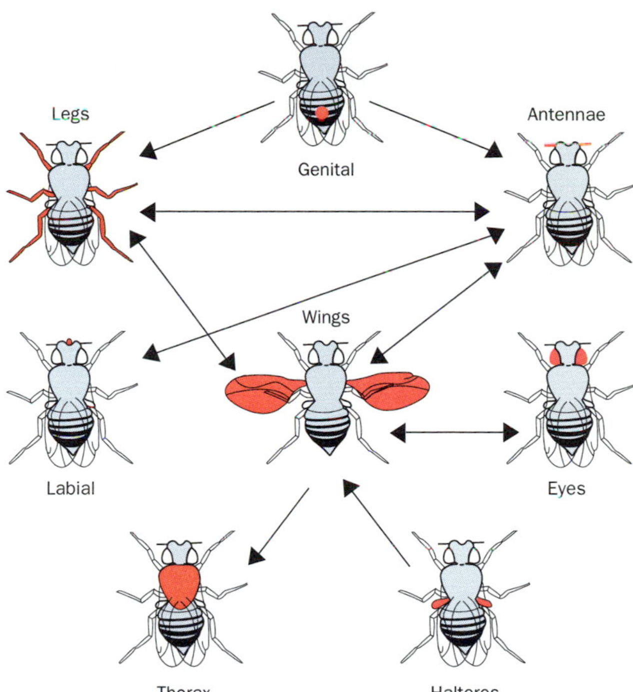

Figure 15.8 Transdetermination events observed in *Drosophila* by E. Hadorn and colleagues. Only transdeterminations of the types indicated by the arrows were observed. (After E. Hadorn, *Sci. Amer.* 219: 110–120, 1968.)

Morphogenesis of a Fly: Sequential Anterior–Posterior Segmentation Events

The dominant feature in the morphogenesis of *Drosophila* is the development of a ***complex pattern of segmentation along the anterior–posterior axis of the body*** (Fig. 15.9). Anterior–posterior polarity is initially established in the egg by gene-products synthesized during oögenesis, the products of so-called ***maternal-effect genes.*** During subsequent stages of embryonic development, finer and finer patterns of segmentation develop through the action of at least 20 different ***segmentation genes.*** After the intricate pattern of segmentation has been established, the ***homeotic genes*** are believed to trigger the expression of the different sets of genes that control the development of the individual organs and tissues of the fly. Superimposed on the development of patterns of segmentation is the ***establishment of dorsal–ventral polarity.*** This occurs very early during embryonic development and is controlled by another set of key morphogenetic genes. Lastly, ***tissue-specific genes*** act to produce the highly differentiated cells of the adult fly.

As mentioned, the anterior–posterior polarity of the embryo is established by maternal-effect genes that are expressed during oögenesis. Eighteen different maternal-effect genes that influence the anterior–posterior polarity and segmentation pattern have been identified. We shall discuss only two of these genes: *bicoid (bcd)* and *oskar (osk).* Females that are homozygous for a *bcd* mutation produce embryos that lack heads and thoracic structures. On the other extreme, females homozygous for an *osk* mutation produce embryos that lack all the abdominal segments. As one might expect from the preceding observations, the *bcd* and *osk* gene-products play key roles in establishing "anterior phenotype" and "posterior phenotype," respectively. In fact, *bcd* embryos can be rescued by injecting them with cytoplasm obtained from the anterior segment of a wild-type embryo. The injected *bcd* embryo will develop fairly normal head and thoracic structures. Likewise, *osk* embryos will develop quite

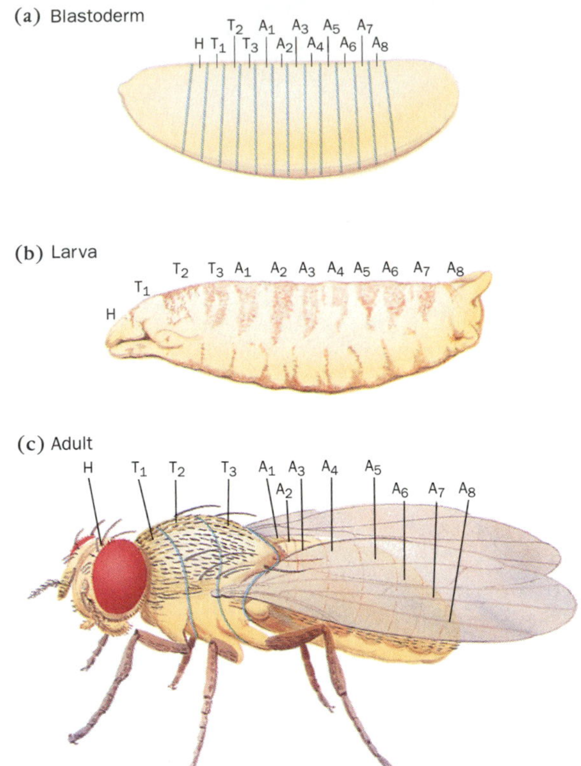

Figure 15.9 Diagram illustrating the major segmentation pattern in *Drosophila* at the (a) blastoderm, (b) larva, and (c) adult stages of development. Although segmentation is not yet visible at the blastoderm stage, the cells are already committed to the formation of the body segments of the larva as shown on the blastoderm fate map (a). [(a) After W. J. Gehring, *Science* 236: 1245–1252, 1987; (b) After G. Struhl, *Nature* 293: 36–41, 1981.]

normally if injected with cytoplasm obtained from the posterior region of a wild-type embryo.

Localization of both the *bcd* gene transcript and the *bcd* protein in the anterior region of the embryo has been demonstrated directly by *in situ* hybridization with a labeled *bcd* cDNA probe and by immunofluorescence microscopy with a *bcd* protein-specific antibody probe (Fig. 15.10), respectively. Gene-products such as the *bcd* protein that establish form-determining gradients in cells and tissues are called ***morphogens.***

Another group of at least 12 genes controls the dorsal–ventral polarity in the embryo. These include the *dorsal (dl)* and *Toll (Tl)* genes, which seem to play important roles in forming a dorsal–ventral gradient of morphogens in embryos. Most of the dorsal–ventral determining genes respond to injections of cytoplasm from wild-type embryos just as *bcd* and *osk* embryos do. The initial dorsal–ventral polarity of the egg appears to be established by the *Toll* gene-product, but the development of normal dorsal–ventral patterns requires the activity of all 12 genes involved in establishing dorsal–ventral polarity.

Segmentation of the embryo, the larva, the pupa, and the adult fly is undoubtedly the most striking feature of morphogenesis in *Drosophila* and related insects. The establishment of this segmentation pattern is under the control of about 20 known genes that act in sequence to produce the final pattern. In fact, it is now clear that the products of the genes that function early often regulate the expression of the next set of segmentation genes, and so on, in a cascade fashion. The products of several of these genes have been shown to be *trans*-acting regulators of the transcrip-

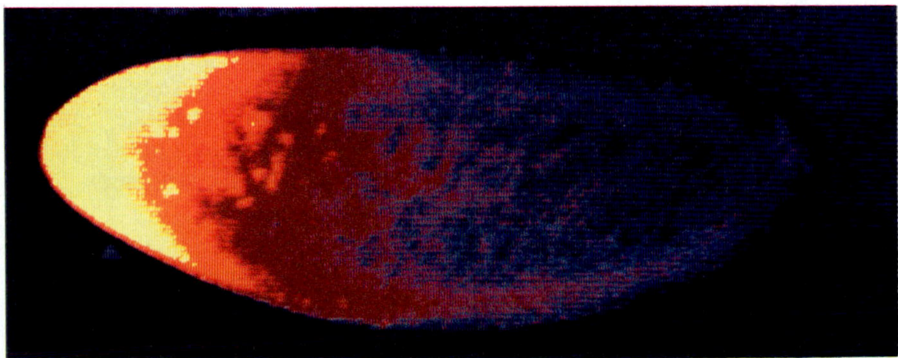

Figure 15.10 Localization of the *bicoid* protein in the anterior zone of a *Drosophila* embryo by immunological staining performed with an antibody to the *bicoid* protein. The intensity of staining has been transformed into arbitrarily selected gradients of color with an image analysis system. Details of the procedures used are given in W. Driever and C. Nüsslein-Volhard, "A Gradient of *Bicoid* Protein in *Drosophila* Embryos." *Cell* 54: 83–93, 1988. (Reproduced with permission from the cover of *Cell,* July 1, 1988. Copyright © 1988 by Cell Press, Cambridge, MA.)

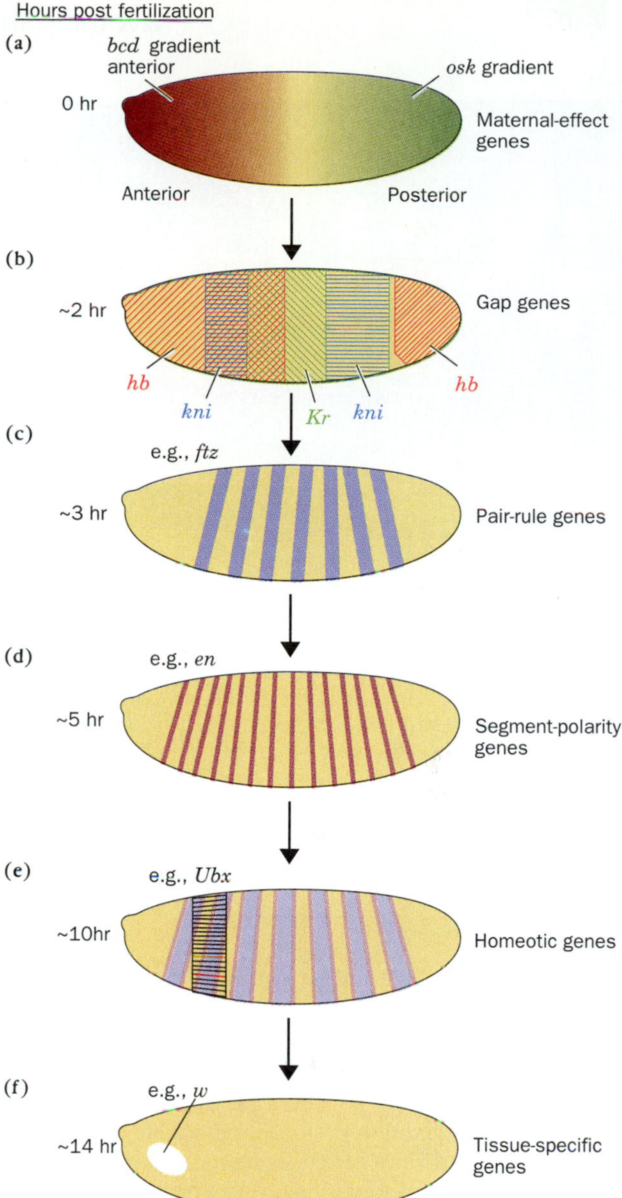

Figure 15.11 Schematic diagram of the cascade of gene expression that produces finer and finer patterns of segmentation and differentiation in *Drosophila* embryos. Times given for expression of specific genes (in hours postfertilization) are only approximate and are accurate for only some of the genes in each class. (a) The initial anterior–posterior polarity is established by maternal-effect genes such as *bicoid (bcd)* and *oskar (osk)* (see text). (b) The expression of the three gap genes (*kr, kni,* and *hb*) produces four broad anterior–posterior zones in the embryo (plus two terminal regions). (c) The pair-rule genes are each expressed in seven bands, as shown for the *ftz* gene along the anterior–posterior axis. Each of the different pair-rule genes is expressed in seven such bands; however, the bands are in different positions along the longitudinal axis for different genes. (d) The segment-polarity genes are expressed in 14 bands or stripes along the anterior–posterior axis. Again, different genes are expressed in different bands along the longitudinal axis. For example, the *engrailed (en)* gene-product is present in bands that are only one cell in width. Only the expression of *en* is shown in (d). (e) The *Ubx* gene is expressed only in thoracic segments T2 and T3 (cross-hatched); expression of *Ubx* is shown superimposed on the combined *ftz* and *en* patterns of (c) and (d). (f) Tissue-specific genes are then expressed in patches of determined cells, such as *white (w)* in the eye progenitors, to yield differentiated groups of cells that will produce specific tissues.

tion of later-acting genes producing an intricate cascade of transcriptional regulators (see the following section).

The segmentation genes have been classified into three groups: (1) **gap genes,** (2) **pair-rule genes,** and (3) **segment-polarity genes.** These genes exert their effects on the segmentation pattern in roughly the sequence in which they are listed (Fig. 15.11).

1. ***Gap Genes*** The gap gene-products are required to form a set of adjacent segments of the larva and fly. Mutants with defects in gap genes are lacking an entire set of contiguous body segments. Three gap genes are well characterized: *Krüppel (Kr), hunchback (hb),* and *knirps (kni). Krüppel* mutant larvae lack eight adjacent segments, thorax segment 1 through abdomen segment 5. When the locations of *Kr* gene transcripts and proteins were examined in cellular blastoderms by *in situ* hybridization with a labeled cDNA probe and with a specific antibody probe, the *Kr* gene was found to be expressed only in the central zone of these embryos. Gap gene *knirps* is expressed in two bands, one on each side of the zone of *Kr* expression, and expression of *kni* only occurs if the maternal-effect *osk* gene-product is present. The *hunchback* gene is required to produce the anterior half and the posterior quarter of the larva, and expression of *hb* requires *bcd* function. Thus, no *hb* protein is made in *bcd* mutants. Acting together, the gap genes establish four broad zones within the fertilized egg (Fig. 15.11).

2. ***Pair-Rule Genes*** The pair-rule genes are each expressed in seven distinct bands or stripes along the anterior–posterior axis (Fig. 15.12), in effect dividing the embryo into 15 segments (Fig. 15.11**c**). Mutants in each of the eight pair-rule genes produce embryos with only half as many segments as wild type. In each mutant,

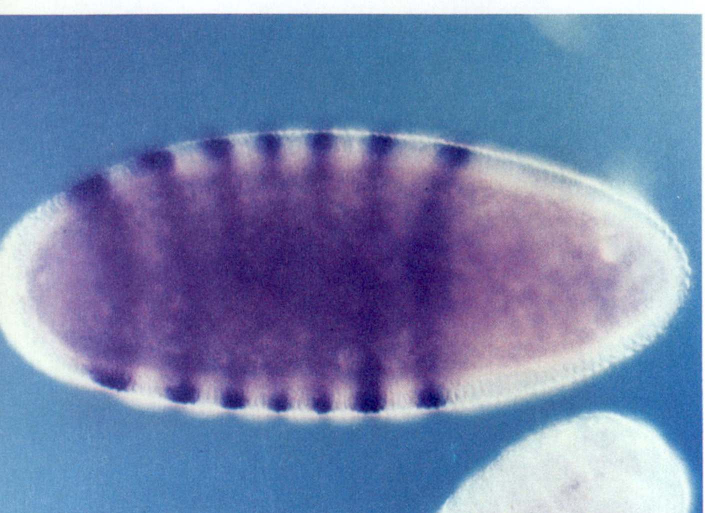

Figure 15.12 The seven-stripe pattern of expression of the pair-rule gene *fushi tarazu (ftz)* in a *Drosophila* blastoderm embryo. The localization of the *ftz* RNA is indicated by the characteristic seven dark stripes detected by *in situ* hybridization with a *ftz* gene-specific probe. The anterior pole of the embryo is to the left; dorsal is at the top. (Courtesy of Matthew Scott, University of Colorado, Boulder.)

every other segment is missing, although the missing segments are not the same in different pair-rule mutants. Depending on how one demarcates the zones along the anterior–posterior axis of an embryo or a larva, one defines either ***segments*** or ***parasegments*** of the body (as defined for use in discussing patterns of development). The relationship between segments and parasegments is shown in Fig. 15.13. Mutants in the pair-rule gene *fushi tarazu (ftz)* lack each of the odd-numbered parasegments (P1 through P13). On the other hand, mutants of the pair-rule gene *even-skipped (eve)* are missing all the even-numbered parasegments.

3. ***Segment-Polarity Genes*** The 10 or more segment-polarity genes function to produce anterior and posterior polarity to the individual segments defined by the expression of the pair-rule genes. At least some of these genes are expressed in 14 zones or stripes perpendicular to the longitudinal axis of the embryo (Fig. 15.11*d*). Mutations in segment-polarity genes result in part of each segment being lost and replaced with a mirror-image copy of the rest of the segment or an adjacent "half-segment." For example, in *gooseberry* mutants, the posterior half of each segment is replaced by a mirror-image copy of the adjacent anterior half-segment. Again, different segment-polarity mutants exhibit subsegment mirror-image replacements of different subsegment or subparasegment compartments, but the overall pattern is the same for all the segment-polarity mutants.

Once the complete segmentation pattern has been established through the sequential action of the maternal-effect genes, gap genes, pair-rule genes, and segment-polarity genes, the homeotic genes act to control the development of organs and tissues of the adult fly. For example, the *Ubx* gene is expressed in thorax segments T2 and T3 (Fig. 15.11*e*), and ultimately tissue-specific genes are expressed to produce the final phenotypes of the highly differentiated cells of the adult (Fig. 15.11*f*).

Clearly, patterns are established in developing embryos in a very precise and ordered manner. This requires very carefully regulated patterns of gene expression. How are these intricate patterns of gene expression regulated? What molecular mechanisms control the precise temporal and spatial patterns of gene expression observed during the morphogenesis of the fruit fly?

A Temporally Ordered Cascade of Regulatory Genes

Recent results indicate that about half of the genes that control pattern formation in *Drosophila,* including maternal-effect genes, gap genes, pair-rule genes, segment-polarity genes, and homeotic genes, encode ***trans-acting, promoter-selective regulators of transcription.*** These results suggest that the temporal and spatial patterns of morphogenesis are governed by the action of a ***cascade of regulatory genes.*** As was mentioned earlier, the gap gene *hunchback (hb)* is not expressed in *bicoid (bcd)* embryos. Recently, the *bcd* protein has been shown to promote transcription *in vitro* from the *hb* promoter. Indeed, the *bcd* protein

H

Segments: | Mn | Mx | La | T1 | T2 | T3 | A1 | A2 | A3 | A4 | A5 | A6 | A7 | A8 | A9/A10

Parasegments: | P1 | P2 | P3 | P4 | P5 | P6 | P7 | P8 | P9 | P10 | P11 | P12 | P13 | P14 |

Figure 15.13 Relative locations of the segments (see Fig. 15.9) and parasegments of *Drosophila* as defined by the actions of the pair-rule and segment-polarity genes. H, head region, includes segments Mn, Mx, and La, the mandibular, maxillary, and labial segments that give rise to the mouthparts; T1–T3, thorax segments 1–3; A1–A9/A10, abdomen segments 1 through 10; P1–P14, parasegments 1 through 14. Note that an anterior half of a segment corresponds to the posterior half of a parasegment.

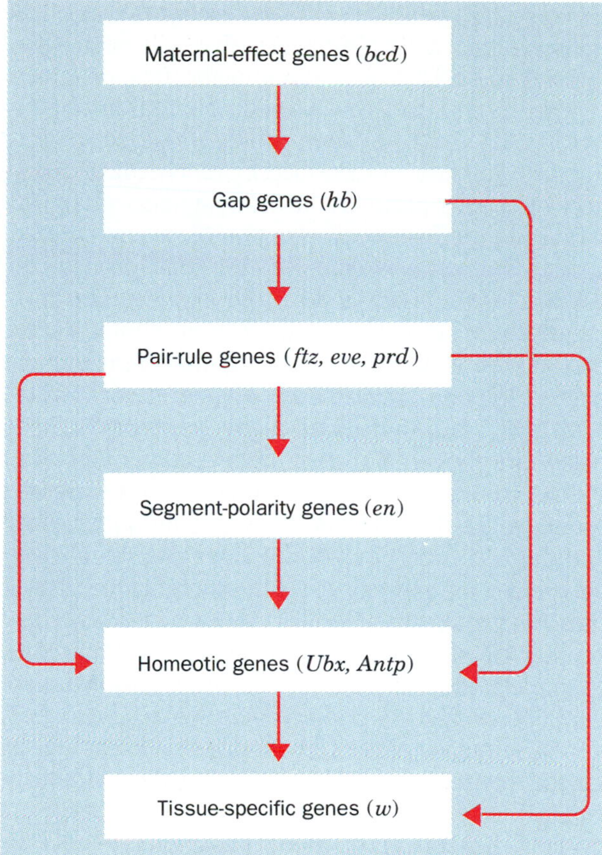

Figure 15.14 Regulatory cascade in the control of development in *Drosophila*. See Fig. 15.11 for the roles of the indicated classes of genes. Each arrow indicates a known pathway of gene regulation in which the product of one or more genes in one class affects the expression of one or more genes in another class. (After M. D. Biggin and R. Tjian, *Trends in Genetics* 5: 377–383, 1989.)

binds to the *hb* promoter in a sequence-specific manner (binding site TCCTAATCC). Other *Drosophila* pattern-determining genes that have been shown to encode proteins with sequence-specific binding properties are *Antennapedia (Antp), engrailed (en), Abdominal-B (Abd-B*^{r}*), even-skipped (eve), fushi tarazu (ftz), paired (prd), Ultrabithorax (Ubx),* and *zerknullt (zen).* In some cases, the gene-products bind to the same promoter sequences. For example, the proteins encoded by *Abd-B*^{r}*, en, eve, ftz, prd,* and *zen* all bind to the sequence TCAATTAAAT. Interestingly, some of these proteins also bind to other related sequences. In most cases, they stimulate transcription; however, in other cases, they repress transcription.

In the sequence of events in anterior–posterior segmentation in *Drosophila* embryos (Fig. 15.11), there are many examples where a gene of a class expressed early produces a sequence-specific transcription factor that controls the expression of a regulator gene that acts later in the pathway. Fig. 15.14 shows some of the known transcriptional regulators that are part of this regulatory cascade. It seems likely that other genes that control pattern formation in *Drosophila* will be shown to be part of this regulatory cascade in the future. If the transcription of regulatory genes in a morphogenetic pathway requires the concerted action of two or more sequence-specific, *trans*-acting regulators, one can easily visualize how complex patterns of gene expression could result (Fig. 15.15).

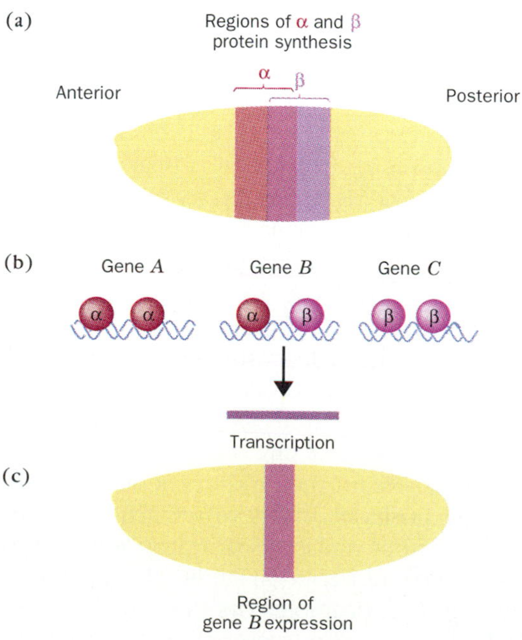

Figure 15.15 Pattern formation based on the effects of two different sequence-specific transcription activators synthesized in overlapping segments of an embryo. The region in which the α activator is synthesized is shown shaded in red. The region containing the β activator is shown shaded in purple. The gene *B* promoter requires both activators, α and β. By adding additional transcription regulators, both activators and repressors of transcription, and additional promoter binding sites, one can produce complex patterns of gene expression. (After M. D. Biggin and R. Tjian, *Trends in Genet.* 5: 377–383, 1989.)

The *Bithorax* Complex

The *bithorax* complex (BX-C) includes a variety of homeotic mutations that affect the second and third segments of the thorax and all the abdominal segments. BX-C contains three complementation groups: *Ubx, abd-A,* and *Abd-B,* each containing several distinct types of homeotic mutants. Mutations in the *Ubx* region of BX-C affect the thoracic region; those in *abd-A* and *Abd-B* alter abdominal segments. The mutations in *abd-A* and *Abd-B* are called *infraabdominals: iab-2, iab-3, iab-4,* to *iab-8.* Interestingly, the mutations are present on the chromosome in the same order as the order of the abdominal segments of the fly that they influence (see Fig. 15.6). Initially, it was thought that each *infraabdominal* mutation might correspond to a gene, and that the abdominal region produced would depend on the *iab* gene-products present. For example, segment A1 would form if no *iab* protein was present, A2 would require the product of the *iab-2* gene, A3 would require *iab-2* and *iab-3* proteins, and abdomen segment A8 would form only if the products of all the *infraabdominal* genes (*iab-2–iab-8*) were present. This model was eliminated when it became clear that *iab-2, iab-3,* and so on were not separate genes and that most of these mutations were ***cis-acting*** (not *trans*-acting).

We still do not understand how the BX-C region functions, but we do know that *cis*-acting regulation is one key component. The BX-C region spans over 300,000 nucleotide-pairs on chromosome 3 (Fig. 15.16). BX-C is clearly large enough to contain many genes. However, BX-C contains only three complementation groups. The entire BX-C region has been cloned and much of it has been sequenced. When the BX-C clones were used to study the structure and expression of the BX-C region, the genes were found to contain huge introns. For example, the major transcription unit of *Ubx* is shown in Fig. 15.16*b*. The major primary transcript of the *Ubx* is over 80,000 nucleotides (80 kb) long. This huge primary transcript is processed and yields two RNAs of sizes 3.2 and 4.3 kb. Both of these RNAs contain common sequences from each end of the primary transcript. Presumably these two RNAs are produced by alternative patterns of transcript splicing. In any case, it seems clear that the vast majority of the primary transcript (about 75 kb) is noncoding and is spliced out during processing.

Many of the classical BX-C homeotic mutations turn out to define *cis*-acting regulatory regions, probably binding sites for the products of other regulatory genes. The positions of known *cis*-acting sequences defined by either mutations or transposon insertions and/or *in vitro* transcription studies are shown in Fig. 15.16*b*. The *Ubx* promoter has been analzyed extensively, by both *in vitro* transcription studies and protein-binding studies. Fig. 15.16*c* shows some of the transcriptional activator proteins that have been found to bind to specific sequences in the *Ubx* promoter. Proteins NTF-1 and GAGA (neurogenic transcription factor-1 and a protein binding to GAGA sequences) are transcription factors identified by biochemical dissection of transcription in *in vitro* systems. They are required for efficient transcription of several different genes. The same is true for the *zeste* protein, the product of the *zeste* gene in *Drosophila.* The most interesting binding sites in the *Ubx* promoter are the two sequences that bind proteins with homeotic domains. Homeotic domains are sequences of 60 amino acids that are found to be highly conserved among many morphogenetic genes of *Drosophila* and other species (see the following section). The *Ubx* promoter sequence TAATAATAATAATAA binds the *Antp* protein and the *Ubx* protein itself; the latter observation suggests that *Ubx* may be, in part, self-regulating, that is, the *Ubx* protein may bind to its own promoter and repress or stimulate its own synthesis. Many such examples of self-regulation or ***autoregulation*** are well documented in prokaryotic systems. The TCAATTAAAT sequence of the *Ubx* promoter is a binding site for the *ftz* protein and the *eve* protein, the products of two pair-rule genes. Thus, *Ubx* appears to be an important component of a morphogenetic regulatory cascade in *Drosophila.*

A second primary transcript is produced from the *pbx* to *bxd* region of *Ubx.* It is about 25 kb in length and gives rise, after processing, to two small RNAs of sizes 0.8 and 1.2 kb. The functions of these small RNAs are unknown. Given the *cis*-acting nature of many of the mutations in BX-C, many geneticists have speculated that there are two important features in regulating the expression of BX-C: (1) many *trans*-acting regulators of transcription (and, possibly, transcript processing) and (2) alternative pathways of transcript splicing. Whether these speculations are correct remains uncertain. What is certain is that we will understand the mechanism(s) of action of these key homeotic genes in the very near future.

Role of the "Homeobox" Domain in Regulation of Transcription

A striking feature of many homeotic genes and other pattern-determining genes such as the gap genes and segmentation genes is the presence of a highly conserved region 180 base-pairs in length. This conserved structure in many morphogenetic genes has been named the ***homeobox.*** At least 20 *Drosophila* genes are known to contain such homeobox regions, which exhibit greater than 70 percent conservation of sequence in most cases. Two segmentation genes, *engrailed (en)* and *invected (inv),* contain a somewhat

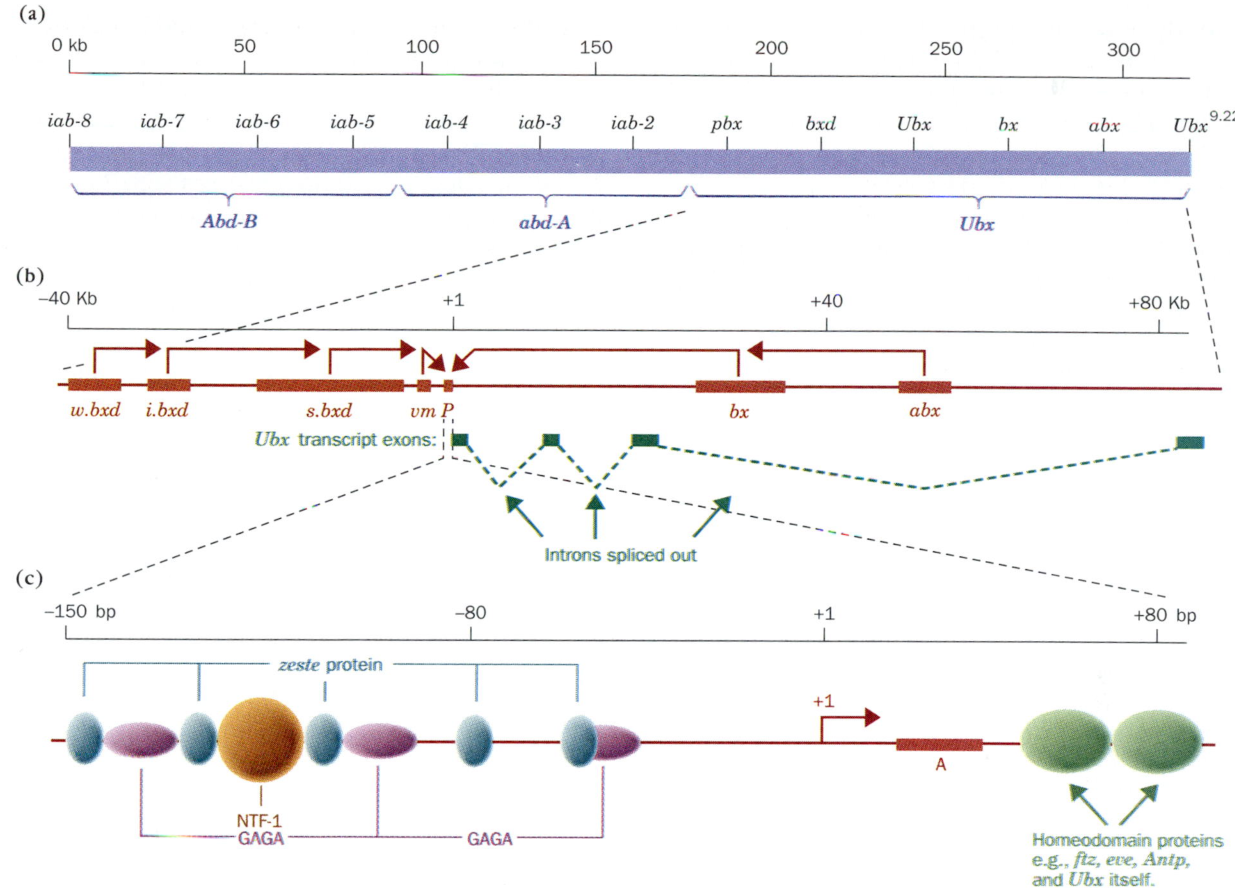

Figure 15.16 Structure of the *bithorax* complex (BX-C) of homeotic mutations. The locus includes three complementation groups and spans 300,000 nucleotide-pairs (300 kb) on chromosome 3 of *D. melanogaster*. Mutations within BX-C affect thorax segments T2 and T3 and abdomen segments A1 through A8 (see Fig. 15.6). (a) Map of BX-C showing the locations of some of the classical homeotic mutations and the positions of three complementation groups: *Ubx, abd-A,* and *Abd-B*. The map is oriented such that transcription of the major *Ubx* transcription unit occurs from left to right. (After E. B. Lewis, *Cold Spring Habor Sympos. Quant. Biol.* 50: 155–164, 1985.) (b) Enlarged map of the *Ubx* major transcription unit showing the positions of *cis*-acting regulatory sequences that have been identified by correlated genetic and *in vitro* transcription studies. Sites *w.bxd, i.bxd, s.bxd,* and *vm* are four distinct *cis*-acting sequences within the *bxd* region; *P* is the *Ubx* major promoter. The exon and intron regions of the primary transcript and the splicing pattern are indicated below the map (in green). (c) Map of known protein binding sites within the *Ubx* major promoter. ***Note*** that the ***scale in (c) is in base-pairs,*** whereas the ***scales in (a) and (b) are in kilobase-pairs.*** Again, +1 represents the 5′ base of the primary transcript. The *zeste* gene has been known to interact with other *Drosophila* genes for several decades. NTF-1 and GAGA are known transcription factors in *Drosophila;* NTF-1 = neurogenic transcription factor 1; GAGA = protein transcription factor that binds to promoter sequence CGAGAGAG. The morphogenic genes *ftz, eve,* and *Antp* are described in the text. The box labeled A is known to be an important *cis*-acting regulatory sequence, but the protein(s) that binds to it is not yet known. [(b, c) After M. D. Biggin and R. Tjian, *Trends in Genet.* 5: 377–383, 1989.]

more divergent homeobox with only 45 percent sequence conservation when compared to the *Antennapedia (Antp)* group of homeoboxes. In all cases, the homeobox lies within the coding region of the gene and specifies a 60-amino acid segment, called the ***homeodomain,*** of the polypeptide product.

Once it became clear that many of the homeodomain-containing proteins were transcription regulators that functioned via sequence-specific binding to DNA *cis*-control elements, attention was directed at the role of the conserved homeodomain regions of the proteins. When *in vitro* studies showed that several different homeobox proteins would bind specifically to the same consensus sequence, namely, TCAGCACCG, researchers began to look for evidence that the homeodomains of these proteins were directly involved. By using site-specific mutagenesis on cloned genes and *in vitro* DNA-binding studies, the homeobox sequences were shown to be essential for DNA-binding activity. By using recombinant DNA techniques, researchers demonstrated that they could replace the homeobox of one gene with a related homeobox from another gene and produce a "hybrid" gene that functioned quite normally. Although the mechanism by which homeodomains allow proteins to bind to DNA in a sequence-specific manner is still the subject of speculation, it seems quite clear that the homeobox sequences of these proteins provide, at least in part, their sequence-specific DNA-binding properties.

Conserved "Homeobox" Sequences in Other Species

By screening cDNA and genomic libraries of different species with *Drosophila* homeobox sequences as hybridization probes, scientists have discovered that these homeoboxes are present in a wide range of higher eukaryotes, including frogs, mice, honeybees, and humans. U. Walldorf, R. Fleig, and W. J. Gehring recently isolated seven homeobox-containing genes from the honeybee (*Apis mellifera*) and found that six of the seven genes contained homeoboxes with greater than 90 percent sequence identity with the homologous homeoboxes in *D. melanogaster*. Two of the homeobox-containing genes of mice and humans, *Hox-1* and *Hox-2,* have homeoboxes with 70 percent amino acid identity with the *Antp* homeobox of *Drosophila.* The conservation of these homeobox sequences among higher animal species suggests that they play important roles, probably in regulating some aspect of morphogenesis. Indeed, recent studies have shown that some homeobox-containing proteins in mammals and in the worm *Caenorhabditis elegans* are sequence-specific DNA-binding proteins that most likely function as transcriptional activators.

CAENORHABDITIS ELEGANS: A MODEL SYSTEM FOR THE STUDY OF DEVELOPMENT

After searching for the ideal animal system to use as a model in dissecting development, S. Brenner chose the small free-living nematode (worm) *Caenorhabditis elegans* (Fig. 15.17) as the species of choice in 1965. Progress in dissecting and understanding the morpho-

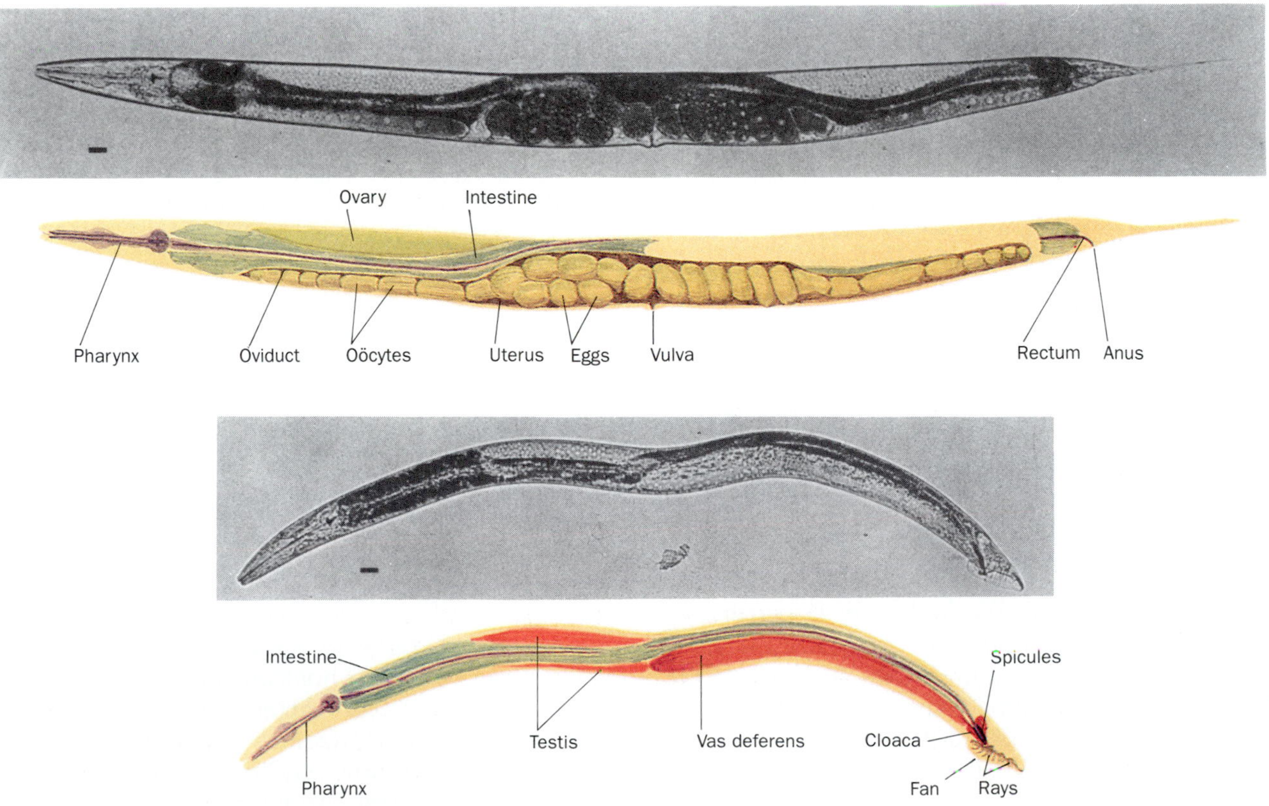

Figure 15.17 Photographs and diagrams showing the phenotypes of a *C. elegans* hermaphrodite (top) and male (below). (Reproduced with permission from J. E. Sulston and H. R. Horvitz, *Develop. Biol.* 56: 110–156, 1977).

genesis of *C. elegans* has been extremely rapid since that time. Only *D. melanogaster* rivals *C. elegans* as a model system for the study of animal development, with each system having some advantages. *Caenorhabditis elegans,* called "the worm" by the large group of *C. elegans* researchers, is particularly suited to correlated genetic and ultrastructural dissection. Its major advantages as a system are given in the following list.

1. ***Small size***— the adult worms are about 1 mm in length.
2. ***Short generation time***—reproduces with a 3-day life cycle under optimal growth conditions.
3. ***Simple growth conditions***—grows on agar plates with *E. coli* as food.
4. ***Two sexes: hermaphrodites and males***—the existence of hermaphrodites (male and female sex organs in the same worm) allows geneticists to produce worms homozygous for newly induced mutations very easily by simply allowing self-fertilization, just as Mendel allowed his pea plants to self-pollinate.
5. ***Large numbers of progeny***—a single hermaphrodite will produce about 300 eggs during its life.
6. ***Small genome***—about 8×10^7 nucleotide-pairs, which greatly facilitates molecular analyses such as cloning, sequencing, and physical mapping of the genome.
7. ***No exoskeleton, transparent bodies***—all cells of the body can be observed directly with the light microscope (Nomarski optics) at all stages of development.
8. ***Precise and invariant developmental program***—each adult hermaphrodite is composed of exactly 945 cells and these are produced from the zygote by precise pathways of cell division, cell migration, cell growth, and cell death; these are called ***invariant cell lineages.***

The transparent body of *C. elegans* has permitted J. E. Sulston and coworkers to establish the entire cell lineage for the adult worm from the single-celled zygote, an accomplishment that would be impossible in almost any other higher animal. Now, a small army of developmental geneticists are dissecting this pathway at the molecular, cellular, and morphological levels. The dominant approach is induce mutations, select those of the desired type, and use biochemical and microscopic tools to define the resulting blocks in the pathway. The final result of their efforts will be a molecular map of the morphogenesis of "the worm" just like that worked out for phage T4 in earlier studies (see Chapter 11, Fig. 11.33).

NEOCLASSICAL EXAMPLES OF DEVELOPMENTALLY REGULATED GENE EXPRESSION

The recent explosion of new information about the regulation of gene expression in eukaryotes that has resulted from the application of recombinant DNA and gene cloning technologies should not cause us to overlook some of the elegant neoclassical studies of developmentally regulated gene expression. The striking pictures of transcription on lampbrush chromosomes and of the amplification of ribosomal RNA genes in amphibian oöcytes are two important examples of developmentally regulated gene expression that predated the recombinant DNA epoch of biology.

Transcription on Lampbrush Chromosomes in Amphibian Oöcytes

In all higher organisms studied so far, the fertilization of a mature egg by a sperm triggers a dramatic increase in protein synthesis, followed by the rapid nuclear and cell divisions of early cleavage stages of embryogenesis. In most eukaryotes, this protein synthesis is not accompanied by RNA synthesis. Instead, all the components required for protein synthesis during early embryogenesis are present in the egg prior to fertilization. Gene transcripts, in the form of mRNA or pre-mRNA molecules, must therefore be stored in the egg in a dormant state. Translation of these preformed mRNA molecules must somehow be triggered by events associated with fertilization.

Therefore, the informational molecules that direct protein synthesis during the early cleavage states following fertilization must be synthesized during oögenesis. Studies of oögenesis in vertebrates, particularly amphibians, reveal that extensive transcription occurs during prophase I (specifically diplotene) of meiosis. During this stage, the chromosomes exist as large ***lampbrush*** structures. (The structure of these oöcyte ***lampbrush chromosomes*** is described in Chapter 6, pp. 134–135).

Most of the DNA in lampbrush chromosomes exists in a highly condensed, transcriptionally inactive state in the so-called axial regions of the chromosomes. Certain segments of the DNA in each lampbrush chromosome, however, exist in highly extended lateral loops (see Figs. 6.6 and 6.7). Each loop consists of a central molecule of DNA that is surrounded by a matrix of newly synthesized RNA and protein. By means of pulse-labeling with [^{3}H]uridine and autoradiography, the loops of lampbrush chromosomes have been shown to be regions of active transcription. The lampbrush chromosomes of oöcytes thus appear to be an excellent example of the correlation between structure and function—the lampbrush morphology being the structural correlate of the transcription of a specific set of chromosomal genes. The transcribed genes of the lampbrush chromosomes apparently are those whose products are required during the early stages of embryogenesis. The gene transcripts synthesized during oögenesis must be stored in an ***inactive but stable form*** (possible in RNA–protein complexes) until fertilization occurs. Clearly, regulatory mechanisms are involved that act at a posttranscriptional (mRNA processing?) or translational level.

In addition, particular gene transcripts and/or other gene-products must become ***localized*** in specific areas of the egg cytoplasm during oögenesis. This is evident from experiments that show that the destiny of a particular cell depends on the section of the egg cytoplasm that the cell receives in the early cleavage divisions.

In amphibians, then, and probably in most vertebrates, the genetic programs controlling early development (up to about the blastula stage) are established during oögenesis. Later stages of development, when cell differentiation begins (from about the gastrula stage on), require new programs of gene expression.

rRNA Gene Amplification in Amphibian Oöcytes

Despite the rapid initiation of protein synthesis following fertilization, no rRNA is synthesized in amphibian embryos until the gastrula stage. This means that large amounts of rRNA must also be synthesized during oögenesis. In fact, the large eggs of amphibians contain vast quantities of ribosomes, on the order of 10^{12} per mature egg. The requirement for the synthesis of such enormous amounts of a particular gene transcript (the 40S amphibian rRNA precursor) in a single cell has resulted in the evolution of a novel mechanism of ***specific gene amplification.*** In amphibian oöcytes, the rRNA genes are selectively amplified about a thousandfold to facilitate the synthesis of the huge quantities of rRNA stored in mature eggs.

As discussed in Chapter 10, the rRNA genes are normally present as tandemly repeated copies located within the nucleolar organizer regions of the chromosomes. D. Brown, J. Gurdon, and colleagues have shown that there are about 500 copies of the rRNA gene in each of the two nucleolar organizer regions of diploid nuclei of *Xenopus laevis.* The rRNA precursors are synthesized and processed in the nucleoli.

Given that about 1000 rRNA genes exist per diploid nucleus, it has been estimated that over 450 years would be required to synthesize the large number of rRNA molecules present in mature *Xenopus* eggs, hardly a plausible situation given the average life expectancy of a toad. This potential dilemma has been resolved by the evolution of a mechanism by which the rRNA genes are selectively replicated in oöcytes.

D. Brown and I. Dawid have shown that the nuclei of oöcytes of *Xenopus laevis* contain hundreds of nucleoli (Fig. 15.18), each containing circular DNA molecules carrying tandemly repeated copies of the rRNA gene. Once formed, these extrachromosomal DNA molecules appear to replicate by the rolling circle mechanism (Chapter 5, pp. 122–124). However, how the nucleolar regions of the chromosomes are selectively replicated to produce the first extrachromosomal DNA molecules is not yet established.

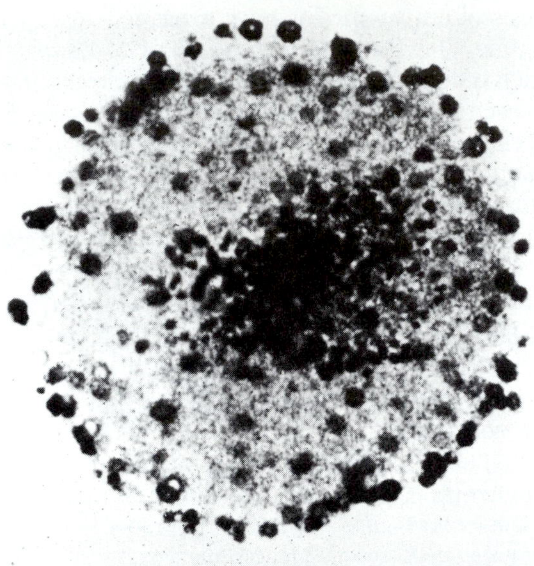

Figure 15.18 Photomicrograph of a nucleus from an oöcyte of the South African clawed toad *Xenopus laevis,* showing a large number of supernumerary nucleoli that contain circular extrachromosomal molecules of DNA carrying tandemly repeated rRNA genes. (From D. W. Brown and I. B. Dawid, *Science* 160: 272–280, 1968. Copyright © 1968 by the American Association for the Advancement of Science.)

The selective replication of the rRNA genes in oöcytes is the best-known example of this type of specific gene amplification. When large quantities of specific protein are required, as in the case of hemoglobin in red blood cells, extensive amplification can be accomplished at the translation level; each mRNA molecule can be translated many times. Of course, this is not possible when the required gene-product is an RNA molecule.

GENE TRANSCRIPT POPULATIONS ARE DIVERGENT IN DIFFERENT CELL TYPES

In higher eukaryotes, only a small proportion of the genome is represented among mRNA molecules in any given cell type. This has been demonstrated by ***RNA–DNA saturation hybridization experiments.*** RNA is extracted from cells of a particular type and allowed to hybridize with total nuclear DNA (denatured). The

RNA is added to the hybridization reaction in large excess (relative to the DNA concentration) so that all DNA sequences complementary to sequences represented in the RNA population will form DNA–RNA hybrids. The proportion of the total genomic DNA present as DNA–RNA hybrids is then determined, providing an estimate of the proportion of the genome represented by sequences in the mRNA population in that particular cell type. (Quantification is usually accomplished by using radioactively labeled DNA, although several other procedures have also been used.)

RNA–DNA saturation hybridization experiments have been done on a number of eukaryotic species using RNA from several different cell types. The results of these experiments show that ***less than 10 percent of the DNA in the genome is represented by mRNA molecules in the cytoplasm of any one cell type.*** In mice, for example, from 2 to 5 percent of the DNA sequences are represented in the mRNA molecules present in liver cells. Brain cells appear to contain the maximum variety of RNA transcripts. In the toad *Xenopus,* 8 percent of the DNA sequences are represented in mRNA from brain cells. The mRNA from *Xenopus* oöcytes, on the other hand, contains sequences homologous to less than 1 percent of the DNA sequences. It is clear, therefore, that the ***majority of the DNA sequences in the genome of a higher eukaryote are not represented among the mRNA populations of any one tissue or cell type.***

Different sets of genes are transcribed and different primary transcripts are processed into mRNA in different types of differentiated cells. Usually, some of the same genes and some different genes are transcribed in different tissues. This can be demonstrated by ***RNA–DNA competitive hybridization experiments.*** In these experiments, one measures the amounts of radioactively labeled RNA from one cell type that hybridize to total genomic DNA (1) in the presence and (2) in the absence of an excess of competing nonradioactive RNA from a second cell type. If the two cell types contain totally different ("nonoverlapping") RNA populations, the same amount of labeled RNA will hybridize with DNA in the presence or absence of the competing RNA. If the two RNA populations overlap (share common sequences), the amount of hybridized labeled RNA will be decreased in proportion to the degree of overlap (decreased by a proportion equal to the proportion of shared RNA sequences).

RNA–DNA competitive hybridization experiments of this type indicate that the RNA sequences present in RNA populations taken from different tissues or different cell types differ by from 10 to almost 100 percent. E. Davidson and colleagues detected no mRNA sequences present in both oöcytes and blastulae of *Xenopus laevis.* Mouse liver, spleen, and kidney cells contain mRNA populations estimated to differ in sequence composition by from 15 to 70 percent. Although these are relatively crude estimates, they do indicate that ***different sets of genes are transcribed and that different transcripts are processed into mRNA in different types of cells.***

Since over 90 percent of the DNA sequences in the genome are ***not*** represented among mRNA populations in any given cell, it has been hypothesized that eukaryotic genes are packaged in chromatin (see Chapter 6, pp. 136–143) in a nonspecifically repressed state, and that the regulation of transcription and/or transcript processing occurs by a ***positive*** mechanism involving specific gene ***activators.*** These activators are proposed to function by somehow turning on or "activating" the transcription of specific genes, or sets of genes, at the proper time in the appropriate cells. What these activators are and how such activation might occur are just beginning to be established. In several cases, the activation is known to involve *cis*-acting regulatory sequences called enhancers. The structure and effects of enhancers are discussed in subsequent sections of this chapter. Evidence indicates that certain nonhistone chromosomal proteins function as specific activators of transcription in some cases. Other evidence suggests that the regulation of RNA transcript processing is important in controlling differentiation in eukaryotes.

The histones are thought to be responsible, at least in part, for the nonspecific repression of eukaryotic genes. The histones, excluding histone H1, have been highly conserved throughout the evolution of eukaryotic organisms, and these histones are known to be tightly complexed with the DNA in nucleosomes. Moreover, histone-complexed DNA is transcribed much less efficiently in *in vitro* transcription systems than is the same DNA after removal of the histones.

The same histones are usually present in the chromatin of various types of differentiated cells, thus many scientists believe that the histones do not function as specific repressors or activators of transcription. Yet other scientists believe that certain modifications of histones, such as phosphorylations and acylations of key amino acids, are involved in the regulation of transcription. Clearly, conclusions regarding the specific roles of histones in the regulation of transcription must await further evidence.

MECHANISMS OF REGULATION OF TRANSCRIPTION IN HIGHER EUKARYOTES

In *E. coli,* the RNA polymerase holoenzyme (with sigma factor) contains all the information needed to accurately initiate transcription given proper promoter

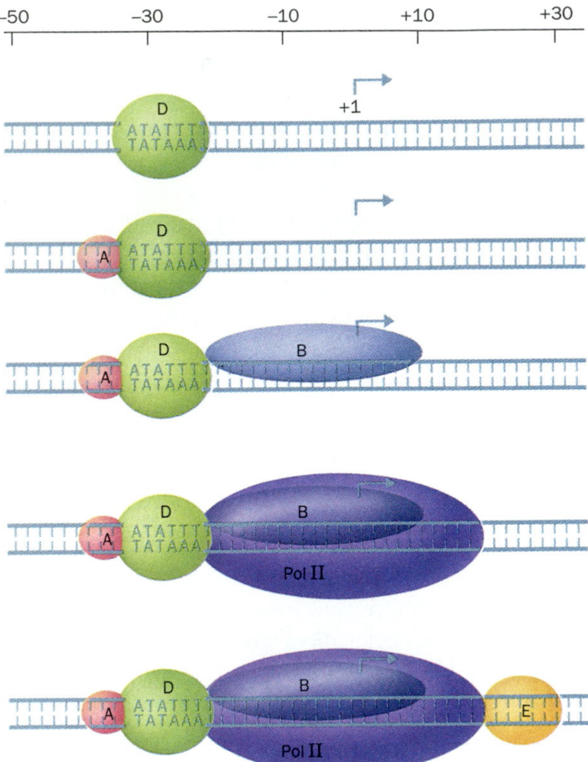

Figure 15.19 Transcription factors required for the accurate initiation of transcription by RNA polymerase II *in vitro.* The model shown provides the best fit to the data available regarding the roles of transcription factors A, B, D, and E. Factor D is believed to interact with the TATA box as the first step in the initiation process. Factor A is believed to act next, forming a complex with factor D and DNA. Factor B is the third factor to join the transcription-initiation complex. After the addition of factor B, the complex partially protects the DNA sequence from −10 to +10 (transcription initiation at +1) from nuclease digestion. RNA polymerase II must be added before factor E will bind. After E is present, sequences from +20 to +30 are protected from nuclease digestion, suggesting that factor E covers this region of the DNA. The complete complex is then capable of initiating transcription accurately from the +1 site. (After S. Buratowski *et al., Cell* 56: 549–561, 1989.)

signals (−35 recognition sequence plus −10 Pribnow box). However, RNA polymerase II of eukaryotes, which transcribes most of the protein-encoding nuclear genes, ***cannot*** initiate transcription accurately *in vitro* without the addition of ***four accessory proteins or general transcription factors.*** The requirement for these transcription factors (Fig. 15.19) provides the potential for additional sites of regulation of transcription. Whether these proteins interact with other transcription factors in a regulatory fashion or whether they are totally autonomous factors remains to be determined. In either case, they yield a slightly more complex picture of the initiation of transcription in eukaryotes.

Most Eukaryotic Transcription Units Are Monogenic

At present, our knowledge about the mechanisms by which gene expression is regulated in eukaryotes is expanding rapidly. We know that different sets of genes are transcribed in different cell types in higher eukaryotes, and we know that the different patterns of gene expression in different tissues are controlled by *trans*-acting proteins encoded by regulatory genes that act in sequence during differentiation. Clearly, regulatory mechanisms acting at the level of transcription are important in cell differentiation. However, the molecular details of these regulatory mechanisms are still being worked out, and many important questions about differentiation promise to challenge geneticists for years to come.

In higher eukaryotes, it does seem very clear that operons are ***not*** important, if they exist at all. Although there is evidence for operons or operonlike units in the lower eukaryotes (e.g., fungi), operons appear to be rare or nonexistent in higher eukaryotes. Most of the mRNAs of higher eukaryotes characterized to date are ***monogenic*** (contain the coding sequence of one structural gene). In a few cases, the primary transcripts are polygenic and are cleaved to produce monogenic mRNAs.

Enhancers and Silencers Modulate Transcription in Eukaryotes

Eukaryotic genes are regulated by ***promoter elements*** located just upstream (5′) from the transcription-initiation sites in a manner quite similar to the regulation of prokaryotic genes (Chapter 14). However, as in the case of the *Drosophila Ubx* gene (see Fig. 15.16), these eukaryotic promoters may be very complex with binding sites for many different regulatory proteins. In addition to the nearby promoters, many eukaryotic genes are also regulated by more distant **cis-*acting elements*** called ***enhancers*** and ***silencers.*** As the names suggest, ***enhancers increase transcription*** and ***silencers decrease transcription*** of the regulated genes. Since enhancers appear to be more common and are much better understood, the following discussion will focus on the properties of enhancer elements.

The basic features of enhancers that distinguish them from promoters are as follows:

1. ***Enhancers can act over relatively large distances***—up to several thousand nucleotide-pairs from the regulated gene(s).

2. ***Enhancers are orientation independent***—they function equally well in either orientation, normal or inverted (turned end-for-end).
3. ***Enhancers are position independent***—they function equally well whether located upstream (5′) from a gene, downstream (3′) from a gene, or present within an intron of a gene.

Enhancers are relatively large elements, up to several hundred nucleotide-pairs in length. They sometimes contain repeated sequences that have partial enhancer activity by themselves. Most enhancer elements function in a complete or partially tissue-specific manner, that is, they frequently will only enhance the transcription of genes in specific target tissues (those tissues where the gene-product is needed).

The most extensively studied enhancer is that present on the minichromosome of simian virus 40 (SV40), a virus of monkeys that can be investigated in cell cultures. The complete ***SV40 enhancer*** is about 220 nucleotide-pairs in length (Fig. 15.20). This region of the SV40 minichromosome is not packaged into nucleosomes (Fig. 15.20*a*). Presumably, the SV40 enhancer is covered with protein transcription factors that prevent it from becoming wrapped into nucleosomes by the histones. The SV40 enhancer contains two 72-nucleotide-pair direct repeats, and deletion of both repeats eliminates enhancer activity. If one of the direct repeats is deleted, the enhancer is still functional. Early experiments demonstrated that the SV40 enhancer could be moved to any other location on the SV40 minichromosome without loss of activity. More-

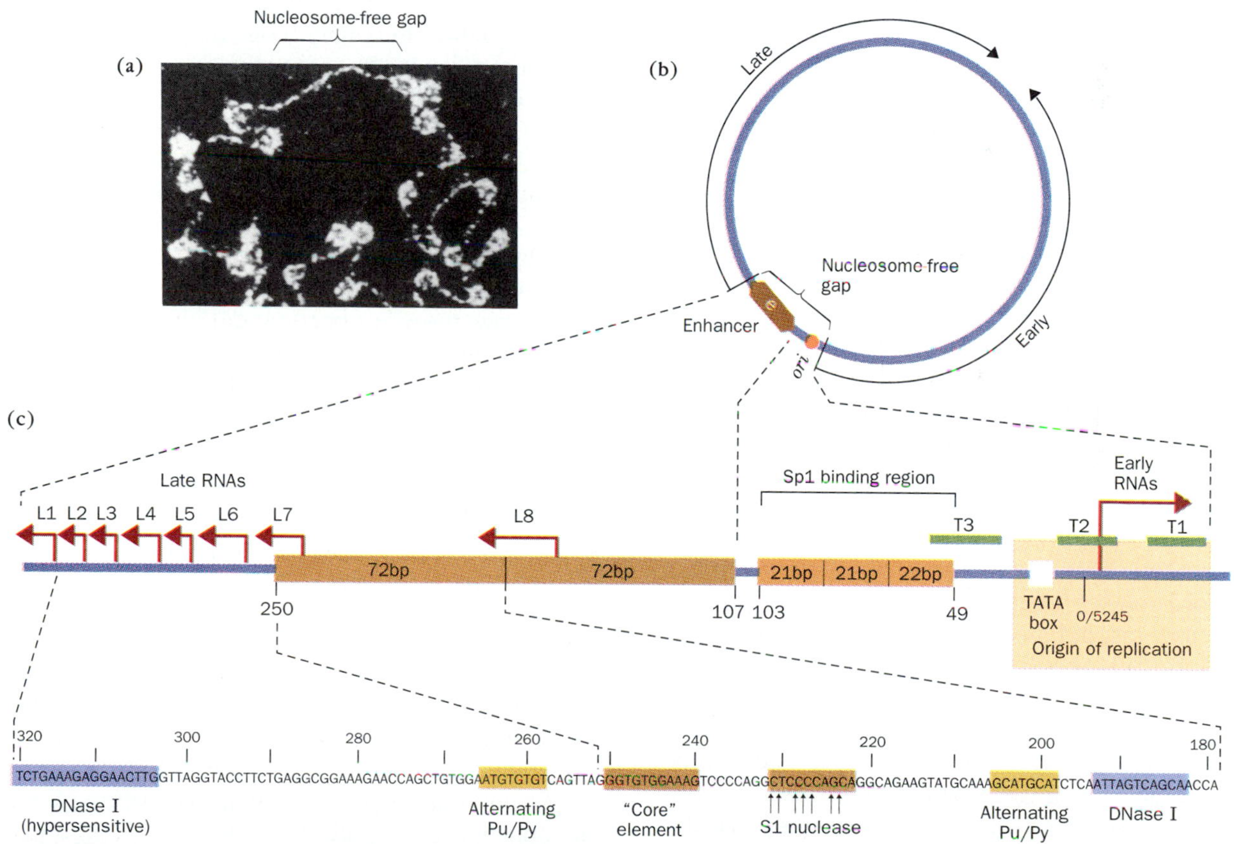

Figure 15.20 Structure of the enhancer of simian virus 40 (SV40). (a) Electron micrograph of an SV40 minichromosome showing nucleosomes except in the region of the enhancer. Within the cell, the enhancer region is probably covered with sequence-specific transcription factors, each bound at its specific recognition site. (b) Diagram of the structure of the SV40 minichromosome showing the location of the enhancer. The enhancer is about 220 nucleotide-pairs in length—covering the region from nucleotide position 100 to about position 320. (c) Diagram showing the components of the enhancer (left) and their relationship to the promoter. The sequence of one of the two 72-base-pair repeats plus flanking nonrepeated enhancer DNA is shown at the bottom. Sequences that are sensitive to nucleases S1 and DNase I are identified at the bottom along with the conserved core element of the enhancer. The regions labeled T1, T2, and T3 are equivalent to the operator regions of prokaryotic operons (see Chapter 14). They bind the SV40 T-antigen (a protein), which then represses transcription. The Sp1 binding sites are sequences to which the mammalian transcription factor Sp1 binds. (Reproduced with permission from E. Serfling, M. Jasin, and W. Schaffner, *Trends in Genet.* 1: 224–230, 1985.)

over, it could be excised from the SV40 chromosome and reinserted in the inverse orientation without loss of function. These early studies established the orientation- and positional-independent nature of enhancer action. Finally, when the SV40 enhancer was placed on either side of any of several different eukaryotic genes (e.g., the human β-globin gene), it sharply increased the rate of transcription from these genes. *In vivo,* the insertion of an SV40 enhancer within a few thousand nucleotide-pairs of a structural gene can increase the rate of transcription by as much as 100-fold.

Many enhancers have now been characterized that play key roles in the regulation of gene expression. A striking feature of the characterized enhancers is that they exhibit tissue specificity. The glucocorticoid hormone-responsive enhancer of mammals is described later in this chapter (see section entitled Glucocorticoid Hormones Act via Enhancer Elements). In Chapter 16 (see Fig. 16.10), the enhancer that regulates immunoglobulin heavy-chain synthesis is described. This enhancer is located within an intron of the complex heavy-chain genes. In the chicken genome, an enhancer is located between the β-hemoglobin gene and the ϵ-hemoglobin gene; it stimulates transcription of the ϵ-globin gene during embryonic development and of the β-globin gene in adults. The tissue specificity of the enhancers is fascinating, however, we do not understand the molecular basis of this specificity. Presumably, the tissue specificity must result from interactions of enhancer sequences with transcriptional activators present only in cells in which a given set of genes is expressed. This, of course, raises another question: What controls the tissue-specific expression of these transcriptional activators?

The ability of enhancers and silencers to act at a distance of 1000 nucleotide-pairs or more is intriguing. By what mechanism can these *cis*-acting regulators influence transcription from promoters that may be more than a kilobase away? We do not know the answer to this question in any detail. However, one point is very clear. The factors bound at the enhancer sequence and the promoter sequence can act in a cooperative manner, either positively or negatively. This indicates that these proteins are probably in contact, at least some of the time, and suggests that the intervening DNA folds or bends to permit this contact (Fig. 15.21). Models of enhancer and silencer action thus resemble the mechanism of regulation of the *ara* operon in *E. coli* (see Chapter 14, Fig. 14.8), except that many more regulatory proteins appear to be involved in higher eukaryotes. For example, in the case of the chicken β-hemoglobin gene, evidence is available that implicates the binding of at least five different proteins to the promoter and at least five different proteins to the enhancer. Since two of the same transcription factors bind to both elements, promoter and enhancer, the data available indicate that at least eight different proteins participate in regulating transcription of the chicken β-globin gene. Obviously, regulation of transcription in higher eukaryotes is complex. Clearly, much work will be required to sort out the specific function(s) of each of these *trans*-acting transcriptional regulators.

Regulation of Levels of Transcription by DNA Methylation

In most higher plants and animals, the DNA is often modified after synthesis by the enzymatic conversion of many cytosine bases to ***5-methylcytosine*** bases (Fig. 15.22). The extent of methylation varies from species to species; in mammals, from 2 to 7 percent of the cytosine residues are methylated.

Methyl groups on the 5-carbons of pyrimidines occupy exposed positions within the major grooves of DNA molecules, thus they have the potential to play influential roles in the interactions of DNA with specific proteins. In fact, studies on the binding of the repressor for the *lac* operon of *E. coli* to the *lac* operator DNA have shown that the addition or removal of a single methyl group can sharply change the affinity of the repressor for the DNA. Thus, the potential regulatory role of 5-methyl groups on pyrimidine bases is well established.

To date, there has been no definitive proof of the role of methylation in the regulation of the expression of any eukaryotic gene. Instead, the arguments for the involvement of methylation in the control of gene expression in eukaryotes are based primarily on three kinds of indirect evidence. (1) Numerous studies have demonstrated a correlation between the level of gene expression and the degree of methylation, such that low methylation → high gene expression, high methylation → low gene expression. (2) Methylation patterns are tissue specific, at least in some cases. (3) The drug (base analog) 5-azacytidine, which can ***not*** be methylated after it is incorporated into DNA, has been shown to result in the expression of genes in tissues where they normally are not expressed.

An important aspect of methylation patterns is that they appear to be clonally heritable. Over 90 percent of the methylation in the DNA of most eukaryotes occurs in CG dinucleotide sequences, and these sequences are symmetrically methylated, that is,

5-methyl
|
—CG—
—GC—
|
5-methyl

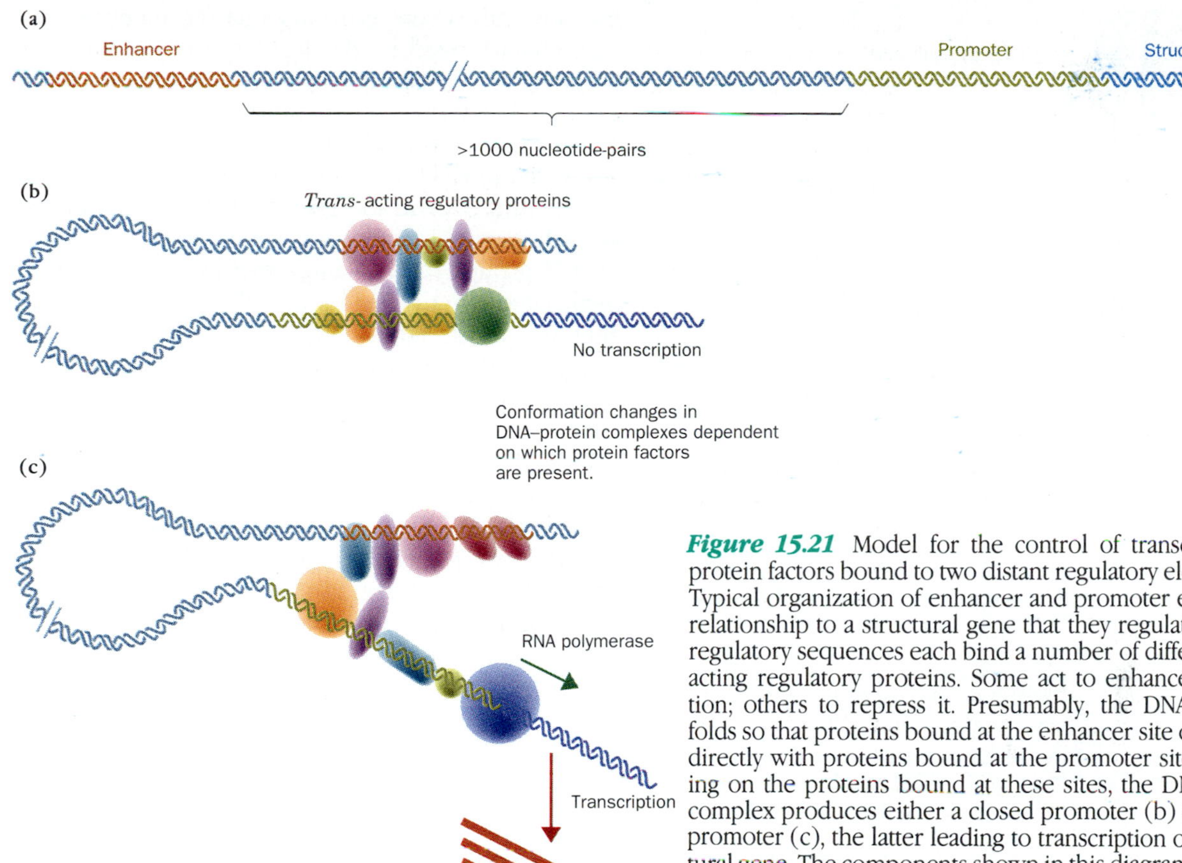

Figure 15.21 Model for the control of transcription by protein factors bound to two distant regulatory elements. (a) Typical organization of enhancer and promoter elements in relationship to a structural gene that they regulate. The two regulatory sequences each bind a number of different *trans*-acting regulatory proteins. Some act to enhance transcription; others to repress it. Presumably, the DNA molecule folds so that proteins bound at the enhancer site can interact directly with proteins bound at the promoter site. Depending on the proteins bound at these sites, the DNA–protein complex produces either a closed promoter (b) or an open promoter (c), the latter leading to transcription of the structural gene. The components shown in this diagram are totally hypothetical; however, the enhancers and eukaryotic promoters that have been studied to date generally contain binding sites for four or more different regulatory proteins.

Semiconservative replication of such a symmetrically methylated sequence will yield two half-methylated sequences, that is,

```
5-methyl
   |
 —CG—                —CG—
 —GC—       and      —GC—
                       |
                    5-methyl
```

In the presence of a DNA methylase that is specific for half-methylated sequences, such methylation patterns will be clonally heritable (Fig. 15.23), that is, once formed, the pattern will be passed on to all the cells of a given lineage. Some bacterial methylases are indeed known to act primarily on half-methylated sites, and considerable evidence points to the existence of eukaryotic methylases with similar properties.

These features of DNA methylation patterns nicely explain their maintenance, once they are formed. However, the key step in any model for the regulation of gene expression or differentiation through DNA methylation involves the ***formation of the tissue-specific methylation patterns***. The most popular hypothesis is that the patterns are formed during development by ***tissue-specific demethylases***, which remove methyl groups from critical sites in genes that are

Cytosine —DNA methylase→ 5-methylcytosine

Figure 15.22 Structures of cytosine and 5-methylcytosine. The DNA methylases that catalyze the addition of the methyl groups act after DNA replication has incorporated cytosine into nascent DNA chains.

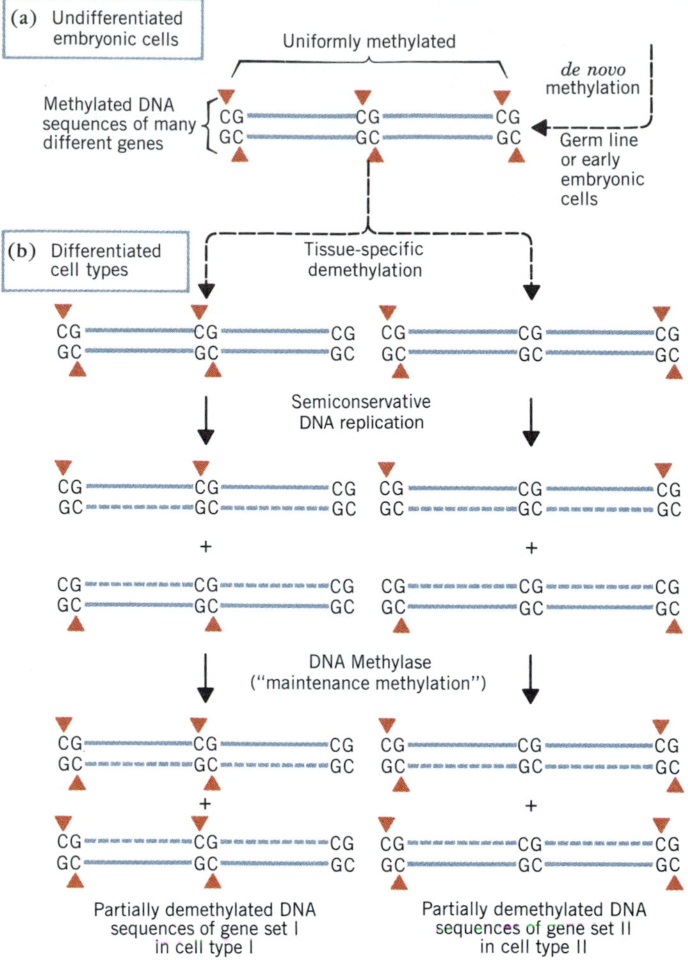

Figure 15.23 Model for the regulation of gene expression by specific patterns of DNA methylation in different cell types. Each red triangle represents a 5-methyl group on cytosine; dashed lines represent nascent DNA chains. (a) The DNA in germ line and early embryonic cells is uniformly methylated. (b) During differentiation, tissue-specific patterns of methylation are established by the action of specific demethylases. These patterns are clonally inherited as a result of semiconservative DNA replication and the subsequent maintenance methylation. Note that the DNA methylase involved acts only on half-methylated sequences. (After A. Razin and A. D. Riggs, *Science* 210: 604–609, 1980.)

scheduled to be expressed in a particular cell type (Fig. 15.23). Although this is an attractive model, we should emphasize that it is just that: a model. No such tissue-specific demethylase has yet been identified. Recently, the methylation-blocking drug 5-azacytidine has been shown to result in the expression of the fetal (γ-hemoglobin) and, to a lesser extent, the embryonic (ε-hemoglobin) β-like hemoglobin genes (see Chapter 13, p. 356) of anemic adult baboons and adult humans with severe β-thalassemia (an inherited disease characterized by the inability to synthesize the β-hemoglobin chain of adult hemoglobin) and with sickle-cell anemia (see Chapter 11, pp. 295–297). These embryonic and fetal genes are normally not expressed in red blood cells of adults. In one of these studies, the DNA in the region of the γ-hemoglobin and ε-hemoglobin genes was shown to contain fewer methyl groups (to be "undermethylated" or "hypomethylated") in the red blood cells of the individuals after treatment. These results not only support the hypothesis that methylation is important in the regulation of gene expression, but also suggest a possible approach to the treatment of these inherited diseases.

Clearly, additional data are needed before any conclusions can be drawn about the role of DNA methylation as a regulatory mechanism in eukaryotes. Nevertheless, the aforementioned and other correlations between methylation patterns and levels of gene expression suggest many interesting questions. For example, of what importance is the stabilizing effect of methylation on the Z-form of DNA, and is the correlation between undermethylation and nuclease sensitivity of DNA in chromatin biologically significant (see the following two sections)?

Does Z-DNA Play a Regulatory Role?

One of the more interesting discoveries of the last decade was that segments of DNA that have sequences in which purines and pyrimidines alternate along each strand can form ***left-handed*** double helices. The Watson–Crick B-form of DNA is a ***right-handed*** double helix. The novel left-handed double-helical form of DNA has been named ***Z-DNA*** for the zigzagged paths of the sugar–phosphate backbones of the molecule (Fig. 15.24).

Normally, the Z-form of alternating purine and pyrimidine DNA sequences occurs only at high salt concentrations. When some of the bases in the potential Z-form sequences are methylated (see the preceding section), however, the Z-conformation is stable at lower salt concentrations. Thus, Z-DNA composed of alternating purine–pyrimidine sequences containing methylated bases may be stable *in vivo*. Moreover, its stability is enhanced by cations, including polyamines such as spermine, by negative supercoiling (see Chapter 5, pp. 124–126), and by DNA binding proteins specific for Z-DNA.

In fact, there is evidence that Z-DNA exists in the interband regions of the giant salivary gland chromosomes of *Drosophila melanogaster* and in the transcriptionally active macronucleus of the ciliated protozoan *Stylonychia mytilus*. A. Rich and colleagues have prepared antibodies specific for Z-DNA; these antibod-

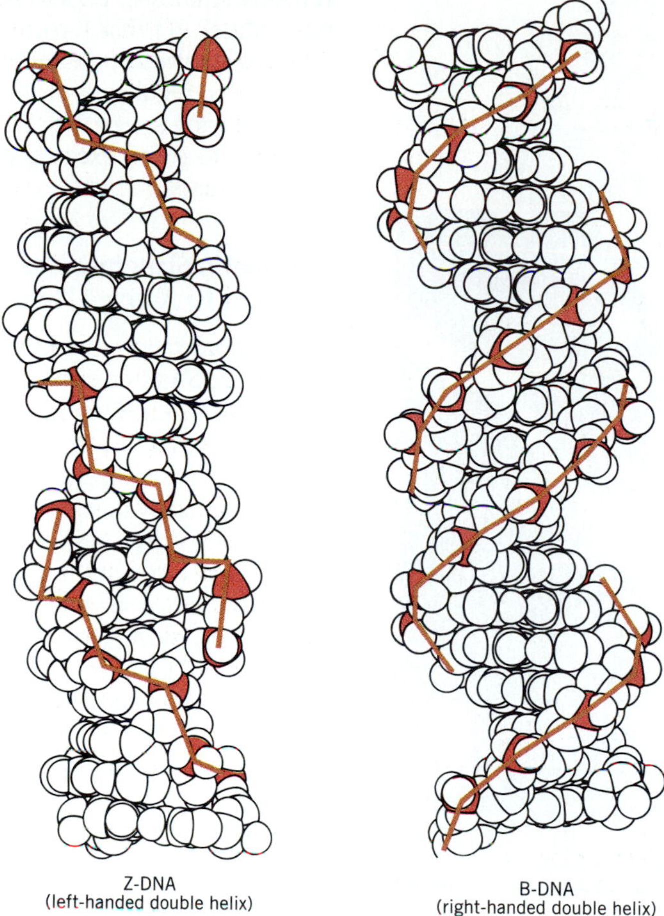

Figure 15.24 Comparison of the structures of DNA in the well-known B-form (right) and the Z-form (left). The heavy lines show the paths of the sugar–phosphate backbones in Z-DNA. (This drawing of the structure of Z-DNA is based on that of A. H.-J. Wang *et al., Science* 211: 171–176, 1981.)

ies do not react with B-form DNA. Rich and coworkers have shown that the Z-DNA-specific antibodies bind to the interband regions of the polytene chromosomes of *D. melanogaster* (Fig. 15.25). It will be of great interest to determine the sequences and methylation patterns of the DNA in the interband regions of these polytene chromosomes. Do these interband sequences control the expression ("puffing") of the genes located in the adjacent bands? Or are these sequences merely structural or involved in synapsis?

Another hint of the possible involvement of Z-DNA in regulating gene expression is that the structures of certain regulatory proteins suggest that they may bind in the major groove of left-handed double helices, but not right-handed helices. In fact, D. B. McKay and T. A. Steitz have proposed that the catabolite activator protein (CAP; see pp. 398–400) stabilizes its CAP-binding sequence in a left-handed conformation. They further propose that this right-handed to left-handed transition in the double helix unwinds the adjacent promoter or RNA polymerase-binding site and thus activates transcription of the adjacent structural genes. Repressor proteins might act in the opposite direction, stabilizing regulatory sequences in the right-handed B-form and preventing transcription. Although their functions are still unknown, Z-DNA-specific binding proteins have been isolated from *Drosophila.*

When alternating purine–pyrimidine sequences are complexed with histones, these sequences do not display B-DNA to Z-DNA transitions. This and other observations suggest that Z-DNA sequences must be stabilized by Z-DNA-specific binding proteins, and that Z-DNA sequences may not be found in nucleosomes. Indications of nuclease sensitivity of B-DNA to Z-DNA junctions has led to speculation that such junctions may be related to the nuclease-sensitive sites near the

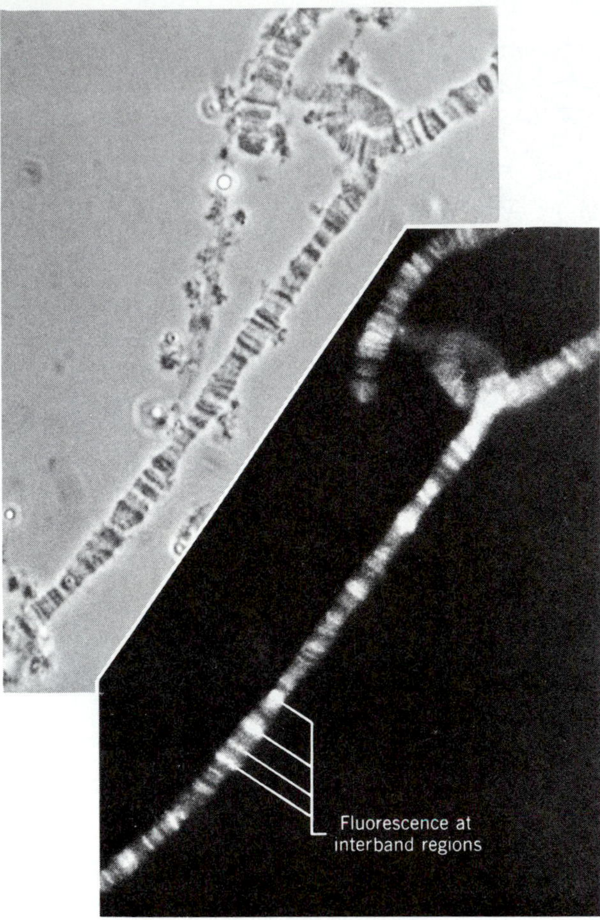

Figure 15.25 Binding of antibodies specific for Z-DNA to the interband regions of the polytene chromosomes of *D. melanogaster*. Phase (top) and fluorescence (bottom) micrographs are shown for a segment of a giant salivary band chromosome. The specific binding of the antibodies to the interband regions of the chromosomes was detected by ultraviolet light fluorescence microscopy. A fluorescent compound was attached to a second antibody (prepared in a goat), which in turn is specific for the original anti-Z-DNA immunoglobulin (prepared in a rabbit). The secondary fluorescein-labeled antibody allows the location of the bound Z-DNA antibody to be visualized as shown in the lower photograph. (For additional details, see A. Nordheim *et al., Nature* 294: 417–422, 1981; photographs courtesy of M. L. Pardue and A. Nordheim.)

promoter regions of transcriptionally active genes (see the following section).

Clearly, much more information is needed before the potential validity of these proposals can be evaluated. Although transitions from sequences of DNA in the B-conformation to sequences in the Z-conformation have been shown to occur within individual plasmids, the biological significance of these transitions remains unknown. Experiments designed to test the possibilities (1) that B-form to Z-form transitions in DNA are involved in the regulation of gene expression and (2) that regulatory proteins may act by binding to and stabilizing one or the other of these conformations may lead to some exciting developments during the next few years.

Chromatin Structure: Nuclease Sensitive Sites Adjacent to Active Genes

The demonstration that much, if not all, of the chromosomal DNA of eukaryotes is packaged into nucleosomes and that the 146-nucleotide-pair length of DNA in the nucleosome core is protected to a considerble degree from nuclease digestion (Chapter 6, pp. 137–138) immediately raised the question of whether transcriptionally active DNA is similarly packaged. Since RNA polymerases are very large, complex enzymes (eukaryotic RNA polymerases are larger than nucleosomes) and since DNA is locally unwound during transcription (Chapter 10, p. 256), it seemed likely that the nucleosomes would have to disassemble or at least undergo conformational changes during transcription of the resident DNA sequences.

Does the DNA of a gene remain packaged in nucleosomes during transcription and, if so, what structural changes, if any, occur in these nucleosomes? Electron microscope and nuclease digestion studies of transcriptionally active genes and chromatin have shown that genes that are being transcribed are indeed packaged into nucleosomes displaying the same frequency and spacing as nucleosomes containing the DNA of genes that are not being transcribed. However, the structures of nucleosomes containing active genes are not identical to those of nucleosomes containing inactive genes. This is shown by the ***increased nuclease sensitivity of transcriptionally active genes.***

In 1976, M. Groudine and H. Weintraub showed that the hemoglobin genes present in chromatin from red blood cells of 18-day-old chickens were more sensitive to degradation by pancreatic deoxyribonuclease I (DNase I) than were the ovalbumin genes (not expressed in red blood cells) in chromatin from these same cells or the hemoglobin genes in chromatin isolated from fibroblasts or brain cells of the same chickens. These experiments were done using globin and ovalbumin cDNAs (DNA sequences synthesized *in vitro* by reverse transcriptase using purified globin and ovalbumin mRNAs as templates; see Chapter 13, p. 354) as hybridization probes to measure the quantities of globin and ovalbumin gene sequences in isolated chromatin before and after partial digestion with DNase I. In these experiments, over 50 percent of the DNA sequences of transcriptionally active genes had already been degraded at a time when only 10 percent of the total DNA had been hydrolyzed by DNase I.

Subsequent studies have demonstrated the nuclease sensitivity of transcriptionally active genes in several other organisms. In addition, the nuclease sensitivity of active genes has been found to depend on the presence of two nonhistone chromosomal proteins called HMG14 and HMG17 (***HMG*** for ***h***igh ***mo***bility ***g***roup; small proteins with high mobility during polyacrylamide gel electrophoresis). When these proteins are removed from active chromatin, nuclease sensitivity is lost. When they are added back, sensitivity is restored.

When isolated chromatin containing transcriptionally active genes is treated with very low concentrations of DNase I, the DNA molecules are cleaved at a few ***specific sites.*** Some of these ***hypersensitive sites*** have been shown to lie "upstream" (adjacent to the end of the gene homologous to the 5′ end of the mRNA) of transcriptionally active genes. In a few cases, these hypersensitive sites have been shown to be located right at the upstream ends of the promoters of the transcribed genes. In other cases, the hypersensitive sites seem to be located in enhancers.

The nature of these hypersensitive sites adjacent to transcriptionally active genes is still unknown. In the case of the active chicken globin genes, however, the hypersensitive sites are cleaved by S1 nuclease, an endonuclease (isolated from *Aspergillus*) that is specific for single-stranded DNA. This suggests that the DNA is not precisely base-paired or has some other structural modification (B-DNA to Z-DNA junction?) at these hypersensitive sites, very possibly in preparation for RNA polymerase binding at the promoter.

HORMONAL CONTROL OF GENE EXPRESSION

Intercellular communication is a very important phenomenon in higher plants and animals. Signals originating in various glands and/or secretory cells somehow stimulate ***target tissues*** or ***target cells*** to undergo dramatic changes in their metabolic patterns. These changes frequently include altered patterns of differentiation that are dependent, at least in some cases, on altered patterns of gene expression. What kinds of molecules carry these signals from one cell to another? How do they trigger the altered patterns of gene expression?

Peptide hormones such as insulin and ***steroid hormones*** such as estrogen and testosterone (Fig. 15.26) represent two types of signal systems utilized in intercellular communications. In higher animals, hormones are synthesized in various specialized secretory cells and are released into the bloodstream. The peptide hormones do not normally enter cells because of their relatively large size. Their effects appear to be mediated by receptor proteins located in target cell membranes and by the intracellular levels of cyclic AMP. The steroid hormones, on the other hand, are small molecules (Fig. 15.26) that readily enter cells through the plasma membranes. Once inside the appropriate target cells, the steroid hormones become tightly bound to ***specific receptor proteins.*** These receptor proteins are present only in the cytoplasm of target cells (an example of cell differentiation at the molecular level).

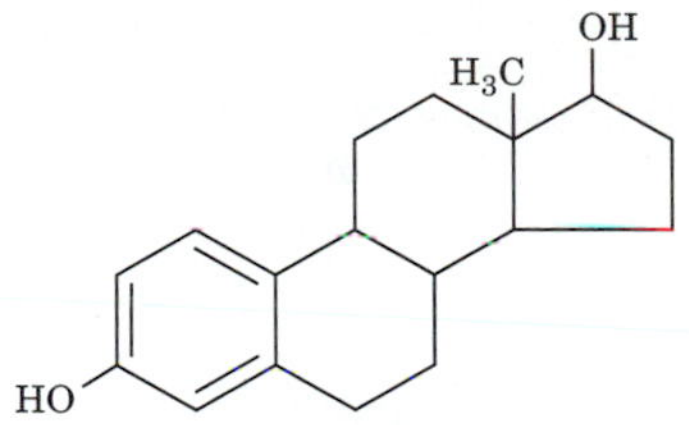

Estrogen

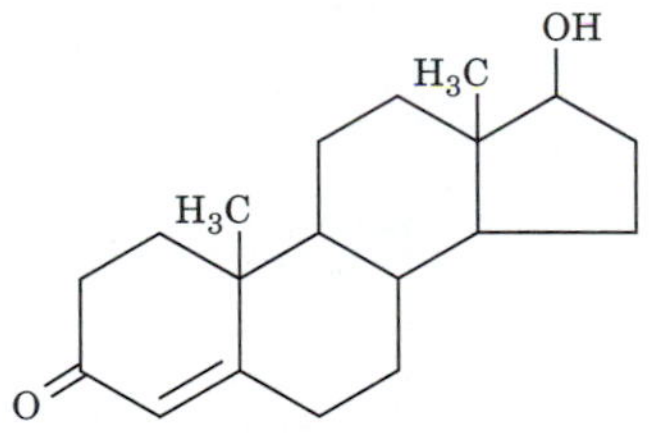

Testosterone

Figure 15.26 Chemical structures of the steroid sex hormones estrogen (female) and testosterone (male). The steroid hormones are relatively small molecules (molecular weights around 300) with a conjugated four-ring structure that is synthesized from cholesterol. The various steroid hormones have different side chains and different bonding patterns within the rings. These differences permit them to be recognized by different receptor proteins that are present in the cytoplasm of various target cells.

Activation of Transcription by Steroid Hormones

Autoradiographic studies using radioactively labeled steroid hormones have shown that the ***hormone–receptor protein complexes rapidly accumulate in the nuclei of target cells.*** Early studies by G. Tomkins and colleagues on mice and by B. W. O'Malley and associates on chickens have provided evidence that these hormone–receptor protein complexes activate the transcription of specific genes or sets of genes (Fig. 15.27). Subsequent studies have shown that at least some of these hormone–receptor protein complexes

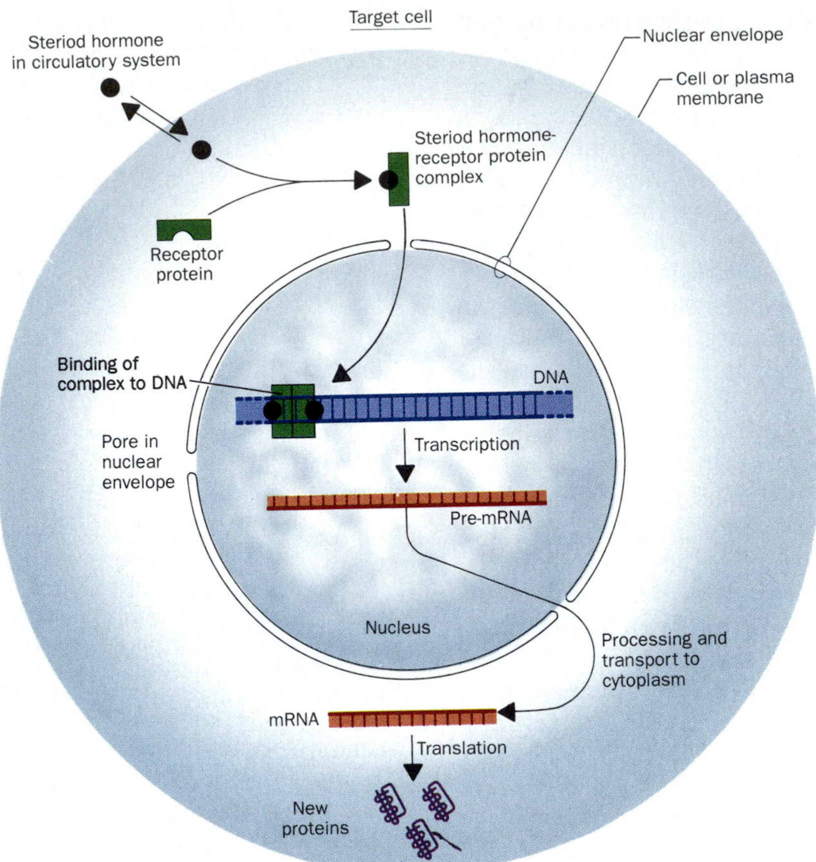

Figure 15.27 Diagram illustrating the effects of steroid hormones on gene expression. Hormones are synthesized in specialized secretory cells and are distributed to the various tissues of the organism through the circulatory system. Their small size (see Fig. 15.26) permits them readily to pass into cells through plasma membranes. ***The target cells*** (the cells responding to the presence of the specific steroid hormone) contain ***receptor proteins*** that specifically bind the hormone molecules. The steroid hormone–receptor protein complexes then pass through the pores in the nuclear envelope and accumulate in target cell nuclei. The hormone–receptor protein complexes then bind (probably as dimers) to a specific DNA sequence in an enhancer or promoter region of each gene that is activated by the hormone. Once bound, the hormone–receptor complex somehow stimulates transcription of the structural genes. Presumably, the bound complex enhances the binding of RNA polymerase to the promoter region; in any case, it functions as an activator of transcription of the regulated gene(s).

activate transcription of target genes by binding to specific DNA sequences present in the *cis*-acting regulatory regions of these genes (see the following section). Another hypothesis is that the hormone–receptor protein complexes interact with specific nonhistone chromosomal proteins (specific nonhistone proteins present only in the chromatin of target cells?) rather than directly with DNA. This interaction would then supposedly stimulate the transcription of the correct genes. In both cases, these hormone–receptor protein complexes would function as positive regulators (or "activators") of transcription, much like the CAP–cAMP complexes in prokaryotes (see Chapter 14, Fig. 14.7).

Early evidence that nonhistone chromosomal proteins can control the transcriptional state of particular genes was obtained by J. Stein, G. Stein, and L. Kleinsmith. Histones are synthesized, like DNA, during the *S* phase of the cell cycle. When chromatin from *S*-phase (DNA synthesis phase) cells is transcribed ***in vitro***, ***histone mRNA*** is synthesized. When chromatin from G_1 phase (the period after the completion of mitosis, but prior to *S*) is used, no histone mRNA is synthesized. When the nonhistones are removed from G_1-phase chromatin and replaced with nonhistone chromosomal proteins from *S*-phase chromatin, and this reconstituted chromatin is transcribed ***in vitro***, histone mRNA is synthesized. On the other hand, when the nonhistones in reconstituted chromatin are from G_1-phase cells and the DNA and histones are from *S*-phase cells, no histone mRNA is synthesized. These results indicate that the nonhistone proteins in chromatin determine whether the genes coding for histones are transcribed. It seems likely, therefore, that the nonhistone chromosomal proteins play important roles in the regulation of gene expression in eukaryotes. Evidence of this type certainly does not exclude the involvement of histones in the regulation of transcription. The regulation of transcription in eukaryotes may well involve specific interactions among DNA, histones, and nonhistone chromosomal proteins.

At present, one cannot exclude the possibility that histone modifications or nonhistone chromosomal proteins are involved in some aspects of hormone-regulated gene expression. On the other hand, the evidence that is available to date strongly suggests that hormone–receptor protein complexes activate gene expression by interacting directly with specific DNA sequences present within the enhancer or promoter regions that regulate the transcription of the target genes. Strong evidence in support of a direct interaction between the complex and *cis*-acting regulatory sequences of target genes is available for the ***glucocorticoids*** (which stimulate increases in blood sugar levels), the ***estrogens*** (which stimulate the development of the female sex phenotype), and the ***thyroid hormones*** (which control basal metabolic rates) in higher animals.

Glucocorticoid Hormones Act via Enhancer Elements

Mammals produce a large number of distinct ***steroid hormones*** that induce a vast array of metabolic changes in different cells of various tissues. Often, a given steroid hormone will have different effects on different cell types. Although the mechanisms by which most of the steroid hormones act are still unknown, specific steroid hormones such as the glucocorticoids (e.g., cortisol) and the estrogens (e.g., β-estradiol) have been shown to activate specific target genes by protein-mediated interactions with *cis*-acting regulatory sequences. These *cis*-acting sequences are usually called ***enhancers*** even though they differ from the classical enhancers in that they affect transcription from nearby promoters ***only when hormone–receptor protein complexes are bound to them.***

The glucocorticoid hormones provide the best-documented example of steroid hormone-activated gene expression. The effects of glucocorticoid hormones have been analyzed by using a synthetic hormone called dexamethasome. The availability of this synthetic hormone has facilitated the preparation of labeled hormone substrates for localization and binding studies and for *in vitro* studies on transcription of cloned target genes. The glucocorticoid hormones act by first binding to a ***receptor protein*** that is present in the cytoplasm of target cells. The hormone–receptor protein complex then accumulates in cell nuclei and binds to specific DNA sequences called ***glucocorticoid response elements (GREs).*** In the absence of hormone, the receptor protein is associated with another cytoplasmic protein and has low affinity for DNA. The available evidence suggests that the associated cytoplasmic protein prevents the receptor protein from forming a dimer, which is believed to be the active, DNA-binding form of the receptor. Presumably, the binding of hormone causes an allosteric conformational change in the receptor protein so that it no longer binds to the cytoplasmic protein. The receptor protein can then dimerize to its active form.

The glucocorticoid hormone–receptor complex activates transcription of target genes by binding to GRE sequences in enhancers located near each of these genes (Fig. 15.28). Binding of the hormone–receptor to the enhancer in turn activates the promoter of the adjacent target genes. Since the mechanism by which enhancers act to stimulate transcription of responding genes is still uncertain, this final stage in hormone activation of gene expression remains to be elucidated. Clearly, the binding of the hormone–receptor complex to the enhancer must somehow produce open promoters that facilitate RNA polymerase loading and transcription. This most likely involves some mechanism of enhancement of localized unwinding of the two DNA strands in the promoter region, but exactly how this occurs is unknown.

The ***hormone response elements*** that bind different steroid hormone–receptor protein complexes contain different DNA sequences as would be expected since distinct sets of genes respond to each hormone. For example, the hormone response element core consensus sequences for glucocorticoid, estrogen, and thyroid hormones are 5′-GGTACANNNTGTTGT-3′, 5′-GGTCANNNTG(A/T)CC-3′, and 5′-CAGGGACGTGACCGCA-3′, respectively. Interestingly, when the amino acid sequences of eight different steroid hormone receptor proteins were compared, they all exhibited a similar organization. The N-terminal regions of these eight proteins are responsible for the activation of gene expression once the hormone–receptor complex has bound to the appropriate hormone response element of the enhancer region. These regions of the receptor proteins are highly variable as one would expect since the different hormones activate different genes. The central regions of the receptor proteins contain the DNA-binding domains, and these regions are highly conserved with from 42 to 94 percent amino

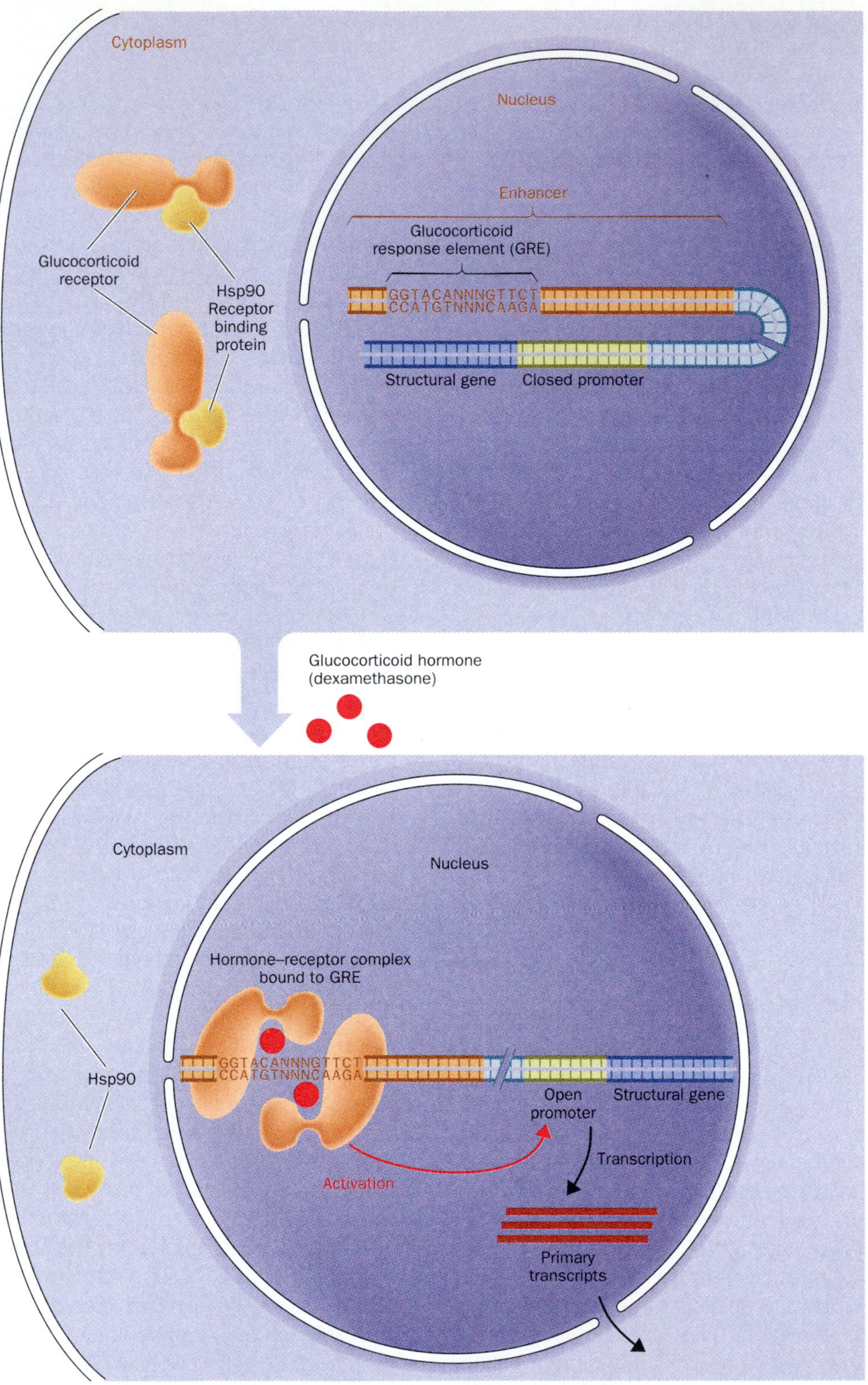

Figure 15.28 Schematic diagram of the mechanism of glucocorticoid hormone-activated gene expression. In the absence of hormone, the glucocorticoid receptor forms a complex with a protein called Hsp90 and remains in the cytoplasm. When hormone is present, it is bound by the receptor protein (Hsp90 is released), and the complex enters the nucleus through pores in the nuclear envelope. The hormone–receptor complex then binds (probably as a dimer) to the glucocorticoid response element (GRE) within an enhancer located near each of the target genes. Binding of the hormone–receptor complex at GRE somehow causes the enhancer to activate the promoter(s) of the target gene(s). The net result is that the level of transcription of the target gene(s) is greatly enhanced in the presence of the hormone. (Based on the data summarized by R. M. Evans, *Science* 240: 889–895, 1988.)

acid identity between different pairs of proteins. The C-terminal regions of the receptor proteins contain the hormone-binding domains; these regions show an intermediate level of conservation with from 15 to 57 percent amino acid identity. Since the steroid hormones all contain a cholesterol core with different side groups, some conservation of structure would be expected in the hormone-binding domains as observed.

As we stated at the beginning of this section, a given steroid hormone activates the expression of particular target genes in specific tissues. Why are these target genes not activated in all tissues and all cells by the presence of the hormone? Are the appropriate receptor proteins only synthesized in cells of the target tissue? If so, what controls the expression of the genes encoding the receptor proteins? If the receptor protein is present in all cells, why is it only active in certain cells? If it is kept inactive by an association with another protein, why isn't this protein present in target cells? If it is present, why doesn't it keep the receptor inactive in target cells? Although much progress has been made in elucidating the modes of action of steroid hormones, there still are many important questions that remain to challenge present and future generations of geneticists.

Ecdysone and Chromosome "Puffs" in Flies

In the giant salivary gland chromosomes of certain dipteran flies, such as *Drosophila* species and *Chironomus tentans,* individual chromosome bands undergo striking morphological changes at particular times during development. The individual bands expand into diffuse, less densely staining structures called ***"puffs"*** (Fig. 15.29); the phenomenon is frequently referred to as "puffing." Each puff almost certainly represents a segment of the chromosome that is in a highly extended state to facilitate the transcription of the resident gene or genes. By means of *in situ* hybridization and autoradiography (see Chapters 5–6, pp. 144 and 107, respectively), the puffs have been shown to contain DNA sequences that are complementary to the RNA sequences present in newly synthesized cytoplasmic mRNA (see Figure 15.29).

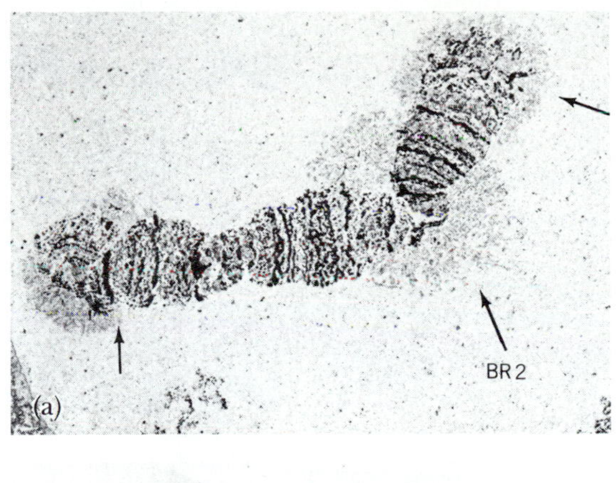

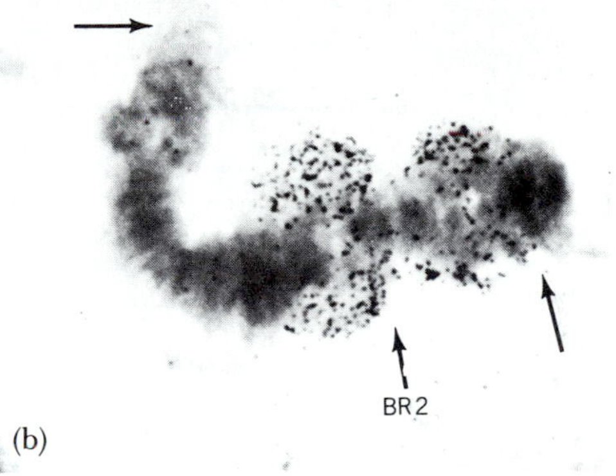

Figure 15.29 Photomicrograph (a) and autoradiograph (b) of the polytene salivary chromosome IV of the fly *Chironomus tentans,* showing the three giant puffs (arrows) characteristic of this chromosome. These three puffs are called Balbiani rings (BR1, BR2, and BR3) after the cytologist who first described them. Balbiani ring 2 (labeled BR2) is a particularly large puff, which has been shown to direct the synthesis of a huge 75S RNA transcript. This giant mRNA molecule is believed to specify the synthesis of one of the large salivary polypeptides. The template for the giant 75S mRNA has been shown to reside in BR2; this was demonstrated by purifying radioactively labeled 75S mRNA from cytoplasmic polyribosomes and using this labeled mRNA to carry out *in situ* hybridization and autoradiography (see Chapters 5–6, p. 144 and pp. 107–109). Note the heavy labeling of BR2 in (b). (From B. Daneholt, S. T. Case, J. Hyde, L. Nelson, and L. Wieslander, *Progr. in Nucleic Acid Res. Molec. Biol.* 19: 319–334, 1976.)

During development of dipteran flies, the steroid hormone ***ecdysone*** is released and triggers molting. Very specific patterns of salivary chromosome puffing occur during these moltings. If larvae of *D. melanogaster* and *C. tentans* are treated with ecdysone at stages of development prior to or between moltings, patterns of chromosome puffing occur that are identical to those occurring during natural moltings. These ***ecdysone-induced patterns of sequential puffing*** provide convincing demonstrations of the ***effect of a steroid hormone on gene expression.*** The patterns of puffing observed are very specific and are completely repeatable from experiment to experiment.

During early larval states of development in *D. melanogaster,* the puffs that existed prior to the ecdysone treatment regress, and a small number of new puffs form within 5 minutes after treatment. These initial puffs regress within a few hours and some 100–125 new puffs appear. By using inhibitors of protein synthesis, such as cycloheximide, the forma-

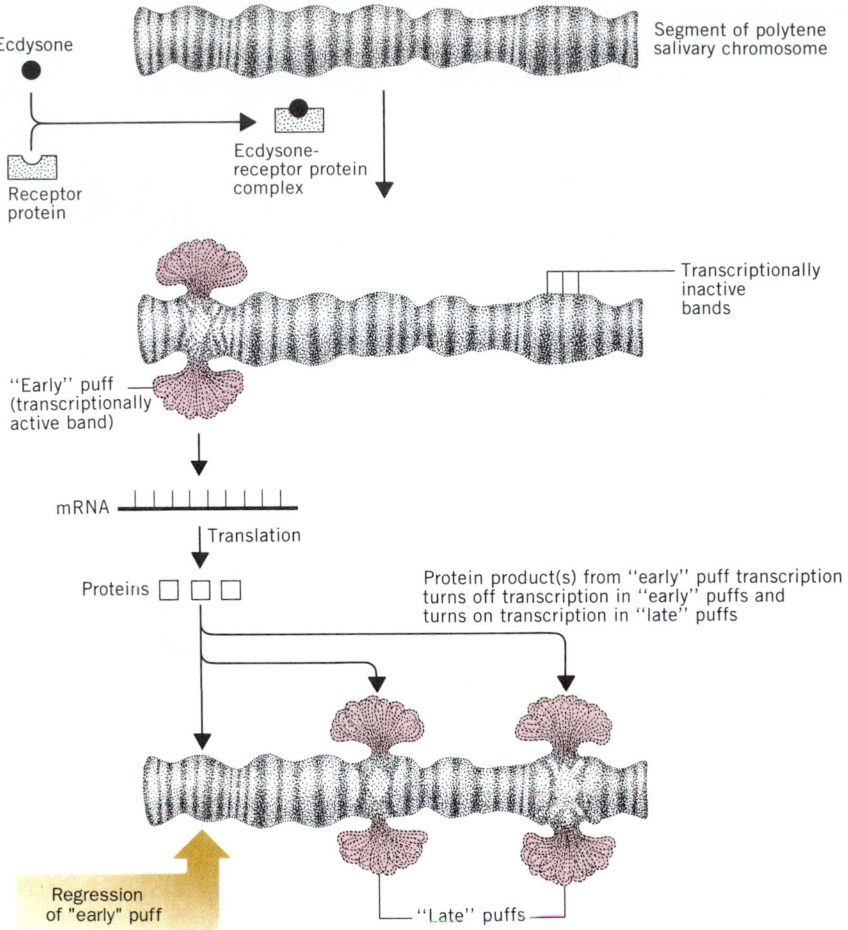

Figure 15.30 Schematic illustration of the ecdysone-induced sequences of chromosome puffing in *Drosophila* larvae. Ecdysone is the steroid hormone responsible for triggering the events associated with molting in many insects. Its putative mode of action is as diagrammed in Figure 15.27. Evidence that the puffs are sites of active transcription is presented in Figure 11.29b. Within about 5 minutes after injection of *Drosophila* larvae with ecdysone, a specific set of bands of the polytene salivary gland chromosomes begin to puff; these are always the same bands. Puffing at these sites terminates within about 4 hours, and new bands begin to develop this puffed morphology. By about 10 hours after the ecdysone injection, some 100–125 bands have formed puffs. The "late" puffs require one or more of the protein products that are synthesized from gene transcripts produced in "early" puffs; these "late" puffs do not form if protein synthesis is inhibited during the ecdysone treatment. Thus, ecdysone is a signal that triggers specific programs of sequential gene expression in target cells.

tion of the "late" puffs has been shown to require protein synthesis after ecdysone treatment. However, "early" puffs form in the absence of posttreatment protein synthesis. This suggests that the "late" ecdysone-induced puffing pattern is triggered by one or more of the proteins encoded by gene transcripts synthesized in "early" puffs (Fig. 15.30).

In addition to illustrating the effect of steroid hormones on gene expression, the ecdysone-induced puffing patterns provide evidence for the existence of preprogrammed patterns of gene expression in eukaryotes.

REGULATION BY ALTERNATE PATHWAYS OF TRANSCRIPT SPLICING

As should be clear from the preceding sections of this chapter, transcriptional regulation plays an important role in controlling development in eukaryotes. However, that does not mean that other modes of regulation are not important. Indeed, the regulation of transcript processing is well documented in many systems and occurs in several ways. Regulation occurs by changing transcript stability, by differential transport

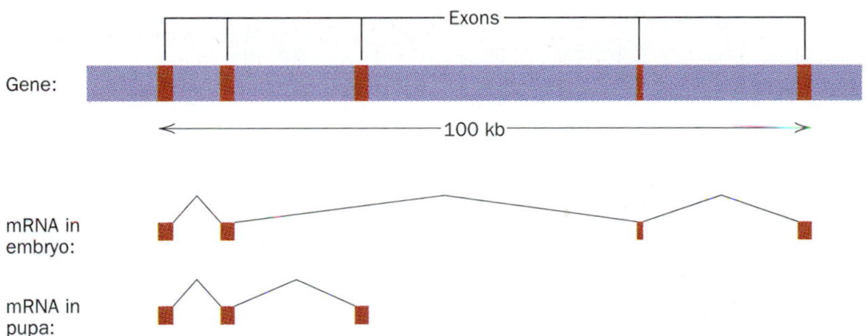

Figure 15.31 Diagram showing the alternate pathways of splicing observed for the *Antennepedia* transcript of *Drosophila* in embryos and pupae. (After R. L. Garber, A. Kuroiwa, and W. J. Gehring, *EMBO J.,* 2:2027–2036, 1983.)

to the cytoplasm, and by differential translation of processed transcripts. In this section, our focus will be on the regulation of gene expression through the use of alternate pathways of splicing primary transcripts. A large number of examples of alternate splicing pathways are known. The alternate splicing of the *Drosophila Ubx* transcript was mentioned earlier in this chapter. Similarly, the *Drosophila Antennapedia (Antp)* transcript is known to undergo alternate pathways of splicing in embryos and pupae (Fig. 15.31), although in this case we do not understand the functions of the different products.

One of the most spectacular examples of alternative modes of transcript splicing occurs in the case of the tropomyosin genes of *Drosophila* and vertebrate animals. Tropomyosins are a family of closely related proteins that mediate the interactions between actin and troponin and thus help regulate muscle contractions. Different tissues, both muscle and nonmuscle, are characterized by the presence of different tropomyosin isoforms. It turns out that many of these isoforms are produced from the same gene by alternative splicing. For example, Fig. 15.32 shows the production of six different tropomyosin isoforms from the two *Drosophila* tropomyosin genes (*TmI* and *TmII*). The tropomyosin transcript splicing patterns appear to be even more complex in mammals, where at least 10 different isoforms are derived from a single gene.

Many other examples of alternative pathways of transcript splicing are known. In Chapter 16, the production of different immunoglobulin heavy chains by alternative splicing of IG heavy chain transcripts is discussed (see Fig. 16.6).

At present, we do not know how important the use of alternative splicing pathways will turn out to be in the overall regulation of gene expression. What we do know is that new examples of the use of alternative splicing pathways are being discovered almost every day. Clearly, the regulation of gene expression by control of splicing pathways is a significant mechanism of regulation in higher eukaryotes.

REGULATION OF COMPLEX CIRCUITS OF GENE EXPRESSION IN EUKARYOTES

In the absence of precise information about the mechanisms that regulate gene expression in eukaryotes, many models and much speculation have ensued. One of the more popular early models was that proposed by R. J. Britten and E. H. Davidson. Britten and Davidson's model (Fig. 15.33) proposes an integrated regulation of sets of structural genes by means of moderately repetitive regulator genes. It therefore takes into account the observed interspersion of single-copy DNA sequences and repetitive DNA sequences (see Chapter 6, pp. 146–148).

According to the Britten–Davidson model, specific ***sensor genes*** represent sequence-specific binding sites (analogous to the CAP–cAMP binding site in the *E. coli lac* promoter?; see pp. 398–400) that respond to specific signals (such as hormone–receptor protein complexes?). When sensor genes receive the appropriate signals, they activate the transcription of the adjacent ***integrator genes.*** The integrator gene-products then interact in a sequence-specific manner with ***receptor genes.*** Britten and Davidson proposed that the integrator gene-products were ***activator RNAs*** that interacted directly with the receptor genes to trigger the transcription of the contiguous ***producer genes*** (analogous to the structural genes in prokaryotic operons). However, they pointed out that, formally, it would make no difference whether the active integrator gene-products were RNA molecules or proteins.

By making either the receptor genes (Fig. 15.33*a*) or the integrator genes (Fig. 15.33*b*) redundant, various combinations of the producer genes could be

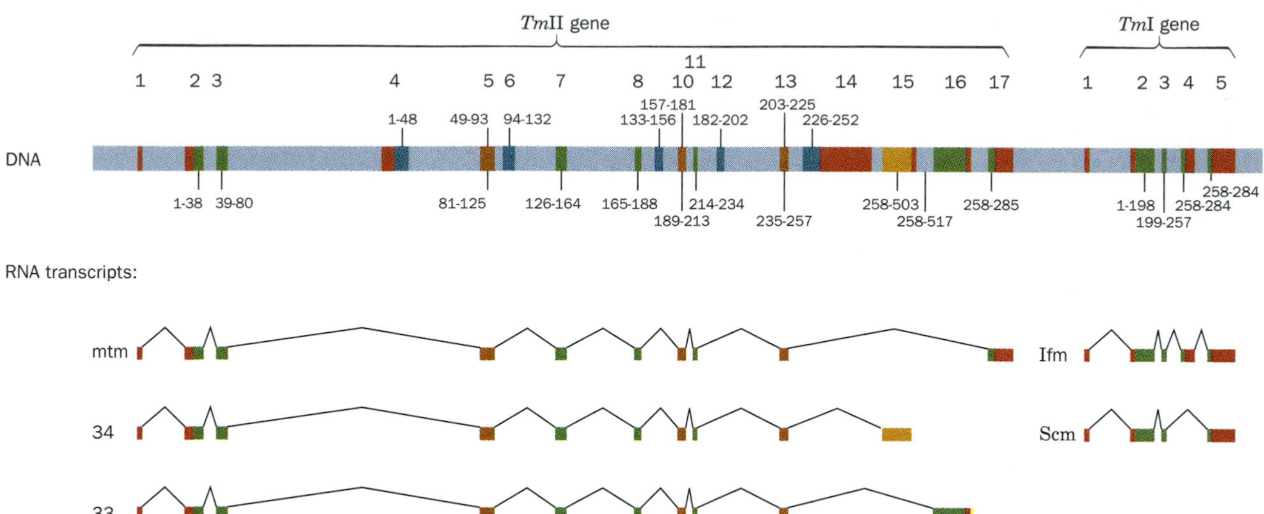

Figure 15.32 Drawing showing the exon–intron structures of the two tropomyosin genes (*TmI* and *TmII*) of *Drosophila* (top) and the alternative pathways of transcript splicing (below) that have been shown to produce different tropomyosin isoforms. The exons are numbered 1 through 17 (*TmII*) and 1 through 5 (*TMI*) at the top; exons are shown as colored boxes. The different colors of the exon boxes are keyed to the type of tropomyosin isoform that contains the amino acid sequence encoded by each exon. The numbers just above and below the *TmII* map indicate the amino acid positions for the cytoskeletal and muscle tropomyosin exons, respectively. Abbreviations: mtm, muscle tropomyosin; ctm, cytoskeletal tropomyosin; Ifm, indirect flight muscle; and Scm, supercontractile muscle (referring to the tissue in which the specific isoform occurs). (After P. D. Hanke and R. V. Storti, *Molec. Cell Biol.* 8: 3591–3602, 1988.)

turned on in response to different signals. If both the integrator genes and the receptor genes are redundant, complex integrated circuits of gene expression can easily be devised. Unfortunately, testing the validity of such models is far more difficult than devising them. The most attractive feature of the Britten and Davidson model is that it provides a plausible reason for the observed pattern of interspersion of moderately repetitive DNA sequences and single-copy DNA sequences. Direct evidence indicates that most structural genes (producer genes) are indeed single-copy DNA sequences. The adjacent moderately repetitive ("middle-repetitive") DNA sequences would contain the various kinds of regulator genes (sensor, integrator, and receptor genes), according to Britten and Davidson's model.

As more studies have been done comparing the complexity of heterogeneous nuclear RNA (hnRNA) populations and mRNA populations in different types of cells, it has become evident that hnRNA populations are almost always more complex (contain more distinct sequences) than mRNA populations. These results suggested that considerable ***regulation is occurring posttranscriptionally during RNA processing***, that is, in the ***hnRNA ⟶ mRNA stage***.

These observations led Davidson and Britten to propose a second model, the ***"Davidson–Britten model,"*** in which gene expression is regulated at the RNA processing level (Fig. 15.34). According to this model, most of the structural genes are located in ***"constitutive transcription units,"*** which are transcribed at basal levels in all cells. These constitutive

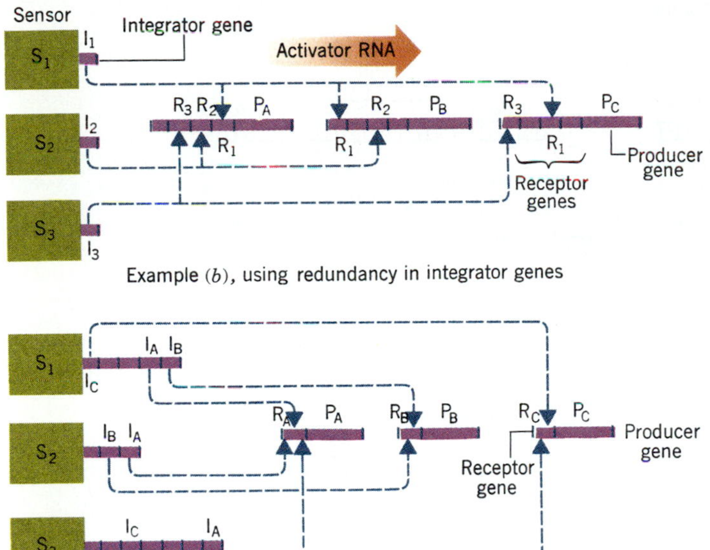

Figure 15.33 The Britten and Davidson model of regulation of gene expression is eukaryotes. Two variations of integrated regulation are shown: (a) a system based on redundancy of "receptor" genes and (b) a system based on the redundancy of "integrator" genes. The three "sensor" genes (S_1, S_2, and S_3) respond to three different signals (perhaps hormone–receptor complexes). The diagrams schematize the events proposed to occur after the sensor genes have triggered the transcription of their respective integrator genes (I_1, I_2, I_3 or I_A, I_B, I_C). The integrator gene-products, "activator RNAs," diffuse (symbolized by dashed lines) from their sites of synthesis (integrator genes) to their sites of action (receptor genes). The binding of the various activator RNAs to the respective receptor genes somehow triggers the transcription of the contiguous producer genes (P_A, P_B, P_C). Depending on which integrator gene (or genes) is activated by its sensor gene (or genes), one, two, or all three of the producer genes (structural genes) may be turned on. (From R. J. Britten and E. H. Davidson, "Gene Regulation for Higher Cells: A Theory," *Science* 165: 349–357. Copyright © 1969 by the American Association for the Advancement of Science.)

transcripts are processed, however, only in cells that contain the appropriate ***"integrating regulatory transcripts."*** The latter are transcribed in a ***cell-specific*** manner and must be present before the ***"constitutive transcripts" of the structural genes can be processed into mRNAs.***

These "integrating regulatory transcripts" contain repetitive sequences that interact with different structural gene transcripts much like the repetitive "integrator" genes interacted with different "receptor" genes in the original Britten–Davidson model. The key difference is that the regulation occurs posttranscriptionally during RNA processing in the new Davidson–Britten model, rather than transcriptionally as in the original model.

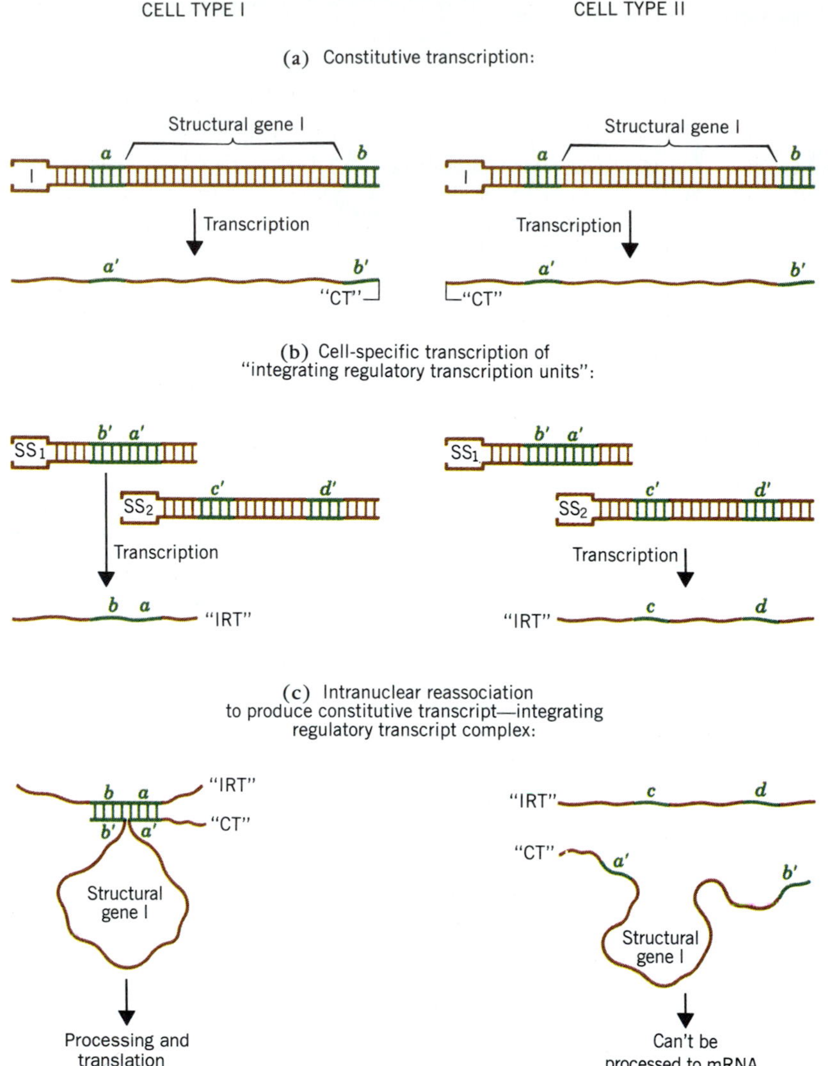

Figure 15.34 The Davidson and Britten model for the regulation of gene expression at the RNA processing level in eukaryotes. (a) The majority of the structural genes are assumed to be located in ***"constitutive transcription units,"*** which are transcribed continually in all cells. *I* denotes the transcrption initiation site. The sequences labeled *a, a', b,* and so on shown in yellow-green are middle-repetitive sequences that are proposed to be involved in regulating gene expression. Sequences *a', b', c',* and *d'* are complementary to sequences *a, b, c,* and *d,* respectively. (b) Structural gene expression is regulated by repetitive RNA sequences transcribed in a tissue- or cell-specific manner as components of ***"integrating regulatory transcription units"*** (*IRTU*s). Transcription of the *IRTU*s is controlled by nucleoprotein ***"sensors"*** (*SS*) that respond to specific external signals. Different *IRTU*s are transcribed in different tissues or cell types. (c) ***A given structural gene constitutive transcript* (CT) *can be processed into mRNA*** and thus expressed ***only if it forms an appropriate complex with an "integrating regulatory transcript"* (IRT*).*** Thus, the expression of a particular structural gene in any given cell requires the presence of the appropriate nuclear *IRT*. Different types of cells will have *IRT*s carrying overlapping populations of repetitive sequences, such that they process some of the same structural gene transcripts and some different structural gene transcripts. No attempt has been made to illustrate the complex nature of the *IRT* populations of different cells in the above diagram. (Modified from E. H. Davidson and R. J. Britten, *Science* 204: 1052–1059, 1979.)

SUMMARY

The process by which a complex, three-dimensional form unfolds during the development of a higher eukaryote such as a fruit fly or a human has fascinated scientists for decades. The genetic control of ***morphogenesis*** is now being dissected in detail in *Drosophila melanogaster* and *Caenorhabditis elegans.* The master plan now seems quite clear: ***development is controlled by a cascade of regulatory genes*** each acting at the proper time and place to trigger the expression of the next set of genes in the cascade.

In *Drosophila,* the basic body plan unfolds through the establishment of ***dorsal–ventral polarity*** and a series of ***anterior–posterior segmentation*** events. These events are controlled by key regulatory genes, at least half of which encode transcriptional activator proteins that function by binding in a sequence-specific manner to enhancers and/or promoters of the next set of regulatory genes in the cascade.

Genes that control pattern formation in *Drosophila* have been classified into five groups: (1) ***maternal-effect genes,*** (2) ***gap genes,*** (3) ***pair-rule genes,*** (4) ***segment-polarity genes,*** and (5) ***homeotic genes.*** The maternal-effect genes are expressed during oögenesis in the female and produce the initial dorsal–ventral and anterior–posterior polarities in the fertilized egg and early embryo. The gap genes, pair-rule genes, and segment-polarity genes then act, roughly in sequence, to subdivide the developing embryo and cellular blastoderm into 15 segments and 14 parasegments along the anterior–posterior axis of the body. Thereafter, the homeotic genes function and control the commitment of groups of cells to differentiate into specific tissues. At least half of these genes produce proteins that function as transcriptional activators.

Groups of cells make commitments to proceed along specific pathways of differentiation long before they actually undergo the differentiation process. Cellular commitments of this type are called ***determination.*** *Drosophila* larvae contain ***imaginal discs*** in which the cells have already undergone determination; each disc will give rise to a specific organ of the adult fly.

Oögenesis in higher animals, especially amphibians, involves extensive transcriptional activity on giant, highly extended "lampbrush chromosomes." In addition, amphibian oöcytes amplify the rRNA genes in order to produce and store the large number of ribosomes in the egg that will be needed for the rapid growth that occurs during embryonic development.

Much of the regulation of gene expression in eukaryotes occurs at the level of transcription. Some occurs at the level of ***transcript processing.*** In some cases, ***alternative pathways of splicing*** a primary transcript give rise to different gene-products. Eukaryotic transcription units are ***monogenic,*** in contrast to the operons of prokaryotes.

Transcription in eukaryotes is often regulated by ***enhancers*** and ***silencers*** that can influence gene expression from distances of greater than 1000 nucleotide-pairs. Enhancers and silencers are orientation and position independent; they work equally well when positioned on either side of a gene or in an intron of the gene and in either forward or reverse (flipped end-for-end) orientation. Considerable evidence suggests that ***methylation*** of the bases in DNA can alter expression and that ***DNA conformation*** plays an important role in the activation of genes.

Steroid hormones are known to be major regulators of transcription in higher animals. These hormones are known, in some cases, to become bound to receptor proteins that then activate transcription via sequence-specific binding to enhancer elements.

REFERENCES

BEERMANN, W., editor. 1972. *Developmental studies on giant chromosomes.* Springer-Verlag, New York.

BIGGIN, M.D., and R. TIJAN. 1989. "Transcription factors and the control of *Drosophila* development." *Trends in Genet.* 5: 377–383.

BRITTEN, R.J., and E.H. DAVIDSON. 1969. "Gene regulation for higher cells: A theory." *Science* 165: 349–357.

BROWN, D.D. 1981. "Gene expression in eukaryotes." *Science* 211: 667–674.

BURATOWSKI, S., S. HAHN, L. GUARENTE, and P.A. SHARP. 1989. "Five intermediate complexes in transcription initiation by RNA polymerase II." *Cell* 56: 549–561.

DAVIDSON, E.H. 1976. *Gene activity in early development.* Academic Press, New York.

DAVIDSON, E.H., and R.J. BRITTEN. 1979. "Regulation of gene expression: Possible role of repetitive sequences." *Science* 204: 1052–1059.

EHRLICH, M., and R.Y.-H. WANG. 1981. "5-Methylcytosine in eukaryotic DNA." *Science* 212: 1350–1357.

EVANS, R.M. 1988. "The steroid and thyroid hormone receptor superfamily." *Science* 240: 889–895.

FREIFELDER, D. 1983. *Molecular biology, a comprehensive introduction to prokaryotes and eukaryotes.* Science Books International, Boston.

FRIGERIO, G., M. BURRI, D. BOPP, S. BAUMGARTNER, and M. NOLL. 1986. "Structure of the segmentation gene *paired* and the *Drosophila* PRD gene set as part of a gene network." *Cell* 47: 735–746.

GARBER, R.L., A. KUROIWA, and W.J. GEHRING. 1983. "Genomic and cDNA clones of the homeotic locus *Antennapedia* in *Drosophila*." *EMBO J.* 2: 2027–2036.

GEHRING, W.J. 1987. "Homeoboxes in the study of development." *Science* 236: 1245–1252.

GROUDINE, M., and H. WEINTRAUB. 1982. "Propagation of globin DNase I-hypersensitive sites in absence of factors required for induction: A possible mechanism for determination." *Cell* 30: 131–139.

HANKE, P.D., and R.V. STORTI. 1988. "The *Drosophila melanogaster* tropomyosin II gene produces multiple proteins by use of alternative tissue-specific promoters and alternative splicing." *Molec. Cell. Biol.* 8: 3591–3602.

HARDING, K., C. WEDEEN, W. MCGINNIS, and M. LEVINE. 1985. "Spatially regulated expression of homeotic genes in *Drosophila*." *Science* 229: 1236–1242.

INGHAM, P.W. 1988. "The molecular genetics of embryonic pattern formation in *Drosophila*." *Nature* 335: 25–34.

JOHNSON, P.F., and S.L. MCKNIGHT. 1989. "Eukaryotic transcriptional regulatory proteins." *Annu. Rev. Biochem.* 58: 799–839.

KARCH, F., B. WEIFFENBACH, M. PEIFER, W. BENDER, I. DUNCAN, S. CELNIKER, M. CROSBY, and E.B. LEWIS. 1985. "The abdominal region of the bithorax complex." *Cell* 43: 81–96.

LEVING, M., and T. HOEY. 1988. "Homeobox proteins as sequence-specific transcription factors." *Cell* 55: 537–540.

LEWIN, B. 1990. *Genes IV.* Cell Press, Cambridge, MA/Oxford Univ. Press, New York.

LEWIS, E.B. 1985. "Regulation of the genes of the bithorax complex in *Drosophila*." *Cold Spring Harbor Sympos. Quant. Biol.* 50: 155–164.

LEY, T.J., J. DESIMONE, N.P. ANAGNOU, G.H. KELLER, R.K. HUMPHRIES, P.H. TURNER, N.S. YOUNG, P. HELLER, and A.W. NIENHUIS. 1982. "5-Azacytidine selectively increases γ-globin synthesis in a patient with β^+ thalassemia." *New England J. Med.* 307: 1469–1475.

LIPPS, H.J., A. NORDHEIM, E.M. LAFER, D. AMMERMANN, B.D. STOLLAR, and A. RICH. 1983. "Antibodies against Z DNA react with the macronucleus but not the micronucleus of the hypotrichous ciliate *Stylonychia mytilus*." *Cell* 32: 435–441.

MANIATIS, T., S. GOODBOURN, and J.A. FISCHER. 1987. "Regulation of inducible and tissue-specific gene expression." *Science* 236: 1237–1245.

MCGINNIS, W., A.W. SHERMOEN, J. HEEMSKERK, and S.K. BECKENDORF. 1983. "DNA sequence changes in an upstream DNase I-hypersensitive region are correlated with reduced gene expression." *Proc. Natl. Acad. Sci. U.S.A.* 80: 1063–1067.

NORDHEIM, A., E.M. LAFER, L.J. PECK, J.C. WANG, B.D. STOLLAR, and A. Rich. 1982. "Negatively supercoiled plasmids contain left-handed Z-DNA segments as detected by specific antibody binding." *Cell* 31: 309–318.

NORDHEIM, A., M.L. PARDUE, E.M. LAFER, A. MOLLER, B.D. STOLLAR, and A. RICH. 1981. "Antibodies to left-handed Z-DNA bind to interband regions of *Drosophila* polytene chromosomes." *Nature* 294: 417–422.

NORDHEIM, A., P. TESSER, F. AZORIN, Y.H. KWON, A. MÖLLER, and A. RICH. 1982. "Isolation of *Drosophila* proteins that bind selectively to left-handed Z-DNA." *Proc. Natl. Acad. Sci. U.S.A.* 79: 7729–7733.

NUSSLEIN-VOLHARD, C., H.G. FROHNHOFER, and R. LEHMANN. 1987. "Determination of anteroposterior polarity in *Drosophila*." *Science* 238: 1675–1681.

O'MALLEY, B.W., and W.T. SCHRADER. 1976. "The receptors of steroid hormones." *Sci. Amer.* 234(2): 32–43.

POSTLETHWAIT, J.H., and H.A. SCHNEIDERMAN. 1971. "Pattern formation and determination in the antenna of the homeotic mutant *Antennapedia* of *Drosophila melanogaster*." *Develop. Biol.* 25: 606–640.

PTASHNE, M. 1988. "How eukaryotic transcriptional activators work." *Nature* 335: 683–689.

______. 1989. "How gene activators work." *Sci. Amer.* 260(1): 40–47.

RAZIN, A., and A.D. RIGGS. 1980. "DNA methylation and gene function." *Science* 210: 604–610.

SERFLING, E., M. JASIN, and W. SCHAFFNER. 1985. "Enhancers and eukaryotic gene transcription." *Trends in Genet.* 1: 224–230.

SMITH, C.W. J., J.G. PATTON, and B. NADAL-GINARD. 1989. "Alternative splicing in the control of gene expression." *Annu. Rev. Genet.* 23: 527–577.

STEIN, G.S., J.S. STEIN, and L.J. KLEINSMITH. 1975. "Chromosomal proteins and gene regulation." *Sci. Amer.* 232(2): 46–57.

STRUHL, G. 1981. "A gene product required for correct initiation of segmental determination in *Drosophila*." *Nature* 293: 36–41.

STRUHL, K. 1989. "Molecular mechanisms of transcriptional regulation in yeast." *Annu. Rev. Biochem.* 58: 1051–1077.

WALLDORF, U., R. FLEIG, and W.J. GEHRING. 1989. "Comparison of homeobox-containing genes of the honeybee and *Drosophila*." *Proc. Natl. Acad. Sci. U.S.A.* 86: 9971–9975.

WOOD, W.B., editor. 1988. *The nematode Caenorhabditis elegans.* Cold Spring Harbor Laboratory, Cold Spring Harbor, NY.

PROBLEMS AND QUESTIONS

15.1 Are operons more common in bacteria or in higher organisms? Why?

15.2 If the operon theory had been established at the time the two classical hypotheses for the mechanics of differentiation were being considered, that is, (1) segregation of nuclear elements and (2) intervention of cytoplasm, which hypothesis would have been most acceptable?

15.3 Many of the RNA transcripts of prokaryotic organisms are polygenic. Are polygenic RNA transcripts common in eukaryotic organisms?

15.4 Why is the Britten and Davidson model more acceptable than the operon model for explaining regulation in cells of higher animals?

15.5 (a) How do histones regulate gene activity? (b) When in the cell cycle are the histones synthesized? (c) How is the synthesis of the histones themselves regulated?

15.6 How can steroid hormones regulate gene activity?

15.7 Why are salivary chromosomes in the larvae of diptera useful in studying hormonal regulation of gene activity?

15.8 (a) Why are certain organisms such as molds, bacteria, and viruses favorable materials for biochemical genetic studies? (b) What type of experimental material would be most suitable for a study of (1) operons, (2) repressing effects of histones, (3) hormonal control of mRNA synthesis, and (4) chromosome puffs?

15.9 At birth, rabbits of the Himalayan breed are all white, but as they grow older the extremities (paws, nose, ears, and tail) become black. When the white fur is shaved from a spot on the body of the adult and the rabbit is kept in a cool place, the new hair that grows in the shaved spot is black. The temperature of the body proper is about 33°C, but in the extremities it is about 27°C. (a) How might genetic and environmental factors be involved? (b) Formulate an explanation for the difference in pigmentation in different areas of the body surface.

15.10 If all cells in a given organism carry the same genes, how can gene expressions that are localized in time and space be explained?

15.11 How are (a) differentiation, (b) organization, and (c) growth involved in the development of an animal?

15.12 The father of two albino children has made widespread inquiries among geneticists and physicians concerning a possible cure for albinism. The steps in pigment production have been elucidated and it seems feasible to him that something might be added to the diet or given by injection that would supply the missing step or steps in melanin production in his children. Evaluate the possibility of such a development.

15.13 Ordinarily, *Drosophila* eye-disc transplants placed in hosts with different genotypes develop according to the genotype of the transplant. For example, transplants from larvae with the genotype for white eye develop white eyes in wild-type hosts, and transplants from wild-type larvae develop wild-type red eyes in hosts with the genotype for white. Beadle and Ephrussi performed transplantation experiments on larvae with vermilion (*v*) and cinnabar (*cn*) eyes. The phenotypes for these two mutants are similar: they have bright red color because they lack the brown pigment that is a part of the wild-type red eye. (1) When discs from wild-type larvae were transplanted into vermilion or cinnabar hosts they developed into wild type, and when *v* or *cn* transplants were placed in wild-type hosts they also developed into wild type. (2) When discs from *cn* larvae were placed in *v* hosts, no brown pigment was formed, and the eyes were bright red. From the reciprocal transplant, that is, *v* in *cn* hosts, brown pigment was produced and the eyes were wild type. Formulate an explanation for these results.

15.14 Why are the homeotic mutants of *Drosophila* particularly interesting?

15.15 How can one determine whether a particular gene is being transcribed in the cells of two different tissues or two different cell cultures derived from the same organism?

15.16 Enhancers and classical promoters both act as positive regulators of gene expression. By what criteria can these two classes of regulatory sequences be distinguished?

15.17 What correlations are apparent in comparing the phenotypic effects of homeotic mutants of *Drosophila* and the transdetermination events observed after serial imaginal disc transplants in *Drosophila* by E. Hadorn? What implications do these correlations have regarding the genetic control of development in fruit flies?

15.18 *Drosophila* females homozygous for mutations in the *bicoid (bcd)* gene produce embryos that lack heads and thoracic structures. Females homozygous for mutations in the *oskar (osk)* gene produce embryos that lack all abdominal segments. (a) What can you conclude about the functions of the *bcd* and *osk* gene-products from these results? (b) Are the products of these genes synthesized before or after the completion of oögenesis? (c) What would you expect the embryos of females homozygous for both *bcd* and *osk* to look like?

15.19 The most striking feature of the morphology of many insects is the intricate pattern of segmentation. How are the anterior–posterior segmentation patterns of *Drosophila* larvae, pupae, and adult flies established?

15.20 Several of the genes that govern segmentation patterns in *Drosophila* have been shown to encode proteins with sequence-specific DNA-binding properties. What does this suggest about the genetic control of development in flies?

15.21 The HOX genes of humans contain sequences with striking similarity to the homeobox domains of homeotic genes of *Drosophila*. What implications does this have regarding the genetic control of development in humans?

15.22 Based on the available evidence, does methylation of DNA sequences in a gene tend to enhance or depress the level of expression of the gene?

15.23 Tropomyosins are proteins that mediate the interaction of actin and troponin and help regulate muscle contractions. In higher animals, tropomyosins exist as a family of closely related proteins that have some amino acid sequences in common and differ in amino acid sequence in other regions. Could such a family of closely related polypeptides be encoded by a single gene, or would a family of related genes be required?

Chapter 16

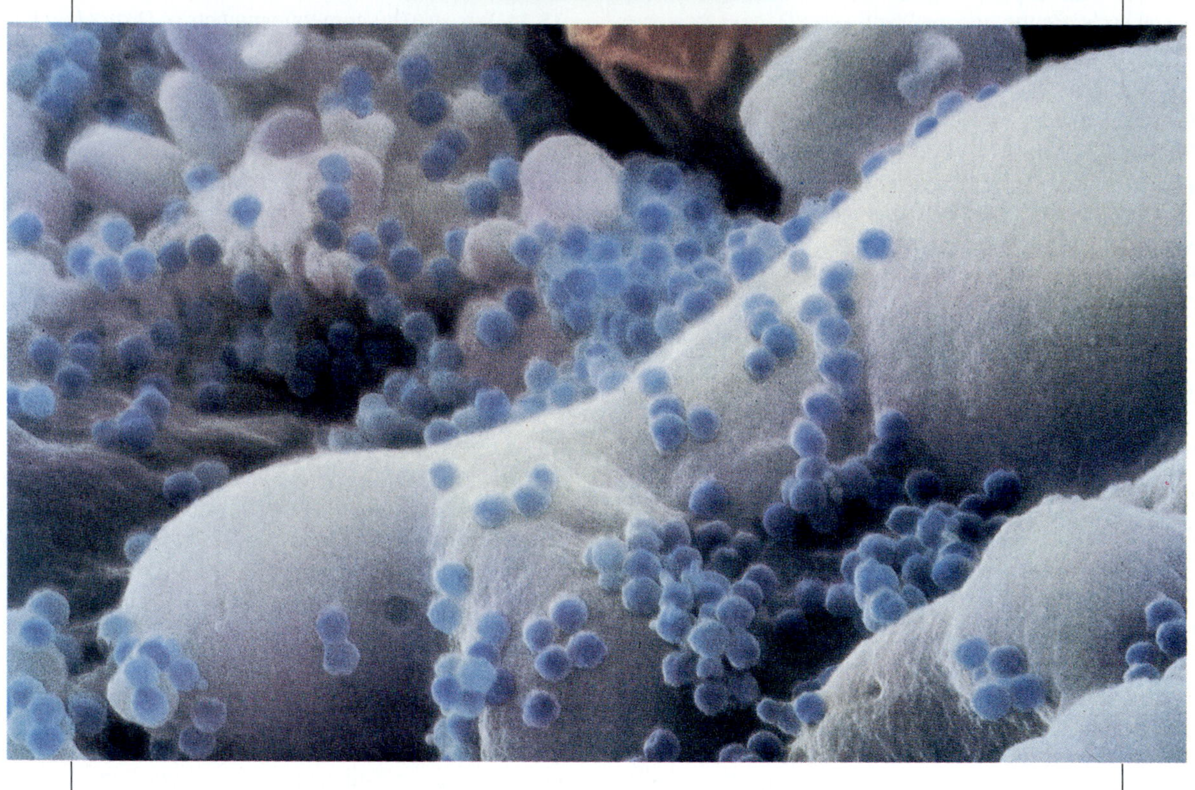

Genetic Control of the Immune Response

WHEN foreign substances called ***antigens*** (e.g., the coat proteins of viruses) enter the bloodstream of a mammal, they trigger a defense mechanism, the ***immune response,*** which results in the synthesis of an extremely important group of proteins called ***antibodies.*** These ***antibodies bind to the antigens with exceptional specificity,*** thus facilitating their removal from the circulatory system. During the last decade, scientists have discovered that the DNA sequences that encode the vast array of antibodies produced by the immune systems of mammals ***are assembled during the differentiation of the antibody-producing cells by the occurrence of a novel set of genome rearrangements.***

COMPONENTS OF THE IMMUNE SYSTEM

Three different types of white blood cells play central roles in the immune response in vertebrates. These cells are (1) ***B lymphocytes*** (called B cells because they are produced in bone marrow), (2) ***T lymphocytes*** (called T cells because they are produced in the thymus gland), and (3) ***macrophages. Antibodies are synthesized by B lymphocytes*** and are either secreted or remain membrane-bound on the surface of the B cell depending on the conditions. During the ***humoral immune response,*** these antibodies bind to free antigens in the circulatory system and agglutinate them. The resulting ***antibody–antigen complexes are then ingested and degraded by macrophages*** (Fig. 16.1, left). ***T lymphocytes mediate the cellular immune response.*** The T cells synthesize antigen receptors that recognize antigens on cell surfaces and trigger the lysis of the antigen-containing cells by the activated T cells (Fig. 16.1, right). Different T lymphocytes perform this function in slightly different ways. However, in general, the attack of the T cell on the antigen-carrying cell requires both the ***specific T cell receptor*** and one or more ***histocompatibility antigen receptors.*** The mechanisms by which antibodies, T cell receptors, and histocompatibility antigen receptors are produced are described in the following sections of this chapter.

VAST REPERTOIRE OF ANTIBODIES

The most remarkable aspect of the immune response, at least from a genetics standpoint, is the ***seemingly infinite variety of antibodies that can be synthesized in response to antigens that the animal has not previously encountered.*** How can an organism have prepared to synthesize an antibody designed to bind very specifically to a particular antigen without ever having made contact with the antigen? Moreover, how can an organism store enough genetic information to code for the amino acid sequences of a virtually unlimited variety of antibodies? These and related questions about the immune response had puzzled geneticists for several decades. Within the last few years, however, the main features of the answers to these questions have become clear.

We do not know how many different antibodies a mouse or a human can produce, but we do know that the number is very large, almost certainly in the millions. This presents a paradox. The complete human genome (i.e., one of each of the 23 pairs of human chromosomes) contains about 3×10^9 nucleotide-pairs. If all of this DNA were in the form of uninterrupted coding sequences of genes each 1000 nucleotide-pairs long (of course, we know that much of it is not), the genome would contain a maximum of about 3 million genes. Since we know that many of these genes code for various RNA molecules, enzymes, and structural proteins, and we know that many of these genes contain long noncoding introns, how can we account for the genetic information needed to code for the plethora of different antibodies?

Hypotheses: Genetic Basis of Antibody Diversity

Past attempts to explain the genetic basis of antibody diversity can be roughly grouped into three different hypotheses.

1. The ***"germ line" hypothesis*** stated that there is a ***separate germ line gene for each antibody.*** (This agreed with our early knowledge about protein synthesis, but presented the paradox of not enough DNA.)
2. The ***"somatic mutation" hypothesis*** stated that there is only one or a few germ line genes specifying each major class of antibodies and that the ***diversity is generated by a high frequency of somatic mutation***—mutation occurring in the antibody-producing somatic cells or in cell lineages leading to antibody-producing cells. (There was no precedent for a high frequency of mutation occurring in only certain genes and in only certain types of cells. By what mechanism could this occur, and how could it be regulated?)
3. The ***"minigene" hypothesis*** stated that the diversity is generated by the ***shuffling of many small segments of a few genes into a multitude of possible combinations.*** The shuffling would occur by recombination processes in somatic cells. (This required totally novel mechanisms for rearranging segments of DNA.)

Scanning electron micrograph of human immunodeficiency virus particles (the causative agent of AIDS) attacking a helper T lymphocyte. (Photo by Lennart Nilsson, Copyright © Boehringer Ingelheim International GmbH.)

We now know that the minigene hypothesis explains a great deal of the observed diversity. However, we also know that somatic mutation contributes additional diversity. Finally, we know that one segment (the "constant" region; see the following discussion) of each antibody chain is specified by a "gene" or "gene segment" that is present in the genome in only a few copies. Thus, all three hypotheses were correct in certain respects.

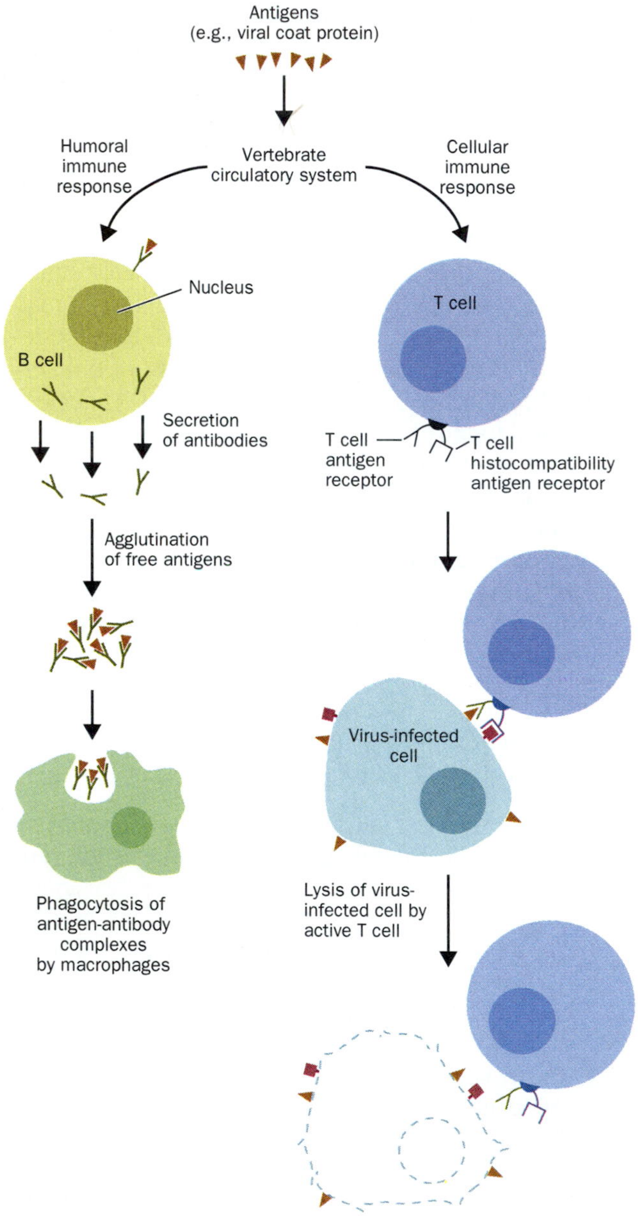

Structure of Antibodies

Antibodies belong to the class of proteins called ***immunoglobulins***. Each antibody is a tetramer composed of four polypeptides, ***two identical light chains and two identical heavy chains***, joined by disulfide bonds (Fig. 16.2). The light chains are about 220 amino acids long, and the heavy chains are about 440–450 amino acids long. Every chain, heavy and light, has an amino-terminal ***variable region***, within which the amino acid sequence varies among antibodies specific for different antigens, and a carboxyl-terminal ***constant region***, within which the amino acid sequence is the same for all antibodies of a given immunoglobulin (Ig) class, regardless of antigen-binding specificity. The variable regions of all antibody chains are about 110 amino acids long.

Regions of proteins that carry out particular functions are called ***domains***. Each antibody has ***two antigen-binding sites or domains***, each of which is formed by the variable regions of one light chain and one heavy chain (Fig. 16.3). In addition, the constant regions of the two heavy chains interact to form a third domain, called the ***effector function domain***, which is responsible for the proper interaction of the antibody with other components of the immune system.

There are five classes of antibodies: IgM, IgD, IgG, IgE, and IgA. The class to which an antibody belongs, and thus the function that it carries out, is determined by the structure of its heavy chain constant region (i.e., the structure of its effector function domain). For example, IgD antibodies usually remain bound to the surfaces of the cells in which they are synthesized, whereas IgG antibodies are usually secreted and circulate through the body in the bloodstream. The light chains of antibodies are of two types, ***kappa*** and

Figure 16.1 Schematic diagram showing the major components of the immune response in vertebrates. A foreign substance such as the coat protein of a virus acts as an antigen to trigger the synthesis of large amounts of antibodies that react specifically with the antigen and remove it from the circulatory system. Two types of immune response occur. (1) B lymphocytes synthesize and secrete antibodies that complex with free antigens in the bloodstream; these complexes are then ingested and destroyed by macrophages. (2) T lymphocytes synthesize antigen receptors that remain bound to the surface of the T cells. These antigen receptors act in concert with histocompatibility antigen receptors to recognize and destroy cells that carry the antigens. Thus, the B cells carry out a humoral immune response, and the T cells carry out a cellular immune response. The vast diversity of both antibodies and T cell receptors is produced by genome rearrangements that take place during the differentiation of B and T lymphocytes from stem cells. The histocompatibility antigens are encoded by a large cluster of genes in the major histocompatibility complex (*MHC*).

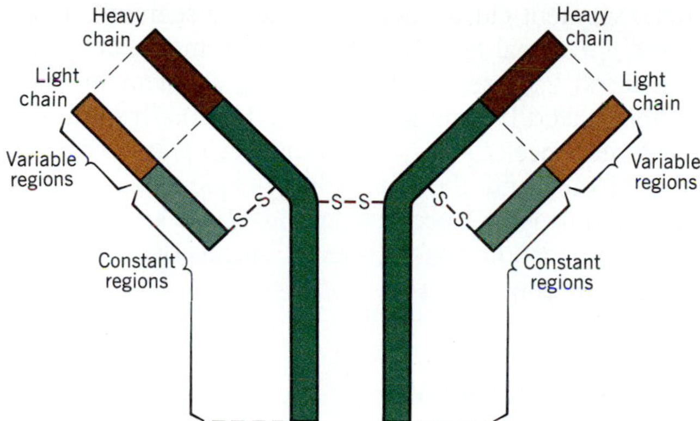

Figure 16.2 Schematic diagram of antibody structure. Each antibody is a tetramer composed of four polypeptide chains: two identical light chains and two identical heavy chains. Each chain consists of a variable region (brown) and a constant region (green). Every antibody has two antigen binding sites, formed by heavy and light chain variable regions (see Fig. 16.3).

lambda, with type being determined by the structure of the light chain constant region. As we shall see, antibodies may have the same antigen-binding specificity, as determined by the variable regions of the four chains, but different immunological functions, as determined by the constant regions of the two heavy chains.

Thus, when we examine the structure of antibodies, we see that their diversity resides almost entirely within the variable regions of the molecules. If these polypeptides were synthesized from colinear nucleotide-pair sequences of genes, one gene per polypeptide chain, the genome would have to contain a vast array of genes with highly variable sequences at one end and essentially identical sequences at the other end. However, this is not the case. Recombinant DNA techniques have made it possible to isolate and sequence many of the segments of chromosomal DNA of mice and humans coding for antibody chains. The results of these studies have provided an elegant explanation for the generation of proteins with great diversity in certain regions and constancy in other regions.

ANTIBODY DIVERSITY: GENOME REARRANGEMENTS DURING B LYMPHOCYTE DIFFERENTIATION

Very simply, the ***genetic information coding for antibody chains is stored in bits and pieces, and these bits and pieces are put together in the appropriate sequences by genome rearrangements occurring during the development of the antibody-producing cells (called B lymphocytes) of the body.*** Each B lymphocyte produces only a single type of antibody, that is, all

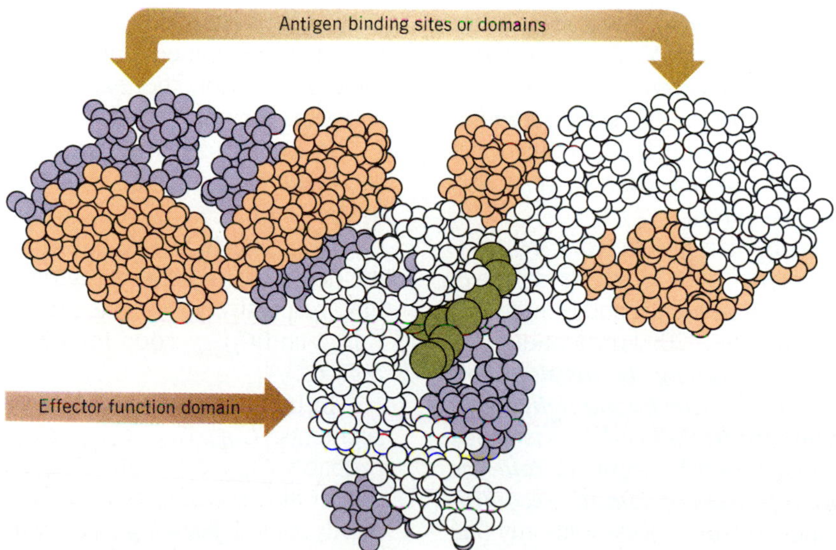

Figure 16.3 Space-filling model of antibody structure showing the three functional domains of immunoglobulins. The two heavy chains are shown in white and blue, respectively. The two light chains are shown in orange. The olive colored components are the associated carbohydrate moieties; their location and structure will vary depending on the immunoglobulin (Ig) class. The structure shown is for a human IgG molecule. (After E. W. Silverton, M. A. Navia, and D. R. Davies, *Proc. Natl. Acad. Sci. U.S.A.* 74: 5140–5144, 1977.)

the antibodies produced by a given B lymphocyte have the same antigen-binding specificity.

Each antibody chain is synthesized using information stored in several different "genes" of "gene segments." (Some geneticists call these DNA sequences "genes," and others prefer to call them "gene segments." We will call them gene segments in the following discussion.) Note that the classical concept of one gene–one polypeptide is not adequate, at least in its simplest form, to explain gene–antibody relationships.

Kappa Light Chains

Synthesis of the kappa light chain is controlled by three different gene segments: (1) a V_κ gene segment, coding for the N-terminal 95 amino acids of the variable region; (2) a J_κ gene segment (J for ***joining segment***), coding for the last (constant region-proximal) 13 amino acids of the variable region; and (3) a C_κ gene segment, coding for the C-terminal constant region. A fourth gene segment, the L_κ segment, codes for an N-terminal ***hydrophobic leader sequence*** 17–20 amino acids long, which is essential for the transport of the antibody chain through the cell membrane. The leader sequence is cleaved off the chain as it passes through the membrane, and thus is not part of the final antibody.

The arrangement of the kappa chain gene segments in germ line cells (in fact, in all cells ***not*** producing antibodies) is shown in Fig. 16.4. In mice and humans, all the kappa chain gene segments are located on the same chromosome (chromosome 2 in humans). The same is true for the lambda gene segments (chromosome 22 in humans) and the heavy chain gene segments (chromosome 14 in humans). ***There are a large number, probably about 300, of V_κ gene segments, each with a nearby L_κ gene segment*** (Fig. 16.4*a*). On the other hand, ***there is only one C_κ gene segment. Five J_κ gene segments*** (one of which is nonfunctional in the mouse) ***are located between the V_κ gene segments and the C_κ gene segment.***

In germ line cells, the five J_κ segments are separated from the V_κ segments by a long (we don't yet know how long) noncoding sequence and from the C_κ segment by an approximately 2000-nucleotide-pair-long noncoding sequence. ***During the development of a B lymphocyte, the particular kappa light chain gene that will be expressed in that cell is assembled from one L_κ–v_κ segment, one J_κ segment, and the single C_κ segment by a process of somatic recombination*** (Figs. 16.4*a* and *b*). This process joins any one of the approximately 300 L_κ–V_κ segments with any one of the five J_κ segments, with the deletion of all intervening DNA (Fig. 16.4*b*). It yields a fused $V_\kappa J_\kappa$ gene segment coding for the entire variable region of the kappa chain. The noncoding sequence between the J_κ gene-segment cluster and the C_κ gene segment, and the C_κ-proximal J_κ segments, if any, remain between the fused $V_\kappa J_\kappa$ segment and the C_κ segment in the differentiated B lymphocytes (Fig. 16.4*b*). This entire DNA sequence (L_κ–$V_\kappa J_\kappa$–noncoding–C_κ) is transcribed (Fig. 16.4*c*) and the noncoding sequences are removed during RNA processing (Figs. 16.4*c* and *d*), just like the noncoding sequences or introns of any other eukaryotic gene.

Lambda Light Chains

Lambda light chain genes are also assembled from separate segments during B lymphocyte development. The major difference is that each J_λ gene segment comes with its own C_λ gene segment, that is, the genome rearrangements required for lambda chain synthesis join L_λ–V_λ segments to J_λ–C_λ segments. Mice have only four J_λ–C_λ gene segments, whereas humans have six. This correlates with the fact that only 5 percent of the antibodies of mice are of the lambda type, whereas 40 percent of the antibodies of humans have lambda light chains.

Heavy Chains

The genetic information coding for antibody heavy chains is organized into L_H–V_H, J_H, and C_H gene segments analogous to those for kappa light chains; but ***there is one additional gene segment,*** called D for diversity, ***that codes for 2–13 amino acids of the variable region.*** The variable region of the heavy chain is thus encoded in three separate gene segments that must be joined during B lymphocyte development. In addition, there are from one to four C_H gene segments for each Ig class.

In the mouse, there are a total of ***eight C_H gene segments,*** all functional, arranged on the chromosome in the sequence $C_{H\mu}$, $C_{H\delta}$, $C_{H\gamma 3}$, $C_{H\gamma 1}$, $C_{H\gamma 2b}$, $C_{H\gamma 2a}$, $C_{H\epsilon}$, $C_{H\alpha}$ (Fig. 16.5*a*). $C_{H\mu}$, $C_{H\delta}$, $C_{H\epsilon}$, and $C_{H\alpha}$ code for the heavy chain constant regions of IgM, IgD, IgE, and IgA, respectively. Four gene segments, $C_{H\gamma 3}$, $C_{H\gamma 1}$, $C_{H\gamma 2b}$, and $C_{H\gamma 2a}$, code for IgG heavy chain constant regions.

In humans, there are 9 or 10 functional C_H gene segments: $C_{H\mu}$, $C_{H\delta}$, $C_{H\gamma 1}$, $C_{H\gamma 2}$, $C_{H\gamma 3}$, $C_{H\gamma 4}$, $C_{H\epsilon 1}$, probably $C_{H\epsilon 2}$, $C_{H\alpha 1}$, and $C_{H\alpha 2}$. The human C_H gene cluster also contains two nonfunctional "genes," commonly called ***pseudogenes,*** with very similar structures. Pseudogenes are partial duplicates of structural genes that have incorporated sufficient changes that they are not biologically active and usually are not transcribed. Pseudogenes are turning out to be quite common in eukaryotes.

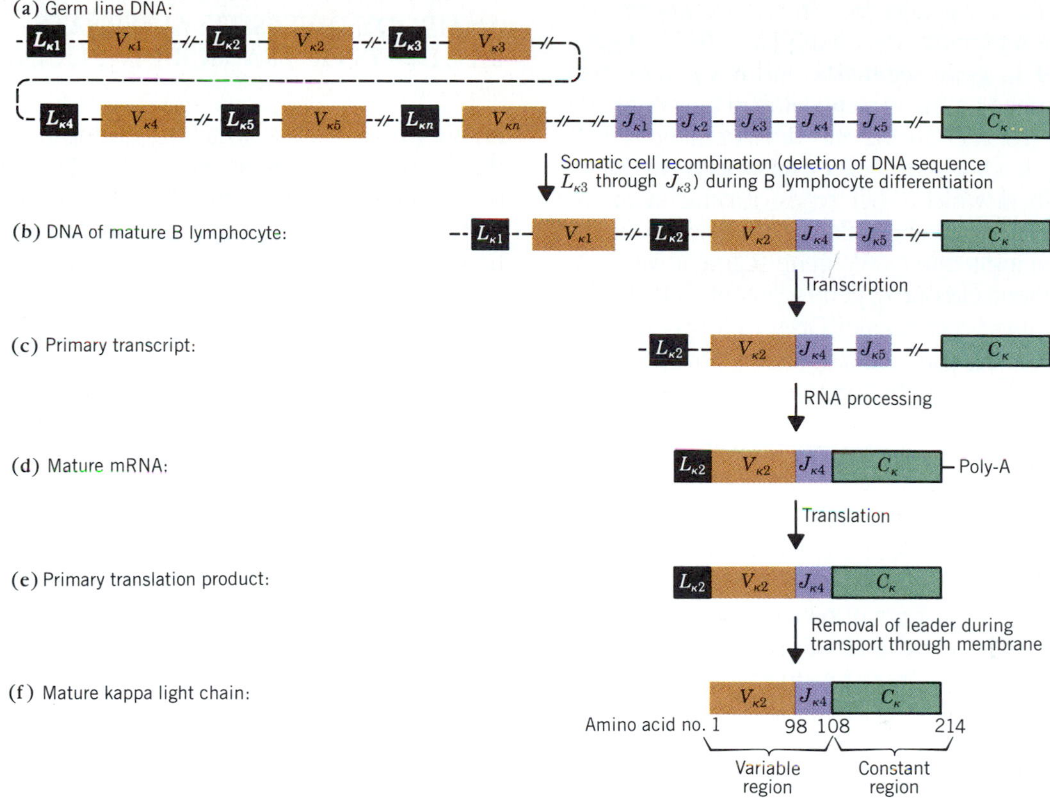

Figure 16.4 The genetic control of human antibody kappa light chains. See the text for a description of the processes involved. (After P. Leder, *Sci. Amer.* 246(5): 102–115, 1982.)

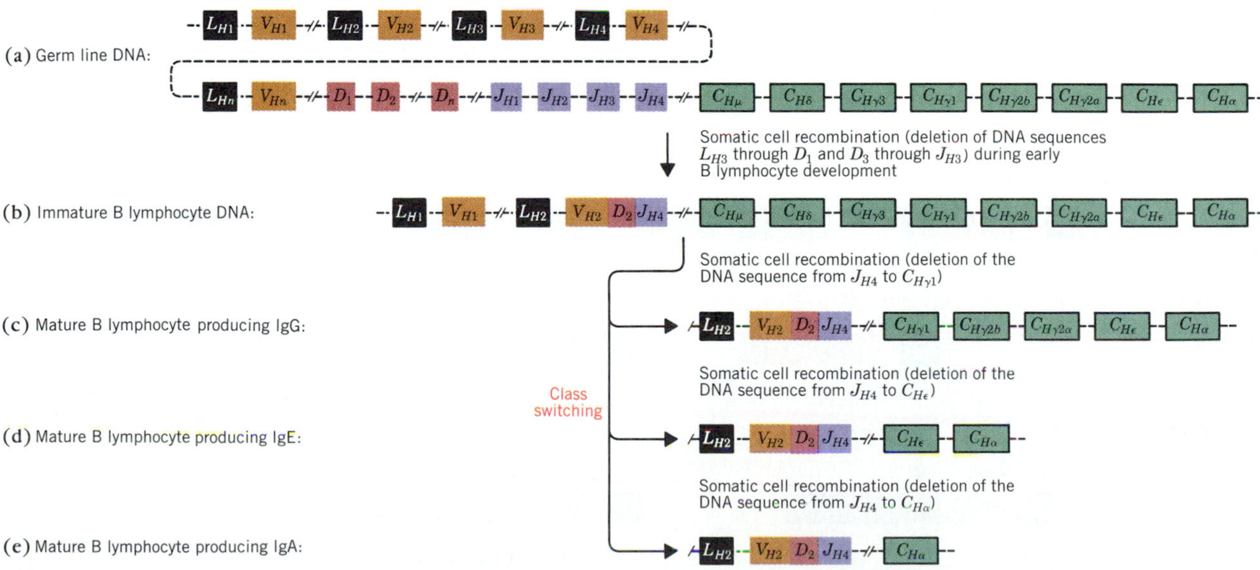

Figure 16.5 The genetic control of mouse antibody heavy chains. $C_{H\mu}$, $C_{H\delta}$, $C_{H\epsilon}$, and $C_{H\alpha}$ encode the heavy chain constant regions of IgM, IgD, IgE, and IgA, respectively. $C_{H\gamma 3}$, $C_{H\gamma 1}$, $C_{H\gamma 2b}$, and $C_{H\gamma 2a}$ encode the heavy chain constant regions of four closely related forms of IgG. See the text for a description of the processes involved. (After M. M. Davis, S. K. Kim, and L. Hood, *Cell* 22: 1–2, 1980.)

In mouse germ line cells, there are ***about 300*** L_H–V_H ***gene segments***, something like ***10–50*** D ***gene segments, 4*** J_H ***gene segments,*** and ***8*** C_H ***gene segments***, arranged on the chromosome in the preceding order (Fig. 16.5*a*). During the development of a B lymphocyte from a stem cell (a mitotically active somatic cell from which other types of cells "stem" or arise by cell division and differentiation), somatic recombination joins one L_H–V_H gene segment with one D gene segment and one J_H gene segment, deleting the two intervening sequences of DNA, to form one continuous DNA sequence (V_HDJ_H) that codes for the entire heavy chain variable region (Figs. 16.5*a* and *b*).

Class Switching

At the time that antibody synthesis begins in the developing B lymphocyte, all the C_H gene segments are still present, separated from the newly formed L_H–V_HDJ_H gene segment by a short noncoding sequence (Fig. 16.5*b*). At this stage, all antibodies synthesized have IgM heavy chains ($C_{H\mu}$ gene-products). If an antigen is recognized and bound to an antibody on the surface of a developing B lymphocyte, however, that cell is stimulated to differentiate into a mature B lymphocyte. During this differentiation, some B lymphocytes will switch from producing antibodies of class IgM to producing antibodies of another class. This phenomenon, called ***class switching***, often involves further genome rearrangements during which the C_H gene segments closest to the previously joined L_H–V_HDJ_H gene segments are deleted (Figs. 16.5*c*–*e*). The class of antibodies produced after class switching is determined by which gene is brought into the closest proximity with the L_H–V_HDJ_H gene segment, as illustrated in Figs. 16.5*c*–*e*).

ANTIBODY DIVERSITY: ALTERNATE PATHWAYS OF TRANSCRIPT SPLICING

Another type of class switching during B lymphocyte differentiation occurs at the level of RNA processing ("splicing"). Certain mature B lymphocytes produce both IgM and IgD antibodies. It should be emphasized, however, that these antibodies differ only in their effector function domains; they have identical antigen-binding domains, specified by the same $V_\kappa J_\kappa$ (or $V_\lambda J_\lambda$) and V_HDJ_H fused gene segments. In these cells, a primary transcript that extends through both the $C_{H\mu}$ and $C_{H\delta}$ gene segments is synthesized. During processing, the V_HDJ_H transcript sequence may be spliced to either the $C_{H\mu}$ sequence or the $C_{H\delta}$ sequence, such that both types of heavy chain are synthesized in the same cell (Fig. 16.6).

A further complexity observed in antibody synthesis is the ***sequential production of membrane-bound and secreted forms of a given antibody***. The first antibodies to appear in developing B lymphocytes are membrane-bound IgM molecules. Subsequently, these cells switch to the production of a secreted form of IgM. These two forms of IgM differ only in the C-terminal portions of their heavy chains. The heavy chain of the membrane-bound form is 21 amino acids longer than that of the secreted form. The ***membrane-bound heavy chain*** has a ***41-amino-acid-long hydrophobic sequence*** at the C terminus that is probably responsible for anchoring it to the cell surface. This hydrophobic sequence is replaced by a ***20-amino-acid-long hydrophilic sequence*** in the ***secreted form***

The coding sequences (exons) of the C_H gene segments are interrupted by noncoding sequences (introns) just like those of many other eukaryotic

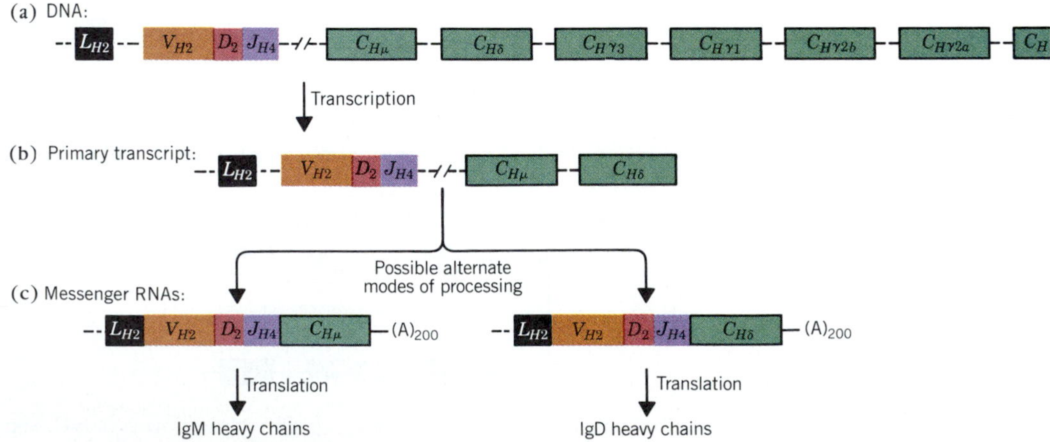

Figure 16.6 Mechanism for the production of IgM and IgD antibody heavy chains in the same B lymphocyte as a result of alternate modes of splicing the heavy chain transcript.

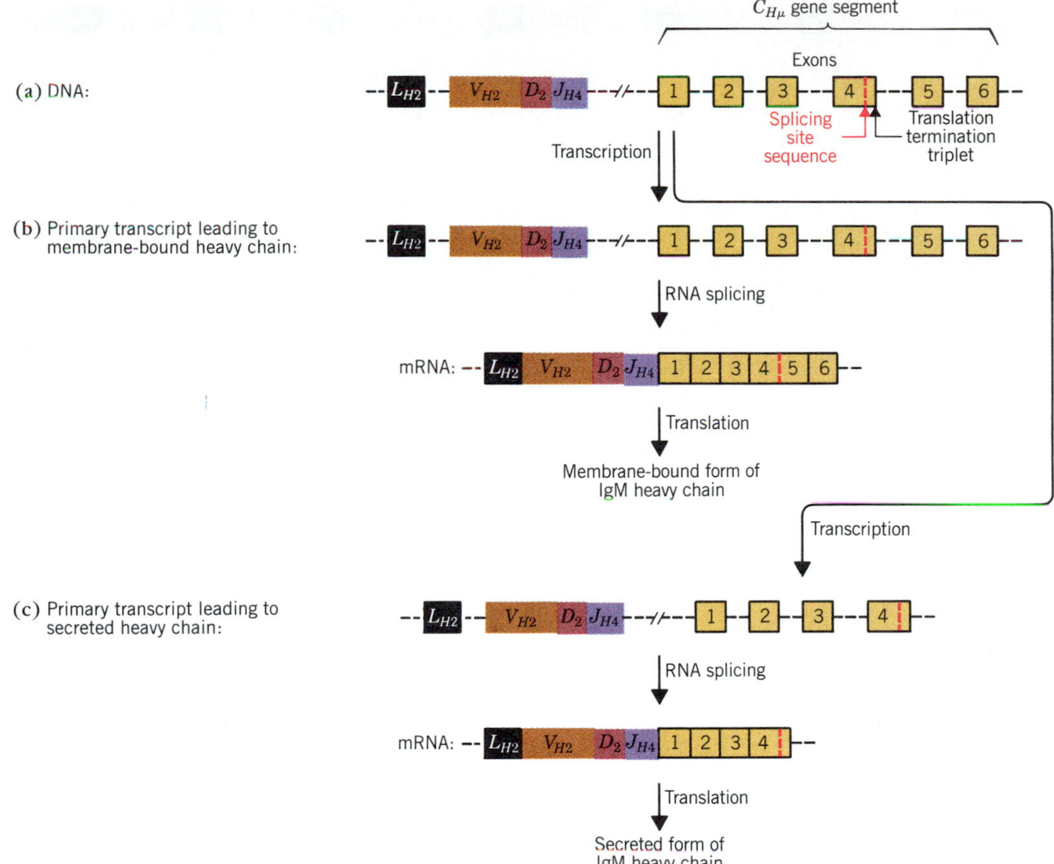

Figure 16.7 Genetic control of the membrane-bound and secreted forms of IgM. Only the heavy chain genes and gene-products are shown since both forms have identical light chains. (After P. Leder, *Sci. Amer.* 246(5): 102–115, 1982.)

genes (Chapter 13, pp. 356–359). The C_H gene segments contain four to six exons and three to five introns (Fig. 16.7). In membrane-bound antibodies, the heavy chain constant regions are produced by splicing all six exons together (Figs. 16.7*a* and *b*). The last two exons code for the hydrophobic tails of the membrane-bound heavy chains. During synthesis of the membrane-bound form, the fifth C_H exon is spliced to a site 20 codons from the end of the fourth exon (Fig. 16.7*b*), thus changing the amino acid sequence of this portion of the heavy chain constant region. In secreted antibodies, the heavy chain constant regions are therefore the product of the first four exons (Fig. 16.7*c*).

The use of alternate pathways of transcription and RNA processing to synthesize membrane-bound and secreted forms has been firmly established for the IgM class of antibodies. Recent evidence suggests that similar alternate pathways of transcription and splicing are responsible for the production of the membrane-bound and secreted forms of the other classes of immunoglobulins as well.

SIGNAL SEQUENCES GOVERN GENOME REARRANGEMENTS

How are the genome rearrangements that occur during B lymphocyte development regulated? What controls the somatic recombination events such that a *V* gene segment is joined to a *J* segment and not to another *V* segment or directly to a *C* segment? Several long segments of chromosomal DNA carrying clusters of *V* gene segments, *D* gene segments, and *J* gene segments of both mice and humans have now been sequenced, and the resulting nucleotide-pair sequences suggest the presence of specific *V–J, V–D,* and *D–J* joining signals. The same ***signal sequences*** are found adjacent to all *V* gene segments. Similarly, all *J* gene segments have identical signal sequences located adjacent to their coding sequences; however, their signal sequence is different from that adjacent to *V* gene segments. Likewise, *D* and *C* gene segments have their own adjacent signal sequences.

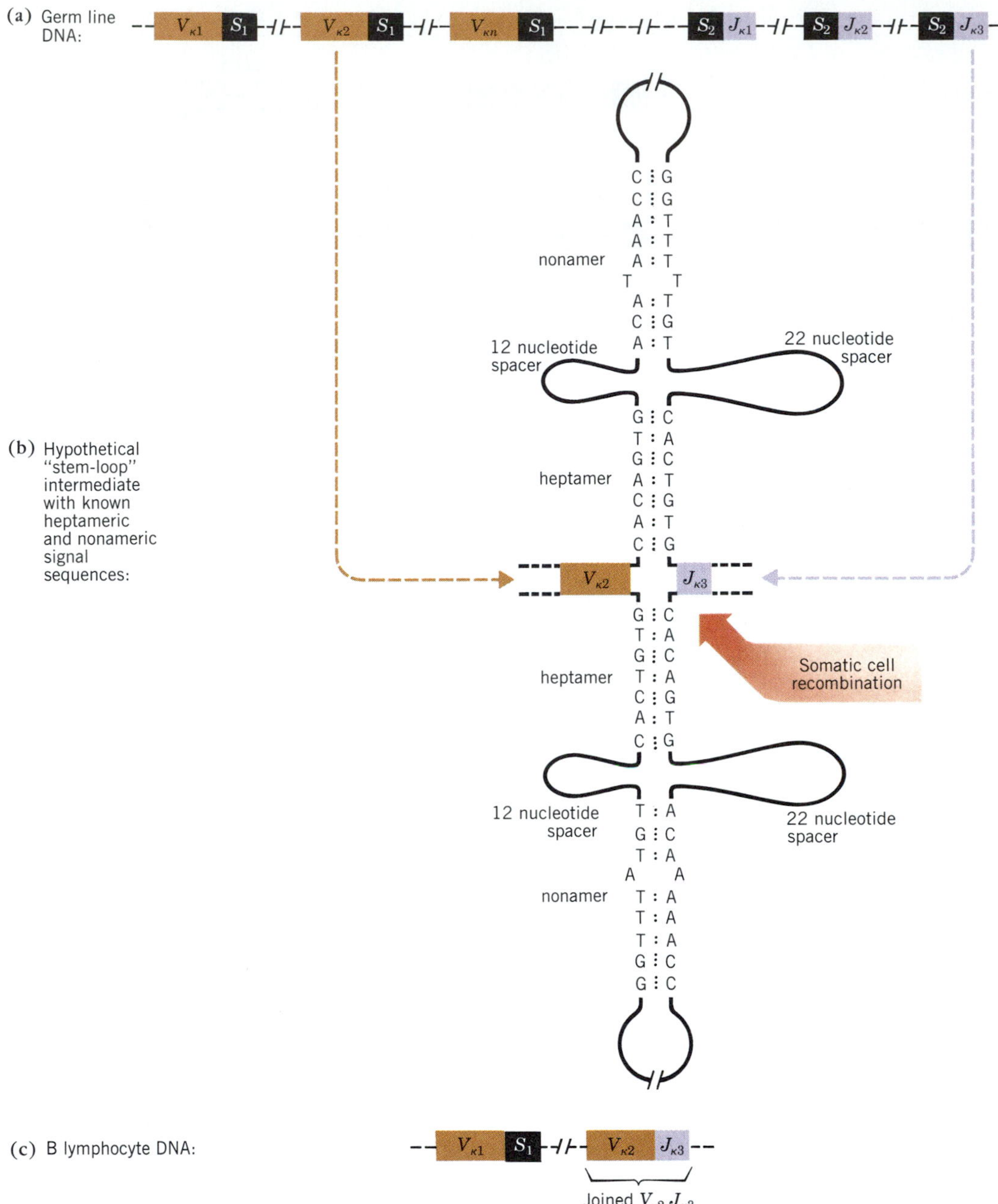

Figure 16.8 Signal sequences and their proposed role in $V_\kappa J_\kappa$ joining. The germ line and B lymphocyte DNA arrangements shown in (a) and (c), respectively, as well as the heptamer and nonamer sequences and spacer sizes shown in (b), are known. The "stem and loop" structure shown in (b) is a hypothetical intermediate.

The signal sequences controlling *V–J, V–D,* and *D–J* joining contain 7-base-pair (heptamer)- and 9-base-pair (nonamer)-long sequences separated by spacers of different, but specific lengths. For V_κ–J_κ joining, the spacer in the V_κ signal sequence is 12 nucleotide-pairs long, whereas that in the J_κ signal sequence is 22 nucleotide-pairs long. The heptamer and nonamer sequences located "after" (to the right as drawn in Figs. 16.4 and 16.8) the V_κ gene segments are complementary (with the exception of one base-pair) to those "preceding" (to the left as drawn in Figs. 16.4 and 16.8) the J_κ gene segments. These signal sequences have the potential to form "stem and loop" structures as diagrammed in Fig. 16.8, thus bringing the V_κ and J_κ gene segments into juxtaposition for joining. Apparently, joining will occur only when one signal sequence contains a 12-base-pair spacer and the other contains a 22-base-pair spacer. This requirement would supposedly be enforced by the specific protein(s) mediating the joining process. Very similar signal sequences appear to control V_H–D and D–J_H joining, whereas somewhat different signal sequences mediate class switching (see P. Leder, *Sci. Amer.* 246(5): 102–115, 1982).

ANTIBODY DIVERSITY: VARIABLE JOINING SITES AND SOMATIC MUTATIONS

A comparison of the diversity of amino acid sequences present in antibody molecules with that predicted from the sequences of gene segments that encode these antibodies revealed that there is more variation in amino acid sequences at the V–J junctions than is predicted by the nucleotide sequences. Subsequent studies showed that ***much of this additional diversity could be explained by variation in the exact site of recombination during the V–J joining events.*** An example of the use of alternate sites of joining of V_κ and J_κ gene segments in the mouse is illustrated in Fig. 16.9. During the joining of gene segments $V_{\kappa 41}$ and J_5, recombination has been shown to occur between four adjacent nucleotide positions at the junction sites. As shown in Fig. 16.9*d*, these recombination events produce four different nucleotide sequences that encode three distinct amino acids at position 96 in the mouse kappa light chain. Since amino acid 96 occurs in a region of the antibody chain that is involved in antigen binding, alternate V–J joining events of this type undoubtedly contribute significantly to the great diversity of antibody specificity that is observed in vertebrates. Similar alternate joining events have been documented for V_λ–J_λ and V_H–D–J_H joining reactions. Thus, ***the use of alternate sites of recombination during the joining events*** that are involved in the assembly of mature antibody genes ***provides an additional mechanism for generating antibody diversity***.

Despite the vast array of antibody diversity produced by (1) the joining of large families of V, D, and J gene segments and (2) the use of alternate positions of recombination during the joining reactions, considerable data demonstrate that still another mechanism must be involved in the generation of antibody diversity. This has been established by comparing (1) the nucleotide-pair sequences of expressed genes with the sequences of germ line gene segments and (2) the actual amino acid sequences of antibody chains with the amino acid sequences predicted from the nucleotide sequences of the genes. For example, when the actual amino acid sequences of different mouse λ_1 chains were compared with the predicted amino acid sequences (based on the nucleotide-pair sequences of λ_1 gene segments) of λ_1 light chains, differences were found in the variable regions away from the junction sites. Similar observations have been made in studies of heavy chain variable regions. In essentially all cases, the changes have resulted from single nucleotide-pair substitutions. Such substitutions may represent 1–2 percent of the nucleotide-pairs of the gene segments encoding the variable regions of antibodies. These

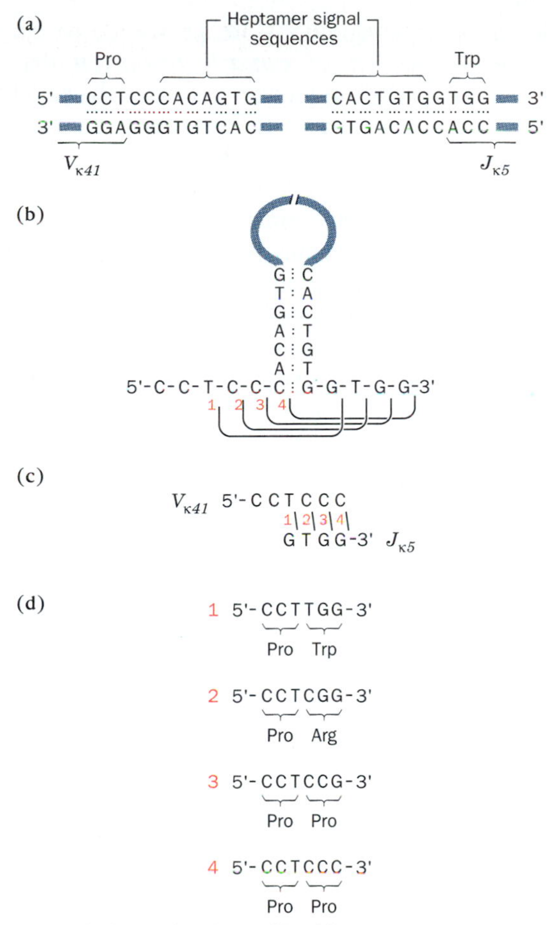

Figure 16.9 Diversity at the $V_\kappa J_\kappa$ junction produced by variation in the exact position of the joining reaction. (a) Sequences of $V_{\kappa 41}$ and $J_{\kappa 5}$ at the sites that will form the junction. The adjacent heptanucleotide signal sequences that are believed to play a central role in bringing the V_κ and J_κ gene segments into juxtaposition are also shown. (b) Putative role of the heptanucleotide signal sequences of $V_{\kappa 41}$ and $J_{\kappa 5}$ in the recombinational joining process (see Fig. 16.8 for a more complete diagram of this event). (c) Four positions of joining that have been documented for the mouse $V_{\kappa 41}$ and $J_{\kappa 5}$ gene segments. (d) The products of the four alternate joining events shown in (c) are listed along with the amino acids that these sequences predict for amino acid residues number 95 and 96 in the resulting kappa light chain. (Based on the results of E. E. Max, J. G. Seidman, and P. Leder, *Proc. Natl. Acad. Sci. U.S.A.* 76: 3450–3454, 1979.)

nucleotide-pair substitutions are presumed to occur by some mechanism of ***somatic mutation*** that is restricted to the DNA sequences encoding the variable regions of antibody chains. Because these changes in the variable segments of antibody genes occur at such a high frequency, the process by which they occur is often called ***somatic hypermutation***. The mechanism by which somatic hypermutation occurs is unknown.

Somatic hypermutation of regions of antibody genes that encode antigen-binding sites may be of great value to the organism. Without this mechanism for generating antibody diversity, the range of available antibody specificity would be fixed in terms of the sequences present in the genome at birth and the combinations that could be produced by the various levels of gene segment joining reactions. Viruses and other pathogens are constantly evolving and producing new variants with new antigenic determinants. To provide an adequate defense against the changing antigenic composition of these viruses and other components of the environment, the immune system must also be capable of rapidly responding to these changes. What better way to provide this safeguard than to endow antibody genes with their own mechanism for rapid adaptation to new antigens that might evolve in the future—somatic hypermutation?

HOW MANY COMBINATIONS?

One can readily see that a large amount of diversity can be generated by the joining of antibody gene segments as just described. For example, consider the number of different kappa light chains possible in humans: 300 V_κ gene segments $\times$ 5 J_κ segments $\approx$1500 fused $V_\kappa J_\kappa$ gene segments. The heavy chain variable region provides even greater diversity because of the multiple D gene segments. If there are 300 V_H gene segments, 25 D gene segments, and 6 J_H gene segments in human germ line cells, 45,000 different heavy chain variable regions could be assembled. Using these estimates, 67,500,000 different antigen-binding sites could be produced using just kappa light chains. Lambda light chains produce another level of diversity.

Clearly, these antibody gene-segment fusions provide for a vast amount of antibody diversity. We now know, however, that further diversity is generated in two additional ways: (1) somatic mutation and (2) variability in the sites at which V–J, V–D, and D–J joining events occur. In total, the possible range of antibody diversity seems almost limitless.

REGULATION OF TRANSCRIPTION: A TISSUE-SPECIFIC ENHANCER

It has been known for several years that germ line antibody genes are not transcribed or are transcribed at very low levels. Yet, in antibody-producing B lymphocytes, 10 to 20 percent of the mRNA molecules are antibody gene transcripts. What, then, is responsible for the activation of transcription of antibody genes that undergo rearrangements and become active? In the case of the heavy chain genes, the answer appears to be that ***the rearrangement process brings the promoters located upstream of L_H–V_H gene segments into the range of influence of a strong enhancer*** element located in the intron between the J_H gene segments and the $C_{H\mu}$ gene segment (Fig. 16.10). Each L_H–V_H gene segment contains an upstream promoter. However, prior to the genomic rearrangement events that lead to heavy chain synthesis, this enhancer is over 100,000 nucleotide-pairs away from the closest L_H–V_H promoter (Fig. 16.10, top). Presumably, this enhancer

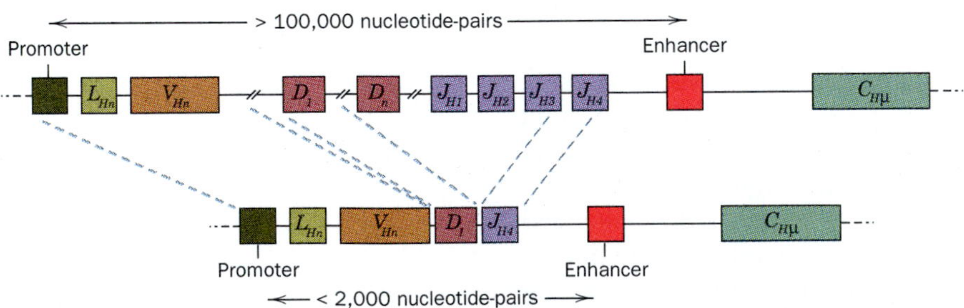

Figure 16.10 Position of the tissue-specific enhancer in the J segment-proximal intron of the heavy chain gene cluster of the mouse relative to the position of the promoter in germ line DNA (top) and after a genomic rearrangement leading to heavy chain synthesis (bottom). Prior to the rearrangement, the enhancer is over 100,000 nucleotide-pairs from the promoter, presumably too distant to have any effect on transcription initiated at the promoter. After the rearrangement, the enhancer is less than 2,000 nucleotide-pairs from the promoter, close enough to activate transcription of the rearranged gene. (After J. Banerji, L. Olson, and W. Schaffner, *Cell* 33: 729–740, 1983.)

cannot activate transcription from promoters that are located that far away. However, rearrangement events occurring during B cell differentiation (see Fig. 16.5) move the promoter of the closest L_H–V_H gene segment to within less than 2000 nucleotide-pairs from the enhancer (Fig. 16.10, bottom). The enhancer can now activate transcription from the promoter located upstream from the L_H–V_H gene segment. The ***enhancer involved in the activation of heavy chain synthesis is tissue specific***; it activates transcription only in lymphocytes and has no effect in cells derived from other tissues. Presumably, the activation process requires the presence of a transcriptional-activating factor that is synthesized in lymphocytes, but not in other types of cells.

A similar enhancer element has been found in the intron between the light chain J_κ gene segment cluster and the C_κ coding sequence. Thus, it seems likely that the movement of antibody gene promoters into the range of influence of tissue-specific enhancers may be a general mechanism of activation of antibody genes during the differentiation of B lymphocytes.

CLONAL SELECTION

Up to this point, we have avoided the question of how an organism initiates the synthesis of antibodies specific to antigens that it has not previously encountered. This is nicely explained by the ***clonal selection theory***. Recall that ***all the antibodies produced by a single B lymphocyte have the same antigen-binding specificity***. But ***different cells*** in a ***population*** of B lymphocytes will have undergone ***different genome rearrangements leading to the production of antibodies with different specificities***. Thus, the population of B lymphocytes in a human or a mouse will be producing a very large variety of antibodies. The clonal selection theory states that the ***binding of a particular foreign antigen to an antibody on the surface of a B lymphocyte stimulates that cell to divide***, producing large numbers of this particular B lymphocyte (a "clone" of identical cells) and thus large amounts of the particular antibody that recognizes the foreign antigen (Fig. 16.11).

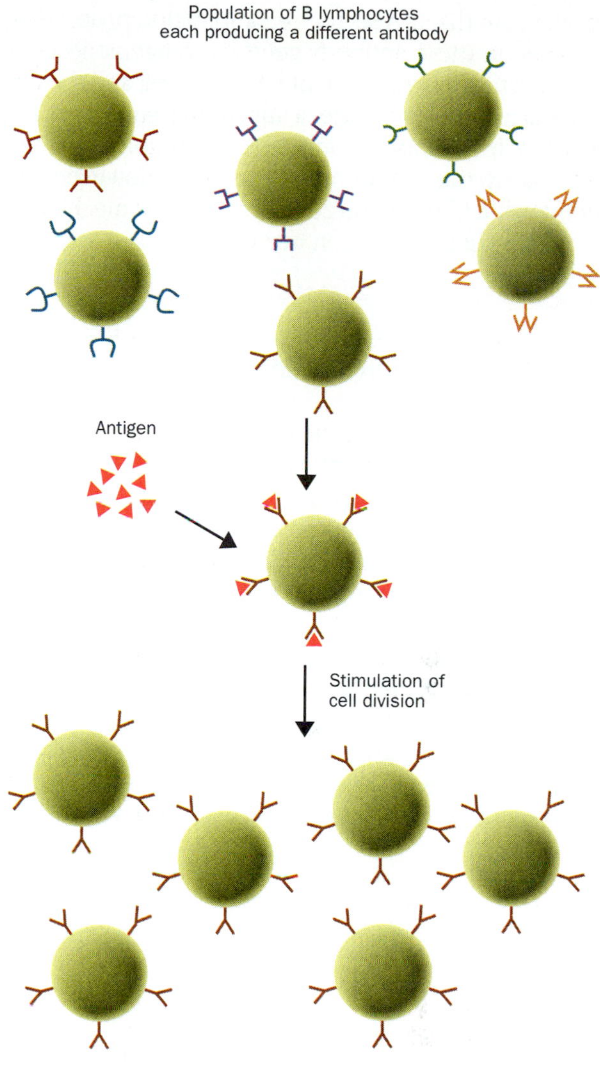

Figure 16.11 Schematic diagram of the role of clonal selection in the immune response. Genomic rearrangements, the use of alternate joining sites, and somatic hypermutation all contribute to the development of a population of B lymphocytes that collectively produce a vast variety of different antibodies (although each cell produces only a single type of antibody). When a particular antibody binds to an antigen, the cell producing that particular antibody is stimulated to divide. After a series of cell divisions, a clonal population of B cells is produced, with all cells producing the same antibody.

ALLELIC EXCLUSION

Consider one final point about the genetic control of antibody synthesis. Each B lymphocyte makes only ***one*** type of antibody. Why? Mammalian cells are diploid; they carry two sets of genetic information coding for each of the antibody chains. But only ***one productive genome rearrangement of light chain coding sequences and one productive genome rearrangement of heavy chain coding sequences occur in each B lymphocyte!*** This phenomenon is called ***allelic exclusion*** because one of the ***"alleles" is excluded from being expressed***. How? Why? At present, we still don't know. Clearly, there must be some type of a feedback mechanism that arrests the recombination process(es) involved in these antibody gene rearrangements once a productive rearrangement has occurred and the cell has started to synthesize a functional antibody. The simplest mechanism would involve inhibition of this process by the mature antibody itself. However, further work will be required to establish the mechanism by which allelic exclusion occurs.

T CELL RECEPTOR VARIABILITY

T lymphocytes mediate the cellular immune response (see Fig. 16.1). The T cells recognize antigens on the surface of cells and kill the cells carrying these antigens. Like the antibodies produced by B lymphocytes, T cells can recognize and destroy cells carrying an amazing variety of antigens. Thus, the T cell response also exhibits a phenomenal degree of specificity. How is this specificity produced? The answer is that ***T cells produce membrane-bound receptors that are very similar to the antibodies*** produced by B lymphocytes. Moreover, ***the diversity of T cell receptor specificity is produced by genome rearrangements analogous to those involved in antibody production*** (see the preceding sections of this chapter). But how do T lymphocytes avoid interacting with free antigens to avoid duplicating the function of B cells in the immune response? As it turns out, T cells must simultaneously recognize both the offending antigen on the cell surface and another protein that occurs only attached to cell surfaces. This second cell-surface protein that the T cell must recognize is the product of one of many genes in the ***major histocompatibility complex (MHC)***. The MHC locus (see the following section) encodes a complex group of proteins that are present on all the cells in the body of a human (or a mouse). Thus, T cells are able to recognize and destroy any cell that is producing a given antigen (e.g., the coat protein of a virus) in any tissue of the body. The interaction of the T cell receptor with the two types of cell-surface antigens (the offending antigen and the histocompatibility antigen) is illustrated in Fig. 16.1.

The T cell receptors are composed of two polypeptide chains, α and β, each encoded by L–V, D, J, and C gene segments just like antibody chains. The α- and β-polypeptides, like antibody chains, contain variable regions that form the antigen-binding sites and constant regions that anchor the receptors on the cell surface (Fig. 16.12*a*). The variable regions of the T cell receptors are encoded by multiple *L–V*, *D*, and *J*

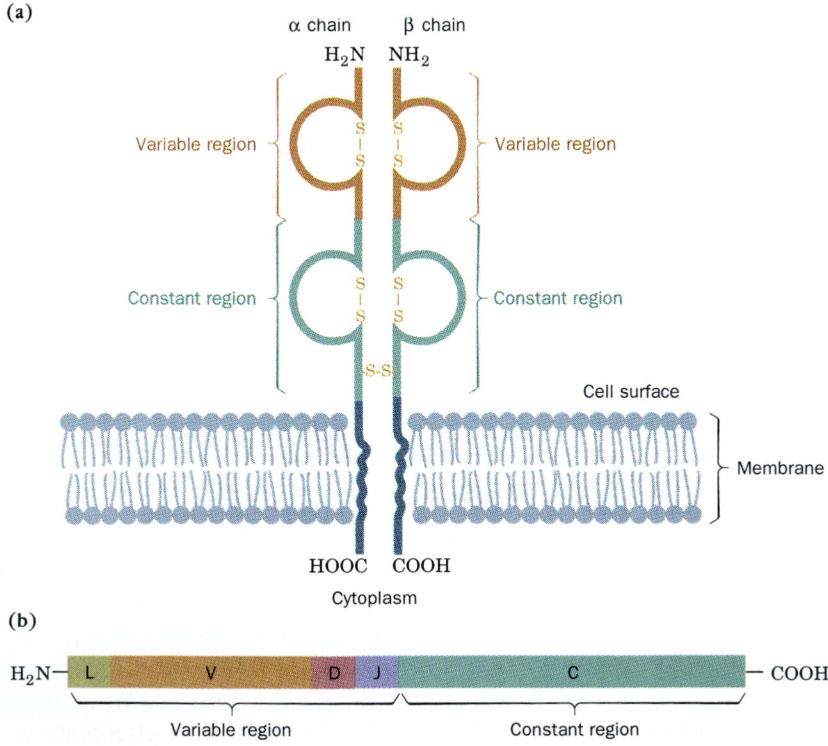

Figure 16.12 Diagrams showing (a) the structure of a T cell receptor anchored in the cell membrane and (b) the regions of the α and β receptor proteins that are encoded by the *L–V*, *D*, *J*, and *C* gene segments, respectively. (After M. K. Davis, Y. Chien, N. R. J. Gascoigne, and S. M. Hedrick, *Immunol. Rev.* 8: 235–258, 1984.)

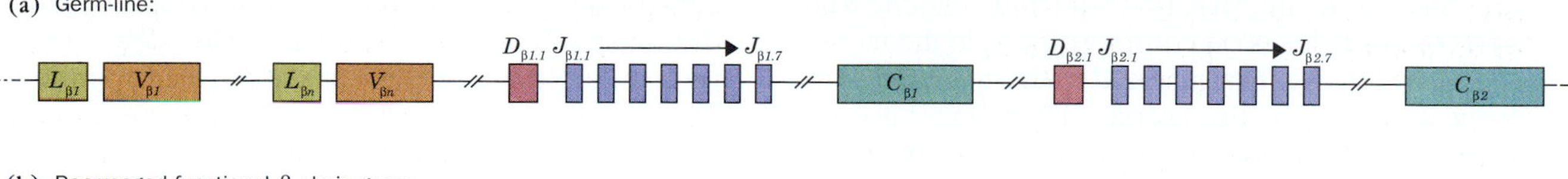

Figure 16.13 Organization of the complex locus in the mouse that encodes T cell β receptor polypeptides. (a) The germ line organization. (b) Structure of an assembled gene in a mature T lymphocyte that is producing a specific β-polypeptide. The arrangement of the receptor β-chain locus in humans is very similar to that in the mouse, however, only 12 of the 14 *J* gene segments are functional in humans. The C_β gene segments each contain three introns; they are not shown in order to minimize complexity. The large intron between the *J* segment and the *C* segment of the assembled gene is removed during processing of the primary transcript just like the three introns within the *C* segment of the gene. The T cell receptor α-polypeptide gene cluster has a similar organization, however, the α-polypeptide gene cluster contains 18 J_α gene segments and only a single C_α sequence. (After A. Winoto, S. Mjolsness, and L. Hood, *Nature* 316: 832–836, 1985.)

gene segments; the constant regions are encoded by a small number of *C* gene segments. The T cell receptor genes are assembled by genomic rearrangements that occur during the differentiation of T lymphocytes from stem cells just as in the case of antibody genes in developing B lymphocytes. Figure 16.12*b* shows the segments of the α and β T cell receptor polypeptides that are encoded by the *L–V*, *D*, *J*, and *C* gene segments. The α and β receptor proteins are encoded by gene segments that are lined up in clusters on chromosomes similar to those encoding antibody chains. In humans, the α and β gene segment clusters are located on chromosomes 14 and 7, respectively. (Another gene segment cluster encodes a third type of T cell receptor polypeptide designated γ that is present on a specific type of T cell. There are several distinct types of T cells that carry out different functions during the immune response.)

The structures of the T cell receptor gene clusters are very similar in humans and in mice. The germ line organization of the gene segments that encode T cell receptor β-polypeptides is shown in Fig. 16.13*a*. The organization of a functional rearranged β-polypeptide gene is shown in Fig. 16.13*b*. Heptamer and nonamer signal sequences that are very similar to those that control antibody gene rearrangements are also present at essentially the same locations in the T cell receptor gene clusters. Their presence in both types of gene clusters suggests that the same mechanism of gene segment joining is employed during rearrangements of both antibody genes and T cell receptor genes. There probably is somewhat less total variability in T cell receptors than in antibodies. T cell receptor variable regions are encoded by only about 30 *V* gene segments, whereas there are about 300 *V* gene segments for both kappa light chains and heavy antibody chains. However, there are more *J* gene segments in the T cell receptor gene clusters. For example, there are 12 functional *J* gene segments for β receptor polypeptides (Fig. 16.13). Moreover, we still do not know whether or not the variable segments of T cell receptor genes undergo somatic hypermutation. In any case, it is clear that ***T cell receptors exhibit a great amount of diversity,*** and that ***this diversity is generated by genome rearrangements during T lymphocyte differentiation*** in a manner analogous to those involved in the production of antibody diversity in B lymphocytes.

As was mentioned earlier, there are several different types of T lymphocytes and they play different roles in the cellular immune response. For an excellent discussion of the different types of T cells and their functions, the reader is referred to Chapter 24, "Immunity," of *Molecular Cell Biology* by J. Darnell, H. Lodish, and D. Baltimore.

MAJOR HISTOCOMPATIBILITY COMPLEX

The immune response in mammals is a very complex process involving a large number of different macromolecules and different cell types. Our discussion to this point has been restricted to the genetic control of the synthesis of antibody chains and T cell receptor proteins. Many of the other components of the immune response, such as the ***transplantation antigens*** that are largely responsible for the rejection of foreign tissues in transplant operations, are controlled by a multigene complex called the ***major histocompatibility complex*** (MHC). In humans, the MHC proteins are encoded by the **HLA** (for ***H****uman* ***L****eukocyte* ***A****nti*gen complex) locus on chromosome 6; in the mouse, the *MHC* locus is designated **H-2** (***H****istocompatibility* locus ***2***) and is on chromosome 17. In both mice and humans, the *MHC* locus is very large ($>2 \times 10^6$ nucleotide-pairs) and contains a large number of genes. Moreover, there is a very large number of distinct alleles for many of these genes such that the probability of any two individuals being identical for all the *MHC* genes is extremely small. The *MHC* genes are said to be ***highly polymorphic*** because of the large number of alleles of individual genes that are usually segregating in a given population.

The *MHC* genes encode three different classes of proteins that are involved in different aspects of the immune response. The structure of the human *MHC* (*HLA*) locus and the relative locations of genes that encode the different classes of histocompatibility antigens are shown in Fig. 16.14. The ***class I genes encode the transplantation antigens*** mentioned. The class I proteins exist as glycoproteins anchored as integral membrane proteins with the antigenic determinants exposed on the outside of cells. They are present on all cells of an organism and permit T lymphocytes to distinguish "self" from "foreign." The MHC class I proteins are the antigens that usually are responsible for the rejection of foreign tissues in tissue and organ transplants. As illustrated in Fig. 16.1, these antigens play a key role in the recognition and destruction of cells carrying foreign antigens by cytotoxic T lymphocytes. A single T cell receptor is believed to recognize both the foreign antigen and the class I histocompatibility antigen during the cytotoxic T cell immune response.

The **MHC *class II genes encode polypeptides that are located primarily on the surfaces of B lymphocytes and macrophages.*** MHC class II proteins provide a special type of T lymphocyte called the "T helper cell" with the capacity for self-recognition and facilitate communication between the different types of cells involved in the immune response. Finally, the **MHC *class III genes encode complement proteins*** that interact with antibody–antigen complexes and induce cell lysis.

The MHC class I and class II antigens are anchored in the cell membrane and have structures very similar

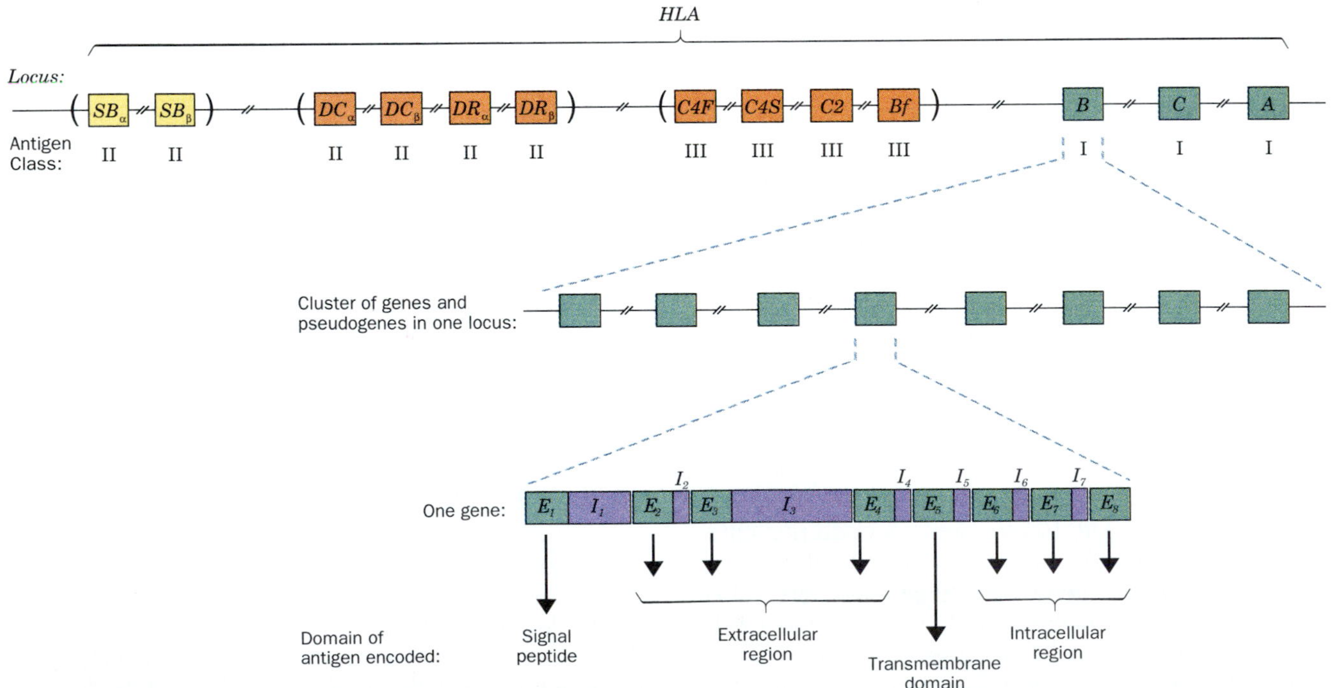

Figure 16.14 Organization of the major histocompatibility complex (*HLA*) on human chromosome 6. The relative positions of mapped loci within this huge gene complex are shown at the top. The order of loci enclosed in parentheses is uncertain. The entire *HLA* complex is over 2×10^6 nucleotide-pairs in length. The class of histocompatibility antigen encoded by genes in each locus is indicated below the map. Note that each of the loci within the *HLA* complex is itself a complex locus containing several genes and pseudogenes (center). The structure of a typical class I gene is shown at the bottom. Note that different exons encode different functional domains of the polypeptide gene-product. (After L. E. Hood, I. L. Weissman, W. B. Wood, and J. H. Wilson, *Immunology*, 2nd ed., Benjamin/Cummings, Menlo Park, California, 1984.)

to the structures of T cell receptors (see Fig. 16.12). However, the diversity of MHC antigens is much less than that of antibodies and T cell receptors, and, so far as is known, no genomic rearrangement is involved in the genetic control of MHC antigen diversity. Instead, the observed diversity results from the presence of a large number of highly polymorphic *MHC* genes.

SUMMARY

The immune system of verebrate animals protects them from invasion by pathogenic microorganisms and other foreign substances. The immune response is very complex. It involves three different types of white blood cells: (1) ***B lymphocytes*** that ***produce antibodies***, (2) ***T lymphocytes*** that produce ***T cell receptors*** and use them to seek out and ***destroy cells carrying foreign antigens***, and (3) ***macrophages*** that carry out ***phagocytosis of antibody–antigen complexes***, viruses, bacteria, and so on. In addition, T cell recognition of cells that are producing foreign antigens requires the presence of specific ***histocompatibility antigens*** encoded by the ***major histocompatibility complex*** (the HLA ***locus*** in humans).

A remarkable feature of the immune response is the ***seemingly limitless variety of antibodies and T cell receptor proteins that can be synthesized in response to antigens that the animal has not previously encountered***. We now understand how this phenomenal diversity of antibody specificity and T cell receptor specificity is produced. The ***genetic information coding for antibody chains is stored in several sets of gene segments***, and the segments are put together in the appropriate sequences by ***genome rearrangements that occur during the development of the antibody-producing cells (the B lymphocytes)***. The same mechanism is used to generate variability in the antigen specificity of T cell receptors. The germ line DNA contains the genes encoding T cell receptor proteins stored in the form of sets of gene segments that are assembled into functional genes during the differentiation of T lymphocytes from stem cells.

Antibodies are composed of ***two light chains*** (either ***kappa*** or ***lambda***) and ***two heavy chains***. Each antibody chain contains a ***variable region that forms the antigen-binding site*** and a ***constant region that anchors the antibody to the cell surface*** in the case of membrane-bound antibodies. The ***variable regions of antibodies are encoded by either two (light chains) or three (heavy chains) gene segments*** that are present in the ***germ line DNA in multiple copies***. The constant regions of antibodies are encoded by gene segments that are present in the genome in only one or a few copies per cell. ***Kappa light chain genes are assembled by genomic rearrangements that occur during B cell differentiation from about 300*** L_κ–V_κ ***gene segments, 5*** J_κ ***gene segments, and one*** C_κ ***gene segment. Functional heavy chain genes are similarly assembled from about 300*** L_H–V_H ***gene segments, 10*** D ***gene segments, 4*** J_H ***gene segments, and 8*** C_H ***gene segments.*** The ***use of alternate sites of gene segment joining*** during the genomic rearrangement events and ***somatic hypermutation*** within the variable region gene segments contribute to the production of additional antibody diversity.

Antibodies are of five different immunoglobulin classes: ***IgM, IgD, IgG, IgE,*** and ***IgA***. The class of antibody is determined by its ***heavy chain constant region***, which in turn is determined by the C ***gene segment*** that was expressed during its synthesis. ***Class switching*** occurs when a B lymphocyte stops synthesizing one class of antibody and begins synthesizing another class of antibody with the same antigen specificity. Class switching involves the expression of the same variable region gene segments but a different heavy chain constant region gene segment. ***Class switching most often occurs by further genomic rearrangements*** similar to those that resulted in the synthesis of the original antibody chains. However, ***class switching can also occur by alternate patterns of transcript splicing***.

Germ line antibody genes are transcribed at a very low rates or not at all. The productive rearrangement of antibody gene segments during B cell differentiation apparently ***activates transcription of the assembled gene by bringing the promoter located upstream from the*** L–V ***gene segment into the range of influence of a tissue-specific enhancer*** located in the intron between the *J* segment cluster and the *C* gene segment.

Clonal selection nicely explains the production of large numbers of B lymphocytes all synthesizing the same antibody with specificity for binding a particular antigen present in the circulatory system of an animal. Only one productive light chain genomic rearrangement and one productive heavy chain genomic rearrangement can occur in a given B lymphocyte. This is known as ***allelic exclusion*** and, clearly, must be controlled by some type of feedback mechanism. However, the molecular basis of allelic exclusion remains unknown.

The class I antigens produced by the major

*histocompatibility complex genes (*HLA *genes* in humans) usually *are responsible for tissue rejections during tissue- and organ-transplant operations.* The genes encoding these class I histocompatibility antigens are *highly polymorphic,* thus different individuals (other than identical twins) are unlikely to carry histocompatibility antigens that are all identical.

REFERENCES

ALBERTS, B., D. BRAY, J. LEWIS, M. RAFF, K. ROBERTS, and J.D. WATSON. 1989. *Molecular biology of the cell,* 2nd ed. Garland Publishing, New York.

BALTIMORE, D. 1981. "Somatic mutation gains its place among the generators of diversity." *Cell* 26: 295–296.

BANERJI, J., L. OLSON, and W. SCHAFFNER. 1983. "A lymphocyte-specific cellular enhancer is located downstream of the joining region in immunological heavy chain genes." *Cell* 33: 729–740.

DARNELL, J., H. LODISH, and D. BALTIMORE. 1986. *Molecular cell biology.* Scientific American Books, New York.

DAVIS, M.M., S.K. KIM, and L. HOOD. 1980. "Immunoglobulin class switching: Developmentally regulated DNA rearrangements during differentiation. *Cell* 22: 1–2.

EARLY, P., and L. HOOD. 1981. "Allelic exclusion and nonproductive immunoglobulin gene rearrangements." *Cell* 24: 1–3.

LEDER, P. 1982. "The genetics of antibody diversity." *Sci. Amer.* 246(5): 102–115.

MAX, E.E., J.G. SEIDMAN, and P. LEDER. 1979. "Sequences of five potential recombination sites encoded close to an immunoglobulin *K* constant region gene." *Proc. Natl. Acad. Sci. U.S.A.* 76: 3450–3454.

WINOTO, A., S. MJOLSNESS, and L. HOOD. 1985. "Genomic organization of the genes encoding mouse T-cell receptor α-chain." *Nature* 316: 832–836.

PROBLEMS AND QUESTIONS

16.1 Is the genetic information specifying antibody chains stored in germ line cells in the same format as that specifying most other polypeptides?

16.2 How many polypeptide chains are present in each antibody molecule? How many antigen-binding sites are present per antibody? How many different antibodies are produced in each mature B lymphocyte?

16.3 What are three different sources of antibody variability?

16.4 Does class switching during B lymphocyte differentiation occur at the DNA level or the RNA level? By what mechanism does it occur?

16.5 Which of the heavy chain antibody gene segments determines whether a given antibody will be secreted or remain membrane-bound?

16.6 (a) Name three different types of white blood cells that play important roles in the immune response of vertebrates. (b) What function(s) does each of these types of cells perform in the immune response?

16.7 Of what importance is the *MHC* locus of humans in medical practice?

16.8 The genes of the *MHC* locus of humans are said to be *highly polymorphic.* (a) What does this mean? (b) Of what significance is this in the field of medicine?

16.9 Mature B lymphocytes make up a population of cells that collectively produce a vast array of different antibodies. Yet each B lymphocyte produces only a single type of antibody (sometimes slightly different antibodies, but all specific for a single antigen). After exposure of a human or other vertebrate to a foreign antigen such as the coat protein of a virus, the individual rapidly initiates the production of large quantities of antibodies specific to the foreign antigen to which she or he was exposed. By what mechanism does this response occur?

16.10 Antibodies are encoded by families of gene segments that are assembled into the genes that direct their synthesis during the differentiation of the cells (B lymphocytes) that produce them. Are any other vertebrate proteins encoded by gene segments that are assembled in a similar manner during cell differentiation?

16.11 In what ways are the structures of antibodies and T-cell antigen receptors similar?

16.12 If the genome of a primate species contains 300 V_κ, 5 J_κ, 1 C_κ, 150 V_λ, 5 J_λ–C_λ, 300 V_H, 4 J_H, 50 D, and 10 C_H gene segments, and these can be joined in all possible appropriate combinations during the genome rearrangements that occur in the development of mature B lymphocytes, how many different antibodies could this primate produce?

16.13 The human immunodeficiency virus (HIV) causes many of the symptoms of AIDS (Acquired Immunodeficiency Syndrome) by infecting and killing specific types of T lymphocytes. HIV gains entry to T cells by binding to a specific cell-surface antigen called CD4. The CD4 antigen is found primarily on T lymphocytes called helper T cells, and the depletion of helper T cells is a major factor in the deterioration of the immune system in individuals with AIDS. How might this knowledge of the process by which HIV infects cells be used to devise a method to prevent or at least retard the effects of HIV?

Chapter 17

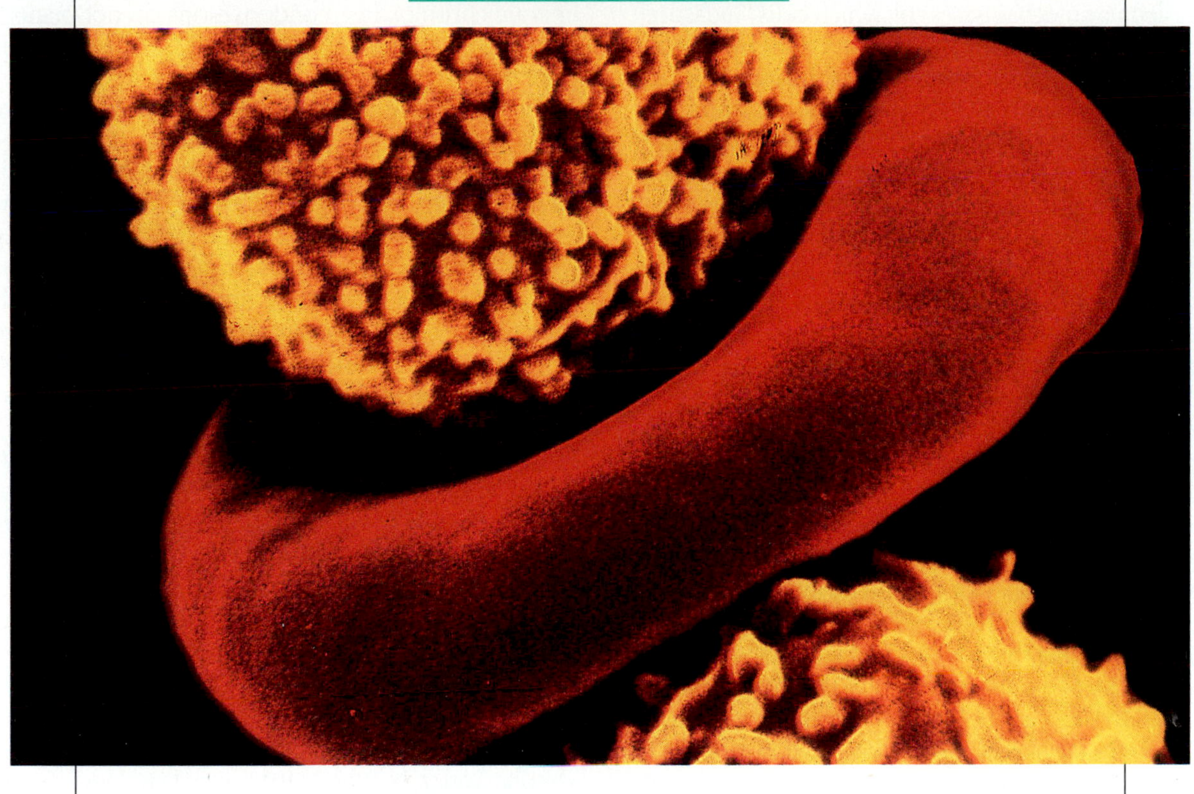

Genetic Control of Cell Division: Oncogenes and Protooncogenes

(Photo above): False-color scanning electron micrograph of abnormal white blood cells (B lymphocytes) along with one normal red blood cell (center) from a patient with chronic lymphocytic leukemia. These malignant B lymphocytes are abnormally small and are covered with microvilli; they are not able to carry out their usual immune functions, leaving the individual more susceptible to infections. (Courtesy of Professor Aaron Pulliack/Science Photo Library/Photo Researchers.)

HOW does a cell know when it has grown large enough such that it should divide and produce progeny cells? Even in the case of *E. coli,* the best known of all cellular organisms, we still do not know the molecular mechanism(s) by which cells "measure" cell mass and initiate cell division at precisely the right time. We know that the ***initiation of rounds of chromosome replication is very precisely controlled irrespective of the growth conditions.*** Moreover, many of the macromolecules that are involved in the initiation of chromosome replication have been identified in *E. coli.* Nevertheless, although several models have been proposed to explain how cells maintain the proper ratio of active origins of replication to cell mass, the details of the mechanism of this precise control of cell division remain unknown.

In eukaryotic cells, the control of cell division is more complex than in prokaryotes because not only must chromosome duplication and cytokinesis be regulated, but, in addition, the ***components of the mitotic apparatus must form and function at the proper time in the cell cycle.*** In the yeast *Saccharomyces cerevisiae,* the process of cell division has been partially dissected by isolating and studying ***temperature-sensitive mutants*** that are blocked at various stages of the cell cycle. The mutant genes are designated ***cdc*** for the ***c***ell ***d***ivision ***c***ycle defects that they cause. By determining the stages in the cell cycle at which the various *cdc* mutants are blocked, geneticists are beginning to define the sequences of events that occur during cell growth and division. If the products of these *cdc* genes can be identified and their functions determined, it will eventually be possible to define the processes that occur during the cell cycle in molecular terms. Although much remains to be learned before we understand the control of cell division at the molecular level, some of the key components of this important regulatory circuitry have been identified.

REGULATION OF THE MITOTIC CELL CYCLE IN EUKARYOTES

As a cell grows, the cell membrane (and cell wall in plant cells) and most components of the cytosol increase in size and/or quantity. When the cell reaches a critical mass, it divides, producing two smaller progeny cells. Under appropriate conditions, the progeny cells will in turn grow and divide. As cells proceed through this ***cell cycle*** (Chapter 3), two key processes must take place in an accurate and coordinated manner. (1) The genetic material must be duplicated (semiconservative replication of DNA; Chapter 5). (2) The two copies of the genetic material must be faithfully distributed to the two progeny cells (mitosis; Chapter 3). How does a cell know when to initiate replication of its DNA? How does a cell know when to begin the chromosomal condensations of mitosis? How are these two processes coordinated so that mitotic chromosome condensations do not begin before the DNA has been replicated? Clearly, a cell must have some mechanism by which it keeps track of where it is in the cell cycle. Although detailed answers to these questions are not yet available, some answers are beginning to emerge, and some of the key players in the circuitry that regulates the cell cycle are known.

Studies of cell division cycle (*cdc*) mutants of the yeasts *Schizosaccharomyces pombe* and *Saccharomyces cerevisiae* and of early embryonic cleavage divisions in the toad *Xenopus laevis* have shown that there are two points at which a cell makes a commitment to proceed through the ensuing stages of the cell cycle (Fig. 17.1). The first point, called ***Start,*** occurs near the end of the G_1 phase. At Start, a cell becomes committed to the initiation of DNA synthesis a short time later at the onset of the *S* phase of the cell cycle. The second point is the commitment to proceed through the chromosomal condensation and chromatid separation events of mitosis; this occurs at the very beginning of the *M* phase of the cell cycle. Recent evidence indicates that certain key regulatory proteins function as signals in both commitment steps.

A ***m***itosis-***p***romoting ***f***actor (***MPF***) was first discovered in *Xenopus.* When MPF was injected into *Xenopus* oöcytes, it stimulated the oöcytes to enter *M* phase. MPF has been shown to contain at least two essential components: (1) proteins called ***cyclins*** that undergo cycles of synthesis and accumulation during G_1 and G_2 and degradation during *M* phase (and probably also after the Start commitment) and (2) a Start- and *M* phase-specific protein kinase called ***pp34*** (***pp*** for ***p***hospho***p***rotein, a protein that may have phosphate groups on the side chains of specific amino acids, and ***34*** for ***34***,000 molecular weight), which is the product of the *cdc2* gene of *S. pombe* and the *CDC28* gene of *S. cerevisiae.* Recent evidence indicates that pp34 is involved in both commitment events: Start and the onset of *M* phase. Moreover, the results show that phosphorylation/dephosphorylation of a single tyrosine residue (Tyr 15—the 15th amino acid from the NH_2 terminus) may directly regulate both the initiation of DNA replication and the onset of *M* phase when pp34 interacts with the appropriate cyclins (Fig. 17.1). Recent evidence indicates that the pp34 protein kinase interacts with two classes of cyclins, one class, ***M-cyclins,*** involved in the *M*-phase commitment and the second class, ***G_1-cyclins,*** involved in the Start decision.

Although additional components are undoubtedly involved in the circuitry that regulates the cell cycle in eukaryotes, the cyclins and the pp34 protein kinase are key components, and homologous proteins have been

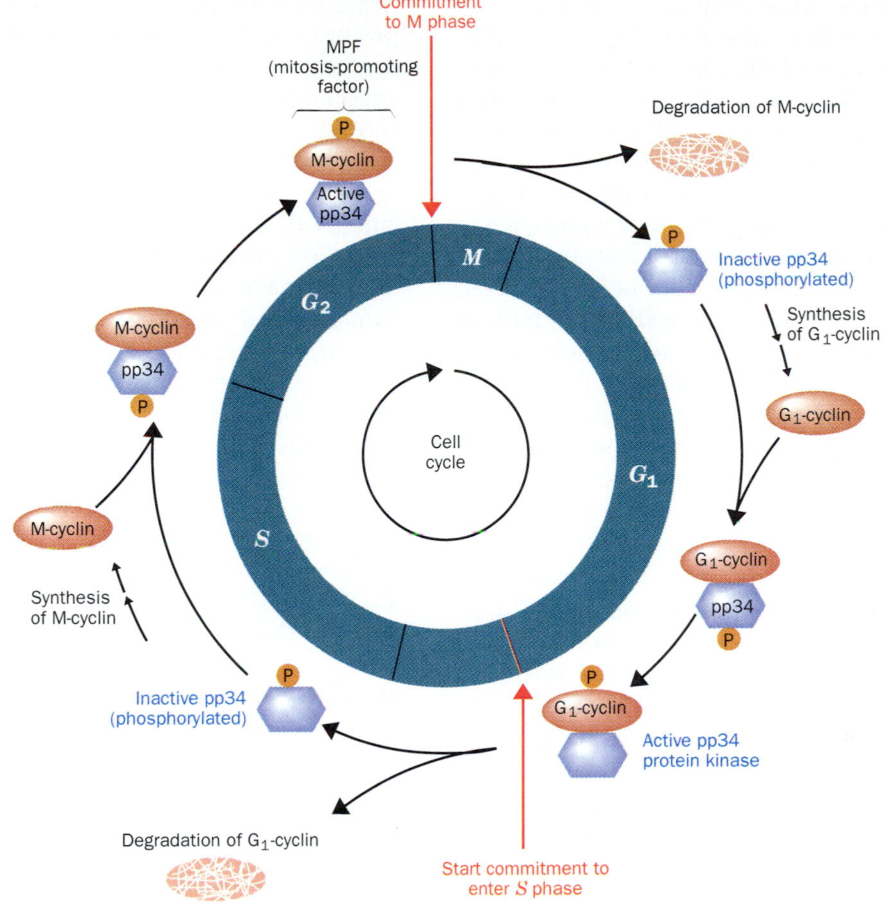

Figure 17.1 Diagram showing some of the factors that regulate the progression of eukaryotic cells through the mitotic cell cycle. Cells make commitments to proceed through the ensuing stages at two points: (1) ***start,*** a point late in G_1 at which a commitment is made to subsequently initiate DNA synthesis (*S* phase), and (2) the beginning of *M* phase, at which time a commitment is made to proceed through the chromosomal condensation and chromatid separation phases of mitosis. Proteins called ***cyclins*** undergo cycles of synthesis and accumulation and then degradation as the cell passes through the cell cycle. The cyclins interact with a key regulatory protein designated pp34 (pp for phosphoprotein, 34 for 34,000 molecular weight). The phosphorylation/dephosphorylation of a single tyrosine residue of pp34 has been shown to determine the inactive/active states of this important regulator. Ⓟ represents a phosphate group on the protein pp34 yielding the inactive state. (For additional details, see the review by A. W. Murray, *Nature* 342: 14–15, 1989.)

identified in several eukaryotes, including humans, frogs, sea urchins, and starfish. One of the attractive aspects of having one or more of the same molecules involved in both commitment steps is that this provides a ***mechanism by which a cell can remember, based on the conformational states of these molecules, where it is in the cell cycle, and thus accurately coordinate Start and the onset of* M *phase.*** Clearly, our knowledge of the molecular mechanisms that control the orderly passage of cells through the cell cycle is still incomplete. Neverthess, new information is accumulating very rapidly, and one can anticipate that a detailed picture of the circuitry controlling cell growth and division will emerge during the next few years—at least for some of the simple eukaryotes such as the yeasts.

INTERCELLULAR COMMUNICATION IN MULTICELLULAR EUKARYOTES

In multicellular eukaryotes, the mechanisms that control cell division and cell growth are even more complex because a new dimension—***intercellular communication***—must play a major role. Each tissue within an organ and each organ within the body of an organism must grow to the proper size for that particular species. The growth of bones, muscles, the liver, the pancreas, and so on must all be correctly coordinated during the growth and development of a mouse, a rabbit, or a human. Clearly, for this to occur, cell division must be under very precise control within each tissue and must be subject to different regulatory signals in different tissues and organs. Because of the intricate interrelationships that exist between the different tissues of a multicellular plant or animal, intercellular communication must play an essential central role in the growth and differentiation of higher plants and animals. How does this intercellular communication occur? What are the mechanisms by which cellular differentiation and cell growth and division are regulated? At present, we know that there are a host of "factors" that stimulate or inhibit the growth and division of specific types of cells. However, we do not understand how any of these factors influence cell division at the molecular level. Enough information has accumulated to indicate that the total picture will be complex, but at present we are just beginning to understand a few of the pieces of this complex picture.

Cell division, like all other biological processes, is under genetic control. Certain genes must regulate the process of cell division in response to intracellular, intercellular, and environmental signals. These regulatory genes are undoubtedly subject to mutation, like all other genes. Mutations that abolish the function of these regulatory genes would be expected to lead to abnormal cell division—in the extreme, either the inability to divide at all or the inability to stop dividing.

To date, we do not know the details of how cell division is controlled for any cell of any higher animal, nor have we identified all of the genes that regulate this process in any higher eukaryote. However, recent studies of viral genes called ***oncogenes*** (from the Greek *onkos,* meaning "tumor"), which can cause a loss of the normal control of cell division, have led to the identification of a set of homologous genes called ***protooncogenes*** in the genomes of normal animals, including humans. These normal cellular protooncogenes can be converted into tumor-causing ***cellular oncogenes*** by mutation or by becoming associated with new regulatory sequences through recombination processes. These and related observations indicate that the normal cellular functions of the protooncogenes involve specific aspects of the control of cell division. In fact, ***it now seems likely that breakthroughs in understanding the normal control of cell division may result from studies on the disruptions of normal control that occur in cancer cells.***

CANCER CELLS: LOSS OF CONTROL OF CELL DIVISION

Cancer is a large class of diverse diseases, all of which exhibit ***uncontrolled cell growth and division.*** In noncirculatory tissues, such uncontrolled cell growth produces cell masses called ***tumors.*** Cancerous or ***malignant tumors*** are those from which cells detach and migrate to other parts of the body, giving rise to secondary tumors (a process called ***metastasis***). Noncancerous or ***benign tumors*** do not metastasize.

Human cancers are responsible for an enormous amount of suffering. Thus, a large amount of money and effort have been directed to the study of these diseases. Although there has been great progress in the detection and treatment of cancers, there has been little progress toward understanding the molecular bases of cancers. However, there is now extensive evidence for the involvement of over 40 different oncogenes in the occurrence of various types of cancer in animals.

TUMOR-INDUCING VIRUSES: VIRAL ONCOGENES

Most of our information about ***oncogenes*** has come from studies of ***RNA tumor viruses*** or ***retroviruses.*** The name retrovirus is derived from the fact that these viruses store their genetic information in a single-stranded RNA genome, and then convert it into a homologous double-stranded DNA form after infecting a host cell. Thus, they use a "backward" ("retro") flow of genetic information. Usually, genetic information flows from DNA to RNA during transcription. The retroviruses code for a special enzyme, called reverse transcriptase, which catalyzes the synthesis of homologous DNA sequences using RNA molecules as templates (Chapter 13, p. 354).

The genomes of DNA tumor viruses such as polyoma virus, SV40 (simian virus 40), and adenoviruses also contain oncogenes that are able to induce the uncontrolled (neoplastic) growth of animal cells growing in culture. However, the oncogenes of these DNA viruses have proven more difficult to study than those of the retroviruses because the tumor-inducing properties of these DNA viruses are not easily distinguished from effects caused by the replication of the viral chromosomes and the expression of other viral genes that eventually lead to the death of the infected host cells. Thus, the discussion of viral oncogenes in this chapter will be restricted to retroviral oncogenes.

Life Cycle of Rous Sarcoma Virus

The best known of the retroviruses is ***Rous sarcoma virus*** ("Rous" for the scientist who discovered it; "sarcoma" for the type of cancer induced), which induces cancer in chicken cells. The life cycle of Rous sarcoma virus is diagrammed in Fig. 17.2. Shortly after Rous sarcoma virus infects a cell, its RNA genome is repli-

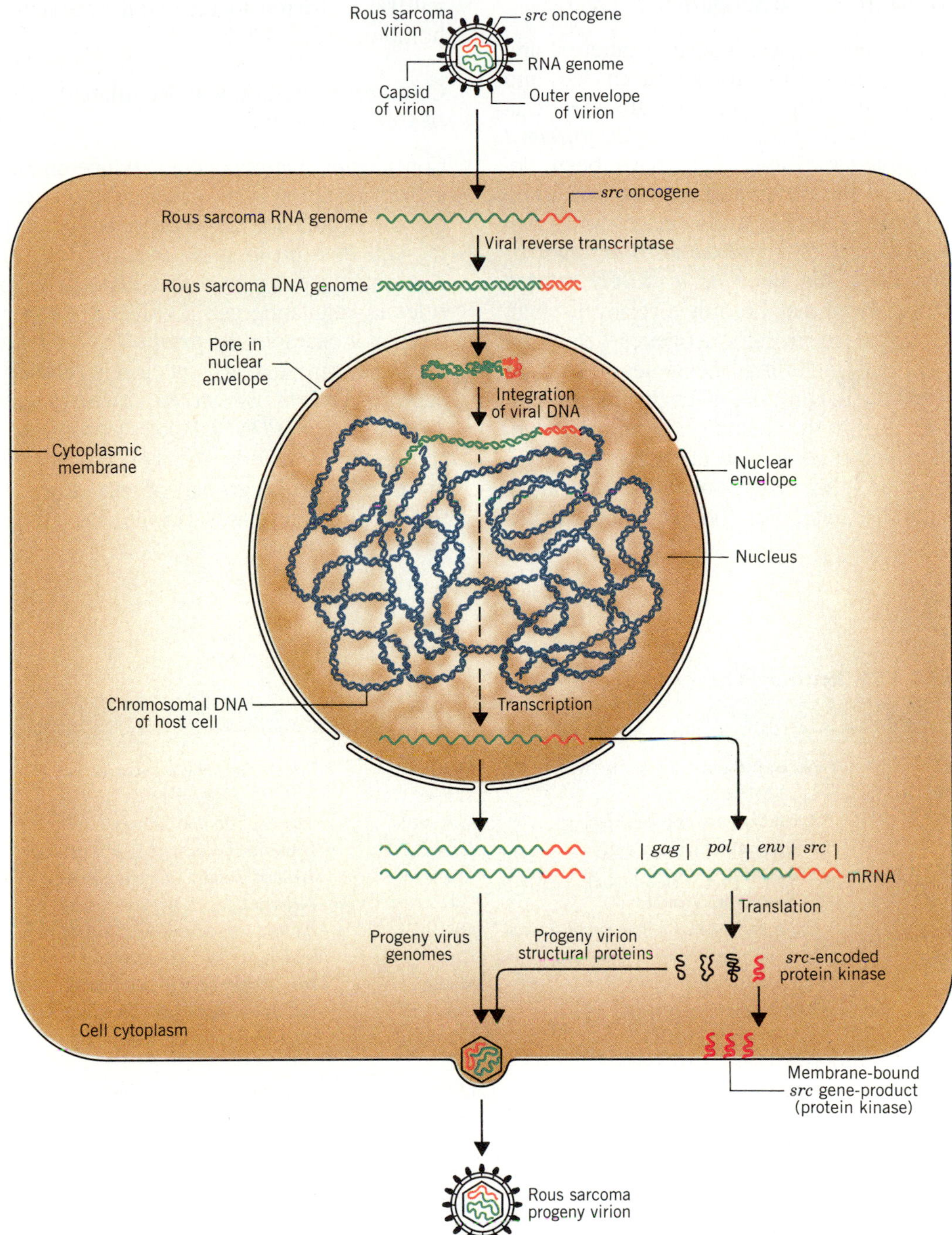

Figure 17.2 Life cycle of the Rous sarcoma RNA tumor virus. The ability of this virus to cause cancer resides in the *src* oncogene (shown in orange). (After J. M. Bishop, *Sci. Amer.* 246(3): 80–92, 1982.)

cated to its DNA form by reverse transcriptase, and the viral DNA is integrated into the chromosomal DNA of the host cell. In this integrated state, it is replicated and transcribed by the host cell's metabolic machinery, just like the normal genes of the host cell.

The genome of Rous sarcoma virus contains just four genes: **gag,** which codes for the capsid protein of the virion; **pol,** which codes for reverse transcriptase; **env,** which codes for the protein spikes of the viral envelope; and the ***oncogene*** **src** (derived from *sar*coma), which codes for a membrane-bound protein kinase. The viral genome also carries its own strong promoter, thus the four viral gene-products are synthesized in large amounts. The **src** ***gene is entirely responsible for the ability of Rous sarcoma virus to cause cancer;*** the deletion of this one gene yields a virus that infects and replicates just like the *src*-containing virus, but one that is totally nononcogenic.

Diversity of Retroviral Oncogenes

Studies of other retroviruses of several different animal species (most notably the mouse, rat, chicken, and certain species of monkey) soon led to the identification of other retroviral oncogenes. Over ***20 different retroviral oncogenes*** (Table 17.1) have been described to date, and more undoubtedly remain to be discovered. However, recent studies have shown that ***different retroviruses that induce similar types of cancers often carry the same or a closely related oncogene.*** Thus, there may be only a relatively small number of unique retroviral oncogenes yet to be discovered. In total, including oncogenes of DNA viruses and of cellular origin, over 40 distinct genes have been identified that fit the definition of an oncogene. However, more information is available for the retroviral oncogenes and the related oncogenes of cellular origin, thus the discussion that follows in this chapter will be restricted to retroviral oncogenes and their homologs.

Oncogene Products as Regulators of Cell Division

Considering that oncogenes induce uncontrolled cell growth resulting in tumors, one might well anticipate that the products of these genes would act by stimulating cell division in some manner. Indeed, it is now clear that the products of these oncogenes play various roles in regulating cell division in one or more cell types. For example, the ***product of the v-sis oncogene*** of simian sarcoma virus is closely ***related to a polypeptide growth hormone called platelet-derived growth factor (PDGF).*** PDGF is produced by platelet cells as the name indicates; it promotes wound healing by stimulating the growth of cells at the wound site. Simian sarcoma viruses carrying the v-*sis* gene induce

TABLE 17.1 Retroviral Oncogenes

ONCOGENE[a]	VIRUS	HOST SPECIES	FUNCTION OF GENE-PRODUCT
abl	Abelson murine leukemia virus	Mouse	Tyrosine-specific protein kinase
cbl	Cas NS-1 virus	Mouse	?
erbA	Avian erythroblastosis virus	Chicken	Analog of thyroid hormone receptor
erbB	Avian erythroblastosis virus	Chicken	Truncated version of EGF receptor[b]
ets	E26 virus	Chicken	Nuclear protein, unknown function
fes	ST feline sarcoma virus	Cat	Tyrosine-specific protein kinase
fgr	Gardner–Rasheed feline sarcoma virus	Cat	Tyrosine-specific protein kinase
fms	McDonough feline sarcoma virus	Cat	Analog of CFS-1 GF receptor[c]
fos	FJB osteosarcoma virus	Mouse	Transcriptional activator protein
fps	Fuginami sarcoma virus	Chicken	Tyrosine-specific protein kinase
jun	Avian sarcoma virus 17	Chicken	Transcriptional activator protein
kit	HZ4 feline sarcoma virus	Cat	?
mil (mht)	MH2 virus	Chicken	Serine/thr-specific protein kinase
mos	Moloney sarcoma virus	Mouse	Serine/thr-specific protein kinase
myb	Avian myeloblastosis virus	Chicken	Nuclear protein, unknown function
myc	MC29 myelocytomatosis virus	Chicken	Probably transcription regulator
raf	3611 murine sarcoma virus	Mouse	Serine/thr-specific protein kinase
H-*ras*	Harvey murine sarcoma virus	Rat	GTP binding protein, GTPase activity
K-*ras*	Kirsten murine sarcoma virus	Rat	GTP binding protein, GTPase activity
rel	Reticuloendotheliosis virus	Turkey	Cytoplasmic protein, unknown function
ros	URII avian sarcoma virus	Chicken	Tyrosine-specific protein kinase
sis	Simian sarcoma virus	Monkey	Analog of platelet-derived GF[c]
ski	Avian SKV770 virus	Chicken	Nuclear protein, unknown function
src	Rous sarcoma virus	Chicken	Tyrosine-specific protein kinase
yes	Y73 sarcoma virus	Chicken	Tyrosine-specific protein kinase

[a]Oncogenes *fes* and *fps*, first isolated from retroviruses of cat and chicken, respectively, are homologous; this was not known when they were assigned different names (see the section of this chapter entitled Conservation of Protooncogenes during Evolution); the same is true for *mil* and *raf*.
[b]EGF = epidermal growth factor.
[c]GF = growth factor; **CSF-1** (**C**olony **S**timulating **F**actor-**1**) is a polypeptide growth factor that specifically stimulates the growth of macrophages; platelet-derived growth factor (secreted by platelet cells during clot formation) promotes growth of cells at wound sites, thereby enhancing the healing process.

sarcomas when injected into woolly monkeys; they also transform fibroblasts growing in culture to a neoplastic or tumorous state. Presumably, this cellular transformation to the cancerous state occurs by a mechanism that is related to the effect of normal PDGF on cells at a wound site.

Other oncogenes encode products that are similar to growth-factor and hormone receptors. ***Oncogenes* erbB *and* fms** encode proteins that are closely ***related to the receptors for epidermal growth factor and CSF-1 growth factor,*** respectively. CSF-1 is a growth factor that stimulates growth and differentiation of macrophages. Both of these growth-factor receptors are transmembrane proteins with growth-factor binding domains on the outside of the cell and protein kinase domains on the inside of the cell. These receptors are key components in transmembrane signaling pathways. Finally, the **erbA *gene-product* is** an ***analog of the nuclear receptor for the thyroid hormone T_3.*** Thus, all of these gene-products are undoubtedly involved in the ***intercellular communication circuitry*** that regulates cell division during the growth and development of highly differentiated tissues and organs in multicellular animals.

Because these transmembrane receptor protein tyrosine kinases are capable of transmitting a mitogenic signal (a signal telling a cell to divide), it is not surprising that alterations in the structure and function of these proteins will sometimes be oncogenic. If they malfunction and transmit a signal telling the cell to divide when it normally should not divide, the result will be tumor formation.

The ***largest group of the oncogenes*** (including **src**) encode ***protein kinases that phosphorylate tyrosine residues.*** Some of these may well be analogous to the receptors for epidermal growth factor and CSF-1 growth factors, but contain receptors for mitogenic factors that have not yet been identified. However, the *src* tyrosine kinase is not a transmembrane protein, but rather is tightly associated with the inner face of the plasma membrane. Although the *src* protein is a very active protein kinase that phosphorylates specific tyrosine residues in proteins, we still do not understand the molecular basis of its oncogenicity or what proteins are the important regulatory factors that are subject to its action.

The *ras* oncogenes encode proteins that bind GTP and exhibit GTPase activity. They may be analogous to proteins called G proteins that have GTPase activity and play a role in the regulation of the enzyme adenylcyclase and, thus, the levels of cyclic AMP in cells. The function(s) of the *ras* gene-products is (are) of particular interest because considerable evidence implicates the involvement of mutant *ras* products in several distinct types of human cancers.

Lastly, other ***oncogenes*** such as **jun, fos, erbA,** and **myc *encode nuclear transcription factors*** that activate the expression of specific genes. Undoubtedly, some of the genes that they activate will prove to encode products that function as positive regulators of cell division.

In summary, oncogene products are simply proteins that play central roles in stimulating cell division in one or more cell types. In some cases, these oncogene products are probably altered or "mutant" proteins that trigger the division of cells that should normally not divide under the existing conditions. In other cases, the oncogene products stimulate abnormal cell division by being overproduced—being synthesized in much larger amounts than in normal cells.

PROTOONCOGENES AND CELLULAR ONCOGENES

Genes with DNA sequences that are very similar to the retroviral oncogenes and that encode proteins with similar properties have been identified in the genomes of higher animals including humans by the use of two distinct experimental approaches. (1) One approach involved looking for ***cellular DNA sequences that would cross-hybridize with the oncogenes*** of animal viruses. (2) The second approach involved looking directly for cancer-causing genes in the genomes of cancer cells by ***transfection experiments,*** experiments in which the DNA of tumor cells is isolated and added to normal tissue culture cells to see if it will convert any of them to the cancerous state. Both approaches have been successful and, in some cases, both have resulted in the identification of the same cellular oncogenes. Genes homologous to some of the viral oncogenes are even present in lower eukaryotes such as the yeasts.

Homology with Viral Oncogenes

As was mentioned earlier, the *src* oncogene was first identified in the genome of Rous sarcoma viruses (RSV) isolated from chickens. When reverse transcriptase was used to convert the *src* oncogene of RSV to a cDNA form (see Fig. 13.3 for an illustration of the action of reverse transcriptase) and this cDNA was labeled with ^{32}P and used as a probe in Southern blot hybridization experiments (Chapter 13) with genomic DNA from normal chickens, the *src* cDNA hybridized with specific restriction fragments of genomic DNA in every experiment. This was true regardless of the source of the chicken genomic DNA. Moreover, similar genomic DNA sequences that hybridize with the viral *src* cDNA probe have been identified in all vertebrate animals and even in the fruit fly *Drosophila melanogaster*.

Subsequent experiments have demonstrated that genomic DNAs from normal (noncancerous) cells of all higher animals contain DNA sequences that hybridize with essentially all the retroviral oncogene sequences. In some cases, sequences homologous to retroviral oncogenes (e.g., *ras*) are even found in lower eukaryotes such as *Saccharomyces cerevisiae*.

One might initially guess that these genomic DNA sequences that hybridize with oncogenes are simply present on integrated proviruses. However, this has proven ***not*** to be the case. Instead, when these sequences were isolated from genomic libraries and characterized, they were found to be normal cellular genes with structures that distinguish them from the homologous viral oncogenes. These ***normal cellular genes with homology to oncogenes*** are now called ***protooncogenes***. In some cases, these ***protooncogenes can mutate to forms that are capable of inducing oncogenesis***—the ability to transform cells to a neoplastic or cancerlike state (see the following section). In the latter form, they are called ***cellular oncogenes*** (abbreviated *c*-onc, e.g., *c*-src, *c*-sis, *c*-myc) to distinguish them from their viral counterparts. This means that we must now denote the ***viral oncogenes*** more precisely as *v*-onc's, for example, *v*-src, *v*-sis, and *v*-myc.

Interestingly, some of the same cellular oncogenes that were identified by their cross-hybridization to viral oncogene sequences have also been identified on the basis of their ability to transform cells growing in culture to neoplastic or tumorlike states in direct DNA transfer studies called transfection experiments.

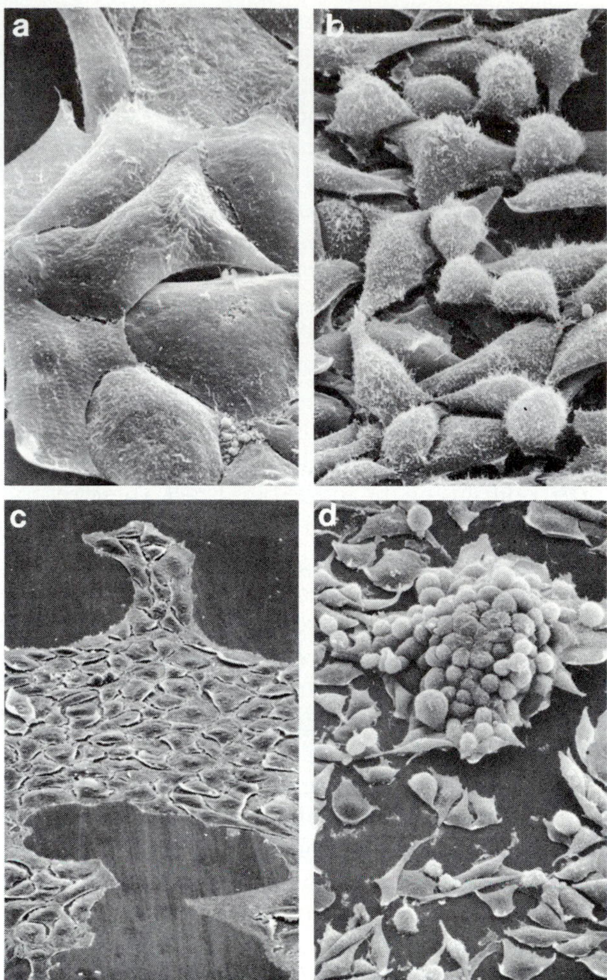

Figure 17.3 Scanning electron micrographs showing the altered morphology and cell–cell interaction characteristics of normal (a and c) and transformed (b and d) rat kidney cells growing in culture. Note that the normal cells adhere to the surface of the culture dish, forming monolayers of flat, extended cells. The transformed cells, on the other hand, tend to overgrow one another, forming clusters or piles of cells. All cells are of the NRK (normal rat kidney) cell culture line. The NRK cells shown in (b) and (d) were transformed by infection with the Prague A strain of avian sarcoma virus. Magnification: (a) ×1100; (b) ×1000; (c) ×200; (d) ×280. (Original photographs courtesy of Mike Atkinson and Chris Frethem, Department of Cell Biology and Neuroanatomy, University of Minnesota.)

Transfection Experiments

The detection of cellular oncogenes by ***transfection experiments*** is based on the ability of the oncogenes to convert noncancerous cells (characterized by controlled cell division) growing in culture to the cancerous state (characterized by uncontrolled cell division). This phenomenon is called ***cell transformation*** or simply ***transformation***. It should not be confused with the recombination process in bacteria called transformation (discussed in Chapters 5 and 8); the two phenomena are totally different processes.

Normal (nontransformed) cells growing in culture will stop dividing when they make contact with neighboring cells (a phenomenon called ***contact inhibition***); they will thus form a monolayer of cells on the surface of the culture flask or petri dish in which they are growing. Transformed cells do not exhibit contact inhibition. They will keep on dividing despite contact with their neighbors and will form piles of cells or "tumors" on the surface of the culture flask (Fig. 17.3).

When DNA from normal cells is used in transfection experiments, a very low, but detectable, level of cell transformation is observed. When the DNA from the transformed cells is used in a second round of transfection experiments, a higher frequency of transformation is sometimes observed. That is, higher frequencies of transformation are observed using DNA isolated from certain transformed cell clones, but not using DNA isolated from other transformed cell clones. This indicates that genetic changes are responsible for the transformed state in the first group of cell clones, but that ***epigenetic changes*** (noninherited developmental changes) are responsible for the transformed state in the second group of cell clones.

Transfection experiments have also been used to demonstrate the presence of cellular oncogenes in cell culture lines derived from various spontaneously occurring and chemically induced animal tumors. Most of the cellular oncogenes detected by transfection experiments have been isolated using recombinant DNA and gene cloning techniques. When these isolated cellular oncogenes were compared with the oncogenes of retroviruses by various procedures (e.g., DNA hybridization, restriction enzyme analysis, DNA sequencing), many of them were found to be homologous to one of the retroviral oncogenes. For example, the c-H-*ras* oncogene identified by transfection experiments in DNA from human bladder carcinoma cells turned out to be homologous to the v-H-*ras* oncogene of Harvey sarcoma virus.

Cellular Oncogenes Contain Introns—Their Viral Homologs Are Single Exons

As mentioned earlier, when viral oncogenes such as *src* are cloned by recombinant DNA techniques and are used as hybridization probes to search for homologous sequences in normal host cells, such sequences are almost always found. These homologous sequences present in the chromosomes of normal cells of normal animals are ***not*** integrated viral oncogenes, because they differ from the viral oncogenes in having interrupted coding sequences, like most other eukaryotic genes. That is, the ***cellular oncogenes and protooncogenes have multiple exons separated by introns*** (see Chapter 13, pp. 356–359), whereas the ***viral oncogenes are single exons.*** For example, the ***chicken cellular* src *protooncogene contains 11 introns separating 12 coding sequences, whereas the RSV v-*src *gene has a single, uninterrupted coding sequence*** (Fig. 17.4).

The v-*src* and c-*src* genes both code for protein kinases that phosphorylate tyrosine residues. Moreover, these two protein kinases are the same size and have very similar structures. In addition, both proteins react with antibodies prepared using the v-*src* protein as antigen. Comparison of the nucleotide sequences of the chicken c-*src* gene and the v-*src* gene of one strain (the Schmid–Ruppin strain) of RSV indicates that the two genes encode very similar proteins. The c-*src* protein is 533 amino acids long; the v-*src* protein is 526 amino acids long. The major difference between these two proteins occurs at the COOH terminus, where the last 12 amino acids of the v-*src* protein are replaced by 19 completely different amino acids at the terminus of the c-*src* protein. In addition, their are 18 single nucleotide-pair differences between the coding sequences of v-*src* and c-*src* that result in 8 amino acid changes in the protein products. These 8 amino acid changes in v-*src* protein of the Schmid–Ruppin strain of RSV do not appear to be involved in the oncogenicity of the v-*src* protein since none of them is found in common in the v-*src* oncogenes that have been sequenced from other RSV strains. Clearly, the major difference between these two genes is the presence of the 11 introns in c-*src* and their absence in v-*src* (Fig. 17.4).

We do not yet know how the protein kinases encoded by these viral oncogenes cause tumors, but the mechanism probably relates to the large quantities of these enzymes made in retrovirus-infected cells. There is 100 times as much v-*src* protein kinase per cell in chicken tumors induced with Rous sarcoma virus as there is c-*src* protein kinase in normal chicken cells.

Conservation of Protooncogenes during Evolution

One argument for the importance of the protooncogenes and the products that they encode in normal cell growth and/or cell division is that the ***protooncogenes have been highly conserved during evolution.*** c-*src* genes are found not only in chickens, but in other birds, in mammals (including humans), in fish, and even in insects such as *Drosophila melanogaster.* Moreover, this picture of the conservation of protooncogenes across a wide range of species is observed for most of the other protooncogenes as well. All vertebrate animals contain protooncogenes that are homologous to essentially all the oncogenes listed in Table 17.1. The fruit fly **Drosophila melanogaster *contains normal cellular genes that exhibit strong homology to the vertebrate cellular oncogenes c-*abl, *c-*erbB, *c-*fps, *c-*raf, *c-*ras, *and c-*myb, *in addition to the c-*src *homolog.*** In fact, the genome of *D. melanogaster* contains two genes with homology to *src* and three genes with homolog to *ras,* just like the genomes of vertebrates. In the case of the *ras* protooncogenes, the genome of the yeast *Saccharomyces cerevisiae* has even been found to contain two homologous sequences. Clearly, then, the various protooncogenes have been widely conserved during the course of evolution.

When the ***sequences of the homologous protooncogenes of different species are compared, the sequences are almost always very highly conserved,*** often differing by less than 15 percent in nucleotide-pair sequence. Even in the case of the relatively distantly related yeast and vertebrate *ras* protooncogenes, the predicted amino acid sequences (predicted from the nucleotide sequences) of the conserved amino-terminal two-thirds of the protein products are identi-

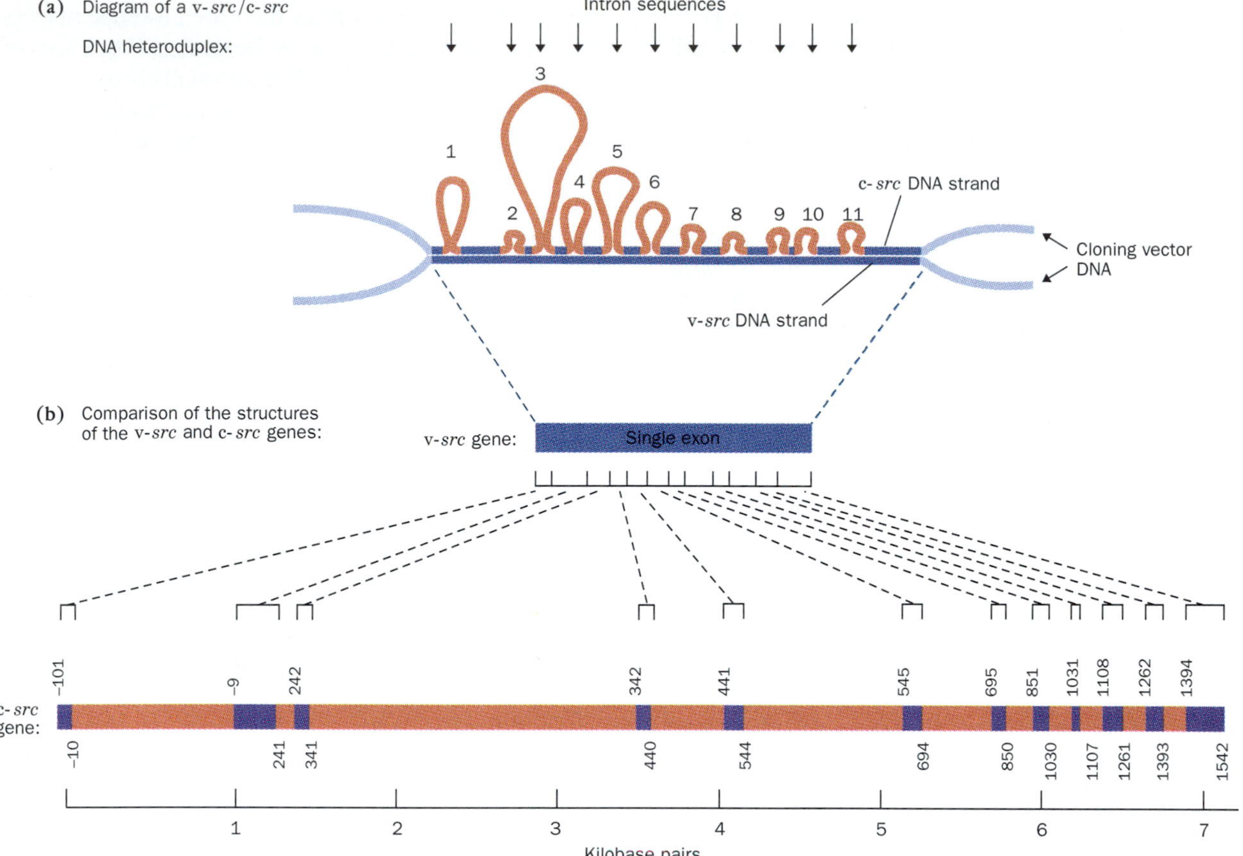

Figure 17.4 Structures of the v-*src* oncogene and the c-*src* protooncogene of the chicken. (a) Diagram of the predicted DNA heteroduplex formed by hybridization of one strand carrying the c-*src* gene (top strand) with a (partially) complementary strand carrying the v-*src* gene (bottom strand) as observed by electron microscopy. The intron segments of the c-*src* strand form single-stranded loops (numbered 1–11 and shown in red). The light blue lines represent the DNA strands of the plasmid in which these genes were cloned. (b) Schematic comparison of the organization of the coding sequences (exons) of the two genes; exon sequences are shown in blue. The v-*src* gene has a single, uninterrupted coding sequence. The c-*src* gene contains 12 exons (shown in blue) separated by 11 introns (shown in red). The starting and ending nucleotide positions of the exons are shown above and below, respectively, the map of the c-*src* gene; position 1 corresponds to the first nucleotide of the coding sequence. The first exon (position −101 to −10) is in the 5′ leader region of the transcript, not in the coding sequence. The dashed lines indicate the approximate correlations between the coding sequences of the two genes. (Based on the data of T. Takeya and H. Hanafusa, *Cell* 32: 881–890, 1983.)

cal at 75 percent of the amino acid positions. Presumably, the highly conserved domains of these homologous protooncogene products carry out the same or very similar functions in all the different species that contain these genes. This high degree of conservation of protooncogene structure in all vertebrate animals and even in many invertebrate species argues strongly that these genes encode important products, and that the sequences of these genes have been conserved by natural selection acting to maintain the functional integrity of the important gene-products that are encoded by these genes. Although we do not yet understand the exact roles that these protooncogene products perform in normal cells, it seems certain that they are directly involved in the control of cell division.

Protooncogene Products: Key Regulators of Cell Division

During the last few years, a wealth of information has accumulated regarding the structure and function of the various protooncogenes. It now seems clear that the only property that unites these genes as a group is that they all play central roles in the control of cell division. When classified according to function, the different protooncogenes appear to fit into four groups: (1) those that encode growth factors (c-*sis*) or growth-factor receptors (c-*fms* and c-*erbB*); (2) those that encode GTP-binding proteins with GTPase activity (c-H-*ras,* c-K-*ras,*, and N-*ras*); (3) those that encode protein kinases, either tyrosine-specific protein kinases (c-*abl,* c-*fes,* c-*fgr,* c-*fps,* c-*ros,* c-*src,* and c-*yes*) or serine/threonine-specific protein kinases (c-*mil,* c-*mos,* and c-*raf*); and (4) those that encode transcriptional regulators (c-*fos,* c-*jun,* c-*erbA,* c-*myc,* and possibly c-*myb* and c-*ets*).

We probably know the most about the function of the protooncogene products that are growth factors or growth-factor receptors because they were studied long before we knew of the existence of protooncogenes. For example, consider the growth-factor receptors encoded by c-*erbB* and c-*fms.* The prototype structure of such growth-factor receptors that have intracellular tyrosine-specific protein kinase activities is shown in Fig. 17.5. Although we still do not know exactly how these proteins function, it seems quite clear that they are involved in the transfer of signals from the cell surface to the cell nucleus. They bind growth factors at their extracellular binding sites and transmit a signal, presumably via an allosteric transition, to the intracellular tyrosine kinase site. Thus, in turn, must activate the kinase and induce phosphorylations of key intracellular proteins. Activation of the tyrosine kinase site may involve autophosphorylation, because these receptor protein kinases have been shown to undergo reversible autophosphorylations of specific tyrosine residues near the intracellular COOH termini of the proteins. The epidermal growth factor receptor is also known to undergo phosphorylation by other cellular protein kinases (e.g., protein kinase C) and to interact with other protein factors that modulate its activity. Thus, an accurate picture of the mode of action of these key regulatory proteins in signal transduction must await the results of further studies. When available, this picture will almost certainly depend on an understanding of the three-dimensional structure and function of the receptor molecule plus all the other macromolecules with which it interacts.

The c-*src* protein and the products of several of the related protooncogenes also have tyrosine-specific protein kinase activities. However, these protein kinases are not transmembrane proteins, but rather are associated with the cytoplasmic face of the plasma membrane. Presumably, these protein kinases also are involved in signal transduction, but we do not know what signal(s) they respond to or how this signal is transmitted. As a working model, it seems reasonable to assume that phosphorylation of key intracellular protein targets is the most likely mode of action of these protooncogene products.

Clearly, the mechanisms of action of the c-*ras* gene-products and the protooncogene products that function as transcription activators are totally distinct from those of the protooncogene products just discussed. The available information about the modes of action of the c-*ras,* c-*fos,* and c-*jun* gene-products is discussed in the next two sections of this chapter.

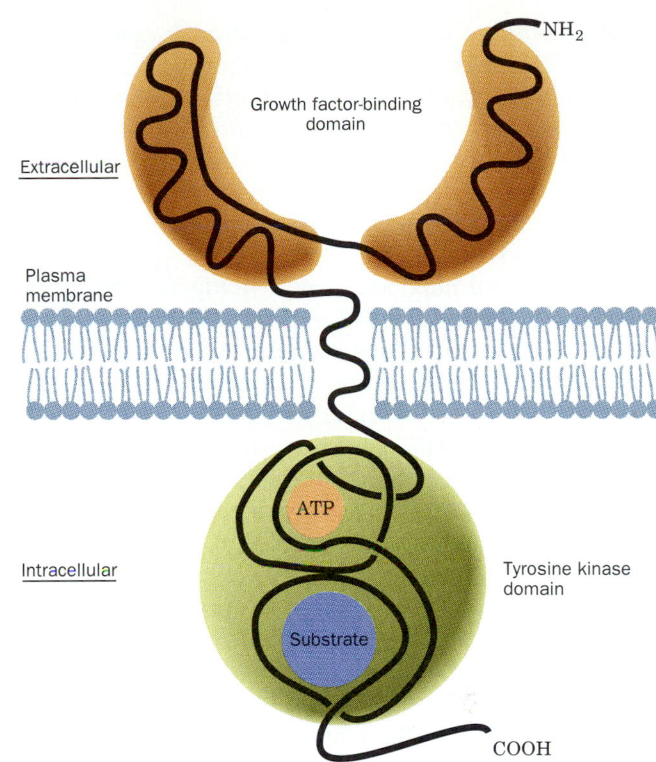

Figure 17.5 Schematic drawing of the prototype structure of transmembrane growth factor receptors with protein tyrosine kinase activities. These receptor proteins play key roles in signal transduction from the cell surface to the cell nucleus; details of the molecular mechanisms by which they transmit these signals are still unknown. (After Y. Yarden and A. Ullrich, *Annu. Rev. Biochem.* 57: 443–478, 1988.)

p*jun* and p*fos* as Activators of Gene Transcription

The products of two protooncogenes, c-*jun* and c-*fos,* have recently been shown to be identical to proteins that had previously been demonstrated to be components of nuclear complexes that activated the transcription of specific genes. The ***product of c-jun is*** now known to be ***transcription factor AP-1,*** which was first identified as a nuclear factor required for transcription induced by certain tumor-promoting compounds. Transcription factor AP-1 (=p*jun*) had been shown to bind specifically to enhancer elements in the simian virus 40 genome and in the human metallothionein IIA gene. ***The DNA-binding sites for***

AP-1 (=pjun) have a core consensus sequence of TGACTCA. Even more recently, the ***product of the c-fos protooncogene has been shown to form a tight complex with the c-jun gene-product.*** Both protooncogene proteins contain leucine-rich motifs that have the potential to form α-helical regions with leucine side chains projecting from the same face of the helix at regular intervals. Such proteins are proposed to interact by forming so-called ***"leucine zippers"*** with the leucine side chains of the two proteins interdigitated. Whether the "leucine zipper" model is correct or not remains to be determined. However, in any case, it is clear that the c-*jun* and c-*fos* products form a tight complex that functions as a *trans*-activator of transcription from enhancer/promoter regions that contain the TGACTCA consensus binding sequence.

A model for the mode of action of the complex containing the protein products of c-*jun* and c-*fos* is shown in Fig. 17.6. Note that the products of these protooncogenes are commonly designated simply c-Jun and c-Fos (the corresponding genes are c-*jun* and c-*fos*). The *trans*-activation of the transcription of responder genes by the c-Jun/c-Fos complex has now been demonstrated in several laboratories. Present research is directed at identifying more of the genes that are regulated by this c-Jun/c-Fos complex and at determining what factors regulate the expression of the c-*jun* and c-*fos* protooncogenes themselves. This c-Jun/c-Fos story is unfolding very rapidly, and our discussion of it promises to be out-of-date long before this textbook is in print.

Mutational Origin of *ras* Cellular Oncogenes

The oncogenes present in cancer cells can sometimes be identified by their ability to transform cells growing in culture (in particular, those of a mouse cell culture line called NIH 3T3) to a neoplastic state by means of transfection experiments as described in a preceding section of this chapter. When such transfection experiments have been used to search for oncogenes in the genomes of various types of human cancer cells, such oncogenes are frequently found. Moreover, when these ***oncogenes from human cancer cells*** are cloned and characterized, they ***are frequently found to be derivatives of c-ras protooncogenes.***

The genomes of all vertebrates contain three distinct, but closely related *ras* protooncogenes. Two of these, ***c-H-ras*** and ***c-K-ras,*** are closely related to the v-*ras* oncogenes of the Harvey and Kirsten strains,

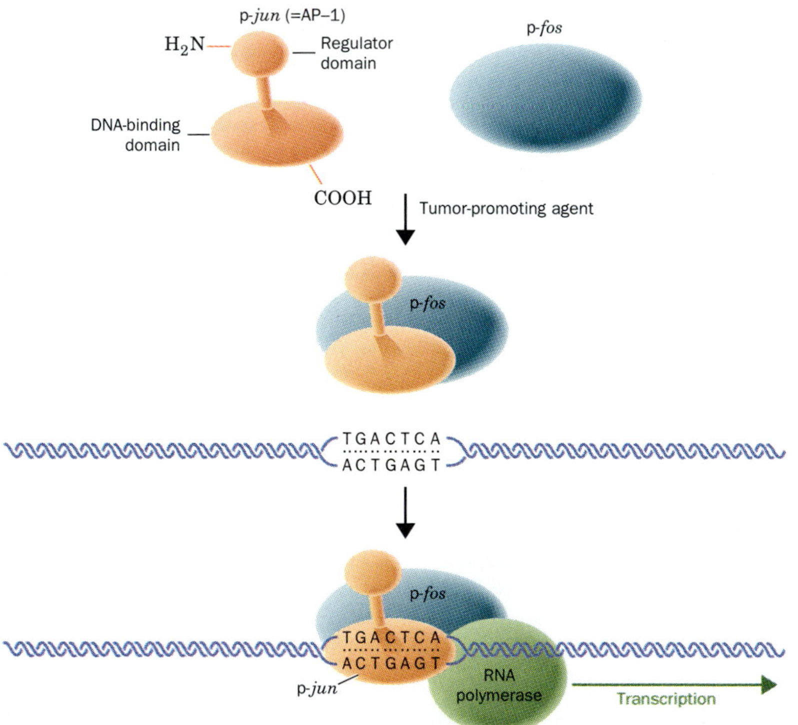

Figure 17.6 Diagrammatic illustration of the interaction of p*jun* and p*fos* and their role in the *trans*-activation of transcription of genes with TGACTCA consensus binding sequences in enhancer/promoter controlling elements. The interaction of p*jun* and p*fos*, their specific binding to DNA at TGACTCA sequences, and their effect on transcription are firmly established; other aspects of the model shown are still speculative. (After R. Chiu, W. J. Boyle, J. Meek, T. Smeal, T. Hunter, and M. Karin, *Cell* 54: 541–552, 1988.)

respectively, of the murine sarcoma virus of the rat (see Table 17.1). The third, designated *N*-ras, has not been found to have a homolog in any retroviral genome to date. All three of these cellular *ras* protooncogenes are known to encode very similar GTP-binding proteins with GTPase activity. Unfortunately, the functions of these *ras* proteins are still unknown.

All three of these cellular *ras* protooncogenes have been shown to undergo mutation to oncogenic derivatives that can be detected by transfection experiments by using the genomic DNAs from spontaneous or carcinogen-induced tumors. In fact, the majority of the NIH 3T3 cell-transforming oncogenes that have been detected in human tumor cells have turned out to be variants of one of the three cellular *ras* protooncogenes. The first cellular oncogene to be characterized in detail was derived from a human bladder carcinoma called EJ. When *the cellular oncogene present in these EJ bladder tumor cells* was cloned and sequenced, it *was found to be a derivative of the c-H-*ras *protooncogene.* Surprisingly, the *oncogenicity of the EJ c-H-*ras *mutant gene was found to result from a single base-pair substitution,* that is, a single base-pair difference was shown to correlate with the ability or inability of the two genes to transform NIH 3T3 cells growing in culture. The oncogene was produced from its protooncogene by a GC ⟶ TA transversion. *This mutation results in the substitution of valine for the glycine present as the twelfth amino acid (from the amino terminus) in the normal c-H-*ras *protein.*

In contrast to many of the retroviral oncogenes, the EJ c-H-*ras* oncogene from the human bladder carcinoma does not result in the synthesis of abnormally large amounts of its protein product. At present, we don't know why such a small change in a protooncogene, a normal cellular gene, should produce an oncogene capable of transforming cells to the cancerous state. Recent studies have indicated that both the normal cells and the tumor cells of some patients with carcinomas are heterozygous for the c-H-*ras* oncogene and the c-H-*ras* protooncogene. These results suggest that the oncogene causes a predisposition, rather than an immediate change, to the cancerous state.

Subsequent to the characterization of the c-H-*ras* oncogene present in the EJ human bladder carcinoma line, *oncogenic variants of the three c-*ras *protooncogenes have been detected and characterized in a large number of different mammalian cancer cell lines.* These have included cells from a variety of *lung, colon, mammary,* and *bladder tumors* plus *neuroblastomas, fibrosarcomas, a teratocarcinoma, a melanoma,* and a *promyelocytic leukemia.* When the oncogenes present in these cancer cells were cloned and sequenced, all were found to be variants of one of the three c-*ras* protooncogenes. Moreover, all the variant *ras* alleles with oncogenic potential as assayed by NIH 3T3 transfection experiments resulted in *amino acid substitutions at one of three amino acid positions in the* ras *gene-product.* All the mutations that conferred oncogenicity on the *ras* genes involved one or more of three codons: *codons number 12, 59, and 61.* Some of the mutations that have been documented in c-*ras* genes in various cancer cells are shown in Table 17.2. Note that the base substitutions that gave rise to the mutant *ras* alleles and the amino acid changes that these mutations produce in the mutant *ras* proteins are given in boldface type in Table 17.2. As would be expected, many of the mutant alleles result from nucleotide-pair substitutions that produce changes of basic amino acids to neutral or acidic amino acids or vice versa and thus change the charge of the mutant *ras* gene-product.

Although we do not know how these mutations in the c-*ras* genes give rise to their oncogenicity, we can

TABLE 17.2 Mutational Origin of c-*ras* Oncogenes in Cancer Cells[a]

TISSUE SOURCE	*ras* ALLELE	CODON 12	CODON 59	CODON 61
Normal human cells	c-H-*ras*	G G C gly	G C C ala	C A G gln
Normal human cells	c-K-*ras*	G G T gly	G C A ala	C A A gln
Normal human cells	N-*ras*	G G T gly	G C T ala	C A A gln
Human bladder cancer	EJ/T24-H-*ras*	C **T** C **val**	G C C ala	C A G gln
Human lung cancer	HS242-H-*ras*	G G C gly	G C C ala	C **T** G **leu**
Human lung cancer	Calu-K-*ras*	**T** G T **cys**	G C A ala	C A A gln
Human colon cancer	SW480-K-*ras*	G **T** T **val**	G C A ala	C A A gln
Human lung cancer	LL-10-K-*ras*	**C** G T **arg**	?	?
Human bladder cancer	A1698-K-*ras*	**C** G T **arg**	?	?
Human lung cancer	PR371-K-*ras*	**T** G T **lys**	?	?
Human lung cancer	PR310-K-*ras*	G G T gly	G C A ala	C A **T** **his**
Human neuroblastoma	SK-N-*ras*	G G T gly	G C T ala	**A** A A **lys**
Human melanoma	Mel-N-*ras*	G G T gly	G C T ala	**A** A A **lys**
Human lung carcinoma	SW1271-N-*ras*	G G T gly	G C T ala	C **G** A **arg**

[a]Note that the nucleotide changes present in the oncogenic *ras* alleles and the corresponding amino acid changes in the mutant *ras* proteins are given in boldface print. (Data are from Table 7 of H. E. Varmus, *Annu. Rev. Genet.* 18:553–612, 1984.)

be quite certain that it involves an activity that is dependent on the amino acids present at positions 12, 59, and 61 in the *ras* polypeptides. Interestingly, these mutations do not seem to alter the GTP-binding properties of the *ras* proteins, but they do reduce or eliminate the GTPase activities of these proteins. This property of the mutant *ras* proteins suggests a possible mode of action. Other GTP-binding proteins with GTPase activity, the so-called G proteins, are known to interact with adenylate cyclase and alter the cAMP levels in cells and, thus, modify the metabolic behavior of these cells. These G proteins are plasma membrane proteins that are inactive except when stimulated by specific hormone–receptor interactions. When stimulated, the G proteins bind GTP and modulate adenylate cyclase activity. However, the GTP bound to G proteins is then hydrolyzed by the resident GTPase activity, returning the protein to the inactive state. If the *ras* proteins act by a similar mechanism, the ***mutational loss of the GTPase activity could lock the* ras *protein in the active form, which might trigger continual cell division and tumor formation.*** Obviously, the results of further studies are needed to evaluate the accuracy of this model and to identify the normal cellular functions of the c-*ras* protooncogene products.

At present, it appears likely that the mutant oncogenic derivatives of the c-*ras* genes present in cancer cells will provide important clues that lead to the discovery of the normal cellular functions of the c-*ras* protooncogenes, once again illustrating the power of the mutational dissection approach in investigating biological processes. Whatever these functions are, it seems safe to conclude that the ***c-*ras *gene-products play key roles in the control of cell division.***

Translocation Breakpoints at Protooncogene Loci

For years, cytogeneticists have documented ***correlations between certain types of cancer and specific changes in chromosome structure.*** In particular, translocations (the breakage and transfer of parts of chromosomes to nonhomologous chromosomes; see Chapter 18, p. 495) and deletions or deficiencies (the breakage and loss of parts of chromosomes; see Chapter 18, p. 490) involving specific chromosomes, and, more importantly, often breakpoints at the same positions on these chromosomes, were repeatedly observed in certain types of cancer cells. The best-known example of this is the so-called ***"Philadelphia" chromosome,*** an altered chromosome 22 that has lost a large segment of its long arm. This abnormal chromosome has been found in various studies in ***up to 90 percent of the patients suffering from*** a specific type of cancer called ***chronic myelogenous leukemia.***

Now that the chromosomal locations of several human cellular protoncogenes have been determined (Fig. 17.7), a striking correlation is evident between their locations and the chromosomal breakpoints of translocations and deficiencies observed in specific types of cancer cells. These results have led to speculation that the observed chromosome breaks and rearrangements may have caused altered expression of protooncogenes or other important regulatory genes present in the vicinity of the breakpoints.

The Philadelphia chromosome (named after the city in which it was discovered) was first detected in patients with chronic myelogenous leukemia in 1960. Initially, the Philadelphia chromosome was thought to be a simple deletion of the distal portion of the long arm of chromosome 22. Subsequently, the ***Philadelphia chromosome was*** shown to have been ***produced by a reciprocal translocation involving the ends of the long arms of chromosomes 9 and 22*** (Fig. 17.8). The ***breakpoints on chromosome 9*** that give rise to these translocations ***occur very close to the c-*abl *protooncogene,*** and the exchange ***transfers the c-*abl *gene to chromosome 22.*** In some patients, abnormal c-*abl* transcripts are produced; in other patients, the breakpoints occur farther from, but always 5′ (relative to the direction of transcription) to c-*abl.* Although we still do not know what is cause and what is effect in this oncogenic pathway, it seems likely that the abnormal expression of c-*abl* resulting from the translocation event is involved in one step of the pathway.

Another type of cancer that is consistently (>80 percent) associated with specific kinds of translocations is ***Burkitt's lymphoma,*** a cancer of the antibody-producing B lymphocytes. This lymphoma is very common in certain regions of Africa and frequently develops in individuals infected with Epstein–Barr virus. The translocations observed in B cells of patients with Burkitt's lymphoma invariably involve chromosome 8 and one of the three chromosomes (2, 14, and 22) that carry genes encoding antibody chains (see Chapter 16). Translocations involving chromosomes 8 and 14 are by far the most common; chromosome 14 carries the heavy chain antibody genes. These B cell lymphomas generally secrete antibodies, thus the heavy chain antibody genes are being expressed in these tumor cells.

The most common sites of the chromosome breaks that give rise to the translocations between chromosomes 8 and 14 in Burkitt's lymphoma cells are in bands q24 and q32, respectively (Fig. 17.9). The ***c-*myc *protooncogene is located in band q24 on chromosome 8,*** and ***c-*myc *is transferred adjacent to the heavy chain antibody genes on chromosome 14 by the translocation events.*** Again, as for the Philadelphia chromosome, we do not understand the consequences of translocating the c-*myc* protooncogene to

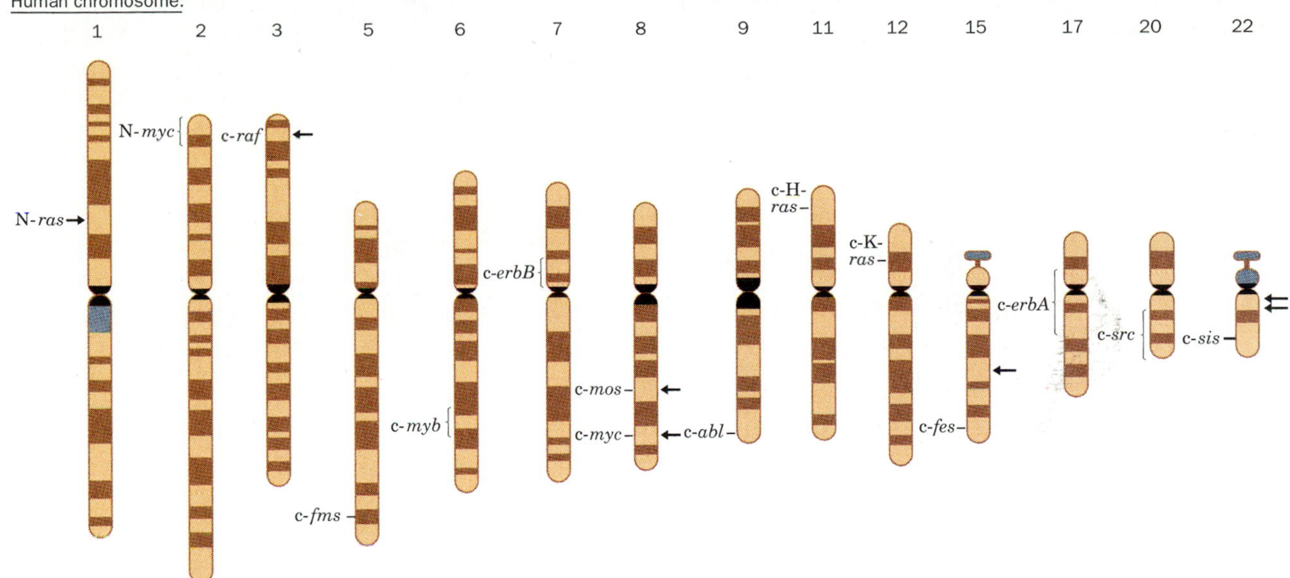

Figure 17.7 Chromosomal locations of fifteen human protooncogenes and of breakpoints (arrows to the right of the chromosomes) observed in translocations found to be consistently associated with specific types of cancer (see text). The symbols for the various protooncogenes are derived from the names of the retroviruses that carry the homologous viral oncogenes (see Table 17.1). (Based on data summarized in Fig. 11 of M. M. Le Beau and J. D. Rowley, *Adv. in Hum. Genet.* 15: 1–54, 1986.)

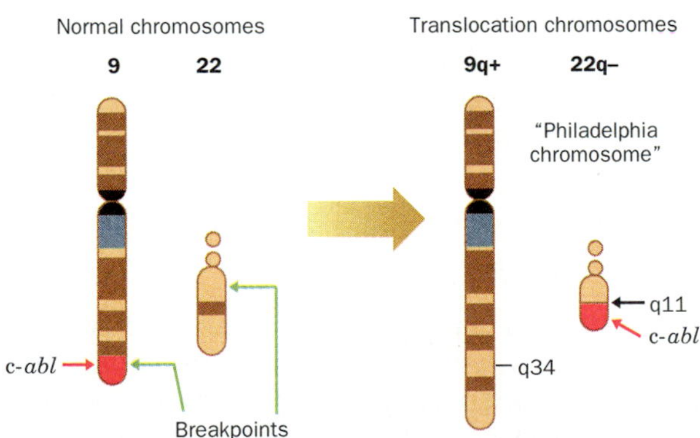

Figure 17.8 Diagram of the reciprocal translocations between chromosomes 9 and 22 that give rise to the Philadelphia chromosomes found in about 90 percent of the patients with chronic myelogenous leukemia. The breakpoints on chromosome 9 occur near the location of the c-*abl* protooncogene, and the resulting translocations transfer the c-*abl* gene to the Philadelphia chromosome (translocation chromosome 22). (After M. M. Le Beau and J. D. Rowley, *Adv. in Hum. Genet.* 15: 1–54, 1986.)

chromosome 14 next to the antibody genes. However, it is generally assumed that the c-*myc* is expressed abnormally in the new location, and that this abnormal expression somehow contributes to the oncogenic transformation of Burkitt's lymphoma cells.

Other chromosomal translocation events have been detected with high frequencies in other types of cancer cells as well, although none with as high an incidence as the translocations associated with chronic myelogenous leukemia and Burkitt's lymphoma. For a discussion of the chromosome aberrations detected in various types of cancer cells, the reader is referred to an excellent review by Le Beau and Rowley (see References).

Insertional Activation of Protooncogenes

Scientists have known for over a decade that the RNA tumor viruses are of two distinct types: (1) the ***acute transforming viruses*** like Rous sarcoma virus that carry oncogenes such as v-*src* and (2) the ***slow transforming viruses*** that ***do not carry oncogenes*** and induce transformations of cells to the neoplastic state only after extended latent periods (usually several

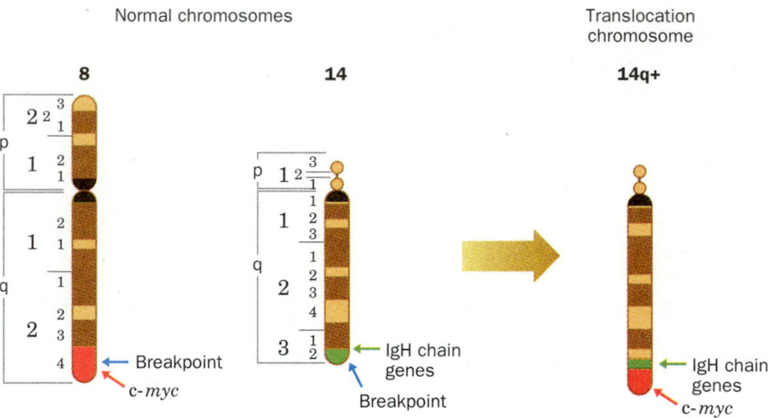

Figure 17.9 Diagram showing the nature of the reciprocal translocation events that give rise to the abnormal 14 $q+$ chromosomes prevalent in tumor cells of patients with Burkitt's lymphoma. Although the translocations usually involve reciprocal exchanges of the ends of the long arms of chromosomes 8 and 14, only the translocation chromosome (14 $q+$) that carries both the c-*myc* protooncogene and the heavy chain antibody genes is shown. (After M. M. Le Beau and J. D. Rowley, *Adv. in Hum. Genet.* 15: 1–54, 1986.)

months). The key question was how these retroviruses that do not carry oncogenes could induce neoplastic transformations. Extensive evidence now indicates that the ***slow transforming viruses most frequently induce cancers by integrating as proviruses adjacent to protooncogenes and, in so doing, activating these protooncogenes to an "overexpressed" state*** (Fig. 17.10). The ***long terminal repeats (LTRs)*** of the DNA proviral forms of RNA tumor viruses (see Chapter 9, Fig. 9.13) ***contain very strong enhancer/promoter elements,*** and integration of these proviruses can lead to increased rates of transcription of adjacent genes.

One of the best-known examples of retroviral activation of a normal cellular protooncogene involves ***B cell lymphomas induced by the avian leukosis virus (ALV).*** The ALV genome does not contain an oncogene. However, ALV is a serious pathogen of chickens; it results in many types of cancers in infected flocks, with lymphomas being the most common. Molecular analysis of genomic DNAs in these lymphomas showed that in many cases an ***ALV provirus had integrated adjacent to the c-*myc* protooncogene*** and had activated transcription such that ***c-*myc* transcript levels were 30- to 100-fold higher than in normal cells.*** Moreover, these transcripts were shown to contain an ALV LTR sequence at the 5′ terminus, indicating that transcription was initiated from the proviral LTR promoter (Fig. 17.10). These results strongly suggest that the lymphomas resulted from the overexpression of c-*myc* caused by the integration of the proviral LTRs with their strong enhancer/promoters adjacent to c-*myc*.

Many similar examples of activation of protooncogenes by the insertion of either intact proviruses or just proviral LTRs have now been described. Early examples that were defined in detail by molecular analyses involved the protooncogenes c-*myc*, c-*myb*, and c-*erbB*. Interestingly, activation may occur by proviral integrations on either side of the protooncogene or within introns of the protooncogene. These results suggest that the enhancer elements present within the proviral LTRs are often responsible for the observed transcriptional activation.

Amplification of Protooncogenes in Cancer Cells

One mechanism that can lead to ***increased levels of a particular gene-product in a cell is to amplify the number of copies of the gene encoding that product.*** Sometimes such an amplification event will occur as a normal component of the process of development as in the case of the amplification of the rRNA genes during oögenesis in animals (see Chapter 15, p. 425). In other cases, amplification can be induced to occur by selecting for increased tolerance to an inhibitor of an essential enzyme. Finally, extensive evidence now indicates that ***specific protooncogenes are frequently amplified in particular types of cancers.*** Although we still do not have proof that these amplification

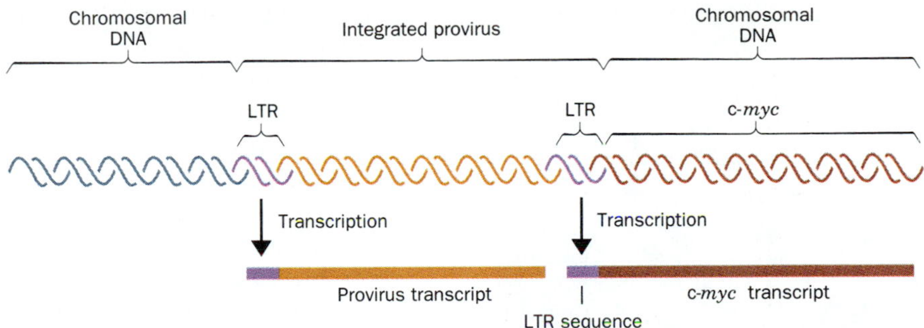

Figure 17.10 Diagram of the process of insertional activation of protooncogenes by RNA tumor proviruses. The example shown is based on the activation of the c-*myc* gene in chicken B cell lymphomas by insertion of proviral genomes of avian leukosis virus adjacent to the c-*myc* locus. Note that the long terminal repeats (LTRs) of the provirus contain a very strong enhancer/promoter element that can lead to greatly elevated levels of transcription of adjacent genes (like c-*myc* in the example shown here). (After W. S. Hayward, B. G. Neel, and S. M. Astrin, *Nature* 290: 475–480, 1981.)

events play a causative role in the oncogenic processes that give rise to these cancer cells, it seems likely that the amplification of a protooncogene and the resulting overexpression of the protooncogene product might well be involved in oncogenesis since these gene-products are known to play a role in the control of cell division.

The best-known example of induced gene amplification involves the tolerance of animal cells growing in culture to the drug methotrexate. ***Methotrexate inhibits the enzyme dihydrofolate reductase,*** an enzyme that catalyzes an essential step in dTMP synthesis (and, thus, in DNA synthesis). Methotrexate binds to the active site of dihydrofolate reductase and prevents it from binding its normal substrate. If one selects for cells with tolerance to gradually increasing concentrations of methotrexate, some of the cells will become tolerant by amplifying the gene that encodes dihydrofolate reductase. The methotrexate-tolerant cells contain more copies of this gene and synthesize more dihydrofolate reductase than normal cells. As a result, they can tolerate higher levels of methotrexate without being killed. Some of the enzyme molecules will bind methotrexate and be inhibited, but with more enzyme, enough free enzyme molecules will exist to allow the cells to survive and grow. The more gene copies that are present (the greater the degree of amplification), the more enzyme that will be synthesized, and the higher the concentration of methotrexate that the cells can "tolerate" and still survive. In some methotrexate-tolerant cell lines, the dihydrofolate reductase gene is highly amplified with over 1000 copies per cell.

The extra copies of the dihydrofolate reductase gene in these methotrexate-tolerant cell lines are present either (1) on extra very small chromosomes called ***"double minutes"*** or ***DMs*** (Fig. 17.11*a*) or (2) as tandemly repeated sequences within so-called ***"homogeneously staining regions"*** or ***HSRs*** (Fig. 17.11*b*) of otherwise normal chromosomes in the genome. The double-minute chromosomes are supernumerary chromosomes that contain the amplified gene and adjacent chromosomal DNA on extrachromosomal circular molecules of DNA. They contain the circular molecules of DNA packaged in nucleosomes and chromatin fibers just like normal chromosomes. These small chromosomes look like two small dots in chromosome spreads (thus the name "double minute"). The circular DNA molecules in these DM chromosomes are predominantly in a postreplication state with the two DNA circles still attached to each other; this explains their bipartite structure. The DNA molecules present in the DM chromosomes range in size from about 50 kilobase-pairs (kb) to several hundred kb. The unit of the chromosome that undergoes the amplification process is often referred to as an ***amplicon.*** In all cases studied to date, the size of the amplicon has been much larger than the size of the gene encoding the enzyme target of the drug used in the selection process. The same amplicon unit that is present in DMs is often present as the tandemly repeated unit within the HSR regions of chromosomes containing amplified genes. A general model of the amplification process is shown in Fig. 17.12; this model has been shown to be accurate on a gross level for the amplification of the dihydrofolate reductase gene in response to selection for tolerance to methotrexate in the protozoan *Leishmania tropica.* Whether or not it provides an accurate picture of the amplification process in higher animals and plants is still uncertain. In no case do we understand the molecular mechanism by which the amplification process occurs.

There is now considerable evidence indicating

(a)

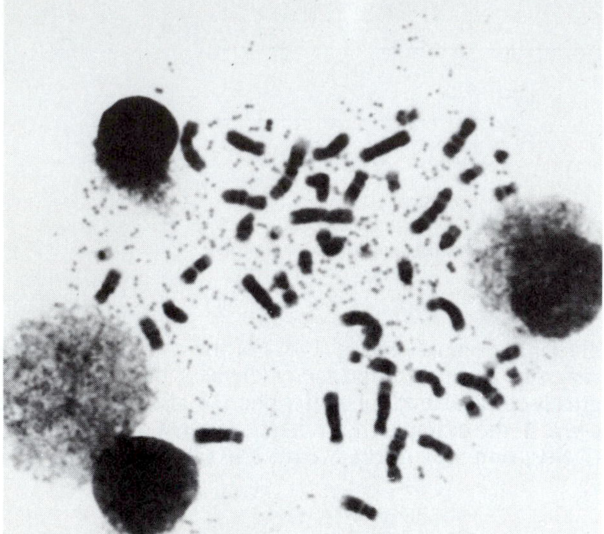

(b)

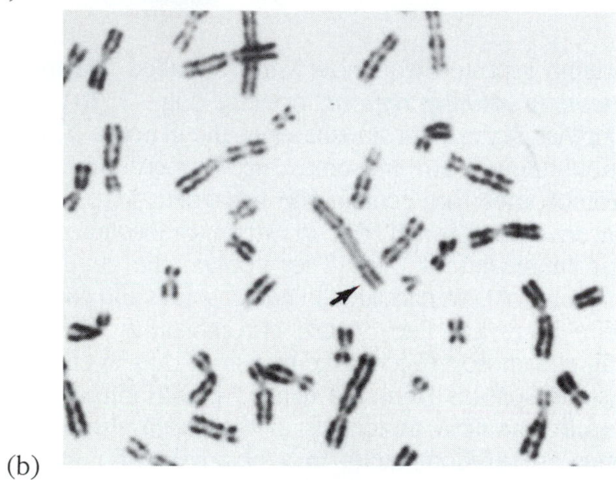

Figure 17.11 Photomicrographs showing the morphology of (a) double-minute chromosomes (DMs) and (b) homogeneously staining regions (HSRs, arrow) of chromosomes carrying amplified genes. The DMs contain circular DNA molecules carrying the amplified gene and adjacent segments of chromosomal DNA, and the HSRs contain tandemly repeated copies of the amplified gene and adjacent chromosomal DNA. The segment of chromosomal DNA that undergoes the amplification process, sometimes called the ***amplicon,*** usually is several hundred kilobase-pairs in length; the amplified unit is thus much larger than a single gene. [(a) DM chromosomes present in an abdominal solid tumor—a stage IV neuroblastoma. Photograph courtesy of B. Streifel and D. Arthur, Dept. of Laboratory Medicine and Pathology, University of Minnesota. (b) HSR region (arrow) on the long arm of the X chromosome detected by replication banding on the mitotic chromosomes from cell line COLO-320 derived from a human colon carcinoma. Photograph courtesy of Jaroslav Cervenka, Division of Cytogenetics and Cell Genetics, Health Sciences, University of Minnesota.]

that the ***amplification of cellular protooncogenes may be directly involved in the progression of oncogenesis in certain types of human cancers.*** In some cases, the amplified protooncogene is present on DM chromosomes; in other cases, the amplified protooncogene is part of a tandemly repeated amplicon present in an HSR of a chromosome. In a few cases, the cancer cells contain both DMs and HSRs. In particular, **c-myc** is found to be ***amplified*** with a very high frequency in small cell carcinomas of the lung and with lower frequencies in some other types of cancer. In an early study, Alitalo and colleagues used *in situ* hybridization and autoradiography to demonstrate the presence of ***multiple copies of c-myc in an HSR*** on an abnormal X chromosome in malignant cells of a patient with a carcinoma of the colon. The malignant cells of this patient also contained DM chromosomes, and these DMs probably carried amplified copies of c-*myc* as well. However, the sensitivity of the autoradiography was not sufficient to establish for certain that the DM chromosomes carried copies of c-*myc.* Many other studies have yielded similar results and demonstrated the amplification of c-*myc* in several different types of human cancer cells, particularly carcinomas of the lung. In addition, two cellular genes that are closely related to c-*myc,* namely, L-*myc* and N-*myc,* are frequently found to be amplified in lung carcinomas and neuroblastomas, respectively. Finally, c-*erbB* is often present in the amplified state in squamous cell carcinomas and glioblastomas.

Presumably, the effect (if indeed there is an effect) ***of amplification of these cellular protooncogenes results from the overproduction of the protooncogene product.*** The amplification events probably are not involved in the initiation of oncogenesis, but may contribute at subsequent steps in the oncogenic pathway. However, we should hasten to add that there is no solid evidence demonstrating that the amplification of any protooncogene plays a causative role in oncogenesis. The amplification events could be nothing more than secondary effects of other steps in oncogenic pathways. Further evidence will be needed before the putative role of protooncogene amplification in oncogenesis can be evaluated. Nevertheless, the ***recurrent amplification of specific protooncogenes in particular types of cancers suggests that this correlation may be more than chance*** and certainly warrants further study. Moreover, given the known central roles of the products of some protooncogenes in the intercellular communications circuitry, it seems likely that the overproduction of certain protooncogene products might well contribute to oncogenesis.

Origin of Viral Oncogenes

Regardless of how oncogenes induce cancers, it now seems clear that ***retroviral oncogenes have evolved from normal cellular protooncogenes.*** Originally, it was thought that the cellular homologs of viral oncogenes might be relics of integrated retroviral proviruses. However, this is clearly not the case. Compari-

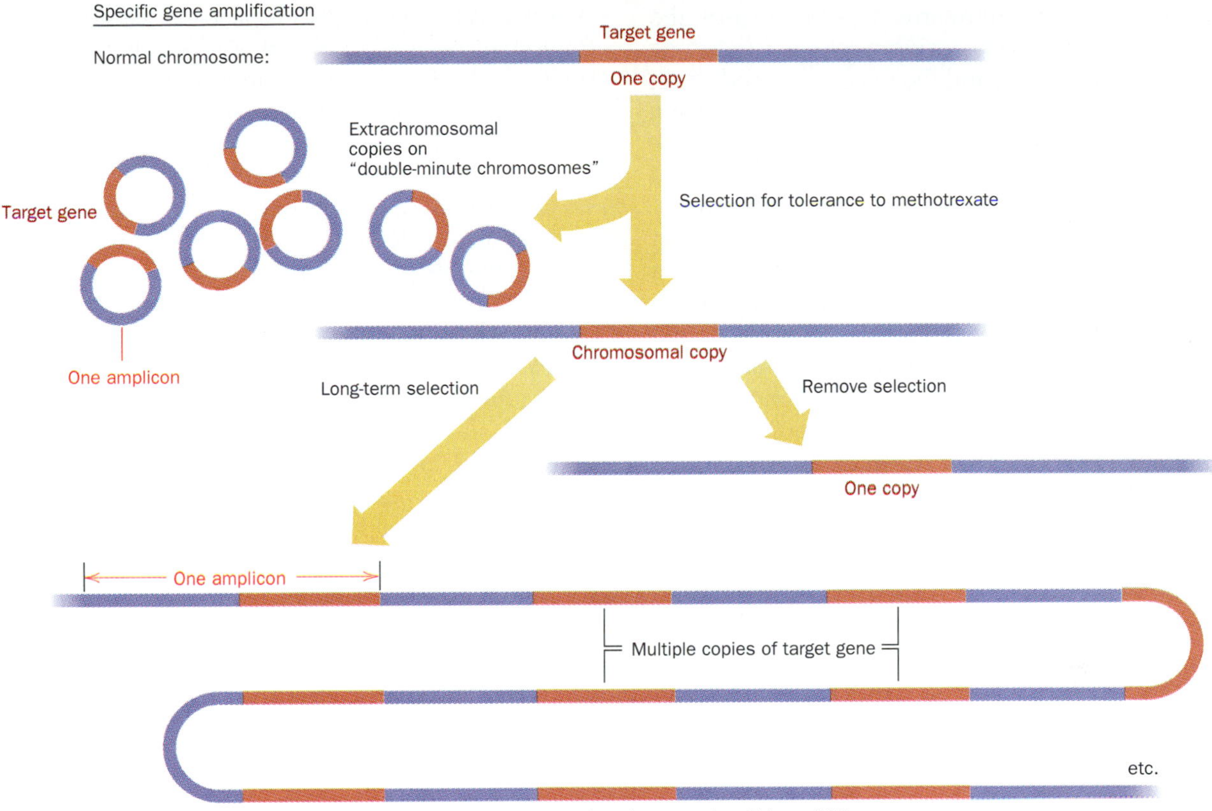

Figure 17.12 Schematic drawing of the mechanism of gene amplification. The model presented here has been shown to be accurate in gross details in the case of the amplification of the dihydrofolate reductase gene of the protozoan *Leishmania tropica* in response to selection for tolerance to the drug methotrexate. Methotrexate specifically inhibits the enzyme dihydrofolate reductase, an essential enzyme in DNA precursor biosynthesis. The structures of the DMs and HSRs that contain amplified copies of protooncogenes (most frequently, c-*myc*) in particular types of human cancer cells appear to be very similar (see Figure 17.11) although molecular details of the structures of the amplicons are not yet available in the latter cases. (After S. M. Beverley, J. A. Coderre, D. V. Santi, and R. T. Schimke, *Cell* 38: 431–439, 1984.)

sons of the nucleotide sequences of viral oncogenes and the homologous cellular protooncogenes have shown that these genes share major regions of sequence identity. The key difference is that the ***cellular protooncogenes contain introns,*** whereas ***viral oncogenes are single exons.*** This is ***not*** consistent with the idea that cellular protooncogenes have evolved from v-oncogenes on integrated proviruses. Instead, it strongly suggests that v-oncogenes are derived from ancestral cellular protooncogenes. This difference is to be expected if the v-oncogenes evolved from cellular protooncogenes. The retroviral genomes are RNA, and the intron sequences of RNA transcripts of protooncogenes should be spliced out during RNA processing (Chapter 10). All that needs to occur is for an mRNA copy of a protooncogene to be ligated into the RNA genome of a retrovirus by a recombination mechanism that preserves the LTR regions of the viral genome. (The viral LTR sequences are required for proviral integration.) The viral reverse transcriptase will then convert the mRNA–viral RNA hybrid into homologous DNA for integration into the host genome. What could be of greater value to a virus than to have a new gene that stimulates increased growth of its host, while its integrated genome goes along for the ride?

In some cases, different retroviruses that infect distantly related species have acquired copies of the same cellular protooncogene. For example, the simian sarcoma virus of the monkey and the P1 feline sarcoma virus of the cat both carry viral oncogenes derived from the c-*sis* protooncogene. In other cases, closely related viruses contain oncogenes derived from completely unrelated cellular protooncogenes.

By comparing the nucleotide sequences of the v-oncogenes and the homologous c-protooncogenes, the sites of breakage and joining in the recombination events that gave rise to the v-oncogenes can sometimes be identified. In other cases, extensive rearrangements have occurred, making it impossible to identify the sites of recombination involved in the acquisition of the oncogene by the retrovirus. In several cases, the viral oncogenes encode fusion proteins containing part of the *gag* protein and the oncogene product. In most cases (Rous sarcoma virus being the best-known exception), the retroviral acquisition of an oncogene has been accompanied by the loss of viral genetic material required for replication. Such defective viruses can integrate normally as proviruses, but can only produce progeny viruses in the presence of a "helper virus" that provides the missing function(s). These defective retroviruses are analogous to the phage lambda defective transducing particles discussed in detail in Chapter 8. Moreover, their ability to transfer cellular genes from one cell (a donor cell) to another cell (a recipient cell) is formally equivalent to ***transduction*** in bacteria (see Chapter 8).

CANCER AS THE END PRODUCT OF A MULTISTEP PROCESS

Finally, we should emphasize that a large amount of data indicates that the ***cancerous state is the end product of a multistep process.*** The cell lines used in transfection experiments are probably already at some intermediate stage in this pathway, possibly simply due to the selection for the ability to grow under cell culture conditions. The oncogene-induced transformation observed in cell cultures is undoubtedly only one part of a more complex pathway.

In fact, there is considerable evidence indicating that certain oncogenes may have cooperative effects in promoting neoplastic transformations. Moreover, different oncogenes seem to play different roles in oncogenic pathways in different cell types. Finally, it seems likely that different molecular events are involved in the acquisition of the enhanced proliferative capacity of cells, in the ability of tumors to invade adjacent tissues, and in the capacity for metastasis. To what extent and in what roles protooncogenes and oncogenes are involved in these processes in human cancers remain to be determined. Regardless of the extent of their involvement in the formation of malignant tumors, the ***ongoing and future investigations of protooncogenes and oncogenes promise to yield important information about the molecular circuitry that controls cell proliferation in higher eukaryotes such as humans.***

SUMMARY

As cells proceed through the ***cell cycle, DNA replication and the chromosome condensation and chromatid separation events of M phase take place in a precise, coordinated manner.*** Cells make commitments to proceed through the cell cycle at two points: (1) ***Start,*** late in G_1, at which the ***decision is made to initiate DNA synthesis*** at the beginning of the subsequent *S* phase; and (2) the ***onset of M phase,*** at which cells ***become committed to the ensuing chromosome condensations and chromatid disjunctions of mitosis.*** Although the complete regulatory circuitry controlling the passage of cells through the cell cycle is still unknown, some of the key components have been identified. Proteins called ***cyclins*** are synthesized and accumulate during interphase and are degraded during *M* phase. A special class of cyclins called ***G_1-cyclins*** also appear to be degraded after the Start commitment.

The ***protein kinase*** designated ***pp34 is a key cell cycle regulator;*** pp34 forms a complex with cyclins that acts as a ***mitosis-promoting factor (MPF).*** Protein kinase pp34 may also control the Start commitment by means of an interaction with the G_1-cyclins. ***The pp34 regulatory protein undergoes inactivation/activation cycles controlled by the phosphorylation/dephosphorylation of a specific tyrosine residue, tyrosine 15.*** A yeast mutant with a phenylalanine residue (not subject to phosphorylation) at position 15 of pp34 exhibits abnormal cell cycle regulation. This mutant initiates cell division prematurely; its phenotype suggests that its cell cycle regulator is locked in the "on" or "divide" state.

The regulation of cell division in ***multicellular eukaryotes*** requires even more complex control mechanisms because each tissue within an organ and each organ in the body must grow to the proper size in the proper position and coordinate its activities with those of the other tissues and organs of the body. Thus, ***intercellular communication*** is of paramount importance in multicellular organisms. Although details of the molecular signals involved in intercellular communication and their modes of action remain poorly understood, a specific set of genes, called ***protooncogenes,*** is known to ***encode proteins that function in various pathways of intercellular communication.***

Retroviral ***oncogenes*** code for products that somehow cause normal cells to undergo ***transformation*** to the cancerous state. Genes that are homologous to the viral oncogenes are found as normal components of the chromosomes of all higher animals. These genes, the ***protooncogenes,*** can be converted to ***cellular oncogenes*** (1) by ***mutation,*** (2) by ***becoming associated with new regulatory sequences,*** for example, by the insertion of a retroviral provirus nearby, or (3) by ***amplification*** resulting in overproduction of the protooncogene product.

Cellular protooncogenes can be detected by ***cross-hybridization to retroviral oncogene probes.*** Oncogenic derivatives of cellular protooncogenes can be identified in ***transfection experiments*** on the basis of their ability to cause cells growing in culture to undergo ***transformation to a neoplastic state*** in which the cells exhibit ***a loss of contact inhibition*** and form "tumors" *in vitro* instead of growing as cellular monolayers.

Over 20 different retroviral oncogenes and ***their homologous cellular protooncogenes*** have been identified and studied. The protooncogenes are highly conserved in all higher animals. Genes with strong homology to several of the protooncogenes are present in *Drosophila,* and homologs to some protooncogenes are even present in the yeast *Saccharomyces cerevisiae.* This conservation of protooncogene structure indicates that these genes encode products that play important roles in the life cycles of a wide variety of species.

The products of some of the cellular protooncogenes have been identified and shown to be well-known growth factors or growth-factor receptors. In addition, the product of one protooncogene, c-erbA, has now been shown to be a ***thyroid hormone receptor protein.*** Finally, the products of two protooncogenes, c-jun and c-fos, have recently been shown to form a complex that is equivalent to a previously characterized trans-*acting transcriptional activator.* The products of other protooncogenes have been found to be ***transmembrane proteins with GTP-binding and GTPase activities.*** Finally, the products of the majority of the protooncogenes are ***protein kinases*** with either ***tyrosine-specific*** or ***serine/threonine-specific kinase activity.*** Although the functions of these proteins are still unknown, they are believed to play central roles in ***signal transduction,*** that is, in transmitting signals from the cell surface to the cell nucleus and the transcriptional machinery within the cell nucleus.

The three members of the c-ras ***protooncogene family (c-K-***ras, ***C-H-***ras, and ***N-***ras***)*** have been shown to ***mutate to allelic cellular oncogenic variants in a variety of different human cancers.*** In almost every case, the c-*ras* oncogene was produced by a ***single nucleotide-pair substitution*** in one of three triplets in the gene, specifically, the ***triplets specifying codons number 12, 59, and 61 in the mRNA.*** Changes in amino acids at positions 12, 59, and 61 in the c-*ras* proteins often seem to result in a loss of the GTPase activity of the gene-product. This loss of GTPase activity, in turn, is believed to cause a loss of the normal regulation of c-*ras* protein activity and, thus, the oncogenicity of the mutant c-*ras* allele.

Cellular protooncogenes contain interrupted coding sequences with exons and introns like most other normal cellular genes. In contrast, ***retroviral oncogenes are single exons.*** Comparisons of the nucleotide sequences of homologous cellular protooncogenes and viral oncogenes indicate that the retroviral oncogenes have originated from processed transcripts of the cellular protooncogenes.

The malignant phenotype of a cancer cell is the end product of a multistep process. The evidence available at present indicates that oncogenes may be involved in promoting certain steps in this pathway, although the exact roles of these oncogenes remain to be determined. One point is very clear: ***studies on the functions of protooncogene products are beginning to provide a wealth of information about the molecular circuitry that regulates cell division in higher eukaryotes.***

REFERENCES

ALITALO, K., M. SCHWAB, C.C. LIN, H.E. VARMUS, and J.M. BISHOP. 1983. "Homogeneously staining chromosomal regions contain amplified copies of an abundantly expressed cellular oncogene (c-*myc*) in malignant neuroendocrine cells from a human colon carcinoma." *Proc. Natl. Acad. Sci. U.S.A.* 80: 1707–1711.

BARBACID, M. 1986. "Mutagens, oncogenes and cancer." *Trends in Genet.* 2: 188–192.

______. 1987. "*ras* genes." *Annu. Rev. Biochem.* 56: 779–827.

BEVERLEY, S.M., J.A. CODERRE, D.V. SANTI, and R.T. SCHIMKE. 1984. "Unstable DNA amplifications in methotrexate-resistant *Leishmania* consist of extrachromosomal circles which relocalize during stabilization." *Cell* 38: 431–439.

BISHOP, J.M. 1981. "Enemies within: The genesis of retrovirus oncogenes." *Cell* 23: 5–6.

______. 1982. "Oncogenes." *Sci. Amer.* 246(3): 80–92.

______. 1985. "Trends in oncogenes." *Trends in Genet.* 1: 245–249.

______. 1985. "Viral oncogenes." *Cell* 42: 23–38.

______. 1987. "The molecular genetics of cancer." *Science* 235: 305–311.

CHIU, R., W.J. BOYLE, J. MEEK, T. SMEAL, T. HUNTER, and M. KARIN. 1988. "The c-Fos protein interacts with c-Jun/AP-1 to stimulate transcription of AP-1 responsive genes." *Cell* 54: 541–552.

COLE, M.D. 1986. "The *myc* oncogene: Its role in transformation and differentiation." *Annu. Rev. Genet.* 20: 361–384.

COOPER, G.M. 1982. "Cellular transforming genes." *Science* 217: 801–806.

GOLDBERG, Y., C. GLINEUR, J.-C. GESQUIERE, A. RICOUART, J. SAP, B. VENNESTROM, and J. GHYSDAEL. 1988. "Activation of protein kinase C or cAMP-dependent protein kinase increases phosphorylation of the c-*erbA*-encoded thyroid hormone receptor and of the v-*erbA*-encoded protein." *EMBO J.* 7: 2425–2433.

GOULD, K.L., and P. NURSE. 1989. "Tyrosine phosphorylation of the fission yeast $cdc2^+$ protein kinase regulates entry into mitosis." *Nature* 342: 39–45.

GREEN, M.R. 1989. "When the products of oncogenes and anti-oncogenes meet." *Cell* 56: 1–3.

HADWIGER, J.A., C. WITTENBERG, H.E. RICHARDSON, M. DE BARROS LOPES, and S. I. REED. 1989. "A family of cyclin homologs that control the G_1 phase in yeast." *Proc. Natl. Acad. Sci. U.S.A.* 86: 6255–6259.

HALAZONETIS, T.D., K. GEORGOPOULOS, M.E. GREENBERG, and P. LEDER. 1988. "c-*jun* dimerizes with itself and with c-*fos*, forming complexes of different DNA binding affinities." *Cell* 55: 917–924.

HARTWELL, L.H., and T.A. WEINER. 1989. "Checkpoints: Controls that ensure the order of cell cycle events." *Science* 246: 629–634.

HAYMAN, M.J. 1986. "*erb*-B: Growth factor receptor turned oncogene." *Trends in Genet.* 2: 260–262.

HAYWARD, W.S., B.G. NEEL, and S.M. ASTRIN. 1981. "Activation of a cellular *onc* gene by promoter insertion in ALV-induced lymphoid leukosis." *Nature* 290: 475–480.

HUNTER, T., and J.A. COOPER. 1985. "Protein-tyrosine kinases." *Annu. Rev. Biochem.* 54: 897–930.

LANGDON, W.Y., J.W. HARTLEY, S.P. KLINKEN, S.K. RUSCETTI, and H.C. MORSE III. 1989. "v-*cbl*, an oncogene from a dual-recombinant murine retrovirus that induces early B-lineage lymphomas." *Proc. Natl. Acad. Sci. U.S.A.* 86: 1168–1172.

LAZAR, M.A., R.A. HODIN, D.S. DARLING, and W.W. CHIN. 1989. "A novel member of the thyroid/steroid hormone receptor family is encoded by the opposite strand of the rat c-*erbA*α transcriptional unit." *Molec. Cell. Biol.* 9: 1128–1136.

LE BEAU, M.M., and J.D. ROWLEY. 1986. "Chromosomal abnormalities in leukemia and lymphoma: Clinical and biological significance." *Adv. Human Genet.* 15:1–54.

LEWIN, B. 1990. *Genes IV.* Cell Press, Cambridge, MA/Oxford Univ. Press, Oxford.

LUGO, T.G., and O.N. WHITE. 1989. "The BCR-ABL oncogene transforms rat-1 cells and cooperates with v-*myc*." *Molec. Cell. Biol.* 9: 1263–1270.

MARX, J.L. 1988. "*jun* is bustin' out all over." *Science* 242: 1377–1378.

MCCORMICK, F. 1989. "*ras* GTPase activating protein: Signal transmitter and signal terminator." *Cell* 56: 5–8.

MURRAY, A.W. 1989. "The cell cycle as a *cdc2* cycle." *Nature* 342: 14–15.

MURRAY, A.W., and M.W. KIRSCHNER. 1989. "Dominoes and clocks: The union of two views of the cell cycle." *Science* 246: 614–621.

NUSSE, R. 1986. "The activation of cellular oncogenes by retroviral insertion." *Trends in Genet.* 2: 244–247.

PARDEE, A.B. 1989. "G_1 events and regulation of cell proliferation." *Science* 246: 603–608.

PARKER, R.C., H.E. VARMUS, and J.M. BISHOP. 1981. "Cellular homologue (*c-src*) of the transforming gene of Rous sarcoma virus: Isolation, mapping, and transcriptional analysis of *c-src* and flanking regions." *Proc. Natl. Acad. Sci. U.S.A.* 78: 5842–5846.

PIECHACZYK, M., J.-M. BLANCHARD, and P. JEANTEUR. 1987. "c-*myc* gene regulation still holds its secret." *Trends in Genet.* 3: 47–51.

RABBITTS, T.H. 1985. "The c-*myc* proto-oncogene: Involvement in chromosomal abnormalities." *Trends in Genet.* 1: 327–331.

REDDY, E.P., R.K. REYNOLDS, E. SANTOS, and M. BARBACID. 1982. "A point mutation is responsible for the acquisition of transforming properties by the T24 human bladder carcinoma oncogene." *Nature* 300: 149–152.

SCHUERMANN, M., M. NEUBERG, J.B. HUNTER, T. JENUWEIN, R.-P. RYSECK, R. BRAVO, and R. MULLER. 1989. "The leucine repeat motif in *fos* protein mediates complex formation with *jun*/AP-1 and is required for transformation." *Cell* 56: 507–516.

SCHWAB, M. 1985. "Amplification of N-*myc* in human neuroblastomas." *Trends in Genet.* 1: 271–275.

SEFTON, B.M. 1985. "Oncogenes encoding protein kineases." *Trends in Genet.* 1: 306–308.

SHILO, B.-Z. 1987. "Proto-oncogenes in *Drosophila melanogaster.*" *Trends in Genet.* 3: 69–72.

STARK, G.R., and G.M. WAHL. 1984. "Gene amplification." *Annu. Rev. Biochem.* 53: 447–491.

TABIN, C.J., S.M. BRADLEY, C.I. BARGMANN, R.A. WEINBERG, A.G. PAPAGEORGE, E.M. SCOLNICK, R. DHAR, D.R. LOWY, and E.H. CHANG. 1982. "Mechanisms of activation of a human oncogene." *Nature* 300: 142–149.

TAYEYA, T., and H. HANAFUSA. 1983. "Structure and sequence of the cellular gene homlgous to the RSV *src* gene and the mechanism for generating the transforming virus." *Cell* 32: 881–890.

VARMUS, H.E. 1982. "Form and function of retroviral proviruses." *Science* 216: 812–820.

______. 1984. "The molecular genetics of cellular oncogenes." *Annu. Rev. Genet.* 18: 553–612.

VERMA, I.M. 1986. "Proto-oncogene *fos:* A multifaceted gene." *Trends in Genet.* 2: 93–96.

WEINBERG, R.A. 1982. "Fewer and fewer oncogenes." *Cell* 30: 3–4.

______. 1985. "The action of oncogenes in the cytoplasm and nucleus." *Science* 230: 770–776.

YARDEN, Y., and A. ULLRICH. 1988. "Growth factor receptor tyrosine kinases." *Annu. Rev. Biochem.* 57: 443–478.

YUNIS, J.J. 1983. "The chromosomal basis of human neoplasia." *Science* 221: 227–236.

PROBLEMS AND QUESTIONS

17.1 What function(s) do cyclins play in the mitotic cell cycle of eukaryotic cells?

17.2 Site-specific mutagenesis was used to replace either tyrosine 15 or tyrosine 19 in the cell cycle regulatory protein pp34, the product of gene *cdc2* of *Schizosaccharomyces pombe,* with a phenylalanine residue. The two *cdc2* mutant alleles that were produced have been referred to as *cdc2.Phe15* and *cdc2.Phe19,* respectively. Phenylalanine is structurally very similar to tyrosine, but is not subject to phosphorylation by protein kinases as is tyrosine. What effect(s), if any, would you expect these modified *cdc2* alleles to have on the cell cycle of cells carrying either (a) *cdc2.Phe15* or (b) *cdc2.Phe19* in the homozygous or hemizygous (haploid) state?

17.3 What evidence is there that the protooncogenes found in chromosomes of normal cells and normal animals are not simply integrated retroviral oncogenes?

17.4 The nucleotide-pair sequences of cellular protooncogenes are highly conserved in a large number of species in the animal kingdom. Does this conservation of protooncogene structure tell us anything about the functions of protooncogenes?

17.5 Which is most likely: (1) that cellular protooncogenes originated from viral oncogenes or (2) that viral oncogenes are derivatives of cellular protooncogenes?

17.6 Variant c-*ras* oncogenes have been detected in a large number of human cancers. What is the origin of these c-*ras* oncogenes?

17.7 The majority of the c-*ras* oncogenes have alterations in one of three codons in the c-*ras* mRNA that lead to substitutions of amino acids in the c-*ras* proteins at positions 12, 59, or 61 from the amino terminus. How can the clustering of mutational events to these three sites in the three c-*ras* genes be explained?

17.8 Describe three different mechanisms by which normal cellular protooncogenes can be converted to cellular oncogenes.

17.9 Provide a molecular mechanism for the conversion of a normal cellular protooncogene to a cellular oncogene by means of the integration of the long terminal repeat (LTR) of a retroviral provirus adjacent to the protooncogene.

17.10 Of what significance are double-minute chromosomes (DMs) and homogeneous staining regions (HSRs) in chromosomes in certain types of cancers?

17.11 How might the amplification of a protooncogene lead to oncogenesis?

17.12 What is the structure of the Philadelphia chromosome? Does the structure of this diagnostic chromosome have any relationship to oncogenes or protooncogenes?

17.13 The products of several of the cellular protooncogenes have been shown to be previously studied components of cells. What are these protooncogene products, and what do their identities tell us about the normal functions of protooncogenes?

Chapter 18

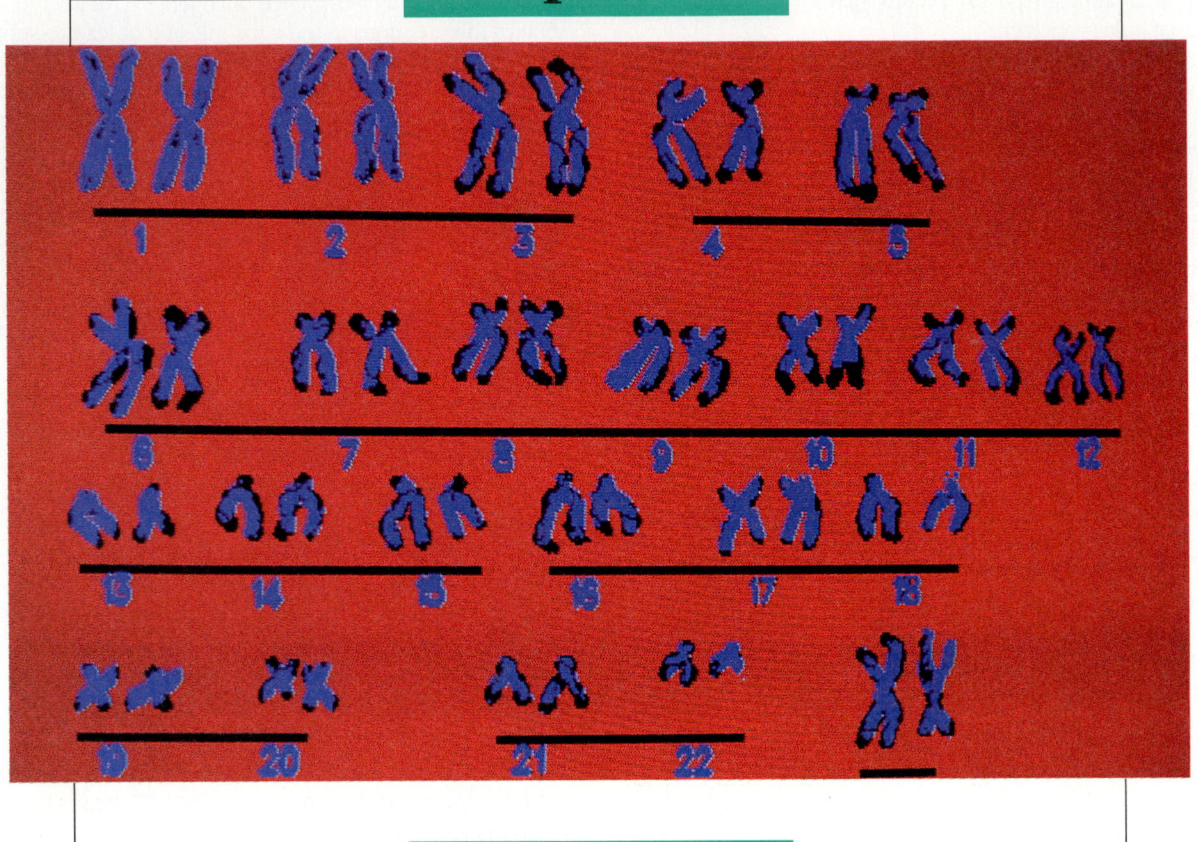

Variations in Chromosome Structure

(Photo above): Computer generated human female karyotype. (Courtesy of Don Kelly, Grant Heilman Photography, Inc.)

STRUCTURAL changes require breaks in the chromosomes. More than one break can occur in a single chromosome or set of chromosomes, and the broken parts may then reunite in new arrangements. Any ***broken end may unite with any other broken end,*** thus potentially resulting in ***new linkage*** arrangements. The loss or addition of a chromosome segment may also occur in the process. More than one type of aberration may occur at the same time. For example, a section may be broken off and lost during the formation of an inversion or translocation; this simultaneously produces a deficiency.

Identifying chromosomal aberrations by observation is a major problem for the cytogeneticist because members of a pair usually lack visible markers for identifying different areas along their length. Primary and secondary constrictions and total length are useful, but more markers are needed. Maize chromosomes are exceptions. They are favored materials for microscopic study because meiotic prophase chromosomes have deep-staining bodies, called ***heteropyknotic knobs*** (Fig. 18.1), distributed along their length. With the aid of these visible markers, many chromosome changes have been detected in maize. The acetocarmine smear method, first applied to other plant materials by J. Belling in 1931, greatly facilitated studies of the maize chromosomes. This technique permits whole chromosomes to be fixed, stained, and spread on a microscope slide in one operation. It provides a way to compare individual chromosomes within a chromosome set or genome, as well as whole sets from different organisms.

The first cytological demonstration of plant chromosomal rearrangements was made in maize by B. McClintock. Working with pachytene and other meiotic prophase stages that present large chromosomes for microscopic observation, she eventually demonstrated that irregular configurations made by chromosome rearrangements in the pairing process led to four different kinds of structural changes: (1) ***deficiencies*** (parts of chromosomes lost or deleted), (2) ***duplications*** (parts added or duplicated), (3) ***inversions*** (sections detached and reunited in reverse order), and (4) ***translocations*** (parts of chromosomes detached and joined to nonhomologous chromosomes). Comparable demonstrations were later made with the giant polytene chromosomes of *Drosophila*.

GIANT POLYTENE CHROMOSOMES IN DIPTERA

Large coiled bodies about 150–200 times as large as gonad-cell chromosomes were observed in the nuclei of glandular tissues of dipterous larvae as early as 1881 by E. G. Balbiani. He described banded structures in the nuclei of cells of larval midges in the genus *Chironomus* but did not attach any genetic significance to the observation. D. Kostoff (in 1930) suggested a relation between the bands of these structures and the linear sequence known to occur among genes. The anatomical significance of the nuclear bodies was further studied by E. Heitz and H. Bauer in the genus *Bibio*, a group of March flies whose larvae feed on the roots of grasses. These authors identified the bodies as ***giant chromosomes*** that occurred ***in pairs.*** They described the morphology in detail and discovered the relation between the salivary gland chromosomes and other somatic- and germ-cell chromosomes. They also demonstrated that comparable elements occurred in the giant chromosomes and in the chromosomes of other cells of the same organism.

It was largely because of the work of T. S. Painter that *Drosophila* salivary gland chromosomes (Fig. 18.2) were first used for the cytological verification of genetic data. Painter related the ***bands*** on the giant chromosomes to genes, but he was more interested in

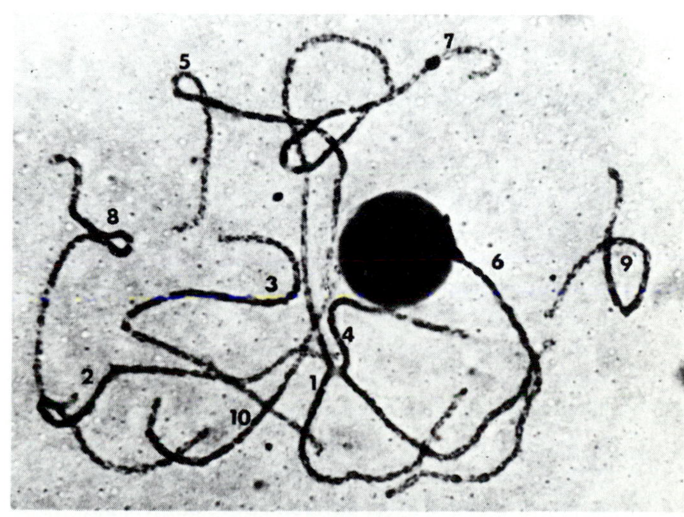

Figure 18.1 The ten pairs of chromosomes in maize, shown in pachytene of prophase I of meiosis. The large nucleolus is associated with the nucleolar organizer region near the end of chromosome 6. The positions of the centromeres and some of the more conspicuous knobs are used to identify specific chromosome pairs (bivalents). The 10 bivalents are numbered 1–10; these correspond with the 10 linkage groups of maize shown in Figure 7.8. (From R. L. Phillips, "Molecular Cytogenetics of the Nucleolus Organizer Region," in *Maize Breeding and Genetics,* pp. 711–741, D. B. Walden, editor. Copyright © 1978 by Wiley, New York. Original photograph courtesy of R. L. Phillips.)

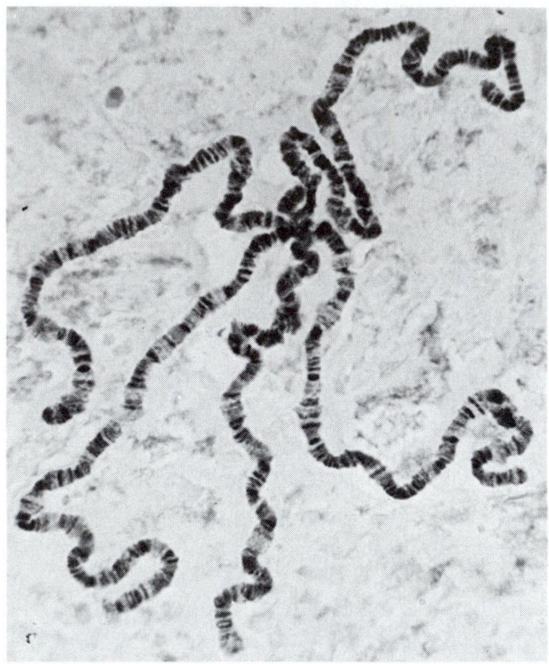

Figure 18.2 Salivary gland chromosomes of *Drosophila melanogaster*. (Courtesy of B. P. Kaufmann, reprinted with permission from *J. Heredity* 30: 179–190, 1939.)

the morphology of the chromosomes and implications concerning speciation than in associating chromosome sections with particular genes. About 25,000 base-pairs are now estimated for each band. C. B. Bridges made extensive and detailed studies of the salivary gland chromosomes and, in the course of his investigations, developed a tool of practical usefulness ***in relating genes to chromosomes***. In applying this method to *Drosophila melanogaster*, he prepared a series of cytological (physical) maps to correspond with the genetic linkage maps already available.

Features of Polytene Chromosomes

The unusual size of the salivary gland chromosomes is a product of the type of growth that occurs in larval glandular tissues of dipterous insects. Salivaries and other glands grow by enlargement rather than by duplication of individual cells. This can be demonstrated by cell counts and measurements taken at different stages in the development of a larva. As a larva develops, the synapsed *Drosophila* chromosomes replicate (usually nine times), giving rise to a polytene strand of about 1000 DNA double-helix molecules. The cross bands are chromomeres, densely packed with chromatin, and the interband regions are chromonemata. Giant chromosomes are thus ***cable-like structures*** with many DNA strands (polytene). Giant polytene chromosomes thus correspond in linear structure with other chromosomes of the same species. The difference is that the ***duplicate strands*** (chromatids) are held together in bundles, and they do not separate out to new cells through cell division.

Another feature that makes the giant salivary chromosomes valuable for study is their continuous state of ***somatic synapsis***. If one member of a homologous pair is altered by deficiency, duplication, inversion, or translocation, an irregularity occurs in pairing. Characteristic and observable irregularities make it possible to recognize different kinds of chromosome modifications and to identify their location on the chromosome.

About 5000 single cross bands have been noted on the four pairs of salivary gland chromosomes in *D. melanogaster*. This number was considered by H. J. Muller to be a minimum approximation of the number of genes in that insect. B. Judd demonstrated by studies of deficiencies that each band corresponds with a unit of genetic function (complementation unit; see Chapter 12). Genes have thus been associated with individual bands.

Linkage data do not correspond exactly with cytological locations, but the linear sequence of genes can be verified from salivary preparations. Not only can the linear sequence be verified, but it agrees generally with the linear sequence determined by cytological mapping. The lack of identical correspondence lies in the relative distances between genes determined by the two methods. Frequency of crossing-over is not constant along the length of the chromosome, but is less near the centromere. The main uses of polytene chromosomes are in (1) ***locating genes*** and (2) ***identifying structural changes*** in chromosomes.

DEFICIENCIES

A single break near the end of a chromosome would be expected to result in a terminal deficiency. If two breaks occur, a section may be deleted and an ***intercalary deficiency*** created. Terminal deficiencies might seem less complicated and more likely to occur than those involving two breaks. Instead, the great majority of deficiencies detected thus far are of the intercalary type within the chromosome. The ends of chromosomes have folded-over "hairpins" protecting them from breaks and rejoining. When a deficiency occurs, the chromosome set is left without having the genes carried in the deleted portion, unless the deleted part becomes fused to a chromosome that has a centromere. Without a centromere, a chromosome section cannot move to the pole of the spindle during cell division, but lags in the dividing cell and is excluded from the chromosome group when the nuclear membrane forms around the chromosomes of a daughter cell.

When an intercalary part of a chromosome is missing, a buckling effect may be observed microscopically in the paired salivary gland chromosomes (Fig. 18.3). Large deficiencies are more readily detected than small ones, but with good optical equipment and patience, an investigator may be able to see single bands of the salivary preparations and thus to identify minute heterozygous deficiencies. By identifying the part of the polytene chromosome in which the buckle occurs and then studying the phenotype of flies carrying a recessive gene in the homologous chromosome opposite the deficiency, the gene can be spatially positioned on the chromosome. Expression of a single recessive opposite a deficiency is called pseudodominance. Chromosome deficiencies have greatly facilitated the ***checking of linkage maps***. The physical location of many genes is now precisely known in *D. melanogaster* and other species of diptera because of the effective use of this technique.

A somatic cell that has lost a small chromosome segment may live and produce other cells heterozygous like itself, each with the deleted section of a chromosome. Phenotypic effects sometimes indicate which cells or portions of the body have descended from the originally deficient cell. If the deficient cell is a gamete that is subsequently fertilized by a gamete carrying a nondeficient homolog, all cells of the resulting organism will carry the deficiency in the heterozygous condition. Recessive genes on the nondeficient chromosome in the region of the deficiency may express themselves (pseudodominance). Heterozygous deficiencies usually decrease the general viability of the flies that carry them. Flies carrying deficiencies in the homozygous condition usually die. Some very small homozygous deficiencies, however, are viable in *Drosophila*. Such deficiencies have been identified in the region in which the *w* (white eye) and *fa* (facet eye) genes are located near the end of the X chromosome.

DUPLICATIONS

Duplications represent additions of chromosome parts. Some chromosome segments behave like dominants and some like recessives with respect to certain phenotypes. Others show intermediate inheritance and still others have cumulative effects. Duplications provide a means for determining effects of chromosome segments when three, four, or ***more similar sections are present*** in individual animals or plants.

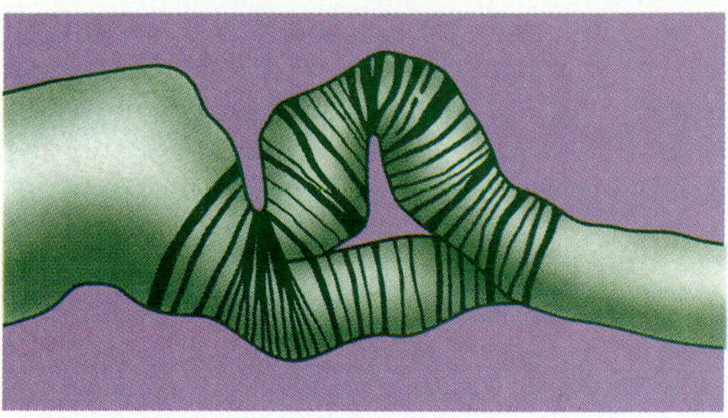

Figure 18.3 Appearance in the *Drosophila* salivary gland chromosomes of an intercalary deficiency showing a buckling effect. A section of about 45 bands is missing in one X chromosome. This deficiency is associated with a notched phenotype at the tips of the wings. (After H. Slizynska, *Genetics* 23: 291–299, 1938.)

The first duplications to be critically examined involved the *B* (bar) locus in the X chromosome of *Drosophila*. The eye of a heterozygous B/B^+ female is somewhat smaller than the normal eye (Fig. 18.4), and the sides are straighter, giving an oblong or bar appearance. In the homozygous or hemizygous condition, the eye is considerably smaller. Bridges and Muller discovered independently that the bar phenotype was the result of a duplication involving a part of the X chromosome already present in the wild-type flies. Both investigators observed not only the effect of a duplication that produced bar eye, but also a duplication that resulted in an extreme decrease in the size of the eye, which was called "double bar." By examining the salivary polytene chromosome, they identified the segments of the chromosome actually involved in the duplication. The different phenotypes and the corresponding segments of the salivary gland chromosome pairs are illustrated in Fig. 18.4. Polytene chromosome section 16A is present in flies with wild-type eyes. When this section is duplicated, it produces the bar phenotype; but when it is represented three times in a single chromosome, the double-bar phenotype results.

Each additional duplicate segment of section 16A makes the eye smaller. Other duplications that have since been found in *Drosophila* work in the opposite direction. These suppress the effects of mutant genes and make the fly appear more nearly normal with respect to certain traits. Further studies demonstrated that duplications need not occur in the immediate vicinity of the section duplicated to exert an influence. Chromosome fragments may become attached to entirely different chromosomes. Through the assortment of such chromosomes in gametes, duplications may be carried to succeeding generations.

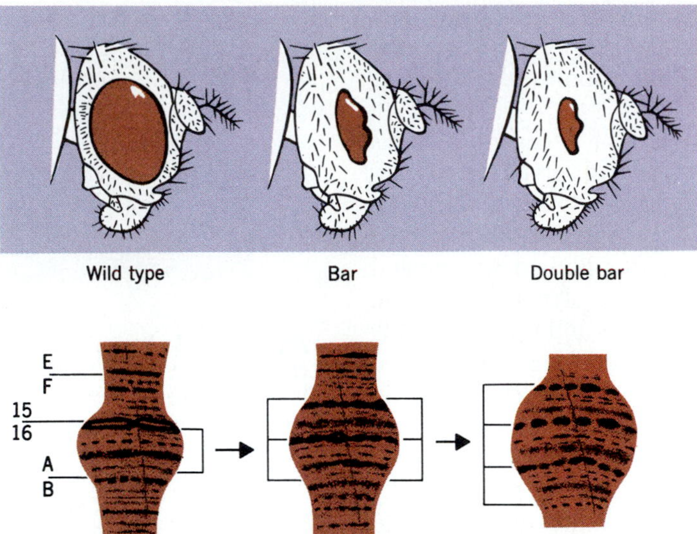

Figure 18.4 Effects on eye size of different arrangements of duplicated sections of 16A in the X chromosome of *D. melanogaster*. The 16A identifies a particular segment on the salivary chromosome following the standard numbering system for *D. melanogaster*. Positions of the 16A segment on a particular chromosome are related to size of the eye. (After C. B. Bridges, *Science* 83: 210–211, 1936.)

INVERSIONS

Inversions occur when parts of chromosomes become detached, turn through 180°, and are reinserted in such a way that the genes are in ***reversed order***. Some inversions presumably result from entanglements of the threads during the meiotic prophase and from the chromosome breaks that occur at that time. For example, a certain segment may be broken in two places, and the two breaks may be in close proximity because of a chance loop in the chromosome. When they rejoin, the wrong ends may become connected. The part on one side of the loop connects with a broken end different from the one with which it was formerly connected. This leaves the other two broken ends to become attached. The part within the loop thus becomes turned around or inverted. It is not known whether all inversions occur in this way, but this is a plausible explanation for many chromosome inversions.

Inversions may survive the meiotic process and segregate into viable gametes. As indicated in earlier chapters, chromosome pairing is essential to the production of fertile gametes. The mechanism by which homologous chromosomes heterozygous for inversions accomplish such pairing in the meiotic sequence is remarkable. The part of the uninverted chromosome corresponding to the inversion forms a ***loop***. A similar loop is formed by the inverted sections of the homologous chromosome, but in reverse direction. If, for example, the loop of the uninverted section is formed with the gene sequence in a clockwise direction, the inverted part will form in a counterclockwise direction. In this way, corresponding parts come together even though one of the sections is inverted, as illustrated diagrammatically in Fig. 18.5.

Inversions have been associated with the ***suppression of crossing-over***. Before *Drosophila* chromosomes were studied extensively, investigators had already identified genetic crossover suppressors in this organism. These were first considered to be genes that somehow interfered with crossing-over. It was shown that the locations of inversions and crossover suppressors coincided and that the apparent suppression of crossing-over was associated directly with inversions. The main process is not a suppression of crossing-over, although the frequency of physical crossing-over may be reduced. Crossover gametes that do occur are not recovered, that is, the zygotes die before they can be detected.

The relation between inversions and crossing-over can be demonstrated from studies of heterozygous inversions. Heterozygous paracentric inversions (Fig. 18.6), those with the centromere outside the inverted area, and heterozygous pericentric inversions (Fig. 18.7), those with the centromere inside the inverted region, can be used to illustrate why crossover gametes lead to inviability.

Chromosomes that carry ***paracentric*** inversions

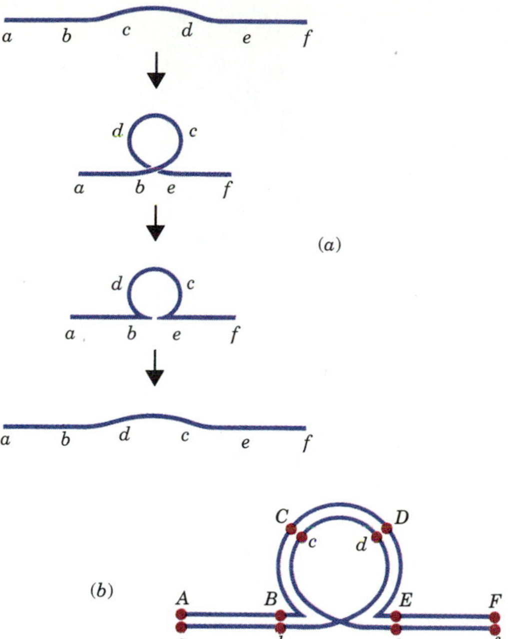

Figure 18.5 (a) Diagram illustrating the mechanism by which some inversions may occur. (b) Pairing of an inverted homolog has occurred with alleles c and d in reverse position. Only two of the four strands are shown. A loop may be observed in salivary gland preparations of *Drosophila* that carry heterozygous inversions.

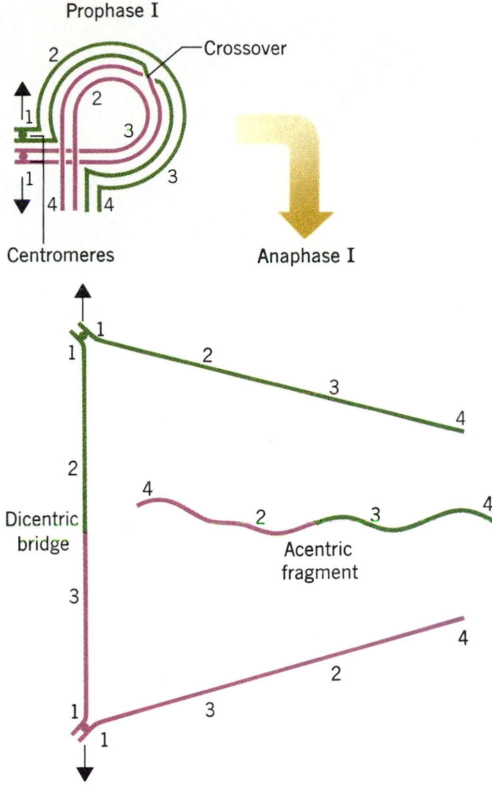

Figure 18.6 Meiotic prophase and anaphase, illustrating the mechanism through which a paracentric inversion acts as a crossover "suppressor." Dicentric (1 2 3 1) and acentric (4 3 2 4) chromosomes, which are unbalanced, are formed from crossover chromatids. Crossover "suppression" results from the unbalanced chromatids and the inability of the chromatids to reach the poles of the cell in division.

cannot cross over within the loop without producing fragments of chromosomes lacking centromeres (***acentric***) and chromosome complexes with two centromeres (***dicentric***). These result either in fragments that lag in the center of the spindle or chromatid bridges that tie together the two homologs involved and interfere with the division process. In either case, the chromosomes do not separate properly to their respective poles. Crossovers within the loops of ***pericentric*** inversions result in ***duplications and deficiencies***. The fate of these cells varies in animals and plants. Following a single crossover within the in-

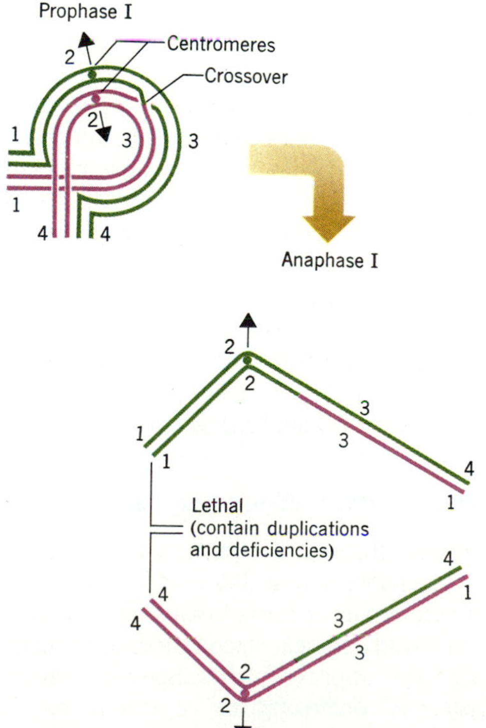

Figure 18.7 Diagram illustrating the crossover "suppressor" action of a pericentric inversion. The resulting chromatids carry both duplications and deficiencies and therefore lead to lethality. Crossover "suppression" results from the loss of the gametes containing chromosomes involved in crossovers.

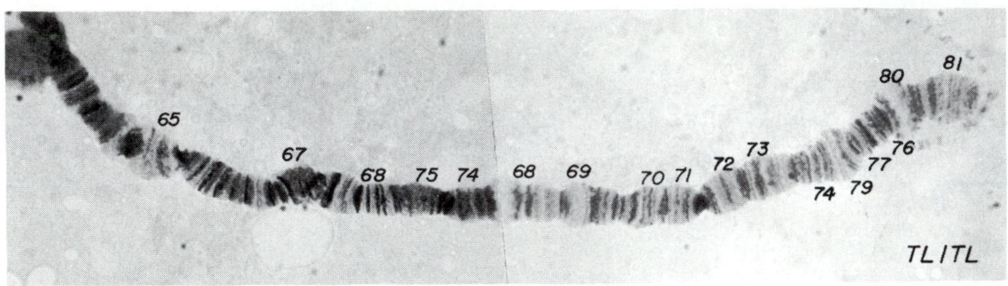

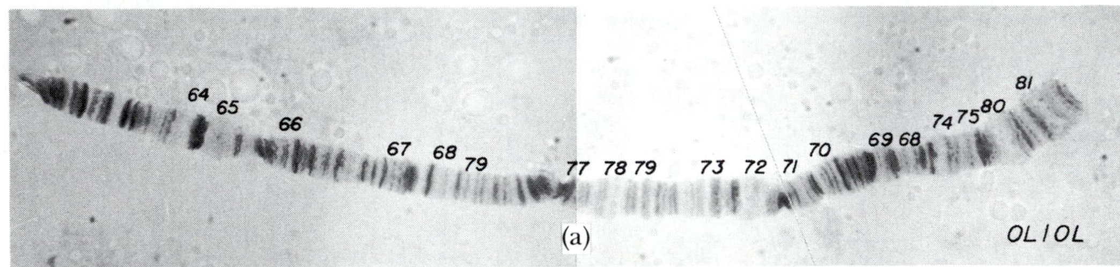

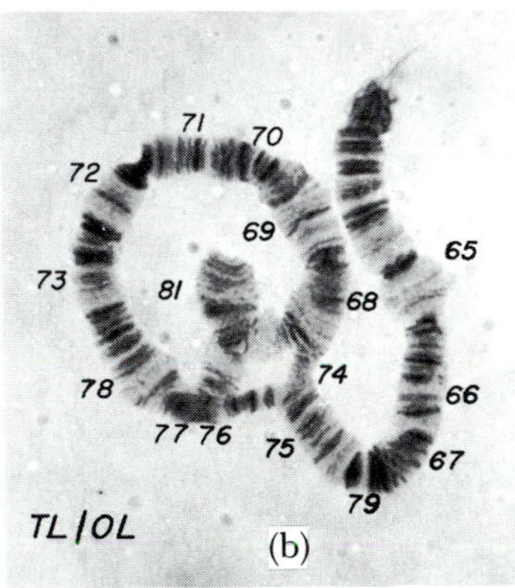

Figure 18.8 Salivary gland chromosomes illustrating an inversion. (a) Homozygous chromosomes *TL/TL* and *OL/OL.* (b) Heterozygote, *TL/OL,* with inversion loop. (Courtesy of C. D. Kastritis and D. W. Crumpaker. Reprinted with permission from *J. Heredity* 58: 113–129, 1967. Copyright © 1967 by the American Genetic Association.)

version loop of a maturing plant cell, the gametes receiving crossover chromatids are inviable; thus, crossing-over is effectively suppressed. In *Drosophila*, unbalanced zygotes produced by abnormal gametes die, thus eliminating crossover chromatids from the population. The apparent suppressing effect of inversions on crossing-over can thus be explained mostly on the basis of secondary results that follow crossing-over in inverted segments.

Polytene chromosomes of *Drosophila* have been especially useful for detecting heterozygous inversions in the flies. A characteristic loop in these chromosomes is illustrated in Fig. 18.8. Loops in giant chromosomes of salivary gland tissues presumably resemble those in meiotic prophase chromosomes, where observation is much more difficult.

Chromosomes with inversions have practical applications in maintaining *Drosophila* stocks. They are used as "balancers," that is, as chromosomes that can be placed opposite homologous chromosomes carrying certain genes that are homozygous lethal. Crossing-over is suppressed in such chromosomes, and it is possible to maintain a gene in the heterozygous state that would cause death in the homozygous condition. Laboratory stocks carrying several mutants are kept intact without crossing-over. Because recessive genes are not expressed when they are in heterozygous condition, however, frequent checks must be made to ensure that the gene is not lost from the stock. Appropriate outcrosses are conducted occasionally to check for the continued presence of the gene or the genes in the stock. A mechanism through which some laboratory stocks are kept balanced and heterozygous involves "balanced lethal mutations."

Inversions and Balanced Lethal Mutations

Sometimes through the process of mutation, two recessive lethals occur at different loci on each member of the same pair of homologous chromosomes. If the loci involved are near each other, or if chromosomal aberrations "suppress" crossing-over between them, an ***enforced heterozygous*** condition may be established. Since individuals homozygous for any lethal die, only those heterozygous for both lethals survive. The first documented case of a balanced lethal was that described by Muller in *Drosophila*. He had been maintaining in his laboratory the *Bd* (beaded) stock, which produced flies with scalloped wings. Since the beaded phenotype was controlled by a dominant homozygous lethal gene (*Bd*), the progeny from the crosses between beaded flies were two-thirds beaded and one-third normal, as illustrated in Fig. 18.9*a*. All those homozygous for beaded were dead; thus, a ***2 : 1 ratio*** replaced the 3:1 ratio, which would otherwise have been expected. This pattern is similar to that described in Chapter 2 for creeper chickens.

Abruptly, with no visible change in the beaded phenotype, all the flies in the culture began to appear beaded. In an attempt to explain the change that had occurred, Muller postulated that a new lethal (*l*) had been created by mutation in the homologous chromosome opposite *Bd*. An inversion that was associated with the gene for beaded "suppressed" crossing-over, and only two kinds of gametes were produced by the beaded flies—one carrying the gene (*Bd*) for beaded and the other carrying the new lethal (*l*). The cross representing the new arrangement is illustrated in Fig. 18.9*b*. When these chromosomes segregated and fertilization occurred, some zygotes became homozygous for *Bd* and were therefore lethal; others became homozygous for *l* and they also died. Only those heterozygous for both *Bd* and *l* survived. Therefore, the heterozygous condition was enforced. This explanation was substantiated, and the term ***"balanced lethal"*** was used to designate such an arrangement involving two lethals in the same chromosome pair in which recovered crossovers are infrequent or entirely absent.

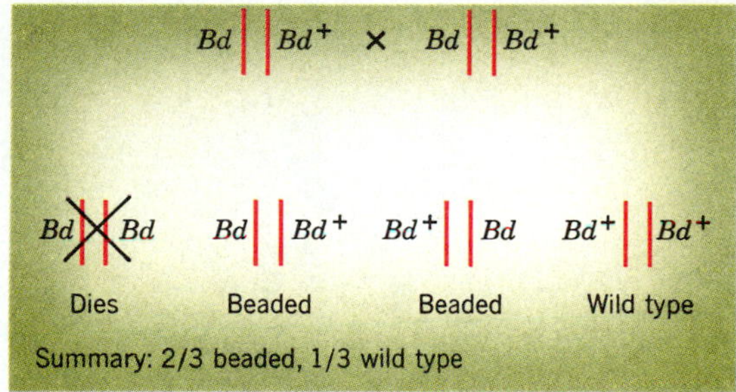

(a)

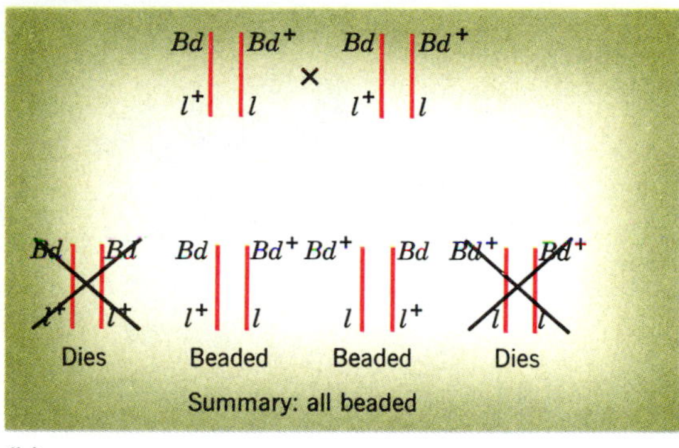

(b)

Figure 18.9 (a) A cross between two *Bd* flies. Since the gene *Bd* is homozygous lethal, all surviving beaded flies are heterozygous and one-fourth of the progeny from beaded parents die before hatching. One-fourth of the total and one-third of the flies that survive are wild type (not beaded). (b) A balanced lethal cross. A cross between two beaded flies after a new lethal mutation (*l*) had occurred on the homolog of the *Bd* chromosome. All progeny homozygous for *Bd* and those homozygous for *l* died. All the survivors expressed the beaded phenotype.

TRANSLOCATIONS

Sometimes a part of a chromosome becomes detached and ***joins*** a part of a ***nonhomologous chromosome***, thus producing a translocation. In heteromorphic sex chromosomes (homologous chromosomes that differ morphologically), exchanges may occur between nonhomologous parts of the same chromosome pair. Reciprocal translocations occur when parts of chromosomes belonging to members of two different pairs become exchanged. Part of chromosome 1, for example, may be detached and become attached to chromosome 2, whereas a section of chromosome 2 becomes attached to chromosome 1. Reciprocal translocations have been described in a number of plants and are important factors in the evolution of certain plant groups such as *Datura* and *Oenothera*. Translocations may not involve a loss or an addition of chromosome material, but frequently they do become associated with deficiencies and duplications, unbalanced combinations of genetic units.

Translocations can be detected from genetic data by noting altered linkage arrangements brought about by exchanges of parts between different chromosomes. Consider, for example, two nonhomologous autosomes number 2 and number 3 in one organism. In number 2, the sequence of genes is *a, b, c, d, e, f, g, h, i*, and the sequence on number 3 is *r, s, t, u, v, w, x, y, z*. Genes *a–i* are in linkage group 2, and genes *r–z* are in linkage group 3. If appropriate testcrosses should show linkage between *d* and *u*, a translocation between chromosome 2 and 3 would be suggested.

Cytological evidence for translocations can be obtained from microscopic studies of polytene chromosomes in heterozygous *Drosophila* and meiotic prophase stages in plant materials. A characteristic cytological pattern for translocations is marked by arrows in Fig. 18.10. It represents the place where the

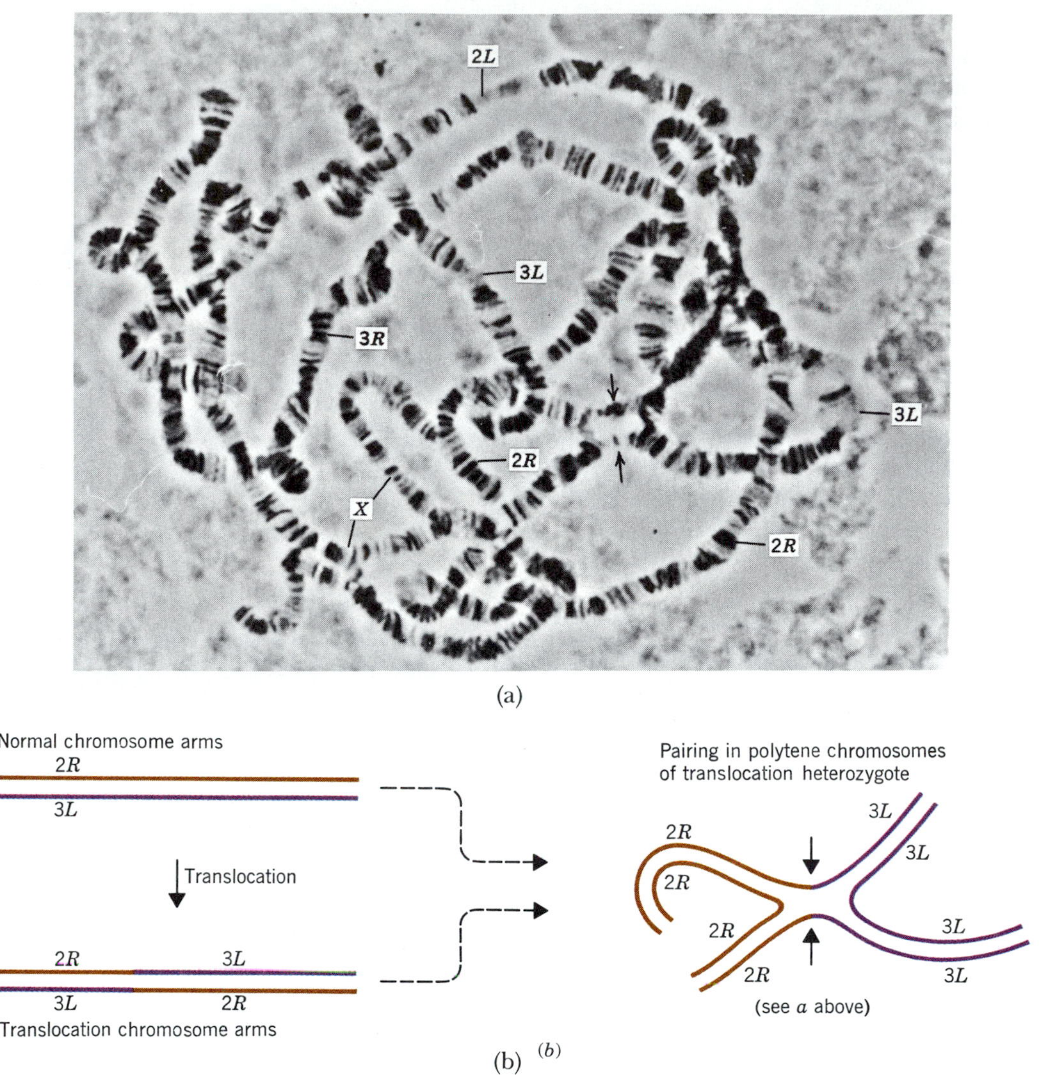

Figure 18.10 (a) Complete complement of chromosomes from a salivary gland preparation of *Drosophila melanogaster,* showing a translocation between the right arm of chromosome 2 and the left arm of chromosome 3. Breakpoints are marked with arrows. Chromosomes are identified as follows: left arm of second, *2L*; right arm of second, *2R*; left arm of third, *3L*; right arm of third, *3R*. X chromosome, *X*. (Courtesy of B. H. Judd.) (b) Interpretive diagram of the translocation chromosome arms shown in the photograph in (a).

right arm of one chromosome 2 homolog has joined the left arm of one chromosome 3 homolog. At this point, the translocation homolog (2 to 3) changes direction, and each part of the homolog is paired with the corresponding part from its homolog. On the other side of the cross, the translocation homolog (3 to 2) changes direction and is paired with its homolog. The diagram with Fig. 18.10 outlines the changes among homologs.

Characteristics of translocations in the germ cells of barley crosses and a method of detecting them are illustrated further from an actual study on barley (Fig. 18.11). The alleles (r^+ and r) for rough and smooth awns and those (s^+ and s) for long and short rachilla hairs are in linkage group 7. Alleles (n^+ and n) for hulled and hull-less and those (l^+ and l) for lax and club head are in group 1. When the progeny of certain plants homozygous for all four genes (r, s, n, and l) were classified, the results indicated that s was linked with n. This genetic evidence suggested that a translocation joining chromosomes 1 and 7 had occurred. When the plants were observed in the field and studied in the laboratory, at least 30 percent of the pollen was found to be sterile. Further investigations showed that many ovules also were sterile. Cytological studies of pollen mother cells (Fig. 18.12) showed one quadripartite ring and five bivalents instead of the seven bivalents usually observed in barley preparation. This

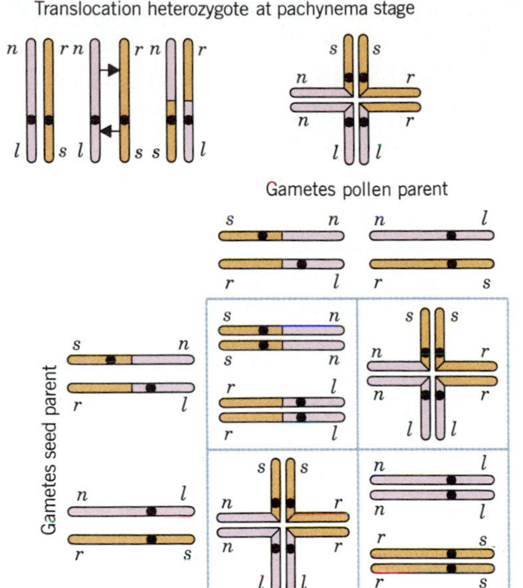

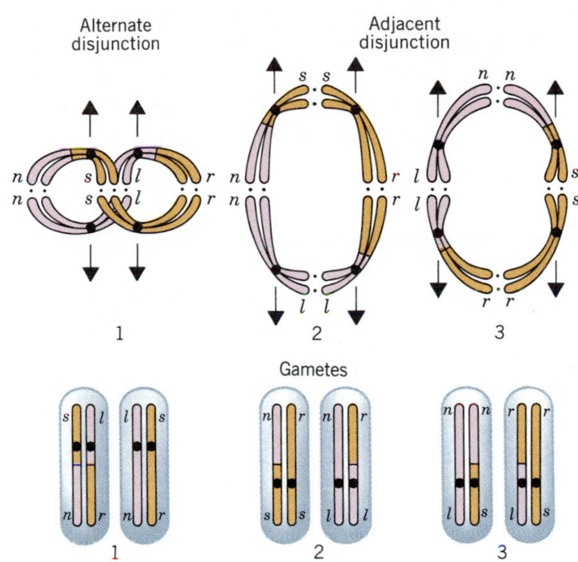

Figure 18.11 Explanation of the consequences of a reciprocal translocation in barley germ cells. A translocation had occurred between chromosomes 1 and 7, bringing the genes *n* and *s* into the same linkage group. The cross configuration at the top of the diagram represents the pairing of the translocated chromosomes (indicated as single chromatids rather than duplicated, for simplicity). Two viable male and two viable female gametes of the selfed plant are represented at the top and left, respectively, of the Punnett square. In the four squares, viable zygotes from the selfed plant are illustrated with the chromosomes as they would appear following synapsis. The two zygotes that have crosslike configurations are heterozygous for the translocation and produce plants that are partially sterile—that is, semisteriles—because of the deficiencies and duplications of chromosome parts occurring in the gametes. Late meiotic prophase configurations of the heterozygous translocation are shown at the right: (1) alternating disjunction resulting in balanced viable gametes; (2) adjacent disjunction resulting in deficiency–duplication, inviable gametes; (3) adjacent disjunction resulting in deficiency–duplication, inviable gametes. Arrows at the centromeres indicate the directions of movement of chromosomes in anaphase. (After W. S. Boyle and A. H. Holmgren, *Genetics* 40: 539–545, 1955.)

cytological evidence demonstrated that a reciprocal translocation had occurred. The explanation for the ***altered linkage*** grouping and the mechanism of gamete and zygote formation that involves chromosomes with translocations are illustrated in Fig. 18.11.

A consequence of heterozygous translocation is ***semisterility*** in pollen or ovules. Some gametes from a plant carrying a heterozygous translocation will be unbalanced and inviable. The imbalance causing the sterility occurs when chromosomes separate to the poles in meiosis. Chance combinations of translocation chromatids leave some gametes deficient in chromosome parts, and some have duplications. Viable and inviable gametes from a plant carrying a translocation heterozygote are shown in Fig. 18.11. Changes in linkage groups were demonstrated by these experiments to be direct consequences of exchanges between nonhomologous chromosomes. Semisterility in barley was shown to result from the irregularities in segregation of heterozygous translocations that developed as deficiencies and duplications.

Figure 18.12 Photograph of meiotic chromosomes in barley, showing a ring of four chromosomes (at left) in a pollen mother cell that resulted from a reciprocal translocation. The arrangement of the four chromosomes is similar to that for adjacent disjunction shown in Fig. 18.11. The quadrivalent ring in the photograph has been broken and opened in the process of preparation. Along with the quadrivalent ring, five bivalents are shown, making a total of 14 barley chromosomes (4 + 2 + 2 + 2 + 2 + 2).

POSITION EFFECTS

When a chromosome rearrangement involves no change in the amount of genetic material, but only in the order of genes, the term ***position effect*** is used to describe any associated phenotypic alteration. Along with gene mutations, position effects represent a source of genetic variation. The extent to which chromosome rearrangements such as ***inversions*** and ***translocations*** are associated with new phenotypic variation is, however, open to question. In this regard, it must be remembered that chromosome modifications, particularly inversions, curtail recombination and thus would be expected generally to restrict genetic variation.

Nevertheless, several well-established position effects are on record. The first example, from the studies of Sturtevant and Bridges on the bar-eye duplication in *Drosophila* (Fig. 18.13), also demonstrated a "dosage" effect. These investigators found a relation between the number of chromosome sections (16*A*) present and the number of facets in the eye. Further critical experiments showed, however, that it is not a strictly proportional relation. The arrangement of the chromosome segments with respect to each other, as well as their presence or absence, influences the size of the eye. The effect of different arrangements was demonstrated by manipulating the chromosomes through appropriate matings and counting the facets in the eyes of the female offspring.

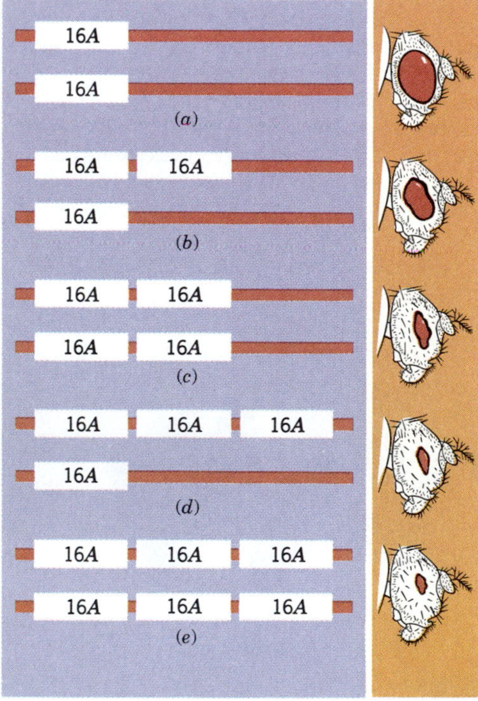

Figure 18.13 Effects on eye size of different arrangements of duplicated sections of 16A in the X chromosome of *D. melanogaster*. The 16A identifies a particular segment on the salivary chromosome following the standard numbering system for *D. melanogaster*. Positions of the 16A segment on a particular chromosome are related to size of eye. (a) Wild-type female (wild-type male would have same phenotype); (b) bar female, heterozygous; (c) bar female, homozygous (hemizygous bar male would have same phenotype); (d) double-bar female, heterozygous, showing position effect as compared with (c); (e) double-bar homozygous. (After A. H. Sturtevant, *Genetics* 10: 117–147, 1925.)

When section 16*A* was duplicated and the extra segment occurred in homozygous conditions with a total of four segments (Fig. 18.13*c*), the number of facets in the eyes averaged 68. But when three 16*A* sections were side by side in one homolog and one section in the other homolog (Fig. 18.13*d*), the eyes averaged 45 facets. Since the same number of 16*A* units is present in the eyes represented by Figs. 18.3*c* and *d*, the difference depends on the arrangement or position of the genes with respect to each other. This phenomenon was interpreted as a position effect.

Several well-established and many possible position effects have now been described in *Drosophila*. E. B. Lewis has shown that position effects fall into two classes: (1) ***stable*** and (2) ***variegated***. Stable position effects are uniform phenotypic effects that result from changes of specific segments of chromosomes. The bar-eye position effect is an example. Variegated position effects result in a mosaic pattern of a trait usually evidenced in a particular structure or area of the body. Specks of different colors, for example, may occur in the eyes of *Drosophila* following rearrangements of the w (white eye) locus. Inversions or translocations that place w^+ in heterochromatin may cause white variegation or mosaicism for eye color.

Variegated position effect in *Drosophila* is associated with chromosome structural changes. Action of a gene is depressed when the gene is transferred to a ***heterochromatic*** region. A position effect usually involves two chromosome breaks, one in the heterochromatic region and one near a euchromatic gene. Regulation of this gene activity occurs at the chromosome rather than the cell level and presumably results from suppression of gene transcription. ***Variegated transcription*** of linear DNA could result in variegated expression or a position effect. Experiments of H. J. Becker on *Drosophila* eye development showed that mutational disturbances in areas of the chromosomes resulted in the same phenotypes as position effects caused by chromosome structural changes. With X rays he induced mitotic exchanges of eye-color genes at time intervals during development. Because each affected sector was made up of the cell progeny of a single cell carrying an exchange product, it was possible to trace the developmental sequence of the tissue composing the eye. Becker then induced structural changes in chromosomes and found that alterations in the position of genes, with reference to heterochromatin, created effects that appeared the same phenotypically as those of X-ray-induced disturbances.

CHROMOSOME ABERRATIONS AND EVOLUTION

A consequence of chromosomal structural aberration in populations is related to evolutionary change including speciation. Chromosomal aberrations are associated with position effects that may be significant in natural selection. More important for evolution is the genetic isolation that is mostly caused by inversions and translocations. Speciation in the *Drosophila* group of dipterous insects, for example, has been related to chromosome inversions. These structural changes occur in chromosomes of individual flies and are carried homozygous in populations. Populations have developed over periods of time with different chromosome inversions. Each may be isolated because matings of flies from a particular population with those of another population carrying a different inversion result in sterile or inviable hybrids. This strengthens the boundaries around a particular population and prevents the exchange of genes between related populations. Speciation in *Drosophila* has been associated with a series of different inversions that occurred by chance in breeding populations and were eventually recognized in different taxonomic groups. Translocations have been shown to occur in certain plant groups and to cause genetic isolation, thus promoting evolutionary stability in populations.

HUMAN CHROMOSOME TECHNIQUES

The 22 pairs of human autosomes were first classified according to length and position of the centromere into ***seven groups***, identified with the letters A–G. All autosomes could be placed satisfactorily within a group, but the numbering within the groups was more or less tentative (Fig. 18.14) until 1970.

Chromosome ***banding techniques***, along with new methods of identification, finally distinguished all 46 human chromosomes. Bands are defined as parts of chromosomes that appear lighter or darker than adjacent regions when treated with particular staining methods. Q-staining methods employ quinacrine compounds and produce fluorescent Q-bands along the chromosomes. G (Giemsa)-staining methods (Fig. 18.15) result in G-bands and, with some Giemsa techniques, the R (reverse)-staining methods result in R-bands. The centric region and other heterochromatin areas are stained by C-banding (Fig. 18.16). Refinements in staining techniques now permit all human chromosomes to be identified (Fig. 18.17). The amount of DNA packed into one G-band is on the order of 10^6–10^7 base-pairs of DNA, and it is clearly not a gene.

In 1971 a conference on the standardization of human cytogenetics was held in Paris. Conferees agreed that to describe extra autosomes, the number of the extra chromosome should be placed after the total number and sex chromosomes with a plus (+) or minus (−) sign before the number of the autosome involved, for example, 47,XX+21 is the karyotype of a female with trisomy-21. A male with an extra X chromosome is symbolized 47,XXY. A plus or minus sign is placed following a chromosome symbol to signify increase or decrease in arm length. The letter *q* symbolizes the long arm and *p* the short arm. For example, 46,XY,1*q*+ indicates an increase in the length of the long arm of chromosome 1. The person represented by this example is a male with 46 chromosomes. The karyotype 47,XY,+14*p*+ symbolizes a male with 47 chromosomes, including an additional chromosome (No. 14), with an increase in the length of its short arm. The types of chromosome aberrations are represented by abbreviations for convenience in presenting chromosome formulas: def (deficiency), dup (duplication), r (ring), inv (inversion), and t (translocation).

CHROMOSOME ABERRATIONS IN HUMANS

Chromosome deletions are usually lethal even as heterozygotes, resulting in zygotic loss, stillbirths, or infant deaths. Sometimes infants with small chromosome deficiencies, however, survive long enough to permit observation of some of the abnormal phenotypes they express. J. Lejeune and his colleagues, for example, discovered a chromosome deficiency in humans that has been associated with the **cri-du-chat** (cat cry) ***syndrome*** (Fig. 18.18). The name of this syndrome came from a plaintive catlike mewing cry from small weak infants with the disorder. Other characteristics are microcephaly (small head), broad face and saddle nose, widely spaced eyes with epicanthic folds, unique facial features, and physical and mental retardation. IQs of *cri-du-chat* children studied are in the range of 20–40. The chromosome deficiency is in the short arm of chromosome 5 and is designated 5*p*−. A karyotype (46,XX,5*p*−) of a newborn is illustrated in Fig. 18.19.

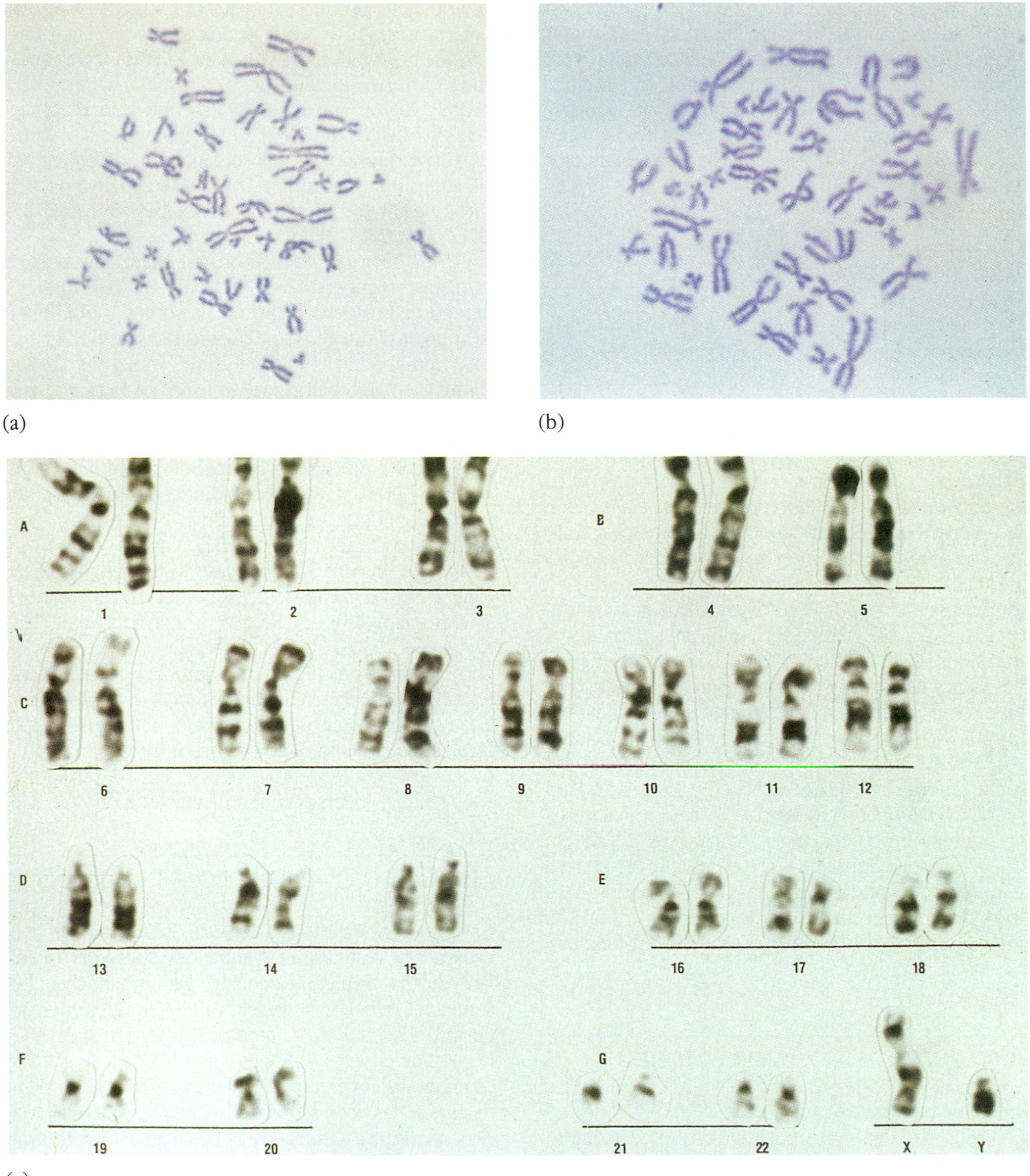

Figure 18.14 Human metaphase chromosomes of (a) a normal female and (b) a normal male as they appear in microscope preparations, and (c) as they are classified (karyotyped) in seven major groups according to criteria established at the Denver (1960), London (1963), and Chicago (1966) conferences. (Photos (a) and (b) by John P. Limbach, Ripon Microslides Laboratory; photo (c) by Peter Arnold, Inc.)

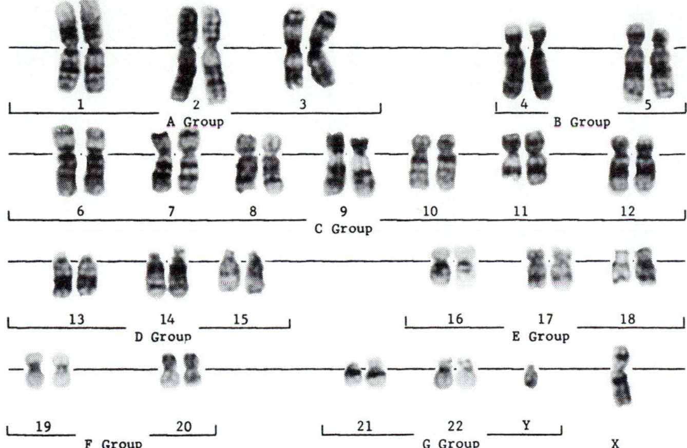

Figure 18.15 Karyotype of a normal male child. Chromosomes were prepared with the G-staining technique. Bands along the chromosomes identify members of chromosome pairs. All 23 pairs of chromosomes can be distinguished by Giemsa-staining methods. (Courtesy of R. M. Fineman, Department of Pediatrics, University of Utah.)

Cri-du-chat patients die in infancy or early childhood and do not transfer the chromosome deletion to offspring. This chromosome deficiency, however, has been shown by Lejeune and others to become involved sometimes in a reciprocal translocation and thus to be transmitted. When the short arm of chromosome 5 became translocated to chromosome 15, the heterozygous translocation was carried in a normal healthy parent. Some gametes, however, carried only the ***deficient member*** of the translocation pair. Children inheriting the $5p^{-}$ chromosome expressed the *cri-du-chat* syndrome.

Another human disorder that is associated with a chromosome abnormality is chronic myelocytic (myelogenous) leukemia. A deletion of chromosome 22 was first described by P.C. Nowell and D.A. Hungerford and was called the Philadelphia (Ph′) chromosome after the city in which the discovery was made. It was observed consistently in bone marrow preparations of patients with chronic myelocytic leukemia. Later, a translocation discovered by J. Rowley in a leukemia patient provided the correct chromosome rearrangement. A part of the long arm of chromosome 22 was translocated to another chromosome, usually chromo-

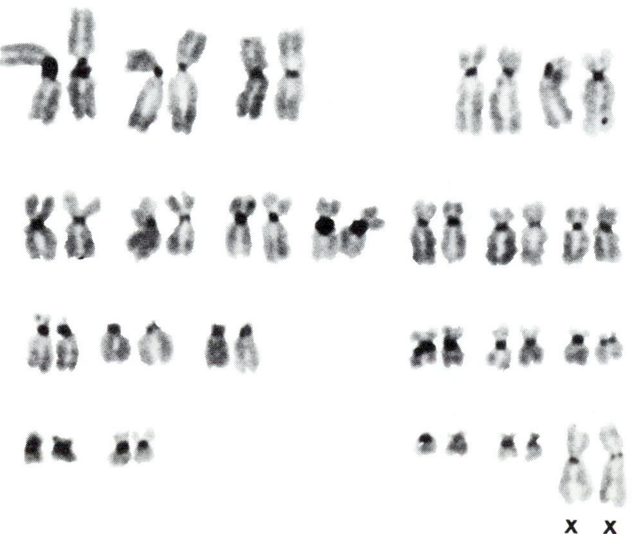

Figure 18.16 Karyotype of a normal young woman. Chromosomes were prepared with the C-staining technique. Heterochromatin regions of chromosomes are darkly stained. (Courtesy of A. D. Stock, Utah State University.)

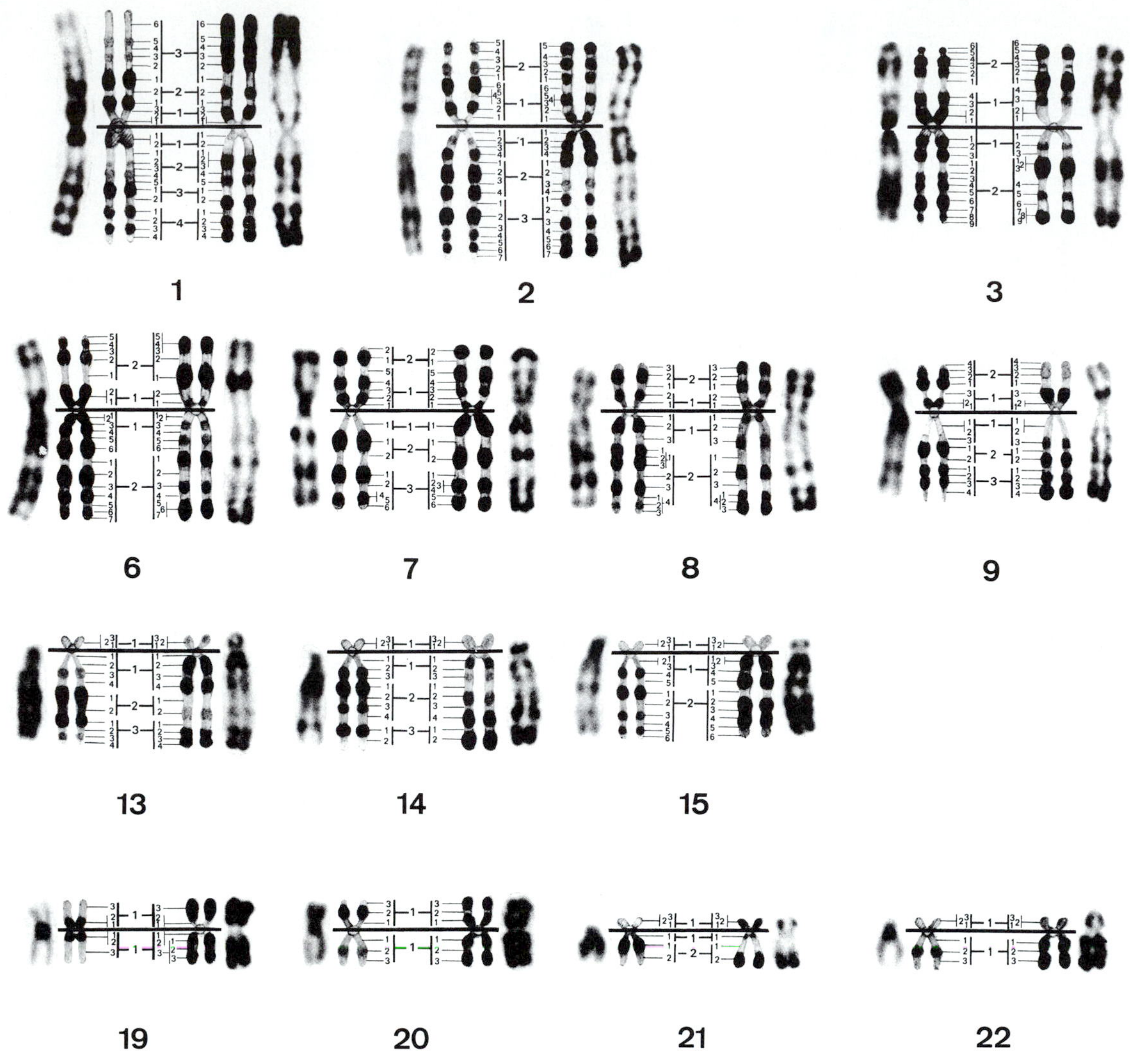

1 2 3

6 7 8 9

13 14 15

19 20 21 22

Figure 18.17 (Top left and right:) Photographs of normal human karyotype G- and R-bands. Autosomes are numbered in order from 1 to 22, and X and Y are included. Short arms of chromosomes (*p*) and long arms (*q*) have regions numbered from 1 (next to centromere) to the distal end of each arm, and bands are numbered within the regions from the centromere to the distal end of each arm. (From "The Normal Human Karyotype." Originally reproduced from "An International System for Human Cytogenetic Nomenclature (1978)," in *Birth Defects: Original Article Series,* XIV: 8, 1978, The National Foundation, New York, and *Cytogenetics and Cell Genetics* 21: 6, 1978, S. Karger AG, Basel, Switzerland, reproduced with permission.)

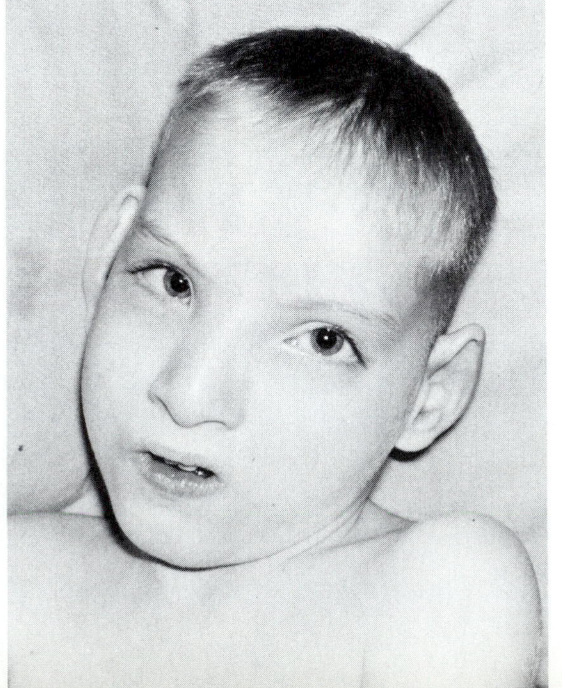

Figure 18.18 (Left:) Patient with *cri-du-chat* syndrome, a malady associated with a deletion of part of the short arm of chromosome 5. The syndrome includes microcephaly and a catlike cry. (Courtesy of I. A. Uchida, Department of Pediatrics, McMaster University.)

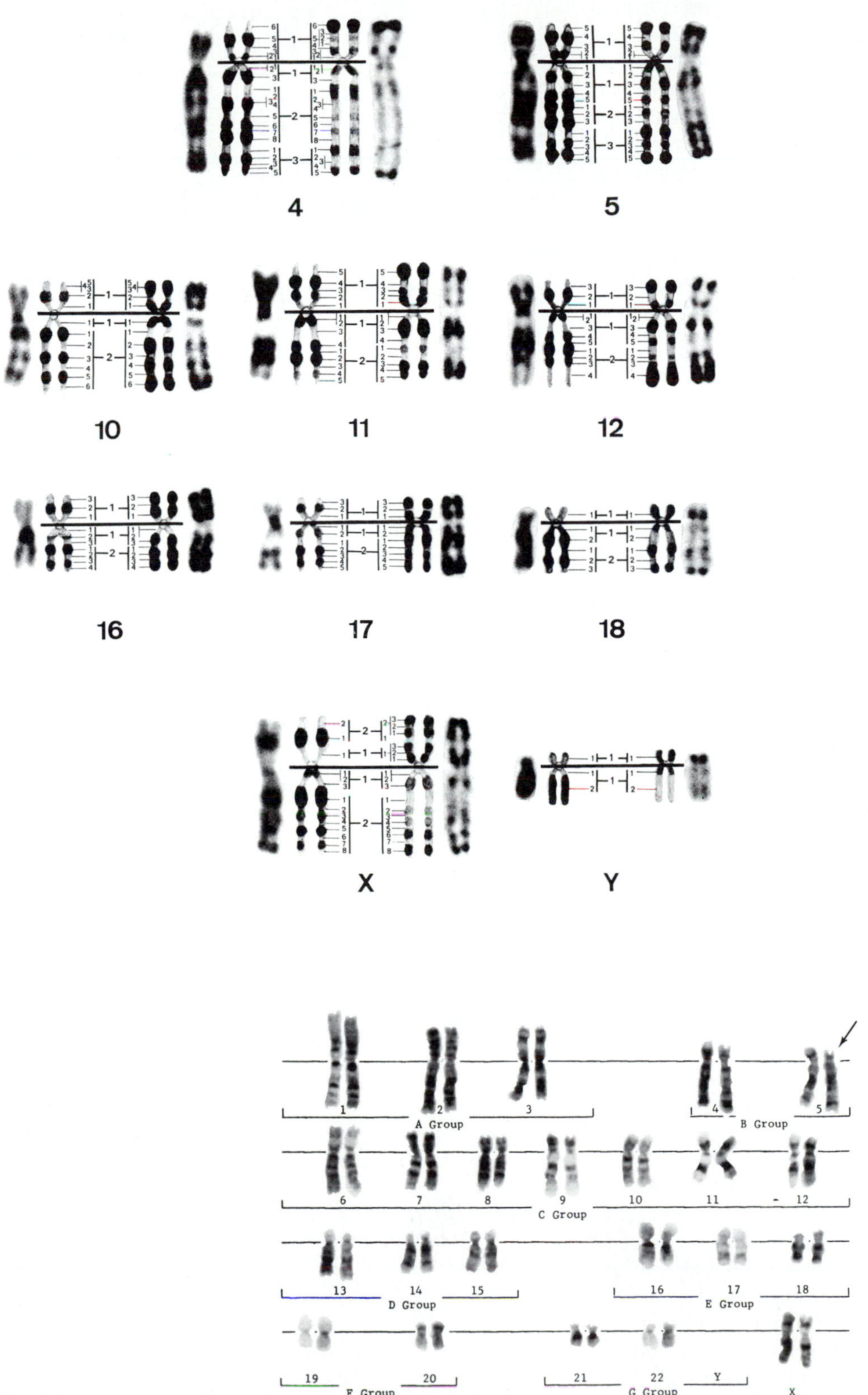

Figure 18.19 Karyotype of newborn with the *cri-du-chat* syndrome. A heterozygous deficiency is shown in chromosome 5 (see arrow). The chromosome arrangement is 46, XX($5p^-$). (Courtesy of R. M. Fineman, Department of Pediatrics, University of Utah.)

some 9 (46,XX,9q+,22q−), leaving a ***deficiency in the long arm of No. 22***. The karyotype prepared from a bone marrow biopsy is presented in Fig. 18.20.

More chromosomal anomalies are now being associated with ***malignancy***. New staining methods permit more precise comparisons of (1) chromosome arms and (2) differentially stained sister chromatids. With banded chromosomes, Burkitt's lymphoma was related to a translocation of chromosome 14 [t(8q−; 14q+)(q24;q32)]. A deletion in chromosome 13q (band 14.1) is associated with the human embryonic tumor retinoblastoma. Wilms' tumor, an embryonic kidney tumor, is associated with a deletion in band 11p13. ***Sister chromatid exchanges*** (Chapter 3), exchanges occurring between sister chromatids, occur with high frequency in somatic cells of patients with Bloom syndrome, indicating chromosome instability. Most people with this syndrome die with some form of cancer before age 30. Irregular chromosome numbers are also observed in malignant cells, particularly in later stages of cancer.

Like deficiencies, chromosome duplications are usually lethal even as heterozygotes, but sometimes they are sufficiently viable to permit observations of the abnormalities they produce. Like deficiencies, they may be associated with translocations and thus transmitted by normal healthy parents. Remember that a translocation is an exchange of parts between nonhomologous chromosomes or a transfer from one chromosome to a nonhomolog. Broken parts may be further divided and some may be lost or gained in a transfer. Thus, deficiencies and duplications may accompany a translocation carried in a parent. Such a parent may be a translocation heterozygote, with the long arm of one chromosome 21 attached at the centromere to the long arm of chromosome 14 along with normal chromosomes 14 and 21. Some gametes will contain a normal 21 and translocation 14q21q. When such a gamete is fertilized with a gamete from a normal individual, a translocated 21, along with two normal chromosomes 21, results in a viable infant with trisomy-21 and Down syndrome.

With fluorescent microscope techniques, I. A. Uchida and C. C. Lin discovered a partial trisomy-12 that could not be identified by conventional methods. A boy (Fig. 18.21) with some clinical features resembling Down syndrome had been studied with conventional chromosome techniques. Because of continued lack of motor development at seven months of age, further chromosomal investigations were carried out on lymphocyte cultures and slides stained with quinacrine dihydrochloride. With this technique, an additional band was identified in the short arm of chromosome 8. The mother's chromosomes were normal; but in the father, the short arm of No. 12 had become

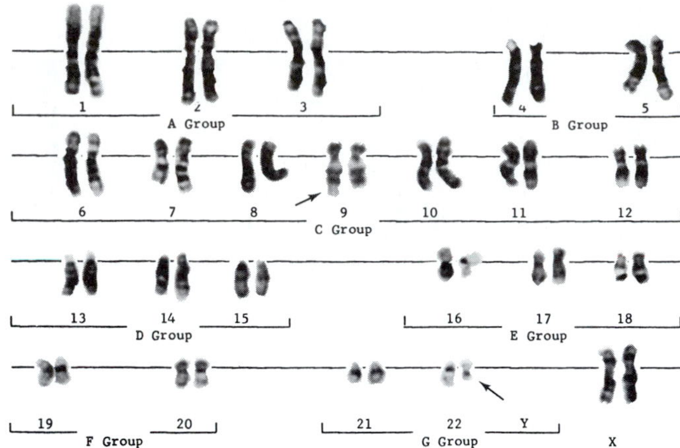

Figure 18.20 Karyotype prepared from the bone marrow biopsy of a patient with chronic myelogenous leukemia. The chromosome arrangement is 46,XX,9q^+,22q^-. (Courtesy of R. M. Fineman, Department of Pediatrics, University of Utah.)

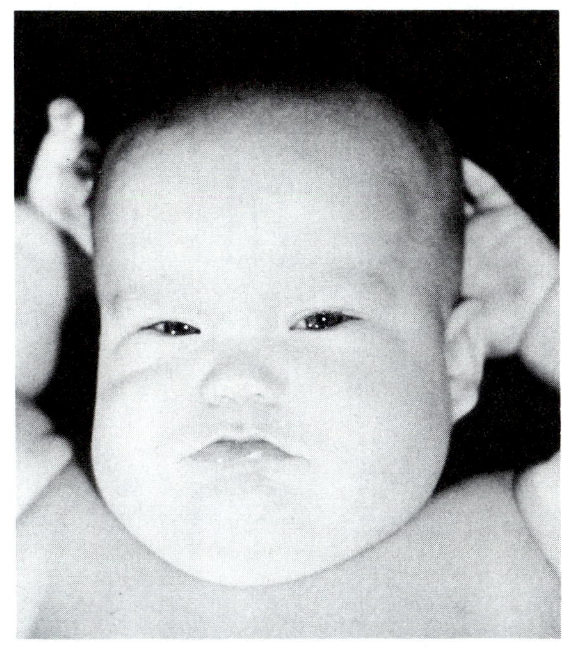

Figure 18.21 A boy with some features of Down syndrome, who was found by quinacrine fluorescence to have partial chromosome trisomy-12 (12p^+) and deficiency of the tip of chromosome 8 (8p^-). There was no apparent involvement of chromosome 21. Down syndrome features were mental retardation and hypotonia; a rather flat facies with prominent epicanthic folds; Brushfield spots in the irides; flat, broad nasal bridge; and ears in the normal position but with small pinnae. The fingers were somewhat spade-shaped with incurving of both fifth digits. Simian creases were present on both palms. (From I. A. Uchida and C. C. Lin, "Identification of Partial 12 Trisomy by Quinacrine Fluorescence," *J. Pediatr.* 82: 269–272, 1973.)

translocated to the short arm of No. 8 [46,XY,(8p+, 12p−)]. The same chromosomal rearrangement was found in his daughter, an older sister of the patient. Since both father and daughter were clinically normal, the translocation was presumed to be reciprocal. The patient's No. 8 pair (Fig. 18.22) consisted of one normal and one translocation chromosome, similar to that of his father and sister (Fig. 18.23), but both of his No. 12 chromosomes were normal (Fig. 18.22). Thus, the boy had a duplication of part of the short arm of No. 12 and a deficiency of the tip of No. 8.

Many translocations, in addition to those cited, have been detected in studies of human chromosomes (see Chapter 17), but most were apparently reciprocal and produced no phenotypic anomalies. At the Yale–New Haven Hospital, for example, cytological studies were conducted on 4500 infants born consecutively during one year. Lymphocytes from umbilical cord blood of each infant were grown *in vitro* and prepared for microscopic observation of chromosomes. Six translocations were detected, and none of these was associated with a distinctive phenotypic anomaly. It is not the translocation process per se that produces abnormalities, but the imbalance of genetic material reflected in chromosome deficiencies and duplications that are produced by segregation of translocated chromosomes.

P. W. Allerdice *et al.* described a syndrome called the ***chromosome 3 duplication–deletion syndrome.*** Phenotypically, this syndrome includes a group of morbid symptoms: stillbirths, neonatal deaths, and spontaneous abortions. Most pregnancies are lost, but two children survived and became probands for the investigation. One of these (Case 1) is shown in Fig. 18.24 at age six. He cannot sit up, turn over, or eat solid food. Facial malformation of the living children includes a distorted head shape; thick, low eyebrows; low hairline; long eyelashes; persistent lanugo; distended veins on scalp; hypertelorism; oblique palpebral fissures; a very short nose with a broad, depressed bridge and anteverted nares; protruding maxilla; thin upper lip; micrognathia; low-set ears; and short, webbed neck. Port-wine stains, congenital glaucoma, cloudy corneas, cleft palate, and harelip also occur frequently. Each infant had difficulty sucking and swallowing. Internal physical abnormalities were noted in infants who died neonatally.

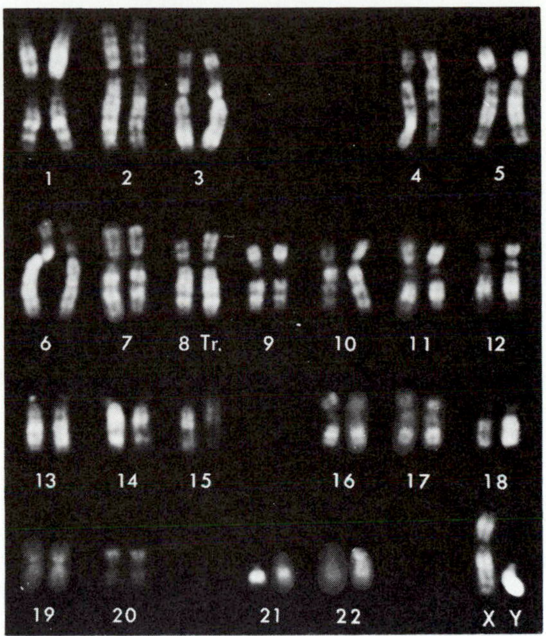

Figure 18.22 Karyotype of patient with abnormal chromosome 8 and two normal No. 12 chromosomes. Note the extra band on the second No. 8 chromosome. (From I. A. Uchida and C. C. Lin, "Identification of Partial 12 Trisomy by Quinacrine Fluorescence," *J. Pediatr.* 82: 269–272, 1973.)

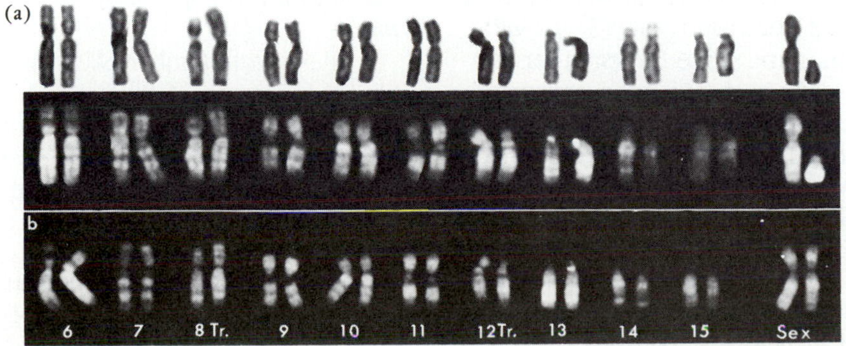

Figure 18.23 Partial karyotypes of the father (a) and sister (b) of the patient whose karyotype is shown in Fig. 18.22. Note translocation of chromosomes 8 and 12. (From I. A. Uchida and C. C. Lin, "Identification of Partial 12 Trisomy by Quinacrine Fluorescence," *J. Pediatr.* 82: 269–272, 1973.)

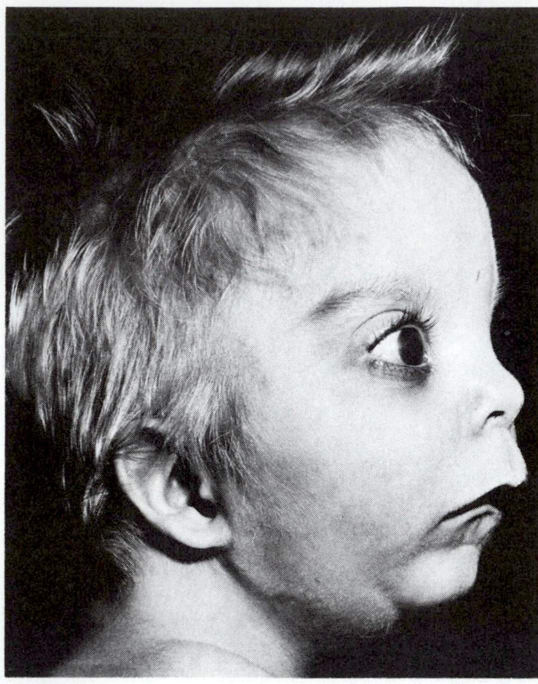

Figure 18.24 Six-year-old child with chromosome 3 duplication–deletion syndrome. The karyotype is 46,XY,rec(3), dup *q* inv(3)(*p*22*q*21). He could never suck adequately, and at six years was still spoon or tube fed. All developmental milestones have been grossly delayed. His functioning as tested on the Denver Developmental Screening Test varied between newborn and four-month level. He has had recurrent urinary, respiratory, and eye infections. (Courtesy of P. W. Allderdice. Copyright © 1975 by the American Society of Human Genetics, University of Chicago Press.)

Giemsa- and quinacrine-banded karyotypes from a parent of an affected child revealed an inversion [inv(3)(*p*25*q*21)]. Fetal cells cultured for prenatal diagnosis from a subsequent pregnancy of this couple carried a recombinant chromosome 3, with the long arm described as rec(3)del *p*, dup *q*, inv(3)(*p*25*q*21). With a banded chromosome 3, the inversion was analyzed. The inverted segment included the centromere (pericentric). Breaks had occurred in the short arm (*p*) at band 25 and in the long arm (*q*) at band 21. The central part of the chromosome including the centromere had rotated 180° and become reinserted into chromosome 3. Inv 3(*p*25*q*21) had been carried in the kindred for at least four generations and 35 kindred members were presumed to carry it. The karyotype of Case 1 is 46,XY,rec(3), dup *q*, inv(3)(*p*25*q*21). He was the second child born to a 23-year-old carrier mother with karyotype 46,XX,inv(3)*p*25*q*21 and a 26-year-old father with karyotype 46,XY.

Crossing-over within the inversion loop is presumed to produce the imbalance of genetic material associated with duplications and deficiencies and to cause the symptoms of the syndrome. Odd numbers of crossovers within the inverted area of a chromosome 3 would be expected to result in genetic imbalance. Earlier in this chapter, crossing-over in *Drosophila* inverted chromosomes was shown to create irregular combinations resulting in lethals. This, in turn, resulted in "suppressing" crossing-over. In humans, unbalanced zygotes from inverted chromosome crossovers sometimes survive. Some infants carrying unbalanced chromosomes are kept alive, but with varying degrees of birth defect handicaps. The degree of imbalance of genetic material may be the determining factor for life or death and degree of abnormality.

BANDED CHROMOSOMES AND PHYLOGENY

Chromosome-banding techniques have now been applied to nonhuman organisms. Comparisons of chromosome-banding patterns within animal groups such as mammals, birds, reptiles, amphibians, and fish, and also between representatives of particular taxonomic groups, have been accomplished. Compared with other phenotypes employed by taxonomists, chromosome structure has proved to be stable and broadly representative of phylogenetic change. Data from comparisons of chromosome-banding patterns in members of different species therefore have broad phylogenetic significance. Relations among different breeding populations can be estimated by comparing chromosomal characteristics of different taxonomic groups. Chimpanzees (*Pan troglodytes*) and human beings (*Homo sapiens*) are placed taxonomically in different genera, *Pan* and *Homo*, respectively. These species are much alike in anatomical and physiological features as well as in DNA and protein composition. Comparisons of chromosomes show ***basic similarities*** with a few ***superimposed structural rearrangements***.

The overall number of chromosomes is comparable (46 for *Homo*, 48 for *Pan*). Homologous pairs can be identified, and general chromosome structure can be matched band for band between the pairs. Two acrocentrics in *P. troglodytes* have the same bands as one submetacentric in *H. sapiens*, to make the equivalent of 46 in both species.

A. D. Stock compared the banded chromosomes of members of the two species (Fig. 18.25) and noted the following structural differences:

1. Chromosome 2 in *H. sapiens* is long and submetacentric, but is comparable with two acrocentric chromo-

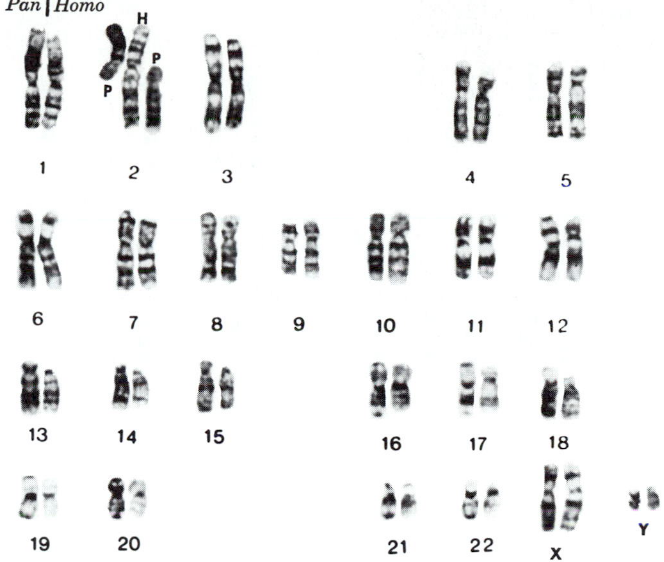

Figure 18.25 Chimpanzee chromosomes compared with human chromosomes. Both species have the same total number if two chimpanzee acrocentrics are joined to correspond with *Homo* No. 2. Six pericentric inversions are recognized in comparing chromosomes of the two species. (Courtesy of A. D. Stock.)

somes in *P. troglodytes*. When these two chromosomes of *P. troglodytes* are fused together near the centromere on the short arm of one chromosome and near the telomere of the short arm of the other, a chromosome similar to *H. sapiens* chromosome 2 is produced.

2. Chromosomes 4, 5, 9, 12, 17, and 18 of the two species each differ by a single pericentric inversion.
3. A probable pericentric inversion distinguishes chromosome 8 of the two species.
4. Chromosomes 1 and 9 of the two species have small, special differences in heterochromatin content.

When chromosome bands are compared and homologies are identified, a close relationship between the two species can be established. In this comparison, at the species level, chromosome morphology is a measure of phylogenetic change. Chromosome number, on the other hand, depends on a more superficial "packaging" process and does not represent a basic criterion for relationship. In this example (Fig. 18.25), one near centric change has made a difference in number, while the basic chromosome units are essentially homologous.

Rhesus and African Green Monkey

Rhesus (*Rhesus macaque*) and African green (*Cercopithecus aethiops*) monkeys are both Old World primates. They are not closely related but they are members of the same family. When Stock and Hsu compared the chromosomes of the two species, differences were observed that could be explained by translocations. Rhesus has a total $2n = 42$ chromosomes, all biarmed, making a total of 84 arms. The chromosomes, in addition to the sex pair, were arranged in four morphological groups; A: seven pairs of medium to large submetacentrics; B: seven pairs of medium submetacentrics with higher arm ratios than those in group A; C: five pairs of small metacentrics and submetacentrics; and D: one pair that carries a pronounced secondary constriction on its short arm. In the sex pair, the X is a large submetacentric, and the Y is the smallest element of the complement. Chromosome-banding demonstrated that the constitutive (stationary) heterochromatin of all chromosomes in this species is centromeric.

The African green monkey has a total of $2n = 60$ biarmed chromosomes (120 arms). These chromosomes were arranged in four morphological groups corresponding with those of the rhesus monkey: A: 11 pairs of subtelocentric; B: 12 pairs of medium to small submetacentrics of varying lengths; C: 5 pairs of small metacentrics and submetacentrics that appeared identical with group C of rhesus; and D: 1 pair of small subtelocentrics with the secondary constriction in the long arm. The long arm is homologous with the short arm of rhesus D chromosomes. In gross morphology,

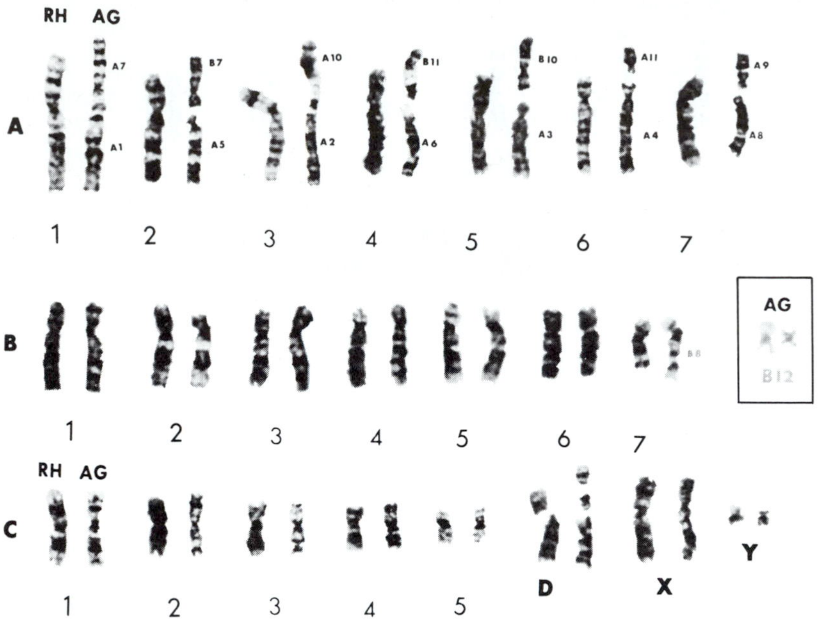

Figure 18.26 Matching between the two monkey species. Rhesus monkey chromosomes, left side of each pair; African green monkey, right side. (From A. D. Stock and T. C. Hsu, *Chromosoma* 43: 211–224, 1973.)

the sex chromosomes were similar to those of rhesus. Constitutive heterochromatin of all members of groups B and C and the X is centromeric. In chromosomes of group A, the short arms of all except one (A2 is euchromatic) are heterochromatic.

When the chromosomes of rhesus and the African green monkey were compared (Fig. 18.26), groups B and C and sex chromosomes had matching elements in the karyotypes of the African green and rhesus. The group A chromosomes in the two species were different, but matching parts were found when chromosomes were compared side by side. Some places that did not fit contained constitutive heterochromatin. When excess heterochromatin was trimmed from the African green monkey chromosomes, and the euchromatin parts remaining on acrocentrics were fused, a nearly perfect fit between the two karyotypes was obtained. The main differences between the chromosomes of the two monkey species were (1) amount and distribution of heterochromatin and (2) translocations. Translocations were whole-arm fusions of centromere to telomere that reduced three arms into two. These species are distantly related. Comparisons of banded chromosomes reveal many ***homologies*** and ***differences***.

SUMMARY

Structural aberrations of chromosomes occur in plants and animals. They result in increased variation and reproductive isolation, thus playing a significant role in the evolution of some groups of living things. Polytene chromosomes in diptera and large meiotic prophase chromosomes in maize have been used extensively for identifying and investigating chromosome structural changes. The mechanism of structural aberration consists of the breaking and rejoining of chromosomes, resulting in ***deficiencies, duplications, inversions***, and ***translocations***. Linear linkage relations among genes in chromosomes are altered by structural aberrations. Both stable and variegated position effects are associated with chromosome structural changes.

All 46 human chromosomes can now be identified with chromosome-banding techniques. Descriptive symbols have been standardized for labeling structural and numerical chromosome changes. Chromosome banding has now been applied to nonhuman as well as human organisms, and comparisons with phylogenetic significance have been established within and among taxonomic groups.

REFERENCES

ALLERDICE, P.W. 1975. "Chromosome 3 duplication $q21 \rightarrow$ qter deletion $p25 \rightarrow$ pter syndrome in children of carriers of a pericentric inversion." *Amer. J. Hum. Genet.* 27: 699–718.

BERGSMA, D., editor. 1972. "Advances in human genetics and their impact on society." *Birth Defects* 8(4): 1–118.

______. 1972. "Paris conference: Standardization in human cytogenetics." *Birth Defects* 8(7): 1–46.

BRIDGES, C.B. 1935. "Salivary chromosome maps with a key to the banding of the chromosomes of *Drosophila melanogaster.*" *J. Hered.* 26: 60–64.

DALLA-FAVERA, R., *et al.* 1982. "Human c-*myc onc* gene is located on the region of chromosome 8 that is translocated in Burkitt lymphoma cells." *Proc. Natl. Acad. Sci. U.S.A.* 79: 7824–7827.

DARLINGTON, C.D., and L.F. LA COUR. 1976. *The handling of chromosomes,* 6th ed. Halsted, New York.

DUTRILLAUX, B., and J. LEJEUNE. 1975. "New techniques in the study of human chromosomes: Methods and applications." In *Advances in human genetics,* Vol. 5, H. Harris and K. Hirschhorn, editors. Plenum, New York.

HSU, T.C. 1972. "Procedures for mammalian chromosome preparation." *Methods in Cell Physiol.* 5: 1–36.

JUDD, B.H., M.W. SHEN, and T.C. KAUFMAN. 1972. "The anatomy and function of a segment of the X chromosome of *Drosophila melanogaster.*" *Genetics* 71: 139–156.

KAISER-MCCAW, B., A.L. EPSTEIN, H.S. KAPLAN, and F. HECHT. 1977. "Chromosome 14 translocation in African and North American Burkitt's lymphoma." *Int. J. Cancer* 19: 482–486.

KLEIN, G. 1981. "The role of gene dosage and genetic transposition in carcinogenesis." *Nature* 294: 313–318.

KLEIN, G., *et al.* 1982. "A cellular oncogene is translocated to the Philadelphia chromosome in chronic myelocytic leukemia." *Nature* 300: 765–767.

LEWIS, E.B. 1950. "The phenomenon of position effect." In *Advances in genetics,* Vol. 3, M. Demerec, editor. Academic Press, New York.

PRIEST, J.H. 1977. *Medical cytogenetics and cell culture,* 2nd ed. Lea & Febiger, New York.

ROWLEY, J. 1973. "A new consistent chromosomal abnormality in chronic myelogenous leukemia identified by quinacrine fluorescence and Giemsa staining." *Nature* 243: 290–293.

______. 1983. "Human oncogene locations and chromosome aberrations." *Nature* 301: 290–291.

STOCK, A.D., and T.C. HSU. 1973. "Evolutionary conservatism in arrangement of genetic material." *Chromosoma* 43: 211–224.

TURLEAU, C., and J. DE GROUCHY. 1973. "New observations on the human and chimpanzee karyotypes." *Humangenetik* 20: 151–157.

TURLEAU, C., J. DE GROUCHY, and M. KLEIN. 1972. "Phylogénie chromosomique de l'homme et des primates hominiens. Essai de reconstitution du caryotype de l'ancetre commun." *Ann. Génét.* 15: 225–240.

TURPIN, R., and J. LEJEUNE. 1969. *Human afflictions and chromosomal aberrations.* Pergamon, Oxford, England.

UCHIDA, I.A., and C.C. LIN. 1973. "Identification of partial 12 trisomy by quinacrine fluorescence." *J. Pediatr.* 82: 269–272.

PROBLEMS AND QUESTIONS

18.1 (a) What genetic evidence first suggested chromosome structural changes? (b) Why did it take so many years to obtain cytological verification for the genetic evidence? (c) How was cytological verification obtained?

18.2 Compare the methods now available for cytogenetic studies of the fruit fly with those available for maize.

18.3 What characteristics of *Drosophila* salivary gland chromosomes make them especially suitable for cytogenetic studies?

18.4 What are the advantages of the acetocarmine smear technique as compared with fixing, sectioning, and staining methods for chromosome studies?

18.5 Formulate a plausible explanation for the origin of the giant salivary gland chromosomes in the developing larva of the fly. What do the cross bands represent?

18.6 What is the difference between a linkage and a cytological chromosome map?

18.7 Describe or illustrate with appropriate sketches how a recessive gene may be expressed through pseudodominance.

18.8 How can the extent of a chromosome deficiency be determined (a) genetically and (b) cytologically?

18.9 If a trait such as "waltzing" in mice, known to depend on a single recessive gene (v), should appear in an animal considered to be heterozygous for the gene, how could it be determined (a) genetically and (b) cytologically whether a mutation had occurred or a deficiency was present in the chromosome opposite v?

18.10 How can (a) paracentric and (b) pericentric inversions act as "crossover suppressors"? Describe or illustrate. (c) Is crossing-over really suppressed?

18.11 In *D. melanogaster*, the gene *Bd* is dominant with respect to a wing abnormality but is homozygous lethal. Another homozygous lethal gene (l) is located on the homologous chromosome, and a crossover "suppressor" prevents crossing-over between *Bd* and l. What results would be expected from a cross between two flies with the genotype $Bd\ l^+/Bd^+\ l$?

18.12 In barley, a_n for white seedlings and x_c for yellow seedlings are on the same chromosomes with a recombination frequency of 10 percent. Plants with the genotype

$a_n{}^+ x_c{}^+$ are green. Homozygous a_n and x_c plants die in the seedling stage. A plant with the genotype $a_n x_c{}^+/a_n{}^+ x_c$ was selfed. Give the expected results.

18.13 Reciprocal crosses were made between flies with striped (*sr*) bodies and wild-type flies (sr^+). No differences were found between results of crosses in either the F_1 or the F_2 progeny. Homozygous *sr* females were then mated to males bearing the balanced curly and plum dominant genes on chromosome 2, and stubble and dichaete dominant genes on chromosome 3. Crossing-over was suppressed. F_1 progeny showing both curly and stubble were mated. Some progeny had striped bodies and curly wings. (a) Is *sr* autosomal or sex-linked? (b) Which chromosome is it on? (Chromosome 4 is very small and may be eliminated from this consideration.)

18.14 Describe or illustrate with sketches the appearance of the following heterozygous chromosome modifications in salivary preparations: (a) deficiency; (b) duplication; (c) inversion; (d) reciprocal translocation.

18.15 What (a) genetic and (b) cytological evidence would indicate that a translocation was present in a plant material such as barley?

18.16 How is pollen sterility associated with translocations? Illustrate.

18.17 In a *Drosophila* salivary chromosome section, the bands have a sequence of 1 2 3 4 5 6 7 8. The homolog with which this chromosome must synapse has a sequence 1 2 3 6 5 4 7 8. (a) What kind of a chromosome change has occurred? (b) Describe or draw a diagram to illustrate the possible pairing arrangement.

18.18 Other chromosomes have sequences as follows: (a) 1 2 5 6 7 8; (b) 1 2 3 4 4 5 6 7 8; (c) 1 2 3 4 5 8 7 6. What kind of chromosome modification is present in each? Illustrate with diagrams the pairing of these chromosomes with their normal homologs in salivary preparations.

18.19 Chromosome 1 in maize has the sequence A B C D E F, whereas chromosome 2 has the sequence M N O P Q R. A reciprocal translocation resulted in the following arrangements: A B C P Q R and M N O D E F. Diagram the expected pachytene configuration and describe the causes of the pollen sterility that might be expected.

18.20 How could a phenotypic effect, such as the number of facets in expressions of bar eye in *Drosophila*, be demonstrated as a position effect?

18.21 (a) How do variegated position effects originate and (b) how can they be explained?

18.22 What is the (a) theoretical and (b) practical significance of the Philadelphia chromosome?

18.23 Describe with standard nomenclature the following human chromosome complements: (a) male karyotype with all normal chromosomes except one missing, No. 21; (b) female karyotype with a translocation between long arms of a No. 14 and a No. 21 chromosome, with the short arms of No. 14 and No. 21 missing; (c) male karyotype with two translocations, involving interchange of both whole arms of chromosomes 5 and 12.

18.24 What is the significance of chromosome inversions and translocations in evolution?

18.25 Why were critical chromosome studies in humans not carried out before 1956?

18.26 Why was it difficult before 1970 to identify individual human chromosomes?

Chapter 19

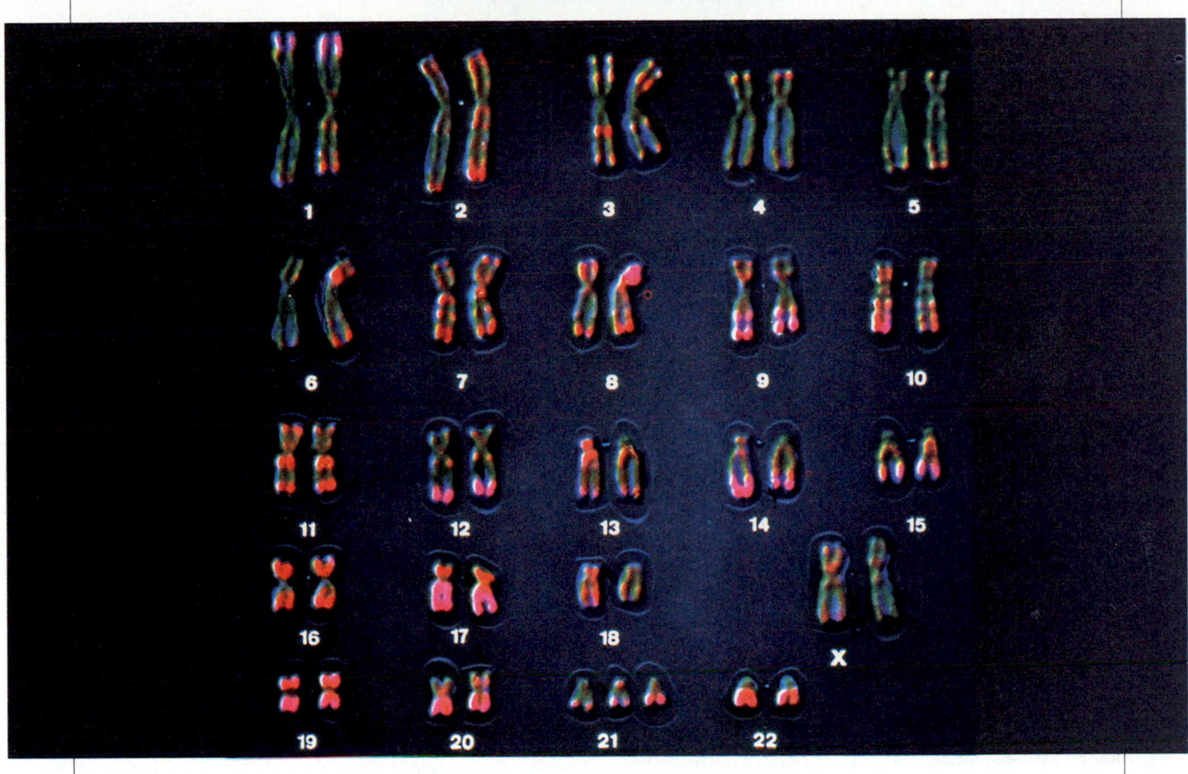

Variations in Chromosome Number

(Photo above): Karyotype of a woman with Down syndrome (trisomy 21). (CNRI/Science Photo Library/Science Source/Photo Researchers.)

SOMATIC cells of higher plants and animals usually have chromosomes in pairs ($2n$), that is, two of each kind of chromosome are present in each cell. Mature germ cells, having undergone reduction division, normally have one member of each pair (n). Many individual plants and animals, however, have local areas of somatic tissue characterized by a multiple of the basic chromosome number. A ***doubling process*** in cell division is the usual explanation for these deviations.

With the exception of sex differences, somatic doubling, and minor variations that occur in natural and experimental populations, all members of a species of plants or animals have the same basic chromosome number. *Ascaris lumbricoides* has only one chromosome pair, at least in early development stages. With this exception, the range of reported chromosome numbers in animals extends from 2 pairs in a rhabdocoel, *Gyratrix hermaphroditus,* and some mites, midges, and scale insects, to more than 100 in some butterflies and Crustacea. The crustacean *Paralithodes camtschatica,* for example, has 208 chromosomes or 104 pairs. The reported range in plants is from 2 pairs in the small composite plant *Haplopappus gracilis* to several hundred in some ferns. A species of fernlike plants of the genus *Ophioglossum* is reported to have 768 chromosomes.

Changes in the number of chromosomes may be reflected in high ***inviability*** and phenotypic ***anomalies*** in those that survive. This constitutes a useful tool for identifying the influence of individual chromosomes. For example, if phenotypically distinguishable individuals with different chromosome numbers can be identified in natural populations or produced experimentally, it is sometimes possible to determine the effect of adding or removing certain chromosomes. Some plants with increased chromosome numbers have phenotypic changes in morphological or physiological characteristics that are of practical importance to humans. Tomato plants, for example, with chromosome numbers above $2n$ are larger and produce more desirable fruit than do corresponding varieties with the usual $2n$ number.

Chromosome changes are classified in terms of additions or eliminations of parts of chromosomes (Chapter 18), whole chromosomes, or whole sets of chromosomes (genomes). Two main classes are ***euploidy*** and ***aneuploidy*** (*ploid,* Greek for "unit"; *eu,* "true or even"; and *aneu,* "uneven"). Euploids have chromosome complements consisting of whole sets, or genomes. The basic chromosome number of euploid organisms is represented by the haploid (n). The symbol **n** represents the ***reduced*** or half the diploid ($2n$) number in a somatic cell. In polyploid plants, the functional reduced haploid number (n) may not represent the smallest number (x) that can make up a chromosome set or genome. Bread wheat, *Triticum aestivum,* for example, has $2n = 42$ chromosomes, $n = 21$. The 21 chromosomes segregate to each gamete in meiosis. Small seedlings that sometimes occur in a wheat field carry one genome of 21 chromosomes. But emmer wheat, *T. dicoccum,* has $n = 14$, and einkorn, *T. monococcum,* has $n = 7$, where $2n = 14$. Seven is an odd number representing a complete genome in this species. It cannot be divided further into equal sets of whole chromosomes. The symbol **x** represents the ***smallest possible*** (monoploid) number in a ***genome.*** Bread wheat is a hexaploid with $6x = 42$. In garden peas, n (haploid) $= x$ (monoploid) $= 7$, and $2x = 2n = 14$ (diploid). Euploids with more than one genome (monoploid) may be diploid ($2n$), triploid ($3n$), tetraploid ($4n$), and so on.

In humans, additions and deletions of chromosomes, particularly the large chromosomes (groups A, B, and C), almost always result in lethals. Some newborns with extra chromosomes of the smaller groups such as G (e.g., chromosome 21) survive but show multiple physical and mental abnormalities.

TRISOMY IN HUMANS

Down Syndrome (47, +21)

The best-known and most common chromosome-related disease syndrome, formely known as "mongolism," is now designated ***Down syndrome,*** after Langdon Down, who first described the clinical signs in 1866. The cost of training and maintaining Down syndrome cases in the United States is estimated at ***$1 billion per year.*** Emotional stress in families with Down syndrome children and adults is also a factor in their care. The need for effective counseling and prevention is readily apparent.

Patients studied by Down, and those with Down syndrome who have since been observed (Fig. 19.1), were short in stature (about four feet tall) and had an epicanthal fold (thus the earlier name "mongolism"), broad short skulls, wide nostrils, large tongues with distinctive furrowing, stubby hands (particularly the fifth digit) with a simian crease (Fig. 19.2) on the palm and a single crease on the fifth digit, and general loose jointedness, observed particularly in the ankles. They are characterized as low in mentality (Table 19.1), but they can be trained in routine mechanical skills. Through the investigations of J. Lejeune in 1959, Down syndrome became the first chromosomal disorder to be described in humans.

Figure 19.1 Facial features of a child with Down syndrome. Down syndrome children are usually sensitive, happy, and lovable. They are skillful and dependable when properly trained and employed. (Copyright © 1988 by Bruce Coleman, Inc., New York, NY. Reproduced with permission.)

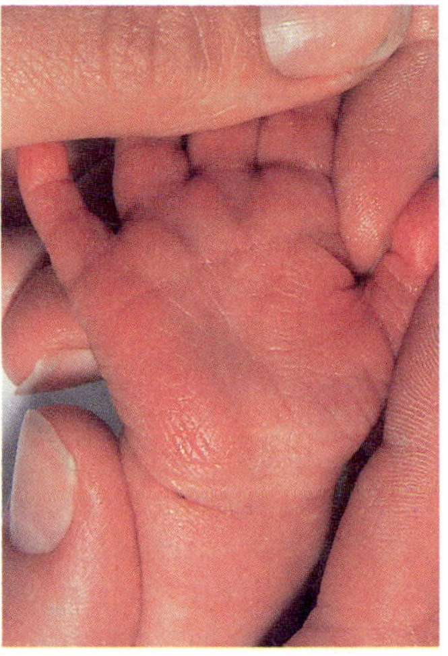

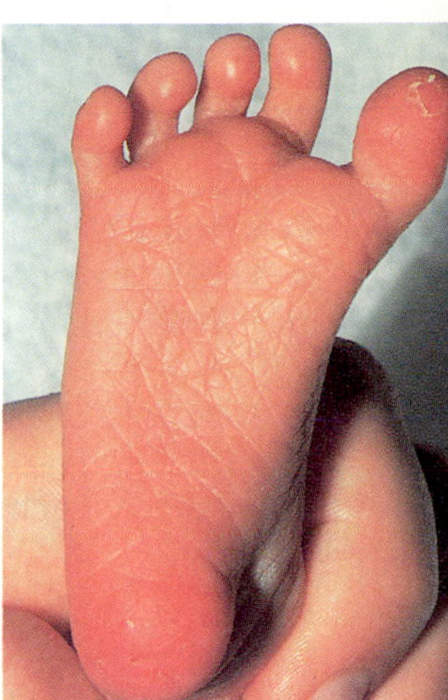

Figure 19.2 Hand and foot of a child with Down syndrome showing abnormal skin creases. (Courtesy of Science Photo Library.)

TABLE 19.1 Aneuploidy Resulting from Nondisjunction in the Human Population

CHROMOSOME NOMENCLATURE	CHROMOSOME FORMULA	CLINICAL SYNDROME	ESTIMATED FREQUENCY AT BIRTH	MAIN PHENOTYPIC CHARACTERISTICS
47,+21	$2n + 1$	Down	1/700	Short broad hands with simian-type palmar crease, short stature, hyperflexibility of joints, mental retardation, broad head with round face, open mouth with large tongue, epicanthal fold.
47,+13	$2n + 1$	Trisomy-13	1/20,000	Mental deficiency and deafness, minor muscle seizures, cleft lip and/or palate, polydactyly, cardiac anomalies, posterior heel prominence.
47,+18	$2n + 1$	Trisomy-18	1/8000	Multiple congenital malformation of many organs; low-set, malformed ears; receding mandible, small mouth and nose with general elfin appearance; mental deficiency; horseshoe or double kidney; short sternum. 90% die in the first 6 months.
45,X	$2n - 1$	Turner	1/2500 female births	Female with retarded sexual development, usually sterile, short stature, webbing of skin in neck region, cardiovascular abnormalities, hearing impairment.
47,XXY 48,XXXY 48,XXYY 49,XXXXY 50,XXXXXY	$2n + 1$ $2n + 2$ $2n + 2$ $2n + 3$ $2n + 4$	Klinefelter	1/500 male births	Male, subfertile with small testes, developed breasts, feminine pitched voice, long limbs, knock knees, rambling talkativeness.
47,XXX	$2n + 1$	Triple X	1/700	Female with usually normal genitalia and limited fertility. Slight mental retardation.

A small autosome, chromosome 21 (Fig. 19.3), added to the normal complement (47, +21) causes Down syndrome. This is a ***trisomic*** for No. ***21***; all other chromosomes are "disomes." Chromosome 21 had been difficult to distinguish from No. 22. Chromosome 21 was considered to be slightly larger than No. 22 when the chromosomes were first arranged (on the basis of size) in a karyotype. Critical study has since shown that No. 22 is larger than No. 21, but the Paris Conference (1971) agreed to leave them in their original positions in the standard karyotype. Both No. 21 and No. 22 usually have satellites—distal chromosome segments that are separated from the rest of the chromosome by a chromatic filament that can be observed in good chromosome preparations. Differential staining and photography through a fluorescent microscope have now distinguished between No. 21 and No. 22 (Fig. 19.4). Trisomy of No. 21 is the result of primary nondisjunction, which can occur at either of the two meiotic divisions and in either parent. Paired chromosomes do not separate properly to the poles at anaphase; one gamete will receive two No. 21 chromosomes and the other none.

Down syndrome occurs once in about 700 live births among European people. Incidence at conception is estimated to be considerably higher (7.3 per 1000), the difference being reflected in fetal loss due to ***spontaneous abortion.*** About one in six of the Down syndrome children born alive dies within the first year. The average expectation of life is about 16.2 years. Some patients with Down syndrome have a total of only 46 instead of 47 chromosomes, but in such cases a translocation has joined the long arm of chromosome 21 with another chromosome in the same complement (Fig. 19.5), most frequently No. 14. When an ***inherited translocation*** is associated with Down syndrome in an infant, the risk figure of a second occurrence in the same family can be calculated. With assurance that the translocated chromosome t(14*q*21*q*) will remain intact and assuming that one parent has a normal chromosome complement, the theoretical risk is one in three that each additional child will have Down syndrome. Karyotypes of parents who have already produced a child with the inherited translocation Down syndrome are 45,XX, t(14*q*21*q*) × 46, XY (normal). Furthermore, gametes from the first parent are expected to be of six different kinds instead of the normal number: (1) normal 14,21; (2) balanced translocation (14*q*21*q*); (3) t(14*q*21*q*),+21; (4) 14; (5) t(14*q*21*q*),+14; (6) 21. If each is fertilized by a normal gamete (14,21) from the second parent, the resulting ratio is 2 normal to 1 trisomy-21 (Down syndrome, Fig. 19.6). All zygotes with ***extra or missing large chromosomes*** (trisomics or monosomics) ***do not survive.*** Only zygotes with the smallest autosomes (No. 21) in trisomy can survive, and this combination produces Down syndrome. Thus, one-third of the viable zygotes from the cross, representing possible future children in the family, are expected to have Down syndrome. The risk factor of 1 in 3 rests on the assumption that the three surviving zygotes are equally viable and that the other three will die. Apparently, this assumption is not

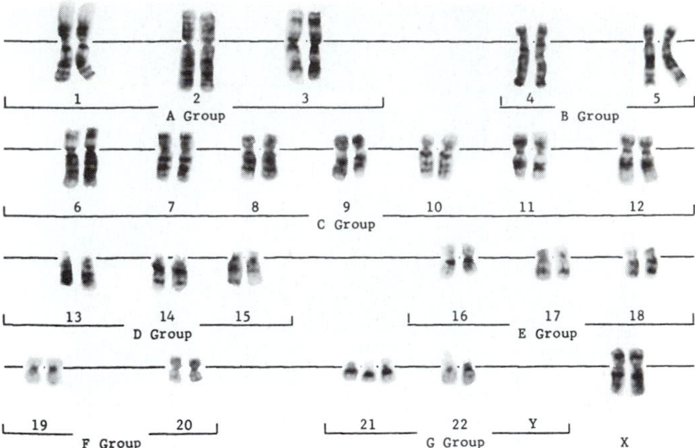

Figure 19.3 Karyotype of a child with Down syndrome, showing trisomy for chromosome 21 (47,XX,+21). (Courtesy of R. M. Fineman.)

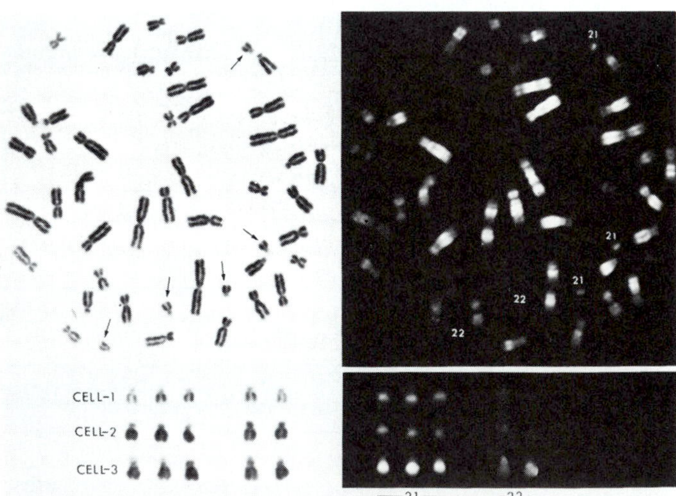

Figure 19.4 Photograph showing the difference between chromosomes 21 and 22 in a trisomic with Down syndrome. The spread on the left is stained with orcein. The one on the right has been stained with quinacrine dihydrochloride and photographed through a fluorescent microscope. This technique shows the brighter fluorescence of chromosome 21. The 21 and 22 chromosomes at the bottom came from three different cells from the same child with Down syndrome. (Courtesy of I. A. Uchida and C. C. Lin.)

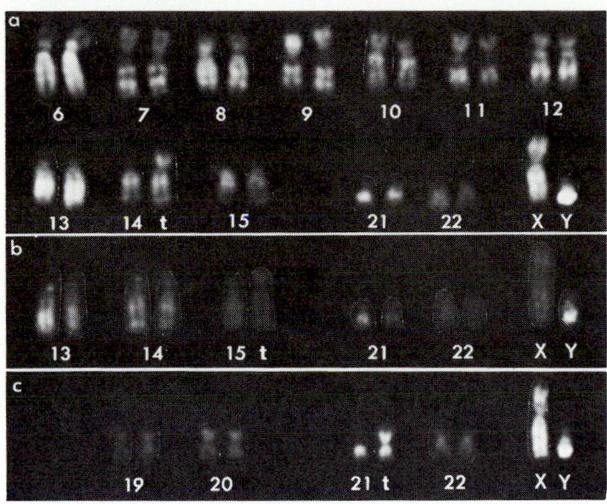

Figure 19.5 Partial karyotypes from three different translocation patients with Down syndrome. (a) The top band is from a child with a translocation of chromosome 21 to a 14 chromosome; (b) the middle band involves Nos. 21 and 15; and (c) the bottom band shows centric fusion of two No. 21 chromosomes. (Courtesy of I. A. Uchida and C. C. Lin.)

valid since the observed frequency for Down syndrome is far lower, about 11 percent, when the mother is a balanced translocation carrier; it is 2 percent when the father is the translocation carrier.

Amniocentesis for Detecting Aneuploidy

Chromosomal abnormalities are sufficiently well understood to permit genetic counseling. A fetus may be checked in early stages of development by karyotyping the cultured cells obtained by a process called ***amniocentesis*** (Figs. 19.7 and 19.8). A sample of fluid is withdrawn with a needle from the amniotic sac. Fetal cells are cultured and, after a period of two to three weeks, chromosomes in dividing cells can be stained and observed. If three ***No. 21*** chromosomes are present, Down syndrome is confirmed. The risk for mothers less than 25 years of age to have the trisomy is about 1 in 1500 births; at 40 years of age, 1 in 100 births; at 45, 1 in 40 births. If all pregnant women of age 45 were checked, about 1 in 40 would be expected to be carrying a child with trisomy-21. Pregnant women over 45 are a special ***high-risk*** group. Some trisomy-21 cases are not related to maternal age. Genetic factors rather than aging mothers are postulated to control nondisjunction in these cases.

Most, and perhaps all, cases of trisomy in humans can be explained by nondisjunction at meiosis, as illustrated in Fig. 19.9. Older women are more likely than younger women to evidence chromosome damage and meiotic irregularities. An ovary may form about 400,000 eggs, but only about 400 will mature. All the eggs remain in prophase I from the time they are formed in the female fetus until the time of life when they become mature and capable of fertilization and development. The longer the time before fertilization, the greater the chance for damage and irregularity. Some drugs and other environmental factors can cause chromosome anomalies.

Trisomy-13 and Trisomy-18

Trisomy-13 syndrome (47,+13; Fig. 19.10), described by K. Patau in 1960, occurs in about 1 in 20,000 newborns. It is rare in children and nonexistent in adults because the severe symptoms result in early death. Most of the deaths occur within the first three months after birth, but a few victims have lived for as long as five years. Symptoms include small brain, apparent mental deficiency, deafness, and numerous other external and internal abnormalities (Table 19.1).

Trisomy-18 syndrome (47,+18; Fig. 19.11), described by J. H. Edwards and his colleagues in 1960,

Cross: 45,XX,t(14*q*21*q*) × 46,XY

GAMETE OF SECOND PARENT	GAMETES OF FIRST PARENTS					
	14,21	t(14*q*21*q*)	t(14*q*21*q*), + 21	14	t(14*q*21*q*), + 14	21
14,21	14,21 14,21	t(14*q*21*q*) 14,21	t(14*q*21*q*), + 21 14,21	14 14,21	t(14*q*21*q*), + 14 14,21	21 14,21
PHENOTYPES OF ZYGOTES	Normal	Normal translocation carrier	Trisomy-21, Down syndrome	Monosomic-21, lethal	Trisomic-14, lethal	Monosomic-14, lethal

Summary: 2 normal, 1 with Down syndrome.

Figure 19.6 Diagram of a cross between an individual with an inherited translocation, carrying chromosome 21, 45, XX, t(14*q*21*q*) and a normal person, 46, XY. The theoretical risk of one-third for Down syndrome among children of a parent carrying the translocation t(14*q*21*q*) and a normal parent is illustrated. The actual risk is far less, presumably because of decreased viability of zygotes carrying the translocation, the only zygotes capable of producing a Down syndrome child.

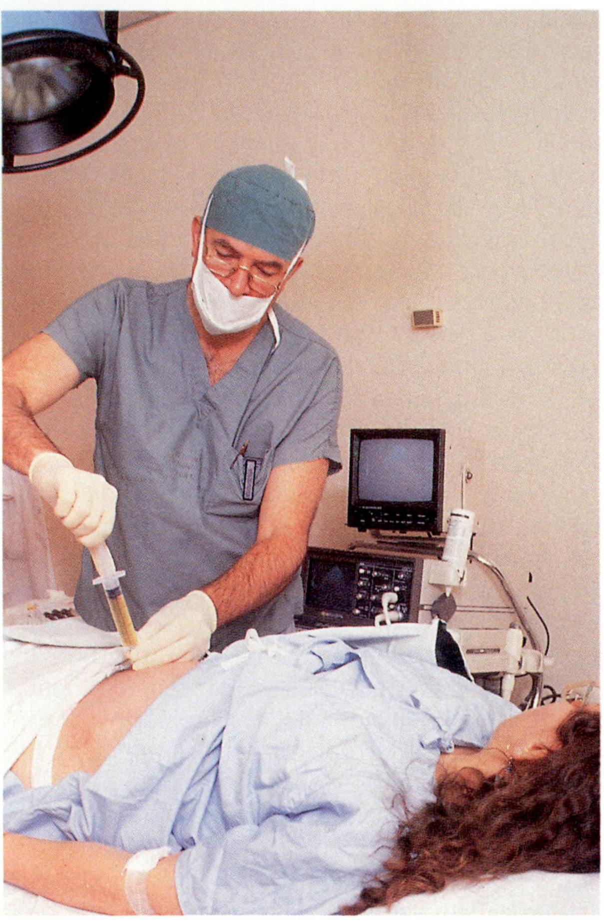

Figure 19.7 A physician taking a sample of fluid from the amniotic sac of a pregnant woman for prenatal diagnosis of a chromosomal or biochemical abnormality. (IMS/Custom Medical Stock Photo.)

includes mental deficiency and multiple congenital malformations (Figs. 19.12 and 19.13) involving virtually every organ system (Table 19.1). Most infants with this syndrome die at an early age, some 90 percent within their first six months. Nearly all are deceased before they reach one year, but a few have been reported to be alive in their teen years.

The incidence of trisomy-18 is about 1 in 8000. Thus far, only a few cases have been observed, and it is not known whether racial or other population groups differ in incidence. In general, the incidence of this deformity is greater among infants of older women, as expected if the cause of this trisomy is ***primary nondisjunction in meiosis.***

Turner Syndrome (45, X)

This monosomic has a chromosome complement of 44 autosomes and one X chromosome (Fig. 19.14). The chromosome anomaly is associated with an abnormal female phenotype described in 1938 by H. H. Turner and associates and known as ***Turner syndrome*** (Fig. 19.15). It occurs in about 1 per 2500 live female births. More than 90 percent abort spontaneously. A rough estimate for 45, X adults in the general population is 1 in 5000. These adults (Fig. 19.16) have virtually no ovaries, have limited secondary sexual characteristics, and are sterile. Microscopic sections of the ovaries show fibrous streaks of tissue representing remnants of ovaries. Affected females have short stature, low-set ears, webbed neck, and a shieldlike chest (Table 19.1). Mental deficiency is not usually associated with this syndrome. Epithelial cells of 45, X patients are ***X chromatin negative*** (Chapter 4), as expected when only one X chromosome is present.

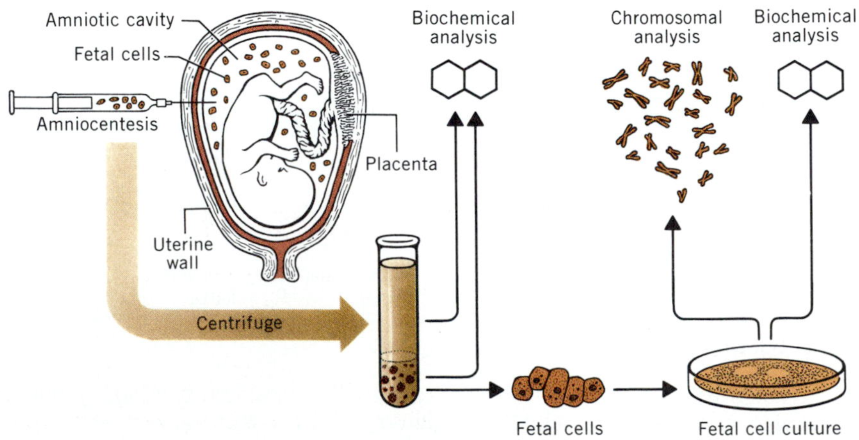

Figure 19.8 Amniocentesis and procedure for prenatal diagnosis of biochemical and chromosomal disorders.

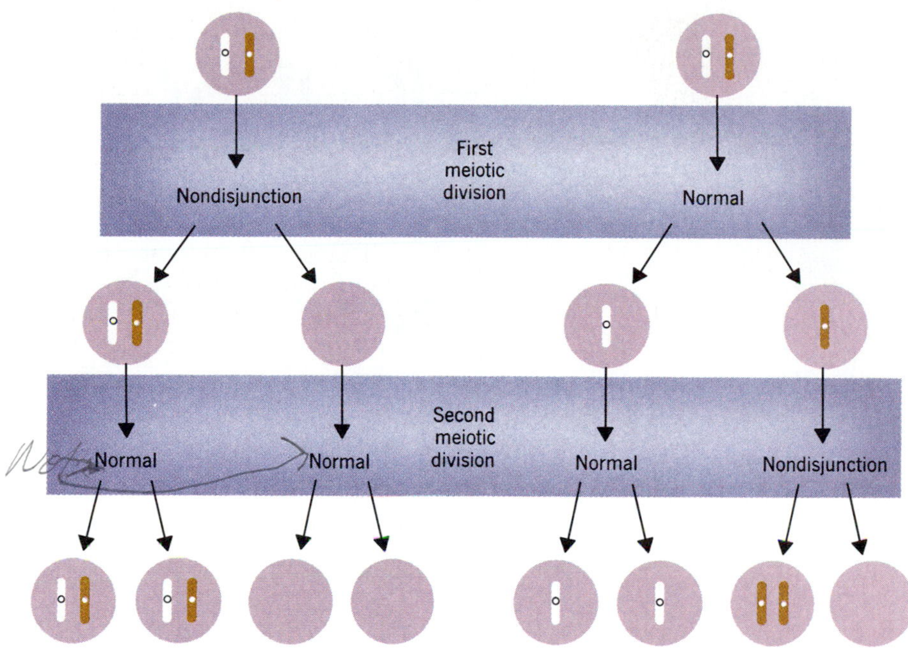

Figure 19.9 Consequences of nondisjunction at the first and second meiotic divisions. Nondisjunction at meiosis I produces no normal gametes. Nondisjunction at meiosis II produces two gametes containing (or lacking) two identical chromosomes, both derived from the same member of the homologous pair, and two normal gametes. It is assumed that other chromosome pairs not shown in this diagram behave normally in the first and second meiotic divisions.

X monosomics probably originate from exceptional eggs or sperm with no sex chromosome or from the loss of a sex chromosome in mitosis during early cleavage stages, after an XX or XY zygote has been formed. This latter possibility is supported by the high frequency of mosaics that result from postzygotic events in patients with Turner syndrome. ***Mosaics*** with X/XX sex chromosomes show symptoms of Turner syndrome but are usually taller than X and have fewer anomalies than nonmosaic 45, X females. They show more feminization, more normal menstruation, and may be fertile. Many cases of the somatic Turner phenotype without the typical 45, X chromosome constitution are now known. Most of these have one

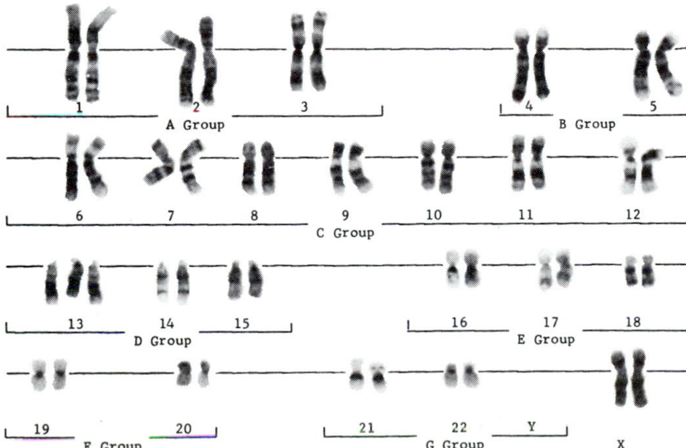

Figure 19.10 Karyotype of a newborn with trisomy-13 syndrome, showing three No. 13 chromosomes: 47, XX,+13. (Courtesy of R. M. Fineman.)

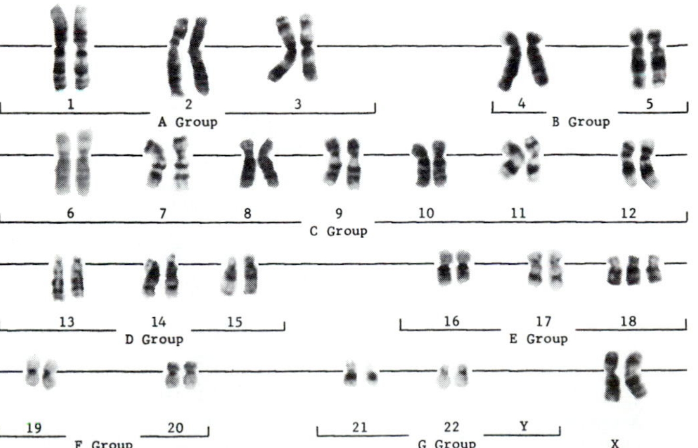

Figure 19.11 Karyotype of a child with trisomy-18 syndrome, showing three No. 18 chromosomes: 47, XX,+18. (Courtesy of R. M. Fineman.)

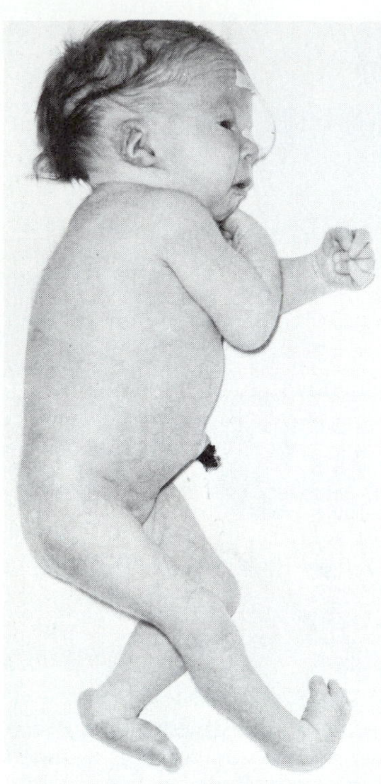

Figure 19.12 Infant with trisomy-X and trisomy-18 (48,XXX,+18). Note abnormal ears, receding mandible, flexion deformity of fingers, small hips, left rockerbottom foot, and right club foot characteristic of trisomy-18. (Courtesy of I. A. Uchida.)

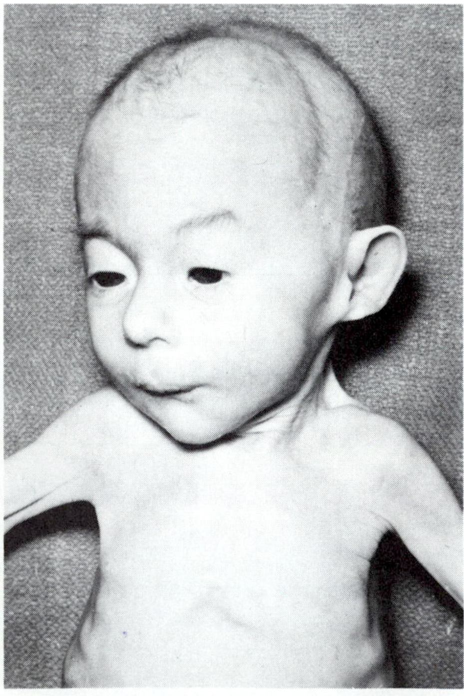

Figure 19.13 Facial features associated with trisomy-18. (Courtesy of I. A. Uchida.)

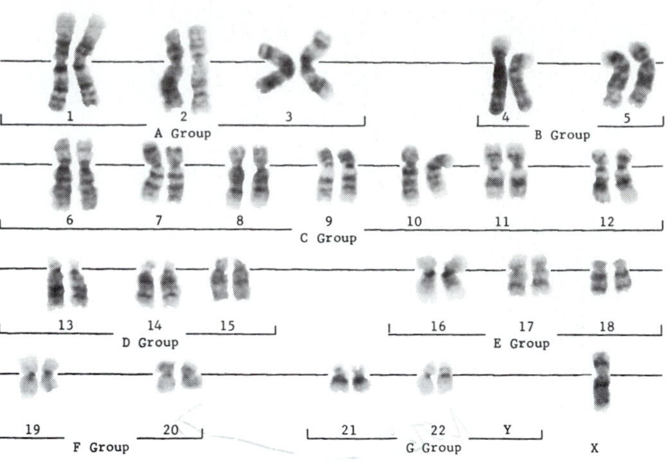

Figure 19.14 Karyotype of a child with Turner syndrome, showing only one sex (X) chromosome (45,X). (Courtesy of R. M. Fineman.)

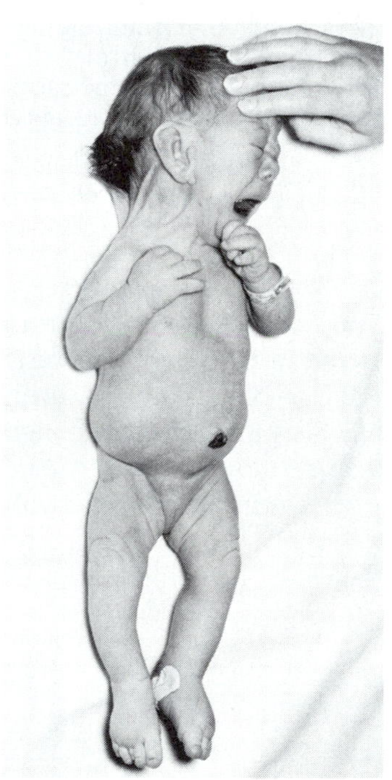

Figure 19.15 Infant with Turner syndrome (45,X). Note webbed neck and puffy hands and feet. (Courtesy of I. A. Uchida.)

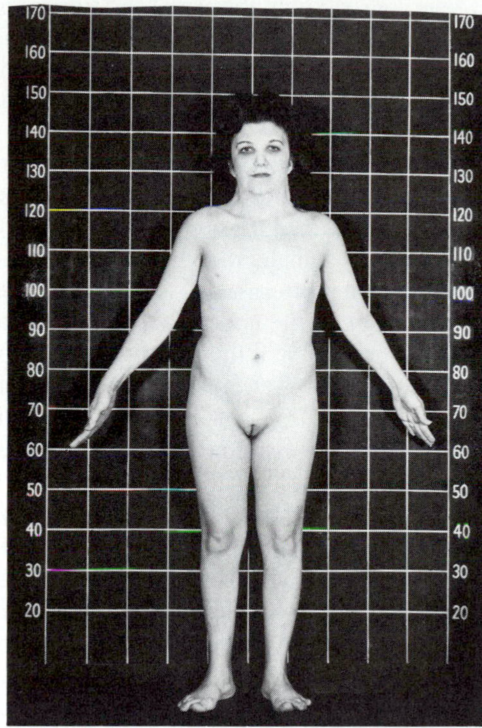

Figure 19.16 Female with Turner syndrome (45,X). She is short in stature, has a short neck, and lacks most female secondary sex characteristics. The breast development shown here has been induced by administering estrogen. Note the broad, shieldlike chest with widely spaced nipples, mild neck webbing, and low-set ears. (Courtesy of V. A. McKusick, Johns Hopkins University.)

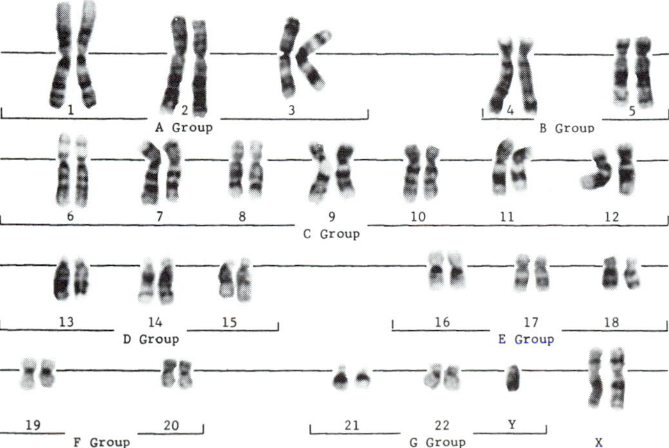

Figure 19.17 Karyotype of a young man with Klinefelter syndrome, showing an extra X chromosome (47,XXY). (Courtesy of R. M. Fineman.)

normal X chromosome and a fragment of a second X chromosome. Both arms of the second X chromosome are apparently necessary for normal ovarian differentiation. Individuals with only the long arm of the second X are short in stature and show other somatic symptoms of Turner syndrome, whereas those with only the short arm of the second X have normal stature and do not show as many signs of Turner syndrome. This indicates that the Turner phenotype is mostly controlled by ***genes on the short arm of the X chromosome.***

Patients with partial deletion of one X chromosome are X chromatin positive and therefore may be misdiagnosed if a buccal smear is the only test used. The deficient X chromosome always forms the X chromatin body. A Y chromosome also occurs in some individuals with the Turner phenotype. These patients are usually mosaic for 45, X/46, XY, with a normal Y. People with one X and a Y fragment, not including the Y short arm, have only streak ovaries but are normal in phenotype. This suggests that ***male-determining genes*** are in the short arm of the ***Y chromosome,*** and those that prevent the Turner phenotype are in the Y long arm as well as the X short arm. Major features of the Turner phenotype occur in some males as well as in females. The male Turner syndrome is characterized by defective development of the testes, sterility, and limited male secondary sexual characteristics, along with somatic features of the Turner phenotype. These people have normal male karyotypes (46, XY).

Klinefelter Syndrome (47, XXY)

An extra X chromosome in addition to the usual male (XY) chromosome complement (47, XXY) has been associated with the abnormal male syndrome described (in 1942) by H. F. Klinefelter and known as ***Klinefelter syndrome*** (Fig. 19.17). It is estimated to occur in 1 per 500 live male births. Individuals with this syndrome (Table 19.1) are phenotypically males but with some tendency toward femaleness, particularly in secondary sex characteristics. Such features as enlarged breasts (Fig. 19.18), underdeveloped body hair, small testes, and small prostate glands are a part of the syndrome. Presumably, the ***XXY*** constitution originates either by fertilization of an exceptional ***XX*** egg by a ***Y*** sperm or of an ***X*** egg by an exceptional ***XY*** sperm. Studies of Klinefelter syndrome and Turner syndrome indicate that the ***Y*** chromosome in human beings, unlike that in *Drosophila,* determines male sex.

The most common karyotype (about three-quarters of the cases) for Klinefelter syndrome is ***47, XXY,*** but the symptoms of the syndrome will usually occur whenever more than one X chromosome is present along with a Y chromosome. More complex karyotypes

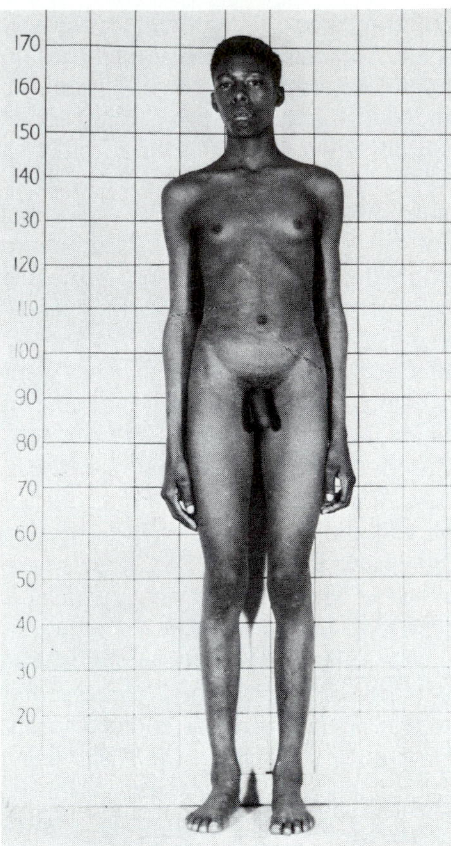

Figure 19.18 Male with Klinefelter syndrome. The syndrome includes small testes and very little body hair. Affected males tend to be long-legged and thin. They often develop breasts (gynecomastia) like females and are sterile. (Courtesy of V. A. McKusick, Johns Hopkins University.)

associated with the Klinefelter syndrome include XXYY, XXXY, XXXYY, XXXXY, XXXXYY, and XXXXXY. All patients with Klinefelter syndrome have one or more X chromatin bodies in their cells. Mental retardation is usually found when there are more than two X chromosomes. The XY/XXY mosaicism in patients with Klinefelter syndrome is associated with less severe ***physical and reproductive anomalies.***

Aneuploidy of X Chromosomes and Mental Deficiency

Other irregular combinations of X chromosomes have also been recognized among females with X chromosome aberrations. About 1 percent of all mentally defective women in institutions have been shown to have one or more extra X chromosomes. This chromosome abnormality occurs in about 1 in 700 live births in the general population. Individuals with ***"triple X syndrome"*** are comparable in some ways with *Drosophila* metafemales (***XXX***). In *Drosophila,* however, such individuals are usually lethal, and those that survive are strikingly abnormal and sterile. Human XXX individuals are sometimes visibly indistinguishable from normal XX females, but there is considerable range in phenotypic expression. They may be mentally abnormal.

The best-known symptoms in this syndrome are abnormalities associated with functional processes such as menstruation. One patient cited by P. A. Jacobs was a 37-year-old female who reported that the first suggestion of an abnormality was highly irregular menstruation. When the abdominal wall was opened, the ovaries appeared as if they were postmenopausal. Microscopically, they showed deficient ovarian follicle formation. Of 63 cells observed, 51 had 47 chromosomes; the extra chromosome was an X. ***Nondisjunction*** in the production of the egg from which this woman developed was postulated as the mechanism for the occurrence of the extra chromosome. Buccal smears showed two sex chromatin bodies in the epithelial cells as expected when three X chromosomes are present. Individuals with tetrasomic X chromosomes (48, ***XXXX***) are all mentally defective. The degree of mental deficiency increases with the number of X chromosomes present.

47, XYY and Behavior

P. A. Jacobs and her associates reported in 1965 that seven ***XYY*** males were detected in a population of 197 male, mentally subnormal inmates of a penal institution in Scotland. The XYY (Fig. 19.19) men were unusually tall, with an average height of 73.1 inches, compared with 64 inches for XY men in the same prison. Numerous other studies, mostly in institutionalized populations, have since confirmed that a high proportion of XYY individuals are tall, subnormal in intelligence (with IQs ranging from 80 to 95), and ***antisocial.*** The aggressive behavior that brought them into conflict with the law was usually against property rather than people.

XYY trisomy occurs about once in 1000 live male births in the general European population. Only a few of these can be accounted for in the criminal population. Furthermore, most XYY men have been described as perfectly normal in behavior. In studies carried out in the 1970s, E.B. Hook showed that only 3.6 percent of all XYY men are institutionalized for any reason.

Environmental factors are presumed to be involved in the development of ***aggressiveness.*** Since some XYY men are subnormal in intelligence and excessively tall in stature, particular environmental situations in childhood or adulthood may lead to withdrawal from society or aggressive behavior. Unfavorable social conditions such as frustration in personal accomplishment and taunting from associates

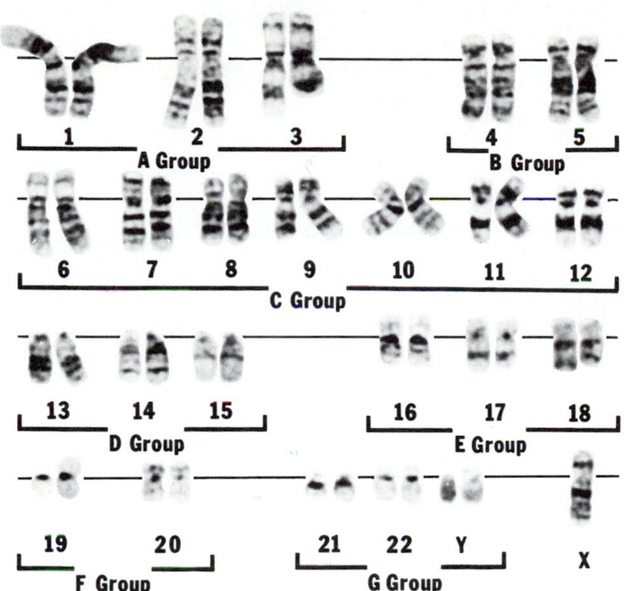

Figure 19.19 Karyotype of a boy showing an extra Y chromosome (47,XYY). (Courtesy of S. W. Rogers.)

may encourage physical aggression as a means of adaptation. Males with this sex trisomy have not been found to transmit the extra Y chromosome to their sons. This extra chromosome seems to be weeded out in gametogenesis. A wide range of physical and mental abnormalities has been detected in institutionalized XYY men, but most of these are irregular in occurrence and do not form a syndrome. Tallness of stature and mental dullness, however, are fairly constant characteristics among institutionalized XYY men.

CHROMOSOMAL MOSAICS

Individuals who have at least two cell lines, with different karyotypes derived from one zygote, originate from nondisjunction in a cleavage division after fertilization. One daughter cell would thus receive one too many chromosomes and the other would be one deficient, as illustrated in Fig. 19.20. Each cell would give rise to a cell line with its irregular chromosome number. Proportions of cells representing the different cell lines would vary in different tissues, making the extent of the mosaicism and the effect on the organism difficult to predict.

Many sex chromosome mosaics have been detected in human beings. The main phenotypic characteristic is ***extreme variability.*** Some sex chromosomes mosaics have been reported—X/XX, X/XY, XX/XY, XXY/XX, XX/XXX, XXX/X, XXX/XXXXY—as well as several other combinations reflecting two or three cell lines. Mild to severe phenotypic symptoms have been associated with these sex chromosome mosaics.

NONDISJUNCTION AND ANEUPLOIDY IN *DROSOPHILA*

C.B. Bridges' example of the trisomic X chromosome (Chapter 4) was explained on the basis of primary nondisjunction of the X chromosomes. When fertilized

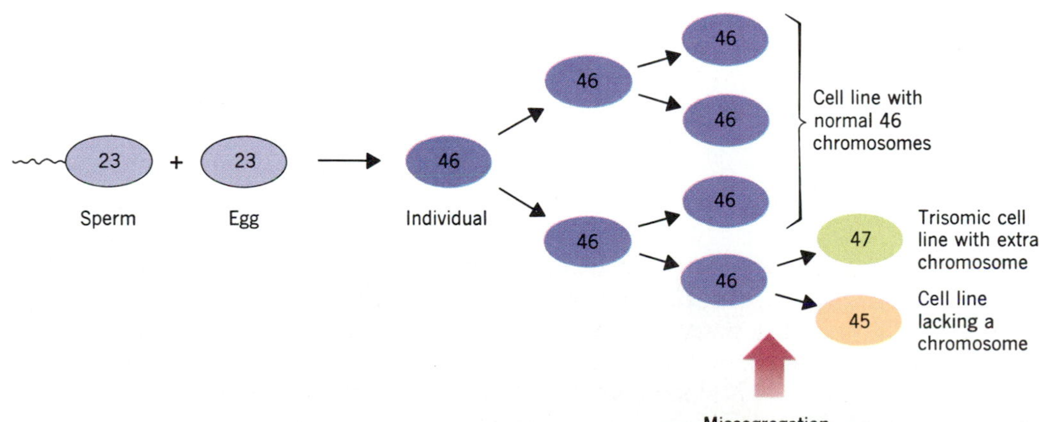

Figure 19.20 Mechanism through which chromosomal mosaicism may occur. A normal individual with 46 chromosomes is formed when sperm and egg unite. During an early cell cleavage, nondisjunction gives rise to a cell with 47 chromosomes and a cell with 45 chromosomes. Each of these cells initiates a cell line, thus forming a mosaic individual with cells carrying different numbers of chromosomes.

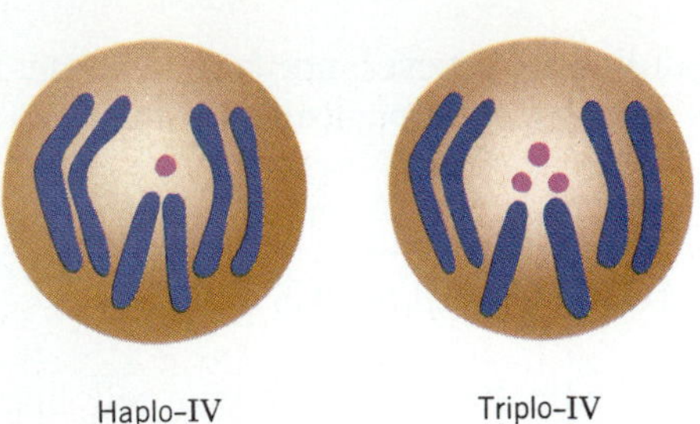

Figure 19.21 Trisomic called "triplo-IV," produced by the addition of a fourth chromosome, and a monosomic called "haplo-IV," which results from the elimination of one fourth chromosome in *Drosophila.*

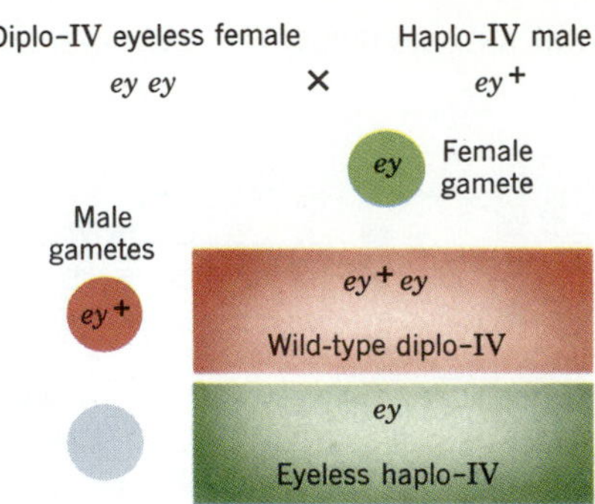

Figure 19.22 Cross between a diplo-IV eyeless female and a haplo-IV male.

with normal sperm, eggs with two X chromosomes produced two kinds of trisomics, ***XXX and XXY.*** The metafemales (***XXX***) were always ***sterile*** and usually ***inviable,*** but the ***XXY*** combinations produced females phenotypically indistinguishable from wild-type females. Zygotes with normal pairs of autosomes but single X or Y chromosomes were monosomics. Although those with a single X were normal in appearance but were sterile, zygotes with a single Y chromosome were inviable. Later investigations by Bridges showed that chromosome IV, which is very small, could be added or eliminated without seriously affecting the viability of the flies. However, when the large II and III chromosomes were lost or added, the resulting cells were always inviable.

A monosomic ($2n - 1$) called ***haplo-IV*** was obtained through the elimination of one IV chromosome. Through the addition of the small fourth chromosome to the wild-type complement (Fig. 19.21), a trisomic ($2n + 1$) called ***triplo-IV*** was obtained. Haplo-IV flies are small with slender bristles and deviate in several minor respects from the wild type. Triplo-IV and diplo-IV flies cannot be separated on the basis of phenotype. Bridges considered triplo-IV to have a darker body than wild type with coarser bristles, smaller eyes, and more elongate wings, but the departure from normal is very slight and the phenotype overlaps with the normal.

Chromosome IV genes, such as *ey* for eyeless (a phenotype characterized by small or missing eyes), segregate differently with different chromosome numbers, as expected. When $2n$ eyeless flies (*ey/ey*) were crossed with haplo-IV (Fig. 19.22), the F_1 generation consisted of normal $2n$ and eyeless haplo-IV flies. When eyeless flies were crossed with triplo-IV, none of the first-generation progeny were eyeless. About half were triplo-IV, and half were normal (diplo-IV), as expected. When F_1 triplo-IV females were crossed with eyeless (*ey/ey*) males, as shown in Fig. 19.23, the progeny again consisted of about half triplo-IV and half diplo-IV. Normal and eyeless phenotypes were present in a ratio of about 5:1, respectively. This ***trisomic ratio*** is explained on the basis of the extra chromosome in the triplo-IV flies and the dominance of 1 ey^+ over 2 *ey*.

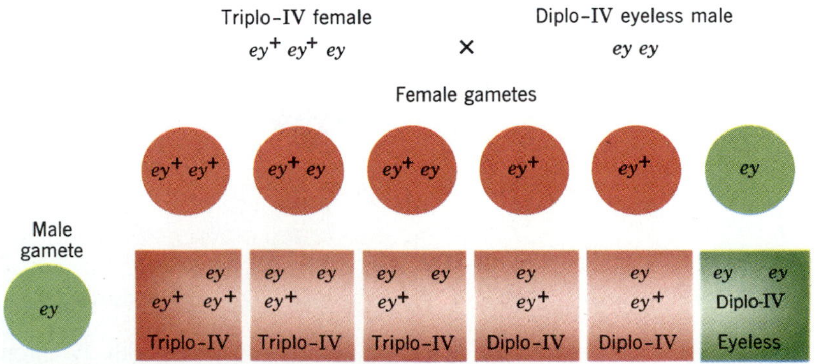

Figure 19.23 Cross between a triplo-IV female and a diplo-IV eyeless male.

ANEUPLOID SEGREGATION IN PLANTS

The first critical study of aneuploid plants was initiated in 1924 by A. F. Blakeslee and J. Belling when they discovered a "mutant type" with 25 chromosomes in the common Jimsonweed, *Datura stramonium,* which normally has 24 chromosomes in the somatic cells. At the meiotic metaphase, one of the 12 pairs was found to have an ***extra member,*** that is, one ***trisome*** ($2n + 1$) was present along with 11 disomes. This trisomic plant differed from wild-type plants in several specific ways, particularly in shape and spine characteristics of seed capsules. Because the complement was composed of 12 chromosome pairs differing in the genes they carried, 12 distinguishable trisomics were possible in Jimsonweed. Through experimental breeding, Blakeslee and his associates succeeded in producing ***all 12 possible trisomics.*** These were grown in Blakeslee's garden, and each was found to have a distinguishable phenotype that was attributed to the extra set of genes contained in one of the 12 chromosomes.

One of the 12 trisomic types, known as poinsettia, had several distinguishing traits, including morphological characteristics of seed capsules, that were attributed to the basic trisomic arrangement. It was also possible to identify some traits that were determined by genes on particular chromosomes by trisomic ratios, that is, 5:1, 17:1 (Fig. 19.24), and 35:1, in contrast to 3:1 and 1:1 expected from monohybrid crosses in regular $2n$ plants. The extra chromosome in poinsettia, for example, was found to carry the locus for alleles p^+ and p, for purple or white flowers, respectively. Any one of three chromosome arrangements with the dominant gene p^+ ($p^+\,p^+\,p^+$, $p^+\,p^+\,p$, or $p^+\,p\,p$) produced poinsettia plants with purple flowers, whereas only one, the fully recessive trisomic ($p\,p\,p$), gave rise to poinsettia plants with white flowers. The $2n$ plants had two chromosome combinations ($p^+\,p^+$ and $p^+\,p$) for purple and one ($p\,p$) for white.

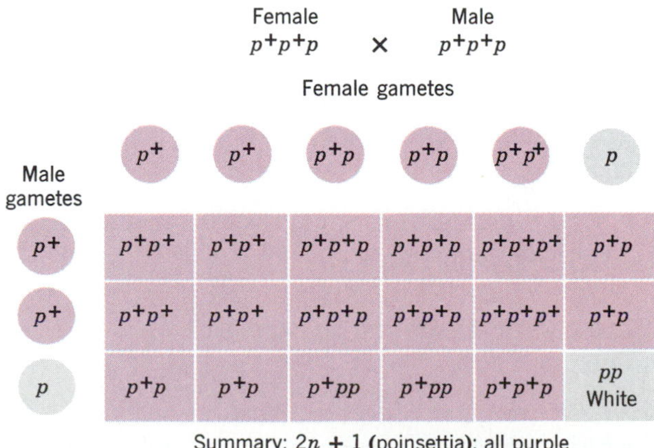

Figure 19.24 Cross between two purple trisomic plants of the genus *Datura,* illustrating a trisomic ratio of 17 : 1.

Plant trisomics experience interesting ***complications*** when undergoing ***meiosis.*** Two chromosomes go to one pole and one goes to the other in the reduction division of megasporogenesis, thus giving rise to different kinds of gametes, some with two and some with one member of the trisome. Trisomic ratios reflect increased proportions of progeny carrying wild-type dominant alleles as compared with diploid ratios. In the Jimsonweed, developing megaspores tolerate extra chromosomes and form gametes with relatively little loss of viability. However, when additional chromosomes above the n number enter developing microspores, the resulting spores cannot successfully compete against those with the normal haploid complement. A cross between two purple poinsettia plants $p^+\,p^+\,p \times p^+\,p^+\,p$ [seed parent (female) written first] may be constructed as illustrated in Fig. 19.24. ***Female gametes*** are of four kinds, two haploid and two carrying an extra chromosome, in the proportion $2p^+$, $2p^+\,p$, $1p^+\,p^+$, $1p$. Because male gametes receiving extra chromosomes are nonfunctional, only two kinds, p^+ and p, occurring in the proportion $2p^+$ and $1p$, were involved in fertilization. All poinsettia ($2n + 1$) that resulted from the cross carried at least one p^+ gene and had purple flowers. The $2n$ plants occurred in the proportion of 8 purple to 1 white. Complete sets of trisomics have since been discovered for other plants, including rice (Fig. 19.25). The different trisomics may be distinguished by characteristics of the grain as well as other features of the plants.

Tetrasomics and Nullisomics

Aneuploids more complex than trisomics have been produced in some species, but ordinarily these are highly inviable. ***Tetrasomics (2n + 2)*** have been identified in wheat, but no phenotypic characteristics were observed that could distinguish them from trisomics. Another variation that could be expected to occur in an aneuploid series is the complete absence of a certain kind of chromosome ($2n - 2$). Plants in which a chromosome pair is completely missing are called ***"nullisomics."*** These may be produced occasionally in nature, but seldom survive long enough to be recognized or to perpetuate the chromosome type. E. R. Sears experimentally produced all the 21 possible nullisomics in wheat, *Triticum aestivum.* By associating certain phenotypes with corresponding chromosome arrangements, nullisomics have been used effectively in locating several different genes in wheat.

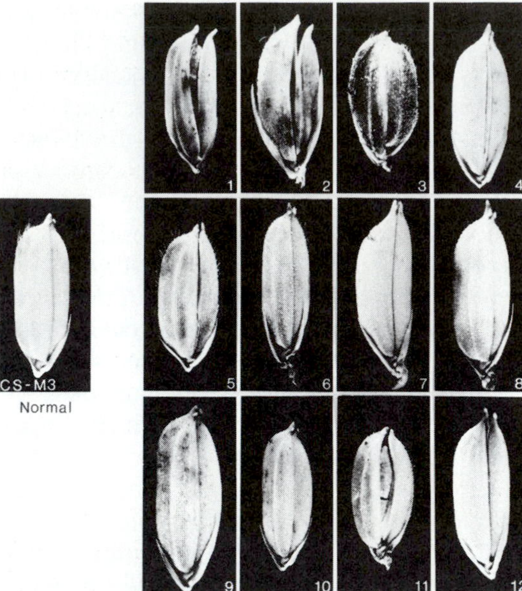

Figure 19.25 A set of 12 primary trisomics of rice, variety CS-M3, showing the grain characteristics of each type along with the normal diploid. The numbers correspond to the extra chromosome in each trisomic in decreasing order of length. (Courtesy of S.H. Khan and J. N. Rutger, University of California, Davis, and ARS, USDA, Davis, CA.)

EUPLOIDY

Whereas aneuploids differ from standard $2n$ chromosome complements in single chromosomes, euploids differ in multiples of n or of x if $n = x$. Monoploids (n) carry one genome. The n or x chromosome number is usual for gametes of diploid animals, but unusual for somatic cells. Monoploidy is seldom observed in animals, but is found in the male honeybee and other insects in which male haploids occur.

By contrast, plants have a gametophyte stage in their cycle that is characterized by the reduced (n) chromosome number. In higher plants, this stage is brief and inconspicuous, but in some lower plant groups it is the major part of the cycle. Occasionally, plants in natural populations or plots can be recognized as monoploids by observation and verified by cytological procedure. These plants are usually frail in structure, with small leaves, low viability, and a high degree of sterility. Sterility is attributed to ***irregularities at meiosis.*** Obviously, no pairing is possible because only one set of chromosomes is present. Therefore, if the meiotic process succeeds at all, the dispersal of chromosomes to the poles is irregular, and the resulting gametes are highly inviable. Because monoploids undergo no segregation and carry a single set of genes, they are valuable experimental tools when they can be produced successfully. Microorganisms that are propagating monoploids are especially useful in genetics (Chapter 8). Diploid plants with two genomes ($2n$) are most common among euploids. Normal chromosome behavior in animals and plants is based on diploids, which are used in the following examples as standards for comparison.

POLYPLOIDY IN ANIMALS

Organisms with three or more genomes are polyploids. Fully one-half of all known plant genera contain polyploids, and about two-thirds of all the grasses are polyploids, but polyploids are rarely seen in animals. One reason is that the sex balance in animals is much more delicate than that in plants. ***Sterility*** in animals is virtually always associated with a departure from the diploid number. The few animals (such as the brine shrimp, *Artemia salina*) that show evidence of polyploidy utilize ***parthenogenesis*** to escape the hazard of anomalous gametes.

An exceptional case of triploidy in salamanders related to *Ambystoma jeffersonianum* has been reported. Female salamanders of this particular group that have large erythrocytes and erythrocyte nuclei produced some triploid larvae with 42 chromosomes, whereas those with small erythrocytes and erythrocyte nuclei produced diploid larvae with 28 chromosomes. Field observations and laboratory studies indicated that distinct, persisting populations of triploid females have become established in parts of the range occupied by this species complex.

Although animals composed entirely of polyploid cells are rare, many diploid animals have polyploid cells within certain tissues of their bodies. For example, in teleostean embryos, giant nuclei (presumably polyploid) have been observed in many species. Giant polyploid nuclei occur in particular tissues of a wide range of diploid animals (e.g., the liver and kidneys in humans).

POLYPLOIDY IN PLANTS

Chromosome Doubling in Somatic and Germ Cells

Two basic irregular processes have been discovered by which polyploids may evolve from diploid plants and become established in nature. (1) With ***somatic doubling,*** cells sometimes undergo irregularities at mitosis and give rise to meristematic cells that perpetuate these irregularities in new generations of plants. (2) Reproductive cells may have an irregular reductional or equational division in which the sets of chromosomes fail to separate completely to the poles in anaphase. Both sets thus become incorporated in the same ***"restitution"*** nucleus, which doubles the chromosome number in the gamete. Both of these irregularities occur in nature. Once polyploidy is established, intercrossing among plants with different chromosome numbers may give rise to numerous chromosome combinations. Most of these are sterile, but some may be fertile and come under the influence of natural selection. All degrees of viability are encountered—from lethal combinations to those that compete favorably with diploids in particular environmental situations.

Two main kinds of polyploids, ***autopolyploids*** and ***allopolyploids,*** may be distinguished on the basis of their source of chromosomes. Autopolyploids occur when the same genome is duplicated (e.g., $n_1 + n_1$). Apparently, this occurs frequently in single cells of many plants, but these cells usually do not survive. Allopolyploids result when different genomes come together through hybridization (e.g., $n_1 + n_2$). Among the surviving plants, it is usually impossible to determine whether the genomes are alike and therefore whether the polyploids are autopolyploids or allopolyploids unless information from ancestral history is available and detailed chromosome studies are performed.

The presence of varying numbers of quadrivalents (i.e., four homologous chromosomes instead of the usual two pairing with one another in meiosis) suggests autopolyploidy. Unequal segregation of chromosomes in quadrivalents is the main reason why ***autopolyploids are sterile*** to varying degrees. In the meiotic prophase, chromosomes must pair with one another throughout their entire length. When four similar chromosomes are present, they usually pair with different chromosomes at different places along their length, complicating the meiotic process and resulting in nonviable gametes or zygotes. Unequal chromosome pairing is not, however, the whole basis for sterility in autopolyploids. ***Unequal disjunction*** occurs, as well as perhaps more important phenomena, none of which is well enough established to permit discussion here. Chromosomes in some autopolyploids appear to pair properly but form bivalents rather than quadrivalents.

Some plant groups have a series of chromosome numbers based on a multiple of a basic number. In the genus *Chrysanthemum,* for example, the basic number is 9, and species are known that have 18, 36, 54, 72, and 90 chromosomes. In *Solanum,* the genus of nightshades including the potato, *S. tuberosum,* the basic number is 12. Members of this genus include species with 24, 36, 48, 60, 72, 96, 108, 120, and 144 chromosomes. In spite of such conspicuous examples where autopolyploidy would appear superficially to be involved in the origin of plants with different chromosome numbers, it is doubtful that autopolyploidy alone has played a major role in the evolution of plant groups. Inviability and sterility would seem to preclude the perpetuation of true autopolyploids in nature. Autopolyploidy combined with allopolyploidy, however, produces ***"autoallopolyploids"*** and has apparently been an important process in the evolution of some plants.

Triploid Plants

Triploids ($3n$) with three complete genomes occur when tetraploid ($4n$) or unreduced diploid ($2n$) plants produce viable $2n$ gametes that unite in fertilization with normal n gametes. Reduction sometimes fails in diploid plants, resulting in gametes with more than a single genome. But some $2n$ gametes are produced and fertilized with n gametes, and viable triploids occur in nature. The triploids, however, do not ordinarily reproduce and become established because of irregularity during meiosis that results in low survival and sterility of the survivors. ***Synapsis*** in meiosis can take place in several ways, but only between two homologous chromosomes in the same region. A $2n + n$ segregation may occur in three ways:

$$\frac{1+2}{3}, \quad \frac{1+3}{2} \text{ and } \frac{2+3}{1}$$

The possibility of getting either a $2n$ or an n chromosome set in a gamete is remote. The few gametes that are functional become fertilized and are mostly ***unbalanced and inviable.*** Meiotic segregation of homologs results in gametes with chromosome number varying from n to $2n$, with all integral values in between. In this respect, triploids are like other polyploids with odd numbers of genomes ($5n$, $7n$, and so on) in sexually reproducing organisms. They are all sterile.

But there is a way for triploids to be propagated. Like other plants, they can be ***propagated vegetatively.*** Bananas are triploids that are propagated through asexual cuttings. Winesap, Gravenstein, and Baldwin apples and European pears are perpetuated by grafting and budding and thus maintain their triploid characteristics. Tulips with three sets of chromosomes ($3n$) are also propagated by vegetative bulbs. Triploids occur in grasses, forest trees, vegetables, and flower garden varieties. ***Seedless watermelons*** and other seedless fruits are often triploids. Among animals, triploids are very rare, but they have been found in *Drosophila,* salamanders, lizards, and the land isopod, *Triconiscus.* On the whole, triploids have been unsuccessful in establishing themselves in natural populations because the usual consequence of triploidy is sterility. Triploids have, however, provided opportunities for human ingenuity in genetic engineering to develop species with odd ploidy for practical use.

Tetraploid Plants

Tetraploids ($4n$) with four genomes frequently originate from a ***doubling of diploids.*** They also may result from the ***duplication*** of somatic chromosomes following irregularities at mitosis. If the spindle does not develop properly in the mitotic sequence of a diploid, and cell division fails to follow chromosome duplication, a single nuclear membrane may develop around the two sets of chromosomes that ordinarily would produce daughter nuclei, forming a single restitution nucleus. If this tetraploid cell perpetuates itself through normal mitosis, the increased chromosome number may become established in a group of cells or tissues within the organism. When such plants are capable of vegetative reproduction, they may be manipulated to produce whole tetraploid plants. Failure at reduction division in the oöcytes of some plants also results in polyploids. Chromosome irregularity in mature pollen is rare because developing male gametes with irregular chromosome numbers do not compete favorably with normal gametes.

Early in the nineteenth century, some seeds of the American saltmarsh grass (*Spartina alterniflora*) were accidentally transported by ship to Bayonne, France, and to Southampton, England. The American species became established in the same localities where a European saltmarsh grass (*S. maritima,* formerly *S. stricta*) was growing. A new saltmarsh grass, *S. townsendii* (currently known as *S. anglica*), commonly called Townsend's grass, was later identified in these localities. By 1907, it had become common along the coast of southern England and northern France. Townsend's grass was more vigorous and aggressive than either the American or the European species and crowded out the native grasses in many places. Therefore, it was intentionally introduced into Holland to support the dikes, and it was also imported into other localities for similar purposes. Townsend's grass was considered to be a hybrid between the American and European species; but, unlike most hybrids, it was fertile and true breeding. The European grass had $2n = 60$ chromosomes, the American species $2n = 62$, and Townsend's grass $2n = 122$; some plants had $2n = 120$ and others 124 (Fig. 19.26). These facts suggested that a cross had occurred in which allopolyploidy was involved. Townsend's grass was an "amphidiploid," with the sum of the diploid chromosomes carried by the two species. This evidence, along with the high fertility and intermediate appearance, indicated that the new plant arose from natural hybridization and doubling of chromosomes. The chromosome doubling had presumably given the hybrid its ***fertility and ability to survive.***

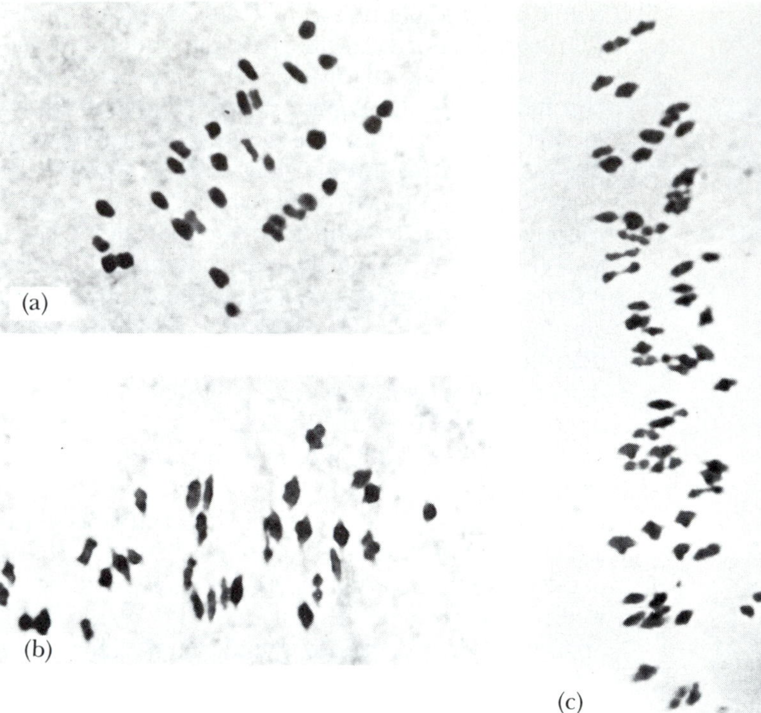

Figure 19.26 Metaphase I (paired) chromosomes of two parent species and the amphidiploid hybrid, Townsend's grass. (a) European marshgrass, *Spartina maritima,* $2n = 60$; (b) American saltmarshgrass, *S. alterniflora,* $2n = 62$; (c) Townsend's grass, *S. townsendii (S. anglica),* $2n = 122$ (some plants had $2n = 120$ and some $2n = 124$). (Courtesy of C. J. Marchant.)

The low fertility and marked phenotypic variation associated with chromosome irregularity are illustrated in Fig. 19.27. A cross was made between Polish wheat, *T. polonicum,* a tetraploid with large, amber kernels, and Marquis, a hexaploid with hard, red kernels. The entire F_2 from the cross is shown in the illustration. Only a few plants were produced from an extensive experiment, and those observed varied widely in phenotype.

Allopolyploidy has occurred in various plant groups, and some present-day plants have resulted from this kind of hybridization. Polyploid species that are established in nature have genomes that correspond more or less completely to the combined chro-

Figure 19.27 Heads of wheat resulting from an F_2 cross between Polish with 28 chromosomes and Marquis with 42 chromosomes. These are the only heads produced from an extensive experiment, indicating the low fertility encountered in crosses between plants with different chromosome numbers.

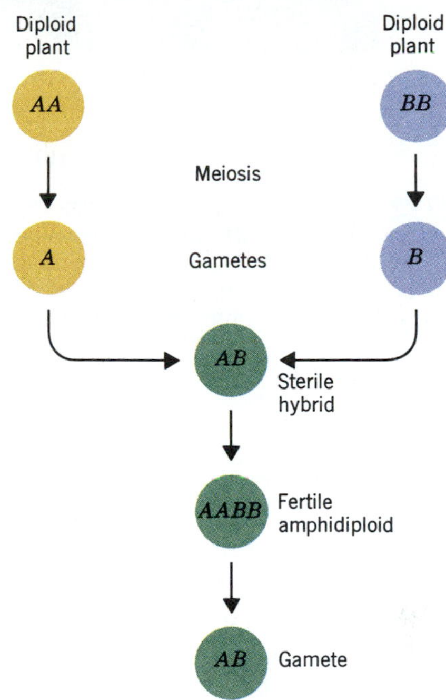

Figure 19.28 The production of a fertile amphidiploid (*AABB*) by doubling of the chromosomes of a sterile hybrid (*AB*) produced by crossing two normal diploid plants (*AA* and *BB*). *A* represents a haploid set of chromosomes from diploid plant *AA*, and *B* represents a haploid set of chromosomes from diploid plant *BB*. Since *A* chromosomes are not homologous with *B* chromosomes, meiosis is highly irregular in hybrid *AB*, resulting in a high degree of sterility. The fertile amphidiploid (*AABB*) may be produced by a rare union of diploid (*AB*) gametes that are produced by the hybrid *AB*. In the amphidiploid, meiosis is normal—*A* chromosomes pair with *A* chromosomes and *B* chromosomes pair with *B* chromosomes.

mosome complements of two different but related diploid plants (Fig. 19.28). These "amphidiploids" or allotetraploids have undergone hybridization somewhere in their ancestral history. Allopolyploidy thus represents a method by which new species may be formed almost immediately, whereas autopolyploidy alone results in meiotic anomalies and "dead ends" with reference to evolution.

INDUCED POLYPLOIDY

Polyploids have been induced experimentally by several methods in various plants. Anything that interferes with spindle formation during mitosis might result in a doubling of the chromosomes. Induced polyploidy was first demonstrated by subjecting growing plants to a higher than usual temperature. Maize and some other plants responded to such treatment with an increase in the chromosome number of certain cells. Some of these cells gave rise to germinal tissue and whole plants were propagated. Other such cells were cultured artificially and polyploid plants were produced. When, for example, the buds of tomato plants were removed, cells of some shoots developing from the ***scar tissue*** were ***tetraploid.*** These were propagated and whole plants with $4n$ chromosomes were produced.

The method of inducing polyploidy in plants that has become most widely used was developed by A. F. Blakeslee, A. G. Avery, and B. R. Nebel. These investigators found that an alkaloid, ***colchicine,*** extracted from the autumn crocus, *Colchicum autumnale,* could ***disturb spindle formation during cell division.*** When root tips or other growing plant parts

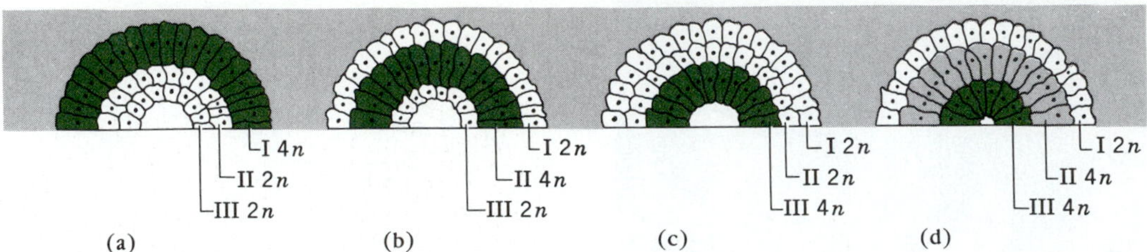

Figure 19.29 Mechanisms through which reproductive cells of a plant may have 4*n* chromosomes and produce tetraploids. Colchicine added to the medium interferes with cell division and transforms 2*n* cells to 4*n* cells. (a) The two inner layers are 2*n* and the outer layer is 4*n*. (b) Horizontal duplication has resulted in a 4*n* cell, which gave rise to a 4*n* middle layer. (c) The inner layer has become 4*n*. (d) The inner layer and middle layer have become 4*n*, and 4*n* reproductive cells may produce 4*n* (tetraploid) plants under experimental treatment.

were placed in appropriate concentrations of colchicine, chromosomes of the treated cells duplicated properly, but spindle formation was inhibited and the cytoplasmic phase of cell division did not occur. Instead, restitution nuclei with different numbers of chromosomes were produced in the treated tissue. Some cells had a completely doubled chromosome number. When these cells were propagated, tetraploid plants were produced and tetraploid seed was obtained.

In some plants, growing areas at the stem tips and lateral buds have three distinct cell layers, as illustrated in Fig. 19.29. Cells from each layer are much alike in early stages of development, but later they give rise to separate tissues in stems, leaves, and other organs. The outer layer (I) becomes the epidermis; the middle layer (II) gives rise to the reproductive cells (eggs and pollen); and the inner layer (III) produces the internal parts of stems and leaves. Colchicine placed in the medium of the growing tips may interfere with division and result in transformation of 2*n* cells into 4*n* cells in one or more layers. Cells usually divide vertically in such a way that the number within a given layer is increased. In examples (*a*), (*b*), and (*c*) in Fig. 19.29, the 4*n* cells are restricted to the first, second, and third layer, respectively. Sometimes cells divide horizontally and the daughter cells enter a new layer. Thus, a doubled (4*n*) cell in layer III may give rise to a 4*n* cell in layer II, as shown in (*d*). This pattern of irregularity may extend 4*n* cells in a new plant. By propagating tetraploid tissue, it is possible to produce ***tetraploid plants***

EXPERIMENTAL PRODUCTION OF POLYPLOIDS

In an early experiment, the Russian cytologist G. D. Karpechenko synthesized polyploids from crosses between two common vegetables that belong to different genera, the radish, *Raphanus sativus,* and the cabbage, *Brassica oleracea.* Although these plants were only distantly related, they were enough alike to be crossed successfully with the intervention of the experimenter. Both had 9 pairs of chromosomes. The diploid hybrid had 18 chromosomes, 9 from each parent, but it was sterile and could not perpetuate itself, largely because of the failure in pairing between the unlike chromosomes in meiosis. Some unreduced gametes were formed, however, and in the F_2 population Karpechenko recovered some tetraploids. When the chromosomes of the F_1 hybrid were doubled in this way, a fertile polyploid named *Raphanobrassica* was produced with 18 radish and 18 cabbage chromosomes. Because two sets of chromosomes were now present from each parental variety, pairing was quite regular. Normal gametes were produced and a high degree of fertility was obtained. This experiment had theoretical significance because it demonstrated a method by which fertile interspecific hybrids could be produced. It also suggested the possibility of incorporating desirable genotypes from two different species into a new polyploid species. Seed capsules of the parent, the sterile hybrid, and the tetraploid plants of Karpechenko's experiment are shown in Fig. 19.30. Unfortunately, from the practical standpoint, *Raphanobrassica* had the foliage of a radish and the root of a cabbage.

Sterile Hybrid Grass Made Fertile by Chromosome Doubling

Interesting experiments that made use of induced polyploidy and cytological analysis in certain grasses have been conducted by W. S. Boyle, A. H. Holmgren, and their associates at Utah State University. A completely sterile perennial grass was tentatively identified as a natural, sterile hybrid between two genera in the

Figure 19.30 Seed pods of (a) radish (*Raphanus*), with 18 *R* chromosomes; (b) cabbage (*Brassica*), with 18 *B* chromosomes; (c) sterile diploid hybrid with 9 *R* + 9 *B* chromosomes; and (d) tetraploid resulting from chromosome doubling, with 18 *R* + 18 *B* chromosomes. (After G. D. Karpechenko.)

Figure 19.31 Parent plants representing different genera and a hybrid produced by these plants. (a) *Hordeum jubatum;* (b) F_1 hybrid; (c) *Agropyron trachycaulum.* (Courtesy of W. S. Boyle.)

tribe Hordeae. Examination of the Hordeae species in the vicinity of the hybrids indicated that the two parents were probably *Agropyron trachycaulum* and *Hordeum jubatum* (Fig. 19.31). Cytological studies on the two presumed parents and the hybrid supported this view. Both *A. trachycaulum* and *H. jubatum* were found to be allotetraploids ($2n = 28$). During meiosis of both species, normal pairing of chromosomes occurred to form 14 paired chromosomes (bivalents). The sterile hybrid was also a tetraploid with the same chromosome number ($2n = 28$), but its chromosome behavior during the meiotic process was highly irregular (Fig. 19.32). Fourteen unpaired chromosomes (univalents) and seven bivalents were frequently observed at metaphase in pollen mother cells. Many lagging chromosomes remained in the center of the spindle during anaphase, and numerous small micronuclei, reflecting chromosome irregularity, were observed following division. No viable pollen was produced. These observations indicated that ***meiotic irregularity*** was a major factor in the sterility.

Sterile hybrids were produced through controlled reciprocal crosses between *A. trachycaulum* and *H. jubatum,* thus confirming the predicted parentage. The hybrids were then treated with colchicine. Some stalks or clumps with double chromosome numbers (octoploids) were produced and set seed. All this seed was viable and produced ***fertile*** plants considered to be autoalloploids. This name was applied because

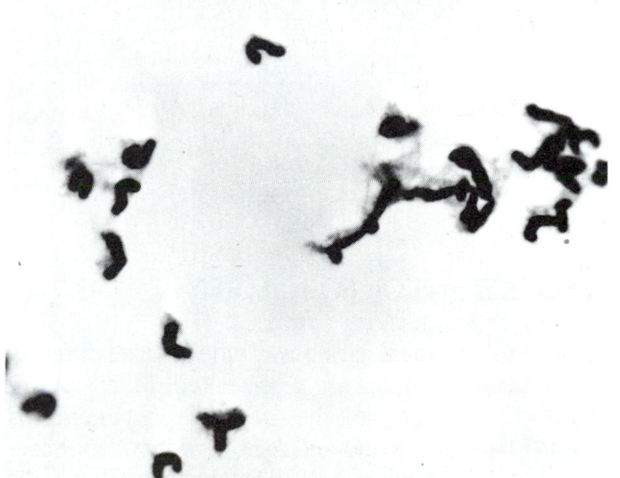

Figure 19.32 Chromosomes of the sterile hybrid between *A. trachycaulum* and *H. jubatum* during meiosis. Seven bivalents and 14 single chromosomes were frequently found as in this photograph. (Courtesy of W. S. Boyle.)

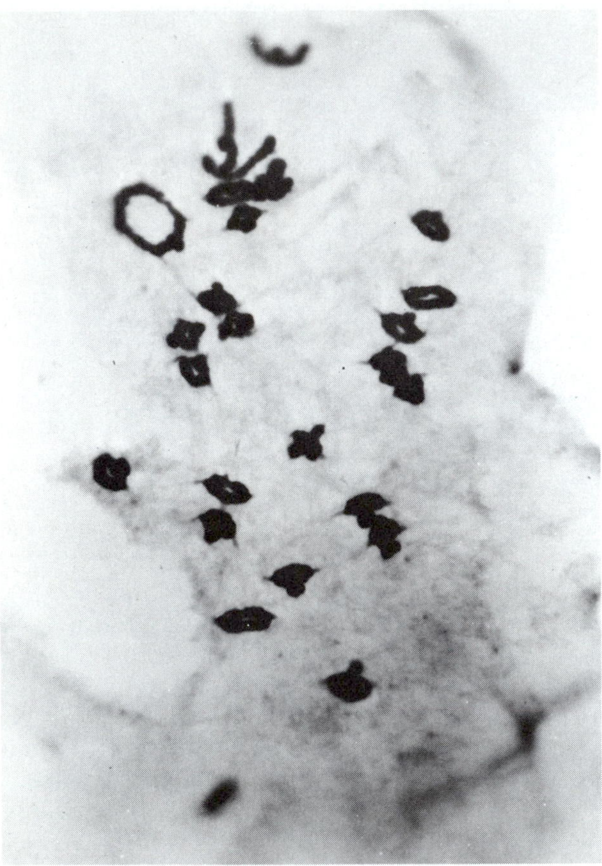

Figure 19.33 Diakinesis stage of colchicine-induced octoploid. In this photograph the following chromosome associations are present: 1 ring of 4; 22 rings of 2; 3 rods of 2; and 2 single chromosomes, making a total of 56 chromosomes. (Courtesy of W. S. Boyle.)

both autopolyploidy and allopolyploidy had entered into the formation of the fertile octoploid. Chromosome studies indicated that the two parents carried a genome in common. Therefore, the genomic formulas of the parents were represented as *AABB* and *AACC.* The sterile hybrid was *AABC* and the colchicine-induced octoploid was *AAAABBCC* (Fig. 19.33).

APPLICATIONS OF POLYPLOIDY

Among the cultivated varieties of wheat, three different chromosome numbers are represented: 14, 28, and 42 ($x = 7$). For example, the primitive small-grained einkorn type of Europe and Asia, *Triticum monococcum,* has 14 chromosomes in its vegetative cells. Its yield is low and it is of comparatively little value. An emmer wheat (durum), *T. dicoccum,* grown chiefly in southern Europe but also in the United States, has 28 chromosomes. It has thick heads with large, hard kernels and is used mainly for macaroni, spaghetti, and stock feed. The ***bread wheats,*** *T. aestivum,* with 42 chromosomes, were postulated by J. Percival in England to have come from a cross between emmer wheat and goat grass (*Aegilops*), both of which are native to the Babylonian region where bread wheat originated.

When techniques for artificial chromosome doubling became established, investigations of the origin of bread wheat confirmed Percival's theory. Experimental evidence obtained by E. S. McFadden and E. R. Sears and separately by H. Kihara traced the pathway for the origin of one type of bread wheat, *T. spelta.*

Aegilops squarrosa ($n = 7$) was found to carry a group of major characteristics that distinguish the hexaploid ($n = 21$) *T. spelta* from the tetraploids ($n = 14$) *T. dicoccum* and *T. dicoccoides.* Hybrids between these tetraploid species of wheat and *A. squarrosa* proved to have all the major taxonomic characters of *T. spelta* but the hybrids were completely or nearly sterile. When the F_1 hybrids of *T. dicoccoides* × *A. squarrosa* were treated with colchicine, highly fertile allopolyploids with 42 chromosomes were obtained. These synthetic hexaploids closely resembled the cultivated *T. spelta,* and they produced highly fertile hybrids with that species and with *T. vulgare,* known to be in the ancestry of the bread wheats. This demonstrated that the genome of the hexaploid wheats contained one chromosome set that originated from *A. squarrosa.* It was postulated that *T. spelta* is the ancestral hexaploid wheat of Europe, having arisen, possibly in fairly recent times, in southeastern Europe or southwestern Asia following chromosome doubling of natural hybrids of *T. dicoccoides* (or its cultivated close relative, *T. dicoccum*) × *A. squarrosa. Triticum spelta* is believed to have been carried over the northerly route into central and western Europe. Experiments of McFadden, Sears, and Kihara reconstructed the pathway through which a moderately useful wheat and a goat grass ***hybridized in nature*** and produced forerunners of a most valuable crop, bread wheat.

New World Cotton

Crosses can be made between distinct species of cotton, members of the genus *Gossypium.* The hybrids show a wide range of vigor and fertility, making the material favorable for studies of origins. Three cytological groups have been found to correspond with the major world distributional areas. Old World cotton has 13 pairs of large chromosomes. American cotton, which originated in Central or South America, has 13 pairs of small chromosomes. New World cotton (the cultivated long-staple type) has 26 pairs, 13 large and

13 small. Evidently, ***hybridization and chromosome duplication*** occurred somewhere in the ancestry of the New World cotton.

J. O. Beasley used the colchicine technique and succeeded in doubling the chromosomes of a hybrid between the Old World and American cotton. The resulting hybrids, with four sets of chromosomes (amphidiploids), crossed readily among themselves and produced fertile plants resembling New World cotton. The process by which the valuable ***polyploid cotton*** may have originated in nature was thus duplicated in the laboratory.

Primrose Hybridization

The primrose, *Primula kewensis,* is an allotetraploid with 36 ($2n$) chromosomes. It was derived from a cross between two diploids, *P. floribunda* ($x = 9$) and *P. verticillata* ($x = 9$). Plants from these two species crossed readily, producing hybrids with 18 chromosomes in their vegetative cells, 9 from *P. floribunda* and 9 from *P. verticillata,* but the hybrids were ***sterile.*** Eventually, however, a branch on a hybrid plant developed from a cell in which the chromosome number was doubled (36), so that each chromosome had a homologous partner. This branch was propagated and gave rise to a fertile primrose plant with cells containing 36 chromosomes. In Fig. 19.34, metaphase chromosomes of the two diploid parents, the sterile diploid hybrid, and the ***fertile allotetraploid*** are shown.

Disease Resistance in Tobacco

Induced polyploidy has been exploited to a great extent. Practical applications may become more common as additional data are accumulated. By artificially induced polyploidy, ***disease resistance*** and other desirable qualities have been incorporated into some commercial crop plants. Tobacco, *Nicotiana tabacum,* for example, is susceptible to the tobacco mosaic virus (TMV), whereas *N. glutinosa* appeared at first observation to be resistant. Further investigation, however, showed that in *N. glutinosa* the virus killed the cells that were invaded and the virus particles became isolated in the dead cell. The apparent resistance thus was attributable to hypersensitivity. When the two tobacco species were crossed, the hybrid was found to be "resistant" to the virus, but totally sterile. When the chromosomes were doubled, it was possible to secure a fertile polyploid "resistant" to the virus.

Polyploid Fruits, Flowers, and Wheat

Some varieties of plants that serve human needs more effectively than others have now been identified as polyploids. Many polyploids were selected and cultivated because of their large size, vigor, and ornamental values, before their chromosome numbers were known. Giant "sports" from twigs of ***McIntosh apple*** trees that were found to be tetraploid ($4n$) were propagated into whole trees, which produce extra-large fruit. The texture of the giant apples is as fine as that of diploids, but the yield is inferior. Mass selection of seedlings may overcome this difficulty. ***Bartlett pears,*** several varieties of ***grapes,*** and ***cranberries*** have also produced sports with ***giant*** fruits. Some of these show promise of practical usefulness. With colchicine treatment, a number of polyploids have been developed artificially. This technique has provided a way to explore the mechanism involved in ***polyploid formation*** and to make use of the good qualities of polyploids. Tetraploid ($4n$) maize is more vigorous than the ordinary diploid and produces some 20 percent more vitamin A. Its fertility is somewhat reduced, but this drawback responds to selection. Polyploid watermelons have been developed from colchicine treatment by Kihara and others. The tetraploid with 44 chromosomes is large and has practical value. Triploid

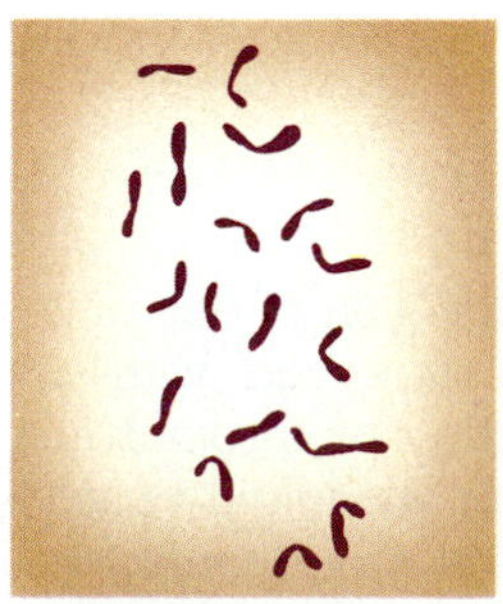

Primula floribunda

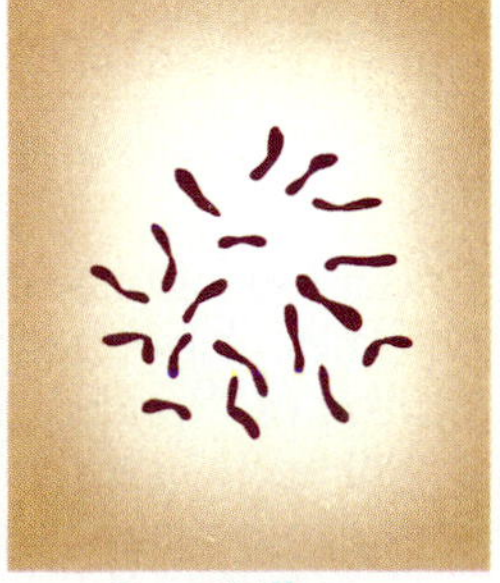
P. verticillata

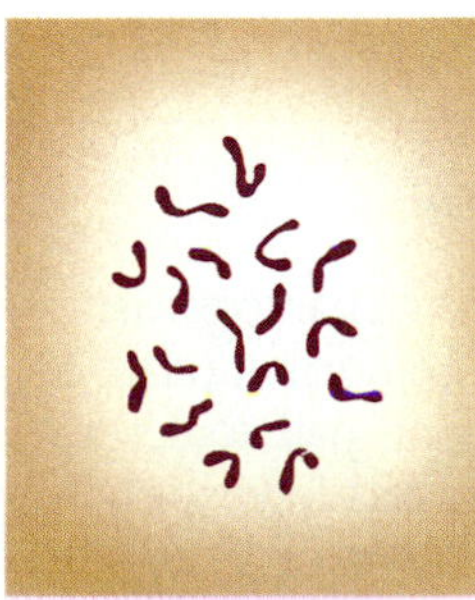
Sterile hybrid

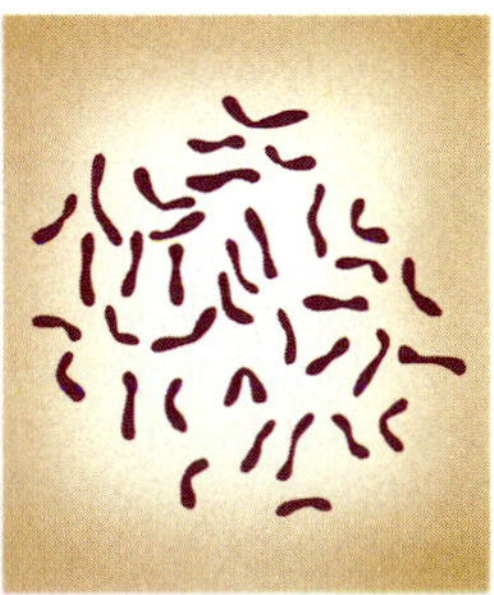
P. kewensis

Figure 19.34 Chromosomes of *Primula floribunda, P. verticillata,* the sterile hybrid, and the allotetraploid, *P. kewensis.* The two parents and the hybrid each have 18 chromosomes, whereas the fertile tetraploid *P. kewensis* has 36 chromosomes.

Figure 19.35 Wheat kernels illustrating the advantage of rust resistance. (Left) Rust-resistant wheat; (right) rust-susceptible wheat infected in nature, similar in other respects to the strain represented at left. (Courtesy of the Utah State Experiment Station.)

watermelons with 33 chromosomes are especially desirable because they are sterile and have no seeds. Among the flower garden varieties, **4n *marigolds*** and ***snapdragons*** are widely cultivated.

Polyploid plants respond to artificial selection and hybridization, as do diploid species. The recent history of plant breeding has been characterized by a marked improvement in many polyploid plant crops. The yield of wheat, for example, has increased appreciably. This has been accomplished by developing disease-resistant strains and breeding for increased hardiness and greater efficiency so that crops may survive under the various environmental conditions found in wheat-growing areas.

A constant threat to the wheat crop is rust—a fungus that attacks the stems and leaves of the growing plants and destroys the ripening grain. Spores are borne by wind and, when conditions are right, they spread like fire through wheat fields. The disease can be combated by developing ***rust-resistant*** strains and by eradicating barberry bushes, which are hosts to the spores during the spring months. But new varieties of rust that destroy previously resistant grain keep evolving, thus perpetuating the job of plant breeders. The advantage of incorporating rust resistance is illustrated in Fig. 19.35. The larger kernels at the left are from a new strain of rust-resistant spring wheat. At the right are shown kernels of wheat, similar in other respects but not resistant, that are dwarfed from infection with stem rust. The number of kernels of grain per plant as well as the size of the kernels is decreased by rust infection. Investigators in agricultural experiment stations are constantly alert for new rusts. When a new one is found, the standard wheat varieties are tested against it. If they are not resistant, breeding programs are initiated immediately to develop ***new strains*** resistant to that particular rust.

SUMMARY

A particular chromosome number is a characteristic of cells of plant and animal species. The normal range in different organisms extends from one pair to several hundred chromosomes. ***Euploid*** plants have a multiple of a basic chromosome number (n or x) for the species ($2n$, $3n$, $4n$, and so on). ***Aneuploids*** deviate from a base number by one, two, or a few chromosomes (e.g., $2n + 1$; $2n - 1$). Human beings have $2n = 46$ chromosomes. Irregularities in number, such as a loss of one or the gain of one or more chromosomes, have been associated with abnormal syndromes. Nondisjunction of chromosomes in meiosis will account for most aneuploid varieties. Irregularities in mitotic division will account for chromosome mosaics in individual organisms. Polyploids, which are multiples of whole sets of chromosomes (e.g., $2n$, $3n$, $4n$), are common among plants but uncommon among animals. In some plant groups, polyploidy has been an important factor in evolution. By using colchicine and some other agents that interfere with cell division, polyploids can be induced. In wheat and cotton, induced polyploidy has paralleled what has occurred in nature; induced polyploidy also has practical value in such plants as apples, grapes, and tobacco.

REFERENCES

BLAKESLEE, A.F. 1941. "Effect of induced polyploidy in plants." *Amer. Natur.* 75: 117–135.

DE GROUCHY, J., and C. TURLEAU. 1977. *Clinical atlas of human chromosomes.* Wiley, New York.

MCFADDEN, E.S., and E.R. SEARS. 1946. "The origin of *Triticum spelta* and its free-threshing hexaploid relatives." *J. Hered.* 37: 107–116.

RAVEN, P.H., and R. EVERT. 1976. *Biology of plants,* 2nd ed. Worth, New York.

SEARS, E.R. 1948. "The cytology and genetics of wheats and their relatives." In *Advances in genetics,* Vol. 2, pp. 240–270, M. Demerec, editor. Academic Press, New York.

SHEPARD, J., D. BIDNEY, T. BARSBY, and R. KEMBLE. 1983. "Genetic transfer in plants through interspecific protoplast fusion." *Science* 219: 683–688.

SMITH, G.F., and J.M. BERG. 1976. *Down's anomaly.* Churchill Livingstone, New York.

SPARKS, R.S., D.E. COMINGS, and C.F. FOX. 1977. *Molecular human cytogenetics.* Academic Press, New York.

SWANSON, C.P., and P. WEBSTER. 1977. *The cell,* 4th ed. Prentice–Hall, Englewood Cliffs, NJ.

UCHIDA, I.A. 1970. "Epidemiology of mongolism: The Manitoba study." *Ann. N.Y. Acad. Sci.* 171: 361–369.

UCHIDA, I.A., and C.C. LIN. 1973. "Identification of partial 12 trisomy by quinacrine fluorescence." *J. Pediatr.* 82(2): 269–272.

PROBLEMS AND QUESTIONS

19.1 According to the Lyon hypothesis (Chapter 4), all but one X chromosome in multi-X individuals are inactivated and form sex chromatin bodies. How many sex chromatin bodies would be expected to occur in a cell from a person with (a) Turner syndrome, (b) trisomic X, and (c) Down syndrome female?

19.2 How can the origin of trisomy in humans be explained?

19.3 (a) If Down syndrome occurs in 1 out of every 700 births in the general population, what is the chance that two cases will be recorded in a city hospital on the same day? (b) If the number of live births for a given year in a country is 42 million, how many would be expected to have Down syndrome? (c) If 40 percent of the babies with Down syndrome are born to mothers over 45 years of age, and mothers in this age group produce 4 percent of all children, what is the chance that a given woman in this age group would have a baby with Down syndrome?

19.4 If nondisjunction of chromosome 21 is known to have occurred in the division of the secondary oöcytes in a particular woman, what is the chance that a mature egg arising from this cell division will receive two No. 21 chromosomes?

19.5 If Down syndrome occurs in about 1 in 700 and Turner syndrome occurs in about 1 in 5000 in the general population, and each is separately and randomly distributed in the population, what is the chance that a baby will be born with both these abnormalities?

19.6 (a) If X chromosome trisomy occurs in 1 in 1000 of the general population and No. 18 trisomy occurs in 1 in 4000, and each is separately and randomly distributed, what is the chance that a baby, such as one described by Uchida, will be born with both abnormalities? (b) If the mother is over 40 years old and if the increased occurrence of nondisjunction makes mothers in this age group 10 times more likely to have babies with each of these abnormalities, what is the probability of the two trisomies occurring in the same baby?

19.7 The poinsettia type of *Datura* carries an extra member of the chromosome set ($2n + 1$) in which the genes for purple (p^+) and white (p) flower color are located. From the following crosses, give the expected proportions of purple and white. (Female parent is always written first. Female gametes may carry either one or two chromosomes of this set, but viable pollen carries only a single chromosome.) (a) $p^+ \, p^+ \, p \times p^+ \, p^+ \, p$; (b) $p^+ \, p^+ \, p \times p^+ \, p$; (c) $p^+ \, p \, p \times p^+ \, p^+ \, p$; and (d) $p^+ \, p \, p \times p^+ \, p$. (e) How do trisomic ratios differ from the usual Mendelian ratios?

19.8 Triplo-IV fruit flies have an extra member of the fourth chromosome in which the gene *ey* is located. Give the expected results from a cross between triplo-IV flies of the genotype $ey^+ \, ey^+ \, ey$ and diplo-IV, $ey^+ \, ey$, flies in terms of (a) $2n$ and $2n + 1$ and (b) eyeless and normal eye phenotypes.

19.9 (a) What evidence concerning the influence of the Y chromosome on sex determination in humans can be obtained by comparing the characteristics of X, XXY, and XXX individuals? (b) Compare the influence of the human Y chromosome on sex determination with that of *Drosophila* and *Melandrium* Y chromosomes (see Chapter 4).

19.10 How can aneuploidy be used as a tool to determine the chromosome on which particular genes are located?

19.11 Why are tetrasomics and nullisomics found in nature less frequently than trisomics?

19.12 What values, potential if not realized at present, could monoploids have for genetic studies?

19.13 Polyploidy is rare in animals, yet some tissues in the bodies of certain diploid animals show evidence of polyploidy. Why do numbers above $2n$ persist in somatic tissues when they do not occur in the whole animal?

19.14 Describe two methods by which polyploidy might occur in nature.

19.15 (a) How may autopolyploidy and allopolyploidy originate? (b) Evaluate the significance that each might have in evolution.

19.16 Why do tetraploids behave more regularly in meiosis than triploids, and perpetuate themselves more readily in populations?

19.17 Why is chromosome irregularity associated with low fertility in plants?

19.18 What is the significance of chromosome number in (a) taxonomy and (b) studies of evolution in plants and animals?

19.19 A plant species A, which has seven chromosomes in its gametes, was crossed with a related species B, which has nine. Hybrids were produced, but they were sterile. Microscopic observation of the pollen mother cells of the F_1 showed no pairing of chromosomes. A section of the hybrid that grew vigorously was propagated vegetatively, and a plant was produced with 32 chromosomes in its somatic cells. What steps might have been involved?

19.20 A plant species A ($n = 5$) was crossed with a related species B with $n = 7$. Only a few pollen grains were produced by the F_1 hybrid. These were used to fertilize the ovules of species B. A few plants were produced with 19 chromosomes. They were highly sterile but following self-fertilization produced a few plants with 24 chromosomes. These plants were different in phenotype from the original parents, and the progeny were fertile. What steps might have been involved?

19.21 How does colchicine treatment result in chromosome doubling in plants?

19.22 What (a) practical and (b) theoretical significance may be associated with colchicine-induced polyploidy?

19.23 How could polyploidy be a significant factor in the evolution of such plants as cotton and wheat?

19.24 How might new species be produced through a combination of polyploidy and hybridization?

19.25 Give a plausible explanation for the origin of (a) *Triticum spelta,* (b) *Raphanobrassica,* (c) *Spartina townsendii,* (d) *Primula kewensis,* and (e) New World cotton.

19.26 What chromosome arrangement is symbolized by each of the following: (a) n; (b) $2n$; (c) $2n + 1$; (d) $2n - 1$; (e) $2n + 2$; (f) $3n$; and (g) $4n$?

Chapter 20

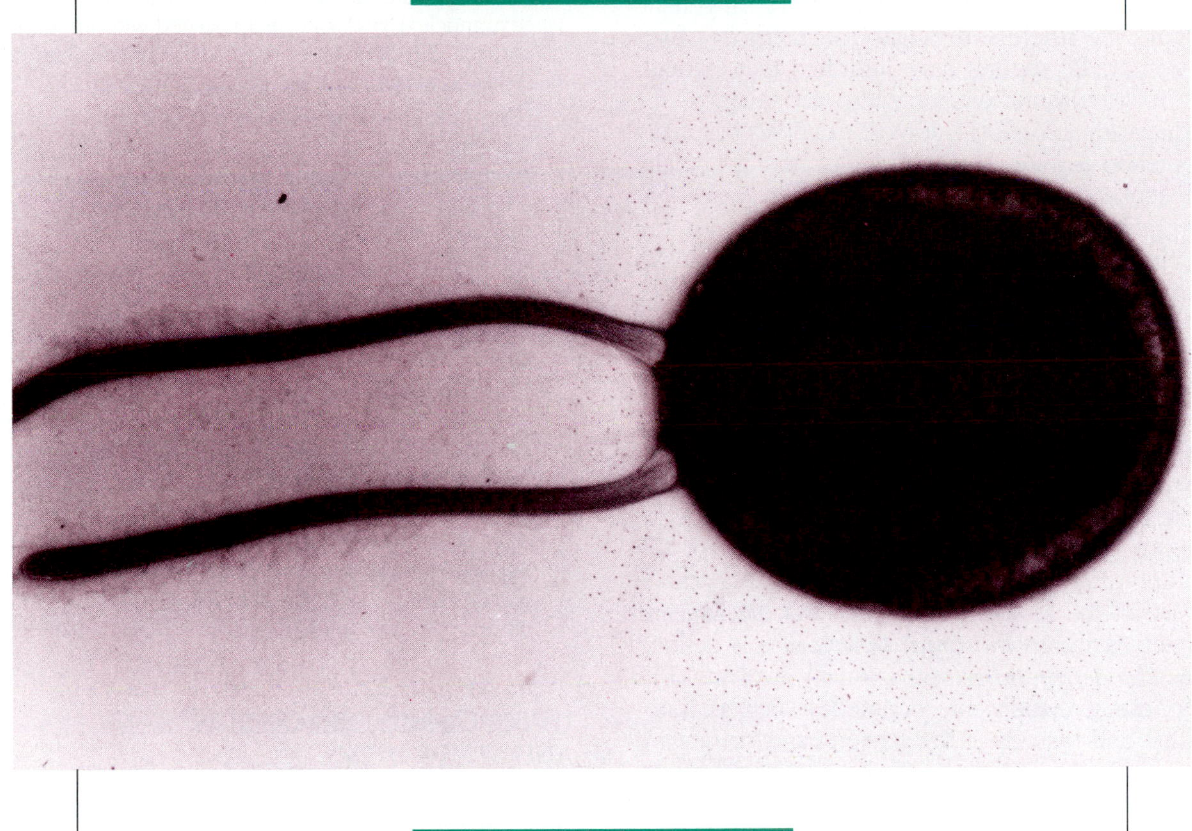

Extrachromosomal Inheritance

(Photo above): Electron micrograph of *Chlamydomonas reinhardtii* stained with uranyl acetate. (From W. J. Snell, *J. Cell Biol.* 68: 48. Copyright ® 1976 by the American Society for Cell Biology/Rockefeller University Press, New York. Original photograph courtesy of Dr. William J. Snell, Department of Cell Biology and Neurosciences, University of Texas Southwestern Medical Center at Dallas.)

THUS far in our treatment of transmission genetics in eukaryotes, we have dealt with nuclear chromosomes and genes. Certainly, nuclear DNA is the most important and very nearly the universal genetic material. Nevertheless, throughout the history of genetics, sporadic reports have indicated that extranuclear or cytoplasmic elements also act as agents for hereditary transmission. But most examples originally attributed to extranuclear inheritance have eventually been explained by nuclear genes. A few cases that appeared at first to depend on cytoplasmic genes and were classified under maternal inheritance were shown by further investigations to be associated with the genes of the mother. Phenotypes were expressed in her progeny, and these cases were reclassified as maternal effects. Since the pattern of transmission for maternal effects is similar to that of cytoplasmic inheritance, maternal effects are discussed at the end of this chapter.

What criteria will distinguish extranuclear from nuclear inheritance? Criteria for identifying this heterogeneous group of traits are consequences of the definition of extranuclear or cytoplasmic inheritance and also of the kinds of organisms and mechanisms involved. ***Extrachromosomal inheritance*** is defined as ***non-Mendelian inheritance***, usually involving DNA in replicating cytoplasmic organelles such as mitochondria and plastids. A few bacteria and viruses are also agents for extranuclear inheritance. General guidelines can be set forth at the outset, and criteria for specific cases can await the content of particular examples. Since the criteria must depend on phenotypes associated with extranuclear DNA, a persistent deviation from Mendelian pattern will suggest extranuclear inheritance.

CRITERIA FOR EXTRANUCLEAR INHERITANCE

Five major criteria can be used to distinguish between traits controlled by nuclear genes and traits controlled by extranuclear genes. These are summarized in the following list.

1. Differences in the results of ***reciprocal crosses*** would suggest a deviation from the pattern of Mendelian autosomal gene transmission. To conduct reciprocal crosses, a female from strain A is mated with a male from strain B and a male from strain A is mated with a female from strain B. If sex linkage were ruled out, differences in results of reciprocal crosses would indicate that one parent (usually maternal) is exerting a greater influence than the other on a particular trait.
2. The female reproductive cell usually carries more cytoplasm and cytoplasmic organelles than the male cell and would be expected to influence non-Mendelian traits. Organelles and symbionts in the cytoplasm might be isolated and analyzed for more specific evidence concerning maternal transmission in inheritance. When extranuclear DNA can be associated with the transmission of particular traits, the case of extranuclear inheritance is established.
3. Chromosomal genes occupy particular loci and map in certain places with respect to other genes. The failure to find linkage to known nuclear genes may rule out chromosomal inheritance and suggest extranuclear inheritance if sufficient data can be obtained.
4. Lack of Mendelian segregation and characteristic Mendelian ratios that depend on chromosomal transmission in meiosis would suggest extrachromosomal transmission.
5. Experimental substitution of nuclei might clarify the relative influence of nucleus and cytoplasm. Transmission of traits without transmission of nuclear genes would suggest extranuclear inheritance. Genes and viruses have much in common. A fine line of distinction may be required to distinguish between persistent infection and ***cytoplasmic DNA***. But phenotypes from either could qualify broadly for extranuclear inheritance.

CYTOPLASMIC ORGANELLES AND SYMBIONTS

Extranuclear inheritance associated with cytoplasmic organelles meets the preceding requirements and deserves an explanation even though organelles represent only a very small proportion of genetic material, perhaps a few hundred genes, based on the amount of functional DNA that is involved. How and why has this independent pocket of extranuclear DNA been preserved in nature? Apparently local, more or less independent cellular control has been advantageous for a few special phenotypes.

It should be noted that ***cytoplasmic organelles*** are especially significant and basic to the functions and, indeed, to the continued existence of living things. Enzymes for cellular respiration and energy production, for example, are located in mitochondria, and foodstuffs are oxidized to produce adenosine triphosphate (ATP), the fuel for biochemical reactions. Chlorophyll and other plant pigments are synthesized in plastids. It is not likely, however, that many, if any, of the autonomous genes in mitochondrial and plastid DNA are directly associated with these basic and vital phenotypes. The fascinating possibility suggested by several earlier investigators, and recently elaborated by Margulis, is that ***mitochondria were once free-living bacteria***. Over long periods of time, they established a hereditary symbiosis with their eukaryote host cells and ultimately evolved into organelles within animal and plant cells. They brought with them from the free-living state their own ***DNA*** and other equipment for genetic mechanisms. Their processing factories, partially independent from control of nuclear genes, must have been favored in evolution and found worthy of continued existence in eukaryote cells.

Similarly, chloroplasts in green plant cells are postulated to have come from free-living algae that established a symbiotic relation with early eukaryote cells. They had much to contribute to their host cells. Chlorophyll—the essential pigment for photosynthesis, with its synthesizing machinery, including specific DNA, mRNA, tRNA, ribosomes, and the machinery for chlorophyll production—was already assembled in the free-living algae. In addition, plastids of green algae are presumed to carry other genetic mechanisms such as streptomycin resistance, which was discovered in the alga *Chlamydomonas*.

Symbiont bacteria have been discovered in the cytoplasm of the protozoan *Paramecium aurelia*, where they produce a toxic substance that kills other susceptible paramecia placed in the same culture medium. This symbiont, now dignified with a specific Latin name, *Caedobacter taeniospiralis*, has worked its way into the genetic system of its host, but it can reproduce only in the presence of a particular host genotype.

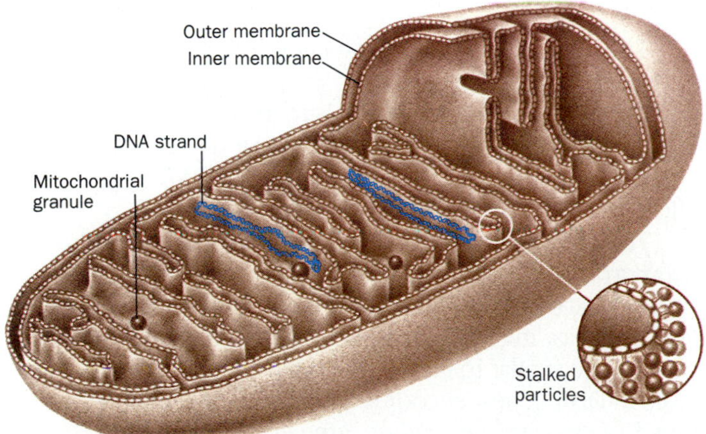

Figure 20.1 A mitochondrion showing a smooth continuous outer membrane and a periodically convoluted inner membrane that forms double membranes called cristae. Mitochondria are the principal energy source in all cells of eukaryotes. They contain a small amount of DNA, RNA polymerase, transfer RNAs, and ribosomes, which are presumably responsible for the extrachromosomal inheritance and protein synthesis in mitochondria.

DNA in Mitochondria

Mitochondria in presently living organisms arise from preexisting mitochondria. They are usually small cytoplasmic organelles (Fig. 20.1) with distinct shelflike internal layers or cristae that arise as invaginations from the inner mitochondrial membrane. They are about the same size as bacteria and occur in cells of eukaryotes but not in bacteria and viruses.

Mitochondria provide higher animals and plants with life-sustaining cellular energy through the oxidative processes of the citric acid and the fatty acid cycles, as well as the coupled processes of oxidative phosphorylation and electron transport. They contain a small amount of unique ***DNA*** that has remained autonomous outside the nuclear genome throughout the long evolutionary history of animals and plants. The mitochondrial genome is small and codes only for a limited number of structures and functions. Mitochondria contain a distinctive ***protein-synthesizing apparatus*** with specific ribosomes, tRNAs, aminoacyl-tRNA synthetases; this apparatus exhibits sensitivity to antibiotics like that of bacteria. Mitochondrial protein synthetic machinery is significantly different from the cytoplasmic machinery for the same purpose. For example, ribosomes in the protein-synthesizing apparatus of mitochondria are like those found in bacteria and yet very different from those in the cytoplasm of eukaryote cells. ***Mitochondrial rRNA*** molecules are the same size as those in bacteria and consistently smaller than those in eukaryote cells.

In yeast cells, 10–20 percent of the cellular DNA is localized in a single mitochondrion. Mitochondrial DNA has properties different from those of nuclear DNA in density and proportion of GC and AT basepairs. One yeast study showed that mitochondrial DNA has a density of 1.683 g/cm^3 and a GC content of 21 percent, whereas nuclear DNA had a density of 1.699 g/cm^3 and a GC content of 40 percent.

The life cycle of normal baker's yeast, *Saccharomyces cerevisiae*, includes a haploid and diploid phase. Mating normally occurs between vegetative haploid cells of opposite mating type (*A* or *a*). These cells fuse to form vegetative diploid cells that divide by mitosis. Cell division is usually unequal, with a small "daughter" cell budding from a larger "mother." Both cells, however, are identical in nuclear composition. Vegetative diploid cells may undergo the complex process of sporulation in which meiosis occurs. The resulting four ascospores divide to form a clone.

The first mutant found in yeast, a small colony type called "petite," has provided the best evidence that now exists for ***mitochondrial mutations***. Petites are defective in their ability to utilize oxygen in the metabolism of carbohydrates. For example, when glucose is in the medium, petite yeast will grow to only small-sized colonies. Enzyme analysis indicates that the mitochondria lack the respiratory enzyme cytochrome oxidase that is normally associated with mitochondria. Not only does this deficiency produce defective growth, but it prevents petites from producing spores. Petite strains that have been analyzed show only a small proportion of G and C and a preponderance of ***repetitive AT base-pairs***. This kind of DNA does not encode meaningful biological information. Absence of cytochrome oxidase from mitochondria does not mean that this enzyme is encoded by mitochondrial DNA, but it does indicate that mutational changes in mitochondrial DNA lead to ***heritable alterations*** in mitochondrial phenotypes.

Mutations other than those causing petites can be induced in yeasts and transmitted by the cytoplasm. For example, resistance to the antibiotics chloramphenicol and erythromycin has been induced. These antibiotics have selective affinity for mitochondrial ribosomal proteins, suggesting that structural genes are present for some ribosomal proteins.

W. L. French has presented evidence that ***sterility*** in hybrid *Culex* mosquitoes is caused by an interaction involving mitochondrial DNA. Several other investigators have shown that mitochondrial DNA is ***inherited maternally in frogs***. J. B. David has compared mitochondrial DNA in cell cultures of different mammals, including rats, mice, and human beings; he has also hybridized cells of different mammals in culture. In hybrid mouse and human cells, for example, he has shown that not only homogeneous mouse and homogeneous human mitochondrial DNA can be detected but also heterogeneous ***hybrid DNA***. In one series of experiments, 20 percent of each circular DNA unit was mouse and 80 percent was human mitochondrial DNA. Heterogeneous DNA was shown to result from mitochondrial DNA recombination in the hybrids.

Yeast chromosomal genes must specify most enzymes associated with mitochondria. Petite yeast strains with damaged DNA continue to synthesize abnormal mitochondrial DNA. This indicates that the proteins needed for mitochondrial DNA replication are not encoded by mitochondrial DNA. Similarly, petite strains continue to synthesize the enzymes of the Krebs cycle that are located in the mitochondria. Control must come from chromosomal genes.

Organization of Mitochondrial Genomes

Although the ***DNA genomes of mitochondria (mtDNA)*** generally make up only a small portion of the total cellular DNA (less than 1 percent in somatic cells of higher animals), this mtDNA usually exists as relatively small circular molecules, which can be easily isolated and characterized. Thus, considerable information is available regarding the structure of mitochondrial genomes. These mtDNAs range in size from about 16 kb in mammals up to several hundred kilobase-pairs in higher plants (e.g., 570 kb in maize). The mtDNAs are usually present in multiple copies per organelle. Human HeLa cells (an extensively studied human cell culture line) contain about 10 copies of the mtDNA per mitochondria and have about 800 mitochondria per cell. Thus, HeLa cells contain about 8000 copies of the mitochondrial genome per cell. Similar estimates of mtDNA copy number have been obtained for other types of mammalian somatic cells. In the giant mammalian egg cells, which contain very large numbers of mitochondria, mtDNA may make up as much as one-third of the total cellular DNA.

Whereas mtDNA structure is highly conserved in higher animals, considerable diversity is observed in plants and especially in lower eukaryotes. The mtDNAs of some ciliated protozoans are linear rather than circular.

The conserved structures of the mitochondrial genomes of related species can be illustrated by comparing the organization of the mtDNAs of humans, mice, and cattle—the first three mtDNAs to be sequenced *in toto*. The sequenced mtDNAs of humans, mice, and cattle were 16,569, 16,275, and 16,338 nucleotide-pairs in length, respectively. Most importantly, all three mtDNAs exhibited the same basic organization of genetic information (Fig. 20.2). Each contained 2 rRNA genes, 22 tRNA genes, and 13 putative protein structural genes. Five genes encode known proteins, but the products and functions of the other putative genes have not yet been identified. The latter open reading frames are designated ***URF***s (for ***u***nassigned ***r***eading ***f***rames).

The entire mammalian mitochondrial genome is transcribed as one unit from a single promoter site, and the giant primary transcript is then cleaved endonucleolytically to produce the individual tRNA, rRNA, and mRNA molecules. Thus, the entire mtDNA is, in effect, equivalent to one operon in bacteria.

The circular mitochondrial genome of the yeast *Saccharomyces cerevisiae* is over five times larger (about 84 kb) than the mtDNAs of mammals. Nevertheless, the yeast mitochondrial genome exhibits organization quite similar to that of mammalian mtDNAs. Two yeast mtDNA genes, encoding cytochrome *b* and subunit 1 of cytochrome oxidase, are very large—almost as large as the entire mtDNA of mammals. These two genes contain several very long intron sequences. The yeast mtDNA contains a cluster of 16 tRNA genes in one short segment of the genome, plus about 10 tRNA genes dispersed throughout the genome. Thus, the yeast mitochondrial tRNA genes are not as uniformly distributed as the mammalian mitochondrial tRNA

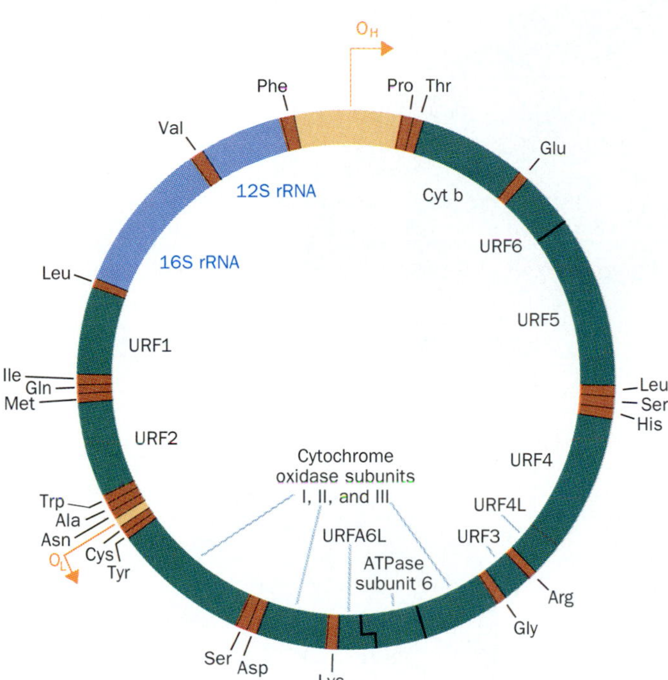

Figure 20.2 Structure of the mammalian mitochondrial genome, based on the established sequences of the human, mouse, and bovine mitochondrial DNAs. The sizes of these three DNAs are 16,569, 16,295, and 16,338 nucleotide-pairs, respectively. The locations of the 22 tRNA genes are shown with the amino acid accepted by each tRNA given outside the circle. The locations of the 12S and 16S rRNA genes and the 13 open reading frames are identified by the designations given inside the circle. URF indicates an unassigned open reading frame. O_H and O_L indicate the positions of the origins of replication for the two complementary strands (designated H for "heavy" and L for "light" based on their band positions in density gradients). Replication of the two strands is initiated asynchronously at these two origins and proceeds in opposite directions as indicated by the arrows. (After D. A. Clayton, "Transcription of the Mammalian Mitochondrial Genome," *Annu. Rev. Biochem.* 53: 573–594, 1984.)

genes. Part of the difference in size between mammalian mtDNAs and yeast mtDNAs is clearly due to the very large introns in the yeast cytochrome *b* and cytochrome oxidase subunit 1 genes. Presumably, the larger yeast mitochondrial genome also encodes more proteins than the mammalian mtDNAs, but this has not yet been established.

Since mitochondria are complex organelles containing a large number of different proteins—far in excess of the 13 potentially encoded by the mammalian mitochondrial genome—the majority of the mitochondrial proteins must be encoded by nuclear genes. Indeed, a large number of soluble proteins such as amino acid biosynthetic enzymes and mitochondrial structural proteins are known to be synthesized on cytoplasmic ribosomes with specifications provided by nuclear gene transcripts. These proteins are then transported into mitochondria, where they function. The transport of cytoplasmic translation products into mitochondria is directed by special ***transit peptides*** at the amino termini of the nascent polypeptides. These transit peptides are usually cleaved off the precursor polypeptides during their transport across the mitochondrial membrane. Thus, mitochondrial genomes specify the rRNA and tRNA molecules needed for mitochondrial protein synthesis and a few proteins, but the vast majority of the mitochondrial proteins are encoded by nuclear genes.

DNA in Plastids

Carl Correns (in 1908) observed a difference in the results of reciprocal crosses and was the first to describe deviations from Mendelian heredity. Different shades of color from white (albino) to dark green in the leaves of some plants were investigated. Instead of equal inheritance from the seed and pollen parent, as demonstrated by Mendel in garden peas, Correns showed in studies of four-o'clock, *Mirabilis jalapa*, that inheritance of certain traits came entirely from the seed parent. Color differences were related to cytoplasmic plastids, most important of which are chloroplasts (Fig. 20.3), which carry chlorophyll. Chloroplasts arise from cytoplasmic particles called ***proplastids that contain DNA*** and duplicate themselves independently of other cell parts. They are distributed more or less equally during cell division. Although some proplastids are transmitted in the cytoplasm of the egg, few if any are transmitted in the pollen of most plants. Thus, some chloroplast characteristics are ***inherited*** from the ***seed-parent cytoplasm***

Many investigators have followed the early lead of Correns, and now almost everything that is known about plastid genetics has come from studies on variegation in seed-bearing plants. Any plant that develops patches of different colors in leaves or other vegetative parts is said to be variegated. Many variegations are not inherited; some are controlled by nuclear genes, and others depend on plastid inheritance. Interactions occur, and it is difficult to distinguish examples that are solely dependent on plastid inheritance. Figurative patterns (marginal bands, diffuse spots, prominent veins and stripes on leaves) that are true-breeding usually represent physiological modifications that influence regulation of normal chloroplast development and not specific gene mutations. The sorting of normal and mutant plastids into color patterns may depend on mutations within plastids.

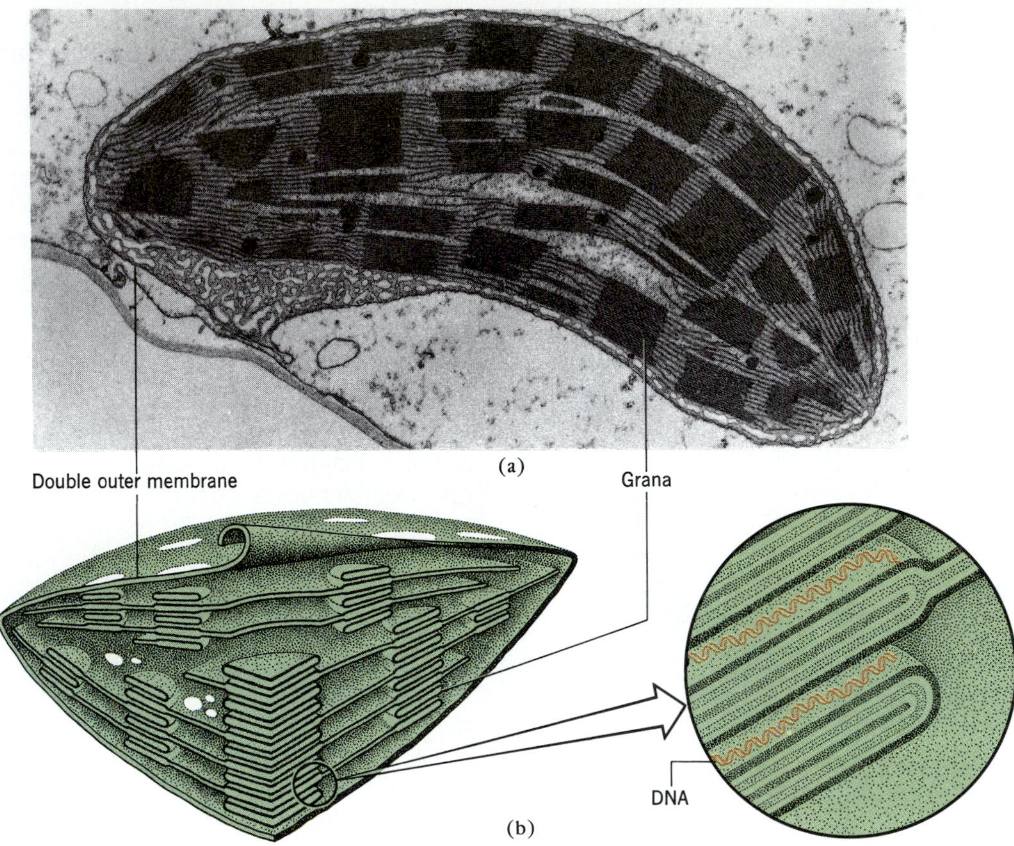

Figure 20.3 Structure of a chloroplast, the cytoplasmic green plant organelle containing DNA. (a) Electron micrograph of a chloroplast; magnification ×15,200. (b) Enlarged diagram of grana stacks of membrane sacs within the chloroplast (magnification ×78,500)—showing the position of DNA. (Photo courtesy of T. E. Weier, University of California, Davis.)

Ovules, as well as somatic cells of mottled plants (e.g., the four-o'clock), may carry both abnormal, nearly ***colorless plastids*** and normal green chloroplasts in their cytoplasm (Fig. 20.4). The mottled effect is transmitted through the maternal line, generation after generation. Because the pollen of the four-o'clock has little if any cytoplasm, its influence on variegation is negligible. A single plant with green, white, and variegated branches or sectors may produce seed that perpetuates each of the three types. Seeds borne on white branches contain only primordia for colorless plastids, those on green branches only green, and those on variegated branches might contain either colorless or green chloroplasts or a combination of both.

In plants such as the primrose, *P. sinensis*, chimeras (sectors containing different plastid types) are sometimes formed, with only part of the plant containing chlorophyll. The areas with abnormal plastids that lack chlorophyll can rely on the green parts of the plant for the products of photosynthesis and therefore can continue to live. Each part of the chimera may produce reproductive cells and thus transmit its type of plastids through female gametes.

Chloroplasts have now been isolated and found to be capable of ***protein synthesis*** in the presence of either adenosine triphosphate or light. The products are identical with authentic chloroplast proteins, demonstrating that isolated chloroplasts have fully functional protein synthesis machinery in which mRNA is translated accurately.

With DNA analysis and the use of restriction endonucleases for DNA fragmentation, much has been learned about plastid DNA. Some 30–60 copies of the chloroplast genome are found in each chloroplast of higher plants; about 100 genome copies occur in each plastid of some algae. Enough unique chloroplast DNA has been discovered to code for about 126 proteins, and about 12 percent of the ***plastid DNA*** sequence ***codes*** for plastid components.

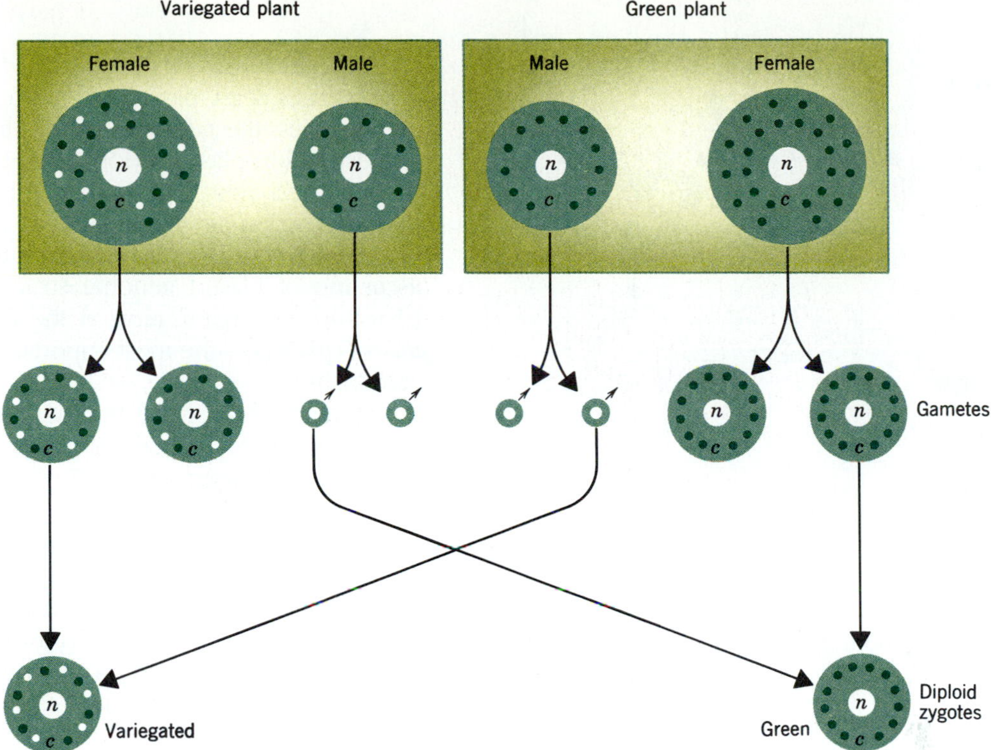

Figure 20.4 Diagrammatic illustration of maternal plastid inheritance in the diploid plant *Mirabilis,* which has abundant cytoplasm in the seed-parent gametes but little or no cytoplasm in pollen-parent gametes. *n*, nucleus; *c*, cytoplasm.

Chloroplast DNA and Drug Resistance

When Ruth Sager placed *Chlamydomonas* algal cells on a culture medium containing the antibiotic streptomycin, most of the cells were killed, but about one per million survived and multiplied, each to form a ***streptomycin-resistant*** colony. Mutants with resistance to streptomycin were being selected from the predominantly streptomycin-susceptible algae. About 90 percent of the mutants involved nuclear genes (*sr*-1), and such mutations were merely being demonstrated by the antibiotic challenge. Approximately 10 percent of the mutations (*sr*-2), however, were uniparental and ***nonchromosomal***. Eventually, nonchromosomal mutants were recovered from almost every colony. Nonchromosomal DNA mutations expressed the same phenotypes as chromosomal DNA mutants. These nonchromosomal genes are presumed to be located in the chloroplasts.

Reciprocal crosses (Fig. 20.5) demonstrated that antibiotic resistance, controlled by nonchromosomal genes, was uniparental in inheritance. On the other hand, mating type in this sexual unicellular alga was controlled by chromosomal genes, which were designated by the investigators mt^+ and mt^- or simply plus (+) and minus (−), instead of female and male. All progeny from each reciprocal mating were like the plus (+) mating type with respect to relative streptomycin resistance thus demonstrating maternal inheritance. When the plus (+; female) mating type was resistant, all progeny were resistant; when the plus (+) mating type was nonresistant, all progeny were nonresistant. These results of reciprocal crosses demonstrated non-Mendelian inheritance, which involved a single pair of contrasting traits. Nonchromosomal genes, *sr* for streptomycin resistance and *ss* for streptomycin sensitive, were postulated to control these two alternative characteristics.

Another mutant, ac_2, which blocked photosynthetic activity, was induced and a pair of nonchromosomal alleles, ac_1, and ac_2, was thus available for study in the same strain of *Chlamydomonas*. The mutant required acetate in the medium for growth. With two pairs of nonchromosomal genes available, a dihybrid cross could be conducted in the same system to check for evidence of recombination. Crosses of the dihybrid type ac_1 *ss* × ac_2 *sr* were prepared, and progeny were allowed to grow for a few vegetative multiplications. Each cell was then classified for its segregating markers, both nonchromosomal and chromosomal (i.e., mating type and others known to be chromosomal). Both the ac_1/ac_2 and the *sr*/*ss* pairs of alleles were observed to segregate, but not always in the same division. After four or five mitotic doublings, both

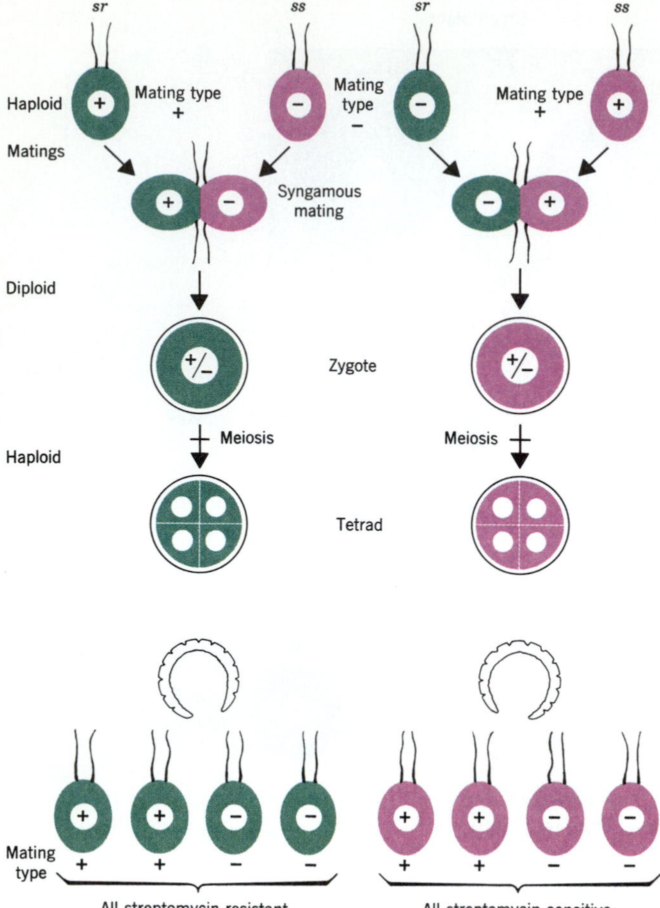

Figure 20.5 Maternal (cytoplasmic) inheritance of resistance to streptomycin and nuclear gene-controlled mating type inheritance in *Chlamydomonas reinhardtii.* The plus and minus signs refer to mating types, which are inherited as single chromosomal gene differences. Progeny are, with rare exceptions, like the plus (maternal) parent in their reaction to streptomycin. In reciprocal crosses, streptomycin resistance shows maternal inheritance, indicating uniparental, cytoplasmic inheritance. (Based on R. Sager and F. J. Ryan, *Cell Heredity,* Wiley, New York, 1961.)

parentals (ac_1 *ss* and ac_2 *sr*) and recombinants (ac_1 *sr* and ac_2 *ss*) had been obtained. The results indicated independent assortment, suggesting that the two pairs of nonchromosomal genes were carried in different plastids. Three- and four-point crosses and reciprocal crosses have been made with the addition of several mutants, which are presumed to be carried in chloroplasts and mitochondria. A ***genetic map of non-Mendelian genes*** in *Chlamydomonas* has been constructed, but uncertainty still exists as to whether some "chloroplast" linkage groups are solely in the chloroplast genome.

Organization of Plastid Genomes

The plastid genomes of over 200 species of higher plants and of many green, blue-green, and red algae have been at least partially characterized. Within a given species, the genomes of the different types of plastids—chloroplast, amyloplasts (plastids that accumulate starch in storage tissues), and chromoplasts (plastids containing pigments)—all are identical in organisms where they have been studied. Thus, our discussion of plastid genome structure will be restricted to the organization of the ***DNAs of chloroplasts (cpDNAs)***—the most important member of the plastid family.

In higher plants, cpDNAs range in size from 120 to 160 kb. In algae, the size range for chloroplast genomes is much larger—from 85 to 292 kb for species known to have circular cpDNAs. In two species of green algae of the genus *Acetabularia*, the cpDNAs appear to be huge, about 2000 kb, and it has not yet been established whether these large chloroplast genomes are linear or circular. As in the case of mitochondrial DNAs, chloroplasts often contain multiple copies of the cpDNA. The large single chloroplast of *Chlamydomonas reinhardtii* contains about 100 copies of the cpDNA. The single-celled flagellate *Euglena gracilis* contains about 15 chloroplasts each with about 40 copies of cpDNA, giving a total of about 600 copies per organism.

All the chloroplast genomes analyzed to date contain basically the same set of genes, but with these genes arranged in very different ways on the cpDNAs. The genes present on cpDNAs can be grouped into two major classes: (1) those that encode components of the chloroplast protein biosynthetic apparatus (RNA polymerase subunits, structural components of chloroplast ribosomes, and a set of tRNAs) and (2) those specifying components of the photosynthetic machinery (photosystems I and II and the electron transport chains).

Chloroplast genomes of higher plants are about one-twentieth to one-thirtieth the size of the genomes of the prokaryotic organisms (blue-green algae or cyanobacteria) from which they are believed to have evolved. Thus, the chloroplasts have lost much of the genetic information of their ancestors and have become very dependent on nuclear genes of the host cell for many essential components. As in the case of mitochondria, the latter components are synthesized on cytoplasmic ribosomes and are imported into chloroplasts with the aid of amino-terminal transit peptides that are cleaved off during transport through the chloroplast membranes.

Comparative studies of chloroplast genomes have provided important new information about evolutionary relationships of plant and algal species. J. D. Palmer has distinguished six major lines of chloroplast evolution (Fig. 20.6). The chloroplast genomes present in different evolutionary lines all contain largely the same genes, but these genes are present in different arrange-

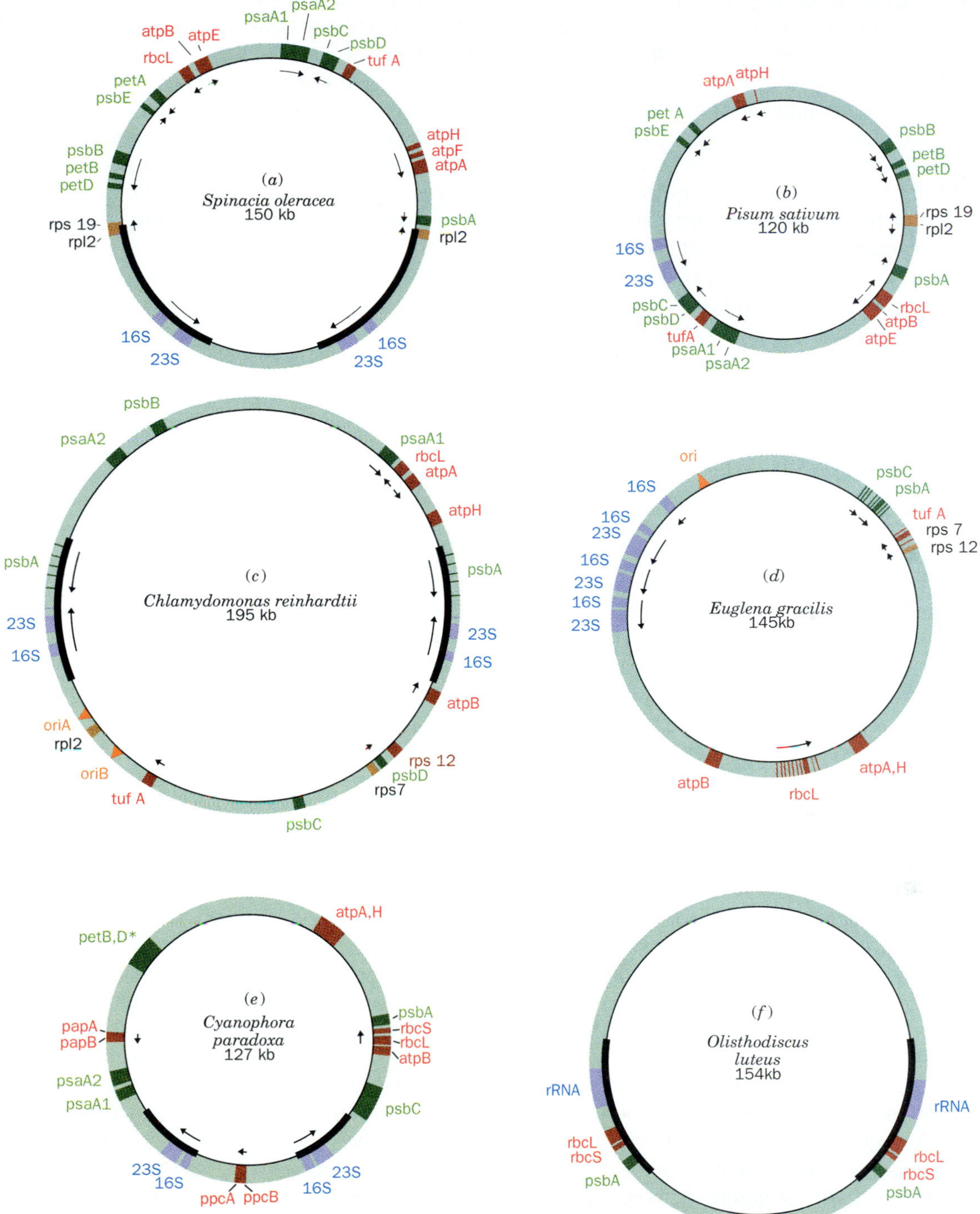

Figure 20.6 Comparative organization of the chloroplast genomes in species representing the six established lines of chloroplast evolution. The sizes and positions of known marker genes are shown for the chloroplast DNAs of (a) spinach, (b) pea, (c) the green alga *Chlamydomonas reinhardtii,* (d) the unicellular green flagellate *Euglena gracilis,* and the blue-green algae (e) *Cyanophora paradoxa* and (f) *Olisthodiscus luteus.* Each arrow inside the circular map indicates the direction of transcription of a gene when known. The symbols used are as follows: *ori, ori*A, and *ori*B, origins of replication; *rbc*L and *rbc*S, genes encoding the large and small subunits, respectively, of ribulose bisphosphate carboxylase; *atp,* genes encoding ATP synthase components; *psa* and *psb,* genes specifying photosystem I and photosystem II components, respectively; *pet,* genes encoding components of the cytochrome b/f complex; *rps* and *rpl,* genes encoding proteins of the small and large subunits, respectively, of chloroplast ribosomes; *tuf*A, gene for elongation factor EF-Tu; *16S, 23S,* genes for the 16S and 23S rRNAs. (After J. D. Palmer, "Comparative Organization of Chloroplast Genomes," *Annu. Rev. Genet.* 19: 325–354, 1985.)

ments on the cpDNA molecules. The rRNA genes are present in duplicate on inverse repeats within the cpDNAs of most species. However, in *Euglena gracilis*, the 16S and 23S rRNA genes are present on three direct tandem repeats, with a separate fourth copy of the 16S rRNA gene located nearby in the genome. By comparing the locations of genes on different chloroplast genomes, Palmer has been able to show that many of the changes in cpDNA organization have resulted from inversions of segments of DNA. In other cases, deletions and insertions of DNA were found to have occurred in intergenic regions and within introns of genes. However, the major documented changes in cpDNA structure seem to have resulted from large inversions.

Photosynthesis occurring within chloroplasts provides the life-sustaining energy sources for all living organisms on planet Earth. Since chloroplast genomes encode many key components of photosystems I and II and the electron transport chains, knowledge of the structure and function of cpDNAs is very important and has received much attention. The complete nucleotide-pair sequences of the cpDNAs of the liverwort *Marchantia polymorpha* and of tobacco (*Nicotiana tobacum*) have been determined. The cpDNAs of *Marchantia* and tobacco are 121,024 and 155,844 nucleotide-pairs in length, respectively. The organization of the single-copy genes in the chloroplast genomes of these two plants is remarkably similar considering that they are evolutionarily very distant from each other. The major difference between these two cpDNAs is that the inverse repeat regions containing the rRNA genes are considerably larger in tobacco. The best estimates of cpDNA gene number are 136 in *Marchantia* and 150 in tobacco. The locations of known genes and open reading frames are shown for the chloroplast genome of *Marchantia* in Fig. 20.7.

Perhaps a complete understanding of the chloroplast genes and the products that they encode will have important practical applications in the future. Information of the exact mechanisms by which photosystems I and II function might someday permit scientists and engineers to "build" a totally synthetic system capable of duplicating the capacity of green plants to capture light energy and convert it to chemical forms useful to living organisms.

Symbiont Bacteria in *Paramecium* Cytoplasm

Paramecia are favored organisms for genetic investigation. They are large, unicellular protozoans that reproduce by both asexual and sexual processes. Asexual reproduction occurs through cell fission to produce clones of genetically identical cells. In the sexual phase, paramecia conjugate periodically and transfer genetic material from one cell to another. Paramecia and other ciliates have two kinds of nuclei: a large vegetative macronucleus and a small micronucleus, which goes through the meiotic sequence and produces haploid gametes. A micronucleus also gives rise to the macronucleus that divides in asexual cell division. It is possible in the laboratory to make sexual crosses through which nuclear DNA is transferred from a donor to a recipient, resulting in heterozygous progeny, that is, $AA \times aa \longrightarrow Aa$. A process of self-fertilization, called ***autogamy***, results in the complete homozygosis of the resulting progeny (Fig. 20.8). Following meiosis, the cells are haploid, but through autogamy they become homozygous diploids. This provides a basis for comparing extranuclear and nuclear inheritance, and thus for demonstrating that progeny can differ from wild type in traits controlled by both nuclear and extranuclear genes.

G. H. Beale discovered that erythromycin resistance in *Paramecium,* like that in yeast, results from non-Mendelian inheritance. A number of additional cytoplasmic and nuclear mutations affecting antibiotic resistance have been studied by both Beale and J. Beisson. These and other investigators made transfers of cytoplasm and also transfers of isolated mitochondria between strains of paramecia and showed that mitochondria (presumably mitochondrial DNA) control the resistance. Studies have also shown that although some mitochondrial traits are determined by the mitochondria themselves, others are dependent on the elements in the protoplasm.

T. M. Sonneborn and others have investigated a persistent extranuclear effect in *Paramecium*. Some strains of *P. aurelia* produce a substance that has a lethal effect on members of other strains of the same species. Paramecia from strains capable of producing the toxic substance are called ***"killers."*** When killers are subject to low temperatures, their killing capacity gradually disappears. The toxic effect also decreases after repeated cell divisions. Separate elements in the cytoplasm were postulated for the production of a toxic substance. From mathematical calculation, it was estimated that about ***400 particles*** are required to make a killer effective. Killers were then observed microscopically and "particles" called "kappa" were observed in the expected numbers. These "particles," shown to be symbiotic bacteria, have been named *Caedobacter taeniospiralis* (the killer bacterium with the spiral ribbon).

A "toxic substance" (paramecin), produced by the killer bacteria, is diffusible in the fluid medium (Fig. 20.9). When killers are allowed to remain in a medium for a time and are then replaced by sensitives, the sensitives are killed. Paramecin, which has no effect on killers, is associated with a particular kind of kappa that

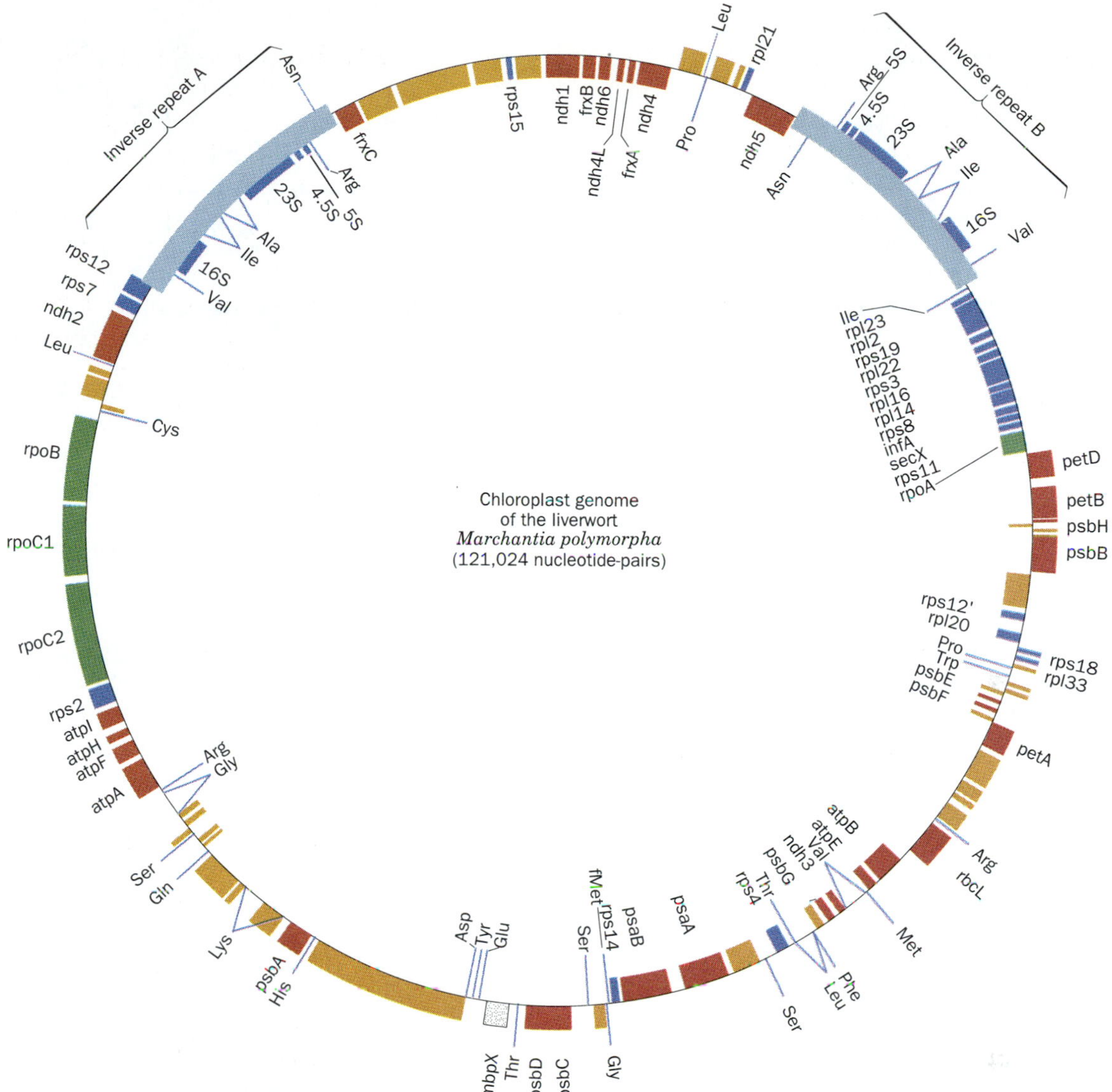

Figure 20.7 Organization of the chloroplast genome of the bryophyte *Marchantia polymorpha.* The complete sequence of this 121,024-nucleotide-pair genome has been determined. The known genes and open reading frames of unknown function are shown as boxes except for the small tRNA genes, which are indicated by single lines. The tRNA genes that contain introns and are thus larger are denoted by triangles extending from the circular map. Genes shown inside the circle are transcribed from one strand in a clockwise direction; those shown outside the circle are transcribed from the other strand in a counterclockwise direction. Genes that encode RNA polymerase subunits are shown in green, whereas those that specify components of the chloroplast translational machinery are shown in blue. Genes that encode proteins involved in photosynthesis and electron transport are shown in red. The *mbpX* gene (stippled box) is believed to specify some type of chloroplast permease. Open reading frames of unknown function are shown in yellow. The tRNA genes are designated by the amino acids that they accept when activated. Symbols used are as follows: rpo, RNA polymerase; rps, ribosomal proteins of small (30S) subunit; rpl, ribosomal proteins of large (50S) subunit; 4.5S, 5S, 16S, 23S, rRNAs of the indicated size; rbc, ribulose bisphosphate carboxylase; psa, photosystem I; psb, photosystem II; pet, cytochrome b/f complex; atp, ATP synthesis; infA, initiation factor A; secX, 50S ribosomal protein; frx, iron–sulfur proteins; ndh, NAD(P)H oxidoreductase. (After K. Umesono and H. Ozeki, "Chloroplast Gene Organization in Plants," *Trends in Genet.* 3: 281–287, 1987.)

occurs in about 20 percent of a kappa population. These kappa bacteria possess a refractile protein-containing "R" body and are called "brights," because they are infected with a virus that controls the synthesis of viral protein. The ***virus is toxic*** to sensitive paramecia but is not toxic in "nonbright" bacteria.

Kappa bacteria are perpetuated only in organisms carrying the dominant ***nuclear allele K***, which establishes the environment necessary for the bacteria to reproduce. When killers conjugate with sensitives under appropriate conditions (to avoid killing the mate) and no cytoplasmic exchange occurs (Fig. 20.10), two

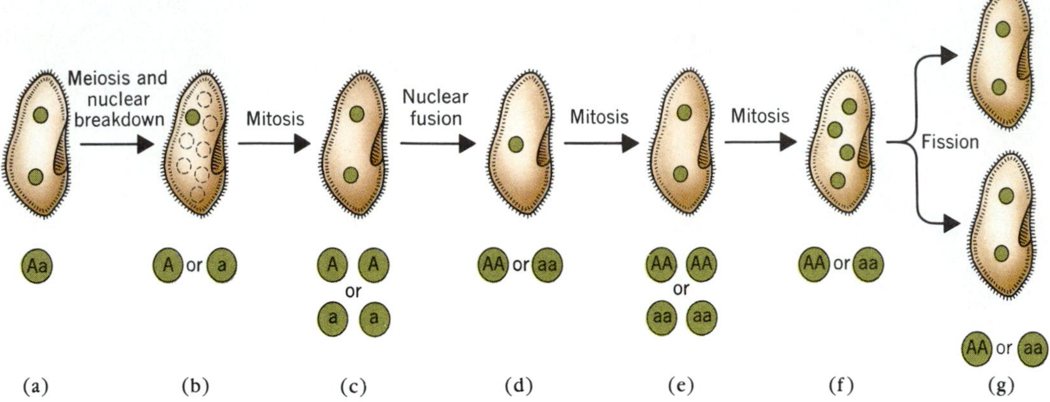

Figure 20.8 Autogamy in *Paramecium* results in homozygosity through the following steps: (a and b) Meiosis of micronuclei resulting in one haploid product that divides mitotically, and of micronuclei that disappear; (c) the haploid micronucleus has divided; and (d) the two micronuclei have fused to form a diploid (2*n*) nucleus. (e and f) Fission of the cell results in two cells, each homozygous as shown in (g).

kinds of clones emerge: one from the original killer cell, which contains allele *K* (*Kk*) and kappa bacteria, and the other from the original sensitive cell, which carries the allele *K* (*Kk*) and lacks kappa. Following autogamy, half the progeny of the killers are killers and half are sensitive paramecia. All progeny of sensitives are sensitive. Since no cytoplasm was transferred in this conjugation, only the cells from original killers inherit kappa bacteria. Kappa cannot reproduce in cells unless a *K* allele is present in the nucleus.

Under some conditions, conjugation persists much longer; a larger connection is established between conjugants, and cytoplasm as well as nuclear genes is exchanged (Fig. 20.11). When the conjugants are *KK* and *kk*, alleles *K* and *k* are exchanged and both exconjugants are *Kk*. ***Cytoplasmic exchange*** has transferred kappa bacteria from the killer to the nonkiller cell.

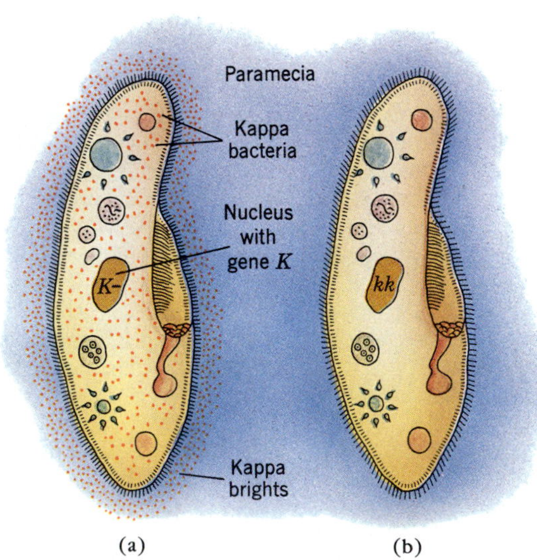

Figure 20.9 Kappa particles in *Paramecium aurelia.* (a) Killer with kappa particles inside the paramecium and paramecin in the liquid medium outside the organism. Gene *K* is present in the nucleus. (b) Sensitive paramecia with no kappa particles, no paramecin, and genes *kk* in the nucleus. (After T. M. Sonneborn.)

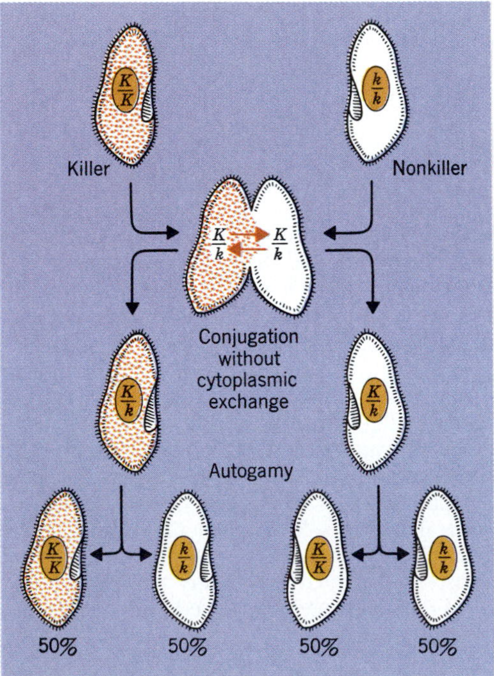

Figure 20.10 Conjugation between a killer paramecium (stippled) and a sensitive paramecium without cytoplasmic exchange but with an exchange of *K* and *k* genes. The autogamy that follows results in *KK* and *kk* homozygotes, which may give rise to *KK* and *kk* clones. (After T. M. Sonneborn.)

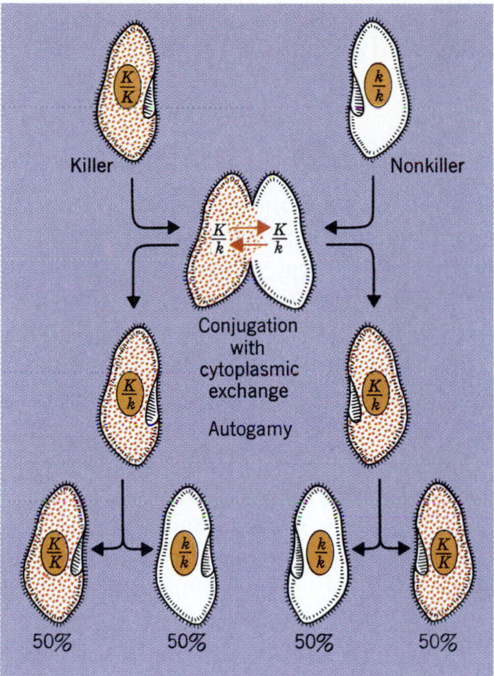

Figure 20.11 Conjugation as in Fig. 20.10, but with cytoplasmic exchange followed by autogamy. (After T. M. Sonneborn.)

Autogamy produces homozygotes *KK* and *kk* cells, which produce clones of killers or nonkillers, respectively.

PLASMID DNA AND TUMOR TRANSFORMATION

Extrachromosomal DNA molecules that replicate independently and maintain themselves in the cytoplasm of plant cells are called ***plasmids*** (Chapter 8). They have much in common with chromosomes of mitochondria and plastids, but they are not organized into organelles that are vital to their host cells. Some plasmids are fragments of a bacterial chromosome and some are recombinants of DNA fragments. Most plasmids are not essential to their host cells, but some control a favorable reaction to antibiotics. Because of their ability to replicate independently, to combine with other DNA, and to carry DNA to cell centers of synthetic activity, they are useful in genetic engineering.

A plasmid called Ti (for tumor-inducing) carries a DNA sequence that transforms cells of dicotyledonous plants (tobacco, sunflower, carrot, tomato, etc.) to tumor cells. Tumor transformation is associated with the crown gall disease. This disease, manifested as a bulbous growth or gall, is induced by the bacterium *Agrobacterium tumefaciens*. The disease is economically important, especially in fruit crops and nursery stocks, and is caused by viable bacteria that enter a wounded surface of a plant, usually at the crown (junction of stem and soil). But the bacteria that initiate the gall disease are not needed to perpetuate the tumor, for they can be killed after a few days and the tumor continues to grow. A fragment of the Ti plasmid carried by the bacterium has been combined with a DNA segment of an infected plant cell. Genes carried by the plasmid, now integrated into plant cells, ***code for enzymes*** that promote continuous and uncontrolled tumor growth, which extends from the bacterial-induced gall. (See Chapter 24 for a discussion of the use of Ti plasmids in plant genetic engineering.)

CYTOPLASMIC MALE STERILITY IN PLANTS

Another example of cytoplasmic inheritance is associated with ***pollen failure***. This occurs in many flowering plants and results in male sterility. In maize, wheat, sugar beets, onions, and some other crop plants, fertility is controlled at least in part by cytoplasmic factors. In other plants, however, male sterility is controlled entirely by nuclear genes. Critical observation and tests must be made in individual cases to determine the mechanism of inheritance. Male sterility has practical importance when crosses are made on a large scale to produce hybrid seed. Hybrid plants are produced commercially in maize, cucumbers, onions, sorghum, and other plants to obtain hybrid vigor (see Chapter 21).

Male Sterility in a Cross-Pollinating Plant

Classical examples of maternal-inheritance mechanisms that transmit male sterility in maize (corn) were discovered and carefully analyzed by M. M. Rhoades. Pollen was aborted in the anthers of certain corn plants, causing them to be male sterile, but female structures and fertility were normal. Nuclear genes did not control this type of sterility, rather, it was transmitted from generation to generation through the ***egg cytoplasm***

A particular male-sterile variety produced only male-sterile progeny when fertilized with pollen from normal maize plants. The male-sterile seed-parent plants were then backcrossed repeatedly with pollen-fertile lines until all chromosomes from the male-sterile line had been exchanged for those of the male-fertile line (Fig. 20.12). In the genetically restored sterile line, male sterility persisted, demonstrating that inheritance was maternal and not controlled by chromosomal genes. As the investigations progressed, a small amount of pollen was obtained from the male-

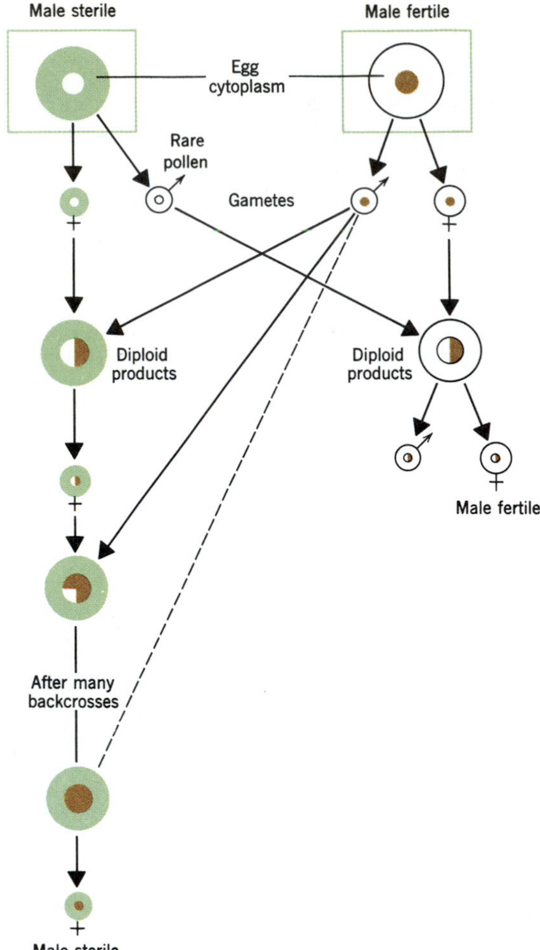

Figure 20.12 Maternal inheritance of male sterility in maize. (After M. M. Rhoades.)

sterile line, making reciprocal crosses possible. These crosses produced progeny from the male-sterile seed plant line that were male-fertile. Thus inheritance of ***male sterility was maternal***, regardless of the direction in which the cross was made. Male sterility in this example was attributed to cytoplasmic genes (plasmagenes) that were transmitted by female gametes.

However, the cytoplasmic effect is not the only factor in male sterility. Specific nuclear genes are now known to suppress maternally inherited sterility in maize. A single dominant chromosomal gene, for example, can restore pollen fertility in the presence of cytoplasm that ordinarily would ensure sterility. In one experiment, pollen abortion occurred only when a specific kind of cytoplasm was present along with a dominant gene for male sterility. The homozygous recessive allele was present at a suppressor locus.

Large-scale use of male-sterile maize for seed production brought disaster to the United States corn crop in 1970. Because of the advantages of uniformity in corn and the great advantage of male sterility in seed production, a single source of cytoplasm, known as Texas (T) male-sterile cytoplasm, had been used in producing seed for most of the corn hybrids planted that year.

Danger of Uniformity

What caused disaster to the corn crop? A new mutant of the fungus *Helminthosporium maydis* (Nisikado and Miyake) became a virulent pathogen on a particular kind of hybrid corn. It was especially destructive on corn with (T) male-sterile cytoplasm. Pathologists and plant breeders met the epidemic by searching for corn varieties that were resistant to the fungus. Because of a previous, less serious yellow leaf blight, some 1970 seed production had been shifted to corn without T cytoplasm. This corn required manual detasseling but was widely used for winter planting in 1971. It also produced some resistant seed for immediate general farm use. Some growers preferred the predictable 20–30 percent loss of yield to the high risk of much larger losses from growing susceptible hybrids.

Most of the 1971 seed production was, therefore, accomplished without the use of male sterility and T cytoplasm. The T race of *H. maydis* was not serious in 1972. Still another race of *H. maydis* may appear, or one of the other corn diseases could become a threat to the highly uniform hybrid corn with T male-sterile cytoplasm. Several varieties of corn that are resistant to the existing T race of *H. maydis* have now been identified and are available for seed production.

This example illustrates the danger of uniformity in germplasm for a crop grown on a large scale. It also indicates that sustained research programs are essential in protecting food supplies from potential losses of catastrophic magnitude.

MATERNAL EFFECTS

Eggs and embryos are expected to be influenced by the maternal environment in which they develop. Even those removed from the body of the mother at an early stage receive cytoplasm and nutrients in the egg from the mother, and special influences on gene action may have already taken effect. Certain potentialities of the egg are known to be determined before fertilization, and, in some cases, these have been influenced by the surrounding maternal environment. Such predetermination by genes of the mother, rather than those of the progeny, is called ***maternal effect***. Existence of a maternal effect is commonly substantiated or disproved by reciprocal crosses. If a maternal effect is involved, results from reciprocal crosses will be different from each other, with genes of the mother being expressed.

Maternal Effect in Snail Shell Coiling

One of the earliest and best-known examples of a maternal effect is that of the direction of coiling in shells of the snail *Limnaea peregra*. Some strains of this species have dextral shells, which coil to the right; other have sinistral shells, which coil to the left. This characteristic is determined by the ***genotype of the mother*** (not her phenotype) rather than by the genes of the developing snail. Allele s^+ for right-handed coiling is dominant over allele s for coiling to the left.

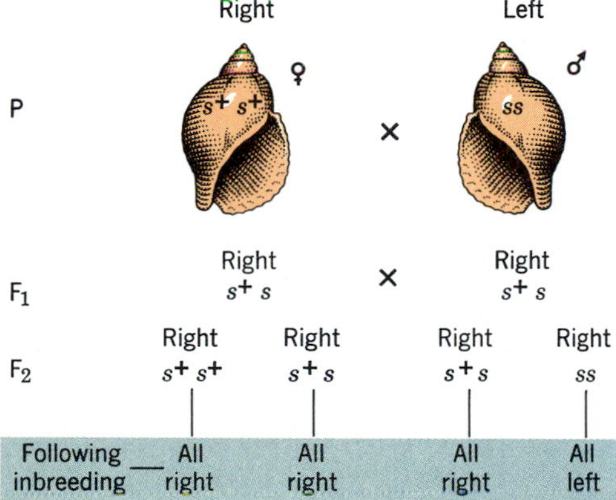

Figure 20.13 A cross illustrating a maternal effect in the snail *Limnaea peregra*. The coiling pattern is controlled by the genes of the mother. Following inbreeding of snails with *ss* genotypes, *ss* mothers produced progeny that coiled to the left. (Data and interpretation from A. E. Boycott *et al.* and A. H. Sturtevant.)

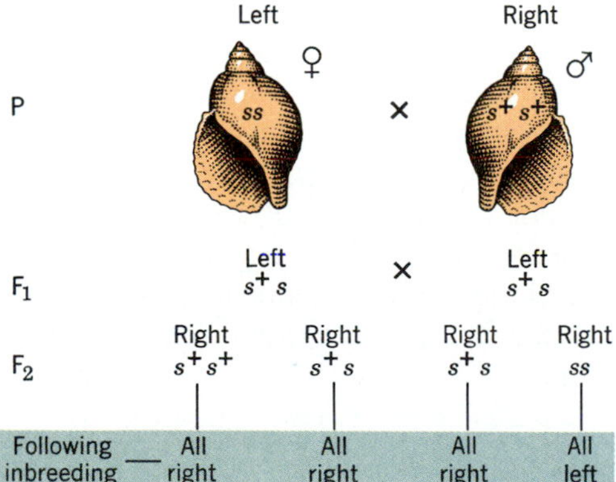

Figure 20.14 Cross reciprocal to that of Fig. 20.13, illustrating a maternal effect. The *ss* (P) mother produced only left-coiling progeny, and the s^+s (F_1) mother produced only right-coiling progeny. Following inbreeding, the *ss* (F_2) mother produced only left-coiling progeny. (Data and interpretation from A. E. Boycott *et al.* and A. H. Sturtevant.)

When crosses (Fig. 20.13) were made between females coiled to the right and males coiled left, the F_1 snails were all coiled to the right. The usual 3:1 ratio was not obtained in the F_2 because the phenotype of *ss* was not expressed. Instead, the pattern determined by the mother's (P) genes (s^+s^+) was expressed in the F_1, and the F_1 mother's genotype (s^+s) was expressed in the F_2. When *ss* individuals were inbred, only progeny that coiled to the left were produced. When the s^+s^+ or s^+s snails were inbred, however, they produced offspring that all coiled to the right. From the reciprocal cross between left-coiling females and right-coiling males (Fig. 20.14), all F_1 progeny were coiled to the left. The F_2 all coiled to the right; but, when each F_2 snail was inbred, those with the genotype *ss* produced progeny that coiled to the left.

Further investigation of coiling in snails has shown that the spindle formed in the metaphase of the first cleavage division influences the direction of coiling. The spindle of potential "dextral" snails is tipped to the right, but that of "sinistral" snails is tipped to the left. This difference in the arrangement of the spindle is controlled by the genes of the mother. They determine the orientation of the spindle, which in turn influences further cell division and results in the adult pattern of coiling. The actual phenotypic characteristic, therefore, is influenced directly by the mother, with no immediate relation to the genes in the egg, sperm, or progeny. However, most other snail traits do not show the maternal-effect pattern. The striping color pattern, for example, is also determined in the early embryo, but it is controlled directly by chromosomal genes of both parents. In this example, comparable color patterns are obtained from the results of reciprocal crosses.

Maternal Effect in *Drosophila*

At the University of Texas abnormal growth in the head region of *Drosophila melanogaster* appeared sporadically in a sample from a wild population collected at Acahuizotla, Mexico. At the University of Utah, these flies were inbred and selected for the abnormal head growths over a period of several years. The proportion of flies expressing the trait, named "tumorous head" (tu-h; Fig. 20.15), was increased to about 76 percent at 22°C when the flies were raised on a cornmeal and molasses medium. When reciprocal crosses were conducted, a maternal effect was indicated.

Tu-h females mated separately with three wild-type males and males from 11 laboratory stocks produced from 14 to 52 percent (average of 30 percent) of abnormal flies in the first generation. From the reciprocal cross between tu-h males and the same three wild-type and the same 11 laboratory stock females,

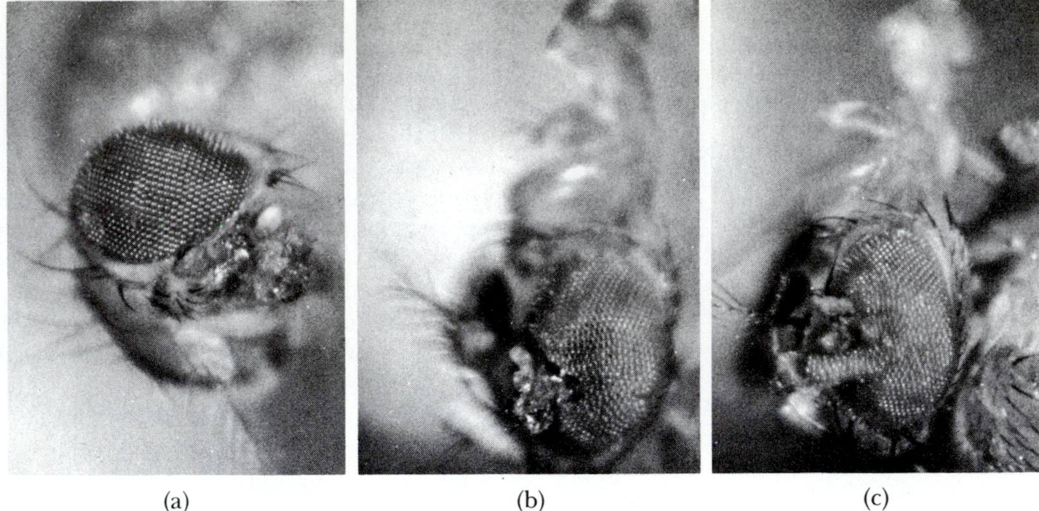

(a) (b) (c)

Figure 20.15 *Drosophila melanogaster* of the tumorous head strain, showing abnormal growths. A sex-linked gene homozygous in the mother controls the proportions of female and male progeny that express the abnormal growths. An autosomal gene on the third chromosome is required in the progeny for the expression, but this gene alone produces the phenotype in only 1 percent or less of the flies. With the maternal effect, 30 to 80 percent of the flies express the trait, depending on the combinations of genes in mothers and progeny.

0–1 percent (average of less than 1 percent) of tumorous-head flies were obtained. Further studies demonstrated the maternal effect. Genes of the mother were exerting an influence in the direction of abnormal growths on the heads of adult progeny during the first 22 hours of development. Two major genes were found to control the tumorous head trait: (1) a sex-linked gene at 64.5 map units on the X chromosome controlling the maternal effect and (2) a structural gene at 58 map units on the third chromosome controlling the tumorous head phenotype.

SUMMARY

Most heritable traits are controlled by nuclear chromosomal genes, but some depend on DNA in cytoplasmic organelles. Mitochondria and chloroplasts carry small amounts of unique DNA that behave independently with respect to nuclear genes. These cytoplasmic organelles have presumably evolved from free-living bacteria and algae, respectively, that entered into symbiotic relationships with eukaryotic cells. Streptomycin resistance in some present-day algae is dependent on plastids carrying DNA. In *Paramecium*, symbionts with their own DNA are established in the cytoplasm, but they can reproduce only in the presence of a particular nuclear genotype (*K*). Cytoplasmic organelles carrying DNA and having developed from prokaryote symbionts have become established through evolution and have retained a limited genetic operation more or less independent of nuclear genes. Plasmids are DNA molecules in the cytoplasm that may transform normal plant cells to tumor cells. Male sterility in maize and some other crop and flower garden plants is controlled by cytoplasmic factors. Maternal effects are controlled by nuclear ***genes*** of the ***mother*** and therefore are not examples of extranuclear inheritance. Extranuclear contents of the egg, however, reflect the influence of the mother's genotype, and the pattern of inheritance is like that of extranuclear inheritance.

REFERENCES

ASHWELL, M., and T.W. WORK. 1970. "The biogenesis of mitochondria." *Annu. Rev. Biochem.* 39: 251–290.

BIRKY, C.W., JR., P.S. PERLMAN, and T.J. BYERS, editors. 1974. *Genetics and biogenesis of mitochondria and chloroplasts.* Ohio State Univ. Press, Columbus.

CLAYTON, D.A. 1984. "Transcription of the mammalian mitochondrial genome." *Annu. Rev. Biochem.* 53: 573–594.

DUVICK, D.N. 1965. "Cytoplasmic pollen sterility in corn." In *Advances in genetics*, Vol. 13, pp. 2–56, E.W. Caspari and J.M. Thoday, editors. Academic Press, New York.

GIBSON, I. 1970. "Interacting genetic systems in *Paramecium*." In *Advances in morphogenesis*, Vol. 9, pp. 159–208, M. Abercrombin, J. Bracker, and T. J. King, editors. Academic Press, New York.

GOODENOUGH, U.W., and R.P. LEVINE. 1970. "The genetic activity of mitochondria and chloroplasts. *Sci. Amer.* 223: 22–29.

KIRK, J.T.O., and R.A.E. TILNEY BASSETT. 1978. *The plastids: Their chemistry, structure, growth and inheritance*, 2nd ed. Elsevier/North-Holland, New York.

KROON, A.M., and C. SACCONE. 1974. *The biogenesis of mitochondria.* Academic Press, New York.

MARGULIS, L. 1971. "Symbiosis and evolution." *Sci. Amer.* 225: 49–57.

NESTER, W.W., and A. MONTOYA. 1979. "Crown gall: A natural case of genetic engineering." *ASM News* 45: 283–287.

PALMER, J.D. 1985. "Comparative organization of chloroplast genomes." *Annu. Rev. Genet.* 19: 325–354.

PREER, J.P., JR. 1969. "Genetics of the protozoa." *Res. Protozool.* 31: 130–278.

———. 1971. "Extrachromosomal inheritance: Hereditary symbionts, mitochondria, chloroplasts." *Annu. Rev. Genet.* 5: 361–406.

SAGER, R., and Z. RAMINIS. 1970. "A genetic map of non-Mendelian genes in *Chlamydomonas*." *Proc. Natl. Acad. Sci. U.S.A.* 65: 593–600.

SWIFT, H., and D.R. WOLSTENHOLME. 1969. "Mitochondria and chloroplasts: Nucleic acids and problems of biogenesis." In *Handbook of molecular cytology*, pp. 972–1046, A. Lima de Faria, editor. North Holland Publishing, Amsterdam.

TATUM, L.A. 1971. "The southern corn leaf blight epidemic." *Science* 171: 1113–1116.

UMESONO, K., and H. OZEKI. 1987. "Chloroplast gene organization in plants." *Trends in Genet.* 3: 281–287.

PROBLEMS AND QUESTIONS

20.1 If a particular trait in a plant could be shown to be inherited solely through mitochondrial DNA, would it be classified as a case of maternal inheritance or maternal effect? Why?

20.2 In most animals, a larger amount of cytoplasm is carried by the egg than by the sperm. Similarly, the egg in plants carries more cytoplasm than the pollen. How could this difference affect the expression of inherited traits (a) dependent on nuclear chromosomal genes and (b) dependent on extranuclear genes?

20.3 Reciprocal crosses with experimental animals or plants sometimes give different results in the F_1. This may be due to (a) sex-linked inheritance, (b) cytoplasmic inheritance, or (c) maternal effects. If such a result were obtained, how could the investigator determine experimentally which category was involved?

20.4 Explain how single plants such as four-o'clocks could have green, pale green, and variegated sectors. If such sectors reached sexual maturity, what color characteristics would each type be expected to transmit through male or female gametes?

20.5 What practical applications could be made with male-sterile lines of maize?

20.6 How could kappa particles have become established in their host organism, *Paramecium*, through evolution?

20.7 O. Renner carried out reciprocal crosses between two types of the evening primrose, *Oenothera hookeri* and *O. muricata*, known to have the same chromosome constitution. When the seed parent was *O. hookeri*, the plastids of the progeny were yellow; but when the seed parent was *O. muricata*, the plastids of the progeny were green. How might this difference in the results of reciprocal crosses be explained?

20.8 In snails of the genus *Limnaea*, coiling is transmitted as a maternal effect. (a) Give the phenotypes that could be associated with the following genotypes in individual snails, and give the reason for each answer: s^+s^+, ss^+, and *ss*. (b) What might be said about the female and male parents and grandparents of snails represented by each of the three genotypes?

20.9 Diagram a cross between a female snail with dextral coiling and the genotype s^+s^+ and an inbred *ss* male with sinistral coiling. Carry the cross to the F_2, and represent the expected results from inbreeding each of the F_2 snails. Explain the results.

20.10 In the beach hopper of the genus *Gammarus*, pigment of the eyes is influenced in early stages by the genotype of the mother, but is later influenced by the genes of the individual hopper. Give the expected results of the following crosses in young and adult stages: (a) dark females (*AA*) × light males (*aa*) and (b) light females (*aa*) × dark males (*AA*). (c) Give a plausible explanation for the change that sometimes occurs from light eyes in young organisms to dark eyes in later stages.

20.11 When ovaries from light-colored (*aa*) flour moths of the genus *Ephestia* are implanted into dark (*AA*) females, which are then mated to *aa* males, the *aa* progeny have dark eyes when first hatched, but the eyes gradually become lighter. Give a plausible explanation for such a change in eye color.

20.12 A female fruit fly known to be heterozygous for *y* (for yellow body color) had patches of yellow on the thorax. Is this expression more likely to be the result of (a) nonhereditary environmental modification, (b) maternal effect, or (c) nondisjunction of chromosomes, resulting in a gynandromorph?

Chapter 21

Quantitative Genetics

Photo above: Contrasting sizes of flour beetles, *Tribolium castaneum*. The smaller beetle is from a standard laboratory stock; the larger beetle was produced over 120 generations of selection for increased body size. (Photograph courtesy of Franklin D. Enfield, Department of Genetics and Cell Biology, University of Minnesota.)

GENETICS is a science of variation. Without variation, genetic analysis at any level—molecular, organismic, or populational—cannot be carried out. In many cases, this variation involves striking differences between organisms. A flower's petals may be red or white; a fruit fly's wings may be long or short; a bacterial virus may be capable or incapable of infecting a particular cell. In other cases, biological variation is difficult to categorize. Height and weight in humans beings, for example, are traits that vary continuously in the population. For such traits, distinct classes cannot be identified, making it difficult to analyze them by conventional genetic methods. These traits are nonetheless important, so special methods have been developed to study them. Quantitative genetics is the discipline that deals with these complex traits.

QUANTITATIVE TRAITS AND QUANTITATIVE GENETICS

Quantitative genetics is based on measurements of individuals within a population of organisms. Many traits, such as height, weight, and yield of agricultural crops, are readily quantified by using a suitable scale (e.g., centimeters, grams, bushels per acre). Other traits, such as intensity of color, can be graded by comparing individuals to defined standards. Sometimes quantitative geneticists convert raw measurements into data that are more suitable for analysis. This is done by using mathematical transformations, such as the logarithm or square root, and has the advantage of smoothing out statistical irregularities.

The methods of quantitative genetics recognize two important facts. First, ***most quantitative traits involve the contributions of many different genes***. Sometimes a single gene may exert primary control over the trait, but usually there are many, more or less equally influential genes. The existence of different alleles at each of these genes will produce genotypic variation in the population and this, in turn, will lead to phenotypic variation for the trait. Second, ***quantitative traits are often influenced by environmental factors***. To the extent that these factors vary in the population, a quantitative trait will show phenotypic variation over and above that caused by genotypic differences. This combination of genotypic and environmental variability is, in fact, what makes quantitative traits difficult to study. Only with special methods is it possible to estimate the relative contributions of these two sources to the total phenotypic variation of the trait.

Like all geneticists, quantitative geneticists try to make predictions about the offspring of crosses. Naturally, their procedures are complicated by the preceding issues, but they are nonetheless effective. The basis for these procedures is a theoretical framework, or ***model***, that takes into account the relative contributions of genes and environment, the number of genes affecting the trait, the nature of individual gene effects, and Mendelian segregation. This model was developed in the first part of the twentieth century, mainly through the work of the British statistician and geneticist ***R. A. Fisher*** (Fig. 21.1). It provided the foundation for modern quantitative genetics by fusing the newly emerged Mendelian theory with a biometrical approach to the study of heredity. This latter approach stressed the importance of statistical analysis and was championed by the British researchers Karl Pearson and Francis Galton. Fisher's work laid to rest a protracted controversy in Britain between the Mendelists and the biometricians, thereby unifying two different points of view.

DESCRIPTIVE STATISTICS

Data on a quantitative trait can be presented graphically as a ***frequency distribution***. The horizontal axis, or ***abscissa***, measures values of the trait that are encountered in a sample from the population; usually the abscissa is subdivided into regular intervals. The vertical axis, or ***ordinate***, then measures the frequency of the observations in each interval.

Figure 21.2 shows the frequency distributions of the time to maturity in four populations of wheat. All

Figure 21.1 Ronald A. Fisher (1890–1962), British statistician and geneticist. (Courtesy of Everett Thorpe.)

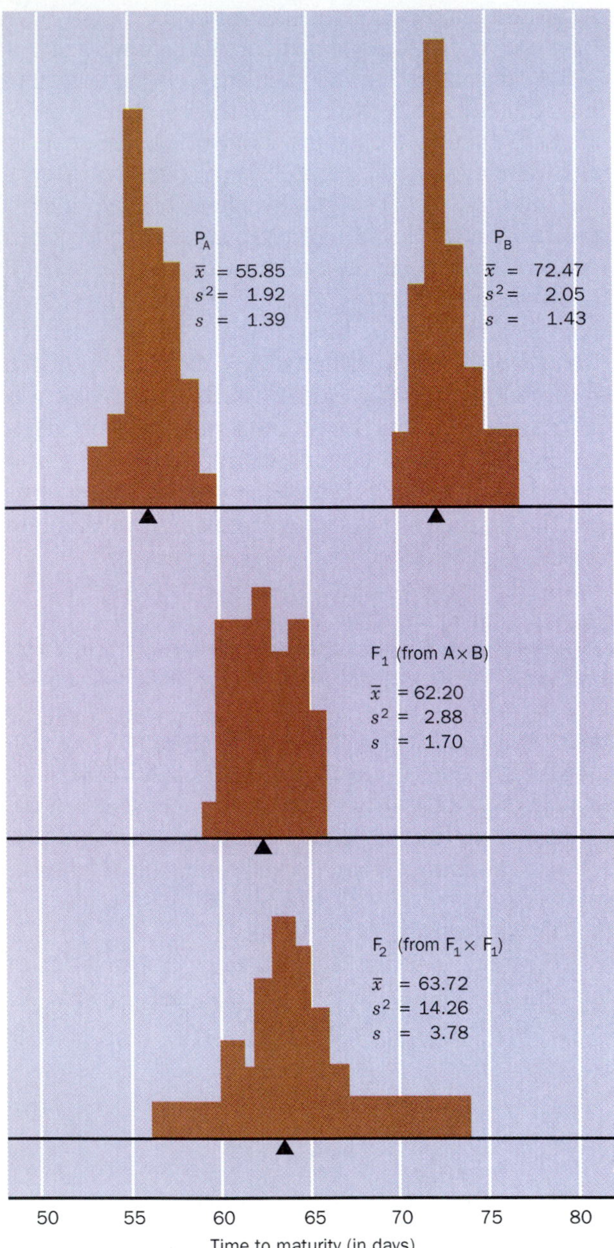

Figure 21.2 Frequency distributions and descriptive statistics of time to maturity in four populations of wheat. P_A and P_B are inbred varieties that were crossed to produce F_1 hybrids. The F_1 plants were then intercrossed to produce an F_2. Seed from all four populations was planted in the same season to determine the time to maturity. In each case, data were obtained from 40 plants. The mean maturation times ($\overline{X}$) are indicated by the triangles; the sample variances (s^2) and standard deviations (s) are also given.

four populations were grown in randomized test plots in the same season, thereby minimizing differences due to environmental variation in soil, moisture, temperature, and other factors. In each population, 40 plants were monitored until the heads of grain matured. The time to maturity for each of the plants was recorded in days, the shortest time being 52 days and the longest being 76. Two of the populations (P_A and P_B) represent inbred varieties of wheat. One matured quickly and the other slowly. In both cases, the frequency distributions are narrow and peaked, indicating that there is relatively little variation among the plants within each of these populations. This is consistent with the fact that both have been inbred to eliminate genetic variability. Notice also that there is no overlap between the two distributions.

An F_1 population was produced by crossing the two inbred varieties with each other. The frequency distribution for this population lies between those of the parents. This distribution is also somewhat broader and less peaked than the distributions of the inbred populations.

An F_2 population was produced by intercrossing the F_1 plants. In this case, the variability in maturation time among the 40 plants in the sample was considerably greater than it was in the inbred or F_1 samples. Some of the F_2 plants were as quick to mature as plants from the P_A variety, whereas others were as slow as plants from the P_B variety.

The Mean

The essential characteristics of a frequency distribution can often be represented by simple descriptive statistics. One of these, called the mean or average, gives the "center" of the distribution. The ***sample mean*** ($\overline{X}$) is calculated by summing all the data (ΣX_i) and dividing by the total number of observations (n); thus, $\overline{X} = (\Sigma X_i)/n$. In this formula, the Greek letter Σ is a mathematical shorthand for $X_1 + X_2 + ... + X_n$. The means of the four distributions in Fig. 21.2 are indicated by triangles; numerical values are given on the right. The means of the F_1 and F_2 samples are 62.20 and 63.72 days, respectively; both are just a little bit less than the average of the means (64.16 days) of the two inbred varieties.

The Variance and Standard Deviation

Another descriptive statistic, called the ***variance***, measures the scatter of the data about the mean. Data that are very dispersed produce a large variance, whereas data that are tightly clustered produce a much smaller value. The sample variance, s^2, is calculated from the formula $s^2 = \Sigma(X_i - \overline{X})^2/(n-1)$. Here $(X_i - \overline{X})^2$ is the squared difference between the ith observation and the sample mean (often called the squared deviation from the mean), and the Greek letter Σ indicates that all such squared deviations must be summed. The sum of the squared deviations is then averaged by dividing by $n-1$. (For technical reasons, the divisor is $n-1$ rather than n.) Because the variance is a squared statistic, it is always positive. Although this feature has useful mathematical properties, it sometimes makes the variance difficult to interpret. Consequently another statistic, called the ***standard deviation***, is used to measure the variability of the sample data. The standard deviation (s) is simply the square root of the sample variance; in mathematical notation, $s = \sqrt{s^2}$.

Figure 21.2 gives the variances and standard deviations of the distributions of the four wheat populations. The F_2 population clearly has the greatest values, corresponding to the large amount of variability that is evident in its frequency distribution. Notice that the variances (and standard deviations) of the two inbred populations are approximately equal to each other. This suggests that both have the same amount of variability for maturation time; since inbred populations are expected to be genetically homogeneous, this variability is probably due to environmental factors.

Correlations

Sometimes one set of measurements can be related to another. For example, researchers might have data on the heights and weights of individuals from a population, or they might have data on only one of these traits, but from pairs of genetically related individuals (e.g., siblings, parent and offspring, twins). In such circumstances, it is sometimes useful to calculate correlations between the matched data. One method is to compute the ***product-moment correlation coefficient***, often simply called the correlation coefficient. The formula is $r = \Sigma[(X_i - \overline{X})(Y_i - \overline{Y})]/[(n-1)\, s_X\, s_Y]$; here X_i and Y_i are the data from the ith pair of observations, $\overline{X}$ and $\overline{Y}$ are the sample means of the two sets of measurements, and s_X and s_Y are their standard deviations. As before, n is the sample size. The correlation coefficient can range from -1 to $+1$, with -1 indicating a perfect negative correlation between the X's and Y's and $+1$ indicating a perfect positive correlation. No correlation is indicated by a value of zero. Figure 21.3 illustrates these correlation concepts. The correlation coefficient often indicates that there is an underlying connection between two traits. Thus, in human beings, height and weight are positively correlated because they are both aspects of overall size. However, in poultry, egg mass and egg number are negatively correlated because a hen can either lay a few large eggs or many small ones.

The term $\Sigma[(X_i - \overline{X})(Y_i - \overline{Y})]/(n-1)$ is called the ***covariance*** between X and Y, symbolized $\mathrm{Cov}(X,Y)$. This is a measure of statistical association between two sets of data; it gives the correlation coefficient its sign. When two sets of data are independently distributed, the covariance between them is zero.

Additivity of Means and Variances

Sometimes it is useful to consider the sums $X_i + Y_i = Z_i$ from a set of paired observations. For instance, X and Y might be the IQ scores of husbands and wives, and Z could be considered a measure of the combined "intelligence" of a couple. From statistical theory, it follows that the mean of Z is equal to the sum of the

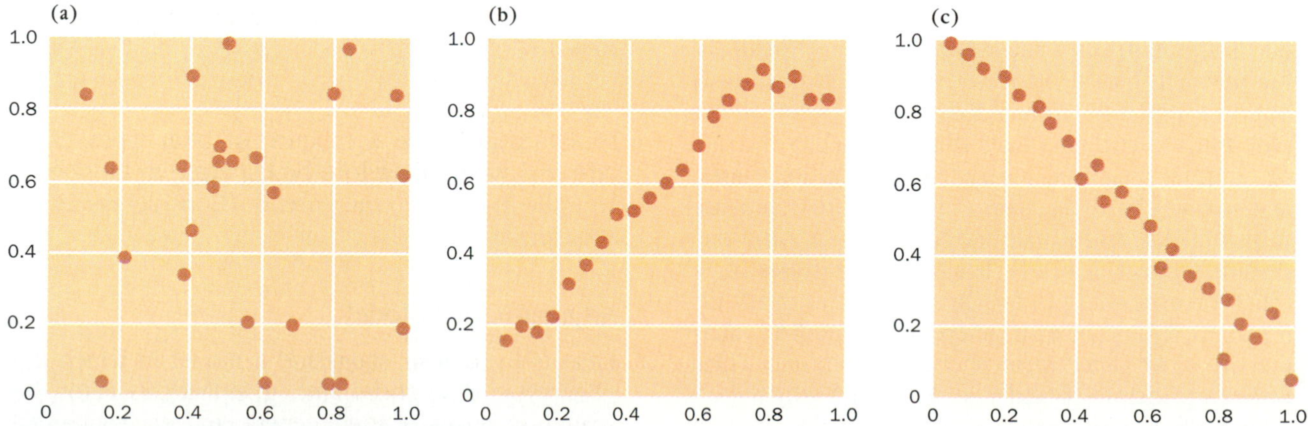

Figure 21.3 Statistical associations and the correlation coefficient. (a) No significant correlation; $r = 0.05$. (b) Strong positive correlation; $r = 0.97$. (c) Strong negative correlation; $r = -0.98$.

means of X and Y; thus, $\bar{Z} = \bar{X} + \bar{Y}$, indicating that means are strictly additive. This principle of additivity also applies to variances, but there is a complication if X and Y are correlated. Letting V stand for variance, the variance of Z can be written as $V_Z = V_X + V_Y + 2\,\mathrm{Cov}(X,Y)$. Clearly, if X and Y are uncorrelated, $\mathrm{Cov}(X,Y) = 0$, and the variance of the sum is simply the sum of the variances.

This principle of additivity does not extend to standard deviations. Perhaps this is why quantitative geneticists prefer to use the variance as a measure of scatter in their data.

THE MULTIPLE FACTOR HYPOTHESIS

Early in this century, R. A. Fisher realized that the study of quantitative traits had to be reconciled with the Mendelian view of heredity. The problem was that no one could show directly that these traits were determined by single Mendelian factors. Fisher appreciated the seriousness of this problem and proposed a solution: quantitative traits were not determined by single genes, but by many, and their alleles each had small and approximately additive effects. Thus, the phenotype of an individual depended on its genotype at all the relevant loci, with each allele adding (or subtracting) a small amount. Traits that have this feature are said to be ***polygenic***. Fisher also proposed that many environmental factors influenced the trait by adding or subtracting effects in a manner similar to that of all the genetic loci. This combination of multiple environmental and genetic factors therefore determined the phenotype of an individual.

Fisher's hypothesis readily explained three features of many quantitative traits. First, the multiplicity of factors contributing to a trait accounted for the variability among phenotypes. This variability arises from the large number of genotypes and environments that are possible when many factors are involved. Second, Fisher's hypothesis suggested an explanation for the bell-shaped nature of many frequency distributions. If a trait is influenced by two types of alleles, plus and minus, defined according to whether they add or subtract an effect, Mendelian segregation should lead to many individuals carrying a mixture of plus and minus alleles, but usually not to many carrying only plus or minus (Fig. 21.4). Thus, extreme values of the trait, corresponding to the tails of the frequency distribution, should be rare, whereas intermediate values that are close to the mean should be common. Third, Fisher's principle of additivity explained why the offspring of crosses between phenotypically different parents often had intermediate phenotypes. One parent might have a large number of plus alleles, the other a large number of minus alleles, and their offspring would have a mixture of both types. Such offspring would therefore be expected to be phenotypically in between the parents. (Consider again the situation in Fig. 21.2.)

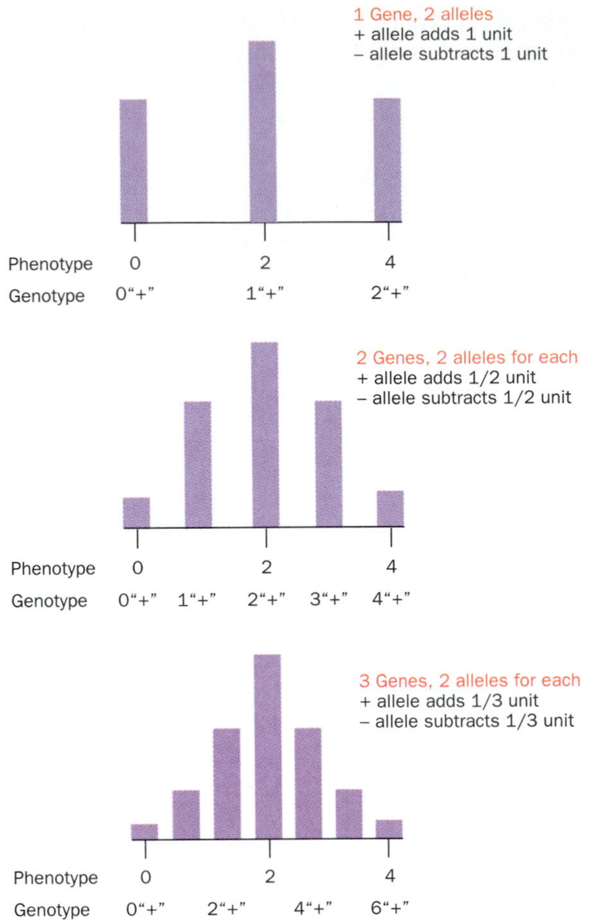

Figure 21.4 Frequency distributions of a quantitative trait influenced by different numbers of genes. For each gene there are two kinds of alleles, plus and minus, with each plus allele adding a constant amount to a basic phenotypic value of 2 and each minus allele subtracting an equal amount. The frequency distributions have been constructed by assuming that the plus and minus alleles are equally frequent, that different genes assort independently, and that the alleles of each gene are combined at random. In each case, the genotype is specified by the number of plus alleles that are present. Notice that as the number of genes influencing the trait increases, the frequency distribution tends to be continuous and bell-shaped.

A Quantitative Model

Fisher's hypothesis suggested a model in which the phenotypic differences among individuals could be written in terms of their genetic and environmental components. Let T be the phenotypic value of an individual and $\bar{T}$ be the overall mean of the population. Then the deviation between T and $\bar{T}$, defined as

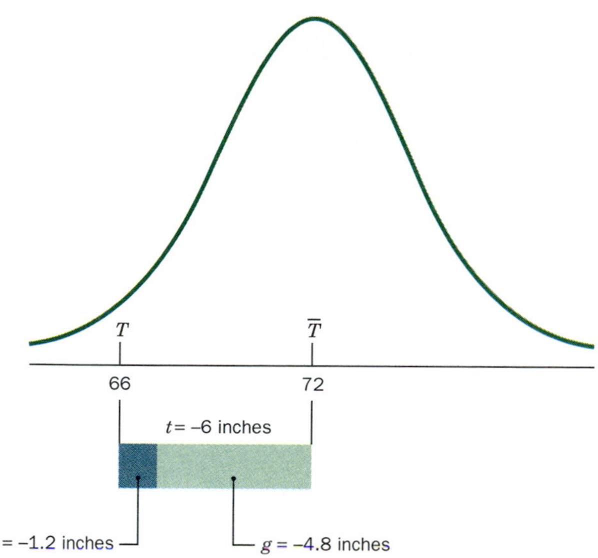

Figure 21.5 Decomposition of the phenotypic deviation into genotypic and environmental components. The distribution of height in a population of human males is shown. The mean ($\overline{T}$) is 72 inches. Individuals with a height (T) of 66 inches deviate from this mean by −6 inches. This deviation can be decomposed into expected genetic (g) and environmental (e) components as shown.

$t = T - \overline{T}$, can be written as the sum of a genotypic (g) and an environmental (e) deviation: $t = g + e$.

An example will clarify the point. In one sample of human males, the mean height was 72 inches. Individuals who were 66 inches tall deviated from this mean by −6 inches. A quantitative geneticist estimated that 80 percent of this deviation, or −4.8 inches, was expected to be due to genotypic differences among individuals, whereas the remaining 20 percent, or −1.2 inches, was expected to be due to environmental differences. Altogether, −4.8 inches (g) − 1.2 inches (e) = −6 inches (t), the total phenotypic deviation (Fig. 21.5).

Analysis of Variance

The quantitative model just described assumes that the phenotypic deviation (t) is equal to the sum of the genotypic (g) and environmental (e) deviations. From a principle given earlier, the variance of a sum is equal to the sum of the variances. Thus, in mathematical notation, $V_t = V_g + V_e$, where V_t is the total phenotypic variance for the trait, V_g is the variance due to genotypic differences, and V_e is the variance due to environmental effects. This formula is correct as long as the covariance between the genotypic and environmental components is zero. If this were not true, the two components would be correlated and their variances would not be strictly additive.

The equation showing that the phenotypic variance is equal to the sum of the genotypic and environmental variances is an example of an approach invented by Fisher. It is called the ***analysis of variance***. In quantitative genetics, the analysis of variance is used to partition the total phenotypic variance into genotypic and environmental components. This technique is useful because it reveals what proportion of the total variance is due to genotypic factors.

As an example, consider a study on the number of abdominal bristles in female *Drosophila* by E. C. R. Reeve and F. W. Robertson. The total variance for bristle number was 5.44. Of this, 2.11 was genotypic and 3.33 was environmental, so 5.44 (V_t) = 2.11 (V_g) + 3.33 (V_e).

Broad-Sense Heritability

The proportion of the total variance that is genotypic is called the ***broad-sense heritability***, usually designated H^2. The exponent 2 is a reminder that the variance is a squared unit; however, it must be remembered that the broad-sense heritability is itself a dimensionless quantity, since $H^2 = V_g/V_t$. In the example just cited, H^2 is 2.11/5.44 = 0.39.

This statistic has two main uses. First, it reveals the relative contributions of genotypic and environmental factors to the variability of the trait. Second, it makes it possible to estimate the expected values of g and e, the genotypic and environmental deviations in the quantitative model described. For human stature, H^2 has been estimated to be 0.8, so a phenotypic deviation of −6 inches from the population mean can be split into an expected genotypic deviation of −4.8 inches (= −6 inches × 0.8) and an expected environmental deviation of −1.2 inches (= −6 inches × 0.2). This separation reflects the relative contributions of the two types of factors to the overall variability.

One method of estimating the broad-sense heritability uses highly inbred strains. Within these strains, the genotypic variance is expected to be zero because all the individuals have essentially the same genotype; any variation must be due to the environment, so the observed variance estimates V_e. In contrast, a randomly mated population is expected to show both genotypic and environmental variability, so the observed variance estimates $V_g + V_e$. The difference between the variances of the randomly mated and inbred populations therefore estimates V_g, and the ratio of this to the variance of the randomly mated population estimates H^2.

Another method utilizes data collected from twins. Identical twins reared apart have the same genotype but different environments. In mathematical notation, this means that the phenotypic deviation of one twin can be written as $t = g + e$, and that of its co-twin can be written as $t' = g + e'$. Consequently, the similarity

TABLE 21.1 Heights in Inches of 31 Pairs of Identical Human Twins[a]

TWIN #1	TWIN #2
63.3	63.0
64.3	65.0
66.8	66.8
63.0	63.5
62.5	62.5
61.8	60.8
68.5	67.3
63.0	65.3
63.5	62.0
62.3	62.3
63.5	63.5
59.0	59.0
62.5	63.0
63.7	63.7
66.0	66.2
65.0	62.7
64.0	63.0
64.8	64.5
64.8	63.0
63.3	65.0
64.0	60.5
63.3	62.0
68.5	67.7
64.2	63.8
61.8	62.6
62.3	64.6
61.2	62.1
65.2	64.9
63.8	64.3
61.5	60.5
66.5	66.5

[a]Data from T. J. Bouchard, University of Minnesota (personal communication, 1989).

between twins depends only on g, the effect of their common genotype, so the covariance between them is $\mathrm{Cov}(t, t') = \Sigma(g_i \times g_i)/(n-1) = \Sigma g_i^2/(n-1) = V_g$, the genotypic variance. Dividing this by the total phenotypic variance gives the broad-sense heritability. Thus, $\mathrm{Cov}(t, t')/V_t = H^2$, which is also the correlation coefficient for a set of identical twin pairs reared in different environments.

As an example, consider the data in Table 21.1, which shows the heights of 31 identical human twins who were separated from each other early in life. Assuming that the separation was early enough to fulfill the requirement for different environments, the correlation coefficient calculated from these data can be used as an estimate of the broad-sense heritability: $r = \mathrm{Cov}(t, t')/V_t = 0.82$, indicating that a high proportion of the population variability is attributable to genotypic differences.

PREDICTING PHENOTYPES

Although the broad-sense heritability provides information about the importance of genetic differences among the individuals in a population, it has very little predictive power. It cannot be used to predict the values of a quantitative trait in the offspring of specific matings, nor can it be used to study the nature of the genes affecting the trait. For both of these purposes, a more refined analysis is needed.

To develop a method for predicting an offspring's phenotype, it is necessary to subdivide the genotypic deviation in the quantitative model $t = g + e$. The reason is that g encompasses all the genetic factors affecting a trait, including interactions among genes. Such interactions have little or no predictive value because Mendelian segregation breaks up gene combinations. As a result, the methods for predicting phenotypes depend primarily on the effects of individual alleles.

The cumulative sum of these effects will be represented by the letter a (for ***additive allelic effects***), whereas the sum of all the ***interactions*** will be represented by the letter i. Thus, $g = a + i$, and $t = a + i + e$. In this quantitative model, the phenotype of an individual is determined by three separate components. Turning to the population, this model implies that the phenotypic variance will also be determined by three components, providing that the terms a, i, and e are uncorrelated with each other. Assuming this to be the case (and it often is), $V_t = V_a + V_i + V_e$.

V_a is called the ***additive genetic variance***, so named because it accounts for the variability caused by the additive effects of alleles. V_i stands for the ***interaction genetic variance***, a term that measures the impact of genetic combinations. Sometimes this term is further subdivided into components for ***dominance*** and ***epistasis***, the former involving combinations of alleles at the same locus and the latter involving combinations at different loci. Because neither of these parts of V_i has much predictive power, quantitative geneticists tend to focus their attention on the additive genetic variance.

Narrow-Sense Heritability

The narrow-sense heritability, often symbolized h^2, is defined as the ratio V_a/V_t, that is, the proportion of the total phenotypic variance that is due to the additive effects of alleles. Clearly, h^2 can never be greater than the broad-sense heritability because the latter also includes interaction effects. Table 21.2 gives some estimates for the narrow-sense heritability for a number of traits. Notice that human stature is highly heritable, whereas litter size in pigs is not. This means that

TABLE 21.2 Estimates of the Narrow-Sense Heritability[a]

TRAIT	h^2
Stature in human beings	0.65
Milk yield in dairy cattle	0.35
Litter size in pigs	0.05
Egg production in poultry	0.10
Tail length in mice	0.40
Body size in *Drosophila*	0.40

[a]From D. S. Falconer, *Introduction to Quantitative Genetics*, 2nd ed., p. 51. Longman, London, 1981.

we would have more luck in predicting the height of a human child than we would in predicting the litter size of a pig, assuming, of course, that we had information about the parents' phenotypes in each case.

Quantitative geneticists often estimate h^2 from correlations between relatives. The method is to compute the correlation coefficient and then divide it by the fraction of genes that the relatives share by virtue of their common ancestry. Half-sibs reared in different environments are particularly useful, especially if they have had different mothers. Such half-sibs share one-fourth of their genes, but none of their environmental or maternal effects. In one study of human half-sibs, the correlation coefficient for height was 0.14, yielding an estimate of $0.14/0.25 = 0.56$ for the narrow-sense heritability.

It should be emphasized that h^2 is a property of a particular population and that different populations can have different values of h^2. This follows because the narrow-sense heritability depends on the amount and nature of the genetic variation that is present in a population, not on some absolute quantity. Care must therefore be taken in using estimates of h^2 to make genetic predictions.

Estimating an Offspring's Phenotype

In the quantitative model $t = a + i + e$, a is the only term that provides much information about the expected phenotype of an offspring. It is sometimes called the ***breeding value*** of the individual. To estimate a for each of an individual's parents, multiply their phenotypic deviations by the narrow-sense heritability. For the father, $a_f = t_f h^2$, and for the mother, $a_m = t_m h^2$. These breeding values then make it possible to predict the phenotype of their offspring. Since the offspring inherits half its genes from each parent, its breeding value is estimated by $a_o = (a_m + a_f)/2 = h^2\,(t_m + a_f)/2$. The predicted phenotypic deviation of the offspring is then $t_o = a_o$ (because i_o and e_o cannot be predicted) and the expected phenotype is $T_o = \overline{T} + t_o = \overline{T} + a_o = \overline{T} + h^2\,(t_m + t_f)/2$. Noting that $(t_m + t_f)/2 = (T_m + T_f)/2 - \overline{T}$, and defining $(T_m + T_f)/2$ as T^*, the average phenotype of the two parents or the ***midparent value***, the final prediction equation is $T_o = \overline{T} + h^2\,(T^* - \overline{T})$.

As an example, consider the situation sketched in Fig. 21.6. A man who has made an IQ score of 110 on a standardized intelligence test has married a woman who has made a score of 120. The mean IQ of the population is 100 and the standard deviation is 15. The narrow-sense heritability of IQ, as measured by this test, is estimated to be 0.4. What is the IQ of a child born to this couple expected to be? To obtain the answer, we must compute the phenotypic deviations of each of the parents. For the father, $t_f = T_f - \overline{T} = 10$,

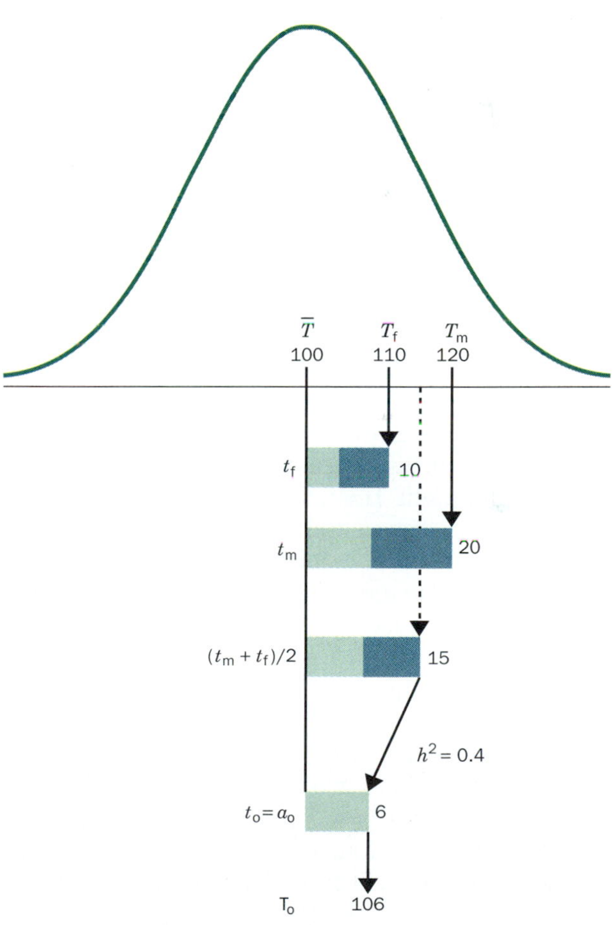

Figure 21.6 Estimating an offspring's phenotype. The distribution of IQ scores from an intelligence test is shown. The mean score ($\overline{T}$) is 100 and the standard deviation is 15. The scores of a particular man (T_f) and a particular woman (T_m) are also shown, along with their phenotypic deviations from the mean (t_f and t_m, respectively). The average of these deviations is $(t_m + t_f)/2 = 15$. By using the narrow-sense heritability ($h^2 = 0.4$), this average allows the phenotypic deviation of a child born to this couple to be predicted; $t_o = h^2\,(t_m + t_f)/2 = 6$. Consequently, the IQ of the child is expected to be $\overline{T} + t_o = 106$.

and for the mother, $t_m = T_m - \overline{T} = 20$. Then we compute the average deviation, which is $(t_m + t_f)/2 = (10 + 20)/2 = 15$. Multiplying this by the narrow-sense heritability, we obtain an estimate for the deviation of the child from the overall mean; this is $t_o = 0.4 \times 15 = 6$. Thus, the expected IQ of the child is $\overline{T} + t_o = 100 + 6 = 106$. Notice that the same result could be obtained by using the formula $T_o = \overline{T} + h^2 (T^* - \overline{T})$. In this case, $T^* = (110 + 120)/2 = 115$, so $100 + 0.4 \times (115 - 100) = 106$. It should be emphasized that this is only an expected value. Because of the uncertainties of genetic segregation and because of random variation in the environment, there is no way to predict the IQ of a child exactly. The value of 106 obtained here is actually the expected, or average, value of the IQs of all the children born to parents whose IQs were 110 and 120. For any particular child, the IQ could deviate significantly from this overall average.

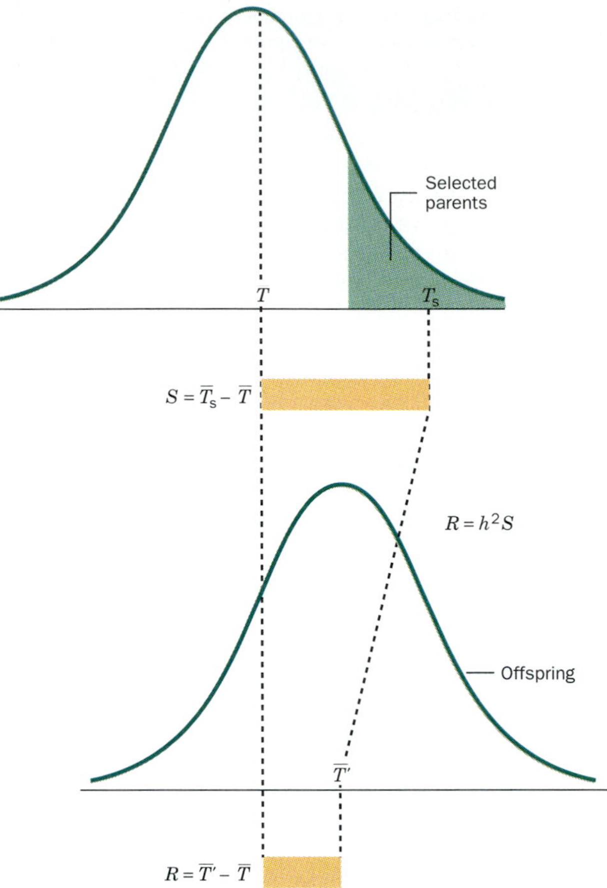

Figure 21.7 Predicting responses to artificial selection. The distribution of a quantitative trait in two generations is shown. In the upper distribution, the individuals that were selected to produce the second generation are indicated by the shaded area. S is the selection differential, defined as the difference between the mean of the selected parents ($\overline{T}_s$) and the mean of the whole population ($\overline{T}$). R is the response to selection, defined as the difference between the mean of the offspring ($\overline{T}'$) and the mean of the original population; h^2 is the narrow-sense heritability.

Predicting Responses to Artificial Selection

The narrow-sense heritability has another important use: to predict changes in the mean of a population under ***artificial selection***. The situation is diagrammed in Fig. 21.7. In each generation, a fraction of the population is selected to produce offspring. The difference between the mean of the selected individuals and the mean of the whole population is $S = \overline{T}_s - \overline{T}$, the ***selection differential***. This measures the strength of selection for extreme values of the trait. The response to this selection is evident in the next generation, where the entire frequency distribution is shifted slightly to the right. The difference between the mean of the next generation and that of the whole population is $R = \overline{T}' - \overline{T}$, the ***selection response***, and the ratio R/S is equal to the narrow-sense heritability. Thus, the response to selection can be predicted by noting that $R = h^2S$.

This relationship turns out to be merely an extension of the formula used to predict the phenotype of an offspring from those of its parents. Substituting $\overline{T}' - \overline{T}$ for R and $\overline{T}_s - \overline{T}$ for S, we have $\overline{T}' - \overline{T} = h^2 [\overline{T}_s - \overline{T}]$, which yields $T' = \overline{T} + h^2 [\overline{T}_s - \overline{T}]$. Here $\overline{T}'$, the mean of the next generation, takes the place of T_o, the expected phenotype of an individual, and $\overline{T}_s$ takes the place of T^*.

To see how the method works in practice, consider an artifical selection experiment that was conducted by F. D. Enfield. In this experiment, Enfield selected for increased body size in the common flour beetle, *Tribolium castaneum*, for 125 generations. Body size was measured by weighing the pupae at 21 days of age. Each generation, the largest pupae were selected to complete development and to serve as parents for the next generation. When the experiment was started, the weight of individual pupae ranged from 1800 to 3000 μg. The mean was 2400 μg with a variance of 40,000 μg^2. After 125 generations of selection, the mean pupa weight had increased to 5800 μg, which is more than twice the mean of the starting population (Fig. 21.8). Moreover, none of the individuals in the selected population was as small as the largest individuals in the starting population (Fig. 21.9). This complete lack of overlap in the frequency distributions indicates that the genetic makeup of the population had been radically changed.

To achieve this stunning result, Enfield utilized a selection differential of 200 μg in each generation in the experiment. Initially, the narrow-sense heritability

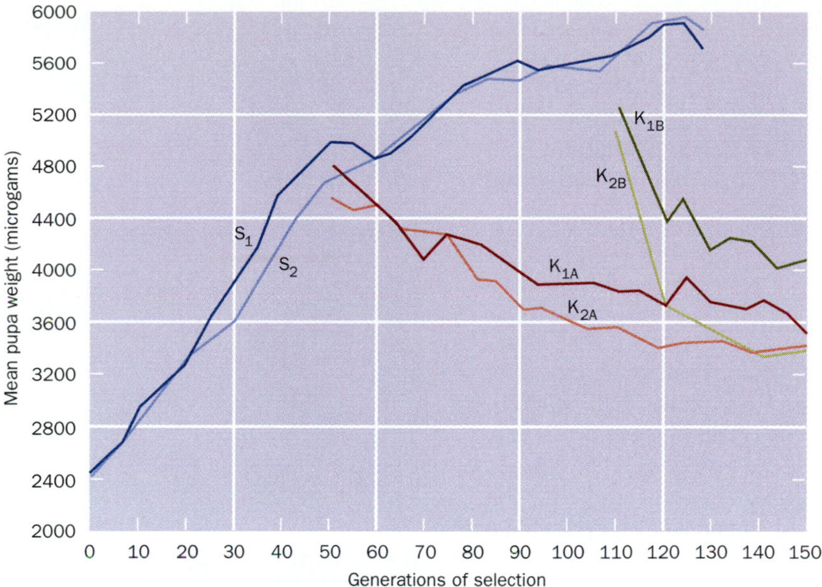

Figure 21.8 Artificial selection for increased size in *Tribolium.* Size was measured by the weight of the pupa at 21 days of age. The curves S_1 and S_2 show the response to selection in two replicate populations. The curves K_{1A}, K_{2A}, K_{1B}, and K_{2B} show what happened when artificial selection was discontinued in subpopulations that were established from the selected populations. (Data courtesy of F. D. Enfield, University of Minnesota.)

for pupa weight was estimated to be about 0.3, so the predicted response to selection was 0.3 × 200 μg = 60 μg per generation. For the first 40 generations, this is approximately what was observed. However, the cumulative response during this time, 4400 μg − 2400 μg = 2000 μg, is a little less than the 2400 μg (=60 μg/generation × 40 generations) that was expected. The discrepancy is due to factors that reduce the efficiency of selection, including such things as infertility among the selected individuals. Thus, although the narrow-sense heritability is a reasonably good predictor of the response to selection over a few generations, it tends to overestimate this response in a long-term experiment.

The later generations of Enfield's project demonstrate this point dramatically. Between generations 40 and 125, the cumulative response was 5800 μg − 4400 μg = 1400 μg, which, although impressive, is very much less than the expected response of 5100 μg (=60 μg/generation × 85 generations). Enfield checked the narrow-sense heritability in these later generations to see if any of the additive genetic variance had been lost during the long selection process. To his surprise, he found that h^2 was only slightly changed, indicating that the population still had selectable genetic variability. A more detailed analysis demonstrated that during these generations, the efficiency of selection was severely reduced by a negative correlation between size and reproductive ability. The larger beetles were less fertile than the smaller ones, thus reducing the effective selection differential and making it very difficult to select for further increases in size. In fact, when selection was relaxed in generation 50, and again in generation 110, mean pupa weight began dropping back to a

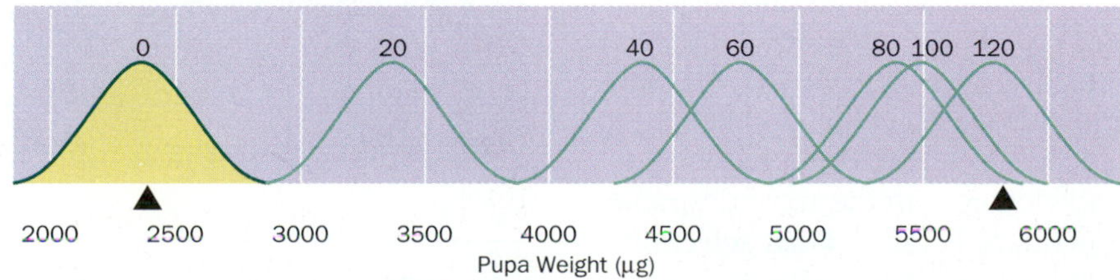

Figure 21.9 Frequency distributions of pupa weight in *Tribolium* populations selected for increased size. The shape of the distributions with the dashed lines is only approximate. Generations are indicated by the numbers at the top.

more nearly normal value. This suggests that a force of ***natural selection*** strongly opposed the artificial selection that was practiced in the main experiment. Enfield's attempts to increase pupa weight beyond 5800 μg failed, not for any lack of genetic variability, but simply because artificial selection had exceeded the natural limits.

Practical Significance and Evolutionary Implications

For tens of thousands of years, human beings have cultivated different varieties of plants and animals. Information on the process of domestication is very sketchy, but many species such as dogs, cats, horses, pigs, and cows were probably domesticated near the end of the last ice age, about 12,000 years ago. With the development of permanent agriculture, about 10,000 years ago, other species were brought under the direct control of human beings. Ancient texts record efforts to breed livestock, indicating that at least some primitive people recognized the possibility of genetic improvement. Moreover, the amazing diversification of species under domestication indicates that even though these people lacked a scientific understanding of the issues, they were quite successful in developing different varieties.

A good example is the dog, which is almost certainly the oldest domesticated species. Figure 21.10 illustrates the tremendous variation that exists among some of the standard breeds. In addition to differences in hair color, length, and texture, and in body size and shape, there are major differences in renal and endocrine physiology, as well as in behavior. The origin of the various breeds of dogs is largely a matter of speculation, but some, like the basenji from central Africa and the chow from China, have probably developed from small, isolated populations. Others, like the golden retriever and the Shetland sheep dog, have been created in recent times by selection from hybrid stock. Without a doubt, the remarkable diversity among dog breeds indicates that human beings have long been aware of the power of artificial selection.

With the advent of Mendelian genetics, artificial selection was given a scientific basis. The formal theory of quantitative genetics establishes procedures for collecting and analyzing data; it also provides guidelines for programs of selective breeding. In the simplest case, a group of individuals from the population is randomly mated to produce the next generation. Usually, these are the individuals with the most desirable phenotypes. This procedure is called ***mass selection***.

(a)

(b)

(c)

Figure 21.10 Variation among breeds of dogs. (a) St. Bernard; (b) bull terrier; (c) shih tzu; (d) chow chow; (e) Dalmation; (f) borzoi. (Courtesy of Gerard Lacz/NHPA.)

(d)

(e)

(f)

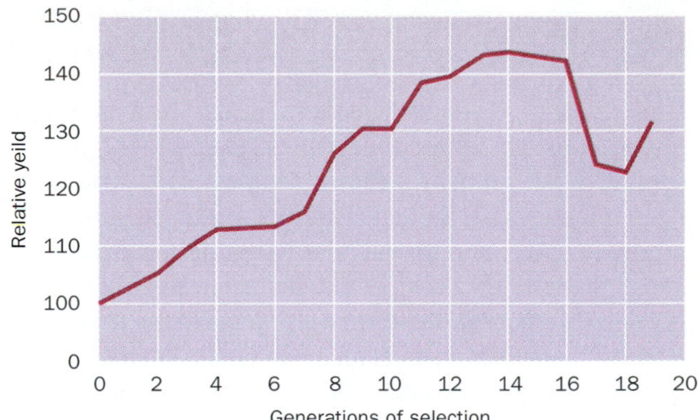

Figure 21.11 Response of a population of maize to mass selection for increased yield. (From C. O. Gardner, "Quantitative Genetic Studies and Population Improvement in Maize and Sorghum," *Proceedings of the International Conference on Quantitative Genetics,* E. Pollak, O. Kempthorne, and T. B. Bailey, Jr., editors, Iowa State Univ. Press, Ames, 1977. Copyright © 1977 the Iowa State University Press, Ames, Iowa. All rights reserved.)

Figure 21.11 illustrates the effect of mass selection to improve grain yield in maize. Relative yield increased steadily at a rate of about 3 percent per generation for about 13 generations; thereafter it declined, possibly because growing conditions had changed. Although mass selection is not the sole reason, it is certainly responsible for some of the improvements that have occurred in crop and livestock species.

Other forms of selection are also used in plant and animal breeding. For instance, in dairy herds, milk production has been improved by selecting for high-yielding cows; however, most of the improvement has actually come from selecting bulls. In practice, this involves assessing the milk yield of a sample of the bull's daughters, a procedure called ***progeny testing***. Bulls with the highest-yielding daughters are likely to carry genes that are favorable for the trait, so by using such bulls to inseminate many cows, the quality of the breed is improved.

Recently, evolutionary biologists have turned their attention to quantitative genetics to elucidate the process of natural selection. Genetic analysis of existing species has been combined with the careful study of fossil forms to determine the rate of phenotypic change and the intensity of natural selection during the evolution of a species.

SUMMARY

Quantitative genetics considers traits in which the phenotypes are determined by the action of ***many genetic and environmental factors***. This complex determination is characteristic of height, weight, and other metric traits. In quantitative genetics, it is standard practice to collect data from individuals in a population in order to obtain the ***frequency distribution*** of the trait. Statistical analyses based on these data are employed to estimate the total amount of phenotypic variation, and also to study the separate contributions of the genetic and environmental factors. The ***mean***, or average, is used to represent the typical value of the trait in the population and the ***variance*** and ***standard deviation*** are used to measure the variation about this value.

The phenotypic deviation of an individual from the mean of the population is equal to the sum of a genotypic and an environmental deviation. In the whole population, the ***phenotypic variance*** of a trait is equal to the sum of the ***genotypic and environmental variances***, providing that these components are uncorrelated. The proportion of the total phenotypic variance that is due to genotypic variance is called the ***broad-sense heritability***.

The genotypic contribution to a trait can be subdivided into ***additive and interaction effects***, each accounting for a portion of the total genotypic variance. The fraction of the total phenotypic variance that is due to additive genetic variance is called the ***narrow-sense heritability***. This quantity is useful in predicting the phenotypes of offspring and in estimating responses to artificial selection. Consequently, it is an important statistic in programs to improve agricultural crops and livestock through selective breeding.

REFERENCES

BOUCHARD, T.J., JR., D.T. LYKKEN, N.L. SEGAL, and K.J. WILCOX. 1986. "Development in twins reared apart: A test of the chronogenetic hypothesis." In *Human growth: A multidisciplinary review,* pp. 299–310, A. Demirjian, editor. Taylor & Francis, London.

CLAYTON, G.A., J.A. MORRIS, and A. ROBERTSON. 1957. "An experimental check on quantitative genetical theory. I. Short-term responses to selection." *J. Genet.* 55: 131–151.

CLAYTON, G.A., and A. ROBERTSON. 1957. "An experimental check on quantitative genetical theory. II. The long-term effects of selection." *J. Genet.* 55: 152–170.

CROW, J.F. 1986. *Basic concepts in population, quantitative and evolutionary genetics.* Freeman, New York.

DUDLEY, J.W. 1977. "76 generations of selection for oil and protein percentage in maize." In *Proceedings, International Conference on Quantitative Genetics*, pp. 459–473. Iowa State Univ. Press, Ames.

ENFIELD, F.D. 1977. "Selection experiments in tribolium designed to look at gene action issues." In *Proceedings, International Conference on Quantitative Genetics*, pp. 177–190. Iowa State Univ. Press, Ames.

FALCONER, D.S. 1981. *Introduction to quantitative genetics,* 2nd ed. Longman, London.

FISHER, R.A. 1918. "The correlation between relatives on the supposition of Mendelian inheritance." *Trans. Roy. Soc. Edinburgh* 52: 399–433.

GARDNER, C.O. 1977. "Quantitative genetic studies and population improvement in maize and sorghum." In *Proceedings, International Conference on Quantitative Genetics,* pp. 475–489. Iowa State Univ. Press, Ames.

REEVE, E.C.R., and F.W. ROBERTSON. 1954. "Studies in quantitative inheritance. VI. Sternite chaeta number in *Drosophila*: A metameric quantitative character. *Z. Ind. Abst. Vererb.* 86: 269–288.

WRIGHT, S. 1968. *Evolution and the genetics of populations,* Vol. 1, *Genetic and biometric foundations.* Univ. Of Chicago Press, Chicago.

______. 1977. *Evolution and the genetics of populations,* Vol. 2, *Experimental results and evolutionary deductions.* Univ. of Chicago Press, Chicago.

PROBLEMS AND QUESTIONS

21.1 A wheat variety with red kernels (genotype $A'A'B'B'$) was crossed with a variety with white kernels (genotype $AABB$). The F_1 were intercrossed to produce an F_2. If each primed allele increases the amount of pigment in the kernel by an equal amount, what phenotypes will be expected in the F_2? Assuming that the A and B loci are unlinked, what will the phenotypic frequencies be?

21.2 A sample of 20 plants from a population was measured in inches as follows: 18, 21, 20, 23, 20, 21, 20, 22, 19, 20, 17, 21, 20, 22, 20, 21, 20, 22, 19, and 23. Calculate (a) the mean, (b) the variance, and (c) the standard deviation.

21.3 Quantitative geneticists use the variance as a measure of scatter in a sample of data; they calculate this statistic by averaging the squared deviations between each measurement and the sample mean. Why don't they simply measure the scatter by computing the average of the deviations without bothering to square them?

21.4 Two inbred strains of corn were crossed to produce an F_1; this was then intercrossed to produce an F_2. Data on ear length from a sample of F_1 and F_2 individuals gave phenotypic variances of 15.2 and 27.6 cm^2, respectively. Why was the phenotypic variance for the F_2 greater than that for the F_1?

21.5 A study of quantitative variation for abdominal bristle number in female *Drosophila* yielded estimates of $V_t = 6.08$, $V_g = 3.17$, and $V_e = 2.91$. What was the broad-sense heritability?

21.6 Measurements on ear length were obtained from three populations of corn—two inbred varieties and a randomly pollinated population derived from a cross between the two inbreds. The phenotypic variances were 9.2 and 9.6 cm^2 for the two inbred varieties and 26.4 cm^2 for the randomly pollinated population. Estimate the broad-sense heritability of ear length for these populations.

21.7 The broad-sense heritability of protein content in a population of soybeans was estimated to be 0.2. In standardized units, the mean protein content was 10. On the same scale, a group of plants in this population had a protein content of 6. (a) In this group, what was the phenotypic deviation from the mean? What were the expected genotypic (b) and environmental (c) deviations?

21.8 The broad-sense heritability can be estimated by computing the correlation coefficient between identical twins reared apart. The formula is $r = \text{Cov}(t, t')/V_t = V_g/V_t = H^2$. Why is $\text{Cov}(t,t')$ equal to V_g?

21.9 A quantitative geneticist claims that the narrow-sense heritability for IQ in human beings is 0.7 whereas the broad-sense heritability is only 0.3. Why must there be an error?

21.10 The mean value of a trait is 100 units and the narrow-sense heritability is 0.4. A male and a female measuring 124 and 126 units, respectively, mate and produce a large number of offspring, which are reared in an average environment. What is the expected value of the trait among these offspring?

21.11 The narrow-sense heritability for abdominal bristle number in a population of *Drosophila* is 0.3. The mean bristle number is 12. A male with 10 bristles is mated to a female with 20 bristles and a large number of progeny are scored for bristle number. What is the expected number of bristles among these progeny?

21.12 A breeder is trying to decrease the maturation time in a population of sunflowers. In this population, the mean time to flowering is 100 days. Plants with a mean flowering time of only 90 days were used to produce the next generation. If the narrow-sense heritability for flowering time is 0.2, what will the average time to flowering be in the next generation?

21.13 A fish breeder wishes to increase the rate of growth in a stock by selecting for increased length at 6 weeks after hatching. The mean length of 6-week-old fingerlings is currently 10 cm. Adult fish that had a mean length of 15 cm at 6 weeks of age were used to produce a new generation of fingerlings. Among these, the mean length was 12.5 cm. Estimate the narrow-sense heritability of fingerling length at 6 weeks of age and advise the breeder about the feasibility of the plan to increase growth rate.

21.14 A selection differential of 40 μg per generation was used in an experiment to select for increased pupa weight in *Tribolium*. The narrow-sense heritability for pupa weight was estimated to be 0.3. If the mean pupa weight was initially 2000 μg and selection was practiced for 10 generations, what was the mean pupa weight expected to become?

Chapter 22

Population and Evolutionary Genetics

Photo above: African cheetahs, *Acinonyx jubatus.* (Courtesy of R. S. Virde/ Heilman.)

GENES do not exist in isolation. Even the simplest cells contain thousands of genes working together to make life possible. Diploid, sexually reproducing species have, in fact, two sets of genes—one derived from the mother and the other from the father. At each locus, there is a possibility for different alleles, creating the potential for an enormous number of distinct genotypes.

The richness of this genetic variation is evident in even the most superficial analyses. Genes in organisms as diverse as primroses, *Drosophila*, mice, and human beings have been shown to have several alternative alleles at many different loci. The nature and extent of this diversity, as well as its biological significance, are subjects that fall within the scope of population genetics. For nearly 80 years, researchers in this area of genetics have been engaged in documenting, analyzing, and interpreting natural genetic variability.

GENETIC VARIATION

One way of detecting genetic variation is to study phenotypic differences. Human beings come in many sizes, shapes, and colors, and experience indicates that each of these traits is at least partially heritable. More definitive studies have focused on simple Mendelian traits, such as those known in *Drosophila*. In one survey, the Russian geneticist N. P. Dubinin found several unusual flies in a collection of 10,000. Breeding tests subsequently demonstrated that some of these flies were homozygous for rare recessive alleles. Other studies have focused on variation in genes that are essential for life. Because mutations in such genes are often lethal in the homozygous condition, these studies have concentrated on detecting heterozygous carriers. Surveys of natural populations of *Drosophila* have shown that almost all the members of a population are heterozygous for at least one recessive, or nearly recessive, lethal mutation.

With the application of biochemical techniques, more sophisticated studies have been possible. ***Gel electrophoresis*** is a technique that can identify variation in the physical and chemical properties of proteins (Fig. 22.1). Protein samples are readily obtained from extracts of tissues such as blood. The individual proteins in these samples can be separated from each other by passing them through a carbohydrate sieve. The carbohydrate is formed into a gel, saturated with a buffering solution, and attached to an electric current. Protein samples inserted into slots at one end of the gel are then moved through the gel by electric force. The rate of movement is determined primarily by the electric charge on the protein, and this, in turn, depends on the protein's amino acid sequence. Thus, if

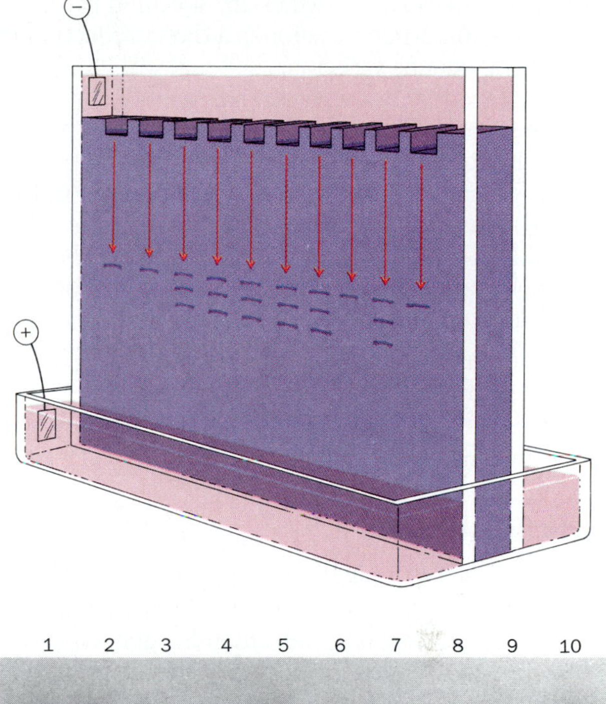

Figure 22.1 Detection of genetic variability by gel electrophoresis. Samples of proteins are inserted into slots at the negative side of the gel and then moved to the positive side by an electric force. Each protein molecule in a sample will move at a rate that depends on its electric charge. When movement is completed, the gel is treated with a stain and individual protein bands become visible. The photograph at the bottom shows the results of an experiment with an enzyme (isocitrate dehydrogenase) that was obtained from a plant, *Trillium pusillum*. The uppermost and lowermost bands represent slow- and fast-moving forms of the enzyme, each consisting of two identical subunits. The band in the middle represents an enzyme that is composed of one fast-moving and one slow-moving subunit; consequently, it has an intermediate electrophoretic mobility. Samples 1, 2, 8, and 10 in the photograph came from plants that were homozygous for the allele for the fast-moving subunit. All the other samples came from plants that were heterozygous for this allele and the allele for the slow-moving subunit. (Data courtesy of Paul R. Cabe, University of Minnesota.)

two forms of a protein have slightly different sequences, and if this causes them to have different charges, they will move at different rates through the gel. When the electric current is turned off, the gel can be treated with a specific stain to reveal mobility differences in the protein molecules. Forms of the same protein that are separated in this way are called

allozymes. Because allozymes are specified by different alleles, allozymic variation is a direct reflection of underlying genetic variation.

The extent of this variation has been documented in numerous studies. For instance, about one-third of all soluble proteins in human beings show allozymic variation. In many other species, the frequency of loci with allozymic variation is 20–30 percent, but in some it is much lower. One example is the African cheetah, in which only 4 percent of enzyme-coding loci are segregating different alleles. This lack of variability is probably the result of a relatively recent bottleneck in the size of the cheetah population. Such bottlenecks are expected to reduce genetic variability by causing the loss of rare alleles in the species.

Allele Frequencies

The frequencies of alleles at segregating loci can be estimated by analyzing a sample of individuals from the population. This is easiest when the genotypes are all phenotypically distinct. As an example, consider the data in Table 22.1, which shows the frequencies of three blood types in a sample from a human population. Each of the phenotypes was identified by testing the blood with immunological sera. One serum detects a blood cell property called the M antigen, and the other detects a related property called the N antigen. Cells that react with both sera possess both antigens.

The genetic basis of these antigenic characteristics is well established. The M and N antigens are encoded by alleles of a gene on chromosome 4. With two alleles, there are three possible genotypes, L^ML^M, L^NL^N, and L^ML^N, corresponding to the M, N, and MN phenotypes. This unambiguous relationship between genotypes and phenotypes permits the frequencies of the two alleles to be estimated directly from the data. The procedure is to count the alleles of each type and then to divide the totals by the number of alleles in the entire sample. Thus, for L^M, there are $2 \times 1787 = 2574$ L^M alleles in the individuals with blood type M and 3039 L^M alleles in those with blood type MN, giving a total of 6613 L^M alleles in the entire sample of $2 \times 6129 = 12{,}258$ alleles. This yields a frequency of $6613/12{,}258 = 0.5395$. For L^N, there are $2 \times 1303 = 2606$ L^N alleles in the individuals with blood type N and 3039 L^N alleles in those with blood type MN, giving a total of 5645 L^N alleles in the sample. The frequency of the L^N allele is therefore $5645/12{,}259 = 0.4605$. Letting p stand for the frequency of L^M and letting q stand for the frequency of L^N, we have $p = 0.5395$ and $q = 0.4605$; notice also that $p + q = 1.0000$.

The L locus is a clear example of ***genetic polymorphism***. By convention, population geneticists recognize a locus as polymorphic if it is segregating two or more alleles and if the frequency of the most common allele is less than 0.99.

RANDOM MATING AND THE HARDY–WEINBERG METHOD

In the example just discussed, the frequencies of the two alleles were estimated from the frequencies of the three genotypes. We might now ask if there is any method of reversing the logic and predicting the genotype frequencies from the estimated allele frequencies.

One procedure is to assume that the population mates randomly with respect to the M–N blood type. We can then think of the diploid genotypes as coming from the random union of haploid eggs and sperm. The calculations of the genotype frequencies are laid out in Table 22.2. Under the assumption of random mating, the probability of producing a zygote that is homozygous for the L^M allele is simply $p \times p = p^2$. This follows because p is the frequency of the L^M allele in both the sperm and the eggs. Likewise, the probability of producing an L^NL^N homozygote is $q \times q = q^2$. For the L^ML^N heterozygotes, two possibilities must be considered. An L^M sperm can unite with an L^N egg, or vice versa, each with a probability of $p \times q$. Since both events are equally likely, the total probability of forming an L^ML^N heterozygote is $2pq$. Under the assumption of random mating, the frequencies of the L^ML^M, L^ML^N, and L^NL^N genotypes are therefore given by p^2, $2pq$, and q^2. These are the terms in the expansion of the expres-

TABLE 22.1 Frequency of M, N, and MN Phenotypes

PHENOTYPE	ANTI-M SERUM[a]	ANTI-N SERUM	GENOTYPE	NUMBER OF INDIVIDUALS
M	+	0	L^ML^M	1787
MN	+	+	L^ML^N	3039
N	0	+	L^NL^N	1303

[a]The + means that the red blood cells of the individual react with the antiserum. The 0 means that no reaction occurs.

TABLE 22.2 Genotype Frequencies with Random Mating

		EGGS	
		L^M ($p = 0.5395$)	L^N ($q = 0.4605$)
SPERM	L^M ($p = 0.5395$)	L^M/L^M ($p^2 = 0.2911$)	L^M/L^N ($pq = 0.2484$)
	L^N ($q = 0.4605$)	L^N/L^M ($pq = 0.2484$)	L^N/L^N ($q^2 = 0.2121$)

Genotypes	L^ML^M	L^ML^N	L^NL^N
Observed frequency	1787	3039	1303
Expected frequencies[a]	$p^2 \times 6129$ = 1784.2	$2pq \times 6129$ = 3044.8	$q^2 \times 6129$ = 1300

[a]Expected frequencies are calculated on the basis of the Hardy–Weinberg genotype frequencies.

sion $(p + q)^2$. Notice that because $p + q = 1$, we also have $p^2 + 2pq + q^2 = 1$.

The genotypic frequencies calculated by this method are called the ***Hardy–Weinberg frequencies***, so named because the method was first described by the British mathematician G. H. Hardy and the German physician W. Weinberg. Notice that the numerical frequencies that were calculated by this method correspond quite closely to the frequencies that were actually observed. This close correspondence suggests that the underlying assumption of random mating is, in fact, justified. Similar results with other cases indicate that the Hardy–Weinberg method is a fairly useful one for predicting genotype frequencies from allele frequencies.

Applications of the Hardy–Weinberg Method

One use of the Hardy–Weinberg method is to predict the frequency of individuals homozygous for a deleterious recessive allele. Consider, for example, the allele that causes phenylketonuria (PKU), a potentially serious disease in the metabolism of aromatic amino acids. In some human populations, the frequency of this allele is 0.01. Assuming random mating, homozygotes should occur with a frequency of $(0.01)^2 = 0.0001$, which is about what is observed. In comparison, the predicted frequency of heterozygotes is $2 \times 0.01 \times 0.99 = 0.0198$, or about 2 percent of the population. This indicates that the vast majority of the alleles for PKU are carried by heterozygous individuals.

When there is dominance, the Hardy–Weinberg method can be turned around to estimate the frequency of the recessive allele. Let C and c stand for the dominant and recessive alleles, with frequencies p and q, respectively. Then the Hardy–Weinberg frequencies of the three genotypes are p^2 (CC), $2pq$ (Cc), and q^2 (cc). However, because CC and Cc have the same phenotype, it is not possible to obtain the allele frequencies by simple enumeration from population data. Fortunately, the observed frequency of the recessive homozygotes provides an estimate of q^2, and using the Hardy–Weinberg method in reverse, the square root of this is an estimate of the frequency of c. For example, consider a population with 0.04 percent phenylketonurics; since $\sqrt{0.0004} = 0.02$, approximately 2 percent of the alleles in this population are recessive.

Cases of multiple alleles can also be handled by the Hardy–Weinberg method. For example, if a gene has alleles A_1, A_2, and A_3, with frequencies p_1, p_2, and p_3, the Hardy–Weinberg proportions of the six different genotypes are p_1^2 (A_1A_1), p_2^2 (A_2A_2), p_3^2 (A_3A_3), $2p_1p_2$ (A_1A_2), $2p_2p_3$ (A_2A_3), and $2p_1p_3$ (A_1A_3).

When two or more loci are considered simultaneously, the situation gets more complicated. Suppose the genotypes AA, Aa, and aa are in Hardy–Weinberg proportions, as are the genotypes BB, Bb, and bb for a gene on the same chromosome. Does this mean that the frequency of $AA\ BB$ is simply the frequency of AA times the frequency of BB? The answer is yes only if the two loci are in a state of ***linkage equilibrium***. This means that recombination has randomized the alleles of one locus with respect to those of the other. Letting p and q be the frequencies of A and a, and letting r and s be the frequencies of B and b, linkage equilibrium implies that the frequency of an AB chromosome is $p \times r$ and that the frequency of an ab chromosome is $q \times s$. Consequently, the frequency of the $AA\ BB$ genotype is $(p \times r)^2$, the frequency of the $aa\ bb$ genotype is $(q \times s)^2$, and so on (Table 22.3). Genes that are far apart on a chromosome attain linkage equilibrium rather quickly, but genes that are close together may remain out of equilibrium for many generations. Tight

TABLE 22.3 Genotype Frequencies at Two Loci with Linkage Equilibrium and Random Mating[a]

		EGGS			
		AB (pr)	*aB* (qr)	*Ab* (ps)	*ab* (qs)
SPERM	*AB* (pr)	*AB/AB* (p^2r^2)	*AB/aB* (pqr^2)	*AB/Ab* (p^2rs)	*AB/ab* ($pqrs$)
	aB (qr)	*aB/AB* (pqr^2)	*aB/aB* (q^2r^2)	*aB/Ab* ($pqrs$)	*aB/ab* (q^2rs)
	Ab (ps)	*Ab/AB* (p^2rs)	*Ab/aB* ($pqrs$)	*Ab/Ab* (p^2s^2)	*Ab/ab* (pqs^2)
	ab (qs)	*ab/AB* ($pqrs$)	*ab/aB* (q^2rs)	*ab/Ab* (pqs^2)	*ab/ab* (q^2s^2)

[a]The frequencies of the gametes and genotypes are shown in parentheses.

linkage therefore makes the prediction of multiple locus genotype frequencies rather difficult.

INBREEDING

Any departure from random mating naturally leads to complications in the relationships between allele frequencies and genotype frequencies. For instance, consider the case of repeated self-fertilization. This occurs in some animal species, such as worms and snails, but it is much more common among plants. If a heterozygote, *Aa*, is self-fertilized, it will produce three kinds of progeny, *AA*, *Aa*, and *aa*, in the ratio 1:2:1. At this stage, the frequency of the heterozygotes is 0.5. If self-fertilization is continued for another generation, the homozygotes will breed true, but the heterozygotes will again segregate, reducing their frequency to 0.25. With every succeeding generation of self-fertilization, the frequency of heterozygotes drops off by 50 percent, reaching 0.008 by generation seven and 0.001 by generation ten. At this stage, the population is 99.9 percent homozygous (Fig. 22.2)

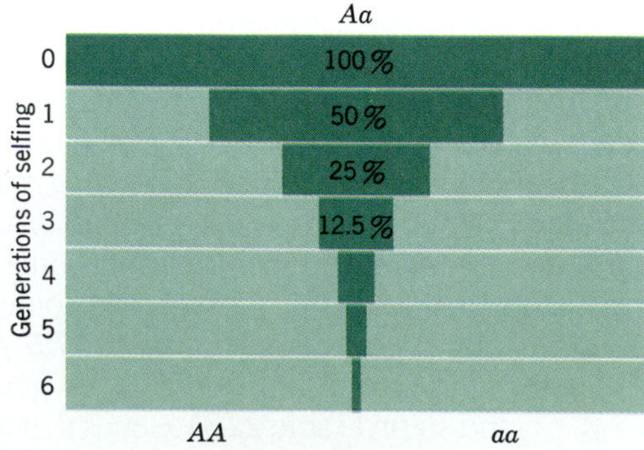

Figure 22.2 Consequences of repeated self-fertilization. In each generation of selfing, the frequency of heterozygotes is reduced by one-half.

The process of repeated self-fertilization illustrates the general effects of a form of nonrandom mating called ***inbreeding***. The allele frequencies remain the same, but the genotype frequencies change. Inbreeding occurs whenever mates are genetically related. Self-fertilization is the most extreme example, but other examples such as the mating of full siblings, first cousins, parents and offspring, and half siblings have the same effect, namely, to increase the frequency of homozygotes and decrease the frequency of heterozygotes. As a result, the Hardy–Weinberg method does not hold.

The Inbreeding Coefficient

The effects of inbreeding can be quantified by using the inbreeding coefficient, symbolized by F. Initially, S. Wright defined F as a correlation coefficient, but later C. W. Cotterman redefined it as a probability. Cotterman's definition will be used here. ***The inbreeding coefficient is the probability that two alleles in an individual are identical by descent from a common ancestor.*** Identity by descent means that two alleles are copies of a particular allele from a previous generation. Obviously, alleles that are identical by descent are also identical in state. Notice that since the inbreeding coefficient is defined as a probability, it must have a value between zero and one. $F = 0$ means that there has been no inbreeding, that is, that mating is random, and $F = 1$ means that inbreeding is complete.

To see the meaning of F, consider the case of self-fertilization illustrated in the pedigree in Fig. 22.3. The individual P is both the mother and father of its offspring O. The inbreeding coefficient of O is therefore the probability that two alleles in O are identical by descent from P. For clarity, each of the alleles at a locus in O (and also in P) has been given a different label. F_O is the probability that a is identical to b, written simply as $F_O = \Pr(a = b)$.

To compute F_O, it is necessary to enumerate all the ways in which a could equal b, Figure 22.3 shows that there are four possibilities. One is for c to be transmitted in both the egg and the sperm, leading to $a = c$ and $b = c$ in the zygote. This has a probability of ¼ because c has a chance of ½ of being included in the egg and another chance of ½ of being included in the sperm and these two events are independent. Another possibility is for d to be transmitted in both the egg and the sperm, leading to $a = d$ and $b = d$; this also has a probability of ¼. The two remaining cases involve the chance that c and d are already identical by virtue of

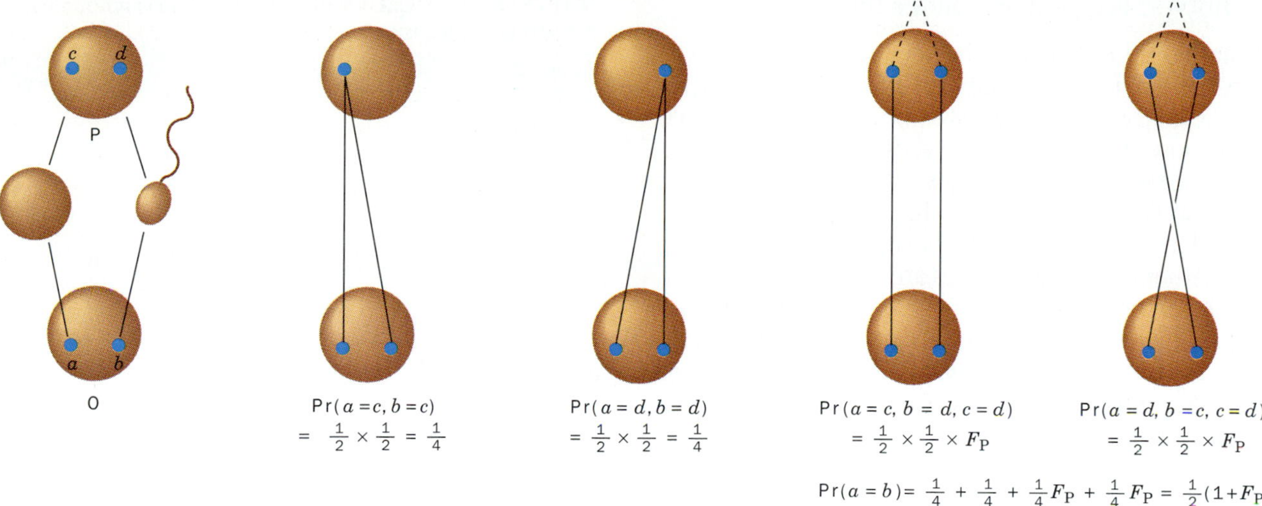

Figure 22.3 Calculating the inbreeding coefficient for the offspring of self-fertilization.

previous, but unstated, inbreeding. The probability of this is simply F_P, the inbreeding coefficient of P; thus, $F_P = \Pr(c = d)$. Then, if c is transmitted in the egg and d is transmitted in the sperm, $a = b$ because $c = d$. This could also happen the other way around, with d passing through the egg and c through the sperm. In both cases, $\Pr(a = b) = \frac{1}{2} \times \frac{1}{2} \times F_P$. Putting all this together, $F_O = \frac{1}{4}\,(a = c, b = c) + \frac{1}{4}\,(a = d, b = d) + \frac{1}{4} \times F_P\,(a = c, b = d, \text{and } c = d) + \frac{1}{4} \times F_P\,(a = d, b = c, \text{and } c = d) = \frac{1}{2}\,(1 + F_P)$. Notice that if $F_P = 0$, then $F_O = \frac{1}{2}$. Thus, when an individual that came from random mating undergoes self-fertilization, its offspring has an inbreeding coefficient of $\frac{1}{2}$.

Calculating *F* from Pedigrees

The coefficients for other cases of inbreeding can be computed by using a simple procedure: (1) Identify the common ancestors of the inbred individual. A common ancestor is connected to the inbred individual through both of the individual's parents. The genetic pathways that descend from the common ancestor to these parents form an inbreeding loop. (2) For each loop, calculate $(\frac{1}{2})^n\,(1 + F_{CA})$, where n is the number of individuals in the loop, including the common ancestor (but excluding the inbred individual), and F_{CA} is the common ancestor's inbreeding coefficient. (3) Sum the results of all the loops.

Three examples will clarify the method. Figure 22.4*a* shows the mating of full-sibs. The individuals in the pedigree are represented by circles and the genetic paths connecting them are indicated by lines. The inbreeding coefficient of the offspring, O, is obtained by dividing the pedigree into two separate inbreeding loops, A X B and A Y B. Each has a common ancestor (X or Y) from which two lines of descent lead to the inbred individual O. Also, for each loop $n = 3$, so $F_O = (\frac{1}{2})^3(1 + F_X) + (\frac{1}{2})^3(1 + F_Y)$. If X and Y were produced by random mating, then $F_X = F_Y = 0$ and $F_O = \frac{1}{8} + \frac{1}{8} = \frac{1}{4}$. Figure 22.4*b* shows the mating of half-sibs. Here, there is only one common ancestor (Z) and only one inbreeding loop, A Z B; $n = 3$, so $F_O = (\frac{1}{2})^3(1 + F_Z)$. Assuming $F_Z = 0$, $F_O = \frac{1}{8}$. The last example is shown in Fig. 22.4*c*. First, we note that P is the offspring of half-sib mating, and assuming no prior inbreeding, $F_P = \frac{1}{8}$. Using this result, we have $F_O = (\frac{1}{8})(1 + F_P) = \frac{9}{64}$.

The inbreeding coefficient of an individual obviously depends on the genetic relationship between its parents. Full-sibs are more closely related than half-sibs, and this is reflected by the inbreeding coefficients of their progeny. ***The general rule is that the fraction of alleles that mates share by virtue of common ancestry is two times the inbreeding coefficient of their progeny.***

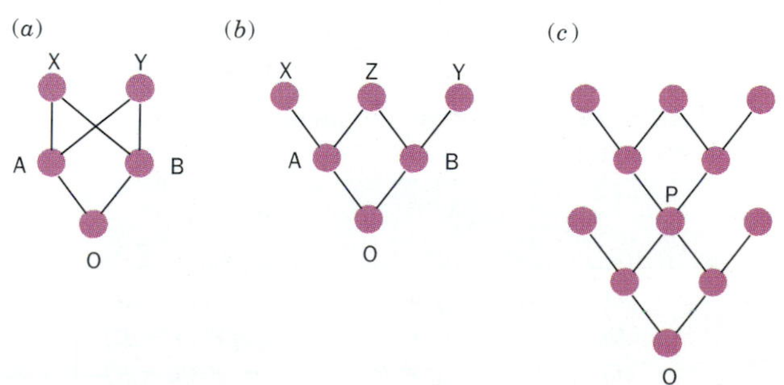

Figure 22.4 Calculation of *F* from pedigrees. (a) Full-sib mating. (b) Half-sib mating. (c) A more complicated pedigree.

Genotype Frequencies under Inbreeding

In an inbred population, the genotypic frequencies of a particular locus will deviate from the frequencies predicted by the Hardy–Weinberg principle. Table 22.4 shows how to calculate these frequencies using the allele frequencies and the inbreeding coefficient. The population in Table 22.4 has been divided in two ways. First, individuals have been classified according to whether or not their alleles are identical by descent. There are two possibilities, corresponding to the genetic terms autozygote and allozygote. ***In an autozygote, the alleles at a locus are identical by descent; in an allozygote, they are not.*** Second, individuals have been classified by genotype; *AA, Aa,* and *aa*. In the population, the intersections of these two classifications create six different groups; however, one group is ruled out by the fact that all autozygotes must also be homozygotes. Table 22.4 therefore gives the population frequencies of five different groups.

These frequencies were obtained by using the allele frequencies (p for A and q for a) and by noting that F is the probability that the alleles in an individual are identical by descent, that is, the probability that an individual is autozygous. It follows, therefore, that $1 - F$ is the probability that an individual is allozygous. Among the autozygotes, a fraction p will be *AA* and a fraction q will be *aa*. Consequently, the population frequencies of these two types will be $p \times F$ and $q \times F$, respectively. Among the allozygotes, there is no identity by descent, meaning that the alleles have been combined at random. The frequencies of the allozygous genotypes can therefore be predicted by the Hardy–Weinberg method. If each of these frequencies is multiplied by $(1 - F)$, that is, the probability of being allozygous, we obtain the frequencies shown in Table 22.4. To determine the overall frequencies of the three genotypes, irrespective of the autozygote/allozygote classification, we sum the entries in each column:

$$AA : pF + p^2(1 - F) = p^2 + pqF$$
$$Aa : 2pq\,(1 - F) = 2pq - 2pqF$$
$$aa : qF + q^2(1 - F) = q^2 + pqF$$

TABLE 22.4 Genotype Frequencies under Inbreeding[a]

GENOTYPES	AUTOZYGOTES (F)	ALLOZYGOTES ($1 - F$)
AA	$p \times F$	$p^2 \times (1 - F)$
Aa	0	$2pq \times (1 - F)$
aa	$q \times F$	$q^2 \times (1 - F)$

[a]The allele frequencies are p (A) and q (a). F is the probability that two alleles in an individual are identical by descent, that is, the frequency of autozygotes in the population.

TABLE 22.5 Frequency of Phenylketonuria among the Offspring of Relatives[a]

PARENTS	F	FREQUENCY OF PKU IN OFFSPRING
Unrelated	0	0.0001
First cousins	1/16	0.0007
Half-sibs	1/8	0.0013
Full-sibs	1/4	0.0026

[a]The frequency of the allele for PKU is assumed to be 0.01.

These are the Hardy–Weinberg frequencies plus or minus a term involving the inbreeding coefficient. To appreciate their usefulness, consider the effect of inbreeding on the incidence of phenylketonuria in the human population. If unrelated parents produce children with PKU at a frequency of about 0.0001, we can estimate the frequency of the deleterious allele as $q = \sqrt{0.0001} = 0.01$. Then, using the formula for the frequency of recessive homozygotes under inbreeding, we can predict the occurrence of PKU among the children of first cousin, half-sib, and full-sib matings. The calculations are given in Table 22.5. Compared to unrelated parents, first cousins have a 7-fold higher risk of having a child with PKU, half-sibs have a 13-fold higher risk, and full-sibs have a 26-fold higher risk. Clearly, the probability of a diseased child increases with the inbreeding coefficient.

Quantitative Effects of Inbreeding

Inbreeding also affects quantitative traits, such as size, vigor, fertility, and the yield of agricultural crops. Detailed studies have indicated that these traits decline more or less linearly with F. Two examples are given in Fig. 22.5. The data in Fig. 22.5 were obtained from a population of maize that had been self-fertilized for several generations. Seed was saved at stages in the inbreeding process and then grown in test plots to determine the values of each trait. It is clear from Fig. 22.5 that both plant height and crop yield declined linearly with F. One explanation is that deleterious recessive alleles were made homozygous by the inbreeding.

OUTBREEDING AND ASSORTATIVE MATING

Outbreeding occurs when matings are preferentially between unrelated individuals. There are many cases in nature, but some of the most striking are found in plants that have developed a system of sexual incompatibility. This incompatibility is based on a genetically complex locus, called S, which has a large number of alleles. In species with ***gametophytic incompatibility***, pollen cannot fertilize an ovule if the pollen grain

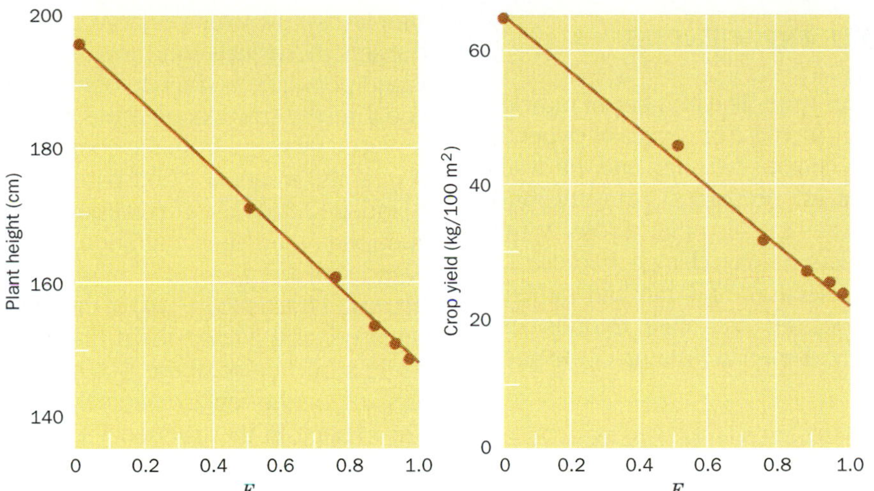

Figure 22.5 Inbreeding decline in maize. (Data from A. R. Hallauer and J. H. Sears, *Crop Sci.* 13: 327–330, 1973.)

carries an allele that is present in the maternal tissue of the style. For example, pollen with the allele S_1 cannot fertilize an ovule produced by an S_1S_2 plant; however, it can fertilize an ovule produced by an S_2S_3 plant. Obviously, in such a system, homozygotes can never be formed, so the *S* locus is usually a highly variable one.

Outbreeding is also a standard feature of crop production and animal husbandry. For example, seed corn is commercially produced by crossing different inbred lines. Although the inbreds themselves are not very productive, their hybrid progeny often yield an abundant crop. This hybrid vigor, or ***heterosis***, is probably due to heterozygosity for the genes that control crop yield.

In conservation biology, there is a need to minimize the deleterious effects of inbreeding in small, managed populations. This is done by selecting the least similar individuals for matings. Prezwalski's wild horse is an example (Fig. 22.6). This species is native to the steppes of central Asia, but because a variety of factors has essentially exterminated the native population, almost all the horses are now kept in zoos. Detailed pedigree records have enabled zoo managers to determine the relationships among these animals. By arranging matings between genetically different individuals, it has been possible to minimize inbreeding effects.

Mating based on phenotypic resemblance is called assortative mating. If the mates are phenotypically similar, we say that there has been ***positive assortative mating***, whereas if they are phenotypically different, we say that there has been ***negative assortative mating***. The effects of positive assortative mating are similar to those of inbreeding (the frequency of homozygotes increases), whereas the effects of negative assortative mating are similar to those of outbreeding (the frequency of heterozygotes increases). These changes occur because individuals that are phenotypically similar tend to have similar alleles. If such individuals mate, they are more likely to produce homozygous offspring than are mates that have been chosen randomly from the population. By the same token, phenotypically different individuals tend not to have the same alleles, so their offspring are more likely to be heterozygous. Human beings probably mate assortatively with respect to several traits, including height, weight, and intelligence. In each case, there is evidence for a positive correlation between mates.

Figure 22.6 Prezwalski's wild horses. (Courtesy of Tom Cajacob, Minnesota Zoo.)

CHANGES IN ALLELE FREQUENCIES

Populations change over time. The number of individuals in a population may increase or decrease, depending on food resources, climate, weather, and the availability of breeding areas; predators, parasites, and disease-causing organisms can also play a role. At the genetic level, a population may change because of mutation, migration, selection, and random genetic drift. These natural processes alter allele frequencies, thereby changing the fundamental makeup of the population.

Mutation

As discussed in Chapter 11, mutation changes one form of a gene into another. Typical values of the mutation rate are 10^{-5} per gene, or less, so the impact of mutation in any one generation is small. Over time, however, a considerable amount of mutational variation may accumulate.

From the standpoint of population genetics, ***mutations are classified as beneficial, harmful, or neutral,*** depending on whether they increase, decrease, or do not change the viability or fertility of their carriers. Viability and fertility are traits associated with ***fitness,*** a term that is used to describe the overall ability of an organism to survive and reproduce. It is obvious that a mutation's effect on fitness will influence its chance of persisting in a population. Many mutations are lost because they reduce the fitness of their carriers. Others are tolerated because they have only negligible effects, and still others spread through the population because they improve fitness.

In some organisms, there are strains that produce mutations at abnormally high rates. This high mutability may be due to error-prone DNA polymerases, defective mutational repair systems, or transposable elements. Such ***mutator strains*** have been used to study the consequences of increasing the mutation rate. In some cases a mutator strain is more fit than a strain with a normal mutation rate. This suggests that a higher rate is sometimes advantageous. However, the mutation rate cannot be too high, for if it is, the frequency of harmful mutations will be too great for the population to survive. A mutation rate of zero is also undesirable, since mutation is the only source of genetic variation. In nature, the mutation rate is probably adjusted so that it provides enough, but not too much, genetic variability for the population.

Migration

One population may contribute to the genetic makeup of another through migration. Figure 22.7 shows a simple case in which a large continental population donates individuals (and therefore genes) to a small island population. The rate of migration is m, which is equal to the fraction of genes on the island that are replaced by genes from the continent each generation. If q_i is the frequency of a particular allele, a, on the island, and q_c is the corresponding frequency on the continent, then after migration, the frequency on the island will be $q_i' = (1 - m)q_i + mq_c$. This follows because a fraction $(1 - m)$ of the island's genes are not replaced, and among them the frequency of a is q_i, while a fraction m of the island's genes are replaced, and among the replacements the frequency of a is q_c. The change in the frequency of a in one generation will therefore be $\Delta q_i = q_i' - q_i = m(q_c - q_i)$; Δq_i will become zero either when migration stops ($m = 0$) or when the frequency of a on the island equals the frequency of a on the continent ($q_i = q_c$).

It is clear from the example in Fig. 22.7 that the effect of migration is to make populations genetically similar. More complicated cases arise when migration goes in both directions, or when there are several populations capable of exchanging genes. Whether or not a population maintains its own genetic identity depends on how many migrants it receives. Theoretical studies have indicated that even a few migrants per generation are sufficient to eliminate the differences among geographically separated populations. Thus, migration can be a powerful homogenizing force in population genetics.

Selection

As mentioned earlier, the term fitness is used to describe the overall ability of an organism to survive and reproduce. In many populations, survival and reproductive ability are variable traits. Some individuals die before they have a chance to reproduce, whereas others leave many progeny. In a population of stable size, the average number of progeny produced by an individual is one.

Variation in fitness is partially explained by underlying genetic differences (Fig. 22.8). We have already seen that mutations can increase or decrease fitness,

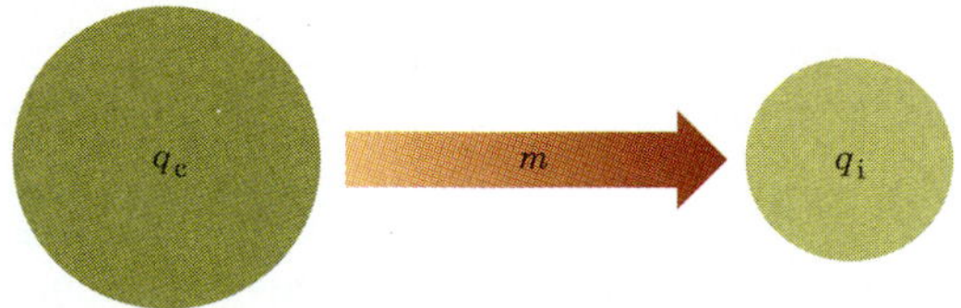

Figure 22.7 Migration between a large continental population and a small island population. The frequencies of the allele a are q_c and q_i on the continent and island, respectively; m is the migration rate.

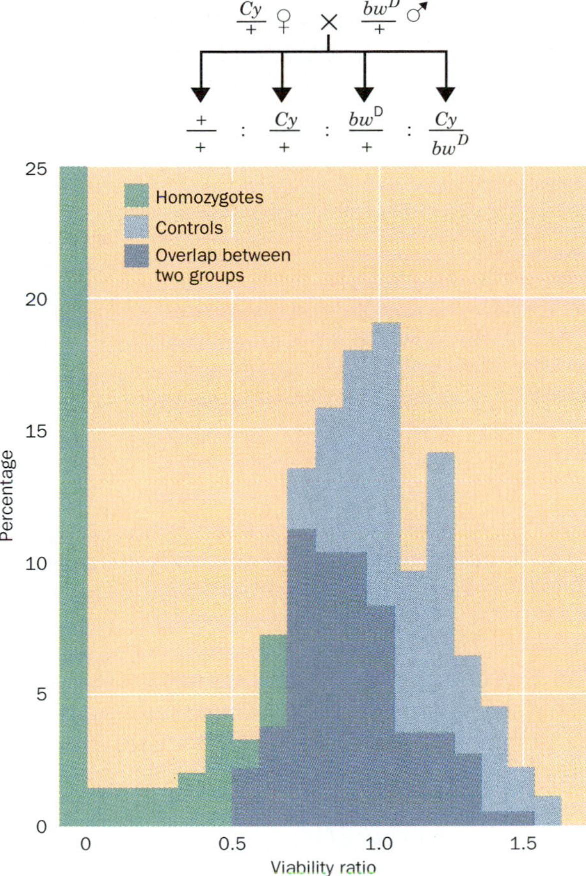

Figure 22.8 Viability of *Drosophila* homozygous for entire second chromosomes derived from a natural population. Chromosomes were manipulated according to the mating scheme shown at the top. The chromosomes that were derived from the natural population are indicated by +. *Cy* is a dominant marker that causes curly wings and bw^D is a dominant marker that causes brown eyes. The *Cy* chromosome also suppresses recombination in females and there is normally no recombination in males. In one set of crosses, *Cy*/+ females and bw^D/+ males that carried the same + chromosome were mated to produce +/+ homozygotes. The ratio $2 \times (+/+)/[(Cy/+) + (bw^D/+)]$ was used to measure the viability of the homozygotes in comparison to the curly and brown heterozygotes. Approximately 25 percent of the + chromosomes tested in this way were lethal in homozygous condition (viability ratio = 0). For controls, *Cy*/+ and bw^D/+ flies carrying different + chromosomes were mated to produce +/+ flies that were not homozygous for the entire second chromosome. Viability ratios were also calculated. From the results, it is clear that the homozygous flies were, on average, less viable than the controls. These results clearly demonstrate that wild second chromosomes can carry mutations that reduce viability in the homozygous condition. (Data from R. G. Temin, *Genetics* 53: 27–46, 1966.)

depending on their effects on viability and fertility. Combinations of mutant alleles can therefore be expected to create genotypes with different fitness values. Among these, the most fit will leave the most offspring. This differential contribution of progeny implies that the alleles that are associated with superior fitness will increase in frequency in the population. When this happens, the population is said to be undergoing ***selection***.

Figure 22.9 illustrates the power of selection in a population of *E. coli* cells. Two different strains were placed in a culture vessel with a steady supply of nutrients. Under such conditions, the two strains will multiply continuously, but each does so at its own rate. Since Fig. 22.9 shows that the Tn5 strain increased in frequency, it must have multiplied more rapidly than the other strain. Thus, in this situation, the Tn5 strain was favored by selection.

Figure 22.10 shows another example. In this case, the frequency of a dominant mutation called Lobe (*L*) was monitored in a population of *Drosophila*. In heterozygous condition, Lobe alters the shape of the eye slightly, but in homozygous condition the effect is more extreme. Figure 22.10 shows that selection decreased the frequency of the Lobe allele from 99 to 12 percent in 120 days, or 8–10 generations. The disappearance of the Lobe allele clearly illustrates the effectiveness of selection in a sexually reproducing population.

Selection Dynamics

The dynamic effects of selection can be studied by using simple quantitative models. First consider a case in which the competing genotypes are haploid and

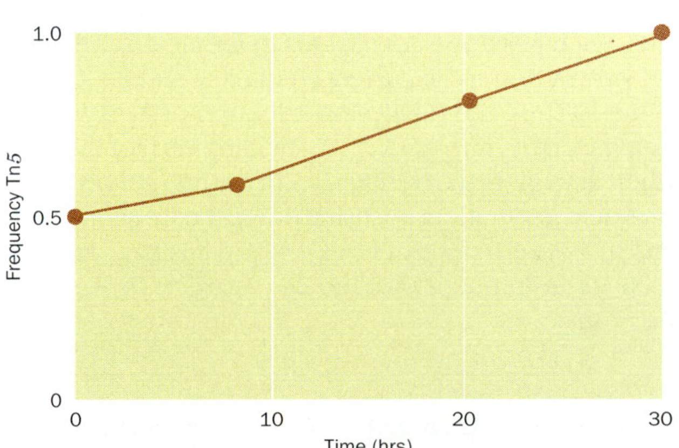

Figure 22.9 Selection in a population of *E. coli* cells. Two strains of *E. coli* were introduced at equal frequencies into a culture vessel that permitted continuous growth. One strain carried the transposable element Tn5 and the other did not. Samples were taken at various times to determine the frequencies of the two strains. The curve indicates that the Tn5 strain was favored by selection. (Data from S. W. Biel and D. L. Hartl, *Genetics* 103: 581–592, 1983.)

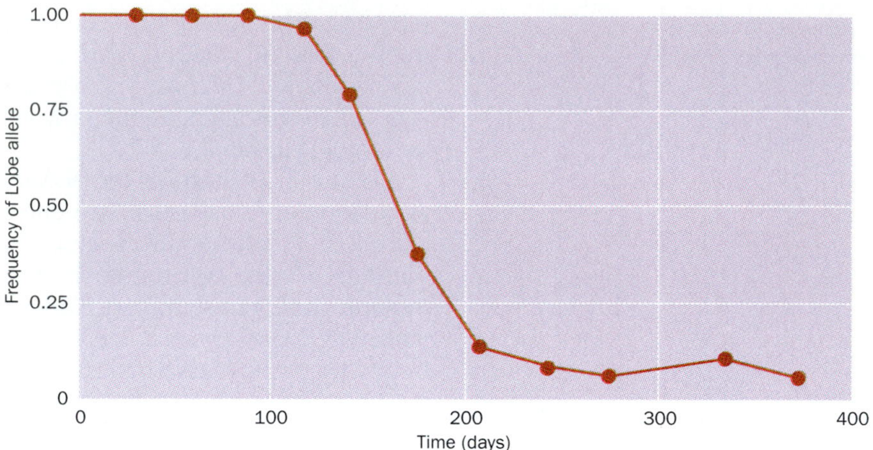

Figure 22.10 Selection in a population of *Drosophila*. A single wild-type male was introduced into a population that was homozygous for the Lobe mutation (L). This mutation affects the shape of the eyes. L/L, $L/+$, and $+/+$ flies are all phenotypically distinguishable. Periodically, samples of flies were taken from the population to determine the frequency of the Lobe allele. (Data from D. J. Merrell, *Genetics* 52: 162–189, 1965.)

reproduction is strictly asexual. Table 22.6 presents the situation.

The frequencies of the two genotypes, A_1 and A_2, are p_1 and p_2, respectively. To determine how these change over time, it is necessary to know the ***relative fitnesses*** of the two competing genotypes. For convenience, one fitness can be used as a standard and the other can be written as a deviation from it. This is done in Table 22.6, where the fitnesses of A_1 and A_2 are 1 and $1 - s$, respectively. The deviation s is called the ***selection coefficient***, and it measures the difference between the relative fitnesses of the two genotypes. If s is positive, A_2 has the lower fitness, whereas if s is negative, it has the higher fitness. In this model, the absolute value of s is assumed to be less than one.

Each genotype contributes to the next generation in proportion to its current frequency and its relative fitness. Thus, the relative contribution of A_1 is $p_1 \times 1$, and that of A_2 is $p_2 \times (1 - s)$. Dividing each of these by their sum gives the frequencies of the two genotypes in the next generation. Focusing on A_1, the frequency will be $p_1' = p_1/[p_1 + p_2(1 - s)] = p_1/[1 - sp_2]$, and the change in frequency will be $\Delta p_1 = p_1' - p_1 = sp_1p_2/[1 - sp_2]$.

This equation reveals that the direction of Δp_1 is determined solely by s; $s > 0$ implies $\Delta p_1 > 0$ and $s < 0$ implies $\Delta p_1 < 0$. However, the size of Δp_1 depends on the current genotype frequencies, as well as on s. In addition, notice that selection is not very effective when either of the genotypes is rare. This last feature follows from the term p_1p_2, which is small if either p_1 or p_2 is small.

A quantitative model of selection dynamics can also be developed for sexually reproducing, diploid organisms. Table 22.7 shows the analysis for a single locus with two alleles under the assumption of random mating. At fertilization, the frequencies of the genotypes can be predicted from the allele frequencies by the Hardy–Weinberg method. However, during development, the allele frequencies can change as a result of differences in survival among the genotypes. Further changes are possible if there are fertility differences among the surviving adults. The combined effects of these changes can be calculated by using the relative fitnesses of the three genotypes. For convenience, the A_1A_1 homozygotes have been chosen as a standard, so their relative fitness is 1. The fitnesses of the other genotypes have been written as deviations from this by

TABLE 22.6 Haploid Selection Dynamics

Genotypes	A_1	A_2
Frequencies	p_1	p_2
Relative fitnesses	1	$1 - s$
Relative contribution to the next generation	$p_1 \times 1$	$p_2 \times (1 - s)$
Frequencies after selection	$p_1' = p_1/[1 - sp_2]$	$p_2' = p_2(1 - s)/[1 - sp_2]$
Change in frequency	$\Delta p_1 = p_1' - p_1$ $= sp_1p_2/[1 - sp_2]$	$\Delta p_2 = p_2' - p_2$ $= -sp_1p_2/[1 - sp_2]$

TABLE 22.7 Diploid Selection Dynamics

Genotypes	A_1A_1	A_1A_2	A_2A_2
Hardy-Weinberg frequencies	p_1^2	$2p_1p_2$	p_2^2
Relative fitnesses	1	$1 - hs$	$1 - s$
Relative contribution to the next generation	$p_1^2 \times 1$	$2p_1p_2 \times (1 - hs)$	$p_2^2 \times (1 - s)$
Average fitness	$\overline{W} = p_1^2 + 2p_1p_2(1 - hs) + p_2^2(1 - s)$		
Frequency of A_1 in next generation	$p_1' = [p_1^2 + (½)2p_1p_2(1 - hs)]/\overline{W} = p_1(1 - p_2hs)/\overline{W}$		
Change in frequency of A_1	$\Delta p_1 = p_1' - p_1 = sp_1p_2(2hp_1 + p_2 - h)/W$		

using selection coefficients. The selection coefficient s measures the deviation between the fitnesses of A_2A_2 and A_1A_1. This coefficient also measures the deviation between the fitnesses of A_1A_2 and A_1A_1, but an adjustment has been included to account for the possibility of dominance. If A_2 is completely dominant to A_1, the fitness of the heterozygote is the same as that of the A_2A_2 homozygote. If A_2 is completely recessive, the fitness of the heterozygote is the same as that of the A_1A_1 homozygote. These two possibilities are accommodated by the quantity h, which measures the expression of A_2 in heterozygous condition. Thus, $h = 1$ corresponds to the case in which A_2 is completely dominant to A_1, whereas $h = 0$ corresponds to the case in which A_2 is completely recessive. A value of h between 0 and 1 means that A_2 has intermediate expression in the heterozygotes.

As in the haploid model, each genotype will contribute to the next generation in proportion to its frequency and relative fitness. The new allele frequencies can then be obtained by summing the contributions of the appropriate genotypes. For example, for A_1, we sum the contributions of A_1A_1 and A_1A_2, but weight the contribution of the heterozygotes by ½ since only half of their alleles are A_1. This sum is then divided by the sum of the contributions of all the genotypes in order to obtain the frequency of A_1 in the next generation. Thus

$$p_1' = \frac{[p_1^2 \times 1 + ½ \times 2p_1p_2 \times (1 - hs)]}{[p_1^2 \times 1 + 2p_1p_2 \times (1 - hs) + p_2^2 \times (1 - s)]}$$

The denominator in this expression is called the average relative fitness; usually it is symbolized by $\overline{W}$. Thus, after simplifying the numerator, we have $p_1' = p_1(1 - p_2hs)/\overline{W}$, and the change in the frequency of A_1 in one generation of selection is $\Delta p_1 = p_1' - p_1 = sp_1p_2[2hp_1 + p_2 - h]/\overline{W}$.

Figure 22.11 illustrates three special cases of selection in sexually reproducing diploids. In all three, s has been set at 0.05, so there is a 5 percent difference between the fitnesses of the two types of homozygotes. The heterozygotes have a fitness that is equal to the fitness of one of the homozygotes, or a fitness that is exactly in between, depending on the value of h.

The case of $h = 0$ illustrates selection against a harmful recessive allele (A_2). Initially, the population has many such alleles but as selection acts on the recessive homozygotes, the frequency of the dominant allele (p_1) increases. After about 500 generations, the allele frequency changes very slowly. This happens because most of the harmful alleles are now carried in heterozygotes, where they are not subject to elimination by selection.

The case of $h = 1$ illustrates selection against a harmful dominant allele (A_2). At the start, this allele is very common, but after a considerable lag, the recessive allele (A_1) takes over. The slow change at the beginning occurs because the favored allele is found mainly in heterozygotes, where it has no selective advantage.

The remaining graph shows what happens when $h = 0.5$, that is, when the heterozygote has a fitness exactly in between the fitnesses of the two homozy-

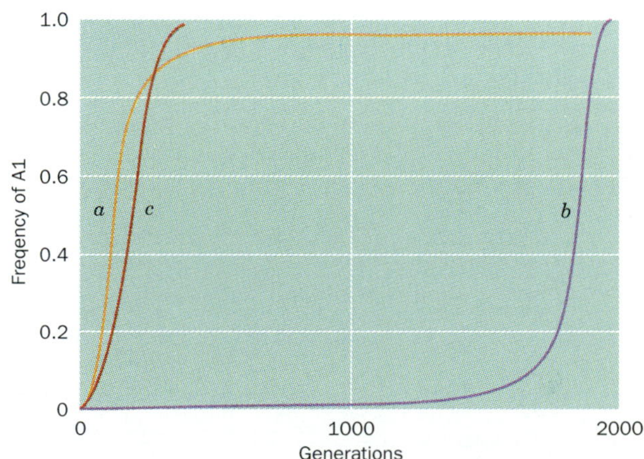

Figure 22.11 Diploid selection dynamics. Frequency of the A_1 allele in three different cases with the selection coefficient $s = 0.05$. (a) Selection against a recessive allele (A_2); $h = 0$. (b) Selection against a dominant allele (A_2); $h = 1$. (c) Selection against an allele (A_2) with partial expression in heterozygotes; $h = 0.5$.

gotes. Starting with a low frequency of the favored allele (A_1), the selection curve rises rather rapidly with an inflection point in the middle. When $h = 0.5$, the process of selection actually works in two phases. In the first phase, most of the competition is between A_1A_2 and A_2A_2 individuals because the A_1 allele is too rare to contribute many homozygotes; however, selection favors A_1 because the A_1A_2 heterozygotes are more fit than the A_2A_2 homozygotes. After a while, A_1 becomes frequent enough to produce A_1A_1 homozygotes and a second phase of selection begins. In this phase, most of the competition is between A_1A_1 and A_1A_2. The A_1A_1 homozygotes are more fit, so the frequency of the A_1 allele continues to increase and eventually it approaches 1.

Random Genetic Drift

During reproduction, allele frequencies can change by chance. Consider, for instance, two human beings heterozygous for the recessive allele that causes phenylketonuria (Fig. 22.12). If they mate and produce a single child, there is a ¼ chance that the child will have the disease. There is also a ½ chance that the child will be heterozygous like its parents, and finally, there is a ¼ chance that it will be homozygous for the normal allele. In the first case, the frequency of the harmful allele has increased from 0.5 in the parents to 1.0 in the child, in the second case it has stayed the same, and in the third case, it has decreased from 0.5 to 0. Overall, the probability of a change in the allele frequency is 0.5.

Such random changes can occur in each family in the population. ***The cumulative effect of these statistical fluctuations is a process called random genetic drift.*** Over many generations, drift can lead to significant changes in the frequency of an allele in the population.

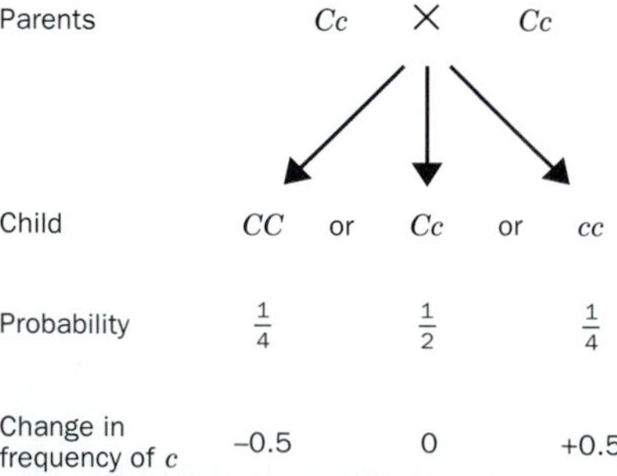

Figure 22.12 Random changes in the frequency of the allele for phenylketonuria in a family where the parents are both heterozygous.

Effects of Random Drift

All the effects of random genetic drift have one feature in common; they involve a ***loss of genetic variability.*** As populations lose their variability, they become more homozygous and, consequently, less heterozygous. The effects of drift can therefore be studied by monitoring the reduction in the frequency of heterozygotes.

The rate of loss of heterozygotes depends on the population size. To see this, consider a single locus with two alleles, A and a, in a population of N individuals. For simplicity, assume that N is constant, that mating is random, and that self-fertilization is allowed. Let H_t be the frequency of heterozygotes in the population at generation t. Since drift reduces the frequency of heterozygotes at a steady rate, we can write $H_t = (1 - c)H_{t-1}$. In this equation, c is the rate at which heterozygotes are lost.

The value of c depends on the probability that two genes in generation t are copies of the same gene in generation $t - 1$. If they are, then obviously they cannot form a heterozygote, since the two genes in a heterozygote cannot be identical by descent. With N individuals, there are $2N$ genes, so the probability that two genes in any individual are direct copies of each other is $1/(2N)$. This follows from the fact that each gene has the same chance of getting into a sperm and an egg and the two kinds of gametes unite randomly to form the next generation (Fig. 22.13). Thus $c = 1/(2N)$ and $H_t = [1 - 1/(2N)]H_{t-1}$. This is a mathematical recursion relationship between H_t and H_{t-1}. What is true of H_t and H_{t-1} is also true of H_{t-1} and H_{t-2}, so $H_{t-1} = [1 - 1/(2N)]H_{t-2}$; substituting the right-hand side of the second equation into the right-hand side of the first, we have $H_t = [1 - 1/(2N)]^2H_{t-2}$, and so on, leading to $H_t = [1 - 1/(2N)]^tH_0$.

This equation shows that in each generation, random genetic drift reduces the frequency of heterozygotes by a fraction 1/(2N). Eventually, the frequency of heterozygotes will approach zero and the population will be completely homozygous. Notice that the rate of loss of genetic variability depends only on the population size. Since this rate is inversely proportional to the size, small populations are expected to lose variation more rapidly than large ones (Fig. 22.14). This loss of variability is sometimes referred to as the ***inbreeding effect of a finite population;*** however, the word inbreeding is not used in the same sense that it was earlier, since in this situation, mating is completely random. In this case, the outcome of the random mating process is similar to that of inbreeding because members of a finite population can share genes that are identical by descent. More complicated mating systems, such as those with sepa-

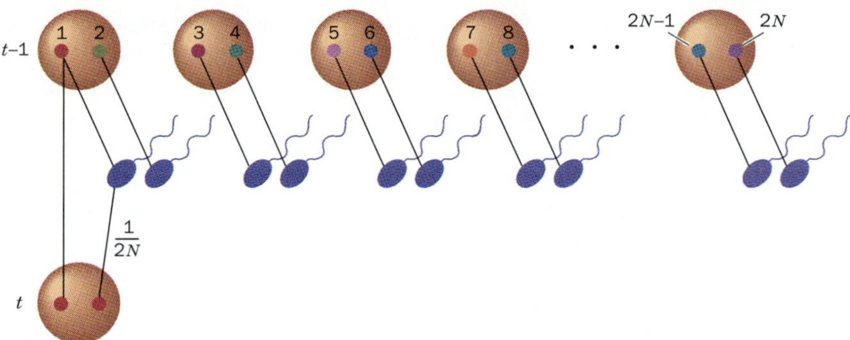

Figure 22.13 The probability that two genes are direct copies of each other in a randomly mating population of N individuals. Each individual is represented by a circle and each gene is identified by a number. The probability that two genes in an individual in generation t are direct copies of a gene in generation $t - 1$ is $1/(2N)$. This can be seen by noting that an egg that carries a copy of a particular gene, for example, gene 1, has a chance of $1/(2N)$ of being fertilized by a sperm that carries a copy of the same gene.

rate sexes, lead to a slightly different mathematical formula, but the basic result is the same.

Another way of studying the effects of drift is to consider the fate of individual alleles. Over time, ***allele frequencies can change randomly***, sometimes increasing and sometimes decreasing (Fig. 22.15). In the absence of other factors, such as mutation, migration, and selection, the cumulative effects of these random changes will produce a state of complete genetic uniformity. When this state is reached, only one allele will remain at each locus in the population.

The chance that a particular allele will become fixed as the only allele at a locus is equal to its current frequency. This can be seen by noting that eventually all the genes at a locus will be descended from a single gene present in the current population. In the absence of mutation, migration and selection, every gene should have the same chance of becoming this great ancestor. Since there are N individuals and $2N$ genes, the chance that any one of the genes will become the great ancestor is $1/(2N)$. Suppose that the frequency of the A allele is p and the frequency of the a allele is q. Then $2N \times p$ of the genes are A, and $2N \times q$ are a, and the chance that one of the A alleles will become the great ancestor is simply $2N \times p \times 1/(2N) = p$. This is the probability of fixation for A; likewise, the probability of fixation for a is $2N \times q \times 1/(2N) = q$.

Fixation (or loss) by random genetic drift is the expected fate of mutations at loci with little or no effect on fitness. Such ***selectively neutral loci will evolve at a rate equal to the mutation rate.*** This follows di-

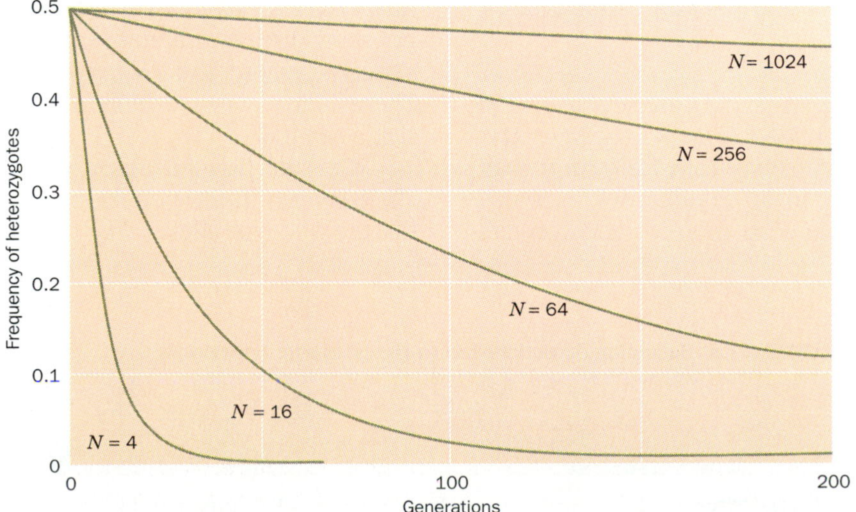

Figure 22.14 Loss of heterozygotes by random genetic drift in populations of different sizes. In each case, the population size is given. Since $p = q = 0.5$, the initial frequency of heterozygotes is $2pq = 0.5$.

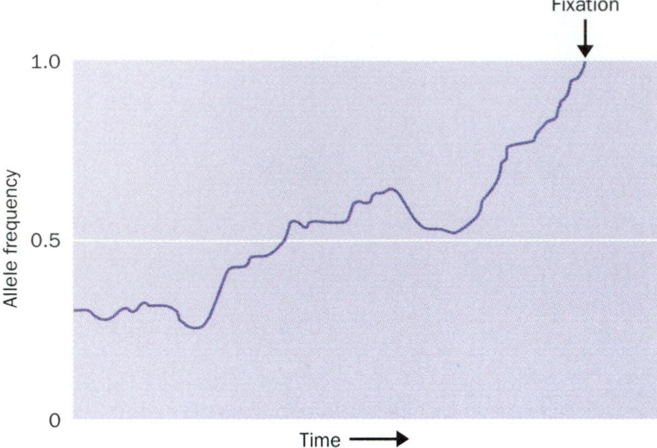

Figure 22.15 Random changes in allele frequency due to genetic drift. When the frequency equals one, the allele has been fixed in the population.

rectly from the principle that the probability of fixation of a neutral allele is equal to its current frequency. In the case of a new mutation, the frequency is simply the mutation rate.

GENETIC EQUILIBRIUM

The genetic characteristics of a population are determined by the combined action of the mating system, mutation, migration, selection, and drift. Sometimes these factors act in opposite ways, leading to a state of genetic equilibrium. ***At equilibrium, the characteristics of a population do not change.*** In the following sections, we consider situations that can lead to such a state.

Balancing Selection

Suppose that two alleles, A_1 and A_2, are segregating in a population, and that ***selection favors the heterozygotes.*** This situation, sometimes called ***selection with overdominance***, leads to an equilibrium in which both alleles are maintained at more or less constant frequencies. The situation is outlined in Table 22.8. Because the heterozygotes are favored, they are given a fitness of one; the other fitnesses, representing the two kinds of homozygotes, are written as deviations from one using the selection coefficients s and t. After selection, the frequency of A_1 is $p_1' = p_1(1 - sp_1)/\overline{W}$, where $\overline{W} = 1 - sp_1^2 - tp_2^2$. From this, we can compute the change in allele frequency due to one generation of selection: $\Delta p_1 = p_1' - p_1 = p_1 p_2 [tp_2 - sp_1]/\overline{W}$. At equilibrium, $\Delta p_1 = 0$, implying that either $p_1 = 0, p_2 = 0$ or $tp_2 - sp_1 = 0$. Under this last condition, $tp_2 = sp_1$, so, noting that $p_1 = 1 - p_2$, we have $p_1 = t/(s + t)$.

This value of p_1 is the frequency of A_1 at genetic equilibrium. Notice that it is strictly a function of the selection coefficients. Thus, if the A_2A_2 homozygotes are lethal ($t = 1$), and the A_1A_1 homozygotes are 50 percent as fit as the heterozygotes ($s = 0.5$), $p_1 = 1/(0.5 + 1) = 2/3$. By subtraction, $p_2 = 1/3$.

This example demonstrates that ***when selection favors the heterozygotes, both of the alleles will remain in the population even though they are deleterious in homozygous condition.*** Such a state of genetic balance is called a ***balanced polymorphism*** and the form of selection that creates it is called ***balancing selection.*** Detailed studies of human populations in Africa have shown that this form of selection operates on the gene that makes the beta polypeptide of hemoglobin. In some African populations, a mutant allele (Hb_β^S) of this gene is found at frequencies as high as 0.1. Homozygotes for this allele suffer from a severe form of anemia in which the hemoglobin crystallizes in the blood, distorting the shape of the red blood cells. Sometimes the cells appear sickle-shaped (Fig. 22.16), giving the disease the name ***sickle-cell anemia.*** Heterozygous carriers of the mutant allele do not show disease symptoms unless they travel to high altitudes, where the oxygen concentration is low. These individuals are therefore said to possess the sickle-cell trait.

With such harmful effects, why does the sickle-cell allele remain in the population at all? The answer is that the wild-type allele (Hb_β^A) is also associated with a

TABLE 22.8 Balancing Selection due to Heterozygote Superiority

Genotypes		A_1A_1	A_1A_2	A_2A_2
Hardy-Weinberg frequencies		p_1^2	$2p_1p_2$	p_2^2
Relative fitnesses		$1 - s$	1	$1 - t$
Average fitness	$\overline{W}$	$= p_1^2(1 - s) + 2p_1p_2 + p_2^2(1 - t)$ $= 1 - sp_1^2 - tp_2^2$		
Frequency of A_1 in next generation	p_1'	$= [p_1^2(1 - s) + p_1p_2]/\overline{W}$ $= p_1(1 - sp_1)/\overline{W}$		
Change in frequency of A_1	Δp_1	$= p_1' - p_1$ $= p_1p_2(tp_2 - sp_1)/\overline{W}$		

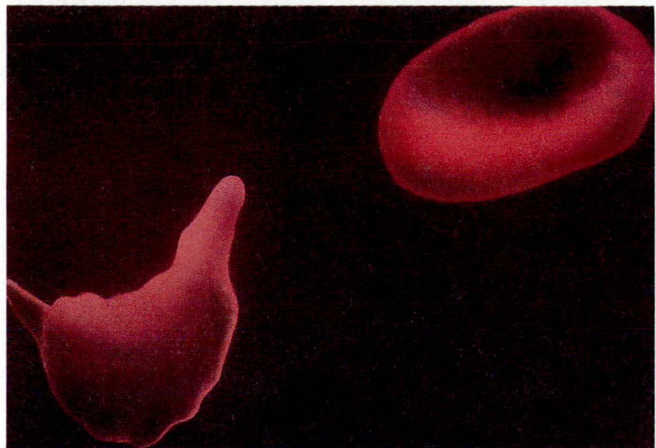

Figure 22.16 A comparison between normal red blood cells and a red blood cell from an individual with sickle-cell anemia. The cells are magnified several thousandfold in this scanning electron micrograph. (Photo by R. Roseman/Custom Medical Stock Agency.)

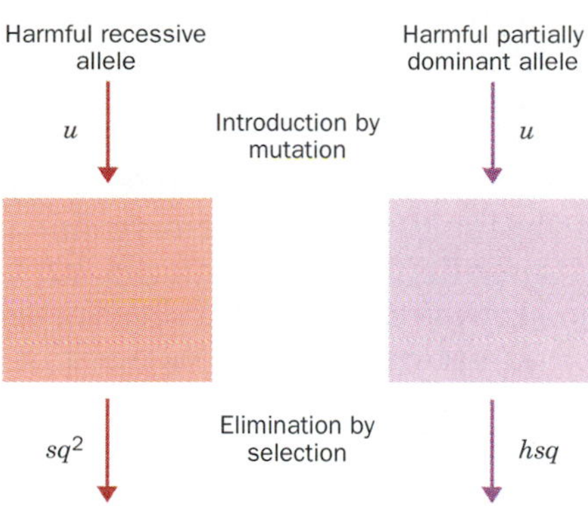

Figure 22.17 Mutation–selection balance. The population, represented by the rectangle, reaches genetic equilibrium when the introduction of harmful alleles by mutation is balanced by the elimination of these alleles through selection.

reduction in fitness. Epidemiological studies have shown that the wild-type homozygotes are more susceptible to infection by malaria parasites than the $Hb_\beta{}^S$ $Hb_\beta{}^A$ heterozygotes are. As a result, both kinds of homozygotes in this system are less fit than the heterozygotes.

If we assume that 0.1 is the equilibrium frequency of the sickle-cell allele, we can calculate the relative fitnesses of the three genotypes. At equilibrium, $0.1 = t/(s + t)$, where t is the selection coefficient for the $Hb_\beta{}^A$ $Hb_\beta{}^A$ homozygotes and s is the selection coefficient for the $Hb_\beta{}^S Hb_\beta{}^S$ homozygotes. Sickle-cell anemia is fatal early in life, so the fitness of the $Hb_\beta{}^S Hb_\beta{}^S$ homozygotes is essentially zero, and $s = 1$. Consequently, we can solve the equation for t, which is the only unknown. $0.1 = t/(1 + t)$, so $t = 0.1/0.9 = 0.11$. Thus, the $Hb_\beta{}^A$ $Hb_\beta{}^A$ homozygotes are about 11 percent less fit than the $Hb_\beta{}^S$ $Hb_\beta{}^A$ heterozygotes.

Mutation–Selection Balance

Harmful alleles are introduced into a population by mutation from the wild-type allele. If they are recessive, these harmful alleles will be eliminated by selection against recessive homozygotes, but if they are partially dominant, they will be eliminated mainly by selection against heterozygotes. In either case, ***mutation–selection equilibrium will be achieved when the rate of introduction is balanced by the rate of elimination*** (Fig. 22.17).

The equilibrium frequency of a harmful recessive allele can be obtained by noting that the rate of elimination by selection is approximately sq^2. This is so because selection acts only against the recessive homozygotes, which under random mating have a frequency q^2, and the intensity of selection is s (see Table 22.7). If the mutation rate is u, the rate of introduction of the recessive allele will be up, which is essentially u since p will be very close to one. Thus, at equilibrium, rate of introduction = rate of elimination implies that $u = sq^2$ so $q^2 = u/s$. The equilibrium frequency of the recessive allele will therefore be $q = \sqrt{u/s}$.

For a partially dominant allele, the rate of elimination can be calculated if we can assume that selection acts mainly against the heterozygotes. With random mating, the heterozygote frequency will be $2pq$, which is approximately $2q$ since p will be very close to one. The strength of selection against the heterozygotes is measured by hs (see Table 21.7). However, only half of the genes that are eliminated in the heterozygotes are harmful. Putting all this together, the rate of elimination by selection is approximately equal to $\frac{1}{2} \times hs \times 2q = hsq$. As before, the rate of introduction by mutation is approximately the mutation rate, u, so at equilibrium, rate of introduction = rate of elimination implies that $u = hsq$. The frequency of the harmful allele will therefore be $q = u/(hs)$.

As an example, consider the population frequencies of lethal alleles on the major autosomes of *Drosophila*. Surveys have indicated that the average frequency of such alleles in natural populations is 2.2×10^{-4}. From laboratory experiments, we know that the per-gene mutation rate for lethals is about 3.0×10^{-6} mutations per generation. If these lethals were completely recessive, the equilibrium frequency would be $\sqrt{u/s} = \sqrt{3 \times 10^{-6}} = 0.0017$, since for lethals $s = 1$. However, this is almost eight times the observed frequency, casting doubt on the assumption of complete recessivity. It seems more likely that the lethals are partially dominant and that they are eliminated primarily through their heterozygous effects. Using the observed equilibrium frequency and the measured mutation rate, we can calculate the magnitude of this effect: $q = u/(hs)$ implies that $h = u/(qs)$, so $h = 3 \times 10^{-6}/2.2 \times 10^{-4} = 0.014$. Thus, lethals appear to reduce the fitness of their heterozygous carriers by about 1.4 percent. This may seem slight, but it is large enough to keep the equilibrium frequency well below that of a completely recessive lethal mutation.

Mutation–Drift Balance

We have seen that random genetic drift eliminates variability from a population. Without any counteracting forces, this process would eventually make all populations completely homozygous. However, ***mutation replenishes the variation that is lost by drift, leading to a state of genetic balance.*** To investigate this balance, consider the situation diagrammed in Fig. 22.18. The population has been divided into heterozygotes and homozygotes, with frequencies H and $1 - H$, respectively. Suppose that the size (N) of the population is fixed, that mating is random, that mutations occur at rate u, and that every mutation creates a new allele. (This last condition will be approximately correct because of differences in molecular structure.) Under the influence of genetic drift, some of the heterozygotes in the population will be lost to the homozygote fraction. As discussed earlier, the rate of loss will be approximately $1/(2N) \times H$ in each generation. At the same time, any mutation that occurs must produce a heterozygote in the next generation. Since the rate of mutation per gene is u, and since each homozygote carries two genes at a locus, the total rate at which heterozygotes will be gained by mutation among the homozygotes is $2u \times (1 - H)$. At equilibrium, the loss of heterozygotes by drift will be balanced by the gain from mutation, so we have $1/(2N) \times H = 2u \times (1 - H)$, which implies that $H = 4Nu/(4Nu + 1)$. This is the equilibrium frequency of heterozygotes in a population where mutation and drift have come into balance.

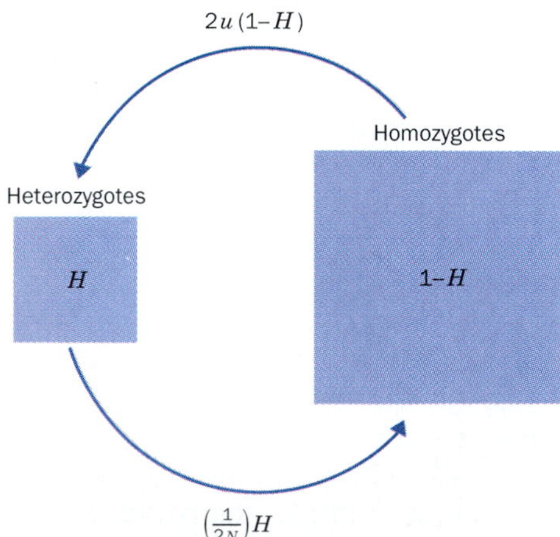

Figure 22.18 Mutation–drift balance in a population. The frequencies of the heterozygotes and homozygotes are H and $1 - H$, respectively. Heterozygotes are lost to the homozygote fraction by drift at a rate of $1/(2N) \times H$ per generation. At the same time, heterozygotes are gained from the homozygote fraction by mutation at a rate of $2u \times (1 - H)$. When these two rates are equal, $H = 4Nu/(4Nu + 1)$ and the population is at equilibrium.

Population surveys using protein electrophoresis have provided estimates of H for many organisms, including human beings. A representative value for the human population is 0.12. Assuming that $u = 10^{-6}$, we can solve for N, the size of the human population: $N = H/[4(1 - H)]u = 0.12/(4 \times 0.88 \times 10^{-6}) = 34{,}091$. Taken at face value, this result is completely ridiculous; the human population is obviously very much larger than 34,091. However, the N that is estimated from this formula is what population geneticists call the ***effective size,*** sometimes symbolized N_e. The effective size is the genetically relevant size, often much less than the actual census size. The reason is that there are often restrictions on mating and reproduction, and these lead to more identity by descent than would otherwise occur. Bottlenecks in the population size are particularly significant, especially if they occur repeatedly during the population's history. This has probably been the case for the human population, in which fluctuations in size and the partial isolation of clans and tribes have created an inbreeding effect. Unequal contributions of the two sexes can also contribute to this effect. This factor is seen clearly in dairy cattle herds where there are frequently many females but few males. This inequality means that half of the genes that are contributed in each generation will be derived from only a few individuals. Consequently, the genetically effective size is much less than the actual population size.

EVOLUTIONARY GENETICS

Evolution is one of the unifying ideas in biology. It emerged in the middle of the nineteenth century, principally through the work of ***Charles Darwin.*** After careful observation and study, Darwin proposed that all living things were descended from a common ancestor and that the differences among them were due to changes that had accumulated slowly over a long period of time. To explain these changes, Darwin invoked a mechanism that he called ***natural selection.***

According to Darwin, not all organisms had the same chances for survival and reproduction. Those with the best chances would leave the greatest numbers of progeny, meaning, of course, that their characteristics would tend to spread in the population. After many generations, such a process might bring about the physical transformation of the species.

Darwin's theory had one glaring defect; there was no satisfactory explanation for heredity. Although Darwin and Mendel were contemporaries, they never met and there is no evidence that Darwin ever read Mendel's paper. Consequently, it was necessary for Darwin's followers to incorporate Mendel's discoveries into the Darwinian theory. This was done in the twentieth century in the decades following the rediscovery of Mendel's work. The principal architects of the new synthetic theory were R. A. Fisher, J. B. S. Haldane, and S. Wright. Numerous other scientists, such as T. Dobzhansky, E. Mayr, H. J. Muller, G. G. Simpson, G. L. Stebbins, and A. H. Sturtevant, also made significant contributions.

The Synthetic Theory of Evolution

The synthetic theory of evolution provides explanations for the transformation of a species by natural selection and for the splitting of a species into reproductively isolated subgroups. The latter process is called ***speciation***. In the synthetic theory, the most fundamental evolutionary event is a change in the frequency of an allele in a population. The body of this theory is therefore built on an analysis of allele frequency changes. All the factors that affect allele frequencies—mutation, migration, selection, and random genetic drift—are taken into account.

One of the most comprehensive analyses was carried out by ***S. Wright***. In Wright's model, each population of a species is symbolized by a point on a topographic map (Fig. 22.19). The contours of the map represent different levels of adaptation to the environment. Populations at high elevations are more adapted than populations at low elevations, so the map depicts the ***adaptive landscape*** of the species. At any one

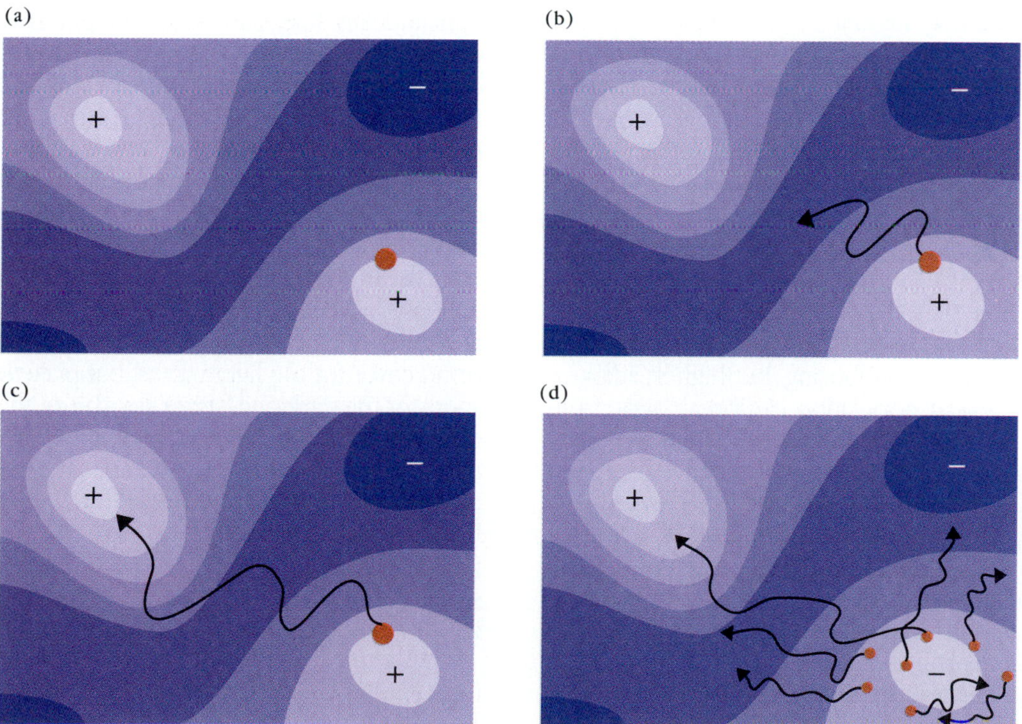

Figure 22.19 Wright's shifting balance theory of evolution. The adaptive landscape is indicated by contour lines; the + and − signs identify adaptive peaks and nonadaptive pits, respectively. The region between the two peaks is a shallow valley. The population is indicated by a red circle. (a) The population hovers around a point of genetic equilibrium near an adaptive peak. (b) Random genetic drift causes the population to move out of its equilibrium position into the valley between the two peaks. (c) Further random changes create a new combination of alleles that is favored by selection; consequently, the population moves up the higher adaptive peak. (d) The shifting balance process in a large, but subdivided, population. Each red circle represents a deme.

time, the position of a population will depend on its genetic makeup. Populations with alleles that improve fitness will be at a higher elevation than populations without these alleles. Consequently, as the genetic makeup of a population changes, so will its position on the adaptive landscape. Mutation and migration will provide a population with genetic variation. Natural selection will sift through this variation, preserving the beneficial mutations and eliminating the harmful ones. As it does this, selection will tend to drive the population uphill. By contrast, random genetic drift will move the population in an unpredictable fashion. The effect of all these forces—mutation, migration, selection, and random genetic drift—will bring the population to a state of genetic equilibrium, corresponding to a point near a peak on the adaptive landscape. In actuality, however, the population will hover around this point because of the fluctuations caused by genetic drift.

In Wright's theory, evolution occurs when a population shifts from one equilibrium position to a higher one. This will happen if the population drops down an adaptive peak, crosses a shallow valley, and scales a neighboring peak of higher adaptive value. Since the initial and final stages of this process involve a balance among several genetic forces, Wright has called this the ***shifting balance theory***.

Genetic drift plays a crucial role in the shift from one equilibrium to another. The random changes brought about by drift can provide a population with the opportunity to test out new combinations of alleles. Many of these will lower fitness, moving the population out of its equilibrium position into a valley on the adaptive landscape. If additional random changes create a favorable allelic combination—what Wright calls a new interaction system—selection will move the population back uphill to another equilibrium point, possibly on a neighboring, but higher, adaptive peak. In this way, a population can break away from one equilibrium position and move to a higher one.

The problem with this theory is that it places great emphasis on random changes, and these are not likely in large populations. To overcome this difficulty, Wright has proposed that evolution occurs in populations that are subdivided into small, partially isolated breeding units called ***demes***. A species might consist of hundreds or even thousands of demes, each represented by a point on the adaptive landscape. Under such conditions, the evolution of the species would depend on the operation of the shifting balance process in each of the demes. As the demes shift from one equilibrium to another, some of them will move to higher adaptive peaks. These populations will then become the most fit, and through migration will spread their highly adaptive interaction systems to other demes in the species. In this way, the whole species will tend to move toward the adaptive peak occupied by its most fit demes.

Evidence for Adaptive Evolution

Some of the best evidence for adaptive evolution comes from species that live in disturbed habitats. In these cases, human activity has altered the environment and there has been a corresponding change in the species. The peppered moth, *Biston betularia*, is a good example. This species is found in wooded areas in Great Britain, where it exists in two color forms, light and dark (see Fig. 1.5). The difference between the two forms involves a single gene, with the dark color allele being dominant. Since 1850, the frequency of the dark form has increased in certain areas in England, particularly in the industrialized Midlands section of the country. Around heavily industrialized cities, such as Manchester and Birmingham, the frequency of the dark form has changed dramatically, from 1 to 90 percent in less than 100 years. In other areas of England, where there is little industrial activity, the dark form has remained very rare.

The rapid spread of the dark form in industrialized areas has been attributed to natural selection. Both light and dark forms are active at night. During the day, the moths remain still, resting on tree trunks and other objects in the woodlands. Since birds may find the moths and eat them during this resting period, camouflage is their only defense against predation. On white or gray tree bark, the light moths are protectively colored, especially if the bark is overgrown with lichens. This is the normal condition in undisturbed habitats. In industrialized areas, however, most of the lichens have been killed by pollution and the tree bark is often blackened by soot. Such conditions offer little or no cover for the light moths, but make ideal resting spots for the dark ones. Predatory birds have a difficult time seeing the dark moths on sooty bark, just as they have a difficult time seeing light moths on lichen-covered bark. Experimental tests by H. B. D. Kettlewell have shown that the two types of moths are selectively favored in these different environments. Thus, the spread of the dark moth appears to have resulted from its selective advantage on a sooty background.

Another example of adaptive evolution comes from studies with the grass *Agrostis tenuis*. This species grows in pastures in Great Britain, where it competes with many other species; however, it is also found around abandoned lead, copper, and zinc mines, where few other species grow. The soil in the vicinity of these mines contains high concentrations of heavy metals. Experimental tests have shown that *Agrostis* plants growing on the contaminated soils are resistant to heavy metal poisoning. Other *Agrostis* plants growing in uncontaminated habitats have no such resistance. This difference indicates that *Agrostis tenuis* has been subdivided into locally adapted races, or ***ecotypes***. The genetic basis of this differentiation is not completely understood, but dominant alleles for metal tolerance seem to be involved. Detailed studies by J. Antonovics, A. D. Bradshaw, S. K. Jain, and T. McNeilly have shown that even though there is hybridization between tolerant and nontolerant populations, natural selection preserves their separate identities.

Molecular Evolution

Evolution is a process of change. At the molecular level, this process involves the insertion, deletion, or substitution of nucleotides in the DNA. If the DNA encodes a polypeptide, these events may cause a change in the amino acid sequence. Over time, such changes can accumulate, leading to a molecule that bears little resemblance to its progenitor. Recent advances in molecular biology have made it possible to determine the nucleotide sequences of DNA and the amino acid sequences of polypeptides. By comparing related sequences, the molecular details of evolution can be analyzed.

Fig. 22.20 shows the results of an analysis of the amino acids in the α-polypeptide chain of vertebrate hemoglobin. All the organisms listed in Fig. 22.20 are derived from an evolutionary line that appeared around 500 million years ago. Since that time, this line has split into many separate branches, each giving rise to a different vertebrate group. Five of the major classes—the cartilaginous fishes, the bony fishes, the amphibians, the marsupials, and the placental mammals—are represented by the branches in the figure. This figure also gives the approximate time for the origin of each branch.

The data in Fig. 22.20 show the number of amino acids that are different when human α-globin is compared to each of the other α-globin molecules. Altogether, there are 141 amino acids in the α-globin chain. The shark α-globin is the most unlike the human α-globin, followed by the carp, newt, kangaroo, and cow α-globins, in that order. Notice that the degree of difference depends on the time that has elapsed since the evolutionary lines diverged from a common ancestor.

The relationship between the degree of difference (D) and the evolutionary time (T) can be used to estimate the rate at which α-globin has evolved. Consider, for instance, the human and carp evolutionary lines. The common ancestor of these organisms existed about 400 million years ago, so the total evolutionary time that has elapsed since the two lines diverged is 800 million years. During this time, human and carp α-globins have accumulated differences in 68 amino acid sites. The degree of difference is therefore $68/141 = 0.482$ differences per amino acid site, suggesting that α-globin has evolved at a rate of $D/T = 0.482$ amino acid substitutions per site over 800 million years, or 0.6×10^{-9} substitutions at each site per year.

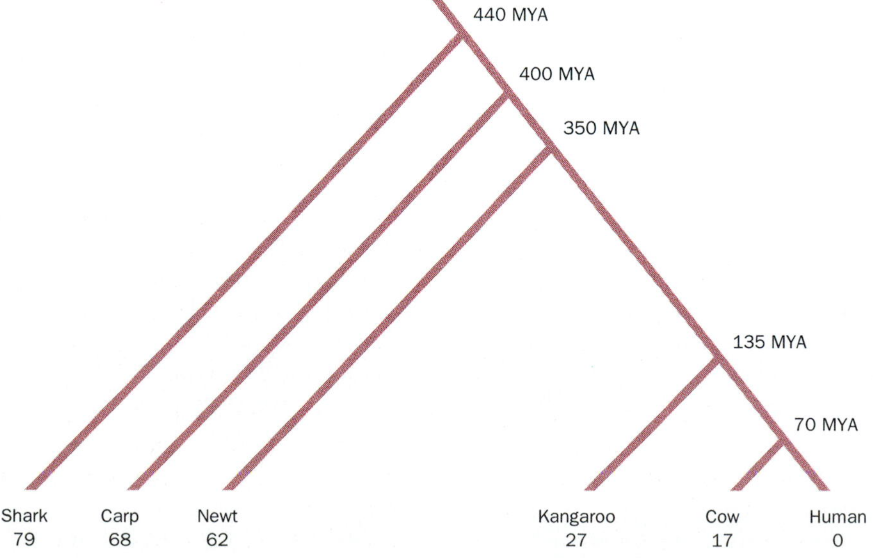

Figure 22.20 Evolution of the α-globin polypeptide of hemoglobin in several vertebrate groups. The data below the evolutionary tree indicate the number of amino acids that are different when human α-globin is compared to each of the other α-globin molecules. MYA = million years ago. (Adapted from M. Kimura, "The Neutral Theory of Molecular Evolution," in *Evolution of Genes and Proteins,* M. Nei, editor, Sinauer, Sonderland, MA, 1983.)

Actually, this value underestimates the true evolutionary rate because some amino acid sites may have changed more than once and there is no way of knowing which ones, or how many times they may have changed. Fortunately, it is possible to correct for these unseen changes and obtain a more realistic value for the evolutionary rate of α-globin; the corrected rate is about one substitution per amino acid site every one billion years.

Similar analyses have been performed with other polypeptides and the results indicate that each polypeptide has its own evolutionary rate. For instance, the fibrinopeptides, which are components of blood clots, are evolving about eight times faster than α-globin, whereas the histones, which bind to DNA, are evolving about a thousand times slower.

How might these observations be explained? One proposal, championed by M. Kimura, is that the great majority of these molecular changes are caused by the random fixation of selectively neutral, or nearly neutral, mutations. In this ***neutral theory of molecular evolution***, mutation and random genetic drift play the critical roles. Recall that the probability of random fixation of a selectively neutral mutation is equal to the mutation rate. ***Thus, in the neutral theory, the rate of evolution is completely determined by the force of mutation.*** Since mutation pressure is more or less constant, this would explain why polypeptides like α-globin evolve at a fairly uniform rate in different taxonomic groups. However, it does not explain why there are different rates for different polypeptides.

To account for such differences, it is necessary to invoke selection pressure. Kimura and others have proposed that some ***polypeptides are functionally constrained***. By this they mean that some sites in a polypeptide cannot be changed without impairing the polypeptide's function. The more constrained a polypeptide, the lower the probability of a neutral amino acid substitution, and the lower the evolutionary rate. In this view, polypeptides such as the histones are highly constrained, whereas those such as the fibrinopeptides are not.

This theory of molecular evolution has been controversial because it places so much emphasis on random processes. Although the neutral theory does admit a role for selection as a purifying agent, that is, as a force that eliminates harmful mutations, it has little or no place for selection in the positive Darwinian sense. In the neutral theory, species do not get better by fixing beneficial mutations; they simply do not get worse by fixing only neutral or nearly neutral mutations. For this reason, the neutral theory is sometimes called the ***non-Darwinian theory of evolution***. Most population geneticists would agree that it has had a major impact on our thinking about the evolutionary process.

Evolution of Multigene Families

Advances in molecular genetics have revealed that many eukaryotic genes belong to ***multigene families***. These are clusters of genes that are often linked together in a tandem array. The genes encoding the β-chain of human hemoglobin are a classic example. As Fig. 22.21 shows, there are six β-globin genes in a region of about 50 kilobases on chromosome 11. The ϵ gene is expressed in the embryo, the two γ genes are expressed in the fetus, and the δ and β genes are expressed after birth and in the adult. In between the fetal and adult genes, there is a pseudogene (ψβ*1*) that has a mutation that prevents its expression. Analysis of the sequences of all six genes indicates that they are structurally related, probably because they are duplicates of an ancestral β-globin gene. This and other multigene families have therefore provided evidence that gene duplication is an important process in eukaryotic evolution.

A duplication event could occur through an error in DNA replication. One possibility is sketched in Fig. 22.22. In this scheme, the segment to be duplicated forms a loop and the DNA polymerase passes around it twice. Although the molecular details are not known, the mechanism is certainly plausible. Another possibility is shown in Fig. 22.23. Here, the duplication event is mediated by transposable elements that are located on either side of the gene sequence. During meiosis, it is possible for these elements to bring the homologous chromosomes together out of register. If displaced pairing is followed by a crossover between the transposable elements, a chromosome with two copies of the gene will be created. There will also be a complementary crossover product that lacks the gene alto-

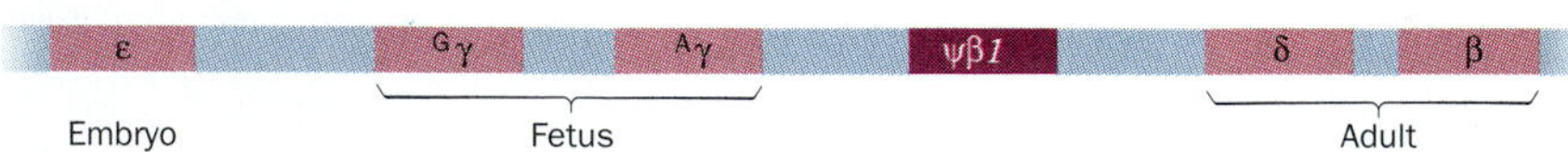

Figure 22.21 The organization of the β-globin genes on human chromosome 11. The genes that are indicated by the open rectangles are expressed at different stages during development. The ψβ1 pseudogene, indicated by the closed rectangle, is never expressed.

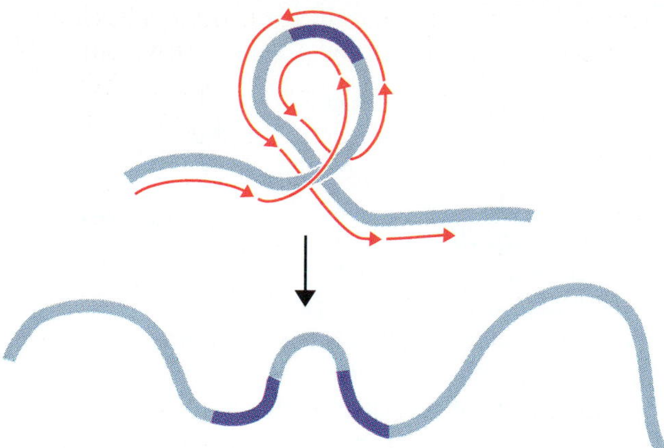

Figure 22.22 Origin of a DNA duplication by a replication error. The DNA polymerase passes over the looped region twice, duplicating the gene that is indicated by the darkened segment.

gether. This process, called ***unequal crossing-over***, can therefore generate chromosomes with zero or two copies of the gene.

Once two genes have been formed by duplication, the system can evolve to a more complex state. Fig. 22.24 shows how duplicate genes can create a gene triplication through unequal crossing-over. The resulting triplication might then go through a similar process, generating a chromosome with additional copies of the gene.

The evolutionary dynamics of multigene families are rather special. First, extra gene copies can evolve under minimal selection pressure. Since the function of the gene family is provided by a single copy, the extra copies are not likely to be subject to rigorous functional constraints. We would therefore expect previously harmful mutations, such as those that create pseudogenes, to be tolerated. Second, the relaxation of functional constraints should provide an opportunity for duplicate genes to diversify. With this diversification, some of the genes might develop new and potentially useful functions. In the β-globin family, this has evidently occurred, since the members of the family are expressed at different times and in different physiological settings. Third, in cases where there is an advantage in maintaining several copies of a specific gene because its function is amplified, the whole multigene family is expected to evolve as a unit. Harmful mutations will not be tolerated and the genes of the unit will tend to have the same sequence. In this situation, sequence similarity will be maintained by selection pressure and by mechanisms such as gene conversion and unequal crossing-over. The overall process of maintaining genetic homogeneity in such a multigene family is called ***concerted evolution***.

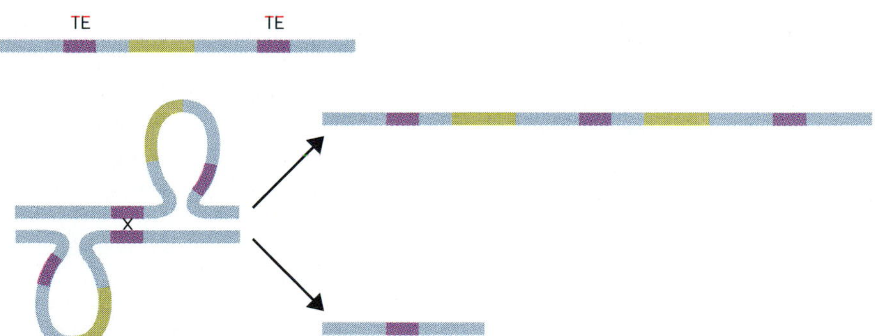

Figure 22.23 Origin of a gene duplication by unequal crossing-over mediated by transposable elements. The transposable elements are indicated in purple.

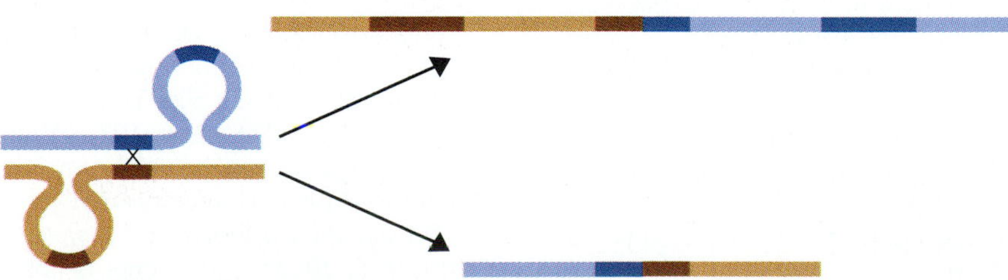

Figure 22.24 Production of a gene triplication by unequal crossing-over between duplicate genes. The genes on homologous chromosomes are distinguished by blue and brown coloring.

SUMMARY

Genetic variability is a feature of many populations of organisms. This variability can be quantified by determining the frequencies of alleles at segregating loci. Loci that segregate two or more alleles are considered to be ***polymorphic*** if the frequency of the most common allele is less than 0.99.

Under the assumption of random mating, allele frequencies can be used to predict genotype frequencies by the ***Hardy–Weinberg method. Inbreeding, outbreeding***, and ***positive*** and ***negative assortative mating*** are examples of nonrandom mating. In each case, the genotype frequencies are altered from the Hardy–Weinberg predictions. The intensity of inbreeding can be measured by using pedigrees to calculate the ***inbreeding coefficient***. This coefficient can also be used along with allele frequencies to predict genotype frequencies in an inbred population. Inbreeding is often associated with a decline in quantitative traits such as height and yield in agricultural crops. Typically, this decline is a linear function of the inbreeding coefficient.

Allele frequencies may change through the action of ***mutation, migration, selection,*** and ***random genetic drift***. Quantitative models have been developed to assess the impact of each of these forces on the genetic variability of a population. Sometimes these forces work in opposite directions, producing a state of ***genetic equilibrium***. In such a state, the genetic characteristics of a population do not change.

Evolution involves the physical transformation of a species. The ***synthetic theory of evolution*** merges Charles Darwin's ideas with the principles of Mendelian genetics and focuses on the factors that change allele frequencies. ***Natural selection*** plays a key role in the synthetic theory. Studies with natural populations have demonstrated that evolution is an ***adaptive process***. However, analyses of the amino acid sequences of polypeptides suggest that some evolutionary changes may be random. The ***neutral theory of evolution*** proposes that the rate of evolution is equal to the rate of mutation. A cluster of genes that has evolved from a single progenitor gene constitutes a ***multigene family***. Such families may expand or contract in size through a process of ***unequal crossing-over***. Sometimes a multigene family evolves as a concerted unit.

REFERENCES

ANTONOVICS, J., A.D. BRADSHAW, and R.G. TURNER. 1971. "Heavy metal tolerance in plants." *Adv. in Ecol. Res.* 7: 1–85.

BIEL, S.W., and D.L. HARTL. 1983. "Evolution of transposons: Natural selection for Tn5 in *Escherichia coli* K12." *Genetics* 103: 581–592.

CAVALLI-SFORZA, L.L., and W.F. BODMER. 1971. *The genetics of human populations.* Freeman, New York.

CROW, J.F. 1986. *Basic concepts in population, quantitative and evolutionary genetics.* Freeman, New York.

CROW, J.F., and M. KIMURA. 1970. *An introduction to population genetics theory.* Harper & Row, New York.

DARWIN, C. 1951. *The origin of species by means of natural selection or the preservation of favoured races in the struggle for life.* Philosophical Library, New York. (Reprint of first edition published November 24, 1859.)

DOBZHANSKY, T. 1951. *Genetics and the origin of species*, 3rd ed. Columbia Univ. Press, New York.

FISHER, R.A. 1958. *The genetical theory of natural selection.* Dover, New York.

HALLAUER, A.R., and J.H. SEARS. 1973. "Changes in quantitative traits associated with inbreeding in a synthetic variety of maize." *Crop Sci.* 13: 327–330.

HARDY, G. 1908. "Mendelian proportions in a mixed population." *Science* 28: 49–50.

JEPSEN, G.L., G.G. SIMPSON, and E. MAYR. 1963. *Genetics, paleontology and evolution.* Atheneum, New York.

KETTLEWELL, H.B.D. 1959. "Darwin's missing evidence." *Sci. Amer.* 200(3): 48–53.

KIMURA, M. 1983. "The neutral theory of molecular evolution." In *Evolution of genes and proteins,* pp. 208–233, M. Nei, editor. Sinauer, Sunderland, MA.

LEWONTIN, R.C. 1974. *The genetic basis of evolutionary change.* Columbia Univ. Press, New York.

MAYR, E. 1963. *Animal species and evolution.* Belknap Press, Cambridge, MA.

MERRELL, D.J. 1965. "Competition involving dominant mutants in experimental populations of *Drosophila melanogaster.*" *Genetics* 52: 165–189.

METTLER, L.E., T.G. GREGG, and H.E. SCHAFFER. 1988. *Population genetics and evolution.* Prentice Hall, Englewood Cliffs, NJ.

O'BRIEN, S.J., D.E. WILDT, M. BUSH, T.M. CARO, C. FITZGIBBON, I. AGGUNDEY, and R.E. LEAKEY. 1987. "East African cheetahs: Evidence for two population bottlenecks." *Proc. Natl. Acad. Sci. U.S.A.* 84: 508–511.

SIMMONS, M.J., and J.F. CROW. 1977. "Mutations affecting fitness in *Drosophila* populations." *Annu. Rev. Genet.* 11: 49–78.

TEMIN, R.G. 1966. "Homozygous viability and fertility loads in *Drosophila melanogaster.*" *Genetics* 53: 27–46.

WRIGHT, S. 1922. "Coefficients of inbreeding and relationship." *Amer. Natur.* 56: 330–338.

———. 1969. *Evolution and the genetics of populations,*

Vol. 2, *The theory of gene frequencies.* Univ. of Chicago Press, Chicago.

———. 1977. *Evolution and the genetics of populations,* Vol. 3, *Experimental results and evolutionary deductions.* Univ. of Chicago Press, Chicago.

———. 1978. *Evolution and the genetics of populations,* Vol. 4, *Variability within and among natural populations.* Univ. of Chicago Press, Chicago.

PROBLEMS AND QUESTIONS

22.1 The frequency of an allele in a large randomly mating population is 0.2. What is the frequency of heterozygous carriers?

22.2 The incidence of recessive albinism is 0.0004 in a human population. What is the frequency of the recessive allele? (Assume random mating.)

22.3 The frequency of newborn infants homozygous for a recessive lethal allele is about 1 in 25,000. What is the expected frequency of the carriers of this allele in the population? (Assume random mating.)

22.4 The following data for the MN blood types were obtained from Indian villages in North and Central America.

GROUP	POPULATION SIZE	M	MN	N
Central American	86	53	29	4
North American	278	78	61	139

Calculate the frequencies of the L^M and L^N alleles for the two groups.

22.5 In a sample from an African population, the frequencies of the L^M and L^N alleles were 0.78 and 0.22, respectively. What are the expected frequencies of the M, MN, and N phenotypes? (Assume random mating.)

22.6 A locus has three alleles $A_1, A_2,$ and A_3, with frequencies of 0.6, 0.3, and 0.1. Assuming random mating, what is the expected frequency of all the heterozygotes in the population?

22.7 Human beings carrying the dominant allele T can taste the substance phenylthiocarbamide. In a randomly mating population in which the frequency of this allele is 0.4, what is the probability that a particular taster is homozygous?

22.8 What is the inbreeding coefficient of the offspring of a mating between (a) uncle and niece? (b) grandfather and granddaughter?

22.9 What fraction of their genes do half-siblings share by reason of common ancestry?

22.10 Assuming that Q and R in the pedigree below arose from random mating, what is the inbreeding coefficient of X?

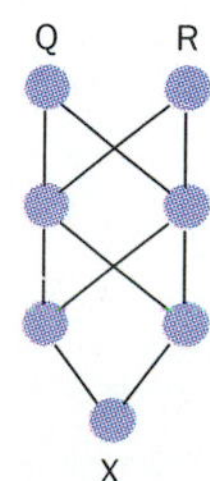

22.11 In the pedigree below, what is the inbreeding coefficient of the common ancestor, C, if the inbreeding coefficient of Y is 1/16?

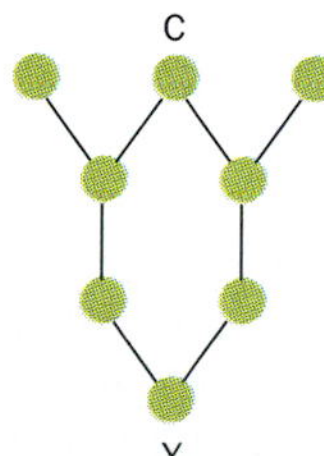

22.12 The A locus is segregating two alleles, A and a, with frequencies of 0.6 and 0.4, respectively. After a long history of random mating, the population goes through one cycle in which all the matings are between full-sibs. What is the expected frequency of heterozygotes among the offspring of these full-sib matings?

22.13 A, B, and C are inbred strains of mice, assumed to be completely homozygous. A is mated to B and B to C. Then the A×B hybrids are mated to C and the offspring are mated to the B×C hybrids. What is the inbreeding coefficient of the offspring of this last mating?

22.14 In mice there are several different loci that segregate recessive alleles causing albinism. A cross between albinos from two highly inbred strains produces normally pigmented F_1 animals. If these F_1 mice are sib-mated, what fraction of their progeny will be albino?

22.15 In a wild prairie grass, the genotypes TT and Tt produce plants that are 100 cm tall. In contrast, the genotype tt produces plants that are only 50 cm tall. In a randomly mating population, the frequency of the t allele is 0.6. (a) What is the average height of plants in this population? (b) What would the average be among the offspring of plants that were self-fertilized? (c) What would the average be among the offspring of plants that were mated to full-sibs? (d) What is the apparent relationship between the average height and the degree of inbreeding?

22.16 In a large, randomly mating population, 0.84 of the individuals express the phenotype of the dominant allele ($A-$) and 0.16 express the phenotype of the recessive allele (aa). (a) What is the frequency of the dominant allele? (b) If the aa individuals are 5 percent less fit than the $A-$ individuals, what will the frequency of the dominant allele be in the next generation?

22.17 A completely recessive allele, g, is lethal in homozygous condition. If G mutates to g at a rate of 10^{-6} per generation, what is the expected frequency of the lethal homozygotes when the population reaches mutation–selection equilibrium?

22.18 Individuals with the genotype *Bb* are 5 percent less fit than those with the genotype *BB*, whereas *bb* individuals die during embryonic development. The allele *B* mutates to *b* at a rate of 10^{-6} per generation. What is the expected equilibrium frequency of *b* in a large, randomly mating population?

22.19 Determine the selection coefficient (s) and the "dominance" parameter (h) for the following sets of relative fitnesses.

SET	A_1A_1	A_1A_2	A_2A_2
(a)	1	1	0.8
(b)	1	1	0.7
(c)	1	0.9	0.8
(d)	1	0.9	0.9

22.20 Mice with the genotype *Hh* are twice as fit as either of the homozygotes. With random mating, what is the expected frequency of the *h* allele in a population at equilibrium?

22.21 A population of size 50 reproduces in such a way that the population size remains constant. If mating is random, how rapidly will genetic variability, as measured by the frequency of heterozygotes at a locus, be lost from this population?

22.22 A population is segregating three alleles, A_1, A_2, and A_3, with respective frequencies of 0.2, 0.5, and 0.3. Assuming these alleles to be selectively neutral, what is the probability that A_2 will ultimately be fixed by genetic drift?

22.23 An island population of human beings consists of 50 males and 50 females. One-fourth of the males have Y chromosomes that are longer than normal. Assuming that the length of the Y has no reproductive significance and that mating is random, what is the probability that the long Y chromosome will ultimately be fixed by genetic drift?

22.24 A transposable element called *roo* is situated at two positions that are 500 kilobases apart in the X chromosome of *Drosophila melanogaster*. Using diagrams, explain how these elements might lead to the formation of chromosomes that are duplicated or deleted for this 500-kilobase region.

22.25 Assume that pattern baldness in humans is controlled by an allele, *B*, of an autosomal gene. Further assume that this allele is dominant in males and recessive in females, that is, is "sex-influenced." If a bald man and a nonbald woman in a randomly mating population in which 51 percent of the men are bald have a child, what is the probability that the child will eventually become bald?

Chapter 23

Genetics of Behavior

Photo above: A cluster of monarch butterflies, *Danaus plexippus,* that have migrated to Morro Bay, California. (Courtesy of Richard R. Hansen/Photo Researchers.)

GENETIC mechanisms for structural and numerical traits have been investigated more successfully than those for behavioral characteristics. The ***complexity of behavioral traits*** and the difficulty of studying them at the molecular level are the main reasons for the difference. Furthermore, behavioral characteristics of any animal develop under the joint, tightly entwined effects of heredity and environment. DNA in the genome determines the individual's physiological, structural, and behavioral potentials, but not all the potentials are realized in the developing individual. ***Behavior genetics is concerned with the effects of genotype on behavior and with the role that genetic differences play in the determination of behavioral differences in a population.***

A fundamental question in the study of the relationship between genes and behavior is whether heredity directly affects behavior or merely defines the stage on which behavioral patterns may be molded by environmental factors. Biologists and psychologists have taken opposite views on this issue, but now the two groups recognize that both heredity and environment are relevant for virtually all behavioral patterns. The question of direct versus indirect effects of heredity can be approached by recognizing the sequential levels of organization of behavior patterns in developing animals. ***Environmental factors are interwoven with inheritance*** mechanisms at every point in development. The problem is to recognize the role of each and to evaluate their relative importance in specific situations.

In some cases, genetic programming simply states that animals have specific learning capabilities at certain developmental periods. Zebra finches, for example, have inherited the physical equipment for the learning process, but when they are isolated from species members early in life, they can never learn to distinguish males and females of their own species. Other aspects of behavior, such as the calls of some birds, are less learned and more inherited. An incubator-raised chicken, for example, having never heard the sound of a hen, will still mature and produce notes typical of other chickens. In other cases, it has been shown that the environment in which an animal matures can drastically affect certain aspects of its adult behavior. H. F. Harlow and his associates, for example, have shown that female rhesus monkeys separated at birth from their own mothers and deprived of early interaction with other monkeys were deficient in basic patterns of maternal behavior, social play, and sexual activity. Again, structural and physiological characteristics that provide the tangible framework for these behavior patterns must follow the DNA "blueprints" for the organism, but an important component of the behavior itself must be learned through contact with the environment.

Basically, then, there are ***ranges of modifiability*** that are inherited, with the segregation of genes and the forces of natural selection accounting for observable individual differences in behavior patterns within a given population.

GENETIC MECHANISMS

In contrast to the more common complex meshing of environment and heredity, the genetic mechanisms of some behavior traits have proved surprisingly simple. Some specific examples depend on a few genes and respond to a limited range of environmental stimuli.

W. C. Rothenbuhler, for example, has found evidence that an interesting behavior pattern in honeybees is controlled by two pairs of recessive alleles in simple Mendelian fashion. Two different races of bees differ in their "hygienic" behavior. Worker bees from the Brown line, a hygienic race, open compartments in the hive that contain pupae dead from American foulbrood and remove the dead (Fig. 23.1*a*). Those from the Van Scoy line, a nonhygienic race, leave the dead pupae in the closed compartments and thus allow the infectious agent for American foulbrood, *Bacillus larvae,* to spread in the colony (Fig. 23.1*b*).

Rothenbuhler crossed the two races and obtained F_1 worker bees, all of which were nonhygienic. When F_1 drones (from unfertilized F_1 gametes) were backcrossed to hygienic queens, four kinds of backcross colonies were obtained in about equal proportions: (1) hygienic bees, (2) bees that opened cells but did not remove the dead pupae, (3) bees that did not open cells but removed dead pupae when cells were opened by the beekeepers, and (4) nonhygienic bees. If a single recessive gene u controls the behavior pattern of uncapping cells and another single recessive gene r controls the behavior pattern for removal of dead pupae, the results obtained by Rothenbuhler can be explained by Mendelian independent assortment. The dihybrid cross is reconstructed in Fig. 23.2. Although many genes and environmental influences may be associated with the complicated neuronal mechanism underlying the behavior patterns of uncapping and removing, the response ***threshold*** is determined primarily by the ***single alleles.***

In this example, two colony-behavior characteristics that appeared together in parental lines separated in the backcross colonies. This manifestation of genetic segregation is expected among individuals but not among colonies. It can occur only when members of a

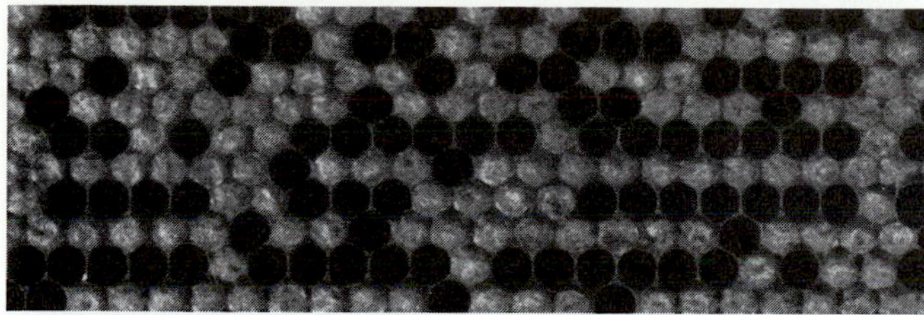

Figure 23.1 Combs of brood from (a) Brown, hygienic colony and (b) Van Scoy, nonhygienic colony. In the Brown colony about two days before brood emergence, many individuals were missing from the spore-inoculated rows, but all brood remaining in the comb were found to be alive. In the Van Scoy colony, most of the brood were present but, when the cells were uncapped in the laboratory, many individuals in the spore-inoculated rows were dead of American foulbrood. (From *Amer. Zool.* 4: 111–123, 1964. Photos courtesy of W. C. Rothenbuhler.)

colony are genetically similar. Colonial bees can be genetically similar when the mother (the queen) is highly inbred and has mated with a single haploid drone to produce the colony. An inbred queen–single drone mating performed by artificial insemination facilitates the study of colony behavior.

In the course of studying the 63 colonies of honeybees in their hygienic behavior experiment, Rothenbuhler and his associates made observations on several other behavioral characteristics of bees. Most obvious and impressive was stinging behavior. In the course of 98 visits to seven Van Scoy colonies, the beekeepers were stung only once, whereas the same number of visits to seven Brown colonies brought 143 stings. The first proposed explanation was that keeping the brood nest free of dead larvae and defending the colony against beekeepers were manifestations of the same general characteristics—a high level of vigor in the worker bees. It was suggested that the same genes might explain the two behavior patterns. If this were the explanation, the hygienic colonies among the backcross colonies would also be stingers. This was not the case, for a very few of the bees in the hygienic backcross colonies were stingers. The stinging behavior in all the 29 backcross colonies indicated that more than two pairs of genes were involved.

P	**Hygienic queens (uncapping and removing)**			**Nonhygienic drones**
P	**Genotypes** $uurr$		×	u^+r^+
P	**Gametes** ur			u^+r^+
F_1	**Genotype**	$u^+u\,r^+r$		
F_1	**Phenotype**	**Nonhygienic**		
F_1	**Gametes (drones)** u^+r^+, ur, u^+r, ur^+ **used to inseminate** $uurr$ **queens**			

From backcross:

FEMALES	$uurr$	uur^+r	u^+urr	u^+ur^+r
Phenotypes	Hygienic	Uncapping, no removal	No uncapping, removal	Nonhygienic
Ratio	1 colony	1 colony	1 colony	1 colony

Figure 23.2 Cross between hygienic and nonhygienic bees and a backcross of four kinds of F_1 drones (from unfertilized haploid F_1 gametes) with *uurr* queens; *u* is the gene for the uncapping and *r* is for removing. The bees (Hymenoptera) used in the P cross are colonial and inbred. Many hygienic worker females and several queens with the same genotype (*uurr*) were produced. Sex is determined by haploidy (nonfertilization) or diploidy (fertilization). By a timing process, queens can determine which eggs are fertilized and thus become females (either workers or queens depending on diet), and which are unfertilized and thus become drones. The nonhygienic drones in the P cross came from a different race. Fully heterozygous (u^+ur^+r) F_1 females were produced by artificial insemination. The four kinds of gametes (u^+r^+, ur, u^+r, ur^+) produced by F_1 queens that were not fertilized became the drones that were used to inseminate P queens (*uurr*). Four kinds of female backcross progeny were produced in equal proportion—1 $uurr$: 1 $uurr^+$: 1 u^+urr : 1 u^+ur^+r, all phenotypically different in behavior characteristics. (From experiments by W. C. Rothenbuhler.)

The genetic aspects of such observable elements of behavior are believed to be determined basically by the same mechanisms that function for the more tangible physical traits that Mendel and others have described. However, the genetic bases for behavior patterns in animals are characteristically difficult to confirm through experimental procedures.

Inheritance and Learning in Bees

W. P. Nye and O. Mackensen have shown that honeybee preference for alfalfa, *Medicago sativa,* pollen depends to a large extent on genetic determinants that respond to ***selective breeding.*** These investigators observed that some colonies of honeybees collected a much higher percentage of alfalfa pollen than did others. Separate inbred lines were developed from colonies that showed either a high or a low preference for collecting alfalfa pollen. At the end of the eighth generation of inbreeding, the high and low lines had been completely separated (Fig. 23.3). Subsequent hybridization of the bees from the two lines produced bees that were intermediate between the high and low lines. Because the preference for alfalfa pollen can be changed markedly by selection, and the trait follows a predictable pattern based on Mendelian inheritance, preference for alfalfa pollen is presumed to have a high hereditary component.

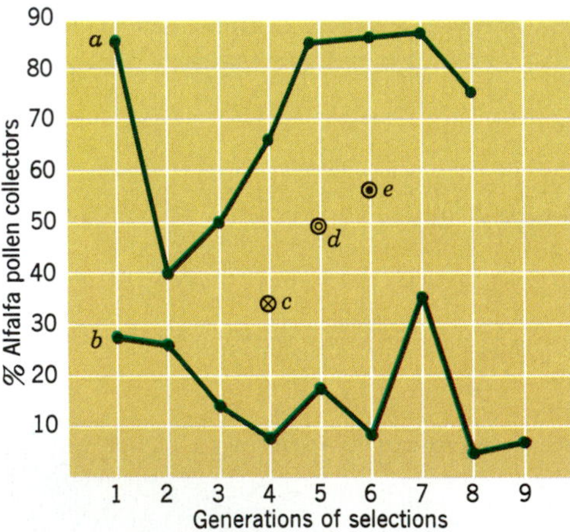

Figure 23.3 Results of selection experiments for high alfalfa pollen collectors (HAPC), low alfalfa pollen collectors (LAPC), and results from outcrosses and backcrosses. (a) HAPC queens and drones selected in each of eight generations. (b) LAPC queens and drones selected in each of nine generations. (c) Hybrid colonies resulting from crosses between LAPC queens and HAPC drones from generation 3. (d) Results from backcrosses between queens from c (hybrid colonies) and drones from generation 4 HAPC queens. (e) Results from outcrosses between HAPC queens from generation 5 and drones from a commercial stock. The dip from 85 to 40 percent between generation 1 and 2 in a is presumed to be the result of heterozygosity of the genes involved in the parents selected to initiate the HAPC line. Both HAPC and LAPC lines are expected to become more homozygous as inbreeding continues. Environmental as well as genetic factors are known to influence pollen collection. In the spring of the seventh year, frosts destroyed most of the early plants on which LAPC bees usually forage, and the bees were forced to move to more abundant alfalfa. (Data from W. P. Nye.)

In later experiments, Nye took high and low alfalfa pollen-collection lines of bees to Donnelly, Idaho, where alsike clover was flowering. Nearly all the collection of all lines was clover pollen, and no significant difference occurred among lines. Inherited behavior for collecting alfalfa pollen was found to be specific to alfalfa pollen.

Both inheritance and learning are involved in the complex social patterns of several hymenoptera, such as the honeybee, *Apis mellifera,* that make use of chemical, optical, and aural signals. Effectiveness of the activities of bees in and around a hive, as well as in foraging areas, depends on the exchange of information among individual bees. Communication symbols for ***distance and direction of food source are mostly learned.*** Even so, in some groups, this behavior is so stereotyped and well established that it can serve as well as a morphological characteristic for distinguishing species.

Although the basic communication system is the same for all *Apis mellifera,* different "dialects" have developed in different races. Members of an Italian race, for example, have a slower dancing rhythm than those of an Austrian race. When bees of these two races are mixed, they misunderstand each other. An Austrian bee receiving information from an Italian bee about food 100 meters from the nest will fly 120 meters because she interprets the "Italian dialect" on the basis of her Austrian knowledge. Conversely, the Italian bee will fly 80 meters when given the information for 100 meters by the Austrian forager. Nevertheless, when M. Lindauer compared the communication systems of three different species of *Apis,* he observed wider differences in behavior as well as in structural characteristics between the species than between races.

EXPERIMENTAL BEHAVIOR GENETICS

Behavior genetics is a challenging field for experimental work. Several investigators, including the late T. Dobzhansky (Fig. 23.4), S. Benzer, J. Adler, C. Kung, M. Nirenberg, G. Stent, M. Levinthal, S. Brenner, and C. M. Woolf, who have already distinguished themselves in other aspects of experimental genetics, have moved into the area of behavior genetics. Each investigator

Figure 23.4 Theodosius Dobzhansky, eminent researcher, writer, and teacher in the fields of genetics and evolution. He made basic contributions to population genetics as well as behavior genetics through his extensive studies on *Drosophila pseudoobscura.*

has an "ideal" experimental material with properties such as large size, rapid reproduction, few neurons, and wide behavior patterns. Organisms such as *E. coli,* phycomycetes, paramecia, nematodes, rotifers, freshwater crustacea (*Daphnia*), leeches, *Drosophila,* and mice have been chosen for basic investigations on behavior. Examples from *E. coli* and *Drosophila* will be discussed here.

Escherichia coli Chemotaxis

In the bacterium *E. coli,* behavior patterns are comparatively simple. ***Response (taxis)*** to chemicals, light, gravity, and temperature have been studied by different investigators. J. Adler, for example, investigating chemotaxis, has shown that the mechanics include detection of the chemical (attractant) by a chemoreceptor and transmission to an effector that produces a flagellar response, the swimming of the organism (Fig. 23.5). By isolating mutants, genes responsible for each step in the process have been postulated. Three *che* mutants, for example, render the bacteria nonchemotactic to specific chemicals: galactose, aspartate, and serine. Eight *fla* (nonflagellar) genes interfere with swimming activity.

Movements of the bacteria are followed with a tracking microscope and the data are fed into the computer. When no chemical gradient is present in the medium, the ***movement is haphazard.*** A bacterium moves in one direction (run) for 2–4 seconds, tumbles around (twiddles), makes another short run, twiddles, and so on, without appreciably changing location (Fig. 23.6). If an ***attractant*** is present, the runs are longer in the ***direction of*** the ***gradient.*** At the end of a run, the organism twiddles, takes a short run, and repeats the activity until the orientation is again in the direction of the gradient. If a ***repellant*** instead of an attractant is present, long runs, twiddles, and short runs occur, but

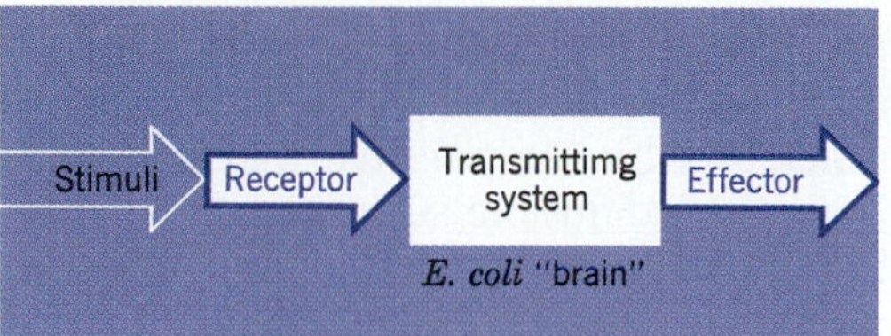

Figure 23.5 Receptor–effector system in *E. coli.*

(a)

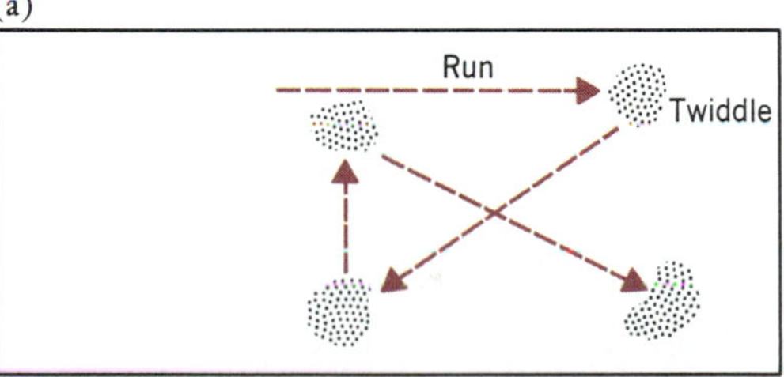

(b)

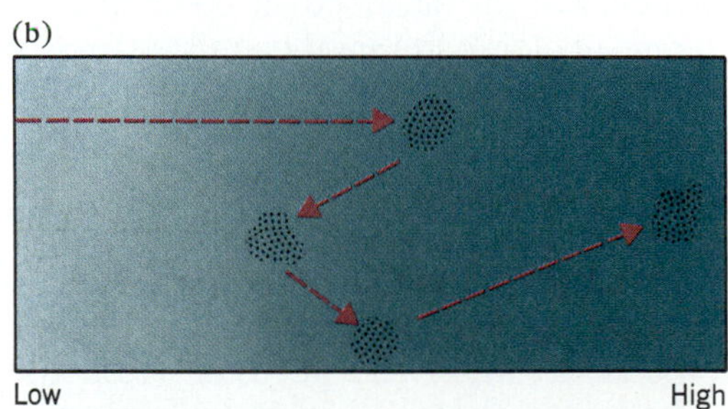

(c)

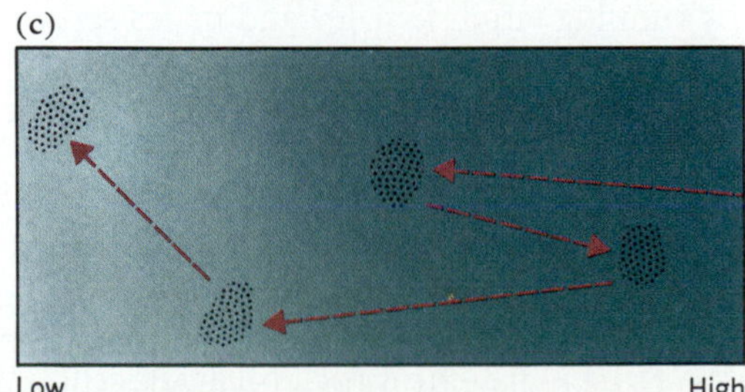

Figure 23.6 Movement of *E. coli* in a liquid medium. (a) No chemical gradient in medium; (b) gradient for attractant; (c) gradient for repellent.

the ***direction*** is toward the ***low concentration*** of the chemical. Several theories, not always exclusive of each other, have been advanced to explain the cause of twiddling: (1) a diffusible substance, (2) a change in membrane potential, (3) a change in membrane configuration, and (4) a general controlled enzyme that influences chemotaxis. Such mutations have now been obtained. Some mutations block the twiddle and others make bacteria twiddle all the time.

Drosophila Behavior Genetics

Drosophila adults are much more complex than *E. coli.* Their brains contain some 10^5 neurons that are arranged in nerve tracts. For such complex animals, the *Drosophila* are small; they also reproduce within a few days, live on simple food in the laboratory, and are well known genetically. Their chromosomes have already been extensively mapped, and behavior genes have been detected in particular areas of the chromosomes. On the X chromosome of *D. melanogaster,* for example, S. Benzer and associates have mapped several behavior loci, including visual receptors, stress receptors, and "wings-up" mutants. Developmental studies by these investigators have related cells in the blastocyst stage with imaginal discs that give rise to eyes, antennae, legs, and wings in the adult flies. Mosaics (gynandromorphs; see Chapter 4) are being used to relate structural parts of the fly to cells in the blastoderm and to genes in the zygote. The wings-up mutation, for example, has been related to a developmental abnormality of the muscles. The "drop-dead" mutation is related to a defect in the brain. Sex and learning behavior are analysed in component parts through mutant blocks and are then related to ***sequential steps in gene activity.***

Response to Light and Gravity in *Drosophila*

C. M. Woolf has carried out hybridization studies with strains of *D. pseudoobscura* that differ in phototactic (i.e., response to light) and geotactic (response to gravity) behavior. These behavior traits were measured by running virgin females and males separately through Hirsch–Hadler classification mazes (Fig. 23.7). Matings were then made among the flies that earned particular classifications, to obtain evidence with respect to ***genetic mechanisms.***

Hirsch–Hadler mazes provide 15 downward or upward choices, that is, toward or away from light. Eventually, a fly enters one of 16 different collecting tubes. The No. 1 collecting tube is entered by flies that make 15 choices upward or away from light. The No. 16 collecting tube will be entered if the 15 opposite choices are made. If an equal number of downward and upward choices are made, the fly enters No. 8 or 9 collecting tube. The number of flies reaching each collecting tube can be used to calculate the reaction of particular flies to light. A completely neutral strain has an expected mean of 8.5. The strongest possible positive strain would have an expected mean of 16. The strongest possible negative strain would have an expected mean of 1.

The strains of flies used by Woolf had been identified through selection as part of the extensive research program of T. Dobzhansky and his associates. Beginning with strains that were phototactically and geotactically neutral, the Dobzhansky group had selected flies with positive and negative response to light and gravity. Strain 25, for example, had become strongly positive. Observations of this kind provided evidence for genetic variance in these strains of *D. pseudoobscura.* The rate of divergence under artificial selection indicated that the genetic component consisted of polygenes. When selection was relaxed, the positive and negative strains reverted to the neutral state, suggesting that the genes responsible for the positive and negative phototactic behavior were maintained in the heterozygous state in natural and laboratory populations by natural selection.

Woolf crossed females from strain 25 with males from strain 27 and also crossed females from strain 27 with males from strain 25. As indicated in Fig. 23.8, the 311 female flies from strain 25 had a mean of 12.28. The 321 male flies from strain 27 chosen for this same experiment had a mean of 4.88. The means of the F_1 progeny and of the F_2 (which were obtained from mating F_1 females with F_1 males) were intermediate between those of the parental strains. The particular arrangement of parental chromosomes in the F_1 and F_2 male and female progeny are indicated in Fig. 23.8, along with the histograms. ***Phototactic behavior*** scores are characterized as being relatively consistent from replication to replication and relatively similar in males and females of the F_1 and F_2 generations.

The position of the F_1 means between the values of the parental means indicates that ***polygenic inheritance*** (groups of genes controlling quantitative traits; see Chapter 21) is involved. A consistent shift, however, was observed in the F_1 and F_2 means toward that of the parental positive strain (strain 25), suggesting that some type of nonadditive influence is present. The similarity of the means of F_1 males and F_1 females indicated that the X chromosome had little or no effect on phototactic behavior. Therefore, the responsible genes are largely located in the autosomes. The variances of the F_1 and F_2 scores are similar. This exception to classical polygenic inheritance, in which a wider spread is usually observed in the F_2 as compared with the F_1, apparently results from the ***low heritabil-***

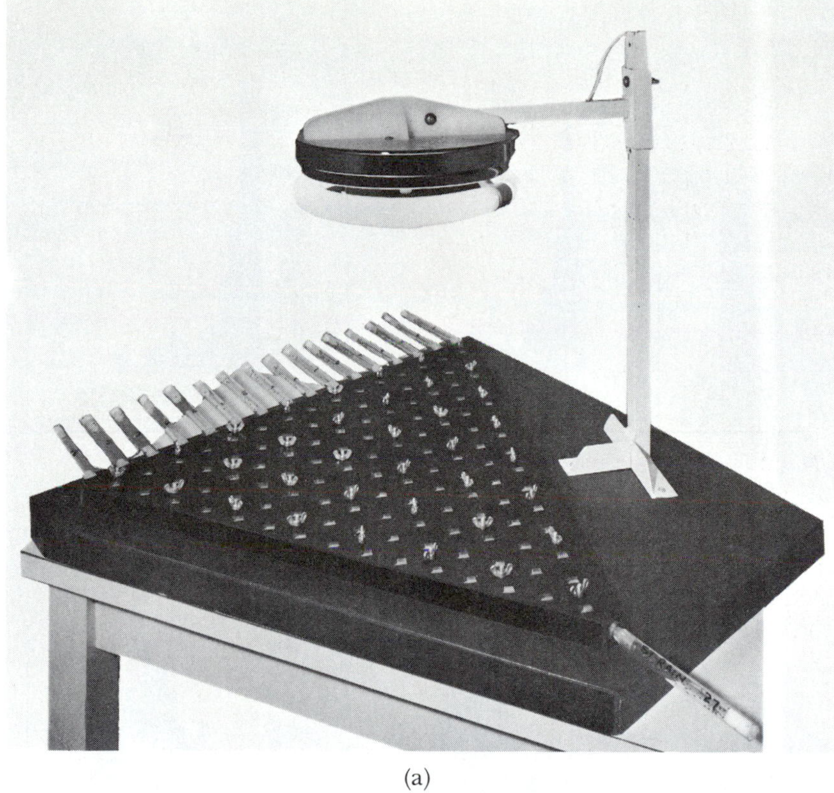

(a)

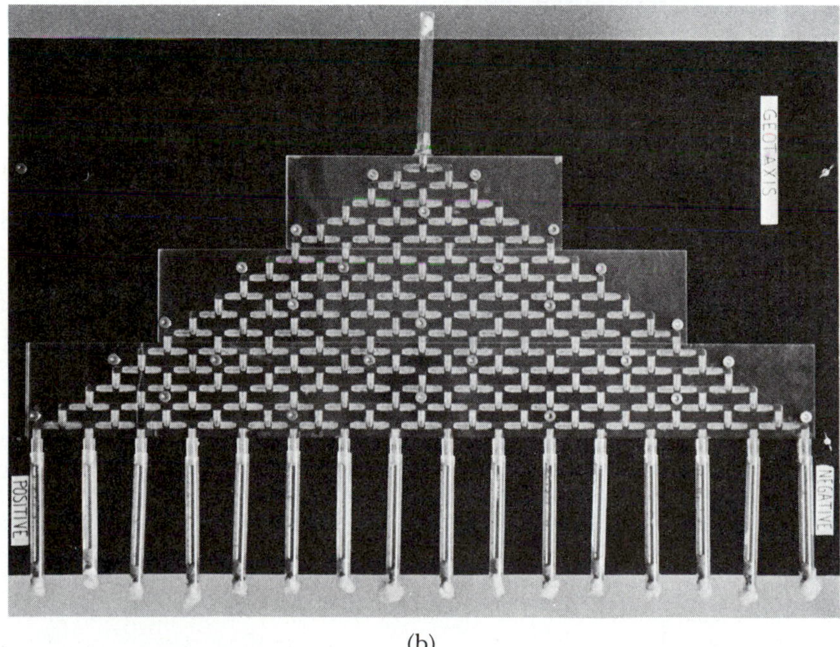

(b)

Figure 23.7 Hirsch–Hadler classification mazes designed to select *Drosophila* for positive and negative geotactic and phototactic behavior. (a) Phototactic behavior maze. (b) Geotactic behavior maze. In each generation, virgin females and males are introduced separately at the maze entrance. The flies have a number of choices, indicating positive or negative behavior in the maze. Those showing each response are collected at the exit and are used as progenitors of the next generation. Beginning with strains that were phototactic and geotactic and neutral, it was possible to obtain positive and negative strains after several generations of selection. (Courtesy of C. M. Woolf, Arizona State University.)

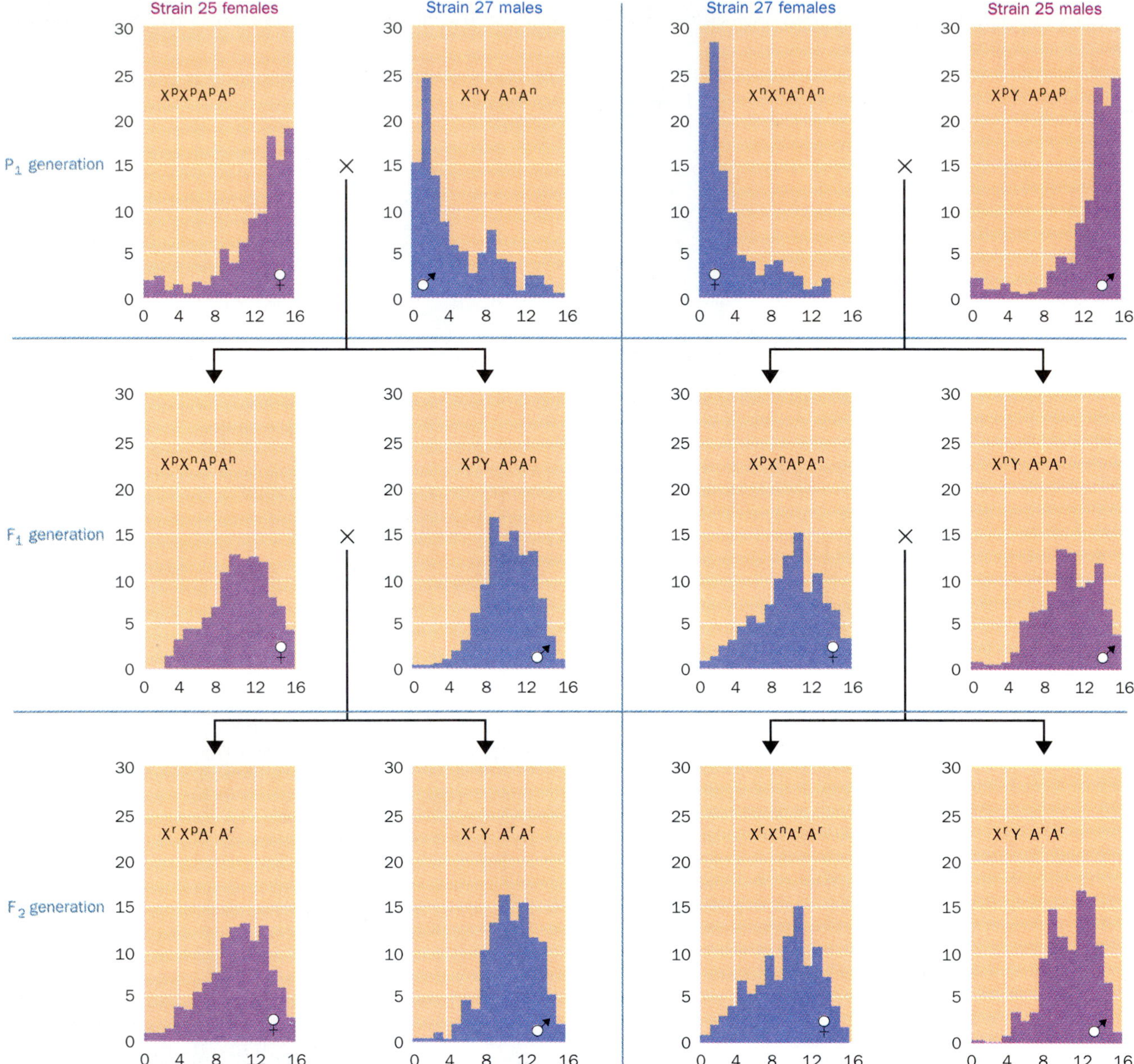

Figure 23.8 Distribution of phototactic scores (pooled data) in the P_1, F_1, and F_2 generations for the parental mating between (left) female strain 25 (positive) × male strain 27 (negative), and (right) female strain 27 (negative) × male strain 25 (positive). In this series of histograms, the area of each rectangle represents the percentage of flies occurring in each of the 16 collecting tubes of the Hirsch–Hadler classification maze. The X chromosome is symbolized by X, the Y chromosome by Y, and the set of autosomes by A. The p and n indicate whether the X chromosomes and autosomes are from the positive or negative strain, respectively. The r designates that the X chromosome and autosomes are recombinant chromosomes as a result of crossing-over and segregation. (Courtesy of C. M. Woolf.)

ity (Chapter 21) of phototactic behavior in *D. pseudoobscura,* the lack of homozygosity of the parental strains, and perhaps some unknown interaction between genotype and environment.

In other comparable experiments carried out in other types of mazes and designed to determine geotactic behavior, Woolf's results were quite different. As in the studies of phototactic behavior, the consistency of the replications indicated the importance of the genotype for these types of behavior under controlled laboratory conditions in spite of the low heritability. Female progeny in the F_1 generation were essentially intermediate between the two parents, but the F_1 males were strongly positive for ***geotactic behavior.***

This is the pattern of crisscross inheritance expected when the ***X chromosome*** carries the gene or genes involved in the transmission of the trait. This same pattern was held through the F_2, F_3, and backcross results, giving evidence that the X chromosome is strongly involved in the transmission of geotactic behavior. The evidence indicated further that the autosomes carried some genes for this trait, but that the great proportion of genes for positive geotactic behavior were located in a region of the X chromosome not readily divisible by crossing-over. Females homozygous or males hemizygous for this region tended to express positive geotactic behavior. Although geotactic behavior in *Drosophila* is not controlled by a few major genes, like those associated with hygienic behavior in bees, particular regions of DNA are involved in the mechanism of this behavior characteristic.

GENETIC AND ENVIRONMENTAL INTERACTIONS IN DOGS

Dogs have personality differences even within a breed. Some are timid and others are confident; some are gentle and others are aggressive. Those that are "socialized" (allowed to interact with people) early in life function as friendly and understanding companions of humans, whereas others of the same breed (or even of the same litter) that are not given similar experiences while young may become fearful or even hostile toward people.

J. P. Scott and J. L. Fuller have made extensive observations on ***genetic and environmental factors*** involved in the ***building*** of ***behavior patterns*** in dogs. In some of their work, daily observations were continued until the dogs were 16 weeks of age. Every effort was made to observe the earliest manifestations of hereditary differences before the effects of experience were noted. Of course, during the first few days after birth, there was very little behavior to observe in the pups.

As soon as recognizable behavior became apparent, interaction between hereditary and environmental influences was already present. The puppies changed markedly in reactions from day to day in response to learning and increasing physical and mental maturity. The environment was optimal for learning, and innate faculties became active at ***progressive developmental stages.*** Genetic control was acting on a very different animal with each new developmental stage, and environmental influence was apparent.

Results obtained by Scott and Fuller were quite unexpected and significant. During the very early stages of development when behavior is minimal, genetic differences had few opportunities to be expressed. When behavior patterns did appear, however, genetically determined differences did not appear all at once early in development, to be modified by later experience. Instead, they developed under the influence of environmental factors. Scott and Fuller concluded that the raising and lowering of response thresholds is one of the most important ways behavior in dogs is affected by heredity. These studies showed that although ***heredity*** is an ***important*** factor in dog behavior, details—and sometimes the actual appearance of specific patterns—depend to varying degrees on the ***individual's experience.*** Furthermore, they demonstrated that at least some genetic behavior difference can be measured and compared as readily as hereditary physical differences.

Dog breeds generally have managed to retain a great deal of genetic flexibility, despite their intensive selection over time. This was borne out by the further studies of Scott and Fuller in which 50 traits were examined in five pure breeds of dogs. Almost all the traits were significantly different among the various breeds. But a very few of them were found to breed true, as would be expected if they were controlled by a single homozygous pair of alleles. In addition, lack of correlation was observed between behavior and phenotypic "type" within each breed. In their crossbreeding work with cocker spaniels and dogs of the Basenji breed, the Basenji personality was often seen in spaniel-appearing dogs, and vice versa. However, selection has apparently produced near homozygosity for certain traits in particular breeds. Fighting behavior, for example, is almost nonexistent in the hound but is well developed in the terrier. But such instances of near homozygosity are rare.

Through selection of genetic qualities and training, remarkable behavior patterns have been made available in some breeds of dogs. The Basenji, for example, commonly known as the African nonbarking hound, is by nature a "scent" hunter. This dog is used in Africa to find and drive wild game. It is basically intelligent and can be taught such feats as the advanced American Kennel Club obedience program, which includes retrieving a dumbbell over a high jump. In this demonstration, the Basenji's inherent intelligence is put to a relatively artificial use.

Retrieving dogs have been selected and trained for more than 100 years in England by enthusiasts of waterfowl hunting. Several breeds exhibited some of the characteristics necessary for an excellent retriever, such as strength, moderate size, endurance, enthusiasm, aquatic ability, keen scenting ability, courage, favorable temperament, and trainability. By intercrossing the most favorable breeds—Newfoundland, setter, and spaniel—and selecting progeny, a litter of four

Figure 23.9 Golden retriever. This breed of dogs has been selected and trained for more than 100 years to produce the characteristics that make good retrievers. (Courtesy of Heather Angel.)

puppies with a favorable combination of traits was obtained in 1868. From this beginning the golden retriever stock was developed. These dogs (Fig. 23.9) are light yellow in coat color. They have aquatic ability, pleasing temperament, keen nose, tracking ability, and tenacity to retrieve under severe conditions. It is natural for a pup from this stock to want to retrieve. Training merely perfects and polishes the performance.

Breed differences in behavior are both ***real*** and ***important.*** The great variability available in dogs means that it is possible to ***modify*** a breed markedly within a few generations ***by careful selection.*** Through crossbreeding, entirely new and unique combinations of behavioral traits can be created and studied.

Sociobiology

Groups of animals have behavior characteristics in common that involve biological mechanisms responsive to selection. Adherence to a particular social order has value when it provides that the group may act as one in behalf of the individual. On the other hand, without a certain amount of social disorder or freedom, the strengths inherent in diversity are lost. A dynamic ***balance*** between ***order*** and ***disorder*** is a healthy condition within living populations.

One aspect of "order" requires respect for one another's "distance" needs. Different species have different ***individual distance*** requirements and differ in their responses to crowding. Black-headed gulls resting in a row will space themselves at about one foot apart. Flamingos maintain about twice that much space between neighbors, and swallows require about half as much space as the gulls (Fig. 23.10). Some tortoises and hedgehogs will crowd together and make an animal pile. Inherent species differences control such behavior patterns. Individual distance or "personal space" requirements, however, may vary with individuals and with seasons. For example, virtually all bird species tend to gather in tight flocks during the winter, but as soon as the breeding season comes in the spring they disperse and strenuously defend their individual territories. ***Social distance*** is the farthest point an animal will stray from the group. A baboon troop is dispersed widely while feeding in the daytime, but at night the group will come together and sleep in a few adjacent trees. In this case, the acceptable social distance varies with environmental conditions.

Overcrowding, whether among human beings or other animals, can lead to a breakdown in social structure and open the gates to overaggressiveness. ***Innate aggressiveness,*** which may have been necessary throughout evolution to help some species to survive, can take a destructive turn under conditions of ***overcrowding.*** In some cases, aggression is directed toward others, but it can also become manifest in ulcers, nervous disorders, and various psychosomatic maladies.

Fortunately, aggression, like many other behavior patterns, is subject to modification—though only within genetic bounds. Aggression can be increased or decreased by purposeful education. If mice or chickens are paired with others of their kind, fighting ensues and one becomes dominant and the other subordinate. If these encounters are arranged so that a particular animal always wins, that animal becomes more and more emphatically dominant over his fellows. In psychological terms, winning reinforces further aggression. On the other hand, repeated losses make an animal submissive.

One can also train animals to be nonaggressive in other ways. Scott prevented puppies from making playful attacks on their handlers by picking them up frequently, thus rendering them helpless with their feet off the ground. This process of ***passive inhibition*** produced nonaggressive adults. He obtained similar results with young male mice by repeatedly stroking them at an early age. Aggressiveness is a product of inheritance, maturation, various endogenous factors, and experience. However, manifestations of aggression among nonhumans depend on the presentation of proper external stimuli, usually specific sign stimuli from other individuals of the same species.

Figure 23.10 Normal spacing of swallows on power lines. (Courtesy of Joseph Van Wormer/Bruce Colman, Inc.)

The Evolutionary Approach

In trying to trace the evolution of behavior in humans, researchers use the same general approaches that have proved productive with other species. It has been found (predictably) that some behavior patterns that serve the function of food intake are phylogenetically quite old—and the human infant shares them with many other mammals. One example is the rhythmic searching for the nipple.

The grasping reflex is also characteristic, perhaps originally serving to keep the baby attached to the mother. In premature infants it is especially strong, and they are able to cling to an outstretched cord. Climbing movements can also be elicited in premature infants.

Observation of congenitally deaf–blind children (in effect deprivation experiments) is giving new insights into situations where human behavior patterns are truly unlearned. Expressions of anger (stamping of feet, facial contortions), high-intensity laughter, and the rejection of strangers seem to be innate. Similarly, components of facial expressions, such as raising the eyebrows when greeting someone, are common to all human cultures that have been studied.

Inborn behavior mechanisms also seem to account for a remarkably universal desire for cover and unobstructed vision, as evidenced in modern-day life by a preference for corner and wall tables in restaurants. Culture imposes specific limits on individual distance requirements, but the basic need for space seems inborn and probably evolved.

The tendencies to seek membership in some sort of group and to accept the group-dictated exclusion of others also seem to be universal human behaviorisms and are therefore likely evolutionary phenomena.

Obviously human beings have many kinds of behavior in common, despite a heavy veneer of individually learned, culturally dictated modifications. It has also been proved (e.g., as in Scott and Fuller's work with puppies) that experiments with other animals can provide valid insights into human behavior.

SUMMARY

DNA in the genome determines the behavioral potentials of individuals, but environmental factors, which are interwoven with inheritance mechanisms, create the setting in which individuals will realize their potentials. Ranges of modifiability are inherited. Segregating genes, forces of natural and artificial selection, and direct environmental influences account for individual differences in behavior patterns within a given population. Although the genetics of most behavior patterns is complex and cannot at this time be reduced to simple Mendelian ratios, the stability and the evolutionary background of these patterns imply genetic mechanisms. Behavior of an animal ***develops*** in response to interactions between inherited limitations and environmental factors.

REFERENCES

ARDREY, R. 1974. *The social contact,* 2nd ed. Atheneum, New York.

______. 1976. *The hunting hypothesis: A personal conclusion concerning the evolutionary nature of man.* Atheneum, New York.

BENZER, S. 1973. "Genetic dissection of behavior." *Sci. Amer.* 229(6): 24–37.

DOBZHANSKY, T., B. SPASSKY, and F. SVED. 1969. "Effects of selection and migration on geotactic and phototactic behavior of *Drosophila,* II." *Proc. Roy. Soc. London B* 173: 191–207.

EHRMAN, L., and P.A. PARSONS. 1976. *The genetics of behavior.* Sinauer, New York.

EIBL-EIBESFELDT, I. 1975. *Ethology, biology of behavior,* 2nd ed. Holt, Rinehart & Winston, New York.

HIRSCH, J., editor. 1969. *Behavior—Genetic analysis.* Brown, New York.

MCCLEARN, G.E., and J.C. DEVRIES. 1973. *Introduction to behavioral genetics.* Freeman, San Francisco.

MCGILL, T., editor. 1977. *Readings in animal behavior,* 3rd ed. Holt, Rinehart & Winston, New York.

NYE, W.P. 1971. "Pollen collection from alsike clover by high and low alfalfa pollen collecting lines and by a commercial line of honeybees." *J. Apicultural Res.* 10: 115–118.

NYE, W.P., and O.MACKENSEN. 1970. "Selective breeding of honeybees for alfalfa pollen collection: With tests on high and low alfalfa collection regions." *J. Apicultural Res.* 9: 61–64.

ROTHENBUHLER, W.C. 1968. "Bee genetics." *Annu. Rev. Genet.* 2: 413–437.

SCOTT, J.P., and J.L. FULLER, editors. 1974. *Dog behavior: The genetic basis.* Univ. of Chicago Press, Chicago.

PROBLEMS AND QUESTIONS

23.1 In general, how is animal behavior related to genetics and environmental factors?

23.2 Why has the genetics of behavior patterns developed more slowly than the genetics of other characteristics, such as size and color patterns?

23.3 Why do certain studies, such as Rothenbuhler's study of hygienic and nonhygienic bees, have particular significance in behavior genetics?

23.4 Why are comparative studies of behavior patterns especially useful for investigating the evolution of behavior?

23.5 What evidence suggests a genetic basis for the preference of some bees for alfalfa pollen?

23.6 (a) How may hormones influence behavior? (b) How and to what extent is hormone production under genetic control?

23.7 What conclusions may be drawn from the studies of Scott and Fuller on dogs concerning the relative influence of heredity and environment on behavior?

23.8 Evaluate the extent of influence of (a) genetic factors and (b) training on retrieving dogs.

23.9 Why are dogs highly reactive to selection?

23.10 Why was the dog one of the first animals to be domesticated and to come into close association with humans?

23.11 How could natural or human selection account for nonbarking in Basenji dogs even though they are capable of barking?

23.12 What are the potential difficulties and dangers in trying to transfer behavioral activities in dogs that have developed in laboratory or wild populations to individual domesticated animals?

23.13 If a behavioral trait has demonstrable survival value, is it more likely to be genetically controlled or a learned phenomenon? Why?

23.14 What characteristics should be considered when choosing suitable material for the following kinds of studies on genetics of behavior: (a) study of simple response to environmental stimuli; (b) study of location of a gene that affects a behavior mechanism?

23.15 If two strains of rats seemed to differ in their ability to solve maze-running problems, what sorts of environmental factors should be considered before concluding that the difference was genetically controlled (whether through general physiology or brain capacity)?

23.16 How may aggression be reduced in animal populations?

23.17 Why are *Drosophila* more suitable than mice as experimental material for maze studies that are designed to determine genetic mechanisms?

23.18 How could it be determined whether sex-linked polygenes control geotactic behavior in *Drosophila?*

23.19 A fully heterozygous ($u^+u\ r^+r$) nonhygienic queen bee was mated with a hygienic drone (ur). (Remember that queen bees are diploid and drones are haploid.) What genotypes and phenotypes (uncapping of cells and removal of dead pupae) might be expected in the first generation and in what proportions?

Preface to Chapter 24

Chapter 24 is written for those students who have mastered the basic principles of genetics and who are anxious to learn more about what is happening in the field of genetics today and what is likely to happen in the future. We do not anticipate that any basic genetics course will cover all the material in Chapter 24 as a core part of the course. Instead, we envision Chapter 24 as optional or possibly assigned reading at the end of the term. However, some classes may wish to cover selected sections of the chapter that relate most directly to the particular emphasis of the course. For example, a class with a human genetics focus might wish to cover the section on somatic-cell human gene therapy, or a class with a plant genetics emphasis might use the sections on gene transfer in plants. ***However, most of all, Chapter 24 is written for those students who have discovered that genetics is a truly fascinating subject and who want to learn more.*** If Chapter 24 can help fulfill the wishes of these students, it will achieve our objective.

Chapter 24

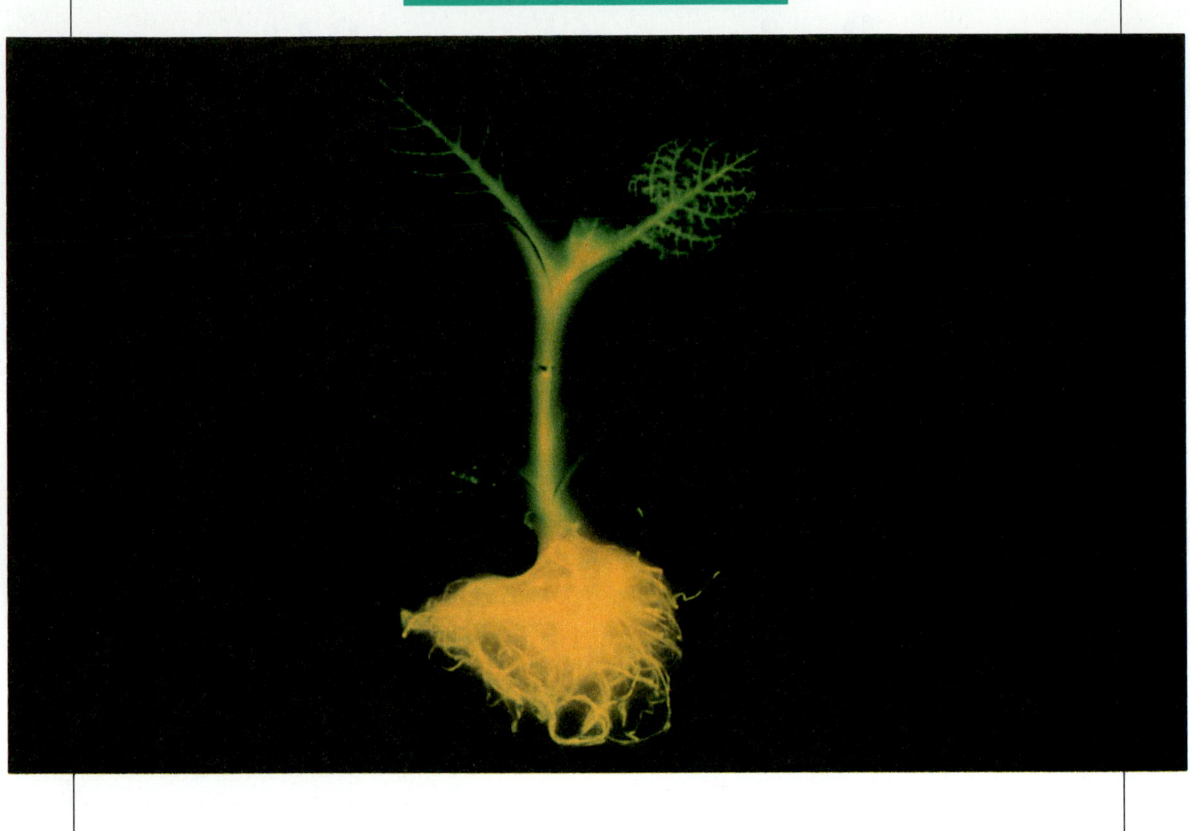

Genetic Engineering and the Future

Photo above: Luminescence from a transgenic tobacco plant harboring the luciferase gene of the firefly. (Reproduced with permission from D. W. Ow, K. V. Wood, M. DeLuca, J. R. de Wet, D. R. Helinski, and S. H. Howell, "Transient and Stable Expression of the Firefly Luciferase Gene in Plant Cells and Transgenic Plants." *Science* 234: 856–859, 1986. Copyright ® 1986 by the American Association for the Advancement of Science. All rights reserved.) Photo by Keith V. Wood. Promega Corp.

R*ECOMBINANT DNA and gene cloning technologies* are, without doubt, the most powerful tools ever developed in the field of biology. Our ability to carry out carefully ***designed genetic engineering*** of living cells has many potential applications. Some of these applications have already reached fruition. If used wisely, genetic engineering promises to enhance the quality of human life. On the other hand, if used haphazardly and carelessly, genetic engineering could have a negative impact on our quality of life. Like any powerful tool, genetic engineering must be used carefully and wisely if we are to avoid unfortunate accidents.

DISSECTION OF BIOLOGICAL PROCESSES

Although some commercial applications of the recombinant DNA and gene cloning technologies have already proven successful, the real importance of these techologies to date has derived from their use in dissecting basic biological processes. These technologies have allowed scientists to identify and isolate genes that encode a great variety of both enzymes and structural proteins, and to dissect these genes, base-pair by base-pair, via site-specific mutagenesis. The functional components of each gene are thus defined in terms of the essential nucleotide-pair sequences. The ultimate product of such an analysis is a catalog giving the complete nucleotide-pair sequence of each genetic element and the functional consequences, if any, of changing the nucleotide-pair present at a given site to each of the other three possible nucleotide-pairs. When combined with classical mutational dissection strategies, such molecular analyses have proven particularly productive in expanding our knowledge of biological processes.

Much of the information in this text owes its existence to the application of recombinant DNA technologies to studies of gene structure and function. A few of the most notable discoveries that resulted directly from gene cloning studies are (1) the presence of noncoding intervening sequences or introns in a few genes of the T-even bacteriophages and in most genes of higher plants and animals, (2) the existence of several "splicing" pathways used in the removal of intron sequences from primary transcripts, (3) the novel arrangement of the genetic material encoding antibodies as tandem arrays of "gene segments" that assort themselves into an enormous variety of combinations via genome "splicing" rearrangements during B lymphocyte differentiation, (4) oncogenes and protooncogenes and their interrelationship, and (5) the presence of conserved "homeobox" sequences in developmentally important regulatory genes.

In the future, we can expect the genetic control of growth and development to be worked out in higher plants and animals. Recent results already indicate that the conserved homeobox sequences probably encode DNA-binding domains in proteins that regulate the cascades of gene expression responsible for cell differentiation in complex eukaryotes. Just as we now know the complete pathway of morphogenesis of bacteriophage T4 (Chapter 10), ***in the future we will know the complete pathway of morphogenesis of a yeast cell, a fruit fly, an* Arabidopsis *plant, or, indeed, even a human being.*** Moreover, although we cannot say when, we can be certain that at some point in the future we will understand the molecular basis of learning and memory and will know what molecular events underlie the aging process. Most importantly, at some future time we will understand the complex mechanisms that regulate cell division in humans and, hopefully, will be able to use this knowledge to prevent and/or cure at least some types of human cancer. And what about AIDS and other serious viral infections? Clearly, much important work remains to be done by the next generation of biologists.

Saccharomyces cerevisiae, a Eukaryotic *Escherichia coli*

Baker's yeast, the budding yeast *Saccharomyces cerevisiae,* has most of the advantages of a prokaryotic system such as *E. coli,* but is a true eukaryote. *Saccharomyces cerevisiae* is the ideal model eukaryotic system for genetic analysis. It can reproduce both asexually (budding) and sexually (mating and meiosis) and can be cultured in either the haploid or the diploid state. Meiosis gives rise to four ascospores that are maintained in an ascus and are easily subjected to unordered tetrad analysis. Large populations of yeast cells can easily be cultured, mutagenized, and screened for phenotypes of interest in a few petri plates. Over 500 genes have already been mapped on the 16 chromosomes of *S. cerevisiae.* Yeast chromosomes undergo high frequencies of recombination, and numerous selection protocols have been devised to introduce foreign genes, to mutationally disrupt endogenous genes, and to retrieve mutant or wild-type genes on specially designed vectors. In fact, yeast currently is an ideal system for genetic analysis and genetic engineering in that most of the desired tools are already in hand.

The "wish list" of any geneticist whose objectives include the genetic manipulation of an organism for either functional dissection or engineering purposes would undoubtedly include the following capabilities:

1. Isolation of any gene or DNA sequence of interest;
2. Site-specific mutagenesis of any isolated DNA sequence;

3. Introduction of any gene or DNA sequence, either native or foreign, of interest into the desired host:
   a. Integrated into random chromosomal sites,
   b. Integrated into homologous chromosomal sites,
   c. Not integrated, on autonomously replicating vectors,
   d. With control of copy number,
   e. With temporal, spatial, or environment-specific control of expression;
4. Targeted disruption of a gene or DNA sequence of interest;
5. Direct retrieval of a mutant gene of interest;
6. Targeted replacement of a mutant gene with its wild-type allele or replacement of a wild-type gene with a specific mutant allele.

All the genetic manipulations listed can be carried out in *S. cerevisiae* using available vectors and established procedures. Although essentially all genes can be isolated from all organisms and subjected to site-specific mutagenesis, these genes can be reintroduced into most higher eukaryotic organisms only at random chromosomal sites of integration. In the past, targeted gene disruptions and replacements have been largely restricted to lower eukaryotes, particularly yeast; however, some successful targeted integration events have recently been reported in animal cells. The direct retrieval of mutant genes has to date only been achieved in yeast.

Yeast cells can be transformed with free DNA molecules in much the same manner as *E. coli* cells. Yeast cells have very thick, largely impenetrable cell walls. Thus, one might correctly anticipate that intact yeast cells will not take up free DNA molecules present in the surrounding milieu. However, by enzymatically removing major components of the rigid cell walls, ***spheroplasts*** can be produced that are permeable to naked DNA molecules in the presence of calcium ions and polyethylene glycol. Several different types of DNA vectors (Fig. 24.1) can then be used to transform these spheroplasts, each type of vector having been designed for a specific type of genetic manipulation. All the plasmid vectors contain a selectable marker gene and restriction enzyme cleavage sites for use in inserting foreign DNAs. Additional components are present or absent depending on the desired application(s).

Because of the great utility of *E. coli* as the host cell of choice for site-specific mutagenesis, isolation of plasmid DNAs for subcloning, construction of nested families of deletions, and for the purification of single-stranded template DNAs for sequencing, some of the most useful cloning vectors are the ***"shuttle vectors" that can replicate in both* E. coli *and in the eukaryotic cell of interest.*** Shuttle vectors designed for use in *S. cerevisiae* contain both an *E. coli* origin of replication and a yeast origin of replication, along with selectable marker genes and a cloning site (Fig. 24.2). Actually, most of the currently used vectors contain a

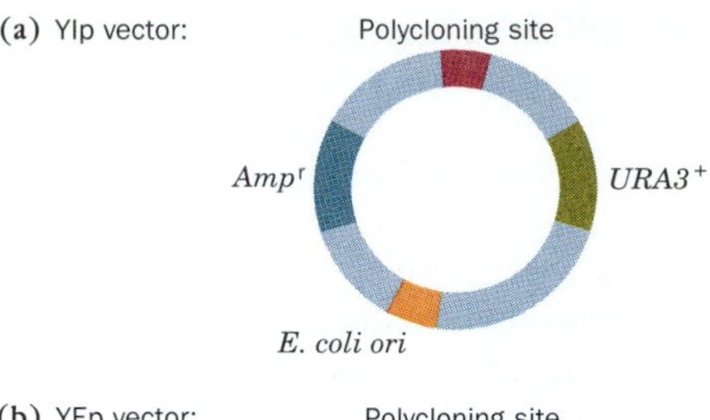

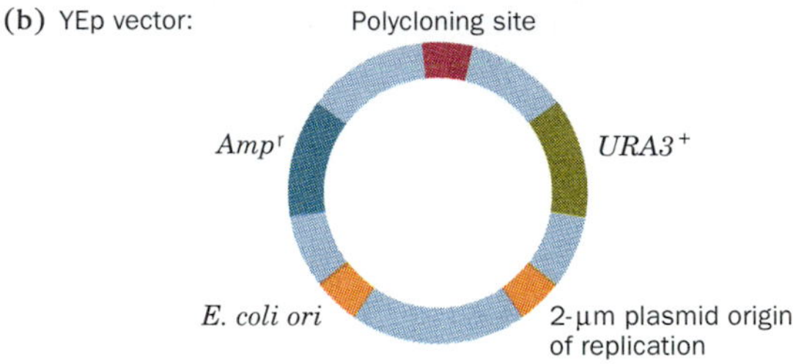

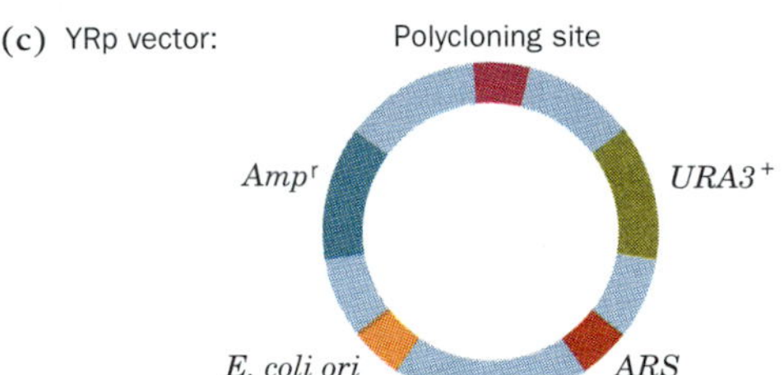

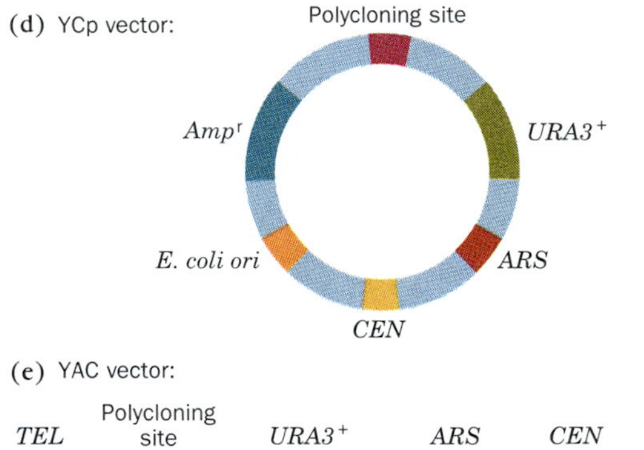

(e) YAC vector:

TEL Polycloning site $URA3^+$ ARS CEN TEL

Figure 24.1 Structures of five classes of yeast vectors used for gene cloning and genetic manipulations of various kinds. YIps, YEps, YRps, and YCps are circular plasmid vectors with different components and different properties as described in detail in the text; YACs are linear artificial chromosomes. Abbreviations are as follows: *ori* = origin of replication functional in *E. coli; ARS* (Autonomously Replicating Sequence) = yeast chromosomal origin of replication; *CEN* = yeast centromere sequence; $URA3^+$ = wild-type allele of yeast *URA3* gene, yeast selectable marker (uracil prototrophy); Amp^r = ampicillin resistance (*E. coli* selectable marker). YIp vectors, no yeast origin of replication, must integrate; YEp and YRp vectors, yeast origin, no centromere, autonomous replication, but unstable without selection; YCp vectors, YRp vectors with centromere added, more stable; YAC vectors, artificial linear yeast "minichromosomes" with centromere and telomeres, accept large (200–400 kb) foreign DNA inserts.

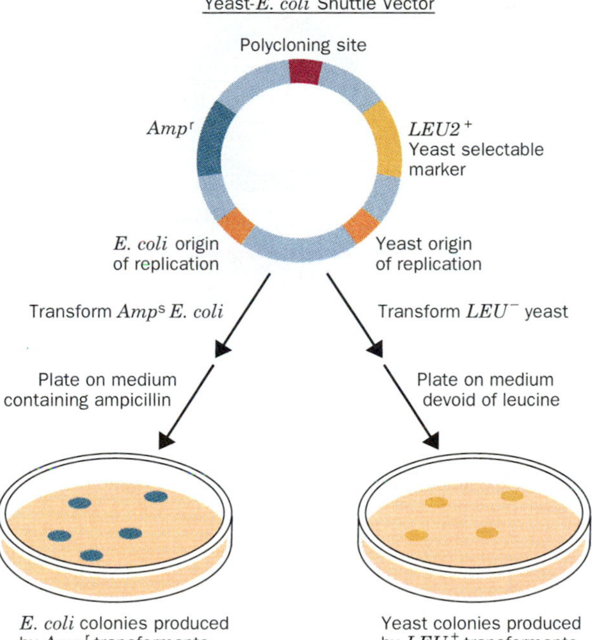

Figure 24.2 Basic structure and utility of an *E. coli*–yeast shuttle vector. These vectors can replicate in both *E. coli* and yeast. Thus, shuttle vectors allow investigators to move genes back and forth between these two organisms and to study the structure and/or function of genes of interest in either host or in both hosts.

polycloning site with recognition sequences for a whole set of restriction endonucleases. The shuttle vectors are extremely useful for genetic dissections. A yeast gene of interest can be subcloned into a shuttle vector, subjected to site-specific mutagenesis in *E. coli,* and then moved back to yeast to examine the effects of the induced modifications on function in native host cells. Similar shuttle vectors are available for use in animal systems; they replicate both in *E. coli* and in animal cells. Unfortunately, no well-established *E. coli*–plant shuttle vectors are yet available. In contrast, four distinct classes of *E. coli*–yeast shuttle vectors are available and are widely used, with the vector of choice depending on the specific experimental objective.

The ***YIp*** (***Y***east ***I***ntegrative ***p***lasmid) vectors contain one or more selectable markers (*URA3* is present on the YIp shown in Fig. 24.1*a*) and one or more unique cloning sites, but ***no*** origin of replication. As a result, YIp vectors must integrate into the host chromosome in order to produce stable transformants. Integration occurs by means of recombination between homologous DNA sequences present in the YIp vector and in a chromosome of the host cell. In the case of the YIp vector shown in Fig. 24.1*a*, integration will occur at the *URA3* locus on host chromosome V since that is the only region of homology between the plasmid and the yeast chromosomes. If a second yeast chromosomal DNA sequence is introduced into a cloning site on the YIp vector, then integration events will occur at both *URA3* and the new region of homology. If the two regions of homology are of approximately the same length, then integration events will be about equally frequent at the two sites.

Many strains of *S. cerevisiae* contain ***a native 2-μm plasmid*** (6300 nucleotide-pairs in length). Although this plasmid is unessential (in fact, it has no known function), it is present as a self-replicating genetic element in the nuclei of numerous natural isolates of *S. cerevisiae.* The yeast 2-μm plasmid was an obvious candidate for use in constructing yeast cloning vectors. Vectors derived from this 2-μm plasmid are called ***YEp*** (***Y***east ***E***pisomal ***p***lasmid) vectors (Fig. 24.1*b*). Their autonomous replication is governed by the origin of replication and two key replication control genes from the 2-μm plasmid. The YEp vectors have the advantage of being present at relatively high copy number, usually about 30 copies per cell. These YEp vectors are unstable. Selectable marker genes must be present on YEp vectors and appropriate selection must be maintained to avoid the loss of YEp plasmids from host cells.

Yeast chromosomes, like other eukaryotic chromosomes, have multiple origins of replication distributed along the giant DNA molecules in these chromosomes. In yeast, these origins of replication are spaced 20,000 to 100,000 nucleotide-pairs apart along each chromosome. Each origin of replication in yeast is called an **ARS** (for ***A***utonomously ***R***eplicating ***S***equence); there are 300 to 400 *ARS*s per haploid yeast cell. The addition of an *ARS* to a YIp vector produces a plasmid that can replicate autonomously. *ARS*-containing plasmids of this type are called ***YRp*** (***Y***east ***R***eplicating ***p***lasmid) vectors (Fig. 24.1*c*). Like YEp vectors, they are unstable and require continual selection for maintenance. YRp vectors replicate once per cell generation, just like the replicons from which the *ARS*s that they carry were isolated. Occasionally, integration events occur and produce stably transformed cell lines.

YRp vectors can be made stable in host cells by adding chromosomal centromere (**CEN**) sequences. Yeast centromeres are 220 to 250 nucleotide-pairs in length; however, all essential sequences reside within a 130-nucleotide-pair-long core. The presence of a *CEN* sequence apparently facilitates the faithful segregation of these "minichromosomes" to daughter nuclei during mitosis and meiosis. Plasmids that contain both *ARS* and *CEN* sequences are called ***YCp*** (***Y***east ***C***entromere-containing ***p***lasmid) vectors (Fig. 24.1*d*). The major advantage of YCp vectors is that they are maintained as stably transmitted, single-copy plasmids or "minichromosomes."

Finally, linear ***YAC*** (***Y****east* ***A****ritificial* ***C****hromosome*) vectors can be constructed from YCp vectors by linearization and the addition of a telomere (**TEL**) sequence to each terminus (Fig. 24.1*e*). The major advantage of YAC vectors is that they permit the insertion of very large sequences of foreign DNA. Small YACs are somewhat unstable, but large YACs (≥300 kb) are just as stable as native yeast chromosomes.

Clearly, the vector of choice will depend on the objectives of the study. YIp vectors should be used whenever integration is desired. Surprisingly, integrative transformation frequencies are increased about 100-fold if YIp vectors are first linearized by cleavage with a site-specific restriction endonuclease. The free ends of DNA molecules are highly recombinogenic and invade homologous chromosomal sequences as the first step in the integrative crossover event. YCp vectors are most appropriate for complementation analyses since the introduced alleles and the endogenous chromosomal alleles will be present in the same copy number. YEp and YRp vectors are the vectors of choice for high-level expression studies since these plasmids are present in high copy number. YAC vectors must be used to study "long-range" (long range at the molecular level, not the genetic level) linkage relationships since they are the only vectors that are stable with 10^5- to 10^6-bp inserts.

For the geneticist, one major advantage of yeast is the ease with which specific gene disruptions, gene replacements, and gene retrievals can be accomplished. For example, if one wants to study the expression of a modified gene in its native host cell, one must first somehow remove or block the expression of the endogenous copy of that gene. In yeast, this is easily done by targeted gene disruptions. Such gene disruptions can also be used to determine whether a gene is essential or nonessential. D. Shortle, J. Haber, and D. Botstein's 1982 demonstration that the actin gene of *S. cerevisiae* is essential provides a good illustration of the use of the specific gene disruption procedure (Fig. 24.3). They inserted the central part of the yeast actin gene into a YIp vector carrying the $URA3^+$ selectable marker gene. They then transformed diploid $URA3^-$ cells with this vector. About half of the integration events occurred at the *URA3* locus (previously mapped); the other half occurred at the actin locus. Recombination between the internal fragment of the actin gene on the YIp vector and the complete actin gene on the recipient chromosome will yield two partial, nonfunctional copies of the actin gene (Fig. 24.3). When diploid cells that contained the YIp DNA inserted at the actin locus were stimulated to go through meiosis and sporulate, only half of the haploid ascospores were viable, and they were all $URA3^-$. No $URA3^+$ spores survived as a result of the absence of a functional actin gene.

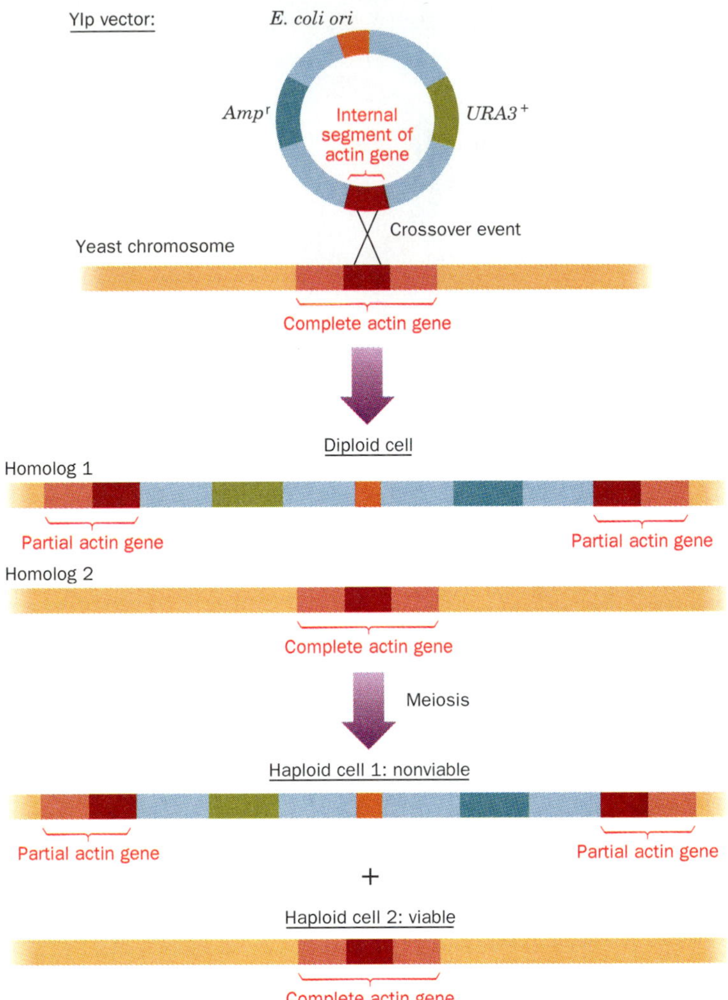

Figure 24.3 Schematic illustration of targeted gene disruption in yeast. Diploid $URA3^-$ cells are transformed with a YIp vector carrying $URA3^+$ as a selectable marker and an internal segment of the actin gene. Integration events at the *URA3* locus are not shown. Homologous recombination between the actin gene segment in the plasmid and the complete actin gene in the chromosome of the recipient cell will yield two incomplete, inactive actin genes. The diploid transformants will be viable because of the complete, functional copy of the actin gene on the homologous chromosome. However, if the actin gene is essential, half of the haploid ascospores produced by meiosis and sporulation will be nonviable due to the absence of a complete, functional actin gene. On the other hand, if the actin gene is nonessential, all the haploid ascospores will be viable. When an experiment of this type was carried out by D. Shortle, J. Haber, and D. Botstein, the results showed that the actin gene was essential; half of the ascospores produced were nonviable.

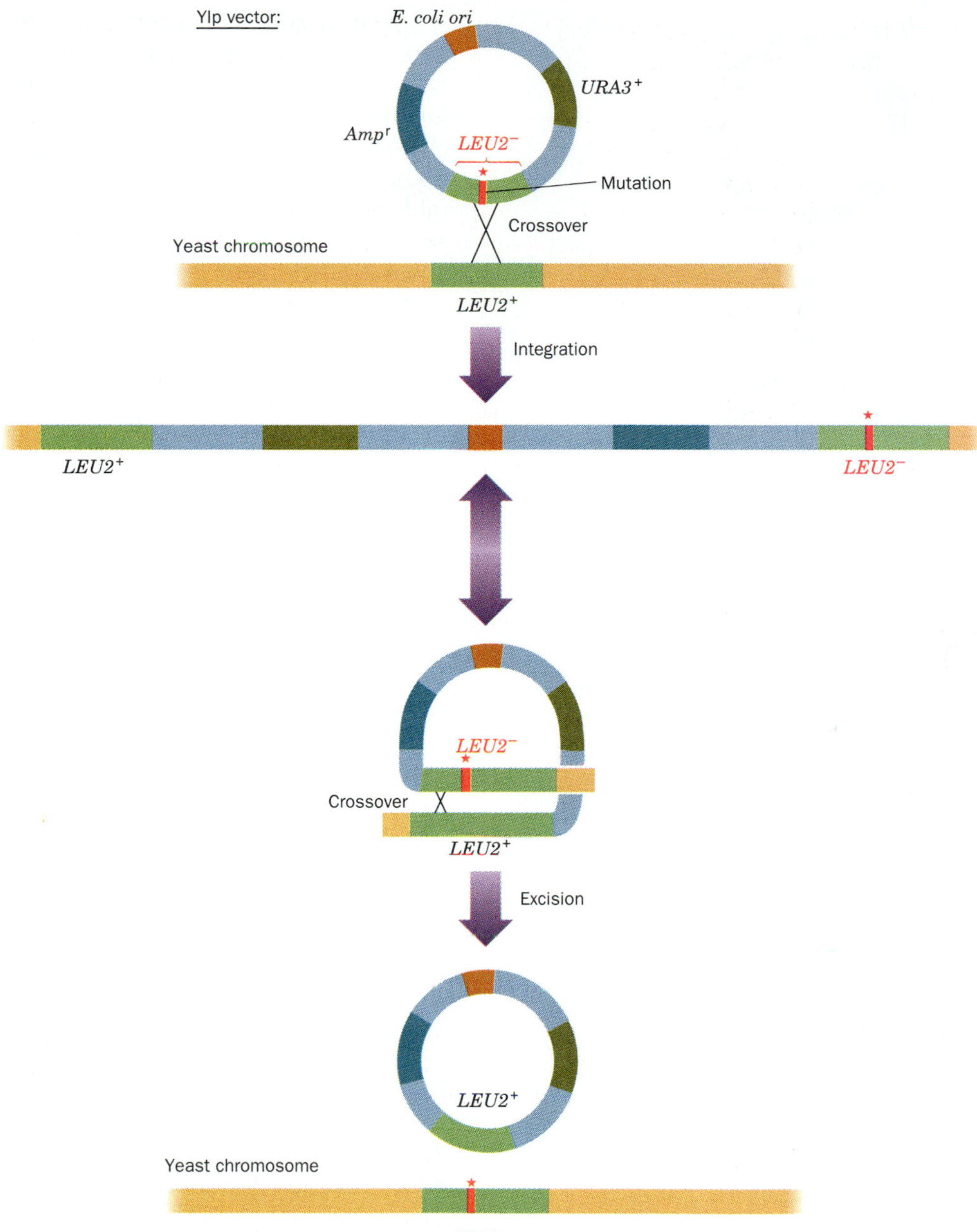

Figure 24.4 Diagram of the use of YIp vectors to perform gene displacements in yeast. In the example shown, the wild-type *LEU2*$^+$ gene is replaced by a specific *LEU2* mutant allele designated *LEU2*$^{*-}$. The mutation indicated by the * might well be a base-pair substitution produced by *in vitro* site-specific mutagenesis.

In addition to specific gene disruptions, gene replacements and gene retrievals can be carried out with YIp vectors and appropriate selection schemes. For example, wild-type genes can be replaced with specific mutant alleles generated *in vitro* by site-specific mutagenesis, and the mutant alleles can then be analyzed for function *in vivo.* Figure 24.4 illustrates the use of a YIp plasmid to introduce a specific *LEU2*$^-$ mutation (symbolized *LEU2*$^{*-}$) into the endogenous *LEU2* gene on chromosome III of yeast. Transformation of haploid *URA3*$^-$ *LEU2*$^+$ yeast cells with a *URA3*$^+$ *LEU2*$^{*-}$ YIp vector will yield transformants about half of which will contain the YIp DNA integrated at the *LEU2* locus. Excision of the YIp plasmid can subsequently occur via recombination between the tandem copies of the *LEU2* gene (Fig. 24.4). Depending on the exact position of the recombination event within the *LEU2* gene, the yeast chromosome will be left with either the original *LEU2*$^+$ allele or a *LEU2*$^-$ allele that carries the *LEU2*$^{*-}$ mutation introduced on the YIp vector.

Specific gene retrievals in yeast are carried out with either YEp or YRp "gapped vectors" (Fig. 24.5).

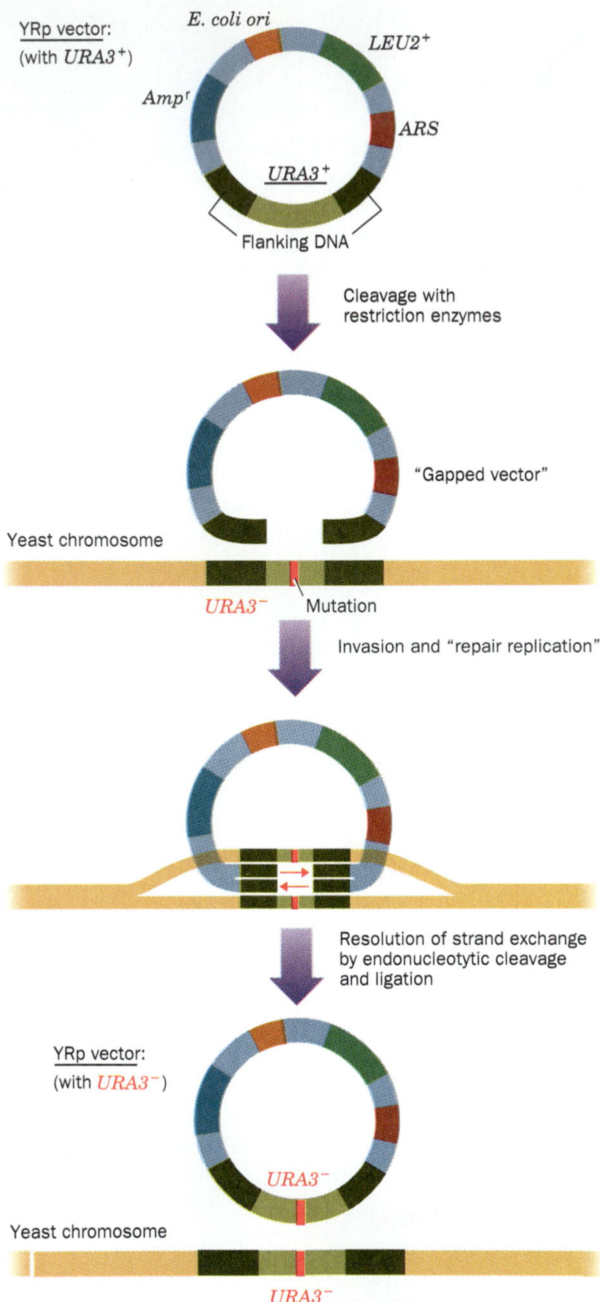

Figure 24.5 Illustration of the use of "gapped YRp vectors" for gene retrievals in yeast. The example shows the retrieval of a *URA3*$^-$ allele from its normal chromosomal location by the "gapped vector." Although YRp vectors contain an origin of replication (*ARS*), they are not stable as autonomous linear macromolecules because of the absence of telomere sequences. The highly recombinogenic ends of the "gapped vectors" invade the homologous flanking sequences in the chromosome and "repair synthesis" occurs using the chromosomal DNA strands as templates. After the strand exchanges are resolved by sequential endonuclease and ligase reactions, the resulting intact plasmid contains a copy of the *URA3* allele that was originally present in the host chromosome.

"Gapped vectors" are constructed by inserting the gene of interest plus flanking sequences into the polycloning site of a YEp or YRp vector, and then excising just the gene by cleavage with restriction endonucleases at sites immediately adjacent to the gene. Thus, the only remaining homology between the plasmid and the yeast chromosomes involves the flanking sequences left in the "gapped vectors." Because of the absence of telomere sequences, these linear "gapped vectors" are unstable in yeast cells unless they integrate or recircularize. Recircularization occurs via "repair of the gap," which involves copying the missing sequence from the host chromosome. Apparently, the highly recombinogenic ends of the "gapped vectors" invade the homologous chromosomal DNA sequences and the 3′-OH termini are covalently extended by repair synthesis with the endogenous DNA strands serving as templates. The resulting strand exchanges are subsequently resolved by endonucleolytic cleavages and ligation reactions yielding intact plasmids and chromosomes (Fig. 24.5). Note that these plasmids now carry a copy of the gene that was originally present in the chromosome of the host cell.

These gene disruption, gene replacement, and gene retrieval procedures are powerful tools for genetic analyses of prokaryotes, yeasts, and other eukaryotic microorganisms. In the future, similar tools will need to be developed that are applicable to higher plants and animals. Some progress in this direction has recently been reported with animal cell-culture systems.

Drosophila, P-Element Transformation

P elements are the transposable genetic elements that cause hybrid dysgenesis in *Drosophila* (Chapter 9). The high frequency of P-element transposition that was observed when P strain males were mated to females that lack P elements (M strains) demonstrated that P elements can integrate efficiently at many sites in *Drosophila* chromosomes. This, in turn, suggested that P elements might be useful as vectors for gene transfer, and, in 1982, A.C. Spradling and G.M. Rubin demonstrated the utility of ***P element-mediated gene introductions in* Drosophila *by injecting P element-containing plasmids into M strain embryos.*** A high frequency (20 to 50 percent) of the surviving embryos produced progeny flies with P elements integrated into random chromosomal sites. Since P-element transposition is restricted to germ line cells (see Chapter 9), the microinjections must be done using very young (0- to 90-minute) embryos.

Complete P elements contain 31-base-pair perfect inverted repeats at their termini, subterminal sequences, and four internal open reading frames (ORFs). The product(s) of all four ORFs is (are) required for transposition. ***Incomplete P elements*** lack one or more of the ORFs. As a result, incomplete P elements undergo transposition only when they are present in the same cell with at least one complete P element. A complete P element can provide the ***transposase*** required for movement of all P elements, complete or incomplete, within a cell, that is, the ***P element transposase acts both in* cis *and in* trans.** Given the presence of P-element transposase in a cell, all that a P element needs to be capable of transposition is intact ends, that is, the terminal repeats and certain adjacent subterminal sequences. This greatly facilitates the construction of P-element vectors for use in gene transfer experiments. Most of the ***internal sequences of the P element can be replaced with foreign DNA,*** and the ***transposase can be supplied in* trans *by a "helper P element."*** Only the DNA that is present between the terminal repeats of a P element is transferred to the host chromosome during integration. Thus, shuttle vectors can be constructed that carry an *E. coli* origin of replication and an *E. coli* selectable marker gene external to the P-element termini in the plasmid (Fig. 24.6). Finally, useful shuttle vectors also contain (1) a ***selectable marker gene for use in* Drosophila** and (2)

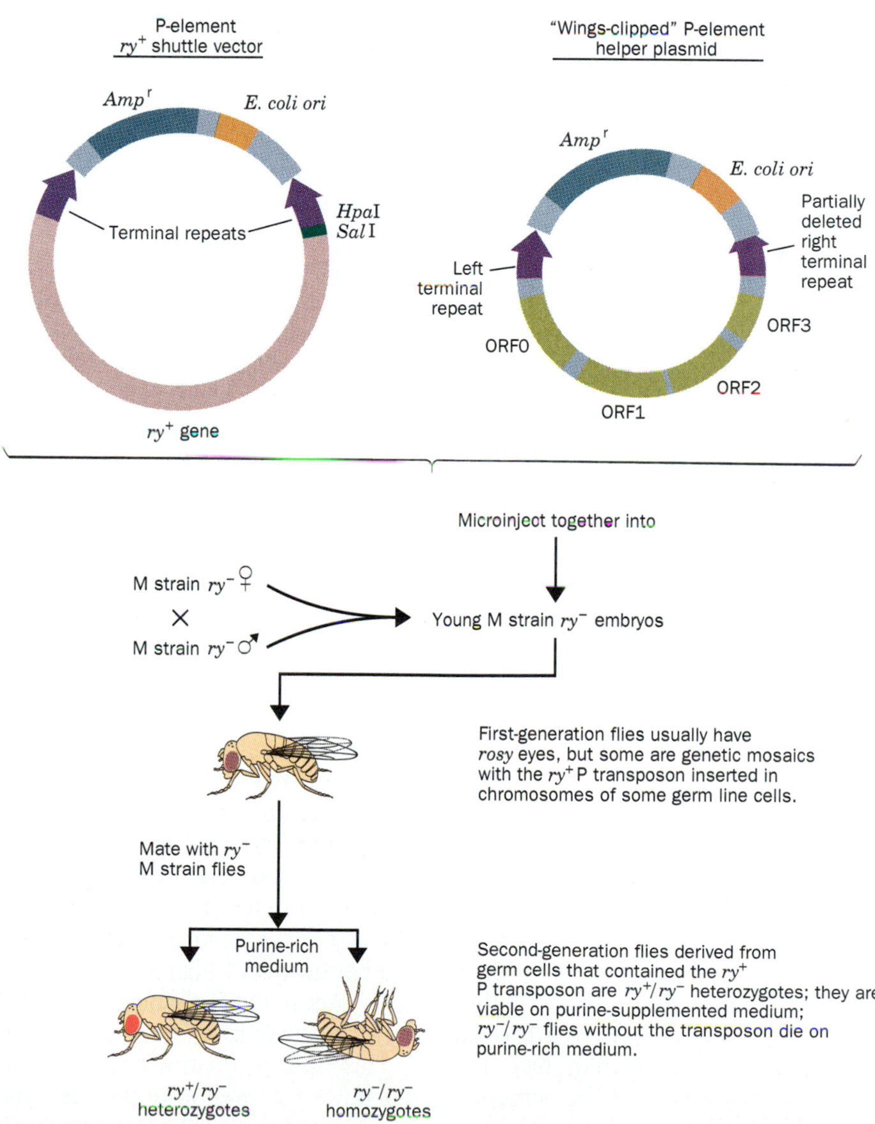

Figure 24.6 Diagram illustrating (1) the structure of a ry^+ P-element shuttle vector, (2) the structure of a "wings-clipped" P-element helper vector, and (3) their use in introducing genes or DNA sequences of interest into *Drosophila.* The helper P element provides transposase activity in *trans* for transposition (integration) of the incomplete P element carrying the ry^+ gene. The helper P element is defective due to the deletion of most (23 of 31 base-pairs) of the right terminal repeat and cannot itself transpose (integrate). *Hpa* I and *Sal* I are two unique restriction enzyme cleavage sites that can be used to insert DNA sequences of interest into the P-element transposon on the ry^+ vector.

one or more ***restriction enzyme cleavage sites for use in inserting DNA sequences*** of interest, all located between the terminal repeats of the P element. The wild-type *rosy* (ry^+) gene is often used as the selectable marker in *Drosophila*. The *rosy* gene encodes the enzyme xanthine dehydrogenase, which is essential for growth of *Drosophila* on media containing high concentrations of purines; ry^- embryos fail to survive to eclosion on purine-supplemented media.

If a plasmid carrying a complete, fully functional P element is used as helper (to provide the transposase), some transformants may carry copies of both the incomplete P element from the vector plasmid and the complete P element from the helper plasmid. This possibility of cointegration events greatly complicates any genetic or molecular studies of the transformants. Such double P-element transformation events can be avoided by using a ***defective helper P element*** that provides transposase activity but cannot integrate itself. R.E. Karess and G.M. Rubin constructed a defective helper P element of this type by deleting 23 base-pairs from the end of the 31-base-pair right terminal repeat. Since intact terminal repeats are required for transposition, this helper P element (called "wings clipped") cannot integrate itself, but it can provide transposase in *trans* for integration of incomplete P elements from vector plasmids. Thus, the use of the "wings clipped" P-element helper plasmid assures that only the incomplete P element carrying the inserted gene or DNA sequence of interest will be transferred to host chromosomes after microinjections of vector and helper P-element plasmids (Fig. 24.6).

P element-mediated gene transfers are performed by injecting young *Drosophila* M strain embryos with both (1) P-element ry^+ vector DNA and (2) "wings-clipped" helper P-element plasmid DNA. The injected embryos are multinucleate single cells with the nuclei destined to form germ cells clustered at the posterior end of the embryo. The DNA is injected through a fine glass needle into the posterior region of the embryo. The flies that develop from the injected embryos usually have rosy-colored eyes, but some are mosaics with the ry^+ P-element transposon integrated into one or more chromosomal sites in some germ line cells. These flies are then mated to homozygous ry^- M strain flies, and the progeny are screened for ry^+ transformants. This mating passes the chromosomes of the flies that developed from the injected embryos through haploid eggs and sperm and, therefore, eliminates the mosaicism. The progeny flies can either be screened visually for wild-type (ry^+) red eyes or selected for the ry^+ phenotype on purine-supplemented medium (Fig. 24.6). ***Offspring that are*** ry^-/ry^- ***die before eclosion on purine-rich media.***

P element-mediated gene introductions have been extremely useful in studying gene structure, function, and regulation in *Drosophila*. This approach has proven particularly fruitful in dissecting the regulatory sequences that control gene expression. It has already been used to precisely define the regulatory sequences of a large number of genes in *D. melanogaster* and promises to continue to be a powerful tool for genetic analysis in the future.

Arabidopsis thaliana, the *Drosophila* of the Plant Kingdom

Despite the fact that genetics originated with Mendel's work on garden peas, our knowledge of the genetics of higher plants has lagged far behind our knowledge of the genetics of higher animals. Our understanding of the genetics of the fruit fly *D. melanogaster* far surpasses our knowledge of the genetics of any other higher organism, although recent work on the nematode *Caenorhabditis elegans* is rapidly closing the gap in the case of this little worm (Chapter 15).

Of the higher plants, maize is probably the best known genetically. It has been extensively studied for three major reasons. (1) Maize is of major economic importance as one of the world's foremost food crops. (2) Being a diploid, maize is subject to standard genetic analyses, whereas many of the other economically important food crops such as oats and wheat are polyploids (Chapter 19). (3) The meiotic chromosomes of maize are excellent subjects for cytogenetic studies (see Fig. 18.1). To date, about 200 genetic loci have been mapped on the linkage groups that correspond to the 10 chromosomes of maize (see Fig. 7.8). The major disadvantages of maize for genetic studies are (1) its relatively long generation time (a maximum of three generations per year) and (2) the large size and the complexity of the maize genome. Whereas the genome of yeast (*S. cerevisiae*) is only 3.5 times the size of the *E. coli* genome, ***the maize genome is 3000 times as large as the*** E. coli ***genome***. In addition, maize contains a large number of repetitive DNA sequences that are widely dispersed throughout the genome. The large genome size and the large amount of dispersed repetitive DNA make molecular genetic analyses much more difficult. These disadvantages of maize for molecular work led geneticists to search for a better model plant system.

Arabidopsis thaliana (Fig. 24.7) is a small, economically unimportant member of the mustard family. This little weed has been used for classical genetic studies for over four decades; its advantages for genetic studies have been extensively documented by G. Rédei, M. Koornneef, A. R. Kranz, and colleagues. Nevertheless, *A. thaliana* did not become a major research subject until the mid-1980s when (1) E. Meyerowitz and R. Pruitt demonstrated that the *A. thaliana* genome is very small and contains almost no inter-

Figure 24.7 Drawing of the phenotype of the small model plant *Arabidopsis thaliana.* Actually, the size and phenotype of this small crucifer vary dramatically depending on the environmental conditions. It will grow over a foot tall and spread about a foot in diameter under optimal conditions. In contrast, it will also grow about a centimeter tall, flower, and set seeds when restricted by growth in a 15 × 60-mm culture dish. In this respect, *Arabidopsis* exhibits all the features of a true weed. (Drawing after M. A. Estelle and C. R. Somerville, *Trends in Genet.* 2:89, 1986.)

spersed repetitive DNA and (2) C. Somerville and coworkers demonstrated the feasibility of mutational dissections in *A. thaliana* by mutagenesis and "brute-force" (no genetic selection) screening. During the last three years, *A. thaliana* has emerged as the plant model system of choice, the *D. melanogaster* or *C. elegans* of the plant kingdom. *Arabidopsis thaliana* is now considered to be the best plant system for genetic dissections of plant development and other biological processes unique to plants (e.g., photosynthesis). ***The major advantages of* A. thaliana *are:***

1. ***small size***—very large numbers of plants can be grown in relatively small growth chambers;
2. ***short generation time***—about 5 weeks;
3. ***high seed production***—up to 40,000 seeds per plant;
4. ***very small genome***—7×10^7 nucleotide-pairs (about 20 times larger than the *E. coli* genome and less than 1/100th the size of the maize genome);
5. ***very little interspersed repetitive DNA***—unique sequences of DNA are on averge 120,000 nucleotide-pairs in length (as compared to about 1000 nucleotide-pairs in length in maize); and
6. ***natural self-pollination***—very low levels of outcrossing (as opposed to natural cross-pollination in maize).

The small size of *A. thaliana* permits one to grow large populations in growth chambers under controlled environmental conditions and greatly facilitates mutational dissections. Large populations of plants can be screened to permit the detection of rare mutant phenotypes, and these screenings can be done at much lower expense than for large plants such as maize. The short generation time allows one to collect eight to nine generations of genetic data per year, thus greatly accelerating progress.

The large seed yield per plant and the natural self-pollination that occur in *Arabidopsis* are very important for mutational-dissection studies. Seeds are mutagenized (usually with ethylmethane sulfonate, EMS; see Chapter 11) and planted. The first-generation plants, called ***M_1 plants*** (for ***1***st generation after ***M***utagenesis), will be heterozygous for the newly induced mutations. However, these M_1 plants will produce large numbers of seeds containing ***M_2*** embryos, and one-fourth of the embryos produced by the self-pollination of a heterozygous M_1 plant will be homozygous for the new mutation. Self-pollination is virtually assured in *Arabidopsis* because pollen is released and fertilization occurs before the flowers even open. Therefore, under natural conditions, receptive stigmas are only accessible to pollen produced in the same flower. The seeds produced by the M_1 plants are planted, and the resulting M_2 plants are screened for the desired mutant phenotype(s). Since most mutations are recessive, self-pollination is the key to rapidly generating homozygous plants expressing the mutant phenotypes of interest.

Because of the small size of the *Arabidopsis* genome, it is possible to identify a mutation in essentially any gene of interest by screening a relatively small number of M_2 plants. The work of C. Somerville and colleagues has shown that screening 2000 to 5000 M_2 plants is often sufficient to permit one to identify a mutant with a defect in a particular gene. This number is low enough that one can frequently identify mutants even when selection is not available and even when the mutation has no visible effect on phenotype. For example, Somerville and coworkers have isolated several mutants with altered membrane composition by using gas chromatography to screen extracts of M_2 plants for altered fatty acid profiles. Given reasonably simple enzyme assays, such direct "brute-force" screening should be applicable to genes encoding enzymes in most, if not all, metabolic pathways of *Arabidopsis*.

A major advantage of *Arabidopsis* is that once a mutation has been identified in a gene of interest, the mutant allele can be mapped and used as a tool in cloning the gene. Because of the small genome size and the rare occurrence of interspersed repetitive DNA sequences in *Arabidopsis*, genes can be cloned by a procedure called ***"chromosome walking."*** The mutant allele of interest is simply mapped relative to the positions of previously mapped molecular markers by standard genetic crosses, and then one "walks" along the chromosome from the closest marker by isolating overlapping (cross-hybridizing) genomic clones until one reaches the gene of interest (see the following discussion for details). Chromosome walking is difficult to impossible in most higher plants because of the huge genomes (one usually has too far to "walk") and abundance of repetitive DNA (each repeated sequence is a potential roadblock to the "walk").

Genetic crosses in *Arabidopsis* are performed by hand under a dissecting scope or magnifying glass by ***emasculating flowers*** (removing the stamens) prior to the release of pollen and then ***applying pollen from the male parent of choice to the stigmas*** of the emasculated female parent. Since the flowers of *Arabidopsis* are quite small, the emasculation procedure is somewhat tedious; however, the other major advantages of *Arabidopsis* far outweight this minor disadvantage.

The molecular markers that are used to initiate chromosome "walks" may be cDNA or genomic clones of genes that have previously been mapped using mutant alleles. However, more frequently, they are ***R***estriction ***F***ragment ***L***ength ***P***olymorphisms (***RFLPs***). RFLPs are simply variations in the lengths of the DNA fragments produced by cleaving DNA molecules with specific restriction endonucleases. Since restriction enzymes cut DNA in a sequence-specific manner (see Chapter 13), every homologous DNA molecule from every cell of a totally (all genes and DNA sequences) homozygous organism will be cut at exactly the same sites. This is what allows one to isolate large quantities of specific DNA fragments for subcloning and sequencing. However, the DNAs isolated from organisms in populations that have evolved independently for substantial periods of time (e.g., different ecotypes or geographical isolates) will usually exhibit sequence divergences that yield RFLPs. That is, base-pair substitutions, deletions, and additions will have occurred, and some of these changes will have altered the positions of restriction enzyme cleavage sites. Many of the changes that accumulate are ***"silent mutations,"*** mutations that occur in intergenic regions, in introns, and at exon positions that correspond to the 3′ bases of codons (the "degeneracy" or "wobble" positions) and have no perceptible effect on phenotype. However, these "silent mutations" will occasionally remove or add restriction enzyme cleavage sites and produce RFLPs (Fig. 24.8).

Not surprisingly, then, the DNAs of different geographical isolates, different ecotypes, and different inbred lines of a species usually contain RFLPs. These RFLPs can be visualized directly when the fragments in DNA digests are separated by agarose gel electrophoresis, stained with ethidium bromide, and viewed under ultraviolet illumination (see Chapter 13). Individual RFLPs can be detected by using specific cDNA or genomic clones as radioactive hybridization probes on genomic Southern blots (see Chapter 13). Since these RFLPs result from specific changes in DNA sequences (mutations), they can be mapped just like any other genetic marker. The RFLPs themselves are the "phenotypes" that are used to classify the progeny of crosses as parental or recombinant. The use of RFLPs as genetic markers in mapping experiments is illustrated in Fig. 24.9.

C. Chang, E.M. Meyerowitz, and colleagues have constructed an ***RFLP linkage map for*** **Arabidopsis thaliana** ***that contains 90 markers distributed over the five small chromosomes*** of this plant (Fig. 24.10). Given the small size of the genome, this places 50 percent of the genome within 1.9 map units or about 270 kilobase-pairs of the mapped sequences, well within the range of reasonable "chromosome walks." This map makes it possible to clone many *A. thaliana* genes that have been identified only by mutant phenotypes resulting from mapped mutations by "chromosome walking" (Fig. 24.11). "Chromosome walking" is done by selecting the molecular marker (RFLP or known gene clone) closest to the gene of interest and using this clone as a hybridization probe to screen a genomic library for overlapping sequences. By isolating a series of overlapping genomic clones, one simply "walks" down the chromosome the required distance to the gene of interest. Unless one has additional information about the orientation of the starting clone on the linkage map, the initial "walk" will have to proceed in both directions until one identifies another RFLP and determines whether it is closer to or farther away from the gene of interest than the starting RFLP. The only major obstacles to such "chromosome walks" are (1) the presence of repeated DNA sequences that lead the "walker" astray (e.g., to copies of the repeated sequence at other sites on the same chromosome or other chromosomes), or (2) the presence of sequences that cannot be cloned because they are lethal when present in the cloning vector being used. Verification that one has indeed isolated a clone

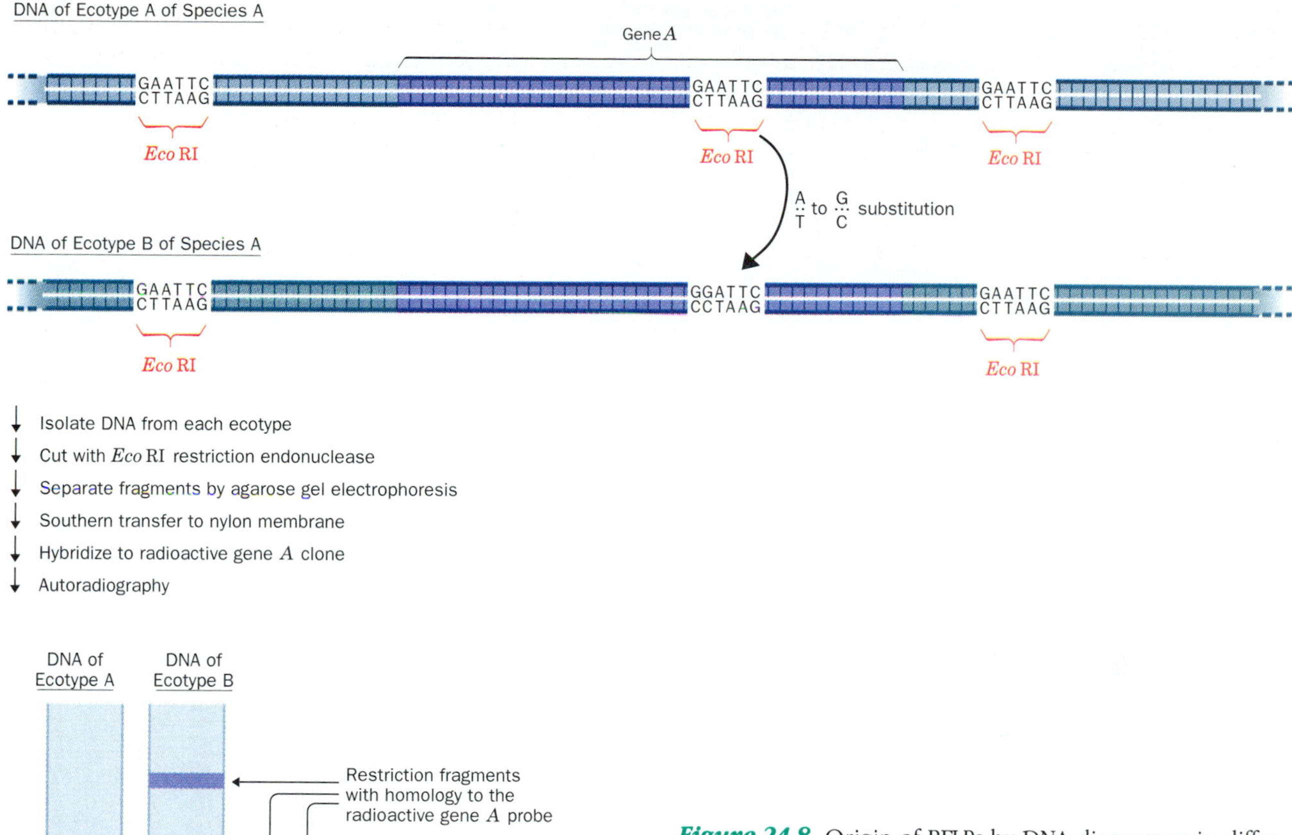

Figure 24.8 Origin of RFLPs by DNA divergence in different ecotypes of a species. In the example shown, an AT → GC base-pair substitution results in the loss of the central *Eco*RI recognition sequence present in gene *A* of the DNA of ecotype A. Presumably, this event occurred in a common progenitor or an ancestor in the early stages of the evolution of ecotype B. In the DNA of individuals of ecotype A, gene *A* sequences are present on two *Eco*RI restriction fragments. In that of individuals of ecotype B, all gene *A* sequences are present on one large *Eco*RI restriction fragment.

of the gene of interest is then accomplished by introducing the wild-type allele of the gene into a mutant plant and showing that the wild-type allele restores the wild-type phenotype (i.e., is dominant to the mutant allele). In *Arabidopsis,* genes can be introduced by using the Ti plasmid of *Agrobacterium tumefaciens* (see the section entitled The Ti Plasmid of *Agrobacterium tumefaciens* later in this chapter).

The small size of the *Arabidopsis thaliana* genome also makes it feasible to construct ***complete physical maps for all five chromosomes.*** By isolating large numbers of genomic clones and constructing detailed restriction maps of all of them, one can identify overlapping genomic clones and construct a physical map of the entire genome. In principle, this procedure is very simple (Fig. 24.12). However, in practice, this is a very formidable task when one is dealing with an entire genome. Indeed, such an undertaking is almost inconceivable in the case of very large genomes with substantial amounts of repetitive DNA (e.g., the maize genome). The construction of physical maps of entire genomes requires that vast amounts of data be searched for the presence of overlapping sets of restriction enzyme cleavage sites. Nevertheless, physical maps are currently being constructed for the relatively small (for higher eukaryotes) genomes of *A. thaliana* and *C. elegans.* The *A. thaliana* map, being constructed by H. Goodman and colleagues, is currently over 90 percent complete. When completed, Goodman and coworkers will have a clone bank containing individual clones that collectively cover every segment of the *A. thaliana* genome. Therefore, if someone needs a clone of a particular segment of the genome, that clone will already be available and catalogued in the

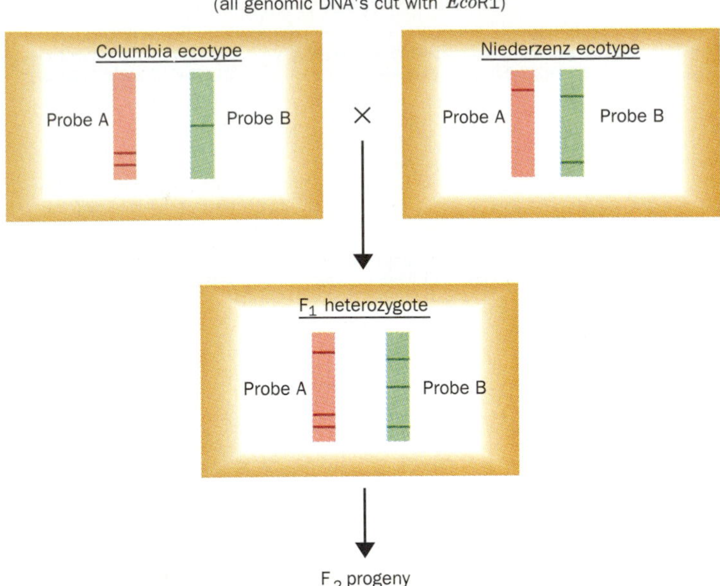

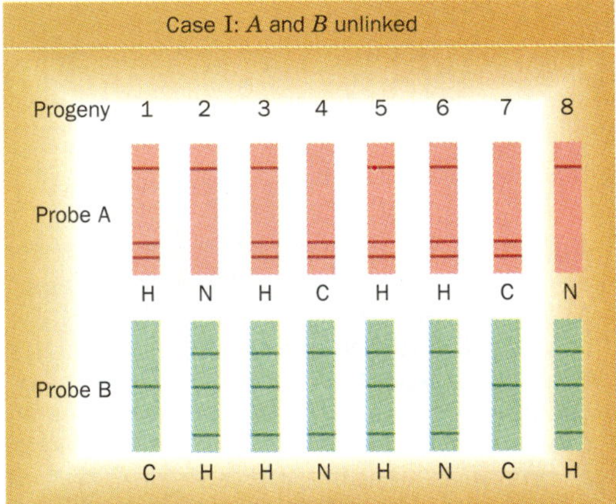

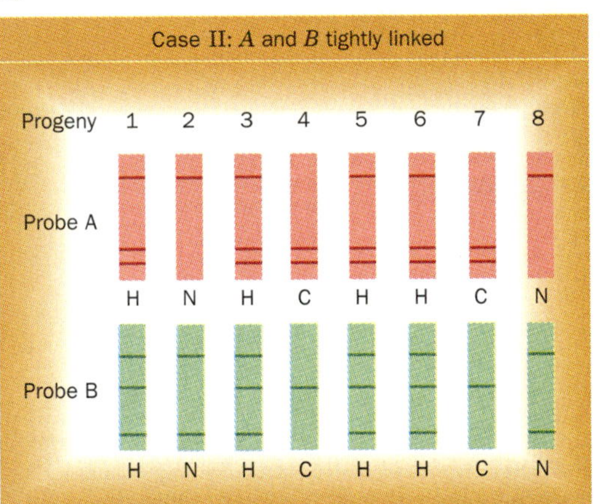

Figure 24.9 Simplified diagram of the technique of RFLP mapping. To minimize complexity, only (1) two ecotypes of *Arabidopsis thaliana,* Columbia and Niederzenz; (2) two probes, gene *A* and gene *B*; and (3) a single restriction endonuclease, *Eco*RI, were used in this example. In reality, several ecotypes, several restriction enzymes, and a large number of probes would be employed (see Fig. 24.10). Thus, all Southern blots shown were for *Eco*RI digests of the respective DNAs, and all blots were hybridized to a radioactive probe for either gene *A* (shown in red) or gene *B* (shown in green). Note that the parental ecotypes are polymorphic for both probes, that is, these ecotypes carry each gene, *A* and *B*, on restriction fragments of different sizes. The F_1 progeny are heterozygous and contain the restriction fragments of both parents as expected. When these F_1 plants are allowed to self-pollinate, one would expect one-fourth of the F_2 progeny to be homozygous for the gene *A* fragments of Columbia, one-fourth to be homozygous for the gene *A* fragments of Niederzenz, and one-half to be heterozygous like the F_1 plants and contain the gene *A* fragments of both Columbia and Niederzenz. The same 1:2:1 segregation pattern would be expected to occur for the gene *B* fragments. However, when the segregation patterns of both genes are examined together, one expects very different results depending on whether the two genes assort independently (Case I) or exhibit linkage (Case II). In the example, Case II shows complete linkage between genes *A* and *B* in order to focus on the difference between independent assortment and linkage. In reality, some recombinants would be expected, the actual number depending on the linkage distance between genes *A* and *B*. Obviously, much larger F_2 populations must be analyzed to establish reliable linkage distances. However, note that in RFLP mapping, linkage distances are calculated as percentage recombination just as they are in crosses involving other types of genetic markers (see Chapter 7). In fact, two "allelic" RFLPs behave genetically in almost all respects just like any other pair of codominant alleles.

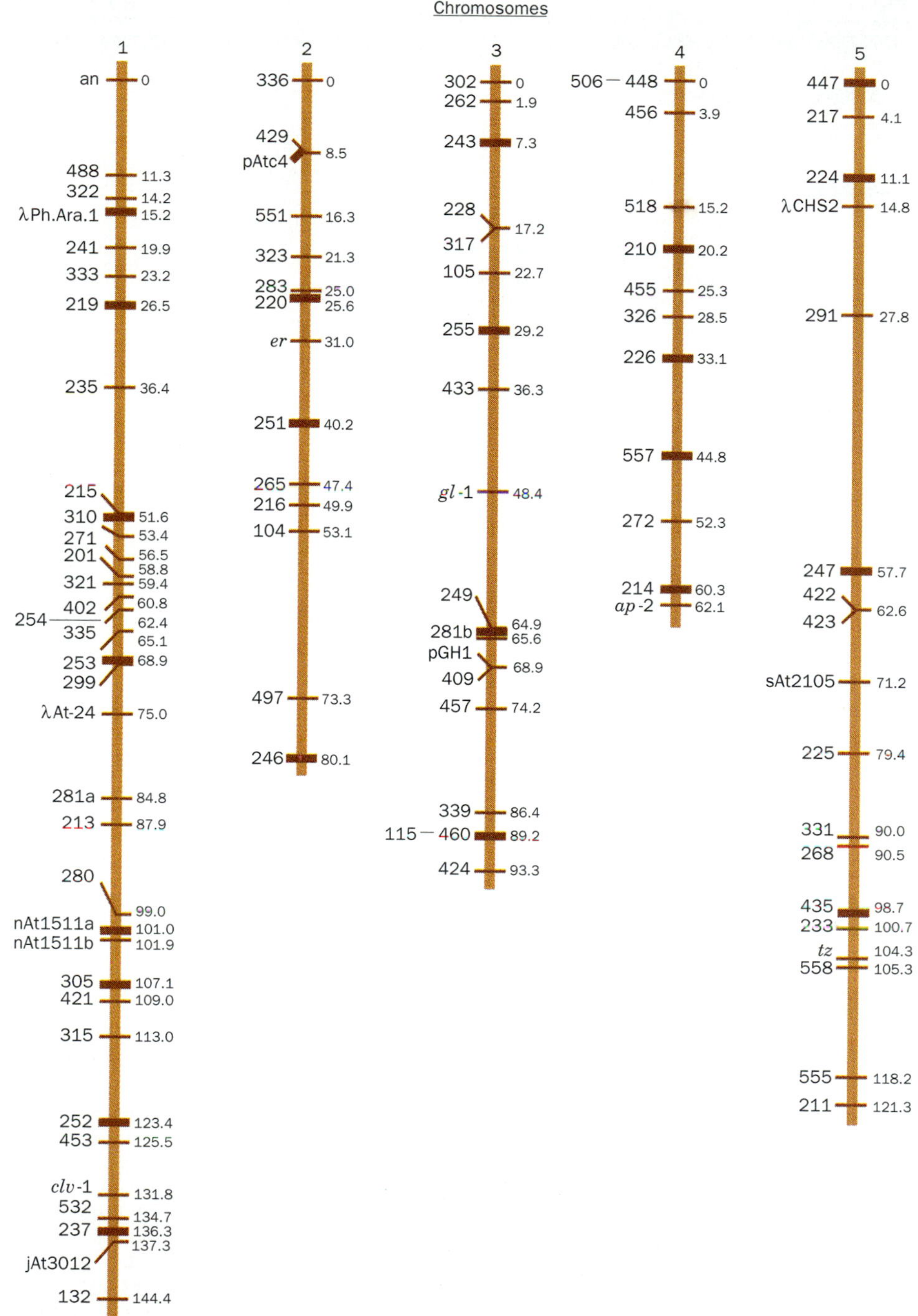

Figure 24.10 RFLP linkage map of the *A. thaliana* genome. The chromosomes are numbered 1 through 5. Clone designations are given to the left of each linkage group. Map designations are given to the right of each linkage group, with the uppermost marker arbitrarily assigned position 0. The three clones offset to the left of linkage groups 1, 3, and 4 (254, 115, and 506) could not be precisely ordered with the data available. Clones nAt1511 and 281 hybridized to sequences at two sites (designated a and b on the map). The following clones encode known products: Ph.Ara.1, phytochrome; At-24, nitrate reductase; nAt1511, small RNA present in seeds; jAt3012, alcohol dehydrogenase; pAtc4, actin; pGH1, acetolactate synthase; CHS2, chalcone synthase; sAt2105, 12S seed storage protein. The remaining numbered clones contain single-copy DNA and were randomly selected from a genomic library. (Data are from C. Chang, J. L. Bowman, A. W. DeJohn, E. S. Lander, and E. M. Meyerowitz, *Proc. Natl. Acad. Sci. U.S.A.* 85: 6856–6860, 1988.)

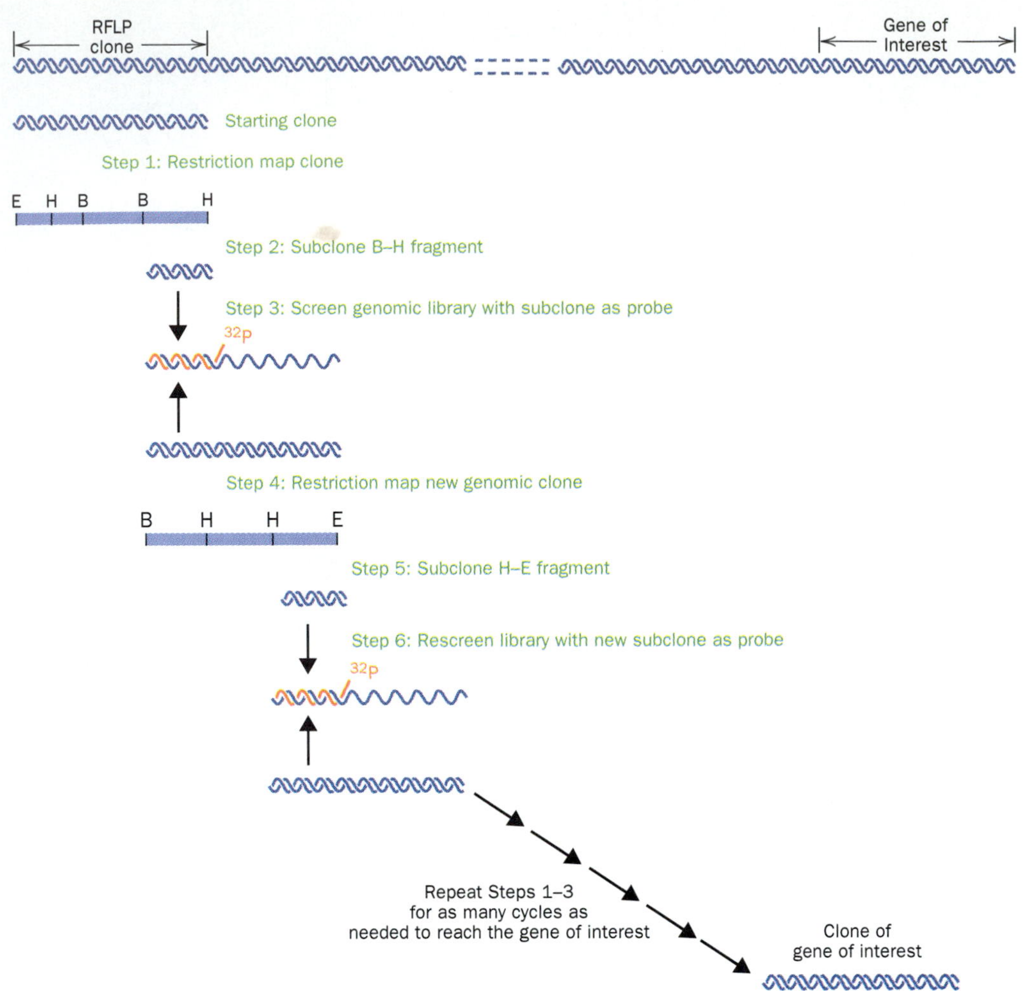

Figure 24.11 Schematic drawing of the procedure used in cloning genes by "chromosome walking." The "walk" starts with the identification of the molecular marker closest to the gene that one wishes to isolate. In the example shown, the closest molecular marker is an RFLP marker clone. A restriction map is prepared for this initial RFLP clone, and the restriction fragment proximal to the gene of interest is subcloned. B, E, and H identify cleavage sites for the enzymes *Bam*HI, *Eco*RI, and *Hin*dIII, respectively. The subclone (a *Bam*HI–*Hin*dIII restriction fragment in the example shown) is used as a radioactive hybridization probe to screen the genomic library for overlapping clones. Restriction maps are then prepared for the new genomic clones (for simplicity, only one is shown) and, again, the restriction fragment proximal to the gene of interest is subcloned. This new subclone (a *Hin*dIII–*Eco*RI fragment in the example shown) is, in turn, used as a probe to isolate a second set of overlapping genomic clones. Restriction maps of the new clones are prepared, and the process is repeated until the "walk" has covered the indicated distance.

clone bank. Obviously, the availability of such a clone bank and the correlated physical map of the entire genome will be of great value to the *A. thaliana* research community.

MICROORGANISMS

Microorganisms have been used to produce important products for humans for decades. Everyone is aware of the impact of the ***penicillin antibiotics*** on human health; fewer of us are aware of the economic importance of penicillins. The wholesale market value of the penicillins in the United States in 1979 was $221 million. The wholesale market value of all antibiotics in the United States in 1979 was over one billion dollars. Microbes also play important roles in the production of many other products, for example, ***antifungal drugs, amino acids, and vitamins.*** Today, thanks to genetic engineering, bacteria are being used to make many important proteins: ***human insulin, human growth hormone, the entire family of human interferons,*** and others.

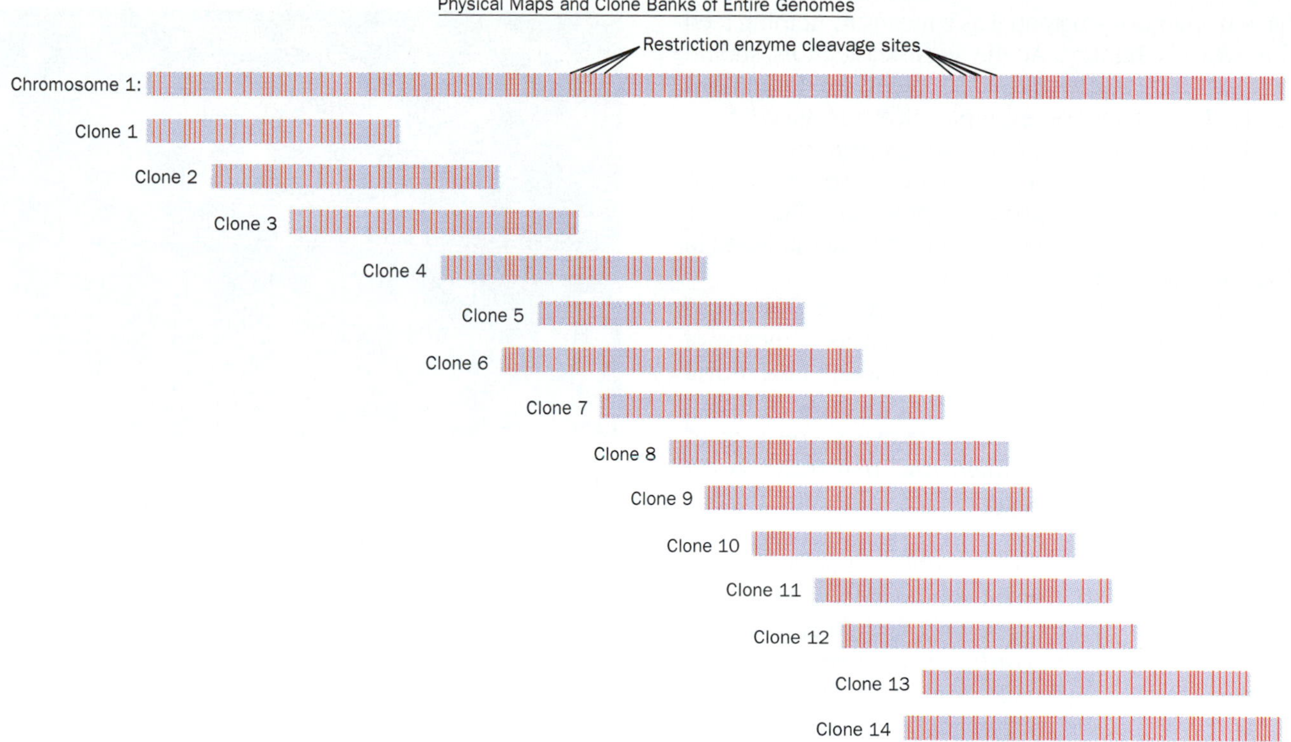

Figure 24.12 Diagram illustrating the procedure for constructing physical maps and complete clone banks for chromosomes and entire genomes. For the sake of simplicity, a small chromosome is shown that is covered by just 14 overlapping genomic clones. Each vertical line represents a restriction enzyme cleavage site. Note how overlapping clones are aligned on the basis of the common patterns of restriction sites. In reality, hundreds to thousands of individual clones would be required to span an entire chromosome of a higher eukaryote. For example, the *A. thaliana* genome contains 7×10^7 nucleotide-pairs. If each clone was 40,000 nucleotide-pairs in length and the clones were evenly distributed with 5000-nucleotide-pair overlaps between adjacent clones, it would still take 2000 clones to cover the entire genome. Since the clones will not be evenly distributed with minimal overlaps, one will actually have to screen several thousand clones to complete the *A. thaliana* physical map. Moreover, certain genomic DNA sequences may not be clonable, that is, their presence may be lethal to the cloning vector being used.

Another extremely important role of microorganisms involves the ***microbial fermentation processes*** used in the production of foods (breads, soda crackers, yogurts, and a vast array of cheeses) and beverages (ales, beers, wines, ciders, etc.). Humans have used microorganisms in fermentation processes for centuries. However, as you might well anticipate, the microbes that carry out these food and beverage fermentations are now being subjected to the restriction enzyme scalpels of genetic engineers, just as *E. coli* has been for the last decade.

Of all living organisms, microbes can be genetically manipulated the most easily. Moreover, microorganisms exhibit a wealth of genetic diversity. As a result, they have great potential to provide useful products such as enzymes and other organic molecules and to provide metabolic machinery for the ***detoxification of pollutants,*** for ***conversion of biomass to combustible compounds,*** and so on. A few present and potential applications of genetically engineered microorganisms are discussed briefly in the next few sections of this chapter.

Single-Cell Proteins

The term ***"single-cell proteins" (SCP)*** refers to the ***dried cells of microorganisms*** (algae, bacteria, yeasts, etc.). These SCPs are produced from cells grown in large-scale culture vats or fermenters. They are then dried and used as ***protein sources in animal feeds*** (including human food supplements). The use of SCP as a food supplement has a long history. The early Aztecs in Mexico harvested algae (genus *Spirulina*) from ponds and used them as a source of food. SCPs have continued to be used as human food in certain regions of the world to the present day, however, they have never become a major food resource for either humans or domestic animals. Yet, SCPs would seem to have tremendous potential as a means to helping feed the world's hungry. At the very least, SCPs should provide important protein supplements for animal feeds. In Britain, for example, SCP produced from the bacterium *Methylophilus methylotrophus* is manufactured and sold under the trade name Pruteen for use as an animal feed protein supplement (Fig. 24.13). Similar products have been produced and marketed in Germany and the Soviet Union.

With the advent of genetic engineering, SCP would seem to have considerable potential as a means of providing improved nutrition in regions of the world where animal proteins are in short supply in human diets. In addition, SCP should prove to be a good ***alternative to soybean meal and fish meal as an animal feed protein supplement.*** The total protein composition of seeds of some of our most important agronomic crops provides inadequate amounts of certain amino acids when used as human food. For example, ***corn seed proteins are very low in lysine, and soybean seed proteins contain suboptimal amounts of methionine.*** Thus, SCPs could be used to supplement the lysine and methionine levels in diets based on corn or soybeans, respectively.

Genetic engineering now allows scientists to manipulate bacteria almost at will. In particular, bacteria can be constructed that greatly overproduce a given protein. ***Up to 50 percent of the protein in a cell may be the product of a single gene.*** This is done by joining the coding sequence of the gene to a strong promoter, inserting the promoter-coding sequence unit into a high-copy-number plasmid, and introducing the recombinant plasmid into an appropriate host cell. Some ***high-copy-number plasmids are present in up to 500 copies per cell.*** Thus, large amounts of protein are produced from genes carried on these plasmids. Given the vast array of genes that are available as starting material, and our ability to change codons at will by site-specific mutagenesis (see Chapter 13), it should be possible to design bacteria with total proteins (SCP) of almost any desired amino acid composition. In any case, it will be interesting to see what roles SCPs play as human food and animal feed supplements in the future.

Figure 24.13 Sample of Pruteen, a single-cell protein made from the bacterium *Methylophilus methylotrophus.* (Photograph courtesy of Imperial Chemical Industries, London.)

Industrial Chemicals

Traditionally, most ***industrial chemicals have been manufactured from fossil fuels.*** However, ***as petroleum costs rise and supplies diminish, the production of chemicals from biomass feedstocks by microbial fermentation processes should become an attractive alternative*** to the use of fossil fuels. Given the tools of genetic engineering, scientists should be able to design microorganisms that can perform specific fermentation reactions with much greater efficiency than their wild-type relatives. The potential for using microorganisms in the production of industrially important chemicals seems almost unlimited and includes a long list of chemicals. A detailed discussion of this topic is far beyond the scope of this text. Industrial molecules and macromolecules that can be produced by microbial fermentation processes have applications as solvents, adhesives, propellants, explosives, alternative fuels, gasoline additives, pesticides, dyes, cosmetics, antifreeze, brake fluids, vitamins, food flavorings, and so on. Major products that are currently manufactured by using microbial production processes include citric acid, lactic acid, ethanol, *n*-butanol, glutamic acid, and lysine. For example, the Kyowa Hakko Kogyo Co., Ltd., in Japan produces 20,000 tons of monosodium glutamate and 10,000 tons of lysine by microbial processes per year. In the future, ***genetic engineering technologies should allow scientists to design microorganisms that synthesize the desired products more efficiently,*** that is, with higher yields per cell or per unit of biomass consumed.

Proteins with Industrial Applications

Some enzymes with important industrial applications have been manufactured by using microorganisms for their synthesis for many years. For example, proteases have been produced from *Bacillus licheniformis* and other bacteria for some time. These proteases have been used extensively as cleaning aids in detergents and in smaller amounts as meat tenderizers and as digestive aids in animal feeds. Amylases have been widely used to break down complex carbohydrates such as starch to glucose. The glucose is then converted to fructose with the enzyme glucose isomerase, and this fructose is used as a food sweetener. The amylases and glucose isomerase are all manufactured by microbiological processes. In 1980, the worldwide production yields of α-amylase and glucose isomerase were 320 and 70 tons, respectively. Clearly, these are major industrial products; they account for millions of dollars in sales each year. The protein rennin is used in making cheeses; 26 tons were used in 1980 (wholesale value $64 million). Prior to the advent of genetic engineering, rennin was extracted from the fourth stomach of cattle. Rennin is now being produced by using genetically engineered bacteria. These examples are all proteins that have had important industrial applications for some time in the past. In the future, we can expect many additional enzymes to be manufactured and used in industrial applications because of the ease of producing these proteins using recombinant microorganisms.

In fact, the major tools of molecular biology today are themselves enzymes, namely, the vast array of restriction endonucleases, ligases, polymerases, and reverse transcriptases, that are used to manipulate DNA molecules *in vitro*. These enzymes are now commonly produced by cloning the respective genes and expressing these genes on high-copy-number plasmids in bacteria. Thus, the development of the recombinant DNA and gene cloning technologies has spawned a whole new industry, a group of companies that manufacture and sell the enzymes, vectors, and so on that are used by genetic engineers. New applications in this field are inevitable. ***It is now possible to produce essentially any enzyme or other protein of interest by cloning the gene that encodes this protein and expressing that gene at very high levels in microorganisms.***

Proteins with Medicinal Applications

As was discussed in Chapter 1, over 3000 human diseases are believed to result from genetic defects. Many of these are known to be caused by mutations in individual genes that result in the synthesis of defective or totally nonfunctional gene-products. The consequences of a deficiency of the enzyme phenylalanine hydroxylase in individuals with phenylketonuria were described in detail in Chapter 1. The abnormalities that occur in individuals with other inherited diseases also frequently result from specific enzyme deficiencies. In principle, one might expect to be able to treat these diseases by supplying the missing enzymes if adequate quantities of pure enzymes were available. In fact, given the tools of genetic engineering, there can be little question about our ability to produce large amounts of any enzyme or other protein of interest. The problem is in delivering these enzymes to the intracellular sites in those tissues where their function is required. One would need a delivery system such that the enzymes retained activity during storage and were gradually and continuously released over an extended period of time. Unfortunately, ***the instability of most enzymes at ambient temperatures makes "enzyme therapy" infeasible as an approach to the treatment of most inherited diseases.*** Indeed, a major advantage of ***"gene therapy"*** is that once the normal gene has been introduced into a cell, that ***gene should be completely stable and should continuously produce the needed gene-product*** for the entire life of the cell. Nevertheless, "enzyme therapy" has proven effective in some cases, and should be included as one approach to the treatment of human diseases. Moreover, enzyme supplies should not be a major factor in the future; it should be possible to program recombinant microorganisms to synthesize large amounts of any desired protein.

Insulin has been used to treat diabetes for several decades. As was mentioned in Chapter 1, insulin has been produced for many years from bovine and porcine pancreas glands. However, in 1982, human insulin produced in recombinant bacteria was approved for sale over the counter of your local drugstore. Thus, ***human insulin was the first commercial success of the new recombinant DNA technologies in the field of pharmaceuticals.*** Other human proteins that have been synthesized in bacteria and have medical roles are ***blood-clotting factor VIII*** (lacking in individuals with one type of hemophilia), ***plasminogen activator*** (a protein that disperses blood clots), and ***human growth hormone*** (a protein deficient in certain types of dwarfism).

As an example of the use of bacteria to synthesize human proteins, let's briefly examine the initial synthesis of human growth hormone (HGH) in *E. coli*. HGH is a single polypeptide chain 191 amino acids in length that is produced in the pituitary gland and is required for normal growth. In contrast to insulin, porcine and bovine pituitary growth hormones do not work in humans. Only growth hormones from humans or from closely related primates function effectively in humans. ***To obtain expression of HGH in*** E. coli, ***it was necessary to place the HGH coding sequence under the control of*** E. coli ***regulatory elements.*** Therefore, the HGH coding sequence was joined to the promoter and ribosome-binding-site sequences of the *E. coli lac* operon (see Chapter 14). To get these elements joined in the proper positions, a *Hae*III cleavage site in the nucleotide-pair triplet specifying codon 24 of HGH was used to fuse a synthetic DNA sequence encoding amino acids 1–23 to a partial cDNA sequence encoding amino acids 24–191. This unit was then inserted into a plasmid carrying the *lac* operon regulatory signals and introduced into *E. coli* by transformation. The structure of the plasmid that was first used to express HGH in *E. coli* is diagrammed in Fig. 24.14.

The HGH produced in *E. coli* in these first experiments contained methionine at the amino terminus (the methionine specified by the ATG initiator codon). Native HGH has an NH_3-terminal phenylalanine; the initiator methionine is cleaved off enzymatically after synthesis. *Escherichia coli* also removes many amino-terminal methionine residues posttranslationally. However, the removal of the terminal methionine is sequence dependent, and *E. coli* does not cleave off the NH_3-terminal methionine from HGH. Nevertheless, this ***HGH synthesized in* E. coli *was found to be fully active in humans*** despite the presence of one extra amino acid. More recently, a DNA sequence encoding a signal sequence (the amino acid sequence required for the transport of proteins across membranes) has been added to an HGH recombinant gene construct similar to that shown in Fig. 24.14. With the signal sequence added, the HGH synthesized in *E. coli* is secreted and is correctly processed, that is, the methionine residue is removed with the rest of the signal peptide when it is cleaved off during transport across the membrane, yielding a 191-amino-acid-long polypeptide identical to native HGH. In 1985, HGH synthesized in *E. coli* became the second genetically engineered pharmaceutical to be approved for use by the U.S. Food and Drug Administration.

Genetic engineering also provides an alternate strategy to the production of vaccines for immunization against viral diseases. Traditionally, vaccines have consisted of virus particles that have been inactivated or of live, but weakened, viral strains that can no longer cause the disease but can still stimulate an immune response against the virulent strains. Although these approaches have proven invaluable in the case of polio, flu, and other viral diseases, there is always the chance that a few virus particles might escape inactivation and remain virulent. In fact, this has occurred on occasion in cattle vaccinated against the virus that causes foot-and-mouth disease. ***Since the coat proteins of viruses are the major antigens that induce immunity, another approach is to clone the gene(s) that encodes the viral coat proteins and to use just the coat proteins in vaccines to induce immunity.*** Such vaccines would be cheaper and safer; there would be no possibility of live virus particles contaminating the vaccines. Although the overall utility of this approach remains to be evaluated, it would

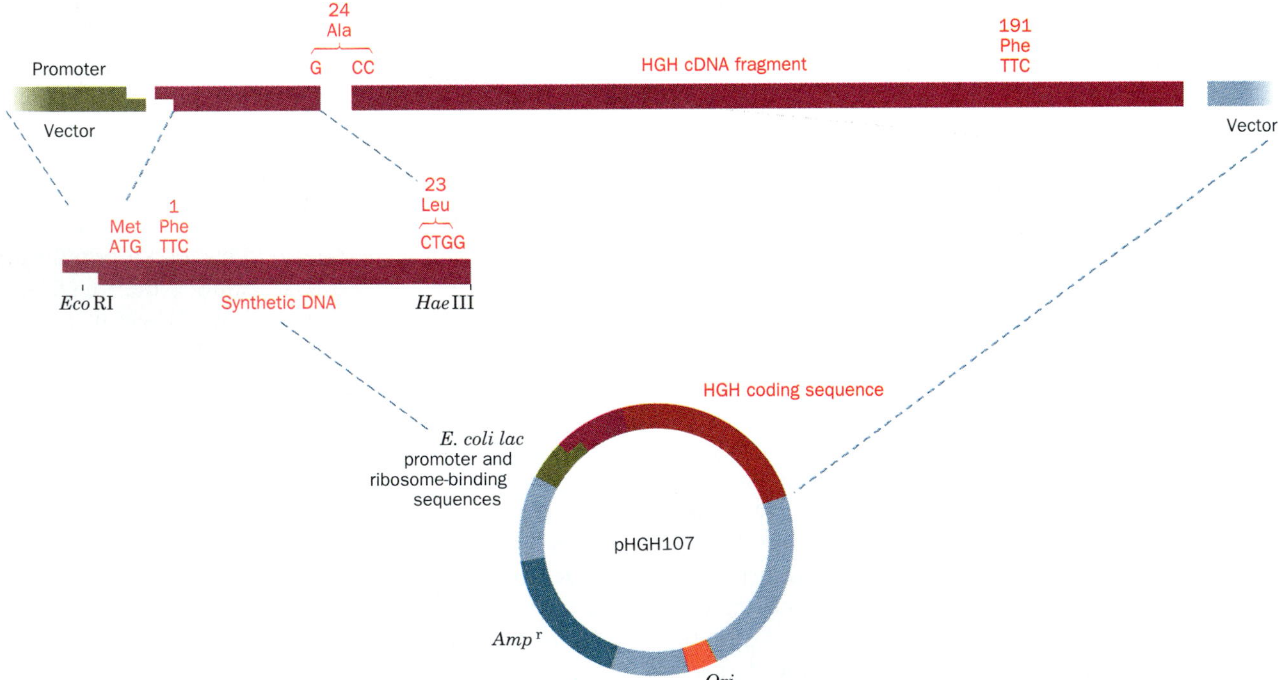

Figure 24.14 Structure of pHGH107, the first vector constructed for the synthesis of human growth hormone in *E. coli.* The DNA segment encoding amino acids 1–23 was assembled *in vitro* from 12 synthetic oligonucleotides (6 for each strand). The rest of the coding sequence (specifying amino acids 24–191) was derived from a cDNA synthesized from human pituitary gland RNA with reverse transcriptase. The desired cDNA fragment was cleaved from the complete cDNA with *Hae*III. (For complete details of the cloning procedure, see P. H. Seeburg, "Human Growth Hormone: From Clone to Clinic," *Cold Spring Harbor Sympos. Quant. Biol.* 51: 669–677, 1986.)

seem to have great potential as a tool for use in our ongoing battle with pathogenic viruses.

Dairy Fermentations

What do Mozzarella, Parmesan, Swiss, Gouda, Colby, Cheddar, Roquefort, Muenster, Monterey, Brick, Blue, Asiago, and Limburger all have in common? The correct answer is *Streptococcus lactis.* All of these ***major cheeses are produced by dairy fermentation processes that utilize the bacterium* S. lactis**. They differ in the presence of other species of bacteria in the starter cultures used and in the specific details of the fermentation processes employed, but they all utilize *S. lactis* as a key component. Dairy fermentations are involved in the manufacture of a vast variety of cheeses and yogurts; collectively, these products are a multibillion dollar industry. *Streptococcus lactis* and *Streptococcus cremoris* are the two most important bacteria used in dairy fermentation starter cultures; other species are used in the production of specific products.

A common scenario in the cheese-making industry is that starter cultures that have been working perfectly and yielding prime cheeses suddenly go bad and begin producing unmarketable products. It is as though the bacteria in these ***starter cultures are somehow unstable and subject to sporadic changes in metabolic properties.*** But why? What is the basis of this apparent instability? For many years this instability remained a complete mystery. However, recently the results of L.L. McKay and colleagues, M.J. Gasson and F.L. Davies, and others have provided at least part of the answer. **Streptococcus lactis *and its relatives are full of plasmids*** (up to 12 different plasmids have been identified in some *S. lactis* strains), and these plasmids (Fig. 24.15) carry genes that encode enzymes with essential roles in the fermentation processes. Lactose metabolism and proteinase activity are absolutely essential for proper cheese ripening, and the three enzymes involved in lactose metabolism as well as an important proteinase are all known to be encoded by genes carried on plasmids. Since most plasmids are by nature unstable, the presence of key genes on plasmids explains, at least in part, the instability of dairy fermentation starter cultures. More importantly, these results suggest an obvious solution to the instability problem, namely, genetic engineering should be employed to incorporate these genes into the main bacterial chromosome in order to create more stable strains.

Another problem that has plagued the dairy fermentation industry for the last 50 years is that of bacteriophage infections of cheese starter cultures. Again, genetic engineering should facilitate the construction of phage-resistant strains of *S. lactis* and related dairy fermentation species. Given the economic impact that the availability of stable, phage-resistant starter culture strains would have in the diary industry, no one should be surprised to discover that the ***genetic engineering of bacteria involved in dairy fermentations is currently an area of intense activity.***

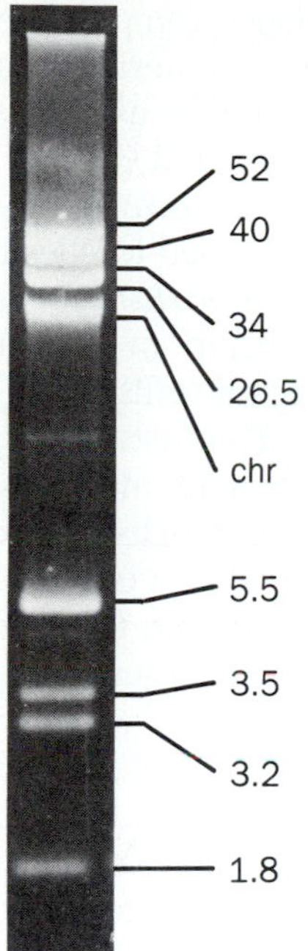

Figure 24.15 Photograph of an agarose gel showing the presence of multiple plasmids in the dairy fermentation bacterium *Streptococcus lactis.* The plasmid profile shown is for *S. lactis* subspecies *diacetylactis* strain DRC3. Plasmid sizes are given in terms of their weights in megadaltons (1 megadalton = 10^6 daltons) on the right. The 52-megadalton (MDa) plasmid carries genes that encode the enzymes responsible for lactose fermentation; the 40-MDa plasmid contains genes that control (1) conjugative transfer functions, (2) a temperature-sensitive resistance to bacteriophage infection, and (3) resistance to a bacterial growth inhibitor protein called nisin; the 34-MDa plasmid encodes proteinase activity, and the 5.5-MDa plasmid encodes enzymes involved in citrate utilization. The other plasmids remain cryptic, that is, they have no known functions. The metabolic functions listed all play important roles in dairy fermentation processes. (Photograph courtesy of L. L. McKay and K. A. Baldwin, Department of Food Science and Nutrition, University of Minnesota.)

Degradation of Pollutants

In 1972, A. Chakrabarty and General Electric Company filed a patent application to protect their interests in a strain of *Pseudomonas* that could degrade several components (octanes, camphors, etc.) of crude petroleum. Chakrabarty's bacterial strain was perceived to have significant value for use in cleaning up oil spills. Actually, Chakrabarty and General Electric filed ***three separate patent applications: (1) a process claim*** to protect the method used to construct the "oil-eating" bacterium, ***(2) an inoculum-for-use claim*** to protect the use of the bacterium plus a carrier substance (straw) in treating oil spills, and ***(3) a patent on the bacterium itself.*** The first two applications were approved by the Patent Examiner; the third was initially denied on the grounds that one could not patent living bacteria, but was subsequently approved by a 5 to 4 vote of the U.S. Supreme Court in 1980 after a long series of appeals. The arguments in this case and the history of the question of whether living cells should be patentable are fascinating topics themselves; unfortunately, they are beyond the scope of this text. The reader is referred to pp. 501–510 in *The DNA Story* by Watson and Tooze for a copy of the text of the Supreme Court's decision. The Supreme Court's decision in the Chakrabarty case was a landmark in patent law because it established for the first time that ***living bacteria could be patented*** if ***they were*** not ***"naturally occurring."***

Chakrabarty's bacterium was constructed by using classical genetic selection to combine genes originally located on four different plasmids onto one compound plasmid that could then be maintained in a single bacterium. The original plasmids were incompatible and, thus, could not coexist in the same cell. The Chakrabarty decision was precedent-setting; it clearly stated that genetically modified bacteria were subject to protection by patent law. Furthermore, ***extrapolation of this decision implied that genetically engineered eukaryotic cells should also be patentable.*** Most certainly, this decision meant that ***self-replicating genetic elements such as plasmids, transposons, and viruses should be patentable.***

The value of being able to patent something is that it protects the capital investments used in the development of the product for commercial application; without patent protection, a competitor could simply copy a product and market it without having made any contribution to the development of the product. Clearly, then, the Chakrabarty decision was a key step in stimulating the use of genetic engineering technologies to produce living cells with novel and useful properties. Numerous examples of recombinant bacteria with useful properties, such as the production of human insulin or human growth hormone, have already been discussed in the preceding sections of this chapter.

Chakrabarty's work itself suggests another area in which ***recombinant microorganisms should be able to make a significant contribution,*** namely, ***in the degradation of chemical pollutants.*** Collectively, microorganisms display an enormous amount of metabolic diversity. When natural populations of microorganisms are screened for species or strains that have the ability to utilize unusual and exotic chemicals as energy sources, the searches are most often successful. In other cases, selection pressure can be applied by requiring the microorganisms to grow with a given chemical as the sole energy source, and, again, variants are frequently obtained that have acquired the ability to catabolize the compound. This is presently a very active field of research, and the microbial degradation of environmental pollutants promises to be of increased importance in the future. Moreover, genetic engineering will undoubtedly play an increasingly important role in this field as it matures. For example, an enzyme that catalyzes the degradation of a pollutant may be produced by a microorganism that one would not want to release in a polluted environment because of other undesirable properties of that organism. In this case, genetic engineering could be used to transfer the gene encoding the enzyme of interest to another species that does not exhibit the undesirable properties of the original host. Alternatively, genetic engineering might be used to eliminate the undesirable characteristics of the original microorganism.

PLANTS

Research on plant genetics has lagged far behind that on animal genetics. This has occurred primarily because the driving force behind most genetic research has been its applicability to human health (and, thus, its "fundability" by the U.S. Public Health Service's National Institutes of Health). Moreover, past research on plants has focused almost exclusively on a few of our most important agronomic crops. More recently, the potential of genetic engineering of plants has been recognized, and research activities (and funding for this research) have increased considerably. Plant systems represent a wealth of genetic diversity, and the increased research activities on plants should yield important information about the basic biology of different plant systems as well as some successful commercial applications of genetically engineered plants. In fact, the first commercial applications of genetic engineering in agronomically important species are currently very close to fruition; a few of these will be discussed in the following sections of this chapter.

Regeneration of Plants from Protoplasts and Callus Tissue

A striking feature of plant cells is their ***totipotency,*** that is, most plant cells retain the ability to give rise to

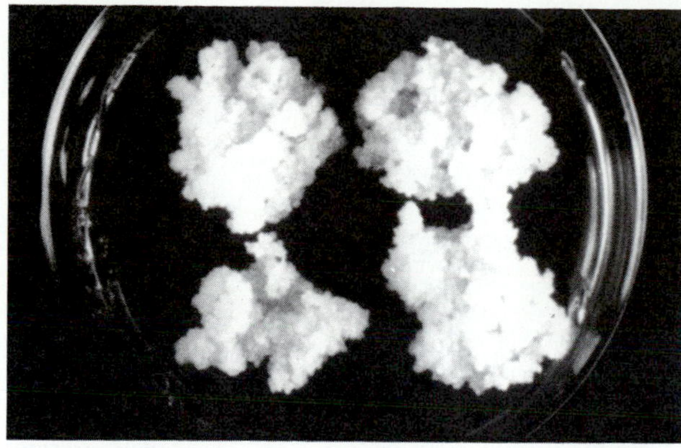

Figure 24.16 Photograph showing the growth of corn cells as undifferentiated callus on synthetic culture medium. (Photograph courtesy of D. A. Somers, Department of Agronomy and Plant Genetics, University of Minnesota.)

Figure 24.17 *Arabidopsis thaliana* plantlets regenerated from callus tissue growing on sterile culture medium. (Photograph courtesy of J. Carpenter and D. P. Snustad, Department of Genetics and Cell Biology, University of Minnesota.)

any differentiated cell type of the mature plant. Many differentiated plant cells are able to dedifferentiate to the embryonic state and to subsequently redifferentiate to new cell types. Thus, there is no separation of germ line cells from somatic cells as in higher animals.

When excised tissues from mature plants are placed in the appropriate sterile tissue culture conditions (notably in the presence of the plant hormone 2,4-dichlorophenoxyacetic acid, 2,4-D), cells in these tissue explants will often dedifferentiate and grow into highly unorganized cell masses called calli (or calluses, singular = callus) (Fig. 24.16). If these undifferentiated callus cell-clumps are subsequently (within not longer than a year or so) transferred to growth medium favoring differentiation (medium lacking 2,4-D, but containing growth hormones such as kinetin), ***plantlets will regenerate*** (Fig. 24.17) in the case of many, but not all, species. Moreover, with some plant species, one can regenerate plants from isolated ***protoplasts*** (single "cells" from which the walls have been removed by enzymatic digestion).

The totipotency of plant cells is a major advantage for genetic engineering since it permits one to regenerate entire plants from individual modified somatic cells. In contrast, one must modify germ line cells to produce nonmosaic, genetically modified higher animals (see the *Drosophila,* P-Element Transformation section earlier in this chapter).

The Ti Plasmid of *Agrobacterium tumefaciens*

To date, ***the most important tool in the genetic engineering of plants has been the Ti plasmid of the soil bacterium*** **Agrobacterium tumefaciens.** This bacterium is the causative agent of ***crown gall disease of dicotyledonous plants.*** The name refers to the "galls" or tumors that often form at the crown (junction between the root and the stem) of infected plants (Fig. 24.18). Since the crown of the plant is usually located at the soil surface, this is where a plant is most likely to be wounded (e.g., due to a soil abrasion from the plant blowing in a strong wind) and infected by a soil bacterium such as *A. tumefaciens.* However, *A. tumefaciens* can infect a plant and induce a tumor at any wound site (Fig. 24.19).

During the infection of a wound site by *A. tumefaciens,* two key events occur: (1) the plant cells begin to proliferate and form ***tumors*** and (2) they begin to synthesize an arginine derivative called an ***opine.*** The opine synthesized is usually either ***nopaline*** or ***octopine*** (Fig. 24.20) depending on the *A. tumefaciens* strain involved. These opines are catabolized and used as energy sources by the infecting bacteria. Interestingly, *A. tumefaciens* strains that induce the synthesis of nopaline can grow on nopaline, but not on octopine, and *vice versa.* Clearly, an interesting interrela-

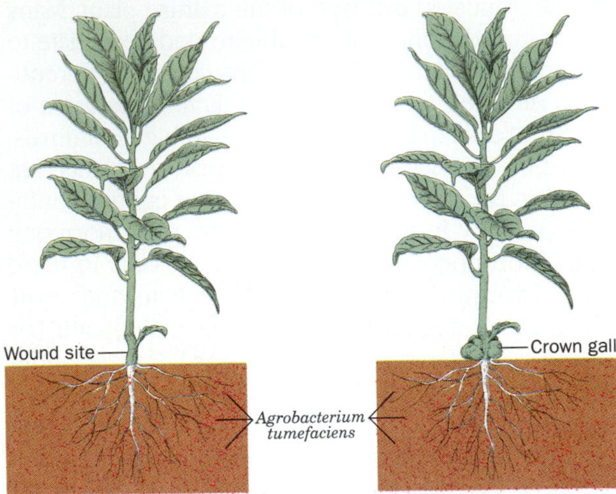

Figure 24.18 The formation of "crown galls" on dicotyledonous plants after infection by the soil bacterium *Agrobacterium tumefaciens.*

(a) Arginine

$$HN{=}C(NH_2)-NH-CH_2-CH_2-CH_2-\underset{\displaystyle NH_2}{\underset{|}{CH}}-COOH$$

(b) Octopine

$$HN{=}C(NH_2)-NH-CH_2-CH_2-CH_2-\underset{\displaystyle \underset{\displaystyle H_3C-CH-COOH}{\underset{|}{NH}}}{\underset{|}{CH}}-COOH$$

(c) Nopaline

$$HN{=}C(NH_2)-NH-CH_2-CH_2-CH_2-\underset{\displaystyle \underset{\displaystyle HOOC-CH_2-CH_2-CH-COOH}{\underset{|}{NH}}}{\underset{|}{CH}}-COOH$$

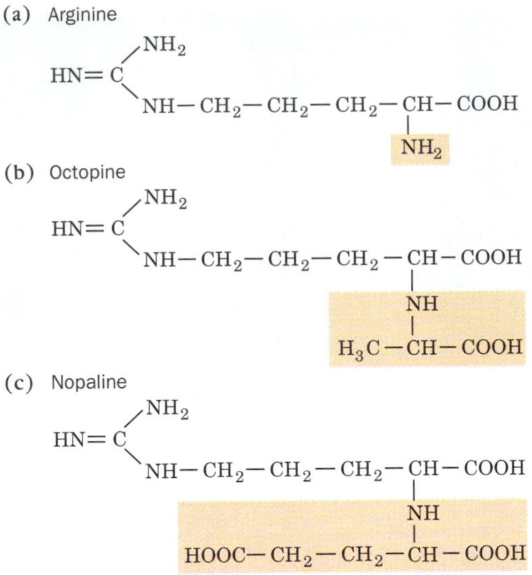

Figure 24.20 Structure of the amino acid arginine (a) and the arginine-derived opines (b) octopine and (c) nopaline synthesized in plant cells transformed with *Agrobacterium tumefaciens* harboring Ti plasmids of the octopine and nopaline types, respectively. These opines are catabolized and used as energy sources by the respective *A. tumefaciens* strains.

tionship has evolved between *A. tumefaciens* strains and their plant hosts. *Agrobacterium tumefaciens* is able to divert the metabolic resources of the host plant to the synthesis of opines, which are of no apparent benefit to the plant, but which provide sustenance to the bacterium.

Figure 24.19 Photograph of tumors on a leaf of *Kalanchoe daigremontiana.* Four abrasions were made on the surface of the leaf and were inoculated with a virulent strain of *Agrobacterium tumefaciens.* (Photograph courtesy of G. Pazour and A. Das, Department of Biochemistry, University of Minnesota.)

The ability of *A. tumefaciens* to induce crown gall disease in plants is controlled by genetic information carried on a large (about 200,000 nucleotide-pairs) plasmid called the ***Ti plasmid*** for its ***T***umor-***i***nducing capacity. ***Two components, the T-DNA and the* vir *region, of the Ti plasmid are essential for the transformation of plant cells.*** During the transformation process, the ***T-DNA*** (for ***T***ransferred ***DNA***) is excised from the Ti plasmid, transferred to a plant cell, and integrated (covalently inserted) into the DNA of the plant cell. The available data indicate that integration of the T-DNA occurs at random chromosomal sites; moreover, in some cases, multiple T-DNA integration events occur in the same cell. In nopaline-type Ti plasmids, the T-DNA is a 23,000-nucleotide-pair segment that carries 13 known genes. In octopine-type Ti plasmids, there are two separate T-DNA segments. For the sake of simplicity, the information presented here and in the following sections will be restricted to nopaline-type Ti plasmids, whose structure is shown in Fig. 24.21. Some of the genes on the T-DNA segment of the Ti plasmid encode enzymes that catalyze the synthesis of phytohormones (the auxin indoleacetic acid and the cytokinin isopentenyl adenosine). These phytohormones are responsible for the tumorous growth of cells in crown galls. The ***T-DNA region is bordered by 25-nucleotide-pair imperfect repeats,*** which are

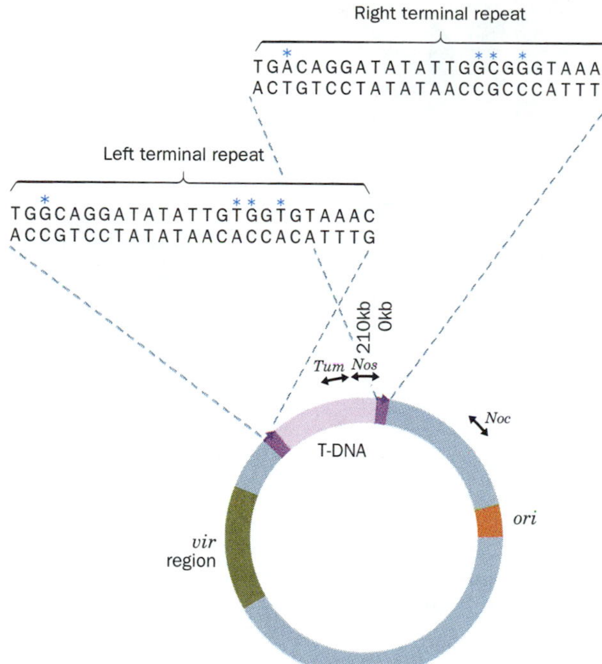

Figure 24.21 Diagram of the structure of the wild-type nopaline Ti plasmid pTi C58. Only the components that are essential to the process of DNA transfer or that are discussed in the text for other reasons are shown. The following descriptive symbols are used: *ori,* plasmid origin of replication; *Tum,* region containing genes responsible for tumor formation (genes encoding phytohormone biosynthetic enzymes); *Nos,* region containing genes involved in nopaline biosynthesis; *Noc,* region containing genes involved in the catabolism of nopaline; *vir,* virulence region containing the genes that are required for T-DNA transfer. The entire plasmid is 210 kilobase-pairs (kb) in length; the diagram is not drawn to scale. The nucleotide-pair sequences of the left and right imperfect direct terminal repeats are shown at the top; the asterisks above the sequences mark the four base-pairs that differ in the two border sequences. (For further details, see the review by M. W. Bevan and M.-D. Chilton, "T-DNA of the *Agrobacterium* Ti and Ri Plasmids," *Annu. Rev. Genet.* 16: 357–384, 1982.)

Acetosyringone

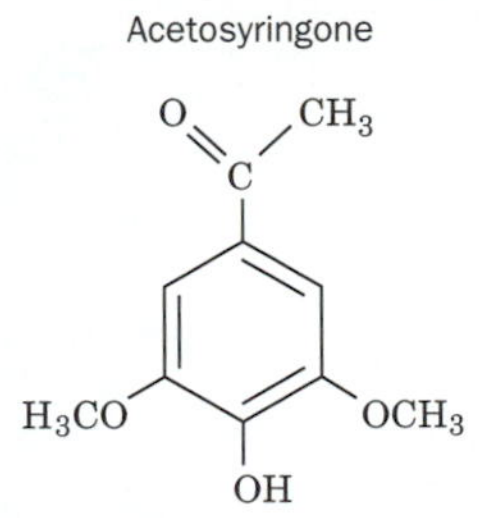

Figure 24.22 Structure of acetosyringone, a phenolic compound in plant cells that induces transcription of the *Agrobacterium tumefaciens* operons containing the *vir* genes.

required in* cis *for T-DNA excision and transfer. The deletion of either border sequence completely blocks the transfer of T-DNA to plant cells.

The **vir** (for ***vir***ulence) region of the Ti plasmid contains the genes required for the T-DNA transfer process. The functions of these genes are just beginning to be worked out; they undoubtedly encode the DNA processing enzymes required for excision, transfer, and integration of the T-DNA segment. An important factor in the construction of Ti gene-transfer vectors is that ***the* vir *genes can supply the functions needed for T-DNA transfer when located either* cis *or* trans *to the T-DNA.*** The *vir* genes are normally expressed at very low levels in *A. tumefaciens* cells growing in soil. However, exposure of the bacteria to plant cells or exudates from plant cells induces increased levels of expression of the operons containing the *vir* genes. Surprisingly, this induction process is very slow for bacteria, taking 10 to 15 hours to reach maximum levels of expression. Recently, certain phenolic compounds, for example, ***acetosyringone*** (Fig. 24.22), have been purified from plant exudates and shown to act as ***inducers of the* vir *operons.*** Thus, the induction process can now be studied *in vitro* by using these purified inducer molecules. Although many details are still unknown, the overall process of transformation of plant cells by the Ti plasmid of *Agrobacterium tumefaciens* is quite clear (Fig. 24.23).

Disarmed Ti Vectors

Once it had been established that the T-DNA region of the Ti plasmid of *Agrobacterium tumefaciens* is transferred to plant cells and becomes integrated in plant chromosomes, the potential use of *A. tumefaciens* in plant genetic engineering was obvious. One could introduce foreign genes into the T-DNA region and, hopefully, these genes would be transferred to the plant with the rest of the T-DNA segment. In fact, this works very well. All the available data indicate that any DNA inserted anywhere between the T-DNA border sequences (see Fig. 24.21) is transferred to plant cells and integrates in plant chromosomes with the rest of the T-DNA. The problem is that the transformed plant cells that have received the T-DNA from a wild-type Ti plasmid lose their normal control of cell division and form tumors. This oncogenicity of T-DNA renders wild-type Ti plasmids incompatible with the goals of most gene-transfer experiments. One usually does not wish to genetically engineer tumor cells, and one certainly cannot study the normal functions of transferred genes in tumor cells. Fortunately, the solution to this problem came early with the identification of the genes in the T-DNA that were responsible for

Figure 24.23 Transformation of plant cells by *Agrobacterium tumefaciens* harboring a wild-type Ti plasmid. Cells in the plant tumor contain the T-DNA segment of the Ti plasmid integrated (covalently inserted) into the DNA of plant chromosomes. The integration site appears to be selected at random or nearly at random. (After M.-D. Chilton, "A Vector for Introducing New Genes into Plants," *Sci. Amer.* 248(6): 51–59, 1983.)

tumor formation (see Fig. 24.21). The deletion of one or more of these genes produces a ***nononcogenic*** or ***disarmed Ti plasmid.*** Unfortunately, the deletion of the tumor-causing genes also makes it extremely difficult to identify plant cells that have received the disarmed T-DNA. With wild-type Ti plasmids, the recipient plant cells form tumors and are easily identified by the tumor phenotype; with disarmed Ti plasmids, the recipient plant cells continue to grow just like their neighbors that don't harbor the T-DNA. Thus, one needs another way to identify plant cells that have received the T-DNA segment of a disarmed Ti plasmid, specifically, a good selectable marker gene located within the T-DNA region of the disarmed Ti plasmid.

Chimeric Selectable Marker Genes

A good selectable marker gene is one that will provide resistance to a drug, antibiotic, or other agent that arrests the growth of normal plant cells (plant cells not harboring the marker gene). The selective agent should inhibit the growth of plant cells or kill them slowly. Agents that kill cells rapidly often result in the release of phenolic compounds and other substances (from the dead cells) that are toxic to the growth of the remaining, otherwise resistant cells. To date, three selectable marker genes have been extensively used in plant systems; they provide resistance to the antibiotics (1) ***kanamycin*** (and the related aminoglycoside ***G418***—often a more effective growth inhibitor in eukaryotes) and (2) ***hygromycin,*** and to the drug (3) ***methotrexate.*** Of these, kanamycin (and G418) has been by far the most widely used with plants.

The **Kan^r *gene from the* E. coli *transposon Tn5*** has been extensively used as a selectable marker in plants; it encodes an enzyme called ***neomycin phosphotransferase type II (NPTII).*** NPTII is one of several prokaryotic enzymes that detoxify the kanamycin family of aminoglycoside antibiotics by phosphorylating them. Since the promoter sequences and transcription-termination signals are different in bacteria and plants, the native Tn5 *Kan^r* gene cannot be used directly in plants. Instead, the NPTII coding sequence must be provided with a plant promoter (5′ to the coding sequence) and plant termination and polyadenylation signals (3′ to the coding sequence). Such constructions with prokaryotic coding sequences flanked by eukaryotic regulatory sequences are called ***chimeric selectable marker genes.*** Regulatory sequences from several different plant genes have been used to construct chimeric marker genes. The two most frequently used promoter sequences are those from (1) the nopaline synthase (*nos*) gene of the Ti plasmid and (2) the 35S transcript of cauliflower mosaic virus (CaMV). The most frequently used termination and polyadenylation sequence has been that from the nopaline synthase gene of the Ti plasmid. One widely used chimeric selectable marker gene has the structure ***CaMV 35S promoter/NPTII coding sequence/Ti* nos *termination sequence;*** this chimeric gene is usually symbolized ***35S/NPTII/*nos.**

An important objective in converting the Ti plasmid to a useful vector for plant gene transfers was to replace the tumor-inducing genes of the T-DNA with a chimeric selectable marker gene such as 35S/NPTII/*nos.* This sounds like a simple task, but it is not. The difficulty results from the large size (about 200 kb) of

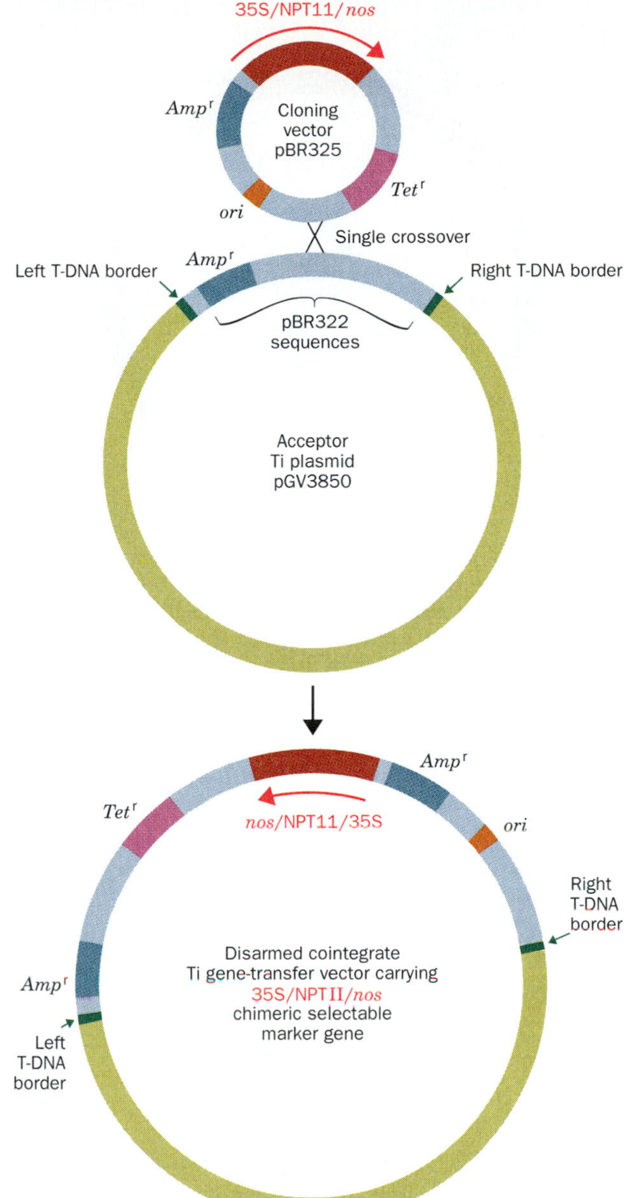

Figure 24.24 Diagram illustrating the use of the *Agrobacterium tumefaciens* Ti acceptor plasmid pGV3850. This plasmid contains a sequence from the *E. coli* cloning vector pBR322 in place of tumor-inducing genes of the T-DNA of a nopaline Ti plasmid. A single crossover between the pBR322 sequence and a homologous sequence on a donor plasmid (such as pBR325) will yield a cointegrate plasmid of the type shown at the bottom. Plasmid pBR325 is a derivative of pBR322; thus, it provides a region of homology to the pBR322 sequence in the acceptor Ti plasmid. This procedure can be used to insert any foreign gene or DNA sequence of interest into the pGV3850 acceptor plasmid. The example diagrammed here shows the introduction of the chimeric selectable marker gene 35S/NPTII/*nos* (see text) into pGV3850. Plasmid pBR325 does not contain an origin of replication that is functional in *Agrobacterium;* thus, pBR325 can persist in *A. tumefaciens* cells only by forming a cointegrate of the type shown. (After P. Zambryski, H. Joos, C. Genetello, J. Leemans, M. Van Montagu, and J. Schell, *EMBO J.* 2: 2143–2150, 1983.).

the Ti plasmid. The Ti plasmid cannot be easily manipulated like the small cloning vectors (3 to 10 kb) described in Chapter 13 and the first two sections of this chapter. The large 200+ kb Ti plasmid will be cut at several sites by almost all restriction enzymes. Thus, one cannot simply open it up (cut the circular DNA molecule once) and insert the chimeric gene as one can with most small cloning vectors. Instead, one has to insert the chimeric gene into a smaller intermediate plasmid containing DNA sequences with homology to T-DNA sequences, put this intermediate plasmid into *Agrobacterium tumefaciens* cells harboring a functional Ti plasmid, and rely on genetic recombination (crossovers within the regions of homology) to insert the chimeric gene into the T-DNA segment of the resident Ti plasmid. This has been accomplished by at least four slightly different procedures. All of these procedures involve putting either T-DNA sequences into smaller plasmids or plasmid sequences into the T-DNA segment of Ti plasmids. Either approach will produce a region of homology between the smaller plasmid and the T-DNA segment of the Ti plasmid and result in insertions of genes on the smaller plasmid via classical genetic crossover events. For the sake of brevity, only one of these approaches is discussed here.

The approach of Zambryski, Van Montagu, Schell, and colleagues was to ***replace the tumor-causing genes of a nopaline Ti plasmid with DNA from the* E. coli *plasmid pBR322.*** This disarmed Ti plasmid functions as an ***acceptor plasmid*** for chimeric genes or other foreign DNA sequences cloned in the small pBR322 plasmid. The pBR322 derivative of interest is introduced into *A. tumefaciens* by conjugative transfer from *E. coli* mediated by *trans*-acting proteins encoded by a broad-host-range mobilizing plasmid (conjugative plasmid) that is a coresident of the *E. coli* donor cell. Since plasmid pBR322 can ***not*** replicate in *A. tumefaciens* (it has no origin of replication that functions in *Agrobacterium*), DNA sequences carried on the pBR322 plasmid will only be maintained in *A. tumefaciens* if they are incorporated into the T-DNA via crossover events (Fig. 24.24).

Recombination between two plasmids to form one larger plasmid is called ***cointegration;*** a compound plasmid produced by a single crossover fusing two smaller plasmids is called a ***cointegrate.*** Cointegrate plasmids are usually unstable. However, they can be stably maintained in cells if each of the original plasmids contains a selectable marker gene and if only one of the plasmids contains an origin of replication that is functional in the host cell.

As mentioned earlier, several disarmed Ti vectors that carry a variety of chimeric selectable marker genes have been constructed. A single example is shown in Fig. 24.24. These cointegrate Ti plasmids have proven to be effective gene-transfer vectors for several dicotyledonous plants. The disadvantage of cointegrate vectors is that they require crossover events to form the cointegrate structures; such events occur with a frequency of 10^{-4} to 10^{-5} in *A. tumefaciens* cells harboring both plasmids. Thus, it would be convenient to have a vector that was not dependent on cointegration.

Binary Ti Vectors

As previously stated, the *vir* genes of *Agrobacterium tumefaciens* can act in *trans* as well as in *cis* to the T-DNA. This has facilitated the construction of bipartite Ti vectors in which the T-DNA segment is on one plasmid and the *vir* genes are on a second plasmid. The plasmid that contains the T-DNA segment, but no *vir* genes, is called a ***binary Ti vector.*** Binary vector plasmids are very similar to the intermediate donor plasmids (Fig. 24.24) used to form cointegrate plasmids. However, ***binary vectors differ from the cointegrate donor plasmids by the presence of (1) T-DNA border sequences*** (sometimes only one) ***and (2) a broad-host-range origin of replication that allows them to replicate autonomously in* A. tumefaciens.** The two major advantages of binary vectors are (1) their small size (usually about 8 to 12 kb), which facilitates structural manipulations *in vitro* (e.g., the insertion of foreign genes), and (2) elimination of the requirement for cointegrate formation prior to T-DNA transfer to plant cells. Although many binary vectors contain both T-DNA border sequences, it turns out that one is sufficient. The small size of the binary plasmid allows a single border to be used twice, essentially functioning like the direct tandem repeats at the ends of the T-DNA of a wild-type Ti plasmid. The structure of a typical binary vector, pMON410, is shown in Fig. 24.25. This vector contains two chimeric selectable marker genes. One, *nos*/NPTII/*nos*, provides resistance to kanamycin (G418); the other, 35S/HPH/*nos* (HPH is the coding sequence for the enzyme hygromycin phosphotransferase), confers resistance to the antibiotic hygromycin.

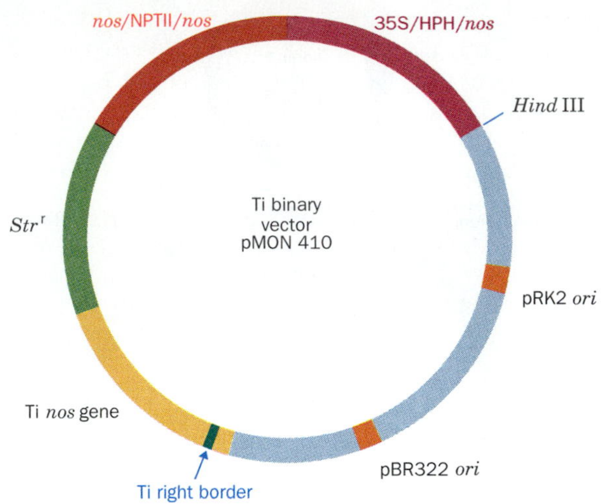

Figure 24.25 Structure of the binary Ti vector pMON410 (constructed at Monsanto Co.). This plasmid contains two origins of replication: the pBR322 *ori* allows replication in *E. coli* and the pRK2 *ori* allows replication in *A. tumefaciens*. The chimeric selectable marker genes *nos*/NPTII/*nos* and 35S/HPH/*nos* confer resistance of recipient plant cells to the antibiotics kanamycin (G418) and hygromycin, respectively. Str^r is an *E. coli* transposon Tn7 gene that provides bacterial cells with resistance to streptomycin. The Ti *nos* (nopaline synthase) gene provides a good marker for use in assaying expression of integrated T-DNAs in transformed plant cells. The *Hin*dIII recognition sequence shown is unique on pMON410; it provides a convenient site for the insertion of foreign genes into the vector. (Simplified version of the map of pMON410 provided courtesy of S. G. Rogers, R. Horsch, and colleagues, Monsanto Company.)

Transformation by Co-cultivation of Tissue Explants with *Agrobacterium*

Transformation is now seldom carried out by inoculating wounded plants with *Agrobacterium tumefaciens*. Instead, either (1) ***tissue explants*** or (2) ***protoplasts*** are ***co-cultivated with* A. tumefaciens** cells harboring the Ti plasmid of choice, and ***plants are regenerated*** from transformed callus cells formed on the cut surfaces of the tissue explants or from transformed protoplasts. The simplest procedure is to co-cultivate ***sterile leaf discs*** or ***root sections*** with *A. tumefaciens* for a few days, and then to transfer the inoculated explants to selection/regeneration medium. The selection/regeneration medium contains (1) an antibiotic (usually carbenicillin) to kill *Agrobacterium* and (2) the appropriate antibiotic (kanamycin, G418, hygromycin, etc.) to select for transformed plant cells. Within two to three weeks, transformed callus tissue has grown on the cut surfaces and has begun to differentiate into shoots. These shoots are excised and transferred to root-inducing medium. After root formation, the plantlets are transplanted to soil. By using the leaf disc or root section co-cultivation procedure, transgenic plants can often be obtained within five to six weeks after co-cultivation is initiated.

Direct Gene Transfer, Electroporation, and Microprojectile Guns

As discussed in the preceding sections, the Ti plasmid of *Agrobacterium tumefaciens* provides an excellent gene-transfer system for use in dicotyledonous plants. Unfortunately, the host range of *Agrobacterium* is limited to dicots, thus Ti vectors cannot be used for gene transfers in monocotyledonous plants such as maize and the cereal crops rice, wheat, oats, and barley. Because so many of the important agronomic plants are monocots, a great amount of effort has been invested in the development of gene-transfer technologies that are applicable to these species. Most of this effort has involved attempts to introduce DNA into plant protoplasts (single cells from which the cell walls have been removed by enzymatic digestions) by transiently perturbing the cell membranes so as to make them permeable to macromolecules such as DNA. One successful approach, called ***direct gene transfer,*** involves the addition of selectable marker genes together with polyethylene glycol (which seems to stimulate membrane fusions). A second successful approach, called ***electroporation,*** employs a short pulse of high-intensity electrical current to disrupt cell membranes and render them temporarily permeable to DNA molecules. Although these procedures routinely yield transiently transformed protoplasts, and yield a reasonable frequency of stably transformed cell lines, monocotyledonous plants can seldom be regenerated from the transformed protoplasts. In fact, to date (May, 1990), there has been ***only one published report of maize plants regenerated from transformed protoplasts.***

Recently, a novel procedure has been developed for delivering ***DNA into intact plant cells.*** This procedure employs a ***microprojectile gun*** (Fig. 24.26) to ***shoot DNA-coated microprojectiles (tungsten particles of diameter 1 to 4 μm) into plant cells,*** either cells growing in culture or epidermal cells of whole plants. This microprojectile approach may well become the method of choice for plant gene transfers. It has the advantage of being technically simple, and it should be applicable to all plant species. The microprojectile procedure is not limited to use in plants; it has also proven effective in transforming fungi and algae that were previously recalcitrant to transformation by other methods.

Herbicide-Tolerant Plants

The scenario is very simple. You plant your tomato seeds in the backyard garden, and once every two weeks you spray the garden lightly with a solution containing a rapidly degradable broad-spectrum herbicide. Your tomato plants carry a gene that makes them tolerant to the herbicide. All other plants ("weeds" when present in your tomato patch) are killed by the herbicide. Your garden is weedless throughout the summer without anyone ever touching a tiller or a hoe. Although this scenario is still fiction today, it may well be reality before this textbook is in print. Of all the proposed applications of plant genetic engineering, the production of commercially available ***herbicide-tolerant plant varieties*** is the closest to fruition.

Although herbicide-tolerant tomato plants for the backyard garden may be of relatively minor importance, the development of herbicide-tolerant varieties of agronomically important plants such as corn, soybeans, and the cereals promises to have a major impact on agriculture, both economically and on production practices. Weeds compete with crops for soil nutrients and routinely lead to significant losses in yield. Modern agriculture makes use of herbicides to control weeds and minimize the losses. Unfortunately, the available herbicides seldom provide the degree of specificity that is desired, and most herbicides will control only certain classes of weeds and not others. Broad-spectrum herbicides may give good weed control, but, in so doing, usually have deleterious effects on the growth of the crop plant as well. As a result, scientists are now evaluating alternate approaches to weed control. The most promising of the alternate approaches is the development of herbicide-tolerant plant varieties for use with broad-spectrum or totally

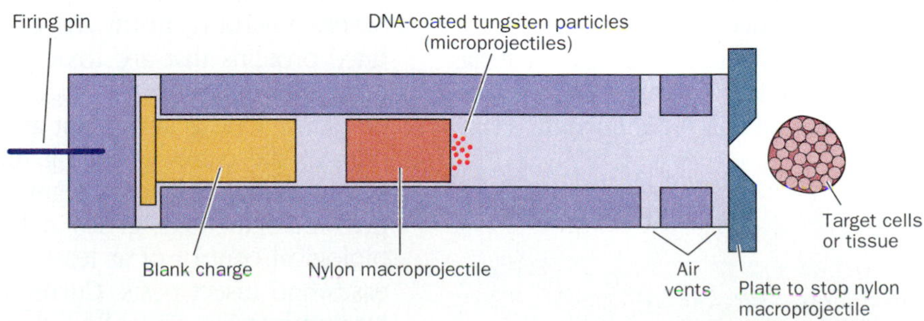

Figure 24.26 Schematic diagram of the structure of a microprojectile gun used in the transformation of plants, fungi, and algae. The gun propels DNA-coated tungsten particles or microprojectiles through the rigid walls of plant, fungal, and algal cells. (After T. M. Klein, E. D. Wolf, R. Wu, and J. C. Sanford, *Nature* 327: 70–73, 1987.)

nonspecific herbicides. Obviously, the potential economic value of herbicide-tolerant plant varieties is substantial; this potential monetary reward has provided the stimulus for the large investment of money and effort that has gone into the production of herbicide-tolerant plants.

Herbicides are simply chemical compounds that kill or inhibit the growth of plants without deleterious effects on animals. Herbicides usually inhibit processes that are unique to plants, for example, photosynthesis. Most frequently, herbicides act as inhibitors of essential enzyme reactions. Thus, anything that diminishes the level of inhibition will provide increased herbicide tolerance. The two most common ***sources of herbicide tolerance are (1) overproduction of the target enzyme and (2) mutations resulting in enzymes that are less sensitive to the inhibitor*** (usually due to a lower affinity of the enzyme for the inhibitor). It seems likely that the most successful strategy for developing herbicide-tolerant plants will be to combine both sources of tolerance, that is, to engineer plants that overproduce herbicide-tolerant mutant enzymes. Although progress has been made in developing plants with tolerance to several different herbicides, the discussion here will be limited to a single example for the sake of brevity.

Glyphosate (Fig. 24.27) is one of the most potent broad-spectrum herbicides known; it is marketed under the trade name Roundup. ***Glyphosate acts by inhibiting the enzyme 5-enolpyruvylshikimate 3-phosphate synthase (EPSP synthase),*** an essential enzyme in the biosynthesis of the aromatic amino acids tyrosine, phenylalanine, and tryptophan. These aromatic amino acids are essential components in the diets of higher animals since the enzymes that catalyze the biosynthesis of these amino acids are not present in higher animals. Therefore, since higher animals contain no EPSP synthase, glyphosate has no toxic effects on animal systems. In this respect, glyphosate is an ideal herbicide.

Glyphosate does inhibit the EPSP synthases of microorganisms as well as those of plants. By selecting for growth in the presence of glyphosate concentrations that inhibit the growth of wild-type bacteria, researchers have been able to isolate ***glyphosate-tolerant mutants of*** **Salmonella typhimurium, *Aerobacter aerogenes, and* Escherichia coli.** In bacteria, EPSP synthase is encoded by the *aro A* gene. When the mutant bacterial *aro A* genes were provided with plant promoters and polyadenylation signals (producing chimeric genes) and were introduced into plants, the transgenic plants exhibited increased tolerance to glyphosate. In plants, aromatic amino acid synthesis takes place in chloroplasts, but the genes encoding the biosynthetic enzymes such as EPSP synthase are nuclear genes. The translation products contain a transit peptide that targets the protein to the chloroplasts. This transit peptide is then cleaved off proteolytically upon entering the chloroplast to yield the active enzyme. Experiments have now shown that the petunia transit peptide will target the *E. coli aroA* gene-product into tobacco chloroplasts and will produce glyphosate tolerance in the recipient cell lines.

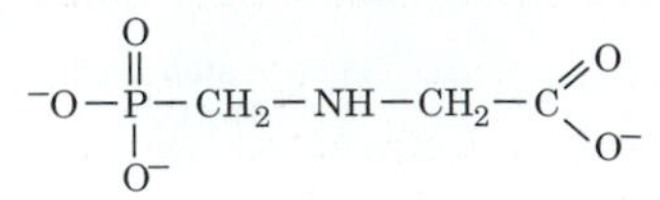

$$^{-}O-\overset{\displaystyle O}{\underset{\displaystyle O^{-}}{\overset{\|}{\underset{|}{P}}}}-CH_2-NH-CH_2-C\begin{matrix}\nearrow O \\ \searrow O^{-}\end{matrix}$$

Figure 24.27 Structure of glyphosate, the active ingredient in the herbicide Roundup.

Glyphosate-tolerant plants have also been produced by using an ***EPSP synthase cDNA isolated from a glyphosate-tolerant petunia cell culture line.*** The glyphosate-tolerant cell line was isolated by selecting for gradually increasing resistance to glyphosate. Molecular analysis of this cell line demonstrated that the glyphosate tolerance resulted from ***amplification*** (an increase in copy number) ***of the EPSP synthase gene*** and, thus, ***overproduction of EPSP synthase*** in these cells. The EPSP synthase cDNA isolated from this cell line was joined to the CaMV 35S promoter and to the Ti *nos* 3′ polyadenylation signal (Fig. 24.28). The strong CaMV 35S promoter results in overproduction of EPSP synthase. When this ***chimeric 35S/EPSP synthase/*nos *gene*** was introduced into petunia plants on a Ti vector (see the preceding section), ***the resulting transgenic plants were tolerant to four times the concentration of glyphosate required to kill control plants*** (Fig. 24.29). The results of these preliminary studies suggest that glyphosate-tolerant varieties of agronomically important plant species will be commercially available in the not-too-distant future.

Disease- and Insect-Resistant Varieties

Several microorganisms and certain native plants produce proteins that are toxic to specific plant pathogens, both microbial pathogens and insects that feed on plants. One goal of plant genetic engineering is to transfer the genes encoding these protein toxins to agronomically important plants with the hope that expression of the toxin genes in these plants will provide biological control of at least some serious plant diseases and insect pests. Currently, plant diseases and insect pests are controlled almost exclusively by the use of broad-spectrum chemical bacteriocides, fungicides, and insecticides. However, there is reason for concern about the potential damage to ecosystems and

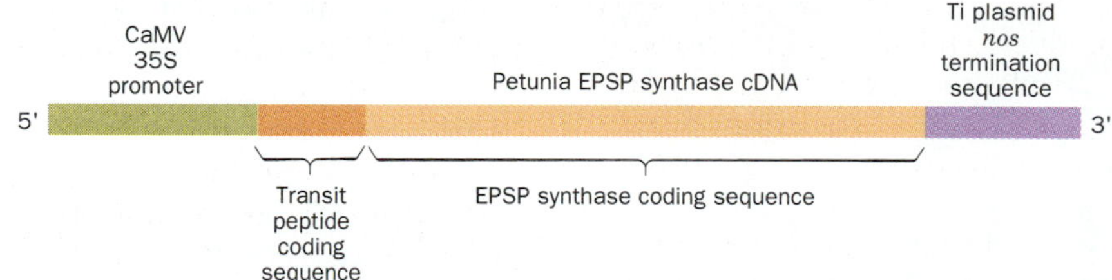

Figure 24.28 Diagram of the structure of the chimeric EPSP synthase gene that was used to produce the glyphosate-tolerant petunia plants shown in Fig. 24.29. The chimeric gene contains the 35S promoter from cauliflower mosaic virus (CaMV), the EPSP synthase coding sequence (cDNA) from petunia (note that this cDNA will also encode the chloroplast transit peptide), and the 3′ polyadenylation sequence from the *nos* (nopaline synthase) gene of an *Agrobacterium tumifaciens* Ti plasmid.

pollution of groundwaters that might result from the widespread use of these chemicals on agricultural crops. Thus, scientists are searching for alternate methods for controlling these pathogens.

The best-known examples of the ***use of natural gene-products to control plant pests are the insect toxins of* Bacillus thuringiensis.** Each of the *B. thuringiensis* toxin genes encodes a large protein that aggregates to form protein crystals in spores, and these protein crystals are highly toxic to certain insects. Some of the insects that are killed by these protein toxins are plant pests of major economic im-

Figure 24.29 Photograph of glyphosate-tolerant petunia plants produced by introducing the chimeric 35S/EPSP synthase/*nos* gene (Fig. 24.28) into plants by means of the leaf disc–*Agrobacterium tumefaciens* co-cultivation procedure (see the preceding section). The plants at the top contained the chimeric gene; the dead plants at the bottom were wild-type controls. The plants were sprayed with Roundup at a dose equivalent to about 0.8 pound per acre. (Photograph courtesy of D. M. Shah. Reproduced with permission from D. M. Shah *et al., Science* 233: 478–481, 1986. Copyright © 1986 by the American Association for the Advancement of Science.)

Figure 24.30 Photograph illustrating the protection of transgenic tomato plants against tobacco hornworm larvae provided by endogenous *B. thuringiensis* insect toxin expressed from a chimeric gene with the structure CaMV 35S promoter/*B. thuringiensis* subspecies *kurstaki* toxin coding sequence/*nos* termination sequence. Ten hornworm larvae were allowed to feed on each plant for seven days. Left, a transgenic plant expressing the chimeric gene; right, a nontransformed control plant. (Photograph courtesy of D. A. Fischhoff and S. R. Sims. Reproduced with permission from D. A. Fischhoff *et al., Bio/Technology* 5: 807–813, 1987; Copyright © 1987 by BIO/TECHNOLOGY.)

portance. Different subspecies of *B. thuringiensis* produce toxins that kill different insects. For example, the toxin produced by *B. thuringiensis* subspecies *kurstaki* kills lepidopteran larvae such as the tobacco hornworm. The gene that encodes this toxin has been isolated and shown to synthesize a functional toxin in *E. coli.* ***A chimeric gene with the structure CaMV 35S promoter/*****B. thuringiensis** ***subsp.*** **kurstaki** ***toxin coding sequence/Ti*** **nos** ***3′ termination sequence was constructed*** (see the preceding sections). This chimeric gene was placed in a Ti vector, and tomato leaf disc cells were transformed by co-cultivation with *A. tumefaciens* harboring the engineered Ti vector–chimeric gene construct. ***Transgenic tomato plants were regenerated and shown to express the chimeric gene.*** The toxicity of the gene-product synthesized in the transgenic plants was tested by allowing tobacco hornworm larvae to feed on the transgenic plants and on control plants (Fig. 24.30). All the ***larvae applied to the transgenic plants died within a few days;*** larvae applied to the control plants remained healthy and eventually consumed the entire plants. These results support the feasibility of using genetic engineering to produce insect- and pathogen-resistant plant varieties. However, the overall potential of this approach remains to be evaluated through future investigations. In particular, the question of how rapidly insects and pathogens evolve resistance to the toxins or other endogenously produced antipathogenic substances is of major concern. If the insects and other pathogens can evolve new modes of resistance rapidly, the utility of this approach may be limited.

"High-Lysine" Corn

The proteins stored in plant seeds function as amino acid reserves for use during seed germination and preemergence growth of the young seedling. Plant seed storage proteins also provide the major source of proteins in the diets of most humans and herbivorous higher animals. Worldwide, the seeds of legumes and cereal grains are estimated to provide humans with 70 percent of their dietary protein requirements. Unfortunately, ***the major seed storage proteins of cereals, called prolamines (zeins in corn), are virtually lacking the amino acid lysine.*** Since prolamines account for about half of the total protein content of cereal seeds, diets based largely on cereal grains will be deficient in lysine. In the case of corn, the seed proteins are also deficient in tryptophan and, to a lesser extent, methionine. Because of the importance of the cereal seeds as human and animal food, plant breeders have attempted for several decades to develop varieties with increased lysine, tryptophan, and methionine content. Although some progress has been made, the successes have had limited applicability in terms of agricultural production. In corn, mutants such as *opaque-2, sugary-1,* and *floury-2* have increased amounts of lysine and/or methionine in seeds, but these mutant strains have undesirable soft kernels and produce lower yields. These "high-lysine" strains all result from mutations that alter the relative proportions of different seed storage proteins. In general, they lower the prolamine (zein) content so that other seed proteins account for a larger proportion of the total seed proteins. This, in turn, increases the relative amounts of lysine and/or methionine (the exact composition depends on the specific mutation) in the seeds.

Several corn genes encoding zeins have now been cloned and sequenced. With this information in hand, researchers have suggested that it might be possible to produce ***"high-lysine" corn by genetic engineering.*** Since the zeins have no known enzymatic functions, one might be able to modify zein genes by mutagenesis without imposing any deleterious effects on function(s). Specifically, ***site-specific mutagenesis could be used to introduce more lysine codons into zein sequences.*** Then, these "high-lysine" zein coding sequences could be joined to strong promoters such as the CaMV 35S promoter and reintroduced into corn plants by transformation by means of electroporation or a microprojectile gun. However, a potential obstacle to engineering "high-lysine" corn by this procedure is that the modified zeins might not package properly in seed storage structures.

The zeins are synthesized on rough endoplasmic reticulum and then aggregate within this membranous structure into dense deposits called ***protein bodies.*** The formation of protein bodies is thought to involve hydrophobic and weak polar interactions between the zein monomers. If so, charged amino acids such as lysine might interfere with proper packaging of zeins during protein body formation. Recently, however, B. A. Larkins and colleagues have ***introduced new lysine and tryptophan codons into a zein cDNA*** by oligonucleotide-directed site-specific mutagenesis (Fig. 24.31). When RNA transcripts of these modified cDNAs were injected into *Xenopus laevis* oöcytes, the RNAs were translated efficiently and the ***"high-lysine" zein products were found to self-aggregate*** into dense structures similar to those present during polar body formation in corn. These results offer encouragement that "high-lysine" corn might indeed be produced by means of genetic engineering. In fact, the same approach can potentially be used to tailor other plant seed storage proteins to the optimal amino acid compositions. Although obstacles are likely to be encountered, the use of genetic engineering to improve the nutritional quality of seed storage proteins would seem to have great potential.

Enhanced Nitrogen Utilization

Plants are only able to utilize nitrogen that has been incorporated into chemical compounds such as ammonia, urea, or nitrates. No green plant is capable of extracting diatomic nitrogen (N_2) molecules directly from the atmosphere. Although plants use only a small fraction of the total nitrogen pool, they are dependent on a continuous supply of nitrogen in usable form (most often called ***"fixed nitrogen"***). Ongoing fixation of atmospheric nitrogen is required because the fixed nitrogen in soil is constantly being depleted by leaching, by utilization for the growth of plants and microorganisms, and by denitrifying bacteria that convert fixed nitrogen back to N_2. As a result, millions of dollars are spent each year on nitrogen fertilizers in order to obtain optimal yields of major crops such as corn and the cereal grains.

Biological nitrogen fixation is the alternative to the use of the industrially fixed nitrogen provided in fertilizers. Several species of bacteria and lower algae are capable of converting N_2 to fixed forms of nitrogen that can be utilized by plants. Because the purchase of nitrogen fertilizers represents one of the major expenses incurred with current agricultural production methods, a major effort has been and continues to be devoted to the development of enhanced methods of biological nitrogen fixation.

Certain free-living soil bacteria such as **Azotobacter vinelandii** and **Klebsiella pneumoniae** ***directly convert atmospheric nitrogen to ammonia.*** These bacteria are an important source of fixed nitrogen and, in addition, have proven to be extremely valuable subjects for studies on the mechanism of nitrogen fixation. In *Klebsiella,* there are 17 **nif** (*ni*trogen *f*ixation) genes organized in seven operons. The complexity of the nitrogen fixation metabolic machinery in these bacteria has important implications for anyone who might aspire to engineer nitrogen-fixing plants. The situation with nitrogen fixation is very different from that of herbicide tolerance (discussed earlier in this chapter). It is one thing to construct a single chimeric gene and transfer that gene to plants, but it is far more difficult to engineer 17 different chimeric genes, to transfer all of them to the same recipient plant, and to coordinate their expression in the plant so that all the components of the complex nitrogen-fixing enzymatic machinery are synthesized in the proper amounts and in the proper cells of the plant. At present, the possibility of engineering nitrogen-fixing plants is largely fantasy, but remember that traveling to the moon was pure science fiction not too many years ago.

Another very important source of biologically fixed nitrogen is that produced through the symbiotic relationship between bacteria of the genus **Rhizobium** and plants of the family ***Leguminosae (the alfalfas, clovers, soybeans, peanuts, peas,*** etc.). This symbiotic nitrogen fixation occurs in highly differentiated ***root nodules*** (Fig. 24.32) that develop when *Rhizobium* bacteria interact with the roots of legumes. Nodule formation is dependent on genetic information of both the plant and the bacterium. ***The nitrogenase that catalyzes N_2 reduction is encoded by the bacterial genome, but the fixed nitrogen is utilized for growth of both the bacteria and the host plants.*** Once the mechanisms responsible for establishing this symbiotic relationship and for nodule formation are known, and the genes that control these processes have been identified, it might be possible to use genetic engineering to modify nonlegume plants (e.g., corn, rice, and wheat) such that they will participate in similar symbiotic relationships with nitrogen-fixing bacteria. However, once again, this will undoubtedly be a formidable task because the genetic control of nodule formation is clearly complex. Nevertheless, experiments are in progress with goals of modifying

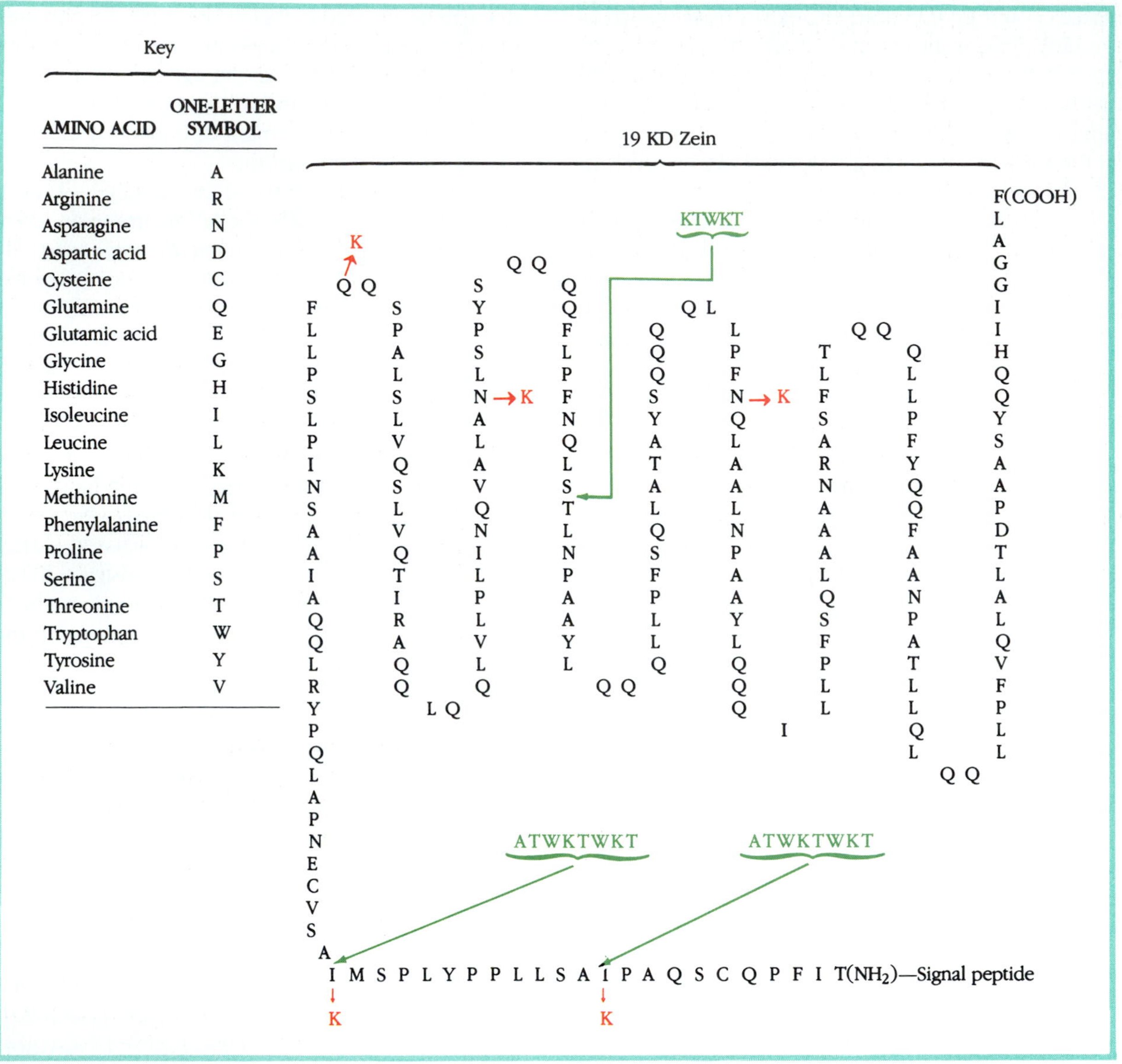

Figure 24.31 Modified zein sequences produced by oligonucleotide-driven site-specific mutagenesis. The amino acid sequence of the zein is shown using the standard one-letter symbols; a key to the one-letter symbols is shown on the left. The amino acid sequence shown is that for the 19,000-Da zein product of maize cDNA clone cZ19c1 (D. M. Marks, J. S. Lindell, and B. A. Larkins, *J. Biol. Chem.* 257: 16445, 1985). The amino acid sequence is drawn in a format representing the proposed secondary structure of the protein with parallel regions corresponding to predicted α-helical segments. The positions of lysine residues introduced by site-specific mutagenesis are shown in red; the positions of oligopeptide insertions produced by inserting synthetic oligonucleotides into restriction enzyme cleavage sites in the cDNA are shown in green. Lysine substitutions were done singly and in pairs. All the modified zeins self-aggregated in structures that appeared similar to those produced by the native zeins. (After J. C. Wallace, G. Galili, E. E. Kawata, R. E. Cuellar, M. Shotwell, and B. A. Larkins, *Science* 240: 662–664, 1988.)

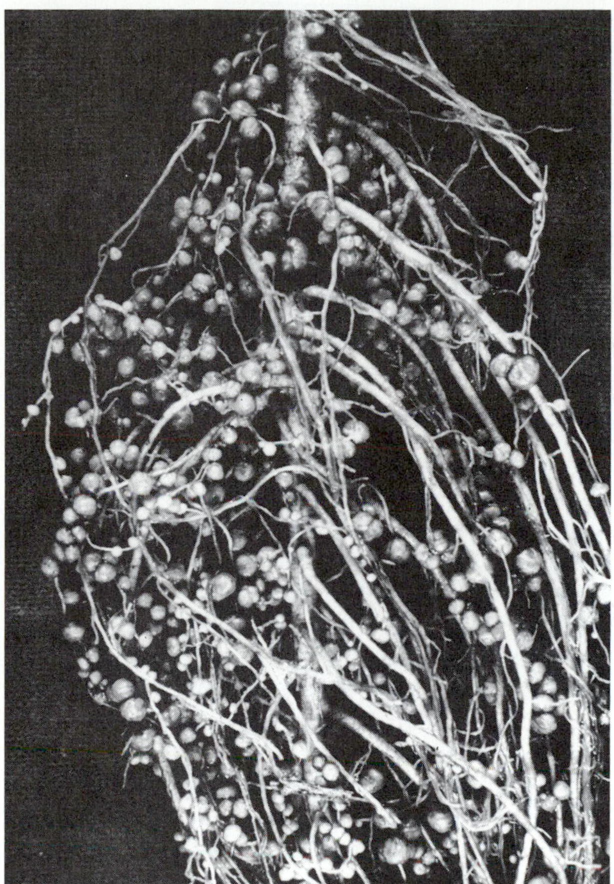

Figure 24.32 Photograph of *Rhizobium*-containing nodules on the root of a soybean plant. Nitrogen fixation that occurs within these nodules provides the major source of ammonia used in the synthesis of all the nitrogen-containing components of the host plant. For the plant breeder, the development of a corn plant with similar nitrogen-fixing root nodules would be a near-Utopian accomplishment. (Photograph courtesy of C. Vance, Department of Agronomy and Plant Genetics, University of Minnesota, and J. Burton, Nitragin Co., Milwaukee, Wisconsin.)

bacteria so as to enhance their nitrogen-fixing capacity and to broaden their host range to include additional plant species.

ANIMALS

The application of genetic engineering technologies to higher animals has three major goals: (1) the development of gene-transfer techniques for future use in treating inherited diseases in humans by "gene therapy," (2) the introduction or enhancement of desirable traits in domestic animals, and (3) the production of valuable products by transgenic animals. The mouse has been the organism of choice for developing gene-transfer methodologies in higher animals, and transgenic mice are now routinely produced and studied in laboratories throughout the world. Over 50 different genes have already been transferred into recipient mouse eggs or embryos and stably maintained in adult mice and their progeny. The foreign or introduced genes are referred to as ***transgenes*** to distinguish them from endogenous genes; the animals that contain the foreign or introduced genes are called ***transgenic animals.*** One of the most encouraging results of these studies on transgenic mice is that the transgenes have usually exhibited their normal patterns of expression (tissue-specific or developmental stage-specific expression) and regulation (induced or constitutive). Transgenic sheep, pigs, rabbits, and chickens have also been produced. Two methods are predominantly used to produce transgenic animals: (1) microinjection of DNA into the pronuclei of fertilized eggs and (2) infection of preimplantation embryos with retroviral vectors.

Microinjection of DNA into Fertilized Eggs

The production of transgenic mice and other higher animals by DNA microinjections is accomplished by essentially the same procedure that is widely used in *Drosophila* (see the section entitled *Drosophila,* P-Element Transformation earlier in this chapter). There are two major differences. (1) P-element transformation in *Drosophila* utilizes the transposase of the P element for integration of the injected DNA molecules. Integration of microinjected DNA in fertilized mouse eggs depends on the recombination mechanisms of the mouse embryo itself. (2) DNA is usually injected into mouse eggs that are at an earlier stage of development than those used in *Drosophila.*

In microinjection, eggs are surgically removed from the female parent and are fertilized with sperm *in vitro.* The ***DNA*** (usually a plasmid vector carrying the gene of interest) ***is microinjected into the male pronucleus*** (the haploid nucleus contributed by the sperm, prior to nuclear fusion) ***of the fertilized egg through a very fine tipped glass needle*** (Fig. 24.33). The injected DNA can be either linear or circular DNA molecules. However, linear DNA integrates with a higher efficiency (about fivefold higher) than circular DNA. Although the reason for this difference is not known, it probably results from the recombinogenicity of ends of DNA molecules owing to their ability to invade homologous DNA molecules (see the section *Saccharomyces cerevisiae,* a Eukaryotic *Escherichia coli* earlier in this chapter). Usually, several hundred to several thousand copies of the gene of interest are injected into each egg, and thus multiple integrations can occur. Surprisingly, when multiple copies do inte-

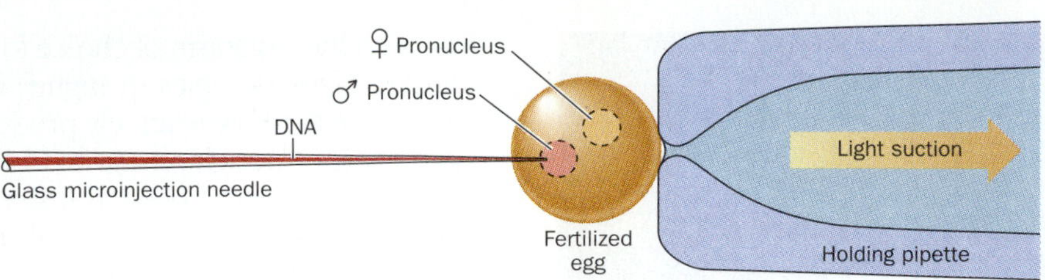

Figure 24.33 Diagram illustrating the procedure used to microinject DNA into fertilized mouse eggs in the production of transgenic mice. A light suction is applied to a blunt-tipped pipette to hold the egg in place without damaging it. DNA is then injected through a very fine glass microneedle into the male pronucleus of the fertilized egg. (Drawing based on the photograph of J. Gordon and F. H. Ruddle, *Methods in Enzymol.* 101: 411–433, 1982.)

grate into the genome, they usually do so as tandem, head-to-tail arrays at a single chromosomal site. These DNA molecules carrying multiple tandem copies of the transgene are believed to form by recombination events between injected DNA molecules prior to integration. The chromosomal sites of integration are apparently selected at random.

Because the DNA is injected so early during development, even before fusion of the haploid nuclei to form the diploid nucleus of the zygote, integration of the injected DNA molecules usually occurs early enough during embryogenesis so that some germ line cells do carry the transgene. As would be expected, the mice that develop from the injected eggs (called the ***G_0 generation***) are almost always ***genetic mosaics*** with some somatic cells carrying the transgene and others not carrying the transgene. The initial (G_0) transgenic mice must be mated and G_1 progeny produced to obtain mice in which all cells carry the transgene. In almost all the cases where progeny studies have been done, the transgenes were stably transmitted to progeny for as many generations as were examined. In a few cases, transgenes have undergone rearrangements in progeny generations. In mammals, transgenes do not seem to be maintained by autonomous replication (with a couple of exceptions). This is in contrast to the autonomous replication of transgenic DNAs observed after microinjection into eggs of other animals such as frogs, sea urchins, and *Caenorhabditis elegans.*

Retroviral Vectors

The ability of retroviral proviruses to integrate into host chromosomes is discussed in some detail in Chapter 17. This capability of retroviruses to incorporate DNA copies of their genomes into the chromosomes of their hosts clearly suggests that one might be able to construct modified ***retroviral gene-transfer vectors*** that have properties analogous to the P-element vectors of *Drosophila.* The key to the development of such retroviral vectors is to modify the viral genome in some manner that will eliminate any chance of pathogenic effects of the vector when present in transgenic animals without changing its capacity to integrate into host chromosomes.

Several retroviral vectors have now been developed and shown to be effective vehicles for gene transfer. These retroviral vectors have proven especially useful in introducing genes into primary and established cell culture lines, although they have also been used to introduce genes into germ line cells by infecting eight-cell embryos. The use of retroviral vectors to introduce genes into primary cell lines may have major applications in somatic-cell "gene therapy." One promising approach is to ***initiate a primary cell line*** by tissue biopsy, to correct the enzyme deficiency in these primary cells by ***introducing a wild-type copy of the defective gene,*** and then to ***reimplant these transgenic cells*** back into the appropriate tissue of the affected individual. Whether or not this approach will prove to be effective in treating human diseases remains to be determined, but current work with mice has yielded very encouraging results. Although several different vectors have been developed, and several different gene-transfer strategies have been employed, a single example will suffice to illustrate the potential use of retroviral vectors as tools in future "gene therapy" treatments of inherited diseases.

As was discussed in Chapter 1, ***phenylketonurea (PKU)*** is an inherited disease in humans caused by a ***deficiency of the enzyme phenylalanine hydroxylase (PAH).*** When untreated, individuals with PKU develop severe mental retardation as a result of the accumulation of toxic derivatives of phenylalanine. Current treatment of PKU involves limiting phenylalanine intake with a protein-restricted, carefully formulated diet. Somatic-cell "gene therapy" might someday provide an alternative method of treatment of PKU.

Both the gene and a cDNA encoding human PAH have been cloned, thus setting the stage for PAH gene-transfer experiments. One complication is that in humans, PAH is synthesized only in the liver. Moreover, the PAH-catalyzed conversion of phenylalanine to tyrosine requires the cofactor tetrahydrobiopterin. This cofactor is present in liver cells, but is not present in most other tissues. Thus, ***corrective "gene therapy" for PKU will probably require expression of the introduced PAH gene in liver cells.*** If so, this excludes the possibility of utilizing the more common bone-marrow transplant technology to treat PKU. For that reason, S. L. C. Woo and colleagues have concentrated on the development of ***chimeric genes and vectors that could be used to express PAH in primary hepatocytes (liver cells).*** Their results to date are most promising; Woo and coworkers have demonstrated the expression of a chimeric human PAH transgene in primary mouse hepatocytes.

The chimeric human PAH gene was introduced into mouse hepatocytes by using a ***retroviral vector called N2*** that was derived in part from the Moloney murine leukemia virus. The N2 vector contains the viral long terminal repeats (LTRs) that are required for integration of the provirus and additional viral sequences that are required for packaging of the vector into enveloped particles that are capable of adsorbing to and entering recipient cells. The coding sequences for the viral structural proteins have been deleted; thus, the vector is totally nonpathogenic. A chimeric selectable marker gene (a derivative of the neomycin-resistance gene of *E. coli* transposon Tn*5*) has been inserted to facilitate the identification of transgenic cells. Into this N2 vector, Woo and colleagues inserted a cassette that contained the ***human PAH cDNA clone and two key regulatory sequences:*** (1) The 5′ regulatory sequence of the human α_1-antitrypsin gene and (2) the promoter sequence from simian virus 40 (SV40). The human α_1-antitrypsin gene is known to be expressed at high levels in hepatocytes, and the 5′ sequence that controls this tissue-specific expression has been localized to the 5′ segment present in the PAH cassette. The structures of the chimeric human PAH "gene" and the N2 vector are diagrammed in Fig. 24.34. Packaged vector is produced by transfection of a cell line that produces viral structural proteins, but no endogenous progeny viral genomes. Thus, only progeny RNA replicas of the vector genome are packaged, and these virions are purified for use in the gene-transfer experiments.

When primary mouse hepatocytes were infected with virions containing the retroviral vector carrying the chimeric human PAH "gene," transgenic cells were identified and were shown to express the chimeric "gene." Transcripts that contained the PAH coding sequence were found to have the predicted structure, and human PAH was shown to be synthesized in the transgenic mouse hepatocytes. Although considerable work remains to be done before somatic-cell "gene therapy" is ready to be tried on PKU patients, these preliminary results with mice suggest that such "gene therapy" will probably be feasible in the future.

Gene Expression in Transgenic Mice

As mentioned earlier in this chapter, ***transgenic mice are now routinely produced*** in laboratories throughout the world. These transgenic mice provide invaluable tools for the study of mammalian gene expression and provide a good model system with which to test

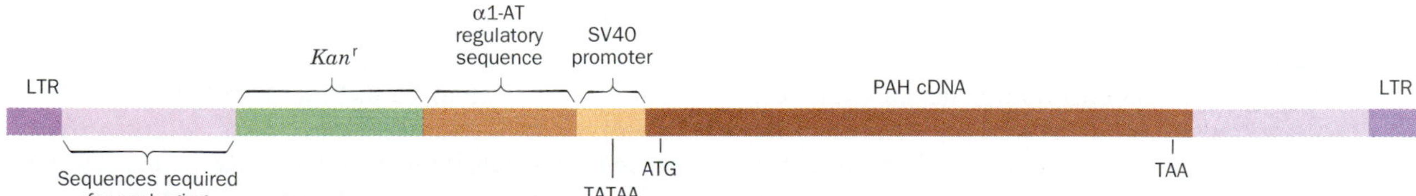

Figure 24.34 Diagram of the structure of the retroviral phenylalanine hydroxylase expression vector pNASPAH (**p**, **p**lasmid; **N**, N2 vector; **A**, α_1-antitrypsin gene tissue-specific regulatory sequence; **S**, SV40 promoter; **PAH**, **p**henylalanine **h**ydroxylase cDNA). N2 vector sequences are shown in purple and green; components of the PAH expression cassette are shown in brown. LTR, long terminal repeat sequences of the Moloney murine leukemia virus; packaging sequences are also those from the Moloney leukemia virus; *Kan^r* (often labeled *NEO* for neomycin resistance), a chimeric *Kan^r* gene containing the coding sequence for neomycin phosphotransferase from the *E. coli* transposon Tn*5*; α_1-AT regulatory sequence, the 5′ regulatory region from the human α_1-antitrypsin gene that dictates the tissue-specific (hepatocyte-specific) expression of this gene; SV40 promoter, the promoter from simian virus 40, but without the SV40 enhancer element; PAH cDNA, a cDNA encoding human phenylalanine hydroxylase; TATAA, position of the "TATA box" of the SV40 promoter; ATG and TAA, translational-initiation and -termination codons, respectively, of the human PAH cDNA. (After H. Peng, D. Armentano, L. MacKenzie-Graham, R.-F. Shen, G. Darlington, F. D. Ledley, and S. L. C. Woo, *Proc. Natl. Acad. Sci. U.S.A.* 85: 8146–8150, 1988.)

the utility of various types of gene-transfer vectors and methodologies. ***In excess of 50 different genes have been shown to exist as transgenes in mice,*** and, surprisingly, most of these transgenes were found to exhibit their normal patterns of expression. This means that the adjacent regulatory sequences transferred with the transgenes were sufficient to produce normal patterns of transcription, processing, and translation in the new host.

The ***elastase-1 gene of the rat*** was one of the first genes to be studied in transgenic mice. The elastase gene is normally expressed only in the pancreas; elastase is a protease that is secreted into the upper intestine to aid digestion. When the rat elastase gene was cloned and microinjected into mouse eggs, the resulting transgenic mice expressed the rat elastase gene only in pancreatic secretory cells. Correct tissue-specific expression of the rat elastase gene was observed in transgenic mice that carried the gene integrated at different chromosomal sites and in their G_1 progeny that carried the gene in all somatic cells. When transgenic mice were produced with rat elastase genes that carried varying amounts of the adjacent 5′ sequence, it was demonstrated that ***only about 200 nucleotide-pairs were necessary to obtain the normal tissue-specific patterns of expression.*** Similar results have been obtained in studies of mammalian protamine genes and gamma-crystalin genes in transgenic mice. All of these genes contain compact promoters, and all the sequences required for the observed tissue-specific expression of these genes are located within a few hundred nucleotide-pairs 5′ to the coding sequences.

Similar studies of other genes in transgenic mice revealed more complex pictures of *cis*-acting regulatory elements, with tissue-specific enhancers located at considerable distances from the coding sequences. For example, the ***liver-specific expression of the mouse albumin gene is completely dependent on an enhancer element located about 10,000 nucleotide-pairs upstream*** (5′ relative to the direction of transcription) from its promoter. Other genes exhibited even more complex regulatory circuits. The yolk sac-specific expression of the mouse alpha-fetoprotein gene was shown to be controlled by three different enhancer elements spread over a 7000-nucleotide-pair region upstream from the coding sequence. Transgenic mice have also been used to identify the sequences that regulate the expression of the embryonic, fetal, and adult hemoglobin genes during development. In short, transgenic mice have been used to study the expression of a large number of mammalian genes, far too many to discuss in detail here. Table 24.1 lists some of the genes that have been studied in transgenic mice. The proper expression of genes from mammals as diverse as humans and rabbits in transgenic mice indicates that many of the regulatory sequences in mammals are quite highly conserved. This conservation of regulatory sequences will make the genetic engineering of animal species less difficult, since each gene of interest will not have to be provided with species-specific regulatory elements. Perhaps the most dramatic phenotypic effect of the expression of heterologous transgenes is ***the increased growth rate that occurs when rat, bovine, or human growth hormone is synthesized in transgenic mice*** (Fig. 24.35).

TABLE 24.1 Expression of Native Genes in Transgenic Mice[a]

GENE	SPECIES OF GENE DONOR	TISSUE IN WHICH TRANSGENE IS EXPRESSED
α-Actin	Rat	Muscle
Amylase-2.2	Mouse	Pancreas
Elastase-1	Rat	Pancreas
α-Fetoprotein	Mouse	Yolk sac and liver
β-Hemoglobin	Mouse	Erythrocytes
β-Hemoglobin	Rabbit	Erythrocytes
β-Hemoglobin	Human	Erythrocytes
Gonadotropin (alpha subunit)	Human	Pituitary
Immunoglobulin κ	Mouse	Lymphocytes
Immunoglobulin μ	Mouse	Lymphocytes
Immunoglobulin μ	Human	Lymphocytes
Insulin	Human	Pancreatic β-cells
α1(I)-Collagen	Human	Various tissues
Myosin (light chain-2)	Rat	Skeletal muscle
Protamine 1	Mouse	Testis
Transferrin	Chicken	Various tissues

[a]Selected data from R. D. Palmiter and R. L. Brinster, "Germ-Line Transformation of Mice," *Annu. Rev. Genet.* 20:465–499, 1986.

As the preceding discussion indicates, the major use of transgenic mice to date has been in defining the *cis*-acting elements that regulate gene expression. Transgenic mice will obviously continue to be invaluable tools for use in dissecting the regulation of gene expression in the future. However, in addition, ***transgenic mice promise to provide tools for the study of animal development, of the immune response, and of the mode of action of oncogenes.*** Finally, there is the hope that appropriately manipulated transgenic mice may provide valuable animal models for the study of inherited human diseases such as PKU and of infectious diseases such as AIDS.

Increased Growth Rates via Amplified Growth Hormone Genes

As noted in the preceding section, transgenic mice that overproduced rat, bovine, or human growth hormone

Figure 24.35 Photograph illustrating the increased growth of a transgenic mouse expressing a chimeric human growth hormone gene. The transgenic mouse on the left is about twice the size of the control mouse on the right. The chimeric human growth hormone gene was microinjected into fertilized eggs as described in Fig. 24.33. This chimeric gene contains the promoter sequence of the metallothionein-I gene (a "housekeeping" gene that is expressed in most tissues) controlling the expression of the human growth hormone coding sequence. Metallothionein is a heavy metal scavenger protein that binds heavy metals and protects cells from the toxic effects of high heavy metal concentrations. Transcription of the metallothionein-I gene (and any foreign coding sequences placed under the control of its promoter) is induced by the presence of heavy metals such as zinc and cadmium. As expected, the presence of zinc or cadmium resulted in increased amounts of human growth hormone in transgenic mice harboring the metallothionein-I promoter/human growth hormone chimeric gene. (Photograph courtesy of R. L. Brinster; for further details, see R. D. Palmiter, G. Norstedt, R. E. Gelinas, R. E. Hammer, and R. L. Brinster, *Science* 222: 809–814, 1983.)

grew to about twice the size of their nontransgenic littermates. This prompted animal breeders to ask whether the introduction of either (1) extra copies of the homologous (same species) growth hormone gene or (2) copies of heterologous growth hormone genes from related species might result in domestic animals with faster than normal growth rates. R. Wall, R. Brinster, and R. Palmiter have collaborated to produce ***transgenic pigs*** (Fig. 24.36) with the hope that ***enhanced growth hormone levels*** might result in leaner pigs with improved meat quality and with faster growth. Other scientists are studying the effects of introduced growth hormone genes in ***transgenic fish*** and ***transgenic chickens.*** The results of the early experiments with transgenic pigs have indicated that growth rates were not increased on standard diets, but were increased on high-protein diets. However, the transgenic pigs were found to be considerably leaner than the controls; this was expected because growth hormone favors the synthesis of proteins instead of fats.

Unfortunately, the transgenic pigs also exhibited several undesirable side effects of the higher growth hormone levels. Most notably, the ***female transgenic pigs were sterile.*** In addition, transgenic animals of both sexes were lethargic with weak muscles and were highly susceptible to arthritis and ulcers. Although scientists are hopeful that ways can be found to overcome these side effects, the potential utility of increased growth hormone levels in transgenic animals as a mechanism to improve growth rates and enhance meat quality in domestic animals remains to be evaluted through further research. The use of transgenic animals carrying appropriately regulated growth hor-

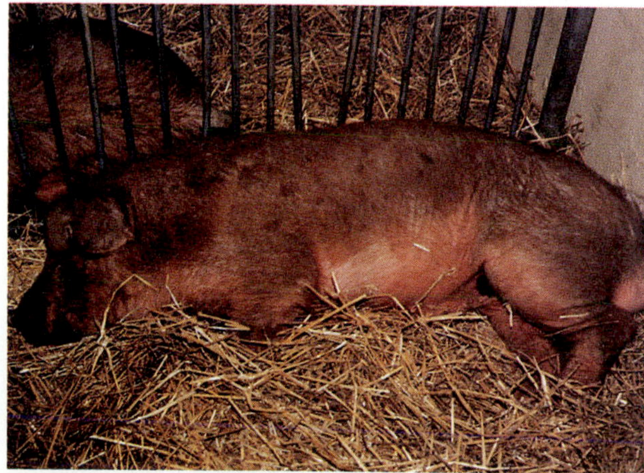

Figure 24.36 Photograph of a transgenic pig containing the chimeric metallothionein-I promoter/growth hormone coding sequence gene. See the legend to Fig. 24.35 for details. (Courtesy Dr. R. L. Brinster, School of Veterinary Medicine, University of Pennsylvania.)

mone genes offers a possible alternative to the use of growth-hormone-releasing ear clips in enhancing growth and meat quality in domestic animals.

Clearly, the genetic control of the native levels of growth hormone represents the end product of many generations of natural selection for optimal fitness in these animals, and any changes in hormone levels induced by transgenes or other manipulations are likely to decrease fitness by upsetting the normal metabolic balance. Thus, deleterious side effects are to be expected. This does not mean that increased growth rates or improved meat quality cannot be achieved without unacceptable decreases in fitness of the transgenic animals. Moderate decreases in vigor may well be offset by highly desirable changes in meat quality (e.g., lower cholesterol levels) or increases in the efficiency of conversion of feed into meat products. After all, domestic animals do not have to survive in the wild; they need only be fit enough to survive and reproduce in the highly protective environment of the farm feedlot.

Virus-Resistant Transgenic Chickens

Cells that are infected with a virus are sometimes resistant to infection by related viruses. In addition, cells that harbor a defective provirus (one that is missing an essential segment of its genome and cannot reproduce) are sometimes resistant to infection by intact, virulent forms of that virus. On the basis of these observations, scientists have proposed that one might be able to genetically engineer domestic animals that are resistant to serious viral pathogens by introducing defective copies of the genome of the virus or a related virus.

One such viral pathogen is the ***avian leukosis virus (ALV).*** ALV causes serious deleterious effects in chicken breeds used in egg production; ALV infections are estimated to cost the poultry industry $50 million to $100 million in losses per year. Thus, the availability of an ***ALV-resistant strain of egg-laying chickens*** would be of major commercial value. D. Salter and L. Crittenden have recently identified transgenic chickens that carry a defective ALV genome. These transgenic chickens produce viral RNA and the viral envelope protein, but no progeny viruses. Most importantly, they are resistant to infection by normally infectious ALV. It appears that the synthesis of large amounts of the retroviral envelope protein somehow blocks the reproductive cycle of intact, pathogenic ALV viruses (at least those of the most closely related ALV subgroup). The ALV resistance has been transmitted to fifth-generation progeny, indicating that the trait is at least reasonably stable. However, much further research is needed to determine whether (1) there are serious side effects that have not yet been detected, (2) the trait can be transferred to economically important egg-laying breeds, and (3) what effect(s) the defective provirus might have, if any, on egg production in the commercially important breeds of chicken. Nevertheless, the results to date with the ALV-resistant transgenic chickens indicate that the use of defective viral genomes may provide one approach to the production of virus-resistant domestic animals.

Secretion of Valuable Proteins in Milk of Transgenic Animals

Many native human proteins contain carbohydrate or lipid side groups that are added posttranslationally. Bacteria usually do not contain the enzymes that catalyze the addition of these side chain moieties to nascent proteins. In such cases, recombinant bacteria cannot be used to synthesize the final product; they will only synthesize the polypeptide gene-product in its unmodified form. For this reason, some researchers have begun to explore alternate methods for producing valuable human proteins, especially glycoproteins and lipoproteins.

One approach that is currently being tested involves the use of transgenic animals to produce human proteins. In particular, the goal is to construct transgenes such that the recipient animals will secrete the proteins in their milk. In principle, this could be accomplished by ***placing the coding sequence of interest under the control of a promoter of a gene that encodes a milk protein such as β-lactoglobulin.*** In addition, the coding sequence of interest would probably need to be joined to the coding sequence for a signal peptide required for secretion of the product from the mammary gland cells in which it is synthesized. However, the construction of chimeric genes of this type is now quite routine and should present no major obstacles.

In fact, A. J. Clark and colleagues in Edinburgh, Great Britain, have produced ***transgenic sheep that secrete either of two human proteins,*** the blood-clotting protein ***factor IX*** or the elastase inhibitor ***alpha$_1$-antitrypsin, in their milk.*** A deficiency of factor IX occurs in one type of hemophilia in humans, and a lack of alpha$_1$-antitrypsin sometimes contributes to the development of emphysema (degenerative lung disease) in humans. Thus, both proteins have important medicinal applications. Clark and co-workers inserted the coding sequences for these proteins into the β-lactoglobulin gene of sheep and microinjected the chimeric constructs into fertilized eggs as described in Fig. 24.33. The injected eggs were then implanted into foster mothers to produce the transgenic lambs. Although the concentrations of human factor IX and human alpha$_1$-antitrypsin were quite low in the milk of the respective transgenic sheep, the researchers believe that the concentrations can be increased by appropriately manipulating the regulatory sequences of the chimeric genes. Importantly, the transgenic sheep exhibited no apparent side effects from the production of either human protein in their milk. Again, further research must be done before the potential applicability of this approach can be adequately evaluated.

"ANTI-SENSE RNA" STRATEGIES FOR REGULATING GENE EXPRESSION

Another new tool that became possible with the advent of recombinant DNA and gene cloning technologies is the use of so-called ***"anti-sense RNA"*** to regulate gene expression. The "anti-sense RNA" method has already proven useful as a tool for studying pathways of gene expression. Presently, "anti-sense RNA" is being touted as a powerful tool for use in pursuing goals as diverse as (1) ***turning off the synthesis of unwanted enzymes during the ripening processes in fruits and vegetables and*** (2) ***controlling viral diseases such as AIDS.***

The "anti-sense RNA" method simply involves the synthesis of RNA molecules that are complementary to the mRNA molecules produced by transcription of a given gene. The normal (*in vivo* synthesized) mRNA of a gene is said to be "sense" because it carries the codons that are "read" during translation to produce the specified sequence of amino acids in the polypeptide gene-product. Normally, the complement to the mRNA "sense" strand will not contain a sequence of codons that can be translated to produce a functional protein; thus, this complementary strand is called "anti-sense RNA." Usually, this "anti-sense RNA" will not contain the regulatory sequences required for translation, but even if it does, it will not usually specify a functional product. Instead, most "anti-sense RNAs" contain multiple termination codons in all three reading frames. However, when "anti-sense RNA" molecules are present in the same cytosol with "sense" (mRNA) molecules of a gene, the ***"anti-sense RNA" and mRNA molecules will anneal to form duplex RNA molecules. These duplex RNA molecules cannot be translated;*** thus, the presence of "anti-sense RNA" will block translation of the mRNA of the affected gene. In fact, recent evidence indicates that these RNA duplexes are often rapidly degraded *in vivo.*

The simplest way to produce "anti-sense RNA" of a gene within a cell or an organism is to (1) clone the gene of interest, (2) separate the coding sequence of the gene from its promoter by the use of appropriately chosen restriction enzymes, (3) religate the coding sequence to its promoter in the inverse orientation (flipped end-for-end), and (4) reintroduce the ***promoter/inverted coding sequence construct ("anti-sense gene")*** into the host cell or organism by transformation. This procedure is illustrated in Fig. 24.37. The net result of this procedure is that ***transcription of the promoter/inverted coding sequence construct will now produce "anti-sense RNA" transcripts.*** These "anti-sense RNA" molecules will hybridize with mRNA ("sense RNA") molecules in cells (form RNA double helices) and block translation of the mRNAs. The final result is that no (or reduced amounts of) protein gene-product will be synthesized in the cells containing the "anti-sense RNA."

"Anti-sense RNA" has been used to block the expression of numerous genes of both prokaryotes and eukaryotes. In addition to its use in blocking the expression of genes with known functions, the "anti-sense RNA" strategy can be used to probe the functions of cloned genes of unknown function. In effect, "anti-sense RNA" can be used to produce a "functional mutation" with the "mutant phenotype" of the affected organism providing important information about the *in vivo* function of the gene involved. To obtain complete blockage of the expression of a gene, one will probably have to either (1) use a very strong promoter to drive the transcription of the inverted coding sequence or (2) reintroduce multiple copies of the "anti-sense gene" into host cells. Both of these approaches can be accomplished with the present tools of genetic engineering.

One potential commercial application of the "anti-sense RNA" technology is already being tested. D. Grierson, in England, has used the ***"anti-sense RNA" technique to reduce the expression of the polygalacturonase gene in tomato to 10 percent of its normal level.*** Polygalacturonase is an enzyme that breaks down cell walls and causes softening in tomatoes as they ripen. As most of us know from first-hand experience, vine-ripened tomatoes are much more flavorful than those that are picked green and allowed to ripen en route to the marketplace. However, vine-ripened tomatoes are too soft to survive the handling required during transport; they simply bruise too easily. Tomato growers hope to enhance the firmness of vine-ripened tomatoes by using the "anti-sense RNA" procedure to lower polygalacturonase levels, and, indeed, the preliminary results are most promising. If successful, this would permit the production of vine-ripened tomatoes that are firm enough to withstand harvesting and transport to the marketplace without major losses due to smashed and bruised fruit.

Another ***potential application of the "anti-sense RNA" technology is in the treatment or prevention of viral diseases.*** In principle, this approach is very sim-

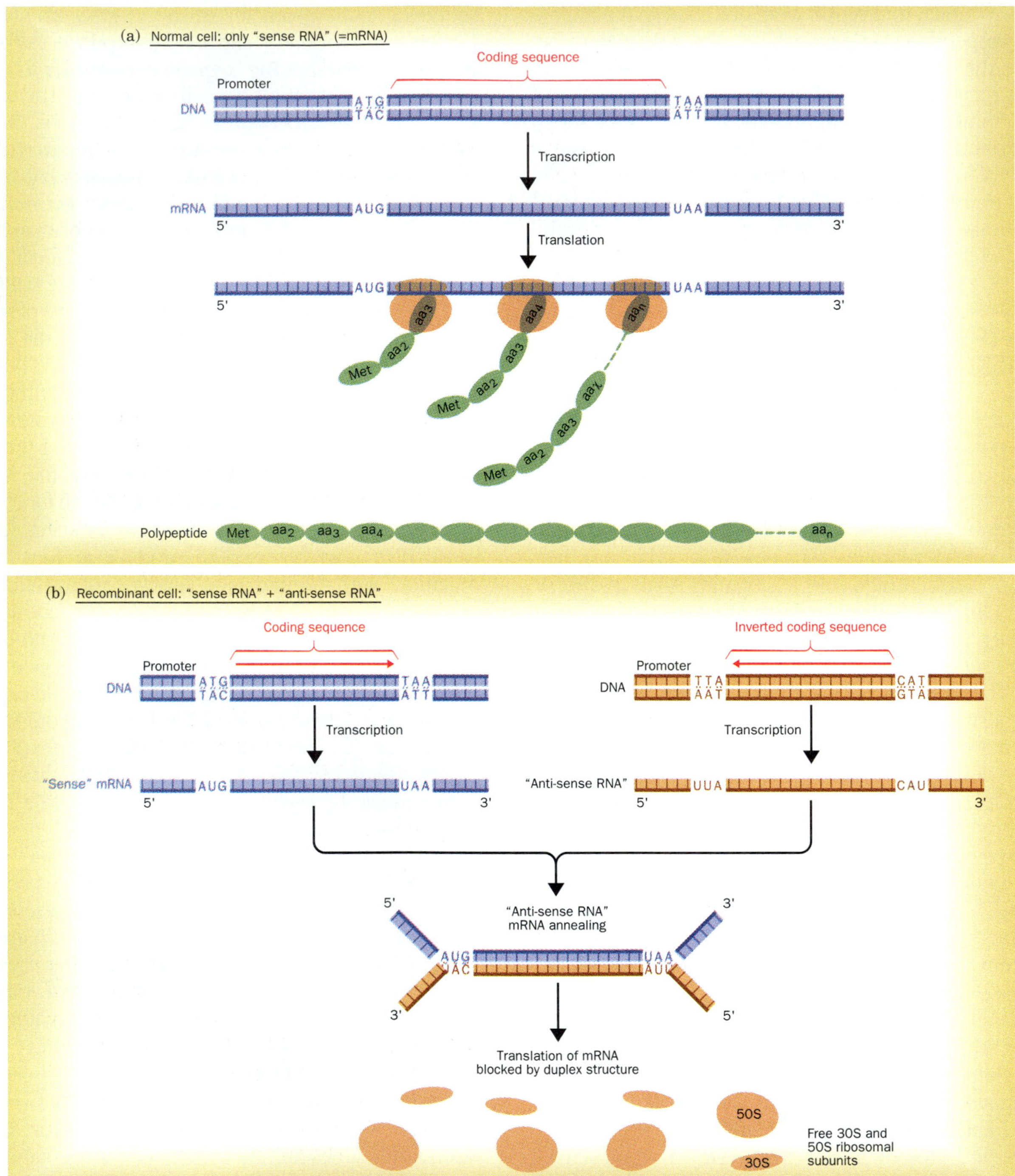

Figure 24.37 Schematic diagram of the use of the "anti-sense RNA" technique to block the expression of a specific gene. (top) A normal cell showing the production of the polypeptide gene-product by transcription and translation. (bottom) A cell containing the homologous "anti-sense gene" in addition to the normal "sense gene" shown in the cell above. The "anti-sense gene" construct was produced by excising the coding sequence (by using appropriate restriction endonucleases), inverting it end-for-end (indicated by the orientation of the arrow above each coding sequence), and rejoining it in the inverted orientation by ligation (catalyzed by DNA ligase). The two nucleotide-pair triplets that gene" and the "anti-sense gene" as markers to use in keeping track of the DNA single strands and their respective orientations. The normal "sense gene" and its mRNA are shown in blue in both cells. The "anti-sense gene" and its "anti-sense RNA" transcript are shown in brown in the cell at the bottom. Note that the mRNA and the "anti-sense RNA" of the gene are complementary to each other; thus, they form an RNA double helix that can**not** be translated by the ribosomes. The net result is that a **reduced level** of polypeptide gene-product is produced in the cell that contains "anti-sense RNA."

ple. One identifies a gene that is completely essential for the reproduction of a virus, and one blocks the expression of this gene with "anti-sense RNA." The difficult step in this approach is in delivering the "anti-sense gene" to the right cells in the desired tissue(s) in order to intercept the reproduction of the target virus. Nevertheless, the "anti-sense RNA" technology has many potential applications. In fact, some scientists believe that this "anti-sense RNA" strategy may be our best hope for treating and/or preventing AIDS in the future. Whether or not the "anti-sense RNA" methodology will live up to the expectations of its most optimistic proponents remains to be seen. In any case, the "anti-sense RNA" approach is one more product of genetic engineering that clearly warrants, and is receiving, further study.

SOMATIC-CELL GENE THERAPY TREATMENT OF HUMAN DISEASES

Of the ***over 3000 inherited human diseases*** catalogued to date, only a few such as phenylketonurea (see Chapter 1) are currently treatable. For many of these diseases, the missing (or defective) gene-product cannot be supplied exogenously as insulin is supplied to diabetics. Most enzymes are unstable and cannot be delivered in functional form to their sites of action in the body, at least not in a form that provides for the ongoing enzyme activity that is usually needed. Cell membranes are not permeable to large macromolecules such as proteins; thus, most enzymes must be synthesized within the cells in which their activity is required. Treatment of inherited diseases is largely restricted to those cases where the product of the defective enzyme is a small molecule that can be distributed to the appropriate tissues of the body through the circulatory system. For most other inherited diseases, somatic-cell gene therapy offers the most promising approach to successful treatment.

Somatic-cell gene therapy involves three sequential steps: (1) the removal of some of the patient's cells, (2) the introduction of normal, functional copies of the gene that is defective in the patient into these cells, and (3) the reintroduction of the transgenic "repaired" cells into the patient. In higher animals such as humans, the gonadal germ line cells are maintained as a cell lineage that is totally separate from all other (somatic) cell lineages. Thus, somatic-cell gene therapy will ***treat*** the disease, ***not cure it.*** In fact, somatic-cell gene therapy experiments will be designed to avoid any chance of affecting germ line cells.

The distinction between somatic-cell gene therapy and germ line gene therapy is pre-eminent when discussing humans. ***All the proposed experiments being discussed here are somatic-cell gene transfers, not germ line modifications.*** The frequently expressed concerns about humankind's "tinkering with nature" or "playing God" apply to germ line gene transfers, not to somatic-cell gene therapy. Major moral and ethical considerations are involved in any decision to perform germ line gene therapy experiments in humans, and ***no germ line gene therapy experiments are currently proposed or are likely to be proposed in the near future.*** Thus, the present discussion will be totally restricted to somatic-cell gene therapy. In terms of moral and ethical considerations, somatic-cell gene therapy is no different than enzyme (gene-product) therapy or cell, tissue, and organ transplants. In transplants, entire organs, with all the foreign genes present in the genome of every cell in the organ, are implanted in patients. With somatic-cell gene therapy, some of the patient's own cells are removed, "repaired," and reimplanted in the patient. Thus, somatic-cell gene therapy is clearly a less extreme manipulation of the patient than an organ transplant.

Several inherited diseases are likely candidates for treatment by somatic-cell gene therapy in the near future. These include ***phenylketonurea,*** the ***thalassemias*** (hemoglobin deficiencies), ***sickle-cell anemia*** (see Mutant Hemoglobins in Chapter 11), ***citrullinemia*** (a life-threatening hyperammonemia caused by a deficiency of the urea-cycle enzyme argininosuccinate synthetase), ***Lesch–Nyhan syndrome*** (see p. 88), and ***combined immunodeficiency disease*** (caused by a deficiency of the enzyme adenosine deaminase).

In an earlier section (Gene Expression in Transgenic Mice) of this chapter, experiments were described in which a retroviral vector (see Fig. 24.34) was used to express a human phenylalanine hydroxylase cDNA in primary mouse hepatocytes. Somatic-cell gene therapy treatment of human phenylketonurea patients would involve the same approach except that (1) primary hepatocytes of the patient would be used and (2) these hepatocytes would be reimplanted in the patient after the introduction of the phenylalanine hydroxylase cDNA expression vector.

Several of the experts in the mammalian gene transplant field believe that human somatic-cell gene therapy will be accomplished first with combined immunodeficiency patients that lack the enzyme adenosine deaminase (ADA). The ADA deficiency in T lymphocytes of patients can be corrected by infusion of normal bone marrow cells from a histocompatible donor. Since ***bone marrow transplant technology is highly advanced,*** obtaining cells for gene transplants and reimplanting the modified cells in the bone mar-

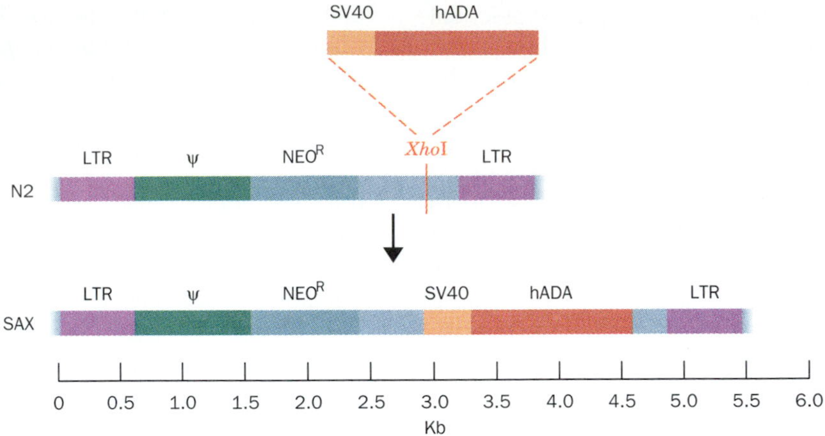

Figure 24.38 Schematic diagram of the construction of the human adenosine deaminase (ADA) gene-transfer vector. A chimeric gene (top) composed of the simian virus 40 early promoter (SV40) and a human ADA cDNA (hADA) was inserted into the *Xho* I site of the retrovirus-derived vector N2 (center) to produce the ADA expression vector **SAX** (bottom; SV40, ADA, *Xho* I site). N2 is the same vector that was used to construct the human phenylalanine hydroxylase expression vector shown in Fig. 24.34. LTR, the long terminal repeats of the Moloney murine leukemia virus; ψ, DNA sequence required for the packaging of vector DNA in virions; NEO^R, neomycin-resistance gene (NEO^R = *Kan^r*, different researchers have adopted the use of different symbols for this gene that provides resistance of host cells to the aminoglycoside antibiotics such as neomycin, kanamycin, and G418). (After W. F. Anderson *et al., Cold Spring Harbor Sympos. Quant. Biol.* 51: 1073–1081, 1986.)

row of the patient should be quite routine. Clearly, graft-versus-host diseases and implant rejections due to histo-incompatibilities should be eliminated by the use of the patient's own cells. The human ADA gene and a human ADA cDNA have been cloned and sequenced, and a retroviral vector that carries a ***strong promoter/ADA cDNA chimeric gene*** has been constructed (Fig. 24.38) for use in gene-transfer experiments. Preliminary experiments with transgenic mice were very encouraging. Therefore, the vector was next tested with ***bone marrow cells of rhesus monkeys*** (Fig. 24.39) in an attempt to more closely simulate an actual human somatic-cell gene transplant. Although human ADA was expressed in the blood cells of some of the monkeys implanted with transgenic bone marrow cells, the levels of expression were lower than expected. Although the results of these first experiments are encouraging, they also indicate that improvements in the structures of the vectors, in the methods used for gene transfers, and/or in the procedures for reimplating transgenic cells will be necessary before high levels of expression of transgenic genes can be obtained in treated patients. This may limit the initial use of somatic-cell gene therapy to the treatment of human diseases in which the symptoms can be alleviated by the synthesis of relatively small amounts of the respective gene-products. Indeed, this is the case with ADA deficiency, and the first somatic-cell gene therapy treatment of an ADA-deficient immunodeficiency patient has just been initiated (September, 1990).

Gene-transfer experiments in nonhuman primates provide the last phase of testing of a somatic-cell gene therapy protocol prior to its use on human patients. The results of these primate experiments are then carefully evaluated by the appropriate committees of the U.S. Public Health Service's National Institutes of Health in arriving at decisions to approve or disapprove applications to perform somatic-cell gene therapy treatments on human patients. Clearly, all possible precautions must be taken to exclude possible side effects or other potential harm to human patients treated by somatic-cell gene therapy, just as in the case of any other experimental treatment procedure.

The somatic-cell gene therapy procedure described simply ***adds functional copies of the gene*** that is defective in the recipient cells to the genomes of those cells (***"gene addition"***); the procedure does ***not replace the defective gene*** with a nondefective copy of the gene (***"gene replacement"***). In fact, the introduced gene is inserted at random (or nearly random) sites in the chromosomes of the transgenic cells. Ideally, one would like to perform somatic-cell gene-replacement therapy analogous to the gene replacements that are routinely carried out in yeast (see Fig. 24.4). Such gene replacements would occur by means of recombination between homologous DNA sequences and would place the introduced copy of the gene at its normal location in the host genome. In the human genetics literature, such gene replacements are usually referred to as ***"targeted gene transfers."*** O. Smithies and coworkers first demonstrated that it was possible to use homologous recombination to target

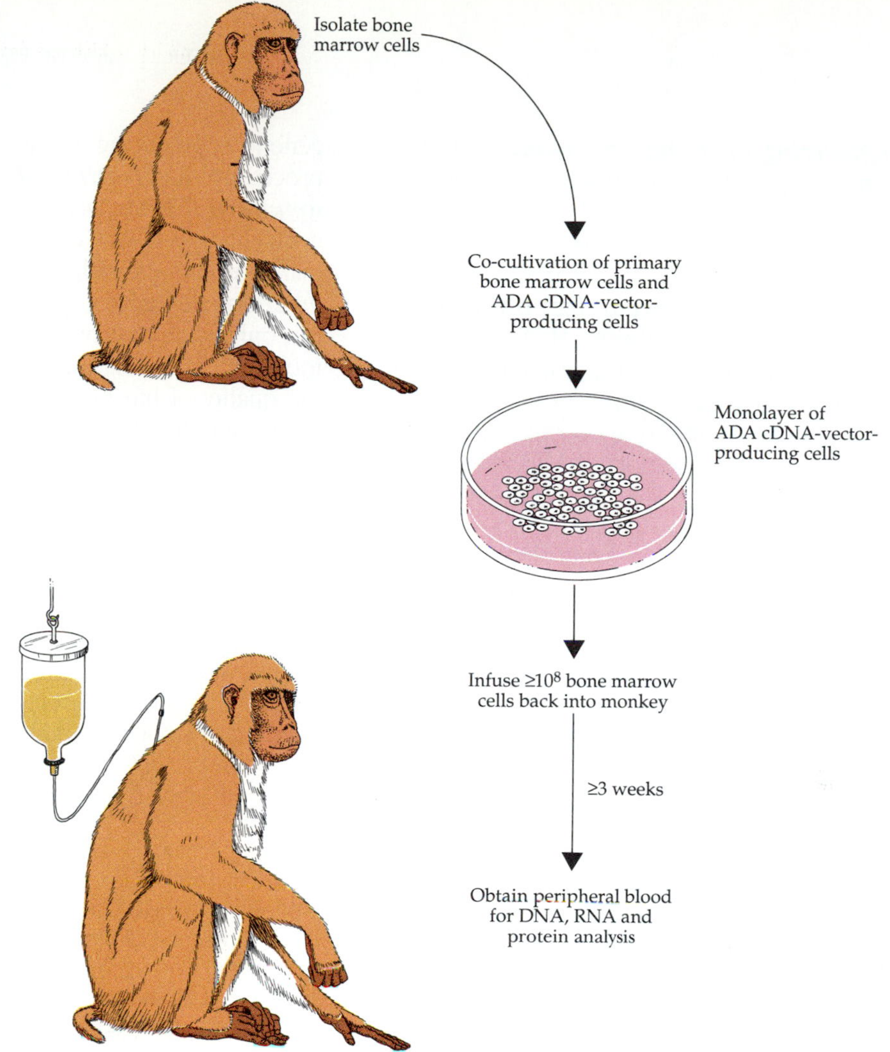

Figure 24.39 Diagram illustrating the application of somatic-cell gene therapy to primary bone marrow cells of rhesus monkeys. In the specific experiment illustrated, the utility of a human adenosine deaminase (ADA) expression vector (see Fig. 24.38) was being tested. The proposed treatment of human diseases by somatic-cell gene therapy will involve the same three steps: (1) the isolation of primary cells from the patient, (2) the introduction of normal, functional copies of the gene that is defective in the patient into these cells, and (3) the reintroduction of the transgenic cells back into the patient. (After W. F. Anderson *et al., Cold Spring Harbor Sympos. Quant. Biol.* 51: 1073–1081, 1986.)

DNA sequences into the β-hemoglobin locus of human tissue-culture cells in 1985. However, the frequency of the targeted gene transfer was very low (about 10^{-5}). More recently, Smithies and colleagues and M. R. Capecchi and colleagues have developed vectors and selection strategies that permit them to carry out targeted gene transfers at much higher frequencies and to select cell clones that carry the desired gene replacements. In the future, targeted gene replacements will probably become the method of choice for somatic-cell gene therapy treatment of human diseases.

SUMMARY

With the development of ***recombinant DNA and gene cloning technologies,*** biologists are now able to isolate and dissect essentially any gene and any segment of any chromosome of any organism. This has resulted in the explosive accumulation of new information in virtually every area of biology. Although some commercial applications of these technologies are in hand, for example, the ***production of human insulin and human growth hormone by bacteria,*** the real value of these technologies has resulted from their use in ***furthering our knowledge of basic biology.***

Biologists can now carry out carefully ***designed***

genetic engineering of living organisms, including ***microorganisms, plants, and animals.*** The goal of this chapter is to provide a survey of some of the ongoing and anticipated applications of genetic engineering. Clearly, a comprehensive discussion of all the potential applications of genetic engineering is far beyond the scope of this textbook. However, an attempt has been made to select representative examples from a variety of areas to illustrate the breadth of the impact of genetic engineering on our lives.

One of the most exciting potential applications of genetic engineering is the development of effective procedures for ***treating inherited human diseases by somatic-cell gene therapy.*** If successful, somatic-cell gene therapy could ease the hardships endured by many of the individuals who were unfortunate in inheriting defective copies of an important gene. One can only hope that humankind will use the powerful tools of genetic engineering in ways that will improve the quality of life on planet Earth for all living creatures, including humans.

REFERENCES

ABELSON, P. H., editor. 1983. "Biotechnology." *Science* 219: 609–746. (An entire issue of *Science* devoted to the applications of biotechnology.)

ANDERSON, W. F. 1984. "Prospects for human gene therapy." *Science* 226: 401–409.

ANDERSON, W. F., P. KANTOFF, M. EGLITIS, J. MCLACHLIN, E. KARSON, J. ZWIEBEL, A. NIENHUIS, S. KARLSSON, R. M. BLAESE, D. KOHN, E. GILBOA, D. ARMENTANO, E. D. ZANJANI, A. FLAKE, M. R. HARRISON, A. GILLIO, C. BORDIGNON, and R. O'REILLY. 1986. "Gene transfer and expression in nonhuman primates using retroviral vectors." *Cold Spring Harbor Sympos. Quant. Biol.* 51: 1073–1081.

BEVAN, M. W., and M.-D. CHILTON. 1982. "T-DNA of the *Agrobacterium* Ti and Ri plasmids," *Annu. Rev. Genet.* 16: 357–384.

BOSSELMAN, R. A., R.-Y. HSU, T. BOGGS, S. HU, J. BRUSZEWSKI, S. OU, L. KOZAR, F. MARTIN, C. GREEN, F. JACOBSEN, M. NICOLSON, J. A. SCHULTZ, K. M. SEMON, W. RISHELL, and R. G. STEWART. 1989. "Germline transmission of exogenous genes in the chicken." *Science* 243: 533–535.

BOTSTEIN, D., and G. R. FINK. 1988. "Yeast: An experimental organism for modern biology." *Science* 240: 1439–1442.

CHANG, C., J. L. BOWMAN, A. W. DEJOHN, E. S. LANDER, and E. M. MEYEROWITZ. 1988. "Restriction fragment length polymorphism linkage map for *Arabidopsis thaliana.*" *Proc. Natl. Acad. Sci. U.S.A.* 85: 6856–6860.

CHILTON, M.-D. 1983. "A vector for introducing new genes into plants." *Sci. Amer.* 246(6): 51–59.

DELISI, C. 1988. "The human genome project." *Amer. Sci.* 76: 488–493.

DZIERZAK, E. A., T. PAPAYANNOPOULOU, and R. C. MULLIGAN. 1988. "Lineage-specific expression of a human β-globin gene in murine bone marrow transplant recipients reconstituted with retrovirus-transduced stem cells." *Nature* 331: 35–41.

ESTELLE, M. A., and C. R. SOMERVILLE. 1986. "The mutants of *Arabidopsis.*" *Trends in Genet.* 2: 89–93.

FISCHHOFF, D. A., K. S. BOWDISH, F. J. PERLAK, P. G. MARRONE, S. M. MCCORMICK, J. G. NIEDERMEYER, D. A. DEAN, K. KUSANO-KRETZMER, E. J. MAYER, D. E. ROCHESTER, S. G. ROGERS, and R. T. FRALEY. 1987. "Insect tolerant transgenic tomato plants." *Bio/Technology* 5: 807–813.

FREELING, M., editor. 1985. *Plant genetics.* Liss, New York.

GREEN, P. J., O. PINES, and M. INOUYE. 1986. "The role of antisense RNA in gene regulation." *Annu. Rev. Biochem.* 55: 569–597.

GREGG, R. G., and O. SMITHIES. 1986. "Targeted modification of human chromosomal genes." *Cold Spring Harbor Sympos. Quant. Biol.* 51: 1093–1099.

JAENISCH, R. 1988. "Transgenic animals." *Science* 240: 1468–1474.

KARESS, R. E., and G. M. RUBIN. 1984. "Analysis of P transposable element functions in *Drosophila.*" *Cell* 38: 135–146.

KISHORE, G. M., and D. M. SHAH. 1988. "Amino acid biosynthesis inhibitors as herbicides." *Annu. Rev. Biochem.* 57: 627–663.

KLEE, H., R. HORSCH, and S. ROGERS. 1987. "*Agrobacterium*-mediated plant transformation and its further applications to plant biology." *Annu. Rev. Plant Physiol.* 38: 467–486.

KLEIN, T. M., E. C. HARPER, Z. SVAB, J. C. SANFORD, M. E. FROMM, and P. MALIGA. 1988. "Stable genetic transformation of intact *Nicotiana* cells by the particle bombardment process." *Proc. Natl. Acad. Sci. U.S.A.* 85: 8502–8505.

LEDLEY, F. D. 1987. "Somatic gene therapy for human disease: A problem of eugenics?" *Trends in Genet.* 3: 112–115.

MCKAY, L. L. 1985. "Role of plasmids in starter cultures." In *Bacterial starter cultures for foods,* pp. 159–174. CRC Press, Boca Raton, FL.

MEYEROWITZ, E. M. 1987. "*Arabidopsis thaliana.*" *Annu. Rev. Genet.* 21: 93–111.

PALMITER, R. D., and R. L. BRINSTER. 1986. "Germ-line transformation of mice." *Annu. Rev. Genet.* 20: 465–499.

PALMITER, R. D., G. NORSTEDT, R. E. GELINAS, R. E. HAMMER, and R. L. BRINSTER. 1983. "Metallothionein–human GH fusion genes stimulate growth of mice." *Science* 222: 809–814.

PENG, H., D. ARMENTANO, L. MACKENZIE-GRAHAM, R.-F. SHEN, G. DARLINGTON, F. D. LEDLEY, and S. L. C. WOO. 1988. "Retroviral-mediated gene transfer and expression of human phenylalanine hydroxylase in primary mouse hepatocytes." *Proc. Natl. Acad. Sci. U.S.A.* 85: 8146–8150.

RHODES, C. A., D. A. PIERCE, I. J. METTLER, D. MASCARENHAS, and J. J. DETMER. 1988. "Genetically transformed maize plants from protoplasts." *Science* 240: 204–207.

RUBIN, G. M., and A. C. SPRADLING. 1982. "Genetic transformation of *Drosophila* with transposable element vectors." *Science* 218: 348–353.

SHAH, D. M., R. B. HORSCH, H. J. KLEE, G. M. KISHORE, J. A.

WINTER, N. E. TUMER, C. M. HIRONAKA, P. R. SANDERS, C. S. GASSER, S. AYKENT, N. R. SIEGEL, S. G. ROGERS, and R. T. FRALEY. 1986. "Engineering herbicide tolerance in transgenic plants." *Science* 233: 478–481.

SHAH, D. M., N. E. TUMER, D. A. FISCHHOFF, R. B. HORSCH, S. G. ROGERS, R. T. FRALEY, and E. G. JAWORSKI. 1987. "The introduction and expression of foreign genes in plants." *Biotechnol. Genet. Eng. Rev.* 5: 81–106.

SIMONS, R. W., and N. KLECKNER. 1988. "Biological regulation by antisense RNA in prokaryotes." *Annu. Rev. Genet.* 22:567–600.

ST. LOUIS, D., and I. M. VERMA. 1988. "An alternative approach to somatic cell gene therapy." *Proc. Natl. Acad. Sci. U.S.A.* 85: 3150–3154.

THOMAS, K. R., and M. R. CAPECCHI. 1986. "Targeting of genes to specific sites in the mammalian genome." *Cold Spring Harbor Sympos. Quant. Biol.* 51: 1101–1113.

VARMUS, H. 1988. "Retroviruses." *Science* 240: 1427–1435.

WALLACE, J. C., G. GALILI, E. E. KAWATA, R. E. CUELLAR, M. A. SHOTWELL, and B. A. LARKINS. 1988. "Aggregation of lysine-containing zeins into protein bodies in *Xenopus oocytes.*" *Science* 240: 662–664.

WALTERS, L. 1986. "The ethics of human gene therapy." *Nature* 320: 225–227.

WEINTRAUB, H. M. 1990. "Antisense RNA and DNA." *Sci. Amer.* 262(1): 40–46.

WEISING, K., J. SCHELL, and G. KAHL. 1988. "Foreign genes in plants: Transfer, structure, expression, and applications." *Annu. Rev. Genet.* 22: 421–477.

WHITE, R., and C. T. CASKEY. 1988. "The human as an experimental system in molecular genetics." *Science* 240: 1483–1488.

ZAMBRYSKI, P. 1988. "Basic processes underlying *Agrobacterium*-mediated DNA transfer to plant cells." *Annu. Rev. Genet.* 22: 1–30.

GLOSSARY

This Glossary provides an introduction to some basic and recurring terms in the text. Names of chemical compounds, definitions of specialized terms, and variants of basic names have been omitted from the Glossary but are given in the Index, where they are identified by page number; the Index specifies where a term first appears and where it is defined in the book. For definitions not in the Glossary, please locate the term by referring to the Index.

Abscissa. The horizontal scale on a graph.
Acentric chromosome. Chromosome fragment lacking a centromere.
Acrocentric. A modifying term for a chromosome or chromatid that has its centromere near the end.
Acrosome. An apical organelle in the head of the sperm.
Adaptation. Adjustment of an organism or a population to an environment.
Adaptive landscape. In the evolutionary theory of Sewall Wright, a representation of the fitness of a population as a topographic map in which contour height (i.e., the fitness value) is a function of allele frequencies at many loci.
Additive allelic effects. Genetic factors that raise or lower the value of a phenotype on a linear scale of measurement.
Adenine. A purine base found in RNA and DNA.
Agglutinin. An antibody in blood plasma that brings about clumping (agglutination) of blood cells that carry an incompatible agglutinogen.
Agglutinogen. An antigen carried by red blood cells that reacts with a specific agglutinin in the plasma and causes clumping of the cells. When a specific antigen is injected into an animal body, it stimulates the production of a corresponding antibody.
Albinism. Absence of pigment in skin, hair, and eyes of an animal. Absence of chlorophyll in plants.
Aleurone. The outermost layer of the endosperm in a seed.
Alkaptonuria. An inherited metabolic disorder. Alkaptonurics excrete excessive amounts of homogentisic acid (alkapton) in the urine.
Allele (allelomorph; *adj.,* allelic, allelomorphic). One of a pair, or series, of alternative forms of a gene that occur at a given locus in a chromosome. Alleles are symbolized with the same basic symbol (e.g., *D* for tall peas and *d* for dwarf). (See ***Multiple alleles.***)
Allele frequency. The proportion of one allele relative to all alleles at a locus in a population.
Allopolyploid. A polyploid having chromosome sets from different species. A polyploid containing genetically different chromosome sets derived from two or more species.
Allosteric effect. Reversible interaction of a small molecule with a protein molecule, causing a change in the shape of the protein and a consequent alteration of the interaction of that protein with a third molecule.
Allotetraploid. An organism with four genomes derived from hybridization of different species. Usually, in forms that become established, two of the four genomes are from one species and two are from another species.
Allozygote. A diploid individual in which the two genes of a particular locus are not identical by descent from a common ancestor. (Cf. ***Autozygote.***)
Amino acid. Any one of a class of organic compounds containing the amino (NH_2) group and the carboxyl (COOH) group. Amino acids are building blocks of proteins. Alanine, proline, threonine, histidine, lysine, glutamine, phenylalanine, tryptophan, valine, arginine, tyrosine, and leucine are among the common amino acids.
Amniocentesis. A procedure for obtaining amniotic fluid from a pregnant woman. Chemical contents of the fluid are studied directly for the diagnosis of some diseases. Cells are cultured, and metaphase chromosomes are examined for irregularities (e.g., Down syndrome).
Amnion. The thin membrane that lines the fluid-filled sac in which the embryo develops in higher vertebrates.
Amniotic fluid. Liquid contents of the amniotic sac of higher vertebrates containing cells of the embryo (not of the mother). Both fluid and cells are used for diagnosis of genetic abnormalities of the embryo or fetus.
Amorph. A mutation that obliterates gene function; a null mutation.
Amphidiploid. A species or type of plant derived from doubling the chromosomes in the F_1 hybrid of two species; an allopolyploid. In an amphidiploid the two species are known, whereas in other allopolyploids they may not be known.
Analysis of variance. Partitioning variation among data into its components by the computation of suitable statistics.
Anaphase. The stage of mitosis or meiosis during which the daughter chromosomes pass from the equatorial plate to opposite poles of the cell (toward the ends of the spindle). Anaphase follows metaphase and precedes telophase.
Androgen. A male hormone that controls sexual activity in vertebrate animals.
Anemia. Abnormal condition characterized by pallor, weakness, and breathlessness, resulting from a deficiency of hemoglobin or a reduced number of red blood cells.
Aneuploid (heteroploid). An organism or cell having a chromosome number that is not an exact multiple of the monoploid (n) with one genome, that is, a hyperploid has a higher number (e.g., $2n + 1$) and a hypoploid has a lower number (e.g., $2n - 1$). Also applied to cases where part of a chromosome is duplicated or deficient.
Antibody. Substance in a tissue or fluid of the body that acts in antagonism to a foreign substance (antigen).
Anticodon. Three bases in a transfer RNA molecule that are complementary to the three bases of a specific codon in messenger RNA.
Antigen. A substance, usually a protein, introduced into a living organism that elicits antibody formation.

Antihemophilic globulin. Blood globulin that reduces the clotting time of hemophilic blood.
Anti-sense RNA. RNA that is made from the DNA strand that is complementary to the sense strand of the DNA.
Artificial selection. The practice of choosing individuals from a population for reproduction, usually because these individuals possess one or more desirable traits.
Ascospore. One of the spores contained in the ascus of certain fungi such as *Neurospora*.
Ascus (*pl.*, asci). Reproductive sac in the sexual stage of a type of fungi (Ascomycetes) in which ascospores are produced.
Asexual reproduction. Any process of reproduction that does not involve the formation and union of gametes from the different sexes or mating types.
Assortative mating. Mating in which the partners are chosen because they are phenotypically similar.
Asynapsis. The failure or partial failure in the pairing of homologous chromosomes during the meiotic prophase.
Atavism. Reappearance of an ancestral trait after several generations because of recessiveness or other masking effects. (See ***Reversion.***)
ATP. Adenosine triphosphate: an energy-rich compound that promotes certain activities in the cell.
Attenuation. A mechanism for controlling gene expression in prokaryotes that involves premature termination of transcription.
Attenuator. A nucleotide sequence in the 5′ region of a prokaryotic gene (or in its RNA) that causes premature termination of transcription, possibly by forming a secondary structure.
Autogamy. A process of self-fertilization within one undivided cell, resulting in homozygosity. This occurs in *Paramecium*.
Autonomous. A term applied to any biological unit that can function on its own, that is, without the help of another unit. For example, a transposable element that encodes an enzyme for its own transposition. (Cf. ***Nonautonomous.***)
Autopolyploid. A polyploid that has multiple and identical or nearly identical sets of chromosomes (genomes). A polyploid species with genomes derived from the same original species.
Autoradiograph. A record or photograph prepared by labeling a substance such as DNA with a radioactive material such as tritiated thymidine and allowing the image produced by decay radiations to develop on a film over a period of time.
Autosome. Any chromosome that is not a sex chromosome.
Autozygote. A diploid individual in which the two genes of a locus are identical by descent from an ancestral gene. (Cf. ***Allozygote.***)
Auxotroph. A mutant microorganism (e.g., bacterium or yeast) that will not grow on a minimal medium but that requires the addition of some compound such as an amino acid or a vitamin.
Backcross. The cross of an F_1 hybrid to one of the parental types. The offspring of such a cross are referred to as the backcross generation or backcross progeny. (See ***Testcross.***)
Bacteriophage. Virus that attacks bacteria. Such viruses are called bacteriophages because they destroy their bacterial hosts.
Balanced lethal. Lethal mutations in different genes on the same pair of chromosomes that remain in repulsion because of close linkage or crossover suppression. In a closed population, only the *trans*-heterozygotes (l_1 +/+ l_2) for the lethal mutations survive.
Balanced polymorphism. Two or more types of individuals maintained in the same breeding population.
Barr body. A condensed mass of chromatin found in the nuclei of placental mammals that contains one or more X chromosomes. Named for its discoverer Murray Barr.
Basal body. Small granule to which a cilium or flagellum is attached.
Base analog. Unnatural purine or pyrimidine base, differing slightly from the normal base, that can be incorporated into nucleic acid.
Binomial expansion. Exponential multiplication of an expression consisting of two terms connected by a plus (+) or minus (−) sign, such as $(a + b)^n$.
Biometry. Application of statistical methods to the study of biological problems.
Biotype. Distinct physiological race or strain within morphological species. A population of individuals with identical genetic constitution. A biotype may be made up of homozygotes or of heterozygotes, of which only the former would be expected to breed true.
Bipartite. Consisting of two identifiable parts.
Bivalent. A pair of synapsed or associated homologous chromosomes that have undergone the duplication process to form a group of four chromatids.
Blastomere. Any one of the cells formed from the first few cleavages in animal embryology.
Blastula. In animals, an early embryo form that follows the morula stage; typically, a single-layered sheet or ball of cells.
Breeding value. In quantitative genetics, the part of the deviation of an individual phenotype from the population mean that is due to the additive effects of alleles.
Broad-sense heritability. In quantitative genetics, the proportion of the total phenotypic variance that is the genotypic variance.
Carcinogen. A substance capable of inducing cancer in an organism.
Carrier. An individual who carries a recessive allele that is not expressed (i.e., obscured by a dominant allele).
Catabolite repression. Glucose-mediated reduction in the rates of transcription of operons that specify enzymes involved in catabolic pathways (such as the *lac* operon).
Centimorgan. See ***Crossover unit.***
Centriole. An organelle in many animal cells that appears to be involved in the formation of the spindle during mitosis.
Centromere. Spindle-fiber attachment region of a chromosome.
Character (*contraction of the word* characteristic). One of the many details of structure, form, substance, or function that make up an individual organism.
Chemotaxis. Attraction or repulsion of organisms by a diffusing substance.
Chiasma (*pl.*, chiasmata). A visible change of partners or a crossover in two of a group of four chromatids during the first meiotic prophase. In the diplotene stage of meiosis, the four chromatids of a bivalent are associated in pairs, but in such a way that one part of two chromatids is exchanged. This point of "change of partner" is the chiasma.
Chimera (animal). Individual derived from two embryos by experimental intervention.
Chimera (plant). Part of a plant with a genetically different constitution as compared with other parts of the same plant. It may result from different zygotes that grow together or from artificial fusion (grafting); it may be either pernical, with parallel layers of genetically different tissues, or sectorial.
Chloroplastid. Green structure in plant cytoplasm that contains chlorophyll and in which starch is synthesized. A

mode of cytoplasmic inheritance, independent of nuclear genes, has been associated with these cytoplasmic structures.

Chromatid. In mitosis or meiosis, one of the two identical strands resulting from self-duplication of a chromosome.

Chromatin. The deoxyribonucleohistone in chromosomes; originally named because of the readiness with which it stains with certain dyes (chromaticity).

Chromatography. A method for separating and identifying the components from mixtures of molecules having similar chemical and physical properties.

Chromocenter. Body produced by fusion of the heterochromatic regions of the chromosomes in the polytene tissues (e.g., the salivary glands) of certain Diptera.

Chromomeres. Small bodies, described by J. Belling, that are identified by their characteristic size and linear arrangement along a chromosome.

Chromonema (*pl.,* chromonemata). An optically single thread forming an axial structure within each chromosome.

Chromosome aberration. Abnormal structure or number of chromosomes; includes deficiency, duplication, inversion, translocation, aneuploidy, polyploidy, or any other change from the normal pattern.

Chromosome banding. Staining of chromosomes in such a way that light and dark areas occur along the length of the chromosomes. Lateral comparisons identify pairs. Each human chromosome can be identified by its banding pattern.

Chromosomes. Darkly staining nucleoprotein bodies that are observed in cells during division. Each chromosome carries a linear array of genes.

Cilium (*pl.,* cilia; *adj.,* ciliate). Hairlike locomotor structure on certain cells; a locomotor structure on a ciliate protozoan.

***cis*-Arrangement.** See *Coupling*.

Cistron. A unit of function in DNA. One DNA cistron specifies one polypeptide chain in protein synthesis.

Clone. All the individuals derived by vegetative propagation from a single original individual. In molecular biology, a strain of organisms that carries a particular DNA sequence.

Codominant alleles. Alleles that produce independent effects when heterozygous.

Codon. A set of three adjacent nucleotides in an mRNA molecule that specifies the incorporation of an amino acid into a polypeptide chain or that signals the end of polypeptide synthesis. Codons with the latter function are called termination codons.

Coefficient. A number expressing the amount of some change or effect under certain conditions (e.g., coefficient of inbreeding).

Coenzyme. A substance necessary for the activity of an enzyme.

Coincidence. The ratio of the observed frequency of double crossovers to the expected frequency, where the latter is calculated by assuming that the two crossover events occur independently of each other.

Cointegrate. A DNA molecule formed by the fusion of two different DNA molecules, usually mediated by a transposable element.

Colchicine. An alkaloid derived from the autumn crocus that is used as an agent to arrest spindle formation and interrupt mitosis.

Colinearity. A relationship in which the units in one molecule occur in the same sequence as the units in another molecule that they specify; for example, the nucleotides in a gene are colinear with the amino acids in the polypeptide encoded by that gene.

Competence. Ability of a bacterial cell to incorporate DNA and become genetically transformed.

Complementation test (*cis–trans* test). Introduction of two recessive mutations into the same cell but on different chromosomes to determine whether they are alleles of the same gene, that is, whether they affect the same genetic function. If the mutations are allelic, the genotye m_1 $+/+$ m_2 will exhibit a mutant phenotype, whereas if they are nonallelic, it will exhibit the wild phenotype.

Concordance. Identity of matched pairs or groups for a given trait; for example, identical twins both expressing the same genetic syndrome.

Conidium (*pl.,* conidia). An asexual spore produced by a specialized hypha in certain fungi.

Conjugation. Union of sex cells (gametes) or unicellular organisms during fertilization; in *Escherichia coli*, a one-way transfer of genetic material from a donor ("male") to a recipient ("female").

Consanguinity. Related by descent from a common ancestor.

Constitutive enzyme. An enzyme that is produced in fixed quantities regardless of need. (Cf. *Inducible enzyme* and *Repressible enzyme.*)

Continuous variation. Variation not represented by distinct classes. Individuals grade into each other and measurement data are required for analysis. (Cf. *Discontinuous variation.*) Multiple genes are usually responsible for this type of variation.

Controlling element. In maize, a transposable element such as *Ac* or *Ds* that is capable of influencing the expression of a nearby gene.

Coordinate repression. Correlated regulation of the structural genes in an operon by a molecule that interacts with the operator sequence.

Copolymers. Mixtures consisting of more than one monomer; for example, polymers of two kinds of organic bases such as uracil and cytosine (poly-UC) have been combined for studies of genetic coding.

Correlation. A statistical association between variables.

Coupling (*cis*-arrangement). The condition in which a double heterozygote has received two linked mutations from one parent and their wild-type alleles from the other parent, e.g., $a\ b/a\ b \times + +/+ +$ produces $a\ b/+ +$. (Cf. *Repulsion.*)

Covalent bond. A bond in which an electron pair is equally shared by protons in two adjacent atoms.

Covariance. A measure of the statistical association between variables.

Crossbreeding. Mating between members of different races or species.

Crossing-over. A process in which chromosomes exchange material through the breakage and reunion of their DNA molecules. (See *Recombination.*)

Crossover unit. A measure of distance on genetic maps that is based on the average number of crossing-over events that take place during meiosis. A map interval that is one crossover unit in length (sometimes called a centimorgan) implies that only one in every hundred chromatids recovered from meiosis will have undergone a crossing-over event in this interval.

Cytogenetics. Area of biology concerned with chromosomes and their implications in genetics.

Cytokinesis. Cytoplasmic division and other changes exclusive of nuclear division that are a part of mitosis or meiosis.

Cytology. The study of the structure and function of cells.

Cytoplasm. The protoplasm of a cell outside the nucleus in which cell organelles (mitochondria, plastids, etc.) reside. All living parts of the cell except the nucleus.

Cytoplasmic inheritance. Hereditary transmission dependent on the cytoplasm or structures in the cytoplasm rather than the nuclear genes; extrachromosomal inheritance. Example: Plastid characteristics in plants may be inherited by a mechanism independent of nuclear genes.
Cystosine. A pyrimidine base found in RNA and DNA.
Cytotype. A maternally inherited cellular condition in *Drosophila* that regulates the activity of transposable *P* elements.
Dalton. The mass of a hydrogen atom.
Deficiency (deletion). Absence of a segment of a chromosome, reducing the number of loci.
Degeneracy. An aspect of the genetic code in which a particular amino acid is encoded by more than one codon.
Deme. A local population of organisms.
Denaturation. Loss of native configuration of a macromolecule, usually accompanied by loss of biological activity. Denatured proteins often unfold their polypeptide chains and express changed properties of solubility.
De novo. From the Latin, meaning arising anew, afresh, once more.
Determination. Process by which undifferentiated cells in an embryo become committed to develop into specific cell types, such as neuron, fibroblast, or muscle cell.
Deviation. As used in statistics, a departure from an expected value.
Diakinesis. A stage of meiosis just before metaphase I in which the bivalents are shortened and thickened.
Dicentric chromosome. One chromosome having two centromeres.
Differentiation. A process in which unspecialized cells develop characteristic structures and functions.
Dihybrid. An individual that is heterozygous for two pairs of alleles. The progeny of a cross between homozygous parents differing in two respects.
Dimer. A compound having the same percentage composition as another but twice the molecular weight; one formed by polymerization.
Dimorphism. Two different forms in a group as determined by such characteristics as sex, size, or coloration.
Diploid. An organism or cell with two sets of chromosomes ($2n$) or two genomes. Somatic tissues of higher plants and animals are ordinarily diploid in chromosome constitution in contrast with the haploid (monoploid) gametes.
Diplonema (*adj.,* diplotene). That stage in prophase of meiosis I following the pachytene stage, but preceding diakinesis, in which the chromosomes of bivalents separate from each other at and around their centromeres.
Discontinuous variation. Phenotypic variability involving distinct classes such as red versus white or tall versus dwarf. (Cf. ***Continuous variation.***)
Discordant. Members of a pair showing different, rather than similar, characteristics.
Disjunction. Separation of homologous chromosomes during anaphase of mitotic or meiotic divisions. (See ***Nondisjunction.***)
Disome. See ***Monosomic.***
Ditype. In fungi, a tetrad that contains two kinds of meiotic products (spores) (e.g., 2*AB* and 2*ab*).
Dizygotic twins. Two-egg or fraternal twins.
DNA. Deoxyribonucleic acid; the information-carrying genetic material that comprises the genes.
DNase. Any enzyme that hydrolyzes DNA.
Dominance. A condition in which one member of an allele pair is manifested to the exclusion of the other.
Dosage compensation. A phenomenon in which the activity of a gene is increased or decreased according to the number of copies of that gene in the cell.
Drift. See ***Random genetic drift.***
Duplication. The occurrence of a segment more than once in the same chromosome or genome. Also, the multiplication of cells.
Ecdysone. A hormone that influences development in insects.
Eclosion. Emergence of an adult insect from the pupal stage.
Ecotype. A population or strain of organisms that is adapted to a particular habitat.
Effector. A molecule that influences the behavior of a regulatory molecule, such as a repressor protein, thereby influencing gene expression.
Egg (ovum). A germ cell produced by a female organism.
Electrophoresis. The migration of suspended particles in an electric field.
Electroporation. A process whereby cell membranes are made permeable to DNA by applying an intense electric current.
Embryo. An organism in the early stages of development; in humans, the first two months in the uterus.
Embryo sac. A large, thin-walled space within the ovule of the seed plant in which the egg and, after fertilization, the embryo develop; the mature female gametophyte in higher plants.
Endogenote. The part of the bacterial chromosome that is homologous to a genome fragment (exogenote) transferred from the donor to the recipient cell in the formation of a merozygote.
Endomitosis. Duplication of chromosomes without division of the nucleus, resulting in increased chromosome number within a cell. Chromosome strands separate but the cell does not divide.
Endonuclease. An enzyme that breaks strands of DNA at internal positions; some are involved in recombination of DNA.
Endoplasmic reticulum. Network of membranes in the cytoplasm to which ribosomes adhere.
Endopolyploidy. A state in which the cells of a diploid organism contain multiples of the diploid chromosome number (i.e., $4n$, $8n$, etc.).
Endosperm. Nutritive tissue that develops in the embryo sac of most angiosperms. It usually forms after the fertilization of the two fused primary endosperm nuclei of the embryo sac with one of the two male gamete nuclei. In most diploid plants, the endosperm is triploid ($3n$), but in some (e.g., the lily) it is $5n$.
Enhancer. A substance or object that increases a chemical activity or a physiological process; a major or modifier gene that increases a physiological process; a DNA sequence that influences transcription of a nearby gene.
Environment. The aggregate of all the external conditions and influences affecting the life and development of an organism.
Enzyme. A protein that accelerates a specific chemical reaction in a living system.
Epigenetic. A term referring to the nongenetic causes of a phenotype.
Episome. A genetic element that may be present or absent in different cells and that may be inserted in a chromosome or independent in the cytoplasm. For example, the fertility factor (F) in *Escherichia coli.*
Epistasis. Interactions between products of nonallelic genes. Genes suppressed are said to be hypostatic. Domi-

nance is associated with members of allelic pairs, whereas epistasis is interaction among products of nonalleles.

Equational (homotypic) division. Mitotic-type division that is usually the second division in the meiotic sequence; somatic mitosis and the nonreductional division of meiosis.

Equatorial plate. The figure formed by the chromosomes in the center (equatorial plane) of the spindle in mitosis.

Equilibrium. A state of dynamical systems in which there is no net change.

Estrogen. Female hormone or estrus-producing compound.

Euchromatin. Genetic material that is not stained so intensely by certain dyes during interphase and that comprises many different kinds of genes. (Cf. ***Heterochromatin.***)

Eugenics. The application of the principles of genetics to the improvement of humankind.

Eukaryote. A member of the large group of organisms that have nuclei enclosed by a membrane within their cells. (Cf. ***Prokaryote.***)

Euploid. An organism or cell having a chromosome number that is an exact multiple of the monoploid (x) or haploid number. Terms used to identify different levels in an euploid series are diploid, triploid, tetraploid, and so on. (Cf. ***Heteroploid*** and ***Aneuploid.***)

Exogenote. Chromosomal fragment homologous to an endogenote and donated to a merozygote.

Exons. The segments of a eukaryotic gene that correspond to the sequences in the final processed RNA transcript of that gene.

Exonuclease. An enzyme that digests DNA or RNA, beginning at the ends of strands.

Expressivity. Degree of expression of a trait controlled by a gene. A particular gene may produce different degrees of expression in different individuals.

Extrachromosomal. Structures that are not a part of the chromosomes; DNA units in the cytoplasm that control cytoplasmic inheritance.

F_1. The first filial generation. The first generation of descent from a given mating.

F_2. The second filial generation produced by crossing *inter se* or by self-pollinating the F_1. The inbred "grandchildren" of a given mating, but in controlled genetic experimentation, self-fertilization of the F_1 (or equivalent) is implied.

F factor. A bacterial plasmid that confers the ability to function as a genetic donor ("male") in conjugation; the fertility factor in bacteria.

Feedback inhibition. The accumulated end product of a biochemical pathway stops synthesis of that product. A late metabolite of a synthetic pathway regulates synthesis at an earlier step of that pathway.

Fertilization. The fusion of a male gamete (sperm) with a female gamete (egg) to form a zygote.

Fetus. Prenatal stage of a viviparous animal between the embryonic stage and the time of birth. In humans, the final seven months before birth.

Filial. See F_1 and F_2.

Fitness. The number of offspring left by an individual, often compared with the average of the population or with some other standard, such as the number left by a particular genotype.

Fixation. An event that occurs when all the alleles at a locus except one are eliminated from a population. The remaining allele, with a frequency of 100 percent, is said to have been fixed.

Flagellum (*pl.*, flagella; *adj.* flagellate). A whiplike organelle of locomotion in certain cells; locomotor structures in flagellate protozoa.

Founder principle. The possibility that a new, small, isolated population may diverge genetically because the founding individuals are a random sample from a large, main population.

Frameshift. A mutation that upsets the reading frame of an mRNA, by either inserting or deleting nucleotides.

Frequency distribution. A graph showing either the relative or absolute incidence of classes in a population. The classes may be defined by either a discrete or a continuous variable; in the latter case, each class represents a different interval on the scale of measurement.

Fusion protein. A polypeptide made from a recombinant gene that contains portions of two or more different genes. The different genes are joined so that their coding sequences are in the same reading frame.

Gall. A tumorous growth in plants.

Gamete. A mature male or female reproductive cell (sperm or egg).

Gametogenesis. The formation of gametes.

Gametophyte. That phase of the plant life cycle that bears the gametes; the cell have n chromosomes.

Gametophytic incompatibility. A botanical phenomenon controlled by the complex S locus in which a pollen grain cannot fertilize an ovule produced by a plant that carries the same S allele as the pollen grain. For example, S_1 pollen cannot fertilize an ovule made by an S_1/S_2 plant.

Gap gene. A gene that controls the formation of adjacent segments in the body of *Drosophila*.

Gastrula. An early animal embryo consisting of two layers of cells; an embryological stage following the blastula.

Gene. A hereditary determinant of a specific biological function; a unit of inheritance (DNA) located in a fixed place on the chromosome.

Gene conversion. A process in which one allele is replicated at the expense of another, leading to non-Mendelian segregation ratios; in whole tetrads, for example, the ratio may be 6 : 2 or 5 : 3 instead of the expected 4 : 4.

Gene flow. The spread of genes from one breeding population to another by migration, possibly leading to allele frequency changes.

Gene pool. The sum total of all different alleles in the breeding members of a population at a given time.

Genetic drift. See ***Random genetic drift.***

Genetic equilibrium. Condition in a group of interbreeding organisms in which the allele frequencies remain constant over time.

Genetic load. The proportion by which the fitness of the optimum genotype is decreased by deleterious genes, as expressed in lethal equivalents or "genetic deaths." Average genetic load of the human population has been estimated to be four to six lethal equivalents.

Genetics. The science of heredity and variation.

Genome. A complete set (n) of chromosomes (hence, of genes) inherited as a unit from one parent.

Genotype. The genetic constitution (gene makeup) of an organism. (Cf. ***Phenotype.***)

Germ cell. A reproductive cell capable when mature of being fertilized and reproducing an entire organism. (Cf. ***Somatic cell.***)

Germ plasm. The hereditary material transmitted to the offspring through the germ cells.

Globulins. Common proteins in the blood that are insoluble in water and soluble in salt solutions. Alpha, beta, and gamma globulins can be distinguished in human blood serum. Gamma globulins are important in developing immunity to diseases.

Glucocorticoid. A steroid hormone that regulates gene expression in higher animals.

Gonad. A sexual organ (i.e., ovary or testis) that produces gametes.

Guanine. A purine base found in DNA and RNA.

Gynandromorph. An individual in which one part of the body is female and another part is male; a sex mosaic.

Haploid (monoploid). An organism or cell having only one complete set (n) of chromosomes or one genome.

Haptoglobin. A serum protein, alpha globulin, in the blood.

Hardy–Weinberg equilibrium. Mathematical relationship that allows the frequencies of genotypes in a population to be predicted from their constituent allele frequencies; a consequence of random mating.

Helix. Any structure with a spiral shape. The Watson and Crick model of DNA is in the form of a double helix.

Hemizygous. The condition in which only one allele of a pair is present, as in sex linkage or as a result of deletion.

Hemoglobin. Conjugated protein compound containing iron, located in erythrocytes of vertebrates; important in the transportation of oxygen to the cells of the body.

Hemolymph. The mixture of blood and other fluids in the body cavity of an invertebrate.

Hemophilia. A bleeder's disease; tendency to bleed freely from even a slight wound; hereditary condition dependent on a sex-linked recessive gene.

Heredity. Resemblance among individuals related by descent; transmission of traits from parents to offspring.

Heritability. Degree to which a given trait is controlled by inheritance. (See ***Broad-sense heritability*** and ***Narrow-sense heritability***.)

Hermaphrodite. An individual with both male and female reproductive organs.

Heterochromatin. Chromatin staining darkly even during interphase, often containing repetitive DNA with few genes.

Heteroduplex. A double-stranded nucleic acid containing one or more mismatched (noncomplementary) base pairs.

Heterogametic sex. Producing unlike gametes with regard to the sex chromosomes. In humans, the XY male is heterogametic and the XX female is homogametic.

Heterokaryon. A cell containing two or more different nuclei.

Heteroploid (aneuploid). An organism characterized by a chromosome number other than the true haploid (monoploid) or diploid number ($2n + 1$ or $2n - 1$). (Cf. ***Euploid***.)

Heteropyknosis (*adj.*, heteropyknotic). Property of certain chromosomes, or of their parts, to remain more dense and to stain more intensely than other chromosomes or parts during the cell cycle.

Heterosis. Superiority of heterozygous genotypes in respect to one or more traits in comparison with corresponding homozygotes.

Heterozygote (*adj.*, heterozygous). An organism with unlike members of any given pair or series of alleles that consequently produces unlike gametes.

Hfr. High-frequency recombination strain of *Escherichia coli*; in such strains, the F episome is integrated into the bacterial chromosome.

Histones. Group of proteins rich in basic amino acids. They function in the coiling of DNA in chromosomes and in the regulation of gene activity.

"Holandric" gene. A gene carried on the Y chromosome and therefore transmitted from father to son.

Homeobox. A DNA sequence found in several genes that are involved in the specification of organs in different body parts in animals; characteristic of genes that influence segmentation in animals. The homeobox corresponds to an amino acid sequence in the polypeptide encoded by these genes; this sequence is called the homeodomain.

Homeotic mutation. A mutation that causes a body part to develop in an inappropriate position in an organism. For example, a mutation in *Drosophila* that causes legs to develop on the head in the place of antennae.

Homogametic sex. Producing like gametes with regard to the sex chromosomes. (Cf. ***Heterogametic sex***.)

Homologous chromosomes. Chromosomes that occur in pairs and are generally similar in size and shape, one having come from the male parent and the other from the female parent. Such chromosomes contain the same array of genes.

Hormone. An organic product of cells of one part of the body that is transported by the body fluids to another part where it influences activity or serves as a coordinating agent.

Hybrid. An offspring of homozygous parents differing in one or more genes; more generally, an offspring of a cross between unrelated strains.

Hybrid dysgenesis. In *Drosophila*, a syndrome of abnormal germ line traits, including mutation, chromosome breakage, and sterility, that results from transposable element activity.

Hybrid vigor (heterosis). Unusual growth, strength, and health of heterozygous hybrids from two less vigorous homozygous parents.

Hybridization. Interbreeding of species, races, varieties, and so on, among plants or animals. A process of forming a hybrid by cross-pollination of plants or by mating animals of different types.

Hypersensitive sites. Regions in the DNA that are highly susceptible to digestion with endonucleases.

Hypomorph. A mutation that reduces but does not completely abolish gene expression.

Hypostasis. See ***Epistasis***.

Imaginal disc. A mass of cells in the larvae of *Drosophila* and other holometabolous insects that gives rise to particular adult organs such as antennae, eyes, and wings.

Immunoglobulin. See ***Globulin***.

Inbreeding. Matings between related individuals.

Incomplete dominance. Expression of two alleles in a heterozygote that allows the heterozygote to be distinguished from either of its homozygous parents.

Independent assortment. The random distribution of alleles to the gametes that occurs when genes are located in different chromosomes. The distribution of one pair of alleles is independent of other genes located in nonhomologous chromosomes.

Inducer. A substance of low molecular weight that inactivates a repressor by combining with it, thereby stimulating gene expression.

Inducible enzyme. A enzyme that is synthesized only in the presence of the substrate that acts as an inducer.

Inhibitor. Any substance or object that retards a chemical reaction; a major or modifier gene that interferes with a reaction.

***In situ*.** From the Latin, meaning "in the natural place." Refers to experimental treatments performed on cells or tissues rather than on extracts from them.

Interaction. In statistics, an effect that cannot be explained by the additive action of contributing factors; a departure from strict additivity.

Interference. Crossing-over at one point that reduces the chance of another crossover nearby; detected by studying the pattern of crossing-over with three or more linked genes.
Interphase. The stage in the cell cycle when the cell is not dividing; the metabolic stage during which DNA replication occurs; the stage following telophase of one division and extending to the beginning of prophase in the next division.
Intersex. An organism displaying secondary sexual characters intermediate between male and female; a type that shows some phenotypic characteristics of both males and females.
Introns. Intervening sequences of DNA bases within eukaryotic genes that are not represented in the mature RNA transcript because they are spliced out of the primary RNA transcript.
Invariant. Constant and unchanging, usually referring to the portion of a molecule that is the same across species.
Inversion. A rearrangement that reverses the order of a linear array of genes in a chromosome.
Inverted repeat. A sequence present twice in a DNA molecule, but in reverse orientation.
In vitro. From the Latin, meaning "within glass"; biological processes made to occur experimentally outside the organism in a test tube or other container.
In vivo. From the Latin, meaning "within the living organism."
IS element (insertion sequence). A short (800–1400 nucleotide-pairs) DNA sequence found in bacteria that is capable of transposing to a new genomic location; other DNA sequences that are bounded by IS elements may also be transposed.
Isoagglutinogen. An antigen, such as A or B blood type factor, that occurs normally (i.e., in an individual, without artificial stimulation).
Isochromosome. A chromosome with two identical arms and identical genes. The arms are mirror images of each another.
Isogenic stocks. Strains of organisms that are genetically uniform; completely homozygous.
Kappa particles. DNA-containing, self-reproducing cytoplasmic particles in certain strains of *Paramecium aurelia*. They control a toxic substance, *paramecin*, that is released into the culture medium and kills sensitive paramecia. Nuclear gene K is required for maintenance of kappa in the cytoplasm of killer strains.
Karyotype. The chromosome constitution of a cell or an individual; chromosomes arranged in order of length and according to position of centromere; also the abbreviated formula for the chromosome constitution, such as 47, XX + 21 for human trisomy-21.
Kinetics. A dynamic process involving motion.
Kinetosome. Granular body at the base of a flagellum or a cilium.
Lamella. A double-membrane structure, plate, or vesicle that is formed by two membranes lying parallel to each other.
Lampbrush chromosomes. Large diplotene chromosomes present in oöcyte nuclei, particularly conspicuous in amphibians. These chromosomes have extended regions called loops, which are active sites of transcription.
Leptonema (*adj.,* leptotene). Stage in meiosis immediately preceding synapsis in which the chromosomes appear as single, fine threadlike structures (but they are really double because DNA replication has already taken place).
Lethal allele. An allele that renders an organism or a cell possessing it inviable.
Ligase. An enzyme that joins the ends of two strands of nucleic acid.
Ligation. The joining of two or more DNA molecules by covalent bonds.
Linkage. A relationship among genes in the same chromosome; such genes tend to be inherited together.
Linkage equilibrium. A state in which the alleles of linked loci are randomized with respect to each other on the chromosomes of a population.
Linkage map. A linear or circular diagram that shows the relative positions of genes on a chromosome as determined by genetic analysis.
Locus (*pl.,* loci). A fixed position on a chromosome that is occupied by a given gene or one of its alleles.
LTR (Long terminal repeat). A DNA sequence present at each of the ends of a retrotransposon.
Lymphocyte. A general class of white blood cells that are important components of the immune system of vertebrate animals.
Lysis. Bursting of a cell by the destruction of the cell membrane following infection by a virus.
Lysogenic bacteria. Bacteria harboring temperate bacteriophages.
Macromolecule. A large molecule; term used to identify molecules of proteins and nucleic acids.
Map unit. See *Crossover unit.*
Mass selection. As practiced in plant and animal breeding, the choosing of individuals for reproduction from the entire population on the basis of the individuals' phenotypes rather than on the phenotypes of their relatives.
Maternal effect. Trait controlled by a gene of the mother but expressed in the progeny.
Maternal inheritance. Inheritance controlled by extrachromosomal (i.e., cytoplasmic) factors that are transmitted through the egg.
Maturation. The formation of gametes or spores.
Mean. The arithmetic average; the sum of all measurements or values in a sample divided by the sample size.
Meiosis. The process by which the chromosome number of a reproductive cell becomes reduced to half the diploid ($2n$) or somatic number; results in the formation of gametes in animals or of spores in plants; important source of variability through recombination.
Meiotic drive. Any mechanism that causes alleles to be recovered unequally in the gametes of a heterozygote.
Melanin. Brown or black pigment.
Mendelian population. A natural interbreeding unit of sexually reproducing plants or animals sharing a common gene pool.
Merozygote. Partial zygote produced by a process of partial genetic exchange, such as transformation in bacteria. An exogenote may be introduced into a bacterial cell in the formation of a merozygote.
Mesoderm. The middle germ layer that forms in the early animal embryo and gives rise to such parts as bone and connective tissue.
Messenger RNA (mRNA). RNA that carries information necessary for protein synthesis from the DNA to the ribosomes.
Metabolic cell. A cell that is not dividing.
Metabolism. Sum total of all chemical processes in living cells by which energy is provided and used.
Metacentric chromosome. A chromosome with the centromere near the middle and two arms of about equal length.
Metafemale (superfemale). In *Drosophila*, abnormal female, usually sterile, with an excess of X chromosomes compared with sets of autosomes (e.g., XXX; AA).
Metaphase. That stage of cell division in which the chromosomes are most discrete and arranged in an equatorial plate; stage following prophase and preceding anaphase.

Metastasis. The spread of cancer cells to previously unaffected organs.
Microtubules. Hollow filaments in the cytoplasm making up a part of the locomotor apparatus of a motile cell; component of the mitotic spindle.
Midparent value. In quantitative genetics, the average of the phenotypes of two mates.
Mitochondria. Organelles in the cytoplasm of plant and animal cells where oxidative phosphorylation takes place to produce ATP.
Mitosis. Disjunction of duplicated chromosomes and division of the cytoplasm to produce two genetically identical daughter cells.
Model. A mathematical description of a biological phenomenon.
Modifier (modifying gene). A gene that affects the expression of some other gene.
Monohybrid. An offspring of two homozygous parents that differ from one another by the alleles present at only one gene locus.
Monohybrid cross. A cross between parents differing in only one trait or in which only one trait is being considered.
Monomer. A single molecular entity that may combine with others to form more complex structures.
Monoploid (haploid). Organism or cell having a single set of chromosomes or one genome (chromosome number n).
Monosomic. A diploid organism lacking one chromosome of its proper complement ($2n - 1$); an aneuploid. Monosome refers to the single chromosome, disome to two chromosomes of a kind, and trisome to three chromosomes of a kind.
Monozygotic twins. One-egg or identical twins.
Morphogen. A substance that stimulates the development of form or structure in an organism.
Morphology. Study of the form of an organism. Developmental history of visible structures and the comparative relation of similar structures in different organisms.
Mosaic. An organism or part of an organism that is composed of cells of different genotypes.
Multigene family. A group of genes that are similar in nucleotide sequence or that produce polypeptides with similar amino acid sequences.
Multiple alleles. A condition in which a particular gene occurs in three or more allelic forms in a population of organisms.
Mutable genes. Genes with an unusually high mutation rate.
Mutagen. An environmental agent, either physical or chemical, that is capable of inducing mutations.
Mutant. A cell or individual organism that shows a change brought about by a mutation; a changed gene.
Mutation. A change in the DNA at a particular locus in an organism. The term is used loosely to include point mutations involving a single gene change as well as a chromosomal change.
Mutation pressure. A constant mutation rate that adds mutant genes to a population; repeated occurrences of mutations in a population.
Mycelium (*pl.,* mycelia). Threadlike filament making up the vegetative portion of thallus fungi.
Narrow-sense heritability. In quantitative genetics, the proportion of the phenotypic variance that is due to the additive effects of alleles.
Natural selection. Differential survival and reproduction in nature that favors individuals that are better adapted to their environment; elimination of less fit organisms.
Neutral theory. The theory that the evolution of traits with little or no effects on fitness is a random process involving mutation and genetic drift.
Nonautonomous. Refers to biological units that cannot function by themselves; such units require the assistance of another unit, or "helper." (Cf. *Autonomous.*)
Nondisjunction. Failure of disjunction or separation of homologous chromosomes in mitosis or meiosis, resulting in too many chromosomes in some daughter cells and too few in others. Examples: In meiosis, both members of a pair of chromosomes go to one pole so that the other pole does not receive either of them; in mitosis, both sister chromatids go to the same pole.
Northern blot. The transfer of RNA molecules from an electrophoretic gel to a cellulose or nylon membrane by capillary action.
Nucleic acid. A macromolecule composed of phosphoric acid, pentose sugar, and organic bases; DNA and RNA.
Nucleolar organizer (NO). A chromosomal segment containing genes that control the synthesis of ribosomal RNA, located at the secondary constriction of some chromosomes.
Nucleolus. An RNA-rich, spherical sack in the nucleus of metabolic cells; associated with the nucleolar organizer; storage place for ribosomes and ribosome precursors.
Nucleoprotein. Conjugated protein composed of nucleic acid and protein; the material of which the chromosomes are made.
Nucleosome (Nu body). Spherical subunits of eukaryotic chromatin that are composed of a core particle consisting of an octamer of histones and 146 nucleotide-pairs.
Nucleotide. A unit of DNA and RNA molecules containing a phosphate, a sugar, and an organic base.
Nucleus. The part of a eukaryotic cell that contains the chromosomes; separated from the cytoplasm by a membrane.
Null mutation. A mutation that abolishes the expression of a gene. (See *Amorph.*)
Nullisomic. An otherwise diploid cell or organism lacking both members of a chromosome pair (chromosome formula $2n-2$).
Octoploid. Cell or organism with eight genomes or sets of chromosomes (chromosome number $8n$).
Oncogene. A gene that can cause neoplastic transformation in animal cells growing in culture and tumor formation in animals themselves.
Oöcyte. The egg-mother cell; the cell that undergoes two meiotic divisions (oögenesis) to form the egg cell. Primary oöcyte—before completion of the first meiotic division; secondary oöcyte—after completion of the first meiotic division.
Oögenesis. The formation of the egg or ovum in animals.
Oögonium (*pl.,* oögonia). A germ cell of the female animal before meiosis begins.
Open reading frame. A sequence of nucleotide triplets that lacks a termination codon
Operator. A part of an operon that controls the activity of one or more structural genes by binding a regulatory protein.
Operon. A group of genes making up a regulatory or control unit. The unit includes an operator, a promoter, and structural genes.
Ordinate. The vertical axis in a graph.
Organelle. Specialized part of a cell with a particular function or functions (e.g., cilium of a protozoan).
Organizer. An inductor; a chemical substance in a living system that determines the fate in development of certain cells or groups of cells.

Otocephaly. Abnormal development of the head of a mammalian fetus.
Outbreeding. Mating of unrelated individuals.
Ovary. The swollen part of the pistil of a plant flower that contains the ovules. The female reproductive organ or gonad in animals.
Overdominance. A condition in which heterozygotes are superior (on some scale of measurement) to either of the associated homozygotes.
Ovule. The macrosporangium of a flowering plant that becomes the seed. It includes the nucellus and the integuments.
P. Symbolizes the parental generation or parents of a given individual.
Pachynema (*adj.,* pachytene). A midprophase stage in meiosis immediately following zygonema and preceding diplonema. In favorable microscopic preparations, the chromosomes are visible as long, paired threads. Rarely, four chromatids are detectable.
Pair-rule gene. A gene that influences the formation of body segements in *Drosophila*.
Palindrome. A segment of DNA in which the base-pair sequence reads the same in both directions from a point of symmetry.
Panmictic population. A population in which mating occurs at random.
Panmixis. Random mating in a population.
Paracentric inversion. An inversion that is entirely within one arm of a chromosome and does not include the centromere.
Parameter. A value or constant based on an entire population. (Cf. *Statistic.*)
Parthenogenesis. The development of a new individual from an egg without fertilization.
Paternal. Pertaining to the father.
Pathogen. An organism that causes a disease.
Pedigree. A table, chart, or diagram representing the ancestry of an individual.
Penetrance. The percentage of individuals that show a particular phenotype among those capable of showing it.
Peptide. A compound containing amino acids. A breakdown or buildup unit in protein metabolism.
Peptide bond. A chemical bond holding amino acid subunits together in proteins.
Pericentric inversion. An inversion including the centromere, hence involving both arms of a chromosome.
Phage. See *Bacteriophage.*
Pharmacogenetics. Biochemical genetics associated with genetically controlled variations in responses to drugs.
Phenocopy. An organism whose phenotype (but not genotype) has been changed by the environment to resemble the phenotype of a different (mutant) organism.
Phenotype. The observable characteristics of an organism.
Phenylalanine. See *Amino acid.*
Phenylketonuria. Metabolic disorder resulting in mental retardation; transmitted as a Mendelian recessive and treated in early childhood by special diet.
Phylogeny. Evolutionary history of populations of related organisms.
Plaque. Clear area on an otherwise opaque culture plate of bacteria where the bacteria have been killed by a virus.
Plasmid. An extrachromosomal hereditary determinant that exists in an autonomous state and is transferred independently of chromosomes. Some plasmids, such as F, are also capable of inserting into a chromosome.
Plastid. A cytoplasmic body found in the cells of plants and some protozoa. Chloroplastids, for example, produce chlorophyll, which is involved in photosynthesis.
Pleiotropy (*adj.,* Pleiotropic). Condition in which a single gene influences more than one trait.
Polar bodies. In female animals, the smaller cells produced at meiosis that do not develop into egg cells. The first polar body is produced at division I and may not go through division II. The second polar body is produced at division II.
Polar mutation. Gene mutation that influences the functioning of genes that are downstream in the same operon.
Pole cells. A group of cells in the posterior of *Drosophila* embryos that are precursors to the adult germ line.
Polycloning region. A segment of DNA in a cloning vector such as a plasmid that contains many different restriction enzyme cleavage sites.
Polydactyly. The occurrence of more than the usual number of fingers or toes.
Polygene (*adj.,* polygenic). One of many genes involved in quantitative inheritance.
Polymer. A compound composed of many smaller subunits; results from the process of polymerization.
Polymerase. An enzyme that catalyzes the formation of DNA or RNA.
Polymerase chain reaction (PCR). A procedure involving multiple cycles of denaturation, renaturation, and polynucleotide synthesis that amplifies a particular DNA sequence.
Polymerization. Chemical union of two or more molecules of the same kind to form a new compound having the same elements in the same proportions but a higher molecular weight and different physical properties.
Polymorphism. Two or more kinds of individuals maintained in a breeding population.
Polynucleotide. A linear sequence of joined nucleotides in DNA or RNA.
Polypeptide. A linear molecule with two or more amino acids and one or more peptide groups. They are called dipeptides, tripeptides, and so on, according to the number of amino acids present.
Polyploid. An organism with more than two sets of chromosomes ($2n$ diploid) or genomes (e.g., triploid ($3n$), tetraploid ($4n$), pentaploid ($5n$), hexaploid ($6n$), heptaploid ($7n$), octoploid ($8n$)).
Polysaccharide capsules. Carbohydrate coverings with antigenic specificity that are present on some types of bacteria.
Polytene chromosomes. Giant chromosomes produced by interphase replication without division and consisting of many identical chromatids arranged side by side in a cable-like pattern.
Population. Entire group of organisms of one kind; an interbreeding group of plants or animals. The extensive group from which a sample might be taken.
Population (effective). Breeding members of the population.
Population genetics. The branch of genetics that deals with frequencies of alleles and genotypes in breeding populations.
Position effect. A difference in phenotype that is dependent on the position of a gene or group of genes, often caused by heterochromatin that is nearby.
Primary oöcyte. See *Oöcyte.*
Primer. A short nucleotide sequence with a reactive 3′-OH that can initiate DNA synthesis along a template.
Progeny testing. The practice of ascertaining the genotype of an individual by mating it to an individual of known genotype and examining the progeny.
Prokaryote. A member of a large group of organisms

(including bacteria and blue-green algae) that lack true nuclei in their cells and that do not undergo meiosis.
Promoter. A nucleotide sequence to which RNA polymerase binds and initiates transcription. Also, a chemical substance that enhances the transformation of benign cells into cancerous cells.
Prophase. The stage of mitosis between interphase and metaphase. During this phase, the centriole divides and the two daughter centrioles move apart. Each sister DNA strand from interphase replication becomes coiled and the chromosome is longitudinally double except in the region of the centromere. Each partially separated chromosome is called a chromatid. The two chromatids of a chromosome are sister chromatids.
Protooncogene. A normal cellular gene that can be changed to an oncogene by mutation.
Protoplast. A plant or bacterial cell from which the wall has been removed.
Prototroph. An organism such as a bacterium that will grow on a minimal medium.
Provirus. A viral chromosome that has integrated into a host genome.
Pseudogene. An inactive but stable component of a genome resembling a gene; apparently derived from active genes by mutation.
Quantitative inheritance. Inheritance of measureable traits (height, weight, color intensity) that depend on the cumulative action of many genes, each producing a small effect on the phenotype.
Race. A distinguishable group of organisms of a particular species.
Radioactive isotope. An unstable isotope (form of an atom) that emits ionizing radiation.
Random genetic drift. Changes in allele frequency in small breeding populations due to chance fluctuations.
Recessive. Applied to one member of an allelic pair lacking the ability to manifest itself when the other or dominant member is present.
Reciprocal crosses. Crosses between different strains with the sexes reversed; for example, female A × male B and male A × female B are reciprocal crosses.
Recombination. The production of gene combinations not found in the parents by the assortment of nonhomologous chromosomes and crossing-over between homologous chromosomes during meiosis. For linked genes, the frequency of recombination can be used to estimate the genetic map distance; however, high frequencies (approaching 50 percent) do not yield accurate estimates.
Reduction division. Phase of meiosis in which the maternal and paternal chromosomes of the bivalent separate. (Cf. *Equational (homotypic) division.*)
Regulator gene. A gene that controls the rate of expression of another gene or genes. Example: The *lac i* gene produces a protein that controls the expression of the structural genes of the *lac* operon in *Escherichia coli*.
Renaturation. The restoration of a molecule to its native form; in nucleic acid biochemistry, this term usually refers to the formation of a double-stranded helix from complementary single-stranded molecules.
Replication. A duplication process that is accomplished by copying from a template (e.g., reproduction at the level of DNA).
Replicon. A unit of replication. In bacteria, replicons are associated with segments of the cell membrane that control replication and coordinate it with cell division.
Repressible enzyme. An enzyme whose synthesis is diminished by a regulatory molecule.
Repulsion (*trans*-arrangement). The condition in which a double heterozygote has received a mutant and a wild-type allele from each parent; for example, $a\,+/a\,+ \times +\,b/+\,b$ produces $a\,+/+\,b$. (Cf. *Coupling*.)
Resistance factor. A plasmid that confers antibiotic resistance to a bacterium.
Restitution nucleus. A nucleus with unreduced or doubled chromosome number that results from the failure of a meiotic or mitotic division.
Restriction enzyme. An endonuclease that recognizes a specific short sequence in DNA and cleaves the DNA molecule.
Restriction map. A linear or circular diagram of a DNA molecule showing the sites that are cleaved by different restriction enzymes.
Reticulocyte. A young red blood cell.
Retroelement. Any of the integrated retroviruses or the transposable elements that resemble them.
Retrotransposon. A transposable element that resembles the integrated form of a retrovirus.
Reverse transcriptase. An enzyme that catalyzes the synthesis of DNA using an RNA template.
Reversion. Restitution of a mutant gene to the wild-type condition, or at least to a form that gives the wild phenotype. More generally, the appearance of a trait expressed by a remote ancestor; a throwback; atavism.
RFLP (Restriction fragment length polymorphism). A genetic difference among individuals that is detected by comparing DNA fragments released by digestion with one or more restriction enzymes.
Rh factor. Antigen in the red blood corpuscles of certain people. A pregnant Rh-negative woman carrying an Rh-positive child may develop antibodies, causing the child to develop a hemolytic disease.
Ribonucleic acid. See *RNA*.
Ribosome. Cytoplasmic organelle on which proteins are synthesized.
RNA. Ribonucleic acid. The information-carrying material in some viruses. More generally, a molecule derived from DNA by transcription that may carry information (messenger or mRNA), provide subcellular structure (ribosomal or rRNA), transport amino acids (transfer or tRNA), or facilitate the biochemical modification of itself or other RNA molecules.
Roentgen (r). Unit of ionizing radiation.
Satellite DNA. A component of the genome that can be isolated from the rest of the DNA by centrifugation; usually it consists of short, highly repetitive sequences.
Secondary oöcyte. See *Oöcyte*.
Secondary spermatocyte. See *Spermatocyte*.
Secretor. A person with a water-soluble form of blood group antigen A or B. In such a person, the antigen may be detected in body fluids (e.g., saliva) as well as on the erythrocytes.
Segment-polarity gene. A gene that functions to define the anterior and posterior components of body segments in *Drosophila*.
Segregation. The separation of paternal and maternal chromosomes from each other at meiosis; the separation of alleles from each other in heterozygotes; the occurrence of different phenotypes among offspring, resulting from chromosome or allele separation in their heterozygous parents. Mendel's first principle of inheritance.
Segregation distorter (*SD*). A factor that alters the segregation ratio in heterozygous *Drosophila*; *SD*+ cells are eliminated during spermiogenesis in *SD*+/*SD* males.
Selection. Differential survival and reproduction among

genotypes. The most important of the factors that change allele frequencies in large populations.

Selection coefficient. A number that measures the fitness of a genotype relative to a standard.

Selection differential. In plant and animal breeding, the difference between the mean of the individuals selected to be parents and the mean of the overall population.

Selection pressure. Effectiveness of differential survival and reproduction in changing the frequency of alleles in a population.

Selection response. In plant and animal breeding, the difference between the mean of the individuals selected to be parents and the mean of their offspring.

Selector gene. A gene that influences the development of specific body segments in *Drosophila*; a homeotic gene.

Self-fertilization. The process by which pollen of a given plant fertilizes the ovules of the same plant. Plants fertilized in this way are said to have been selfed. An analogous process may occur in some animals, such as nematodes and molluscs.

Semisterility. A condition of only partial fertility in plant zygotes (e.g., maize); usually associated with translocations.

Serology. (*adj.,* serological). The study of interactions between antigens and antibodies.

Sex chromosomes. Chromosomes that are connected with the determination of sex.

Sexduction. The incorporation of bacterial genes into F factors and their subsequent transfer by conjugation to a recipient cell.

Sex factor. A bacterial episome (e.g., the F plasmid in *E. coli*) that enables the cell to be a donor of genetic material. The sex factor may be propagated in the cytoplasm, or it may be integrated into the bacterial chromosome.

Sex-influenced dominance. A dominant expression that depends on the sex of the individual (e.g., horns in sheep are dominant in males and recessive in females).

Sex-limited. Expression of a trait in only one sex. Examples: Milk production in mammals; horns in Rambouillet sheep; egg production in chickens.

Sex linkage. Association or linkage of a hereditary trait with sex; the gene is in a sex chromosome, usually the X; often used synonymously with X-linkage.

Sex mosaic. See ***Gynandromorph***.

Sexual reproduction. Reproduction involving the formation of mature germ cells (i.e., eggs and sperm).

Shuttle vector. A plasmid capable of replicating in two different organisms, such as yeast and *E. coli*.

Sib-mating (crossing of siblings). Matings involving two individuals of the same parentage; brother–sister matings.

Silencer. A DNA sequence that helps to reduce or shut off the expression of a nearby gene.

Somatic cell. A cell that is a component of the body, in contrast with a germ cell that is capable, when fertilized, of reproducing the organism.

Southern blot. The transfer of DNA fragments from an electrophoretic gel to a cellulose or nylon membrane by capillary action.

Species. Interbreeding, natural populations that are reproductively isolated from other such groups.

Sperm (*abbreviation of* spermatozoon; *pl.,* spermatozoa). A mature male germ cell.

Spermatids. The four cells formed by the meiotic divisions in spermatogenesis. Spermatids become mature spermatozoa or sperm.

Spermatocyte (sperm mother cell). The cell that undergoes two meiotic divisions (spermatogenesis) to form four spermatids; the *primary* spermatocyte before completion of the first meiotic division; the *secondary* spermatocyte after completion of the first meiotic division.

Spermatogenesis. The process by which maturation of the gametes (sperm) of the male takes place.

Spermatogonium (*pl.,* spermatogonia). Primordial male germ cell that may divide by mitosis to produce more spermatogonia. A spermatogonium may enter a growth phase and give rise to a primary spermatocyte.

Spermiogenesis. Formation of sperm from spermatids; the part of spermatogenesis that follows the meiotic divisions of spermatocytes.

Spheroplast. A plant or bacterial cell from which the wall has been removed. (See ***Protoplast.***)

Splicing. The process that covalently joins exon sequences of RNA and eliminates the intervening intron sequences.

Standard deviation. A measure of variability in a set of data. The square root of the variance.

Standard error. A measure of variation among a population of means.

Statistic. A value based on a sample or samples of a population from which estimates of a population value or parameter may be obtained.

Stem cell. A mitotically active somatic cell from which other cells arise by differentiation.

Sterility. Inability to produce offspring.

Structural gene. A gene that specifies the synthesis of a polypeptide.

Supercoil. A DNA molecule that has been twisted.

Suppressor mutation. A mutation that partially or completely cancels the phenotypic effect of another mutation.

Symbiont. An organism living in intimate association with another, dissimilar organism.

Synapsis. The pairing of homologous chromosomes in the meiotic prophase.

Synaptinemal complex. A ribbonlike structure formed between synapsed homologs at the end of the first meiotic prophase, binding the chromatids along their length and facilitating chromatid exchange.

Syndrome. A group of symptoms that occur together and represent a particular disease.

Synkaryon. A nucleus formed by the fusion of nuclei from two different somatic cells during somatic-cell hybridization.

Synteny. The occurrence of two loci on the same chromosome, without regard to the distance between them.

Target site duplication. A sequence of DNA that is duplicated when a transposable element inserts, usually found at each end of the insertion.

Telophase. The last stage in each mitotic or meiotic division in which the chromosomes are assembled at the poles of the division spindle.

Temperate phage. A phage (virus) that invades but may not destroy (lyse) the host (bacterial cell) (cf. ***Virulent phage.***); however, it may continue into the lytic cycle.

Template. A pattern or mold. DNA stores coded information and acts as a model or template from which information is transcribed into messenger RNA.

Terminalization. Repelling movement of the centromeres of bivalents in the diplotene stages of the meiotic prophase that tends to move the visible chiasmata toward the ends of the bivalents.

Testcross. Backcross to the recessive parental type, or a cross between genetically unknown individuals with a fully recessive tester to determine whether an individual in ques-

tion is heterozygous or homozygous for a certain allele. Also used as a test for linkage.
Tetrad. The four cells arising from the second meiotic division in plants (pollen tetrads) or fungi (ascospores). The term is also used to identify the quadruple group of chromatids that is formed by the association of duplicated homologous chromosomes during meiosis.
Tetraploid. An organism whose cells contain four haploid ($4n$) sets of chromosomes or genomes.
Tetrasomic (*noun,* tetrasome). Pertaining to a nucleus or an organism with four members of one of its chromosomes whereas the remainder of its chromosome complement is diploid. (Chromosome formula: $2n + 2$.)
Tetratype. In fungi, a tetrad of spores that contains four different types; for example, *AB*, *aB*, *Ab*, and *ab*.
Thymine. A pyrimidine base found in DNA. The other three organic bases—adenine, cytosine, and guanine—are found in both RNA and DNA, but in RNA, thymine is replaced by uracil.
Totipotent cell (*or* nucleus). An undifferentiated cell (or nucleus) such as a blastomere that when isolated or suitably transplanted can develop into a complete embryo.
***trans*-Acting.** A term describing substances that are diffusable and that can affect spatially separated entities within cells.
***trans*-Arrangement.** See *Repulsion.*
Transcription. Process through which messenger RNA is formed along a DNA template. The enzyme RNA polymerase catalyzes the formation of RNA from ribonucleotide triphosphates.
Transdetermination. The alteration of the developmental fate of a group of cells, causing them to differentiate in an anomalous way. For example, in *Drosophila*, an experimentally manipulated wing imaginal disc may differentiate into eye structures.
Transduction. Genetic recombination in bacteria mediated by bacteriophage. Abortive: Bacterial DNA is injected by a phage into a bacterium but it does not replicate. Generalized: Any bacterial gene may be transferred by a phage to a recipient bacterium. Restricted: Transfer of bacterial DNA by a temperate phage is restricted to only one site on the bacterial chromosome.
Transfection. The uptake of DNA by a eukaryotic cell, followed by the incorporation of genetic markers present in the DNA into the cell's genome.
Transfer RNA (tRNA). RNA that transports amino acids to the ribosomes, where the amino acids are assembled into proteins.
Transferrin. Blood serum protein, beta globulin. (See Globulin.)
Transformation (bacteria). Genetic alteration of bacteria brought about by the incorporation of foreign DNA in the bacterial cells.
Transformation (eukaryotic cells). The conversion of eukaryotic cells growing in culture to a state of uncontrolled cell growth (similar to tumor cell growth).
Transgenic. A term applied to organisms that have been altered by introducing DNA molecules into them.
Transgressive variation. The appearance in the F_2 (or a later) generation of individuals showing more extreme development of a trait than either of the original parents.
Transition. A mutation caused by the substitution of one purine by another purine or one pyrimidine by another pyrimidine in DNA or RNA.
Translation. Protein (polypeptide) synthesis directed by a specific messenger RNA; occurs on ribosomes.
Translocation. Change in position of a segment of a chromosome to another part of the same chromosome or to a different chromosome.
Transposase. An enzyme that catalyzes the movement of a DNA sequence to a different site in a DNA molecule.
Transposons. DNA elements that can move ("transpose themselves") from one position in a DNA molecule to another.
Transposon tagging. The insertion of a transposable element into or nearby a gene, thereby marking that gene with a known DNA sequence.
Transversion. A mutation caused by the substitution of a purine for a pyrimidine or a pyrimidine for a purine in DNA or RNA.
Trihybrid. The offspring from homozygous parents differing in three pairs of genes.
Trisomic. An otherwise diploid cell or organism that has an extra chromosome of one pair. (Chromosome formula: $2n + 1$.)
Tryptophan. See *Amino acids.*
Unequal crossing-over. Crossing-over between repeated DNA sequences that have paired out of register, creating duplicated and deficient products.
Unipartite structures. Single units.
Univalent. An unpaired chromosome at meiosis.
Uracil. A pyrimidine base found in RNA but not in DNA. In DNA, uracil is replaced by thymine.
Variance. A measure of variation in a population. The square of the standard deviation.
Variation. In biology, the occurrence of differences among individuals.
Vector. A plasmid or viral chromosome that may be used to construct recombinant DNA molecules for introduction into living cells.
Viability. The capability to live and develop normally.
Virulent phage. A phage (virus) that destroys the host (bacterial) cell. (Cf. *Temperate phage.*)
Viscoelastometry. A method to study the physical properties of molecules in solution.
Western blot. The transfer of proteins from an electrophoretic gel to a cellulose or nylon membrane by means of an electric force.
Wild type. The customary phenotype or standard for comparison.
Wobble hypothesis. Hypothesis to explain how one tRNA may recognize two codons. The first two bases of the mRNA codon and anticodon pair properly, but the third base in the anticodon has some play (or wobble) that permits it to pair with more than one base.
X chromosome. A chromosome associated with sex determination. In most animals, the female has two and the male has one X chromosome.
Xenia. Hereditary influence from the pollen parent to the endosperm in the seed in certain plants.
Y chromosome. The partner of the X chromosome in the male of many animal species.
Zygonema (*adj.,* zygotene). Stage in meiosis during which synapsis occurs; after the leptotene stage and before the pachytene stage in the meiotic prophase.
Zygote. The cell produced by the union of two mature sex cells (gametes) in reproduction; also used in genetics to designate the individual developing from such a cell.

ANSWERS TO PROBLEMS

Chapter 2

2.1 (a) All tall; (b) ¾ tall, ¼ dwarf; (c) all tall; (d) ½ tall, ½ dwarf

2.2 (a) *WW* × *ww*; (b) Ⓦ and ⓦ; (c) *Ww*; (d) *Ww* × *Ww*; (e) Ⓦ, ⓦ and Ⓦ, ⓦ; (f) see the following table:

PHENOTYPES	GENOTYPES	GENOTYPIC FREQUENCY	PHENOTYPIC RATIO
Round	*WW*	1	3
	Ww	2	
Wrinkled	*ww*	1	1

2.3 (a) The 3 : 1 ratio suggests a single pair of genes, with the gene for color dominant over that for white.

(b)

CC × *cc*	P
Ⓒ ⓒ	Gametes
Cc	F_1
Cc × *Cc*	$F_1 \times F_1$
Ⓒⓒ Ⓒⓒ	F_1 gametes
1*CC* : 2*Cc* : 1*cc*	F_2

PHENOTYPES	OBSERVED	CALCULATED	DEVIATION
Colored	198	202.5	−4.5
White	72	67.5	4.5

2.4 (a) Woman, *Pp*; her father, *Pp*; her mother, *pp*. (b) Half of her children are expected to have ptosis.

2.5

CCBB × *ccbb*	P
(*CB*) (*cb*)	Gametes
CcBb × *CcBb*	$F_1 \times F_1$

GAMETES	(*CB*)	(*Cb*)	(*cB*)	(*cb*)
(*CB*)	*CCBB*	*CCBb*	*CcBB*	*CcBb*
(*Cb*)	*CCBb*	*CCbb*	*CcBb*	*Ccbb*
(*cB*)	*CcBB*	*CcBb*	*ccBB*	*ccBb*
(*cb*)	*CcBb*	*Ccbb*	*ccBb*	*ccbb*

Summary of F_2:

PHENOTYPES	GENO-TYPES	GENOTYPIC FREQUENCY	PHENO-TYPIC RATIO
Checkered red	*CCBB*	1	9
	CCBb	2	
	CcBB	2	
	CcBb	4	

Summary of F_2: *(continued)*

PHENOTYPES	GENO-TYPES	GENOTYPIC FREQUENCY	PHENO-TYPIC RATIO
Checkered brown	*CCbb*	1	3
	Ccbb	2	
Plain red	*ccBB*	1	3
	ccBb	2	
Plain brown	*ccbb*	1	1

2.6 (a) *CCVv* × *ccVv*; (b) *CcVv* × *CcVv*; (c) *CcVv* × *ccvv*.

2.7 (a)

BBRR × *bbrr*	P
(*BR*) (*br*)	Gametes
BbRr	F_1
BbRr × *BbRr*	$F_1 \times F_1$

GAMETES	(*BR*)	(*Br*)	(*bR*)	(*br*)
(*BR*)	*BBRR*	*BBRr*	*BbRR*	*BbRr*
(*Br*)	*BBRr*	*BBrr*	*BbRr*	*Bbrr*
(*bR*)	*BbRR*	*BbRr*	*bbRR*	*bbRr*
(*br*)	*BbRr*	*Bbrr*	*bbRr*	*bbrr*

Summary of F_2 phenotypes: 9 black, long : 3 black, rex : 3 brown, long : 1 brown, rex.

(b) 1/9.

(c) *BbRr* × *bbrr*.

GAMETES	(*BR*)	(*Br*)	(*bR*)	(*br*)
(*br*)	*BbRr*	*Bbrr*	*bbRr*	*bbrr*

Summary of backcross results: 1 black, long : 1 black, rex : 1 brown, long : 1 brown, rex.

2.8 (a) All red; (b) ½ red : ½ roan; (c) all roan; (d) ¼ red : ½ roan : ¼ white; (e) ½ roan : ½ white; (f) all white; (g) red. Mate red × red and all progeny will be red. If roan animals are mated together, red and white as well as roan will be produced.

2.9 (a) ½; (b) ½; (c) ⅜.

2.10 (a) 81/256; (b) 108/256; (c) 54/256; (d) 12/256.

2.11 (a)

Dd	×	*Dd*	P
Ⓓⓓ		Ⓓⓓ	Gametes
1*DD*	2*Dd*	1*dd*	Progeny
dies	dichaete	wild type	

Summary: 2 dichaete, 1 wild type.

(b) *Dd* × *dd* — P

Ⓓ Ⓓ Ⓓ (gametes: *D*, *d* × *d*) — Gametes

1*Dd* 1*dd* — Progeny

Summary: 1 dichaete, 1 wild-type.

2.12 (a) ¼; (b) ¼; (c) ¼; (d) ¼.

2.13 (a) Single recessive gene. (b) 1. Parents, *aa* × *Aa;* progeny, *Aa, aa, Aa, aa*. 2. Parents, *Aa* × *Aa;* progeny, *Aa* or *AA, aa, Aa* or *AA, aa*. 3. Parents, *Aa* × *Aa;* progeny, *aa, Aa* or *AA, Aa* or *AA, Aa* or *AA, aa*. 4. Parents, *aa* × *aa;* progeny, *aa, aa, aa, aa.*

2.14 (a) *DDggWW* × *ddGGww* — P

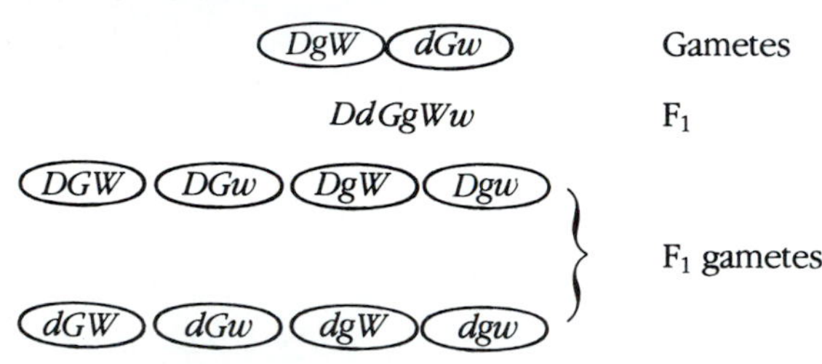

(b) *DdGgWw* × *DdGgWw*

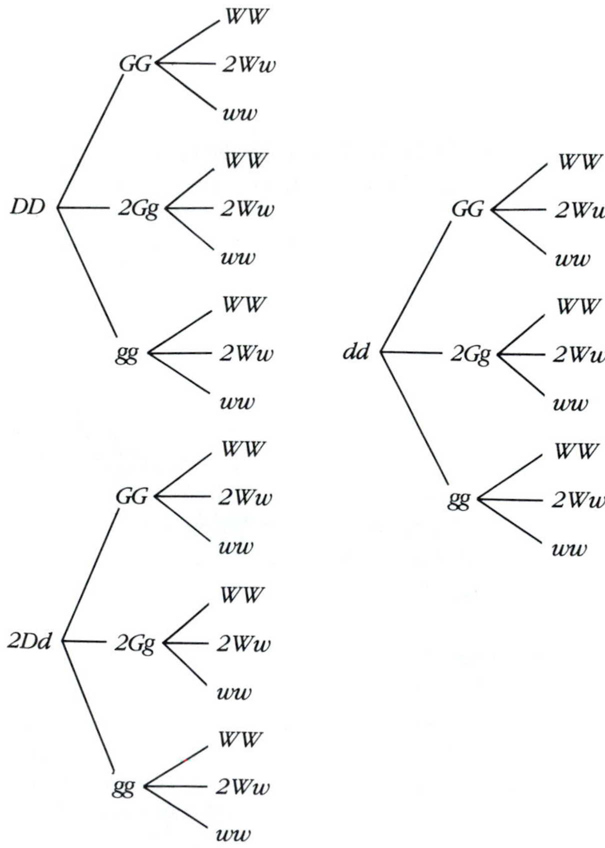

SUMMARY OF F_2 PHENOTYPES:

27 tall, yellow, round
9 tall, yellow, wrinkled
9 tall, green, round
9 dwarf, yellow, round
3 tall, green, wrinkled
3 dwarf, yellow, wrinkled
3 dwarf, green, round
1 dwarf, green, wrinkled

(c) *DdGgWw* × *ddggww*
Summary of backcross results:

	PHENOTYPES	GENOTYPES	GENO-TYPIC RATIO	PHENO-TYPIC RATIO
Dd – *Gg* – *Ww*	Tall, yellow, round	*DdGgWw*	1	1
Dd – *Gg* – *ww*	Tall, yellow, wrinkled	*DdGgww*	1	1
Dd – *gg* – *Ww*	Tall, green, round	*DdggWw*	1	1
Dd – *gg* – *ww*	Tall, green, wrinkled	*Ddggww*	1	1
dd – *Gg* – *Ww*	Dwarf, yellow, round	*ddGgWw*	1	1
dd – *Gg* – *ww*	Dwarf, yellow, wrinkled	*ddGgww*	1	1
dd – *gg* – *Ww*	Dwarf, green, round	*ddggWw*	1	1
dd – *gg* – *ww*	Dwarf, green, wrinkled	*ddggww*	1	1

2.15

P CROSS	*AA* × *aa*	*AABB* × *aabb*	*AABBCC* × *aabbcc*	GENERAL FORMULA
F_1 gametes	2	4	8	2^n[a]
F_2 genotypes	3	9	27	3^n
F_2 phenotypes[b]	2	4	8	2^n

[a] *n* = number of segregating pairs of alleles.
[b] Under complete dominance of *A, B,* and *C.*

2.16 The Punnett square method is illustrated as follows:

RRLL × *R′R′L′L′* — P

(*RL*) (*R′L′*) — Gametes

RR′LL′ — F_1

RR′LL′ × *RR′LL′* — $F_1 \times F_1$

GAMETES	(*RL*)	(*RL′*)	(*R′L*)	(*R′L′*)
(*RL*)	*RRLL*	*RRLL′*	*RR′LL*	*RR′LL′*
(*RL′*)	*RRLL′*	*RRL′L′*	*RR′LL′*	*RR′L′L′*
(*R′L*)	*RR′LL*	*RR′LL′*	*R′R′LL*	*R′R′LL′*
(*R′L′*)	*RR′LL′*	*RR′L′L′*	*R′R′LL′*	*R′R′L′L′*

PHENOTYPES	GENOTYPES	GENOTYPIC FREQUENCY	PHENOTYPIC RATIO
Red, long	*RRLL*	1	1
Red, oval	*RRLL'*	2	2
Red, round	*RRL'L'*	1	1
Purple, long	*RR'LL*	2	2
Purple, oval	*RR'LL'*	4	4
Purple, round	*RR'L'L'*	2	2
White, long	*R'R'LL*	1	1
White, oval	*R'R'L'L*	2	2
White, round	*R'R'L'L'*	1	1

2.17 (a) 3 walnut : 1 rose; (b) 1 walnut : 1 pea; (c) 3 walnut : 3 rose : 1 pea : 1 single; (d) 1 rose : 1 single.

2.18 *Rrpp* × *RrPp*.

2.19 12 white : 3 yellow : 1 green.

2.20 13 white : 3 colored.

2.21 (a) 3 : 1; (b) 3 : 1; (c) 1 : 1 : 2; (d) 2 : 1 : 1; (e) 1 : 1 : 1: 1.

2.22 (a) χ^2 = .75, P = .20 – .50, not significant; (b) χ^2 = 8.39, P = .004, significant.

2.23 (a) χ^2 = .246, P = .80 – .50, not significant; (b) χ^2 = .390, P = .80 – .50, not significant; (c) χ^2 = .450, P = .80 – .50, not significant; (d) χ^2 = .618, P = .95 – .80, not significant; (e) χ^2 = .530, P = .95 – .80, not significant.

2.24 (a) 1/16; (b) 1/16; (c) 4/16; (d) 4/16; (e) 6/16.

2.25 (a) 6/16 or 3/8; (b) 1/16; (c) 2 boys and 2 girls, because more combinations of the four independent events result in 2 and 2 than in any other arrangement; (d) 1/2.

2.26 (a) 6/64; (b) 20/64; (c) 1/64.

2.27 (a) 0; (b) 1/2.

2.28 (a) 1/2 × 1 × 1/4 = 1/8; (b) 1/16; (c) 1/16; (d) 1/24.

2.29 (a) 243/1024; (b) 405/1024; (c) 270/1024; (d) 90/1024; (e) 15/1024; (f) 1/1024.

2.30 (a) Environmental factors may influence developing phenotypes in the same way that genes do. If so, studies of phenocopies might suggest mechanisms of gene action. (b) Phenocopies provide one approach to the study of physiological genetics. Such studies are indirect and can only provide suggestions about the mechanisms of gene action. (c) Mutations are inherited; phenocopies are not inherited.

2.31 Phenocopies in humans are often very difficult to distinguish from inherited abnormalities in which the mutant gene has variable expressivity, such as variable age of onset.

2.32 70 percent.

2.33 (a) By independent changes at various sites within an ancestral gene. Such changes may accumulate over time, resulting in the divergence of different alleles. (b) The same basic symbol is used conventionally for all members of a group of multiple alleles, with superscripts to identify particular alleles. (c) Most series of multiple alleles are associated with gradations in the same phenotype. Some produce quite different end results, such as legs in the place of antennae in *Drosophila*.

2.34 (a) All colored; (b) 3 colored : 1 albino; (c) 3 colored : 1 chinchilla; (d) 1 chinchilla : 1 albino; (e) 3 colored : 1 himalayan; (f) 1 himalayan : 1 albino.

2.35 (a) Yellow and yellow; 2 yellow : 1 agouti light belly (A^yA^y is lethal); (b) yellow and agouti light belly; 2 yellow : 1 agouti light belly : 1 black and tan; (c) black and tan and yellow; 2 yellow : 1 black and tan : 1 black; (d) agouti light belly and agouti light belly; all agouti light belly; (e) agouti light belly and yellow; 1 yellow : 1 agouti light belly; (f) agouti and black and tan; 1 agouti : 1 black and tan; (g) black and tan and black; 1 black and tan : 1 black; (h) yellow and agouti; 1 yellow : 1 agouti light belly; (i) yellow and yellow; 2 yellow : 1 agouti.

2.36 (a) 1; (b) 2; (c) 2; (d) 10.

2.37 30.

2.38 (a) 28; (b) by crossing fish that carry the different alleles, intercrossing the F_1 progeny, and checking for monohybrid F_2 ratios.

2.39 (a) S_1S_3 and S_2S_3; (b) S_1S_3, S_1S_4, S_2S_3, and S_2S_4; (c) none; (d) S_3S_5, S_3S_6, S_4S_5, and S_4S_6.

2.40 (a) all AB; (b) 1 A : 1 B; (c) 1 AB : 1 A : 1 B : 1 O-type; (d) 1 A : 1 O-type.

2.41 It is ***extremely unlikely*** (approximately 10^{-5} to 10^{-6}) that the man with AB-type blood was the father of the child with O-type blood. This could happen only if a change (new mutation) had occurred in one of the A or B blood-type alleles of the man's reproductive cells.

2.42 It is ***extremely unlikely*** (see answer to Problem 2.41) that the baby with O-type blood was the daughter of the woman with AB-type blood.

2.43 (a) All children would be heterozygous (*Rr*) and Rh-positive. The mother became immunized during the first pregnancy and the next child was affected. (b) All future children would be expected to be affected.

2.44 Half of the children carried by the *rr* mother would be *Rr* (Rh-positive) and half would be *rr* (rh-negative), like the mother. Because the woman was sensitized at the second pregnancy, the *Rr* children would be expected to be erythroblastotic and the *rr* children normal.

2.45 9 black : 3 gray : 52 white.

2.46 9 black : 39 gray : 16 white.

2.47 (a) $(3/4)^3(1/4)$ = 27/256; (b) $(3/4)^4 + (3/4)^2(1/4)$ = 117/256; (c) 256/256 − 144/256 = 112/256.

Chapter 3

3.1 (a) 0; (b) 0; (c) 0; (d) 0; (e) 0; (f) +; (g) +.

3.2 (a) 23; (b) 23; (c) 23; (d) 23; (e) 46; (f) 46; (g) 23; (h) 46.

3.3 (a) 200; (b) 50.

3.4 $\frac{M}{m} \times \frac{m}{m}$ P

Ⓜ ⓜ ⓜ Gametes

$\frac{M}{m}$ $\frac{m}{m}$ F_1, half of the progeny are expected to have myopia

3.5 Model in text, Fig. 3.6, right.

3.6 The chromosome mechanism is similar in animals and plants. Division of the cytoplasmic part of the flexible animal cell is accomplished by constriction (cytokinesis), whereas the rigid plant cell forms a partition or cell plate.

3.7 Meiosis includes a pairing (synapsis) of corresponding maternal and paternal chromosomes. In the cell division that follows, the chromosomes that have previously paired separate. This results in a reduction of chromosome number from $2n$ (diploid) to n (haploid).

3.8 (a) Many plants have male and female parts on the same plant or in the same flower. Unlike animals, plants have a gametophyte stage that consists (in higher plants) of a few cell divisions. (b) The chromosome mechanism is essentially the same during the formation of gametes in plants and animals.

3.9 An egg and two polar nuclei are developed in the ovule. Two haploid nuclei are introduced by the pollen tube. One nucleus fuses with the egg and the other with the two polar nuclei to produce the $3n$ endosperm nucleus. The fertilized egg is the zygote that develops into an embryo. The endosperm forms the nutrient material that supports the developing embryo.

3.10 Model for diagram, Fig. 2.11. About ¼ of all children would be expected to have only the gene for intestinal polyposis; ¼ to have only the gene for Huntington's chorea; ¼ to have neither; and about ¼ to have both.

3.11 Model in text, Fig. 3.6, right.

3.12 Model in text, Fig. 3.6, left.

3.13 *AaBb*.

3.14 Bisexual organisms. Asexual reproduction provides for no genetic variation, except that of rare mutations. Self-fertilization tends toward homozygosity or pure lines. Bisexual reproduction in higher organisms is associated with great hereditary variation through recombination.

3.15 (a) Early primary oöcyte; (b) prophase, first meiotic division; (c) suspended prophase; (d) first meiotic division is completed just before ovulation of each egg.

Chapter 4

4.1 Male Y and female X bearing sperm.

4.2 (a) Female (tetraploid); (b) intersex; (c) intersex; (d) metamale; (e) female (diploid); (f) male (sterile if no Y chromosome is present in *Drosophila*).

4.3 Female gametes would be (2X 2A), (2X A), (X 2A), and (X A). Zygotes and sex would be 3X 3A female (triploid), 3X 2A metafemale, 2X 3A intersex, 2X 2A female (diploid), 2XY 3A intersex, 2XY 2A female, XY 3A metamale, XY 2A male.

4.4 (a) Male; (b) female; (c) bisex; (d) female.

4.5 (a) Male; (b) female; (c) male; (d) female.

4.6 The single gene (*ba*) removes the female part of the monoecious plant and makes the stalk only staminate. Another gene (*ts*) transforms the tassel into a pistillate structure. A plant of the genotype *ba ba ts ts* would be only pistillate (female), whereas a plant with the genotype $ba\ ba\ ts^+\ ts^+$ would be only staminate (male).

4.7 (a) 0; (b) 1; (c) 1; (d) 2; (e) 3; (f) 0.

4.8 (a) ¾; (b) ¼.

4.9 (a) 3 hen-feathered : 1 cock-feathered; (b) all hen-feathered.

4.10 2 rods; 1 hook and 1 rod; 1 knob and 1 rod; 1 hook and 1 knob.

4.11 (a) ½ $X^{bb}X^{bb}$ bobbed females, ½ $X^{bb}Y^{bb+}$ wild males; (b) ½ $X^{bb}X^{bb+}$ wild females, ½ $X^{bb}Y^{bb}$ bobbed males; (c) ½ $X^{bb+}X^{bb+}$ and $X^{bb}X^{bb+}$ wild females, ¼ $X^{bb+}Y^{bb}$ wild males, ¼ $X^{bb}Y^{bb}$ bobbed males; (d) ¼ $X^{bb+}X^{bb}$ wild females, ¼ $X^{bb}X^{bb}$ bobbed females, ½ $X^{bb+}Y^{bb+}$ and $X^{bb}Y^{bb+}$ wild males.

4.12 (a) Neither. Both the dominant and the recessive would be expressed with equal frequency in males. (b) The dominant would be expressed more frequently in females.

4.13 (a) If it breeds true in succeeding generations, it is probably hereditary and may be assumed to have arisen through mutation. (b) The crisscross pattern of inheritance is evidence for sex linkage. (c) Females whose father had white eyes could be crossed with white-eyed males. Half of the females would be expected to have white eyes. This would conform to genetic theory based on sex linkage.

4.14

(a) $\overset{w}{\rightharpoondown}$, $\overset{w^+}{\rightharpoondown}$, $\frac{w^+}{w^+}$, $\frac{w^+}{w}$, and $\frac{w}{w}$

(b) $\frac{w^+}{w^+} \times \overset{w}{\rightharpoondown}$ P

$\underline{w^+}$ and $\underline{w}$, $\rightharpoondown$ Gametes

$\frac{w}{w^+} \times \overset{w^+}{\rightharpoondown}$ $F_1 \times F_1$ (all red-eyed)

$\underline{w}$, $\underline{w^+}$ and $\underline{w^+}$, $\rightharpoondown$ F_1 gametes

$\frac{w}{w^+}$ and $\frac{w^+}{w^+}$ F_2 females: all red-eyed

$\overset{w}{\rightharpoondown}$ and $\overset{w^+}{\rightharpoondown}$ F_2 males: ½ red, ½ white

(c) (i) $\frac{w}{w^+} \times \overset{w}{\rightharpoondown}$ Females: ½ red, ½ white; Males: ½ red, ½ white

(ii) $\frac{w}{w^+} \times \overset{w^+}{\rightharpoondown}$ Females: all red; Males: ½ red, ½ white

4.15 (1) $\overset{g^+}{\rightharpoondown}$ (2) $\frac{g^+}{g}$ (3) $\overset{g}{\rightharpoondown}$ (4) $\frac{g^+}{g}$ (5) Probability is 63/64 for $\frac{g^+}{g^+}$ and 1/64 for $\frac{g^+}{g}$ (6) $\frac{g^+}{g}$ (7) $\overset{g}{\rightharpoondown}$ (8) $\overset{g^+}{\rightharpoondown}$ (9) $\overset{g^+}{\rightharpoondown}$

4.16 Half of their sons would be color-blind, and their daughters would be normal.

4.17 No. A son receives his X chromosome from his mother. The father contributes a Y chromosome.

4.18

$\dfrac{g^+}{g} \times \dfrac{g}{\rightharpoonup}$	P
$\underline{g^+}$, $\underline{g}$, $\underline{g}$, $\rightharpoonup$	Gametes
$\dfrac{g^+}{g}$, $\dfrac{g}{g}$, $\dfrac{g^+}{\rightharpoonup}$, $\dfrac{g}{\rightharpoonup}$	Sons and daughters: ½ normal, ½ color-blind

4.19

(a) $\dfrac{h}{h} \times \dfrac{h^+}{\rightharpoonup}$	P
$\underline{h}$ and $\underline{h^+}$, $\rightharpoonup$	Gametes
$\dfrac{h}{h^+}$ and $\dfrac{h}{\rightharpoonup}$	F_1: normal daughters, hemophiliac sons
(b) $\dfrac{h^+}{h} \times \dfrac{h}{\rightharpoonup}$	P
$\underline{h^+}$, $\underline{h}$ and $\underline{h}$, $\rightharpoonup$	Gametes
$\dfrac{h^+}{h}$, $\dfrac{h}{h}$, $\dfrac{h^+}{\rightharpoonup}$, $\dfrac{h}{\rightharpoonup}$	F_1 daughters and sons: ½ normal, ½ hemophiliac
(c) $\dfrac{h^+}{h^+} \times \dfrac{h}{\rightharpoonup}$	P
$\underline{h^+}$, $\underline{h}$, $\rightharpoonup$	Gametes
$\dfrac{h}{h^+}$, $\dfrac{h^+}{\rightharpoonup}$	F_1 all normal

4.20 ½ chance for each son or ¼ for each child (½ for male × ½ affected).

4.21 (a) ½; (b) ½.

Chapter 5

5.1 (a) Griffith's *in vivo* experiments demonstrated the occurrence of transformation in pneumococcus. They provided no indication as to the molecular basis of the transformation phenomenon. Avery and colleagues carried out *in vitro* experiments, employing biochemical analyses to demonstrate that transformation was mediated by DNA. (b) Griffith showed that a transforming substance existed; Avery *et al.* defined it as DNA. (c) Griffith's experiments did not include any attempt to characterize the substance responsible for transformation. Avery *et al.* isolated DNA in "pure" form and demonstrated that it could mediate transformation.

5.2 (a) No effect; (b) no effect; (c) DNase will destroy the capacity of the extract to transform Type IIR cells to Type IIIS by degrading the DNA in the extract. Protease and RNase will degrade the proteins and RNA, respectively, in the extract. They will have no effect, since the proteins and RNA are not involved in transformation.

5.3 Purified DNA from Type III cells was shown to be sufficient to transform Type II cells. This occurred in the absence of any dead Type III cells.

5.4 About ½ protein, ½ DNA. A single long molecule of DNA is enclosed within a complex "coat" that is composed of many proteins.

5.5 DNA contains phosphorus (normally ^{31}P), but no sulfur; it can be labeled with ^{32}P. Proteins contain sulfur (normally ^{32}S), but usually no phosphorus; they can be labeled with ^{35}S.

5.6 (a) The objective was to determine whether the genetic material was DNA or protein. (b) By labeling phosphorus, a constituent of DNA, and sulfur, a constituent of protein, in a virus, it was possible to demonstrate that only the labeled phosphorus was introduced into the host cell during the viral reproductive cycle. This was enough to produce new phages. (c) Therefore DNA, not protein, is the genetic material.

5.7 (a) The ladderlike pattern was known from X-ray diffraction studies. Chemical analyses had shown that a 1 : 1 relationship existed between the organic bases adenine and thymine and between cytosine and guanine. Physical data concerning the length of each spiral and the stacking of bases were also available. (b) Watson and Crick developed the model of a double helix, with the rigid strands of sugar and phosphorous forming spirals around an axis, and hydrogen bonds connecting the complementary bases in base-pairs.

5.8 (a) A multistranded, spiral structure was suggested by the X-ray diffraction patterns. A double-stranded helix with specific base-pairing nicely fits the 1 : 1 stoichiometry observed for A : T and G : C in DNA. (b) Use of the known hydrogen-bonding potential of the bases provided a means of holding the two complementary strands in a stable configuration in such a double helix.

5.9 (a) 400,000 nucleotides; (b) 20,000 complete spirals; (c) 400,000 phosphorus atoms; (d) 680,000 Å.

5.10 3′-C A G T A C T G-5′.

5.11 (a) DNA has one atom less of oxygen than RNA in the sugar part of the molecule. In DNA, thymine replaces the uracil that is present in RNA. (In certain bacteriophages, DNA-containing uracil is present.) DNA is most frequently double-stranded, but bacteriophages such as ΦX174 contain single-stranded DNA. RNA is most frequently single-stranded. Some viruses, such as the reoviruses, however, contain double-stranded RNA chromosomes.

5.12 13 percent.

5.13 No. TMV RNA is single-stranded. Thus, the base-pair stoichiometry of DNA does not apply.

5.14 (a) False; (b) false; (c) true; (d) true; (e) true; (f) true; (g) false; (h) true; (i) true; (j) false; (k) true; (l) false; (m) true.

5.15 (a) (1) One-half of the DNA molecules with ^{15}N in both strands and ½ with ^{14}N in both strands; (2) all DNA molecules with one strand containing ^{15}N and the complementary strand containing ^{14}N; (3) all DNA molecules with both strands containing roughly equal amounts of ^{15}N and ^{14}N. (b) (1) ¼ of the DNA molecules with ^{15}N in both strands and ¾ with ^{14}N in both strands; (2) ½ of the DNA molecules with one strand containing ^{15}N and the complementary strand containing ^{14}N and the other ½ with ^{14}N in both strands; (3) all DNA molecules with both strands containing about ¼ ^{15}N and ¾ ^{14}N. See Fig. 5.13.

5.16 One-half of the DNA molecules fully heavy (^{15}N in both strands); the other half of the molecules "hybrid" (^{15}N in one strand, ^{14}N in the complementary strand).

5.17 (a) Both 3′⟶5′ and 5′⟶3′ exonuclease activities. (b) The 3′⟶5′ exonuclease "proofreads" the nascent DNA strand during its synthesis. If a mismatched base-pair occurs at the 3′-OH end of the primer, the 3′⟶5′ exonuclease removes the incorrect terminal nucleotide before polymerization proceeds again. The 5′⟶3′ exonuclease is responsible for the removal of RNA primers during DNA replication and functions in pathways involved in the repair of damaged DNA (see Chapter 11). (c) Yes, both exonuclease activities appear to be very important. Without the 3′⟶5′ proofreading activity during replication, an intolerable mutation frequency would probably occur. The 5′⟶3′ exonuclease activity is essential to the survival of the cell. Conditional mutations (see Chapter 11) that affect the 5′⟶3′ exonuclease activity of DNA polymerase I are lethal to the cell under conditions where the exonuclease is nonfunctional.

5.18 (a)

3′ Ⓟ-TGCGAATTAGCGACAT-Ⓟ 5′
5′ Ⓟ-ATCGGTACGÄC̈G̈C̈T̈T̈ÄÄT̈C̈G̈C̈T̈G̈T̈Ä- OH 3′;

note that DNA synthesis will ***not*** occur on the left end since the 3′ terminus of the potential primer strand is blocked with a phosphate group—all DNA polymerases require a free 3′-OH terminus. (b) The first step will be the removal of the mismatched C (exiting as dCMP) from the 3′-OH primer terminus by the 3′⟶5′ exonuclease ("proofreading") activity.

5.19 If nascent DNA is labeled by exposure to [^{3}H]thymidine for very short periods of time, continuous replication predicts that the label would be incorporated into chromosome-size DNA molecules, whereas discontinuous replication predicts that the label would first appear in small pieces of nascent DNA (prior to covalent joining, catalyzed by polynucleotide ligase).

5.20

PROTEINS	FUNCTIONS
1. DNA polymerase III	1. (a) Catalyzes polymerization (covalent extension) of new DNA chains. (b) The 3′ → 5′ exonuclease activity "proofreads" the product, removing any mismatched base-pairs at the 3′ end of the primer strand.
2. DNA polymerase I	2. Removes the RNA primers (5′ → 3′ exonuclease activity) and replaces them with new DNA strands (5′ → 3′ polymerase activity).
3. DNA ligase	3. Catalyzes covalent joining of "Okazaki fragments."
4. Primase (*dnaG* protein)	4. Catalyzes RNA primer synthesis.
5. DNA gyrase	5. Catalyzes the formation of negative supercoils; facilitates unwinding?
6. DNA helicase	6. Catalyzes unwinding.
7. DNA single-strand binding protein	7. Maintains an "extended" single-stranded template; aids unwinding?
8. *dnaB* protein	8. Required for initiation of replication.
9. proteins i, n, n′, *dnaC* protein	9. "Prepriming"—required prior to initiation of primer synthesis
10. *dnaI, dnaJ,* etc., proteins	10. Required; but functions unknown.

Chapter 6

6.1 (1) The nucleosome level; the core containing an octamer of histones plus 146 nucleotide-pairs of DNA arranged as 1¾ turns of a supercoil (see Fig. 6.12), yielding an approximately 100-Å-diameter spherical body; or juxtaposed, a roughly 100-Å-diameter fiber. (2) The 300 Å fiber observed in condensed mitotic and meiotic chromosomes; structure unknown, apparently a second level of folding or coiling. (3) The highly condensed mitotic and meiotic chromosomes (e.g., metaphase chromosomes); the tight folding or coiling maintained by a "scaffold" composed of nonhistone chromosomal proteins (see Fig. 6.17).

6.2 In the diploid nucleus of *D. melanogaster*, 10^6 nucleosomes would be present; these would contain 2×10^6 molecules of each histone, H2a, H2b, H3, and H4.

6.3 The satellite DNA fragments would renature much more rapidly than the main-band DNA fragments. In *D. virilus* satellite DNAs, all three have repeating heptanucleotide-pair sequences. Thus, essentially every 40-nucleotide-long (average) single-stranded fragment from one strand will have a sequence complementary (in part) with every single-stranded fragment from the complementary strand. Many of the nucleotide-pair sequences in main-band DNA will be unique sequences (present only once in the genome).

6.4 It indicates that highly repetitive DNA sequences do not contain structural genes specifying RNA and polypeptide gene-products.

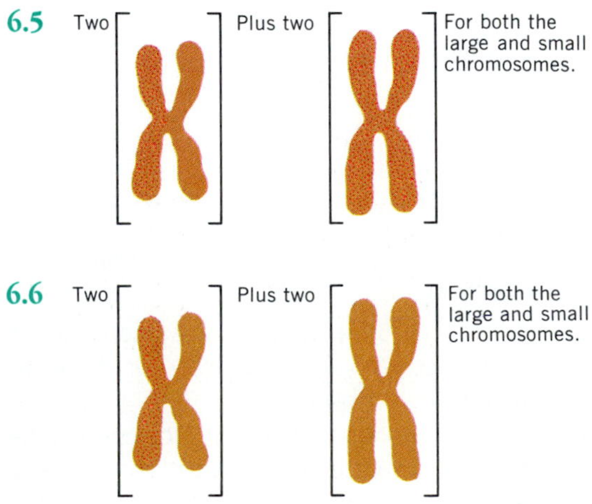

6.7 That DNA replication was unidirectional rather than bidirectional. As the intracellular pools of radioactive [^{3}H]thymidine are gradually diluted after transfer to nonradioactive medium, less and less [^{3}H]thymidine will be incorporated into DNA at each replicating fork. This will produce autoradiograms with tails of decreasing grain density at each growing point. Since such tails appear at only one end of each track, replication must be unidirectional. Bidirectional replication would produce such tails at both ends of an autoradiographic track (see Fig. 6.25).

6.8 (a) (1) Euchromatin; (2) euchromatin; (3) heterochromatin. (b) (1) Yes; (2) no. (c) (1) Most of the single-copy DNA sequences are believed to be structural genes encoding structural proteins and the vast repertoire of enzymes employed by living organisms; (2) essentially nothing is known regarding the functions of the highly repetitive DNA sequences—your hypotheses are probably as valid as anyone else's. (d) Some middle-repetitive DNA sequences specify products such as ribosomal RNA molecules that are required by cells in large quantities. Others are believed to be binding sites for proteins that regulate gene expression and replication of the multiple replicons of the giant DNA molecules in eukaryotic chromosomes. Some middle-repetitive sequences probably play structural roles in chromosomes, especially during the condensations of mitosis and meiosis. Others undoubtedly carry out functions that are still unknown.

6.9 Interphase. Chromosomes are for the most part metabolically inactive (very little transcription) during the various stages of condensation in mitosis and meiosis.

6.10 (a) Telomeres provide at least three important functions: (1) prevention of exonucleolytic degradation of the ends of the linear DNA molecules in eukaryotic chromosomes, (2) prevention of the fusion of ends of DNA molecules of different chromosomes, and (3) provision of a mechanism for replication of the distal tips of linear DNA molecules in eukaryotic chromosomes. (b) Yes. All telomeres studied to date contain DNA sequence repeat units (e.g., TTAGGG in human chromosomes), and, at least in some species, telomeres terminate with single-stranded 3′ overhangs that form "hairpin" structures. The bases in these "hairpins" exhibit unique patterns of methylation that presumably contribute to the structure and stability of telomeres. (c) The enzyme telomere transferase catalyzes the addition of telomere repeat sequences to the ends of the giant DNA molecules present in chromosomes of eukaryotes. (d) The broken ends resulting from irradiation will not contain telomeres; as a result the free ends of the DNA molecules are apparently subject to the activities of enzymes such as exonucleases, ligases, etc., that modify the ends. They can regain stability by fusing to broken ends of other DNA molecules that contain terminal telomere sequences.

6.11 Viscoelastometry is a procedure used to measure the viscosity of molecules in solution. In addition, viscoelastometric methods can be used to estimate the sizes of the largest DNA molecules present in aqueous solutions. By using viscoelastometry, scientists have obtained evidence (see Fig. 6.8) that indicates that all the DNA present in chromosomes of eukaryotes exists as giant, "chromosome-size" DNA molecules (one huge DNA molecule per chromosome). This eliminated early models of chromosome structure that proposed multiple DNA molecules joined end-to-end by protein or RNA "linkers."

6.12 (a) Two. The axial region of a "lampbrush" chromosome contains the two chromatids of one homologous chromosome (postreplication). (b) One. Each lateral loop of a "lampbrush" chromosome is a segment of a single chromatid.

6.13 Nonhistone chromosomal proteins. The "scaffold" structures of metaphase chromosomes can be observed by light microscopy after removal of the histones by differential extraction procedures.

6.14 (a) 23; (b) 23; (c) 46; (d) 92; (e) 92.

6.15 (a) Histones have been highly conserved throughout the evolution of eukaryotes. A major function of histones is to package DNA into nucleosomes and chromatin fibers. Since DNA is composed of the same four nucleotides and has the same basic structure in all eukaryotes, one might expect that the proteins that play a structural role in packaging this DNA would be similarly conserved. (b) The nonhistone chromosomal proteins exhibit the greater heterogeneity in chromatin from different tissues and cell types of an organism. The histone composition is largely the same in all cell types within a given species—consistent with the role of histones in packaging DNA into nucleosomes. The nonhistone chromosomal proteins include proteins that regulate gene expression. Since different sets of genes are transcribed in different cell types, one would expect heterogeneity in some of the nonhistone chromosomal proteins of different tissues.

6.16 The rate of DNA renaturation is proportional to the square of the concentration of single strands (C^2) because renaturation requires a collision between and a reassociation of two complementary molecules (single strands) that are present in the reaction mixture in ***identical concentrations*** since they were originally present in the same double helix. Thus, the reaction depends on the concentrations of both strands (both concentrations equal, therefore, $C \times C = C^2$), whereas the rates of most other bimolecular reactions are a function of the concentrations (often different, therefore, $C_1 \times C_2$) of both reacting substances. The fact that DNA renaturation rates are proportional to C^2 is thus a direct consequence of the double-helix structure of DNA with each double helix composed of two complementary single strands.

6.17 The 6 percent of the human DNA that is already renatured at $t=0$ in standard DNA renaturation experiments results from the presence of single strands that themselves contain complementary sequences with opposite chemical polarity (one sequence reading 5′ to 3′ complementary to another sequence reading 3′ to 5′). Single strands containing such complementary sequences form double-stranded "hairpin" or "foldback" structures. Such reactions are concentration independent since collisions and reassociations between two molecules are not required for the renaturation events to occur. Thus, they occur very fast (too fast to be measured in the standard renaturation experiments) with unimolecular reaction kinetics. Some DNA sequences that exhibit unimolecular renaturation kinetics are called palindromes; such DNAs contain sequences that are the same when read in opposite direction starting from a central point of symmetry (see Chapter 13).

6.18 DNA polymerases α, β, and δ are located in the nuclei of cells; polymerase γ is located in mitochondria and chloroplasts. Current evidence suggests that polymerases α and δ are both required for the replication of nuclear DNA. Polymerase δ is thought to catalyze the continuous synthesis of the leading strand, and polymerase α is believed to catalyze discontinuous synthesis of the lagging strand since it contains the primase activity required for the repeated initiation of "Okazaki fragments." Polymerase γ presumably catalyzes replication of organellar chromosomes. The function of polymerase β remains unknown; perhaps it functions in DNA repair pathways like DNA polymerase I of *E. coli*.

6.19 (a) Given bidirectional replication of a single replicon, each replication fork must traverse 2×10^6 nucleotide-pairs in *E. coli* and 3×10^7 nucleotide-pairs in the largest *Drosophila* chromosome. If the rates were the same in both species, it would take 15 times ($3 \times 10^7/2 \times 10^6$) as long to replicate the *Drosophila* chromosome or 10 hours (40 minutes $\times$ 15 = 600 minutes). (b) If replication forks in *E. coli* move 20 times as fast as replication forks in *Drosophila* (100,000 nucleotide-pairs per minute/5000 nucleotide-pairs per minute), the largest *Drosophila* chromosome would require 8.3 days (10 hours $\times$ 20 = 200 hours) to complete one round of replication. (c) Each *Drosophila* chromosome must contain many replicons to complete replication in less than 10 minutes.

6.20 (a) One microgram of human DNA will contain, on average, 3.04×10^5 copies of the genome. Using an average molecular weight per nucleotide-pair of 660, the "molecular" weight of the entire human genome is 1.98×10^{12} ($3 \times 10^9 \times 660$). Thus, 1.98×10^{12} g (1 "mole" = number of grams equivalent to the "molecular" weight) of human DNA will contain, on average, 6.02×10^{23} molecules [Avogadro's number = number of molecules (here, copies of the genome) present in one "mole" of a substance]. One gram will contain, on average, 3.04×10^{11} ($6.02 \times 10^{23}/1.98 \times 10^{12}$) copies of the genome; thus, 1 μg will contain, on average, 3.04×10^5 copies of the human genome. (b) One copy of the human genome weighs approximately 3.3×10^{-12} g (1.98×10^{12} g per "mole"/6.02×10^{23} molecules per "mole") or 3.3×10^{-6} μg. (c) By analogous calculations, 1 μg of *Arabidopsis thaliana* DNA contains, on average, 1.18×10^7 copies of the genome. (d) Similarly, one copy of the *A. thaliana* genome weighs approximately 8.4×10^{-8} μg. (e) In carrying out molecular analyses of the structures of genomes, geneticists frequently need to know how many copies of a genome are present, on average, in a given quantity of DNA.

Chapter 7

7.1 (a) Prepare a testcross and compare results with those expected from the hypothesis of independent assortment. If they do not fit, linkage may be the next hypothesis. (b) The parental combinations should occur in greater proportion in the progeny. If the parental combinations are not known and only progeny data are available, determine whether the expressions controlled by the two dominants or those controlled by a dominant and recessive are in greater proportion. (c) With most materials, a testcross is easily prepared, and the results can be compared with the simple 1 : 1 : 1 : 1 ratio rather than with the more complex 9 : 3 : 3 : 1 F_2 ratio.

7.2 The one class that is represented by 351/1000 is out of proportion for independent assortment. At least two of the three gene pairs must be on the same chromosome.

7.3 The three classes given fit the hypothesis of independent assortment, suggesting that the gene pairs are on three different chromosome pairs.

7.4 20 percent.

7.5 (a) $\dfrac{a^+b^+}{a^+b^+} \times \dfrac{a\ b}{a\ b}$ P

(a^+b^+) $(a\ b)$ Gametes

$\dfrac{a^+b^+}{a\ b}$ F_1

(b)

(a^+b^+) 40%

(a^+b) 10%

$(a\ b^+)$ 10%

$(a\ b)$ 40%

(c)

$\dfrac{a^+b^+}{a\ b}$ 40%

$\dfrac{a^+b}{a\ b}$ 10%

$\dfrac{a\ b^+}{a\ b}$ 10%

$\dfrac{a\ b}{a\ b}$ 40%

(d) Coupling.

7.6

(a) $\dfrac{a^+b}{a\ b^+}$ F_1

(b)

(a^+b) 40%

(a^+b^+) 10%

$(a\ b)$ 10%

$(a\ b^+)$ 40%

(c)

$\dfrac{a^+b}{a\ b}$ 40%

$\dfrac{a^+b^+}{a\ b}$ 10%

$\dfrac{a\ b}{a\ b}$ 10%

$\dfrac{a\ b^+}{a\ b}$ 40%

(d) Repulsion.

7.7 The parental gametes would be in the proportion of 30 percent each and the recombinants in the proportion of 20 percent each. The zygotes from the testcross would be in the same proportion as the F_1 gametes.

7.8

	(*a*)	(*b*)
Two dominant expressions (a^+-b^+-)	66%	51%
One dominant and one recessive (a^+-bb)	9%	24%
Other dominant and other recessive (aab^+-)	9%	24%
Two recessives ($aabb$)	16%	1%
	100%	100%

7.9 (a) No; (b) —.

(c) Parentals, frequency = .496

$\dfrac{b^+}{b}\ \dfrac{ts^+}{ts}$; red, normal; 124/498 = .249

$\dfrac{b}{b}\ \dfrac{ts}{ts}$; green, tassel; 123/498 = .247

Recombinants, frequency = .504

$\dfrac{b^+}{b}\ \dfrac{ts}{ts}$; red, tassel; 126/498 = .253

$\dfrac{b}{b}\ \dfrac{ts^+}{ts}$; green, normal; 125/498 = .251

(d) $\dfrac{b}{b^+}\ \dfrac{ts^+}{ts^+} \times \dfrac{b}{b}\ \dfrac{ts}{ts}$

7.10 (a) Yes; (b) 16.7 percent; (c) *b* and *vg* are linked; see (b).

(d) $\dfrac{b^+vg^+}{b\ vg} \times \dfrac{b\ vg}{b\ vg}$

7.11 (a) Yes; (b) 16.3 percent; (c) $\dfrac{b^+vg}{b\ vg^+} \times \dfrac{b\ vg}{b\ vg}$

7.12 (a) Yes; (b) 34 percent; (c) $\dfrac{c^+b}{c^+b} \times \dfrac{c\ b^+}{c\ b^+}$

7.13 (a) $\dfrac{d^+p^+}{d\,p}$; (b) $\dfrac{d^+p}{d\,p^+}$

(c) 54% d^+-p^+- : 21% d^+-pp : 21% ddp^+- : 4% $ddpp$.

7.14 (a) $\dfrac{a^+b^+}{a^+b^+}\ \dfrac{c^+d^+}{c^+d^+} \times \dfrac{a\,b}{a\,b}\ \dfrac{c\,d}{c\,d}$ P

(b) $\dfrac{a^+b^+}{a\,b}\ \dfrac{c^+d^+}{c\,d}$ F_1

(c)

12%	a^+b^+	c^+d^+	8%	a^+b^+	c^+d
12%	a^+b^+	$c\,d$	8%	a^+b^+	$c\,d^+$
12%	$a\,b$	c^+d^+	8%	$a\,b$	c^+d
12%	$a\,b$	$c\,d$	8%	$a\,b$	$c\,d^+$
3%	a^+b	c^+d^+	2%	a^+b	c^+d
3%	a^+b	$c\,d$	2%	$a\,b^+$	$c\,d^+$
3%	$a\,b^+$	c^+d^+	2%	a^+b	$c\,d^+$
3%	$a\,b^+$	$c\,d$	2%	$a\,b^+$	c^+d

7.15 (a) Either cataract or polydactyly. The genes would be in repulsion. In the separation at meiosis, each gamete would get one or the other. A crossover would be required to produce a gamete with both or neither.

7.16 (a) 46% $\underline{sr\ e^+}$
46% $\underline{sr^+\ e}$
4% $\underline{sr^+\ e^+}$
4% $\underline{sr\ e}$

(b) 46% striped, gray
46% not striped, ebony
4% not striped, gray
4% striped, ebony

7.17 (a) Four kinds in the proportion of 5, 45, 45, and 5 percent. (Repulsion.) (b) Wild type: 5 percent; vestigial, red: 45 percent; long wing, cinnabar: 45 percent, vestigial, cinnabar: 5 percent.

7.18 Map distances may be converted to probabilities as follows: the chance that crossing-over will occur between *st* and *ss* is .14 and the chance that crossing-over will not occur is .86; the chance that crossing-over will occur between *ss* and *e* is .12 and the chance that it will not occur is .88. By applying the multiplication procedure, the probabilities for the different combinations expected to be represented in the gametes can be calculated. Only one kind of gamete is produced by the male, so the proportion of zygotes will be the same as the proportion of gametes. Phenotypes may be expected in the following percentages:

Red eyes, normal bristles, gray body:
$$\frac{(.12 \times .86 \times 100)}{2} = 5.16 \text{ percent}$$

Red, normal, ebony:
$$\frac{(.86 \times .88 \times 100)}{2} = 37.84 \text{ percent}$$

Red, spineless, gray:
$$\frac{(.14 \times .88 \times 100)}{2} = 6.16 \text{ percent}$$

Red, spineless, ebony:
$$\frac{(.14 \times .12 \times 100)}{2} = .84 \text{ percent}$$

Scarlet, normal, gray:
$$\frac{(.14 \times .12 \times 100)}{2} = .84 \text{ percent}$$

Scarlet, normal, ebony:
$$\frac{(.14 \times .88 \times 100)}{2} = 6.16 \text{ percent}$$

Scarlet, spineless, gray:
$$\frac{(.86 \times .88 \times 100)}{2} = 37.84 \text{ percent}$$

Scarlet, spineless, ebony:
$$\frac{(.12 \times .86 \times 100)}{2} = 5.16 \text{ percent}$$

7.19 Purple, salmon silk, pigmy:

$\frac{.81}{2} = .405 = 40.5$ percent

Green, yellow silk, normal:

$\frac{.81}{2} = .405 = 40.5$ percent

Purple, yellow silk, normal:

$\frac{.09}{2} = .045 = 4.5$ percent

Green, salmon silk, pigmy:

$\frac{.09}{2} = .045 = 4.5$ percent

Purple, salmon silk, normal:

$\frac{.09}{2} = .045 = 4.5$ percent

Green, yellow silk, pigmy:

$\frac{.09}{2} = .045 = 4.5$ percent

Purple, yellow silk, pigmy:

$\frac{.01}{2} = .005 = .5$ percent

Green, salmon silk, normal:

$\frac{.01}{2} = .005 = .5$ percent

These predictions are based entirely on probability, assuming equal crossing-over in all areas along the chromosome. Interference could reduce the frequency of the double crossover classes.

7.20 (Tu is a dominant mutant; Tu^+ is recessive.)

$$Tu^+\ j_2\ gl_3 : \frac{.893}{2} = .4465 = 44.65 \text{ percent}$$

$$Tu\ j_2^+\ gl_3^+ : \frac{.893}{2} = .4465 = 44.65 \text{ percent}$$

$$Tu\ j_2\ gl_3 : \frac{.047}{2} = .00235 = 2.35 \text{ percent}$$

$$Tu^+\ j_2^+\ gl_3^+ : \frac{.047}{2} = .0235 = 2.35 \text{ percent}$$

$$Tu^+\ j_2\ gl_3^+ : \frac{.057}{2} = .0285 = 2.85 \text{ percent}$$

$$Tu\ j_2^+\ gl_3 : \frac{.057}{2} = .0285 = 2.85 \text{ percent}$$

$$Tu\ j_2\ gl_3^+ : \frac{.0030}{2} = .0015 = .15 \text{ percent}$$

$$Tu^+\ j_2^+\ gl_3 : \frac{.0030}{2} = .0015 = .15 \text{ percent}$$

7.21 The classes with the smallest numbers ($+ + w$ and $y\ ec\ +$) must be double crossovers. The gene in these classes that differs from the parentals must be in the center, that is, $+\ w +$ and $y +\ ec$. With the sequence established, single crossovers can be identified and percentages calculated.

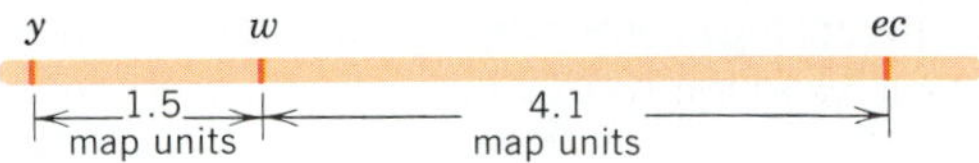

Or with y arbitrarily set at position 0 on the linkage map,

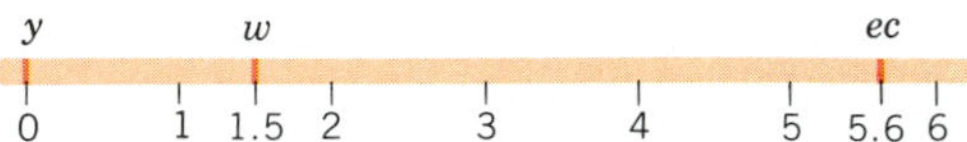

7.22 The smallest class is the double crossover class, and the v locus must be in the center. The double crossover value must be added to each single crossover value to obtain each map distance.

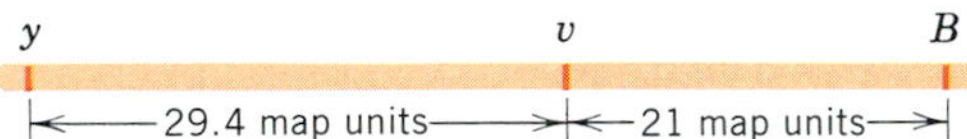

Or with y arbitrarily set at position 0 on the linkage map,

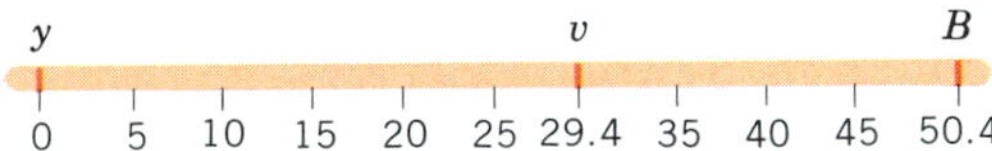

7.23 (a) Yes. The testcross results do not follow the 1 : 1 : 1 : 1 : 1 : 1 : 1 : 1 pattern expected from a trihybrid cross with independent assortment.

(b) s ss e

(c) Parental cross: $\frac{s^+\ ss\ e}{s^+\ ss\ e} \times \frac{s\ ss^+\ e^+}{s\ ss^+\ e^+}$

Testcross: $\frac{s^+\ ss\ e}{s\ ss^+\ e^+} \times \frac{s\ ss\ e}{s\ ss\ e}$

(d) 30.3 units.

(e) 14.0 units.

(f) $\frac{18}{.14 \times .163\ (1000)} = .79.$

7.24 80 percent (30 percent w dor^+ + 30 percent w^+ dor + 20 percent w dor); only the 20 percent w^+ dor^+ recombinants will exhibit a wild-type phenotype.

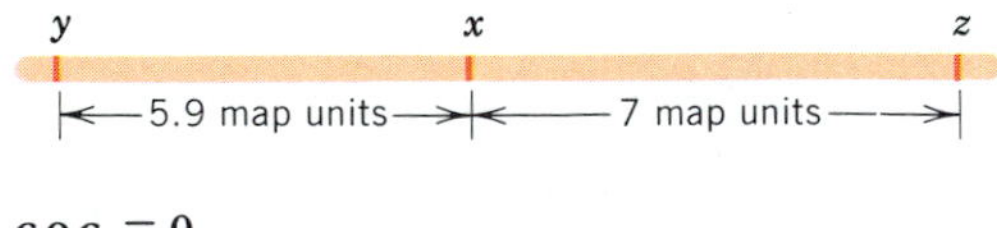

c.o.c. = 0.

Note that these genes are sex-linked; thus male progeny are hemizygous for X chromosome markers that are transmitted through the eggs (from their female parents). All female progeny will necessarily carry the wild-type alleles of all three genes, namely, wild-type alleles transmitted through the sperm (from their male parents). Therefore, linkage relationships can be determined only by analyzing the male progeny of the cross. Only six genotypes are represented among the progeny; thus, no double crossover types were present among the progeny analyzed. However, the missing double crossover genotypes can be determined by elimination.

7.26 $P^2/4$.

7.27 $\dfrac{(.5)(.02)(1000)}{2} = 5.$

7.28

b a c

|←20.6 map units→|←14.0 map units→|

7.29 cv sn v 7.2 map units; 13.2 map units; 20.4 map units;

$$\text{c.o.c.} = \frac{.006}{(.072)(.132)} \approx .63.$$

7.30 2A- : 1*aa*.

7.31 From the frequency of the homozygous recessive genotype *aabb*; this genotype can be formed only when an *ab* ovule is fertilized by an *ab* pollen. The frequency of the *ab* gamete type is therefore equal to the square root of the frequency of the double homozygous recessive genotype. In our example, the frequency of the *aabb* genotype equals 2/200 or .01. Thus, the frequency of the *ab* gamete equals $\sqrt{.01}$ or .1. Since the cross was done in repulsion, this is a recombinant gamete type. Therefore, the recombination frequency equals 2(.1) or .2, and the *A* and *B* loci are approximately 20 map units apart.

7.32 21 map units.

7.33 It gives definitive evidence that crossing-over occurs in the postreplication four-strand or tetrad stage rather than in the prereplication two-strand stage.

7.34

A Centromere B

|←5.5 map units→|←0.35 map units→|

7.35

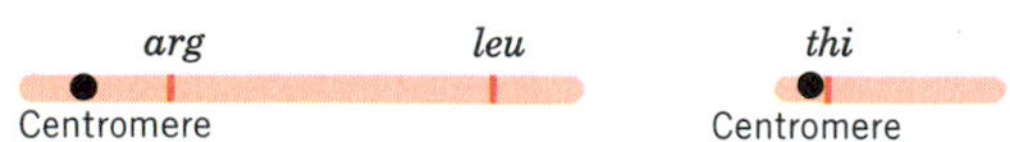

0 map units *thi* to centromere.

0.17 map units *arg* to centromere; from 1/2(1)/300 × 100.

11 map units *arg* to *leu*; from 1/2(44)/200 × 100.

7.36 All loci are unlinked; the parental ditype to nonparental ditype ratios are 52 : 46, 50 : 40, and 46 : 46 for gene pairs *xy, xz,* and *yz*, respectively. Distances from genes to centromeres cannot be determined from unordered tetrad data by the simple procedure discussed in this text. In some cases, these distances can be calculated by a more complex procedure (see H. L. K. Whitehouse, *J. Genet.* 55: 348, 1957).

7.37 The two genes are unlinked, but each is very tightly linked to its respective centromere.

7.38 The ascus pattern shown is most simply explained by a four-strand double crossover, with one crossover between *x* and *y* and the second between *y* and *z*.

7.39 *kar-ry*^{402}*-ry*^{8}.

7.40 Chromosome 17.

Chapter 8

8.1 Recombination has occurred, producing wild-type or prototrophic bacteria.

8.2 Perform two experiments: (1) determine whether the process is sensitive to DNase, and (2) determine whether cell contact is required for the process to take place. The cell contact requirement can be tested by a U-tube experiment (see Fig. 8.1). If the process is sensitive to DNase, it is similar to transformation. If cell contact is required, it is similar to conjugation. If it is neither sensitive to DNase nor requires cell contact, it is similar to transduction.

8.3

RECOMBINATION PROCESS	AGENT MEDIATING DNA TRANSFER	SIZE OF DNA UNITS TRANSFERRED	STATE OF DONOR DNA IN RECOMBINANT CELL
Transformation	Active uptake of free DNA	Small (about 1/200 to 1/100 of a chromosome)	Single strand integrated (see Fig. 8.19)
Transduction	Bacteriophage	Small (about 1/100 to 1/50 of a chromosome)	Integrated into host chromosome (except autonomous in abortive transduction)
Sexduction	F factor	Variable (see Fig. 8.16)	Initially added to the host cell as separate plasmid; may undergo recombination with host chromosome to yield stable sexductant

8.4 F^- cells, no F factor present; F^+ cells, autonomous F factor; Hfr cells, integrated F factor (see Fig. 8.9). (b) F^+ and Hfr cells have F pili; F^- cells do not (see Fig. 8.8). (c) F^- cells are converted to F^+ cells by the conjugative transfer of F factors from F^+ cells; Hfr cells are formed when F factors in F^+ cells become integrated into the chromosomes of these cells.

8.5 F′ factors are useful for genetic mapping and, particularly, for *cis–trans* or complementation tests in bacteria (see Chapter 12). (b) F′ factors are formed by abnormal excision of F factors from Hfr chromosomes (see Fig. 8.15). (c) By the conjugative transfer of an F′ factor from a donor cell to a recipient (F^-) cell.

8.6 Generalized transduction: (1) transducing particles often contain only host DNA; (2) transducing particles may carry any segment of the host chromosome. Thus, all host genes are transduced. Specialized transduction: (1) transducing particles carry a recombinant chromosome, which contains both phage DNA and host DNA; (2) only host genes that are adjacent to the prophage integration site are transduced.

8.7 (a) A prophage is a phage chromosome that has become integrated into the host chromosome. The prophage is dormant in the sense that the phage genes involved in lytic development (replication and maturation) are repressed (see Chapter 14). The prophage replicates during host chromosome replication just as any other segment of the host chromosome. (b) No. (c) The mature (packaged) lambda chromosome and the lambda prophage are circular permutations of one another (see Fig. 8.4).

8.8 IS elements (or insertion sequences) are short (800–1400 nucleotide-pairs) DNA sequences that are transposable, that is, capable of moving from one position in a chromosome to another position or from one chromosome to another chromosome. (b) IS elements appear to mediate recombination between nonhomologous DNA molecules or sequences (so-called "illegitimate recombination"). (c) IS elements are implicated in the evolution of episomes and plasmids by mediating recombination between otherwise nonhomologous genetic elements.

8.9 By interrupting conjugation at various times after the donor and recipient cells are mixed (using a blender or other form of agitation), one can determine the length of time required to transfer a given genetic marker from an Hfr cell to an F^- cell. Since the chromosome is transferred in a linear sequence, the positions of genetic markers can be ordered relative to each other.

8.10 Cotransduction refers to the simultaneous transduction of two different genetic markers to a single recipient cell. Since bacteriophage particles can package only 1/100 to 1/50 of the total bacterial chromosome, only markers that are relatively closely linked can be cotransduced. The frequency of cotransduction of any two markers will be an inverse function of the distance between them on the chromosome. As such, this frequency can be used as an estimate of the linkage distance. Specific cotransduction-linkage functions must be prepared for each phage–host system studied.

8.11

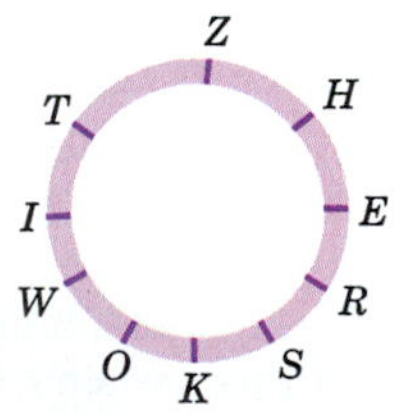

Z
T *H*
I *E*
W *R*
O *K* *S*

8.12 *lacY–lacZ–proC.*

8.13 *anth–A34–A223–A46.*

8.14 *anth–A487–A223–A58.*

8.15 *am* A453–*nrd*11–*nd*28–*am* M69–*am* N54.

Chapter 9

9.1 Same orientation leads to deletion; opposite orientation to inversion.

9.2 M cytotype and P transposase.

9.3 Transposon-induced mutation or loss of a *Bz* allele in the endosperm.

9.4 The pair in (d) are inverted repeats and could therefore qualify.

9.5 Through crossing-over between the LTRs of a *gypsy* element.

9.6 A transposon carrying the kanamycin resistance gene transposed from the original resistant cells to some of the infecting phage and subsequently conferred resistance on some of the lysogenized C107 cells produced by the second infection.

9.7 The pair in (a) are direct repeats and could therefore qualify.

Chapter 10

10.1 (a) RNA contains the sugar ribose, which has a hydroxyl (OH) group on the 2-carbon; DNA contains the sugar 2-deoxyribose, with hydrogens only on the 2-carbon. RNA usually contains the base uracil at positions where thymine is present in DNA. However, some DNAs contain uracil, and some RNAs contain thymine. DNA exists most frequently as a double helix (double-stranded molecule); RNA exists more frequently as a single-stranded molecule; but some DNAs are single-stranded, and some RNAs are double-stranded. (b) The main function of DNA is to store genetic information and to transmit that information from cell to cell and from generation to generation. RNA stores and transmits genetic information in some viruses that contain no DNA. In cells with both DNA and RNA, (1) mRNA acts as an intermediary in protein synthesis, carrying the information from DNA in the chromosomes to the ribosomes (sites at which proteins are synthesized), (2) tRNAs carry amino acids to the ribosomes and function in codon recognition during the synthesis of polypeptides, and (3) rRNA molecules are essential components of the ribosomes. (c) DNA is located primarily in the chromosomes (with some in cytoplasmic organelles, such as mitochondria and chloroplasts), whereas RNA is located throughout cells.

10.2 3′—ACGUCUGU—5′.

10.3 3′—GACTA—5′.

10.4 The genetic information of cells is stored in DNA, which is located predominantly in the chromosomes. The gene-products (polypeptides) are synthesized primarily in the cytoplasm on ribosomes. Some intermediate must therefore carry the genetic information from the chromosomes to the ribosomes. RNA molecules (mRNAs) were shown to perform this function by means of RNA pulse-labeling and pulse-chase experiments combined with autoradiography (see Fig. 10.9). The enzyme RNA polymerase was subsequently shown to catalyze the synthesis of mRNA using chromosomal DNA as a template. Finally, the mRNA molecules synthesized by RNA polymerase were shown to faithfully direct the synthesis of specific polypeptides when used in *in vitro* protein synthesis systems.

10.5 Proteins are long, chainlike molecules made up of amino acids linked together by peptide bonds. Proteins are composed of carbon, hydrogen, nitrogen, oxygen, and usually sulfur. They provide the enzymatic capacity and much of the structure of living organisms. DNA is composed of phosphate, the pentose sugar 2-deoxyribose, and four nitrogen-containing organic bases (adenine, cytosine, guanine, and thymine). DNA stores and transmits the genetic information in most living organisms. Protein synthesis is of particular interest to geneticists because proteins are the primary gene-products—the key intermediates through which genes control the phenotypes of living organisms.

10.6 Protein synthesis occurs on ribosomes. In eukaryotes, most of the ribosomes are located in the cytoplasm and are attached to the extensive membranous network of the endoplasmic reticulum (see Fig. 10.12). Some protein synthesis also occurs in cytoplasmic organelles such as chloroplasts and mitochondria.

10.7 Ribosomes are from 100 to 200 Å in diameter. They are located primarily in the cytoplasm of cells. In bacteria, they are largely free in the cytoplasm. In eukaryotes, many of the ribosomes are attached to the endoplasmic reticulum. Ribosomes are complex structures composed of over 50 different polypeptides and 3–5 different RNA molecules.

10.8 (a) The nucleus, specifically the nucleoli. (b) The cytoplasm.

10.9 Messenger RNA molecules carry genetic information from the chromosomes (where the information is stored) to the ribosomes in the cytoplasm (where the information is expressed during protein synthesis). The linear sequence of triplet codons in an mRNA molecule specifies the linear sequence of amino acids in the polypeptide(s) produced during translation of that mRNA. Transfer RNA molecules are small (about 80 nucleotides long) molecules that carry amino acids to the ribosomes and provide the codon-recognition specificity during translation. Ribosomal RNA molecules provide part of the structure and function of ribosomes; they represent an important part of the machinery required for the synthesis of polypeptides.

10.10 (a) Polysomes are formed when two or more ribosomes are simultaneously translating the same mRNA molecule. Ribosomes are usually spaced about 90 nucleotides apart on an mRNA molecule. Thus, polysome size is determined by mRNA size. (b) A ribosome, containing its three rRNA molecules, can participate in the synthesis of any polypeptide specified by the ribosome-associated mRNA. In that sense, rRNA is ***nonspecific.*** Messenger RNAs and tRNAs, in contrast, are ***specific,*** in directing the synthesis of a particular polypeptide or set of polypeptides (mRNA) or in attaching to a particular amino acid (tRNA). (c) Transfer RNA molecules are much smaller (about 80 nucleotides) than DNA or mRNA molecules. They are single-stranded molecules but have complex secondary structures because of the base-pairing between different segments of the molecules (see Figs. 10.15 and 10.16).

10.11 A specific aminoacyl–tRNA synthetase catalyzes the formation of an amino acid–AMP complex from the appropriate amino acid and ATP (with the release of pyrophosphate). The same enzyme then catalyzes the formation of the aminoacyl–tRNA complex, with the release of AMP. The amino acid–AMP and aminoacyl–tRNA linkages are both high-energy phosphate bonds.

10.12 (a) Synthetic RNA molecules (polyuridylic acid molecules) containing only the base uracil were prepared. When these synthetic molecules were used to activate *in vitro* protein synthesis systems, small polypeptides containing only the amino acid phenylalanine (polyphenylalanine molecules) were synthesized. Codons composed only of uracil were therefore shown to specify phenylalanine. Similar experiments were carried out using synthetic RNA molecules with different base compositions. (b) Better *in vitro* systems activated with synthetic RNA molecules with known repeating base sequences were developed. Ultimately, *in vitro* systems in which specific aminoacyl–tRNAs were shown to bind to ribosomes activated with specific mini-mRNAs, which were trinucleotides of known base sequence, were developed and used in codon identification.

10.13 (a) The genetic code is degenerate in that all but 2 of the 20 amino acids are specified by two or more codons. Some amino acids are specified by six different codons. The degeneracy occurs largely at the third or 3′ base of the codons. "Partial degeneracy" occurs where the third base of the codon may be either of the two purines or either of the two pyrimidines and the codon still specifies the same amino acid. "Complete degeneracy" occurs where the third base of the codon may be any one of the four bases and the codon still specifies the same amino acid. (b) The code appears to be almost completely universal. Known exceptions to universality include strains carrying suppressor mutations that alter the reading of certain codons (with low efficiencies in most cases) and the use of UGA as a tryptophan codon in yeast and human mitochondria.

10.14 Colinearity received strong support from studies that showed a direct correlation between the linear sequence of mutational sites in a gene (established by genetic mapping experiments) and the linear sequence of mutational defects (amino acid substitutions or chain terminations) in the polypeptide gene-product (established by purification and amino acid sequencing, etc., of mutant polypeptides). See Figs. 10.33 and 12.19.

10.15 Colinearity is an important prediction of our present concepts of transcription and translation. The presence of noncoding sequences ("introns" or "intervening sequences") within eukaryotic genes does not violate the concept of colinearity; their presence simply means that the colinear structures frequently contain interruptions.

10.16 Blueprints transcribed into building instructions and translated into structures composed of boards, bricks, and mortar by skilled craftsmen and craftswomen may be likened to DNA, mRNA, and tRNA functions in the assembly of amino acids into polypeptides by ribosomes and other required factors.

10.17 (a) Met ⟶ Val. This substitution occurs as a result of a transition. All other amino acid substitutions listed would require transversions.

10.18 His ⟶ Arg results from a transition; His ⟶ Pro would require a transversion (not induced by 5-bromouracil).

10.19 (a) By RNA polymerase binding to ***promoter sequences*** (see Chapter 14). (b) By specific ***terminator sequences*** (specific AT-rich sequences of DNA) that are recognized by RNA polymerase or the transcription termination protein ***rho.*** (c) By a complex reaction involving mRNA, ribosomes, initiation factors (IF-1, IF-2, and IF-3), GTP, the initiator codon AUG, and a special initiator tRNA ($tRNA_f^{Met}$). It also appears to involve a base-pairing interaction between a base sequence near the 3′ end of the 16S rRNA and a base sequence in the "leader sequence" of the mRNA. (d) By recognition of one or more of the chain-termination codons (UAG, UAA, and UGA) by the appropriate release factors (RF-1 and/or RF-2).

10.20 At least 813 nucleotides.

10.21 By changing the anticodon or by changing the sequence recognized by the aminoacyl-tRNA synthetase. Yes. Mutations in genes coding for aminoacyl-tRNA synthetases. Possibly also mutations in genes coding for certain components of the ribosome.

10.22 3′-TAC ACC-5′ (transcribed strand)
5′-ATGTGG-3′ (nontranscribed strand)

10.23 Incorporation of alanine into polypeptide chains.

10.24 *Amber* (UAG). This is the only nonsense codon that is related to tryptophan, serine, tyrosine, leucine, glutamic acid, glutamine, and lysine codons by a single base-pair substitution in each case.

10.25 (1) The promoters of eukaryotic and prokaryotic organisms are different. Thus, the eukaryotic genes will probably have to be joined to bacterial promoters. (2) Eukaryotic genes often contain introns. Bacteria do not have the enzymatic machinery needed to remove eukaryotic introns. Thus, unless the introns are "self-splicing," they will probably have to be removed from the eukaryotic genes before their insertion into bacterial plasmids in order to obtain expression in bacteria.

10.26 The entire nucleotide-pair sequences—including the introns—of the genes are transcribed by RNA polymerase to produce primary transcripts that still contain the intron sequences. The intron sequences are then "spliced out" of the primary transcripts to produce the mature, functional RNA molecules. In the case of protein-encoding nuclear genes of higher eukaryotes, the introns are "spliced out" by complex ribosomelike macromolecular structures called ***spliceosomes*** (see Fig. 10.26).

10.27 "Self-splicing" of RNA precursors demonstrates that RNA molecules can also contain catalytic sites; this property is not totally restricted to proteins.

10.28 Ribosomes and spliceosomes both play essential roles in gene expression, and both are complex macromolecular structures composed of RNA and protein molecules. Ribosomes are located in the cytoplasm, and spliceosomes in the nucleus. Ribosomes are larger and more complex than spliceosomes.

10.29 The introns of protein-encoding nuclear genes of higher eukaryotes almost invariably begin (5′) with GT and end (3′) with AG. In addition, the 3′ subterminal A in the "TACTAAC box" is completely conserved; this A is involved in bond formation during intron excision (see Fig. 10.25).

10.30 (a) Sequence 5. It contains the conserved intron sequences: a 5′ GU, a 3′ AG, and a UACUAAC internal sequence providing a potential bonding site for intron excision. Sequence 4 has a 5′ GU and a 3′ AG, but contains no internal A for the bonding site during intron excision. (b) 5′—UAGUCUCAA—3′; the putative intron from the 5′ GU through the 3′ AG has been removed.

Chapter 11

11.1 (a) Transition; (b) transition; (c) transversion; (d) transversion; (e) frameshift; (f) transition.

11.2 (a) *CℓB* method; (b) attached-X method (see Figs. 11.16 and 11.17).

11.3 Bacteria treated with a mutagen or expected to carry mutations may be introduced into media with particular drugs in appropriate concentrations. Colonies that appear have originated from cells carrying preexisting mutations for resistance. This may be verified by the replica-plating technique (see Fig. 11.1). The frequency of mutations in wild-type (drug-sensitive) cells to drug resistance can be measured in the presence or the absence of the drug.

11.4 Probably not. A human is larger than a bacterium, with more cells and a longer life span. If mutation frequencies are calculated in terms of cell generations, the rates for human cells and bacterial cells are similar.

11.5 A dominant mutation presumably occurred in the woman in whom the condition was first known.

11.6 The sex-linked gene is carried by mothers, and the disease is expressed in half of their sons. Such a disease is difficult to follow in pedigree studies because of the recessive nature of the gene, the tendency for the expression to skip generations in a family line, and the loss of the males who carry the gene. One explanation for the sporadic occurrence and tendency for the gene to persist is that by mutation, new defective genes are constantly being added to the load already present in the population.

11.7 Plants can be propagated vegetatively, but no such methods are available for widespread use in animals.

11.8 The sheep with short legs could be mated to unrelated animals with long legs. If the trait is expressed in the first generation, it could be presumed to be inherited and to depend on a dominant gene. On the other hand, if it does not appear in the first generation, F_1 sheep could be crossed back to the short-legged parent. If the trait is expressed in one-half of the backcross progeny, it might be presumed to be inherited as a simple recessive. If two short-legged sheep of different sex could be obtained, they could be mated repeatedly to test the hypothesis of dominance. In the event that the trait is not transmitted to the progeny that result from these matings, it might be considered to be environmental or dependent on some complex genetic mechanism that could not be identified by the simple test used in the experiments.

11.9 Enzymes may discriminate among the different nucleotides that are being incorporated. Some enzymes may utilize a higher proportion of incorrect nucleotides, whereas others may select fewer incorrect bases in DNA replication. In the case of the phage T4 DNA polymerase, the relative efficiencies of polymerization and proofreading by the polymerase's 3′ ⟶ 5′ exonuclease activity play key roles in determining the mutation rate.

11.10 If both mutators and antimutators operate in the same living system, an optimum mutation rate for a particular organism in a given environment may result from natural selection.

11.11 *Dt* is a mutator gene that induces somatic mutations in developing kernels.

11.12 (a) Yes. (b) A block would result in the accumulation of phenylalanine and a decrease in the amount of tyrosine, which would be expected to result in several different phenotypic expressions.

11.13 These hemoglobins can be distinguished by mobility of molecules in an electric field (electrophoretic mobility) and by determining the amino acid sequences of their β-polypeptides.

11.14

AMINO ACID	mRNA	DNA
Glutamic acid	—GAA→	—GAA→ ←CTT— ←Transcribed strand ↓ Mutation
Valine	—GUA→	—GTA→ ←CAT— Mutation
Lysine	—AAA→	—AAA→ ←TTT—

11.15 The label "molecular disease" became common in speaking of sickle-cell anemia because its molecular basis (the substitution of a valine residue for the glutamic acid residue at amino acid position number 6 in the β-chain) was recognized quite early during the emergence of the science of molecular biology. Actually, most if not all inherited diseases probably have very similar molecular bases. We just don't know what the molecular defects are in most instances.

11.16 Mutations: transitions, transversions, and frameshifts.

11.17 Irradiate the nonresistant strain and plate the irradiated organisms on a medium containing streptomycin. Those that survive and produce colonies are resistant. They could then be replicated to a medium without streptomycin. Those that survive would be of the first type; those that can live with streptomycin but not without it would be the second type.

11.18 3 percent; 4 percent; 6 percent.

11.19 Each quantum of energy from the X rays that is absorbed in a cell has a certain probability of hitting and breaking a chromosome. Hence, the greater the number of quanta of energy or dosage, the more likely that breaks will occur. The rate at which this dosage is delivered does not change the probability of each quantum inducing a break.

11.20 The person receiving a total of 100 r would be expected to have twice as many mutations as the one receiving 50 r.

11.21 During the replicating process, ultraviolent light produces mispairing alterations mostly in pyrimidines (e.g., cytosine to thymine transitions). Thymine may be altered to cytosine (or a modified pyrimidine with the base-pairing potential of cytosine), which pairs with guanine. A reverse mutation may occur when cytosine is changed to thymine (or a derivative of cytosine with the hydrogen-bonding potential of thymine), which pairs with adenine. The T-A base-pair may thus be changed to a C-G, and the reverse mutation may occur from C-G to T-A.

11.22 Nitrous acid brings about a substitution of an OH group for an NH_2 group in those bases (A, C, and G) having NH_2 side groups. In so doing, adenine is converted to hypoxanthine, which base-pairs with cytosine, and cytosine is converted to uracil, which base-pairs with adenine. The net effects are GC ⟷ AT base-pair substitutions (see Fig. 11.27).

11.23 Transitions.

11.24 Nitrous acid acts as a mutagen on either replicating or nonreplicating DNA and produces transitions from A to G or C to T, whereas 5-bromouracil does ***not*** affect nonreplicating DNA but acts during the replication process to cause GC ⟷ AT transitions. 5-Bromouracil must be incorporated into DNA during the replication process to induce mispairing of bases and thus mutations.

11.25 Large numbers of seeds have been given massive X-ray doses and screened to see if induced mutations have improved any of the resulting plants for desired traits in particular environments. In one study of oats, a few seeds produced plants with enhanced disease resistance.

11.26 The immediate effects are surface and internal burns. Irradiation is known to induce leukemia and, in general, to shorten the life span—an effect that is difficult to evaluate. Genetic effects are insidious; small doses produce mutations and have a cumulative effect. The number of mutations generally is proportional to total dosage. This is a potential danger to future generations. Clearly, great care should be taken to prevent humans from being exposed to increased levels of irradiation.

11.27 Mutations induced by acridine dyes are primarily insertions or deletions of single base-pairs. Such mutations alter the reading frame (the in-phase triplets specifying mRNA codons) for that portion of the gene distal (relative to the direction of transcription and translation) to the mutation (see Fig. 11.14). This would be expected to totally change the amino acid sequences of polypeptides distal to the mutation site and produce inactive polypeptides. In addition, such frameshift mutations frequently produce in-frame termination codons that result in truncated proteins.

11.28 Proline and serine.

11.29 No. Leucine ⟶ proline would occur more frequently. Leu (CUA) $\xrightarrow{\text{5-BU}}$ Pro (CCA) occurs by a single base-pair transition, whereas Leu (**CU**A) $\xrightarrow{\text{5-BU}}$ Ser (**UC**A) requires two base-pair transitions. Recall that 5-bromouracil (5-BU) induces only transitions (see Figs. 11.25 and 11.26).

11.30 No. 5-Bromouracil is mutagenic only to replicating nucleic acids.

11.31 Yes.

DNA:	←GGX— —CCX′→	$\xrightarrow{HNO_2}$	←GGX— —UCX′→
	↓		↓
mRNA:	GGX		AGX
	↓		↓
Polypeptide:	Gly		Ser or Arg
	or	$\xrightarrow{HNO_2}$	←GGX— —CUX′→
			↓
			GAX
			↓
			Asp or Glu (depending on X)

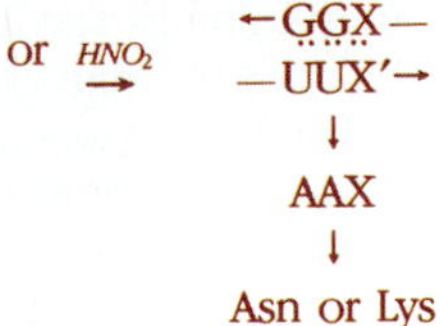

Note: The X at the third position in each codon in mRNA and in each triplet of base-pairs in DNA refers to the fact that there is complete degeneracy at the third base in the glycine codon. Any base may be present in the codon, and it will still specify glycine.

11.32 No. The glycine codon is GGX, where X can be any one of the four bases. Because of this complete degeneracy at the third position of the glycine codon, changing X to any other base will have no effect (i.e., the codon will still specify glycine). Nitrous acid deaminates guanine (G) to xanthine, but xanthine still base-pairs with cytosine. Thus guanine is not a target for mutagenesis by nitrous acid.

11.33 Tyr ⟶ Cys substitutions; Tyr to Cys requires a transition, which is induced by nitrous acid. Tyr to Ser would require a transversion, and nitrous acid is not expected to induce transversions.

11.34 (b) Met ⟶ Thr. 5-Bromouracil induces transitions, not transversions. All other changes listed require transversions.

11.35 5′-AUGCCCUUUGGG**GAAAGG**UUUCCCUAA-3′.

Chapter 12

12.1 Prior to 1940, the gene was considered a "bead-on-a-string," not subdivisible by recombination or mutation. Today, the gene is considered to be the unit of genetic material that codes for one polypeptide. The unit of structure, not subdivisible by recombination or mutation, is known to be the single nucleotide-pair.

12.2 The recombination observed between lz^s and lz^g, two functionally allelic mutations at the *lozenge* locus of *Drosophila.*

12.3 The *cis–trans* test, which defines the unit of genetic material specifying the amino acid sequence of one polypeptide.

12.4 They provide powerful selective sieves for identifying rare recombinants. This is accomplished by using the restrictive environmental conditions to select wild-type recombinant progency from crosses between pairs of conditional lethal mutants.

12.5 7-5-6-4-1-2-3.

12.6 Two genes; mutations 1, 2, 3, 4, 5, 6, and 8 are in one gene; mutation 7 is in a second gene.

12.7 Four genes; mutations 1 and 2 in one gene; mutations 3 and 4 in a second gene; mutations 5 and 6 in a third gene; mutations 7 and 8 in a fourth gene.

12.8 The size of the gene (assuming that all nucleotide-pairs in the gene are capable of undergoing base-pair substitutions, as seems highly probable). Dominant lethal alleles and recessive lethal alleles in haploids will (under normal conditions) exist only transiently, of course.

12.9 Homoalleles are structurally and functionally allelic; they are not separable by recombination. Heteroalleles are functionally allelic (based on *cis–trans* tests), but are structurally nonallelic (based on recombination tests). Heteroalleles thus result from mutations occurring at different sites within a gene.

12.10 (1) Cross the two white-flowered varieties. The F_1 plants will be *trans*-heterozygotes. If the F_1 plants have white flowers, the two varieties probably carry mutations in the same gene, causing white flowers. (2) Cross white-flowered varieties with red-flowered varieties and self-pollinate or intercross the F_1 plants. If alleles of a single gene are involved, monohybrid F_2 ratios should be observed in all cases.

12.11 In most cases, the positions of loci on chromosomes can be determined only by genetic mapping experiments. Such experiments require at least one mutant allele of a gene, so that the segregation of the pairs of alleles can be followed in crosses. The recombinant DNA, cloning, restriction endonuclease mapping, and nucleic acid sequencing techniques developed in the last few years now allow some wild-type genes to be analyzed and localized in the chromosome without genetic mapping data.

12.12 (a) The amino acid sequences of polypeptides that are specified by overlapping genes or "genes-within-genes" would not be able to evolve independently. However, the degeneracy of the code (see Chapter 10) will allow for some independence. In addition, a single mutation within a shared nucleotide sequence could result in nonfunctional products for both genes. (b) 3. (c) 6.

12.13 (a) Five genes. (b) Mutations 1, 3, and 5 are in one gene; mutations 7 and 8 are in a second gene; mutations 2, 4, and 6 identify genes 3, 4, and 5, respectively.

12.14 Yes. The female progeny are *trans*-heterozygotes. The results indicate that *white* and *cherry* are functionally allelic.

12.15 (a) True. (b) False. (c) True. (d) True. (e) True.

12.16 The observed complementation between ry^2 and ry^{42} is **intra**genic complementation. Xanthine dehydrogenase is a dimeric protein, and dimers that contain one polypeptide encoded by the ry^2 allele and one polypeptide encoded by the ry^{42} allele are partially active. Presumably, the wild-type segment of the ry^2 polypeptide somehow stabilizes the mutant segment of the ry^{42} polypeptide, and vice versa, yielding a dimer with enzymatic activity.

12.17 *Am* mutations result in UAG chain-termination codons within the coding sequence of the mRNA product of a gene; they thus produce truncated polypeptide gene-products. Since all *am* mutant alleles of a gene will produce polypeptides lacking the COOH terminus, they would not be expected to exhibit intragenic complementation except in very rare cases. In contrast, most *ts* mutations are caused by missense mutations that change the amino acid sequence of the polypeptide gene-product, making it more heat-labile. However, most *ts* mutant alleles produce a complete, although altered, gene-product. As a result, *ts* mutant alleles often exhibit intragenic complementation when the active form of the protein gene-product is a homomultimer. For this reason, *cis–trans* tests carried out with *am* mutants, not *ts* mutants, have been used whenever possible to operationally define the genes of phage T4.

12.18 (a) The seven *sus* mutants are located in three different genes. (b) Mutant strains 1, 3, and 7 contain mutations in one gene; 4, 5, and 6 carry mutations in a second gene; 2 has a mutation in a third gene.

12.19 Three: (1) A/T ⟶ T/A, (2) A/T ⟶ G/C, and (3) A/T ⟶ C/G.

12.20

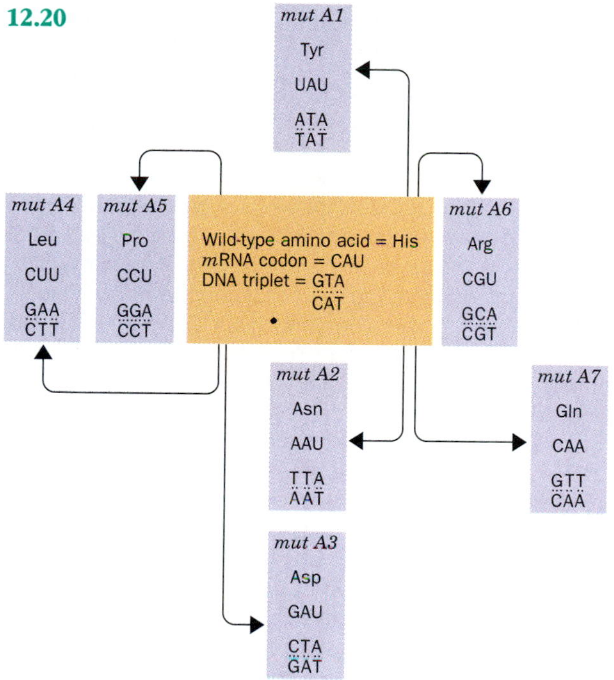

DEDUCTIONS

(a) Wild-type His codon must be CAU based on DNA sequence of gene given.

(b) The codons for the seven amino acids found at position 6 in the mutant polypeptides must be connected to CAU by a single base change since the mutants were all derived from wild type by a single nucleotide-pair substitution. Thus, the degeneracy of the genetic code is not a factor in deducing specific codon assignments.

(c) Because of the nature of the genetic code—specifically the degeneracy at the third (3′) position in each codon—there are three possible amino acid substitutions due to single base substitutions at each of the first two positions (the 5′ base and the middle base), but only one possible amino acid change due to a single base change at position 3 (the 3′ base in the codon).

Note: For ease of discussion, the three nucleotide-pair positions in the triplet of nucleotide-pairs in the gene specifying the CAU codon in mRNA and the amino acid His in the wild-type polypeptide will be referred to as position 1 (corresponding to the 5′ base in the codon), position 2 (the middle nucleotide-pair), and position 3 (corresponding to the 3′ base in the mRNA codon).

(d) Since *A*1, *A*2, and *A*3 do not recombine with each other, they must all result from base-pair substitutions at the same position in the triplet, at either position 1 or position 2. The same is true for *A*4, *A*5, and *A*6. Since *A*7 recombines with each of the other six mutant alleles, it must result from the single base-pair substitution at position 3 that leads to an amino acid change.

(e) The only amino acid with a codon connected to the His codon CAU by a single base change at position 3 is Gln (codon CAA). Thus, the *mut A*7 polypeptide must have glutamine as the sixth amino acid.

(f) Since *mut A*7 (the third position substitution) yields about twice as many wild-type recombinants in crosses with *mut A*1 as in crosses with *mut A*6, the *A*1 substitution must be at position 1 and the *mut A*6 substitution must be at position 2. Combined with (d) above, this places the *A*2 and *A*3 substitutions at position 1 and the *A*4 and *A*5 substitutions at position 2.

(g) Since *mut A*1 and *mut A*6 are induced to revert to wild type by 5-BU, they must be connected to the His triplet of nucleotide-pairs by transition mutations, that is,

$$(mut\ A1)\ \frac{ATA}{TAT} \xleftrightarrow{\text{5-BU}} \frac{GTA}{CAT} \xleftrightarrow{\text{5-BU}} \frac{GCA}{CGT}\ (mut\ A6)$$

(h) Since *mut A*3 and *mut A*5 are induced to mutate to *mut A*2 and *mut A*4, respectively, by hydroxylamine, *A*3 must be connected to *A*2 and *A*5 to *A*4 specifically by GC ⟶ AT transitions, that is,

$$(mut\ A3)\ \frac{CTA}{GAT} \xrightarrow{\text{HA}} \frac{TTA}{AAT}\ (mut\ A2)$$

and

$$(mut\ A5)\ \frac{GGA}{CCT} \xrightarrow{\text{HA}} \frac{GGA}{CTT}\ (mut\ A4)$$

Chapter 13

13.1 (a) Both introduce new genetic variability into the cell. In both cases, only one gene or a small segment of DNA representing a small fraction of the total genome is changed or added to the genome. The vast majority of the genes of the organism remain the same. (b) The introduction of recombinant DNA molecules, if introducing genes from a very different species, is more likely to result in a novel, functional gene-product in the cell, ***if the introduced gene (or genes) is capable of being expressed in the foreign protoplasm***. The introduction of recombinant DNA molecules is more analogous to duplication mutations (see Chapter 18) than to other types of mutations.

13.2 This is a wide-open question at present! There is much speculation (see p. 359), but little hard evidence. In one yeast mitochondrial gene, the introns contain open reading frames that encode "maturases" that splice out these introns—a neat negative feedback control. Other introns may be merely relics of evolution.

13.3 Recombinant DNA and gene cloning techniques allow geneticists to isolate essentially any gene or DNA sequence of interest and to characterize it structurally and functionally. Large quantities of a given gene can be obtained in pure form, which permits one to determine its nucleotide-pair sequence (to "sequence it" in common lab jargon). From the nucleotide sequence and our knowledge of the genetic code, geneticists can predict the amino acid sequence of any polypeptide encoded by the gene. By using an appropriate subclone of the gene as a hybridization probe in northern blot analyses, geneticists can identify the tissues in which the gene is expressed. On the basis of the predicted amino acid sequence of a polypeptide encoded by a gene, geneticists can synthesize oligopeptides and use these to raise antibodies that, in turn, can be used to identify the actual product of the gene and localize it within cells or tissues of the organism. Thus recombinant DNA and gene cloning technologies provide very powerful tools with which to study the genetic control of essentially all biological processes. These tools have played major roles in the explosive progress in knowledge in the field of biology during the last two decades.

13.4 The nucleotide-pair sequence. Restriction endonucleases recognize a specific nucleotide-pair sequence in DNA regardless of the source of the DNA. In most cases, this is a 4- or 6-nucleotide-pair sequence; in a few cases, the recognition sequence is longer (e.g., 8 nucleotide-pairs). Most restriction enzymes cleave the two strands of the DNA at a specific position (between the same two adjacent nucleotides in each strand) within the recognition sequence. A few restriction enzymes bind at a specific recognition sequence, but cut the DNA at a nearby site outside of the recognition sequence. Some restriction endonucleases cut both strands between the same two nucleotide-pairs ("blunt end" cutters), whereas others cut the two strands at different positions and yield complementary single-stranded ends ("sticky or staggered end" cutters). See Table 13.1 for examples.

13.5 Restriction endonucleases are believed to provide a kind of primitive "immune system" to the microorganisms that produce them by protecting their genetic material from "invasion" by foreign DNAs from viruses or other pathogens or just DNA in the environment that might be taken up by the microorganism. Obviously, these microorganisms do not have a sophisticated immune system like that of higher animals (Chapter 16).

13.6 Microorganisms that produce restriction endonucleases also produce enzymes that modify one or more bases in the recognition sequence for that endonuclease so that it can no longer cleave the DNA at that site. In most cases, the modifying enzyme is a methylase that attaches a methyl group to one or more of the bases in the recognition sequence. For example, *E. coli* strains that produce the restriction endonuclease *Eco*RI also produce *Eco*RI methylase, an enzyme that transfers a methyl group from *S*-adenosylmethionine to the 3′ adenine residue in each strand of the recognition sequence (5′-GAATCC-3′), thus producing N^6-methyladenines at these positions. *Eco*RI can ***not*** cleave DNA that contains N^6-methyladenine at these positions even if the *Eco*RI recognition sequence is present in this DNA. Thus, if one wishes to digest DNA with *Eco*RI, that DNA must ***not*** be isolated from an *E. coli* strain that is producing *Eco*RI methylase.

13.7 A foreign DNA cloned using an enzyme that produces single-stranded complementary ends can always be excised from the cloning vector by cleavage with the same restriction enzyme that was originally used to clone it. For example, if a *Hin*dIII fragment from the human genome is cloned into *Hin*dIII-cleaved pUC119, the human *Hin*dIII fragment can be excised from a plasmid DNA preparation of this clone by cleavage with restriction endonuclease *Hin*dIII. The human *Hin*dIII fragment will be flanked in the recombinant plasmid DNA clone by *Hin*dIII cleavage sites. When terminal transferase is used to add complementary single-stranded ends during cloning, the original restriction endonuclease cleavage sites are destroyed. Thus, the restriction enzyme used to generate the fragment for cloning can ***not*** be used to excise the original fragment from the cloning vector.

13.8 Maize genomic DNA would be purified and digested with *Eco*RI. Plasmid vector DNA would be similarly purified and digested with *Eco*RI. The *Eco*RI-cut plasmid DNA would then be subjected to alkaline phosphatase digestion, which will remove the terminal phosphate groups and prevent recircularization of nonrecombinant plasmid DNA molecules during the subsequent ligation step. The maize *Eco*RI restriction fragments and the *Eco*RI-cut plasmid DNA molecules will now all have complementary single-stranded ends (5′-AATT-3′). The maize restriction fragments would next be mixed with the *Eco*RI-cut, phosphatased plasmid molecules and covalently inserted into the linearized vector molecules in an ATP-dependent ligation reaction catalyzed by the enzyme DNA ligase. The ligation reaction will produce circular recombinant plasmids containing maize *Eco*RI fragment inserts. Insertion of maize DNA fragments into the *Eco*RI site of the plasmid disrupts the tet^r gene so that the resulting recombinant plasmids will no longer confer tetracycline resistance to host cells. Amp^s tet^s *E. coli* cells would then be transformed with the recombinant plasmid DNAs, and the transformed (and nontransformed) cells would be plated on medium containing ampicillin to select for transformed cells harboring plasmids. The majority of the cells plated will ***not*** be transformed; however, these nontransformed cells will ***not*** be able to grow in the presence of ampicillin. The cells that grow on ampicillin-containing medium would then be pooled and frozen at −80°C in 20 percent glycerol for subsequent screening. This collection of cells harboring different *Eco*RI fragments of the maize genome represents a clone library that should contain clones with intact *gln2* genes since the *gln2* gene contains no *Eco*RI cleavage site. Note that the *Hin*dIII site of the vector could be used to construct a similar maize genomic *Hin*dIII fragment library, but such a library would not contain intact *gln2* genes because of the *Hin*dIII cleavage site in the *gln2* gene.

13.9 Most genes of higher plants and animals contain noncoding intron sequences. These intron sequences will be present in genomic clones, but not in cDNA clones since cDNAs are synthesized using mRNA templates and intron sequences are removed during the processing of the primary transcripts to produce mature mRNAs.

13.10 Higher eukaryotes have very large genomes; for example, the genomes of mammals contain approximately 3×10^9 nucleotide pairs. Thus, trying to identify a particular single-copy gene from a clone library is like looking for the proverbial "needle in a haystack." To accomplish this, one needs a nucleic acid hybridization probe specific for the gene or an antibody probe specific for the gene-product.

Given a specific cell or tissue type producing the mRNA and/or the protein gene-product in large amounts, it was relatively easy to obtain pure mRNA or pure protein to use in making a hybridization or antibody probe, respectively, with which to screen a library for the gene or cDNA of interest. These approaches are much more difficult for the majority of the genes that encode products that represent only a small proportion of the total gene-products in any given cell type.

13.11 The maize *gln2* gene contains many introns, and one of the introns contains an *Hin*dIII cleavage site. The intron sequences (and thus the *Hin*dIII cleavage site) are ***not*** present in mRNA sequences and thus are also ***not*** present in full-length *gln2* cDNA clones.

13.12 By oligonucleotide-directed site-specific mutagenesis, you can change each of the eight nucleotide-pairs in the protein binding site to each of the other three possible nucleotide-pairs. You can then examine the binding of the regulatory protein to each of the mutant sequences (24 single nucleotide-pair changes are possible at the eight positions), and you can examine the ability of each of the mutant sequences to mediate induction of transcription of the gene at 45°C *in vivo*. You could study the binding of the protein using synthetic oligonucleotide sequences more easily, but that approach would not let you study induction of transcription of the gene *in vivo*. The site-specific mutagenesis approach will allow you to carry out a saturation mutagenesis analysis of the octameric regulatory protein binding sequence.

13.13 The PCR technique has much greater ***sensitivity*** than any other method available for analyzing nucleic acids. Thus, PCR procedures allow one to analyze nucleic acid structure given extremely minute amounts of starting material. DNA sequences can be amplified and structurally analyzed from very small amounts of a tissue like blood or sperm in assault and rape cases. In addition, PCR protocols permit investigators to detect the presence of rare gene transcripts (e.g., in specific types of cells) that could not be detected by less sensitive procedures such as northern blot analyses or *in situ* hybridization studies.

13.14 Some bacterial viruses package single-stranded DNA rather than double-stranded DNA. These bacteriophages employ a double-stranded DNA intermediate for replication, but then switch to the production of single-stranded DNAs for packaging during phage maturation (see Chapter 5, Fig. 5.32, for details). When a foreign DNA segment is cloned into the replicative form of the chromosome of one of these phages, only one of the two strands gets packaged during the subsequent maturation processes. In the case of the filamentous single-stranded DNA phages such as M13 and f2, the progeny phage particles are simply extruded through the cell membrane and wall without lysing the host bacterium. Because the phage are so small relative to the size of the bacteria, the phage particles can be separated from the bacteria by a simple low-speed centrifugation step. The bacteria form a pellet at the bottom of the centrifuge tube; the phage remain in the supernatant suspension. The bacterial pellets are discarded, and the phage particles can be collected by high-speed centrifugation or by precipitation with polyethylene glycol and low-speed centrifugation. Pure single-stranded DNA can then be separated from the phage proteins by phenol–chloroform extractions. Many cloning vectors like pUC118 and pUC119 (see Fig. 13.18) are phage–plasmid hybrids. They have both plasmid and M13 phage origins of replication as well as M13 phage packaging signals so that they can replicate either as plasmids or as phages and can package single-stranded phage DNA when in the phage mode of replication. The switch from the plasmid mode of replication to the phage mode is accomplished by superinfecting the host bacteria with a mutant "helper" phage.

13.15 All modern cloning vectors contain a "polycloning site"—a cluster of cleavage sites for a number of different restriction endonucleases in a nonessential region of the vector into which foreign DNAs can be inserted. In general, the greater the complexity of the polycloning site, that is, the more restriction endonuclease cleavage sites that are present, the greater the utility of the vector for cloning a wide variety of different restriction fragments. For example, see the polycloning site present in pUC118 and pUC119 shown in Fig. 13.18.

13.16 Transcription vectors permit one to synthesize high-specific-activity radioactive RNA transcripts for use as hybridization probes in northern blot and *in situ* hybridization studies. In addition, *in vitro* translation systems can be coupled to the *in vitro* transcription systems and used to synthesize radioactively labeled polypeptide gene-products encoded by the cloned gene or cDNA. These polypeptide products can be analyzed by polyacrylamide gel electrophoresis and autoradiography. In some cases, this approach can be used to identify the product of the cloned gene or cDNA.

13.17 Restriction endonuclease *Hpa*I produces fragments with blunt ends. The simplest way to clone the desired *Hpa*I fragment into the *Hin*dIII site of pUC119 would be to use *Hin*dIII linkers. To obtain the desired clone, the *Hpa*I fragment of interest would be isolated by agarose gel electrophoresis, extracted from the agarose in the slice cut out of the gel, ligated to *Hin*dIII linkers, digested with *Hin*dIII restriction enzyme, separated from the remaining linker fragments by agarose gel electrophoresis and reextraction, and ligated into *Hin*dIII-cut pUC119. *Amp*^{s} *E. coli* cells would then be transformed with the ligation products and plated on medium containing ampicillin and X-gal (5-bromo-4-chloro-3-indolyl-β-D-galactoside). Only bacteria containing pUC119 plasmids will be able to grow on the ampicillin-containing medium. Two kinds of colonies will be present: blue colonies that harbor pUC119 plasmids with no foreign DNA inserts and white colonies that harbor plasmids with inserts. A white colony should be used to inoculate rich broth medium containing ampicillin to grow a culture of cells containing the desired clone. Plasmid DNA should then be isolated from the resulting cell culture and used to verify the presence of the *Hpa*I fragment of interest. The entire procedure can be carried out in two or three days.

13.18 The name "western blot" has no literal significance; it is pure laboratory jargon. In 1975, E. M. Southern published a procedure for the transfer of DNA molecules that had been separated by agarose gel electrophoresis to nitrocellulose membranes and their subsequent detection by hybridization to radioactive probes and autoradiography. The resulting autoradiographs showing the positions of the DNA bands that had hybridized to the probes were called Southern blots in reference to E. M. Southern, who had developed the technique. When similar procedures were developed for analyzing RNA molecules that had been separated by gel electrophoresis, researchers started calling them northern blots because of their similarity to Southern blots. When the methodology was extended to proteins, the term western blot supposedly was the logical extension of the Southern/northern laboratory jargon. Or was it? Why wasn't the western blot called an eastern blot?

13.19 Because the nucleotide-pair sequences of both the normal CF gene and the CFΔ508 mutant gene are known, labeled oligonucleotides can be synthesized and used as hybridization probes to detect the presence of each allele (normal and Δ508). Under high stringency hybridization conditions, each probe will hybridize only with the CF allele that exhibits perfect complementarity to itself. Since the sequences of the CF gene flanking the Δ508 site are known, oligonucleotide PCR primers can be synthesized and used to amplify this segment of the DNA obtained from small tissue explants of putative CF patients and their relatives by PCR. The amplified DNAs can then be separated by agarose gel electrophoresis, transferred to nylon membranes, and hybridized to the respective labeled oligonucleotide probes, and the presence of each CF allele detected by autoradiography. For a demonstration of the utility of this procedure, see B. Kerem *et al.*, "Identification of the Cystic Fibrosis Gene: Genetic Analysis," *Science* 245: 1073–1080, 1989. Kerem and coworkers used two synthetic oligonucleotide probes (oligo-N = 3′-CTTTTATAGTAGAAACCAC-5′ and oligo-ΔF = 3′-TTCTTTTATAGTA--ACCACAA-5′; the dashes indicate the deleted nucleotides in the CFΔ508 mutant allele) to analyze the DNA of CF patients and their parents. For confirmed CF families, the results of these Southern blot hybridizations with the oligo-N (normal) and oligo-ΔF (CFΔ508) labeled probes were often as follows:

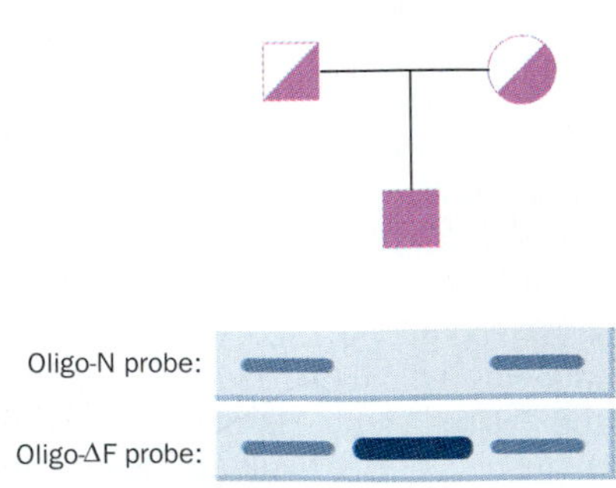

Both parents were heterozygous for the normal CF allele and the mutant CFΔ508 allele as would be expected for a rare recessive trait, and the CF patient was homozygous for the CFΔ508 allele. In such families, one-fourth of the children would be expected to be homozygous for the Δ508 mutant allele and exhibit the symptoms of CF, whereas three-fourths would be normal (not have CF). However, two-thirds of these normal children would be expected to be heterozygous and transmit the allele to their children. Only one-fourth of the children of this family would be homozygous for the normal CF allele and have no chance of transmitting the mutant CF gene to their offspring. Note that the screening procedure described here can be used to determine which of the normal children are carriers of the CFΔ508 allele, that is, the mutant gene can be detected in heterozygotes as well as homozygotes.

13.20 You could attempt to isolate either a dihydropicolinate synthase (DHPS) cDNA clone or a DHPS genomic clone. Once you have isolated either DHPS clone, it can be used as a hybridization probe to isolate the other (genomic or cDNA) by screening an appropriate library by *in situ* colony hybridization. Four approaches that have proven effective in isolating other eukaryotic coding sequences of interest are the following. (1) You could obtain a clone of the DHPS gene of a lower eukaryote (a clone of the DHPS gene of *Saccharomyces cerevisiae* is available) or even a prokaryote and use it as a heterologous hybridization probe to screen a maize cDNA library using low stringency conditions. Sometimes this approach is successful; sometimes it is not successful. Whether or not this approach works depends on how similar the coding sequences of the specific gene of interest are in the two species. (2) You could purify the DHPS enzyme from corn and use the purified protein to produce an antibody to DHPS. This DHPS-specific antibody could then be used to screen a maize cDNA expression library by a protocol analogous to the western blot procedure. (An expression library contains the cDNA coding sequences fused to appropriate transcription and translation signals so that they are expressed in *E. coli* or other host cells in which the cDNA library is prepared.) (3) You could purify the DHPS enzyme from maize and determine the amino acid sequence of its NH_2 terminus by microsequencing techniques. From the amino acid sequence and the known genetic code, you could predict the possible nucleotide sequences encoding this segment of the protein. Because of the degeneracy in the code, there would be a set of nucleotide sequences that would all specify the same amino acid sequence, and you would not know which one was present in the maize DHPS gene. However, the synthesis of oligonucleotides is now routine and quite inexpensive. Thus, you could synthesize a mixture of oligonucleotides containing all possible coding sequences and use this mixture as a set of hybridization probes to screen an appropriate library by *in situ* colony hybridization. (4) Finally, you might try a simple and very quick genetic approach based on the ability of cDNAs in an expression library to rescue DHPS mutants of *E. coli* or other species that can be transformed at high frequencies. You would obtain a DHPS-deficient mutant of *E. coli* (available from the *E. coli* Genetics Stock Center at Yale University), transform it with your cDNA expression library, and plate the transformed cells on medium lacking diaminopimelic acid (the product of DHPS). DHPS-deficient *E. coli* mutants can ***not*** grow in the absence of diaminopimelic acid; thus, any colonies that grow on your selection plates should be the result of rescue of the DHPS mutant bacteria by corn DHPS encoded by cDNAs in the library. This entire screening procedure can be carried out in three or four days; thus, it is much simpler than the preceding approaches. In fact, D. A. Frisch has recently isolated a maize DHPS cDNA by this simple but powerful genetic approach. However, note that this approach would only be expected to work in the case of enzymes that are active as monomers or homomultimers; it is not applicable when the active form of the enzyme is a heteromultimer.

13.21 (a) Southern, northern, and western blot procedures all share one common step, namely, the transfer of macromolecules (DNAs, RNAs, and proteins, respectively) that have been separated by gel electrophoresis to a solid support—usually a nitrocellulose or nylon membrane—for further analysis. (b) The major difference between these techniques is the class of macromolecules that are separated during the electrophoresis step: DNA for Southern blots, RNA for northern blots, and protein for western blots.

13.22 No. The genome of the species in question may contain one, two, or three copies of the gene (or family of closely related genes) encoding this protein. The possibilities are: (1) one copy of the gene with two *Eco*RI cleavage sites located within intron sequences; (2) two copies of the gene with one *Eco*RI cleavage site located within an intron sequence of one of the copies; and (3) three copies of the gene with no *Eco*RI cleavage site in any of the copies, that is, each copy present on a single *Eco*RI restriction fragment.

13.23

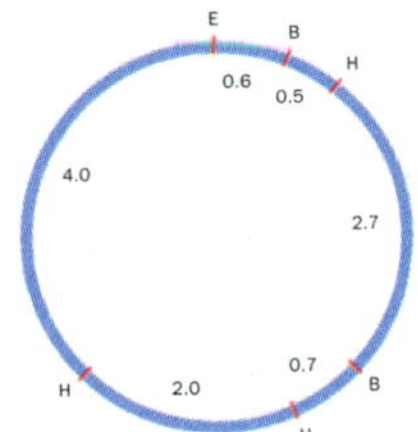

Restriction enzyme cleavage sites for *Bam*HI, *Eco*RI, and *Hin*dIII are denoted by B, E, and H, respectively. The numbers give distances in kilobase-pairs.

13.24

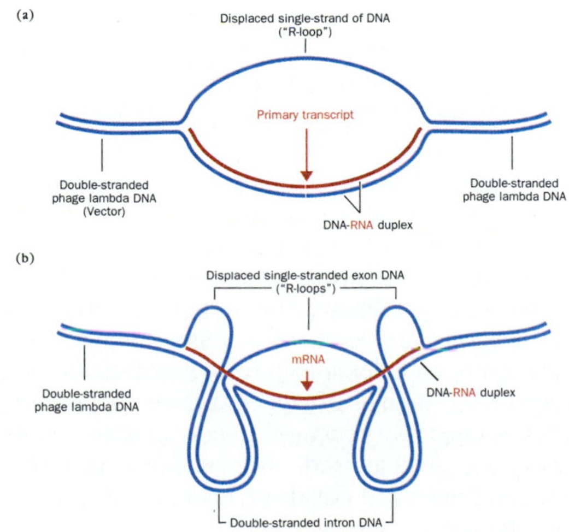

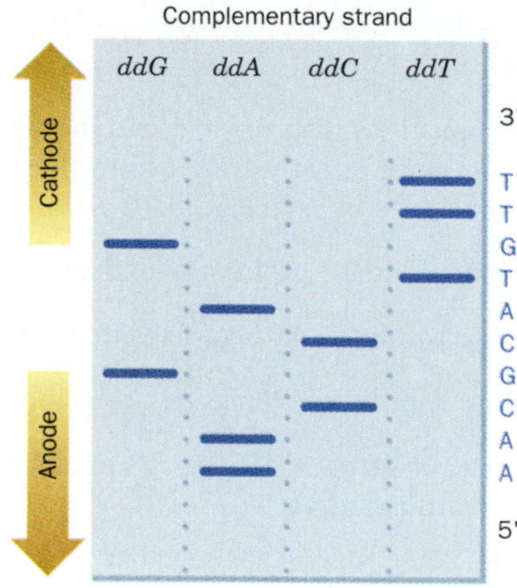

Chapter 14

14.1 By studying the synthesis or lack of synthesis of the enzyme in cells grown on chemically defined media. If the enzyme is synthesized only in the presence of a certain metabolite or a particular set of metabolites, it is probably inducible. If it is synthesized in the absence but not in the presence of a particular metabolite or group of metabolites, it is probably repressible.

14.2 Repression occurs at the level of transcription during enzyme synthesis. The end product, or a derivative of the end product, of a repressible system acts as an effector molecule that usually, if not always, combines with the product of one or more regulator genes to turn off the ***synthesis*** of the enzymes in the biosynthetic pathway. Feedback inhibition occurs at the level of enzyme ***activity***; it usually involves the first enzyme of the biosynthetic pathway. Feedback inhibition thus brings about an immediate arrest of the biosynthesis of the end product. Together, feedback inhibition and repression rapidly and efficiently turn off the synthesis of both the enzymes and the end products that no longer need to be synthesized by the cell.

14.3

GENE OR REGULATORY ELEMENT	FUNCTION
(1) Regulator gene	Codes for repressor
(2) Operator	Binding site of repressor
(3) Promoter	Binding site of RNA polymerase and CAP–cAMP complex
(4) Structural gene *z*	Encodes β-galactosidase
(5) Structural gene *y*	Encodes β-galactoside permease

14.4 (1) Constitutive synthesis of the *lac* enzymes. (2) Constitutive synthesis of the *lac* enzymes. (3) Uninducibility of the *lac* enzymes. (4) No β-galactosidase activity. (5) No β-galactoside permease activity.

14.5 (a) 1, 2, 3*, and 5; (b) 2, 3*, and 5; 3* may be either noninducible or constitutive depending on whether the specific o^c mutation eliminates binding of the i^s "superrepressor."

14.6 (a) Inducible; (b) constitutive; (c) constitutive; (d) inducible; (e) constitutive.

14.7 (a) $\dfrac{i^+\ o^c\ z^+\ y^-}{i^+\ o^+\ z^-\ y^+}$; (b) $\dfrac{i^s\ o^c\ z^+\ y^-}{i^s\ o^+\ z^-\ y^+}$

14.8 (a) The o^c mutations map very close to the *z* structural gene; i^- mutations map slightly farther from the structural gene (but still very close by; see Fig. 14.7).

(b) An $\dfrac{i^+\ o^c\ z^+\ y^+}{i^+\ o^+\ z^+\ y^+}$ partial diploid would exhibit constitutive synthesis of β-galactosidase and β-galactoside permease; an $\dfrac{i^+\ o^+\ z^+\ y^+}{i^-\ o^+\ z^+\ y^+}$ partial diploid would be inducible for the synthesis of these enzymes. (c) The o^c mutation is *cis*-dominant; i^- is *trans*-recessive.

14.9 The system could have developed from a series of tandem duplications of a single ancestral gene. Mutational changes that make the system more efficient and, therefore, favored in selection could have brought the system to its present level of efficiency.

14.10 Catabolite repression has apparently evolved to assure the use of glucose as a carbon source when this carbohydrate is available, rather than less efficient energy sources.

14.11 Possibly by directly or indirectly inhibiting the enzyme adenylcyclase, which catalyzes the synthesis of cyclic AMP from ATP.

14.12 Positive regulation; the CAP–cAMP complex has a "positive" effect on the expression of the *lac* operon. It functions in ***turning on*** the transcription of the structural genes in the operon.

14.13 Yes, in the gene coding for CAP; some mutations in this gene might result in a CAP that binds to the promoter in the absence of cAMP. Also, mutations in the gene (or genes) coding for the protein (or proteins) that regulate the cAMP level as a function of glucose concentration.

14.14 Negative regulatory mechanisms such as that involving the repressor in the lactose operon block the transcription of the structural genes of the operon, whereas positive mechanisms such as the activator in the arabinose operon or the CAP–cAMP effect on the *lac* operon promote the transcription of the structural genes of the operon.

14.15 Uninducible, but with a higher level of baseline synthesis of the arabinose enzymes; or, stated differently, a low level of constitutive synthesis of the enzymes. The product of the arabinose regulator gene is required for induction; in its absence, induction could not occur. However, in the absence of arabinose, the regulator gene-product represses the level of *ara* operon transcription. This effect would also be eliminated in such a deletion mutant, resulting in a higher baseline level of synthesis (or a low level of constitutive synthesis).

14.16 0.1, 100, 25.1, 200.

14.17 Constitutive synthesis of the *araB*, *araA*, and *araD* proteins would occur at basal or noninduced levels because the *araC* protein could no longer bind at $araO_2$ and form the loop structure (see Fig. 14.8*c*).

14.18 Repression/derepression of the *trp* operon occurs at the level of transcript initiation, modulating the frequency at which RNA polymerase initiates transcription from the *trp* operon promoters. Attenuation modulates *trp* transcript levels by altering the frequency of termination of transcription within the *trp* operon leader region (*trpL*).

14.19 Deletion of the *trpL* region would result in the levels of the tryptophan biosynthetic enzymes in cells growing in the presence of tryptophan being increased about 10-fold because attenuation would no longer occur if this region were absent.

14.20 First, remember that ***transcription and translation are coupled in prokaryotes***. When tryptophan is present in cells, tryptophan-charged $tRNA_{trp}$ is produced. This allows translation of the *trp* leader sequence through the two UGG Trp codons to the *trp* leader sequence UGA termination codon. This translation of the *trp* leader region prevents base-pairing between the partially complementary mRNA leader sequences 74–85 and 108–119 (see Fig. 14.11*b*), which in turn permits formation of the transcription-termination "hairpin" involving leader sequences 108–119 and 126–134 (see Fig. 14.11*c*).

14.21 No. Attenuation of the *trp* operon would now be controlled by the presence or absence of Gly–$tRNA_{gly}$.

14.22 No. Since transcription (nucleus) and translation (cytoplasm) are ***not*** coupled in eukaryotes, attenuation of the type occurring in prokaryotes would ***not*** be possible.

Chapter 15

15.1 Operons are common in prokaryotes. Gene clusters that resemble operons, if they are not in fact operons, are known to exist in eukaryotic microorganisms such as yeasts and fungi. No operons are known in higher eukaryotes. Why operons are common in prokaryotes but are rare, if present at all, in higher eukaryotes is unknown. More knowledge about the organization of genes and modes of regulation of gene expression in eukaryotes will be required before this question can be answered.

15.2 Probably the hypothesis most acceptable would be (2) the intervention of cytoplasm. The cytoplasmic repressor or activator substances involved in the process of differentiation according to this hypothesis fit rather nicely with the operon concept.

15.3 No. The vast majority of the RNA transcripts that have been studied in higher eukaryotes are monogenic. However, exceptions are quite prevalent in certain viruses of eukaryotic organisms.

15.4 In prokaryotes, the structural genes specifying the enzymes in a metabolic pathway are often arranged as clusters of contiguous genes. This facilitates regulation by the operon mechanism. In higher animals, such genes are usually not in clusters, often being unlinked. The complex patterns of gene expression during development in higher animals almost certainly require complex integrated controls that can govern expression. Genomes of higher animals contain single-copy DNA sequences (structural genes, in some cases at least) that are interspersed with middle-repetitive sequences. The Britten and Davidson model is consistent with these observations.

15.5 (a) We really don't know yet. Histones are thought to be nonspecific repressors or inhibitors of transcription. This is true *in vitro*, where the rate of RNA synthesis from chromatin (native or reconstituted) is inversely related to the amount of histone present, given a constant amount of DNA. Whether this is true *in vivo* is not known. Some researchers believe that histone modifications, such as phosphorylations, acetylations, and so on, play an important role in the regulation of gene expression in eukaryotes. (b) Histones are synthesized along with DNA during the *S* phase of the cell cycle (c) Again, we don't know how the synthesis of histones is regulated. However, nonhistone chromosomal proteins apparently play a key role.

15.6 We do not know the details of the mechanism(s) by which steroid hormones regulate gene expression in very many cases. These hormones form complexes with specific receptor proteins in target cells. In some way, these steroid hormone–receptor protein complexes trigger the transcription of the appropriate genes in target cell nuclei (see Fig. 15.27). In the case of the glucocorticoid hormones, the hormone–receptor protein complex is known to activate gene expression by binding to a specific enhancer element (see Fig. 15.28).

15.7 When salivary chromosomes are followed in their developmental sequence, puffs that are controlled by a hormone occur in specific regions in a regular pattern. Puffs have been interpreted as regions of RNA synthesis and have been related to particular gene loci. The chromosomes are large and easily observed. Staining techniques are available to identify areas of RNA synthesis, and the gene sequences along the chromosomes are known and identifiable with phenotypes in the flies. *In situ* DNA–RNA hybridization and autoradiography can be used to determine the chromosomal location of genes transcribed in response to a particular hormone.

15.8 (a) These organisms can be cultured readily and investigated in the laboratory. Techniques are available for studies of biochemical pathways. (b) (1) The material most suitable for experiments would be organisms such as *E. coli* and other bacteria that have a single linkage group, many biochemical mutants, and, for the most part, organisms that reproduce asexually; these organisms would be most likely to develop clusters of coordinately regulated genes. (2) The effects of histones might be studied using *in vitro* chromatin-reconstitution (fractionation and reassociation) experiments in combination with *in vitro* transcription systems. These *in vitro* systems might be derived from cells such as chicken erythrocytes, which have large amounts of histones. (3) Insects or higher animals provide excellent systems with which to study the effects of steroid hormones on gene expression. These organisms use many different steroid hormones as signals that control the metabolic patterns in different tissues. (4) *Drosophila* and other Diptera exhibit puffing of bands in their giant salivary chromosomes.

15.9 (a) Genes controlling pigment production are present in all cells of the rabbit. The chemical reaction required for black pigment can proceed at a temperature of 27°C, but it is blocked at higher temperatures. (b) While the rabbits are developing in the body of the mother, the temperature is about 33°C, and after birth the main part of the body is maintained at that temperature. Extremities and treated areas in this experiment were coöler as the reaction proceeded.

15.10 Although the genomes remain largely the same in all cells of an organism, the molecular composition of the cytoplasms of cells is constantly changing in response to both the genome and the environment. Many of the metabolic processes occurring in cells take place in the cytoplasm. Thus changes in the cytoplasm may induce or repress the expression of genes. In some cases, these repress the expression of metabolism and appear to be genetically preprogrammed. In other cases, they occur in response to environmental stimuli.

15.11 Differentiation provides the necessary cell forms and functions; organization results in appropriate groupings of cells and tissues; and growth, through cell division and cell expansion, determines the ultimate size and shape of the whole organism as well as its various parts.

15.12 At present, there is no known artificial method for producing and supplying to the cells the necessary enzymes for pigment production. Enzymes that operate in cells must be synthesized in these cells. However, gene transplant experiments now make it feasible to introduce wild-type copies of the genes encoding enzymes into mutant cells or tissues of animals. In the future, it will almost certainly become possible to perform somatic-cell gene therapy in humans and thus treat inherited metabolic abnormalities such as albinism (see Chapter 24).

15.13 (a) Wild-type host tissue apparently can compensate for the chemical blocks induced by the *v* and *cn* genes. (b) Two steps in the production of wild-type pigment are involved. Discs with *v* can synthesize a material needed for pigment production when supplemented with a substance from the *cn* host. The *cn* discs, however, had their pigment production blocked at a later state and required a substance that could not be supplied by the *v* host.

15.14 Homeotic mutants provide direct evidence that patterns of gene expression occurring during differentiation are under genetic control (at least in part). Ongoing studies of these mutants and of the wild-type genes in which the mutations have occurred are providing new information about the mechanisms that regulate the pathways of development in *Drosophila*. Many of the genes involved in the control of development in *Drosophila* are now known to encode *trans*-acting proteins that regulate the expression of genes controlling subsequent steps in the developmental pathway. Current evidence indicates that development is controlled by a cascade of regulatory genes that act sequentially during the process of differentiation (e.g., see Fig. 15.14).

15.15 If purified mRNA transcribed from the gene can be obtained, it can be used to synthesize a radioactive cDNA probe (a complementary strand of DNA) with the enzyme "reverse transcriptase" (RNA-dependent DNA polymerase). The radioactive cDNA can then be allowed to hybridize with total RNA extracted from each of the two tissues. After hybridization, all single-stranded molecules can be degraded with a single-strand-specific nuclease. The amount of radioactivity remaining in macromolecular form (precipitable by trichloroacetic acid) is a measure of the amount of gene-specific mRNA present in the two tissues. If a clone of the gene is available, it can be used as a hybridization probe to look for complementary RNA sequences in RNA populations purified from the cells or tissues of interest. Northern blot hybridization experiments (see Chapter 13) provide a simple procedure for the detection of specific mRNAs; however, northern blots have limited sensitivity and often will not permit one to detect rare mRNAs. The polymerase chain reaction (PCR) procedure is much more sensitive and is the method of choice for detecting rare mRNAs. If the gene is known to be the structural gene for a specific enzyme or polypeptide, the synthesis of this enzyme or polypeptide can be used as an indirect assay for the expression of the gene. To be certain that synthesis of the protein accurately reflects what is occurring at the transcription level, one would have to make sure that it is not synthesized from a stable mRNA.

15.16 Classical promoters can regulate the transcription of genes ***only*** when they are located just 5′ (upstream) to the gene and have the proper orientation (i.e., they are inactive when inverted relative to the gene). Enhancers are (1) orientation independent (i.e., they are active in either orientation), (2) position independent (i.e., they are active whether 5′ to the gene, within the gene, or 3′ to the gene), and (3) can function over relatively large distances (>1000 nucleotide-pairs).

15.17 Some of the same changes in the determined states of cells (cells destined to form specific organs) are observed in the phenotypes of flies expressing homeotic mutant alleles as were observed in Hadorn's serial transplantations of larval imaginal discs to adult flies. For example, transdeterminations of antennae imaginal discs to leg discs were observed by Hadorn. This phenotypic change is analogous to that produced by certain *Antennapedia* homeotic mutations. In contrast, no transdetermination of an antenna imaginal disc to an eye disc was ever observed. Similarly, no homeotic mutation has been identified that converts antennae to eyes. These results suggest that the programs of gene expression that give rise to certain states of differentiation are more closely related than others. Apparently, a mutation in a single gene (master regulator or "selector" gene) can change the pattern of gene expression from that required to produce antennae to that required to produce legs but not to that required to produce eyes.

15.18 (a) The *bcd* gene-product is required for the formation of the head and thoracic (anterior) segments of the fly; the *osk* gene-product is required for the development of the abdominal (posterior) segments of the fly. (b) These genes must be expressed before the completion of oögenesis because the phenotype of the embryo is controlled by the genotype of the mother rather than the genotype of the embryo itself. (c) The growth of such embryos would be arrested very early in development with essentially no body segments being formed.

15.19 The anterior–posterior segmentation pattern in *Drosophila* is established by the sequential expression of sets of regulator genes. The initial anterior–posterior polarity is established by the expression of maternal-effect genes such as *bcd* and *osk* (see Problem 15.18). These genes are expressed during oögenesis with the *bcd* transcript being synthesized at the anterior cortex of the oöcyte and the *osk* transcript being synthesized at the posterior cortex of the oöcyte. The *bcd* and *osk* proteins represent morphogens and their gradients within the embryo establish the anterior–posterior axis. Finer anterior–posterior gradients are established by the products of the gap genes, the pair-rule genes, and the segment-polarity genes acting in a roughly sequential manner. The final pattern of segmentation requires the expression of homeotic genes and the tissue-specific genes that produce the final differentiated phenotypes of individual tissues and cells of the adult fly. See Fig. 15.11 for a schematic overview of this developmental pathway. Many of these genes in this pathway have been shown to encode *trans*-acting regulatory proteins that turn on the expression of subsequent regulatory genes in the pathway—creating a sort of cascade of regulatory controls.

15.20 The sequence-specific DNA-binding properties of these proteins provide support for the hypothesis that development in *Drosophila* is controlled by a cascade of regulatory genes—with maternal-effect genes regulating the expression of gap genes, and gap genes regulating the expression of pair-rule genes, and so on.

15.21 The conserved homeobox sequences of HOX genes of humans (and other higher animals as well) suggest that some of the key regulatory controls may be conserved in many different eukaryotic species.

15.22 Methylation has been found to decrease the level of expression of those genes where its effect has been studied.

15.23 A family of closely related proteins such as the tropomyosins could either be encoded by a family of related genes or be produced by differential RNA or protein processing of a single gene-product. In the case of the tropomyosins of *Drosophila* and vertebrate animals, both of these possibilities are correct. There are families of closely related tropomyosin genes and, in addition, the transcripts of some of these genes undergo different pathways of exon splicing to produce multiple protein species from a single gene (see Fig. 15.31).

Chapter 16

16.1 The genetic information specifying antibody chains is stored in sequences of nucleotide-pairs encoding sequences of amino acids, just like the genetic information specifying other polypeptides. However, the information specifying an antibody chain is stored in bits and pieces that are assembled into functional genes encoding antibody chains by genome rearrangements (somatic-cell recombination events) occurring during the development of the B lymphocytes (the antibody-producing cells). See Figs. 16.4 and 16.5.

16.2 4; 2; 1.

16.3 (1) The joining of different *V*, *D*, and *J* gene segments by somatic recombination during B lymphocyte development (see Figs. 16.4 and 16.5); (2) variability in the exact location of the joining reaction during *V–D–J* joining events (see Fig. 16.9); and (3) somatic mutation.

16.4 At the DNA level; class switching occurs by somatic recombination during B lymphocyte differentiation (see Fig. 16.5).

16.5 The C_H gene segments that are used to direct the synthesis of the individual antibodies determine whether a given antibody will be secreted or membrane bound. In some cases the pattern of splicing of the exons of a C_H gene segment will determine whether the antibody is soluble or membrane bound (see Fig. 16.7).

16.6 (a) B lymphocytes, T lymphocytes, and macrophages. (b) B lymphocytes synthesize the ***antibodies*** responsible for the ***humoral immune response***; T lymphocytes synthesize the ***T cell antigen receptors*** responsible for the ***cellular immune response***; and macrophages ingest and ***degrade antigen–antibody complexes***.

16.7 The *MHC* locus encodes the ***transplantation antigens*** that play a major role in the rejection of foreign tissues after transplant operations.

16.8 (a) The *MHC* genes are said to be highly polymorphic because each is represented by a large number of alleles segregating in human populations. (b) The polymorphic nature of the *MHC* genes makes it more difficult to match the tissue types of potential organ donors and organ recipients so as to avoid rejection of the foreign organ following transplant surgery.

16.9 This response occurs by a process called ***clonal selection*** during which antibodies on the surface of the B lymphocyte producing them bind to the antigen and this in turn stimulates the cell to divide and produce a population (clone) of B cells all producing the same antibody.

16.10 Yes, the T lymphocyte antigen receptor proteins (see Fig. 16.13).

16.11 Both contain polypeptides with variable regions that form the antigen-binding domains and constant regions that facilitate interactions with cell membranes and other components of the immune response.

16.12 1.35 billion (composed of 2250 different light chains and 600,000 different heavy chains).

16.13 If one could prevent HIV from entering T lymphocytes, one could eliminate many of the symptoms of AIDS. Since HIV must first bind to the CD4 surface antigen of T cells, one could prevent entry of the virus by blocking this binding step. One possible way to block this binding step would be to introduce an altered, soluble form of CD4 into the circulatory systems of HIV-positive individuals. This soluble CD4 would bind to the virus and block its interaction with authentic CD4 on the surface of T cells. Alternatively, one might implant transgenic white blood cells carrying a modified gene that would produce a secreted, soluble form of CD4. However, a potential problem with these approaches is that the altered CD4 might cause serious side effects by interfering with the normal function of the CD4 antigen. If the virus binds to a region in CD4 that is not essential to its normal function, one might be able to use an oligopeptide that contains the HIV-binding domain, but not other regions of the CD4 protein required for its normal function(s). Alternatively, one might be able to produce a monoclonal antibody that binds to the HIV-binding domain of CD4 without interfering with normal CD4 function(s). Unfortunately, there are many unknowns in these potential strategies for treating and preventing AIDS that must be investigated before their feasibility can be evaluated. Nevertheless, one must continue to hope that successful methods for combating HIV will be forthcoming in the near future.

Chapter 17

17.1 *M*-cyclins are positive regulators of cell division; they are components of the mitosis-inducing factor (MPF) that stimulates the onset of the *M* phase of mitosis. G_1-cyclins are probably components of a similar complex that provides the signal committing cells to initiate DNA replication in the *S* phase of the cell cycle.

17.2 *Schizosaccharomyces pombe* mutants carrying the *cdc*2 · Phe15 and *cdc* · Phe19 alleles have in fact been constructed by site-specific mutagenesis (K. L. Gould and P. Nurse, *Nature* 342: 39–45, 1989). The *cdc*2 · Phe19 allele had no effect on phenotype; the cell cycle of this mutant was normal, indicating that phosphorylation of tyrosine 19 is not involved in cell cycle regulation. The *cdc*2 · Phe15 mutant strain was found to initiate mitosis while the cells were still very small—dividing at about half the size of the cdc^+ strain. Presumably, the inability to inactivate the *cdc* · Phe15 pp34 by phosphorylation leaves this regulatory protein locked in the active form, which stimulates cell division prematurely.

17.3 The protooncogenes and retroviral oncogenes have very different structures. The viral oncogenes contain a single uninterrupted coding sequence or exon, whereas the protooncogenes contain several coding sequences interrupted by several introns (see Fig. 17.4).

17.4 The conservation of the sequences of protooncogenes across the animal kingdom indicates that these genes encode products that perform essential functions that must be preserved in all species. In fact, the protooncogene products that have been identified to date play central roles in cellular metabolism and in intercellular communication. They clearly provide functions that are essential to the normal growth and development of the host organism.

17.5 Comparisons of the sequences of cellular protooncogenes and related viral oncogenes indicate that viral oncogenes have evolved from cellular protooncogenes. Many of the viral oncogenes have nucleotide sequences suggesting that they have been derived from processed (introns spliced out) mRNA products of the related cellular protooncogenes.

17.6 Most of the *c-ras* oncogenes that have been cloned and sequenced differ from the homologous cellular *ras* protooncogenes by single nucleotide-pair substitutions. These results strongly indicate that the c-*ras* oncogenes have been produced from the homologous c-*ras* protooncogenes by single base-pair mutations (see Table 17.2).

17.7 The fact that most of the mutations that convert c-*ras* protooncogenes to c-*ras* oncogenes alter three specific codons causing amino acid substitutions at positions 12, 59, and 61 of the c-*ras* proteins indicates that the amino acids at these sites (glycine, alanine, and glutamine, respectively) are critical to the normal functioning of the c-*ras* protooncogene products. It seems likely that the observed amino acid substitutions at these sites (see Table 17.2) cause the proteins to be "locked" in a form that stimulates cell division.

17.8 (1) Base-pair substitution mutations such as those observed in the c-*ras* genes (see Table 17.2) producing altered gene-products. (2) Insertion of a strong enhancer or promoter sequence adjacent to the protooncogene, thus leading to overproduction of its product. (3) Amplification of the protooncogene leading to overproduction of its product.

17.9 The LTRs of retroviral proviruses contain very strong enhancer/promoter elements. If an LTR integrates adjacent to a protooncogene, it may cause overproduction of the protooncogene product. The increased level of the gene-product may in turn cause the host cells to divide when they would not normally be undergoing cell divisions.

17.10 DMs and HSRs represent regions of amplification of genetic material. They contain multiple tandem copies of a unit of DNA called an "amplicon." Long HSRs may contain hundreds to thousands of copies of a specific amplicon. Such DMs and HSRs are frequently observed in certain types of cancers. If the amplicon contains a protooncogene, the increase in copy number might cause a loss of the normal control of cell division and tumor formation. However, we still do not know what is "cause" and what is "effect." Is the amplification event a cause of the cancer or is it an effect of some earlier event that caused both the malignancy and the amplification?

17.11 The amplification of a protooncogene might result in overproduction of its product—each copy of the gene contributing a certain quantity of product. If the protooncogene product is a positive regulator of cell division, its overproduction might stimulate precocious cell divisions.

17.12 The "Philadelphia" chromosome, which is diagnostic for chronic myelogenous leukemia, is an altered chromosome 22 resulting from a reciprocal translocation with chromosome 9. The Philadelphia chromosome has lost much of its long arm, which has been replaced with a short segment from the tip of the long arm of chromosome 9. The breakpoints on chromosome 9 that are involved in the translocations producing the Philadelphia chromosome occur very close to the c-*abl* protooncogene and transfer it to chromosome 22. Although the relationship between c-*abl* in the Philadelphia chromosome and chronic myelogenous leukemia is still not understood, it seems likely that the translocation event somehow causes inappropriate expression of the c-*abl* protooncogene, which in turn leads to the cancerous state.

17.13 The products of several cellular protooncogenes are known to be involved in normal pathways of intercellular communication and in the normal control of cell division (see Table 17.1). Thus, it should not be surprising that the mutations in these genes may result in a loss of the normal control of cell division, just as mutations in other genes lead to abnormal phenotypes related to their functions.

Chapter 18

18.1 (a) A recessive gene presumed to be carried in heterozygous condition unexpectedly came to expression. Bridges postulated that a section in the homologous chromosome carrying the wild-type allele was missing (i.e., there was a chromosome deficiency). On another occasion, Bridges found that a recessive gene presumed to be homozygous did not come to expression. He postulated that a gene acting as a dominant allele must be present elsewhere in the chromosome set (i.e., there was a duplication). (b) It was impossible to distinguish microscopically between the structural parts of homologous chromosomes at the time these genetic results were obtained. (c) The discovery of attached-X chromosomes in *Drosophila*, meiotic configurations in maize, and salivary chromosomes in *Drosophila* provided tools for cytological verification.

18.2 Salivary preparations in *Drosophila* provide larger chromosomes to work with than those of meiotic stages in maize. Chromosome parts can be identified with the aid the Bridges' chromosome maps. Salivary chromosome studies are performed with somatic cells, whereas cytogenetic studies in maize are done with germ line cells.

18.3 Enlarged size, somatic pairing, identifiable bands, and distinguishable anatomical features along the length of the different chromosomes make salivary gland chromosomes especially useful.

18.4 Chromosomes can be spread on a slide, fixed, and stained in a single operation with a single solution. When they are well spread out on the slide, the linear sequence of morphological markers on an individual chromosome can be followed. On the other hand, it would be necessary to reconstruct the chromosomes if sections were employed for study.

18.5 The chromonemata duplicate themselves many times, but the chromosomes and cells do not divide. Bundles of chromonemata are thus produced. The cross bands represent groups of identical chromomeres.

18.6 Linkage maps are constructed by placing the relative gene positions calculated from crossover data along a line representing a chromosome. Cytological maps are constructed by microscopic observations of chromosomes from actual cell preparations. The salivary chromosome maps prepared by Bridges are linkage maps superimposed on cytological maps.

18.7 When a section of a chromosome carrying a dominant gene is deleted, a recessive allele carried on the homologous chromosome will be expressed.

18.8 (a) The extent of a deficiency can be determined by testing the genes on either side of the point known to be in the deficiency to see if they behave as pseudodominants. (b) The determination can be made cytologically by observing microscopically the extent of the deleted segment in a suitable chromosome preparation (e.g., a salivary chromosome preparation with chromosomes heterozygous for the deficiency).

18.9 (a) This is not an easy determination. New mutants of a particular kind may occur infrequently at a particular locus and produce phenotypic changes like those associated with a deficiency. A mutation v^+ to v in this case would appear unlikely, but, if it did occur, the pair of recessive genes would behave in a regular Mendelian pattern. Testcross or F_2 results could be predicted and checked. In animals such as *Dro-*

sophila, for which more complete data are available, deficiencies are usually homozygous lethal, resulting in modified F_2 ratios. If other recessives in the same chromosome, presumed to be heterozygous, came to expression, the case for a deficiency would be strengthened. (b) Cytological determination in mice is also difficult. There are no enlarged salivary or meiotic chromosomes available for study. If a chromosome could be shown to be shorter or structurally different from its homolog, the deficiency hypothesis would gain support. Banded chromosomes can be used to identify large deficiencies in mouse chromosomes.

18.10 (a) When crossing-over occurs in the area of a paracentric inversion, acentric and dicentric chromosomes occur that do not separate properly to the poles during meiotic divisions. Gametes carrying crossover chromatids are often abnormal and inviable. (b) Crossovers within pericentric inversions also result in unbalanced chromosome arrangements that make the crossover gametes or zygotes inviable. (c) Crossing-over is reduced to some extent, but the main "suppression" results from inviable gametes or zygotes.

18.11 All flies would be beaded with the genotype $Bd\ \ell^+/Bd^+\ell$.

18.12 The seedlings would be approximately 1 white : 2 green : 1 yellow; white and yellow seedlings would die, and all mature plants would be green.

18.13 (a) Autosomal; (b) chromosome III.

18.14 (a) A loop is formed or a buckling occurs in the unpaired normal chromosome segment, which corresponds in size with the deficiency. (b) A loop similar in appearance to that described for a deficiency is formed. (c) A loop is formed by the synapsed chromosomes; the orientation of one member of the pair is reversed, making it possible for the corresponding segments of the inverted and uninverted parts to pair. (d) A cross-shaped structure is formed by the paired chromosomes (see Figs. 18.3, 18.5, 18.6, 18.7, 18.10, and 18.11).

18.15 (a) Altered linkage groups, pollen or ovule sterility, and position effects suggest the presence of translocations. (b) Cross configurations plus rings and chains of chromosomes that can be seen microscopically during meiotic prophase provide cytological evidence.

18.16 Unbalanced chromosome arrangements occur during gamete formation, giving rise to inviable pollen (see Fig. 18.11).

18.17 (a) An inversion has occurred involving section 4 5 6. (b) A loop would be formed by the paired chromosomes, and the elements of the inverted segment would be in reverse order (i.e., 6 5 4).

18.18 (a) Segment 3 4 has been deleted. (b) Segment 4 has been duplicated. (c) Segment 8 7 6 has been inverted (see Fig. 18.5 for a model).

18.19 See Fig. 18.11 for a model.

18.20 If a structural alteration in a chromosome can be definitively related to a phenotypic change, a position effect would be indicated. The phenotypic effect is caused by the change in position of genetic material rather than by an addition or deletion of genetic material.

18.21 (a) When a section of euchromatin is moved by a structural change to a location within or near heterochromatin, a variegated position effect sometimes occurs. (b) Apparently genes moved to a different chromosome location behave differently from those remaining in the genetic environment to which they are adjusted, presumably because they are moved into proximity with different regulatory elements such as enhancers and silencers.

18.22 (a) The Philadelphia chromosome is a fragment of a No. 22 chromosome that is associated with a particular type of cancer. (b) It is a diagnostic factor for chronic myelocytic leukemia.

18.23 (a) 45, XY, −21; (b) 45, XX, −14, −21, +t(14*q*21*q*); (c) 46, XY, −5, −12, +t(5*p*12*p*), +t(5*q*12*q*).

18.24 Inversions and translocations act as isolating factors among individuals in populations, thus preventing normal gene exchange and promoting speciation.

18.25 Techniques for preparing slides for critical microscopic observation were not effectively applied to human chromosome studies until 1956. Better sources of human material from surgical procedures and tissue-culture techniques are now available.

18.26 Human chromosomes are small, quite uniform, and, using older techniques, difficult to distinguish. They could be separated into seven groups following the Denver conference criteria, but enough overlap occurred among members of the same group (partly because of variations in techniques) to make it impossible to consistently identify individual chromosomes.

Chapter 19

19.1 (a) 0; (b) 2; (c) 1.

19.2 Nondisjunction of chromosomes during the production of gametes (eggs and sperm) seems to be the explanation for most cases of trisomy in humans.

19.3 (a) 1/490,000; (b) 60,000; (c) 1/70.

19.4 ½.

19.5 1/3,500,000.

19.6 (a) 1/4,000,000; (b) 1/40,000.

19.7 (a) 17 purple : 1 white; (b) 11 purple : 1 white; (c) 15 purple : 3 white reduced to 5 purple : 1 white; (d) 3 purple : 1 white. (e) The mutant trait occurs less frequently than expected in the results of crosses because of the presence of an extra chromosome, usually carrying a wild-type allele.

19.8 (a) ½ ($2n$) and ½ ($2n + 1$); (b) 11 normal : 1 eyeless.

19.9 (a) XO is basically female; XXY male; and XXX metafemale. The human Y carries the male-determining capacity. (b) The *Drosophila* Y chromosome has no influence on sex determination. The *Melandrium* Y chromosome carries male-determining genes.

19.10 Some plants and animals with a particular extra chromosome can be recognized phenotypically. Genes that give trisomic ratios (e.g., genes for purple or white in *Datura*) are located on the particular chromosome that makes the trisome. Monosomics are useful because recessive genes on a chromosome that has no homolog (e.g., a monosome) express themselves and can thus be associated with a particular chromosome.

19.11 Tetrasomics and nullisomics apparently occur in nature, but they are less viable and usually die before they are detected.

19.12 Monoploids have only one set of genes and, therefore, no gene segregation. They could be used for experimental work where it is desirable to relate genes with phenotypes or traits. In fungi such as *Neurospora*, monoploids represent the most important part of the life cycle and are used extensively for genetic studies.

19.13 Polyploid tissues grow through cell divisions that can occur regularly among polyploid cells. Sexual reproduction of animals requires gamete formation, fertilization, and sex determination. Irregularities associated with polyploidy nearly always result in inviability and sterility in animals.

19.14 Polyploidy might occur through a doubling of the chromosomes in the somatic cells or the failure of the reduction division of meiosis to produce gametes with the haploid number of chromosomes.

19.15 (a) Autopolyploidy is produced by the doubling of a $2n$ chromosome complement, resulting in four similar genomes per cell. Allopolyploidy results from hybridizations in which two $2n$ complements are combined. (b) Because autopolyploidy results in four sets of chromosomes, pairing is irregular; parts of all four similar chromosomes may be paired in different places, thus interfering with normal reduction divisions. This and more subtle genetic factors make autopolyploids sterile and incapable of perpetuating themselves. On the other hand, allopolyploids contain two diploid sets of at least partially distinct chromosomes. Thus, the chromosomes can pair properly (two-by-two) and produce gametes with complete sets of chromosomes. Allopolyploidy appears to have played an important role in the evolution of many plant groups.

19.16 Tetraploids have an even number of chromosome sets or genomes. If they have arisen through hybridization and the four genomes are represented as two pairs, synapsis—an essential part of meiosis—can be accomplished normally. Triploids with three genomes have inherent difficulties in pairing and gamete formation, and they usually do not perpetuate themselves through normal sexual reproduction, although they may arise through normal sexual reproduction. They may arise through fusion of unreduced ($2n$) and reduced (n) gametes, but usually they do not reproduce themselves as triploids unless apomixis or some other deviation from sexual reproduction occurs.

19.17 Improper pairing, the most obvious cause of sterility, is often associated with chromosome irregularities. More subtle genetic factors may also be involved in some cases.

19.18 (a) Chromosome number is an important criterion in the classification of living organisms. It is at least as significant as any well-defined morphological characteristic. (b) Numerical as well as structural differences in chromosomes have been significant in the mechanics of evolution. Analysis and comparison of chromosome numbers within and between taxonomic groups have aided in solving questions about evolutionary pathways.

19.19 (1) Doubling of chromosome number in the section of the hybrid that grew vigorously. (2) Propagation of cells from the polyploid tissue to reproduce a plant with two sets of chromosomes from each parent species.

19.20 (1) A few cells of the hybrid were apparently polyploid with 24 chromosomes. (2) These gave rise to pollen grains with 12 chromosomes. (3) When these pollen grains were used to fertilize eggs of species B, a few plants with 19 chromosomes were produced. (4) These produced gametes with different numbers of chromosomes including some gametes with 12 chromosomes. (5) Gametes with 12 chromosomes fused to form 24-chromosome plants with the characteristics of a new polyploid species.

19.21 Colchicine treatment interferes with spindle formation during cell division. Irregular numbers of chromosomes are included in the daughter cells. Sometimes the nuclear envelope forms around the two sets of chromosomes that normally would separate to two progeny nuclei, and the number of chromosomes per cell is doubled.

19.22 (a) Colchicine-induced polyploidy provides a means of duplicating in the laboratory a process that has apparently occurred naturally in many important plant groups. Some polyploids have qualities that give them practical advantages. New polyploids are being produced and compared with established commercial varieties of grapes, tomatoes, and other plants. Some have been found to possess valuable characteristics. (b) Induced polyploidy has theoretical significance as a tool for studying the mechanisms that have occurred in the evolution of some plant groups. A better understanding of the origins of modern varieties of cotton, wheat, and other valuable and interesting polyploids presents a challenge for present and future investigations.

19.23 Polyploidy that is associated with hybridization could account for the differences among some known strains of cotton and wheat. The processes through which some modern polyploids may have developed in nature have been reconstructed in the laboratory (see text for details).

19.24 Progeny of interspecific crosses are usually sterile. If hybridization is combined with chromosome doubling, fertile hybrids may occur and perpetuate their new chromosome arrangement. They do not cross with the diploids and, if they are perpetuated in nature, they might eventually develop into a new species.

19.25 (a) *Triticum spelta* has been reproduced experimentally by doubling the chromosomes of emmer wheat and goat grass and crossing the two polyploids. (b) *Raphanobrassica* was developed experimentally by crossing the radish and the cabbage. The F_1 progeny were sterile, but some unreduced gametes were obtained, and the F_2 tetraploid (*Raphanobrassica*) was produced from them. (c) *Spartina townsendii* apparently occurred in nature from the crossing of *Spartina alterniflora* with *S. stricta* and the doubling of the chromosome number of the hybrid. (d) *Primula kewensis* was produced by the crossing of *Primula floribunda* and *P.*

verticillata and the doubling of the chromosome number of the hybrid. (e) The origin of New World cotton has been reconstructed experimentally by crossing Old World cotton with upland cotton and doubling the chromosome number of the hybrid.

19.26 (a) Haploid; (b) diploid; (c) trisomic; (d) monosomic; (e) tetrasomic; (f) triploid; (g) tetraploid.

Chapter 20

20.1 Maternal inheritance. It is inherited through the maternal line and does not involve nuclear genes of the mother.

20.2 (a) The cytoplasm provides the environment in which the genes act. Therefore, the mother would be expected to influence secondary actions of certain genes more than the father. (b) Nonchromosomal genes would be inherited through the maternal line.

20.3 (a) Sex-linked genes are located on the X chromosome, and characteristic crisscross inheritance could be detected from appropriate crosses if sex linkage were involved. (b) Cytoplasmic inheritance would be transmitted through the maternal line because most of the cytoplasm of the zygote comes from the egg. A series of backcrosses could be made from F_1 males and females to the appropriate females and males of the two parent types. If the trait is transmitted repeatedly for several generations from the maternal parent to her progeny and not through the paternal parent, it may be cytoplasmic. (c) If the trait was transmitted from mother to progeny but did not persist in the maternal line, it might be attributed to the influence of the mother's genes on the developing egg or embryo, that is, a maternal effect. In this case, nuclear genes would be involved, but they would be the genes of the mother rather than those of the individual itself.

20.4 Inherited changes in the proplastids and the multiplication of cells in which the changes occurred resulted in sectors that showed different color characteristics, as in the four-o'clocks. If the plastid characteristics are determined solely by eggs, only female gametes would determine green, pale green, or variegated plants. In examples from other species, a small amount of cytoplasm is carried with sperm, which also results in variegated plants.

20.5 Male sterility facilitates crosses involving plants that are ordinarily self-fertilized. Large-scale crossing for obtaining hybrid vigor is accomplished more economically if the plants are male sterile.

20.6 Kappa particles are bacteria that have developed an intimate symbiotic relationship with paramecia of a particular genotype.

20.7 Maternal inheritance. *Oenothera hookeri* must have yellow plastids, and *O. muricata* must have green plastids.

20.8 (a) Because this trait is transmitted as a maternal effect, all three genotypes could have dextral or sinistral coiling, depending on the genes of the mother in question. (b) The mother and grandmother might be expected to determine the coiling characteristics in their immediate progeny. An *ss* mother would produce sinistral progeny. Male parents have no immediate effect on this trait.

20.9

$s^+ s^+ \times s\ s$	P
(s^+) (s)	Gametes
$s^+ s$	F_1
$1s^+ s^+ : 2s^+ s : 1s\ s$	F_2 genotypes; all F_2 snails would coil right because the F_1 mother was $s^+ s$.
right, right, left	Phenotypic results from inbreeding the F_2 snails.

Phenotype depends on the mother's genotype. All F_2 snails would coil to the right because the F_1 mothers were $s^+ s$. Following inbreeding, progeny of $s\ s$ snails would coil to the left.

20.10 (a) Dark young, dark adult; (b) light young, dark adult. (c) The pigment condition is at first influenced by the cells of the mother. When the hoppers grow older, the condition determined by the genes of the individual hopper replaces the one that originated from the mother (if there is a difference in genotype between mother and young).

20.11 The eyes became lighter as the kynurenine that diffused from *AA* host into the egg was metabolized and broken down by the *aa* individuals that were unable to manufacture more kynurenine.

20.12 (c) A gynandromorph.

Chapter 21

21.1 The phenotypes are determined by the number of primed alleles (x) in the genotype. Their frequencies are given by the formula $f = \{4!/[x!(4-x)!]\}\,(1/2)^x\,(1/2)^{4-x}$; for $x = 0, f = 1/16$; $x = 1, f = 4/16$; $x = 2, f = 6/16$; $x = 3, f = 4/16$; $x = 4, f = 1/16$.

21.2 (a) $\bar{x} = 20.45$; (b) $s^2 = 2.37$; (c) $s = 1.54$.

21.3 $\Sigma(x_i - \bar{x}) = 0$.

21.4 For the F_1, $V_g = 0$ since they are all genetically identical; for the F_2, $V_g > 0$ since genetic differences result from segregation.

21.5 3.17/6.08 = 0.52.

21.6 V_e can be estimated by the average of the variances of the F_1; V_g is obtained as the difference between the variances of the F_2 and the F_1; thus, the broad-sense heritability is $(26.4 - 9.4)/26.4 = 0.64$.

21.7 (a) −4; (b) −0.8; (c) −3.2.

21.8 $t = g + e$ and $t' = g + e'$; therefore, $\text{Cov}(t,t') = \Sigma(t)(t')/(n-1) = \Sigma(g + e)(g + e')/(n-1) = \Sigma(g^2 + ge + ge' + ee')/(n-1) = [\Sigma g^2 + \Sigma ge + \Sigma ge' + \Sigma ee']/(n-1) = V_g + \text{Cov}(g,e) + \text{Cov}(g,e') + \text{Cov}(e,e')$. If g is uncorrelated with e and e', and if e is uncorrelated with e', then the last three terms are each zero, and $\text{Cov}(t,t') = V_g$.

21.9 Broad-sense heritability must be greater than narrow-sense heritability.

21.10 (125 − 100)(0.4) + 100 = 110.

21.11 (15 − 12) (0.3) + 12 = 12.9.

21.12 (90 − 100) (0.2) + 100 = 98.

21.13 (12.5 − 10)/(15 − 10) = 0.5.

21.14 (40)(0.3)(10) + 2000 = 2120.

Chapter 22

22.1 (2)(0.2)(0.8) = 0.32.

22.2 $\sqrt{0.0004} = 0.02$.

22.3 $\sim 2\sqrt{0.00004} = 0.012$.

22.4 Central American: L^M (0.78), L^N (0.22); North American: L^M (0.39), L^N (0.61).

22.5 M (0.61), MN (0.34), N (0.05).

22.6 Compute the probability of not being homozygous: 1 − 0.36 −0.09 − 0.01 = 0.54.

22.7 The frequency of homozygous tasters is $(0.4)^2 = 0.16$ and the frequency of heterozygous tasters is (2)(0.4)(0.6) = 0.48; thus, 0.16/(0.16 + 0.48) = 0.25.

22.8 (a) ⅛; (b) ⅛.

22.9 ¼.

22.10 There are six inbreeding loops; assuming that Q and R came from random mating, $F_X = (4)(½)^5 + 2(½)^3 = ⅜$.

22.11 $F_C = 1$ to satisfy the equation $(½)^5(1 + F_C) = 1/16$.

22.12 $F = 0.25$ and the allele frequencies are 0.6 and 0.4; therefore, (2)(0.6)(0.4)(1 − 0.25) = 0.36.

22.13 Note that for the inbred strains $F = 1$; thus, $(½)^3(1 + 1) + (½)^4(1 + 1) = ⅜$.

22.14 Assuming two unlinked loci are involved, ¼ + ¼ − 1/16 = 7/16.

22.15 For each answer, sum the phenotypic values weighted by their frequencies in the population; these frequencies are obtained by computing the frequencies of the corresponding genotypes using the allele frequencies ($T = 0.4$ and $t = 0.6$) and the inbreeding coefficient: (a) $F = 0$ and mean = (0.16)(100) + (0.48)(100) + (0.36)(50) = 82; (b) $F = ½$ and mean = (0.28)(100) + (0.24)(100) + (0.48)(50) = 76; (c) $F = ¼$ and mean = (0.22)(100) + (0.36)(100) + (0.42)(50) = 79; (d) the mean decreases linearly with F.

22.16 (a) $p_1 = 1 - p_2 = 1 - \sqrt{0.16} = 0.6$; (b) see the formula for p_1' in Table 22.7. Here, $p_1 = 0.6$, $hs = 0$, and $\overline{W} = 0.992$; thus, $p_1' = 0.6048$.

22.17 $\sqrt{10^{-6}} = 0.001$.

22.18 $q = \sqrt{u/hs}$; here $hs = 0.05$, so $q = 0.00447$.

22.19 (a) $s = 0.2$, $h = 0$; (b) $s = 0.3$, $h = 0$; (c) $s = 0.2$, $h = 0.5$; (d) $s = 0.1$, $h = 1$.

22.20 Apply the formula for a balanced polymorphism; 0.5.

22.21 Heterozygosity will be lost at a rate of $1/(2N)$, or 1 percent per generation.

22.22 0.5.

22.23 0.25.

22.24 Duplications and deficiencies could be produced by unequal crossing-over.

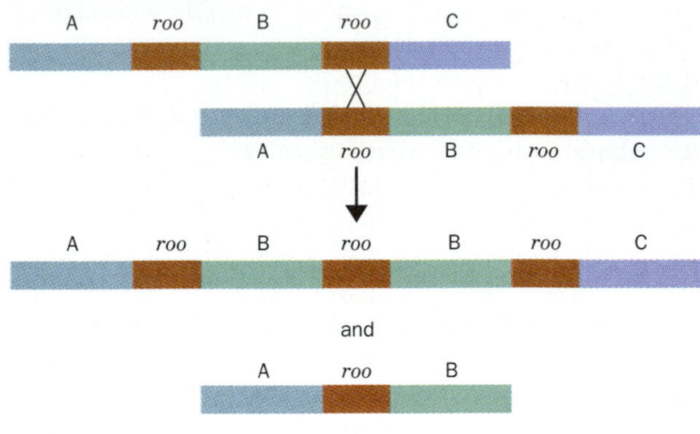

22.25

♂♂		♀♀	
BB, BB'	bald	BB	bald
$B'B'$	nonbald	BB', $B'B'$	nonbald

Total population

p = freq. B allele
q = freq. B' allele
$p^2 + 2pq = 0.51$
$q^2 = 0.49$
$q = .7$; $p = .3$

Mating subpopulation

bald men:

$$BB\text{:}\ \frac{p^2}{p^2 + 2pq} = \frac{.09}{.51} = .176$$

$$BB'\text{:}\ \frac{2pq}{p^2 + 2pq} = \frac{.42}{.51} = .824$$

nonbald women:

$$BB'\text{:}\ \frac{2pq}{2pq + q^2} = \frac{.42}{.91} = .462$$

$$B'B'\text{:}\ \frac{q^2}{2pq + q^2} = \frac{.49}{.91} = .538$$

♀ gametes \ ♂ gametes	p_m 0.588	q_m 0.412
p_f 0.231	0.1358	0.0952
q_f 0.769	0.4522	0.3168

$$p_{male} = .176 + \frac{.824}{2} = .588$$

$$q_{male} = \frac{.824}{2} = .412$$

$$p_{female} = \frac{.462}{2} = .231$$

$$q_{female} = \frac{.462}{2} + .538 = .769$$

$$\text{Probability bald child} = p_f p_m + \frac{p_f q_m}{2} + \frac{p_m q_f}{2}$$

$$= 0.1358 + 0.0476 + 0.2261 \cong 0.41$$

Chapter 23

23.1 Animal behavior, the sum total of the animal's responses to its environment, has a hereditary basis. The structural and physiological characteristics on which behavior depends are inherited. Now some behavioral patterns are shown to have a genetic basis, but most of these are also strongly influenced by environmental factors.

23.2 Behavior is more difficult to explain in terms of specific genetic mechanisms than tangible, structural characteristics because behavior traits are complex and the genetic control is intimately interwoven with environmental influences. It is difficult to determine exactly what is inherited, but the general genetic basis as indicated by response to selection is well established.

23.3 Such studies indicate that simple Mendelian inheritance can be applied to behavior traits. They suggest that other behavior patterns that now seem complex and involved with environmental factors may be reduced to common denominators when the mechanics are disentangled.

23.4 Such studies identify differences in the behavior of different species and provide specific areas where hypotheses might be developed and experiments designed to investigate mechanisms that have evolved in these species.

23.5 High and low lines have been selected for preference to alfalfa pollen. A trait that responds to selection and becomes established in a line of bees must have a genetic basis.

23.6 (a) Hormones provide chemical coordination and thus influence the general health and temperament as well as the mating behavior of an animal. (b) Glands that produce hormones develop under genetic control, and hormone production in mature animals can be altered by genetic factors.

23.7 Scott and Fuller showed that genetically determined behavior differences develop under the influence of environmental factors.

23.8 The dogs are innately intelligent, and they have been selected for desired physical and mental traits. Training guides and perfects the particular skills for which the dogs are genetically qualified.

23.9 Dogs have a great amount of variability, even within breeds, that responds readily to selection.

23.10 The dog is an intelligent and highly adaptive animal. Humans and dogs apparently adopted each other at an early period in their known cultural history. Humans and dogs work together in many ways for mutual advantage. Dogs live in the same shelters with people, feed on their leftovers, and adapt to the requirements for human companionship.

23.11 In their African environment, noisy dogs would call attention to themselves and their masters and either humans or nature (in the form of carnivores) would tend to remove the conspicuous.

23.12 A large component of behavior in a population having undergone long periods of selection is genetic. Basic differences in motivation, cultural overlay, hormonal response, and physical variations make it difficult and unsettling to transfer into a population behavioral insights that have not developed in that population.

23.13 A trait with survival value is likely to be predominantly and basically genetic. Selection in nature tends to preserve individuals and populations with traits that have survival value.

23.14 (a) Size, simplicity of nervous system, mutability, generation time, response to culture methods, defined behavior traits, and observability; (b) size, generation time, culture requirements, defined simple behavior traits, and abundant linkage data.

23.15 To be sure that the experiments were properly controlled and environmental factors were ruled out, questions such as the following must be considered: Are diets exactly alike? Do mothers differ in treatment of young? Does early environment provide more diversity for one strain than the other? Are both strains run through the same maze? Does time of day when tested alter results? Are temperature and light and possible disturbing factors controlled?

23.16 Aggression can be decreased in animal populations by training, by reducing stimuli that arouse aggressive tendencies, and by avoiding overcrowding.

23.17 *Drosophila* are better known genetically than mice. All chromosomes have been mapped and numerous genes have been located on chromosomes. It is possible not only to detect behavioral traits, but to determine some basic genetic mechanisms that are involved.

23.18 From the results of reciprocal crosses between geotactic positive and geotactic negative strains of *Drosophila*, it can be shown whether the F_1 and F_2 plotted pattern is that of quantitative inheritance (polygenes) and whether a criss-cross pattern (from mother to son, sex linkage) can be detected.

23.19

Genotype	$u^+u\ r^+r$	$u^+u\ rr$	$uu\ r^+r$	$uu\ rr$
Phenotype	nonhygienic	no uncapping, but removal	uncapping, no removal	hygienic
Proportion	1	1	1	1

INDEX